The National Hockey League

Official Guide & Record Book

2003

THE NATIONAL HOCKEY LEAGUE
Official Guide & Record Book/2003

Compiled by the NHL Public Relations Department and the 30 NHL Club Public Relations Directors.

Published in Canada by:
Dan Diamond and Associates, Inc, 194 Dovercourt Road, Toronto, Ontario M6J 3C8 Canada
ISBN in Canada 0-920445-79-9

Published in the United States by:
Triumph Books, 601 South LaSalle, Suite 500, Chicago, Illinois 60605
ISBN in USA 1-57243-500-3

Staff

For the NHL: Dave McCarthy, Denise Gomez; Supervising Editor: Greg Inglis; Statistician: Benny Ercolani;
Editorial Staff: Dave Baker, David Keon, Jackie Rinaldi, Kelley Rosset, Chris Tredree, Julie Young.

Managing Editor: Ralph Dinger
Player Register Editor: James Duplacey
Photo Editor: Eric Zweig
Production Editors: John Pasternak, Alex Dubiel
Records Editor and Fact Checker: Paul Bontje
Assistant Editor: Jonathan Zweig

Contributors: Jarvis Abela, Ken Anderson, Eddie Bajus, Heiko Behrens, Steven M. Black, Bob Borgen, Paul R. Carroll Jr., Steve Cherwonak (CHL), Diana Danforth (ECHL), Denis Demers (QMJHL), Dennis Demeule, Mark DeWitt, Richard A. Donnan, Bob Duff, D.A. England, Peter Fillman, Ernie Fitzsimmons, Mel Foster, Pierre Genest, Dan Gognavic, Hockey Hall of Fame, Patrick Houda, Mary Hutchinson (WCHL), Seppo Kittila, James Kochias, Len Kotylo, Dana Lapierre, Eric Lavigne, Roger Leblond, Manon Gagnon Leroux (QMJHL), Al Mason, Christopher McDonald, Penny McEwen, Leroy McKinnon (WHL), Herb Morell (OHL), Isabella Murphy, NHL Broadcasters' Association, NHL Central Registry, NHL Officiating, NHL Players' Association, Greg Nesteroff, Joseph Nieforth, John Norlin, Jeff Obermeyer, Becky Pasternak, Brenda Pasternak, Stephanie Pasternak, John Paton, Gary J. Pearce, Lisa Pepin (UHL), Valentina Riazanova, Mrs. Claude Rompré (QMJHL), Ron Rumball, Minako Saki, Robert Schulz, Ralph Slate, David Stewart-Candy, Bret Stothart (AHL), Andrew Szabo, U.S. Hockey Hall of Fame, Drew White, Ian Wilson.

Special thanks to NCAA Conference and School Sports Information Departments.

Publisher: Dan Diamond

Data Management and Typesetting: Caledon Data Management, Hillsburgh, Ontario
Film Output and Scanning: Stafford Graphics, Toronto, Ontario
Printing: Fidelity National Information Solutions Canada, Scarborough, Ontario
Production Management: Dan Diamond and Associates, Inc., Toronto, Ontario

Photo Credits

NHL Images: Anita Cechowski.

Photographers: Graig Abel, Toronto; Scott Audette, Tampa Bay; Bruce Bennett Studios; Andrew D. Bernstein/Andrew Bernstein Associates, Los Angeles; David Bier; Joe Black; Marc Buckner, St. Louis; Scott Cunningham, Atlanta; Tim DeFrisco, Colorado; Gregg Forwerck, Carolina; Barry Gossage, Phoenix; J. Henson Photographics; Hockey Hall of Fame Collections; Glenn James, Dallas; Ladislas Kadyszewski, Montreal; Bruce Kluckhohn, Minnesota; David E. Klutho; Steve Kovich, Detroit; Mitchell Layton, Washington; Jim Leary; Richard C. Lewis, Florida; NHL Images/Allsport; Andre Pichette; Matt Polk, Pittsburgh; Debora Robinson, Anaheim; Len Redkoles, Philadelphia; A. Ringuette and P. MacCallum/Freestyle Photography; Ottawa; John Russell, Nashville; Sports Action Photography; Boston; Jamie Sabau, Columbus; Harry Scull, Jr.; Robert Shaver; Bill Smith, Chicago; Don Smith, San Jose; Gerry Thomas, Calgary and Edmonton; Jeff Vinnick, Vancouver; Rocky Widner, San Jose; Bill Wippert, Buffalo.

Distribution

Trade sales and distribution in Canada by:
North 49 Books, 35 Prince Andrew Drive, Toronto, Ontario M3C 2H2
416/449-4000; FAX 416/449-9924

Dan Diamond and Associates, Inc., 194 Dovercourt Road, Toronto, Ontario M6J 3C8
416/531-6535; FAX 416/531-3939 e-mail: dda.nhl@sympatico.ca

Trade sales and distribution in the United States by:
Triumph Books, 601 South LaSalle, Suite 500, Chicago, Illinois 60605
312/939-3330; FAX 312/663-3557

International representatives:
Barkers Worldwide Publications, Unit 6/7 The Elms Centre, Glaziers Lane, Normandy, Guildford, Surrey GU3 2DF England
Tel: 011/441/483/811-971 and FAX: 011/441/483/811-972 e-mail: sales@bwpu.demon.co.uk website: www.bwpu.demon.co.uk

Dan Diamond and Associates books may be purchased for educational, business or sales promotional use.
For information please write to: Dan Diamond and Associates, 194 Dovercourt Road, Toronto, Ontario M6J 3C8 Canada
e-mail: dda.nhl@sympatico.ca

The National Hockey League
1251 Avenue of the Americas, 47th Floor, New York, New York 10020-1198
1800 McGill College Ave., Suite 2600, Montreal, Quebec H3A 3J6
50 Bay Street, 11th Floor, Toronto, Ontario M5J 2X8

Table of Contents

Table of Contents *continued*

(2002-03 NHL Schedule begins inside front cover)

Introduction

WELCOME TO *THE NHL OFFICIAL GUIDE & RECORD BOOK 2003.* This is the 71st edition of a book that has been published annually and without interruption since 1932-33, spanning almost two-thirds of the history of organized hockey and more than three-quarters of the history of the NHL. During these years, the National Hockey League has seen its megastars leave their imprint on the game. The achievements of every one of them, from Howie Morenz, who was called the "Babe Ruth of hockey," through Maurice Richard, Gordie Howe, Bobby Orr, Wayne Gretzky and Mario Lemieux, are recorded in these pages, as are the next level of star players such as Syl Apps, Terry Sawchuk, Jean Beliveau, Bobby and Brett Hull, Raymond Bourque, Patrick Roy and Steve Yzerman. In fact, no player, whether one-game wonder or multiple trophy winner and First Team All-Star, is omitted. All 5,500 forwards, defensemen and goaltenders who have played in the NHL since its inception in 1917-18 can be found herein. As a spry 71-year-old, *The NHL Official Guide & Record Book*'s task remains unchanged: to provide comprehensive statistical coverage of the National Hockey League, its players and top prospects in minor pro, European, junior or college leagues and conferences. Our readers range from general managers to scouts, reporters, broadcasters and, of greatest importance, fans. Every fall, books are delivered coast-to-coast in North America and overseas, both Atlantic and Pacific. Readers reward us with solid suggestions, intelligent questions and more than a few pieces of missing information. Your support is much appreciated.

The 2001-02 NHL season was a cascade of great stories and milestones achieved. The Detroit Red Wings were deserving champions, icing three 600-goal scorers (Hull, Yzerman and Luc Robitaille) and a fourth 500-goal man in Brendan Shanahan, multiple Norris Trophy winners on defense (Nicklas Lidstrom and Chris Chelios) and the game's only goaltender (Dominik Hasek) to win the Hart Trophy as the NHL's most valuable player more than once.

Other club's posted greatly improved records. Chicago, Montreal and the New York Islanders returned to postseason play after long absences. Carolina, a team that had missed the playoffs in seven of the past ten seasons, won the Prince of Wales Trophy and capably represented the Eastern Conference in the Stanley Cup Finals.

It was a season in which honors seemed to be distributed in pairs, beginning with Canada's double Olympic gold medals in men's and women's hockey in Salt Lake City. This trend continued at the League's end-of-season awards gala when captain Ron Francis of the Hurricanes won the Lady Byng Trophy for gentlemanly conduct and the King Clancy Trophy for leadership and contribution to his community. Nicklas Lidstrom won the Norris Trophy as the League's top defenseman to add to his Conn Smythe Trophy win as playoff MVP. Calgary right wing Jarome Iginla became the first player to win the Art Ross Trophy as the NHL's top scorer with 96 points and the Maurice Richard Trophy as its top goal-getter with 52. Montreal goaltender Jose Theodore won the Vezina as top goaltender and the Hart Trophy as the League's most valuable player. In both cases, Theodore finished tied for first place in award balloting, but was declared the winner on the basis of most first-place votes. Voting for an NHL individual award had never before yielded a tie; in Theodore's case it did so twice in a row. Both Iginla and Theodore were first-time NHL award winners. In addition, the Atlanta Thrashers featured two multi-talented rookie forwards, Dany Heatley, the top rookie of 2001-02, and Ilya Kovalchuk, the first player selected in the 2001 Entry Draft. The *Guide & Record Book*'s complete award coverage begins on page 201. The Entry Draft begins on page 209.

Other special features for 2003 include a review of the 2002 Olympic men's and women's hockey competition (page 13). Olympic participants are also noted in the 2002-03 player personnel panel in each club's four-page section. In addition, in response to reader suggestions, we have added goaltenders' scoring points and penalty minutes for 2001-02. These can be found in the 2001-02 scoring panel that is located on the second page of each club. Club pages begin with the Mighty Ducks of Anaheim on page 15. The Washington Capitals complete club coverage on page 134.

As first presented in 2002, overtime losses (abbreviated OL or, if space permits, OTL) are not included in a team's loss total. Therefore W+L+T+OTL=GP. In 1999-2000, the first season in which a loser in overtime received a point in the standings, this was not the case. The 1999-2000 standings (page 155) are presented using today's standard. Also note that overtime losses are only reflected in team statistics. Goaltender and coaching statistics do not include OTLs.

This edition's Player Register begins on page 269 with a Prospect Register made up of active forwards and defensemen who have yet to play in the NHL. Players in the Prospect Register either have been recently drafted, signed as free agents or invited to training camp by NHL clubs.

The NHL Player Register begins on page 338. It includes active forwards and defensemen who have appeared in an NHL regular-season or playoff game at any time. In addition to the standard GP-G-A-Pts-PIM, an NHLer's player panel includes the following statistical categories, listed from left to right as they appear in the book: power-play goals (PP), shorthand goals (SH), game-winning goals (GW), shots on goal (S), percentage of shots that score (%), plus-minus rating (+/–), total face-offs taken (TF*), face-off winning percentage (F%*), hits (H*), shots blocked (SB*) and average time-on-ice per game played (Min*). Categories marked with an asterisk (*) are NHL Real-Time statistics gathered by teams of trained spotters who, working with laptop computers and custom software, record hits, shots blocked, face-off wins, etc. "on-the-fly" at each game. These statistics were kept officially for the first time in 1998-99, so no player in this year's *NHL Guide* has more than four years of Real-Time statistics.

The order of the Registers is as follows: Prospect, NHL Player, Goaltender (page 576), Retired Player (600) and Retired Goaltender (632).

A key to the abbreviations and symbols used in individual player and goaltender data panels, along with useful information on how to use the Registers, is found on page 268. Late additions to the Registers are found on page 337 along with a list of abbreviations used for league names. Each NHL club's minor-pro affiliates are found on page 14.

As always, our thanks to readers, correspondents and members of the media who take the time to comment on the *Guide & Record Book*. Thanks as well to the people working in the communications departments of the NHL's member clubs and to their counterparts in minor pro, junior, college and European hockey.

Best wishes for an enjoyable 2002-03 NHL season.

ACCURACY REMAINS THE *GUIDE & RECORD BOOK*'S TOP PRIORITY.

We appreciate comments and clarification from our readers. Please direct these to:

- James Duplacey — Player Register Editor, 194 Dovercourt Road, Toronto, Ontario M6J 3C8. e-mail: jj.nhl@sympatico.ca.
- Greg Inglis — 47th floor, 1251 Avenue of the Americas, New York, New York 10020-1198 . . . or . . .
- David Keon — 50 Bay Street, 11th Floor, Toronto, Ontario, M5J 2X8

Your involvement makes a better book.

NATIONAL HOCKEY LEAGUE

Established November 22, 1917

New York, 1251 Avenue of the Americas, 47th Floor, New York, NY 10020-1198, 212/789-2000, Fax: 212/789-2020, PR Fax: 212/789-2080
Montréal, 1800 McGill College Avenue, Suite 2600, Montréal, Québec, H3A 3J6, 514/841-9220, Fax: 514/841-1070
Toronto, 50 Bay Street, 11th Floor, Toronto, Ontario, M5J 2X8, 416/981-2777, Fax: 416/981-2779
NHL Enterprises, L.P. — 1251 Avenue of the Americas, 47th Floor, New York, NY 10020-1198, 212/789-2000, Fax: 212/789-2020
NHL Enterprises Canada, L.P. — 50 Bay Street, 11th Floor, Toronto, Ontario, M5J 2X8, 416/981-2777, Fax: 416/981-2779
NHL Productions — 183 Oak Tree Road, Tappan, NY 10983-2809, 845/365-6701, Fax: 845/365-6010

Executive

Commissioner .... Gary B. Bettman
Executive Vice President & Chief Legal Officer .... William Daly
Executive Vice President & Director of Hockey Operations .... Colin Campbell
Executive Vice President & Chief Operating Officer .... Jon Litner
Executive Vice President & Chief Financial Officer .... Craig Harnett
Director, Administration & Executive Assistant to the Commissioner .... Debbie Jordan

Administration

Director of Administration .... Debbie Jordan
Director, Human Resources .... Janet Meyers
Director, Offices & Facilities .... Andrew Crawford
Manager, Human Resources .... Patrice Distler

Broadcasting/Scheduling

Vice President, Broadcasting & Programming .... Adam Acone
Director, Television Production & Technology .... Onnie Bose
Director, NHL Radio .... Brian G. Hamilton
Manager, Business & Special Events .... Phyllis DeCongilio
Manager, Television Production & Operations .... Stacie Watkins
Vice President, Scheduling, Operations & Research (Montreal) .... Steve Hatze Petros
Director, Research & Scheduling .... Mark Erlichson
Manager, Scheduling & Operations .... William Bredin

NHL Productions

Executive Producer .... Ken Rosen
Vice President .... Patti Fallick
Coordinating Producer .... Darryl Lepik
Director, Operations/Footage .... Peg Walsh
Producers .... Janice Arbour, Michele Giordano-Moore, Robert Lekhwani, Gary Waksman
Senior Editor .... Chip Swain
Associate Producer .... Nick Mascolo
Sr. Production Manager .... Christine Cortez
Manager, Video Services .... Chris Cesa

NHL Images
Director .... Anita Cechowski

Communications

Group Vice President, Communications .... Bernadette Mansur
Vice President, Media Relations .... Frank Brown
Vice President, Public Relations & Media Services (Toronto) .... Gary Meagher
Chief Statistician (Toronto) .... Benny Ercolani
Director, Communications .... Jamey Horan
Director, Community and Diversity Programming .... Ken Martin
Director, Media Relations .... Amy Early
Director, News Services .... Greg Inglis
Director, Player Publicity .... Sandra Carreon
Director, Youth Development, NHL Diversity .... Willie O'Ree
Senior Manager, Corporate Communications .... Brian Walker
Manager, Community Relations .... Ann Marie Lynch
Manager, NHL Diversity .... Nirva Milord
Manager, News Services .... Adam Schwartz
Managers, Public Relations (Toronto) .... David Keon, Chris Tredree, Julie Young

Events and Entertainment

Group Vice President .... Frank Supovitz
Vice President .... Ken Chin
Directors .... Sammy Choi, Katherine Krautter, Bill Miller
Senior Managers .... Susan Aglietti, Danny Frank, Dean Matsuzaki, Greta Palmer
Managers .... Eileen Murphy, Chie Sakuma

Finance

Executive Vice President & Chief Financial Officer .... Craig Harnett
Senior Vice President, Finance .... Joseph DeSousa
Vice President, Finance and Office Manager (Montreal) .... Olivia Pietrantonio
Director, Financial Systems .... Belinda Haeberlein
Director, Finance .... Lowell Heit
Corporate Controller .... Kenneth Cartisano

Hockey Operations

Executive Vice President & Director of Hockey Operations .... Colin Campbell
Senior Vice President, Hockey Operations (Toronto) .... Jim Gregory
Vice President, Hockey Operations (Toronto) .... Mike Murphy
Director of Officiating (Toronto) .... Andy VanHellemond
Associate Director of Hockey Operations .... Claude Loiselle
Consultant (Toronto) .... Kris King
Vice President & Managing Director, Central Registry (Toronto) .... Stephen Pellegrini
Assistant Director, Central Registry (Montreal) .... Madeleine Supino
Project Manager, Central Registry (Toronto) .... Sean MacLeod
Director, Central Scouting (Toronto) .... Frank Bonello
Director of Alumni Relations (Toronto) .... Patrick Flatley
Video Director .... Damian Echevarrieta
Video Technologies Consultant .... Jed Dole
Video Coordinator (Toronto) .... Paul Brighty
Facilities Operation Manager .... Dan Craig
Consultant (Toronto) .... Dave Dryden

Information Technology

Group Vice President, Information Technology .... Peter DelGiacco
Assistant Director (Montreal) .... Luc Coulombe
Senior Director .... Carol Dann
Director, Network Services .... Patrick Powers
Director, Technical Services .... John Ho
Manager, Technical Support .... Dan O'Neill

Legal

Executive Vice President & Chief Legal Officer .... William Daly
Senior Vice President, General Counsel .... David Zimmerman
Vice President, Deputy General Counsel .... Julie Grand
Associate Counsel .... Daniel Ages

Pension

Vice President and Managing Director, Pension (Montreal) .... Yvon Chamberland
Controller, Pension (Montreal) .... Mary Skiadopoulos
Manager, Pension (Montreal) .... Lise de Jocas

Security

Senior Vice President, Security .... Dennis Cunningham
Senior Director, Security .... Joseph Caporicci
Manager, Security .... Al Young

Television and Media Ventures

Senior Vice President, Television & Media Ventures .... Doug Perlman
Vice President, Television & Business Affairs .... Leslie Gittess
Director, Team Television & Business Affairs .... John Tortora
Director, NHL Center Ice & Program Development .... Ken Gelman
Manager, NHL Center Ice & Media Ventures .... Peter Aquilone
Manager, Television & Business Affairs .... Bridget DeMouy

NHL Interactive CyberEnterprises (NHL ICE)
President, NHL ICE & Senior Vice President, New Business Development .... Keith Ritter
Vice President, Editorial & Production .... Richard Libero
Vice President, Revenue & Administration .... Ken Nova
Director, Technology & Operations .... Grant Nodine

NHL Enterprises

President, NHL Enterprises .... Ed Horne

Consumer Products Marketing

Group Vice President, Consumer Products Marketing .... Brian Jennings
Vice President, Consumer Products Marketing (Toronto) .... Glenn Wakefield
Senior Director, Consumer Products Marketing .... James Haskins
Senior Director, Retail Sales & Marketing, Canada (Toronto) .... Barry Monaghan
Director, Center Ice Program and Sporting Goods .... Lloyd Haymes
Director, Consumer Products Marketing, Canada (Toronto) .... Karen Hanson
Director, Entertainment Products .... Dave McCarthy
Director, Non-Apparel .... Judith Salsberg
Director, Retail Sales & Marketing .... Cathy Groves
Manager, Youth Licensing .... Rachel Podradchik
Manager, Printed Products & Publishing .... Denise Gomez
Manager, Apparel, Non-Apparel, & Trade Shows .... John Gulla
Manager, Entertainment Products .... Linda M. Santiago
Manager, Center Ice & Sporting Goods .... Richard Villani

Club Marketing

Vice President, Club Marketing .... Scott Carmichael
Managers .... Tammy Levine, Maryann Thorgrimson

Corporate Marketing

Group Vice President, Corporate Marketing .... Andrew Judelson
Senior Director, Canada (Toronto) .... Laurie Kepron
Directors .... Eustace King, Susan Rosenfeld
Managers .... Jean Marie Cesare, David Levy, Chris Long, Lauren Ordower
Manager, Canada (Toronto) .... Jeff Rockwell

Creative Services

Associate Director, Creative Services .... Kathy Drew

Fan Development

Vice President, Fan Development .... Alysse Soll
Managers .... Felicia Sass, Suzanne Sherman

Finance

Vice President, Finance – NHL Enterprises .... Mary McCarthy
Director, Finance .... Scott Weinfeld
Director, Accounting Operations .... Deborah Corletta

International

Group Vice President & Managing Director, NHL International .... Ken Yaffe
Senior Director, International Business Operations .... Frank Nakano
Director, International Broadcasting .... Susanna Mandel-Mantello
Director, International Marketing .... Kamini Sharma
Director, International Licensing & Special Projects .... Lynn White
Manager, International Marketing & Special Projects .... Michael Rolnick

NHLE Legal and Business Affairs

Executive Vice President & General Counsel .... Richard Zahnd
Group Vice President & Associate General Counsel .... Mary Sotis
Vice President, Licensing and Trademark Compliance .... Ruth Gruhin
Vice President & Corporate Counsel .... Robert Hawkins
Senior Counsel .... Tom Prochnow
Associate Counsels .... Jason Camhi, Michael Gold, Matthew Kline
Staff Attorney .... Lisa Stancati
Director, Contract Administration .... Heather Atria
Director, Quality Control .... Catherine O'Brien
Senior Manager, Intellectual Property .... Alison Nunez

Strategic Development

Vice President, Strategic Development .... Susan Cohig

BOARD OF GOVERNORS

Chairman of the Board – Harley N. Hotchkiss

Mighty Ducks of Anaheim

Paul Pressler....Governor
Michael D. Eisner....Alternate Governor
Rick Schlesinger....Alternate Governor

Atlanta Thrashers

Stan Kasten....Governor
Don Waddell....Alternate Governor

Boston Bruins

Jeremy M. Jacobs....Governor
Louis Jacobs....Alternate Governor
Harry J. Sinden....Alternate Governor
Jeremy M. Jacobs, Jr.....Alternate Governor
Charles M. Jacobs....Alternate Governor
Mike O'Connell....Alternate Governor

Buffalo Sabres

Darcy Regier....Governor
Kevin Billet....Alternate Governor

Calgary Flames

Harley N. Hotchkiss....Governor
Ken King....Alternate Governor
N. Murray Edwards....Alternate Governor
Alvin Libin....Alternate Governor

Carolina Hurricanes

Peter Karmanos, Jr.....Governor
Jim Rutherford....Alternate Governor
Jason Karmanos....Alternate Governor
Jim Cain....Alternate Governor

Chicago Blackhawks

William W. Wirtz....Governor
John A. Ziegler, Jr.....Alternate Governor
Robert J. Pulford....Alternate Governor
Peter R. Wirtz....Alternate Governor

Colorado Avalanche

E. Stanley Kroenke....Governor
Donald M. Elliman, Jr.....Alternate Governor
Pierre Lacroix....Alternate Governor

Columbus Blue Jackets

John H. McConnell....Governor
John P. McConnell....Alternate Governor
John S. Christie....Alternate Governor
Doug MacLean....Alternate Governor

Dallas Stars

Thomas O. Hicks....Governor
Doug Armstrong....Alternate Governor
Mike Cramer....Alternate Governor

Detroit Red Wings

Michael Ilitch....Governor
Jay A. Bielfield....Alternate Governor
Jim Devellano....Alternate Governor
Christopher Ilitch....Alternate Governor
Denise Ilitch....Alternate Governor
Ken Holland....Alternate Governor

Edmonton Oilers

Cal Nichols....Governor
Kevin Lowe....Alternate Governor
Gordon Buchanan....Alternate Governor
Patrick R. LaForge....Alternate Governor

Florida Panthers

William A. Torrey....Governor
Alan Cohen....Alternate Governor
Jordan Zimmerman....Alternate Governor
Jeff Cogen....Alternate Governor
Steven Cohen....Alternate Governor

Los Angeles Kings

Timothy J. Leiweke....Governor
Philip F. Anschutz....Alternate Governor
David Taylor....Alternate Governor

Minnesota Wild

Robert O. Naegele, Jr.....Governor
Jac Sperling....Alternate Governor
Tod Leiweke....Alternate Governor
Doug Risebrough....Alternate Governor

Montréal Canadiens

George N. Gillett, Jr.....Governor
Pierre Boivin....Alternate Governor
Fred Steer....Alternate Governor
Jeff Joyce....Alternate Governor
Foster Gillett....Alternate Governor

Nashville Predators

Craig Leipold....Governor
David Poile....Alternate Governor
Jack Diller....Alternate Governor

New Jersey Devils

Lou Lamoriello....Governor
Michael Gilfillan....Alternate Governor

New York Islanders

Charles B. Wang....Governor
Sanjay Kumar....Alternate Governor
Michael J. Picker....Alternate Governor
Roy E. Reichbach....Alternate Governor

New York Rangers

James L. Dolan....Governor
Glen Sather....Alternate Governor
Steve Mills....Alternate Governor

Ottawa Senators

Roderick M. Bryden....Governor
Roy Mlakar....Alternate Governor

Philadelphia Flyers

Edward M. Snider....Governor
Bob Clarke....Alternate Governor
Ronald K. Ryan....Alternate Governor
Philip I. Weinberg....Alternate Governor

Phoenix Coyotes

Steve Ellman....Governor
Wayne Gretzky....Alternate Governor
Mike Barnett....Alternate Governor
James R. Lites....Alternate Governor

Pittsburgh Penguins

Kenneth Sawyer....Governor
Craig Patrick....Alternate Governor
Thomas J. Rooney....Alternate Governor
Ronald Burkle....Alternate Governor
Anthony Liberati....Alternate Governor

St. Louis Blues

William J. Laurie....Governor
Richard C. Thomas....Alternate Governor
Mark Sauer....Alternate Governor
Larry Pleau....Alternate Governor
Brent P. Karasiuk....Alternate Governor

San Jose Sharks

Greg Jamison....Governor
Dean Lombardi....Alternate Governor
Kevin Compton....Alternate Governor

Tampa Bay Lightning

Thomas S. Wilson....Governor
Ronald J. Campbell....Alternate Governor
Jay H. Feaster....Alternate Governor

Toronto Maple Leafs

Steve A. Stavro....Governor
Brian P. Bellmore....Alternate Governor
Ken Dryden....Alternate Governor
Richard Peddie....Alternate Governor

Vancouver Canucks

John E. McCaw, Jr.....Governor
Stanley B. McCammon....Alternate Governor
Brian P. Burke....Alternate Governor
David M. Nonis....Alternate Governor
David Cobb....Alternate Governor

Washington Capitals

Richard M. Patrick....Governor
Ted Leonsis....Alternate Governor
George McPhee....Alternate Governor

Commissioner and League Presidents

Gary B. Bettman

Gary B. Bettman took office as the NHL's first Commissioner on February 1, 1993. Since the League was formed in 1917, there have been five League Presidents.

NHL President	Years in Office
Frank Calder	1917-1943
Mervyn "Red" Dutton	1943-1946
Clarence Campbell	1946-1977
John A. Ziegler, Jr.	1977-1992
Gil Stein	1992-1993

Hockey Hall of Fame

BCE Place
30 Yonge Street
Toronto, Ontario M5E 1X8
Phone: 416/360-7735
Executive Fax: 416/360-1501
Resource Center/Retail Fax: 416/360-1316
www.hhof.com

Bill Hay – Chairman and Chief Executive Officer
Jeff Denomme – President, Chief Operating Officer and Treasurer
Craig Baines – Director, Marketing and Facilities Services
Ron Ellis – Director, Public Affairs and Assistant to the President
Ray Paquet – Creative Director, Exhibit Development
Phil Pritchard – Director, Hockey Operations and Curator
Craig Campbell – Manager, Resource Center and Archives
Peter Jagla – Producer, New Media and E-Business
Jan Barrina – Manager, Special Events and Hospitality
Kelly Massé – Manager, Corporate and Media Relations
Craig Beckim – Manager, Merchandising and Retail Operations
Jackie Boughazaje – Manager, Promotions and Attractions Services
Anthony Fusco – Manager, Information Systems
Sandra Walters – Controller and Office Manager
Pearl Rajwanth – Executive Assistant to the President

National Hockey League Players' Association

777 Bay Street, Suite 2400
Toronto, Ontario M5G 2C8
Phone: 416/313-2300
Fax: 416/313-2301
www.nhlpa.com

Robert W. Goodenow – Executive Director and General Counsel
Ted Saskin – Senior Director, Business Affairs and Licensing
Mike Gartner – Director, Business Relations
Kenneth Kim – Director, Marketing
Ian Pulver, Ian Penny, Roland Lee – Associate Counsel, Labour
Mike Ouellet – Associate Counsel, Licensing
Eric Weisz – Manager, Licensing and International Business
Steve Larmer – Player Relations
Greg Dick – Senior Manager, Finance and Business Administration
Kim Murdoch – Manager, Pensions and Benefits
Devin Smith – Program Manager, Goals & Dreams Fund
Dave Tredgett – Executive Producer-Television
Jonathan Weatherdon – Media Relations

NHL On-Ice Officials

Total NHL Games and 2001-02 Games columns count regular-season games only.

Referees

#	Name	Birthplace	Birthdate	First NHL Game	Total NHL Games	2001-02 Games
9	Blaine Angus	Shawville, Que.	9/25/61	10/17/92	319	72
15	Stephane Auger	Montreal, Que.	12/9/70	4/1/00	98	71
10	Paul Devorski	Guelph, Ont.	8/18/58	10/14/89	751	74
44	Harry Dumas	Mount Laurel, N.J.	7/7/73	12/27/00	8	7
11	Mark Faucette	Springfield, MA	6/9/58	12/23/87	846	72
2	Kerry Fraser	Sarnia, Ont.	5/30/52	4/6/75	1407	72
27	Eric Furlatt	Cap de la Madelaine, Que.	12/2/71	10/8/01	30	30
4	Terry Gregson	Erin, Ont.	11/7/53	12/19/81	1290	72
30	Mike Hasenfratz	Regina, Sask.	7/19/66	10/21/00	106	72
17	Shane Heyer	Summerland, B.C.	2/7/64	**10/1/99	*955	73
46	Scott Hoberg	Windsor, Ont.	1/23/71			0
8	Dave Jackson	Montreal, Que.	11/28/64	12/23/90	610	72
25	Marc Joannette	Verdun, Que.	11/3/68	10/27/99	145	73
18	Greg Kimmerly	Toronto, Ont.	12/8/64	11/30/96	232	72
12	Don Koharski	Halifax, N.S.	12/2/55	10/14/77	*1435	73
48	Tom Kowal	Vernon, B.C.	11/2/67	10/29/99	140	69
37	Bob Langdon	Woodstock, Ont.	3/11/71	11/11/01	10	10
14	Dennis LaRue	Savannah, GA	7/14/59	3/26/91	433	71
28	Chris Lee	Saint John, N.B.	7/7/70	4/2/00	22	17
3	Mike Leggo	North Bay, Ont.	10/7/64	3/3/98	227	73
6	Dan Marouelli	Edmonton, Alta.	7/16/55	11/2/84	1122	61
26	Rob Martell	Winnipeg, Man.	10/21/63	3/14/84	*145	68
41	Wes McCauley	Georgetown, Ont.	1/11/72			0
7	Bill McCreary	Guelph, Ont.	11/17/55	11/3/84	1161	73
19	Mick McGeough	Regina, Sask.	6/20/57	1/19/89	739	74
34	Brad Meier	Dayton, OH	4/11/67	10/23/99	150	73
36	Dean Morton	Peterborough, Ont.	2/27/68	11/11/00	1	0
13	Dan O'Halloran	Essex, Ont.	3/25/64	10/14/95	238	8
42	Dan O'Rourke	Calgary, Alta.	8/31/72	10/2/99	*120	0
20	Tim Peel	Toronto, Ont.	4/27/66	10/21/99	154	73
43	Brian Pochmara	Detroit, MI	11/27/76			0
33	Kevin Pollock	Kincardine, Ont.	2/7/70	3/28/00	151	74
21	Chris Rooney	Boston, MA	5/26/74	11/22/00	53	40
40	Jay Sharrers	Jamaica, West Indies	7/3/67	**4/3/01	*672	29
16	Rob Shick	Port Alberni, B.C.	12/4/57	4/6/86	930	72
49	Jeff Smith	Hamilton, Ont.	9/2/69			0
38	Craig Spada	Welland, Ont.	9/7/71	3/28/02	6	6
22	Paul Stewart	Boston, MA	3/21/55	3/27/87	939	62
31	Kelly Sutherland	Victoria, B.C.	4/18/71	12/19/00	67	67
5	Don Van Massenhoven	London, Ont.	7/17/60	11/11/93	548	72
24	Stephen Walkom	North Bay, Ont.	8/8/63	10/18/92	549	73
45	Ian Walsh	Philadelphia, PA	5/9/72	10/14/00	27	27
35	Dean Warren	Toronto, Ont.	7/22/63	10/8/99	149	72
23	Brad Watson	Regina, Sask.	10/4/61	2/5/94	252	73
29	Scott Zelkin	Wilmette, IL	9/12/68	4/13/97	236	72

* Includes some games worked as a linesman. ** First game as an NHL referee. Previously worked as a linesman.

Linesmen

#	Name	Birthplace	Birthdate	First NHL Game	Total NHL Games	2001-02 Games
75	Derek Amell	Port Colborne, Ont.	9/16/68	10/13/97	290	75
59	Steve Barton	Ottawa, Ont.	12/27/71	11/1/00	71	48
94	Wayne Bonney	Ottawa, Ont.	5/27/53	10/10/79	1598	71
96	David Brisebois	Sudbury, Ont.	4/14/76	10/11/99	116	46
74	Lonnie Cameron	Victoria, B.C.	7/15/64	10/5/96	404	74
67	Pierre Champoux	Ville St-Pierre, Que.	4/18/63	10/8/88	874	42
50	Kevin Collins	Springfield, MA	12/15/50	10/13/77	1828	0
88	Mike Cvik	Calgary, Alta.	7/6/62	10/8/87	1001	74
83	Angelo D'Amico	Etobicoke, Ont.	5/29/74	11/27/00	51	43
60	Pat Dapuzzo	Hoboken, NJ	12/29/58	12/5/84	1272	73
54	Greg Devorski	Guelph, Ont.	8/3/69	10/9/93	577	74
68	Scott Driscoll	Seaforth, Ont.	5/2/68	10/10/92	646	73
82	Ryan Galloway	Winnipeg, Man.	7/12/72			0
63	Gerard Gauthier	Montreal, Que.	9/5/48	10/16/71	2278	72
66	Darren Gibbs	Edmonton, Alta.	9/30/66	10/1/97	274	44
91	Don Henderson	Calgary, Alta.	9/23/68	3/10/95	396	74
71	Brad Kovachik	Woodstock, Ont.	3/7/71	10/10/96	379	74
86	Brad Lazarowich	Vancouver, B.C.	8/4/62	10/9/86	1087	74
78	Brian Mach	Little Falls, MN	4/15/74	10/7/00	137	74
51	Dan McCourt	Falconbridge, Ont.	8/14/54	12/27/80	1488	74
90	Andy McElman	Chicago Heights, IL	8/4/61	10/7/93	579	74
89	Steve Miller	Stratford, Ont.	6/22/72	10/7/00	134	69
98	Randy Mitton	Fredericton, N.B.	9/22/50	2/2/74	1973	74
97	Jean Morin	Sorel, Que.	8/10/63	10/5/91	702	73
93	Brian Murphy	Dover, NH	12/13/64	10/7/88	*875	73
95	Jonny Murray	Beauport, Que.	8/10/74	10/7/00	139	73
70	Derek Nansen	Ottawa, Ont.	12/6/71			0
80	Thor Nelson	Westminister, CA	1/6/68	2/16/95	313	74
77	Tim Nowak	Buffalo, NY	9/6/67	10/8/93	589	74
79	Mark Paré	Windsor, Ont.	7/26/57	10/11/79	1684	74
72	Stephane Provost	Montreal, Que.	5/5/67	1/25/95	557	74
65	Pierre Racicot	Verdun, Que.	2/15/67	10/12/93	614	74
73	Vaughan Rody	Winnipeg, Man.	12/13/68	10/8/00	144	74
81	Troy Sartison	Swift Current, Sask.	2/25/70	10/6/99	195	72
53	Ray Scapinello	Guelph, Ont.	11/5/46	10/17/71	2366	74
52	Dan Schachte	Madison, WI	7/13/58	10/6/82	1398	74
61	Lyle Seitz	Brooks, Alta.	1/22/69	10/6/92	*268	24
84	Anthony Sericolo	Troy, NY	7/17/68	10/21/98	235	74
56	Mark Wheler	North Battleford, Sask.	9/20/65	10/10/92	669	74

* Includes some games worked as a referee.

NHL History

1917 — National Hockey League organized November 22 in Montreal following suspension of operations by the National Hockey Association of Canada Limited (NHA). Montreal Canadiens, Montreal Wanderers, Ottawa Senators and Quebec Bulldogs attended founding meeting. Delegates decided to use NHA rules.

Toronto Arenas were later admitted as fifth team; Quebec decided not to operate during the first season. Quebec players allocated to remaining four teams.

Frank Calder elected president and secretary-treasurer.

First NHL games played December 19, with Toronto only arena with artificial ice. Clubs played 22-game split schedule.

1918 — Emergency meeting held January 3 due to destruction by fire of Montreal Arena which was home ice for both Canadiens and Wanderers.

Wanderers withdrew, reducing the NHL to three teams; Canadiens played remaining home games at 3,250-seat Jubilee rink.

Quebec franchise sold to P.J. Quinn of Toronto on October 18 on the condition that the team operate in Quebec City for 1918-19 season. Quinn did not attend the November League meeting and Quebec did not play in 1918-19.

1919-20 — NHL reactivated Quebec Bulldogs franchise. Former Quebec players returned to the club. New Mount Royal Arena became home of Canadiens. Toronto Arenas changed name to St. Patricks. Clubs played 24-game split schedule.

1920-21 — H.P. Thompson of Hamilton, Ontario made application for the purchase of an NHL franchise. Quebec franchise shifted to Hamilton with other NHL teams providing players to strengthen the club.

1921-22 — Split schedule abandoned. First and second place teams at the end of full schedule to play for championship.

1922-23 — Clubs agreed that players could not be sold or traded to clubs in any other league without first being offered to all other clubs in the NHL. In March, Foster Hewitt broadcasts radio's first hockey game.

1923-24 — Ottawa's new 10,000-seat arena opened. First U.S. franchise granted to Boston for following season.

Dr. Cecil Hart Trophy donated to NHL to be awarded to the player judged most useful to his team.

1924-25 — Canadian Arena Company of Montreal granted a franchise to operate Montreal Maroons. NHL now six team league with two clubs in Montreal. Inaugural game in new Montreal Forum played November 29, 1924 as Canadiens defeated Toronto 7-1. Forum was home rink for the Maroons, but no ice was available in the Canadiens arena November 29, resulting in shift to Forum.

Hamilton finished first in the standings, receiving a bye into the finals. But Hamilton players, demanding $200 each for additional games in the playoffs, went on strike. The NHL suspended all players, fining them $200 each. Stanley Cup finalist to be the winner of NHL semi-final between Toronto and Canadiens.

Prince of Wales and Lady Byng trophies donated to NHL.

Clubs played 30-game schedule.

1925-26 — Hamilton club dropped from NHL. Players signed by new New York Americans franchise. Pittsburgh Pirates granted franchise.

Clubs played 36-game schedule.

1926-27 — New York Rangers granted franchise May 15, 1926. Chicago Black Hawks and Detroit Cougars granted franchises September 25, 1926. NHL now ten-team league with an American and a Canadian Division.

Stanley Cup came under the control of NHL. In previous seasons, winners of the now-defunct Western or Pacific Coast leagues would play NHL champion in Cup finals.

Toronto franchise sold to a new company controlled by Hugh Aird and Conn Smythe. Name changed from St. Patricks to Maple Leafs.

Clubs played 44-game schedule.

The Montreal Canadiens donated the Vezina Trophy to be awarded to the team allowing the fewest goals-against in regular season play. The winning team would, in turn, present the trophy to the goaltender playing in the greatest number of games during the season.

1930-31 — Detroit franchise changed name from Cougars to Falcons. Pittsburgh transferred to Philadelphia for one season. Pirates changed name to Philadelphia Quakers. Trading deadline for teams set at February 15 of each year. NHL approved operation of farm teams by Rangers, Americans, Falcons and Bruins. Four-sided electric arena clock first demonstrated.

1931-32 — Philadelphia dropped out. Ottawa withdrew for one season. New Maple Leaf Gardens completed.

Clubs played 48-game schedule

1932-33 — Detroit franchise changed name from Falcons to Red Wings. Franchise application received from St. Louis but refused because of additional travel costs. Ottawa team resumed play.

1933-34 — First All-Star Game played as a benefit for injured player Ace Bailey. Leafs defeated All-Stars 7-3 in Toronto.

1934-35 — Ottawa franchise transferred to St. Louis. Team called St. Louis Eagles and consisted largely of Ottawa's players.

1935-36 — Ottawa-St. Louis franchise terminated. Montreal Canadiens finished season with very poor record. To strengthen the club, NHL gave Canadiens first call on the services of all French-Canadian players for three seasons.

1937-38 — Second benefit All-Star game staged November 2 in Montreal in aid of the family of the late Canadiens star Howie Morenz.

Montreal Maroons withdrew from the NHL on June 22, 1938, leaving seven clubs in the League.

1938-39 — Expenses for each club regulated at $5 per man per day for meals and $2.50 per man per day for accommodation.

1939-40 — Benefit All-Star Game played October 29, 1939 in Montreal for the children of the late Albert (Babe) Siebert.

1940-41 — Ross-Tyer puck adopted as the official puck of the NHL. Early in the season it was apparent that this puck was too soft. The Spalding puck was adopted in its place.

On May 16, 1941, Arthur Ross, NHL governor from Boston, donated a perpetual trophy to be awarded annually to the player voted outstanding in the league. Due to wartime restrictions, the trophy was never awarded.

1941-42 — New York Americans changed name to Brooklyn Americans.

1942-43 — Brooklyn Americans withdrew from NHL, leaving six teams: Boston, Chicago, Detroit, Montreal, New York and Toronto. Playoff format saw first-place team play third-place team and second play fourth.

Clubs played 50-game schedule.

Frank Calder, president of the NHL since its inception, died in Montreal. Meryn "Red" Dutton, former manager of the New York Americans, became president. The NHL commissioned the Calder Memorial Trophy to be awarded to the League's outstanding rookie each year.

1945-46 — Philadelphia, Los Angeles and San Francisco applied for NHL franchises.

The Philadelphia Arena Company of the American Hockey League applied for an injunction to prevent the possible operation of an NHL franchise in that city.

1946-47 — Mervyn Dutton retired as president of the NHL prior to the start of the season. He was succeeded by Clarence S. Campbell.

Individual trophy winners and all-star team members to receive $1,000 awards.

Playoff guarantees for players introduced.

Clubs played 60-game schedule.

1947-48 — The first annual All-Star Game for the benefit of the players' pension fund was played when the All-Stars defeated the Stanley Cup Champion Toronto Maple Leafs 4-3 in Toronto on October 13, 1947.

Criteria for awarding Art Ross Trophy changed. Now awarded to top scorer. Elmer Lach was its first winner.

Philadelphia and Los Angeles franchise applications refused.

National Hockey League Pension Society formed.

1949-50 — Clubs played 70-game schedule.

First intra-league draft held April 30, 1950. Clubs allowed to protect 30 players. Remaining players available for $25,000 each.

1951-52 — Referees included in the League's pension plan.

1952-53 — In May of 1952, City of Cleveland applied for NHL franchise. Application denied. In March of 1953, the Cleveland Barons of the AHL challenged the NHL champions for the Stanley Cup. The NHL governors did not accept this challenge.

1953-54 — The James Norris Memorial Trophy presented to the NHL for annual presentation to the League's best defenseman.

Intra-league draft rules amended to allow teams to protect 18 skaters and two goaltenders, claiming price reduced to $15,000.

1954-55 — Each arena to operate an "out-of-town" scoreboard. Referees and linesmen to wear shirts of black and white vertical stripes.

1956-57 — Standardized signals for referees and linesmen introduced.

1960-61 — Canadian National Exhibition, City of Toronto and NHL reach agreement for the construction of a Hockey Hall of Fame on the CNE grounds. Hall opens on August 26, 1961.

1963-64 — Player development league established with clubs operated by NHL franchises located in Minneapolis, St. Paul, Indianapolis, Omaha and, beginning in 1964-65, Tulsa. First universal amateur draft took place. All players of qualifying age (17) unaffected by sponsorship of junior teams available to be drafted.

1964-65 — Conn Smythe Trophy presented to the NHL to be awarded annually to the outstanding player in the Stanley Cup playoffs.

Minimum age of players subject to amateur draft changed to 18.

1965-66 — NHL announced expansion plans for a second six-team division to begin play in 1967-68.

1966-67 — Fourteen applications for NHL franchises received.

Lester Patrick Trophy presented to the NHL to be awarded annually for outstanding service to hockey in the United States.

NHL sponsorship of junior teams ceased, making all players of qualifying age not already on NHL-sponsored lists eligible for the amateur draft.

1967-68 — Six new teams added: California Seals, Los Angeles Kings, Minnesota North Stars, Philadelphia Flyers, Pittsburgh Penguins, St. Louis Blues. New teams to play in West Division. Remaining six teams to play in East Division.

Minimum age of players subject to amateur draft changed to 20.

Clubs played 74-game schedule.

Clarence S. Campbell Trophy awarded to team finishing the regular season in first place in West Division.

California Seals change name to Oakland Seals on December 8, 1967.

1968-69 — Clubs played 76-game schedule.

Amateur draft expanded to cover any amateur player of qualifying age throughout the world.

1970-71 — Two new teams added: Buffalo Sabres and Vancouver Canucks. These teams joined East Division: Chicago switched to West Division. Oakland Seals change name to California Golden Seals prior to season.

Clubs played 78-game schedule.

1971-72 — Playoff format amended. In each division, first to play fourth; second to play third.

1972-73 — Soviet Nationals and Canadian NHL stars play eight-game pre-season series. Canadians win 4-3-1.

Two new teams added. Atlanta Flames join West Division; New York Islanders join East Division.

1974-75 — Two new teams added: Kansas City Scouts and Washington Capitals. Teams realigned into two nine-team conferences, the Prince of Wales made up of the Norris and Adams Divisions, and the Clarence Campbell made up of the Smythe and Patrick Divisions.

Clubs played 80-game schedule.

1976-77 — California franchise transferred to Cleveland. Team named Cleveland Barons. Kansas City franchise transferred to Denver. Team named Colorado Rockies.

1977-78 — Clarence S. Campbell retires as NHL president. Succeeded by John A. Ziegler, Jr.

1978-79 — Cleveland and Minnesota franchises merge, leaving NHL with 17 teams. Merged team placed in Adams Division, playing home games in Minnesota.

Minimum age of players subject to amateur draft changed to 19.

1979-80 — Four new teams added: Edmonton Oilers, Hartford Whalers, Quebec Nordiques and Winnipeg Jets.

Minimum age of players subject to entry draft changed to 18.

1980-81 — Atlanta franchise shifted to Calgary, retaining "Flames" name.

1981-82 — Teams realigned within existing divisions. New groupings based on geographical areas. Unbalanced schedule adopted.

1982-83 — Colorado Rockies franchise shifted to East Rutherford, New Jersey. Team named New Jersey Devils. Franchise moved to Patrick Division from Smythe; Winnipeg moved to Smythe Division from Norris.

NHL History — *continued*

1991-92 — San Jose Sharks added, making the NHL a 22-team league. NHL celebrates 75th Anniversary Season. The 1991-92 regular season suspended due to a strike by members of the NHL Players' Association on April 1, 1992. Play resumed April 12, 1992.

1992-93 — Gil Stein named NHL president (October, 1992). Gary Bettman named first NHL Commissioner (February, 1993). Ottawa Senators and Tampa Bay Lightning added, making the NHL a 24-team league. NHL celebrates Stanley Cup Centennial. Clubs played 84-game schedule.

1993-94 — Mighty Ducks of Anaheim and Florida Panthers added, making the NHL a 26-team league. Minnesota franchise shifted to Dallas, team named Dallas Stars. Prince of Wales and Clarence Campbell Conferences renamed Eastern and Western. Adams, Patrick, Norris and Smythe Divisions renamed Northeast, Atlantic, Central and Pacific. Winnipeg moved to Central Division from Pacific; Tampa Bay moved to Atlantic Division from Central; Pittsburgh moved to Northeast Division from Atlantic.

1994-95 — A labor disruption forced the cancellation of 468 games from October 1, 1994 to January 19, 1995. Clubs played a 48-game schedule that began January 20, 1995 and ended May 3, 1995. No inter-conference games were played.

1995-96 — Quebec franchise transferred to Denver. Team named Colorado Avalanche and placed in Pacific Division of Western Conference. Clubs to play 82-game schedule.

1996-97 — Winnipeg franchise transferred to Phoenix. Team named Phoenix Coyotes and placed in Central Division of Western Conference.

1997-98 — Hartford franchise transferred to Raleigh. Team named Carolina Hurricanes and remains in Northeast Division of Eastern Conference.

1998-99 — The addition of the Nashville Predators made the NHL a 27-team league and brought about the creation of two new divisions and a League-wide realignment in preparation for further expansion to 30 teams by 2000-2001. Nashville was added to the Central Division of the Western Conference, while Toronto moved into the Northeast Division of the Eastern Conference. Pittsburgh was shifted from the Northeast to the Atlantic, while Carolina left the Northeast for the newly created Southeast Division of the Eastern Conference. Florida, Tampa Bay and Washington also joined the Southeast. In the Western Conference, Calgary, Colorado, Edmonton and Vancouver make up the new Northwest Division. Dallas and Phoenix moved from the Central to the Pacific Division.

The NHL retired uniform number 99 in honor of all-time scoring leader Wayne Gretzky who retired at the end of the season.

1999-2000 — Atlanta Thrashers added, making the NHL a 28-team league.

2000-01 — Columbus Blue Jackets and Minnesota Wild added, making the NHL a 30-team league.

NHL Attendance

Season	Regular Season Games	Regular Season Attendance	Playoffs Games	Playoffs Attendance	Total Attendance
1960-61	210	2,317,142	17	242,000	2,559,142
1961-62	210	2,435,424	18	277,000	2,712,424
1962-63	210	2,590,574	16	220,906	2,811,480
1963-64	210	2,732,642	21	309,149	3,041,791
1964-65	210	2,822,635	20	303,859	3,126,494
1965-66	210	2,941,164	16	249,000	3,190,184
1966-67	210	3,084,759	16	248,336	3,333,095
1967-68[1]	444	4,938,043	40	495,089	5,433,132
1968-69	456	5,550,613	33	431,739	5,982,352
1969-70	456	5,992,065	34	461,694	6,453,759
1970-71[2]	546	7,257,677	43	707,633	7,965,310
1971-72	546	7,609,368	36	582,666	8,192,034
1972-73[3]	624	8,575,651	38	624,637	9,200,288
1973-74	624	8,640,978	38	600,442	9,241,420
1974-75[4]	720	9,521,536	51	784,181	10,305,717
1975-76	720	9,103,761	48	726,279	9,830,040
1976-77	720	8,563,890	44	646,279	9,210,169
1977-78	720	8,526,564	45	686,634	9,213,198
1978-79	680	7,758,053	45	694,521	8,452,574
1979-80[5]	840	10,533,623	63	976,699	11,510,322
1980-81	840	10,726,198	68	966,390	11,692,588
1981-82	840	10,710,894	71	1,058,948	11,769,842
1982-83	840	11,020,610	66	1,088,222	12,028,832
1983-84	840	11,359,386	70	1,107,400	12,466,786
1984-85	840	11,633,730	70	1,107,500	12,741,230
1985-86	840	11,621,000	72	1,152,503	12,773,503
1986-87	840	11,855,880	87	1,383,967	13,239,847
1987-88	840	12,117,512	83	1,336,901	13,454,413
1988-89	840	12,417,969	83	1,327,214	13,745,183
1989-90	840	12,579,651	85	1,355,593	13,935,244
1990-91	840	12,343,897	92	1,442,203	13,786,100
1991-92[6]	880	12,769,676	86	1,327,920	14,097,596
1992-93[7]	1,008	14,158,177[8]	83	1,346,034	15,504,211
1993-94[9]	1,092	16,105,604[10]	90	1,440,095	17,545,699
1994-95	624[11]	9,233,884	81	1,329,130	10,563,014
1995-96	1,066	17,041,614	86	1,540,140	18,581,754
1996-97	1,066	17,640,529	82	1,494,878	19,135,407
1997-98	1,066	17,264,678	82	1,507,416	18,772,094
1998-99[12]	1,107	18,001,741	86	1,509,411	19,511,152
1999-2000[13]	1,148	18,800,139	83	1,524,629	20,324,768
2000-01[14]	1,230	20,373,379	86	1,584,011	21,957,390
2001-02	1,230	20,614,613	90	1,691,174	22,305,787

[1] First expansion: Los Angeles, Pittsburgh, California (Cleveland),Philadelphia, St. Louis and Minnesota (Dallas)
[2] Second expansion: Buffalo and Vancouver
[3] Third expansion: Atlanta (Calgary) and New York Islanders
[4] Fourth expansion: Kansas City (Colorado, New Jersey) and Washington
[5] Fifth expansion: Edmonton, Hartford, Quebec (Colorado) and Winnipeg
[6] Sixth expansion: San Jose
[7] Seventh expansion: Ottawa and Tampa Bay
[8] Includes 24 neutral site games
[9] Eighth expansion: Anaheim and Florida
[10] Includes 26 neutral site games
[11] Lockout resulted in the cancellation of 468 regular-season games.
[12] Ninth expansion: Nashville
[13] Tenth expansion: Atlanta
[14] Eleventh expansion: Columbus and Minnesota

Major Rule Changes

1910-11 — Game changed from two 30-minute periods to three 20-minute periods.

1911-12 — National Hockey Association (forerunner of the NHL) originated six-man hockey, replacing seven-man game.

1917-18 — Goalies permitted to fall to the ice to make saves. Previously a goaltender was penalized for dropping to the ice.

1918-19 — Penalty rules amended. For minor fouls, substitutes not allowed until penalized player had served three minutes. For major fouls, no substitutes for five minutes. For match fouls, no substitutes allowed for the remainder of the game.

With the addition of two lines painted on the ice twenty feet from center, three playing zones were created, producing a forty-foot neutral center ice area in which forward passing was permitted. Kicking the puck was permitted in this neutral zone.

Tabulation of assists began.

1921-22 — Goaltenders allowed to pass the puck forward up to their own blue line.

Overtime limited to twenty minutes.

Minor penalties changed from three minutes to two minutes.

1923-24 — Match foul defined as actions deliberately injuring or disabling an opponent. For such actions, a player was fined not less than $50 and ruled off the ice for the balance of the game. A player assessed a match penalty may be replaced by a substitute at the end of 20 minutes. Match penalty recipients must meet with the League president who can assess additional punishment.

1925-26 — Delayed penalty rules introduced. Each team must have a minimum of four players on the ice at all times.

Two rules were amended to encourage offense: No more than two defensemen permitted to remain inside a team's own blue line when the puck has left the defensive zone. A faceoff to be called for ragging the puck unless short-handed.

Team captains only players allowed to talk to referees.

Goaltender's leg pads limited to 12-inch width.

Timekeeper's gong to mark end of periods rather than referee's whistle. Teams to dress a maximum of 12 players for each game from a roster of no more than 14 players.

1926-27 — Blue lines repositioned to sixty feet from each goal-line, thereby enlarging the neutral zone and standardizing distance from blueline to goal.

Uniform goal nets adopted throughout NHL with goal posts securely fastened to the ice.

1927-28 — To further encourage offense, forward passes allowed in defending and neutral zones and goaltender's pads reduced in width from 12 to 10 inches.

Game standardized at three twenty-minute periods of stop-time separated by ten-minute intermissions.

Teams to change ends after each period.

Ten minutes of sudden-death overtime to be played if the score is tied after regulation time.

Minor penalty to be assessed to any player other than a goaltender for deliberately picking up the puck while it is in play. Minor penalty to be assessed for deliberately shooting the puck out of play.

The Art Ross goal net adopted as the official net of the NHL.

Maximum length of hockey sticks limited to 53 inches measured from heel of blade to end of handle. No minimum length stipulated.

Home teams given choice of end to defend at start of game.

1928-29 — Forward passing permitted in defensive and neutral zones and into attacking zone if pass receiver is in neutral zone when pass is made. No forward passing allowed inside attacking zone.

Minor penalty to be assessed to any player who delays the game by passing the puck back into his defensive zone.

Ten-minute overtime without sudden-death provision to be played in games tied after regulation time. Games tied after this overtime period declared a draw.

Exclusive of goaltenders, team to dress at least 8 and no more than 12 skaters.

Major Rule Changes — *continued*

1929-30 — Forward passing permitted inside all three zones but not permitted across either blue line.

Kicking the puck allowed, but a goal cannot be scored by kicking the puck in.

No more than three players including the goaltender may remain in their defensive zone when the puck has gone up ice. Minor penalties to be assessed for the first two violations of this rule in a game; major penalties thereafter.

Goaltenders forbidden to hold the puck. Pucks caught must be cleared immediately. For infringement of this rule, a faceoff to be taken ten feet in front of the goal with no player except the goaltender standing between the faceoff spot and the goal-line.

Highsticking penalties introduced.

Maximum number of players in uniform increased from 12 to 15.

December 21, 1929 — Forward passing rules instituted at the beginning of the 1929-30 season more than doubled number of goals scored. Partway through the season, these rules were further amended to read, "No attacking player allowed to precede the play when entering the opposing defensive zone." This is similar to modern offside rule.

1930-31 — A player without a complete stick ruled out of play and forbidden from taking part in further action until a new stick is obtained. A player who has broken his stick must obtain a replacement at his bench.

A further refinement of the offside rule stated that the puck must first be propelled into the attacking zone before any player of the attacking side can enter that zone; for infringement of this rule a faceoff to take place at the spot where the infraction took place.

1931-32 — Though there is no record of a team attempting to play with two goaltenders on the ice, a rule was instituted which stated that each team was allowed only one goaltender on the ice at one time.

Attacking players forbidden to impede the movement or obstruct the vision of opposing goaltenders.

Defending players with the exception of the goaltender forbidden from falling on the puck within 10 feet of the net.

1932-33 — Each team to have captain on the ice at all times.

If the goaltender is removed from the ice to serve a penalty, the manager of the club to appoint a substitute.

Match penalty with substitution after five minutes instituted for kicking another player.

1933-34 — Number of players permitted to stand in defensive zone restricted to three including goaltender.

Visible time clocks required in each rink.

Two referees replace one referee and one linesman.

1934-35 — Penalty shot awarded when a player is tripped and thus prevented from having a clear shot on goal, having no player to pass to other than the offending player. Shot taken from inside a 10-foot circle located 38 feet from the goal. The goaltender must not advance more than one foot from his goal-line when the shot is taken.

1937-38 — Rules introduced governing icing the puck.

Penalty shot awarded when a player other than a goaltender falls on the puck within 10 feet of the goal.

1938-39 — Penalty shot modified to allow puck carrier to skate in before shooting.

One referee and one linesman replace two referee system.

Blue line widened to 12 inches.

Maximum number of players in uniform increased from 14 to 15.

1939-40 — A substitute replacing a goaltender removed from ice to serve a penalty may use a goaltender's stick and gloves but no other goaltending equipment.

1940-41 — Flooding ice surface between periods made obligatory.

1941-42 — Penalty shots classified as minor and major. Minor shot to be taken from a line 28 feet from the goal. Major shot, awarded when a player is tripped with only the goaltender to beat, permits the player taking the penalty shot to skate right into the goalkeeper and shoot from point-blank range.

One referee and two linesmen employed to officiate games.

For playoffs, standby minor league goaltenders employed by NHL as emergency substitutes.

1942-43 — Because of wartime restrictions on train scheduling, regular-season overtime was discontinued on November 21, 1942.

Player limit reduced from 15 to 14. Minimum of 12 men in uniform abolished.

1943-44 — Red line at center ice introduced to speed up the game and reduce offside calls. This rule is considered to mark the beginning of the modern era in the NHL.

1945-46 — Goal indicator lights synchronized with official time clock required at all rinks.

1946-47 — System of signals by officials to indicate infractions introduced.

Linesmen from neutral cities employed for all games.

1947-48 — Goal awarded when a player with the puck has an open net to shoot at and a thrown stick prevents the shot on goal. Major penalty to any player who throws his stick in any zone other than defending zone. If a stick is thrown by a player in his defending zone but the thrown stick is not considered to have prevented a goal, a penalty shot is awarded.

All playoff games played until a winner determined, with 20-minute sudden-death overtime periods separated by 10-minute intermissions.

1949-50 — Ice surface painted white.

Clubs allowed to dress 17 players exclusive of goaltenders.

Major penalties incurred by goaltenders served by a member of the goaltender's team instead of resulting in a penalty shot.

1950-51 — Each team required to provide an emergency goaltender in attendance with full equipment at each game for use by either team in the event of illness or injury to a regular goaltender.

1951-52 — Home teams to wear basic white uniforms; visiting teams basic colored uniforms.

Goal crease enlarged from 3 × 7 feet to 4 × 8 feet.

Number of players in uniform reduced to 15 plus goaltenders.

Faceoff circles enlarged from 10-foot to 15-foot radius.

1952-53 — Teams permitted to dress 15 skaters on the road and 16 at home.

1953-54 — Number of players in uniform set at 16 plus goaltenders.

1954-55 — Number of players in uniform set at 18 plus goaltenders up to December 1 and 16 plus goaltenders thereafter. Teams agree to wear colored uniforms at home and white uniforms on the road.

1956-57 — Player serving a minor penalty allowed to return to ice when a goal is scored by opposing team.

1959-60 — Players prevented from leaving their benches to enter into an altercation. Substitutions permitted providing substitutes do not enter into altercation.

1960-61 — Number of players in uniform set at 16 plus goaltenders.

1961-62 — Penalty shots to be taken by the player against whom the foul was committed. In the event of a penalty shot called in a situation where a particular player hasn't been fouled, the penalty shot to be taken by any player on the ice when the foul was committed.

1964-65 — No bodily contact on faceoffs.

In playoff games, each team to have its substitute goaltender dressed in his regular uniform except for leg pads and body protector. All previous rules governing standby goaltenders terminated.

1965-66 — Teams required to dress two goaltenders for each regular-season game. Maximum stick length increased to 55 inches.

1966-67 — Substitution allowed on coincidental major penalties.

Between-periods intermissions fixed at 15 minutes.

1967-68 — If a penalty incurred by a goaltender is a co-incident major, the penalty to be served by a player of the goaltender's team on the ice at the time the penalty was called. Limit of curvature of hockey stick blade set at 1-1/2 inches.

1969-70 — Limit of curvature of hockey stick blade set at 1 inch.

1970-71 — Home teams to wear basic white uniforms; visiting teams basic colored uniforms.

Limit of curvature of hockey stick blade set at 1/2 inch.

Minor penalty for deliberately shooting the puck out of the playing area.

1971-72 — Number of players in uniform set at 17 plus 2 goaltenders.

Third man to enter an altercation assessed an automatic game misconduct penalty.

1972-73 — Minimum width of stick blade reduced to 2 inches from 2-1/2 inches.

1974-75 — Bench minor penalty imposed if a penalized player does not proceed directly and immediately to the penalty box.

1976-77 — Rule dealing with fighting amended to provide a major and game misconduct penalty for any player who is clearly the instigator of a fight.

1977-78 — Teams requesting a stick measurement to be assessed a minor penalty in the event that the measured stick does not violate the rules.

1979-80 — Wearing of helmets made mandatory for players entering the NHL.

1980-81 — Maximum stick length increased to 58 inches.

1981-82 — If both of a team's listed goaltenders are incapacitated, the team can dress and play any eligible goaltender who is available.

1982-83 — Number of players in uniform set at 18 plus 2 goaltenders.

1983-84 — Five-minute sudden-death overtime to be played in regular-season games that are tied at the end of regulation time.

1985-86 — Substitutions allowed in the event of co-incidental minor penalties. Maximum stick length increased to 60 inches.

1986-87 — Delayed off-side is no longer in effect once the players of the offending team have cleared the opponents' defensive zone.

1990-91 — The goal lines, blue lines, defensive zone face-off circles and markings all moved one foot out from the end boards, creating 11 feet of room behind the nets and shrinking the neutral zone from 60 to 58 feet.

1991-92 — Video replays employed to assist referees in goal/no goal situations. Size of goal crease increased. Crease changed to semi-circular configuration. Time clock to record tenths of a second in last minute of each period and overtime. Major and game misconduct penalty for checking from behind into boards. Penalties added for crease infringement and unnecessary contact with goaltender. Goal disallowed if puck enters net while a player of the attacking team is standing on the goal crease line, is in the goal crease or places his stick in the goal crease.

1992-93 — No substitutions allowed in the event of coincidental minor penalties called when both teams are at full strength. Wearing of helmets made optional for forwards and defensemen. Minor penalty for attempting to draw a penalty ("diving"). Major and game misconduct penalty for checking from behind into goal frame. Game misconduct penalty for instigating a fight. Highsticking redefined to include any use of the stick above waist-height. Previous rule stipulated shoulder-height.

1993-94 — High sticking redefined to allow goals scored with a high stick below the height of the crossbar of the goal frame.

1996-97 — Maximum stick length increased to 63 inches.

1998-99 — The league instituted a two-referee system with each team to play 20 regular-season games with two referees and a pair of linesmen. Also, the goal lines, blue lines, defensive zone face-off circles and markings all moved two feet closer to center, creating 13 feet of room behind the nets and cutting the neutral zone from 58 to 54 feet. The goal crease was altered so that it extends only one foot beyond each goal post (eight feet across in total) and has square sides for the first 4'6". Only the top of the crease remains rounded.

1999-2000 — Each team to play 25 home and 25 road games using the two-referee system. Crease rule revised to implement a "no harm, no foul, no video review" standard. An attacking player's position, whether inside or outside the crease, does not, in itself, determine whether a goal should be allowed or disallowed. The on-ice judgement of the referee(s) — instead of video review — will determine if a goal is "good" or not. Also, regular-season games tied at the end of three periods will result in each team being awarded one point in the standings. As before, there will be a five-minute sudden death overtime when the score is tied after three periods, but each team will play "four on four," with four skaters and a goalkeeper. In the event that penalties dictate that one team has a two-man advantage, the penalized team plays with three skaters while the team with the two-man advantage adds a fifth skater. A team that scores a goal in regular-season overtime is credited with a win and earns two points in the standings. A team scored upon in regular-season overtime is credited with an overtime loss and earns one point in the standings.

2000-01 — All games to be played using the two-referee system.

2002-03 — "Hurry-up" faceoff and line-change rules implemented.

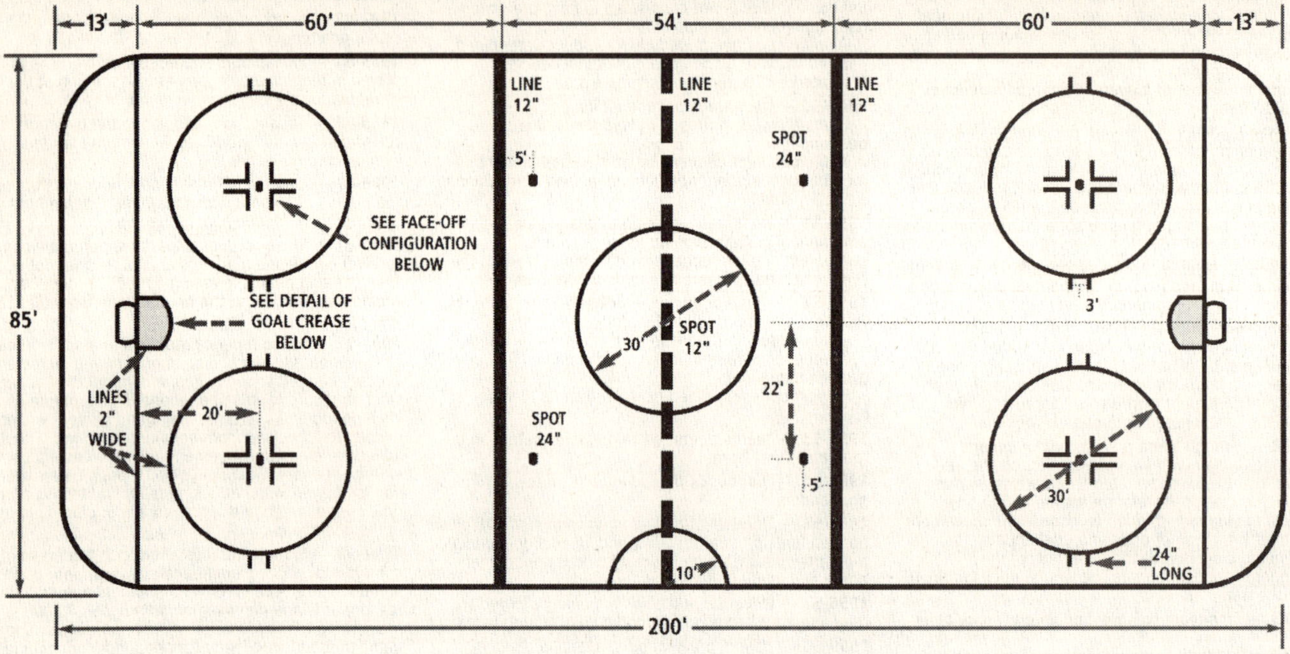

NHL RINK DIMENSIONS

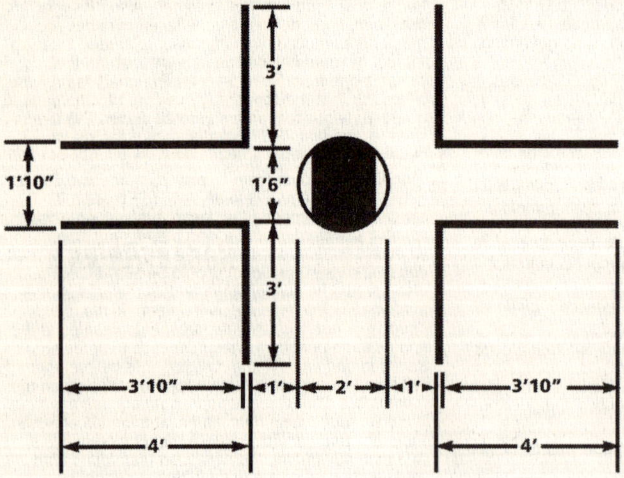

ALL LINES ARE 2" IN WIDTH

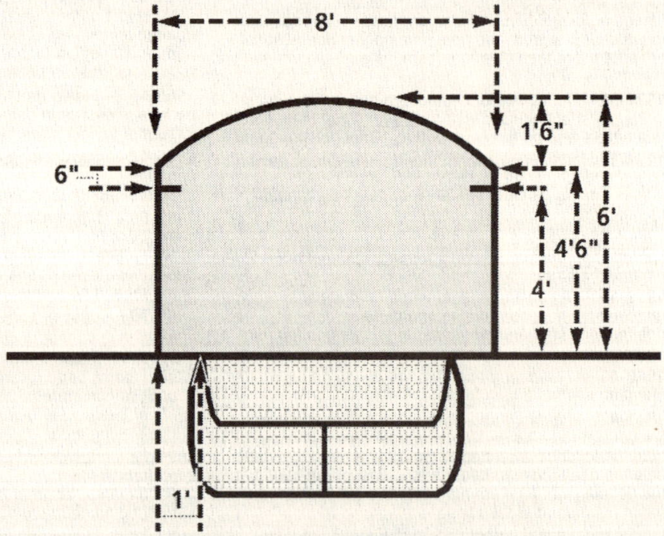

FACEOFF CONFIGURATION

CREASE DIMENSIONS

Double Gold for Canada at 2002 Olympic Winter Games

The NHL's top players represented their countries at the 2002 Olympic Winter Games in Salt Lake City, Utah. The tournament was played in two phases. The preliminary round saw eight teams play three games each (one against each of the other teams in their group), with the winners of Group A (Germany) and Group B (Belarus) advancing to join Canada, Czech Republic, Finland, Russia, Sweden and USA in the final round. Again, eight teams were divided into two goups with each team playing three games.

Sweden and the United States topped the round-robin phase, but the Swedes were upset by Belarus in the quarter-finals. The Americans defeated Germany and edged Russia to reach the final. Russia downed Belarus to claim the bronze.

After struggling in early games, Canada defeated Finland in the quarter-finals and Belarus in the semis before playing its best game of the tournament, winning 5-2 over Team USA in the gold medal game. It was the Canadian men's first hockey gold medal since 1952, but was the country's second hockey gold at Salt Lake as the Canadian women's team defeated their U.S. rivals 3-2, reversing the result from Nagano in 1998. Sweden upset favored Finland to win the women's bronze.

2002 Men's Olympic Hockey Results

Preliminary Round (round robin)

Feb. 9	Belarus	1	Ukraine	0
Feb. 9	Slovakia	0	Germany	3
Feb. 9	Austria	2	Latvia	4
Feb. 9	Switzerland	3	France	3
Feb. 10	Austria	2	Germany	3
Feb. 10	Latvia	6	Slovakia	6
Feb. 11	Ukraine	6	Switzerland	2
Feb. 11	Belarus	3	France	1
Feb. 12	Slovakia	2	Austria	3
Feb. 12	Germany	4	Latvia	1
Feb. 13	Switzerland	2	Belarus	1
Feb. 13	France	2	Ukraine	4

13th-Place Game

Feb. 14	Slovakia	7	France	1

11th-Place Game

Feb. 14	Switzerland	4	Austria	1

Ninth-Place Game

Feb. 14	Latvia	9	Ukraine	2

Final Round (round robin)

Feb. 15	Russia	6	Belarus	4
Feb. 15	Canada	2	Sweden	5
Feb. 15	Czech Republic	8	Germany	2
Feb. 15	Finland	0	United States	6
Feb. 16	Finland	7	Belarus	1
Feb. 16	United States	2	Russia	2
Feb. 17	Sweden	2	Czech Republic	1
Feb. 17	Canada	3	Germany	2
Feb. 18	Belarus	1	United States	8
Feb. 18	Russia	1	Finland	3
Feb. 18	Czech Republic	3	Canada	3
Feb. 18	Germany	1	Sweden	7

Quarter-final Games

Feb. 20	Belarus	4	Sweden	3
Feb. 20	Russia	2	Czech Republic	1
Feb. 20	United States	5	Germany	0
Feb. 20	Canada	2	Finland	1

Semi-final Games

Feb. 22	Canada	7	Belarus	1
Feb. 22	United States	3	Russia	2

Bronze Medal Game

Feb. 23	Russia	7	Belarus	2

Gold Medal Game

Feb. 24	Canada	5	United States	2

2002 Women's Olympic Hockey Results

Feb. 11	Canada	7	Kazakhstan	0
Feb. 11	Sweden	3	Russia	2
Feb. 12	United States	10	Germany	0
Feb. 12	Finland	4	China	0
Feb. 13	Russia	0	Canada	7
Feb. 13	Sweden	7	Kazakhstan	0
Feb. 14	Finland	3	Germany	1
Feb. 14	China	1	United States	12
Feb. 15	Kazakhstan	1	Russia	4
Feb. 16	United States	5	Finland	0
Feb. 16	Germany	5	China	5
Feb. 16	Canada	11	Sweden	0

Classification Round

Feb. 17	Russia	5	China	0
Feb. 17	Germany	4	Kazakhstan	0

Semi-final Games

Feb. 19	Canada	7	Finland	3
Feb. 19	United States	4	Sweden	0

Seventh-Place Game

Feb. 19	China	2	Kazakhstan	1 (OT)

Fifth-Place Game

Feb. 20	Russia	5	Germany	0

Bronze Medal Game

Feb. 21	Sweden	2	Finland	0

Gold Medal Game

Feb. 21	Canada	3	United States	2

Cumulative Medal Standings, Men's Olympic Hockey, 1924-2002

		G	S	B	Total	Last Medal
1.	USSR/Russia*	8	2	2	12	Bronze 02
2.	Canada	6	4	2	12	Gold 02
3.	USA	2	6	1	9	Silver 02
4.	Czechoslovakia/ Czech Republic	1	4	3	8	Gold 98
5.	Sweden	1	2	4	7	Gold 94
6.	Great Britain	1	0	1	2	Gold 36
7.	Finland	0	1	2	3	Bronze 98
8.	W. Germany	0	0	2	2	Bronze 76
9.	Switzerland	0	0	2	2	Bronze 48

** Soviet Union/Russia played as the Unified Team in 1992.*

Men's Standings • 2002

Preliminary Round

Group A

Team	GP	W	L	T	GF	GA	Pts
Germany	3	3	0	0	10	3	6
Latvia	3	1	1	1	11	12	3
Austria	3	1	2	0	7	9	2
Slovakia	3	0	2	1	8	12	1

Group B

Team	GP	W	L	T	GF	GA	Pts
Belarus	3	2	1	0	5	3	4
Ukraine	3	2	1	0	9	5	4
Switzerland	3	1	1	1	7	9	3
France	3	0	2	1	6	10	1

Final Round

Group C

Team	GP	W	L	T	GF	GA	Pts
Sweden	3	3	0	0	14	4	6
Czech Rep.	3	1	1	1	12	7	3
Canada	3	1	1	1	8	10	3
Germany	3	0	3	0	5	18	0

Group D

Team	GP	W	L	T	GF	GA	Pts
USA	3	2	0	1	16	3	5
Finland	3	2	1	0	11	8	4
Russia	3	1	1	1	9	9	3
Belarus	3	0	3	0	6	22	0

Final Rankings

Rank	Team	GP	W	L	T	GF	GA
1.	Canada	6	4	1	1	22	14
2.	USA	6	4	1	1	26	10
3.	Russia	6	3	2	1	20	15
4.	Belarus	9	3	6	0	18	42
5.	Czech R.	4	1	2	1	13	9
	Finland	4	2	2	0	12	10
	Germany	7	3	4	0	15	26
	Sweden	4	3	1	0	17	8
9.	Latvia	4	2	1	1	20	14
10.	Ukraine	4	2	2	0	11	14
11.	Switz.	4	2	1	1	11	10
12.	Austria	4	1	3	0	8	13
13.	Slovakia	4	1	2	1	15	13
14.	France	4	0	3	1	7	17

2002 Scoring Leaders

Player	Team	GP	G	A	PTS	PIM
Mats Sundin	Sweden	4	5	4	9	10
Brett Hull	USA	6	3	5	8	6
John LeClair	USA	6	6	1	7	2
Joe Sakic	Canada	6	4	3	7	0
Marian Hossa	Slovakia	2	4	2	6	0
J-J Aeschlimann	Switzerland	4	3	3	6	2
Philippe Bozon	France	4	3	3	6	2
Len Soccio	Germany	7	3	3	6	8
Mario Lemieux	Canada	5	2	4	6	0
Steve Yzerman	Canada	6	2	4	6	2
Nicklas Lidstrom	Sweden	4	1	5	6	0
Mike Modano	USA	6	0	6	6	0
M. Rozenthal	France	4	4	1	5	2
Klaus Kathan	Germany	7	3	2	5	0
A. Macijevskis	Latvia	4	2	3	5	0
A. Nizivijs	Latvia	4	2	3	5	2
Jaromir Jagr	Czech Rep.	4	2	3	5	4
R. Pavlikovsky	Slovakia	4	2	3	5	6
D. Alfredsson	Sweden	4	1	4	5	2
Phil Housley	USA	6	1	4	5	0

2002 Goaltending Leaders

Player	Team	GP	Min	GA	SO	GAA
Mike Dunham	USA	1	60	0	1	0.00
Tom Barrasso	USA	1	60	1	0	1.00
Pasi Nurminen	Finland	1	60	1	0	1.00
Rastislav Stana	Slovakia	1	60	1	0	1.00
Johan Hedberg	Sweden	1	60	1	0	1.00
Martin Gerber	Switzerland	3	157	4	0	1.52
Martin Brodeur	Canada	5	300	9	0	1.80
Dominik Hasek	Czech Rep.	4	239	8	0	2.01
Mike Richter	USA	4	240	9	1	2.34
N. Khabibulin	Russia	6	359	14	1	2.34
Tommy Salo	Sweden	3	179	7	0	2.35
Marc Selinger	Germany	6	302	15	1	2.89
Jani Hurme	Finland	3	179	9	0	3.01

Women's Standings • 2002

First Round

Group A

Team	GP	W	L	T	GF	GA	PTS
Canada	3	3	0	0	25	0	6
Sweden	3	2	1	0	10	13	4
Russia	3	1	2	0	6	11	2
Kazakhstan	3	0	0	0	1	18	0

Group B

Team	GP	W	L	T	GF	GA	PTS
USA	3	3	0	0	27	1	6
Finland	3	2	1	0	7	6	4
Germany	3	0	2	1	6	18	1
China	3	0	2	1	6	21	1

Final Rankings

Rank	Team	GP	W	L	T	GF	GA
1.	Canada	5	5	0	0	35	5
2.	USA	5	4	1	0	33	4
3.	Sweden	5	3	2	0	12	17
4.	Finland	5	2	3	0	10	15
5.	Russia	5	3	2	0	16	11
6.	Germany	5	1	3	1	10	23
7.	China	5	1	3	1	8	27
8.	Kazakhstan	5	0	5	0	2	24

2002 Scoring Leaders

Player	Team	GP	G	A	PTS	PIM
H. Wickenheiser	Canada	5	7	3	10	2
Cammi Granato	USA	5	6	4	10	0
Danielle Goyette	Canada	5	3	7	10	0
Natalie Darwitz	USA	5	7	1	8	2
Katie King	USA	5	4	3	7	4
Jenna Hefford	Canada	5	3	4	7	2
Jenny Potter	USA	5	1	6	7	2
Tara Mounsey	USA	5	0	7	7	4
Vicky Sunohora	Canada	5	4	2	6	6
Jennifer Botterill	Canada	5	3	3	6	8
Katja Riipi	Finland	5	3	3	6	6
Karyn Bye	USA	5	3	3	6	0
Caroline Oullette	Canada	5	2	4	6	6
Krissy Wendell	USA	5	1	5	6	6
Cherie Piper	Canada	5	3	2	5	0
Maritta Becker	Germany	5	3	2	5	8
E. Pachkevitch	Russia	5	3	2	5	2
Laurie Baker	USA	5	3	2	5	4
Therese Brisson	Canada	5	2	3	5	6
Erika Holst	Sweden	5	2	3	5	10

2002 Goaltending Leaders

Player	Team	GP	Min	GA	GAA
Sami Jo Small	Canada	1	60	0	0.00
Sarah Tueting	USA	2	120	1	0.50
Sara DeCosta	USA	3	180	3	1.00
Kim St-Pierre	Canada	4	240	5	1.25
Kim Martin	Sweden	3	180	5	1.67
I. Gachennikova	Russia	5	300	12	2.40
Tuula Puputti	Finland	5	299	15	3.01
S. Wartosh-Kurten	Germany	5	228	14	3.67
Natalya Trunova	Kazak.	5	301	24	4.77
Hong Guo	China	5	301	26	5.17

Cumulative Medal Standings, Women's Olympic Hockey, 1998-2002

		G	S	B	Total	Last Medal
1.	Canada	1	1	0	2	Gold 02
2.	USA	1	1	0	2	Silver 02
3.	Sweden	0	0	1	1	Bronze 02
4.	Finland	0	0	1	1	Bronze 98

NHL Clubs' Minor-League Affiliations, 2002-03

NHL Club	Minor-League Affiliates
Anaheim	Cincinnati Mighty Ducks (AHL)
Atlanta	Chicago Wolves (AHL) Greenville Grrrowl (ECHL)
Boston	Providence Bruins (AHL)
Buffalo	Rochester Americans (AHL) South Carolina Stingrays (ECHL)
Calgary	Saint John Flames (AHL) Johnstown Chiefs (ECHL)
Carolina	Lowell Lock Monsters (AHL) Florida Everblades (ECHL)
Chicago	Norfolk Admirals (AHL) Roanoke Express (ECHL)
Colorado	Hershey Bears (AHL)
Columbus	Syracuse Crunch (AHL) Dayton Bombers (ECHL) Elmira Jackals (UHL)
Dallas	Utah Grizzlies (AHL) Fort Worth Brahmas (CHL)
Detroit	Grand Rapids Griffins (AHL) Toledo Storm (ECHL)
Edmonton	Hamilton Bulldogs (AHL) Columbus Cottonmouths (ECHL) Odessa Jackalopes (CHL)
Florida	San Antonio Rampage (AHL)
Los Angeles	Manchester Monarchs (AHL) Reading Royals (ECHL)
Minnesota	Houston Aeros (AHL) Louisiana IceGators (ECHL)
Montreal	Hamilton Bulldogs (AHL) Utah Grizzlies (AHL)
Nashville	Milwaukee Admirals (AHL) Toledo Storm (ECHL)
New Jersey	Albany River Rats (AHL)
NY Islanders	Bridgeport Sound Tigers (AHL) Trenton Titans (ECHL)
NY Rangers	Hartford Wolf Pack (AHL)
Ottawa	Binghamton Senators (AHL)
Philadelphia	Philadelphia Phantoms (AHL) Trenton Titans (ECHL)
Phoenix	Springfield Falcons (AHL) Augusta Lynx (ECHL)
Pittsburgh	Wilkes-Barre/Scranton Penguins (AHL) Wheeling Nailers (ECHL)
St. Louis	Worcester IceCats (AHL) Peoria Rivermen (ECHL)
San Jose	Cleveland Barons (AHL) Richmond Renegades (ECHL) Cincinnati Cyclones (ECHL)
Tampa Bay	Springfield Falcons (AHL)) Pensacola Ice Pilots (ECHL)
Toronto	St. John's Maple Leafs (AHL) Memphis RiverKings (CHL)
Vancouver	Manitoba Moose (AHL) Columbia Inferno (ECHL)
Washington	Portland Pirates (AHL) Richmond Renegades (ECHL) Quad City Mallards (UHL)

Mighty Ducks of Anaheim

2001-02 Results: 29W-42L-8T-3OTL 69PTS.
Fifth, Pacific Division

2002-03 Schedule

Oct.	Thu.	10	at St. Louis
	Fri.	11	at Dallas
	Sun.	13	Detroit
	Wed.	16	Los Angeles
	Fri.	18	Vancouver
	Sun.	20	Colorado*
	Thu.	24	at Vancouver
	Sat.	26	at Edmonton
	Mon.	28	at Toronto
	Tue.	29	at Montreal
	Thu.	31	at Boston
Nov.	Sun.	3	San Jose*
	Wed.	6	Nashville
	Fri.	8	at Colorado
	Sun.	10	Minnesota*
	Tue.	12	at New Jersey
	Thu.	14	at Columbus
	Fri.	15	at Detroit
	Sun.	17	at Atlanta*
	Tue.	19	at NY Rangers
	Fri.	22	Dallas
	Sun.	24	Florida*
	Wed.	27	Phoenix
	Fri.	29	Los Angeles*
Dec.	Sun.	1	Chicago*
	Tue.	3	at Detroit
	Wed.	4	at Buffalo
	Fri.	6	at Chicago
	Sun.	8	Nashville*
	Wed.	11	Washington
	Sun.	15	Pittsburgh*
	Wed.	18	St. Louis
	Thu.	19	at Los Angeles
	Sun.	22	Phoenix*
	Thu.	26	at San Jose
	Sat.	28	at Vancouver
	Sun.	29	at Calgary
	Tue.	31	at Minnesota*
Jan.	Fri.	3	Philadelphia
	Sun.	5	Dallas*
	Wed.	8	Edmonton
	Thu.	9	at Colorado
	Sun.	12	St. Louis*
	Wed.	15	at Columbus
	Thu.	16	at Ottawa
	Sat.	18	at Minnesota
	Mon.	20	Minnesota*
	Wed.	22	Los Angeles
	Fri.	24	New Jersey
	Wed.	29	Ottawa
	Thu.	30	at San Jose
Feb.	Tue.	4	at Calgary
	Wed.	5	at Edmonton
	Fri.	7	Phoenix
	Sun.	9	Carolina*
	Wed.	12	Calgary
	Fri.	14	at Dallas
	Sat.	15	at Nashville
	Mon.	17	NY Islanders
	Wed.	19	Columbus
	Fri.	21	NY Rangers
	Sun.	23	at Carolina*
	Tue.	25	at Tampa Bay
	Wed.	26	at Florida
	Fri.	28	at Phoenix
Mar.	Sun.	2	Atlanta*
	Tue.	4	at Los Angeles
	Wed.	5	Montreal
	Fri.	7	Edmonton
	Sun.	9	Detroit*
	Wed.	12	Chicago
	Sat.	15	at Phoenix
	Sun.	16	Calgary*
	Wed.	19	at Chicago
	Thu.	20	at St. Louis
	Sat.	22	at San Jose
	Mon.	24	Columbus
	Wed.	26	San Jose
	Sun.	30	Vancouver*
Apr.	Tue.	1	at Nashville
	Wed.	2	at Dallas
	Fri.	4	Colorado

* Denotes afternoon game.

Year-by-Year Record

		Home				Road				Overall								
Season	GP	W	L	T	OL	W	L	T	OL	W	L	T	OL	GF	GA	Pts.	Finished	Playoff Result
2001-02	82	15	19	5	2	14	23	3	1	29	42	8	3	175	198	69	5th, Pacific Div.	Out of Playoffs
2000-01	82	15	20	4	2	10	21	7	3	25	41	11	5	188	245	66	5th, Pacific Div.	Out of Playoffs
1999-2000	82	19	13	7	2	15	20	5	1	34	33	12	3	217	227	83	5th, Pacific Div.	Out of Playoffs
1998-99	82	21	14	6	...	14	20	7	...	35	34	13	...	215	206	83	3rd, Pacific Div.	Lost Conf. Quarter-Final
1997-98	82	12	23	6	...	14	20	7	...	26	43	13	...	205	261	65	6th, Pacific Div.	Out of Playoffs
1996-97	82	23	12	6	...	13	21	7	...	36	33	13	...	245	233	85	2nd, Pacific Div.	Lost Conf. Semi-Final
1995-96	82	22	15	4	...	13	24	4	...	35	39	8	...	234	247	78	4th, Pacific Div.	Out of Playoffs
1994-95	48	11	9	4	...	5	18	1	...	16	27	5	...	125	164	37	6th, Pacific Div.	Out of Playoffs
1993-94	84	14	26	2	...	19	20	3	...	33	46	5	...	229	251	71	4th, Pacific Div.	Out of Playoffs

After 10 seasons in the NHL, Keith Carney added a veteran presence to the Anaheim defense corps in 2001-02. Though he collected just 14 points in 60 games, his plus-minus rating of +14 was by far the best on the team last year.

Franchise date: June 15, 1993

PACIFIC DIVISION

10th NHL Season

2002-03 Player Personnel

FORWARDS	HT	WT	S	Place of Birth	Date	2001-02 Club
BELANGER, Francis	6-3	228	L	Bellefeuille, Que.	1/15/78	Quebec (AHL)
BRIGLEY, Travis	6-1	200	L	Coronation, Alta.	6/16/77	Macon-Cincinnati
BYLSMA, Dan	6-2	212	L	Grand Haven, MI	9/19/70	Anaheim
CHISTOV, Stanislav	5-10	178	R	Chelyabinsk, USSR	4/17/83	Omsk-CSKA Moscow 2
CHOUINARD, Marc	6-5	210	R	Charlesbourg, Que.	5/6/77	Anaheim
CULLEN, Matt	6-0	205	L	Virginia, MN	11/2/76	Anaheim
GORNICK, Brian	6-5	210	L	St. Paul, MN	3/17/80	Air Force (CHA)
GUITE, Ben	6-1	205	R	Montreal, Que.	7/17/78	Bridgeport-Cincinnati
HEDSTROM, Jonathan	6-0	200	L	Skelleftea, Sweden	12/27/77	Lulea
KARIYA, Paul	5-10	176	L	Vancouver, B.C.	10/16/74	Anaheim-Canada
KJELLBERG, Patric	6-2	210	L	Trelleborg, Sweden	6/17/69	Nashville-Anaheim
KROG, Jason	5-11	191	R	Fernie, B.C.	10/9/75	NY Islanders-Bridgeport
LAMBERT, Denny	5-11	211	L	Wawa, Ont.	1/7/70	Anaheim
LECLERC, Mike	6-2	208	L	Winnipeg, Man.	11/10/76	Anaheim
MARTENSSON, Tony	6-0	189	L	Upplands Vasby, Sweden	6/23/80	Brynas
McDONALD, Andy	5-10	186	L	Strathroy, Ont.	8/25/77	Anaheim-Cincinnati
OATES, Adam	5-11	190	R	Weston, Ont.	8/27/62	Washington-Philadelphia
PAHLSSON, Sami	5-11	212	L	Ornskoldsvik, Sweden	12/17/77	Anaheim
PECKER, Cory	6-0	195	R	Montreal, Que.	3/20/81	Erie (OHL)
RUCCHIN, Steve	6-2	211	L	Thunder Bay, Ont.	7/4/71	Anaheim
SAWYER, Kevin	6-2	212	L	Christina Lake, B.C.	2/21/74	Anaheim
SEVERSON, Cam	6-1	215	L	Canora, Sask.	1/15/78	Hartford
SMIRNOV, Alexei	6-3	211	L	Tver, USSR	1/28/82	CSKA Moscow 2-CSKA Moscow
SMITH, Jarrett	6-2	204	L	Edmonton, Alta.	6/15/79	Cincinnati
SMITH, Nick	6-2	196	L	Hamilton, Ont.	3/23/79	Fla-Bridgeport-Saint John
SYKORA, Petr	6-0	190	L	Plzen, Czech.	11/19/76	New Jersey-Czech Republic
TITOV, German	6-1	203	L	Moscow, USSR	10/16/65	Anaheim
VALICEVIC, Rob	6-1	198	R	Detroit, MI	1/6/71	Los Angeles-Manchester
DEFENSEMEN						
CARNEY, Keith	6-2	211	L	Providence, RI	2/3/70	Anaheim
COMMODORE, Mike	6-4	230	R	Fort Saskatchewan, Alta.	11/7/79	New Jersey-Albany
DeWOLF, Josh	6-2	203	L	Bloomington, MN	7/25/77	Cincinnati
HAVELID, Niclas	5-11	196	L	Stockholm, Sweden	4/12/73	Anaheim
OLAUSSON, Fredrik	6-2	198	R	Dadesjo, Sweden	10/5/66	Detroit-Sweden
O'SULLIVAN, Chris	6-2	205	L	Dorchester, MA	5/15/74	Kloten
PODHRADSKY, Peter	6-2	204	R	Bratislava, Czech.	12/10/79	Cincinnati
POPOVIC, Mark	6-1	191	L	Stoney Creek, Ont.	10/11/82	St. Michael's
REIRDEN, Todd	6-5	225	L	Deerfield, IL	6/25/71	Atlanta
SALEI, Ruslan	6-1	205	L	Minsk, USSR	11/2/74	Anaheim-Belarus
SAUER, Kurt	6-4	225	L	St. Cloud, MN	1/16/81	Spokane
TABACEK, Jan	5-11	169	L	Martin, Czech.	4/7/80	Slov. Bratislava
TRNKA, Pavel	6-2	200	L	Plzen, Czech.	7/27/76	Anaheim
VISHNEVSKI, Vitaly	6-2	206	L	Kharkov, USSR	3/18/80	Anaheim
YORK, Jason	6-1	208	R	Nepean, Ont.	5/20/70	Anaheim

GOALTENDERS	HT	WT	C	Place of Birth	Date	2001-02 Club
BRYZGALOV, Ilja	6-3	198	L	Togliatti, USSR	6/22/80	Cincinnati-Ana-Russia
DAMPHOUSSE, Jean-Francois	6-0	180	L	St-Alexis-des-Monts, Que.	7/21/79	New Jersey-Albany
GERBER, Martin	6-0	185	L	Burgdorf, Switz.	9/3/74	Farjestad-Switzerland
GIGUERE, Jean-Sebastien	6-1	199	L	Montreal, Que.	5/16/77	Anaheim

In his first full season as the club's top goaltender, Jean-Sebastien Giguere put up stellar numbers in 2001-02. His 2.13 goals-against average was fifth in the NHL and his .920 save percentage also ranked among the league leaders.

2001-02 Scoring

* – rookie

Regular Season

Pos	#	Player	Team	GP	G	A	Pts	+/–	PIM	PP	SH	GW	GT	S	%
L	9	Paul Kariya	ANA	82	32	25	57	–15	28	11	0	8	1	289	11.1
C	17	Matt Cullen	ANA	79	18	30	48	–1	24	3	1	4	0	164	11.0
L	12	Mike Leclerc	ANA	82	20	24	44	–12	107	8	0	4	0	178	11.2
L	11	Jeff Friesen	ANA	81	17	26	43	–1	44	1	1	0	0	161	10.6
D	10	Oleg Tverdovsky	ANA	73	6	26	32	0	31	2	0	1	0	147	4.1
C	19 *	Andy McDonald	ANA	53	7	21	28	2	10	2	0	3	0	79	8.9
L	13	German Titov	ANA	66	13	14	27	4	36	1	0	2	0	63	20.6
D	33	Jason York	ANA	74	5	20	25	–11	60	3	0	2	0	104	4.8
C	20	Steve Rucchin	ANA	38	7	16	23	–3	6	4	0	1	0	57	12.3
C	26	Sami Pahlsson	ANA	80	6	14	20	–16	26	1	1	0	0	99	6.1
R	18	Patric Kjellberg	NSH	12	1	3	4	–3	6	0	0	0	0	19	5.3
			ANA	65	7	8	15	–9	10	4	0	0	0	69	10.1
			TOTAL	77	8	11	19	–12	16	4	0	0	0	88	9.1
R	21	Dan Bylsma	ANA	77	8	9	17	5	28	0	1	2	0	72	11.1
D	3	Keith Carney	ANA	60	5	9	14	14	30	0	0	1	0	66	7.6
D	7	Pavel Trnka	ANA	71	2	11	13	–5	66	1	0	0	0	78	2.6
D	24	Ruslan Salei	ANA	82	4	7	11	–10	97	0	0	1	0	96	4.2
C	32	Marc Chouinard	ANA	45	4	5	9	2	10	0	0	0	0	40	10.0
L	27	Denny Lambert	ANA	73	2	5	7	1	213	0	0	0	0	51	3.9
R	23	Sergei Krivokrasov	MIN	9	1	1	2	–1	17	0	0	1	0	13	7.7
			ANA	17	1	2	3	–1	19	0	0	0	0	38	2.6
			TOTAL	26	2	3	5	–2	36	0	0	1	0	51	3.9
D	28	Niclas Havelid	ANA	52	1	2	3	–13	40	0	0	0	0	45	2.2
L	29 *	Timo Parssinen	ANA	17	0	3	3	0	2	0	0	0	0	16	0.0
D	5	Vitaly Vishnevski	ANA	74	0	3	3	–10	60	0	0	0	0	54	0.0
L	25	Kevin Sawyer	ANA	57	1	1	2	–4	221	0	0	0	0	29	3.4
D	36	Drew Bannister	ANA	1	0	0	0	0	0	0	0	0	0	1	0.0
D	4	Antti-Jussi Niemi	ANA	1	0	0	0	–1	0	0	0	0	0	0	0.0
D	37	Aris Brimanis	ANA	5	0	0	0	–1	9	0	0	0	0	2	0.0

Goaltending

No.	Goaltender	GPI	Mins	Avg	W	L	T	EN	SO	GA	SA	S%	G	A	PIM
30	* Ilja Bryzgalov	1	32	1.88	0	0	0	0	0	1	12	.917	0	0	0
35	J-S Giguere	53	3127	2.13	20	25	6	3	4	111	1384	.920	0	0	28
31	Steve Shields	33	1777	2.67	9	20	2	4	0	79	850	.907	0	0	4
	Totals	**82**	**4976**	**2.39**	**29**	**45**	**8**	**7**	**5**	**198**	**2253**	**.912**			

Coach

BABCOCK, MIKE
Coach, Mighty Ducks of Anaheim.
Born in Manitouwadge, Ont., April 29, 1963.

The Mighty Ducks of Anaheim announced Mike Babcock as the club's head coach on May 22, 2002. Babcock spent the previous two seasons as head coach of the Cincinnati Mighty Ducks, Anaheim's primary development affiliate in the American Hockey League. While with Cincinnati, he led the club to a franchise-best 41 wins and 95 points in 2000-01. The team qualified for the Calder Cup playoffs in each of his two seasons.

Babcock earned the honor of coaching the Canadian World Junior Team in 1997, leading the club to its fifth consecutive gold medal in the tournament. Prior to joining the Mighty Ducks, Babcock had a successful six-year run as the head coach of the Spokane Chiefs of the Western Hockey League. While with Spokane, he had a regular-season record of 224-175-29 (.557 winning percentage, the highest in the WHL in that span). He was twice named WHL coach of the year (1996 & 2000) after taking the franchise to the league finals both seasons. Additionally, he was the head coach of the 2000 WHL West Division All-Star Team.

In 1988, Babcock was named head coach at Red Deer College in Red Deer, Alberta. He spent three seasons at the school, winning the Alberta college championship and coach of the year honors in 1989. Babcock won a national championship and was again named the coach of the year while with the University of Lethbridge in 1993-94. He began his WHL career as head coach of the Moose Jaw Warriors from 1991 to 1993.

Coaching Record

		Regular Season				Playoffs		
Season	Team	Games	W	L	T	Games	W	L
1991-92	Moose Jaw (WHL)	72	33	36	3	4	0	4
1992-93	Moose Jaw (WHL)	72	27	42	3	....	....	....
1993-94	U. of Lethbridge (CIAU)	28	19	7	2	....	....	....
1994-95	Spokane (WHL)	72	32	36	4	11	6	5
1995-96	Spokane (WHL)	72	50	18	4	9	3	6
1996-97	Spokane (WHL)	65	31	30	4	9	4	5
1997-98	Spokane (WHL)	72	45	23	4	18	10	8
1998-99	Spokane (WHL)	72	19	44	9	....	....	....
1999-2000	Spokane (WHL)	72	47	21	4	20	15	5
2000-01	Cincinnati (WHL)	80	41	26	13	4	1	3
2001-02	Cincinnati (WHL)	80	33	33	14	3	1	2

Club Records

Team

(Figures in brackets for season records are games played; records for fewest points, wins, ties, losses, goals, goals against are for 70 or more games)

Record		
Most Points	**85**	1996-97 (82)
Most Wins	**36**	1996-97 (82)
Most Ties	**13**	1996-97 (82); 1997-98 (82); 1998-99 (82)
Most Losses	**46**	1993-94 (84)
Most Goals	**245**	1996-97 (82)
Most Goals Against	**261**	1997-98 (82)
Fewest Points	**65**	1997-98 (82)
Fewest Wins	**25**	2000-01 (82)
Fewest Ties	**5**	1993-94 (84)
Fewest Losses	**33**	1996-97 (82)
Fewest Goals	**175**	2001-02 (82)
Fewest Goals Against	**198**	2001-02 (82)
Longest Winning Streak		
Overall	**7**	Feb. 20-Mar. 7/99
Home	**5**	Three times
Away	**5**	Nov. 26-Dec. 26/99
Longest Undefeated Streak		
Overall	**12**	Feb. 22-Mar. 19/97 (7 wins, 5 ties)
Home	**14**	Feb. 12-Apr. 9/97 (10 wins, 4 ties)
Away	**5**	Five times
Longest Losing Streak		
Overall	**8**	Oct. 12-30/96
Home	**8**	Jan. 10-Feb. 9/01
Away	**6**	Three times
Longest Winless Streak		
Overall	**9**	Twice
Home	**11**	Jan. 5-Feb. 14/01 (8 losses, 3 ties)
Away	**10**	Mar. 26-Oct. 11/95 (9 losses, 1 tie)
Most Shutouts, Season	**7**	1998-99 (82)
Most PIM, Season	**1,843**	1997-98 (82)
Most Goals, Game	**8**	Jan. 21/98 (Ana. 8, Fla. 3)

Individual

Most Seasons	**8**	Guy Hebert, Paul Kariya
Most Games	**524**	Paul Kariya
Most Goals, Career	**275**	Paul Kariya
Most Assists, Career	**313**	Paul Kariya
Most Points, Career	**588**	Paul Kariya (275G, 313A)
Most PIM, Career	**788**	Dave Karpa
Most Shutouts, Career	**27**	Guy Hebert
Longest Consecutive Games Streak	**237**	Oleg Tverdovsky (Oct. 2/99-Mar. 24/02)
Most Goals, Season	**52**	Teemu Selanne (1997-98)
Most Assists, Season	**62**	Paul Kariya (1998-99)
Most Points, Season	**109**	Teemu Selanne (1996-97; 51G, 58A)
Most PIM, Season	**285**	Todd Ewen (1995-96)
Most Points, Defenseman, Season	**56**	Fredrik Olausson (1998-99; 16G, 40A)
Most Points, Center, Season	**67**	Steve Rucchin (1996-97; 19G, 48A)
Most Points, Right Wing, Season	**109**	Teemu Selanne (1996-97; 51G, 58A)
Most Points, Left Wing, Season	**108**	Paul Kariya (1995-96; 50G, 58A)
Most Points, Rookie, Season	**39**	Paul Kariya (1994-95; 18G, 21A)
Most Shutouts, Season	**6**	Guy Hebert (1998-99)
Most Goals, Game	**3**	Thirteen times
Most Assists, Game	**5**	Dmitri Mironov (Dec. 12/97)
Most Points, Game	**5**	Six times

General Managers' History

Jack Ferreira, 1993-94 to 1997-98; Pierre Gauthier, 1998-99 to 2001-02; Bryan Murray, 2002-03.

Coaching History

Ron Wilson, 1993-94 to 1996-97; Pierre Page, 1997-98; Craig Hartsburg, 1998-99, 1999-2000; Craig Hartsburg and Guy Charron, 2000-01; Bryan Murray, 2001-02; Mike Babcock, 2002-03.

Captains' History

Troy Loney, 1993-94; Randy Ladouceur, 1994-95, 1995-96; Paul Kariya, 1996-97; Paul Kariya and Teemu Selanne, 1997-98; Paul Kariya, 1998-99 to date.

All-time Record vs. Other Clubs

Regular Season

	At Home								On Road								Total							
	GP	W	L	T	OL	GF	GA	PTS	GP	W	L	T	OL	GF	GA	PTS	GP	W	L	T	OL	GF	GA	PTS
Atlanta	3	2	1	0	0	11*	5	4	2	2	0	0	0	9	3	4	5	4	1	0	0	20	8	8
Boston	8	2	3	2	1	15	20	7	7	3	4	0	0	20	22	6	15	5	7	2	1	35	42	13
Buffalo	8	2	6	0	0	13	25	4	7	2	2	3	0	18	16	7	15	4	8	3	0	31	41	11
Calgary	22	10	7	5	0	71	58	25	21	8	12	1	0	51	58	17	43	18	19	6	0	122	116	42
Carolina	7	3	3	1	0	24	23	7	7	1	5	1	0	12	22	3	14	4	8	2	0	36	45	10
Chicago	18	9	7	2	0	45	39	20	20	7	11	2	0	44	56	16	38	16	18	4	0	89	95	36
Colorado	17	4	10	3	0	37	44	11	17	4	9	4	0	43	55	12	34	8	19	7	0	80	99	23
Columbus	4	2	2	0	0	12	13	4	4	0	4	0	0	6	13	0	8	2	6	0	0	18	26	4
Dallas	22	9	12	1	0	51	58	19	21	4	15	1	1	41	82	10	43	13	27	2	1	92	140	29
Detroit	18	5	10	3	0	36	52	13	18	2	12	3	1	44	69	8	36	7	22	6	1	80	121	21
Edmonton	22	14	6	2	0	64	54	30	21	8	12	0	1	45	48	17	43	22	18	2	1	109	102	47
Florida	8	3	5	0	0	23	26	6	6	1	2	2	1	13	19	5	14	4	7	2	1	36	45	11
Los Angeles	24	11	5	5	3	77	58	30	25	8	13	4	0	62	72	20	49	19	18	9	3	139	130	50
Minnesota	4	2	2	0	0	9	13	4	4	2	1	1	0	10	6	5	8	4	3	1	0	19	19	9
Montreal	6	2	4	0	0	20	21	4	7	2	4	1	0	17	22	5	13	4	8	1	0	37	43	9
Nashville	8	7	1	0	0	23	12	14	8	3	3	2	0	20	17	8	16	10	4	2	0	43	29	22
New Jersey	8	4	4	0	0	22	20	8	7	1	6	0	0	12	28	2	15	5	10	0	0	34	48	10
NY Islanders	7	2	3	2	0	15	19	6	7	3	3	1	0	22	20	7	14	5	6	3	0	37	39	13
NY Rangers	7	6	0	0	1	30	19	13	7	4	2	1	0	21	19	9	14	10	2	1	1	51	38	22
Ottawa	7	3	2	2	0	18	13	8	7	3	3	1	0	18	21	7	14	6	5	3	0	36	34	15
Philadelphia	7	3	2	2	0	26	24	8	8	2	3	3	0	17	22	7	15	5	5	5	0	43	46	15
Phoenix	21	11	8	2	0	58	57	24	21	12	8	1	0	67	61	25	42	23	16	3	0	125	118	49
Pittsburgh	7	4	3	0	0	24	24	8	8	2	4	2	0	26	27	6	15	6	7	2	0	50	51	14
St. Louis	18	4	13	1	0	42	58	9	18	6	9	3	0	47	56	15	36	10	22	4	0	89	114	24
San Jose	25	9	14	2	0	66	85	20	24	11	10	2	1	69	71	25	49	20	24	4	1	135	156	45
Tampa Bay	8	4	3	1	0	24	20	9	7	4	3	0	0	21	15	8	15	8	6	1	0	45	35	17
Toronto	10	4	5	1	0	29	27	9	14	2	8	4	0	28	45	8	24	6	13	5	0	57	72	17
Vancouver	21	6	8	6	1	51	64	19	22	6	15	1	0	50	80	13	43	12	23	7	1	101	144	32
Washington	8	5	2	1	0	26	22	11	8	4	4	0	0	18	14	8	16	9	6	1	0	44	36	19
Totals	**353**	**152**	**151**	**44**	**6**	**962**	**973**	**354**	**353**	**117**	**187**	**44**	**5**	**871**	**1059**	**283**	**706**	**269**	**338**	**88**	**11**	**1833**	**2032**	**637**

Playoffs

	Series	W	L	GP	W	L	T	GF	GA	Last Mtg.	Rnd.	Result
Detroit	2	0	2	8	0	8	0	14	30	1999	CQF	L 0-4
Phoenix	1	1	0	7	4	3	0	17	17	1997	CQF	W 4-3
Totals	**3**	**1**	**2**	**15**	**4**	**11**	**0**	**31**	**47**			

Carolina totals include Hartford, 1993-94 to 1996-97.
Colorado totals include Quebec, 1993-94 to 1994-95.
Phoenix totals include Winnipeg, 1993-94 to 1995-96.

Playoff Results 2002-1998

Year	Round	Opponent	Result	GF	GA
1999	CQF	Detroit	L 0-4	6	17

Abbreviations: Round: CQF - conference quarter-final

2001-02 Results

Oct.	4	at Boston	2-4		31	at Columbus	1-3
	6	at Pittsburgh	4-2	**Jan.**	2	at Detroit	3-5
	8	at Toronto	1-6		4	Florida	1-2
	9	at Montreal	1-3		9	St. Louis	2-3
	12	Washington	2-1		11	at Minnesota	2-2
	14	Tampa Bay	2-3		12	at Nashville	1-2
	17	Boston	2-2		14	Nashville	5-3
	18	at Los Angeles	1-4		16	Buffalo	1-3
	21	Vancouver	3-1		18	at Edmonton	1-3
	24	at Phoenix	3-2*		19	at Calgary	1-2
	28	Colorado	2-3		21	Los Angeles	2-4
	31	San Jose	2-4		23	Minnesota	3-2*
Nov.	2	Chicago	5-2		25	at Dallas	6-1
	4	Atlanta	5-0		26	at Nashville	3-1
	7	Calgary	3-3		28	at St. Louis	1-0
	9	Detroit	0-1		30	Columbus	3-1
	11	Dallas	2-2	**Feb.**	6	Philadelphia	5-4
	14	San Jose	2-4		8	Carolina	1-4
	16	at Columbus	2-3		10	Dallas	1-5
	17	at Washington	1-4		13	Calgary	3-2
	20	at Tampa Bay	2-3		27	Minnesota	3-5
	21	at Florida	0-6	**Mar.**	3	at Chicago	1-2
	24	at NY Islanders	3-5		6	at Atlanta	4-1
	25	at NY Rangers	3-2		8	New Jersey	2-1
	28	Edmonton	0-2		10	Ottawa	2-4
	30	San Jose	2-5		13	Pittsburgh	4-2
Dec.	2	Nashville	4-2		15	Chicago	1-1
	5	at Edmonton	2-3*		17	St. Louis	2-3
	6	at Vancouver	2-3		19	at Detroit	2-1
	8	at Calgary	4-0		21	at Philadelphia	1-2
	10	at Colorado	1-1		22	at St. Louis	2-3
	12	Vancouver	0-1*		24	at Dallas	2-1
	14	Columbus	2-3		27	Phoenix	2-4
	16	Los Angeles	2-3*		28	at Phoenix	2-1*
	18	at Minnesota	5-1		30	at Vancouver	1-4
	19	at Colorado	1-2	**Apr.**	2	at San Jose	1-3
	21	Phoenix	2-1		3	Detroit	1-1
	23	at Phoenix	4-0		5	Edmonton	2-0
	26	at San Jose	2-1		7	Dallas	4-1
	27	at Los Angeles	2-2		12	Colorado	1-3
	30	at Chicago	1-2		14	at Los Angeles	0-1

* – Overtime

Entry Draft Selections 2002-1993

2002

Pick	
7	Joffrey Lupul
37	Tim Brent
71	Brian Lee
103	Joonas Vihko
140	George Davis
173	Luke Fritshaw
261	Francois Caron
267	Chris Petrow

2001

Pick	
5	Stanislav Chistov
35	Mark Popovic
69	Joel Stepp
102	Timo Parssinen
105	Vladimir Korsunov
118	Brandon Rogers
137	Joel Perreault
170	Jan Tabacek
224	Tony Martensson
232	Martin Gerber
264	Pierre Parenteau

2000

Pick	
12	Alexei Smirnov
44	Ilja Bryzgalov
98	Jonas Ronnqvist
134	Peter Podhradsky
153	Bill Cass

1999

Pick	
44	Jordan Leopold
83	Niclas Havelid
105	Alexandr Chagodayev
141	Maxim Rybin
173	Jan Sandstrom
230	Petr Tenkrat
258	Brian Gornick

1998

Pick	
5	Vitaly Vishnevski
32	Stephen Peat
112	Viktor Wallin
150	Trent Hunter
178	Jesse Fibiger
205	David Bernier
205	David Bernier
233	Pelle Prestberg
245	Andreas Andersson

1997

Pick	
18	Mikael Holmqvist
45	Maxim Balmochnykh
72	Jay Legault
125	Luc Vaillancourt
178	Tony Mohagen
181	Mat Snesrud
209	Rene Stussi
235	Tommi Degerman

1996

Pick	
9	Ruslan Salei
35	Matt Cullen
117	Brendan Buckley
149	Blaine Russell
172	Timo Ahmaoja
198	Kevin Kellett
224	Tobias Johansson

1995

Pick	
4	Chad Kilger
29	Brian Wesenberg
55	Mike Leclerc
107	Igor Nikulin
133	Peter LeBoutillier
159	Mike LaPlante
185	Igor Karpenko

1994

Pick	
2	Oleg Tverdovsky
28	Johan Davidsson
67	Craig Reichert
80	Byron Briske
106	Pavel Trnka
132	Bates Battaglia
158	Rocky Welsing
184	Brad Englehart
236	Tommi Miettinen
262	Jeremy Stevenson

1993

Pick	
4	Paul Kariya
30	Nikolai Tsulygin
56	Valeri Karpov
82	Joel Gagnon
108	Mikhail Shtalenkov
134	Antti Aalto
160	Matt Peterson
186	Tom Askey
212	Vitali Kozel
238	Anatoli Fedotov
264	David Penney

General Manager

MURRAY, BRYAN
General Manager, Mighty Ducks of Anaheim.
Born in Shawville, Que., December 5, 1942.

After serving behind the bench during the 2001-02 season, Bryan Murray stepped down as head coach on May 2, 2002 in order to become the team's new general manager. Prior to joining the Mighty Ducks in 2001, Murray had served most recently as the Florida Panthers' vice president and general manager from 1994 to 2001. He also assumed head coaching duties with Florida during the 1997-98 season, prior to naming his brother Terry Murray as head coach before the 1998-99 season. Bryan joined the Panthers on August 1, 1994 and assembled a team that went to the Stanley Cup Finals in just its third year of existence (1996).

Prior to joining the Panthers, Murray was the general manager of the Detroit Red Wings from 1990 to 1994, also taking on head coaching duties for the first three seasons. He earned his first NHL head coaching job with the Washington Capitals, taking over on November 11, 1981. He spent the next eight and a half seasons with the Capitals, winning the Jack Adams Award as coach of the year in 1983-84. Murray led Washington to the postseason seven times during his tenure, including the club's first division title in 1988-89.

A graduate of McGill University, Murray spent four years as the athletic director and coach at the school. He left that post to become the coach of the Regina Pats (WHL) in 1979-80. Murray took over as coach of the Hershey Bears (AHL) the next season and was named the Minor League Coach of the Year by *The Hockey News* after leading Hershey to its best mark in 40 years.

NHL Coaching Record

		Regular Season				Playoffs		
Season	**Team**	**Games**	**W**	**L**	**T**	**Games**	**W**	**L**
1981-82	Washington	66	25	28	13	....	....	....
1982-83	Washington	80	39	25	16	4	1	3
1983-84	Washington	80	48	27	5	8	4	4
1984-85	Washington	80	46	25	9	5	2	3
1985-86	Washington	80	50	23	7	9	5	4
1986-87	Washington	80	38	32	10	7	3	4
1987-88	Washington	80	38	33	9	14	7	7
1988-89	Washington	80	41	29	10	6	2	4
1989-90	Washington	46	18	24	4	....	....	....
1990-91	Detroit	80	34	38	8	7	3	4
1991-92	Detroit	80	43	25	12	11	4	7
1992-93	Detroit	84	47	28	9	7	3	4
1997-98	Florida	59	17	31	11	....	....	....
2001-02	Anaheim	82	29	45	8	....	....	....
	NHL Totals	**1057**	**513**	**413**	**131**	**78**	**34**	**44**

Club Directory

Arrowhead Pond of Anaheim

Mighty Ducks of Anaheim
Arrowhead Pond of Anaheim
2695 Katella Ave.
Anaheim, CA 92806
Phone **714/940-2900**
FAX 714/940-2953
Ticket Information 877/WILDWING
www.mightyducks.com
Capacity: 17,174

Executive Management

Chairman and Governor	Paul Pressler
Senior Vice President and General Manager	Bryan Murray
Assistant General Manager	David McNab
Senior Vice President, Business Operations	Douglas Moss
Vice President, Finance and Administration	Andy Roundtree
Vice President, Business and Legal Affairs	Rick Schlesinger
Administrative Assistant, Senior Vice President and General Manager	Maureen Nyeholt
Administrative Assistant, Senior Vice President, Business Operations	Pat Navarro
Administrative Assistant, Finance and Administration	Meta Maynard
Sr. Paralegal	Tia Wood

Coaching Staff

Head Coach	Mike Babcock
Assistant Coaches	Lorne Henning, Paul MacLean
Goaltending Consultant	François Allaire

Hockey Club Operations

Director of Hockey Operations	Chuck Fletcher
Player Personnel Director	Tim Murray
Director of Amateur Scouting	Alain Chainey
Scouting Staff	Jan-Åke Danielson, Richard Green, Todd Hearty, Konstantin Krylov, Donald Marier, Wayne Meier, Tomas Prucha, Floyd Smith, Tom Watt
Head Athletic Trainer	Chris Phillips
Strength and Conditioning Coach	Sean Skahan
Equipment Manager	Mark O'Neill
Assistant Equipment Manager	John Allaway
Cincinnati Mighty Ducks (AHL) Head Coach	Brad Shaw
Cincinnati Assistant Coach	Darryl Williams
Team Physicians	Dr. Ronald Glousman, Dr. Craig Milhouse
Oral Surgeon	Dr. Jeff Pulver
Visiting Team Equipment Attendant	Chris Kincaid

Communications Department

Communications and Team Services Manager	Alex Gilchrist
Publications Manager	Doug Ward
Sr. Media Relations Representative	Merit Tully
Team Photographer	V.J. Lovero (Lovero Group)

Finance and Administration Department

Director, Finance	Molly Taylor
Sr. Financial Analyst	Amy Langdale
Director, Human Resources	Jenny Price
Manager, Information Services	Al Castro
Sr. Network Engineer	Neil Fariss
Sr. End User Analyst	Phil Alger
Sr. Desktop Support Analyst	David Yun
Assistant Controller	Melody Martin
Accountant	Rosanna Sitzman
Accounting Assistants	Linda Chubak, Rob Dumlao
Administrative Assistant, Human Resources	Cindy Williams
Director, Ballpark Operations	John Drum
Assistant Operations Manager	Sam Maida
Sr. Travel Consultant	Chantelle Ball
General Manager, Disney ICE	Art Trottier
Receptionist	Jeannette Radillo

Sales and Marketing Department

Director, Ticket Sales & Customer Service	Bill Chapin
Director, Marketing and Promotions	TBA
Premium Ticket Services Specialist	Lisa Yamamoto
Sr. Marketing Representative	Jared Rice
Telemarketing Supervisor	TBA
Account Executives	Ron Campbell, Bob Ruiz, William Schaeffer
Group Sales Account Executives	Ken Bamberg, Patricia Perez-Freund, Derek Wilson
Ticketing Representative	Mandi Van Eps
Executive Secretary	Roxanne Gandara

Ticketing Department

Ticketing Operations Manager	Christa Richards
Ticketing Supervisor	Jonas Calicdan

Publicity and Community Development

Director of Publicity and Community Development	Charles Harris
Publicity and Broadcast Manager	Aaron Tom
Community Development Manager	Erin Bickmeier
Website Editor	Terry Crowley

Broadcasting and Entertainment

Telecast Producer	Aaron Teats
Telecast Director	Mike Levy
Entertainment Manager	Rod Murray
Producer, Video & Scoreboard Operations	Robert Castillo
Associate Producer, Video & Scoreboard Operations	David Tsuruda
Television, KCAL (Ch. 9) & Fox Sports West 2 (Cable)	John Ahlers, Brian Hayward
Radio, Flagship TBA & Mighty Ducks Radio Network	Steve Carroll

Advertising Sales and Broadcasting Department

Director, Advertising Sales	John Covarrubias
Sponsorship Services Manager	Tamara Goddard
Advertising Sales Managers	Andrew Campagnone, Alexander Grant
Executive Secretary, Advertising Sales	Jacklyn Perkins

Miscellaneous

Practice Facilities	Disney ICE (300 W. Lincoln Ave.) and the Arrowhead Pond (2695 Katella Ave.)
Primary Development Affiliate	Cincinnati Mighty Ducks (AHL)

Atlanta Thrashers

2001-02 Results: 19W-47L-11T-5OTL 54PTS.
Fifth, Southeast Division

Year-by-Year Record

		Home				Road				Overall								
Season	GP	W	L	T	OL	W	L	T	OL	W	L	T	OL	GF	GA	Pts.	Finished	Playoff Result
2001-02	82	11	21	9	0	8	26	2	5	19	47	11	5	187	288	54	5th, Southeast Div.	Out of Playoffs
2000-01	82	10	23	6	2	13	22	6	0	23	45	12	2	211	289	60	4th, Southeast Div.	Out of Playoffs
1999-2000	82	9	26	3	3	5	31	4	1	14	57	7	4	170	313	39	5th, Southeast Div.	Out of Playoffs

2002-03 Schedule

Oct.	Fri.	11	at Carolina
	Sat.	12	Florida
	Wed.	16	at Pittsburgh
	Fri.	18	at Tampa Bay
	Sat.	19	NY Islanders
	Mon.	21	at Florida
	Wed.	23	New Jersey
	Sat.	26	at Boston
	Tue.	29	Los Angeles
	Thu.	31	at Toronto
Nov.	Sat.	2	at Florida
	Thu.	7	at Chicago
	Sat.	9	at Buffalo
	Mon.	11	Calgary*
	Wed.	13	San Jose
	Fri.	15	Phoenix
	Sun.	17	Anaheim*
	Tue.	19	Florida
	Fri.	22	Pittsburgh
	Sat.	23	at Washington
	Tue.	26	at Montreal
	Thu.	28	NY Rangers
Dec.	Sun.	1	Washington*
	Thu.	5	at Boston
	Fri.	6	at Washington
	Sun.	8	Edmonton*
	Wed.	11	at Phoenix
	Fri.	13	at Dallas
	Sat.	14	at St. Louis
	Mon.	16	Toronto
	Wed.	18	Philadelphia
	Fri.	20	Carolina
	Mon.	23	at Toronto
	Fri.	27	at Carolina
	Sat.	28	Boston
	Mon.	30	at Pittsburgh
Jan.	Thu.	2	at Ottawa
	Fri.	3	Pittsburgh
	Sun.	5	Philadelphia*
	Tue.	7	Carolina
	Thu.	9	at Tampa Bay
	Sat.	11	at NY Islanders
	Mon.	13	at Philadelphia
	Wed.	15	Montreal
	Fri.	17	Boston
	Sun.	19	NY Islanders*
	Tue.	21	St. Louis
	Thu.	23	Ottawa
	Sat.	25	at NY Rangers*
	Tue.	28	NY Rangers
	Thu.	30	Toronto
Feb.	Tue.	4	at Montreal
	Fri.	7	at New Jersey
	Sat.	8	at Ottawa
	Wed.	12	Washington
	Fri.	14	Tampa Bay
	Sat.	15	Detroit
	Mon.	17	Buffalo*
	Wed.	19	at Tampa Bay
	Sun.	23	at Edmonton
	Tue.	25	at Vancouver
	Thu.	27	at Colorado
Mar.	Sat.	1	at Los Angeles*
	Sun.	2	at Anaheim*
	Thu.	6	at Washington
	Fri.	7	Florida
	Sun.	9	Minnesota*
	Tue.	11	at New Jersey
	Thu.	13	Montreal
	Sat.	15	Buffalo
	Mon.	17	Columbus
	Wed.	19	Dallas
	Fri.	21	Ottawa
	Sat.	22	at Columbus
	Mon.	24	at Philadelphia
	Wed.	26	Carolina
	Fri.	28	New Jersey
	Sat.	29	at Nashville
	Mon.	31	at NY Rangers
Apr.	Wed.	2	at Buffalo
	Sat.	5	at NY Islanders*
	Sun.	6	Tampa Bay*

* Denotes afternoon game.

Franchise date: June 25, 1997

SOUTHEAST DIVISION

4th NHL Season

Atlanta's first choice, second overall, in the 2000 NHL Entry Draft, Dany Heatley won the Calder Trophy as rookie of the year in 2001-02. The NHL's top-scoring freshman also led the Thrashers with 67 points (26 goals, 41 assists).

2002-03 Player Personnel

FORWARDS	HT	WT	S	Place of Birth	Date	2001-02 Club
BARTECKO, Lubos	5-11	200	L	Kezmarok, Czech.	7/14/76	Atlanta-Slovakia
BLATNY, Zdenek	6-1	190	L	Brno, Czech.	1/14/81	Chi (AHL)-Greenville
BUTSAYEV, Yuri	6-0	195	L	Togliatti, USSR	10/11/78	Det-Cin-Atl-Chi (AHL)
COWAN, Jeff	6-2	215	L	Scarborough, Ont.	9/27/76	Calgary-Atlanta
GAMACHE, Simon	5-9	185	L	Montreal, Que.	1/3/81	Chi (AHL)-Greenville
HARTIGAN, Mark	6-0	200	L	Fort St. John, B.C.	10/15/77	St. Cloud State-Atlanta
HEATLEY, Dany	6-3	210	L	Freiburg, West Germany	1/21/81	Atlanta
HERPERGER, Chris	6-0	190	L	Esterhazy, Sask.	2/24/74	Ottawa
HRKAC, Tony	5-10	190	L	Thunder Bay, Ont.	7/7/66	Atlanta
KACZOWKA, David	6-3	220	L	Regina, Sask.	7/5/81	Greenville-Chi (AHL)
KALLIO, Tomi	6-0	190	L	Turku, Finland	1/27/77	Atlanta-Finland
KARLSSON, Andreas	6-4	205	L	Ludvika, Sweden	8/19/75	Atlanta-Chicago (AHL)
KOVALCHUK, Ilya	6-1	220	R	Tver, USSR	4/15/83	Atlanta-Russia
KOZLOV, Vyacheslav	5-10	185	L	Voskresensk, USSR	5/3/72	Buffalo
MacKENZIE, Derek	5-11	175	L	Sudbury, Ont.	6/11/81	Atlanta-Chicago (AHL)
McEACHERN, Shawn	5-11	200	L	Waltham, MA	2/28/69	Ottawa
ODGERS, Jeff	5-11	200	R	Spy Hill, Sask.	5/31/69	Atlanta
PIROS, Kamil	6-0	200	L	Most, Czech.	11/20/78	Atlanta-Chicago (AHL)
RHEAUME, Pascal	6-1	210	L	Quebec City, Que.	6/21/73	Chicago (AHL)-Atlanta
SIMON, Ben	6-0	195	L	Shaker Heights, OH	6/14/78	Atlanta-Chicago (AHL)
SNYDER, Dan	6-0	185	L	Elmira, Ont.	2/23/78	Atlanta-Chicago (AHL)
STEFAN, Patrik	6-3	205	L	Pribram, Czech.	9/16/80	Atlanta-Chicago (AHL)
SVARTVADET, Per	6-1	195	L	Solleftea, Sweden	5/17/75	Atlanta
TAPPER, Brad	6-0	185	R	Scarborough, Ont.	4/28/78	Atlanta-Chicago (AHL)
VIGIER, Jean-Pierre	6-1	190	R	Notre Dame de Lourdes, Man.	9/11/76	Atlanta-Chicago (AHL)
DEFENSEMEN						
DiPENTA, Joe	6-2	235	L	Barrie, Ont.	2/25/79	Phi (AHL)-Chi (AHL)
EAKINS, Dallas	6-2	195	L	Dade City, FL	2/27/67	Calgary-Chicago (AHL)
EXELBY, Garnet	6-1	210	L	Craik, Sask.	8/16/81	Chicago (AHL)
FLACHE, Paul	6-5	215	R	Toronto, Ont.	3/4/82	Brampton
FOSTER, Kurtis	6-5	230	R	Carp, Ont.	11/24/81	Peterborough-Chi (AHL)
KABERLE, Frantisek	6-0	190	L	Kladno, Czech.	11/8/73	Atlanta
KRUPP, Uwe	6-6	235	R	Cologne, West Germany	6/24/65	Detroit
LESSARD, Francis	6-2	220	R	Montreal, Que.	5/30/79	Phi (AHL)-Atl-Chi (AHL)
LEVOKARI, Pauli	6-7	260	L	Luvia, Finland	4/7/79	HIFK Helsinki
SAFRONOV, Kirill	6-1	210	L	Leningrad, USSR	2/26/81	Phx-Sprfld-Atl-Chi (AHL)
SELLARS, Luke	6-1	205	L	Toronto, Ont.	5/21/81	Atl-Chi (AHL)-Greenville
SMEHLIK, Richard	6-4	222	L	Ostrava, Czech.	1/23/70	Buffalo-Czech Republic
SUTTON, Andy	6-6	245	L	Edmonton, Alta.	3/10/75	Minnesota-Atlanta
TAMER, Chris	6-2	205	L	Dearborn, MI	11/17/70	Atlanta
TJARNQVIST, Daniel	6-2	190	L	Umea, Sweden	10/14/76	Atlanta
TREMBLAY, Yannick	6-2	200	R	Pointe-aux-Trembles, Que.	11/15/75	Atlanta
USTRNUL, Libor	6-5	230	L	Steruberk, Czech.	2/20/82	Plymouth-Chi (AHL)
WEAVER, Mike	5-9	180	R	Bramalea, Ont.	5/2/78	Atlanta-Chicago (AHL)

GOALTENDERS	HT	WT	C	Place of Birth	Date	2001-02 Club
CASSIVI, Frederic	6-4	220	L	Sorel, Que.	6/12/75	Hershey-Atl-Chi (AHL)
GARNETT, Michael	6-1	185	L	Saskatoon, Sask.	11/25/82	Saskatoon
HNILICKA, Milan	5-11	190	L	Kladno, Czech.	6/25/73	Atlanta
MARACLE, Norm	5-9	195	L	Belleville, Ont.	10/2/74	Atlanta-Chicago (AHL)
NURMINEN, Pasi	5-10	190	L	Lahti, Finland	12/17/75	Atl-Chi (AHL)-Finland

General Managers' History

Don Waddell, 1999-2000 to date.

General Manager

WADDELL, DON

General Manager, Atlanta Thrashers. Born in Detroit, MI, August 19, 1958.

Don Waddell serves as vice president and general manager of the Atlanta Thrashers. He came to the Thrashers on June 23, 1998, — almost a year to the day after the NHL granted Atlanta a franchise — bringing with him more than 20 years experience in professional hockey as a player, coach and general manager. During his tenure in Atlanta, Waddell has used a combination of intelligent draft picks, shrewd trades and astute free agent signings to work towards the long-term success of the Thrashers. Previously, he built two professional hockey franchises, the San Diego Gulls and the Orlando Solar Bears of the International Hockey League. He's also no stranger to winning through his role as assistant general manager for the NHL's Stanley Cup champion Detroit Red Wings during the 1997-98 season.

Prior to Detroit, Waddell was vice president of RDV Sports, where he served on the Executive Committee which oversaw operations of the National Basketball Association's Orlando Magic, the International Hockey League's Orlando Solar Bears, Magic Fanattics (retail) and Magic Carpet Aviation. While at RDV Sports, Waddell was vice president and general manager of the IHL's Orlando Solar Bears from 1995 to 1997. Prior to the Solar Bears, he held the same role with the IHL's San Diego Gulls from 1990 to 1995. He also served as the club's head coach for the 1991-92 season, guiding the team to the franchise's first playoff berth. He spent two seasons with the IHL's Flint Spirits where he served as head coach and general manager in 1988-89, and general manager in 1989-90.

Waddell's playing experience includes being player/coach for the Flint Spirits from 1986 to 1988, and the Goaldiggers Hockey Club in Toledo, Ohio for the 1985-86 season. He was drafted by the NHL's Los Angeles Kings back in 1978, and spent three years with the organization from 1980 to 1983. He was a member of the 1983 U.S. national team and had been a member of the 1980 gold medal Olympic hockey team, but was injured prior to play.

Waddell played Division I hockey at Northern Michigan University from 1976 to 1980, where he majored in business management. He was inducted into the Northern Michigan University Sports Hall of Fame in 1992.

2001-02 Scoring

* - rookie

Regular Season

Pos	#	Player	Team	GP	G	A	Pts	+/–	PIM	PP	SH	GW	GT	S	%
R	15	* Dany Heatley	ATL	82	26	41	67	–19	56	7	0	4	0	202	12.9
R	17	* Ilya Kovalchuk	ATL	65	29	22	51	–19	28	7	0	4	1	184	15.8
C	12	Tony Hrkac	ATL	80	18	26	44	–12	12	5	1	2	0	101	17.8
R	18	Lubos Bartecko	ATL	71	13	14	27	–15	30	1	0	0	0	96	13.5
D	8	Frantisek Kaberle	ATL	61	5	20	25	–11	24	1	0	0	0	82	6.1
D	38	Yannick Tremblay	ATL	66	9	15	24	–15	47	1	0	1	1	115	7.8
C	13	Patrik Stefan	ATL	59	7	16	23	–4	22	0	1	0	0	67	10.4
C	26	Pascal Rheaume	CHI	19	0	2	2	–1	4	0	0	0	0	19	0.0
			ATL	42	11	9	20	–3	25	6	0	2	0	61	18.0
			TOTAL	61	11	11	22	–4	29	6	0	2	0	80	13.8
L	14	Tomi Kallio	ATL	60	8	14	22	–8	12	1	0	0	0	102	7.8
D	36	* Daniel Tjarnqvist	ATL	75	2	16	18	–22	14	1	0	0	0	68	2.9
C	39	Per Svartvadet	ATL	78	3	12	15	–12	24	0	0	0	0	80	3.8
D	25	Andy Sutton	MIN	19	2	4	6	–4	35	1	0	0	0	21	9.5
			ATL	24	0	4	4	0	46	0	0	0	0	20	0.0
			TOTAL	43	2	8	10	–4	81	1	0	0	0	41	4.9
D	3	* Brian Pothier	ATL	33	3	6	9	–19	22	1	0	1	0	65	4.6
R	20	Jeff Odgers	ATL	46	4	4	8	–3	135	0	0	1	0	34	11.8
D	28	Todd Reirden	ATL	65	3	5	8	–25	82	1	0	0	0	85	3.5
C	24	Andreas Karlsson	ATL	42	1	7	8	–8	20	0	0	0	0	41	2.4
L	16	Jeff Cowan	CGY	19	1	0	1	–3	40	0	0	1	0	13	7.7
			ATL	38	4	1	5	–11	50	0	0	1	0	51	7.8
			TOTAL	57	5	1	6	–14	90	0	0	2	0	64	7.8
L	16	Jeff Cowan	ATL	57	5	1	6	–14	90	0	0	2	0	64	7.8
D	4	Chris Tamer	ATL	78	3	3	6	–11	111	0	1	0	0	66	4.5
R	19	* Brad Tapper	ATL	20	2	4	6	–3	43	0	0	0	0	34	5.9
R	47	* Jean-Pierre Vigier	ATL	15	4	1	5	–5	4	0	0	0	0	18	22.2
C	10	Yuri Butsayev	DET	3	0	0	0	–1	0	0	0	0	0	4	0.0
			ATL	8	2	0	2	1	4	0	0	0	0	6	33.3
			TOTAL	11	2	0	2	0	4	0	0	0	0	10	20.0
C	37	* Dan Snyder	ATL	11	1	1	2	–3	30	0	0	0	0	7	14.3
C	22	* Kamil Piros	ATL	8	0	1	1	–2	4	0	0	0	0	4	0.0
D	43	* Mike Weaver	ATL	16	0	1	1	0	10	0	0	0	0	9	0.0
D	6	David Harlock	ATL	19	0	1	1	–2	18	0	0	0	0	12	0.0
D	40	* Luke Sellars	ATL	1	0	0	0	0	2	0	0	0	0	0	0.0
L	23	* Derek Mackenzie	ATL	1	0	0	0	–1	2	0	0	0	0	1	0.0
D	23	Brett Clark	ATL	2	0	0	0	–3	0	0	0	0	0	0	0.0
C	9	* Mark Hartigan	ATL	2	0	0	0	–2	2	0	0	0	0	3	0.0
D	29	* Kiril Safronov	PHX	1	0	0	0	–2	0	0	0	0	0	0	0.0
			ATL	2	0	0	0	–3	2	0	0	0	0	2	0.0
			TOTAL	3	0	0	0	–5	2	0	0	0	0	2	0.0
D	6	* Francis Lessard	ATL	5	0	0	0	0	26	0	0	0	0	2	0.0
C	45	* Ben Simon	ATL	6	0	0	0	1	6	0	0	0	0	7	0.0

Goaltending

No.	Goaltender	GPI	Mins	Avg	W	L	T	EN	SO	GA	SA	S%	G	A	PIM
34	Norm Maracle	1	60	3.00	0	1	0	0	0	3	20	.850	0	0	0
33	Milan Hnilicka	60	3367	3.19	13	33	10	9	3	179	1956	.908	0	2	8
35	Frederic Cassivi	6	307	3.32	2	3	0	1	0	17	207	.918	0	0	0
31	* Pasi Nurminen	9	465	3.61	2	5	0	1	0	28	275	.898	0	0	0
1	Damian Rhodes	15	769	3.67	2	10	1	3	0	47	439	.893	0	0	0
	Totals	**82**	**4994**	**3.46**	**19**	**52**	**11**	**14**	**3**	**288**	**2911**	**.901**			

Captains' History

Kelly Buchberger, 1999-2000; Steve Staios, 2000-01; Ray Ferraro, 2001-02.

Coaching History

Curt Fraser, 1999-2000 to date.

Club Records

Team

(Figures in brackets for season records are games played.)

Record		
Most Points	60	2000-01 (82)
Most Wins	23	2000-01 (82)
Most Ties	12	2000-01 (82)
Most Losses	61	1999-2000 (82)
Most Goals	211	2000-01 (82)
Most Goals Against	313	1999-2000 (82)
Fewest Points	39	1999-2000 (82)
Fewest Wins	14	1999-2000 (82)
Fewest Ties	7	1999-2000 (82)
Fewest Losses	45	2000-01 (82)
Fewest Goals	170	1999-2000 (82)
Fewest Goals Against	288	2001-02 (82)
Longest Winning Streak		
Overall	3	Dec. 1-4/00
Home	2	Six times
Away	2	Four times
Longest Undefeated Streak		
Overall	4	Oct. 21-28/00 (1 win, 3 ties), Jan. 9-15/02 (3 wins, 1 tie)
Home	3	Feb. 13-Mar. 3/01 (1 win, 2 ties), Jan. 9-13/02 (2 wins, 1 tie)
Away	7	Oct. 21-Nov. 13/00 (3 wins, 4 ties)
Longest Losing Streak		
Overall	12	Jan. 24-Feb. 20/00
Home	*11	Jan. 24-Mar. 16/00
Away	10	Oct. 6-Nov. 18/01
Longest Winless Streak		
Overall	16	Jan. 16-Feb. 20/00 (2 ties, 14 losses)
Home	*17	Jan. 19-Mar. 29/00 (2 ties, 15 losses)
Away	10	Oct. 6-Nov. 18/01 (10 losses)
Most Shutouts, Season	3	2001-02 (82)
Most PIM, Season	1,500	2000-01 (82)
Most Goals, Game	6	Four times

Individual

Most Seasons	3	Many players
Most Games	229	Chris Tamer
Most Goals, Career	56	Ray Ferraro
Most Assists, Career	91	Ray Ferraro
Most Points, Career	147	Ray Ferraro (56G, 91A)
Most PIM, Career	434	Denny Lambert
Most Shutouts, Career	5	Milan Hnilicka
Longest Consecutive Games Streak	103	Chris Tamer (Mar. 16/00-Oct. 23/01)
Most Goals, Season	32	Donald Audette (2000-01)
Most Assists, Season	47	Ray Ferraro (2000-01)
Most Points, Season	76	Ray Ferraro (2000-01; 29G, 47A)
Most PIM, Season	226	Jeff Odgers (2000-01)
Most Points, Defenseman, Season	31	Yannick Tremblay (1999-2000; 10G, 21A)
Most Points, Center, Season	76	Ray Ferraro (2000-01; 29G, 47A)
Most Points, Right Wing, Season	71	Donald Audette (2000-01; 32G, 39A)
Most Points, Left Wing, Season	59	Andrew Brunette (2000-01; 15G, 44A)
Most Points, Rookie, Season	67	Dany Heatley (2001-02; 26G, 41A)
Most Shutouts, Season	3	Milan Hnilicka (2001-02)
Most Goals, Game	4	Pascal Rheaume (Jan. 19/02)
Most Assists, Game	4	Andrew Brunette (Dec. 19/00), Ilya Kovalchuk (Jan. 19/02)
Most Points, Game	5	Ilya Kovalchuk (Jan. 19/02; 1G, 4A), Pascal Rheaume (Jan. 19/02; 4G, 1A)

* NHL Record.

Milan Hnilicka was the top goaltender for the Czech Republic teams that won the World Championships in 1999 and 2001. He played 60 games in goal for the Thrashers last season.

2001-02 Results

Month	Date	Opponent	Score
Oct.	4	at Buffalo	2-1
	6	at Boston	3-4*
	13	Carolina	2-5
	16	Philadelphia	3-3
	19	NY Rangers	3-4
	20	at Carolina	1-2*
	23	Pittsburgh	2-4
	26	Washington	1-0
	27	Tampa Bay	4-3
	30	Ottawa	3-6
Nov.	1	at San Jose	2-5
	3	at Los Angeles	1-4
	4	at Anaheim	0-5
	7	at New Jersey	2-3
	8	at Buffalo	0-8
	10	at Washington	0-3
	13	at Minnesota	2-4
	16	Nashville	4-4
	18	at NY Rangers	2-6
	19	Buffalo	3-2
	22	Montreal	2-5
	24	at Ottawa	6-3
	27	at Montreal	1-5
	29	at Tampa Bay	2-5
Dec.	1	at Florida	5-2
	4	Boston	2-3
	6	Washington	3-3
	8	at Pittsburgh	3-6
	10	Philadelphia	1-3
	12	Montreal	3-3
	14	Chicago	1-3
	15	at Washington	2-5
	18	at Boston	3-2*
	19	San Jose	2-4
	21	at Carolina	4-5*
	23	Dallas	1-4
	26	Florida	3-3
	28	Toronto	5-4
	31	at Florida	3-4
Jan.	2	at Dallas	1-2
	3	at Phoenix	1-2
	6	NY Islanders	2-3
	8	at Philadelphia	4-7
	9	Ottawa	4-3*
	11	Calgary	1-0
	13	Tampa Bay	2-2
	15	at Toronto	3-2
	17	at Philadelphia	3-6
	19	at Florida	6-1
	22	Washington	0-3
	24	New Jersey	4-2
	26	at Pittsburgh	2-3*
	28	Phoenix	2-4
	30	Toronto	0-6
Feb.	5	Edmonton	3-2
	7	at New Jersey	3-3
	8	NY Rangers	1-2
	11	at Toronto	4-5
	12	at St. Louis	0-3
	26	Buffalo	2-1
Mar.	1	NY Islanders	4-3
	2	at NY Islanders	1-4
	4	at Montreal	3-5
	6	Anaheim	1-4
	8	Boston	0-3
	10	at NY Islanders	1-6
	12	Tampa Bay	4-4
	14	Colorado	2-0
	16	Vancouver	2-4
	18	Pittsburgh	2-4
	20	at Tampa Bay	2-4
	22	at NY Rangers	5-2
	23	at Ottawa	3-2*
	27	Minnesota	2-4
	30	at Detroit	1-4
Apr.	2	at Calgary	2-4
	3	at Colorado	0-6
	5	New Jersey	1-3
	7	at Carolina	1-1
	10	Florida	4-4
	12	at Columbus	4-5*
	14	Carolina	2-2

* – Overtime

All-time Record vs. Other Clubs

Regular Season

	At Home								On Road								Total							
	GP	W	L	T	OL	GF	GA	PTS	GP	W	L	T	OL	GF	GA	PTS	GP	W	L	T	OL	GF	GA	PTS
Anaheim	2	0	2	0	0	3	9	0	3	1	2	0	0	5	11	2	5	1	4	0	0	8	20	2
Boston	6	2	4	0	0	16	18	4	6	3	1	1	1	26	21	8	12	5	5	1	1	42	39	12
Buffalo	6	3	2	1	0	15	20	7	6	2	4	0	0	11	24	4	12	5	6	1	0	26	44	11
Calgary	3	2	0	1	0	6	4	5	2	0	2	0	0	4	9	0	5	2	2	1	0	10	13	5
Carolina	7	0	5	2	0	14	24	2	8	0	5	1	2	15	28	3	15	0	10	3	2	29	52	5
Chicago	3	1	2	0	0	7	8	2	1	0	1	0	0	0	1	0	4	1	3	0	0	7	9	2
Colorado	3	1	1	0	1	6	7	3	3	1	2	0	0	6	14	2	6	2	3	0	1	12	21	5
Columbus	1	0	1	0	0	0	3	0	2	1	0	0	1	6	6	3	3	1	1	0	1	6	9	3
Dallas	3	0	3	0	0	5	11	0	2	0	2	0	0	2	4	0	5	0	5	0	0	7	15	0
Detroit	2	0	2	0	0	5	13	0	3	0	2	0	1	3	11	1	5	0	4	0	1	8	24	1
Edmonton	2	1	1	0	0	3	5	2	2	1	1	0	0	7	6	2	4	2	2	0	0	10	11	4
Florida	7	2	1	4	0	21	23	8	8	4	3	1	0	26	21	9	15	6	4	5	0	47	44	17
Los Angeles	2	0	2	0	0	3	9	0	3	1	2	0	0	8	14	2	5	1	4	0	0	11	23	2
Minnesota	1	0	1	0	0	2	4	0	2	0	1	1	0	3	5	1	3	0	2	1	0	5	9	1
Montreal	6	0	4	2	0	7	23	2	6	1	5	0	0	13	22	2	12	1	9	2	0	20	45	4
Nashville	3	1	0	1	1	8	8	4	2	0	2	0	0	3	10	0	5	1	2	1	1	11	18	4
New Jersey	6	1	4	1	0	11	25	3	6	0	5	1	0	8	24	1	12	1	9	2	0	19	49	4
NY Islanders	6	2	3	1	0	20	23	5	6	3	3	0	0	15	20	6	12	5	6	1	0	35	43	11
NY Rangers	6	0	6	0	0	13	24	0	6	3	3	0	0	22	21	6	12	3	9	0	0	35	45	6
Ottawa	6	2	4	0	0	21	26	4	6	2	3	1	0	19	29	5	12	4	7	1	0	40	55	9
Philadelphia	6	1	3	1	1	14	20	4	6	0	4	2	0	15	27	2	12	1	7	3	1	29	47	6
Phoenix	2	0	2	0	0	4	7	0	3	0	3	0	0	4	11	0	5	0	5	0	0	8	18	0
Pittsburgh	6	0	5	0	1	14	26	1	6	0	5	0	1	10	26	1	12	0	10	0	2	24	52	2
St. Louis	2	0	2	0	0	5	11	0	2	0	2	0	0	1	7	0	4	0	4	0	0	6	18	0
San Jose	2	0	2	0	0	2	7	0	3	0	2	1	0	5	11	1	5	0	4	1	0	7	18	1
Tampa Bay	8	5	1	2	0	33	24	12	7	1	5	1	0	15	28	3	15	6	6	3	0	48	52	15
Toronto	5	2	2	0	1	11	19	5	5	2	3	0	0	12	19	4	10	4	5	0	1	23	38	9
Vancouver	3	1	2	0	0	11	12	2	1	0	0	1	0	1	1	1	4	1	2	1	0	12	13	3
Washington	8	3	3	2	0	14	20	8	7	0	6	1	0	9	26	1	15	3	9	3	0	23	46	9
Totals	**123**	**30**	**70**	**18**	**5**	**294**	**433**	**83**	**123**	**26**	**79**	**12**	**6**	**274**	**457**	**70**	**246**	**56**	**149**	**30**	**11**	**568**	**890**	**153**

Entry Draft Selections 2002-1999

2002 Pick		2001 Pick		2000 Pick		1999 Pick	
2	Kari Lehtonen	1	Ilya Kovalchuk	2	Dany Heatley	1	Patrik Stefan
30	Jim Slater	80	Michael Garnett	31	Ilja Nikulin	30	Luke Sellars
116	Patrick Dwyer	100	Brian Sipotz	42	Libor Ustrnul	68	Zdenek Blatny
124	Lane Manson	112	Milan Gajic	107	Carl Mallette	98	David Kaczowka
144	Paul Flache	135	Colin Stuart	108	Blake Robson	99	Rob Zepp
167	Brad Schell	189	Pasi Nurminen	147	Matt McRae	128	Derek MacKenzie
198	Nathan Oystrick	199	Matt Suderman	168	Zdenek Smid	159	Yuri Dobryshkin
230	Colton Fretter	201	Colin FitzRandolph	178	Jeff Dwyer	188	Stephan Baby
236	Tyler Boldt	262	Mario Cartelli	180	Darcy Hordichuk	217	Garnet Exelby
257	Pauli Levokari			230	Samu Isosalo	245	Tommi Santala
				242	Evan Nielsen	246	Raymond DiLauro
				244	Eric Bowen		
				288	Mark McRae		
				290	Simon Gamache		

Ilya Kovalchuk stepped directly into the NHL after the Thrashers selected him first overall in the 2001 NHL Entry Draft. His 29 goals last season led all rookies and he finished second to teammate Dany Heatley in Calder Trophy voting.

Coach

FRASER, CURT
Coach, Atlanta Thrashers. Born in Cincinnati, OH, January 12, 1958.

Curt Fraser became the first head coach in the history of the Atlanta Thrashers on July 14, 1999. Fraser had spent the previous four seasons as the head coach of the IHL's Orlando Solar Bears, where he worked with Thrashers g.m. Don Waddell from 1995 to 1997.

During Fraser's four years with Orlando, the Solar Bears posted a 192-111-25 record with four consecutive playoff appearances, including reaching the Eastern Conference Finals on three occasions and two Turner Cup Finals (1995-96 and 1998-99). He won eight of 12 playoff rounds and posted a 17-4 record in playoff elimination games during that span.

In 1996-97, Fraser led the club to its second consecutive 50-plus win season and eclipsed the 100-point plateau for the second time with Orlando and the third time in his IHL career. He led the club to a 16-game winning streak, the third longest in professional hockey history. He was also named as co-coach for the Eastern Conference at the 1997 IHL All-Star Game for the second consecutive season (with Orlando) and for the third time in his IHL career. He also represented Milwaukee in 1993.

During their 1995-96 inaugural season, the Solar Bears clinched the Central Division championship and became the first Eastern Conference expansion team to reach the Turner Cup Finals in the history of the IHL.

Fraser came to Orlando in 1995 after 12 years as a player in the NHL and five years as a coach in the professional ranks. He spent the 1994-95 season as an associate coach with the AHL's Syracuse Crunch, the top affiliate for the Vancouver Canucks. Before joining Syracuse, Fraser served as an assistant coach and later head coach of the IHL's Milwaukee Admirals.

Prior to reaching the coaching ranks, Fraser was a highly respected left winger in the NHL with Vancouver, Chicago and Minnesota. In 704 career games, he scored 193 goals, notched 433 points and amassed 1,306 penalty minutes. Originally a second round pick of Vancouver in 1978, Fraser established himself as a hard working and fearless player, who combined toughness with the ability to score.

Coaching Record

		Regular Season				Playoffs		
Season	**Team**	**Games**	**W**	**L**	**T**	**Games**	**W**	**L**
1992-93	Milwaukee (IHL)	82	49	23	10	6	2	4
1993-94	Milwaukee (IHL)	81	40	24	17	4	0	4
1995-96	Orlando (IHL)	82	52	24	6	23	11	12
1996-97	Orlando (IHL)	82	53	24	5	10	4	6
1997-98	Orlando (IHL)	82	42	30	10	17	9	8
1998-99	Orlando (IHL)	82	45	33	4	17	10	7
1999-2000	**Atlanta (NHL)**	**82**	**14**	**61**	**7**	....	....	....
2000-01	**Atlanta (NHL)**	**82**	**23**	**47**	**12**	....	....	....
2001-02	**Atlanta (NHL)**	**82**	**19**	**52**	**11**	....	....	....
	NHL Totals	**246**	**56**	**160**	**30**	....	....	....

Club Directory

Philips Arena

Atlanta Thrashers
One CNN Center
12 South
Atlanta, GA 30303
Phone **404/827-5300**
FAX 404/827-5769
www.atlantathrashers.com
Capacity: 18,545

Executive Management
President and Governor . . . . Stan Kasten
VP & Gen. Manager (Alt. Governor) . . . . Don Waddell
VP of Sales and Marketing . . . . Derek Schiller
Senior Vice President of Public Relations . . . . Greg Hughes
Team Counsel . . . . John Cooper

Hockey Operations
Dir. of Player Personnel . . . . Jack Ferreira
Dir. of Player Development & Evaluation . . . . Bob Owen
Head Coach . . . . Curt Fraser
Assistant Coaches . . . . Tim Bothwell, Steve Weeks
Chief Scout . . . . Dan Marr
Professional Scouts . . . . Mark Dobson, Peter Mahovlich
Amateur Scouts . . . . Bernd Freimuller, John Perpich, Normand Poisson, Marcel Comeau
Part-Time Scouts . . . . Evgeny Bogdanovich, Terry Brennan, Pat Carmichael, Pentti Katainen, Tehro Koskela, Jan Lindegren
Dir. of Team Services . . . . Michele Zarzaca
Director of Hockey Administration . . . . Larry Simmons
Strength and Conditioning Coach . . . . Ray Bear
Head Athletic Trainer . . . . Scott Green
Assistant Athletic Trainer . . . . Craig Brewer
Manager of Scouting/Video . . . . Jon Barkan
Massage Therapist . . . . Inar Treiguts
Head Equipment Manager . . . . Bobby Stewart
Assistant Equipment Managers . . . . Joe Guilmet, Rob Thomson
Team Physician . . . . Dr. Scott Gillogly
Team Internist . . . . Dr. William Whaley
Team Dentists . . . . Dr. Gary Saban, Dr. Lawrence Saltzman, Dr. Brett Silverman
Hockey Operations Coordinator . . . . Tony Borgford

Administration
Assistant to Stan Kasten/Office Manager . . . . Carole Harding
Assistant to Don Waddell/ Practice Facility Office Manager . . . . Leisa Strickland

Corporate Sales (Philips Arena Sports Marketing)
V.P. of Broadcast and Corp. Sales . . . . Tracy White
Dirs. of Broadcast and Corp. Sales . . . . Terri Cameron, Bill Abercrombie
Dir. of Sponsor Services . . . . Joan Lanier-Heath
Mgrs. of Broadcast and Corp. Sales . . . . Stewart Tanner, Arden Robbins
Mgr. of Broadcast Operations . . . . Diana Corbin
Mgr. Of Promotions/Sponsor Services . . . . Chris Carter
Account Executive . . . . Chris Beaudin
Sponsor Services Coordinators . . . . Robin Halliburton, Amanda Loftus, Cari Pawlicki, Adam Ragsdale

Finance/Accounting
Controller . . . . David Kane
Assistant Controller . . . . Darius Nixon
Senior Financial Analyst . . . . Lonna Donaldson
Financial Analyst . . . . Ryan Floyd

Human Resources
V.P. of Management Company/Turner Sports/ Sports Teams/Philips Arena/Ad Sales . . . . Tim Goodly
HR Director of Sports/Teams/Philips Arena . . . . Michelle Golden
Human Resources Advisor . . . . Nikki Tinsley
Human Resources Coordinator . . . . Stacy Higginbotham

Marketing
Director of Marketing . . . . Jim Pfeifer
Senior Manager of Marketing . . . . Rob Preiditsch
Manager of Community Relations . . . . Terri Hickman
Senior Manager of Game Presentation and Special Events . . . . Peter Sorckoff
Manager of Special Events . . . . Connie Zaleski
Promotions Coordinator . . . . Cameron Brent
Mascot Coordinator . . . . Javier Presas
Marketing Coordinator . . . . Ralph Humphlett
Fan Development Coordinator . . . . David Porter

Media Relations
Director of Media Relations . . . . Tom Hughes
Senior Manager of Media Relations . . . . Rob Koch
Manager of Publications . . . . Matt Musgrove
Multimedia Specialist . . . . John Heid
Website Specialist . . . . Kevin McCormack
Media Relations Assistant . . . . Katie McLennan

Television/Radio Broadcasting
Coordinating Television Producer . . . . Tim Kiely
Television Producers . . . . Howard Zalkowitz, Glenn Diamond
Television Director . . . . Jim Allen
Sr. Remote Operations Managers . . . . Tom Sahara, Tom Cox
Director of Radio Operations/ Radio Play by Play Broadcaster . . . . Dan Kamal
Radio Analyst . . . . Billy Jaffe
Broadcast Operations Associate . . . . Mary Moran
TV Broadcasters . . . . Matt McConnell, Darren Eliot

Ticket Sales
Director of Ticket Sales . . . . Dan Froehlich
Director of Ticket Operations . . . . Wendell Byrne
New Account Sales, Sr. Manager . . . . Keith Brennan
Group Sales Managers . . . . Evan Kellner, Sasha Trendley
New Account Sales Managers . . . . John Farrell, David Forrest, Larry Jones, John Morgan, Rob Stephenson
Client Services, Sr. Manager . . . . Scott Fillmore
Client Services Managers . . . . Renee Carter, Grady Landis, Jennifer Tibbitts

Joe Thornton had 68 points in just 66 games.

Boston Bruins

2001-02 Results: 43W-24L-6T-9OTL 101PTS.

First, Northeast Division

2002-03 Schedule

Month	Day	Date	Opponent	Month	Day	Date	Opponent
Oct.	Fri.	11	at Minnesota		Sat.	11	Toronto
	Mon.	14	at Colorado		Mon.	13	Pittsburgh
	Wed.	16	at Vancouver		Wed.	15	at Florida
	Thu.	17	at Calgary		Fri.	17	at Atlanta
	Sat.	19	at Edmonton		Sat.	18	Columbus
	Mon.	21	at Toronto		Mon.	20	Washington*
	Thu.	24	Ottawa		Thu.	23	at Pittsburgh
	Sat.	26	Atlanta		Sat.	25	Philadelphia
	Wed.	30	at Washington		Tue.	28	Nashville
	Thu.	31	Anaheim		Thu.	30	Chicago
Nov.	Sat.	2	NY Rangers	**Feb.**	Tue.	4	Colorado
	Thu.	7	at Detroit		Thu.	6	Montreal
	Sat.	9	Ottawa		Sat.	8	Pittsburgh*
	Mon.	11	Edmonton*		Tue.	11	at Montreal
	Tue.	12	at Buffalo		Fri.	14	at Florida
	Thu.	14	NY Islanders		Sat.	15	at Tampa Bay
	Sat.	16	at Philadelphia		Mon.	17	at Nashville
	Tue.	19	at Toronto		Wed.	19	at Carolina
	Thu.	21	Carolina		Fri.	21	at New Jersey
	Sat.	23	Buffalo		Sun.	23	at NY Islanders*
	Tue.	26	Calgary		Tue.	25	Dallas
	Fri.	29	Montreal*		Thu.	27	at NY Rangers
	Sat.	30	at Pittsburgh	**Mar.**	Sat.	1	Philadelphia*
Dec.	Tue.	3	St. Louis		Mon.	3	Vancouver
	Thu.	5	Atlanta		Tue.	4	at Carolina
	Sat.	7	Tampa Bay*		Thu.	6	NY Islanders
	Sun.	8	at NY Rangers*		Sat.	8	Washington*
	Tue.	10	Montreal		Sun.	9	at Chicago*
	Thu.	12	Ottawa		Tue.	11	at Ottawa
	Sat.	14	at Montreal		Thu.	13	New Jersey
	Wed.	18	at Buffalo		Sat.	15	Florida*
	Thu.	19	at Washington		Tue.	18	at Phoenix
	Sat.	21	Florida		Fri.	21	at San Jose
	Mon.	23	San Jose		Sat.	22	at Los Angeles
	Fri.	27	at Tampa Bay		Mon.	24	Toronto
	Sat.	28	at Atlanta		Thu.	27	at Philadelphia
	Mon.	30	New Jersey		Sat.	29	NY Rangers*
Jan.	Fri.	3	at NY Islanders		Mon.	31	Tampa Bay
	Sat.	4	Carolina	**Apr.**	Tue.	1	at Ottawa
	Tue.	7	at Toronto		Thu.	3	at New Jersey
	Fri.	10	at Buffalo		Sat.	5	Buffalo*

* Denotes afternoon game.

Franchise date: November 1, 1924

NORTHEAST DIVISION

79th NHL Season

Year-by-Year Record

		Home				Road				Overall								
Season	**GP**	**W**	**L**	**T**	**OL**	**W**	**L**	**T**	**OL**	**W**	**L**	**T**	**OL**	**GF**	**GA**	**Pts.**	**Finished**	**Playoff Result**
2001-02	82	23	11	2	5	20	13	4	4	43	24	6	9	236	201	101	1st, Northeast Div.	Lost Conf. Quarter-Final
2000-01	82	21	12	5	3	15	18	3	5	36	30	8	8	227	249	88	4th, Northeast Div.	Out of Playoffs
1999-2000	82	12	17	11	1	12	16	8	5	24	33	19	6	210	248	73	5th, Northeast Div.	Out of Playoffs
1998-99	82	22	10	9	...	17	20	4	...	39	30	13	...	214	181	91	3rd, Northeast Div.	Lost Conf. Semi-Final
1997-98	82	19	16	6	...	20	14	7	...	39	30	13	...	221	194	91	2nd, Northeast Div.	Lost Conf. Quarter-Final
1996-97	82	14	20	7	...	12	27	2	...	26	47	9	...	234	300	61	6th, Northeast Div.	Out of Playoffs
1995-96	82	22	14	5	...	18	17	6	...	40	31	11	...	282	269	91	2nd, Northeast Div.	Lost Conf. Quarter-Final
1994-95	48	15	7	2	...	12	11	1	...	27	18	3	...	150	127	57	3rd, Northeast Div.	Lost Conf. Quarter-Final
1993-94	84	20	14	8	...	22	15	5	...	42	29	13	...	289	252	97	2nd, Northeast Div.	Lost Conf. Semi-Final
1992-93	84	29	10	3	...	22	16	4	...	51	26	7	...	332	268	109	1st, Adams Div.	Lost Div. Semi-Final
1991-92	80	23	11	6	...	13	21	6	...	36	32	12	...	270	275	84	2nd, Adams Div.	Lost Conf. Championship
1990-91	80	26	9	5	...	18	15	7	...	44	24	12	...	299	264	100	1st, Adams Div.	Lost Conf. Championship
1989-90	80	23	13	4	...	23	12	5	...	46	25	9	...	289	232	101	1st, Adams Div.	Lost Final
1988-89	80	17	15	8	...	20	14	6	...	37	29	14	...	289	256	88	2nd, Adams Div.	Lost Div. Final
1987-88	80	24	13	3	...	20	17	3	...	44	30	6	...	300	251	94	2nd, Adams Div.	Lost Final
1986-87	80	25	11	4	...	14	23	3	...	39	34	7	...	301	276	85	3rd, Adams Div.	Lost Div. Semi-Final
1985-86	80	24	9	7	...	13	22	5	...	37	31	12	...	311	288	86	3rd, Adams Div.	Lost Div. Semi-Final
1984-85	80	21	15	4	...	15	19	6	...	36	34	10	...	303	287	82	4th, Adams Div.	Lost Div. Semi-Final
1983-84	80	25	12	3	...	24	13	3	...	49	25	6	...	336	261	104	1st, Adams Div.	Lost Div. Semi-Final
1982-83	80	28	6	6	...	22	14	4	...	50	20	10	...	327	228	110	1st, Adams Div.	Lost Conf. Championship
1981-82	80	24	12	4	...	19	15	6	...	43	27	10	...	323	285	96	2nd, Adams Div.	Lost Div. Final
1980-81	80	26	10	4	...	11	20	9	...	37	30	13	...	316	272	87	2nd, Adams Div.	Lost Prelim. Round
1979-80	80	27	9	4	...	19	12	9	...	46	21	13	...	310	234	105	2nd, Adams Div.	Lost Quarter-Final
1978-79	80	25	10	5	...	18	13	9	...	43	23	14	...	316	270	100	1st, Adams Div.	Lost Semi-Final
1977-78	80	29	6	5	...	22	12	6	...	51	18	11	...	333	218	113	1st, Adams Div.	Lost Final
1976-77	80	27	7	6	...	22	16	2	...	49	23	8	...	312	240	106	1st, Adams Div.	Lost Final
1975-76	80	27	5	8	...	21	10	9	...	48	15	17	...	313	237	113	1st, Adams Div.	Lost Semi-Final
1974-75	80	29	5	6	...	11	21	8	...	40	26	14	...	345	245	94	2nd, Adams Div.	Lost Prelim. Round
1973-74	78	33	4	2	...	19	13	7	...	52	17	9	...	349	221	113	1st, East Div.	Lost Final
1972-73	78	27	10	2	...	24	12	3	...	51	22	5	...	330	235	107	2nd, East Div.	Lost Quarter-Final
1971-72	**78**	**28**	**4**	**7**	**...**	**26**	**9**	**4**	**...**	**54**	**13**	**11**	**...**	**330**	**204**	**119**	**1st, East Div.**	**Won Stanley Cup**
1970-71	78	33	4	2	...	24	10	5	...	57	14	7	...	399	207	121	1st, East Div.	Lost Quarter-Final
1969-70	**76**	**27**	**3**	**8**	**...**	**13**	**14**	**11**	**...**	**40**	**17**	**19**	**...**	**277**	**216**	**99**	**2nd, East Div.**	**Won Stanley Cup**
1968-69	76	29	3	6	...	13	15	10	...	42	18	16	...	303	221	100	2nd, East Div.	Lost Semi-Final
1967-68	74	22	9	6	...	15	18	4	...	37	27	10	...	259	216	84	3rd, East Div.	Lost Quarter-Final
1966-67	70	10	21	4	...	7	22	6	...	17	43	10	...	182	253	44	6th,	Out of Playoffs
1965-66	70	15	17	3	...	6	26	3	...	21	43	6	...	174	275	48	5th,	Out of Playoffs
1964-65	70	12	17	6	...	9	26	0	...	21	43	6	...	166	253	48	6th,	Out of Playoffs
1963-64	70	13	15	7	...	5	25	5	...	18	40	12	...	170	212	48	6th,	Out of Playoffs
1962-63	70	7	18	10	...	7	21	7	...	14	39	17	...	198	281	45	6th,	Out of Playoffs
1961-62	70	9	22	4	...	6	25	4	...	15	47	8	...	177	306	38	6th,	Out of Playoffs
1960-61	70	13	17	5	...	2	25	8	...	15	42	13	...	176	254	43	6th,	Out of Playoffs
1959-60	70	21	11	3	...	7	23	5	...	28	34	8	...	220	241	64	5th,	Out of Playoffs
1958-59	70	21	11	3	...	11	18	6	...	32	29	9	...	205	215	73	2nd,	Lost Semi-Final
1957-58	70	15	14	6	...	12	14	9	...	27	28	15	...	199	194	69	4th,	Lost Final
1956-57	70	20	9	6	...	14	15	6	...	34	24	12	...	195	174	80	3rd,	Lost Final
1955-56	70	14	14	7	...	9	20	6	...	23	34	13	...	147	185	59	5th,	Out of Playoffs
1954-55	70	16	10	9	...	7	16	12	...	23	26	21	...	169	188	67	4th,	Lost Semi-Final
1953-54	70	22	8	5	...	10	20	5	...	32	28	10	...	177	181	74	4th,	Lost Semi-Final
1952-53	70	19	10	6	...	9	19	7	...	28	29	13	...	152	172	69	3rd,	Lost Final
1951-52	70	15	12	8	...	10	17	8	...	25	29	16	...	162	176	66	4th,	Lost Semi-Final
1950-51	70	13	12	10	...	9	18	8	...	22	30	18	...	178	197	62	4th,	Lost Semi-Final
1949-50	70	15	12	8	...	7	20	8	...	22	32	16	...	198	228	60	5th,	Out of Playoffs
1948-49	60	18	10	2	...	11	13	6	...	29	23	8	...	178	163	66	2nd,	Lost Semi-Final
1947-48	60	12	8	10	...	11	16	3	...	23	24	13	...	167	168	59	3rd,	Lost Semi-Final
1946-47	60	18	7	5	...	8	16	6	...	26	23	11	...	190	175	63	3rd,	Lost Semi-Final
1945-46	50	11	5	4	...	13	13	4	...	24	18	8	...	167	156	56	2nd,	Lost Final
1944-45	50	11	12	2	...	5	18	2	...	16	30	4	...	179	219	36	4th,	Lost Semi-Final
1943-44	50	15	8	2	...	4	18	3	...	19	26	5	...	223	268	43	5th,	Out of Playoffs
1942-43	50	17	3	5	...	7	14	4	...	24	17	9	...	195	176	57	2nd,	Lost Final
1941-42	48	17	4	3	...	8	13	3	...	25	17	6	...	160	118	56	3rd,	Lost Semi-Final
1940-41	**48**	**15**	**4**	**5**	**...**	**12**	**4**	**8**	**...**	**27**	**8**	**13**	**...**	**168**	**102**	**67**	**1st,**	**Won Stanley Cup**
1939-40	48	20	3	1	...	11	9	4	...	31	12	5	...	170	98	67	1st,	Lost Semi-Final
1938-39	**48**	**20**	**2**	**2**	**...**	**16**	**8**	**0**	**...**	**36**	**10**	**2**	**...**	**156**	**76**	**74**	**1st,**	**Won Stanley Cup**
1937-38	48	18	3	3	...	12	8	4	...	30	11	7	...	142	89	67	1st, Amn. Div.	Lost Semi-Final
1936-37	48	9	11	4	...	14	7	3	...	23	18	7	...	120	110	53	2nd, Amn. Div.	Lost Quarter-Final
1935-36	48	15	8	1	...	7	12	5	...	22	20	6	...	92	83	50	2nd, Amn. Div.	Lost Quarter-Final
1934-35	48	17	7	0	...	9	9	6	...	26	16	6	...	129	112	58	1st, Amn. Div.	Lost Semi-Final
1933-34	48	11	11	2	...	7	14	3	...	18	25	5	...	111	130	41	4th, Amn. Div.	Out of Playoffs
1932-33	48	19	2	3	...	6	13	5	...	25	15	8	...	124	88	58	1st, Amn. Div.	Lost Semi-Final
1931-32	48	11	10	3	...	4	11	9	...	15	21	12	...	122	117	42	4th, Amn. Div.	Out of Playoffs
1930-31	44	16	1	5	...	12	9	1	...	28	10	6	...	143	90	62	1st, Amn. Div.	Lost Semi-Final
1929-30	44	21	1	0	...	17	4	1	...	38	5	1	...	179	98	77	1st, Amn. Div.	Lost Final
1928-29	**44**	**15**	**6**	**1**	**...**	**11**	**7**	**4**	**...**	**26**	**13**	**5**	**...**	**89**	**52**	**57**	**1st, Amn. Div.**	**Won Stanley Cup**
1927-28	44	13	4	5	...	7	9	6	...	20	13	11	...	77	70	51	1st, Amn. Div.	Lost Semi-Final
1926-27	44	15	7	0	...	6	13	3	...	21	20	3	...	97	89	45	2nd, Amn. Div.	Lost Final
1925-26	36	10	7	1	...	7	8	3	...	17	15	4	...	92	85	38	4th,	Out of Playoffs
1924-25	30	3	12	0	...	3	12	0	...	6	24	0	...	49	119	12	6th,	Out of Playoffs

2002-03 Player Personnel

FORWARDS	HT	WT	S	Place of Birth	Date	2001-02 Club
AXELSSON, P.J.	6-1	175	L	Kungalv, Sweden	2/26/75	Boston-Sweden
BOWLER, Bill	5-9	180	L	Toronto, Ont.	9/25/74	Milwaukee-Norfolk
CORAZZINI, Carl	5-10	170	R	Framingham, MA	4/21/79	Providence (AHL)
GELLARD, Mike	6-1	193	L	Markham, Ont.	10/10/78	did not play
GOREN, Lee	6-3	205	R	Winnipeg, Man.	12/26/77	Providence (AHL)
GROSEK, Michal	6-2	207	R	Vyskov, Czech.	6/1/75	NY Rangers-Hartford
HENDERSON, Jay	5-11	190	L	Edmonton, Alta.	9/17/78	Boston
HERR, Matt	6-2	204	L	Hackensack, NJ	5/26/76	Florida-Hershey
HILBERT, Andy	5-11	190	L	Howell, MI	2/6/81	Boston-Prov (AHL)
HUML, Ivan	6-2	195	L	Kladno, Czech.	9/6/81	Boston-Prov (AHL)
KNUBLE, Mike	6-3	208	R	Toronto, Ont.	7/4/72	Boston
LAPOINTE, Martin	5-11	200	R	Ville St-Pierre, Que.	9/12/73	Boston
McINNIS, Marty	5-11	187	R	Hingham, MA	6/2/70	Anaheim-Boston
MURRAY, Glen	6-3	225	R	Halifax, N.S.	11/1/72	Los Angeles-Boston
ORR, Colton	6-2	210	R	Winnipeg, Man.	3/3/82	Kamloops
PARADISE, Chris	6-2	200	R	St. Paul, MN	8/6/77	U. of Denver
ROLSTON, Brian	6-2	205	L	Flint, MI	2/21/73	Boston-United States
SAMSONOV, Sergei	5-8	180	R	Moscow, USSR	10/27/78	Boston-Russia
SAMUELSSON, Martin	6-2	194	L	Upplands Vasby, Sweden	1/25/82	Hammarby Jr.-Hammarby
STOCK, P.J.	5-10	190	L	Victoriaville, Que.	5/26/75	Boston
STUMPEL, Jozef	6-3	225	R	Nitra, Czech.	7/20/72	L.A.-Bos-Slovakia
THORNTON, Joe	6-4	220	L	London, Ont.	7/2/79	Boston
VAN OENE, Darren	6-4	216	L	Edmonton, Alta.	1/18/78	Rochester
VERNARSKY, Kris	6-3	201	L	Detroit, MI	4/5/82	Plymouth
ZAMUNER, Rob	6-3	203	L	Oakville, Ont.	9/17/69	Boston
DEFENSEMEN						
BERARD, Bryan	6-2	195	L	Woonsocket, RI	3/5/77	NY Rangers
BOYNTON, Nick	6-2	210	R	Nobleton, Ont.	1/14/79	Boston
BRENNAN, Rich	6-2	200	R	Schenectady, NY	11/26/72	Nashville-Milwaukee-Manchester
BROWN, Sean	6-3	205	L	Oshawa, Ont.	11/5/76	Edmonton-Boston
DALLMAN, Kevin	5-11	195	R	Niagara Falls, Ont.	2/26/81	Guelph
GILL, Hal	6-7	230	L	Concord, MA	4/6/75	Boston
GIRARD, Jonathan	5-11	192	R	Joliette, Que.	5/27/80	Boston-Prov (AHL)
KELLEHER, Chris	6-1	210	L	Cambridge, MA	3/23/75	Boston-Prov (AHL)
KULTANEN, Jarno	6-2	198	L	Luumaki, Finland	1/8/73	Boston
KUTLAK, Zdenek	6-3	207	L	Budejovice, Czech.	2/13/80	Providence (AHL)
McLAREN, Kyle	6-4	230	L	Humboldt, Sask.	6/18/77	Boston
METCALF, Peter	6-0	200	L	Colorado Springs, CO	2/25/79	U. of Maine
MORRISONN, Shaone	6-3	205	L	Vancouver, B.C.	12/23/82	Kamloops
O'DONNELL, Sean	6-3	230	L	Ottawa, Ont.	10/13/71	Boston
SWEENEY, Don	5-10	185	L	St. Stephen, N.B.	8/17/66	Boston

GOALTENDERS	HT	WT	C	Place of Birth	Date	2001-02 Club
GRAHAME, John	6-2	214	L	Denver, CO	8/31/75	Boston
RAYCROFT, Andrew	6-0	174	L	Belleville, Ont.	5/4/80	Boston-Prov (AHL)
SHIELDS, Steve	6-3	215	L	Toronto, Ont.	7/19/72	Anaheim
THOMAS, Tim	5-11	181	L	Flint, MI	4/15/74	Karpat

General Manager

O'CONNELL, MIKE
General Manager, Boston Bruins. Born in Chicago, IL, November 25, 1955.

Mike O'Connell was named the general manager of the Boston Bruins on November 1, 2000, becoming just the sixth man in club history to hold that position. He was involved in all aspects of the on-ice operation of the hockey team over the previous six seasons as the team's assistant general manager and was instrumental in bringing much of the young talent into the organization.

O'Connell's experience as both a player and assistant coach in the National Hockey League, and as a head coach in both the American and International Hockey Leagues, dates back to the 1977-78 season. Raised in Cohasset, MA, he played two years of high school hockey at Archbishop Williams High School in Braintree, MA. He then made what at the time was an unusual move for an American player, jumping to the Ontario Hockey League to play Canadian major junior hockey at the suggestion of Harry Sinden and Tom Johnson. The move proved beneficial as, after two seasons with Kingston of the OHL, he was drafted by Chicago 43rd overall in the 1975 NHL Amateur Draft.

He turned professional with the Blackhawks organization in 1975 and played five-plus seasons with Chicago and their Central Hockey League affiliate in Dallas before coming to Boston on December 18, 1980, in a trade for Al Secord. He enjoyed his best NHL seasons during his six years in a Bruins uniform, recording 50+ point campaigns for three straight years from 1982 to 1985 and representing the team in the 1984 NHL All-Star Game in New Jersey. He was traded to Detroit for Reed Larson on March 10, 1986 and concluded his playing career with the Red Wings at the end of the 1989-90 season.

O'Connell then moved into the coaching ranks, assuming the head coaching position for the IHL's San Diego Gulls in 1990-91. He then returned to the NHL, with Boston as an assistant coach. On June 12, 1992, he was named as the head coach of Boston's American Hockey League affiliate in Providence. Working with many players who also wore a Boston uniform during his tenure, he compiled a 74-71-15 record over a two-year span and won a Northern Division title in 1992-93. He then returned to Boston when he was named the club's assistant general manager on July 5, 1994. He was named as a vice president of the team in 1998.

2001-02 Scoring

* - rookie

Regular Season

Pos	#	Player	Team	GP	G	A	Pts	+/–	PIM	PP	SH	GW	GT	S	%
R	27	Glen Murray	L.A.	9	6	5	11	5	0	4	0	2	0	34	17.6
			BOS	73	35	25	60	26	40	5	0	7	0	212	16.5
			TOTAL	82	41	30	71	31	40	9	0	9	0	246	16.7
L	14	Sergei Samsonov	BOS	74	29	41	70	21	27	3	0	4	0	192	15.1
C	19	Joe Thornton	BOS	66	22	46	68	7	127	6	0	5	1	152	14.5
R	13	Bill Guerin	BOS	78	41	25	66	–1	91	10	1	7	0	355	11.5
C	12	Brian Rolston	BOS	82	31	31	62	11	30	6	9	7	0	331	9.4
C	16	Jozef Stumpel	L.A.	9	1	3	4	1	4	0	0	0	0	7	14.3
			BOS	72	7	47	54	21	14	1	0	3	0	93	7.5
			TOTAL	81	8	50	58	22	18	1	0	3	0	100	8.0
R	20	Martin Lapointe	BOS	68	17	23	40	12	101	4	0	2	1	141	12.1
R	10	Marty McInnis	ANA	60	9	14	23	–14	25	2	0	0	0	131	6.9
			BOS	19	2	3	5	–1	8	0	0	1	0	26	7.7
			TOTAL	79	11	17	28	–15	33	2	0	1	0	157	7.0
L	17	Rob Zamuner	BOS	66	12	13	25	6	24	1	2	0	0	98	12.2
D	21	Sean O'Donnell	BOS	80	3	22	25	27	89	1	0	2	0	112	2.7
L	11	P.J. Axelsson	BOS	78	7	17	24	6	16	0	2	0	0	127	5.5
D	25	Hal Gill	BOS	79	4	18	22	16	77	0	0	0	0	137	2.9
D	44 *	Nick Boynton	BOS	80	4	14	18	18	107	0	0	1	0	136	2.9
D	32	Don Sweeney	BOS	81	3	15	18	22	35	1	0	0	0	70	4.3
R	26	Mike Knuble	BOS	54	8	6	14	9	42	0	0	2	0	77	10.4
D	23	Sean Brown	EDM	61	6	4	10	8	127	3	0	1	0	58	10.3
			BOS	12	0	1	1	–1	47	0	0	0	0	6	0.0
			TOTAL	73	6	5	11	7	174	3	0	1	0	64	9.4
D	18	Kyle McLaren	BOS	38	0	8	8	–4	19	0	0	0	0	57	0.0
D	31	Jamie Rivers	OTT	2	0	0	0	–3	4	0	0	0	0	3	0.0
			BOS	64	4	2	6	6	45	1	0	1	0	48	8.3
			TOTAL	66	4	2	6	3	49	1	0	1	0	51	7.8
D	33	Jeff Norton	FLA	29	0	4	4	–5	8	0	0	0	0	13	0.0
			BOS	3	0	1	1	0	2	0	0	0	0	2	0.0
			TOTAL	32	0	5	5	–5	10	0	0	0	0	15	0.0
R	22	Dennis Bonvie	BOS	23	1	2	3	3	84	0	0	0	0	5	20.0
D	55	Jonathan Girard	BOS	20	0	3	3	0	9	0	0	0	0	28	0.0
D	64	Jarno Kultanen	BOS	38	0	3	3	–1	33	0	0	0	0	31	0.0
C	42	P.J. Stock	BOS	58	0	3	3	–2	122	0	0	0	0	12	0.0
D	6	Gord Murphy	BOS	15	0	2	2	2	13	0	0	0	0	10	0.0
C	23	John Emmons	BOS	22	0	2	2	–4	16	0	0	0	0	20	0.0
C	29 *	Andy Hilbert	BOS	6	1	0	1	–2	2	0	0	0	0	11	9.1
L	36 *	Ivan Huml	BOS	1	0	1	1	2	0	0	0	0	0	2	0.0
D	48	Chris Kelleher	BOS	1	0	0	0	0	0	0	0	0	0	0	0.0
R	46 *	Tony Tuzzolino	BOS	2	0	0	0	–1	0	0	0	0	0	1	0.0
D	43	Richard Jackman	BOS	2	0	0	0	–1	2	0	0	0	0	4	0.0
C	57	Eric Manlow	BOS	3	0	0	0	0	0	0	0	0	0	2	0.0
D	72	Pavel Kolarik	BOS	13	0	0	0	0	6	0	0	0	0	6	0.0

Goaltending

No.	Goaltender	GPI	Mins	Avg	W	L	T	EN	SO	GA	SA	S%	G	A	PIM
34	Byron Dafoe	64	3827	2.21	35	26	3	5	4	141	1520	.907	0	1	27
1	* Andrew Raycroft	1	65	2.77	0	0	1	0	0	3	29	.897	0	0	0
47	John Grahame	19	1079	2.89	8	7	2	0	1	52	503	.897	0	1	6
	Totals	**82**	**4993**	**2.42**	**43**	**33**	**6**	**5**	**6**	**201**	**2057**	**.902**			

Playoffs

Pos	#	Player	Team	GP	G	A	Pts	+/–	PIM	PP	SH	GW	GT	S	%
R	13	Bill Guerin	BOS	6	4	2	6	–2	6	3	0	0	0	25	16.0
C	19	Joe Thornton	BOS	6	2	4	6	0	10	0	0	0	0	14	14.3
C	12	Brian Rolston	BOS	6	4	1	5	–1	0	1	1	0	0	33	12.1
R	27	Glen Murray	BOS	6	1	4	5	–1	4	0	0	0	0	22	4.5
L	14	Sergei Samsonov	BOS	6	2	2	4	–2	0	0	0	0	0	26	7.7
L	11	P.J. Axelsson	BOS	6	2	1	3	2	6	0	1	1	0	9	22.2
R	20	Martin Lapointe	BOS	6	1	2	3	–2	12	1	0	1	0	11	9.1
D	44 *	Nick Boynton	BOS	6	1	2	3	2	8	0	0	0	0	20	5.0
L	17	Rob Zamuner	BOS	6	0	2	2	–2	4	0	0	0	0	8	0.0
C	16	Jozef Stumpel	BOS	6	0	2	2	–2	0	0	0	0	0	10	0.0
D	21	Sean O'Donnell	BOS	6	0	2	2	–2	4	0	0	0	0	6	0.0
C	42	P.J. Stock	BOS	6	1	0	1	0	19	0	0	0	0	2	50.0
D	32	Don Sweeney	BOS	6	0	1	1	–3	2	0	0	0	0	2	0.0
R	10	Marty McInnis	BOS	6	0	1	1	0	0	0	0	0	0	6	0.0
D	25	Hal Gill	BOS	6	0	1	1	1	2	0	0	0	0	10	0.0
R	22	Dennis Bonvie	BOS	1	0	0	0	0	0	0	0	0	0	0	0.0
D	55	Jonathan Girard	BOS	1	0	0	0	0	2	0	0	0	0	1	0.0
R	26	Mike Knuble	BOS	2	0	0	0	0	0	0	0	0	0	2	0.0
D	33	Jeff Norton	BOS	3	0	0	0	1	5	0	0	0	0	1	0.0
D	31	Jamie Rivers	BOS	3	0	0	0	0	0	0	0	0	0	1	0.0
D	18	Kyle McLaren	BOS	4	0	0	0	2	20	0	0	0	0	2	0.0
D	23	Sean Brown	BOS	4	0	0	0	0	2	0	0	0	0	1	0.0

Goaltending

No.	Goaltender	GPI	Mins	Avg	W	L	EN	SO	GA	SA	S%	G	A	PIM
34	Byron Dafoe	6	358	3.18	2	4	1	0	19	141	.865	0	0	0
	Totals	**6**	**360**	**3.33**	**2**	**4**	**1**	**0**	**20**	**142**	**.859**			

Captains' History

No captain, 1924-25 to 1926-27; Lionel Hitchman, 1927-28 to 1930-31; George Owen, 1931-32; Dit Clapper, 1932-33 to 1937-38; Cooney Weiland, 1938-39; Dit Clapper, 1939-40 to 1945-46; Dit Clapper and John Crawford, 1946-47; John Crawford 1947-48 to 1949-50; Milt Schmidt, 1950-51 to 1953-54; Milt Schmidt, Ed Sanford, 1954-55; Fern Flaman, 1955-56 to 1960-61; Don McKenney, 1961-62, 1962-63; Leo Boivin, 1963-64 to 1965-66; John Bucyk, 1966-67; no captain, 1967-68 to 1972-73; John Bucyk, 1973-74 to 1976-77; Wayne Cashman, 1977-78 to 1982-83; Terry O'Reilly, 1983-84, 1984-85; Raymond Bourque, Rick Middleton (co-captains) 1985-86 to 1987-88; Raymond Bourque, 1988-89 to 1999-2000; Jason Allison, 2000-01; no captain, 2001-02.

Club Records

Team

(Figures in brackets for season records are games played; records for fewest points, wins, ties, losses, goals, goals against are for 70 or more games)

Record		
Most Points	**121**	1970-71 (78)
Most Wins	**57**	1970-71 (78)
Most Ties	**21**	1954-55 (70)
Most Losses	**47**	1961-62 (70), 1996-97 (82)
Most Goals	**399**	1970-71 (78)
Most Goals Against	**306**	1961-62 (70)
Fewest Points	**38**	1961-62 (70)
Fewest Wins	**14**	1962-63 (70)
Fewest Ties	**5**	1972-73 (78)
Fewest Losses	**13**	1971-72 (78)
Fewest Goals	**147**	1955-56 (70)
Fewest Goals Against	**172**	1952-53 (70)
Longest Winning Streak		
Overall	**14**	Dec. 3/29-Jan. 9/30
Home	***20**	Dec. 3/29-Mar. 18/30
Away	**8**	Feb. 17-Mar. 8/72, Mar. 15-Apr. 14/93
Longest Undefeated Streak		
Overall	**23**	Dec. 22/40-Feb. 23/41 (15 wins, 8 ties)
Home	**27**	Nov. 22/70-Mar. 20/71 (26 wins, 1 tie)
Away	**15**	Dec. 22/40-Mar. 16/41 (9 wins, 6 ties)
Longest Losing Streak		
Overall	**11**	Dec. 3/24-Jan. 5/25
Home	***11**	Dec. 8/24-Feb. 17/25
Away	**14**	Dec. 27/64-Feb. 21/65
Longest Winless Streak		
Overall	**20**	Jan. 28-Mar. 11/62 (16 losses, 4 ties)
Home	**11**	Dec. 8/24-Feb. 17/25 (11 losses)
Away	**14**	Three times
Most Shutouts, Season	**15**	1927-28 (44)
Most PIM, Season	**2,443**	1987-88 (80)
Most Goals, Game	**14**	Jan. 21/45 (NYR 3 at Bos. 14)

Individual

Most Seasons	**21**	John Bucyk, Raymond Bourque
Most Games	**1,518**	Raymond Bourque
Most Goals, Career	**545**	John Bucyk
Most Assists, Career	**1,111**	Raymond Bourque
Most Points, Career	**1,506**	Raymond Bourque (395G, 1,111A)
Most PIM, Career	**2,095**	Terry O'Reilly
Most Shutouts, Career	**74**	Tiny Thompson
Longest Consecutive Games Streak	**418**	John Bucyk (Jan. 23/69-Mar. 2/75)
Most Goals, Season	**76**	Phil Esposito (1970-71)
Most Assists, Season	**102**	Bobby Orr (1970-71)
Most Points, Season	**152**	Phil Esposito (1970-71; 76G, 76A)
Most PIM, Season	**302**	Jay Miller (1987-88)
Most Points, Defenseman, Season	***139**	Bobby Orr (1970-71; 37G, 102A)
Most Points, Center, Season	**152**	Phil Esposito (1970-71; 76G, 76A)
Most Points, Right Wing, Season	**105**	Ken Hodge (1970-71; 43G, 62A), (1973-74; 50G, 55A), Rick Middleton (1983-84; 47G, 58A)
Most Points, Left Wing, Season	**116**	John Bucyk (1970-71; 51G, 65A)
Most Points, Rookie, Season	**102**	Joe Juneau (1992-93; 32G, 70A)
Most Shutouts, Season	**15**	Hal Winkler (1927-28)
Most Goals, Game	**4**	Twenty times
Most Assists, Game	**6**	Ken Hodge (Feb. 9/71), Bobby Orr (Jan. 1/73)
Most Points, Game	**7**	Bobby Orr (Nov. 15/73; 3G, 4A), Phil Esposito (Dec. 19/74; 3G, 4A), Barry Pederson (Apr. 4/82; 3G, 4A), Cam Neely (Oct. 16/88; 3G, 4A)

* NHL Record.

Retired Numbers

2	Eddie Shore	1926-1940
3	Lionel Hitchman	1925-1934
4	Bobby Orr	1966-1976
5	Dit Clapper	1927-1947
7	Phil Esposito	1967-1975
9	John Bucyk	1957-1978
15	Milt Schmidt	1936-1955
77	Raymond Bourque	1979-2000

All-time Record vs. Other Clubs

Regular Season

	At Home								On Road								Total							
	GP	W	L	T	OL	GF	GA	PTS	GP	W	L	T	OL	GF	GA	PTS	GP	W	L	T	OL	GF	GA	PTS
Anaheim	7	4	3	0	0	22	20	8	8	4	2	2	0	20	15	10	15	8	5	2	0	42	35	18
Atlanta	6	2	2	1	1	21	26	6	6	4	2	0	0	18	16	8	12	6	4	1	1	39	42	14
Buffalo	102	57	31	14	0	379	297	128	102	36	50	15	1	304	371	88	204	93	81	29	1	683	668	216
Calgary	45	27	11	6	1	158	123	61	43	22	18	3	0	151	157	47	88	49	29	9	1	309	280	108
Carolina	74	45	22	7	0	268	198	97	72	34	31	7	0	252	244	75	146	79	53	14	0	520	442	172
Chicago	283	161	88	34	0	1022	805	356	284	94	144	45	1	761	916	234	567	255	232	79	1	1783	1721	590
Colorado	61	31	21	9	0	238	189	71	64	34	24	6	0	265	232	74	125	65	45	15	0	503	421	145
Columbus	1	0	1	0	0	1	5	0	2	1	0	0	1	10	3	3	3	1	1	0	1	11	8	3
Dallas	58	40	9	9	0	249	140	89	60	29	17	13	1	218	175	72	118	69	26	22	1	467	315	161
Detroit	286	153	89	43	1	1005	760	350	283	78	153	52	0	716	950	208	569	231	242	95	1	1721	1710	558
Edmonton	28	19	6	3	0	116	76	41	28	14	11	3	0	93	95	31	56	33	17	6	0	209	171	72
Florida	18	7	8	3	0	48	49	17	17	9	7	0	1	50	47	19	35	16	15	3	1	98	96	36
Los Angeles	61	44	11	6	0	287	169	94	59	31	21	7	0	216	204	69	120	75	32	13	0	503	373	163
Minnesota	2	0	2	0	0	4	11	0	1	0	1	0	0	1	2	0	3	0	3	0	0	5	13	0
Montreal	329	152	121	56	0	973	889	360	329	96	187	46	0	783	1112	238	658	248	308	102	0	1756	2001	598
Nashville	3	1	1	1	0	11	6	3	4	3	0	0	1	12	6	7	7	4	1	1	1	23	12	10
New Jersey	52	30	13	7	2	209	159	69	49	25	12	10	2	161	132	62	101	55	25	17	4	370	291	131
NY Islanders	55	29	15	10	1	207	158	69	57	27	23	7	0	189	188	61	112	56	38	17	1	396	346	130
NY Rangers	294	157	94	42	1	1064	826	357	298	114	129	55	0	842	911	283	592	271	223	97	1	1906	1737	640
Ottawa	27	17	7	3	0	106	77	37	26	14	6	3	3	83	62	34	53	31	13	6	3	189	139	71
Philadelphia	72	44	18	10	0	276	205	98	69	30	31	8	0	201	231	68	141	74	49	18	0	477	436	166
Phoenix	29	22	4	3	0	133	89	47	29	14	12	3	0	101	99	31	58	36	16	6	0	234	188	78
Pittsburgh	74	54	14	6	0	333	209	114	76	30	31	15	0	271	264	75	150	84	45	21	0	604	473	189
St. Louis	58	35	13	9	1	247	157	80	58	23	24	9	2	195	184	57	116	58	37	18	3	442	341	137
San Jose	8	6	0	2	0	29	19	14	10	5	3	2	0	36	27	12	18	11	3	4	0	65	46	26
Tampa Bay	19	14	1	4	0	73	40	32	19	10	6	3	0	59	51	23	38	24	7	7	0	132	91	55
Toronto	293	158	87	47	1	955	771	364	293	89	153	51	0	754	991	229	586	247	240	98	1	1709	1762	593
Vancouver	50	37	6	7	0	211	117	81	50	26	16	8	0	203	163	60	100	63	22	15	0	414	280	141
Washington	52	30	15	7	0	195	139	67	51	24	14	12	1	173	141	61	103	54	29	19	1	368	280	128
Defunct Clubs	164	112	39	13	0	525	306	237	164	79	67	18	0	496	440	176	328	191	106	31	0	1021	746	413
Totals	**2611**	**1488**	**752**	**362**	**9**	**9365**	**7035**	**3347**	**2611**	**999**	**1195**	**403**	**14**	**7634**	**8429**	**2415**	**5222**	**2487**	**1947**	**765**	**23**	**16999**	**15464**	**5762**

Playoffs

	Series	W	L	GP	W	L	T	GF	GA	Last Mtg.	Rnd.	Result
Buffalo	7	5	2	39	21	18	0	139	130	1999	CSF	L 2-4
Carolina	3	3	0	19	12	7	0	63	48	1999	CQF	W 4-2
Chicago	6	5	1	22	16	5	1	97	63	1978	QF	W 4-0
Colorado	2	1	1	11	6	5	0	37	36	1983	DSF	W 3-1
Dallas	1	0	1	3	0	3	0	13	20	1981	PRE	L 0-3
Detroit	7	4	3	33	19	14	0	96	98	1957	SF	W 4-1
Edmonton	2	0	2	9	1	8	0	20	41	1990	F	L 1-4
Florida	1	0	1	5	1	4	0	16	22	1996	CQF	L 1-4
Los Angeles	2	2	0	13	8	5	0	56	38	1977	QF	W 4-2
Montreal	29	7	22	145	54	91	0	357	450	2002	CQF	L 2-4
New Jersey	3	1	2	18	7	11	0	52	55	1995	CQF	L 1-4
NY Islanders	2	0	2	11	3	8	0	35	49	1983	CF	L 2-4
NY Rangers	9	6	3	42	22	18	2	114	104	1973	QF	L 1-4
Philadelphia	4	2	2	20	11	9	0	60	57	1978	SF	W 4-1
Pittsburgh	4	2	2	19	9	10	0	62	67	1992	CF	L 0-4
St. Louis	2	2	0	8	8	0	0	48	15	1972	SF	W 4-0
Toronto	13	5	8	62	30	31	1	153	150	1974	QF	W 4-0
Washington	2	1	1	10	6	4	0	28	21	1998	CQF	L 2-4
Defunct Clubs	3	1	2	11	4	5	2	20	20			
Totals	**102**	**47**	**55**	**500**	**238**	**256**	**6**	**1466**	**1484**			

Playoff Results 2002-1998

Year	Round	Opponent	Result	GF	GA
2002	CQF	Montreal	L 2-4	18	20
1999	CSF	Buffalo	L 2-4	14	17
	CQF	Carolina	W 4-2	16	10
1998	CQF	Washington	L 2-4	13	15

Abbreviations: Round: F - Final; **CF** - conference final; **CSF** - conference semi-final; **CQF** - conference quarter-final; **DSF** - division semi-final; **SF** - semi-final; **QF** - quarter-final; **PRE** - preliminary round.

Calgary totals include Atlanta Flames, 1972-73 to 1979-80. Carolina totals include Hartford, 1979-80 to 1996-97.
Colorado totals include Quebec, 1979-80 to 1994-95. Dallas totals include Minnesota North Stars, 1967-68 to 1992-93.
New Jersey totals include Kansas City, 1974-75 to 1975-76, and Colorado Rockies, 1976-77 to 1981-82.
Phoenix totals include Winnipeg, 1979-80 to 1995-96.

2001-02 Results

Oct.	4	Anaheim	4-2		5	Washington	7-4
	6	Atlanta	4-3*		8	at Pittsburgh	3-2
	8	Washington	4-0		10	Los Angeles	5-0
	10	at Minnesota	1-2		12	NY Islanders	4-5
	13	at San Jose	2-3		14	at Washington	0-1*
	16	at Phoenix	1-1		17	Ottawa	5-2
	17	at Anaheim	2-2		19	at St. Louis	1-2*
	20	at Nashville	2-1		21	St. Louis	3-4*
	23	at Toronto	0-2		23	at NY Rangers	4-8
	25	Toronto	2-1*		24	at Ottawa	3-4
	27	NY Rangers	1-2*		26	Florida	4-2
	28	at Chicago	3-3		28	Chicago	2-1
	30	New Jersey	3-4*		30	at Montreal	4-3*
Nov.	3	at New Jersey	2-1	**Feb.**	4	at Columbus	8-0
	6	Edmonton	1-0*		5	Buffalo	2-2
	8	Minnesota	3-5		9	Florida	4-1
	10	Columbus	1-5		11	at Colorado	2-5
	13	Montreal	5-3		12	at Vancouver	2-1*
	15	New Jersey	5-4*		26	at NY Islanders	3-3
	17	Buffalo	3-1		28	Carolina	2-6
	20	at Montreal	2-3	**Mar.**	1	at Buffalo	3-4
	23	Vancouver	3-2		4	Philadelphia	1-4
	24	at Toronto	0-2		6	at Montreal	3-5
	27	Tampa Bay	6-3		8	at Atlanta	3-0
	29	at Philadelphia	3-2*		9	Calgary	3-2
Dec.	1	at Ottawa	1-2*		13	at NY Rangers	3-1
	4	at Atlanta	3-2		14	Toronto	1-2
	6	Pittsburgh	1-4		16	Detroit	2-1
	8	Buffalo	4-2		19	Phoenix	4-2
	12	at Pittsburgh	4-2		21	at Buffalo	2-1
	13	at Washington	1-2		23	at Florida	3-1
	15	Philadelphia	2-5		24	at Tampa Bay	4-3*
	18	Atlanta	2-3*		26	at Carolina	3-2
	20	Montreal	5-0		30	Carolina	0-2
	22	at NY Islanders	4-2	**Apr.**	2	at Philadelphia	4-2
	26	Ottawa	3-2		4	NY Islanders	1-2*
	28	at Florida	7-1		6	NY Rangers	4-6
	29	at Tampa Bay	5-4*		7	at New Jersey	2-3*
	31	at Dallas	1-2		9	Tampa Bay	2-2
Jan.	2	at Carolina	6-3		11	at Ottawa	0-4
	3	Toronto	1-2		13	Pittsburgh	7-1

* – Overtime

Entry Draft Selections 2002-1988

2002
Pick
29 Hannu Toivonen
56 Vladislav Yevseyev
130 Jan Kubista
153 Peter Hamerlik
228 Dmitri Utkin
259 Yan Stastny
290 Pavel Frolov

2001
Pick
19 Shaone Morrisonn
77 Darren McLachlan
111 Matti Kaltiainen
147 Jiri Jakes
179 Andrew Alberts
209 Jordan Sigalet
241 Milan Jurcina
282 Marcel Rodman

2000
Pick
7 Lars Jonsson
27 Martin Samuelsson
37 Andy Hilbert
59 Ivan Huml
66 Tuukka Makela
73 Sergei Zinovjev
102 Brett Nowak
174 Jarno Kultanen
204 Chris Berti
237 Zdenek Kutlak
268 Pavel Kolarik
279 Andreas Lindstrom

1999
Pick
21 Nick Boynton
56 Matt Zultek
89 Kyle Wanvig
118 Jaakko Harikkala
147 Seamus Kotyk
179 Donald Choukalos
207 Greg Barber
236 John Cronin
247 Mikko Eloranta
264 Georgy Pujacs

1998
Pick
48 Jonathan Girard
52 Bobby Allen
78 Peter Nordstrom
135 Andrew Raycroft
165 Ryan Milanovic

1997
Pick
1 Joe Thornton
8 Sergei Samsonov
27 Ben Clymer
54 Mattias Karlin
63 Lee Goren
81 Karol Bartanus
135 Denis Timofeev
162 Joel Trottier
180 Jim Baxter
191 Antti Laaksonen
218 Eric Van Acker
246 Jay Henderson

1996
Pick
8 Johnathan Aitken
45 Henry Kuster
53 Eric Naud
80 Jason Doyle
100 Trent Whitfield
132 Elias Abrahamsson
155 Chris Lane
182 Thomas Brown
208 Bob Prier
234 Anders Soderberg

1995
Pick
9 Kyle McLaren
21 Sean Brown
47 Paxton Schafer
73 Bill McCauley
99 Cameron Mann
151 Yevgeny Shaldybin
177 P.J. Axelsson
203 Sergei Zhukov
229 Jonathon Murphy

1994
Pick
21 Evgeni Ryabchikov
47 Daniel Goneau
99 Eric Nickulas
125 Darren Wright
151 Andre Roy
177 Jeremy Schaefer
229 John Grahame
255 Neil Savary
281 Andrei Yakhanov

1993
Pick
25 Kevyn Adams
51 Matt Alvey
88 Charles Paquette
103 Shawn Bates
129 Andrei Sapozhnikov
155 Milt Mastad
181 Ryan Golden
207 Hal Gill
233 Joel Prpic
259 Joakim Persson

1992
Pick
16 Dmitri Kvartalnov
55 Sergei Zholtok
112 Scott Bailey
133 Jiri Dopita
136 Grigori Panteleev
184 Kurt Seher
208 Mattias Timander
232 Chris Crombie
256 Denis Chervyakov
257 Evgeny Pavlov

1991
Pick
18 Glen Murray
40 Jozef Stumpel
62 Marcel Cousineau
84 Brad Tiley
106 Mariusz Czerkawski
150 Gary Golczewski
172 Jay Moser
194 Daniel Hodge
216 Steve Norton
238 Stephen Lombardi
260 Torsten Kienass

1990
Pick
21 Bryan Smolinski
63 Cam Stewart
84 Jerome Buckley
105 Mike Bales
126 Mark Woolf
147 Jim Mackey
168 John Gruden
189 Darren Wetherill
210 Dean Capuano
231 Andy Bezeau
252 Ted Miskolczi

1989
Pick
17 Shayne Stevenson
38 Mike Parson
57 Wes Walz
80 Jackson Penney
101 Mark Montanari
122 Stephen Foster
143 Otto Hascak
164 Rick Allain
185 James Lavish
206 Geoff Simpson
227 David Franzosa

1988
Pick
18 Rob Cimetta
60 Steve Heinze
81 Joe Juneau
102 Daniel Murphy
123 Derek Geary
165 Mark Krys
186 Jon Rohloff
228 Eric Reisman
249 Doug Jones

Coach

FTOREK, ROBBIE
Coach, Boston Bruins. Born in Needham, MA, January 2, 1952.

Robbie Ftorek was named the head coach of the Bruins on May 9, 2001. In his first season behind the bench in Boston in 2001-02, he led the club to the best record in the Eastern Conference and back into the playoffs after a two-year absence. He was rewarded with a nomination for the Jack Adams Award as coach of the year. Ftorek began his coaching career as head coach of the Los Angeles Kings' American Hockey League affiliate in New Haven in 1985 and was named as the head coach in Los Angeles on December 9, 1987. In 1989-90 he joined the Quebec Nordiques organization when he was named head coach of their AHL affiliate in Halifax. After 48 games, he was called to Quebec as an assistant coach through the 1990-91 season.

Ftorek went on to spend 10 seasons in the New Jersey Devils organization. He won a Calder Cup championship with Albany in 1994-95 and was named the AHL's outstanding coach in both 1994-95 and 1995-96. He moved up to New Jersey as an assistant coach in 1996-97 before being named the club's head coach in 1998.

A native of Needham, Massachusetts, Ftorek was a schoolboy standout in both hockey and soccer and is still regarded as the top high school hockey player in state history. He joined the U.S. national team in 1971 and won a silver medal with the 1972 U.S. Olympic team in Japan before signing with the Detroit Red Wings. Ftorek later starred in the WHA from 1974-75 to 1978-79 before returning to the NHL for six more seasons. Ftorek was honored with induction into the U. S. Hockey Hall of Fame in 1991.

Coaching Record

		Regular Season				Playoffs		
Season	Team	Games	W	L	T	Games	W	L
1985-86	New Haven (AHL)	80	36	37	7	5	1	4
1986-87	New Haven (AHL)	80	44	25	11	7	3	4
1987-88	New Haven (AHL)	27	16	8	3	....	....	....
	Los Angeles (NHL)	**52**	**23**	**25**	**4**	**5**	**1**	**4**
1988-89	**Los Angeles (NHL)**	**80**	**42**	**31**	**7**	**11**	**4**	**7**
1989-90	Halifax (AHL)	48	25	19	4	....	....	....
1992-93	Utica (AHL)	80	33	36	11	5	1	4
1993-94	Albany (AHL)	80	38	34	8	5	1	4
1994-95	Albany (AHL)	80	46	17	17	14	12	2
1995-96	Albany (AHL)	80	54	19	17	4	1	3
1998-99	**New Jersey (NHL)**	**82**	**47**	**24**	**11**	**7**	**3**	**4**
1999-2000	**New Jersey (NHL)**	**74**	**41**	**25**	**8**	....	....	....
2001-02	**Boston (NHL)**	**82**	**43**	**33**	**6**	**6**	**4**	**2**
	NHL Totals	**370**	**196**	**138**	**36**	**29**	**10**	**19**

Club Directory

FleetCenter

Boston Bruins
One FleetCenter Place, Suite 250
Boston, MA 02114
Phone **617/624-1900**
FAX 617/523-7184
www.bostonbruins.com
Capacity: 17,565

Executive
Owner and Governor .... Jeremy M. Jacobs
Alternate Governors .... Louis Jacobs, Charles Jacobs, Jeremy Jacobs, Jr., Mike O'Connell
President and Alternate Governor .... Harry Sinden
Senior Assistant to the President .... Nate Greenberg
Chief Legal Officer .... Michael Wall
Chief Financial Officer .... Jessica Rahuba
Vice President and General Manager .... Mike O'Connell
Assistant General Manager .... Jeff Gorton
Executive Vice President .... Charles Jacobs
Executive Vice President .... Richard Krezwick
Director of Administration .... Dale Hamilton-Powers
Assistant to the President .... Joe Curnane
Asst. Dir. of Administration/Travel Coordinator .... Carol Gould
Receptionist .... Karen Ondo

Coaching Staff
Head Coach .... Robbie Ftorek
Assistant Coaches .... Wayne Cashman, Jim Hughes
Video Coordinator .... Nickolai Bobrov
Team Road Services Coordinator .... John Bucyk
Coach, Providence Bruins .... Mike Sullivan

Scouting Staff
Director of Pro Scouting & Player Development .... Sean Coady
Director of Amateur Scouting .... Scott Bradley
Scouting Staff .... Adam Creighton, Gerry Cheevers, Daniel Dore, Don Matheson, Mike McGraw, Tom McVie, Tom Songin, Svenake Svensson

Medical & Training Staff
Strength and Conditioning Coach .... John Whitesides
Athletic Trainer .... Don Del Negro
Physical Therapist .... Scott Waugh
Equipment Manager .... Peter Henderson
Assistant Equipment Managers .... Chris "Muggsy" Aldrich, Keith Robinson

Communications & Marketing Staff
Director of Media Relations .... Heidi Holland
Media Relations Manager .... Mark Awdycki
Director of Marketing & Community Relations .... Sue Byrne
Promotions Manager .... Dave Murray
Community Relations Coordinator .... Heather Riva
Game Presentation and Marketing Coordinator .... Mike Burns
Administrative Assistant, Alumni Office .... Mal Viola

Ticketing & Finance Staff
Director of Ticket Operations .... Matt Brennan
Assistant Director of Ticket Operations .... Jim Foley
Ticket Office Receptionist .... Jo-Ann Connolly-White
Accounting Manager .... Rick McGlinchey
Payroll Manager .... Rita Brandano
Accounts Payable .... Linda Bartlett

Television & Radio
TV Outlet .... New England Sports Network (NESN)
Radio Station .... WBZ Radio (1030 AM) and Bruins Radio Network
Television Broadcasters .... Dave Shea & Dale Arnold (play-by-play); Gord Kluzak & Andy Brickley (color)
Radio Broadcasters .... Dave Goucher (play-by-play) & Bob Beers (color)

Miscellaneous
Club Colors .... Gold, Black and White
Ice Surface .... 200 feet by 85 feet

General Managers' History

Art Ross, 1924-25 to 1953-54; Lynn Patrick, 1954-55 to 1964-65; Hap Emms, 1965-66, 1966-67; Milt Schmidt, 1967-68 to 1971-72; Harry Sinden, 1972-73 to 1999-2000; Harry Sinden and Mike O'Connell, 2000-01; Mike O'Connell, 2001-02 to date.

Coaching History

Art Ross, 1924-25 to 1927-28; Cy Denneny, 1928-29; Art Ross, 1929-30 to 1933-34; Frank Patrick, 1934-35, 1935-36; Art Ross, 1936-37 to 1938-39; Cooney Weiland, 1939-40, 1940-41; Art Ross, 1941-42 to 1944-45; Dit Clapper, 1945-46 to 1948-49; Georges Boucher, 1949-50; Lynn Patrick, 1950-51 to 1953-54; Lynn Patrick and Milt Schmidt, 1954-55; Milt Schmidt, 1955-56 to 1960-61; Phil Watson, 1961-62; Phil Watson and Milt Schmidt, 1962-63; Milt Schmidt, 1963-64 to 1965-66; Harry Sinden, 1966-67 to 1969-70; Tom Johnson, 1970-71, 1971-72; Tom Johnson and Bep Guidolin, 1972-73; Bep Guidolin, 1973-74; Don Cherry, 1974-75 to 1978-79; Fred Creighton and Harry Sinden, 1979-80; Gerry Cheevers, 1980-81 to 1983-84; Gerry Cheevers and Harry Sinden, 1984-85; Butch Goring, 1985-86; Butch Goring and Terry O'Reilly, 1986-87; Terry O'Reilly, 1987-88, 1988-89; Mike Milbury, 1989-90, 1990-91; Rick Bowness, 1991-92; Brian Sutter, 1992-93 to 1994-95; Steve Kasper, 1995-96, 1996-97; Pat Burns, 1997-98 to 1999-2000; Pat Burns and Mike Keenan, 2000-01; Robbie Ftorek, 2001-02 to date.

Buffalo Sabres

2001-02 Results: 35W-35L-11T-1OTL 82PTS.
Fifth, Northeast Division

2002-03 Schedule

Month	Day	Date	Opponent	Month	Day	Date	Opponent
Oct.	Thu.	10	NY Islanders		Sat.	11	at Montreal
	Sat.	12	at Montreal		Tue.	14	at Minnesota
	Sun.	13	at Chicago		Thu.	16	at San Jose
	Thu.	17	NY Rangers		Sat.	18	at Phoenix
	Sat.	19	Phoenix		Tue.	21	Pittsburgh
	Tue.	22	Philadelphia		Fri.	24	Toronto
	Fri.	25	New Jersey		Sat.	25	at Ottawa
	Sat.	26	at Pittsburgh		Mon.	27	Nashville
	Tue.	29	at Vancouver		Thu.	30	at St. Louis
	Thu.	31	at Calgary	**Feb.**	Tue.	4	at New Jersey
Nov.	Fri.	1	at Edmonton		Fri.	7	Vancouver
	Sun.	3	at Columbus		Sat.	8	at NY Islanders
	Thu.	7	at Carolina		Tue.	11	St. Louis
	Sat.	9	Atlanta		Thu.	13	at Detroit
	Tue.	12	Boston		Sat.	15	NY Rangers
	Fri.	15	Toronto		Mon.	17	at Atlanta*
	Sat.	16	at Ottawa		Wed.	19	Montreal
	Tue.	19	at New Jersey		Fri.	21	Los Angeles
	Fri.	22	Columbus		Sun.	23	at Tampa Bay*
	Sat.	23	at Boston		Mon.	24	at Florida
	Wed.	27	Tampa Bay		Wed.	26	at Washington
	Fri.	29	Pittsburgh		Fri.	28	Dallas
	Sat.	30	at Toronto	**Mar.**	Sat.	1	at NY Islanders
Dec.	Wed.	4	Anaheim		Tue.	4	Washington
	Fri.	6	at NY Rangers		Thu.	6	Toronto
	Sat.	7	Washington		Sat.	8	at Florida
	Tue.	10	Ottawa		Sun.	9	at Tampa Bay*
	Fri.	13	Chicago		Wed.	12	Carolina
	Sat.	14	at Philadelphia		Fri.	14	Tampa Bay
	Wed.	18	Boston		Sat.	15	at Atlanta
	Fri.	20	Florida		Tue.	18	Philadelphia
	Sat.	21	at Montreal		Wed.	19	at NY Rangers
	Mon.	23	at Pittsburgh		Sat.	22	at Toronto
	Thu.	26	Ottawa		Mon.	24	Colorado
	Sat.	28	Minnesota		Wed.	26	Florida
	Mon.	30	at Washington		Fri.	28	Montreal
	Tue.	31	NY Islanders		Sat.	29	at Carolina
Jan.	Fri.	3	Carolina		Mon.	31	at Dallas
	Sat.	4	at Ottawa	**Apr.**	Wed.	2	Atlanta
	Tue.	7	at Philadelphia		Sat.	5	at Boston*
	Fri.	10	Boston		Sun.	6	New Jersey*

* Denotes afternoon game.

With 23 goals in each of the last two seasons, only Miroslav Satan has scored more goals for the Sabres than J-P Dumont over that stretch. Dumont's shooting percentage of 14.9 was tops on the team for all players with more than 10 goals.

Year-by-Year Record

		Home				Road				Overall								
Season	**GP**	**W**	**L**	**T**	**OL**	**W**	**L**	**T**	**OL**	**W**	**L**	**T**	**OL**	**GF**	**GA**	**Pts.**	**Finished**	**Playoff Result**
2001-02	82	20	16	5	0	15	19	6	1	35	35	11	1	213	200	82	5th, Northeast Div.	Out of Playoffs
2000-01	82	26	12	3	0	20	18	2	1	46	30	5	1	218	184	98	2nd, Northeast Div.	Lost Conf. Semi-Final
1999-2000	82	21	14	5	1	14	18	6	3	35	32	11	4	213	204	85	3rd, Northeast Div.	Lost Conf. Quarter-Final
1998-99	82	23	12	6	...	14	16	11	...	37	28	17	...	207	175	91	4th, Northeast Div.	Lost Final
1997-98	82	20	13	8	...	16	16	9	...	36	29	17	...	211	187	89	3rd, Northeast Div.	Lost Conf. Final
1996-97	82	24	11	6	...	16	19	6	...	40	30	12	...	237	208	92	1st, Northeast Div.	Lost Conf. Semi-Final
1995-96	82	19	17	5	...	14	25	2	...	33	42	7	...	247	262	73	5th, Northeast Div.	Out of Playoffs
1994-95	48	15	8	1	...	7	11	6	...	22	19	7	...	130	119	51	4th, Northeast Div.	Lost Conf. Quarter-Final
1993-94	84	22	17	3	...	21	15	6	...	43	32	9	...	282	218	95	4th, Northeast Div.	Lost Conf. Quarter-Final
1992-93	84	25	15	2	...	13	21	8	...	38	36	10	...	335	297	86	4th, Adams Div.	Lost Div. Final
1991-92	80	22	13	5	...	9	24	7	...	31	37	12	...	289	299	74	3rd, Adams Div.	Lost Div. Semi-Final
1990-91	80	15	13	12	...	16	17	7	...	31	30	19	...	292	278	81	3rd, Adams Div.	Lost Div. Semi-Final
1989-90	80	27	11	2	...	18	16	6	...	45	27	8	...	286	248	98	2nd, Adams Div.	Lost Div. Semi-Final
1988-89	80	25	12	3	...	13	23	4	...	38	35	7	...	291	299	83	3rd, Adams Div.	Lost Div. Semi-Final
1987-88	80	19	14	7	...	18	18	4	...	37	32	11	...	283	305	85	3rd, Adams Div.	Lost Div. Semi-Final
1986-87	80	18	18	4	...	10	26	4	...	28	44	8	...	280	308	64	5th, Adams Div.	Out of Playoffs
1985-86	80	23	16	1	...	14	21	5	...	37	37	6	...	296	291	80	5th, Adams Div.	Out of Playoffs
1984-85	80	23	10	7	...	15	18	7	...	38	28	14	...	290	237	90	3rd, Adams Div.	Lost Div. Semi-Final
1983-84	80	25	9	6	...	23	16	1	...	48	25	7	...	315	257	103	2nd, Adams Div.	Lost Div. Semi-Final
1982-83	80	25	7	8	...	13	22	5	...	38	29	13	...	318	285	89	3rd, Adams Div.	Lost Div. Final
1981-82	80	23	8	9	...	16	18	6	...	39	26	15	...	307	273	93	3rd, Adams Div.	Lost Div. Semi-Final
1980-81	80	21	7	12	...	18	13	9	...	39	20	21	...	327	250	99	1st, Adams Div.	Lost Quarter-Final
1979-80	80	27	5	8	...	20	12	8	...	47	17	16	...	318	201	110	1st, Adams Div.	Lost Semi-Final
1978-79	80	19	13	8	...	17	15	8	...	36	28	16	...	280	263	88	2nd, Adams Div.	Lost Prelim. Round
1977-78	80	25	7	8	...	19	12	9	...	44	19	17	...	288	215	105	2nd, Adams Div.	Lost Quarter-Final
1976-77	80	27	8	5	...	21	16	3	...	48	24	8	...	301	220	104	2nd, Adams Div.	Lost Quarter-Final
1975-76	80	28	7	5	...	18	14	8	...	46	21	13	...	339	240	105	2nd, Adams Div.	Lost Quarter-Final
1974-75	80	28	6	6	...	21	10	9	...	49	16	15	...	354	240	113	1st, Adams Div.	Lost Final
1973-74	78	23	10	6	...	9	24	6	...	32	34	12	...	242	250	76	5th, East Div.	Out of Playoffs
1972-73	78	30	6	3	...	7	21	11	...	37	27	14	...	257	219	88	4th, East Div.	Lost Quarter-Final
1971-72	78	11	19	9	...	5	24	10	...	16	43	19	...	203	289	51	6th, East Div.	Out of Playoffs
1970-71	78	16	13	10	...	8	26	5	...	24	39	15	...	217	291	63	5th, East Div.	Out of Playoffs

Franchise date: May 22, 1970

NORTHEAST DIVISION

33rd NHL Season

2002-03 Player Personnel

FORWARDS	HT	WT	S	Place of Birth	Date	2001-02 Club
ADDUONO, Jeremy	6-0	182	R	Thunder Bay, Ont.	8/4/78	Rochester
AFINOGENOV, Maxim	6-0	190	L	Moscow, USSR	9/4/79	Buffalo-Russia
BARNES, Stu	5-11	180	R	Spruce Grove, Alta.	12/25/70	Buffalo
BOTTERILL, Jason	6-4	220	L	Edmonton, Alta.	5/19/76	Calgary-Saint John
BOULTON, Eric	6-0	222	L	Halifax, N.S.	8/17/76	Buffalo
BROWN, Curtis	6-0	197	L	Unity, Sask.	2/12/76	Buffalo
CONNOLLY, Tim	6-1	182	R	Syracuse, NY	5/7/81	Buffalo
DUMONT, J-P	6-1	205	L	Montreal, Que.	4/1/78	Buffalo
GRATTON, Chris	6-4	225	L	Brantford, Ont.	7/5/75	Buffalo
HAMEL, Denis	6-1	201	L	Lachute, Que.	5/10/77	Buffalo
HECHT, Jochen	6-1	200	L	Mannheim, West Germany	6/21/77	Edmonton-Germany
KOTALIK, Ales	6-1	217	R	Jindrichuv Hradec, Czech.	12/23/78	Buffalo-Rochester
MAIR, Adam	6-2	215	R	Hamilton, Ont.	2/15/79	Los Angeles-Manchester
METHOT, Francois	6-0	203	R	Montreal, Que.	4/26/78	Rochester
MILLEY, Norm	6-0	200	R	Toronto, Ont.	2/14/80	Buffalo-Rochester
PYATT, Taylor	6-4	222	L	Thunder Bay, Ont.	8/19/81	Buffalo-Rochester
RAY, Rob	6-0	217	L	Stirling, Ont.	6/8/68	Buffalo
SATAN, Miroslav	6-3	190	L	Topolcany, Czech.	10/22/74	Buffalo-Slovakia
VARADA, Vaclav	6-0	208	L	Vsetin, Czech.	4/26/76	Buffalo
DEFENSEMEN						
CAMPBELL, Brian	6-0	190	L	Strathroy, Ont.	5/23/79	Buffalo-Rochester
HECL, Radoslav	6-1	196	L	Partizanske, Czech.	10/11/74	Slovan Bratislava
KALININ, Dmitri	6-3	215	L	Chelyabinsk, USSR	7/22/80	Buffalo
McKEE, Jay	6-4	212	L	Kingston, Ont.	9/8/77	Buffalo
PATRICK, James	6-2	202	R	Winnipeg, Man.	6/14/63	Buffalo
RATCHUK, Peter	6-1	185	L	Buffalo, NY	9/10/77	Wilkes-Barre
WARRENER, Rhett	6-2	217	R	Shaunavon, Sask.	1/27/76	Buffalo
WOOLLEY, Jason	6-0	203	L	Toronto, Ont.	7/27/69	Buffalo
ZHITNIK, Alexei	5-11	215	L	Kiev, USSR	10/10/72	Buffalo

GOALTENDERS	HT	WT	C	Place of Birth	Date	2001-02 Club
BIRON, Martin	6-2	168	L	Lac-St-Charles, Que.	8/15/77	Buffalo
NORONEN, Mika	6-2	196	L	Tampere, Finland	6/17/79	Buffalo-Rochester

Coaching History

Punch Imlach, 1970-71; Punch Imlach, Floyd Smith and Joe Crozier, 1971-72; Joe Crozier, 1972-73, 1973-74; Floyd Smith, 1974-75 to 1976-77; Marcel Pronovost, 1977-78; Marcel Pronovost and Billy Inglis, 1978-79; Scotty Bowman, 1979-80; Roger Neilson, 1980-81; Jim Roberts and Scotty Bowman, 1981-82; Scotty Bowman 1982-83 to 1984-85; Jim Schoenfeld and Scotty Bowman, 1985-86; Scotty Bowman, Craig Ramsay and Ted Sator, 1986-87; Ted Sator, 1987-88, 1988-89; Rick Dudley, 1989-90, 1990-91; Rick Dudley and John Muckler, 1991-92; John Muckler, 1992-93 to 1994-95; Ted Nolan, 1995-96, 1996-97; Lindy Ruff, 1997-98 to date.

Head Coach

RUFF, LINDY
Head Coach, Buffalo Sabres. Born in Warburg, Alta., February, 17, 1960.

A former captain of the Sabres, Lindy Ruff was appointed as the club's 15th head coach on July 21, 1997. In 1999, he led the Sabres to the Stanley Cup Finals for just the second time in club history. As a player, Ruff was drafted 32nd overall by the Sabres in the 1979 Entry Draft. He played both defense and left wing in an NHL career that spanned 12 seasons including 608 regular-season games with Buffalo. He became a playing assistant coach with Rochester of the AHL in 1991-92 and San Diego of the IHL in 1992-93. Ruff's San Diego club set a pro hockey record with 62 wins. In 1993-94 he became an NHL assistant coach with the Florida Panthers.

Coaching Record

		Regular Season				Playoffs		
Season	Team	Games	W	L	T	Games	W	L
1997-98	Buffalo (NHL)	82	36	29	17	15	10	5
1998-99	Buffalo (NHL)	82	37	28	17	21	14	7
1999-2000	Buffalo (NHL)	82	35	36	11	5	1	4
2000-01	Buffalo (NHL)	82	46	31	5	13	7	6
2001-02	Buffalo (NHL)	82	35	36	11	....	....	....
	NHL Totals	410	189	160	61	54	32	22

2001-02 Scoring

* - rookie

Regular Season

Pos	#	Player	Team	GP	G	A	Pts	+/–	PIM	PP	SH	GW	GT	S	%
L	81	Miroslav Satan	BUF	82	37	36	73	14	33	15	5	5	0	267	13.9
C	41	Stu Barnes	BUF	68	17	31	48	6	26	5	0	4	1	127	13.4
C	18	Tim Connolly	BUF	82	10	35	45	4	34	3	0	3	0	126	7.9
R	17	Jean-Pierre Dumont	BUF	76	23	21	44	–10	42	7	0	3	1	154	14.9
R	61	Maxim Afinogenov	BUF	81	21	19	40	–9	69	3	1	0	0	234	9.0
C	77	Chris Gratton	BUF	82	15	24	39	0	75	1	0	5	0	139	10.8
C	37	Curtis Brown	BUF	82	20	17	37	–4	32	4	1	5	0	171	11.7
D	44	Alexei Zhitnik	BUF	82	1	33	34	–1	80	1	0	0	0	150	0.7
D	5	Jason Woolley	BUF	59	8	20	28	–6	34	6	0	2	0	90	8.9
R	25	Vaclav Varada	BUF	76	7	16	23	–7	82	1	0	1	0	138	5.1
L	13	Vyacheslav Kozlov	BUF	38	9	13	22	0	16	3	0	1	1	68	13.2
L	24	Taylor Pyatt	BUF	48	10	10	20	4	35	0	0	0	1	61	16.4
C	9	Erik Rasmussen	BUF	69	8	11	19	–1	34	0	0	2	0	89	9.0
D	3	James Patrick	BUF	56	5	8	13	3	16	1	0	0	0	45	11.1
D	45	Dmitri Kalinin	BUF	58	2	11	13	–6	26	0	0	0	0	67	3.0
D	74	Jay McKee	BUF	81	2	11	13	18	43	0	0	1	0	50	4.0
D	4	Rhett Warrener	BUF	65	5	5	10	15	113	0	0	1	1	66	7.6
D	42	Richard Smehlik	BUF	60	3	6	9	–9	22	0	0	1	1	52	5.8
C	29	Bob Corkum	ATL	65	3	4	7	–30	16	0	0	0	0	70	4.3
			BUF	10	0	1	1	–2	4	0	0	0	0	10	0.0
			TOTAL	75	3	5	8	–32	20	0	0	0	0	80	3.8
R	55	Denis Hamel	BUF	61	2	6	8	–1	28	0	0	0	0	80	2.5
D	51	Brian Campbell	BUF	29	3	3	6	0	12	0	0	0	0	30	10.0
L	26	Eric Boulton	BUF	35	2	3	5	–1	129	0	0	1	0	21	9.5
R	32	Rob Ray	BUF	71	2	3	5	–3	200	0	0	0	0	23	8.7
R	12	* Ales Kotalik	BUF	13	1	3	4	–1	2	0	0	0	0	21	4.8
R	19	* Norman Milley	BUF	5	0	1	1	0	0	0	0	0	0	10	0.0
D	10	* Henrik Tallinder	BUF	2	0	0	0	–1	0	0	0	0	0	4	0.0
D	8	Rory Fitzpatrick	BUF	5	0	0	0	–2	4	0	0	0	0	2	0.0

Goaltending

No.	Goaltender	GPI	Mins	Avg	W	L	T	EN	SO	GA	SA	S%	G	A	PIM
43	Martin Biron	72	4085	2.22	31	28	10	6	4	151	1781	.915	0	1	8
35	* Mika Noronen	10	518	2.66	4	3	1	0	0	23	217	.894	0	0	2
31	Bob Essensa	9	350	2.91	0	5	0	3	0	17	113	.850	0	0	0
	Totals	**82**	**4986**	**2.41**	**35**	**36**	**11**	**9**	**4**	**200**	**2120**	**.906**			

Stu Barnes was named captain of the Sabres last season. His 48 points in 2001-02 represented the second-best offensive total of his 11-year career.

Captains' History

Floyd Smith, 1970-71; Gerry Meehan, 1971-72 to 1973-74; Gerry Meehan and Jim Schoenfeld, 1974-75; Jim Schoenfeld, 1975-76, 1976-77; Danny Gare, 1977-78 to 1980-81; Danny Gare and Gilbert Perreault, 1981-82; Gilbert Perreault, 1982-83 to 1985-86; Gilbert Perreault and Lindy Ruff, 1986-87; Lindy Ruff, 1987-88; Lindy Ruff and Mike Foligno, 1988-89; Mike Foligno, 1989-90; Mike Foligno and Mike Ramsey, 1990-91; Mike Ramsey, 1991-92; Mike Ramsey and Pat LaFontaine, 1992-93; Pat LaFontaine and Alexander Mogilny, 1993-94; Pat LaFontaine, 1994-95 to 1996-97; Donald Audette and Michael Peca, 1997-98; Michael Peca, 1998-99, 1999-2000; no captain, 2000-01; Stu Barnes. 2001-02 to date.

Club Records

Team

(Figures in brackets for season records are games played; records for fewest points, wins, ties, losses, goals, goals against are for 70 or more games)

Most Points 113 1974-75 (80)
Most Wins 49 1974-75 (80)
Most Ties 21 1980-81 (80)
Most Losses 44 1986-87 (80)
Most Goals 354 1974-75 (80)
Most Goals Against 308 1986-87 (80)
Fewest Points 51 1971-72 (78)
Fewest Wins 16 1971-72 (78)
Fewest Ties 5 2000-01 (82)
Fewest Losses 16 1974-75 (80)
Fewest Goals 203 1971-72 (78)
Fewest Goals Against 175 1998-99 (82)
Longest Winning Streak
Overall .... 10 Jan. 4-23/84
Home .... 12 Nov. 12/72-Jan. 7/73, Oct. 13-Dec. 10/89
Away .... *10 Dec. 10/83-Jan. 23/84
Longest Undefeated Streak
Overall .... 14 Mar. 6-Apr. 6/80 (8 wins, 6 ties)
Home .... 21 Oct. 8/72-Jan. 7/73 (18 wins, 3 ties)
Away .... 10 Dec. 10/83-Jan. 23/84 (10 wins)
Longest Losing Streak
Overall .... 7 Oct. 25-Nov. 8/70, Apr. 3-15/93, Oct. 9-22/93
Home .... 6 Oct. 10-Nov. 10/93, Mar. 3-Apr. 3/96
Away .... 7 Oct. 14-Nov. 7/70, Feb. 6-27/71, Jan. 10-Feb. 3/96
Longest Winless Streak
Overall .... 12 Nov. 23-Dec. 20/91 (8 losses, 4 ties)
Home .... 12 Jan. 27-Mar. 10/91 (7 losses, 5 ties)
Away .... 23 Oct. 30/71-Feb. 19/72 (15 losses, 8 ties)
Most Shutouts, Season 13 1997-98 (82)
Most PIM, Season *2,713 1991-92 (80)
Most Goals, Game .... 14 Jan. 21/75 (Wsh. 2 at Buf. 14), Mar. 19/81 (Tor. 4 at Buf. 14)

Individual

Most Seasons .... 17 Gilbert Perreault
Most Games .... 1,191 Gilbert Perreault
Most Goals, Career .... 512 Gilbert Perreault
Most Assists, Career .... 814 Gilbert Perreault
Most Points, Career .... 1,326 Gilbert Perreault (512G, 814A)
Most PIM, Career .... 3,097 Rob Ray
Most Shutouts, Career .... 55 Dominik Hasek
Longest Consecutive Games Streak .... 776 Craig Ramsay (Mar. 27/73-Feb. 10/83)
Most Goals, Season .... 76 Alexander Mogilny (1992-93)
Most Assists, Season .... 95 Pat LaFontaine (1992-93)
Most Points, Season .... 148 Pat LaFontaine (1992-93; 53G, 95A)
Most PIM, Season .... 354 Rob Ray (1991-92)
Most Points, Defenseman, Season .... 81 Phil Housley (1989-90; 21G, 60A)
Most Points, Center, Season .... 148 Pat LaFontaine (1992-93; 53G, 95A)
Most Points, Right Wing, Season .... 127 Alexander Mogilny (1992-93; 76G, 51A)
Most Points, Left Wing, Season .... 95 Rick Martin (1974-75; 52G, 43A)
Most Points, Rookie, Season .... 74 Rick Martin (1971-72; 44G, 30A)
Most Shutouts, Season .... 13 Dominik Hasek (1997-98)
Most Goals, Game .... 5 Dave Andreychuk (Feb. 6/86)
Most Assists, Game .... 5 Gilbert Perreault (Feb. 1/76, Mar. 9/80, Jan. 4/84), Dale Hawerchuk (Jan. 15/92), Pat LaFontaine (Dec. 31/92, Feb. 10/93)
Most Points, Game .... 7 Gilbert Perreault (Feb. 1/76; 2G, 5A)

* NHL Record.

Retired Numbers

2	Tim Horton	1972-1974
7	Rick Martin	1971-1981
11	Gilbert Perreault	1970-1987
14	Rene Robert	1971-1979

All-time Record vs. Other Clubs

Regular Season

	At Home								On Road								Total							
	GP	W	L	T	OL	GF	GA	PTS	GP	W	L	T	OL	GF	GA	PTS	GP	W	L	T	OL	GF	GA	PTS
Anaheim	7	2	2	3	0	16	18	7	8	6	2	0	0	25	13	12	15	8	4	3	0	41	31	19
Atlanta	6	4	2	0	0	24	11	8	6	2	3	1	0	20	15	5	12	6	5	1	0	44	26	13
Boston	102	51	36	15	0	371	304	117	102	31	57	14	0	297	379	76	204	82	93	29	0	668	683	193
Calgary	45	27	13	5	0	189	131	59	44	17	16	11	0	144	149	45	89	44	29	16	0	333	280	104
Carolina	73	43	23	7	0	293	219	93	74	33	30	11	0	222	220	77	147	76	53	18	0	515	439	170
Chicago	52	32	14	6	0	198	137	70	49	17	26	6	0	136	159	40	101	49	40	12	0	334	296	110
Colorado	62	35	18	9	0	243	201	79	63	21	31	11	0	194	226	53	125	56	49	20	0	437	427	132
Columbus	2	0	2	0	0	2	5	0	1	0	0	1	0	2	2	1	3	0	2	1	0	4	7	1
Dallas	50	26	13	11	0	179	133	63	53	21	26	6	0	156	170	48	103	47	39	17	0	335	303	111
Detroit	51	33	10	8	0	224	146	74	54	18	30	5	1	157	199	42	105	51	40	13	1	381	345	116
Edmonton	30	10	13	7	0	109	112	27	27	5	20	2	0	72	115	12	57	15	33	9	0	181	227	39
Florida	19	13	4	2	0	57	29	28	17	8	9	0	0	54	54	16	36	21	13	2	0	111	83	44
Los Angeles	51	27	15	9	0	208	150	63	53	22	22	9	0	182	184	53	104	49	37	18	0	390	334	116
Minnesota	1	1	0	0	0	3	1	2	1	1	0	0	0	4	1	2	2	2	0	0	0	7	2	4
Montreal	97	49	29	19	0	303	263	117	97	32	53	12	0	291	371	76	194	81	82	31	0	594	634	193
Nashville	3	0	2	1	0	8	11	1	4	3	1	0	0	8	6	6	7	3	3	1	0	16	17	7
New Jersey	50	31	13	6	0	206	154	68	50	26	15	9	0	170	142	61	100	57	28	15	0	376	296	129
NY Islanders	57	31	17	9	0	190	152	71	57	25	23	9	0	161	159	59	114	56	40	18	0	351	311	130
NY Rangers	64	37	18	9	0	266	200	83	62	20	26	15	1	169	206	56	126	57	44	24	1	435	406	139
Ottawa	26	19	5	2	0	86	34	40	27	14	7	6	0	76	58	34	53	33	12	8	0	162	92	74
Philadelphia	59	29	23	7	0	195	170	65	63	15	35	12	1	159	219	43	122	44	58	19	1	354	389	108
Phoenix	29	20	4	5	0	123	74	45	28	13	13	2	0	91	87	28	57	33	17	7	0	214	161	73
Pittsburgh	67	33	17	16	1	267	183	83	67	17	33	17	0	210	252	51	134	50	50	33	1	477	435	134
St. Louis	50	29	15	6	0	197	156	64	49	14	28	7	0	124	178	35	99	43	43	13	0	321	334	99
San Jose	10	10	0	0	0	50	26	20	9	1	4	3	1	31	34	6	19	11	4	3	1	81	60	26
Tampa Bay	19	12	6	1	0	55	52	25	19	13	4	2	0	63	42	28	38	25	10	3	0	118	94	53
Toronto	63	40	17	6	0	259	168	86	62	25	25	11	1	211	189	62	125	65	42	17	1	470	357	148
Vancouver	51	26	17	8	0	184	148	60	50	16	24	10	0	160	186	42	101	42	41	18	0	344	334	102
Washington	52	32	14	6	0	199	138	70	52	30	13	9	0	182	132	69	104	62	27	15	0	381	270	139
Defunct Clubs	23	13	5	5	0	94	63	31	23	12	8	3	0	97	76	27	46	25	13	8	0	191	139	58
Totals	**1271**	**715**	**367**	**188**	**1**	**4798**	**3589**	**1619**	**1271**	**478**	**584**	**204**	**5**	**3868**	**4223**	**1165**	**2542**	**1193**	**951**	**392**	**6**	**8666**	**7812**	**2784**

Playoffs

	Series	W	L	GP	W	L	T	GF	GA	Last Mtg.	Rnd.	Result
Boston	7	2	5	39	18	21	0	130	139	1999	CSF	W 4-2
Chicago	2	2	0	9	8	1	0	36	17	1980	QF	W 4-0
Colorado	2	0	2	8	2	6	0	27	35	1985	DSF	L 2-3
Dallas	3	1	2	13	5	8	0	37	39	1999	F	L 2-4
Montreal	7	3	4	35	17	18	0	111	124	1998	CSF	W 4-0
New Jersey	1	0	1	7	3	4	0	14	14	1994	CQF	L 3-4
NY Islanders	3	0	3	16	4	12	0	45	59	1980	SF	L 2-4
NY Rangers	1	1	0	3	2	1	0	11	6	1978	PRE	W 2-1
Ottawa	2	2	0	11	8	3	0	26	19	1999	CQF	W 4-0
Philadelphia	7	2	5	37	14	23	0	96	110	2001	CQF	W 4-2
Pittsburgh	2	0	2	10	4	6	0	26	26	2001	CSF	L 3-4
St. Louis	1	1	0	3	2	1	0	7	8	1976	PRE	W 2-1
Toronto	1	1	0	5	4	1	0	21	16	1999	CF	W 4-1
Vancouver	2	2	0	7	6	1	0	28	14	1981	PRE	W 3-0
Washington	1	0	1	6	2	4	0	11	13	1998	CF	L 2-4
Totals	**42**	**17**	**25**	**209**	**99**	**110**	**0**	**626**	**639**			

Playoff Results 2002-1998

Year	Round	Opponent	Result	GF	GA
2001	CSF	Pittsburgh	L 3-4	17	17
	CQF	Philadelphia	W 4-2	21	13
2000	CQF	Philadelphia	L 1-4	8	14
1999	F	Dallas	L 2-4	9	13
	CF	Toronto	W 4-1	21	16
	CSF	Boston	W 4-2	17	14
	CQF	Ottawa	W 4-0	12	6
1998	CF	Washington	L 2-4	11	13
	CSF	Montreal	W 4-0	17	10
	CQF	Philadelphia	W 4-1	18	9

Abbreviations: Round: F - final; **CF** - conference final; **CSF** - conference semi-final; **CQF** - conference quarter-final; **DSF** - division semi-final; **SF** - semi-final; **QF** - quarter-final; **PRE** - preliminary round.

Calgary totals include Atlanta Flames, 1972-73 to 1979-80. Carolina totals include Hartford, 1979-80 to 1996-97.
Colorado totals include Quebec, 1979-80 to 1994-95. Dallas totals include Minnesota North Stars, 1970-71 to 1992-93.
New Jersey totals include Kansas City, 1974-75 to 1975-76, and Colorado Rockies, 1976-77 to 1981-82.
Phoenix totals include Winnipeg, 1979-80 to 1995-96.

2001-02 Results

Month	Day	Opponent	Score
Oct.	4	Atlanta	1-2
	6	Ottawa	3-2
	7	at NY Rangers	4-5*
	10	Philadelphia	2-1
	12	at Detroit	2-4
	14	Pittsburgh	4-1
	16	Nashville	3-3
	19	Columbus	1-3
	20	at Montreal	3-1
	23	San Jose	4-1
	26	Montreal	2-5
	27	at New Jersey	1-3
	30	Phoenix	2-3
Nov.	2	Tampa Bay	4-1
	3	at Ottawa	0-3
	8	Atlanta	8-0
	10	NY Rangers	2-4
	12	at Florida	5-3
	13	at Nashville	1-4
	16	Florida	0-2
	17	at Boston	1-3
	19	at Atlanta	2-3
	21	Toronto	4-2
	23	Calgary	5-2
	24	at Pittsburgh	1-3
	27	NY Rangers	2-2
	28	at Washington	5-2
Dec.	1	at NY Islanders	4-2
	4	at Carolina	4-2
	7	Colorado	1-4
	8	at Boston	2-4
	12	at Dallas	3-4
	14	Carolina	3-2*
	15	at NY Rangers	2-4
	19	Chicago	5-6
	21	Toronto	3-3
	22	at Toronto	2-3
	26	Montreal	3-1
	29	at Columbus	2-2
	31	Carolina	4-5
Jan.	3	at Calgary	1-3
	6	at Minnesota	4-1
	8	Vancouver	3-2
	10	Pittsburgh	0-2
	12	New Jersey	2-1
	16	at Anaheim	3-1
	17	at Los Angeles	2-4
	19	at Phoenix	3-1
	21	at Colorado	2-3
	23	St. Louis	2-5
	25	Tampa Bay	4-1
	27	at Washington	3-2*
	29	at Carolina	2-2
Feb.	5	at Boston	2-2
	8	Ottawa	3-2*
	10	at New Jersey	1-4
	12	New Jersey	2-2
	26	at Atlanta	1-2
Mar.	1	Boston	4-3
	2	at Toronto	3-3
	4	Edmonton	0-3
	7	at NY Islanders	5-0
	8	Montreal	3-0
	10	Detroit	5-1
	12	NY Islanders	0-3
	14	at Philadelphia	3-1
	15	at Florida	2-5
	17	at Tampa Bay	2-2
	19	Ottawa	5-1
	21	Boston	1-2
	23	at Toronto	0-2
	24	at Ottawa	3-2*
	26	Washington	3-4
	28	at St. Louis	1-4
	30	at Philadelphia	3-1
Apr.	1	Philadelphia	3-1
	3	NY Islanders	1-1
	5	Florida	3-1
	7	at Tampa Bay	5-3
	10	at Pittsburgh	4-4
	12	Washington	1-3
	13	at Montreal	3-0

* – Overtime

Entry Draft Selections 2002-1988

2002

Pick	
11	Keith Ballard
20	Dan Paille
76	Michael Tessier
82	John Adams
108	Jakub Hulva
121	Marty Magers
178	Maxim Schevjev
208	Radoslav Hecl
241	Dennis Wideman
271	Martin Cizek

2001

Pick	
22	Jiri Novotny
32	Derek Roy
50	Chris Thorburn
55	Jason Pominville
155	Michal Vondrka
234	Calle Aslund
247	Marek Dubec
279	Ryan Jorde

2000

Pick	
15	Artem Kryukov
48	Gerard Dicaire
111	Ghyslain Rousseau
149	Denis Denisov
213	Vasili Bizyayev
220	Paul Gaustad
258	Sean McMorrow
277	Ryan Courtney

1999

Pick	
20	Barrett Heisten
35	Milan Bartovic
55	Doug Janik
64	Mike Zigomanis
73	Tim Preston
117	Karel Mosovsky
138	Ryan Miller
146	Matt Kinch
178	Seneque Hyacinthe
206	Bret DeCecco
235	Brad Self
263	Craig Brunel

1998

Pick	
18	Dmitri Kalinin
34	Andrew Peters
47	Norm Milley
50	Jaroslav Kristek
77	Mike Pandolfo
137	Aaron Goldade
164	Ales Kotalik
191	Brad Moran
218	David Moravec
249	Edo Terglav

1997

Pick	
21	Mika Noronen
48	Henrik Tallinder
69	Maxim Afinogenov
75	Jeff Martin
101	Luc Theoret
128	Torrey DiRoberto
156	Brian Campbell
184	Jeremy Adduono
212	Kamil Piros
238	Dylan Kemp

1996

Pick	
7	Erik Rasmussen
27	Cory Sarich
33	Darren Van Oene
54	Francois Methot
87	Kurt Walsh
106	Mike Martone
115	Alexei Tezikov
142	Ryan Davis
161	Darren Mortier
222	Scott Buhler

1995

Pick	
14	Jay McKee
16	Martin Biron
42	Mark Dutiaume
68	Mathieu Sunderland
94	Matt Davidson
111	Marian Menhart
119	Kevin Popp
123	Daniel Bienvenue
172	Brian Scott
198	Mike Zanutto
224	Rob Skrlac

1994

Pick	
17	Wayne Primeau
43	Curtis Brown
69	Rumun Ndur
121	Sergei Klimentiev
147	Cal Benazic
168	Steve Plouffe
173	Shane Hnidy
176	Steve Webb
199	Bob Westerby
225	Craig Millar
251	Mark Polak
277	Shayne Wright

1993

Pick	
38	Denis Tsygurov
64	Ethan Philpott
116	Richard Safarik
142	Kevin Pozzo
168	Sergei Petrenko
194	Mike Barrie
220	Barrie Moore
246	Chris Davis
272	Scott Nichol

1992

Pick	
11	David Cooper
35	Jozef Cierny
59	Ondrej Steiner
80	Dean Melanson
83	Matthew Barnaby
107	Markus Ketterer
108	Yuri Khmylev
131	Paul Rushforth
179	Dean Tiltgen
203	Todd Simon
227	Rick Kowalsky
251	Chris Clancy

1991

Pick	
13	Philippe Boucher
35	Jason Dawe
57	Jason Young
72	Peter Ambroziak
101	Steve Shields
123	Sean O'Donnell
124	Brian Holzinger
145	Chris Snell
162	Jiri Kuntos
189	Tony Iob
211	Spencer Meany
233	Mikhail Volkov
255	Michael Smith

1990

Pick	
14	Brad May
82	Brian McCarthy
97	Richard Smehlik
100	Todd Bojcun
103	Brad Pascall
142	Viktor Gordiouk
166	Milan Nedoma
187	Jason Winch
208	Sylvain Naud
229	Kenneth Martin
250	Brad Rubachuk

1989

Pick	
14	Kevin Haller
56	Scott Thomas
77	Doug MacDonald
98	Ken Sutton
107	Bill Pye
119	Mike Barkley
161	Derek Plante
183	Donald Audette
194	Mark Astley
203	John Nelson
224	Todd Henderson
245	Michael Bavis

1988

Pick	
13	Joel Savage
55	Darcy Loewen
76	Keith Carney
89	Alexander Mogilny
97	Rob Ray
106	David Di Vita
118	Mike McLaughlin
139	Mike Griffith
160	Daniel Ruoho
181	Wade Flaherty
223	Thomas Nieman
243	Michael Pohl

General Managers' History

Punch Imlach, 1970-71 to 1977-78; John Anderson, 1978-79; Scotty Bowman, 1979-80 to 1985-86; Scotty Bowman and Gerry Meehan, 1986-87; Gerry Meehan, 1987-88 to 1992-93; John Muckler, 1993-94 to 1996-97; Darcy Regier, 1997-98 to date.

General Manager

REGIER, DARCY

General Manager, Buffalo Sabres. Born in Swift Current, Sask., Nov. 27, 1957.

Darcy Regier became the sixth general manager of the Buffalo Sabres on June 11, 1997 after a lengthy management apprenticeship in the New York Islanders organization. As a player, Regier played eight pro seasons, including part of the 1977-78 season with the Cleveland Barons and parts of the 1982-83 and 1983-84 campaigns with the New York Islanders.

He began his career as an administrator with the Islanders in 1984-85 and went on to serve in a variety of capacities including director of administration, assistant director of hockey operations, assistant coach and assistant general manager. He also served as an assistant coach with Hartford in 1991-92.

While with the Islanders, Regier benefitted from working with talented managers and coaches including Bill Torrey and Al Arbour. As a minor pro player with Indianapolis of the CHL he became associated with another important influence on his hockey career, current Detroit Red Wing executive Jim Devellano.

Club Directory

HSBC Arena

Buffalo Sabres
HSBC Arena
One Seymour H. Knox III Plaza
Buffalo, NY 14203
Phone **716/855-4100**
Fax 716/855-4110
Tickets, U.S.: 716/223-6000
Tickets, Canada: 888/669-GOAL
Capacity: 18,690

Executive

Executive Vice President/Administration	Ron Bertovich
Senior Vice President/ Sales	Kerry Atkinson
Senior Vice President/Legal & Business Affairs	Kevin Billet
Senior Vice President/Marketing	Christye Peterson
Vice President/Communications	Michael Gilbert
Vice President/Corporate Relations	Seymour H. Knox, IV
Vice President/Ticket Sales & Operations	John Sinclair
Senior Director of Sports & Arena Planning	Chris Schoepflin
Special Consultant	Joe Crozier
Executive Assistants	Eleanore MacKenzie, Donna Webb-Smith

Hockey Department

General Manager	Darcy Regier
Assistant to the General Manager	Larry Carriere
Director of Player Personnel	Don Luce
Executive Assistant	Elaine Burzynski
On-Site Travel Coordinator	Kim Christiano
Professional Scouts	Kevin Devine, Terry Martin
Scouting Staff	Don Barrie, Jim Benning, Bo Berglund, Iouri Khmylev, Paul Merritt, Rudy Migay, Darryl Plandowski, Mike Racicot, David Volek
Head Coach	Lindy Ruff
Assistant Coach	Brian McCutcheon
Assistant Coach	Scott Arniel
Strength & Conditioning Coach	Doug McKenney
Assistant Strength Coach	Dennis Cole
Goaltender Coach	Jim Corsi
Administrative Assistant, Coaches	Jeff Holbrook
Head Trainer/Massage Therapist	Jim Pizzutelli
Head Equipment Manager	Rip Simonick
Assistant Equipment Manager	George Babcock
Equipment Assistant	Encil "Porky" Palmer

Medical

Club Doctor	Les Bisson, M.D.
Doctors	Nicholas Aquino, M.D., William Hartrich, M.D.
Oral Surgeon	Steven Jenson, DDS
Club Dentist	Daniel Yustin, DDS, M.S.
Physical Therapist	Joe Aquino
Club Doctor Emeritus	John L. Butsch, M.D.

Legal

Associate Counsel	Richard Mugel

Administration

Human Resources Coordinator	Mary Jones

Broadcast Production

Senior Director of Broadcast & Production Services	Joe Guarnieri
Producer	Lowell MacDonald
Director	TBA
Broadcast Team	Rick Jeanneret (play-by-play), Jim Lorentz (color commentary), Danny Gare (reporter)

Merchandise

Director of Merchandise	Mike Kaminska
Store Manager	Tammy Preteroti

Communications

Director of Communications	Gregg Huller
Communications Coordinator – Buffalo Sabres	Kevin Wiles
Team Photographer	Bill Wippert
Director of Alumni Relations	Larry Playfair
Corporate & Community Relations Liaison	Gilbert Perreault

Empire Sports Sales

General Sales Manager	Dan Rozanski
National Sales Manager	Mark Kennedy
Radio General Sales Manager	Steve Cuccia

Finance

Controller	John Marsh
Accounting Manager – Buffalo Sabres	Christine Ivansitz

Marketing

Director of Advertising and Promotions	Robert Kopacz
Director of Game Presentation & Special Events	Matt Copolla

Community Development

Director of Community Development	Peter Hassen

Canadian Sales & Marketing

Director of Canadian Sales & Marketing	Steve Katzman

Ticket Sales & Operations

Ticket Sales Manager	Dan Carroll
Box Office Manager	Michael Tout

HSBC Arena

Senior Director of Facilities Management	Stan Makowski
Director of Event Booking	Jennifer Van Rysdam
Director of Event Services	John Faso
Director of Premium Sales	Nick Turano
Director of Suite Services	Natalie DeSilva

General Information

Practice Site	Pepsi Center, Amherst, NY
TV Station	TBA
Radio Flagship Station	TBA

Calgary Flames

2001-02 Results: 32W-35L-12T-3OTL 79PTS.
Fourth, Northwest Division

2002-03 Schedule

Month	Day	Date	Opponent	Month	Day	Date	Opponent
Oct.	Thu.	10	Vancouver		Thu.	9	Ottawa
	Sat.	12	Philadelphia		Sat.	11	Columbus
	Mon.	14	at Vancouver		Mon.	13	at Montreal
	Thu.	17	Boston		Tue.	14	at Toronto
	Sat.	19	at Chicago		Thu.	16	Nashville
	Mon.	21	at Detroit		Sat.	18	Los Angeles
	Tue.	22	at Minnesota		Mon.	20	Edmonton
	Thu.	24	Dallas		Thu.	23	Phoenix
	Sat.	26	St. Louis		Sat.	25	Detroit
	Thu.	31	Buffalo		Tue.	28	at Phoenix
Nov.	Sat.	2	Colorado		Wed.	29	at Dallas
	Mon.	4	at NY Islanders	**Feb.**	Tue.	4	Anaheim
	Tue.	5	at New Jersey		Thu.	6	Chicago
	Thu.	7	at NY Rangers		Fri.	7	at Edmonton
	Sat.	9	at Florida		Sun.	9	at Colorado
	Mon.	11	at Atlanta*		Wed.	12	at Anaheim
	Thu.	14	NY Rangers		Thu.	13	at Los Angeles
	Sat.	16	St. Louis		Sat.	15	Vancouver
	Tue.	19	Detroit		Mon.	17	at St. Louis*
	Thu.	21	Edmonton		Wed.	19	at Dallas
	Sat.	23	Chicago		Thu.	20	at Nashville
	Tue.	26	at Boston		Sun.	23	at Phoenix*
	Wed.	27	at Washington		Mon.	24	at San Jose
	Fri.	29	at St. Louis	**Mar.**	Sat.	1	San Jose
Dec.	Sun.	1	at Detroit		Wed.	5	New Jersey
	Tue.	3	at Colorado		Fri.	7	at Chicago
	Thu.	5	Minnesota		Sat.	8	at Columbus
	Mon.	9	at Vancouver		Tue.	11	Edmonton
	Thu.	12	Carolina		Thu.	13	Toronto
	Sat.	14	Colorado		Sat.	15	at San Jose
	Sun.	15	at Vancouver		Sun.	16	at Anaheim*
	Tue.	17	at Nashville		Tue.	18	at Los Angeles
	Thu.	19	at Columbus		Thu.	20	Washington
	Sat.	21	at Pittsburgh		Sat.	22	Nashville
	Mon.	23	at Minnesota		Mon.	24	Phoenix
	Fri.	27	Toronto		Thu.	27	Dallas
	Sun.	29	Anaheim		Sat.	29	Columbus
	Tue.	31	Montreal		Mon.	31	at Minnesota
Jan.	Thu.	2	Tampa Bay	**Apr.**	Wed.	2	San Jose
	Sat.	4	Minnesota		Fri.	4	Los Angeles
	Tue.	7	at Colorado		Sat.	5	at Edmonton

* Denotes afternoon game.

Year-by-Year Record

		Home				Road				Overall								
Season	**GP**	**W**	**L**	**T**	**OL**	**W**	**L**	**T**	**OL**	**W**	**L**	**T**	**OL**	**GF**	**GA**	**Pts.**	**Finished**	**Playoff Result**
2001-02	82	20	14	5	2	12	21	7	1	32	35	12	3	201	220	79	4th, Northwest Div.	Out of Playoffs
2000-01	82	12	18	9	2	15	18	6	2	27	36	15	4	197	236	73	4th, Northwest Div.	Out of Playoffs
1999-2000	82	20	14	6	1	11	22	4	4	31	36	10	5	211	256	77	4th, Northwest Div.	Out of Playoffs
1998-99	82	15	20	6	...	15	20	6	...	30	40	12	...	211	234	72	3rd, Northwest Div.	Out of Playoffs
1997-98	82	18	17	6	...	8	24	9	...	26	41	15	...	217	252	67	5th, Pacific Div.	Out of Playoffs
1996-97	82	21	18	2	...	11	23	7	...	32	41	9	...	214	239	73	5th, Pacific Div.	Out of Playoffs
1995-96	82	18	18	5	...	16	19	6	...	34	37	11	...	241	240	79	2nd, Pacific Div.	Lost Conf. Quarter-Final
1994-95	48	15	7	2	...	9	10	5	...	24	17	7	...	163	135	55	1st, Pacific Div.	Lost Conf. Quarter-Final
1993-94	84	25	12	5	...	17	17	8	...	42	29	13	...	302	256	97	1st, Pacific Div.	Lost Conf. Quarter-Final
1992-93	84	23	14	5	...	20	16	6	...	43	30	11	...	322	282	97	2nd, Smythe Div.	Lost Div. Semi-Final
1991-92	80	19	14	7	...	12	23	5	...	31	37	12	...	296	305	74	5th, Smythe Div.	Out of Playoffs
1990-91	80	29	8	3	...	17	18	5	...	46	26	8	...	344	263	100	2nd, Smythe Div.	Lost Div. Semi-Final
1989-90	80	28	7	5	...	14	16	10	...	42	23	15	...	348	265	99	1st, Smythe Div.	Lost Div. Semi-Final
1988-89	**80**	**32**	**4**	**4**	**...**	**22**	**13**	**5**	**...**	**54**	**17**	**9**	**...**	**354**	**226**	**117**	**1st, Smythe Div.**	**Won Stanley Cup**
1987-88	80	26	11	3	...	22	12	6	...	48	23	9	...	397	305	105	1st, Smythe Div.	Lost Div. Final
1986-87	80	25	13	2	...	21	18	1	...	46	31	3	...	318	289	95	2nd, Smythe Div.	Lost Div. Semi-Final
1985-86	80	23	11	6	...	17	20	3	...	40	31	9	...	354	315	89	2nd, Smythe Div.	Lost Final
1984-85	80	23	11	6	...	18	16	6	...	41	27	12	...	363	302	94	3rd, Smythe Div.	Lost Div. Semi-Final
1983-84	80	22	11	7	...	12	21	7	...	34	32	14	...	311	314	82	2nd, Smythe Div.	Lost Div. Final
1982-83	80	21	12	7	...	11	22	7	...	32	34	14	...	321	317	78	2nd, Smythe Div.	Lost Div. Final
1981-82	80	20	11	9	...	9	23	8	...	29	34	17	...	334	345	75	3rd, Smythe Div.	Lost Div. Semi-Final
1980-81	80	25	5	10	...	14	22	4	...	39	27	14	...	329	298	92	3rd, Patrick Div.	Lost Semi-Final
1979-80*	80	18	15	7	...	17	17	6	...	35	32	13	...	282	269	83	4th, Patrick Div.	Lost Prelim. Round
1978-79*	80	25	11	4	...	16	20	4	...	41	31	8	...	327	280	90	4th, Patrick Div.	Lost Prelim. Round
1977-78*	80	20	13	7	...	14	14	12	...	34	27	19	...	274	252	87	3rd, Patrick Div.	Lost Prelim. Round
1976-77*	80	22	11	7	...	12	23	5	...	34	34	12	...	264	265	80	3rd, Patrick Div.	Lost Prelim. Round
1975-76*	80	19	14	7	...	16	19	5	...	35	33	12	...	262	237	82	3rd, Patrick Div.	Lost Prelim. Round
1974-75*	80	24	9	7	...	10	22	8	...	34	31	15	...	243	233	83	4th, Patrick Div.	Out of Playoffs
1973-74*	78	17	15	7	...	13	19	7	...	30	34	14	...	214	238	74	4th, West Div.	Lost Quarter-Final
1972-73*	78	16	16	7	...	9	22	8	...	25	38	15	...	191	239	65	7th, West Div.	Out of Playoffs

* Atlanta Flames

Long considered one of the game's best young players, Jarome Iginla truly arrived in 2001-02, leading the NHL with 52 goals and 96 points. He was the runner-up for the Hart Trophy as MVP and won an Olympic gold medal with Team Canada.

Franchise date: June 6, 1972
Transferred from Atlanta to Calgary, June 24, 1980.

NORTHWEST DIVISION

31st NHL Season

2002-03 Player Personnel

FORWARDS	HT	WT	S	Place of Birth	Date	2001-02 Club
BEGIN, Steve	5-11	190	L	Trois-Rivieres, Que.	6/14/78	Calgary
BERUBE, Craig	6-1	210	L	Calahoo, Alta.	12/17/65	Calgary
BETTS, Blair	6-1	200	L	Edmonton, Alta.	2/16/80	Calgary-Saint John
CHRISTIE, Ryan	6-3	200	L	Beamsville, Ont.	7/3/78	Calgary-Saint John
CLARK, Chris	6-0	200	R	South Windsor, CT	3/8/76	Calgary
CONROY, Craig	6-2	197	R	Potsdam, NY	9/4/71	Calgary
DOME, Robert	6-0	210	L	Skalica, Czech.	1/29/79	Wilkes-Barre
GELINAS, Martin	5-11	195	L	Shawinigan, Que.	6/5/70	Carolina
HAY, Dwayne	6-1	203	L	London, Ont.	2/11/77	Saint John
IGINLA, Jarome	6-1	207	R	Edmonton, Alta.	7/1/77	Calgary-Canada
JOHANSSON, Mathias	6-2	185	L	Oskarshamn, Sweden	2/22/74	Farjestad-Sweden
KOBASEW, Chuck	5-11	195	R	Osoyoos, B.C.	4/17/82	Kelowna
LOMBARDI, Matthew	5-11	191	L	Montreal, Que.	3/18/82	Victoriaville
LOWRY, Dave	6-1	195	L	Sudbury, Ont.	2/14/65	Calgary
McAMMOND, Dean	5-11	193	L	Grand Cache, Alta.	6/15/73	Calgary
MORGAN, Jason	6-1	200	L	St. John's, Nfld.	10/9/76	Saint John
NICHOL, Scott	5-8	173	R	Edmonton, Alta.	12/31/74	Calgary
NIEDERMAYER, Rob	6-2	205	L	Cassiar, B.C.	12/28/74	Calgary
NYSTROM, Eric	6-1	195	L	Syosset, NY	2/14/83	U. of Michigan
PETROVICKY, Ronald	5-11	190	R	Zilina, Czech.	2/15/77	Calgary
SAPRYKIN, Oleg	6-0	195	L	Moscow, USSR	2/12/81	Calgary-Saint John
SAVARD, Marc	5-10	188	L	Ottawa, Ont.	7/17/77	Calgary
SHANTZ, Jeff	6-0	195	R	Duchess, Alta.	10/10/73	Calgary-Saint John
SLOAN, Blake	5-10	196	R	Park Ridge, IL	7/27/75	Columbus-Calgary
SONNENBERG, Martin	6-0	197	L	Wetaskiwin, Alta.	1/23/78	Wilkes-Barre
VEROT, Darcy	6-0	202	L	Radville, Sask.	7/13/76	Wilkes-Barre
WRIGHT, Jamie	6-0	195	L	Kitchener, Ont.	5/13/76	Calgary-Saint John
DEFENSEMEN						
BOUGHNER, Bob	6-0	203	R	Windsor, Ont.	3/8/71	Calgary
BUZEK, Petr	6-0	210	L	Jihlava, Czech.	4/26/77	Atlanta-Calgary
DuPONT, Micki	5-9	186	R	Calgary, Alta.	4/15/80	Calgary-Saint John
GAUTHIER, Denis	6-2	224	L	Montreal, Que.	10/1/76	Calgary
LEOPOLD, Jordan	6-0	193	L	Golden Valley, MN	8/3/80	U. of Minnesota
LYDMAN, Toni	6-1	202	L	Lahti, Finland	9/25/77	Calgary
MARTIN, Mike	6-2	205	R	Stratford, Ont.	10/27/76	Saint John
MONTADOR, Steve	6-0	210	R	Vancouver, B.C.	12/21/79	Calgary-Saint John
MORRIS, Derek	5-11	210	R	Edmonton, Alta.	8/24/78	Calgary
MROZIK, Rick	6-2	214	L	Duluth, MN	1/2/75	Saint John
REGEHR, Robyn	6-2	226	L	Recife, Brazil	4/19/80	Calgary

GOALTENDERS	HT	WT	C	Place of Birth	Date	2001-02 Club
McLENNAN, Jamie	6-0	190	L	Edmonton, Alta.	6/30/71	Houston
SABOURIN, Dany	6-2	182	L	Val d'Or, Que.	9/2/80	Johnstown (ECHL)
SZUPER, Levente	5-11	180	L	Budapest, Hungary	6/11/80	Saint John
TUREK, Roman	6-3	220	R	Strakonice, Czech.	5/21/70	Calgary
WHITMORE, Kay	5-11	175	L	Sudbury, Ont.	4/10/67	Calgary-Saint John

Roman Turek posted shutouts in his first two games with the Flames to open the 2001-02 campaign. He was named NHL player of the week for the first week of the season.

General Manager

BUTTON, CRAIG
Vice President/General Manager, Calgary Flames.
Born in Montreal, Que., January 3, 1963.

Craig Button was named vice president and general manager of the Flames on June 6, 2000. Button spent the previous 12 seasons with the Dallas Stars organization, serving as the director of player personnel for the last two seasons after spending six years as the director of scouting. Button oversaw the Stars' top minor league team in Kalamazoo, aiding in the evaluation and development of the Stars' minor league prospects. His other responsibilities included the management and development of the Stars' amateur and professional scouting program, the evaluation of players as it relates to movement within the organization, including the Entry Draft, trades and free agent signings, and the player development program for their amateur prospects.

A native of Montreal, Button graduated from Concordia University in Montreal in 1987 with a BA in Economics with an emphasis in international finance.

Button's family has a long history in hockey. His late father, Jack, was a former NHL general manager and a highly respected 34-year veteran of NHL management. His mother, Bridget, worked for Punch Imlach and the Toronto Maple Leafs, and his brother Tod is a pro scout with the Flames.

2001-02 Scoring

* - rookie

Regular Season

Pos	#	Player	Team	GP	G	A	Pts	+/–	PIM	PP	SH	GW	GT	S	%
R	12	Jarome Iginla	CGY	82	52	44	96	27	77	16	1	7	2	311	16.7
C	22	Craig Conroy	CGY	81	27	48	75	24	32	7	2	4	1	146	18.5
C	37	Dean McAmmond	CGY	73	21	30	51	2	60	7	0	4	1	152	13.8
D	53	Derek Morris	CGY	61	4	30	34	–4	88	2	0	1	0	166	2.4
C	27	Marc Savard	CGY	56	14	19	33	–18	48	7	0	3	1	140	10.0
D	32	Toni Lydman	CGY	79	6	22	28	–8	52	1	0	0	0	126	4.8
D	25	Igor Kravchuk	CGY	78	4	22	26	3	19	1	0	1	0	135	3.0
C	44	Rob Niedermayer	CGY	57	6	14	20	–15	49	1	2	1	1	87	6.9
C	23	Clarke Wilm	CGY	66	4	14	18	–1	61	0	1	0	0	83	4.8
R	17	Chris Clark	CGY	64	10	7	17	–12	79	2	1	4	0	109	9.2
C	40	Scott Nichol	CGY	60	8	9	17	–9	107	2	1	0	0	49	16.3
L	18	Jamie Wright	CGY	44	4	12	16	6	20	0	0	0	0	64	6.3
L	10	Dave Lowry	CGY	62	7	6	13	–20	51	2	1	1	1	74	9.5
D	3	Denis Gauthier	CGY	66	5	8	13	9	91	0	1	2	0	76	6.6
C	26 *	Steve Begin	CGY	51	7	5	12	–3	79	1	0	0	1	65	10.8
L	36	Ronald Petrovicky	CGY	77	5	7	12	0	85	1	0	1	0	78	6.4
R	24	Blake Sloan	CBJ	60	2	7	9	–18	46	0	0	0	0	49	4.1
			CGY	7	0	2	2	1	4	0	0	0	0	7	0.0
			TOTAL	67	2	9	11	–17	50	0	0	0	0	56	3.6
D	28	Robyn Regehr	CGY	77	2	6	8	–24	93	0	0	0	0	82	2.4
C	11	Jeff Shantz	CGY	40	3	3	6	–3	23	2	0	0	0	37	8.1
D	6	Bob Boughner	CGY	79	2	4	6	9	170	0	0	0	0	58	3.4
L	16	Craig Berube	CGY	66	3	1	4	–2	164	1	0	0	0	34	8.8
D	8	Petr Buzek	ATL	9	0	0	0	–4	13	0	0	0	0	2	0.0
			CGY	32	1	3	4	4	14	0	0	0	0	34	2.9
			TOTAL	41	1	3	4	0	27	0	0	0	0	36	2.8
D	58 *	Steve Montador	CGY	11	1	2	3	–2	26	0	0	0	0	10	10.0
L	20	Jason Botterill	CGY	4	1	0	1	–3	2	1	0	1	0	4	25.0
C	15 *	Blair Betts	CGY	6	1	0	1	–1	2	0	0	1	0	4	25.0
D	2	Alan Letang	CGY	2	0	0	0	–2	0	0	0	0	0	0	0.0
L	39 *	Ryan Christie	CGY	2	0	0	0	–1	0	0	0	0	0	0	0.0
D	42 *	Micki DuPont	CGY	2	0	0	0	0	2	0	0	0	0	2	0.0
D	4	Dallas Eakins	CGY	3	0	0	0	1	4	0	0	0	0	0	0.0
C	19	Oleg Saprykin	CGY	3	0	0	0	–2	0	0	0	0	0	9	0.0

Goaltending

No.	Goaltender	GPI	Mins	Avg	W	L	T	EN	SO	GA	SA	S%	G	A	PIM
1	Roman Turek	69	4081	2.53	30	28	11	6	5	172	1839	.906	0	5	4
29	Mike Vernon	18	825	2.76	2	9	1	1	1	38	375	.899	0	3	0
35	Kay Whitmore	1	58	3.10	0	1	0	0	0	3	21	.857	0	0	0
	Totals	**82**	**4990**	**2.65**	**32**	**38**	**12**	**7**	**7**	**220**	**2242**	**.902**			

Captains' History

Keith McCreary, 1972-73 to 1974-75; Pat Quinn, 1975-76, 1976-77; Tom Lysiak, 1977-78, 1978-79; Jean Pronovost, 1979-80; Brad Marsh, 1980-81; Phil Russell, 1981-82, 1982-83; Lanny McDonald, Doug Risebrough (co-captains), 1983-84; Lanny McDonald, Doug Risebrough, Jim Peplinski (tri-captains), 1984-85 to 1986-87; Lanny McDonald, Jim Peplinski (co-captains), 1987-88; Lanny McDonald, Jim Peplinski, Tim Hunter (tri-captains), 1988-89; Brad McCrimmon, 1989-90; alternating captains, 1990-91; Joe Nieuwendyk, 1991-92 to 1994-95; Theoren Fleury, 1995-96, 1996-97; Todd Simpson, 1997-98, 1998-99; Steve Smith, 1999-2000; Steve Smith and Dave Lowry, 2000-01; Dave Lowry, Bob Boughner and Craig Conroy (co-captains), 2001-02; Bob Boughner and Craig Conroy (co-captains), 2002-03.

General Managers' History

Cliff Fletcher, 1972-73 to 1990-91; Doug Risebrough, 1991-92 to 1994-95; Doug Risebrough and Al Coates, 1995-96; Al Coates, 1996-97 to 1999-2000; Craig Button, 2000-01 to date.

Coaching History

Bernie Geoffrion, 1972-73, 1973-74; Bernie Geoffrion and Fred Creighton, 1974-75; Fred Creighton, 1975-76 to 1978-79; Al MacNeil, 1979-80 to 1981-82; Bob Johnson, 1982-83 to 1986-87; Terry Crisp, 1987-88 to 1989-90; Doug Risebrough, 1990-91; Doug Risebrough and Guy Charron, 1991-92; Dave King, 1992-93 to 1994-95; Pierre Page, 1995-96, 1996-97; Brian Sutter, 1997-98 to 1999-2000; Don Hay and Greg Gilbert, 2000-01; Greg Gilbert, 2001-02 to date.

Club Records

Team

(Figures in brackets for season records are games played; records for fewest points, wins, ties, losses, goals, goals against are for 70 or more games)

Record		
Most Points	**117**	1988-89 (80)
Most Wins	**54**	1988-89 (80)
Most Ties	**19**	1977-78 (80)
Most Losses	**41**	1996-97 (82), 1997-98 (82), 1999-2000 (82)
Most Goals	**397**	1987-88 (80)
Most Goals Against	**345**	1981-82 (80)
Fewest Points	**65**	1972-73 (78)
Fewest Wins	**25**	1972-73 (78)
Fewest Ties	**3**	1986-87 (80)
Fewest Losses	**17**	1988-89 (80)
Fewest Goals	**191**	1972-73 (78)
Fewest Goals Against	**220**	2001-02 (82)
Longest Winning Streak		
Overall	**10**	Oct. 14-Nov. 3/78
Home	**9**	Oct. 17-Nov. 15/78, Jan. 3-Feb. 5/89, Mar. 3-Apr. 1/90, Feb. 21-Mar. 14/91
Away	**7**	Nov. 10-Dec. 4/88
Longest Undefeated Streak		
Overall	**13**	Nov. 10-Dec. 8/88 (12 wins, 1 tie)
Home	**18**	Dec. 29/90-Mar. 14/91 (17 wins, 1 tie)
Away	**9**	Feb. 20-Mar. 21/88 (6 wins, 3 ties), Nov. 11-Dec. 16/90 (6 wins, 3 ties)
Longest Losing Streak		
Overall	**11**	Dec. 14/85-Jan. 7/86
Home	**6**	Dec. 5-31/98
Away	**9**	Dec. 1/85-Jan. 12/86
Longest Winless Streak		
Overall	**11**	Dec. 14/85-Jan. 7/86 (11 losses), Jan. 5-26/93 (9 losses, 2 ties)
Home	**10**	Oct. 21-Dec. 4/00 (6 losses, 4 ties)
Away	**13**	Feb. 3-Mar. 29/73 (10 losses, 3 ties)
Most Shutouts, Season	**8**	1974-75 (80), 2000-01 (82)
Most PIM, Season	**2,643**	1991-92 (80)
Most Goals, Game	**13**	Feb. 10/93 (S.J. 1 at Cgy. 13)

Individual

Most Seasons	**13**	Al MacInnis
Most Games	**803**	Al MacInnis
Most Goals, Career	**364**	Theoren Fleury
Most Assists, Career	**609**	Al MacInnis
Most Points, Career	**830**	Theoren Fleury (364G, 466A)
Most PIM, Career	**2,405**	Tim Hunter
Most Shutouts, Career	**20**	Dan Bouchard
Longest Consecutive Games Streak	**257**	Brad Marsh (Oct. 11/78-Nov. 10/81)
Most Goals, Season	**66**	Lanny McDonald (1982-83)
Most Assists, Season	**82**	Kent Nilsson (1980-81)
Most Points, Season	**131**	Kent Nilsson (1980-81; 49G, 82A)
Most PIM, Season	**375**	Tim Hunter (1988-89)
Most Points, Defenseman, Season	**103**	Al MacInnis (1990-91; 28G, 75A)
Most Points, Center, Season	**131**	Kent Nilsson (1980-81; 49G, 82A)
Most Points, Right Wing, Season	**110**	Joe Mullen (1988-89; 51G, 59A)
Most Points, Left Wing, Season	**90**	Gary Roberts (1991-92; 53G, 37A)
Most Points, Rookie, Season	**92**	Joe Nieuwendyk (1987-88; 51G, 41A)
Most Shutouts, Season	**5**	Dan Bouchard (1973-74), Phil Myre (1974-75), Fred Brathwaite (1999-2000, 2000-01), Roman Turek (2001-02)
Most Goals, Game	**5**	Joe Nieuwendyk (Jan. 11/89)
Most Assists, Game	**6**	Guy Chouinard (Feb. 25/81), Gary Suter (Apr. 4/86)
Most Points, Game	**7**	Sergei Makarov (Feb. 25/90; 2G, 5A)

Records include Atlanta Flames, 1972-73 through 1979-80.

Retired Numbers

9	Lanny McDonald	1981-1989

All-time Record vs. Other Clubs

Regular Season

	At Home								On Road								Total							
	GP	W	L	T	OL	GF	GA	PTS	GP	W	L	T	OL	GF	GA	PTS	GP	W	L	T	OL	GF	GA	PTS
Anaheim	21	12	8	1	0	58	51	25	22	7	9	5	1	58	71	20	43	19	17	6	1	116	122	45
Atlanta	2	2	0	0	0	9	4	4	3	0	2	1	0	4	6	1	5	2	2	1	0	13	10	5
Boston	43	18	22	3	0	157	151	39	45	12	27	6	0	123	158	30	88	30	49	9	0	280	309	69
Buffalo	44	16	17	11	0	149	144	43	45	13	26	5	1	131	189	32	89	29	43	16	1	280	333	75
Carolina	27	20	5	2	0	136	88	42	28	13	10	5	0	103	91	31	55	33	15	7	0	239	179	73
Chicago	59	27	20	12	0	191	179	66	57	19	25	13	0	165	189	51	116	46	45	25	0	356	368	117
Colorado	42	20	15	7	0	148	125	47	41	14	15	11	1	135	153	40	83	34	30	18	1	283	278	87
Columbus	4	3	1	0	0	13	5	6	4	0	4	0	0	8	16	0	8	3	5	0	0	21	21	6
Dallas	58	32	13	13	0	201	144	77	58	21	27	10	0	190	214	52	116	53	40	23	0	391	358	129
Detroit	56	33	17	6	0	223	168	72	55	17	28	10	0	167	204	44	111	50	45	16	0	390	372	116
Edmonton	75	40	27	8	0	313	262	88	76	25	41	10	0	250	297	60	151	65	68	18	0	563	559	148
Florida	7	3	3	1	0	17	18	7	8	4	2	2	0	22	18	10	15	7	5	3	0	39	36	17
Los Angeles	90	52	27	11	0	402	302	115	87	34	43	9	1	306	327	78	177	86	70	20	1	708	629	193
Minnesota	5	3	0	1	1	15	12	8	5	2	2	1	0	8	10	5	10	5	2	2	1	23	22	13
Montreal	47	15	26	6	0	143	162	36	44	12	24	8	0	109	155	32	91	27	50	14	0	252	317	68
Nashville	8	5	1	1	1	27	20	12	9	3	6	0	0	16	27	6	17	8	7	1	1	43	47	18
New Jersey	41	27	6	8	0	179	107	62	43	26	14	3	0	158	123	55	84	53	20	11	0	337	230	117
NY Islanders	49	24	14	11	0	172	145	59	49	15	25	9	0	136	187	39	98	39	39	20	0	308	332	98
NY Rangers	48	27	10	10	1	215	146	65	50	22	22	5	1	179	177	50	98	49	32	15	2	394	323	115
Ottawa	9	5	3	1	0	34	22	11	9	2	4	3	0	23	25	7	18	7	7	4	0	57	47	18
Philadelphia	51	25	17	9	0	204	167	59	50	14	33	3	0	134	197	31	101	39	50	12	0	338	364	90
Phoenix	67	36	21	9	1	287	218	82	66	22	32	11	1	225	257	56	133	58	53	20	2	512	475	138
Pittsburgh	45	26	11	8	0	198	139	60	43	10	23	10	0	133	165	30	88	36	34	18	0	331	304	90
St. Louis	58	29	23	5	1	196	171	64	60	22	29	9	0	183	215	53	118	51	52	14	1	379	386	117
San Jose	28	15	10	3	0	104	79	33	30	17	9	4	0	95	85	38	58	32	19	7	0	199	164	71
Tampa Bay	8	5	3	0	0	26	17	10	10	4	5	1	0	32	31	9	18	9	8	1	0	58	48	19
Toronto	58	32	21	5	0	229	186	69	51	18	26	7	0	186	195	43	109	50	47	12	0	415	381	112
Vancouver	93	57	23	13	0	385	269	127	93	41	34	17	1	311	322	100	186	98	57	30	1	696	591	227
Washington	37	24	6	7	0	156	89	55	39	14	20	5	0	134	146	33	76	38	26	12	0	290	235	88
Defunct Clubs	13	8	4	1	0	51	34	17	13	7	3	3	0	43	33	17	26	15	7	4	0	94	67	34
Totals	**1193**	**641**	**374**	**173**	**5**	**4638**	**3624**	**1460**	**1193**	**430**	**570**	**186**	**7**	**3767**	**4283**	**1053**	**2386**	**1071**	**944**	**359**	**12**	**8405**	**7907**	**2513**

Playoffs

	Series	W	L	GP	W	L	T	GF	GA	Last Mtg.	Rnd.	Result
Chicago	3	2	1	12	7	5	0	37	33	1996	CQF	L 0-4
Dallas	1	0	1	6	2	4	0	18	25	1981	SF	L 2-4
Detroit	1	0	1	2	0	2	0	5	8	1978	PRE	L 0-2
Edmonton	5	1	4	30	11	19	0	96	132	1991	DSF	L 3-4
Los Angeles	6	2	4	26	13	13	0	102	105	1993	DSF	L 2-4
Montreal	2	1	1	11	5	6	0	32	31	1989	F	W 4-2
NY Rangers	1	0	1	4	1	3	0	8	14	1980	PRE	L 1-3
Philadelphia	2	1	1	11	4	7	0	28	43	1981	QF	W 4-3
St. Louis	1	1	0	7	4	3	0	28	22	1986	CF	W 4-3
San Jose	1	0	1	7	3	4	0	35	26	1995	CQF	L 3-4
Toronto	1	0	1	2	0	2	0	5	9	1979	PRE	L 0-2
Vancouver	5	3	2	25	13	12	0	82	80	1994	CQF	L 3-4
Winnipeg	3	1	2	13	6	7	0	43	45	1987	DSF	L 2-4
Totals	**32**	**12**	**20**	**156**	**69**	**87**	**0**	**519**	**573**			

Playoff Results 2002-1998

(Last playoff appearance: 1996)

Abbreviations: Round: F - Final; **CF** - conference final; **CQF** - conference quarter-final; **DSF** - division semi-final; **SF** - semi-final; **QF** - quarter-final; **PRE** - preliminary round.

Carolina totals include Hartford, 1979-80 to 1996-97.
Colorado totals include Quebec, 1979-80 to 1994-95.
Dallas totals include Minnesota North Stars, 1972-73 to 1992-93.
New Jersey totals include Kansas City, 1974-75 to 1975-76, and Colorado Rockies, 1976-77 to 1981-82.
Phoenix totals include Winnipeg, 1979-80 to 1995-96.

2001-02 Results

Month	Day	Opponent	Score
Oct.	3	Edmonton	1-0
	6	Chicago	4-0
	8	Phoenix	1-2*
	10	at Detroit	4-2
	11	at Nashville	0-1
	13	at Dallas	4-3*
	18	Florida	3-1
	20	Toronto	4-1
	22	at St. Louis	3-2
	23	at Chicago	3-6
	25	Nashville	4-5*
	27	Minnesota	4-2
Nov.	1	Columbus	2-1
	3	Montreal	6-2
	7	at Anaheim	3-3
	8	at Los Angeles	3-2
	10	Colorado	2-0
	15	Chicago	2-2
	17	St. Louis	2-0
	20	Los Angeles	5-5
	22	at Ottawa	4-4
	23	at Buffalo	2-5
	25	at Columbus	3-4
	27	at Detroit	2-4
	29	Dallas	0-3
Dec.	1	Colorado	2-2
	3	at Los Angeles	2-0
	4	at San Jose	2-2
	6	San Jose	1-3
	8	Anaheim	0-4
	10	Detroit	2-0
	12	Tampa Bay	1-3
	14	at Dallas	4-3
	15	at St. Louis	0-4
	19	at Phoenix	3-6
	21	at Colorado	2-2
	26	at Edmonton	2-3
	27	at Vancouver	2-4
	29	Minnesota	4-3
	31	Edmonton	2-2
Jan.	3	Buffalo	3-1
	5	Montreal	2-4
	8	at NY Islanders	5-2
	9	at New Jersey	1-5
	11	at Atlanta	0-1
	15	NY Islanders	1-3
	17	Pittsburgh	4-6
	19	Anaheim	2-1
	22	Toronto	1-6
	24	Colorado	0-2
	26	Vancouver	0-2
	28	at Minnesota	3-2*
	30	Detroit	4-3
Feb.	6	at San Jose	0-2
	8	Vancouver	1-4
	9	at Vancouver	4-3
	12	at Phoenix	3-4*
	13	at Anaheim	2-3
	26	at Colorado	2-2
	28	St. Louis	3-2
Mar.	2	Nashville	5-2
	4	at NY Rangers	5-3
	6	at Washington	2-3
	7	at Philadelphia	4-2
	9	at Boston	2-3
	11	at Carolina	3-3
	13	at Florida	3-3
	14	at Tampa Bay	2-3
	16	at Columbus	1-3
	18	at Minnesota	2-4
	21	San Jose	1-4
	23	at Edmonton	1-3
	25	Columbus	6-1
	28	Dallas	2-2
	30	Los Angeles	5-3
Apr.	2	Atlanta	4-2
	4	Minnesota	4-3
	6	at Nashville	1-3
	7	at Chicago	2-3
	9	Phoenix	2-4
	12	at Edmonton	2-0
	13	Vancouver	1-4

* – Overtime

Entry Draft Selections 2002-1988

2002
Pick
10 Eric Nystrom
39 Brian McConnell
90 Matthew Lombardi
112 Yuri Artemenkov
141 Jiri Cetkovsky
142 Emanuel Peter
146 Victor Bobrov
159 Kristofer Persson
176 Curtis McElhinney
206 David Van Der Gulik
207 Pierre Johnsson
238 Jyri Marttinen

2001
Pick
14 Chuck Kobasew
41 Andrei Taratukhin
56 Andrei Medvedev
108 Tomi Maki
124 Yegor Shastin
145 James Hakewill
164 Yuri Trubachev
207 Garrett Bembridge
220 David Moss
233 Joe Campbell
251 Ville Hamalainen

2000
Pick
9 Brent Krahn
40 Kurtis Foster
46 Jarret Stoll
116 Levente Szuper
141 Wade Davis
155 Travis Moen
176 Jukka Hentunen
239 David Hajek
270 Micki DuPont

1999
Pick
11 Oleg Saprykin
38 Dan Cavanaugh
77 Craig Andersson
106 Rail Rozakov
135 Matt Doman
153 Jesse Cook
166 Cory Pecker
170 Matt Underhill
190 Blair Stayzer
252 Dmitri Kirilenko

1998
Pick
6 Rico Fata
33 Blair Betts
62 Paul Manning
102 Shaun Sutter
108 Dany Sabourin
120 Brent Gauvreau
192 Radek Duda
206 Jonas Frogren
234 Kevin Mitchell

1997
Pick
6 Daniel Tkaczuk
32 Evan Lindsay
42 John Tripp
51 Dimitri Kokorev
60 Derek Schutz
70 Erik Andersson
92 Chris St. Croix
100 Ryan Ready
113 Martin Moise
140 Ilja Demidov
167 Jeremy Rondeau
223 Dustin Paul

1996
Pick
13 Derek Morris
39 Travis Brigley
40 Steve Begin
73 Dmitri Vlasenkov
89 Toni Lydman
94 Christian Lefebvre
122 Josef Straka
202 Ryan Wade
228 Ronald Petrovicky

1995
Pick
20 Denis Gauthier
46 Pavel Smirnov
72 Rocky Thompson
98 Jan Labraaten
150 Clarke Wilm
176 Ryan Gillis
233 Steve Shirreffs

1994
Pick
19 Chris Dingman
45 Dmitri Ryabykin
77 Chris Clark
91 Ryan Duthie
97 Johan Finnstrom
107 Nils Ekman
123 Frank Appel
149 Patrick Haltia
175 Ladislav Kohn
201 Keith McCambridge
227 Jorgen Jonsson
253 Mike Peluso
279 Pavel Torgaev

1993
Pick
18 Jesper Mattsson
44 Jamie Allison
70 Dan Tompkins
95 Jason Smith
96 Marty Murray
121 Darryl Lafrance
122 John Emmons
148 Andreas Karlsson
200 Derek Sylvester
252 German Titov
278 Burke Murphy

1992
Pick
6 Cory Stillman
30 Chris O'Sullivan
54 Mathias Johansson
78 Robert Svehla
102 Sami Helenius
126 Ravil Yakubov
129 Joel Bouchard
150 Pavel Rajnoha
174 Ryan Mulhern
198 Brandon Carper
222 Jonas Hoglund
246 Andrei Potaichuk

1991
Pick
19 Niklas Sundblad
41 Francois Groleau
52 Sandy McCarthy
63 Brian Caruso
85 Steven Magnusson
107 Jerome Butler
129 Bobby Marshall
140 Matt Hoffman
151 Kelly Harper
173 David St-Pierre
195 David Struch
217 Sergei Zolotov
239 Marko Jantunen
261 Andrei Trefilov

1990
Pick
11 Trevor Kidd
26 Nicolas Perreault
32 Vesa Viitakoski
41 Etienne Belzile
62 Glen Mears
83 Paul Kruse
125 Chris Tschupp
146 Dimitri Frolov
167 Shawn Murray
188 Mike Murray
209 Rob Sumner
230
251 Leo Gudas

1989
Pick
24 Kent Manderville
42 Ted Drury
50 Veli-Pekka Kautonen
63 Corey Lyons
70 Robert Reichel
84 Ryan O'Leary
105 Toby Kearney
147 Alex Nikolic
168 Kevin Wortman
189 Sergei Gomolyako
210 Dan Sawyer
231 Alexander Yudin
252 Kenneth Kennholt

1988
Pick
21 Jason Muzzatti
42 Todd Harkins
84 Gary Socha
85 Tomas Forslund
90 Scott Matusovich
126 Jonas Bergqvist
168 Troy Kennedy
189 Brett Peterson
210 Guy Darveau
231 Dave Tretowicz
252 Sergei Priakin

Coach

GILBERT, GREG
Coach, Calgary Flames. Born in Mississauga, Ont., January 22, 1962.

Greg Gilbert joined the Calgary Flames as an assistant coach on August 11, 2000 and assumed the head coaching duties on March 14, 2001. He was formally announced as the club's head coach on May 1, 2001.

Before joining the Flames, Gilbert spent four seasons as the bench boss of the St. Louis Blues' American Hockey League affiliate in Worcester, Massachusetts. In his first season as head coach of the IceCats, Gilbert guided the 1996-97 club to a first place finish in the New England Division. He was subsequently named the AHL coach of the year and *The Sporting News* Minor League coach of the year. Under Gilbert's direction, the IceCats made the playoffs every season.

Gilbert began coaching immediately following a successful 15-year National Hockey League playing career with the New York Islanders, Chicago Blackhawks, New York Rangers and St. Louis Blues. Drafted 80th overall by the Islanders in the 1980 NHL Entry Draft, the 1981-82 season was to be the first of his three Stanley Cup championships. Gilbert was also a member of the Islanders' Stanley Cup team the following season and celebrated his third and final Cup win with the Rangers in 1994. Statistically, his finest season was the 1983-84 campaign with the Islanders when he scored 31 goals and added 35 assists. In 837 career games, Gilbert recorded 150 goals and 228 assists for 378 points. He retired after the 1995-96 season.

Coaching Record

		Regular Season				Playoffs		
Season	Team	Games	W	L	T	Games	W	L
1996-97	Worcester (AHL)	80	43	23	14	5	2	3
1997-98	Worcester (AHL)	80	34	31	15	11	6	5
1998-99	Worcester (AHL)	80	34	36	10	4	1	3
1999-2000	Worcester (AHL)	80	34	41	15	9	4	5
2000-01	**Calgary (NHL)**	**14**	**4**	**8**	**2**	....	....	....
2001-02	**Calgary (NHL)**	**82**	**32**	**38**	**12**	....	....	....
	NHL Total	**96**	**36**	**46**	**14**	....	....	....

Club Directory

Pengrowth Saddledome

Calgary Flames
Pengrowth Saddledome
P.O. Box 1540 Station M
Calgary, Alberta T2P 3B9
Phone **403/777-2177**
FAX 403/777-2199
www.calgaryflames.com
Capacity: 17,409

Owners . . . . N. Murray Edwards, Harley N. Hotchkiss, Alvin G. Libin, Allan P. Markin, J.R. (Bud) McCaig, Byron J. Seaman, Daryl K. Seaman

Executive
President & Chief Executive Officer . . . . Ken King
Vice-President/General Manager . . . . Craig Button
Vice President, Finance & Administration. . . . . Michael Holditch
Vice-President, Building Operations . . . . Libby Raines
Vice-President, Business Development . . . . Jim Peplinski
Vice President, Advertising,
Sponsorship & Marketing . . . . Jim Bagshaw
Vice-President, Sales & Ticket Operations . . . . Rollie Cyr

Hockey Club Personnel
Vice-President/General Manager . . . . Craig Button
Assistant to the General Manager . . . . Dan Stuchal
Director, Hockey Administration . . . . Mike Burke
Special Assistant to the GM. . . . . Al MacNeil
Executive Advisor to Hockey Operations . . . . Lanny McDonald
Head Coach . . . . Greg Gilbert
Assistant Coaches . . . . Brad McCrimmon, Brian Skrudland, Rob Cookson
Development Coach . . . . Jamie Hislop
Goaltending Coach . . . . Wendell Young
Exec. Asst. to GM and Hockey Operations. . . . . Brenda Koyich
Team Services Manager. . . . . Kelly Chesla
Director of Scouting . . . . Tod Button
Director of Amateur Scouting . . . . Mike Sands
Pro Scout . . . . Ron Sutter
Scouts. . . . . Bob Atrill, Jeff Crisp, Steve Graves, Pertti Hasanen, Tomas Jelinek, Larry Johnston, Bob Richardson, Sergei Samoylov, Al Tuer
Saint John Flames Head Coach . . . . Jim Playfair
Saint John Flames Asst. Coach . . . . Ron Wilson

Medical/Training Staff
Athletic Therapist . . . . Morris Boyer
Assistant Athletic Therapist . . . . Terence "TD" Forss
Strength & Conditioning Coach . . . . Rich Hesketh
Equipment Manager . . . . Gus Thorson
Assistant Equipment Manager. . . . . Les Jarvis
Head Physician . . . . Dr. Nicholas Mohtadi
Sport Medicine Physician. . . . . Dr. Kelly Brett
Internal Medicine . . . . Dr. Terry Groves
Team Dentist. . . . . Dr. Bill Blair
Dressing Room Attendant . . . . Jules Carriere

Communications
Director, Communications. . . . . Peter Hanlon
Manager, Media Relations. . . . . Sean O'Brien
Administrative Assistant, Communications . . . . Bernie Hargrave
Community Relations Ambassador . . . . Jim "Bearcat" Murray
Community Relations Coordinator . . . . Trevor Elgar
Flames Foundation/Community
Relations Coordinator . . . . Lauren Smith

Administration
Controller . . . . Jackie Manwaring
Assistant Controller. . . . . Karen Kingham
Interim Assistant Controller . . . . Trudy McInnes
Exec. Asst. to President/CEO . . . . Gita Nayak
Exec. Asst. to VP, Finance & Administration. . . . . . . Christine Macri

Marketing
Director of Sponsorship. Al Molnar
Director, Advertising . . . . Pat Halls
Director, Retail/FanAttic. Kip Reghenas
Director, Executive Suites . . . . Bob White
Business Development Manager . . . . Kevin Gross
Manager, Ticket Operations . . . . Mike Franco
Director, Game Presentation . . . . Dave Imbach
Director/Producer, Jumbotron . . . . Carlo Petrini
Desktop Publishing . . . . Laurie Wheeler
Exec. Asst. to VP, Advertising/Marketing . . . . Yvette Mutcheson
Mascot . . . . Harvey the Hound

Pengrowth Saddledome
Operations Manager . . . . George Greenwood
Food Services Manager . . . . Art Hernandez
Concessions Manager . . . . Sheila Parisien
Security/Parking Manager . . . . Bob Godun

Calgary Hitmen
Governor . . . . Michael Holditch
General Manager . . . . Kelly Kisio
Asst. General Manager . . . . Blaine Forsythe
Head Coach . . . . Richard Kromm
Assistant Coach . . . . Bruno Campese

Miscellaneous Data
Practice Facility . . . . Pengrowth Saddledome
Training Camp . . . . Banff, Alberta
Club Colours. . . . . Red, white, gold and black
Radio Affiliate . . . . The Team 960 (960 AM)
TV Affiliate . . . . Rogers Sportsnet, CBC-TV, TSN

Carolina Hurricanes

2001-02 Results: 35W-26L-16T-5OTL 91PTS.
First, Southeast Division

2002-03 Schedule

Oct.	Wed.	9	NY Rangers		Wed.	8	at NY Rangers
	Fri.	11	Atlanta		Fri.	10	Washington
	Sat.	12	at Tampa Bay		Sun.	12	Colorado*
	Tue.	15	at St. Louis		Wed.	15	Pittsburgh
	Thu.	17	Washington		Fri.	17	New Jersey
	Sat.	19	New Jersey		Sat.	18	at New Jersey
	Tue.	22	at NY Islanders		Mon.	20	St. Louis
	Wed.	23	at Ottawa		Wed.	22	at Washington
	Sat.	26	Chicago		Fri.	24	Florida
	Tue.	29	at New Jersey		Sat.	25	at Florida
	Wed.	30	NY Islanders		Wed.	29	Toronto
Nov.	Fri.	1	Montreal		Thu.	30	at Tampa Bay
	Tue.	5	Philadelphia	**Feb.**	Wed.	5	at San Jose
	Thu.	7	Buffalo		Fri.	7	at Los Angeles
	Sat.	9	Pittsburgh		Sun.	9	at Anaheim*
	Tue.	12	Phoenix		Tue.	11	at Dallas
	Fri.	15	Philadelphia		Fri.	14	Washington
	Sun.	17	Tampa Bay*		Sat.	15	at Philadelphia
	Tue.	19	Ottawa		Tue.	18	at Toronto
	Thu.	21	at Boston		Wed.	19	Boston
	Sat.	23	at Montreal		Fri.	21	Tampa Bay
	Mon.	25	at NY Rangers		Sun.	23	Anaheim*
	Wed.	27	Vancouver		Wed.	26	at Phoenix
	Fri.	29	Detroit	**Mar.**	Sat.	1	at Toronto
	Sat.	30	at Columbus		Sun.	2	at Washington
Dec.	Tue.	3	at Nashville		Tue.	4	Boston
	Wed.	4	at Florida		Thu.	6	at Pittsburgh
	Fri.	6	Florida		Fri.	7	Minnesota
	Sat.	7	at Ottawa		Mon.	10	Columbus
	Wed.	11	at Edmonton		Wed.	12	at Buffalo
	Thu.	12	at Calgary		Thu.	13	at Philadelphia
	Sun.	15	at Minnesota*		Sat.	15	Los Angeles
	Wed.	18	Tampa Bay		Tue.	18	Ottawa
	Fri.	20	at Atlanta		Sat.	22	at Montreal
	Sun.	22	Dallas*		Tue.	25	Toronto
	Fri.	27	Atlanta		Wed.	26	at Atlanta
	Sat.	28	at NY Islanders		Sat.	29	Buffalo
	Tue.	31	NY Rangers		Mon.	31	Montreal
Jan.	Fri.	3	at Buffalo	**Apr.**	Wed.	2	at Pittsburgh
	Sat.	4	at Boston		Fri.	4	at Florida
	Tue.	7	at Atlanta		Sun.	6	NY Islanders*

* Denotes afternoon game.

Year-by-Year Record

		Home				Road				Overall								
Season	**GP**	**W**	**L**	**T**	**OL**	**W**	**L**	**T**	**OL**	**W**	**L**	**T**	**OL**	**GF**	**GA**	**Pts.**	**Finished**	**Playoff Result**
2001-02	82	15	13	11	2	20	13	5	3	35	26	16	5	217	217	91	1st, Southeast Div.	Lost Final
2000-01	82	23	15	3	0	15	17	6	3	38	32	9	3	212	225	88	2nd, Southeast Div.	Lost Conf. Quarter-Final
1999-2000	82	20	16	5	0	17	19	5	0	37	35	10	0	217	216	84	3rd, Southeast Div.	Out of Playoffs
1998-99	82	20	12	9	...	14	18	9	...	34	30	18	...	210	202	86	1st, Southeast Div.	Lost Conf. Quarter-Final
1997-98	82	16	18	7	...	17	23	1	...	33	41	8	...	200	219	74	6th, Northeast Div.	Out of Playoffs
1996-97*	82	23	15	3	...	9	24	8	...	32	39	11	...	226	256	75	5th, Northeast Div.	Out of Playoffs
1995-96*	82	22	15	4	...	12	24	5	...	34	39	9	...	237	259	77	4th, Northeast Div.	Out of Playoffs
1994-95*	48	12	10	2	...	7	14	3	...	19	24	5	...	127	141	43	5th, Northeast Div.	Out of Playoffs
1993-94*	84	14	22	6	...	13	26	3	...	27	48	9	...	227	288	63	6th, Northeast Div.	Out of Playoffs
1992-93*	84	12	25	5	...	14	27	1	...	26	52	6	...	284	369	58	5th, Adams Div.	Out of Playoffs
1991-92*	80	13	17	10	...	13	24	3	...	26	41	13	...	247	283	65	4th, Adams Div.	Lost Div. Semi-Final
1990-91*	80	18	16	6	...	13	22	5	...	31	38	11	...	238	276	73	4th, Adams Div.	Lost Div. Semi-Final
1989-90*	80	17	18	5	...	21	15	4	...	38	33	9	...	275	268	85	4th, Adams Div.	Lost Div. Semi-Final
1988-89*	80	21	17	2	...	16	21	3	...	37	38	5	...	299	290	79	4th, Adams Div.	Lost Div. Semi-Final
1987-88*	80	21	14	5	...	14	24	2	...	35	38	7	...	249	267	77	4th, Adams Div.	Lost Div. Semi-Final
1986-87*	80	26	9	5	...	17	21	2	...	43	30	7	...	287	270	93	1st, Adams Div.	Lost Div. Semi-Final
1985-86*	80	21	17	2	...	19	19	2	...	40	36	4	...	332	302	84	4th, Adams Div.	Lost Div. Final
1984-85*	80	17	18	5	...	13	23	4	...	30	41	9	...	268	318	69	5th, Adams Div.	Out of Playoffs
1983-84*	80	19	16	5	...	9	26	5	...	28	42	10	...	288	320	66	5th, Adams Div.	Out of Playoffs
1982-83*	80	13	22	5	...	6	32	2	...	19	54	7	...	261	403	45	5th, Adams Div.	Out of Playoffs
1981-82*	80	13	17	10	...	8	24	8	...	21	41	18	...	264	351	60	5th, Adams Div.	Out of Playoffs
1980-81*	80	14	17	9	...	7	24	9	...	21	41	18	...	292	372	60	4th, Norris Div.	Out of Playoffs
1979-80*	80	22	12	6	...	5	22	13	...	27	34	19	...	303	312	73	4th, Norris Div.	Lost Prelim. Round

* Hartford Whalers

The Hurricanes picked up Bret Hedican from the Panthers, along with Kevyn Adams, for Sandis Ozolinsh and Byron Ritchie. A defensive defenseman, he became a key part of the Carolina blueline corps during their surprising playoff run.

Franchise date: June 22, 1979
Transferred from Hartford to Carolina, June 25, 1997.

24th NHL Season

SOUTHEAST DIVISION

2002-03 Player Personnel

FORWARDS	HT	WT	S	Place of Birth	Date	2001-02 Club
ADAMS, Craig	6-0	200	R	Seria, Brunei	4/26/77	Carolina-Lowell
ADAMS, Kevyn	6-1	195	R	Washington, DC	10/8/74	Florida-Carolina
ASTASHENKO, Kaspars	6-2	183	L	Riga, Latvia	2/17/75	Springfield-Lowell-Latvia
BATTAGLIA, Bates	6-2	205	L	Chicago, IL	12/13/75	Carolina
BAYDA, Ryan	5-11	185	L	Saskatoon, Sask.	12/9/80	North Dakota-Lowell
BOULERICE, Jesse	6-1	215	R	Plattsburgh, NY	8/10/78	Phi-Phi (AHL)-Lowell (AHL)
BRIND'AMOUR, Rod	6-1	202	L	Ottawa, Ont.	8/9/70	Carolina
COLE, Erik	6-1	200	L	Oswego, NY	11/6/78	Carolina
DANIELS, Jeff	6-1	200	L	Oshawa, Ont.	6/24/68	Carolina
DEFAUW, Brad	6-2	210	L	Edina, MN	11/10/77	Lowell
FRANCIS, Ron	6-3	200	L	Sault Ste. Marie, Ont.	3/1/63	Carolina
HEEREMA, Jeff	6-1	190	R	Thunder Bay, Ont.	1/17/80	Lowell
KAPANEN, Sami	5-10	195	L	Vantaa, Finland	6/14/73	Carolina-Finland
KURKA, Tomas	5-11	190	L	Most, Czech.	12/14/81	Lowell
LANGDON, Darren	6-1	205	L	Deer Lake, Nfld.	1/8/71	Carolina
MacDONALD, Craig	6-2	195	L	Antigonish, N.S.	4/7/77	Carolina-Lowell
O'NEILL, Jeff	6-1	190	R	Richmond Hill, Ont.	2/23/76	Carolina
SVOBODA, Jaroslav	6-2	190	L	Cervenka, Czech.	6/1/80	Carolina-Lowell
VASICEK, Josef	6-4	200	L	Havlickuv Brod, Czech.	9/12/80	Carolina
WATT, Mike	6-2	212	L	Seaforth, Ont.	3/31/76	Philadelphia (AHL)
WESTLUND, Tommy	6-0	210	R	Fors, Sweden	12/29/74	Carolina
ZIGOMANIS, Mike	6-1	189	R	North York, Ont.	1/17/81	Lowell
DEFENSEMEN						
BAST, Ryan	6-2	190	L	Spruce Grove, Alta.	8/27/75	Pee Dee-Lowell
HALKO, Steven	6-1	200	R	Etobicoke, Ont.	3/8/74	Carolina-Worcester
HEDICAN, Bret	6-2	205	L	St. Paul, MN	8/10/70	Florida-Carolina
HILL, Sean	6-0	203	R	Duluth, MN	2/14/70	St. Louis-Carolina
KNYAZEV, Igor	6-0	191	L	Elektrostal, USSR	1/27/83	Spartak Moscow-Spartak Mos. 2-Kazan
MALEC, Tomas	6-2	193	L	Skalica, Czech.	5/13/82	Rimouski-Lowell
MALIK, Marek	6-5	215	L	Ostrava, Czech.	6/24/75	Carolina
TANABE, David	6-1	190	R	White Bear Lake, MN	7/19/80	Carolina
TSELIOS, Nikos	6-5	210	L	Oak Park, IL	1/20/79	Carolina-Lowell
WALLIN, Niclas	6-3	220	L	Boden, Sweden	2/20/75	Carolina
WARD, Aaron	6-2	200	R	Windsor, Ont.	1/17/73	Carolina
WESLEY, Glen	6-1	205	L	Red Deer, Alta.	10/2/68	Carolina

GOALTENDERS	HT	WT	C	Place of Birth	Date	2001-02 Club
IRBE, Arturs	5-8	190	L	Riga, Latvia	2/2/67	Carolina-Latvia
PELLETIER, Jean-Marc	6-3	200	L	Atlanta, GA	3/4/78	Lowell
PETRUK, Randy	5-9	175	R	Cranbrook, B.C.	4/23/78	Florida
WEEKES, Kevin	6-0	195	L	Toronto, Ont.	4/4/75	Tampa Bay-Carolina
ZEPP, Rob	6-1	181	L	Scarborough, Ont.	9/7/81	Florida

Coaching History

Don Blackburn, 1979-80; Don Blackburn and Larry Pleau, 1980-81; Larry Pleau, 1981-82; Larry Kish, Larry Pleau and John Cuniff, 1982- 83; Jack Evans, 1983-84 to 1986-87; Jack Evans and Larry Pleau, 1987-88; Larry Pleau, 1988-89; Rick Ley, 1989-90, 1990-91; Jim Roberts, 1991-92; Paul Holmgren, 1992-93; Paul Holmgren and Pierre Maguire, 1993-94; Paul Holmgren, 1994-95; Paul Holmgren and Paul Maurice, 1995-96; Paul Maurice, 1996-97 to date.

Coach

MAURICE, PAUL
Coach, Carolina Hurricanes. Born in Sault Ste. Marie, Ont., January 30, 1967.

Paul Maurice became the tenth coach in franchise history on November 6, 1995 and is the only man to serve as head coach since the club moved to Carolina in 1997. He was the youngest coach in the NHL when he first stepped behind the bench 12 games into the 1995-96 season and remains the youngest despite having the longest tenure among current NHL head coaches. He ranks as the club's all-time leader in wins and games coached, and guided the team to the Stanley Cup Finals for the first time in franchise history in 2002.

Maurice joined the Whalers in June of 1995 as an assistant coach after serving as the head coach of the Detroit Junior Red Wings for two seasons. The Junior Wings won the OHL Western Division regular season title and played for the 1995 Memorial Cup by winning the OHL playoffs. The Wings lost in the Cup finals to Kamloops. For his efforts, Maurice was the runner-up for OHL coach of the year honors in 1995. In the 1993-94 season, Maurice's squad won the OHL Hap Emms Division title and advanced to the finals of the OHL playoffs before losing in seven games to North Bay.

Maurice began his coaching career in 1986 as an assistant coach for the Detroit Junior Red Wings after an eye injury ended his junior playing career. He served six seasons in that capacity before taking over the head coaching responsibilities in the 1993-94 season.

Coaching Record

		Regular Season				Playoffs		
Season	Team	Games	W	L	T	Games	W	L
1993-94	Detroit (OHL)	66	42	20	4	17	11	6
1994-95	Detroit (OHL)	66	44	18	4	21	16	5
1995-96	**Hartford (NHL)**	**70**	**29**	**33**	**8**	....	....	....
1996-97	**Hartford (NHL)**	**82**	**32**	**39**	**11**	....	....	....
1997-98	**Carolina (NHL)**	**82**	**33**	**41**	**8**	....	....	....
1998-99	**Carolina (NHL)**	**82**	**34**	**30**	**18**	**6**	**2**	**4**
1999-2000	**Carolina (NHL)**	**82**	**37**	**35**	**10**	....	....	....
2000-01	**Carolina (NHL)**	**82**	**38**	**35**	**9**	**6**	**2**	**4**
2001-02	**Carolina (NHL)**	**82**	**35**	**31**	**16**	**23**	**13**	**10**
	NHL Totals	**562**	**238**	**244**	**80**	**35**	**17**	**18**

2001-02 Scoring

* - rookie

Regular Season

Pos	#	Player	Team	GP	G	A	Pts	+/–	PIM	PP	SH	GW	GT	S	%
C	10	Ron Francis	CAR	80	27	50	77	4	18	14	0	5	2	165	16.4
R	24	Sami Kapanen	CAR	77	27	42	69	9	23	11	0	4	1	248	10.9
C	92	Jeff O'Neill	CAR	76	31	33	64	–5	63	11	0	6	1	272	11.4
C	17	Rod Brind'Amour	CAR	81	23	32	55	3	40	5	2	5	1	162	14.2
L	13	Bates Battaglia	CAR	82	21	25	46	–6	44	5	1	2	3	167	12.6
L	26 *	Erik Cole	CAR	81	16	24	40	–10	35	3	0	2	0	159	10.1
D	22	Sean Hill	STL	23	0	3	3	1	28	0	0	0	0	29	0.0
			CAR	49	7	23	30	–1	61	4	0	2	0	116	6.0
			TOTAL	72	7	26	33	0	89	4	0	2	0	145	4.8
C	63	Josef Vasicek	CAR	78	14	17	31	–7	53	3	0	3	0	117	12.0
L	23	Martin Gelinas	CAR	72	13	16	29	–1	30	3	0	1	1	121	10.7
D	5	Marek Malik	CAR	82	4	19	23	8	88	0	0	0	0	91	4.4
D	2	Glen Wesley	CAR	77	5	13	18	–8	56	1	0	0	1	88	5.7
C	15	Kevyn Adams	FLA	44	4	8	12	–3	28	0	0	1	0	71	5.6
			CAR	33	2	3	5	–2	15	0	0	1	0	37	5.4
			TOTAL	77	6	11	17	–5	43	0	0	2	0	108	5.6
D	6	Bret Hedican	FLA	31	3	7	10	–4	12	0	0	0	0	46	6.5
			CAR	26	2	4	6	3	10	0	0	1	0	39	5.1
			TOTAL	57	5	11	16	–1	22	0	0	1	0	85	5.9
D	45	David Tanabe	CAR	78	1	15	16	–13	35	0	0	0	0	113	0.9
D	4	Aaron Ward	CAR	79	3	11	14	0	74	0	0	2	1	69	4.3
L	11	Jeff Daniels	CAR	65	4	1	5	–6	12	0	1	0	0	40	10.0
L	62 *	Jaroslav Svoboda	CAR	10	2	2	4	0	2	0	0	0	0	12	16.7
L	20	Darren Langdon	CAR	58	2	1	3	2	106	0	0	1	0	12	16.7
D	7	Niclas Wallin	CAR	52	1	2	3	1	36	0	0	0	0	33	3.0
C	12 *	Craig Macdonald	CAR	12	1	1	2	–1	0	0	0	0	0	15	6.7
L	16	Tommy Westlund	CAR	40	0	2	2	–8	6	0	0	0	0	29	0.0
D	14	Steven Halko	CAR	5	0	1	1	3	6	0	0	0	0	0	0.0
R	27	Craig Adams	CAR	33	0	1	1	2	38	0	0	0	0	17	0.0
D	48 *	Nikos Tselios	CAR	2	0	0	0	–2	6	0	0	0	0	3	0.0
C	21	Josh Holden	CAR	8	0	0	0	0	2	0	0	0	0	3	0.0

Goaltending

No.	Goaltender	GPI	Mins	Avg	W	L	T	EN	SO	GA	SA	S%	G	A	PIM
80	Kevin Weekes	2	120	1.50	2	0	0	0	0	3	41	.927	0	0	0
1	Arturs Irbe	51	2974	2.54	20	19	11	3	3	126	1282	.902	0	1	10
30	Tom Barrasso	34	1908	2.61	13	12	5	2	2	83	886	.906	0	0	4
	Totals	**82**	**5021**	**2.59**	**35**	**31**	**16**	**5**	**5**	**217**	**2214**	**.902**			

Playoffs

Pos	#	Player	Team	GP	G	A	Pts	+/–	PIM	PP	SH	GW	GT	S	%
C	10	Ron Francis	CAR	23	6	10	16	–2	6	4	0	3	1	51	11.8
L	13	Bates Battaglia	CAR	23	5	9	14	2	14	1	0	1	1	44	11.4
C	92	Jeff O'Neill	CAR	22	8	5	13	2	27	3	0	1	1	72	11.1
C	17	Rod Brind'Amour	CAR	23	4	8	12	–3	16	2	1	1	0	48	8.3
L	26 *	Erik Cole	CAR	23	6	3	9	–2	30	1	0	1	0	68	8.8
R	24	Sami Kapanen	CAR	23	1	8	9	–2	6	0	0	0	0	50	2.0
D	22	Sean Hill	CAR	23	4	4	8	0	20	4	0	1	0	57	7.0
L	23	Martin Gelinas	CAR	23	3	4	7	6	10	0	0	1	1	33	9.1
C	63	Josef Vasicek	CAR	23	3	2	5	6	12	0	0	1	1	32	9.4
D	6	Bret Hedican	CAR	23	1	4	5	0	20	0	0	0	0	39	2.6
L	62 *	Jaroslav Svoboda	CAR	23	1	4	5	2	28	1	0	1	0	26	3.8
D	7	Niclas Wallin	CAR	23	2	1	3	4	12	0	0	2	2	17	11.8
D	5	Marek Malik	CAR	23	0	3	3	3	18	0	0	0	0	24	0.0
D	4	Aaron Ward	CAR	23	1	1	2	0	22	0	0	0	0	23	4.3
D	2	Glen Wesley	CAR	22	0	2	2	0	12	0	0	0	0	21	0.0
L	16	Tommy Westlund	CAR	19	1	0	1	2	0	0	0	0	0	6	16.7
C	15	Kevyn Adams	CAR	23	1	0	1	–1	4	0	0	0	0	23	4.3
D	45	David Tanabe	CAR	1	0	1	1	0	0	0	0	0	0	3	0.0
L	11	Jeff Daniels	CAR	23	0	1	1	–1	0	0	0	0	0	7	0.0
R	27	Craig Adams	CAR	1	0	0	0	0	0	0	0	0	0	0	0.0
C	12 *	Craig MacDonald	CAR	4	0	0	0	–1	2	0	0	0	0	1	0.0

Goaltending

No.	Goaltender	GPI	Mins	Avg	W	L	EN	SO	GA	SA	S%	G	A	PIM
80	Kevin Weekes	8	408	1.62	3	2	1	2	11	180	.939	0	0	0
1	Arturs Irbe	18	1078	1.67	10	8	1	1	30	480	.938	0	0	0
	Totals	**23**	**1493**	**1.73**	**13**	**10**	**2**	**3**	**43**	**662**	**.935**			

General Managers' History

Jack Kelly, 1979-80, 1980-81; Larry Pleau, 1981-82, 1982-83; Emile Francis, 1983-84 to 1988-89; Eddie Johnston, 1989-90 to 1991-92; Brian Burke, 1992-93; Paul Holmgren, 1993-94; Jim Rutherford, 1994-95 to date.

Captains' History

Rick Ley, 1979-80; Rick Ley and Mike Rogers, 1980-81; Dave Keon, 1981-82; Russ Anderson, 1982-83; Mark Johnson, 1983-84; Mark Johnson and Ron Francis, 1984-85; Ron Francis, 1985-86 to 1990-91; Randy Ladouceur, 1991-92; Pat Verbeek, 1992-93 to 1994-95; Brendan Shanahan, 1995-96; Kevin Dineen, 1996-97, 1997-98; Keith Primeau, 1998-99; Keith Primeau and Ron Francis, 1999-2000; Ron Francis, 2000-01 to date.

Club Records

Team

(Figures in brackets for season records are games played; records for fewest points, wins, ties, losses, goals, goals against are for 70 or more games)

Most Points **93** 1986-87 (80)
Most Wins **43** 1986-87 (80)
Most Ties **19** 1979-80 (80)
Most Losses **54** 1982-83 (80)
Most Goals **332** 1985-86 (80)
Most Goals Against **403** 1982-83 (80)
Fewest Points **45** 1982-83 (80)
Fewest Wins **19** 1982-83 (80)
Fewest Ties **4** 1985-86 (80)
Fewest Losses **30** 1986-87 (80); 1998-99 (82)
Fewest Goals **200** 1997-98 (82)
Fewest Goals Against **202** 1998-99 (82)
Longest Winning Streak
  Overall .... **7** Mar. 16-29/85
  Home .... **5** Mar. 17-29/85
  Away .... **6** Nov. 10-Dec. 7/90
Longest Undefeated Streak
  Overall .... **10** Jan. 20-Feb. 10/82 (6 wins, 4 ties)
  Home .... **9** Dec. 15/00-Jan. 18/01 (8 wins, 1 tie)
  Away .... **8** Nov. 11-Dec. 5/96 (4 wins, 4 ties)
Longest Losing Streak
  Overall .... **9** Feb. 19-Mar. 8/83
  Home .... **6** Feb. 19-Mar. 12/83, Feb. 10-Mar. 3/85
  Away .... **13** Dec. 18/82-Feb. 5/83
Longest Winless Streak
  Overall .... **14** Jan. 4-Feb. 9/92 (8 losses, 6 ties)
  Home .... **13** Jan. 15-Mar. 10/85 (11 losses, 2 ties)
  Away .... **15** Nov. 11/79-Jan. 9/80 (11 losses, 4 ties)
Most Shutouts, Season **8** 1998-99 (82)
Most PIM, Season **2,354** 1992-93 (84)
Most Goals, Game .... **11** Feb. 12/84 (Edm. 0 at Hfd. 11), Oct. 19/85 (Mtl. 6 at Hfd. 11), Jan. 17/86 (Que. 6 at Hfd. 11), Mar. 15/86 (Chi. 4 at Hfd. 11)

Individual

Most Seasons .... **14** Ron Francis
Most Games .... **1,036** Ron Francis
Most Goals, Career .... **350** Ron Francis
Most Assists, Career .... **738** Ron Francis
Most Points, Career .... **1,088** Ron Francis (350G, 738A)
Most PIM, Career .... **1,439** Kevin Dineen
Most Shutouts, Career .... **20** Arturs Irbe
Longest Consecutive Games Streak .... **419** Dave Tippett (Mar. 3/84-Oct. 7/89)
Most Goals, Season .... **56** Blaine Stoughton (1979-80)
Most Assists, Season .... **69** Ron Francis (1989-90)
Most Points, Season .... **105** Mike Rogers (1979-80; 44G, 61A), (1980-81; 40G, 65A)
Most PIM, Season .... **358** Torrie Robertson (1985-86)
Most Points, Defenseman, Season .... **69** Dave Babych (1985-86; 14G, 55A)
Most Points, Center, Season .... **105** Mike Rogers (1979-80; 44G, 61A), (1980-81; 40G, 65A)
Most Points, Right Wing, Season .... **100** Blaine Stoughton (1979-80; 56G, 44A)
Most Points, Left Wing, Season .... **89** Geoff Sanderson (1992-93; 46G, 43A)
Most Points, Rookie, Season .... **72** Sylvain Turgeon (1983-84; 40G, 32A)
Most Shutouts, Season .... **6** Arturs Irbe (1998-99, 2000-01)
Most Goals, Game .... **4** Jordy Douglas (Feb. 3/80), Ron Francis (Feb. 12/84)
Most Assists, Game .... **6** Ron Francis (Mar. 5/87)
Most Points, Game .... **6** Paul Lawless (Jan. 4/87; 2G, 4A), Ron Francis (Mar. 5/87; 6A) (Oct. 8/89; 3G, 3A)

Records include Hartford Whalers, 1979-80 through 1996-97.

All-time Record vs. Other Clubs

Regular Season

	At Home								On Road								Total							
	GP	W	L	T	OL	GF	GA	PTS	GP	W	L	T	OL	GF	GA	PTS	GP	W	L	T	OL	GF	GA	PTS
Anaheim	7	5	1	1	0	22	12	11	7	3	3	1	0	23	24	7	14	8	4	2	0	45	36	18
Atlanta	8	7	0	1	0	28	15	15	7	5	0	2	0	24	14	12	15	12	0	3	0	52	29	27
Boston	72	31	34	7	0	244	252	69	74	22	45	7	0	198	268	51	146	53	79	14	0	442	520	120
Buffalo	74	30	33	11	0	220	222	71	73	23	42	7	1	219	293	54	147	53	75	18	1	439	515	125
Calgary	28	10	13	5	0	91	103	25	27	5	20	2	0	88	136	12	55	15	33	7	0	179	239	37
Chicago	29	14	12	3	0	94	90	31	28	9	16	3	0	80	115	21	57	23	28	6	0	174	205	52
Colorado	61	24	25	12	0	201	211	60	64	17	38	9	0	190	271	43	125	41	63	21	0	391	482	103
Columbus	2	2	0	0	0	7	3	4	1	0	1	0	0	1	3	0	3	2	1	0	0	8	6	4
Dallas	30	12	14	4	0	99	107	28	28	10	16	2	0	85	113	22	58	22	30	6	0	184	220	50
Detroit	28	16	11	1	0	97	78	33	30	7	16	6	1	82	115	21	58	23	27	7	1	179	193	54
Edmonton	29	11	11	7	0	112	98	29	29	6	18	5	0	88	115	17	58	17	29	12	0	200	213	46
Florida	20	10	8	2	0	62	55	22	20	7	6	7	0	42	51	21	40	17	14	9	0	104	106	43
Los Angeles	29	14	11	4	0	111	113	32	29	10	16	3	0	111	121	23	58	24	27	7	0	222	234	55
Minnesota	1	1	0	0	0	2	0	2	2	1	0	1	0	8	4	3	3	2	0	1	0	10	4	5
Montreal	74	28	35	11	0	219	260	67	71	17	47	7	0	202	297	41	145	45	82	18	0	421	557	108
Nashville	4	2	1	1	0	13	11	5	3	0	3	0	0	5	8	0	7	2	4	1	0	18	19	5
New Jersey	40	17	15	8	0	131	123	42	41	14	22	4	1	133	147	33	81	31	37	12	1	264	270	75
NY Islanders	41	20	15	5	1	141	135	46	40	17	18	4	1	112	126	39	81	37	33	9	2	253	261	85
NY Rangers	39	22	14	3	0	135	126	47	41	14	24	3	0	110	158	31	80	36	38	6	0	245	284	78
Ottawa	23	16	5	2	0	69	50	34	25	12	9	4	0	71	64	28	48	28	14	6	0	140	114	62
Philadelphia	40	13	20	7	0	134	149	33	39	9	25	3	2	96	147	23	79	22	45	10	2	230	296	56
Phoenix	28	13	9	6	0	100	84	32	30	15	13	2	0	110	106	32	58	28	22	8	0	210	190	64
Pittsburgh	44	19	20	5	0	169	169	43	42	16	21	5	0	160	177	37	86	35	41	10	0	329	346	80
St. Louis	29	11	16	2	0	87	92	24	30	9	18	3	0	93	117	21	59	20	34	5	0	180	209	45
San Jose	10	5	5	0	0	31	23	10	10	4	6	0	0	29	44	8	20	9	11	0	0	60	67	18
Tampa Bay	21	15	3	3	0	74	55	33	21	8	10	3	0	53	55	19	42	23	13	6	0	127	110	52
Toronto	33	18	10	5	0	137	109	41	32	16	11	5	0	118	111	37	65	34	21	10	0	255	220	78
Vancouver	28	12	11	5	0	92	97	29	29	10	13	6	0	80	103	26	57	22	24	11	0	172	200	55
Washington	43	14	19	9	1	116	135	38	42	13	25	4	0	111	144	30	85	27	44	13	1	227	279	68
Totals	**915**	**412**	**371**	**130**	**2**	**3038**	**2977**	**956**	**915**	**299**	**502**	**108**	**6**	**2722**	**3447**	**712**	**1830**	**711**	**873**	**238**	**8**	**5760**	**6424**	**1668**

Playoffs

	Series	W	L	GP	W	L	T	GF	GA	Last Mtg.	Rnd.	Result
Boston	3	0	3	19	7	12	0	48	63	1999	CQF	L 2-4
Colorado	2	1	1	9	5	4	0	35	34	1987	DSF	L 2-4
Detroit	1	0	1	5	1	4	0	7	14	2002	F	L 1-4
Montreal	6	1	5	33	12	21	0	91	108	2002	CSF	W 4-2
New Jersey	2	1	1	12	6	6	0	17	31	2002	CQF	W 4-2
Toronto	1	1	0	6	4	2	0	10	6	2002	CF	W 4-2
Totals	**15**	**4**	**11**	**84**	**35**	**49**	**0**	**208**	**256**			

Playoff Results 2002-1998

Year	Round	Opponent	Result	GF	GA
2002	F	Detroit	L 1-4	7	14
	CF	Toronto	W 4-2	10	6
	CSF	Montreal	W 4-2	21	12
	CQF	New Jersey	W 4-2	9	11
2001	CQF	New Jersey	L 2-4	8	20
1999	CQF	Boston	L 2-4	10	16

Abbreviations: Round: F - Final; **CF** - conference final; **CSF** - conference semi-final; **CQF** - conference quarter-final; **DSF** - division semi-final.

Calgary totals include Atlanta Flames, 1979-80.
Dallas totals include Minnesota North Stars, 1979-80 to 1992-93.
Phoenix totals include Winnipeg, 1979-80 to 1995-96.

Colorado totals include Quebec, 1979-80 to 1994-95.
New Jersey totals include Colorado Rockies, 1979-80 to 1981-82.

2001-02 Results

Month	Date	Opponent	Score
Oct.	5	NY Rangers	3-1
	7	Dallas	3-0
	9	Ottawa	2-6
	11	Toronto	2-3
	13	at Atlanta	5-2
	17	NY Islanders	0-4
	18	at NY Islanders	1-2*
	20	Atlanta	2-1*
	23	at Colorado	1-5
	24	at Minnesota	7-3
	26	NY Islanders	2-3*
	28	Los Angeles	3-2*
	30	Detroit	2-5
Nov.	1	at St. Louis	3-4
	2	NY Rangers	3-2
	4	Phoenix	1-0*
	6	Pittsburgh	2-2
	8	at Washington	3-2
	9	San Jose	3-2
	11	Edmonton	1-1
	13	at Detroit	3-4
	15	at Ottawa	1-1
	17	at Tampa Bay	0-2
	19	Columbus	5-2
	21	at Dallas	4-4
	25	Tampa Bay	0-4
	27	at Toronto	5-2
	29	at NY Rangers	0-5
	30	at Washington	2-6
Dec.	2	Washington	3-4*
	4	Buffalo	2-4
	8	at Florida	3-2
	10	at NY Rangers	4-3*
	12	Florida	4-1
	14	at Buffalo	2-3*
	16	at Pittsburgh	7-0
	18	Ottawa	1-5
	21	Atlanta	5-4*
	22	at Philadelphia	3-4*
	26	Toronto	4-3
	27	at Tampa Bay	3-2
	30	at Washington	5-5
	31	at Buffalo	5-4
Jan.	2	Boston	3-6
	5	New Jersey	1-2
	6	Philadelphia	3-4
	10	at Edmonton	4-1
	12	at Vancouver	1-7
	15	Minnesota	2-0
	17	Montreal	1-1
	19	at New Jersey	3-3
	21	Vancouver	5-7
	23	Nashville	2-2
	25	Florida	1-1
	26	at Philadelphia	2-4
	29	Buffalo	2-2
	30	at Tampa Bay	3-1
Feb.	5	Pittsburgh	3-3
	7	at Los Angeles	1-2
	8	at Anaheim	4-1
	10	at San Jose	0-4
	26	at Toronto	1-4
	28	at Boston	6-2
Mar.	2	at Montreal	4-3
	5	at Chicago	2-1*
	7	at Pittsburgh	3-1
	8	Washington	2-2
	11	Calgary	3-3
	16	at Montreal	2-3
	18	Montreal	1-1
	21	Florida	3-2
	23	at New Jersey	4-2
	26	Boston	2-3
	28	Philadelphia	4-1
	30	at Boston	2-0
Apr.	2	at Ottawa	3-4
	3	New Jersey	2-3
	7	Atlanta	1-1
	8	at NY Islanders	2-1
	10	Tampa Bay	4-2
	12	at Florida	3-1
	14	at Atlanta	2-2

* – Overtime

Entry Draft Selections 2002-1988

2002
Pick
25 Cam Ward
91 Jesse Lane
160 Daniel Manzato
224 Adam Taylor

2001
Pick
15 Igor Knyazev
91 Kevin Estrada
181 Daniel Boisclair
211 Sean Curry
244 Carter Trevisani

2000
Pick
32 Tomas Kurka
80 Ryan Bayda
97 Niclas Wallin
110 Jared Newman
181 J.D. Forrest
212 Magnus Kahnberg
235 Craig Kowalski
276 Troy Ferguson

1999
Pick
16 David Tanabe
49 Brett Lysak
84 Brad Fast
113 Ryan Murphy
174 Damian Surma
231 David Evans
237 Antti Jokela
259 Yevgeny Kurilin

1998
Pick
11 Jeff Heerema
70 Kevin Holdridge
71 Erik Cole
91 Josef Vasicek
93 Tommy Westlund
97 Chris Madden
184 Don Smith
208 Jaroslav Svoboda
211 Mark Kosick
239 Brent McDonald

1997
Pick
22 Nikos Tselios
28 Brad DeFauw
80 Francis Lessard
142 Kyle Dafoe
169 Andrew Merrick
195 Niklas Nordgren
199 Randy Fitzgerald
225 Kent McDonell

1996
Pick
34 Trevor Wasyluk
61 Andrei Petrunin
88 Craig MacDonald
104 Steve Wasylko
116 Mark McMahon
143 Aaron Baker
171 Greg Kuznik
197 Kevin Marsh
223 Craig Adams
231 Ashkat Rakhmatullin

1995
Pick
13 Jean-Sebastien Giguere
35 Sergei Fedotov
85 Ian MacNeil
87 Sami Kapanen
113 Hugh Hamilton
165 Byron Ritchie
191 Milan Kostolny
217 Mike Rucinski

1994
Pick
5 Jeff O'Neill
83 Hnat Domenichelli
109 Ryan Risidore
187 Tom Buckley
213 Ashlin Halfnight
230 Matt Ball
239 Brian Regan
265 Steve Nimigon

1993
Pick
2 Chris Pronger
72 Marek Malik
84 Trevor Roenick
115 Nolan Pratt
188 Manny Legace
214 Dmitri Gorenko
240 Wes Swinson
266 Igor Chibirev

1992
Pick
9 Robert Petrovicky
47 Andrei Nikolishin
57 Jan Vopat
79 Kevin Smyth
81 Jason McBain
143 Jarrett Reid
153 Ken Belanger
177 Konstantin Korotkov
201 Greg Zwakman
225 Steven Halko
249 Joacim Esbjors

1991
Pick
9 Patrick Poulin
31 Martin Hamrlik
53 Todd Hall
59 Michael Nylander
75 Jim Storm
119 Mike Harding
141 Brian Mueller
163 Steve Yule
185 Chris Belanger
207 Jason Currie
229 Mike Santonelli
251 Rob Peters

1990
Pick
15 Mark Greig
36 Geoff Sanderson
57 Mike Lenarduzzi
78 Chris Bright
120 Cory Keenan
141 Jergus Baca
162 Martin D'Orsonnens
183 Corey Osmak
204 Espen Knutsen
225 Tommie Eriksen
246 Denis Chalifoux

1989
Pick
10 Bobby Holik
52 Blair Atcheynum
73 Jim McKenzie
94 James Black
115 Jerome Bechard
136 Scott Daniels
157 Raymond Saumier
178 Michel Picard
199 Trevor Buchanan
220 John Battice
241 Peter Kasowski

1988
Pick
11 Chris Govedaris
32 Barry Richter
74 Dean Dyer
95 Scott Morrow
116 Corey Beaulieu
137 Kerry Russell
158 Jim Burke
179 Mark Hirth
200 Wayde Bucsis
221 Rob White
242 Dan Slatalla

President and General Manager

RUTHERFORD, JIM
President and General Manager, Carolina Hurricanes.
Born in Beeton, Ont., February 17, 1949.

Jim Rutherford, a former NHL goaltender, is the franchise's seventh general manager and the only general manager of the Carolina Hurricanes. Named to his position on June 28, 1994, Rutherford has taken an aggressive approach towards improving the fortunes of the franchise through trades and the NHL draft. In 2002, the team reached the Stanley Cup Finals for the first time in history.

A veteran of 13 NHL seasons, Rutherford began his professional goaltending career in 1969 as a first-round selection of the Detroit Red Wings. While playing for Detroit, Pittsburgh, Toronto and Los Angeles, Rutherford collected 14 career shutouts. For five seasons he also served as the Red Wings' player representative. Rutherford also played for Team Canada in the IIHF World Championships in Vienna in 1977 and Moscow in 1979.

After his playing days with the Red Wings, Rutherford joined Compuware to serve as the director of hockey operations for Compuware Sports Corporation. Rutherford gained a wealth of experience in youth hockey and junior programs. As a former player, coach, and general manager, his ability to develop players and produce winning programs is widely respected throughout the hockey community.

He started his management career by guiding Compuware Sports Corporation's purchase of the Windsor Spitfires of the Ontario Hockey League in April of 1984. During the next four years, Rutherford acted as general manager of the Spitfires. After the Spitfires advanced to the 1988 Memorial Cup finals, Rutherford led Compuware's efforts to bring the first American-based OHL franchise to Detroit on December 11, 1989. Rutherford was voted the 1987 executive of the year in both the OHL and the Canadian Hockey League and won the OHL executive of the year award again in 1988.

Club Directory

Entertainment and Sports Arena

Carolina Hurricanes
1400 Edwards Mill Rd.
Raleigh, NC 27607
Phone **919/467-7825**
FAX 919/462-0123
www.carolinahurricanes.com
Capacity: 18,730

Carolina Hurricanes Directory

Owner	Peter Karmanos Jr.
Chief Executive Officer/General Manager	Jim Rutherford
President/Chief Operating Officer	Jim Cain
Vice President/Assistant General Manager	Jason Karmanos
Head Coach	Paul Maurice
Assistant Coaches	Randy Ladouceur, Kevin McCarthy
Goaltending Consultant	Don Edwards
Director of Public and Media Relations	Jerry Higgins
Media Relations Manager	Mike Sundheim
Head Athletic Therapist/ Strength Conditioning Coach	Peter Friesen
Associate Athletic Therapist	Stu Lempke
Equipment Managers	Wally Tatomir, Bob Gorman, Skip Cunningham
Entertainment & Sports Arena Capacity	18,730
Public Relations Phone	(919) 861-5429
Public Relations Fax	(919) 462-0123
Press Box Phone	(919) 861-2300
ext. 6531, 6563, 6564	
Website	www.carolinahurricanes.com
Radio	WKXU-FM (101.1)
	WKIX-FM (102.3)
	WDTF-AM (570)
TV	FOX Sports South

Centering the "BBC Line" with Bate Battaglia and Erik Cole, Rod Brind'Amour played his best hockey of the season during Carolina's second-round playoff victory over the Montreal Canadiens.

Eric Daze scored 38 goals last season.

Chicago Blackhawks

2001-02 Results: 41W-27L-13T-1OTL 96PTS.
Third, Central Division

2002-03 Schedule

Oct.	Thu.	10	at Columbus
	Sun.	13	Buffalo
	Thu.	17	Florida
	Sat.	19	Calgary
	Thu.	24	Minnesota
	Sat.	26	at Carolina
	Sun.	27	San Jose
	Tue.	29	Columbus
	Thu.	31	Los Angeles
Nov.	Sat.	2	at New Jersey*
	Sun.	3	Edmonton
	Tue.	5	at Detroit
	Thu.	7	Atlanta
	Sat.	9	at Tampa Bay
	Mon.	11	at Florida
	Fri.	15	Washington
	Sun.	17	Nashville
	Tue.	19	at Edmonton
	Wed.	20	at Vancouver
	Sat.	23	at Calgary
	Mon.	25	at Colorado
	Thu.	28	at Phoenix
	Sat.	30	at Los Angeles
Dec.	Sun.	1	at Anaheim*
	Wed.	4	Ottawa
	Fri.	6	Anaheim
	Sun.	8	Tampa Bay
	Tue.	10	at NY Islanders
	Wed.	11	at NY Rangers
	Fri.	13	at Buffalo
	Sun.	15	Dallas
	Tue.	17	Vancouver
	Fri.	20	Columbus
	Sun.	22	Los Angeles
	Thu.	26	Minnesota
	Sat.	28	at San Jose
	Mon.	30	at Los Angeles
Jan.	Thu.	2	at St. Louis
	Sat.	4	at Nashville
	Sun.	5	Detroit
	Wed.	8	Phoenix
	Thu.	9	at Dallas
	Sun.	12	Nashville
	Mon.	13	at Detroit
	Wed.	15	Detroit
	Fri.	17	Vancouver
	Sat.	18	at St. Louis
	Mon.	20	at Columbus
	Thu.	23	St. Louis
	Sat.	25	at Pittsburgh*
	Sun.	26	at Montreal*
	Thu.	30	at Boston
Feb.	Wed.	5	at Minnesota
	Thu.	6	at Calgary
	Sat.	8	at Edmonton
	Mon.	10	at Vancouver
	Wed.	12	Toronto
	Fri.	14	San Jose
	Sat.	15	at Columbus
	Mon.	17	Colorado
	Thu.	20	Phoenix
	Sun.	23	Dallas*
	Tue.	25	Philadelphia
	Thu.	27	at Philadelphia
Mar.	Sat.	1	at Nashville*
	Sun.	2	Colorado
	Wed.	5	at Dallas
	Fri.	7	Calgary
	Sun.	9	Boston*
	Wed.	12	at Anaheim
	Fri.	14	at Phoenix
	Mon.	17	at San Jose
	Wed.	19	Anaheim
	Sat.	22	at Colorado*
	Sun.	23	Pittsburgh*
	Tue.	25	NY Islanders
	Thu.	27	Nashville
	Fri.	28	at Minnesota
	Sun.	30	Edmonton*
Apr.	Thu.	3	at St. Louis
	Fri.	4	St. Louis
	Sun.	6	Detroit*

* Denotes afternoon game.

Franchise date: September 25, 1926

77th NHL Season

CENTRAL DIVISION

Year-by-Year Record

		Home				Road				Overall								
Season	**GP**	**W**	**L**	**T**	**OL**	**W**	**L**	**T**	**OL**	**W**	**L**	**T**	**OL**	**GF**	**GA**	**Pts.**	**Finished**	**Playoff Result**
2001-02	82	28	7	5	1	13	20	8	0	41	27	13	1	216	207	96	3rd, Central Div.	Lost Conf. Quarter-Final
2000-01	82	14	21	4	2	15	19	4	3	29	40	8	5	210	246	71	4th, Central Div.	Out of Playoffs
1999-2000	82	16	19	5	1	17	18	5	1	33	37	10	2	242	245	78	3rd, Central Div.	Out of Playoffs
1998-99	82	20	17	4	...	9	24	8	...	29	41	12	...	202	248	70	3rd, Central Div.	Out of Playoffs
1997-98	82	14	19	8	...	16	20	5	...	30	39	13	...	192	199	73	5th, Central Div.	Out of Playoffs
1996-97	82	16	21	4	...	18	14	9	...	34	35	13	...	223	210	81	5th, Central Div.	Lost Conf. Quarter-Final
1995-96	82	22	13	6	...	18	15	8	...	40	28	14	...	273	220	94	2nd, Central Div.	Lost Conf. Semi-Final
1994-95	48	11	10	3	...	13	9	2	...	24	19	5	...	156	115	53	3rd, Central Div.	Lost Conf. Championship
1993-94	84	21	16	5	...	18	20	4	...	39	36	9	...	254	240	87	5th, Central Div.	Lost Conf. Quarter-Final
1992-93	84	25	11	6	...	22	14	6	...	47	25	12	...	279	230	106	1st, Norris Div.	Lost Div. Semi-Final
1991-92	80	23	9	8	...	13	20	7	...	36	29	15	...	257	236	87	2nd, Norris Div.	Lost Final
1990-91	80	28	8	4	...	21	15	4	...	49	23	8	...	284	211	106	1st, Norris Div.	Lost Div. Semi-Final
1989-90	80	25	13	2	...	16	20	4	...	41	33	6	...	316	294	88	1st, Norris Div.	Lost Conf. Championship
1988-89	80	16	14	10	...	11	27	2	...	27	41	12	...	297	335	66	4th, Norris Div.	Lost Conf. Championship
1987-88	80	21	17	2	...	9	24	7	...	30	41	9	...	284	328	69	3rd, Norris Div.	Lost Div. Semi-Final
1986-87	80	18	13	9	...	11	24	5	...	29	37	14	...	290	310	72	3rd, Norris Div.	Lost Div. Semi-Final
1985-86	80	23	12	5	...	16	21	3	...	39	33	8	...	351	349	86	1st, Norris Div.	Lost Div. Semi-Final
1984-85	80	22	16	2	...	16	19	5	...	38	35	7	...	309	299	83	2nd, Norris Div.	Lost Conf. Championship
1983-84	80	25	13	2	...	5	29	6	...	30	42	8	...	277	311	68	4th, Norris Div.	Lost Div. Semi-Final
1982-83	80	29	8	3	...	18	15	7	...	47	23	10	...	338	268	104	1st, Norris Div.	Lost Conf. Championship
1981-82	80	20	13	7	...	10	25	5	...	30	38	12	...	332	363	72	4th, Norris Div.	Lost Conf. Championship
1980-81	80	21	11	8	...	10	22	8	...	31	33	16	...	304	315	78	2nd, Smythe Div.	Lost Prelim. Round
1979-80	80	21	12	7	...	13	15	12	...	34	27	19	...	241	250	87	1st, Smythe Div.	Lost Quarter-Final
1978-79	80	18	12	10	...	11	24	5	...	29	36	15	...	244	277	73	1st, Smythe Div.	Lost Quarter-Final
1977-78	80	20	9	11	...	12	20	8	...	32	29	19	...	230	220	83	1st, Smythe Div.	Lost Quarter-Final
1976-77	80	19	16	5	...	7	27	6	...	26	43	11	...	240	298	63	3rd, Smythe Div.	Lost Prelim. Round
1975-76	80	17	15	8	...	15	15	10	...	32	30	18	...	254	261	82	1st, Smythe Div.	Lost Quarter-Final
1974-75	80	24	12	4	...	13	23	4	...	37	35	8	...	268	241	82	3rd, Smythe Div.	Lost Quarter-Final
1973-74	78	20	6	13	...	21	8	10	...	41	14	23	...	272	164	105	2nd, West Div.	Lost Semi-Final
1972-73	78	26	9	4	...	16	18	5	...	42	27	9	...	284	225	93	1st, West Div.	Lost Final
1971-72	78	28	3	8	...	18	14	7	...	46	17	15	...	256	166	107	1st, West Div.	Lost Semi-Final
1970-71	78	30	6	3	...	19	14	6	...	49	20	9	...	277	184	107	1st, West Div.	Lost Final
1969-70	76	26	7	5	...	19	15	4	...	45	22	9	...	250	170	99	1st, East Div.	Lost Semi-Final
1968-69	76	20	14	4	...	14	19	5	...	34	33	9	...	280	246	77	6th, East Div.	Out of Playoffs
1967-68	74	20	13	4	...	12	13	12	...	32	26	16	...	212	222	80	4th, East Div.	Lost Semi-Final
1966-67	70	24	5	6	...	17	12	6	...	41	17	12	...	264	170	94	1st,	Lost Semi-Final
1965-66	70	21	8	6	...	16	17	2	...	37	25	8	...	240	187	82	2nd,	Lost Semi-Final
1964-65	70	20	13	2	...	14	15	6	...	34	28	8	...	224	176	76	3rd,	Lost Final
1963-64	70	26	4	5	...	10	18	7	...	36	22	12	...	218	169	84	2nd,	Lost Semi-Final
1962-63	70	17	9	9	...	15	12	8	...	32	21	17	...	194	178	81	2nd,	Lost Semi-Final
1961-62	70	20	10	5	...	11	16	8	...	31	26	13	...	217	186	75	3rd,	Lost Final
1960-61	**70**	**20**	**6**	**9**	**...**	**9**	**18**	**8**	**...**	**29**	**24**	**17**	**...**	**198**	**180**	**75**	**3rd,**	**Won Stanley Cup**
1959-60	70	18	11	6	...	10	18	7	...	28	29	13	...	191	180	69	3rd,	Lost Semi-Final
1958-59	70	14	12	9	...	14	17	4	...	28	29	13	...	197	208	69	3rd,	Lost Semi-Final
1957-58	70	15	17	3	...	9	22	4	...	24	39	7	...	163	202	55	5th,	Out of Playoffs
1956-57	70	12	15	8	...	4	24	7	...	16	39	15	...	169	225	47	6th,	Out of Playoffs
1955-56	70	9	19	7	...	10	20	5	...	19	39	12	...	155	216	50	6th,	Out of Playoffs
1954-55	70	6	21	8	...	7	19	9	...	13	40	17	...	161	235	43	6th,	Out of Playoffs
1953-54	70	8	21	6	...	4	30	1	...	12	51	7	...	133	242	31	6th,	Out of Playoffs
1952-53	70	14	11	10	...	13	17	5	...	27	28	15	...	169	175	69	4th,	Lost Semi-Final
1951-52	70	9	19	7	...	8	25	2	...	17	44	9	...	158	241	43	6th,	Out of Playoffs
1950-51	70	8	22	5	...	5	25	5	...	13	47	10	...	171	280	36	6th,	Out of Playoffs
1949-50	70	13	18	4	...	9	20	6	...	22	38	10	...	203	244	54	6th,	Out of Playoffs
1948-49	60	13	12	5	...	8	19	3	...	21	31	8	...	173	211	50	5th,	Out of Playoffs
1947-48	60	10	17	3	...	10	17	3	...	20	34	6	...	195	225	46	6th,	Out of Playoffs
1946-47	60	10	17	3	...	9	20	1	...	19	37	4	...	193	274	42	6th,	Out of Playoffs
1945-46	50	15	5	5	...	8	15	2	...	23	20	7	...	200	178	53	3rd,	Lost Semi-Final
1944-45	50	9	14	2	...	4	16	5	...	13	30	7	...	141	194	33	5th,	Out of Playoffs
1943-44	50	15	6	4	...	7	17	1	...	22	23	5	...	178	187	49	4th,	Lost Final
1942-43	50	14	3	8	...	3	15	7	...	17	18	15	...	179	180	49	5th,	Out of Playoffs
1941-42	48	15	8	1	...	7	15	2	...	22	23	3	...	145	155	47	4th,	Lost Quarter-Final
1940-41	48	11	10	3	...	5	15	4	...	16	25	7	...	112	139	39	5th,	Lost Semi-Final
1939-40	48	15	7	2	...	8	12	4	...	23	19	6	...	112	120	52	4th,	Lost Quarter-Final
1938-39	48	5	13	6	...	7	15	2	...	12	28	8	...	91	132	32	7th,	Out of Playoffs
1937-38	**48**	**10**	**10**	**4**	**...**	**4**	**15**	**5**	**...**	**14**	**25**	**9**	**...**	**97**	**139**	**37**	**3rd, Amn. Div.**	**Won Stanley Cup**
1936-37	48	8	13	3	...	6	14	4	...	14	27	7	...	99	131	35	4th, Amn. Div.	Out of Playoffs
1935-36	48	15	7	2	...	6	12	6	...	21	19	8	...	93	92	50	3rd, Amn. Div.	Lost Quarter-Final
1934-35	48	12	9	3	...	14	8	2	...	26	17	5	...	118	88	57	2nd, Amn. Div.	Lost Quarter-Final
1933-34	**48**	**13**	**4**	**7**	**...**	**7**	**13**	**4**	**...**	**20**	**17**	**11**	**...**	**88**	**83**	**51**	**2nd, Amn. Div.**	**Won Stanley Cup**
1932-33	48	12	7	5	...	4	13	7	...	16	20	12	...	88	101	44	4th, Amn. Div.	Out of Playoffs
1931-32	48	13	5	6	...	5	14	5	...	18	19	11	...	86	101	47	2nd, Amn. Div.	Lost Quarter-Final
1930-31	44	13	8	1	...	11	9	2	...	24	17	3	...	108	78	51	2nd, Amn. Div.	Lost Final
1929-30	44	12	9	1	...	9	9	4	...	21	18	5	...	117	111	47	2nd, Amn. Div.	Lost Quarter-Final
1928-29	44	3	13	6	...	4	16	2	...	7	29	8	...	33	85	22	5th, Amn. Div.	Out of Playoffs
1927-28	44	2	18	2	...	5	16	1	...	7	34	3	...	68	134	17	5th, Amn. Div.	Out of Playoffs
1926-27	44	12	8	2	...	7	14	1	...	19	22	3	...	115	116	41	3rd, Amn. Div.	Lost Quarter-Final

2002-03 Player Personnel

FORWARDS	HT	WT	S	Place of Birth	Date	2001-02 Club
ARNASON, Tyler	5-11	207	L	Oklahoma City, OK	3/16/79	Chicago-Norfolk
BAINES, Ajay	5-10	178	L	Kamloops, B.C.	3/25/78	Norfolk
BELL, Mark	6-3	198	L	St. Paul's, Ont.	8/5/80	Chicago
BEREZIN, Sergei	5-10	200	R	Voskresensk, USSR	11/5/71	Phoenix-Montreal
CALDER, Kyle	5-11	180	L	Mannville, Alta.	1/5/79	Chicago
DAZE, Eric	6-6	234	L	Montreal, Que.	7/2/75	Chicago
FLEURY, Theoren	5-6	180	R	Oxbow, Sask.	6/29/68	NY Rangers-Canada
HANKINSON, Casey	6-1	187	L	Edina, MN	5/8/76	Chicago-Norfolk
KOROLEV, Igor	6-1	190	L	Moscow, USSR	9/6/70	Chicago
McLEAN, Brett	5-11	194	L	Comox, B.C.	8/14/78	Houston
NYLANDER, Michael	6-1	195	L	Stockholm, Sweden	10/3/72	Chicago-Sweden
PELUSO, Mike	6-1	208	R	Bismark, ND	9/2/74	Chicago-Norfolk
PROBERT, Bob	6-3	225	L	Windsor, Ont.	6/5/65	Chicago
RADULOV, Igor	6-0	194	L	Nizhny Tagil, USSR	8/23/82	Mississauga
RUUTU, Tuomo	6-0	201	L	Vantaa, Finland	2/16/83	Jokerit
SOUZA, Mike	6-1	210	L	Melrose, MA	1/28/78	Norfolk
SULLIVAN, Steve	5-9	160	R	Timmins, Ont.	7/6/74	Chicago
THOMAS, Steve	5-10	185	L	Stockport, England	7/15/63	Chicago
THORNTON, Shawn	6-1	196	R	Oshawa, Ont.	7/23/77	Norfolk
VANDENBUSSCHE, Ryan	6-0	200	R	Simcoe, Ont.	2/28/73	Chicago
VON ARX, Reto	5-10	190	L	Egerkingen, Switz.	9/13/76	Davos-Switzerland
VOROBIEV, Pavel	6-0	183	L	Karaganda, USSR	5/5/82	Yaroslavl
WHITE, Peter	5-11	200	L	Montreal, Que.	3/15/69	Chicago-Norfolk
YAKUBOV, Mikhail	6-3	208	L	Barnaul, USSR	2/16/82	Red Deer
ZHAMNOV, Alexei	6-1	200	L	Moscow, USSR	10/1/70	Chicago-Russia
DEFENSEMEN						
BALAN, Scott	6-3	191	R	Medicine Hat, Alta.	5/29/82	Saskatoon
DEMPSEY, Nathan	6-0	190	R	Spruce Grove, Alta.	7/14/74	Toronto-St. John's
HOUSLEY, Phil	5-10	185	L	St. Paul, MN	3/9/64	Chicago-United States
HUSKINS, Kent	6-2	190	L	Ottawa, Ont.	5/4/79	Norfolk
KARPOVTSEV, Alexander	6-3	215	R	Moscow, USSR	4/7/70	Chicago
KLEMM, Jon	6-2	200	R	Cranbrook, B.C.	1/8/70	Chicago
McCARTHY, Steve	6-0	197	L	Trail, B.C.	2/3/81	Chicago-Norfolk
MIRONOV, Boris	6-3	223	R	Moscow, USSR	3/21/72	Chicago-Russia
ODELEIN, Lyle	5-11	210	R	Quill Lake, Sask.	7/21/68	Columbus-Chicago
POAPST, Steve	6-0	200	L	Cornwall, Ont.	1/3/69	Chicago
STRUDWICK, Jason	6-3	215	L	Edmonton, Alta.	7/17/75	Vancouver
TOLKUNOV, Dmitri	6-2	200	R	Kiev, USSR	5/5/79	Norfolk

GOALTENDERS	HT	WT	C	Place of Birth	Date	2001-02 Club
ANDERSSON, Craig	6-2	174	L	Park Ridge, IL	5/21/81	Norfolk
LEIGHTON, Michael	6-2	175	L	Petrolia, Ont.	5/19/81	Norfolk
MUNRO, Adam	6-1	194	L	St. George, Ont.	11/12/82	Erie (OHL)
PASSMORE, Steve	5-9	165	L	Thunder Bay, Ont.	1/29/73	Chicago-Norfolk
THIBAULT, Jocelyn	5-11	170	L	Montreal, Que.	1/12/75	Chicago

Coach

SUTTER, BRIAN
Coach, Chicago Blackhawks. Born in Viking, Alta., October 7, 1956.

Brian Sutter was hired as the head coach in Chicago on May 3, 2001. In his first year behind the bench in 2001-02, Sutter guided the club into the playoffs for the first time since 1997 and was rewarded with a nomination for the Jack Adams Award as coach of the year. He is the second member of this storied family to have coached the Blackhawks (Darryl Sutter was the head coach of the Blackhawks from 1992 to 1995). The Sutter name has always been synonymous with intensity, honesty, tenacity, and hard work.

Following a 12-year playing career with the St. Louis Blues from 1976 to 1988, Sutter immediately joined the NHL coaching ranks by taking over the reigns of the team he had captained for nine of his 12 seasons. He spent four seasons as the coach of the Blues, posting a mark of 153-124-43 which then exceeded Scotty Bowman's club record for coaching victories. Sutter captured the Jack Adams Award in 1990-91.

Sutter became the head coach of the Boston Bruins in the 1992-93 season and immediately led the club to its first 50-win season in ten years. In his three seasons as the Bruins' chief mentor, Sutter had a coaching record of 120-73-23. After leaving the coaching ranks for two seasons, Sutter returned behind the bench in his native province of Alberta as the head coach of the Calgary Flames for the 1997-98 season. He coached the Flames for three seasons.

As a player, Sutter was drafted by the St. Louis Blues with their second pick, 20th overall, in the 1976 Entry Draft. He wound up playing his entire 12-year NHL career with the Blues. A three time NHL All-Star Game selection, Sutter played 779 games and had 303 goals and 333 assists for 636 points. A great leader and an outstanding performer on the ice, Sutter's #11 was retired by the Blues on December 30, 1988.

Coaching Record

		Regular Season				Playoffs		
Season	Team	Games	W	L	T	Games	W	L
1988-89	St. Louis (NHL)	80	33	35	12	10	5	5
1989-90	St. Louis (NHL)	80	37	34	9	12	7	5
1990-91	St. Louis (NHL)	80	47	22	11	13	6	7
1991-92	St. Louis (NHL)	80	36	33	11	6	2	4
1992-93	Boston (NHL)	84	51	26	7	4	0	4
1993-94	Boston (NHL)	84	42	29	13	13	6	7
1994-95	Boston (NHL)	48	27	18	3	5	1	4
1997-98	Calgary (NHL)	82	26	41	15	....	....	....
1998-99	Calgary (NHL)	82	30	40	12	....	....	....
1999-2000	Calgary (NHL)	82	31	41	10	....	....	....
2001-02	Chicago (NHL)	82	41	28	13	5	1	4
	NHL Totals	864	401	347	116	68	28	40

2001-02 Scoring

* - rookie

Regular Season

Pos	#	Player	Team	GP	G	A	Pts	+/–	PIM	PP	SH	GW	GT	S	%
L	55	Eric Daze	CHI	82	38	32	70	17	36	12	0	5	1	264	14.4
C	13	Alexei Zhamnov	CHI	77	22	45	67	8	67	6	0	3	0	173	12.7
R	10	Tony Amonte	CHI	82	27	39	66	11	67	6	1	4	0	232	11.6
C	92	Michael Nylander	CHI	82	15	46	61	28	50	6	0	2	0	158	9.5
R	26	Steve Sullivan	CHI	78	21	39	60	23	67	3	0	8	2	155	13.5
C	19	Kyle Calder	CHI	81	17	36	53	8	47	6	0	3	0	133	12.8
D	6	Phil Housley	CHI	80	15	24	39	–3	34	8	0	6	1	218	6.9
C	22	Igor Korolev	CHI	82	9	20	29	–5	20	0	1	1	0	78	11.5
L	28 *	Mark Bell	CHI	80	12	16	28	–6	124	1	0	1	0	120	10.0
R	12	Tom Fitzgerald	NSH	63	7	9	16	–4	33	0	1	0	0	101	6.9
			CHI	15	1	3	4	–3	6	0	1	0	0	24	4.2
			TOTAL	78	8	12	20	–7	39	0	2	0	0	125	6.4
D	42	Jon Klemm	CHI	82	4	16	20	–3	42	2	0	1	0	111	3.6
D	2	Boris Mironov	CHI	64	4	14	18	15	68	0	0	1	0	129	3.1
D	7	Lyle Odelein	CBJ	65	2	14	16	–28	89	0	0	0	0	76	2.6
			CHI	12	0	2	2	0	4	0	0	0	0	10	0.0
			TOTAL	77	2	16	18	–28	93	0	0	0	0	86	2.3
L	32	Steve Thomas	CHI	34	11	4	15	0	17	3	0	0	0	66	16.7
D	25	Alexander Karpovtsev	CHI	65	1	9	10	10	40	0	1	0	0	40	2.5
D	23	Joe Reekie	WSH	38	2	4	6	–7	41	0	0	0	0	26	7.7
			CHI	17	0	2	2	2	28	0	0	0	0	7	0.0
			TOTAL	55	2	6	8	–5	69	0	0	0	0	33	6.1
D	8	Steve Poapst	CHI	56	1	7	8	6	30	0	0	0	0	48	2.1
L	17	Mike Peluso	CHI	37	4	2	6	–3	19	0	0	1	1	45	8.9
C	11	Peter White	CHI	48	3	3	6	–8	10	1	0	1	0	21	14.3
C	39 *	Tyler Arnason	CHI	21	3	1	4	–3	4	0	0	0	0	19	15.8
L	24	Bob Probert	CHI	61	1	3	4	–9	176	0	0	1	0	33	3.0
R	14	Ryan Vandenbussche	CHI	50	1	2	3	–10	103	0	0	0	0	22	4.5
D	4	Chris McAlpine	CHI	40	0	3	3	8	36	0	0	0	0	40	0.0
R	15	Jim Campbell	CHI	9	1	1	2	–3	4	0	0	0	0	12	8.3
D	45	Vlad Chebaturkin	CHI	13	0	2	2	0	6	0	0	0	0	8	0.0
R	44	Aaron Downey	CHI	36	1	0	1	–2	76	0	0	1	0	10	10.0
L	16	Matt Henderson	CHI	4	0	1	1	–1	0	0	0	0	0	3	0.0
L	20 *	Casey Hankinson	CHI	3	0	0	0	–2	0	0	0	0	0	1	0.0
D	5	Steve McCarthy	CHI	3	0	0	0	–1	2	0	0	0	0	2	0.0

Goaltending

No.	Goaltender	GPI	Mins	Avg	W	L	T	EN	SO	GA	SA	S%	G	A	PIM
29	Steve Passmore	23	1142	2.26	8	5	4	3	0	43	446	.904	0	1	2
41	Jocelyn Thibault	67	3838	2.49	33	23	9	2	6	159	1626	.902	0	0	2
	Totals	**82**	**4996**	**2.49**	**41**	**28**	**13**	**5**	**6**	**207**	**2077**	**.900**			

Playoffs

Pos	#	Player	Team	GP	G	A	Pts	+/–	PIM	PP	SH	GW	GT	S	%
C	92	Michael Nylander	CHI	5	0	3	3	2	2	0	0	0	0	6	0.0
C	19	Kyle Calder	CHI	5	2	0	2	0	2	1	0	0	0	10	20.0
L	32	Steve Thomas	CHI	5	1	1	2	–3	0	0	0	0	0	8	12.5
D	25	Alexander Karpovtsev	CHI	5	1	0	1	–1	0	0	0	1	0	6	16.7
R	26	Steve Sullivan	CHI	5	1	0	1	–1	2	0	0	0	0	7	14.3
D	7	Lyle Odelein	CHI	4	0	1	1	–1	25	0	0	0	0	3	0.0
D	6	Phil Housley	CHI	5	0	1	1	–1	4	0	0	0	0	13	0.0
R	10	Tony Amonte	CHI	5	0	1	1	1	4	0	0	0	0	9	0.0
D	42	Jon Klemm	CHI	5	0	1	1	0	4	0	0	0	0	6	0.0
C	22	Igor Korolev	CHI	5	0	1	1	–3	0	0	0	0	0	3	0.0
D	23	Joe Reekie	CHI	1	0	0	0	0	2	0	0	0	0	1	0.0
D	4	Chris McAlpine	CHI	1	0	0	0	0	0	0	0	0	0	1	0.0
D	2	Boris Mironov	CHI	1	0	0	0	0	2	0	0	0	0	0	0.0
R	14	Ryan Vandenbussche	CHI	1	0	0	0	0	0	0	0	0	0	1	0.0
L	24	Bob Probert	CHI	2	0	0	0	0	0	0	0	0	0	0	0.0
D	45	Vlad Chebaturkin	CHI	3	0	0	0	–2	2	0	0	0	0	1	0.0
C	39 *	Tyler Arnason	CHI	3	0	0	0	–1	0	0	0	0	0	2	0.0
R	44	Aaron Downey	CHI	4	0	0	0	–1	8	0	0	0	0	0	0.0
R	12	Tom Fitzgerald	CHI	5	0	0	0	–2	4	0	0	0	0	2	0.0
C	13	Alexei Zhamnov	CHI	5	0	0	0	–1	0	0	0	0	0	14	0.0
D	8	Steve Poapst	CHI	5	0	0	0	–2	0	0	0	0	0	8	0.0
L	55	Eric Daze	CHI	5	0	0	0	–2	2	0	0	0	0	13	0.0
L	28 *	Mark Bell	CHI	5	0	0	0	–2	8	0	0	0	0	7	0.0

Goaltending

No.	Goaltender	GPI	Mins	Avg	W	L	EN	SO	GA	SA	S%	G	A	PIM
29	Steve Passmore	3	138	2.61	0	2	0	0	6	62	.903	0	0	0
41	Jocelyn Thibault	3	159	2.64	1	2	0	0	7	77	.909	0	0	0
	Totals	**5**	**300**	**2.60**	**1**	**4**	**0**	**0**	**13**	**139**	**.906**			

Captains' History

Dick Irvin, 1926-27 to 1928-29; Duke Dukowski, 1929-30; Ty Arbour, 1930-31; Cy Wentworth, 1931-32; Helge Bostrom, 1932-33; Chuck Gardiner, 1933-34; no captain, 1934-35; Johnny Gottselig, 1935-36 to 1939-40; Earl Seibert, 1940-41, 1941-42; Doug Bentley, 1942-43, 1943-44; Clint Smith 1944-45; John Mariucci, 1945-46; Red Hamill, 1946-47; John Mariucci, 1947-48; Gaye Stewart, 1948-49; Doug Bentley, 1949-50; Jack Stewart, 1950-51, 1951-52; Bill Gadsby, 1952-53, 1953-54; Gus Mortson, 1954-55 to 1956-57; no captain, 1957-58; Ed Litzenberger, 1958-59 to 1960-61; Pierre Pilote, 1961-62 to 1967-68, no captain, 1968-69; Pat Stapleton, 1969-70; no captain, 1970-71 to 1974-75; Stan Mikita and Pit Martin, 1975-76; Stan Mikita, Pit Martin and Keith Magnuson, 1976-77; Keith Magnuson, 1977-78, 1978-79; Keith Magnuson and Terry Ruskowski, 1979-80; Terry Ruskowski, 1980-81, 1981-82; Darryl Sutter, 1982-83 to 1984-85; Darryl Sutter and Bob Murray, 1985-86; Darryl Sutter, 1986-87; no captain, 1987-88; Denis Savard and Dirk Graham, 1988-89; Dirk Graham, 1989-90 to 1994-95; Chris Chelios, 1995-96 to 1998-99; Doug Gilmour, 1999-2000; Tony Amonte, 2000-01, 2001-02.

Club Records

Team

(Figures in brackets for season records are games played; records for fewest points, wins, ties, losses, goals, goals against are for 70 or more games)

Most Points **107** 1970-71 (78), 1971-72 (78)
Most Wins **49** 1970-71 (78), 1990-91 (80)
Most Ties **23** 1973-74 (78)
Most Losses **51** 1953-54 (70)
Most Goals **351** 1985-86 (80)
Most Goals Against **363** 1981-82 (80)
Fewest Points **31** 1953-54 (70)
Fewest Wins **12** 1953-54 (70)
Fewest Ties **6** 1989-90 (80)
Fewest Losses **14** 1973-74 (78)
Fewest Goals ***133** 1953-54 (70)
Fewest Goals Against **164** 1973-74 (78)
Longest Winning Streak
Overall .... **8** Dec. 9-26/71, Jan. 4-21/81
Home .... **13** Nov. 11-Dec. 20/70
Away .... **7** Dec. 9-29/64
Longest Undefeated Streak
Overall .... **15** Jan. 14-Feb. 16/67 (12 wins, 3 ties)
Home .... **18** Oct. 11-Dec. 20/70 (16 wins, 2 ties)
Away .... **12** Nov. 2-Dec. 16/67 (6 wins, 6 ties)
Longest Losing Streak
Overall .... **12** Feb. 25-Mar. 25/51
Home .... **9** Feb. 8-Mar. 21/28
Away .... **16** Jan. 2-Mar. 21/54
Longest Winless Streak
Overall .... **21** Dec. 17/50-Jan. 28/51 (18 losses, 3 ties)
Home .... **15** Dec. 16/28-Feb. 28/29 (11 losses, 4 ties)
Away .... **22** Dec. 19/50-Mar. 25/51 (20 losses, 2 ties)
Most Shutouts, Season **15** 1969-70 (76)
Most PIM, Season **2,663** 1991-92 (80)
Most Goals, Game .... **12** Jan. 30/69 (Chi. 12 at Phi. 0)

Individual

Most Seasons .... **22** Stan Mikita
Most Games .... **1,394** Stan Mikita
Most Goals, Career .... **604** Bobby Hull
Most Assists, Career .... **926** Stan Mikita
Most Points, Career .... **1,467** Stan Mikita (541G, 926A)
Most PIM, Career .... **1,495** Chris Chelios
Most Shutouts, Career .... **74** Tony Esposito
Longest Consecutive Games Streak .... **884** Steve Larmer (Oct. 6/82-Apr. 15/93)
Most Goals, Season .... **58** Bobby Hull (1968-69)
Most Assists, Season .... **87** Denis Savard (1981-82, 1987-88)
Most Points, Season .... **131** Denis Savard (1987-88; 44G, 87A)
Most PIM, Season .... **408** Mike Peluso (1991-92)
Most Points, Defenseman, Season .... **85** Doug Wilson (1981-82; 39G, 46A)
Most Points, Center, Season .... **131** Denis Savard (1987-88; 44G, 87A)
Most Points, Right Wing, Season .... **101** Steve Larmer (1990-91; 44G, 57A)
Most Points, Left Wing, Season .... **107** Bobby Hull (1968-69; 58G, 49A)
Most Points, Rookie, Season .... **90** Steve Larmer (1982-83; 43G, 47A)
Most Shutouts, Season .... **15** Tony Esposito (1969-70)
Most Goals, Game .... **5** Grant Mulvey (Feb. 3/82)
Most Assists, Game .... **6** Pat Stapleton (Mar. 30/69)
Most Points, Game .... **7** Max Bentley (Jan. 28/43; 4G, 3A), Grant Mulvey (Feb. 3/82; 5G, 2A)

* NHL Record.

Retired Numbers

1	Glenn Hall	1957-1967
9	Bobby Hull	1957-1972
18	Denis Savard	1980-1990, 1995-1997
21	Stan Mikita	1958-1980
35	Tony Esposito	1969-1984

All-time Record vs. Other Clubs

Regular Season

	At Home								On Road								Total							
	GP	W	L	T	OL	GF	GA	PTS	GP	W	L	T	OL	GF	GA	PTS	GP	W	L	T	OL	GF	GA	PTS
Anaheim	20	11	7	2	0	56	44	24	18	7	9	2	0	39	45	16	38	18	16	4	0	95	89	40
Atlanta	1	1	0	0	0	1	0	2	3	2	1	0	0	8	7	4	4	3	1	0	0	9	7	6
Boston	284	145	94	45	0	916	761	335	283	88	161	34	0	805	1022	210	567	233	255	79	0	1721	1783	545
Buffalo	49	26	16	6	1	159	136	59	52	14	32	6	0	137	198	34	101	40	48	12	1	296	334	93
Calgary	57	25	19	13	0	189	165	63	59	20	26	12	1	179	191	53	116	45	45	25	1	368	356	116
Carolina	28	16	8	3	1	115	80	36	29	12	14	3	0	90	94	27	57	28	22	6	1	205	174	63
Colorado	37	21	13	3	0	131	113	45	35	12	18	5	0	121	142	29	72	33	31	8	0	252	255	74
Columbus	5	4	0	1	0	13	4	9	5	2	1	1	1	17	15	6	10	6	1	2	1	30	19	15
Dallas	106	64	28	14	0	411	276	142	108	44	48	16	0	331	360	104	214	108	76	30	0	742	636	246
Detroit	332	151	130	51	0	994	936	353	330	98	199	33	0	823	1129	229	662	249	329	84	0	1817	2065	582
Edmonton	40	21	13	6	0	156	136	48	41	17	19	5	0	139	152	39	81	38	32	11	0	295	288	87
Florida	8	4	3	1	0	28	27	9	7	5	2	0	0	28	16	10	15	9	5	1	0	56	43	19
Los Angeles	71	33	29	9	0	253	213	75	70	31	31	8	0	239	236	70	141	64	60	17	0	492	449	145
Minnesota	4	2	2	0	0	10	12	4	4	2	2	0	0	10	11	4	8	4	4	0	0	20	23	8
Montreal	273	93	125	55	0	731	761	241	274	54	171	48	1	646	1057	157	547	147	296	103	1	1377	1818	398
Nashville	11	5	4	1	1	30	30	12	11	5	5	1	0	30	32	11	22	10	9	2	1	60	62	23
New Jersey	45	24	12	9	0	174	125	57	45	16	18	11	0	137	136	43	90	40	30	20	0	311	261	100
NY Islanders	47	26	16	5	0	160	153	57	45	13	17	15	0	136	156	41	92	39	33	20	0	296	309	98
NY Rangers	285	128	115	42	0	868	790	298	284	112	117	55	0	803	838	279	569	240	232	97	0	1671	1628	577
Ottawa	7	3	2	2	0	17	18	8	9	6	3	0	0	29	25	12	16	9	5	2	0	46	43	20
Philadelphia	59	26	14	19	0	205	168	71	60	16	33	11	0	159	196	43	119	42	47	30	0	364	364	114
Phoenix	44	26	11	7	0	179	122	59	46	16	26	4	0	143	158	36	90	42	37	11	0	322	280	95
Pittsburgh	58	39	10	9	0	235	155	87	57	23	27	7	0	187	204	53	115	62	37	16	0	422	359	140
St. Louis	113	63	34	16	0	421	338	142	109	38	54	17	0	334	366	93	222	101	88	33	0	755	704	235
San Jose	21	11	7	2	1	66	64	25	22	9	11	1	1	59	61	20	43	20	18	3	2	125	125	45
Tampa Bay	12	7	3	2	0	39	29	16	10	3	4	3	0	25	24	9	22	10	7	5	0	64	53	25
Toronto	317	156	119	42	0	967	828	354	314	96	164	54	0	817	1070	246	631	252	283	96	0	1784	1898	600
Vancouver	67	45	15	7	0	255	151	97	68	22	31	15	0	202	205	59	135	67	46	22	0	457	356	156
Washington	38	22	11	5	0	149	114	49	40	14	21	5	0	123	142	33	78	36	32	10	0	272	256	82
Defunct Clubs	139	79	40	20	0	408	268	178	140	52	67	21	0	316	346	125	279	131	107	41	0	724	614	303
Totals	**2578**	**1277**	**900**	**397**	**4**	**8336**	**7017**	**2955**	**2578**	**849**	**1332**	**393**	**4**	**7112**	**8634**	**2095**	**5156**	**2126**	**2232**	**790**	**8**	**15448**	**15651**	**5050**

Playoffs

	Series	W	L	GP	W	L	T	GF	GA	Last Mtg.	Rnd.	Result
Boston	6	1	5	22	5	16	1	63	97	1978	QF	L 0-4
Buffalo	2	0	2	9	1	8	0	17	36	1980	QF	L 0-4
Calgary	3	1	2	12	5	7	0	33	37	1996	CQF	W 4-0
Colorado	2	0	2	12	4	8	0	28	49	1997	CQF	L 2-4
Dallas	6	4	2	33	19	14	0	120	118	1991	DSF	L 2-4
Detroit	14	8	6	69	38	31	0	210	190	1995	CF	L 1-4
Edmonton	4	1	3	20	8	12	0	77	102	1992	CF	W 4-0
Los Angeles	1	1	0	5	4	1	0	10	7	1974	QF	W 4-1
Montreal	17	5	12	81	29	50	2	185	261	1976	QF	L 0-4
NY Islanders	2	0	2	6	0	6	0	6	21	1979	QF	L 0-4
NY Rangers	5	4	1	24	14	10	0	66	54	1973	SF	W 4-1
Philadelphia	1	1	0	4	4	0	0	20	8	1971	QF	W 4-0
Pittsburgh	2	1	1	8	4	4	0	24	23	1992	F	L 0-4
St. Louis	10	7	3	50	28	22	0	171	142	2002	CQF	L 1-4
Toronto	9	3	6	38	15	22	1	89	111	1995	CQF	W 4-3
Vancouver	2	1	1	9	5	4	0	24	24	1995	CSF	W 4-0
Defunct Clubs	4	2	2	9	5	3	1	16	15			
Totals	**90**	**40**	**50**	**411**	**188**	**218**	**5**	**1159**	**1295**			

Calgary totals include Atlanta Flames, 1972-73 to 1979-80. Carolina totals include Hartford, 1979-80 to 1996-97.
Colorado totals include Quebec, 1979-80 to 1994-95. Dallas totals include Minnesota North Stars, 1967-68 to 1992-93.
New Jersey totals include Kansas City, 1974-75 to 1975-76, and Colorado Rockies, 1976-77 to 1981-82.
Phoenix totals include Winnipeg, 1979-80 to 1995-96.

Playoff Results 2002-1998

Year	Round	Opponent	Result	GF	GA
2002	CQF	St. Louis	L 1-4	5	13

Abbreviations: Round: F - Final; **CF** - conference final; **CSF** - conference semi-final; **CQF** - conference quarter-final; **DSF** - division semi-final; **SF** - semi-final; **QF** - quarter-final.

2001-02 Results

Oct.	4	at Vancouver	5-4		30	Anaheim	2-1
	6	at Calgary	0-4		31	at Ottawa	5-4*
	9	at Edmonton	0-1	**Jan.**	4	Tampa Bay	2-0
	11	Phoenix	3-0		6	Pittsburgh	2-0
	12	at Minnesota	4-6		9	at Colorado	3-7
	14	Columbus	2-2		10	Columbus	2-1
	18	at Nashville	5-3		12	at Columbus	4-5
	20	at Dallas	2-2		14	Edmonton	2-1
	21	Colorado	4-2		16	at Florida	3-0
	23	Calgary	6-3		18	at Tampa Bay	2-2
	25	San Jose	4-2		20	Dallas	3-2
	28	Boston	3-3		23	Phoenix	1-4
	30	Los Angeles	5-1		25	St. Louis	2-1
Nov.	1	at Los Angeles	3-2		28	at Boston	1-2
	2	at Anaheim	2-5		30	at New Jersey	1-3
	4	Detroit	5-4	**Feb.**	6	at Phoenix	5-2
	6	Philadelphia	2-1		8	at San Jose	2-4
	9	Vancouver	3-1		9	at Colorado	3-2
	11	San Jose	3-2*		13	Florida	5-4
	13	at Vancouver	2-3		26	at Philadelphia	4-5
	15	at Calgary	2-2		27	Montreal	2-3
	16	at Edmonton	1-7	**Mar.**	3	Anaheim	2-1
	21	at Nashville	3-4		5	Carolina	1-2*
	23	at Columbus	2-2		7	NY Rangers	5-1
	25	at Detroit	4-4		11	at Los Angeles	1-2
	28	Vancouver	3-3		12	at Phoenix	1-3
	30	Toronto	1-2		15	at Anaheim	1-1
Dec.	1	at Toronto	1-4		16	at San Jose	2-2
	3	at Montreal	3-2		18	Dallas	2-2
	5	Minnesota	4-2		20	New Jersey	1-3
	7	NY Islanders	4-3		24	St. Louis	4-3*
	9	Los Angeles	2-5		27	Nashville	4-1
	12	St. Louis	2-2		29	at Minnesota	3-1
	14	at Atlanta	3-1		31	Minnesota	2-1
	15	at Nashville	2-5	**Apr.**	3	Nashville	1-3
	17	at Detroit	2-0		5	at St. Louis	1-5
	19	at Buffalo	6-5		7	Calgary	3-2
	21	Edmonton	5-1		9	at Washington	1-3
	23	Detroit	0-5		10	at Detroit	3-3
	26	at St. Louis	3-1		12	at Dallas	1-3
	27	Colorado	3-1		14	Columbus	2-0

* – Overtime

Entry Draft Selections 2002-1988

2002

Pick	
21	Anton Babchuk
54	Duncan Keith
93	Alexander Kozhevnikov
128	Matt Ellison
156	James Wisniewski
188	Kevin Kantee
219	Tyson Kellerman
251	Jason Kostadine
282	Adam Burish

2001

Pick	
9	Tuomo Ruutu
29	Adam Munro
59	Matt Keith
73	Craig Andersson
104	Brent MacLellan
115	Vladimir Gusev
119	Alexei Zotkin
142	Tommi Jaminki
174	Alexander Golovin
186	Petr Puncochar
205	Teemu Jaaskelainen
216	Oleg Minakov
268	Jeff Miles

2000

Pick	
10	Mikhail Yakoubov
11	Pavel Vorobiev
49	Jonas Nordqvist
74	Igor Radulov
106	Scott Balan
117	Olli Malmivaara
151	Alexander Barkunov
177	Michael Ayers
193	Joey Martin
207	Cliff Loya
225	Vladislav Luchkin
240	Adam Berkhoel
262	Peter Flache
271	Reto Von Arx
291	Arne Ramholt

1999

Pick	
23	Steve McCarthy
46	Dimitri Levinski
63	Stepan Mokhov
134	Michael Jacobsen
165	Michael Leighton
194	Mattias Wennerberg
195	Yorick Treille
223	Andrew Carver

1998

Pick	
8	Mark Bell
94	Matthias Trattnig
156	Kent Huskins
158	Jari Viuhkola
166	Jonathan Pelletier
183	Tyler Arnason
210	Sean Griffin
238	Alexandre Couture
240	Andrei Yershov

1997

Pick	
13	Daniel Cleary
16	Ty Jones
39	Jeremy Reich
67	Mike Souza
110	Ben Simon
120	Peter Gardiner
130	Kyle Calder
147	Heath Gordon
174	Jerad Smith
204	Sergei Shikhanov
230	Chris Feil

1996

Pick	
31	Remi Royer
42	Jeff Paul
46	Geoff Peters
130	Andy Johnson
184	Mike Vellinga
210	Chris Twerdun
236	Andrei Kozyrev

1995

Pick	
19	Dmitri Nabokov
45	Christian Laflamme
71	Kevin McKay
82	Chris Van Dyk
97	Pavel Kriz
146	Marc Magliarditi
149	Marty Wilford
175	Steve Tardif
201	Casey Hankinson
227	Mike Pittman

1994

Pick	
14	Ethan Moreau
40	Jean-Yves Leroux
85	Steve McLaren
118	Marc Dupuis
144	Jim Enson
170	Tyler Prosofsky
196	Mike Josephson
222	Lubomir Jandera
248	Lars Weibel
263	Rob Mara

1993

Pick	
24	Eric Lecompte
50	Eric Manlow
54	Bogdan Savenko
76	Ryan Huska
90	Eric Daze
102	Patrik Pysz
128	Jonni Vauhkonen
180	Tom White
206	Sergei Petrov
232	Mike Rusk
258	Mike McGhan
284	Tom Noble

1992

Pick	
12	Sergei Krivokrasov
36	Jeff Shantz
41	Sergei Klimovich
89	Andy MacIntyre
113	Tim Hogan
137	Gerry Skrypec
161	Mike Prokopec
185	Layne Roland
209	David Hymovitz
233	Richard Raymond

1991

Pick	
22	Dean McAmmond
39	Michael Pomichter
44	Jamie Matthews
66	Bobby House
71	Igor Kravchuk
88	Zac Boyer
110	Maco Balkovec
112	Kevin St. Jacques
132	Jacques Auger
154	Scott Kirton
176	Roch Belley
198	Scott MacDonald
220	Alexander Andrievski
242	Mike Larkin
264	Scott Dean

1990

Pick	
16	Karl Dykhuis
37	Ivan Droppa
79	Chris Tucker
121	Brett Stickney
124	Derek Edgerly
163	Hugo Belanger
184	Owen Lessard
205	Erik Peterson
226	Steve Dubinsky
247	Dino Grossi

1989

Pick	
6	Adam Bennett
27	Michael Speer
48	Bob Kellogg
111	Tommi Pullola
132	Tracy Egeland
153	Milan Tichy
174	Jason Greyerbiehl
195	Matt Saunders
216	Mike Kozak
237	Michael Doneghey

1988

Pick	
8	Jeremy Roenick
50	Trevor Dam
71	Stefan Elvenas
92	Joe Cleary
113	Justin Lafayette
134	Craig Woodcroft
155	Jon Pojar
176	Mathew Hentges
197	Daniel Maurice
218	Dirk Tenzer
239	Andreas Lupzig

Coaching History

Pete Muldoon, 1926-27; Barney Stanley and Hugh Lehman, 1927-28; Herb Gardiner and Dick Irvin, 1928-29; Tom Shaughnessy and Bill Tobin, 1929-30; Dick Irvin, 1930-31; Bill Tobin, 1931-32; Emil Iverson, Godfrey Matheson and Tommy Gorman, 1932-33; Tommy Gorman, 1933-34; Clem Loughlin, 1934-35 to 1936-37; Bill Stewart, 1937-38; Bill Stewart and Paul Thompson, 1938-39; Paul Thompson, 1939-40 to 1943-44; Paul Thompson and Johnny Gottselig, 1944-45; Johnny Gottselig, 1945-46, 1946-47; Johnny Gottselig and Charlie Conacher, 1947-48; Charlie Conacher, 1948-49, 1949-50; Ebbie Goodfellow, 1950-51, 1951-52; Sid Abel, 1952-53, 1953-54; Frank Eddolls, 1954-55; Dick Irvin, 1955-56; Tommy Ivan, 1956-57; Tommy Ivan and Rudy Pilous, 1957-58; Rudy Pilous, 1958-59 to 1962-63; Billy Reay, 1963-64 to 1975-76; Billy Reay and Bill White, 1976-77; Bob Pulford, 1977-78, 1978-79; Eddie Johnston, 1979-80; Keith Magnuson, 1980-81; Keith Magnuson and Bob Pulford, 1981-82; Orval Tessier, 1982-83, 1983-84; Orval Tessier and Bob Pulford, 1984-85; Bob Pulford, 1985-86, 1986-87; Bob Murdoch, 1987-88; Mike Keenan, 1988-89 to 1991-92; Darryl Sutter, 1992-93 to 1994-95; Craig Hartsburg, 1995-96 to 1997-98; Dirk Graham and Lorne Molleken, 1998-99; Lorne Molleken and Bob Pulford, 1999-2000; Alpo Suhonen, 2000-01; Brian Sutter, 2001-02 to date.

General Manager

SMITH, MIKE
General Manager, Chicago Blackhawks.
Born in Potsdam, NY, August 31, 1945.

Mike Smith joined the Chicago Blackhawks as manager of hockey operations on December 12, 1999, and was officially named the club's general manager on September 22, 2000. Smith served as associate general manager of the Toronto Maple Leafs for two seasons (1997-98 and 1998-99) before joining the Blackhawks. Previously, he had served as a consultant under Bob Pulford with Chicago from 1995 to 1997. Smith held a variety of positions with the Winnipeg Jets from 1979 to 1994, including general manager. Under Smith, the Jets entered into a formal agreement with Sokol Kiev in 1989, the first of its kind for any NHL team.

Smith has a doctorate in Political Science and Russian Studies from Syracuse University. He has authored 10 books, mostly on coaching hockey.

Club Directory

United Center

Chicago Blackhawks
United Center
1901 W. Madison Street
Chicago, IL 60612
Phone **312/455-7000**
FAX 312/455-7041
www.chicagoblackhawks.com
Capacity: 20,500

President	William W. Wirtz
Senior Vice President	Robert J. Pulford
Vice President	Jack Davison
Vice President	Peter R. Wirtz
General Manager	Mike Smith
Assistant General Manager	Nick Beverley
Director of Professional Scouting	Joe Yannetti
Director of Amateur Scouting	Bill Lesuk
Director of Player Evaluation	Marshall Johnston
Head Coach	Brian Sutter
Assistant Coach	Denis Savard
Assistant Coach	Al Mac Adam
Asst. Coach, Strength & Cond.	Phil Walker
Goaltending Consultant	Vladislav Tretiak
Chief Amateur Scout	Michel Dumas
Amateur Scouts	Bruce Franklin, Tim Higgins, Ron Anderson, Gord Donnelly
European Amateur Scouting Coord.	Sakari Pietela
European Amateur Scouts	Ruslan Shabonov, Karl Pavlik, Matti Kautto
Amateur Scouting Analyst	Brad Hornung
Executive Assistant	Cindy Brueck
Manager of Team Services	Matt Colleran
Video Coordinator	Ike Rhodes
Medical Staff	
Club Doctors	Mark Bowen, Gordon Nuber, Greg Ewert
Team Dentist	Dr. Daniel Mackey, Dr. Dean Sana
Oral Surgeon	Dr. Eric Pulver
Eye Doctor	Dr. Robert Stein
Head Trainer	Michael Gapski
Equipment Manager	Troy Parchman
Asst. Equipment Mgr.	Bill Stehle
Asst. Equipment Mgr.	Lou Varga
Equipment Assistant	Mark DePasquale
Massage Therapist	Pawel Prylinski
Public Relations/Marketing	
Exec. Dir. of Communications	Jim De Maria
Dir. of Comm. Relations/PR Asst	Barbara Davidson
Manager of Public Relations	Tony Ommen
Exec. Dir. of Marketing and New Business Development	Jim Sofranko
Dir. of Corporate Sponsorships	Steve Waight
Acct. Exec., Corp. Sponsorship	David Stensby
Manager, Client Services	Kelly Bodnarchuk
Ex. Dir. of Fan Development, Operations and Charities	Carol Czaplicki
Mgr., Youth and Fan Development	Drew Stevenson
Manager, Game Operations	Mike Sullivan
Web Producer	TBD
Marketing Coordinator	Maxine Ohlava
Marketing Associate	Alison Tragesser
Administrative Assistant	Angela Armbruster
Finance	
Controller	Tracy Hernandez
Treasurer	Robert Rinkus
Accounting Manager	Deb Kulir
Accounting Clerk	Rita Loretto
Ticketing	
Director, Ticket Operations	James K. Bare
Director, Ticket Sales	Doug Ryan
Account Executives	Brad Bober, Dustin Corey, Adam Collopy, Jocelyn Gay, Katie Golem
Season Tickets Sales Manager	Steve Risnoy
Ticket Operations Manager	Kathie Raimondi
Customer Service Rep.	Holly Manthei
Miscellaneous Information	
Team Photographer	Bill Smith
Organist	Frank Pellico
Public Address Announcer	Gene Honda
Executive Offices/Home Ice	United Center
Location of Press Box	South Side of United Center
Dimensions of Rink	200 feet by 85 feet
Club Colors	Red, White & Black
Radio Station	WSCR (AM 670)
Television Station	Fox Sports Net Chicago
Broadcasters	Pat Foley, Dale Tallon

General Managers' History

Major Frederic McLaughlin, 1926-27 to 1941-42; Bill Tobin, 1942-43 to 1953-54; Tommy Ivan, 1954-55 to 1976-77; Bob Pulford, 1977-78 to 1989-90; Mike Keenan, 1990-91, 1991-92; Mike Keenan and Bob Pulford, 1992-93; Bob Pulford, 1993-94 to 1996-97; Bob Murray, 1997-98, 1998-99; Bob Murray and Bob Pulford, 1999-2000; Mike Smith, 2000-01 to date.

NHL Coaching Record

		Regular Season				Playoffs		
Season	Team	Games	W	L	T	Games	W	L
1980-81	Winnipeg	23	2	17	4	....	....	....
	NHL Totals	**23**	**2**	**17**	**4**	....	....	....

Colorado Avalanche

2001-02 Results: 45W-28L-8T-1OTL 99PTS.
First, Northwest Division

In his first full season with the Avalanche, Rob Blake finished third in scoring among NHL defenseman (16-40-56) and was fourth in the league in average ice time (27:34 minutes per game). Blake was a Norris Trophy finalist for the third time.

2002-03 Schedule

Month	Day	Date	Opponent
Oct.	Wed.	9	Dallas
	Mon.	14	Boston
	Thu.	17	at Los Angeles
	Sat.	19	at San Jose
	Sun.	20	at Anaheim*
	Tue.	22	Edmonton
	Thu.	24	at Phoenix
	Sun.	27	Minnesota
	Tue.	29	at Minnesota
	Thu.	31	at Vancouver
Nov.	Sat.	2	at Calgary
	Mon.	4	Vancouver
	Wed.	6	Ottawa
	Fri.	8	Anaheim
	Sun.	10	Nashville
	Tue.	12	Columbus
	Thu.	14	at Nashville
	Fri.	15	at Dallas
	Sun.	17	at Phoenix*
	Thu.	21	Nashville
	Sat.	23	at St. Louis
	Mon.	25	Chicago
	Wed.	27	St. Louis
	Fri.	29	at Minnesota*
	Sat.	30	at Edmonton
Dec.	Tue.	3	Calgary
	Fri.	6	Montreal
	Wed.	11	at Vancouver
	Fri.	13	at Edmonton
	Sat.	14	at Calgary
	Mon.	16	Washington
	Thu.	19	Edmonton
	Sat.	21	Minnesota
	Mon.	23	Vancouver
	Thu.	26	at St. Louis
	Fri.	27	Philadelphia
	Sun.	29	Los Angeles*
Jan.	Wed.	1	at Nashville*
	Thu.	2	Florida
	Sat.	4	at San Jose
	Tue.	7	Calgary
	Thu.	9	Anaheim
	Sat.	11	at Dallas*
	Sun.	12	at Carolina*
	Thu.	16	Detroit
	Mon.	20	Dallas
	Thu.	23	Columbus
	Sat.	25	at Toronto
	Tue.	28	at Columbus
	Thu.	30	at NY Rangers
Feb.	Tue.	4	at Boston
	Thu.	6	at Detroit
	Sat.	8	Detroit*
	Sun.	9	Calgary
	Tue.	11	New Jersey
	Thu.	13	at Vancouver
	Sat.	15	Minnesota
	Mon.	17	at Chicago
	Thu.	20	at Pittsburgh
	Fri.	21	at NY Islanders
	Sun.	23	NY Rangers*
	Tue.	25	Edmonton
	Thu.	27	Atlanta
Mar.	Sat.	1	Pittsburgh*
	Sun.	2	at Chicago
	Wed.	5	at Florida
	Fri.	7	at Tampa Bay
	Sat.	8	at Philadelphia
	Mon.	10	Phoenix
	Thu.	13	at Columbus
	Sat.	15	at Detroit*
	Sun.	16	at Washington
	Wed.	19	San Jose
	Sat.	22	Chicago*
	Mon.	24	at Buffalo
	Tue.	25	at Ottawa
	Thu.	27	Los Angeles
	Sat.	29	Phoenix*
	Mon.	31	San Jose
Apr.	Wed.	2	at Los Angeles
	Fri.	4	at Anaheim
	Sun.	6	St. Louis*

* Denotes afternoon game.

Franchise date: June 22, 1979
Transferred from Quebec to Denver, June 21, 1995

NORTHWEST DIVISION

24th NHL Season

Year-by-Year Record

Season	GP	Home W	Home L	Home T	Home OL	Road W	Road L	Road T	Road OL	Overall W	Overall L	Overall T	Overall OL	GF	GA	Pts.	Finished	Playoff Result
2001-02	82	24	12	4	1	21	16	4	0	45	28	8	1	212	169	99	1st, Northwest Div.	Lost Conf. Championship
2000-01	**82**	**28**	**6**	**5**	**2**	**24**	**10**	**5**	**2**	**52**	**16**	**10**	**4**	**270**	**192**	**118**	**1st, Northwest Div.**	**Won Stanley Cup**
1999-2000	82	25	12	4	0	17	16	7	1	42	28	11	1	233	201	96	1st, Northwest Div.	Lost Conf. Championship
1998-99	82	21	14	6	...	23	14	4	...	44	28	10	...	239	205	98	1st, Northwest Div.	Lost Conf. Championship
1997-98	82	21	10	10	...	18	16	7	...	39	26	17	...	231	205	95	1st, Pacific Div.	Lost Conf. Quarter-Final
1996-97	82	26	10	5	...	23	14	4	...	49	24	9	...	277	205	107	1st, Pacific Div.	Lost Conf. Championship
1995-96	**82**	**24**	**10**	**7**	**...**	**23**	**15**	**3**	**...**	**47**	**25**	**10**	**...**	**326**	**240**	**104**	**1st, Pacific Div.**	**Won Stanley Cup**
1994-95*	48	19	1	4	...	11	12	1	...	30	13	5	...	185	134	65	1st, Northeast Div.	Lost Conf. Quarter-Final
1993-94*	84	19	17	6	...	15	25	2	...	34	42	8	...	277	292	76	5th, Northeast Div.	Out of Playoffs
1992-93*	84	23	17	2	...	24	10	8	...	47	27	10	...	351	300	104	2nd, Adams Div.	Lost Div. Semi-Final
1991-92*	80	18	19	3	...	2	29	9	...	20	48	12	...	255	318	52	5th, Adams Div.	Out of Playoffs
1990-91*	80	9	23	8	...	7	27	6	...	16	50	14	...	236	354	46	5th, Adams Div.	Out of Playoffs
1989-90*	80	8	26	6	...	4	35	1	...	12	61	7	...	240	407	31	5th, Adams Div.	Out of Playoffs
1988-89*	80	16	20	4	...	11	26	3	...	27	46	7	...	269	342	61	5th, Adams Div.	Out of Playoffs
1987-88*	80	15	23	2	...	17	20	3	...	32	43	5	...	271	306	69	5th, Adams Div.	Out of Playoffs
1986-87*	80	20	13	7	...	11	26	3	...	31	39	10	...	267	276	72	4th, Adams Div.	Lost Div. Final
1985-86*	80	23	13	4	...	20	18	2	...	43	31	6	...	330	289	92	1st, Adams Div.	Lost Div. Semi-Final
1984-85*	80	24	12	4	...	17	18	5	...	41	30	9	...	323	275	91	2nd, Adams Div.	Lost Conf. Championship
1983-84*	80	24	11	5	...	18	17	5	...	42	28	10	...	360	278	94	3th, Adams Div.	Lost Div. Final
1982-83*	80	23	10	7	...	11	24	5	...	34	34	12	...	343	336	80	4th, Adams Div.	Lost Div. Semi-Final
1981-82*	80	24	13	3	...	9	18	13	...	33	31	16	...	356	345	82	4th, Adams Div.	Lost Conf. Championship
1980-81*	80	18	11	11	...	12	21	7	...	30	32	18	...	314	318	78	4th, Adams Div.	Lost Prelim. Round
1979-80*	80	17	16	7	...	8	28	4	...	25	44	11	...	248	313	61	5th, Adams Div.	Out of Playoffs

* Quebec Nordiques

2002-03 Player Personnel

FORWARDS	HT	WT	S	Place of Birth	Date	2001-02 Club
BRULE, Steve	6-0	200	R	Montreal, Que.	1/15/75	Cincinnati
DRURY, Chris	5-10	180	R	Trumbull, CT	8/20/76	Colorado-United States
FORSBERG, Peter	6-0	205	L	Ornskoldsvik, Sweden	7/20/73	Colorado
HAHL, Riku	6-0	190	L	Hameenlinna, Finland	11/1/80	Colorado-Hershey
HEJDUK, Milan	5-11	185	R	Usti-nad-Labem, Czech.	2/14/76	Col-Czech Republic
HINOTE, Dan	6-0	190	R	Leesburg, FL	1/30/77	Colorado
KEANE, Mike	5-10	185	R	Winnipeg, Man.	5/29/67	St. Louis-Colorado
KRESTANOVICH, Jordan	6-1	170	L	Langley, B.C.	6/14/81	Colorado-Hershey
KULESHOV, Mikhail	6-2	205	R	Perm, USSR	1/7/81	Hershey
LARSEN, Brad	6-0	200	L	Nakusp, B.C.	6/28/77	Colorado
LAZAREV, Yevgeny	6-2	205	L	Kharkov, USSR	4/25/80	Hershey
MESSIER, Eric	6-2	200	L	Drummondville, Que.	10/29/73	Colorado
MOORE, Steve	6-2	205	R	Windsor, Ont.	9/22/78	Colorado-Hershey
NEDOROST, Vaclav	6-1	190	L	Budejovice, Czech.	3/16/82	Colorado-Hershey
PARKER, Scott	6-5	230	R	Hanford, CA	1/29/78	Colorado
REINPRECHT, Steve	6-0	190	L	Edmonton, Alta.	5/7/76	Colorado
SAKIC, Joe	5-11	195	L	Burnaby, B.C.	7/7/69	Colorado-Canada
TANGUAY, Alex	6-0	190	L	Ste-Justine, Que.	11/21/79	Colorado
TIMMONS, K.C.	6-4	215	L	Victoria, B.C.	4/6/80	Hershey
VRBATA, Radim	6-1	185	R	Boleslav, Czech.	6/13/81	Colorado-Hershey
WILLSIE, Brian	6-1	195	R	London, Ont.	3/16/78	Colorado
YELLE, Stephane	6-1	190	L	Ottawa, Ont.	5/9/74	Colorado
DEFENSEMEN						
BLAKE, Rob	6-4	225	R	Simcoe, Ont.	12/10/69	Colorado-Canada
CLARK, Brett	6-1	195	L	Wapella, Sask.	12/23/76	Atl-Chi (AHL)-Hershey
de VRIES, Greg	6-3	215	L	Sundridge, Ont.	1/4/73	Colorado
FOOTE, Adam	6-2	215	R	Toronto, Ont.	7/10/71	Colorado-Canada
MUIR, Bryan	6-4	220	L	Winnipeg, Man.	6/8/73	Colorado-Hershey
PAUL, Jeff	6-3	200	R	London, Ont.	3/1/78	Hershey
RIAZANTSEV, Alexander	6-0	210	R	Moscow, USSR	3/15/80	Hershey
SKOULA, Martin	6-2	195	L	Litomerice, Czech.	10/28/79	Col-Czech Republic
SMITH, D.J.	6-2	205	L	Windsor, Ont.	5/13/77	St. John's-Hershey
THOMPSON, Brent	6-2	205	L	Calgary, Alta.	1/9/71	Hershey

GOALTENDERS	HT	WT	C	Place of Birth	Date	2001-02 Club
AEBISCHER, David	6-1	190	L	Fribourg, Switz.	2/7/78	Colorado-Switzerland
ROY, Patrick	6-2	185	L	Quebec, Que.	10/5/65	Colorado
SAUVE, Philippe	6-0	180	L	Buffalo, NY	2/27/80	Hershey

Captains' History

Marc Tardif, 1979-80, 1980-81; Robbie Ftorek and Andre Dupont, 1981-82; Mario Marois, 1982-83 to 1984-85; Mario Marois and Peter Stastny, 1985-86; Peter Stastny, 1986-87 to 1989-90; Joe Sakic and Steven Finn, 1990-91; Mike Hough, 1991-92; Joe Sakic, 1992-93 to date.

Coaching History

Jacques Demers, 1979-80; Maurice Filion and Michel Bergeron, 1980-81; Michel Bergeron, 1981-82 to 1986-87; Andre Savard and Ron Lapointe, 1987-88; Ron Lapointe and Jean Perron, 1988-89; Michel Bergeron, 1989-90; Dave Chambers, 1990-91; Dave Chambers and Pierre Page, 1991-92; Pierre Page, 1992-93, 1993-94; Marc Crawford, 1994-95 to 1997-98; Bob Hartley, 1998-99 to date.

Coach

HARTLEY, BOB
Coach, Colorado Avalanche. Born in Hawkesbury, Ont., September 7, 1960.

Bob Hartley became the second coach of the Colorado Avalanche and the 11th coach in franchise history when he was named to the position on June 30, 1998. In 2001, he led the team to its second Stanley Cup title. Before joining the Avalanche, Hartley spent four years as a head coach with the organization's American Hockey League affiliates in Cornwall and Hershey compiling a record of 151-136-33.

Hartley began his coaching career with the Hawksbury Hawks, where he won two Central Ontario Junior A championships in four seasons. In 1991 he became head coach of the Laval Titans of the Quebec Major Junior Hockey League, where he won another championship prior to becoming an assistant coach with the Cornwall Aces in 1993. He became head coach in Cornwall the following year and remained with the Avalanche affiliate after it relocated to Hershey for the 1996-97 season. Hartley coached Hershey to the Calder Cup championship that year. In addition to his on-ice success in Hershey, Hartley was known for his summer hockey camps and volunteer work within the community.

Coaching Record

		Regular Season				Playoffs		
Season	Team	Games	W	L	T	Games	W	L
1991-92	Laval (QMJHL)	70	38	27	5	10	4	6
1992-93	Laval (QMJHL)	70	43	25	2	13	12	1
1994-95	Cornwall (AHL)	80	38	33	9	15	8	7
1995-96	Cornwall (AHL)	80	34	39	7	8	3	5
1996-97	Hershey (AHL)	80	43	27	10	23	15	8
1997-98	Hershey (AHL)	80	36	37	7	7	3	4
1998-99	**Colorado (NHL)**	**82**	**44**	**28**	**10**	**19**	**11**	**8**
1999-2000	**Colorado (NHL)**	**82**	**42**	**29**	**11**	**17**	**11**	**6**
2000-01	**Colorado (NHL)**	**82**	**52**	**20**	**10**	**23**	**16**	**7***
2001-02	**Colorado (NHL)**	**82**	**45**	**29**	**8**	**21**	**11**	**10**
	NHL Totals	**328**	**183**	**106**	**39**	**80**	**49**	**31**

* Stanley Cup win.

2001-02 Scoring

* - rookie

Regular Season

Pos	#	Player	Team	GP	G	A	Pts	+/–	PIM	PP	SH	GW	GT	S	%
C	19	Joe Sakic	COL	82	26	53	79	12	18	9	1	4	1	260	10.0
D	4	Rob Blake	COL	75	16	40	56	16	58	10	0	2	0	229	7.0
L	40	Alex Tanguay	COL	70	13	35	48	8	36	7	0	2	0	90	14.4
C	18	Chris Drury	COL	82	21	25	46	1	38	5	0	6	0	236	8.9
C	28	Steve Reinprecht	COL	67	19	27	46	14	18	4	0	3	1	111	17.1
R	23	Milan Hejduk	COL	62	21	23	44	0	24	7	1	5	1	139	15.1
D	41	Martin Skoula	COL	82	10	21	31	–3	42	5	0	1	0	100	10.0
R	17 *	Radim Vrbata	COL	52	18	12	30	7	14	6	0	3	0	112	16.1
D	52	Adam Foote	COL	55	5	22	27	7	55	1	1	0	0	85	5.9
D	7	Greg de Vries	COL	82	8	12	20	18	57	1	1	3	0	148	5.4
L	12	Mike Keane	STL	56	4	6	10	–2	22	1	0	0	1	47	8.5
			COL	22	2	5	7	–2	16	0	0	0	0	26	7.7
			TOTAL	78	6	11	17	–4	38	1	0	0	1	73	8.2
C	26	Stephane Yelle	COL	73	5	12	17	1	48	0	1	1	0	71	7.0
L	29	Eric Messier	COL	74	5	10	15	–5	26	0	0	3	0	84	6.0
R	50 *	Brian Willsie	COL	56	7	7	14	4	14	2	0	1	1	66	10.6
D	11	Darius Kasparaitis	PIT	69	2	12	14	–1	123	0	0	0	0	75	2.7
			COL	11	0	0	0	1	19	0	0	0	0	6	0.0
			TOTAL	80	2	12	14	0	142	0	0	0	0	81	2.5
D	3	Pascal Trepanier	COL	74	4	9	13	4	59	2	0	0	0	87	4.6
R	13	Dan Hinote	COL	58	6	6	12	8	39	0	1	3	0	75	8.0
L	9 *	Brad Larsen	COL	50	2	7	9	4	47	1	0	0	0	38	5.3
C	32 *	Riku Hahl	COL	22	2	3	5	1	14	0	0	1	0	17	11.8
R	27	Scott Parker	COL	63	1	4	5	0	154	0	0	0	0	32	3.1
C	22 *	Vaclav Nedorost	COL	25	2	2	4	–4	2	1	0	0	0	22	9.1
D	5	Todd Gill	COL	36	0	4	4	3	25	0	0	0	0	26	0.0
C	20	Kelly Fairchild	COL	10	2	0	2	1	2	0	0	0	0	6	33.3
D	2	Bryan Muir	COL	22	1	1	2	1	9	0	0	0	0	26	3.8
L	37 *	Jordan Krestanovich	COL	8	0	2	2	1	0	0	0	0	0	6	0.0
C	45	Jeff Daw	COL	1	0	1	1	0	0	0	0	0	0	2	0.0
D	43 *	Jaroslav Obsut	COL	3	0	0	0	0	0	0	0	0	0	3	0.0
C	36 *	Steve Moore	COL	8	0	0	0	–4	4	0	0	0	0	5	0.0

Goaltending

No.	Goaltender	GPI	Mins	Avg	W	L	T	EN	SO	GA	SA	S%	G	A	PIM
1	David Aebischer	21	1184	1.88	13	6	0	4	2	37	538	.931	0	0	4
33	Patrick Roy	63	3773	1.94	32	23	8	6	9	122	1629	.925	0	3	26
	Totals	**82**	**4979**	**2.04**	**45**	**29**	**8**	**10**	**11**	**169**	**2177**	**.922**			

Playoffs

Pos	#	Player	Team	GP	G	A	Pts	+/–	PIM	PP	SH	GW	GT	S	%
C	21	Peter Forsberg	COL	20	9	18	27	8	20	0	0	4	2	35	25.7
C	19	Joe Sakic	COL	21	9	10	19	–2	4	4	0	1	0	76	11.8
L	40	Alex Tanguay	COL	19	5	8	13	–8	0	3	0	0	0	20	25.0
D	7	Greg de Vries	COL	21	4	9	13	1	2	0	0	1	0	41	9.8
C	28	Steve Reinprecht	COL	21	7	5	12	7	8	0	0	2	0	29	24.1
D	4	Rob Blake	COL	20	6	6	12	–1	16	1	0	0	0	67	9.0
C	18	Chris Drury	COL	21	5	7	12	4	10	1	0	3	1	56	8.9
D	52	Adam Foote	COL	21	1	6	7	–2	28	0	0	0	0	26	3.8
R	23	Milan Hejduk	COL	16	3	3	6	–4	4	1	0	0	0	45	6.7
D	41	Martin Skoula	COL	21	0	6	6	–5	2	0	0	0	0	13	0.0
L	12	Mike Keane	COL	18	1	4	5	0	8	0	0	0	0	24	4.2
R	13	Dan Hinote	COL	19	1	2	3	–3	9	0	0	0	0	16	6.3
L	29	Eric Messier	COL	21	1	2	3	–4	0	0	0	0	0	18	5.6
C	32 *	Riku Hahl	COL	21	1	2	3	–3	0	0	0	0	0	17	5.9
D	11	Darius Kasparaitis	COL	21	0	3	3	10	18	0	0	0	0	18	0.0
L	9 *	Brad Larsen	COL	21	1	1	2	–4	13	0	0	0	0	8	12.5
C	26	Stephane Yelle	COL	20	0	2	2	–3	14	0	0	0	0	21	0.0
R	50 *	Brian Willsie	COL	4	0	1	1	0	2	0	0	0	0	4	0.0
D	3	Pascal Trepanier	COL	2	0	0	0	–1	0	0	0	0	0	0	0.0
R	17 *	Radim Vrbata	COL	9	0	0	0	–5	0	0	0	0	0	19	0.0
D	2	Bryan Muir	COL	21	0	0	0	–3	2	0	0	0	0	8	0.0

Goaltending

No.	Goaltender	GPI	Mins	Avg	W	L	EN	SO	GA	SA	S%	G	A	PIM
1	David Aebischer	1	34	1.76	0	0	0	0	1	14	.929	0	0	2
33	Patrick Roy	21	1241	2.51	11	10	3	3	52	572	.909	0	2	0
	Totals	**21**	**1287**	**2.61**	**11**	**10**	**3**	**3**	**56**	**589**	**.905**			

Club Records

Team

(Figures in brackets for season records are games played; records for fewest points, wins, ties, losses, goals, goals against are for 70 or more games)

Record		
Most Points	**118**	2000-01 (82)
Most Wins	**52**	2000-01 (82)
Most Ties	**18**	1980-81 (80)
Most Losses	**61**	1989-90 (80)
Most Goals	**360**	1983-84 (80)
Most Goals Against	**407**	1989-90 (80)
Fewest Points	**31**	1989-90 (80)
Fewest Wins	**12**	1989-90 (80)
Fewest Ties	**5**	1987-88 (80)
Fewest Losses	**16**	2000-01 (82)
Fewest Goals	**212**	2001-02 (82)
Fewest Goals Against	**169**	2001-02 (82)
Longest Winning Streak		
Overall	**12**	Jan. 10-Feb. 7/99
Home	**10**	Nov. 26/83-Jan. 10/84, Mar. 6-Apr. 16/95
Away	**7**	Jan. 10-Feb. 7/99
Longest Undefeated Streak		
Overall	**12**	Dec. 23/96-Jan. 20/97 (9 wins, 3 ties), Jan. 10-Feb. 7/99 (12 wins)
Home	**14**	Nov. 19/83-Jan. 21/84 (11 wins, 3 ties)
Away	**10**	Jan. 10-Mar. 3/99 (8 wins, 2 ties)
Longest Losing Streak		
Overall	**14**	Oct. 21-Nov. 19/90
Home	**8**	Oct. 21-Nov. 24/90
Away	**18**	Jan. 18-Apr. 1/90
Longest Winless Streak		
Overall	**17**	Oct. 21-Nov. 25/90 (15 losses, 2 ties)
Home	**11**	Nov. 14-Dec. 26/89 (7 losses, 4 ties)
Away	**33**	Oct. 8/91-Feb. 27/92 (25 losses, 8 ties)
Most Shutouts, Season	**11**	2001-02 (82)
Most PIM, Season	**2,104**	1989-90 (80)
Most Goals, Game	**12**	Three times

Individual

Most Seasons	**14**	Joe Sakic
Most Games	**1,016**	Joe Sakic
Most Goals, Career	**483**	Joe Sakic
Most Assists, Career	**774**	Joe Sakic
Most Points, Career	**1,257**	Joe Sakic (483G, 774A)
Most PIM, Career	**1,562**	Dale Hunter
Most Shutouts, Career	**32**	Patrick Roy
Longest Consecutive Games Streak	**312**	Dale Hunter (Oct. 9/80-Mar. 13/84)
Most Goals, Season	**57**	Michel Goulet (1982-83)
Most Assists, Season	**93**	Peter Stastny (1981-82)
Most Points, Season	**139**	Peter Stastny (1981-82; 46G, 93A)
Most PIM, Season	**301**	Gord Donnelly (1987-88)
Most Points, Defenseman, Season	**82**	Steve Duchesne (1992-93; 20G, 62A)
Most Points, Center, Season	**139**	Peter Stastny (1981-82; 46G, 93A)
Most Points, Right Wing, Season	**103**	Jacques Richard (1980-81; 52G, 51A)
Most Points, Left Wing, Season	**121**	Michel Goulet (1983-84; 56G, 65A)
Most Points, Rookie, Season	**109**	Peter Stastny (1980-81; 39G, 70A)
Most Shutouts, Season	**9**	Patrick Roy (2001-02)
Most Goals, Game	**5**	Mats Sundin (Mar. 5/92), Mike Ricci (Feb. 17/94)
Most Assists, Game	**5**	Six times
Most Points, Game	**8**	Peter Stastny (Feb. 22/81; 4G, 4A), Anton Stastny (Feb. 22/81; 3G, 5A)

Records include Quebec Nordiques, 1979-80 through 1994-95.

Retired Numbers

3	J.C. Tremblay*	1972-1979
8	Marc Tardif*	1979-1983
16	Michel Goulet*	1979-1990
77	Raymond Bourque	2000-2001

* Quebec Nordiques

All-time Record vs. Other Clubs

Regular Season

	At Home								On Road								Total							
	GP	W	L	T	OL	GF	GA	PTS	GP	W	L	T	OL	GF	GA	PTS	GP	W	L	T	OL	GF	GA	PTS
Anaheim	17	9	4	4	0	55	43	22	17	10	4	3	0	44	37	23	34	19	8	7	0	99	80	45
Atlanta	3	2	1	0	0	14	6	4	3	2	1	0	0	7	6	4	6	4	2	0	0	21	12	8
Boston	64	24	34	6	0	232	265	54	61	21	31	9	0	189	238	51	125	45	65	15	0	421	503	105
Buffalo	63	31	20	11	1	226	194	74	62	18	35	9	0	201	243	45	125	49	55	20	1	427	437	119
Calgary	41	16	14	11	0	153	135	43	42	15	20	7	0	125	148	37	83	31	34	18	0	278	283	80
Carolina	64	38	17	9	0	271	190	85	61	25	24	12	0	211	201	62	125	63	41	21	0	482	391	147
Chicago	35	18	12	5	0	142	121	41	37	13	21	3	0	113	131	29	72	31	33	8	0	255	252	70
Columbus	4	4	0	0	0	17	5	8	4	4	0	0	0	16	3	8	8	8	0	0	0	33	8	16
Dallas	37	21	11	5	0	141	100	47	37	14	17	5	1	105	114	34	74	35	28	10	1	246	214	81
Detroit	38	19	15	4	0	138	129	42	36	13	21	1	1	111	133	28	74	32	36	5	1	249	262	70
Edmonton	41	20	18	3	0	160	156	43	41	15	22	4	0	124	172	34	82	35	40	7	0	284	328	77
Florida	10	4	3	3	0	29	23	11	9	8	1	0	0	39	24	16	19	12	4	3	0	68	47	27
Los Angeles	38	20	15	3	0	158	136	43	39	12	24	3	0	124	164	27	77	32	39	6	0	282	300	70
Minnesota	5	5	0	0	0	17	7	10	5	4	1	0	0	20	8	8	10	9	1	0	0	37	15	18
Montreal	62	31	26	5	0	209	216	67	63	16	37	10	0	196	257	42	125	47	63	15	0	405	473	109
Nashville	8	6	1	1	0	25	14	13	8	3	2	3	0	27	25	9	16	9	3	4	0	52	39	22
New Jersey	33	17	13	3	0	120	96	37	35	13	18	4	0	118	143	30	68	30	31	7	0	238	239	67
NY Islanders	33	20	11	2	0	122	96	42	31	13	17	1	0	108	125	27	64	33	28	3	0	230	221	69
NY Rangers	33	17	13	3	0	135	126	37	32	9	19	4	0	92	129	22	65	26	32	7	0	227	255	59
Ottawa	14	12	1	1	0	68	36	25	16	7	7	2	0	62	49	16	30	19	8	3	0	130	85	41
Philadelphia	33	11	10	12	0	120	118	34	33	9	21	2	1	89	121	21	66	20	31	14	1	209	239	55
Phoenix	37	18	15	4	0	128	128	40	36	15	15	6	0	130	133	36	73	33	30	10	0	258	261	76
Pittsburgh	31	16	13	2	0	138	121	34	35	15	15	5	0	142	140	35	66	31	28	7	0	280	261	69
St. Louis	37	19	12	5	1	126	101	44	36	12	21	3	0	112	141	27	73	31	33	8	1	238	242	71
San Jose	19	11	3	4	1	71	38	27	20	13	7	0	0	74	58	26	39	24	10	4	1	145	96	53
Tampa Bay	12	8	2	2	0	50	25	18	10	2	7	1	0	24	31	5	22	10	9	3	0	74	56	23
Toronto	29	17	7	5	0	111	87	39	33	14	15	4	0	128	113	32	62	31	22	9	0	239	200	71
Vancouver	42	21	14	7	0	143	117	49	41	20	15	6	0	160	142	46	83	41	29	13	0	303	259	95
Washington	32	14	14	4	0	99	113	32	32	11	17	4	0	104	129	26	64	25	31	8	0	203	242	58
Totals	**915**	**469**	**319**	**124**	**3**	**3418**	**2942**	**1065**	**915**	**346**	**455**	**111**	**3**	**2995**	**3358**	**806**	**1830**	**815**	**774**	**235**	**6**	**6413**	**6300**	**1871**

Playoffs

	Series	W	L	GP	W	L	T	GF	GA	Last Mtg.	Rnd.	Result
Boston	2	1	1	11	5	6	0	36	37	1983	DSF	L 1-3
Buffalo	2	2	0	8	6	2	0	35	27	1985	DSF	W 3-2
Chicago	2	2	0	12	8	4	0	49	28	1997	CQF	W 4-2
Dallas	2	0	2	14	6	8	0	29	37	2000	CF	L 3-4
Detroit	5	3	2	30	17	13	0	79	76	2002	CF	L 3-4
Edmonton	2	1	1	12	7	5	0	35	30	1998	CQF	L 3-4
Florida	1	1	0	4	4	0	0	15	4	1996	F	W 4-0
Hartford	2	1	1	9	4	5	0	34	35	1987	DSF	W 4-2
Los Angeles	2	2	0	14	8	6	0	33	23	2002	CQF	W 4-3
Montreal	5	2	3	31	14	17	0	85	105	1993	DSF	L 2-4
New Jersey	1	1	0	7	4	3	0	19	11	2001	F	W 4-3
NY Islanders	1	0	1	4	0	4	0	9	18	1982	CF	L 0-4
NY Rangers	1	0	1	6	2	4	0	19	25	1995	CQF	L 2-4
Philadelphia	2	0	2	11	4	7	0	29	39	1985	CF	L 2-4
Phoenix	1	1	0	5	4	1	0	17	10	2000	CQF	W 4-1
St. Louis	1	1	0	5	4	1	0	17	11	2001	CF	W 4-1
San Jose	2	2	0	13	8	5	0	44	38	2002	CSF	W 4-3
Vancouver	2	2	0	10	8	2	0	40	26	2001	CQF	W 4-0
Totals	**36**	**22**	**14**	**206**	**113**	**93**	**0**	**624**	**580**			

Playoff Results 2002-1998

Year	Round	Opponent	Result	GF	GA
2002	CF	Detroit	L 3-4	13	22
	CSF	San Jose	W 4-3	25	21
	CQF	Los Angeles	W 4-3	16	13
2001	**F**	**New Jersey**	**W 4-3**	**19**	**11**
	CF	St. Louis	W 4-1	17	11
	CSF	Los Angeles	W 4-3	17	10
	CQF	Vancouver	W 4-0	16	9
2000	CF	Dallas	L 3-4	13	14
	CSF	Detroit	W 4-1	13	8
	CQF	Phoenix	W 4-1	17	10
1999	CF	Dallas	L 3-4	16	23
	CSF	Detroit	W 4-2	21	14
	CQF	San Jose	W 4-2	19	17
1998	CQF	Edmonton	L 3-4	16	19

Abbreviations: Round: F - Final; **CF** - conference final; **CSF** - conference semi-final; **CQF** - conference quarter-final; **DSF** - division semi-final.

Calgary totals include Atlanta Flames, 1979-80.
Dallas totals include Minnesota North Stars, 1979-80 to 1992-93.
Phoenix totals include Winnipeg, 1979-80 to 1995-96.
Carolina totals include Hartford, 1979-80 to 1996-97.
New Jersey totals include Colorado Rockies, 1979-80 to 1981-82.

2001-02 Results

Oct.	3	at Pittsburgh	3-1	**Jan.**	1	at Nashville	4-4
	9	Vancouver	5-4		3	NY Rangers	3-2*
	11	at Edmonton	3-5		5	at Detroit	1-3
	13	at Vancouver	0-4		9	Chicago	7-3
	16	Tampa Bay	2-1		12	at Edmonton	2-2
	18	Edmonton	1-4		15	San Jose	0-1*
	20	at Columbus	5-0		17	Phoenix	3-2*
	21	at Chicago	2-4		19	at San Jose	3-1
	23	Carolina	5-1		21	Buffalo	3-2
	25	Vancouver	4-1		23	at Edmonton	4-2
	27	at Phoenix	0-1		24	at Calgary	2-0
	28	at Anaheim	3-2		26	at Los Angeles	4-2
	31	St. Louis	0-1		28	Los Angeles	4-6
Nov.	2	at Minnesota	2-4		30	Nashville	2-5
	3	at Toronto	1-4	**Feb.**	4	Detroit	1-3
	6	at Montreal	1-1		8	at Minnesota	6-0
	8	at Ottawa	0-1		9	Chicago	2-3
	10	at Calgary	0-2		11	Boston	5-2
	14	Minnesota	1-0		13	St. Louis	3-1
	16	NY Islanders	1-0		26	Calgary	2-2
	18	at New Jersey	2-0		28	Phoenix	2-1*
	20	at NY Rangers	3-5	**Mar.**	2	Dallas	1-2
	21	at NY Islanders	4-5		4	New Jersey	2-0
	24	Edmonton	2-0		6	Columbus	4-1
	27	Florida	4-1		9	Los Angeles	4-3
	30	at Vancouver	5-2		11	at St. Louis	3-2
Dec.	1	at Calgary	2-2		14	at Atlanta	0-2
	3	Ottawa	4-2		16	at Philadelphia	2-1
	5	at Detroit	4-1		17	at Nashville	5-4
	7	at Buffalo	4-1		19	Washington	0-3
	8	at Columbus	2-0		21	at Los Angeles	1-3
	10	Anaheim	1-1		23	Detroit	0-2
	12	Columbus	5-1		28	at San Jose	3-2
	14	San Jose	0-3		30	at Phoenix	3-5
	16	at Minnesota	3-2*	**Apr.**	1	Nashville	5-1
	19	Anaheim	2-1		3	Atlanta	6-0
	21	Calgary	2-2		5	at Dallas	1-3
	23	Minnesota	6-3		7	at St. Louis	4-2
	26	at Dallas	2-0		9	Vancouver	1-2
	27	at Chicago	1-3		12	at Anaheim	3-1
	29	Philadelphia	2-5		14	Dallas	2-2

* – Overtime

Entry Draft Selections 2002-1988

2002
Pick
28 Jonas Johansson
61 Johnny Boychuk
94 Eric Lundberg
107 Mikko Kalteva
129 Tom Gilbert
164 Tyler Weiman
195 Taylor Christie
227 Ryan Steeves
258 Sergei Shemetov
289 Sean Collins

2001
Pick
63 Peter Budaj
97 Danny Bois
130 Colt King
143 Frantisek Skladany
144 Cody McCormick
149 Mikko Viitanen
165 Pierre-Luc Emond
184 Scott Horvath
196 Charlie Stephens
227 Marek Svatos

2000
Pick
14 Vaclav Nedorost
47 Jared Aulin
50 Sergei Soin
63 Agris Saviels
88 Kurt Sauer
92 Sergei Klyazmin
119 Brian Fahey
159 John-Michael Liles
189 Chris Bahen
221 Aaron Molnar
252 Darryl Bootland
266 Sean Kotary
285 Blake Ward

1999
Pick
25 Mikhail Kuleshov
45 Martin Grenier
93 Branko Radivojevic
112 Sanny Lindstrom
122 Kristian Kovac
142 Will Magnuson
152 Jordan Krestanovich
158 Anders Lovdahl
183 Riku Hahl
212 Radim Vrbata
240 Jeff Finger

1998
Pick
12 Alex Tanguay
17 Martin Skoula
19 Robyn Regehr
20 Scott Parker
28 Ramzi Abid
38 Philippe Sauve
53 Steve Moore
79 Yevgeny Lazarev
141 K.C. Timmons
167 Alexander Riazantsev

1997
Pick
26 Kevin Grimes
53 Graham Belak
55 Rick Berry
78 Ville Nieminen
87 Brad Larsen
133 Aaron Miskovich
161 David Aebischer
217 Doug Schmidt
243 Kyle Kidney
245 Stephen Lafleur

1996
Pick
25 Peter Ratchuk
51 Yuri Babenko
79 Mark Parrish
98 Ben Storey
107 Randy Petruk
134 Luke Curtin
146 Brian Willsie
160 Kai Fischer
167 Dan Hinote
176 Sami Pahlsson
188 Roman Pylner
214 Matt Scorsune
240 Justin Clark

1995
Pick
25 Marc Denis
51 Nic Beaudoin
77 John Tripp
81 Tomi Kallio
129 Brent Johnson
155 John Cirjak
181 Dan Smith
207 Tomi Hirvonen
228 Chris George

1994
Pick
12 Wade Belak
22 Jeffrey Kealty
35 Josef Marha
61 Sebastien Bety
72 Chris Drury
87 Milan Hejduk
113 Tony Tuzzolino
139 Nicholas Windsor
165 Calvin Elfring
191 Jay Bertsch
217 Tim Thomas
243 Chris Pittman
285 Steven Low

1993
Pick
10 Jocelyn Thibault
14 Adam Deadmarsh
49 Ashley Buckberger
75 Bill Pierce
101 Ryan Tocher
127 Anders Myrvold
137 Nicholas Checco
153 Christian Matte
179 David Ling
205 Petr Franek
231 Vincent Auger
257 Mark Pivetz
283 John Hillman

1992
Pick
4 Todd Warriner
28 Paul Brousseau
29 Tuomas Gronman
52 Manny Fernandez
76 Ian McIntyre
100 Charlie Wasley
124 Paxton Schulte
148 Martin Lepage
172 Mike Jickling
196 Steve Passmore
220 Anson Carter
244 Aaron Ellis

1991
Pick
1 Eric Lindros
24 Rene Corbet
46 Rich Brennan
68 Dave Karpa
90 Patrick Labrecque
103 Bill Lindsay
134 Mikael Johansson
156 Janne Laukkanen
157 Aaron Asp
178 Adam Bartell
188 Brent Brekke
200 Paul Koch
222 Doug Friedman
244 Eric Meloche

1990
Pick
1 Owen Nolan
22 Ryan Hughes
43 Brad Zavisha
106 Jeff Parrott
127 Dwayne Norris
148 Andrei Kovalenko
158 Alexander Karpovtsev
169 Pat Mazzoli
190 Scott Davis
211 Mika Stromberg
232 Wade Klippenstein

1989
Pick
1 Mats Sundin
22 Adam Foote
43 Stephane Morin
54 John Tanner
68 Niklas Andersson
76 Eric Dubois
85 Kevin Kaiser
106 Dan Lambert
127 Sergei Mylnikov
148 Paul Krake
169 Vyacheslav Bykov
190 Andrei Khomutov
211 Byron Witkowski
232 Noel Rahn

1988
Pick
3 Curtis Leschyshyn
5 Daniel Dore
24 Stephane Fiset
45 Petri Aaltonen
66 Darin Kimble
87 Stephane Venne
108 Ed Ward
129 Valeri Kamensky
150 Sakari Lindfors
171 Dan Wiebe
213 Alexei Gusarov
234 Claude Lapointe

General Managers' History

Maurice Filion, 1979-80 to 1987-88; Martin Madden, 1988-89; Martin Madden and Maurice Filion, 1989-90; Pierre Page, 1990-91 to 1993-94; Pierre Lacroix, 1994-95 to date.

President and General Manager

LACROIX, PIERRE
President and General Manager, Colorado Avalanche.
Born in Montreal, Que., August 3, 1948.

Pierre Lacroix was appointed to the general manager's post on May 24, 1994 after 21 years as a respected player agent. In his first season as general manager, his leadership was instrumental in moving the team from 11th to second place in the NHL. Lacroix's second season began with the club's move to Denver. The revamped Avs finished atop the Pacific Division and went on to win the Stanley Cup. He was named NHL executive of the year by *The Hockey News* and became president of the club's hockey operations in August, 1995. The Avalanche have continued to rank among the NHL's top teams, and won the Stanley Cup again in 2001.

Club Directory

Pepsi Center

Colorado Avalanche
Pepsi Center
1000 Chopper Circle
Denver, CO 80204
Phone **303/405-1100**
FAX 303/893-0614
Press Box 303/575-1926
www.coloradoavalanche.com
Capacity: 18,007

Owner and Governor	E. Stanley Kroenke
Alternate Governor, President & General Manager	Pierre Lacroix
Head Coach	Bob Hartley
Assistant Coach	Jacques Cloutier
Assistant Coach	Tony Granato
Assistant Coach, Video	Paul Jerrard
Vice President of Player Personnel	Michel Goulet
Assistant to the General Manager	Greg Sherman
Director of Hockey Operations	Eric Lacroix
Director of Hockey Administration	Charlotte Grahame
Team Services Assistant	Ronnie Jameson
Hockey Administration Assistant	Andrea Furness
Chief Scout	Jim Hammett
Pro Scouts	Brad Smith, Garth Joy
Scouts	Yvon Gendron, Alan Hepple, Steve Lyons, Don Paarup, Richard Pracey
European Scout	Joni Lehto
Computer Research Consultant	John Donohue
Strength and Conditioning Coach	Paul Goldberg
Head Athletic Trainer	Pat Karns
Kinesiologist	Matt Sokolowski
Massage Therapist	Gregorio Pradera
Inventory Manager	Wayne Flemming
Head Equipment Manager	Mark Miller
Assistant Equipment Managers	Dave Randolph, Terry Geer
Equipment Assistant	Cliff Halstead
Communications Department	
Vice President of Communications & Team Services	Jean Martineau
Director of Special Projects and New Media	Hayne Ellis
Assistant Director of Media Relations	Damen Zier
Hershey Bears Staff	
Head Coach	Mike Foligno
Assistant Coach	Paul Fixter
Trainer	Dan "Beaker" Stuck
Equipment Manager	Brian Waselko
Rink Dimensions	200′ × 85′
Team Colors	Burgundy, Silver, Blue and Black
Press Box Location	West Side – Level 6
Practice Facility	South Suburban Family Sports Center, Englewood, CO
Minor League Affiliate	Hershey Bears (AHL)
Television Outlets	FOX Sports Net Rocky Mountain, KTVD UPN-20
Radio Flagship	KKFN AM-950

Patrick Roy established career bests with a 1.94 goals-against average and nine shutouts in 2001-02. Both numbers led the NHL, placing him atop the leaderboard in those categories for the first time since the early 1990s in Montreal.

Columbus Blue Jackets

2001-02 Results: 22W-47L-8T-5OTL 57PTS.
Fifth, Central Division

Year-by-Year Record

Season	GP	Home W	Home L	Home T	Home OL	Road W	Road L	Road T	Road OL	Overall W	Overall L	Overall T	Overall OL	GF	GA	Pts.	Finished	Playoff Result
2001-02	82	14	18	5	4	8	29	3	1	22	47	8	5	164	255	57	5th, Central Div.	Out of Playoffs
2000-01	82	19	15	4	3	9	24	5	3	28	39	9	6	190	233	71	5th, Central Div.	Out of Playoffs

2002-03 Schedule

Month	Day	Date	Opponent
Oct.	Thu.	10	Chicago
	Sat.	12	at New Jersey
	Mon.	14	Phoenix
	Thu.	17	at St. Louis
	Sat.	19	Florida
	Wed.	23	Tampa Bay
	Fri.	25	San Jose
	Sun.	27	Los Angeles
	Tue.	29	at Chicago
Nov.	Fri.	1	Dallas
	Sun.	3	Buffalo
	Tue.	5	Washington
	Thu.	7	at St. Louis
	Sat.	9	NY Rangers
	Tue.	12	at Colorado
	Thu.	14	Anaheim
	Sat.	16	at Nashville
	Sun.	17	at Dallas
	Wed.	20	St. Louis
	Fri.	22	at Buffalo
	Sat.	23	at Ottawa
	Wed.	27	Edmonton
	Fri.	29	at NY Islanders
	Sat.	30	Carolina
Dec.	Tue.	3	at NY Rangers
	Fri.	6	at San Jose
	Sat.	7	at Los Angeles
	Mon.	9	at Phoenix
	Thu.	12	New Jersey
	Sat.	14	at Detroit
	Thu.	19	Calgary
	Fri.	20	at Chicago
	Mon.	23	Detroit
	Thu.	26	at Detroit
	Sat.	28	St. Louis
	Sun.	29	at St. Louis
	Tue.	31	Pittsburgh
Jan.	Fri.	3	at Washington
	Sat.	4	Phoenix
	Mon.	6	Nashville
	Wed.	8	at Minnesota
	Fri.	10	at Vancouver
	Sat.	11	at Calgary
	Mon.	13	at Edmonton
	Wed.	15	Anaheim
	Sat.	18	at Boston
	Mon.	20	Chicago
	Wed.	22	at Dallas
	Thu.	23	at Colorado
	Sat.	25	NY Islanders
	Tue.	28	Colorado
	Thu.	30	Nashville
Feb.	Wed.	5	Vancouver
	Sat.	8	at Nashville
	Wed.	12	San Jose
	Thu.	13	at Montreal
	Sat.	15	Chicago
	Tue.	18	at Phoenix
	Wed.	19	at Anaheim
	Fri.	21	at San Jose
	Sun.	23	at Vancouver
	Tue.	25	at Nashville
	Thu.	27	Los Angeles
Mar.	Sat.	1	Edmonton
	Mon.	3	Detroit
	Thu.	6	Vancouver
	Sat.	8	Calgary
	Mon.	10	at Carolina
	Tue.	11	Dallas
	Thu.	13	Colorado
	Sat.	15	Minnesota
	Mon.	17	at Atlanta
	Thu.	20	Toronto
	Sat.	22	Atlanta
	Mon.	24	at Anaheim
	Tue.	25	at Los Angeles
	Fri.	28	at Edmonton
	Sat.	29	at Calgary
Apr.	Tue.	1	at Philadelphia
	Wed.	2	Minnesota
	Fri.	4	Detroit
	Sun.	6	at Minnesota*

* Denotes afternoon game.

Acquired by Columbus at the 2001 NHL trade deadline, Ray Whitney bounced back from a back injury to lead the Blue Jackets in goals (21), assists (40) and points (61) during the 2001-02 season.

Franchise date: June 25, 1997

CENTRAL DIVISION

3rd NHL Season

2002-03 Player Personnel

FORWARDS	HT	WT	S	Place of Birth	Date	2001-02 Club
BELLEFEUILLE, Blake	5-10	208	R	Framingham, MA	12/27/77	Columbus-Syracuse
CASSELS, Andrew	6-1	185	L	Bramalea, Ont.	7/23/69	Vancouver
DARCHE, Mathieu	6-1	210	L	St-Laurent, Que.	11/26/76	Columbus-Syracuse
DAVIDSON, Matt	6-3	196	R	Flin Flon, Man.	8/9/77	Columbus-Syracuse
DINEEN, Kevin	5-11	198	R	Quebec City, Que.	10/28/63	Columbus
JACKMAN, Tim	6-3	190	R	Minot, ND	11/14/81	Minnesota State
KNOPP, Ben	6-1	190	R	Calgary, Alta.	4/8/82	Moose Jaw-Kamloops
KNUTSEN, Espen	5-11	188	L	Oslo, Norway	1/12/72	Columbus
LING, David	5-10	204	R	Halifax, N.S.	1/9/75	Columbus-Syracuse
MacLEAN, Don	6-2	199	L	Sydney, N.S.	1/14/77	St. John's-Toronto
MARSHALL, Grant	6-1	200	R	Mississauga, Ont.	6/9/73	Columbus
McDONELL, Kent	6-2	205	R	Williamstown, Ont.	3/1/79	Syracuse
McLEOD, Kiel	6-6	229	R	Ft. Saskatchewan, Alta.	12/30/82	Kelowna
MORAN, Brad	5-11	187	L	Abbotsford, B.C.	3/20/79	Columbus-Syracuse
NASH, Rick	6-3	188	L	Brampton, Ont.	6/16/84	London
NEDOROST, Andrej	6-0	192	L	Trencin, Czech.	4/30/80	Columbus-Syracuse
NIELSEN, Chris	6-2	204	R	Moshi, Tanzania	2/16/80	Columbus-Syracuse
PANDOLFO, Mike	6-3	221	L	Winchester, MA	9/15/79	Boston University
PAROULEK, Martin	6-0	193	R	Uherske Hradiste, Czech.	11/4/79	Syracuse
PIRJETA, Lasse	6-3	222	L	Oulu, Finland	4/4/74	Karpat
PRONGER, Sean	6-3	209	L	Thunder Bay, Ont.	11/30/72	Columbus-Syracuse
REICH, Jeremy	6-1	204	L	Craik, Sask.	2/11/79	Syracuse
SANDERSON, Geoff	6-0	190	L	Hay River, N.W.T.	2/1/72	Columbus
SHELLEY, Jody	6-4	225	L	Thompson, Man.	2/7/76	Columbus-Syracuse
SILLINGER, Mike	5-11	196	R	Regina, Sask.	6/29/71	Columbus
VYBORNY, David	5-10	189	L	Jihlava, Czech.	6/2/75	Columbus
WHITNEY, Ray	5-10	175	R	Fort Saskatchewan, Alta.	5/8/72	Columbus
WRIGHT, Tyler	6-0	190	R	Kamsack, Sask.	4/6/73	Columbus
DEFENSEMEN						
ALLISON, Jamie	6-1	200	L	Lindsay, Ont.	5/13/75	Calgary-Columbus
BICANEK, Radim	6-1	209	L	Uherske Hradiste, Czech.	1/18/75	Columbus
GRAND-PIERRE, Jean-Luc	6-3	223	R	Montreal, Que.	2/2/77	Columbus
KLESLA, Rostislav	6-3	206	L	Novy Jicin, Czech.	3/21/82	Columbus
LACHANCE, Scott	6-1	215	L	Charlottesville, VA	10/22/72	Vancouver
MANNING, Paul	6-4	205	L	Red Deer, Alta.	4/15/79	Syracuse-Elmira
RICHARDSON, Luke	6-4	210	L	Ottawa, Ont.	3/26/69	Philadelphia
SCOVILLE, Darrel	6-3	215	L	Swift Current, Sask.	10/13/75	Syracuse
SPACEK, Jaroslav	5-11	206	L	Rokycany, Czech.	2/11/74	Chi-Czech Republic-CBJ
WALSER, Derrick	5-10	196	L	New Glasgow, N.S.	5/12/78	Columbus-Syracuse
WESTCOTT, Duvie	5-11	192	R	Winnipeg, Man.	10/30/77	Columbus-Syracuse

GOALTENDERS	HT	WT	C	Place of Birth	Date	2001-02 Club
DENIS, Marc	6-1	190	L	Montreal, Que.	8/1/77	Columbus
LABBE, Jean-Francois	5-10	175	L	Sherbrooke, Que.	6/15/72	Columbus-Syracuse
LECLAIRE, Pascal	6-2	185	L	Repentigny, Que.	11/7/82	Montreal

The club leader in virtually every offensive category during the Blue Jackets' inaugural season, injuries limited Geoff Sanderson to just 11 goals and 5 assists in 2001-02.

2001-02 Scoring

* - rookie

Regular Season

Pos	#	Player	Team	GP	G	A	Pts	+/–	PIM	PP	SH	GW	GT	S	%
L	14	Ray Whitney	CBJ	67	21	40	61	–22	12	6	0	3	0	210	10.0
C	16	Mike Sillinger	CBJ	80	20	23	43	–35	54	8	0	5	0	150	13.3
C	21	Espen Knutsen	CBJ	77	11	31	42	–28	47	5	2	1	0	102	10.8
R	29	Grant Marshall	CBJ	81	15	18	33	–20	86	6	0	4	1	152	9.9
R	9	David Vyborny	CBJ	75	13	18	31	–14	6	6	0	2	0	103	12.6
D	7	Deron Quint	CBJ	75	7	18	25	–34	26	3	0	1	0	169	4.1
C	28	Tyler Wright	CBJ	77	13	11	24	–40	100	4	0	1	1	120	10.8
D	3	Jaroslav Spacek	CHI	60	3	10	13	5	29	0	0	1	0	64	4.7
			CBJ	14	2	3	5	–9	24	1	1	1	0	29	6.9
			TOTAL	74	5	13	18	–4	53	1	1	2	0	93	5.4
L	8	Geoff Sanderson	CBJ	42	11	5	16	–15	12	5	0	2	0	112	9.8
L	10	Serge Aubin	CBJ	71	8	8	16	–20	32	1	0	1	0	86	9.3
D	44	* Rostislav Klesla	CBJ	75	8	8	16	–6	74	1	0	0	1	102	7.8
R	18	Robert Kron	CBJ	59	4	11	15	–14	4	1	0	0	0	92	4.3
C	42	Brett Harkins	CBJ	25	2	12	14	–5	8	2	0	0	0	9	22.2
R	11	Kevin Dineen	CBJ	59	5	8	13	–6	62	0	0	0	0	73	6.8
D	37	Mattias Timander	CBJ	78	4	7	11	–34	44	1	0	0	0	68	5.9
D	34	J-Luc Grand-Pierre	CBJ	81	2	6	8	–28	90	0	0	0	0	62	3.2
L	45	* Jody Shelley	CBJ	52	3	3	6	1	206	0	0	0	0	35	8.6
D	32	Radim Bicanek	CBJ	60	1	5	6	–15	34	0	0	0	0	43	2.3
R	22	Chris Nielsen	CBJ	23	2	3	5	–3	4	0	0	1	0	28	7.1
C	12	Sean Pronger	CBJ	26	3	1	4	–4	4	0	0	0	0	25	12.0
R	41	* Matt Davidson	CBJ	17	1	2	3	–11	10	0	0	0	0	18	5.6
D	6	Jamie Heward	CBJ	28	1	2	3	–9	7	0	0	0	0	38	2.6
L	19	* Mathieu Darche	CBJ	14	1	1	2	–5	6	0	0	0	0	15	6.7
C	26	* Andrej Nedorost	CBJ	7	0	2	2	–3	2	0	0	0	0	12	0.0
D	33	Jamie Allison	CGY	37	0	2	2	–3	24	0	0	0	0	14	0.0
			CBJ	7	0	0	0	–4	28	0	0	0	0	2	0.0
			TOTAL	44	0	2	2	–7	52	0	0	0	0	16	0.0
D	23	* Derrick Walser	CBJ	2	1	0	1	–2	0	0	0	0	0	2	50.0
L	20	* Martin Spanhel	CBJ	4	1	0	1	–2	2	0	0	0	0	6	16.7
C	38	* Blake Bellefeuille	CBJ	2	0	1	1	1	0	0	0	0	0	2	0.0
C	40	* Brad Moran	CBJ	3	0	0	0	0	0	0	0	0	0	2	0.0
D	40	* Duvie Westcott	CBJ	4	0	0	0	–2	2	0	0	0	0	3	0.0
R	43	David Ling	CBJ	5	0	0	0	–1	7	0	0	0	0	5	0.0

Goaltending

No.	Goaltender	GPI	Mins	Avg	W	L	T	EN	SO	GA	SA	S%	G	A	PIM
31	Ron Tugnutt	44	2502	2.85	12	27	3	4	2	119	1195	.900	0	1	0
1	Jean-Francois Labbe	3	117	3.08	1	1	0	0	0	6	68	.912	0	0	0
30	Marc Denis	42	2335	3.11	9	24	5	5	1	121	1197	.899	0	1	2
	Totals	**82**	**4979**	**3.07**	**22**	**52**	**8**	**9**	**3**	**255**	**2469**	**.897**			

Coaching History

Dave King, 2000-01 to date.

Coach

KING, DAVE

Coach, Columbus Blue Jackets. Born in Saskatoon, Sask., December 22, 1947.

Former Canadian national team coach Dave King was named the first head coach of the Columbus Blue Jackets on July 5, 2000. King joined the Blue Jackets after spending three seasons with the Montreal Canadiens organization.

King enjoyed a successful three-year stint as the head coach of the NHL's Calgary Flames from 1992 to 1995, guiding the club to consecutive Pacific Division titles in 1994 and 1995. He joined the Flames after spending nine seasons with the Canadian national hockey program. King coached Canada to the gold medal at the 1982 World Junior Championships and served as an assistant coach with the bronze medal-winning Team Canada at the 1982 World Championships. He later coached Canada at the Olympics in 1984, 1988 and 1992 and at the World Championships from 1989 to 1992.

King began his coaching career at the University of Saskatchewan in 1972-73. He then coached the Saskatoon Junior B Quakers to a pair of provincial and divisional championships from 1974 to 1976. After splitting the 1976-77 season between the Tier II Saskatoon Olympiques and Saskatoon Blades of the Western Hockey League, he joined the Billings Bighorns in 1977 and captured WHL coach of the year honors after leading the club to the 1978 WHL Finals. He then returned to the University of Saskatchewan, where he led the Huskies to three conference championships and the 1983 CIAU national title.

Coaching Record

		Regular Season or World Championships				Playoffs or Olympics			
Year	Team	Games	W	L	T	Games	W	L	T
1984	Canadian National	....	....	....	....	7	4	3	0
1987	Canadian National	10	3	5	2	....	....	....	....
1988	Canadian National	....	....	....	....	8	5	2	1
1989	Canadian National	10	7	3	0	....	....	....	....
1990	Canadian National	10	6	3	1	....	....	....	....
1991	Canadian National	10	5	2	3	....	....	....	....
1992	Canadian National	6	2	3	1	8	6	2	0
1992-93	**Calgary (NHL)**	**84**	**43**	**30**	**11**	**6**	**2**	**4**	....
1993-94	**Calgary (NHL)**	**84**	**42**	**29**	**13**	**7**	**3**	**4**	....
1994-95	**Calgary (NHL)**	**48**	**24**	**17**	**7**	**7**	**3**	**4**	....
2000-01	**Columbus (NHL)**	**82**	**28**	**45**	**9**	....	....	....	....
2001-02	**Columbus (NHL)**	**82**	**22**	**52**	**8**	....	....	....	....
	NHL Totals	**380**	**159**	**173**	**48**	**20**	**8**	**12**	....

Club Records

Team

(Figures in brackets for season records are games played.)

Most Points 71 2000-01 (82)
Most Wins 28 2000-01 (82)
Most Ties 9 2000-01 (82)
Most Losses 47 2001-02 (82)
Most Goals 190 2000-01 (82)
Most Goals Against 255 2001-02 (82)
Fewest Points 57 2001-02 (82)
Fewest Wins 22 2001-02 (82)
Fewest Ties 8 2001-02 (82)
Fewest Losses 39 2000-01 (82)
Fewest Goals 164 2001-02 (82)
Fewest Goals Against 233 2000-01 (82)

Longest Winning Streak
Overall . . . . 4 Nov. 9-Nov. 16/00
Home . . . . 4 Mar. 24-Apr. 8/01, Dec. 31/01-Jan. 16/02
Away . . . . 2 Five times

Longest Undefeated Streak
Overall . . . . 4 Nov. 9-Nov. 16/00 (4 wins)
Home . . . . 5 Mar. 21-Apr. 8/01 (4 wins, 1 tie), Dec. 29/01-Jan. 16/02 (4 wins, 1 tie)
Away . . . . 2 Eight times

Longest Losing Streak
Overall . . . . 8 Nov. 17-Dec. 3/00
Home . . . . 6 Oct. 12-Nov. 9/01
Away . . . . 8 Mar. 25-Apr. 14/02

Longest Winless Streak
Overall . . . . 8 Nov. 17-Dec. 3/00 (8 losses)
Home . . . . 8 Oct. 4-Nov. 9/01 (6 losses, 2 ties)
Away . . . . 10 Dec. 15/01-Jan. 30/02 (9 losses, 1 tie)

Most Shutouts, Season 4 2000-01 (82)
Most PIM, Season 1,234 2000-01 (82)
Most Goals, Game . . . . 7 Dec. 23/00 (CBJ 7 at NYI 5), Mar. 9/01 (CBJ 7 at Fla. 6)

Individual

Most Seasons . . . . 2 Many players
Most Games . . . . 154 Mattias Timander, David Vyborny
Most Goals, Career . . . . 41 Geoff Sanderson
Most Assists, Career . . . . 73 Espen Knutsen
Most Points, Career . . . . 95 Espen Knutsen (22G, 73A)
Most PIM, Career . . . . 240 Tyler Wright
Most Shutouts, Career . . . . 6 Ron Tugnutt
Longest Consecutive Games Streak . . . . 75 Jamie Pushor (Oct. 7/00-Mar. 26/01)

Most Goals, Season . . . . 30 Geoff Sanderson (2000-01)
Most Assists, Season . . . . 42 Espen Knutsen (2000-01)
Most Points, Season . . . . 61 Ray Whitney (2001-02; 21G, 40A)
Most PIM, Season . . . . 206 Jody Shelley (2001-02)
Most Points, Defenseman, Season . . . . 27 Jamie Heward (2000-01; 11G, 16A)
Most Points, Center, Season . . . . 53 Epsen Knutsen (2000-01; 11G, 42A)
Most Points, Right Wing, Season . . . . 42 Steve Heinze (2000-01; 22G, 20A)
Most Points, Left Wing, Season . . . . 61 Ray Whitney (2001-02; 21G, 40A)
Most Points, Rookie, Season . . . . 32 David Vyborny (2000-01; 13G, 19A)
Most Shutouts, Season . . . . 4 Ron Tugnutt (2000-01)
Most Goals, Game . . . . 3 Four times
Most Assists, Game . . . . 5 Espen Knutsen (Mar. 24/01)
Most Points, Game . . . . 5 Espen Knutsen (Mar. 24/01; 5A)

With the departure of Ron Tugnutt to Dallas, the way is clear for Marc Denis to establish himself as the top goaltender in Columbus. Denis was named the Canadian Major Junior Goaltender of the Year back in 1997.

Captains' History

Lyle Odelein, 2000-01, 2001-02.

All-time Record vs. Other Clubs

Regular Season

	At Home								On Road								Total							
	GP	W	L	T	OL	GF	GA	PTS	GP	W	L	T	OL	GF	GA	PTS	GP	W	L	T	OL	GF	GA	PTS
Anaheim	4	4	0	0	0	13	6	8	4	2	1	0	1	13	12	5	8	6	1	0	1	26	18	13
Atlanta	2	1	1	0	0	6	6	2	1	1	0	0	0	3	0	2	3	2	1	0	0	9	6	4
Boston	2	1	1	0	0	3	10	2	1	1	0	0	0	5	1	2	3	2	1	0	0	8	11	4
Buffalo	1	0	0	1	0	2	2	1	2	2	0	0	0	5	2	4	3	2	0	1	0	7	4	5
Calgary	4	4	0	0	0	16	8	8	4	1	3	0	0	5	13	2	8	5	3	0	0	21	21	10
Carolina	1	1	0	0	0	3	1	2	2	0	2	0	0	3	7	0	3	1	2	0	0	6	8	2
Chicago	5	2	2	1	0	15	17	5	5	0	4	1	0	4	13	1	10	2	6	2	0	19	30	6
Colorado	4	0	4	0	0	3	16	0	4	0	4	0	0	5	17	0	8	0	8	0	0	8	33	0
Dallas	4	1	3	0	0	10	15	2	4	0	4	0	0	3	14	0	8	1	7	0	0	13	29	2
Detroit	5	1	0	0	4	7	10	6	5	1	4	0	0	10	19	2	10	2	4	0	4	17	29	8
Edmonton	4	1	3	0	0	9	14	2	4	1	3	0	0	7	12	2	8	2	6	0	0	16	26	4
Florida	1	0	1	0	0	0	3	0	2	1	1	0	0	7	8	2	3	1	2	0	0	7	11	2
Los Angeles	4	1	3	0	0	7	20	2	4	1	3	0	0	5	9	2	8	2	6	0	0	12	29	4
Minnesota	3	1	1	1	0	7	6	3	4	1	2	0	1	7	12	3	7	2	3	1	1	14	18	6
Montreal	1	0	1	0	0	1	3	0	2	1	1	0	0	3	4	2	3	1	2	0	0	4	7	2
Nashville	5	3	1	0	1	10	9	7	5	2	3	0	0	14	11	4	10	5	4	0	1	24	20	11
New Jersey	2	1	1	0	0	6	7	2	1	0	0	1	0	2	2	1	3	1	1	1	0	8	9	3
NY Islanders	2	1	0	1	0	7	5	3	1	1	0	0	0	7	5	2	3	2	0	1	0	14	10	5
NY Rangers	2	1	1	0	0	5	4	2	1	0	0	1	0	2	2	1	3	1	1	1	0	7	6	3
Ottawa	2	0	1	1	0	7	9	1	1	0	0	1	0	2	2	1	3	0	1	2	0	9	11	2
Philadelphia	2	0	1	1	0	5	6	1	1	0	0	1	0	3	3	1	3	0	1	2	0	8	9	2
Phoenix	4	3	1	0	0	12	6	6	4	0	3	1	0	5	10	1	8	3	4	1	0	17	16	7
Pittsburgh	2	0	0	0	2	5	7	2	2	1	1	0	0	6	7	2	4	1	1	0	2	11	14	4
St. Louis	5	2	1	2	0	11	11	6	5	0	4	1	0	8	19	1	10	2	5	3	0	19	30	7
San Jose	4	2	2	0	0	15	12	4	4	0	3	0	1	5	17	1	8	2	5	0	1	20	29	5
Tampa Bay	1	1	0	0	0	3	1	2	2	0	2	0	0	1	5	0	3	1	2	0	0	4	6	2
Toronto	0	0	0	0	0	0	0	0	2	0	1	1	0	3	6	1	2	0	1	1	0	3	6	1
Vancouver	4	0	3	1	0	5	15	1	4	0	3	0	1	10	15	1	8	0	6	1	1	15	30	2
Washington	2	1	1	0	0	5	6	2	1	0	1	0	0	3	6	0	3	1	2	0	0	8	12	2
Totals	**82**	**33**	**33**	**9**	**7**	**198**	**235**	**82**	**82**	**17**	**53**	**8**	**4**	**156**	**253**	**46**	**164**	**50**	**86**	**17**	**11**	**354**	**488**	**128**

2001-02 Results

Month	Date	Opponent	Score
Oct.	4	St. Louis	3-3
	6	at Philadelphia	3-3
	8	Philadelphia	2-2
	12	Montreal	1-3
	14	at Chicago	2-2
	16	at Detroit	3-4
	19	at Buffalo	3-1
	20	Colorado	0-5
	23	Los Angeles	1-7
	25	Edmonton	2-5
	27	at San Jose	0-2
	30	at Vancouver	1-3
Nov.	1	at Calgary	1-2
	2	at Edmonton	2-1
	6	Vancouver	2-3
	9	Edmonton	0-3
	10	at Boston	5-1
	13	St. Louis	3-2*
	16	Anaheim	3-2
	17	at Nashville	2-3
	19	at Carolina	2-5
	21	Detroit	0-1*
	23	Chicago	2-2
	25	Calgary	4-3
	27	Phoenix	3-0
	29	St. Louis	1-3
Dec.	1	at St. Louis	3-4
	5	at Florida	0-2
	6	at Tampa Bay	0-1
	8	Colorado	0-2
	10	New Jersey	3-1
	12	at Colorado	1-5
	14	at Anaheim	3-2
	15	at Los Angeles	2-3
	17	at Phoenix	1-4
	22	Dallas	2-4
	27	at Detroit	1-5
	29	Buffalo	2-2
	31	Anaheim	3-1
Jan.	3	at St. Louis	2-4
	6	Nashville	4-3
	9	at Washington	3-6
	10	at Chicago	1-2
	12	Chicago	5-4
	14	at NY Rangers	2-2
	16	NY Rangers	2-0
	18	Minnesota	1-3
	19	at Nashville	1-2
	21	Dallas	3-5
	24	San Jose	6-2
	26	Phoenix	2-3
	28	at Dallas	2-4
	30	at Anaheim	1-3
Feb.	4	Boston	0-8
	6	Ottawa	4-6
	8	at Detroit	3-2
	9	Nashville	0-1*
	12	Minnesota	3-3
	26	Los Angeles	1-5
	28	Pittsburgh	3-4*
Mar.	2	at Los Angeles	2-0
	3	at Phoenix	1-2
	6	at Colorado	1-4
	8	NY Islanders	4-2
	10	at Minnesota	0-5
	11	at Pittsburgh	4-2
	14	Vancouver	1-5
	16	Calgary	3-1
	20	at Minnesota	3-1
	21	Detroit	2-3*
	23	Washington	2-5
	25	at Calgary	1-6
	26	at Edmonton	1-3
	28	at Vancouver	3-4*
	30	at San Jose	2-10
Apr.	1	at Dallas	1-3
	4	Nashville	2-1
	6	at Montreal	1-4
	8	at Toronto	1-4
	10	San Jose	3-5
	12	Atlanta	5-4*
	14	at Chicago	0-2

* – Overtime

Entry Draft Selections 2002-2000

2002 Pick		2001 Pick		2000 Pick	
1	Rick Nash	8	Pascal Leclaire	4	Rostislav Klesla
41	Joakim Lindstrom	38	Tim Jackman	69	Ben Knopp
65	Ole-Kristian Tollefsen	53	Kiel McLeod	133	Petteri Nummelin
96	Jeff Genovy	85	Aaron Johnson	138	Scott Heffernan
98	Ivan Tkachenko	87	Per Mars	150	Tyler Kolarik
119	Jekabs Redlihs	141	Cole Jarrett	169	Shane Bendera
133	Lasse Pirjeta	173	Justin Aikins	200	Janne Jokila
168	Tim Konsorada	187	Artem Vostrikov	231	Peter Zingoni
184	Jaroslav Balastik	204	Raffaele Sannitz	278	Martin Paroulek
199	Greg Mauldin	236	Ryan Bowness	286	Andrej Nedorost
225	Steve Goertzen	242	Andrew Murray	292	Louis Mandeville
231	Jaroslav Kracik				
263	Sergei Mozyakin				

TM

General Managers' History

Doug MacLean, 2000-01 to date.

General Manager

MacLEAN, DOUG
General Manager, Columbus Blue Jackets.
Born in Summerside, P.E.I., April 12, 1954.

Doug MacLean joined the Columbus Blue Jackets on February 11, 1998, when he was named the club's first general manager. A month later, he was named team president. As the organization's top hockey executive, he holds the dual role of overseeing both the business and on-ice operations of the franchise, as well as the management of Nationwide Arena. MacLean joined the Blue Jackets after a successful stint with the Florida Panthers that saw him lead the team to the Stanley Cup Finals in 1996 during his first season as head coach. He was selected as coach of the year by *The Hockey News* and was runner-up for the Jack Adams Award that same season.

MacLean began his NHL coaching career in 1986 as an assistant to Jacques Martin in St. Louis. He spent two seasons with the Blues before joining the Washington Capitals in 1988, assisting Bryan Murray behind the bench. He was named coach of the Capitals' American Hockey League affiliate in Baltimore for the final 35 games of the 1989-90 season.

The following season, MacLean joined Murray on the Detroit Red Wings, serving as an assistant coach for two years. In 1992, MacLean was named assistant general manager of the Red Wings and also served as general manager of the team's AHL affiliate in Adirondack for two years. MacLean followed Murray to the Panthers in 1994, becoming the expansion club's director of player development. He was named head coach on July 24, 1995.

A collegiate hockey player at the University of Prince Edward Island, MacLean graduated with a bachelor's degree in education. He also played for the Montreal Jr. Canadiens and was invited to training camp with the St. Louis Blues in 1974. Following his playing career, MacLean enrolled at the University of Western Ontario, where he received a master's degree in educational psychology. While attending Western, MacLean began his coaching career as an assistant with London of the Ontario Hockey League.

NHL Coaching Record

		Regular Season				Playoffs		
Season	**Team**	**Games**	**W**	**L**	**T**	**Games**	**W**	**L**
1995-96	Florida	82	41	31	10	22	12	10
1996-97	Florida	82	35	28	19	5	1	4
1997-98	Florida	23	7	12	4	....	....	....
	NHL Totals	**187**	**83**	**71**	**33**	**27**	**13**	**14**

Club Directory

Nationwide Arena

Columbus Blue Jackets
Nationwide Arena
200 W. Nationwide Blvd.
Columbus, Ohio 43215
Phone **614/246-4625**
FAX 614/246-4007
www.BlueJackets.com
Capacity: 18,136

Ownership

Majority Owner/Governor	John H. McConnell
Alternate Governor	John P. McConnell

Executive Staff

President/General Manager/Alternate Governor	Doug MacLean
Executive Vice-President/Assistant General Manager	Jim Clark
Senior Vice-President of Business Operations	Michael Humes
Vice-President of Marketing	David Paitson
Vice-President of Ticket Sales	Andy Silverman
Chief Financial Officer	T.J. LaMendola
General Counsel	Greg Kirstein

Hockey Operations

Head Coach	Dave King
Associate Coach	Newell Brown
Assistant Coaches	Gerard Gallant, Gord Murphy
Goaltending Coach, Pro Scout	Rick Wamsley
Director of Amateur Scouting	Don Boyd
Director of Pro Scouting	Bob Strumm
Director of Player Development	Paul Castron
Manager of Hockey Operations	Chris MacFarland
Manager of Team Services	Jim Rankin
Video Coordinator	Dan Singleton
Administrative Assistant, Hockey Operations	Julie Uhler
Amateur Scouts	Sam McMaster,Wayne Smith, John Williams
Pro Scout	Peter Dineen
European Scout	Kjell Larsson
Regional Scouts	Brian Bates, Scott Fitzgerald, Jukka Holtari, Denis LeBlanc, John McNamara, Artem Telepin Nicholaevich, Bryan Raymond, Andrew Shaw, Milan Tichy
Head Athletic Trainer	Chris Mizer
Strength and Conditioning Coach	Mark Casterline
Equipment Manager	Tim LeRoy
Assistant Equipment Manager	Jamie Healy
Equipment Assistant	Andre Szucko

Business Operations

Executive Director of Sales	Paul D'Aiuto
Director of Communications	Todd Sharrock
Director of Advertising and Promotions	Marc Gregory
Director of Client Services	Carson Woods
Director of Game Operations	Kimberly Kershaw
Director of Fan Development	J.D. Kershaw
Director of Community Development	Wendy Peterson
Business Development Manager	Scott Klein
Business Development Manager	Scott Shepherd
Assistant Director of Communications	Jason Rothwell
Manager of Multimedia	Steve Ostaszewicz
Graphic Designer/Manager of Print Production	Will Bennett
Client Services Managers	Brent Baker, John Sass
Manager of Advertising and Promotions	Chris Sprague
Manager of Video Production	David Bakalik
Game Operations Assistant	Matt Bettinger
Fan Development Coordinator	Joel Siegman
Mascot Coordinator	Jason Zumpano
Community Development Coordinator	Tracey Vogelpohl
Coordinator of Video Production	Jonny Greco
Corporate Development Representative	Brice Clark
Executive Assistant to Doug MacLean	Kari Pitzer
Administrative Assistant to Michael Humes	Michelle LeVeque
Administrative Assistant to Greg Kirstein	Nikki Ward
Administrative Assistant to David Paitson	Jennifer Pritz
Legal/Immigration Associate	Kelley Walton

Finance

Controller	Rich Gross
Financial Analyst	Dana Fletcher
Staff Accountants	Nora Ludwig, Pete Nyikes
Accounts Payable	Rose Phillips, Malika Dickerson
Accounts Receivable	Shelly Phillips
MIS Manager	Jim Connolly
Human Resources Manager	Harry Coder
Payroll Administrator	Vangie Tavella
General Manager, Facility Merchandising, Inc.	Jennifer Davis
Team Store Manager, Facility Merchandising, Inc.	Krista Sheridan
Warehouse Manager, Facility Merchandising, Inc.	Bill Bellville
Office Manager	Rachel Durham
Receptionist	Beth Carlisle

Ticket Operations

Director of Ticket Operations/Customer Service	Mark Morris
Director of Ticket Sales	Todd Taylor
Manager of Premium and Suite Services	Derrill Smith
Manager of Ticket Operations/Customer Service	Karen Bierley
Asst. Manager of Ticket Operations/ Customer Service	Mark Metz
Account Executives – PSL	Celeste Leadingham, Ted Hritz, David Melfi
Ticket Sales Development Manager	John Motto
Account Executives – Group Sales	Heather Bardocz, Kristie Miller
Premium and Suite Services Coordinator	Melissa DeGraw
Database Coordinator	Krista Vicars
Season Ticket Service Coordinator	Liz Burri

Broadcasting

Director of Broadcasting	Russ Mollohan
Fox Sports Net Play-By-Play Announcer	Dan Kelly
Fox Sports Net Color Analyst	Steve Konroyd
Radio Play-By-Play Announcer	George Matthews
Radio Color Analyst	Bill Davidge

Dallas Stars

2001-02 Results: 36W-28L-13T-5OTL 90PTS.
Fourth, Pacific Division

Consistently ranking among the NHL's best offensive defensemen, Sergei Zubov skates and passes well and has a good slap shot. Defensively sound as well, Zubov was sixth in the NHL in ice time per game last season.

2002-03 Schedule

Month	Day	Date	Opponent
Oct.	Wed.	9	at Colorado
	Fri.	11	Anaheim
	Sat.	12	at Phoenix
	Tue.	15	Edmonton
	Thu.	17	at Minnesota
	Sat.	19	at St. Louis
	Sun.	20	Washington
	Thu.	24	at Calgary
	Sat.	26	at Vancouver
	Mon.	28	at Edmonton
	Wed.	30	Florida
Nov.	Fri.	1	at Columbus
	Sun.	3	at Detroit
	Wed.	6	Vancouver
	Fri.	8	Toronto
	Sun.	10	at NY Islanders*
	Tue.	12	at Montreal
	Wed.	13	at Washington
	Fri.	15	Colorado
	Sun.	17	Columbus
	Wed.	20	at Phoenix
	Fri.	22	at Anaheim
	Sat.	23	at Los Angeles
	Mon.	25	Phoenix
	Wed.	27	Minnesota
	Fri.	29	NY Rangers
	Sat.	30	at Nashville
Dec.	Wed.	4	Montreal
	Fri.	6	Detroit
	Wed.	11	Los Angeles
	Fri.	13	Atlanta
	Sun.	15	at Chicago
	Tue.	17	at Philadelphia
	Thu.	19	at Detroit
	Sat.	21	at New Jersey*
	Sun.	22	at Carolina*
	Thu.	26	at Nashville
	Fri.	27	at Florida
	Sun.	29	Detroit
	Tue.	31	Edmonton
Jan.	Thu.	2	at San Jose
	Sat.	4	at Los Angeles
	Sun.	5	at Anaheim*
	Tue.	7	Los Angeles
	Thu.	9	Chicago
	Sat.	11	Colorado*
	Sat.	18	at San Jose
	Mon.	20	at Colorado
	Wed.	22	Columbus
	Fri.	24	Tampa Bay
	Sat.	25	at St. Louis
	Mon.	27	Ottawa
	Wed.	29	Calgary
Feb.	Wed.	5	St. Louis
	Sat.	8	at Phoenix
	Sun.	9	Los Angeles
	Tue.	11	Carolina
	Fri.	14	Anaheim
	Sun.	16	San Jose*
	Wed.	19	Calgary
	Fri.	21	Phoenix
	Sun.	23	at Chicago*
	Tue.	25	at Boston
	Thu.	27	at Ottawa
	Fri.	28	at Buffalo
Mar.	Sun.	2	Pittsburgh*
	Wed.	5	Chicago
	Fri.	7	Nashville
	Sun.	9	San Jose
	Tue.	11	at Columbus
	Wed.	12	at Minnesota
	Sat.	15	at Edmonton
	Mon.	17	Vancouver
	Wed.	19	at Atlanta
	Fri.	21	Minnesota
	Sun.	23	St. Louis
	Tue.	25	at Vancouver
	Thu.	27	at Calgary
	Sat.	29	at San Jose
	Mon.	31	Buffalo
Apr.	Wed.	2	Anaheim
	Sun.	6	Nashville*

* Denotes afternoon game.

Franchise date: June 5, 1967
Transferred from Minnesota to Dallas, June 9, 1993.

WESTERN NHL CONFERENCE
PACIFIC DIVISION

36th NHL Season

Year-by-Year Record

		Home				Road				Overall								
Season	**GP**	**W**	**L**	**T**	**OL**	**W**	**L**	**T**	**OL**	**W**	**L**	**T**	**OL**	**GF**	**GA**	**Pts.**	**Finished**	**Playoff Result**
2001-02	82	18	13	6	4	18	15	7	1	36	28	13	5	215	213	90	4th, Pacific Div.	Out of Playoffs
2000-01	82	26	10	5	0	22	14	3	2	48	24	8	2	241	187	106	1st, Pacific Div.	Lost Conf. Semi-Final
1999-2000	82	21	11	5	4	22	12	5	2	43	23	10	6	211	184	102	1st, Pacific Div.	Lost Final
1998-99	**82**	**29**	**8**	**4**	**...**	**22**	**11**	**8**	**...**	**51**	**19**	**12**	**...**	**236**	**168**	**114**	**1st, Pacific Div.**	**Won Stanley Cup**
1997-98	82	26	8	7	...	23	14	4	...	49	22	11	...	242	167	109	1st, Central Div.	Lost Conf. Final
1996-97	82	25	13	3	...	23	13	5	...	48	26	8	...	252	198	104	1st, Central Div.	Lost Conf. Quarter-Final
1995-96	82	14	18	9	...	12	24	5	...	26	42	14	...	227	280	66	6th, Central Div.	Out of Playoffs
1994-95	48	9	10	5	...	8	13	3	...	17	23	8	...	136	135	42	5th, Central Div.	Lost Conf. Quarter-Final
1993-94	84	23	12	7	...	19	17	6	...	42	29	13	...	286	265	97	3rd, Central Div.	Lost Conf. Semi-Final
1992-93*	84	18	17	7	...	18	21	3	...	36	38	10	...	272	293	82	5th, Norris Div.	Out of Playoffs
1991-92*	80	20	16	4	...	12	26	2	...	32	42	6	...	246	278	70	4th, Norris Div.	Lost Div. Semi-Final
1990-91*	80	19	15	6	...	8	24	8	...	27	39	14	...	256	266	68	4th, Norris Div.	Lost Final
1989-90*	80	26	12	2	...	10	28	2	...	36	40	4	...	284	291	76	4th, Norris Div.	Lost Div. Semi-Final
1988-89*	80	17	15	8	...	10	22	8	...	27	37	16	...	258	278	70	3rd, Norris Div.	Lost Div. Semi-Final
1987-88*	80	10	24	6	...	9	24	7	...	19	48	13	...	242	349	51	5th, Norris Div.	Out of Playoffs
1986-87*	80	17	20	3	...	13	20	7	...	30	40	10	...	296	314	70	5th, Norris Div.	Out of Playoffs
1985-86*	80	21	15	4	...	17	18	5	...	38	33	9	...	327	305	85	2nd, Norris Div.	Lost Div. Semi-Final
1984-85*	80	14	19	7	...	11	24	5	...	25	43	12	...	268	321	62	4th, Norris Div.	Lost Div. Final
1983-84*	80	22	14	4	...	17	17	6	...	39	31	10	...	345	344	88	1st, Norris Div.	Lost Conf. Championship
1982-83*	80	23	6	11	...	17	18	5	...	40	24	16	...	321	290	96	2nd, Norris Div.	Lost Div. Final
1981-82*	80	21	7	12	...	16	16	8	...	37	23	20	...	346	288	94	1st, Norris Div.	Lost Div. Semi-Final
1980-81*	80	23	10	7	...	12	18	10	...	35	28	17	...	291	263	87	3rd, Adams Div.	Lost Final
1979-80*	80	25	8	7	...	11	20	9	...	36	28	16	...	311	253	88	3rd, Adams Div.	Lost Semi-Final
1978-79*	80	19	15	6	...	9	25	6	...	28	40	12	...	257	289	68	4th, Adams Div.	Out Of Playoffs
1977-78*	80	12	24	4	...	6	29	5	...	18	53	9	...	218	325	45	5th, Smythe Div.	Out of Playoffs
1976-77*	80	17	14	9	...	6	25	9	...	23	39	18	...	240	310	64	2nd, Smythe Div.	Lost Prelim. Round
1975-76*	80	15	22	3	...	5	31	4	...	20	53	7	...	195	303	47	4th, Smythe Div.	Out of Playoffs
1974-75*	80	17	20	3	...	6	30	4	...	23	50	7	...	221	341	53	4th, Smythe Div.	Out of Playoffs
1973-74*	78	18	15	6	...	5	23	11	...	23	38	17	...	235	275	63	7th, West Div.	Out of Playoffs
1972-73*	78	26	8	5	...	11	22	6	...	37	30	11	...	254	230	85	3rd, West Div.	Lost Quarter-Final
1971-72*	78	22	11	6	...	15	18	6	...	37	29	12	...	212	191	86	2nd, West Div.	Lost Quarter-Final
1970-71*	78	16	15	8	...	12	19	8	...	28	34	16	...	191	223	72	4th, West Div.	Lost Semi-Final
1969-70*	76	11	16	11	...	8	19	11	...	19	35	22	...	224	257	60	3rd, West Div.	Lost Quarter-Final
1968-69*	76	11	21	6	...	7	22	9	...	18	43	15	...	189	270	51	6th, West Div.	Out of Playoffs
1967-68*	74	17	12	8	...	10	20	7	...	27	32	15	...	191	226	69	4th, West Div.	Lost Semi-Final

* Minnesota North Stars

2002-03 Player Personnel

FORWARDS	HT	WT	S	Place of Birth	Date	2001-02 Club
ARNOTT, Jason	6-4	225	R	Collingwood, Ont.	10/11/74	New Jersey-Dallas
BATEMAN, Jeff	5-11	184	L	Belleville, Ont.	8/29/81	Utah-Fort Worth (CHL)
COX, Justin	6-0	173	R	Merritt, B.C.	3/13/81	Utah-Fort Worth (CHL)
DAHLEN, Ulf	6-2	199	L	Ostersund, Sweden	1/12/67	Washington-Sweden
DiMAIO, Rob	5-10	190	R	Calgary, Alta.	2/19/68	Dallas-Utah
DOWNEY, Aaron	6-1	216	R	Shelburne, Ont.	8/27/74	Chicago-Norfolk
DRANEY, Brett	6-1	195	L	Merritt, B.C.	3/12/81	Medicine Hat-Ft. Worth (CHL)-Utah
GAINEY, Steve	6-1	192	L	Montreal, Que.	1/26/79	Dallas-Utah
GOSSELIN, David	6-1	205	R	Levis, Que.	6/22/77	Nashville-Milwaukee
GUERIN, Bill	6-2	210	R	Worcester, MA	11/9/70	Boston-United States
HEISTEN, Barrett	6-1	200	L	Anchorage, AK	3/19/80	NYR-Hartford-Utah
KAPANEN, Niko	5-9	180	L	Hattula, Finland	4/29/78	Dallas-Utah
KRISTOFFERSSON, Marcus	6-3	217	L	Ostersund, Sweden	1/22/79	Utah
LEHTINEN, Jere	6-0	200	R	Espoo, Finland	6/24/73	Dallas-Finland
MALHOTRA, Manny	6-2	215	L	Mississauga, Ont.	5/18/80	NY Rangers-Dallas
MODANO, Mike	6-3	205	L	Livonia, MI	6/7/70	Dallas-United States
MONTGOMERY, Jim	5-10	180	R	Montreal, Que.	6/30/69	Dallas-Utah
MORGAN, Gavin	5-11	191	R	Scarborough, Ont.	7/9/76	Utah
MORROW, Brenden	5-11	200	L	Carlyle, Sask.	1/16/79	Dallas
MULLER, Kirk	6-0	205	L	Kingston, Ont.	2/8/66	Dallas
OLIVER, David	6-0	190	R	Sechelt, B.C.	4/17/71	Munchen
OTT, Steve	6-0	160	L	Summerside, P.E.I.	8/19/82	Windsor
PELLERIN, Scott	5-11	190	L	Shediac, N.B.	1/9/70	Boston-Dallas
SIM, Jonathan	5-10	190	L	New Glasgow, N.S.	9/29/77	Dallas-Utah
TURGEON, Pierre	6-1	199	L	Rouyn, Que.	8/28/69	Dallas
YOUNG, Scott	6-1	200	R	Clinton, MA	10/1/67	St. Louis-United States
DEFENSEMEN						
BOUCHER, Philippe	6-2	221	R	Ste-Apollinaire, Que.	3/24/73	Los Angeles
ERSKINE, John	6-4	215	L	Kingston, Ont.	6/26/80	Dallas-Utah
HATCHER, Derian	6-5	235	L	Sterling Hts., MI	6/4/72	Dallas
HAWGOOD, Greg	5-10	190	L	Edmonton, Alta.	8/10/68	Dallas-Utah
HELENIUS, Sami	6-6	230	L	Helsinki, Finland	1/22/74	Dallas
JANCEVSKI, Dan	6-3	212	L	Windsor, Ont.	6/15/81	Utah
KOMAROV, Alexei	6-4	194	L	Moscow, USSR	6/11/78	Spartak Moscow 2-Spartak Moscow
MacMILLAN, Jeff	6-3	206	L	Durham, Ont.	3/30/79	Utah
MATVICHUK, Richard	6-2	215	L	Edmonton, Alta.	2/5/73	Dallas
SYDOR, Darryl	6-1	205	L	Edmonton, Alta.	5/13/72	Dallas
WOTTON, Mark	6-1	195	L	Foxwarren, Man.	11/16/73	Utah
ZUBOV, Sergei	6-1	200	R	Moscow, USSR	7/22/70	Dallas

GOALTENDERS	HT	WT	C	Place of Birth	Date	2001-02 Club
BACASHIHUA, Jason	5-11	175	L	Garden City, MI	9/20/82	Plymouth-Utah
HIRSCH, Corey	5-10	175	L	Medicine Hat, Alta.	8/10/72	Port (AHL)-Phi (AHL)
SMITH, Mike	6-3	189	L	Kingston, Ont.	3/22/82	Sudbury
TUGNUTT, Ron	5-11	160	L	Scarborough, Ont.	10/22/67	Columbus
TURCO, Marty	5-11	183	L	Sault Ste. Marie, Ont.	8/13/75	Dallas

2001-02 Scoring

* - rookie

Regular Season

Pos	#	Player	Team	GP	G	A	Pts	+/–	PIM	PP	SH	GW	GT	S	%
C	9	Mike Modano	DAL	78	34	43	77	14	38	6	2	5	0	219	15.5
R	26	Jere Lehtinen	DAL	73	25	24	49	27	14	7	1	4	2	198	12.6
C	77	Pierre Turgeon	DAL	66	15	32	47	–4	16	7	0	1	1	121	12.4
C	44	Jason Arnott	N.J.	63	22	19	41	3	59	8	0	1	0	169	13.0
			DAL	10	3	1	4	–1	6	2	0	2	0	28	10.7
			TOTAL	73	25	20	45	2	65	10	0	3	0	197	12.7
D	56	Sergei Zubov	DAL	80	12	32	44	–4	22	8	0	2	0	198	6.1
L	10	Brenden Morrow	DAL	72	17	18	35	12	109	4	0	3	0	102	16.7
D	5	Darryl Sydor	DAL	78	4	29	33	3	50	2	0	0	0	183	2.2
L	22	Kirk Muller	DAL	78	10	20	30	–12	28	4	0	1	0	111	9.0
D	2	Derian Hatcher	DAL	80	4	21	25	12	87	1	0	0	2	111	3.6
D	24	Richard Matvichuk	DAL	82	9	12	21	11	52	4	0	2	0	109	8.3
R	16	Pat Verbeek	DAL	64	7	13	20	–4	72	3	0	1	0	81	8.6
R	12	Randy McKay	N.J.	55	6	7	13	2	65	3	0	1	0	63	9.5
			DAL	14	1	4	5	2	7	0	0	0	0	10	10.0
			TOTAL	69	7	11	18	4	72	3	0	1	0	73	9.6
C	27	Manny Malhotra	NYR	56	7	6	13	–1	42	0	1	1	0	41	17.1
			DAL	16	1	0	1	–3	5	0	0	0	0	19	5.3
			TOTAL	72	8	6	14	–4	47	0	1	1	0	60	13.3
L	33	Scott Pellerin	BOS	35	1	5	6	–6	6	0	0	0	0	41	2.4
			DAL	33	3	5	8	–5	15	0	0	0	0	22	13.6
			TOTAL	68	4	10	14	–11	21	0	0	0	0	63	6.3
R	18	Rob DiMaio	DAL	61	6	6	12	–2	25	0	2	2	0	63	9.5
L	41	Brent Gilchrist	DET	19	1	1	2	–3	8	0	0	1	0	24	4.2
			DAL	26	2	5	7	–6	6	0	0	0	0	29	6.9
			TOTAL	45	3	6	9	–9	14	0	0	1	0	53	5.7
D	37	Brad Lukowich	DAL	66	1	6	7	–1	40	0	0	0	1	56	1.8
R	51	John MacLean	DAL	20	3	3	6	–1	17	0	0	1	0	40	7.5
C	14	Jonathan Sim	DAL	26	3	0	3	–3	10	1	0	0	0	43	7.0
C	32	Jim Montgomery	DAL	8	0	2	2	–1	0	0	0	0	0	8	0.0
D	4	Dave Manson	TOR	13	0	1	1	3	10	0	0	0	0	12	0.0
			DAL	34	0	1	1	–1	23	0	0	0	0	24	0.0
			TOTAL	47	0	2	2	2	33	0	0	0	0	36	0.0
L	43	* Steve Gainey	DAL	5	0	1	1	–1	7	0	0	0	0	1	0.0
C	39	* Niko Kapanen	DAL	9	0	1	1	–1	2	0	0	0	0	3	0.0
D	3	* John Erskine	DAL	33	0	1	1	–8	62	0	0	0	0	16	0.0
D	28	Greg Hawgood	DAL	2	0	0	0	0	2	0	0	0	0	1	0.0
D	6	Sami Helenius	DAL	39	0	0	0	–4	58	0	0	0	0	18	0.0

Goaltending

No.	Goaltender	GPI	Mins	Avg	W	L	T	EN	SO	GA	SA	S%	G	A	PIM
35	Marty Turco	31	1519	2.09	15	6	2	4	2	53	670	.921	0	0	10
20	Ed Belfour	60	3467	2.65	21	27	11	3	1	153	1458	.895	0	5	12
	Totals	**82**	**5008**	**2.55**	**36**	**33**	**13**	**7**	**3**	**213**	**2135**	**.900**			

Coach

TIPPETT, DAVE

Coach, Dallas Stars. Born in Moosomin, Sask., August 25, 1961.

Dallas Stars general manager Doug Armstrong announced the hiring of Dave Tippett as the club's head coach on May 16, 2002. Tippett had spent the previous three seasons as an assistant coach with the Los Angeles Kings. He served a five-game stint as interim head coach in 2002 while head coach Andy Murray recovered from an auto accident. In all three seasons Tippett was in Los Angeles the Kings qualified for the playoffs. They had reached the postseason just once out of the previous six seasons.

Under Tippett's direction, the Kings power-play led the NHL in 2001-02 with a 20.7 percent success rate. The year before Tippett came aboard the Kings, in 1998-99, the Kings power-play unit ranked 24th in the league. As a highly regarded minor league coach with tremendous work ethic, Tippett posted two 50-win seasons at Houston (International Hockey League) and led the Aeros to the 1999 Turner Cup championship while serving as general manager/head coach. He was also named IHL coach of the year.

Prior to becoming a coach, Tippett played 11 years as a forward in the National Hockey League with the Hartford Whalers, the Washington Capitals, the Pittsburgh Penguins and the Philadelphia Flyers. He ended his playing career in 1995 as a player-assistant coach with the Houston Aeros (IHL). Internationally, he captained the 1984 Canadian Olympic team in Sarajevo, Yugoslavia, and he earned a silver medal as a member of the Canadian Olympic team in Albertville, France, in 1992. He was a member of the 1982 NCAA Division I championship squad at the University of North Dakota with former Stars defenseman Craig Ludwig.

Coaching Record

		Regular Season				Playoffs		
Season	Team	Games	W	L	T	Games	W	L
1995-96	Houston (IHL)	42	17	18	7	....	....	....
1996-97	Houston (IHL)	82	44	30	8	13	8	5
1997-98	Houston (IHL)	82	50	22	10	4	1	3
1998-99	Houston (IHL)	82	54	15	13	19	11	8

Coaching History

Wren Blair, 1967-68; Wren Blair and John Muckler, 1968-69; Wren Blair and Charlie Burns, 1969-70; Jack Gordon, 1970-71 to 1972-73; Jack Gordon and Parker MacDonald, 1973-74; Jack Gordon and Charlie Burns, 1974-75; Ted Harris, 1975-76, 1976-77; Ted Harris, André Beaulieu and Lou Nanne, 1977-78; Harry Howell and Glen Sonmor, 1978-79; Glen Sonmor, 1979-80 to 1981-82; Glen Sonmor and Murray Oliver, 1982-83; Bill Mahoney, 1983-84, 1984-85; Lorne Henning, 1985-86; Lorne Henning and Glen Sonmor, 1986-87; Herb Brooks, 1987-88; Pierre Page, 1988-89, 1989-90; Bob Gainey, 1990-91 to 1994-95; Bob Gainey and Ken Hitchcock, 1995-96; Ken Hitchcock, 1996-97 to 2000-01; Ken Hitchcock and Rick Wilson, 2001-02; Dave Tippett, 2002-03.

Club Records

Team

(Figures in brackets for season records are games played; records for fewest points, wins, ties, losses, goals, goals against are for 70 or more games)

Record		
Most Points	**114**	1998-99 (82)
Most Wins	**51**	1998-99 (82)
Most Ties	**22**	1969-70 (76)
Most Losses	**53**	1975-76, 1977-78 (80)
Most Goals	**346**	1981-82 (80)
Most Goals Against	**349**	1987-88 (80)
Fewest Points	**45**	1977-78 (80)
Fewest Wins	**18**	1968-69 (76), 1977-78 (80)
Fewest Ties	**4**	1989-90 (80)
Fewest Losses	**19**	1998-99 (82)
Fewest Goals	**189**	1968-69 (76)
Fewest Goals Against	**167**	1997-98 (82)
Longest Winning Streak		
Overall	**7**	Mar. 16-28/80, Mar. 16-Apr. 2/97, Nov. 22-Dec. 5/97
Home	**11**	Nov. 4-Dec. 27/72
Away	**7**	Three times
Longest Undefeated Streak		
Overall	**15**	Dec. 6/98-Jan. 6/99 (12 wins, 3 ties)
Home	**13**	Oct. 28-Dec. 27/72 (12 wins, 1 tie), Nov. 21/79-Jan. 9/80 (10 wins, 3 ties), Jan. 17-Mar. 17/91 (11 wins, 2 ties)
Away	**10**	Jan. 12-Mar. 4/99 (8 wins, 2 ties)
Longest Losing Streak		
Overall	**10**	Feb. 1-20/76
Home	**6**	Jan. 17-Feb. 4/70
Away	**8**	Oct. 19-Nov. 13/75, Jan. 28-Mar. 3/88
Longest Winless Streak		
Overall	**20**	Jan. 15-Feb. 28/70 (15 losses, 5 ties)
Home	**12**	Jan. 17-Feb. 25/70 (8 losses, 4 ties)
Away	**23**	Oct. 25/74-Jan. 28/75 (19 losses, 4 ties)
Most Shutouts, Season	**11**	2000-01 (82)
Most PIM, Season	**2,313**	1987-88 (80)
Most Goals, Game	**15**	Nov. 11/81 (Wpg. 2 at Min. 15)

Individual

Most Seasons	**16**	Neal Broten
Most Games	**992**	Neal Broten
Most Goals, Career	**416**	Mike Modano
Most Assists, Career	**593**	Neal Broten
Most Points, Career	**977**	Mike Modano (416G, 561A)
Most PIM, Career	**1,883**	Shane Churla
Most Shutouts, Career	**27**	Ed Belfour
Longest Consecutive Games Streak	**442**	Danny Grant (Dec. 4/68-Apr. 7/74)
Most Goals, Season	**55**	Dino Ciccarelli (1981-82), Brian Bellows (1989-90)
Most Assists, Season	**76**	Neal Broten (1985-86)
Most Points, Season	**114**	Bobby Smith (1981-82; 43G, 71A)
Most PIM, Season	**382**	Basil McRae (1987-88)
Most Points, Defenseman, Season	**77**	Craig Hartsburg (1981-82; 17G, 60A)
Most Points, Center, Season	**114**	Bobby Smith (1981-82; 43G, 71A)
Most Points, Right Wing, Season	**106**	Dino Ciccarelli (1981-82; 55G, 51A)
Most Points, Left Wing, Season	**99**	Brian Bellows (1989-90; 55G, 44A)
Most Points, Rookie, Season	**98**	Neal Broten (1981-82; 38G, 60A)
Most Shutouts, Season	**9**	Ed Belfour (1997-98)
Most Goals, Game	**5**	Tim Young (Jan. 15/79)
Most Assists, Game	**5**	Murray Oliver (Oct. 24/71), Larry Murphy (Oct. 17/89)
Most Points, Game	**7**	Bobby Smith (Nov. 11/81; 4G, 3A)

Records include Minnesota North Stars, 1967-68 through 1992-93.

Retired Numbers

7	Neal Broten	1980-1995, 1996-1997
8	Bill Goldsworthy*	1967-1976
19	Bill Masterton*	1967-1968

* Minnesota North Stars

All-time Record vs. Other Clubs

Regular Season

	At Home								On Road								Total							
	GP	W	L	T	OL	GF	GA	PTS	GP	W	L	T	OL	GF	GA	PTS	GP	W	L	T	OL	GF	GA	PTS
Anaheim	21	16	4	1	0	82	41	33	22	12	9	1	0	58	51	25	43	28	13	2	0	140	92	58
Atlanta	2	2	0	0	0	4	2	4	3	3	0	0	0	11	5	6	5	5	0	0	0	15	7	10
Boston	60	18	29	13	0	175	218	49	58	9	40	9	0	140	249	27	118	27	69	22	0	315	467	76
Buffalo	53	26	21	6	0	170	156	58	50	13	26	11	0	133	179	37	103	39	47	17	0	303	335	95
Calgary	58	27	20	10	1	214	190	65	58	13	31	13	1	144	201	40	116	40	51	23	2	358	391	105
Carolina	28	16	10	2	0	113	85	34	30	14	12	4	0	107	99	32	58	30	22	6	0	220	184	66
Chicago	108	48	43	16	1	360	331	113	106	28	64	14	0	276	411	70	214	76	107	30	1	636	742	183
Colorado	37	18	12	5	2	114	105	43	37	11	20	5	1	100	141	28	74	29	32	10	3	214	246	71
Columbus	4	4	0	0	0	14	3	8	4	3	1	0	0	15	10	6	8	7	1	0	0	29	13	14
Detroit	102	51	35	15	1	361	306	118	102	37	51	14	0	332	395	88	204	88	86	29	1	693	701	206
Edmonton	41	21	13	7	0	150	118	49	40	13	18	8	1	134	163	35	81	34	31	15	1	284	281	84
Florida	7	3	2	2	0	23	19	8	8	4	3	1	0	24	21	9	15	7	5	3	0	47	40	17
Los Angeles	78	49	17	12	0	308	206	110	77	26	33	18	0	225	265	70	155	75	50	30	0	533	471	180
Minnesota	4	1	2	1	0	14	13	3	4	3	1	0	0	11	11	6	8	4	3	1	0	25	24	9
Montreal	57	16	30	11	0	147	201	43	57	11	37	9	0	140	248	31	114	27	67	20	0	287	449	74
Nashville	8	7	1	0	0	24	8	14	8	4	4	0	0	16	18	8	16	11	5	0	0	40	26	22
New Jersey	44	25	13	6	0	162	117	56	42	19	20	3	0	129	141	41	86	44	33	9	0	291	258	97
NY Islanders	46	18	20	7	1	135	166	44	46	14	23	8	1	131	169	37	92	32	43	15	2	266	335	81
NY Rangers	60	20	30	10	0	184	218	50	61	15	35	11	0	165	210	41	121	35	65	21	0	349	428	91
Ottawa	9	5	4	0	0	34	22	10	8	5	3	0	0	22	19	10	17	10	7	0	0	56	41	20
Philadelphia	65	27	23	15	0	214	210	69	65	9	42	14	0	146	251	32	130	36	65	29	0	360	461	101
Phoenix	50	25	18	7	0	184	152	57	48	24	21	3	0	164	156	51	98	49	39	10	0	348	308	108
Pittsburgh	63	36	21	6	0	243	212	78	62	19	37	6	0	178	232	44	125	55	58	12	0	421	444	122
St. Louis	111	50	39	21	1	373	330	122	113	30	61	21	1	319	413	82	224	80	100	42	2	692	743	204
San Jose	24	11	10	3	0	65	59	25	24	14	9	1	0	70	59	29	48	25	19	4	0	135	118	54
Tampa Bay	10	7	2	1	0	38	23	15	12	9	1	2	0	35	19	20	22	16	3	3	0	73	42	35
Toronto	95	49	35	11	0	362	302	109	101	35	49	17	0	319	356	87	196	84	84	28	0	681	658	196
Vancouver	67	34	21	12	0	247	206	80	67	27	30	10	0	205	243	64	134	61	51	22	0	452	449	144
Washington	39	19	11	8	1	143	106	47	39	16	15	8	0	123	119	40	78	35	26	16	1	266	225	87
Defunct Clubs	33	19	8	6	0	123	86	44	32	10	16	6	0	84	105	26	65	29	24	12	0	207	191	70
Totals	**1384**	**668**	**494**	**214**	**8**	**4780**	**4211**	**1558**	**1384**	**450**	**712**	**217**	**5**	**3956**	**4959**	**1122**	**2768**	**1118**	**1206**	**431**	**13**	**8736**	**9170**	**2680**

Playoffs

	Series	W	L	GP	W	L	T	GF	GA	Last Mtg.	Rnd.	Result
Boston	1	1	0	3	3	0	0	20	13	1981	PRE	W 3-0
Buffalo	3	2	1	13	8	5	0	39	37	1999	F	W 4-2
Calgary	1	1	0	6	4	2	0	25	18	1981	SF	W 4-2
Chicago	6	2	4	33	14	19	0	118	120	1991	DSF	W 4-2
Colorado	2	2	0	14	8	6	0	37	29	2000	CF	W 4-3
Detroit	3	0	3	18	6	12	0	40	55	1998	CF	L 2-4
Edmonton	7	5	2	36	23	13	0	98	93	2001	CQF	W 4-2
Los Angeles	1	1	0	7	4	3	0	26	21	1968	QF	W 4-3
Montreal	2	1	1	13	6	7	0	37	48	1980	QF	W 4-3
New Jersey	1	0	1	6	2	4	0	9	15	2000	F	L 2-4
NY Islanders	1	0	1	5	1	4	0	16	26	1981	F	L 1-4
Philadelphia	2	0	2	11	3	8	0	26	41	1980	SF	L 1-4
Pittsburgh	1	0	1	6	2	4	0	16	28	1991	F	L 2-4
St. Louis	12	6	6	66	34	32	0	197	187	2001	CSF	L 0-4
San Jose	2	2	0	11	8	3	0	31	19	2000	CSF	W 4-1
Toronto	2	2	0	7	6	1	0	35	26	1983	DSF	W 3-1
Vancouver	1	0	1	5	1	4	0	11	18	1994	CSF	L 1-4
Totals	**48**	**25**	**23**	**260**	**133**	**127**	**0**	**781**	**794**			

Playoff Results 2002-1998

Year	Round	Opponent	Result	GF	GA
2001	CSF	St. Louis	L 0-4	6	13
	CQF	Edmonton	W 4-2	16	13
2000	F	New Jersey	L 2-4	9	15
	CF	Colorado	W 4-3	14	13
	CSF	San Jose	W 4-1	15	7
	CQF	Edmonton	W 4-1	14	11
1999	**F**	**Buffalo**	**W 4-2**	**13**	**9**
	CF	Colorado	W 4-3	23	16
	CSF	St. Louis	W 4-2	17	12
	CQF	Edmonton	W 4-0	11	7
1998	CF	Detroit	L 2-4	11	15
	CSF	Edmonton	W 4-1	9	5
	CQF	San Jose	W 4-2	16	12

Abbreviations: Round: F - Final; **CF** - conference final; **CSF** - conference semi-final; **CQF** - conference quarter-final; **DSF** - division semi-final; **SF** - semi-final; **QF** - quarter-final; **PRE** - preliminary round.

Calgary totals include Atlanta Flames, 1972-73 to 1979-80. Carolina totals include Hartford, 1979-80 to 1996-97.
Colorado totals include Quebec, 1979-80 to 1994-95.
New Jersey totals include Kansas City, 1974-75 to 1975-76, and Colorado Rockies, 1976-77 to 1981-82.
Phoenix totals include Winnipeg, 1979-80 to 1995-96.

2001-02 Results

Month	Date	Opponent	Score
Oct.	5	Nashville	4-1
	7	at Carolina	0-3
	9	Los Angeles	2-1
	11	Vancouver	1-4
	13	Calgary	3-4*
	17	at St. Louis	2-2
	18	Phoenix	3-1
	20	Chicago	2-2
	24	at Pittsburgh	2-3
	26	at Detroit	5-3
	28	at NY Islanders	2-3*
	29	at NY Rangers	2-4
	31	Detroit	3-4*
Nov.	2	Nashville	3-0
	3	at Nashville	1-4
	7	San Jose	2-2
	9	Phoenix	1-5
	11	at Anaheim	2-2
	15	at Los Angeles	4-3
	17	at San Jose	3-2*
	19	NY Islanders	2-3
	21	Carolina	4-4
	23	Philadelphia	3-3
	25	at Minnesota	4-3
	29	at Calgary	3-0
Dec.	1	at Edmonton	6-4
	2	at Vancouver	4-2
	5	Ottawa	3-6
	7	Edmonton	5-0
	12	Buffalo	4-3
	14	Calgary	3-4
	15	at Phoenix	6-2
	17	San Jose	1-4
	20	at Philadelphia	1-2
	22	at Columbus	4-2
	23	at Atlanta	4-1
	26	Colorado	0-2
	28	Washington	2-3*
	31	Boston	2-1
Jan.	2	Atlanta	2-1
	5	at St. Louis	2-5
	8	at Tampa Bay	2-1
	9	at Florida	3-2
	12	at Detroit	2-5
	13	at Minnesota	3-1
	16	Detroit	3-2
	18	Florida	2-3
	20	at Chicago	2-3
	21	at Columbus	5-3
	23	Vancouver	2-4
	25	Anaheim	1-6
	28	Columbus	4-2
Feb.	6	at Nashville	1-2
	8	Edmonton	1-1
	10	at Anaheim	5-1
	11	at Los Angeles	1-2
	13	NY Rangers	4-2
	26	at Phoenix	1-5
	28	at Vancouver	4-3*
Mar.	2	at Colorado	2-1
	3	San Jose	4-1
	6	Los Angeles	3-2*
	8	Minnesota	3-5
	10	New Jersey	0-3
	12	at Washington	5-2
	14	at Montreal	3-3
	16	at Toronto	5-5
	18	at Chicago	2-2
	20	St. Louis	3-2
	22	Phoenix	4-3
	24	Anaheim	1-2
	26	at San Jose	2-3
	28	at Calgary	2-2
	30	at Edmonton	1-3
Apr.	1	Columbus	3-1
	3	St. Louis	1-2*
	5	Colorado	3-1
	7	at Anaheim	1-4
	8	at Los Angeles	0-3
	10	Minnesota	4-4
	12	Chicago	3-1
	14	at Colorado	2-2

* – Overtime

Entry Draft Selections 2002-1988

2002

Pick	
26	Martin Vagner
32	Janos Vas
34	Tobias Stephan
42	Marius Holtet
43	Trevor Daley
78	Geoff Waugh
110	Jarkko A. Immonen
147	David Bararuk
180	Kirill Sidorenko
210	Bryan Hamm
243	Tuomas Mikkonen
273	Ned Havern

2001

Pick	
26	Jason Bacashihua
70	Yared Hagos
92	Anthony Aquino
126	Daniel Volrab
161	Mike Smith
167	Michal Blazek
192	Jussi Jokinen
255	Marco Rosa
265	Dale Sullivan
285	Marek Tomica

2000

Pick	
25	Steve Ott
60	Dan Ellis
68	Joel Lundqvist
91	Alexei Tereschenko
123	Vadim Khomitsky
139	Ruslan Bernikov
162	Artem Chernov
192	Ladislav Vlcek
219	Marco Tuokko
224	Antti Miettinen

1999

Pick	
32	Mike Ryan
66	Dan Jancevski
96	Mathias Tjarnqvist
126	Jeff Bateman
156	Gregor Baumgartner
184	Justin Cox
186	Brett Draney
215	Jeff MacMillan
243	Brian Sullivan
265	Jamie Chamberlain
272	Mikhail Donika

1998

Pick	
39	John Erskine
57	Tyler Bouck
86	Gabriel Karlsson
153	Pavel Patera
173	Niko Kapanen
200	Scott Perry

1997

Pick	
25	Brenden Morrow
52	Roman Lyashenko
77	Steve Gainey
105	Marcus Kristoffersson
132	Teemu Elomo
160	Alexei Timkin
189	Jeff McKercher
216	Alexei Komarov
242	Brett McLean

1996

Pick	
5	Richard Jackman
70	Jonathan Sim
90	Mike Hurley
112	Ryan Christie
113	Yevgeny Tsybuk
166	Eoin McInerney
194	Joel Kwiatkowski
220	Nick Bootland

1995

Pick	
11	Jarome Iginla
37	Patrick Cote
63	Petr Buzek
69	Sergey Gusev
115	Wade Strand
141	Dominic Marleau
173	Jeff Dewar
193	Anatoli Koveshnikov
202	Sergei Luchinkin
219	Stephen Lowe

1994

Pick	
20	Jason Botterill
46	Lee Jinman
98	Jamie Wright
124	Marty Turco
150	Evgeny Petrochinin
228	Marty Flichel
254	Jimmy Roy
280	Chris Szysky

1993

Pick	
9	Todd Harvey
35	Jamie Langenbrunner
87	Chad Lang
136	Rick Mrozik
139	Per Svartvadet
165	Jeremy Stasiuk
191	Rob Lurtsema
243	Jordan Willis
249	Bill Lang
269	Cory Peterson

1992

Pick	
34	Jarkko Varvio
58	Jeff Bes
88	Jere Lehtinen
130	Michael Johnson
154	Kyle Peterson
178	Juha Lind
202	Lars Edstrom
226	Jeff Romfo
250	Jeffrey Moen

1991

Pick	
8	Richard Matvichuk
74	Mike Torchia
97	Mike Kennedy
118	Mark Lawrence
137	Geoff Finch
174	Michael Burkett
184	Derek Herlofsky
206	Tom Nemeth
228	Shayne Green
250	Jukka Suomalainen

1990

Pick	
8	Derian Hatcher
50	Laurie Billeck
70	Cal McGowan
71	Frank Kovacs
92	Enrico Ciccone
113	Roman Turek
134	Jeff Levy
155	Doug Barrault
176	Joe Biondi
197	Troy Binnie
218	Ole-Eskild Dahlstrom
239	John McKersie

1989

Pick	
7	Doug Zmolek
28	Mike Craig
60	Murray Garbutt
75	Jean-Francois Quintin
87	Pat MacLeod
91	Bryan Schoen
97	Rhys Hollyman
112	Scott Cashman
154	Jonathon Pratt
175	Kenneth Blum
196	Arturs Irbe
217	Tom Pederson
238	Helmut Balderis

1988

Pick	
1	Mike Modano
40	Link Gaetz
43	Shaun Kane
64	Jeffrey Stolp
148	Ken MacArthur
169	Travis Richards
190	Ari Matilainen
211	Grant Bischoff
232	Trent Andison

General Managers' History

Wren Blair, 1967-68 to 1973-74; Jack Gordon, 1974-75 to 1976-77; Lou Nanne, 1977-78 to 1987-88; Jack Ferreira, 1988-89, 1989-90; Bob Clarke 1990-91, 1991-92; Bob Gainey, 1992-93 to 2000-01; Bob Gainey and Doug Armstrong, 2001-02; Doug Armstrong, 2002-03.

General Manager

ARMSTRONG, DOUG
General Manager, Dallas Stars. Born in Sarnia, Ont., September 24, 1964.

Doug Armstrong was in his ninth season as an assistant to Bob Gainey when he was elevated to the position of general manager on January 25, 2002. Armstrong originally joined the club in 1991. As Gainey's assistant, he worked on contract information and season scheduling and handled the day-to-day operations of the hockey department. In five seasons from 1996 to 2001, he helped Gainey build a team that won five straight division championships, as well as the Presidents' Trophy for the best regular-season record in the NHL twice, and the 1999 Stanley Cup. At the international level, Armstrong served as Team Canada's assistant general manger at the 2002 World Championships in Sweden.

A native of Sarnia, Ontario, Armstrong attended Western Michigan University for two years before transferring to Florida State University in Tallahassee, where he earned his B.S. in Business Administration with a major in marketing.

Club Directory

American Airlines Center

Dallas Stars
Office Address:
Dr Pepper StarCenter
211 Cowboys Parkway
Irving, TX 75063
Phone **972/831-2401**
FAX 972/868-2860
Ticket Information 214/GO STARS
www.dallasstars.com
Capacity: 18,532

Dallas Stars Front Office

Chairman of the Board/Owner	Thomas O. Hicks
President/Chief Operating Officer, Southwest Sports Group	Michael J. Cramer
General Manager	Doug Armstrong
Assistant General Manager	Francois Giguere
Director of Hockey Operations	Les Jackson
Special Assistant to the General Manager	Guy Carbonneau
Head Coach	Dave Tippett
Associate Coach	Rick Wilson
Assistant Coach	Mark Lamb
Assistant/Goaltending Coach	Andy Moog
Director, Amateur Scouting	Tim Bernhardt
Director, Professional Scouting	Doug Overton
Scout	Bob Gernander
Professional Scout	Paul McIntosh
Regional Scouts	Shannon Currie, Hans Edlund, Jack Foley, Jiri Hrdina, Dennis Holland, Jimmy Johnston, Jim Pedersen, Brad Robson, Karri Takko
Senior Director of Hockey Communications	Rob Scichili
Director of Media Relations	Mark Janko

Two-way terror Jere Lehtinen scored a career-high 25 goals last season while maintaining his defensive excellence. He was a finalist for the Selke Trophy for the fourth time. It's an award he has won twice (1998 and 1999).

Captains' History

Bob Woytowich, 1967-68; Moose Vasko, 1968-69; Claude Larose, 1969-70; Ted Harris, 1970-71 to 1973-74; Bill Goldsworthy, 1974-75, 1975-76; Bill Hogaboam, 1976-77; Nick Beverley, 1977-78; J.P. Parise, 1978-79; Paul Shmyr, 1979-80, 1980-81; Tim Young, 1981-82; Craig Hartsburg, 1982-83; Craig Hartsburg and Brian Bellows, 1983-84; Craig Hartsburg, 1984-85 to 1987-88; Curt Fraser, Bob Rouse and Curt Giles, 1988-89; Curt Giles, 1989-90, 1990-91; Mark Tinordi, 1991-92 to 1993-94; Neal Broten and Derian Hatcher, 1994-95; Derian Hatcher, 1995-96 to date.

Sergei Fedorov scored 31 goals last season.

Detroit Red Wings

2001-02 Results: 51W-17L-10T-4OTL 116PTS.
First, Central Division

2002-03 Schedule

Month	Day	Date	Opponent
Oct.	Thu.	10	at San Jose
	Sat.	12	at Los Angeles
	Sun.	13	at Anaheim
	Thu.	17	Montreal
	Sat.	19	at Minnesota
	Mon.	21	Calgary
	Wed.	23	Los Angeles
	Fri.	25	Pittsburgh
	Sat.	26	at Nashville
	Tue.	29	San Jose
Nov.	Sat.	2	at Ottawa
	Sun.	3	Dallas
	Tue.	5	Chicago
	Thu.	7	Boston
	Tue.	12	Nashville
	Fri.	15	Anaheim
	Sat.	16	at Toronto
	Tue.	19	at Calgary
	Fri.	22	at Vancouver
	Sat.	23	at Edmonton
	Mon.	25	Edmonton
	Wed.	27	New Jersey
	Fri.	29	at Carolina
Dec.	Sun.	1	Calgary
	Tue.	3	Anaheim
	Thu.	5	at Phoenix
	Fri.	6	at Dallas
	Sun.	8	St. Louis
	Thu.	12	Minnesota
	Sat.	14	Columbus
	Tue.	17	at NY Islanders
	Thu.	19	Dallas
	Sat.	21	NY Rangers
	Mon.	23	at Columbus
	Thu.	26	Columbus
	Sat.	28	at Nashville
	Sun.	29	at Dallas
	Tue.	31	St. Louis
Jan.	Fri.	3	Phoenix
	Sun.	5	at Chicago
	Tue.	7	at Tampa Bay
	Wed.	8	at Florida
	Sat.	11	at Philadelphia*
	Mon.	13	Chicago
	Wed.	15	at Chicago
	Thu.	16	at Colorado
	Sun.	19	Vancouver
	Wed.	22	at Edmonton
	Fri.	24	at Vancouver
	Sat.	25	at Calgary
	Tue.	28	at New Jersey
	Thu.	30	Florida
Feb.	Tue.	4	Nashville
	Thu.	6	Colorado
	Sat.	8	at Colorado*
	Mon.	10	San Jose
	Thu.	13	Buffalo
	Sat.	15	at Atlanta
	Tue.	18	Vancouver
	Thu.	20	Edmonton
	Sat.	22	at Washington
	Mon.	24	Los Angeles
	Thu.	27	Toronto
Mar.	Sun.	2	Phoenix
	Mon.	3	at Columbus
	Wed.	5	Tampa Bay
	Fri.	7	St. Louis
	Sun.	9	at Anaheim*
	Mon.	10	at Los Angeles
	Wed.	12	at Phoenix
	Sat.	15	Colorado*
	Sun.	16	Ottawa
	Tue.	18	at Pittsburgh
	Sat.	22	at St. Louis*
	Sun.	23	at Minnesota
	Tue.	25	Minnesota
	Thu.	27	at San Jose
	Sat.	29	at St. Louis*
	Mon.	31	Nashville
Apr.	Thu.	3	NY Islanders
	Fri.	4	at Columbus
	Sun.	6	at Chicago*

* Denotes afternoon game.

Franchise date: September 25, 1926

CENTRAL DIVISION

77th NHL Season

Year-by-Year Record

		Home				Road				Overall								
Season	GP	W	L	T	OL	W	L	T	OL	W	L	T	OL	GF	GA	Pts.	Finished	Playoff Result
2001-02	**82**	**28**	**7**	**5**	**1**	**23**	**10**	**5**	**3**	**51**	**17**	**10**	**4**	**251**	**187**	**116**	**1st, Central Div.**	**Won Stanley Cup**
2000-01	82	27	9	3	2	22	11	6	2	49	20	9	4	253	202	111	1st, Central Div.	Lost Conf. Quarter-Final
1999-2000	82	28	9	3	1	20	13	7	1	48	22	10	2	278	210	108	2nd, Central Div.	Lost Conf. Semi-Final
1998-99	82	27	12	2	...	16	20	5	...	43	32	7	...	245	202	93	1st, Central Div.	Lost Conf. Semi-Final
1997-98	**82**	**25**	**8**	**8**	**...**	**19**	**15**	**7**	**...**	**44**	**23**	**15**	**...**	**250**	**196**	**103**	**2nd, Central Div.**	**Won Stanley Cup**
1996-97	**82**	**20**	**12**	**9**	**...**	**18**	**14**	**9**	**...**	**38**	**26**	**18**	**...**	**253**	**197**	**94**	**2nd, Central Div.**	**Won Stanley Cup**
1995-96	82	36	3	2	...	26	10	5	...	62	13	7	...	325	181	131	1st, Central Div.	Lost Conf. Championship
1994-95	48	17	4	3	...	16	7	1	...	33	11	4	...	180	117	70	1st, Central Div.	Lost Final
1993-94	84	23	13	6	...	23	17	2	...	46	30	8	...	356	275	100	1st, Central Div.	Lost Conf. Quarter-Final
1992-93	84	25	14	3	...	22	14	6	...	47	28	9	...	369	280	103	2nd, Norris Div.	Lost Div. Semi-Final
1991-92	80	24	12	4	...	19	13	8	...	43	25	12	...	320	256	98	1st, Norris Div.	Lost Div. Final
1990-91	80	26	14	0	...	8	24	8	...	34	38	8	...	273	298	76	3rd, Norris Div.	Lost Div. Semi-Final
1989-90	80	20	14	6	...	8	24	8	...	28	38	14	...	288	323	70	5th, Norris Div.	Out of Playoffs
1988-89	80	20	14	6	...	14	20	6	...	34	34	12	...	313	316	80	1st, Norris Div.	Lost Div. Semi-Final
1987-88	80	24	10	6	...	17	18	5	...	41	28	11	...	322	269	93	1st, Norris Div.	Lost Conf. Championship
1986-87	80	20	14	6	...	14	22	4	...	34	36	10	...	260	274	78	2nd, Norris Div.	Lost Conf. Championship
1985-86	80	10	26	4	...	7	31	2	...	17	57	6	...	266	415	40	5th, Norris Div.	Out of Playoffs
1984-85	80	19	14	7	...	8	27	5	...	27	41	12	...	313	357	66	3rd, Norris Div.	Lost Div. Semi-Final
1983-84	80	18	20	2	...	13	22	5	...	31	42	7	...	298	323	69	3rd, Norris Div.	Lost Div. Semi-Final
1982-83	80	14	19	7	...	7	25	8	...	21	44	15	...	263	344	57	5th, Norris Div.	Out of Playoffs
1981-82	80	15	19	6	...	6	28	6	...	21	47	12	...	270	351	54	6th, Norris Div.	Out of Playoffs
1980-81	80	16	15	9	...	3	28	9	...	19	43	18	...	252	339	56	5th, Norris Div.	Out of Playoffs
1979-80	80	14	21	5	...	12	22	6	...	26	43	11	...	268	306	63	5th, Norris Div.	Out of Playoffs
1978-79	80	15	17	8	...	8	24	8	...	23	41	16	...	252	295	62	5th, Norris Div.	Out of Playoffs
1977-78	80	22	11	7	...	10	23	7	...	32	34	14	...	252	266	78	2nd, Norris Div.	Lost Quarter-Final
1976-77	80	12	22	6	...	4	33	3	...	16	55	9	...	183	309	41	5th, Norris Div.	Out of Playoffs
1975-76	80	17	15	8	...	9	29	2	...	26	44	10	...	226	300	62	4th, Norris Div.	Out of Playoffs
1974-75	80	17	17	6	...	6	28	6	...	23	45	12	...	259	335	58	4th, Norris Div.	Out of Playoffs
1973-74	78	21	12	6	...	8	27	4	...	29	39	10	...	255	319	68	6th, East Div.	Out of Playoffs
1972-73	78	22	12	5	...	15	17	7	...	37	29	12	...	265	243	86	5th, East Div.	Out of Playoffs
1971-72	78	25	11	3	...	8	24	7	...	33	35	10	...	261	262	76	5th, East Div.	Out of Playoffs
1970-71	78	17	15	7	...	5	30	4	...	22	45	11	...	209	308	55	7th, East Div.	Out of Playoffs
1969-70	76	20	11	7	...	20	10	8	...	40	21	15	...	246	199	95	3rd, East Div.	Lost Quarter-Final
1968-69	76	23	8	7	...	10	23	5	...	33	31	12	...	239	221	78	5th, East Div.	Out of Playoffs
1967-68	74	18	15	4	...	9	20	8	...	27	35	12	...	245	257	66	6th, East Div.	Out of Playoffs
1966-67	70	21	11	3	...	6	28	1	...	27	39	4	...	212	241	58	5th,	Out of Playoffs
1965-66	70	20	8	7	...	11	19	5	...	31	27	12	...	221	194	74	4th,	Lost Final
1964-65	70	25	7	3	...	15	16	4	...	40	23	7	...	224	175	87	1st,	Lost Semi-Final
1963-64	70	23	9	3	...	7	20	8	...	30	29	11	...	191	204	71	4th,	Lost Final
1962-63	70	19	10	6	...	13	15	7	...	32	25	13	...	200	194	77	4th,	Lost Final
1961-62	70	17	11	7	...	6	22	7	...	23	33	14	...	184	219	60	5th,	Out of Playoffs
1960-61	70	15	13	7	...	10	16	9	...	25	29	16	...	195	215	66	4th,	Lost Final
1959-60	70	18	14	3	...	8	15	12	...	26	29	15	...	186	197	67	4th,	Lost Semi-Final
1958-59	70	13	17	5	...	12	20	3	...	25	37	8	...	167	218	58	6th,	Out of Playoffs
1957-58	70	16	11	8	...	13	18	4	...	29	29	12	...	176	207	70	3rd,	Lost Semi-Final
1956-57	70	23	7	5	...	15	13	7	...	38	20	12	...	198	157	88	1st,	Lost Semi-Final
1955-56	70	21	6	8	...	9	18	8	...	30	24	16	...	183	148	76	2nd,	Lost Final
1954-55	**70**	**25**	**5**	**5**	**...**	**17**	**12**	**6**	**...**	**42**	**17**	**11**	**...**	**204**	**134**	**95**	**1st,**	**Won Stanley Cup**
1953-54	**70**	**24**	**4**	**7**	**...**	**13**	**15**	**7**	**...**	**37**	**19**	**14**	**...**	**191**	**132**	**88**	**1st,**	**Won Stanley Cup**
1952-53	70	20	5	10	...	16	11	8	...	36	16	18	...	222	133	90	1st,	Lost Semi-Final
1951-52	**70**	**24**	**7**	**4**	**...**	**20**	**7**	**8**	**...**	**44**	**14**	**12**	**...**	**215**	**133**	**100**	**1st,**	**Won Stanley Cup**
1950-51	70	25	3	7	...	19	10	6	...	44	13	13	...	236	139	101	1st,	Lost Semi-Final
1949-50	**70**	**19**	**9**	**7**	**...**	**18**	**10**	**7**	**...**	**37**	**19**	**14**	**...**	**229**	**164**	**88**	**1st,**	**Won Stanley Cup**
1948-49	60	21	6	3	...	13	13	4	...	34	19	7	...	195	145	75	1st,	Lost Final
1947-48	60	16	9	5	...	14	9	7	...	30	18	12	...	187	148	72	2nd,	Lost Final
1946-47	60	14	10	6	...	8	17	5	...	22	27	11	...	190	193	55	4th,	Lost Semi-Final
1945-46	50	16	5	4	...	4	15	6	...	20	20	10	...	146	159	50	4th,	Lost Semi-Final
1944-45	50	19	5	1	...	12	9	4	...	31	14	5	...	218	161	67	2nd,	Lost Final
1943-44	50	18	5	2	...	8	13	4	...	26	18	6	...	214	177	58	2nd,	Lost Semi-Final
1942-43	**50**	**16**	**4**	**5**	**...**	**9**	**10**	**6**	**...**	**25**	**14**	**11**	**...**	**169**	**124**	**61**	**1st,**	**Won Stanley Cup**
1941-42	48	14	7	3	...	5	18	1	...	19	25	4	...	140	147	42	5th,	Lost Final
1940-41	48	14	5	5	...	7	11	6	...	21	16	11	...	112	102	53	3rd,	Lost Final
1939-40	48	11	10	3	...	5	16	3	...	16	26	6	...	90	126	38	5th,	Lost Semi-Final
1938-39	48	14	8	2	...	4	16	4	...	18	24	6	...	107	128	42	5th,	Lost Semi-Final
1937-38	48	8	10	6	...	4	15	5	...	12	25	11	...	99	133	35	4th, Amn. Div.	Out of Playoffs
1936-37	**48**	**14**	**5**	**5**	**...**	**11**	**9**	**4**	**...**	**25**	**14**	**9**	**...**	**128**	**102**	**59**	**1st, Amn. Div.**	**Won Stanley Cup**
1935-36	**48**	**14**	**5**	**5**	**...**	**10**	**11**	**3**	**...**	**24**	**16**	**8**	**...**	**124**	**103**	**56**	**1st, Amn. Div.**	**Won Stanley Cup**
1934-35	48	11	8	5	...	8	14	2	...	19	22	7	...	127	114	45	4th, Amn. Div.	Out of Playoffs
1933-34	48	15	5	4	...	9	9	6	...	24	14	10	...	113	98	58	1st, Amn. Div.	Lost Final
1932-33*	48	17	3	4	...	8	12	4	...	25	15	8	...	111	93	58	2nd, Amn. Div.	Lost Semi-Final
1931-32	48	15	3	6	...	3	17	4	...	18	20	10	...	95	108	46	3rd, Amn. Div.	Lost Quarter-Final
1930-31**	44	10	7	5	...	6	14	2	...	16	21	7	...	102	105	39	4th, Amn. Div.	Out of Playoffs
1929-30	44	9	10	3	...	5	14	3	...	14	24	6	...	117	133	34	4th, Amn. Div.	Out of Playoffs
1928-29	44	11	6	5	...	8	10	4	...	19	16	9	...	72	63	47	3rd, Amn. Div.	Lost Quarter-Final
1927-28	44	9	10	3	...	10	9	3	...	19	19	6	...	88	79	44	4th, Amn. Div.	Out of Playoffs
1926-27***	44	5	16	0	...	7	12	4	...	12	28	4	...	76	105	28	5th, Amn. Div.	Out of Playoffs

* Team name changed to Red Wings. ** Team name changed to Falcons. *** Team named Cougars.

2002-03 Player Personnel

FORWARDS	HT	WT	S	Place of Birth	Date	2001-02 Club
ADAMS, Bryan	6-0	185	L	Fort St. James, B.C.	3/20/77	Chicago
AVERY, Sean	5-10	185	L	North York, Ont.	4/10/80	Detroit-Cincinnati
BARNES, Ryan	6-1	201	L	Dunnville, Ont.	1/30/80	Cincinnati-Toledo
DANDENAULT, Mathieu	6-0	200	R	Sherbrooke, Que.	2/3/76	Detroit
DATSYUK, Pavel	5-11	180	L	Sverdlovsk, USSR	7/20/78	Detroit-Russia
DEVEREAUX, Boyd	6-2	195	L	Seaforth, Ont.	4/16/78	Detroit
DRAPER, Kris	5-11	190	L	Toronto, Ont.	5/24/71	Detroit
FEDOROV, Sergei	6-1	200	L	Pskov, USSR	12/13/69	Detroit-Russia
HOLMSTROM, Tomas	6-0	200	L	Pitea, Sweden	1/23/73	Detroit-Sweden
HULL, Brett	5-11	203	R	Belleville, Ont.	8/9/64	Detroit-United States
KING, Derek	6-1	203	L	Hamilton, Ont.	2/11/67	Munchen
KOPECKY, Tomas	6-3	187	L	Ilava, Czech.	2/5/82	Lethbridge-Cincinnati
LARIONOV, Igor	5-9	170	L	Voskresensk, USSR	12/3/60	Detroit-Russia
MALTBY, Kirk	6-0	180	R	Guelph, Ont.	12/22/72	Detroit
McCARTY, Darren	6-1	210	R	Burnaby, B.C.	4/1/72	Detroit
MOWERS, Mark	5-11	187	R	Whitesboro, NY	2/16/74	Nashville-Milwaukee
PICARD, Michel	5-11	190	L	Beauport, Que.	11/7/69	Mannheim
ROBITAILLE, Luc	6-1	215	L	Montreal, Que.	2/17/66	Detroit
SHANAHAN, Brendan	6-3	218	R	Mimico, Ont.	1/23/69	Detroit-Canada
WILLIAMS, Jason	5-11	185	R	London, Ont.	8/11/80	Detroit-Cincinnati
YZERMAN, Steve	5-11	185	R	Cranbrook, B.C.	5/9/65	Detroit-Canada
ZETTERBERG, Henrik	5-11	176	L	Njurunda, Sweden	10/9/80	Timra-Sweden
DEFENSEMEN						
BALLANTYNE, Paul	6-3	200	R	Waterloo, Ont.	7/16/82	Sault Ste. Marie
BOILEAU, Patrick	6-0	202	R	Montreal, Que.	2/22/75	Washington-Port (AHL)
BYKOV, Dmitri	5-10	169	L	Izhevsk, USSR	5/5/77	Kazan
CAMPBELL, Eddy	6-2	212	L	Worcester, MA	11/26/74	Worcester
CHELIOS, Chris	6-1	190	R	Chicago, IL	1/25/62	Detroit-United States
FISCHER, Jiri	6-5	225	L	Horovice, Czech.	7/31/80	Detroit
GROULX, Danny	6-0	205	L	LaSalle, Que.	6/23/81	Victoriaville
KUZNETSOV, Maxim	6-5	198	L	Pavlodar, USSR	3/24/77	Detroit-Cincinnati
LIDSTROM, Nicklas	6-2	185	L	Vasteras, Sweden	4/28/70	Detroit-Sweden
WALLIN, Jesse	6-2	190	L	Saskatoon, Sask.	3/10/78	Detroit-Cincinnati
WIKSTROM, John	6-3	200	L	Lulea, Sweden	1/30/79	Toledo-Cincinnati

GOALTENDERS	HT	WT	C	Place of Birth	Date	2001-02 Club
JOSEPH, Curtis	5-11	190	L	Keswick, Ont.	4/29/67	Toronto-Canada
LAMOTHE, Marc	6-2	210	L	New Liskeard, Ont.	2/27/74	Hamilton
LEGACE, Manny	5-9	162	L	Toronto, Ont.	2/4/73	Detroit

Coach

LEWIS, DAVE

Coach, Detroit Red Wings. Born in Kindersley, Sask,. July 3, 1953.

A member of the Red Wings coaching staff since retiring as a player on November 6, 1987, Dave Lewis was officially named to replace Scotty Bowman as Detroit's head coach on July 17, 2002. In his 14 seasons as an assistant coach, Lewis worked under Jacques Demers, Bryan Murray and Bowman. He served as an associate coach alongside Barry Smith during Bowman's nine-year tenure as head coach. Lewis excelled as both a motivator and a tactician. Besides the ability to shape the young talent on the Red Wings roster, Lewis also earned the respect of the Wings veterans like Steve Yzerman, Chris Chelios, Nicklas Lidstrom and Brett Hull. His primary focus was the team's defensive corps. His other duties included extensive video work used in scouting opponents.

Lewis joined the Red Wings organization as a player when he was signed as a free agent on July 27, 1986. He played his 1,000th NHL game with Detroit on April 1, 1987. Lewis was originally selected 33rd overall by the New York Islanders in the 1973 Amateur Draft and entered the NHL for the 1973-74 season directly out of junior hockey with the Saskatoon Blades. He never played a game in the minor leagues. In all, Lewis played 1,008 games with the Islanders, Los Angeles, New Jersey and Detroit. He recorded 36 goals, 187 assists and 953 penalty minutes. He was never a Stanley Cup winner during 15 years as a player, but he helped the Red Wings win the championship three times (1997, 1998 and 2002) as an assistant coach.

Off the ice, Lewis has been actively involved with the Make-A-Wish Foundation. He has organized the Dave Lewis Detroit Red Wings Fantasy Camp and celebrity auctions to raise funds for the charitable organization.

Coaching Record

		Regular Season				Playoffs		
Season	Team	Games	W	L	T	Games	W	L
1998-99	Detroit (NHL)	5	4	1	0	....	....	....
	NHL Totals	5	4	1	0	....	....	....

Shared a 4-1-0 record with associate coach Barry Smith while serving as co-head coaches until Scotty Bowman received medical clearance and returned to coaching on October 23, 1998.

2001-02 Scoring

* - rookie

Regular Season

Pos	#	Player	Team	GP	G	A	Pts	+/–	PIM	PP	SH	GW	GT	S	%
L	14	Brendan Shanahan	DET	80	37	38	75	23	118	12	3	7	3	277	13.4
C	91	Sergei Fedorov	DET	81	31	37	68	20	36	10	0	6	0	256	12.1
R	17	Brett Hull	DET	82	30	33	63	18	35	7	1	4	2	247	12.1
D	5	Nicklas Lidstrom	DET	78	9	50	59	13	20	6	0	0	0	215	4.2
L	20	Luc Robitaille	DET	81	30	20	50	–2	38	13	0	5	2	190	15.8
C	19	Steve Yzerman	DET	52	13	35	48	11	18	5	1	5	0	104	12.5
C	8	Igor Larionov	DET	70	11	32	43	–5	50	4	0	1	0	50	22.0
D	24	Chris Chelios	DET	79	6	33	39	40	126	1	0	1	0	128	4.7
C	13	* Pavel Datsyuk	DET	70	11	24	35	4	4	2	0	1	0	79	13.9
C	33	Kris Draper	DET	82	15	15	30	26	56	0	2	3	0	137	10.9
L	96	Tomas Holmstrom	DET	69	8	18	26	–12	58	6	0	1	0	79	10.1
C	21	Boyd Devereaux	DET	79	9	16	25	9	24	0	0	2	0	116	7.8
L	18	Kirk Maltby	DET	82	9	15	24	15	40	0	1	5	0	108	8.3
D	11	Mathieu Dandenault	DET	81	8	12	20	–5	44	2	0	3	0	97	8.2
D	28	Steve Duchesne	DET	64	3	15	18	3	28	1	0	2	0	70	4.3
D	27	Fredrik Olausson	DET	47	2	13	15	9	22	0	1	1	0	61	3.3
R	25	Darren McCarty	DET	62	5	7	12	2	98	0	0	1	2	74	6.8
C	29	* Jason Williams	DET	25	8	2	10	2	4	4	0	0	0	32	25.0
D	2	Jiri Fischer	DET	80	2	8	10	17	67	0	0	1	0	103	1.9
D	71	Jiri Slegr	ATL	38	3	5	8	–21	51	1	0	0	0	56	5.4
			DET	8	0	1	1	1	8	0	0	0	0	11	0.0
			TOTAL	46	3	6	9	–20	59	1	0	0	0	67	4.5
C	42	* Sean Avery	DET	36	2	2	4	1	68	0	0	1	0	30	6.7
D	32	* Maxim Kuznetsov	DET	39	1	2	3	0	40	0	0	0	0	27	3.7
D	4	Uwe Krupp	DET	8	0	1	1	–1	8	0	0	0	0	9	0.0
D	3	* Jesse Wallin	DET	15	0	1	1	–1	13	0	0	0	0	8	0.0
R	15	Ladislav Kohn	DET	4	0	0	0	0	4	0	0	0	0	2	0.0

Goaltending

No.	Goaltender	GPI	Mins	Avg	W	L	T	EN	SO	GA	SA	S%	G	A	PIM
39	Dominik Hasek	65	3872	2.17	41	15	8	2	5	140	1654	.915	0	1	8
34	Manny Legace	20	1117	2.42	10	6	2	0	1	45	503	.911	0	1	0
	Totals	**82**	**5008**	**2.24**	**51**	**21**	**10**	**2**	**7**	**187**	**2159**	**.913**			

Playoffs

Pos	#	Player	Team	GP	G	A	Pts	+/–	PIM	PP	SH	GW	GT	S	%
C	19	Steve Yzerman	DET	23	6	17	23	4	10	4	0	2	0	52	11.5
L	14	Brendan Shanahan	DET	23	8	11	19	5	20	1	0	2	0	78	10.3
C	91	Sergei Fedorov	DET	23	5	14	19	4	20	2	1	0	0	88	5.7
R	17	Brett Hull	DET	23	10	8	18	1	4	3	2	2	0	61	16.4
D	5	Nicklas Lidstrom	DET	23	5	11	16	6	2	2	1	2	0	41	12.2
D	24	Chris Chelios	DET	23	1	13	14	15	44	1	0	0	0	28	3.6
L	96	Tomas Holmstrom	DET	23	8	3	11	7	8	3	0	2	0	34	23.5
C	8	Igor Larionov	DET	18	5	6	11	5	4	0	0	1	1	24	20.8
L	20	Luc Robitaille	DET	23	4	5	9	4	10	1	0	1	0	43	9.3
R	25	Darren McCarty	DET	23	4	4	8	5	34	0	0	1	0	26	15.4
C	13	* Pavel Datsyuk	DET	21	3	3	6	1	2	1	0	1	0	20	15.0
D	2	Jiri Fischer	DET	22	3	3	6	6	30	0	0	1	0	24	12.5
L	18	Kirk Maltby	DET	23	3	3	6	7	32	0	2	0	0	36	8.3
D	27	Fredrik Olausson	DET	21	2	4	6	3	10	1	0	1	1	23	8.7
C	21	Boyd Devereaux	DET	21	2	4	6	5	4	0	0	0	0	29	6.9
D	28	Steve Duchesne	DET	23	0	6	6	6	24	0	0	0	0	15	0.0
C	33	Kris Draper	DET	23	2	3	5	4	20	0	0	0	0	44	4.5
D	11	Mathieu Dandenault	DET	23	1	2	3	7	8	0	1	0	0	11	9.1
D	71	Jiri Slegr	DET	1	0	0	0	2	2	0	0	0	0	2	0.0
D	4	Uwe Krupp	DET	2	0	0	0	–5	2	0	0	0	0	1	0.0
C	29	* Jason Williams	DET	9	0	0	0	–1	2	0	0	0	0	7	0.0

Goaltending

No.	Goaltender	GPI	Mins	Avg	W	L	EN	SO	GA	SA	S%	G	A	PIM
39	Dominik Hasek	23	1455	1.86	16	7	1	6	45	562	.920	0	1	8
34	Manny Legace	1	11	5.45	0	0	0	0	1	2	.500	0	0	0
	Totals	**23**	**1471**	**1.92**	**16**	**7**	**1**	**6**	**47**	**565**	**.917**			

Coaching History

Art Duncan, 1926-27; Jack Adams, 1927-28 to 1946-47; Tommy Ivan, 1947-48 to 1953-54; Jimmy Skinner, 1954-55 to 1956-57; Jimmy Skinner and Sid Abel, 1957-58; Sid Abel, 1958-59 to 1967-68; Bill Gadsby, 1968-69; Bill Gadsby and Sid Abel, 1969-70; Ned Harkness and Doug Barkley, 1970-71; Doug Barkley and Johnny Wilson, 1971-72; Johnny Wilson, 1972-73; Ted Garvin and Alex Delvecchio, 1973-74; Alex Delvecchio, 1974-75; Doug Barkley and Alex Delvecchio, 1975-76; Alex Delvecchio and Larry Wilson, 1976-77; Bobby Kromm, 1977-78, 1978-79; Bobby Kromm and Ted Lindsay, 1979-80; Ted Lindsay and Wayne Maxner, 1980-81; Wayne Maxner and Billy Dea, 1981-82; Nick Polano, 1982-83 to 1984-85; Harry Neale and Brad Park, 1985-86; Jacques Demers, 1986-87 to 1989-90; Bryan Murray, 1990-91 to 1992-93; Scotty Bowman, 1993-94 to 1997-98; Dave Lewis, Barry Smith (co-coaches) and Scotty Bowman, 1998-99; Scotty Bowman, 1999-2000 to 2001-02; Dave Lewis, 2002-03.

Club Records

Team

(Figures in brackets for season records are games played; records for fewest points, wins, ties, losses, goals, goals against are for 70 or more games)

Most Points .... 131 — 1995-96 (82)
Most Wins .... *62 — 1995-96 (82)
Most Ties .... 18 — 1952-53 (70), 1980-81 (80), 1996-97 (82)
Most Losses .... 57 — 1985-86 (80)
Most Goals .... 369 — 1992-93 (84)
Most Goals Against .... 415 — 1985-86 (80)
Fewest Points .... 40 — 1985-86 (80)
Fewest Wins .... 16 — 1976-77 (80)
Fewest Ties .... 4 — 1966-67 (70)
Fewest Losses .... 13 — 1950-51 (70), 1995-96 (82)
Fewest Goals .... 167 — 1958-59 (70)
Fewest Goals Against .... 132 — 1953-54 (70)
Longest Winning Streak
Overall .... 9 — Mar. 3-21/51, Feb. 27-Mar. 20/55, Dec. 12-31/95, Mar. 3-22/96
Home .... 14 — Jan. 21-Mar. 25/65
Away .... 7 — Mar. 25-Apr. 14/95, Feb. 18-Mar. 20/96
Longest Undefeated Streak
Overall .... 15 — Nov. 27-Dec. 28/52 (8 wins, 7 ties)
Home .... 19 — Dec. 31/00-Apr.7/01 (17 wins, 2 ties)
Away .... 15 — Oct. 18-Dec. 20/51 (10 wins, 5 ties)
Longest Losing Streak
Overall .... 14 — Feb. 24-Mar. 25/82
Home .... 7 — Feb. 20-Mar. 25/82
Away .... 14 — Oct. 19-Dec. 21/66
Longest Winless Streak
Overall .... 19 — Feb. 26-Apr. 3/77 (18 losses, 1 tie)
Home .... 10 — Dec. 11/85-Jan. 18/86 (9 losses, 1 tie)
Away .... 26 — Dec. 15/76-Apr. 3/77 (23 losses, 3 ties)
Most Shutouts, Season .... 13 — 1953-54 (70)
Most. PIM, Season .... 2,393 — 1985-86 (80)
Most Goals, Game .... 15 — Jan. 23/44 (NYR 0 at Det. 15)

Individual

Most Seasons .... 25 — Gordie Howe
Most Games .... 1,687 — Gordie Howe
Most Goals, Career .... 786 — Gordie Howe
Most Assists, Career ...... 1,023 — Gordie Howe
Most Points, Career .... 1,809 — Gordie Howe (786G, 1,023A)
Most PIM, Career .... 2,090 — Bob Probert
Most Shutouts, Career .... 85 — Terry Sawchuk
Longest Consecutive Games Streak .... 548 — Alex Delvecchio (Dec. 13/56-Nov. 11/64)
Most Goals, Season .... 65 — Steve Yzerman (1988-89)
Most Assists, Season .... 90 — Steve Yzerman (1988-89)
Most Points, Season .... 155 — Steve Yzerman (1988-89; 65G, 90A)
Most PIM, Season .... 398 — Bob Probert (1987-88)
Most Points, Defenseman, Season .... 77 — Paul Coffey (1993-94; 14G, 63A)
Most Points, Center, Season .... 155 — Steve Yzerman (1988-89; 65G, 90A)
Most Points, Right Wing, Season .... 103 — Gordie Howe (1968-69; 44G, 59A)
Most Points, Left Wing, Season .... 105 — John Ogrodnick (1984-85; 55G, 50A)
Most Points, Rookie, Season .... 87 — Steve Yzerman (1983-84; 39G, 48A)
Most Shutouts, Season .... 12 — Terry Sawchuk (1951-52, 1953-54, 1954-55), Glenn Hall (1955-56)
Most Goals, Game .... 6 — Syd Howe (Feb. 3/44)
Most Assists, Game .... *7 — Billy Taylor (Mar. 16/47)
Most Points, Game .... 7 — Carl Liscombe (Nov. 5/42; 3G, 4A), Don Grosso (Feb. 3/44; 1G, 6A), Billy Taylor (Mar. 16/47; 7A)

* NHL Record.

Retired Numbers

No.	Player	Years
1	Terry Sawchuk	1949-55, 57-64, 68-69
7	Ted Lindsay	1944-57, 64-65
9	Gordie Howe	1946-1971
10	Alex Delvecchio	1951-1973
12	Sid Abel	1938-43, 45-52

All-time Record vs. Other Clubs

Regular Season

	At Home								On Road								Total							
	GP	W	L	T	OL	GF	GA	PTS	GP	W	L	T	OL	GF	GA	PTS	GP	W	L	T	OL	GF	GA	PTS
Anaheim	18	13	2	3	0	69	44	29	18	10	5	3	0	52	36	23	36	23	7	6	0	121	80	52
Atlanta	3	3	0	0	0	11	3	6	2	2	0	0	0	13	5	4	5	5	0	0	0	24	8	10
Boston	283	153	78	52	0	950	716	358	286	90	153	43	0	760	1005	223	569	243	231	95	0	1710	1721	581
Buffalo	54	31	18	5	0	199	157	67	51	10	33	8	0	146	224	28	105	41	51	13	0	345	381	95
Calgary	55	28	17	10	0	204	167	66	56	17	33	6	0	168	223	40	111	45	50	16	0	372	390	106
Carolina	30	17	7	6	0	115	82	40	28	11	16	1	0	78	97	23	58	28	23	7	0	193	179	63
Chicago	330	199	97	33	1	1129	823	432	332	130	150	51	1	936	994	312	662	329	247	84	2	2065	1817	744
Colorado	36	22	13	1	0	133	111	45	38	15	19	4	0	129	138	34	74	37	32	5	0	262	249	79
Columbus	5	4	1	0	0	19	10	8	5	4	1	0	0	10	7	8	10	8	2	0	0	29	17	16
Dallas	102	51	37	14	0	395	332	116	102	36	51	15	0	306	361	87	204	87	88	29	0	701	693	203
Edmonton	40	22	15	3	0	153	137	47	40	14	19	7	0	146	158	35	80	36	34	10	0	299	295	82
Florida	6	3	1	2	0	24	18	8	8	5	1	2	0	25	16	12	14	8	2	4	0	49	34	20
Los Angeles	75	33	30	12	0	286	260	78	76	21	40	14	1	232	310	57	151	54	70	26	1	518	570	135
Minnesota	4	3	1	0	0	19	12	6	4	2	0	1	1	10	8	6	8	5	1	1	1	29	20	12
Montreal	279	130	96	53	0	803	714	313	281	67	171	43	0	635	992	177	560	197	267	96	0	1438	1706	490
Nashville	11	9	0	1	1	45	26	20	11	5	3	2	1	33	28	13	22	14	3	3	2	78	54	33
New Jersey	38	23	13	2	0	156	125	48	39	10	20	9	0	103	137	29	77	33	33	11	0	259	262	77
NY Islanders	43	24	17	2	0	154	133	50	45	19	23	3	0	135	162	41	88	43	40	5	0	289	295	91
NY Rangers	284	163	76	45	0	1001	697	371	283	92	133	58	0	737	865	242	567	255	209	103	0	1738	1562	613
Ottawa	8	5	3	0	0	27	17	10	8	5	2	1	0	24	21	11	16	10	5	1	0	51	38	21
Philadelphia	58	30	18	10	0	206	180	70	57	13	33	11	0	166	227	37	115	43	51	21	0	372	407	107
Phoenix	47	23	17	7	0	186	159	53	45	16	16	13	0	142	138	45	92	39	33	20	0	328	297	98
Pittsburgh	64	39	13	12	0	246	175	90	63	16	43	4	0	187	276	36	127	55	56	16	0	433	451	126
St. Louis	104	47	40	17	0	376	322	111	105	30	54	19	2	289	364	81	209	77	94	36	2	665	686	192
San Jose	21	18	2	1	0	89	35	37	22	13	6	3	0	91	70	29	43	31	8	4	0	180	105	66
Tampa Bay	10	9	1	0	0	43	18	18	13	9	3	1	0	61	42	19	23	18	4	1	0	104	60	37
Toronto	321	167	106	46	2	961	790	382	314	104	163	47	0	842	1039	255	635	271	269	93	2	1803	1829	637
Vancouver	62	39	15	8	0	265	177	86	61	25	26	10	0	201	220	60	123	64	41	18	0	466	397	146
Washington	46	21	14	11	0	160	131	53	44	18	21	5	0	137	166	41	90	39	35	16	0	297	297	94
Defunct Clubs	141	76	40	25	0	430	307	177	141	49	63	29	0	364	375	127	282	125	103	54	0	794	682	304
Totals	**2578**	**1405**	**788**	**381**	**4**	**8854**	**6878**	**3195**	**2578**	**858**	**1301**	**413**	**6**	**7158**	**8704**	**2135**	**5156**	**2263**	**2089**	**794**	**10**	**16012**	**15582**	**5330**

Playoffs

	Series	W	L	GP	W	L	T	GF	GA	Last Mtg.	Rnd.	Result
Anaheim	2	2	0	8	8	0	0	30	14	1999	CQF	W 4-0
Boston	7	3	4	33	14	19	0	98	96	1957	SF	L 1-4
Calgary	1	1	0	2	2	0	0	8	5	1978	PRE	W 2-0
Carolina	1	1	0	5	4	1	0	14	7	2002	F	W 4-1
Chicago	14	6	8	69	31	38	0	190	210	1995	CF	W 4-1
Colorado	5	2	3	30	13	17	0	76	79	2002	CF	W 4-3
Dallas	3	3	0	18	12	6	0	55	40	1998	CF	W 4-2
Edmonton	2	0	2	10	2	8	0	26	39	1988	CF	L 1-4
Los Angeles	2	1	1	10	6	4	0	32	21	2001	CQF	L 2-4
Montreal	12	7	5	62	29	33	0	149	161	1978	QF	L 1-4
New Jersey	1	0	1	4	0	4	0	7	16	1995	F	L 0-4
NY Rangers	5	4	1	23	13	10	0	57	49	1950	F	W 4-3
Philadelphia	1	1	0	4	4	0	0	16	6	1997	F	W 4-0
Phoenix	2	2	0	12	8	4	0	44	28	1998	CQF	W 4-2
St. Louis	7	5	2	40	24	16	0	125	103	2002	CSF	W 4-1
San Jose	2	1	1	11	7	4	0	51	27	1995	CSF	W 4-0
Toronto	23	11	12	117	59	58	0	321	311	1993	DSF	L 3-4
Vancouver	1	1	0	6	4	2	0	22	16	2002	CQF	W 4-2
Washington	1	1	0	4	4	0	0	13	7	1998	F	W 4-0
Defunct Clubs	4	3	1	10	7	2	1	21	13			
Totals	**96**	**55**	**41**	**478**	**251**	**226**	**1**	**1355**	**1248**			

Playoff Results 2002-1998

Year	Round	Opponent	Result	GF	GA
2002	**F**	**Carolina**	**W 4-1**	**14**	**7**
	CF	Colorado	W 4-3	22	13
	CSF	St. Louis	W 4-1	14	11
	CQF	Vancouver	W 4-2	22	16
2001	CQF	Los Angeles	L 2-4	17	15
2000	CSF	Colorado	L 1-4	8	13
	CQF	Los Angeles	W 4-0	15	6
1999	CSF	Colorado	L 2-4	14	21
	CQF	Anaheim	W 4-0	17	6
1998	**F**	**Washington**	**W 4-0**	**13**	**7**
	CF	Dallas	W 4-2	15	11
	CSF	St. Louis	W 4-2	23	13
	CQF	Phoenix	W 4-2	24	18

Abbreviations: Round: F - Final; **CF** - conference final; **CSF** - conference semi-final; **CQF** - conference quarter-final; **DSF** - division semi-final; **SF** - semi-final; **QF** - quarter-final; **PRE** - preliminary round.

Calgary totals include Atlanta Flames, 1972-73 to 1979-80. Carolina totals include Hartford, 1979-80 to 1996-97.
Colorado totals include Quebec, 1979-80 to 1994-95. Dallas totals include Minnesota North Stars, 1967-68 to 1992-93.
New Jersey totals include Kansas City, 1974-75 to 1975-76, and Colorado Rockies, 1976-77 to 1981-82.
Phoenix totals include Winnipeg, 1979-80 to 1995-96.

2001-02 Results

Month	Day	Opponent	Score	Month	Day	Opponent	Score
Oct.	4	at San Jose	4-3*	**Jan.**	2	Anaheim	5-3
	6	at Vancouver	4-1		5	Colorado	3-1
	10	Calgary	2-4		9	Vancouver	5-4*
	12	Buffalo	4-2		12	Dallas	5-2
	13	at NY Islanders	5-4*		15	at Phoenix	2-2
	16	Columbus	4-3		16	at Dallas	2-3
	18	Philadelphia	3-2		18	Washington	3-1
	20	Los Angeles	3-2		20	Ottawa	3-2*
	24	Edmonton	4-1		23	San Jose	2-2
	26	Dallas	3-5		25	Phoenix	4-1
	27	at Nashville	1-0		26	at St. Louis	5-2
	30	at Carolina	5-2		28	at Edmonton	1-1
	31	at Dallas	4-3*		30	at Calgary	3-4
Nov.	2	NY Islanders	2-1	**Feb.**	4	at Colorado	3-1
	4	at Chicago	4-5		6	NY Rangers	3-1
	7	at Phoenix	3-1		8	Columbus	2-3
	9	at Anaheim	1-0		9	at Ottawa	3-2
	10	at Los Angeles	2-3*		11	at Montreal	3-2
	13	Carolina	4-3		13	at Minnesota	2-0
	16	Minnesota	8-3		26	at Tampa Bay	4-3*
	17	Los Angeles	4-2		27	at Florida	3-2*
	20	Nashville	6-3	**Mar.**	2	at Pittsburgh	4-2
	21	at Columbus	1-0*		6	Toronto	6-2
	23	St. Louis	3-1		9	at St. Louis	5-2
	25	Chicago	4-4		10	at Buffalo	1-5
	27	Calgary	4-2		13	Edmonton	4-3*
	30	New Jersey	4-2		16	at Boston	1-2
Dec.	1	at New Jersey	1-4		17	at NY Rangers	5-3
	5	Colorado	1-4		19	Anaheim	1-2
	7	Phoenix	1-1		21	at Columbus	3-2*
	10	at Calgary	0-2		23	at Colorado	2-0
	13	at Edmonton	2-1		25	at Nashville	3-3
	15	at Vancouver	0-3		28	Nashville	3-3
	17	Chicago	0-2		30	Atlanta	4-1
	19	Vancouver	4-1	**Apr.**	1	Toronto	4-5*
	21	San Jose	3-0		3	at Anaheim	1-1
	23	at Chicago	5-0		4	at Los Angeles	0-3
	26	at Minnesota	3-3		6	at San Jose	3-6
	27	Columbus	5-1		10	Chicago	3-3
	29	at Nashville	2-3*		13	at St. Louis	2-3*
	31	Minnesota	4-2		14	St. Louis	3-5

* – Overtime

Entry Draft Selections 2002-1988

2002

Pick	
58	Jiri Hudler
63	Tomas Fleischmann
95	Valtteri Filppula
131	Johan Berggren
166	Logan Koopmans
197	James Cuddihy
229	Derek Meech
260	Pierre-Olivier Beaulieu
262	Christian Soderstrom
291	Jonathan Ericsson

2001

Pick	
62	Igor Grigorenko
121	Drew MacIntyre
129	Miroslav Blatak
157	Andreas Jamtin
195	Nick Pannoni
258	Dmitri Bykov
288	Francois Senez

2000

Pick	
29	Niklas Kronwall
38	Tomas Kopecky
102	Stefan Liv
127	Dmitri Semenov
128	Alexander Seluyanov
130	Aaron Van Leusen
187	Per Backer
196	Paul Ballantyne
228	Jimmie Svensson
251	Todd Jackson
260	Yevgeny Bumagin

1999

Pick	
120	Jari Tolsa
149	Andrei Maximenko
181	Kent McDonell
210	Henrik Zetterberg
238	Anton Borodkin
266	Ken Davis

1998

Pick	
25	Jiri Fischer
55	Ryan Barnes
56	Tomek Valtonen
84	Jake McCracken
111	Brent Hobday
142	Calle Steen
151	Adam DeLeeuw
171	Pavel Datsyuk
198	Jeremy Goetzinger
226	David Petrasek
256	Petja Pietilainen

1997

Pick	
49	Yuri Butsayev
76	Petr Sykora
102	Quintin Laing
129	John Wikstrom
157	B.J. Young
186	Mike Laceby
213	Steve Willejto
239	Greg Willers

1996

Pick	
26	Jesse Wallin
52	Aren Miller
108	Johan Forsander
135	Michal Podolka
144	Magnus Nilsson
162	Alexandre Jacques
189	Colin Beardsmore
215	Craig Stahl
241	Eugeny Afanasiev

1995

Pick	
26	Maxim Kuznetsov
52	Philippe Audet
58	Darryl Laplante
104	Anatoli Ustyugov
125	Chad Wilchynski
126	David Arsenault
156	Tyler Perry
182	Per Eklund
208	Andrei Samokhvalov
234	David Engblom

1994

Pick	
23	Yan Golubovsky
49	Mathieu Dandenault
75	Sean Gillam
114	Frederic Deschenes
127	Doug Battaglia
153	Pavel Agarkov
205	Jason Elliot
231	Jeff Mikesch
257	Tomas Holmstrom
283	Toivo Suursoo

1993

Pick	
5	Benoit Larose
22	Anders Eriksson
48	Jon Coleman
74	Kevin Hilton
97	John Jakopin
126	Norm Maracle
152	Tim Spitzig
178	Yuri Yeresko
204	Vitezslav Skuta
230	Ryan Shanahan
256	James Kosecki
282	Gordon Hunt

1992

Pick	
22	Curtis Bowen
46	Darren McCarty
70	Sylvain Cloutier
118	Mike Sullivan
142	Jason MacDonald
166	Greg Scott
183	Justin Krall
189	C. J. Denomme
214	Jeff Walker
238	Dan McGillis
262	Ryan Bach

1991

Pick	
10	Martin Lapointe
32	Jamie Pushor
54	Chris Osgood
76	Mike Knuble
98	Dimitri Motkov
142	Igor Malykhin
186	Jim Bermingham
208	Jason Firth
230	Bart Turner
252	Andrew Miller

1990

Pick	
3	Keith Primeau
45	Vyacheslav Kozlov
66	Stewart Malgunas
87	Tony Burns
108	Claude Barthe
129	Jason York
150	Wes McCauley
171	Anthony Gruba
192	Travis Tucker
213	Brett Larson
234	John Hendry

1989

Pick	
11	Mike Sillinger
32	Bob Boughner
53	Nicklas Lidstrom
74	Sergei Fedorov
95	Shawn McCosh
116	Dallas Drake
137	Scott Zygulski
158	Andy Suhy
179	Bob Jones
200	Greg Bignell
204	Rick Judson
221	Vladimir Konstantinov
242	Joseph Frederick
246	Jason Glickman

1988

Pick	
17	Kory Kocur
38	Serge Anglehart
47	Guy Dupuis
59	Petr Hrbek
80	Sheldon Kennedy
143	Kelly Hurd
164	Brian McCormack
185	Jody Praznik
206	Glen Goodall
227	Darren Colbourne
248	Donald Stone

Club Directory

Joe Louis Arena

Detroit Red Wings
Joe Louis Arena
600 Civic Center Drive
Detroit, MI 48226
Phone **313/396-7544**
FAX PR: 313/567-0296
Media Hotline: 313/396-7599
www.detroitredwings.com
Capacity: 20,053

Owner/Governor	Mike Ilitch
Owner/Secretary-Treasurer	Marian Ilitch
Senior Vice-President/Alternate Governor	Jim Devellano
Vice-President, Red Wings/President, Ilitch Holdings, Inc./Alternate Governor	Christopher Ilitch
President, Ilitch Holdings, Inc./Alternate Governor	Denise Ilitch
General Counsel	Rob Carr
General Manager/Alternate Governor	Ken Holland
Assistant General Manager	Jim Nill
Head Coach	Dave Lewis
Associate Coach	Barry Smith
Assistant Coach	Joe Kocur
Goaltending Coach	Jim Bedard
NHL Scout	Dan Belisle
Pro Scout	Mark Howe
Pro Scout	Bob McCammon
Amateur Scout	Glenn Merkosky
Amateur Scout	Joe McDonnell
Amateur Scout	Bruce Haralson
Amateur Scout	Mark Leach
Director of European Scouting	Hakan Andersson
European Scout	Vladimir Havluj
Part-Time European Scout	Evgeni Erfilov
Part-Time Scout	Marty Stein
Senior Director of Finance	Paul MacDonald
Executive Assistant	Nancy Beard
Administrative and Scouting Coordinator	David Kolb
Accounting Assistant	Bridget Merritt
Athletic Therapist	Piet Van Zant
Assistant Athletic Therapist	Russ Baumann
Equipment Manager	Paul Boyer
Assistant Equipment Manager	Tim Abbott
Senior Director of Communications	John Hahn
Media Relations Manager	Michael Kuta
Community Relations Manager	Anne Marie Krappmann
Public Relations Assistant	Jennie Hagler
Team Physicians	John Finley, D.O.; David Collon, M.D.
Team Dentist	C.J. Regula, D.M.D.
Radio Broadcasters, Team 1270 WXYT	Ken Kal, Paul Woods
Television Broadcasters, WKBD UPN-50 & FOX Sports Net Detroit	Ken Daniels, Mickey Redmond

General Manager

HOLLAND, KEN
General Manager, Detroit Red Wings. Born in Vernon, B.C., Nov. 10, 1955.

Ken Holland is entering his sixth season as a general manager and his 20th year with the Red Wings organization. In his five seasons as Detroit's general manager, Holland has established himself as one of the most innovative and aggressive GMs in the National Hockey League. Detroit's Stanley Cup victory in 2002 marked the team's second championship under his leadership. Holland began his tenure as the club's general manager after serving as assistant general manager for the previous three seasons. He was elevated to his present position July 18, 1997.

Holland oversees all aspects of hockey operations including all matters relating to player personnel, development, contract negotiations and player movements. He also continues to be Detroit's point person at the NHL Entry Draft, as he has been for the past 12 years.

Holland has deftly handled several different front-office duties for the club over the past 19 years. At the conclusion of his playing days as a goaltender, spending most of his pro career at the American Hockey League level, Holland began his off-ice career in 1985 as a western Canada scout followed by five years as amateur scouting director before promotions leading to his current position as general manager.

A native of Vernon, BC, Holland played in the junior ranks for Medicine Hat (WHL) in 1974-75. He was Toronto's 13th pick (188th overall) in the 1975 draft but never saw action with the Maple Leafs. Holland twice signed with NHL teams as a free agent — in 1980 with Hartford and 1983 with Detroit. He spent most of his pro career with AHL clubs in Binghamton and Springfield, along with Adirondack, but did appear in four NHL games, making his debut with Hartford in 1980-81 and playing three contests for Detroit in 1983-84.

General Managers' History

Art Duncan and Duke Keats, 1926-27; Jack Adams, 1927-28 to 1961-62; Sid Abel, 1962-63 to 1969-70; Sid Abel and Ned Harkness, 1970-71; Ned Harkness, 1971-72 to 1973-74; Alex Delvecchio, 1974-75, 1975-76; Alex Delvecchio and Ted Lindsay, 1976-77; Ted Lindsay, 1977-78 to 1979-80; Jimmy Skinner, 1980-81, 1981-82; Jim Devellano, 1982-83 to 1989-90; Bryan Murray, 1990-91 to 1993-94; Jim Devellano (Senior Vice President), 1994-95 to 1996-97; Ken Holland, 1997-98 to date.

Captains' History

Art Duncan, 1926-27; Reg Noble, 1927-28 to 1929-30; George Hay, 1930-31; Carson Cooper, 1931-32; Larry Aurie, 1932-33; Herbie Lewis, 1933-34; Ebbie Goodfellow, 1934-35; Doug Young, 1935-36 to 1937-38; Ebbie Goodfellow, 1938-39 to 1940-41; Ebbie Goodfellow and Syd Howe, 1941-42; Sid Abel, 1942-43; Mud Bruneteau, Flash Hollett (co-captains), 1943-44; Flash Hollett, 1944-45; Flash Hollett and Sid Abel, 1945-46; Sid Abel, 1946-47 to 1951-52; Ted Lindsay, 1952-53 to 1955-56; Red Kelly, 1956-57, 1957-58; Gordie Howe, 1958-59 to 1961-62; Alex Delvecchio, 1962-63 to 1972-73; Alex Delvecchio, Nick Libett, Red Berenson, Gary Bergman, Ted Harris, Mickey Redmond and Larry Johnston, 1973-74; Marcel Dionne, 1974-75; Danny Grant and Terry Harper, 1975-76; Danny Grant and Dennis Polonich, 1976-77; Dan Maloney and Dennis Hextall, 1977-78; Dennis Hextall, Nick Libett and Paul Woods, 1978-79; Dale McCourt, 1979-80; Errol Thompson and Reed Larson, 1980-81; Reed Larson, 1981-82; Danny Gare, 1982-83 to 1985-86; Steve Yzerman, 1986-87 to date.

Edmonton Oilers

2001-02 Results: 38W-28L-12T-4OTL 92PTS.
Third, Northwest Division

2002-03 Schedule

Oct.	Thu.	10	Philadelphia
	Sat.	12	at Nashville
	Tue.	15	at Dallas
	Thu.	17	at San Jose
	Sat.	19	Boston
	Tue.	22	at Colorado
	Thu.	24	St. Louis
	Sat.	26	Anaheim
	Mon.	28	Dallas
Nov.	Fri.	1	Buffalo
	Sun.	3	at Chicago
	Tue.	5	at NY Rangers
	Fri.	8	at NY Islanders
	Sat.	9	at New Jersey
	Mon.	11	at Boston*
	Tue.	12	at Minnesota
	Fri.	15	St. Louis
	Sat.	16	Los Angeles
	Tue.	19	Chicago
	Thu.	21	at Calgary
	Sat.	23	Detroit
	Mon.	25	at Detroit
	Wed.	27	at Columbus
	Sat.	30	Colorado
Dec.	Tue.	3	Minnesota
	Thu.	5	at Tampa Bay
	Sat.	7	at Florida
	Sun.	8	at Atlanta*
	Wed.	11	Carolina
	Fri.	13	Colorado
	Sat.	14	Vancouver
	Tue.	17	at Minnesota
	Thu.	19	at Colorado
	Sat.	21	at Vancouver
	Thu.	26	Vancouver
	Sat.	28	Toronto
	Mon.	30	at Phoenix
	Tue.	31	at Dallas
Jan.	Thu.	2	Minnesota
	Sat.	4	Montreal
	Mon.	6	at San Jose
	Wed.	8	at Anaheim
	Thu.	9	at Los Angeles
	Sat.	11	Ottawa
	Mon.	13	Columbus
	Thu.	16	Los Angeles
	Sat.	18	Nashville
	Mon.	20	at Calgary
	Wed.	22	Detroit
	Fri.	24	Phoenix
	Wed.	29	Minnesota
	Thu.	30	at Vancouver
Feb.	Wed.	5	Anaheim
	Fri.	7	Calgary
	Sat.	8	Chicago
	Tue.	11	at Toronto
	Thu.	13	at Ottawa
	Sat.	15	at Montreal*
	Tue.	18	at Pittsburgh
	Thu.	20	at Detroit
	Sat.	22	Vancouver
	Sun.	23	Atlanta
	Tue.	25	at Colorado
	Thu.	27	at St. Louis
Mar.	Sat.	1	at Columbus
	Tue.	4	San Jose
	Thu.	6	at Los Angeles
	Fri.	7	at Anaheim
	Mon.	10	Toronto
	Tue.	11	at Calgary
	Thu.	13	NY Islanders
	Sat.	15	Dallas
	Mon.	17	at Nashville
	Thu.	20	at Phoenix
	Sat.	22	Washington
	Sun.	23	Nashville
	Wed.	26	Phoenix
	Fri.	28	Columbus
	Sun.	30	at Chicago*
	Mon.	31	at St. Louis
Apr.	Thu.	3	San Jose
	Sat.	5	Calgary

* Denotes afternoon game.

Year-by-Year Record

		Home				Road				Overall								
Season	GP	W	L	T	OL	W	L	T	OL	W	L	T	OL	GF	GA	Pts.	Finished	Playoff Result
2001-02	82	23	14	4	0	15	14	8	4	38	28	12	4	205	182	92	3rd, Northwest Div.	Out of Playoffs
2000-01	82	23	9	7	2	16	19	5	1	39	28	12	3	243	222	93	2nd, Northwest Div.	Lost Conf. Quarter-Final
1999-2000	82	18	11	9	3	14	15	7	5	32	26	16	8	226	212	88	2nd, Northwest Div.	Lost Conf. Quarter-Final
1998-99	82	17	19	5	...	16	18	7	...	33	37	12	...	230	226	78	2nd, Northwest Div.	Lost Conf. Quarter-Final
1997-98	82	20	16	5	...	15	21	5	...	35	37	10	...	215	224	80	3rd, Pacific Div.	Lost Conf. Semi-Final
1996-97	82	21	16	4	...	15	21	5	...	36	37	9	...	252	247	81	3rd, Pacific Div.	Lost Conf. Semi-Final
1995-96	82	15	21	5	...	15	23	3	...	30	44	8	...	240	304	68	5th, Pacific Div.	Out of Playoffs
1994-95	48	11	12	1	...	6	15	3	...	17	27	4	...	136	183	38	5th, Pacific Div.	Out of Playoffs
1993-94	84	17	22	3	...	8	23	11	...	25	45	14	...	261	305	64	6th, Pacific Div.	Out of Playoffs
1992-93	84	16	21	5	...	10	29	3	...	26	50	8	...	242	337	60	5th, Smythe Div.	Out of Playoffs
1991-92	80	22	13	5	...	14	21	5	...	36	34	10	...	295	297	82	3rd, Smythe Div.	Lost Conf. Championship
1990-91	80	22	15	3	...	15	22	3	...	37	37	6	...	272	272	80	3rd, Smythe Div.	Lost Conf. Championship
1989-90	**80**	**23**	**11**	**6**	**...**	**15**	**17**	**8**	**...**	**38**	**28**	**14**	**...**	**315**	**283**	**90**	**2nd, Smythe Div.**	**Won Stanley Cup**
1988-89	80	21	16	3	...	17	18	5	...	38	34	8	...	325	306	84	3rd, Smythe Div.	Lost Div. Semi-Final
1987-88	**80**	**28**	**8**	**4**	**...**	**16**	**17**	**7**	**...**	**44**	**25**	**11**	**...**	**363**	**288**	**99**	**2nd, Smythe Div.**	**Won Stanley Cup**
1986-87	**80**	**29**	**6**	**5**	**...**	**21**	**18**	**1**	**...**	**50**	**24**	**6**	**...**	**372**	**284**	**106**	**1st, Smythe Div.**	**Won Stanley Cup**
1985-86	80	32	6	2	...	24	11	5	...	56	17	7	...	426	310	119	1st, Smythe Div.	Lost Div. Final
1984-85	**80**	**26**	**7**	**7**	**...**	**23**	**13**	**4**	**...**	**49**	**20**	**11**	**...**	**401**	**298**	**109**	**1st, Smythe Div.**	**Won Stanley Cup**
1983-84	**80**	**31**	**5**	**4**	**...**	**26**	**13**	**1**	**...**	**57**	**18**	**5**	**...**	**446**	**314**	**119**	**1st, Smythe Div.**	**Won Stanley Cup**
1982-83	80	25	9	6	...	22	12	6	...	47	21	12	...	424	315	106	1st, Smythe Div.	Lost Final
1981-82	80	31	5	4	...	17	12	11	...	48	17	15	...	417	295	111	1st, Smythe Div.	Lost Div. Semi-Final
1980-81	80	17	13	10	...	12	22	6	...	29	35	16	...	328	327	74	4th, Smythe Div.	Lost Quarter-Final
1979-80	80	17	14	9	...	11	25	4	...	28	39	13	...	301	322	69	4th, Smythe Div.	Lost Prelim. Round

Jason Smith wraps up Blake Sloan of the Columbus Blue Jackets. The Oilers captain ranked 17th in the NHL with 220 hits last season and led club defensemen with a plus-minus rating of +14.

Franchise date: June 22, 1979

NORTHWEST DIVISION

24th NHL Season

2002-03 Player Personnel

FORWARDS	HT	WT	S	Place of Birth	Date	2001-02 Club
BISHAI, Mike	5-11	185	L	Edmonton, Alta.	5/30/79	W. Michigan-Hamilton
CARTER, Anson	6-1	200	R	Toronto, Ont.	6/6/74	Edmonton
CHIMERA, Jason	6-0	215	L	Edmonton, Alta.	5/2/79	Edmonton-Hamilton
CLEARY, Daniel	6-0	203	L	Carbonear, Nfld.	12/18/78	Edmonton
COMRIE, Mike	5-9	175	L	Edmonton, Alta.	9/11/80	Edmonton
DiCASMIRRO, Nate	5-11	205	L	Burnsville, MN	9/27/78	St. Cloud State-Hamilton
DOPITA, Jiri	6-4	210	L	Sumperk, Czech.	12/2/68	Phi-Czech Republic
GREEN, Josh	6-4	212	L	Camrose, Alta.	11/16/77	Edmonton
GRIER, Mike	6-1	227	R	Detroit, MI	1/5/75	Edmonton
HEMSKY, Ales	6-0	191	R	Pardubice, Czech.	8/13/83	Hull
HENRICH, Michael	6-2	206	R	Thornhill, Ont.	3/3/80	Hamilton
HINZ, Chad	5-10	190	R	Saskatoon, Sask.	3/21/79	Hamilton
HORCOFF, Shawn	6-1	202	L	Trail, B.C.	9/17/78	Edmonton-Hamilton
LARAQUE, Georges	6-3	240	R	Montreal, Que.	12/7/76	Edmonton
MARCHANT, Todd	5-10	178	L	Buffalo, NY	8/12/73	Edmonton
McASLAN, Sean	6-1	190	R	Okootoks, Alta.	1/12/80	Columbus (ECHL)
MOREAU, Ethan	6-2	211	L	Huntsville, Ont.	9/22/75	Edmonton
PISANI, Fernando	6-1	185	L	Edmonton, Alta.	12/27/76	Hamilton
REASONER, Marty	6-1	190	L	Rochester, NY	2/26/77	Edmonton
REICHERT, Craig	6-1	200	R	Winnipeg, Man.	5/11/74	Hamilton
RITA, Jani	6-1	206	L	Helsinki, Finland	7/25/81	Edmonton-Hamilton
SALMELAINEN, Tony	5-9	176	R	Espoo, Finland	8/8/81	Ilves Jr.-Ilves
SARNO, Peter	5-11	185	L	Toronto, Ont.	7/26/79	Hamilton
SMYTH, Ryan	6-1	195	L	Banff, Alta.	2/21/76	Edmonton-Canada
STOLL, Jarret	6-1	199	R	Melville, Sask.	6/25/82	Kootenay
SWANSON, Brian	5-10	185	L	Eagle River, AK	3/24/76	Edmonton-Hamilton
YORK, Mike	5-10	185	R	Waterford, MI	1/3/78	NYR-United States-Edm
DEFENSEMEN						
ALLEN, Bobby	6-1	205	L	Braintree, MA	11/14/78	Providence (AHL)-Hamilton
BERGERON, Marc-Andre	5-9	185	L	St-Louis-de-France, Que.	10/13/80	Hamilton
BREWER, Eric	6-3	220	L	Vernon, B.C.	4/17/79	Edmonton-Canada
FERGUSON, Scott	6-1	195	L	Camrose, Alta.	1/6/73	Edmonton
HAAKANA, Kari	6-1	222	L	Outokumpu, Finland	11/8/73	Hamilton-Jokerit
HENRY, Alex	6-5	220	L	Elliot Lake, Ont.	10/18/79	Hamilton
LIUBIMOV, Alexander	6-3	196	L	Ust-Kamenogorsk, USSR	2/15/80	Odessa
NIINIMAA, Janne	6-1	220	L	Raahe, Finland	5/22/75	Edmonton-Finland
PISA, Ales	6-0	195	L	Pardibuce, Czech.	1/2/77	Edmonton-Hamilton
SEMENOV, Alexei	6-6	210	L	Murmansk, USSR	4/10/81	Hamilton
SMITH, Jason	6-3	210	R	Calgary, Alta.	11/2/73	Edmonton
STAIOS, Steve	6-1	200	R	Hamilton, Ont.	7/28/73	Edmonton

GOALTENDERS	HT	WT	C	Place of Birth	Date	2001-02 Club
ANTILA, Kristian	6-3	207	L	Vammala, Finland	1/10/80	Assat
CONKLIN, Ty	6-0	180	L	Anchorage, AK	3/30/76	Edmonton-Hamilton
MARKKANEN, Jussi	5-11	183	L	Imatra, Finland	5/8/75	Edmonton-Hamilton-Finland
MORRISON, Mike	6-3	194	R	Medford, MA	7/11/79	U. of Maine
SALO, Tommy	5-11	173	L	Surahammar, Sweden	2/1/71	Edmonton-Sweden

Coaching History

Glen Sather, 1979-80; Bryan Watson and Glen Sather, 1980-81; Glen Sather, 1981-82 to 1988-89; John Muckler, 1989-90, 1990-91; Ted Green, 1991-92, 1992-93; Ted Green and Glen Sather, 1993-94; George Burnett and Ron Low, 1994-95; Ron Low, 1995-96 to 1998-99; Kevin Lowe, 1999-2000; Craig MacTavish, 2000-01 to date.

Coach

MacTAVISH, CRAIG
Coach, Edmonton Oilers. Born in London, Ont., August 15, 1958.

The Edmonton Oilers named Craig MacTavish as their head coach on June 22, 2000. He became the eighth person in the club's NHL history to hold the position. MacTavish joined Kevin Lowe and Glen Sather as head coaches who were former captains of the Oilers.

MacTavish played for 18 seasons in the NHL, including eight-and-three-quarter campaigns with the Oilers. He was instrumental in helping his teams win four Stanley Cup titles; three with Edmonton and one with the New York Rangers. Although he was the last player in the NHL to play without a helmet, MacTavish was known for his aggressive style, combined with above average skills.

MacTavish retired as a player in 1997 and was immediately named an assistant coach with the New York Rangers. He was with the Rangers for two seasons prior to joining the Oilers' coaching staff as an assistant under Kevin Lowe in 1999-2000.

Coaching Record

		Regular Season				Playoffs		
Season	Team	Games	W	L	T	Games	W	L
2000-01	Edmonton (NHL)	82	39	31	12	6	2	4
2001-02	Edmonton (NHL)	82	38	32	12	....	....	....
	NHL Totals	**164**	**77**	**63**	**24**	**6**	**2**	**4**

2001-02 Scoring

* - rookie

Regular Season

Pos	#	Player	Team	GP	G	A	Pts	+/–	PIM	PP	SH	GW	GT	S	%
C	16	Mike York	NYR	69	18	39	57	8	16	2	0	5	1	188	9.6
			EDM	12	2	2	4	–1	0	1	0	1	0	30	6.7
			TOTAL	81	20	41	61	7	16	3	0	6	1	218	9.2
C	89	Mike Comrie	EDM	82	33	27	60	16	45	8	0	5	3	170	19.4
R	22	Anson Carter	EDM	82	28	32	60	3	25	12	0	6	0	181	15.5
L	94	Ryan Smyth	EDM	61	15	35	50	7	48	7	1	5	1	150	10.0
D	44	Janne Niinimaa	EDM	81	5	39	44	13	80	1	0	2	0	119	4.2
C	20	Jochen Hecht	EDM	82	16	24	40	4	60	5	0	3	1	211	7.6
C	26	Todd Marchant	EDM	82	12	22	34	7	41	0	3	1	0	124	9.7
R	7	Daniel Cleary	EDM	65	10	19	29	–1	51	2	1	1	0	75	13.3
R	25	Mike Grier	EDM	82	8	17	25	1	32	0	2	3	0	112	7.1
D	2	Eric Brewer	EDM	81	7	18	25	–5	45	6	0	2	0	165	4.2
C	10	Shawn Horcoff	EDM	61	8	14	22	3	18	0	0	0	0	57	14.0
R	27	Georges Laraque	EDM	80	5	14	19	6	157	1	0	1	0	95	5.3
D	21	Jason Smith	EDM	74	5	13	18	14	103	0	1	1	0	85	5.9
L	18	Ethan Moreau	EDM	80	11	5	16	4	81	0	2	1	1	129	8.5
L	12	Josh Green	EDM	61	10	5	15	9	52	1	0	1	0	78	12.8
C	15	Marty Reasoner	EDM	52	6	5	11	0	41	3	0	2	0	66	9.1
D	24	Steve Staios	EDM	73	5	5	10	10	108	0	0	1	0	101	5.0
C	14	Domenic Pittis	EDM	22	0	6	6	–2	8	0	0	0	0	18	0.0
D	32	Scott Ferguson	EDM	50	3	2	5	11	75	0	0	0	0	27	11.1
C	37	* Brian Swanson	EDM	8	1	1	2	–1	0	0	0	0	0	7	14.3
C	28	* Jason Chimera	EDM	3	1	0	1	–3	0	0	0	0	0	3	33.3
R	46	* Jani Rita	EDM	1	0	0	0	0	0	0	0	0	0	0	0.0
D	45	* Ales Pisa	EDM	2	0	0	0	0	2	0	0	0	0	3	0.0
D	19	Sven Butenschon	EDM	14	0	0	0	0	4	0	0	0	0	8	0.0

Goaltending

No.	Goaltender	GPI	Mins	Avg	W	L	T	EN	SO	GA	SA	S%	G	A	PIM
1	* Ty Conklin	4	148	1.62	2	0	0	0	0	4	66	.939	0	0	0
30	Jussi Markkanen	14	784	1.84	6	4	2	0	2	24	336	.929	0	0	0
35	Tommy Salo	69	4035	2.22	30	28	10	5	6	149	1713	.913	0	1	2
	Totals	**82**	**4996**	**2.19**	**38**	**32**	**12**	**5**	**8**	**182**	**2120**	**.914**			

Edmonton native Mike Comrie was a finalist for the Hobey Baker Award as the top NCAA player in 1999-2000. Playing his first full season with the Oilers last year, Comrie ranked among the NHL leaders with 33 goals.

Club Records

Team

(Figures in brackets for season records are games played; records for fewest points, wins, ties, losses, goals, goals against are for 70 or more games)

Record		
Most Points	**119**	1983-84 (80), 1985-86 (80)
Most Wins	**57**	1983-84 (80)
Most Ties	**16**	1980-81 (80), 1999-2000 (82)
Most Losses	**50**	1992-93 (84)
Most Goals	***446**	1983-84 (80)
Most Goals Against	**337**	1992-93 (84)
Fewest Points	**60**	1992-93 (84)
Fewest Wins	**25**	1993-94 (84)
Fewest Ties	**5**	1983-84 (80)
Fewest Losses	**17**	1981-82 (80), 1985-86 (80)
Fewest Goals	**205**	2001-02 (82)
Fewest Goals Against	**182**	2001-02 (82)
Longest Winning Streak		
Overall	**9**	Feb. 20-Mar. 13/01
Home	**8**	Jan. 19-Feb. 22/85, Feb. 24-Apr. 2/86
Away	**8**	Dec. 9/86-Jan. 17/87
Longest Undefeated Streak		
Overall	**15**	Oct. 11-Nov. 9/84 (12 wins, 3 ties)
Home	**14**	Nov. 15/89-Jan. 6/90 (11 wins, 3 ties)
Away	**9**	Jan. 17-Mar. 2/82 (6 wins, 3 ties), Nov. 23/82-Jan. 18/83 (7 wins, 2 ties)
Longest Losing Streak		
Overall	**11**	Oct. 16-Nov. 7/93
Home	**9**	Oct. 16-Nov. 24/93
Away	**9**	Nov. 25-Dec. 30/80
Longest Winless Streak		
Overall	**14**	Oct. 11-Nov. 7/93 (13 losses, 1 tie)
Home	**9**	Oct. 16-Nov. 24/93 (9 losses)
Away	**11**	Dec. 18/01-Feb. 8/02 (7 losses, 4 ties)
Most Shutouts, Season	**8**	1997-98 (82); 2000-01 (82); 2001-02 (82)
Most PIM, Season	**2,173**	1987-88 (80)
Most Goals, Game	**13**	Nov. 19/83 (N.J. 4 at Edm. 13), Nov. 8/85 (Van. 0 at Edm. 13)

Individual

Most Seasons	**15**	Kevin Lowe
Most Games	**1,037**	Kevin Lowe
Most Goals, Career	**583**	Wayne Gretzky
Most Assists, Career	**1,086**	Wayne Gretzky
Most Points, Career	**1,669**	Wayne Gretzky (583G, 1,086A)
Most PIM, Career	**1,747**	Kelly Buchberger
Most Shutouts, Career	**16**	Tommy Salo
Longest Consecutive Games Streak	**519**	Craig MacTavish (Oct. 11/86-Jan. 2/93)
Most Goals, Season	***92**	Wayne Gretzky (1981-82)
Most Assists, Season	***163**	Wayne Gretzky (1985-86)
Most Points, Season	***215**	Wayne Gretzky (1985-86; 52G, 163A)
Most PIM, Season	**286**	Steve Smith (1987-88)
Most Points, Defenseman, Season	**138**	Paul Coffey (1985-86; 48G, 90A)
Most Points, Center, Season	***215**	Wayne Gretzky (1985-86; 52G, 163A)
Most Points, Right Wing, Season	**135**	Jari Kurri (1984-85; 71G, 64A)
Most Points, Left Wing, Season	**106**	Mark Messier (1982-83; 48G, 58A)
Most Points, Rookie, Season	**75**	Jari Kurri (1980-81; 32G, 43A)
Most Shutouts, Season	**8**	Curtis Joseph (1997-98), Tommy Salo (2000-01)
Most Goals, Game	**5**	Wayne Gretzky (Feb. 18/81, Dec. 30/81, Dec. 15/84, Dec. 6/87), Jari Kurri (Nov. 19/83), Pat Hughes (Feb. 3/84)
Most Assists, Game	***7**	Wayne Gretzky (Feb. 15/80, Dec. 11/85, Feb. 14/86)
Most Points, Game	**8**	Wayne Gretzky (Nov. 19/83; 3G, 5A), (Jan. 4/84; 4G, 4A), Paul Coffey (Mar. 14/86; 2G, 6A)

* NHL Record.

Retired Numbers

3	Al Hamilton	1972-1980
17	Jari Kurri	1980-1990
99	Wayne Gretzky	1979-1988

Captains' History

Ron Chipperfield, 1979-80; Blair MacDonald and Lee Fogolin Jr., 1980-81; Lee Fogolin Jr., 1981-82, 1982-83; Wayne Gretzky, 1983-84 to 1987-88; Mark Messier, 1988-89 to 1990-91; Kevin Lowe, 1991-92; Craig MacTavish, 1992-93, 1993-94; Shayne Corson, 1994-95; Kelly Buchberger, 1995-96 to 1998-99; Doug Weight, 1999-2000, 2000-01; Jason Smith, 2001-02 to date.

All-time Record vs. Other Clubs

Regular Season

	At Home								On Road								Total							
	GP	W	L	T	OL	GF	GA	PTS	GP	W	L	T	OL	GF	GA	PTS	GP	W	L	T	OL	GF	GA	PTS
Anaheim	21	13	8	0	0	48	45	26	22	6	14	2	0	54	64	14	43	19	22	2	0	102	109	40
Atlanta	2	1	1	0	0	6	7	2	2	1	1	0	0	5	3	2	4	2	2	0	0	11	10	4
Boston	28	11	14	3	0	95	93	25	28	6	18	3	1	76	116	16	56	17	32	6	1	171	209	41
Buffalo	27	20	5	2	0	115	72	42	30	13	10	7	0	112	109	33	57	33	15	9	0	227	181	75
Calgary	76	41	24	10	1	297	250	93	75	27	40	8	0	262	313	62	151	68	64	18	1	559	563	155
Carolina	29	18	6	5	0	115	88	41	29	11	11	7	0	98	112	29	58	29	17	12	0	213	200	70
Chicago	41	19	17	5	0	152	139	43	40	13	21	6	0	136	156	32	81	32	38	11	0	288	295	75
Colorado	41	22	15	4	0	172	124	48	41	18	20	3	0	156	160	39	82	40	35	7	0	328	284	87
Columbus	4	3	1	0	0	12	7	6	4	3	1	0	0	14	9	6	8	6	2	0	0	26	16	12
Dallas	40	19	13	8	0	163	134	46	41	13	21	7	0	118	150	33	81	32	34	15	0	281	284	79
Detroit	40	19	14	7	0	158	146	45	40	15	20	3	2	137	153	35	80	34	34	10	2	295	299	80
Florida	6	3	2	1	0	20	14	7	8	1	5	2	0	20	24	4	14	4	7	3	0	40	38	11
Los Angeles	73	37	21	15	0	338	266	89	73	31	26	15	1	307	288	78	146	68	47	30	1	645	554	167
Minnesota	5	4	0	1	0	13	6	9	5	4	0	1	0	19	12	9	10	8	0	2	0	32	18	18
Montreal	33	17	16	0	0	109	105	34	28	9	15	4	0	89	100	22	61	26	31	4	0	198	205	56
Nashville	8	4	3	0	1	23	21	9	9	4	3	2	0	26	23	10	17	8	6	2	1	49	44	19
New Jersey	31	14	10	6	1	136	114	35	31	15	13	3	0	104	105	33	62	29	23	9	1	240	219	68
NY Islanders	28	16	7	5	0	105	82	37	29	7	13	9	0	106	121	23	57	23	20	14	0	211	203	60
NY Rangers	28	12	13	3	0	101	94	27	28	13	8	6	1	106	103	33	56	25	21	9	1	207	197	60
Ottawa	9	6	1	2	0	36	22	14	8	5	2	1	0	23	14	11	17	11	3	3	0	59	36	25
Philadelphia	27	14	8	5	0	96	81	33	30	8	20	2	0	82	128	18	57	22	28	7	0	178	209	51
Phoenix	68	43	19	6	0	299	220	92	67	36	26	4	1	299	272	77	135	79	45	10	1	598	492	169
Pittsburgh	29	21	7	1	0	144	95	43	29	12	14	3	0	124	113	27	58	33	21	4	0	268	208	70
St. Louis	40	21	15	4	0	146	133	46	40	16	18	5	1	146	146	38	80	37	33	9	1	292	279	84
San Jose	29	17	6	6	0	101	63	40	28	10	13	3	2	85	99	25	57	27	19	9	2	186	162	65
Tampa Bay	9	7	2	0	0	24	18	14	10	6	2	2	0	33	28	14	19	13	4	2	0	57	46	28
Toronto	40	21	12	6	1	170	131	49	35	14	19	2	0	148	146	30	75	35	31	8	1	318	277	79
Vancouver	75	48	19	7	1	342	233	104	77	37	28	11	1	306	278	86	152	85	47	18	2	648	511	190
Washington	28	14	10	4	0	115	88	32	28	9	17	2	0	93	117	20	56	23	27	6	0	208	205	52
Totals	**915**	**505**	**289**	**116**	**5**	**3651**	**2891**	**1131**	**915**	**363**	**419**	**123**	**10**	**3284**	**3462**	**859**	**1830**	**868**	**708**	**239**	**15**	**6935**	**6353**	**1990**

Playoffs

	Series	W	L	GP	W	L	T	GF	GA	Last Mtg.	Rnd.	Result
Boston	2	2	0	9	8	1	0	41	20	1990	F	W 4-1
Calgary	5	4	1	30	19	11	0	132	96	1991	DSF	W 4-3
Chicago	4	3	1	20	12	8	0	102	77	1992	CF	L 0-4
Colorado	2	1	1	12	5	7	0	30	35	1998	CQF	W 4-3
Dallas	7	2	5	36	13	23	0	93	98	2001	CQF	L 2-4
Detroit	2	2	0	10	8	2	0	39	26	1988	CF	W 4-1
Los Angeles	7	5	2	36	24	12	0	154	127	1992	DSF	W 4-2
Montreal	1	1	0	3	3	0	0	15	6	1981	PRE	W 3-0
NY Islanders	3	1	2	15	6	9	0	47	58	1984	F	W 4-1
Philadelphia	3	2	1	15	8	7	0	49	44	1987	F	W 4-3
Vancouver	2	2	0	9	7	2	0	35	20	1992	DF	W 4-2
Winnipeg	6	6	0	26	22	4	0	120	75	1990	DSF	W 4-3
Totals	**44**	**31**	**13**	**221**	**135**	**86**	**0**	**857**	**682**			

Playoff Results 2002-1998

Year	Round	Opponent	Result	GF	GA
2001	CQF	Dallas	L 2-4	13	16
2000	CQF	Dallas	L 1-4	11	14
1999	CQF	Dallas	L 0-4	7	11
1998	CSF	Dallas	L 1-4	5	9
	CQF	Colorado	W 4-3	19	16

Abbreviations: Round: F - Final; **CF** - conference final; **CSF** - conference semi-final; **CQF** - conference quarter-final; **DF** - division final; **DSF** - division semi-final; **PRE** - preliminary round.

Calgary totals include Atlanta Flames, 1979-80.
Colorado totals include Quebec, 1979-80 to 1994-95.
New Jersey totals include Colorado Rockies, 1979-80 to 1981-82.
Carolina totals include Hartford, 1979-80 to 1996-97.
Dallas totals include Minnesota North Stars, 1979-80 to 1992-93.
Phoenix totals include Winnipeg, 1979-80 to 1995-96.

2001-02 Results

Month	Day	Opponent	Score
Oct.	3	at Calgary	0-1
	6	Phoenix	6-2
	9	Chicago	1-0
	11	Colorado	5-3
	13	at Nashville	4-3
	14	at Minnesota	3-3
	16	Toronto	1-4
	18	at Colorado	4-1
	20	Florida	6-2
	22	Nashville	2-4
	24	at Detroit	1-4
	25	at Columbus	5-2
	27	Vancouver	3-2
	30	Montreal	3-1
Nov.	2	Columbus	1-2
	4	at Minnesota	2-0
	6	at Boston	0-1*
	9	at Columbus	3-0
	11	at Carolina	1-1
	13	at Phoenix	5-4*
	16	Chicago	7-1
	17	at Vancouver	2-2
	20	St. Louis	2-0
	22	Los Angeles	2-4
	24	at Colorado	0-2
	28	at Anaheim	2-0
	29	at Los Angeles	3-1
Dec.	1	Dallas	4-6
	5	Anaheim	3-2*
	7	at Dallas	0-5
	8	at Nashville	2-2
	11	at San Jose	4-5*
	13	Detroit	1-2
	14	Tampa Bay	2-1
	16	at Philadelphia	3-2
	18	at NY Islanders	1-4
	20	at New Jersey	3-3
	21	at Chicago	1-5
	26	Calgary	3-2
	28	Minnesota	3-2
	30	New Jersey	1-2
	31	at Calgary	2-2
Jan.	2	NY Rangers	4-1
	5	Vancouver	3-4
	6	Montreal	7-6
	10	Carolina	1-4
	12	Colorado	2-2
	14	at Chicago	1-2
	15	at St. Louis	2-3
	18	Anaheim	3-1
	19	Pittsburgh	0-1
	21	at San Jose	3-4
	23	Colorado	2-4
	26	Toronto	4-1
	28	Detroit	1-1
	30	at Vancouver	2-2
Feb.	5	at Atlanta	2-3
	7	at St. Louis	1-3
	8	at Dallas	1-1
	10	at Phoenix	4-3
	12	San Jose	2-3
	28	Nashville	2-3
Mar.	2	St. Louis	1-1
	4	at Buffalo	3-0
	6	at Tampa Bay	3-2
	8	at Florida	4-5
	10	at Washington	2-4
	13	at Detroit	3-4*
	14	at Ottawa	4-1
	16	Washington	4-1
	20	San Jose	2-1
	23	Calgary	3-1
	24	at Vancouver	2-0
	26	Columbus	3-1
	28	Los Angeles	2-2
	30	Dallas	3-1
Apr.	2	Minnesota	2-1*
	5	at Anaheim	0-2
	6	at Los Angeles	3-4*
	10	Phoenix	3-0
	12	Calgary	0-2
	14	at Minnesota	4-2

* – Overtime

Entry Draft Selections 2002-1988

2002
Pick	
15	Jesse Niinimaki
31	Jeff Deslauriers
36	Jarret Stoll
44	Matt Greene
79	Brock Radunske
106	Ivan Koltsov
111	Jonas Almtorp
148	Glenn Fisher
181	Mikko Luoma
205	J.F. Dufort
211	Patrick Murphy
244	Dwight Helminen
245	Tomas Micka
274	Fredrik Johansson

2001
Pick	
13	Ales Hemsky
43	Doug Lynch
52	Ed Caron
84	Kenny Smith
133	Jussi Markkanen
154	Jake Brenk
185	Mikael Svensk
215	Dan Baum
248	Kari Haakana
272	Ales Pisa
278	Shay Stephenson

2000
Pick	
17	Alexei Mikhnov
35	Brad Winchester
83	Alexander Liubimov
113	Lou Dickenson
152	Paul Flache
184	Shaun Norrie
211	Joe Cullen
215	Matthew Lombardi
247	Jason Platt
274	Yevgeny Muratov

1999
Pick	
13	Jani Rita
36	Alexei Semenov
41	Tony Salmelainen
81	Adam Hauser
91	Mike Comrie
139	Jonathan Fauteux
171	Chris Legg
199	Chris Chartier
256	Tamas Groschl

1998
Pick	
13	Michael Henrich
67	Alex Henry
99	Shawn Horcoff
113	Kristian Antila
128	Paul Elliott
144	Oleg Smirnov
159	Trevor Ettinger
186	Mike Morrison
213	Christian Lefebvre
241	Maxim Spiridonov

1997
Pick	
14	Michel Riesen
41	Patrick Dovigi
68	Sergei Yerkovich
94	Jonas Elofsson
121	Jason Chimera
141	Peter Sarno
176	Kevin Bolibruck
187	Chad Hinz
205	Chris Kerr
231	Alexander Fomitchev

1996
Pick	
6	Boyd Devereaux
19	Matthieu Descoteaux
32	Chris Hajt
59	Tom Poti
114	Brian Urick
141	Bryan Randall
168	David Bernier
168	David Bernier
170	Brandon Lafrance
195	Fernando Pisani
221	John Hultberg

1995
Pick	
6	Steve Kelly
31	Georges Laraque
57	Lukas Zib
83	Mike Minard
109	Jan Snopek
161	Martin Cerven
187	Stephen Douglas
213	Jiri Antonin

1994
Pick	
4	Jason Bonsignore
6	Ryan Smyth
32	Mike Watt
53	Corey Neilson
60	Brad Symes
79	Adam Copeland
95	Jussi Tarvainen
110	Jon Gaskins
136	Terry Marchant
160	Curtis Sheptak
162	Dmitri Shulga
179	Chris Wickenheiser
185	Rob Guinn
188	Jason Reid
214	Jeremy Jablonski
266	Ladislav Benysek

1993
Pick	
7	Jason Arnott
16	Nick Stajduhar
33	David Vyborny
59	Kevin Paden
60	Alexander Kerch
111	Miroslav Satan
163	Alexander Zhurik
189	Martin Bakula
215	Brad Norton
241	Oleg Maltsev
267	Ilja Byakin

1992
Pick	
13	Joe Hulbig
37	Martin Reichel
61	Simon Roy
65	Kirk Maltby
96	Ralph Intranuovo
109	Joaquin Gage
157	Steve Gibson
181	Kyuin Shim
190	Colin Schmidt
205	Marko Tuomainen
253	Bryan Rasmussen

1991
Pick	
12	Tyler Wright
20	Martin Rucinsky
34	Andrew Verner
56	George Breen
78	Mario Nobili
93	Ryan Haggerty
144	David Oliver
166	Gary Kitching
210	Vegar Barlie
232	Yevgeny Belosheiken
254	Juha Riihijarvi

1990
Pick	
17	Scott Allison
38	Alexandre Legault
59	Joe Crowley
67	Joel Blain
101	Greg Louder
122	Keijo Sailynoja
143	Mike Power
164	Roman Mejzlik
185	Richard Zemlicka
206	Petr Korinek
227	
248	Sami Nuutinen

1989
Pick	
15	Jason Soules
36	Richard Borgo
78	Josef Beranek
92	Peter White
120	Anatoli Semenov
140	Davis Payne
141	Sergei Yashin
162	Darcy Martini
225	Roman Bozek

1988
Pick	
19	Francois Leroux
39	Petro Koivunen
53	Trevor Sim
61	Collin Bauer
82	Cam Brauer
103	Don Martin
124	Len Barrie
145	Mike Glover
166	Shjon Podein
187	Tim Cole
208	Vladimir Zubkov
229	Darin MacDonald
250	Tim Tisdale

General Managers' History

Larry Gordon, 1979-80; Glen Sather, 1980-81 to 1999-2000; Kevin Lowe, 2000-01 to date.

General Manager

LOWE, KEVIN
General Manager, Edmonton Oilers. Born in Lachute, Que., April 15, 1959.

The Edmonton Oilers named Kevin Lowe as their general manager on June 9, 2000, filling the position left vacant when Glen Sather resigned on May 19th. Lowe moved into the front office after spending the 1999-2000 season as coach of the Oilers.

After a brilliant 19-year playing career with the Edmonton Oilers and New York Rangers, Kevin Lowe announced his retirement on July 30, 1998 and joined the Edmonton Oilers coaching staff. He replaced Ron Low as head coach on June 18, 1999.

Lowe was the Oilers' first-ever draft pick when he was selected 21st overall in the 1979 NHL Amateur Draft. He went on to play in 1,254 regular season games and 214 playoff games, winning six Stanley Cup championships; the first five with Edmonton (1984, 1985, 1987, 1988, 1990) followed by a sixth title with the Rangers in 1994.

Besides being the first draft choice in Oilers history, Lowe also scored the first goal in team history on October 10, 1979. He holds the Oilers' record for most games played in both the regular season (1,037) and playoffs (172), and became the sixth captain in team history in 1990-91. He was no less a leader off the ice, becoming the only player to win the King Clancy Memorial Trophy and the Budweiser/NHL Man of the Year Award in the same season (1989-90). Both awards are presented for leadership qualities and humanitarian contributions. His work with the Edmonton Christmas Bureau has set the standard for the Oilers' commitment to community involvement.

NHL Coaching Record

		Regular Season				Playoffs		
Season	Team	Games	W	L	T	Games	W	L
1999-2000	Edmonton	82	32	34	16	5	1	4
	NHL Totals	**82**	**32**	**34**	**16**	**5**	**1**	**4**

Club Directory

Skyreach Centre

Edmonton Oilers
11230 – 110 Street
Edmonton, Alberta T5G 3H7
Phone **780/414-4000**
Press Box 780/414-4235
Ticketing 780/414-4400
Media Lounge 780/414-4173
FAX 780/414-4659
www.edmontonoilers.com
Capacity: 16,839

Owner	Edmonton Investors Group Ltd.
Governor	Cal Nichols
Alternate Governors	Patrick R. LaForge, Kevin Lowe & Gordon Buchanan
President & Chief Executive Officer	Patrick R. LaForge
Executive Vice-President & General Manager	Kevin Lowe
Vice-President, Hockey Operations	Kevin Prendergast
Assistant General Manager	Scott Howson
Head Coach	Craig MacTavish
Assistant Coaches	Charlie Huddy, Bill Moores
Goaltending Coach	Pete Peeters
Video Coach	Brian Ross
European Scout/Development Coach	Frank Musil
Vice President, Public Relations	Bill Tuele
Information Coordinator	Steve Knowles
Public Relations Coordinator, Hockey	Warren Suitor
Director of Research, Analysis and Software Development	Sean Draper
Scouting Staff	Bob Brown, Bill Dandy, Brad Davis, Lorne Davis, Morey Gare, Stu MacGregor, Bob Mancini, Chris McCarthy, Kent Nilsson, Gord Pell, Dave Semenko, John Stevenson
Executive Assistant to the President	Donna Perman
Executive Assistant to the General Manager	Valerie Rendell
Security Advisor	Gary Goulet
Medical and Training Staff	
Head Medical Trainer	Ken Lowe
Assistant Medical Trainer	Ryan McInnes
Head Equipment Manager	Barrie Stafford
Equipment Manager	Lyle Kulchisky
Assistant Equipment Manager	Chris Delorey
Massage Therapist	Stewart Poirier
Team Medical Chief of Staff/Director of Glen Sather Sports Medicine Clinic	Dr. David C. Reid
Team Physicians	Dr. Boris Boyko, Dr. Paul Paludet
Team Dermatologist	Dr. Don Groot
Team Dentists	Dr. Tony Sneazwell, Dr. Ben Eastwood
Fitness Consultants	Dr. Art Quinney, Dr. Gordon Bell
Physical Therapy Consultant	Dr. Dave Magee
Team Optometrist	Dr. Brent Saik
Strength & Conditioning Consultant	Daryl Duke
Finance & Administration	
Vice-President of Finance and CFO	Darryl Boessenkool
Controller	Jason Quilley
Senior Accounting Manager	Colleen Rolston
Facilities Manager	Craig Tkachuk
Systems Administrators	Terry Rhoades, Rod Pruden
Payroll & Benefits Manager	Michelle Schwendeman
Payroll & Benefits Supervisor	Shawna Quigley
Finance Staff	Lynn Schmidl, Donna Chizen, Corinne McGregor, Cheryl Thomas
Marketing & Communications	
Vice-President, Marketing & Communications	Allan Watt
Director of Broadcast	Don Metz
Director, Corporate Communications & Marketing	Natalie Minckler
Marketing & Promotions Coordinator	Trena Vik
Marketing & Communications Coordinator	Darren Krill
New Media Production Manager	Andreas Schwabe
Game Night Operations	Glenn Wiun
Game Night Supervisor	Marilyn Riddell
Publications Coordinator	Steve Sandor
Community Relations	
Director Community Relations & Executive Director Edmonton Oilers Community Foundation	Gillian Andries
Community Relations Adminstrator	Heidi Lippert
Community Relations Coordinator	Kel Parry
Sales	
Director of Sales	Eric Upton
Sponsorships Sales	
Manager, Corporate Sponsorships	Brad MacGregor
National Accounts Managers	Matt Cummings, Michael Lake, Sean Price
Broadcast Sponsorship Coordinator	Kristie Brown
Suite Sales	
Suite Manager	Cathy Cookson
Ticket Sales	
Director of Ticket Sales	Ken Brown
Ticket Sales Department Coordinator	Erin Lewyk
Senior Account Executive	Bob Haromy
Corporate Account Executives	Bruce Rakoczy, Damon Bunting, Scott Jacques, Jeff Tetz
Group Account Executives	Marran Vogelesang, Tom Tobin
Ticket Operations	
Manager, Ticket Operations	John Yeomans
Ticketing Services Representative	Sandy Langley
Database Administrator	Tuan Nguyen
Team Information	
Training Camp Site	Skyreach Centre; Edmonton, Alberta & Millennium Place; Sherwood Park, Alberta
Television Outlets	SportsNet, CBXT TV & TSN
Radio Flagship Station	630 CHED (AM); Rod Phillips (Play-by-play) & Morley Scott (colour)

Florida Panthers

2001-02 Results: 22W-44L-10T-6OTL 60PTS.
Fourth, Southeast Division

Year-by-Year Record

		Home				Road				Overall								
Season	GP	W	L	T	OL	W	L	T	OL	W	L	T	OL	GF	GA	Pts.	Finished	Playoff Result
2001-02	82	11	23	3	4	11	21	7	2	22	44	10	6	180	250	60	4th, Southeast Div.	Out of Playoffs
2000-01	82	12	18	7	4	10	20	6	5	22	38	13	9	200	246	66	3rd, Southeast Div.	Out of Playoffs
1999-2000	82	26	9	4	2	17	18	2	4	43	27	6	6	244	209	98	2nd, Southeast Div.	Lost Conf. Quarter-Final
1998-99	82	17	17	7	...	13	17	11	...	30	34	18	...	210	228	78	2nd, Southeast Div.	Out of Playoffs
1997-98	82	11	24	6	...	13	19	9	...	24	43	15	...	203	256	63	6th, Atlantic Div.	Out of Playoffs
1996-97	82	21	12	8	...	14	16	11	...	35	28	19	...	221	201	89	3rd, Atlantic Div.	Lost Conf. Quarter-Final
1995-96	82	25	12	4	...	16	19	6	...	41	31	10	...	254	234	92	3rd, Atlantic Div.	Lost Final
1994-95	48	9	12	3	...	11	10	3	...	20	22	6	...	115	127	46	5th, Atlantic Div.	Out of Playoffs
1993-94	84	15	18	9	...	18	16	8	...	33	34	17	...	233	233	83	5th, Atlantic Div.	Out of Playoffs

2002-03 Schedule

Oct.	Thu.	10	Tampa Bay
	Sat.	12	at Atlanta
	Tue.	15	at Minnesota
	Thu.	17	at Chicago
	Sat.	19	at Columbus
	Mon.	21	Atlanta
	Wed.	23	at Toronto
	Thu.	24	at NY Islanders
	Sat.	26	Washington
	Mon.	28	Tampa Bay
	Wed.	30	at Dallas
Nov.	Sat.	2	Atlanta
	Wed.	6	Pittsburgh
	Thu.	7	at Washington
	Sat.	9	Calgary
	Mon.	11	Chicago
	Wed.	13	at Philadelphia
	Thu.	14	at Ottawa
	Sat.	16	San Jose
	Tue.	19	at Atlanta
	Wed.	20	NY Islanders
	Fri.	22	at Phoenix
	Sun.	24	at Anaheim*
	Wed.	27	at Los Angeles
	Sat.	30	Vancouver
Dec.	Wed.	4	Carolina
	Fri.	6	at Carolina
	Sat.	7	Edmonton
	Tue.	10	Philadelphia
	Fri.	13	NY Islanders
	Wed.	18	Toronto
	Fri.	20	at Buffalo
	Sat.	21	at Boston
	Mon.	23	Nashville
	Fri.	27	Dallas
	Sat.	28	NY Rangers
	Mon.	30	at NY Islanders
Jan.	Wed.	1	at New Jersey*
	Thu.	2	at Colorado
	Sat.	4	at Vancouver
	Wed.	8	Detroit
	Fri.	10	New Jersey
	Sat.	11	at Washington
	Mon.	13	at New Jersey
	Wed.	15	Boston
	Sat.	18	Pittsburgh
	Mon.	20	Montreal*
	Wed.	22	Ottawa
	Fri.	24	at Carolina
	Sat.	25	Carolina
	Tue.	28	at Montreal
	Thu.	30	at Detroit
Feb.	Wed.	5	Toronto
	Thu.	6	at Pittsburgh
	Sat.	8	Tampa Bay
	Wed.	12	NY Rangers
	Fri.	14	Boston
	Sat.	15	Washington
	Tue.	18	at Montreal
	Thu.	20	at Ottawa
	Sat.	22	at Philadelphia
	Mon.	24	Buffalo
	Wed.	26	Anaheim
	Thu.	27	at Tampa Bay
Mar.	Sat.	1	at NY Rangers
	Mon.	3	at Toronto
	Wed.	5	Colorado
	Fri.	7	at Atlanta
	Sat.	8	Buffalo
	Mon.	10	at NY Rangers
	Wed.	12	Montreal
	Sat.	15	at Boston*
	Sun.	16	at Pittsburgh*
	Wed.	19	Minnesota
	Sat.	22	Ottawa
	Mon.	24	New Jersey
	Wed.	26	at Buffalo
	Thu.	27	at St. Louis
	Sat.	29	at Tampa Bay
Apr.	Tue.	1	at Washington
	Fri.	4	Carolina
	Sun.	6	Philadelphia*

* Denotes afternoon game.

Franchise date: June 14, 1993

SOUTHEAST DIVISION

10th NHL Season

Kristian Huselius scored a goal in his first NHL game during the Panthers' 2001-02 season opener versus Philadelphia. His nine goals last October were the most by an NHL rookie in the season's opening month since 1992.

2002-03 Player Personnel

FORWARDS	HT	WT	S	Place of Birth	Date	2001-02 Club
BEAUDOIN, Eric	6-5	204	L	Ottawa, Ont.	5/3/80	Florida-Utah
BURE, Valeri	5-10	185	R	Moscow, USSR	6/13/74	Florida-Russia
CAMPBELL, Jim	6-2	205	R	Worcester, MA	4/3/73	Chicago-Norfolk
DAGENAIS, Pierre	6-5	215	L	Blainville, Que.	3/4/78	New Jersey-Albany-Florida-Utah
HAGMAN, Niklas	6-0	200	L	Espoo, Finland	12/5/79	Florida-Finland
HUSELIUS, Kristian	6-1	190	L	Osterhaninge, Sweden	11/10/78	Florida
HYVONEN, Hannes	6-2	200	R	Oulu, Finland	8/29/75	San Jose-Cleveland
JOHNSON, Ryan	6-1	200	L	Thunder Bay, Ont.	6/14/76	Florida
JOKINEN, Olli	6-3	205	L	Kuopio, Finland	12/5/78	Florida-Finland
KOZLOV, Viktor	6-5	225	R	Togliatti, USSR	2/14/75	Florida
LUNDBOHM, Andy	6-3	225	L	Roseau, MN	3/24/77	Cleveland
MATTEAU, Stephane	6-4	215	L	Rouyn-Noranda, Que.	9/2/69	San Jose
NILSON, Marcus	6-2	195	R	Balsta, Sweden	3/1/78	Florida
NOVOSELTSEV, Ivan	6-1	210	L	Golitsino, USSR	1/23/79	Florida
PAYER, Serge	6-0	203	L	Rockland, Ont.	5/7/79	Utah
RITCHIE, Byron	5-10	195	L	Burnaby, B.C.	4/24/77	Carolina-Lowell-Florida
SHVIDKI, Denis	6-2	215	L	Kharkov, USSR	11/21/80	Florida-Utah
TATICEK, Petr	6-2	188	L	Rakovnik, Czech.	9/22/83	Sault Ste. Marie
TETARENKO, Joey	6-2	215	R	Prince Albert, Sask.	3/3/78	Florida
THOMPSON, Rocky	6-2	205	R	Calgary, Alta.	8/8/77	Florida-Hershey
TOMS, Jeff	6-5	200	L	Swift Current, Sask.	6/4/74	NYR-Hartford-Pit
WEISS, Stephen	5-11	183	L	Toronto, Ont.	4/3/83	Florida-Plymouth
WORRELL, Peter	6-6	235	L	Pierrefonds, Que.	8/18/77	Florida
DEFENSEMEN						
BOUWMEESTER, Jay	6-4	210	L	Edmonton, Alta.	9/27/83	Medicine Hat
BUTENSCHON, Sven	6-4	215	L	Itzehoe, West Germany	3/22/76	Edmonton-Hamilton
FERENCE, Brad	6-3	210	R	Calgary, Alta.	4/2/79	Florida
KRAJICEK, Lukas	6-2	182	L	Prostejov, Czech.	3/11/83	Florida-Peterborough
LAUS, Paul	6-1	215	R	Beamsville, Ont.	9/26/70	Florida
MEZEI, Branislav	6-5	236	L	Nitra, Czech.	10/8/80	NY Islanders-Bridgeport
MYRVOLD, Anders	6-2	200	L	Lorenskog, Norway	8/12/75	Hartford-Fribourg
NOVAK, Filip	6-1	185	L	Ceske Budejovice, Czech.	5/7/82	Regina
OZOLINSH, Sandis	6-3	215	L	Riga, Latvia	8/3/72	Carolina-Florida-Latvia
ROSSITER, Kyle	6-3	217	L	Edmonton, Alta.	6/9/80	Florida-Utah
ULANOV, Igor	6-3	220	L	Krasnokamsk, USSR	10/1/69	NYR-Hartford-Fla
WARD, Lance	6-3	220	L	Lloydminster, Alta.	6/2/78	Florida
YUSHKEVICH, Dmitry	5-11	208	R	Cherepovets, USSR	11/19/71	Toronto

GOALTENDERS	HT	WT	C	Place of Birth	Date	2001-02 Club
FLAHERTY, Wade	6-0	170	L	Terrace, B.C.	1/11/68	Florida-Utah
KIDD, Trevor	6-2	210	L	Dugald, Man.	3/26/72	Florida
LUONGO, Roberto	6-3	205	L	Montreal, Que.	4/4/79	Florida

2001-02 Scoring

* - rookie

Regular Season

Pos	#	Player	Team	GP	G	A	Pts	+/–	PIM	PP	SH	GW	GT	S	%
D	44	Sandis Ozolinsh	CAR	46	4	19	23	–4	34	1	0	0	0	71	5.6
			FLA	37	10	19	29	–3	24	2	0	1	0	101	9.9
			TOTAL	83	14	38	52	–7	58	3	0	1	0	172	8.1
R	22 *	Kristian Huselius	FLA	79	23	22	45	–4	14	6	1	3	0	169	13.6
L	18	Marcus Nilson	FLA	81	14	19	33	–14	55	6	1	2	0	147	9.5
C	28	Jason Wiemer	FLA	70	11	20	31	–4	178	5	1	1	0	115	9.6
R	39	Ivan Novoseltsev	FLA	70	13	16	29	–10	44	1	1	5	0	109	11.9
C	12	Olli Jokinen	FLA	80	9	20	29	–16	98	3	1	0	1	153	5.9
D	24	Robert Svehla	FLA	82	7	22	29	–19	87	3	0	0	0	119	5.9
L	14 *	Niklas Hagman	FLA	78	10	18	28	–6	8	0	1	2	0	134	7.5
C	25	Viktor Kozlov	FLA	50	9	18	27	–16	20	6	0	1	0	143	6.3
R	20	Valeri Bure	FLA	31	8	10	18	–3	12	2	0	1	0	100	8.0
D	45	Brad Ference	FLA	80	2	15	17	–13	254	0	0	0	0	65	3.1
L	26 *	Pierre Dagenais	N.J.	16	3	3	6	–5	4	1	0	1	0	30	10.0
			FLA	26	7	1	8	–5	4	2	0	0	2	47	14.9
			TOTAL	42	10	4	14	–10	8	3	0	1	2	77	13.0
C	43	Byron Ritchie	CAR	4	0	0	0	0	2	0	0	0	0	5	0.0
			FLA	31	5	6	11	–2	34	2	0	0	1	55	9.1
			TOTAL	35	5	6	11	–2	36	2	0	0	1	60	8.3
D	55	Igor Ulanov	NYR	39	0	6	6	–4	53	0	0	0	0	17	0.0
			FLA	14	0	4	4	–3	11	0	0	0	0	9	0.0
			TOTAL	53	0	10	10	–7	64	0	0	0	0	26	0.0
L	8	Peter Worrell	FLA	79	4	5	9	–15	354	0	0	1	0	65	6.2
D	3	Paul Laus	FLA	45	4	3	7	1	157	0	1	0	0	39	10.3
D	49	Lance Ward	FLA	68	1	4	5	–20	131	0	0	0	0	39	2.6
L	38 *	Eric Beaudoin	FLA	8	1	3	4	–2	4	0	0	1	0	11	9.1
C	17	Ryan Johnson	FLA	29	1	3	4	–5	10	0	0	0	0	24	4.2
R	21	Denis Shvidki	FLA	8	1	2	3	–4	2	0	0	0	0	11	9.1
C	19 *	Stephen Weiss	FLA	7	1	1	2	0	0	1	0	0	0	15	6.7
D	2	Lance Pitlick	FLA	35	1	1	2	–14	12	0	0	0	0	15	6.7
L	32 *	Ryan Jardine	FLA	8	0	2	2	0	2	0	0	0	0	6	0.0
L	42	Brad Norton	FLA	22	0	2	2	–2	45	0	0	0	0	6	0.0
D	36	Joey Tetarenko	FLA	38	1	0	1	–5	123	0	0	0	0	10	10.0
D	23 *	Kyle Rossiter	FLA	2	0	0	0	–1	2	0	0	0	0	0	0.0
C	42	Matt Herr	FLA	3	0	0	0	–2	0	0	0	0	0	1	0.0
R	33 *	Dave Morisset	FLA	4	0	0	0	–7	5	0	0	0	0	2	0.0
D	29 *	Lukas Krajicek	FLA	5	0	0	0	0	0	0	0	0	0	3	0.0
D	23 *	Rocky Thompson	FLA	6	0	0	0	0	12	0	0	0	0	1	0.0
C	41 *	Nick Smith	FLA	15	0	0	0	–1	0	0	0	0	0	2	0.0

Goaltending

No.	Goaltender	GPI	Mins	Avg	W	L	T	EN	SO	GA	SA	S%	G	A	PIM
1	Roberto Luongo	58	3030	2.77	16	33	4	4	4	140	1653	.915	0	1	2
31	Wade Flaherty	4	245	2.94	2	1	1	0	0	12	148	.919	0	0	0
37	Trevor Kidd	33	1683	3.21	4	16	5	4	1	90	857	.895	0	1	0
	Totals	**82**	**4987**	**3.01**	**22**	**50**	**10**	**8**	**5**	**250**	**2666**	**.906**			

General Manager

DUDLEY, RICK
General Manager, Florida Panthers. Born in Toronto, Ont., January 31, 1949.

The Florida Panthers named Rick Dudley as their general manager on May 10, 2002. Dudley's career in hockey has been successful at all levels as a player, coach and general manager. He most recently served as senior vice president of hockey operations and general manager of the Tampa Bay Lightning from 1999-2000 to the 2001-02 season. Prior to joining the Lightning, Dudley served one season as the general manager of the Ottawa Senators in 1998-99, where his team finished with a mark of 44-23-15 and 103 points. The Senators improved by 20 points in the standings and finished with the third best overall record in the NHL during Dudley's tenure.

Before serving with Ottawa, Dudley spent four successful seasons with the Detroit Vipers of the International Hockey League. He began his tenure in Detroit as general manager and head coach in 1994-95, the team's inaugural year. Midway through the 1995-96 season, he stepped down as coach to concentrate on his g.m. duties. Under Dudley's direction, the Vipers finished with more than 100 points in each of their first four seasons and won the 1996-97 Turner Cup championship. While with Detroit, Dudley developed future NHL stars such as Petr Sykora and Miroslav Satan and imported a 17-year-old Sergei Samsonov.

Dudley's minor league coaching career also includes stints with the IHL's Phoenix Roadrunners (1993-94), San Diego Gulls (1992-93) and Flint Spirits (1986-88), the AHL's New Haven Nighthawks (1988-89) and the ECHL's Carolina Thunderbirds (1981-86). He spent two-and-a-half seasons as an NHL coach with the Buffalo Sabres between 1989 and 1992. As a coach, Dudley amassed a lifetime record of 476-196-51, and while with San Diego he set marks for the best start in professional hockey (25-0-1) and the most wins in a season in professional hockey history (62). As a general manager, he has led his teams to the finals eight times in three different leagues (IHL, American Hockey League and East Coast Hockey League), winning four championships.

As a player, Dudley skated six seasons in the NHL with the Sabres from 1972-73 to 1974-75 and 1978-79 to 1980-81. He finished ninth in Hart Trophy voting after posting 70 points (31 goals, 39 assists) in the 1974-75 season. Dudley posted back-to-back 40-goal seasons with Cincinnati of the World Hockey Association in 1975-76 and 1976-77 and played briefly with the Winnipeg Jets and Fredericton of the AHL before retiring early in the 1981-82 season.

NHL Coaching Record

		Regular Season				Playoffs		
Season	Team	Games	W	L	T	Games	W	L
1989-90	Buffalo	80	45	27	8	6	2	4
1990-91	Buffalo	80	31	30	9	6	2	4
1991-92	Buffalo	28	9	15	4	....	....	....
	NHL Totals	**188**	**85**	**72**	**31**	**12**	**4**	**8**

General Managers' History

Bob Clarke, 1993-94; Bryan Murray, 1994-95 to 1999-2000; Bryan Murray and Bill Torrey, 2000-01; Bill Torrey and Chuck Fletcher, 2001-02; Rick Dudley, 2002-03.

Acquired from Carolina on January 16, 2002, defenseman Sandis Ozolinsh wound up playing 83 games last season. He scored 10 goals for Florida in just 37 games.

Club Records

Team

(Figures in brackets for season records are games played; records for fewest points, wins, ties, losses, goals, goals against are for 70 or more games)

Record		
Most Points	**98**	1999-2000 (82)
Most Wins	**43**	1999-2000 (82)
Most Ties	**19**	1996-97 (82)
Most Losses	**44**	2001-02 (82)
Most Goals	**254**	1995-96 (82)
Most Goals Against	**256**	1997-98 (82)
Fewest Points	**60**	2001-02 (82)
Fewest Wins	**22**	2000-01 (82), 2001-02 (82)
Fewest Ties	**6**	1999-2000 (82)
Fewest Losses	**27**	1999-2000 (82)
Fewest Goals	**180**	2001-02 (82)
Fewest Goals Against	**201**	1996-97 (82)
Longest Winning Streak		
Overall	**7**	Nov. 2-14/95
Home	**5**	Nov. 5-14/95
Away	**4**	Dec. 2-12/95, Nov. 13-Dec 1/96, Oct. 25-Nov. 22/97
Longest Undefeated Streak		
Overall	**12**	Oct. 5-30/96 (8 wins, 4 ties)
Home	**8**	Nov. 5-26/95 (7 wins, 1 tie)
Away	**7**	Twice
Longest Losing Streak		
Overall	**13**	Feb. 7-Mar. 23/98
Home	**6**	Feb. 25-Mar. 23/98
Away	**9**	Feb. 9-Mar. 25/02
Longest Winless Streak		
Overall	**15**	Feb. 1-Mar. 23/98 (14 losses, 1 tie)
Home	**8**	Feb. 1-Mar. 23/98 (7 losses, 1 tie)
Away	**16**	Jan. 2-Mar. 21/98 (12 losses, 4 ties)
Most Shutouts, Season	**6**	1994-95 (48), 2000-01 (82)
Most PIM, Season	**1,994**	2001-02 (82)
Most Goals, Game	**10**	Nov. 26/97 (Bos. 5 at Fla. 10)

Individual

Most Seasons	**9**	Paul Laus
Most Games	**573**	Robert Svehla
Most Goals, Career	**157**	Scott Mellanby
Most Assists, Career	**229**	Robert Svehla
Most Points, Career	**354**	Scott Mellanby (157G, 197A)
Most PIM, Career	**1,702**	Paul Laus
Most Shutouts, Career	**13**	John Vanbiesbrouck
Longest Consecutive Games Streak	**300**	Robert Svehla (Dec. 23/98-Apr. 14/02)
Most Goals, Season	**59**	Pavel Bure (2000-01)
Most Assists, Season	**53**	Viktor Kozlov (1999-2000)
Most Points, Season	**94**	Pavel Bure (1999-2000; 58G, 36A)
Most PIM, Season	**354**	Peter Worrell (2001-02)
Most Points, Defenseman, Season	**57**	Robert Svehla (1995-96; 8G, 49A)
Most Points, Center, Season	**70**	Viktor Kozlov (1999-2000; 17G, 53A)
Most Points, Right Wing, Season	**94**	Pavel Bure (1999-2000; 58G, 36A)
Most Points, Left Wing, Season	**71**	Ray Whitney (1999-2000; 29G, 42A)
Most Points, Rookie, Season	**50**	Jesse Belanger (1993-94; 17G, 33A)
Most Shutouts, Season	**5**	Roberto Luongo (2000-01)
Most Goals, Game	**4**	Mark Parrish (Oct. 30/98); Pavel Bure (Jan. 1/00, Feb. 10/01)
Most Assists, Game	**4**	Scott Mellanby (Nov. 26/97); Ray Whitney (Oct. 30/00)
Most Points, Game	**5**	Pavel Bure (Feb. 10/01; 4G, 1A)

Coaching History

Roger Neilson, 1993-94, 1994-95; Doug MacLean, 1995-96, 1996-97; Doug MacLean and Bryan Murray, 1997-98; Terry Murray, 1998-99, 1999-2000; Terry Murray and Duane Sutter, 2000-01; Duane Sutter and Mike Keenan, 2001-02; Mike Keenan, 2002-03.

Captains' History

Brian Skrudland, 1993-94 to 1996-97; Scott Mellanby, 1997-98 to 2000-01; Pavel Bure, 2001-02.

All-time Record vs. Other Clubs

Regular Season

	At Home								On Road								Total							
	GP	W	L	T	OL	GF	GA	PTS	GP	W	L	T	OL	GF	GA	PTS	GP	W	L	T	OL	GF	GA	PTS
Anaheim	6	3	1	2	0	19	13	8	8	5	2	0	1	26	23	11	14	8	3	2	1	45	36	19
Atlanta	8	3	4	1	0	21	26	7	7	1	1	4	1	23	21	7	15	4	5	5	1	44	47	14
Boston	17	8	9	0	0	47	50	16	18	8	7	3	0	49	48	19	35	16	16	3	0	96	98	35
Buffalo	17	9	8	0	0	54	54	18	19	4	12	2	1	29	57	11	36	13	20	2	1	83	111	29
Calgary	8	2	3	2	1	18	22	7	7	3	3	1	0	18	17	7	15	5	6	3	1	36	39	14
Carolina	20	6	5	7	2	51	42	21	20	8	9	2	1	55	62	19	40	14	14	9	3	106	104	40
Chicago	7	2	5	0	0	16	28	4	8	3	4	1	0	27	28	7	15	5	9	1	0	43	56	11
Colorado	9	1	8	0	0	24	39	2	10	3	4	3	0	23	29	9	19	4	12	3	0	47	68	11
Columbus	2	1	0	0	1	8	7	3	1	1	0	0	0	3	0	2	3	2	0	0	1	11	7	5
Dallas	8	3	4	1	0	21	24	7	7	2	3	2	0	19	23	6	15	5	7	3	0	40	47	13
Detroit	8	1	4	2	1	16	25	5	6	1	3	2	0	18	24	4	14	2	7	4	1	34	49	9
Edmonton	8	5	1	2	0	24	20	12	6	2	3	1	0	14	20	5	14	7	4	3	0	38	40	17
Los Angeles	7	4	0	3	0	21	11	11	8	3	5	0	0	24	25	6	15	7	5	3	0	45	36	17
Minnesota	1	1	0	0	0	2	1	2	2	0	1	1	0	0	6	1	3	1	1	1	0	2	7	3
Montreal	18	9	6	3	0	60	51	21	17	8	6	2	1	39	45	19	35	17	12	5	1	99	96	40
Nashville	3	2	0	1	0	9	5	5	4	2	1	1	0	8	6	5	7	4	1	2	0	17	11	10
New Jersey	21	7	10	4	0	46	51	18	20	5	10	3	2	42	61	15	41	12	20	7	2	88	112	33
NY Islanders	21	10	7	4	0	63	60	24	21	11	8	2	0	58	51	24	42	21	15	6	0	121	111	48
NY Rangers	21	9	11	1	0	56	59	19	20	7	9	4	0	52	60	18	41	16	20	5	0	108	119	37
Ottawa	18	9	7	1	1	58	53	20	18	8	8	2	0	49	50	18	36	17	15	3	1	107	103	38
Philadelphia	20	5	14	0	1	49	72	11	21	7	9	5	0	50	54	19	41	12	23	5	1	99	126	30
Phoenix	7	3	4	0	0	22	19	6	8	3	2	2	1	23	18	9	15	6	6	2	1	45	37	15
Pittsburgh	18	9	8	1	0	46	43	19	19	4	10	3	2	53	68	13	37	13	18	4	2	99	111	32
St. Louis	8	2	4	2	0	16	18	6	7	1	5	1	0	11	21	3	15	3	9	3	0	27	39	9
San Jose	7	2	0	5	0	20	14	9	8	2	4	2	0	17	23	6	15	4	4	7	0	37	37	15
Tampa Bay	22	14	4	3	1	64	42	32	23	12	7	4	0	68	47	28	45	26	11	7	1	132	89	60
Toronto	13	4	6	3	0	38	39	11	11	2	7	2	0	26	41	6	24	6	13	5	0	64	80	17
Vancouver	7	3	2	1	1	19	22	8	7	1	2	4	0	15	20	6	14	4	4	5	1	34	42	14
Washington	23	10	10	2	1	61	61	23	22	6	11	4	1	52	65	17	45	16	21	6	2	113	126	40
Totals	**353**	**147**	**145**	**51**	**10**	**969**	**971**	**355**	**353**	**123**	**156**	**63**	**11**	**891**	**1013**	**320**	**706**	**270**	**301**	**114**	**21**	**1860**	**1984**	**675**

Playoffs

	Series	W	L	GP	W	L	T	GF	GA	Last Mtg.	Rnd.	Result
Boston	1	1	0	5	4	1	0	22	16	1996	CQF	W 4-1
Colorado	1	0	1	4	0	4	0	4	15	1996	F	L 0-4
New Jersey	1	0	1	4	0	4	0	6	12	2000	CQF	L 0-4
NY Rangers	1	0	1	5	1	4	0	10	13	1997	CQF	L 1-4
Philadelphia	1	1	0	6	4	2	0	15	11	1996	CSF	W 4-2
Pittsburgh	1	1	0	7	4	3	0	20	15	1996	CF	W 4-3
Totals	**6**	**3**	**3**	**31**	**13**	**18**	**0**	**77**	**82**			

Playoff Results 2002-1998

Year	Round	Opponent	Result	GF	GA
2000	CQF	New Jersey	L 0-4	6	12

Abbreviations: Round: F - Final; **CF** - conference final; **CSF** - conference semi-final; **CQF** - conference quarter-final.

Colorado totals include Quebec, 1993-94 to 1994-95.
Phoenix totals include Winnipeg, 1993-94 to 1995-96.

Carolina totals include Hartford, 1993-94 to 1996-97.

2001-02 Results

Oct.	4	at Philadelphia	2-5		7	at Washington	2-1
	6	NY Islanders	0-3		9	Dallas	2-3
	7	at Tampa Bay	5-0		11	Ottawa	2-4
	10	Ottawa	0-2		12	Washington	0-1
	13	Philadelphia	2-5		16	Chicago	0-3
	16	at Vancouver	2-2		18	at Dallas	3-2
	18	at Calgary	1-3		19	Atlanta	1-6
	20	at Edmonton	2-6		21	Montreal	7-5
	24	Washington	3-4*		23	New Jersey	1-3
	26	Los Angeles	3-2		25	at Carolina	1-1
	28	at Pittsburgh	2-2		26	at Boston	2-4
	30	at NY Islanders	2-3		30	Phoenix	1-3
	31	at NY Rangers	3-1	**Feb.**	4	NY Islanders	6-6
Nov.	3	NY Rangers	3-5		6	Tampa Bay	2-3
	7	Pittsburgh	2-0		7	at Tampa Bay	3-1
	10	Philadelphia	2-3*		9	at Boston	1-4
	12	Buffalo	3-5		12	at Nashville	0-1
	14	Toronto	2-3		13	at Chicago	4-5
	16	at Buffalo	2-0		26	at Washington	3-4
	17	at Montreal	0-1*		27	Detroit	2-3*
	19	at Toronto	1-5	**Mar.**	2	at Tampa Bay	2-3
	21	Anaheim	6-0		5	at Pittsburgh	5-6*
	24	New Jersey	1-5		8	Edmonton	5-4
	27	at Colorado	1-4		9	Nashville	2-2
	29	at Minnesota	0-6		13	Calgary	3-3
Dec.	1	Atlanta	2-5		15	Buffalo	5-2
	5	Columbus	2-0		17	at Ottawa	0-2
	8	Carolina	2-3		20	Montreal	1-4
	12	at Carolina	1-4		21	at Carolina	2-3
	14	at New Jersey	3-2		23	Boston	1-3
	15	at NY Islanders	3-1		25	at New Jersey	1-3
	17	at NY Rangers	2-4		26	at Montreal	2-1
	19	Washington	5-2		28	at Ottawa	4-3
	22	St. Louis	0-2		30	NY Rangers	2-4
	26	at Atlanta	3-3	**Apr.**	3	Pittsburgh	3-2
	28	Boston	1-7		5	at Buffalo	1-3
	29	Toronto	4-2		6	at Toronto	2-2
	31	Atlanta	4-3		8	at Philadelphia	4-4
Jan.	2	at Los Angeles	1-3		10	at Atlanta	4-4
	4	at Anaheim	2-1		12	Carolina	1-3
	5	at San Jose	0-6		14	Tampa Bay	2-3*

* – Overtime

Entry Draft Selections 2002-1993

2002

Pick	
3	Jay Bouwmeester
9	Petr Taticek
40	Rob Globke
67	Gregory Campbell
134	Topi Jaakola
158	Vince Bellissimo
169	Jeremy Swanson
196	Mikael Vuorio
200	Denis Yachmenev
232	Peter Hafner

2001

Pick	
4	Stephen Weiss
24	Lukas Krajicek
34	Greg Watson
64	Tomas Malec
68	Grant McNeill
117	Mike Woodford
136	Billy Thompson
169	Dustin Johner
200	Toni Koivisto
231	Kyle Bruce
263	Jan Blanar
267	Ivan Majesky

2000

Pick	
58	Vladimir Sapozhnikov
77	Robert Fried
82	Sean O'Connor
115	Chris Eade
120	Davis Parley
190	Josh Olson
234	Janis Sprukts
253	Mathew Sommerfeld

1999

Pick	
12	Denis Shvidki
40	Alexander Auld
70	Niklas Hagman
80	Jean-Francois Laniel
103	Morgan McCormick
109	Rod Sarich
169	Brad Woods
198	Travis Eagles
227	Jonathon Charron

1998

Pick	
30	Kyle Rossiter
61	Joe DiPenta
63	Lance Ward
89	Ryan Jardine
117	Jaroslav Spacek
148	Chris Ovington
176	B.J. Ketcheson
203	Ian Jacobs
231	Adrian Wichser

1997

Pick	
20	Mike Brown
47	Kristian Huselius
56	Vratislav Cech
74	Nick Smith
95	Ivan Novoseltsev
127	Pat Parthenais
155	Keith Delaney
183	Tyler Palmer
211	Doug Schueller
237	Benoit Cote

1996

Pick	
20	Marcus Nilson
60	Chris Allen
65	Oleg Kvasha
82	Joey Tetarenko
129	Andrew Long
156	Gaetan Poirier
183	Alexandre Couture
209	Denis Khloptonov
235	Russell Smith

1995

Pick	
10	Radek Dvorak
36	Aaron MacDonald
62	Mike O'Grady
80	Dave Duerden
88	Daniel Tjarnqvist
114	Francois Cloutier
166	Peter Worrell
192	Filip Kuba
218	David Lemanowicz

1994

Pick	
1	Ed Jovanovski
27	Rhett Warrener
31	Jason Podollan
36	Ryan Johnson
84	David Nemirovsky
105	Dave Geris
157	Matt O'Dette
183	Jason Boudrias
235	Tero Lehtera
261	Per Gustafsson

1993

Pick	
5	Rob Niedermayer
41	Kevin Weekes
57	Chris Armstrong
67	Mikael Tjallden
78	Steve Washburn
83	Bill McCauley
109	Todd MacDonald
135	Alain Nasreddine
161	Trevor Doyle
187	Briane Thompson
213	Chad Cabana
239	John Demarco
265	Eric Montreuil

Coach

KEENAN, MIKE
Coach, Florida Panthers. Born in Bowmanville, Ont., October 21, 1949.

Mike Keenan became the sixth head coach in Panthers history on December 3, 2001. The veteran coach has guided six other NHL franchises — the Philadelphia Flyers (1984-88), the Chicago Blackhawks (1988-92), the New York Rangers (1993-94), the St. Louis Blues (1994-96), the Vancouver Canucks (1997-98) and the Boston Bruins (2000-01) — and ranks among the NHL's all-time coaching leaders in both games coached and victories. In addition to his coaching duties, Keenan has acted as interim general manager for Vancouver, and as general manager in Chicago and St. Louis.

Keenan led the New York Rangers to the Stanley Cup championship in 1994, the team's first championship in 54 years. The Rangers went 52-24-8 under Keenan in 1993-94, going from a non-playoff team to the Presidents' Trophy winner and Stanley Cup champion in just one year. The Keenan-led Rangers won a thrilling seven-game Stanley Cup final over the Vancouver Canucks that year. During his four years in Chicago, the team made the playoffs every year and reached the Stanley Cup finals in 1992. Keenan's NHL coaching career began in Philadelphia, where his team made two appearances in the Stanley Cup finals (1985 and 1987) in four years. He also won the Jack Adams Award as the NHL's coach of the year in 1985. His resume includes three Presidents' Trophy wins (1985, 1991 and 1994), and six division titles (1985, 1986, 1987, 1990, 1991 and 1994). He led Team Canada to victory at the Canada Cup in 1987 and 1991. His international career also includes head coaching jobs with Team Canada at the 1980 World Junior Championships and the World Championships in 1993.

An accomplished coach at all levels of hockey, Keenan guided the Peterborough Petes (OHL) to the Memorial Cup finals in 1980. He took the Rochester Americans (AHL) from a non-playoff team to a Calder Cup winner in just three seasons, taking over in 1980-81 and winning the AHL championship in 1983. He also led the University of Toronto to a CIAU championship, winning the University Cup in 1984.

Coaching Record

		Regular Season				Playoffs		
Season	Team	Games	W	L	T	Games	W	L
1979-80	Peterborough (OHL)	68	47	20	1	18	15	3
1980-81	Rochester (AHL)	80	30	42	8	....	....	....
1981-82	Rochester (AHL)	80	40	31	9	9	4	5
1982-83	Rochester (AHL)	80	46	25	9	16	12	4
1983-84	U. of Toronto (CIAU)	49	41	5	3	....	....	....
1984-85	**Philadelphia (NHL)**	**80**	**53**	**20**	**7**	**19**	**12**	**7**
1985-86	**Philadelphia (NHL)**	**80**	**53**	**23**	**4**	**5**	**2**	**3**
1986-87	**Philadelphia (NHL)**	**80**	**46**	**26**	**8**	**26**	**15**	**11**
1987-88	**Philadelphia (NHL)**	**80**	**38**	**33**	**9**	**7**	**3**	**4**
1988-89	**Chicago (NHL)**	**80**	**27**	**41**	**12**	**16**	**9**	**7**
1989-90	**Chicago (NHL)**	**80**	**41**	**33**	**6**	**20**	**10**	**10**
1990-91	**Chicago (NHL)**	**80**	**49**	**23**	**8**	**6**	**2**	**4**
1991-92	**Chicago (NHL)**	**80**	**36**	**29**	**15**	**18**	**12**	**6**
1993-94	**NY Rangers (NHL)**	**84**	**52**	**24**	**8**	**23**	**16**	**7***
1994-95	**St. Louis (NHL)**	**48**	**28**	**15**	**5**	**7**	**3**	**4**
1995-96	**St. Louis (NHL)**	**82**	**32**	**34**	**16**	**13**	**7**	**6**
1996-97	**St. Louis (NHL)**	**33**	**15**	**17**	**1**	....	....	....
1997-98	**Vancouver (NHL)**	**63**	**21**	**30**	**12**	....	....	....
1998-99	**Vancouver (NHL)**	**45**	**15**	**24**	**6**	....	....	....
2000-01	**Boston (NHL)**	**74**	**33**	**34**	**7**	....	....	....
2001-02	**Florida (NHL)**	**56**	**16**	**32**	**8**	....	....	....
	NHL Totals	**1125**	**555**	**438**	**132**	**160**	**91**	**69**

* Stanley Cup win.

Club Directory

Office Depot Center

Florida Panthers
Office Depot Center
One Panthers Parkway
Sunrise, FL 33323
Phone **954/835-7000**
FAX 954/835-7600
www.floridapanthers.com
Capacity: 19,250

Executive

General Partner and Chairman of the Board/ Chief Executive Officer . . . . . Alan Cohen
Partner/President, Panthers Hockey LLLP . . . . . Jordan Zimmerman
Partners . . . . . Steve Cohen, David Epstein, Dr. Elliott Hahn, H. Wayne Huizenga, Bernie Kosar, Al Maroone, Michael Maroone
Governor . . . . . William A. Torrey
Chief Operating Officer . . . . . Jeff Cogen
Chief Financial Officer . . . . . Bill Duffy
Senior Vice President . . . . . Steve Dangerfield
Executive Assistant to Chairman and CEO . . . . . Athena Melfi
Executive Assistant to COO & Sr. Vice President . . . Janine Shea
Executive Assistant to Governor & CFO . . . . . Cathy Stevenson
Executive Assistant to Director of Finance/Controller and Director of Human Resources/Payroll . . . . . Cathy Cuffe
Administrative Assistant . . . . . Jackie Smid-Cortez

Hockey Operations

General Manager . . . . . Rick Dudley
Head Coach . . . . . Mike Keenan
Assistant Coaches . . . . . Paul Baxter, George Kingston
Goaltending Instructor . . . . . Clint Malarchuk
Dir. of Hockey Operations/Corporate Counsel . . . . . Mike Santos
Assistant to the General Manager . . . . . Grant Sonier
Director of Player Development . . . . . Duane Sutter
Director of Amateur Scouting . . . . . Scott Luce
Director of Professional Player Evaluation . . . . . Michael Abbamont
Skating & Skills Instructor and Scout . . . . . Paul Vincent
Head Amateur Scout . . . . . Darwin Bennett
Amateur Scouts . . . . . Erin Ginnell, Ron Harris, Sean O'Brien
Part-time Scouts . . . . . Dale Degray, Richard Rothermel, Buck Steele
Pro Scout . . . . . Billy Dea
European Scouts . . . . . Niklas Blomgren, Pavel Routa
Executive Assistant to the General Manager . . . . . Vanessa Rey-Fischel
Strength & Conditioning Coach . . . . . Chris Reichart
Head Medical Trainer . . . . . Dave Boyer
Head Equipment Manager . . . . . Mark Brennan
Associate Equipment Manager . . . . . Scott Tinkler
Assistant Equipment Manager . . . . . Jon Korman
Team Services Coordinator . . . . . Marni S. Bloomster
Massage Therapist . . . . . Mikhail Manchik
Video Coordinator . . . . . Scott Masters
Orthopaedic Surgeon . . . . . Lex Simpson, M.D.
Internist . . . . . Howard Bush, M.D.
Team Dentist . . . . . Martin Robins, D.D.S.
Neuropsychologist . . . . . Kathleen Knee, Psy.D.
Laser Eye Surgeon . . . . . Cory Lessner, M.D.

Communications

Director of Media Relations . . . . . Randy Sieminski
Dir. of Community Relations & Youth Hockey . . . . . Randy Moller
Media Relations Manager . . . . . Ryan Nadeau
Community Development Manager . . . . . Jean Marshall
Communications/Publications Coordinator . . . . . Michael Citro
Community Development Coordinator . . . . . Robyn Fink
Youth & Amateur Hockey Coordinator . . . . . Keith Martin
Website Coordinator . . . . . Mary Lou Veroline
Administrative Assistant . . . . . Giselle Seoane

Corporate Sales

Director of Corporate Sales . . . . . Jason Camp
Corporate Marketing Manager . . . . . Susan Ferro
Corporate Sales Manager . . . . . Bob Ohrablo

Marketing Partnerships

Director of Marketing Partnerships . . . . . Brette Sadler
Marketing Partnerships Managers . . . . . Heather Germano, Kathy Stock

Finance and Administration

Director of Finance/Controller . . . . . Evelyn Lopez
Director of Human Resources/Payroll . . . . . Carol Duncanson
Director of Information Technology . . . . . Kelly Moyer
Assistant Controller . . . . . Michele Gilbert
Office Manager . . . . . Laura Barrera

Game Presentation/Promotions

Senior Manager of Event Presentation . . . . . Sean Collins
Game Presentation Producer . . . . . Marc Bick
Special Events and Mascot Manager . . . . . Phil Crowhurst
Fan Interaction Coordinator . . . . . Angela Carrasco
Special Events Coordinator . . . . . Eric Wasser

Ticket Operations and Sales

Senior Director of Ticket Sales . . . . . Scott Wampold
Director of Season Ticket Sales . . . . . John Borozzi
Director of Suite & Club Level Services . . . . . Steve Woznick
Manager of Ticket Operations . . . . . Matt Coyne
Manager of Suite & Club Level Services . . . . . Peter Cameron
Group Sales Manager . . . . . Steve Golub

Miscellaneous

Television . . . . . Fox Sports Net
Television Announcers . . . . . Jeff Rimer, Denis Potvin
Radio Flagship . . . . . WQAM (560 AM)
Radio Announcers . . . . . Jiggs McDonald, Randy Moller, Steve Goldstein
Practice Facility . . . . . incredible Ice
General Manager, incredible Ice . . . . . Jeff Campol

Los Angeles Kings

2001-02 Results: 40W-27L-11T-4OTL 95PTS.
Third, Pacific Division

2002-03 Schedule

Oct.	Wed.	9	Phoenix
	Sat.	12	Detroit
	Wed.	16	at Anaheim
	Thu.	17	Colorado
	Sat.	19	Vancouver
	Wed.	23	at Detroit
	Fri.	25	at NY Rangers
	Sun.	27	at Columbus
	Tue.	29	at Atlanta
	Thu.	31	at Chicago
Nov.	Sat.	2	Nashville
	Mon.	4	Minnesota
	Tue.	5	at San Jose
	Fri.	8	at Ottawa
	Sat.	9	at Montreal
	Tue.	12	at Toronto
	Thu.	14	at Vancouver
	Sat.	16	at Edmonton
	Tue.	19	at Minnesota
	Thu.	21	at St. Louis
	Sat.	23	Dallas
	Wed.	27	Florida
	Fri.	29	at Anaheim*
	Sat.	30	Chicago
Dec.	Thu.	5	Nashville
	Sat.	7	Columbus
	Tue.	10	at Nashville
	Wed.	11	at Dallas
	Sat.	14	Pittsburgh
	Sun.	15	at Phoenix*
	Tue.	17	St. Louis
	Thu.	19	Anaheim
	Sun.	22	at Chicago
	Mon.	23	at St. Louis
	Thu.	26	Phoenix
	Sun.	29	at Colorado*
	Mon.	30	Chicago
Jan.	Thu.	2	Philadelphia
	Sat.	4	Dallas
	Mon.	6	at Minnesota
	Tue.	7	at Dallas
	Thu.	9	Edmonton
	Sat.	11	St. Louis
	Mon.	13	San Jose
	Thu.	16	at Edmonton
	Sat.	18	at Calgary
	Wed.	22	at Anaheim
	Thu.	23	Minnesota
	Sat.	25	New Jersey
	Mon.	27	San Jose
	Tue.	28	at San Jose
	Thu.	30	Ottawa
Feb.	Wed.	5	Phoenix
	Fri.	7	Carolina
	Sun.	9	at Dallas
	Tue.	11	at Nashville
	Thu.	13	Calgary
	Sat.	15	NY Islanders*
	Mon.	17	San Jose
	Thu.	20	at Philadelphia
	Fri.	21	at Buffalo
	Mon.	24	at Detroit
	Tue.	25	at Pittsburgh
	Thu.	27	at Columbus
Mar.	Sat.	1	Atlanta*
	Tue.	4	Anaheim
	Thu.	6	Edmonton
	Sat.	8	Montreal
	Mon.	10	Detroit
	Wed.	12	at Tampa Bay
	Fri.	14	at Washington
	Sat.	15	at Carolina
	Tue.	18	Calgary
	Thu.	20	Tampa Bay
	Sat.	22	Boston
	Tue.	25	Columbus
	Thu.	27	at Colorado
	Sat.	29	Vancouver
	Mon.	31	at Phoenix
Apr.	Wed.	2	Colorado
	Fri.	4	at Calgary
	Sun.	6	at Vancouver*

* Denotes afternoon game.

With 32 goals in just 63 games last season, Ziggy Palffy was one of the NHL's most efficient scorers. His shooting percentage of 19.9 was one of the best in the NHL, as were his 15 power-play goals.

Year-by-Year Record

		Home				Road				Overall								
Season	GP	W	L	T	OL	W	L	T	OL	W	L	T	OL	GF	GA	Pts.	Finished	Playoff Result
2001-02	82	22	12	6	1	18	15	5	3	40	27	11	4	214	190	95	3rd, Pacific Div.	Lost Conf. Quarter-Final
2000-01	82	20	12	8	1	18	16	5	2	38	28	13	3	252	228	92	3rd, Pacific Div.	Lost Conf. Semi-Final
1999-2000	82	21	13	5	2	18	14	7	2	39	27	12	4	245	228	94	2nd, Pacific Div.	Lost Conf. Quater-Final
1998-99	82	18	20	3	...	14	25	2	...	32	45	5	...	189	222	69	5th, Pacific Div.	Out of Playoffs
1997-98	82	22	16	3	...	16	17	8	...	38	33	11	...	227	225	87	2nd, Pacific Div.	Lost Conf. Quater-Final
1996-97	82	18	16	7	...	10	27	4	...	28	43	11	...	214	268	67	6th, Pacific Div.	Out of Playoffs
1995-96	82	16	16	9	...	8	24	9	...	24	40	18	...	256	302	66	6th, Pacific Div.	Out of Playoffs
1994-95	48	7	11	6	...	9	12	3	...	16	23	9	...	142	174	41	4th, Pacific Div.	Out of Playoffs
1993-94	84	18	19	5	...	9	26	7	...	27	45	12	...	294	322	66	5th, Pacific Div.	Out of Playoffs
1992-93	84	22	15	5	...	17	20	5	...	39	35	10	...	338	340	88	3rd, Smythe Div.	Lost Final
1991-92	80	20	11	9	...	15	20	5	...	35	31	14	...	287	296	84	2nd, Smythe Div.	Lost Div. Semi-Final
1990-91	80	26	9	5	...	20	15	5	...	46	24	10	...	340	254	102	1st, Smythe Div.	Lost Div. Final
1989-90	80	21	16	3	...	13	23	4	...	34	39	7	...	338	337	75	4th, Smythe Div.	Lost Div. Final
1988-89	80	25	12	3	...	17	19	4	...	42	31	7	...	376	335	91	2nd, Smythe Div.	Lost Div. Final
1987-88	80	19	18	3	...	11	24	5	...	30	42	8	...	318	359	68	4th, Smythe Div.	Lost Div. Semi-Final
1986-87	80	20	17	3	...	11	24	5	...	31	41	8	...	318	341	70	4th, Smythe Div.	Lost Div. Semi-Final
1985-86	80	9	27	4	...	14	22	4	...	23	49	8	...	284	389	54	5th, Smythe Div.	Out of Playoffs
1984-85	80	20	14	6	...	14	18	8	...	34	32	14	...	339	326	82	4th, Smythe Div.	Lost Div. Semi-Final
1983-84	80	13	19	8	...	10	25	5	...	23	44	13	...	309	376	59	5th, Smythe Div.	Out of Playoffs
1982-83	80	20	13	7	...	7	28	5	...	27	41	12	...	308	365	66	5th, Smythe Div.	Out of Playoffs
1981-82	80	19	15	6	...	5	26	9	...	24	41	15	...	314	369	63	4th, Smythe Div.	Lost Div. Final
1980-81	80	22	11	7	...	21	13	6	...	43	24	13	...	337	290	99	2nd, Norris Div.	Lost Prelim. Round
1979-80	80	18	13	9	...	12	23	5	...	30	36	14	...	290	313	74	2nd, Norris Div.	Lost Prelim. Round
1978-79	80	20	13	7	...	14	21	5	...	34	34	12	...	292	286	80	3rd, Norris Div.	Lost Prelim. Round
1977-78	80	18	16	6	...	13	18	9	...	31	34	15	...	243	245	77	3rd, Norris Div.	Lost Prelim. Round
1976-77	80	20	13	7	...	14	18	8	...	34	31	15	...	271	241	83	2nd, Norris Div.	Lost Quarter-Final
1975-76	80	22	13	5	...	16	20	4	...	38	33	9	...	263	265	85	2nd, Norris Div.	Lost Quarter-Final
1974-75	80	22	7	11	...	20	10	10	...	42	17	21	...	269	185	105	2nd, Norris Div.	Lost Prelim. Round
1973-74	78	22	13	4	...	11	20	8	...	33	33	12	...	233	231	78	3rd, West Div.	Lost Quarter-Final
1972-73	78	21	11	7	...	10	25	4	...	31	36	11	...	232	245	73	6th, West Div.	Out of Playoffs
1971-72	78	14	23	2	...	6	26	7	...	20	49	9	...	206	305	49	7th, West Div.	Out of Playoffs
1970-71	78	17	14	8	...	8	26	5	...	25	40	13	...	239	303	63	5th, West Div.	Out of Playoffs
1969-70	76	12	22	4	...	2	30	6	...	14	52	10	...	168	290	38	6th, West Div.	Out of Playoffs
1968-69	76	19	14	5	...	5	28	5	...	24	42	10	...	185	260	58	4th, West Div.	Lost Semi-Final
1967-68	74	20	13	4	...	11	20	6	...	31	33	10	...	200	224	72	2nd, West Div.	Lost Quarter-Final

Franchise date: June 5, 1967

PACIFIC DIVISION

36th NHL Season

2002-03 Player Personnel

FORWARDS	HT	WT	S	Place of Birth	Date	2001-02 Club
ALLISON, Jason	6-3	215	R	North York, Ont.	5/29/75	Los Angeles
ARMSTRONG, Derek	5-11	188	R	Ottawa, Ont.	4/23/73	Bern
AULIN, Jared	6-0	180	R	Calgary, Alta.	3/15/82	Kamloops
BEDNAR, Jaroslav	5-11	198	R	Prague, Czech.	11/8/76	Los Angeles-Manchester
BEKAR, Derek	6-2	205	L	Burnaby, B.C.	9/15/75	Manchester
BELANGER, Eric	6-0	185	L	Sherbrooke, Que.	12/16/77	Los Angeles
BELANGER, Ken	6-4	225	L	Sault Ste. Marie, Ont.	5/14/74	Los Angeles
BRENNAN, Kip	6-4	210	L	Kingston, Ont.	8/27/80	Los Angeles-Manchester
CAMMALLERI, Mike	5-9	180	L	Richmond Hill, Ont.	6/8/82	U. of Michigan
CHARTRAND, Brad	5-11	191	L	Winnipeg, Man.	12/14/74	Los Angeles-Manchester
DEADMARSH, Adam	6-0	195	R	Trail, B.C.	5/10/75	Los Angeles-United States
ELORANTA, Mikko	6-0	190	L	Turku, Finland	8/24/72	Boston-Los Angeles-Finland
FLINN, Ryan	6-5	223	L	Halifax, N.S.	4/20/80	Reading-Los Angeles-Manchester
HEINZE, Steve	5-11	202	R	Lawrence, MA	1/30/70	Los Angeles
JOHNSON, Craig	6-2	200	L	St. Paul, MN	3/18/72	Los Angeles
KELLY, Steve	6-2	210	L	Vancouver, B.C.	10/26/76	Los Angeles-Manchester
LAPERRIERE, Ian	6-1	201	R	Montreal, Que.	1/19/74	Los Angeles
LEHOUX, Yanick	6-0	170	R	Montreal, Que.	4/8/82	Baie-Comeau-Manchester
PALFFY, Ziggy	5-10	183	L	Skalica, Czech.	5/5/72	Los Angeles-Slovakia
RASMUSSEN, Erik	6-3	208	L	Minneapolis, MN	3/28/77	Buffalo
ROSA, Pavel	6-0	195	R	Most, Czech.	6/7/77	Jokerit
SCHMIDT, Chris	6-3	212	L	Beaverlodge, Alta.	3/1/76	Manchester
SMITHSON, Jerred	6-2	190	R	Vernon, B.C.	2/4/79	Manchester
SMOLINSKI, Bryan	6-1	208	R	Toledo, OH	12/27/71	Los Angeles
DEFENSEMEN						
CORVO, Joe	6-0	205	R	Oak Park, IL	6/20/77	Manchester
HOLLAND, Jason	6-3	209	R	Morinville, Alta.	4/30/76	Los Angeles-Manchester
LILJA, Andreas	6-3	222	L	Landskrona, Sweden	7/13/75	Los Angeles-Manchester
MILLER, Aaron	6-3	200	R	Buffalo, NY	8/11/71	Los Angeles-United States
MODRY, Jaroslav	6-2	220	L	Ceske Budejovice, Czech.	2/27/71	Los Angeles
NORSTROM, Mattias	6-2	211	L	Stockholm, Sweden	1/2/72	Los Angeles-Sweden
PUDLICK, Michael	6-3	190	L	Blaine, MN	2/24/78	Manchester
RULLIER, Joe	6-3	200	R	Montreal, Que.	1/28/80	Manchester
SCHNEIDER, Mathieu	5-10	192	L	New York, NY	6/12/69	Los Angeles
SEELEY, Richard	6-2	205	L	Powell River, B.C.	4/30/79	Manchester
VISNOVSKY, Lubomir	5-10	183	L	Topolcany, Czech.	8/11/76	Los Angeles-Slovakia
ZIZKA, Tomas	6-1	198	L	Sternberk, Czech.	10/10/79	Manchester

GOALTENDERS	HT	WT	C	Place of Birth	Date	2001-02 Club
HUET, Cristobal	6-0	194	L	St-Martin-d'Heres, France	9/3/75	Lugano-France
POTVIN, Felix	6-1	190	L	Anjou, Que.	6/23/71	Los Angeles
SCOTT, Travis	6-2	185	L	Kanata, Ont.	9/14/75	Manchester
STORR, Jamie	6-2	195	L	Brampton, Ont.	12/28/75	Los Angeles

General Manager

TAYLOR, DAVE
General Manager, Los Angeles Kings. Born in Levack, Ont., December 4, 1955.

No player in the history of the Kings ever wore the uniform with more distinction and class than Dave Taylor. For 17 seasons, Taylor gave his all, both on and off the ice, receiving All-Star status for his outstanding play.

Fittingly, after finishing his illustrious career during the 1993-94 season, Taylor remains a key part of the Kings organization, now serving as vice president and general manager for the NHL club. Taylor assumed his current responsibilities on April 22, 1997, becoming the seventh g.m. in team history. He joined the Kings front office four years earlier as an assistant to his predecessor, Sam McMaster.

An All-American hockey player while at Clarkson College, Taylor was relatively unknown when the Kings picked him in the 15th round of the 1975 draft. His grit and work ethic kept him around long enough to hook up with a center named Marcel Dionne, who virtually ignited Taylor's career. As a member of the renowned Triple Crown line with Dionne and left winger Charlie Simmer, Taylor became a prolific scorer and a fearsome checker. Taylor's NHL career stats include a Kings-record 1,111 games, 431 goals, 638 assists and 1,069 points.

A four-time NHL All-Star Game selection, Taylor served as the Kings captain for four seasons (1985-89). After posting career highs in goals (47) and points (112) during the 1980-81 season, Taylor earned a spot on the NHL Second All-Star Team. On April 3, 1995, Taylor's jersey No. 18 was retired, joining Rogie Vachon (No. 30) and Marcel Dionne (No. 16). For all his individual accomplishments in hockey, his crowning glory was reaching the Stanley Cup Finals with the 1992-93 Kings.

Away from the ice, Taylor has worked tirelessly for numerous charities throughout the years. Each year he hosts the Dave Taylor Golf Classic benefiting the Cystic Fibrosis Foundation, which annually raises more than $125,000. In 1991, the NHL honored Taylor's contributions to hockey and the community by awarding him both the Bill Masterton and King Clancy trophies.

2001-02 Scoring

* - rookie

Regular Season

Pos	#	Player	Team	GP	G	A	Pts	+/–	PIM	PP	SH	GW	GT	S	%
C	41	Jason Allison	L.A.	73	19	55	74	2	68	5	0	2	2	139	13.7
R	28	Adam Deadmarsh	L.A.	76	29	33	62	8	71	12	0	5	0	139	20.9
R	33	Ziggy Palffy	L.A.	63	32	27	59	5	26	15	1	6	1	161	19.9
C	7	Cliff Ronning	NSH	67	18	31	49	0	24	4	0	0	2	164	11.0
			L.A.	14	1	4	5	0	8	1	0	0	0	35	2.9
			TOTAL	81	19	35	54	0	32	5	0	0	2	199	9.5
D	44	Jaroslav Modry	L.A.	80	4	38	42	–4	65	4	0	0	0	119	3.4
C	21	Bryan Smolinski	L.A.	80	13	25	38	7	56	4	1	0	2	187	7.0
R	57	Steve Heinze	L.A.	73	15	16	31	–15	46	8	0	4	1	123	12.2
D	10	Mathieu Schneider	L.A.	55	7	23	30	3	68	4	0	0	0	123	5.7
D	43	Philippe Boucher	L.A.	80	7	23	30	0	94	4	0	2	0	198	3.5
L	23	Craig Johnson	L.A.	72	13	14	27	14	24	4	1	3	1	102	12.7
C	25	Eric Belanger	L.A.	53	8	16	24	2	21	2	1	1	0	67	11.9
C	22	Ian Laperriere	L.A.	81	8	14	22	5	125	0	0	3	0	89	9.0
D	17	Lubomir Visnovsky	L.A.	72	4	17	21	–5	14	1	0	2	0	95	4.2
L	42	Mikko Eloranta	BOS	6	0	0	0	–1	2	0	0	0	0	15	0.0
			L.A.	71	9	9	18	0	54	1	0	2	0	121	7.4
			TOTAL	77	9	9	18	–1	56	1	0	2	0	136	6.6
D	3	Aaron Miller	L.A.	74	5	12	17	14	54	0	1	3	0	75	6.7
R	29	Brad Chartrand	L.A.	46	7	9	16	5	40	0	0	1	0	49	14.3
R	9	Kelly Buchberger	L.A.	74	6	7	13	–13	105	0	0	1	0	39	15.4
D	14	Mattias Norstrom	L.A.	79	2	9	11	–2	38	0	0	0	0	42	4.8
R	19	Nelson Emerson	L.A.	41	5	2	7	–8	25	0	0	1	0	40	12.5
C	27 *	Jaroslav Bednar	L.A.	22	4	2	6	–4	8	1	0	2	0	20	20.0
D	6	Andreas Lilja	L.A.	26	1	4	5	3	22	1	0	0	0	12	8.3
L	12	Ken Belanger	L.A.	43	2	0	2	–5	85	0	0	0	0	22	9.1
R	38	Rob Valicevic	L.A.	17	1	1	2	–4	8	0	0	0	0	9	11.1
C	24	Adam Mair	L.A.	18	1	1	2	1	57	0	0	0	0	10	10.0
C	11	Steve Kelly	L.A.	8	0	1	1	–1	2	0	0	0	0	0	0.0
D	53	Jason Holland	L.A.	3	0	0	0	–1	0	0	0	0	0	1	0.0
D	37 *	Kip Brennan	L.A.	4	0	0	0	1	22	0	0	0	0	0	0.0
L	49 *	Ryan Flinn	L.A.	10	0	0	0	0	51	0	0	0	0	2	0.0

Goaltending

No.	Goaltender	GPI	Mins	Avg	W	L	T	EN	SO	GA	SA	S%	G	A	PIM
1	Jamie Storr	19	886	1.90	9	4	3	2	2	28	360	.922	0	0	4
39	Felix Potvin	71	4071	2.31	31	27	8	3	6	157	1686	.907	0	1	19
	Totals	**82**	**4989**	**2.29**	**40**	**31**	**11**	**5**	**8**	**190**	**2051**	**.907**			

Playoffs

Pos	#	Player	Team	GP	G	A	Pts	+/–	PIM	PP	SH	GW	GT	S	%
R	33	Ziggy Palffy	L.A.	7	4	5	9	4	0	0	0	0	0	23	17.4
C	41	Jason Allison	L.A.	7	3	3	6	2	4	0	0	1	0	15	20.0
R	28	Adam Deadmarsh	L.A.	4	1	3	4	0	2	0	0	0	0	9	11.1
L	23	Craig Johnson	L.A.	7	1	2	3	3	2	0	0	1	1	15	6.7
C	21	Bryan Smolinski	L.A.	7	2	0	2	–1	2	1	0	0	0	13	15.4
R	29	Brad Chartrand	L.A.	7	1	1	2	0	2	0	0	1	0	6	16.7
L	42	Mikko Eloranta	L.A.	7	1	1	2	–1	2	0	0	0	0	14	7.1
D	44	Jaroslav Modry	L.A.	7	0	2	2	1	0	0	0	0	0	10	0.0
C	7	Cliff Ronning	L.A.	4	0	1	1	–1	2	0	0	0	0	3	0.0
D	17	Lubomir Visnovsky	L.A.	4	0	1	1	1	0	0	0	0	0	2	0.0
R	19	Nelson Emerson	L.A.	5	0	1	1	1	2	0	0	0	0	9	0.0
D	43	Philippe Boucher	L.A.	5	0	1	1	2	2	0	0	0	0	4	0.0
D	10	Mathieu Schneider	L.A.	7	0	1	1	–8	18	0	0	0	0	12	0.0
C	22	Ian Laperriere	L.A.	7	0	1	1	–5	9	0	0	0	0	5	0.0
C	11	Steve Kelly	L.A.	1	0	0	0	0	0	0	0	0	0	0	0.0
C	27 *	Jaroslav Bednar	L.A.	3	0	0	0	0	0	0	0	0	0	6	0.0
R	57	Steve Heinze	L.A.	4	0	0	0	–1	2	0	0	0	0	7	0.0
D	6	Andreas Lilja	L.A.	5	0	0	0	–1	6	0	0	0	0	2	0.0
R	9	Kelly Buchberger	L.A.	7	0	0	0	–3	7	0	0	0	0	2	0.0
D	3	Aaron Miller	L.A.	7	0	0	0	1	0	0	0	0	0	5	0.0
D	14	Mattias Norstrom	L.A.	7	0	0	0	0	4	0	0	0	0	7	0.0
C	25	Eric Belanger	L.A.	7	0	0	0	–5	4	0	0	0	0	7	0.0

Goaltending

No.	Goaltender	GPI	Mins	Avg	W	L	EN	SO	GA	SA	S%	G	A	PIM
1	Jamie Storr	1	0	0.00	0	0	1	0	0	0	.000	0	0	0
39	Felix Potvin	7	417	2.16	3	4	0	1	15	201	.925	0	0	0
	Totals	**7**	**422**	**2.27**	**3**	**4**	**1**	**1**	**16**	**202**	**.921**			

General Managers' History

Larry Regan, 1967-68 to 1972-73; Larry Regan and Jake Milford, 1973-74; Jake Milford, 1974-75 to 1976-77; George Maguire, 1977-78 to 1982-83; George Maguire and Rogie Vachon, 1983-84; Rogie Vachon, 1984-85 to 1991-92; Nick Beverley, 1992-93, 1993-94; Sam McMaster, 1994-95 to 1996-97; Dave Taylor, 1997-98 to date.

Captains' History

Bob Wall, 1967-68, 1968-69; Larry Cahan, 1969-70, 1970-71; Bob Pulford, 1971-72, 1972-73; Terry Harper, 1973-74, 1974-75; Mike Murphy, 1975-76 to 1980-81; Dave Lewis, 1981-82, 1982-83; Terry Ruskowski, 1983-84, 1984-85; Dave Taylor, 1985-86 to 1988-89; Wayne Gretzky, 1989-90 to 1991-92; Wayne Gretzky and Luc Robitaille, 1992-93; Wayne Gretzky, 1993-94, 1994-95; Wayne Gretzky and Rob Blake, 1995-96; Rob Blake, 1996-97 to 2000-01; Mattias Norstrom, 2001-02 to date.

Club Records

Team

(Figures in brackets for season records are games played; records for fewest points, wins, ties, losses, goals, goals against are for 70 or more games)

Record		
Most Points	**105**	1974-75 (80)
Most Wins	**46**	1990-91 (80)
Most Ties	**21**	1974-75 (80)
Most Losses	**52**	1969-70 (76)
Most Goals	**376**	1988-89 (80)
Most Goals Against	**389**	1985-86 (80)
Fewest Points	**38**	1969-70 (76)
Fewest Wins	**14**	1969-70 (76)
Fewest Ties	**5**	1998-99 (82)
Fewest Losses	**17**	1974-75 (80)
Fewest Goals	**168**	1969-70 (76)
Fewest Goals Against	**185**	1974-75 (80)
Longest Winning Streak		
Overall	**8**	Oct. 21-Nov. 7/72, Feb. 23-Mar. 9/92
Home	**12**	Oct. 10-Dec. 5/92
Away	**8**	Dec. 18/74-Jan. 16/75
Longest Undefeated Streak		
Overall	**11**	Feb. 28-Mar. 24/74 (9 wins, 2 ties)
Home	**13**	Oct. 10-Dec. 8/92 (12 wins, 1 tie)
Away	**11**	Oct. 10-Dec. 11/74 (6 wins, 5 ties)
Longest Losing Streak		
Overall	**10**	Feb. 22-Mar. 9/84
Home	**9**	Feb. 8-Mar. 12/86
Away	**12**	Jan. 11-Feb. 15/70
Longest Winless Streak		
Overall	**17**	Jan. 29-Mar. 5/70 (13 losses, 4 ties)
Home	**9**	Jan. 29-Mar. 5/70 (8 losses, 1 tie), Feb. 8-Mar. 12/86 (9 losses)
Away	**21**	Jan. 11-Apr. 3/70 (17 losses, 4 ties)
Most Shutouts, Season	**10**	2000-01 (82)
Most PIM, Season	**2,247**	1992-93 (84)
Most Goals, Game	**12**	Nov. 29/84 (Van. 1 at L.A. 12)

Individual

Most Seasons	**17**	Dave Taylor
Most Games	**1,111**	Dave Taylor
Most Goals, Career	**550**	Marcel Dionne
Most Assists, Career	**757**	Marcel Dionne
Most Points Career	**1,307**	Marcel Dionne (550G, 757A)
Most PIM, Career	**1,846**	Marty McSorley
Most Shutouts, Career	**32**	Rogie Vachon
Longest Consecutive Games Streak	**324**	Marcel Dionne (Jan. 7/78-Jan. 9/82)
Most Goals, Season	**70**	Bernie Nicholls (1988-89)
Most Assists, Season	**122**	Wayne Gretzky (1990-91)
Most Points, Season	**168**	Wayne Gretzky (1988-89; 54G, 114A)
Most PIM, Season	**399**	Marty McSorley (1992-93)
Most Points, Defenseman, Season	**76**	Larry Murphy (1980-81; 16G, 60A)
Most Points, Center, Season	**168**	Wayne Gretzky (1988-89; 54G, 114A)
Most Points, Right Wing, Season	**112**	Dave Taylor (1980-81; 47G, 65A)
Most Points, Left Wing, Season	***125**	Luc Robitaille (1992-93; 63G, 62A)
Most Points, Rookie, Season	**84**	Luc Robitaille (1986-87; 45G, 39A)
Most Shutouts, Season	**8**	Rogie Vachon (1976-77)
Most Goals, Game	**4**	Sixteen times
Most Assists, Game	**6**	Bernie Nicholls (Dec. 1/88), Tomas Sandstrom (Oct. 9/93)
Most Points, Game	**8**	Bernie Nicholls (Dec. 1/88; 2G, 6A)

* NHL Record.

Coaching History

Red Kelly, 1967-68, 1968-69; Hal Laycoe and Johnny Wilson, 1969-70; Larry Regan, 1970-71; Larry Regan and Fred Glover, 1971-72; Bob Pulford, 1972-73 to 1976-77; Ron Stewart, 1977-78; Bob Berry, 1978-79 to 1980-81; Parker MacDonald and Don Perry, 1981-82; Don Perry, 1982-83; Don Perry, Rogie Vachon and Roger Neilson, 1983-84; Pat Quinn, 1984-85, 1985-86; Pat Quinn and Mike Murphy 1986-87; Mike Murphy, Rogie Vachon and Robbie Ftorek, 1987-88; Robbie Ftorek, 1988-89; Tom Webster, 1989-90 to 1991-92; Barry Melrose, 1992-93, 1993-94; Barry Melrose and Rogie Vachon, 1994-95; Larry Robinson, 1995-96 to 1998-99; Andy Murray, 1999-2000 to date.

Retired Numbers

16	Marcel Dionne	1975-1987
18	Dave Taylor	1977-1994
30	Rogie Vachon	1971-1978

All-time Record vs. Other Clubs

Regular Season

	At Home								On Road								Total							
	GP	W	L	T	OL	GF	GA	PTS	GP	W	L	T	OL	GF	GA	PTS	GP	W	L	T	OL	GF	GA	PTS
Anaheim	25	13	8	4	0	72	62	30	24	8	11	5	0	58	77	21	49	21	19	9	0	130	139	51
Atlanta	3	2	0	0	1	14	8	5	2	2	0	0	0	9	3	4	5	4	0	0	1	23	11	9
Boston	59	21	31	7	0	204	216	49	61	11	44	6	0	169	287	28	120	32	75	13	0	373	503	77
Buffalo	53	22	22	9	0	184	182	53	51	15	27	9	0	150	208	39	104	37	49	18	0	334	390	92
Calgary	87	44	34	9	0	327	306	97	90	27	52	11	0	302	402	65	177	71	86	20	0	629	708	162
Carolina	29	16	10	3	0	121	111	35	29	11	13	4	1	113	111	27	58	27	23	7	1	234	222	62
Chicago	70	31	31	8	0	236	239	70	71	29	33	9	0	213	253	67	141	60	64	17	0	449	492	137
Colorado	39	24	12	3	0	164	124	51	38	15	20	3	0	136	158	33	77	39	32	6	0	300	282	84
Columbus	4	3	1	0	0	9	5	6	4	3	1	0	0	20	7	6	8	6	2	0	0	29	12	12
Dallas	77	33	26	18	0	265	225	84	78	17	47	12	2	206	308	48	155	50	73	30	2	471	533	132
Detroit	76	41	21	14	0	310	232	96	75	30	33	12	0	260	286	72	151	71	54	26	0	570	518	168
Edmonton	73	27	31	15	0	288	307	69	73	21	37	15	0	266	338	57	146	48	68	30	0	554	645	126
Florida	8	5	3	0	0	25	24	10	7	0	4	3	0	11	21	3	15	5	7	3	0	36	45	13
Minnesota	4	2	2	0	0	11	11	4	4	2	1	1	0	11	6	5	8	4	3	1	0	22	17	9
Montreal	63	18	36	9	0	195	251	45	63	8	44	11	0	159	286	27	126	26	80	20	0	354	537	72
Nashville	8	4	3	0	1	22	20	9	8	5	1	2	0	21	13	12	16	9	4	2	1	43	33	21
New Jersey	40	27	7	6	0	199	125	60	42	19	18	5	0	146	139	43	82	46	25	11	0	345	264	103
NY Islanders	44	21	16	7	0	161	140	49	43	15	24	4	0	122	155	34	87	36	40	11	0	283	295	83
NY Rangers	59	23	26	10	0	197	213	56	57	16	35	6	0	166	231	38	116	39	61	16	0	363	444	94
Ottawa	8	6	1	1	0	39	18	13	8	3	4	1	0	25	28	7	16	9	5	2	0	64	46	20
Philadelphia	64	20	36	8	0	188	219	48	62	16	39	7	0	156	239	39	126	36	75	15	0	344	458	87
Phoenix	68	23	31	13	1	268	272	60	71	25	34	11	1	237	284	62	139	48	65	24	2	505	556	122
Pittsburgh	68	43	17	8	0	262	181	94	71	23	38	10	0	225	262	56	139	66	55	18	0	487	443	150
St. Louis	74	34	28	12	0	251	216	80	74	18	46	10	0	191	279	46	148	52	74	22	0	442	495	126
San Jose	31	21	7	3	0	103	69	45	32	12	15	3	2	95	108	29	63	33	22	6	2	198	177	74
Tampa Bay	10	1	8	1	0	22	34	3	8	4	4	0	0	19	19	8	18	5	12	1	0	41	53	11
Toronto	64	34	21	9	0	230	187	77	67	22	34	11	0	220	262	55	131	56	55	20	0	450	449	132
Vancouver	95	51	29	15	0	388	297	117	93	30	47	15	1	292	357	76	188	81	76	30	1	680	654	193
Washington	46	26	13	6	1	180	141	59	44	19	18	7	0	165	182	45	90	45	31	13	1	345	323	104
Defunct Clubs	35	27	6	2	0	141	76	56	34	11	14	9	0	91	109	31	69	38	20	11	0	232	185	87
Totals	**1384**	**663**	**517**	**200**	**4**	**5076**	**4511**	**1530**	**1384**	**437**	**738**	**202**	**7**	**4254**	**5418**	**1083**	**2768**	**1100**	**1255**	**402**	**11**	**9330**	**9929**	**2613**

Playoffs

	Series	W	L	GP	W	L	T	GF	GA	Last Mtg.	Rnd.	Result
Boston	2	0	2	13	5	8	0	38	56	1977	QF	L 2-4
Calgary	6	4	2	26	13	13	0	105	102	1993	DSF	W 4-2
Chicago	1	0	1	5	1	4	0	7	10	1974	QF	L 1-4
Colorado	2	0	2	14	6	8	0	23	33	2002	CQF	L 3-4
Dallas	1	0	1	7	3	4	0	21	26	1968	QF	L 3-4
Detroit	2	1	1	10	4	6	0	21	32	2001	CQF	W 4-2
Edmonton	7	2	5	36	12	24	0	127	154	1992	DSF	L 2-4
Montreal	1	0	1	5	1	4	0	12	15	1993	F	L 1-4
NY Islanders	1	0	1	4	1	3	0	10	21	1980	PRE	L 1-3
NY Rangers	2	0	2	6	1	5	0	14	32	1981	PRE	L 1-3
St. Louis	2	0	2	8	0	8	0	13	32	1998	CQF	L 0-4
Toronto	3	1	2	12	5	7	0	31	41	1993	CF	W 4-3
Vancouver	3	2	1	17	9	8	0	66	60	1993	DF	W 4-2
Defunct Clubs	1	1	0	7	4	3	0	23	25			
Totals	**34**	**11**	**23**	**170**	**65**	**105**	**0**	**511**	**639**			

Playoff Results 2002-1998

Year	Round	Opponent	Result	GF	GA
2002	CQF	Colorado	L 3-4	13	16
2001	CSF	Colorado	L 3-4	10	17
	CQF	Detroit	W 4-2	15	17
2000	CQF	Detroit	L 0-4	6	15
1998	CQF	St. Louis	L 0-4	8	16

Abbreviations: Round: F - Final; **CF** - conference final; **CSF** - conference semi-final; **CQF** - conference quarter-final; **DF** - division final; **DSF** - division semi-final; **QF** - quarter-final; **PRE** - preliminary round.

Calgary totals include Atlanta Flames, 1972-73 to 1979-80. Carolina totals include Hartford, 1979-80 to 1996-97.
Colorado totals include Quebec, 1979-80 to 1994-95. Dallas totals include Minnesota North Stars, 1967-68 to 1992-93.
New Jersey totals include Kansas City, 1974-75 to 1975-76, and Colorado Rockies, 1976-77 to 1981-82.
Phoenix totals include Winnipeg, 1979-80 to 1995-96.

2001-02 Results

Month	Date	Opponent	Score
Oct.	4	Phoenix	2-2
	7	Minnesota	3-4
	9	at Dallas	1-2
	11	at St. Louis	6-5
	13	Tampa Bay	0-1
	16	Washington	2-3*
	18	Anaheim	4-1
	20	at Detroit	2-3
	23	at Columbus	7-1
	25	at Tampa Bay	0-3
	26	at Florida	2-3
	28	at Carolina	2-3*
	30	at Chicago	1-5
Nov.	1	Chicago	2-3
	3	Atlanta	4-1
	8	Calgary	2-3
	10	Detroit	3-2*
	15	Dallas	3-4
	17	at Detroit	2-4
	18	at Minnesota	2-2
	20	at Calgary	5-5
	22	at Edmonton	4-2
	24	San Jose	3-1
	29	Edmonton	1-3
Dec.	1	Nashville	4-2
	3	Calgary	0-2
	6	St. Louis	1-1
	8	at St. Louis	0-2
	9	at Chicago	5-2
	11	at Nashville	1-1
	13	Vancouver	6-3
	15	Columbus	3-2
	16	at Anaheim	3-2*
	18	at Toronto	3-1
	20	at Ottawa	4-2
	22	at Montreal	1-2
	26	at Phoenix	1-1
	27	Anaheim	2-2
	29	NY Rangers	4-5
Jan.	2	Florida	3-1
	5	at NY Islanders	3-0
	7	at New Jersey	3-2
	9	at NY Rangers	4-0
	10	at Boston	0-5
	12	at San Jose	3-2
	15	Nashville	2-0
	17	Buffalo	4-2
	19	NY Islanders	2-3
	21	at Anaheim	4-2
	24	Minnesota	4-1
	26	Colorado	2-4
	28	at Colorado	6-4
	30	at Minnesota	2-0
Feb.	4	Philadelphia	1-3
	7	Carolina	2-1
	8	at Phoenix	5-6*
	11	Dallas	2-1
	13	Phoenix	2-2
	26	at Columbus	5-1
	27	at Pittsburgh	5-4
Mar.	2	Columbus	0-2
	4	Ottawa	1-1
	6	at Dallas	2-3*
	7	at Nashville	3-2
	9	at Colorado	3-4
	11	Chicago	2-1
	14	St. Louis	2-1
	16	Pittsburgh	4-3
	18	at San Jose	3-2
	21	Colorado	3-1
	23	San Jose	3-0
	24	at Phoenix	0-4
	26	at Vancouver	0-4
	28	at Edmonton	2-2
	30	at Calgary	3-5
Apr.	2	Vancouver	4-4
	4	Detroit	3-0
	6	Edmonton	4-3*
	8	Dallas	3-0
	11	at Vancouver	2-5
	13	at San Jose	1-3
	14	Anaheim	1-0

* – Overtime

Entry Draft Selections 2002-1988

2002
Pick
18 Denis Grebeshkov
50 Sergei Anshakov
66 Petr Kanko
104 Aaron Rome
115 Mark Rooneem
152 Greg Hogeboom
157 Joel Andresen
185 Ryan Murphy
215 Mikhail Lyubushin
248 Tuukka Pulliainen
279 Connor James

2001
Pick
18 Jens Karlsson
30 Dave Steckel
49 Mike Cammalleri
51 Jaroslav Bednar
83 Henrik Juntunen
116 Richard Petiot
152 Terry Denike
153 Tuukka Mantyla
214 Cristobal Huet
237 Mike Gabinet
277 Sebastien Laplante

2000
Pick
20 Alexander Frolov
54 Andreas Lilja
86 Yanick Lehoux
118 Lubomir Visnovsky
165 Nathan Marsters
201 Yevgeny Fedorov
206 Tim Eriksson
218 Craig Olynick
245 Dan Welch
250 Flavien Conne
282 Carl Grahn

1999
Pick
43 Andrei Shefer
74 Jason Crain
76 Frantisek Kaberle
92 Cory Campbell
104 Brian McGrattan
125 Daniel Johansson
133 Jean-Francois Nogues
193 Kevin Baker
222 George Parros
250 Noah Clarke

1998
Pick
21 Mathieu Biron
46 Justin Papineau
76 Alexei Volkov
103 Kip Brennan
133 Joe Rullier
163 Tomas Zizka
190 Tommi Hannus
217 Jim Henkel
248 Matthew Yeats

1997
Pick
3 Olli Jokinen
15 Matt Zultek
29 Scott Barney
83 Joe Corvo
99 Sean Blanchard
137 Richard Seeley
150 Jeff Katcher
193 Jay Kopischke
220 Konrad Brand

1996
Pick
30 Josh Green
37 Marian Cisar
57 Greg Phillips
84 Mikael Simons
96 Eric Belanger
120 Jesse Black
123 Peter Hogan
190 Stephen Valiquette
193 Kai Nurminen
219 Sebastien Simard

1995
Pick
3 Aki Berg
33 Don MacLean
50 Pavel Rosa
59 Vladimir Tsyplakov
118 Jason Morgan
137 Igor Melyakov
157 Benoit Larose
163 Juha Vuorivirta
215 Brian Stewart

1994
Pick
7 Jamie Storr
33 Matt Johnson
59 Vitali Yachmenev
111 Chris Schmidt
163 Luc Gagne
189 Andrew Dale
215 Jan Nemecek
241 Sergei Shalomai

1993
Pick
42 Shayne Toporowski
68 Jeff Mitchell
94 Bob Wren
105 Frederick Beaubien
117 Jason Saal
120 Tomas Vlasak
146 Jere Karalahti
172 Justin Martin
198 John-Tra Dillabough
224 Martin Strbak
250 Kimmo Timonen
276 Patrick Howald

1992
Pick
39 Justin Hocking
63 Sandy Allan
87 Kevin Brown
111 Jeff Shevalier
135 Rem Murray
207 Magnus Wernblom
231 Ryan Pisiak
255 Jukka Tiilikainen

1991
Pick
42 Guy Leveque
79 Keith Redmond
81 Alexei Zhitnik
108 Pauli Jaks
130 Brett Seguin
152 Kelly Fairchild
196 Craig Brown
218 Mattias Olsson
240 Andre Bouliane
262 Mike Gaul

1990
Pick
7 Darryl Sydor
28 Brandy Semchuk
49 Bill Berg
91 David Goverde
112 Erik Andersson
133 Robert Lang
154 Dean Hulett
175 Denis Leblanc
196 Patrik Ross
217 K.J.(Kevin) White
238 Troy Mohns

1989
Pick
39 Brent Thompson
81 Jim Maher
102 Eric Ricard
103 Thomas Newman
123 Daniel Rydmark
144 Ted Kramer
165 Sean Whyte
182 Jim Giacin
186 Martin Maskarinec
207 Jim Hiller
228 Steve Jaques
249 Kevin Sneddon

1988
Pick
7 Martin Gelinas
28 Paul Holden
49 John Van Kessel
70 Rob Blake
91 Jeff Robison
109 Micah Aivazoff
112 Robert Larsson
133 Jeff Kruesel
154 Timo Peltomaa
175 Jim Larkin
196 Brad Hyatt
217 Doug Laprade
238 Joe Flanagan

Coach

MURRAY, ANDY
Coach, Los Angeles Kings. Born in Gladstone, Man., March 3, 1951.

Andy Murray became the 19th head coach in Kings history on June 14, 1999. His coaching experience dates back to 1974 and includes seven seasons as an NHL assistant or associate coach with the Winnipeg Jets (1993 to 1995), Minnesota North Stars (1990 to 1992) and Philadelphia Flyers (1988 to 1990). As an assistant coach in Minnesota, Murray reached the Stanley Cup Finals in 1991.

In addition to his NHL service, Murray brings to the Kings a tremendous amount of international coaching experience. As head coach of the Canadian national team, he guided his team to a 77-29-14 record and the gold medal in the 1997 World Hockey Championships.

From 1976 to 1978, Murray served his first head coaching position with the Brandon Travelers of the Manitoba Junior Hockey League. He moved on to become head coach for Brandon University from 1978 to 1981, leading the Bobcats to the #1 ranking in Canadian university hockey during his final year. In 1981-82, Murray moved to Switzerland, where for the next seven years he coached several Swiss-A Division teams.

Murray returned to North America as an assistant coach for the Hershey Bears of the American Hockey League in 1987 and helped guide the Bears to the 1988 Calder Cup championship. In 1992, Murray returned to Europe to coach Lugano in Switzerland and then Eisbaren Berlin in Germany a year later. Most recently, Murray served as the head coach for Shattuck-St. Mary's in Faribault, Minnesota, where he led the prep school to a 70-9-2 record and the Midget Triple A USA Hockey national championship in 1998-99.

Coaching Record

		Regular Season				Playoffs		
Season	**Team**	**Games**	**W**	**L**	**T**	**Games**	**W**	**L**
1999-2000	**Los Angeles (NHL)**	**82**	**39**	**31**	**12**	**4**	**0**	**4**
2000-01	**Los Angeles (NHL)**	**82**	**38**	**31**	**13**	**13**	**7**	**6**
2001-02	**Los Angeles (NHL)**	**82**	**40**	**31**	**11**	**7**	**3**	**4**
	NHL Totals	**246**	**117**	**93**	**36**	**24**	**10**	**14**

Assistant coach Dave Tippett posted a 2-2-1 record as replacement coach when Murray was sidelined following a car accident, February 26 to March 6, 2002. All games are credited to Murray's coaching record.

Club Directory

STAPLES Center

Los Angeles Kings
STAPLES Center
1111 South Figueroa Street
Los Angeles, CA 90015
Phone **213/742-7100**
GM FAX 310/535-4507
www.lakings.com
Capacity: 18,118

Executive
Owner . . . . . Philip F. Anschutz
Owner . . . . . Edward P. Roski
President . . . . . Timothy J. Leiweke

Hockey Operations
Senior Vice President/General Manager . . . . . Dave Taylor
Vice President of Hockey Operations,
Assistant General Manager . . . . . Kevin Gilmore
Director, Player Personnel . . . . . Bill O'Flaherty
Director, Amateur Scouting . . . . . Al Murray
Assistant to the General Manager . . . . . John Wolf
Executive Assistant to the General Manager . . . . . Marcia Galloway
Head Coach . . . . . Andy Murray
Assistant Coach . . . . . TBA, Mark Hardy, Ray Bennett
Goaltending Consultant . . . . . Andy Nowicki
Video Coordinator . . . . . Bill Gurney
Pro Scout – Director of European Evaluation . . . . . Rob Laird
Scout . . . . . Greg Drechsel, Vaclav Nedomansky, Brian Putnam, Parry Shockey, John Stanton, Jan Vopat, Ari Vuori, Glen Williamson, Michel Boucher, Jim Cassidy, Mike Donnelly, Viacheslav Golovin, Gary Harker, Jerry Sodomlak, Victor Tjumenev

Medical
Athletic Trainer . . . . . Peter Demers, ATC
Assistant Athletic Trainer . . . . . Rick Burrill, ATC
Rehabilitation Trainer . . . . . Robert Zolg, MPT, ATC
Head Speed-Strength and Conditioning Coach . . . . . Joseph Horrigan, DC, CSCS
Assistant Speed-Strength and Conditioning Coach . Dave Good, CSCS, SSC
Nutrition Consultant . . . . . Doug Andersen, DC, CCN
Team Physician . . . . . Dr. Ronald Kvitne (Kerlan-Jobe Orthopaedic Clinic)
Internist . . . . . Dr. Michael Mellman
Dentist . . . . . Dr. Jeffrey Hoy
Opthamologist . . . . . Dr. Howard Lazerson

Equipment Staff
Equipment Manager . . . . . Peter Millar
Assistant Equipment Manager . . . . . Rick Garcia, Dan Del Vecchio

Media Relations/Team Services
Director, Media Relations/Team Services . . . . . Mike Altieri
Manager, Media Relations/Team Services . . . . . Jeff Moeller
Media Relations Assistant . . . . . Lee Callans

Broadcasters
TV Play-by-Play Announcer . . . . . Bob Miller
Radio Play-by-Play Announcer . . . . . Nick Nickson
TV Color Commentator . . . . . Jim Fox
Radio Color Commentator . . . . . Daryl Evans
Training Center . . . . . HealthSouth Training Center
Rink Dimensions . . . . . 200 feet by 85 feet
Team Colors . . . . . Purple, Silver, Black, White
Television . . . . . FOX Sports Net
Radio FlagshipESPN Radio 1110-KSPN
Minor League Affiliates . . . . . Manchester Monarchs (AHL); Reading Royals (ECHL)

Center Ian Laperriere appeared in 81 regular-season and seven playoff games for the Kings in 2001-02, leading the club with a total of 134 penalty minutes. He is entering his eighth season in Los Angeles.

Minnesota Wild

2001-02 Results: 26W-35L-12T-9OTL 73PTS.
Fifth, Northwest Division

Year-by-Year Record

Season	GP	Home W	Home L	Home T	Home OL	Road W	Road L	Road T	Road OL	Overall W	Overall L	Overall T	Overall OL	GF	GA	Pts.	Finished	Playoff Result
2001-02	82	14	14	8	5	12	21	4	4	26	35	12	9	195	238	73	5th, Northwest Div.	Out of Playoffs
2000-01	82	14	13	10	4	11	26	3	1	25	39	13	5	168	210	68	5th, Northwest Div.	Out of Playoffs

2002-03 Schedule

Month	Day	Date	Opponent
Oct.	Fri.	11	Boston
	Sat.	12	at St. Louis
	Tue.	15	Florida
	Thu.	17	Dallas
	Sat.	19	Detroit
	Tue.	22	Calgary
	Thu.	24	at Chicago
	Sat.	26	at Phoenix
	Sun.	27	at Colorado
	Tue.	29	Colorado
	Thu.	31	San Jose
Nov.	Sat.	2	Vancouver
	Mon.	4	at Los Angeles
	Thu.	7	at Phoenix
	Sat.	9	at San Jose
	Sun.	10	at Anaheim*
	Tue.	12	Edmonton
	Thu.	14	Pittsburgh
	Sat.	16	Washington
	Tue.	19	Los Angeles
	Thu.	21	at Washington
	Sat.	23	Nashville*
	Mon.	25	Vancouver
	Wed.	27	at Dallas
	Fri.	29	Colorado*
Dec.	Tue.	3	at Edmonton
	Thu.	5	at Calgary
	Sat.	7	at Vancouver
	Tue.	10	Tampa Bay
	Thu.	12	at Detroit
	Sat.	14	at Nashville
	Sun.	15	Carolina*
	Tue.	17	Edmonton
	Thu.	19	NY Islanders
	Sat.	21	at Colorado
	Mon.	23	Calgary
	Thu.	26	at Chicago
	Sat.	28	at Buffalo
	Tue.	31	Anaheim*
Jan.	Thu.	2	at Edmonton
	Sat.	4	at Calgary
	Mon.	6	Los Angeles
	Wed.	8	Columbus
	Fri.	10	Phoenix
	Tue.	14	Buffalo
	Thu.	16	Vancouver
	Sat.	18	Anaheim
	Mon.	20	at Anaheim*
	Thu.	23	at Los Angeles
	Sat.	25	at San Jose
	Tue.	28	at Vancouver
	Wed.	29	at Edmonton
Feb.	Wed.	5	Chicago
	Fri.	7	San Jose
	Sun.	9	at New Jersey*
	Mon.	10	at Philadelphia
	Wed.	12	Philadelphia
	Fri.	14	Phoenix
	Sat.	15	at Colorado
	Wed.	19	NY Rangers
	Sun.	23	St. Louis*
	Tue.	25	at Ottawa
	Thu.	27	at Montreal
Mar.	Sat.	1	at St. Louis
	Tue.	4	New Jersey
	Thu.	6	at Nashville
	Fri.	7	at Carolina
	Sun.	9	at Atlanta*
	Wed.	12	Dallas
	Fri.	14	Nashville
	Sat.	15	at Columbus
	Mon.	17	at Tampa Bay
	Wed.	19	at Florida
	Fri.	21	at Dallas
	Sun.	23	Detroit
	Tue.	25	at Detroit
	Wed.	26	St. Louis
	Fri.	28	Chicago
	Mon.	31	Calgary
Apr.	Wed.	2	at Columbus
	Thu.	3	at Toronto
	Sun.	6	Columbus*

* Denotes afternoon game.

Franchise date: June 25, 1997

NORTHWEST DIVISION

3rd NHL Season

A 19-year-old rookie with the Wild last season, Nick Schultz showed great promise as one of only three Minnesota regulars not to register a negative plus-minus rating. He also played for Team Canada at the World Junior Championships last year.

2002-03 Player Personnel

FORWARDS	HT	WT	S	Place of Birth	Date	2001-02 Club
BLOUIN, Sylvain	6-2	207	L	Montreal, Que.	5/21/74	Minnesota
BRUNETTE, Andrew	6-1	210	L	Sudbury, Ont.	8/24/73	Minnesota
CAVANAUGH, Dan	6-1	190	R	Springfield, MA	3/3/80	Houston
CAVOSIE, Marc	6-0	173	L	Albany, NY	8/6/81	RPI Engineers
CROZIER, Greg	6-3	200	L	Calgary, Alta.	7/6/76	Prov (AHL)-Houston
CULLEN, Mark	5-11	175	L	Moorhead, MN	10/28/78	Colorado College
DOMENICHELLI, Hnat	6-0	195	L	Edmonton, Alta.	2/17/76	Atlanta-Minnesota
DOWD, Jim	6-1	190	R	Brick, NJ	12/25/68	Minnesota
DUPUIS, Pascal	6-0	195	R	Laval, Que.	4/7/79	Minnesota
GABORIK, Marian	6-1	183	L	Trencin, Czech.	2/14/82	Minnesota
HENDRICKSON, Darby	6-1	195	L	Richfield, MN	8/28/72	Minnesota
HOGGAN, Jeff	6-0	200	R	Hope, B.C.	2/1/78	Nebraska-Omaha
JOHNSON, Matt	6-5	232	L	Welland, Ont.	11/23/75	Minnesota
LAAKSONEN, Antti	6-0	180	L	Tammela, Finland	10/3/73	Minnesota
LAROSE, Cory	6-0	188	L	Campbellton, N.B.	5/14/75	Houston
MUCKALT, Bill	6-1	200	R	Surrey, B.C.	7/15/74	Ottawa
PARK, Richard	5-11	190	R	Seoul, South Korea	5/27/76	Minnesota-Houston
PAVLIKOVSKY, Rastislav	6-1	180	L	Dubnica, Czech.	3/22/77	HV 71 Jonkoping-Slovakia
RONNING, Cliff	5-8	165	L	Burnaby, B.C.	10/1/65	Nashville-Los Angeles
TRUDEL, Jean-Guy	5-11	202	L	Sudbury, Ont.	10/18/75	Phoenix-Springfield
TUZZOLINO, Tony	6-2	208	R	Buffalo, NY	10/9/75	Boston-Prov (AHL)
VEILLEUX, Stephane	6-1	187	L	Beaureville, Que.	11/16/81	Houston
VIRTA, Tony	5-10	187	L	Hameenlinna, Finland	6/28/72	Minnesota-Houston
WALLIN, Rickard	6-2	185	L	Stockholm, Sweden	4/19/80	Farjestad
WALZ, Wes	5-10	180	R	Calgary, Alta.	5/15/70	Minnesota
WANVIG, Kyle	6-2	219	R	Calgary, Alta.	1/29/81	Houston
ZHOLTOK, Sergei	6-2	191	R	Riga, Latvia	12/2/72	Minnesota
DEFENSEMEN						
BENYSEK, Ladislav	6-2	190	L	Olomouc, Czech.	3/24/75	Minnesota
BOMBARDIR, Brad	6-1	205	L	Powell River, B.C.	5/5/72	Minnesota
BROWN, Brad	6-4	220	R	Baie Verte, Nfld.	12/27/75	Minnesota
CROWLEY, Mike	5-11	190	L	Bloomington, MN	7/4/75	Houston
CULLEN, David	6-2	209	R	St. Catharines, Ont.	12/30/76	Phoenix-Springfield-Minnesota-Houston
DYMENT, Chris	6-3	210	R	Stoneham, MA	10/24/79	Boston University
KUBA, Filip	6-3	205	L	Ostrava, Czech.	12/29/76	Minnesota
MARSHALL, Jason	6-2	200	R	Cranbrook, B.C.	2/22/71	Minnesota
MICHALEK, Zbynek	6-1	176	R	Jindrchuv Hradec, Czech.	12/23/82	Shawinigan
MITCHELL, Willie	6-3	205	L	Port McNeill, B.C.	4/23/77	Minnesota
MURPHY, Curtis	5-8	185	R	Kerrobert, Sask.	12/3/75	Houston
NYCHOLAT, Lawrence	6-0	192	L	Calgary, Alta.	5/7/79	Houston
REITZ, Erik	6-0	192	R	Detroit, MI	7/29/82	Barrie
ROCHE, Travis	6-1	190	R	Grand Cache, Alta	6/17/78	Minnesota-Houston
SCHULTZ, Nick	6-0	187	L	Regina, Sask.	8/25/82	Minnesota-Houston
SEKERAS, Lubomir	6-0	183	L	Trencin, Czech.	11/18/68	Minnesota

GOALTENDERS	HT	WT	C	Place of Birth	Date	2001-02 Club
CLOUTIER, Frederic	6-0	165	R	St-Georges, Que.	5/14/81	Louisiana
FERNANDEZ, Manny	6-0	180	L	Etobicoke, Ont.	8/27/74	Minnesota
GUSTAFSON, Derek	5-11	210	L	Gresham, OR	6/21/79	Minnesota-Houston
KETTLES, Kyle	6-3	180	L	Lac du Bonnet, Man.	2/19/81	Medicine Hat-Moose Jaw
KOCHAN, Dieter	6-1	180	L	Saskatoon, Sask.	5/11/74	Tampa Bay-Springfield
ROLOSON, Dwayne	6-1	178	L	Simcoe, Ont.	10/12/69	Minnesota

2001-02 Scoring

* - rookie

Regular Season

Pos	#	Player	Team	GP	G	A	Pts	+/–	PIM	PP	SH	GW	GT	S	%
L	15	Andrew Brunette	MIN	81	21	48	69	–4	18	10	0	2	1	106	19.8
L	10	Marian Gaborik	MIN	78	30	37	67	0	34	10	0	4	1	221	13.6
C	34	Jim Dowd	MIN	82	13	30	43	–14	54	5	0	1	0	111	11.7
R	33	Sergei Zholtok	MIN	73	19	20	39	–10	28	10	0	2	1	146	13.0
L	24	Antti Laaksonen	MIN	82	16	17	33	–5	22	0	0	1	0	104	15.4
C	37	Wes Walz	MIN	64	10	20	30	0	43	0	2	5	0	97	10.3
L	11	* Pascal Dupuis	MIN	76	15	12	27	–10	16	3	2	0	0	154	9.7
C	18	Richard Park	MIN	63	10	15	25	–1	10	2	1	2	0	115	8.7
L	9	Hnat Domenichelli	ATL	40	8	11	19	–18	34	1	0	1	1	87	9.2
			MIN	27	1	5	6	–5	10	0	0	0	0	57	1.8
			TOTAL	67	9	16	25	–23	44	1	0	1	1	144	6.3
C	14	Darby Hendrickson	MIN	68	9	15	24	–22	50	2	2	1	0	79	11.4
D	17	Filip Kuba	MIN	62	5	19	24	–6	32	3	0	1	0	101	5.0
D	77	Lubomir Sekeras	MIN	69	4	20	24	–7	38	4	0	1	0	82	4.9
R	22	Stacy Roest	MIN	58	10	11	21	–3	8	1	4	2	1	98	10.2
C	44	Aaron Gavey	MIN	71	6	11	17	–21	38	1	0	0	0	75	8.0
D	2	Willie Mitchell	MIN	68	3	10	13	–16	68	0	0	1	0	67	4.5
D	23	Jason Marshall	MIN	80	5	6	11	–8	148	1	0	0	0	73	6.8
D	55	* Nick Schultz	MIN	52	4	6	10	0	14	1	0	1	0	47	8.5
D	3	Ladislav Benysek	MIN	74	1	7	8	–12	28	0	0	0	0	44	2.3
R	26	Tony Virta	MIN	8	2	3	5	0	0	0	0	0	0	15	13.3
L	12	Matt Johnson	MIN	60	4	0	4	–13	183	0	0	1	0	23	17.4
D	4	Brad Brown	MIN	51	0	4	4	–11	123	0	0	0	0	23	0.0
D	5	Brad Bombardir	MIN	28	1	2	3	–6	14	1	0	0	0	24	4.2
C	31	Roman Simicek	MIN	6	2	0	2	1	8	0	0	0	0	4	50.0
L	36	Sylvain Blouin	MIN	43	0	2	2	–11	130	0	0	0	0	28	0.0
D	6	Mike Matteucci	MIN	3	0	0	0	1	2	0	0	0	0	0	0.0
D	39	* Travis Roche	MIN	4	0	0	0	–1	2	0	0	0	0	1	0.0
D	19	* David Cullen	PHX	14	0	0	0	–5	6	0	0	0	0	3	0.0
			MIN	3	0	0	0	–3	0	0	0	0	0	0	0.0
			TOTAL	17	0	0	0	–8	6	0	0	0	0	3	0.0

Goaltending

No.	Goaltender	GPI	Mins	Avg	W	L	T	EN	SO	GA	SA	S%	G	A	PIM
31	* Derek Gustafson	1	26	0.00	0	0	0	0	0	0	7	1.000	0	0	0
30	Dwayne Roloson	45	2506	2.68	14	20	7	0	5	112	1132	.901	0	0	8
35	Manny Fernandez	44	2463	3.05	12	24	5	1	1	125	1157	.892	0	0	4
	Totals	**82**	**5004**	**2.85**	**26**	**44**	**12**	**1**	**6**	**238**	**2297**	**.896**			

Coach

LEMAIRE, JACQUES

Coach, Minnesota Wild. Born in LaSalle, Que., September 7, 1945.

The Minnesota Wild announced the signing of Jacques Lemaire as the club's first head coach on June 19, 2000. Lemaire had spent parts of the previous two seasons as a senior consultant to the general manager for the Montreal Canadiens, the franchise with which he captured eight Stanley Cup championships as a player.

Lemaire spent five seasons behind the New Jersey Devils bench and compiled a 199-122-57 mark. In 1994-95, he coached the Devils to their first Stanley Cup championship. In his first season with the team (1993-94), he was awarded the Jack Adams Award as the NHL's outstanding coach.

Lemaire began his NHL coaching career with the Montreal Canadiens in 1983-84. The next year, he coached Montreal to the Adams Division championship. He stepped aside as head coach following the 1984-85 campaign and moved to the front office where he held the position of assistant to the managing director for seven of his last eight years with the Canadiens. During that time, Lemaire played a role in Montreal's Stanley Cup championships of 1986 and 1993.

Lemaire spent his entire NHL playing career with Montreal from 1967 to 1979. He then began his coaching career in Switzerland where he served as player/coach of the Sierre club. He returned to North America in 1981 and was named the first head coach of the Quebec Major Junior Hockey League's expansion Longueuil Chevaliers. In his only season at the helm (1982-83), Lemaire guided the team to the QMJHL finals.

Coaching Record

			Regular Season			Playoffs		
Season	**Team**	**Games**	**W**	**L**	**T**	**Games**	**W**	**L**
1979-80	Sierre (Switzerland)				UNAVAILABLE			
1980-81	Sierre (Switzerland)				UNAVAILABLE			
1982-83	Longueuil (QMJHL)	70	37	29	4	15	9	6
1983-84	**Montreal (NHL)**	**17**	**7**	**10**	**0**	**15**	**9**	**6**
1984-85	**Montreal (NHL)**	**80**	**41**	**27**	**12**	**12**	**6**	**6**
1993-94	**New Jersey (NHL)**	**84**	**47**	**25**	**12**	**20**	**11**	**9**
1994-95	**New Jersey (NHL)**	**48**	**22**	**18**	**8**	**20**	**16**	**4***
1995-96	**New Jersey (NHL)**	**82**	**37**	**33**	**12**	....	....	....
1996-97	**New Jersey (NHL)**	**82**	**45**	**23**	**14**	**10**	**5**	**5**
1997-98	**New Jersey (NHL)**	**82**	**48**	**23**	**11**	**6**	**2**	**4**
2000-01	**Minnesota (NHL)**	**82**	**25**	**44**	**13**	....	....	....
2001-02	**Minnesota (NHL)**	**82**	**26**	**44**	**12**	....	....	....
	NHL Totals	**639**	**298**	**247**	**94**	**83**	**49**	**34**

* Stanley Cup win.

Club Records

Team

(Figures in brackets for season records are games played.)

Most Points **73** 2001-02 (82)
Most Wins **26** 2001-02 (82)
Most Ties **13** 2000-01 (82)
Most Losses **39** 2000-01 (82)
Most Goals **195** 2001-02 (82)
Most Goals Against **238** 2001-02 (82)
Fewest Points **68** 2000-01 (82)
Fewest Wins **25** 2000-01 (82)
Fewest Ties **12** 2001-02 (82)
Fewest Losses **35** 2001-02 (82)
Fewest Goals **168** 2000-01 (82)
Fewest Goals Against **210** 2000-01 (82)
Longest Winning Streak
Overall .... **3** Oct. 7-12/01, Mar. 7-10/02
Home .... **2** Five times
Away .... **2** Four times
Longest Undefeated Streak
Overall .... **8** Dec. 17/00-Jan. 5/01 (5 wins, 3 ties)
Home .... **9** Dec. 13/00-Jan. 10/01 (5 wins, 4 ties)
Away .... **2** Six times
Longest Losing Streak
Overall .... **5** Mar. 11-19/01, Jan. 28-Feb. 8/02, Mar. 29-Apr. 5/02
Home .... **4** Oct. 29-Nov. 15/00
Away .... **5** Mar. 15-Apr. 2/01, Jan. 19-Feb. 6/02
Longest Winless Streak
Overall .... **12** Mar. 11-Apr. 4/01 (9 losses, 3 ties)
Home .... **8** Feb. 26-Mar. 28/01 (5 losses, 3 ties)
Away .... **6** Five times
Most Shutouts, Season **6** 2000-01 (82), 2001-02 (82)
Most PIM, Season **1,200** 2000-01 (82)
Most Goals, Game .... **6** Four times

Individual

Most Seasons .... **2** Numerous players
Most Games .... **164** Antti Laaksonen
Most Goals, Career .... **48** Marian Gaborik
Most Assists, Career .... **55** Marian Gaborik
Most Points, Career .... **103** Marian Gaborik (48G, 55A)
Most PIM, Career .... **320** Matt Johnson
Most Shutouts, Career .... **5** Manny Fernandez, Dwayne Roloson
Longest Consecutive Games Streak .... **164** Antti Laaksonen (Oct. 6/00-to date)
Most Goals, Season .... **30** Marian Gaborik (2001-02)
Most Assists, Season .... **48** Andrew Brunette (2001-02)
Most Points, Season .... **69** Andrew Brunette (2001-02; 21G, 48A)
Most PIM, Season .... **183** Matt Johnson (2001-02)
Most Points, Defenseman, Season .... **34** Lubomir Sekeras (2000-01; 11G, 23A)
Most Points, Center, Season .... **43** Jim Dowd (2001-02; 13G, 30A)
Most Points, Right Wing, Season .... **22** Sergei Krivokrasov (2000-01; 7G, 15A)
Most Points, Left Wing, Season .... **69** Andrew Brunette (2001-02; 21G, 48A)
Most Points, Rookie, Season .... **36** Marian Gaborik (2000-01; 18G, 18A)
Most Shutouts, Season .... **5** Dwayne Roloson (2001-02)
Most Goals, Game .... **3** Antti Laaksonen (Nov. 26/00), Marian Gaborik (Nov. 13/01, Mar. 10/02)
Most Assists, Game .... **4** Andrew Brunette (Mar. 10/02)
Most Points, Game .... **4** Scott Pellerin (Oct. 18/00; 1G, 3A), Antti Laaksonen (Nov. 26/00; 3G, 1A), Andrew Brunette (Mar. 10/02; 4A), Marion Gaborik (Mar. 10/02; 3G, 1A)

General Managers' History

Doug Risebrough, 2000-01 to date.

Coaching History

Jacques Lemaire, 2000-01 to date.

Captains' History

Sean O'Donnell, Scott Pellerin, Wes Walz, Brad Bombardir, Darby Hendrickson, 2000-01; Jim Dowd, Filip Kuba, Brad Brown, Andrew Brunette, 2001-02.

After spending four years in Europe, Wes Walz scored a career-high 18 goals for Minnesota in their inaugural season and had a career-best 20 assists last season.

All-time Record vs. Other Clubs

Regular Season

	At Home								On Road								Total							
	GP	W	L	T	OL	GF	GA	PTS	GP	W	L	T	OL	GF	GA	PTS	GP	W	L	T	OL	GF	GA	PTS
Anaheim	4	1	1	1	1	6	10	4	4	2	1	0	1	13	9	5	8	3	2	1	2	19	19	9
Atlanta	2	1	0	1	0	5	3	3	1	1	0	0	0	4	2	2	3	2	0	1	0	9	5	5
Boston	1	1	0	0	0	2	1	2	2	2	0	0	0	11	4	4	3	3	0	0	0	13	5	6
Buffalo	1	0	1	0	0	1	4	0	1	0	1	0	0	1	3	0	2	0	2	0	0	2	7	0
Calgary	5	2	1	1	1	10	8	6	5	1	3	1	0	12	15	3	10	3	4	2	1	22	23	9
Carolina	2	0	1	1	0	4	8	1	1	0	1	0	0	0	2	0	3	0	2	1	0	4	10	1
Chicago	4	2	2	0	0	11	10	4	4	2	2	0	0	12	10	4	8	4	4	0	0	23	20	8
Colorado	5	1	3	0	1	8	20	3	5	0	5	0	0	7	17	0	10	1	8	0	1	15	37	3
Columbus	4	3	1	0	0	12	7	6	3	1	1	1	0	6	7	3	7	4	2	1	0	18	14	9
Dallas	4	1	3	0	0	11	11	2	4	2	1	1	0	13	14	5	8	3	4	1	0	24	25	7
Detroit	4	1	1	1	1	8	10	4	4	1	3	0	0	12	19	2	8	2	4	1	1	20	29	6
Edmonton	5	0	3	1	1	12	19	2	5	0	3	1	1	6	13	2	10	0	6	2	2	18	32	4
Florida	2	1	0	1	0	6	0	3	1	0	1	0	0	1	2	0	3	1	1	1	0	7	2	3
Los Angeles	4	1	2	1	0	6	11	3	4	2	2	0	0	11	11	4	8	3	4	1	0	17	22	7
Montreal	1	1	0	0	0	4	2	2	2	0	1	1	0	2	6	1	3	1	1	1	0	6	8	3
Nashville	4	1	2	1	0	7	9	3	4	1	2	1	0	8	10	3	8	2	4	2	0	15	19	6
New Jersey	2	0	1	1	0	4	6	1	1	0	1	0	0	2	6	0	3	0	2	1	0	6	12	1
NY Islanders	2	2	0	0	0	7	5	4	2	1	1	0	0	5	3	2	4	3	1	0	0	12	8	6
NY Rangers	2	0	1	0	1	4	6	1	2	0	2	0	0	3	7	0	4	0	3	0	1	7	13	1
Ottawa	2	0	0	1	1	5	6	2	1	0	1	0	0	1	4	0	3	0	1	1	1	6	10	2
Philadelphia	1	0	0	1	0	3	3	1	2	0	2	0	0	1	8	0	3	0	2	1	0	4	11	1
Phoenix	4	1	1	2	0	9	8	4	4	0	3	0	1	3	15	1	8	1	4	2	1	12	23	5
Pittsburgh	1	1	0	0	0	4	2	2	2	1	1	0	0	6	4	2	3	2	1	0	0	10	6	4
St. Louis	4	0	0	2	2	9	11	4	4	2	2	0	0	6	8	4	8	2	2	2	2	15	19	8
San Jose	4	2	1	1	0	10	8	5	4	1	2	1	0	5	8	3	8	3	3	2	0	15	16	8
Tampa Bay	2	2	0	0	0	8	5	4	1	1	0	0	0	4	2	2	3	3	0	0	0	12	7	6
Toronto	0	0	0	0	0	0	0	0	2	0	2	0	0	2	9	0	2	0	2	0	0	2	9	0
Vancouver	5	2	2	1	0	10	15	5	5	2	1	0	2	15	18	6	10	4	3	1	2	25	33	11
Washington	1	1	0	0	0	3	0	2	2	0	2	0	0	2	4	0	3	1	2	0	0	5	4	2
Totals	**82**	**28**	**27**	**18**	**9**	**189**	**208**	**83**	**82**	**23**	**47**	**7**	**5**	**174**	**240**	**58**	**164**	**51**	**74**	**25**	**14**	**363**	**448**	**141**

2001-02 Results

Month	Day	Opponent	Score	Month	Day	Opponent	Score
Oct.	6	at San Jose	0-0		6	Buffalo	1-4
	7	at Los Angeles	4-3		8	Montreal	4-2
	10	Boston	2-1		10	at Nashville	2-2
	12	Chicago	6-4		11	Anaheim	2-2
	14	Edmonton	3-3		13	Dallas	1-3
	16	San Jose	3-3		15	at Carolina	0-2
	19	St. Louis	2-3*		18	at Columbus	3-1
	24	Carolina	3-7		19	at Ottawa	1-4
	27	at Calgary	2-4		23	at Anaheim	2-3*
	30	at Nashville	4-2		24	at Los Angeles	1-4
	31	Nashville	4-6		26	New Jersey	2-2
Nov.	2	Colorado	4-2		28	Calgary	2-3*
	4	Edmonton	0-2		30	Los Angeles	0-2
	6	at NY Rangers	1-3	**Feb.**	5	at Toronto	1-3
	8	at Boston	5-3		6	at Washington	1-2
	11	Vancouver	0-5		8	Colorado	0-6
	13	Atlanta	4-2		10	NY Islanders	4-3
	14	at Colorado	0-1		12	at Columbus	3-3
	16	at Detroit	3-8		13	Detroit	0-2
	18	Los Angeles	2-2		27	at Anaheim	5-3
	20	at Phoenix	1-2*	**Mar.**	2	at Vancouver	3-6
	21	at San Jose	2-0		5	NY Rangers	2-3*
	23	Phoenix	5-2		7	at St. Louis	3-0
	25	Dallas	3-4		8	at Dallas	5-3
	27	Vancouver	2-1		10	Columbus	5-0
	29	Florida	6-0		12	Ottawa	3-4*
Dec.	2	St. Louis	4-4		17	Phoenix	2-2
	5	at Chicago	2-4		18	Calgary	4-2
	8	at Philadelphia	1-5		20	Columbus	1-3
	10	at Montreal	0-4		23	at NY Islanders	1-2
	14	at Pittsburgh	5-2		26	at St. Louis	2-1
	16	Colorado	2-3*		27	at Atlanta	4-2
	18	Anaheim	1-5		29	Chicago	1-3
	22	at Vancouver	2-1		31	at Chicago	1-2
	23	at Colorado	3-6	**Apr.**	2	at Edmonton	1-2*
	26	Detroit	3-3		4	at Calgary	3-4
	28	at Edmonton	2-3		5	at Vancouver	4-5*
	29	at Calgary	3-4		8	San Jose	3-1
	31	at Detroit	2-4		10	at Dallas	4-4
Jan.	2	Tampa Bay	2-0		12	at Phoenix	1-7
	4	Nashville	2-1		14	Edmonton	2-4

* – Overtime

Entry Draft Selections 2002-2000

2002 Pick		2001 Pick		2000 Pick	
8	Pierre-Marc Bouchard	6	Mikko Koivu	3	Marian Gaborik
38	Josh Harding	36	Kyle Wanvig	33	Nick Schultz
72	Mike Erickson	74	Chris Heid	99	Marc Cavosie
73	Barry Brust	93	Stephane Veilleux	132	Maxim Sushinsky
155	Armands Berzins	103	Tony Virta	170	Erik Reitz
175	Matt Foy	202	Derek Boogaard	199	Brian Passmore
204	Niklas Eckerblom	239	Jake Riddle	214	Peter Bartos
237	Christoph Brandner			232	Lubomir Sekeras
268	Mikhail Tyulyapkin			255	Eric Johansson
269	Mika Hannula				

Andrew Brunette played his first full NHL season with the Nashville Predators in 1998-99. After spending the next two years in Atlanta, he joined the Wild in 2001-02 and led the team with a career-high 48 assists and 69 points.

General Manager

RISEBROUGH, DOUG
Executive Vice President and General Manager, Minnesota Wild.
Born in Guelph, Ont., January 29, 1954.

Doug Risebrough was hired as the first executive vice president and general manager of the Minnesota Wild on September 2, 1999. He is responsible for the club's overall hockey operations. His efforts to build a winner through the draft has been exemplified by the success of Martin Gaborik, the club's first-round choice in 2000.

After ending his 13-year NHL playing career with the Flames in 1987, Risebrough was named as assistant coach with Calgary and joined Terry Crisp behind the bench. Risebrough was appointed head coach of the Flames on May 18, 1990 and on May 16, 1991, he also assumed the role of general manager. Late in the 1991-92 campaign he directed his energies full-time to general manager, handing the coaching responsibilities over to Guy Charron for the balance of the season. Risebrough served as g.m. in Calgary through the start of the 1995-96 season. He was vice president of hockey operations for the Edmonton Oilers from 1996 to 1999.

Risebrough was Montreal's first selection, seventh overall, in the 1974 Amateur Draft. During his nine years with the Canadiens, he helped his club to four consecutive Stanley Cup championships between 1976 and 1979. He joined the Flames prior to the start of the club's 1982 training camp. During his NHL career, his clubs have won five Stanley Cup titles (1976-1979 as a player and 1989 as an assistant coach with Calgary) and two Presidents' Trophies (1987-88 and 1988-89 as an assistant coach).

NHL Coaching Record

		Regular Season				Playoffs		
Season	Team	Games	W	L	T	Games	W	L
1990-91	Calgary	80	46	26	8	7	3	4
1991-92	Calgary	64	25	30	9	....	....	....
	NHL Totals	**144**	**71**	**56**	**17**	**7**	**3**	**4**

Club Directory

Xcel Energy Center

Minnesota Wild
317 Washington Street
St. Paul, MN 55102
Phone **651/602-6000**
FAX 651/293-9574
Tickets 651/222-9453
www.wild.com
Capacity: 18,064

Executive Management
Chairman . . . . Bob Naegele, Jr.
Chief Executive Officer . . . . Jac Sperling
President and Chief Operating Officer. . . . . Tod Leiweke
Senior Vice President and Chief Financial Officer . . . Pamela Wheelock
Executive Vice President/General Manager . . . . Doug Risebrough
Vice President of Finance and Controller . . . . Mike Nealy
Vice President of Administration . . . . Mike Reeves
Vice President of Information Technology/ Corporate Projects. . . . . Brian Jore
Vice President of Customer Sales and Service . . . . . Steve Griggs
Vice President/General Manager of Xcel Energy Center . . . . Jack Larson
Vice President/General Manager of RiverCentre. . . . Jim Ibister
Vice President of Marketing . . . . Matt Majka
Vice President of Communications and Broadcast . . Bill Robertson
Vice President of Corporate Partnerships. . . . . Laura Day
Manager of Special Projects . . . . Kris Parod
Executive Assistant to Jac Sperling. . . . . Heather Bernier
Executive Assistant to Tod Leiweke . . . . Stephanie Huseby

Coaching Staff
Head Coach . . . . Jacques Lemaire
Assistant Coaches. . . . . Mike Ramsey, Mario Tremblay
Strength and Conditioning Coach. . . . . George Kinnear

Hockey Operations
Director of Hockey Administration and Legal Affairs . . Tom Lynn
Chief Amateur Scout. . . . . Tom Thompson
Coordinator of Amateur Scouting. . . . . Guy Lapointe
Scouts. . . . . Marc Chamard, Paul Charles, Frank Effinger, Branislav Gaborik, Ken Hoodikoff, Patrick Kipler, Jiri Koluch, Barry MacKenzie, Doug Mosher, Noel Rahn, Glen Sonmor, Bruce Southern, Thomas Steen, Rich Sutter, Tim Sweeney, Matti Vaisanen
Head Athletic Therapist. . . . . Don Fuller
Head Equipment Manager . . . . Tony DaCosta
Assistant Athletic Trainer. . . . . Mike Vogt
Assistant Equipment Manager. . . . . Brent Proulx
Assistant Equipment Manager. . . . . Matt Benz
Video Coordinator . . . . Todd Woodcroft
Hockey Operations Assistant. . . . . Tobin Wright
Hockey Operations Administrator . . . . Cindy Sweiger
Hockey Operations . . . . Denny Scanlon
Medical Director . . . . Dr. Sheldon Burns
Orthopedic Surgeon . . . . Dr. Joel Boyd

Customer Sales and Service
Director of Customer Sales and Service . . . . Jamie Spencer
Director of Group and Event Suite Sales . . . . Kelly Harens
Director of Ticket Operations. . . . . Holly Cedarblade

Corporate Partnerships
Director of Premium Service/Operation . . . . Rachael Johnson
Director of Corporate Services. . . . . Carin Anderson

Communications and Broadcasting
Manager of Media Relations and Team Services . . . Brad Smith
Director of Internet Services . . . . Brian Hutchinson
Director of Broadcasting . . . . Pat O'Connor
Manager of Arena Communications . . . . Chris Kelleher
Manager of Corporate Communications and Internet Services . . . . Aaron Sickman

Finance
Vice President of Finance and Controller . . . . Mike Nealy
Director of Accounting Operations . . . . Dean Harris
Manager of Payroll . . . . Jackie Ryan
Senior Staff Accountant . . . . Mindee Mills

Marketing
Director of Retail Operations . . . . Chris Poitras
Director of SPAC Marketing . . . . Peter Johns
Director of Advertising and Promotion . . . . Wayne Petersen
Executive Director of 10K Rinks Foundation. . . . . . . Heather McGinty
Senior Director of Creative Services . . . . John Maher
Manager of Retail Operations . . . . Dave Krolow
Manager of SPAC Marketing. . . . . Anette Andruss
Director of Community Relations. . . . . Marlene Wall
Director of Graphic Services and Publications . . . . . Brian Israel

Information Technology
Vice President of Information Technology & Corporate Projects. . . . . Brian Jore
Director of Technology Services. . . . . Chris Monicatti

Administration
Vice President of Administration . . . . Mike Reeves
Employment and Benefits Manager. . . . . Timothy Case
317 Facilities Manager . . . . Tim Wolfgram
Construction Manager . . . . Mark Anger
317 Receptionist . . . . Christine Haas
Mail Clerk . . . . Al Hilton

Miscellaneous
Training Site . . . . Parade Ice Garden and Xcel Energy Center
Primary Affiliate . . . . Houston Aeros (AHL)
Radio Network Flagship. . . . . WCCO Radio (830 AM)
TV Networks . . . . KMSP 9 (over-air), Fox Sports Net
Radio Play-By-Play Announcer . . . . Bob Kurtz
Radio Analyst . . . . Tom Reid
Television Play-By-Play Announcer. . . . . Mike Goldberg
Television Analyst . . . . Mike Greenlay
Team Photographer . . . . Bruce Kluckhohn

Richard Zednik had 22 goals and 22 assists.

Montreal Canadiens

2001-02 Results: 36W-31L-12T-3OTL 87PTS.
Fourth, Northeast Division

2002-03 Schedule

Month	Day	Date	Opponent
Oct.	Fri.	11	at NY Rangers
	Sat.	12	Buffalo
	Tue.	15	Philadelphia
	Thu.	17	at Detroit
	Sat.	19	Toronto
	Tue.	22	Pittsburgh
	Thu.	24	at Philadelphia
	Sat.	26	Ottawa
	Tue.	29	Anaheim
Nov.	Fri.	1	at Carolina
	Sat.	2	at Toronto
	Tue.	5	St. Louis
	Thu.	7	NY Islanders
	Sat.	9	Los Angeles
	Tue.	12	Dallas
	Fri.	15	at New Jersey
	Sat.	16	New Jersey
	Mon.	18	Pittsburgh
	Wed.	20	at Pittsburgh
	Thu.	21	at Ottawa
	Sat.	23	Carolina
	Tue.	26	Atlanta
	Fri.	29	at Boston*
	Sat.	30	Philadelphia
Dec.	Wed.	4	at Dallas
	Fri.	6	at Colorado
	Sat.	7	at Phoenix
	Tue.	10	at Boston
	Thu.	12	Tampa Bay
	Sat.	14	Boston
	Mon.	16	at Ottawa
	Tue.	17	San Jose
	Thu.	19	at NY Rangers
	Sat.	21	Buffalo
	Mon.	23	at NY Islanders
	Fri.	27	at Ottawa
	Sat.	28	at Pittsburgh
	Tue.	31	at Calgary
Jan.	Thu.	2	at Vancouver
	Sat.	4	at Edmonton
	Tue.	7	at New Jersey
	Thu.	9	NY Rangers
	Sat.	11	Buffalo
	Mon.	13	Calgary
	Wed.	15	at Atlanta
	Thu.	16	at Philadelphia
	Sat.	18	Toronto
	Mon.	20	at Florida*
	Wed.	22	at Tampa Bay
	Sat.	25	Washington*
	Sun.	26	Chicago*
	Tue.	28	Florida
	Thu.	30	at NY Islanders
Feb.	Tue.	4	Atlanta
	Thu.	6	at Boston
	Sat.	8	at Toronto
	Sun.	9	at Washington
	Tue.	11	Boston
	Thu.	13	Columbus
	Sat.	15	Edmonton*
	Tue.	18	Florida
	Wed.	19	at Buffalo
	Sat.	22	Toronto
	Mon.	24	at Washington
	Thu.	27	Minnesota
Mar.	Sat.	1	Vancouver
	Wed.	5	at Anaheim
	Thu.	6	at San Jose
	Sat.	8	at Los Angeles
	Mon.	10	at Nashville
	Wed.	12	at Florida
	Thu.	13	at Atlanta
	Sat.	15	Tampa Bay
	Tue.	18	New Jersey
	Thu.	20	NY Islanders
	Sat.	22	Carolina
	Tue.	25	Washington
	Fri.	28	at Buffalo
	Sat.	29	Ottawa
	Mon.	31	at Carolina
Apr.	Wed.	2	at Tampa Bay
	Sat.	5	NY Rangers

* Denotes afternoon game.

Franchise date: November 22, 1917

EASTERN CONFERENCE

NORTHEAST DIVISION

86th NHL Season

Year-by-Year Record

		Home				Road				Overall								
Season	GP	W	L	T	OL	W	L	T	OL	W	L	T	OL	GF	GA	Pts.	Finished	Playoff Result
2001-02	82	21	13	6	1	15	18	6	2	36	31	12	3	207	209	87	4th, Northeast Div.	Lost Conf. Semi-Final
2000-01	82	15	20	4	2	13	20	4	4	28	40	8	6	206	232	70	5th, Northeast Div.	Out of Playoffs
1999-2000	82	18	17	5	1	17	17	4	3	35	34	9	4	196	194	83	4th, Northeast Div.	Out of Playoffs
1998-99	82	21	15	5	...	11	24	6	...	32	39	11	...	184	209	75	5th, Northeast Div.	Out of Playoffs
1997-98	82	15	17	9	...	22	15	4	...	37	32	13	...	235	208	87	4th, Northeast Div.	Lost Conf. Semi-Final
1996-97	82	17	17	7	...	14	19	8	...	31	36	15	...	249	276	77	4th, Northeast Div.	Lost Conf. Quarter-Final
1995-96	82	23	12	6	...	17	20	4	...	40	32	10	...	265	248	90	3rd, Northeast Div.	Lost Conf. Quarter-Final
1994-95	48	15	5	4	...	3	18	3	...	18	23	7	...	125	148	43	6th, Northeast Div.	Out of Playoffs
1993-94	84	26	12	4	...	15	17	10	...	41	29	14	...	283	248	96	3rd, Northeast Div.	Lost Conf. Quarter-Final
1992-93	**84**	**27**	**13**	**2**	**...**	**21**	**17**	**4**	**...**	**48**	**30**	**6**	**...**	**326**	**280**	**102**	**3rd, Adams Div.**	**Won Stanley Cup**
1991-92	80	27	8	5	...	14	20	6	...	41	28	11	...	267	207	93	1st, Adams Div.	Lost Div. Final
1990-91	80	23	12	5	...	16	18	6	...	39	30	11	...	273	249	89	2nd, Adams Div.	Lost Div. Final
1989-90	80	26	8	6	...	15	20	5	...	41	28	11	...	288	234	93	3rd, Adams Div.	Lost Div. Final
1988-89	80	30	6	4	...	23	12	5	...	53	18	9	...	315	218	115	1st, Adams Div.	Lost Final
1987-88	80	26	8	6	...	19	14	7	...	45	22	13	...	298	238	103	1st, Adams Div.	Lost Div. Final
1986-87	80	27	9	4	...	14	20	6	...	41	29	10	...	277	241	92	2nd, Adams Div.	Lost Conf. Championship
1985-86	**80**	**25**	**11**	**4**	**...**	**15**	**22**	**3**	**...**	**40**	**33**	**7**	**...**	**330**	**280**	**87**	**2nd, Adams Div.**	**Won Stanley Cup**
1984-85	80	24	10	6	...	17	17	6	...	41	27	12	...	309	262	94	1st, Adams Div.	Lost Div. Final
1983-84	80	19	19	2	...	16	21	3	...	35	40	5	...	286	295	75	4th, Adams Div.	Lost Conf. Championship
1982-83	80	25	6	9	...	17	18	5	...	42	24	14	...	350	286	98	2nd, Adams Div.	Lost Div. Semi-Final
1981-82	80	25	6	9	...	21	11	8	...	46	17	17	...	360	223	109	1st, Adams Div.	Lost Div. Semi-Final
1980-81	80	31	7	2	...	14	15	11	...	45	22	13	...	332	232	103	1st, Norris Div.	Lost Prelim. Round
1979-80	80	30	7	3	...	17	13	10	...	47	20	13	...	328	240	107	1st, Norris Div.	Lost Quarter-Final
1978-79	**80**	**29**	**6**	**5**	**...**	**23**	**11**	**6**	**...**	**52**	**17**	**11**	**...**	**337**	**204**	**115**	**1st, Norris Div.**	**Won Stanley Cup**
1977-78	**80**	**32**	**4**	**4**	**...**	**27**	**6**	**7**	**...**	**59**	**10**	**11**	**...**	**359**	**183**	**129**	**1st, Norris Div.**	**Won Stanley Cup**
1976-77	**80**	**33**	**1**	**6**	**...**	**27**	**7**	**6**	**...**	**60**	**8**	**12**	**...**	**387**	**171**	**132**	**1st, Norris Div.**	**Won Stanley Cup**
1975-76	**80**	**32**	**3**	**5**	**...**	**26**	**8**	**6**	**...**	**58**	**11**	**11**	**...**	**337**	**174**	**127**	**1st, Norris Div.**	**Won Stanley Cup**
1974-75	80	27	8	5	...	20	6	14	...	47	14	19	...	374	225	113	1st, Norris Div.	Lost Semi-Final
1973-74	78	24	12	3	...	21	12	6	...	45	24	9	...	293	240	99	2nd, East Div.	Lost Quarter-inal
1972-73	**78**	**29**	**4**	**6**	**...**	**23**	**6**	**10**	**...**	**52**	**10**	**16**	**...**	**329**	**184**	**120**	**1st, East Div.**	**Won Stanley Cup**
1971-72	78	29	3	7	...	17	13	9	...	46	16	16	...	307	205	108	3rd, East Div.	Lost Quarter-Final
1970-71	**78**	**29**	**7**	**3**	**...**	**13**	**16**	**10**	**...**	**42**	**23**	**13**	**...**	**291**	**216**	**97**	**3rd, East Div.**	**Won Stanley Cup**
1969-70	76	21	9	8	...	17	13	8	...	38	22	16	...	244	201	92	5th, East Div.	Out of Playoffs
1968-69	**76**	**26**	**7**	**5**	**...**	**20**	**12**	**6**	**...**	**46**	**19**	**11**	**...**	**271**	**202**	**103**	**1st, East Div.**	**Won Stanley Cup**
1967-68	**74**	**26**	**5**	**6**	**...**	**16**	**17**	**4**	**...**	**42**	**22**	**10**	**...**	**236**	**167**	**94**	**1st, East Div.**	**Won Stanley Cup**
1966-67	70	19	9	7	...	13	16	6	...	32	25	13	...	202	188	77	2nd,	Lost Final
1965-66	**70**	**23**	**11**	**1**	**...**	**18**	**10**	**7**	**...**	**41**	**21**	**8**	**...**	**239**	**173**	**90**	**1st,**	**Won Stanley Cup**
1964-65	**70**	**20**	**8**	**7**	**...**	**16**	**15**	**4**	**...**	**36**	**23**	**11**	**...**	**211**	**185**	**83**	**2nd,**	**Won Stanley Cup**
1963-64	70	22	7	6	...	14	14	7	...	36	21	13	...	209	167	85	1st,	Lost Semi-Final
1962-63	70	15	10	10	...	13	9	13	...	28	19	23	...	225	183	79	3rd,	Lost Semi-Final
1961-62	70	26	2	7	...	16	12	7	...	42	14	14	...	259	166	98	1st,	Lost Semi-Final
1960-61	70	24	6	5	...	17	13	5	...	41	19	10	...	254	188	92	1st,	Lost Semi-Final
1959-60	**70**	**23**	**4**	**8**	**...**	**17**	**14**	**4**	**...**	**40**	**18**	**12**	**...**	**255**	**178**	**92**	**1st,**	**Won Stanley Cup**
1958-59	**70**	**21**	**8**	**6**	**...**	**18**	**10**	**7**	**...**	**39**	**18**	**13**	**...**	**258**	**158**	**91**	**1st,**	**Won Stanley Cup**
1957-58	**70**	**23**	**8**	**4**	**...**	**20**	**9**	**6**	**...**	**43**	**17**	**10**	**...**	**250**	**158**	**96**	**1st,**	**Won Stanley Cup**
1956-57	**70**	**23**	**6**	**6**	**...**	**12**	**17**	**6**	**...**	**35**	**23**	**12**	**...**	**210**	**155**	**82**	**2nd,**	**Won Stanley Cup**
1955-56	**70**	**29**	**5**	**1**	**...**	**16**	**10**	**9**	**...**	**45**	**15**	**10**	**...**	**222**	**131**	**100**	**1st,**	**Won Stanley Cup**
1954-55	70	26	5	4	...	15	13	7	...	41	18	11	...	228	157	93	2nd,	Lost Final
1953-54	70	27	5	3	...	8	19	8	...	35	24	11	...	195	141	81	2nd,	Lost Final
1952-53	**70**	**18**	**12**	**5**	**...**	**10**	**11**	**14**	**...**	**28**	**23**	**19**	**...**	**155**	**148**	**75**	**2nd,**	**Won Stanley Cup**
1951-52	70	22	8	5	...	12	18	5	...	34	26	10	...	195	164	78	2nd,	Lost Final
1950-51	70	17	10	8	...	8	20	7	...	25	30	15	...	173	184	65	3rd,	Lost Final
1949-50	70	17	8	10	...	12	14	9	...	29	22	19	...	172	150	77	2nd,	Lost Semi-Final
1948-49	60	19	8	3	...	9	15	6	...	28	23	9	...	152	126	65	3rd,	Lost Semi-Final
1947-48	60	13	13	4	...	7	16	7	...	20	29	11	...	147	169	51	5th,	Out of Playoffs
1946-47	60	19	6	5	...	15	10	5	...	34	16	10	...	189	138	78	1st,	Lost Final
1945-46	**50**	**16**	**6**	**3**	**...**	**12**	**11**	**2**	**...**	**28**	**17**	**5**	**...**	**172**	**134**	**61**	**1st,**	**Won Stanley Cup**
1944-45	50	21	2	2	...	17	6	2	...	38	8	4	...	228	121	80	1st,	Lost Semi-Final
1943-44	**50**	**22**	**0**	**3**	**...**	**16**	**5**	**4**	**...**	**38**	**5**	**7**	**...**	**234**	**109**	**83**	**1st,**	**Won Stanley Cup**
1942-43	50	14	4	7	...	5	15	5	...	19	19	12	...	181	191	50	4th,	Lost Semi-Final
1941-42	48	12	10	2	...	6	17	1	...	18	27	3	...	134	173	39	6th,	Lost Quarter-Final
1940-41	48	11	9	4	...	5	17	2	...	16	26	6	...	121	147	38	6th,	Lost Quarter-Final
1939-40	48	5	14	5	...	5	19	0	...	10	33	5	...	90	167	25	7th,	Out of Playoffs
1938-39	48	8	11	5	...	7	13	4	...	15	24	9	...	115	146	39	6th,	Lost Quarter-Final
1937-38	48	13	4	7	...	5	13	6	...	18	17	13	...	123	128	49	3rd, Cdn. Div.	Lost Quarter-Final
1936-37	48	16	8	0	...	8	10	6	...	24	18	6	...	115	111	54	1st, Cdn. Div.	Lost Semi-Final
1935-36	48	5	11	8	...	6	15	3	...	11	26	11	...	82	123	33	4th, Cdn. Div.	Out of Playoffs
1934-35	48	11	11	2	...	8	12	4	...	19	23	6	...	110	145	44	3rd, Cdn. Div.	Lost Quarter-Final
1933-34	48	16	6	2	...	6	14	4	...	22	20	6	...	99	101	50	2nd, Cdn. Div.	Lost Quarter-Final
1932-33	48	15	5	4	...	3	20	1	...	18	25	5	...	92	115	41	3rd, Cdn. Div.	Lost Quarter-Final
1931-32	48	18	3	3	...	7	13	4	...	25	16	7	...	128	111	57	1st, Cdn. Div.	Lost Semi-Final
1930-31	**44**	**15**	**3**	**4**	**...**	**11**	**7**	**4**	**...**	**26**	**10**	**8**	**...**	**129**	**89**	**60**	**1st, Cdn. Div.**	**Won Stanley Cup**
1929-30	**44**	**13**	**5**	**4**	**...**	**8**	**9**	**5**	**...**	**21**	**14**	**9**	**...**	**142**	**114**	**51**	**2nd, Cdn. Div.**	**Won Stanley Cup**
1928-29	44	12	4	6	...	10	3	9	...	22	7	15	...	71	43	59	1st, Cdn. Div.	Lost Semi-Final
1927-28	44	12	7	3	...	14	4	4	...	26	11	7	...	116	48	59	1st, Cdn. Div.	Lost Semi-Final
1926-27	44	15	5	2	...	13	9	0	...	28	14	2	...	99	67	58	2nd, Cdn. Div.	Lost Semi-Final
1925-26	36	5	12	1	...	6	12	0	...	11	24	1	...	79	108	23	7th,	Out of Playoffs
1924-25	30	10	5	0	...	7	6	2	...	17	11	2	...	93	56	36	3rd,	Lost Final
1923-24	**24**	**10**	**2**	**0**	**...**	**3**	**9**	**0**	**...**	**13**	**11**	**0**	**...**	**59**	**48**	**26**	**2nd,**	**Won Stanley Cup**
1922-23	24	10	2	0	...	3	7	2	...	13	9	2	...	73	61	28	2nd,	Lost NHL Final
1921-22	24	8	3	1	...	4	8	0	...	12	11	1	...	88	94	25	3rd,	Out of Playoffs
1920-21	24	9	3	0	...	4	8	0	...	13	11	0	...	112	99	26	3rd and 2nd*	Out of Playoffs
1919-20	24	8	4	0	...	5	7	0	...	13	11	0	...	129	113	26	2nd and 3rd*	Out of Playoffs
1918-19	18	7	2	0	...	3	6	0	...	10	8	0	...	88	78	20	1st and 2nd*	Cup Final but no Decision
1917-18	22	8	3	0	...	5	6	0	...	13	9	0	...	115	84	26	1st and 3rd*	Lost NHL Final

* Season played in two halves with no combined standing at end.
From 1917-18 through 1925-26, NHL champions played against PCHA/WCHL champions for Stanley Cup.

2002-03 Player Personnel

FORWARDS	HT	WT	S	Place of Birth	Date	2001-02 Club
AUDETTE, Donald	5-8	190	R	Laval, Que.	9/23/69	Dallas-Montreal
BULIS, Jan	6-2	201	L	Pardubice, Czech.	3/18/78	Montreal
CHOUINARD, Eric	6-3	205	L	Atlanta, GA	7/8/80	Quebec
CZERKAWSKI, Mariusz	6-0	200	L	Radomsko, Poland	4/13/72	NY Islanders
DACKELL, Andreas	5-11	194	R	Gavle, Sweden	12/29/72	Montreal
GILMOUR, Doug	5-11	177	L	Kingston, Ont.	6/25/63	Montreal
GRATTON, Benoit	5-11	194	L	Montreal, Que.	12/28/76	Montreal-Quebec
HOSSA, Marcel	6-2	211	L	Ilava, Czech.	10/12/81	Montreal-Quebec
JUNEAU, Joe	6-0	195	L	Pont-Rouge, Que.	1/5/68	Montreal
KILGER, Chad	6-4	224	L	Cornwall, Ont.	11/27/76	Montreal
KOIVU, Saku	5-10	181	L	Turku, Finland	11/23/74	Montreal
LANDRY, Eric	5-10	184	L	Gatineau, Que.	1/20/75	Montreal-Quebec
LINDSAY, Bill	6-0	195	L	Fernie, B.C.	5/17/71	Florida-Montreal
McKAY, Randy	6-2	210	R	Montreal, Que.	1/25/67	New Jersey-Dallas
ODJICK, Gino	6-3	224	L	Maniwaki, Que.	9/7/70	Montreal-Quebec
PERREAULT, Yanic	5-11	185	L	Sherbrooke, Que.	4/4/71	Montreal
PETROV, Oleg	5-9	172	L	Moscow, USSR	4/18/71	Montreal
RIBEIRO, Mike	6-0	177	L	Montreal, Que.	2/10/80	Montreal-Quebec
RYDER, Michael	6-1	195	R	St. John's, Nfld.	3/31/80	Mississippi-Quebec
WARD, Jason	6-3	200	R	Chapleau, Ont.	1/16/79	Quebec
ZEDNIK, Richard	6-0	200	L	Bystrica, Czech.	1/6/76	Montreal
DEFENSEMEN						
BOUILLON, Francis	5-8	194	L	New York, NY	10/17/75	Montreal-Quebec
BRISEBOIS, Patrice	6-2	203	R	Montreal, Que.	1/27/71	Montreal
DESCOTEAUX, Matthieu	6-3	216	L	Pierreville, Que.	9/23/77	Quebec
DYKHUIS, Karl	6-3	214	L	Sept-Iles, Que.	7/8/72	Montreal
HAINSEY, Ron	6-3	200	L	Bolton, CT	3/24/81	Quebec
MARKOV, Andrei	6-0	208	L	Voskresensk, USSR	12/20/78	Montreal-Quebec
QUINTAL, Stephane	6-3	231	R	Boucherville, Que.	10/22/68	Montreal
RIVET, Craig	6-2	207	R	North Bay, Ont.	9/13/74	Montreal
ROBIDAS, Stephane	5-11	189	R	Sherbrooke, Que.	3/3/77	Montreal
SOURAY, Sheldon	6-4	223	L	Elk Point, Alta.	7/13/76	Montreal
TRAVERSE, Patrick	6-4	207	L	Montreal, Que.	3/14/74	Montreal-Quebec

GOALTENDERS	HT	WT	C	Place of Birth	Date	2001-02 Club
GARON, Mathieu	6-2	192	R	Chandler, Que.	1/9/78	Montreal-Quebec
HACKETT, Jeff	6-1	198	L	London, Ont.	6/1/68	Montreal
MICHAUD, Olivier	5-11	160	L	Beloeil, Que.	9/14/83	Montreal-Shawinigan
THEODORE, Jose	5-11	182	R	Laval, Que.	9/13/76	Montreal

Coaching History

Jack Laviolette, 1909-10; Adolphe Lecours, 1910-11; Napoleon Dorval, 1911-12, 1912-13; Jimmy Gardner, 1913-14, 1914-15; Newsy Lalonde, 1915-16 to 1920-21; Newsy Lalonde and Léo Dandurand, 1921-22; Léo Dandurand, 1922-23 to 1925-26; Cecil Hart, 1926-27 to 1931-32; Newsy Lalonde, 1932-33, 1933-34; Newsy Lalonde and Léo Dandurand, 1934-35; Sylvio Mantha, 1935-36; Cecil Hart, 1936-37, 1937-38; Cecil Hart and Jules Dugal, 1938-39; Babe Siebert, 1939*; Pit Lepine, 1939-40; Dick Irvin 1940-41 to 1954-55; Toe Blake, 1955-56 to 1967-68; Claude Ruel, 1968-69, 1969-70; Claude Ruel and Al MacNeil, 1970-71; Scotty Bowman, 1971-72 to 1978-79; Bernie Geoffrion and Claude Ruel, 1979-80; Claude Ruel, 1980-81; Bob Berry, 1981-82, 1982-83; Bob Berry and Jacques Lemaire, 1983-84; Jacques Lemaire, 1984-85; Jean Perron, 1985-86 to 1987-88; Pat Burns, 1988-89 to 1991-92; Jacques Demers, 1992-93 to 1994-95; Jacques Demers and Mario Tremblay, 1995-96; Mario Tremblay, 1996-97; Alain Vigneault, 1997-98 to 1999-2000; Alain Vigneault and Michel Therrien, 2000 -01; Michel Therrien, 2001-02 to date.

* Named coach in summer but died before 1939-40 season began.

Coach

THERRIEN, MICHEL
Coach, Montreal Canadiens. Born in Montreal, Que., November 4, 1963.

Michel Therrien was named head coach of the Montreal Canadiens on November 20, 2000. He had joined the organization on June 10, 1997, as head coach of the AHL Fredericton Canadiens. In his first full season behind the bench in 2001-02, he guided the club back into the playoffs after the Canadiens had missed the postseason for three straight years.

In 2000-01, he had coached the AHL's Quebec Citadelles to a 12-6-1-0 record, when he was hired to coach the Canadiens. His career record in the AHL is 115-108-36 in 259 regular-season games.

Therrien previously coached the Laval Titan and the Granby Predateurs in the QMJHL, winning the Memorial Cup with Granby in 1996. He also reached the league finals twice with Laval in 1993-94 and 1994-95.

Coaching Record

		Regular Season				Playoffs		
Season	Team	Games	W	L	T	Games	W	L
1993-94	Laval (QMJHL)	72	49	22	1	21	14	7
1994-95	Laval (QMJHL)	72	48	22	2	20	14	6
1995-96	Granby (QMJHL)	70	56	12	2	21	17	4
1996-97	Granby (QMJHL)	70	44	20	6	5	1	4
1997-98	Fredericton (AHL)	80	33	32	15	4	1	3
1998-99	Fredericton (AHL)	80	33	36	11	15	9	6
1999-2000	Quebec (AHL)	80	37	34	9	3	0	3
2000-01	Quebec (AHL)	19	12	6	1	....	....	....
	Montreal (NHL)	**62**	**23**	**33**	**6**	....	....	....
2001-02	**Montreal (NHL)**	**82**	**36**	**34**	**12**	**12**	**6**	**6**
	NHL Totals	**144**	**59**	**67**	**18**	**12**	**6**	**6**

2001-02 Scoring

* - rookie

Regular Season

Pos	#	Player	Team	GP	G	A	Pts	+/–	PIM	PP	SH	GW	GT	S	%
C	94	Yanic Perreault	MTL	82	27	29	56	–3	40	6	0	7	2	156	17.3
L	20	Richard Zednik	MTL	82	22	22	44	–3	59	4	0	3	0	249	8.8
R	14	Oleg Petrov	MTL	75	24	17	41	–4	12	3	1	6	0	152	15.8
C	93	Doug Gilmour	MTL	70	10	31	41	–7	48	5	0	2	0	78	12.8
C	90	Joe Juneau	MTL	70	8	28	36	–3	10	1	0	1	1	96	8.3
R	24	Andreas Dackell	MTL	79	15	18	33	–3	24	2	3	2	0	83	18.1
D	43	Patrice Brisebois	MTL	71	4	29	33	9	25	2	1	1	0	95	4.2
L	95	Sergei Berezin	PHX	41	7	9	16	–1	4	1	0	4	0	120	5.8
			MTL	29	4	6	10	3	4	3	0	1	0	80	5.0
			TOTAL	70	11	15	26	2	8	4	0	5	0	200	5.5
D	52	Craig Rivet	MTL	82	8	17	25	1	76	0	0	0	0	90	8.9
D	79	Andrei Markov	MTL	56	5	19	24	–1	24	2	0	1	0	73	6.8
L	25	Chad Kilger	MTL	75	8	15	23	–7	27	0	1	2	0	87	9.2
C	27	Shaun Van Allen	DAL	19	2	4	6	–5	6	0	0	0	0	20	10.0
			MTL	54	6	9	15	5	20	0	1	1	0	28	21.4
			TOTAL	73	8	13	21	0	26	0	1	1	0	48	16.7
C	38	Jan Bulis	MTL	53	9	10	19	–2	8	1	0	3	0	87	10.3
C	71	* Mike Ribeiro	MTL	43	8	10	18	–11	12	3	0	0	0	48	16.7
R	82	Donald Audette	DAL	20	4	8	12	2	12	3	0	2	0	49	8.2
			MTL	13	1	5	6	1	8	0	0	1	0	33	3.0
			TOTAL	33	5	13	18	3	20	3	0	3	0	82	6.1
D	5	Stephane Quintal	MTL	75	6	10	16	–7	87	1	0	1	0	85	7.1
L	22	Bill Lindsay	FLA	63	4	7	11	–11	117	0	0	1	0	63	6.3
			MTL	13	1	3	4	0	23	0	0	0	0	14	7.1
			TOTAL	76	5	10	15	–11	140	0	0	1	0	77	6.5
D	28	Karl Dykhuis	MTL	80	5	7	12	16	32	0	0	1	0	85	5.9
D	56	Stephane Robidas	MTL	56	1	10	11	–25	14	1	0	0	0	68	1.5
C	45	Arron Asham	MTL	35	5	4	9	7	55	0	0	0	0	30	16.7
L	29	Gino Odjick	MTL	36	4	4	8	3	104	0	0	0	0	40	10.0
D	44	Sheldon Souray	MTL	34	3	5	8	–5	62	1	0	0	1	56	5.4
D	54	Patrick Traverse	MTL	25	2	3	5	–7	14	2	0	0	0	24	8.3
L	37	Patrick Poulin	MTL	28	0	5	5	6	6	0	0	0	0	15	0.0
D	51	Francis Bouillon	MTL	28	0	5	5	–5	33	0	0	0	0	24	0.0
C	36	* Marcel Hossa	MTL	10	3	1	4	2	2	0	0	0	0	20	15.0
C	11	Saku Koivu	MTL	3	0	2	2	0	0	0	0	0	0	2	0.0
C	46	Benoit Gratton	MTL	8	1	0	1	–1	8	0	0	0	0	8	12.5
C	78	Eric Landry	MTL	2	0	1	1	2	0	0	0	0	0	0	0.0
D	59	* Martti Jarventie	MTL	1	0	0	0	2	0	0	0	0	0	0	0.0
C	63	Craig Darby	MTL	2	0	0	0	0	0	0	0	0	0	0	0.0

Goaltending

No.	Goaltender	GPI	Mins	Avg	W	L	T	EN	SO	GA	SA	S%	G	A	PIM
95	* Olivier Michaud	1	18	0.00	0	0	0	0	0	0	14	1.000	0	0	0
60	Jose Theodore	67	3864	2.11	30	24	10	9	7	136	1972	.931	0	2	2
31	Jeff Hackett	15	717	3.18	5	5	2	0	0	38	395	.904	0	0	2
35	Stephane Fiset	2	109	3.85	0	1	0	0	0	7	60	.883	0	0	0
30	* Mathieu Garon	5	261	4.37	1	4	0	0	0	19	147	.871	0	0	0
	Totals	**82**	**4989**	**2.51**	**36**	**34**	**12**	**9**	**7**	**209**	**2597**	**.920**			

Playoffs

Pos	#	Player	Team	GP	G	A	Pts	+/–	PIM	PP	SH	GW	GT	S	%
R	82	Donald Audette	MTL	12	6	4	10	–2	10	2	0	2	1	25	24.0
C	93	Doug Gilmour	MTL	12	4	6	10	–2	16	1	0	0	0	17	23.5
C	11	Saku Koivu	MTL	12	4	6	10	2	4	1	0	1	0	16	25.0
R	20	Richard Zednik	MTL	4	4	4	8	3	6	2	0	0	0	22	18.2
C	94	Yanic Perreault	MTL	11	3	5	8	0	0	2	0	1	0	19	15.8
R	14	Oleg Petrov	MTL	12	1	5	6	2	2	0	0	1	0	30	3.3
C	90	Joe Juneau	MTL	12	1	4	5	–6	6	0	0	0	0	18	5.6
L	22	Bill Lindsay	MTL	11	2	2	4	1	2	0	0	0	0	3	66.7
D	5	Stephane Quintal	MTL	12	1	3	4	–2	12	0	0	0	0	9	11.1
D	79	Andrei Markov	MTL	12	1	3	4	–2	8	0	0	1	0	8	12.5
R	24	Andreas Dackell	MTL	12	1	2	3	–4	6	0	0	0	0	11	9.1
D	52	Craig Rivet	MTL	12	0	3	3	2	4	0	0	0	0	19	0.0
L	95	Sergei Berezin	MTL	6	1	1	2	2	0	1	0	0	0	11	9.1
D	43	Patrice Brisebois	MTL	10	1	1	2	–4	2	0	0	0	0	16	6.3
D	28	Karl Dykhuis	MTL	12	1	1	2	2	8	0	0	0	0	11	9.1
L	29	Gino Odjick	MTL	12	1	0	1	–5	47	0	0	0	0	13	7.7
C	45	Arron Asham	MTL	3	0	1	1	1	0	0	0	0	0	3	0.0
C	27	Shaun Van Allen	MTL	7	0	1	1	–3	2	0	0	0	0	1	0.0
D	44	Sheldon Souray	MTL	12	0	1	1	–4	16	0	0	0	0	10	0.0
L	25	Chad Kilger	MTL	12	0	1	1	1	9	0	0	0	0	17	0.0
D	56	Stephane Robidas	MTL	2	0	0	0	–1	4	0	0	0	0	2	0.0
C	38	Jan Bulis	MTL	6	0	0	0	–6	6	0	0	0	0	6	0.0

Goaltending

No.	Goaltender	GPI	Mins	Avg	W	L	EN	SO	GA	SA	S%	G	A	PIM
60	Jose Theodore	12	686	3.06	6	6	1	0	35	413	.915	0	0	0
35	Stephane Fiset	1	38	4.74	0	0	0	0	3	19	.842	0	0	0
	Totals	**12**	**726**	**3.22**	**6**	**6**	**1**	**0**	**39**	**433**	**.910**			

Captains' History

Jack Laviolette, 1909-10; Newsy Lalonde, 1910-11; Jack Laviolette, 1911-12; Newsy Lalonde, 1912-13; Jimmy Gardner, 1913-14, 1914-15; Howard McNamara, 1915-16; Newsy Lalonde, 1916-17 to 1921-22; Sprague Cleghorn, 1922-23 to 1924-25; Bill Coutu, 1925-26; Sylvio Mantha, 1926-27 to 1931-32; George Hainsworth, 1932-33; Sylvio Mantha, 1933-34 to 1935-36; Babe Siebert, 1936-37 to 1938-39; Walt Buswell, 1939-40; Toe Blake, 1940-41 to 1946-47; Toe Blake and Bill Durnan, 1947-48; Butch Bouchard, 1948-49 to 1955-56; Maurice Richard, 1956-57 to 1959-60; Doug Harvey, 1960-61; Jean Béliveau, 1961-62 to 1970-71; Henri Richard, 1971-72 to 1974-75; Yvan Cournoyer, 1975-76 to 1978-79; Serge Savard, 1979-80, 1980-81; Bob Gainey, 1981-82 to 1988-89; Guy Carbonneau and Chris Chelios (co-captains), 1989-90; Guy Carbonneau, 1990-91 to 1993-94; Kirk Muller and Mike Keane, 1994-95; Mike Keane and Pierre Turgeon, 1995-96; Pierre Turgeon and Vincent Damphousse, 1996-97; Vincent Damphousse, 1997-98, 1998-99; Saku Koivu, 1999-2000 to date.

Club Records

Team

(Figures in brackets for season records are games played; records for fewest points, wins, ties, losses, goals, goals against are for 70 or more games)

Record		
Most Points	***132**	1976-77 (80)
Most Wins	**60**	1976-77 (80)
Most Ties	**23**	1962-63 (70)
Most Losses	**40**	1983-84 (80), 2000-01 (82)
Most Goals	**387**	1976-77 (80)
Most Goals Against	**295**	1983-84 (80)
Fewest Points	**65**	1950-51 (70)
Fewest Wins	**25**	1950-51 (70)
Fewest Ties	**5**	1983-84 (80)
Fewest Losses	***8**	1976-77 (80)
Fewest Goals	**155**	1952-53 (70)
Fewest Goals Against	***131**	1955-56 (70)
Longest Winning Streak		
Overall	**12**	Jan. 6-Feb. 3/68
Home	**13**	Nov. 2/43-Jan. 8/44, Jan. 30-Mar. 26/77
Away	**8**	Dec. 18/77-Jan. 18/78, Jan. 21-Feb. 21/82
Longest Undefeated Streak		
Overall	**28**	Dec. 18/77-Feb. 23/78 (23 wins, 5 ties)
Home	***34**	Nov. 1/76-Apr. 2/77 (28 wins, 6 ties)
Away	***23**	Nov. 27/74-Mar. 12/75 (14 wins, 9 ties)
Longest Losing Streak		
Overall	**12**	Feb. 13-Mar. 13/26
Home	**7**	Dec. 16/39-Jan. 18/40, Oct. 28-Nov. 25/00
Away	**10**	Jan. 16-Mar. 13/26
Longest Winless Streak		
Overall	**12**	Feb. 13-Mar. 13/26 (12 losses), Nov. 28-Dec. 29/35 (8 losses, 4 ties)
Home	**15**	Dec. 16/39-Mar. 7/40 (12 losses, 3 ties)
Away	**12**	Nov. 26/33-Jan. 28/34 (8 losses, 4 ties), Oct. 20/50-Dec. 13/51 (8 losses, 4 ties)
Most Shutouts, Season	***22**	1928-29 (44)
Most PIM, Season	**1,847**	1995-96 (82)
Most Goals, Game	***16**	Mar. 3/20 (Mtl. 16 at Que. 3)

Individual

Most Seasons	**20**	Henri Richard, Jean Béliveau
Most Games	**1,256**	Henri Richard
Most Goals, Career	**544**	Maurice Richard
Most Assists, Career	**728**	Guy Lafleur
Most Points, Career	**1,246**	Guy Lafleur (518G, 728A)
Most PIM, Career	**2,248**	Chris Nilan
Most Shutouts, Career	**75**	George Hainsworth
Longest Consecutive Games Streak	**560**	Doug Jarvis (Oct. 8/75-Apr. 4/82)
Most Goals, Season	**60**	Steve Shutt (1976-77), Guy Lafleur (1977-78)
Most Assists, Season	**82**	Pete Mahovlich (1974-75)
Most Points, Season	**136**	Guy Lafleur (1976-77; 56G, 80A)
Most PIM, Season	**358**	Chris Nilan (1984-85)
Most Points, Defenseman, Season	**85**	Larry Robinson (1976-77; 19G, 66A)
Most Points, Center, Season	**117**	Pete Mahovlich (1974-75; 35G, 82A)
Most Points, Right Wing, Season	**136**	Guy Lafleur (1976-77; 56G, 80A)
Most Points, Left Wing, Season	**110**	Mats Naslund (1985-86; 43G, 67A)
Most Points, Rookie, Season	**71**	Mats Naslund (1982-83; 26G, 45A), Kjell Dahlin (1985-86; 32G, 39A)
Most Shutouts, Season	***22**	George Hainsworth (1928-29)
Most Goals, Game	**6**	Newsy Lalonde (Jan. 10/20)
Most Assists, Game	**6**	Elmer Lach (Feb. 6/43)
Most Points, Game	**8**	Maurice Richard (Dec. 28/44; 5G, 3A), Bert Olmstead (Jan. 9/54; 4G, 4A)

* NHL Record.

Retired Numbers

1	Jacques Plante	1952-1963
2	Doug Harvey	1947-1961
4	Jean Béliveau	1950-1971
7	Howie Morenz	1923-1937
9	Maurice Richard	1942-1960
10	Guy Lafleur	1971-1984
16	Henri Richard	1955-1975

All-time Record vs. Other Clubs

Regular Season

	At Home								On Road								Total							
	GP	W	L	T	OL	GF	GA	PTS	GP	W	L	T	OL	GF	GA	PTS	GP	W	L	T	OL	GF	GA	PTS
Anaheim	7	4	2	1	0	22	17	9	6	4	2	0	0	21	20	8	13	8	4	1	0	43	37	17
Atlanta	6	5	1	0	0	22	13	10	6	4	0	2	0	23	7	10	12	9	1	2	0	45	20	20
Boston	329	187	95	46	1	1112	783	421	329	121	150	56	2	889	973	300	658	308	245	102	3	2001	1756	721
Buffalo	97	53	32	12	0	371	291	118	97	29	49	19	0	263	303	77	194	82	81	31	0	634	594	195
Calgary	44	24	12	8	0	155	109	56	47	26	14	6	1	162	143	59	91	50	26	14	1	317	252	115
Carolina	71	47	16	7	1	297	202	102	74	35	27	11	1	260	219	82	145	82	43	18	2	557	421	184
Chicago	274	172	54	48	0	1057	646	392	273	125	93	55	0	761	731	305	547	297	147	103	0	1818	1377	697
Colorado	63	37	15	10	1	257	196	85	62	26	31	5	0	216	209	57	125	63	46	15	1	473	405	142
Columbus	2	1	1	0	0	4	3	2	1	1	0	0	0	3	1	2	3	2	1	0	0	7	4	4
Dallas	57	37	11	9	0	248	140	83	57	30	16	11	0	201	147	71	114	67	27	20	0	449	287	154
Detroit	281	171	67	43	0	992	635	385	279	96	129	53	1	714	803	246	560	267	196	96	1	1706	1438	631
Edmonton	28	15	8	4	1	100	89	35	33	16	16	0	1	105	109	33	61	31	24	4	2	205	198	68
Florida	17	7	8	2	0	45	39	16	18	6	9	3	0	51	60	15	35	13	17	5	0	96	99	31
Los Angeles	63	44	8	11	0	286	159	99	63	36	18	9	0	251	195	81	126	80	26	20	0	537	354	180
Minnesota	2	1	0	1	0	6	2	3	1	0	1	0	0	2	4	0	3	1	1	1	0	8	6	3
Nashville	3	3	0	0	0	10	5	6	3	0	2	1	0	4	13	1	6	3	2	1	0	14	18	7
New Jersey	50	30	14	6	0	174	129	66	50	25	22	3	0	186	149	53	100	55	36	9	0	360	278	119
NY Islanders	56	33	14	9	0	206	161	75	56	24	24	6	2	163	173	56	112	57	38	15	2	369	334	131
NY Rangers	286	188	59	39	0	1114	656	415	286	114	117	54	1	831	831	283	572	302	176	93	1	1945	1487	698
Ottawa	28	15	9	4	0	85	77	34	25	13	11	1	0	73	70	27	53	28	20	5	0	158	147	61
Philadelphia	70	34	22	14	0	249	216	82	69	27	26	16	0	208	207	70	139	61	48	30	0	457	423	152
Phoenix	29	24	3	2	0	142	66	50	27	11	9	7	0	104	89	29	56	35	12	9	0	246	155	79
Pittsburgh	78	59	10	9	0	367	194	127	78	38	27	13	0	273	230	89	156	97	37	22	0	640	424	216
St. Louis	57	40	10	7	0	248	154	87	57	28	14	15	0	195	147	71	114	68	24	22	0	443	301	158
San Jose	10	7	1	2	0	35	17	16	9	4	3	2	0	22	25	10	19	11	4	4	0	57	42	26
Tampa Bay	18	10	7	1	0	52	42	21	19	8	7	4	0	50	44	20	37	18	14	5	0	102	86	41
Toronto	327	197	88	42	0	1150	809	436	328	114	169	45	0	849	993	273	655	311	257	87	0	1999	1802	709
Vancouver	51	38	9	4	0	241	129	80	53	33	12	8	0	196	138	74	104	71	21	12	0	437	267	154
Washington	56	33	16	7	0	216	120	73	55	22	24	9	0	167	153	53	111	55	40	16	0	383	273	126
Defunct Clubs	231	148	58	25	0	779	469	321	230	98	97	35	0	586	606	231	461	246	155	60	0	1365	1075	552
Totals	**2691**	**1664**	**650**	**373**	**4**	**10042**	**6568**	**3705**	**2691**	**1114**	**1119**	**449**	**9**	**7829**	**7792**	**2686**	**5382**	**2778**	**1769**	**822**	**13**	**17871**	**14360**	**6391**

Playoffs

	Series	W	L	GP	W	L	T	GF	GA	Last Mtg.	Rnd.	Result
Boston	29	22	7	145	91	54	0	450	357	2002	CQF	W 4-2
Buffalo	7	4	3	35	18	17	0	124	111	1998	CSF	L 0-4
Calgary	2	1	1	11	6	5	0	31	32	1989	F	L 2-4
Carolina	6	5	1	33	21	12	0	108	91	2002	CSF	L 2-4
Chicago	17	12	5	81	50	29	2	261	185	1976	QF	W 4-0
Colorado	5	3	2	31	17	14	0	105	85	1993	DSF	W 4-2
Dallas	2	1	1	13	7	6	0	48	37	1980	QF	L 3-4
Detroit	12	5	7	62	33	29	0	161	149	1978	QF	W 4-1
Edmonton	1	0	1	3	0	3	0	6	15	1981	PRE	L 0-3
Los Angeles	1	1	0	5	4	1	0	15	12	1993	F	W 4-1
New Jersey	1	0	1	5	1	4	0	11	22	1997	CQF	L 1-4
NY Islanders	4	3	1	22	14	8	0	64	55	1993	CF	W 4-1
NY Rangers	14	7	7	61	34	25	2	188	158	1996	CQF	L 2-4
Philadelphia	4	3	1	21	14	7	0	72	52	1989	CF	W 4-2
Pittsburgh	1	1	0	6	4	2	0	18	15	1998	CQF	W 4-2
St. Louis	3	3	0	12	12	0	0	42	14	1977	QF	W 4-0
Toronto	15	8	7	71	42	29	0	215	160	1979	QF	W 4-0
Vancouver	1	1	0	5	4	1	0	20	9	1975	QF	W 4-1
Defunct Clubs	11*	6	4	28	15	9	4	70	71			
Totals	**136***	**86**	**49**	**650**	**387**	**255**	**8**	**2009**	**1630**			

* 1919 Final incomplete due to influenza epidemic.

Playoff Results 2002-1998

Year	Round	Opponent	Result	GF	GA
2002	CSF	Carolina	L 2-4	12	21
	CQF	Boston	W 4-2	20	18
1998	CSF	Buffalo	L 0-4	10	17
	CQF	Pittsburgh	W 4-2	18	15

Abbreviations: Round: F - Final; **CF** - conference final; **CSF** - conference semi-final; **CQF** - conference quarter-final; **DSF** - division semi-final; **QF** - quarter-final; **PRE** - preliminary round.

Calgary totals include Atlanta Flames, 1972-73 to 1979-80. Carolina totals include Hartford, 1979-80 to 1996-97.
Colorado totals include Quebec, 1979-80 to 1994-95. Dallas totals include Minnesota North Stars, 1967-68 to 1992-93.
New Jersey totals include Kansas City, 1974-75 to 1975-76, and Colorado Rockies, 1976-77 to 1981-82.
Phoenix totals include Winnipeg, 1979-80 to 1995-96.

2001-02 Results

Month	Day	Opponent	Score
Oct.	4	at Ottawa	6-4
	6	Toronto	2-2
	9	Anaheim	3-1
	12	at Columbus	3-1
	13	New Jersey	3-1
	15	NY Rangers	1-2
	19	at Washington	1-4
	20	Buffalo	1-3
	26	at Buffalo	5-2
	27	Philadelphia	1-5
	30	at Edmonton	1-3
Nov.	1	at Vancouver	0-4
	3	at Calgary	2-6
	6	Colorado	1-1
	8	Nashville	3-1
	10	NY Islanders	3-2
	11	at NY Rangers	2-3*
	13	at Boston	3-5
	17	Florida	1-0*
	20	Boston	3-2
	22	at Atlanta	5-2
	24	Washington	5-3
	27	Atlanta	5-1
	29	at NY Islanders	1-1
Dec.	1	NY Rangers	1-3
	3	Chicago	2-3
	5	New Jersey	1-2
	8	Phoenix	3-3
	10	Minnesota	4-0
	12	at Atlanta	3-3
	13	at Philadelphia	3-2
	15	at Toronto	4-6
	17	Tampa Bay	3-4
	19	at Pittsburgh	3-1
	20	at Boston	0-5
	22	Los Angeles	2-1
	26	at Buffalo	1-3
	28	at St. Louis	0-3
	29	at NY Islanders	5-6*
Jan.	3	at Vancouver	2-5
	5	at Calgary	4-2
	6	at Edmonton	6-7
	8	at Minnesota	2-4
	10	NY Islanders	4-0
	12	at Toronto	1-1
	14	Philadelphia	3-5
	16	Washington	2-0
	17	at Carolina	1-1
	19	at Tampa Bay	5-1
	21	at Florida	5-7
	23	at Washington	5-3
	26	Ottawa	1-1
	27	San Jose	3-1
	30	Boston	3-4*
Feb.	5	at New Jersey	1-0
	7	Pittsburgh	1-0
	9	at Toronto	1-4
	11	Detroit	2-3
	26	Ottawa	2-5
	27	at Chicago	3-2
Mar.	2	Carolina	3-4
	4	Atlanta	5-3
	6	Boston	5-3
	8	at Buffalo	0-3
	9	Toronto	1-1
	11	at NY Rangers	1-2
	14	Dallas	3-3
	16	Carolina	3-2
	18	at Carolina	1-1
	20	at Florida	4-1
	22	at Tampa Bay	3-3
	23	at Nashville	1-5
	26	Florida	1-2
	28	Tampa Bay	2-1
	30	Pittsburgh	2-1*
Apr.	1	at Pittsburgh	3-0
	4	at Philadelphia	3-1
	6	Columbus	4-1
	7	at Ottawa	3-1
	9	Ottawa	4-3
	12	at New Jersey	2-5
	13	Buffalo	0-3

* – Overtime

Entry Draft Selections 2002-1988

2002
Pick
14 Christopher Higgins
45 Tomas Linhart
99 Michael Lambert
182 Andre Deveaux
212 Jonathan Ferland
275 Konstantin Korneyev

2001
Pick
7 Mike Komisarek
25 Alexander Perezhogin
37 Duncan Milroy
71 Tomas Plekanec
109 Martti Jarventie
171 Eric Himelfarb
266 Viktor Ujcik

2000
Pick
13 Ron Hainsey
16 Marcel Hossa
78 Josef Balej
79 Tyler Hanchuck
109 Johan Eneqvist
114 Christian Larrivee
145 Ryan Glenn
172 Scott Selig
182 Petr Chvojka
243 Joni Puurula
275 Jonathan Gauthier

1999
Pick
39 Alexander Buturlin
58 Matt Carkner
97 Chris Dyment
107 Evan Lindsay
136 Dusty Jamieson
145 Marc-Andre Thinel
150 Matt Shasby
167 Sean Dixon
196 Vadim Tarasov
225 Mikko Hyytia
253 Jerome Marois

1998
Pick
16 Eric Chouinard
45 Mike Ribeiro
75 Francois Beauchemin
132 Andrei Bashkirov
152 Gordie Dwyer
162 Andrei Markov
189 Andrei Kruchinin
201 Craig Murray
216 Michael Ryder
247 Darcy Harris

1997
Pick
11 Jason Ward
37 Gregor Baumgartner
65 Ilkka Mikkola
91 Daniel Tetrault
118 Konstantin Sidulov
122 Gennady Razin
145 Jonathan Desroches
172 Ben Guite
197 Petr Kubos
202 Andrei Sidyakin
228 Jarl Espen Ygranes

1996
Pick
18 Matt Higgins
44 Mathieu Garon
71 Arron Asham
92 Kim Staal
99 Etienne Drapeau
127 Daniel Archambault
154 Brett Clark
181 Timo Vertala
207 Mattia Baldi
233 Michel Tremblay

1995
Pick
8 Terry Ryan
60 Miloslav Guren
74 Martin Hohenberger
86 Jonathan Delisle
112 Niklas Anger
138 Boyd Olson
164 Stephane Robidas
190 Greg Hart
216 Eric Houde

1994
Pick
18 Brad Brown
44 Jose Theodore
54 Chris Murray
70 Marko Kiprusoff
74 Martin Belanger
96 Arto Kuki
122 Jimmy Drolet
148 Joel Irving
174 Jessie Rezansoff
200 Peter Strom
226 Tomas Vokoun
252 Chris Aldous
278 Ross Parsons

1993
Pick
21 Saku Koivu
47 Rory Fitzpatrick
73 Sebastien Bordeleau
85 Adam Wiesel
99 Jean-Francois Houle
113 Jeff Lank
125 Dion Darling
151 Darcy Tucker
177 David Ruhly
203 Alan Letang
229 Alexandre Duchesne
255 Brian Larochelle
281 Russell Guzior

1992
Pick
20 David Wilkie
33 Valeri Bure
44 Keli Corpse
68 Craig Rivet
82 Louis Bernard
92 Marc Lamothe
116 Don Chase
140 Martin Sychra
164 Christian Proulx
188 Michael Burman
212 Earl Cronan
236 Trent Cavicchi
260 Hiroyuki Miura

1991
Pick
17 Brent Bilodeau
28 Jim Campbell
43 Craig Darby
61 Yves Sarault
73 Vladimir Vujtek
83 Sylvain Lapointe
100 Brad Layzell
105 Tony Prpic
127 Oleg Petrov
149 Brady Kramer
171 Brian Savage
193 Scott Fraser
215 Greg MacEachern
237 Paul Lepler
259 Dale Hooper

1990
Pick
12 Turner Stevenson
39 Ryan Kuwabara
58 Charles Poulin
60 Robert Guillet
81 Gilbert Dionne
102 Paul Di Pietro
123 Craig Conroy
144 Stephen Rohr
165 Brent Fleetwood
186 Derek Maguire
207 Mark Kettelhut
228 John Uniac
249 Sergei Martynyuk

1989
Pick
13 Lindsay Vallis
30 Patrice Brisebois
41 Steve Larouche
51 Pierre Sevigny
83 Andre Racicot
104 Marc Deschamps
146 Craig Ferguson
167 Patrick Lebeau
188 Roy Mitchell
209 Ed Henrich
230 Justin Duberman
251 Steve Cadieux

1988
Pick
20 Eric Charron
34 Martin St. Amour
46 Neil Carnes
83 Patric Kjellberg
93 Peter Popovic
104 Jean-Claude Bergeron
125 Patrik Carnback
146 Tim Chase
167 Sean Hill
188 Harijs Vitolinsh
209 Yuri Krivokhizha
230 Kevin Dahl
251 Dave Kunda

General Managers' History

Jack Laviolette and Joseph Cattarinich, 1909-1910; George Kennedy, 1910-11 to 1920-21; Leo Dandurand, 1921-22 to 1934-35; Ernest Savard, 1935-36; Cecil Hart, 1936-37 to 1938-39; Jules Dugal, 1939-40; Tom P. Gorman, 1940-41 to 1945-46; Frank J. Selke, 1946-47 to 1963-64; Sam Pollock, 1964-65 to 1977-78; Irving Grundman, 1978-79 to 1982-83; Serge Savard, 1983-84 to 1994-95; Serge Savard and Réjean Houle, 1995-96; Réjean Houle, 1996-97 to 1999-2000; Réjean Houle and Andre Savard, 2000-01; Andre Savard, 2001-02 to date.

General Manager

SAVARD, ANDRE
General Manager, Montreal Canadiens.
Born in Témiscamingue, Que., February 9, 1953.

André Savard was named general manager of the Montreal Canadiens on November 20, 2000 after having joined the organization as director of hockey personnel. He has tremendous experience in scouting and player development.

Savard has been associated with the NHL since 1973-74. Before joining the Canadiens, he spent five seasons with the Ottawa Senators organization, including the first four as head scout. In June of 1987, he became the first former Nordiques player to assume the Quebec head coaching position.

Savard played 12 seasons in the NHL. The Boston Bruins' first-round pick, sixth overall in 1973, he also played for the Buffalo Sabres and the Nordiques. In 790 NHL regular season games, he totalled 482 points (211 goals, 271 assists). He also added 31 points in 85 playoff games (13 goals, 18 assists). The former QMJHL star retired as a player after the 1984-85 season following a serious knee injury.

NHL Coaching Record

		Regular Season				Playoffs		
Season	Team	Games	W	L	T	Games	W	L
1987-88	Quebec	24	10	13	1	....	....	....

Club Directory

Bell Centre

Bell Centre
1260 de La Gauchetière Street W.
Montréal, QC H3B 5E8
Phone: **514/932-2582**
Fax Lines (all area code 514):
Hockey 932-8736
Media Relations 932-8285
Marketing 925-2145
Community Relations 925-2144
www.canadiens.com
Capacity: 21,273

Executive Management

Chairman and Governor	George N. Gillett, Jr.
Vice-Chairman	Jeff Joyce
President of Club de Hockey Canadien and L'Aréna des Canadiens and Alternate Governor	Pierre Boivin
Assistant to the President	Foster Gillett
Administrative Assistant to the President	Lise Beaudry
Executive Vice-President Hockey and General Manager	André Savard
Chief Financial Officer and Alternate Governor	Fred Steer
Vice-President, Marketing and Sales	Ray Lalonde
Vice-President, Communications and Community Relations	Donald Beauchamp
Vice-President, Operations, Bell Centre	Alain Gauthier
President of the Gillett Entertainment Group	Aldo Giampaolo

Hockey Operations

Assistant General Manager	Martin Madden
Head Coach	Michel Therrien
Assistant Coaches	Roland Melanson, Rick Green, Clément Jodoin, Guy Charron
Director of Player Personnel	Trevor Timmins
Director of Legal Affairs	Julien BriseBois
Amateur Scouting Coordinator	Pierre Dorion
Director of Team Services	Michèle Lapointe
Professional Scouts	Pierre Mondou, Gordie Roberts
Amateur Scouts	Elmer Benning, William A. Berglund, Hannu Laine, Dave Mayville, Trent McCleary, Gerry O'Flaherty, Antonin Routa, Claude Ruel, Nikolai Vakourov
Administrative Assistant to the General Manager	Donna Stuart
Administrative Assistant, Team Services	Claudine Crépin

Medical and Training Staff

Club Physician and Chief Surgeon	Dr. David Mulder
Orthopaedist	Dr. Eric Lenczner
General Physician	Dr. Vincent Lacroix
Dentist	Dr. Pierre Desautels
Ophthalmologist	Dr. John Little
Head Athletic Therapist	Graham Rynbend
Assistant to the Athletic Therapist	Luc LeBlanc
Strength & Conditioning Coordinator	Scott Livingston
Equipment Manager	Pierre Gervais
Assistants to the Equipment Manager	Robert Boulanger, Pierre Ouellette
Video Supervisor	Mario Leblanc

Communications and Community Relations

Director of Media Relations	Dominick Saillant
Administrative Assistant, Communications	Sylvie Lambert
Community Relations Manager	Frédérique Cardinal
Community Relations Coordinator	Geneviève Paquette

Marketing

Administrative Assistant to the Vice President, Marketing and Sales	Nicole Malboeuf
Manager, Creative Services	Jean-Pierre Lacombe
Manager, Game Operations	Chantal Bunnett
Manager, Editorial	Carl Lavigne
Manager, Consumer Products	Luc Rocheleau

Ticket Sales

Executive Director, Premium Sales and Services	Richard Primeau
Group Manager, Game Day Sales and Promotions	Vincent Lussier
Manager, Group Sales	Pierre Constant
Coordinator, Luxury Suites Services	Sabina D'Ascoli

Advertising and Sponsorship Sales

EFFIX Inc.	François-Xavier Seigneur

Finance

Executive Director of Finance	Jacques Aubé
Controller, Budgeting & Analysis	Dennis McKinley
Controller, Financial Reporting	Françoise Brault
Director of Information Technology	Pierre-Éric Belzile
Administrative Assistant to the CFO	Christine Ouellette

Building Operations

Director, Ticket Office	Cathy D'Ascoli
Assistants to Director of Ticket Office	Jocelyne Rocan, Lucie Masse
Director, Building Operations	Xavier Ludyin
Administrative Assistant to the V.P., Operations	Maryse Cartwright

Entertainment

Executive Director, Events	Louise Laliberté
Administrative Assistant to the President of Gillett Entertainment Group	Vicki Mercuri

AHL Affiliation

Hamilton Bulldogs	Copps Coliseum - 85, York Blvd. - Hamilton, ONT, L8R 3L4 - Tel. : (905) 529-8500
Governor	Scott Hawson
Head Coach	Claude Julien
Assistant Coach	Geoff Ward
Athletic Therapist	Chris Davie
Equipment Managers	Murray Stevens, Patrick Langlois
Manager of Public Relations	Craig Downey

Team Information and Rightsholders

Location of Press Box	Suspended above ice - East Side
Location of Radio and TV booth	Suspended above ice - West side
Club trains at	Centre Bell
Play-by-play - Radio/TV	Pierre Houde (RDS), Pierre Rinfret (CKAC) French, Dino Sisto (CJAD-English), TBA (TSN-English)
TV Channels	CBFT (2) (French)
Cable TV	RDS (33) (French), TSN (28) (English)
Radio Stations	CKAC (730) (French), CJAD (800) (English)

Nashville Predators

2001-02 Results: 28W-41L-13T-0OTL 69PTS.
Fourth, Central Division

Year-by-Year Record

Season	GP	Home W	Home L	Home T	Home OL	Road W	Road L	Road T	Road OL	Overall W	Overall L	Overall T	Overall OL	GF	GA	Pts.	Finished	Playoff Result
2001-02	82	17	16	8	0	11	25	5	0	28	41	13	0	196	230	69	4th, Central Div.	Out of Playoffs
2000-01	82	16	18	7	0	18	18	2	3	34	36	9	3	186	200	80	3rd, Central Div.	Out of Playoffs
1999-2000	82	15	21	3	2	13	19	4	5	28	40	7	7	199	240	70	4th, Central Div.	Out of Playoffs
1998-99	82	15	22	4	...	13	25	3	...	28	47	7	...	190	261	63	4th, Central Div.	Out of Playoffs

2002-03 Schedule

Month	Day	Date	Opponent
Oct.	Fri.	11	at Washington
	Sat.	12	Edmonton
	Tue.	15	at NY Islanders
	Fri.	18	at New Jersey
	Sat.	19	at NY Rangers
	Tue.	22	Phoenix
	Thu.	24	San Jose
	Sat.	26	Detroit
	Wed.	30	at St. Louis
Nov.	Sat.	2	at Los Angeles
	Sun.	3	at Phoenix*
	Wed.	6	at Anaheim
	Thu.	7	at San Jose
	Sun.	10	at Colorado
	Tue.	12	at Detroit
	Thu.	14	Colorado
	Sat.	16	Columbus
	Sun.	17	at Chicago
	Thu.	21	at Colorado
	Sat.	23	at Minnesota*
	Wed.	27	San Jose
	Fri.	29	New Jersey
	Sat.	30	Dallas
Dec.	Tue.	3	Carolina
	Thu.	5	at Los Angeles
	Sat.	7	at San Jose
	Sun.	8	at Anaheim*
	Tue.	10	Los Angeles
	Thu.	12	St. Louis
	Sat.	14	Minnesota
	Tue.	17	Calgary
	Thu.	19	Vancouver
	Sat.	21	at Tampa Bay
	Mon.	23	at Florida
	Thu.	26	Dallas
	Sat.	28	Detroit
	Mon.	30	Ottawa
Jan.	Wed.	1	Colorado*
	Sat.	4	Chicago
	Mon.	6	at Columbus
	Tue.	7	St. Louis
	Sat.	11	Phoenix
	Sun.	12	at Chicago
	Tue.	14	at Vancouver
	Thu.	16	at Calgary
	Sat.	18	at Edmonton
	Tue.	21	Vancouver
	Thu.	23	NY Rangers
	Sat.	25	Tampa Bay
	Mon.	27	at Buffalo
	Tue.	28	at Boston
	Thu.	30	at Columbus
Feb.	Tue.	4	at Detroit
	Sat.	8	Columbus
	Tue.	11	Los Angeles
	Thu.	13	NY Islanders
	Sat.	15	Anaheim
	Mon.	17	Boston
	Thu.	20	Calgary
	Sat.	22	at Ottawa
	Sun.	23	at Toronto
	Tue.	25	Columbus
	Thu.	27	Pittsburgh
Mar.	Sat.	1	Chicago*
	Tue.	4	at St. Louis
	Thu.	6	Minnesota
	Fri.	7	at Dallas
	Mon.	10	Montreal
	Wed.	12	at Pittsburgh
	Fri.	14	at Minnesota
	Sat.	15	St. Louis
	Mon.	17	Edmonton
	Thu.	20	at Vancouver
	Sat.	22	at Calgary
	Sun.	23	at Edmonton
	Tue.	25	Philadelphia
	Thu.	27	at Chicago
	Sat.	29	Atlanta
	Mon.	31	at Detroit
Apr.	Tue.	1	Anaheim
	Fri.	4	at Phoenix
	Sun.	6	at Dallas*

* Denotes afternoon game.

Franchise date: June 25, 1997

CENTRAL DIVISION

5th NHL Season

With the trade of Cliff Ronning to Los Angeles, Greg Johnson emerged as Nashville's scoring leader last season. An original member of the Predators, his 18 goals in 2001-02 equalled a career-high established in Detroit back in 1995-96.

2002-03 Player Personnel

FORWARDS	HT	WT	S	Place of Birth	Date	2001-02 Club
ANDERSON, Erik	5-9	190	L	Plymouth, MI	3/6/78	Cincinnati-Milwaukee
ANDERSSON, Jonas	6-3	202	L	Stockholm, Sweden	2/24/81	Nashville-Milwaukee
ARKHIPOV, Denis	6-3	214	L	Kazan, USSR	5/19/79	Nashville
BARTEK, Martin	6-1	210	L	Kindgseed Jill, Czech.	7/17/80	Milwaukee-Cincinnati
CLASSEN, Greg	6-1	200	L	Aylsham, Sask.	8/24/77	Nashville-Milwaukee
ERAT, Martin	6-0	195	L	Trebic, Czech.	8/28/81	Nashville
FIDDLER, Vernon	5-11	195	L	Edmonton, Alta.	5/9/80	Roanoke-Norfolk
GILCHRIST, Brent	5-11	180	L	Moose Jaw, Sask.	4/3/67	Detroit-Dallas
GRIMSON, Stu	6-4	240	L	Kamloops, B.C.	5/20/65	Nashville
HALL, Adam	6-3	205	R	Kalamazoo, MI	8/14/80	Michigan State-Nashville-Milwaukee
HARTNELL, Scott	6-2	208	L	Regina, Sask.	4/18/82	Nashville
HAYDAR, Darren	5-9	170	L	Toronto, Ont.	10/22/79	New Hampshire
HUBACEK, Petr	6-2	183	R	Brno, Czech.	9/2/79	Phi (AHL)-Milwaukee
JOHNSON, Greg	5-11	202	L	Thunder Bay, Ont.	3/16/71	Nashville
KOEHLER, Greg	6-2	195	L	Scarborough, Ont.	2/27/75	Lowell-Phi (AHL)
LEGWAND, David	6-2	190	L	Detroit, MI	8/17/80	Nashville
LUNDBOHM, Bryan	5-10	184	L	Roseau, MN	8/24/77	Milwaukee
MANN, Cameron	6-0	195	R	Thompson, Man.	4/20/77	Utah
ORSZAGH, Vladimir	5-11	193	L	Banska Bystrica, Czech.	5/24/77	Nashville
PANOV, Konstantin	6-0	195	L	Chelyabinsk, USSR	6/29/80	Milwaukee
PAVLOV, Yevgeny	6-1	205	R	Togliatti, USSR	1/10/81	Cincinnati-Milwaukee
PEDERSON, Denis	6-2	205	R	Prince Albert, Sask.	9/10/75	Vancouver-Phoenix
PERROTT, Nathan	6-0	225	R	Owen Sound, Ont.	12/8/76	Norfolk-Nashville-Milwaukee
PITTIS, Domenic	5-11	190	L	Calgary, Alta.	10/1/74	Edmonton
SMITH, Wyatt	5-11	208	L	Thief River Falls, MN	2/13/77	Phoenix-Springfield
UPSHALL, Scottie	6-0	184	L	Fort McMurray, Alta.	10/7/83	Kamloops
WALKER, Scott	5-10	196	R	Cambridge, Ont.	7/19/73	Nashville
WILM, Clarke	6-0	202	L	Central Butte, Sask.	10/24/76	Calgary
YACHMENEV, Vitali	5-11	200	L	Chelyabinsk, USSR	1/8/75	Nashville
DEFENSEMEN						
BECKETT, Jason	6-3	208	R	Lethbridge, Alta.	7/23/80	Philadelphia (AHL)-Milwaukee-Trenton
BERENZWEIG, Bubba	6-1	217	L	Arlington Heights, IL	8/8/77	Nashville-Milwaukee
DELMORE, Andy	6-1	200	R	LaSalle, Ont.	12/26/76	Nashville
DURAK, Miroslav	6-4	212	R	Topolcany, Czech.	6/9/81	Acadie-Bathurst
EATON, Mark	6-2	205	L	Wilmington, DE	5/6/77	Nashville
HAMHUIS, Dan	6-0	208	L	Smithers, B.C.	12/13/82	Prince George
HELBLING, Timo	6-3	209	R	Basel, Switz.	7/21/81	Milwaukee
HOULDER, Bill	6-2	217	L	Thunder Bay, Ont.	3/11/67	Nashville
HULSE, Cale	6-3	220	R	Edmonton, Alta.	11/10/73	Nashville
HUTCHINSON, Andrew	6-2	198	R	Evanston, IL	3/24/80	Michigan State-Milwaukee
KARALAHTI, Jere	6-2	210	R	Helsinki, Finland	3/25/75	Los Angeles-Nashville
SCHNABEL, Robert	6-5	230	L	Prague, Czech.	11/10/78	Nashville-Milwaukee
SKRASTINS, Karlis	6-1	212	L	Riga, USSR	7/9/74	Nashville-Latvia
SMREK, Peter	6-1	215	L	Martin, Czech.	2/16/79	NYR-Hart-Slovakia-Milwaukee
TIMONEN, Kimmo	5-10	196	L	Kuopio, Finland	3/18/75	Nashville-Finland
TREPANIER, Pascal	6-0	210	R	Gaspe, Que.	9/4/73	Colorado

GOALTENDERS	HT	WT	C	Place of Birth	Date	2001-02 Club
DUNHAM, Mike	6-3	200	L	Johnson City, NY	6/1/72	Nashville-United States
FINLEY, Brian	6-3	205	R	Sault Ste. Marie, Ont.	7/13/81	Did not play – injured
LASAK, Jan	6-1	204	L	Zvolen, Czech.	4/10/79	Nashville-Milwaukee-Slovakia
VOKOUN, Tomas	6-0	195	R	Karlovy Vary, Czech.	7/2/76	Nashville

In his first full NHL season, Denis Arkhipov led the Predators with 20 goals in 2001-02. A product of the Ak Bars Kazan system, Arkhipov won a gold medal with the Russian team at the 1999 World Junior Championships.

2001-02 Scoring

* - rookie

Regular Season

Pos	#	Player	Team	GP	G	A	Pts	+/–	PIM	PP	SH	GW	GT	S	%
C	22	Greg Johnson	NSH	82	18	26	44	–14	38	3	0	2	1	145	12.4
L	25	Denis Arkhipov	NSH	82	20	22	42	–18	16	7	0	6	1	118	16.9
D	44	Kimmo Timonen	NSH	82	13	29	42	2	28	9	0	1	0	154	8.4
L	17	Scott Hartnell	NSH	75	14	27	41	5	111	3	0	4	0	162	8.6
D	5	Andy Delmore	NSH	73	16	22	38	–13	22	11	0	3	0	175	9.1
R	33	Vladimir Orszagh	NSH	79	15	21	36	–15	56	5	0	3	0	113	13.3
L	19 *	Martin Erat	NSH	80	9	24	33	–11	32	2	0	2	0	84	10.7
C	11	David Legwand	NSH	63	11	19	30	1	54	1	1	1	0	121	9.1
L	43	Vitali Yachmenev	NSH	75	11	16	27	–16	14	1	2	0	0	103	10.7
R	15	Petr Tenkrat	ANA	9	0	0	0	–6	6	0	0	0	0	13	0.0
			NSH	58	8	16	24	–4	28	0	1	2	1	82	9.8
			TOTAL	67	8	16	24	–10	34	0	1	2	1	95	8.4
D	3	Karlis Skrastins	NSH	82	4	13	17	–12	36	0	0	1	1	84	4.8
C	9	Greg Classen	NSH	55	5	6	11	1	30	0	1	0	0	32	15.6
D	26 *	Bubba Berenzweig	NSH	26	3	7	10	–3	14	0	0	1	0	27	11.1
R	24	Scott Walker	NSH	28	4	5	9	–13	18	1	0	0	0	46	8.7
R	27	Jukka Hentunen	CGY	28	2	3	5	–9	4	1	0	0	0	38	5.3
			NSH	10	2	2	4	0	0	0	0	1	0	12	16.7
			TOTAL	38	4	5	9	–9	4	1	0	1	0	50	8.0
C	14	Steve Dubinsky	CHI	3	1	0	1	1	4	0	0	1	0	6	16.7
			NSH	26	5	2	7	–2	10	0	0	0	0	42	11.9
			TOTAL	29	6	2	8	–1	14	0	0	1	0	48	12.5
D	4	Mark Eaton	NSH	58	3	5	8	–12	24	0	0	0	0	52	5.8
D	23	Bill Houlder	NSH	82	0	8	8	–1	40	0	0	0	0	44	0.0
L	12	Reid Simpson	MTL	25	1	1	2	0	63	0	0	1	0	9	11.1
			NSH	26	5	0	5	–1	69	0	0	0	0	13	38.5
			TOTAL	51	6	1	7	–1	132	0	0	1	0	22	27.3
R	39	Marian Cisar	NSH	10	1	2	3	–3	8	1	0	0	0	16	6.3
R	18	Mark Mowers	NSH	14	1	2	3	–2	2	0	0	0	0	5	20.0
R	20 *	Nathan Perrott	NSH	22	1	2	3	–1	74	0	0	1	0	7	14.3
L	8	Stu Grimson	NSH	30	1	1	2	0	76	0	0	0	0	5	20.0
D	13	Jere Karalahti	L.A.	30	0	1	1	–5	29	0	0	0	0	19	0.0
			NSH	15	0	1	1	–1	12	0	0	0	0	8	0.0
			TOTAL	45	0	2	2	–6	41	0	0	0	0	27	0.0
D	32	Cale Hulse	NSH	63	0	2	2	–18	121	0	0	0	0	70	0.0
R	34	Adam Hall	NSH	1	0	1	1	0	0	0	0	0	0	2	0.0
C	12	Yves Sarault	NSH	1	0	0	0	0	0	0	0	0	0	0	0.0
D	42 *	Robert Schnabel	NSH	1	0	0	0	0	0	0	0	0	0	0	0.0
D	14	Brett Hauer	NSH	3	0	0	0	–3	6	0	0	0	0	2	0.0
R	40 *	David Gosselin	NSH	3	0	0	0	–1	5	0	0	0	0	0	0.0
D	42 *	Pavel Skrbek	NSH	3	0	0	0	–2	2	0	0	0	0	0	0.0
D	14	Rich Brennan	NSH	4	0	0	0	0	2	0	0	0	0	0	0.0
L	28	Jeremy Stevenson	NSH	4	0	0	0	0	9	0	0	0	0	0	0.0
R	48 *	Jonas Andersson	NSH	5	0	0	0	–2	2	0	0	0	0	4	0.0

Goaltending

No.	Goaltender	GPI	Mins	Avg	W	L	T	EN	SO	GA	SA	S%	G	A	PIM
1	Mike Dunham	58	3316	2.61	23	24	9	2	3	144	1525	.906	0	2	2
29	Tomas Vokoun	29	1471	2.69	5	14	4	4	2	66	678	.903	0	0	2
35	* Jan Lasak	3	177	4.41	0	3	0	1	0	13	104	.875	0	0	2
	Totals	**82**	**4995**	**2.76**	**28**	**41**	**13**	**7**	**5**	**230**	**2314**	**.901**			

General Manager

POILE, DAVID
General Manager, Nashville Predators.
Born in Toronto, Ont., February 14, 1949.

Since joining the Predators as general manager on July 9, 1997, David Poile has made a commitment to building for the future, surrounding himself with one of the youngest and most talented staffs in the National Hockey League. Poile has an impressive reputation as an NHL leader and in 2001 he received the Lester Patrick Trophy for his contributions to hockey in the United States. His father, Norman "Bud" Poile, had won the honor in 1989.

Prior to joining Nashville, Poile spent 15 seasons as vice president/general manager of the Washington Capitals. During his tenure in Washington, the Capitals made 14 postseason appearances, winning their only Patrick Division title in 1989 and advancing to the Conference Finals in 1990. During Poile's 15 years in Washington, the Capitals compiled a record of 594-454-132, finished second in the Patrick Division seven times and recorded 90-or-more points seven different seasons.

Poile started his professional hockey career as an administrative assistant for the Atlanta Flames in 1972, shortly after graduating from Northeastern University in Boston. At Northeastern, he was hockey team captain, leading scorer and most valuable player for two years.

In 1977, he was named assistant general manager of the Atlanta Flames (moved to Calgary in 1980), serving as the manager and coordinator of the Flames farm club.

Poile is a member of the NHL's general managers committee and was instrumental in the NHL's adoption of the instant replay rule in 1991. He was awarded *Inside Hockey*'s Man of the Year for his leadership on the issue. He was also twice honored as *The Sporting News* NHL Executive of the Year following the 1982-83 and 1983-84 seasons. Poile served as general manager of the 1998 and 1999 U.S. National team for the International Ice Hockey Federation World Championships.

Poile was introduced to hockey by watching his father play seven seasons in the NHL. Bud Poile later became general manager for the Vancouver Canucks and the Philadelphia Flyers, both NHL expansion franchises at the time. He was inducted into the Hockey Hall of Fame in 1990.

Club Records

Team

(Figures in brackets for season records are games played; records for fewest points, wins, ties, losses, goals, goals against are for 70 or more games)

Most Points **80** 2000-01 (82)
Most Wins **34** 2000-01 (82)
Most Ties **13** 2001-02 (82)
Most Losses **47** 1998-99 (82)
Most Goals **199** 1999-2000 (82)
Most Goals Against **261** 1998-99 (82)
Fewest Points **63** 1998-99 (82)
Fewest Wins **28** 1998-99 (82), 1999-2000 (82), 2001-02 (82)
Fewest Ties **7** 1998-99 (82) 1999-2000 (82)
Fewest Losses **36** 2000-01 (82)
Fewest Goals **186** 2000-01 (82)
Fewest Goals Against **200** 2000-01 (82)
Longest Winning Streak
Overall .... **4** Dec. 26/99-Jan. 1/00, Jan. 29-Feb. 8/01
Home .... **6** Nov. 3-Dec. 6/01
Away .... **3** Feb. 12-24/99, Jan. 29-Feb. 1/01, Jan. 30-Feb. 28/02
Longest Undefeated Streak
Overall .... **8** Dec. 18/99-Jan. 1/00 (5 wins, 3 ties)
Home .... **11** Nov. 3-Dec. 23/01 (8 wins, 3 ties)
Away .... **3** Eight times
Longest Losing Streak
Overall .... **7** Nov. 20-Dec. 2/99
Home .... **6** Jan. 21-Feb. 15/99, Feb. 26-Mar. 21/02
Away .... **5** Four times
Longest Winless Streak
Overall .... **8** Mar. 7-21/02 (7 losses, 1 tie)
Home .... **9** Jan. 21-Mar. 2/99 (8 losses, 1 tie)
Away .... **9** Nov. 2-Dec. 2/01 (8 losses, 1 tie)
Most Shutouts, Season **6** 2000-01 (82)
Most PIM, Season **1,420** 1998-99 (82)
Most Goals, Game .... **7** Nov. 26/00 (Nsh. 7 at Car. 4)

Individual

Most Seasons .... **4** Many players
Most Games .... **314** Greg Johnson
Most Goals, Career .... **81** Cliff Ronning
Most Assists, Career .... **145** Cliff Ronning
Most Points, Career .... **226** Cliff Ronning (81G, 145A)
Most PIM, Career .... **327** Drake Berehowsky
Most Shutouts, Career .... **8** Mike Dunham
Longest Consecutive Games Streak .... **246** Greg Johnson (Oct. 2/99-to date)
Most Goals, Season .... **26** Cliff Ronning (1999-2000)
Most Assists, Season .... **43** Cliff Ronning (2000-01)
Most Points, Season .... **62** Cliff Ronning (1999-2000; 26G, 36A) (2000-01; 19G, 43A)
Most PIM, Season .... **242** Patrick Cote (1999-2000)
Most Points, Defenseman, Season .... **42** Kimmo Timonen (2001-02; 13G, 29A)
Most Points, Center, Season .... **62** Cliff Ronning (1999-2000; 26G, 36A) (2000-01; 19G, 43A)
Most Points, Right Wing, Season .... **54** Scott Walker (2000-01; 25G, 29A)
Most Points, Left Wing, Season .... **41** Scott Hartnell (2001-02; 14G, 27A)
Most Points, Rookie, Season .... **33** Martin Erat (2001-02; 9G, 24A)
Most Shutouts, Season .... **4** Mike Dunham (2000-01)
Most Goals, Game .... **3** Rob Valicevic (Nov. 10/99), Scott Walker (Dec. 26/00), Petr Tenkrat (Dec. 15/01)
Most Assists, Game .... **3** Nine times
Most Points, Game .... **4** Six times

General Managers' History

David Poile, 1998-99 to date.

Coaching History

Barry Trotz, 1998-99 to date.

Captains' History

Tom Fitzgerald, 1998-99 to 2001-02.

2001-02 Results

Oct.	5	at Dallas	1-4		8	at Toronto	3-4
	6	St. Louis	2-2		10	Minnesota	2-2
	11	Calgary	1-0		12	Anaheim	2-1
	13	Edmonton	3-4		14	at Anaheim	3-5
	16	at Buffalo	3-3		15	at Los Angeles	0-2
	18	Chicago	3-5		17	Toronto	3-2*
	20	Boston	1-2		19	Columbus	2-1
	22	at Edmonton	4-2		21	Phoenix	0-1
	23	at Vancouver	2-4		23	at Carolina	2-2
	25	at Calgary	5-4*		24	at Philadelphia	3-2*
	27	Detroit	0-1		26	Anaheim	1-3
	30	Minnesota	2-4		28	at Vancouver	1-5
	31	at Minnesota	6-4		30	at Colorado	5-2
Nov.	2	at Dallas	0-3	**Feb.**	6	Dallas	2-1
	3	Dallas	4-1		8	Washington	3-3
	8	at Montreal	1-3		9	at Columbus	1-0*
	10	at Ottawa	2-3		12	Florida	1-0
	13	Buffalo	4-1		26	San Jose	1-5
	16	at Atlanta	4-4		28	at Edmonton	3-2
	17	Columbus	3-2	**Mar.**	2	at Calgary	2-5
	20	at Detroit	3-6		5	at San Jose	2-0
	21	Chicago	4-3		7	Los Angeles	2-3
	23	Pittsburgh	5-0		9	at Florida	2-2
	27	at San Jose	2-3		10	at Tampa Bay	1-5
	29	at Phoenix	0-1		12	Vancouver	0-5
Dec.	1	at Los Angeles	2-4		15	Phoenix	2-3
	2	at Anaheim	2-4		17	Colorado	4-5
	6	Ottawa	4-5		19	at St. Louis	1-5
	8	Edmonton	2-2		21	New Jersey	3-4
	11	Los Angeles	1-1		23	Montreal	5-1
	12	at NY Rangers	4-2		25	Detroit	3-3
	15	Chicago	5-2		27	at Chicago	1-4
	20	Vancouver	6-2		28	at Detroit	3-3
	23	San Jose	1-1		30	St. Louis	2-4
	26	Tampa Bay	0-1	**Apr.**	1	at Colorado	1-5
	29	Detroit	3-2*		3	at Chicago	3-1
	30	at St. Louis	2-7		4	at Columbus	1-2
Jan.	1	Colorado	4-4		6	Calgary	3-1
	3	at New Jersey	4-3		9	at St. Louis	2-3
	4	at Minnesota	1-2		11	NY Islanders	2-5
	6	at Columbus	3-4		14	at Phoenix	4-6

* – Overtime

All-time Record vs. Other Clubs

Regular Season

	At Home								On Road								Total							
	GP	W	L	T	OL	GF	GA	PTS	GP	W	L	T	OL	GF	GA	PTS	GP	W	L	T	OL	GF	GA	PTS
Anaheim	8	3	3	2	0	17	20	8	8	1	6	0	1	12	23	3	16	4	9	2	1	29	43	11
Atlanta	2	2	0	0	0	10	3	4	3	1	1	1	0	8	8	3	5	3	1	1	0	18	11	7
Boston	4	1	3	0	0	6	12	2	3	1	1	1	0	6	11	3	7	2	4	1	0	12	23	5
Buffalo	4	1	2	0	1	6	8	3	3	2	0	1	0	11	8	5	7	3	2	1	1	17	16	8
Calgary	9	6	3	0	0	27	16	12	8	2	3	1	2	20	27	7	17	8	6	1	2	47	43	19
Carolina	3	3	0	0	0	8	5	6	4	1	2	1	0	11	13	3	7	4	2	1	0	19	18	9
Chicago	11	5	5	1	0	32	30	11	11	5	5	1	0	30	30	11	22	10	10	2	0	62	60	22
Colorado	8	2	3	3	0	25	27	7	8	1	5	1	1	14	25	4	16	3	8	4	1	39	52	11
Columbus	5	3	2	0	0	11	14	6	5	2	3	0	0	9	10	4	10	5	5	0	0	20	24	10
Dallas	8	4	4	0	0	18	16	8	8	1	6	0	1	8	24	3	16	5	10	0	1	26	40	11
Detroit	11	4	5	2	0	28	33	10	11	1	8	1	1	26	45	4	22	5	13	3	1	54	78	14
Edmonton	9	3	4	2	0	23	26	8	8	4	4	0	0	21	23	8	17	7	8	2	0	44	49	16
Florida	4	1	2	1	0	6	8	3	3	0	2	1	0	5	9	1	7	1	4	2	0	11	17	4
Los Angeles	8	1	5	2	0	13	21	4	8	4	3	0	1	20	22	9	16	5	8	2	1	33	43	13
Minnesota	4	2	1	1	0	10	8	5	4	2	1	1	0	9	7	5	8	4	2	2	0	19	15	10
Montreal	3	2	0	1	0	13	4	5	3	0	3	0	0	5	10	0	6	2	3	1	0	18	14	5
New Jersey	3	0	3	0	0	5	10	0	4	3	1	0	0	12	12	6	7	3	4	0	0	17	22	6
NY Islanders	3	1	2	0	0	8	13	2	3	2	1	0	0	9	7	4	6	3	3	0	0	17	20	6
NY Rangers	2	1	1	0	0	9	9	2	4	2	2	0	0	9	14	4	6	3	3	0	0	18	23	6
Ottawa	3	1	2	0	0	5	9	2	3	1	2	0	0	5	7	2	6	2	4	0	0	10	16	4
Philadelphia	3	0	2	1	0	3	6	1	4	2	1	1	0	7	12	5	7	2	3	2	0	10	18	6
Phoenix	8	3	4	1	0	20	23	7	8	3	5	0	0	23	27	6	16	6	9	1	0	43	50	13
Pittsburgh	4	2	2	0	0	12	8	4	3	1	1	1	0	6	8	3	7	3	3	1	0	18	16	7
St. Louis	11	2	7	2	0	23	36	6	11	3	7	0	1	20	38	7	22	5	14	2	1	43	74	13
San Jose	8	2	5	1	0	14	25	5	8	4	4	0	0	22	18	8	16	6	9	1	0	36	43	13
Tampa Bay	4	1	3	0	0	7	9	2	3	1	1	1	0	8	9	3	7	2	4	1	0	15	18	5
Toronto	1	1	0	0	0	3	2	2	4	2	1	1	0	12	9	5	5	3	1	1	0	15	11	7
Vancouver	9	4	3	1	1	27	29	10	8	2	6	0	0	19	31	4	17	6	9	1	1	46	60	14
Washington	4	2	1	1	0	11	9	5	3	1	2	0	0	4	5	2	7	3	3	1	0	15	14	7
Totals	**164**	**63**	**77**	**22**	**2**	**400**	**439**	**150**	**164**	**55**	**87**	**14**	**8**	**371**	**492**	**132**	**328**	**118**	**164**	**36**	**10**	**771**	**931**	**282**

Entry Draft Selections 2002-1998

2002

Pick	
6	Scottie Upshall
102	Brandon Segal
138	Patrick Jarrett
172	Mike McKenna
203	Josh Morrow
235	Kaleb Betts
264	Matt Davis
266	Steve Spencer

2001

Pick	
12	Dan Hamhuis
33	Timofei Shishkanov
42	Tomas Slovak
75	Denis Platonov
76	Oliver Setzinger
98	Jordin Tootoo
178	Anton Lavrentjev
240	Gustav Grasberg
271	Mikko Lehtonen

2000

Pick	
6	Scott Hartnell
36	Daniel Widing
72	Mattias Nilsson
89	Libor Pivko
131	Matt Hendricks
137	Mike Stuart
154	Matt Koalska
173	Tomas Harant
197	Zbynek Irgl
203	Jure Penko
236	Mats Christeen
284	Martin Hohener

1999

Pick	
6	Brian Finley
33	Jonas Andersson
52	Adam Hall
54	Andrew Hutchinson
61	Ed Hill
65	Jan Lasak
72	Brett Angel
121	Yevgeny Pavlov
124	Alexandre Krevsun
131	Konstantin Panov
162	Timo Helbling
191	Martin Erat
205	Kyle Kettles
220	Miroslav Durak
248	Darren Haydar

1998

Pick	
2	David Legwand
60	Denis Arkhipov
85	Geoff Koch
88	Kent Sauer
138	Martin Beauchesne
147	Craig Brunel
202	Martin Bartek
230	Karlis Skrastins

Scott Hartnell had almost as many goals (14) for Nashville in 2001-02 as he had points (16) as a rookie in 2000-01. He also led all team forwards in hits (144) for the second year in a row.

Coach

TROTZ, BARRY
Coach, Nashville Predators. Born in Winnipeg, Man., July 15, 1962.

Barry Trotz realized his dream of becoming an NHL head coach on August 6, 1997, after serving four seasons as head coach and director of hockey operations for the American Hockey League's Portland Pirates. He and assistant Paul Gardner spent the 1997-98 season scouting in preparation for the inaugural season of the Nashville Predators.

Trotz began his coaching career in 1984 as assistant coach with the University of Manitoba for one season, before serving two seasons as the head coach and general manager of the Dauphin Kings Junior Hockey Club from 1985 to 1987. He became head coach of the University of Manitoba during the 1987 season and also served as a scout for the Spokane Chiefs of the Western Hockey League that season. Trotz joined the Washington Capitals organization as their chief western scout during the 1988 season. The Winnipeg, Manitoba native was appointed an assistant coach of the Capitals' American Hockey League affiliate in Baltimore prior to the 1991 season before being named head coach prior to the 1992 season. When the franchise relocated to Portland, he guided the Pirates to two AHL Calder Cup Final appearances in the club's first four seasons. He led the Pirates to a league-best 43-27-10 record, captured the Calder Cup championship and was named the American Hockey League coach of the year following the 1994-95 season.

In 1995, Trotz guided Portland to a new North American professional hockey league record 17-game unbeaten streak (14-0-3) to start the season. He was named head coach for the U.S. team at the American Hockey League All-Star Game in 1996.

Prior to his coaching career, Trotz played junior hockey for the Western Hockey League's Regina Pats from 1979-83. During that time, he recorded 39 goals, 121 assists for 160 points, along with 490 penalty minutes in 204 games.

Coaching Record

		Regular Season				Playoffs		
Season	Team	Games	W	L	T	Games	W	L
1992-93	Baltimore (AHL)	80	28	40	12	7	3	4
1993-94	Portland (AHL)	80	43	27	10	8	6	2
1994-95	Portland (AHL)	80	46	22	12	7	3	4
1995-96	Portland (AHL)	80	32	38	10	24	14	10
1996-97	Portland (AHL)	80	37	33	10	5	2	3
1998-99	**Nashville (NHL)**	**82**	**28**	**47**	**7**	....	....	....
1999-2000	**Nashville (NHL)**	**82**	**28**	**47**	**7**	....	....	....
2000-01	**Nashville (NHL)**	**82**	**34**	**39**	**9**	....	....	....
2001-02	**Nashville (NHL)**	**82**	**28**	**41**	**13**	....	....	....
	NHL Totals	**328**	**118**	**174**	**36**	....	....	....

Club Directory

Gaylord Entertainment Center

Nashville Predators
Gaylord Entertainment Center
501 Broadway
Nashville, TN 37203
Phone **615/770-2300**
FAX 615/770-2309
Ticket Information 615/770-PUCK
www.nashvillepredators.com
Capacity: 17,113

Owner, Chairman and Governor	Craig Leipold
General Partner	Nashville Predators, LLC
Limited Partner	Gaylord Entertainment Company
President, COO and Alternate Governor	Jack Diller
Exec. V.P./G.M. and Alternate Governor	David Poile
Executive Vice President/Business Operations	Tom Ward
Sr. V.P. of Finance & Administration/CFO	Ed Lang
Vice President/Communications & Development	Gerry Helper

Hockey Operations

Assistant General Manager	Ray Shero
Head Coach	Barry Trotz
Assistant Coaches	Paul Gardner, Brent Peterson
Strength and Conditioning Coach	Mark Nemish
Goaltending Coach	Mitch Korn
Video Coach	Robert Bouchard
Director of Player Personnel	Paul Fenton
Director of Amateur Scouting	Craig Channell
Assistant Directors of Amateur Scouting	Rick Knickle, Greg Royce
Professional Scout	Dan MacKinnon
Amateur Scouts – Europe	Lucas Bergman, Alexei Dementiev, Martin Divis, Janne Kekalainen
Amateur Scouts – North America	Luc Gauthier, Jeff Kealty, Mike Rooney, Dennis Schueller
Head Athletic Trainer	Dan Redmond
Equipment Manager	Pete Rogers
Assistant Equipment Manager	Chris Scoppetto
Equipment Assistant	Chris Moody
Locker Room Attendant	Craig "Partner" Baugh
Massage Therapist	Anthony Garrett
Manager of Team Services	Greg Harvey
Executive Assistant	Kalli Quinn
Hockey Operations Coordinator	Mike Corbett
Hockey Operations Assistant	Brandon Walker

Team Doctors
Dr. Michael J. Pagnani, MD, Dr. Blake Garside, MD, Dr. James W. McPherson Jr., DDS, Dr. Daniel Weikert, MD, Dr. Bryan D. Oslin, MD, Dr. Donald Griffin, MD, Dr. Gary S. Solomon, Ph. D., Dr. Carl Hampf, MD, Dr. Richard W. Garman, MD

Communications/Development

Communications Manager	Ken Anderson
Communications Coordinator	Tim Darling
Corporate Communications Coordinator	Cathy Lewandowski
Director of Community Relations/ Executive Director of Predators Foundation	Polly Pearce
Community Relations Manager	Alexis Herbster
Community Relations Assistant	Angel Winter
Manager, Amateur Hockey	Marc Spigel
Graphic Artist, Communications & Development	Maggie Bizwell
Team Photographer	John Russell

Business/Marketing/Corporate Sales

Vice President, Corporate Services	Susie Masotti
Director of Corporate Partnerships	David Nivison
Director of National Sales	Jim Gibson
Corporate Sponsorship Sales, Account Executives	Allison Gay, David Morse
Sponsor Services, Account Managers	Evelyn Finch, Kristin Fricke, Tom Moulton
Premium Seating Manager	Britt Kincheloe
Account Manager, Premium Seating	Myron Murray
Director of Marketing	Randy Campbell
Marketing and Special Events Manager	Christel Foley
Advertising Manager	Julia Robinson
Marketing and Presentation Coordinator	Carrie Poss
Entertainment Coordinator	Adam DeVault
Database Marketing Manager	Michael Vivelo
Promotions Manager	Mark Iralson
Graphic Artist, Marketing	Jennifer Sheets
Executive Assistant	Linda Adams
Sponsor Sales Administrative Assistant	Kelly Preuett

Finance/Human Resources

Director of Finance	Beth Snider
Director of Human Resources	Stephanie Ditenhafer
Administrator, Payroll & Accounting	Susan Charnley
Accountants	Tracy Hardes, Sjar Toney
Accounts Payable Clerk	Carter Lynch
Financial Project Coordinator	Jonathan Norris
Manager, Internet Development	Scott Pilkinton
Information Systems Manager	Jeff Beck
Computer Support Tech	Wesley Green
Administrative Assistant	Elaine Lewis
Receptionist	Erin Hart

Broadcast/Game Presentation Department

Vice President, Broadcasting	John Guagliano
Producer	Erik Barnhart
Operations Manager	Blake Grant
Game Presentation Manager	Bryan Shaffer
Broadcast and Entertainment Manager	Susan Morgan
Production Assistant	Robert Hill
Play-by-Play Announcer	Pete Weber
Color Analyst	Terry Crisp

Ticket Operations

Ticket Operations Manager	Jamie Hall
Ticket Operations Coordinator	Cordell Johnson
Ticket Operations Coordinator	Cindy Yother
Business Development Manager	Geoff Dunnuck
Club/Suite Sales Executive	Tom Phillips
Corporate Account Executives	Jake Bye, Dan Bauchiero, Nat Harden, Bob Milhizer, Jonathan Tuschl, Bill Walker
Ticket Sales Administrative Assistant	Annie Snelgrove
Fan Relations Manager	Gene Connelly
Fan Relations Account Service Representatives	Brad Gillispie, Elizabeth Mitchell, Tiffany Williams
Radio Flagship	WTN-FM (99.7 FM)
TV Flagship	FOX Sports Net

New Jersey Devils

®

2001-02 Results: 41W-28L-9T-4OTL 95PTS.
Third, Atlantic Division

2002-03 Schedule

Month	Day	Date	Opponent
Oct.	Thu.	10	at Ottawa
	Sat.	12	Columbus
	Fri.	18	Nashville
	Sat.	19	at Carolina
	Wed.	23	at Atlanta
	Fri.	25	at Buffalo
	Sat.	26	Tampa Bay
	Tue.	29	Carolina
Nov.	Sat.	2	Chicago*
	Tue.	5	Calgary
	Thu.	7	at Philadelphia
	Sat.	9	Edmonton
	Tue.	12	Anaheim
	Fri.	15	Montreal
	Sat.	16	at Montreal
	Tue.	19	Buffalo
	Thu.	21	NY Rangers
	Sat.	23	Tampa Bay
	Wed.	27	at Detroit
	Fri.	29	at Nashville
	Sat.	30	at St. Louis
Dec.	Mon.	2	at Philadelphia
	Wed.	4	Vancouver
	Fri.	6	Pittsburgh
	Sat.	7	at Toronto
	Tue.	10	St. Louis
	Thu.	12	at Columbus
	Sat.	14	at Ottawa
	Wed.	18	Ottawa
	Thu.	19	at Pittsburgh
	Sat.	21	Dallas*
	Mon.	23	at NY Rangers
	Fri.	27	at Washington
	Sat.	28	Washington
	Mon.	30	at Boston
Jan.	Wed.	1	Florida*
	Fri.	3	Toronto
	Sat.	4	at Toronto
	Tue.	7	Montreal
	Fri.	10	at Florida
	Sat.	11	at Tampa Bay
	Mon.	13	Florida
	Wed.	15	NY Islanders
	Fri.	17	at Carolina
	Sat.	18	Carolina
	Wed.	22	at San Jose
	Fri.	24	at Anaheim
	Sat.	25	at Los Angeles
	Tue.	28	Detroit
	Thu.	30	Philadelphia
Feb.	Tue.	4	Buffalo
	Wed.	5	at Washington
	Fri.	7	Atlanta
	Sun.	9	Minnesota*
	Tue.	11	at Colorado
	Wed.	12	at Phoenix
	Sat.	15	Pittsburgh
	Tue.	18	at Philadelphia
	Wed.	19	Ottawa
	Fri.	21	Boston
	Sun.	23	at Pittsburgh*
	Tue.	25	NY Rangers
	Thu.	27	at NY Islanders
Mar.	Sat.	1	Washington
	Tue.	4	at Minnesota
	Wed.	5	at Calgary
	Sat.	8	at NY Islanders
	Tue.	11	Atlanta
	Thu.	13	at Boston
	Sat.	15	NY Rangers*
	Mon.	17	Philadelphia
	Tue.	18	at Montreal
	Fri.	21	Pittsburgh
	Sat.	22	at NY Islanders
	Mon.	24	at Florida
	Thu.	27	at Tampa Bay
	Fri.	28	at Atlanta
	Sun.	30	NY Islanders*
Apr.	Tue.	1	Toronto
	Thu.	3	Boston
	Fri.	4	at NY Rangers
	Sun.	6	at Buffalo*

* Denotes afternoon game.

Year-by-Year Record

		Home				Road				Overall								
Season	**GP**	**W**	**L**	**T**	**OL**	**W**	**L**	**T**	**OL**	**W**	**L**	**T**	**OL**	**GF**	**GA**	**Pts.**	**Finished**	**Playoff Result**
2001-02	82	22	13	4	2	19	15	5	2	41	28	9	4	205	187	95	3rd, Atlantic Div.	Lost Conf. Quarter-Final
2000-01	82	24	11	6	0	24	8	6	3	48	19	12	3	295	195	111	1st, Atlantic Div.	Lost Final
1999-2000	**82**	**28**	**9**	**3**	**1**	**17**	**15**	**5**	**4**	**45**	**24**	**8**	**5**	**251**	**203**	**103**	**2nd, Atlantic Div.**	**Won Stanley Cup**
1998-99	82	19	14	8	...	28	10	3	...	47	24	11	...	248	196	105	1st, Atlantic Div.	Lost Conf. Quarter-Final
1997-98	82	29	10	2	...	19	13	9	...	48	23	11	...	225	166	107	1st, Atlantic Div.	Lost Conf. Quarter-Final
1996-97	82	23	9	9	...	22	14	5	...	45	23	14	...	231	182	104	1st, Atlantic Div.	Lost Conf. Semi-Final
1995-96	82	22	17	2	...	15	16	10	...	37	33	12	...	215	202	86	6th, Atlantic Div.	Out of Playoffs
1994-95	**48**	**14**	**4**	**6**	**...**	**8**	**14**	**2**	**...**	**22**	**18**	**8**	**...**	**136**	**121**	**52**	**2nd, Atlantic Div.**	**Won Stanley Cup**
1993-94	84	29	11	2	...	18	14	10	...	47	25	12	...	306	220	106	2nd, Atlantic Div.	Lost Conf. Championship
1992-93	84	24	14	4	...	16	23	3	...	40	37	7	...	308	299	87	4th, Patrick Div.	Lost Div. Semi-Final
1991-92	80	24	12	4	...	14	19	3	...	38	31	11	...	289	259	87	4th, Patrick Div.	Lost Div. Semi-Final
1990-91	80	23	10	7	...	9	23	8	...	32	33	15	...	272	264	79	4th, Patrick Div.	Lost Div. Semi-Final
1989-90	80	22	15	3	...	15	19	6	...	37	34	9	...	295	288	83	2nd, Patrick Div.	Lost Div. Semi-Final
1988-89	80	17	18	5	...	10	23	7	...	27	41	12	...	281	325	66	5th, Patrick Div.	Out of Playoffs
1987-88	80	23	16	1	...	15	20	5	...	38	36	6	...	295	296	82	4th, Patrick Div.	Lost Conf. Championship
1986-87	80	20	17	3	...	9	28	3	...	29	45	6	...	293	368	64	6th, Patrick Div.	Out of Playoffs
1985-86	80	17	21	2	...	11	28	1	...	28	49	3	...	300	374	59	6th, Patrick Div.	Out of Playoffs
1984-85	80	13	21	6	...	9	27	4	...	22	48	10	...	264	346	54	5th, Patrick Div.	Out of Playoffs
1983-84	80	10	28	2	...	7	28	5	...	17	56	7	...	231	350	41	5th, Patrick Div.	Out of Playoffs
1982-83	80	11	20	9	...	6	29	5	...	17	49	14	...	230	338	48	5th, Patrick Div.	Out of Playoffs
1981-82**	80	14	21	5	...	4	28	8	...	18	49	13	...	241	362	49	5th, Smythe Div.	Out of Playoffs
1980-81**	80	15	16	9	...	7	29	4	...	22	45	13	...	258	344	57	5th, Smythe Div.	Out of Playoffs
1979-80**	80	12	20	8	...	7	28	5	...	19	48	13	...	234	308	51	6th, Smythe Div.	Out of Playoffs
1978-79**	80	8	24	8	...	7	29	4	...	15	53	12	...	210	331	42	4th, Smythe Div.	Out of Playoffs
1977-78**	80	17	14	9	...	2	26	12	...	19	40	21	...	257	305	59	2nd, Smythe Div.	Lost Prelim. Round
1976-77**	80	12	20	8	...	8	26	6	...	20	46	14	...	226	307	54	5th, Smythe Div.	Out of Playoffs
1975-76*	80	8	24	8	...	4	32	4	...	12	56	12	...	190	351	36	5th, Smythe Div.	Out of Playoffs
1974-75*	80	12	20	8	...	3	34	3	...	15	54	11	...	184	328	41	5th, Smythe Div.	Out of Playoffs

* Kansas City Scouts. ** Colorado Rockies.

Though he slipped into second place last year after leading all NHL goaltenders in wins for four straight seasons, Martin Brodeur's 38 victories in 2001-02 still ran his string of 30-win seasons to an NHL-record tying seven straight years.

Franchise date: June 11, 1974

Transferred from Denver to New Jersey, June 30, 1982.
Previously transferred from Kansas City to Denver.

29th NHL Season

ATLANTIC DIVISION

2002-03 Player Personnel

FORWARDS	HT	WT	S	Place of Birth	Date	2001-02 Club
BERGLUND, Christian	5-11	195	L	Orebro, Sweden	3/12/80	New Jersey-Albany
BICEK, Jiri	5-10	190	L	Kosice, Czech.	12/3/78	New Jersey-Albany
BIRBRAER, Max	6-2	195	L	Ust-Kamenogorsk, USSR	12/15/80	Albany
BRYLIN, Sergei	5-10	190	L	Moscow, USSR	1/13/74	New Jersey
CAMERON, Scott	6-0	190	L	Sudbury, Ont.	4/11/81	Albany
CLOUTHIER, Brett	6-5	245	L	Ottawa, Ont.	6/9/81	Albany
DANTON, Mike	5-9	190	R	Brampton, Ont.	10/21/80	did not play
DARBY, Craig	6-3	200	R	Oneida, NY	9/26/72	Montreal-Quebec
ELIAS, Patrik	6-1	195	L	Trebic, Czech.	4/13/76	New Jersey-Czech Republic
FOSTER, Adrian	6-1	200	L	Lethbridge, Alta.	1/15/82	Saskatoon-Brandon
FRIESEN, Jeff	6-0	215	L	Meadow Lake, Sask.	8/5/76	Anaheim
GIONTA, Brian	5-7	175	R	Rochester, NY	1/18/79	New Jersey-Albany
GOMEZ, Scott	5-11	200	L	Anchorage, AK	12/23/79	New Jersey
GUOLLA, Stephen	6-0	190	L	Scarborough, Ont.	3/15/73	Albany
HARTSBURG, Chris	6-0	190	R	Edina, MN	5/30/80	Colorado (WCHA)
JANSSEN, Cam	5-11	200	R	St. Louis, MO	4/15/84	Windsor
JOHANSSON, Eric	6-0	190	L	Edmonton, Alta.	1/7/82	Tri-City
KINKEL, Bill	6-5	230	L	Buffalo, NY	2/27/84	Kitchener
LANGENBRUNNER, Jamie	6-1	200	R	Duluth, MN	7/24/75	Dallas-New Jersey
LEBLANC, Robin	6-1	175	R	Chur, Switz.	1/11/83	Baie-Comeau
LEHOUX, Jason	6-2	220	L	Ste-Marie-Beauce, Que.	7/21/79	Albany
MADDEN, John	5-11	190	L	Barrie, Ont.	5/4/75	New Jersey
McKENZIE, Jim	6-4	230	L	Gull Lake, Sask.	11/3/69	New Jersey
NIEUWENDYK, Joe	6-1	205	L	Oshawa, Ont.	9/10/66	Dallas-Canada-New Jersey
NITTEL, Ahren	6-3	215	L	Waterloo, Ont.	12/6/84	Windsor
NOLAN, Brandon	6-0	180	L	Sault Ste. Marie, Ont.	7/18/83	Oshawa
PANDOLFO, Jay	6-1	190	L	Winchester, MA	12/27/74	New Jersey
ROCHE, Dave	6-4	230	L	Lindsay, Ont.	6/13/75	Cin-NYI-Bridgeport
RUPP, Mike	6-5	235	L	Cleveland, OH	1/13/80	Albany
SALOMONSSON, Andreas	6-1	200	L	Ornskoldsvik, Sweden	12/19/73	New Jersey-Albany
SKRLAC, Rob	6-5	245	L	Port McNeill, B.C.	6/10/76	Alb-Mississippi-Port (AHL)
STEVENSON, Turner	6-3	230	R	Prince George, B.C.	5/18/72	New Jersey
DEFENSEMEN						
ALBELIN, Tommy	6-2	195	L	Stockholm, Sweden	5/21/64	New Jersey
ANDREWS, Daryl	6-3	215	L	Campbell River, B.C.	4/27/77	Albany
BROOKS, Alex	6-2	200	R	Madison, WI	8/21/76	Jokerit
COLE, Phil	6-4	205	L	Winnipeg, Man.	9/6/82	Leth-Vancouver-M.Hat
DANEYKO, Ken	6-1	215	L	Windsor, Ont.	4/17/64	New Jersey
DEZAINDE, Joel	6-0	200	L	Simcoe, Ont.	11/2/78	Albany
GIROUX, Raymond	6-1	190	L	North Bay, Ont.	7/20/76	NY Islanders-Bridgeport
JOKELA, Mikko	6-1	210	R	Lappeenranta, Finland	3/4/80	Albany
KADEIKIN, Anton	6-2	180	L	Elektrostal, USSR	5/17/84	Elektrostal 2-Elektrostal
MATTEUCCI, Mike	6-2	210	L	Trail, B.C.	12/27/71	Minnesota-Houston
NIEDERMAYER, Scott	6-1	200	L	Edmonton, Alta.	8/31/73	New Jersey-Canada
RAFALSKI, Brian	5-9	190	R	Dearborn, MI	9/28/73	New Jersey-United States
REDLIHS, Krisjanis	6-2	185	L	Riga, Latvia	1/15/81	Liepaja
STEVENS, Scott	6-2	215	L	Kitchener, Ont.	4/1/64	New Jersey
TVERDOVSKY, Oleg	6-1	205	L	Donetsk, USSR	5/18/76	Anaheim-Russia
UCHEVATOV, Victor	6-4	215	L	Angarsk, USSR	2/10/83	Albany
WHITE, Colin	6-4	215	L	New Glasgow, N.S.	12/12/77	New Jersey
ZYUZIN, Andrei	6-1	215	R	Ufa, USSR	1/21/78	Tampa Bay-New Jersey-Albany

GOALTENDERS	HT	WT	C	Place of Birth	Date	2001-02 Club
AHONEN, Ari	6-1	190	L	Jyvaskyla, Finland	2/6/81	Albany
BRODEUR, Martin	6-2	210	L	Montreal, Que.	5/6/72	New Jersey-Canada
CLEMMENSEN, Scott	6-2	205	L	Des Moines, IA	7/23/77	New Jersey-Albany
KOSTUR, Matus	6-1	190	L	Banska Bystrica, Czech.	3/28/80	Zvolen
SCHWAB, Corey	6-0	180	L	North Battleford, Sask.	11/4/70	Toronto

General Managers' History

Sid Abel, 1974-75, 1975-76; Ray Miron, 1976-77 to 1980-81; Bill MacMillan, 1981-82, 1982-83; Bill MacMillan and Max McNab, 1983-84; Max McNab 1984-85 to 1986-87; Lou Lamoriello, 1987-88 to date.

General Manager

LAMORIELLO, LOU
CEO/President/General Manager, New Jersey Devils.
Born in Providence, RI, October 21, 1942.

Lou Lamoriello's life-long dedication to the game of hockey was rewarded in 1992 when he was named a recipient of the Lester Patrick Trophy for outstanding service to hockey in the United States. Lamoriello is entering his 16th season as president and general manager of the Devils following more than 20 years with Providence College as a player, coach and administrator. His trades, signings and draft choices helped lead the Devils to their first Stanley Cup championship in 1995 and another in 2000. A member of the varsity hockey Friars during his undergraduate days, he became an assistant coach with the college club after graduating in 1963. Lamoriello was later named head coach and in the ensuing 15 years, led his teams to a 248-179-13 record and appearances in 10 post-season tournaments, including the 1983 NCAA Final Four. Lamoriello also served a five-year term as athletic director at Providence and was a co-founder of Hockey East, one of the strongest collegiate hockey conferences in the U.S. He remained as athletic director until he was hired as president of the Devils on April 30, 1987. He assumed the responsibility of general manager on September 10, 1987. He was g.m. of Team USA for the first World Cup of Hockey in 1996 as the U.S. captured the championship. He was also the g.m. for the 1998 U.S. Olympic team.

2001-02 Scoring

* - rookie

Regular Season

Pos	#	Player	Team	GP	G	A	Pts	+/–	PIM	PP	SH	GW	GT	S	%
L	26	Patrik Elias	N.J.	75	29	32	61	4	36	8	1	8	0	199	14.6
C	25	Joe Nieuwendyk	DAL	67	23	24	47	–2	18	6	0	5	1	157	14.6
			N.J.	14	2	9	11	2	4	0	0	1	0	32	6.3
			TOTAL	81	25	33	58	0	22	6	0	6	1	189	13.2
C	16	Bobby Holik	N.J.	81	25	29	54	7	97	6	0	3	2	270	9.3
R	17	Petr Sykora	N.J.	73	21	27	48	12	44	4	0	4	0	194	10.8
C	23	Scott Gomez	N.J.	76	10	38	48	–4	36	1	0	1	0	156	6.4
D	28	Brian Rafalski	N.J.	76	7	40	47	15	18	2	0	4	1	125	5.6
C	18	Sergei Brylin	N.J.	76	16	28	44	21	10	5	0	3	0	133	12.0
D	27	Scott Niedermayer	N.J.	76	11	22	33	12	30	2	0	6	0	129	8.5
R	15	Jamie Langenbrunner	DAL	68	10	16	26	–11	54	0	1	2	0	132	7.6
			N.J.	14	3	3	6	2	23	0	0	2	0	31	9.7
			TOTAL	82	13	19	32	–9	77	0	1	4	0	163	8.0
R	44	Stephane Richer	PIT	58	13	12	25	–8	14	1	0	2	0	107	12.1
			N.J.	10	1	2	3	–1	0	0	0	0	0	16	6.3
			TOTAL	68	14	14	28	–9	14	1	0	2	0	123	11.4
C	11	John Madden	N.J.	82	15	8	23	6	25	0	0	2	1	170	8.8
L	22	Valeri Kamensky	DAL	24	3	6	9	3	2	0	0	1	0	32	9.4
			N.J.	30	4	8	12	–2	18	0	0	0	0	40	10.0
			TOTAL	54	7	14	21	1	20	0	0	1	0	72	9.7
D	4	Scott Stevens	N.J.	82	1	16	17	15	44	0	0	1	0	121	0.8
L	20	Jay Pandolfo	N.J.	65	4	10	14	12	15	0	1	0	0	72	5.6
R	14 *	Brian Gionta	N.J.	33	4	7	11	10	8	0	0	0	0	58	6.9
L	12	Sergei Nemchinov	N.J.	68	5	5	10	–9	10	0	0	1	0	49	10.2
L	21	Andreas Salomonsson	N.J.	39	4	5	9	–12	22	1	0	1	0	58	6.9
C	10 *	Christian Berglund	N.J.	15	2	7	9	–3	8	0	0	0	0	22	9.1
L	19	Jim McKenzie	N.J.	67	3	5	8	0	123	1	0	0	0	33	9.1
D	3	Ken Daneyko	N.J.	67	0	6	6	2	60	0	0	0	0	44	0.0
D	5	Colin White	N.J.	73	2	3	5	6	133	0	0	0	0	81	2.5
D	7	Andrei Zyuzin	T.B.	9	0	2	2	–6	6	0	0	0	0	14	0.0
			N.J.	38	1	2	3	1	25	1	0	0	0	47	2.1
			TOTAL	47	1	4	5	–5	31	1	0	0	0	61	1.6
D	6	Tommy Albelin	N.J.	42	1	3	4	0	4	0	0	1	0	33	3.0
R	9	Bruce Gardiner	N.J.	7	2	1	3	–1	2	1	0	0	0	10	20.0
R	24	Turner Stevenson	N.J.	21	0	2	2	–3	25	0	0	0	0	33	0.0
D	29	Joel Bouchard	N.J.	1	0	1	1	1	0	0	0	0	0	0	0.0
D	2 *	Mike Commodore	N.J.	37	0	1	1	–12	30	0	0	0	0	22	0.0
L	9 *	Jiri Bicek	N.J.	1	0	0	0	–1	0	0	0	0	0	2	0.0

Goaltending

No.	Goaltender	GPI	Mins	Avg	W	L	T	EN	SO	GA	SA	S%	G	A	PIM
34	John Vanbiesbrouck	5	300	2.00	2	3	0	2	0	10	117	.915	0	0	4
30	Martin Brodeur	73	4347	2.15	38	26	9	5	4	156	1655	.906	0	4	8
1	* J-F Damphousse	6	294	2.45	1	3	0	1	0	12	115	.896	0	0	0
40	* Scott Clemmensen	2	20	3.00	0	0	0	0	0	1	5	.800	0	0	0
	Totals	**82**	**4988**	**2.25**	**41**	**32**	**9**	**8**	**4**	**187**	**1900**	**.902**			

Playoffs

Pos	#	Player	Team	GP	G	A	Pts	+/–	PIM	PP	SH	GW	GT	S	%
L	26	Patrik Elias	N.J.	6	2	4	6	–1	6	2	0	0	0	16	12.5
C	16	Bobby Holik	N.J.	6	4	1	5	4	2	1	0	0	0	28	14.3
D	28	Brian Rafalski	N.J.	6	3	2	5	–2	4	3	0	0	0	19	15.8
R	14 *	Brian Gionta	N.J.	6	2	2	4	–1	0	0	1	2	0	16	12.5
D	27	Scott Niedermayer	N.J.	6	0	2	2	–1	6	0	0	0	0	6	0.0
C	18	Sergei Brylin	N.J.	6	0	2	2	3	2	0	0	0	0	15	0.0
R	17	Petr Sykora	N.J.	4	0	1	1	1	0	0	0	0	0	6	0.0
L	21	Andreas Salomonsson	N.J.	4	0	1	1	1	0	0	0	0	0	5	0.0
C	25	Joe Nieuwendyk	N.J.	5	0	1	1	–2	0	0	0	0	0	8	0.0
R	15	Jamie Langenbrunner	N.J.	5	0	1	1	–2	8	0	0	0	0	14	0.0
R	24	Turner Stevenson	N.J.	1	0	0	0	0	4	0	0	0	0	3	0.0
L	22	Valeri Kamensky	N.J.	2	0	0	0	0	0	0	0	0	0	2	0.0
R	44	Stephane Richer	N.J.	3	0	0	0	–1	0	0	0	0	0	6	0.0
L	12	Sergei Nemchinov	N.J.	3	0	0	0	–2	0	0	0	0	0	0	0.0
C	10 *	Christian Berglund	N.J.	3	0	0	0	0	2	0	0	0	0	6	0.0
D	6	Tommy Albelin	N.J.	6	0	0	0	–4	0	0	0	0	0	6	0.0
D	3	Ken Daneyko	N.J.	6	0	0	0	3	8	0	0	0	0	4	0.0
L	19	Jim McKenzie	N.J.	6	0	0	0	0	2	0	0	0	0	2	0.0
D	4	Scott Stevens	N.J.	6	0	0	0	5	4	0	0	0	0	7	0.0
L	20	Jay Pandolfo	N.J.	6	0	0	0	0	0	0	0	0	0	6	0.0
D	5	Colin White	N.J.	6	0	0	0	–1	2	0	0	0	0	6	0.0
C	11	John Madden	N.J.	6	0	0	0	–1	0	0	0	0	0	16	0.0

Goaltending

No.	Goaltender	GPI	Mins	Avg	W	L	EN	SO	GA	SA	S%	G	A	PIM
30	Martin Brodeur	6	381	1.42	2	4	0	1	9	145	.938	0	0	0
	Totals	**6**	**384**	**1.41**	**2**	**4**	**0**	**1**	**9**	**145**	**.938**			

Coaching History

Bep Guidolin, 1974-75; Bep Guidolin, Sid Abel and Eddie Bush, 1975-76; Johnny Wilson, 1976-77; Pat Kelly, 1977-78; Pat Kelly and Aldo Guidolin, 1978-79; Don Cherry, 1979-80; Bill MacMillan, 1980-81; Bert Marshall and Marshall Johnston, 1981-82; Bill MacMillan, 1982-83; Bill MacMillan and Tom McVie, 1983-84; Doug Carpenter, 1984-85 to 1986-87; Doug Carpenter and Jim Schoenfeld, 1987-88; Jim Schoenfeld, 1988-89; Jim Schoenfeld and John Cunniff, 1989-90; John Cunniff and Tom McVie, 1990-91; Tom McVie, 1991-92; Herb Brooks, 1992-93; Jacques Lemaire, 1993-94 to 1997-98; Robbie Ftorek, 1998-99; Robbie Ftorek and Larry Robinson, 1999-2000; Larry Robinson, 2000-01; Larry Robinson and Kevin Constantine, 2001-02; Pat Burns, 2002-03.

Club Records

Team

(Figures in brackets for season records are games played; records for fewest points, wins, ties, losses, goals, goals against are for 70 or more games)

Most Points	111	2000-01 (82)
Most Wins	48	1997-98 (82), 2000-01 (82)
Most Ties	21	1977-78 (80)
Most Losses	56	1983-84 (80), 1975-76 (80)
Most Goals	308	1992-93 (84)
Most Goals Against	374	1985-86 (80)
Fewest Points	*36	1975-76 (80)
	41	1983-84 (80)
Fewest Wins	*12	1975-76 (80)
	17	1982-83 (80), 1983-84 (80)
Fewest Ties	3	1985-86 (80)
Fewest Losses	19	2000-01 (82)
Fewest Goals	*184	1974-75 (80)
	205	2001-02 (82)
Fewest Goals Against	166	1997-98 (82)
Longest Winning Streak		
Overall	13	Feb. 26-Mar. 23/01
Home	8	Oct. 9-Nov. 7/87
Away	**10	Feb. 27-Apr. 7/01
Longest Undefeated Streak		
Overall	13	Three times
Home	15	Jan. 8-Mar. 15/97 (9 wins, 6 ties)
Away	10	Feb. 27-Apr. 7/01 (10 wins)
Longest Losing Streak		
Overall	*14	Dec. 30/75-Jan. 29/76
	10	Oct. 14-Nov. 4/83
Home	9	Dec. 22/85-Feb. 6/86
Away	12	Oct. 19-Dec. 1/83
Longest Winless Streak		
Overall	*27	Feb. 12-Apr. 4/76 (21 losses, 6 ties)
	18	Oct. 20-Nov. 26/82 (14 losses 4 ties)
Home	*14	Feb. 12-Mar. 30/76 (10 losses, 4 ties), Feb. 4-Mar. 31/79 (12 losses, 2 ties)
	9	Dec. 22/85-Feb. 6/86 (9 losses)
Away	*32	Nov. 12/77-Mar. 15/78 (22 losses, 10 ties)
	14	Dec. 26/82-Mar. 5/83 (13 losses, 1 tie)
Most Shutouts, Season	13	1996-97 (82)
Most PIM, Season	2,494	1988-89 (80)
Most Goals, Game	9	Nine times

Individual

Most Seasons	19	Ken Daneyko
Most Games	1,214	Ken Daneyko
Most Goals, Career	347	John MacLean
Most Assists, Career	354	John MacLean
Most Points, Career	701	John MacLean (347G, 354A)
Most PIM, Career	2,486	Ken Daneyko
Most Shutouts, Career	55	Martin Brodeur
Longest Consecutive Games Streak	388	Ken Daneyko (Nov. 4/89-Mar. 29/94)
Most Goals, Season	46	Pat Verbeek (1987-88)
Most Assists, Season	60	Scott Stevens (1993-94)
Most Points, Season	96	Patrik Elias (2000-01; 40G, 56A)
Most PIM, Season	295	Krzysztof Oliwa (1997-98)
Most Points, Defenseman, Season	78	Scott Stevens (1993-94; 18G, 60A)
Most Points, Center, Season	94	Kirk Muller (1987-88; 37G, 57A)
Most Points, Right Wing, Season	*87	Wilf Paiement (1977-78; 31G, 56A)
	87	John MacLean (1988-89; 42G, 45A)
Most Points, Left Wing, Season	96	Patrik Elias (2000-01; 40G, 56A)
Most Points, Rookie, Season	70	Scott Gomez (1999-2000; 19G, 51A)
Most Shutouts, Season	10	Martin Brodeur (1996-97, 1997-98)
Most Goals, Game	4	Four times
Most Assists, Game	5	Greg Adams (Oct. 10/85), Kirk Muller (Mar. 25/87), Tom Kurvers (Feb. 13/89)
Most Points, Game	6	Kirk Muller (Nov. 29/86; 3G, 3A)

* Records include Kansas City Scouts and Colorado Rockies, 1974-75 through 1981-82.
** NHL Record.

Captains' History

Simon Nolet, 1974-75 to 1976-77; Wilf Paiement, 1977-78; Gary Croteau, 1978-79; Mike Christie, Rene Robert and Lanny McDonald, 1979-80; Lanny McDonald, 1980-81; Lanny McDonald and Rob Ramage, 1981-82; Don Lever, 1982-83; Don Lever and Mel Bridgman, 1983-84; Mel Bridgman, 1984-85 to 1986-87; Kirk Muller, 1987-88 to 1990-91; Bruce Driver, 1991-92; Scott Stevens, 1992-93 to date.

All-time Record vs. Other Clubs

Regular Season

	At Home								On Road								Total							
	GP	W	L	T	OL	GF	GA	PTS	GP	W	L	T	OL	GF	GA	PTS	GP	W	L	T	OL	GF	GA	PTS
Anaheim	7	6	1	0	0	28	12	12	8	4	4	0	0	20	22	8	15	10	5	0	0	48	34	20
Atlanta	6	5	0	1	0	24	8	11	6	4	1	1	0	25	11	9	12	9	1	2	0	49	19	20
Boston	49	14	25	10	0	132	161	38	52	15	28	7	2	159	209	39	101	29	53	17	2	291	370	77
Buffalo	50	15	26	9	0	142	170	39	50	13	31	6	0	154	206	32	100	28	57	15	0	296	376	71
Calgary	43	14	26	3	0	123	158	31	41	6	27	8	0	107	179	20	84	20	53	11	0	230	337	51
Carolina	41	23	14	4	0	147	133	50	40	15	16	8	1	123	131	39	81	38	30	12	1	270	264	89
Chicago	45	18	16	11	0	136	137	47	45	12	24	9	0	125	174	33	90	30	40	20	0	261	311	80
Colorado	35	18	13	4	0	143	118	40	33	13	17	3	0	96	120	29	68	31	30	7	0	239	238	69
Columbus	1	0	0	1	0	2	2	1	2	1	1	0	0	7	6	2	3	1	1	1	0	9	8	3
Dallas	42	20	19	3	0	141	129	43	44	13	24	6	1	117	162	33	86	33	43	9	1	258	291	76
Detroit	39	20	10	9	0	137	103	49	38	13	23	2	0	125	156	28	77	33	33	11	0	262	259	77
Edmonton	31	13	15	3	0	105	104	29	31	11	14	6	0	114	136	28	62	24	29	9	0	219	240	57
Florida	20	12	5	3	0	61	42	27	21	10	7	4	0	51	46	24	41	22	12	7	0	112	88	51
Los Angeles	42	18	19	5	0	139	146	41	40	7	27	6	0	125	199	20	82	25	46	11	0	264	345	61
Minnesota	1	1	0	0	0	6	2	2	2	1	0	1	0	6	4	3	3	2	0	1	0	12	6	5
Montreal	50	22	25	3	0	149	186	47	50	14	29	6	1	129	174	35	100	36	54	9	1	278	360	82
Nashville	4	1	3	0	0	12	12	2	3	3	0	0	0	10	5	6	7	4	3	0	0	22	17	8
NY Islanders	79	32	36	11	0	263	283	75	79	16	53	10	0	227	340	42	158	48	89	21	0	490	623	117
NY Rangers	80	42	33	5	0	280	269	89	79	21	39	18	1	238	313	61	159	63	72	23	1	518	582	150
Ottawa	19	12	5	2	0	61	41	26	20	12	5	3	0	52	41	27	39	24	10	5	0	113	82	53
Philadelphia	78	40	31	7	0	269	272	87	79	21	49	9	0	206	314	51	157	61	80	16	0	475	586	138
Phoenix	28	12	10	6	0	94	84	30	30	6	21	3	0	78	114	15	58	18	31	9	0	172	198	45
Pittsburgh	75	35	27	13	0	278	257	83	74	30	39	4	1	255	284	65	149	65	66	17	1	533	541	148
St. Louis	45	21	17	7	0	144	128	49	44	10	26	7	1	134	187	28	89	31	43	14	1	278	315	77
San Jose	11	6	4	1	0	41	22	13	9	5	2	1	1	27	20	12	20	11	6	2	1	68	42	25
Tampa Bay	22	17	2	2	1	84	33	37	21	12	6	3	0	67	49	27	43	29	8	5	1	151	82	64
Toronto	42	16	13	13	0	148	130	45	44	11	28	5	0	136	176	27	86	27	41	18	0	284	306	72
Vancouver	47	20	20	6	1	149	152	47	47	9	27	11	0	130	175	29	94	29	47	17	1	279	327	76
Washington	75	35	32	7	1	231	223	78	75	23	47	5	0	214	299	51	150	58	79	12	1	445	522	129
Defunct Clubs	8	4	2	2	0	25	19	10	8	2	3	3	0	19	27	7	16	6	5	5	0	44	46	17
Totals	**1115**	**512**	**449**	**151**	**3**	**3694**	**3536**	**1178**	**1115**	**333**	**618**	**155**	**9**	**3276**	**4279**	**830**	**2230**	**845**	**1067**	**306**	**12**	**6970**	**7815**	**2008**

Playoffs

	Series	W	L	GP	W	L	T	GF	GA	Last Mtg.	Rnd.	Result
Boston	3	2	1	18	11	7	0	55	52	1995	CQF	W 4-1
Buffalo	1	1	0	7	4	3	0	14	14	1994	CQF	W 4-3
Carolina	2	1	1	12	6	6	0	31	17	2002	CQF	L 2-4
Colorado	1	0	1	7	3	4	0	11	19	2001	F	L 3-4
Dallas	1	1	0	6	4	2	0	15	9	2000	F	W 4-2
Detroit	1	1	0	4	4	0	0	16	7	1995	F	W 4-0
Florida	1	1	0	4	4	0	0	12	6	2000	CQF	W 4-0
Montreal	1	1	0	5	4	1	0	22	11	1997	CQF	W 4-1
NY Islanders	1	1	0	6	4	2	0	23	18	1988	DSF	W 4-2
NY Rangers	3	0	3	19	7	12	0	46	56	1997	CSF	L 1-4
Ottawa	1	0	1	6	2	4	0	12	13	1998	CQF	L 2-4
Philadelphia	3	2	1	15	8	7	0	41	35	2000	CF	W 4-3
Pittsburgh	5	2	3	29	15	14	0	86	80	2001	CF	W 4-1
Toronto	2	2	0	13	8	5	0	37	27	2001	CSF	W 4-3
Washington	2	1	1	13	6	7	0	43	44	1990	DSF	L 2-4
Totals	**28**	**16**	**12**	**164**	**90**	**74**	**0**	**464**	**408**			

Playoff Results 2002-1998

Year	Round	Opponent	Result	GF	GA
2002	CQF	Carolina	L 2-4	11	9
2001	F	Colorado	L 3-4	11	19
	CF	Pittsburgh	W 4-1	17	7
	CSF	Toronto	W 4-3	21	18
	CQF	Carolina	W 4-2	20	8
2000	**F**	**Dallas**	**W 4-2**	**15**	**9**
	CF	Philadelphia	W 4-3	18	15
	CSF	Toronto	W 4-2	16	9
	CQF	Florida	W 4-0	12	6
1999	CQF	Pittsburgh	L 3-4	18	21
1998	CQF	Ottawa	L 2-4	12	13

Abbreviations: Round: F – Final; **CF** – conference final; **CSF** – conference semi-final; **CQF** – conference quarter-final; **DSF** – division semi-final.

Calgary totals include Atlanta Flames, 1974-75 to 1979-80.
Colorado totals include Quebec, 1979-80 to 1994-95.
Phoenix totals include Winnipeg, 1979-80 to 1995-96.
Carolina totals include Hartford, 1979-80 to 1996-97.
Dallas totals include Minnesota North Stars, 1974-75 to 1992-93.

2001-02 Results

Month	Day	Opponent	Score
Oct.	6	at Washington	1-6
	11	NY Islanders	4-6
	13	at Montreal	1-3
	17	at NY Rangers	3-4*
	18	San Jose	6-1
	20	Ottawa	3-2
	23	at Ottawa	2-1
	27	Buffalo	3-1
	30	at Boston	4-3*
Nov.	1	Phoenix	5-2
	3	Boston	1-2
	7	Atlanta	3-2
	9	Toronto	3-2*
	10	at Toronto	1-1
	13	Pittsburgh	1-5
	15	at Boston	4-5*
	17	Philadelphia	1-3
	18	Colorado	0-2
	20	at Philadelphia	3-3
	23	at Tampa Bay	0-2
	24	at Florida	5-1
	27	at Pittsburgh	0-6
	30	at Detroit	2-4
Dec.	1	Detroit	4-1
	4	Tampa Bay	1-1
	5	at Montreal	2-1
	8	Washington	3-1
	10	at Columbus	1-3
	12	NY Islanders	3-2*
	14	Florida	2-3
	15	at Ottawa	2-0
	19	at NY Rangers	2-2
	20	Edmonton	3-3
	22	Ottawa	0-1
	26	Pittsburgh	4-0
	29	at Vancouver	2-4
	30	at Edmonton	2-1
Jan.	1	St. Louis	2-1
	3	Nashville	3-4
	5	at Carolina	2-1
	7	Los Angeles	2-3
	9	Calgary	5-1
	10	at Philadelphia	2-3
	12	at Buffalo	1-2
	15	Tampa Bay	4-5*
	17	NY Rangers	6-4
	19	Carolina	3-3
	21	at Tampa Bay	2-3
	23	at Florida	3-1
	24	at Atlanta	2-4
	26	at Minnesota	2-2
	29	at NY Islanders	3-1
	30	Chicago	3-1
Feb.	5	Montreal	0-1
	7	Atlanta	3-3
	9	at Pittsburgh	2-1*
	10	Buffalo	4-1
	12	at Buffalo	2-2
	26	at NY Rangers	4-3
	27	Philadelphia	0-1
Mar.	1	Toronto	4-2
	4	at Colorado	0-2
	5	at Phoenix	1-4
	8	at Anaheim	1-2
	10	at Dallas	3-0
	13	NY Islanders	3-2
	16	NY Rangers	3-1
	17	Vancouver	2-3*
	20	at Chicago	3-1
	21	at Nashville	4-3
	23	Carolina	2-4
	25	Florida	3-1
	27	at Pittsburgh	4-3
	29	Washington	1-3
	30	at Toronto	3-1
Apr.	1	at NY Islanders	2-4
	3	at Carolina	3-2
	5	at Atlanta	3-1
	7	Boston	3-2*
	10	Philadelphia	1-0
	12	Montreal	5-2
	13	at Washington	4-3*

* – Overtime

Entry Draft Selections 2002-1988

2002
Pick
51 Anton Kadeikin
53 Barry Tallackson
64 Jason Ryznar
84 Marek Chvatal
85 Ahren Nittel
117 Cam Janssen
154 Krisjanis Redlihs
187 Eric Johansson
218 Ilkka Pikkarainen
250 Dan Glover
281 Bill Kinkel

2001
Pick
28 Adrian Foster
44 Igor Pohanka
48 Thomas Pihlman
60 Victor Uchevatov
67 Robin Leblanc
72 Brandon Nolan
128 Andrei Posnov
163 Andreas Salomonsson
194 James Massen
229 Aaron Voros
257 Yevgeny Gamalei

2000
Pick
22 David Hale
39 Teemu Laine
56 Alexander Suglobov
57 Matt DeMarchi
62 Paul Martin
67 Max Birbraer
76 Mike Rupp
125 Phil Cole
135 Mike Danton
164 Matus Kostur
194 Deryk Engelland
198 Ken Magowan
257 Warren McCutcheon

1999
Pick
27 Ari Ahonen
42 Mike Commodore
50 Brett Clouthier
95 Andre Lakos
100 Teemu Kesa
185 Scott Cameron
214 Chris Hartsburg
242 Justin Dziama

1998
Pick
26 Mike Van Ryn
27 Scott Gomez
37 Christian Berglund
82 Brian Gionta
96 Mikko Jokela
105 Pierre Dagenais
119 Anton But
143 Ryan Flinn
172 Jacques Lariviere
199 Erik Jensen
227 Marko Ahosilta
257 Ryan Held

1997
Pick
24 Jean-Francois Damphousse
38 Stanislav Gron
104 Lucas Nehrling
131 Jiri Bicek
159 Sascha Goc
188 Mathieu Benoit
215 Scott Clemmensen
241 Jan Srdinko

1996
Pick
10 Lance Ward
38 Wes Mason
41 Josh DeWolf
47 Pierre Dagenais
49 Colin White
63 Scott Parker
91 Josef Boumedienne
101 Josh MacNevin
118 Glenn Crawford
145 Sean Ritchlin
173 Daryl Andrews
199 Willie Mitchell
205 Jay Bertsch
225 Pasi Petrilainen

1995
Pick
18 Petr Sykora
44 Nathan Perrott
70 Sergei Vyshedkevich
78 David Gosselin
79 Alyn McCauley
96 Henrik Rehnberg
122 Chris Mason
148 Adam Young
174 Richard Rochefort
200 Frederic Henry
226 Colin O'Hara

1994
Pick
25 Vadim Sharifijanov
51 Patrik Elias
71 Sheldon Souray
103 Zdenek Skorepa
129 Christian Gosselin
134 Ryan Smart
155 Luciano Caravaggio
181 Jeff Williams
207 Eric Bertrand
233 Steve Sullivan
259 Scott Swanjord
269 Mike Hanson

1993
Pick
13 Denis Pederson
32 Jay Pandolfo
39 Brendan Morrison
65 Krzysztof Oliwa
110 John Guirestante
143 Steve Brule
169 Nikolai Zavarukhin
195 Thomas Cullen
221 Judd Lambert
247 Jimmy Provencher
273 Mike Legg

1992
Pick
18 Jason Smith
42 Sergei Brylin
66 Cale Hulse
90 Vitali Tomilin
94 Scott McCabe
114 Ryan Black
138 Dan Trebil
162 Geordie Kinnear
186 Stephane Yelle
210 Jeff Toms
234 Heath Weenk
258 Vladislav Yakovenko

1991
Pick
3 Scott Niedermayer
11 Brian Rolston
33 Donevan Hextall
55 Fredrik Lindquist
77 Bradley Willner
121 Curt Regnier
143 David Craievich
165 Paul Wolanski
187 Daniel Reimann
231 Kevin Riehl
253 Jason Hehr

1990
Pick
20 Martin Brodeur
24 David Harlock
29 Chris Gotziaman
53 Mike Dunham
56 Brad Bombardir
64 Mike Bodnarchuk
95 Dean Malkoc
104 Petr Kuchyna
116 Lubomir Kolnik
137 Chris McAlpine
179 Jaroslav Modry
200 Corey Schwab
221 Valeri Zelepukin
242 Todd Reirden

1989
Pick
5 Bill Guerin
18 Jason Miller
26 Jarrod Skalde
47 Scott Pellerin
89 Mike Heinke
110 David Emma
152 Sergei Starikov
173 Andre Faust
215 Jason Simon
236 Peter Larsson

1988
Pick
12 Corey Foster
23 Jeff Christian
54 Zdeno Ciger
65 Matt Ruchty
75 Scott Luik
96 Chris Nelson
117 Chad Johnson
138 Chad Erickson
159 Bryan Lafort
180 Sergei Svetlov
201 Bob Woods
207 Alexander Semak
222 Charles Hughes
244 Robert Wallwork

Coach

BURNS, PAT

Coach, New Jersey Devils. Born in St-Henri, Que., April 4, 1952.

Pat Burns was hired as head coach of the New Jersey Devils on June 13, 2002. His most recent coaching position was with the Boston Bruins, whom he guided from 1997-98 through the first eight games of the 2000-01 season. Burns earned the Jack Adams Award as coach of the year in 1998 after his Bruins showed a 30-point improvement over the previous season. He became the first man in NHL history to win the award three times, having won it previously with Toronto (1993) and Montreal (1989).

Burns began his coaching career with the Hull Olympiques of the QMJHL in 1983. He spent four seasons with the club, guiding them to a berth in the Memorial Cup finals in 1986. He moved into the professional ranks in 1987 with Montreal's AHL affiliate and took over the Canadiens the following year. Burns was the winningest coach in the NHL during his four-year tenure the Canadiens, posting a record of 174-104-42. He was hired by the Toronto Maple Leafs on May 29, 1992, and promptly led the team to a club-record 32-point improvement on their 1991-92 record with a mark of 44-29-11 and 99 points. Both the Maple Leafs' win and point totals that season represented club records at the time.

Coaching Record

		Regular Season				Playoffs		
Season	Team	Games	W	L	T	Games	W	L
1983-84	Hull (QMJHL)	70	25	45	0	....	....	....
1984-85	Hull (QMJHL)	68	33	34	1	5	1	4
1985-86	Hull (QMJHL)	72	54	18	0	15	15	0
1986-87	Hull (QMJHL)	70	26	39	5	8	4	4
1987-88	Sherbrooke (AHL)	80	42	34	4	6	2	4
1988-89	**Montreal (NHL)**	**80**	**53**	**18**	**9**	**21**	**14**	**7**
1989-90	**Montreal (NHL)**	**80**	**41**	**28**	**11**	**11**	**5**	**6**
1990-91	**Montreal (NHL)**	**80**	**39**	**30**	**11**	**13**	**7**	**6**
1991-92	**Montreal (NHL)**	**80**	**41**	**28**	**11**	**11**	**4**	**7**
1992-93	**Toronto (NHL)**	**84**	**44**	**29**	**11**	**21**	**11**	**10**
1993-94	**Toronto (NHL)**	**84**	**43**	**29**	**12**	**18**	**9**	**9**
1994-95	**Toronto (NHL)**	**48**	**21**	**19**	**8**	**7**	**3**	**4**
1995-96	**Toronto (NHL)**	**65**	**25**	**30**	**10**	....	....	....
1997-98	**Boston (NHL)**	**82**	**39**	**30**	**13**	**6**	**2**	**4**
1998-99	**Boston (NHL)**	**82**	**39**	**30**	**13**	**12**	**6**	**6**
1999-2000	**Boston (NHL)**	**82**	**24**	**39**	**19**	....	....	....
2000-01	**Boston (NHL)**	**8**	**3**	**4**	**1**	....	....	....
	NHL Totals	**855**	**412**	**314**	**129**	**120**	**61**	**59**

Club Directory

Continental Airlines Arena

New Jersey Devils
Continental Airlines Arena
50 Route 120 North
P.O. Box 504
East Rutherford, NJ 07073
Phone **201/935-6050**
FAX 201/935-2127
www.newjerseydevils.com
Capacity: 19,040

Chairman	Raymond G. Chambers
CEO/President/General Manager	Louis A. Lamoriello
Executive Vice President	Peter S. McMullen
Executive Vice President	Chris Modrzynski
Vice President, Community Development/ Broadcasting	Glenn Adamo
Vice President/General Counsel	Joseph C. Benedetti
Vice President, Ticket Operations	Terry Farmer
Vice President, Corporate Partnerships	Kenneth F. Ferriter
Vice President, Sales/Marketing	Jason Siegel
Vice President, Finance	Scott Struble

Hockey Club Personnel

Head Coach	Pat Burns
Assistant Coach	Bob Carpenter
Goaltending Coach	Jacques Caron
Special Assignment Coach	Larry Robinson
Director, Scouting	David Conte
Assistant Director, Scouting	Claude Carrier
Scouting Staff	Glen Dirk, Milt Fisher, Ferny Flaman, Dan Labraaten, Chris Lamoriello, Vladimir Lokotko, Joe Mahoney, Larry Perris, Marcel Pronovost, Lou Reycroft, Vaclav Slansky, Jr., Geoff Stevens, Ed Thomlinson, Les Widdifield
Pro Scouting Staff	Andre Boudrias, Bob Hoffmeyer, Jan Ludvig
Special Assignment Scout	Kurt Kleinendorst
Hockey Operations Video Coordinator	Taran Singleton
Scouting Staff Assistant	Callie A. Smith
Medical Trainer	Bill Murray
Strength/Conditioning Coordinator	Michael Vasalani
Equipment Manager	Rich Matthews
Assistant Equipment Managers	Alex Abasto, Joe Murray
Massage Therapist	Juergen Merz
Team Cardiologist	Dr. Joseph Niznik
Team Dentist	Dr. H. Hugh Gardy
Team Optometrist	Dr. Paul Berman
Team Orthopedists	Dr. Barry Fisher, Dr. Len Jaffe
Fitness Consultant	Vladimir Bure
Exercise Physiologist	Dr. Garret Caffrey
Physical Therapist	David Feniger
Video Consultant	Mitch Kaufman
Head Coach, Albany	Dennis Gendron
Assistant Coaches, Albany	Chris Terreri, Geordie Kinnear, Gates Orlando
Athletic Trainer, Albany	Curtis Bell
Equipment Manager, Albany	Jason McGrath

Administration

Hockey Operations Executive Assistant to the CEO/President/General Manager	Marie Carnevale
Corporate Executive Assistant to the CEO/President/General Manager	Mary K. Morrison
Receptionists	Jelsa Belotta, Pat Maione
Corporate Staff Assistant	Christie Zdanowicz
Operations Staff Assistant	John Gerba

Ticket Operations

Director, Ticket Operations	Tom Bates
Director, Customer Service/Season Ticket Accounts	Dave Beck
Customer Service Representative	Andrea Marchesani
Director, Group Sales	Neil Desormeaux
Group Account Managers	Rich Davis, Christie Freid

Corporate Partnerships

Director, Corporate Accounts	Michael DeMartino
Account Manager, Corporate Partnerships	Michael Merolla
Staff Assistant, Corporate Partner Services	Matt Dugan

Marketing

Director, Season Ticket Sales	Todd Hyland
Account Managers	Erica Brask, Chris Brehm, Kyra Coots, Michael Greenberg, Scott Hollingshead, Zachary W. Holmes, Jason Lee, Louie Leone, James Winters
Assistant Director, Community Development	Paul Viola
Coordinator, Game Entertainment	Bruce Cohn
Merchandise Manager	David Perricone
Merchandise Assistant	Adam Manger

Communications

Director, Information/Publications	Mike Levine
Director, Public Relations	Jeff Altstadter
Staff Assistants	Erica Luthman, Peter M. Albietz

Finance

Assistant Controller	Craig Wolman
Staff Accountants	Jill Bach, Matt Courtney
Administrative Assistant	Eileen Howell

Computer Operations

Director, Programming/Computer Operations	Jack Skelley
Systems Administrator	Mike Tukes
Director, Website Operations	Antonio Barrera
Coordinator, Website Operations	Anthony Bovasso

Nets & Devils Foundation

Executive Director	Shane Harris
Program/Grants Coordinator	Milagros Rodriguez

Television/Radio

Television Outlet	FOX Sports Net
Broadcasters	Mike Emrick, Play-by-Play; Glenn Resch, Color
Radio Outlet	WABC 770 AM
Broadcasters	John Hennessy, Play-by-Play; Randy Velischek, Color

New York Islanders

2001-02 Results: 42W-28L-8T-4OTL 96PTS.
Second, Atlantic Division

2002-03 Schedule

Month	Day	Date	Opponent
Oct.	Thu.	10	at Buffalo
	Sat.	12	Washington
	Tue.	15	Nashville
	Thu.	17	at Philadelphia
	Sat.	19	at Atlanta
	Tue.	22	Carolina
	Thu.	24	Florida
	Sat.	26	Philadelphia
	Tue.	29	Phoenix
	Wed.	30	at Carolina
Nov.	Sat.	2	St. Louis
	Mon.	4	Calgary
	Thu.	7	at Montreal
	Fri.	8	Edmonton
	Sun.	10	Dallas*
	Tue.	12	Ottawa
	Thu.	14	at Boston
	Sat.	16	at Pittsburgh
	Wed.	20	at Florida
	Thu.	21	at Tampa Bay
	Sat.	23	at NY Rangers*
	Wed.	27	Ottawa
	Fri.	29	Columbus
	Sat.	30	at Ottawa
Dec.	Tue.	3	Vancouver
	Fri.	6	Toronto
	Sat.	7	at Pittsburgh
	Tue.	10	Chicago
	Fri.	13	at Florida
	Sat.	14	at Tampa Bay
	Tue.	17	Detroit
	Thu.	19	at Minnesota
	Sat.	21	Washington
	Mon.	23	Montreal
	Sat.	28	Carolina
	Mon.	30	Florida
	Tue.	31	at Buffalo
Jan.	Fri.	3	Boston
	Sat.	4	at Pittsburgh
	Tue.	7	Pittsburgh
	Thu.	9	Philadelphia
	Sat.	11	Atlanta
	Mon.	13	at Washington
	Wed.	15	at New Jersey
	Thu.	16	at St. Louis
	Sun.	19	at Atlanta*
	Tue.	21	NY Rangers
	Fri.	24	at Philadelphia
	Sat.	25	at Columbus
	Tue.	28	Pittsburgh
	Thu.	30	Montreal
Feb.	Tue.	4	Philadelphia
	Fri.	7	at Washington
	Sat.	8	Buffalo
	Tue.	11	Tampa Bay
	Thu.	13	at Nashville
	Sat.	15	at Los Angeles*
	Mon.	17	at Anaheim
	Wed.	19	at San Jose
	Fri.	21	Colorado
	Sun.	23	Boston*
	Tue.	25	at Toronto
	Thu.	27	New Jersey
Mar.	Sat.	1	Buffalo
	Mon.	3	at NY Rangers
	Tue.	4	Tampa Bay
	Thu.	6	at Boston
	Sat.	8	New Jersey
	Tue.	11	at Vancouver
	Thu.	13	at Edmonton
	Sat.	15	at Ottawa
	Mon.	17	at NY Rangers
	Tue.	18	at Toronto
	Thu.	20	at Montreal
	Sat.	22	New Jersey
	Tue.	25	at Chicago
	Fri.	28	Toronto
	Sun.	30	at New Jersey*
Apr.	Tue.	1	NY Rangers
	Thu.	3	at Detroit
	Sat.	5	Atlanta*
	Sun.	6	at Carolina*

* Denotes afternoon game.

Franchise date: June 6, 1972

31st NHL Season

ATLANTIC DIVISION

Mark Parrish began the 2001-02 season with at least two goals in each of the Islanders' first four games, including his first career hat trick on October 11. He finished the season with career highs in goals (30), assists (30) and points (60).

Year-by-Year Record

		Home				Road				Overall								
Season	**GP**	**W**	**L**	**T**	**OL**	**W**	**L**	**T**	**OL**	**W**	**L**	**T**	**OL**	**GF**	**GA**	**Pts.**	**Finished**	**Playoff Result**
2001-02	82	21	13	5	2	21	15	3	2	42	28	8	4	239	220	96	2nd, Atlantic Div.	Lost Conf. Quarter-Final
2000-01	82	12	27	1	1	9	24	6	2	21	51	7	3	185	268	52	5th, Atlantic Div.	Out of Playoffs
1999-2000	82	10	25	5	1	14	23	4	0	24	48	9	1	194	275	58	5th, Atlantic Div.	Out of Playoffs
1998-99	82	11	23	7	...	13	25	3	...	24	48	10	...	194	244	58	5th, Atlantic Div.	Out of Playoffs
1997-98	82	17	20	4	...	13	21	7	...	30	41	11	...	212	225	71	4th, Atlantic Div.	Out of Playoffs
1996-97	82	19	18	4	...	10	23	8	...	29	41	12	...	240	250	70	7th, Atlantic Div.	Out of Playoffs
1995-96	82	14	21	6	...	8	29	4	...	22	50	10	...	229	315	54	7th, Atlantic Div.	Out of Playoffs
1994-95	48	10	11	3	...	5	17	2	...	15	28	5	...	126	158	35	7th, Atlantic Div.	Out of Playoffs
1993-94	84	23	15	4	...	13	21	8	...	36	36	12	...	282	264	84	4th, Atlantic Div.	Lost Conf. Quarter-Final
1992-93	84	20	19	3	...	20	18	4	...	40	37	7	...	335	297	87	3rd, Patrick Div.	Lost Conf. Championship
1991-92	80	20	15	5	...	14	20	6	...	34	35	11	...	291	299	79	5th, Patrick Div.	Out of Playoffs
1990-91	80	15	19	6	...	10	26	4	...	25	45	10	...	223	290	60	6th, Patrick Div.	Out of Playoffs
1989-90	80	15	17	8	...	16	21	3	...	31	38	11	...	281	288	73	4th, Patrick Div.	Lost Div. Semi-Final
1988-89	80	19	18	3	...	9	29	2	...	28	47	5	...	265	325	61	6th, Patrick Div.	Out of Playoffs
1987-88	80	24	10	6	...	15	21	4	...	39	31	10	...	308	267	88	1st, Patrick Div.	Lost Div. Semi-Final
1986-87	80	20	15	5	...	15	18	7	...	35	33	12	...	279	281	82	3rd, Patrick Div.	Lost Div. Final
1985-86	80	22	11	7	...	17	18	5	...	39	29	12	...	327	284	90	3rd, Patrick Div.	Lost Div. Semi-Final
1984-85	80	26	11	3	...	14	23	3	...	40	34	6	...	345	312	86	3rd, Patrick Div.	Lost Div. Final
1983-84	80	28	11	1	...	22	15	3	...	50	26	4	...	357	269	104	1st, Patrick Div.	Lost Final
1982-83	**80**	**26**	**11**	**3**	**...**	**16**	**15**	**9**	**...**	**42**	**26**	**12**	**...**	**302**	**226**	**96**	**2nd, Patrick Div.**	**Won Stanley Cup**
1981-82	**80**	**33**	**3**	**4**	**...**	**21**	**13**	**6**	**...**	**54**	**16**	**10**	**...**	**385**	**250**	**118**	**1st, Patrick Div.**	**Won Stanley Cup**
1980-81	**80**	**23**	**6**	**11**	**...**	**25**	**12**	**3**	**...**	**48**	**18**	**14**	**...**	**355**	**260**	**110**	**1st, Patrick Div.**	**Won Stanley Cup**
1979-80	**80**	**26**	**9**	**5**	**...**	**13**	**19**	**8**	**...**	**39**	**28**	**13**	**...**	**281**	**247**	**91**	**2nd, Patrick Div.**	**Won Stanley Cup**
1978-79	80	31	3	6	...	20	12	8	...	51	15	14	...	358	214	116	1st, Patrick Div.	Lost Semi-Final
1977-78	80	29	3	8	...	19	14	7	...	48	17	15	...	334	210	111	1st, Patrick Div.	Lost Quarter-Final
1976-77	80	24	11	5	...	23	10	7	...	47	21	12	...	288	193	106	2nd, Patrick Div.	Lost Semi-Final
1975-76	80	24	8	8	...	18	13	9	...	42	21	17	...	297	190	101	2nd, Patrick Div.	Lost Semi-Final
1974-75	80	22	6	12	...	11	19	10	...	33	25	22	...	264	221	88	3rd, Patrick Div.	Lost Semi-Final
1973-74	78	13	17	9	...	6	24	9	...	19	41	18	...	182	247	56	8th, East Div.	Out of Playoffs
1972-73	78	10	25	4	...	2	35	2	...	12	60	6	...	170	347	30	8th, East Div.	Out of Playoffs

2002-03 Player Personnel

FORWARDS	HT	WT	S	Place of Birth	Date	2001-02 Club
ASHAM, Arron	5-11	209	R	Portage La Prairie, Man.	4/13/78	Montreal-Quebec
BATES, Shawn	6-0	205	R	Melrose, MA	4/3/75	NY Islanders
BLAKE, Jason	5-10	180	L	Moorhead, MN	9/2/73	NY Islanders
ISBISTER, Brad	6-4	227	R	Edmonton, Alta.	5/7/77	NY Islanders
KVASHA, Oleg	6-5	230	R	Moscow, USSR	7/26/78	NY Islanders-Russia
LAPOINTE, Claude	5-9	188	L	Lachine, Que.	10/11/68	NY Islanders
PARRISH, Mark	5-11	200	R	Edina, MN	2/2/77	NY Islanders
PECA, Michael	5-11	190	R	Toronto, Ont.	3/26/74	NY Islanders-Canada
SCATCHARD, Dave	6-2	224	R	Hinton, Alta.	2/20/76	NY Islanders
WEBB, Steve	6-0	211	R	Peterborough, Ont.	4/30/75	NY Islanders
WIEMER, Jason	6-1	225	L	Kimberley, B.C.	4/14/76	Florida
YASHIN, Alexei	6-3	225	R	Sverdlovsk, USSR	11/5/73	NY Islanders-Russia
DEFENSEMEN						
AUCOIN, Adrian	6-2	214	R	Ottawa, Ont.	7/3/73	NY Islanders
CAIRNS, Eric	6-6	230	L	Oakville, Ont.	6/27/74	NY Islanders
HALLER, Kevin	6-2	199	L	Trochu, Alta.	12/5/70	NY Islanders
HAMRLIK, Roman	6-2	200	L	Zlin, Czech.	4/12/74	NY Islanders-Czech Republic
JONSSON, Kenny	6-3	217	L	Angelholm, Sweden	10/6/74	NY Islanders-Sweden
KOROLEV, Evgeny	6-1	214	L	Moscow, USSR	7/24/78	NY Islanders-Bridgeport
MARTINEK, Radek	6-1	200	R	Havlickuv Brod, Czech.	8/31/76	NY Islanders
TIMANDER, Mattias	6-2	230	L	Solleftea, Sweden	4/16/74	Columbus

GOALTENDERS	HT	WT	C	Place of Birth	Date	2001-02 Club
OSGOOD, Chris	5-10	175	L	Peace River, Alta.	11/26/72	NY Islanders
SNOW, Garth	6-3	200	L	Wrentham, MA	7/28/69	NY Islanders

Coach

LAVIOLETTE, PETER
Coach, New York Islanders. Born in Norwood, MA, December 7, 1964.

Peter Laviolette was named the head coach of the New York Islanders on May 23, 2001, after having served as an assistant coach with the Boston Bruins under Pat Burns and Mike Keenan during the 2000-01 season. In his first season behind the Islanders bench in 2001-02, he led the club into the playoffs for the first time since 1994.

Prior to joining the Bruins, Laviolette had spent two seasons as the head coach of Boston's AHL affiliate in Providence. In 1998-99, he led that club to the winningest season in AHL history as they captured the Calder Cup championship. At the conclusion of that season, he was named the AHL's outstanding coach. In 1999-2000, he again led the AHL Bruins to a playoff berth, despite using over 80 players due to injuries and recalls to Boston.

Laviolette began his coaching career in 1997-98 with the ECHL's Wheeling Thunderbirds and led them to the playoff semifinals. Laviolette played four seasons of college hockey at Westfield (MA) State College. He turned professional in 1986-87 and spent most of his 11 pro seasons in the International and American Hockey Leagues, but played 12 NHL games with the New York Rangers in the 1988-89 season. Laviolette represented the United States on two Olympic teams — playing in the 1988 Games in Calgary and captaining the 1994 squad in Lillehammer. He concluded his playing career with Providence, becoming the team's first captain in 1992-93.

Coaching Record

		Regular Season				Playoffs		
Season	Team	Games	W	L	T	Games	W	L
1997-98	Wheeling (ECHL)	70	37	24	9	15	8	7
1998-99	Providence (AHL)	80	56	16	8	19	15	4
1999-2000	Providence (AHL)	80	33	38	9	14	10	4
2001-02	**NY Islanders (NHL)**	**82**	**42**	**32**	**8**	**7**	**3**	**4**
	NHL Totals	**82**	**42**	**32**	**8**	**7**	**3**	**4**

2001-02 Scoring

* - rookie

Regular Season

Pos	#	Player	Team	GP	G	A	Pts	+/–	PIM	PP	SH	GW	GT	S	%
C	79	Alexei Yashin	NYI	78	32	43	75	–3	25	15	0	5	0	239	13.4
R	37	Mark Parrish	NYI	78	30	30	60	10	32	9	1	6	0	162	18.5
C	27	Michael Peca	NYI	80	25	35	60	19	62	3	6	5	1	168	14.9
C	17	Shawn Bates	NYI	71	17	35	52	18	30	1	4	4	1	150	11.3
R	21	Mariusz Czerkawski	NYI	82	22	29	51	–8	48	6	0	6	1	169	13.0
R	15	Brad Isbister	NYI	79	17	21	38	1	113	4	0	2	0	142	12.0
L	12	Oleg Kvasha	NYI	71	13	25	38	–4	80	2	0	3	0	119	10.9
D	4	Roman Hamrlik	NYI	70	11	26	37	7	78	4	1	1	0	169	6.5
D	3	Adrian Aucoin	NYI	81	12	22	34	23	62	7	0	1	0	232	5.2
D	29	Kenny Jonsson	NYI	76	10	22	32	15	26	2	1	0	0	107	9.3
C	38	Dave Scatchard	NYI	80	12	15	27	–4	111	3	1	4	0	117	10.3
C	11	Kip Miller	NYI	37	7	17	24	2	6	2	0	1	0	52	13.5
C	13	Claude Lapointe	NYI	80	9	12	21	–9	60	0	3	0	0	74	12.2
D	8	Dick Tarnstrom	NYI	62	3	16	19	–12	38	0	0	0	1	59	5.1
C	55	Jason Blake	NYI	82	8	10	18	–11	36	0	0	1	0	136	5.9
L	10	Mats Lindgren	NYI	59	3	12	15	0	16	0	0	1	0	35	8.6
D	28	Darren Van Impe	NYR	17	1	0	1	3	12	1	0	0	0	19	5.3
			FLA	36	1	6	7	3	31	1	0	1	0	27	3.7
			NYI	14	1	2	3	6	16	0	0	0	0	14	7.1
			TOTAL	67	3	8	11	12	59	2	0	1	0	60	5.0
D	33	Eric Cairns	NYI	74	2	5	7	–2	176	0	0	1	0	34	5.9
R	20	Steve Webb	NYI	60	2	4	6	0	104	0	0	0	0	31	6.5
D	45	Marko Kiprusoff	NYI	27	0	6	6	0	4	0	0	0	0	17	0.0
D	24 *	Radek Martinek	NYI	23	1	4	5	5	16	0	0	1	0	25	4.0
R	25	Juraj Kolnik	NYI	7	2	0	2	–2	0	1	0	0	0	10	20.0
D	36	Evgeny Korolev	NYI	17	0	2	2	0	6	0	0	0	0	9	0.0
D	6	Ken Sutton	NYI	21	0	2	2	–5	8	0	0	0	0	22	0.0
D	2	Branislav Mezei	NYI	24	0	2	2	2	12	0	0	0	0	4	0.0
L	16 *	Raffi Torres	NYI	14	0	1	1	2	6	0	0	0	0	9	0.0
D	7	Kevin Haller	NYI	1	0	0	0	–1	2	0	0	0	0	2	0.0
R	46	Marko Tuomainen	NYI	1	0	0	0	–1	0	0	0	0	0	0	0.0
L	51	Dave Roche	NYI	1	0	0	0	0	0	0	0	0	0	0	0.0
C	56	Jason Podollan	NYI	1	0	0	0	0	2	0	0	0	0	1	0.0
D	41 *	Raymond Giroux	NYI	2	0	0	0	–1	2	0	0	0	0	2	0.0
D	39	Ray Schultz	NYI	2	0	0	0	–1	5	0	0	0	0	0	0.0
C	28	Jason Krog	NYI	2	0	0	0	0	0	0	0	0	0	0	0.0
R	14	Alexander Kharitonov	NYI	5	0	0	0	–1	4	0	0	0	0	5	0.0
R	18	Jim Cummins	ANA	2	0	0	0	–1	0	0	0	0	0	0	0.0
			NYI	10	0	0	0	–5	31	0	0	0	0	3	0.0
			TOTAL	12	0	0	0	–6	31	0	0	0	0	3	0.0

Goaltending

No.	Goaltender	GPI	Mins	Avg	W	L	T	EN	SO	GA	SA	S%	G	A	PIM
35	Chris Osgood	66	3743	2.50	32	25	6	6	4	156	1727	.910	0	4	10
30	Garth Snow	25	1217	2.71	10	7	2	3	2	55	549	.900	0	0	14
	Totals	**82**	**4987**	**2.65**	**42**	**32**	**8**	**9**	**6**	**220**	**2285**	**.904**			

Playoffs

Pos	#	Player	Team	GP	G	A	Pts	+/–	PIM	PP	SH	GW	GT	S	%
C	79	Alexei Yashin	NYI	7	3	4	7	–2	2	1	0	0	0	29	10.3
D	3	Adrian Aucoin	NYI	7	2	5	7	–1	4	2	0	0	0	16	12.5
D	4	Roman Hamrlik	NYI	7	1	6	7	–6	6	0	0	0	0	21	4.8
C	11	Kip Miller	NYI	7	4	2	6	–2	2	2	0	1	0	8	50.0
C	17	Shawn Bates	NYI	7	2	4	6	–5	11	1	0	1	0	18	11.1
R	21	Mariusz Czerkawski	NYI	7	2	2	4	–3	4	1	0	0	0	11	18.2
D	28	Darren Van Impe	NYI	7	0	4	4	–5	8	0	0	0	0	8	0.0
R	37	Mark Parrish	NYI	7	2	1	3	–6	6	2	0	0	0	12	16.7
D	29	Kenny Jonsson	NYI	5	1	2	3	–2	4	1	0	0	0	8	12.5
L	15	Brad Isbister	NYI	3	1	1	2	0	17	1	0	1	0	4	25.0
R	43 *	Trent Hunter	NYI	4	1	1	2	1	2	0	0	0	0	11	9.1
C	38	Dave Scatchard	NYI	7	1	1	2	–2	22	0	0	0	0	7	14.3
C	27	Michael Peca	NYI	5	1	0	1	–5	2	0	0	0	0	5	20.0
L	12	Oleg Kvasha	NYI	7	0	1	1	1	6	0	0	0	0	12	0.0
C	55	Jason Blake	NYI	7	0	1	1	–2	13	0	0	0	0	15	0.0
R	18	Jim Cummins	NYI	1	0	0	0	0	9	0	0	0	0	0	0.0
D	39	Ray Schultz	NYI	2	0	0	0	2	2	0	0	0	0	0	0.0
D	36	Evgeny Korolev	NYI	2	0	0	0	0	0	0	0	0	0	0	0.0
D	8	Dick Tarnstrom	NYI	5	0	0	0	–2	2	0	0	0	0	2	0.0
C	13	Claude Lapointe	NYI	7	0	0	0	–2	14	0	0	0	0	4	0.0
D	33	Eric Cairns	NYI	7	0	0	0	–2	15	0	0	0	0	4	0.0
R	20	Steve Webb	NYI	7	0	0	0	1	12	0	0	0	0	3	0.0

Goaltending

No.	Goaltender	GPI	Mins	Avg	W	L	EN	SO	GA	SA	S%	G	A	PIM
35	Chris Osgood	7	392	2.60	3	4	3	0	17	193	.912	0	0	4
30	Garth Snow	1	26	4.62	0	0	0	0	2	19	.895	0	0	0
	Totals	**7**	**420**	**3.14**	**3**	**4**	**3**	**0**	**22**	**215**	**.898**			

Coaching History

Phil Goyette and Earl Ingarfield, 1972-73; Al Arbour, 1973-74 to 1985-86; Terry Simpson, 1986-87, 1987-88; Terry Simpson and Al Arbour, 1988-89; Al Arbour, 1989-90 to 1993-94; Lorne Henning, 1994-95; Mike Milbury, 1995-96; Mike Milbury and Rick Bowness, 1996-97; Rick Bowness and Mike Milbury, 1997-98; Mike Milbury and Bill Stewart, 1998-99; Butch Goring, 1999-2000; Butch Goring and Lorne Henning, 2000-01; Peter Laviolette, 2001-02 to date.

Club Records

Team

(Figures in brackets for season records are games played; records for fewest points, wins, ties, losses, goals, goals against are for 70 or more games)

Most Points	**118**	1981-82 (80)
Most Wins	**54**	1981-82 (80)
Most Ties	**22**	1974-75 (80)
Most Losses	**60**	1972-73 (78)
Most Goals	**385**	1981-82 (80)
Most Goals Against	**347**	1972-73 (78)
Fewest Points	**30**	1972-73 (78)
Fewest Wins	**12**	1972-73 (78)
Fewest Ties	**4**	1983-84 (80)
Fewest Losses	**15**	1978-79 (80)
Fewest Goals	**170**	1972-73 (78)
Fewest Goals Against	**190**	1975-76 (80)
Longest Winning Streak		
Overall	**15**	Jan. 21-Feb. 20/82
Home	**14**	Jan. 2-Feb. 25/82
Away	**8**	Feb. 27-Mar. 29/81
Longest Undefeated Streak		
Overall	**15**	Three times
Home	**23**	Oct. 17/78-Jan. 27/79 (19 wins, 4 ties), Jan. 2-Apr. 3/82 (21 wins, 2 ties)
Away	**8**	Three times
Longest Losing Streak		
Overall	**12**	Dec. 27/72-Jan. 16/73, Nov. 22-Dec. 15/88
Home	**7**	Nov. 13-Dec. 14/99
Away	**15**	Jan. 20-Mar. 31/73
Longest Winless Streak		
Overall	**15**	Nov. 22-Dec. 21/72 (12 losses, 3 ties)
Home	**9**	Mar. 2-Apr. 6/99 (7 losses, 2 ties)
Away	**20**	Nov. 3/72-Jan. 13/73 (19 losses, 1 tie)
Most Shutouts, Season	**10**	1975-76 (80)
Most PIM, Season	**1,857**	1986-87 (80)
Most Goals, Game	**11**	Dec. 20/83 (Pit. 3 at NYI 11), Mar. 3/84 (NYI 11 at Tor. 6)

Individual

Most Seasons	**17**	Billy Smith
Most Games	**1,123**	Bryan Trottier
Most Goals, Career	**573**	Mike Bossy
Most Assists, Career	**853**	Bryan Trottier
Most Points, Career	**1,353**	Bryan Trottier (500G, 853A)
Most PIM, Career	**1,879**	Mick Vukota
Most Shutouts, Career	**25**	Glenn Resch
Longest Consecutive Games Streak	**576**	Billy Harris (Oct. 7/72-Nov. 30/79)
Most Goals, Season	**69**	Mike Bossy (1978-79)
Most Assists, Season	**87**	Bryan Trottier (1978-79)
Most Points, Season	**147**	Mike Bossy (1981-82; 64G, 83A)
Most PIM, Season	**356**	Brian Curran (1986-87)
Most Points, Defenseman, Season	**101**	Denis Potvin (1978-79; 31G, 70A)
Most Points, Center, Season	**134**	Bryan Trottier (1978-79; 47G, 87A)
Most Points, Right Wing, Season	**147**	Mike Bossy (1981-82; 64G, 83A)
Most Points, Left Wing, Season	**100**	John Tonelli (1984-85; 42G, 58A)
Most Points, Rookie, Season	**95**	Bryan Trottier (1975-76; 32G, 63A)
Most Shutouts, Season	**7**	Glenn Resch (1975-76)
Most Goals, Game	**5**	Bryan Trottier (Dec. 23/78, Feb. 13/82), John Tonelli (Jan. 6/81)
Most Assists, Game	**6**	Mike Bossy (Jan. 6/81)
Most Points, Game	**8**	Bryan Trottier (Dec. 23/78; 5G, 3A)

Retired Numbers

5	Denis Potvin	1973-1988
9	Clark Gillies	1974-1986
19	Bryan Trottier	1975-1990
22	Mike Bossy	1977-1987
23	Bob Nystrom	1972-1986
31	Billy Smith	1972-1989

Captains' History

Ed Westfall, 1972-73 to 1975-76; Ed Westfall and Clark Gillies, 1976-77; Clark Gillies, 1977-78, 1978-79; Denis Potvin, 1979-80 to 1986-87; Brent Sutter, 1987-88 to 1990-91; Brent Sutter and Pat Flatley, 1991-92; Pat Flatley, 1992-93 to 1995-96; no captain, 1996-97; Bryan McCabe and Trevor Linden, 1997-98; Trevor Linden, 1998-99; Kenny Jonsson, 1999-2000, 2000-01; Michael Peca, 2001-02 to date.

All-time Record vs. Other Clubs

Regular Season

	At Home								On Road								Total							
	GP	W	L	T	OL	GF	GA	PTS	GP	W	L	T	OL	GF	GA	PTS	GP	W	L	T	OL	GF	GA	PTS
Anaheim	7	3	3	1	0	20	22	7	7	3	2	2	0	19	15	8	14	6	5	3	0	39	37	15
Atlanta	6	3	3	0	0	20	15	6	6	3	2	1	0	23	20	7	12	6	5	1	0	43	35	13
Boston	57	23	27	7	0	188	189	53	55	16	29	10	0	158	207	42	112	39	56	17	0	346	396	95
Buffalo	57	23	25	9	0	159	161	55	57	17	30	9	1	152	190	44	114	40	55	18	1	311	351	99
Calgary	49	25	15	9	0	187	136	59	49	14	24	11	0	145	172	39	98	39	39	20	0	332	308	98
Carolina	40	19	17	4	0	126	112	42	41	16	20	5	0	135	141	37	81	35	37	9	0	261	253	79
Chicago	45	17	13	15	0	156	136	49	47	16	26	5	0	153	160	37	92	33	39	20	0	309	296	86
Colorado	31	17	13	1	0	125	108	35	33	11	20	2	0	96	122	24	64	28	33	3	0	221	230	59
Columbus	1	0	1	0	0	5	7	0	2	0	1	1	0	5	7	1	3	0	2	1	0	10	14	1
Dallas	46	24	14	8	0	169	131	56	46	21	18	7	0	166	135	49	92	45	32	15	0	335	266	105
Detroit	45	23	18	3	1	162	135	50	43	17	24	2	0	133	154	36	88	40	42	5	1	295	289	86
Edmonton	29	13	7	9	0	121	106	35	28	7	16	5	0	82	105	19	57	20	23	14	0	203	211	54
Florida	21	8	11	2	0	51	58	18	21	7	10	4	0	60	63	18	42	15	21	6	0	111	121	36
Los Angeles	43	24	15	4	0	155	122	52	44	16	21	7	0	140	161	39	87	40	36	11	0	295	283	91
Minnesota	2	1	1	0	0	3	5	2	2	0	2	0	0	5	7	0	4	1	3	0	0	8	12	2
Montreal	56	26	24	6	0	173	163	58	56	14	33	9	0	161	206	37	112	40	57	15	0	334	369	95
Nashville	3	1	2	0	0	7	9	2	3	2	1	0	0	13	8	4	6	3	3	0	0	20	17	6
New Jersey	79	53	16	10	0	340	227	116	79	36	31	11	1	283	263	84	158	89	47	21	1	623	490	200
NY Rangers	91	54	29	7	1	362	286	116	90	29	51	10	0	274	341	68	181	83	80	17	1	636	627	184
Ottawa	20	3	11	5	1	62	77	12	19	4	11	4	0	51	67	12	39	7	22	9	1	113	144	24
Philadelphia	92	49	29	14	0	345	267	112	90	27	53	10	0	256	329	64	182	76	82	24	0	601	596	176
Phoenix	29	13	8	8	0	111	88	34	29	14	11	4	0	100	94	32	58	27	19	12	0	211	182	66
Pittsburgh	81	42	30	8	1	324	272	93	82	31	39	12	0	286	313	74	163	73	69	20	1	610	585	167
St. Louis	47	25	11	11	0	180	121	61	45	19	17	9	0	148	161	47	92	44	28	20	0	328	282	108
San Jose	11	5	4	2	0	40	35	12	10	5	4	1	0	33	25	11	21	10	8	3	0	73	60	23
Tampa Bay	21	10	10	1	0	63	62	21	22	10	9	2	1	65	57	23	43	20	19	3	1	128	119	44
Toronto	49	27	19	3	0	194	145	57	51	23	24	3	1	177	173	50	100	50	43	6	1	371	318	107
Vancouver	45	24	11	10	0	167	124	58	46	21	22	3	0	150	151	45	91	45	33	13	0	317	275	103
Washington	77	41	34	2	0	292	246	84	77	29	37	11	0	242	250	69	154	70	71	13	0	534	496	153
Defunct Clubs	13	11	0	2	0	75	33	24	13	4	5	4	0	35	41	12	26	15	5	6	0	110	74	36
Totals	**1193**	**607**	**421**	**161**	**4**	**4382**	**3598**	**1379**	**1193**	**432**	**593**	**164**	**4**	**3746**	**4138**	**1032**	**2386**	**1039**	**1014**	**325**	**8**	**8128**	**7736**	**2411**

Playoffs

	Series	W	L	GP	W	L	T	GF	GA	Last Mtg.	Rnd.	Result
Boston	2	2	0	11	8	3	0	49	35	1983	CF	W 4-2
Buffalo	3	3	0	16	12	4	0	59	45	1980	SF	W 4-2
Chicago	2	2	0	6	6	0	0	21	6	1979	QF	W 4-0
Colorado	1	1	0	4	4	0	0	18	9	1982	CF	W 4-0
Dallas	1	1	0	5	4	1	0	26	16	1981	F	W 4-1
Edmonton	3	2	1	15	9	6	0	58	47	1984	F	L 1-4
Los Angeles	1	1	0	4	3	1	0	21	10	1980	PRE	W 3-1
Montreal	4	1	3	22	8	14	0	55	64	1993	CF	L 1-4
New Jersey	1	0	1	6	2	4	0	18	23	1988	DSF	L 2-4
NY Rangers	8	5	3	39	20	19	0	129	132	1994	CQF	L 0-4
Philadelphia	4	1	3	25	11	14	0	69	83	1987	DF	L 3-4
Pittsburgh	3	3	0	19	11	8	0	67	58	1993	DF	W 4-3
Toronto	3	1	2	17	9	8	0	54	42	2002	CQF	L 3-4
Vancouver	2	2	0	6	6	0	0	26	14	1982	F	W 4-0
Washington	6	5	1	30	18	12	0	99	88	1993	DSF	W 4-2
Totals	**44**	**30**	**14**	**225**	**131**	**94**	**0**	**769**	**672**			

Playoff Results 2002-1998

Year	Round	Opponent	Result	GF	GA
2002	CQF	Toronto	L 3-4	21	22

Abbreviations: Round: F – Final; **CF** – conference final; **CQF** – conference quarter-final; **DF** – division final; **DSF** – division semi-final; **SF** – semi-final; **QF** – quarter-final; **PRE** – preliminary round.

Calgary totals include Atlanta Flames, 1972-73 to 1979-80. Carolina totals include Hartford, 1979-80 to 1996-97.
Colorado totals include Quebec, 1979-80 to 1994-95. Dallas totals include Minnesota North Stars, 1972-73 to 1992-93.
New Jersey totals include Kansas City, 1974-75 to 1975-76, and Colorado Rockies, 1976-77 to 1981-82.
Phoenix totals include Winnipeg, 1979-80 to 1995-96.

2001-02 Results

Month	Day	Opponent	Score
Oct.	5	at Tampa Bay	3-2
	6	at Florida	3-0
	10	at Pittsburgh	6-3
	11	at New Jersey	6-4
	13	Detroit	4-5*
	17	at Carolina	4-0
	18	Carolina	2-1*
	20	San Jose	2-2
	26	at Carolina	3-2*
	28	Dallas	3-2*
	30	Florida	3-2
Nov.	2	at Detroit	1-2
	3	at Philadelphia	2-1
	6	Tampa Bay	3-0
	8	NY Rangers	2-6
	10	at Montreal	2-3
	14	at Pittsburgh	3-3
	16	at Colorado	0-1
	17	at Phoenix	1-6
	19	at Dallas	3-2
	21	Colorado	5-4
	23	Toronto	3-1
	24	Anaheim	5-3
	27	Washington	5-5
	29	Montreal	1-1
Dec.	1	Buffalo	2-4
	4	Philadelphia	2-3
	6	at Philadelphia	2-0
	7	at Chicago	3-4
	11	Ottawa	2-2
	12	at New Jersey	2-3*
	15	Florida	1-3
	18	Edmonton	4-1
	21	at NY Rangers	2-1
	22	Boston	2-4
	27	at Ottawa	2-5
	29	Montreal	6-5*
Jan.	1	at Washington	2-3
	3	Pittsburgh	4-2
	5	Los Angeles	0-3
	6	at Atlanta	3-2
	8	Calgary	2-5
	10	at Montreal	0-4
	12	at Boston	5-4
	15	at Calgary	3-1
	17	at San Jose	2-3
	19	at Los Angeles	3-2
	22	NY Rangers	4-5
	24	Pittsburgh	4-5*
	26	Tampa Bay	6-2
	29	New Jersey	1-3
	30	at NY Rangers	6-3
Feb.	4	at Florida	6-6
	5	St. Louis	4-3
	7	Toronto	4-1
	10	at Minnesota	3-4
	12	at Philadelphia	1-0*
	26	Boston	3-3
Mar.	1	at Atlanta	3-4
	2	Atlanta	4-1
	4	Pittsburgh	2-4
	7	Buffalo	0-5
	8	at Columbus	2-4
	10	Atlanta	6-1
	12	at Buffalo	3-0
	13	at New Jersey	2-3
	16	at Ottawa	3-4
	19	at Toronto	2-3*
	21	Vancouver	3-2
	23	Minnesota	2-1
	25	NY Rangers	4-2
	27	Ottawa	1-4
	28	at Toronto	5-4
	30	at Washington	2-4
Apr.	1	New Jersey	4-2
	3	at Buffalo	1-1
	4	at Boston	2-1*
	6	Washington	5-4
	8	Carolina	1-2
	11	at Nashville	5-2
	12	at Tampa Bay	3-1
	14	Philadelphia	3-1

* – Overtime

Entry Draft Selections 2002-1988

2002
Pick
22 Sean Bergenheim
87 Frans Nielsen
149 Marcus Paulsson
189 Alexei Stonkus
220 Brad Topping
252 Martin Chabada
283 Per Braxenholm

2001
Pick
101 Cory Stillman
132 Dusan Salficky
166 Andy Chiodo
197 Jan Holub
228 Mike Bray
260 Bryan Perez
280 Roman Kukhtinov
287 Juha-Pekka Ketola

2000
Pick
1 Rick DiPietro
5 Raffi Torres
101 Arto Tukio
105 Vladimir Gorbunov
136 Dmitri Upper
148 Kristofer Ottosson
202 Ryan Caldwell
264 Dmitri Altarev
267 Tomi Pettinen

1999
Pick
5 Tim Connolly
8 Taylor Pyatt
10 Branislav Mezei
28 Kristian Kudroc
78 Mattias Weinhandl
87 Brian Collins
101 Juraj Kolnik
102 Johan Halvardsson
130 Justin Mapletoft
140 Adam Johnson
163 Bjorn Melin
228 Radek Martinek
255 Brett Henning
268 Tyler Scott

1998
Pick
9 Mike Rupp
36 Chris Nielsen
95 Andy Burnham
123 Jiri Dopita
155 Kevin Clauson
182 Evgeny Korolev
209 Frederik Brindamour
237 Ben Blais
242 Jason Doyle
250 Radek Matejovsky

1997
Pick
4 Roberto Luongo
5 Eric Brewer
31 Jeff Zehr
59 Jarrett Smith
79 Robert Schnabel
85 Petr Mika
115 Adam Edinger
139 Bobby Leavins
166 Kris Knoblauch
196 Jeremy Symington
222 Ryan Clark

1996
Pick
3 J-P Dumont
29 Dan Lacouture
56 Zdeno Chara
83 Tyrone Garner
109 Bubba Berenzweig
128 Petr Sachl
138 Todd Miller
165 J.R. Prestifilippo
192 Evgeny Korolev
218 Mike Muzechka

1995
Pick
2 Wade Redden
28 Jan Hlavac
41 D.J. Smith
106 Vladimir Orszagh
158 Andrew Taylor
210 David MacDonald
211 Mike Broda

1994
Pick
9 Brett Lindros
38 Jason Holland
63 Jason Strudwick
90 Brad Lukowich
112 Mark McArthur
116 Albert O'Connell
142 Jason Stewart
194 Mike Loach
203 Peter Hogardh
220 Gord Walsh
246 Kirk Dewaele
272 Dick Tarnstrom

1993
Pick
23 Todd Bertuzzi
40 Bryan McCabe
66 Vladimir Chebaturkin
92 Warren Luhning
118 Tommy Salo
144 Peter LeBoutillier
170 Darren Van Impe
196 Rod Hinks
222 Daniel Johansson
248 Stephane Larocque
274 Carl Charland

1992
Pick
5 Darius Kasparaitis
56 Jarrett Deuling
104 Thomas Klimt
105 Ryan Duthie
128 Derek Armstrong
152 Vladimir Grachev
159 Steve O'Rourke
176 Jason Widmer
200 Daniel Paradis
224 David Wainwright
248 Andrei Vasilyev

1991
Pick
4 Scott Lachance
26 Ziggy Palffy
48 Jamie McLennan
70 Milan Hnilicka
92 Steve Junker
114 Rob Valicevic
136 Andreas Johansson
158 Todd Sparks
180 John Johnson
202 Robert Canavan
224 Marcus Thuresson
246 Marty Schriner

1990
Pick
6 Scott Scissons
27 Chris Taylor
48 Dan Plante
90 Chris Marinucci
111 Joni Lehto
132 Michael Guilbert
153 Sylvain Fleury
174 John Joyce
195 Richard Enga
216 Martin Lacroix
237 Andy Shier

1989
Pick
2 Dave Chyzowski
23 Travis Green
44 Jason Zent
65 Brent Grieve
86 Jace Reed
90 Steve Young
99 Kevin O'Sullivan
128 Jon Larson
133 Brett Harkins
149 Phil Huber
170 Matthew Robbins
191 Vladimir Malakhov
212 Kelly Ens
233 Iain Fraser

1988
Pick
16 Kevin Cheveldayoff
29 Wayne Doucet
37 Sean Lebrun
58 Danny Lorenz
79 Andre Brassard
100 Paul Rutherford
111 Pavel Gross
121 Jason Rathbone
142 Yves Gaucher
163 Marty McInnis
173 Shorty Forrest
184 Jeff Blumer
205 Jeff Kampersal
226 Phillip Neururer
247 Joe Caprinni

General Managers' History

Bill Torrey, 1972-73 to 1991-92; Don Maloney, 1992-93 to 1994-95; Don Maloney and Mike Milbury, 1995-96; Mike Milbury, 1996-97 to date.

General Manager

MILBURY, MIKE

General Manager, New York Islanders. Born in Walpole, MA, June 17, 1952.

Mike Milbury came to the Islanders with 20 years of professional hockey experience with the Boston Bruins — as a player, assistant coach, assistant general manager, general manager and coach on both the NHL and AHL levels. Milbury took over as general manager from Don Maloney on December 12, 1995.

His recent trades have brought the Islanders established stars like Alexei Yashin and Michael Peca as well as an abundance of young talent. In 2001-02, the Islanders returned to the playoffs for the first time since 1994.

Milbury joined the Boston organization after graduating from Colgate University with a degree in urban sociology and enjoyed a 10-year playing career with the team. He retired May 6, 1985 and took over as assistant coach. He returned to the ice late in the 1985-86 season when injuries decimated the Bruins defense.

Milbury's playing career concluded after the 1986-87 season and on July 16, 1987 he took over as coach of the Maine Mariners, Boston's top AHL affiliate. In his first year with the team he guided the Mariners to the AHL's Northern Division title and was named both AHL coach of the year and *The Hockey News* minor league coach of the year.

NHL Coaching Record

			Regular Season				Playoffs	
Season	Team	Games	W	L	T	Games	W	L
1989-90	Boston	80	46	25	9	21	13	8
1990-91	Boston	80	44	24	12	19	10	9
1995-96	NY Islanders	82	22	50	10	....	....	....
1996-97	NY Islanders	45	13	23	9	....	....	....
1997-98	NY Islanders	19	8	9	2	....	....	....
	NHL Totals	**306**	**133**	**131**	**42**	**40**	**23**	**17**

Club Directory

Nassau Veterans' Memorial Coliseum

New York Islanders
Nassau Veterans' Memorial Coliseum
Uniondale, NY 11553
Phone **516/501-6700**
FAX 516/501-6746
www.newyorkislanders.com
Capacity: 16,234

Team Information	
Colors	Orange, Blue, White
Television Coverage	FSNY
TV Announcers	Howie Rose, Joe Micheletti
Radio	1050 AM ESPN RADIO
Radio Announcers	John Wiedeman, Chris King
Executive	
Owner and Governor	Charles B. Wang
Owner and Alternate Governor	Sanjay Kumar
Senior VP of Operations and Alternate Governor	Michael J. Picker
Senior Vice President of Sales and Marketing	Paul Lancey
Alternate Governor and General Counsel	Roy E. Reichbach
Executive Assistant	Theresa Dewar
Hockey Staff	
General Manager and Alternate Governor	Mike Milbury
Manager, Hockey Administration	Joanne Holewa
Assistant Manager of Player Contracts	Pam Genzardi
Player Liaison/Travel Secretary	Kerry Gwydir
Head Coach	Peter Laviolette
Assistant Coaches	Jacques Laperriere, Kelly Miller, Greg Cronin
Goaltending Consultant/Scout	Bill Smith
Head Amateur Scout	Tony Feltrin
Western Scout	Earl Ingarfield
Sweden/Finland Amateur Scout	Anders Kallur
Ontario Scout	Doug Gibson
Russian Amateur Scout	Yuri Karmanov
U.S. Amateur Scout	Jay Heinbuck
Czech Republic Amateur Scout	Karel Pavlik
Director of Pro Scouting	Ken Morrow
Assistant Director of Pro Scouting	Kevin Maxwell
Scouting Staff	Jim Madigan, Mario Saraceno, Brian Hunter, Greg Morrow, Harri Rindell, Harkie Singh
Video Coordinator	Bob Smith
Medical Staff	
Director of Medical Services	Dr. Elliot Pellman
Internist	Dr. Clifford Cooper
Team Orthopedists	Dr. Elliott Hershman, Dr, Kenneth Montgomery, Dr. David Gazzaniga
Team Dentists	Dr. Bruce Michnick, Dr. Jan Sherman
Training/Equipment Staff	
Head Trainer	Rich Campbell
Strength and Conditioning Coach	Sean Donellan
Assistant Athletic Trainer	Andy Wetstein
Head Equipment Manager	Joe McMahon
Assistant Equipment Manager	Bill Nichols
Lockerroom Attendants	Charles E. Nass, Matt Brager, Robert Dobrzeniecki, Arthur Verdi
Communications	
Vice President of Communications	Chris Botta
Manager of Media Relations	Jamie Fabos
Media Relations Coordinator	Howie Wirtheim
Manager of Broadcasting	Alice Vanderveldt
Manager of Community Relations	Heather Umen
Community Relations Coordinators	Heather Cozzens, Erin Leavy
Sales and Administration	
Senior Vice President/CFO	Arthur McCarthy
Vice President of Administration	Janet L. Kask
Vice President of Corporate and Community Relations	Bill Kain
Director of Corporate Relations	Bob Nystrom
Director of Merchandise	Chris DiPierri
Vice President of Ticket Sales/Customer Service	Larry Fitzpatrick
Director of Game Operations	Tim Beach
Director of Corporate Sponsorships	Ted Van Zelst
Director of Executive Suites	Mary Dolan Grippo
Controller	Ralph Sellitti
Assistant Controller	Ginna Cotton
Customer Service Manager	Kerry Cornils
Payroll Manager	Christine Bowler
Staff Accountants	Laura Ferretti, Heather Jabick
Manager, Ticket Sales	Brian Reynolds
Manager of Corporate Ticket Sales	Erik Scheibe
Arena Manager, Merchandise	Danny DiPierri
Creative Services Manager	Timothy Gilroy
Manager of Executive Suite Services	Jennifer Meilan
Marketing Manager	Jessica Rotoli
Manager of Game Operations and Events	Brad Preston
Team Store Manager	MaryAnne Steves
Group Sales Manager	Cliff Gault
Corporate Account Executives	Anthony Mercogliano, Rob Olenchak, Steven Bromberg
Customer Service Representative	Jesse Mones
Group Ticket Sales Executives	Emily Derkasch, Kevin Schwab
Corporate Sales Account Executive	Larry Sragow
Account Executives	Mike Bellinzoni, Mike Clough, Steven Beisel
Creative Services Coordinator	Elizabeth McFadden
Creative Services Avid Editor	Nima Forough
Ticket Coordinator	Maria Corvino
Assistant Ticket Coordinator	Adam Ortiz
Sponsor Services Coordinators	Lorraine Bittles, Rainbow Kirby
Administrative Services Coordinator	Sheriene Ahmed
Accounts Payable Bookkeeper	Janet Nelson
Receptionists	Chere O'Neill, Bonnie Dreher
Office Attendant	Todd Aronovitch

New York Rangers

Eric Lindros played 72 games last season.

2001-02 Results: 36W-38L-4T-4OTL 80PTS.
Fourth, Atlantic Division

Year-by-Year Record

Season	GP	Home W	Home L	Home T	Home OL	Road W	Road L	Road T	Road OL	Overall W	Overall L	Overall T	Overall OL	GF	GA	Pts.	Finished	Playoff Result
2001-02	82	19	19	2	1	17	19	2	3	36	38	4	4	227	258	80	4th, Atlantic Div.	Out of Playoffs
2000-01	82	17	20	3	1	16	23	2	0	33	43	5	1	250	290	72	4th, Atlantic Div.	Out of Playoffs
1999-2000	82	15	20	5	1	14	18	7	2	29	38	12	3	218	246	73	4th, Atlantic Div.	Out of Playoffs
1998-99	82	17	19	5	...	16	19	6	...	33	38	11	...	217	227	77	4th, Atlantic Div.	Out of Playoffs
1997-98	82	14	18	9	...	11	21	9	...	25	39	18	...	197	231	68	5th, Atlantic Div.	Out of Playoffs
1996-97	82	21	14	6	...	17	20	4	...	38	34	10	...	258	231	86	4th, Atlantic Div.	Lost Conf. Final
1995-96	82	22	10	9	...	19	17	5	...	41	27	14	...	272	237	96	2nd, Atlantic Div.	Lost Conf. Semi-Final
1994-95	48	11	10	3	...	11	13	0	...	22	23	3	...	139	134	47	4th, Atlantic Div.	Lost Conf. Semi-Final
1993-94	**84**	**28**	**8**	**6**	**...**	**24**	**16**	**2**	**...**	**52**	**24**	**8**	**...**	**299**	**231**	**112**	**1st, Atlantic Div.**	**Won Stanley Cup**
1992-93	84	20	17	5	...	14	22	6	...	34	39	11	...	304	308	79	6th, Patrick Div.	Out of Playoffs
1991-92	80	28	8	4	...	22	17	1	...	50	25	5	...	321	246	105	1st, Patrick Div.	Lost Div. Final
1990-91	80	22	11	7	...	14	20	6	...	36	31	13	...	297	265	85	2nd, Patrick Div.	Lost Div. Semi-Final
1989-90	80	20	11	9	...	16	20	4	...	36	31	13	...	279	267	85	1st, Patrick Div.	Lost Div. Final
1988-89	80	21	17	2	...	16	18	6	...	37	35	8	...	310	307	82	3rd, Patrick Div.	Lost Div. Semi-Final
1987-88	80	22	13	5	...	14	21	5	...	36	34	10	...	300	283	82	5th, Patrick Div.	Out of Playoffs
1986-87	80	18	18	4	...	16	20	4	...	34	38	8	...	307	323	76	4th, Patrick Div.	Lost Div. Semi-Final
1985-86	80	20	18	2	...	16	20	4	...	36	38	6	...	280	276	78	4th, Patick Div.	Lost Conf. Championship
1984-85	80	16	18	6	...	10	26	4	...	26	44	10	...	295	345	62	4th, Patrick Div.	Lost Div. Semi-Final
1983-84	80	27	12	1	...	15	17	8	...	42	29	9	...	314	304	93	4th, Patrick Div.	Lost Div. Semi-Final
1982-83	80	24	13	3	...	11	22	7	...	35	35	10	...	306	287	80	4th, Patrick Div.	Lost Div. Final
1981-82	80	19	15	6	...	20	12	8	...	39	27	14	...	316	306	92	2nd, Patrick Div.	Lost Div. Final
1980-81	80	17	13	10	...	13	23	4	...	30	36	14	...	312	317	74	4th, Patrick Div.	Lost Semi-Final
1979-80	80	22	10	8	...	16	22	2	...	38	32	10	...	308	284	86	3rd, Patrick Div.	Lost Quarter-Final
1978-79	80	19	13	8	...	21	16	3	...	40	29	11	...	316	292	91	3rd, Patrick Div.	Lost Final
1977-78	80	18	15	7	...	12	22	6	...	30	37	13	...	279	280	73	4th, Patrick Div.	Lost Prelim. Round
1976-77	80	17	18	5	...	12	19	9	...	29	37	14	...	272	310	72	4th, Patrick Div.	Out of Playoffs
1975-76	80	16	16	8	...	13	26	1	...	29	42	9	...	262	333	67	4th, Patrick Div.	Out of Playoffs
1974-75	80	21	11	8	...	16	18	6	...	37	29	14	...	319	276	88	2nd, Patrick Div.	Lost Prelim. Round
1973-74	78	26	7	6	...	14	17	8	...	40	24	14	...	300	251	94	3rd, East Div.	Lost Semi-Final
1972-73	78	26	8	5	...	21	15	3	...	47	23	8	...	297	208	102	3rd, East Div.	Lost Semi-Final
1971-72	78	26	6	7	...	22	11	6	...	48	17	13	...	317	192	109	2nd, East Div.	Lost Final
1970-71	78	30	2	7	...	19	16	4	...	49	18	11	...	259	177	109	2nd, East Div.	Lost Semi-Final
1969-70	76	22	8	8	...	16	14	8	...	38	22	16	...	246	189	92	4th, East Div.	Lost Quarter-Final
1968-69	76	27	7	4	...	14	19	5	...	41	26	9	...	231	196	91	3rd, East Div.	Lost Quarter-Final
1967-68	74	22	8	7	...	17	15	5	...	39	23	12	...	226	183	90	2nd, East Div.	Lost Quarter-Final
1966-67	70	18	12	5	...	12	16	7	...	30	28	12	...	188	189	72	4th,	Lost Semi-Final
1965-66	70	12	16	7	...	6	25	4	...	18	41	11	...	195	261	47	6th,	Out of Playoffs
1964-65	70	8	19	8	...	12	19	4	...	20	38	12	...	179	246	52	5th,	Out of Playoffs
1963-64	70	14	13	8	...	8	25	2	...	22	38	10	...	186	242	54	5th,	Out of Playoffs
1962-63	70	12	17	6	...	10	19	6	...	22	36	12	...	211	233	56	5th,	Out of Playoffs
1961-62	70	16	11	8	...	10	21	4	...	26	32	12	...	195	207	64	4th,	Lost Semi-Final
1960-61	70	15	15	5	...	7	23	5	...	22	38	10	...	204	248	54	5th,	Out of Playoffs
1959-60	70	10	15	10	...	7	23	5	...	17	38	15	...	187	247	49	6th,	Out of Playoffs
1958-59	70	14	16	5	...	12	16	7	...	26	32	12	...	201	217	64	5th,	Out of Playoffs
1957-58	70	14	15	6	...	18	10	7	...	32	25	13	...	195	188	77	2nd,	Lost Semi-Final
1956-57	70	15	12	8	...	11	18	6	...	26	30	14	...	184	227	66	4th,	Lost Semi-Final
1955-56	70	20	7	8	...	12	21	2	...	32	28	10	...	204	203	74	3rd,	Lost Semi-Final
1954-55	70	10	12	13	...	7	23	5	...	17	35	18	...	150	210	52	5th,	Out of Playoffs
1953-54	70	18	12	5	...	11	19	5	...	29	31	10	...	161	182	68	5th,	Out of Playoffs
1952-53	70	11	14	10	...	6	23	6	...	17	37	16	...	152	211	50	6th,	Out of Playoffs
1951-52	70	16	13	6	...	7	21	7	...	23	34	13	...	192	219	59	5th,	Out of Playoffs
1950-51	70	14	11	10	...	6	18	11	...	20	29	21	...	169	201	61	5th,	Out of Playoffs
1949-50	70	19	12	4	...	9	19	7	...	28	31	11	...	170	189	67	4th,	Lost Final
1948-49	60	13	12	5	...	5	19	6	...	18	31	11	...	133	172	47	6th,	Out of Playoffs
1947-48	60	11	12	7	...	10	14	6	...	21	26	13	...	176	201	55	4th,	Lost Semi-Final
1946-47	60	11	14	5	...	11	18	1	...	22	32	6	...	167	186	50	5th,	Out of Playoffs
1945-46	50	8	12	5	...	5	16	4	...	13	28	9	...	144	191	35	6th,	Out of Playoffs
1944-45	50	7	11	7	...	4	18	3	...	11	29	10	...	154	247	32	6th,	Out of Playoffs
1943-44	50	4	17	4	...	2	22	1	...	6	39	5	...	162	310	17	6th,	Out of Playoffs
1942-43	50	7	13	5	...	4	18	3	...	11	31	8	...	161	253	30	6th,	Out of Playoffs
1941-42	48	15	8	1	...	14	9	1	...	29	17	2	...	177	143	60	1st,	Lost Semi-Final
1940-41	48	13	7	4	...	8	12	4	...	21	19	8	...	143	125	50	4th,	Lost Quarter-Final
1939-40	**48**	**17**	**4**	**3**	**...**	**10**	**7**	**7**	**...**	**27**	**11**	**10**	**...**	**136**	**77**	**64**	**2nd,**	**Won Stanley Cup**
1938-39	48	13	8	3	...	13	8	3	...	26	16	6	...	149	105	58	2nd,	Lost Semi-Final
1937-38	48	15	5	4	...	12	10	2	...	27	15	6	...	149	96	60	2nd, Amn. Div.	Lost Quarter-Final
1936-37	48	9	7	8	...	10	13	1	...	19	20	9	...	117	106	47	3rd, Amn. Div.	Lost Final
1935-36	48	11	6	7	...	8	11	5	...	19	17	12	...	91	96	50	4th, Amn. Div.	Out of Playoffs
1934-35	48	11	8	5	...	11	12	1	...	22	20	6	...	137	139	50	3rd, Amn. Div.	Lost Semi-Final
1933-34	48	11	7	6	...	10	12	2	...	21	19	8	...	120	113	50	3rd, Amn. Div.	Lost Quarter-Final
1932-33	**48**	**12**	**7**	**5**	**...**	**11**	**10**	**3**	**...**	**23**	**17**	**8**	**...**	**135**	**107**	**54**	**3rd, Amn. Div.**	**Won Stanley Cup**
1931-32	48	13	7	4	...	10	10	4	...	23	17	8	...	134	112	54	1st, Amn. Div.	Lost Final
1930-31	44	10	9	3	...	9	7	6	...	19	16	9	...	106	87	47	3rd, Amn. Div.	Lost Semi-Final
1929-30	44	11	5	6	...	6	12	4	...	17	17	10	...	136	143	44	3rd, Amn. Div.	Lost Semi-Final
1928-29	44	12	6	4	...	9	7	6	...	21	13	10	...	72	65	52	2nd, Amn. Div.	Lost Final
1927-28	**44**	**10**	**8**	**4**	**...**	**9**	**8**	**5**	**...**	**19**	**16**	**9**	**...**	**94**	**79**	**47**	**2nd, Amn. Div.**	**Won Stanley Cup**
1926-27	44	13	5	4	...	12	8	2	...	25	13	6	...	95	72	56	1st, Amn. Div.	Lost Quarter-Final

2002-03 Schedule

Month	Day	Date	Opponent	Month	Day	Date	Opponent
Oct.	Wed.	9	at Carolina	**Jan.**	Sat.	4	Washington
	Fri.	11	Montreal		Mon.	6	Ottawa
	Sat.	12	at Pittsburgh		Wed.	8	Carolina
	Tue.	15	Toronto		Thu.	9	at Montreal
	Thu.	17	at Buffalo		Sat.	11	at Pittsburgh*
	Sat.	19	Nashville		Mon.	13	Toronto
	Mon.	21	Tampa Bay		Wed.	15	at Washington
	Wed.	23	Washington		Sun.	19	Philadelphia
	Fri.	25	Los Angeles		Tue.	21	at NY Islanders
	Sat.	26	at Toronto		Thu.	23	at Nashville
	Mon.	28	Phoenix		Sat.	25	Atlanta*
	Wed.	30	at Tampa Bay		Sun.	26	at Washington*
Nov.	Sat.	2	at Boston		Tue.	28	at Atlanta
	Sun.	3	St. Louis*		Thu.	30	Colorado
	Tue.	5	Edmonton	**Feb.**	Wed.	5	Ottawa
	Thu.	7	Calgary		Thu.	6	at St. Louis
	Sat.	9	at Columbus		Sat.	8	at Philadelphia*
	Mon.	11	at San Jose		Wed.	12	at Florida
	Thu.	14	at Calgary		Fri.	14	Pittsburgh
	Sat.	16	at Vancouver		Sat.	15	at Buffalo
	Tue.	19	Anaheim		Mon.	17	at Ottawa
	Thu.	21	at New Jersey		Wed.	19	at Minnesota
	Sat.	23	NY Islanders*		Fri.	21	at Anaheim
	Mon.	25	Carolina		Sun.	23	at Colorado*
	Thu.	28	at Atlanta		Tue.	25	at New Jersey
	Fri.	29	at Dallas		Thu.	27	Boston
Dec.	Sun.	1	Tampa Bay*	**Mar.**	Sat.	1	Florida
	Tue.	3	Columbus		Mon.	3	NY Islanders
	Thu.	5	at Philadelphia		Fri.	7	Philadelphia
	Fri.	6	Buffalo		Mon.	10	Florida
	Sun.	8	Boston*		Thu.	13	at Ottawa
	Wed.	11	Chicago		Sat.	15	at New Jersey*
	Sat.	14	at Toronto		Mon.	17	NY Islanders
	Mon.	16	San Jose		Wed.	19	Buffalo
	Thu.	19	Montreal		Sat.	22	at Philadelphia*
	Sat.	21	at Detroit		Wed.	26	Pittsburgh
	Mon.	23	New Jersey		Sat.	29	at Boston*
	Thu.	26	Pittsburgh		Mon.	31	Atlanta
	Sat.	28	at Florida	**Apr.**	Tue.	1	at NY Islanders
	Sun.	29	at Tampa Bay*		Fri.	4	New Jersey
	Tue.	31	at Carolina		Sat.	5	at Montreal

* Denotes afternoon game.

Franchise date: May 15, 1926

ATLANTIC DIVISION

77th NHL Season

2002-03 Player Personnel

FORWARDS	HT	WT	S	Place of Birth	Date	2001-02 Club
ANDREWS, Bobby	6-1	200	L	Birtle, Man.	1/5/78	Alaska-Fairbanks-Hartford
BARNABY, Matthew	6-0	189	L	Ottawa, Ont.	5/4/73	Tampa Bay-NY Rangers
BURE, Pavel	5-10	189	L	Moscow, USSR	3/31/71	Florida-Russia-NY Rangers
DONATO, Ted	5-10	178	L	Boston, MA	4/28/69	NYI-Bridgeport-StL-L.A.-Manchester
DUSABLON, Benoit	6-1	207	L	Ste Anne de la Perard, Que.	8/1/79	Charlotte-Hartford
DVORAK, Radek	6-1	194	R	Tabor, Czech.	3/9/77	NY Rangers-Czech Republic
EKMAN, Nils	5-11	185	L	Stockholm, Sweden	3/11/76	Djurgarden
FATA, Rico	5-11	200	L	Sault Ste. Marie, Ont.	2/12/80	NY Rangers-Hartford
GERNANDER, Ken	5-10	175	L	Coleraine, MN	6/30/69	Hartford
HOLIK, Bobby	6-4	230	R	Jihlava, Czech.	1/1/71	New Jersey
KANE, Boyd	6-2	218	L	Swift Current, Sask.	4/18/78	Hartford
LINDROS, Eric	6-4	240	R	London, Ont.	2/28/73	NY Rangers-Canada
LUNDMARK, Jamie	6-0	174	R	Edmonton, Alta.	1/16/81	Hartford
LYASHENKO, Roman	6-0	189	R	Murmansk, Russia	5/2/79	Dal-Utah-NYR-Hart
McCARTHY, Sandy	6-3	225	R	Toronto, Ont.	6/15/72	NY Rangers
MESSIER, Mark	6-1	210	L	Edmonton, Alta.	1/18/61	NY Rangers
MURRAY, Garth	6-1	205	L	Regina, Sask.	9/17/82	Regina-Hartford
MURRAY, Rem	6-2	195	L	Stratford, Ont.	10/9/72	Edmonton-NY Rangers
NEDVED, Petr	6-3	195	L	Liberec, Czech.	12/9/71	Liberec-NY Rangers
OLIWA, Krzysztof	6-5	235	L	Tychy, Poland	4/12/73	Pittsburgh
SAMUELSSON, Mikael	6-1	195	L	Mariefred, Sweden	12/23/76	NY Rangers-Hartford
SCOTT, Richard	6-2	195	L	Orillia, Ont.	8/1/78	NY Rangers-Hartford
ULMER, Layne	6-1	205	L	North Battleford, Sask.	9/14/80	Hartford-Charlotte
DEFENSEMEN						
BOUCHARD, Joel	6-1	209	L	Montreal, Que.	1/23/74	New Jersey-Albany
CHEBATURKIN, Vladimir	6-2	226	L	Tyumen, USSR	4/23/75	Chicago-Norfolk
KARPA, Dave	6-1	210	R	Regina, Sask.	5/7/71	NY Rangers
KASPARAITIS, Darius	5-11	212	L	Elektrenai, USSR	10/16/72	Pit-Russia-Col
KINCH, Matt	6-0	195	L	Red Deer, Alta.	2/17/80	Hartford-Charlotte
KLOUCEK, Tomas	6-3	203	L	Prague, Czech.	3/7/80	NY Rangers-Hartford
LEETCH, Brian	6-1	190	L	Corpus Christi, TX	3/3/68	NYR-United States
LEFEBVRE, Sylvain	6-2	205	L	Richmond, Que.	10/14/67	NY Rangers-Hartford
LINTNER, Richard	6-3	212	R	Trencin, Czech.	11/15/77	Dukla Trencin-MoDo-Slovakia
MALAKHOV, Vladimir	6-4	230	L	Sverdlovsk, USSR	8/30/68	NY Rangers-Russia
MOTTAU, Mike	6-0	192	L	Quincy, MA	3/19/78	NY Rangers-Hartford
POTI, Tom	6-3	215	L	Worcester, MA	3/22/77	Edm-United States-NYR
PURINTON, Dale	6-3	214	L	Fort Wayne, IN	10/11/76	NY Rangers
STATE, Jeff	6-6	235	R	Tonowanda, NY	9/17/79	Merrimack
TUTIN, Fedor	6-2	196	L	Izhevsk, USSR	7/19/83	Guelph
WELLER, Craig	6-3	195	R	Calgary, Alta.	1/17/81	Kootenay

GOALTENDERS	HT	WT	C	Place of Birth	Date	2001-02 Club
BLACKBURN, Dan	6-0	180	L	Montreal, Que.	5/20/83	NY Rangers-Hartford
HOLMQVIST, Johan	6-3	190	L	Tolfta, Sweden	5/24/78	NY Rangers-Hartford
LABARBERA, Jason	6-2	205	L	Prince George, B.C.	1/18/80	Hartford-Charlotte
MEYER, Scott	6-0	185	L	White Bear Lake, MN	4/10/76	Charlotte-Hartford
RICHTER, Mike	5-11	185	L	Abington, PA	9/22/66	NYR-United States

2001-02 Scoring

* - rookie

Regular Season

Pos	#	Player	Team	GP	G	A	Pts	+/–	PIM	PP	SH	GW	GT	S	%
C	88	Eric Lindros	NYR	72	37	36	73	19	138	12	1	4	0	196	18.9
R	9	Pavel Bure	FLA	56	22	27	49	–14	56	9	1	1	1	238	9.2
			NYR	12	12	8	20	9	6	3	0	1	0	49	24.5
			TOTAL	68	34	35	69	–5	62	12	1	2	1	287	11.8
R	14	Theoren Fleury	NYR	82	24	39	63	0	216	7	0	5	0	267	9.0
D	2	Brian Leetch	NYR	82	10	45	55	14	28	1	0	3	0	202	5.0
C	93	Petr Nedved	NYR	78	21	25	46	–8	36	6	1	3	1	175	12.0
L	19	Martin Rucinsky	MTL	18	2	6	8	–1	12	1	0	0	0	41	4.9
			DAL	42	6	11	17	3	24	2	0	1	0	63	9.5
			NYR	15	3	10	13	6	6	0	0	1	0	24	12.5
			TOTAL	75	11	27	38	8	42	3	0	2	0	128	8.6
R	20	Radek Dvorak	NYR	65	17	20	37	–20	14	3	3	1	0	210	8.1
D	23	Vladimir Malakhov	NYR	81	6	22	28	10	83	1	0	0	0	145	4.1
C	17	Rem Murray	EDM	69	7	17	24	5	14	0	2	1	1	84	8.3
			NYR	11	1	2	3	–9	4	0	0	0	0	14	7.1
			TOTAL	80	8	19	27	–4	18	0	2	1	1	98	8.2
D	3	Tom Poti	EDM	55	1	16	17	–6	42	1	0	0	0	100	1.0
			NYR	11	1	7	8	–4	2	1	0	1	0	9	11.1
			TOTAL	66	2	23	25	–10	44	2	0	1	0	109	1.8
L	26	Andreas Johansson	NYR	70	14	10	24	6	46	3	0	1	0	108	13.0
R	10	Sandy McCarthy	NYR	82	10	13	23	–8	171	1	0	3	0	90	11.1
C	11	Mark Messier	NYR	41	7	16	23	–1	32	2	0	2	0	69	10.1
D	34	Bryan Berard	NYR	82	2	21	23	–1	60	0	0	0	0	132	1.5
R	36	Matthew Barnaby	T.B.	29	0	0	0	–7	70	0	0	0	0	13	0.0
			NYR	48	8	13	21	–3	144	0	0	1	0	56	14.3
			TOTAL	77	8	13	21	–10	214	0	0	1	0	69	11.6
C	37	* Mikael Samuelsson	NYR	67	6	10	16	10	23	1	2	1	0	94	6.4
D	33	Dave Karpa	NYR	75	1	10	11	–9	131	0	0	1	0	53	1.9
L	8	Michal Grosek	NYR	15	3	2	5	–3	12	0	0	0	0	23	13.0
D	24	Sylvain Lefebvre	NYR	41	0	5	5	–3	23	0	0	0	0	20	0.0
D	22	Tomas Kloucek	NYR	52	1	3	4	–2	137	0	0	0	0	21	4.8
D	5	Dale Purinton	NYR	40	0	4	4	4	113	0	0	0	0	11	0.0
L	21	Steve McKenna	NYR	54	2	1	3	0	144	1	0	1	0	17	11.8
C	29	Roman Lyashenko	DAL	4	0	0	0	–2	0	0	0	0	0	3	0.0
			NYR	15	2	0	2	0	0	0	0	0	0	13	15.4
			TOTAL	19	2	0	2	–2	0	0	0	0	0	16	12.5
D	25	* Peter Smrek	NYR	8	0	1	1	–7	4	0	0	0	0	2	0.0
C	39	Trent Whitfield	WSH	24	0	1	1	–3	28	0	0	0	0	15	0.0
			NYR	1	0	0	0	1	0	0	0	0	0	0	0.0
			TOTAL	25	0	1	1	–2	28	0	0	0	0	15	0.0
R	27	Jason Dawe	NYR	1	0	0	0	–1	0	0	0	0	0	1	0.0
D	4	* Mike Mottau	NYR	1	0	0	0	0	0	0	0	0	0	0	0.0
R	13	* Richard Scott	NYR	5	0	0	0	0	5	0	0	0	0	1	0.0
R	38	* Rico Fata	NYR	10	0	0	0	–2	0	0	0	0	0	8	0.0
L	47	* Barrett Heisten	NYR	10	0	0	0	–4	2	0	0	0	0	7	0.0

Goaltending

No.	Goaltender	GPI	Mins	Avg	W	L	T	EN	SO	GA	SA	S%	G	A	PIM
40	* Johan Holmqvist	1	9	0.00	0	0	0	0	0	0	2	1.000	0	0	0
35	Mike Richter	55	3195	2.95	24	26	4	3	2	157	1675	.906	0	0	4
31	* Dan Blackburn	31	1737	3.28	12	16	0	3	0	95	935	.898	0	0	10
	Totals	**82**	**4963**	**3.12**	**36**	**42**	**4**	**6**	**2**	**258**	**2618**	**.901**			

Captains' History

Bill Cook, 1926-27 to 1936-37; Art Coulter, 1937-38 to 1941-42; Ott Heller, 1942-43 to 1944-45; Neil Colville 1945-46 to 1948-49; Buddy O'Connor, 1949-50; Frank Eddolls, 1950-51; Frank Eddolls and Allan Stanley, 1951-52; Allan Stanley, 1952-53; Allan Stanley and Don Raleigh, 1953-54; Don Raleigh, 1954-55; Harry Howell, 1955-56, 1956-57; Red Sullivan, 1957-58 to 1960-61; Andy Bathgate, 1961-62, 1962-63; Andy Bathgate and Camille Henry, 1963-64; Camille Henry and Bob Nevin, 1964-65; Bob Nevin 1965-66 to 1970-71; Vic Hadfield, 1971-72 to 1973-74; Brad Park, 1974-75; Brad Park and Phil Esposito, 1975-76; Phil Esposito, 1976-77, 1977-78; Dave Maloney, 1978-79, 1979-80; Dave Maloney, Walt Tkaczuk and Barry Beck, 1980-81; Barry Beck, 1981-82 to 1985-86; Ron Greschner, 1986-87; Ron Greschner and Kelly Kisio, 1987-88; Kelly Kisio, 1988-89 to 1990-91; Mark Messier, 1991-92 to 1996-97; Brian Leetch, 1997-98 to 1999-2000; Mark Messier, 2000-01 to date.

Coaching History

Lester Patrick, 1926-27 to 1938-39; Frank Boucher, 1939-40 to 1947-48; Frank Boucher and Lynn Patrick, 1948-49; Lynn Patrick, 1949-50; Neil Colville, 1950-51; Neil Colville and Bill Cook, 1951-52; Bill Cook, 1952-53; Frank Boucher and Muzz Patrick, 1953-54; Phil Watson, 1954-55; Phil Watson, 1955-56 to 1958-59; Phil Watson and Alf Pike, 1959-60; Alf Pike, 1960-61; Doug Harvey, 1961-62; Muzz Patrick and Red Sullivan, 1962-63; Red Sullivan, 1963-64, 1964-65; Red Sullivan and Emile Francis, 1965-66; Emile Francis, 1966-67, 1967-68; Bernie Geoffrion and Emile Francis, 1968-69; Emile Francis, 1969-70 to 1972-73; Larry Popein and Emile Francis, 1973-74; Emile Francis, 1974-75; Ron Stewart and John Ferguson, 1975-76; John Ferguson, 1976-77; Jean-Guy Talbot, 1977-78; Fred Shero, 1978-79, 1979-80; Fred Shero and Craig Patrick, 1980-81; Herb Brooks, 1981-82 to 1983-84; Herb Brooks and Craig Patrick, 1984-85; Ted Sator, 1985-86; Ted Sator, Tom Webster and Phil Esposito, 1986-87; Michel Bergeron, 1987-88; Michel Bergeron and Phil Esposito, 1988-89; Roger Neilson, 1989-90 to 1991-92; Roger Neilson and Ron Smith, 1992-93; Mike Keenan, 1993-94; Colin Campbell, 1994-95 to 1996-97; Colin Campbell and John Muckler, 1997-98; John Muckler, 1998-99; John Muckler and John Tortorella, 1999-2000; Ron Low, 2000-01, 2001-02; Bryan Trottier, 2002-03.

General Managers' History

Lester Patrick, 1927-28 to 1945-46; Frank Boucher, 1946-47 to 1954-55; Muzz Patrick, 1955-56 to 1963-64; Emile Francis, 1964-65 to 1974-75; Emile Francis and John Ferguson, 1975-76; John Ferguson, 1976-77, 1977-78; John Ferguson and Fred Shero, 1978-79; Fred Shero, 1979-80; Fred Shero and Craig Patrick, 1980-81; Craig Patrick, 1981-82 to 1985-86; Phil Esposito, 1986-87 to 1988-89; Neil Smith, 1989-90 to 1999-2000; Glen Sather, 2000-01 to date.

Coach

TROTTIER, BRYAN
Coach, New York Rangers. Born in Val Marie, Sask., July 17, 1956.

Former New York Islanders superstar Bryan Trottier was named head coach of the New York Rangers on June 6, 2002. Trottier joined the Rangers after spending the previous four seasons in the Colorado Avalanche organization, serving as an assistant coach. Trottier was responsible for overseeing the specialty team units, which ranked consistently among the NHL's best. As a member of the Avalanche's coaching staff, Trottier was instrumental in keeping the team in the upper echelon of the NHL, including helping the team to the 2001 Stanley Cup championship.

Prior to joining the Colorado organization, Trottier served as head coach and director of hockey operations for the Portland Pirates, American Hockey League affiliate of the Washington Capitals, during the 1997-98 season. Trottier began his coaching career as an assistant coach for the Pittsburgh Penguins from 1994-95 through 1996-97.

During his playing career, Trottier was regarded as one of the best two-way centers in the history of the NHL. He was inducted into the Hockey Hall of Fame in 1997. An 18-year NHL veteran, he appeared in 1,279 regular-season games with the Islanders and Penguins. He scored 524 goals and added 901 assists for 1,425 points. Trottier's playing career is highlighted with six Stanley Cup championships; four with the Islanders (1980, 1981, 1982 and 1983) and two with the Penguins (1991 and 1992). His individual honors include the Calder Trophy as NHL rookie of the year in 1976, the Art Ross Trophy (leading scorer) and the Hart Trophy (MVP) in 1979 and the Conn Smythe Trophy as the 1980 playoff MVP. Trottier was also honored with the King Clancy Memorial Trophy in 1989 for his noteworthy humanitarian contributions.

Coaching Record

		Regular Season				Playoffs		
Season	Team	Games	W	L	T	Games	W	L
1997-98	Portland (AHL)	80	33	33	14	10	5	5

Club Records

Team

(Figures in brackets for season records are games played; records for fewest points, wins, ties, losses, goals, goals against are for 70 or more games)

Most Points **112** 1993-94 (84)
Most Wins **52** 1993-94 (84)
Most Ties **21** 1950-51 (70)
Most Losses **44** 1984-85 (80)
Most Goals **321** 1991-92 (80)
Most Goals Against **345** 1984-85 (80)
Fewest Points **47** 1965-66 (70)
Fewest Wins **17** 1952-53 (70), 1954-55 (70), 1959-60 (70)
Fewest Ties **4** 2001-02 (82)
Fewest Losses **17** 1971-72 (78)
Fewest Goals **150** 1954-55 (70)
Fewest Goals Against **177** 1970-71 (78)

Longest Winning Streak
Overall .... **10** Dec. 19/39-Jan. 13/40, Jan. 19-Feb. 10/73
Home .... **14** Dec. 19/39-Feb. 25/40
Away .... **7** Jan. 12-Feb. 12/35, Oct. 28-Nov. 29/78

Longest Undefeated Streak
Overall .... **19** Nov. 23/39-Jan. 13/40 (14 wins, 5 ties)
Home .... **26** Mar. 29/70-Jan. 31/71 (19 wins, 7 ties)
Away .... **11** Nov. 5/39-Jan. 13/40 (6 wins, 5 ties)

Longest Losing Streak
Overall .... **11** Oct. 30-Nov. 27/43
Home .... **7** Oct. 20-Nov. 14/76, Mar. 24-Apr. 14/93
Away .... **10** Oct. 30-Dec. 23/43, Feb. 2-Mar. 15/61

Longest Winless Streak
Overall .... **21** Jan. 23-Mar. 19/44 (17 losses, 4 ties)
Home .... **10** Jan. 30-Mar. 19/44 (7 losses, 3 ties)
Away .... **16** Oct. 9-Dec. 20/52 (12 losses, 4 ties)

Most Shutouts, Season **13** 1928-29 (44)
Most PIM, Season **2,018** 1989-90 (80)
Most Goals, Game .... **12** Nov. 21/71 (Cal. 1 at NYR 12)

Individual

Most Seasons .... **18** Rod Gilbert
Most Games .... **1,160** Harry Howell
Most Goals, Career .... **406** Rod Gilbert
Most Assists, Career .... **700** Brian Leetch
Most Points, Career .... **1,021** Rod Gilbert (406G, 615A)
Most PIM, Career .... **1,226** Ron Greschner
Most Shutouts, Career .... **49** Ed Giacomin
Longest Consecutive Games Streak .... **560** Andy Hebenton (Oct. 7/55-Mar. 24/63)
Most Goals, Season .... **52** Adam Graves (1993-94)
Most Assists, Season .... **80** Brian Leetch (1991-92)
Most Points, Season .... **109** Jean Ratelle (1971-72; 46G, 63A)
Most PIM, Season .... **305** Troy Mallette (1989-90)
Most Points, Defenseman, Season .... **102** Brian Leetch (1991-92; 22G, 80A)
Most Points, Center, Season .... **109** Jean Ratelle (1971-72; 46G, 63A)
Most Points, Right Wing, Season .... **97** Rod Gilbert (1971-72; 43G, 54A), (1974-75; 36G, 61A)
Most Points, Left Wing, Season .... **106** Vic Hadfield (1971-72; 50G, 56A)
Most Points, Rookie, Season .... **76** Mark Pavelich (1981-82; 33G, 43A)
Most Shutouts, Season .... **13** John Ross Roach (1928-29)
Most Goals, Game .... **5** Don Murdoch (Oct. 12/76), Mark Pavelich (Feb. 23/83)
Most Assists, Game .... **5** Walt Tkaczuk (Feb. 12/72), Rod Gilbert (Mar. 2/75, Mar. 30/75, Oct. 8/76), Don Maloney (Jan. 3/87), Brian Leetch (Apr. 18/95), Wayne Gretzky (Feb. 15/99)
Most Points, Game .... **7** Steve Vickers (Feb. 18/76; 3G, 4A)

Retired Numbers

1	Ed Giacomin	1965-1976
7	Rod Gilbert	1960-1978

All-time Record vs. Other Clubs

Regular Season

	At Home								On Road								Total							
	GP	W	L	T	OL	GF	GA	PTS	GP	W	L	T	OL	GF	GA	PTS	GP	W	L	T	OL	GF	GA	PTS
Anaheim	7	2	4	1	0	19	21	5	7	1	6	0	0	19	30	2	14	3	10	1	0	38	51	7
Atlanta	6	3	3	0	0	21	22	6	6	6	0	0	0	24	13	12	12	9	3	0	0	45	35	18
Boston	298	129	114	55	0	911	842	313	294	95	157	42	0	826	1064	232	592	224	271	97	0	1737	1906	545
Buffalo	62	27	20	15	0	206	169	69	64	18	37	9	0	200	266	45	126	45	57	24	0	406	435	114
Calgary	50	23	22	5	0	177	179	51	48	11	27	10	0	146	215	32	98	34	49	15	0	323	394	83
Carolina	41	24	13	3	1	158	110	52	39	14	22	3	0	126	135	31	80	38	35	6	1	284	245	83
Chicago	284	117	112	55	0	838	803	289	285	115	128	42	0	790	868	272	569	232	240	97	0	1628	1671	561
Colorado	32	19	9	4	0	129	92	42	33	13	16	3	1	126	135	30	65	32	25	7	1	255	227	72
Columbus	1	0	0	1	0	2	2	1	2	1	1	0	0	4	5	2	3	1	1	1	0	6	7	3
Dallas	61	35	15	11	0	210	165	81	60	30	19	10	1	218	184	71	121	65	34	21	1	428	349	152
Detroit	283	133	92	58	0	865	737	324	284	76	163	45	0	697	1001	197	567	209	255	103	0	1562	1738	521
Edmonton	28	9	13	6	0	103	106	24	28	13	12	3	0	94	101	29	56	22	25	9	0	197	207	53
Florida	20	9	7	4	0	60	52	22	21	11	9	1	0	59	56	23	41	20	16	5	0	119	108	45
Los Angeles	57	35	16	6	0	231	166	76	59	26	23	10	0	213	197	62	116	61	39	16	0	444	363	138
Minnesota	2	2	0	0	0	7	3	4	2	2	0	0	0	6	4	4	4	4	0	0	0	13	7	8
Montreal	286	118	114	54	0	831	831	290	286	59	188	39	0	656	1114	157	572	177	302	93	0	1487	1945	447
Nashville	4	2	1	0	1	14	9	5	2	1	1	0	0	9	9	2	6	3	2	0	1	23	18	7
New Jersey	79	40	21	18	0	313	238	98	80	33	42	5	0	269	280	71	159	73	63	23	0	582	518	169
NY Islanders	90	51	29	10	0	341	274	112	91	30	54	7	0	286	362	67	181	81	83	17	0	627	636	179
Ottawa	19	10	9	0	0	62	54	20	19	11	5	3	0	59	51	25	38	21	14	3	0	121	105	45
Philadelphia	105	46	36	23	0	342	308	115	103	37	52	14	0	287	339	88	208	83	88	37	0	629	647	203
Phoenix	28	17	9	2	0	125	100	36	30	13	13	4	0	98	105	30	58	30	22	6	0	223	205	66
Pittsburgh	95	49	37	9	0	378	323	107	95	40	39	14	2	351	346	96	190	89	76	23	2	729	669	203
St. Louis	59	44	9	6	0	243	140	94	61	28	24	9	0	194	179	65	120	72	33	15	0	437	319	159
San Jose	9	6	2	1	0	38	28	13	11	8	2	1	0	42	25	17	20	14	4	2	0	80	53	30
Tampa Bay	23	12	9	2	0	78	73	26	21	10	8	3	0	76	73	23	44	22	17	5	0	154	146	49
Toronto	279	118	105	56	0	856	821	292	278	83	155	39	1	731	959	206	557	201	260	95	1	1587	1780	498
Vancouver	53	37	11	5	0	233	136	79	50	33	14	3	0	203	160	69	103	70	25	8	0	436	296	148
Washington	78	37	32	8	1	297	273	83	80	31	40	9	0	263	301	71	158	68	72	17	1	560	574	154
Defunct Clubs	139	87	30	22	0	460	290	196	139	82	34	23	0	441	291	187	278	169	64	45	0	901	581	383
Totals	**2578**	**1241**	**894**	**440**	**3**	**8548**	**7367**	**2925**	**2578**	**931**	**1291**	**351**	**5**	**7513**	**8868**	**2218**	**5156**	**2172**	**2185**	**791**	**8**	**16061**	**16235**	**5143**

Playoffs

	Series	W	L	GP	W	L	T	GF	GA	Last Mtg.	Rnd.	Result
Boston	9	3	6	42	18	22	2	104	114	1973	QF	W 4-1
Buffalo	1	0	1	3	1	2	0	6	11	1978	PRE	L 1-2
Calgary	1	1	0	4	3	1	0	14	8	1980	PRE	W 3-1
Chicago	5	1	4	24	10	14	0	54	66	1973	SF	L 1-4
Colorado	1	1	0	6	4	2	0	25	19	1995	CQF	W 4-2
Detroit	5	1	4	23	10	13	0	49	57	1950	F	L 3-4
Florida	1	1	0	5	4	1	0	13	10	1997	CQF	W 4-1
Los Angeles	2	2	0	6	5	1	0	32	14	1981	PRE	W 3-1
Montreal	14	7	7	61	25	34	2	158	188	1996	CQF	W 4-2
New Jersey	3	3	0	19	12	7	0	56	46	1997	CSF	W 4-1
NY Islanders	8	3	5	39	19	20	0	132	129	1994	CQF	W 4-0
Philadelphia	10	4	6	47	20	27	0	153	157	1997	CF	L 1-4
Pittsburgh	3	0	3	15	3	12	0	45	64	1996	CSF	L 1-4
St. Louis	1	1	0	6	4	2	0	29	22	1981	QF	W 4-2
Toronto	8	5	3	35	19	16	0	86	86	1971	QF	W 4-2
Vancouver	1	1	0	7	4	3	0	21	19	1994	F	W 4-3
Washington	4	2	2	22	11	11	0	71	75	1994	CSF	W 4-1
Defunct Clubs	9	6	3	22	11	7	4	43	29			
Totals	**86**	**42**	**44**	**386**	**183**	**195**	**8**	**1091**	**1114**			

Calgary totals include Atlanta Flames, 1972-73 to 1979-80. Carolina totals include Hartford, 1979-80 to 1996-97.
Colorado totals include Quebec, 1979-80 to 1994-95. Dallas totals include Minnesota North Stars, 1967-68 to 1992-93.
New Jersey totals include Kansas City, 1974-75 to 1975-76, and Colorado Rockies, 1976-77 to 1981-82.
Phoenix totals include Winnipeg, 1979-80 to 1995-96.

Playoff Results 2002-1998

(Last playoff appearance: 1997)

Abbreviations: Round: F – Final; **CF** – conference final; **CSF** – conference semi-final; **CQF** – conference quarter-final; **SF** – semi-final; **QF** – quarter-final; **PRE** – preliminary round.

2001-02 Results

Oct.	5	at Carolina	1-3		31	at Phoenix	0-5
	7	Buffalo	5-4*	**Jan.**	2	at Edmonton	1-4
	10	Washington	2-5		3	at Colorado	2-3*
	13	at Ottawa	2-2		5	at Pittsburgh	1-4
	15	at Montreal	2-1		9	Los Angeles	0-4
	17	New Jersey	4-3*		12	at Philadelphia	2-4
	19	at Atlanta	4-3		14	Columbus	2-2
	20	at Tampa Bay	2-5		16	at Columbus	0-2
	22	San Jose	1-5		17	at New Jersey	4-6
	25	at St. Louis	1-5		22	at NY Islanders	5-4
	27	at Boston	2-1*		23	Boston	8-4
	29	Dallas	4-2		26	Washington	6-3
	31	Florida	1-3		28	Tampa Bay	0-1
Nov.	2	at Carolina	2-3		30	NY Islanders	3-6
	3	at Florida	5-3	**Feb.**	6	at Detroit	1-3
	6	Minnesota	3-1		8	at Atlanta	2-1
	8	at NY Islanders	6-2		10	Pittsburgh	4-3
	10	at Buffalo	4-2		13	at Dallas	2-4
	11	Montreal	3-2*		26	New Jersey	3-4
	14	Philadelphia	4-2		28	Ottawa	0-3
	17	at Pittsburgh	0-1*	**Mar.**	2	Philadelphia	6-5
	18	Atlanta	6-2		4	Calgary	3-5
	20	Colorado	5-3		5	at Minnesota	3-2*
	23	at Washington	2-6		7	at Chicago	1-5
	25	Anaheim	2-3		9	at Pittsburgh	2-3*
	27	at Buffalo	2-2		11	Montreal	2-1
	29	Carolina	5-0		13	Boston	1-3
Dec.	1	at Montreal	3-1		16	at New Jersey	1-3
	2	Tampa Bay	1-0		17	Detroit	3-5
	4	at Washington	2-5		19	Vancouver	1-3
	6	Toronto	3-6		21	at Ottawa	5-2
	8	at Toronto	3-4		22	Atlanta	2-5
	10	Carolina	3-4*		25	at NY Islanders	2-4
	12	Nashville	2-4		27	Philadelphia	2-4
	15	Buffalo	4-2		30	at Florida	4-2
	17	Florida	4-2	**Apr.**	1	at Tampa Bay	6-4
	19	New Jersey	2-2		4	at Toronto	4-2
	21	NY Islanders	1-2		6	at Boston	6-4
	23	Ottawa	3-2		8	Pittsburgh	3-2
	28	at San Jose	5-3		10	Toronto	2-7
	29	at Los Angeles	5-4		13	at Philadelphia	1-2

* – Overtime

Entry Draft Selections 2002-1988

2002
Pick
33 Lee Falardeau
81 Marcus Jonasen
127 Nate Guenin
143 Mike Walsh
177 Jake Taylor
194 Kim Hirschovits
226 Joseph Crabb
240 Petr Prucha
270 Rob Flynn

2001
Pick
10 Dan Blackburn
40 Fedor Tutin
79 Garth Murray
113 Bryce Lampman
139 Shawn Collymore
176 Marek Zidlicky
206 Petr Preucil
226 Pontus Petterstrom
230 Leonid Zhvachkin
238 Ryan Hollweg
269 Juris Stals

2000
Pick
64 Filip Novak
95 Dominic Moore
112 Premysl Duben
140 Nathan Martz
143 Brandon Snee
175 Sven Helfenstein
205 Henrik Lundqvist
238 Dan Eberly
269 Martin Richter

1999
Pick
4 Pavel Brendl
9 Jamie Lundmark
59 David Inman
79 Johan Asplund
90 Patrick Aufiero
137 Garrett Bembridge
177 Jay Dardis
197 Arto Laatikainen
226 Yevgeny Gusakov
251 Petter Henning
254 Alexei Bulatov

1998
Pick
7 Manny Malhotra
40 Randy Copley
66 Jason Labarbera
114 Boyd Kane
122 Patrick Leahy
131 Tomas Kloucek
180 Stefan Lundqvist
207 Johan Witehall
235 Jan Mertzig

1997
Pick
19 Stefan Cherneski
46 Wes Jarvis
73 Burke Henry
93 Tomi Kallarsson
126 Jason McLean
134 Johan Lindbom
136 Mike York
154 Shawn Degagne
175 Johan Holmqvist
182 Mike Mottau
210 Andrew Proskurnicki
236 Richard Miller

1996
Pick
22 Jeff Brown
48 Daniel Goneau
76 Dmitri Subbotin
131 Colin Pepperall
158 Ola Sandberg
185 Jeff Dessner
211 Ryan McKie
237 Ronnie Sundin

1995
Pick
39 Christian Dube
65 Mike Martin
91 Marc Savard
110 Alexei Vasiliev
117 Dale Purinton
143 Peter Slamiar
169 Jeff Heil
195 Ilja Gorokhov
221 Bob Maudie

1994
Pick
26 Dan Cloutier
52 Rudolf Vercik
78 Adam Smith
100 Alexander Korobolin
104 Sylvain Blouin
130 Martin Ethier
135 Yuri Litvinov
156 David Brosseau
182 Alexei Lazarenko
208 Craig Anderson
209 Vitali Yeremeyev
234 Eric Boulton
260 Radoslav Kropac
267 Jamie Butt
286 Kim Johnsson

1993
Pick
8 Niklas Sundstrom
34 Lee Sorochan
61 Maxim Galanov
86 Sergei Olimpiyev
112 Gary Roach
138 Dave Trofimenkoff
162 Sergei Kondrashkin
164 Todd Marchant
190 Eddy Campbell
216 Ken Shepard
242 Andrei Kudinov
261 Pavel Komarov
268 Maxim Smelnitsky

1992
Pick
24 Peter Ferraro
48 Mattias Norstrom
72 Eric Cairns
85 Chris Ferraro
120 Dmitri Starostenko
144 David Dal Grande
168 Matt Oates
192 Mickey Elick
216 Daniel Brierley
240 Vladimir Vorobiev

1991
Pick
15 Alexei Kovalev
37 Darcy Werenka
96 Corey Machanic
125 Fredrik Jax
128 Barry Young
147 John Rushin
169 Corey Hirsch
191 Vyacheslav Uvayev
213 Jamie Ram
235 Vitali Chinakhov
257 Brian Wiseman

1990
Pick
13 Michael Stewart
34 Doug Weight
55 John Vary
69 Jeff Nielsen
76 Rick Willis
85 Sergei Zubov
99 Lubos Rob
118 Jason Weinrich
139 Brian Lonsinger
160 Todd Hedlund
181 Andrew Silverman
202 Jon Hillebrandt
223 Brett Lievers
244 Sergei Nemchinov

1989
Pick
20 Steven Rice
40 Jason Prosofsky
45 Rob Zamuner
49 Louie DeBrusk
67 Jim Cummins
88 Aaron Miller
118 Joby Messier
139 Greg Leahy
160 Greg Spenrath
181 Mark Bavis
202 Roman Oksiuta
223 Steve Locke
244 Ken MacDermid

1988
Pick
22 Troy Mallette
26 Murray Duval
68 Tony Amonte
99 Martin Bergeron
110 Dennis Vial
131 Mike Rosati
152 Eric Couvrette
194 Paul Cain
202 Eric Fenton
215 Peter Fiorentino
236 Keith Slifstein

President and General Manager

SATHER, GLEN
President and General Manager, New York Rangers
Born in High River, Alta., Sept. 2, 1943.

Glen Sather, who spent parts of four seasons with the New York Rangers as a player from 1970 to 1974, became the franchise's 12th president and tenth general manager on June 1, 2000. He joined the club following a 24-year career with the Edmonton Oilers, where he was the architect of five Stanley Cup championships between 1984 and 1990. One of the most respected executives in the National Hockey League, Sather was honored for his tremendous achievements in 1997 by becoming the first member of the Oilers organization to be selected to the Hockey Hall of Fame.

Named coach and vice president of hockey operations for the Oilers when the franchise joined the NHL in June of 1979, Sather became general manager and club president in May of 1980. He coached through the 1988-89 season and also returned for 60 games behind the bench in 1993-94. Sather-coached teams won the Stanley Cup four times in the 1980s. As general manager, Sather was instrumental in the Oilers' fifth Cup triumph in 1990.

He played for six different teams during a 10-year NHL career. He scored 80 goals in 658 games.

NHL Coaching Record

		Regular Season				Playoffs		
Season	**Team**	**Games**	**W**	**L**	**T**	**Games**	**W**	**L**
1979-80	Edmonton	80	28	39	13	3	0	3
1980-81	Edmonton	62	25	26	11	9	5	4
1981-82	Edmonton	80	48	17	15	5	2	3
1982-83	Edmonton	80	47	21	12	16	11	5
1983-84	Edmonton	80	57	18	5	19	15	4*
1984-85	Edmonton	80	49	20	11	18	15	3*
1985-86	Edmonton	80	56	17	7	10	6	4
1986-87	Edmonton	80	50	24	6	21	16	5*
1987-88	Edmonton	80	44	25	11	18	16	2*
1988-89	Edmonton	80	38	34	8	7	3	4
1993-94	Edmonton	60	22	27	11	....	....	....
	NHL Totals	**842**	**464**	**268**	**110**	**126**	**89**	**37**

* Stanley Cup win.

Club Directory

Madison Square Garden

New York Rangers
14th Floor
2 Pennsylvania Plaza
New York, New York 10121
Phone **212/465-6000**
PR FAX 212/465-6494
www.newyorkrangers.com
Capacity: 18,200

Office of the Chairman, Madison Square Garden
President & CEO, Cablevision Systems Corporation; Chairman, Madison Square Garden . . . . James L. Dolan
Vice Chairman, Cablevision Systems Corporation, Vice Chairman, Madison Square Garden . . . . Robert S. Lemle
President and General Manager, New York Rangers . Glen Sather
President, Sports Team Operations . . . . Steve Mills
President, MSG/Radio City Entertainment . . . . Seth Abraham
President and General Manager, New York Knicks . Scott Layden

Team Executive Management
Governor . . . . James L. Dolan
President & General Manager, Alternate Governor . Glen Sather
President, Sports Team Operations, Alternate Governor . . . . Steve Mills
Senior Vice President, Legal Affairs, MSG . . . . Marc Schoenfeld
Senior Vice President, Business Operations . . . . Mark Piazza
Vice President, Marketing . . . . Jeanie Baumgartner
Vice President, Controller . . . . John Cudmore
Vice President, Public Relations . . . . John Rosasco

Madison Square Garden Executive Management
President, MSG Facilities . . . . Robert Russo
Executive Vice President, Advertising Sales . . . . Joe Gangone
Executive Vice President, MSG Networks . . . . Mike McCarthy
Executive Vice President, Finance . . . . Robert Pollichino
Senior Vice President, Marketing . . . . Betsy Bruce
Senior Vice President, Sports & Facility Event Sales . . Joel Fisher
Senior Vice President, Communications . . . . Barry Watkins

Hockey Club Personnel
Vice President, Player Personnel and Assistant General Manager . . . . Don Maloney
Vice President, Player Development . . . . Tom Renney
Head Coach . . . . Bryan Trottier
Assistant Coaches . . . . Ted Green, Terry O'Reilly and Jim Schoenfeld
Vice President, Hockey Administration and Scouting . . Peter Stephan
Goaltending Analyst . . . . Sam St. Laurent
Amateur Scouting Staff . . . . Rich Brown, Ray Clearwater, Andre Beaulieu, Jan Gajdosik, Ernie Gare, Martin Madden Jr., Christer Rockstom, Bob Crocker, Jamie McDonald
Pro Scouting Staff . . . . Dave Brown, Harry Howell, Gilles Leger, Ron Low, Brad Park, Shanon Sather
Medical Trainer . . . . Jim Ramsay
Equipment Manager . . . . Acacio Marques
Assistant Equipment Manager . . . . James Johnson
Massage Therapist . . . . Bruce Lifrieri
Strength and Conditioning Coordinator . . . . Reg Grant
Video Analyst . . . . Jerry Dineen
Manager, Practice Facility . . . . Pat Boller
Locker Room Assistant . . . . Jason Devenney

Operations
Director, Business Operations . . . . Barbara Dand
Director, Team Operations . . . . Darren Blake
Executive Assistant to the President and GM . . . . Sara Adamson
Manager, Accounting . . . . Nicole Florit
Operations Assistant . . . . Victor Saljanin
Operations Assistant . . . . Chris Smith

Public Relations
Director, Public Relations . . . . Jason Vogel
Public Relations Coordinator . . . . Keith Soutar
Public Relations Coordinator . . . . Jennifer Schoenfeld

Marketing
Director, Marketing Partnerships . . . . Rob Scolaro
Manager, Marketing Partnerships . . . . Kelly Jutras
Manager, Game Presentation . . . . Ryan Halkett
Manager, Website . . . . Jeff Schwartzenberg
Manager, Marketing . . . . Janet Duch
Marketing Assistant . . . . Adam Evert

Community Development
Director, Community Development . . . . Rob Capilli
Director, Special Projects/Community Relations Representative . . . . Rod Gilbert
Manager, Community Development . . . . Anthony Triano
Community Development Assistant . . . . Jan Greenberg

Medical/Training Staff
Team Physician and Orthopedic Surgeon . . . . Dr. Andrew Feldman
Assistant Team Physician . . . . Dr. Anthony Maddalo
Medical Consultant . . . . Dr. Ronald Weissman
Team Dentists . . . . Drs. Irwin Miller, Don Soloman, Jeff Shapiro

Additional Information
Television Network . . . . MSG Network
Radio Network . . . . MSG Radio
Practice Facility . . . . TBA

Ottawa Senators

2001-02 Results: 39W-27L-9T-7OTL 94PTS.
Third, Northeast Division

Year-by-Year Record

Season	GP	Home W	Home L	Home T	Home OL	Road W	Road L	Road T	Road OL	Overall W	Overall L	Overall T	Overall OL	GF	GA	Pts.	Finished	Playoff Result
2001-02	82	21	13	3	4	18	14	6	3	39	27	9	7	243	208	94	3rd, Northeast Div.	Lost Conf. Semi-Final
2000-01	82	26	7	5	3	22	14	4	1	48	21	9	4	274	205	109	1st, Northeast Div.	Lost Conf. Quarter-Final
1999-2000	82	24	10	5	2	17	18	6	0	41	28	11	2	244	210	95	2nd, Northeast Div.	Lost Conf. Quarter-Final
1998-99	82	22	11	8	...	22	12	7	...	44	23	15	...	239	179	103	1st, Northeast Div.	Lost Conf. Quarter-Final
1997-98	82	18	16	7	...	16	17	8	...	34	33	15	...	193	200	83	5th, Northeast Div.	Lost Conf. Semi-Final
1996-97	82	16	17	8	...	15	19	7	...	31	36	15	...	226	234	77	3rd, Northeast Div.	Lost Conf. Quarter-Final
1995-96	82	8	28	5	...	10	31	0	...	18	59	5	...	191	291	41	6th, Northeast Div.	Out of Playoffs
1994-95	48	5	16	3	...	4	18	2	...	9	34	5	...	117	174	23	7th, Northeast Div.	Out of Playoffs
1993-94	84	8	30	4	...	6	31	5	...	14	61	9	...	201	397	37	7th, Northeast Div.	Out of Playoffs
1992-93	84	9	29	4	...	1	41	0	...	10	70	4	...	202	395	24	6th, Adams Div.	Out of Playoffs

2002-03 Schedule

Month	Day	Date	Opponent	Month	Day	Date	Opponent
Oct.	Thu.	10	New Jersey		Thu.	9	at Calgary
	Sat.	12	at Toronto		Sat.	11	at Edmonton
	Tue.	15	Phoenix		Tue.	14	Tampa Bay
	Wed.	23	Carolina		Thu.	16	Anaheim
	Thu.	24	at Boston		Sat.	18	Washington
	Sat.	26	at Montreal		Mon.	20	at Tampa Bay
	Tue.	29	at Philadelphia		Wed.	22	at Florida
	Wed.	30	Pittsburgh		Thu.	23	at Atlanta
Nov.	Sat.	2	Detroit		Sat.	25	Buffalo
	Wed.	6	at Colorado		Mon.	27	at Dallas
	Fri.	8	Los Angeles		Wed.	29	at Anaheim
	Sat.	9	at Boston		Thu.	30	at Los Angeles
	Tue.	12	at NY Islanders	Feb.	Wed.	5	at NY Rangers
	Thu.	14	Florida		Thu.	6	Philadelphia
	Sat.	16	Buffalo		Sat.	8	Atlanta
	Tue.	19	at Carolina		Wed.	12	at Pittsburgh
	Thu.	21	Montreal		Thu.	13	Edmonton
	Sat.	23	Columbus		Sat.	15	at Toronto
	Mon.	25	Toronto		Mon.	17	NY Rangers
	Wed.	27	at NY Islanders		Wed.	19	at New Jersey
	Fri.	29	at Washington		Thu.	20	Florida
	Sat.	30	NY Islanders		Sat.	22	Nashville
Dec.	Wed.	4	at Chicago		Tue.	25	Minnesota
	Thu.	5	at St. Louis		Thu.	27	Dallas
	Sat.	7	Carolina	Mar.	Sat.	1	Tampa Bay
	Tue.	10	at Buffalo		Tue.	4	Toronto
	Thu.	12	at Boston		Sat.	8	at Pittsburgh
	Sat.	14	New Jersey		Sun.	9	Pittsburgh
	Mon.	16	Montreal		Tue.	11	Boston
	Wed.	18	at New Jersey		Thu.	13	NY Rangers
	Thu.	19	San Jose		Sat.	15	NY Islanders
	Sat.	21	at Philadelphia*		Sun.	16	at Detroit
	Mon.	23	Philadelphia		Tue.	18	at Carolina
	Thu.	26	at Buffalo		Fri.	21	at Atlanta
	Fri.	27	Montreal		Sat.	22	at Florida
	Mon.	30	at Nashville		Tue.	25	Colorado
	Tue.	31	at Tampa Bay		Fri.	28	Washington
Jan.	Thu.	2	Atlanta		Sat.	29	at Montreal
	Sat.	4	Buffalo	Apr.	Tue.	1	Boston
	Mon.	6	at NY Rangers		Thu.	3	at Washington
	Wed.	8	at Vancouver		Sat.	5	at Toronto

* Denotes afternoon game.

With a career-high 37 goals last season, Daniel Alfredsson was the Senators' leading goal scorer for the first time since his rookie season. His 71 points tied a career high and was also the team's top total.

Franchise date: December 16, 1991

NORTHEAST DIVISION

11th NHL Season

2002-03 Player Personnel

FORWARDS	HT	WT	S	Place of Birth	Date	2001-02 Club
ALFREDSSON, Daniel	5-11	195	R	Goteborg, Sweden	12/11/72	Ottawa-Sweden
ARVEDSON, Magnus	6-2	198	L	Karlstad, Sweden	11/25/71	Ottawa-Sweden
BALA, Chris	6-1	180	L	Alexandria, VA	9/24/78	Ottawa-Grand Rapids
BONK, Radek	6-3	210	L	Krnov, Czech.	1/9/76	Ottawa
DAHLMAN, Toni	6-0	193	R	Helsinki, Finland	9/3/79	Ottawa-Grand Rapids
FISHER, Mike	6-1	193	R	Peterborough, Ont.	6/5/80	Ottawa
GIROUX, Alexandre	6-3	182	L	Quebec, Que.	6/16/81	Grand Rapids
HAVLAT, Martin	6-1	190	L	Mlada Boleslav, Czech.	4/19/81	Ottawa-Czech Republic
HOSSA, Marian	6-1	199	L	Stara Lubovna, Czech.	1/12/79	Dukla Trencin-Ottawa-Slovakia
HYMOVITZ, David	5-11	170	L	Randolph, MA	5/30/74	Grand Rapids
KELLY, Chris	6-0	190	L	Toronto, Ont.	11/11/80	Muskegon-Grand Rapids
LANGFELD, Josh	6-3	216	R	Fridley, MN	7/17/77	Ottawa-Grand Rapids
MARTINS, Steve	5-9	175	L	Gatineau, Que.	4/13/72	Ottawa-Grand Rapids
MURPHY, Joe	6-0	200	R	Didsbury, Alta.	1/21/75	Grand Rapids
NEIL, Christopher	6-0	213	R	Markdale, Ont.	6/18/79	Ottawa
SCHASTLIVY, Petr	6-1	204	L	Angarsk, USSR	4/18/79	Ottawa-Grand Rapids
SMYTH, Brad	6-0	195	R	Ottawa, Ont.	3/13/73	Hartford
SPEZZA, Jason	6-2	214	R	Mississauga, Ont.	6/13/83	Windsor-Belleville-Grand Rapids
ULMER, Jeff	5-11	195	R	Wilcox, Sask.	4/27/77	Grand Rapids
VAN ALLEN, Shaun	6-1	205	L	Calgary, Alta.	8/29/67	Dallas-Montreal
VERMETTE, Antoine	6-0	184	L	St-Agapit, Que.	7/20/82	Victoriaville
WHITE, Todd	5-10	194	L	Kanata, Ont.	5/21/75	Ottawa
DEFENSEMEN						
BOUMEDIENNE, Josef	6-1	200	L	Stockholm, Sweden	1/12/78	N.J.-Alb-T.B.-Sprfld
BROOKBANK, Wade	6-4	219	L	Lanigan, Sask.	9/29/77	Grand Rapids
CHARA, Zdeno	6-9	255	L	Trencin, Czech.	3/18/77	Dukla Trencin-Ottawa
DEMIDOV, Ilja	6-3	185	L	Moscow, USSR	4/14/79	Grand Rapids-Mobile
GRUDEN, John	6-0	203	L	Virginia, MN	6/4/70	Grand Rapids
HNIDY, Shane	6-2	210	R	Neepawa, Man.	11/8/75	Ottawa
KWIATKOWSKI, Joel	6-2	210	L	Kindersley, Sask.	3/22/77	Ottawa-Grand Rapids
LESCHYSHYN, Curtis	6-1	220	L	Thompson, Man.	9/21/69	Ottawa
PHILLIPS, Chris	6-3	215	L	Calgary, Alta.	3/9/78	Ottawa
POTHIER, Brian	6-0	195	R	New Bedford, MA	4/15/77	Atlanta-Chicago
RACHUNEK, Karel	6-2	202	R	Gottwaldov, Czech.	8/27/79	Ottawa
REDDEN, Wade	6-2	205	L	Lloydminster, Sask.	6/12/77	Ottawa
SALO, Sami	6-3	215	R	Turku, Finland	9/2/74	Ottawa-Finland
SCHUBERT, Christoph	6-3	198	L	Munich, West Germany	2/5/82	Munchen
VAUCLAIR, Julien	6-0	198	L	Delemont, Switz.	10/2/79	Grand Rapids-Switzerland
VOLCHENKOV, Anton	6-0	209	L	Moscow, USSR	2/25/82	Krylja Sovetov 2-Krylja Sovetov

GOALTENDERS	HT	WT	C	Place of Birth	Date	2001-02 Club
CHOUINARD, Mathieu	6-1	211	L	Laval, Que.	4/11/80	Grand Rapids
EMERY, Ray	6-3	192	L	Cayuga, Ont.	9/28/82	Sault Ste. Marie
HURME, Jani	6-0	187	L	Turku, Finland	1/7/75	Ottawa-Finland
LAJEUNESSE, Simon	6-0	175	L	Quebec, Que.	1/22/81	Ottawa-Mobile-Grand Rapids
LALIME, Patrick	6-3	185	L	St-Bonaventure, Que.	7/7/74	Ottawa
PRUSEK, Martin	6-1	176	L	Ostrava, Czech.	12/11/75	Ottawa-Grand Rapids

General Manager

MUCKLER, JOHN
General Manager, Ottawa Senators. Born in Midland, Ont., April 3, 1934.

John Muckler was named the sixth general manager in Senators history on June 12, 2002. Most recently, Muckler had served as coach of the New York Rangers from 1997-98 to 1999-2000. Previously, he was general manager of the Buffalo Sabres from 1993 to 1997, and was named NHL executive of the year by *The Sporting News* for the 1996-97 season. Muckler is the first g.m. hired by the Senators to have previous NHL experience as a general manager.

Working for Glen Sather, Muckler enjoyed Edmonton's great 1980s run. He was an assistant coach with the Stanley Cup winners in 1984 and 1985, and designated co-coach during the 1987 and 1988 championship seasons. When Sather gave up the Oilers' coaching reins in 1989, Muckler stepped in and led the team to its fifth Stanley Cup in seven years. In 1991, he left the Oilers for the Buffalo Sabres.

Muckler has been involved in professional hockey since the 1949-50 season. He was a defenseman in the minor leagues for 13 seasons, playing the bulk of his career in the old Eastern Hockey League. His professional coaching career began while he was still a player in 1959 when he took over the New York Rovers of the EHL. He had great success with the team in the 1960s when they were known as the Long Island Ducks. Muckler joined the Minnesota North Stars after NHL expansion in 1967 and spent six seasons in the organization, mostly as a coach and g.m. in the minor leagues. His first NHL coaching job came with the North Stars midway through the 1968-69 season. He later worked in the Rangers and Canucks organizations before joining the Oilers as coach of their Wichita farm club in 1981.

NHL Coaching Record

		Regular Season				Playoffs		
Season	Team	Games	W	L	T	Games	W	L
1968-69	Minnesota	35	6	23	6	....	....	....
1989-90	Edmonton	80	38	28	14	22	16	6*
1990-91	Edmonton	80	37	37	6	18	9	9
1991-92	Buffalo	52	22	22	8	7	3	4
1992-93	Buffalo	84	38	36	10	8	4	4
1993-94	Buffalo	84	43	32	9	7	3	4
1994-95	Buffalo	48	22	19	7	5	1	4
1997-98	NY Rangers	25	8	15	2	....	....	....
1998-99	NY Rangers	82	33	38	11	....	....	....
1999-2000	NY Rangers	78	29	38	11	....	....	....
	NHL Totals	**648**	**276**	**288**	**84**	**67**	**36**	**31**

* Stanley Cup win.

2001-02 Scoring

* - rookie

Regular Season

Pos	#	Player	Team	GP	G	A	Pts	+/–	PIM	PP	SH	GW	GT	S	%
R	11	Daniel Alfredsson	OTT	78	37	34	71	3	45	9	1	4	2	243	15.2
C	14	Radek Bonk	OTT	82	25	45	70	3	52	6	2	5	0	170	14.7
R	18	Marian Hossa	OTT	80	31	35	66	11	50	9	1	4	1	278	11.2
R	9	Martin Havlat	OTT	72	22	28	50	–7	66	9	0	6	1	145	15.2
C	28	Todd White	OTT	81	20	30	50	12	24	4	0	1	0	147	13.6
L	15	Shawn McEachern	OTT	80	15	31	46	9	52	5	0	3	0	196	7.7
L	20	Magnus Arvedson	OTT	74	12	27	39	27	35	0	0	1	0	121	9.9
D	6	Wade Redden	OTT	79	9	25	34	22	48	4	1	1	0	156	5.8
C	12	Mike Fisher	OTT	58	15	9	24	8	55	0	3	4	0	123	12.2
D	3	Zdeno Chara	OTT	75	10	13	23	30	156	4	1	2	0	105	9.5
L	26	Benoit Brunet	MTL	16	0	2	2	–4	4	0	0	0	0	19	0.0
			DAL	32	4	9	13	5	8	0	1	1	0	33	12.1
			OTT	13	5	3	8	–4	0	1	0	2	0	20	25.0
			TOTAL	61	9	14	23	–3	12	1	1	3	0	72	12.5
D	4	Chris Phillips	OTT	63	6	16	22	5	29	1	0	1	0	103	5.8
D	5	Sami Salo	OTT	66	4	14	18	1	14	1	1	2	0	122	3.3
D	23	Karel Rachunek	OTT	51	3	15	18	7	24	1	0	2	0	55	5.5
R	25	* Chris Neil	OTT	72	10	7	17	5	231	1	0	0	0	56	17.9
C	36	Juha Ylonen	T.B.	65	3	10	13	–10	8	0	0	0	0	75	4.0
			OTT	15	1	1	2	–1	2	0	1	0	0	22	4.5
			TOTAL	80	4	11	15	–11	10	0	1	0	0	97	4.1
L	33	Chris Herperger	OTT	72	4	9	13	4	43	0	0	0	0	81	4.9
D	7	Curtis Leschyshyn	OTT	79	1	9	10	–5	44	0	0	0	0	59	1.7
D	27	Ricard Persson	OTT	34	2	7	9	3	42	0	0	0	0	35	5.7
R	17	Bill Muckalt	OTT	70	0	8	8	–3	46	0	0	0	0	73	0.0
R	16	Jody Hull	OTT	24	2	2	4	0	6	0	0	1	0	12	16.7
D	34	Shane Hnidy	OTT	33	1	1	2	–10	57	0	0	0	0	34	2.9
C	21	Steve Martins	OTT	14	1	0	1	1	4	0	0	0	0	11	9.1
L	13	Petr Schastlivy	OTT	1	0	1	1	1	0	0	0	0	0	0	0.0
L	52	* Chris Bala	OTT	6	0	1	1	1	0	0	0	0	0	2	0.0
R	10	* Toni Dahlman	OTT	10	0	1	1	–1	0	0	0	0	0	5	0.0
R	39	* Josh Langfeld	OTT	1	0	0	0	0	2	0	0	0	0	5	0.0
D	29	* Joel Kwiatkowski	OTT	11	0	0	0	5	12	0	0	0	0	9	0.0

Goaltending

No.	Goaltender	GPI	Mins	Avg	W	L	T	EN	SO	GA	SA	S%	G	A	PIM
1	* Simon Lajeunesse	1	24	0.00	0	0	0	0	0	0	9	1.000	0	0	0
35	Jani Hurme	25	1309	2.48	12	9	1	2	3	54	578	.907	0	0	17
40	Patrick Lalime	61	3583	2.48	27	24	8	1	7	148	1521	.903	0	1	19
31	* Martin Prusek	1	62	2.90	0	1	0	0	0	3	15	.800	0	0	0
	Totals	**82**	**4991**	**2.50**	**39**	**34**	**9**	**3**	**10**	**208**	**2126**	**.902**			

Playoffs

Pos	#	Player	Team	GP	G	A	Pts	+/–	PIM	PP	SH	GW	GT	S	%
R	11	Daniel Alfredsson	OTT	12	7	6	13	6	4	3	0	3	0	55	12.7
R	18	Marian Hossa	OTT	12	4	6	10	2	2	1	0	0	0	42	9.5
C	14	Radek Bonk	OTT	12	3	7	10	–3	6	2	0	1	0	26	11.5
R	9	Martin Havlat	OTT	12	2	5	7	0	14	2	0	2	1	20	10.0
D	6	Wade Redden	OTT	12	3	2	5	4	6	1	0	1	0	8	37.5
C	36	Juha Ylonen	OTT	12	0	5	5	2	2	0	0	0	0	8	0.0
C	28	Todd White	OTT	12	2	2	4	4	6	0	0	0	0	29	6.9
L	15	Shawn McEachern	OTT	12	0	4	4	3	2	0	0	0	0	15	0.0
C	12	Mike Fisher	OTT	10	2	1	3	4	0	0	0	0	0	25	8.0
D	5	Sami Salo	OTT	12	2	1	3	4	4	0	0	0	0	19	10.5
L	20	Magnus Arvedson	OTT	12	2	1	3	–2	4	0	0	0	0	16	12.5
L	26	Benoit Brunet	OTT	12	0	3	3	8	0	0	0	0	0	12	0.0
R	16	Jody Hull	OTT	12	1	1	2	2	2	0	0	0	0	8	12.5
D	34	Shane Hnidy	OTT	12	1	1	2	–1	12	0	0	0	0	7	14.3
D	3	Zdeno Chara	OTT	10	0	1	1	4	12	0	0	0	0	17	0.0
D	7	Curtis Leschyshyn	OTT	12	0	1	1	2	0	0	0	0	0	13	0.0
D	27	Ricard Persson	OTT	2	0	0	0	0	15	0	0	0	0	1	0.0
C	21	Steve Martins	OTT	2	0	0	0	0	0	0	0	0	0	0	0.0
D	4	Chris Phillips	OTT	12	0	0	0	3	12	0	0	0	0	23	0.0
R	25	* Chris Neil	OTT	12	0	0	0	–1	12	0	0	0	0	6	0.0

Goaltending

No.	Goaltender	GPI	Mins	Avg	W	L	EN	SO	GA	SA	S%	G	A	PIM
40	Patrick Lalime	12	778	1.39	7	5	0	4	18	332	.946	0	0	0
	Totals	**12**	**780**	**1.38**	**7**	**5**	**0**	**4**	**18**	**332**	**.946**			

General Managers' History

Mel Bridgman, 1992-93; Randy Sexton, 1993-94, 1994-95; Randy Sexton and Pierre Gauthier, 1995-96; Pierre Gauthier, 1996-97, 1997-98; Rick Dudley, 1998-99; Marshall Johnston, 1999-2000 to 2001-02; John Muckler, 2002-03.

Club Records

Team

(Figures in brackets for season records are games played; records for fewest points, wins, ties, losses, goals, goals against are for 70 or more games)

Record		
Most Points	**109**	2000-01 (82)
Most Wins	**48**	2000-01 (82)
Most Ties	**15**	1996-97 (82), 1997-98 (82), 1998-99 (82)
Most Losses	**70**	1992-93 (84)
Most Goals	**274**	2000-01 (82)
Most Goals Against	**397**	1993-94 (84)
Fewest Points	**24**	1992-93 (84)
Fewest Wins	**10**	1992-93 (84)
Fewest Ties	**4**	1992-93 (84)
Fewest Losses	**21**	2000-01 (82)
Fewest Goals	**191**	1995-96 (82)
Fewest Goals Against	**179**	1998-99 (82)
Longest Winning Streak		
Overall	**7**	Oct. 25-Nov. 13/01
Home	**7**	Feb. 13-Mar. 8/99
Away	**5**	Apr. 3-19/98
Longest Undefeated Streak		
Overall	**11**	Dec. 28/98-Jan. 16/99 (8 wins, 3 ties), Oct. 25-Nov. 22/01 (9 wins, 2 ties)
Home	**8**	Feb. 5-Mar. 20/98 (5 wins, 3 ties), Oct. 27-Nov. 22/01 (7 wins, 1 tie)
Away	**7**	Twice
Longest Losing Streak		
Overall	**14**	Mar. 2-Apr. 7/93
Home	***11**	Oct. 27-Dec. 8/93
Away	***38**	Oct. 10/92-Apr. 3/93**
Longest Winless Streak		
Overall	**21**	Oct. 10-Nov. 23/92 (20 losses, 1 tie)
Home	***17**	Oct. 28/95-Jan. 27/96 (15 losses, 2 ties)
Away	***38**	Oct. 10/92-Apr. 3/93 (38 losses)
Most Shutouts, Season	**10**	2001-02 (82)
Most PIM, Season	**1,716**	1992-93 (84)
Most Goals, Game	**11**	Nov. 13/01 (Ott. 11 at Wsh. 5)

** NHL records do not include neutral site games

Individual

Record		
Most Seasons	**8**	Radek Bonk
Most Games, Career	**553**	Radek Bonk
Most Goals, Career	**218**	Alexei Yashin
Most Assists, Career	**273**	Alexei Yashin
Most Points, Career	**491**	Alexei Yashin (218G, 273A)
Most PIM, Career	**625**	Dennis Vial
Most Shutouts, Career	**17**	Patrick Lalime
Longest Consecutive Games Streak	**292**	Alexei Yashin (Dec. 31/95-Apr. 17/99)
Most Goals, Season	**44**	Alexei Yashin (1998-99)
Most Assists, Season	**50**	Alexei Yashin (1998-99)
Most Points, Season	**94**	Alexei Yashin (1998-99; 44G, 50A)
Most PIM, Season	**318**	Mike Peluso (1992-93)
Most Points, Defenseman, Season	**63**	Norm Maciver (1992-93; 17G, 46A)
Most Points, Center, Season	**94**	Alexei Yashin (1998-99; 44G, 50A)
Most Points, Right Wing, Season	**75**	Marian Hossa (2000-01; 32G, 43A)
Most Points, Left Wing, Season	**72**	Shawn McEachern (2000-01; 32G, 40A)
Most Points, Rookie, Season	**79**	Alexei Yashin (1993-94; 30G, 49A)
Most Shutouts, Season	**7**	Patrick Lalime (2000-01, 2001-02)
Most Goals, Game	**3**	Twenty-five times
Most Assists, Game	**5**	Marian Hossa (Jan. 4/01)
Most Points, Game	**6**	Dan Quinn (Oct. 15/95; 3G, 3A), Radek Bonk (Jan. 4/01; 3G, 3A)

* NHL Record.

Coaching History

Rick Bowness, 1992-93 to 1994-95; Rick Bowness, Dave Allison and Jacques Martin, 1995-96; Jacques Martin, 1996-97 to date.

Retired Numbers

8	Frank Finnigan	1924-1934

Captains' History

Laurie Boschman, 1992-93; Brad Shaw, Mark Lamb and Gord Dineen, 1993-94; Randy Cunneyworth, 1994-95 to 1997-98; Alexei Yashin, 1998-99; Daniel Alfredsson, 1999-2000 to date.

All-time Record vs. Other Clubs

Regular Season

	At Home								On Road								Total							
	GP	W	L	T	OL	GF	GA	PTS	GP	W	L	T	OL	GF	GA	PTS	GP	W	L	T	OL	GF	GA	PTS
Anaheim	7	3	3	1	0	21	18	7	7	2	3	2	0	13	18	6	14	5	6	3	0	34	36	13
Atlanta	6	3	1	1	1	29	19	8	6	4	1	0	1	26	21	9	12	7	2	1	2	55	40	17
Boston	26	9	14	3	0	62	83	21	27	7	17	3	0	77	106	17	53	16	31	6	0	139	189	38
Buffalo	27	7	12	6	2	58	76	22	26	5	18	2	1	34	86	13	53	12	30	8	3	92	162	35
Calgary	9	4	1	3	1	25	23	12	9	3	5	1	0	22	34	7	18	7	6	4	1	47	57	19
Carolina	25	9	11	4	1	64	71	23	23	5	16	2	0	50	69	12	48	14	27	6	1	114	140	35
Chicago	9	3	5	0	1	25	29	7	7	2	3	2	0	18	17	6	16	5	8	2	1	43	46	13
Colorado	16	7	7	2	0	49	62	16	14	1	12	1	0	36	68	3	30	8	19	3	0	85	130	19
Columbus	1	0	0	1	0	2	2	1	2	1	0	1	0	9	7	3	3	1	0	2	0	11	9	4
Dallas	8	3	5	0	0	19	22	6	9	4	5	0	0	22	34	8	17	7	10	0	0	41	56	14
Detroit	8	2	5	1	0	21	24	5	8	3	4	0	1	17	27	7	16	5	9	1	1	38	51	12
Edmonton	8	2	5	1	0	14	23	5	9	1	6	2	0	22	36	4	17	3	11	3	0	36	59	9
Florida	18	8	8	2	0	50	49	18	18	8	9	1	0	53	58	17	36	16	17	3	0	103	107	35
Los Angeles	8	4	2	1	1	28	25	10	8	1	6	1	0	18	39	3	16	5	8	2	1	46	64	13
Minnesota	1	1	0	0	0	4	1	2	2	1	0	1	0	6	5	3	3	2	0	1	0	10	6	5
Montreal	25	11	13	1	0	70	73	23	28	9	15	4	0	77	85	22	53	20	28	5	0	147	158	45
Nashville	3	2	1	0	0	7	5	4	3	2	1	0	0	9	5	4	6	4	2	0	0	16	10	8
New Jersey	20	5	11	3	1	41	52	14	19	5	11	2	1	41	61	13	39	10	22	5	2	82	113	27
NY Islanders	19	11	4	4	0	67	51	26	20	12	3	5	0	77	62	29	39	23	7	9	0	144	113	55
NY Rangers	19	5	11	3	0	51	59	13	19	9	10	0	0	54	62	18	38	14	21	3	0	105	121	31
Philadelphia	20	7	10	3	0	56	66	17	19	6	11	2	0	53	64	14	39	13	21	5	0	109	130	31
Phoenix	10	3	6	1	0	27	33	7	9	4	4	1	0	34	33	9	19	7	10	2	0	61	66	16
Pittsburgh	23	6	13	4	0	56	73	16	23	3	16	4	0	49	91	10	46	9	29	8	0	105	164	26
St. Louis	9	3	6	0	0	20	35	6	8	3	4	1	0	23	25	7	17	6	10	1	0	43	60	13
San Jose	8	3	1	4	0	27	23	10	8	3	5	0	0	12	17	6	16	6	6	4	0	39	40	16
Tampa Bay	19	12	7	0	0	72	44	24	19	10	7	2	0	63	56	22	38	22	14	2	0	135	100	46
Toronto	15	10	4	1	0	45	39	21	16	7	8	1	0	44	47	15	31	17	12	2	0	89	86	36
Vancouver	9	5	3	1	0	22	21	11	9	4	4	1	0	20	26	9	18	9	7	2	0	42	47	20
Washington	19	9	8	1	1	67	62	20	20	6	11	3	0	52	71	15	39	15	19	4	1	119	133	35
Totals	**395**	**157**	**177**	**52**	**9**	**1099**	**1163**	**375**	**395**	**131**	**215**	**45**	**4**	**1031**	**1330**	**311**	**790**	**288**	**392**	**97**	**13**	**2130**	**2493**	**686**

Playoffs

	Series	W	L	GP	W	L	T	GF	GA	Last Mtg.	Rnd.	Result
Buffalo	2	0	2	11	3	8	0	19	26	1999	CQF	L 0-4
New Jersey	1	1	0	6	4	2	0	13	12	1998	CQF	W 4-2
Philadelphia	1	1	0	5	4	1	0	11	2	2002	CQF	W 4-1
Toronto	3	0	3	17	5	12	0	31	43	2002	CSF	L 3-4
Washington	1	0	1	5	1	4	0	7	18	1998	CSF	L 1-4
Totals	**8**	**2**	**6**	**44**	**17**	**27**	**0**	**81**	**101**			

Playoff Results 2002-1998

Year	Round	Opponent	Result	GF	GA
2002	CSF	Toronto	L 3-4	18	16
	CQF	Philadelphia	W 4-1	11	2
2001	CQF	Toronto	L 0-4	3	10
2000	CQF	Toronto	L 2-4	10	17
1999	CQF	Buffalo	L 0-4	6	12
1998	CSF	Washington	L 1-4	7	18
	CQF	New Jersey	W 4-2	13	12

Abbreviations: Round: CSF – conference semi-final; **CQF** – conference quarter-final.

Colorado totals include Quebec, 1992-93 to 1994-95.
Dallas totals include Minnesota North Stars, 1992-93.
Carolina totals include Hartford, 1992-93 to 1996-97.
Phoenix totals include Winnipeg, 1992-93 to 1995-96.

2001-02 Results

Month	Day	Opponent	Score	Month	Day	Opponent	Score
Oct.	3	at Toronto	5-4		7	Toronto	4-3
	4	Montreal	4-6		9	at Atlanta	3-4*
	6	at Buffalo	2-3		11	at Florida	4-2
	9	at Carolina	6-2		12	at Tampa Bay	2-1
	10	at Florida	2-0		15	Philadelphia	1-4
	13	NY Rangers	2-2		17	at Boston	2-5
	16	at Pittsburgh	2-5		19	Minnesota	4-1
	18	Pittsburgh	0-3		20	at Detroit	2-3*
	20	at New Jersey	2-3		22	at Philadelphia	1-1
	23	New Jersey	1-2		24	Boston	4-3
	25	at Philadelphia	7-2		26	at Montreal	1-1
	27	St. Louis	4-1		30	Philadelphia	3-1
	30	at Atlanta	6-3	**Feb.**	4	at Tampa Bay	4-4
Nov.	3	Buffalo	3-0		6	at Columbus	6-4
	8	Colorado	1-0		8	at Buffalo	2-3*
	10	Nashville	3-2		9	Detroit	2-3
	13	at Washington	11-5		12	Pittsburgh	5-1
	15	Carolina	1-1		26	at Montreal	5-2
	17	Toronto	2-1*		28	at NY Rangers	3-0
	20	Vancouver	3-0	**Mar.**	2	Washington	2-3*
	22	Calgary	4-4		4	at Los Angeles	1-1
	24	Atlanta	3-6		7	at San Jose	2-5
	27	at St. Louis	2-4		9	at Phoenix	2-3
Dec.	1	Boston	2-1*		10	at Anaheim	4-2
	3	at Colorado	2-4		12	at Minnesota	4-3*
	5	at Dallas	6-3		14	Edmonton	1-4
	6	at Nashville	2-4		16	NY Islanders	4-3
	8	Tampa Bay	5-2		17	Florida	2-0
	11	at NY Islanders	2-2		19	at Buffalo	1-5
	13	Phoenix	6-0		21	NY Rangers	2-5
	15	New Jersey	0-2		23	Atlanta	2-3*
	18	at Carolina	5-1		24	Buffalo	2-3*
	20	Los Angeles	2-4		27	at NY Islanders	4-1
	22	at New Jersey	1-0		28	Florida	3-4
	23	at NY Rangers	2-3		30	Tampa Bay	3-1
	26	at Boston	2-3	**Apr.**	2	Carolina	4-3
	27	NY Islanders	5-2		5	at Washington	0-0
	29	at Pittsburgh	5-2		7	Montreal	1-3
	31	Chicago	4-5*		9	at Montreal	3-4
Jan.	3	Washington	4-1		11	Boston	4-0
	5	at Toronto	1-3		13	Toronto	2-5

* – Overtime

Entry Draft Selections 2002-1992

2002

Pick	
16	Jakub Klepis
47	Alexei Kaigorodov
75	Arttu Luttinen
113	Scott Dobben
125	Johan Bjork
150	Brock Hooton
246	Josef Vavra
276	Vitali Atyushov

2001

Pick	
2	Jason Spezza
23	Tim Gleason
81	Neil Komadoski
99	Ray Emery
127	Christoph Schubert
162	Stefan Schauer
193	Brooks Laich
218	Jan Platil
223	Brandon Bochenski
235	Neil Petruic
256	Gregg Johnson
286	Toni Dahlman

2000

Pick	
21	Anton Volchenkov
45	Mathieu Chouinard
55	Antoine Vermette
87	Jan Bohac
122	Derrick Byfuglien
156	Greg Zanon
157	Grant Potulny
158	Sean Connolly
188	Jason Maleyko
283	James Demone

1999

Pick	
26	Martin Havlat
48	Simon Lajeunesse
62	Teemu Sainomaa
94	Chris Kelly
154	Andrew Ianiero
164	Martin Prusek
201	Mikko Ruutu
209	Layne Ulmer
213	Alexandre Giroux
269	Konstantin Gorovikov

1998

Pick	
15	Mathieu Chouinard
44	Mike Fisher
58	Chris Bala
74	Julien Vauclair
101	Petr Schastlivy
130	Gavin McLeod
161	Christopher Neil
188	Michel Periard
223	Sergei Verenikin
246	Rastislav Pavlikovsky

1997

Pick	
12	Marian Hossa
58	Jani Hurme
66	Josh Langfeld
119	Magnus Arvedson
146	Jeff Sullivan
173	Robin Bacul
203	Nick Gillis
229	Karel Rachunek

1996

Pick	
1	Chris Phillips
81	Antti-Jussi Niemi
136	Andreas Dackell
163	Francois Hardy
212	Erich Goldmann
216	Ivan Ciernik
239	Sami Salo

1995

Pick	
1	Bryan Berard
27	Marc Moro
53	Brad Larsen
89	Kevin Bolibruck
89	Kevin Bolibruck
103	Kevin Boyd
131	David Hruska
183	Kaj Linna
184	Ray Schultz
231	Erik Kaminski

1994

Pick	
3	Radek Bonk
29	Stan Neckar
81	Bryan Masotta
131	Mike Gaffney
133	Daniel Alfredsson
159	Doug Sproule
210	Frederic Cassivi
211	Danny Dupont
237	Stephen MacKinnon
274	Antti Tormanen

1993

Pick	
1	Alexandre Daigle
27	Radim Bicanek
53	Patrick Charbonneau
91	Cosmo Dupaul
131	Rick Bodkin
157	Sergei Poleschuk
183	Jason Disher
209	Toby Kvalevog
227	Pavol Demitra
235	Rick Schuwerk

1992

Pick	
2	Alexei Yashin
25	Chad Penney
50	Patrick Traverse
73	Radek Hamr
98	Daniel Guerard
121	Al Sinclair
146	Jaroslav Miklenda
169	Jay Kenney
194	Claude Savoie
217	Jake Grimes
242	Tomas Jelinek
264	Petter Ronnqvist

Club Directory

Corel Centre

Ottawa Senators
Corel Centre
1000 Palladium Drive
Ottawa, Ontario
K2V 1A5
Phone **613/599-0250**
FAX 613/599-5562
www.ottawasenators.com
Capacity: 18,500

Chairman and Governor	Rod Bryden
President and CEO	Roy Mlakar
General Manager	John Muckler
Assistant to the General Manager	Allison Vaughan
Director of Hockey Operations	Trevor Timmins
Director of Player Personnel	Jarmo Kekalainen
Director of Legal Relations	Peter Chiarelli
Chief Amateur Scout	Frank Jay
Head Coach	Jacques Martin
Assistant Coaches	Perry Pearn, Roger Neilson, Don Jackson
Strength & Conditioning & Video Coach	Randy Lee
Pro Scout/Goaltending Coach	Phil Myre
VP, Communications	Phil Legault
Manager, Communications	Ian Mendes
Coordinator, Communications	Tim Pattyson
Communications/Hockey Operations Assistant	Jennifer Vuong
Head Athletic Trainer	Kevin Wagner
Head Equipment Manager	John Gervais
Assistant Equipment Manager	Chris Cook
Massage Therapist	Brad Joyal
Professional scout	Bob Janecyk
Mental skills coach	John Phelan
Scouts	Dale Engel, George Fargher, Ken Williamson, Lewis Mongelluzzo, Patrick Savard, Boris Shagas, Ales Volek, Ilkka Ikonen
Radio	Sports Radio 1200 The Team (English), Radio 1150 (French)
Television	Sportsnet, New RO

Coach

MARTIN, JACQUES
Coach, Ottawa Senators. Born in St. Pascal, Ont., October 1, 1952.

Jacques Martin led the Ottawa Senators to their best regular season in team history in 2000-01, breaking team records for wins (48) and points (109) the club had established under his leadership two years before.

When appointed the Senators' third head coach on January 24, 1996, Martin brought 10 years of NHL coaching experience, including five with the Quebec Nordiques, an organization often compared with the Senators, in that both teams were built around young, talented players requiring patience and teaching.

Martin's coaching career began at the collegiate level in 1976. He was appointed head coach of the Guelph Platers (now Storm) in 1985, winning the OHL title, the Memorial Cup and being named the OHL coach of the year. That summer, Martin became head coach of the St. Louis Blues. In his NHL rookie year, he led the Blues to the Norris Division championship and, in two seasons with the Blues, posted a 66-71-23 record. He then spent two seasons as an assistant to Chicago's head coach Mike Keenan, before joining the Nordiques in 1990. With Quebec, he worked four years as assistant coach and one year (1993-94) as both head coach and general manager of the AHL Cornwall Aces.

Coaching Record

		Regular Season				Playoffs		
Season	**Team**	**Games**	**W**	**L**	**T**	**Games**	**W**	**L**
1983-84	Peterborough (OHL)	70	43	23	4	....	....	....
1984-85	Peterborough (OHL)	66	42	20	4	....	....	....
1985-86	Guelph (OHL)	66	41	23	2	....	....	....
1986-87	**St. Louis (NHL)**	**80**	**32**	**33**	**15**	**6**	**2**	**4**
1987-88	**St. Louis (NHL)**	**80**	**34**	**38**	**8**	**10**	**5**	**5**
1993-94	Cornwall (AHL)	80	33	36	11	13	8	5
1995-96	**Ottawa (NHL)**	**38**	**10**	**24**	**4**	....	....	....
1996-97	**Ottawa (NHL)**	**82**	**31**	**36**	**15**	**7**	**3**	**4**
1997-98	**Ottawa (NHL)**	**82**	**34**	**33**	**15**	**11**	**5**	**6**
1998-99	**Ottawa (NHL)**	**82**	**44**	**23**	**15**	**4**	**0**	**4**
1999-2000	**Ottawa (NHL)**	**82**	**41**	**30**	**11**	**6**	**2**	**4**
2000-01	**Ottawa (NHL)**	**82**	**48**	**25**	**9**	**4**	**0**	**4**
2001-02	**Ottawa (NHL)**	**80**	**38**	**33**	**9**	**12**	**7**	**5**
	NHL Totals	**688**	**312**	**275**	**101**	**60**	**24**	**36**

Martin stepped aside (with NHL permission) during the final two games of the 2001-02 season in order to allow assistant coach Roger Neilson to reach the 1,000-game plateau, April 11 and 13, 2002.

After splitting the previous four seasons between NHL teams and the minor leagues, Todd White finally spent his first full season at the NHL level in 2001-02. He played 81 games and collected 20 goals and 30 assists.

Philadelphia Flyers

2001-02 Results: 42W-27L-10T-3OTL 97PTS.
First, Atlantic Division

2002-03 Schedule

Oct.	Thu.	10	at Edmonton		Sat.	11	Detroit*
	Sat.	12	at Calgary		Mon.	13	Atlanta
	Tue.	15	at Montreal		Thu.	16	Montreal
	Thu.	17	NY Islanders		Sat.	18	Tampa Bay*
	Sat.	19	Washington*		Sun.	19	at NY Rangers
	Tue.	22	at Buffalo		Tue.	21	at Toronto
	Thu.	24	Montreal		Fri.	24	NY Islanders
	Sat.	26	at NY Islanders		Sat.	25	at Boston
	Tue.	29	Ottawa		Tue.	28	Tampa Bay
	Thu.	31	Phoenix		Thu.	30	at New Jersey
Nov.	Sat.	2	Washington	**Feb.**	Tue.	4	at NY Islanders
	Tue.	5	at Carolina		Thu.	6	at Ottawa
	Thu.	7	New Jersey		Sat.	8	NY Rangers*
	Sat.	9	at Washington		Mon.	10	Minnesota
	Wed.	13	Florida		Wed.	12	at Minnesota
	Fri.	15	at Carolina		Thu.	13	at St. Louis
	Sat.	16	Boston		Sat.	15	Carolina
	Tue.	19	at Tampa Bay		Tue.	18	New Jersey
	Thu.	21	San Jose		Thu.	20	Los Angeles
	Sat.	23	at Toronto		Sat.	22	Florida
	Wed.	27	at Pittsburgh		Tue.	25	at Chicago
	Fri.	29	Toronto*		Thu.	27	Chicago
	Sat.	30	at Montreal	**Mar.**	Sat.	1	at Boston*
Dec.	Mon.	2	New Jersey		Tue.	4	Vancouver
	Thu.	5	NY Rangers		Fri.	7	at NY Rangers
	Sat.	7	St. Louis*		Sat.	8	Colorado
	Tue.	10	at Florida		Mon.	10	at Washington
	Thu.	12	Toronto		Thu.	13	Carolina
	Sat.	14	Buffalo		Sat.	15	at Pittsburgh*
	Tue.	17	Dallas		Mon.	17	at New Jersey
	Wed.	18	at Atlanta		Tue.	18	at Buffalo
	Sat.	21	Ottawa*		Thu.	20	Pittsburgh
	Mon.	23	at Ottawa		Sat.	22	NY Rangers*
	Fri.	27	at Colorado		Mon.	24	Atlanta
	Sat.	28	at Phoenix		Tue.	25	at Nashville
	Mon.	30	at San Jose		Thu.	27	Boston
Jan.	Thu.	2	at Los Angeles		Sat.	29	Pittsburgh
	Fri.	3	at Anaheim		Mon.	31	at Pittsburgh
	Sun.	5	at Atlanta*	**Apr.**	Tue.	1	Columbus
	Tue.	7	Buffalo		Fri.	4	at Tampa Bay
	Thu.	9	at NY Islanders		Sun.	6	at Florida*

* Denotes afternoon game.

Franchise date: June 5, 1967

ATLANTIC DIVISION

36th NHL Season

Year-by-Year Record

		Home				Road				Overall								
Season	**GP**	**W**	**L**	**T**	**OL**	**W**	**L**	**T**	**OL**	**W**	**L**	**T**	**OL**	**GF**	**GA**	**Pts.**	**Finished**	**Playoff Result**
2001-02	82	20	13	5	3	22	14	5	0	42	27	10	3	234	192	97	1st, Atlantic Div.	Lost Conf. Quarter-Final
2000-01	82	26	11	4	0	17	14	7	3	43	25	11	3	240	207	100	2nd, Atlantic Div.	Lost Conf. Quarter-Final
1999-2000	82	25	6	7	3	20	16	5	0	45	22	12	3	237	179	105	1st, Atlantic Div.	Lost Conf. Championship
1998-99	82	21	9	11	...	16	17	8	...	37	26	19	...	231	196	93	2nd, Atlantic Div.	Lost Conf. Quarter-Final
1997-98	82	24	11	6	...	18	18	5	...	42	29	11	...	242	193	95	2nd, Atlantic Div.	Lost Conf. Quarter-Final
1996-97	82	23	12	6	...	22	12	7	...	45	24	13	...	274	217	103	2nd, Atlantic Div.	Lost Final
1995-96	82	27	9	5	...	18	15	8	...	45	24	13	...	282	208	103	1st, Atlantic Div.	Lost Conf. Semi-Final
1994-95	48	16	7	1	...	12	9	3	...	28	16	4	...	150	132	60	1st, Atlantic Div.	Lost Conf. Championship
1993-94	84	19	20	3	...	16	19	7	...	35	39	10	...	294	314	80	6th, Atlantic Div.	Out of Playoffs
1992-93	84	23	14	5	...	13	23	6	...	36	37	11	...	319	319	83	5th, Patrick Div.	Out of Playoffs
1991-92	80	22	11	7	...	10	26	4	...	32	37	11	...	252	273	75	6th, Patrick Div.	Out of Playoffs
1990-91	80	18	16	6	...	15	21	4	...	33	37	10	...	252	267	76	5th, Patrick Div.	Out of Playoffs
1989-90	80	17	19	4	...	13	20	7	...	30	39	11	...	290	297	71	6th, Patrick Div.	Out of Playoffs
1988-89	80	22	15	3	...	14	21	5	...	36	36	8	...	307	285	80	4th, Patrick Div.	Lost Conf. Championship
1987-88	80	20	14	6	...	18	19	3	...	38	33	9	...	292	292	85	3rd, Patrick Div.	Lost Div. Semi-Final
1986-87	80	29	9	2	...	17	17	6	...	46	26	8	...	310	245	100	1st, Patrick Div.	Lost Final
1985-86	80	33	6	1	...	20	17	3	...	53	23	4	...	335	241	110	1st, Patrick Div.	Lost Div. Semi-Final
1984-85	80	32	4	4	...	21	16	3	...	53	20	7	...	348	241	113	1st, Patrick Div.	Lost Final
1983-84	80	25	10	5	...	19	16	5	...	44	26	10	...	350	290	98	3rd, Patrick Div.	Lost Div. Semi-Final
1982-83	80	29	8	3	...	20	15	5	...	49	23	8	...	326	240	106	1st, Patrick Div.	Lost Div. Semi-Final
1981-82	80	25	10	5	...	13	21	6	...	38	31	11	...	325	313	87	3rd. Patrick Div.	Lost Div. Semi-Final
1980-81	80	23	9	8	...	18	15	7	...	41	24	15	...	313	249	97	2nd, Patrick Div.	Lost Quarter-Final
1979-80	80	27	5	8	...	21	7	12	...	48	12	20	...	327	254	116	1st, Patrick Div.	Lost Final
1978-79	80	26	10	4	...	14	15	11	...	40	25	15	...	281	248	95	2nd, Patrick Div.	Lost Quarter-Final
1977-78	80	29	6	5	...	16	14	10	...	45	20	15	...	296	200	105	2nd, Patrick Div.	Lost Semi-Final
1976-77	80	33	6	1	...	15	10	15	...	48	16	16	...	323	213	112	1st, Patrick Div.	Lost Semi-Final
1975-76	80	36	2	2	...	15	11	14	...	51	13	16	...	348	209	118	1st, Patrick Div.	Lost Final
1974-75	**80**	**32**	**6**	**2**	**...**	**19**	**12**	**9**	**...**	**51**	**18**	**11**	**...**	**293**	**181**	**113**	**1st, Patrick Div.**	**Won Stanley Cup**
1973-74	**78**	**28**	**6**	**5**	**...**	**22**	**10**	**7**	**...**	**50**	**16**	**12**	**...**	**273**	**164**	**112**	**1st, West Div.**	**Won Stanley Cup**
1972-73	78	27	8	4	...	10	22	7	...	37	30	11	...	296	256	85	2nd, West Div.	Lost Semi-Final
1971-72	78	19	13	7	...	7	25	7	...	26	38	14	...	200	236	66	5th, West Div.	Out of Playoffs
1970-71	78	20	10	9	...	8	23	8	...	28	33	17	...	207	225	73	3rd, West Div.	Lost Quarter-Final
1969-70	76	11	14	13	...	6	21	11	...	17	35	24	...	197	225	58	5th, West Div.	Out of Playoffs
1968-69	76	14	16	8	...	6	19	13	...	20	35	21	...	174	225	61	3rd, West Div.	Lost Quarter-Final
1967-68	74	17	13	7	...	14	19	4	...	31	32	11	...	173	179	73	1st, West Div.	Lost Quarter-Final

Philadelphia acquired Kim Johnsson as part of the package of players they received from the Rangers for the rights to Eric Lindros. He emerged as the club's top offensive defenseman last season with 11 goals and 30 assists.

2002-03 Player Personnel

FORWARDS	HT	WT	S	Place of Birth	Date	2001-02 Club
BRASHEAR, Donald	6-2	225	L	Bedford, IN	1/7/72	Vancouver-Philadelphia
BRENDL, Pavel	6-1	204	R	Opocno, Czech.	3/23/81	Phi-Phi (AHL)
FEDORUK, Todd	6-2	235	L	Redwater, Alta.	2/13/79	Phi-Phi (AHL)
GAGNE, Simon	6-0	190	L	Ste-Foy, Que.	2/29/80	Philadelphia-Canada
GREIG, Mark	5-11	190	R	High River, Alta.	1/25/70	Philadelphia (AHL)
HANDZUS, Michal	6-5	210	L	Banska Bystrica, Czech.	3/11/77	Phoenix-Slovakia
LAW, Kirby	6-1	185	R	McCreary, Man.	3/11/77	Philadelphia (AHL)
LeCLAIR, John	6-3	226	L	St. Albans, VT	7/5/69	Phi-United States
LEFEBVRE, Guillaume	6-1	195	L	Amos, Que.	5/7/81	Phi-Phi (AHL)
MacNEIL, Ian	6-2	190	L	Halifax, N.S.	4/27/77	Lowell
MURRAY, Marty	5-9	180	L	Deloraine, Man.	2/16/75	Phi-Phi (AHL)
PLETKA, Vaclav	5-11	182	L	Mlada Boleslav, Czech.	6/8/79	Phi-Phi (AHL)
PRIMEAU, Keith	6-5	220	L	Toronto, Ont.	11/24/71	Philadelphia
RANHEIM, Paul	6-1	210	R	St. Louis, MO	1/25/66	Philadelphia
RECCHI, Mark	5-10	185	L	Kamloops, B.C.	2/1/68	Philadelphia
ROENICK, Jeremy	6-1	207	R	Boston, MA	1/17/70	Phi-United States
SAVAGE, Andre	6-0	195	R	Ottawa, Ont.	5/27/75	Manitoba
SHARP, Patrick	6-0	188	R	Thunder Bay, Ont.	12/27/81	U. of Vermont
WILLIAMS, Justin	6-1	190	R	Cobourg, Ont.	10/4/81	Philadelphia
DEFENSEMEN						
DESJARDINS, Eric	6-1	205	R	Rouyn, Que.	6/14/69	Philadelphia
HARLOCK, David	6-2	215	L	Toronto, Ont.	3/16/71	Atl-Chi (AHL)-Phi (AHL)
JOHNSSON, Kim	6-1	205	L	Malmo, Sweden	3/16/76	Philadelphia-Sweden
McALLISTER, Chris	6-8	240	L	Saskatoon, Sask.	6/16/75	Philadelphia
McGILLIS, Dan	6-2	230	L	Hawkesbury, Ont.	7/1/72	Philadelphia
ST-JACQUES, Bruno	6-2	210	L	Montreal, Que.	8/22/80	Phi-Phi (AHL)
SLANEY, John	6-0	189	L	St. John's, Nfld.	2/2/72	Phi-Phi (AHL)
THERIEN, Chris	6-5	235	L	Ottawa, Ont.	12/14/71	Philadelphia
TILEY, Brad	6-1	199	L	Markdale, Ont.	7/5/71	Philadelphia (AHL)
WEINRICH, Eric	6-1	213	L	Roanoke, VA	12/19/66	Philadelphia
WOYWITKA, Jeff	6-2	209	L	Vermilion, Alta.	9/1/83	Red Deer

GOALTENDERS	HT	WT	C	Place of Birth	Date	2001-02 Club
CECHMANEK, Roman	6-3	187	L	Gottwaldov, Czech.	3/2/71	Phi-Czech Republic
ESCHE, Robert	6-1	210	L	Whitesboro, NY	1/22/78	Phoenix-Springfield
LITTLE, Neil	6-1	193	L	Medicine Hat, Alta.	12/18/71	Phi-Phi (AHL)
NIITTYMAKI, Antero	6-0	176	L	Turku, Finland	6/18/80	TPS Turku

Coach

HITCHCOCK, KEN
Coach, Philadelphia Flyers. Born in Edmonton, Alta., December 17, 1951.

The Philadelphia Flyers named Ken Hitchcock as their head coach on May 14, 2002.

Hitchcock is the 15th head coach in Flyers history. Prior to joining the Flyers, he won a gold medal as an associate coach with Team Canada at the 2002 Winter Olympic Games. Hitchcock served as head coach of the Dallas Stars for parts of seven seasons (1995-96 to 2001-02), compiling a 277-166-60 record in 503 regular season games for a .610 winning percentage.

Hitchcock served as head coach of Dallas' International Hockey League affiliate, the Kalamazoo Wings/Michigan K-Wings for three seasons, from the 1993-94 season until being named Stars' head coach on January 8, 1996. Prior to joining the Stars' organization, Hitchcock served three seasons as an assistant coach with the Flyers (1990-91 through 1992-93).

Hitchcock joined the Flyers after six seasons as head coach of the Kamloops Blazers of the Western Hockey League from 1984-85 through 1989-90. His .693 winning percentage as head coach at Kamloops is the second highest in the history of the WHL (291-125-15). His international experience also includes serving as an assistant coach for the Team Canada team that captured the gold medal at the 1987 World Junior Championships.

Coaching Record

		Regular Season				Playoffs		
Season	Team	Games	W	L	T	Games	W	L
1984-85	Kamloops (WHL)	71	52	17	2	15	10	5
1985-86	Kamloops (WHL)	72	49	19	4	16	14	2
1986-87	Kamloops (WHL)	72	55	14	3	13	8	5
1987-88	Kamloops (WHL)	72	45	26	1	18	12	6
1988-89	Kamloops (WHL)	72	34	33	5	16	8	8
1989-90	Kamloops (WHL)	72	56	16	0	17	14	3
1993-94	Kalamazoo (IHL)	81	48	26	7	5	1	4
1994-95	Kalamazoo (IHL)	81	43	24	14	16	10	6
1995-96	Michigan (IHL)	40	19	10	11	....	....	....
	Dallas (NHL)	**43**	**15**	**23**	**5**	....	....	....
1996-97	**Dallas (NHL)**	**82**	**48**	**26**	**8**	**7**	**3**	**4**
1997-98	**Dallas (NHL)**	**82**	**49**	**22**	**11**	**17**	**10**	**7**
1998-99	**Dallas (NHL)**	**82**	**51**	**19**	**12**	**23**	**16**	**7***
1999-2000	**Dallas (NHL)**	**82**	**43**	**29**	**10**	**23**	**14**	**9**
2000-01	**Dallas (NHL)**	**82**	**48**	**26**	**8**	**10**	**4**	**6**
2001-02	**Dallas (NHL)**	**50**	**23**	**21**	**6**	....	....	....
	NHL Totals	**503**	**277**	**166**	**60**	**80**	**47**	**33**

* Stanley Cup win.

2001-02 Scoring

* - rookie

Regular Season

Pos	#	Player	Team	GP	G	A	Pts	+/–	PIM	PP	SH	GW	GT	S	%
C	77	Adam Oates	WSH	66	11	57	68	–2	22	3	0	1	0	85	12.9
			PHI	14	3	7	10	–2	6	0	0	0	0	17	17.6
			TOTAL	80	14	64	78	–4	28	3	0	1	0	102	13.7
C	97	Jeremy Roenick	PHI	75	21	46	67	32	74	5	0	3	0	167	12.6
L	12	Simon Gagne	PHI	79	33	33	66	31	32	4	1	7	0	199	16.6
R	8	Mark Recchi	PHI	80	22	42	64	5	46	7	2	4	0	205	10.7
L	10	John LeClair	PHI	82	25	26	51	5	30	4	0	6	1	220	11.4
C	25	Keith Primeau	PHI	75	19	29	48	–3	128	5	0	3	0	151	12.6
D	5	Kim Johnsson	PHI	82	11	30	41	12	42	5	0	1	0	150	7.3
R	14	Justin Williams	PHI	75	17	23	40	11	32	0	0	1	1	162	10.5
L	87	Donald Brashear	VAN	31	5	8	13	–8	90	1	0	0	0	45	11.1
			PHI	50	4	15	19	0	109	0	0	2	0	62	6.5
			TOTAL	81	9	23	32	–8	199	1	0	2	0	107	8.4
C	39	Marty Murray	PHI	74	12	15	27	10	10	1	1	2	0	109	11.0
C	20	Jiri Dopita	PHI	52	11	16	27	9	8	3	0	2	0	79	13.9
L	26	Ruslan Fedotenko	PHI	78	17	9	26	15	43	0	1	3	0	121	14.0
D	37	Eric Desjardins	PHI	65	6	19	25	–1	24	2	1	0	0	117	5.1
D	2	Eric Weinrich	PHI	80	4	20	24	27	26	0	0	2	0	102	3.9
D	3	Daniel McGillis	PHI	75	5	14	19	17	46	2	0	1	0	147	3.4
D	6	Chris Therien	PHI	77	4	10	14	16	30	0	2	3	0	105	3.8
L	19	Paul Ranheim	PHI	79	5	4	9	5	36	1	0	0	0	75	6.7
D	22	Luke Richardson	PHI	72	1	8	9	18	102	0	0	0	0	65	1.5
L	29	Todd Fedoruk	PHI	55	3	4	7	–2	141	0	0	0	0	21	14.3
R	17	Billy Tibbetts	PIT	33	1	5	6	–13	109	0	0	1	0	42	2.4
			PHI	9	0	1	1	–3	69	0	0	0	0	6	0.0
			TOTAL	42	1	6	7	–16	178	0	0	1	0	48	2.1
D	24	Chris McAllister	PHI	42	0	5	5	–7	113	0	0	0	0	26	0.0
R	92	Rick Tocchet	PHI	14	0	2	2	–2	28	0	0	0	0	10	0.0
R	18 *	Tomas Divisek	PHI	3	1	0	1	1	0	0	0	1	0	3	33.3
R	55 *	Pavel Brendl	PHI	8	1	0	1	–1	2	0	0	0	0	6	16.7
C	15	Jarrod Skalde	PHI	1	0	0	0	0	2	0	0	0	0	5	0.0
D	15	John Slaney	PHI	1	0	0	0	2	0	0	0	0	0	1	0.0
L	11 *	Vaclav Pletka	PHI	1	0	0	0	0	0	0	0	0	0	2	0.0
R	21 *	Jesse Boulerice	PHI	3	0	0	0	–1	5	0	0	0	0	1	0.0
L	23 *	Guillaume Lefebvre	PHI	3	0	0	0	–1	0	0	0	0	0	3	0.0
D	42 *	Bruno St. Jacques	PHI	7	0	0	0	4	2	0	0	0	0	4	0.0

Goaltending

No.	Goaltender	GPI	Mins	Avg	W	L	T	EN	SO	GA	SA	S%	G	A	PIM
32	Roman Cechmanek	46	2603	2.05	24	13	6	2	4	89	1131	.921	0	0	10
33	Brian Boucher	41	2295	2.41	18	16	4	5	2	92	972	.905	0	0	4
35	Neil Little	1	60	4.00	0	1	0	0	0	4	29	.862	0	0	10
	Totals	**82**	**4986**	**2.31**	**42**	**30**	**10**	**7**	**7**	**192**	**2139**	**.910**			

Playoffs

Pos	#	Player	Team	GP	G	A	Pts	+/–	PIM	PP	SH	GW	GT	S	%
C	77	Adam Oates	PHI	5	0	2	2	–1	0	0	0	0	0	4	0.0
D	3	Daniel McGillis	PHI	5	1	0	1	–1	8	1	0	0	0	10	10.0
L	26	Ruslan Fedotenko	PHI	5	1	0	1	0	2	0	0	1	1	11	9.1
D	37	Eric Desjardins	PHI	5	0	1	1	–3	2	0	0	0	0	6	0.0
C	39	Marty Murray	PHI	5	0	1	1	–2	0	0	0	0	0	2	0.0
D	15	John Slaney	PHI	1	0	0	0	–1	0	0	0	0	0	1	0.0
R	55 *	Pavel Brendl	PHI	2	0	0	0	0	0	0	0	0	0	2	0.0
L	29	Todd Fedoruk	PHI	3	0	0	0	0	0	0	0	0	0	0	0.0
R	8	Mark Recchi	PHI	4	0	0	0	–1	2	0	0	0	0	5	0.0
L	10	John LeClair	PHI	5	0	0	0	–2	2	0	0	0	0	7	0.0
C	25	Keith Primeau	PHI	5	0	0	0	–3	6	0	0	0	0	8	0.0
L	19	Paul Ranheim	PHI	5	0	0	0	–2	0	0	0	0	0	1	0.0
D	22	Luke Richardson	PHI	5	0	0	0	–1	4	0	0	0	0	11	0.0
C	97	Jeremy Roenick	PHI	5	0	0	0	–3	14	0	0	0	0	14	0.0
D	2	Eric Weinrich	PHI	5	0	0	0	0	4	0	0	0	0	6	0.0
D	6	Chris Therien	PHI	5	0	0	0	–3	2	0	0	0	0	9	0.0
L	87	Donald Brashear	PHI	5	0	0	0	–1	19	0	0	0	0	2	0.0
D	5	Kim Johnsson	PHI	5	0	0	0	–2	2	0	0	0	0	12	0.0
L	12	Simon Gagne	PHI	5	0	0	0	–3	2	0	0	0	0	11	0.0
R	14	Justin Williams	PHI	5	0	0	0	–3	4	0	0	0	0	15	0.0

Goaltending

No.	Goaltender	GPI	Mins	Avg	W	L	EN	SO	GA	SA	S%	G	A	PIM
33	Brian Boucher	2	88	1.36	0	1	0	0	2	33	.939	0	0	0
32	Roman Cechmanek	4	227	1.85	1	3	2	1	7	109	.936	0	0	0
	Totals	**5**	**315**	**2.10**	**1**	**4**	**2**	**1**	**11**	**144**	**.924**			

Captains' History

Lou Angotti, 1967-68; Ed Van Impe, 1968-69 to 1971-72; Ed Van Impe and Bobby Clarke, 1972-73; Bobby Clarke, 1973-74 to 1978-79; Mel Bridgman, 1979-80, 1980-81; Bill Barber, 1981-82; Bill Barber and Bobby Clarke, 1982-83; Bobby Clarke, 1983-84; Dave Poulin, 1984-85 to 1988-89; Dave Poulin and Ron Sutter, 1989-90; Ron Sutter, 1990-91; Rick Tocchet, 1991-92; no captain, 1992-93; Kevin Dineen, 1993-94; Eric Lindros, 1994-95 to 1998-99; Eric Lindros and Eric Desjardins, 1999-2000; Eric Desjardins, 2000-01; Eric Desjardins and Keith Primeau, 2001-02; Keith Primeau, 2002-03.

Coaching History

Keith Allen, 1967-68, 1968-69; Vic Stasiuk, 1969-70, 1970-71; Fred Shero, 1971-72 to 1977-78; Bob McCammon and Pat Quinn, 1978-79; Pat Quinn, 1979-80, 1980-81; Pat Quinn and Bob McCammon, 1981-82; Bob McCammon, 1982-83, 1983-84; Mike Keenan, 1984-85 to 1987-88; Paul Holmgren, 1988-89 to 1990-91; Paul Holmgren and Bill Dineen, 1991-92; Bill Dineen, 1992-93; Terry Simpson, 1993-94; Terry Murray, 1994-95 to 1996-97; Wayne Cashman and Roger Neilson, 1997-98; Roger Neilson, 1998-99, 1999-2000; Craig Ramsay and Bill Barber, 2000-01; Bill Barber, 2001-02; Ken Hitchcock, 2002-03.

Club Records

Team

(Figures in brackets for season records are games played; records for fewest points, wins, ties, losses, goals, goals against are for 70 or more games)

Most Points **118** 1975-76 (80)
Most Wins **53** 1984-85 (80), 1985-86 (80)
Most Ties ***24** 1969-70 (76)
Most Losses **39** 1993-94 (84)
Most Goals **350** 1983-84 (80)
Most Goals Against **319** 1992-93 (84)
Fewest Points **58** 1969-70 (76)
Fewest Wins **17** 1969-70 (76)
Fewest Ties **4** 1985-86 (80)
Fewest Losses **12** 1979-80 (80)
Fewest Goals **173** 1967-68 (74)
Fewest Goals Against **164** 1973-74 (78)
Longest Winning Streak
Overall .... **13** Oct. 19-Nov. 17/85
Home .... ***20** Jan. 4-Apr. 3/76
Away .... **8** Dec. 22/82-Jan. 16/83
Longest Undefeated Streak
Overall .... ***35** Oct. 14/79-Jan. 6/80 (25 wins, 10 ties)
Home .... **26** Oct. 11/79-Feb. 3/80 (19 wins, 7 ties)
Away .... **16** Oct. 20/79-Jan. 6/80 (11 wins, 5 ties)
Longest Losing Streak
Overall .... **6** Mar. 25-Apr. 4/70, Dec. 5-Dec. 17/92, Jan. 25-Feb. 5/94
Home .... **5** Jan. 30-Feb. 15/69, Dec. 19/89-Jan. 23/90
Away .... **8** Oct. 25-Nov. 26/72, Mar. 3-29/88
Longest Winless Streak
Overall .... **12** Feb. 24-Mar. 16/99 (8 losses, 4 ties)
Home .... **8** Dec. 19/68-Jan. 18/69 (4 losses, 4 ties), Nov. 17-Dec. 14/91 (4 losses, 4 ties)
Away .... **19** Oct. 23/71-Jan. 27/72 (15 losses, 4 ties)
Most Shutouts, Season **13** 1974-75 (80)
Most PIM, Season **2,621** 1980-81 (80)
Most Goals, Game .... **13** Mar. 22/84 (Pit. 4 at Phi. 13), Oct. 18/84 (Van. 2 at Phi. 13)

Individual

Most Seasons .... **15** Bobby Clarke
Most Games .... **1,144** Bobby Clarke
Most Goals, Career .... **420** Bill Barber
Most Assists, Career .... **852** Bobby Clarke
Most Points, Career .... **1,210** Bobby Clarke (358G, 852A)
Most PIM, Career .... **1,817** Rick Tocchet
Most Shutouts, Career .... **50** Bernie Parent
Longest Consecutive Game Streak .... **484** Rod Brind'Amour (Feb. 24/93-Apr. 18/99)
Most Goals, Season .... **61** Reggie Leach (1975-76)
Most Assists, Season .... **89** Bobby Clarke (1974-75, 1975-76)
Most Points, Season .... **123** Mark Recchi (1992-93; 53G, 70A)
Most PIM, Season .... ***472** Dave Schultz (1974-75)
Most Points, Defenseman, Season .... **82** Mark Howe (1985-86; 24G, 58A)
Most Points, Center, Season .... **119** Bobby Clarke (1975-76; 30G, 89A)
Most Points, Right Wing, Season .... **123** Mark Recchi (1992-93; 53G, 70A)
Most Points, Left Wing, Season .... **112** Bill Barber (1975-76; 50G, 62A)
Most Points, Rookie, Season .... **82** Mikael Renberg (1993-94; 38G, 44A)
Most Shutouts, Season .... **12** Bernie Parent (1973-74, 1974-75)
Most Goals, Game .... **4** Fifteen times
Most Assists, Game .... **6** Eric Lindros (Feb. 26/97)
Most Points, Game .... **8** Tom Bladon (Dec. 11/77; 4G, 4A)

* NHL Record.

Retired Numbers

1	Bernie Parent	1967-1971, 1973-1979
4	Barry Ashbee	1970-1974
7	Bill Barber	1972-1985
16	Bobby Clarke	1969-1984

All-time Record vs. Other Clubs

Regular Season

	At Home								On Road								Total							
	GP	W	L	T	OL	GF	GA	PTS	GP	W	L	T	OL	GF	GA	PTS	GP	W	L	T	OL	GF	GA	PTS
Anaheim	8	3	2	3	0	22	17	9	7	2	3	2	0	24	26	6	15	5	5	5	0	46	43	15
Atlanta	6	4	0	2	0	27	15	10	6	4	1	1	0	20	14	9	12	8	1	3	0	47	29	19
Boston	69	31	29	8	1	231	201	71	72	18	44	10	0	205	276	46	141	49	73	18	1	436	477	117
Buffalo	63	36	15	12	0	219	159	84	59	23	29	7	0	170	195	53	122	59	44	19	0	389	354	137
Calgary	50	33	14	3	0	197	134	69	51	17	25	9	0	167	204	43	101	50	39	12	0	364	338	112
Carolina	39	27	9	3	0	147	96	57	40	20	13	7	0	149	134	47	79	47	22	10	0	296	230	104
Chicago	60	33	16	11	0	196	159	77	59	14	26	19	0	168	205	47	119	47	42	30	0	364	364	124
Colorado	33	22	9	2	0	121	89	46	33	10	11	12	0	118	120	32	66	32	20	14	0	239	209	78
Columbus	1	0	0	1	0	3	3	1	2	1	0	1	0	6	5	3	3	1	0	2	0	9	8	4
Dallas	65	42	9	14	0	251	146	98	65	23	27	15	0	210	214	61	130	65	36	29	0	461	360	159
Detroit	57	33	13	11	0	227	166	77	58	18	30	10	0	180	206	46	115	51	43	21	0	407	372	123
Edmonton	30	20	8	2	0	128	82	42	27	8	14	5	0	81	96	21	57	28	22	7	0	209	178	63
Florida	21	9	7	5	0	54	50	23	20	15	5	0	0	72	49	30	41	24	12	5	0	126	99	53
Los Angeles	62	39	15	7	1	239	156	86	64	36	20	8	0	219	188	80	126	75	35	15	1	458	344	166
Minnesota	2	2	0	0	0	8	1	4	1	0	0	1	0	3	3	1	3	2	0	1	0	11	4	5
Montreal	69	26	26	16	1	207	208	69	70	22	33	14	1	216	249	59	139	48	59	30	2	423	457	128
Nashville	4	1	1	1	1	12	7	4	3	2	0	1	0	6	3	5	7	3	1	2	1	18	10	9
New Jersey	79	49	21	9	0	314	206	107	78	31	40	7	0	272	269	69	157	80	61	16	0	586	475	176
NY Islanders	90	53	25	10	2	329	256	118	92	29	48	14	1	267	345	73	182	82	73	24	3	596	601	191
NY Rangers	103	52	37	14	0	339	287	118	105	36	45	23	1	308	342	96	208	88	82	37	1	647	629	214
Ottawa	19	11	6	2	0	64	53	24	20	10	7	3	0	66	56	23	39	21	13	5	0	130	109	47
Phoenix	29	21	8	0	0	125	79	42	29	15	12	2	0	101	91	32	58	36	20	2	0	226	170	74
Pittsburgh	101	78	16	7	0	430	248	163	100	35	45	20	0	325	357	90	201	113	61	27	0	755	605	253
St. Louis	66	45	11	10	0	263	150	100	66	33	26	7	0	208	189	73	132	78	37	17	0	471	339	173
San Jose	9	6	2	1	0	32	20	13	10	7	2	1	0	30	20	15	19	13	4	2	0	62	40	28
Tampa Bay	21	13	1	7	0	71	34	33	22	14	7	1	0	68	57	29	43	27	8	8	0	139	91	62
Toronto	63	41	14	8	0	246	146	90	63	27	22	14	0	211	202	68	126	68	36	22	0	457	348	158
Vancouver	52	35	16	1	0	227	156	71	51	29	10	12	0	203	144	70	103	64	26	13	0	430	300	141
Washington	79	49	24	6	0	298	214	104	76	33	30	13	0	252	252	79	155	82	54	19	0	550	466	183
Defunct Clubs	34	24	4	6	0	137	67	54	35	13	14	8	0	102	89	34	69	37	18	14	0	239	156	88
Totals	**1384**	**838**	**358**	**182**	**6**	**5164**	**3605**	**1864**	**1384**	**545**	**589**	**247**	**3**	**4427**	**4600**	**1340**	**2768**	**1383**	**947**	**429**	**9**	**9591**	**8205**	**3204**

Playoffs

	Series	W	L	GP	W	L	T	GF	GA	Last Mtg.	Rnd.	Result
Boston	4	2	2	20	9	11	0	57	60	1978	SF	L 1-4
Buffalo	7	5	2	37	23	14	0	110	96	2001	CQF	L 2-4
Calgary	2	1	1	11	7	4	0	43	28	1981	QF	L 3-4
Chicago	1	0	1	4	0	4	0	8	20	1971	QF	L 0-4
Colorado	2	2	0	11	7	4	0	39	29	1985	CF	W 4-2
Dallas	2	2	0	11	8	3	0	41	26	1980	SF	W 4-1
Detroit	1	0	1	4	0	4	0	6	16	1997	F	L 0-4
Edmonton	3	1	2	15	7	8	0	44	49	1987	F	L 3-4
Florida	1	0	1	6	2	4	0	11	15	1996	CSF	L 2-4
Montreal	4	1	3	21	7	14	0	52	72	1989	CF	L 2-4
New Jersey	3	1	2	15	7	8	0	35	41	2000	CF	L 3-4
NY Islanders	4	3	1	25	14	11	0	83	69	1987	DF	W 4-3
NY Rangers	10	6	4	47	27	20	0	157	153	1997	CF	W 4-1
Ottawa	1	0	1	5	1	4	0	2	11	2002	CQF	L 1-4
Pittsburgh	3	3	0	18	12	6	0	66	51	2000	CSF	W 4-2
St. Louis	2	0	2	11	3	8	0	20	34	1969	QF	L 0-4
Tampa Bay	1	1	0	6	4	2	0	26	13	1996	CQF	W 4-2
Toronto	4	3	1	23	14	9	0	78	56	1999	CQF	L 2-4
Vancouver	1	1	0	3	2	1	0	15	9	1979	PRE	W 2-1
Washington	3	1	2	16	7	9	0	55	65	1989	DSF	W 4-2
Totals	**59**	**33**	**26**	**309**	**161**	**148**	**0**	**948**	**913**			

Calgary totals include Atlanta Flames, 1972-73 to 1979-80. Carolina totals include Hartford, 1979-80 to 1996-97.
Colorado totals include Quebec, 1979-80 to 1994-95. Dallas totals include Minnesota North Stars, 1967-68 to 1992-93.
New Jersey totals include Kansas City, 1974-75 to 1975-76, and Colorado Rockies, 1976-77 to 1981-82.
Phoenix totals include Winnipeg, 1979-80 to 1995-96.

Playoff Results 2002-1998

Year	Round	Opponent	Result	GF	GA
2002	CQF	Ottawa	L 1-4	2	11
2001	CQF	Buffalo	L 2-4	13	21
2000	CF	New Jersey	L 3-4	15	18
	CSF	Pittsburgh	W 4-2	15	14
	CQF	Buffalo	W 4-1	14	8
1999	CQF	Toronto	L 2-4	11	9
1998	CQF	Buffalo	L 1-4	9	18

Abbreviations: Round: F – Final; **CF** – conference final; **CSF** – conference semi-final; **CQF** – conference quarter-final; **DF** – division final; **DSF** – division semi-final; **SF** – semi-final; **QF** – quarter-final; **PRE** – preliminary round.

2001-02 Results

Oct.	4	Florida	5-2
	6	Columbus	3-3
	8	at Columbus	2-2
	10	at Buffalo	1-2
	13	at Florida	5-2
	16	at Atlanta	3-3
	18	at Detroit	2-3
	20	Washington	6-3
	25	Ottawa	2-7
	27	at Montreal	5-1
	30	at Washington	3-0
	31	Pittsburgh	3-0
Nov.	3	NY Islanders	1-2
	6	at Chicago	1-2
	8	at Tampa Bay	2-1
	10	at Florida	3-2*
	14	at NY Rangers	2-4
	15	Washington	5-0
	17	at New Jersey	3-1
	20	New Jersey	3-3
	23	at Dallas	3-3
	25	Vancouver	1-4
	29	Boston	2-3*
Dec.	1	Tampa Bay	2-0
	4	at NY Islanders	3-2
	6	NY Islanders	0-2
	8	Minnesota	5-1
	10	at Atlanta	3-1
	13	Montreal	2-3
	15	at Boston	5-2
	16	Edmonton	2-3
	18	St. Louis	6-3
	20	Dallas	2-1
	22	Carolina	4-3*
	26	at Washington	4-1
	28	at Phoenix	2-4
	29	at Colorado	5-2
	31	at Vancouver	2-1
Jan.	2	at San Jose	2-5
	6	at Carolina	4-3
	8	Atlanta	7-4
	10	New Jersey	3-2
	12	NY Rangers	4-2
	14	at Montreal	5-3
	15	at Ottawa	4-1
	17	Atlanta	6-3
	19	at Toronto	3-0
	21	at Pittsburgh	2-5
	22	Ottawa	1-1
	24	Nashville	2-3*
	26	Carolina	4-2
	29	Pittsburgh	3-2*
	30	at Ottawa	1-3
Feb.	4	at Los Angeles	3-1
	6	at Anaheim	4-5
	9	at St. Louis	5-0
	12	NY Islanders	0-1*
	26	Chicago	5-4
	27	at New Jersey	1-0
Mar.	2	at NY Rangers	5-6
	4	at Boston	4-1
	7	Calgary	2-4
	8	at Tampa Bay	4-2
	10	Toronto	1-3
	12	at Toronto	1-1
	14	Buffalo	1-3
	16	Colorado	1-2
	18	Tampa Bay	3-3
	21	Anaheim	2-1
	23	at Pittsburgh	4-4
	25	Toronto	4-1
	27	at NY Rangers	4-2
	28	at Carolina	1-4
	30	Buffalo	1-3
Apr.	1	at Buffalo	1-3
	2	Boston	2-4
	4	Montreal	1-3
	6	Pittsburgh	3-1
	8	Florida	4-4
	10	at New Jersey	0-1
	13	NY Rangers	2-1
	14	at NY Islanders	1-3

* – Overtime

Entry Draft Selections 2002-1988

2002
Pick
4 Joni Pitkanen
105 Rosario Ruggeri
126 Konstantin Baranov
161 Dov Grumet-Morris
192 Nikita Korovkin
193 Joey Mormina
201 Mathieu Brunelle

2001
Pick
27 Jeff Woywitka
95 Patrick Sharp
146 Jussi Timonen
150 Bernd Bruckler
158 Roman Malek
172 Denis Seidenberg
177 Andrei Razin
208 Thierry Douville
225 David Printz

2000
Pick
28 Justin Williams
94 Alexander Drozdetsky
171 Roman Cechmanek
195 Colin Shields
210 John Eichelberger
227 Guillaume Lefebvre
259 Regan Kelly
287 Milan Kopecky

1999
Pick
22 Maxime Ouellet
119 Jeff Feniak
160 Konstantin Rudenko
200 Pavel Kasparik
208 Vaclav Pletka
224 David Nystrom

1998
Pick
22 Simon Gagne
42 Jason Beckett
51 Ian Forbes
109 Jean-Philippe Morin
124 Francis Belanger
139 Garrett Prosofsky
168 Antero Niittymaki
175 Cam Ondrik
195 Tomas Divisek
222 Lubomir Pistek
243 Petr Hubacek
253 Bruno St-Jacques
258 Sergei Skrobot

1997
Pick
30 Jean-Marc Pelletier
50 Pat Kavanagh
62 Kris Mallette
103 Mikhail Chernov
158 Jordon Flodell
164 Todd Fedoruk
214 Marko Kauppinen
240 Par Styf

1996
Pick
15 Dainius Zubrus
64 Chester Gallant
124 Per-Ragna Bergqvist
133 Jesse Boulerice
187 Roman Malov
213 Jeff Milleker

1995
Pick
22 Brian Boucher
48 Shane Kenny
100 Radovan Somik
132 Dmitri Tertyshny
135 Jamie Sokolsky
152 Martin Spanhel
178 Martin Streit
204 Ruslan Shafikov
230 Jeff Lank

1994
Pick
62 Artem Anisimov
88 Adam Magarrell
101 Sebastien Vallee
140 Alex Selivanov
166 Colin Forbes
192 Derek Diener
202 Ray Giroux
218 Johan Hedberg
244 Andre Payette
270 Jan Lipiansky

1993
Pick
36 Janne Niinimaa
71 Vaclav Prospal
77 Milos Holan
114 Vladimir Krechin
140 Mike Crowley
166 Aaron Israel
192 Paul Healey
218 Tripp Tracy
226 E.J. Bradley
244 Jeff Staples
270 Ken Hemenway

1992
Pick
7 Ryan Sittler
15 Jason Bowen
31 Denis Metlyuk
103 Vladislav Buljin
127 Roman Zolotov
151 Kirk Daubenspeck
175 Claude Jr. Jutras
199 Jonas Hakansson
223 Chris Herperger
247 Patrice Paquin

1991
Pick
6 Peter Forsberg
50 Yanick Dupre
86 Aris Brimanis
94 Yanick Degrace
116 Clayton Norris
122 Dmitry Yushkevich
138 Andrei Lomakin
182 James Bode
204 Josh Bartell
226 Neil Little
248 John Porco

1990
Pick
4 Mike Ricci
25 Chris Simon
40 Mikael Renberg
44 Kimbi Daniels
46 Bill Armstrong
47 Chris Therien
52 Al Kinisky
88 Dan Kordic
109 Viacheslav Butsayev
151 Patrik Englund
172 Toni Porkka
193 Greg Hanson
214 Tommy Soderstrom
235 William Lund

1989
Pick
33 Greg Johnson
34 Patrik Juhlin
72 Reid Simpson
117 Niklas Eriksson
138 John Callahan Jr.
159 Sverre Sears
180 Glen Wisser
201 Al Kummu
222 Matt Brait
243 James Pollio

1988
Pick
14 Claude Boivin
35 Pat Murray
56 Craig Fisher
63 Dominic Roussel
77 Scott Lagrand
98 Edward O'Brien
119 Gordie Frantti
140 Jamie Cooke
161 Johan Salle
182 Brian Arthur
203 Jeff Dandretta
224 Scott Billey
245 Drahomir Kadlec

General Managers' History

Bud Poile, 1967-68, 1968-69; Bud Poile and Keith Allen, 1969-70; Keith Allen, 1970-71 to 1982-83; Bob McCammon, 1983-84; Bob Clarke, 1984-85 to 1989-90; Russ Farwell, 1990-91 to 1993-94; Bob Clarke, 1994-95 to date.

President and General Manager

CLARKE, BOB
President/General Manager, Philadelphia Flyers.
Born in Flin Flon, Man., August 13, 1949.

Bob Clarke was named president and general manager of the Philadelphia Flyers on June 15, 1994. Clarke's appointment marked the second time he has served as the Flyers' general manager. The Flin Flon native was the Flyers' vice president and general manager from 1984 to 1990. During his 14 years as the team's general manager, the Flyers have won six divisional titles, three conference championships, reached the Stanley Cup semifinals six times and the finals three times.

Prior to re-joining the Flyers' family in 1994, Clarke served as vice president and general manager of the Florida Panthers. In 1993-94, their first season in the NHL, the Panthers established NHL records for wins (33) and points (83) by an expansion franchise. Clarke also served as the vice president and general manager of the Minnesota North Stars from 1990 to 1992, guiding the team to the Stanley Cup Finals in 1991.

As a player, the former Philadelphia captain led his club to Stanley Cup championships in 1974 and 1975 and captured numerous individual awards, including the Hart Trophy as the league's most valuable player in 1973, 1975 and 1976. The four-time All-Star also received the Bill Masterton Memorial Trophy (perseverance and dedication) in 1972 and the Frank J. Selke Trophy (top defensive forward) in 1983. He appeared in eight All-Star Games and was elected to the Hockey Hall of Fame in 1987. He was awarded the Lester Patrick Trophy in 1979-80 in recognition of his contribution to hockey in the United States. Clarke appeared in 1,144 regular season games, recording 358 goals and 852 assists for 1,210 points. He also added 119 points in 136 playoff games.

Club Directory

First Union Center

Philadelphia Flyers
First Union Center
3601 South Broad Street
Philadelphia, PA 19148-5290
Phone **215/465-4500**
PR FAX 215/389-9403
www.philadelphiaflyers.com
Capacity: 19,523

Executive Management

Chairman	Ed Snider
Limited Partners	Pat Croce, Jay Snider, Sylvan and Fran Tobin
President and General Manager	Bob Clarke
Executive Vice President and Chief Operating Officer	Ron Ryan
Executive Vice President	Keith Allen
Governor	Ed Snider
Alternate Governors	Bob Clarke, Ron Ryan, Phil Weinberg
Executive Assistants	Lisa D'Aprile, Gina Pelle
Receptionist	Maureen McGuckin

Hockey Club Personnel

Assistant General Manager	Paul Holmgren
Head Coach	Ken Hitchcock
Assistant Coaches	Wayne Fleming, Craig Hartsburg
Goaltending Coach	Rejean Lemelin
Skating Coach	David Roy
Director of Pro Hockey Personnel	Ron Hextall
Chief Scout	Dennis Patterson
Scouting Staff	Serge Boudreault, John Chapman, Inge Hammarstrom, Simon Nolet, Chris Pryor, Vaclav Slansky, Evgeny Zimin
Pro Scouts	Al Hill, Terry Murray
Assistant to the President	Barry Hanrahan
Video Coordinator	Steve Romanowski
Scouting Information Coordinator	Bryan Hardenbergh
Executive Assistant	Dianna Taylor
Receptionist	Sharon Allison

Medical/Training Staff

Team Physicians	Arthur Bartolozzi, M.D.; Gary Dorshimer, M.D.; Jeff Hartzell, M.D.; Guy Lanzi, D.M.D.
Athletic Trainer	John Worley
Strength and Conditioning Coach	Jim McCrossin
Massage Therapist	Tom D'Ancona
Head Equipment Manager	Jim Evers
Equipment Managers	Anthony Oratorio, Harry Bricker, Luke Clarke
Training Center Maintenance	Mike Craytor

Public Relations Department

Director of Public Relations	Zack Hill
Assistant Director of Public Relations	Jill Lipson
Director of Media Services and Publications	Joe Klueg
Director of Fan Services	Joe Kadlec
Archivist and Special Projects Manager	Kerrianne Brady
Public Relations Assistant	Kevin Kurz

Fan Development Department

Executive Director of Youth Hockey and Fan Development	Eric Turner
Director of Youth Hockey and Fan Development	Melissa Wilson

Sales/Marketing Department

Vice President, Sales	Jack Betson
Director of Community Relations	Linda Panasci
Director of Ticket Operations	Cecilia Baker
Ticket Office Administration	Joan Kadlec
Manager, Sales and Services	Nicole Allison
Marketing Assistant	Kevin Morley
Assistant Manager, Sales and Services	Diane Smith
Account Executive, Sales and Marketing	Edwin Gregory
Manager of Game Presentation and Special Events	Linda Held
Assistant to the VP, Sales and Marketing	Debbie Brown

Finance Department

Director of Finance	Dave Jablonski
Controller	Lisa Cataldo
Payroll Accountant	Susann Schaffer
Accounts Payable	Marilyn Trout

Advertising Sales Department

Vice President, Sales	Joe Croce
Director, Advertising Sales	Jeffrey Kirk
General Sales Manager	Brian Monihan
National Sales Manager	Lee Stein
Senior Account Executive	Joe Watson
Account Executives	Stephanie Bennett, Mike Garrity, Joe Heyer, Andrew Humphreys, Steve Jeffries, Traci Kloss, Bo Koelle, Ray Lyons, Steve Rex, Jon Roche, Peter Schwartz
Sponsorship Manager	Maura Hood
Television Coordinator	Shannan Archer
Manager of Contracts and Client Services	Thea Vogel
Senior Account Coordinator	Colleen Molloy
Account Coordinator	Kim Windt
Director of Premium Seating	Rick Campbell
Manager of Finance and Inventory	Jimmy Dunk
Sales Executives	Chris Genther, Tara Ritting, Pete Seelaus, Dennis Shea
Administrative Assistant	Tarah Linus

Broadcast Department

TV Play-by-Play, Analyst, Color Commentary	Jim Jackson, Gary Dornhoefer, Steve Coates
Radio Play-by-Play, Color Analyst	Tim Saunders, Brian Propp
Executive Producer/Director of Broadcasting	Bryan Cooper
Associate Producer	Jennifer Roman
Director, Flyers Game Operations	Brian Mantai
Public Address Announcer	Lou Nolan
TV Rightsholders	Comcast SportsNet, UPN-57 WPSG-TV
Radio Rightsholder	SportsRadio 610 WIP (610 AM)

Flyers Wives Charities

Executive Director	Fran Tobin
Director	Rita Johanson
Event Coordinator	Roseanne Uhl, Susan Wechsler

Phoenix Coyotes

2001-02 Results: 40W-27L-9T-6OTL 95PTS.
Second, Pacific Division

2002-03 Schedule

Oct.	Wed.	9	at Los Angeles		Wed.	8	at Chicago
	Sat.	12	Dallas		Fri.	10	at Minnesota
	Mon.	14	at Columbus		Sat.	11	at Nashville
	Tue.	15	at Ottawa		Tue.	14	St. Louis
	Thu.	17	at Toronto		Sat.	18	Buffalo
	Sat.	19	at Buffalo		Mon.	20	San Jose
	Tue.	22	at Nashville		Thu.	23	at Calgary
	Thu.	24	Colorado		Fri.	24	at Edmonton
	Sat.	26	Minnesota		Sun.	26	at Vancouver
	Mon.	28	at NY Rangers		Tue.	28	Calgary
	Tue.	29	at NY Islanders	**Feb.**	Wed.	5	at Los Angeles
	Thu.	31	at Philadelphia		Fri.	7	at Anaheim
Nov.	Sun.	3	Nashville*		Sat.	8	Dallas
	Thu.	7	Minnesota		Wed.	12	New Jersey
	Sat.	9	Vancouver		Fri.	14	at Minnesota
	Mon.	11	at Tampa Bay		Sat.	15	at St. Louis
	Tue.	12	at Carolina		Tue.	18	Columbus
	Fri.	15	at Atlanta		Thu.	20	at Chicago
	Sun.	17	Colorado*		Fri.	21	at Dallas
	Wed.	20	Dallas		Sun.	23	Calgary*
	Fri.	22	Florida		Wed.	26	Carolina
	Mon.	25	at Dallas		Fri.	28	Anaheim
	Wed.	27	at Anaheim	**Mar.**	Sun.	2	at Detroit
	Thu.	28	Chicago		Tue.	4	at Pittsburgh
	Sat.	30	at San Jose		Thu.	6	at St. Louis
Dec.	Tue.	3	San Jose		Sat.	8	San Jose
	Thu.	5	Detroit		Mon.	10	at Colorado
	Sat.	7	Montreal		Wed.	12	Detroit
	Mon.	9	Columbus		Fri.	14	Chicago
	Wed.	11	Atlanta		Sat.	15	Anaheim
	Fri.	13	Washington		Tue.	18	Boston
	Sun.	15	Los Angeles*		Thu.	20	Edmonton
	Tue.	17	Pittsburgh		Sat.	22	Tampa Bay
	Fri.	20	St. Louis		Mon.	24	at Calgary
	Sun.	22	at Anaheim*		Wed.	26	at Edmonton
	Thu.	26	at Los Angeles		Thu.	27	at Vancouver
	Sat.	28	Philadelphia		Sat.	29	at Colorado*
	Mon.	30	Edmonton		Mon.	31	Los Angeles
Jan.	Wed.	1	at Washington*	**Apr.**	Wed.	2	Vancouver
	Fri.	3	at Detroit		Fri.	4	Nashville
	Sat.	4	at Columbus		Sun.	6	at San Jose*

* Denotes afternoon game.

Year-by-Year Record

		Home				Road				Overall								
Season	**GP**	**W**	**L**	**T**	**OL**	**W**	**L**	**T**	**OL**	**W**	**L**	**T**	**OL**	**GF**	**GA**	**Pts.**	**Finished**	**Playoff Result**
2001-02	82	27	8	3	3	13	19	6	3	40	27	9	6	228	210	95	2nd, Pacific Div.	Lost Conf. Quarter-Final
2000-01	82	21	11	7	2	14	16	10	1	35	27	17	3	214	212	90	4th, Pacific Div.	Out of Playoffs
1999-2000	82	22	16	2	1	17	15	6	3	39	31	8	4	232	228	90	3rd, Pacific Div.	Lost Conf. Quarter-Final
1998-99	82	23	13	5	...	16	18	7	...	39	31	12	...	205	197	90	2nd, Pacific Div.	Lost Conf. Quarter-Final
1997-98	82	19	16	6	...	16	19	6	...	35	35	12	...	224	227	82	4th, Central Div.	Lost Conf. Quarter-Final
1996-97	82	15	19	7	...	23	18	0	...	38	37	7	...	240	243	83	3rd, Central Div.	Lost Conf. Quarter-Final
1995-96*	82	22	16	3	...	14	24	3	...	36	40	6	...	275	291	78	5th, Central Div.	Lost Conf. Quarter-Final
1994-95*	48	10	10	4	...	6	15	3	...	16	25	7	...	157	177	39	6th, Central Div.	Out of Playoffs
1993-94*	84	15	23	4	...	9	28	5	...	24	51	9	...	245	344	57	6th, Central Div.	Out of Playoffs
1992-93*	84	23	16	3	...	17	21	4	...	40	37	7	...	322	320	87	4th, Smythe Div.	Lost Div. Semi-Final
1991-92*	80	20	14	6	...	13	18	9	...	33	32	15	...	251	244	81	4th, Smythe Div.	Lost Div. Semi-Final
1990-91*	80	17	18	5	...	9	25	6	...	26	43	11	...	260	288	63	5th, Smythe Div.	Out of Playoffs
1989-90*	80	22	13	5	...	15	19	6	...	37	32	11	...	298	290	85	3rd, Smythe Div.	Lost Div. Semi-Final
1988-89*	80	17	18	5	...	9	24	7	...	26	42	12	...	300	355	64	5th, Smythe Div.	Out of Playoffs
1987-88*	80	20	14	6	...	13	22	5	...	33	36	11	...	292	310	77	3rd, Smythe Div.	Lost Div. Semi-Final
1986-87*	80	25	12	3	...	15	20	5	...	40	32	8	...	279	271	88	3rd, Smythe Div.	Lost Div. Final
1985-86*	80	18	19	3	...	8	28	4	...	26	47	7	...	295	372	59	3rd, Smythe Div.	Lost Div. Semi-Final
1984-85*	80	21	13	6	...	22	14	4	...	43	27	10	...	358	332	96	2nd, Smythe Div.	Lost Div. Final
1983-84*	80	17	15	8	...	14	23	3	...	31	38	11	...	340	374	73	4th, Smythe Div.	Lost Div. Semi-Final
1982-83*	80	22	16	2	...	11	23	6	...	33	39	8	...	311	333	74	4th, Smythe Div.	Lost Div. Semi-Final
1981-82*	80	18	13	9	...	15	20	5	...	33	33	14	...	319	332	80	2nd, Norris Div.	Lost Div. Semi-Final
1980-81*	80	7	25	8	...	2	32	6	...	9	57	14	...	246	400	32	6th, Smythe Div.	Out of Playoffs
1979-80*	80	13	19	8	...	7	30	3	...	20	49	11	...	214	314	51	5th, Smythe Div.	Out of Playoffs

* Winnipeg Jets

Despite the departure of several key players, the Phoenix Coyotes enjoyed a strong season in 2001-02. Daymond Langkow was one of the reasons why. Acquired from the Flyers, Langkow led Phoenix in scoring with 62 points.

Franchise date: June 22, 1979

Transferred from Winnipeg to Phoenix, July 1, 1996

PACIFIC DIVISION

24th NHL Season

2002-03 Player Personnel

FORWARDS	HT	WT	S	Place of Birth	Date	2001-02 Club
ABID, Ramzi	6-2	210	L	Montreal, Que.	3/24/80	Springfield
AMONTE, Tony	6-0	200	L	Hingham, MA	8/2/70	Chicago-United States
BRIERE, Daniel	5-10	178	R	Gatineau, Que.	10/6/77	Phoenix
BUCHBERGER, Kelly	6-2	210	L	Langenburg, Sask.	12/2/66	Los Angeles
DOAN, Shane	6-2	228	R	Halkirk, Alta.	10/10/76	Phoenix
HORDICHUK, Darcy	6-1	215	L	Kamsack, Sask.	8/10/80	Atl-Chi (AHL)-Phx
JASPERS, Jason	5-11	204	L	Thunder Bay, Ont.	4/8/81	Phoenix-Springfield
JOHNSON, Mike	6-2	201	R	Scarborough, Ont.	10/3/74	Phoenix
KOLANOS, Krys	6-3	201	R	Calgary, Alta.	7/27/81	Phoenix
LANGKOW, Daymond	5-11	183	L	Edmonton, Alta	9/27/76	Phoenix
LEMIEUX, Claude	6-1	227	R	Buckingham, Que.	7/16/65	Phoenix
MAY, Brad	6-1	217	L	Toronto, Ont.	11/29/71	Phoenix
NAGY, Ladislav	5-11	186	L	Saca, Czech.	6/1/79	Phoenix
NAZAROV, Andrei	6-5	241	R	Chelyabinsk, USSR	5/22/74	Boston-Phoenix
PODLESAK, Martin	6-6	218	L	Melnik, Czech.	9/26/82	Lethbridge
RADIVOJEVIC, Branko	6-1	209	R	Piestany, Czech.	11/24/80	Phoenix-Springfield
SAVAGE, Brian	6-1	205	L	Sudbury, Ont.	2/24/71	Montreal-Phoenix
SJOSTROM, Fredrik	6-1	210	L	Fargelanda, Sweden	5/6/83	Calgary (WHL)
TAFFE, Jeff	6-3	195	L	Hastings, MN	2/19/81	U. of Minnesota
WESTRUM, Erik	6-0	204	L	Minneapolis, MN	7/26/79	Springfield
WILSON, Landon	6-3	232	R	St. Louis, MO	3/13/75	Phoenix-Springfield
DEFENSEMEN						
BEREHOWSKY, Drake	6-2	225	R	Toronto, Ont.	1/3/72	Vancouver-Phoenix
BEZINA, Goran	6-2	220	L	Split, Yugoslavia	3/21/80	Springfield
FOCHT, Dan	6-6	242	L	Regina, Sask.	12/31/77	Phoenix-Springfield
GRENIER, Martin	6-5	245	L	Laval, Que.	11/2/80	Phoenix-Springfield
MARA, Paul	6-4	217	L	Ridgewood, NJ	9/7/79	Phoenix
MARKOV, Danny	6-1	190	L	Moscow, USSR	7/30/76	Phoenix-Russia
NUMMINEN, Teppo	6-2	197	R	Tampere, Finland	7/3/68	Phoenix-Finland
SCHUTTE, Michael	6-2	199	L	Burlington, Ont.	7/28/79	U. of Maine
SIMPSON, Todd	6-3	218	L	North Vancouver, B.C.	5/28/73	Phoenix
SUCHY, Radoslav	6-2	198	L	Kezmarok, Czech.	4/7/76	Phoenix
VAANANEN, Ossi	6-4	215	L	Vantaa, Finland	8/18/80	Phoenix-Finland

GOALTENDERS	HT	WT	C	Place of Birth	Date	2001-02 Club
BIERK, Zac	6-4	205	L	Peterborough, Ont.	9/17/76	Augusta-Springfield
BLACKBURN, Josh	6-0	203	L	Delrio, TX	11/13/78	U. of Michigan
BOUCHER, Brian	6-2	190	L	Woonsocket, RI	1/2/77	Philadelphia
BURKE, Sean	6-4	211	L	Windsor, Ont.	1/29/67	Phoenix
DesROCHERS, Patrick	6-3	209	L	Penetanguishene, Ont.	10/27/79	Phoenix-Springfield

Coach

FRANCIS, BOB
Coach, Phoenix Coyotes. Born in North Battleford, Sask., December 5, 1958.

The Phoenix Coyotes named Bob Francis as the team's head coach on June 16, 1999. Francis became the 14th head coach in franchise history and the third since moving to Phoenix in 1996. During the 2001-02 season, he led a rebuilding Coyotes club to an impressive 95-point season and was rewarded with the Jack Adams Award as coach of the year.

Francis joined the Coyotes after two successful seasons as an assistant coach with the Boston Bruins. The son of former NHL coaching great Emile Francis joined the Bruins as an assistant on June 27, 1997. He had previously spent two seasons in the Boston organization as head coach of the Bruins' AHL affiliate in Providence.

Francis began his coaching career with the Calgary Flames organization in the 1986-87 season, first as a player/assistant coach with Calgary's IHL affiliate in Salt Lake City. Francis helped guide the Golden Eagles to the IHL championship, winning the Turner Cup that season and successfully defending its title the following year with Francis serving as a full-time assistant coach. In 1989-90, Francis became the Golden Eagles' head coach and held that position for four seasons. The highlight of his coaching career at Salt Lake City was a 50-win season during the 1990-91 campaign. When Calgary moved their development team to Saint John (AHL) in 1993-94, Francis moved as well and served as their head coach before joining Providence.

Before his move to the coaching ranks, Francis played four years of college hockey at the University of New Hampshire (ECAC). Francis spent most of his professional career at the minor-league level though he did play 14 NHL games with Detroit during the 1982-83 season.

Coaching Record

		Regular Season				Playoffs		
Season	Team	Games	W	L	T	Games	W	L
1989-90	Salt Lake (IHL)	82	37	36	9	10	5	5
1990-91	Salt Lake (IHL)	83	50	28	5	4	0	4
1991-92	Salt Lake (IHL)	82	33	40	9	5	1	4
1992-93	Salt Lake (IHL)	82	38	39	5	....	....	....
1993-94	Saint John (AHL)	80	37	33	10	7	3	4
1994-95	Saint John (AHL)	80	27	40	13	5	1	4
1995-96	Providence (AHL)	80	30	40	10	4	1	3
1996-97	Providence (AHL)	80	35	40	5	10	4	6
1999-2000	**Phoenix (NHL)**	**82**	**39**	**35**	**8**	**5**	**1**	**4**
2000-01	**Phoenix (NHL)**	**82**	**35**	**30**	**17**	....	....	....
2001-02	**Phoenix (NHL)**	**82**	**40**	**33**	**9**	**5**	**1**	**4**
	NHL Totals	**246**	**114**	**98**	**34**	**10**	**2**	**8**

2001-02 Scoring

* - rookie

Regular Season

Pos	#	Player	Team	GP	G	A	Pts	+/–	PIM	PP	SH	GW	GT	S	%
C	11	Daymond Langkow	PHX	80	27	35	62	18	36	6	3	2	1	171	15.8
C	8	Daniel Briere	PHX	78	32	28	60	6	52	12	0	5	1	149	21.5
R	19	Shane Doan	PHX	81	20	29	49	11	61	6	0	2	0	205	9.8
D	27	Teppo Numminen	PHX	76	13	35	48	13	20	4	0	6	0	117	11.1
C	16	Michal Handzus	PHX	79	15	30	45	–8	34	3	1	1	0	94	16.0
C	17	Ladislav Nagy	PHX	74	23	19	42	6	50	5	0	5	0	187	12.3
L	49	Brian Savage	MTL	47	14	15	29	–14	30	7	0	2	0	117	12.0
			PHX	30	6	6	12	1	8	2	0	2	1	47	12.8
			TOTAL	77	20	21	41	–13	38	9	0	4	1	164	12.2
R	22	Claude Lemieux	PHX	82	16	25	41	–5	70	4	1	3	0	174	9.2
D	55	Danny Markov	PHX	72	6	30	36	–7	67	4	0	1	0	103	5.8
R	12	Mike Johnson	PHX	57	5	22	27	14	28	1	2	0	0	73	6.8
D	23	Paul Mara	PHX	75	7	17	24	–6	58	2	0	0	1	112	6.3
C	36 *	Krys Kolanos	PHX	57	11	11	22	6	48	0	0	5	0	81	13.6
L	32	Brad May	PHX	72	10	12	22	11	95	1	0	3	1	105	9.5
R	28	Landon Wilson	PHX	47	7	12	19	4	46	1	0	0	0	100	7.0
D	15	Radoslav Suchy	PHX	81	4	13	17	25	10	1	0	0	0	49	8.2
D	2	Todd Simpson	PHX	67	2	13	15	20	152	0	0	0	0	51	3.9
D	4	Ossi Vaananen	PHX	76	2	12	14	6	74	0	1	0	0	41	4.9
L	44	Andrei Nazarov	BOS	47	0	2	2	–2	164	0	0	0	0	18	0.0
			PHX	30	6	3	9	7	51	0	0	0	0	38	15.8
			TOTAL	77	6	5	11	5	215	0	0	0	0	56	10.7
C	10	Denis Pederson	VAN	29	1	5	6	–2	31	0	0	0	0	24	4.2
			PHX	19	1	1	2	–2	20	0	0	0	0	18	5.6
			TOTAL	48	2	6	8	–4	51	0	0	0	0	42	4.8
D	5	Drake Berehowsky	VAN	25	1	2	3	–5	18	0	0	1	0	15	6.7
			PHX	32	1	4	5	5	42	0	0	0	0	23	4.3
			TOTAL	57	2	6	8	0	60	0	0	1	0	38	5.3
R	29 *	Branko Radivojevic	PHX	18	4	2	6	1	4	0	0	1	0	19	21.1
C	18	Sebastien Bordeleau	MIN	14	1	4	5	–1	8	0	0	0	0	25	4.0
			PHX	6	0	0	0	–1	2	0	0	0	0	3	0.0
			TOTAL	20	1	4	5	–2	10	0	0	0	0	28	3.6
C	26	Mike Sullivan	PHX	42	1	2	3	–3	16	0	0	0	0	28	3.6
L	24 *	Darcy Hordichuk	ATL	33	1	1	2	–5	127	0	0	0	0	8	12.5
			PHX	1	0	0	0	0	14	0	0	0	0	0	0.0
			TOTAL	34	1	1	2	–5	141	0	0	0	0	8	12.5
C	45 *	Jason Jaspers	PHX	4	0	1	1	–1	4	0	0	0	0	1	0.0
L	31 *	Jean-Guy Trudel	PHX	3	0	0	0	0	2	0	0	0	0	1	0.0
D	38 *	Martin Grenier	PHX	5	0	0	0	0	5	0	0	0	0	1	0.0
R	14	Tyler Bouck	PHX	7	0	0	0	–1	4	0	0	0	0	3	0.0
D	37 *	Dan Focht	PHX	8	0	0	0	0	11	0	0	0	0	5	0.0
C	20	Wyatt Smith	PHX	10	0	0	0	–5	0	0	0	0	0	4	0.0

Goaltending

No.	Goaltender	GPI	Mins	Avg	W	L	T	EN	SO	GA	SA	S%	G	A	PIM
1	Sean Burke	60	3587	2.29	33	21	6	3	5	137	1711	.920	0	1	14
42	Robert Esche	22	1145	2.72	6	10	2	3	1	52	533	.902	0	0	16
40	* Patrick DesRochers	5	243	3.70	1	2	1	0	0	15	99	.848	0	0	2
	Totals	**82**	**4993**	**2.52**	**40**	**33**	**9**	**6**	**6**	**210**	**2349**	**.911**			

Playoffs

Pos	#	Player	Team	GP	G	A	Pts	+/–	PIM	PP	SH	GW	GT	S	%
R	19	Shane Doan	PHX	5	2	2	4	–2	6	0	0	0	0	14	14.3
C	8	Daniel Briere	PHX	5	2	1	3	–4	2	1	0	1	0	12	16.7
R	12	Mike Johnson	PHX	5	1	1	2	1	6	0	0	0	0	7	14.3
C	10	Denis Pederson	PHX	5	0	2	2	2	0	0	0	0	0	6	0.0
D	2	Todd Simpson	PHX	5	0	2	2	–1	6	0	0	0	0	7	0.0
C	11	Daymond Langkow	PHX	5	1	0	1	–2	0	0	0	0	0	11	9.1
D	15	Radoslav Suchy	PHX	5	1	0	1	–5	0	0	0	0	0	5	20.0
D	37 *	Dan Focht	PHX	1	0	1	1	1	0	0	0	0	0	1	0.0
D	5	Drake Berehowsky	PHX	5	0	1	1	–3	4	0	0	0	0	4	0.0
R	29 *	Branko Radivojevic	PHX	1	0	0	0	–1	2	0	0	0	0	0	0.0
C	36 *	Krys Kolanos	PHX	2	0	0	0	–1	6	0	0	0	0	1	0.0
L	44	Andrei Nazarov	PHX	3	0	0	0	0	2	0	0	0	0	2	0.0
D	27	Teppo Numminen	PHX	4	0	0	0	0	2	0	0	0	0	6	0.0
R	28	Landon Wilson	PHX	4	0	0	0	–2	12	0	0	0	0	2	0.0
R	22	Claude Lemieux	PHX	5	0	0	0	–2	2	0	0	0	0	13	0.0
L	32	Brad May	PHX	5	0	0	0	0	0	0	0	0	0	5	0.0
L	49	Brian Savage	PHX	5	0	0	0	–1	0	0	0	0	0	8	0.0
C	16	Michal Handzus	PHX	5	0	0	0	–2	2	0	0	0	0	7	0.0
D	23	Paul Mara	PHX	5	0	0	0	–4	4	0	0	0	0	7	0.0
C	17	Ladislav Nagy	PHX	5	0	0	0	–2	21	0	0	0	0	12	0.0
D	4	Ossi Vaananen	PHX	5	0	0	0	0	6	0	0	0	0	4	0.0

Goaltending

No.	Goaltender	GPI	Mins	Avg	W	L	EN	SO	GA	SA	S%	G	A	PIM
1	Sean Burke	5	297	2.63	1	4	0	0	13	133	.902	0	2	0
	Totals	**5**	**300**	**2.60**	**1**	**4**	**0**	**0**	**13**	**133**	**.902**			

Coaching History

Tom McVie and Bill Sutherland, 1979-80; Tom McVie, Bill Sutherland and Mike Smith, 1980-81; Tom Watt, 1981-82, 1982-83; Tom Watt and Barry Long, 1983-84; Barry Long, 1984-85; Barry Long and John Ferguson, 1985-86; Dan Maloney, 1986-87, 1987-88; Dan Maloney and Rick Bowness, 1988-89; Bob Murdoch, 1989-90, 1990-91; John Paddock, 1991-92 to 1993-94; John Paddock and Terry Simpson, 1994-95; Terry Simpson, 1995-96; Don Hay, 1996-97; Jim Schoenfeld, 1997-98, 1998-99; Bob Francis, 1999-2000 to date.

Club Records

Team

(Figures in brackets for season records are games played; records for fewest points, wins, ties, losses, goals, goals against are for 70 or more games)

Record		
Most Points	96	1984-85 (80)
Most Wins	43	1984-85 (80)
Most Ties	17	2000-01 (82)
Most Losses	57	1980-81 (80)
Most Goals	358	1984-85 (80)
Most Goals Against	400	1980-81 (80)
Fewest Points	32	1980-81 (80)
Fewest Wins	9	1980-81 (80)
Fewest Ties	6	1995-96 (82)
Fewest Losses	27	1984-85 (80), 2000-01 (82), 2001-02 (82)
Fewest Goals	205	1998-99 (82)
Fewest Goals Against	197	1998-99 (82)
Longest Winning Streak		
Overall	9	Mar. 8-27/85
Home	9	Dec. 27/92-Jan. 23/93
Away	8	Feb. 25-Apr. 6/85
Longest Undefeated Streak		
Overall	14	Oct. 25-Nov. 28/98 (12 wins, 2 ties)
Home	11	Dec. 23/83-Feb. 5/84 (6 wins, 5 ties), Oct. 15-Dec. 20/98 (10 wins, 1 tie)
Away	9	Feb. 25-Apr. 7/85 (8 wins, 1 tie)
Longest Losing Streak		
Overall	10	Nov. 30-Dec. 20/80, Feb. 6-25/94
Home	5	Oct. 29-Nov. 13/93, Mar. 13-23/00
Away	13	Jan. 26-Apr. 14/94
Longest Winless Streak		
Overall	*30	Oct. 19-Dec. 20/80 (23 losses, 7 ties)
Home	14	Oct. 19-Dec. 14/80 (9 losses, 5 ties)
Away	18	Oct. 10-Dec. 20/80 (16 losses, 2 ties)
Most Shutouts, Season	9	1998-99 (82)
Most PIM, Season	2,278	1987-88 (80)
Most Goals, Game	12	Feb. 25/85 (Wpg. 12 at NYR 5)

Individual

Most Seasons	14	Thomas Steen, Teppo Numminen
Most Games	1,020	Teppo Numminen
Most Goals, Career	379	Dale Hawerchuk
Most Assists, Career	553	Thomas Steen
Most Points, Career	929	Dale Hawerchuk (379G, 550A)
Most PIM, Career	1,508	Keith Tkachuk
Most Shutouts, Career	21	Nikolai Khabibulin
Longest Consecutive Games Streak	475	Dale Hawerchuk (Dec. 19/82-Dec. 10/88)
Most Goals, Season	76	Teemu Selanne (1992-93)
Most Assists, Season	79	Phil Housley (1992-93)
Most Points, Season	132	Teemu Selanne (1992-93; 76G, 56A)
Most PIM, Season	347	Tie Domi (1993-94)
Most Points, Defenseman, Season	97	Phil Housley (1992-93; 18G, 79A)
Most Points, Center, Season	130	Dale Hawerchuk (1984-85; 53G, 77A)
Most Points, Right Wing, Season	132	Teemu Selanne (1992-93; 76G, 56A)
Most Points, Left Wing, Season	98	Keith Tkachuk (1995-96; 50G, 48A)
Most Points, Rookie, Season	*132	Teemu Selanne (1992-93; 76G, 56A)
Most Shutouts, Season	8	Nikolai Khabibulin (1998-99)
Most Goals, Game	5	Willy Lindstrom (Mar. 2/82), Alexei Zhamnov (Apr. 1/95)
Most Assists, Game	5	Dale Hawerchuk (Mar. 6/84, Mar. 18/89, Mar. 4/90), Phil Housley (Jan. 18/93), Keith Tkachuk (Feb. 23/01)
Most Points, Game	6	Willy Lindstrom (Mar. 2/82; 5G, 1A), Dale Hawerchuk (Dec. 14/83; 3G, 3A, Mar. 5/88; 2G, 4A, Mar. 18/89; 1G, 5A), Thomas Steen (Oct. 24/84; 2G, 4A), Ed Olczyk (Dec. 21/91; 2G, 4A)

* NHL Record.

Records include Winnipeg Jets, 1979-80 through 1995-96.

Winnipeg Jets Retired Numbers

9	Bobby Hull	1972-1980
25	Thomas Steen	1981-1995

Captains' History

Lars-Erik Sjoberg, 1979-80; Morris Lukowich, 1980-81; Dave Christian, 1981-82; Dave Christian and Lucien DeBlois, 1982-83; Lucien DeBlois, 1983-84; Dale Hawerchuk, 1984-85 to 1988-89; Randy Carlyle, Dale Hawerchuk and Thomas Steen (tri-captains), 1989-90; Randy Carlyle and Thomas Steen (co-captains), 1990-91; Troy Murray, 1991-92; Troy Murray and Dean Kennedy, 1992-93; Dean Kennedy and Keith Tkachuk, 1993-94; Keith Tkachuk, 1994-95; Kris King, 1995-96; Keith Tkachuk, 1996-97 to 2000-01; Teppo Numminen, 2001-02 to date.

All-time Record vs. Other Clubs

Regular Season

	At Home								On Road								Total							
	GP	W	L	T	OL	GF	GA	PTS	GP	W	L	T	OL	GF	GA	PTS	GP	W	L	T	OL	GF	GA	PTS
Anaheim	21	8	10	1	2	61	67	19	21	8	10	2	1	57	58	19	42	16	20	3	3	118	125	38
Atlanta	3	3	0	0	0	11	4	6	2	2	0	0	0	7	4	4	5	5	0	0	0	18	8	10
Boston	29	12	14	3	0	99	101	27	29	4	22	3	0	89	133	11	58	16	36	6	0	188	234	38
Buffalo	28	13	13	2	0	87	91	28	29	4	20	5	0	74	123	13	57	17	33	7	0	161	214	41
Calgary	66	33	22	11	0	257	225	77	67	22	36	9	0	218	287	53	133	55	58	20	0	475	512	130
Carolina	30	13	14	2	1	106	110	29	28	9	12	6	1	84	100	25	58	22	26	8	2	190	210	54
Chicago	46	26	16	4	0	158	143	56	44	11	26	7	0	122	179	29	90	37	42	11	0	280	322	85
Colorado	36	15	15	6	0	133	130	36	37	15	16	4	2	128	128	36	73	30	31	10	2	261	258	72
Columbus	4	3	0	1	0	10	5	7	4	1	3	0	0	6	12	2	8	4	3	1	0	16	17	9
Dallas	48	21	24	3	0	156	164	45	50	18	25	7	0	152	184	43	98	39	49	10	0	308	348	88
Detroit	45	16	16	13	0	138	142	45	47	17	23	7	0	159	186	41	92	33	39	20	0	297	328	86
Edmonton	67	27	35	4	1	272	299	59	68	19	42	6	1	220	299	45	135	46	77	10	2	492	598	104
Florida	8	3	3	2	0	18	23	8	7	4	3	0	0	19	22	8	15	7	6	2	0	37	45	16
Los Angeles	71	35	25	11	0	284	237	81	68	32	23	13	0	272	268	77	139	67	48	24	0	556	505	158
Minnesota	4	4	0	0	0	15	3	8	4	1	1	2	0	8	9	4	8	5	1	2	0	23	12	12
Montreal	27	9	11	7	0	89	104	25	29	3	24	2	0	66	142	8	56	12	35	9	0	155	246	33
Nashville	8	5	2	0	1	27	23	11	8	4	2	1	1	23	20	10	16	9	4	1	2	50	43	21
New Jersey	30	21	6	3	0	114	78	45	28	10	12	6	0	84	94	26	58	31	18	9	0	198	172	71
NY Islanders	29	11	14	4	0	94	100	26	29	8	13	8	0	88	111	24	58	19	27	12	0	182	211	50
NY Rangers	30	13	13	4	0	105	98	30	28	9	17	2	0	100	125	20	58	22	30	6	0	205	223	50
Ottawa	9	4	4	1	0	33	34	9	10	6	3	1	0	33	27	13	19	10	7	2	0	66	61	22
Philadelphia	29	12	15	2	0	91	101	26	29	8	21	0	0	79	125	16	58	20	36	2	0	170	226	42
Pittsburgh	29	13	13	3	0	108	101	29	29	9	20	0	0	82	119	18	58	22	33	3	0	190	220	47
St. Louis	47	24	17	6	0	155	144	54	46	12	24	10	0	125	168	34	93	36	41	16	0	280	312	88
San Jose	29	16	9	3	1	95	79	36	27	11	14	2	0	87	95	24	56	27	23	5	1	182	174	60
Tampa Bay	10	6	4	0	0	28	22	12	8	5	3	0	0	30	24	10	18	11	7	0	0	58	46	22
Toronto	38	20	12	6	0	157	137	46	42	21	19	2	0	158	152	44	80	41	31	8	0	315	289	90
Vancouver	65	33	23	9	0	249	229	75	68	19	40	9	0	196	257	47	133	52	63	18	0	445	486	122
Washington	29	15	7	7	0	109	101	37	29	6	17	5	1	80	118	18	58	21	24	12	1	189	219	55
Totals	**915**	**434**	**357**	**118**	**6**	**3259**	**3095**	**992**	**915**	**298**	**491**	**119**	**7**	**2846**	**3569**	**722**	**1830**	**732**	**848**	**237**	**13**	**6105**	**6664**	**1714**

Playoffs

	Series	W	L	GP	W	L	T	GF	GA	Last Mtg.	Rnd.	Result
Anaheim	1	0	1	7	3	4	0	17	17	1997	CQF	L 3-4
Calgary	3	2	1	13	7	6	0	45	43	1987	DSF	W 4-2
Colorado	1	0	1	5	1	4	0	10	17	2000	CQF	L 1-4
Detroit	2	0	2	12	4	8	0	28	44	1998	CQF	L 2-4
Edmonton	6	0	6	26	4	22	0	75	120	1990	DSF	L 3-4
St. Louis	2	0	2	11	4	7	0	29	39	1999	CQF	L 3-4
San Jose	1	0	1	5	1	4	0	7	13	2002	CQF	L 1-4
Vancouver	2	0	2	13	5	8	0	34	50	1993	DSF	L 2-4
Totals	**18**	**2**	**16**	**92**	**29**	**63**	**0**	**245**	**343**			

Playoff Results 2002-1998

Year	Round	Opponent	Result	GF	GA
2002	CQF	San Jose	L 1-4	7	13
2000	CQF	Colorado	L 1-4	10	17
1999	CQF	St. Louis	L 3-4	16	19
1998	CQF	Detroit	L 2-4	18	24

Abbreviations: Round: CQF – conference quarter-final; **DSF** – division semi-final.

Calgary totals include Atlanta Flames, 1979-80.
Colorado totals include Quebec, 1979-80 to 1994-95.
New Jersey totals include Colorado Rockies, 1979-80 to 1981-82.
Carolina totals include Hartford, 1979-80 to 1996-97.
Dallas totals include Minnesota North Stars, 1979-80 to 1992-93.

2001-02 Results

Oct.	4	at Los Angeles	2-2
	6	at Edmonton	2-6
	8	at Calgary	2-1*
	11	at Chicago	0-3
	13	Washington	5-2
	16	Boston	1-1
	18	at Dallas	1-3
	20	Vancouver	5-2
	24	Anaheim	2-3*
	27	Colorado	1-0
	30	at Buffalo	3-2
Nov.	1	at New Jersey	2-5
	2	at Washington	2-2
	4	at Carolina	0-1*
	7	Detroit	1-3
	9	at Dallas	5-1
	10	at St. Louis	1-4
	13	Edmonton	4-5*
	15	San Jose	5-3
	17	NY Islanders	6-1
	20	Minnesota	2-1*
	23	at Minnesota	2-5
	24	at St. Louis	3-5
	27	at Columbus	0-3
	29	Nashville	1-0
Dec.	1	Pittsburgh	5-2
	5	St. Louis	3-0
	7	at Detroit	1-1
	8	at Montreal	3-3
	11	at Toronto	3-6
	13	at Ottawa	0-6
	15	Dallas	2-6
	17	Columbus	4-1
	19	Calgary	6-3
	21	at Anaheim	1-2
	23	Anaheim	0-4
	26	Los Angeles	1-1
	28	Philadelphia	4-2
	30	at San Jose	4-2
	31	NY Rangers	5-0
Jan.	3	Atlanta	2-1
	4	at San Jose	3-5
	6	Tampa Bay	0-3
	9	San Jose	5-6
	15	Detroit	2-2
	17	at Colorado	2-3*
	19	Buffalo	1-3
	21	at Nashville	1-0
	23	at Chicago	4-1
	25	at Detroit	1-4
	26	at Columbus	3-2
	28	at Atlanta	4-2
	30	at Florida	3-1
Feb.	4	at Vancouver	2-4
	6	Chicago	2-5
	8	Los Angeles	6-5*
	10	Edmonton	3-4
	12	Calgary	4-3*
	13	at Los Angeles	2-2
	26	Dallas	5-1
	28	at Colorado	1-2*
Mar.	3	Columbus	2-1
	5	New Jersey	4-1
	7	Vancouver	6-1
	9	Ottawa	3-2
	12	Chicago	3-1
	15	at Nashville	3-2
	17	at Minnesota	2-2
	19	at Boston	2-4
	20	at Pittsburgh	3-1
	22	at Dallas	3-4
	24	Los Angeles	4-0
	27	at Anaheim	4-2
	28	Anaheim	1-2*
	30	Colorado	5-3
Apr.	1	St. Louis	5-3
	4	at San Jose	2-5
	7	at Vancouver	3-4
	9	at Calgary	4-2
	10	at Edmonton	0-3
	12	Minnesota	7-1
	14	Nashville	6-4

* – Overtime

Entry Draft Selections 2002-1988

2002

Pick	
19	Jakub Koreis
23	Ben Eager
46	David Leneveu
70	Joe Callahan
80	Matt Jones
97	Lance Monych
132	John Zeiler
186	Jeff Pietrasiak
216	Ladislav Kouba
249	Marcus Smith
280	Russell Spence

2001

Pick	
11	Fredrik Sjostrom
31	Matthew Spiller
45	Martin Podlesak
78	Beat Forster
148	David Klema
180	Scott Polaski
210	Steve Belanger
243	Frantisek Lukes
273	Severin Blindenbacher

2000

Pick	
19	Krys Kolanos
53	Alexander Tatarinov
85	Ramzi Abid
160	Nate Kiser
186	Brent Gauvreau
217	Igor Samoilov
249	Sami Venalainen
281	Peter Fabus

1999

Pick	
15	Scott Kelman
19	Kirill Safronov
53	Brad Ralph
71	Jason Jaspers
116	Ryan Lauzon
123	Preston Mizzi
168	Erik Lewerstrom
234	Goran Bezina
262	Alexei Litvinenko

1998

Pick	
14	Patrick DesRochers
43	Ossi Vaananen
73	Pat O'Leary
100	Ryan Vanbuskirk
115	Jay Leach
116	Josh Blackburn
129	Robert Schnabel
160	Rickard Wallin
187	Erik Westrum
214	Justin Hansen

1997

Pick	
43	Juha Gustafsson
96	Scott McCallum
123	Curtis Suter
151	Robert Francz
207	Alexander Andreyev
233	Wyatt Smith

1996

Pick	
11	Dan Focht
24	Daniel Briere
62	Per-Anton Lundstrom
119	Richard Lintner
139	Robert Esche
174	Trevor Letowski
200	Nicholas Lent
226	Marc-Etienne Hubert

1995

Pick	
7	Shane Doan
32	Marc Chouinard
34	Jason Doig
67	Brad Isbister
84	Justin Kurtz
121	Brian Elder
136	Sylvain Daigle
162	Paul Traynor
188	Jaroslav Obsut
189	Fredrik Loven
214	Rob Deciantis

1994

Pick	
30	Deron Quint
56	Dorian Anneck
58	Tavis Hansen
82	Steve Cheredaryk
108	Craig Mills
143	Steve Vezina
146	Chris Kibermanis
186	Ramil Saifullin
212	Henrik Smangs
238	Mike Mader
264	Jason Issel

1993

Pick	
15	Mats Lindgren
31	Scott Langkow
43	Alexei Budayev
79	Ruslan Batyrshin
93	Ravil Gusmanov
119	Larry Courville
145	Michal Grosek
171	Martin Woods
197	Adrian Murray
217	Vladimir Potapov
223	Ilja Stashenkov
228	Harijs Vitolinsh
285	Russ Hewson

1992

Pick	
17	Sergei Bautin
27	Boris Mironov
60	Jeremy Stevenson
84	Mark Visheau
132	Alexander Alexeyev
155	Artur Oktyabrev
156	Andrei Raisky
204	Nikolai Khabibulin
228	Yevgeny Garanin
229	Teemu Numminen
252	Andrei Karpovstev
254	Ivan Vologzhaninov

1991

Pick	
5	Aaron Ward
49	Dmitri Filimonov
91	Juha Ylonen
99	Yan Kaminsky
115	Jeff Sebastian
159	Jeff Ricciardi
181	Sean Gauthier
203	Igor Ulanov
225	Jason Jennings
247	Sergei Sorokin

1990

Pick	
19	Keith Tkachuk
35	Mike Muller
74	Roman Meluzin
75	Scott Levins
77	Alexei Zhamnov
98	Craig Martin
119	Daniel Jardemyr
140	John Lilley
161	Henrik Andersson
182	Rauli Raitanen
203	Mika Alatalo
224	Sergei Selyanin
245	Keith Morris

1989

Pick	
4	Stu Barnes
25	Dan Ratushny
46	Jason Cirone
62	Kris Draper
64	Mark Brownschidle
69	Allain Roy
109	Dan Bylsma
130	Pekka Peltola
131	Doug Evans
151	Jim Solly
172	Stephane Gauvin
193	Joe Larson
214	Bradley Podiak
235	Evgeny Davydov
240	Sergei Kharin

1988

Pick	
10	Teemu Selanne
31	Russell Romaniuk
52	Stephane Beauregard
73	Brian Hunt
94	Tony Joseph
101	Benoit Lebeau
115	Ronald Jones
127	Markus Akerblom
136	Jukka Marttila
157	Mark Smith
178	Mike Helber
199	Pavel Kostichkin
220	Kevin Heise
241	Kyle Galloway

General Managers' History

John Ferguson, 1979-80 to 1987-88; John Ferguson and Mike Smith, 1988-89; Mike Smith, 1989-90 to 1992-93; Mike Smith and John Paddock, 1993-94; John Paddock, 1994-95, 1995-96; John Paddock and Bobby Smith, 1996-97; Bobby Smith, 1997-98 to 1999-2000; Bobby Smith and Cliff Fletcher, 2000-01; Cliff Fletcher and Mike Barnett, 2001-02; Mike Barnett, 2002-03.

General Manager

BARNETT, MIKE

General Manager, Phoenix Coyotes. Born in Olds, Alta., October 9, 1948.

Mike Barnett joined the Coyotes as vice president and general manager on August 28, 2001 after serving as president of International Management Group's (IMG) hockey division since 1990. Barnett is the sixth general manager in franchise history and follows in the footsteps of Brian Burke (Vancouver Canucks), Pierre Lacroix (Colorado Avalanche) and Dean Lombardi (San Jose Sharks) as former player agents who have become NHL general managers.

With over 20 years of experience in the game prior to joining the Coyotes, Barnett left IMG as one of hockey's most distinguished and well-respected player agents. Over the years, Barnett earned acclaim for his integrity, vision and success as a negotiator. He developed a reputation within the NHL as one of the most creative and well-informed agents in the industry. He is reunited in Phoenix with his longtime friend Wayne Gretzky, the Coyotes' managing partner. Barnett served as Gretzky's agent for 20 years. He also represented some of the NHL's most high-profile players including Jaromir Jagr, Brett Hull, Paul Coffey, Alexander Mogilny, Owen Nolan, Mats Sundin and Joe Thornton.

Barnett actually began his career in hockey as a player. He played hockey at St. Lawrence University in Canton, New York and later attended the University of Calgary, where he played both intercollegiate hockey and football for three years. In 1973-74, he turned professional with the Chicago Cougars (WHA) playing left wing for their minor league affiliate, the Long Island Cougars (NAHL). The following season (1974-75), while playing for the Roanoke-Valley Rebels (SHL) — the Houston Aeros' (WHA) minor league affiliate — Barnett suffered a career ending eye injury.

In 1980, Barnett opened a Western Canadian sports management agency and began his long-lasting relationship with Gretzky by signing him on as his top client. In 1990, Barnett merged his company with Mark McCormack's IMG and became president of IMG hockey operations.

Club Directory

America West Arena

Phoenix Coyotes
ALLTEL Ice Den
9375 E. Bell Road
Scottsdale, AZ 85260
Phone **480/473-5600**
FAX 480/473-5699
www.PhoenixCoyotes.com
Capacity: 16,210

Chairman, CEO & Governor	Steve Ellman
Co-Owner	Jerry Moyes
Managing Partner & Alternate Governor	Wayne Gretzky
President, COO & Alternate Governor	Jim Lites
Senior Exec. V.P. of Hockey Operations	Cliff Fletcher
Vice President, G.M. & Alternate Governor	Michael Barnett
Senior Consultant	Bill Strong
Senior Vice President, Sales and Marketing	Brian Byrnes
Sr. V.P., Corporate Sales & Broadcasting	Dave Groff
Senior Vice President, Finance & Administration	Vaibhav Gupta
Executive Assistant to the President	Cheryl Hocker
Hockey Operations	
Assistant General Manager	Laurence Gilman
Head Coach	Bob Francis
Assistant Coaches	Rick Bowness, Pat Conacher
Goaltending Coach	Benoit Allaire
Vice President, Scouting & Player Personnel	Dave Draper
Director of Amateur Scouting	Vaughn Karpan
Professional Scouts	Tom Kurvers, Warren Rychel
Co-ordinator, Hockey Operations	Igor Kuperman
Strength & Conditioning Coordinator	Stieg Theander
Video Coordinator	Steve Peters
Amateur Scouts	Connie Broden, Shane Churla, Pelle Eklund, Keith Gretzky, Paul Henry, Blair Reid, Evzen Slansky, Boris Yemeljanov
Player Development	Charles Henry
Athletic Therapist	Gord Hart
Massage Therapist	Jukka Nieminen
Equipment Manager	Stan Wilson
Assistant Equipment Managers	Tony Silva, Jason Rudee
Manager of Team Services	Lesa Guth
Administrative Assistant, Hockey Operations	Maryjane DeBiasio
Team Doctors	Matt Maddox, D.O., Robert Luberto, D.O.
Team Dentists	Dr. Rick Lawson, Dr. Lawrence Emmott
Springfield Falcons (AHL) Head Coach	Marty McSorley
Communications	
Vice President of Media & Player Relations	Richard Nairn
Director of Media Relations	Rick Braunstein
Manager of Publications & Media Relations	Ryan Lichtenfels
Broadcasting	
TV/Radio Play-by-Play	Curt Keilback
TV/Radio Color Analyst	Charlie Simmer
Manager of Broadcasting	Graham Taylor
Community Relations	
V.P. of Comm./Exec. Dir. of Coyotes Charities	Susan Kricun
Community Relations Manager	Heather Bennett
Player Relations Coordinator	Adam Cresswell
Community Relations Coordinator	Melissa Doyle
Corporate Sales & Service	
Vice President of Corporate Sales	Cullen Maxey
Manager of Corporate Sales	John Allen
Manager of Corporate Promotions	Jason Levy
Manager of Corporate Sales Services	Ashley Ritt
Manager of Sponsorship Sales	Jeff Tummonds
Corporate Sales Coordinator	Stacie Nelson
Creative Services	
V.P., Strategic Mktg. & Creative Services	Becky Thielen
Lead Graphic Designer	A. Brad Hazelton
Graphic Designer	Julia Theile Elefson
Database Intern	Scott Neidig
Finance & Administration	
Vice President and Controller	Joe Leibfried
Assistant Controller	Larry Silver
Payroll Administrator	Cheri Sedor
Accounting Assistant	Julie deWit
Human Resources Assistant	Sarah Delp
Receptionist	Tomi Stern
Marketing	
Vice President of Marketing	Brett Rogers
Director of Game Operations	Greg Hanover
Marketing Manager	Jason Shughart
Manager of Fan Development	Amy Robertson
Promotions & Events Coordinator	Leigh Goldstein
Hockey Programs Coordinator	Ben Weber
Ticket Sales & Service	
Vice President, Ticket Sales & Service	Augie Manfredo
Director of Ticket Sales	Jim Willits
Manager of Individual Sales	Scott Newhouse
Account Executives	Lisa Anderson, Dean Blixt, Randy Just, Neils Lund, Alexia Matak, E.A.McDonough, Brian Powell, Brian Woods
Customer Service Representatives	Alyssa VanKlaveren, Tudor Wadell, Kelley Dilworth
Director of Ticket Operations	David Drake
Manager of Ticket Operations	Kevin Prebil
Ticket Operations Assistant	Adam Somers
Security	
Director of Security	Jim O'Neal
Suite Sales	
Vice President of Suite Sales	Mike McCoy
Luxury Suite Co-ordinator	Kristin Anderson
Team Information	
Training Camp	Scottsdale, Arizona
Television Stations	Fox Sports Net, KTVK-3TV, KASW TV-WB61
Radio Stations	KDKB 93.3 FM, KDUS 1060 AM

Pittsburgh Penguins

2001-02 Results: 28W-41L-8T-5OTL 69PTS.
Fifth, Atlantic Division

2002-03 Schedule

Oct.	Thu.	10	Toronto
	Sat.	12	NY Rangers
	Mon.	14	at Toronto
	Wed.	16	Atlanta
	Sat.	19	Tampa Bay
	Tue.	22	at Montreal
	Fri.	25	at Detroit
	Sat.	26	Buffalo
	Mon.	28	Washington
	Wed.	30	at Ottawa
Nov.	Sat.	2	Tampa Bay
	Wed.	6	at Florida
	Fri.	8	at Tampa Bay
	Sat.	9	at Carolina
	Thu.	14	at Minnesota
	Sat.	16	NY Islanders
	Mon.	18	at Montreal
	Wed.	20	Montreal
	Fri.	22	at Atlanta
	Sat.	23	San Jose
	Wed.	27	Philadelphia
	Fri.	29	at Buffalo
	Sat.	30	Boston
Dec.	Tue.	3	Washington
	Fri.	6	at New Jersey
	Sat.	7	NY Islanders
	Tue.	10	at Toronto
	Thu.	12	at San Jose
	Sat.	14	at Los Angeles
	Sun.	15	at Anaheim*
	Tue.	17	at Phoenix
	Thu.	19	New Jersey
	Sat.	21	Calgary
	Mon.	23	Buffalo
	Thu.	26	at NY Rangers
	Sat.	28	Montreal
	Mon.	30	Atlanta
	Tue.	31	at Columbus
Jan.	Fri.	3	at Atlanta
	Sat.	4	NY Islanders
	Tue.	7	at NY Islanders
	Thu.	9	Toronto
	Sat.	11	NY Rangers*
	Mon.	13	at Boston
	Wed.	15	at Carolina
	Fri.	17	at Tampa Bay
	Sat.	18	at Florida
	Tue.	21	at Buffalo
	Thu.	23	Boston
	Sat.	25	Chicago*
	Tue.	28	at NY Islanders
	Thu.	30	at Washington
Feb.	Tue.	4	Vancouver
	Thu.	6	Florida
	Sat.	8	at Boston*
	Wed.	12	Ottawa
	Fri.	14	at NY Rangers
	Sat.	15	at New Jersey
	Tue.	18	Edmonton
	Thu.	20	Colorado
	Sat.	22	St. Louis*
	Sun.	23	New Jersey*
	Tue.	25	Los Angeles
	Thu.	27	at Nashville
Mar.	Sat.	1	at Colorado*
	Sun.	2	at Dallas*
	Tue.	4	Phoenix
	Thu.	6	Carolina
	Sat.	8	Ottawa
	Sun.	9	at Ottawa
	Wed.	12	Nashville
	Sat.	15	Philadelphia*
	Sun.	16	Florida*
	Tue.	18	Detroit
	Thu.	20	at Philadelphia
	Fri.	21	at New Jersey
	Sun.	23	at Chicago*
	Wed.	26	at NY Rangers
	Sat.	29	at Philadelphia
	Mon.	31	Philadelphia
Apr.	Wed.	2	Carolina
	Sat.	5	at Washington

* Denotes afternoon game.

Though injuries cut short his 2001-02 campaign, Mario Lemieux still collected 31 points in just 24 games. He also led Team Canada to an Olympic gold medal at the 2002 Salt Lake City Winter Games.

Year-by-Year Record

		Home				Road				Overall								
Season	GP	W	L	T	OL	W	L	T	OL	W	L	T	OL	GF	GA	Pts.	Finished	Playoff Result
2001-02	82	16	20	4	1	12	21	4	4	28	41	8	5	198	249	69	5th, Atlantic Div.	Out of Playoffs
2000-01	82	24	15	2	0	18	13	7	3	42	28	9	3	281	256	96	3rd, Atlantic Div.	Lost Conf. Championship
1999-2000	82	23	11	7	0	14	20	1	6	37	31	8	6	241	236	88	3rd, Atlantic Div.	Lost Conf. Semi-Final
1998-99	82	21	10	10	...	17	20	4	...	38	30	14	...	242	225	90	3rd, Atlantic Div.	Lost Conf. Semi-Final
1997-98	82	21	10	10	...	19	14	8	...	40	24	18	...	228	188	98	1st, Northeast Div.	Lost Conf. Quarter-Final
1996-97	82	25	11	5	...	13	25	3	...	38	36	8	...	285	280	84	2nd, Northeast Div.	Lost Conf. Quarter-Final
1995-96	82	32	9	0	...	17	20	4	...	49	29	4	...	362	284	102	1st, Northeast Div.	Lost Conf. Championship
1994-95	48	18	5	1	...	11	11	2	...	29	16	3	...	181	158	61	2nd, Northeast Div.	Lost Conf. Semi-Final
1993-94	84	25	9	8	...	19	18	5	...	44	27	13	...	299	285	101	1st, Northeast Div.	Lost Conf. Quarter-Final
1992-93	84	32	6	4	...	24	15	3	...	56	21	7	...	367	268	119	1st, Patrick Div.	Lost Div. Final
1991-92	**80**	**21**	**13**	**6**	**...**	**18**	**19**	**3**	**...**	**39**	**32**	**9**	**...**	**343**	**308**	**87**	**3rd, Patrick Div.**	**Won Stanley Cup**
1990-91	**80**	**25**	**12**	**3**	**...**	**16**	**21**	**3**	**...**	**41**	**33**	**6**	**...**	**342**	**305**	**88**	**1st, Patrick Div.**	**Won Stanley Cup**
1989-90	80	22	15	3	...	10	25	5	...	32	40	8	...	318	359	72	5th, Patrick Div.	Out of Playoffs
1988-89	80	24	13	3	...	16	20	4	...	40	33	7	...	347	349	87	2nd, Patrick Div.	Lost Div. Final
1987-88	80	22	12	6	...	14	23	3	...	36	35	9	...	319	316	81	6th, Patrick Div.	Out of Playoffs
1986-87	80	19	15	6	...	11	23	6	...	30	38	12	...	297	290	72	5th, Patrick Div.	Out of Playoffs
1985-86	80	20	15	5	...	14	23	3	...	34	38	8	...	313	305	76	5th, Patrick Div.	Out of Playoffs
1984-85	80	17	20	3	...	7	31	2	...	24	51	5	...	276	385	53	6th, Patrick Div.	Out of Playoffs
1983-84	80	7	29	4	...	9	29	2	...	16	58	6	...	254	390	38	6th, Patrick Div.	Out of Playoffs
1982-83	80	14	22	4	...	4	31	5	...	18	53	9	...	257	394	45	6th, Patrick Div.	Out of Playoffs
1981-82	80	21	11	8	...	10	25	5	...	31	36	13	...	310	337	75	4th, Patrick Div.	Lost Div. Semi-Final
1980-81	80	21	16	3	...	9	21	10	...	30	37	13	...	302	345	73	3rd, Norris Div.	Lost Prelim. Round
1979-80	80	20	13	7	...	10	24	6	...	30	37	13	...	251	303	73	3rd, Norris Div.	Lost Prelim. Round
1978-79	80	23	12	5	...	13	19	8	...	36	31	13	...	281	279	85	2nd, Norris Div.	Lost Quarter-Final
1977-78	80	16	15	9	...	9	22	9	...	25	37	18	...	254	321	68	4th, Norris Div.	Out of Playoffs
1976-77	80	22	12	6	...	12	21	7	...	34	33	13	...	240	252	81	3rd, Norris Div.	Lost Prelim. Round
1975-76	80	23	11	6	...	12	22	6	...	35	33	12	...	339	303	82	3rd, Norris Div.	Lost Prelim. Round
1974-75	80	25	5	10	...	12	23	5	...	37	28	15	...	326	289	89	3rd, Norris Div.	Lost Quarter-Final
1973-74	78	15	18	6	...	13	23	3	...	28	41	9	...	242	273	65	5th, West Div.	Out of Playoffs
1972-73	78	24	11	4	...	8	26	5	...	32	37	9	...	257	265	73	5th, West Div.	Out of Playoffs
1971-72	78	18	15	6	...	8	23	8	...	26	38	14	...	220	258	66	4th, West Div.	Lost Quarter-Final
1970-71	78	18	12	9	...	3	25	11	...	21	37	20	...	221	240	62	6th, West Div.	Out of Playoffs
1969-70	76	17	13	8	...	9	25	4	...	26	38	12	...	182	238	64	2nd, West Div.	Lost Semi-Final
1968-69	76	12	20	6	...	8	25	5	...	20	45	11	...	189	252	51	5th, West Div.	Out of Playoffs
1967-68	74	15	12	10	...	12	22	3	...	27	34	13	...	195	216	67	5th, West Div.	Out of Playoffs

Franchise date: June 5, 1967

ATLANTIC DIVISION

36th NHL Season

2002-03 Player Personnel

FORWARDS	HT	WT	S	Place of Birth	Date	2001-02 Club
ARMSTRONG, Colby	5-11	185	R	Lloydminster, Sask.	11/23/82	Red Deer
BEECH, Kris	6-3	199	L	Salmon Arm, B.C.	2/5/81	Pittsburgh
CRAMPTON, Steve	6-2	200	R	Winnipeg, Man.	4/12/82	Moose Jaw
DONOVAN, Shean	6-2	200	R	Timmins, Ont.	1/22/75	Atlanta-Pittsburgh
ENDICOTT, Shane	6-4	200	L	Saskatoon, Sask.	12/21/81	Pittsburgh-Wilkes-Barre
FADRNY, Jan	6-0	195	R	Brno, Czech.	6/14/80	Wilkes-Barre
HRDINA, Jan	6-0	206	R	Hradec Kralove, Czech.	2/5/76	Pit-Czech Republic
KOLTSOV, Konstantin	6-0	190	L	Minsk, USSR	4/17/81	Kazan-Spartak Mos. 2-Spartak Moscow
KOSTOPOULOS, Tom	6-0	200	R	Mississauga, Ont.	1/24/79	Pittsburgh-Wilkes-Barre
KOVALEV, Alexei	6-1	220	L	Togliatti, USSR	2/24/73	Pittsburgh-Russia
KRAFT, Milan	6-3	211	R	Plzen, Czech.	1/7/80	Pittsburgh-Wilkes-Barre
LaCOUTURE, Dan	6-2	208	L	Hyannis, MA	4/18/77	Pittsburgh
LEMIEUX, Mario	6-4	230	R	Montreal, Que.	10/5/65	Pittsburgh-Canada
MacDONALD, Jason	5-11	205	R	Charlottetown, P.E.I.	4/1/74	Wilkes-Barre
MANDERVILLE, Kent	6-3	200	L	Edmonton, Alta.	4/12/71	Philadelphia-Pittsburgh
McKENNA, Steve	6-8	255	L	Toronto, Ont.	8/21/73	NY Rangers-Hartford
MELOCHE, Eric	5-10	195	R	Montreal, Que.	5/1/76	Pittsburgh-Wilkes-Barre
MOROZOV, Aleksey	6-1	202	L	Moscow, USSR	2/16/77	Pittsburgh
MURLEY, Matt	6-1	192	L	Troy, NY	12/17/79	RPI Engineers
NIEMINEN, Ville	6-0	200	L	Tampere, Finland	4/6/77	Col-Finland-Pit
OUELLET, Michel	6-0	190	R	Rimouski, Que.	3/5/82	Rimouski
PETERSEN, Toby	5-9	197	L	Minneapolis, MN	10/27/78	Pittsburgh
PRIMEAU, Wayne	6-3	220	L	Scarborough, Ont.	6/4/76	Pittsburgh
ROBITAILLE, Randy	5-11	196	L	Ottawa, Ont.	10/12/75	L.A.-Manchester-Pit
SIVEK, Michal	6-3	209	L	Nachod, Czech.	1/21/81	Wilkes-Barre-Sparta Praha
STRAKA, Martin	5-9	178	L	Plzen, Czech.	9/3/72	Pittsburgh
SUROVY, Tomas	6-1	187	L	Banska Bystrica, Czech.	9/24/81	Wilkes-Barre
VUJTEK, Vladimir	6-2	200	L	Ostrava, Czech.	2/17/72	HPK
ZEVAKHIN, Alexander	5-11	208	L	Perm, USSR	6/4/80	Wilkes-Barre
DEFENSEMEN						
BERGEVIN, Marc	6-1	214	L	Montreal, Que.	8/11/65	St. Louis-Worcester
BERRY, Rick	6-2	210	L	Birtle, Man.	11/4/78	Colorado-Pittsburgh
BUCKLEY, Brendan	6-1	200	R	Boston, MA	2/26/77	Wilkes-Barre
FERENCE, Andrew	5-10	196	L	Edmonton, Alta.	3/17/79	Pittsburgh
JONSSON, Hans	6-1	205	L	Jarved, Sweden	8/2/73	Pittsburgh
KOCI, David	6-6	225	L	Prague, Czech.	5/12/81	Wheeling-Wilkes-Barre
LAUKKANEN, Janne	6-1	196	L	Lahti, Finland	3/19/70	Pittsburgh
LEROUX, Francois	6-6	247	L	Ste-Adele, Que.	4/18/70	Berlin Capitals
LUPASCHUK, Ross	6-1	217	R	Edmonton, Alta.	1/19/81	Wilkes-Barre
MELICHAR, Josef	6-2	221	L	Ceske Budejovice, Czech.	1/20/79	Pittsburgh
MORAN, Ian	6-0	200	R	Cleveland, OH	8/24/72	Pittsburgh
ORPIK, Brooks	6-2	222	L	San Francisco, CA	9/26/80	Wilkes-Barre
PUSHOR, Jamie	6-3	218	R	Lethbridge, Alta.	2/11/73	Columbus-Pittsburgh
ROBINSON, Darcy	6-3	221	R	Kamloops, B.C.	5/3/81	Wheeling-Wilkes-Barre
ROZSIVAL, Michal	6-1	208	R	Vlasim, Czech.	9/3/78	Pittsburgh
SCUDERI, Rob	6-0	208	L	Syosset, NY	12/30/78	Wilkes-Barre
TARNSTROM, Dick	6-2	200	L	Sundbyberg, Sweden	1/20/75	NY Islanders-Bridgeport
WILSON, Mike	6-6	229	L	Brampton, Ont.	2/26/75	Pittsburgh-Wilkes-Barre

GOALTENDERS	HT	WT	C	Place of Birth	Date	2001-02 Club
AUBIN, Jean-Sebastien	5-11	180	R	Montreal, Que.	7/19/77	Pittsburgh
CARON, Sebastian	6-1	167	L	Amqui, Que.	6/25/80	Wilkes-Barre
HEDBERG, Johan	6-0	184	L	Leksand, Sweden	5/3/73	Pittsburgh-Sweden
TALLAS, Robbie	6-0	170	L	Edmonton, Alta.	3/20/73	Wilkes-Barre

General Managers' History

Jack Riley, 1967-68 to 1969-70; Red Kelly, 1970-71; Red Kelly and Jack Riley, 1971-72; Jack Riley, 1972-73; Jack Riley and Jack Button, 1973-74; Jack Button, 1974-75; Wren Blair, 1975-76; Wren Blair and Baz Bastien, 1976-77; Baz Bastien, 1977-78 to 1982-83; Eddie Johnston, 1983-84 to 1987-88; Tony Esposito, 1988-89; Tony Esposito and Craig Patrick, 1989-90; Craig Patrick, 1990-91 to date.

Coach

KEHOE, RICK
Coach, Pittsburgh Penguins. Born in Windsor, Ont., July 15, 1951.

Rick Kehoe was in his 27th season with the Penguins organization when he was named head coach on October 15, 2001. Kehoe had served as an assistant under nine Penguins head coaches: Pierre Creamer, Gene Ubriaco, Craig Patrick (twice), Bob Johnson, Scotty Bowman, Ed Johnston, Kevin Constantine, Herb Brooks and Ivan Hlinka.

A talented right winger in his playing days, "Chico" played 11 of his 14 NHL seasons with the Penguins. He scored 312 goals in 722 games with the Penguins and won the Lady Byng Trophy for sportsmanship in 1981 after setting career highs with 55 goals and 88 points while recording just six penalty minutes. He appeared in the NHL All-Star Game in 1981 and 1983 before retiring in 1985. Kehoe was named a charter member of the Penguins' Hall of Fame in 1993 and is also a member of the Western Pennsylvania Sports Hall of Fame.

After retiring, Kehoe was a Penguins scout for two seasons, then began his coaching career as an assistant on Pierre Creamer's staff in 1987. He served as an assistant until the 1988-89 season when he served as the Penguins' professional scout. Kehoe went back behind the bench as an assistant to Herb Brooks on December 9, 1999.

Coaching Record

		Regular Season				Playoffs		
Season	Team	Games	W	L	T	Games	W	L
2001-02	Pittsburgh (NHL)	78	28	42	8	....	....	....
	NHL Totals	78	28	42	8	....	....	....

2001-02 Scoring

* - rookie

Regular Season

Pos	#	Player	Team	GP	G	A	Pts	+/–	PIM	PP	SH	GW	GT	S	%
R	27	Alexei Kovalev	PIT	67	32	44	76	2	80	8	1	3	2	266	12.0
C	38	Jan Hrdina	PIT	79	24	33	57	–7	50	6	0	6	0	115	20.9
C	20	Robert Lang	PIT	62	18	32	50	9	16	5	1	3	0	175	10.3
R	95	Alexei Morozov	PIT	72	20	29	49	–7	16	7	0	3	0	162	12.3
C	22	Randy Robitaille	L.A.	18	4	3	7	–9	17	2	0	0	0	30	13.3
			PIT	40	10	20	30	–14	16	3	0	1	0	91	11.0
			TOTAL	58	14	23	37	–23	33	5	0	1	0	121	11.6
C	66	Mario Lemieux	PIT	24	6	25	31	0	14	2	0	0	0	75	8.0
D	28	Michal Rozsival	PIT	79	9	20	29	–6	47	4	0	4	0	89	10.1
L	10	Ville Nieminen	COL	53	10	14	24	1	30	1	0	5	0	72	13.9
			PIT	13	1	2	3	–2	8	0	0	0	0	11	9.1
			TOTAL	66	11	16	27	–1	38	1	0	5	0	83	13.3
C	16	* Kris Beech	PIT	79	10	15	25	–25	45	2	0	0	0	126	7.9
C	17	* Toby Petersen	PIT	79	8	10	18	–15	4	1	1	0	0	116	6.9
L	33	Dan LaCouture	PIT	82	6	11	17	–19	71	0	1	0	0	77	7.8
C	14	Milan Kraft	PIT	68	8	8	16	–9	16	1	0	2	1	103	7.8
R	18	Shean Donovan	ATL	48	6	6	12	–16	40	1	0	2	0	64	9.4
			PIT	13	2	1	3	–5	4	0	0	0	0	18	11.1
			TOTAL	61	8	7	15	–21	44	1	0	2	0	82	9.8
L	9	Jeff Toms	NYR	38	7	4	11	–4	10	2	0	0	0	62	11.3
			PIT	14	2	1	3	–5	4	0	0	0	1	22	9.1
			TOTAL	52	9	5	14	–9	14	2	0	0	1	84	10.7
D	5	Janne Laukkanen	PIT	47	6	7	13	–18	28	3	0	1	0	66	9.1
D	7	Andrew Ference	PIT	75	4	7	11	–12	73	1	0	0	1	82	4.9
C	15	Wayne Primeau	PIT	33	3	7	10	–1	18	0	1	0	0	28	10.7
R	24	Ian Moran	PIT	64	2	8	10	–11	54	0	0	1	0	94	2.1
C	82	Martin Straka	PIT	13	5	4	9	3	0	1	0	1	0	33	15.2
C	26	Kent Manderville	PHI	34	2	5	7	2	8	0	0	0	0	39	5.1
			PIT	4	1	0	1	1	4	0	0	0	0	4	25.0
			TOTAL	38	3	5	8	3	12	0	0	0	0	43	7.0
D	3	Jamie Pushor	CBJ	61	0	6	6	–10	54	0	0	0	0	46	0.0
			PIT	15	0	2	2	–3	30	0	0	0	0	14	0.0
			TOTAL	76	0	8	8	–13	84	0	0	0	0	60	0.0
D	8	Hans Jonsson	PIT	53	2	5	7	–12	22	2	0	0	0	37	5.4
L	25	Kevin Stevens	PIT	32	1	4	5	–9	25	0	0	0	0	34	2.9
D	23	John Jakopin	PIT	19	0	4	4	2	42	0	0	0	0	3	0.0
R	37	* Tom Kostopoulos	PIT	11	1	2	3	–1	9	0	0	0	0	8	12.5
D	2	* Josef Melichar	PIT	60	0	3	3	–1	68	0	0	0	0	46	0.0
D	4	Mike Wilson	PIT	21	1	1	2	–12	17	0	0	0	0	14	7.1
L	29	Krzysztof Oliwa	PIT	57	0	2	2	–5	150	0	0	0	0	31	0.0
D	6	* Rick Berry	COL	57	0	0	0	1	60	0	0	0	0	29	0.0
			PIT	13	0	2	2	–4	21	0	0	0	0	20	0.0
			TOTAL	70	0	2	2	–3	81	0	0	0	0	49	0.0
C	41	* Shane Endicott	PIT	4	0	1	1	–1	4	0	0	0	0	2	0.0
R	72	* Eric Meloche	PIT	23	0	1	1	–7	8	0	0	0	0	29	0.0

Goaltending

No.	Goaltender	GPI	Mins	Avg	W	L	T	EN	SO	GA	SA	S%	G	A	PIM
1	Johan Hedberg	66	3877	2.75	25	34	7	3	6	178	1851	.904	0	1	22
30	Jean-Sebastien Aubin	21	1094	3.56	3	12	1	3	0	65	537	.879	0	0	0
	Totals	**82**	**4994**	**2.99**	**28**	**46**	**8**	**6**	**6**	**249**	**2394**	**.896**			

Alexei Kovalev enjoyed another fine season in 2001-02, leading the Penguins in goals (32), assists (44) and points (76) and finishing 11th in the NHL scoring race.

Coaching History

Red Sullivan, 1967-68, 1968-69; Red Kelly, 1969-70 to 1971-72; Red Kelly and Ken Schinkel, 1972-73; Ken Schinkel and Marc Boileau, 1973-74; Marc Boileau, 1974-75; Marc Boileau and Ken Schinkel, 1975-76; Ken Schinkel, 1976-77; Johnny Wilson, 1977-78 to 1979-80; Eddie Johnston, 1980-81 to 1982-83; Lou Angotti, 1983-84; Bob Berry, 1984-85 to 1986-87; Pierre Creamer, 1987-88; Gene Ubriaco, 1988-89; Gene Ubriaco and Craig Patrick, 1989-90; Bob Johnson, 1990-91, 1991-92; Scotty Bowman, 1991-92, 1992-93; Eddie Johnston, 1993-94 to 1995-96; Eddie Johnston and Craig Patrick, 1996-97; Kevin Constantine, 1997-98, 1998-99; Kevin Constantine and Herb Brooks, 1999-2000; Ivan Hlinka, 2000-01; Ivan Hlinka and Rick Kehoe, 2001-02; Rick Kehoe, 2002-03.

Club Records

Team

(Figures in brackets for season records are games played; records for fewest points, wins, ties, losses, goals, goals against are for 70 or more games)

Record		
Most Points	**119**	1992-93 (84)
Most Wins	**56**	1992-93 (84)
Most Ties	**20**	1970-71 (78)
Most Losses	**58**	1983-84 (80)
Most Goals	**367**	1992-93 (84)
Most Goals Against	**394**	1982-83 (80)
Fewest Points	**38**	1983-84 (80)
Fewest Wins	**16**	1983-84 (80)
Fewest Ties	**4**	1995-96 (82)
Fewest Losses	**21**	1992-93 (84)
Fewest Goals	**182**	1969-70 (76)
Fewest Goals Against	**188**	1997-98 (82)
Longest Winning Streak		
Overall	***17**	Mar. 9-Apr. 10/93
Home	**11**	Jan. 5-Mar. 7/91
Away	**7**	Mar. 14-Apr. 9/93
Longest Undefeated Streak		
Overall	**18**	Mar. 9-Apr. 14/93 (17 wins, 1 tie)
Home	**20**	Nov. 30/74-Feb. 22/75 (12 wins, 8 ties)
Away	**8**	Mar. 14-Apr. 14/93 (7 wins, 1 tie)
Longest Losing Streak		
Overall	**11**	Jan. 22-Feb. 10/83
Home	**7**	Oct. 8-29/83
Away	**18**	Dec. 23/82-Mar. 4/83
Longest Winless Streak		
Overall	**18**	Jan. 2-Feb. 10/83 (17 losses, 1 tie)
Home	**11**	Oct. 8-Nov. 19/83 (9 losses, 2 ties)
Away	**18**	Oct. 25/70-Jan. 14/71 (11 losses, 7 ties), Dec. 23/82-Mar. 4/83 (18 losses)
Most Shutouts, Season	**9**	1998-99 (82)
Most PIM, Season	**2,670**	1988-89 (80)
Most Goals, Game	**12**	Mar. 15/75 (Wsh. 1 at Pit. 12), Dec. 26/91 (Tor. 1 at Pit. 12)

Individual

Most Seasons	**14**	Mario Lemieux
Most Games	**812**	Mario Lemieux
Most Goals, Career	**654**	Mario Lemieux
Most Assists, Career	**947**	Mario Lemieux
Most Points, Career	**1,601**	Mario Lemieux (654G, 947A)
Most PIM, Career	**1,023**	Kevin Stevens
Most Shutouts, Career	**22**	Tom Barrasso
Longest Consecutive Games Streak	**320**	Ron Schock (Oct. 24/73-Apr. 3/77)
Most Goals, Season	**85**	Mario Lemieux (1988-89)
Most Assists, Season	**114**	Mario Lemieux (1988-89)
Most Points, Season	**199**	Mario Lemieux (1988-89; 85G, 114A)
Most PIM, Season	**409**	Paul Baxter (1981-82)
Most Points, Defenseman, Season	**113**	Paul Coffey (1988-89; 30G, 83A)
Most Points, Center, Season	**199**	Mario Lemieux (1988-89; 85G, 114A)
Most Points, Right Wing, Season	***149**	Jaromir Jagr (1995-96; 62G, 87A)
Most Points, Left Wing, Season	**123**	Kevin Stevens (1991-92; 54G, 69A)
Most Points, Rookie, Season	**100**	Mario Lemieux (1984-85; 43G, 57A)
Most Shutouts, Season	**7**	Tom Barrasso (1997-98)
Most Goals, Game	**5**	Mario Lemieux (Three times)
Most Assists, Game	**6**	Ron Stackhouse (Mar. 8/75), Greg Malone (Nov. 28/79), Mario Lemieux (Three times)
Most Points, Game	**8**	Mario Lemieux (Oct. 15/88; 2G, 6A, Dec. 31/88; 5G, 3A)

* NHL Record.

Captains' History

Ab McDonald, 1967-68; no captain, 1968-69 to 1972-73; Ron Schock, 1973-74 to 1976-77; Jean Pronovost, 1977-78; Orest Kindrachuk, 1978-79 to 1980-81; Randy Carlyle, 1981-82 to 1983-84; Mike Bullard, 1984-85, 1985-86; Mike Bullard and Terry Ruskowski, 1986-87; Dan Frawley and Mario Lemieux, 1987-88; Mario Lemieux, 1988-89 to 1993-94; Ron Francis, 1994-95; Mario Lemieux, 1995-96, 1996-97; Ron Francis, 1997-98; Jaromir Jagr, 1998-99 to 2000-01; Mario Lemieux, 2001-02 to date.

Retired Numbers

21	Michel Brière	1969-1970

All-time Record vs. Other Clubs

Regular Season

	At Home								On Road								Total							
	GP	W	L	T	OL	GF	GA	PTS	GP	W	L	T	OL	GF	GA	PTS	GP	W	L	T	OL	GF	GA	PTS
Anaheim	8	4	2	2	0	27	26	10	7	3	3	0	1	24	24	7	15	7	5	2	1	51	50	17
Atlanta	6	6	0	0	0	26	10	12	6	6	0	0	0	26	14	12	12	12	0	0	0	52	24	24
Boston	76	31	30	15	0	264	271	77	74	14	54	6	0	209	333	34	150	45	84	21	0	473	604	111
Buffalo	67	33	17	17	0	252	210	83	67	18	33	16	0	183	267	52	134	51	50	33	0	435	477	135
Calgary	43	23	10	10	0	165	133	56	45	11	26	8	0	139	198	30	88	34	36	18	0	304	331	86
Carolina	42	21	16	5	0	177	160	47	44	20	19	5	0	169	169	45	86	41	35	10	0	346	329	92
Chicago	57	27	23	7	0	204	187	61	58	10	39	9	0	155	235	29	115	37	62	16	0	359	422	90
Colorado	35	15	15	5	0	140	142	35	31	13	15	2	1	121	138	29	66	28	30	7	1	261	280	64
Columbus	2	1	1	0	0	7	6	2	2	2	0	0	0	7	5	4	4	3	1	0	0	14	11	6
Dallas	62	37	19	6	0	232	178	80	63	21	35	6	1	212	243	49	125	58	54	12	1	444	421	129
Detroit	63	43	16	4	0	276	187	90	64	13	38	12	1	175	246	39	127	56	54	16	1	451	433	129
Edmonton	29	14	12	3	0	113	124	31	29	7	21	1	0	95	144	15	58	21	33	4	0	208	268	46
Florida	19	12	4	3	0	68	53	27	18	8	8	1	1	43	46	18	37	20	12	4	1	111	99	45
Los Angeles	71	38	23	10	0	262	225	86	68	17	43	8	0	181	262	42	139	55	66	18	0	443	487	128
Minnesota	2	1	1	0	0	4	6	2	1	0	1	0	0	2	4	0	3	1	2	0	0	6	10	2
Montreal	78	27	38	13	0	230	273	67	78	10	58	9	1	194	367	30	156	37	96	22	1	424	640	97
Nashville	3	1	1	1	0	8	6	3	4	2	2	0	0	8	12	4	7	3	3	1	0	16	18	7
New Jersey	74	40	29	4	1	284	255	85	75	27	35	13	0	257	278	67	149	67	64	17	1	541	533	152
NY Islanders	82	39	31	12	0	313	286	90	81	31	42	8	0	272	324	70	163	70	73	20	0	585	610	160
NY Rangers	95	41	40	14	0	346	351	96	95	37	49	9	0	323	378	83	190	78	89	23	0	669	729	179
Ottawa	23	16	3	4	0	91	49	36	23	13	6	4	0	73	56	30	46	29	9	8	0	164	105	66
Philadelphia	100	45	35	20	0	357	325	110	101	16	75	7	3	248	430	42	201	61	110	27	3	605	755	152
Phoenix	29	20	9	0	0	119	82	40	29	13	13	3	0	101	108	29	58	33	22	3	0	220	190	69
St. Louis	63	31	20	12	0	236	189	74	63	15	41	6	1	169	244	37	126	46	61	18	1	405	433	111
San Jose	8	3	4	1	0	37	31	7	11	6	3	2	0	50	25	14	19	9	7	3	0	87	56	21
Tampa Bay	19	14	3	2	0	79	44	30	19	8	8	2	1	55	55	19	38	22	11	4	1	134	99	49
Toronto	65	36	23	6	0	272	205	78	63	22	29	11	1	201	250	56	128	58	52	17	1	473	455	134
Vancouver	49	33	9	7	0	225	168	73	49	23	22	4	0	184	176	50	98	56	31	11	0	409	344	123
Washington	79	44	28	7	0	310	251	95	82	32	40	9	1	303	343	74	161	76	68	16	1	613	594	169
Defunct Clubs	35	22	6	7	0	148	93	51	34	13	10	11	0	108	101	37	69	35	16	18	0	256	194	88
Totals	**1384**	**718**	**468**	**197**	**1**	**5272**	**4526**	**1634**	**1384**	**431**	**768**	**172**	**13**	**4287**	**5475**	**1047**	**2768**	**1149**	**1236**	**369**	**14**	**9559**	**10001**	**2681**

Playoffs

	Series	W	L	GP	W	L	T	GF	GA	Last Mtg.	Rnd.	Result
Boston	4	2	2	19	10	9	0	67	62	1992	CF	W 4-0
Buffalo	2	2	0	10	6	4	0	26	26	2001	CSF	W 4-3
Chicago	2	1	1	8	4	4	0	23	24	1992	F	W 4-0
Dallas	1	1	0	6	4	2	0	28	16	1991	F	W 4-2
Florida	1	0	1	7	3	4	0	15	20	1996	CF	L 3-4
Montreal	1	0	1	6	2	4	0	15	18	1998	CQF	L 2-4
New Jersey	5	3	2	29	14	15	0	80	86	2001	CF	L 1-4
NY Islanders	3	0	3	19	8	11	0	58	67	1993	DF	L 3-4
NY Rangers	3	3	0	15	12	3	0	64	45	1996	CSF	W 4-1
Philadelphia	3	0	3	18	6	12	0	51	66	2000	CSF	L 2-4
St. Louis	3	1	2	13	6	7	0	40	45	1981	PRE	L 2-3
Toronto	3	0	3	12	4	8	0	27	39	1999	CSF	L 2-4
Washington	7	6	1	42	26	16	0	137	121	2001	CQF	W 4-2
Defunct Clubs	1	1	0	4	4	0	0	13	6			
Totals	**39**	**20**	**19**	**208**	**109**	**99**	**0**	**644**	**641**			

Playoff Results 2002-1998

Year	Round	Opponent	Result	GF	GA
2001	CF	New Jersey	L 1-4	7	17
	CSF	Buffalo	W 4-3	17	17
	CQF	Washington	W 4-2	14	10
2000	CSF	Philadelphia	L 2-4	14	15
	CQF	Washington	W 4-1	17	8
1999	CSF	Toronto	L 2-4	14	18
	CQF	New Jersey	W 4-3	21	18
1998	CQF	Montreal	L 2-4	15	18

Abbreviations: Round: F – Final; **CF** – conference final; **CSF** – conference semi-final; **CQF** – conference quarter-final; **DF** – division final; **PRE** – preliminary round.

Calgary totals include Atlanta Flames, 1972-73 to 1979-80. Carolina totals include Hartford, 1979-80 to 1996-97.
Colorado totals include Quebec, 1979-80 to 1994-95. Dallas totals include Minnesota North Stars, 1967-68 to 1992-93.
New Jersey totals include Kansas City, 1974-75 to 1975-76, and Colorado Rockies, 1976-77 to 1981-82.
Phoenix totals include Winnipeg, 1979-80 to 1995-96.

2001-02 Results

Month	Day	Opponent	Score
Oct.	3	Colorado	1-3
	6	Anaheim	2-4
	10	NY Islanders	3-6
	14	at Buffalo	1-4
	16	Ottawa	5-2
	18	at Ottawa	3-0
	20	at St. Louis	1-2*
	23	at Atlanta	4-2
	24	Dallas	3-2
	27	at Toronto	0-4
	28	Florida	2-2
	31	at Philadelphia	0-3
Nov.	1	Toronto	3-1
	3	Tampa Bay	2-1*
	6	at Carolina	2-2
	7	at Florida	0-2
	10	at Tampa Bay	2-3*
	13	at New Jersey	5-1
	14	NY Islanders	3-3
	17	NY Rangers	1-0*
	21	Vancouver	1-4
	23	at Nashville	0-5
	24	Buffalo	3-1
	27	New Jersey	6-0
	29	at San Jose	0-5
Dec.	1	at Phoenix	2-5
	4	at Toronto	1-0
	6	at Boston	4-1
	8	Atlanta	6-3
	11	at Washington	2-2
	12	Boston	2-4
	14	Minnesota	2-5
	16	Carolina	0-7
	19	Montreal	1-3
	21	Washington	4-3
	22	at Washington	4-4
	26	at New Jersey	0-4
	29	Ottawa	2-5
Jan.	3	at NY Islanders	2-4
	5	NY Rangers	4-1
	6	at Chicago	0-2
	8	Boston	2-3
	10	at Buffalo	2-0
	12	St. Louis	1-4
	15	at Vancouver	2-5
	17	at Calgary	6-4
	19	at Edmonton	1-0
	21	Philadelphia	5-2
	23	Tampa Bay	5-1
	24	at NY Islanders	5-4*
	26	Atlanta	3-2*
	29	at Philadelphia	2-3*
	30	San Jose	3-6
Feb.	5	at Carolina	3-3
	7	at Montreal	0-1
	9	New Jersey	1-2*
	10	at NY Rangers	3-4
	12	at Ottawa	1-5
	27	Los Angeles	4-5
	28	at Columbus	4-3*
Mar.	2	Detroit	2-4
	4	at NY Islanders	4-2
	5	Florida	6-5*
	7	Carolina	1-3
	9	NY Rangers	3-2*
	11	Columbus	2-4
	13	at Anaheim	2-4
	16	at Los Angeles	3-4
	18	at Atlanta	4-2
	20	Phoenix	1-3
	23	Philadelphia	4-4
	24	Washington	6-2
	27	New Jersey	3-4
	30	at Montreal	1-2*
Apr.	1	Montreal	0-3
	3	at Florida	2-3
	4	at Tampa Bay	2-4
	6	at Philadelphia	1-3
	8	at NY Rangers	2-3
	10	Buffalo	4-4
	12	Toronto	2-5
	13	at Boston	1-7

* – Overtime

Entry Draft Selections 2002-1988

2002

Pick	
5	Ryan Whitney
35	Ondrej Nemec
69	Erik Christensen
101	Daniel Fernholm
136	Andrew Sertich
137	Cam Paddock
171	Robert Goepfert
202	Patrik Bartschi
234	Maxime Talbot
239	Ryan Lannon
265	Dwight Labrosse

2001

Pick	
21	Colby Armstrong
54	Noah Welch
86	Drew Fata
96	Alexandre Rouleau
120	Tomas Surovy
131	Ben Eaves
156	Andrew Schneider
217	Tomas Duba
250	Brandon Crawford-West

2000

Pick	
18	Brooks Orpik
52	Shane Endicott
84	Peter Hamerlik
124	Michel Ouellet
146	David Koci
185	Patrick Foley
216	Jim Abbott
248	Steven Crampton
273	Roman Simicek
280	Nick Boucher

1999

Pick	
18	Konstantin Koltsov
51	Matt Murley
57	Jeremy Van Hoof
86	Sebastian Caron
115	Ryan Malone
144	Tomas Skvaridlo
157	Vladimir Malenkikh
176	Doug Meyer
204	Tom Kostopoulos
233	Darcy Robinson
261	Andrew McPherson

1998

Pick	
23	Milan Kraft
54	Alexander Zevakhin
80	David Cameron
110	Scott Myers
134	Rob Scuderi
169	Jan Fadrny
196	Joel Scherban
224	Mika Lehto
244	Toby Petersen
254	Matt Hussey

1997

Pick	
17	Robert Dome
44	Brian Gaffaney
71	Josef Melichar
97	Alexandre Mathieu
124	Harlan Pratt
152	Petr Havelka
179	Mark Moore
208	Andrew Ference
234	Eric Lind

1996

Pick	
23	Craig Hillier
28	Pavel Skrbek
72	Boyd Kane
77	Boris Protsenko
105	Michal Rozsival
150	Peter Bergman
186	Eric Meloche
238	Timo Seikkula

1995

Pick	
24	Aleksey Morozov
76	Jean-Sebastien Aubin
102	Oleg Belov
128	Jan Hrdina
154	Alexei Kolkunov
180	Derrick Pyke
206	Sergei Voronov
232	Frank Ivankovic

1994

Pick	
24	Chris Wells
50	Richard Park
57	Sven Butenschon
73	Greg Crozier
76	Alexei Krivchenkov
102	Tom O'Connor
128	Clint Johnson
154	Valentin Morozov
161	Serge Aubin
180	Drew Palmer
206	Boris Zelenko
232	Jason Godbout
258	Mikhail Kazakevich
284	Brian Leitza

1993

Pick	
26	Stefan Bergkvist
52	Domenic Pittis
62	Dave Roche
104	Jonas Andersson-Junkka
130	Chris Kelleher
156	Patrick Lalime
182	Sean Selmser
208	Larry McMorran
234	Timothy Harberts
260	Leonid Toropchenko
286	Hans Jonsson

1992

Pick	
19	Martin Straka
43	Marc Hussey
67	Travis Thiessen
91	Todd Klassen
115	Philippe DeRouville
139	Artem Kopot
163	Jan Alinc
187	Fran Bussey
211	Brian Bonin
235	Brian Callahan

1991

Pick	
16	Markus Naslund
38	Rusty Fitzgerald
60	Shane Peacock
82	Joe Tamminen
104	Robert Melanson
126	Brian Clifford
148	Ed Patterson
170	Peter McLaughlin
192	Jeff Lembke
214	Chris Tok
236	Paul Dyck
258	Pasi Huura

1990

Pick	
5	Jaromir Jagr
61	Joe Dziedzic
68	Chris Tamer
89	Brian Farrell
107	Ian Moran
110	Denis Casey
130	Mika Valila
131	Ken Plaquin
145	Pat Neaton
152	Petteri Koskimaki
173	Ladislav Karabin
194	Timothy Fingerhut
215	Michael Thompson
236	Brian Bruininks

1989

Pick	
16	Jamie Heward
37	Paul Laus
58	John Brill
79	Todd Nelson
100	Tom Nevers
121	Mike Markovich
126	Mike Needham
142	Patrick Schafhauser
163	Dave Shute
184	Andrew Wolf
205	Greg Hagen
226	Scott Farrell
247	Jason Smart

1988

Pick	
4	Darrin Shannon
25	Mark Major
62	Daniel Gauthier
67	Mark Recchi
88	Greg Andrusak
130	Troy Mick
151	Jeff Blaeser
172	Rob Gaudreau
193	David Pancoe
214	Cory Laylin
235	Darren Stolk

General Manager

PATRICK, CRAIG

General Manager, Pittsburgh Penguins. Born in Detroit, MI, May 20, 1946.

Known for his calm and patient management style, Craig Patrick has led the Penguins to two Stanley Cup championships, one Presidents' Trophy title and five division championships since taking over as general manager on December 5, 1989. In 2000, he and Mario Lemieux were recipients of the Lester Patrick Trophy for their contributions to hockey in the United States. He was elected to the Hockey Hall of Fame in 2001.

A member of one of hockey's most famous families — including grandfather Lester, father Lynn and uncle Muzz — Patrick played collegiate hockey at the University of Denver and captained the Pioneers to the NCAA championship in 1969. He played eight NHL seasons with four different teams, registering 72 goals and 163 points in 401 games before retiring in 1979. He made the transition to management and coaching when he landed the dual role of assistant coach and assistant g.m. of the 1980 U.S. Olympic team that won the gold medal at Lake Placid.

Patrick joined the New York Rangers organization as director of operations in 1980 and became the youngest general manager in club history one year later. He served in that capacity through the 1985-86 season, leading his team to the playoffs every year.

Prior to joining the Penguins, Patrick spent two years as director of athletics and recreation at the University of Denver.

NHL Coaching Record

		Regular Season				Playoffs		
Season	Team	Games	W	L	T	Games	W	L
1980-81	NY Rangers	60	26	23	11	14	7	7
1984-85	NY Rangers	35	11	22	2	3	0	3
1989-90	Pittsburgh	54	22	26	6	....	....	....
1996-97	Pittsburgh	20	7	10	3	5	1	4
	NHL Totals	**169**	**66**	**81**	**22**	**22**	**8**	**14**

Club Directory

Mellon Arena

Pittsburgh Penguins
Mellon Arena
66 Mario Lemieux Place
Pittsburgh, PA 15219
Phone **412/642-1300**
FAX 412/642-1859
Media Relations FAX 412/642-1322
Capacity: 16,958

Ownership
Mario Lemieux and The Lemieux Group

Administration
Chairman/CEO .... Mario Lemieux
President, Lemieux Group LP .... Ken Sawyer
President, Team Lemieux LLC .... Tom Rooney
Executive Vice President/General Manager .... Craig Patrick
Vice President & General Counsel .... Ted Black
Executive Assistants .... Elaine Heufelder, Fay McNamara
Receptionist .... Kelly Hart
Mailroom Supervisor .... Brett Hart

Hockey Operations
General Manager .... Craig Patrick
Assistant General Manager .... Ed Johnston
Director of Player Development .... Herb Brooks
Head Coach .... Rick Kehoe
Assistant Coaches .... Joe Mullen, Randy Hillier
Goaltending Coach and Scout .... Gilles Meloche
Head Scout .... Greg Malone
Scouts .... Wayne Daniels, Charlie Hodge, Mark Kelley, Neil Shea, Chuck Grillo, Richard Rose
Strength and Conditioning Coach .... John Welday
Equipment Manager .... Steve Latin
Assistant Equipment Manager .... Paul Flati
Equipment Staff .... Paul DeFazio
Trainers .... Mark Mortland, Scott Johnson, Tom Plasko
Team Physician .... Dr. Charles Burke
Executive Assistant .... Tracey Botsford
Team Staff .... Mike Lang, Jr.
Head Coach, Wilkes-Barre/Scranton Penguins (AHL) .... Glenn Patrick
Head Coach, Wheeling Nailers (ECHL) .... John Brophy

Communications and Marketing
Vice President, Communications/Marketing .... Tom McMillan
Director of Media Relations .... Steve Bovino
Manager, Media Relations .... Keith Wehner
Director of Public & Alumni Relations .... Cindy Himes
Director, Community Relations .... Renee Petrichevich
Director of Marketing .... Brian Magness
Director, Video Production Department .... Paul Barto
Manager, Game Presentation .... Mike Wurman
Manager, Video Production .... Joe Hale
Manager, Art and Graphics .... Doreen Minnis
Director of Operations, Lemieux Hockey Development .... Mark Shuttleworth
Director of Alumni Relations .... Jack Riley

Finance
Vice President/Controller .... Kevin Hart
Assistant Controller .... Michael McCullough
Accounting Staff .... Tawni Love, Troy Ussack, Andrea Winschel

Properties
Vice President, Business Development & Properties .. Mike Lee
Director of Publications .... Brian Coe
Creative Director .... Barb Pilarski
Executive Producer, Penguins Radio Network/PensNet .... Ray Walker
Multi Media Manager .... Chris DeVivo

Ticketing
Vice President, Ticketing .... Mark Anderson
Director, Premium Seating .... Terri Smith
Manager, Premium Services .... Michelle Follen
Assistant Manager, Premium Services .... Sherry Huggins
Senior Director, Ticketing .... James Santilli
Senior Director, Ticketing .... Chad Slencak
Manager, Ticket Sales .... Mike Guiffre
Director, Ticket Operations .... Laura Bryer
Box Office Manager .... Carol Coulson
Premium Seating Account Representative .... Bonnie Golinski
Box Office Staff .... Kelly Gabany, Jason Onufer, Jenn Tuite, Dana Kirkpatrick
Ticket Sales Representatives .... Peter Barakat, George Birman, Ted Miller, Chuck Pukansky, Mike McLaughlin, Joe Traynor, Craig Wheeler, Jason Florian, George Murphy, Dave Benson, Jill Shaw, Ross Miller
Customer Service Representatives .... Lynda Gladdng, Jill Weisbrod, Kathy Davis, Erin Exley, Jennifer Snyder, Erica Myones

Corporate Sales
Vice President, Corporate Sales .... David Soltesz
Senior Director, Corporate Sponsorships .... Kimberly Bogesdorfer
Directors, Corporate Sponsorships .... Carl D'Alicandro, Mark DeAndrea
Director, Corporate Sales, Lemieux Hockey Development .... Terri Dobos Young
Manager, Sales Service .... Marie Mays
Assistant Manager, Sales Service .... Darren Morchesky
Corporate Sales Liaison .... Pierre Larouche

General Information
Team Colors .... Black, gold and white
TV Station .... Fox Sports Net Pittsburgh
TV Announcers .... Mike Lange, Eddie Olczyk
Flagship Radio Station .... 3WS (94.5 FM, 970 AM)
Radio Announcers .... Paul Steigerwald, Bob Errey
Minor League Affiliates .... Wilkes-Barre/Scranton (AHL)
Wheeling (ECHL)

St. Louis Blues

2001-02 Results: 43W-27L-8T-4OTL 98PTS.
Second, Central Division

With 46 points last season (11 goals, 35 assists), Al MacInnis became the active leader in points by a defenseman with 1,204. He enters the 2002-03 season one point ahead of Phil Housley and 12 back of Larry Murphy for third place all-time.

2002-03 Schedule

Month	Day	Date	Opponent	Month	Day	Date	Opponent
Oct.	Thu.	10	Anaheim		Thu.	9	at San Jose
	Sat.	12	Minnesota		Sat.	11	at Los Angeles
	Tue.	15	Carolina		Sun.	12	at Anaheim*
	Thu.	17	Columbus		Tue.	14	at Phoenix
	Sat.	19	Dallas		Thu.	16	NY Islanders
	Thu.	24	at Edmonton		Sat.	18	Chicago
	Sat.	26	at Calgary		Mon.	20	at Carolina
	Wed.	30	Nashville		Tue.	21	at Atlanta
Nov.	Sat.	2	at NY Islanders		Thu.	23	at Chicago
	Sun.	3	at NY Rangers*		Sat.	25	Dallas
	Tue.	5	at Montreal		Tue.	28	at Washington
	Thu.	7	Columbus		Thu.	30	Buffalo
	Sat.	9	Toronto	Feb.	Wed.	5	at Dallas
	Tue.	12	at Vancouver		Thu.	6	NY Rangers
	Fri.	15	at Edmonton		Sat.	8	San Jose
	Sat.	16	at Calgary		Tue.	11	at Buffalo
	Wed.	20	at Columbus		Thu.	13	Philadelphia
	Thu.	21	Los Angeles		Sat.	15	Phoenix
	Sat.	23	Colorado		Mon.	17	Calgary*
	Mon.	25	San Jose		Thu.	20	Vancouver
	Wed.	27	at Colorado		Sat.	22	at Pittsburgh*
	Fri.	29	Calgary		Sun.	23	at Minnesota*
	Sat.	30	New Jersey		Thu.	27	Edmonton
Dec.	Tue.	3	at Boston	Mar.	Sat.	1	Minnesota
	Thu.	5	Ottawa		Tue.	4	Nashville
	Sat.	7	at Philadelphia*		Thu.	6	Phoenix
	Sun.	8	at Detroit		Fri.	7	at Detroit
	Tue.	10	at New Jersey		Tue.	11	at San Jose
	Thu.	12	at Nashville		Thu.	13	at Vancouver
	Sat.	14	Atlanta		Sat.	15	at Nashville
	Tue.	17	at Los Angeles		Tue.	18	Vancouver
	Wed.	18	at Anaheim		Thu.	20	Anaheim
	Fri.	20	at Phoenix		Sat.	22	Detroit*
	Mon.	23	Los Angeles		Sun.	23	at Dallas
	Thu.	26	Colorado		Wed.	26	at Minnesota
	Sat.	28	at Columbus		Thu.	27	Florida
	Sun.	29	Columbus		Sat.	29	Detroit*
	Tue.	31	at Detroit		Mon.	31	Edmonton
Jan.	Thu.	2	Chicago	Apr.	Thu.	3	Chicago
	Sat.	4	Tampa Bay		Fri.	4	at Chicago
	Tue.	7	at Nashville		Sun.	6	at Colorado*

* Denotes afternoon game.

Franchise date: June 5, 1967

WESTERN CONFERENCE

CENTRAL DIVISION

36th NHL Season

Year-by-Year Record

		Home				Road				Overall								
Season	**GP**	**W**	**L**	**T**	**OL**	**W**	**L**	**T**	**OL**	**W**	**L**	**T**	**OL**	**GF**	**GA**	**Pts.**	**Finished**	**Playoff Result**
2001-02	82	27	12	1	1	16	15	7	3	43	27	8	4	227	188	98	2nd, Central Div.	Lost Conf. Semi-Final
2000-01	82	28	5	5	3	15	17	7	2	43	22	12	5	249	195	103	2nd, Central Div.	Lost Conf. Championship
1999-2000	82	24	9	7	1	27	10	4	0	51	19	11	1	248	165	114	1st, Central Div.	Lost Conf. Quarter-Final
1998-99	82	18	17	6	...	19	15	7	...	37	32	13	...	237	209	87	2nd, Central Div.	Lost Conf. Semi-Final
1997-98	82	26	10	5	...	19	19	3	...	45	29	8	...	256	204	98	3rd, Central Div.	Lost Conf. Semi-Final
1996-97	82	17	20	4	...	19	15	7	...	36	35	11	...	236	239	83	4th, Central Div.	Lost Conf. Quarter-Final
1995-96	82	15	17	9	...	17	17	7	...	32	34	16	...	219	248	80	4th, Central Div.	Lost Conf. Semi-Final
1994-95	48	16	6	2	...	12	9	3	...	28	15	5	...	178	135	61	2nd, Central Div.	Lost Conf. Quarter-Final
1993-94	84	23	11	8	...	17	22	3	...	40	33	11	...	270	283	91	4th, Central Div.	Lost Conf. Quarter-Final
1992-93	84	22	13	7	...	15	23	4	...	37	36	11	...	282	278	85	4th, Norris Div.	Lost Div. Final
1991-92	80	25	12	3	...	11	21	8	...	36	33	11	...	279	266	83	3rd, Norris Div.	Lost Div. Semi-Final
1990-91	80	24	9	7	...	23	13	4	...	47	22	11	...	310	250	105	2nd, Norris Div.	Lost Div. Final
1989-90	80	20	15	5	...	17	19	4	...	37	34	9	...	295	279	83	2nd, Norris Div.	Lost Div. Final
1988-89	80	22	11	7	...	11	24	5	...	33	35	12	...	275	285	78	2nd, Norris Div.	Lost Div. Final
1987-88	80	18	17	5	...	16	21	3	...	34	38	8	...	278	294	76	2nd, Norris Div.	Lost Div. Final
1986-87	80	21	12	7	...	11	21	8	...	32	33	15	...	281	293	79	1st, Norris Div.	Lost Div. Semi-Final
1985-86	80	23	11	6	...	14	23	3	...	37	34	9	...	302	291	83	3rd, Norris Div.	Lost Conf. Championship
1984-85	80	21	12	7	...	16	19	5	...	37	31	12	...	299	288	86	1st, Norris Div.	Lost Div. Semi-Final
1983-84	80	23	14	3	...	9	27	4	...	32	41	7	...	293	316	71	2nd, Norris Div.	Lost Div. Final
1982-83	80	16	16	8	...	9	24	7	...	25	40	15	...	285	316	65	4th, Norris Div.	Lost Div. Semi-Final
1981-82	80	22	14	4	...	10	26	4	...	32	40	8	...	315	349	72	3rd Norris Div.	Lost Div. Final
1980-81	80	29	7	4	...	16	11	13	...	45	18	17	...	352	281	107	1st, Smythe Div.	Lost Quarter-Final
1979-80	80	20	13	7	...	14	21	5	...	34	34	12	...	266	278	80	2nd, Smythe Div.	Lost Prelim. Round
1978-79	80	14	20	6	...	4	30	6	...	18	50	12	...	249	348	48	3rd, Smythe Div.	Out of Playoffs
1977-78	80	12	20	8	...	8	27	5	...	20	47	13	...	195	304	53	4th, Smythe Div.	Out of Playoffs
1976-77	80	22	13	5	...	10	26	4	...	32	39	9	...	239	276	73	1st, Smythe Div.	Lost Quarter-Final
1975-76	80	20	12	8	...	9	25	6	...	29	37	14	...	249	290	72	3rd, Smythe Div.	Lost Prelim. Round
1974-75	80	23	13	4	...	12	18	10	...	35	31	14	...	269	267	84	2nd, Smythe Div.	Lost Prelim. Round
1973-74	78	16	16	7	...	10	24	5	...	26	40	12	...	206	248	64	6th, West Div.	Out of Playoffs
1972-73	78	21	11	7	...	11	23	5	...	32	34	12	...	233	251	76	4th, West Div.	Lost Quarter-Final
1971-72	78	17	17	5	...	11	22	6	...	28	39	11	...	208	247	67	3rd, West Div.	Lost Semi-Final
1970-71	78	23	7	9	...	11	18	10	...	34	25	19	...	223	208	87	2nd, West Div.	Lost Quarter-Final
1969-70	76	24	9	5	...	13	18	7	...	37	27	12	...	224	179	86	1st, West Div.	Lost Final
1968-69	76	21	8	9	...	16	17	5	...	37	25	14	...	204	157	88	1st, West Div.	Lost Final
1967-68	74	18	12	7	...	9	19	9	...	27	31	16	...	177	191	70	3rd, West Div.	Lost Final

2002-03 Player Personnel

FORWARDS	HT	WT	S	Place of Birth	Date	2001-02 Club
BOGUNIECKI, Eric	5-8	192	R	New Haven, CT	5/6/75	St. Louis-Worcester
CAJANEK, Petr	5-11	176	L	Gottwaldov, Czech.	8/18/75	Zlin-Czech Republic
CORSO, Daniel	5-10	187	L	Montreal, Que.	4/3/78	St. Louis
DAWE, Jason	5-10	189	L	North York, Ont.	5/29/73	NY Rangers-Hartford
DEMITRA, Pavol	5-11	203	L	Dubnica, Czech.	11/29/74	St. Louis-Slovakia
DRAKE, Dallas	6-1	187	L	Trail, B.C.	2/4/69	St. Louis
DUBINSKY, Steve	6-0	190	L	Montreal, Que.	7/9/70	Chi-Norfolk-Nsh-Mil
EASTWOOD, Mike	6-3	213	R	Ottawa, Ont.	7/1/67	St. Louis
LOW, Reed	6-3	222	R	Moose Jaw, Sask.	6/21/76	St. Louis
MAYERS, Jamal	6-1	212	R	Toronto, Ont.	10/24/74	St. Louis
MELLANBY, Scott	6-1	205	R	Montreal, Que.	6/11/66	St. Louis
NASH, Tyson	6-0	185	L	Edmonton, Alta.	3/11/75	St. Louis
PODEIN, Shjon	6-2	200	L	Rochester, MN	3/5/68	Colorado-St. Louis
STILLMAN, Cory	6-0	194	L	Peterborough, Ont.	12/20/73	St. Louis
TKACHUK, Keith	6-2	225	L	Melrose, MA	3/28/72	St. Louis-United States
VARLAMOV, Sergei	5-11	195	L	Kiev, USSR	7/21/78	St. Louis-Ukraine
WEIGHT, Doug	5-11	200	L	Warren, MI	1/21/71	St. Louis-United States
DEFENSEMEN						
BACKMAN, Christian	6-4	198	L	Alingsas, Sweden	4/28/80	Vastra Frolunda
BANCROFT, Steve	6-1	214	L	Toronto, Ont.	10/6/70	San Jose-Cleveland
BRIMANIS, Aris	6-3	210	R	Cleveland, OH	3/14/72	Anaheim-Cincinnati
FINLEY, Jeff	6-2	205	L	Edmonton, Alta.	4/14/67	St. Louis
JACKMAN, Barret	6-1	200	L	Trail, B.C.	3/5/81	St. Louis-Worcester
KHAVANOV, Alexander	6-0	187	L	Ryazan, USSR	1/30/72	St. Louis
KOIVISTO, Tom	5-10	194	R	Turku, Finland	6/4/74	Jokerit Helsinki
LAFLAMME, Christian	6-1	210	R	St-Charles, Que.	11/24/76	St. Louis-Worcester
MacINNIS, Al	6-2	209	R	Inverness, N.S.	7/11/63	St. Louis-Canada
PILON, Rich	6-2	220	L	Saskatoon, Sask.	4/30/68	St. Louis
PRONGER, Chris	6-6	220	L	Dryden, Ont.	10/10/74	St. Louis-Canada
SALVADOR, Bryce	6-2	215	L	Brandon, Man.	2/11/76	St. Louis
VAN RYN, Mike	6-1	190	R	London, Ont.	5/14/79	St. Louis-Worcester

GOALTENDERS	HT	WT	C	Place of Birth	Date	2001-02 Club
BRATHWAITE, Fred	5-7	175	L	Ottawa, Ont.	11/24/72	St. Louis
JOHNSON, Brent	6-2	200	L	Farmington, MI	3/12/77	St. Louis

Coaching History

Lynn Patrick and Scotty Bowman, 1967-68; Scotty Bowman, 1968-69, 1969-70; Al Arbour and Scotty Bowman, 1970-71; Sid Abel, Bill McCreary and Al Arbour, 1971-72; Al Arbour and Jean-Guy Talbot, 1972-73; Jean-Guy Talbot and Lou Angotti, 1973-74; Lou Angotti, Lynn Patrick and Garry Young, 1974-75; Garry Young, Lynn Patrick and Leo Boivin, 1975-76; Emile Francis, 1976-77; Leo Boivin and Barclay Plager, 1977-78; Barclay Plager, 1978-79; Barclay Plager and Red Berenson, 1979-80; Red Berenson, 1980-81; Red Berenson and Emile Francis, 1981-82; Emile Francis and Barclay Plager, 1982-83; Jacques Demers, 1983-84 to 1985-86; Jacques Martin, 1986-87, 1987-88; Brian Sutter, 1988-89 to 1991-92; Bob Plager and Bob Berry, 1992-93; Bob Berry, 1993-94; Mike Keenan, 1994-95, 1995-96; Mike Keenan, Jim Roberts and Joel Quenneville, 1996-97; Joel Quenneville, 1997-98 to date.

Coach

QUENNEVILLE, JOEL
Head Coach, St. Louis Blues. Born in Windsor, Ont., September 15, 1958.

Joel Quenneville was named head coach on January 6, 1997, becoming the 20th head coach in Blues history. His first game in St. Louis was on January 7, 1997. He won the Jack Adams Award as coach of the year in 1999-2000 after leading the Blues to the Presidents' Trophy with a club record 51 wins and 114 points. In 2000-01, he led the team to the Western Conference Finals.

Prior to joining the Blues the former NHL defenseman spent three seasons with the Colorado Avalanche organization as an assistant coach. He was instrumental in the Avalanche's drive for their first Stanley Cup championship during the 1995-96 season.

Prior to joining the Avalanche he was head coach for the Springfield Indians of the American Hockey League during the 1993-94 season. He retired from hockey after the 1991-92 season after serving the St. John's Maple Leafs (AHL) as a player/coach. Quenneville played 13 NHL seasons and finished with 803 career games played, 54 goals, 136 assists and 705 penalty minutes. His best years on the ice were spent with Hartford where he earned most valuable defenseman honors in 1984 and 1985. He played an integral part in helping Hartford win a divisional championship in 1986-87.

Coaching Record

		Regular Season				Playoffs		
Season	Team	Games	W	L	T	Games	W	L
1993-94	Springfield (AHL)	80	29	38	13	6	2	4
1996-97	**St. Louis (NHL)**	**40**	**18**	**15**	**7**	**6**	**2**	**4**
1997-98	**St. Louis (NHL)**	**82**	**45**	**29**	**8**	**10**	**6**	**4**
1998-99	**St. Louis (NHL)**	**82**	**37**	**32**	**13**	**13**	**6**	**7**
1999-2000	**St. Louis (NHL)**	**82**	**51**	**20**	**11**	**7**	**3**	**4**
2000-01	**St. Louis (NHL)**	**82**	**43**	**27**	**12**	**15**	**9**	**6**
2001-02	**St. Louis (NHL)**	**82**	**43**	**31**	**8**	**10**	**5**	**5**
	NHL Totals	**450**	**237**	**154**	**59**	**61**	**31**	**30**

2001-02 Scoring

* - rookie

Regular Season

Pos	#	Player	Team	GP	G	A	Pts	+/–	PIM	PP	SH	GW	GT	S	%
R	38	Pavol Demitra	STL	82	35	43	78	13	46	11	0	10	0	212	16.5
L	7	Keith Tkachuk	STL	73	38	37	75	21	117	13	0	7	1	244	15.6
C	39	Doug Weight	STL	61	15	34	49	20	40	3	0	1	0	131	11.5
D	44	Chris Pronger	STL	78	7	40	47	23	120	4	1	3	0	204	3.4
D	2	Al MacInnis	STL	71	11	35	46	3	52	6	0	4	1	231	4.8
L	61	Cory Stillman	STL	80	23	22	45	8	36	6	0	4	0	140	16.4
R	48	Scott Young	STL	67	19	22	41	11	26	5	0	1	0	210	9.0
R	19	Scott Mellanby	STL	64	15	26	41	–5	93	8	0	2	0	137	10.9
C	13	Ray Ferraro	ATL	61	8	19	27	–32	66	2	0	0	1	72	11.1
			STL	15	6	4	10	2	8	2	0	1	0	27	22.2
			TOTAL	76	14	23	37	–30	74	4	0	1	1	99	14.1
R	10	Dallas Drake	STL	80	11	15	26	8	87	1	3	2	0	116	9.5
D	29	Alexander Khavanov	STL	81	3	21	24	9	55	0	0	0	0	87	3.4
L	25	Shjon Podein	COL	41	6	6	12	0	39	0	1	2	0	43	14.0
			STL	23	2	4	6	2	2	0	0	0	0	24	8.3
			TOTAL	64	8	10	18	2	41	0	1	2	0	67	11.9
R	21	Jamal Mayers	STL	77	9	8	17	9	99	0	1	0	0	105	8.6
C	32	Mike Eastwood	STL	71	7	10	17	–2	41	0	0	2	0	60	11.7
L	9	Tyson Nash	STL	64	6	7	13	2	100	0	0	1	0	66	9.1
L	17 *	Sergei Varlamov	STL	52	5	7	12	4	26	0	0	0	0	83	6.0
D	27	Bryce Salvador	STL	66	5	7	12	3	78	1	0	2	0	37	13.5
C	15	Daniel Corso	STL	41	4	7	11	3	6	1	0	2	0	25	16.0
D	43 *	Mike Van Ryn	STL	48	2	8	10	10	18	0	0	1	0	52	3.8
D	37	Jeff Finley	STL	78	0	6	6	12	30	0	0	0	0	39	0.0
R	34	Reed Low	STL	58	0	5	5	–3	160	0	0	0	0	25	0.0
R	42 *	Mark Rycroft	STL	9	0	3	3	0	4	0	0	0	0	14	0.0
D	4	Marc Bergevin	STL	30	0	3	3	6	2	0	0	0	0	13	0.0
D	47	Richard Pilon	STL	8	0	2	2	–1	9	0	0	0	0	3	0.0
C	14	Eric Boguniecki	STL	8	0	1	1	–2	4	0	0	0	0	10	0.0
D	46	Christian Laflamme	STL	8	0	1	1	3	4	0	0	0	0	6	0.0
C	26 *	Justin Papineau	STL	1	0	0	0	–2	0	0	0	0	0	0	0.0
D	5 *	Barret Jackman	STL	1	0	0	0	0	0	0	0	0	0	1	0.0
L	6	Ted Donato	NYI	1	0	0	0	–1	0	0	0	0	0	1	0.0
			L.A.	2	0	0	0	–2	2	0	0	0	0	1	0.0
			STL	2	0	0	0	–2	2	0	0	0	0	0	0.0
			TOTAL	5	0	0	0	–5	4	0	0	0	0	2	0.0

Goaltending

No.	Goaltender	GPI	Mins	Avg	W	L	T	EN	SO	GA	SA	S%	G	A	PIM
30	Reinhard Divis	1	25	0.00	0	0	0	1	0	0	4	1.000	0	0	0
35	Brent Johnson	58	3491	2.18	34	20	4	5	5	127	1293	.902	0	4	2
40	Fred Brathwaite	25	1446	2.24	9	11	4	1	2	54	543	.901	0	0	0
	Totals	**82**	**4990**	**2.26**	**43**	**31**	**8**	**7**	**7**	**188**	**1847**	**.898**			

Playoffs

Pos	#	Player	Team	GP	G	A	Pts	+/–	PIM	PP	SH	GW	GT	S	%
R	38	Pavol Demitra	STL	10	4	7	11	3	6	2	1	1	0	26	15.4
R	19	Scott Mellanby	STL	10	7	3	10	1	18	4	0	1	0	25	28.0
L	7	Keith Tkachuk	STL	10	5	5	10	1	18	1	0	0	0	35	14.3
D	44	Chris Pronger	STL	9	1	7	8	5	24	0	0	0	0	16	6.3
D	2	Al MacInnis	STL	10	0	7	7	3	4	0	0	0	0	30	0.0
R	48	Scott Young	STL	10	3	0	3	–1	2	1	1	0	0	22	13.6
R	21	Jamal Mayers	STL	10	3	0	3	–1	2	0	0	2	0	11	27.3
C	13	Ray Ferraro	STL	10	0	3	3	–2	4	0	0	0	0	7	0.0
C	39	Doug Weight	STL	10	1	1	2	0	4	1	0	1	0	22	4.5
L	61	Cory Stillman	STL	9	0	2	2	–1	2	0	0	0	0	15	0.0
C	33	Eric Boguniecki	STL	1	0	1	1	1	0	0	0	0	0	1	0.0
L	9	Tyson Nash	STL	9	0	1	1	0	20	0	0	0	0	6	0.0
D	27	Bryce Salvador	STL	10	0	1	1	–1	4	0	0	0	0	6	0.0
L	17 *	Sergei Varlamov	STL	1	0	0	0	0	2	0	0	0	0	0	0.0
D	5 *	Barret Jackman	STL	1	0	0	0	0	2	0	0	0	0	2	0.0
C	15	Daniel Corso	STL	2	0	0	0	0	0	0	0	0	0	1	0.0
D	29	Alexander Khavanov	STL	4	0	0	0	0	2	0	0	0	0	5	0.0
D	4	Marc Bergevin	STL	7	0	0	0	0	4	0	0	0	0	2	0.0
R	10	Dallas Drake	STL	8	0	0	0	0	8	0	0	0	0	8	0.0
D	43 *	Mike Van Ryn	STL	9	0	0	0	–2	0	0	0	0	0	12	0.0
D	37	Jeff Finley	STL	10	0	0	0	–1	8	0	0	0	0	3	0.0
C	32	Mike Eastwood	STL	10	0	0	0	–1	6	0	0	0	0	5	0.0
L	25	Shjon Podein	STL	10	0	0	0	0	6	0	0	0	0	9	0.0

Goaltending

No.	Goaltender	GPI	Mins	Avg	W	L	EN	SO	GA	SA	S%	G	A	PIM
40	Fred Brathwaite	1	0	0.00	0	0	0	0	0	0	.000	0	0	0
35	Brent Johnson	10	590	1.83	5	5	1	3	18	252	.929	0	0	4
	Totals	**10**	**600**	**1.90**	**5**	**5**	**1**	**3**	**19**	**253**	**.925**			

Captains' History

Al Arbour, 1967-68 to 1969-70; Red Berenson and Barclay Plager, 1970-71; Barclay Plager, 1971-72 to 1975-76; no captain, 1976-77; Red Berenson, 1977-78; Barry Gibbs, 1978-79; Brian Sutter, 1979-80 to 1987-88; Bernie Federko, 1988-89; Rick Meagher, 1989-90; Scott Stevens, 1990-91; Garth Butcher, 1991-92; Brett Hull, 1992-93 to 1994-95; Brett Hull, Shayne Corson and Wayne Gretzky, 1995-96; no captain, 1996-97; Chris Pronger, 1997-98 to date.

Club Records

Team

(Figures in brackets for season records are games played; records for fewest points, wins, ties, losses, goals, goals against are for 70 or more games)

Record		
Most Points	**114**	1999-2000 (82)
Most Wins	**51**	1999-2000 (82)
Most Ties	**19**	1970-71 (78)
Most Losses	**50**	1978-79 (80)
Most Goals	**352**	1980-81 (80)
Most Goals Against	**349**	1981-82 (80)
Fewest Points	**48**	1978-79 (80)
Fewest Wins	**18**	1978-79 (80)
Fewest Ties	**7**	1983-84 (80)
Fewest Losses	**18**	1980-81 (80)
Fewest Goals	**177**	1967-68 (74)
Fewest Goals Against	**157**	1968-69 (76)
Longest Winning Streak		
Overall	**10**	Jan. 3-23/02
Home	**9**	Jan. 26-Feb. 26/91
Away	***10**	Jan. 21-Mar. 2/00
Longest Undefeated Streak		
Overall	**12**	Nov. 10-Dec. 8/68 (5 wins, 7 ties), Nov. 24-Dec. 26/00 (11 wins, 1 tie)
Home	**11**	Four times
Away	**11**	Jan. 21-Mar. 4/00 (10 wins, 1 tie)
Longest Losing Streak		
Overall	**7**	Nov. 12-26/67, Feb. 12-25/89
Home	**6**	Nov. 23-Dec. 19/96
Away	**10**	Jan. 20-Mar. 8/82
Longest Winless Streak		
Overall	**12**	Jan. 17-Feb. 15/78 (10 losses, 2 ties)
Home	**7**	Dec. 28/82-Jan. 25/83 (5 losses, 2 ties)
Away	**17**	Jan. 23-Oct. 9/74 (13 losses, 4 ties)
Most Shutouts, Season	**13**	1968-69 (76)
Most PIM, Season	**2,041**	1990-91 (80)
Most Goals, Game	**11**	Feb. 26/94 (St.L. 11 at Ott. 1)

Individual

Most Seasons	**13**	Bernie Federko
Most Games	**927**	Bernie Federko
Most Goals, Career	**527**	Brett Hull
Most Assists, Career	**721**	Bernie Federko
Most Points, Career	**1,073**	Bernie Federko (352G, 721A)
Most PIM, Career	**1,786**	Brian Sutter
Most Shutouts, Career	**16**	Glenn Hall
Longest Consecutive Games Streak	**662**	Garry Unger (Feb. 7/71-Apr. 8/79)
Most Goals, Season	**86**	Brett Hull (1990-91)
Most Assists, Season	**90**	Adam Oates (1990-91)
Most Points, Season	**131**	Brett Hull (1990-91) (86G, 45A)
Most PIM, Season	**306**	Bob Gassoff (1975-76)
Most Points, Defenseman, Season	**78**	Jeff Brown (1992-93; 25G, 53A)
Most Points, Center, Season	**115**	Adam Oates (1990-91; 25G, 90A)
Most Points, Right Wing, Season	**131**	Brett Hull (1990-91; 86G, 45A)
Most Points, Left Wing, Season	**102**	Brendan Shanahan (1993-94; 52G, 50A)
Most Points, Rookie, Season	**73**	Jorgen Pettersson (1980-81; 37G, 36A)
Most Shutouts, Season	**8**	Glenn Hall (1968-69)
Most Goals, Game	**6**	Red Berenson (Nov. 7/68)
Most Assists, Game	**5**	Brian Sutter (Nov. 22/83), Bernie Federko (Feb. 27/88), Adam Oates (Jan. 26/91)
Most Points, Game	**7**	Red Berenson (Nov. 7/68; 6G, 1A), Garry Unger (Mar. 13/71; 3G, 4A)

* NHL Record.

Retired Numbers

No.	Player	Years
3	Bob Gassoff	1973-1977
8	Barclay Plager	1967-1977
11	Brian Sutter	1976-1988
24	Bernie Federko	1976-1989

All-time Record vs. Other Clubs

Regular Season

	At Home								On Road								Total							
	GP	W	L	T	OL	GF	GA	PTS	GP	W	L	T	OL	GF	GA	PTS	GP	W	L	T	OL	GF	GA	PTS
Anaheim	18	9	6	3	0	56	47	21	18	13	4	1	0	58	42	27	36	22	10	4	0	114	89	48
Atlanta	2	2	0	0	0	7	1	4	2	2	0	0	0	11	5	4	4	4	0	0	0	18	6	8
Boston	58	26	23	9	0	184	195	61	58	14	35	9	0	157	247	37	116	40	58	18	0	341	442	98
Buffalo	49	28	14	7	0	178	124	63	50	15	29	6	0	156	197	36	99	43	43	13	0	334	321	99
Calgary	60	29	22	9	0	215	183	67	58	24	28	5	1	171	196	54	118	53	50	14	1	386	379	121
Carolina	30	18	9	3	0	117	93	39	29	16	11	2	0	92	87	34	59	34	20	5	0	209	180	73
Chicago	109	54	37	17	1	366	334	126	113	34	62	16	1	338	421	85	222	88	99	33	2	704	755	211
Colorado	36	21	12	3	0	141	112	45	37	13	19	5	0	101	126	31	73	34	31	8	0	242	238	76
Columbus	5	4	0	1	0	19	8	9	5	1	1	2	1	11	11	5	10	5	1	3	1	30	19	14
Dallas	113	62	30	21	0	413	319	145	111	40	50	21	0	330	373	101	224	102	80	42	0	743	692	246
Detroit	105	56	30	19	0	364	289	131	104	40	47	17	0	322	376	97	209	96	77	36	0	686	665	228
Edmonton	40	19	16	5	0	146	146	43	40	15	21	4	0	133	146	34	80	34	37	9	0	279	292	77
Florida	7	5	1	1	0	21	11	11	8	4	2	2	0	18	16	10	15	9	3	3	0	39	27	21
Los Angeles	74	46	18	10	0	279	191	102	74	28	34	12	0	216	251	68	148	74	52	22	0	495	442	170
Minnesota	4	2	2	0	0	8	6	4	4	2	0	2	0	11	9	6	8	4	2	2	0	19	15	10
Montreal	57	14	28	15	0	147	195	43	57	10	40	7	0	154	248	27	114	24	68	22	0	301	443	70
Nashville	11	8	3	0	0	38	20	16	11	7	2	2	0	36	23	16	22	15	5	2	0	74	43	32
New Jersey	44	27	10	7	0	187	134	61	45	17	21	7	0	128	144	41	89	44	31	14	0	315	278	102
NY Islanders	45	17	18	9	1	161	148	44	47	11	25	11	0	121	180	33	92	28	43	20	1	282	328	77
NY Rangers	61	24	28	9	0	179	194	57	59	9	44	6	0	140	243	24	120	33	72	15	0	319	437	81
Ottawa	8	4	3	1	0	25	23	9	9	6	3	0	0	35	20	12	17	10	6	1	0	60	43	21
Philadelphia	66	26	32	7	1	189	208	60	66	11	45	10	0	150	263	32	132	37	77	17	1	339	471	92
Phoenix	46	24	12	10	0	168	125	58	47	17	24	6	0	144	155	40	93	41	36	16	0	312	280	98
Pittsburgh	63	42	15	6	0	244	169	90	63	20	31	12	0	189	236	52	126	62	46	18	0	433	405	142
San Jose	24	16	6	1	1	83	55	34	20	16	3	1	0	73	44	33	44	32	9	2	1	156	99	67
Tampa Bay	9	8	1	0	0	35	20	16	12	5	4	2	1	40	36	13	21	13	5	2	1	75	56	29
Toronto	101	57	29	14	1	342	280	129	98	29	58	11	0	289	367	69	199	86	87	25	1	631	647	198
Vancouver	67	39	19	9	0	252	191	87	68	33	26	8	1	221	199	75	135	72	45	17	1	473	390	162
Washington	40	19	13	8	0	161	126	46	38	14	20	4	0	112	133	32	78	33	33	12	0	273	259	78
Defunct Clubs	32	25	4	3	0	131	55	53	33	11	10	12	0	95	100	34	65	36	14	15	0	226	155	87
Totals	**1384**	**731**	**441**	**207**	**5**	**4856**	**4002**	**1674**	**1384**	**477**	**699**	**203**	**5**	**4052**	**4894**	**1162**	**2768**	**1208**	**1140**	**410**	**10**	**8908**	**8896**	**2836**

Playoffs

	Series	W	L	GP	W	L	T	GF	GA	Last Mtg.	Rnd.	Result
Boston	2	0	2	8	0	8	0	15	48	1972	SF	L 0-4
Buffalo	1	0	1	3	1	2	0	8	7	1976	PRE	L 1-2
Calgary	1	0	1	7	3	4	0	22	28	1986	CF	L 3-4
Chicago	10	3	7	50	22	28	0	142	171	2002	CQF	W 4-1
Colorado	1	0	1	5	1	4	0	11	17	2001	CF	L 1-4
Dallas	12	6	6	66	32	34	0	187	197	2001	CSF	W 4-0
Detroit	7	2	5	40	16	24	0	103	125	2002	CSF	L 1-4
Los Angeles	2	2	0	8	8	0	0	32	13	1998	CQF	W 4-0
Montreal	3	0	3	12	0	12	0	14	42	1977	QF	L 0-4
NY Rangers	1	0	1	6	2	4	0	22	29	1981	QF	L 2-4
Philadelphia	2	2	0	11	8	3	0	34	20	1969	QF	W 4-0
Phoenix	2	2	0	11	7	4	0	39	29	1999	CQF	W 4-3
Pittsburgh	3	2	1	13	7	6	0	45	40	1981	PRE	W 3-2
San Jose	2	1	1	13	7	6	0	38	31	2001	CQF	W 4-2
Toronto	5	3	2	31	17	14	0	88	90	1996	CQF	W 4-2
Vancouver	1	0	1	7	3	4	0	27	27	1995	CQF	L 3-4
Totals	**55**	**23**	**32**	**291**	**134**	**157**	**0**	**827**	**914**			

Playoff Results 2002-1998

Year	Round	Opponent	Result	GF	GA
2002	CSF	Detroit	L 1-4	11	14
	CQF	Chicago	W 4-1	13	5
2001	CF	Colorado	L 1-4	11	17
	CSF	Dallas	W 4-0	13	6
	CQF	San Jose	W 4-2	16	11
2000	CQF	San Jose	L 3-4	22	20
1999	CSF	Dallas	L 2-4	12	17
	CQF	Phoenix	W 4-3	19	16
1998	CSF	Detroit	L 2-4	13	23
	CQF	Los Angeles	W 4-0	16	8

Abbreviations: Round: CF – conference final; **CSF** – conference semi-final; **CQF** – conference quarter-final; **DSF** – division semi-final; **SF** – semi-final; **QF** – quarter-final; **PRE** – preliminary round.

Calgary totals include Atlanta Flames, 1972-73 to 1979-80. Carolina totals include Hartford, 1979-80 to 1996-97.
Colorado totals include Quebec, 1979-80 to 1994-95. Dallas totals include Minnesota North Stars, 1967-68 to 1992-93.
New Jersey totals include Kansas City, 1974-75 to 1975-76, and Colorado Rockies, 1976-77 to 1981-82.
Phoenix totals include Winnipeg, 1979-80 to 1995-96.

2001-02 Results

Oct.	4	at Columbus	3-3		8	at San Jose	6-2
	6	at Nashville	2-2		9	at Anaheim	3-2
	11	Los Angeles	5-6		12	at Pittsburgh	4-1
	13	at Toronto	5-2		15	Edmonton	3-2
	17	Dallas	2-2		17	Vancouver	5-4
	19	at Minnesota	3-2*		19	Boston	2-1*
	20	Pittsburgh	2-1*		21	at Boston	4-3*
	22	Calgary	2-3		23	at Buffalo	5-2
	25	NY Rangers	5-1		25	at Chicago	1-2
	27	at Ottawa	1-4		26	Detroit	2-5
	31	at Colorado	1-0		28	Anaheim	0-1
Nov.	1	Carolina	4-3		30	at Washington	4-1
	3	Washington	4-1	**Feb.**	5	at NY Islanders	3-4
	6	San Jose	1-4		7	Edmonton	3-1
	8	Vancouver	3-1		9	Philadelphia	0-5
	10	Phoenix	4-1		12	Atlanta	3-0
	13	at Columbus	2-3*		13	at Colorado	1-3
	15	at Vancouver	1-2		26	at Vancouver	4-4
	17	at Calgary	0-2		28	at Calgary	2-3
	20	at Edmonton	0-2	**Mar.**	2	at Edmonton	1-1
	23	at Detroit	1-3		7	Minnesota	0-3
	24	Phoenix	5-3		9	Detroit	2-5
	27	Ottawa	4-2		11	Colorado	2-3
	29	at Columbus	3-1		13	at San Jose	2-0
Dec.	1	Columbus	4-3		14	at Los Angeles	1-2
	2	at Minnesota	4-4		17	at Anaheim	3-2
	5	at Phoenix	0-3		19	Nashville	5-1
	6	at Los Angeles	1-1		20	at Dallas	2-3
	8	Los Angeles	2-0		22	Anaheim	3-2
	12	at Chicago	2-2		24	at Chicago	3-4*
	13	Toronto	3-4*		26	Minnesota	1-2
	15	Calgary	4-0		28	Buffalo	4-1
	18	at Philadelphia	3-6		30	at Nashville	4-2
	21	at Tampa Bay	3-4*	**Apr.**	1	at Phoenix	3-5
	22	at Florida	2-0		3	at Dallas	2-1*
	26	Chicago	1-3		5	Chicago	5-1
	28	Montreal	3-0		7	Colorado	2-4
	30	Nashville	7-2		9	Nashville	3-2
Jan.	1	at New Jersey	1-2		11	San Jose	4-1
	3	Columbus	4-2		13	Detroit	3-2*
	5	Dallas	5-2		14	at Detroit	5-3

* – Overtime

Entry Draft Selections 2002-1988

2002
Pick
48 Alexei Shkotov
62 Andrei Mikhnov
89 Tomas Troliga
120 Robin Jonsson
165 Justin Maiser
191 D.J. King
221 Jonas Johnson
253 Tom Koivisto
284 Ryan MacMurchy

2001
Pick
57 Jay McClement
89 Tuomas Nissinen
122 Igor Valeyev
159 Dmitri Semin
190 Brett Scheffelmaier
253 Petr Cajanek
270 Grant Jacobsen
283 Simon Skoog

2000
Pick
30 Jeff Taffe
65 Dave Morisset
75 Justin Papineau
96 Antoine Bergeron
129 Troy Riddle
167 Craig Weller
229 Brett Lutes
261 Reinhard Divis
293 Lauri Kinos

1999
Pick
17 Barret Jackman
85 Peter Smrek
114 Chad Starling
143 Trevor Byrne
180 Tore Vikingstad
203 Phil Osaer
221 Colin Hemingway
232 Alexander Khavanov
260 Brian McMeekin
270 James Desmarais

1998
Pick
24 Christian Backman
41 Maxim Linnik
83 Matt Walker
157 Brad Voth
170 Andrei Troschinsky
197 Brad Twordik
225 Yevgeny Pastukh
255 John Pohl

1997
Pick
40 Tyler Rennette
86 Didier Tremblay
98 Jan Horacek
106 Jame Pollock
149 Nicholas Bilotto
177 Ladislav Nagy
206 Bobby Haglund
232 Dmitri Plekhanov
244 Marek Ivan

1996
Pick
14 Marty Reasoner
67 Gordie Dwyer
95 Jonathan Zukiwsky
97 Andrei Petrakov
159 Stephen Wagner
169 Daniel Corso
177 Reed Low
196 Andrej Podkonicky
203 Tony Hutchins
229 Konstantin Shafranov

1995
Pick
49 Jochen Hecht
75 Scott Roche
101 Michal Handzus
127 Jeff Ambrosio
153 Denis Hamel
179 Jean-Luc Grand-Pierre
205 Derek Bekar
209 Libor Zabransky

1994
Pick
68 Stephane Roy
94 Tyler Harlton
120 Edvin Frylen
172 Roman Vopat
198 Steve Noble
224 Marc Stephan
250 Kevin Harper
276 Scott Fankhouser

1993
Pick
37 Maxim Bets
63 Jamie Rivers
89 Jamal Mayers
141 Todd Kelman
167 Mike Buzak
193 Eric Boguniecki
219 Mike Grier
245 Libor Prochazka
271 Alexander Vasilevski
275 Christer Olsson

1992
Pick
38 Igor Korolev
62 Vitali Karamnov
64 Vitali Prokhorov
86 Lee Leslie
134 Bob Lachance
158 Ian Laperriere
160 Lance Burns
180 Igor Boldin
182 Nick Naumenko
206 Todd Harris
230 Yuri Gunko
259 Wade Salzman

1991
Pick
27 Steve Staios
64 Kyle Reeves
65 Nathan LaFayette
87 Grayden Reid
109 Jeff Callinan
131 Bruce Gardiner
153 Terry Hollinger
175 Chris Kenady
197 Jed Fiebelkorn
219 Chris MacKenzie
241 Kevin Rappana
263 Mike Veisor

1990
Pick
33 Craig Johnson
54 Patrice Tardif
96 Jason Ruff
117 Kurtis Miller
138 Wayne Conlan
180 Parris Duffus
201 Steve Widmeyer
222 Joe Hawley
243 Joe Fleming

1989
Pick
9 Jason Marshall
31 Rick Corriveau
55 Denny Felsner
93 Daniel Laperriere
114 David Roberts
124 Derek Frenette
135 Jeff Batters
156 Kevin Plager
177 John Roderick
198 John Valo
219 Brian Lukowski

1988
Pick
9 Rod Brind'Amour
30 Adrien Plavsic
51 Rob Fournier
72 Jaan Luik
105 Dave Lacouture
114 Dan Fowler
135 Matt Hayes
156 John McCoy
177 Tony Twist
198 Bret Hedican
219 Heath DeBoer
240 Michael Francis

Club Directory

Savvis Center

St. Louis Blues
Savvis Center
1401 Clark Avenue
St. Louis, MO 63103
Phone **314/622-2500**
FAX 314/622-2533
www.stlouisblues.com
Capacity: 19,022

Owner and Chairman . . . . Bill Laurie
President and CEO . . . . Mark Sauer
Sr. Vice President and General Manager . . . . Larry Pleau
Vice President and Director of Hockey Operations . . John Ferguson, Jr.
Head Coach . . . . Joel Quenneville
Assistant Coaches . . . . Mike Kitchen, Don Lever
Goaltending Coach . . . . Keith Allain
Video Coach . . . . Jamie Kompon
Athletic Trainer . . . . Ray Barile
Equipment Manager . . . . Bert Godin
Assistant Equipment Manager . . . . Eric Bechtol
Massage Therapist . . . . Jeff Wright
Exercise Physiologist . . . . Dr. Howie Wenger
Sr. Vice President pf Marketing
and Communications . . . . Jim Woodcock
Director of Team Services . . . . Mike Caruso
Director of Communications . . . . Frank Buonomo
Communications Assistants . . . . Scott Bonanni, Rich Jankowski
Team Photographer . . . . Mark Buckner
Radio Station . . . . KTRS 550 AM
Radio Broadcasters . . . . Chris Kerber, Kelly Chase
Television Station . . . . KPLR-TV WB 11
Television Broadcasters . . . . Ken Wilson, Bernie Federko, Dan McLaughlin
Regional Sports Network . . . . Fox Sports Net (Midwest)

Though Chris Pronger only scored seven goals last season, three of them were game winners. His 40 assists tied him for third among all NHL defensemen.

General Manager

PLEAU, LARRY
General Manager, St. Louis Blues. Born in Lynn, MA, June 29, 1947.

Larry Pleau was named general manager on June 9, 1997, becoming the tenth person to hold that position in team history. He has built the Blues into one of the NHL's top teams, winning the President's Trophy in 1999-2000 and reaching the Western Conference Finals in 2000-01.

Pleau joined the Blues after spending eight seasons with the New York Rangers organization, most recently as vice president of player personnel. He joined the Rangers in 1989 as assistant general manager of player development. During Pleau's tenure in New York, the Rangers drafted NHL stars Sergei Zubov, Doug Weight, Alexei Kovalev and Niklas Sundstrom. Prior to joining the Rangers, Pleau spent 17 seasons with the Hartford Whalers organization as a player, assistant coach, head coach, general manager and minor league general manager and head coach. He was also instrumental in drafting Ray Ferraro, Ron Francis, Kevin Dineen and Ulf Samuelsson while a member of the Whalers organization.

Pleau played three seasons with the Montreal Canadiens (1969-1972) in the National Hockey League before being the first player signed by the Hartford Whalers of the World Hockey Association. He was a center/left wing for the Whalers from 1972 until his retirement in 1979. He played in 468 regular season games for Hartford, accumulating 157 goals and 215 assists for 372 points. He also played for the 1968 United States Olympic team, the 1969 U.S. national team and went to training camp with Team USA for the 1976 Canada Cup tournament.

NHL Coaching Record

		Regular Season				Playoffs		
Season	Team	Games	W	L	T	Games	W	L
1980-81	Hartford	20	6	12	2	....	....	....
1981-82	Hartford	80	21	41	18	....	....	....
1982-83	Hartford	18	4	13	1	....	....	....
1987-88	Hartford	26	13	13	0	6	2	4
1988-89	Hartford	80	37	38	5	4	0	4
	NHL Totals	**224**	**81**	**117**	**26**	**10**	**2**	**8**

General Managers' History

Lynn Patrick, 1967-68; Scotty Bowman, 1968-69 to 1970-71; Lynn Patrick, 1971-72; Sid Abel, 1972-73; Charles Catto, 1973-74; Gerry Ehman, 1974-75; Dennis Ball, 1975-76; Emile Francis, 1976-77 to 1982-83; Ron Caron, 1983-84 to 1993-94; Mike Keenan, 1994-95, 1995-96; Mike Keenan and Ron Caron, 1996-97; Larry Pleau, 1997-98 to date.

San Jose Sharks

2001-02 Results: 44W-27L-8T-3OTL 99PTS.
First, Pacific Division

Year-by-Year Record

		Home				Road				Overall								
Season	GP	W	L	T	OL	W	L	T	OL	W	L	T	OL	GF	GA	Pts.	Finished	Playoff Result
2001-02	82	25	11	3	2	19	16	5	1	44	27	8	3	248	199	99	1st, Pacific Div.	Lost Conf. Semi-Final
2000-01	82	22	14	4	1	18	13	8	2	40	27	12	3	217	192	95	2nd, Pacific Div.	Lost Conf. Quarter-Final
1999-2000	82	21	14	3	3	14	16	7	4	35	30	10	7	225	214	87	4th, Pacific Div.	Lost Conf. Semi-Final
1998-99	82	17	15	9	...	14	18	9	...	31	33	18	...	196	191	80	4th, Pacific Div.	Lost Conf. Quarter-Final
1997-98	82	17	19	5	...	17	19	5	...	34	38	10	...	210	216	78	4th, Pacific Div.	Lost Conf. Quarter-Final
1996-97	82	14	23	4	...	13	24	4	...	27	47	8	...	211	278	62	7th, Pacific Div.	Out of Playoffs
1995-96	82	12	26	3	...	8	29	4	...	20	55	7	...	252	357	47	7th, Pacific Div.	Out of Playoffs
1994-95	48	10	13	1	...	9	12	3	...	19	25	4	...	129	161	42	3rd, Pacific Div.	Lost Conf. Semi-Final
1993-94	84	19	13	10	...	14	22	6	...	33	35	16	...	252	265	82	3rd, Pacific Div.	Lost Conf. Semi-Final
1992-93	84	8	33	1	...	3	38	1	...	11	71	2	...	218	414	24	6th, Smythe Div.	Out of Playoffs
1991-92	80	14	23	3	...	3	35	2	...	17	58	5	...	219	359	39	6th, Smythe Div.	Out of Playoffs

2002-03 Schedule

Oct.	Thu.	10	Detroit
	Sat.	12	at Vancouver
	Thu.	17	Edmonton
	Sat.	19	Colorado
	Mon.	21	Vancouver
	Thu.	24	at Nashville
	Fri.	25	at Columbus
	Sun.	27	at Chicago
	Tue.	29	at Detroit
	Thu.	31	at Minnesota
Nov.	Sun.	3	at Anaheim*
	Tue.	5	Los Angeles
	Thu.	7	Nashville
	Sat.	9	Minnesota
	Mon.	11	NY Rangers
	Wed.	13	at Atlanta
	Fri.	15	at Tampa Bay
	Sat.	16	at Florida
	Tue.	19	at Washington
	Thu.	21	at Philadelphia
	Sat.	23	at Pittsburgh
	Mon.	25	at St. Louis
	Wed.	27	at Nashville
	Sat.	30	Phoenix
Dec.	Tue.	3	at Phoenix
	Fri.	6	Columbus
	Sat.	7	Nashville
	Thu.	12	Pittsburgh
	Sat.	14	Washington
	Mon.	16	at NY Rangers
	Tue.	17	at Montreal
	Thu.	19	at Ottawa
	Sat.	21	at Toronto
	Mon.	23	at Boston
	Thu.	26	Anaheim
	Sat.	28	Chicago
	Mon.	30	Philadelphia
Jan.	Thu.	2	Dallas
	Sat.	4	Colorado
	Mon.	6	Edmonton
	Thu.	9	St. Louis
	Sat.	11	Vancouver
	Mon.	13	at Los Angeles
	Thu.	16	Buffalo
	Sat.	18	Dallas
	Mon.	20	at Phoenix
	Wed.	22	New Jersey
	Sat.	25	Minnesota
	Mon.	27	at Los Angeles
	Tue.	28	Los Angeles
	Thu.	30	Anaheim
Feb.	Wed.	5	Carolina
	Fri.	7	at Minnesota
	Sat.	8	at St. Louis
	Mon.	10	at Detroit
	Wed.	12	at Columbus
	Fri.	14	at Chicago
	Sun.	16	at Dallas*
	Mon.	17	at Los Angeles
	Wed.	19	NY Islanders
	Fri.	21	Columbus
	Mon.	24	Calgary
	Thu.	27	at Vancouver
Mar.	Sat.	1	at Calgary
	Tue.	4	at Edmonton
	Thu.	6	Montreal
	Sat.	8	at Phoenix
	Sun.	9	at Dallas
	Tue.	11	St. Louis
	Sat.	15	Calgary
	Mon.	17	Chicago
	Wed.	19	at Colorado
	Fri.	21	Boston
	Sat.	22	Anaheim
	Mon.	24	Tampa Bay
	Wed.	26	at Anaheim
	Thu.	27	Detroit
	Sat.	29	Dallas
	Mon.	31	at Colorado
Apr.	Wed.	2	at Calgary
	Thu.	3	at Edmonton
	Sun.	6	Phoenix*

* Denotes afternoon game.

Though his goal-scoring production has dropped since he scored 44 two years ago, Owen Nolan established a career high with 43 assists last season and led the Sharks in scoring (66 points) for the fourth time in his seven seasons in San Jose.

Franchise date: May 9, 1990

PACIFIC DIVISION

12th NHL Season

2002-03 Player Personnel

FORWARDS	HT	WT	S	Place of Birth	Date	2001-02 Club
BRADLEY, Matt	6-2	195	R	Stittsville, Ont.	6/13/78	San Jose
CHEECHOO, Jonathan	6-0	205	R	Moose Factory, Ont.	7/15/80	Cleveland
DAMPHOUSSE, Vincent	6-1	200	L	Montreal, Que.	12/17/67	San Jose
DIMITRAKOS, Nico	5-11	190	R	Boston, MA	5/21/79	U. of Maine
GRAVES, Adam	6-0	205	L	Toronto, Ont.	4/12/68	San Jose
HANSEN, Tavis	6-1	205	R	Prince Albert, Sask.	6/17/75	Hershey
HARVEY, Todd	6-0	200	R	Hamilton, Ont.	2/17/75	San Jose
KOROLYUK, Alexander	5-9	195	L	Moscow, USSR	1/15/76	San Jose
KRAFT, Ryan	5-9	190	L	Bottineau, ND	11/7/75	Cleveland
LAPLANTE, Eric	6-0	185	L	St-Maurice, Que.	12/1/79	Cleveland
LEVESQUE, Willie	6-0	195	R	Oak Bluff, MA	1/22/80	Northeastern U.
MARLEAU, Patrick	6-2	210	L	Aneroid, Sask.	9/15/79	San Jose
MISCHLER, Greg	6-3	174	L	Holbrook, NY	9/15/78	Cleveland
NELSON, Jeff	5-11	190	L	Prince Albert, Sask.	12/18/72	Portland (AHL)
NOLAN, Owen	6-1	210	R	Belfast, Ireland	2/12/72	San Jose-Canada
RICCI, Mike	6-0	185	L	Scarborough, Ont.	10/27/71	San Jose
SELANNE, Teemu	6-0	204	R	Helsinki, Finland	7/3/70	San Jose-Finland
SMITH, Mark	5-10	205	L	Edmonton, Alta.	10/24/77	San Jose
STURM, Marco	6-0	195	L	Dingolfing, West Germany	9/8/78	San Jose-Germany
SUNDSTROM, Niklas	6-0	190	L	Ornskoldsvik, Sweden	6/6/75	San Jose-Sweden
THOMAS, Scott	6-2	200	R	Buffalo, NY	1/18/70	Manchester
THORNTON, Scott	6-3	220	L	London, Ont.	1/9/71	San Jose
WISEMAN, Chad	6-0	190	L	Burlington, Ont.	3/25/81	Cleveland
ZALESAK, Miroslav	6-0	185	L	Skalica, Czech.	1/2/80	Cleveland

DEFENSEMEN	HT	WT	S	Place of Birth	Date	2001-02 Club
CARKNER, Matt	6-4	229	R	Winchester, Ont.	11/3/80	Cleveland
CLOUTIER, David	6-2	200	R	Quebec City, Que.	12/17/81	Val-d'Or-Cape Breton
DAVISON, Rob	6-2	220	L	St. Catharines, Ont.	5/1/80	Cleveland
FAHEY, Jim	6-0	215	R	Boston, MA	5/11/79	Northeastern U.
FIBIGER, Jesse	6-3	210	L	Victoria, B.C.	4/4/78	Cleveland
HANNAN, Scott	6-2	220	L	Richmond, B.C.	1/23/79	San Jose
HEINS, Shawn	6-4	210	L	Eganville, Ont.	12/24/73	San Jose
JAKOPIN, John	6-5	240	R	Toronto, Ont.	5/16/75	Pittsburgh-Wilkes-Barre
JILLSON, Jeff	6-3	220	R	North Smithfield, RI	7/24/80	San Jose-Cleveland
MARCHMENT, Bryan	6-1	200	L	Scarborough, Ont.	5/1/69	San Jose
MULICK, Robert	6-2	210	R	Toronto, Ont.	10/23/79	Cleveland
RAGNARSSON, Marcus	6-1	215	L	Ostervala, Sweden	8/13/71	San Jose-Sweden
RATHJE, Mike	6-5	245	L	Mannville, Alta.	5/11/74	San Jose
STUART, Brad	6-2	215	L	Rocky Mountain House, Alta.	11/6/79	San Jose
SUTER, Gary	6-0	215	L	Madison, WI	6/24/64	San Jose-United States

GOALTENDERS	HT	WT	C	Place of Birth	Date	2001-02 Club
KIELKUCKI, Marc	6-4	195	L	Brooklyn Park, MN	6/5/79	Dayton
KIPRUSOFF, Miikka	6-2	190	L	Turku, Finland	10/26/76	San Jose-Cleveland
KOTYK, Seamus	5-11	190	L	London, Ont.	10/7/80	Cleveland
NABOKOV, Evgeni	6-0	200	L	Ust-Kamenogorsk, USSR	7/25/75	San Jose
TOSKALA, Vesa	5-10	190	L	Tampere, Finland	5/20/77	San Jose-Cleveland

Coach

SUTTER, DARRYL
Coach, San Jose Sharks. Born in Viking, Alta., August 19, 1958.

Darryl Sutter became the Sharks' fifth head coach on June 9, 1997. San Jose's point total has improved during each of Sutter's first five seasons and the team set club records with 44 wins and 99 points in 2001-02 en route to the first division title in team history. As a player, Sutter spent eight seasons in the NHL, all with the Chicago Blackhawks (1979-87). He began his coaching career as an assistant in Chicago in 1987-88 before taking over as head coach of the Blackhawks' IHL affiliate that played in Saginaw (1988-89) and in Indianapolis (1989-90). His club won the IHL Turner Cup championship in 1990 and Sutter was named coach of the year.

He later served as an associate coach under Mike Keenan in Chicago in 1990-91 and 1991-92 and began a three-year tenure as head coach of the Blackhawks in 1992-93. As coach of Chicago, Sutter's teams reached the playoffs in all three seasons. He resigned as head coach following the 1994-95 season to spend more time with his family and worked as a consultant to the Blackhawks for special assignments in 1995-96 and 1996-97.

In 20 years of hockey as a player and coach, Sutter has never failed to qualify for post-season play. During his eight-year playing career, he scored 161 goals and added 118 assists in 406 regular-season games. He added 24 goals and 19 assists in 51 playoff games. Drafted 179th overall by Chicago in the 1978 NHL Entry Draft, he scored a remarkable 40 goals during his rookie season. The left winger served as team captain from 1982-83 until injuries forced his retirement after the 1986-87 season.

He is one of six brothers to play in the NHL. The others are Brian, Brent, Duane, Rich and Ron. All are involved in the Sutter Foundation which raises raises money for non-profit organizations in their home province of Alberta.

Coaching Record

		Regular Season				Playoffs		
Season	Team	Games	W	L	T	Games	W	L
1988-89	Saginaw (IHL)	82	46	26	10	6	2	4
1989-90	Indianapolis (IHL)	82	53	21	8	14	12	2
1992-93	**Chicago (NHL)**	**84**	**47**	**25**	**12**	**4**	**0**	**4**
1993-94	**Chicago (NHL)**	**84**	**39**	**36**	**9**	**6**	**2**	**4**
1994-95	**Chicago (NHL)**	**48**	**24**	**19**	**5**	**16**	**9**	**7**
1997-98	**San Jose (NHL)**	**82**	**34**	**38**	**10**	**6**	**2**	**4**
1998-99	**San Jose (NHL)**	**82**	**31**	**33**	**18**	**6**	**2**	**4**
1999-2000	**San Jose (NHL)**	**82**	**35**	**37**	**10**	**12**	**5**	**7**
2000-01	**San Jose (NHL)**	**82**	**40**	**30**	**12**	**6**	**2**	**4**
2001-02	**San Jose (NHL)**	**82**	**44**	**30**	**8**	**12**	**7**	**5**
	NHL Totals	**626**	**294**	**248**	**84**	**68**	**29**	**39**

2001-02 Scoring

* - rookie

Regular Season

Pos	#	Player	Team	GP	G	A	Pts	+/–	PIM	PP	SH	GW	GT	S	%
R	11	Owen Nolan	S.J.	75	23	43	66	7	93	8	2	2	1	217	10.6
C	25	Vincent Damphousse	S.J.	82	20	38	58	8	60	7	2	4	1	172	11.6
R	8	Teemu Selanne	S.J.	82	29	25	54	–11	40	9	1	8	0	202	14.4
C	18	Mike Ricci	S.J.	79	19	34	53	9	44	5	2	0	0	115	16.5
C	12	Patrick Marleau	S.J.	79	21	23	44	9	40	3	0	5	0	121	17.4
L	17	Scott Thornton	S.J.	77	26	16	42	11	116	6	0	5	0	144	18.1
L	19	Marco Sturm	S.J.	77	21	20	41	23	32	4	3	5	0	174	12.1
R	24	Niklas Sundstrom	S.J.	73	9	30	39	7	50	0	1	0	1	74	12.2
D	20	Gary Suter	S.J.	82	6	27	33	13	57	3	0	1	1	174	3.4
L	9	Adam Graves	S.J.	81	17	14	31	11	51	1	3	1	0	139	12.2
D	7	Brad Stuart	S.J.	82	6	23	29	13	39	2	0	2	0	96	6.3
R	28	* Matt Bradley	S.J.	54	9	13	22	22	43	0	0	2	0	63	14.3
R	13	Todd Harvey	S.J.	69	9	13	22	16	73	0	0	1	0	66	13.6
D	27	Bryan Marchment	S.J.	72	2	20	22	22	178	0	0	0	0	68	2.9
D	10	Marcus Ragnarsson	S.J.	70	5	15	20	4	44	2	0	3	0	68	7.4
D	5	* Jeff Jillson	S.J.	48	5	13	18	2	29	3	0	2	0	47	10.6
D	2	Mike Rathje	S.J.	52	5	12	17	23	48	4	0	0	0	56	8.9
D	22	Scott Hannan	S.J.	75	2	12	14	10	57	0	0	1	0	68	2.9
L	32	Stephane Matteau	S.J.	55	7	4	11	4	15	1	1	0	0	43	16.3
R	15	Alexander Korolyuk	S.J.	32	3	7	10	2	14	0	0	1	0	49	6.1
C	16	Mark Smith	S.J.	49	3	3	6	–1	72	0	0	1	0	40	7.5
D	23	Shawn Heins	S.J.	17	0	2	2	1	24	0	0	0	0	20	0.0
D	3	Steve Bancroft	S.J.	5	0	1	1	–2	2	0	0	0	0	5	0.0
R	26	Mike Craig	S.J.	2	0	0	0	0	2	0	0	0	0	2	0.0
R	50	Hannes Hyvonen	S.J.	6	0	0	0	–2	0	0	0	0	0	4	0.0

Goaltending

No.	Goaltender	GPI	Mins	Avg	W	L	T	EN	SO	GA	SA	S%	G	A	PIM
29	* Vesa Toskala	1	10	0.00	0	0	0	0	0	0	2	1.000	0	0	0
35	Evgeni Nabokov	67	3901	2.29	37	24	5	5	7	149	1818	.918	1	3	14
37	* Miikka Kiprusoff	20	1037	2.49	7	6	3	2	2	43	508	.915	0	0	4
	Totals	**82**	**4973**	**2.40**	**44**	**30**	**8**	**7**	**9**	**199**	**2335**	**.915**			

Playoffs

Pos	#	Player	Team	GP	G	A	Pts	+/–	PIM	PP	SH	GW	GT	S	%
C	12	Patrick Marleau	S.J.	12	6	5	11	3	6	1	0	3	0	21	28.6
C	18	Mike Ricci	S.J.	12	4	6	10	6	4	0	0	1	0	17	23.5
R	11	Owen Nolan	S.J.	12	3	6	9	–2	8	0	0	0	0	46	6.5
R	8	Teemu Selanne	S.J.	12	5	3	8	–3	2	2	0	1	0	35	14.3
C	25	Vincent Damphousse	S.J.	12	2	6	8	–1	12	1	0	0	0	17	11.8
R	24	Niklas Sundstrom	S.J.	12	1	6	7	3	6	0	0	0	0	17	5.9
L	17	Scott Thornton	S.J.	12	3	3	6	3	6	0	0	0	0	23	13.0
L	19	Marco Sturm	S.J.	12	3	2	5	–1	2	0	0	0	0	31	9.7
L	9	Adam Graves	S.J.	12	3	1	4	–2	6	0	0	2	0	15	20.0
D	2	Mike Rathje	S.J.	12	1	3	4	5	6	1	0	0	0	10	10.0
D	10	Marcus Ragnarsson	S.J.	12	1	3	4	–1	12	0	0	0	0	12	8.3
D	20	Gary Suter	S.J.	12	0	4	4	2	8	0	0	0	0	22	0.0
L	32	Stephane Matteau	S.J.	10	1	2	3	1	2	0	0	0	0	5	20.0
D	7	Brad Stuart	S.J.	12	0	3	3	1	8	0	0	0	0	12	0.0
D	27	Bryan Marchment	S.J.	12	1	1	2	–1	10	0	0	0	0	11	9.1
R	13	Todd Harvey	S.J.	12	0	2	2	3	12	0	0	0	0	13	0.0
D	22	Scott Hannan	S.J.	12	0	2	2	–1	12	0	0	0	0	10	0.0
D	5	* Jeff Jillson	S.J.	4	0	0	0	0	0	0	0	0	0	2	0.0
R	28	* Matt Bradley	S.J.	10	0	0	0	0	0	0	0	0	0	4	0.0

Goaltending

No.	Goaltender	GPI	Mins	Avg	W	L	EN	SO	GA	SA	S%	G	A	PIM
37	* Miikka Kiprusoff	1	8	0.00	0	0	0	0	0	2	1.000	0	0	0
35	Evgeni Nabokov	12	712	2.61	7	5	1	0	31	322	.904	0	2	2
	Totals	**12**	**723**	**2.66**	**7**	**5**	**1**	**0**	**32**	**325**	**.902**			

Captains' History

Doug Wilson, 1991-92, 1992-93; Bob Errey, 1993-94; Bob Errey and Jeff Odgers, 1994-95; Jeff Odgers, 1995-96; Todd Gill, 1996-97, 1997-98; Owen Nolan, 1998-99 to date.

Coaching History

George Kingston, 1991-92, 1992-93; Kevin Constantine, 1993-94, 1994-95; Kevin Constantine and Jim Wiley, 1995-96; Al Sims, 1996-97; Darryl Sutter, 1997-98 to date.

Club Records

Team

(Figures in brackets for season records are games played; records for fewest points, wins, ties, losses, goals, goals against are for 70 or more games)

Record		
Most Points	99	2001-02 (82)
Most Wins	44	2001-02 (82)
Most Ties	18	1998-99 (82)
Most Losses	*71	1992-93 (84)
Most Goals	252	1993-94 (84), 1995-96 (82)
Most Goals Against	414	1992-93 (84)
Fewest Points	24	1992-93 (84)
Fewest Wins	11	1992-93 (84)
Fewest Ties	*2	1992-93 (84)
Fewest Losses	33	1998-99 (82)
Fewest Goals	196	1998-99 (82)
Fewest Goals Against	191	1998-99 (82)
Longest Winning Streak		
Overall	7	Mar. 24-Apr. 5/94, Jan. 30-Feb. 28/02
Home	5	Jan. 21-Feb. 15/95, Oct. 11-Nov. 3/01
Away	6	Nov. 30-Dec. 19/01
Longest Undefeated Streak		
Overall	10	Nov. 27-Dec. 19/01 (9 wins, 1 tie)
Home	7	Oct. 12-Nov. 22/00 (6 wins, 1 tie)
Away	10	Dec. 26/00-Feb. 16/01 (6 wins, 4 ties)
Longest Losing Streak		
Overall	*17	Jan. 4-Feb. 12/93
Home	9	Nov. 19-Dec. 19/92
Away	19	Nov. 27/92-Feb. 12/93
Longest Winless Streak		
Overall	20	Dec. 29/92-Feb. 12/93 (19 losses, 1 tie)
Home	9	Nov. 19-Dec. 19/92 (9 losses)
Away	19	Nov. 27/92-Feb. 12/93 (19 losses)
Most Shutouts, Season	9	2000-01 (82), 2001-02 (82)
Most PIM, Season	2,134	1992-93 (84)
Most Goals, Game	10	Jan. 13/96 (S.J. 10 at Pit. 8)

Individual

Most Seasons	9	Mike Rathje
Most Games, Career	512	Jeff Friesen
Most Goals, Career	184	Owen Nolan
Most Assists, Career	225	Owen Nolan
Most Points, Career	409	Owen Nolan (184G, 225A)
Most PIM, Career	1,001	Jeff Odgers
Most Shutouts, Career	14	Evgeni Nabokov
Longest Consecutive Games Streak	228	Mike Ricci (Nov. 22/97-Oct. 20/00)
Most Goals, Season	44	Owen Nolan (1999-2000)
Most Assists, Season	52	Kelly Kisio (1992-93)
Most Points, Season	84	Owen Nolan (1999-2000; 44G, 40A)
Most PIM, Season	326	Link Gaetz (1991-92)
Most Points, Defenseman, Season	64	Sandis Ozolnish (1993-94; 26G, 38A)
Most Points, Center, Season	78	Kelly Kisio (1992-93; 26G, 52A)
Most Points, Right Wing, Season	84	Owen Nolan (1999-2000; 44G, 40A)
Most Points, Left Wing, Season	66	Johan Garpenlov (1992-93; 22G, 44A)
Most Points, Rookie, Season	59	Pat Falloon (1991-92; 25G, 34A)
Most Shutouts, Season	7	Evgeni Nabokov (2001-02)
Most Goals, Game	4	Owen Nolan (Dec. 19/95)
Most Assists, Game	4	Three times
Most Points, Game	6	Owen Nolan (Oct. 4/99; 3G, 3A)

* NHL Record.

All-time Record vs. Other Clubs

Regular Season

	At Home								On Road								Total							
	GP	W	L	T	OL	GF	GA	PTS	GP	W	L	T	OL	GF	GA	PTS	GP	W	L	T	OL	GF	GA	PTS
Anaheim	24	11	11	2	0	71	69	24	25	14	9	2	0	85	66	30	49	25	20	4	0	156	135	54
Atlanta	3	2	0	1	0	11	5	5	2	2	0	0	0	7	2	4	5	4	0	1	0	18	7	9
Boston	10	3	5	2	0	27	36	8	8	0	6	2	0	19	29	2	18	3	11	4	0	46	65	10
Buffalo	9	5	1	3	0	34	31	13	10	0	10	0	0	26	50	0	19	5	11	3	0	60	81	13
Calgary	30	9	17	4	0	85	95	22	28	10	14	3	1	79	104	24	58	19	31	7	1	164	199	46
Carolina	10	6	4	0	0	44	29	12	10	5	5	0	0	23	31	10	20	11	9	0	0	67	60	22
Chicago	22	12	9	1	0	61	59	25	21	8	9	2	2	64	66	20	43	20	18	3	2	125	125	45
Colorado	20	7	13	0	0	58	74	14	19	4	11	4	0	38	71	12	39	11	24	4	0	96	145	26
Columbus	4	4	0	0	0	17	5	8	4	2	2	0	0	12	15	4	8	6	2	0	0	29	20	12
Dallas	24	9	12	1	2	59	70	21	24	10	11	3	0	59	65	23	48	19	23	4	2	118	135	44
Detroit	22	6	12	3	1	70	91	16	21	2	18	1	0	35	89	5	43	8	30	4	1	105	180	21
Edmonton	28	15	10	3	0	99	85	33	29	6	17	6	0	63	101	18	57	21	27	9	0	162	186	51
Florida	8	4	2	2	0	23	17	10	7	0	2	5	0	14	20	5	15	4	4	7	0	37	37	15
Los Angeles	32	17	12	3	0	108	95	37	31	7	20	3	1	69	103	18	63	24	32	6	1	177	198	55
Minnesota	4	2	1	1	0	8	5	5	4	1	2	1	0	8	10	3	8	3	3	2	0	16	15	8
Montreal	9	3	3	2	1	25	22	9	10	1	7	2	0	17	35	4	19	4	10	4	1	42	57	13
Nashville	8	4	4	0	0	18	22	8	8	5	2	1	0	25	14	11	16	9	6	1	0	43	36	19
New Jersey	9	3	5	1	0	20	27	7	11	4	6	1	0	22	41	9	20	7	11	2	0	42	68	16
NY Islanders	10	4	4	1	1	25	33	10	11	4	5	2	0	35	40	10	21	8	9	3	1	60	73	20
NY Rangers	11	2	8	1	0	25	42	5	9	2	6	1	0	28	38	5	20	4	14	2	0	53	80	10
Ottawa	8	5	3	0	0	17	12	10	8	1	3	4	0	23	27	6	16	6	6	4	0	40	39	16
Philadelphia	10	2	7	1	0	20	30	5	9	2	6	1	0	20	32	5	19	4	13	2	0	40	62	10
Phoenix	27	14	10	2	1	95	87	31	29	10	15	3	1	79	95	24	56	24	25	5	2	174	182	55
Pittsburgh	11	3	6	2	0	25	50	8	8	4	3	1	0	31	37	9	19	7	9	3	0	56	87	17
St. Louis	20	3	16	1	0	44	73	7	24	7	15	1	1	55	83	16	44	10	31	2	1	99	156	23
Tampa Bay	9	3	5	1	0	31	32	7	10	4	5	0	1	27	26	9	19	7	10	1	1	58	58	16
Toronto	14	5	7	2	0	30	38	12	17	4	12	1	0	44	66	9	31	9	19	3	0	74	104	21
Vancouver	30	11	14	5	0	89	96	27	28	8	16	4	0	74	104	20	58	19	30	9	0	163	200	47
Washington	9	5	3	1	0	27	26	11	10	5	5	0	0	30	30	10	19	10	8	1	0	57	56	21
Totals	**435**	**179**	**204**	**46**	**6**	**1266**	**1356**	**410**	**435**	**132**	**242**	**54**	**7**	**1111**	**1490**	**325**	**870**	**311**	**446**	**100**	**13**	**2377**	**2846**	**735**

Playoffs

	Series	W	L	GP	W	L	T	GF	GA	Last Mtg.	Rnd.	Result
Calgary	1	1	0	7	4	3	0	26	35	1995	CQF	W 4-3
Colorado	2	0	2	13	5	8	0	38	44	2002	CSF	L 3-4
Dallas	2	0	2	11	3	8	0	19	31	2000	CSF	L 1-4
Detroit	2	1	1	11	4	7	0	27	51	1995	CSF	L 0-4
Phoenix	1	1	0	5	4	1	0	13	7	2002	CQF	W 4-1
St. Louis	2	1	1	13	6	7	0	31	38	2001	CQF	L 2-4
Toronto	1	0	1	7	3	4	0	21	26	1994	CSF	L 3-4
Totals	**11**	**4**	**7**	**67**	**29**	**38**	**0**	**175**	**232**			

Playoff Results 2002-1998

Year	Round	Opponent	Result	GF	GA
2002	CSF	Colorado	L 3-4	21	25
	CQF	Phoenix	W 4-1	13	7
2001	CQF	St. Louis	L 2-4	11	16
2000	CSF	Dallas	L 1-4	7	15
	CQF	St. Louis	W 4-3	20	22
1999	CQF	Colorado	L 2-4	17	19
1998	CQF	Dallas	L 2-4	12	16

Abbreviations: Round: CSF – conference semi-final; **CQF** – conference quarter-final.

Carolina totals include Hartford, 1991-92 to 1996-97.
Dallas totals include Minnesota North Stars, 1991-92 to 1992-93.
Colorado totals include Quebec, 1991-92 to 1994-95.
Phoenix totals include Winnipeg, 1991-92 to 1995-96.

2001-02 Results

Month	Day	Opponent	Score
Oct.	4	Detroit	3-4*
	6	Minnesota	0-0
	11	Tampa Bay	4-3
	13	Boston	3-2
	16	at Minnesota	3-3
	18	at New Jersey	1-6
	20	at NY Islanders	2-2
	22	at NY Rangers	5-1
	23	at Buffalo	1-4
	25	at Chicago	2-4
	27	Columbus	2-0
	31	at Anaheim	4-2
Nov.	1	Atlanta	5-2
	3	Vancouver	5-1
	6	at St. Louis	4-1
	7	at Dallas	2-2
	9	at Carolina	2-3
	11	at Chicago	2-3*
	14	at Anaheim	4-2
	15	at Phoenix	3-5
	17	Dallas	2-3*
	21	Minnesota	0-2
	24	at Los Angeles	1-3
	27	Nashville	3-2
	29	Pittsburgh	5-0
	30	at Anaheim	5-2
Dec.	4	Calgary	2-2
	6	at Calgary	3-1
	8	at Vancouver	5-3
	11	Edmonton	5-4*
	14	at Colorado	3-0
	17	at Dallas	4-1
	19	at Atlanta	4-2
	21	at Detroit	0-3
	23	at Nashville	1-1
	26	Anaheim	1-2
	28	NY Rangers	3-5
	30	Phoenix	2-4
Jan.	2	Philadelphia	5-2
	4	Phoenix	5-3
	5	Florida	6-0
	8	St. Louis	2-6
	9	at Phoenix	6-5
	12	Los Angeles	2-3
	15	at Colorado	1-0*
	17	NY Islanders	3-2
	19	Colorado	1-3
	21	Edmonton	4-3
	23	at Detroit	2-2
	24	at Columbus	2-6
	27	at Montreal	1-3
	29	at Toronto	3-4
	30	at Pittsburgh	6-3
Feb.	6	Calgary	2-0
	8	Chicago	4-2
	10	Carolina	4-0
	12	at Edmonton	3-2
	26	at Nashville	5-1
	28	at Washington	5-2
Mar.	1	at Tampa Bay	2-4
	3	at Dallas	1-4
	5	Nashville	0-2
	7	Ottawa	5-2
	9	Vancouver	2-0
	10	at Vancouver	7-4
	13	St. Louis	0-2
	15	Washington	5-4
	16	Chicago	2-2
	18	Los Angeles	2-3
	20	at Edmonton	1-2
	21	at Calgary	4-1
	23	at Los Angeles	0-3
	26	Dallas	3-2
	28	Colorado	2-3
	30	Columbus	10-2
Apr.	2	Anaheim	3-1
	4	Phoenix	5-2
	6	Detroit	6-3
	8	at Minnesota	1-3
	10	at Columbus	5-3
	11	at St. Louis	1-4
	13	Los Angeles	3-1

* – Overtime

Entry Draft Selections 2002-1991

2002
Pick
27 Mike Morris
52 Dan Spang
86 Jonas Fidler
139 Kris Newbury
163 Tom Walsh
217 Tim Conboy
288 Michael Hutchins

2001
Pick
20 Marcel Goc
106 Christian Ehrhoff
107 Dimitri Patzold
140 Tomas Plihal
175 Ryan Clowe
182 Tom Cavanagh

2000
Pick
41 Tero Maatta
104 Jon Disalvatore
142 Michal Pinc
166 Nolan Schaefer
183 Michal Macho
246 Chad Wiseman
256 Pasi Saarinen

1999
Pick
14 Jeff Jillson
82 Mark Concannon
111 Willie Levesque
155 Nico Dimitrakos
229 Eric Betournay
241 Doug Murray
257 Hannes Hyvonen

1998
Pick
3 Brad Stuart
29 Jonathan Cheechoo
65 Eric Laplante
98 Rob Davison
104 Miroslav Zalesak
127 Brandon Coalter
145 Mikael Samuelsson
185 Robert Mulick
212 Jim Fahey

1997
Pick
2 Patrick Marleau
23 Scott Hannan
82 Adam Colagiacomo
107 Adam Nittel
163 Joe Dusbabek
192 Cam Severson
219 Mark Smith

1996
Pick
2 Andrei Zyuzin
21 Marco Sturm
55 Terry Friesen
102 Matt Bradley
137 Michel Larocque
164 Jake Deadmarsh
191 Cory Cyrenne
217 David Thibeault

1995
Pick
12 Teemu Riihijarvi
38 Peter Roed
64 Marko Makinen
90 Vesa Toskala
116 Miikka Kiprusoff
130 Michal Bros
140 Timo Hakanen
142 Jaroslav Kudrna
167 Brad Mehalko
168 Robert Jindrich
194 Ryan Kraft
220 Mikko Markkanen

1994
Pick
11 Jeff Friesen
37 Angel Nikolov
66 Alexei Yegorov
89 Vaclav Varada
115 Brian Swanson
141 Alexander Korolyuk
167 Sergei Gorbachev
193 Eric Landry
219 Evgeni Nabokov
240 Tomas Pisa
245 Aniket Dhadphale
271 David Beauregard

1993
Pick
6 Viktor Kozlov
28 Shean Donovan
45 Vlastimil Kroupa
58 Ville Peltonen
80 Alexander Osadchy
106 Andrei Buschan
132 Petri Varis
154 Fredrik Oduya
158 Anatoli Filatov
184 Todd Holt
210 Jonas Forsberg
236 Jeff Salajko
262 Jamie Matthews

1992
Pick
3 Mike Rathje
10 Andrei Nazarov
51 Alexander Cherbayev
75 Jan Caloun
99 Marcus Ragnarsson
123 Michal Sykora
147 Eric Bellerose
171 Ryan Smith
195 Chris Burns
219 Alexander Kholomeyev
243 Victor Ignatjev

1991
Pick
2 Pat Falloon
23 Ray Whitney
30 Sandis Ozolinsh
45 Dody Wood
67 Kerry Toporowski
89 Dan Ryder
111 Frank Nilsson
133 Jaroslav Otevrel
155 Dean Grillo
177 Corwin Saurdiff
199 Dale Craigwell
221 Aaron Kriss
243 Mikhail Kravets

Adam Graves added a veteran presence to the Sharks lineup. He contributed three short-handed goals to one of the NHL's most dangerous penalty killing units and had 17 goals overall.

General Managers' History

Jack Ferreira, 1991-92; Chuck Grillo (V.P. Director of Player Personnel), 1992-93 to 1995-96; Dean Lombardi, 1996-97 to date.

General Manager

LOMBARDI, DEAN
Executive Vice President and General Manager, San Jose Sharks.
Born in Holyoke, MA, March 5, 1958.

Dean Lombardi is the architect who built the San Jose hockey club. A charter member of the Sharks management team, Lombardi joined the club in 1990 as assistant general manager after having served in a similar capacity with the Minnesota North Stars. He was named San Jose's director of hockey operations on June 26, 1992 and became the club's general manager on March 6, 1996. As the team's top hockey executive, he oversees player personnel decisions, negotiates player contracts and coordinates the efforts of the Sharks' scouting and player evaluation departments.

Lombardi has spent considerable effort in building a professional scouting staff, reorganizing the amateur scouting department, and establishing a system to evaluate pro players at all levels. He has stood firm in building the Sharks through the draft, though trades have brought players like Owen Nolan, Vincent Damphousse and Adam Graves.

After the 1993-94 season that saw the Sharks post an NHL record single-season improvement of 58 points, Lombardi finished third in *The Hockey News* award voting for executive of the year.

Club Directory

Compaq Center at San Jose

San Jose Sharks
Compaq Center at San Jose
525 West Santa Clara Street
San Jose, CA 95113
Phone **408/287-7070**
FAX 408/999-5797
www.sjsharks.com
Capacity: 17,496

San Jose Sports & Entertainment Enterprises

Board Members	Greg Jamison, Kevin Compton, Greg Reyes, Tom McEnery, Brent Jones

Investors In SJSEE Include (listed in alphabetical order)

Investors	Blue Line Associates (Harvey Armstrong, Kevin Compton, Hasso Plattner, Greg Reyes, Stratton Sclavos, Gary Valenzuela), William DelBiaggio, George Gund III, Greg Jamison, Floyd Kvamme, Tom McEnery, Gordon Russell, Rudy Staedler

Executive Staff

President & Chief Executive Officer	Greg Jamison
Executive Vice President of Business Operations	Malcolm Bordelon
Executive Vice President & GM (HP Pavilion at San Jose)	Jim Goddard
Executive Vice President & General Counsel	Don Gralnek
Executive Vice President & General Manager (Sharks)	Dean Lombardi
Executive Vice President & Chief Financial Officer	Gregg Olson
Vice President of Finance	Ken Caveney
Vice President of Corporate Partnerships	Greg Elliott
Vice President of Sales & Marketing	Kent Russell
Vice President of Building Operations	Rich Sotelo
Vice President and Assistant General Manager (Sharks)	Wayne Thomas
Executive Assistants	Tricia Sullivan, Michelle Simmons, Kristen Fuce

Hockey Operations

Head Coach	Darryl Sutter
Assistant Coach	Rich Preston
Assistant Coach	Lorne Molleken
Goaltender Coach	Warren Strelow
Director of Pro Development	Doug Wilson
Special Consultant to the General Manager	John Ferguson
Professional Scout	Barry Long, Cap Raeder
Director of Amateur Scouting	Tim Burke
Chief Scout	Ray Payne
Assistant to the General Manager	Joe Will
Scouts	Ilkka Sinisalo, Pat Funk, Rob Grillo, Brian Gross, Karel Masopust
Executive Assistant	Brenda Will
Video Scouting Coordinator	Bob Friedlander
Team Services Coordinator	Marshall Dickerson
Head Athletic Trainer	Ray Tufts, A.T.,C
Athletic Trainer	Tom Woodcock, A.T.,C,L
Strength & Conditioning Coordinator	Mac Read
Massage Therapist	Wes Howard
Equipment Manager	Mike Aldrich
Assistant Equipment Manager	Kurt Harvey
Equipment Assistant & Equipment Transportation	Roy Sneesby
Administrative Assistant	Cathy Hancock
President and CEO, Cleveland Barons (AHL)	Michael T. Lehr
V.P. of Hockey and Business Ops, Cleveland Barons (AHL)	Michael Mudd
Head Coach, Cleveland Barons (AHL)	Roy Sommer
Assistant Coach, Cleveland Barons (AHL)	TBA
Head Trainer, Cleveland Barons (AHL)	Dave Zenobi
Equipment Manager, Cleveland Barons (AHL)	Steve Wissman
Assistant Equipment Manager, Cleveland Barons (AHL)	Kenny Bufton
Team Physician	Arthur J. Ting, M.D.
Team Dentist	Robert Bonahoom, D.D.S.
Team Vision Specialist	Vincent S. Zuccaro, O.D., F.A.A.O.
Medical Staff	Warren King, M.D., Mark Sontag, M.D., Will Straw, M.D.

Silicon Valley Sports & Entertainment/Business Operations

Senior Director of Media Relations & Publishing	Ken Arnold
Director of Broadcasting	Frank Albin
Director of Marketing	Beth Brigino
Director of Ticket Operations	Mary Enriquez
Director of Ticket Sales	Andy Fiske
Director of Fan Development/The Sharks Foundation	Rob Jaynes
Director of Event Presentation	Jason Minsky
Director of Suite Hospitality	Jay O'Sullivan
Director of Internet Services	Roger Ross
Media Relations Manager	Scott Emmert
Media Relations Coordinator	Ben Stephenson

The Sharks Foundation

Manager, The Sharks Foundation	Jackie Fuce

Finance

Director of Information Technology	James Struckle
Human Resources Manager	Cathy Chandler
Accounting Manager	Tina Park

Building Operations

Director of Ticket Operations	Daniel DeBoer
Director of Booking & Events	Steve Kirsner, Chuck Ryder
Director of Guest Services	Ken Sweezey
Facilities Technical Director	Greg Carrolan
Director of Building Services	Monte Chavez
Chief Engineer	Mark Mullins

Miscellaneous

Television Station	Fox Sports Net
Radio Network Flagship	KFOX 98.5 (KUFX FM)
Television Play-By-Play Broadcaster	Randy Hahn
Television Color Analyst	Drew Remenda
Radio Play-By-Play Broadcaster	Dan Rusanowsky
Radio Color Analyst	Pete Stemkowski
Radio Hockey Analyst	Rob Zettler
Team Photographer	Don Smith, Rocky Widner
P.A. Announcer	Joe Ike
In Game Host	Danny Miller
Mascot	S.J. Sharkie
Organist	James Day

Tampa Bay Lightning

2001-02 Results: 27W-40L-11T-4OTL 69PTS.
Third, Southeast Division

2002-03 Schedule

Oct.	Thu.	10	at Florida
	Sat.	12	Carolina
	Fri.	18	Atlanta
	Sat.	19	at Pittsburgh
	Mon.	21	at NY Rangers
	Wed.	23	at Columbus
	Fri.	25	Washington
	Sat.	26	at New Jersey
	Mon.	28	at Florida
	Wed.	30	NY Rangers
Nov.	Fri.	1	at Washington
	Sat.	2	at Pittsburgh
	Tue.	5	at Toronto
	Fri.	8	Pittsburgh
	Sat.	9	Chicago
	Mon.	11	Phoenix
	Fri.	15	San Jose
	Sun.	17	at Carolina*
	Tue.	19	Philadelphia
	Thu.	21	NY Islanders
	Sat.	23	at New Jersey
	Wed.	27	at Buffalo
	Fri.	29	Vancouver
Dec.	Sun.	1	at NY Rangers*
	Tue.	3	at Toronto
	Thu.	5	Edmonton
	Sat.	7	at Boston*
	Sun.	8	at Chicago
	Tue.	10	at Minnesota
	Thu.	12	at Montreal
	Sat.	14	NY Islanders
	Wed.	18	at Carolina
	Thu.	19	Toronto
	Sat.	21	Nashville
	Mon.	23	at Washington
	Fri.	27	Boston
	Sun.	29	NY Rangers*
	Tue.	31	Ottawa
Jan.	Thu.	2	at Calgary
	Sat.	4	at St. Louis
	Tue.	7	Detroit
	Thu.	9	Atlanta
	Sat.	11	New Jersey
	Tue.	14	at Ottawa
	Fri.	17	Pittsburgh
	Sat.	18	at Philadelphia*
	Mon.	20	Ottawa
	Wed.	22	Montreal
	Fri.	24	at Dallas
	Sat.	25	at Nashville
	Tue.	28	at Philadelphia
	Thu.	30	Carolina
Feb.	Tue.	4	Washington
	Thu.	6	Toronto
	Sat.	8	at Florida
	Tue.	11	at NY Islanders
	Fri.	14	at Atlanta
	Sat.	15	Boston
	Mon.	17	Washington
	Wed.	19	Atlanta
	Fri.	21	at Carolina
	Sun.	23	Buffalo*
	Tue.	25	Anaheim
	Thu.	27	Florida
Mar.	Sat.	1	at Ottawa
	Tue.	4	at NY Islanders
	Wed.	5	at Detroit
	Fri.	7	Colorado
	Sun.	9	Buffalo*
	Wed.	12	Los Angeles
	Fri.	14	at Buffalo
	Sat.	15	at Montreal
	Mon.	17	Minnesota
	Thu.	20	at Los Angeles
	Sat.	22	at Phoenix
	Mon.	24	at San Jose
	Thu.	27	New Jersey
	Sat.	29	Florida
	Mon.	31	at Boston
Apr.	Wed.	2	Montreal
	Fri.	4	Philadelphia
	Sun.	6	at Atlanta*

* Denotes afternoon game.

Year-by-Year Record

Season	GP	Home W	Home L	Home T	Home OL	Road W	Road L	Road T	Road OL	Overall W	Overall L	Overall T	Overall OL	GF	GA	Pts.	Finished	Playoff Result
2001-02	82	16	17	5	3	11	23	6	1	27	40	11	4	178	219	69	3rd, Southeast Div.	Out of Playoffs
2000-01	82	17	19	3	2	7	28	3	3	24	47	6	5	201	280	59	5th, Southeast Div.	Out of Playoffs
1999-2000	82	13	20	4	4	6	27	5	3	19	47	9	7	204	310	54	4th, Southeast Div.	Out of Playoffs
1998-99	82	12	25	4	...	7	29	5	...	19	54	9	...	179	292	47	4th, Southeast Div.	Out of Playoffs
1997-98	82	11	23	7	...	6	32	3	...	17	55	10	...	151	269	44	7th, Atlantic Div.	Out of Playoffs
1996-97	82	15	18	8	...	17	22	2	...	32	40	10	...	217	247	74	6th, Atlantic Div.	Out of Playoffs
1995-96	82	22	14	5	...	16	18	7	...	38	32	12	...	238	248	88	5th, Atlantic Div.	Lost Conf. Quarter-Final
1994-95	48	10	14	0	...	7	14	3	...	17	28	3	...	120	144	37	6th, Atlantic Div.	Out of Playoffs
1993-94	84	14	22	6	...	16	21	5	...	30	43	11	...	224	251	71	7th, Atlantic Div.	Out of Playoffs
1992-93	84	12	27	3	...	11	27	4	...	23	54	7	...	245	332	53	6th, Norris Div.	Out of Playoffs

In his first full season with Tampa Bay, Nikolai Khabibulin was once again one of the NHL's busiest goaltenders. He was one of only five netminders to play as many as 70 games and his .920 save percentage was fifth best in the league.

Franchise date: December 16, 1991

EASTERN CONFERENCE
SOUTHEAST DIVISION

11th NHL Season

2002-03 Player Personnel

FORWARDS	HT	WT	S	Place of Birth	Date	2001-02 Club
AFANASENKOV, Dmitry	6-2	200	R	Arkhangelsk, USSR	5/12/80	T.B.-Springfield-Grand Rapids
ALEXEEV, Nikita	6-5	210	L	Murmansk, USSR	12/27/81	Tampa Bay-Springfield
ANDREYCHUK, Dave	6-4	220	R	Hamilton, Ont.	9/29/63	Tampa Bay
CIBAK, Martin	6-1	195	L	Liptovmikulas, Czech.	5/17/80	Tampa Bay-Springfield
CLYMER, Ben	6-1	199	R	Edina, MN	4/11/78	Tampa Bay
DINGMAN, Chris	6-4	225	L	Edmonton, Alta.	7/6/76	Carolina-Tampa Bay
DWYER, Gordie	6-3	215	L	Dalhousie, N.B.	1/25/78	Tampa Bay-Springfield
FEDOTENKO, Ruslan	6-2	195	L	Kiev, Ukraine	1/18/79	Philadelphia-Ukraine
HOLZINGER, Brian	5-11	190	R	Parma, OH	10/10/72	Tampa Bay
KEEFE, Sheldon	5-11	185	R	Brampton, Ont.	9/17/80	Tampa Bay-Springfield
LECAVALIER, Vincent	6-4	205	L	Ile Bizard, Que.	4/21/80	Tampa Bay
MODIN, Fredrik	6-4	225	L	Sundsvall, Sweden	10/8/74	Tampa Bay
OLVESTAD, Jimmie	6-1	189	L	Stockholm, Sweden	2/16/80	Tampa Bay
PROSPAL, Vaclav	6-2	195	L	Ceske Budejovice, Czech.	2/17/75	Tampa Bay
RICHARDS, Brad	6-1	198	L	Montague, P.E.I.	5/2/80	Tampa Bay
ROY, Andre	6-4	213	L	Port Chester, NY	2/8/75	Ottawa-Tampa Bay
ST-LOUIS, Martin	5-9	185	L	Laval, Que.	6/18/75	Tampa Bay
SVITOV, Alexander	6-3	198	L	Omsk, USSR	11/3/82	CSKA Moscow 2-Omsk
TAYLOR, Tim	6-1	189	L	Stratford, Ont.	2/6/69	Tampa Bay
TOBLER, Ryan	6-3	227	L	Calgary, Alta.	5/13/76	Tampa Bay-Springfield
WILLIS, Shane	6-1	190	R	Edmonton, Alta.	6/13/77	Carolina-Tampa Bay
DEFENSEMEN						
BIRON, Mathieu	6-6	226	R	Lac-St-Charles, Que.	4/29/80	Tampa Bay-Springfield
BOYLE, Dan	5-11	190	R	Ottawa, Ont.	7/12/76	Florida-Tampa Bay
CULLIMORE, Jassen	6-5	244	L	Simcoe, Ont.	12/4/72	Tampa Bay
GOC, Sascha	6-2	225	R	Calw, West Germany	4/17/79	N.J.-Alb-T.B.-Sprfld
JONES, Mike	6-3	190	L	Toledo, OH	5/18/76	Pensacola
KUBINA, Pavel	6-4	230	R	Celadna, Czech.	4/15/77	Tampa Bay-Czech Republic
KUDROC, Kristian	6-6	255	R	Michalovce, Czech.	5/21/81	T.B.-Sprfld-Phi (AHL)
LUKOWICH, Brad	6-1	200	L	Cranbrook, B.C.	8/12/76	Dallas
NECKAR, Stan	6-1	214	L	Ceske Budejovice, Czech.	12/22/75	Tampa Bay
PRATT, Harlan	6-1	195	L	Fort McMurray, Alta.	12/10/78	Lowell (AHL)-Fla (ECHL)-Springfield-Pensacola
PRATT, Nolan	6-3	200	L	Fort McMurray, Alta.	8/14/75	Tampa Bay
SARICH, Cory	6-3	204	R	Saskatoon, Sask.	8/16/78	Tampa Bay-Springfield

GOALTENDERS	HT	WT	C	Place of Birth	Date	2001-02 Club
HODSON, Kevin	6-0	182	L	Winnipeg, Man.	3/27/72	Sault Ste. Marie
KHABIBULIN, Nikolai	6-1	203	L	Sverdlovsk, USSR	1/13/73	Tampa Bay-Russia
KONSTANTINOV, Evgeny	6-0	176	L	Kazan, USSR	3/29/81	Pensacola-Springfield

Coaching History

Terry Crisp, 1992-93 to 1996-97; Terry Crisp, Rick Paterson and Jacques Demers, 1997-98; Jacques Demers, 1998-99; Steve Ludzik, 1999-2000; Steve Ludzik and John Tortorella, 2000-01; John Tortorella, 2001-02 to date.

Coach

TORTORELLA, JOHN
Coach, Tampa Bay Lightning. Born in Boston, MA, June 24, 1958.

After finishing out the 1999-2000 season as the interim coach of the New York Rangers, John Tortorella joined the Tampa Bay Lightning as an associate coach on July 7, 2000. He took over head coaching duties on January 6, 2001.

Prior to his season with the Rangers, Tortorella had spent two years as an assistant coach with the Phoenix Coyotes and eight years in the Buffalo Sabres organization. He was an assistant coach in Buffalo from 1989-90 to 1994-95. Tortorella served as the head coach of the Rochester Americans, Buffalo's AHL affiliate, in 1995-96 and 1996-97. He guided Rochester to the Calder Cup championship during his first season behind the bench.

Tortorella starred at the University of Maine for three seasons and was twice named a Conference All-Star. After playing hockey in Sweden, Tortorella played in the Atlantic Coast Hockey League with Virginia, Hampton Roads and Erie. He later spent two seasons (1986-87 and 1987-88) as the coach and general manager of the Virginia Lancers, compiling a record of 87-31 and winning the league championship and coach of the year honors during both campaigns. Following the 1987-88 season, Tortorella joined the Fort Wayne Komets of the IHL for their 1988 playoff run. He was an assistant coach with the New Haven Nighthawks of the AHL in 1988-89.

Coaching Record

			Regular Season				Playoffs	
Season	Team	Games	W	L	T	Games	W	L
1995-96	Rochester (AHL)	80	37	38	5	19	15	4
1996-97	Rochester (AHL)	80	40	30	9	10	6	4
1999-2000	**NY Rangers (NHL)**	**4**	**0**	**3**	**1**	....	....	....
2000-01	**Tampa Bay (NHL)**	**43**	**12**	**30**	**1**	....	....	....
2001-02	**Tampa Bay (NHL)**	**82**	**27**	**44**	**11**	....	....	....
	NHL Totals	**129**	**39**	**77**	**13**	....	....	....

2001-02 Scoring

* - rookie

Regular Season

Pos	#	Player	Team	GP	G	A	Pts	+/–	PIM	PP	SH	GW	GT	S	%
L	19	Brad Richards	T.B.	82	20	42	62	–18	13	5	0	0	1	251	8.0
C	20	Vaclav Prospal	T.B.	81	18	37	55	–11	38	7	0	2	0	166	10.8
L	25	Dave Andreychuk	T.B.	82	21	17	38	–12	109	9	1	5	0	161	13.0
C	4	Vincent Lecavalier	T.B.	76	20	17	37	–18	61	5	0	3	1	164	12.2
C	26	Martin St. Louis	T.B.	53	16	19	35	4	20	6	1	2	2	105	15.2
D	7	Ben Clymer	T.B.	81	14	20	34	–10	36	4	0	2	0	151	9.3
D	13	Pavel Kubina	T.B.	82	11	23	34	–22	106	5	2	3	0	189	5.8
L	33	Fredrik Modin	T.B.	54	14	17	31	0	27	2	0	4	1	141	9.9
D	22	Dan Boyle	FLA	25	3	3	6	–1	12	1	0	0	0	31	9.7
			T.B.	41	5	15	20	–15	27	2	0	1	1	68	7.4
			TOTAL	66	8	18	26	–16	39	3	0	1	1	99	8.1
L	18	Zdeno Ciger	NYR	29	6	7	13	–3	16	1	0	1	0	48	12.5
			T.B.	27	6	6	12	–12	10	0	0	2	0	44	13.6
			TOTAL	56	12	13	25	–15	26	1	0	3	0	92	13.0
R	24	Shane Willis	CAR	59	7	10	17	–8	24	2	0	0	1	126	5.6
			T.B.	21	4	3	7	0	6	0	0	0	0	29	13.8
			TOTAL	80	11	13	24	–8	30	2	0	0	1	155	7.1
L	36	Andre Roy	OTT	56	6	8	14	3	148	0	0	0	0	60	10.0
			T.B.	9	1	1	2	–5	63	0	0	0	0	6	16.7
			TOTAL	65	7	9	16	–2	211	0	0	0	0	66	10.6
L	41	* Jimmie Olvestad	T.B.	74	3	11	14	3	24	0	0	0	0	99	3.0
R	28	Sheldon Keefe	T.B.	39	6	7	13	–11	16	0	0	1	2	52	11.5
D	5	Jassen Cullimore	T.B.	78	4	9	13	–1	58	0	0	1	0	84	4.8
D	21	Cory Sarich	T.B.	72	0	11	11	–4	105	0	0	0	0	55	0.0
R	15	* Nikita Alexeev	T.B.	44	4	4	8	–9	8	1	0	1	0	47	8.5
C	27	Tim Taylor	T.B.	48	4	4	8	–2	25	0	1	0	0	50	8.0
D	2	Stan Neckar	T.B.	77	1	7	8	–18	24	0	1	0	0	38	2.6
C	46	* Martin Cibak	T.B.	26	1	5	6	–6	8	0	0	0	0	22	4.5
L	11	Chris Dingman	CAR	30	0	1	1	–2	77	0	0	0	0	17	0.0
			T.B.	14	0	4	4	–8	26	0	0	0	0	24	0.0
			TOTAL	44	0	5	5	–10	103	0	0	0	0	41	0.0
D	3	Grant Ledyard	T.B.	53	1	3	4	–5	12	0	0	0	0	27	3.7
C	9	Brian Holzinger	T.B.	23	1	2	3	–4	4	0	0	0	0	20	5.0
D	44	Nolan Pratt	T.B.	46	0	3	3	–4	51	0	0	0	0	38	0.0
L	34	Gordie Dwyer	T.B.	26	0	2	2	–4	60	0	0	0	0	6	0.0
D	55	* Josef Boumedienne	N.J.	1	1	0	1	–1	2	0	0	0	0	1	100.0
			T.B.	3	0	0	0	–1	4	0	0	0	0	0	0.0
			TOTAL	4	1	0	1	–2	6	0	0	0	0	1	100.0
D	11	* Kristian Kudroc	T.B.	2	0	0	0	0	0	0	0	0	0	0	0.0
R	54	* Gaetan Royer	T.B.	3	0	0	0	–1	2	0	0	0	0	0	0.0
L	17	* Ryan Tobler	T.B.	4	0	0	0	–2	5	0	0	0	0	1	0.0
L	29	* Dimitry Afanasenkov	T.B.	5	0	0	0	–1	0	0	0	0	0	1	0.0
D	6	* Sasha Goc	N.J.	2	0	0	0	–2	0	0	0	0	0	2	0.0
			T.B.	9	0	0	0	0	0	0	0	0	0	2	0.0
			TOTAL	11	0	0	0	–2	0	0	0	0	0	4	0.0
D	43	Mathieu Biron	T.B.	36	0	0	0	–16	12	0	0	0	0	35	0.0

Goaltending

No.	Goaltender	GPI	Mins	Avg	W	L	T	EN	SO	GA	SA	S%	G	A	PIM
35	Nikolai Khabibulin	70	3896	2.36	24	32	10	7	7	153	1914	.920	0	2	6
80	Kevin Weekes	19	830	2.89	3	9	0	3	2	40	470	.915	0	0	0
31	Dieter Kochan	5	237	4.05	0	3	1	0	0	16	129	.876	0	0	0
	Totals	**82**	**4997**	**2.63**	**27**	**44**	**11**	**10**	**9**	**219**	**2523**	**.913**			

Showing no signs of a sophomore jinx, Brad Richards led the Lightning in scoring for the second straight season and posted numbers (20 goals, 42 assists) that were virtually identical to the totals that led all rookies during the 2000-01 season.

Club Records

Team

(Figures in brackets for season records are games played; records for fewest points, wins, ties, losses, goals, goals against are for 70 or more games)

Most Points **88** 1995-96 (82)
Most Wins **38** 1995-96 (82)
Most Ties **12** 1995-96 (82)
Most Losses **55** 1997-98 (82)
Most Goals **245** 1992-93 (84)
Most Goals Against **332** 1992-93 (84)
Fewest Points **44** 1997-98 (82)
Fewest Wins **17** 1997-98 (82)
Fewest Ties **6** 2000-01 (82)
Fewest Losses **32** 1995-96 (82)
Fewest Goals **151** 1997-98 (82)
Fewest Goals Against **219** 2001-02 (82)
Longest Winning Streak
Overall .... **5** Twice
Home .... **6** Feb. 15-Mar. 10/96, Nov. 17-Dec. 21/01
Away .... **4** Jan. 6-13/97
Longest Undefeated Streak
Overall .... **7** Feb. 28-Mar. 13/96 (5 wins, 2 ties)
Home .... **8** Twice
Away .... **6** Dec. 28/93-Jan. 12/94 (5 wins, 1 tie)
Longest Losing Streak
Overall .... **13** Jan. 3-Feb. 2/98
Home .... **10** Jan. 3-Feb. 26/98
Away .... **11** Oct. 24-Dec. 10/97
Longest Winless Streak
Overall .... **16** Twice
Home .... **11** Jan. 2-Feb. 26/98 (10 losses, 1 tie)
Away .... **17** Dec. 2/99-Feb. 19/00 (14 losses, 3 ties)
Most Shutouts, Season **9** 2001-02 (82)
Most PIM, Season **1,823** 1997-98 (82)
Most Goals, Game .... **8** Nov. 22/00 (Atl. 2 at T.B. 8)

Individual

Most Seasons .... **7** Mikael Andersson, Rob Zamuner, Daren Puppa
Most Games, Career .... **475** Rob Zamuner
Most Goals, Career .... **111** Brian Bradley
Most Assists, Career .... **189** Brian Bradley
Most Points, Career .... **300** Brian Bradley (111G, 189A)
Most PIM, Career .... **782** Chris Gratton
Most Shutouts, Career .... **12** Daren Puppa
Longest Consecutive Games Streak .... **226** Rob Zamuner (Nov. 1/95-Mar. 30/98)
Most Goals, Season .... **42** Brian Bradley (1992-93)
Most Assists, Season .... **56** Brian Bradley (1995-96)
Most Points, Season .... **86** Brian Bradley (1992-93; 42G, 44A)
Most PIM, Season .... **258** Enrico Ciccone (1995-96)
Most Points, Defenseman, Season .... **65** Roman Hamrlik (1995-96; 16G, 49A)
Most Points, Center, Season .... **86** Brian Bradley (1992-93; 42G, 44A)
Most Points, Right Wing, Season .... **60** Dino Ciccarelli (1996-97; 35G, 25A)
Most Points, Left Wing, Season .... **62** Brad Richards (2001-02; 20G, 42A)
Most Points, Rookie, Season .... **62** Brad Richards (2000-01; 21G, 41A)
Most Shutouts, Season .... **7** Nikolai Khabibulin (2001-02)
Most Goals, Game .... **4** Chris Kontos (Oct. 7/92)
Most Assists, Game .... **4** Four times
Most Points, Game .... **6** Doug Crossman (Nov. 7/92; 3G, 3A)

Captains' History

No captain, 1992-93 to 1994-95; Paul Ysebaert, 1995-96, 1996-97; Paul Ysebaert and Mikael Renberg, 1997-98; Rob Zamuner, 1998-99; Bill Houlder, Chris Gratton and Vincent Lecavalier, 1999-2000; Vincent Lecavalier, 2000-01; no captain, 2001-02.

All-time Record vs. Other Clubs

Regular Season

	At Home								On Road								Total							
	GP	W	L	T	OL	GF	GA	PTS	GP	W	L	T	OL	GF	GA	PTS	GP	W	L	T	OL	GF	GA	PTS
Anaheim	7	3	4	0	0	15	21	6	8	3	4	1	0	20	24	7	15	6	8	1	0	35	45	13
Atlanta	7	5	1	1	0	28	15	11	8	1	5	2	0	24	33	4	15	6	6	3	0	52	48	15
Boston	19	6	8	3	2	51	59	17	19	1	14	4	0	40	73	6	38	7	22	7	2	91	132	23
Buffalo	19	4	12	2	1	42	63	11	19	6	12	1	0	52	55	13	38	10	24	3	1	94	118	24
Calgary	10	5	4	1	0	31	32	11	8	3	4	0	1	17	26	7	18	8	8	1	1	48	58	18
Carolina	21	10	8	3	0	55	53	23	21	3	14	3	1	55	74	10	42	13	22	6	1	110	127	33
Chicago	10	4	3	3	0	24	25	11	12	3	7	2	0	29	39	8	22	7	10	5	0	53	64	19
Colorado	10	7	2	1	0	31	24	15	12	2	8	2	0	25	50	6	22	9	10	3	0	56	74	21
Columbus	2	2	0	0	0	5	1	4	1	0	1	0	0	1	3	0	3	2	1	0	0	6	4	4
Dallas	12	1	9	2	0	19	35	4	10	2	7	1	0	23	38	5	22	3	16	3	0	42	73	9
Detroit	13	3	8	1	1	42	61	8	10	1	9	0	0	18	43	2	23	4	17	1	1	60	104	10
Edmonton	10	2	5	2	1	28	33	7	9	2	7	0	0	18	24	4	19	4	12	2	1	46	57	11
Florida	23	7	12	4	0	47	68	18	22	5	12	3	2	42	64	15	45	12	24	7	2	89	132	33
Los Angeles	8	4	4	0	0	19	19	8	10	8	1	1	0	34	22	17	18	12	5	1	0	53	41	25
Minnesota	1	0	1	0	0	2	4	0	2	0	2	0	0	5	8	0	3	0	3	0	0	7	12	0
Montreal	19	7	7	4	1	44	50	19	18	7	10	1	0	42	52	15	37	14	17	5	1	86	102	34
Nashville	3	1	1	1	0	9	8	3	4	3	1	0	0	9	7	6	7	4	2	1	0	18	15	9
New Jersey	21	6	12	3	0	49	67	15	22	3	17	2	0	33	84	8	43	9	29	5	0	82	151	23
NY Islanders	22	10	10	2	0	57	65	22	21	10	9	1	1	62	63	22	43	20	19	3	1	119	128	44
NY Rangers	21	8	9	3	1	73	76	20	23	9	11	2	1	73	78	21	44	17	20	5	2	146	154	41
Ottawa	19	7	10	2	0	56	63	16	19	7	12	0	0	44	72	14	38	14	22	2	0	100	135	30
Philadelphia	22	7	13	1	1	57	68	16	21	1	13	7	0	34	71	9	43	8	26	8	1	91	139	25
Phoenix	8	3	5	0	0	24	30	6	10	4	6	0	0	22	28	8	18	7	11	0	0	46	58	14
Pittsburgh	19	9	8	2	0	55	55	20	19	3	13	2	1	44	79	9	38	12	21	4	1	99	134	29
St. Louis	12	5	5	2	0	36	40	12	9	1	8	0	0	20	35	2	21	6	13	2	0	56	75	14
San Jose	10	6	4	0	0	26	27	12	9	5	3	1	0	32	31	11	19	11	7	1	0	58	58	23
Toronto	16	2	14	0	0	31	59	4	17	6	10	1	0	42	63	13	33	8	24	1	0	73	122	17
Vancouver	8	3	4	0	1	28	32	7	8	0	6	2	0	13	35	2	16	3	10	2	1	41	67	9
Washington	23	5	16	2	0	49	77	12	24	5	15	4	0	51	88	14	47	10	31	6	0	100	165	26
Totals	**395**	**142**	**199**	**45**	**9**	**1033**	**1230**	**338**	**395**	**104**	**241**	**43**	**7**	**924**	**1362**	**258**	**790**	**246**	**440**	**88**	**16**	**1957**	**2592**	**596**

Playoffs

	Series	W	L	GP	W	L	T	GF	GA	Last Mtg.	Rnd.	Result
Philadelphia	1	0	1	6	2	4	0	13	26	1996	CQF	L 2-4
Totals	**1**	**0**	**1**	**6**	**2**	**4**	**0**	**13**	**26**			

Carolina totals include Hartford, 1992-93 to 1996-97.
Dallas totals include Minnesota North Stars, 1992-93.

Playoff Results 2002-1998

(Last playoff appearance: 1996)

Abbreviations: Round: CQF – conference quarter-final.

Colorado totals include Quebec, 1992-93 to 1994-95.
Phoenix totals include Winnipeg, 1992-93 to 1995-96.

2001-02 Results

Oct.	5	NY Islanders	2-3
	7	Florida	0-5
	11	at San Jose	3-4
	13	at Los Angeles	1-0
	14	at Anaheim	3-2
	16	at Colorado	1-2
	20	NY Rangers	5-2
	23	Washington	1-1
	25	Los Angeles	3-0
	27	at Atlanta	3-4
	30	at Toronto	2-3
Nov.	2	at Buffalo	1-4
	3	at Pittsburgh	1-2*
	6	at NY Islanders	0-3
	8	Philadelphia	1-2
	10	Pittsburgh	3-2*
	15	Toronto	2-3
	17	Carolina	2-0
	20	Anaheim	3-2
	21	at Washington	2-3
	23	New Jersey	2-0
	25	at Carolina	4-0
	27	at Boston	3-6
	29	Atlanta	5-2
Dec.	1	at Philadelphia	0-2
	2	at NY Rangers	0-1
	4	at New Jersey	1-1
	6	Columbus	1-0
	8	at Ottawa	2-5
	10	at Vancouver	1-1
	12	at Calgary	3-1
	14	at Edmonton	1-2
	17	at Montreal	4-3
	21	St. Louis	4-3*
	26	at Nashville	1-0
	27	Carolina	2-3
	29	Boston	4-5*
	31	Toronto	1-4
Jan.	2	at Minnesota	0-2
	4	at Chicago	0-2
	6	at Phoenix	3-0
	8	Dallas	1-2
	12	Ottawa	1-2
	13	at Atlanta	2-2
	15	at New Jersey	5-4*
	18	Chicago	2-2
	19	Montreal	1-5
	21	New Jersey	3-2
	23	at Pittsburgh	1-5
	25	at Buffalo	1-4
	26	at NY Islanders	2-6
	28	at NY Rangers	1-0
	30	Carolina	1-3
Feb.	4	Ottawa	4-4
	6	at Florida	3-2
	7	Florida	1-3
	9	Washington	2-4
	11	at Washington	1-3
	26	Detroit	3-4*
Mar.	1	San Jose	4-2
	2	Florida	3-2
	6	Edmonton	2-3
	8	Philadelphia	2-4
	10	Nashville	5-1
	12	at Atlanta	4-4
	14	Calgary	3-2
	17	Buffalo	2-2
	18	at Philadelphia	3-3
	20	Atlanta	4-2
	22	Montreal	3-3
	24	Boston	3-4*
	26	at Toronto	2-7
	28	at Montreal	1-2
	30	at Ottawa	1-3
Apr.	1	NY Rangers	4-6
	3	at Washington	1-4
	4	Pittsburgh	4-2
	7	Buffalo	3-5
	9	at Boston	2-2
	10	at Carolina	2-4
	12	NY Islanders	1-3
	14	at Florida	3-2*

* – Overtime

Entry Draft Selections 2002-1992

2002

Pick	
60	Adam Henrich
135	Joseph Pearce
162	Gerard Dicaire
170	P.J. Atherton
174	Karri Akkanen
183	Paul Ranger
213	Fredrik Norrena
233	Vasili Koshechkin
255	Ryan Craig
256	Darren Reid
286	Alexei Glukhov
287	John Toffey

2001

Pick	
3	Alexander Svitov
47	Alexander Polushin
61	Andreas Holmqvist
94	Evgeni Artukhin
123	Aaron Lobb
138	Paul Lynch
188	Arthur Femenella
219	Dennis Packard
222	Jeremy Van Hoof
252	J.F. Soucy
259	Dmitri Bezrukov
261	Vitali Smolyaninov
281	Ilja Solarev
289	Henrik Bergfors

2000

Pick	
8	Nikita Alexeev
34	Ruslan Zainullin
81	Alexander Kharitonov
126	Johan Hagglund
161	Pavel Sedov
191	Aaron Gionet
222	Marek Priechodsky
226	Brian Eklund
233	Alexander Polukeyev
263	Thomas Ziegler

1999

Pick	
47	Sheldon Keefe
67	Evgeny Konstantinov
75	Brett Scheffelmaier
88	Jimmie Olvestad
127	Kaspars Astashenko
148	Michal Lanicek
182	Fedor Fedorov
187	Ivan Rachunek
216	Erkki Rajamaki
244	Mikko Kuparinen

1998

Pick	
1	Vincent Lecavalier
64	Brad Richards
72	Dmitry Afanasenkov
92	Eric Beaudoin
121	Curtis Rich
146	Sergei Kuznetsov
174	Brett Allan
194	Oak Hewer
221	Daniel Hulak
229	Chris Lyness
252	Martin Cibak

1997

Pick	
7	Paul Mara
33	Kyle Kos
61	Matt Elich
108	Mark Thompson
109	Jan Sulc
112	Karel Betik
153	Andrei Skopintsev
168	Justin Jack
170	Eero Somervuori
185	Samuel St-Pierre
198	Shawn Skolney
224	Paul Comrie

1996

Pick	
16	Mario Larocque
69	Curtis Tipler
125	Jason Robinson
152	Nikolai Ignatov
157	Xavier Delisle
179	Pavel Kubina

1995

Pick	
5	Daymond Langkow
30	Mike McBain
56	Shane Willis
108	Konstantin Golokhvastov
134	Eduard Pershin
160	Cory Murphy
186	Joe Cardarelli
212	Zac Bierk

1994

Pick	
8	Jason Wiemer
34	Colin Cloutier
55	Vadim Epanchintsev
86	Dmitri Klevakin
137	Daniel Juden
138	Bryce Salvador
164	Chris Maillet
190	Alexei Baranov
216	Yuri Smirnov
242	Shawn Gervais
268	Brian White

1993

Pick	
3	Chris Gratton
29	Tyler Moss
55	Allan Egeland
81	Marian Kacir
107	Ryan Brown
133	Kiley Hill
159	Matthieu Raby
185	Ryan Nauss
211	Alexandre Laporte
237	Brett Duncan
263	Mark Szoke

1992

Pick	
1	Roman Hamrlik
26	Drew Bannister
49	Brent Gretzky
74	Aaron Gavey
97	Brantt Myhres
122	Martin Tanguay
145	Derek Wilkinson
170	Dennis Maxwell
193	Andrew Kemper
218	Marc Tardif
241	Tom MacDonald

General Managers' History

Phil Esposito, 1992-93 to 1997-98; Jacques Demers, 1998-99; Rick Dudley, 1999-2000 to 2000-01; Rick Dudley and Jay Feaster, 2001-02; Jay Feaster, 2002-03.

General Manager

FEASTER, JAY
General Manager, Tampa Bay Lightning.
Born in Williamstown, PA, July 30, 1962.

Jay Feaster was named general manager of the Tampa Bay Lightning on February 10, 2002. He had joined the Lightning on October 20, 1998, from the Hershey Bears of the AHL and spent three-plus seasons as Tampa Bay's assistant general manager, overseeing all contractual, collective bargaining and National Hockey League legal issues, as well as the organization's scouting department and its minor league affiliates in Springfield (American Hockey League) and Pensacola (East Coast Hockey League).

Feaster spent nine years with the Hershey Bears, leading the team to a division title (1993-94) and a Calder Cup championship (1997), while establishing three consecutive single-season attendance records (1991-92 to 1993-94) and entering into a five-year affiliation agreement with the NHL's Colorado Avalanche. For his work, Feaster was named the AHL's executive of the year in 1997. In 1996, Feaster negotiated the purchase of an A-League American Professional Soccer League (APSL) Division II franchise by Herco, the Hershey Wildcats, and then served on the A-League committee responsible for finalizing the merger between the APSL and USISL, Inc.

Prior to joining the Hershey Company, Feaster practiced law in Harrisburg, Pennsylvania. He is a Summa Cum Laude graduate of Susquehanna University and a Cum Laude graduate of The Georgetown Law Center in Washington, D.C.

While in Hershey, Feaster spent time on the advisory boards of the Big 33 Scholarship Foundation, the Four Diamonds Fund at the Pennsylvania State University Milton S. Hershey Medical Center, and the Central PA Chapter of the National Multiple Sclerosis Society. He also taught business law and hotel law as a visiting faculty member at the Lebanon Valley College in Annville, PA.

Club Directory

Ice Palace

Tampa Bay Lightning
Ice Palace
401 Channelside Drive
Tampa, FL 33602
Phone **813/301-6500**
FAX 813/301-1480
Ticket Info. 813/301-6600
www.icepalace.com
Capacity: 19,758

Executive Staff

Owner	Palace Sports & Entertainment, Bill Davidson
President of Palace Sports & Entertainment/Governor	Tom Wilson
President of Tampa Bay Lightning/ Alternate Governor	Ron Campbell
Senior Vice President of Sales and Marketing	Michael Yormark
Vice President of Administration	Sean Henry
Vice President of Community Affairs	Bill Newton

Hockey Operations

General Manager/Alternate Governor	Jay H. Feaster
Director of Player Personnel	Bill Barber
Hockey Operations Assistant	Kathy Paterson
Assistant to the General Manager	Ryan Belec
Head Coach	John Tortorella
Associate Coach	Craig Ramsay
Assistant Coach	Jeff Reese
Strength & Conditioning Coach	Eric Lawson
Video Coach	Nigel Kirwan
Head Scout	Jake Goertzen
Chief Professional Scout	Rick Paterson
Scouting Staff	Stephen Baker, Dave Heitz, Karri Kettunen, Craig Muni, Luke Williams, Yuri Yanchenkov, Darrell Young, Glen Zacharias
Director of Team Services	Phil Thibodeau
Head Medical Trainer	Thomas Mulligan
Massage Therapist	Mike Griebel
Equipment Manager	Ray Thill
Assistant Equipment Managers	Dana Heinze, Jim Pickard
Team Physician	Dr. Ira Guttentag
Assistant Coach, Springfield Falcons	Norm Maciver
Director of Alumni	John Tucker

Executive Suites/Premium Seating

Director of Premium Seating	RJ Martino
Suite Sales Manager	Bill Bullock
Director of Premium Services	Karrie Yager

Finance

Chief Financial Officer	Joe Fada
Finance Manager	Michelle Ekiss
Senior Accountants	Doug Riefler, Dave Weber
Accounts Payable	Tim Winans
Staff Accountants	Jane Sheill, Joey Armbruster

Internal Support Staff

Senior Systems Analyst	Dave Everett
Assistant Information Services Managers	Roberto Camejo, Ed Belzer

Box Office

Director of Ticket Operations	Jim Mannino
Box Office Manager	Aaron Corso
Assistant Box Office Manager	Alex Bohne

Sales

Vice President of Sales	Dave Bullock
Director of Corporate Sales	Todd Lambert
Director of Group Events	Brent Stehlik
Director of Outside Sales	Chad Johnson
Corporate Account Managers	Ross Bley, James Piatt, Ben Milsom, Paul Wallace, Brad Lott, Joe Ondrejko, Mike Janowicz, Eric Campailla, Alex English, Andrew Hammer

Sales & Marketing

Vice President of Integrated Sales	Pedro Goncalves
Vice President of Marketing	Sean Flynn
Director of Promotions	Mark Gullett
Entertainment Marketing Manager	Jason Franke
Marketing Coordinator	Sandi Lundin
Director of Client Services	Katherine Lesinski
Client Services Managers	Alaina Miller, Arleen Zulawski, Sean McHale
Client Services Coordinator	Kelly McCoy
Executive Assistant	Julie Stein
Corporate Marketing Sales Executive	Tom DeCaprio
Corporate Marketing Managers	Chris Hibbs, Ted Major, Tim Zulawski, Dan Heiserer, Aaron Cohn, RJ Martino, Tim Zulawski
Director of Event Marketing	Holly Brown
Director of Sports Marketing	Bina Kumar
Director of Broadcast Production & Game Ops.	Jim Ciotoli
Video Production Coordinators	Terry Levandoski, JC Kent, Mike Cooley, Chad Collins
Director of Fan Development	David Cole
Director of Web Services	Martin Quessenberry
Web Audio Reporter	Tom Gilbert

Public Relations

Vice President of Public Relations	Bill Wickett
Director of Public Relations	Jay Preble
Public Relations Manager	Jay Levin
Community Relations Manager	Stephanie Hanchey
Public Relations Assistant	Mary China

Business Relations

Director of Business Relations	Donna Ferris

Broadcast Information

Director of Broadcasting & Programming	Jason Dixon
Television	Sunshine Network
Television Broadcasters	Rick Peckham, Bobby Taylor, Mike Nabors
Radio	WDAE 620 AM
Broadcasters	TBD, Phil Esposito

Maple Leafs captain Mats Sundin.

Toronto Maple Leafs

2001-02 Results: 43W-25L-10T-4OTL 100PTS.
Second, Northeast Division

2002-03 Schedule

Oct.	Thu.	10	at Pittsburgh
	Sat.	12	Ottawa
	Mon.	14	Pittsburgh
	Tue.	15	at NY Rangers
	Thu.	17	Phoenix
	Sat.	19	at Montreal
	Mon.	21	Boston
	Wed.	23	Florida
	Sat.	26	NY Rangers
	Mon.	28	Anaheim
	Thu.	31	Atlanta
Nov.	Sat.	2	Montreal
	Tue.	5	Tampa Bay
	Fri.	8	at Dallas
	Sat.	9	at St. Louis
	Tue.	12	Los Angeles
	Fri.	15	at Buffalo
	Sat.	16	Detroit
	Tue.	19	Boston
	Sat.	23	Philadelphia
	Mon.	25	at Ottawa
	Tue.	26	Washington
	Fri.	29	at Philadelphia*
	Sat.	30	Buffalo
Dec.	Tue.	3	Tampa Bay
	Fri.	6	at NY Islanders
	Sat.	7	New Jersey
	Tue.	10	Pittsburgh
	Thu.	12	at Philadelphia
	Sat.	14	NY Rangers
	Mon.	16	at Atlanta
	Wed.	18	at Florida
	Thu.	19	at Tampa Bay
	Sat.	21	San Jose
	Mon.	23	Atlanta
	Fri.	27	at Calgary
	Sat.	28	at Edmonton
	Tue.	31	at Vancouver
Jan.	Fri.	3	at New Jersey
	Sat.	4	New Jersey
	Tue.	7	Boston
	Thu.	9	at Pittsburgh
	Sat.	11	at Boston
	Mon.	13	at NY Rangers
	Tue.	14	Calgary
	Fri.	17	at Washington
	Sat.	18	at Montreal
	Tue.	21	Philadelphia
	Fri.	24	at Buffalo
	Sat.	25	Colorado
	Wed.	29	at Carolina
	Thu.	30	at Atlanta
Feb.	Wed.	5	at Florida
	Thu.	6	at Tampa Bay
	Sat.	8	Montreal
	Tue.	11	Edmonton
	Wed.	12	at Chicago
	Sat.	15	Ottawa
	Tue.	18	Carolina
	Thu.	20	at Washington
	Sat.	22	at Montreal
	Sun.	23	Nashville
	Tue.	25	NY Islanders
	Thu.	27	at Detroit
Mar.	Sat.	1	Carolina
	Mon.	3	Florida
	Tue.	4	at Ottawa
	Thu.	6	at Buffalo
	Sat.	8	Vancouver
	Mon.	10	at Edmonton
	Thu.	13	at Calgary
	Sat.	15	at Vancouver*
	Tue.	18	NY Islanders
	Thu.	20	at Columbus
	Sat.	22	Buffalo
	Mon.	24	at Boston
	Tue.	25	at Carolina
	Fri.	28	at NY Islanders
	Sat.	29	Washington
Apr.	Tue.	1	at New Jersey
	Thu.	3	Minnesota
	Sat.	5	Ottawa

* Denotes afternoon game.

Franchise date: November 22, 1917

EASTERN CONFERENCE

NORTHEAST DIVISION

86th NHL Season

Year-by-Year Record

		Home				Road				Overall								
Season	GP	W	L	T	OL	W	L	T	OL	W	L	T	OL	GF	GA	Pts.	Finished	Playoff Result
2001-02	82	24	11	6	0	19	14	4	4	43	25	10	4	249	207	100	2nd, Northeast Div.	Lost Conf. Championship
2000-01	82	19	11	7	4	18	18	4	1	37	29	11	5	232	207	90	3rd, Northeast Div.	Lost Conf. Semi-Final
1999-2000	82	24	12	5	0	21	15	2	3	45	27	7	3	246	222	100	1st, Northeast Div.	Lost Conf. Semi-Final
1998-99	82	23	13	5	...	22	17	2	...	45	30	7	...	268	231	97	2nd, Northeast Div.	Lost Conf. Championship
1997-98	82	16	20	5	...	14	23	4	...	30	43	9	...	194	237	69	6th, Central Div.	Out of Playoffs
1996-97	82	18	20	3	...	12	24	5	...	30	44	8	...	230	273	68	6th, Central Div.	Out of Playoffs
1995-96	82	19	15	7	...	15	21	5	...	34	36	12	...	247	252	80	3rd, Central Div.	Lost Conf. Quarter-Final
1994-95	48	15	7	2	...	6	12	6	...	21	19	8	...	135	146	50	4th, Central Div.	Lost Conf. Quarter-Final
1993-94	84	23	15	4	...	20	14	8	...	43	29	12	...	280	243	98	2nd, Central Div.	Lost Conf. Championship
1992-93	84	25	11	6	...	19	18	5	...	44	29	11	...	288	241	99	3rd, Norris Div.	Lost Conf. Championship
1991-92	80	21	16	3	...	9	27	4	...	30	43	7	...	234	294	67	5th, Norris Div.	Out of Playoffs
1990-91	80	15	21	4	...	8	25	7	...	23	46	11	...	241	318	57	5th, Norris Div.	Out of Playoffs
1989-90	80	24	14	2	...	14	24	2	...	38	38	4	...	337	358	80	3rd, Norris Div.	Lost Div. Semi-Final
1988-89	80	15	20	5	...	13	26	1	...	28	46	6	...	259	342	62	5th, Norris Div.	Out of Playoffs
1987-88	80	14	20	6	...	7	29	4	...	21	49	10	...	273	345	52	4th, Norris Div.	Lost Div. Semi-Final
1986-87	80	22	14	4	...	10	28	2	...	32	42	6	...	286	319	70	4th, Norris Div.	Lost Div. Final
1985-86	80	16	21	3	...	9	27	4	...	25	48	7	...	311	386	57	4th, Norris Div.	Lost Div. Final
1984-85	80	10	28	2	...	10	24	6	...	20	52	8	...	253	358	48	5th, Norris Div.	Out of Playoffs
1983-84	80	17	16	7	...	9	29	2	...	26	45	9	...	303	387	61	5th, Norris Div.	Out of Playoffs
1982-83	80	20	15	5	...	8	25	7	...	28	40	12	...	293	330	68	3rd, Norris Div.	Lost Div. Semi-Final
1981-82	80	12	20	8	...	8	24	8	...	20	44	16	...	298	380	56	5th, Norris Div.	Out of Playoffs
1980-81	80	14	21	5	...	14	16	10	...	28	37	15	...	322	367	71	5th, Adams Div.	Lost Prelim. Round
1979-80	80	17	19	4	...	18	21	1	...	35	40	5	...	304	327	75	4th, Adams Div.	Lost Prelim. Round
1978-79	80	20	12	8	...	14	21	5	...	34	33	13	...	267	252	81	3rd, Adams Div.	Lost Quarter-Final
1977-78	80	21	13	6	...	20	16	4	...	41	29	10	...	271	237	92	3rd, Adams Div.	Lost Semi-Final
1976-77	80	18	13	9	...	15	19	6	...	33	32	15	...	301	285	81	3rd, Adams Div.	Lost Quarter-Final
1975-76	80	23	12	5	...	11	19	10	...	34	31	15	...	294	276	83	3rd, Adams Div.	Lost Quarter-Final
1974-75	80	19	12	9	...	12	21	7	...	31	33	16	...	280	309	78	3rd, Adams Div.	Lost Quarter-Final
1973-74	78	21	11	7	...	14	16	9	...	35	27	16	...	274	230	86	4th, East Div.	Lost Quarter-Final
1972-73	78	20	12	7	...	7	29	3	...	27	41	10	...	247	279	64	6th, East Div.	Out of Playoffs
1971-72	78	21	11	7	...	12	20	7	...	33	31	14	...	209	208	80	4th, East Div.	Lost Quarter-Final
1970-71	78	24	9	6	...	13	24	2	...	37	33	8	...	248	211	82	4th, East Div.	Lost Quarter-Final
1969-70	76	18	13	7	...	11	21	6	...	29	34	13	...	222	242	71	6th, East Div.	Out of Playoffs
1968-69	76	20	8	10	...	15	18	5	...	35	26	15	...	234	217	85	4th, East Div.	Lost Quarter-Final
1967-68	74	24	9	4	...	9	22	6	...	33	31	10	...	209	176	76	5th, East Div.	Out of Playoffs
1966-67	**70**	**21**	**8**	**6**	**...**	**11**	**19**	**5**	**...**	**32**	**27**	**11**	**...**	**204**	**211**	**75**	**3rd,**	**Won Stanley Cup**
1965-66	70	22	9	4	...	12	16	7	...	34	25	11	...	208	187	79	3rd,	Lost Semi-Final
1964-65	70	17	15	3	...	13	11	11	...	30	26	14	...	204	173	74	4th,	Lost Semi-Final
1963-64	**70**	**22**	**7**	**6**	**...**	**11**	**18**	**6**	**...**	**33**	**25**	**12**	**...**	**192**	**172**	**78**	**3rd,**	**Won Stanley Cup**
1962-63	**70**	**21**	**8**	**6**	**...**	**14**	**15**	**6**	**...**	**35**	**23**	**12**	**...**	**221**	**180**	**82**	**1st,**	**Won Stanley Cup**
1961-62	**70**	**25**	**5**	**5**	**...**	**12**	**17**	**6**	**...**	**37**	**22**	**11**	**...**	**232**	**180**	**85**	**2nd,**	**Won Stanley Cup**
1960-61	70	21	6	8	...	18	13	4	...	39	19	12	...	234	176	90	2nd,	Lost Semi-Final
1959-60	70	20	9	6	...	15	17	3	...	35	26	9	...	199	195	79	2nd,	Lost Final
1958-59	70	17	13	5	...	10	19	6	...	27	32	11	...	189	201	65	4th,	Lost Final
1957-58	70	12	16	7	...	9	22	4	...	21	38	11	...	192	226	53	6th,	Out of Playoffs
1956-57	70	12	16	7	...	9	18	8	...	21	34	15	...	174	192	57	5th,	Out of Playoffs
1955-56	70	19	10	6	...	5	23	7	...	24	33	13	...	153	181	61	4th,	Lost Semi-Final
1954-55	70	14	10	11	...	10	14	11	...	24	24	22	...	147	135	70	3rd,	Lost Semi-Final
1953-54	70	22	6	7	...	10	18	7	...	32	24	14	...	152	131	78	3rd,	Lost Semi-Final
1952-53	70	17	12	6	...	10	18	7	...	27	30	13	...	156	167	67	5th,	Out of Playoffs
1951-52	70	17	10	8	...	12	15	8	...	29	25	16	...	168	157	74	3rd,	Lost Semi-Final
1950-51	**70**	**22**	**8**	**5**	**...**	**19**	**8**	**8**	**...**	**41**	**16**	**13**	**...**	**212**	**138**	**95**	**2nd,**	**Won Stanley Cup**
1949-50	70	18	9	8	...	13	18	4	...	31	27	12	...	176	173	74	3rd,	Lost Semi-Final
1948-49	**60**	**12**	**8**	**10**	**...**	**10**	**17**	**3**	**...**	**22**	**25**	**13**	**...**	**147**	**161**	**57**	**4th,**	**Won Stanley Cup**
1947-48	**60**	**22**	**3**	**5**	**...**	**10**	**12**	**8**	**...**	**32**	**15**	**13**	**...**	**182**	**143**	**77**	**1st,**	**Won Stanley Cup**
1946-47	**60**	**20**	**8**	**2**	**...**	**11**	**11**	**8**	**...**	**31**	**19**	**10**	**...**	**209**	**172**	**72**	**2nd,**	**Won Stanley Cup**
1945-46	50	10	13	2	...	9	11	5	...	19	24	7	...	174	185	45	5th,	Out of Playoffs
1944-45	**50**	**13**	**9**	**3**	**...**	**11**	**13**	**1**	**...**	**24**	**22**	**4**	**...**	**183**	**161**	**52**	**3rd,**	**Won Stanley Cup**
1943-44	50	13	11	1	...	10	12	3	...	23	23	4	...	214	174	50	3rd,	Lost Semi-Final
1942-43	50	17	6	2	...	5	13	7	...	22	19	9	...	198	159	53	3rd,	Lost Semi-Final
1941-42	**48**	**18**	**6**	**0**	**...**	**9**	**12**	**3**	**...**	**27**	**18**	**3**	**...**	**158**	**136**	**57**	**2nd,**	**Won Stanley Cup**
1940-41	48	16	5	3	...	12	9	3	...	28	14	6	...	145	99	62	2nd,	Lost Semi-Final
1939-40	48	15	3	6	...	10	14	0	...	25	17	6	...	134	110	56	3rd,	Lost Final
1938-39	48	13	8	3	...	6	12	6	...	19	20	9	...	114	107	47	3rd,	Lost Final
1937-38	48	13	6	5	...	11	9	4	...	24	15	9	...	151	127	57	1st, Cdn. Div.	Lost Final
1936-37	48	14	9	1	...	8	12	4	...	22	21	5	...	119	115	49	3rd, Cdn. Div.	Lost Quarter-Final
1935-36	48	15	4	5	...	8	15	1	...	23	19	6	...	126	106	52	2nd, Cdn. Div.	Lost Final
1934-35	48	16	6	2	...	14	8	2	...	30	14	4	...	157	111	64	1st, Cdn. Div.	Lost Final
1933-34	48	19	2	3	...	7	11	6	...	26	13	9	...	174	119	61	1st, Cdn. Div.	Lost Semi-Final
1932-33	48	16	4	4	...	8	14	2	...	24	18	6	...	119	111	54	1st, Cdn. Div.	Lost Final
1931-32	**48**	**17**	**4**	**3**	**...**	**6**	**14**	**4**	**...**	**23**	**18**	**7**	**...**	**155**	**127**	**53**	**2nd, Cdn. Div.**	**Won Stanley Cup**
1930-31	44	15	4	3	...	7	9	6	...	22	13	9	...	118	99	53	2nd, Cdn. Div.	Lost Quarter-Final
1929-30	44	10	8	4	...	7	13	2	...	17	21	6	...	116	124	40	4th, Cdn. Div.	Out of Playoffs
1928-29	44	15	5	2	...	6	13	3	...	21	18	5	...	85	69	47	3rd, Cdn. Div.	Lost Semi-Final
1927-28	44	9	8	5	...	9	10	3	...	18	18	8	...	89	88	44	4th, Cdn. Div.	Out of Playoffs
1926-27*	44	10	10	2	...	5	14	3	...	15	24	5	...	79	94	35	5th, Cdn. Div.	Out of Playoffs
1925-26	36	11	5	2	...	1	16	1	...	12	21	3	...	92	114	27	6th,	Out of Playoffs
1924-25	30	10	5	0	...	9	6	0	...	19	11	0	...	90	84	38	2nd,	Lost NHL S-Final
1923-24	24	7	5	0	...	3	9	0	...	10	14	0	...	59	85	20	3rd,	Out of Playoffs
1922-23	24	10	1	1	...	3	9	0	...	13	10	1	...	82	88	27	3rd,	Out of Playoffs
1921-22	**24**	**8**	**4**	**0**	**...**	**5**	**6**	**1**	**...**	**13**	**10**	**1**	**...**	**98**	**97**	**27**	**2nd,**	**Won Stanley Cup**
1920-21	24	9	3	0	...	6	6	0	...	15	9	0	...	105	100	30	2nd and 1st***	Lost NHL Final
1919-20**	24	8	4	0	...	4	8	0	...	12	12	0	...	119	106	24	3rd and 2nd***	Out of Playoffs
1918-19	18	5	4	0	...	0	9	0	...	5	13	0	...	64	92	10	3rd and 3rd***	Out of Playoffs
1917-18	**22**	**10**	**1**	**0**	**...**	**3**	**8**	**0**	**...**	**13**	**9**	**0**	**...**	**108**	**109**	**26**	**2nd and 1st*** **	**Won Stanley Cup**

* Name changed from St. Patricks to Maple Leafs. ** Name changed from Arenas to St. Patricks.
*** Season played in two halves with no combined standing at end.
From 1917-18 through 1925-26, NHL champions played against PCHA/WCHL champions for Stanley Cup.

2002-03 Player Personnel

FORWARDS	HT	WT	S	Place of Birth	Date	2001-02 Club
ANTROPOV, Nik	6-5	203	L	Vost, USSR	2/18/80	Toronto-St. John's
BARRETT, Nathan	5-11	192	L	Vancouver, B.C.	8/3/81	Lethbridge
BOYES, Brad	6-0	181	R	Mississauga, Ont.	4/17/82	Erie (OHL)
CORSON, Shayne	6-1	202	L	Barrie, Ont.	8/13/66	Toronto
DOMI, Tie	5-10	200	R	Windsor, Ont.	11/1/69	Toronto
DOULL, Doug	6-2	216	L	Glace Bay, N.S.	5/31/74	St. John's
FITZGERALD, Tom	6-0	195	R	Billerica, MA	8/28/68	Nashville-Chicago
GAVEY, Aaron	6-2	200	L	Sudbury, Ont.	2/22/74	Minnesota
GREEN, Travis	6-2	200	R	Castlegar, B.C.	12/20/70	Toronto
HEALEY, Paul	6-2	198	R	Edmonton, Alta.	3/20/75	Toronto-St. John's
HOGLUND, Jonas	6-3	215	R	Hammaro, Sweden	8/29/72	Toronto
HOLDEN, Josh	6-0	190	L	Calgary, Alta.	1/18/78	Carolina-Manitoba
HOUSE, Bobby	6-1	205	R	Whitehorse, Yukon	1/7/73	St. John's
McCAULEY, Alyn	5-11	190	L	Brockville, Ont.	5/29/77	Toronto
MILLS, Craig	6-0	190	R	Toronto, Ont.	8/27/76	St. John's
MOGILNY, Alexander	6-0	200	L	Khabarovsk, USSR	2/18/69	Toronto
PONIKAROVSKY, Alexei	6-4	196	L	Kiev, USSR	4/9/80	Toronto-St.J's-Ukraine
REICHEL, Robert	5-10	185	L	Litvinov, Czech.	6/25/71	Toronto-Czech Republic
RENBERG, Mikael	6-2	218	L	Pitea, Sweden	5/5/72	Toronto-Sweden
ROBERTS, Gary	6-1	190	L	North York, Ont.	5/23/66	Toronto
SUNDIN, Mats	6-4	220	R	Bromma, Sweden	2/13/71	Toronto-Sweden
TUCKER, Darcy	5-11	185	L	Castor, Alta.	3/15/75	Toronto
WREN, Bob	5-10	185	L	Preston, Ont.	9/16/74	Toronto-St. John's
DEFENSEMEN						
BELAK, Wade	6-4	225	R	Saskatoon, Sask.	7/3/76	Toronto
BERG, Aki	6-3	220	L	Turku, Finland	7/28/77	Toronto-Finland
BONNI, Ryan	6-4	190	L	Winnipeg, Man.	2/18/79	Manitoba-Columbia
COLAIACOVO, Carlo	6-1	184	L	Toronto, Ont.	1/27/83	Erie (OHL)
ERIKSSON, Anders	6-2	220	L	Bollnas, Sweden	1/9/75	Toronto-St. John's
JACKMAN, Richard	6-2	192	R	Toronto, Ont.	6/28/78	Boston-Prov (AHL)
KABERLE, Tomas	6-2	200	L	Rakovnik, Czech.	3/2/78	Kladno-Toronto-Czech Republic
LUMME, Jyrki	6-1	209	L	Tampere, Finland	7/16/66	Dallas-Toronto-Finland
McCABE, Bryan	6-2	213	L	St. Catharines, Ont.	6/8/75	Toronto
MORO, Marc	6-1	220	L	Toronto, Ont.	7/17/77	Nsh-Mil-Tor-St.J's
PILAR, Karel	6-3	210	R	Prague, Czech.	12/23/77	Toronto-St. John's
SVEHLA, Robert	6-1	210	R	Martin, Czech.	1/2/69	Florida
SVOBODA, Petr	6-3	200	R	Jihlava, Czech.	6/20/80	St. John's
YAKUSHIN, Dmitri	6-0	200	L	Kharkov, USSR	1/21/78	Kiev

GOALTENDERS	HT	WT	C	Place of Birth	Date	2001-02 Club
BELFOUR, Ed	5-11	192	L	Carman, Man.	4/21/65	Dallas-Canada
CENTOMO, Sebastien	6-1	193	R	Montreal, Que.	3/26/81	Toronto-Mem-St.J's
TELLQVIST, Mikael	5-11	185	L	Sundbyberg, Sweden	9/19/79	St. John's-Sweden

Coach and General Manager

QUINN, PAT
Coach and General Manager, Toronto Maple Leafs.
Born in Hamilton, Ont., January 29, 1943.

Pat Quinn became the 25th head coach of the Toronto Maple Leafs on June 26, 1998 and quickly turned the club's fortunes around. He earned a nomination for the Jack Adams Award as coach of the year in his first year behind the bench in Toronto after guiding the Leafs to a club-record 45 victories. After the season, Quinn was named general manager on July 14, 1999. He is the first man since Punch Imlach in the 1960s to serve the dual role in Toronto. Quinn served as both coach and general manager during much of his time with the Vancouver Canucks. He led the Leafs to their first 100-point season in 1999-2000. The Leafs reached the 100-point plateau again in 2001-02 and advanced to the Conference Finals for the second time under Quinn's leadership.

Quinn joined the Canucks as president and general manager in 1987 and took over the coaching reigns on January 31, 1991. He guided the Canucks to single-season records for wins (46) and points (101) in 1992-93, and led the team to the Stanley Cup Finals in 1994. Quinn had previously guided Philadelphia to the Stanley Cup Finals in 1980. He won the Jack Adams Award with the Flyers in 1979-80 and in Vancouver in 1992-93. He also coached in Los Angeles from 1984 to 1987.

Quinn spent the 1968-69 and 1969-70 seasons in a Maple Leafs uniform. He played 99 games for Toronto and later played for Vancouver and the Atlanta Flames. He holds a law degree from Widener University, Delaware School of Law.

NHL Coaching Record

		Regular Season				Playoffs		
Season	Team	Games	W	L	T	Games	W	L
1978-79	Philadelphia	30	18	8	4	8	3	5
1979-80	Philadelphia	80	48	12	20	19	13	6
1980-81	Philadelphia	80	41	24	15	12	6	6
1981-82	Philadelphia	72	34	29	9	....	....	....
1984-85	Los Angeles	80	34	32	14	3	0	3
1985-86	Los Angeles	80	23	49	8	....	....	....
1986-87	Los Angeles	42	18	20	4	....	....	....
1990-91	Vancouver	26	9	13	4	6	2	4
1991-92	Vancouver	80	42	26	12	13	6	7
1992-93	Vancouver	84	46	29	9	12	6	6
1993-94	Vancouver	84	41	40	3	24	15	9
1995-96	Vancouver	6	3	3	0	6	2	4
1998-99	Toronto	82	45	30	7	17	9	8
1999-2000	Toronto	82	45	30	7	12	6	6
2000-01	Toronto	82	37	34	11	11	7	4
2001-02	Toronto	82	43	29	10	20	10	10
	NHL Totals	**1072**	**527**	**408**	**137**	**163**	**85**	**78**

Assistant coach Rick Ley posted a 1-1 record as replacement coach when Quinn was sidelined with heart arrythmia, May 21 and 25, 2002. Both games are credited to Quinn's coaching record.

2001-02 Scoring

* - rookie

Regular Season

Pos	#	Player	Team	GP	G	A	Pts	+/–	PIM	PP	SH	GW	GT	S	%
C	13	Mats Sundin	TOR	82	41	39	80	6	94	10	2	9	2	262	15.6
C	16	Darcy Tucker	TOR	77	24	35	59	24	92	7	0	5	0	124	19.4
R	89	Alexander Mogilny	TOR	66	24	33	57	1	8	5	0	4	0	188	12.8
R	19	Mikael Renberg	TOR	71	14	38	52	11	36	4	0	3	0	130	10.8
C	21	Robert Reichel	TOR	78	20	31	51	7	26	1	0	3	0	152	13.2
L	7	Gary Roberts	TOR	69	21	27	48	–4	63	6	2	2	1	122	17.2
R	14	Jonas Hoglund	TOR	82	13	34	47	11	26	1	1	4	0	199	6.5
D	24	Bryan McCabe	TOR	82	17	26	43	16	129	8	0	1	0	157	10.8
D	15	Tomas Kaberle	TOR	69	10	29	39	5	2	5	0	3	1	85	11.8
C	39	Travis Green	TOR	82	11	23	34	13	61	3	0	2	0	119	9.2
L	27	Shayne Corson	TOR	74	12	21	33	11	120	0	1	1	0	111	10.8
R	28	Tie Domi	TOR	74	9	10	19	3	157	0	0	2	0	93	9.7
D	36	Dmitry Yushkevich	TOR	55	6	13	19	14	26	3	0	0	0	79	7.6
C	18	Alyn McCauley	TOR	82	6	10	16	10	18	0	1	1	0	95	6.3
R	10	Garry Valk	TOR	63	5	10	15	2	28	0	0	0	0	80	6.3
D	25	Jyrki Lumme	DAL	15	0	1	1	–5	4	0	0	0	0	12	0.0
			TOR	51	4	8	12	13	18	1	2	1	0	61	6.6
			TOTAL	66	4	9	13	8	22	1	2	1	0	73	5.5
D	4	Cory Cross	TOR	50	3	9	12	11	54	0	0	1	0	39	7.7
D	8	Aki Berg	TOR	81	1	10	11	14	46	0	0	0	0	66	1.5
R	26	Paul Healey	TOR	21	3	7	10	7	2	0	0	0	0	29	10.3
D	29	* Karel Pilar	TOR	23	1	3	4	3	8	0	0	0	0	32	3.1
D	2	Wade Belak	TOR	63	1	3	4	2	142	0	0	0	0	47	2.1
R	22	* Alexei Ponikarovsky	TOR	8	2	0	2	2	0	0	0	1	0	8	25.0
C	11	Nik Antropov	TOR	11	1	1	2	–1	4	0	0	0	0	12	8.3
C	20	* Jeff Farkas	TOR	6	0	2	2	1	4	0	0	0	0	3	0.0
D	44	Anders Eriksson	TOR	34	0	2	2	–1	12	0	0	0	0	31	0.0
L	33	Bob Wren	TOR	1	0	0	0	1	0	0	0	0	0	0	0.0
L	43	Nathan Dempsey	TOR	3	0	0	0	1	0	0	0	0	0	3	0.0
D	3	Marc Moro	NSH	13	0	0	0	–3	23	0	0	0	0	7	0.0
			TOR	2	0	0	0	0	2	0	0	0	0	0	0.0
			TOTAL	15	0	0	0	–3	25	0	0	0	0	7	0.0

Goaltending

No.	Goaltender	GPI	Mins	Avg	W	L	T	EN	SO	GA	SA	S%	G	A	PIM
31	Curtis Joseph	51	3065	2.23	29	17	5	4	4	114	1210	.906	0	1	10
35	Corey Schwab	30	1646	2.73	12	10	5	1	1	75	707	.894	0	0	2
30	Tom Barrasso	4	219	2.74	2	2	0	0	0	10	110	.909	0	0	0
30	* Sebastien Centomo	1	40	4.50	0	0	0	0	0	3	12	.750	0	0	0
	Totals	**82**	**4986**	**2.49**	**43**	**29**	**10**	**5**	**5**	**207**	**2044**	**.899**			

Playoffs

Pos	#	Player	Team	GP	G	A	Pts	+/–	PIM	PP	SH	GW	GT	S	%
L	7	Gary Roberts	TOR	19	7	12	19	6	56	3	0	1	1	47	14.9
C	18	Alyn McCauley	TOR	20	5	10	15	3	4	1	0	2	0	48	10.4
R	89	Alexander Mogilny	TOR	20	8	3	11	1	8	2	0	2	0	59	13.6
D	24	Bryan McCabe	TOR	20	5	5	10	4	30	3	0	1	0	64	7.8
R	14	Jonas Hoglund	TOR	20	4	6	10	–3	2	3	0	1	0	50	8.0
D	15	Tomas Kaberle	TOR	20	2	8	10	7	16	0	0	0	0	29	6.9
C	39	Travis Green	TOR	20	3	6	9	5	34	0	0	1	0	35	8.6
C	16	Darcy Tucker	TOR	17	4	4	8	–1	38	1	0	1	0	31	12.9
C	13	Mats Sundin	TOR	8	2	5	7	5	4	0	0	0	0	21	9.5
L	27	Shayne Corson	TOR	19	1	6	7	–2	33	0	0	0	0	37	2.7
R	28	Tie Domi	TOR	19	1	3	4	0	61	0	0	1	0	25	4.0
D	29	* Karel Pilar	TOR	11	0	4	4	–3	12	0	0	0	0	12	0.0
C	21	Robert Reichel	TOR	18	0	3	3	–6	4	0	0	0	0	22	0.0
L	43	Nathan Dempsey	TOR	6	0	2	2	1	0	0	0	0	0	7	0.0
R	10	Garry Valk	TOR	11	1	0	1	–1	4	0	0	0	0	9	11.1
D	2	Wade Belak	TOR	16	1	0	1	4	18	0	0	0	0	6	16.7
R	26	Paul Healey	TOR	18	0	1	1	2	2	0	0	0	0	15	0.0
D	8	Aki Berg	TOR	20	0	1	1	0	37	0	0	0	0	10	0.0
L	33	Bob Wren	TOR	1	0	0	0	0	0	0	0	0	0	0	0.0
C	20	* Jeff Farkas	TOR	2	0	0	0	0	0	0	0	0	0	0	0.0
R	19	Mikael Renberg	TOR	3	0	0	0	1	2	0	0	0	0	6	0.0
C	37	* Donald MacLean	TOR	3	0	0	0	0	0	0	0	0	0	0	0.0
D	44	Anders Eriksson	TOR	10	0	0	0	–3	0	0	0	0	0	11	0.0
R	22	* Alexei Ponikarovsky	TOR	10	0	0	0	0	4	0	0	0	0	7	0.0
D	4	Cory Cross	TOR	12	0	0	0	–1	8	0	0	0	0	8	0.0
D	25	Jyrki Lumme	TOR	14	0	0	0	–2	4	0	0	0	0	9	0.0

Goaltending

No.	Goaltender	GPI	Mins	Avg	W	L	EN	SO	GA	SA	S%	G	A	PIM
35	Corey Schwab	1	12	0.00	0	0	0	0	0	5	1.000	0	0	0
31	Curtis Joseph	20	1253	2.30	10	10	1	3	48	557	.914	0	0	4
	Totals	**20**	**1272**	**2.31**	**10**	**10**	**1**	**3**	**49**	**563**	**.913**			

Coaching History

Conn Smythe, 1927-28 to 1929-30; Conn Smythe and Art Duncan, 1930-31; Art Duncan and Dick Irvin, 1931-32; Dick Irvin, 1932-33 to 1939-40; Hap Day, 1940-41 to 1949-50; Joe Primeau, 1950-51 to 1952-53; King Clancy, 1953-54 to 1955-56; Howie Meeker, 1956-57; Billy Reay, 1957-58; Billy Reay and Punch Imlach, 1958-59; Punch Imlach, 1959-60 to 1968-69; John McLellan, 1969-70 to 1972-73; Red Kelly, 1973-74 to 1976-77; Roger Neilson, 1977-78, 1978-79; Floyd Smith, Dick Duff and Punch Imlach, 1979-80; Punch Imlach, Joe Crozier and Mike Nykoluk, 1980-81; Mike Nykoluk, 1981-82 to 1983-84; Dan Maloney, 1984-85, 1985-86; John Brophy, 1986-87, 1987-88; John Brophy and George Armstrong, 1988-89; Doug Carpenter, 1989-90; Doug Carpenter and Tom Watt, 1990-91; Tom Watt, 1991-92; Pat Burns, 1992-93 to 1994-95; Pat Burns and Nick Beverley, 1995-96; Mike Murphy, 1996-97, 1997-98; Pat Quinn, 1998-99 to date.

Club Records

Team

(Figures in brackets for season records are games played; records for fewest points, wins, ties, losses, goals, goals against are for 70 or more games)

Record		
Most Points	100	1999-2000 (82), 2001-02 (82)
Most Wins	45	1998-99 (82), 1999-2000 (82)
Most Ties	22	1954-55 (70)
Most Losses	52	1984-85 (80)
Most Goals	337	1989-90 (80)
Most Goals Against	387	1983-84 (80)
Fewest Points	48	1984-85 (80)
Fewest Wins	20	1981-82 (80), 1984-85 (80)
Fewest Ties	4	1989-90 (80)
Fewest Losses	16	1950-51 (70)
Fewest Goals	147	1954-55 (70)
Fewest Goals Against	*131	1953-54 (70)
Longest Winning Streak		
Overall	10	Oct. 7-28/93
Home	9	Nov. 11-Dec. 26/53
Away	7	Nov. 14-Dec. 15/40, Dec. 4/60-Jan. 5/61
Longest Undefeated Streak		
Overall	11	Oct. 15-Nov. 8/50 (8 wins, 3 ties), Jan. 6-Feb. 1/94 (7 wins, 4 ties)
Home	18	Nov. 28/33-Mar. 10/34 (15 wins, 3 ties), Oct. 31/53-Jan. 23/54 (16 wins, 2 ties)
Away	9	Nov. 30/47-Jan. 11/48 (4 wins, 5 ties)
Longest Losing Streak		
Overall	10	Jan. 15-Feb. 8/67
Home	7	Nov. 11-Dec. 5/84
Away	11	Feb. 20-Apr. 1/88
Longest Winless Streak		
Overall	15	Dec. 26/87-Jan. 25/88 (11 losses, 4 ties)
Home	11	Dec. 19/87-Jan. 25/88 (7 losses, 4 ties)
Away	18	Oct. 6/82-Jan. 5/83 (13 losses, 5 ties)
Most Shutouts, Season	13	1953-54 (70)
Most PIM, Season	2,419	1989-90 (80)
Most Goals, Game	14	Mar. 16/57 (NYR 1 at Tor. 14)

Individual

Most Seasons	21	George Armstrong
Most Games	1,187	George Armstrong
Most Goals, Career	389	Darryl Sittler
Most Assists, Career	620	Borje Salming
Most Points, Career	916	Darryl Sittler (389G, 527A)
Most PIM, Career	1,777	Tie Domi
Most Shutouts, Career	62	Turk Broda
Longest Consecutive Games Streak	486	Tim Horton (Feb. 11/61-Feb. 4/68)
Most Goals, Season	54	Rick Vaive (1981-82)
Most Assists, Season	95	Doug Gilmour (1992-93)
Most Points, Season	127	Doug Gilmour (1992-93; 32G, 95A)
Most PIM, Season	365	Tie Domi (1997-98)
Most Points, Defenseman, Season	79	Ian Turnbull (1976-77; 22G, 57A)
Most Points, Center, Season	127	Doug Gilmour (1992-93; 32G, 95A)
Most Points, Right Wing, Season	97	Wilf Paiement (1980-81; 40G, 57A)
Most Points, Left Wing, Season	99	Dave Andreychuk (1993-94; 53G, 46A)
Most Points, Rookie, Season	66	Peter Ihnacak (1982-83; 28G, 38A)
Most Shutouts, Season	13	Harry Lumley (1953-54)
Most Goals, Game	6	Corb Denneny (Jan. 26/21), Darryl Sittler (Feb. 7/76)
Most Assists, Game	6	Babe Pratt (Jan. 8/44), Doug Gilmour (Feb. 13/93)
Most Points, Game	*10	Darryl Sittler (Feb. 7/76; 6G, 4A)

* NHL Record.

Retired Numbers

No.	Player	Years
5	Bill Barilko	1946-1951
6	Ace Bailey	1926-1934

Honored Numbers

No.	Player	Years
1	Turk Broda	1936-43, 45-52
	Johnny Bower	1958-1970
7	King Clancy	1930-1937
	Tim Horton	1949-50, 51-70
9	Charlie Conacher	1929-1938
	Ted Kennedy	1942-55, 56-57
10	Syl Apps	1936-43, 45-48
	George Armstrong	1949-50, 51-71
27	Frank Mahovlich	1956-1968

All-time Record vs. Other Clubs

Regular Season

	At Home								On Road								Total							
	GP	W	L	T	OL	GF	GA	PTS	GP	W	L	T	OL	GF	GA	PTS	GP	W	L	T	OL	GF	GA	PTS
Anaheim	14	8	2	4	0	45	28	20	10	5	4	1	0	27	29	11	24	13	6	5	0	72	57	31
Atlanta	5	3	2	0	0	19	12	6	5	3	2	0	0	19	11	6	10	6	4	0	0	38	23	12
Boston	293	153	89	51	0	991	754	357	293	88	156	47	2	771	955	225	586	241	245	98	2	1762	1709	582
Buffalo	62	26	25	11	0	189	211	63	63	17	40	6	0	168	259	40	125	43	65	17	0	357	470	103
Calgary	51	26	17	7	1	195	186	60	58	21	32	5	0	186	229	47	109	47	49	12	1	381	415	107
Carolina	32	11	16	5	0	111	118	27	33	10	18	5	0	109	137	25	65	21	34	10	0	220	255	52
Chicago	314	164	96	54	0	1070	817	382	317	119	156	42	0	828	967	280	631	283	252	96	0	1898	1784	662
Colorado	33	15	14	4	0	113	128	34	29	7	17	5	0	87	111	19	62	22	31	9	0	200	239	53
Columbus	2	1	0	1	0	6	3	3	0	0	0	0	0	0	0	0	2	1	0	1	0	6	3	3
Dallas	101	49	35	17	0	356	319	115	95	35	49	11	0	302	362	81	196	84	84	28	0	658	681	196
Detroit	314	163	104	47	0	1039	842	373	321	108	167	46	0	790	961	262	635	271	271	93	0	1829	1803	635
Edmonton	35	19	14	2	0	146	148	40	40	13	21	6	0	131	170	32	75	32	35	8	0	277	318	72
Florida	11	7	2	2	0	41	26	16	13	6	4	3	0	39	38	15	24	13	6	5	0	80	64	31
Los Angeles	67	34	22	11	0	262	220	79	64	21	34	9	0	187	230	51	131	55	56	20	0	449	450	130
Minnesota	2	2	0	0	0	9	2	4	0	0	0	0	0	0	0	0	2	2	0	0	0	9	2	4
Montreal	328	169	114	45	0	993	849	383	327	88	197	42	0	809	1150	218	655	257	311	87	0	1802	1999	601
Nashville	4	1	2	1	0	9	12	3	1	0	0	0	1	2	3	1	5	1	2	1	1	11	15	4
New Jersey	44	28	11	5	0	176	136	61	42	13	15	13	1	130	148	40	86	41	26	18	1	306	284	101
NY Islanders	51	25	23	3	0	173	177	53	49	19	27	3	0	145	194	41	100	44	50	6	0	318	371	94
NY Rangers	278	156	83	39	0	959	731	351	279	105	116	56	2	821	856	268	557	261	199	95	2	1780	1587	619
Ottawa	16	8	6	1	1	47	44	18	15	4	9	1	1	39	45	10	31	12	15	2	2	86	89	28
Philadelphia	63	22	26	14	1	202	211	59	63	14	40	8	1	146	246	37	126	36	66	22	2	348	457	96
Phoenix	42	19	21	2	0	152	158	40	38	12	20	6	0	137	157	30	80	31	41	8	0	289	315	70
Pittsburgh	63	30	22	11	0	250	201	71	65	23	36	6	0	205	272	52	128	53	58	17	0	455	473	123
St. Louis	98	58	28	11	1	367	289	128	101	30	57	14	0	280	342	74	199	88	85	25	1	647	631	202
San Jose	17	12	4	1	0	66	44	25	14	7	5	2	0	38	30	16	31	19	9	3	0	104	74	41
Tampa Bay	17	10	6	1	0	63	42	21	16	14	2	0	0	59	31	28	33	24	8	1	0	122	73	49
Vancouver	58	27	21	10	0	214	191	64	61	21	29	11	0	205	219	53	119	48	50	21	0	419	410	117
Washington	44	23	16	5	0	196	153	51	46	15	28	3	0	129	175	33	90	38	44	8	0	325	328	84
Defunct Clubs	232	158	53	21	0	860	515	337	233	84	120	29	0	607	745	197	465	242	173	50	0	1467	1260	534
Totals	**2691**	**1427**	**874**	**386**	**4**	**9319**	**7567**	**3244**	**2691**	**902**	**1401**	**380**	**8**	**7396**	**9072**	**2192**	**5382**	**2329**	**2275**	**766**	**12**	**16715**	**16639**	**5436**

Playoffs

	Series	W	L	GP	W	L	T	GF	GA	Last Mtg.	Rnd.	Result
Boston	13	8	5	62	31	30	1	150	153	1974	QF	L 0-4
Buffalo	1	0	1	5	1	4	0	16	21	1999	CF	L 1-4
Calgary	1	1	0	2	2	0	0	9	5	1979	PRE	W 2-0
Carolina	1	0	1	6	2	4	0	6	10	2002	CF	L 2-4
Chicago	9	6	3	38	22	15	1	111	89	1995	CQF	L 3-4
Dallas	2	0	2	7	1	6	0	26	35	1983	DSF	L 1-3
Detroit	23	12	11	117	58	59	0	311	321	1993	DSF	W 4-3
Los Angeles	3	2	1	12	7	5	0	41	31	1993	CF	L 3-4
Montreal	15	7	8	71	29	42	0	160	215	1979	QF	L 0-4
New Jersey	2	0	2	13	5	8	0	27	37	2001	CSF	L 3-4
NY Islanders	3	2	1	17	8	9	0	42	54	2002	CQF	W 4-3
NY Rangers	8	3	5	35	16	19	0	86	86	1971	QF	L 2-4
Ottawa	3	3	0	17	12	5	0	43	31	2002	CSF	W 4-3
Philadelphia	4	1	3	23	9	14	0	56	78	1999	CQF	W 4-2
Pittsburgh	3	3	0	12	8	4	0	39	27	1999	CSF	W 4-2
St. Louis	5	2	3	31	14	17	0	90	88	1996	CQF	L 2-4
San Jose	1	1	0	7	4	3	0	26	21	1994	CSF	W 4-3
Vancouver	1	0	1	5	1	4	0	9	16	1994	CF	L 1-4
Defunct Clubs	8	6	2	24	12	10	2	59	57			
Totals	**106**	**57**	**49**	**504**	**242**	**258**	**4**	**1307**	**1375**			

Calgary totals include Atlanta Flames, 1972-73 to 1979-80.
Carolina totals include Hartford, 1979-80 to 1996-97.
Colorado totals include Quebec, 1979-80 to 1994-95.
Dallas totals include Minnesota North Stars, 1967-68 to 1992-93.
New Jersey totals include Kansas City, 1974-75 to 1975-76, and Colorado Rockies, 1976-77 to 1981-82.
Phoenix totals include Winnipeg, 1979-80 to 1995-96.

Playoff Results 2002-1998

Year	Round	Opponent	Result	GF	GA
2002	CF	Carolina	L 2-4	6	10
	CSF	Ottawa	W 4-3	16	18
	CQF	NY Islanders	W 4-3	22	21
2001	CSF	New Jersey	L 3-4	18	21
	CQF	Ottawa	W 4-0	10	3
2000	CSF	New Jersey	L 2-4	9	16
	CQF	Ottawa	W 4-2	17	10
1999	CF	Buffalo	L 1-4	16	21
	CSF	Pittsburgh	W 4-2	18	14
	CQF	Philadelphia	W 4-2	9	11

Abbreviations: Round: CF – conference final; **CSF** – conference semi-final; **CQF** – conference quarter-final; **DSF** – division semi-final; **QF** – quarter-final; **PRE** – preliminary round.

2001-02 Results

Month	Day	Opponent	Score	Month	Day	Opponent	Score
Oct.	3	Ottawa	4-5		5	Ottawa	3-1
	6	at Montreal	2-2		7	at Ottawa	3-4
	8	Anaheim	6-1		8	Nashville	4-3
	11	at Carolina	3-2		11	at Washington	3-3
	13	St. Louis	2-5		12	Montreal	1-1
	16	at Edmonton	4-1		15	Atlanta	2-3
	18	at Vancouver	6-5		17	at Nashville	2-3*
	20	at Calgary	1-4		19	Philadelphia	0-3
	23	Boston	2-0		22	at Calgary	6-1
	25	at Boston	1-2*		25	at Vancouver	1-6
	27	Pittsburgh	4-0		26	at Edmonton	1-4
	30	Tampa Bay	3-2		29	San Jose	4-3
Nov.	1	at Pittsburgh	1-3		30	at Atlanta	6-0
	3	Colorado	4-1	Feb.	5	Minnesota	3-1
	6	Washington	4-2		7	at NY Islanders	1-4
	9	at New Jersey	2-3*		9	Montreal	4-1
	10	New Jersey	1-1		11	Atlanta	5-4
	14	at Florida	3-2		26	Carolina	4-1
	15	at Tampa Bay	3-2	Mar.	1	at New Jersey	2-4
	17	at Ottawa	1-2*		2	Buffalo	3-3
	19	Florida	5-1		4	at Washington	3-2
	21	at Buffalo	2-4		6	at Detroit	2-6
	23	at NY Islanders	1-3		9	at Montreal	1-1
	24	Boston	2-0		10	at Philadelphia	3-1
	27	Carolina	2-5		12	Philadelphia	1-1
	30	at Chicago	2-1		14	at Boston	2-1
Dec.	1	Chicago	4-1		16	Dallas	5-5
	4	Pittsburgh	0-1		19	NY Islanders	3-2*
	6	at NY Rangers	6-3		21	Washington	3-4
	8	NY Rangers	4-3		23	Buffalo	2-0
	11	Phoenix	6-3		25	at Philadelphia	1-4
	13	at St. Louis	4-3*		26	Tampa Bay	7-2
	15	Montreal	6-4		28	NY Islanders	4-5
	18	Los Angeles	1-3		30	New Jersey	1-3
	21	at Buffalo	3-3	Apr.	1	at Detroit	5-4*
	22	Buffalo	3-2		4	NY Rangers	2-4
	26	at Carolina	3-4		6	Florida	2-2
	28	at Atlanta	4-5		8	Columbus	4-1
	29	at Florida	2-4		10	at NY Rangers	7-2
	31	at Tampa Bay	4-1		12	at Pittsburgh	5-2
Jan.	3	at Boston	2-1		13	at Ottawa	5-2

* – Overtime

Entry Draft Selections 2002-1988

2002

Pick	
24	Alexander Steen
57	Matthew Stajan
74	Todd Ford
88	Dominic D'Amour
122	David Turon
191	Ian White
222	Scott May
254	Jarkko Immonen
285	Staffan Kronvall

2001

Pick	
17	Carlo Colaiacovo
39	Karel Pilar
65	Brendan Bell
82	Jay Harrison
88	Nicolas Corbeil
134	Kyle Wellwood
168	Maxim Kondratjev
183	Jaroslav Sklenar
198	Ivan Kolozvary
213	Jan Chovan
246	Tomas Mojzis
276	Mike Knoepfli

2000

Pick	
24	Brad Boyes
51	Kris Vernarsky
70	Mikael Tellqvist
90	Jean-Francois Racine
100	Miguel Delisle
179	Vadim Sozinov
209	Markus Seikola
223	Lubos Velebny
254	Alexander Shinkar
265	Jean-Philippe Cote

1999

Pick	
24	Luca Cereda
60	Peter Reynolds
108	Mirko Murovic
110	Jon Zion
151	Vaclav Zavoral
161	Jan Sochor
211	Vladimir Kulikov
239	Pierre Hedin
267	Peter Metcalf

1998

Pick	
10	Nik Antropov
35	Petr Svoboda
69	Jamie Hodson
87	Alexei Ponikarovsky
126	Morgan Warren
154	Allan Rourke
181	Jonathan Gagnon
215	Dwight Wolfe
228	Michal Travnicek
236	Sergei Rostov

1997

Pick	
57	Jeff Farkas
84	Adam Mair
111	Frantisek Mrazek
138	Eric Gooldy
165	Hugo Marchand
190	Shawn Thornton
194	Russ Bartlett
221	Jonathan Hedstrom

1996

Pick	
36	Marek Posmyk
50	Francis Larivee
66	Mike Lankshear
68	Konstantin Kalmikov
86	Jason Sessa
103	Vladimir Antipov
110	Peter Cava
111	Brandon Sugden
140	Dmitri Yakushin
148	Chris Bogas
151	Lucio DeMartinis
178	Reggie Berg
204	Tomas Kaberle
230	Jared Hope

1995

Pick	
15	Jeff Ware
54	Ryan Pepperall
139	Doug Bonner
145	Yannick Tremblay
171	Marek Melenovsky
197	Mark Murphy
223	Danny Markov

1994

Pick	
16	Eric Fichaud
48	Sean Haggerty
64	Fredrik Modin
126	Mark Deyell
152	Karri White
178	Tommi Rajamaki
204	Rob Butler
256	Sergei Berezin
282	Doug Nolan

1993

Pick	
12	Kenny Jonsson
19	Landon Wilson
123	Zdenek Nedved
149	Paul Vincent
175	Jeff Andrews
201	David Brumby
253	Kyle Ferguson
279	Mikhail Lapin

1992

Pick	
8	Brandon Convery
23	Grant Marshall
77	Nikolai Borschevsky
95	Mark Raiter
101	Janne Gronvall
106	Chris Deruiter
125	Mikael Hakansson
149	Patrik Augusta
173	Ryan Vandenbussche
197	Wayne Clarke
221	Sergei Simonov
245	Nathan Dempsey

1991

Pick	
47	Yanic Perreault
69	Terry Chitaroni
102	Alexei Kudashov
113	Jeff Perry
120	Alexander Kuzminsky
135	Martin Prochazka
160	Dmitri Mironov
164	Robb McIntyre
167	Tomas Kucharcik
179	Guy Lehoux
201	Gary Miller
223	Johnathon Kelley
245	Chris O'Rourke

1990

Pick	
10	Drake Berehowsky
31	Felix Potvin
73	Darby Hendrickson
80	Greg Walters
115	Alexander Godynyuk
136	Eric Lacroix
157	Dan Stiver
178	Robert Horyna
199	Rob Chebator
220	Scott Malone
241	Nick Vachon

1989

Pick	
3	Scott Thornton
12	Rob Pearson
21	Steve Bancroft
66	Matt Martin
96	Keith Carney
108	David Burke
125	Michael Doers
129	Keith Merkler
150	Derek Langille
171	Jeffrey St. Laurent
192	Justin Tomberlin
213	Mike Jackson
234	Steve Chartrand

1988

Pick	
6	Scott Pearson
27	Tie Domi
48	Peter Ing
69	Ted Crowley
86	Len Esau
132	Matt Mallgrave
153	Peter Elvenas
174	Mike Delay
195	David Sacco
216	Mike Gregorio
237	Peter DeBoer

General Managers' History

Conn Smythe, 1927-28 to 1956-57; Hap Day, 1957-58; Punch Imlach, 1958-59 to 1968-69; Jim Gregory, 1969-70 to 1978-79; Punch Imlach, 1979-80, 1980-81; Punch Imlach and Gerry McNamara, 1981-82; Gerry McNamara, 1982-83 to 1987-88; Gord Stellick, 1988-89; Floyd Smith, 1989-90, 1990-91; Cliff Fletcher, 1991-92 to 1996-97; Ken Dryden, 1997-98, 1998-99; Pat Quinn, 1999-2000 to date.

Captains' History

Hap Day, 1927-28 to 1936-37; Charlie Conacher, 1937-38; Red Horner, 1938-39, 1939-40; Syl Apps, 1940-41 to 1942-43; Bob Davidson, 1943-44, 1944-45; Syl Apps, 1945-46 to 1947-48; Ted Kennedy, 1948-49 to 1954-55; Sid Smith, 1955-56; Jimmy Thomson, Ted Kennedy, 1956-57; George Armstrong, 1957-58 to 1968-69; Dave Keon, 1969-70 to 1974-75; Darryl Sittler, 1975-76 to 1980-81; Rick Vaive, 1981-82 to 1985-86; no captain, 1986-87 to 1988-89; Rob Ramage, 1989-90, 1990-91; Wendel Clark, 1991-92 to 1993-94; Doug Gilmour, 1994-95 to 1996-97; Mats Sundin, 1997-98 to date.

Club Directory

Air Canada Centre

Toronto Maple Leafs
Air Canada Centre
40 Bay St., Suite 400
Toronto, Ontario M5J 2X2
Phone **416/815-5700**
FAX 416/359-9331
www.mapleleafs.com
Capacity: 18,819

Board of Directors
Steve A. Stavro (Chairman of the Board and NHL Governor), Brian P. Bellmore (Alternate NHL Governor), Dale Lastman, Robert G. Bertram, John MacIntyre, Dean Metcalf

Maple Leaf Sports & Entertainment Ltd.

Chairman of the Board and NHL Governor	Steve A. Stavro
Alternate NHL Governor	Brian P. Bellmore
President, Chief Executive Officer and Alternate NHL Governor	Richard Peddie
Executive Vice-President and Alternate NHL Governor	Ken Dryden
Sr. Vice-President and General Manager, Air Canada Centre	Bob Hunter
Sr. Vice-President, Business, Chief Marketing Officer	Tom Anselmi
Sr. Vice-President, Chief Financial Officer	Ian Clarke
Vice-President, Sports Communications and Community Development	John Lashway
Vice-President, People	Mardi Walker
Vice-President, Regulatory Affairs and General Counsel	Robin Brudner
Vice-President, Sales and Service	Chris Overholt
Corporate Secretary	Paul Perantinos
Corporate Legal Counsel	David Matheson

Maple Leafs Management

President and Alternate NHL Governor	Ken Dryden
General Manager and Head Coach	Pat Quinn
Assistant to the General Manager	Bill Watters
Assistant General Manager & Director of Player Personnel	Mike Penny
Assistant Coaches	Keith Acton, Rick Ley
Player Development Coach	Paul Dennis
Video Coach	Reid Mitchell
Strength & Conditioning Coach	Matt Nichol
Community Representatives	Wendel Clark, Darryl Sittler
Director, Amateur Scouting	Barry Trapp
Pro Scout	Murray Oliver
Scouts	George Armstrong, Bob Johnson, Garth Malarchuk, Mike Palmateer, Mark Yannetti
European Scouts	Thommie Bergman, Jan Kovac, Nikolai Ladygin
Director, Team Services	Casey Vanden Heuvel
Travel Coordinator	Mary Speck
Executive Assistant to the President	Ann Clark
Executive Assistant to the General Manager	Maria Tomasevic

Maple Leafs Communications and Community Development

Vice-President, Sports Communications and Community Development	John Lashway
Director, Media Relations	Pat Park
Coordinators, Media Relations	Matthew Frost, Dave Griffiths
Manager, Corporate Communications	Tara McCarthy
Director, Community Relations	Kristy Fletcher
Director, Go Kids Go! The Leaf Fund	Angela McManus
Coordinators, Go Kids Go! The Leaf Fund	Paula Dal Maso, Ted Warner
Coordinators, Community Relations	Sefu Bernard, Paulette Minard
Manager, Game Presentation	Mike Ferriman
Manager, Game Operations	Nancy Gilks
Assistant, Game Operations	Stephenie Summerhill
Production, Video and Scoreboard	Ian MacMillan
Head Audio Engineer	Courtney Ross
Coordinator, Youth Hockey Development	Greg Schell
Alumni Relations	Susanna Tyson
Executive Assistant, Communications	Laura Leite

Maple Leafs Medical and Training Staff

Head Athletic Therapist	Chris Broadhurst
Athletic Therapist	Brent Smith
Equipment Manager	Brian Papineau
Assistant Equipment Managers	Bobby Hastings, Scott McKay
Team Doctors	Dr. Michael Clarfield, Dr. Darrell Ogilvie-Harris, Dr. Leith Douglas, Dr. Rob Devenyi, Dr. Simon McGrail
Team Dentist	Dr. Allan Hawryluk

Broadcast Information

Radio Play-By-Play	Joe Bowen, Dennis Beyak
Radio Analyst	Jim Ralph
Television Play-By-Play	Bob Cole, Joe Bowen
Television Analyst	Harry Neale

Air Canada Centre

Director, Building Operations	Diego Roccasalva
Director, Event Operations and Production	Jim Roe
Director, Guest Service	Chris Gibbs
Director, Programming and Event Marketing	Patti-Anne Tarlton
Director, Marketing	Beth Robertson
Director, Information Technology	Sasha Puric
Director, Ticket Operations	Donna Henderson
Manager, Video and Scoreboard Production	Curtis Emerson
Director, Executive Suites Services	Nancy Read
Director, Consumer Products	Jeff Newman
Director of Restaurant Operations, Executive Chef	Brad Long
Executive Producer, New Media	John Shannon
Director, Business, New Media	Frank Bertolas
Director, Marketing Media	Alon Marcovici
Director, Corporate Sales	Dave Hopkinson
Director, Service and Ticket Administration	Paul Beirne
Director, Food and Beverage Finance	Alldrick Britto
Director, Quick Service, Food & Beverage Operations	Michael Doyle

Vancouver Canucks

2001-02 Results: 42W-30L-7T-3OTL 94PTS.
Second, Northwest Division

2002-03 Schedule

Oct.	Thu.	10	at Calgary
	Sat.	12	San Jose
	Mon.	14	Calgary
	Wed.	16	Boston
	Fri.	18	at Anaheim
	Sat.	19	at Los Angeles
	Mon.	21	at San Jose
	Thu.	24	Anaheim
	Sat.	26	Dallas
	Tue.	29	Buffalo
	Thu.	31	Colorado
Nov.	Sat.	2	at Minnesota
	Mon.	4	at Colorado
	Wed.	6	at Dallas
	Sat.	9	at Phoenix
	Tue.	12	St. Louis
	Thu.	14	Los Angeles
	Sat.	16	NY Rangers
	Wed.	20	Chicago
	Fri.	22	Detroit
	Mon.	25	at Minnesota
	Wed.	27	at Carolina
	Fri.	29	at Tampa Bay
	Sat.	30	at Florida
Dec.	Tue.	3	at NY Islanders
	Wed.	4	at New Jersey
	Sat.	7	Minnesota
	Mon.	9	Calgary
	Wed.	11	Colorado
	Sat.	14	at Edmonton
	Sun.	15	Calgary
	Tue.	17	at Chicago
	Thu.	19	at Nashville
	Sat.	21	Edmonton
	Mon.	23	at Colorado
	Thu.	26	at Edmonton
	Sat.	28	Anaheim
	Tue.	31	Toronto
Jan.	Thu.	2	Montreal
	Sat.	4	Florida
	Wed.	8	Ottawa
	Fri.	10	Columbus
	Sat.	11	at San Jose
	Tue.	14	Nashville
	Thu.	16	at Minnesota
	Fri.	17	at Chicago
	Sun.	19	at Detroit
	Tue.	21	at Nashville
	Fri.	24	Detroit
	Sun.	26	Phoenix
	Tue.	28	Minnesota
	Thu.	30	Edmonton
Feb.	Tue.	4	at Pittsburgh
	Wed.	5	at Columbus
	Fri.	7	at Buffalo
	Mon.	10	Chicago
	Thu.	13	Colorado
	Sat.	15	at Calgary
	Tue.	18	at Detroit
	Thu.	20	at St. Louis
	Sat.	22	at Edmonton
	Sun.	23	Columbus
	Tue.	25	Atlanta
	Thu.	27	San Jose
Mar.	Sat.	1	at Montreal
	Mon.	3	at Boston
	Tue.	4	at Philadelphia
	Thu.	6	at Columbus
	Sat.	8	at Toronto
	Tue.	11	NY Islanders
	Thu.	13	St. Louis
	Sat.	15	Toronto*
	Mon.	17	at Dallas
	Tue.	18	at St. Louis
	Thu.	20	Nashville
	Sun.	23	Washington
	Tue.	25	Dallas
	Thu.	27	Phoenix
	Sat.	29	at Los Angeles
	Sun.	30	at Anaheim*
Apr.	Wed.	2	at Phoenix
	Sun.	6	Los Angeles*

* Denotes afternoon game.

Year-by-Year Record

		Home				Road				Overall								
Season	GP	W	L	T	OL	W	L	T	OL	W	L	T	OL	GF	GA	Pts.	Finished	Playoff Result
2001-02	82	23	11	5	2	19	19	2	1	42	30	7	3	254	211	94	2nd, Northwest Div.	Lost Conf. Quarter-Final
2000-01	82	21	12	5	3	15	16	6	4	36	28	11	7	239	238	90	3rd, Northwest Div.	Lost Conf. Quarter-Final
1999-2000	82	16	14	5	6	14	15	10	2	30	29	15	8	227	237	83	3rd, Northwest Div.	Out of Playoffs
1998-99	82	14	21	6	...	9	26	6	...	23	47	12	...	192	258	58	4th, Northwest Div.	Out of Playoffs
1997-98	82	15	22	4	...	10	21	10	...	25	43	14	...	224	273	64	7th, Pacific Div.	Out of Playoffs
1996-97	82	20	17	4	...	15	23	3	...	35	40	7	...	257	273	77	4th, Pacific Div.	Out of Playoffs
1995-96	82	15	19	7	...	17	16	8	...	32	35	15	...	278	278	79	3rd, Pacific Div.	Lost Conf. Quarter-Final
1994-95	48	10	8	6	...	8	10	6	...	18	18	12	...	153	148	48	2nd, Pacific Div.	Lost Conf. Semi-Final
1993-94	84	20	19	3	...	21	21	0	...	41	40	3	...	279	276	85	2nd, Pacific Div.	Lost Final
1992-93	84	27	11	4	...	19	18	5	...	46	29	9	...	346	278	101	1st, Smythe Div.	Lost Div. Final
1991-92	80	23	10	7	...	19	16	5	...	42	26	12	...	285	250	96	1st, Smythe Div.	Lost Div. Final
1990-91	80	18	17	5	...	10	26	4	...	28	43	9	...	243	315	65	4th, Smythe Div.	Lost Div. Semi-Final
1989-90	80	13	16	11	...	12	25	3	...	25	41	14	...	245	306	64	5th, Smythe Div.	Out of Playoffs
1988-89	80	19	15	6	...	14	24	2	...	33	39	8	...	251	253	74	4th, Smythe Div.	Lost Div. Semi-Final
1987-88	80	15	20	5	...	10	26	4	...	25	46	9	...	272	320	59	5th, Smythe Div.	Out of Playoffs
1986-87	80	17	19	4	...	12	24	4	...	29	43	8	...	282	314	66	5th, Smythe Div.	Out of Playoffs
1985-86	80	17	18	5	...	6	26	8	...	23	44	13	...	282	333	59	4th, Smythe Div.	Lost Div. Semi-Final
1984-85	80	15	21	4	...	10	25	5	...	25	46	9	...	284	401	59	5th, Smythe Div.	Out of Playoffs
1983-84	80	20	16	4	...	12	23	5	...	32	39	9	...	306	328	73	3rd, Smythe Div.	Lost Div. Semi-Final
1982-83	80	20	12	8	...	10	23	7	...	30	35	15	...	303	309	75	3rd, Smythe Div.	Lost Div. Semi-Final
1981-82	80	20	8	12	...	10	25	5	...	30	33	17	...	290	286	77	2nd, Smythe Div.	Lost Final
1980-81	80	17	12	11	...	11	20	9	...	28	32	20	...	289	301	76	3rd, Smythe Div.	Lost Prelim. Round
1979-80	80	14	17	9	...	13	20	7	...	27	37	16	...	256	281	70	3rd, Smythe Div.	Lost Prelim. Round
1978-79	80	15	18	7	...	10	24	6	...	25	42	13	...	217	291	63	2nd, Smythe Div.	Lost Prelim. Round
1977-78	80	13	15	12	...	7	28	5	...	20	43	17	...	239	320	57	3rd, Smythe Div.	Out of Playoffs
1976-77	80	13	21	6	...	12	21	7	...	25	42	13	...	235	294	63	4th, Smythe Div.	Out of Playoffs
1975-76	80	22	11	7	...	11	21	8	...	33	32	15	...	271	272	81	2nd, Smythe Div.	Lost Prelim. Round
1974-75	80	23	12	5	...	15	20	5	...	38	32	10	...	271	254	86	1st, Smythe Div.	Lost Quarter-Final
1973-74	78	14	18	7	...	10	25	4	...	24	43	11	...	224	296	59	7th, East Div.	Out of Playoffs
1972-73	78	17	18	4	...	5	29	5	...	22	47	9	...	233	339	53	7th, East Div.	Out of Playoffs
1971-72	78	14	20	5	...	6	30	3	...	20	50	8	...	203	297	48	7th, East Div.	Out of Playoffs
1970-71	78	17	18	4	...	7	28	4	...	24	46	8	...	229	296	56	6th, East Div.	Out of Playoffs

In his first full season with the Canucks in 2001-02, Dan Cloutier shattered all his previous career bests with 62 games played, 31 wins and a 2.43 goals-against average. His seven shutouts tied him for second in the NHL behind Patrick Roy's nine.

Franchise date: May 22, 1970

NORTHWEST DIVISION

33rd NHL Season

2002-03 Player Personnel

FORWARDS	HT	WT	S	Place of Birth	Date	2001-02 Club
BERTUZZI, Todd	6-3	235	L	Sudbury, Ont.	2/2/75	Vancouver
BOUCK, Tyler	6-0	196	L	Camrose, Alta.	1/13/80	Phoenix-Springfield-Manitoba
BROWN, Mike	6-5	185	L	Surrey, B.C.	4/27/79	Vancouver-Manitoba
CHUBAROV, Artem	6-1	189	L	Gorky, USSR	12/12/79	Vancouver-Manitoba
COOKE, Matt	5-11	205	L	Belleville, Ont.	9/7/78	Vancouver
DAVIDSSON, Johan	6-1	190	R	Jonkoping, Sweden	1/6/76	HV 71 Jonkoping
DRUKEN, Harold	6-0	205	L	St. John's, Nfld.	1/26/79	Vancouver-Manitoba
FARKAS, Jeff	6-0	185	L	Amherst, MA	1/24/78	Toronto-St. John's
HLAVAC, Jan	6-0	185	L	Prague, Czech.	9/20/76	Philadelphia-Vancouver
KARIYA, Steve	5-8	170	R	North Vancouver, B.C.	12/22/77	Vancouver-Manitoba
KLATT, Trent	6-1	210	R	Robbinsdale, MN	1/30/71	Vancouver
LEEB, Brad	5-11	180	R	Red Deer, Alta.	8/27/79	Vancouver-Manitoba
LETOWSKI, Trevor	5-10	176	R	Thunder Bay, Ont.	4/5/77	Phoenix-Vancouver
LINDEN, Trevor	6-4	215	R	Medicine Hat, Alta.	4/11/70	Washington-Vancouver
MORRISON, Brendan	5-11	190	L	Pitt Meadows, B.C.	8/15/75	Vancouver
MORRISON, Justin	6-3	205	R	Los Angeles, CA	9/10/79	Manitoba
NASLUND, Markus	5-11	195	L	Ornskoldsvik, Sweden	7/30/73	Vancouver-Sweden
REID, Brandon	5-8	165	R	Kirkland, Que.	3/9/81	Manitoba
RUUTU, Jarkko	6-2	194	L	Vantaa, Finland	8/23/75	Vancouver-Finland
SAVAGE, Andre	6-0	195	R	Ottawa, Ont.	5/27/75	Manitoba
SCHAEFER, Peter	5-11	195	L	Yellow Grass, Sask.	7/12/77	TPS Turku
SEDIN, Daniel	6-1	200	L	Ornskoldsvik, Sweden	9/26/80	Vancouver
SEDIN, Henrik	6-2	200	L	Ornskoldsvik, Sweden	9/26/80	Vancouver
SHARIFIJANOV, Vadim	6-0	205	L	Ufa, USSR	12/23/75	Togliatti-Cherepovets
SMITH, Tim	5-9	160	L	Whitecourt, Alta.	7/21/81	Swift Current
VASILJEVS, Herbert	5-11	180	R	Riga, Latvia	5/27/76	Vancouver-Manitoba
WARRINER, Todd	6-1	200	L	Blenheim, Ont.	1/3/74	Phoenix-Springfield-Vancouver-Manitoba

DEFENSEMEN	HT	WT	S	Place of Birth	Date	2001-02 Club
ALLEN, Bryan	6-4	215	L	Kingston, Ont.	8/21/80	Vancouver-Manitoba
BARON, Murray	6-3	215	L	Prince George, B.C.	6/1/67	Vancouver
BAUMGARTNER, Nolan	6-2	205	R	Calgary, Alta.	3/23/76	Norfolk
HELMER, Bryan	6-1	200	R	Sault Ste. Marie, Ont.	7/15/72	Vancouver-Manitoba
JOVANOVSKI, Ed	6-2	210	L	Windsor, Ont.	6/26/76	Vancouver-Canada
KOMARNISKI, Zenith	6-0	200	L	Edmonton, Alta.	8/13/78	Manitoba
KURTZ, Justin	6-0	188	L	Winnipeg, Manitoba	1/14/77	Vancouver-Manitoba
OBSUT, Jaroslav	6-1	200	L	Presov, Czech.	9/3/76	Colorado-Hershey-Slovakia
OHLUND, Mattias	6-2	220	L	Pitea, Sweden	9/9/76	Vancouver-Sweden
SOPEL, Brent	6-1	205	R	Calgary, Alta.	1/7/77	Vancouver
TEZIKOV, Alexei	6-1	208	L	Togliatti, USSR	6/22/78	Vancouver-Manitoba

GOALTENDERS	HT	WT	C	Place of Birth	Date	2001-02 Club
BROCHU, Martin	6-0	199	L	Anjou, Que.	3/10/73	Vancouver-Manitoba
CLOUTIER, Dan	6-1	182	L	Mont-Laurier, Que.	4/22/76	Vancouver
MICHAUD, Alfie	5-10	177	L	Selkirk, Man.	11/6/76	Manitoba-Reading
MOSS, Tyler	6-0	185	R	Ottawa, Ont.	6/29/75	Lowell
SKUDRA, Peter	6-1	189	L	Riga, Latvia	4/24/73	Vancouver-Hartford

Coaching History

Hal Laycoe, 1970-71, 1971-72; Vic Stasiuk, 1972-73; Bill McCreary and Phil Maloney, 1973-74; Phil Maloney, 1974-75, 1975-76; Phil Maloney and Orland Kurtenbach, 1976-77; Orland Kurtenbach, 1977-78; Harry Neale, 1978-79 to 1980-81; Harry Neale and Roger Neilson, 1981-82; Roger Neilson, 1982-83; Roger Neilson and Harry Neale, 1983-84; Bill Laforge and Harry Neale, 1984-85; Tom Watt, 1985-86, 1986-87; Bob McCammon, 1987-88 to 1989-90; Bob McCammon and Pat Quinn, 1990-91; Pat Quinn, 1991-92 to 1993-94; Rick Ley, 1994-95; Rick Ley and Pat Quinn, 1995-96; Tom Renney, 1996-97; Tom Renney and Mike Keenan, 1997-98; Mike Keenan and Marc Crawford, 1998-99; Marc Crawford, 1999-2000 to date.

Coach

CRAWFORD, MARC
Coach, Vancouver Canucks. Born in Belleville, Ont., February 13, 1961.

Marc Crawford became the 15th head coach in Canucks history on January 24, 1999. Crawford began his NHL coaching career with the Quebec Nordiques in 1994 and won a Stanley Cup in 1996 when the team moved to Denver to become the Colorado Avalanche. With the win, Crawford became the third-youngest coach in NHL history to win a Stanley Cup. Crawford coached the Avalanche for two seasons after winning the Cup before leaving following the 1997-98 season. He began the 1998-99 season as a colour commentator for CBC's Hockey Night in Canada before joining the Canucks. He led the team to 83 points in his first full season behind the bench in 1999-2000, then guided the Canucks back into the playoffs in 2000-01 and 2001-02.

Crawford was the head coach for Team Canada at the 1998 Olympic Winter Games in Nagano, Japan and he was an assistant coach with Canada's silver medal-winning team in the 1996 World Cup of Hockey. He began his coaching career when he was hired by Brian Burke as a playing assistant with Fredericton (AHL) for the 1987-88 season. At the end of the year he moved to Milwaukee where he served as an assistant coach for the Canucks' IHL minor league affiliate for the 1988-89 campaign. He then moved to Cornwall where he served as the Royals' general manager and head coach in 1989-90.

After two seasons with Cornwall, Crawford went on to coach the St. John's Maple Leafs of the AHL before joining the Nordiques in 1994. He received the 1995 Jack Adams Award as the NHL coach of the year, becoming the first rookie coach to win the award since it was inaugurated in 1974.

Crawford played every game of his six-year NHL career with the Vancouver Canucks, recording 19 goals and 31 assists in 176 games. He was a rookie on the Canucks team that made a run to the Stanley Cup finals to face the New York Islanders in 1982.

2001-02 Scoring

* - rookie

Regular Season

Pos	#	Player	Team	GP	G	A	Pts	+/–	PIM	PP	SH	GW	GT	S	%
L	19	Markus Naslund	VAN	81	40	50	90	22	50	8	0	6	1	302	13.2
L	44	Todd Bertuzzi	VAN	72	36	49	85	21	110	14	0	3	0	203	17.7
C	7	Brendan Morrison	VAN	82	23	44	67	18	26	6	0	4	0	183	12.6
C	25	Andrew Cassels	VAN	53	11	39	50	5	22	7	0	1	0	64	17.2
D	55	Ed Jovanovski	VAN	82	17	31	48	–7	101	7	1	3	1	202	8.4
C	16	Trevor Linden	WSH	16	1	2	3	–2	6	1	0	0	0	19	5.3
			VAN	64	12	22	34	–3	65	2	0	2	0	122	9.8
			TOTAL	80	13	24	37	–5	71	3	0	2	0	141	9.2
C	33	Henrik Sedin	VAN	82	16	20	36	9	36	3	0	1	1	78	20.5
D	2	Mattias Ohlund	VAN	81	10	26	36	16	56	4	1	3	0	193	5.2
C	24	Matt Cooke	VAN	82	13	20	33	4	111	1	0	2	0	103	12.6
L	22	Daniel Sedin	VAN	79	9	23	32	1	32	4	0	2	0	117	7.7
L	17	Jan Hlavac	PHI	31	7	3	10	5	8	0	0	1	1	62	11.3
			VAN	46	9	12	21	4	10	1	0	2	0	70	12.9
			TOTAL	77	16	15	31	9	18	1	0	3	1	132	12.1
C	10	Trevor Letowski	PHX	33	2	6	8	2	4	0	0	0	0	43	4.7
			VAN	42	7	10	17	2	15	1	0	0	0	65	10.8
			TOTAL	75	9	16	25	4	19	1	0	0	0	108	8.3
D	3	Brent Sopel	VAN	66	8	17	25	21	44	1	0	3	0	116	6.9
R	26	Trent Klatt	VAN	34	8	7	15	9	10	2	1	3	0	67	11.9
D	14	Scott Lachance	VAN	81	1	10	11	15	50	0	0	0	0	48	2.1
D	28	Bryan Helmer	VAN	40	5	5	10	10	53	2	0	1	1	43	11.6
C	13	Artem Chubarov	VAN	51	5	5	10	–3	10	0	0	3	0	73	6.8
L	15	Todd Warriner	PHX	18	0	3	3	–3	8	0	0	0	0	10	0.0
			VAN	14	2	4	6	4	12	0	0	0	0	18	11.1
			TOTAL	32	2	7	9	1	20	0	0	0	0	28	7.1
L	37	Jarkko Ruutu	VAN	49	2	7	9	–1	74	0	0	0	0	37	5.4
C	9	Harold Druken	VAN	27	4	4	8	–1	6	1	0	2	0	33	12.1
D	4	* Justin Kurtz	VAN	27	3	5	8	–4	14	2	0	0	0	30	10.0
D	23	Murray Baron	VAN	61	1	6	7	8	68	0	0	0	0	38	2.6
D	34	Jason Strudwick	VAN	44	2	4	6	4	96	0	0	0	0	13	15.4
R	29	Herbert Vasiljevs	VAN	18	3	2	5	0	2	2	0	0	0	17	17.6
L	18	Steve Kariya	VAN	3	0	1	1	–2	2	0	0	0	0	0	0.0
D	6	* Alexei Tezikov	VAN	2	0	0	0	2	0	0	0	0	0	0	0.0
R	38	* Brad Leeb	VAN	2	0	0	0	1	0	0	0	0	0	1	0.0
D	5	* Bryan Allen	VAN	11	0	0	0	1	6	0	0	0	0	4	0.0
L	27	* Mike Brown	VAN	15	0	0	0	1	72	0	0	0	0	2	0.0

Goaltending

No.	Goaltender	GPI	Mins	Avg	W	L	T	EN	SO	GA	SA	S%	G	A	PIM
35	* Alexander Auld	1	60	2.00	1	0	0	0	0	2	22	.909	0	1	0
1	Peter Skudra	23	1166	2.42	10	8	2	2	1	47	508	.907	0	0	2
39	Dan Cloutier	62	3502	2.43	31	22	5	2	7	142	1440	.901	0	0	20
30	Martin Brochu	6	216	4.17	0	3	0	1	0	15	104	.856	0	0	0
	Totals	**82**	**4978**	**2.54**	**42**	**33**	**7**	**5**	**8**	**211**	**2079**	**.899**			

Playoffs

Pos	#	Player	Team	GP	G	A	Pts	+/–	PIM	PP	SH	GW	GT	S	%
C	24	Matt Cooke	VAN	6	3	2	5	3	0	1	0	0	0	11	27.3
C	16	Trevor Linden	VAN	6	1	4	5	1	0	0	0	0	0	10	10.0
D	55	Ed Jovanovski	VAN	6	1	4	5	1	8	1	0	0	0	10	10.0
L	44	Todd Bertuzzi	VAN	6	2	2	4	–1	14	1	0	0	0	16	12.5
C	33	Henrik Sedin	VAN	6	3	0	3	0	0	0	0	1	1	7	42.9
C	25	Andrew Cassels	VAN	6	2	1	3	–4	0	1	0	0	0	12	16.7
D	14	Scott Lachance	VAN	6	1	1	2	2	4	0	0	1	0	4	25.0
L	19	Markus Naslund	VAN	6	1	1	2	–1	2	0	0	0	0	18	5.6
D	2	Mattias Ohlund	VAN	6	1	1	2	–3	6	0	0	0	0	10	10.0
C	7	Brendan Morrison	VAN	6	0	2	2	–2	6	0	0	0	0	4	0.0
D	3	Brent Sopel	VAN	6	0	2	2	–1	2	0	0	0	0	9	0.0
L	15	Todd Warriner	VAN	6	1	0	1	0	2	0	0	0	0	5	20.0
L	17	Jan Hlavac	VAN	5	0	1	1	–3	0	0	0	0	0	5	0.0
D	23	Murray Baron	VAN	6	0	1	1	–3	10	0	0	0	0	2	0.0
C	10	Trevor Letowski	VAN	6	0	1	1	–2	8	0	0	0	0	5	0.0
C	13	Artem Chubarov	VAN	6	0	1	1	0	0	0	0	0	0	8	0.0
L	22	Daniel Sedin	VAN	6	0	1	1	0	0	0	0	0	0	6	0.0
L	37	Jarkko Ruutu	VAN	1	0	0	0	–1	0	0	0	0	0	0	0.0
D	28	Bryan Helmer	VAN	6	0	0	0	–1	0	0	0	0	0	5	0.0

Goaltending

No.	Goaltender	GPI	Mins	Avg	W	L	EN	SO	GA	SA	S%	G	A	PIM
1	Peter Skudra	2	96	3.13	0	1	0	0	5	46	.891	0	0	0
39	Dan Cloutier	6	273	3.52	2	3	1	0	16	123	.870	0	0	0
	Totals	**6**	**374**	**3.53**	**2**	**4**	**1**	**0**	**22**	**170**	**.871**			

Coaching Record

		Regular Season				Playoffs		
Season	Team	Games	W	L	T	Games	W	L
1989-90	Cornwall (OHL)	66	24	38	4	6	2	4
1990-91	Cornwall (OHL)	66	23	42	1	....	....	....
1991-92	St. John's (AHL)	80	39	29	12	16	11	5
1992-93	St. John's (AHL)	80	41	26	13	9	4	5
1993-94	St. John's (AHL)	80	45	23	12	11	6	5
1994-95	**Quebec (NHL)**	**48**	**30**	**13**	**5**	**6**	**2**	**4**
1995-96	**Colorado (NHL)**	**82**	**47**	**25**	**10**	**22**	**16**	**6***
1996-97	**Colorado (NHL)**	**82**	**49**	**24**	**9**	**17**	**10**	**7**
1997-98	**Colorado (NHL)**	**82**	**39**	**26**	**17**	**7**	**3**	**4**
1998-99	**Vancouver (NHL)**	**37**	**8**	**23**	**6**	....	....	....
1999-2000	**Vancouver (NHL)**	**82**	**30**	**37**	**15**	....	....	....
2000-01	**Vancouver (NHL)**	**82**	**36**	**35**	**11**	**4**	**0**	**4**
2001-02	**Vancouver (NHL)**	**82**	**42**	**33**	**7**	**6**	**2**	**4**
	NHL Totals	**577**	**281**	**216**	**80**	**62**	**33**	**29**

* Stanley Cup win.

Club Records

Team

(Figures in brackets for season records are games played; records for fewest points, wins, ties, losses, goals, goals against are for 70 or more games)

Record		
Most Points	**101**	1992-93 (84)
Most Wins	**46**	1992-93 (84)
Most Ties	**20**	1980-81 (80)
Most Losses	**50**	1971-72 (78)
Most Goals	**346**	1992-93 (84)
Most Goals Against	**401**	1984-85 (80)
Fewest Points	**48**	1971-72 (78)
Fewest Wins	**20**	1971-72 (78), 1977-78 (80)
Fewest Ties	**3**	1993-94 (84)
Fewest Losses	**26**	1991-92 (80)
Fewest Goals	**192**	1998-99 (82)
Fewest Goals Against	**211**	2001-02 (82)
Longest Winning Streak		
Overall	**7**	Feb. 10-23/89
Home	**9**	Nov. 6-Dec. 9/92
Away	**5**	Four times
Longest Undefeated Streak		
Overall	**10**	Mar. 5-25/77 (5 wins, 5 ties)
Home	**18**	Nov. 4/92-Jan. 16/93 (16 wins, 2 ties)
Away	**5**	Eight times
Longest Losing Streak		
Overall	**10**	Oct. 23-Nov. 11/97
Home	**6**	Dec. 18/70-Jan. 20/71
Away	**12**	Nov. 28/81-Feb. 6/82
Longest Winless Streak		
Overall	**13**	Nov. 9-Dec. 7/73 (10 losses, 3 ties)
Home	**11**	Dec. 18/70-Feb. 6/71 (10 losses, 1 tie)
Away	**20**	Jan. 2-Apr. 2/86 (14 losses, 6 ties)
Most Shutouts, Season	**8**	1974-75 (80), 2001-02 (82)
Most PIM, Season	**2,326**	1992-93 (84)
Most Goals, Game	**11**	Mar. 28/71 (Cal. 5 at Van. 11), Nov. 25/86 (L.A. 5 at Van. 11), Mar. 1/92 (Cgy. 0 at Van. 11)

Individual

Most Seasons	**13**	Stan Smyl
Most Games	**896**	Stan Smyl
Most Goals, Career	**262**	Stan Smyl
Most Assists, Career	**411**	Stan Smyl
Most Points, Career	**673**	Stan Smyl (262G, 411A)
Most PIM, Career	**2,127**	Gino Odjick
Most Shutouts, Career	**20**	Kirk McLean
Longest Consecutive Games Streak	**482**	Trevor Linden (Oct. 4/90-Dec. 7/96)
Most Goals, Season	**60**	Pavel Bure (1992-93, 1993-94)
Most Assists, Season	**62**	André Boudrias (1974-75)
Most Points, Season	**110**	Pavel Bure (1992-93; 60G, 50A)
Most PIM, Season	**372**	Donald Brashear (1997-98)
Most Points, Defenseman, Season	**63**	Doug Lidster (1986-87; 12G, 51A)
Most Points, Center, Season	**91**	Patrik Sundstrom (1983-84; 38G, 53A)
Most Points, Right Wing, Season	**110**	Pavel Bure (1992-93; 60G, 50A)
Most Points, Left Wing, Season	**81**	Darcy Rota (1982-83; 42G, 39A)
Most Points, Rookie, Season	**60**	Ivan Hlinka (1981-82; 23G, 37A), Pavel Bure (1991-92; 34G, 26A)
Most Shutouts, Season	**7**	Dan Cloutier (2001-02)
Most Goals, Game	**4**	Nine times
Most Assists, Game	**6**	Patrik Sundstrom (Feb. 29/84)
Most Points, Game	**7**	Patrik Sundstrom (Feb. 29/84; 1G, 6A)

Retired Numbers

12	Stan Smyl	1978-1991

Captains' History

Orland Kurtenbach, 1970-71 to 1973-74; no captain, 1974-75; Andre Boudrias, 1975-76; Chris Oddleifson, 1976-77; Don Lever, 1977-78; Don Lever and Kevin McCarthy, 1978-79; Kevin McCarthy, 1979-80 to 1981-82; Stan Smyl, 1982-83 to 1989-90; Dan Quinn, Doug Lidster and Trevor Linden, 1990-91; Trevor Linden, 1991-92 to 1996-97; Mark Messier, 1997-98 to 1999-2000; Markus Naslund, 2000-01 to date.

General Managers' History

Bud Poile, 1970-71 to 1972-73; Hal Laycoe, 1973-74; Phil Maloney, 1974-75 to 1976-77; Jake Milford, 1977-78 to 1981-82; Harry Neale, 1982-83 to 1984-85; Jack Gordon, 1985-86, 1986-87; Pat Quinn, 1987-88 to 1997-98; Brian Burke, 1998-99 to date.

All-time Record vs. Other Clubs

Regular Season

	At Home								On Road								Total							
	GP	W	L	T	OL	GF	GA	PTS	GP	W	L	T	OL	GF	GA	PTS	GP	W	L	T	OL	GF	GA	PTS
Anaheim	22	15	6	1	0	80	50	31	21	9	6	6	0	64	51	24	43	24	12	7	0	144	101	55
Atlanta	1	0	0	1	0	1	1	1	3	2	1	0	0	12	11	4	4	2	1	1	0	13	12	5
Boston	50	16	25	8	1	163	203	41	50	6	37	7	0	117	211	19	100	22	62	15	1	280	414	60
Buffalo	50	24	16	10	0	186	160	58	51	17	26	8	0	148	184	42	101	41	42	18	0	334	344	100
Calgary	93	35	40	17	1	322	311	88	93	23	57	13	0	269	385	59	186	58	97	30	1	591	696	147
Carolina	29	13	10	6	0	103	80	32	28	11	12	5	0	97	92	27	57	24	22	11	0	200	172	59
Chicago	68	31	22	15	0	205	202	77	67	15	44	7	1	151	255	38	135	46	66	22	1	356	457	115
Colorado	41	15	19	6	1	142	160	37	42	14	20	7	1	117	143	36	83	29	39	13	2	259	303	73
Columbus	4	4	0	0	0	15	10	8	4	3	0	1	0	15	5	7	8	7	0	1	0	30	15	15
Dallas	67	30	26	10	1	243	205	71	67	21	34	12	0	206	247	54	134	51	60	22	1	449	452	125
Detroit	61	26	25	10	0	220	201	62	62	15	38	8	1	177	265	39	123	41	63	18	1	397	466	101
Edmonton	77	29	36	11	1	278	306	70	75	20	47	7	1	233	342	48	152	49	83	18	2	511	648	118
Florida	7	2	1	4	0	20	15	8	7	3	3	1	0	22	19	7	14	5	4	5	0	42	34	15
Los Angeles	93	48	30	15	0	357	292	111	95	29	50	15	1	297	388	74	188	77	80	30	1	654	680	185
Minnesota	5	3	1	0	1	18	15	7	5	2	2	1	0	15	10	5	10	5	3	1	1	33	25	12
Montreal	53	12	33	8	0	138	196	32	51	9	38	4	0	129	241	22	104	21	71	12	0	267	437	54
Nashville	8	6	2	0	0	31	19	12	9	4	4	1	0	29	27	9	17	10	6	1	0	60	46	21
New Jersey	47	27	9	11	0	175	130	65	47	21	20	6	0	152	149	48	94	48	29	17	0	327	279	113
NY Islanders	46	22	21	3	0	151	150	47	45	11	24	10	0	124	167	32	91	33	45	13	0	275	317	79
NY Rangers	50	14	33	3	0	160	203	31	53	11	37	5	0	136	233	27	103	25	70	8	0	296	436	58
Ottawa	9	4	4	1	0	26	20	9	9	3	5	1	0	21	22	7	18	7	9	2	0	47	42	16
Philadelphia	51	10	28	12	1	144	203	33	52	16	35	1	0	156	227	33	103	26	63	13	1	300	430	66
Phoenix	68	40	18	9	1	257	196	90	65	23	32	9	1	229	249	56	133	63	50	18	2	486	445	146
Pittsburgh	49	22	23	4	0	176	184	48	49	9	33	7	0	168	225	25	98	31	56	11	0	344	409	73
St. Louis	68	27	33	8	0	199	221	62	67	19	39	9	0	191	252	47	135	46	72	17	0	390	473	109
San Jose	28	16	8	4	0	104	74	36	30	14	11	5	0	96	89	33	58	30	19	9	0	200	163	69
Tampa Bay	8	6	0	2	0	35	13	14	8	5	3	0	0	32	28	10	16	11	3	2	0	67	41	24
Toronto	61	29	19	11	2	219	205	71	58	21	27	10	0	191	214	52	119	50	46	21	2	410	419	123
Washington	38	17	15	5	1	130	123	40	39	13	21	4	1	114	131	31	77	30	36	9	2	244	254	71
Defunct Clubs	19	14	3	2	0	82	48	30	19	10	8	1	0	71	68	21	38	24	11	3	0	153	116	51
Totals	**1271**	**557**	**506**	**197**	**11**	**4380**	**4196**	**1322**	**1271**	**379**	**714**	**171**	**7**	**3779**	**4930**	**936**	**2542**	**936**	**1220**	**368**	**18**	**8159**	**9126**	**2258**

Playoffs

	Series	W	L	GP	W	L	T	GF	GA	Last Mtg.	Rnd.	Result
Buffalo	2	0	2	7	1	6	0	14	28	1981	PRE	L 0-3
Calgary	5	2	3	25	12	13	0	80	82	1994	CQF	W 4-3
Chicago	2	1	1	9	4	5	0	24	24	1995	CSF	L 0-4
Colorado	2	0	2	10	2	8	0	26	40	2001	CQF	L 0-4
Dallas	1	1	0	5	4	1	0	18	11	1994	CSF	W 4-1
Detroit	1	0	1	6	2	4	0	16	22	2002	CQF	L 2-4
Edmonton	2	0	2	9	2	7	0	20	35	1992	DF	L 2-4
Los Angeles	3	1	2	17	8	9	0	60	66	1993	DF	L 2-4
Montreal	1	0	1	5	1	4	0	9	20	1975	QF	L 1-4
NY Islanders	2	0	2	6	0	6	0	14	26	1982	F	L 0-4
NY Rangers	1	0	1	7	3	4	0	19	21	1994	F	L 3-4
Philadelphia	1	0	1	3	1	2	0	9	15	1979	PRE	L 1-2
St. Louis	1	1	0	7	4	3	0	27	27	1995	CQF	W 4-3
Toronto	1	1	0	5	4	1	0	16	9	1994	CF	W 4 1
Winnipeg	2	2	0	13	8	5	0	50	34	1993	DSF	W 4-2
Totals	**27**	**9**	**18**	**134**	**56**	**78**	**0**	**402**	**460**			

Playoff Results 2002-1998

Year	Round	Opponent	Result	GF	GA
2002	CQF	Detroit	L 2-4	16	22
2001	CQF	Colorado	L 0-4	9	16

Abbreviations: Round: F – Final; **CF** – conference final; **CSF** – conference semi-final; **CQF** – conference quarter-final; **DF** – division final; **DSF** – division semi-final; **QF** – quarter-final; **PRE** – preliminary round.

Calgary totals include Atlanta Flames, 1972-73 to 1979-80.
Carolina totals include Hartford, 1979-80 to 1996-97.
Colorado totals include Quebec, 1979-80 to 1994-95.
Dallas totals include Minnesota North Stars, 1970-71 to 1992-93.
New Jersey totals include Kansas City, 1974-75 to 1975-76, and Colorado Rockies, 1976-77 to 1981-82.
Phoenix totals include Winnipeg, 1979-80 to 1995-96.

2001-02 Results

Oct.	4	Chicago	4-5		31	Philadelphia	1-2
	6	Detroit	1-4	**Jan.**	3	Montreal	5-2
	9	at Colorado	4-5		5	at Edmonton	4-3
	11	at Dallas	4-1		8	at Buffalo	2-3
	13	Colorado	4-0		9	at Detroit	4-5*
	16	Florida	2-2		12	Carolina	7-1
	18	Toronto	5-6		15	Pittsburgh	5-2
	20	at Phoenix	2-5		17	at St. Louis	4-5
	21	at Anaheim	1-3		19	at Washington	5-1
	23	Nashville	4-2		21	at Carolina	7-5
	25	at Colorado	1-4		23	at Dallas	4-2
	27	at Edmonton	2-3		25	Toronto	6-1
	30	Columbus	3-1		26	at Calgary	2-0
Nov.	1	Montreal	4-0		28	Nashville	5-1
	3	at San Jose	1-5		30	Edmonton	2-2
	6	at Columbus	3-2	**Feb.**	4	Phoenix	4-2
	8	at St. Louis	1-3		8	at Calgary	4-1
	9	at Chicago	1-3		9	Calgary	3-4
	11	at Minnesota	5-0		12	Boston	1-2*
	13	Chicago	3-2		26	St. Louis	4-4
	15	St. Louis	2-1		28	Dallas	3-4*
	17	Edmonton	2-2	**Mar.**	2	Minnesota	6-3
	20	at Ottawa	0-3		7	at Phoenix	1-6
	21	at Pittsburgh	4-1		9	at San Jose	0-2
	23	at Boston	2-3		10	San Jose	4-7
	25	at Philadelphia	4-1		12	at Nashville	5-0
	27	at Minnesota	1-2		14	at Columbus	5-1
	28	at Chicago	3-3		16	at Atlanta	4-2
	30	Colorado	2-5		17	at New Jersey	3-2*
Dec.	2	Dallas	2-4		19	at NY Rangers	3-1
	6	Anaheim	3-2		21	at NY Islanders	2-3
	8	San Jose	3-5		24	Edmonton	0-2
	10	Tampa Bay	1-1		26	Los Angeles	4-0
	12	at Anaheim	1-0*		28	Columbus	4-3*
	13	at Los Angeles	3-6		30	Anaheim	4-1
	15	Detroit	3-0	**Apr.**	2	at Los Angeles	4-4
	19	at Detroit	1-4		5	Minnesota	5-4*
	20	at Nashville	2-6		7	Phoenix	4-3
	22	Minnesota	1-2		9	at Colorado	2-1
	27	Calgary	4-2		11	Los Angeles	5-2
	29	New Jersey	4-2		13	at Calgary	4-1

* – Overtime

Entry Draft Selections 2002-1988

2002

Pick	
49	Kirill Koltsov
55	Denis Grot
68	Brett Skinner
83	Lukas Mensator
114	John Laliberte
151	Rob McVicar
214	Marc-Andre Roy
223	Ilja Krikunov
247	Matt Violin
277	Thomas Nussli
278	Matt Gens

2001

Pick	
16	R.J. Umberger
66	Fedor Fedorov
114	Evgeny Gladskikh
151	Kevin Bieksa
212	Jason King
245	Konstantin Mikhailov

2000

Pick	
23	Nathan Smith
71	Thatcher Bell
93	Tim Branham
144	Pavel Duma
208	Brandon Reid
241	Nathan Barrett
272	Tim Smith

1999

Pick	
2	Daniel Sedin
3	Henrik Sedin
69	Rene Vydareny
129	Ryan Thorpe
172	Josh Reed
189	Kevin Swanson
218	Markus Kankaanpera
271	Darrell Hay

1998

Pick	
4	Bryan Allen
31	Artem Chubarov
68	Jarkko Ruutu
81	Justin Morrison
90	Regan Darby
136	David Ytfeldt
140	Rick Bertran
149	Paul Cabana
177	Vincent Malts
204	Greg Mischler
219	Curtis Valentine
232	Jason Metcalfe

1997

Pick	
10	Brad Ference
34	Ryan Bonni
36	Harold Druken
64	Kyle Freadrich
90	Chris Stanley
114	David Darguzas
117	Matt Cockell
144	Matt Cooke
148	Larry Shapley
171	Rod Leroux
201	Denis Martynyuk
227	Peter Brady

1996

Pick	
12	Josh Holden
75	Zenith Komarniski
93	Jonas Soling
121	Tyler Prosofsky
147	Nolan McDonald
175	Clint Cabana
201	Jeff Scissons
227	Lubomir Vaic

1995

Pick	
40	Chris McAllister
61	Larry Courville
66	Peter Schaefer
92	Lloyd Shaw
120	Todd Norman
144	Brent Sopel
170	Stewart Bodtker
196	Tyler Willis
222	Jason Cugnet

1994

Pick	
13	Mattias Ohlund
39	Robb Gordon
42	Dave Scatchard
65	Chad Allan
92	Mike Dubinsky
117	Yanick Dube
169	Yuri Kuznetsov
195	Rob Trumbley
221	Bill Muckalt
247	Tyson Nash
273	Robert Longpre

1993

Pick	
20	Mike Wilson
46	Rick Girard
98	Dieter Kochan
124	Scott Walker
150	Troy Creurer
176	Yevgeni Babariko
202	Sean Tallaire
254	Bert Robertsson
280	Sergei Tkachenko

1992

Pick	
21	Libor Polasek
40	Michael Peca
45	Mike Fountain
69	Jeff Connolly
93	Brent Tully
110	Brian Loney
117	Adrian Aucoin
141	Jason Clark
165	Scott Hollis
213	Sonny Mignacca
237	Mark Wotton
261	Aaron Boh

1991

Pick	
7	Alek Stojanov
29	Jassen Cullimore
51	Sean Pronger
95	Dan Kesa
117	John Namestnikov
139	Brent Thurston
161	Eric Johnson
183	David Neilson
205	Brad Barton
227	Jason Fitzsimmons
249	Xavier Majic

1990

Pick	
2	Petr Nedved
18	Shawn Antoski
23	Jiri Slegr
65	Darin Bader
86	Gino Odjick
128	Daryl Filipek
149	Paul O'Hagan
170	Mark Cipriano
191	Troy Neumier
212	Tyler Ertel
233	Karri Kivi

1989

Pick	
8	Jason Herter
29	Robert Woodward
71	Brett Hauer
113	Pavel Bure
134	James Revenberg
155	Rob Sangster
176	Sandy Moger
197	Gus Morschauser
218	Hayden O'Rear
239	Darcy Cahill
248	Jan Bergman

1988

Pick	
2	Trevor Linden
33	Leif Rohlin
44	Dane Jackson
107	Corrie D'Alessio
122	Phil Von Stefenelli
128	Dixon Ward
149	Greg Geldart
170	Roger Akerstrom
191	Paul Constantin
212	Chris Wolanin
233	Steffan Nilsson

President and General Manager

BURKE, BRIAN
President and General Manager, Vancouver Canucks.
Born in Providence, RI, June 30, 1955.

The Vancouver Canucks announced the appointment of Brian Burke to the position of president and general manager on June 22, 1998. Burke became the eighth general manager in Canucks history after serving as the National Hockey League's senior vice president and director of hockey operations for five years.

Burke's prior experience with the Canucks began when he was named vice president and director of hockey operations on June 2, 1987. Burke worked with former Canucks president and general manager Pat Quinn for five seasons and assisted in rebuilding Vancouver's team through his contract negotiation skills and his overseeing of the club's scouting systems and its minor league affiliates. Burke helped reshape the Canucks from a 59-point team in 1987-88, to a 96-point team in his final season of 1991-92. It was the first time since the 1974-75 regular season that the Canucks finished first in the Smythe Division.

Brian Burke was appointed general manager of the Hartford Whalers on May 26, 1992. In his only season in Hartford, Burke made a number of player moves, changed the team's uniform and completed a major draft-day trade in 1993. After acquiring the second overall selection from San Jose, Burke selected Chris Pronger who has developed into one of the NHL's premier defencemen.

Burke joined the NHL front office in September of 1993. In five years as NHL senior vice president, Burke was most visible in his role as the league's chief disciplinarian. He spent much of his time overseeing the league's on-ice officials and was responsible for many disciplinary decisions handed down by the NHL based on his interpretation of league rules. Brian worked closely with NHL commissioner Gary Bettman on the direction of the league and was a key member of the group that introduced NHL excitement to Japan when the Vancouver Canucks and Mighty Ducks of Anaheim opened the 1997-98 regular season in Tokyo.

Club Directory

General Motors Place

Vancouver Canucks
General Motors Place
800 Griffiths Way
Vancouver, B.C. V6B 6G1
Phone **604/899-4600**
FAX 604/899-4640
www.canucks.com
Capacity: 18,422

Executive	
Chairman, OBSE; Governor, NHL	John E. McCaw Jr.
President, Chief Executive Officer, OBSE	Stanley B. McCammon
President & General Manager, Vancouver Canucks, Alternate Governor, NHL	Brian P. Burke
Chief Operating Officer, Alternate Governor, NHL	David Cobb
Senior Vice-President, Sales and Marketing	John Rizzardini
Vice President & General Manager, Arena Operations	Harvey Jones
Vice President, Finance	Victor de Bonis
Vice President, Broadcast and New Media	Chris Hebb
Vice President, Business Development	Ric Thomsen
Vice-President, Customer Sales & Service	John Rocha
Vice President, People Development	Susanne Haine
Hockey Operations	
Senior Vice-President, Director Hockey Operations	David M. Nonis
Vice President, Player Personnel	Steve Tambellini
Head Coach	Marc Crawford
Assistant Coaches	Jack McIlhargey, Mike Johnston
General Manager, Manitoba Moose	Randy Carlyle
Head Coach, Manitoba Moose	Stan Smyl
Assistant Coach, Manitoba Moose	Barry Smith
Strength & Conditioning Coach	Peter Twist
Goaltending Consultant	Andy Moog
Assistant Coach, Video	Eric Crawford
Senior Editor, Alumni Liaison	Norm Jewison
Manager, Media Relations	Chris Brumwell
Coordinator, Media Relations	T.C. Carling
Assistant, Media Relations	Rob Viccars
Director, Community Relations	Veronica Varhaug
Coordinator, Community Relations	Allanah Mooney
Assistant, Community Relations	Regan McDonald
Executive Assistant to Brian Burke	Patti Timms
Executive Assistant to David Nonis	Chris Stephens
Scouting	
Professional Scouts	Bob Murray, Shawn Dineen
European Scout	Thomas Gradin
Russian Scout	Sergei Chibisov
Amateur Scouts	Ron Delorme, Ken Slater, Jack McCartan, Barry Dean, Dave Morrison, Daryl Stanley, Jim Eagle, Mike McHugh, Tim Lenardon, Mario Marois
Scouting Information Coordinator	Jonathan Wall
Medical and Training Staff	
Medical Trainer	Mike Burnstein
Assistant Medical Trainers	Jon Sanderson, Marty Dudgeon
Equipment Manager	Pat O'Neill
Assistant Equipment Manager	Darren Granger
Assistant Equipment Trainer	Jamie Hendricks
Game Dressing Room Attendants	Ron Shute, John Jukitch
Team Doctors	Dr. Rui Avelar, Dr. Bill Regan
Team Dentist	Dr. David Lawson
Team Chiropractor	Dr. Sid Sheard
Team Optometrist	Dr. Alan R. Boyco
Corporate Communications	
Manager, Creative Services	Jackie Boucher
Photo Editor/Librarian	Kathy McAdam
Broadcasting	
Director, Facilities and In-house Productions	Paul Brettell
Director, Production Services	Mike Hall
Business Development	
Directors, Business Development	David Altman, Dave Cannon, Tom Mauthe
Director, Executive Suite Sales & Marketing	Chris Bradley
Sr. Manager, Suite and Sponsorship Services	Darren Moscovitch
Manager, Suite and Sponsorship Services	Shannon Soper
Manager, Hospitality Suite Sales & Service	Lara Grescoe
Legal	
Corporate Counsel	James Conrad
Customer Sales & Service	
Executive Assistant	Annabelle Kroes
Director, Customer Sales & Service	Caley Denton
Director, Customer Sales	Jordan Thorsteinson
Senior Managers, Customer Sales	Sharon Butler, Graham Wall
Managers, Customer Sales	Andrew Merai, Josh Bender, Marc Tourigny, Terry Craig, John Bellefeuille
Account Managers	Kevin Yeung , Paul Maaker, Andrew Marchand
Manager, Sales & Marketing	Rick Ramsbottom
Marketing and Game Operations	
Director, Marketing	Paul Dal Monte
Manager, Game Presentation & Events	Karen Brydon
Finance, Administration and People Development	
Director of Finance	Chris Samis
Executive Assistant to Victor de Bonis	Denise Ouang
Manager, People Development	Tracey Arnish
Assistant Controller	Patricia Bigonzi
Authentix, Fan Apparel and Collectibles	
Senior Manager, Retail Operations	Dennis Kim
Merchandise Manager	Karen Saunders-Smith
Arena Store Manager	Alan Cook
Gate 5 Receptionist, OBSE	Lynn Bradley

Washington Capitals

2001-02 Results: 36W-33L-11T-2OTL 85PTS.
Second, Southeast Division

2002-03 Schedule

Month	Day	Date	Opponent	Month	Day	Date	Opponent
Oct.	Fri.	11	Nashville		Fri.	10	at Carolina
	Sat.	12	at NY Islanders		Sat.	11	Florida
	Thu.	17	at Carolina		Mon.	13	NY Islanders
	Sat.	19	at Philadelphia*		Wed.	15	NY Rangers
	Sun.	20	at Dallas		Fri.	17	Toronto
	Wed.	23	at NY Rangers		Sat.	18	at Ottawa
	Fri.	25	at Tampa Bay		Mon.	20	at Boston*
	Sat.	26	at Florida		Wed.	22	Carolina
	Mon.	28	at Pittsburgh		Sat.	25	at Montreal*
	Wed.	30	Boston		Sun.	26	NY Rangers*
Nov.	Fri.	1	Tampa Bay		Tue.	28	St. Louis
	Sat.	2	at Philadelphia		Thu.	30	Pittsburgh
	Tue.	5	at Columbus	Feb.	Tue.	4	at Tampa Bay
	Thu.	7	Florida		Wed.	5	New Jersey
	Sat.	9	Philadelphia		Fri.	7	NY Islanders
	Wed.	13	Dallas		Sun.	9	Montreal
	Fri.	15	at Chicago		Wed.	12	at Atlanta
	Sat.	16	at Minnesota		Fri.	14	at Carolina
	Tue.	19	San Jose		Sat.	15	at Florida
	Thu.	21	Minnesota		Mon.	17	at Tampa Bay
	Sat.	23	Atlanta		Thu.	20	Toronto
	Tue.	26	at Toronto		Sat.	22	Detroit
	Wed.	27	Calgary		Mon.	24	Montreal
	Fri.	29	Ottawa		Wed.	26	Buffalo
Dec.	Sun.	1	at Atlanta*	Mar.	Sat.	1	at New Jersey
	Tue.	3	at Pittsburgh		Sun.	2	Carolina
	Fri.	6	Atlanta		Tue.	4	at Buffalo
	Sat.	7	at Buffalo		Thu.	6	Atlanta
	Wed.	11	at Anaheim		Sat.	8	at Boston*
	Fri.	13	at Phoenix		Mon.	10	Philadelphia
	Sat.	14	at San Jose		Fri.	14	Los Angeles
	Mon.	16	at Colorado		Sun.	16	Colorado
	Thu.	19	Boston		Thu.	20	at Calgary
	Sat.	21	at NY Islanders		Sat.	22	at Edmonton
	Mon.	23	Tampa Bay		Sun.	23	at Vancouver
	Fri.	27	New Jersey		Tue.	25	at Montreal
	Sat.	28	at New Jersey		Fri.	28	at Ottawa
	Mon.	30	Buffalo		Sat.	29	at Toronto
Jan.	Wed.	1	Phoenix*	Apr.	Tue.	1	Florida
	Fri.	3	Columbus		Thu.	3	Ottawa
	Sat.	4	at NY Rangers		Sat.	5	Pittsburgh

* Denotes afternoon game.

Year-by-Year Record

		Home				Road				Overall								
Season	GP	W	L	T	OL	W	L	T	OL	W	L	T	OL	GF	GA	Pts.	Finished	Playoff Result
2001-02	82	21	12	6	2	15	21	5	0	36	33	11	2	228	240	85	2nd, Southeast Div.	Out of Playoffs
2000-01	82	24	9	6	2	17	18	4	2	41	27	10	4	233	211	96	1st, Southeast Div.	Lost Conf. Quarter-Final
1999-2000	82	26	5	8	2	18	19	4	0	44	24	12	2	227	194	102	1st, Southeast Div.	Lost Conf. Quarter-Final
1998-99	82	16	23	2	...	15	22	4	...	31	45	6	...	200	218	68	3rd, Southeast Div.	Out of Playoffs
1997-98	82	23	12	6	...	17	18	6	...	40	30	12	...	219	202	92	3rd, Atlantic Div.	Lost Final
1996-97	82	19	17	5	...	14	23	4	...	33	40	9	...	214	231	75	5th, Atlantic Div.	Out of Playoffs
1995-96	82	21	15	5	...	18	17	6	...	39	32	11	...	234	204	89	4th, Atlantic Div.	Lost Conf. Quarter-Final
1994-95	48	15	6	3	...	7	12	5	...	22	18	8	...	136	120	52	3rd, Atlantic Div.	Lost Conf. Quarter-Final
1993-94	84	17	16	9	...	22	19	1	...	39	35	10	...	277	263	88	3rd, Atlantic Div.	Lost Conf. Semi-Final
1992-93	84	21	15	6	...	22	19	1	...	43	34	7	...	325	286	93	2nd, Patrick Div.	Lost Div. Semi-Final
1991-92	80	25	12	3	...	20	15	5	...	45	27	8	...	330	275	98	2nd, Patrick Div.	Lost Div. Semi-Final
1990-91	80	21	14	5	...	16	22	2	...	37	36	7	...	258	258	81	3rd, Patrick Div.	Lost Div. Final
1989-90	80	19	18	3	...	17	20	3	...	36	38	6	...	284	275	78	3rd, Patrick Div.	Lost Conf. Championship
1988-89	80	25	12	3	...	16	17	7	...	41	29	10	...	305	259	92	1st, Patrick Div.	Lost Div. Semi-Final
1987-88	80	22	14	4	...	16	19	5	...	38	33	9	...	281	249	85	2nd, Patrick Div.	Lost Div. Final
1986-87	80	22	15	3	...	16	17	7	...	38	32	10	...	285	278	86	2nd, Patrick Div.	Lost Div. Semi-Final
1985-86	80	30	8	2	...	20	15	5	...	50	23	7	...	315	272	107	2nd, Patrick Div.	Lost Div. Final
1984-85	80	27	11	2	...	19	14	7	...	46	25	9	...	322	240	101	2nd, Patrick Div.	Lost Div. Semi-Final
1983-84	80	26	11	3	...	22	16	2	...	48	27	5	...	308	226	101	2nd, Patrick Div.	Lost Div. Final
1982-83	80	22	12	6	...	17	13	10	...	39	25	16	...	306	283	94	3rd, Patrick Div.	Lost Div. Semi-Final
1981-82	80	16	16	8	...	10	25	5	...	26	41	13	...	319	338	65	5th, Patrick Div.	Out of Playoffs
1980-81	80	16	17	7	...	10	19	11	...	26	36	18	...	286	317	70	5th, Patrick Div.	Out of Playoffs
1979-80	80	20	14	6	...	7	26	7	...	27	40	13	...	261	293	67	5th, Patrick Div.	Out of Playoffs
1978-79	80	15	19	6	...	9	22	9	...	24	41	15	...	273	338	63	4th, Norris Div.	Out of Playoffs
1977-78	80	10	23	7	...	7	26	7	...	17	49	14	...	195	321	48	5th, Norris Div.	Out of Playoffs
1976-77	80	17	15	8	...	7	27	6	...	24	42	14	...	221	307	62	4th, Norris Div.	Out of Playoffs
1975-76	80	6	26	8	...	5	33	2	...	11	59	10	...	224	394	32	5th, Norris Div.	Out of Playoffs
1974-75	80	7	28	5	...	1	39	0	...	8	67	5	...	181	446	21	5th, Norris Div.	Out of Playoffs

Limited to just 69 games due to injuries and struggling at times to acclimatize himself to a new situation, Jaromir Jagr still managed to rank fifth in the NHL in scoring with 79 points (31 goals, 48 assists).

Franchise date: June 11, 1974

SOUTHEAST DIVISION

29th NHL Season

2002-03 Player Personnel

FORWARDS	HT	WT	S	Place of Birth	Date	2001-02 Club
BONDRA, Peter	6-0	200	L	Luck, USSR	2/7/68	Washington
CIERNIK, Ivan	6-1	234	L	Levice, Czech.	10/30/77	Ott-Grand Rapids-Wsh-Port (AHL)
CORRINET, Chris	6-3	220	R	Derby, CT	10/29/78	Washington-Port (AHL)
FARRELL, Michael	6-0	222	R	Edina, MN	10/20/78	Washington-Portland Pirates
FERRARO, Peter	5-10	180	R	Port Jefferson, NY	1/24/73	Washington-Port (AHL)
FORBES, Colin	6-3	205	L	New Westminster, B.C.	2/16/76	Utah-Washington-Portland (AHL)
HALPERN, Jeff	6-0	201	R	Potomac, MD	5/3/76	Washington
JAGR, Jaromir	6-2	234	L	Kladno, Czech.	2/15/72	Washington-Czech Republic
KONOWALCHUK, Steve	6-1	207	L	Salt Lake City, UT	11/11/72	Washington
LANG, Robert	6-2	216	R	Teplice, Czech.	12/19/70	Pittsburgh-Czech Republic
METROPOLIT, Glen	5-10	200	R	Toronto, Ont.	6/25/74	T.B.-Wsh-Port (AHL)
MILLER, Kip	5-10	190	L	Lansing, MI	6/11/69	Grand Rapids-NY Islanders
MINK, Graham	6-3	210	R	Stowe, VT	5/12/79	Richmond-Port (AHL)
MURPHY, Mark	5-11	200	L	Stoughton, MA	8/6/76	Portland (AHL)
NIKOLISHIN, Andrei	6-0	213	L	Vorkuta, USSR	3/25/73	Washington-Russia
PETTINGER, Matt	6-1	205	L	Edmonton, Alta	10/22/80	Washington-Port (AHL)
SIMON, Chris	6-4	235	L	Wawa, Ont.	1/30/72	Washington
SUTHERBY, Brian	6-2	180	L	Edmonton, Alta.	3/1/82	Washington-Moose Jaw
TVRDON, Roman	6-1	189	L	Trencin, Czech.	1/29/81	Portland (AHL)
WHITFIELD, Trent	5-11	204	L	Estevan, Sask.	6/17/77	Wsh-Port (AHL)-NYR-Port (AHL)
ZUBRUS, Dainius	6-4	231	L	Elektrenai, USSR	6/16/78	Washington
DEFENSEMEN						
COTE, Sylvain	5-11	201	R	Quebec City, Que.	1/19/66	Washington
CUTTA, Jakub	6-3	217	L	Jablonec nad Nisou, Czech.	12/29/81	Washington-Port (AHL)
FORTIN, Jean-Francois	6-2	205	R	Laval, Que.	3/15/79	Washington-Port (AHL)
GONCHAR, Sergei	6-2	208	L	Chelyabinsk, USSR	4/13/74	Washington-Russia
HAJT, Chris	6-3	206	L	Saskatoon, Sask.	7/5/78	Hamilton
JOHANSSON, Calle	5-11	203	L	Goteborg, Sweden	2/14/67	Washington
KLEE, Ken	6-0	210	R	Indianapolis, IN	4/24/71	Washington
MELANSON, Dean	5-11	190	R	Antigonish, N.S.	11/19/73	Washington-Port (AHL)
PEAT, Stephen	6-3	210	R	Princeton, B.C.	3/10/80	Washington-Port (AHL)
ROHLOFF, Todd	6-3	213	L	Grand Rapids, IL	1/16/74	Washington-Port (AHL)
WITT, Brendan	6-2	229	L	Humboldt, Sask.	2/20/75	Washington
YONKMAN, Nolan	6-6	236	R	Punnicht, Sask.	4/1/81	Washington-Port (AHL)
ZINGER, Dwayne	6-4	225	L	Coronation, Alta.	7/5/76	Cincinnati

GOALTENDERS	HT	WT	C	Place of Birth	Date	2001-02 Club
BILLINGTON, Craig	5-10	170	L	London, Ont.	9/11/66	Washington
CHARPENTIER, Sebastien	5-9	177	L	Drummondville, Que.	4/18/77	Washington-Port (AHL)
KOLZIG, Olaf	6-3	225	L	Johannesburg, South Africa	4/9/70	Washington
OUELLET, Maxime	6-2	195	L	Beauport, Que.	6/17/81	Phi (AHL)-Port (AHL)

General Manager

McPHEE, GEORGE
General Manager, Washington Capitals. Born in Guelph, Ont., July 2, 1958.

On June 9, 1997, George McPhee became the fifth general manager of the Washington Capitals. In his first year on the job, McPhee led the Caps to the Stanley Cup Finals for the first time in franchise history. He provides the Capitals with the leadership and knowledge to bring the Stanley Cup Finals back to Washington in the years to come.

A back injury forced McPhee to retire as an active player at the conclusion of the 1988-89 season, after a seven year playing career with the New York Rangers and New Jersey Devils. McPhee originally signed as a free agent with the Rangers in July, 1982, after graduating from Bowling Green State University with a business degree. McPhee did not waste any time in college, tallying 40 goals and 48 assists in his freshman season and easily winning CCHA rookie of the year honors. His outstanding collegiate hockey career was capped off when he was named the recipient of the Hobey Baker Award as the top U.S. collegiate player in his senior season. McPhee also earned All-America honors as a senior and finished his career at Bowling Green as the CCHA's all-time leading scorer with 114-153-267. He was the first player in CCHA history to make the Conference's all-academic team three straight seasons.

2001-02 Scoring

* - rookie

Regular Season

Pos	#	Player	Team	GP	G	A	Pts	+/–	PIM	PP	SH	GW	GT	S	%
R	68	Jaromir Jagr	WSH	69	31	48	79	0	30	10	0	3	0	197	15.7
R	12	Peter Bondra	WSH	77	39	31	70	–2	80	17	1	8	1	333	11.7
D	55	Sergei Gonchar	WSH	76	26	33	59	–1	58	7	0	2	0	216	12.0
R	10	Ulf Dahlen	WSH	69	23	29	52	–5	8	7	0	4	1	141	16.3
R	9	Dainius Zubrus	WSH	71	17	26	43	5	38	4	0	3	0	138	12.3
C	13	Andrei Nikolishin	WSH	80	13	23	36	–1	40	1	0	0	0	143	9.1
L	17	Chris Simon	WSH	82	14	17	31	–8	137	1	0	1	1	121	11.6
R	8	Dmitri Khristich	WSH	61	9	12	21	2	12	3	0	2	0	54	16.7
C	11	Jeff Halpern	WSH	48	5	14	19	–9	29	0	0	4	0	74	6.8
C	20	Glen Metropolit	T.B.	2	0	0	0	–2	0	0	0	0	0	1	0.0
			WSH	33	1	16	17	3	6	0	0	0	0	51	2.0
			TOTAL	35	1	16	17	1	6	0	0	0	0	52	1.9
D	2	Ken Klee	WSH	68	8	8	16	4	38	2	0	3	0	85	9.4
L	33	Benoit Hogue	DAL	32	3	3	6	–4	24	0	0	0	0	20	15.0
			BOS	17	4	4	8	–3	9	0	0	1	0	17	23.5
			WSH	9	0	1	1	2	4	0	0	0	0	5	0.0
			TOTAL	58	7	8	15	–5	37	0	0	1	0	42	16.7
D	3	Sylvain Cote	WSH	70	3	11	14	–15	26	1	0	2	0	101	3.0
L	22	Steve Konowalchuk	WSH	28	2	12	14	–2	23	0	0	0	0	36	5.6
D	4	Frantisek Kucera	WSH	56	1	13	14	7	12	0	0	0	0	67	1.5
L	18 *	Matt Pettinger	WSH	61	7	3	10	–8	44	1	0	1	0	73	9.6
D	19	Brendan Witt	WSH	68	3	7	10	–1	78	0	0	0	0	81	3.7
C	36	Colin Forbes	WSH	38	5	3	8	–2	15	0	1	1	0	49	10.2
R	14	Joe Sacco	WSH	65	0	7	7	–13	51	0	0	0	0	57	0.0
D	24	Rob Zettler	WSH	49	1	4	5	3	56	0	0	0	0	28	3.6
R	51 *	Stephen Peat	WSH	38	2	2	4	–1	85	0	0	0	0	11	18.2
L	23 *	Ivan Ciernik	OTT	23	1	2	3	0	4	0	0	0	0	18	5.6
			WSH	6	0	1	1	0	2	0	0	0	0	5	0.0
			TOTAL	29	1	3	4	0	6	0	0	0	0	23	4.3
D	58 *	Jean-Francois Fortin	WSH	36	1	3	4	–1	20	0	0	0	0	24	4.2
D	6	Calle Johansson	WSH	11	2	0	2	–4	8	0	0	1	0	18	11.1
D	40 *	Nolan Yonkman	WSH	11	1	0	1	3	4	0	0	0	0	7	14.3
R	27	Chris Ferraro	WSH	1	0	1	1	0	0	0	0	0	0	4	0.0
R	28	Peter Ferraro	WSH	4	0	1	1	–1	0	0	0	0	0	3	0.0
R	48 *	Chris Corrinet	WSH	8	0	1	1	–4	6	0	0	0	0	8	0.0
D	38	Todd Rohloff	WSH	16	0	1	1	–2	14	0	0	0	0	6	0.0
D	39	Patrick Boileau	WSH	2	0	0	0	–1	2	0	0	0	0	0	0.0
D	34 *	Jakub Cutta	WSH	2	0	0	0	–3	0	0	0	0	0	2	0.0
D	57	Dean Melanson	WSH	4	0	0	0	1	4	0	0	0	0	6	0.0
C	41 *	Brian Sutherby	WSH	7	0	0	0	–3	2	0	0	0	0	3	0.0
D	46 *	Michael Farrell	WSH	8	0	0	0	–1	0	0	0	0	0	1	0.0

Goaltending

No.	Goaltender	GPI	Mins	Avg	W	L	T	EN	SO	GA	SA	S%	G	A	PIM
35	* Sebastien Charpentier	2	122	2.46	1	1	0	0	0	5	78	.936	0	0	0
37	Olaf Kolzig	71	4131	2.79	31	29	8	5	6	192	1977	.903	0	1	8
1	Craig Billington	17	710	3.04	4	5	3	2	0	36	295	.878	0	0	0
	Totals	**82**	**4987**	**2.89**	**36**	**35**	**11**	**7**	**6**	**240**	**2357**	**.898**			

With 71 games played for the Capitals last season, only New Jersey's Martin Brodeur (73) and Buffalo's Martin Biron (72) spent more games in goal than Olaf Kolzig.

General Managers' History

Milt Schmidt, 1974-75; Milt Schmidt and Max McNab, 1975-76; Max McNab, 1976-77 to 1980-81; Max McNab and Roger Crozier, 1981-82; David Poile, 1982-83 to 1996-97; George McPhee, 1997-98 to date.

Club Records

Team

(Figures in brackets for season records are games played; records for fewest points, wins, ties, losses, goals, goals against are for 70 or more games)

Most Points .... 107 — 1985-86 (80)
Most Wins .... 50 — 1985-86 (80)
Most Ties .... 18 — 1980-81 (80)
Most Losses .... 67 — 1974-75 (80)
Most Goals .... 330 — 1991-92 (80)
Most Goals Against .... *446 — 1974-75 (80)
Fewest Points .... *21 — 1974-75 (80)
Fewest Wins .... *8 — 1974-75 (80)
Fewest Ties .... 5 — 1974-75 (80), 1983-84 (80)
Fewest Losses .... 23 — 1985-86 (80)
Fewest Goals .... 181 — 1974-75 (80)
Fewest Goals Against .... 202 — 1997-98 (82)

Longest Winning Streak
Overall .... 10 — Jan. 27-Feb. 18/84
Home .... 10 — Jan. 4-Feb. 23/00
Away .... 6 — Feb. 26-Apr. 1/84

Longest Undefeated Streak
Overall .... 14 — Nov. 24-Dec. 23/82 (9 wins, 5 ties), Jan. 17-Feb. 18/84 (13 wins, 1 tie)
Home .... 13 — Nov. 25/92-Jan. 31/93 (9 wins, 4 ties), Dec. 27/99-Feb. 23/00 (11 wins, 2 ties)
Away .... 10 — Nov. 24/82-Jan. 8/83 (6 wins, 4 ties)

Longest Losing Streak
Overall .... *17 — Feb. 18-Mar. 26/75
Home .... *11 — Feb. 18-Mar. 30/75
Away .... 37 — Oct. 9/74-Mar. 26/75

Longest Winless Streak
Overall .... 25 — Nov. 29/75-Jan. 21/76 (22 losses, 3 ties)
Home .... 14 — Dec. 3/75-Jan. 21/76 (11 losses, 3 ties)
Away .... 37 — Oct. 9/74-Mar. 26/75 (37 losses)

Most Shutouts, Season .... 9 — 1995-96 (82)
Most PIM, Season .... 2,204 — 1989-90 (80)
Most Goals, Game .... 12 — Feb. 6/90 (Que. 2 at Wsh. 12)

Individual

Most Seasons .... 14 — Calle Johansson
Most Games .... 940 — Kelly Miller
Most Goals, Career .... 421 — Peter Bondra
Most Assists, Career .... 418 — Michal Pivonka
Most Points, Career .... 789 — Mike Gartner (397G, 392A)
Most PIM, Career .... 2,003 — Dale Hunter
Most Shutouts, Career .... 27 — Olaf Kolzig
Longest Consecutive Games Streak .... 422 — Bob Carpenter (Oct. 7/81-Nov. 22/86)
Most Goals, Season .... 60 — Dennis Maruk (1981-82)
Most Assists, Season .... 76 — Dennis Maruk (1981-82)
Most Points, Season .... 136 — Dennis Maruk (1981-82; 60G, 76A)
Most PIM, Season .... 339 — Alan May (1989-90)
Most Points, Defenseman, Season .... 81 — Larry Murphy (1986-87; 23G, 58A)
Most Points, Center, Season .... 136 — Dennis Maruk (1981-82; 60G, 76A)
Most Points, Right Wing, Season .... 102 — Mike Gartner (1984-85; 50G, 52A)
Most Points, Left Wing, Season .... 87 — Ryan Walter (1981-82; 38G, 49A)
Most Points, Rookie, Season .... 67 — Bob Carpenter (1981-82; 32G, 35A), Chris Valentine (1981-82; 30G, 37A)
Most Shutouts, Season .... 9 — Jim Carey (1995-96)
Most Goals, Game .... 5 — Bengt Gustafsson (Jan. 8/84), Peter Bondra (Feb. 5/94)
Most Assists, Game .... 6 — Mike Ridley (Jan. 7/89)
Most Points, Game .... 7 — Dino Ciccarelli (Mar. 18/89; 4G, 3A)

* NHL Record.

Retired Numbers

5	Rod Langway	1982-1993
7	Yvon Labre	1974-1981
32	Dale Hunter	1987-1999

Coaching History

Jim Anderson, Red Sullivan and Milt Schmidt, 1974-75; Milt Schmidt and Tom McVie, 1975-76; Tom McVie, 1976-77, 1977-78; Danny Belisle, 1978-79; Danny Belisle and Gary Green, 1979-80; Gary Green, 1980-81; Gary Green, Roger Crozier and Bryan Murray, 1981-82; Bryan Murray, 1982-83 to 1988-89; Bryan Murray and Terry Murray, 1989-90; Terry Murray, 1990-91 to 1992-93; Terry Murray and Jim Schoenfeld, 1993-94; Jim Schoenfeld, 1994-95 to 1996-97; Ron Wilson, 1997-98 to 2001-02; Bruce Cassidy, 2002-03.

Captains' History

Doug Mohns, 1974-75; Bill Clement and Yvon Labre, 1975-76; Yvon Labre, 1976-77, 1977-78; Guy Charron, 1978-79; Ryan Walter, 1979-80 to 1981-82; Rod Langway, 1982-83 to 1991-92; Rod Langway and Kevin Hatcher, 1992-93; Kevin Hatcher, 1993-94; Dale Hunter, 1994-95 to 1998-99; Adam Oates, 1999-2000 to 2001-02.

All-time Record vs. Other Clubs

Regular Season

	At Home								On Road								Total							
	GP	W	L	T	OL	GF	GA	PTS	GP	W	L	T	OL	GF	GA	PTS	GP	W	L	T	OL	GF	GA	PTS
Anaheim	8	4	4	0	0	14	18	8	8	2	5	1	0	22	26	5	16	6	9	1	0	36	44	13
Atlanta	7	6	0	1	0	26	9	13	8	3	3	2	0	20	14	8	15	9	3	3	0	46	23	21
Boston	51	15	24	12	0	141	173	42	52	15	30	7	0	139	195	37	103	30	54	19	0	280	368	79
Buffalo	52	13	29	9	1	132	182	36	52	14	32	6	0	138	199	34	104	27	61	15	1	270	381	70
Calgary	39	20	14	5	0	146	134	45	37	6	24	7	0	89	156	19	76	26	38	12	0	235	290	64
Carolina	42	25	13	4	0	144	111	54	43	20	13	9	1	135	116	50	85	45	26	13	1	279	227	104
Chicago	40	21	14	5	0	142	123	47	38	11	22	5	0	114	149	27	78	32	36	10	0	256	272	74
Colorado	32	17	10	4	1	129	104	39	32	14	14	4	0	113	99	32	64	31	24	8	1	242	203	71
Columbus	1	1	0	0	0	6	3	2	2	1	1	0	0	6	5	2	3	2	1	0	0	12	8	4
Dallas	39	15	16	8	0	119	123	38	39	12	19	8	0	106	143	32	78	27	35	16	0	225	266	70
Detroit	44	21	18	5	0	166	137	47	46	14	20	11	1	131	160	40	90	35	38	16	1	297	297	87
Edmonton	28	17	9	2	0	117	93	36	28	10	14	4	0	88	115	24	56	27	23	6	0	205	208	60
Florida	22	12	6	4	0	65	52	28	23	11	10	2	0	61	61	24	45	23	16	6	0	126	113	52
Los Angeles	44	18	19	7	0	182	165	43	46	14	26	6	0	141	180	34	90	32	45	13	0	323	345	77
Minnesota	2	2	0	0	0	4	2	4	1	0	1	0	0	0	3	0	3	2	1	0	0	4	5	4
Montreal	55	24	22	9	0	153	167	57	56	16	33	7	0	120	216	39	111	40	55	16	0	273	383	96
Nashville	3	2	1	0	0	5	4	4	4	1	2	1	0	9	11	3	7	3	3	1	0	14	15	7
New Jersey	75	47	22	5	1	299	214	100	75	33	35	7	0	223	231	73	150	80	57	12	1	522	445	173
NY Islanders	77	37	29	11	0	250	242	85	77	34	41	2	0	246	292	70	154	71	70	13	0	496	534	155
NY Rangers	80	40	30	9	1	301	263	90	78	33	37	8	0	273	297	74	158	73	67	17	1	574	560	164
Ottawa	20	11	6	3	0	71	52	25	19	9	9	1	0	62	67	19	39	20	15	4	0	133	119	44
Philadelphia	76	30	33	13	0	252	252	73	79	24	49	6	0	214	298	54	155	54	82	19	0	466	550	127
Phoenix	29	18	6	5	0	118	80	41	29	7	15	7	0	101	109	21	58	25	21	12	0	219	189	62
Pittsburgh	82	41	31	9	1	343	303	92	79	28	44	7	0	251	310	63	161	69	75	16	1	594	613	155
St. Louis	38	20	14	4	0	133	112	44	40	13	19	8	0	126	161	34	78	33	33	12	0	259	273	78
San Jose	10	5	5	0	0	30	30	10	9	3	5	1	0	26	27	7	19	8	10	1	0	56	57	17
Tampa Bay	24	15	5	4	0	88	51	34	23	16	5	2	0	77	49	34	47	31	10	6	0	165	100	68
Toronto	46	28	14	3	1	175	129	60	44	16	23	5	0	153	196	37	90	44	37	8	1	328	325	97
Vancouver	39	22	13	4	0	131	114	48	38	16	17	5	0	123	130	37	77	38	30	9	0	254	244	85
Defunct Clubs	10	2	8	0	0	28	42	4	10	4	5	1	0	30	39	9	20	6	13	1	0	58	81	13
Totals	**1115**	**549**	**415**	**145**	**6**	**3910**	**3484**	**1249**	**1115**	**400**	**573**	**140**	**2**	**3337**	**4054**	**942**	**2230**	**949**	**988**	**285**	**8**	**7247**	**7538**	**2191**

Playoffs

	Series	W	L	GP	W	L	T	GF	GA	Last Mtg.	Rnd.	Result
Boston	2	1	1	10	4	6	0	21	28	1998	CQF	W 4-2
Buffalo	1	1	0	6	4	2	0	13	11	1998	CF	W 4-2
Detroit	1	0	1	4	0	4	0	7	13	1998	F	L 0-4
New Jersey	2	1	1	13	7	6	0	44	43	1990	DSF	W 4-2
NY Islanders	6	1	5	30	12	18	0	88	99	1993	DSF	L 2-4
NY Rangers	4	2	2	22	11	11	0	75	71	1994	CSF	L 1-4
Ottawa	1	1	0	5	4	1	0	18	7	1998	CSF	W 4-1
Philadelphia	3	2	1	16	9	7	0	65	55	1989	DSF	L 2-4
Pittsburgh	7	1	6	42	16	26	0	121	137	2001	CQF	L 2-4
Totals	**27**	**10**	**17**	**148**	**67**	**81**	**0**	**452**	**464**			

Playoff Results 2002-1998

Year	Round	Opponent	Result	GF	GA
2001	CQF	Pittsburgh	L 2-4	10	14
2000	CQF	Pittsburgh	L 1-4	8	17
1998	F	Detroit	L 0-4	7	13
	CF	Buffalo	W 4-2	13	11
	CSF	Ottawa	W 4-1	18	7
	CQF	Boston	W 4-2	15	13

Abbreviations: Round: F – Final; **CF** – conference final; **CSF** – conference semi-final; **CQF** – conference quarter-final; **DSF** – division semi-final.

Calgary totals include Atlanta Flames, 1974-75 to 1979-80. Carolina totals include Hartford, 1979-80 to 1996-97.
Colorado totals include Quebec, 1979-80 to 1994-95. Dallas totals include Minnesota North Stars, 1974-75 to 1992-93.
New Jersey totals include Kansas City, 1974-75 to 1975-76, and Colorado Rockies, 1976-77 to 1981-82.
Phoenix totals include Winnipeg, 1979-80, 1995-96.

2001-02 Results

Month	Day	Opponent	Score	Month	Day	Opponent	Score
Oct.	6	New Jersey	6-1		5	at Boston	4-7
	8	at Boston	0-4		7	Florida	1-2
	10	at NY Rangers	5-2		9	Columbus	6-3
	12	at Anaheim	1-2		11	Toronto	3-3
	13	at Phoenix	2-5		12	at Florida	1-0
	16	at Los Angeles	3-2*		14	Boston	1-0*
	19	Montreal	4-1		16	at Montreal	0-2
	20	at Philadelphia	3-6		18	at Detroit	1-3
	23	at Tampa Bay	1-1		19	Vancouver	1-5
	24	at Florida	4-3*		22	at Atlanta	3-0
	26	at Atlanta	0-1		23	Montreal	3-5
	30	Philadelphia	0-3		26	at NY Rangers	3-6
Nov.	2	Phoenix	2-2		27	Buffalo	2-3*
	3	at St. Louis	1-4		30	St. Louis	1-4
	6	at Toronto	2-4	**Feb.**	6	Minnesota	2-1
	8	Carolina	2-3		8	at Nashville	3-3
	10	Atlanta	3-0		9	at Tampa Bay	4-2
	13	Ottawa	5-11		11	Tampa Bay	3-1
	15	at Philadelphia	0-5		26	Florida	4-3
	17	Anaheim	4-1		28	San Jose	2-5
	21	Tampa Bay	3-2	**Mar.**	2	at Ottawa	3-2*
	23	NY Rangers	6-2		4	Toronto	2-3
	24	at Montreal	3-5		6	Calgary	3-2
	27	at NY Islanders	5-5		8	at Carolina	2-2
	28	Buffalo	2-5		10	Edmonton	4-2
	30	Carolina	6-2		12	Dallas	2-5
Dec.	2	at Carolina	4-3*		15	at San Jose	4-5
	4	NY Rangers	5-2		16	at Edmonton	1-4
	6	at Atlanta	3-3		19	at Colorado	3-0
	8	at New Jersey	1-3		21	at Toronto	4-3
	11	Pittsburgh	2-2		23	at Columbus	5-2
	13	Boston	2-1		24	at Pittsburgh	2-6
	15	Atlanta	5-2		26	at Buffalo	4-3
	19	at Florida	2-5		29	at New Jersey	3-1
	21	at Pittsburgh	3-4		30	NY Islanders	4-2
	22	Pittsburgh	4-4	**Apr.**	3	Tampa Bay	4-1
	26	Philadelphia	1-4		5	Ottawa	0-0
	28	at Dallas	3-2*		6	at NY Islanders	4-5
	30	Carolina	5-5		9	Chicago	3-1
Jan.	1	NY Islanders	3-2		12	at Buffalo	3-1
	3	at Ottawa	1-4		13	New Jersey	3-4*

* – Overtime

Entry Draft Selections 2002-1988

2002
Pick
12 Steve Eminger
13 Alexander Semin
17 Boyd Gordon
59 Maxime Daigneault
77 Patrick Wellar
92 Derek Krestanovich
109 Jevon Desautels
118 Petr Dvorak
145 Rob Gherson
179 Marian Havel
209 Joni Lindlof
242 Igor Ignatushkin
272 Patric Blomdahl

2001
Pick
58 Nathan Paetsch
90 Owen Fussey
125 Jeff Lucky
160 Artem Ternavsky
191 Zbynek Novak
221 Johnny Oduya
249 Matt Maglione
254 Peter Polcik
275 Robert Muller
284 Viktor Hubl

2000
Pick
26 Brian Sutherby
43 Matt Pettinger
61 Jakub Cutta
121 Ryan Vanbuskirk
163 Ivan Nepryayev
289 Bjorn Nord

1999
Pick
7 Kris Beech
29 Michal Sivek
31 Charlie Stephens
34 Ross Lupaschuk
37 Nolan Yonkman
132 Roman Tvrdon
175 Kyle Clark
192 David Bornhammar
219 Maxim Orlov
249 Igor Schadilov

1998
Pick
49 Jomar Cruz
59 Todd Hornung
106 Krys Barch
107 Chris Corrinet
118 Mike Siklenka
125 Erik Wendell
179 Nate Forster
193 Ratislav Stana
220 Michael Farrell
251 Blake Evans

1997
Pick
9 Nick Boynton
35 Jean-Francois Fortin
89 Curtis Cruickshank
116 Kevin Caulfield
143 Henrik Petre
200 Pierre-Luc Therrien
226 Matt Oikawa

1996
Pick
4 Alexandre Volchkov
17 Jaroslav Svejkovsky
43 Jan Bulis
58 Sergei Zimakov
74 Dave Weninger
78 Shawn McNeil
85 Justin Davis
126 Matthew Lahey
153 Andrew Van Bruggen
180 Michael Anderson
206 Oleg Orekhovsky
232 Chad Cavanagh

1995
Pick
17 Brad Church
23 Miika Elomo
43 Dwayne Hay
93 Sebastien Charpentier
95 Joel Theriault
105 Benoit Gratton
124 Joel Cort
147 Frederick Jobin
199 Vasili Turkovsky
225 Scott Swanson

1994
Pick
10 Nolan Baumgartner
15 Alexander Kharlamov
41 Scott Cherrey
93 Matt Herr
119 Yanick Jean
145 Dmitri Mekeshkin
171 Daniel Reja
197 Chris Patrick
223 John Tuohy
249 Richard Zednik
275 Sergei Tertyshny

1993
Pick
11 Brendan Witt
17 Jason Allison
69 Patrick Boileau
147 Frank Banham
173 Daniel Hendrickson
174 Andrew Brunette
199 Joel Poirier
225 Jason Gladney
251 Mark Seliger
277 Dany Bousquet

1992
Pick
14 Sergei Gonchar
32 Jim Carey
53 Stefan Ustorf
71 Martin Gendron
119 John Varga
167 Mark Matier
191 Mike Mathers
215 Brian Stagg
239 Gregory Callahan
263 Billy Jo MacPherson

1991
Pick
14 Pat Peake
21 Trevor Halverson
25 Eric Lavigne
36 Jeff Nelson
58 Steve Konowalchuk
80 Justin Morrison
146 Dave Morissette
168 Rick Corriveau
190 Trevor Duhaime
209 Rob Leask
212 Carl Leblanc
234 Rob Puchniak
256 Bill Kovacs

1990
Pick
9 John Slaney
30 Rod Pasma
51 Chris Longo
72 Randy Pearce
93 Brian Sakic
94 Mark Ouimet
114 Andrei Kovalev
135 Roman Kontsek
156 Peter Bondra
159 Steve Martell
177 Ken Klee
198 Michael Boback
219 Alan Brown
240 Todd Hlushko

1989
Pick
19 Olaf Kolzig
35 Byron Dafoe
59 Jim Mathieson
61 Jason Woolley
82 Trent Klatt
145 Dave Lorentz
166 Dean Holoien
187 Victor Gervais
208 Jiri Vykoukal
229 Sidorov Sidorov
250 Ken House

1988
Pick
15 Reggie Savage
36 Tim Taylor
41 Wade Bartley
57 Duane Derksen
78 Bob Krauss
120 Dmitri Khristich
141 Keith Jones
144 Brad Schlegel
162 Todd Hilditch
183 Petr Pavlas
192 Mark Sorensen
204 Claudio Scremin
225 Chris Venkus
246 Ron Pascucci

Coach

CASSIDY, BRUCE
Coach, Washington Capitals. Born in Ottawa, Ont., May 20, 1965.

A former first-round draft pick (18th overall) by the Chicago Blackhawks in 1983, Bruce Cassidy became the 12th head coach of the Washington Capitals on June 25, 2002. The hiring came shortly after Cassidy was named American Hockey League coach of the year for the 2001-02 season. He is the second-youngest coach in the NHL behind only Carolina's Paul Maurice.

Cassidy played just 36 games over parts of six NHL seasons following a severe knee injury suffered while he was still a teenager. His pro career spanned 14 seasons, several leagues and four countries before another knee injury a decade later essentially ended his playing career. He played on championship clubs (including the 1984 Memorial Cup and the 1990 Turner Cup), made all-star teams and developed a reputation as a guy with a good head for the game. His coaching career began early in the 1996-97 season when he decided to give up playing at the age of 31. Cassidy stepped directly from the ice to the back of the bench as the head man of one of the worst teams in hockey, the Jacksonville Lizard Kings of the East Coast Hockey League. He spent six seasons as a minor league coach while also doubling as the director of hockey operations for Indianapolis of the IHL in 1998-99 and Trenton of the ECHL in 1999-2000. He guided Grand Rapids to the best record in the IHL in 2000-01 and had another successful season when the Griffins moved into the AHL for the 2001-02 season.

Known as "Butch" since he was 10 years old, Cassidy cites Darryl Sutter and Mike Keenan among those who have had an impact on his career to date.

Coaching Record

		Regular Season				Playoffs		
Season	Team	Games	W	L	T	Games	W	L
1996-97	Jacksonville (ECHL)	50	15	25	10	....	....	....
1997-98	Jacksonville (ECHL)	70	35	29	6	....	....	....
1998-99	Indianapolis (IHL)	82	33	37	12	7	3	4
1999-00	Trenton (ECHL)	70	37	29	4	14	8	6
2000-01	Grand Rapids (IHL)	82	53	22	7	10	6	4
2001-02	Grand Rapids (AHL)	82	42	27	11	5	2	3

Club Directory

MCI Center

Washington Capitals
401 Ninth Street, NW, Suite 750
Washington, DC 20004
Phone **202/226-2200**
PR FAX 202/266-2360
www.washingtoncaps.com
Capacity: 18,277

Executive Management
Chairman & Majority Owner . . . Ted Leonsis
President & Owner . . . Dick Patrick
Owners . . . Raul Fernandez, Jack Davies, Richard Kay, George Stamas, Richard Fairbank, Jeong Kim
Executive Assistant . . . Michelle Trostle

Hockey Operations
Vice President and General Manager . . . George McPhee
Director of Hockey Operations . . . Shawn Simpson
Assistant to the General Manager . . . Frank Provenzano
Head Coach . . . Bruce Cassidy
Assistant Coaches . . . Randy Carlyle, Glen Hanlon
Goaltending Coach . . . Dave Prior
Strength/Conditioning Coach . . . Frank Costello
Director of Team Services . . . Todd Warren
Scouting Coordinator . . . Kris Wagner
Video Coordinator . . . Ted Dent
Piney Orchard Staff . . . Alex Walker
Security Representative . . . James Wiseman
Executive Assistant . . . Katy Headman

Scouting Staff
Director, Amateur Scouting . . . Ross Mahoney
Pro Scouts . . . Archie Henderson, Brian MacLellan, Mike Backman
Ontario Scout . . . Steve Bowman
Western Scout . . . Dale Derkatch
Quebec Scout . . . Martin Pouliot
European Scouts . . . Gleb Chistyakov, Vojtech Kucera
Eastern U.S. Scout . . . Ed McColgan
Western U.S. Scout . . . Steve Richmond

Medical Staff
Head Athletic Trainer . . . Greg Smith
Assistant Athletic Trainer . . . Tim Clark
Massage Therapist . . . Curt Millar
Team Physician . . . Ben Shaffer, MD
Team Internist . . . Richard Feldman, MD
Team Ophthalmologist . . . Michael Herr, MD
Team Dentist . . . Howard Salob, DDS
Team Nutritionist . . . Tom Fox

Training Staff
Head Equipment Manager . . . Doug Shearer
Assistant Equipment Manager . . . Craig Leydig
Equipment Assistant . . . Brian Metzger

Miscellaneous
Television Rightsholder . . . Comcast SportsNet
Radio Flagship . . . Sportstalk 980, WTEM
Television Play-by-Play . . . Joe Beninati
Television Color Analyst . . . Craig Laughlin
Television Producer . . . Bill Bell
Radio Play-by-Play . . . Steve Kolbe

Business Operations
Sr. Vice President, Business Operations . . . Declan J. Bolger
Director, Operations . . . George Parr
Director, Sponsorship Partnerships . . . Chris Hudgins
IT Manager . . . Kevin McDermott
Executive Assistant . . . Piper Sammons
Receptionist . . . Deborah Anderson
Mail Room Coordinator . . . Jennifer Whittington

Communications
Senior Director, Public Relations . . . Kurt Kehl
Manager, Media Relations . . . Brian Potter
Manager, Information . . . Kyle Hanlin
Manager, Community Development . . . Stephanie Boyer

New Media
New Media Manager . . . Brett Robinson
Senior Sports Media Producer . . . Mike Vogel
Sports Media Producer . . . Jon Carr

Finance
Controller . . . Keith Burrows
Senior Accountant . . . Michael Mercer
Accounts Payable Manager . . . Jennifer Simpson
Staff Accountant . . . Jill Shannon

Marketing and Advertising
Senior Director, Marketing . . . John Vidalin
Director, Game Operations . . . Mark Tamar
Promotions & Advertising Manager . . . Missy Rentz
Fan Development & Alumni Relations Manager . . . Chris Lewis
Fan Development Coordinator . . . Ryan Ahern
Promotions Coordinator . . . Lindsay Donald
Mascot Coordinator . . . Desi Deceder
Sr. Administrative Assistant . . . Kirsten Bergman

Sales
Vice President, Sales . . . Kevin Morgan
Group Sales Director . . . Brian Simpson
Senior Regional Sales Managers . . . Darren Bruening, Tim Bronaugh
Regional Sales Managers . . . Letitia Petrillo, Greg Voss, Jyermal Jones, David Dzwonkowski, Doug Pristach, Scott Borden, Jay Wheeler, Jill Colby, Ryan Smith, Rich Little, Josh Havey, Meghan Crowley
Executive Assistant . . . Ingrid Harrell-Lee

Ticket Operations & Guest Services
Director, Ticket Operations & Guest Services . . . Laini Samuels Delawter
Assistant Director, Guest Services . . . Greg Monares
Manager, Ticket Operations . . . Chris Turns

2001-2002 Final Statistics

Standings

Abbreviations: GP – games played; **W** – wins; **L** – losses; **T** – ties; **OTL** – overtime losses; **GF** – goals for; **GA** – goals against; **PTS** – points.

EASTERN CONFERENCE

Northeast Division

	GP	W	L	T	OTL	GF	GA	PTS
Boston	82	43	24	6	9	236	201	101
Toronto	82	43	25	10	4	249	207	100
Ottawa	82	39	27	9	7	243	208	94
Montreal	82	36	31	12	3	207	209	87
Buffalo	82	35	35	11	1	213	200	82

Atlantic Division

	GP	W	L	T	OTL	GF	GA	PTS
Philadelphia	82	42	27	10	3	234	192	97
NY Islanders	82	42	28	8	4	239	220	96
New Jersey	82	41	28	9	4	205	187	95
NY Rangers	82	36	38	4	4	227	258	80
Pittsburgh	82	28	41	8	5	198	249	69

Southeast Division

	GP	W	L	T	OTL	GF	GA	PTS
Carolina	82	35	26	16	5	217	217	91
Washington	82	36	33	11	2	228	240	85
Tampa Bay	82	27	40	11	4	178	219	69
Florida	82	22	44	10	6	180	250	60
Atlanta	82	19	47	11	5	187	288	54

WESTERN CONFERENCE

Central Division

	GP	W	L	T	OTL	GF	GA	PTS
Detroit	82	51	17	10	4	251	187	116
St. Louis	82	43	27	8	4	227	188	98
Chicago	82	41	27	13	1	216	207	96
Nashville	82	28	41	13	0	196	230	69
Columbus	82	22	47	8	5	164	255	57

Pacific Division

	GP	W	L	T	OTL	GF	GA	PTS
San Jose	82	44	27	8	3	248	199	99
Phoenix	82	40	27	9	6	228	210	95
Los Angeles	82	40	27	11	4	214	190	95
Dallas	82	36	28	13	5	215	213	90
Anaheim	82	29	42	8	3	175	198	69

Northwest Division

	GP	W	L	T	OTL	GF	GA	PTS
Colorado	82	45	28	8	1	212	169	99
Vancouver	82	42	30	7	3	254	211	94
Edmonton	82	38	28	12	4	205	182	92
Calgary	82	32	35	12	3	201	220	79
Minnesota	82	26	35	12	9	195	238	73

A junior scoring sensation, Daniel Briere was considered too small for the NHL. He split four seasons between Phoenix and the minors before getting a chance in 2001-02. He responded with 32 goals and a league-leading 21.5 percent shooting percentage.

INDIVIDUAL LEADERS

Goal Scoring

Player	Team	GP	G
Jarome Iginla	Calgary	82	52
Bill Guerin	Boston	78	41
Mats Sundin	Toronto	82	41
Glen Murray	L.A., Bos.	82	41
Markus Naslund	Vancouver	81	40
Peter Bondra	Washington	77	39
Keith Tkachuk	St. Louis	73	38
Eric Daze	Chicago	82	38

Assists

Player	Team	GP	A
Adam Oates	Wsh., Phi.	80	64
Jason Allison	Los Angeles	73	55
Joe Sakic	Colorado	82	53
Nicklas Lidstrom	Detroit	78	50
Ron Francis	Carolina	80	50
Markus Naslund	Vancouver	81	50
Jozef Stumpel	L.A., Bos.	81	50
Todd Bertuzzi	Vancouver	72	49
Jaromir Jagr	Washington	69	48
Craig Conroy	Calgary	81	48
Andrew Brunette	Minnesota	81	48

Power-play Goals

Player	Team	GP	PP
Peter Bondra	Washington	77	17
Jarome Iginla	Calgary	82	16
Ziggy Palffy	Los Angeles	63	15
Alexei Yashin	NY Islanders	78	15
Miroslav Satan	Buffalo	82	15

Short-handed Goals

Player	Team	GP	SH
Brian Rolston	Boston	82	9
Michael Peca	NY Islanders	80	6
Miroslav Satan	Buffalo	82	5
Stacy Roest	Minnesota	58	4
Shawn Bates	NY Islanders	71	4

Game-winning Goals

Player	Team	GP	GW
Pavol Demitra	St. Louis	82	10
Mats Sundin	Toronto	82	9
Glen Murray	L.A., Bos.	82	9
Patrik Elias	New Jersey	75	8
Peter Bondra	Washington	77	8
Steve Sullivan	Chicago	78	8
Teemu Selanne	San Jose	82	8
Paul Kariya	Anaheim	82	8

Game-tying Goals

Player	Team	GP	GT
Brendan Shanahan	Detroit	80	3
Bates Battaglia	Carolina	82	3
Mike Comrie	Edmonton	82	3

Shots

Player	Team	GP	S
Bill Guerin	Boston	78	355
Peter Bondra	Washington	77	333
Brian Rolston	Boston	82	331
Jarome Iginla	Calgary	82	311
Markus Naslund	Vancouver	81	302

Shooting Percentage

(minimum 82 shots)

Player	Team	GP	G	S	%
Daniel Briere	Phoenix	78	32	149	21.5
Adam Deadmarsh	Los Angeles	76	29	139	20.9
Jan Hrdina	Pittsburgh	79	24	115	20.9
Ziggy Palffy	Los Angeles	63	32	161	19.9
Andrew Brunette	Minnesota	81	21	106	19.8

Penalty Minutes

Player	Team	GP	PIM
Peter Worrell	Florida	79	354
Brad Ference	Florida	80	254
*Chris Neil	Ottawa	72	231
Kevin Sawyer	Anaheim	57	221
Theoren Fleury	NY Rangers	82	216

Plus/Minus

Player	Team	GP	+/–
Chris Chelios	Detroit	79	40
Jeremy Roenick	Philadelphia	75	32
Simon Gagne	Philadelphia	79	31
Glen Murray	L.A., Bos.	82	31
Zdeno Chara	Ottawa	75	30

* rookie

Individual Leaders

Abbreviations: * – rookie eligible for Calder Trophy; **A** – assists; **G** – goals; **GP** – games played; **GT** – game-tying goals; **GW** – game-winning goals; **PIM** – penalties in minutes; **PP** – power play goals; **Pts** – points; **S** – shots on goal; **SH** – short-handed goals; **%** – percentage of shots on goal resulting in goals; **+/–** – difference between Goals For (**GF**) scored when a player is on the ice with his team at even strength or short-handed and Goals Against (**GA**) scored when the same player is on the ice with his team at even strength or on a power play.

Individual Scoring Leaders for Art Ross Trophy

Player	Team	GP	G	A	Pts	+/–	PIM	PP	SH	GW	GT	S	%
Jarome Iginla	Calgary	82	52	44	96	27	77	16	1	7	2	311	16.7
Markus Naslund	Vancouver	81	40	50	90	22	50	8	0	6	1	302	13.2
Todd Bertuzzi	Vancouver	72	36	49	85	21	110	14	0	3	0	203	17.7
Mats Sundin	Toronto	82	41	39	80	6	94	10	2	9	2	262	15.6
Jaromir Jagr	Washington	69	31	48	79	0	30	10	0	3	0	197	15.7
Joe Sakic	Colorado	82	26	53	79	12	18	9	1	4	1	260	10.0
Pavol Demitra	St. Louis	82	35	43	78	13	46	11	0	10	0	212	16.5
Adam Oates	Wsh., Phi.	80	14	64	78	–4	28	3	0	1	0	102	13.7
Mike Modano	Dallas	78	34	43	77	14	38	6	2	5	0	219	15.5
Ron Francis	Carolina	80	27	50	77	4	18	14	0	5	2	165	16.4
Alexei Kovalev	Pittsburgh	67	32	44	76	2	80	8	1	3	2	266	12.0
Keith Tkachuk	St. Louis	73	38	37	75	21	117	13	0	7	1	244	15.6
Brendan Shanahan	Detroit	80	37	38	75	23	118	12	3	7	3	277	13.4
Alexei Yashin	NY Islanders	78	32	43	75	–3	25	15	0	5	0	239	13.4
Craig Conroy	Calgary	81	27	48	75	24	32	7	2	4	1	146	18.5
Jason Allison	Los Angeles	73	19	55	74	2	68	5	0	2	2	139	13.7
Eric Lindros	NY Rangers	72	37	36	73	19	138	12	1	4	0	196	18.9
Miroslav Satan	Buffalo	82	37	36	73	14	33	15	5	5	0	267	13.9
Glen Murray	L.A., Bos.	82	41	30	71	31	40	9	0	9	0	246	16.7
Daniel Alfredsson	Ottawa	78	37	34	71	3	45	9	1	4	2	243	15.2
Peter Bondra	Washington	77	39	31	70	–2	80	17	1	8	1	333	11.7
Eric Daze	Chicago	82	38	32	70	17	36	12	0	5	1	264	14.4
Sergei Samsonov	Boston	74	29	41	70	21	27	3	0	4	0	192	15.1
Radek Bonk	Ottawa	82	25	45	70	3	52	6	2	5	0	170	14.7

Defencemen Scoring Leaders

Player	Team	GP	G	A	Pts	+/–	PIM	PP	SH	GW	GT	S	%
Sergei Gonchar	Washington	76	26	33	59	–1	58	7	0	2	0	216	12.0
Nicklas Lidstrom	Detroit	78	9	50	59	13	20	6	0	0	0	215	4.2
Rob Blake	Colorado	75	16	40	56	16	58	10	0	2	0	229	7.0
Brian Leetch	NY Rangers	82	10	45	55	14	28	1	0	3	0	202	5.0
Sandis Ozolinsh	Car., Fla.	83	14	38	52	–7	58	3	0	1	0	172	8.1
Ed Jovanovski	Vancouver	82	17	31	48	–7	101	7	1	3	1	202	8.4
Teppo Numminen	Phoenix	76	13	35	48	13	20	4	0	6	0	117	11.1
Brian Rafalski	New Jersey	76	7	40	47	15	18	2	0	4	1	125	5.6
Chris Pronger	St. Louis	78	7	40	47	23	120	4	1	3	0	204	3.4
Al MacInnis	St. Louis	71	11	35	46	3	52	6	0	4	1	231	4.8

Jarome Iginla scored 18 goals in his first 20 games last season. He also finished the season on a tear with nine goals in his last 10 games en route to topping the 50-goal plateau for the first time in his career.

CONSECUTIVE SCORING STREAKS

Goals

Games	Player	Team	G
7	Luc Robitaille	Detroit	7
6	Mike Modano	Dallas	9
6	Patrick Marleau	San Jose	9
6	Jarome Iginla	Calgary	8
6	Mike Modano	Dallas	7
5	Keith Tkachuk	St. Louis	7
5	Jarome Iginla	Calgary	7
5	*Kristian Huselius	Florida	7
5	Mats Sundin	Toronto	6
5	Rob Zamuner	Boston	6
5	Chris Drury	Colorado	6
5	Sergei Fedorov	Detroit	5
5	Brian Rolston	Boston	5
5	Glen Murray	L.A., Bos	5
5	Ray Whitney	Columbus	5
5	Pavol Demitra	St. Louis	5
5	Jan Hrdina	Pittsburgh	5
5	Daniel Briere	Phoenix	5

Assists

Games	Player	Team	A
11	Adam Oates	Washington	13
8	Jim Dowd	Minnesota	9
6	Nicklas Lidstrom	Detroit	10
6	Jarome Iginla	Calgary	10
6	Jason Allison	Los Angeles	9
6	Sami Kapanen	Carolina	9
6	Brian Leetch	NY Rangers	8
6	Adam Oates	Washington	8
6	Joe Sakic	Colorado	8
6	Brendan Morrison	Vancouver	8
6	Daniel Briere	Phoenix	8
6	Sylvain Cote	Washington	6
6	Theoren Fleury	NY Rangers	6
6	Anson Carter	Edmonton	6
6	Miroslav Satan	Buffalo	6
6	Joe Thornton	Boston	6

Points

Games	Player	Team	G	A	PTS
15	Jarome Iginla	Calgary	18	13	31
15	Todd Bertuzzi	Vancouver	7	13	20
12	Mark Parrish	NY Islanders	5	8	13
11	Adam Oates	Washington	3	13	16
10	Brad Richards	Tampa Bay	6	12	18
10	Andrew Brunette	Minnesota	5	12	17
9	Mike Modano	Dallas	8	9	17
9	Markus Naslund	Vancouver	9	8	17
9	B. Shanahan	Detroit	8	7	15
9	Doug Weight	St. Louis	4	9	13
9	Jason Allison	Los Angeles	2	10	12
9	Eric Daze	Chicago	7	5	12
9	Pavol Demitra	St. Louis	7	5	12

Vancouver's Ed Jovanovski established personal bests in goals and points for the second straight season. His 17 goals ranked second among NHL blueliners, while his 48 points placed him sixth in defenseman scoring.

Atlanta's Dany Heatley led all rookie scorers during the 2001-02 season. He ranked first or second among first-year players in many major offensive categories and was rewarded with the Calder Trophy.

Individual Rookie Scoring Leaders

Rookie	Team	GP	G	A	Pts	+/–	PIM	PP	SH	GW	GT	S	%
Dany Heatley	Atlanta	82	26	41	67	–19	56	7	0	4	0	202	12.9
Ilya Kovalchuk	Atlanta	65	29	22	51	–19	28	7	0	4	1	184	15.8
Kristian Huselius	Florida	79	23	22	45	–4	14	6	1	3	0	169	13.6
Erik Cole	Carolina	81	16	24	40	–10	35	3	0	2	0	159	10.1
Pavel Datsyuk	Detroit	70	11	24	35	4	4	2	0	1	0	79	13.9
Martin Erat	Nashville	80	9	24	33	–11	32	2	0	2	0	84	10.7
Radim Vrbata	Colorado	52	18	12	30	7	14	6	0	3	0	112	16.1
Mark Bell	Chicago	80	12	16	28	–6	124	1	0	1	0	120	10.0
Niklas Hagman	Florida	78	10	18	28	–6	8	0	1	2	0	134	7.5
Andy McDonald	Anaheim	53	7	21	28	2	10	2	0	3	0	79	8.9

Goal Scoring

Name	Team	GP	G
Ilya Kovalchuk	Atlanta	65	29
Dany Heatley	Atlanta	82	26
Kristian Huselius	Florida	79	23
Radim Vrbata	Colorado	52	18
Erik Cole	Carolina	81	16
Pascal Dupuis	Minnesota	76	15
Mark Bell	Chicago	80	12
Krys Kolanos	Phoenix	57	11
Pavel Datsyuk	Detroit	70	11
Pierre Dagenais	N.J., Fla.	42	10
Chris Neil	Ottawa	72	10
Niklas Hagman	Florida	78	10
Kris Beech	Pittsburgh	79	10
Matt Bradley	San Jose	54	9
Martin Erat	Nashville	80	9

Assists

Name	Team	GP	A
Dany Heatley	Atlanta	82	41
Pavel Datsyuk	Detroit	70	24
Martin Erat	Nashville	80	24
Erik Cole	Carolina	81	24
Ilya Kovalchuk	Atlanta	65	22
Kristian Huselius	Florida	79	22
Andy McDonald	Anaheim	53	21
Niklas Hagman	Florida	78	18
Daniel Tjarnqvist	Atlanta	75	16
Mark Bell	Chicago	80	16
Kris Beech	Pittsburgh	79	15
Nicholas Boynton	Boston	80	14
Jeff Jillson	San Jose	48	13
Matt Bradley	San Jose	54	13

Power-play Goals

Name	Team	GP	PP
Ilya Kovalchuk	Atlanta	65	7
Dany Heatley	Atlanta	82	7
Radim Vrbata	Colorado	52	6
Kristian Huselius	Florida	79	6
Jason Williams	Detroit	25	4
Pierre Dagenais	N.J., Fla.	42	3
Mike Ribeiro	Montreal	43	3
Jeff Jillson	San Jose	48	3
Pascal Dupuis	Minnesota	76	3
Erik Cole	Carolina	81	3

Short-handed Goals

Name	Team	GP	SH
Mikael Samuelsson	NY Rangers	67	2
Pascal Dupuis	Minnesota	76	2
Niklas Hagman	Florida	78	1
Kristian Huselius	Florida	79	1
Toby Petersen	Pittsburgh	79	1

Game-winning Goals

Name	Team	GP	GW
Krys Kolanos	Phoenix	57	5
Ilya Kovalchuk	Atlanta	65	4
Dany Heatley	Atlanta	82	4
Radim Vrbata	Colorado	52	3
Andy McDonald	Anaheim	53	3
Kristian Huselius	Florida	79	3

Game-tying Goals

Name	Team	GP	GT
Pierre Dagenais	N.J., Fla.	42	2
Steve Begin	Calgary	51	1
Brian Willsie	Colorado	56	1
Ilya Kovalchuk	Atlanta	65	1
Rostislav Klesla	Columbus	75	1

Shots

Name	Team	GP	S
Dany Heatley	Atlanta	82	202
Ilya Kovalchuk	Atlanta	65	184
Kristian Huselius	Florida	79	169
Erik Cole	Carolina	81	159
Pascal Dupuis	Minnesota	76	154

Shooting Percentage

(minimum 82 shots)

Name	Team	GP	G	S	%
Radim Vrbata	Colorado	52	18	112	16.1
Ilya Kovalchuk	Atlanta	65	29	184	15.8
Kristian Huselius	Florida	79	23	169	13.6
Dany Heatley	Atlanta	82	26	202	12.9
Martin Erat	Nashville	80	9	84	10.7

Penalty Minutes

Name	Team	GP	PIM
Chris Neil	Ottawa	72	231
Jody Shelley	Columbus	52	206
Darcy Hordichuk	Atl., Phx.	34	141
Mark Bell	Chicago	80	124
Nick Boynton	Boston	80	107

Plus/Minus

Name	Team	GP	+/–
Matt Bradley	San Jose	54	22
Nicholas Boynton	Boston	80	18
Brian Gionta	New Jersey	33	10
Mike Van Ryn	St. Louis	48	10
Mikael Samuelsson	NY Rangers	67	10

Three-or-More-Goal Games

Player	Team	Date	Final Score	G
Daniel Alfredsson	Ottawa	Oct. 27	St.L. 1 Ott. 4	3
Daniel Alfredsson	Ottawa	Nov. 13	Ott. 11 Wsh. 5	3
Jason Arnott	New Jersey	Feb. 10	Buf. 1 N.J. 4	3
Todd Bertuzzi	Vancouver	Mar. 19	Van. 3 NYR 1	3
Peter Bondra	Washington	Mar. 10	Edm. 2 Wsh. 4	3
Zdeno Ciger	Tampa Bay	Jan. 15	T.B. 5 N.J. 4	3
*Erik Cole	Carolina	Dec. 21	Atl. 4 Car. 5	3
Mariusz Czerkawski	NY Islanders	Nov. 27	Wsh. 5 NYI 5	3
Eric Daze	Chicago	Dec. 07	NYI 3 Chi. 4	3
Eric Daze	Chicago	Dec. 31	Chi. 5 Ott. 4	3
Jiri Dopita	Philadelphia	Jan. 08	Atl. 4 Phi. 7	**4**
J.P. Dumont	Buffalo	Nov. 12	Buf. 5 Fla. 3	3
Patrik Elias	New Jersey	Oct. 18	S.J. 1 N.J. 6	3
Patrik Elias	New Jersey	Nov. 09	Tor. 2 N.J. 3	3
Marian Gaborik	Minnesota	Nov. 13	Atl. 2 Min. 4	3
Marian Gaborik	Minnesota	Mar. 10	CBJ 0 Min. 5	3
Simon Gagne	Philadelphia	Mar. 02	Phi. 5 NYR 6	3
Bill Guerin	Boston	Jan. 02	Bos. 6 Car. 3	3
Bill Guerin	Boston	Feb. 04	Bos. 8 CBJ 0	3
Martin Havlat	Ottawa	Dec. 13	Phx. 0 Ott. 6	3
Brett Hull	Detroit	Jan. 26	Det. 5 St.L. 2	3
Jarome Iginla	Calgary	Mar. 04	Cgy. 5 NYR 3	3
Sami Kapanen	Carolina	Nov. 27	Car. 5 Tor. 2	3
Paul Kariya	Anaheim	Jan. 14	Nsh. 3 Ana. 5	3
Paul Kariya	Anaheim	Feb. 06	Phi. 4 Ana. 5	3
Espen Knutsen	Columbus	Jan. 24	S.J. 2 CBJ 6	3
Alexei Kovalev	Pittsburgh	Nov. 13	Pit. 5 N.J. 1	3
Alexei Kovalev	Pittsburgh	Nov. 14	NYI 3 Pit. 3	3
Alexei Kovalev	Pittsburgh	Jan. 05	NYR 1 Pit. 4	3
Daymond Langkow	Phoenix	Oct. 13	Wsh. 2 Phx. 5	3
Eric Lindros	NY Rangers	Nov. 18	Atl. 2 NYR 6	3
Eric Lindros	NY Rangers	Mar. 02	Phi. 5 NYR 6	3
Patrick Marleau	San Jose	Apr. 06	Det. 3 S.J. 6	3
Alexei Morozov	Pittsburgh	Jan. 17	Pit. 6 Cgy. 4	3
Glen Murray	Boston	Apr. 13	Pit. 1 Bos. 7	3
Markus Naslund	Vancouver	Oct. 18	Tor. 6 Van. 5	3
Markus Naslund	Vancouver	Jan. 28	Nsh. 1 Van. 5	3
Markus Naslund	Vancouver	Apr. 11	L.A. 2 Van. 5	3
Ziggy Palffy	Los Angeles	Feb. 08	L.A. 5 Phx. 6	3
Mark Parrish	NY Islanders	Oct. 11	NYI 6 N.J. 4	3
Mark Parrish	NY Islanders	Nov. 24	Ana. 3 NYI 5	3
Yanic Perreault	Montreal	Jan. 21	Mtl. 5 Fla. 7	3
*Toby Petersen	Pittsburgh	Oct. 16	Ott. 2 Pit. 5	3
Oleg Petrov	Montreal	Jan. 14	Phi. 5 Mtl. 3	3
Mark Recchi	Philadelphia	Oct. 20	Wsh. 3 Phi. 6	3
Steve Reinprecht	Colorado	Apr. 03	Atl. 0 Col. 6	3
Pascal Rheaume	Atlanta	Jan. 19	Atl. 6 Fla. 1	**4**
Gary Roberts	Toronto	Dec. 06	Tor. 6 NYR 3	3
Brian Savage	Montreal	Oct. 26	Mtl. 5 Buf. 2	3
Marc Savard	Calgary	Jan. 30	Det. 3 Cgy. 4	3
Brendan Shanahan	Detroit	Oct. 04	Det. 4 S.J. 3	3
Cory Stillman	St Louis	Jan. 05	Dal. 2 St.L. 5	3
Petr Tenkrat	Nashville	Dec. 15	Chi. 2 Nsh. 5	3
Joe Thornton	Boston	Jan. 05	Wsh. 4 Bos. 7	3
Jeff Toms	NY Rangers	Jan. 23	Bos. 4 NYR 8	3
*Radim Vrbata	Colorado	Feb. 08	Col. 6 Min. 0	3
Alexei Yashin	NY Islanders	Jan. 30	NYI 6 NYR 3	3

* indicates rookie

Building on an impressive rookie season, Minnesota's Marian Gaborik collected 30 goals in 2001-02, including the first two hat tricks of his career.

2001-02 Penalty Shots

Scored

Sebastien Bordeleau (Minnesota) scored against Felix Potvin (Los Angeles), October 7. Final score: Minnesota 4 at Los Angeles 3.

Markus Naslund (Vancouver) scored against Curtis Joseph (Toronto), October 18. Final score: Toronto 6 at Vancouver 5.

Wayne Primeau (Pittsburgh) scored against Nikolai Khabibulin (Tampa Bay), November 10. Final score: Pittsburgh 2 at Tampa Bay 3.

Tim Connolly (Buffalo) scored against Milan Hnilicka (Atlanta), November 19. Final score: Buffalo 2 at Atlanta 3.

Mike Fisher (Ottawa) scored against Marty Turco (Dallas), December 5. Final score: Ottawa 6 at Dallas 3.

Stacy Roest (Minnesota) scored against Dan Cloutier (Vancouver), March 2. Final score: Minnesota 3 at Vancouver 6.

Antti Laaksonen (Minnesota) scored against Brent Johnson (St. Louis), March 26. Final score: Minnesota 2 at St. Louis 1.

Krys Kolanos (Phoenix) scored against Patrick Roy (Colorado), March 30. Final score: Colorado 3 at Phoenix 5.

Scott Nichol (Calgary) scored against Milan Hnilicka (Atlanta), April 2. Final score: Atlanta 2 at Calgary 4.

Tony Hrkac (Atlanta) scored against Marc Denis (Columbus), April 12. Final score: Atlanta 4 at Columbus 5.

Chris Drury (Colorado) scored against Jean-Sebastien Giguere (Anaheim), April 12. Final score: Colorado 3 at Anaheim 1.

Stopped

Fred Brathwaite (St. Louis) stopped Vladimir Orszagh (Nashville), October 6. Final score: St. Louis 2 at Nashville 2.

Manny Fernandez (Minnesota) stopped Ziggy Palffy (Los Angeles), October 7. Final score: Minnesota 4 at Los Angeles 3.

Johan Hedberg (Pittsburgh) stopped Marian Hossa (Ottawa), October 16. Final score: Ottawa 2 at Pittsburgh 5.

Olaf Kolzig (Washington) stopped Chad Kilger (Montreal), October 19. Final score: Montreal 1 at Washington 4.

Johan Hedberg (Pittsburgh) stopped Mike Modano (Dallas), October 24. Final score: Dallas 2 at Pittsburgh 3.

Sean Burke (Phoenix) stopped Greg de Vries (Colorado), October 27. Final score: Colorado 0 at Phoenix 1.

Curtis Joseph (Toronto) stopped Vincent Lecavalier (Tampa Bay), October 30. Final score: Tampa Bay 2 at Toronto 3.

Patrick Lalime (Ottawa) stopped Miroslav Satan (Buffalo), November 3. Final score: Buffalo 0 at Ottawa 3.

Milan Hnilicka (Atlanta) stopped Matt Cullen (Anaheim), November 4. Final score: Atlanta 0 at Anaheim 5.

Curtis Joseph (Toronto) stopped Joe Sacco (Washington), November 6. Final score: Washington 2 at Toronto 4.

Tom Barrasso (Carolina) stopped Anson Carter (Edmonton), November 11. Final score: Edmonton 1 at Carolina 1.

Jamie Storr (Los Angeles) stopped Alexander Korolyuk (San Jose), November 24. Final score: San Jose 1 at Los Angeles 3.

Curtis Joseph (Toronto) stopped Kyle Calder (Chicago), December 1. Final score: Chicago 1 at Toronto 4.

Jean-Sebastien Giguere (Anaheim) stopped Jarome Iginla (Calgary), December 8. Final score: Anaheim 4 at Calgary 0.

Marc Denis (Columbus) stopped Shjon Podein (Colorado), December 12. Final score: Columbus 1 at Colorado 5.

Dan Cloutier (Vancouver) stopped Dan Bylsma (Anaheim), December 12. Final score: Vancouver 1 at Anaheim 0.

Mike Richter (NY Rangers) stopped Colin White (New Jersey), December 19. Final score: New Jersey 2 at NY Rangers 2.

Jose Theodore (Montreal) stopped Ian Laperriere (Los Angeles), December 22. Final score: Los Angeles 1 at Montreal 2.

Steve Shields (Anaheim) stopped Alexei Zhamnov (Chicago), December 30. Final score: Anaheim 1 at Chicago 2.

Evgeni Nabokov (San Jose) stopped Sergei Varlamov (St. Louis), January 8. Final score: St. Louis 6 at San Jose 2.

Roman Turek (Calgary) stopped Jeff Cowan (Atlanta), January 11. Final score: Calgary 0 at Atlanta 1.

Jani Hurme (Ottawa) stopped Martin St. Louis (Tampa Bay), January 12. Final score: Ottawa 2 at Tampa Bay 1.

Marc Denis (Columbus) stopped Manny Malhotra (NY Rangers), January 14. Final score: Columbus 2 at NY Rangers 2.

Curtis Joseph (Toronto) stopped Jarome Iginla (Calgary), January 22. Final score: Toronto 6 at Calgary 1.

Arturs Irbe (Carolina) stopped Pavel Bure (Florida), January 25. Final score: Florida 1 at Carolina 1.

Byron Dafoe (Boston) stopped Pavel Bure (Florida), February 9. Final score: Florida 1 at Boston 4.

Dwayne Roloson (Minnesota) stopped Brad Isbister (NY Islanders), February 10. Final score: NY Islanders 3 at Minnesota 4.

Dominik Hasek (Detroit) stopped Dan Boyle (Tampa Bay), February 26. Final score: Detroit 4 at Tampa Bay 3.

Milan Hnilicka (Atlanta) stopped Mike Ribeiro (Montreal), March 4. Final score: Atlanta 3 at Montreal 5.

Dan Cloutier (Vancouver) stopped Mike Sillinger (Columbus), March 14. Final score: Vancouver 5 at Columbus 1.

Jocelyn Thibault (Chicago) stopped Sergei Brylin (New Jersey), March 20. Final score: New Jersey 3 at Chicago 1.

Chris Osgood (NY Islanders) stopped Markus Naslund (Vancouver), March 21. Final score: Vancouver 2 at NY Islanders 3.

Nikolai Khabibulin (Tampa Bay) stopped Richard Zednik (Montreal), March 22. Final score: Montreal 3 at Tampa Bay 3.

Jan Lasak (Nashville) stopped Keith Tkachuk (St. Louis), April 9. Final score: Nashville 2 at St. Louis 3.

Roman Turek (Calgary) stopped Michal Handzus (Phoenix), April 9. Final score: Phoenix 4 at Calgary 2.

Summary

46 penalty shots resulted in 11 goals

Goaltending Leaders

Minimum 25 games

Goals Against Average

Goaltender	Team	GPI	MINS	GA	Avg
Patrick Roy	Colorado	63	3773	122	1.94
Roman Cechmanek	Philadelphia	46	2603	89	2.05
Marty Turco	Dallas	31	1519	53	2.09
Jose Theodore	Montreal	67	3864	136	2.11
Jean-Sebastien Giguere	Anaheim	53	3127	111	2.13

Wins

Goaltender	Team	GPI	MINS	W	L	T
Dominik Hasek	Detroit	65	3872	41	15	8
Martin Brodeur	New Jersey	73	4347	38	26	9
Evgeni Nabokov	San Jose	67	3901	37	24	5
Byron Dafoe	Boston	64	3827	35	26	3
Brent Johnson	St. Louis	58	3491	34	20	4

Save Percentage

Goaltender	Team	GPI	MINS	GA	SA	S%	W	L	T
Jose Theodore	Montreal	67	3864	136	1972	.931	30	24	10
Patrick Roy	Colorado	63	3773	122	1629	.925	32	23	8
Roman Cechmanek	Philadelphia	46	2603	89	1131	.921	24	13	6
Marty Turco	Dallas	31	1519	53	670	.921	15	6	2
Nikolai Khabibulin	Tampa Bay	70	3896	153	1914	.920	24	32	10

Shutouts

Goaltender	Team	GPI	MINS	SO	W	L	T
Patrick Roy	Colorado	63	3773	9	32	23	8
Dan Cloutier	Vancouver	62	3502	7	31	22	5
Patrick Lalime	Ottawa	61	3583	7	27	24	8
Jose Theodore	Montreal	67	3864	7	30	24	10
Nikolai Khabibulin	Tampa Bay	70	3896	7	24	32	10
Evgeni Nabokov	San Jose	67	3901	7	37	24	5

Team-by-Team Point Totals

1997-98 to 2001-02

(Ranked by five-year point %)

	01-02	00-01	99-00	98-99	97-98	Pts%
Detroit	116	111	108	93	103	.648
New Jersey	95	111	103	105	107	.635
Dallas	90	106	102	114	109	.635
Colorado	99	118	96	98	95	.617
St. Louis	98	103	114	87	98	.610
Philadelphia	97	100	105	93	95	.598
Ottawa	94	109	95	103	83	.590
Toronto	100	90	100	97	69	.556
Phoenix	95	90	90	90	82	.545
Buffalo	82	98	85	91	89	.543
Boston	101	88	73	91	91	.541
Washington	85	96	102	68	92	.540
Pittsburgh	69	96	88	90	98	.538
San Jose	99	95	87	80	78	.535
Los Angeles	95	92	94	69	87	.533
Edmonton	92	93	88	78	80	.526
Carolina	91	88	84	86	74	.516
Montreal	87	70	83	75	87	.490
Vancouver	94	90	83	58	64	.474
Chicago	96	71	78	70	73	.473
NY Rangers	80	72	73	77	68	.451
Calgary	79	73	77	72	67	.449
Anaheim	69	66	83	83	65	.446
Florida	60	66	98	78	63	.445
Nashville	69	80	70	63	—	.430
Minnesota	73	68	—	—	—	.430
NY Islanders	96	52	58	58	71	.409
Columbus	57	71	—	—	—	.390
Tampa Bay	69	59	54	47	44	.333
Atlanta	54	60	39	—	—	.311

Team Record When Scoring First Goal of a Game

Team	GP	FG	W	L	T
Anaheim	82	39	21	13	5
Atlanta	82	34	12	16	6
Boston	82	48	**34**	10	4
Buffalo	82	44	28	9	7
Calgary	82	40	23	11	6
Carolina	82	41	25	10	6
Chicago	82	40	27	6	7
Colorado	82	38	27	8	3
Columbus	82	37	17	13	7
Dallas	82	45	25	11	9
Detroit	82	46	**34**	7	5
Edmonton	82	38	24	8	6
Florida	82	44	19	21	4
Los Angeles	82	42	27	10	5
Minnesota	82	38	19	12	7
Montreal	82	45	28	9	8
Nashville	82	39	20	12	7
New Jersey	82	42	30	10	2
NY Islanders	82	**49**	31	14	4
NY Rangers	82	32	25	6	1
Ottawa	82	45	28	13	4
Philadelphia	82	47	29	12	6
Phoenix	82	45	30	10	5
Pittsburgh	82	31	16	11	4
San Jose	82	46	**34**	8	4
St. Louis	82	38	28	8	2
Tampa Bay	82	32	17	13	2
Toronto	82	44	30	10	4
Vancouver	82	47	32	12	3
Washington	82	32	20	8	4

Team Plus/Minus Differential

Team	GF	PPGF	Net GF	GA	PPGA	Net GA	Goal Differential
San Jose	248	59	189	199	54	145	+44
Philadelphia	234	39	195	192	40	152	+43
Boston	236	39	197	201	45	156	+41
Toronto	249	54	195	207	51	156	+39
Detroit	251	73	178	187	48	139	+39
St. Louis	227	62	165	188	56	132	+33
Vancouver	254	69	185	211	57	154	+31
Ottawa	243	55	188	208	48	160	+28
Phoenix	228	53	175	210	60	150	+25
Edmonton	205	51	154	182	50	132	+22
Colorado	212	62	150	169	41	128	+22
Chicago	216	54	162	207	63	144	+18
New Jersey	205	44	161	187	43	144	+17
NY Islanders	239	59	180	220	52	168	+12
Buffalo	213	50	163	200	41	159	+4
NY Rangers	227	48	179	258	80	178	+1
Dallas	215	60	155	213	59	154	+1
Los Angeles	214	73	141	190	49	141	0
Calgary	201	55	146	220	66	154	-8
Montreal	207	45	162	209	38	171	-9
Carolina	217	63	154	217	54	163	-9
Washington	228	58	170	240	58	182	-12
Anaheim	175	43	132	198	46	152	-20
Nashville	196	48	148	230	50	180	-32
Minnesota	195	55	140	238	62	176	-36
Pittsburgh	198	47	151	249	57	192	-41
Florida	180	50	130	250	76	174	-44
Tampa Bay	178	46	132	219	40	179	-47
Atlanta	187	37	150	288	65	223	-73
Columbus	164	50	114	255	52	203	-89

Team Record When Leading, Trailing, Tied

Team	Leading after 1 period			Leading after 2 periods			Trailing after 1 period			Trailing after 2 periods			Tied after 1 period			Tied after 2 periods		
	W	L	T	W	L	T	W	L	T	W	L	T	W	L	T	W	L	T
Anaheim	14	6	3	22	1	2	3	22	1	2	27	1	12	17	4	5	17	5
Atlanta	9	9	2	12	4	4	1	27	5	0	40	5	9	16	4	7	8	2
Boston	25	6	0	30	7	2	5	14	0	3	20	0	13	13	5	10	6	4
Buffalo	23	7	5	29	2	4	2	17	2	2	28	2	10	12	3	4	6	5
Calgary	14	4	3	21	1	5	7	20	2	3	31	2	11	14	5	8	6	5
Carolina	17	8	4	22	2	2	6	19	5	4	21	5	12	4	8	9	8	9
Chicago	21	3	2	32	0	4	4	16	3	3	23	3	16	9	9	6	5	6
Colorado	19	4	2	34	0	1	10	13	3	5	22	3	16	12	3	6	7	4
Columbus	11	4	3	17	2	3	2	30	1	0	42	2	9	18	4	5	8	3
Dallas	21	5	3	22	1	2	8	15	4	1	23	7	7	13	6	13	9	4
Detroit	23	5	2	37	2	2	7	9	3	5	15	4	21	7	5	9	4	4
Edmonton	20	6	5	26	3	4	7	15	2	6	20	4	11	11	5	6	9	4
Florida	13	11	2	17	5	1	2	22	4	0	30	5	7	17	4	5	15	4
Los Angeles	20	4	3	26	2	1	8	16	3	4	21	4	12	11	5	10	8	6
Minnesota	12	6	4	20	5	3	3	25	2	0	36	1	11	13	6	6	3	8
Montreal	19	6	2	26	5	3	6	20	2	2	22	2	11	8	8	8	7	7
Nashville	15	6	5	21	4	7	6	20	5	3	33	3	7	15	3	4	4	3
New Jersey	18	5	1	24	4	3	7	14	3	5	23	3	16	13	5	12	5	3
NY Islanders	23	8	1	30	1	4	7	13	3	4	20	3	12	11	4	8	11	1
NY Rangers	16	4	1	24	1	1	3	28	2	3	31	2	17	10	1	9	10	1
Ottawa	22	4	1	30	2	1	3	18	0	2	23	2	14	12	8	7	9	6
Philadelphia	21	7	2	29	4	6	6	13	2	2	21	3	15	10	6	11	5	1
Phoenix	21	6	3	28	4	2	4	17	2	4	25	1	15	10	4	8	4	6
Pittsburgh	11	5	3	20	1	4	9	23	2	2	32	4	8	18	3	6	13	0
San Jose	24	2	3	33	2	4	7	14	1	2	19	2	13	14	4	9	9	2
St. Louis	19	5	2	30	2	4	8	16	2	2	21	1	16	10	4	11	8	3
Tampa Bay	11	8	1	16	3	2	8	24	1	1	28	5	8	12	9	10	13	4
Toronto	24	6	3	33	4	2	7	13	2	4	21	2	12	10	5	6	4	6
Vancouver	27	8	1	35	4	1	4	14	1	2	23	3	11	11	5	5	6	3
Washington	14	2	3	21	2	4	10	20	5	4	26	4	12	13	3	11	7	3

Evgeni Nabokov enjoyed another stellar season in 2001-02, ranking among the NHL leaders in shutouts and wins. The Sharks rarely let a lead slip away last year and posted the NHL's best plus-minus differential.

Team Statistics

TEAMS' HOME AND ROAD RECORD

Eastern Conference

	Home								Road							
	GP	W	L	T	OTL	GF	GA	PTS	GP	W	L	T	OTL	GF	GA	PTS
BOS	41	23	11	2	5	126	104	53	41	20	13	4	4	110	97	48
TOR	41	24	11	6	0	130	93	54	41	19	14	4	4	119	114	46
PHI	41	20	13	5	3	116	100	48	41	22	14	5	0	118	92	49
NYI	41	21	13	5	2	124	115	49	41	21	15	3	2	115	105	47
N.J.	41	22	13	4	2	112	88	50	41	19	15	5	2	93	99	45
OTT	41	21	13	3	4	114	98	49	41	18	14	6	3	129	110	45
CAR	41	15	13	11	2	98	105	43	41	20	13	5	3	119	112	48
MTL	41	21	13	6	1	102	86	49	41	15	18	6	2	105	123	38
WSH	41	21	12	6	2	124	110	50	41	15	21	5	0	104	130	35
BUF	41	20	16	5	0	111	93	45	41	15	19	6	1	102	107	37
NYR	41	19	19	2	1	119	129	41	41	17	19	2	3	108	129	39
PIT	41	16	20	4	1	114	128	37	41	12	21	4	4	84	121	32
T.B.	41	16	17	5	3	103	109	40	41	11	23	6	1	75	110	29
FLA	41	11	23	3	4	96	131	29	41	11	21	7	2	84	119	31
ATL	41	11	21	9	0	93	129	31	41	8	26	2	5	94	159	23
Total	**615**	**281**	**228**	**76**	**30**	**1682**	**1618**	**668**	**615**	**243**	**266**	**70**	**36**	**1559**	**1727**	**592**

Western Conference

	GP	W	L	T	OTL	GF	GA	PTS	GP	W	L	T	OTL	GF	GA	PTS
DET	41	28	7	5	1	143	95	62	41	23	10	5	3	108	92	54
COL	41	24	12	4	1	109	80	53	41	21	16	4	0	103	89	46
S.J.	41	25	11	3	2	131	89	55	41	19	16	5	1	117	110	44
ST.L.	41	27	12	1	1	126	90	56	41	16	15	7	3	101	98	42
CHI	41	28	7	5	1	114	81	62	41	13	20	8	0	102	126	34
PHX	41	27	8	3	3	139	92	60	41	13	19	6	3	89	118	35
L.A.	41	22	12	6	1	103	79	51	41	18	15	5	3	111	111	44
VAN	41	23	11	5	2	139	100	53	41	19	19	2	1	115	111	41
EDM	41	23	14	4	0	110	84	50	41	15	14	8	4	95	98	42
DAL	41	18	13	6	4	104	105	46	41	18	15	7	1	111	108	44
CGY	41	20	14	5	2	103	100	47	41	12	21	7	1	98	120	32
MIN	41	14	14	8	5	102	114	41	41	12	21	4	4	93	124	32
ANA	41	15	19	5	2	93	101	37	41	14	23	3	1	82	97	32
NSH	41	17	16	8	0	101	95	42	41	11	25	5	0	95	135	27
CBJ	41	14	18	5	4	93	127	37	41	8	29	3	1	71	128	20
Total	**615**	**325**	**188**	**73**	**29**	**1710**	**1432**	**752**	**615**	**232**	**278**	**79**	**26**	**1491**	**1665**	**569**
	1230	**606**	**416**	**149**	**59**	**3392**	**3050**	**1420**	**1230**	**475**	**544**	**149**	**62**	**3050**	**3392**	**1161**

Kyle Calder (far left) and the Blackhawks were tough to beat at home last season. Curtis Leschyshyn (left) and the Ottawa Senators posted the league's longest undefeated streak with a run of nine wins and two ties early in the season.

TEAMS' DIVISIONAL RECORD

Northeast Division

	Against Own Division								Against Other Divisions							
	GP	W	L	T	OTL	GF	GA	PTS	GP	W	L	T	OTL	GF	GA	PTS
BOS	20	9	9	1	1	49	47	20	62	34	15	5	8	187	154	81
TOR	20	10	3	5	2	52	39	27	62	33	22	5	2	197	168	73
OTT	20	8	9	1	2	52	58	19	62	31	18	8	5	191	150	75
MTL	20	6	9	4	1	46	61	17	62	30	22	8	2	161	148	70
BUF	20	10	7	3	0	50	44	23	62	25	28	8	1	163	156	59
Total	**100**	**43**	**37**	**14**	**6**	**249**	**249**	**106**	**310**	**153**	**105**	**34**	**18**	**899**	**776**	**358**

Atlantic Division

	GP	W	L	T	OTL	GF	GA	PTS	GP	W	L	T	OTL	GF	GA	PTS
PHI	20	10	7	2	1	47	44	23	62	32	20	8	2	187	148	74
NYI	20	11	6	1	2	62	54	25	62	31	22	7	2	177	166	71
N.J.	20	10	7	2	1	51	54	23	62	31	21	7	3	154	133	72
NYR	20	7	10	1	2	56	66	17	62	29	28	3	2	171	192	63
PIT	20	8	8	2	2	57	55	20	62	20	33	6	3	141	194	49
Total	**100**	**46**	**38**	**8**	**8**	**273**	**273**	**108**	**310**	**143**	**124**	**31**	**12**	**830**	**833**	**329**

Southeast Division

	GP	W	L	T	OTL	GF	GA	PTS	GP	W	L	T	OTL	GF	GA	PTS
CAR	20	11	3	5	1	54	47	28	62	24	23	11	4	163	170	63
WSH	20	12	4	4	0	60	41	28	62	24	29	7	2	168	199	57
T.B.	20	7	10	3	0	46	53	17	62	20	30	8	4	132	166	52
FLA	20	5	10	3	2	48	57	15	62	17	34	7	4	132	193	45
ATL	20	4	7	7	2	51	61	17	62	15	40	4	3	136	227	37
Total	**100**	**39**	**34**	**22**	**5**	**259**	**259**	**105**	**310**	**100**	**156**	**37**	**17**	**731**	**955**	**254**

Central Division

	GP	W	L	T	OTL	GF	GA	PTS	GP	W	L	T	OTL	GF	GA	PTS
DET	20	10	4	4	2	64	48	26	62	41	13	6	2	187	139	90
STL	20	10	5	3	2	62	51	25	62	33	22	5	2	165	137	73
CHI	20	9	6	5	0	53	54	23	62	32	21	8	1	163	153	73
NSH	20	7	10	3	0	47	60	17	62	21	31	10	0	149	170	52
CBJ	20	5	9	3	3	40	53	16	62	17	38	5	2	124	202	41
Total	**100**	**41**	**34**	**18**	**7**	**266**	**266**	**107**	**310**	**144**	**125**	**34**	**7**	**788**	**801**	**329**

Pacific Division

	GP	W	L	T	OTL	GF	GA	PTS	GP	W	L	T	OTL	GF	GA	PTS
S.J.	20	10	8	1	1	58	53	22	62	34	19	7	2	190	146	77
PHX	20	7	8	3	2	58	59	19	62	33	19	6	4	170	151	76
L.A.	20	10	4	4	2	48	40	26	62	30	23	7	2	166	150	69
DAL	20	9	9	2	0	47	54	20	62	27	19	11	5	168	159	70
ANA	20	8	9	2	1	44	49	19	62	21	33	6	2	131	149	50
Total	**100**	**44**	**38**	**12**	**6**	**255**	**255**	**106**	**310**	**145**	**113**	**37**	**15**	**825**	**755**	**342**

Northwest Division

	GP	W	L	T	OTL	GF	GA	PTS	GP	W	L	T	OTL	GF	GA	PTS
COL	20	10	6	4	0	53	43	24	62	35	22	4	1	159	126	75
VAN	20	10	8	2	0	58	46	22	62	32	22	5	3	196	165	72
EDM	20	10	5	5	0	47	38	25	62	28	23	7	4	158	144	67
CGY	20	8	8	4	0	41	47	20	62	24	27	8	3	160	173	59
MIN	20	4	11	1	4	42	67	13	62	22	24	11	5	153	171	60
Total	**100**	**42**	**38**	**16**	**4**	**241**	**241**	**104**	**310**	**141**	**118**	**35**	**16**	**826**	**779**	**333**

TEAM STREAKS

Consecutive Wins

Games	Team	From	To
10	St. Louis	Jan. 3	Jan. 23
8	Philadelphia	Jan. 6	Jan. 19
8	Detroit	Feb. 9	Mar. 9
7	Ottawa	Oct. 25	Nov. 13
7	San Jose	Jan. 30	Feb. 28
7	Montreal	Mar. 28	Apr. 9

Consecutive Home Wins

Games	Team	From	To
10	Detroit	Dec. 19	Jan. 20
8	Phoenix	Feb. 12	Mar. 24
7	Philadelphia	Dec. 18	Jan. 17
7	Chicago	Dec. 27	Jan. 20
7	St. Louis	Dec. 28	Jan. 19

Consecutive Road Wins

Games	Team	From	To
8	Detroit	Feb. 4	Mar. 9
7	Dallas	Nov. 15	Dec. 15
7	Boston	Mar. 8	Apr. 2
6	Detroit	Oct. 4	Oct. 31
6	NY Islanders	Oct. 5	Oct. 26
6	San Jose	Nov. 30	Dec. 19
6	Colorado	Jan. 19	Mar. 11

Consecutive Undefeated

Games	Team	W	T	From	To
11	Ottawa	9	2	Oct. 25	Nov. 22
10	Calgary	6	4	Oct. 27	Nov. 22
10	Colorado	8	2	Nov. 24	Dec. 12
10	San Jose	9	1	Nov. 27	Dec. 19
10	St. Louis	10	0	Jan. 3	Jan. 23
9	Chicago	6	3	Oct. 14	Nov. 1
9	Detroit	8	1	Nov. 13	Nov. 30
9	Vancouver	8	1	Jan. 19	Feb. 8
9	Edmonton	8	1	Mar. 14	Apr. 2
9	Vancouver	8	1	Mar. 26	Apr. 13

Consecutive Home Undefeated

Games	Team	W	T	From	To
13	Detroit	12	1	Dec. 19	Feb. 6
12	Chicago	9	3	Oct. 11	Nov. 28
11	Nashville	8	3	Nov. 3	Dec. 23
11	Los Angeles	9	2	Mar. 4	Apr. 14
9	Detroit	8	1	Nov. 2	Nov. 30
9	Toronto	6	3	Jan. 29	Mar. 19
9	Edmonton	7	2	Mar. 2	Apr. 10

Consecutive Road Undefeated

Games	Team	W	T	From	To
8	Dallas	7	1	Nov. 11	Dec. 15
8	Detroit	8	0	Feb. 4	Mar. 9
7	Colorado	6	1	Nov. 30	Dec. 26
7	Colorado	6	1	Jan. 12	Mar. 11
7	Boston	7	0	Mar. 8	Apr. 2

TEAM PENALTIES

Abbreviations: GP – games played; **PEN** – total penalty minutes including bench minutes; **BMI** – total bench minor minutes; **AVG** – average penalty minutes/game calculated by dividing total penalty minutes by games played

Team	GP	PEN	BMI	AVG	Team	GP	PEN	BMI	AVG
DAL	82	959	12	11.7	PIT	82	1248	16	15.2
MTL	82	974	24	11.9	S.J.	82	1249	8	15.2
COL	82	1007	10	12.3	ANA	82	1254	14	15.3
N.J.	82	1010	18	12.3	NYI	82	1255	14	15.3
CAR	82	1022	12	12.5	EDM	82	1267	6	15.5
WSH	82	1043	6	12.7	ATL	82	1290	22	15.7
DET	82	1053	10	12.8	L.A.	82	1348	16	16.4
NSH	82	1071	20	13.1	OTT	82	1347	16	16.4
T.B.	82	1072	22	13.1	ST.L.	82	1343	26	16.4
PHX	82	1154	16	14.1	VAN	82	1342	28	16.4
CBJ	82	1198	14	14.6	BOS	82	1454	12	17.7
MIN	82	1209	16	14.7	CGY	82	1586	12	19.3
BUF	82	1217	16	14.8	NYR	82	1753	10	21.4
TOR	82	1212	12	14.8	FLA	82	1994	26	24.3
CHI	82	1234	10	15.0	**Total**	**1230**	**37407**	**460**	
PHI	82	1242	16	15.1	**Two-Team Avg. PIM/GP**				**30.4**

Michael Peca was a key reason why the New York Islanders returned to the postseason in 2001-02 for the first time in eight years. Peca had six of the Islanders' league-leading 17 short-handed goals and a league-best 10 short-handed points.

TEAMS' POWER-PLAY RECORD

Abbreviations: ADV – total advantages; **PPGF** – power-play goals for;
% – calculated by dividing number of power-play goals by total advantages.

	Home					Road					Overall				
	Team	GP	ADV	PPGF	%	Team	GP	ADV	PPGF	%	Team	GP	ADV	PPGF	%
1	DET	41	181	44	24.3	L.A.	41	168	41	24.4	L.A.	82	353	73	20.7
2	ST.L.	41	182	38	20.9	NYI	41	187	37	19.8	DET	82	359	73	20.3
3	WSH	41	145	29	20.0	COL	41	177	34	19.2	WSH	82	299	58	19.4
4	N.J.	41	136	27	19.9	OTT	41	159	30	18.9	VAN	82	372	69	18.5
5	FLA	41	187	35	18.7	CHI	41	149	28	18.8	COL	82	344	62	18.0
6	VAN	41	191	35	18.3	WSH	41	154	29	18.8	CHI	82	302	54	17.9
7	S.J.	41	195	35	17.9	VAN	41	181	34	18.8	ST.L.	82	356	62	17.4
8	NYR	41	176	31	17.6	DAL	41	156	28	17.9	DAL	82	350	60	17.1
9	TOR	41	178	31	17.4	DET	41	178	29	16.3	N.J.	82	261	44	16.9
10	L.A.	41	185	32	17.3	CAR	41	190	29	15.3	OTT	82	331	55	16.6
11	PHX	41	187	32	17.1	BOS	41	122	18	14.8	NYI	82	363	59	16.3
12	CHI	41	153	26	17.0	EDM	41	163	24	14.7	CAR	82	391	63	16.1
13	CAR	41	201	34	16.9	MTL	41	145	21	14.5	S.J.	82	369	59	16.0
14	COL	41	167	28	16.8	NSH	41	155	22	14.2	PHX	82	340	53	15.6
15	CGY	41	187	31	16.6	PIT	41	151	21	13.9	TOR	82	349	54	15.5
16	DAL	41	194	32	16.5	ST.L.	41	174	24	13.8	EDM	82	333	51	15.3
17	MIN	41	198	32	16.2	S.J.	41	174	24	13.8	CGY	82	364	55	15.1
18	T.B.	41	170	27	15.9	PHX	41	153	21	13.7	MTL	82	301	45	15.0
19	EDM	41	170	27	15.9	N.J.	41	125	17	13.6	FLA	82	341	50	14.7
20	CBJ	41	188	29	15.4	CGY	41	177	24	13.6	NYR	82	326	48	14.7
21	MTL	41	156	24	15.4	TOR	41	171	23	13.5	T.B.	82	312	46	14.7
22	OTT	41	172	25	14.5	T.B.	41	142	19	13.4	MIN	82	373	55	14.7
23	NSH	41	182	26	14.3	BUF	41	181	24	13.3	NSH	82	337	48	14.2
24	PIT	41	184	26	14.1	MIN	41	175	23	13.1	PIT	82	335	47	14.0
25	BUF	41	194	26	13.4	PHI	41	140	18	12.9	CBJ	82	356	50	14.0
26	PHI	41	160	21	13.1	CBJ	41	168	21	12.5	BOS	82	283	39	13.8
27	BOS	41	161	21	13.0	ATL	41	147	17	11.6	BUF	82	375	50	13.3
28	ANA	41	181	23	12.7	NYR	41	150	17	11.3	PHI	82	300	39	13.0
29	ATL	41	159	20	12.6	ANA	41	192	20	10.4	ATL	82	306	37	12.1
30	NYI	41	176	22	12.5	FLA	41	154	15	9.7	ANA	82	373	43	11.5
TOTAL		**1230**	**5296**	**869**	**16.4**		**1230**	**4858**	**732**	**15.1**		**1230**	**10154**	**1601**	**15.8**

TEAMS' PENALTY KILLING RECORD

Abbreviations: TSH – total times short-handed; **PPGA** – power-play goals against;
% – calculated by dividing times short minus power-play goals against by times short.

	Home					Road					Overall				
	Team	GP	TSH	PPGA	%	Team	GP	TSH	PPGA	%	Team	GP	TSH	PPGA	%
1	BOS	41	177	18	89.8	DET	41	173	20	88.4	BOS	82	349	45	87.1
2	NSH	41	176	18	89.8	ANA	41	171	21	87.7	COL	82	315	41	87.0
3	MTL	41	125	13	89.6	COL	41	167	21	87.4	L.A.	82	366	49	86.6
4	TOR	41	156	17	89.1	NYI	41	204	28	86.3	BUF	82	304	41	86.5
5	PHI	41	133	15	88.7	L.A.	41	190	26	86.3	MTL	82	277	38	86.3
6	S.J.	41	170	22	87.1	BUF	41	150	21	86.0	PHI	82	293	40	86.3
7	BUF	41	154	20	87.0	T.B.	41	142	20	85.9	ANA	82	330	46	86.1
8	L.A.	41	176	23	86.9	ST.L.	41	204	29	85.8	DET	82	343	48	86.0
9	WSH	41	151	20	86.8	PIT	41	181	27	85.1	NSH	82	354	50	85.9
10	EDM	41	174	23	86.8	S.J.	41	211	32	84.8	S.J.	82	381	54	85.8
11	VAN	41	186	25	86.6	N.J.	41	136	21	84.6	NYI	82	366	52	85.8
12	COL	41	148	20	86.5	EDM	41	174	27	84.5	EDM	82	348	50	85.6
13	PHX	41	163	24	85.3	PHI	41	160	25	84.4	T.B.	82	273	40	85.3
14	NYI	41	162	24	85.2	BOS	41	172	27	84.3	ST.L.	82	379	56	85.2
15	CAR	41	158	24	84.8	OTT	41	157	25	84.1	VAN	82	379	57	85.0
16	OTT	41	150	23	84.7	CBJ	41	163	26	84.0	TOR	82	328	51	84.5
17	T.B.	41	131	20	84.7	MTL	41	152	25	83.6	OTT	82	307	48	84.4
18	ST.L.	41	175	27	84.6	VAN	41	193	32	83.4	PIT	82	352	57	83.8
19	DAL	41	147	23	84.4	FLA	41	189	32	83.1	N.J.	82	265	43	83.8
20	ANA	41	159	25	84.3	CAR	41	174	30	82.8	CAR	82	332	54	83.7
21	ATL	41	169	27	84.0	CHI	41	171	30	82.5	PHX	82	356	60	83.1
22	DET	41	170	28	83.5	MIN	41	199	35	82.4	CBJ	82	306	52	83.0
23	MIN	41	163	27	83.4	NSH	41	178	32	82.0	MIN	82	362	62	82.9
24	CGY	41	184	31	83.2	PHX	41	193	36	81.3	WSH	82	325	58	82.2
25	N.J.	41	129	22	82.9	NYR	41	213	40	81.2	ATL	82	358	65	81.8
26	PIT	41	171	30	82.5	TOR	41	172	34	80.2	CGY	82	356	66	81.5
27	CBJ	41	143	26	81.8	ATL	41	189	38	79.9	CHI	82	340	63	81.5
28	CHI	41	169	33	80.5	CGY	41	172	35	79.7	DAL	82	319	59	81.5
29	FLA	41	204	44	78.4	DAL	41	172	36	79.1	FLA	82	393	76	80.7
30	NYR	41	185	40	78.4	WSH	41	174	38	78.2	NYR	82	398	80	79.9
TOTAL		**1230**	**4858**	**732**	**84.9**		**1230**	**5296**	**869**	**83.6**		**1230**	**10154**	**1601**	**84.2**

SHORT-HANDED GOALS FOR

	Home			Road			Overall		
	Team	GP	SHGF	Team	GP	SHGF	Team	GP	SHGF
1	NYI	41	10	S.J.	41	11	NYI	82	17
2	CGY	41	7	BOS	41	9	S.J.	82	15
3	EDM	41	7	OTT	41	7	BOS	82	14
4	COL	41	7	NYI	41	7	EDM	82	12
5	PHX	41	7	MIN	41	7	OTT	82	11
6	BOS	41	5	EDM	41	5	MIN	82	11
7	BUF	41	5	DET	41	5	CGY	82	10
8	PHI	41	5	NYR	41	5	TOR	82	9
9	MTL	41	5	TOR	41	5	DET	82	9
10	NSH	41	5	FLA	41	4	PHI	82	8
11	TOR	41	4	ST.L.	41	4	FLA	82	8
12	OTT	41	4	DAL	41	3	PHX	82	8
13	S.J.	41	4	PHI	41	3	NYR	82	8
14	DAL	41	4	L.A.	41	3	COL	82	7
15	FLA	41	4	CGY	41	3	BUF	82	7
16	DET	41	4	T.B.	41	3	MTL	82	7
17	MIN	41	4	ANA	41	2	DAL	82	7
18	CHI	41	4	MTL	41	2	T.B.	82	6
19	PIT	41	3	BUF	41	2	NSH	82	6
20	T.B.	41	3	CAR	41	2	PIT	82	5
21	NYR	41	3	PIT	41	2	ST.L.	82	5
22	CBJ	41	2	ATL	41	1	L.A.	82	5
23	VAN	41	2	PHX	41	1	ANA	82	4
24	L.A.	41	2	NSH	41	1	CHI	82	4
25	CAR	41	2	CBJ	41	1	CAR	82	4
26	WSH	41	2	VAN	41	1	ATL	82	3
27	N.J.	41	2	COL	41	0	CBJ	82	3
28	ANA	41	2	CHI	41	0	VAN	82	3
29	ATL	41	2	N.J.	41	0	WSH	82	2
30	ST.L.	41	1	WSH	41	0	N.J.	82	2
TOTAL		**1230**	**121**		**1230**	**99**		**1230**	**220**

SHORT-HANDED GOALS AGAINST

	Home			Road			Overall		
	Team	GP	SHGA	Team	GP	SHGA	Team	GP	SHGA
1	BUF	41	0	CHI	41	1	CHI	82	2
2	MTL	41	0	S.J.	41	1	MTL	82	4
3	CGY	41	1	COL	41	2	OTT	82	5
4	OTT	41	1	PHX	41	2	CGY	82	5
5	CHI	41	1	NYR	41	2	BOS	82	5
6	ST.L.	41	1	EDM	41	3	NYR	82	5
7	TOR	41	2	ANA	41	3	TOR	82	6
8	BOS	41	2	FLA	41	3	BUF	82	6
9	CBJ	41	3	WSH	41	3	L.A.	82	6
10	VAN	41	3	L.A.	41	3	PHX	82	6
11	L.A.	41	3	BOS	41	3	S.J.	82	6
12	NSH	41	3	OTT	41	4	COL	82	7
13	NYR	41	3	MTL	41	4	EDM	82	7
14	DET	41	3	DET	41	4	FLA	82	7
15	MIN	41	3	TOR	41	4	ST.L.	82	7
16	N.J.	41	3	DAL	41	4	DET	82	7
17	T.B.	41	4	CGY	41	4	CBJ	82	8
18	DAL	41	4	PIT	41	4	DAL	82	8
19	FLA	41	4	NYI	41	5	NSH	82	8
20	EDM	41	4	NSH	41	5	WSH	82	8
21	PHX	41	4	CBJ	41	5	N.J.	82	8
22	PHI	41	5	N.J.	41	5	ANA	82	9
23	S.J.	41	5	PHI	41	5	VAN	82	9
24	CAR	41	5	T.B.	41	5	T.B.	82	9
25	WSH	41	5	ST.L.	41	6	MIN	82	9
26	COL	41	5	MIN	41	6	PIT	82	10
27	NYI	41	5	BUF	41	6	PHI	82	10
28	ATL	41	5	VAN	41	6	NYI	82	10
29	PIT	41	6	CAR	41	6	CAR	82	11
30	ANA	41	6	ATL	41	7	ATL	82	12
TOTAL		**1230**	**99**		**1230**	**121**		**1230**	**220**

Regular-Season Overtime Results

1983-84 to 2001-02

	2001-02				2000-01				1999-2000				1998-99				1997-98			
Team	GP	W	L	T	GP	W	L	T	GP	W	L	T	GP	W	L	T	GP	W	L	T
ANA	14	3	3	8	20	4	5	11	18	3	3	12	17	1	3	13	20	3	4	13
ATL	19	3	5	11	16	2	2	12	11	0	4	7	...	...	...	...	...	...	...	...
BOS	24	9	9	6	20	4	8	8	26	1	6	19	17	2	2	13	17	3	1	13
BUF	16	4	1	11	10	4	1	5	20	5	4	11	23	3	3	17	21	3	1	17
CGY	17	2	3	12	22	3	4	15	26	11	5	10	16	3	1	12	22	4	3	15
CAR/HFD	27	6	5	16	18	6	3	9	14	4	0	10	24	1	5	18	12	2	2	8
CHI	17	3	1	13	15	2	5	8	17	5	2	10	15	1	2	12	18	1	4	13
COL/QUE	13	4	1	8	20	6	4	10	17	5	1	11	12	2	0	10	22	2	3	17
CBJ	15	2	5	8	18	3	6	9	...	...	...	...	...	...	...	...	...	...	...	...
DAL/MIN	21	3	5	13	16	6	2	8	19	3	6	10	16	3	1	12	17	5	1	11
DET	24	10	4	10	23	10	4	9	16	4	2	10	10	2	1	7	15	0	0	15
EDM	19	3	4	12	20	5	3	12	27	3	8	16	20	3	5	12	15	3	2	10
FLA	16	0	6	10	24	2	9	13	15	3	6	6	21	1	2	18	20	3	2	15
L.A.	18	3	4	11	19	3	3	13	21	5	4	12	12	5	2	5	16	3	2	11
MIN	21	0	9	12	22	4	5	13	...	...	...	...	...	...	...	...	...	...	...	...
MTL	17	2	3	12	16	2	6	8	17	4	4	9	15	0	4	11	20	3	4	13
NSH	18	5	0	13	17	5	3	9	18	4	7	7	10	1	2	7	...	...	...	...
N.J.	19	6	4	9	20	5	3	12	16	3	5	8	15	3	1	11	16	2	3	11
NYI	18	6	4	8	12	2	3	7	15	5	1	9	17	1	6	10	13	0	2	11
NYR	13	5	4	4	11	5	1	5	21	6	3	12	19	5	3	11	24	2	4	18
OTT	19	3	7	9	16	3	4	9	15	2	2	11	18	1	2	15	17	2	0	15
PHI	16	3	3	10	19	5	3	11	21	6	3	12	24	2	3	19	15	3	1	11
PHX/WPG	19	4	6	9	23	3	3	17	16	4	4	8	15	2	1	12	14	0	2	12
PIT	20	7	5	8	15	3	3	9	17	3	6	8	22	7	1	14	23	3	2	18
ST.L.	18	6	4	8	23	6	5	12	17	5	1	11	15	1	1	13	12	2	2	8
S.J.	13	2	3	8	22	7	3	12	21	4	7	10	21	1	2	18	12	0	2	10
T.B.	19	4	4	11	13	2	5	6	16	0	7	9	12	1	2	9	13	0	3	10
TOR	17	3	4	10	19	3	5	11	17	7	3	7	14	6	1	7	10	1	0	9
VAN	14	4	3	7	23	5	7	11	27	4	8	15	13	0	1	12	17	0	3	14
WSH	19	6	2	11	16	2	4	10	19	5	2	12	11	2	3	6	17	4	1	12
Totals	**270**	**121**		**149**	**274**	**122**		**152**	**260**	**114**		**146**	**222**	**60**		**162**	**219**	**54**		**165**

	1996-97				1995-96				1994-95				1993-94				1992-93			
Team	GP	W	L	T	GP	W	L	T	GP	W	L	T	GP	W	L	T	GP	W	L	T
ANA	16	3	0	13	16	6	2	8	7	2	0	5	12	2	5	5	...	...	...	...
ATL	...	...	...	...	...	...	...	...	...	...	...	...	...	...	...	...	...	...	...	...
BOS	15	3	3	9	19	2	6	11	8	2	3	3	17	2	2	13	15	5	3	7
BUF	21	5	4	12	15	2	6	7	9	1	1	7	13	0	4	9	18	4	4	10
CGY	16	3	4	9	16	2	3	11	9	1	1	7	18	3	2	13	19	4	4	11
CAR/HFD	18	3	4	11	14	2	3	9	9	1	1	7	14	4	1	9	18	3	9	6
CHI	19	1	5	13	19	1	4	14	7	2	0	5	16	2	5	9	16	1	3	12
COL/QUE	15	2	3	10	6	1	0	5	8	0	0	8	15	3	3	9	15	4	1	10
CBJ	...	...	...	...	...	...	...	...	...	...	...	...	...	...	...	...	...	...	...	...
DAL/MIN	15	4	3	8	15	1	0	14	9	0	1	8	22	6	3	13	10	0	0	10
DET	27	7	2	18	11	3	1	7	4	0	0	4	15	5	2	8	11	2	0	9
EDM	16	1	6	9	14	4	2	8	7	1	2	4	21	1	6	14	17	5	4	8
FLA	26	3	4	19	13	0	3	10	9	0	3	6	24	2	5	17	...	...	...	...
L.A.	14	0	3	11	23	3	2	18	9	0	0	9	18	3	3	12	13	2	1	10
MIN	...	...	...	...	...	...	...	...	...	...	...	...	...	...	...	...	...	...	...	...
MTL	21	2	4	15	15	2	3	10	10	1	2	7	19	3	2	14	14	5	3	6
NSH	...	...	...	...	...	...	...	...	...	...	...	...	...	...	...	...	...	...	...	...
N.J.	17	1	2	14	19	7	0	12	11	1	2	8	14	1	1	12	11	4	0	7
NYI	17	3	2	12	17	2	5	10	7	1	1	5	19	5	2	12	13	3	3	7
NYR	13	3	0	10	17	2	1	14	3	0	0	3	12	3	1	8	17	2	4	11
OTT	17	0	2	15	8	0	3	5	7	1	1	5	17	4	4	9	10	0	6	4
PHI	18	3	2	13	20	4	3	13	8	3	1	4	18	3	5	10	17	4	2	11
PHX/WPG	16	5	4	7	8	2	0	6	9	0	2	7	15	1	5	9	11	2	2	7
PIT	13	1	4	8	9	3	2	4	5	1	1	3	19	4	2	13	10	3	0	7
ST.L.	13	1	1	11	18	1	1	16	7	1	1	5	17	4	2	11	17	2	4	11
S.J.	12	3	1	8	9	1	1	7	5	1	0	4	19	2	1	16	10	3	5	2
T.B.	16	4	2	10	18	3	3	12	7	2	2	3	18	3	4	11	14	3	4	7
TOR	10	1	1	8	18	4	2	12	8	0	0	8	17	4	1	12	13	1	1	11
VAN	14	5	2	7	20	1	4	15	13	0	1	12	12	5	4	3	10	1	0	9
WSH	13	2	2	9	16	4	1	11	9	0	1	8	14	2	2	10	11	2	2	7
Totals	**214**	**70**		**144**	**201**	**64**		**137**	**101**	**26**		**75**	**214**	**74**		**140**	**165**	**65**		**100**

2001-02
Home Team Wins: 62
Visiting Team Wins: 59

	1991-92				1990-91				1989-90				1988-89				1987-88			
Team	GP	W	L	T	GP	W	L	T	GP	W	L	T	GP	W	L	T	GP	W	L	T
ANA	...	...	...	...	...	...	...	...	...	...	...	...	...	...	...	...	...	...	...	...
ATL	...	...	...	...	...	...	...	...	...	...	...	...	...	...	...	...	...	...	...	...
BOS	20	6	2	12	17	5	0	12	14	3	2	9	19	3	2	14	14	4	4	6
BUF	16	2	2	12	24	3	2	19	15	4	3	8	13	2	4	7	12	0	1	11
CGY	19	2	5	12	15	3	4	8	21	3	3	15	17	5	3	9	15	2	4	9
CAR/HFD	18	2	3	13	9	1	1	7	9	0	0	9	10	1	4	5	12	3	2	7
CHI	19	2	2	15	12	3	1	8	10	2	2	6	17	2	3	12	15	4	2	9
COL/QUE	17	0	5	12	18	1	3	14	8	0	1	7	10	2	1	7	9	2	2	5
CBJ	...	...	...	...	...	...	...	...	...	...	...	...	...	...	...	...	...	...	...	...
DAL/MIN	8	0	2	6	17	0	3	14	11	3	4	4	17	0	1	16	16	1	2	13
DET	16	3	1	12	14	2	4	8	17	2	1	14	16	3	1	12	16	2	3	11
EDM	12	0	2	10	15	4	5	6	20	5	1	14	15	4	3	8	16	3	2	11
FLA	...	...	...	...	...	...	...	...	...	...	...	...	...	...	...	...	...	...	...	...
L.A.	16	1	1	14	16	4	2	10	12	3	2	7	14	6	1	7	12	1	3	8
MIN	...	...	...	...	...	...	...	...	...	...	...	...	...	...	...	...	...	...	...	...
MTL	20	6	3	11	17	3	3	11	17	4	2	11	11	2	0	9	16	1	2	13
NSH	...	...	...	...	...	...	...	...	...	...	...	...	...	...	...	...	...	...	...	...
N.J.	17	2	4	11	17	1	1	15	16	3	4	9	17	1	4	12	12	4	2	6
NYI	16	3	2	11	15	2	3	10	16	3	2	11	11	3	3	5	13	3	0	10
NYR	11	5	1	5	16	1	2	13	17	2	2	13	10	1	1	8	11	0	1	10
OTT	...	...	...	...	...	...	...	...	...	...	...	...	...	...	...	...	...	...	...	...
PHI	17	2	4	11	11	1	0	10	18	2	5	11	14	1	5	8	13	1	3	9
PHX/WPG	20	1	4	15	14	1	2	11	19	4	4	11	20	6	2	12	21	8	2	11
PIT	12	2	1	9	12	4	2	6	14	3	3	8	10	2	1	7	16	5	2	9
ST.L.	15	2	2	11	18	3	4	11	15	2	4	9	16	3	1	12	14	2	4	8
S.J.	9	1	3	5	...	...	...	...	...	...	...	...	...	...	...	...	...	...	...	...
T.B.	...	...	...	...	...	...	...	...	...	...	...	...	...	...	...	...	...	...	...	...
TOR	11	4	0	7	17	4	2	11	11	3	4	4	11	1	4	6	13	1	2	10
VAN	17	4	1	12	15	3	3	9	21	2	5	14	14	2	4	8	11	0	2	9
WSH	12	2	2	8	14	4	3	7	9	2	1	6	16	2	4	10	15	2	4	9
Totals	**169**	**52**		**117**	**166**	**54**		**112**	**155**	**55**		**100**	**149**	**52**		**97**	**146**	**49**		**97**

	1986-87				1985-86				1984-85				1983-84			
Team	GP	W	L	T	GP	W	L	T	GP	W	L	T	GP	W	L	T
ANA	...	...	...	...	...	...	...	...	...	...	...	...	...	...	...	...
ATL	...	...	...	...	...	...	...	...	...	...	...	...	...	...	...	...
BOS	12	2	3	7	17	2	3	12	18	4	4	10	7	1	0	6
BUF	13	1	4	8	9	1	2	6	17	0	3	14	13	5	1	7
CGY	4	1	0	3	12	1	2	9	14	1	1	12	18	4	0	14
CAR/HFD	9	2	0	7	7	1	2	4	17	4	4	9	15	2	3	10
CHI	15	1	0	14	12	3	1	8	12	2	3	7	9	0	1	8
COL/QUE	14	0	4	10	11	4	1	6	14	3	2	9	15	0	5	10
CBJ	...	...	...	...	...	...	...	...	...	...	...	...	...	...	...	...
DAL/MIN	14	2	2	10	15	4	2	9	15	1	2	12	18	5	3	10
DET	17	2	5	10	13	2	5	6	14	0	2	12	11	3	1	7
EDM	14	5	3	6	14	5	2	7	12	0	1	11	9	4	0	5
FLA	...	...	...	...	...	...	...	...	...	...	...	...	...	...	...	...
L.A.	12	2	2	8	14	3	3	8	19	3	2	14	17	1	3	13
MIN	...	...	...	...	...	...	...	...	...	...	...	...	...	...	...	...
MTL	16	2	4	10	14	1	6	7	18	3	3	12	7	1	1	5
NSH	...	...	...	...	...	...	...	...	...	...	...	...	...	...	...	...
N.J.	13	3	4	6	10	4	3	3	12	0	2	10	15	1	7	7
NYI	19	4	3	12	17	4	1	12	15	1	8	6	10	3	3	4
NYR	19	5	6	8	13	0	7	6	17	2	5	10	17	5	3	9
OTT	...	...	...	...	...	...	...	...	...	...	...	...	...	...	...	...
PHI	10	1	1	8	9	4	1	4	9	1	1	7	14	3	1	10
PHX/WPG	11	2	1	8	8	0	1	7	14	3	1	10	24	7	6	11
PIT	21	5	4	12	14	3	3	8	8	3	0	5	12	1	5	6
ST.L.	21	4	2	15	17	5	3	9	15	2	1	12	11	3	1	7
S.J.	...	...	...	...	...	...	...	...	...	...	...	...	...	...	...	...
T.B.	...	...	...	...	...	...	...	...	...	...	...	...	...	...	...	...
TOR	13	3	4	6	17	4	6	7	15	5	2	8	13	1	3	9
VAN	10	2	0	8	16	1	2	13	17	7	1	9	16	3	4	9
WSH	17	5	2	10	11	4	0	7	12	3	0	9	9	1	3	5
Totals	**147**	**54**		**93**	**135**	**56**		**79**	**152**	**48**		**104**	**140**	**54**		**86**

NHL Record Book

Year-By-Year Final Standings & Leading Scorers

*Stanley Cup winner

1917-18

First Half

Team	GP	W	L	T	GF	GA	PTS
Montreal	14	10	4	0	81	47	20
Toronto	14	8	6	0	71	75	16
Ottawa	14	5	9	0	67	79	10
**Mtl. Wanderers	6	1	5	0	17	35	2

**Montreal Arena burned down and Wanderers forced to withdraw from League. Montreal Canadiens and Toronto each counted a win for defaulted games with Wanderers.

Second Half

Team	GP	W	L	T	GF	GA	PTS
*Toronto	8	5	3	0	37	34	10
Ottawa	8	4	4	0	35	35	8
Montreal	8	3	5	0	34	37	6

Leading Scorers

Player	Club	GP	G	A	PTS	PIM
Malone, Joe	Montreal	20	44	4	48	30
Denneny, Cy	Ottawa	20	36	10	46	80
Noble, Reg	Toronto	20	30	10	40	35
Lalonde, Newsy	Montreal	14	23	7	30	51
Denneny, Corb	Toronto	21	20	9	29	14
Cameron, Harry	Toronto	21	17	10	27	28
Pitre, Didier	Montreal	20	17	6	23	29
Gerard, Eddie	Ottawa	20	13	7	20	26
Darragh, Jack	Ottawa	18	14	5	19	26
Nighbor, Frank	Ottawa	10	11	8	19	6
Meeking, Harry	Toronto	21	10	9	19	28

1918-19

First Half

Team	GP	W	L	T	GF	GA	PTS
• Montreal	10	7	3	0	57	50	14
Ottawa	10	5	5	0	39	39	10
Toronto	10	3	7	0	42	49	6

Second Half

Team	GP	W	L	T	GF	GA	PTS
Ottawa	8	7	1	0	32	14	14
Montreal	8	3	5	0	31	28	6
Toronto	8	2	6	0	22	43	4

• NHL Champion. Stanley Cup not awarded due to influenza epidemic.

Leading Scorers

Player	Club	GP	G	A	PTS	PIM
Lalonde, Newsy	Montreal	17	22	10	32	40
Cleghorn, Odie	Montreal	17	22	6	28	22
Nighbor, Frank	Ottawa	18	19	9	28	27
Denneny, Cy	Ottawa	18	18	4	22	58
Pitre, Didier	Montreal	17	14	5	19	12
Skinner, Alf	Toronto	17	12	4	16	26
Cameron, Harry	Tor., Ott.	14	11	3	14	35
Darragh, Jack	Ottawa	14	11	3	14	33
Randall, Ken	Toronto	15	8	6	14	27
Cleghorn, Sprague	Ottawa	18	7	6	13	27

All-Time Standings of NHL Teams

(ranked by percentage)

Active Clubs

Team	Games	Wins	Losses	Ties	OT Losses	Goals For	Goals Against	Points	Pts %	First Season
Montreal	5382	2778	1769	822	13	17871	14360	6391	.593	1917-18
Philadelphia	2768	1383	947	429	9	9591	8205	3204	.577	1967-68
Boston	5222	2487	1947	765	23	16999	15464	5762	.550	1924-25
Buffalo	2542	1193	951	392	6	8666	7812	2784	.546	1970-71
Edmonton	1830	868	708	239	15	6935	6353	1990	.540	1979-80
Calgary	2386	1071	944	359	12	8405	7907	2513	.524	1972-73
Detroit	5156	2263	2089	794	10	16012	15582	5330	.515	1926-27
St. Louis	2768	1208	1140	410	10	8908	8896	2836	.510	1967-68
Colorado	1830	815	774	235	6	6413	6300	1871	.508	1979-80
Toronto	5382	2329	2275	766	12	16715	16639	5436	.504	1917-18
NY Islanders	2386	1039	1014	325	8	8128	7736	2411	.503	1972-73
NY Rangers	5156	2172	2185	791	8	16061	16235	5143	.497	1926-27
Chicago	5156	2126	2232	790	8	15448	15651	5050	.489	1926-27
Washington	2230	949	988	285	8	7247	7538	2191	.489	1974-75
Dallas	2768	1118	1206	431	13	8736	9170	2680	.482	1967-68
Pittsburgh	2768	1149	1236	369	14	9559	10001	2681	.482	1967-68
Florida	706	270	301	114	21	1860	1984	675	.471	1993-94
Los Angeles	2768	1100	1255	402	11	9330	9929	2613	.470	1967-68
Phoenix	1830	732	848	237	13	6105	6664	1714	.465	1979-80
Carolina	1830	711	873	238	8	5760	6424	1668	.452	1979-80
New Jersey	2230	845	1067	306	12	6970	7815	2008	.447	1974-75
Anaheim	706	269	338	88	11	1833	2032	637	.444	1993-94
Vancouver	2542	936	1220	368	18	8159	9126	2258	.441	1970-71
Ottawa	790	288	392	97	13	2130	2493	686	.428	1992-93
San Jose	870	311	446	100	13	2377	2846	735	.415	1991-92
Nashville	328	118	164	36	10	771	931	282	.411	1998-99
Minnesota	164	51	74	25	14	363	448	141	.408	2000-01
Tampa Bay	790	246	440	88	16	1957	2592	596	.371	1992-93
Columbus	164	50	86	17	11	354	488	128	.369	2000-01
Atlanta	246	56	149	30	11	568	890	153	.297	99-2000

Defunct Clubs

Team	Games	Wins	Losses	Ties	Goals For	Goals Against	Points	Pts %	First Season	Last Season
Ottawa Senators	542	258	221	63	1458	1333	579	.534	1917-18	1933-34
Montreal Maroons	622	271	260	91	1474	1405	633	.509	1924-25	1937-38
NY/Brooklyn Americans	784	255	402	127	1643	2182	637	.406	1925-26	1941-42
Hamilton Tigers	126	47	78	1	414	475	95	.377	1920-21	1924-25
Cleveland Barons	160	47	87	26	470	617	120	.375	1976-77	1977-78
Pittsburgh Pirates	212	67	122	23	376	519	157	.370	1925-26	1929-30
Calif./Oakland Seals	698	182	401	115	1826	2580	479	.343	1967-68	1975-76
St. Louis Eagles	48	11	31	6	86	144	28	.292	1934-35	1934-35
Quebec Bulldogs	24	4	20	0	91	177	8	.167	1919-20	1919-20
Montreal Wanderers	6	1	5	0	17	35	2	.167	1917-18	1917-18
Philadelphia Quakers	44	4	36	4	76	184	12	.136	1930-31	1930-31

Calgary totals include Atlanta Flames, 1972-73 to 1979-80.
Carolina totals include Hartford, 1979-80 to 1996-97.
Colorado totals include Quebec, 1979-80 to 1994-95.
Dallas totals include Minnesota North Stars, 1967-68 to 1992-93.
Detroit totals include Cougars, 1926-27 to 1929-30, and Falcons, 1930-31 to 1931-32.
New Jersey totals include Kansas City, 1974-75 to 1975-76, and Colorado Rockies, 1976-77 to 1981-82.
Phoenix totals include Winnipeg, 1979-80 to 1995-96.
Toronto totals include Arenas, 1917-18 to 1918-19, and St. Patricks, 1919-20 to 1925-26.

1919-20

First Half

Team	GP	W	L	T	GF	GA	PTS
Ottawa	12	9	3	0	59	23	18
Montreal	12	8	4	0	62	51	16
Toronto	12	5	7	0	52	62	10
Quebec	12	2	10	0	44	81	4

Second Half

Team	GP	W	L	T	GF	GA	PTS
*Ottawa	12	10	2	0	62	41	20
Montreal	12	5	7	0	67	62	10
Toronto	12	7	5	0	67	44	14
Quebec	12	2	10	0	47	96	4

Leading Scorers

Player	Club	GP	G	A	PTS	PIM
Malone, Joe	Quebec	24	39	10	49	12
Lalonde, Newsy	Montreal	23	37	9	46	34
Nighbor, Frank	Ottawa	23	26	15	41	18
Denneny, Corb	Toronto	24	24	12	36	20
Darragh, Jack	Ottawa	23	22	14	36	22
Noble, Reg	Toronto	24	24	9	33	52
Arbour, Amos	Montreal	22	21	5	26	13
Wilson, Cully	Toronto	23	20	6	26	86
Pitre, Didier	Montreal	22	14	12	26	6
Broadbent, Punch	Ottawa	21	19	6	25	40

1920-21

First Half

Team	GP	W	L	T	GF	GA	PTS
*Ottawa	10	8	2	0	49	23	16
Toronto	10	5	5	0	39	47	10
Montreal	10	4	6	0	37	51	8
Hamilton	10	3	7	0	34	38	6

Second Half

Team	GP	W	L	T	GF	GA	PTS
Toronto	14	10	4	0	66	53	20
Montreal	14	9	5	0	75	48	18
Ottawa	14	6	8	0	48	52	12
Hamilton	14	3	11	0	58	94	6

Leading Scorers

Player	Club	GP	G	A	PTS	PIM
Lalonde, Newsy	Montreal	24	33	10	43	36
Dye, Babe	Ham., Tor.	24	35	5	40	32
Denneny, Cy	Ottawa	24	34	5	39	10
Malone, Joe	Hamilton	20	28	9	37	6
Nighbor, Frank	Ottawa	24	19	10	29	10
Noble, Reg	Toronto	24	19	8	27	54
Cameron, Harry	Toronto	24	18	9	27	35
Prodgers, Goldie	Hamilton	24	18	9	27	8
Denneny, Corb	Toronto	20	19	7	26	29
Darragh, Jack	Ottawa	24	11	15	26	20

1921-22

Team	GP	W	L	T	GF	GA	PTS
Ottawa	24	14	8	2	106	84	30
*Toronto	24	13	10	1	98	97	27
Montreal	24	12	11	1	88	94	25
Hamilton	24	7	17	0	88	105	14

Leading Scorers

Player	Club	GP	G	A	PTS	PIM
Broadbent, Punch	Ottawa	24	32	14	46	28
Denneny, Cy	Ottawa	22	27	12	39	20
Dye, Babe	Toronto	24	31	7	38	39
Cameron, Harry	Toronto	24	18	17	35	22
Malone, Joe	Hamilton	24	24	7	31	4
Denneny, Corb	Toronto	24	19	9	28	28
Noble, Reg	Toronto	24	17	11	28	19
Cleghorn, Sprague	Montreal	24	17	9	26	80
Boucher, Georges	Ottawa	23	13	12	25	12
Cleghorn, Odie	Montreal	23	21	3	24	26

1922-23

Team	GP	W	L	T	GF	GA	PTS
*Ottawa	24	14	9	1	77	54	29
Montreal	24	13	9	2	73	61	28
Toronto	24	13	10	1	82	88	27
Hamilton	24	6	18	0	81	110	12

Leading Scorers

Player	Club	GP	G	A	PTS	PIM
Dye, Babe	Toronto	22	26	11	37	19
Denneny, Cy	Ottawa	24	23	11	34	28
Boucher, Billy	Montreal	24	24	7	31	55
Adams, Jack	Toronto	23	19	9	28	42
Roach, Mickey	Hamilton	24	17	10	27	8
Cleghorn, Odie	Montreal	24	19	6	25	18
Boucher, Georges	Ottawa	24	14	9	23	58
Noble, Reg	Toronto	24	12	11	23	47
Wilson, Cully	Hamilton	23	16	5	21	46
Joliat, Aurel	Montreal	24	12	9	21	37

1923-24

Team	GP	W	L	T	GF	GA	PTS
Ottawa	24	16	8	0	74	54	32
*Montreal	24	13	11	0	59	48	26
Toronto	24	10	14	0	59	85	20
Hamilton	24	9	15	0	63	68	18

Leading Scorers

Player	Club	GP	G	A	PTS	PIM
Denneny, Cy	Ottawa	22	22	2	24	10
Boucher, Georges	Ottawa	21	13	10	23	38
Boucher, Billy	Montreal	23	16	6	22	48
Burch, Billy	Hamilton	24	16	6	22	6
Joliat, Aurel	Montreal	24	15	5	20	27
Dye, Babe	Toronto	19	16	3	19	23
Adams, Jack	Toronto	22	14	4	18	51
Noble, Reg	Toronto	23	12	5	17	79
Morenz, Howie	Montreal	24	13	3	16	20
Clancy, King	Ottawa	24	8	8	16	26

1924-25

Team	GP	W	L	T	GF	GA	PTS
Hamilton	30	19	10	1	90	60	39
Toronto	30	19	11	0	90	84	38
• Montreal	30	17	11	2	93	56	36
Ottawa	30	17	12	1	83	66	35
Mtl. Maroons	30	9	19	2	45	65	20
Boston	30	6	24	0	49	119	12

• NHL Champion (Stanley Cup won by Victoria Cougars, WCHL)

Leading Scorers

Player	Club	GP	G	A	PTS	PIM
Dye, Babe	Toronto	29	38	8	46	41
Denneny, Cy	Ottawa	29	27	15	42	16
Joliat, Aurel	Montreal	25	30	11	41	85
Morenz, Howie	Montreal	30	28	11	39	46
Green, Red	Hamilton	30	19	15	34	81
Adams, Jack	Toronto	27	21	10	31	67
Boucher, Billy	Montreal	30	17	13	30	92
Burch, Billy	Hamilton	27	20	7	27	10
Herberts, Jimmy	Boston	30	17	7	24	55
Smith, Hooley	Ottawa	30	10	13	23	81

1925-26

Team	GP	W	L	T	GF	GA	PTS
Ottawa	36	24	8	4	77	42	52
*Mtl. Maroons	36	20	11	5	91	73	45
Pittsburgh	36	19	16	1	82	70	39
Boston	36	17	15	4	92	85	38
NY Americans	36	12	20	4	68	89	28
Toronto	36	12	21	3	92	114	27
Montreal	36	11	24	1	79	108	23

Leading Scorers

Player	Club	GP	G	A	PTS	PIM
Stewart, Nels	Mtl. Maroons	36	34	8	42	119
Denneny, Cy	Ottawa	36	24	12	36	18
Cooper, Carson	Boston	36	28	3	31	10
Herberts, Jimmy	Boston	36	26	5	31	47
Morenz, Howie	Montreal	31	23	3	26	39
Adams, Jack	Toronto	36	21	5	26	52
Joliat, Aurel	Montreal	35	17	9	26	52
Burch, Billy	NY Americans	36	22	3	25	33
Smith, Hooley	Ottawa	28	16	9	25	53
Nighbor, Frank	Ottawa	35	12	13	25	40

1926-27

Canadian Division

Team	GP	W	L	T	GF	GA	PTS
*Ottawa	44	30	10	4	86	69	64
Montreal	44	28	14	2	99	67	58
Mtl. Maroons	44	20	20	4	71	68	44
NY Americans	44	17	25	2	82	91	36
Toronto	44	15	24	5	79	94	35

American Division

Team	GP	W	L	T	GF	GA	PTS
NY Rangers	44	25	13	6	95	72	56
Boston	44	21	20	3	97	89	45
Chicago	44	19	22	3	115	116	41
Pittsburgh	44	15	26	3	79	108	33
Detroit	44	12	28	4	76	105	28

Leading Scorers

Player	Club	GP	G	A	PTS	PIM
Cook, Bill	NY Rangers	44	33	4	37	58
Irvin, Dick	Chicago	43	18	18	36	34
Morenz, Howie	Montreal	44	25	7	32	49
Fredrickson, Frank	Det., Bos.	41	18	13	31	46
Dye, Babe	Chicago	41	25	5	30	14
Bailey, Ace	Toronto	42	15	13	28	82
Boucher, Frank	NY Rangers	44	13	15	28	17
Burch, Billy	NY Americans	43	19	8	27	40
Oliver, Harry	Boston	42	18	6	24	17
Keats, Duke	Bos., Det.	42	16	8	24	52

1927-28

Canadian Division

Team	GP	W	L	T	GF	GA	PTS
Montreal	44	26	11	7	116	48	59
Mtl. Maroons	44	24	14	6	96	77	54
Ottawa	44	20	14	10	78	57	50
Toronto	44	18	18	8	89	88	44
NY Americans	44	11	27	6	63	128	28

American Division

Team	GP	W	L	T	GF	GA	PTS
Boston	44	20	13	11	77	70	51
*NY Rangers	44	19	16	9	94	79	47
Pittsburgh	44	19	17	8	67	76	46
Detroit	44	19	19	6	88	79	44
Chicago	44	7	34	3	68	134	17

Leading Scorers

Player	Club	GP	G	A	PTS	PIM
Morenz, Howie	Montreal	43	33	18	51	66
Joliat, Aurel	Montreal	44	28	11	39	105
Boucher, Frank	NY Rangers	44	23	12	35	15
Hay, George	Detroit	42	22	13	35	20
Stewart, Nels	Mtl. Maroons	41	27	7	34	104
Gagne, Art	Montreal	44	20	10	30	75
Cook, Bun	NY Rangers	44	14	14	28	45
Carson, Bill	Toronto	32	20	6	26	36
Finnigan, Frank	Ottawa	38	20	5	25	34
Cook, Bill	NY Rangers	43	18	6	24	42
Keats, Duke	Det., Chi.	38	14	10	24	60

1928-29

Canadian Division

Team	GP	W	L	T	GF	GA	PTS
Montreal	44	22	7	15	71	43	59
NY Americans	44	19	13	12	53	53	50
Toronto	44	21	18	5	85	69	47
Ottawa	44	14	17	13	54	67	41
Mtl. Maroons	44	15	20	9	67	65	39

American Division

Team	GP	W	L	T	GF	GA	PTS
*Boston	44	26	13	5	89	52	57
NY Rangers	44	21	13	10	72	65	52
Detroit	44	19	16	9	72	63	47
Pittsburgh	44	9	27	8	46	80	26
Chicago	44	7	29	8	33	85	22

Leading Scorers

Player	Club	GP	G	A	PTS	PIM
Bailey, Ace	Toronto	44	22	10	32	78
Stewart, Nels	Mtl. Maroons	44	21	8	29	74
Cooper, Carson	Detroit	43	18	9	27	14
Morenz, Howie	Montreal	42	17	10	27	47
Blair, Andy	Toronto	44	12	15	27	41
Boucher, Frank	NY Rangers	44	10	16	26	8
Oliver, Harry	Boston	43	17	6	23	24
Cook, Bill	NY Rangers	43	15	8	23	41
Ward, Jimmy	Mtl. Maroons	43	14	8	22	46

Seven players tied with 19 points

1929-30

Canadian Division

Team	GP	W	L	T	GF	GA	PTS
Mtl. Maroons	44	23	16	5	141	114	51
*Montreal	44	21	14	9	142	114	51
Ottawa	44	21	15	8	138	118	50
Toronto	44	17	21	6	116	124	40
NY Americans	44	14	25	5	113	161	33

American Division

Team	GP	W	L	T	GF	GA	PTS
Boston	44	38	5	1	179	98	77
Chicago	44	21	18	5	117	111	47
NY Rangers	44	17	17	10	136	143	44
Detroit	44	14	24	6	117	133	34
Pittsburgh	44	5	36	3	102	185	13

Leading Scorers

Player	Club	GP	G	A	PTS	PIM
Weiland, Cooney	Boston	44	43	30	73	27
Boucher, Frank	NY Rangers	42	26	36	62	16
Clapper, Dit	Boston	44	41	20	61	48
Cook, Bill	NY Rangers	44	29	30	59	56
Kilrea, Hec	Ottawa	44	36	22	58	72
Stewart, Nels	Mtl. Maroons	44	39	16	55	81
Morenz, Howie	Montreal	44	40	10	50	72
Himes, Normie	NY Americans	44	28	22	50	15
Lamb, Joe	Ottawa	44	29	20	49	119
Gainor, Dutch	Boston	42	18	31	49	39

1930-31

Canadian Division

Team	GP	W	L	T	GF	GA	PTS
*Montreal	44	26	10	8	129	89	60
Toronto	44	22	13	9	118	99	53
Mtl. Maroons	44	20	18	6	105	106	46
NY Americans	44	18	16	10	76	74	46
Ottawa	44	10	30	4	91	142	24

American Division

Team	GP	W	L	T	GF	GA	PTS
Boston	44	28	10	6	143	90	62
Chicago	44	24	17	3	108	78	51
NY Rangers	44	19	16	9	106	87	47
Detroit	44	16	21	7	102	105	39
Philadelphia	44	4	36	4	76	184	12

Leading Scorers

Player	Club	GP	G	A	PTS	PIM
Morenz, Howie	Montreal	39	28	23	51	49
Goodfellow, Ebbie	Detroit	44	25	23	48	32
Conacher, Charlie	Toronto	37	31	12	43	78
Cook, Bill	NY Rangers	43	30	12	42	39
Bailey, Ace	Toronto	40	23	19	42	46
Primeau, Joe	Toronto	38	9	32	41	18
Stewart, Nels	Mtl. Maroons	42	25	14	39	75
Boucher, Frank	NY Rangers	44	12	27	39	20
Weiland, Cooney	Boston	44	25	13	38	14
Cook, Bun	NY Rangers	44	18	17	35	72
Joliat, Aurel	Montreal	43	13	22	35	73

1931-32

Canadian Division

Team	GP	W	L	T	GF	GA	PTS
Montreal	48	25	16	7	128	111	57
*Toronto	48	23	18	7	155	127	53
Mtl. Maroons	48	19	22	7	142	139	45
NY Americans	48	16	24	8	95	142	40

American Division

Team	GP	W	L	T	GF	GA	PTS
NY Rangers	48	23	17	8	134	112	54
Chicago	48	18	19	11	86	101	47
Detroit	48	18	20	10	95	108	46
Boston	48	15	21	12	122	117	42

Leading Scorers

Player	Club	GP	G	A	PTS	PIM
Jackson, Busher	Toronto	48	28	25	53	63
Primeau, Joe	Toronto	46	13	37	50	25
Morenz, Howie	Montreal	48	24	25	49	46
Conacher, Charlie	Toronto	44	34	14	48	66
Cook, Bill	NY Rangers	48	34	14	48	33
Trottier, Dave	Mtl. Maroons	48	26	18	44	94
Smith, Hooley	Mtl. Maroons	43	11	33	44	49
Siebert, Babe	Mtl. Maroons	48	21	18	39	64
Clapper, Dit	Boston	48	17	22	39	21
Joliat, Aurel	Montreal	48	15	24	39	46

1932-33

Canadian Division

Team	GP	W	L	T	GF	GA	PTS
Toronto	48	24	18	6	119	111	54
Mtl. Maroons	48	22	20	6	135	119	50
Montreal	48	18	25	5	92	115	41
NY Americans	48	15	22	11	91	118	41
Ottawa	48	11	27	10	88	131	32

American Division

Team	GP	W	L	T	GF	GA	PTS
Boston	48	25	15	8	124	88	58
Detroit	48	25	15	8	111	93	58
*NY Rangers	48	23	17	8	135	107	54
Chicago	48	16	20	12	88	101	44

Leading Scorers

Player	Club	GP	G	A	PTS	PIM
Cook, Bill	NY Rangers	48	28	22	50	51
Jackson, Busher	Toronto	48	27	17	44	43
Northcott, Baldy	Mtl. Maroons	48	22	21	43	30
Smith, Hooley	Mtl. Maroons	48	20	21	41	66
Haynes, Paul	Mtl. Maroons	48	16	25	41	18
Joliat, Aurel	Montreal	48	18	21	39	53
Barry, Marty	Boston	48	24	13	37	40
Cook, Bun	NY Rangers	48	22	15	37	35
Stewart, Nels	Boston	47	18	18	36	62
Morenz, Howie	Montreal	46	14	21	35	32
Gagnon, Johnny	Montreal	48	12	23	35	64
Shore, Eddie	Boston	48	8	27	35	102
Boucher, Frank	NY Rangers	46	7	28	35	4

1933-34

Canadian Division

Team	GP	W	L	T	GF	GA	PTS
Toronto	48	26	13	9	174	119	61
Montreal	48	22	20	6	99	101	50
Mtl. Maroons	48	19	18	11	117	122	49
NY Americans	48	15	23	10	104	132	40
Ottawa	48	13	29	6	115	143	32

American Division

Team	GP	W	L	T	GF	GA	PTS
Detroit	48	24	14	10	113	98	58
*Chicago	48	20	17	11	88	83	51
NY Rangers	48	21	19	8	120	113	50
Boston	48	18	25	5	111	130	41

Leading Scorers

Player	Club	GP	G	A	PTS	PIM
Conacher, Charlie	Toronto	42	32	20	52	38
Primeau, Joe	Toronto	45	14	32	46	8
Boucher, Frank	NY Rangers	48	14	30	44	4
Barry, Marty	Boston	48	27	12	39	12
Dillon, Cecil	NY Rangers	48	13	26	39	10
Stewart, Nels	Boston	48	21	17	38	68
Jackson, Busher	Toronto	38	20	18	38	38
Joliat, Aurel	Montreal	48	22	15	37	27
Smith, Reg	Mtl. Maroons	47	18	19	37	58
Thompson, Paul	Chicago	48	20	16	36	17

1934-35

Canadian Division

Team	GP	W	L	T	GF	GA	PTS
Toronto	48	30	14	4	157	111	64
*Mtl. Maroons	48	24	19	5	123	92	53
Montreal	48	19	23	6	110	145	44
NY Americans	48	12	27	9	100	142	33
St. Louis	48	11	31	6	86	144	28

American Division

Team	GP	W	L	T	GF	GA	PTS
Boston	48	26	16	6	129	112	58
Chicago	48	26	17	5	118	88	57
NY Rangers	48	22	20	6	137	139	50
Detroit	48	19	22	7	127	114	45

Leading Scorers

Player	Club	GP	G	A	PTS	PIM
Conacher, Charlie	Toronto	47	36	21	57	24
Howe, Syd	St.L., Det.	50	22	25	47	34
Aurie, Larry	Detroit	48	17	29	46	24
Boucher, Frank	NY Rangers	48	13	32	45	2
Jackson, Busher	Toronto	42	22	22	44	27
Lewis, Herbie	Detroit	47	16	27	43	26
Chapman, Art	NY Americans	47	9	34	43	4
Barry, Marty	Boston	48	20	20	40	33
Schriner, Sweeney	NY Americans	48	18	22	40	6
Stewart, Nels	Boston	47	21	18	39	45
Thompson, Paul	Chicago	48	16	23	39	20

1935-36

Canadian Division

Team	GP	W	L	T	GF	GA	PTS
Mtl. Maroons	48	22	16	10	114	106	54
Toronto	48	23	19	6	126	106	52
NY Americans	48	16	25	7	109	122	39
Montreal	48	11	26	11	82	123	33

American Division

Team	GP	W	L	T	GF	GA	PTS
*Detroit	48	24	16	8	124	103	56
Boston	48	22	20	6	92	83	50
Chicago	48	21	19	8	93	92	50
NY Rangers	48	19	17	12	91	96	50

Leading Scorers

Player	Club	GP	G	A	PTS	PIM
Schriner, Sweeney	NY Americans	48	19	26	45	8
Barry, Marty	Detroit	48	21	19	40	16
Thompson, Paul	Chicago	45	17	23	40	19
Thoms, Bill	Toronto	48	23	15	38	29
Conacher, Charlie	Toronto	44	23	15	38	74
Smith, Hooley	Mtl. Maroons	47	19	19	38	75
Romnes, Doc	Chicago	48	13	25	38	6
Chapman, Art	NY Americans	47	10	28	38	14
Lewis, Herbie	Detroit	45	14	23	37	25
Northcott, Baldy	Mtl. Maroons	48	15	21	36	41

1936-37

Canadian Division

Team	GP	W	L	T	GF	GA	PTS
Montreal	48	24	18	6	115	111	54
Mtl. Maroons	48	22	17	9	126	110	53
Toronto	48	22	21	5	119	115	49
NY Americans	48	15	29	4	122	161	34

American Division

Team	GP	W	L	T	GF	GA	PTS
*Detroit	48	25	14	9	128	102	59
Boston	48	23	18	7	120	110	53
NY Rangers	48	19	20	9	117	106	47
Chicago	48	14	27	7	99	131	35

Leading Scorers

Player	Club	GP	G	A	PTS	PIM
Schriner, Sweeney	NY Americans	48	21	25	46	17
Apps, Syl	Toronto	48	16	29	45	10
Barry, Marty	Detroit	48	17	27	44	6
Aurie, Larry	Detroit	45	23	20	43	20
Jackson, Busher	Toronto	46	21	19	40	12
Gagnon, Johnny	Montreal	48	20	16	36	38
Gracie, Bob	Mtl. Maroons	47	11	25	36	18
Stewart, Nels	Bos., NYA	43	23	12	35	37
Thompson, Paul	Chicago	47	17	18	35	28
Cowley, Bill	Boston	46	13	22	35	4

1937-38

Canadian Division

Team	GP	W	L	T	GF	GA	PTS
Toronto	48	24	15	9	151	127	57
NY Americans	48	19	18	11	110	111	49
Montreal	48	18	17	13	123	128	49
Mtl. Maroons	48	12	30	6	101	149	30

American Division

Team	GP	W	L	T	GF	GA	PTS
Boston	48	30	11	7	142	89	67
NY Rangers	48	27	15	6	149	96	60
*Chicago	48	14	25	9	97	139	37
Detroit	48	12	25	11	99	133	35

Leading Scorers

Player	Club	GP	G	A	PTS	PIM
Drillon, Gordie	Toronto	48	26	26	52	4
Apps, Syl	Toronto	47	21	29	50	9
Thompson, Paul	Chicago	48	22	22	44	14
Mantha, Georges	Montreal	47	23	19	42	12
Dillon, Cecil	NY Rangers	48	21	18	39	6
Cowley, Bill	Boston	48	17	22	39	8
Schriner, Sweeney	NY Americans	49	21	17	38	22
Thoms, Bill	Toronto	48	14	24	38	14
Smith, Clint	NY Rangers	48	14	23	37	0
Stewart, Nels	NY Americans	48	19	17	36	29
Colville, Neil	NY Rangers	45	17	19	36	11

1938-39

Team	GP	W	L	T	GF	GA	PTS
*Boston	48	36	10	2	156	76	74
NY Rangers	48	26	16	6	149	105	58
Toronto	48	19	20	9	114	107	47
NY Americans	48	17	21	10	119	157	44
Detroit	48	18	24	6	107	128	42
Montreal	48	15	24	9	115	146	39
Chicago	48	12	28	8	91	132	32

Leading Scorers

Player	Club	GP	G	A	PTS	PIM
Blake, Toe	Montreal	48	24	23	47	10
Schriner, Sweeney	NY Americans	48	13	31	44	20
Cowley, Bill	Boston	34	8	34	42	2
Smith, Clint	NY Rangers	48	21	20	41	2
Barry, Marty	Detroit	48	13	28	41	4
Apps, Syl	Toronto	44	15	25	40	4
Anderson, Tom	NY Americans	48	13	27	40	14
Gottselig, Johnny	Chicago	48	16	23	39	15
Haynes, Paul	Montreal	47	5	33	38	27
Conacher, Roy	Boston	47	26	11	37	12
Carr, Lorne	NY Americans	46	19	18	37	16
Colville, Neil	NY Rangers	48	18	19	37	12
Watson, Phil	NY Rangers	48	15	22	37	42

1939-40

Team	GP	W	L	T	GF	GA	PTS
Boston	48	31	12	5	170	98	67
*NY Rangers	48	27	11	10	136	77	64
Toronto	48	25	17	6	134	110	56
Chicago	48	23	19	6	112	120	52
Detroit	48	16	26	6	90	126	38
NY Americans	48	15	29	4	106	140	34
Montreal	48	10	33	5	90	168	25

Leading Scorers

Player	Club	GP	G	A	PTS	PIM
Schmidt, Milt	Boston	48	22	30	52	37
Dumart, Woody	Boston	48	22	21	43	16
Bauer, Bobby	Boston	48	17	26	43	2
Drillon, Gordie	Toronto	43	21	19	40	13
Cowley, Bill	Boston	48	13	27	40	24
Hextall, Bryan	NY Rangers	48	24	15	39	52
Colville, Neil	NY Rangers	48	19	19	38	22
Howe, Syd	Detroit	46	14	23	37	17
Blake, Toe	Montreal	48	17	19	36	48
Armstrong, Murray	NY Americans	48	16	20	36	12

A slick playmaker, Syl Apps led the NHL in assists his first two seasons in the league (1936-37 and 1937-38) and finished second in scoring both years. The long-time captain of the Maple Leafs was a top-10 scorer six times in his career.

Brought along slowly by his Hall of Fame father Lester, Lynn Patrick helped the Rangers win the Stanley Cup in 1940. He led the NHL with 32 goals in 1941-42 and finished second in scoring to linemate Bryan Hextall that year.

1940-41

Team	GP	W	L	T	GF	GA	PTS
*Boston	48	27	8	13	168	102	67
Toronto	48	28	14	6	145	99	62
Detroit	48	21	16	11	112	102	53
NY Rangers	48	21	19	8	143	125	50
Chicago	48	16	25	7	112	139	39
Montreal	48	16	26	6	121	147	38
NY Americans	48	8	29	11	99	186	27

Leading Scorers

Player	Club	GP	G	A	PTS	PIM
Cowley, Bill	Boston	46	17	45	62	16
Hextall, Bryan	NY Rangers	48	26	18	44	16
Drillon, Gordie	Toronto	42	23	21	44	2
Apps, Syl	Toronto	41	20	24	44	6
Patrick, Lynn	NY Rangers	48	20	24	44	12
Howe, Syd	Detroit	48	20	24	44	8
Colville, Neil	NY Rangers	48	14	28	42	28
Wiseman, Eddie	Boston	48	16	24	40	10
Bauer, Bobby	Boston	48	17	22	39	2
Schriner, Sweeney	Toronto	48	24	14	38	6
Conacher, Roy	Boston	40	24	14	38	7
Schmidt, Milt	Boston	44	13	25	38	23

1941-42

Team	GP	W	L	T	GF	GA	PTS
NY Rangers	48	29	17	2	177	143	60
*Toronto	48	27	18	3	158	136	57
Boston	48	25	17	6	160	118	56
Chicago	48	22	23	3	145	155	47
Detroit	48	19	25	4	140	147	42
Montreal	48	18	27	3	134	173	39
Brooklyn	48	16	29	3	133	175	35

Leading Scorers

Player	Club	GP	G	A	PTS	PIM
Hextall, Bryan	NY Rangers	48	24	32	56	30
Patrick, Lynn	NY Rangers	47	32	22	54	18
Grosso, Don	Detroit	48	23	30	53	13
Watson, Phil	NY Rangers	48	15	37	52	48
Abel, Sid	Detroit	48	18	31	49	45
Blake, Toe	Montreal	47	17	28	45	19
Thoms, Bill	Chicago	47	15	30	45	8
Drillon, Gordie	Toronto	48	23	18	41	6
Apps, Syl	Toronto	38	18	23	41	0
Anderson, Tom	Brooklyn	48	12	29	41	54

1942-43

Team	GP	W	L	T	GF	GA	PTS
*Detroit	50	25	14	11	169	124	61
Boston	50	24	17	9	195	176	57
Toronto	50	22	19	9	198	159	53
Montreal	50	19	19	12	181	191	50
Chicago	50	17	18	15	179	180	49
NY Rangers	50	11	31	8	161	253	30

Leading Scorers

Player	Club	GP	G	A	PTS	PIM
Bentley, Doug	Chicago	50	33	40	73	18
Cowley, Bill	Boston	48	27	45	72	10
Bentley, Max	Chicago	47	26	44	70	2
Patrick, Lynn	NY Rangers	50	22	39	61	28
Carr, Lorne	Toronto	50	27	33	60	15
Taylor, Billy	Toronto	50	18	42	60	2
Hextall, Bryan	NY Rangers	50	27	32	59	28
Blake, Toe	Montreal	48	23	36	59	28
Lach, Elmer	Montreal	45	18	40	58	14
O'Connor, Buddy	Montreal	50	15	43	58	2

1943-44

Team	GP	W	L	T	GF	GA	PTS
*Montreal	50	38	5	7	234	109	83
Detroit	50	26	18	6	214	177	58
Toronto	50	23	23	4	214	174	50
Chicago	50	22	23	5	178	187	49
Boston	50	19	26	5	223	268	43
NY Rangers	50	6	39	5	162	310	17

Leading Scorers

Player	Club	GP	G	A	PTS	PIM
Cain, Herb	Boston	48	36	46	82	4
Bentley, Doug	Chicago	50	38	39	77	22
Carr, Lorne	Toronto	50	36	38	74	9
Liscombe, Carl	Detroit	50	36	37	73	17
Lach, Elmer	Montreal	48	24	48	72	23
Smith, Clint	Chicago	50	23	49	72	4
Cowley, Bill	Boston	36	30	41	71	12
Mosienko, Bill	Chicago	50	32	38	70	10
Jackson, Art	Boston	49	28	41	69	8
Bodnar, Gus	Toronto	50	22	40	62	18

1944-45

Team	GP	W	L	T	GF	GA	PTS
Montreal	50	38	8	4	228	121	80
Detroit	50	31	14	5	218	161	67
*Toronto	50	24	22	4	183	161	52
Boston	50	16	30	4	179	219	36
Chicago	50	13	30	7	141	194	33
NY Rangers	50	11	29	10	154	247	32

Leading Scorers

Player	Club	GP	G	A	PTS	PIM
Lach, Elmer	Montreal	50	26	54	80	37
Richard, Maurice	Montreal	50	50	23	73	36
Blake, Toe	Montreal	49	29	38	67	15
Cowley, Bill	Boston	49	25	40	65	2
Kennedy, Ted	Toronto	49	29	25	54	14
Mosienko, Bill	Chicago	50	28	26	54	0
Carveth, Joe	Detroit	50	26	28	54	6
DeMarco, Ab	NY Rangers	50	24	30	54	10
Smith, Clint	Chicago	50	23	31	54	0
Howe, Syd	Detroit	46	17	36	53	6

1945-46

Team	GP	W	L	T	GF	GA	PTS
*Montreal	50	28	17	5	172	134	61
Boston	50	24	18	8	167	156	56
Chicago	50	23	20	7	200	178	53
Detroit	50	20	20	10	146	159	50
Toronto	50	19	24	7	174	185	45
NY Rangers	50	13	28	9	144	191	35

Leading Scorers

Player	Club	GP	G	A	PTS	PIM
Bentley, Max	Chicago	47	31	30	61	6
Stewart, Gaye	Toronto	50	37	15	52	8
Blake, Toe	Montreal	50	29	21	50	2
Smith, Clint	Chicago	50	26	24	50	2
Richard, Maurice	Montreal	50	27	21	48	50
Mosienko, Bill	Chicago	40	18	30	48	12
DeMarco, Ab	NY Rangers	50	20	27	47	20
Lach, Elmer	Montreal	50	13	34	47	34
Kaleta, Alex	Chicago	49	19	27	46	17
Taylor, Billy	Toronto	48	23	18	41	14
Horeck, Pete	Chicago	50	20	21	41	34

1946-47

Team	GP	W	L	T	GF	GA	PTS
Montreal	60	34	16	10	189	138	78
*Toronto	60	31	19	10	209	172	72
Boston	60	26	23	11	190	175	63
Detroit	60	22	27	11	190	193	55
NY Rangers	60	22	32	6	167	186	50
Chicago	60	19	37	4	193	274	42

Leading Scorers

Player	Club	GP	G	A	PTS	PIM
Bentley, Max	Chicago	60	29	43	72	12
Richard, Maurice	Montreal	60	45	26	71	69
Taylor, Billy	Detroit	60	17	46	63	35
Schmidt, Milt	Boston	59	27	35	62	40
Kennedy, Ted	Toronto	60	28	32	60	27
Bentley, Doug	Chicago	52	21	34	55	18
Bauer, Bobby	Boston	58	30	24	54	4
Conacher, Roy	Detroit	60	30	24	54	6
Mosienko, Bill	Chicago	59	25	27	52	2
Dumart, Woody	Boston	60	24	28	52	12

1947-48

Team	GP	W	L	T	GF	GA	PTS
*Toronto	60	32	15	13	182	143	77
Detroit	60	30	18	12	187	148	72
Boston	60	23	24	13	167	168	59
NY Rangers	60	21	26	13	176	201	55
Montreal	60	20	29	11	147	169	51
Chicago	60	20	34	6	195	225	46

Leading Scorers

Player	Club	GP	G	A	PTS	PIM
Lach, Elmer	Montreal	60	30	31	61	72
O'Connor, Buddy	NY Rangers	60	24	36	60	8
Bentley, Doug	Chicago	60	20	37	57	16
Stewart, Gaye	Tor., Chi.	61	27	29	56	83
Bentley, Max	Chi., Tor.	59	26	28	54	14
Poile, Bud	Tor., Chi.	58	25	29	54	17
Richard, Maurice	Montreal	53	28	25	53	89
Apps, Syl	Toronto	55	26	27	53	12
Lindsay, Ted	Detroit	60	33	19	52	95
Conacher, Roy	Chicago	52	22	27	49	4

1948-49

Team	GP	W	L	T	GF	GA	PTS
Detroit	60	34	19	7	195	145	75
Boston	60	29	23	8	178	163	66
Montreal	60	28	23	9	152	126	65
*Toronto	60	22	25	13	147	161	57
Chicago	60	21	31	8	173	211	50
NY Rangers	60	18	31	11	133	172	47

Leading Scorers

Player	Club	GP	G	A	PTS	PIM
Conacher, Roy	Chicago	60	26	42	68	8
Bentley, Doug	Chicago	58	23	43	66	38
Abel, Sid	Detroit	60	28	26	54	49
Lindsay, Ted	Detroit	50	26	28	54	97
Conacher, Jim	Det., Chi.	59	26	23	49	43
Ronty, Paul	Boston	60	20	29	49	11
Watson, Harry	Toronto	60	26	19	45	0
Reay, Billy	Montreal	60	22	23	45	33
Bodnar, Gus	Chicago	59	19	26	45	14
Peirson, Johnny	Boston	59	22	21	43	45

1949-50

Team	GP	W	L	T	GF	GA	PTS
*Detroit	70	37	19	14	229	164	88
Montreal	70	29	22	19	172	150	77
Toronto	70	31	27	12	176	173	74
NY Rangers	70	28	31	11	170	189	67
Boston	70	22	32	16	198	228	60
Chicago	70	22	38	10	203	244	54

Leading Scorers

Player	Club	GP	G	A	PTS	PIM
Lindsay, Ted	Detroit	69	23	55	78	141
Abel, Sid	Detroit	69	34	35	69	46
Howe, Gordie	Detroit	70	35	33	68	69
Richard, Maurice	Montreal	70	43	22	65	114
Ronty, Paul	Boston	70	23	36	59	8
Conacher, Roy	Chicago	70	25	31	56	16
Bentley, Doug	Chicago	64	20	33	53	28
Peirson, Johnny	Boston	57	27	25	52	49
Prystai, Metro	Chicago	65	29	22	51	31
Guidolin, Bep	Chicago	70	17	34	51	42

Detroit's Bob Goldham wraps up a young Jean Beliveau during his three-game stint with the Canadiens in 1952-53. Beliveau joined the Canadiens full-time the following season and quickly established himself as one of the NHL's top players.

1950-51

Team	GP	W	L	T	GF	GA	PTS
Detroit	70	44	13	13	236	139	101
*Toronto	70	41	16	13	212	138	95
Montreal	70	25	30	15	173	184	65
Boston	70	22	30	18	178	197	62
NY Rangers	70	20	29	21	169	201	61
Chicago	70	13	47	10	171	280	36

Leading Scorers

Player	Club	GP	G	A	PTS	PIM
Howe, Gordie	Detroit	70	43	43	86	74
Richard, Maurice	Montreal	65	42	24	66	97
Bentley, Max	Toronto	67	21	41	62	34
Abel, Sid	Detroit	69	23	38	61	30
Schmidt, Milt	Boston	62	22	39	61	33
Kennedy, Ted	Toronto	63	18	43	61	32
Lindsay, Ted	Detroit	67	24	35	59	110
Sloan, Tod	Toronto	70	31	25	56	105
Kelly, Red	Detroit	70	17	37	54	24
Smith, Sid	Toronto	70	30	21	51	10
Gardner, Cal	Toronto	66	23	28	51	42

1951-52

Team	GP	W	L	T	GF	GA	PTS
*Detroit	70	44	14	12	215	133	100
Montreal	70	34	26	10	195	164	78
Toronto	70	29	25	16	168	157	74
Boston	70	25	29	16	162	176	66
NY Rangers	70	23	34	13	192	219	59
Chicago	70	17	44	9	158	241	43

Leading Scorers

Player	Club	GP	G	A	PTS	PIM
Howe, Gordie	Detroit	70	47	39	86	78
Lindsay, Ted	Detroit	70	30	39	69	123
Lach, Elmer	Montreal	70	15	50	65	36
Raleigh, Don	NY Rangers	70	19	42	61	14
Smith, Sid	Toronto	70	27	30	57	6
Geoffrion, Bernie	Montreal	67	30	24	54	66
Mosienko, Bill	Chicago	70	31	22	53	10
Abel, Sid	Detroit	62	17	36	53	32
Kennedy, Ted	Toronto	70	19	33	52	33
Schmidt, Milt	Boston	69	21	29	50	57
Peirson, Johnny	Boston	68	20	30	50	30

1952-53

Team	GP	W	L	T	GF	GA	PTS
Detroit	70	36	16	18	222	133	90
*Montreal	70	28	23	19	155	148	75
Boston	70	28	29	13	152	172	69
Chicago	70	27	28	15	169	175	69
Toronto	70	27	30	13	156	167	67
NY Rangers	70	17	37	16	152	211	50

Leading Scorers

Player	Club	GP	G	A	PTS	PIM
Howe, Gordie	Detroit	70	49	46	95	57
Lindsay, Ted	Detroit	70	32	39	71	111
Richard, Maurice	Montreal	70	28	33	61	112
Hergesheimer, Wally	NY Rangers	70	30	29	59	10
Delvecchio, Alex	Detroit	70	16	43	59	28
Ronty, Paul	NY Rangers	70	16	38	54	20
Prystai, Metro	Detroit	70	16	34	50	12
Kelly, Red	Detroit	70	19	27	46	8
Olmstead, Bert	Montreal	69	17	28	45	83
Mackell, Fleming	Boston	65	27	17	44	63
McFadden, Jim	Chicago	70	23	21	44	29

1953-54

Team	GP	W	L	T	GF	GA	PTS
*Detroit	70	37	19	14	191	132	88
Montreal	70	35	24	11	195	141	81
Toronto	70	32	24	14	152	131	78
Boston	70	32	28	10	177	181	74
NY Rangers	70	29	31	10	161	182	68
Chicago	70	12	51	7	133	242	31

Leading Scorers

Player	Club	GP	G	A	PTS	PIM
Howe, Gordie	Detroit	70	33	48	81	109
Richard, Maurice	Montreal	70	37	30	67	112
Lindsay, Ted	Detroit	70	26	36	62	110
Geoffrion, Bernie	Montreal	54	29	25	54	87
Olmstead, Bert	Montreal	70	15	37	52	85
Kelly, Red	Detroit	62	16	33	49	18
Reibel, Dutch	Detroit	69	15	33	48	18
Sandford, Ed	Boston	70	16	31	47	42
Mackell, Fleming	Boston	67	15	32	47	60
Mosdell, Ken	Montreal	67	22	24	46	64
Ronty, Paul	NY Rangers	70	13	33	46	18

1954-55

Team	GP	W	L	T	GF	GA	PTS
*Detroit	70	42	17	11	204	134	95
Montreal	70	41	18	11	228	157	93
Toronto	70	24	24	22	147	135	70
Boston	70	23	26	21	169	188	67
NY Rangers	70	17	35	18	150	210	52
Chicago	70	13	40	17	161	235	43

Leading Scorers

Player	Club	GP	G	A	PTS	PIM
Geoffrion, Bernie	Montreal	70	38	37	75	57
Richard, Maurice	Montreal	67	38	36	74	125
Béliveau, Jean	Montreal	70	37	36	73	58
Reibel, Dutch	Detroit	70	25	41	66	15
Howe, Gordie	Detroit	64	29	33	62	68
Sullivan, Red	Chicago	69	19	42	61	51
Olmstead, Bert	Montreal	70	10	48	58	103
Smith, Sid	Toronto	70	33	21	54	14
Mosdell, Ken	Montreal	70	22	32	54	82
Lewicki, Danny	NY Rangers	70	29	24	53	8

1955-56

Team	GP	W	L	T	GF	GA	PTS
*Montreal	70	45	15	10	222	131	100
Detroit	70	30	24	16	183	148	76
NY Rangers	70	32	28	10	204	203	74
Toronto	70	24	33	13	153	181	61
Boston	70	23	34	13	147	185	59
Chicago	70	19	39	12	155	216	50

Leading Scorers

Player	Club	GP	G	A	PTS	PIM
Béliveau, Jean	Montreal	70	47	41	88	143
Howe, Gordie	Detroit	70	38	41	79	100
Richard, Maurice	Montreal	70	38	33	71	89
Olmstead, Bert	Montreal	70	14	56	70	94
Sloan, Tod	Toronto	70	37	29	66	100
Bathgate, Andy	NY Rangers	70	19	47	66	59
Geoffrion, Bernie	Montreal	59	29	33	62	66
Reibel, Dutch	Detroit	68	17	39	56	10
Delvecchio, Alex	Detroit	70	25	26	51	24
Creighton, Dave	NY Rangers	70	20	31	51	43
Gadsby, Bill	NY Rangers	70	9	42	51	84

1956-57

Team	GP	W	L	T	GF	GA	PTS
Detroit	70	38	20	12	198	157	88
*Montreal	70	35	23	12	210	155	82
Boston	70	34	24	12	195	174	80
NY Rangers	70	26	30	14	184	227	66
Toronto	70	21	34	15	174	192	57
Chicago	70	16	39	15	169	225	47

Leading Scorers

Player	Club	GP	G	A	PTS	PIM
Howe, Gordie	Detroit	70	44	45	89	72
Lindsay, Ted	Detroit	70	30	55	85	103
Béliveau, Jean	Montreal	69	33	51	84	105
Bathgate, Andy	NY Rangers	70	27	50	77	60
Litzenberger, Ed	Chicago	70	32	32	64	48
Richard, Maurice	Montreal	63	33	29	62	74
McKenney, Don	Boston	69	21	39	60	31
Moore, Dickie	Montreal	70	29	29	58	56
Richard, Henri	Montreal	63	18	36	54	71
Ullman, Norm	Detroit	64	16	36	52	47

1957-58

Team	GP	W	L	T	GF	GA	PTS
*Montreal	70	43	17	10	250	158	96
NY Rangers	70	32	25	13	195	188	77
Detroit	70	29	29	12	176	207	70
Boston	70	27	28	15	199	194	69
Chicago	70	24	39	7	163	202	55
Toronto	70	21	38	11	192	226	53

Leading Scorers

Player	Club	GP	G	A	PTS	PIM
Moore, Dickie	Montreal	70	36	48	84	65
Richard, Henri	Montreal	67	28	52	80	56
Bathgate, Andy	NY Rangers	65	30	48	78	42
Howe, Gordie	Detroit	64	33	44	77	40
Horvath, Bronco	Boston	67	30	36	66	71
Litzenberger, Ed	Chicago	70	32	30	62	63
Mackell, Fleming	Boston	70	20	40	60	72
Béliveau, Jean	Montreal	55	27	32	59	93
Delvecchio, Alex	Detroit	70	21	38	59	22
McKenney, Don	Boston	70	28	30	58	22

1958-59

Team	GP	W	L	T	GF	GA	PTS
*Montreal	70	39	18	13	258	158	91
Boston	70	32	29	9	205	215	73
Chicago	70	28	29	13	197	208	69
Toronto	70	27	32	11	189	201	65
NY Rangers	70	26	32	12	201	217	64
Detroit	70	25	37	8	167	218	58

Leading Scorers

Player	Club	GP	G	A	PTS	PIM
Moore, Dickie	Montreal	70	41	55	96	61
Béliveau, Jean	Montreal	64	45	46	91	67
Bathgate, Andy	NY Rangers	70	40	48	88	48
Howe, Gordie	Detroit	70	32	46	78	57
Litzenberger, Ed	Chicago	70	33	44	77	37
Geoffrion, Bernie	Montreal	59	22	44	66	30
Sullivan, Red	NY Rangers	70	21	42	63	56
Hebenton, Andy	NY Rangers	70	33	29	62	8
McKenney, Don	Boston	70	32	30	62	20
Sloan, Tod	Chicago	59	27	35	62	79

1959-60

Team	GP	W	L	T	GF	GA	PTS
*Montreal	70	40	18	12	255	178	92
Toronto	70	35	26	9	199	195	79
Chicago	70	28	29	13	191	180	69
Detroit	70	26	29	15	186	197	67
Boston	70	28	34	8	220	241	64
NY Rangers	70	17	38	15	187	247	49

Leading Scorers

Player	Club	GP	G	A	PTS	PIM
Hull, Bobby	Chicago	70	39	42	81	68
Horvath, Bronco	Boston	68	39	41	80	60
Béliveau, Jean	Montreal	60	34	40	74	57
Bathgate, Andy	NY Rangers	70	26	48	74	28
Richard, Henri	Montreal	70	30	43	73	66
Howe, Gordie	Detroit	70	28	45	73	46
Geoffrion, Bernie	Montreal	59	30	41	71	36
McKenney, Don	Boston	70	20	49	69	28
Stasiuk, Vic	Boston	69	29	39	68	121
Prentice, Dean	NY Rangers	70	32	34	66	43

1960-61

Team	GP	W	L	T	GF	GA	PTS
Montreal	70	41	19	10	254	188	92
Toronto	70	39	19	12	234	176	90
*Chicago	70	29	24	17	198	180	75
Detroit	70	25	29	16	195	215	66
NY Rangers	70	22	38	10	204	248	54
Boston	70	15	42	13	176	254	43

Leading Scorers

Player	Club	GP	G	A	PTS	PIM
Geoffrion, Bernie	Montreal	64	50	45	95	29
Béliveau, Jean	Montreal	69	32	58	90	57
Mahovlich, Frank	Toronto	70	48	36	84	131
Bathgate, Andy	NY Rangers	70	29	48	77	22
Howe, Gordie	Detroit	64	23	49	72	30
Ullman, Norm	Detroit	70	28	42	70	34
Kelly, Red	Toronto	64	20	50	70	12
Moore, Dickie	Montreal	57	35	34	69	62
Richard, Henri	Montreal	70	24	44	68	91
Delvecchio, Alex	Detroit	70	27	35	62	26

1961-62

Team	GP	W	L	T	GF	GA	PTS
Montreal	70	42	14	14	259	166	98
*Toronto	70	37	22	11	232	180	85
Chicago	70	31	26	13	217	186	75
NY Rangers	70	26	32	12	195	207	64
Detroit	70	23	33	14	184	219	60
Boston	70	15	47	8	177	306	38

Leading Scorers

Player	Club	GP	G	A	PTS	PIM
Hull, Bobby	Chicago	70	50	34	84	35
Bathgate, Andy	NY Rangers	70	28	56	84	44
Howe, Gordie	Detroit	70	33	44	77	54
Mikita, Stan	Chicago	70	25	52	77	97
Mahovlich, Frank	Toronto	70	33	38	71	87
Delvecchio, Alex	Detroit	70	26	43	69	18
Backstrom, Ralph	Montreal	66	27	38	65	29
Ullman, Norm	Detroit	70	26	38	64	54
Hay, Bill	Chicago	60	11	52	63	34
Provost, Claude	Montreal	70	33	29	62	22

1962-63

Team	GP	W	L	T	GF	GA	PTS
*Toronto	70	35	23	12	221	180	82
Chicago	70	32	21	17	194	178	81
Montreal	70	28	19	23	225	183	79
Detroit	70	32	25	13	200	194	77
NY Rangers	70	22	36	12	211	233	56
Boston	70	14	39	17	198	281	45

Leading Scorers

Player	Club	GP	G	A	PTS	PIM
Howe, Gordie	Detroit	70	38	48	86	100
Bathgate, Andy	NY Rangers	70	35	46	81	54
Mikita, Stan	Chicago	65	31	45	76	69
Mahovlich, Frank	Toronto	67	36	37	73	56
Richard, Henri	Montreal	67	23	50	73	57
Béliveau, Jean	Montreal	69	18	49	67	68
Bucyk, John	Boston	69	27	39	66	36
Delvecchio, Alex	Detroit	70	20	44	64	8
Hull, Bobby	Chicago	65	31	31	62	27
Oliver, Murray	Boston	65	22	40	62	38

1963-64

Team	GP	W	L	T	GF	GA	PTS
Montreal	70	36	21	13	209	167	85
Chicago	70	36	22	12	218	169	84
*Toronto	70	33	25	12	192	172	78
Detroit	70	30	29	11	191	204	71
NY Rangers	70	22	38	10	186	242	54
Boston	70	18	40	12	170	212	48

Leading Scorers

Player	Club	GP	G	A	PTS	PIM
Mikita, Stan	Chicago	70	39	50	89	146
Hull, Bobby	Chicago	70	43	44	87	50
Béliveau, Jean	Montreal	68	28	50	78	42
Bathgate, Andy	NYR, Tor.	71	19	58	77	34
Howe, Gordie	Detroit	69	26	47	73	70
Wharram, Kenny	Chicago	70	39	32	71	18
Oliver, Murray	Boston	70	24	44	68	41
Goyette, Phil	NY Rangers	67	24	41	65	15
Gilbert, Rod	NY Rangers	70	24	40	64	62
Keon, Dave	Toronto	70	23	37	60	6

1964-65

Team	GP	W	L	T	GF	GA	PTS
Detroit	70	40	23	7	224	175	87
*Montreal	70	36	23	11	211	185	83
Chicago	70	34	28	8	224	176	76
Toronto	70	30	26	14	204	173	74
NY Rangers	70	20	38	12	179	246	52
Boston	70	21	43	6	166	253	48

Leading Scorers

Player	Club	GP	G	A	PTS	PIM
Mikita, Stan	Chicago	70	28	59	87	154
Ullman, Norm	Detroit	70	42	41	83	70
Howe, Gordie	Detroit	70	29	47	76	104
Hull, Bobby	Chicago	61	39	32	71	32
Delvecchio, Alex	Detroit	68	25	42	67	16
Provost, Claude	Montreal	70	27	37	64	28
Gilbert, Rod	NY Rangers	70	25	36	61	52
Pilote, Pierre	Chicago	68	14	45	59	162
Bucyk, John	Boston	68	26	29	55	24
Backstrom, Ralph	Montreal	70	25	30	55	41
Esposito, Phil	Chicago	70	23	32	55	44

1965-66

Team	GP	W	L	T	GF	GA	PTS
*Montreal	70	41	21	8	239	173	90
Chicago	70	37	25	8	240	187	82
Toronto	70	34	25	11	208	187	79
Detroit	70	31	27	12	221	194	74
Boston	70	21	43	6	174	275	48
NY Rangers	70	18	41	11	195	261	47

Leading Scorers

Player	Club	GP	G	A	PTS	PIM
Hull, Bobby	Chicago	65	54	43	97	70
Mikita, Stan	Chicago	68	30	48	78	58
Rousseau, Bobby	Montreal	70	30	48	78	20
Béliveau, Jean	Montreal	67	29	48	77	50
Howe, Gordie	Detroit	70	29	46	75	83
Ullman, Norm	Detroit	70	31	41	72	35
Delvecchio, Alex	Detroit	70	31	38	69	16
Nevin, Bob	NY Rangers	69	29	33	62	10
Richard, Henri	Montreal	62	22	39	61	47
Oliver, Murray	Boston	70	18	42	60	30

1966-67

Team	GP	W	L	T	GF	GA	PTS
Chicago	70	41	17	12	264	170	94
Montreal	70	32	25	13	202	188	77
*Toronto	70	32	27	11	204	211	75
NY Rangers	70	30	28	12	188	189	72
Detroit	70	27	39	4	212	241	58
Boston	70	17	43	10	182	253	44

Leading Scorers

Player	Club	GP	G	A	PTS	PIM
Mikita, Stan	Chicago	70	35	62	97	12
Hull, Bobby	Chicago	66	52	28	80	52
Ullman, Norm	Detroit	68	26	44	70	26
Wharram, Kenny	Chicago	70	31	34	65	21
Howe, Gordie	Detroit	69	25	40	65	53
Rousseau, Bobby	Montreal	68	19	44	63	58
Esposito, Phil	Chicago	69	21	40	61	40
Goyette, Phil	NY Rangers	70	12	49	61	6
Mohns, Doug	Chicago	61	25	35	60	58
Richard, Henri	Montreal	65	21	34	55	28
Delvecchio, Alex	Detroit	70	17	38	55	10

1967-68

East Division

Team	GP	W	L	T	GF	GA	PTS
*Montreal	74	42	22	10	236	167	94
NY Rangers	74	39	23	12	226	183	90
Boston	74	37	27	10	259	216	84
Chicago	74	32	26	16	212	222	80
Toronto	74	33	31	10	209	176	76
Detroit	74	27	35	12	245	257	66

West Division

Team	GP	W	L	T	GF	GA	PTS
Philadelphia	74	31	32	11	173	179	73
Los Angeles	74	31	33	10	200	224	72
St. Louis	74	27	31	16	177	191	70
Minnesota	74	27	32	15	191	226	69
Pittsburgh	74	27	34	13	195	216	67
Oakland	74	15	42	17	153	219	47

Leading Scorers

Player	Club	GP	G	A	PTS	PIM
Mikita, Stan	Chicago	72	40	47	87	14
Esposito, Phil	Boston	74	35	49	84	21
Howe, Gordie	Detroit	74	39	43	82	53
Ratelle, Jean	NY Rangers	74	32	46	78	18
Gilbert, Rod	NY Rangers	73	29	48	77	12
Hull, Bobby	Chicago	71	44	31	75	39
Ullman, Norm	Det., Tor.	71	35	37	72	28
Delvecchio, Alex	Detroit	74	22	48	70	14
Bucyk, John	Boston	72	30	39	69	8
Wharram, Kenny	Chicago	74	27	42	69	18

1968-69

East Division

Team	GP	W	L	T	GF	GA	PTS
*Montreal	76	46	19	11	271	202	103
Boston	76	42	18	16	303	221	100
NY Rangers	76	41	26	9	231	196	91
Toronto	76	35	26	15	234	217	85
Detroit	76	33	31	12	239	221	78
Chicago	76	34	33	9	280	246	77

West Division

Team	GP	W	L	T	GF	GA	PTS
St. Louis	76	37	25	14	204	157	88
Oakland	76	29	36	11	219	251	69
Philadelphia	76	20	35	21	174	225	61
Los Angeles	76	24	42	10	185	260	58
Pittsburgh	76	20	45	11	189	252	51
Minnesota	76	18	43	15	189	270	51

Leading Scorers

Player	Club	GP	G	A	PTS	PIM
Esposito, Phil	Boston	74	49	77	126	79
Hull, Bobby	Chicago	74	58	49	107	48
Howe, Gordie	Detroit	76	44	59	103	58
Mikita, Stan	Chicago	74	30	67	97	52
Hodge, Ken	Boston	75	45	45	90	75
Cournoyer, Yvan	Montreal	76	43	44	87	31
Delvecchio, Alex	Detroit	72	25	58	83	8
Berenson, Red	St. Louis	76	35	47	82	43
Béliveau, Jean	Montreal	69	33	49	82	55
Mahovlich, Frank	Detroit	76	49	29	78	38
Ratelle, Jean	NY Rangers	75	32	46	78	26

1969-70

East Division

Team	GP	W	L	T	GF	GA	PTS
Chicago	76	45	22	9	250	170	99
*Boston	76	40	17	19	277	216	99
Detroit	76	40	21	15	246	199	95
NY Rangers	76	38	22	16	246	189	92
Montreal	76	38	22	16	244	201	92
Toronto	76	29	34	13	222	242	71

West Division

Team	GP	W	L	T	GF	GA	PTS
St. Louis	76	37	27	12	224	179	86
Pittsburgh	76	26	38	12	182	238	64
Minnesota	76	19	35	22	224	257	60
Oakland	76	22	40	14	169	243	58
Philadelphia	76	17	35	24	197	225	58
Los Angeles	76	14	52	10	168	290	38

Leading Scorers

Player	Club	GP	G	A	PTS	PIM
Orr, Bobby	Boston	76	33	87	120	125
Esposito, Phil	Boston	76	43	56	99	50
Mikita, Stan	Chicago	76	39	47	86	50
Goyette, Phil	St. Louis	72	29	49	78	16
Tkaczuk, Walt	NY Rangers	76	27	50	77	38
Ratelle, Jean	NY Rangers	75	32	42	74	28
Berenson, Red	St. Louis	67	33	39	72	38
Parise, Jean-Paul	Minnesota	74	24	48	72	72
Howe, Gordie	Detroit	76	31	40	71	58
Mahovlich, Frank	Detroit	74	38	32	70	59
Balon, Dave	NY Rangers	76	33	37	70	100
McKenzie, John	Boston	72	29	41	70	114

1970-71

East Division

Team	GP	W	L	T	GF	GA	PTS
Boston	78	57	14	7	399	207	121
NY Rangers	78	49	18	11	259	177	109
*Montreal	78	42	23	13	291	216	97
Toronto	78	37	33	8	248	211	82
Buffalo	78	24	39	15	217	291	63
Vancouver	78	24	46	8	229	296	56
Detroit	78	22	45	11	209	308	55

West Division

Team	GP	W	L	T	GF	GA	PTS
Chicago	78	49	20	9	277	184	107
St. Louis	78	34	25	19	223	208	87
Philadelphia	78	28	33	17	207	225	73
Minnesota	78	28	34	16	191	223	72
Los Angeles	78	25	40	13	239	303	63
Pittsburgh	78	21	37	20	221	240	62
California	78	20	53	5	199	320	45

Leading Scorers

Player	Club	GP	G	A	PTS	PIM
Esposito, Phil	Boston	78	76	76	152	71
Orr, Bobby	Boston	78	37	102	139	91
Bucyk, John	Boston	78	51	65	116	8
Hodge, Ken	Boston	78	43	62	105	113
Hull, Bobby	Chicago	78	44	52	96	32
Ullman, Norm	Toronto	73	34	51	85	24
Cashman, Wayne	Boston	77	21	58	79	100
McKenzie, John	Boston	65	31	46	77	120
Keon, Dave	Toronto	76	38	38	76	4
Béliveau, Jean	Montreal	70	25	51	76	40
Stanfield, Fred	Boston	75	24	52	76	12

1971-72

East Division

Team	GP	W	L	T	GF	GA	PTS
*Boston	78	54	13	11	330	204	119
NY Rangers	78	48	17	13	317	192	109
Montreal	78	46	16	16	307	205	108
Toronto	78	33	31	14	209	208	80
Detroit	78	33	35	10	261	262	76
Buffalo	78	16	43	19	203	289	51
Vancouver	78	20	50	8	203	297	48

West Division

Team	GP	W	L	T	GF	GA	PTS
Chicago	78	46	17	15	256	166	107
Minnesota	78	37	29	12	212	191	86
St. Louis	78	28	39	11	208	247	67
Pittsburgh	78	26	38	14	220	258	66
Philadelphia	78	26	38	14	200	236	66
California	78	21	39	18	216	288	60
Los Angeles	78	20	49	9	206	305	49

Leading Scorers

Player	Club	GP	G	A	PTS	PIM
Esposito, Phil	Boston	76	66	67	133	76
Orr, Bobby	Boston	76	37	80	117	106
Ratelle, Jean	NY Rangers	63	46	63	109	4
Hadfield, Vic	NY Rangers	78	50	56	106	142
Gilbert, Rod	NY Rangers	73	43	54	97	64
Mahovlich, Frank	Montreal	76	43	53	96	36
Hull, Bobby	Chicago	78	50	43	93	24
Cournoyer, Yvan	Montreal	73	47	36	83	15
Bucyk, John	Boston	78	32	51	83	4
Clarke, Bobby	Philadelphia	78	35	46	81	87
Lemaire, Jacques	Montreal	77	32	49	81	26

1972-73

East Division

Team	GP	W	L	T	GF	GA	PTS
*Montreal	78	52	10	16	329	184	120
Boston	78	51	22	5	330	235	107
NY Rangers	78	47	23	8	297	208	102
Buffalo	78	37	27	14	257	219	88
Detroit	78	37	29	12	265	243	86
Toronto	78	27	41	10	247	279	64
Vancouver	78	22	47	9	233	339	53
NY Islanders	78	12	60	6	170	347	30

West Division

Team	GP	W	L	T	GF	GA	PTS
Chicago	78	42	27	9	284	225	93
Philadelphia	78	37	30	11	296	256	85
Minnesota	78	37	30	11	254	230	85
St. Louis	78	32	34	12	233	251	76
Pittsburgh	78	32	37	9	257	265	73
Los Angeles	78	31	36	11	232	245	73
Atlanta	78	25	38	15	191	239	65
California	78	16	46	16	213	323	48

Leading Scorers

Player	Club	GP	G	A	PTS	PIM
Esposito, Phil	Boston	78	55	75	130	87
Clarke, Bobby	Philadelphia	78	37	67	104	80
Orr, Bobby	Boston	63	29	72	101	99
MacLeish, Rick	Philadelphia	78	50	50	100	69
Lemaire, Jacques	Montreal	77	44	51	95	16
Ratelle, Jean	NY Rangers	78	41	53	94	12
Redmond, Mickey	Detroit	76	52	41	93	24
Bucyk, John	Boston	78	40	53	93	12
Mahovlich, Frank	Montreal	78	38	55	93	51
Pappin, Jim	Chicago	76	41	51	92	82

1973-74

East Division

Team	GP	W	L	T	GF	GA	PTS
Boston	78	52	17	9	349	221	113
Montreal	78	45	24	9	293	240	99
NY Rangers	78	40	24	14	300	251	94
Toronto	78	35	27	16	274	230	86
Buffalo	78	32	34	12	242	250	76
Detroit	78	29	39	10	255	319	68
Vancouver	78	24	43	11	224	296	59
NY Islanders	78	19	41	18	182	247	56

West Division

Team	GP	W	L	T	GF	GA	PTS
*Philadelphia	78	50	16	12	273	164	112
Chicago	78	41	14	23	272	164	105
Los Angeles	78	33	33	12	233	231	78
Atlanta	78	30	34	14	214	238	74
Pittsburgh	78	28	41	9	242	273	65
St. Louis	78	26	40	12	206	248	64
Minnesota	78	23	38	17	235	275	63
California	78	13	55	10	195	342	36

Leading Scorers

Player	Club	GP	G	A	PTS	PIM
Esposito, Phil	Boston	78	68	77	145	58
Orr, Bobby	Boston	74	32	90	122	82
Hodge, Ken	Boston	76	50	55	105	43
Cashman, Wayne	Boston	78	30	59	89	111
Clarke, Bobby	Philadelphia	77	35	52	87	113
Martin, Rick	Buffalo	78	52	34	86	38
Apps Jr., Syl	Pittsburgh	75	24	61	85	37
Sittler, Darryl	Toronto	78	38	46	84	55
MacDonald, Lowell	Pittsburgh	78	43	39	82	14
Park, Brad	NY Rangers	78	25	57	82	148
Hextall, Dennis	Minnesota	78	20	62	82	138

1974-75

PRINCE OF WALES CONFERENCE

Norris Division

Team	GP	W	L	T	GF	GA	PTS
Montreal	80	47	14	19	374	225	113
Los Angeles	80	42	17	21	269	185	105
Pittsburgh	80	37	28	15	326	289	89
Detroit	80	23	45	12	259	335	58
Washington	80	8	67	5	181	446	21

Adams Division

Team	GP	W	L	T	GF	GA	PTS
Buffalo	80	49	16	15	354	240	113
Boston	80	40	26	14	345	245	94
Toronto	80	31	33	16	280	309	78
California	80	19	48	13	212	316	51

CLARENCE CAMPBELL CONFERENCE

Patrick Division

Team	GP	W	L	T	GF	GA	PTS
*Philadelphia	80	51	18	11	293	181	113
NY Rangers	80	37	29	14	319	276	88
NY Islanders	80	33	25	22	264	221	88
Atlanta	80	34	31	15	243	233	83

Smythe Division

Team	GP	W	L	T	GF	GA	PTS
Vancouver	80	38	32	10	271	254	86
St. Louis	80	35	31	14	269	267	84
Chicago	80	37	35	8	268	241	82
Minnesota	80	23	50	7	221	341	53
Kansas City	80	15	54	11	184	328	41

Leading Scorers

Player	Club	GP	G	A	PTS	PIM
Orr, Bobby	Boston	80	46	89	135	101
Esposito, Phil	Boston	79	61	66	127	62
Dionne, Marcel	Detroit	80	47	74	121	14
Lafleur, Guy	Montreal	70	53	66	119	37
Mahovlich, Pete	Montreal	80	35	82	117	64
Clarke, Bobby	Philadelphia	80	27	89	116	125
Robert, Rene	Buffalo	74	40	60	100	75
Gilbert, Rod	NY Rangers	76	36	61	97	22
Perreault, Gilbert	Buffalo	68	39	57	96	36
Martin, Rick	Buffalo	68	52	43	95	72

One of the all-time great defensive forwards, Dave Keon played tough but clean. A four-time Stanley Cup champion in the 1960s, Keon enjoyed his best offensive season in 1970-71 when he had 38 goals and 38 assists to crack the top 10 in scoring.

1975-76

PRINCE OF WALES CONFERENCE

Norris Division

Team	GP	W	L	T	GF	GA	PTS
*Montreal	80	58	11	11	337	174	127
Los Angeles	80	38	33	9	263	265	85
Pittsburgh	80	35	33	12	339	303	82
Detroit	80	26	44	10	226	300	62
Washington	80	11	59	10	224	394	32

Adams Division

Team	GP	W	L	T	GF	GA	PTS
Boston	80	48	15	17	313	237	113
Buffalo	80	46	21	13	339	240	105
Toronto	80	34	31	15	294	276	83
California	80	27	42	11	250	278	65

CLARENCE CAMPBELL CONFERENCE

Patrick Division

Team	GP	W	L	T	GF	GA	PTS
Philadelphia	80	51	13	16	348	209	118
NY Islanders	80	42	21	17	297	190	101
Atlanta	80	35	33	12	262	237	82
NY Rangers	80	29	42	9	262	333	67

Smythe Division

Team	GP	W	L	T	GF	GA	PTS
Chicago	80	32	30	18	254	261	82
Vancouver	80	33	32	15	271	272	81
St. Louis	80	29	37	14	249	290	72
Minnesota	80	20	53	7	195	303	47
Kansas City	80	12	56	12	190	351	36

Leading Scorers

Player	Club	GP	G	A	PTS	PIM
Lafleur, Guy	Montreal	80	56	69	125	36
Clarke, Bobby	Philadelphia	76	30	89	119	13
Perreault, Gilbert	Buffalo	80	44	69	113	36
Barber, Bill	Philadelphia	80	50	62	112	104
Larouche, Pierre	Pittsburgh	76	53	58	111	33
Ratelle, Jean	Bos., NYR	80	36	69	105	18
Mahovlich, Pete	Montreal	80	34	71	105	76
Pronovost, Jean	Pittsburgh	80	52	52	104	24
Sittler, Darryl	Toronto	79	41	59	100	90
Apps Jr., Syl	Pittsburgh	80	32	67	99	24

1976-77

PRINCE OF WALES CONFERENCE

Norris Division

Team	GP	W	L	T	GF	GA	PTS
*Montreal	80	60	8	12	387	171	132
Los Angeles	80	34	31	15	271	241	83
Pittsburgh	80	34	33	13	240	252	81
Washington	80	24	42	14	221	307	62
Detroit	80	16	55	9	183	309	41

Adams Division

Team	GP	W	L	T	GF	GA	PTS
Boston	80	49	23	8	312	240	106
Buffalo	80	48	24	8	301	220	104
Toronto	80	33	32	15	301	285	81
Cleveland	80	25	42	13	240	292	63

CLARENCE CAMPBELL CONFERENCE

Patrick Division

Team	GP	W	L	T	GF	GA	PTS
Philadelphia	80	48	16	16	323	213	112
NY Islanders	80	47	21	12	288	193	106
Atlanta	80	34	34	12	264	265	80
NY Rangers	80	29	37	14	272	310	72

Smythe Division

Team	GP	W	L	T	GF	GA	PTS
St. Louis	80	32	39	9	239	276	73
Minnesota	80	23	39	18	240	310	64
Chicago	80	26	43	11	240	298	63
Vancouver	80	25	42	13	235	294	63
Colorado	80	20	46	14	226	307	54

Leading Scorers

Player	Club	GP	G	A	PTS	PIM
Lafleur, Guy	Montreal	80	56	80	136	20
Dionne, Marcel	Los Angeles	80	53	69	122	12
Shutt, Steve	Montreal	80	60	45	105	28
MacLeish, Rick	Philadelphia	79	49	48	97	42
Perreault, Gilbert	Buffalo	80	39	56	95	30
Young, Tim	Minnesota	80	29	66	95	58
Ratelle, Jean	Boston	78	33	61	94	22
McDonald, Lanny	Toronto	80	46	44	90	77
Sittler, Darryl	Toronto	73	38	52	90	89
Clarke, Bobby	Philadelphia	80	27	63	90	71

1977-78

PRINCE OF WALES CONFERENCE

Norris Division

Team	GP	W	L	T	GF	GA	PTS
*Montreal	80	59	10	11	359	183	129
Detroit	80	32	34	14	252	266	78
Los Angeles	80	31	34	15	243	245	77
Pittsburgh	80	25	37	18	254	321	68
Washington	80	17	49	14	195	321	48

Adams Division

Team	GP	W	L	T	GF	GA	PTS
Boston	80	51	18	11	333	218	113
Buffalo	80	44	19	17	288	215	105
Toronto	80	41	29	10	271	237	92
Cleveland	80	22	45	13	230	325	57

CLARENCE CAMPBELL CONFERENCE

Patrick Division

Team	GP	W	L	T	GF	GA	PTS
NY Islanders	80	48	17	15	334	210	111
Philadelphia	80	45	20	15	296	200	105
Atlanta	80	34	27	19	274	252	87
NY Rangers	80	30	37	13	279	280	73

Smythe Division

Team	GP	W	L	T	GF	GA	PTS
Chicago	80	32	29	19	230	220	83
Colorado	80	19	40	21	257	305	59
Vancouver	80	20	43	17	239	320	57
St. Louis	80	20	47	13	195	304	53
Minnesota	80	18	53	9	218	325	45

Leading Scorers

Player	Club	GP	G	A	PTS	PIM
Lafleur, Guy	Montreal	78	60	72	132	26
Trottier, Bryan	NY Islanders	77	46	77	123	46
Sittler, Darryl	Toronto	80	45	72	117	100
Lemaire, Jacques	Montreal	76	36	61	97	14
Potvin, Denis	NY Islanders	80	30	64	94	81
Bossy, Mike	NY Islanders	73	53	38	91	6
O'Reilly, Terry	Boston	77	29	61	90	211
Perreault, Gilbert	Buffalo	79	41	48	89	20
Clarke, Bobby	Philadelphia	71	21	68	89	83
McDonald, Lanny	Toronto	74	47	40	87	54
Paiement, Wilf	Colorado	80	31	56	87	114

1978-79

PRINCE OF WALES CONFERENCE

Norris Division

Team	GP	W	L	T	GF	GA	PTS
*Montreal	80	52	17	11	337	204	115
Pittsburgh	80	36	31	13	281	279	85
Los Angeles	80	34	34	12	292	286	80
Washington	80	24	41	15	273	338	63
Detroit	80	23	41	16	252	295	62

Adams Division

Team	GP	W	L	T	GF	GA	PTS
Boston	80	43	23	14	316	270	100
Buffalo	80	36	28	16	280	263	88
Toronto	80	34	33	13	267	252	81
Minnesota	80	28	40	12	257	289	68

CLARENCE CAMPBELL CONFERENCE

Patrick Division

Team	GP	W	L	T	GF	GA	PTS
NY Islanders	80	51	15	14	358	214	116
Philadelphia	80	40	25	15	281	248	95
NY Rangers	80	40	29	11	316	292	91
Atlanta	80	41	31	8	327	280	90

Smythe Division

Team	GP	W	L	T	GF	GA	PTS
Chicago	80	29	36	15	244	277	73
Vancouver	80	25	42	13	217	291	63
St. Louis	80	18	50	12	249	348	48
Colorado	80	15	53	12	210	331	42

Leading Scorers

Player	Club	GP	G	A	PTS	PIM
Trottier, Bryan	NY Islanders	76	47	87	134	50
Dionne, Marcel	Los Angeles	80	59	71	130	30
Lafleur, Guy	Montreal	80	52	77	129	28
Bossy, Mike	NY Islanders	80	69	57	126	25
MacMillan, Bob	Atlanta	79	37	71	108	14
Chouinard, Guy	Atlanta	80	50	57	107	14
Potvin, Denis	NY Islanders	73	31	70	101	58
Federko, Bernie	St. Louis	74	31	64	95	14
Taylor, Dave	Los Angeles	78	43	48	91	124
Gillies, Clark	NY Islanders	75	35	56	91	68

1979-80

PRINCE OF WALES CONFERENCE

Norris Division

Team	GP	W	L	T	GF	GA	PTS
Montreal	80	47	20	13	328	240	107
Los Angeles	80	30	36	14	290	313	74
Pittsburgh	80	30	37	13	251	303	73
Hartford	80	27	34	19	303	312	73
Detroit	80	26	43	11	268	306	63

Adams Division

Team	GP	W	L	T	GF	GA	PTS
Buffalo	80	47	17	16	318	201	110
Boston	80	46	21	13	310	234	105
Minnesota	80	36	28	16	311	253	88
Toronto	80	35	40	5	304	327	75
Quebec	80	25	44	11	248	313	61

CLARENCE CAMPBELL CONFERENCE

Patrick Division

Team	GP	W	L	T	GF	GA	PTS
Philadelphia	80	48	12	20	327	254	116
*NY Islanders	80	39	28	13	281	247	91
NY Rangers	80	38	32	10	308	284	86
Atlanta	80	35	32	13	282	269	83
Washington	80	27	40	13	261	293	67

Smythe Division

Team	GP	W	L	T	GF	GA	PTS
Chicago	80	34	27	19	241	250	87
St. Louis	80	34	34	12	266	278	80
Vancouver	80	27	37	16	256	281	70
Edmonton	80	28	39	13	301	322	69
Winnipeg	80	20	49	11	214	314	51
Colorado	80	19	48	13	234	308	51

Leading Scorers

Player	Club	GP	G	A	PTS	PIM
Dionne, Marcel	Los Angeles	80	53	84	137	32
Gretzky, Wayne	Edmonton	79	51	86	137	21
Lafleur, Guy	Montreal	74	50	75	125	12
Perreault, Gilbert	Buffalo	80	40	66	106	57
Rogers, Mike	Hartford	80	44	61	105	10
Trottier, Bryan	NY Islanders	78	42	62	104	68
Simmer, Charlie	Los Angeles	64	56	45	101	65
Stoughton, Blaine	Hartford	80	56	44	100	16
Sittler, Darryl	Toronto	73	40	57	97	62
MacDonald, Blair	Edmonton	80	46	48	94	6
Federko, Bernie	St. Louis	79	38	56	94	24

1980-81

PRINCE OF WALES CONFERENCE

Norris Division

Team	GP	W	L	T	GF	GA	PTS
Montreal	80	45	22	13	332	232	103
Los Angeles	80	43	24	13	337	290	99
Pittsburgh	80	30	37	13	302	345	73
Hartford	80	21	41	18	292	372	60
Detroit	80	19	43	18	252	339	56

Adams Division

Team	GP	W	L	T	GF	GA	PTS
Buffalo	80	39	20	21	327	250	99
Boston	80	37	30	13	316	272	87
Minnesota	80	35	28	17	291	263	87
Quebec	80	30	32	18	314	318	78
Toronto	80	28	37	15	322	367	71

CLARENCE CAMPBELL CONFERENCE

Patrick Division

Team	GP	W	L	T	GF	GA	PTS
*NY Islanders	80	48	18	14	355	260	110
Philadelphia	80	41	24	15	313	249	97
Calgary	80	39	27	14	329	298	92
NY Rangers	80	30	36	14	312	317	74
Washington	80	26	36	18	286	317	70

Smythe Division

Team	GP	W	L	T	GF	GA	PTS
St. Louis	80	45	18	17	352	281	107
Chicago	80	31	33	16	304	315	78
Vancouver	80	28	32	20	289	301	76
Edmonton	80	29	35	16	328	327	74
Colorado	80	22	45	13	258	344	57
Winnipeg	80	9	57	14	246	400	32

Leading Scorers

Player	Club	GP	G	A	PTS	PIM
Gretzky, Wayne	Edmonton	80	55	109	164	28
Dionne, Marcel	Los Angeles	80	58	77	135	70
Nilsson, Kent	Calgary	80	49	82	131	26
Bossy, Mike	NY Islanders	79	68	51	119	32
Taylor, Dave	Los Angeles	72	47	65	112	130
Stastny, Peter	Quebec	77	39	70	109	37
Simmer, Charlie	Los Angeles	65	56	49	105	62
Rogers, Mike	Hartford	80	40	65	105	32
Federko, Bernie	St. Louis	78	31	73	104	47
Richard, Jacques	Quebec	78	52	51	103	39
Middleton, Rick	Boston	80	44	59	103	16
Trottier, Bryan	NY Islanders	73	31	72	103	74

1981-82

CLARENCE CAMPBELL CONFERENCE

Norris Division

Team	GP	W	L	T	GF	GA	PTS
Minnesota	80	37	23	20	346	288	94
Winnipeg	80	33	33	14	319	332	80
St. Louis	80	32	40	8	315	349	72
Chicago	80	30	38	12	332	363	72
Toronto	80	20	44	16	298	380	56
Detroit	80	21	47	12	270	351	54

Smythe Division

Team	GP	W	L	T	GF	GA	PTS
Edmonton	80	48	17	15	417	295	111
Vancouver	80	30	33	17	290	286	77
Calgary	80	29	34	17	334	345	75
Los Angeles	80	24	41	15	314	369	63
Colorado	80	18	49	13	241	362	49

PRINCE OF WALES CONFERENCE

Adams Division

Team	GP	W	L	T	GF	GA	PTS
Montreal	80	46	17	17	360	223	109
Boston	80	43	27	10	323	285	96
Buffalo	80	39	26	15	307	273	93
Quebec	80	33	31	16	356	345	82
Hartford	80	21	41	18	264	351	60

Patrick Division

Team	GP	W	L	T	GF	GA	PTS
*NY Islanders	80	54	16	10	385	250	118
NY Rangers	80	39	27	14	316	306	92
Philadelphia	80	38	31	11	325	313	87
Pittsburgh	80	31	36	13	310	337	75
Washington	80	26	41	13	319	338	65

Leading Scorers

Player	Club	GP	G	A	PTS	PIM
Gretzky, Wayne	Edmonton	80	92	120	212	26
Bossy, Mike	NY Islanders	80	64	83	147	22
Stastny, Peter	Quebec	80	46	93	139	91
Maruk, Dennis	Washington	80	60	76	136	128
Trottier, Bryan	NY Islanders	80	50	79	129	88
Savard, Denis	Chicago	80	32	87	119	82
Dionne, Marcel	Los Angeles	78	50	67	117	50
Smith, Bobby	Minnesota	80	43	71	114	82
Ciccarelli, Dino	Minnesota	76	55	51	106	138
Taylor, Dave	Los Angeles	78	39	67	106	130

1982-83

CLARENCE CAMPBELL CONFERENCE

Norris Division

Team	GP	W	L	T	GF	GA	PTS
Chicago	80	47	23	10	338	268	104
Minnesota	80	40	24	16	321	290	96
Toronto	80	28	40	12	293	330	68
St. Louis	80	25	40	15	285	316	65
Detroit	80	21	44	15	263	344	57

Smythe Division

Team	GP	W	L	T	GF	GA	PTS
Edmonton	80	47	21	12	424	315	106
Calgary	80	32	34	14	321	317	78
Vancouver	80	30	35	15	303	309	75
Winnipeg	80	33	39	8	311	333	74
Los Angeles	80	27	41	12	308	365	66

PRINCE OF WALES CONFERENCE

Adams Division

Team	GP	W	L	T	GF	GA	PTS
Boston	80	50	20	10	327	228	110
Montreal	80	42	24	14	350	286	98
Buffalo	80	38	29	13	318	285	89
Quebec	80	34	34	12	343	336	80
Hartford	80	19	54	7	261	403	45

Patrick Division

Team	GP	W	L	T	GF	GA	PTS
Philadelphia	80	49	23	8	326	240	106
*NY Islanders	80	42	26	12	302	226	96
Washington	80	39	25	16	306	283	94
NY Rangers	80	35	35	10	306	287	80
New Jersey	80	17	49	14	230	338	48
Pittsburgh	80	18	53	9	257	394	45

Leading Scorers

Player	Club	GP	G	A	PTS	PIM
Gretzky, Wayne	Edmonton	80	71	125	196	59
Stastny, Peter	Quebec	75	47	77	124	78
Savard, Denis	Chicago	78	35	86	121	99
Bossy, Mike	NY Islanders	79	60	58	118	20
Dionne, Marcel	Los Angeles	80	56	51	107	22
Pederson, Barry	Boston	77	46	61	107	47
Messier, Mark	Edmonton	77	48	58	106	72
Goulet, Michel	Quebec	80	57	48	105	51
Anderson, Glenn	Edmonton	72	48	56	104	70
Nilsson, Kent	Calgary	80	46	58	104	10
Kurri, Jari	Edmonton	80	45	59	104	22

1983-84

CLARENCE CAMPBELL CONFERENCE

Norris Division

Team	GP	W	L	T	GF	GA	PTS
Minnesota	80	39	31	10	345	344	88
St. Louis	80	32	41	7	293	316	71
Detroit	80	31	42	7	298	323	69
Chicago	80	30	42	8	277	311	68
Toronto	80	26	45	9	303	387	61

Smythe Division

Team	GP	W	L	T	GF	GA	PTS
*Edmonton	80	57	18	5	446	314	119
Calgary	80	34	32	14	311	314	82
Vancouver	80	32	39	9	306	328	73
Winnipeg	80	31	38	11	340	374	73
Los Angeles	80	23	44	13	309	376	59

PRINCE OF WALES CONFERENCE

Adams Division

Team	GP	W	L	T	GF	GA	PTS
Boston	80	49	25	6	336	261	104
Buffalo	80	48	25	7	315	257	103
Quebec	80	42	28	10	360	278	94
Montreal	80	35	40	5	286	295	75
Hartford	80	28	42	10	288	320	66

Patrick Division

Team	GP	W	L	T	GF	GA	PTS
NY Islanders	80	50	26	4	357	269	104
Washington	80	48	27	5	308	226	101
Philadelphia	80	44	26	10	350	290	98
NY Rangers	80	42	29	9	314	304	93
New Jersey	80	17	56	7	231	350	41
Pittsburgh	80	16	58	6	254	390	38

Leading Scorers

Player	Club	GP	G	A	PTS	PIM
Gretzky, Wayne	Edmonton	74	87	118	205	39
Coffey, Paul	Edmonton	80	40	86	126	104
Goulet, Michel	Quebec	75	56	65	121	76
Stastny, Peter	Quebec	80	46	73	119	73
Bossy, Mike	NY Islanders	67	51	67	118	8
Pederson, Barry	Boston	80	39	77	116	64
Kurri, Jari	Edmonton	64	52	61	113	14
Trottier, Bryan	NY Islanders	68	40	71	111	59
Federko, Bernie	St. Louis	79	41	66	107	43
Middleton, Rick	Boston	80	47	58	105	14

1984-85

CLARENCE CAMPBELL CONFERENCE

Norris Division

Team	GP	W	L	T	GF	GA	PTS
St. Louis	80	37	31	12	299	288	86
Chicago	80	38	35	7	309	299	83
Detroit	80	27	41	12	313	357	66
Minnesota	80	25	43	12	268	321	62
Toronto	80	20	52	8	253	358	48

Smythe Division

Team	GP	W	L	T	GF	GA	PTS
*Edmonton	80	49	20	11	401	298	109
Winnipeg	80	43	27	10	358	332	96
Calgary	80	41	27	12	363	302	94
Los Angeles	80	34	32	14	339	326	82
Vancouver	80	25	46	9	284	401	59

PRINCE OF WALES CONFERENCE

Adams Division

Team	GP	W	L	T	GF	GA	PTS
Montreal	80	41	27	12	309	262	94
Quebec	80	41	30	9	323	275	91
Buffalo	80	38	28	14	290	237	90
Boston	80	36	34	10	303	287	82
Hartford	80	30	41	9	268	318	69

Patrick Division

Team	GP	W	L	T	GF	GA	PTS
Philadelphia	80	53	20	7	348	241	113
Washington	80	46	25	9	322	240	101
NY Islanders	80	40	34	6	345	312	86
NY Rangers	80	26	44	10	295	345	62
New Jersey	80	22	48	10	264	346	54
Pittsburgh	80	24	51	5	276	385	53

Leading Scorers

Player	Club	GP	G	A	PTS	PIM
Gretzky, Wayne	Edmonton	80	73	135	208	52
Kurri, Jari	Edmonton	73	71	64	135	30
Hawerchuk, Dale	Winnipeg	80	53	77	130	74
Dionne, Marcel	Los Angeles	80	46	80	126	46
Coffey, Paul	Edmonton	80	37	84	121	97
Bossy, Mike	NY Islanders	76	58	59	117	38
Ogrodnick, John	Detroit	79	55	50	105	30
Savard, Denis	Chicago	79	38	67	105	56
Federko, Bernie	St. Louis	76	30	73	103	27
Gartner, Mike	Washington	80	50	52	102	71

1985-86

CLARENCE CAMPBELL CONFERENCE

Norris Division

Team	GP	W	L	T	GF	GA	PTS
Chicago	80	39	33	8	351	349	86
Minnesota	80	38	33	9	327	305	85
St. Louis	80	37	34	9	302	291	83
Toronto	80	25	48	7	311	386	57
Detroit	80	17	57	6	266	415	40

Smythe Division

Team	GP	W	L	T	GF	GA	PTS
Edmonton	80	56	17	7	426	310	119
Calgary	80	40	31	9	354	315	89
Winnipeg	80	26	47	7	295	372	59
Vancouver	80	23	44	13	282	333	59
Los Angeles	80	23	49	8	284	389	54

PRINCE OF WALES CONFERENCE

Adams Division

Team	GP	W	L	T	GF	GA	PTS
Quebec	80	43	31	6	330	289	92
*Montreal	80	40	33	7	330	280	87
Boston	80	37	31	12	311	288	86
Hartford	80	40	36	4	332	302	84
Buffalo	80	37	37	6	296	291	80

Patrick Division

Team	GP	W	L	T	GF	GA	PTS
Philadelphia	80	53	23	4	335	241	110
Washington	80	50	23	7	315	272	107
NY Islanders	80	39	29	12	327	284	90
NY Rangers	80	36	38	6	280	276	78
Pittsburgh	80	34	38	8	313	305	76
New Jersey	80	28	49	3	300	374	59

Leading Scorers

Player	Club	GP	G	A	PTS	PIM
Gretzky, Wayne	Edmonton	80	52	163	215	52
Lemieux, Mario	Pittsburgh	79	48	93	141	43
Coffey, Paul	Edmonton	79	48	90	138	120
Kurri, Jari	Edmonton	78	68	63	131	22
Bossy, Mike	NY Islanders	80	61	62	123	14
Stastny, Peter	Quebec	76	41	81	122	60
Savard, Denis	Chicago	80	47	69	116	111
Naslund, Mats	Montreal	80	43	67	110	16
Hawerchuk, Dale	Winnipeg	80	46	59	105	44
Broten, Neal	Minnesota	80	29	76	105	47

1986-87

CLARENCE CAMPBELL CONFERENCE

Norris Division

Team	GP	W	L	T	GF	GA	PTS
St. Louis	80	32	33	15	281	293	79
Detroit	80	34	36	10	260	274	78
Chicago	80	29	37	14	290	310	72
Toronto	80	32	42	6	286	319	70
Minnesota	80	30	40	10	296	314	70

Smythe Division

Team	GP	W	L	T	GF	GA	PTS
*Edmonton	80	50	24	6	372	284	106
Calgary	80	46	31	3	318	289	95
Winnipeg	80	40	32	8	279	271	88
Los Angeles	80	31	41	8	318	341	70
Vancouver	80	29	43	8	282	314	66

PRINCE OF WALES CONFERENCE

Adams Division

Team	GP	W	L	T	GF	GA	PTS
Hartford	80	43	30	7	287	270	93
Montreal	80	41	29	10	277	241	92
Boston	80	39	34	7	301	276	85
Quebec	80	31	39	10	267	276	72
Buffalo	80	28	44	8	280	308	64

Patrick Division

Team	GP	W	L	T	GF	GA	PTS
Philadelphia	80	46	26	8	310	245	100
Washington	80	38	32	10	285	278	86
NY Islanders	80	35	33	12	279	281	82
NY Rangers	80	34	38	8	307	323	76
Pittsburgh	80	30	38	12	297	290	72
New Jersey	80	29	45	6	293	368	64

Leading Scorers

Player	Club	GP	G	A	PTS	PIM
Gretzky, Wayne	Edmonton	79	62	121	183	28
Kurri, Jari	Edmonton	79	54	54	108	41
Lemieux, Mario	Pittsburgh	63	54	53	107	57
Messier, Mark	Edmonton	77	37	70	107	73
Gilmour, Doug	St. Louis	80	42	63	105	58
Ciccarelli, Dino	Minnesota	80	52	51	103	92
Hawerchuk, Dale	Winnipeg	80	47	53	100	54
Goulet, Michel	Quebec	75	49	47	96	61
Kerr, Tim	Philadelphia	75	58	37	95	57
Bourque, Raymond	Boston	78	23	72	95	36

1987-88

CLARENCE CAMPBELL CONFERENCE

Norris Division

Team	GP	W	L	T	GF	GA	PTS
Detroit	80	41	28	11	322	269	93
St. Louis	80	34	38	8	278	294	76
Chicago	80	30	41	9	284	328	69
Toronto	80	21	49	10	273	345	52
Minnesota	80	19	48	13	242	349	51

Smythe Division

Team	GP	W	L	T	GF	GA	PTS
Calgary	80	48	23	9	397	305	105
*Edmonton	80	44	25	11	363	288	99
Winnipeg	80	33	36	11	292	310	77
Los Angeles	80	30	42	8	318	359	68
Vancouver	80	25	46	9	272	320	59

PRINCE OF WALES CONFERENCE

Adams Division

Team	GP	W	L	T	GF	GA	PTS
Montreal	80	45	22	13	298	238	103
Boston	80	44	30	6	300	251	94
Buffalo	80	37	32	11	283	305	85
Hartford	80	35	38	7	249	267	77
Quebec	80	32	43	5	271	306	69

Patrick Division

Team	GP	W	L	T	GF	GA	PTS
NY Islanders	80	39	31	10	308	267	88
Washington	80	38	33	9	281	249	85
Philadelphia	80	38	33	9	292	292	85
New Jersey	80	38	36	6	295	296	82
NY Rangers	80	36	34	10	300	283	82
Pittsburgh	80	36	35	9	319	316	81

Leading Scorers

Player	Club	GP	G	A	PTS	PIM
Lemieux, Mario	Pittsburgh	77	70	98	168	92
Gretzky, Wayne	Edmonton	64	40	109	149	24
Savard, Denis	Chicago	80	44	87	131	95
Hawerchuk, Dale	Winnipeg	80	44	77	121	59
Robitaille, Luc	Los Angeles	80	53	58	111	82
Stastny, Peter	Quebec	76	46	65	111	69
Messier, Mark	Edmonton	77	37	74	111	103
Carson, Jimmy	Los Angeles	80	55	52	107	45
Loob, Hakan	Calgary	80	50	56	106	47
Goulet, Michel	Quebec	80	48	58	106	56

1988-89

CLARENCE CAMPBELL CONFERENCE

Norris Division

Team	GP	W	L	T	GF	GA	PTS
Detroit	80	34	34	12	313	316	80
St. Louis	80	33	35	12	275	285	78
Minnesota	80	27	37	16	258	278	70
Chicago	80	27	41	12	297	335	66
Toronto	80	28	46	6	259	342	62

Smythe Division

Team	GP	W	L	T	GF	GA	PTS
*Calgary	80	54	17	9	354	226	117
Los Angeles	80	42	31	7	376	335	91
Edmonton	80	38	34	8	325	306	84
Vancouver	80	33	39	8	251	253	74
Winnipeg	80	26	42	12	300	355	64

PRINCE OF WALES CONFERENCE

Adams Division

Team	GP	W	L	T	GF	GA	PTS
Montreal	80	53	18	9	315	218	115
Boston	80	37	29	14	289	256	88
Buffalo	80	38	35	7	291	299	83
Hartford	80	37	38	5	299	290	79
Quebec	80	27	46	7	269	342	61

Patrick Division

Team	GP	W	L	T	GF	GA	PTS
Washington	80	41	29	10	305	259	92
Pittsburgh	80	40	33	7	347	349	87
NY Rangers	80	37	35	8	310	307	82
Philadelphia	80	36	36	8	307	285	80
New Jersey	80	27	41	12	281	325	66
NY Islanders	80	28	47	5	265	325	61

Leading Scorers

Player	Club	GP	G	A	PTS	PIM
Lemieux, Mario	Pittsburgh	76	85	114	199	100
Gretzky, Wayne	Los Angeles	78	54	114	168	26
Yzerman, Steve	Detroit	80	65	90	155	61
Nicholls, Bernie	Los Angeles	79	70	80	150	96
Brown, Rob	Pittsburgh	68	49	66	115	118
Coffey, Paul	Pittsburgh	75	30	83	113	193
Mullen, Joe	Calgary	79	51	59	110	16
Kurri, Jari	Edmonton	76	44	58	102	69
Carson, Jimmy	Edmonton	80	49	51	100	36
Robitaille, Luc	Los Angeles	78	46	52	98	65

1989-90

CLARENCE CAMPBELL CONFERENCE

Norris Division

Team	GP	W	L	T	GF	GA	PTS
Chicago	80	41	33	6	316	294	88
St. Louis	80	37	34	9	295	279	83
Toronto	80	38	38	4	337	358	80
Minnesota	80	36	40	4	284	291	76
Detroit	80	28	38	14	288	323	70

Smythe Division

Team	GP	W	L	T	GF	GA	PTS
Calgary	80	42	23	15	348	265	99
*Edmonton	80	38	28	14	315	283	90
Winnipeg	80	37	32	11	298	290	85
Los Angeles	80	34	39	7	338	337	75
Vancouver	80	25	41	14	245	306	64

PRINCE OF WALES CONFERENCE

Adams Division

Team	GP	W	L	T	GF	GA	PTS
Boston	80	46	25	9	289	232	101
Buffalo	80	45	27	8	286	248	98
Montreal	80	41	28	11	288	234	93
Hartford	80	38	33	9	275	268	85
Quebec	80	12	61	7	240	407	31

Patrick Division

Team	GP	W	L	T	GF	GA	PTS
NY Rangers	80	36	31	13	279	267	85
New Jersey	80	37	34	9	295	288	83
Washington	80	36	38	6	284	275	78
NY Islanders	80	31	38	11	281	288	73
Pittsburgh	80	32	40	8	318	359	72
Philadelphia	80	30	39	11	290	297	71

Leading Scorers

Player	Club	GP	G	A	PTS	PIM
Gretzky, Wayne	Los Angeles	73	40	102	142	42
Messier, Mark	Edmonton	79	45	84	129	79
Yzerman, Steve	Detroit	79	62	65	127	79
Lemieux, Mario	Pittsburgh	59	45	78	123	78
Hull, Brett	St. Louis	80	72	41	113	24
Nicholls, Bernie	L.A., NYR	79	39	73	112	86
Turgeon, Pierre	Buffalo	80	40	66	106	29
LaFontaine, Pat	NY Islanders	74	54	51	105	38
Coffey, Paul	Pittsburgh	80	29	74	103	95
Sakic, Joe	Quebec	80	39	63	102	27
Oates, Adam	St. Louis	80	23	79	102	30

1990-91

CLARENCE CAMPBELL CONFERENCE

Norris Division

Team	GP	W	L	T	GF	GA	PTS
Chicago	80	49	23	8	284	211	106
St. Louis	80	47	22	11	310	250	105
Detroit	80	34	38	8	273	298	76
Minnesota	80	27	39	14	256	266	68
Toronto	80	23	46	11	241	318	57

Smythe Division

Team	GP	W	L	T	GF	GA	PTS
Los Angeles	80	46	24	10	340	254	102
Calgary	80	46	26	8	344	263	100
Edmonton	80	37	37	6	272	272	80
Vancouver	80	28	43	9	243	315	65
Winnipeg	80	26	43	11	260	288	63

PRINCE OF WALES CONFERENCE

Adams Division

Team	GP	W	L	T	GF	GA	PTS
Boston	80	44	24	12	299	264	100
Montreal	80	39	30	11	273	249	89
Buffalo	80	31	30	19	292	278	81
Hartford	80	31	38	11	238	276	73
Quebec	80	16	50	14	236	354	46

Patrick Division

Team	GP	W	L	T	GF	GA	PTS
*Pittsburgh	80	41	33	6	342	305	88
NY Rangers	80	36	31	13	297	265	85
Washington	80	37	36	7	258	258	81
New Jersey	80	32	33	15	272	264	79
Philadelphia	80	33	37	10	252	267	76
NY Islanders	80	25	45	10	223	290	60

Leading Scorers

Player	Club	GP	G	A	PTS	PIM
Gretzky, Wayne	Los Angeles	78	41	122	163	16
Hull, Brett	St. Louis	78	86	45	131	22
Oates, Adam	St. Louis	61	25	90	115	29
Recchi, Mark	Pittsburgh	78	40	73	113	48
Cullen, John	Pit., Hfd.	78	39	71	110	101
Sakic, Joe	Quebec	80	48	61	109	24
Yzerman, Steve	Detroit	80	51	57	108	34
Fleury, Theoren	Calgary	79	51	53	104	136
MacInnis, Al	Calgary	78	28	75	103	90
Larmer, Steve	Chicago	80	44	57	101	79

1991-92

CLARENCE CAMPBELL CONFERENCE

Norris Division

Team	GP	W	L	T	GF	GA	PTS
Detroit	80	43	25	12	320	256	98
Chicago	80	36	29	15	257	236	87
St. Louis	80	36	33	11	279	266	83
Minnesota	80	32	42	6	246	278	70
Toronto	80	30	43	7	234	294	67

Smythe Division

Team	GP	W	L	T	GF	GA	PTS
Vancouver	80	42	26	12	285	250	96
Los Angeles	80	35	31	14	287	296	84
Edmonton	80	36	34	10	295	297	82
Winnipeg	80	33	32	15	251	244	81
Calgary	80	31	37	12	296	305	74
San Jose	80	17	58	5	219	359	39

PRINCE OF WALES CONFERENCE

Adams Division

Team	GP	W	L	T	GF	GA	PTS
Montreal	80	41	28	11	267	207	93
Boston	80	36	32	12	270	275	84
Buffalo	80	31	37	12	289	299	74
Hartford	80	26	41	13	247	283	65
Quebec	80	20	48	12	255	318	52

Patrick Division

Team	GP	W	L	T	GF	GA	PTS
NY Rangers	80	50	25	5	321	246	105
Washington	80	45	27	8	330	275	98
*Pittsburgh	80	39	32	9	343	308	87
New Jersey	80	38	31	11	289	259	87
NY Islanders	80	34	35	11	291	299	79
Philadelphia	80	32	37	11	252	273	75

Leading Scorers

Player	Club	GP	G	A	PTS	PIM
Lemieux, Mario	Pittsburgh	64	44	87	131	94
Stevens, Kevin	Pittsburgh	80	54	69	123	254
Gretzky, Wayne	Los Angeles	74	31	90	121	34
Hull, Brett	St. Louis	73	70	39	109	48
Robitaille, Luc	Los Angeles	80	44	63	107	95
Messier, Mark	NY Rangers	79	35	72	107	76
Roenick, Jeremy	Chicago	80	53	50	103	23
Yzerman, Steve	Detroit	79	45	58	103	64
Leetch, Brian	NY Rangers	80	22	80	102	26
Oates, Adam	St.L., Bos.	80	20	79	99	22

1992-93

CLARENCE CAMPBELL CONFERENCE

Norris Division

Team	GP	W	L	T	GF	GA	PTS
Chicago	84	47	25	12	279	230	106
Detroit	84	47	28	9	369	280	103
Toronto	84	44	29	11	288	241	99
St. Louis	84	37	36	11	282	278	85
Minnesota	84	36	38	10	272	293	82
Tampa Bay	84	23	54	7	245	332	53

Smythe Division

Team	GP	W	L	T	GF	GA	PTS
Vancouver	84	46	29	9	346	278	101
Calgary	84	43	30	11	322	282	97
Los Angeles	84	39	35	10	338	340	88
Winnipeg	84	40	37	7	322	320	87
Edmonton	84	26	50	8	242	337	60
San Jose	84	11	71	2	218	414	24

PRINCE OF WALES CONFERENCE

Adams Division

Team	GP	W	L	T	GF	GA	PTS
Boston	84	51	26	7	332	268	109
Quebec	84	47	27	10	351	300	104
*Montreal	84	48	30	6	326	280	102
Buffalo	84	38	36	10	335	297	86
Hartford	84	26	52	6	284	369	58
Ottawa	84	10	70	4	202	395	24

Patrick Division

Team	GP	W	L	T	GF	GA	PTS
Pittsburgh	84	56	21	7	367	268	119
Washington	84	43	34	7	325	286	93
NY Islanders	84	40	37	7	335	297	87
New Jersey	84	40	37	7	308	299	87
Philadelphia	84	36	37	11	319	319	83
NY Rangers	84	34	39	11	304	308	79

Leading Scorers

Player	Club	GP	G	A	PTS	PIM
Lemieux, Mario	Pittsburgh	60	69	91	160	38
LaFontaine, Pat	Buffalo	84	53	95	148	63
Oates, Adam	Boston	84	45	97	142	32
Yzerman, Steve	Detroit	84	58	79	137	44
Selanne, Teemu	Winnipeg	84	76	56	132	45
Turgeon, Pierre	NY Islanders	83	58	74	132	26
Mogilny, Alexander	Buffalo	77	76	51	127	40
Gilmour, Doug	Toronto	83	32	95	127	100
Robitaille, Luc	Los Angeles	84	63	62	125	100
Recchi, Mark	Philadelphia	84	53	70	123	95

1993-94

EASTERN CONFERENCE

Northeast Division

Team	GP	W	L	T	GF	GA	PTS
Pittsburgh	84	44	27	13	299	285	101
Boston	84	42	29	13	289	252	97
Montreal	84	41	29	14	283	248	96
Buffalo	84	43	32	9	282	218	95
Quebec	84	34	42	8	277	292	76
Hartford	84	27	48	9	227	288	63
Ottawa	84	14	61	9	201	397	37

Atlantic Division

Team	GP	W	L	T	GF	GA	PTS
*NY Rangers	84	52	24	8	299	231	112
New Jersey	84	47	25	12	306	220	106
Washington	84	39	35	10	277	263	88
NY Islanders	84	36	36	12	282	264	84
Florida	84	33	34	17	233	233	83
Philadelphia	84	35	39	10	294	314	80
Tampa Bay	84	30	43	11	224	251	71

WESTERN CONFERENCE

Central Division

Team	GP	W	L	T	GF	GA	PTS
Detroit	84	46	30	8	356	275	100
Toronto	84	43	29	12	280	243	98
Dallas	84	42	29	13	286	265	97
St. Louis	84	40	33	11	270	283	91
Chicago	84	39	36	9	254	240	87
Winnipeg	84	24	51	9	245	344	57

Pacific Division

Team	GP	W	L	T	GF	GA	PTS
Calgary	84	42	29	13	302	256	97
Vancouver	84	41	40	3	279	276	85
San Jose	84	33	35	16	252	265	82
Anaheim	84	33	46	5	229	251	71
Los Angeles	84	27	45	12	294	322	66
Edmonton	84	25	45	14	261	305	64

Leading Scorers

Player	Club	GP	G	A	PTS	PIM
Gretzky, Wayne	Los Angeles	81	38	92	130	20
Fedorov, Sergei	Detroit	82	56	64	120	34
Oates, Adam	Boston	77	32	80	112	45
Gilmour, Doug	Toronto	83	27	84	111	105
Bure, Pavel	Vancouver	76	60	47	107	86
Roenick, Jeremy	Chicago	84	46	61	107	125
Recchi, Mark	Philadelphia	84	40	67	107	46
Shanahan, Brendan	St. Louis	81	52	50	102	211
Andreychuk, Dave	Toronto	83	53	46	99	98
Jagr, Jaromir	Pittsburgh	80	32	67	99	61

1994-95

EASTERN CONFERENCE

Northeast Division

Team	GP	W	L	T	GF	GA	PTS
Quebec	48	30	13	5	185	134	65
Pittsburgh	48	29	16	3	181	158	61
Boston	48	27	18	3	150	127	57
Buffalo	48	22	19	7	130	119	51
Hartford	48	19	24	5	127	141	43
Montreal	48	18	23	7	125	148	43
Ottawa	48	9	34	5	117	174	23

Atlantic Division

Team	GP	W	L	T	GF	GA	PTS
Philadelphia	48	28	16	4	150	132	60
*New Jersey	48	22	18	8	136	121	52
Washington	48	22	18	8	136	120	52
NY Rangers	48	22	23	3	139	134	47
Florida	48	20	22	6	115	127	46
Tampa Bay	48	17	28	3	120	144	37
NY Islanders	48	15	28	5	126	158	35

WESTERN CONFERENCE

Central Division

Team	GP	W	L	T	GF	GA	PTS
Detroit	48	33	11	4	180	117	70
St. Louis	48	28	15	5	178	135	61
Chicago	48	24	19	5	156	115	53
Toronto	48	21	19	8	135	146	50
Dallas	48	17	23	8	136	135	42
Winnipeg	48	16	25	7	157	177	39

Pacific Division

Team	GP	W	L	T	GF	GA	PTS
Calgary	48	24	17	7	163	135	55
Vancouver	48	18	18	12	153	148	48
San Jose	48	19	25	4	129	161	42
Los Angeles	48	16	23	9	142	174	41
Edmonton	48	17	27	4	136	183	38
Anaheim	48	16	27	5	125	164	37

Leading Scorers

Player	Club	GP	G	A	PTS	PIM
Jagr, Jaromir	Pittsburgh	48	32	38	70	37
Lindros, Eric	Philadelphia	46	29	41	70	60
Zhamnov, Alexei	Winnipeg	48	30	35	65	20
Sakic, Joe	Quebec	47	19	43	62	30
Francis, Ron	Pittsburgh	44	11	48	59	18
Fleury, Theoren	Calgary	47	29	29	58	112
Coffey, Paul	Detroit	45	14	44	58	72
Renberg, Mikael	Philadelphia	47	26	31	57	20
LeClair, John	Mtl., Phi.	46	26	28	54	30
Messier, Mark	NY Rangers	46	14	39	53	40
Oates, Adam	Boston	48	12	41	53	8

1995-96

EASTERN CONFERENCE

Northeast Division

Team	GP	W	L	T	GF	GA	PTS
Pittsburgh	82	49	29	4	362	284	102
Boston	82	40	31	11	282	269	91
Montreal	82	40	32	10	265	248	90
Hartford	82	34	39	9	237	259	77
Buffalo	82	33	42	7	247	262	73
Ottawa	82	18	59	5	191	291	41

Atlantic Division

Team	GP	W	L	T	GF	GA	PTS
Philadelphia	82	45	24	13	282	208	103
NY Rangers	82	41	27	14	272	237	96
Florida	82	41	31	10	254	234	92
Washington	82	39	32	11	234	204	89
Tampa Bay	82	38	32	12	238	248	88
New Jersey	82	37	33	12	215	202	86
NY Islanders	82	22	50	10	229	315	54

WESTERN CONFERENCE

Central Division

Team	GP	W	L	T	GF	GA	PTS
Detroit	82	62	13	7	325	181	131
Chicago	82	40	28	14	273	220	94
Toronto	82	34	36	12	247	252	80
St. Louis	82	32	34	16	219	248	80
Winnipeg	82	36	40	6	275	291	78
Dallas	82	26	42	14	227	280	66

Pacific Division

Team	GP	W	L	T	GF	GA	PTS
*Colorado	82	47	25	10	326	240	104
Calgary	82	34	37	11	241	240	79
Vancouver	82	32	35	15	278	278	79
Anaheim	82	35	39	8	234	247	78
Edmonton	82	30	44	8	240	304	68
Los Angeles	82	24	40	18	256	302	66
San Jose	82	20	55	7	252	357	47

Leading Scorers

Player	Club	GP	G	A	PTS	PIM
Lemieux, Mario	Pittsburgh	70	69	92	161	54
Jagr, Jaromir	Pittsburgh	82	62	87	149	96
Sakic, Joe	Colorado	82	51	69	120	44
Francis, Ron	Pittsburgh	77	27	92	119	56
Forsberg, Peter	Colorado	82	30	86	116	47
Lindros, Eric	Philadelphia	73	47	68	115	163
Kariya, Paul	Anaheim	82	50	58	108	20
Selanne, Teemu	Wpg., Ana.	79	40	68	108	22
Mogilny, Alexander	Vancouver	79	55	52	107	16
Fedorov, Sergei	Detroit	78	39	68	107	48

One of the top-scoring left wingers in NHL history, Dave Andreychuk enjoyed his most productive seasons playing alongside Doug Gilmour with the Maple Leafs in the mid 1990s. He ranked among the top 10 scorers with 99 points in 1993-94.

1996-97

EASTERN CONFERENCE

Northeast Division

Team	GP	W	L	T	GF	GA	PTS
Buffalo	82	40	30	12	237	208	92
Pittsburgh	82	38	36	8	285	280	84
Ottawa	82	31	36	15	226	234	77
Montreal	82	31	36	15	249	276	77
Hartford	82	32	39	11	226	256	75
Boston	82	26	47	9	234	300	61

Atlantic Division

Team	GP	W	L	T	GF	GA	PTS
New Jersey	82	45	23	14	231	182	104
Philadelphia	82	45	24	13	274	217	103
Florida	82	35	28	19	221	201	89
NY Rangers	82	38	34	10	258	231	86
Washington	82	33	40	9	214	231	75
Tampa Bay	82	32	40	10	217	247	74
NY Islanders	82	29	41	12	240	250	70

WESTERN CONFERENCE

Central Division

Team	GP	W	L	T	GF	GA	PTS
Dallas	82	48	26	8	252	198	104
*Detroit	82	38	26	18	253	197	94
Phoenix	82	38	37	7	240	243	83
St. Louis	82	36	35	11	236	239	83
Chicago	82	34	35	13	223	210	81
Toronto	82	30	44	8	230	273	68

Pacific Division

Team	GP	W	L	T	GF	GA	PTS
Colorado	82	49	24	9	277	205	107
Anaheim	82	36	33	13	245	233	85
Edmonton	82	36	37	9	252	247	81
Vancouver	82	35	40	7	257	273	77
Calgary	82	32	41	9	214	239	73
Los Angeles	82	28	43	11	214	268	67
San Jose	82	27	47	8	211	278	62

Leading Scorers

Player	Club	GP	G	A	PTS	PIM
Lemieux, Mario	Pittsburgh	76	50	72	122	65
Selanne, Teemu	Anaheim	78	51	58	109	34
Kariya, Paul	Anaheim	69	44	55	99	6
LeClair, John	Philadelphia	82	50	47	97	58
Gretzky, Wayne	NY Rangers	82	25	72	97	28
Jagr, Jaromir	Pittsburgh	63	47	48	95	40
Sundin, Mats	Toronto	82	41	53	94	59
Palffy, Ziggy	NY Islanders	80	48	42	90	43
Francis, Ron	Pittsburgh	81	27	63	90	20
Shanahan, Brendan	Hfd., Det.	81	47	41	88	131

1997-98

EASTERN CONFERENCE

Northeast Division

Team	GP	W	L	T	GF	GA	PTS
Pittsburgh	82	40	24	18	228	188	98
Boston	82	39	30	13	221	194	91
Buffalo	82	36	29	17	211	187	89
Montreal	82	37	32	13	235	208	87
Ottawa	82	34	33	15	193	200	83
Carolina	82	33	41	8	200	219	74

Atlantic Division

Team	GP	W	L	T	GF	GA	PTS
New Jersey	82	48	23	11	225	166	107
Philadelphia	82	42	29	11	242	193	95
Washington	82	40	30	12	219	202	92
NY Islanders	82	30	41	11	212	225	71
NY Rangers	82	25	39	18	197	231	68
Florida	82	24	43	15	203	256	63
Tampa Bay	82	17	55	10	151	269	44

WESTERN CONFERENCE

Central Division

Team	GP	W	L	T	GF	GA	PTS
Dallas	82	49	22	11	242	167	109
*Detroit	82	44	23	15	250	196	103
St. Louis	82	45	29	8	256	204	98
Phoenix	82	35	35	12	224	227	82
Chicago	82	30	39	13	192	199	73
Toronto	82	30	43	9	194	237	69

Pacific Division

Team	GP	W	L	T	GF	GA	PTS
Colorado	82	39	26	17	231	205	95
Los Angeles	82	38	33	11	227	225	87
Edmonton	82	35	37	10	215	224	80
San Jose	82	34	38	10	210	216	78
Calgary	82	26	41	15	217	252	67
Anaheim	82	26	43	13	205	261	65
Vancouver	82	25	43	14	224	273	64

Leading Scorers

Player	Club	GP	G	A	PTS	PIM
Jagr, Jaromir	Pittsburgh	77	35	67	102	64
Forsberg, Peter	Colorado	72	25	66	91	94
Bure, Pavel	Vancouver	82	51	39	90	48
Gretzky, Wayne	NY Rangers	82	23	67	90	28
LeClair, John	Philadelphia	82	51	36	87	32
Palffy, Ziggy	NY Islanders	82	45	42	87	34
Francis, Ron	Pittsburgh	81	25	62	87	20
Selanne, Teemu	Anaheim	73	52	34	86	30
Allison, Jason	Boston	81	33	50	83	60
Stumpel, Jozef	Los Angeles	77	21	58	79	53

1998-99

EASTERN CONFERENCE

Northeast Division

Team	GP	W	L	T	GF	GA	PTS
Ottawa	82	44	23	15	239	179	103
Toronto	82	45	30	7	268	231	97
Boston	82	39	30	13	214	181	91
Buffalo	82	37	28	17	207	175	91
Montreal	82	32	39	11	184	209	75

Atlantic Division

Team	GP	W	L	T	GF	GA	PTS
New Jersey	82	47	24	11	248	196	105
Philadelphia	82	37	26	19	231	196	93
Pittsburgh	82	38	30	14	242	225	90
NY Rangers	82	33	38	11	217	227	77
NY Islanders	82	24	48	10	194	244	58

Southeast Division

Team	GP	W	L	T	GF	GA	PTS
Carolina	82	34	30	18	210	202	86
Florida	82	30	34	18	210	228	78
Washington	82	31	45	6	200	218	68
Tampa Bay	82	19	54	9	179	292	47

WESTERN CONFERENCE

Central Division

Team	GP	W	L	T	GF	GA	PTS
Detroit	82	43	32	7	245	202	93
St Louis	82	37	32	13	237	209	87
Chicago	82	29	41	12	202	248	70
Nashville	82	28	47	7	190	261	63

Pacific Division

Team	GP	W	L	T	GF	GA	PTS
*Dallas	82	51	19	12	236	168	114
Phoenix	82	39	31	12	205	197	90
Anaheim	82	35	34	13	215	206	83
San Jose	82	31	33	18	196	191	80
Los Angeles	82	32	45	5	189	222	69

Northwest Division

Team	GP	W	L	T	GF	GA	PTS
Colorado	82	44	28	10	239	205	98
Edmonton	82	33	37	12	230	226	78
Calgary	82	30	40	12	211	234	72
Vancouver	82	23	47	12	192	258	58

Leading Scorers

Player	Club	GP	G	A	PTS	PIM
Jagr, Jaromir	Pittsburgh	81	44	83	127	66
Selanne, Teemu	Anaheim	75	47	60	107	30
Kariya, Paul	Anaheim	82	39	62	101	40
Forsberg, Peter	Colorado	78	30	67	97	108
Sakic, Joe	Colorado	73	41	55	96	29
Yashin, Alexei	Ottawa	82	44	50	94	54
Lindros, Eric	Philadelphia	71	40	53	93	120
Fleury, Theoren	Cgy., Col.	75	40	53	93	86
LeClair, John	Philadelphia	76	43	47	90	30
Demitra, Pavol	St Louis	82	37	52	89	16

Now playing in Los Angeles, Jason Allison was a top-10 scorer for the Bruins in 2000-01 and just missed cracking the top 10 again with the Kings last season. His 55 assists ranked second in the league, while his 74 points placed him 16th.

Though he had a 50-goal season early in his career, Mike Modano did not crack the top 10 in scoring until 1999-2000. He ranked among the NHL's elite scorers again in 2001-02 with 77 points (34 goals, 43 assists).

1999-2000

EASTERN CONFERENCE

Northeast Division

Team	GP	W	L	T	OTL	GF	GA	PTS
Toronto	82	45	27	7	3	246	222	100
Ottawa	82	41	28	11	2	244	210	95
Buffalo	82	35	32	11	4	213	204	85
Montreal	82	35	34	9	4	196	194	83
Boston	82	24	33	19	6	210	248	73

Atlantic Division

Team	GP	W	L	T	OTL	GF	GA	PTS
Philadelphia	82	45	22	12	3	237	179	105
*New Jersey	82	45	24	8	5	251	203	103
Pittsburgh	82	37	31	8	6	241	236	88
NY Rangers	82	29	38	12	3	218	246	73
NY Islanders	82	24	48	9	1	194	275	58

Southeast Division

Team	GP	W	L	T	OTL	GF	GA	PTS
Washington	82	44	24	12	2	227	194	102
Florida	82	43	27	6	6	244	209	98
Carolina	82	37	35	10	0	217	216	84
Tampa Bay	82	19	47	9	7	204	310	54
Atlanta	82	14	57	7	4	170	313	39

WESTERN CONFERENCE

Central Division

Team	GP	W	L	T	OTL	GF	GA	PTS
St. Louis	82	51	19	11	1	248	165	114
Detroit	82	48	22	10	2	278	210	108
Chicago	82	33	37	10	2	242	245	78
Nashville	82	28	40	7	7	199	240	70

Pacific Division

Team	GP	W	L	T	OTL	GF	GA	PTS
Dallas	82	43	23	10	6	211	184	102
Los Angeles	82	39	27	12	4	245	228	94
Phoenix	82	39	31	8	4	232	228	90
San Jose	82	35	30	10	7	225	214	87
Anaheim	82	34	33	12	3	217	227	83

Northwest Division

Team	GP	W	L	T	OTL	GF	GA	PTS
Colorado	82	42	28	11	1	233	201	96
Edmonton	82	32	26	16	8	226	212	88
Vancouver	82	30	29	15	8	227	237	83
Calgary	82	31	36	10	5	211	256	77

Leading Scorers

Player	Club	GP	G	A	PTS	PIM
Jagr, Jaromir	Pittsburgh	63	42	54	96	50
Bure, Pavel	Florida	74	58	36	94	16
Recchi, Mark	Philadelphia	82	28	63	91	50
Kariya, Paul	Anaheim	74	42	44	86	24
Selanne, Teemu	Anaheim	79	33	52	85	12
Nolan, Owen	San Jose	78	44	40	84	110
Amonte, Tony	Chicago	82	43	41	84	48
Modano, Mike	Dallas	77	38	43	81	48
Sakic, Joe	Colorado	60	28	53	81	28
Yzerman, Steve	Detroit	78	35	44	79	34

2000-2001

EASTERN CONFERENCE

Northeast Division

Team	GP	W	L	T	OTL	GF	GA	PTS
Ottawa	82	48	21	9	4	274	205	109
Buffalo	82	46	30	5	1	218	184	98
Toronto	82	37	29	11	5	232	207	90
Boston	82	36	30	8	8	227	249	88
Montreal	82	28	40	8	6	206	232	70

Atlantic Division

Team	GP	W	L	T	OTL	GF	GA	PTS
New Jersey	82	48	19	12	3	295	195	111
Philadelphia	82	43	25	11	3	240	207	100
Pittsburgh	82	42	28	9	3	281	256	96
NY Rangers	82	33	43	5	1	250	290	72
NY Islanders	82	21	51	7	3	185	268	52

Southeast Division

Team	GP	W	L	T	OTL	GF	GA	PTS
Washington	82	41	27	10	4	233	211	96
Carolina	82	38	32	9	3	212	225	88
Florida	82	22	38	13	9	200	246	66
Atlanta	82	23	45	12	2	211	289	60
Tampa Bay	82	24	47	6	5	201	280	59

WESTERN CONFERENCE

Central Division

Team	GP	W	L	T	OTL	GF	GA	PTS
Detroit	82	49	20	9	4	253	202	111
St. Louis	82	43	22	12	5	249	195	103
Nashville	82	34	36	9	3	186	200	80
Chicago	82	29	40	8	5	210	246	71
Columbus	82	28	39	9	6	190	233	71

Pacific Division

Team	GP	W	L	T	OTL	GF	GA	PTS
Dallas	82	48	24	8	2	241	187	106
San Jose	82	40	27	12	3	217	192	95
Los Angeles	82	38	28	13	3	252	228	92
Phoenix	82	35	27	17	3	214	212	90
Anaheim	82	25	41	11	5	188	245	66

Northwest Division

Team	GP	W	L	T	OTL	GF	GA	PTS
*Colorado	82	52	16	10	4	270	192	118
Edmonton	82	39	28	12	3	243	222	93
Vancouver	82	36	28	11	7	239	238	90
Calgary	82	27	36	15	4	197	236	73
Minnesota	82	25	39	13	5	168	210	68

Leading Scorers

Player	Club	GP	G	A	PTS	PIM
Jagr, Jaromir	Pittsburgh	81	52	69	121	42
Sakic, Joe	Colorado	82	54	64	118	30
Elias, Patrik	New Jersey	82	40	56	96	51
Kovalev, Alexei	Pittsburgh	79	44	51	95	96
Allison, Jason	Boston	82	36	59	95	85
Straka, Martin	Pittsburgh	82	27	68	95	38
Bure, Pavel	Florida	82	59	33	92	58
Weight, Doug	Edmonton	82	25	65	90	91
Palffy, Ziggy	Los Angeles	73	38	51	89	20
Forsberg, Peter	Colorado	73	27	62	89	54

2001-2002

EASTERN CONFERENCE

Northeast Division

Team	GP	W	L	T	OTL	GF	GA	PTS
Boston	82	43	24	6	9	236	201	101
Toronto	82	43	25	10	4	249	207	100
Ottawa	82	39	27	9	7	243	208	94
Montreal	82	36	31	12	3	207	209	87
Buffalo	82	35	35	11	1	213	200	82

Atlantic Division

Team	GP	W	L	T	OTL	GF	GA	PTS
Philadelphia	82	42	27	10	3	234	192	97
NY Islanders	82	42	28	8	4	239	220	96
New Jersey	82	41	28	9	4	205	187	95
NY Rangers	82	36	38	4	4	227	258	80
Pittsburgh	82	28	41	8	5	198	249	69

Southeast Division

Team	GP	W	L	T	OTL	GF	GA	PTS
Carolina	82	35	26	16	5	217	217	91
Washington	82	36	33	11	2	228	240	85
Tampa Bay	82	27	40	11	4	178	219	69
Florida	82	22	44	10	6	180	250	60
Atlanta	82	19	47	11	5	187	288	54

WESTERN CONFERENCE

Central Division

Team	GP	W	L	T	OTL	GF	GA	PTS
*Detroit	82	51	17	10	4	251	187	116
St. Louis	82	43	27	8	4	227	188	98
Chicago	82	41	27	13	1	216	207	96
Nashville	82	28	41	13	0	196	230	69
Columbus	82	22	47	8	5	164	255	57

Pacific Division

Team	GP	W	L	T	OTL	GF	GA	PTS
San Jose	82	44	27	8	3	248	199	99
Phoenix	82	40	27	9	6	228	210	95
Los Angeles	82	40	27	11	4	214	190	95
Dallas	82	36	28	13	5	215	213	90
Anaheim	82	29	42	8	3	175	198	69

Northwest Division

Team	GP	W	L	T	OTL	GF	GA	PTS
Colorado	82	45	28	8	1	212	169	99
Vancouver	82	42	30	7	3	254	211	94
Edmonton	82	38	28	12	4	205	182	92
Calgary	82	32	35	12	3	201	220	79
Minnesota	82	26	35	12	9	195	238	73

Leading Scorers

Player	Club	GP	G	A	PTS	PIM
Iginla, Jarome	Calgary	82	52	44	96	77
Naslund, Markus	Vancouver	81	40	50	90	50
Bertuzzi, Todd	Vancouver	72	36	49	85	110
Sundin, Mats	Toronto	82	41	39	80	94
Jagr, Jaromir	Washington	69	31	48	79	30
Sakic, Joe	Colorado	82	26	53	79	18
Demitra, Pavol	St. Louis	82	35	43	78	46
Oates, Adam	Wsh., Phi.	80	14	64	78	28
Modano, Mike	Dallas	78	34	43	77	38
Francis, Ron	Carolina	80	27	50	77	18

Note: Detailed statistics for 2001-2002 are listed in the Final Statistics, 2001-2002 section of the *NHL Guide & Record Book*. **See page 135.**

Todd Bertuzzi has been a solid performer since returning from a leg injury that sidelined him for most of the 1998-99 season. Last year, he exploded onto the scene with 85 points, trailing only Jarome Iginla and Canucks teammate Markus Naslund.

Team Records

Regular Season

FINAL STANDINGS

MOST POINTS, ONE SEASON:
132 — Montreal Canadiens, 1976-77. 60W-8L-12T. 80GP
131 — Detroit Red Wings, 1995-96. 62W-13L-7T. 82GP
129 — Montreal Canadiens, 1977-78. 59W-10L-11T. 80GP

BEST POINTS PERCENTAGE, ONE SEASON:
.875 — Boston Bruins, 1929-30. 38W-5L-1T. 77PTS in 44GP
.830 — Montreal Canadiens, 1943-44. 38W-5L-7T. 83PTS in 50GP
.825 — Montreal Canadiens, 1976-77. 60W-8L-12T. 132PTS in 80GP
.806 — Montreal Canadiens, 1977-78. 59W-10L-11T. 129PTS in 80GP
.800 — Montreal Canadiens, 1944-45. 38W-8L-4T. 80PTS in 50GP

FEWEST POINTS, ONE SEASON:
8 — Quebec Bulldogs, 1919-20. 4W-20L-0T. 24GP
10 — Toronto Arenas, 1918-19. 5W-13L-0T. 18GP
12 — Hamilton Tigers, 1920-21. 6W-18L-0T. 24GP
— Hamilton Tigers, 1922-23. 6W-18L-0T. 24GP
— Boston Bruins, 1924-25. 6W-24L-0T. 30GP
— Philadelphia Quakers, 1930-31. 4W-36L-4T. 44GP

FEWEST POINTS, ONE SEASON (MINIMUM 70-GAME SCHEDULE):
21 — Washington Capitals, 1974-75. 8W-67L-5T. 80GP
24 — Ottawa Senators, 1992-93. 10W-70L-4T. 84GP
— San Jose Sharks, 1992-93. 11W-71L-2T. 84GP
30 — NY Islanders, 1972-73. 12W-60L-6T. 78GP

WORST POINTS PERCENTAGE, ONE SEASON:
.131 — Washington Capitals, 1974-75. 8W-67L-5T. 21PTS in 80GP
.136 — Philadelphia Quakers, 1930-31. 4W-36L-4T. 12PTS in 44GP
.143 — Ottawa Senators, 1992-93. 10W-70L-4T. 24PTS in 84GP
.143 — San Jose Sharks, 1992-93. 11W-71L-2T. 24PTS in 84GP
.148 — Pittsburgh Pirates, 1929-30. 5W-36L-3T. 13PTS in 44GP

TEAM WINS

Most Wins

MOST WINS, ONE SEASON:
62 — Detroit Red Wings, 1995-96. 82GP
60 — Montreal Canadiens, 1976-77. 80GP
59 — Montreal Canadiens, 1977-78. 80GP

MOST HOME WINS, ONE SEASON:
36 — Philadelphia Flyers, 1975-76. 40GP
— Detroit Red Wings, 1995-96. 41GP
33 — Boston Bruins, 1970-71. 39GP
— Boston Bruins, 1973-74. 39GP
— Montreal Canadiens, 1976-77. 40GP
— Philadelphia Flyers, 1976-77. 40GP
— NY Islanders, 1981-82. 40GP
— Philadelphia Flyers, 1985-86. 40GP

MOST ROAD WINS, ONE SEASON:
28 — New Jersey Devils, 1998-99. 41GP
27 — Montreal Canadiens, 1976-77. 40GP
— Montreal Canadiens, 1977-78. 40GP
— St. Louis Blues, 1999-2000. 41GP
26 — Boston Bruins, 1971-72. 39GP
— Montreal Canadiens, 1975-76. 40GP
— Edmonton Oilers, 1983-84. 40GP
— Detroit Red Wings, 1995-96. 41GP

Fewest Wins

FEWEST WINS, ONE SEASON:
4 — Quebec Bulldogs, 1919-20. 24GP
— Philadelphia Quakers, 1930-31. 44GP
5 — Toronto Arenas, 1918-19. 18GP
— Pittsburgh Pirates, 1929-30. 44GP

FEWEST WINS, ONE SEASON (MINIMUM 70-GAME SCHEDULE):
8 — Washington Capitals, 1974-75. 80GP
9 — Winnipeg Jets, 1980-81. 80GP
10 — Ottawa Senators, 1992-93. 84GP

FEWEST HOME WINS, ONE SEASON:
2 — Chicago Blackhawks, 1927-28. 22GP
3 — Boston Bruins, 1924-25. 15GP
— Chicago Blackhawks, 1928-29. 22GP
— Philadelphia Quakers, 1930-31. 22GP

FEWEST HOME WINS, ONE SEASON (MINIMUM 70-GAME SCHEDULE):
6 — Chicago Blackhawks, 1954-55. 35GP
— Washington Capitals, 1975-76. 40GP
7 — Boston Bruins, 1962-63. 35GP
— Washington Capitals, 1974-75. 40GP
— Winnipeg Jets, 1980-81. 40GP
— Pittsburgh Penguins, 1983-84. 40GP

FEWEST ROAD WINS, ONE SEASON:
0 — Toronto Arenas, 1918-19. 9GP
— Quebec Bulldogs, 1919-20. 12GP
— Pittsburgh Pirates, 1929-30. 22GP
1 — Hamilton Tigers, 1921-22. 12GP
— Toronto St. Patricks, 1925-26. 18GP
— Philadelphia Quakers, 1930-31. 22GP
— NY Americans, 1940-41. 24GP
— Washington Capitals, 1974-75. 40GP
* — Ottawa Senators, 1992-93. 41GP

FEWEST ROAD WINS, ONE SEASON (MINIMUM 70-GAME SCHEDULE):
1 — Washington Capitals, 1974-75. 40GP
*** — Ottawa Senators,** 1992-93. 41GP
2 — Boston Bruins, 1960-61. 35GP
— Los Angeles Kings, 1969-70. 38GP
— NY Islanders, 1972-73. 39GP
— California Golden Seals, 1973-74. 39GP
— Colorado Rockies, 1977-78. 40GP
— Winnipeg Jets, 1980-81. 40GP
— Quebec Nordiques, 1991-92. 40GP

TEAM LOSSES

Fewest Losses

FEWEST LOSSES, ONE SEASON:
5 — Ottawa Senators, 1919-20. 24GP
— Boston Bruins, 1929-30. 44GP
— Montreal Canadiens, 1943-44. 50GP

FEWEST HOME LOSSES, ONE SEASON:
0 — Ottawa Senators, 1922-23. 12GP
— Montreal Canadiens, 1943-44. 25GP
1 — Toronto Arenas, 1917-18. 11GP
— Ottawa Senators, 1918-19. 9GP
— Ottawa Senators, 1919-20. 12GP
— Toronto St. Patricks, 1922-23. 12GP
— Boston Bruins, 1929-30. 22GP
— Boston Bruins, 1930-31. 22GP
— Montreal Canadiens, 1976-77. 40GP
— Quebec Nordiques, 1994-95. 24GP

FEWEST ROAD LOSSES, ONE SEASON:
3 — Montreal Canadiens, 1928-29. 22GP
4 — Ottawa Senators, 1919-20. 12GP
— Montreal Canadiens, 1927-28. 22GP
— Boston Bruins, 1929-30. 20GP
— Boston Bruins, 1940-41. 24GP

FEWEST LOSSES, ONE SEASON (MINIMUM 70-GAME SCHEDULE):
8 — Montreal Canadiens, 1976-77. 80GP
10 — Montreal Canadiens, 1972-73. 78GP
— Montreal Canadiens, 1977-78. 80GP
11 — Montreal Canadiens, 1975-76. 80GP

FEWEST HOME LOSSES, ONE SEASON (MINIMUM 70-GAME SCHEDULE):
1 — Montreal Canadiens, 1976-77. 40GP
2 — Montreal Canadiens, 1961-62. 35GP
— NY Rangers, 1970-71. 39GP
— Philadelphia Flyers, 1975-76. 40GP

FEWEST ROAD LOSSES, ONE SEASON (MINIMUM 70-GAME SCHEDULE):
6 — Montreal Canadiens, 1972-73. 39GP
— Montreal Canadiens, 1974-75. 40GP
— Montreal Canadiens, 1977-78. 40GP
7 — Detroit Red Wings, 1951-52. 35GP
— Montreal Canadiens, 1976-77. 40GP
— Philadelphia Flyers, 1979-80. 40GP

Most Losses

MOST LOSSES, ONE SEASON:
71 — San Jose Sharks, 1992-93. 84GP
70 — Ottawa Senators, 1992-93. 84GP
67 — Washington Capitals, 1974-75. 80GP
61 — Quebec Nordiques, 1989-90. 80GP
—Ottawa Senators, 1993-94. 84GP

MOST HOME LOSSES, ONE SEASON:
***32 — San Jose Sharks,** 1992-93. 41GP
29 — Pittsburgh Penguins, 1983-84. 40GP
* — Ottawa Senators, 1993-94. 41GP

MOST ROAD LOSSES, ONE SEASON:
***40 — Ottawa Senators,** 1992-93. 41GP
39 — Washington Capitals, 1974-75. 40GP
37 — California Seals, 1973-74. 39GP
* — San Jose Sharks, 1992-93. 41GP

* — Does not include neutral site games

TEAM TIES

Most Ties

MOST TIES, ONE SEASON:
24 — Philadelphia Flyers, 1969-70. 76GP
23 — Montreal Canadiens, 1962-63. 70GP
— Chicago Blackhawks, 1973-74. 78GP

MOST HOME TIES, ONE SEASON:
13 — NY Rangers, 1954-55. 35GP
— Philadelphia Flyers, 1969-70. 38GP
— California Golden Seals, 1971-72. 39GP
— California Golden Seals, 1972-73. 39GP
— Chicago Blackhawks, 1973-74. 39GP

MOST ROAD TIES, ONE SEASON:
15 — Philadelphia Flyers, 1976-77. 40GP
14 — Montreal Canadiens, 1952-53. 35GP
— Montreal Canadiens, 1974-75. 40GP
— Philadelphia Flyers, 1975-76. 40GP

Fewest Ties

FEWEST TIES, ONE SEASON (Since 1926-27):
1 — Boston Bruins, 1929-30. 44GP
2 — Montreal Canadians, 1926-27. 44GP
— NY Americans, 1926-27. 44GP
— Boston Bruins, 1938-39. 48GP
— NY Rangers, 1941-42. 48GP
— San Jose Sharks, 1992-93. 84GP

FEWEST TIES, ONE SEASON (MINIMUM 70-GAME SCHEDULE):
2 — San Jose Sharks, 1992-93. 84GP
3 — New Jersey Devils, 1985-86. 80GP
— Calgary Flames, 1986-87. 80GP
— Vancouver Canucks, 1993-94. 84GP

WINNING STREAKS

LONGEST WINNING STREAK, ONE SEASON:
17 Games — Pittsburgh Penguins, Mar. 9 - Apr. 10, 1993.
15 Games — NY Islanders, Jan. 21 - Feb. 20, 1982.
14 Games — Boston Bruins, Dec. 3, 1929 - Jan. 9, 1930.

LONGEST HOME WINNING STREAK, ONE SEASON:
20 Games — Boston Bruins, Dec. 3, 1929 - Mar. 18, 1930.
— Philadelphia Flyers, Jan. 4 - Apr. 3, 1976.

LONGEST ROAD WINNING STREAK, ONE SEASON:
10 Games — Buffalo Sabres, Dec. 10, 1983 - Jan. 23, 1984.
— St. Louis Blues, Jan. 21 - Mar. 2, 2000.
— New Jersey Devils, Feb. 27 - Apr. 7, 2001.
8 Games — Boston Bruins, Feb. 17 - Mar. 8, 1972.
— Los Angeles Kings, Dec. 18, 1974 - Jan. 16, 1975.
— Montreal Canadiens, Dec. 18, 1977 - Jan. 18, 1978.
— NY Islanders, Feb. 27 - Mar. 29, 1981.
— Montreal Canadiens, Jan. 21 - Feb. 21, 1982.
— Philadelphia Flyers, Dec. 22, 1982 - Jan. 16, 1983.
— Winnipeg Jets, Feb. 25 - Apr. 6, 1985.
— Edmonton Oilers, Dec. 9, 1986 - Jan. 17, 1987.
— Boston Bruins, Mar. 15 - Apr. 14, 1993.
— Detroit Red Wings, Feb. 4 - Mar. 9, 2002

LONGEST WINNING STREAK FROM START OF SEASON:
10 Games — Toronto Maple Leafs, 1993-94.
8 Games — Toronto Maple Leafs, 1934-35.
— Buffalo Sabres, 1975-76.
7 Games — Edmonton Oilers, 1983-84.
— Quebec Nordiques, 1985-86.
— Pittsburgh Penguins, 1986-87.
— Pittsburgh Penguins, 1994-95.

LONGEST HOME WINNING STREAK FROM START OF SEASON:
11 Games — Chicago Blackhawks, 1963-64.
10 Games — Ottawa Senators, 1925-26.
9 Games — Montreal Canadiens, 1953-54.
— Chicago Blackhawks, 1971-72.

LONGEST ROAD WINNING STREAK FROM START OF SEASON:
7 Games — Toronto Maple Leafs, Nov. 14 - Dec. 15, 1940.
— Philadelphia Flyers, Oct. 12 - Nov. 16, 1985

LONGEST WINNING STREAK, INCLUDING PLAYOFFS:
15 Games — Detroit Red Wings, Feb. 27 - Apr. 5, 1955. Nine regular-season games, six playoff games.

LONGEST HOME WINNING STREAK, INCLUDING PLAYOFFS:
24 Games — Philadelphia Flyers, Jan. 4 - Apr. 25, 1976. Twenty regular-season games, four playoff games.

LONGEST ROAD WINNING STREAK, INCLUDING PLAYOFFS:
11 Games — New Jersey Devils, Feb. 27 - Apr. 17, 2001. Ten regular season games, one playoff game.

UNDEFEATED STREAKS

LONGEST UNDEFEATED STREAK, ONE SEASON:
35 Games — Philadelphia Flyers, Oct. 14, 1979 - Jan. 6, 1980. 25W-10T.
28 Games — Montreal Canadiens, Dec. 18, 1977 - Feb. 23, 1978. 23W-5T.

LONGEST HOME UNDEFEATED STREAK, ONE SEASON:
34 Games — Montreal Canadiens, Nov. 1, 1976 - Apr. 2, 1977. 28W-6T.
27 Games — Boston Bruins, Nov. 22, 1970 - Mar. 20, 1971. 26W-1T.

LONGEST ROAD UNDEFEATED STREAK, ONE SEASON:
23 Games — Montreal Canadiens, Nov. 27, 1974 - Mar. 12, 1975. 14W-9T.
17 Games — Montreal Canadiens, Dec. 18, 1977 - Mar. 1, 1978. 14W-3T.

LONGEST UNDEFEATED STREAK FROM START OF SEASON:
15 Games — Edmonton Oilers, 1984-85. 12W-3T.
14 Games — Montreal Canadiens, 1943-44. 11W-3T.

LONGEST HOME UNDEFEATED STREAK FROM START OF SEASON:
25 Games — Montreal Canadiens, Oct. 30, 1943 - Mar. 18, 1944. 22W-3T.

LONGEST ROAD UNDEFEATED STREAK FROM START OF SEASON:
15 Games — Detroit Red Wings, Oct. 18 - Dec. 20, 1951. 10W-5T.

LONGEST UNDEFEATED STREAK, INCLUDING PLAYOFFS:
21 Games — Pittsburgh Penguins, Mar. 9 - Apr. 22, 1993. 17W-1T in regular season and 3W in playoffs.

LONGEST HOME UNDEFEATED STREAK, INCLUDING PLAYOFFS:
38 Games — Montreal Canadiens, Nov. 1, 1976 - Apr. 26, 1977. 28W-6T in regular season and 4W in playoff.

LONGEST ROAD UNDEFEATED STREAK, INCLUDING PLAYOFFS:
13 Games — Montreal Canadiens, Feb. 26 - Apr. 20, 1980. 6W-4T in regular season and 3W in playoffs.
— NY Islanders, Mar. 16 - May 1, 1980. 3W-3T in regular season and 7W in playoffs.
— Philadelphia Flyers, Feb. 26 - Apr. 21, 1977. 6W-4T in regular season and 3W in playoffs.

LOSING STREAKS

LONGEST LOSING STREAK, ONE SEASON:
17 Games — Washington Capitals, Feb. 18 - Mar. 26, 1975.
— San Jose Sharks, Jan. 4 - Feb. 12, 1993.
15 Games — Philadelphia Quakers, Nov. 29, 1930 - Jan. 8, 1931.

LONGEST HOME LOSING STREAK, ONE SEASON:
11 Games — Boston Bruins, Dec. 8, 1924 - Feb. 17, 1925.
— Washington Capitals, Feb. 18 - Mar. 30, 1975.
— Ottawa Senators, Oct. 27 - Dec. 8, 1993.
— Atlanta Thrashers, Jan. 24-Mar. 16, 2000.

LONGEST ROAD LOSING STREAK, ONE SEASON:
***38 Games — Ottawa Senators,** Oct. 10, 1992 - Apr. 3, 1993.
37 Games — Washington Capitals, Oct. 9, 1974 - Mar. 26, 1975.

LONGEST LOSING STREAK FROM START OF SEASON:
11 Games — NY Rangers, 1943-44.
7 Games — Montreal Canadiens, 1938-39.
— Chicago Blackhawks, 1947-48.
— Washington Capitals, 1983-84.
— Chicago Blackhawks, 1997-98.

LONGEST HOME LOSING STREAK FROM START OF SEASON:
8 Games — Los Angeles Kings, Oct. 13 - Nov. 6, 1971.

LONGEST ROAD LOSING STREAK FROM START OF SEASON:
***38 Games — Ottawa Senators,** Oct. 10, 1992 - Apr. 3, 1993.

WINLESS STREAKS

LONGEST WINLESS STREAK, ONE SEASON:
30 Games — Winnipeg Jets, Oct. 19 - Dec. 20, 1980. 23L-7T.
27 Games — Kansas City Scouts, Feb. 12 - Apr. 4, 1976. 21L-6T.
25 Games — Washington Capitals, Nov. 29, 1975 - Jan. 21, 1976. 22L-3T.

LONGEST HOME WINLESS STREAK, ONE SEASON:
17 Games — Ottawa Senators, Oct. 28, 1995 - Jan. 27, 1996. 15L-2T.
— Atlanta Thrashers, Jan. 19 - Mar. 29, 2000. 15L-2T.
15 Games — Chicago Blackhawks, Dec. 16, 1928 - Feb. 28, 1929. 11L-4T.
— Montreal Canadiens, Dec. 16, 1939 - Mar. 7, 1940. 12L-3T.

LONGEST ROAD WINLESS STREAK, ONE SEASON:
***38 Games — Ottawa Senators,** Oct. 10, 1992 - Apr. 3, 1993. 38L-0T.
37 Games — Washington Capitals, Oct. 9, 1974 - Mar. 26, 1975. 37L-0T.

LONGEST WINLESS STREAK FROM START OF SEASON:
15 Games — NY Rangers, 1943-44. 14L-1T.
11 Games — Pittsburgh Pirates, 1927-28. 8L-3T.
— Minnesota North Stars, 1973-74. 5L-6T.
— San Jose Sharks, 1995-96. 7L-4T.

LONGEST HOME WINLESS STREAK FROM START OF SEASON:
11 Games — Pittsburgh Penguins, Oct. 8 - Nov. 19, 1983. 9L-2T.

LONGEST ROAD WINLESS STREAK FROM START OF SEASON:
***38 Games — Ottawa Senators,** Oct. 10, 1992 - Apr. 3, 1993. 38L-0T.

NON-SHUTOUT STREAKS

LONGEST NON-SHUTOUT STREAK:
264 Games — Calgary Flames, Nov. 12, 1981 - Jan. 9, 1985.
261 Games — Los Angeles Kings, Mar. 15, 1986 - Oct. 22, 1989.
244 Games — Washington Capitals, Oct. 31, 1989 - Nov. 11, 1993.
236 Games — NY Rangers, Dec. 20, 1989 - Dec. 13, 1992.
230 Games — Quebec Nordiques, Feb. 10, 1980 - Jan. 12, 1983.

LONGEST NON-SHUTOUT STREAK INCLUDING PLAYOFFS:
264 Games — Los Angeles Kings, Mar. 15, 1986 - Apr. 6, 1989. (5 playoff games in 1987; 5 in 1988; 2 in 1989).
262 Games — Chicago Blackhawks, Mar. 14, 1970 - Feb. 21, 1973. (8 playoff games in 1970; 18 in 1971; 8 in 1972).
251 Games — Quebec Nordiques, Feb. 10, 1980 - Jan. 12, 1983. (5 playoff games in 1981; 16 in 1982).
246 Games — Pittsburgh Penguins, Jan. 7, 1989 - Oct. 26, 1991. (11 playoff games in 1989; 24 in 1991).

TEAM GOALS

Most Goals

MOST GOALS, ONE SEASON:
446 — Edmonton Oilers, 1983-84. 80GP
426 — Edmonton Oilers, 1985-86. 80GP
424 — Edmonton Oilers, 1982-83. 80GP
417 — Edmonton Oilers, 1981-82. 80GP
401 — Edmonton Oilers, 1984-85. 80GP

MOST GOALS, ONE TEAM, ONE GAME:
16 — Montreal Canadiens, Mar. 3, 1920, at Quebec. Defeated Que. Bulldogs 16-3.

MOST GOALS, BOTH TEAMS, ONE GAME:
21 — Montreal Canadiens, Toronto St. Patricks, at Montreal, Jan. 10, 1920. Montreal won 14-7.
— **Edmonton Oilers, Chicago Blackhawks,** at Chicago, Dec. 11, 1985. Edmonton won 12-9.
20 — Edmonton Oilers, Minnesota North Stars, at Edmonton, Jan. 4, 1984. Edmonton won 12-8.
— Toronto Maple Leafs, Edmonton Oilers, at Toronto, Jan. 8, 1986. Toronto won 11-9.
19 — Montreal Wanderers, Toronto Arenas, at Montreal, Dec. 19, 1917. Montreal won 10-9.
— Montreal Canadiens, Quebec Bulldogs, at Quebec, Mar. 3, 1920. Montreal won 16-3.
— Montreal Canadiens, Hamilton Tigers, at Montreal, Feb. 26, 1921. Montreal won 13-6.
— Boston Bruins, NY Rangers, at Boston, Mar. 4, 1944. Boston won 10-9.
— Boston Bruins, Detroit Red Wings, at Detroit, Mar. 16, 1944. Detroit won 10-9.
— Vancouver Canucks, Minnesota North Stars, at Vancouver, Oct. 7, 1983. Vancouver won 10-9.

MOST GOALS, ONE TEAM, ONE PERIOD:
9 — Buffalo Sabres, Mar. 19, 1981, at Buffalo, second period during 14-4 win over Toronto.
8 — Detroit Red Wings, Jan. 23, 1944, at Detroit, third period during 15-0 win over NY Rangers.
— Boston Bruins, Mar. 16, 1969, at Boston, second period during 11-3 win over Toronto.
— NY Rangers, Nov. 21, 1971, at NY Rangers, third period during 12-1 win over California.
— Philadelphia Flyers, Mar. 31, 1973, at Philadelphia, second period during 10-2 win over NY Islanders.
— Buffalo Sabres, Dec. 21, 1975, at Buffalo, third period during 14-2 win over Washington.
— Minnesota North Stars, Nov. 11, 1981, at Minnesota, second period during 15-2 win over Winnipeg.
— Pittsburgh Penguins, Dec. 17, 1991, at Pittsburgh, second period during 10-2 win over San Jose.
— Washington Capitals, Feb. 3, 1999, at Washington, second period during 10-1 win over Tampa Bay.

MOST GOALS, BOTH TEAMS, ONE PERIOD:
12 — Buffalo Sabres, Toronto Maple Leafs, at Buffalo, March 19, 1981, second period. Buffalo scored 9 goals, Toronto 3. Buffalo won 14-4.
— **Edmonton Oilers, Chicago Blackhawks,** at Chicago, Dec. 11, 1985, second period. Edmonton scored 6 goals, Chicago 6. Edmonton won 12-9.
10 — NY Rangers, NY Americans, at NY Americans, March 16, 1939, third period. NY Rangers scored 7 goals, NY Americans 3. NY Rangers won 11-5.
— Toronto Maple Leafs, Detroit Red Wings, at Detroit, March 17, 1946, third period. Toronto scored 6 goals, Detroit 4. Toronto won 11-7.
— Vancouver Canucks, Buffalo Sabres, at Buffalo, Jan. 8, 1976, third period. Buffalo scored 6 goals, Vancouver 4. Buffalo won 8-5.
— Buffalo Sabres, Montreal Canadiens, at Montreal, Oct. 26, 1982, first period. Montreal scored 5 goals, Buffalo 5. 7-7 tie.
— Boston Bruins, Quebec Nordiques, at Quebec, Dec. 7, 1982, second period. Quebec scored 6 goals, Boston 4. Quebec won 10-5.
— Calgary Flames, Vancouver Canucks, at Vancouver, Jan. 16, 1987, first period. Vancouver scored 6 goals, Calgary 4. Vancouver won 9-5.
— Winnipeg Jets, Detroit Red Wings, at Detroit, Nov. 25, 1987, third period. Detroit scored 7 goals, Winnipeg 3. Detroit won 10-8.
— Chicago Blackhawks, St. Louis Blues, at St. Louis, March 15, 1988, third period. Chicago scored 5 goals, St. Louis 5. 7-7 tie.

MOST CONSECUTIVE GOALS, ONE TEAM, ONE GAME:
15 — Detroit Red Wings, Jan. 23, 1944, at Detroit. Defeated NY Rangers 15-0.

Fewest Goals

FEWEST GOALS, ONE SEASON:
33 — Chicago Blackhawks, 1928-29. 44GP
45 — Montreal Maroons, 1924-25. 30GP
46 — Pittsburgh Pirates, 1928-29. 44GP

FEWEST GOALS, ONE SEASON (MINIMUM 70-GAME SCHEDULE):
133 — Chicago Blackhawks, 1953-54. 70GP
147 — Toronto Maple Leafs, 1954-55. 70GP
— Boston Bruins, 1955-56. 70GP
150 — NY Rangers, 1954-55. 70GP

TEAM POWER-PLAY GOALS

MOST POWER-PLAY GOALS, ONE SEASON:
119 — Pittsburgh Penguins, 1988-89. 80GP
113 — Detroit Red Wings, 1992-93. 84GP
111 — NY Rangers, 1987-88. 80GP
110 — Pittsburgh Penguins, 1987-88. 80GP
— Winnipeg Jets, 1987-88, 80GP

TEAM SHORTHAND GOALS

MOST SHORTHAND GOALS, ONE SEASON:
36 — Edmonton Oilers, 1983-84. 80GP
28 — Edmonton Oilers, 1986-87. 80GP
27 — Edmonton Oilers, 1985-86. 80GP
— Edmonton Oilers, 1988-89. 80GP

TEAM GOALS-PER-GAME

HIGHEST GOALS-PER-GAME AVERAGE, ONE SEASON:
5.58 — Edmonton Oilers, 1983-84. 446G in 80GP
5.38 — Montreal Canadiens, 1919-20. 129G in 24GP
5.33 — Edmonton Oilers, 1985-86. 426G in 80GP
5.30 — Edmonton Oilers, 1982-83. 424G in 80GP
5.23 — Montreal Canadiens, 1917-18. 115G in 22GP

LOWEST GOALS-PER-GAME AVERAGE, ONE SEASON:
.75 — Chicago Blackhawks, 1928-29, 33G in 44GP
1.05 — Pittsburgh Pirates, 1928-29. 46G in 44GP
1.20 — NY Americans, 1928-29. 53G in 44GP

TEAM ASSISTS

MOST ASSISTS, ONE SEASON:
737 — Edmonton Oilers, 1985-86. 80GP
736 — Edmonton Oilers, 1983-84. 80GP
706 — Edmonton Oilers, 1981-82. 80GP

FEWEST ASSISTS, ONE SEASON (Since 1926-27):
45 — NY Rangers, 1926-27. 44GP

FEWEST ASSISTS, ONE SEASON (MINIMUM 70-GAME SCHEDULE):
206 — Chicago Blackhawks, 1953-54. 70GP

TEAM TOTAL POINTS

MOST SCORING POINTS, ONE SEASON:
1,182 — Edmonton Oilers, 1983-84. 80GP
1,163 — Edmonton Oilers, 1985-86. 80GP
1,123 — Edmonton Oilers, 1981-82. 80GP

MOST SCORING POINTS, ONE TEAM, ONE GAME:
40 — Buffalo Sabres, Dec. 21, 1975, at Buffalo. Buffalo defeated Washington 14-2, receiving 26A.
39 — Minnesota North Stars, Nov. 11, 1981, at Minnesota. Minnesota defeated Winnipeg 15-2, receiving 24A.
37 — Detroit Red Wings, Jan. 23, 1944, at Detroit. Detroit defeated NY Rangers 15-0, receiving 22A.
— Toronto Maple Leafs, Mar. 16, 1957, at Toronto. Toronto defeated NY Rangers 14-1, receiving 23A.
— Buffalo Sabres, Feb. 25, 1978, at Cleveland. Buffalo defeated Cleveland 13-3, receiving 24A.
— Calgary Flames, Feb. 10, 1993, at Calgary. Calgary defeated San Jose 13-1, receiving 24A.

MOST SCORING POINTS, BOTH TEAMS, ONE GAME:
62 — Edmonton Oilers, Chicago Blackhawks, at Chicago, Dec. 11, 1985. Edmonton won 12-9. Edmonton had 24A, Chicago, 17.
53 — Quebec Nordiques, Washington Capitals, at Washington, Feb. 22, 1981. Quebec won 11-7. Quebec had 22A, Washington, 13.
— Edmonton Oilers, Minnesota North Stars, at Edmonton, Jan. 4, 1984. Edmonton won 12-8. Edmonton had 20A, Minnesota 13.
— Minnesota North Stars, St. Louis Blues, at St. Louis, Jan. 27, 1984. Minnesota won 10-8. Minnesota had 19A, St. Louis 16.
— Toronto Maple Leafs, Edmonton Oilers, at Toronto, Jan. 8, 1986. Toronto won 11-9. Toronto had 17A, Edmonton 16.
52 — Mtl. Maroons, NY Americans, at NY Americans, Feb. 18, 1936. 8-8 tie. NY Americans had 20A, Montreal 16. (3A allowed for each goal.)
— Vancouver Canucks, Minnesota North Stars, at Vancouver, Oct. 7, 1983. Vancouver won 10-9. Vancouver had 16A, Minnesota 17.

MOST SCORING POINTS, ONE TEAM, ONE PERIOD:
23 — NY Rangers, Nov. 21, 1971, at NY Rangers, third period during 12-1 win over California. NY Rangers scored 8G and 15A.
— Buffalo Sabres, Dec. 21, 1975, at Buffalo, third period during 14-2 win over Washington. Buffalo scored 8G and 15A.
— Buffalo Sabres, March 19, 1981, at Buffalo, second period during 14-4 win over Toronto. Buffalo scored 9G and 14A.
22 — Detroit Red Wings, Jan. 23, 1944, at Detroit, third period during 15-0 win over NY Rangers. Detroit scored 8G and 14A.
— Boston Bruins, March 16, 1969, at Boston, second period during 11-3 win over Toronto Maple Leafs. Boston scored 8G and 14A.
— Minnesota North Stars, Nov. 11, 1981, at Minnesota, second period during 15-2 win over Winnipeg. Minnesota scored 8G and 14A.
— Pittsburgh Penguins, Dec. 17, 1991, at Pittsburgh, second period during 10-2 win over San Jose. Pittsburgh scored 8G and 14A.
— Washington Capitals, Feb. 3, 1999, at Washington, second period during 10-1 win over Tampa Bay. Washington scored 8G and 14A.

MOST SCORING POINTS, BOTH TEAMS, ONE PERIOD:
35 — Edmonton, Oilers, Chicago Blackhawks, at Chicago, Dec. 11, 1985, second period. Edmonton had 6G, 12A; Chicago, 6G, 11A. Edmonton won 12-9.
31 — Buffalo Sabres, Toronto Maple Leafs, at Buffalo, March 19, 1981, second period. Buffalo had 9G, 14A; Toronto, 3G, 5A. Buffalo won 14-4.
29 — Winnipeg Jets, Detroit Red Wings, at Detroit, Nov. 25, 1987, third period. Detroit had 7G, 13A; Winnipeg had 3G, 6A. Detroit won 10-8.
— Chicago Blackhawks, St. Louis Blues, at St. Louis, March 15, 1988, third period. St. Louis had 5G, 10A; Chicago had 5G, 9A. 7-7 tie.

FASTEST GOALS

FASTEST SIX GOALS, BOTH TEAMS
3 Minutes — Quebec Nordiques, Washington Capitals, at Washington, Feb. 22, 1981, second and third periods. Quebec scored 5G, Washington 1. Quebec won 11-7.
3 Minutes, 15 Seconds — Montreal Canadiens, Toronto Maple Leafs, at Montreal, Jan. 4, 1944, first period. Montreal scored 4G, Toronto 2. Montreal won 6-3.

FASTEST FIVE GOALS, BOTH TEAMS:
1 Minute, 24 Seconds — Chicago Blackhawks, Toronto Maple Leafs, at Toronto, Oct. 15, 1983, second period. Scorers: Gaston Gingras, Toronto, 16:49; Denis Savard, Chicago, 17:12; Steve Larmer, Chicago, 17:27; Savard, 17:42; John Anderson, Toronto, 18:13. Toronto won 10-8.
1 Minute, 39 Seconds — Detroit Red Wings, Toronto Maple Leafs, at Toronto, Nov. 15, 1944, third period. Scorers: Ted Kennedy, Toronto, 10:36 and 10:55; Hal Jackson, Detroit, 11:48; Steve Wochy, Detroit, 12:02; Don Grosso, Detroit, 12:15. Detroit won 8-4.

FASTEST FIVE GOALS, ONE TEAM:
2 Minutes, 7 Seconds — Pittsburgh Penguins, at Pittsburgh, Nov. 22, 1972, third period. Scorers: Bryan Hextall, 12:00; Jean Pronovost, 12:18; Al McDonough, 13:40; Ken Schinkel, 13:49; Ron Schock, 14:07. Pittsburgh defeated St. Louis 10-4.
2 Minutes, 37 Seconds — NY Islanders, at NY Islanders, Jan. 26, 1982, first period. Scorers: Duane Sutter, 1:31; John Tonelli, 2:30; Bryan Trottier, 2:46; Bryan Trottier, 3:31; Duane Sutter, 4:08. NY Islanders defeated Pittsburgh 9-2.
2 Minutes, 55 Seconds — Boston Bruins, at Boston, Dec. 19, 1974. Scorers: Bobby Schmautz, 19:13 (first period); Ken Hodge, 0:18; Phil Esposito, 0:43; Don Marcotte, 0:58; John Bucyk, 2:08 (second period). Boston defeated NY Rangers 11-3.

FASTEST FOUR GOALS, BOTH TEAMS:
53 Seconds — Chicago Blackhawks, Toronto Maple Leafs, at Toronto, Oct. 15, 1983, second period. Scorers: Gaston Gingras, Toronto, 16:49; Denis Savard, Chicago, 17:12; Steve Larmer, Chicago, 17:27; and Savard, 17:42. Toronto won 10-8.
57 Seconds — Quebec Nordiques, Detroit Red Wings, at Quebec, Jan. 27, 1990, first period. Scorers: Paul Gillis, Quebec, 18:01; Claude Loiselle, Quebec, 18:12; Joe Sakic, Quebec, 18:27; and Jimmy Carson, Detroit, 18:58. Detroit won 8-6.
1 Minute, 1 Second — Colorado Rockies, NY Rangers, at NY Rangers, Jan. 15, 1980, first period. Scorers: Doug Sulliman, NY Rangers, 7:52; Ed Johnstone, NY Rangers, 7:57; Warren Miller, NY Rangers, 8:20; Rob Ramage, Colorado, 8:53. 6-6 tie.
— Chicago Blackhawks, Toronto Maple Leafs, at Toronto, Oct. 15, 1983, second period. Scorers: Denis Savard, Chicago, 17:12; Steve Larmer, Chicago, 17:27; Savard, 17:42; John Anderson, Toronto, 18:13. Toronto won 10-8.

FASTEST FOUR GOALS, ONE TEAM:
1 Minute, 20 Seconds — Boston Bruins, at Boston, Jan. 21, 1945, second period. Scorers: Bill Thoms, 6:34; Frank Mario, 7:08 and 7:27; and Ken Smith, 7:54. Boston defeated NY Rangers 14-3.

FASTEST THREE GOALS, BOTH TEAMS:
15 Seconds — Minnesota North Stars, NY Rangers, at Minnesota, Feb. 10, 1983, second period. Scorers: Mark Pavelich, NY Rangers, 19:18; Ron Greschner, NY Rangers, 19:27; Willi Plett, Minnesota, 19:33. Minnesota won 7-5.
18 Seconds — Montreal Canadiens, NY Rangers, at Montreal, Dec. 12, 1963, first period. Scorers: Dave Balon, Montreal, 0:58; Gilles Tremblay, Montreal, 1:04; Camille Henry, NY Rangers, 1:16. Montreal won 6-4.
— California Golden Seals, Buffalo Sabres, at California, Feb. 1, 1976, third period. Scorers: Jim Moxey, California, 19:38; Wayne Merrick, California, 19:45; Danny Gare, Buffalo, 19:56. Buffalo won 9-5.

FASTEST THREE GOALS, ONE TEAM:
20 Seconds — Boston Bruins, at Boston, Feb. 25, 1971, third period. Scorers: John Bucyk, 4:50; Ed Westfall, 5:02; Ted Green, 5:10. Boston defeated Vancouver 8-3.
21 Seconds — Chicago Blackhawks, at New York, Mar. 23, 1952, third period. Bill Mosienko scored all three goals, at 6:09, 6:20 and 6:30. Chicago defeated NY Rangers 7-6.
— Washington Capitals, at Washington, Nov. 23, 1990, first period. Scorers: Michal Pivonka, 16:18; Stephen Leach, 16:29 and 16:39. Washington defeated Pittsburgh 7-3.

FASTEST THREE GOALS FROM START OF PERIOD, BOTH TEAMS:
1 Minute, 5 Seconds — Hartford Whalers, Montreal Canadiens, at Montreal, March 11, 1989, second period. Scorers: Kevin Dineen, Hartford, 0:11; Guy Carbonneau, Montreal, 0:36; Petr Svoboda, Montreal, 1:05. Montreal won 5-3.

FASTEST THREE GOALS FROM START OF PERIOD, ONE TEAM:
53 Seconds — Calgary Flames, at Calgary, Feb. 10, 1993, third period. Scorers: Gary Suter, 0:17; Chris Lindbergh, 0:40; Ron Stern, 0:53. Calgary defeated San Jose 13-1.

FASTEST TWO GOALS, BOTH TEAMS:
2 Seconds — St. Louis Blues, Boston Bruins, at Boston, Dec. 19, 1987, third period. Scorers: Ken Linseman, Boston, 19:50; Doug Gilmour, St. Louis, 19:52. St. Louis won 7-5.
3 Seconds — Chicago Blackhawks, Minnesota North Stars, at Minnesota, Nov. 5, 1988, third period. Scorers: Steve Thomas, Chicago, 6:03; Dave Gagner, Minnesota, 6:06. 5-5 tie.

FASTEST TWO GOALS, ONE TEAM:
4 Seconds — Montreal Maroons, at Montreal, Jan. 3, 1931, third period. Nels Stewart scored both goals, at 8:24 and 8:28. Mtl. Maroons defeated Boston 5-3.
— Buffalo Sabres, at Buffalo, Oct. 17, 1974, third period. Scorers: Lee Fogolin, 14:55; Don Luce, 14:59. Buffalo defeated California 6-1.
— Toronto Maple Leafs, at Quebec, Dec. 29, 1988, third period. Scorers: Ed Olczyk, 5:24; Gary Leeman, 5:28. Toronto defeated Quebec 6-5.
— Calgary Flames, at Quebec, Oct. 17, 1989, third period. Scorers: Doug Gilmour, 19:45; Paul Ranheim, 19:49. Calgary and Quebec tied 8-8.
— Winnipeg Jets, at Winnipeg, Dec. 15, 1995, second period. Deron Quint scored both goals, at 7:51 and 7:55. Winnipeg defeated Edmonton 9-4.

FASTEST TWO GOALS FROM START OF GAME, ONE TEAM:
24 Seconds — Edmonton Oilers, Mar. 28, 1982, at Los Angeles. Scorers: Mark Messier, 0:14; Dave Lumley, 0:24. Edmonton defeated Los Angeles 6-2.
29 Seconds — Pittsburgh Penguins, Dec. 6, 1980, at Pittsburgh. Scorers: George Ferguson, 0:17; Greg Malone, 0:29. Pittsburgh defeated Chicago 6-4.
32 Seconds — Calgary Flames, Mar. 11, 1987, at Hartford. Scorers: Doug Risebrough, 0:09; Colin Patterson, 0:32. Calgary defeated Hartford 6-1.

FASTEST TWO GOALS FROM START OF PERIOD, BOTH TEAMS:
14 Seconds — NY Rangers, Quebec Nordiques, at Quebec, Nov. 5, 1983, third period. Scorers: Andre Savard, Quebec, 0:08; Pierre Larouche, NY Rangers, 0:14. 4-4 tie.
26 Seconds — Buffalo Sabres, St. Louis Blues, at Buffalo, Jan. 3, 1993, third period. Scorers: Alexander Mogilny, Buffalo, 0:08; Phillippe Bozon, St. Louis, 0:26. Buffalo won 6-5.
28 Seconds — Boston Bruins, Montreal Canadiens, at Montreal, Oct. 11, 1989, third period. Scorers: Jim Wiemer, Boston 0:10; Tom Chorske, Montreal 0:28. Montreal won 4-2.

FASTEST TWO GOALS FROM START OF PERIOD, ONE TEAM:
21 Seconds — Chicago Blackhawks, Nov. 5, 1983, at Minnesota, second period. Scorers: Ken Yaremchuk, 0:12; Darryl Sutter, 0:21. Minnesota defeated Chicago 10-5.
30 Seconds — Washington Capitals, Jan. 27, 1980, at Washington, second period. Scorers: Mike Gartner, 0:08; Bengt Gustafsson, 0:30. Washington defeated NY Islanders 7-1.
31 Seconds —Buffalo Sabres, Jan. 10, 1974, at Buffalo, third period. Scorers: Rene Robert, 0:21; Rick Martin, 0:31. Buffalo defeated NY Rangers 7-2.
— NY Islanders, Feb. 22, 1986, at NY Islanders, third period. Scorers: Roger Kortko, 0:10; Bob Bourne, 0:31. NY Islanders defeated Detroit 5-2.

Mario Lemieux was the newly named captain of the Penguins when he exploded for 85 goals and 114 assists for 199 points in 1988-89. His 31 goals with a man advantage led Pittsburgh to an NHL-record 119 power-play goals that season.

50, 40, 30, 20-GOAL SCORERS

MOST 50-OR-MORE-GOAL SCORERS, ONE SEASON:

3 — **Edmonton Oilers,** 1983-84. Wayne Gretzky, 87; Glenn Anderson, 54; Jari Kurri, 52. 80GP
— **Edmonton Oilers,** 1985-86. Jari Kurri, 68; Glenn Anderson, 54; Wayne Gretzky, 52. 80GP
2 — Boston Bruins, 1970-71. Phil Esposito, 76; John Bucyk, 51. 78GP
— Boston Bruins, 1973-74. Phil Esposito, 68; Ken Hodge, 50. 78GP
— Philadelphia Flyers, 1975-76. Reggie Leach, 61; Bill Barber, 50. 80GP
— Pittsburgh Penguins, 1975-76. Pierre Larouche, 53; Jean Pronovost, 52. 80GP
— Montreal Canadiens, 1976-77. Steve Shutt, 60; Guy Lafleur, 56. 80GP
— Los Angeles Kings, 1979-80. Charlie Simmer, 56; Marcel Dionne, 53. 80GP
— Montreal Canadiens, 1979-80. Pierre Larouche, 50; Guy Lafleur, 50. 80GP
— Los Angeles Kings, 1980-81. Marcel Dionne, 58; Charlie Simmer, 56. 80GP
— Edmonton Oilers, 1981-82. Wayne Gretzky, 92; Mark Messier, 50. 80GP
— NY Islanders, 1981-82. Mike Bossy, 64; Bryan Trottier, 50. 80GP
— Edmonton Oilers, 1984-85. Wayne Gretzky, 73; Jari Kurri, 71. 80GP
— Washington Capitals, 1984-85. Bob Carpenter, 53; Mike Gartner, 50. 80GP
— Edmonton Oilers, 1986-87. Wayne Gretzky, 62; Jari Kurri, 54. 80GP
— Calgary Flames, 1987-88. Joe Nieuwendyk, 51; Hakan Loob, 50. 80GP
— Los Angeles Kings, 1987-88. Jimmy Carson, 55; Luc Robitaille, 53. 80GP
— Los Angeles Kings, 1988-89. Bernie Nicholls, 70; Wayne Gretzky, 54. 80GP
— Calgary Flames, 1988-89. Joe Nieuwendyk, 51; Joe Mullen, 51. 80GP
— Buffalo Sabres, 1992-93. Alexander Mogilny, 76; Pat LaFontaine, 53. 84GP
— Pittsburgh Penguins, 1992-93. Mario Lemieux, 69; Kevin Stevens, 55. 84GP
— St. Louis Blues, 1992-93. Brett Hull, 54; Brendan Shanahan, 51. 84GP
— St. Louis Blues, 1993-94. Brett Hull, 57; Brendan Shanahan, 52. 84GP
— Detroit Red Wings, 1993-94. Sergei Fedorov, 56; Ray Sheppard, 52. 84GP
— Pittsburgh Penguins, 1995-96. Mario Lemieux, 69; Jaromir Jagr, 62. 82GP

MOST 40-OR-MORE-GOAL SCORERS, ONE SEASON:

4 — **Edmonton Oilers,** 1982-83. Wayne Gretzky, 71; Glenn Anderson, 48; Mark Messier, 48; Jari Kurri, 45. 80GP
— **Edmonton Oilers,** 1983-84. Wayne Gretzky, 87; Glenn Anderson, 54; Jari Kurri, 52; Paul Coffey, 40. 80GP
— **Edmonton Oilers,** 1984-85. Wayne Gretzky, 73; Jari Kurri, 71; Mike Krushelnyski, 43; Glenn Anderson, 42. 80GP
— **Edmonton Oilers,** 1985-86. Jari Kurri, 68; Glenn Anderson, 54; Wayne Gretzky, 52; Paul Coffey, 48. 80GP
— **Calgary Flames,** 1987-88. Joe Nieuwendyk, 51; Hakan Loob, 50; Mike Bullard, 48; Joe Mullen, 40. 80GP
3 — Boston Bruins, 1970-71. Phil Esposito, 76; John Bucyk, 51; Ken Hodge, 43. 78GP
— NY Rangers, 1971-72. Vic Hadfield, 50; Jean Ratelle, 46; Rod Gilbert, 43. 78GP
— Buffalo Sabres, 1975-76. Danny Gare, 50; Rick Martin, 49; Gilbert Perreault, 44. 80GP
— Montreal Canadiens, 1979-80. Guy Lafleur, 50; Pierre Larouche, 50; Steve Shutt, 47. 80GP
— Buffalo Sabres, 1979-80. Danny Gare, 56; Rick Martin, 45; Gilbert Perreault, 40. 80GP
— Los Angeles Kings, 1980-81. Marcel Dionne, 58; Charlie Simmer, 56; Dave Taylor, 47. 80GP
— Los Angeles Kings, 1984-85. Marcel Dionne, 46; Bernie Nicholls, 46; Dave Taylor, 41. 80GP
— NY Islanders, 1984-85. Mike Bossy, 58; Brent Sutter, 42; John Tonelli, 42. 80GP
— Chicago Blackhawks, 1985-86. Denis Savard, 47; Troy Murray, 45; Al Secord, 40. 80GP
— Chicago Blackhawks, 1987-88. Denis Savard, 44; Rick Vaive, 43; Steve Larmer, 41. 80GP
— Edmonton Oilers, 1987-88. Craig Simpson, 43; Jari Kurri, 43; Wayne Gretzky, 40. 80GP
— Los Angeles Kings, 1988-89. Bernie Nicholls, 70; Wayne Gretzky, 54; Luc Robitaille, 46. 80GP
— Los Angeles Kings, 1990-91. Luc Robitaille, 45; Tomas Sandstrom, 45; Wayne Gretzky 41. 80GP
— Pittsburgh Penguins, 1991-92. Kevin Stevens, 54; Mario Lemieux, 44; Joe Mullen, 42. 80GP
— Pittsburgh Penguins, 1992-93. Mario Lemieux, 69; Kevin Stevens, 55; Rick Tocchet, 48. 84GP
— Calgary Flames, 1993-94. Gary Roberts, 41; Robert Reichel, 40; Theoren Fleury, 40. 84GP
— Pittsburgh Penguins, 1995-96. Mario Lemieux, 69; Jaromir Jagr, 62; Petr Nedved, 45. 82GP

MOST 30-OR-MORE GOAL SCORERS, ONE SEASON:

6 — **Buffalo Sabres,** 1974-75. Rick Martin, 52; Rene Robert, 40; Gilbert Perreault, 39; Don Luce, 33; Rick Dudley, Danny Gare, 31 each. 80GP
— **NY Islanders,** 1977-78. Mike Bossy, 53; Bryan Trottier, 46; Clark Gillies, 35; Denis Potvin, Bob Nystrom, Bob Bourne, 30 each. 80GP
— **Winnipeg Jets,** 1984-85. Dale Hawerchuk, 53; Paul MacLean, 41; Laurie Boschman, Brian Mullen, 32 each; Doug Smail, 31; Thomas Steen, 30. 80GP
5 — Chicago Blackhawks, 1968-69. 76GP
— Boston Bruins, 1970-71. 78GP
— Montreal Canadiens, 1971-72. 78GP
— Philadelphia Flyers, 1972-73. 78GP
— Boston Bruins, 1973-74. 78GP
— Montreal Canadiens, 1974-75. 80GP
— Montreal Canadiens, 1975-76. 80GP
— Pittsburgh Penguins, 1975-76. 80GP
— NY Islanders, 1978-79. 80GP
— Detroit Red Wings, 1979-80. 80GP
— Philadelphia Flyers, 1979-80. 80GP
— NY Islanders, 1980-81. 80GP
— St. Louis Blues, 1980-81. 80GP
— Chicago Blackhawks, 1981-82. 80GP
— Edmonton Oilers, 1981-82. 80GP
— Montreal Canadiens, 1981-82. 80GP
— Quebec Nordiques, 1981-82. 80GP
— Washington Capitals, 1981-82. 80GP
— Edmonton Oilers, 1982-83. 80GP
— Edmonton Oilers, 1983-84. 80GP
— Edmonton Oilers, 1984-85. 80GP
— Los Angeles Kings, 1984-85. 80GP
— Edmonton Oilers, 1985-86. 80GP
— Edmonton Oilers, 1986-87. 80GP
— Edmonton Oilers, 1987-88. 80GP
— Edmonton Oilers, 1988-89. 80GP
— Detroit Red Wings, 1991-92. 80GP
— NY Rangers, 1991-92. 80GP
— Pittsburgh Penguins, 1991-92. 80GP
— Detroit Red Wings, 1992-93. 84GP
— Pittsburgh Penguins, 1992-93. 84GP

MOST 20-OR-MORE GOAL SCORERS, ONE SEASON:

11 — **Boston Bruins,** 1977-78; Peter McNab, 41; Terry O'Reilly, 29; Bobby Schmautz, Stan Jonathan, 27 each; Jean Ratelle, Rick Middleton, 25 each; Wayne Cashman, 24; Gregg Sheppard, 23; Brad Park, 22; Don Marcotte, Bob Miller, 20 each. 80GP
10 — Boston Bruins, 1970-71. 78GP
— Montreal Canadiens, 1974-75. 80GP
— St. Louis Blues, 1980-81. 80GP

Alex Connell posted an NHL-record six consecutive shutouts in 1927-28. He had 15 shutouts overall that year. Connell had the same number back in 1925-26 when the Ottawa Senators allowed a record-low 42 goals during the 36-game season.

100-POINT SCORERS

MOST 100 OR-MORE-POINT SCORERS, ONE SEASON:

4 — Boston Bruins, 1970-71, Phil Esposito, 76G-76A-152PTS; Bobby Orr, 37G-102A-139PTS; John Bucyk, 51G-65A-116PTS; Ken Hodge, 43G-62A-105PTS. 78GP
— Edmonton Oilers, 1982-83, Wayne Gretzky, 71G-125A-196PTS; Mark Messier, 48G-58A-106PTS; Glenn Anderson, 48G-56A-104PTS; Jari Kurri, 45G-59A-104PTS. 80GP
— Edmonton Oilers, 1983-84, Wayne Gretzky, 87G-118A-205PTS; Paul Coffey, 40G-86A-126PTS; Jari Kurri, 52G-61A-113PTS; Mark Messier, 37G-64A-101PTS. 80GP
— Edmonton Oilers,1985-86, Wayne Gretzky, 52G-163A-215PTS; Paul Coffey, 48G-90A-138PTS; Jari Kurri, 68G-63A-131PTS; Glenn Anderson, 54G-48A-102PTS. 80GP
— Pittsburgh Penguins,1992-93, Mario Lemieux, 69G-91A-160PTS; Kevin Stevens, 55G-56A-111PTS; Rick Tocchet, 48G-61A-109PTS; Ron Francis, 24G-76A-100PTS. 84GP
3 — Boston Bruins, 1973-74, Phil Esposito, 68G-77A-145PTS; Bobby Orr, 32G-90A-122PTS; Ken Hodge, 50G-55A-105PTS. 78GP
— NY Islanders, 1978-79, Bryan Trottier, 47G-87A-134PTS; Mike Bossy, 69G-57A-126PTS; Denis Potvin, 31G-70A-101PTS. 80GP
— Los Angeles Kings, 1980-81, Marcel Dionne, 58G-77A-135PTS; Dave Taylor, 47G-65A-112PTS; Charlie Simmer, 56G-49A-105PTS. 80GP
— Edmonton Oilers, 1984-85, Wayne Gretzky, 73G-135A-208PTS; Jari Kurri, 71G-64A-135PTS; Paul Coffey, 37G-84A-121PTS. 80GP
— NY Islanders, 1984-85. Mike Bossy, 58G-59A-117PTS; Brent Sutter, 42G-60A-102PTS; John Tonelli, 42G-58A-100PTS. 80GP
— Edmonton Oilers, 1986-87, Wayne Gretzky, 62G-121A-183PTS; Jari Kurri, 54G-54A-108PTS; Mark Messier, 37G-70A-107PTS. 80GP
— Pittsburgh Penguins, 1988-89, Mario Lemieux, 85G-114A-199PTS; Rob Brown, 49G-66A-115PTS; Paul Coffey, 30G-83A-113PTS. 80GP
— Pittsburgh Penguins, 1995-96, Mario Lemieux, 69G-92A-161PTS; Jaromir Jagr, 62G-87A-149PTS; Ron Francis, 27G-92A-119PTS. 82GP

SHOTS ON GOAL

MOST SHOTS, BOTH TEAMS, ONE GAME:

141 — NY Americans, Pittsburgh Pirates, Dec. 26, 1925, at NY Americans. NY Americans, who won game 3-1, had 73 shots; Pit. Pirates, 68 shots.

MOST SHOTS, ONE TEAM, ONE GAME:

83 — Boston Bruins, March 4, 1941, at Boston. Boston defeated Chicago 3-2.
73 — NY Americans, Dec. 26, 1925, at NY Americans. NY Americans defeated Pit. Pirates 3-1.
— Boston Bruins, March 21, 1991, at Boston. Boston tied Quebec 3-3.
72 — Boston Bruins, Dec. 10, 1970, at Boston. Boston defeated Buffalo 8-2.

MOST SHOTS, ONE TEAM, ONE PERIOD:

33 — Boston Bruins, March 4, 1941, at Boston, second period. Boston defeated Chicago 3-2.

TEAM GOALS AGAINST

Fewest Goals Against

FEWEST GOALS AGAINST, ONE SEASON:

42 — Ottawa Senators, 1925-26. 36GP
43 — Montreal Canadiens, 1928-29. 44GP
48 — Montreal Canadiens, 1923-24. 24GP
— Montreal Canadiens, 1927-28. 44GP

FEWEST GOALS AGAINST, ONE SEASON (MINIMUM 70-GAME SCHEDULE):

131 — Toronto Maple Leafs, 1953-54. 70GP
— Montreal Canadiens, 1955-56. 70GP
132 — Detroit Red Wings, 1953-54. 70GP
133 — Detroit Red Wings, 1951-52. 70GP
— Detroit Red Wings, 1952-53. 70GP

LOWEST GOALS-AGAINST-PER-GAME AVERAGE, ONE SEASON:

.98 — Montreal Canadiens, 1928-29. 43GA in 44GP.
1.09 — Montreal Canadiens, 1927-28. 48GA in 44GP.
1.17 — Ottawa Senators, 1925-26. 42GA in 36GP.

Most Goals Against

MOST GOALS AGAINST, ONE SEASON:

446 — Washington Capitals, 1974-75. 80GP
415 — Detroit Red Wings, 1985-86. 80GP
414 — San Jose Sharks, 1992-93. 84GP
407 — Quebec Nordiques, 1989-90. 80GP
403 — Hartford Whalers, 1982-83. 80GP

HIGHEST GOALS-AGAINST-PER-GAME AVERAGE, ONE SEASON:

7.38 — Quebec Bulldogs, 1919-20, 177GA in 24GP.
6.20 — NY Rangers, 1943-44, 310GA in 50GP.
5.58 — Washington Capitals, 1974-75, 446GA in 80GP.

MOST POWER-PLAY GOALS AGAINST, ONE SEASON:

122 — Chicago Blackhawks, 1988-89. 80GP
120 — Pittsburgh Penguins, 1987-88. 80GP
115 — New Jersey Devils, 1988-89. 80GP
— Ottawa Senators, 1992-93. 84GP
114 — Los Angeles Kings, 1992-93. 84GP

MOST SHORTHAND GOALS AGAINST, ONE SEASON:

22 — Pittsburgh Penguins, 1984-85. 80GP
— Minnesota North Stars, 1991-92. 80GP
— Colorado Avalanche, 1995-96. 82GP
21 — Calgary Flames, 1984-85. 80GP
— Pittsburgh Penguins, 1989-90. 80GP

SHUTOUTS

MOST SHUTOUTS, ONE SEASON:

22 — Montreal Canadiens, 1928-29. All by George Hainsworth. 44GP
16 — NY Americans, 1928-29. Roy Worters had 13; Flat Walsh 3. 44GP
15 — Ottawa Senators, 1925-26. All by Alex Connell. 36GP
— Ottawa Senators, 1927-28. All by Alex Connell. 44GP
— Boston Bruins, 1927-28. All by Hal Winkler. 44GP
— Chicago Blackhawks, 1969-70. All by Tony Esposito. 76GP

MOST CONSECUTIVE SHUTOUTS, ONE SEASON:

6 — Ottawa Senators, Jan. 31 - Feb. 18, 1928.

MOST CONSECUTIVE SHUTOUTS TO START SEASON:

5 — Toronto Maple Leafs, Nov. 13 - 22, 1930.

MOST GAMES SHUTOUT, ONE SEASON:

20 — Chicago Blackhawks, 1928-29. 44GP

MOST CONSECUTIVE GAMES SHUTOUT:

8 — Chicago Blackhawks, Feb. 7 - 28, 1929.

MOST CONSECUTIVE GAMES SHUTOUT TO START SEASON:

3 — Montreal Maroons, Nov. 11 - 18, 1930.

TEAM PENALTIES

MOST PENALTY MINUTES, ONE SEASON:

2,713 — Buffalo Sabres, 1991-92. 80GP
2,670 — Pittsburgh Penguins, 1988-89. 80GP
2,663 — Chicago Blackhawks, 1991-92. 80GP
2,643 — Calgary Flames, 1991-92. 80GP
2,621 — Philadelphia Flyers, 1980-81. 80GP

MOST PENALTIES, BOTH TEAMS, ONE GAME:

85 Penalties — Edmonton Oilers (44), Los Angeles Kings (41) at Los Angeles, Feb. 28, 1990. Edmonton received 26 minors, 7 majors, 6 10-minute misconducts, 4 game misconducts and 1 match penalty; Los Angeles received 26 minors, 9 majors, 3 10-minute misconducts and 3 game misconducts.

MOST PENALTY MINUTES, BOTH TEAMS, ONE GAME:

406 Minutes — Minnesota North Stars, Boston Bruins at Boston, Feb. 26, 1981. Minnesota received 18 minors, 13 majors, 4 10-minute misconducts and 7 game misconducts; a total of 211PIM. Boston received 20 minors, 13 majors, 3 10-minute misconducts and six game misconducts; a total of 195PIM.

MOST PENALTIES, ONE TEAM, ONE GAME:

44 — Edmonton Oilers, Feb. 28, 1990, at Los Angeles. Edmonton received 26 minors, 7 majors, 6 10-minute misconducts, 4 game misconducts and 1 match penalty.
42 — Minnesota North Stars, Feb. 26, 1981, at Boston. Minnesota received 18 minors, 13 majors, 4 10-minute misconducts and 7 game misconducts.
— Boston Bruins, Feb. 26, 1981, at Boston vs. Minnesota. Boston received 20 minors, 13 majors, 3 10-minute misconducts and 6 game misconducts.

MOST PENALTY MINUTES, ONE TEAM, ONE GAME:

211 — Minnesota North Stars, Feb. 26, 1981, at Boston. Minnesota received 18 minors, 13 majors, 4 10-minute misconducts and 7 game misconducts.

MOST PENALTIES, BOTH TEAMS, ONE PERIOD:

67 — Minnesota North Stars, Boston Bruins, at Boston, Feb. 26, 1981, first period. Minnesota received 15 minors, 8 majors, 4 10-minute misconducts and 7 game misconducts, a total of 34 penalties. Boston had 16 minors, 8 majors, 3 10-minute misconducts and 6 game misconducts, a total of 33 penalties.

MOST PENALTY MINUTES, BOTH TEAMS, ONE PERIOD:

372 — Los Angeles Kings, Philadelphia Flyers, at Philadelphia, March 11, 1979, first period. Philadelphia received 4 minors, 8 majors, 6 10-minute misconducts and 8 game misconducts for 188 minutes. Los Angeles received 2 minors, 8 majors, 6 10-minute misconducts and 8 game misconducts for 184 minutes.

MOST PENALTIES, ONE TEAM, ONE PERIOD:

34 — Minnesota North Stars, Feb. 26, 1981, at Boston, first period. 15 minors, 8 majors, 4 10-minute misconducts, 7 game misconducts.

MOST PENALTY MINUTES, ONE TEAM, ONE PERIOD:

190 — Calgary Flames, Dec. 8, 2001, at Calgary vs. Anaheim, third period. Flames received 10 minors, 10 majors, 8 10-minute misconducts and 4 game misconducts.
188 — Philadelphia Flyers, March 11, 1979, at Philadelphia vs. Los Angeles, first period. Flyers received 4 minors, 8 majors, 6 10-minute misconducts and 8 game misconducts.

NHL Individual Scoring Records - History

Six individual scoring records stand as benchmarks in the history of the game: most goals, single-season and career; most assists, single-season and career; and most points, single-season and career. The evolution of these six records is traced here, beginning with 1917-18, the NHL's first season. New research has resulted in changes to scoring records in the NHL's first nine seasons.

MOST GOALS, ONE SEASON

44 —Joe Malone, Montreal, 1917-18.
Scored goal #44 against Toronto's Harry Holmes on March 2, 1918 and finished season with 44 goals.
50 —Maurice Richard, Montreal, 1944-45.
Scored goal #45 against Toronto's Frank McCool on February 25, 1945 and finished the season with 50 goals.
50 —Bernie Geoffrion, Montreal, 1960-61.
Scored goal #50 against Toronto's Cesare Maniago on March 16, 1961 and finished the season with 50 goals.
50 —Bobby Hull, Chicago, 1961-62.
Scored goal #50 against NY Rangers' Gump Worsley on March 25, 1962 and finished the season with 50 goals.
54 —Bobby Hull, Chicago, 1965-66.
Scored goal #51 against NY Rangers' Cesare Maniago on March 12, 1966 and finished the season with 54 goals.
58 —Bobby Hull, Chicago, 1968-69.
Scored goal #55 against Boston's Gerry Cheevers on March 20, 1969 and finished the season with 58 goals.
76 —Phil Esposito, Boston, 1970-71.
Scored goal #59 against Los Angeles' Denis DeJordy on March 11, 1971 and finished the season with 76 goals.
92 —Wayne Gretzky, Edmonton, 1981-82.
Scored goal #77 against Buffalo's Don Edwards on February 24, 1982 and finished the season with 92 goals.

MOST ASSISTS, ONE SEASON

10 —Cy Denneny, Ottawa, 1917-18.
—Reg Noble, Toronto, 1917-18.
—Harry Cameron, Toronto, 1917-18.
—Newsy Lalonde, Montreal, 1918-19.
15 —Frank Nighbor, Ottawa, 1919-20.
—Jack Darragh, Ottawa, 1920-21.
17 —Harry Cameron, Toronto, 1921-22.
18 —Dick Irvin, Chicago, 1926-27.
18 —Howie Morenz, Montreal, 1927-28.
36 —Frank Boucher, NY Rangers, 1929-30.
37 —Joe Primeau, Toronto, 1931-32.
45 —Bill Cowley, Boston, 1940-41.
45 —Bill Cowley, Boston, 1942-43.
49 —Clint Smith, Chicago, 1943-44.
54 —Elmer Lach, Montreal, 1944-45.
55 —Ted Lindsay, Detroit, 1949-50.
56 —Bert Olmstead, Montreal, 1955-56.
58 —Jean Beliveau, Montreal, 1960-61.
58 —Andy Bathgate, NY Rangers/Toronto, 1963-64.
59 —Stan Mikita, Chicago, 1964-65.
62 —Stan Mikita, Chicago, 1966-67.
77 —Phil Esposito, Boston, 1968-69.
87 —Bobby Orr, Boston, 1969-70.
102 —Bobby Orr, Boston, 1970-71.
109 —Wayne Gretzky, Edmonton, 1980-81.
120 —Wayne Gretzky, Edmonton, 1981-82.
125 —Wayne Gretzky, Edmonton, 1982-83.
135 —Wayne Gretzky, Edmonton, 1984-85.
163 —Wayne Gretzky, Edmonton, 1985-86.

MOST POINTS, ONE SEASON

48 —Joe Malone, Montreal, 1917-18.
49 —Joe Malone, Montreal, 1919-20.
51 —Howie Morenz, Montreal, 1927-28.
73 —Cooney Weiland, Boston, 1929-30.
73 —Doug Bentley, Chicago, 1942-43.
82 —Herb Cain, Boston, 1943-44.
86 —Gordie Howe, Detroit, 1950-51.
95 —Gordie Howe, Detroit, 1952-53.
96 —Dickie Moore, Montreal, 1958-59.
97 —Bobby Hull, Chicago, 1965-66.
97 —Stan Mikita, Chicago, 1966-67.
126 —Phil Esposito, Boston, 1968-69.
152 —Phil Esposito, Boston, 1970-71.
164 —Wayne Gretzky, Edmonton, 1980-81.
212 —Wayne Gretzky, Edmonton, 1981-82.
215 —Wayne Gretzky, Edmonton, 1985-86.

MOST REGULAR-SEASON GOALS, CAREER

44 —Joe Malone, 1917-18, Montreal.
Malone led the NHL in goals in the league's first season and finished with 44 goals in 22 games in 1917-18.
54 —Cy Denneny, 1918-19, Ottawa.
Denneny passed Malone during the 1918-19 season, finishing the year with a two-year total of 54 goals. He held the career goal- scoring mark until 1919-20.
143 —Joe Malone, Montreal, Quebec Bulldogs, Hamilton.
Malone passed Denneny in 1919-20 and remained the NHL's career goal-scoring leader until 1922-23.
248 —Cy Denneny, Ottawa, Boston.
Denneny passed Malone with goal #144 in 1922-23 and remained the NHL's career goal-scoring leader until his retirement. He finished with a career total of 248 goals.
271 —Howie Morenz, Montreal, Chicago, NY Rangers.
Morenz passed Denneny with goal #249 in 1933-34 and finished his career with 271 goals.
324 —Nels Stewart, Montreal Maroons, Boston, NY Americans.
Stewart passed Morenz with goal #272 in 1936-37 and remained the NHL's career goal-scoring leader until his retirement. He finished his career with 324 goals.
544 —Maurice Richard, Montreal.
Richard passed Nels Stewart with goal #325 on Nov. 8, 1952 and remained the NHL's career goal-scoring leader until his retirement. He finished his career with 544 goals.
801 —Gordie Howe, Detroit, Hartford.
Howe passed Richard with goal #545 on Nov. 10, 1963 and remained the NHL's career goal-scoring leader until his retirement. He finished his career with 801 goals.
894 —Wayne Gretzky, Edmonton, Los Angeles, St. Louis, NY Rangers.
Gretzky passed Gordie Howe with goal #802 on March 23, 1994. He retired as the NHL's current goal-scoring leader with 894.

Briefly the NHL's all-time scoring leader during the mid 1950s, Elmer Lach (facing page) was the first player in NHL history to have more than 50 assists in a single season. He had 54 helpers in 1944-45, the year in which Rocket Richard collected 50 goals in 50 games. The first player to score more than 50 goals in a single season was Bobby Hull (above). Hull had become the NHL's third 50-goal scorer in the 70th and final game of the 1961-62 season. He reached the 50-goal plateau again in just 57 games during the 1965-66 campaign, but was then shut out for three-straight games before netting his record-breaking 51st. He finished the season with 54. Hull was behind his scoring pace in 1967-68, but used the extra six games of the expanded schedule to set a new record with 58 goals.

MOST REGULAR-SEASON ASSISTS, CAREER

(minimum 100 assists)

100 —Frank Boucher, Ottawa, NY Rangers.
In 1930-31, Boucher became the first NHL player to reach the 100-assist milestone.

263 —Frank Boucher, Ottawa, NY Rangers.
Boucher retired as the NHL's career assist leader in 1938 with 253. He returned to the NHL in 1943-44 and remained the NHL's career assist leader until he was overtaken by Bill Cowley in 1943-44. He finished his career with 263 assists.

353 —Bill Cowley, St. Louis Eagles, Boston.
Cowley passed Boucher with assist #264 in 1943-44. He retired as the NHL's career assist leader in 1947 with 353.

408 —Elmer Lach, Montreal.
Lach passed Cowley with assist #354 in 1951-52. He retired as the NHL's career assist leader in 1954 with 408.

1,049 —Gordie Howe, Detroit, Hartford.
Howe passed Lach with assist #409 in 1957-58. He retired as the NHL's career assist leader in 1980 with 1,049.

1,963 —Wayne Gretzky, Edmonton, Los Angeles, St. Louis, NY Rangers.
Gretzky passed Howe with assist #1,050 in 1988-89. He retired as the NHL's current career assist leader with 1,963.

MOST REGULAR-SEASON POINTS, CAREER

(minimum 100 points)

100 —Joe Malone, Montreal, Quebec Bulldogs, Hamilton.
In 1919-20, Malone became the first player in NHL history to record 100 points.

200 —Cy Denneny, Ottawa.
In 1923-24, Denneny became the first player in NHL history to record 200 points.

300 —Cy Denneny, Ottawa.
In 1926-27, Denneny became the first player in NHL history to record 300 points.

333 —Cy Denneny, Ottawa, Boston.
Denneny retired as the NHL's career point-scoring leader in 1929 with 333 points.

472 —Howie Morenz, Montreal, Chicago, NY Rangers.
Morenz passed Cy Denneny with point #334 in 1931-32. At the time his career ended in 1937, he was the NHL's career point- scoring leader with 472 points.

515 —Nels Stewart, Montreal Maroons, Boston, NY Americans.
Stewart passed Morenz with point #473 in 1938-39. He retired as the NHL's career point-scoring leader in 1940 with 515 points.

528 —Syd Howe, Ottawa, Philadelphia Quakers, Toronto, St. Louis Eagles, Detroit.
Howe passed Nels Stewart with point #516 on March 8, 1945. He retired as the NHL's career point-scoring leader in 1946 with 528 points.

548 —Bill Cowley, St. Louis Eagles, Boston.
Cowley passed Syd Howe with point #529 on Feb. 12, 1947. He retired as the NHL's career point-scoring leader in 1947 with 548 points.

610 —Elmer Lach, Montreal.
Lach passed Bill Cowley with point #549 on Feb. 23, 1952. He remained the NHL's career point-scoring leader until he was overtaken by Maurice Richard in 1953-54. He finished his career with 623 points.

946 —Maurice Richard, Montreal.
Richard passed teammate Elmer Lach with point #611 on Dec. 12, 1953. He remained the NHL's career point-scoring leader until he was overtaken by Gordie Howe in 1959-60. He finished his career with 965 points.

1,850 —Gordie Howe, Detroit, Hartford.
Howe passed Richard with point #947 on Jan. 16, 1960. He retired as the NHL's career point-scoring leader in 1980 with 1,850 points.

2,857 —Wayne Gretzky, Edmonton, Los Angeles, St. Louis, NY Rangers.
Gretzky passed Howe with point #1,851 on Oct. 15, 1989. He retired as the NHL's current career points leader with 2,857.

Individual Records

Regular Season

SEASONS

MOST SEASONS:
26 — Gordie Howe, Detroit, 1946-47 – 1970-71; Hartford, 1979-80.
24 — Alex Delvecchio, Detroit, 1950-51 – 1973-74.
— Tim Horton, Toronto, NY Rangers, Pittsburgh, Buffalo, 1949-50, 1951-52 – 1973-74.
23 — John Bucyk, Detroit, Boston, 1955-56 – 1977-78.
— Mark Messier, Edmonton, NY Rangers, Vancouver, 1979-80 – 2001-02.
22 — Dean Prentice, NY Rangers, Boston, Detroit, Pittsburgh, Minnesota, 1952-53 – 1973-74.
— Doug Mohns, Boston, Chicago, Minnesota, Atlanta, Washington, 1953-54 – 1974-75.
— Stan Mikita, Chicago, 1958-59 – 1979-80.
— Raymond Bourque, Boston, Colorado, 1979-80 – 2000-01.

GAMES

MOST GAMES:
1,767 — Gordie Howe, Detroit, 1946-47 – 1970-71; Hartford, 1979-80.
1,616 — Larry Murphy, Los Angeles, Washington, Minnesota, Pittsburgh, Toronto, Detroit, 1980-81 – 2000-01.
1,612 — Raymond Bourque, Boston, Colorado, 1979-80 – 2000-01.
1,602 — Mark Messier, Edmonton, NY Rangers, Vancouver, 1979-80 – 2001-02.
1,569 — Ron Francis, Hartford, Pittsburgh, Carolina, 1981-82 – 2001-02.

MOST GAMES, INCLUDING PLAYOFFS:
1,924 — Gordie Howe, Detroit, Hartford, 1,767 regular-season and 157 playoff games.
1,838 — Mark Messier, Edmonton, NY Rangers, Vancouver, 1,602 regular-season and 236 playoff games.
1,830 — Larry Murphy, Los Angeles, Washington, Minnesota, Pittsburgh, Toronto, Detroit, 1,615 regular-season and 215 playoff games.
1,826 — Raymond Bourque, Boston, Colorado, 1,612 regular-season and 214 playoff games.
1,728 — Ron Francis, Hartford, Pittsburgh, Carolina, 1,569 regular-season and 159 playoff games.

MOST CONSECUTIVE GAMES:
964 — Doug Jarvis, Montreal, Washington, Hartford, from Oct. 8, 1975 – Oct. 10, 1987.
914 — Garry Unger, Toronto, Detroit, St. Louis, Atlanta, from Feb. 24, 1968 – Dec. 21, 1979.
884 — Steve Larmer, Chicago, from Oct. 6, 1982 – Apr. 15, 1993.
776 — Craig Ramsay, Buffalo, from Mar. 27, 1973 – Feb. 10, 1983.
630 — Andy Hebenton, NY Rangers, Boston, from Oct. 7, 1955 – Mar. 22, 1964.

GOALS

MOST GOALS:
894 — Wayne Gretzky, Edmonton, Los Angeles, St. Louis, NY Rangers, in 20 seasons, 1,487GP.
801 — Gordie Howe, Detroit, Hartford, in 26 seasons, 1,767GP.
731 — Marcel Dionne, Detroit, Los Angeles, NY Rangers, in 18 seasons, 1,348GP.
717 — Phil Esposito, Chicago, Boston, NY Rangers, in 18 seasons, 1,282GP.
708 — Mike Gartner, Washington, Minnesota, NY Rangers, Toronto, Phoenix, in 19 seasons, 1,432GP.

MOST GOALS, INCLUDING PLAYOFFS:
1,016 — Wayne Gretzky, Edmonton, Los Angeles, St. Louis, NY Rangers, 894 regular-season and 122 playoff goals.
869 — Gordie Howe, Detroit, Hartford, 801 regular-season and 68 playoff goals.
779 — Brett Hull, Calgary, St. Louis, Dallas, Detroit, 679 regular-season and 100 playoff goals.
778 — Phil Esposito, Chicago, Boston, NY Rangers, 717 regular-season and 61 playoff goals.
767 — Mark Messier, Edmonton, NY Rangers, Vancouver, 658 regular-season and 109 playoff goals.

MOST GOALS, ONE SEASON:
92 — Wayne Gretzky, Edmonton, 1981-82. 80 game schedule.
87 — Wayne Gretzky, Edmonton, 1983-84. 80 game schedule.
86 — Brett Hull, St. Louis, 1990-91. 80 game schedule.
85 — Mario Lemieux, Pittsburgh, 1988-89. 80 game schedule.
76 — Phil Esposito, Boston, 1970-71. 78 game schedule.
— Alexander Mogilny, Buffalo, 1992-93. 84 game schedule.
— Teemu Selanne, Winnipeg, 1992-93. 84 game schedule.
73 — Wayne Gretzky, Edmonton, 1984-85. 80 game schedule.
72 — Brett Hull, St. Louis, 1989-90. 80 game schedule.
71 — Wayne Gretzky, Edmonton, 1982-83. 80 game schedule.
— Jari Kurri, Edmonton, 1984-85. 80 game schedule.
70 — Mario Lemieux, Pittsburgh, 1987-1988. 80 game schedule.
— Bernie Nicholls, Los Angeles, 1988-89. 80 game schedule.
— Brett Hull, St. Louis, 1991-92. 80 game schedule.

MOST GOALS, ONE SEASON, INCLUDING PLAYOFFS:
100 — Wayne Gretzky, Edmonton, 1983-84, 87G in 74 regular-season games and 13G in 19 playoff games.
97 — Wayne Gretzky, Edmonton, 1981-82, 92G in 80 regular-season games and 5G in 5 playoff games.
— Mario Lemieux, Pittsburgh, 1988-89, 85G in 76 regular-season games and 12G in 11 playoff games.
— Brett Hull, St. Louis, 1990-91, 86G in 78 regular-season games and 11G in 13 playoff games.
90 — Wayne Gretzky, Edmonton, 1984-85, 73G in 80 regular-season games and 17G in 18 playoff games.
— Jari Kurri, Edmonton, 1984-85, 71G in 80 regular-season games and 19G in 18 playoff games.
85 — Mike Bossy, NY Islanders, 1980-81, 68G in 79 regular-season games and 17G in 18 playoff games.
— Brett Hull, St. Louis, 1989-90, 72G in 80 regular-season games and 13G in 12 playoff games.
83 — Wayne Gretzky, Edmonton, 1982-83, 71G in 73 regular-season games and 12G in 16 playoff games.
— Alexander Mogilny, Buffalo, 1992-93, 76G in 77 regular-season games and 7G in 7 playoff games.

MOST GOALS, 50 GAMES FROM START OF SEASON:
61 — Wayne Gretzky, Edmonton, 1981-82. Oct. 7, 1981 - Jan. 22, 1982. (80-game schedule)
— Wayne Gretzky, Edmonton, 1983-84. Oct. 5, 1983 - Jan. 25, 1984. (80-game schedule)
54 — Mario Lemieux, Pittsburgh, 1988-89. Oct. 7, 1988 - Jan. 31, 1989. (80-game schedule)
53 — Wayne Gretzky, Edmonton, 1984-85. Oct. 11, 1984 - Jan. 28, 1985. (80-game schedule).
52 — Brett Hull, St. Louis, 1990-91. Oct. 4, 1990 - Jan. 26, 1991. (80-game schedule).
50 — Maurice Richard, Montreal, 1944-45. Oct. 28, 1944 - March 18, 1945. (50-game schedule)
— Mike Bossy, NY Islanders, 1980-81. Oct. 11, 1980 - Jan. 24, 1981. (80-game schedule)
— Brett Hull, St. Louis, 1991-92. Oct. 5, 1991 – Jan 28, 1992. (80 game schedule)

MOST GOALS, ONE GAME:
7 — Joe Malone, Que. Bulldogs, Jan. 31, 1920, at Quebec. Quebec 10, Toronto 6.
6 — Newsy Lalonde, Montreal, Jan. 10, 1920, at Montreal. Montreal 14, Toronto 7.
— Joe Malone, Que. Bulldogs, March 10, 1920, at Quebec. Quebec 10, Ottawa 4.
— Corb Denneny, Toronto, Jan. 26, 1921, at Toronto. Toronto 10, Hamilton 3.
— Cy Denneny, Ottawa, Mar. 7, 1921, at Ottawa. Ottawa 12, Hamilton 5.
— Syd Howe, Detroit, Feb. 3, 1944, at Detroit. Detroit 12, NY Rangers 2.
— Red Berenson, St. Louis, Nov. 7, 1968, at Philadelphia. St. Louis 8, Philadelphia 0.
— Darryl Sittler, Toronto, Feb. 7, 1976, at Toronto. Toronto 11, Boston 4.

With 30 goals in his first season as a Red Wing in 2001-02, plus a league-leading 10 in the playoffs, Brett Hull moved past Phil Esposito into third place all-time for the most goals scored in a career including playoffs with 779 (679 + 100).

Red Green of the Hamilton Tigers (lower left) was the second player in NHL history to score five goals in a road game. The Tigers staged the NHL's first players' strike in 1925 and the team became the New York Americans in 1925-26.

MOST GOALS, ONE ROAD GAME:

6 — Red Berenson, St. Louis, Nov. 7, 1968, at Philadelphia. St. Louis 8, Philadelphia 0.
5 — Joe Malone, Montreal, Dec. 19, 1917, at Ottawa. Montreal 9, Ottawa 4.
— Red Green, Hamilton, Dec. 5, 1924, at Toronto. Hamilton 10, Toronto 3.
— Babe Dye, Toronto, Dec. 22, 1924, at Boston. Toronto 10, Boston 1.
— Punch Broadbent, Mtl. Maroons, Jan. 7, 1925, at Hamilton. Mtl. Maroons 6, Hamilton 2.
— Don Murdoch, NY Rangers, Oct. 12, 1976, at Minnesota. NY Rangers 10, Minnesota 4.
— Tim Young, Minnesota, Jan. 15, 1979, at NY Rangers. Minnesota 8, NY Rangers 1.
— Willy Lindstrom, Winnipeg, Mar. 2, 1982, at Philadelphia. Winnipeg 7, Philadelphia 6.
— Bengt Gustafsson, Washington, Jan. 8, 1984, at Philadelphia. Washington 7, Philadelphia 1.
— Wayne Gretzky, Edmonton, Dec. 15, 1984, at St. Louis. Edmonton 8, St. Louis 2.
— Dave Andreychuk, Buffalo, Feb. 6, 1986, at Boston. Buffalo 8, Boston 6.
— Mats Sundin, Quebec, Mar. 5, 1992, at Hartford. Quebec 10, Hartford 4.
— Mario Lemieux, Pittsburgh, Apr. 9, 1993, at NY Rangers. Pittsburgh 10, NY Rangers 4.
— Mike Ricci, Quebec, Feb. 17, 1994, at San Jose. Quebec 8, San Jose 2.
— Alexei Zhamnov, Winnipeg, Apr. 1, 1995, at Los Angeles. Winnipeg 7, Los Angeles 7.

MOST GOALS, ONE PERIOD:

4 — Busher Jackson, Toronto, Nov. 20, 1934, at St. Louis, third period. Toronto 5, St. Louis Eagles 2.
— Max Bentley, Chicago, Jan. 28, 1943, at Chicago, third period. Chicago 10, NY Rangers 1.
— Clint Smith, Chicago, Mar. 4, 1945, at Chicago, third period. Chicago 6, Montreal 4.
— Red Berenson, St. Louis, Nov. 7, 1968, at Philadelphia, second period. St. Louis 8, Philadelphia 0.
— Wayne Gretzky, Edmonton, Feb. 18, 1981, at Edmonton, third period. Edmonton 9, St. Louis 2.
— Grant Mulvey, Chicago, Feb. 3, 1982, at Chicago, first period. Chicago 9, St. Louis 5.
— Bryan Trottier, NY Islanders, Feb. 13, 1982, at NY Islanders, second period. NY Islanders 8, Philadelphia 2.
— Al Secord, Chicago, Jan. 7, 1987, at Chicago, second period. Chicago 6, Toronto 4.
— Joe Nieuwendyk, Calgary, Jan. 11, 1989, at Calgary, second period. Calgary 8, Winnipeg 3.
— Peter Bondra, Washington, Feb. 5, 1994, at Washington, first period. Washington 6, Tampa Bay 3.
— Mario Lemieux, Pittsburgh, Jan. 26, 1997, at Montreal, third period. Pittsburgh 5, Montreal 2.

ASSISTS

MOST ASSISTS:

1,963 — Wayne Gretzky, Edmonton, Los Angeles, St. Louis, NY Rangers, in 20 seasons, 1,487GP.
1,187 — Ron Francis, Hartford, Pittsburgh, Carolina, in 21 seasons, 1,569GP.
1,169 — Raymond Bourque, Boston, Colorado, in 22 seasons, 1,612GP.
1,146 — Mark Messier, Edmonton, NY Rangers, Vancouver, in 23 seasons, 1,602GP.
1,135 — Paul Coffey, Edmonton, Pittsburgh, Los Angeles, Detroit, Hartford, Philadelphia, Chicago, Carolina, Boston, in 21 seasons, in 1,409GP.

MOST ASSISTS, INCLUDING PLAYOFFS:

2,223 — Wayne Gretzky, Edmonton, Los Angeles, St. Louis, NY Rangers, 1,963 regular-season and 260 playoff assists.
1,332 — Mark Messier, Edmonton, NY Rangers, Vancouver, 1,146 regular-season and 186 playoff assists.
1,308 — Raymond Bourque, Boston, Colorado, 1,169 regular season and 139 playoff assists.
1,280 — Ron Francis, Hartford, Pittsburgh, Carolina, 1,187 regular-season and 93 playoff assists.
1,272 — Paul Coffey, Edmonton, Pittsburgh, Los Angeles, Detroit, Hartford, Philadelphia, Chicago, Carolina, Boston, 1,135 regular-season and 137 playoff assists.

MOST ASSISTS, ONE SEASON:

163 — Wayne Gretzky, Edmonton, 1985-86. 80 game schedule.
135 — Wayne Gretzky, Edmonton, 1984-85. 80 game schedule.
125 — Wayne Gretzky, Edmonton, 1982-83. 80 game schedule.
122 — Wayne Gretzky, Los Angeles, 1990-91. 80 game schedule.
121 — Wayne Gretzky, Edmonton, 1986-87. 80 game schedule.
120 — Wayne Gretzky, Edmonton, 1981-82. 80 game schedule.
118 — Wayne Gretzky, Edmonton, 1983-84. 80 game schedule.
114 — Wayne Gretzky, Los Angeles, 1988-89. 80 game schedule.
— Mario Lemieux, Pittsburgh, 1988-89. 80 game schedule.
109 — Wayne Gretzky, Edmonton, 1980-81. 80 game schedule.
— Wayne Gretzky, Edmonton, 1987-88. 80 game schedule.
102 — Bobby Orr, Boston, 1970-71. 78 game schedule.
— Wayne Gretzky, Los Angeles, 1989-90. 80 game schedule.

MOST ASSISTS, ONE SEASON, INCLUDING PLAYOFFS:

174 — Wayne Gretzky, Edmonton, 1985-86, 163A in 80 regular-season games and 11A in 10 playoff games.
165 — Wayne Gretzky, Edmonton, 1984-85, 135A in 80 regular-season games and 30A in 18 playoff games.
151 — Wayne Gretzky, Edmonton, 1982-83, 125A in 80 regular-season games and 26A in 16 playoff games.
150 — Wayne Gretzky, Edmonton, 1986-87, 121A in 79 regular-season games and 29A in 21 playoff games.
140 — Wayne Gretzky, Edmonton, 1983-84, 118A in 74 regular-season games and 22A in 19 playoff games.
— Wayne Gretzky, Edmonton, 1987-88, 109A in 64 regular-season games and 31A in 19 playoff games.
133 — Wayne Gretzky, Los Angeles, 1990-91, 122A in 78 regular-season games and 11A in 12 playoff games.
131 — Wayne Gretzky, Los Angeles, 1988-89, 114A in 78 regular-season games and 17A in 11 playoff games.
127 — Wayne Gretzky, Edmonton, 1981-82, 120A in 80 regular-season games and 7A in 5 playoff games.
123 — Wayne Gretzky, Edmonton, 1980-81, 109A in 80 regular-season games and 14A in 9 playoff games.
121 — Mario Lemieux, Pittsburgh, 1988-89, 114A in 76 regular-season games and 7A in 11 playoff games.

MOST ASSISTS, ONE GAME:

7 — Billy Taylor, Detroit, Mar. 16, 1947, at Chicago. Detroit 10, Chicago 6.
— **Wayne Gretzky,** Edmonton, Feb. 15, 1980, at Edmonton. Edmonton 8, Washington 2.
— **Wayne Gretzky,** Edmonton, Dec. 11, 1985, at Chicago. Edmonton 12, Chicago 9.
— **Wayne Gretzky,** Edmonton, Feb. 14, 1986, at Edmonton. Edmonton 8, Quebec 2.
6 — Six assists have been recorded in one game on 24 occasions since Elmer Lach of Montreal first accomplished the feat vs. Boston on Feb. 6, 1943. The most recent player is Eric Lindros of Philadelphia (Feb. 26, 1997 at Ottawa).

MOST ASSISTS, ONE ROAD GAME:

7 — Billy Taylor, Detroit, Mar. 16, 1947, at Chicago. Detroit 10, Chicago 6.
— **Wayne Gretzky,** Edmonton, Dec. 11, 1985, at Chicago. Edmonton 12, Chicago 9.
6 — Bobby Orr, Boston, Jan. 1, 1973, at Vancouver. Boston 8, Vancouver 2.
— Patrik Sundstrom, Vancouver, Feb. 29, 1984, at Pittsburgh. Vancouver 9, Pittsburgh 5.
— Mario Lemieux, Pittsburgh, Dec. 5, 1992, at San Jose. Pittsburgh 9, San Jose 4.
— Eric Lindros, Philadelphia, Feb. 26, 1997, at Ottawa. Philadelphia 8, Ottawa 5.

MOST ASSISTS, ONE PERIOD:

5 — Dale Hawerchuk, Winnipeg, Mar. 6, 1984, at Los Angeles, second period. Winnipeg 7, Los Angeles 3.
4 — Four assists have been recorded in one period on 63 occasions since Mickey Roach of Hamilton first accomplished the feat vs. Toronto St. Pats on Feb. 23, 1921. Most recent player, Paul Kariya of Anaheim, (Dec. 16, 1998 vs Nashville).

POINTS

MOST POINTS:

2,857 — Wayne Gretzky, Edmonton, Los Angeles, St. Louis, NY Rangers, in 20 seasons, 1,487GP (894G-1963A).
1,850 — Gordie Howe, Detroit, Hartford, in 26 seasons, 1,767GP (801G-1049A).
1,804 — Mark Messier, Edmonton, NY Rangers, Vancouver, in 23 seasons, 1,602GP (658G-1,146A).
1,771 — Marcel Dionne, Detroit, Los Angeles, NY Rangers, in 18 seasons, 1,348GP (731G-1,040A).
1,701 — Ron Francis, Hartford, Pittsburgh, Carolina, in 21 seasons, 1,569GP (514G-1,187A).

MOST POINTS, INCLUDING PLAYOFFS:

3,239 — Wayne Gretzky, Edmonton, Los Angeles, St. Louis, NY Rangers, 2,857 regular-season and 382 playoff points.
2,099 — Mark Messier, Edmonton, NY Rangers, Vancouver, 1,804 regular-season and 295 playoff points.
2,010 — Gordie Howe, Detroit, Hartford, 1,850 regular-season and 160 playoff points.
1,840 — Ron Francis, Hartford, Pittsburgh, Carolina, 1,701 regular-season and 139 playoff points
1,837 — Steve Yzerman, Detroit, 1,662 regular-season and 175 playoff points.

MOST POINTS, ONE SEASON:

215 — Wayne Gretzky, Edmonton, 1985-86. 80 game schedule.
212 — Wayne Gretzky, Edmonton, 1981-82. 80 game schedule.
208 — Wayne Gretzky, Edmonton, 1984-85. 80 game schedule.
205 — Wayne Gretzky, Edmonton, 1983-84. 80 game schedule.
199 — Mario Lemieux, Pittsburgh, 1988-89. 80 game schedule.
196 — Wayne Gretzky, Edmonton, 1982-83. 80 game schedule.
183 — Wayne Gretzky, Edmonton, 1986-87. 80 game schedule.
168 — Mario Lemieux, Pittsburgh, 1987-88, 80 game schedule.
— Wayne Gretzky, Los Angeles, 1988-89. 80 game schedule.
164 — Wayne Gretzky, Edmonton, 1980-81. 80 game schedule.
163 — Wayne Gretzky, Los Angeles, 1990-91. 80 game schedule.
161 — Mario Lemieux, Pittsburgh, 1995-96. 82 game schedule.
160 — Mario Lemieux, Pittsburgh, 1992-93. 84 game schedule.

MOST POINTS, ONE SEASON, INCLUDING PLAYOFFS:

255 — Wayne Gretzky, Edmonton, 1984-85, 208PTS in 80 regular-season games and 47PTS in 18 playoff games.
240 — Wayne Gretzky, Edmonton, 1983-84, 205PTS in 74 regular-season games and 35PTS in 19 playoff games.
234 — Wayne Gretzky, Edmonton, 1982-83, 196PTS in 80 regular-season games and 38PTS in 16 playoff games.
— Wayne Gretzky, Edmonton, 1985-86, 215PTS in 80 regular-season games and 19PTS in 10 playoff games.
224 — Wayne Gretzky, Edmonton, 1981-82, 212PTS in 80 regular-season games and 12PTS in 5 playoff games.
218 — Mario Lemieux, Pittsburgh, 1988-89, 199PTS in 76 regular-season games and 19PTS in 11 playoff games.
217 — Wayne Gretzky, Edmonton, 1986-87, 183PTS in 79 regular-season games and 34PTS in 21 playoff games.
192 — Wayne Gretzky, Edmonton, 1987-88, 149PTS in 64 regular-season games and 43PTS in 19 playoff games.
190 — Wayne Gretzky, Los Angeles, 1988-89, 168PTS in 78 regular-season games and 22PTS in 11 playoff games.
188 — Mario Lemieux, Pittsburgh, 1995-96, 161PTS in 70 regular-season games and 27PTS in 18 playoff games.
185 — Wayne Gretzky, Edmonton, 1980-81, 164PTS in 80 regular-season games and 21PTS in 9 playoff games.

Originally a Toronto Maple Leaf, Billy Taylor was a member of the Detroit Red Wings when he had a league-leading 46 assists – including a record seven in one game – in 1946-47.

MOST POINTS, ONE GAME:

10 — Darryl Sittler, Toronto, Feb. 7, 1976, at Toronto, 6G-4A. Toronto 11, Boston 4.
8 — Maurice Richard, Montreal, Dec. 28, 1944, at Montreal, 5G-3A. Montreal 9, Detroit 1.
— Bert Olmstead, Montreal, Jan. 9, 1954, at Montreal, 4G-4A. Montreal 12, Chicago 1.
— Tom Bladon, Philadelphia, Dec. 11, 1977, at Philadelphia, 4G-4A. Philadelphia 11, Cleveland 1.
— Bryan Trottier, NY Islanders, Dec. 23, 1978, at NY Islanders, 5G-3A. NY Islanders 9, NY Rangers 4.
— Peter Stastny, Quebec, Feb. 22, 1981, at Washington, 4G-4A. Quebec 11, Washington 7.
— Anton Stastny, Quebec, Feb. 22, 1981, at Washington, 3G-5A. Quebec 11, Washington 7.
— Wayne Gretzky, Edmonton, Nov. 19, 1983, at Edmonton, 3G-5A. Edmonton 13, New Jersey 4.
— Wayne Gretzky, Edmonton, Jan. 4, 1984, at Edmonton, 4G-4A. Edmonton 12, Minnesota 8.
— Paul Coffey, Edmonton, Mar. 14, 1986, at Edmonton, 2G-6A. Edmonton 12, Detroit 3.
— Mario Lemieux, Pittsburgh, Oct. 15, 1988, at Pittsburgh, 2G-6A. Pittsburgh 9, St. Louis 2.
— Bernie Nicholls, Los Angeles, Dec. 1, 1988, at Los Angeles, 2G-6A. Los Angeles 9, Toronto 3.
— Mario Lemieux, Pittsburgh, Dec. 31, 1988, at Pittsburgh, 5G-3A. Pittsburgh 8, New Jersey 6.

MOST POINTS, ONE ROAD GAME:

8 — Peter Stastny, Quebec, Feb. 22, 1981, at Washington, 4G-4A. Quebec 11, Washington 7.
— **Anton Stastny,** Quebec, Feb. 22, 1981, at Washington, 3G-5A. Quebec 11, Washington 7.
7 — Red Green, Hamilton, Dec. 5, 1924, at Toronto, 5G-2A. Hamilton 10, Toronto 3.
— Billy Taylor, Detroit, Mar. 16, 1947, at Chicago, 7A. Detroit 10, Chicago 6.
— Red Berenson, St. Louis, Nov. 7, 1968, at Philadelphia, 6G-1A. St. Louis 8, Philadelphia 0.
— Gilbert Perreault, Buffalo, Feb. 1, 1976, at California, 2G-5A. Buffalo 9, California 5.
— Peter Stastny, Quebec, Apr. 1, 1982, at Boston, 3G-4A. Quebec 8, Boston 5.
— Wayne Gretzky, Edmonton, Nov. 6, 1983, at Winnipeg, 4G-3A. Edmonton 8, Winnipeg 5.
— Patrik Sundstrom, Vancouver, Feb. 29, 1984, at Pittsburgh, 1G-6A. Vancouver 9, Pittsburgh 5.
— Wayne Gretzky, Edmonton, Dec. 11, 1985, at Chicago, 7A, Edmonton 12, Chicago 9.
— Cam Neely, Boston, Oct. 16, 1988, at Chicago, 3G-4A. Boston 10, Chicago 3.
— Mario Lemieux, Pittsburgh, Jan. 21, 1989, at Edmonton, 2G-5A. Pittsburgh 7, Edmonton 4.
— Dino Ciccarelli, Washington, Mar. 18, 1989, at Hartford, 4G-3A. Washington 8, Hartford 2.
— Mats Sundin, Quebec, Mar. 5, 1992, at Hartford, 5G-2A. Quebec 10, Hartford 4.
— Mario Lemieux, Pittsburgh, Dec. 5, 1992, at San Jose, 1G-6A. Pittsburgh 9, San Jose 4.
— Eric Lindros, Philadelphia, Feb. 26, 1997, at Ottawa, 1G-6A. Philadelphia 8, Ottawa 5.

MOST POINTS, ONE PERIOD:

6 — Bryan Trottier, NY Islanders, Dec. 23, 1978, at NY Islanders, second period. 3G-3A. NY Islanders 9, NY Rangers 4.
5 — Les Cunningham, Chicago, Jan. 28, 1940, at Chicago, third period. 2G-3A. Chicago 8, Montreal 1.
— Max Bentley, Chicago, Jan. 28, 1943, at Chicago, third period. 4G-1A, Chicago 10, NY Rangers 1.
— Leo Labine, Boston, Nov. 28, 1954, at Boston, second period. 3G-2A. Boston 6, Detroit 2.
— Darryl Sittler, Toronto, Feb. 7, 1976, at Toronto, second period. 3G-2A. Toronto 11, Boston 4.
— Grant Mulvey, Chicago, Feb. 3, 1982, at Chicago, first period. 4G-1A. Chicago 9, St. Louis 5.
— Dale Hawerchuk, Winnipeg, Mar. 6, 1984, at Los Angeles, second period. 5A. Winnipeg 7, Los Angeles 3.
— Jari Kurri, Edmonton, Oct. 26, 1984, at Edmonton, second period. 2G-3A. Edmonton 8, Los Angeles 2.
— Pat Elynuik, Winnipeg, Jan. 20, 1989, at Winnipeg, second period. 2G-3A. Winnipeg 7, Pittsburgh 3.
— Ray Ferraro, Hartford, Dec. 9, 1989, at Hartford, first period. 3G-2A. Hartford 7, New Jersey 3.
— Stephane Richer, Montreal, Feb. 14, 1990, at Montreal, first period. 2G-3A. Montreal 10, Vancouver 1.
— Cliff Ronning, Vancouver, Apr. 15, 1993, at Los Angeles, third period. 3G-2A. Vancouver 8, Los Angeles 6.
— Peter Forsberg, Colorado, Mar. 3, 1999, at Florida, third period. 2G-3A. Colorado 7, Florida 5.

POWER-PLAY and SHORTHAND GOALS

MOST POWER-PLAY GOALS, ONE SEASON:

34 — Tim Kerr, Philadelphia, 1985-86. 80 game schedule.
32 — Dave Andreychuk, Buffalo, Toronto, 1992-93. 84 game schedule.
31 — Joe Nieuwendyk, Calgary, 1987-88. 80 game schedule.
— Mario Lemieux, Pittsburgh, 1988-89. 80 game schedule.
— Mario Lemieux, Pittsburgh, 1995-96. 82 game schedule.
29 — Michel Goulet, Quebec, 1987-88. 80 game schedule.
— Brett Hull, St. Louis, 1990-91. 80 game schedule.
— Brett Hull, St. Louis, 1992-93. 84 game schedule.

MOST SHORTHAND GOALS, ONE SEASON:

13 — Mario Lemieux, Pittsburgh, 1988-89. 80 game schedule.
12 — Wayne Gretzky, Edmonton, 1983-84. 80 game schedule.
11 — Wayne Gretzky, Edmonton, 1984-85. 80 game schedule.
10 — Marcel Dionne, Detroit, 1974-75. 80 game schedule.
— Mario Lemieux, Pittsburgh, 1987-88. 80 game schedule.
— Dirk Graham, Chicago, 1988-89. 80 game schedule.

MOST SHORTHAND GOALS, ONE GAME:

3 —Theoren Fleury, Calgary, Mar. 9, 1991, at St. Louis. Calgary 8, St. Louis 4.

OVERTIME SCORING

MOST OVERTIME GOALS, CAREER:

11 — Steve Thomas, Toronto, Chicago, NY Islanders, New Jersey.
10 — Mario Lemieux, Pittsburgh.
— Jaromir Jagr, Pittsburgh.
— Sergei Fedorov, Detroit.
— Theoren Fleury, Calgary, Colorado, NY Rangers.
— Mats Sundin, Quebec, Toronto.

MOST OVERTIME ASSISTS, CAREER:

16 — Adam Oates, Detroit, St. Louis, Boston, Washington, Philadelphia.
15 — Wayne Gretzky, Edmonton, Los Angeles, St. Louis, NY Rangers.
— Mark Messier, Edmonton, NY Rangers, Vancouver.
14 — Doug Gilmour, St. Louis, Calgary, Toronto, New Jersey, Chicago, Buffalo, Montreal.
13 — Raymond Bourque, Boston, Colorado.

MOST OVERTIME POINTS, CAREER:

23 — Mark Messier, Edmonton, NY Rangers, Vancouver. 8G-15A
21 — Steve Thomas, Toronto, Chicago, NY Islanders, New Jersey. 11G-10A
— Adam Oates, Detroit, St. Louis, Boston, Washington, Philadelphia. 5G-16A
20 — Mario Lemieux, Pittsburgh. 10G-10A
— Steve Yzerman, Detroit. 9G-11A

SCORING BY A CENTER

MOST GOALS BY A CENTER, CAREER

894 — Wayne Gretzky, Edmonton, Los Angeles, St. Louis, NY Rangers, in 20 seasons.
731 — Marcel Dionne, Detroit, Los Angeles, NY Rangers, in 18 seasons.
717 — Phil Esposito, Chicago, Boston, NY Rangers, in 18 seasons.
658 — Mark Messier, Edmonton, NY Rangers, Vancouver, in 23 seasons.
— Steve Yzerman, Detroit, in 19 seasons.

MOST GOALS BY A CENTER, ONE SEASON:

92 — Wayne Gretzky, Edmonton, 1981-82. 80 game schedule.
87 — Wayne Gretzky, Edmonton, 1983-84. 80 game schedule.
85 — Mario Lemieux, Pittsburgh, 1988-89. 80 game schedule.
76 — Phil Esposito, Boston, 1970-71. 78 game schedule.
73 — Wayne Gretzky, Edmonton, 1984-85. 80 game schedule.

A six-time Stanley Cup champion and one of the game's all-time great clutch players, it's no surprise that Mark Messier is the NHL's all-time leading scorer in regular-season overtime games.

MOST ASSISTS BY A CENTER, CAREER:
1,963 — Wayne Gretzky, Edmonton, Los Angeles, St. Louis, NY Rangers, in 20 seasons.
1,187 — Ron Francis, Hartford, Pittsburgh, Carolina, in 21 seasons.
1,146 — Mark Messier, Edmonton, NY Rangers, Vancouver, in 23 seasons.
1,040 — Marcel Dionne, Detroit, Los Angeles, NY Rangers, in 18 seasons.
1,027 — Adam Oates, Detroit, St. Louis, Boston, Washington, Philadelphia, in 17 seasons.

MOST ASSISTS BY A CENTER, ONE SEASON:
163 — Wayne Gretzky, Edmonton, 1985-86. 80 game schedule.
135 — Wayne Gretzky, Edmonton, 1984-85. 80 game schedule.
125 — Wayne Gretzky, Edmonton, 1982-83. 80 game schedule.
122 — Wayne Gretzky, Los Angeles, 1990-91. 80 game schedule.
121 — Wayne Gretzky, Edmonton, 1986-87. 80 game schedule.

MOST POINTS BY A CENTER, CAREER:
2,857 — Wayne Gretzky, Edmonton, Los Angeles, St. Louis, NY Rangers, in 20 seasons.
1,804 — Mark Messier, Edmonton, NY Rangers, Vancouver, in 23 seasons.
1,771 — Marcel Dionne, Detroit, Los Angeles, NY Rangers, in 18 seasons.
1,701 — Ron Francis, Hartford, Pittsburgh, Carolina, in 21 seasons.
1,662 — Steve Yzerman, Detroit, in 19 seasons.

MOST POINTS BY A CENTER, ONE SEASON:
215 — Wayne Gretzky, Edmonton, 1985-86. 80 game schedule.
212 — Wayne Gretzky, Edmonton, 1981-82. 80 game schedule.
208 — Wayne Gretzky, Edmonton, 1984-85. 80 game schedule.
205 — Wayne Gretzky, Edmonton, 1983-84. 80 game schedule.
199 — Mario Lemieux, Pittsburgh, 1988-89. 80 game schedule.

SCORING BY A LEFT WING

MOST GOALS BY A LEFT WING, CAREER:
620 — Luc Robitaille, Los Angeles, Pittsburgh, NY Rangers, Detroit, in 16 seasons.
610 — Bobby Hull, Chicago, Winnipeg, Hartford, in 16 seasons.
593 — Dave Andreychuk, Buffalo, Toronto, New Jersey, Boston, Colorado, Tampa Bay, in 20 seasons.
556 — John Bucyk, Detroit, Boston, in 23 seasons.
548 — Michel Goulet, Quebec, Chicago, in 15 seasons.

MOST GOALS BY A LEFT WING, ONE SEASON:
63 — Luc Robitaille, Los Angeles, 1992-93. 84 game schedule.
60 — Steve Shutt, Montreal, 1976-77. 80 game schedule.
58 — Bobby Hull, Chicago, 1968-69. 76 game schedule.
57 — Michel Goulet, Quebec, 1982-83. 80 game schedule.
56 — Charlie Simmer, Los Angeles, 1979-80. 80 game schedule.
— Charlie Simmer, Los Angeles, 1980-81. 80 game schedule.
— Michel Goulet, Quebec, 1983-84. 80 game schedule.

MOST ASSISTS BY A LEFT WING, CAREER:
813 — John Bucyk, Detroit, Boston, in 23 seasons.
668 — Luc Robitaille, Los Angeles, Pittsburgh, NY Rangers, Detroit, in 16 seasons.
654 — Dave Andreychuk, Buffalo, Toronto, New Jersey, Boston, Colorado, Tampa Bay, in 20 seasons.
604 — Michel Goulet, Quebec, Chicago, in 15 seasons.
579 — Brian Propp, Philadelphia, Boston, Minnesota, Hartford, in 15 seasons.

MOST ASSISTS BY A LEFT WING, ONE SEASON:
70 — Joe Juneau, Boston, 1992-93. 84 game schedule.
69 — Kevin Stevens, Pittsburgh, 1991-92. 80 game schedule.
67 — Mats Naslund, Montreal, 1985-86. 80 game schedule.
65 — John Bucyk, Boston, 1970-71. 78 game schedule.
— Michel Goulet, Quebec, 1983-84. 80 game schedule.
64 — Mark Messier, Edmonton, 1983-84. 80 game schedule.

MOST POINTS BY A LEFT WING, CAREER:
1,369 — John Bucyk, Detroit, Boston, in 23 seasons.
1,288 — Luc Robitaille, Los Angeles, Pittsburgh, NY Rangers, Detroit, in 16 seasons.
1,247 — Dave Andreychuk, Buffalo, Toronto, New Jersey, Boston, Colorado, Tampa Bay, in 20 seasons.
1,170 — Bobby Hull, Chicago, Winnipeg, Hartford, in 16 seasons.
1,152 — Michel Goulet, Quebec, Chicago, in 15 seasons.

MOST POINTS BY A LEFT WING, ONE SEASON:
125 — Luc Robitaille, Los Angeles, 1992-93. 84 game schedule.
123 — Kevin Stevens, Pittsburgh, 1991-92. 80 game schedule.
121 — Michel Goulet, Quebec, 1983-84. 80 game schedule.
116 — John Bucyk, Boston, 1970-71. 78 game schedule.
112 — Bill Barber, Philadelphia, 1975-76. 80 game schedule.

SCORING BY A RIGHT WING

MOST GOALS BY A RIGHT WING, CAREER:
801 — Gordie Howe, Detroit, Hartford, in 26 seasons.
708 — Mike Gartner, Washington, Minnesota, NY Rangers, Toronto, Phoenix, in 19 seasons.
679 — Brett Hull, Calgary, St. Louis, Dallas, Detroit, in 17 seasons.
608 — Dino Ciccarelli, Minnesota, Washington, Detroit, Tampa Bay, Florida, in 19 seasons.
601 — Jari Kurri, Edmonton, Los Angeles, NY Rangers, Anaheim, Colorado, in 17 seasons.

MOST GOALS BY A RIGHT WING, ONE SEASON:
86 — Brett Hull, St. Louis, 1990-91. 80 game schedule.
76 — Alexander Mogilny, Buffalo, 1992-93. 84 game schedule.
— Teemu Selanne, Winnipeg, 1992-93. 84 game schedule.
72 — Brett Hull, St. Louis, 1989-90. 80 game schedule.
71 — Jari Kurri, Edmonton, 1984-85. 80 game schedule.
70 — Brett Hull, St. Louis, 1991-92. 80 game schedule.

MOST ASSISTS BY A RIGHT WING, CAREER:
1,049 — Gordie Howe, Detroit, Hartford, in 26 seasons.
797 — Jari Kurri, Edmonton, Los Angeles, NY Rangers, Anaheim, Colorado, in 17 seasons.
793 — Guy Lafleur, Montreal, NY Rangers, Quebec, in 17 seasons.
688 — Jaromir Jagr, Pittsburgh, Washington, in 12 seasons.
664 — Mark Recchi, Pittsburgh, Philadelphia, Montreal, in 14 seasons..

MOST ASSISTS BY A RIGHT WING, ONE SEASON:
87 — Jaromir Jagr, Pittsburgh, 1995-96. 82 game schedule.
83 — Mike Bossy, NY Islanders, 1981-82. 80 game schedule.
— Jaromir Jagr, Pittsburgh, 1998-99. 82 game schedule.
80 — Guy Lafleur, Montreal, 1976-77. 80 game schedule.
77 — Guy Lafleur, Montreal, 1978-79. 80 game schedule.

Though no one is going to catch Wayne Gretzky, Steve Yzerman is one of several veterans jockeying for position among the all-time leaders at center. Yzerman, Mark Messier, Ron Francis, Mario Lemieux and Adam Oates are all near the top.

MOST POINTS BY A RIGHT WING, CAREER:
1,850 — Gordie Howe, Detroit, Hartford, in 26 seasons.
1,398 — Jari Kurri, Edmonton, Los Angeles, NY Rangers, Anaheim, Colorado, in 17 seasons.
1,353 — Guy Lafleur, Montreal, NY Rangers, Quebec, in 17 seasons.
1,335 — Mike Gartner, Washington, Minnesota, NY Rangers, Toronto, Phoenix, in 19 seasons.
1,246 — Brett Hull, Calgary, St. Louis, Dallas, Detroit, in 17 seasons.

MOST POINTS BY A RIGHT WING, ONE SEASON:
149 — Jaromir Jagr, Pittsburgh, 1995-96. 82 game schedule.
147 — Mike Bossy, NY Islanders, 1981-82. 80 game schedule.
136 — Guy Lafleur, Montreal, 1976-77. 80 game schedule.
135 — Jari Kurri, Edmonton, 1984-85. 80 game schedule.
132 — Guy Lafleur, Montreal, 1977-78. 80 game schedule.
— Teemu Selanne, Winnipeg, 1992-93. 84 game schedule.

SCORING BY A DEFENSEMAN

MOST GOALS BY A DEFENSEMAN, CAREER:
410 — Raymond Bourque, Boston, Colorado, in 22 seasons.
396 — Paul Coffey, Edmonton, Pittsburgh, Los Angeles, Detroit, Hartford, Philadelphia, Chicago, Carolina, Boston, in 21 seasons.
332 — Phil Housley, Buffalo, Winnipeg, St. Louis, Calgary, New Jersey, Washington, Chicago, in 20 seasons.
324 — Al MacInnis, Calgary, St. Louis, in 21 seasons.
310 — Denis Potvin, NY Islanders, in 15 seasons.

MOST GOALS BY A DEFENSEMAN, ONE SEASON:
48 — Paul Coffey, Edmonton, 1985-86. 80 game schedule.
46 — Bobby Orr, Boston, 1974-75. 80 game schedule.
40 — Paul Coffey, Edmonton, 1983-84. 80 game schedule.
39 — Doug Wilson, Chicago, 1981-82. 80 game schedule.
37 — Bobby Orr, Boston, 1970-71. 78 game schedule.
— Bobby Orr, Boston, 1971-72. 78 game schedule.
— Paul Coffey, Edmonton, 1984-85. 80 game schedule.

MOST GOALS BY A DEFENSEMAN, ONE GAME:
5 — Ian Turnbull, Toronto, Feb. 2, 1977, at Toronto. Toronto 9, Detroit 1.
4 — Harry Cameron, Toronto, Dec. 26, 1917, at Toronto. Toronto 7, Montreal 5.
— Harry Cameron, Montreal, Mar. 3, 1920, at Quebec City. Montreal 16, Que. Bulldogs 3.
— Sprague Cleghorn, Montreal, Jan. 14, 1922, at Montreal. Montreal 10, Hamilton 6.
— Johnny McKinnon, Pit. Pirates, Nov. 19, 1929, at Pittsburgh. Pit. Pirates 10, Toronto 5.
— Hap Day, Toronto, Nov. 19, 1929, at Pittsburgh. Pit. Pirates 10, Toronto 5.
— Tom Bladon, Philadelphia, Dec. 11, 1977, at Philadelphia. Philadelphia 11, Cleveland 1.
— Ian Turnbull, Los Angeles, Dec. 12, 1981, at Los Angeles. Los Angeles 7, Vancouver 5.
— Paul Coffey, Edmonton, Oct. 26, 1984, at Calgary. Edmonton 6, Calgary 5.

MOST ASSISTS BY A DEFENSEMAN, CAREER:
1,169 — Raymond Bourque, Boston, Colorado, in 22 seasons.
1,135 — Paul Coffey, Edmonton, Pittsburgh, Los Angeles, Detroit, Hartford, Philadelphia, Chicago, Carolina, Boston, in 21 seasons.
929 — Larry Murphy, Los Angeles, Washington, Pittsburgh, Toronto, Detroit, in 21 seasons.
880 — Al MacInnis, Calgary, St. Louis, in 21 seasons.
871 — Phil Housley, Buffalo, Winnipeg, St. Louis, Calgary, New Jersey, Washington, Chicago, in 20 seasons.

MOST ASSISTS BY A DEFENSEMAN, ONE SEASON:
102 — Bobby Orr, Boston, 1970-71. 78 game schedule.
90 — Bobby Orr, Boston, 1973-74. 78 game schedule.
— Paul Coffey, Edmonton, 1985-86. 80 game schedule.
89 — Bobby Orr, Boston, 1974-75. 80 game schedule.

MOST ASSISTS BY A DEFENSEMAN, ONE GAME:
6 — Babe Pratt, Toronto, Jan. 8, 1944, at Toronto. Toronto 12, Boston 3.
— **Pat Stapleton,** Chicago, Mar. 30, 1969, at Chicago. Chicago 9, Detroit 5.
— **Bobby Orr,** Boston, Jan. 1, 1973, at Vancouver. Boston 8, Vancouver 2.
— **Ron Stackhouse,** Pittsburgh, Mar. 8, 1975, at Pittsburgh. Pittsburgh 8, Philadelphia 2.
— **Paul Coffey,** Edmonton, Mar. 14, 1986, at Edmonton. Edmonton 12, Detroit 3.
— **Gary Suter,** Calgary, Apr. 4, 1986, at Calgary. Calgary 9, Edmonton 3.

MOST POINTS BY A DEFENSEMAN, CAREER:
1,579 — Raymond Bourque, Boston, Colorado, in 22 seasons.
1,531 — Paul Coffey, Edmonton, Pittsburgh, Los Angeles, Detroit, Hartford, Philadelphia, Chicago, Carolina, Boston, in 21 seasons.
1,216 — Larry Murphy, Los Angeles, Washington, Pittsburgh, Toronto, Detroit, in 21 seasons.
1,204 — Al MacInnis, Calgary, St. Louis, in 21 seasons.
1,203 — Phil Housley, Buffalo, Winnipeg, St. Louis, Calgary, New Jersey, Washington, Chicago, in 20 seasons.

MOST POINTS BY A DEFENSEMAN, ONE SEASON:
139 — Bobby Orr, Boston, 1970-71. 78 game schedule.
138 — Paul Coffey, Edmonton, 1985-86. 80 game schedule.
135 — Bobby Orr, Boston, 1974-75. 80 game schedule.
126 — Paul Coffey, Edmonton, 1983-84. 80 game schedule.
122 — Bobby Orr, Boston, 1973-74. 78 game schedule.

MOST POINTS BY A DEFENSEMAN, ONE GAME:
8 — Tom Bladon, Philadelphia, Dec. 11, 1977, at Philadelphia. 4G-4A. Philadelphia 11, Cleveland 1.
— **Paul Coffey,** Edmonton, Mar. 14, 1986, at Edmonton. 2G-6A. Edmonton 12, Detroit 3.
7 — Bobby Orr, Boston, Nov. 15, 1973, at Boston. 3G-4A. Boston 10, NY Rangers 2.

SCORING BY A GOALTENDER

MOST POINTS BY A GOALTENDER, CAREER:
48 — Tom Barrasso, Buffalo, Pittsburgh, in 17 seasons. (48A)
46 — Grant Fuhr, Edmonton, Toronto, Buffalo, Los Angeles, St. Louis, in 19 seasons. (46A)

MOST POINTS BY A GOALTENDER, ONE SEASON:
14 — Grant Fuhr, Edmonton, 1983-84. (14A)
9 — Curtis Joseph, St. Louis, 1991-92. (9A)
8 — Mike Palmateer, Washington, 1980-81. (8A)
— Grant Fuhr, Edmonton, 1987-88. (8A)
— Ron Hextall, Philadelphia, 1988-89. (8A)
— Tom Barrasso, Pittsburgh, 1992-93. (8A)
7 — Ron Hextall, Philadelphia, 1987-88. (1G-6A)
— Mike Vernon, Calgary, 1987-88. (7A)

MOST POINTS BY A GOALTENDER, ONE GAME:
3 — Jeff Reese, Calgary, Feb. 10, 1993, at Calgary. Calgary 13, San Jose 1. (3A)

Defenseman Babe Pratt joined the Rangers in 1935-36 and won the Stanley Cup in New York in 1940, but his most productive seasons came in Toronto. His career-high 40 assists in 1943-44 included six in one game, and he won the Hart Trophy.

SCORING BY A ROOKIE

MOST GOALS BY A ROOKIE, ONE SEASON:
76 — Teemu Selanne, Winnipeg, 1992-93. 84 game schedule.
53 — Mike Bossy, NY Islanders, 1977-78. 80 game schedule.
51 — Joe Nieuwendyk, Calgary, 1987-88. 80 game schedule.
45 — Dale Hawerchuk, Winnipeg, 1981-82. 80 game schedule.
— Luc Robitaille, Los Angeles, 1986-87. 80 game schedule.

MOST GOALS BY A PLAYER IN HIS FIRST NHL SEASON, ONE GAME:
5 — Howie Meeker, Toronto, Jan. 8, 1947, at Toronto. Toronto 10, Chicago 4.
— Don Murdoch, NY Rangers, Oct. 12, 1976, at Minnesota. NY Rangers 10, Minnesota 4.

MOST GOALS BY A PLAYER IN HIS FIRST NHL GAME:
3 — Alex Smart, Montreal, Jan. 14, 1943, at Montreal. Montreal 5, Chicago 1.
— Real Cloutier, Quebec, Oct. 10, 1979, at Quebec. Atlanta 5, Quebec 3.

MOST ASSISTS BY A ROOKIE, ONE SEASON:
70 — Peter Stastny, Quebec, 1980-81. 80 game schedule.
— Joe Juneau, Boston, 1992-93. 84 game schedule.
63 — Bryan Trottier, NY Islanders, 1975-76. 80 game schedule.
62 — Sergei Makarov, Calgary, 1989-90. 80 game schedule.
60 — Larry Murphy, Los Angeles, 1980-81. 80 game schedule.

MOST ASSISTS BY A PLAYER IN HIS FIRST NHL SEASON, ONE GAME:
7 — Wayne Gretzky, Edmonton, Feb. 15, 1980, at Edmonton. Edmonton 8, Washington 2.
6 — Gary Suter, Calgary, Apr. 4, 1986, at Calgary. Calgary 9, Edmonton 3.

MOST ASSISTS BY A PLAYER IN HIS FIRST NHL GAME:
4 — Dutch Reibel, Detroit, Oct. 8, 1953, at Detroit. Detroit 4, NY Rangers 1.
— Roland Eriksson, Minnesota, Oct. 6, 1976, at NY Rangers. NY Rangers 6, Minnesota 5.
3 — Al Hill, Philadelphia, Feb. 14, 1977, at Philadelphia. Philadelphia 6, St. Louis 4.
— Jarno Kultanen, Boston, Oct. 5, 2000, at Boston. Boston 4, Ottawa 4.

MOST POINTS BY A ROOKIE, ONE SEASON:
132 — Teemu Selanne, Winnipeg, 1992-93. 84 game schedule.
109 — Peter Stastny, Quebec, 1980-81. 80 game schedule.
103 — Dale Hawerchuk, Winnipeg, 1981-82. 80 game schedule.
102 — Joe Juneau, Boston, 1992-93. 84 game schedule.
100 — Mario Lemieux, Pittsburgh, 1984-85. 80 game schedule.

MOST POINTS BY A PLAYER IN HIS FIRST NHL SEASON, ONE GAME:
8 — Peter Stastny, Quebec, Feb. 22, 1981, at Washington. 4G-4A. Quebec 11, Washington 7.
— Anton Stastny, Quebec, Feb. 22, 1981, at Washington. 3G-5A. Quebec 11, Washington 7.
7 — Wayne Gretzky, Edmonton, Feb. 15, 1980, at Edmonton. 7A. Edmonton 8, Washington 2.
— Sergei Makarov, Calgary, Feb. 25, 1990, at Calgary. 2G-5A. Calgary 10, Edmonton 4.
6 — Wayne Gretzky, Edmonton, Mar. 29, 1980, at Toronto. 2G-4A. Edmonton 8, Toronto 5.
— Gary Suter, Calgary, Apr. 4, 1986, at Calgary. 6A. Calgary 9, Edmonton 3.

MOST POINTS BY A PLAYER IN HIS FIRST NHL GAME:
5 — Al Hill, Philadelphia, Feb. 14, 1977, at Philadelphia. 2G-3A. Philadelphia 6, St. Louis 4.
4 — Alex Smart, Montreal, Jan. 14, 1943, at Montreal. 3G-1A. Montreal 5, Chicago 1.
— Dutch Reibel, Detroit, Oct. 8, 1953, at Detroit. 4A. Detroit 4, NY Rangers 1.
— Roland Eriksson, Minnesota, Oct. 6, 1976 at NY Rangers. 4A. NY Rangers 6, Minnesota 5.

SCORING BY A ROOKIE DEFENSEMAN

MOST GOALS BY A ROOKIE DEFENSEMAN, ONE SEASON:
23 — Brian Leetch, NY Rangers, 1988-89. 80 game schedule.
22 — Barry Beck, Col. Rockies, 1977-78. 80 game schedule.
19 — Reed Larson, Detroit, 1977-78. 80 game schedule.
— Phil Housley, Buffalo, 1982-83. 80 game schedule.

MOST ASSISTS BY A ROOKIE DEFENSEMAN, ONE SEASON:
60 — Larry Murphy, Los Angeles, 1980-81. 80 game schedule.
55 — Chris Chelios, Montreal, 1984-85. 80 game schedule.
50 — Stefan Persson, NY Islanders, 1977-78. 80 game schedule.
— Gary Suter, Calgary, 1985-86. 80 game schedule.
49 — Nicklas Lidstrom, Detroit, 1991-92. 80 game schedule.

MOST POINTS BY A ROOKIE DEFENSEMAN, ONE SEASON:
76 — Larry Murphy, Los Angeles, 1980-81. 80 game schedule.
71 — Brian Leetch, NY Rangers, 1988-89. 80 game schedule.
68 — Gary Suter, Calgary, 1985-86. 80 game schedule.
66 — Phil Housley, Buffalo, 1982-83. 80 game schedule.
65 — Raymond Bourque, Boston, 1979-80. 80 game schedule.

A veteran of international hockey, Sergei Makarov had a stellar first season in the NHL in 1989-90 and was named rookie of the year. Before the next season, the rules were changed regarding eligibility for the Calder Trophy.

PER-GAME SCORING AVERAGES

HIGHEST GOALS-PER-GAME AVERAGE, CAREER (AMONG PLAYERS WITH 200 OR MORE GOALS):

.805 — Mario Lemieux, Pittsburgh, 654G, 812GP, from 1984-85 – 1996-97, 2000-01 – 2001-02.
.762 — Mike Bossy, NY Islanders, 573G, 752GP, from 1977-78 – 1986-87.
.756 — Cy Denneny, Ottawa, Boston, 248G, 328GP, from 1917-18 – 1928-29.
.742 — Babe Dye, Toronto, Hamilton, Chicago, NY Americans, 201G, 271GP, from 1919-20 – 1930-31.
.630 — Pavel Bure, Vancouver, Florida, NY Rangers, 418G, 663GP, from 1991-92 – 2001-02.

HIGHEST GOALS-PER-GAME AVERAGE, ONE SEASON (AMONG PLAYERS WITH 20-OR-MORE GOALS):

2.20 — Joe Malone, Montreal, 1917-18, with 44G in 20GP.
1.80 — Cy Denneny, Ottawa, 1917-18, with 36G in 20GP.
1.64 — Newsy Lalonde, Montreal, 1917-18, with 23G in 14GP.
1.63 — Joe Malone, Quebec, 1919-20, with 39G in 24GP.
1.61 — Newsy Lalonde, Montreal, 1919-20, with 37G in 23GP.

HIGHEST GOALS-PER-GAME AVERAGE, ONE SEASON (AMONG PLAYERS WITH 50-OR-MORE GOALS):

1.18 — Wayne Gretzky, Edmonton, 1983-84, with 87G in 74GP.
1.15 — Wayne Gretzky, Edmonton, 1981-82, with 92G in 80GP.
— Mario Lemieux, Pittsburgh, 1992-93, with 69G in 60GP.
1.12 — Mario Lemieux, Pittsburgh, 1988-89, with 85G in 76GP.
1.10 — Brett Hull, St. Louis, 1990-91, with 86G in 78GP.
1.02 — Cam Neely, Boston, 1993-94, with 50G in 49GP.
1.00 — Maurice Richard, Montreal, 1944-45, with 50G in 50GP.

HIGHEST ASSISTS-PER-GAME AVERAGE, CAREER (AMONG PLAYERS WITH 300 OR MORE ASSISTS):

1.320 — Wayne Gretzky, Edmonton, Los Angeles, St. Louis, NY Rangers, 1,963A, 1,487GP from 1979-80 – 1998-99.
1.166 — Mario Lemieux, Pittsburgh, 947A, 812GP, from 1984-85 – 1996-97, 2000-01 – 2001-02.
.982 — Bobby Orr, Boston, Chicago, 645A, 657GP from 1966-67 – 1978-79.
.882 — Peter Forsberg, Quebec, Colorado, 411A, 466GP, from 1994-95 – 2000-01.
.849 — Adam Oates, Detroit, St. Louis, Boston, Washington, Philadelphia, 1,027A, 1,210GP, from 1984-85 – 2001-02.

HIGHEST ASSISTS-PER-GAME AVERAGE, ONE SEASON (AMONG PLAYERS WITH 35-OR-MORE ASSISTS):

2.04 — Wayne Gretzky, Edmonton, 1985-86, with 163A in 80GP.
1.70 — Wayne Gretzky, Edmonton, 1987-88, with 109A in 64GP.
1.69 — Wayne Gretzky, Edmonton, 1984-85, with 135A in 80GP.
1.59 — Wayne Gretzky, Edmonton, 1983-84, with 118A in 74GP.
1.56 — Wayne Gretzky, Edmonton, 1982-83, with 125A in 80GP.
1.56 — Wayne Gretzky, Los Angeles, 1990-91, with 122A in 78GP.
1.53 — Wayne Gretzky, Edmonton, 1986-87, with 121A in 79GP.
1.52 — Mario Lemieux, Pittsburgh, 1992-93, with 91A in 60GP.
1.50 — Wayne Gretzky, Edmonton, 1981-82, with 120A in 80GP.
1.50 — Mario Lemieux, Pittsburgh, 1988-89, with 114A in 76GP.

HIGHEST POINTS-PER-GAME AVERAGE, CAREER: (AMONG PLAYERS WITH 500 OR MORE POINTS):

1.972 — Mario Lemieux, Pittsburgh, 1,601PTS (654G-947A), 812GP, from 1984-85 – 1996-97, 2000-01 – 2001-02.
1.921 — Wayne Gretzky, Edmonton, Los Angeles, St. Louis, NY Rangers, 2,857PTS (894G-1,963A), 1,487GP, from 1979-80 – 1998-99.
1.497 — Mike Bossy, NY Islanders, 1,126PTS (573G-553A), 752GP, from 1977-78 – 1986-87.
1.393 — Bobby Orr, Boston, Chicago, 915PTS (270G-645A), 657GP, from 1966-67 – 1978-79.
1.323 — Jaromir Jagr, Pittsburgh, Washington, 1,158PTS (470G-688A), 875GP, from 1990-91 – 2001-02.

HIGHEST POINTS-PER-GAME AVERAGE, ONE SEASON (AMONG PLAYERS WITH 50-OR-MORE POINTS):

2.77 — Wayne Gretzky, Edmonton, 1983-84, with 205PTS in 74GP.
2.69 — Wayne Gretzky, Edmonton, 1985-86, with 215PTS in 80GP.
2.67 — Mario Lemieux, Pittsburgh, 1992-93, with 160PTS in 60GP.
2.65 — Wayne Gretzky, Edmonton, 1981-82, with 212PTS in 80GP.
2.62 — Mario Lemieux, Pittsburgh, 1988-89, with 199PTS in 76GP.
2.60 — Wayne Gretzky, Edmonton, 1984-85, with 208PTS in 80GP.
2.45 — Wayne Gretzky, Edmonton, 1982-83, with 196PTS in 80GP.
2.33 — Wayne Gretzky, Edmonton, 1987-88, with 149PTS in 64GP.
2.32 — Wayne Gretzky, Edmonton, 1986-87, with 183PTS in 79GP.
2.30 — Mario Lemieux, Pittsburgh, 1995-96 with 161PTS in 70GP.
2.18 — Mario Lemieux, Pittsburgh, 1987-88 with 168PTS in 77GP.
2.15 — Wayne Gretzky, Los Angeles, 1988-89, with 168PTS in 78GP.
2.09 — Wayne Gretzky, Los Angeles, 1990-91, with 163 PTS in 78GP.
2.08 — Mario Lemieux, Pittsburgh, 1989-90, with 123 PTS in 59GP.

SCORING PLATEAUS

MOST 20-OR-MORE GOAL SEASONS:

22 — Gordie Howe, Detroit, Hartford, in 26 seasons.
19 — Ron Francis, Hartford, Pittsburgh, Carolina, in 21 seasons.
17 — Marcel Dionne, Detroit, Los Angeles, NY Rangers, in 18 seasons.
— Mike Gartner, Washington, Minnesota, NY Rangers, Toronto, Phoenix, in 19 seasons.
— Wayne Gretzky, Edmonton, Los Angeles, St. Louis, NY Rangers, in 20 seasons.
— Mark Messier, Edmonton, NY Rangers, Vancouver, in 23 seasons.
— Dave Andreychuk, Buffalo, Toronto, New Jersey, Boston, Colorado, Tampa Bay, in 20 seasons.

MOST CONSECUTIVE 20-OR-MORE GOAL SEASONS:

22 — Gordie Howe, Detroit, 1949-50 – 1970-71.
17 — Marcel Dionne, Detroit, Los Angeles, NY Rangers, 1971-72 – 1987-88.
16 — Phil Esposito, Chicago, Boston, NY Rangers, 1964-65 – 1979-80.
15 — Mike Gartner, Washington, Minnesota, NY Rangers, Toronto, 1979-80 – 1993-94.
— Brett Hull, Calgary, St. Louis, Dallas, Detroit, 1987-88 – 2001-02.

MOST 30-OR-MORE GOAL SEASONS:

17 — Mike Gartner, Washington, Minnesota, NY Rangers, Toronto, Phoenix, in 19 seasons.
14 — Gordie Howe, Detroit, Hartford, in 26 seasons.
— Marcel Dionne, Detroit, Los Angeles, NY Rangers, in 18 seasons.
— Wayne Gretzky, Edmonton, Los Angeles, St. Louis, NY Rangers, in 20 seasons.
13 — Bobby Hull, Chicago, Winnipeg, Hartford, in 16 seasons.
— Phil Esposito, Chicago, Boston, NY Rangers, in 18 seasons.

Not only does Wayne Gretzky hold the record for the highest assists-per-game average in one season (2.04 in 1985-86), he holds each of the first seven spots and eight of the top ten. He also has 11 of the top 15 spots in points-per-game average.

MOST CONSECUTIVE 30-OR-MORE GOAL SEASONS:
15 — Mike Gartner, Washington, Minnesota, NY Rangers, Toronto, 1979-80 – 1993-94.
13 — Bobby Hull, Chicago, 1959-60 – 1971-72.
— Phil Esposito, Boston, NY Rangers, 1967-68 – 1979-80.
— Wayne Gretzky, Edmonton, Los Angeles, 1979-80 – 1991-92.
12 — Marcel Dionne, Detroit, Los Angeles,1974-75 – 1985-86.

MOST 40-OR-MORE GOAL SEASONS:
12 — Wayne Gretzky, Edmonton, Los Angeles, St. Louis, NY Rangers, in 20 seasons.
10 — Marcel Dionne, Detroit, Los Angeles, NY Rangers, in 18 seasons.
— Mario Lemieux, Pittsburgh, in 14 seasons.
9 — Mike Bossy, NY Islanders, in 10 seasons.
— Mike Gartner, Washington, Minnesota, NY Rangers, Toronto, Phoenix, in 19 seasons.

MOST CONSECUTIVE 40-OR-MORE GOAL SEASONS:
12 — Wayne Gretzky, Edmonton, Los Angeles, 1979-80 – 1990-91.
9 — Mike Bossy, NY Islanders, 1977-78 – 1985-86.
8 — Luc Robitaille, Los Angeles, 1986-87 – 1993-94.
7 — Phil Esposito, Boston, 1968-69 – 1974-75.
— Michel Goulet, Quebec, 1981-82 – 1987-88.
— Jari Kurri, Edmonton, 1982-83 – 1988-89.

MOST 50-OR-MORE GOAL SEASONS:
9 — Mike Bossy, NY Islanders, in 10 seasons.
— Wayne Gretzky, Edmonton, Los Angeles, St. Louis, NY Rangers, in 20 seasons.
6 — Guy Lafleur, Montreal, NY Rangers, Quebec, in 17 seasons.
— Marcel Dionne, Detroit, Los Angeles, NY Rangers, in 18 seasons.
— Mario Lemieux, Pittsburgh, in 14 seasons.
5 — Bobby Hull, Chicago, Winnipeg, Hartford, in 16 seasons.
— Phil Esposito, Chicago, Boston, NY Rangers, in 18 seasons.
— Brett Hull, Calgary, St. Louis, Dallas, in 16 seasons.
— Steve Yzerman, Detroit, in 18 seasons.
— Pavel Bure, Vancouver, Florida, in 10 seasons.

MOST CONSECUTIVE 50-OR-MORE GOAL SEASONS:
9 — Mike Bossy, NY Islanders, 1977-78 – 1985-86.
8 — Wayne Gretzky, Edmonton, 1979-80 – 1986-87.
6 — Guy Lafleur, Montreal, 1974-75 – 1979-80.
5 — Phil Esposito, Boston, 1970-71 – 1974-75.
— Marcel Dionne, Los Angeles, 1978-79 – 1982-83.
— Brett Hull, St. Louis, 1989-90 – 1993-94.

MOST 60-OR-MORE GOAL SEASONS:
5 — Mike Bossy, NY Islanders, in 10 seasons.
— Wayne Gretzky, Edmonton, Los Angeles, St. Louis, NY Rangers, in 20 seasons.
4 — Phil Esposito, Chicago, Boston, NY Rangers, in 18 seasons.
— Mario Lemieux, Pittsburgh, in 14 seasons.

MOST CONSECUTIVE 60-OR-MORE GOAL SEASONS:
4 — Wayne Gretzky, Edmonton, 1981-82 – 1984-85.
3 — Mike Bossy, NY Islanders, 1980-81 – 1982-83.
— Brett Hull, St. Louis, 1989-90 – 1991-92.
2 — Phil Esposito, Boston, 1970-71 – 1971-72, 1973-74 – 1974-75.
— Jari Kurri, Edmonton, 1984-85 – 1985-86.
— Mario Lemieux, Pittsburgh, 1987-88 – 1988-89.
— Steve Yzerman, Detroit, 1988-89 – 1989-90.
— Pavel Bure, Vancouver, 1992-93 – 1993-94.

MOST 100-OR-MORE POINT SEASONS:
15 — Wayne Gretzky, Edmonton, Los Angeles, St. Louis, NY Rangers, in 20 seasons.
10 — Mario Lemieux, Pittsburgh, in 14 seasons.
8 — Marcel Dionne, Detroit, Los Angeles, NY Rangers, in 18 seasons.
7 — Mike Bossy, NY Islanders, in 10 seasons.
— Peter Stastny, Quebec, New Jersey, St. Louis, in 15 seasons.

MOST CONSECUTIVE 100-OR-MORE POINT SEASONS:
13 — Wayne Gretzky, Edmonton, Los Angeles, 1979-80 – 1991-92.
6 — Bobby Orr, Boston, 1969-70 – 1974-75.
— Guy Lafleur, Montreal, 1974-75 – 1979-80.
— Mike Bossy, NY Islanders,1980-81 – 1985-86.
— Peter Stastny, Quebec, 1980-81 – 1985-86.
— Mario Lemieux, Pittsburgh, 1984-85 – 1989-90.
— Steve Yzerman, Detroit, 1987-88 – 1992-93.

THREE-OR-MORE-GOAL GAMES

MOST THREE-OR-MORE GOAL GAMES, CAREER:
50 — Wayne Gretzky, Edmonton, Los Angeles, St. Louis, NY Rangers, in 20 seasons, 37 three-goal games, 9 four-goal games, 4 five-goal games.
40 — Mario Lemieux, Pittsburgh, in 14 seasons, 27 three-goal games, 10 four-goal games and 3 five-goal games.
39 — Mike Bossy, NY Islanders, in 10 seasons, 30 three-goal games, 9 four-goal games.
32 — Phil Esposito, Chicago, Boston, NY Rangers, in 18 seasons, 27 three-goal games, 5 four-goal games.
31 — Brett Hull, Calgary, St. Louis, Dallas, Detroit, in 17 seasons, 28 three-goal games, 3 four-goal games.
28 — Bobby Hull, Chicago, Winnipeg, Hartford, in 16 seasons, 24 three-goal games, 4 four-goal games.
— Marcel Dionne, Detroit, Los Angeles, NY Rangers, in 18 seasons, 25 three-goal games, 3 four-goal games.

MOST THREE-OR-MORE GOAL GAMES, ONE SEASON:
10 — Wayne Gretzky, Edmonton, 1981-82. 6 three-goal games, 3 four-goal games, 1 five-goal game.
— Wayne Gretzky, Edmonton, 1983-84. 6 three-goal games, 4 four-goal games.
9 — Mike Bossy, NY Islanders, 1980-81. 6 three-goal games, 3 four-goal games.
— Mario Lemieux, Pittsburgh, 1988-89. 7 three-goal games, 1 four-goal game, 1 five-goal game.
8 — Brett Hull, St. Louis, 1991-92. 8 three-goal games.
7 — Joe Malone, Montreal, 1917-18. 2 three-goal games, 2 four-goal games, 3 five-goal games.
— Phil Esposito, Boston, 1970-71. 7 three-goal games.
— Rick Martin, Buffalo, 1975-76. 6 three-goal games, 1 four-goal game.
— Alexander Mogilny, Buffalo, 1992-93. 5 three-goal games, 2 four-goal games.

Peter Stastny is one of only five players in NHL history to have at least seven 100-point seasons, including a streak of six years in a row. He also finished among the top 10 in scoring on six occasions, placing second to Wayne Gretzky in 1982-83.

SCORING STREAKS

LONGEST CONSECUTIVE GOAL-SCORING STREAK:
16 Games — Punch Broadbent, Ottawa, 1921-22. 27 goals during streak.
14 Games — Joe Malone, Montreal, 1917-18. 35 goals during streak.
13 Games — Newsy Lalonde, Montreal, 1920-21. 24 goals during streak.
— Charlie Simmer, Los Angeles, 1979-80. 17 goals during streak.
12 Games — Cy Denneny, Ottawa, 1917-18. 23 goals during streak.
— Dave Lumley, Edmonton, 1981-82. 15 goals during streak.
— Mario Lemieux, Pittsburgh, 1992-93. 18 goals during streak.

LONGEST CONSECUTIVE ASSIST-SCORING STREAK:
23 Games — Wayne Gretzky, Los Angeles, 1990-91. 48A during streak.
18 Games — Adam Oates, Boston, 1992-93. 28A during streak.
17 Games — Wayne Gretzky, Edmonton, 1983-84. 38A during streak.
— Paul Coffey, Edmonton, 1985-86. 27A during streak.
— Wayne Gretzky, Los Angeles, 1989-90. 35A during streak.
16 Games — Jaromir Jagr, Pittsburgh, 2000-01. 24A during streak.

LONGEST CONSECUTIVE POINT SCORING STREAK:
51 Games — Wayne Gretzky, Edmonton, 1983-84. 61G-92A-153PTS during streak.
46 Games — Mario Lemieux, Pittsburgh, 1989-90. 39G-64A-103PTS during streak.
39 Games — Wayne Gretzky, Edmonton, 1985-86. 33G-75A-108PTS during streak.
30 Games — Wayne Gretzky, Edmonton, 1982-83. 24G-52A-76PTS during streak.
— Mats Sundin, Quebec, 1992-93. 21G-25A-46PTS during streak.
28 Games — Guy Lafleur, Montreal, 1976-77. 19G-42A-61PTS during streak.
— Wayne Gretzky, Edmonton, 1984-85. 20G-43A-63PTS during streak.
— Mario Lemieux, Pittsburgh, 1985-86. 21G-38A-59PTS during streak.
— Paul Coffey, Edmonton, 1985-86. 16G-39A-55PTS during streak.
— Steve Yzerman, Detroit, 1988-89. 29G-36A-65PTS during streak.

LONGEST CONSECUTIVE POINT-SCORING STREAK FROM START OF SEASON:
51 Games — Wayne Gretzky, Edmonton, 1983-84. 61G-92A-153PTS during streak which was stopped by goaltender Markus Mattsson and Los Angeles on Jan. 28, 1984.

LONGEST CONSECUTIVE POINT-SCORING STREAK BY A DEFENSEMAN:
28 Games — Paul Coffey, Edmonton, 1985-86. 16G-39A-55PTS during streak.
19 Games — Raymond Bourque, Boston, 1987-88. 6G-21A-27PTS during streak.
17 Games — Raymond Bourque, Boston, 1984-85. 4G-24A-28PTS during streak.
— Brian Leetch, NY Rangers, 1991-92. 5G-24A-29PTS during streak.
16 Games — Gary Suter, Calgary, 1987-88. 8G-17A-25PTS during streak.
15 Games — Bobby Orr, Boston, 1970-71. 10G-23A-33PTS during streak.
— Bobby Orr, Boston, 1973-74. 8G-15A-23PTS during streak.
— Steve Duchesne, Quebec, 1992-93. 4G-17A-21PTS during streak.
— Chris Chelios, Chicago, 1995-96. 4G-16A-20PTS during streak.

FASTEST GOALS AND ASSISTS

FASTEST GOAL FROM START OF A GAME:
5 Seconds — Doug Smail, Winnipeg, Dec. 20, 1981, at Winnipeg. Winnipeg 5, St. Louis 4.
— Bryan Trottier, NY Islanders, Mar. 22, 1984, at Boston. NY Islanders 3, Boston 3.
— Alexander Mogilny, Buffalo, Dec. 21, 1991, at Toronto. Buffalo 4, Toronto 1.
6 Seconds — Henry Boucha, Detroit, Jan. 28, 1973, at Montreal. Detroit 4, Montreal 2.
— Jean Pronovost, Pittsburgh, Mar. 25, 1976, at St. Louis. St. Louis 5, Pittsburgh 2.
7 Seconds — Charlie Conacher, Toronto, Feb. 6, 1932, at Toronto. Toronto 6, Boston 0.
— Danny Gare, Buffalo, Dec. 17, 1978, at Buffalo. Buffalo 6, Vancouver 3.
— Dave Williams, Los Angeles, Feb. 14, 1987 at Los Angeles. Los Angeles 5, Harford 2.
8 Seconds — Ron Martin, NY Americans, Dec. 4, 1932, at NY Americans. NY Americans 4, Montreal 2.
— Chuck Arnason, Col. Rockies, Jan. 28, 1977, at Atlanta. Col. Rockies 3, Atlanta 3.
— Wayne Gretzky, Edmonton, Dec. 14, 1983, at NY Rangers. Edmonton 9, NY Rangers 4.
— Gaetan Duchesne, Washington, Mar. 14, 1987, at St. Louis. Washington 3, St. Louis 3.
— Tim Kerr, Philadelphia, Mar. 7, 1989, at Philadelphia. Philadelphia 4, Edmonton 4.
— Grant Ledyard, Buffalo, Dec. 4, 1991, at Winnipeg. Buffalo 4, Winnipeg 4.
— Brent Sutter, Chicago, Feb. 5, 1995, at Vancouver. Chicago 9, Vancouver 4.
— Paul Kariya, Anaheim, Mar. 9, 1997, at Colorado. Anaheim 2, Colorado 2.
— Tony Hrkac, Dallas, Nov. 7, 1998, at Los Angeles. Dallas 4, Los Angeles 3.
— Sergei Fedorov, Detroit, Nov. 21, 1998, at Vancouver. Detroit 4, Vancouver 2.

FASTEST GOAL FROM START OF A PERIOD:
4 Seconds — Claude Provost, Montreal, Nov. 9, 1957, at Montreal, second period. Montreal 4, Boston 2.
— Denis Savard, Chicago, Jan. 12, 1986, at Chicago, third period. Chicago 4, Hartford 2.

FASTEST GOAL BY A PLAYER IN HIS FIRST NHL GAME:
15 Seconds — Gus Bodnar, Toronto, Oct. 30, 1943. Toronto 5, NY Rangers 2.
18 Seconds — Danny Gare, Buffalo, Oct. 10, 1974. Buffalo 9, Boston 5.
20 Seconds — Alexander Mogilny, Buffalo, Oct. 5, 1989. Buffalo 4, Quebec 3.

FASTEST TWO GOALS:
4 Seconds — Nels Stewart, Mtl. Maroons, Jan. 3, 1931, at Montreal at 8:24 and 8:28, third period. Mtl. Maroons 5, Boston 3.
— Deron Quint, Winnipeg, Dec. 15, 1995, at Winnipeg at 7:51 and 7:55, second period. Winnipeg 9, Edmonton 4.
5 Seconds — Pete Mahovlich, Montreal, Feb. 20, 1971, at Montreal at 12:16 and 12:21, third period. Montreal 7, Chicago 1.
6 Seconds — Jim Pappin, Chicago, Feb. 16, 1972, at Chicago at 2:57 and 3:03, third period. Chicago 3, Philadelphia 3.
— Ralph Backstrom, Los Angeles, Nov. 2, 1972, at Los Angeles at 8:30 and 8:36, third period. Los Angeles 5, Boston 2.
— Lanny McDonald, Calgary, Mar. 22, 1984, at Calgary at 16:23 and 16:29, first period. Detroit 6, Calgary 4.
— Sylvain Turgeon, Hartford, Mar. 28, 1987, at Hartford at 13:59 and 14:05, second period. Hartford 5, Pittsburgh 4.

FASTEST THREE GOALS:
21 Seconds — Bill Mosienko, Chicago, Mar. 23, 1952, at NY Rangers, against goaltender Lorne Anderson. Mosienko scored at 6:09, 6:20 and 6:30 of third period, all with both teams at full strength. Chicago 7, NY Rangers 6.
44 Seconds — Jean Béliveau, Montreal, Nov. 5, 1955, at Montreal, against goaltender Terry Sawchuk. Béliveau scored at :42, 1:08 and 1:26 of second period, all with Montreal holding a 6-4 man advantage. Montreal 4, Boston 2.

FASTEST THREE ASSISTS:
21 Seconds — Gus Bodnar, Chicago, Mar. 23, 1952, at NY Rangers, Bodnar assisted on Bill Mosienko's three goals at 6:09, 6:20 and 6:30 of third period. Chicago 7, NY Rangers 6.
44 Seconds — Bert Olmstead, Montreal, Nov. 5, 1955, at Montreal, Olmstead assisted on Jean Béliveau's three goals at :42, 1:08 and 1:26 of second period. Montreal 4, Boston 2.

SHOTS ON GOAL

MOST SHOTS ON GOAL, ONE SEASON:
550 — Phil Esposito, Boston, 1970-71. 78 game schedule.
429 — Paul Kariya, Anaheim, 1998-99. 82 game schedule.
426 — Phil Esposito, Boston, 1971-72. 78 game schedule.
414 — Bobby Hull, Chicago, 1968-69. 76 game schedule.

PENALTIES

MOST PENALTY MINUTES, CAREER:
3,966 — Dave Williams, Toronto, Vancouver, Detroit, Los Angeles, Hartford, in 14 seasons, 962GP.
3,565 — Dale Hunter, Quebec, Washington, Colorado, in 19 seasons, 1,407GP.
3,381 — Marty McSorley, Pittsburgh, Edmonton, Los Angeles, NY Rangers, San Jose, Boston, in 17 seasons, 961GP.
3,300 — Bob Probert, Detroit, Chicago, in 17 seasons, 935GP.
3,146 — Tim Hunter, Calgary, Quebec, Vancouver, San Jose, in 16 seasons, 815GP.

MOST PENALTY MINUTES, CAREER, INCLUDING PLAYOFFS:
4,421 — Dave Williams, Toronto, Vancouver, Detroit, Los Angeles, Hartford, 3,966 in regular-season; 455 in playoffs.
4,294 — Dale Hunter, Quebec, Washington, Colorado, 3,565 in regular-season; 729 in playoffs.
3,755 — Marty McSorley, Pittsburgh, Edmonton, Los Angeles, NY Rangers, San Jose, Boston, 3,381 in regular-season; 374 in playoffs.
3,584 — Chris Nilan, Montreal, NY Rangers, Boston, 3,043 in regular-season; 541 in playoffs.
3,574 — Bob Probert, Detroit, Chicago, 3,300 in regular season; 274 in playoffs.

MOST PENALTY MINUTES, ONE SEASON:
472 — Dave Schultz, Philadelphia, 1974-75.
409 — Paul Baxter, Pittsburgh, 1981-82.
408 — Mike Peluso, Chicago, 1991-92.
405 — Dave Schultz, Los Angeles, Pittsburgh, 1977-78.

MOST PENALTIES, ONE GAME:
10 — Chris Nilan, Boston, Mar. 31, 1991, at Boston against Hartford. 6 minors, 2 majors, 1 10-minute misconduct, 1 game misconduct.
9 — Jim Dorey, Toronto, Oct. 16, 1968, at Toronto against Pittsburgh. 4 minors, 2 majors, 2 10-minute misconducts, 1 game misconduct.
— Dave Schultz, Pittsburgh, Apr. 6, 1978, at Detroit. 5 minors, 2 majors, 2 10-minute misconducts.
— Randy Holt, Los Angeles, Mar. 11, 1979, at Philadelphia. 1 minor, 3 majors, 2 10-minute misconducts, 3 game misconducts.
— Russ Anderson, Pittsburgh, Jan. 19, 1980, at Pittsburgh. 3 minors, 3 majors, 3 game misconducts.
— Kim Clackson, Quebec, Mar. 8, 1981, at Quebec. 4 minors, 3 majors, 2 game misconducts.
— Terry O'Reilly, Boston, Dec. 19, 1984, at Hartford. 5 minors, 3 majors, 1 game misconduct.
— Larry Playfair, Los Angeles, Dec. 9, 1986, at NY Islanders. 6 minors, 2 majors, 1 10-minute misconduct.
— Marty McSorley, Los Angeles, Apr. 14, 1992, at Vancouver. 5 minors, 2 majors, 1 10-minute misconduct, 1 game misconduct.

MOST PENALTY MINUTES, ONE GAME:
67 — Randy Holt, Los Angeles, Mar. 11, 1979, at Philadelphia. 1 minor, 3 majors, 2 10-minute misconducts, 3 game misconducts.
57 — Brad Smith, Toronto, Nov. 15, 1986, at Toronto vs. Detroit. 1 minor, 3 majors, 2 10-minute misconducts, 2 game misconducts.
— Reed Low, St. Louis, Feb. 28, 2002, at St. Louis vs. Calgary. 1 minor, 3 majors, 1 10-minute misconduct, 3 game misconducts.

MOST PENALTIES, ONE PERIOD:
9 — Randy Holt, Los Angeles, Mar. 11, 1979, at Philadelphia, first period. 1 minor, 3 majors, 2 10-minute misconducts, 3 game misconducts.

MOST PENALTY MINUTES, ONE PERIOD:
67 — Randy Holt, Los Angeles, Mar. 11, 1979, at Philadelphia, first period. 1 minor, 3 majors, 2 10-minute misconducts, 3 game misconducts.

GOALTENDING

MOST GAMES APPEARED IN BY A GOALTENDER, CAREER:
971 — Terry Sawchuk, Detroit, Boston, Toronto, Los Angeles, NY Rangers from 1949-50 – 1969-70.
966 — Patrick Roy, Montreal, Colorado from 1984-85 – 2001-02.
906 — Glenn Hall, Detroit, Chicago, St. Louis from 1952-53 – 1970-71.
886 — Tony Esposito, Montreal, Chicago from 1968-69 – 1983-84.
882 — John Vanbiesbrouck, NY Rangers, Florida, Philadelphia, NY Islanders, New Jersey, from 1981-82 – 2001-02.

MOST CONSECUTIVE COMPLETE GAMES BY A GOALTENDER:
502 — Glenn Hall, Detroit, Chicago. Played 502 games from beginning of 1955-56 season - first 12 games of 1962-63. In his 503rd straight game, Nov. 7, 1962, at Chicago, Hall was removed from the game against Boston with a back injury in the first period.

MOST GAMES APPEARED IN BY A GOALTENDER, ONE SEASON:
79 — Grant Fuhr, St. Louis, 1995-96.
77 — Martin Brodeur, New Jersey, 1995-96.
— Bill Ranford, Edmonton, Boston, 1995-96.
— Arturs Irbe, Carolina, 2000-01.
75 — Grant Fuhr, Edmonton, 1987-88.
— Arturs Irbe, Carolina, 1999-2000.

MOST MINUTES PLAYED BY A GOALTENDER, CAREER:
57,194 — Terry Sawchuk, Detroit, Boston, Toronto, Los Angeles, NY Rangers, from 1949-50 – 1969-70.
56,466 — Patrick Roy, Montreal, Colorado, from 1984-85 – 2001-02.

MOST MINUTES PLAYED BY A GOALTENDER, ONE SEASON:
4,433 — Martin Brodeur, New Jersey, 1995-96.

MOST SHUTOUTS, CAREER:
103 — Terry Sawchuk, Detroit, Boston, Toronto, Los Angeles, NY Rangers in 21 seasons.
94 — George Hainsworth, Montreal, Toronto in 10 seasons.
84 — Glenn Hall, Detroit, Chicago, St. Louis in 16 seasons.

MOST SHUTOUTS, ONE SEASON:
22 — George Hainsworth, Montreal, 1928-29. 44GP
15 — Alex Connell, Ottawa, 1925-26. 36GP
— Alex Connell, Ottawa, 1927-28. 44GP
— Hal Winkler, Boston, 1927-28. 44GP
— Tony Esposito, Chicago, 1969-70. 63GP
14 — George Hainsworth, Montreal, 1926-27. 44GP

LONGEST SHUTOUT SEQUENCE BY A GOALTENDER:
461 Minutes, 29 Seconds — Alex Connell, Ottawa, 1927-28, six consecutive shutouts. (Forward passing not permitted in attacking zones in 1927-28.)
343 Minutes, 5 Seconds — George Hainsworth, Montreal, 1928-29, four consecutive shutouts. (Forward passing not permitted in attacking zones in 1928-29.)
324 Minutes, 40 Seconds — Roy Worters, NY Americans, 1930-31, four consecutive shutouts.
309 Minutes, 21 Seconds — Bill Durnan, Montreal, 1948-49, four consecutive shutouts.

MOST WINS BY A GOALTENDER, CAREER:
516 — Patrick Roy, Montreal, Colorado, in 18 seasons. 966GP
447 — Terry Sawchuk, Detroit, Boston, Toronto, Los Angeles, NY Rangers, in 21 seasons. 972GP
435 — Jacques Plante, Montreal, NY Rangers, St. Louis, Toronto, Boston, in 18 seasons. 837GP
423 — Tony Esposito, Montreal, Chicago, in 16 seasons. 886GP

MOST WINS BY A GOALTENDER, ONE SEASON:
47 — Bernie Parent, Philadelphia, 1973-74. 73GP
44 — Bernie Parent, Philadelphia, 1974-75. 68GP
— Terry Sawchuk, Detroit, 1950-51. 70GP
— Terry Sawchuk, Detroit, 1951-52. 70GP

LONGEST WINNING STREAK BY A GOALTENDER, ONE SEASON:
17 — Gilles Gilbert, Boston, 1975-76.
14 — Tiny Thompson, Boston, 1929-30.
— Ross Brooks, Boston, 1973-74.
— Don Beaupre, Minnesota, 1985-86.
— Tom Barrasso, Pittsburgh, 1992-93.

LONGEST UNDEFEATED STREAK BY A GOALTENDER, ONE SEASON:
32 Games — Gerry Cheevers, Boston, 1971-72. 24W-8T
31 Games — Pete Peeters, Boston, 1982-83. 26W-5T
27 Games — Pete Peeters, Philadelphia, 1979-80. 22W-5T
23 Games — Frank Brimsek, Boston, 1940-41. 15W-8T
— Chico Resch, NY Islanders, 1978-79. 15W-8T
— Grant Fuhr, Edmonton, 1981-82. 15W-8T

LONGEST UNDEFEATED STREAK BY A GOALTENDER IN HIS FIRST NHL SEASON:
23 Games — Grant Fuhr, 1981-82. 15W-8T.

LONGEST UNDEFEATED STREAK BY A GOALTENDER FROM START OF CAREER:
16 Games — Patrick Lalime, Pittsburgh, 1996-97. 14W-2T.

MOST 40-OR-MORE WIN SEASONS BY A GOALTENDER:
3 — Terry Sawchuk, Detroit, Boston, Toronto, Los Angeles, NY Rangers, in 21 seasons.
— Jacques Plante, Montreal, NY Rangers, St. Louis, Toronto, Boston, in 18 seasons.
— Martin Brodeur, New Jersey, in 10 seasons.
2 — Bernie Parent, Boston, Philadelphia, Toronto, in 13 seasons.
— Ken Dryden, Montreal, in 8 seasons.
— Ed Belfour, Chicago, San Jose, Dallas, in 13 seasons.

MOST CONSECUTIVE 40-OR-MORE WIN SEASONS BY A GOALTENDER:
2 — Terry Sawchuk, Detroit, 1950-51 – 1951-52.
— Bernie Parent, Philadelphia, 1973-74 – 1974-75.
— Ken Dryden, Montreal, 1975-76 – 1976-77.
— Martin Brodeur, New Jersey, 1999-2000 – 2000-01.

MOST 30-OR-MORE WIN SEASONS BY A GOALTENDER:
12 — Patrick Roy, Montreal, Colorado, in 18 seasons.
8 — Tony Esposito, Montreal, Chicago, in 16 seasons.
7 — Jacques Plante, Montreal, NY Rangers, St. Louis, Toronto, Boston, in 18 seasons.
— Ken Dryden, Montreal, in 8 seasons.
— Ed Belfour, Chicago, San Jose, Dallas, in 13 seasons.
— Martin Brodeur, New Jersey, in 10 seasons

MOST CONSECUTIVE 30-OR-MORE WIN SEASONS BY A GOALTENDER:
7 — Tony Esposito, Chicago, 1969-70 – 1975-76.
— Patrick Roy, Montreal, Colorado, 1995-96 – 2001-02.
— Martin Brodeur, New Jersey, 1995-96 – 2001-02.
6 — Jacques Plante, Montreal, 1954-55 – 1959-60.
5 — Terry Sawchuk, Detroit, 1950-51 – 1954-55.
— Ken Dryden, Montreal, 1974-75 – 1978-79.

MOST LOSSES BY A GOALTENDER, CAREER:
352 — Gump Worsley, NY Rangers, Montreal, Minnesota, in 21 seasons. 861GP
351 — Gilles Meloche, Chicago, California, Cleveland, Minnesota, Pittsburgh, in 18 seasons. 788GP
346 — John Vanbiesbrouck, NY Rangers, Florida, Philadelphia, NY Islanders, New Jersey, in 20 seasons. 882GP
332 — Terry Sawchuk, Detroit, Boston, Toronto, Los Angeles, NY Rangers, in 21 seasons. 972GP

MOST LOSSES BY A GOALTENDER, ONE SEASON:
48 — Gary Smith, California, 1970-71.
47 — Al Rollins, Chicago, 1953-54.

Though still a long way from the all-time leaders, Felix Potvin topped the 500-game plateau early in the 2001-02 season. His 71 games played for the Kings last year was the most for Potvin since leading the NHL with 74 games for Toronto in 1996-97.

Active NHL Players' Three-or-More-Goal Games

Regular Season

Teams named are the ones the players were with at the time of their multiple-scoring games. Players listed alphabetically.

Player	Team	3-Goals	4-Goals	5-Goals
Alfredsson, Daniel	Ottawa	4	—	—
Allison, Jason	Boston	4	—	—
Amonte, Tony	NYR, Chi.	7	—	—
Andersson, Niklas	NY Islanders	1	—	—
Andreychuk, Dave	Buf., Tor., Bos.	7	3	1
Antropov, Nik	Toronto	1	—	—
Arnott, Jason	Edm., N.J.	3	—	—
Arvedson, Magnus	Ottawa	1	—	—
Audette, Donald	Buf., Atl.	4	—	—
Barnes, Stu	Wpg., Pit.	3	—	—
Battaglia, Bates	Carolina	1	—	—
Beranek, Josef	Philadelphia	1	—	—
Berezin, Sergei	Toronto	2	—	—
Bertuzzi, Todd	Vancouver	2	—	—
Blake, Rob	Los Angeles	1	—	—
Bondra, Peter	Washington	12	5	1
Bonk, Radek	Ottawa	1	—	—
Brind'Amour, Rod	Phi., Car.	2	—	—
Brown, Rob	Pittsburgh	7	—	—
Buchberger, Kelly	Edmonton	1	—	—
Bure, Pavel	Van., Fla.	16	3	—
Bure, Valeri	Calgary	1	—	—
Butsayev, Viacheslav	Philadelphia	1	—	—
Carter, Anson	Boston	1	—	—
Cassels, Andrew	Vancouver	1	—	—
Ciger, Zdeno	Tampa Bay	1	—	—
Cole, Erik	Carolina	1	—	—
Conroy, Craig	St. Louis	1	—	—
Corson, Shayne	Mtl., Edm.	3	—	—
Czerkawski, Mariusz	Edm., NYI	4	—	—
Dackell, Andreas	Ottawa	1	—	—
Dahlen, Ulf	NYR, Min., S.J.	4	—	—
Daigle, Alexandre	Ott., Phi.	2	—	—
Damphousse, Vincent	Tor., Edm., Mtl., S.J.	11	1	—
Dawe, Jason	Buffalo	2	—	—
Daze, Eric	Chicago	3	1	—
Deadmarsh, Adam	Colorado	1	—	—
Demitra, Pavol	St. Louis	2	—	—
Devereaux, Boyd	Edmonton	1	—	—
Dineen, Kevin	Hfd., Phi.	9	1	—
Donovan, Shean	Atlanta	1	—	—
Dopita, Jiri	Philadelphia	—	1	—
Druken, Harold	Vancouver	1	—	—
Duchesne, Steve	L.A., Phi., St.L.	3	—	—
Dumont, Jean-Pierre	Chi., Buf.	3	—	—
Dvorak, Radek	NY Rangers	1	1	—
Eastwood, Mike	St. Louis	1	—	—
Elias, Patrik	New Jersey	5	—	—
Emerson, Nelson	Winnipeg	1	—	—
Fedorov, Sergei	Detroit	2	1	1
Fleury, Theoren	Cgy., Col., NYR	15	—	—
Forsberg, Peter	Colorado	4	—	—
Francis, Ron	Hfd., Pit.	10	1	—
Friesen, Jeff	San Jose	2	—	—
Gaborik, Marian	Minnesota	2	—	—
Gagne, Simon	Philadelphia	1	—	—
Garpenlov, Johan	Det., S.J., Fla.	2	1	—
Gelinas, Martin	Edm., Van.	2	1	—
Gilchrist, Brent	Montreal	1	—	—
Gilmour, Doug	St.L., Tor.	3	—	—
Gomez, Scott	New Jersey	1	—	—
Gonchar, Sergei	Washington	1	—	—
Gratton, Chris	Tampa Bay	1	—	—
Graves, Adam	Edm., NYR	6	—	—
Green, Travis	NY Islanders	1	—	—
Grier, Mike	Edmonton	1	—	—
Grosek, Michal	Buffalo	1	—	—
Guerin, Bill	N.J., Bos.	3	—	—
Handzus, Michal	St. Louis	1	—	—
Harvey, Todd	Dal., S.J.	2	—	—
Havlat, Martin	Ottawa	2	—	—
Heinze, Steve	Bos., Buf.	5	—	—
Hlavac, Jan	NY Rangers	2	—	—
Hoglund, Jonas	Toronto	1	—	—
Hogue, Benoit	NY Islanders	1	—	—
Holik, Bobby	New Jersey	3	—	—
Holmstrom, Tomas	Detroit	1	—	—
Hossa, Marian	Ottawa	1	—	—
Housley, Phil	Buffalo	2	—	—
Hull, Brett	Cgy., St.L., Dal., Det.	28	3	—
Hull, Jody	Hartford	1	—	—
Iginla, Jarome	Calgary	1	—	—
Jagr, Jaromir	Pittsburgh	8	1	—
Juneau, Joe	Bos., Wsh.	2	—	—
Kallio, Tomi	Atlanta	1	—	—

Vancouver's Markus Naslund was second in the NHL with 90 points last season and fifth in goals with 40. He and Alexei Kovalev of the Penguins each had a league-leading three hat tricks.

Player	Team	3-Goals	4-Goals	5-Goals
Kamensky, Valeri	Colorado	5	—	—
Kapanen, Sami	Carolina	3	—	—
Kariya, Paul	Anaheim	7	—	—
Khristich, Dimitri	Wsh., Bos.	3	—	—
King, Derek	NYI, Tor.	6	1	—
Klatt, Trent	Philadelphia	1	—	—
Knutsen, Espen	Columbus	1	—	—
Konowalchuk, Steve	Washington	3	—	—
Korolev, Igor	Winnipeg	1	—	—
Kovalenko, Andrei	Que., Bos.	2	—	—
Kovalev, Alexei	NYR, Pit.	9	—	—
Kozlov, Viktor	Florida	1	—	—
Kozlov, Vyacheslav	Detroit	2	1	—
Laaksonen, Antti	Minnesota	1	—	—
Langkow, Daymond	Phoenix	1	—	—
Laperriere, Ian	Los Angeles	1	—	—
Lapointe, Martin	Detroit	1	—	—
Laraque, Georges	Edmonton	1	—	—
Larionov, Igor	Van., S.J.	4	—	—
LeClair, John	Philadelphia	8	2	—
Lehtinen, Jere	Dallas	1	—	—
Lemieux, Claude	Mtl., N.J., Col.	7	—	—
Lemieux, Mario	Pittsburgh	27	10	3
Linden, Trevor	Van., Mtl.	5	—	—
Lindros, Eric	Phi., NYR	12	1	—
MacInnis, Al	Cgy., St.L.	3	—	—
Madden, John	New Jersey	—	1	—
Malakhov, Vladimir	Montreal	1	—	—
Maltby, Kirk	Detroit	1	—	—
Manderville, Kent	Hartford	1	—	—
Marleau, Patrick	San Jose	1	—	—
McEachern, Shawn	Ottawa	1	—	—
McInnis, Marty	Cgy., Ana.	2	—	—
McKay, Randy	New Jersey	1	1	—
McKenzie, Jim	Phoenix	1	—	—
Messier, Mark	Edm., NYR	15	4	—
Miller, Kevin	Det., St.L., S.J.	4	—	—
Modano, Mike	Min., Dal.	6	1	—
Modin, Fredrik	Tampa Bay	1	—	—
Mogilny, Alexander	Buf., Van., N.J.	13	2	—
Morozov, Alexei	Pittsburgh	2	—	—
Muller, Kirk	N.J., Mtl., Tor.	7	—	—
Murray, Glen	L.A., Bos.	3	—	—
Murray, Rem	Edmonton	1	—	—
Naslund, Markus	Pit., Van.	7	—	—
Nedved, Petr	Pit., NYR	5	1	—
Nemchinov, Sergei	NY Rangers	1	—	—
Nieuwendyk, Joe	Cgy., Dal.	9	3	1
Nolan, Owen	Que., S.J.	9	1	—
Nylander, Michael	Hfd., Chi.	1	1	—
Oates, Adam	Bos., Wsh.	6	1	—
Odelein, Lyle	Montreal	1	—	—
Oliver, David	Edmonton	1	—	—
O'Neill, Jeff	Hartford	1	—	—
Ozolinsh, Sandis	Col., Car.	2	—	—
Palffy, Ziggy	NYI, L.A.	8	—	—
Parrish, Mark	Fla., NYI	2	1	—
Peca, Michael	Buffalo	1	—	—
Perreault, Yanic	L.A., Tor., Mtl.	3	1	—
Petersen, Toby	Pittsburgh	1	—	—
Petrov, Oleg	Montreal	1	—	—
Plante, Derek	Buffalo	1	—	—
Podein, Shjon	Colorado	1	—	—
Primeau, Keith	Philadelphia	1	—	—
Probert, Bob	Detroit	1	—	—

Player	Team	3-Goals	4-Goals	5-Goals
Quint, Deron	Columbus	1	—	—
Ranheim, Paul	Calgary	1	—	—
Recchi, Mark	Pit., Mtl., Phi.	4	—	—
Reichel, Robert	Cgy., NYI	5	—	—
Reinprecht, Steve	Colorado	1	—	—
Renberg, Mikael	Phi., T.B.	2	—	—
Rheaume, Pascal	Atlanta	—	1	—
Ricci, Mike	Que., S.J.	1	—	1
Roberts, Gary	Cgy., Car., Tor.	12	1	—
Robitaille, Luc	L.A., Pit.	11	3	—
Roenick, Jeremy	Chi., Phx.	7	2	—
Rolston, Brian	New Jersey	1	—	—
Ronning, Cliff	St.L., Van.	3	—	—
Rucinsky, Martin	Montreal	2	—	—
Sakic, Joe	Que., Col.	10	1	—
Salo, Sami	Ottawa	1	—	—
Samsonov, Sergei	Boston	1	—	—
Sanderson, Geoff	Har., Buf., CBJ	7	—	—
Satan, Miroslav	Buffalo	3	—	—
Savage, Brian	Montreal	6	1	—
Savard, Marc	Calgary	1	1	—
Selanne, Teemu	Wpg., Ana., S.J.	16	2	—
Selivanov, Alexander	Edmonton	2	1	—
Shanahan, Brendan	N.J., St.L., Hfd., Det.	13	1	—
Smolinski, Bryan	Bos., L.A.	2	—	—
Smyth, Ryan	Edmonton	4	—	—
Stillman, Cory	Cgy., St.L.	3	—	—
Straka, Martin	Pittsburgh	4	—	—
Stumpel, Jozef	Bos., L.A.	2	—	—
Sturm, Marco	San Jose	1	—	—
Sullivan, Steve	Tor., Chi.	1	1	—
Sundin, Mats	Que., Tor.	5	—	1
Sydor, Darryl	Dallas	1	—	—
Sylvester, Dean	Atlanta	1	—	—
Tenkrat, Petr	Nashville	1	—	—
Thomas, Steve	Chi., NYI	4	2	—
Thornton, Joe	Boston	2	—	—
Thornton, Scott	San Jose	1	—	—
Titov, German	Calgary	2	—	—
Tkachuk, Keith	Phoenix	7	2	—
Tocchet, Rick	Phi., Pit., L.A., Bos.	12	2	—
Toms, Jeff	NY Rangers	1	—	—
Turgeon, Pierre	Buf., NYI, Mtl., St.L.	15	—	—
Valicevic, Robert	Nashville	1	—	—
Valk, Garry	Anaheim	1	—	—
Verbeek, Pat	N.J., Hfd., NYR, Dal.	11	1	—
Vrbata, Radim	Colorado	1	—	—
Walker, Scott	Nashville	1	—	—
Ward, Dixon	Buffalo	1	—	—
Weight, Doug	Edmonton	1	—	—
Wesley, Glen	Boston	1	—	—
Wiemer, Jason	Tampa Bay	1	—	—
Willis, Shane	Carolina	1	—	—
Wright, Tyler	Columbus	1	—	—
Yachmenev, Vitali	Los Angeles	1	—	—
Yake, Terry	Anaheim	1	—	—
Yashin, Alexei	Ott., NYI	7	—	—
Young, Scott	Que., Col.	4	—	—
Yzerman, Steve	Detroit	17	1	—
Zamuner, Rob	Tampa Bay	1	—	—
Zednik, Richard	Washington	1	—	—
Zhamnov, Alexei	Wpg., Chi.	5	—	1
Zubrus, Dainus	Montreal	1	—	—

Top 100 All-Time Goal-Scoring Leaders

* active player

	Player	Seasons	Games	Goals	Goals per game
1.	**Wayne Gretzky**, Edm., L.A., St.L., NYR	20	1487	**894**	.601
2.	**Gordie Howe**, Det., Hfd.	26	1767	**801**	.453
3.	**Marcel Dionne**, Det., L.A., NYR	18	1348	**731**	.542
4.	**Phil Esposito**, Chi., Bos., NYR	18	1282	**717**	.559
5.	**Mike Gartner**, Wsh., Min., NYR, Tor., Phx.	19	1432	**708**	.494
* 6.	**Brett Hull**, Cgy., St.L., Dal., Det.	17	1101	**679**	.617
* 7.	**Steve Yzerman**, Det.	19	1362	**658**	.483
* 8.	**Mark Messier**, Edm., NYR, Van.	23	1602	**658**	.411
* 9.	**Mario Lemieux**, Pit.	15	812	**654**	.805
* 10.	**Luc Robitaille**, L.A., Pit., NYR, Det.	16	1205	**620**	.515
11.	**Bobby Hull**, Chi., Wpg., Hfd.	16	1063	**610**	.574
12.	**Dino Ciccarelli**, Min., Wsh., Det., T.B., Fla.	19	1232	**608**	.494
13.	**Jari Kurri**, Edm., L.A., NYR, Ana., Col.	17	1251	**601**	.480
* 14.	**Dave Andreychuk**, Buf., Tor., N.J., Bos., Col., T.B.	20	1443	**593**	.411
15.	**Mike Bossy**, NYI	10	752	**573**	.762
16.	**Guy Lafleur**, Mtl., NYR, Que.	17	1126	**560**	.497
17.	**John Bucyk**, Det., Bos.	23	1540	**556**	.361
18.	**Michel Goulet**, Que., Chi.	15	1089	**548**	.503
19.	**Maurice Richard**, Mtl.	18	978	**544**	.556
20.	**Stan Mikita**, Chi.	22	1394	**541**	.388
21.	**Frank Mahovlich**, Tor., Det., Mtl.	18	1181	**533**	.451
22.	**Bryan Trottier**, NYI, Pit.	18	1279	**524**	.410
* 23.	**Pat Verbeek**, N.J., Hfd., NYR, Dal., Det.	20	1424	**522**	.367
24.	**Dale Hawerchuk**, Wpg., Buf., St.L., Phi.	16	1188	**518**	.436
* 25.	**Ron Francis**, Hfd., Pit., Car.	21	1569	**514**	.328
26.	**Gilbert Perreault**, Buf.	17	1191	**512**	.430
27.	**Jean Beliveau**, Mtl.	20	1125	**507**	.451
* 28.	**Brendan Shanahan**, N.J., St.L., Hfd., Det.	15	1108	**503**	.454
29.	**Joe Mullen**, St.L., Cgy., Pit., Bos.	17	1062	**502**	.473
30.	**Lanny McDonald**, Tor., Col., Cgy.	16	1111	**500**	.450
31.	**Glenn Anderson**, Edm., Tor., NYR, St.L.	16	1129	**498**	.441
* 32.	**Joe Nieuwendyk**, Cgy., Dal., N.J.	16	1033	**494**	.478
33.	**Jean Ratelle**, NYR, Bos.	21	1281	**491**	.383
34.	**Norm Ullman**, Det., Tor.	20	1410	**490**	.348
35.	**Brian Bellows**, Min., Mtl., T.B., Ana., Wsh.	17	1188	**485**	.408
36.	**Darryl Sittler**, Tor., Phi., Det.	15	1096	**484**	.442
* 37.	**Joe Sakic**, Que., Col.	14	1016	**483**	.475
38.	**Bernie Nicholls**, L.A., NYR, Edm., N.J., Chi., S.J.	18	1127	**475**	.421
39.	**Denis Savard**, Chi., Mtl., T.B.	17	1196	**473**	.395
* 40.	**Jaromir Jagr**, Pit., Wsh.	12	875	**470**	.537
41.	**Pat LaFontaine**, NYI, Buf., NYR	15	865	**468**	.541
* 42.	**Pierre Turgeon**, Buf., NYI, Mtl., St.L., Dal.	15	1074	**468**	.436
43.	**Alex Delvecchio**, Det.	24	1549	**456**	.294
44.	**Peter Stastny**, Que., N.J., St.L.	15	977	**450**	.461
45.	**Rick Middleton**, NYR, Bos.	14	1005	**448**	.446
* 46.	**Theoren Fleury**, Cgy., Col., NYR	14	1030	**443**	.430
47.	**Steve Larmer**, Chi., NYR	15	1006	**441**	.438
48.	**Rick Vaive**, Van., Tor., Chi., Buf.	13	876	**441**	.503
* 49.	**Rick Tocchet**, Phi., Pit., L.A., Bos., Wsh., Phx.	18	1144	**440**	.385
* 50.	**Doug Gilmour**, St.L., Cgy., Tor., N.J., Chi., Buf., Mtl.	19	1412	**439**	.311
51.	**Dave Taylor**, L.A.	17	1111	**431**	.388
* 52.	**Jeremy Roenick**, Chi., Phx., Phi.	14	983	**429**	.436
53.	**Yvan Cournoyer**, Mtl.	16	968	**428**	.442
54.	**Brian Propp**, Phi., Bos., Min., Hfd.	15	1016	**425**	.418
55.	**Steve Shutt**, Mtl., L.A.	13	930	**424**	.456
* 56.	**Peter Bondra**, Wsh.	12	831	**421**	.507
* 57.	**Stephane Richer**, Mtl., N.J., T.B., St.L., Pit.	17	1054	**421**	.399
58.	**Bill Barber**, Phi.	12	903	**420**	.465
* 59.	**Alexander Mogilny**, Buf., Van., N.J., Tor.	13	846	**420**	.496
* 60.	**Pavel Bure**, Van., Fla., NYR	11	663	**418**	.630
* 61.	**Mike Modano**, Min., Dal.	14	946	**416**	.440
62.	**John MacLean**, N.J., S.J., NYR, Dal.	18	1194	**413**	.346
63.	**Garry Unger**, Tor., Det., St.L., Atl., L.A., Edm.	16	1105	**413**	.374
64.	**Raymond Bourque**, Bos., Col.	22	1612	**410**	.254
* 65.	**Mark Recchi**, Pit., Phi., Mtl.	14	1012	**410**	.405
66.	**Ray Ferraro**, Hfd., NYI, NYR, L.A., Atl., St.L.	18	1258	**408**	.324
* 67.	**Teemu Selanne**, Wpg., Ana., S.J.	10	719	**408**	.567
68.	**Rod Gilbert**, NYR	18	1065	**406**	.381
69.	**John Ogrodnick**, Det., Que., NYR	14	928	**402**	.433
* 70.	**Vincent Damphousse**, Tor., Edm., Mtl., S.J.	16	1214	**397**	.327
* 71.	**Steve Thomas**, Tor., Chi., NYI, N.J.	18	1110	**397**	.358
* 72.	**Mats Sundin**, Que., Tor.	12	930	**397**	.427
73.	**Paul Coffey**, Edm., Pit., L.A., Det., Hfd., Phi., Chi., Car., Bos.	21	1409	**396**	.281
74.	**Dave Keon**, Tor., Hfd.	18	1296	**396**	.306
75.	**Pierre Larouche**, Pit., Mtl., Hfd., NYR	14	812	**395**	.486
76.	**Cam Neely**, Van., Bos.	13	726	**395**	.544
77.	**Tomas Sandstrom**, NYR, L.A., Pit., Det., Ana.	15	983	**394**	.401
78.	**Bernie Geoffrion**, Mtl., NYR	16	883	**393**	.445
79.	**Dean Prentice**, NYR, Bos., Det., Pit., Min.	22	1378	**391**	.284
80.	**Jean Pronovost**, Pit., Atl., Wsh.	14	998	**391**	.392
81.	**Rick Martin**, Buf., L.A.	11	685	**384**	.561
82.	**Reggie Leach**, Bos., Cal., Phi., Det.	13	934	**381**	.408
83.	**Ted Lindsay**, Det., Chi.	17	1068	**379**	.355
84.	**Butch Goring**, L.A., NYI, Bos.	16	1107	**375**	.339
* 85.	**Claude Lemieux**, Mtl., N.J., Col., Phx.	19	1129	**371**	.329
86.	**Rick Kehoe**, Tor., Pit.	14	906	**371**	.409
87.	**Tim Kerr**, Phi., NYR, Hfd.	13	655	**370**	.565
88.	**Bernie Federko**, St.L., Det.	14	1000	**369**	.369
* 89.	**Keith Tkachuk**, Wpg., Phx., St.L.	11	725	**367**	.506
90.	**Geoff Courtnall**, Bos., Edm., Wsh., St.L., Van.	17	1048	**367**	.350
91.	**Jacques Lemaire**, Mtl.	12	853	**366**	.429
* 92.	**Gary Roberts**, Cgy., Car., Tor.	15	943	**364**	.386
* 93.	**Sergei Fedorov**, Det.	12	828	**364**	.440
94.	**Peter McNab**, Buf., Bos., Van., N.J.	14	954	**363**	.381
95.	**Brent Sutter**, NYI, Chi.	18	1111	**363**	.327
96.	**Ivan Boldirev**, Bos., Cal., Chi., Atl., Van., Det.	15	1052	**361**	.343
97.	**Bobby Clarke**, Phi.	15	1144	**358**	.313
98.	**Henri Richard**, Mtl.	20	1256	**358**	.285
99.	**Bobby Smith**, Min., Mtl.	15	1077	**357**	.331
100.	**Ray Sheppard**, Buf., NYR, Det., S.J., Fla., Car.	13	817	**357**	.437

With his 34th goal of the season on March 23, 2002, Brendan Shanahan became the 30th member of the NHL's 500-goal club. He finished the season with 37 goals, topping the Red Wings and ranking among the league leaders.

Top 100 Active Goal-Scoring Leaders

	Player	Seasons	Games	Goals	Goals per game
1.	**Brett Hull**, Cgy., St.L., Dal., Det.	17	1101	**679**	.617
2.	**Mark Messier**, Edm., NYR, Van.	23	1602	**658**	.411
3.	**Steve Yzerman**, Det.	19	1362	**658**	.483
4.	**Mario Lemieux**, Pit.	15	812	**654**	.805
5.	**Luc Robitaille**, L.A., Pit., NYR, Det.	16	1205	**620**	.515
6.	**Dave Andreychuk**, Buf., Tor., N.J., Bos., Col., T.B.	20	1443	**593**	.411
7.	**Pat Verbeek**, N.J., Hfd., NYR, Dal., Det.	20	1424	**522**	.367
8.	**Ron Francis**, Hfd., Pit., Car.	21	1569	**514**	.328
9.	**Brendan Shanahan**, N.J., St.L., Hfd., Det.	15	1108	**503**	.454
10.	**Joe Nieuwendyk**, Cgy., Dal., N.J.	16	1033	**494**	.478
11.	**Joe Sakic**, Que., Col.	14	1016	**483**	.475
12.	**Jaromir Jagr**, Pit., Wsh.	12	875	**470**	.537
13.	**Pierre Turgeon**, Buf., NYI, Mtl., St.L., Dal.	15	1074	**468**	.436
14.	**Theoren Fleury**, Cgy., Col., NYR	14	1030	**443**	.430
15.	**Rick Tocchet**, Phi., Pit., L.A., Bos., Wsh., Phx.	18	1144	**440**	.385
16.	**Doug Gilmour**, St.L., Cgy., Tor., N.J., Chi., Buf., Mtl.	19	1412	**439**	.311
17.	**Jeremy Roenick**, Chi., Phx., Phi.	14	983	**429**	.436
18.	**Peter Bondra**, Wsh.	12	831	**421**	.507
19.	**Stephane Richer**, Mtl., N.J., T.B., St.L., Pit.	17	1054	**421**	.399
20.	**Alexander Mogilny**, Buf., Van., N.J., Tor.	13	846	**420**	.496
21.	**Pavel Bure**, Van., Fla., NYR	11	663	**418**	.630
22.	**Mike Modano**, Min., Dal.	14	946	**416**	.440
23.	**Mark Recchi**, Pit., Phi., Mtl.	14	1012	**410**	.405
24.	**Teemu Selanne**, Wpg., Ana., S.J.	10	719	**408**	.567
25.	**Vincent Damphousse**, Tor., Edm., Mtl., S.J.	16	1214	**397**	.327
26.	**Steve Thomas**, Tor., Chi., NYI, N.J.	18	1110	**397**	.358
27.	**Mats Sundin**, Que., Tor.	12	930	**397**	.427
28.	**Claude Lemieux**, Mtl., N.J., Col., Phx.	19	1129	**371**	.329
29.	**Keith Tkachuk**, Wpg., Phx., St.L.	11	725	**367**	.506
30.	**Sergei Fedorov**, Det.	12	828	**364**	.440
31.	**Gary Roberts**, Cgy., Car., Tor.	15	943	**364**	.386
32.	**Kirk Muller**, N.J., Mtl., NYI, Tor., Fla., Dal.	18	1294	**356**	.275
33.	**Kevin Dineen**, Hfd., Phi., Car., Ott., CBJ	18	1184	**355**	.300
34.	**Tony Amonte**, NYR, Chi.	12	861	**352**	.409
35.	**John LeClair**, Mtl., Phi.	12	763	**341**	.447
36.	**Phil Housley**, Buf., Wpg., St.L., Cgy., N.J., Wsh., Chi.	20	1437	**332**	.231
37.	**Adam Oates**, Det., St.L., Bos., Wsh., Phi.	17	1210	**330**	.273
38.	**Kevin Stevens**, Pit., Bos., L.A., NYR, Phi.	15	874	**329**	.376
39.	**Eric Lindros**, Phi., NYR	9	558	**327**	.586
40.	**Rod Brind'Amour**, St.L., Phi., Car.	14	983	**325**	.331
41.	**Al MacInnis**, Cgy., St.L.	21	1333	**324**	.243
42.	**Adam Graves**, Det., Edm., NYR, S.J.	15	1070	**320**	.299
43.	**Trevor Linden**, Van., NYI, Mtl., Wsh.	14	1008	**316**	.313
44.	**Owen Nolan**, Que., Col., S.J.	12	775	**301**	.388
45.	**Scott Mellanby**, Phi., Edm., Fla., St.L.	17	1143	**300**	.262
46.	**Scott Young**, Hfd., Pit., Que., Col., Ana., St.L.	14	970	**293**	.302
47.	**Ulf Dahlen**, NYR, Min., Dal., S.J., Chi., Wsh.	13	903	**284**	.315
48.	**Cliff Ronning**, St.L., Van., Phx., Nsh., L.A.	16	1017	**280**	.275
49.	**Paul Kariya**, Ana.	8	524	**275**	.525
50.	**Geoff Sanderson**, Hfd., Car., Van., Buf., CBJ	12	766	**266**	.347
51.	**Ziggy Palffy**, NYI, L.A.	9	531	**265**	.499
52.	**Derek King**, NYI, Hfd., Tor., St.L.	14	830	**261**	.314
53.	**Shayne Corson**, Mtl., Edm., St.L., Tor.	17	1093	**261**	.239
54.	**Dmitri Khristich**, Wsh., L.A., Bos., Tor.	12	811	**259**	.319
55.	**Bill Guerin**, N.J., Edm., Bos.	11	733	**256**	.349
56.	**Petr Nedved**, Van., St.L., NYR, Pit.	11	730	**255**	.349
57.	**Alexei Yashin**, Ott., NYI	8	582	**250**	.430
58.	**Alexei Kovalev**, NYR, Pit.	10	693	**241**	.348
59.	**Donald Audette**, Buf., L.A., Atl., Dal., Mtl.	13	630	**240**	.381
60.	**Bobby Holik**, Hfd., N.J.	12	878	**240**	.273
61.	**Keith Primeau**, Det., Hfd., Car., Phi.	12	766	**239**	.312
62.	**Martin Gelinas**, Edm., Que., Van., Car.	14	895	**231**	.258
63.	**Robert Reichel**, Cgy., NYI, Phx., Tor.	9	680	**229**	.337
64.	**Shawn McEachern**, Pit., L.A., Bos., Ott.	11	755	**227**	.301
65.	**Steve Duchesne**, L.A., Phi., Que., St.L., Ott., Det.	16	1113	**227**	.204
66.	**Alexei Zhamnov**, Wpg., Chi.	10	666	**222**	.333
67.	**Benoit Hogue**, Buf., NYI, Tor., Dal., T.B., Phx., Bos., Wsh.	15	863	**222**	.257
68.	**Mike Ricci**, Phi., Que., Col., S.J.	12	868	**215**	.248
69.	**Brian Leetch**, NYR	15	1021	**215**	.211
70.	**Vyacheslav Kozlov**, Det., Buf.	11	645	**211**	.327

Alexander Mogilny scored twice in his first game as a Maple Leaf on October 3, 2001. Five nights later, he scored two more goals to reach 400 for his career. He finished the season with 24 goals in just 66 games.

	Player	Seasons	Games	Goals	Goals per game
71.	**Markus Naslund**, Pit., Van.	9	630	**207**	.329
72.	**Ron Sutter**, Phi., St.L., Que., NYI, Bos., S.J., Cgy.	19	1093	**205**	.188
73.	**Miroslav Satan**, Edm., Buf.	7	543	**204**	.376
74.	**Gary Suter**, Cgy., Chi., S.J.	17	1145	**203**	.177
75.	**Eric Daze**, Chi.	8	527	**200**	.380
76.	**Valeri Kamensky**, Que., Col., NYR, Dal., N.J.	11	637	**200**	.314
77.	**Jason Arnott**, Edm., N.J., Dal.	9	598	**200**	.334
78.	**Stu Barnes**, Wpg., Fla., Pit., Buf.	11	739	**197**	.267
79.	**Nelson Emerson**, St.L., Wpg., Hfd., Car., Chi., Ott., Atl., L.A.	12	771	**195**	.253
80.	**Doug Weight**, NYR, Edm., St.L.	12	767	**195**	.254
81.	**Glen Murray**, Bos., Pit., L.A.	11	660	**192**	.291
82.	**Bryan Smolinski**, Bos., Pit., NYI, L.A.	10	681	**191**	.280
83.	**Rob Brown**, Pit., Hfd., Chi., Dal., L.A.	11	543	**190**	.350
84.	**Scott Stevens**, Wsh., St.L., N.J.	20	1516	**189**	.125
85.	**Martin Rucinsky**, Edm., Que., Col., Mtl., Dal., NYR	11	674	**178**	.264
86.	**Mariusz Czerkawski**, Bos., Edm., NYI	9	586	**177**	.302
87.	**Andrew Cassels**, Mtl., Hfd., Cgy., Van.	13	847	**174**	.205
88.	**Chris Chelios**, Mtl., Chi., Det.	19	1260	**174**	.138
89.	**Jarome Iginla**, Cgy.	7	470	**174**	.370
90.	**Steve Heinze**, Bos., CBJ, Buf., L.A.	11	667	**173**	.259
91.	**Andrei Kovalenko**, Que., Col., Mtl., Edm., Phi., Car., Bos.	9	620	**173**	.279
92.	**Adam Deadmarsh**, Que., Col., L.A.	8	547	**171**	.313
93.	**Peter Forsberg**, Que., Col.	8	466	**169**	.363
94.	**Jeff Friesen**, S.J., Ana.	8	608	**168**	.276
95.	**Ray Whitney**, S.J., Edm., Fla., CBJ	11	552	**167**	.303
96.	**Martin Straka**, Pit., Ott., NYI, Fla.	10	616	**164**	.266
97.	**Mikael Renberg**, Phi., T.B., Phx., Tor.	8	535	**164**	.307
98.	**Bob Probert**, Det., Chi.	16	935	**163**	.174
99.	**Marty McInnis**, NYI, Cgy., Ana., Bos.	11	719	**161**	.224
100.	**Brian Savage**, Mtl., Phx.	9	491	**161**	.328

Top 100 All-Time Assist Leaders

* active player

	Player	Seasons	Games	Assists	Assists per game
1.	**Wayne Gretzky**, Edm., L.A., St.L., NYR	20	1487	**1963**	1.320
* 2.	**Ron Francis**, Hfd., Pit., Car.	21	1569	**1187**	.757
3.	**Raymond Bourque**, Bos., Col.	22	1612	**1169**	.725
* 4.	**Mark Messier**, Edm., NYR, Van.	23	1602	**1146**	.715
5.	**Paul Coffey**, Edm., Pit., L.A., Det., Hfd., Phi., Chi., Car., Bos.	21	1409	**1135**	.806
6.	**Gordie Howe**, Det., Hfd.	26	1767	**1049**	.594
7.	**Marcel Dionne**, Det., L.A., NYR	18	1348	**1040**	.772
* 8.	**Adam Oates**, Det., St.L., Bos., Wsh., Phi.	17	1210	**1027**	.849
* 9.	**Steve Yzerman**, Det.	19	1362	**1004**	.737
* 10.	**Mario Lemieux**, Pit.	15	812	**947**	1.166
* 11.	**Doug Gilmour**, St.L., Cgy., Tor., N.J., Chi., Buf., Mtl.	19	1412	**945**	.669
12.	**Larry Murphy**, L.A., Wsh., Min., Pit., Tor., Det.	21	1615	**929**	.575
13.	**Stan Mikita**, Chi.	22	1394	**926**	.664
14.	**Bryan Trottier**, NYI, Pit.	18	1279	**901**	.704
15.	**Dale Hawerchuk**, Wpg., Buf., St.L., Phi.	16	1188	**891**	.750
* 16.	**Al MacInnis**, Cgy., St.L.	21	1333	**880**	.660
17.	**Phil Esposito**, Chi., Bos., NYR	18	1282	**873**	.681
* 18.	**Phil Housley**, Buf., Wpg., St.L., Cgy., N.J., Wsh., Chi.	20	1437	**871**	.606
19.	**Denis Savard**, Chi., Mtl., T.B.	17	1196	**865**	.723
20.	**Bobby Clarke**, Phi.	15	1144	**852**	.745
21.	**Alex Delvecchio**, Det.	24	1549	**825**	.533
22.	**Gilbert Perreault**, Buf.	17	1191	**814**	.683
23.	**John Bucyk**, Det., Bos.	23	1540	**813**	.528
24.	**Jari Kurri**, Edm., L.A., NYR, Ana., Col.	17	1251	**797**	.637
25.	**Guy Lafleur**, Mtl., NYR, Que.	17	1126	**793**	.704
26.	**Peter Stastny**, Que., N.J., St.L.	15	977	**789**	.808
27.	**Jean Ratelle**, NYR, Bos.	21	1281	**776**	.606
* 28.	**Joe Sakic**, Que., Col.	14	1016	**774**	.762
29.	**Bernie Federko**, St.L., Det.	14	1000	**761**	.761
30.	**Larry Robinson**, Mtl., L.A.	20	1384	**750**	.542
31.	**Denis Potvin**, NYI	15	1060	**742**	.700
32.	**Norm Ullman**, Det., Tor.	20	1410	**739**	.524
33.	**Bernie Nicholls**, L.A., NYR, Edm., N.J., Chi., S.J.	18	1127	**734**	.651
* 34.	**Pierre Turgeon**, Buf., NYI, Mtl., St.L., Dal.	15	1074	**724**	.674
35.	**Jean Beliveau**, Mtl.	20	1125	**712**	.633
* 36.	**Vincent Damphousse**, Tor., Edm., Mtl., S.J.	16	1214	**706**	.582
* 37.	**Chris Chelios**, Mtl., Chi., Det.	19	1260	**700**	.556
* 38.	**Brian Leetch**, NYR	15	1021	**700**	.686
39.	**Dale Hunter**, Que., Wsh., Col.	19	1407	**697**	.495
* 40.	**Jaromir Jagr**, Pit., Wsh.	12	875	**688**	.786
41.	**Henri Richard**, Mtl.	20	1256	**688**	.548
* 42.	**Scott Stevens**, Wsh., St.L., N.J.	20	1516	**687**	.453
43.	**Brad Park**, NYR, Bos., Det.	17	1113	**683**	.614
44.	**Bobby Smith**, Min., Mtl.	15	1077	**679**	.630
* 45.	**Luc Robitaille**, L.A., Pit., NYR, Det.	16	1205	**668**	.554
* 46.	**Mark Recchi**, Pit., Phi., Mtl.	14	1012	**664**	.656
* 47.	**Dave Andreychuk**, Buf., Tor., N.J., Bos., Col., T.B.	20	1443	**654**	.453
48.	**Bobby Orr**, Bos., Chi.	12	657	**645**	.982
* 49.	**Gary Suter**, Cgy., Chi., S.J.	17	1145	**641**	.560
50.	**Dave Taylor**, L.A.	17	1111	**638**	.574
51.	**Darryl Sittler**, Tor., Phi., Det.	15	1096	**637**	.581
52.	**Borje Salming**, Tor., Det.	17	1148	**637**	.555
53.	**Neal Broten**, Min., Dal., N.J., L.A.	17	1099	**634**	.577
54.	**Mike Gartner**, Wsh., Min., NYR, Tor., Phx.	19	1432	**627**	.438
55.	**Andy Bathgate**, NYR, Tor., Det., Pit.	17	1069	**624**	.584
56.	**Rod Gilbert**, NYR	18	1065	**615**	.577
* 57.	**Theoren Fleury**, Cgy., Col., NYR	14	1030	**612**	.594
58.	**Michel Goulet**, Que., Chi.	15	1089	**604**	.555
59.	**Glenn Anderson**, Edm., Tor., NYR, St.L.	16	1129	**601**	.532
* 60.	**Kirk Muller**, N.J., Mtl., NYI, Tor., Fla., Dal.	18	1294	**597**	.461
61.	**Dino Ciccarelli**, Min., Wsh., Det., T.B., Fla.	19	1232	**592**	.481
62.	**Dave Keon**, Tor., Hfd.	18	1296	**590**	.455
63.	**Doug Wilson**, Chi., S.J.	16	1024	**590**	.576
* 64.	**Jeremy Roenick**, Chi., Phx., Phi.	14	983	**585**	.595
65.	**Dave Babych**, Wpg., Hfd., Van., Phi., L.A.	19	1195	**581**	.486
66.	**Brian Propp**, Phi., Bos., Min., Hfd.	15	1016	**579**	.570
67.	**Steve Larmer**, Chi., NYR	15	1006	**571**	.568
68.	**Frank Mahovlich**, Tor., Det., Mtl.	18	1181	**570**	.483
* 69.	**Brett Hull**, Cgy., St.L., Dal., Det.	17	1101	**567**	.515
70.	**Craig Janney**, Bos., St.L., S.J., Wpg., Phx., T.B., NYI	12	760	**563**	.741
71.	**Joe Mullen**, St.L., Cgy., Pit., Bos.	17	1062	**561**	.528

Many people thought Henri Richard was too small to play in the NHL. He would last for 20 seasons and win the Stanley Cup a record 11 times. He is the Canadiens' all-time leader with 1,256 games played and ranks third with 688 assists.

	Player	Seasons	Games	Assists	Assist per game
* 72.	**Mike Modano**, Min., Dal.	14	946	**561**	.593
73.	**Bobby Hull**, Chi., Wpg., Hfd.	16	1063	**560**	.527
74.	**Thomas Steen**, Wpg.	14	950	**553**	.582
75.	**Mike Bossy**, NYI	10	752	**553**	.735
76.	**Ken Linseman**, Phi., Edm., Bos., Tor.	14	860	**551**	.641
77.	**Tom Lysiak**, Atl., Chi.	13	919	**551**	.600
78.	**Mark Howe**, Hfd., Phi., Det.	16	929	**545**	.587
* 79.	**Mats Sundin**, Que., Tor.	12	930	**545**	.586
80.	**Pat LaFontaine**, NYI, Buf., NYR	15	865	**545**	.630
81.	**Red Kelly**, Det., Tor.	20	1316	**542**	.412
* 82.	**Pat Verbeek**, N.J., Hfd., NYR, Dal., Det.	20	1424	**541**	.380
83.	**Rick Middleton**, NYR, Bos.	14	1005	**540**	.537
84.	**Brian Bellows**, Min., Mtl., T.B., Ana., Wsh.	17	1188	**537**	.452
* 85.	**Brendan Shanahan**, N.J., St.L., Hfd., Det.	15	1108	**527**	.476
* 86.	**Steve Duchesne**, L.A., Phi., Que., St.L., Ott., Det.	16	1113	**525**	.472
87.	**Dennis Maruk**, Cal., Cle., Min., Wsh.	14	888	**522**	.588
* 88.	**Cliff Ronning**, St.L., Van., Phx., Nsh., L.A.	16	1017	**517**	.508
89.	**Wayne Cashman**, Bos.	17	1027	**516**	.502
90.	**Butch Goring**, L.A., NYI, Bos.	16	1107	**513**	.463
* 91.	**Rick Tocchet**, Phi., Pit., L.A., Bos., Wsh., Phx.	18	1144	**512**	.448
* 92.	**Rod Brind'Amour**, St.L., Phi., Car.	14	983	**511**	.520
93.	**John Tonelli**, NYI, Cgy., L.A., Chi., Que.	14	1028	**511**	.497
* 94.	**Sergei Fedorov**, Det.	12	828	**507**	.612
95.	**Lanny McDonald**, Tor., Col., Cgy.	16	1111	**506**	.455
96.	**Ivan Boldirev**, Bos., Cal., Chi., Atl., Van., Det.	15	1052	**505**	.480
* 97.	**Doug Weight**, NYR, Edm., St.L.	12	767	**501**	.653
98.	**Randy Carlyle**, Tor., Pit., Wpg.	17	1055	**499**	.473
99.	**Murray Craven**, Det., Phi., Hfd., Van., Chi., S.J.	18	1071	**493**	.460
100.	**Ray Ferraro**, Hfd., NYI, NYR, L.A., Atl., St.L.	18	1258	**490**	.390

Top 100 Active Assist Leaders

Player	Seasons	Games	Assists	Assists per game
1. **Ron Francis**, Hfd., Pit., Car.	21	1569	**1187**	.757
2. **Mark Messier**, Edm., NYR, Van.	23	1602	**1146**	.715
3. **Adam Oates**, Det., St.L., Bos., Wsh., Phi.	17	1210	**1027**	.849
4. **Steve Yzerman**, Det.	19	1362	**1004**	.737
5. **Mario Lemieux**, Pit.	15	812	**947**	1.166
6. **Doug Gilmour**, St.L., Cgy., Tor., N.J., Chi., Buf., Mtl.	19	1412	**945**	.669
7. **Al MacInnis**, Cgy., St.L.	21	1333	**880**	.660
8. **Phil Housley**, Buf., Wpg., St.L., Cgy., N.J., Wsh., Chi.	20	1437	**871**	.606
9. **Joe Sakic**, Que., Col.	14	1016	**774**	.762
10. **Pierre Turgeon**, Buf., NYI, Mtl., St.L., Dal.	15	1074	**724**	.674
11. **Vincent Damphousse**, Tor., Edm., Mtl., S.J.	16	1214	**706**	.582
12. **Chris Chelios**, Mtl., Chi., Det.	19	1260	**700**	.556
13. **Brian Leetch**, NYR	15	1021	**700**	.686
14. **Jaromir Jagr**, Pit., Wsh.	12	875	**688**	.786
15. **Scott Stevens**, Wsh., St.L., N.J.	20	1516	**687**	.453
16. **Luc Robitaille**, L.A., Pit., NYR, Det.	16	1205	**668**	.554
17. **Mark Recchi**, Pit., Phi., Mtl.	14	1012	**664**	.656
18. **Dave Andreychuk**, Buf., Tor., N.J., Bos., Col., T.B.	20	1443	**654**	.453
19. **Gary Suter**, Cgy., Chi., S.J.	17	1145	**641**	.560
20. **Theoren Fleury**, Cgy., Col., NYR	14	1030	**612**	.594
21. **Kirk Muller**, N.J., Mtl., NYI, Tor., Fla., Dal.	18	1294	**597**	.461
22. **Jeremy Roenick**, Chi., Phx., Phi.	14	983	**585**	.595
23. **Brett Hull**, Cgy., St.L., Dal., Det.	17	1101	**567**	.515
24. **Mike Modano**, Min., Dal.	14	946	**561**	.593
25. **Mats Sundin**, Que., Tor.	12	930	**545**	.586
26. **Pat Verbeek**, N.J., Hfd., NYR, Dal., Det.	20	1424	**541**	.380
27. **Brendan Shanahan**, N.J., St.L., Hfd., Det.	15	1108	**527**	.476
28. **Steve Duchesne**, L.A., Phi., Que., St.L., Ott., Det.	16	1113	**525**	.472
29. **Cliff Ronning**, St.L., Van., Phx., Nsh., L.A.	16	1017	**517**	.508
30. **Rick Tocchet**, Phi., Pit., L.A., Bos., Wsh., Phx.	18	1144	**512**	.448
31. **Rod Brind'Amour**, St.L., Phi., Car.	14	983	**511**	.520
32. **Sergei Fedorov**, Det.	12	828	**507**	.612
33. **Doug Weight**, NYR, Edm., St.L.	12	767	**501**	.653
34. **Steve Thomas**, Tor., Chi., NYI, N.J.	18	1110	**484**	.436
35. **Nicklas Lidstrom**, Det.	11	853	**481**	.564
36. **Alexander Mogilny**, Buf., Van., N.J., Tor.	13	846	**478**	.565
37. **Joe Nieuwendyk**, Cgy., Dal., N.J.	16	1033	**473**	.458
38. **James Patrick**, NYR, Hfd., Cgy., Buf.	19	1156	**471**	.407
39. **Andrew Cassels**, Mtl., Hfd., Cgy., Van.	13	847	**452**	.534
40. **Teemu Selanne**, Wpg., Ana., S.J.	10	719	**447**	.622
41. **Igor Larionov**, Van., S.J., Det., Fla.	12	798	**432**	.541
42. **Fredrik Olausson**, Wpg., Edm., Ana., Pit., Det.	15	978	**428**	.438
43. **Trevor Linden**, Van., NYI, Mtl., Wsh.	14	1008	**421**	.418
44. **Peter Forsberg**, Que., Col.	8	466	**411**	.882
45. **Shayne Corson**, Mtl., Edm., St.L., Tor.	17	1093	**407**	.372
46. **Eric Lindros**, Phi., NYR	9	558	**405**	.726
47. **Sergei Zubov**, NYR, Pit., Dal.	10	697	**405**	.581
48. **Kevin Dineen**, Hfd., Phi., Car., Ott., CBJ	18	1184	**405**	.342
49. **Teppo Numminen**, Wpg., Phx.	14	1020	**402**	.394
50. **Calle Johansson**, Buf., Wsh.	15	1019	**398**	.391
51. **Stephane Richer**, Mtl., N.J., T.B., St.L., Pit.	17	1054	**398**	.378
52. **Kevin Stevens**, Pit., Bos., L.A., NYR, Phi.	15	874	**397**	.454
53. **Claude Lemieux**, Mtl., N.J., Col., Phx.	19	1129	**394**	.349
54. **Alexei Zhamnov**, Wpg., Chi.	10	666	**393**	.590
55. **Joe Juneau**, Bos., Wsh., Buf., Ott., Phx., Mtl.	11	686	**390**	.569
56. **Gary Roberts**, Cgy., Car., Tor.	15	943	**386**	.409
57. **Eric Desjardins**, Mtl., Phi.	14	971	**384**	.395
58. **Scott Mellanby**, Phi., Edm., Fla., St.L.	17	1143	**382**	.334
59. **Tony Amonte**, NYR, Chi.	12	861	**372**	.432
60. **Glen Wesley**, Bos., Hfd., Car.	15	1103	**366**	.332
61. **Scott Young**, Hfd., Pit., Que., Col., Ana., St.L.	14	970	**357**	.368
62. **Derek King**, NYI, Hfd., Tor., St.L.	14	830	**351**	.423
63. **Jyrki Lumme**, Mtl., Van., Phx., Dal., Tor.	14	912	**343**	.376
64. **Petr Svoboda**, Mtl., Buf., Phi., T.B.	17	1028	**341**	.332
65. **Rob Blake**, L.A., Col.	13	750	**339**	.452
66. **Keith Tkachuk**, Wpg., Phx., St.L.	11	725	**339**	.468
67. **John LeClair**, Mtl., Phi.	12	763	**337**	.442
68. **Dmitri Khristich**, Wsh., L.A., Bos., Tor.	12	811	**337**	.416
69. **Ulf Dahlen**, NYR, Min., Dal., S.J., Chi., Wsh.	13	903	**334**	.370

With 38 assists for San Jose last season, Vincent Damphousse improved to 706 for his career. He also has 397 goals and 1,103 points. Quietly consistent, Damphousse ranks highly among active players.

Player		Games	Assists	Assists per game
70. **Jeff Norton**, NYI, S.J., St.L., Edm., T.B., Fla., Pit., Bos.	15	799	**332**	.416
71. **Owen Nolan**, Que., Col., S.J.	12	775	**332**	.428
72. **Pavel Bure**, Van., Fla., NYR	11	663	**331**	.499
73. **Jozef Stumpel**, Bos., L.A.	11	616	**331**	.537
74. **Robert Reichel**, Cgy., NYI, Phx., Tor.	9	680	**329**	.484
75. **Sandis Ozolinsh**, S.J., Col., Car., Fla.	10	661	**324**	.490
76. **Petr Nedved**, Van., St.L., NYR, Pit.	11	730	**321**	.440
77. **Benoit Hogue**, Buf., NYI, Tor., Dal., T.B., Phx., Bos., Wsh.	15	863	**321**	.372
78. **Mathieu Schneider**, Mtl., NYI, Tor., NYR, L.A.	14	836	**318**	.380
79. **Alexei Kovalev**, NYR, Pit.	10	693	**317**	.457
80. **Alexei Yashin**, Ott., NYI	8	582	**316**	.543
81. **Peter Bondra**, Wsh.	12	831	**313**	.377
82. **Paul Kariya**, Ana.	8	524	**313**	.597
83. **Sylvain Cote**, Hfd., Wsh., Tor., Chi., Dal.	18	1170	**313**	.268
84. **Mike Ricci**, Phi., Que., Col., S.J.	12	868	**313**	.361
85. **Bobby Holik**, Hfd., N.J.	12	878	**311**	.354
86. **Keith Primeau**, Det., Hfd., Car., Phi.	12	766	**305**	.398
87. **Valeri Kamensky**, Que., Col., NYR, Dal., N.J.	11	637	**301**	.473
88. **Scott Niedermayer**, N.J.	11	730	**296**	.405
89. **Martin Straka**, Pit., Ott., NYI, Fla.	10	616	**294**	.477
90. **Nelson Emerson**, St.L., Wpg., Hfd., Car., Chi., Ott., Atl., L.A.	12	771	**293**	.380
91. **Darryl Sydor**, L.A., Dal.	11	782	**292**	.373
92. **Mike Keane**, Mtl., Col., NYR, Dal., St.L.	14	1032	**288**	.279
93. **Dave Manson**, Chi., Edm., Wpg., Phx., Mtl., Dal., Tor.	16	1103	**288**	.261
94. **Ziggy Palffy**, NYI, L.A.	9	531	**280**	.527
95. **Adam Graves**, Det., Edm., NYR, S.J.	15	1070	**278**	.260
96. **Grant Ledyard**, NYR, L.A., Wsh., Buf., Dal., Van., Bos., Ott., T.B.	18	1028	**276**	.268
97. **Alexei Zhitnik**, L.A., Buf.	10	744	**273**	.367
98. **Todd Gill**, Tor., S.J., St.L., Det., Phx., Col.	18	1002	**271**	.270
99. **Roman Hamrlik**, T.B., Edm., NYI	10	719	**270**	.376
100. **Eric Weinrich**, N.J., Hfd., Chi., Mtl., Bos., Phi.	14	921	**269**	.292

Top 100 All-Time Point Leaders

* active player

Elected to the Hockey Hall of Fame in 2002, Bernie Federko played exactly 1,000 games in his career and had 1,130 points. He was the first player in NHL history to record at least 50 assists in 10 straight seasons.

	Player	Seasons	Games	Goals	Assists	Points	Points per game
1.	**Wayne Gretzky**, Edm., L.A., St.L., NYR	20	1487	894	1963	**2857**	1.921
2.	**Gordie Howe**, Det., Hfd.	26	1767	801	1049	**1850**	1.047
* 3.	**Mark Messier**, Edm., NYR, Van.	23	1602	658	1146	**1804**	1.126
4.	**Marcel Dionne**, Det., L.A., NYR	18	1348	731	1040	**1771**	1.314
* 5.	**Ron Francis**, Hfd., Pit., Car.	21	1569	514	1187	**1701**	1.084
* 6.	**Steve Yzerman**, Det.	19	1362	658	1004	**1662**	1.220
* 7.	**Mario Lemieux**, Pit.	15	812	654	947	**1601**	1.972
8.	**Phil Esposito**, Chi., Bos., NYR	18	1282	717	873	**1590**	1.240
9.	**Raymond Bourque**, Bos., Col.	22	1612	410	1169	**1579**	.980
10.	**Paul Coffey**, Edm., Pit., L.A., Det., Hfd., Phi., Chi., Car., Bos.	21	1409	396	1135	**1531**	1.087
11.	**Stan Mikita**, Chi.	22	1394	541	926	**1467**	1.052
12.	**Bryan Trottier**, NYI, Pit.	18	1279	524	901	**1425**	1.114
13.	**Dale Hawerchuk**, Wpg., Buf., St.L., Phi.	16	1188	518	891	**1409**	1.186
14.	**Jari Kurri**, Edm., L.A., NYR, Ana., Col.	17	1251	601	797	**1398**	1.118
* 15.	**Doug Gilmour**, St.L., Cgy., Tor., N.J., Chi., Buf., Mtl.	19	1412	439	945	**1384**	.980
16.	**John Bucyk**, Det., Bos.	23	1540	556	813	**1369**	.889
* 17.	**Adam Oates**, Det., St.L., Bos., Wsh., Phi.	17	1210	330	1027	**1357**	1.121
18.	**Guy Lafleur**, Mtl., NYR, Que.	17	1126	560	793	**1353**	1.202
19.	**Denis Savard**, Chi., Mtl., T.B.	17	1196	473	865	**1338**	1.119
20.	**Mike Gartner**, Wsh., Min., NYR, Tor., Phx.	19	1432	708	627	**1335**	.932
21.	**Gilbert Perreault**, Buf.	17	1191	512	814	**1326**	1.113
* 22.	**Luc Robitaille**, L.A., Pit., NYR, Det.	16	1205	620	668	**1288**	1.069
23.	**Alex Delvecchio**, Det.	24	1549	456	825	**1281**	.827
24.	**Jean Ratelle**, NYR, Bos.	21	1281	491	776	**1267**	.989
* 25.	**Joe Sakic**, Que., Col.	14	1016	483	774	**1257**	1.237
* 26.	**Dave Andreychuk**, Buf., Tor., N.J., Bos., Col., T.B.	20	1443	593	654	**1247**	.864
* 27.	**Brett Hull**, Cgy., St.L., Dal., Det.	17	1101	679	567	**1246**	1.132
28.	**Peter Stastny**, Que., N.J., St.L.	15	977	450	789	**1239**	1.268
29.	**Norm Ullman**, Det., Tor.	20	1410	490	739	**1229**	.872
30.	**Jean Beliveau**, Mtl.	20	1125	507	712	**1219**	1.084
31.	**Larry Murphy**, L.A., Wsh., Min., Pit., Tor., Det.	21	1615	287	929	**1216**	.753
32.	**Bobby Clarke**, Phi.	15	1144	358	852	**1210**	1.058
33.	**Bernie Nicholls**, L.A., NYR, Edm., N.J., Chi., S.J.	18	1127	475	734	**1209**	1.073
* 34.	**Al MacInnis**, Cgy., St.L.	21	1333	324	880	**1204**	.903
* 35.	**Phil Housley**, Buf., Wpg., St.L., Cgy., N.J., Wsh., Chi.	20	1437	332	871	**1203**	.837
36.	**Dino Ciccarelli**, Min., Wsh., Det., T.B., Fla.	19	1232	608	592	**1200**	.974
* 37.	**Pierre Turgeon**, Buf., NYI, Mtl., St.L., Dal.	15	1074	468	724	**1192**	1.110
38.	**Bobby Hull**, Chi., Wpg., Hfd.	16	1063	610	560	**1170**	1.101
* 39.	**Jaromir Jagr**, Pit., Wsh.	12	875	470	688	**1158**	1.323
40.	**Michel Goulet**, Que., Chi.	15	1089	548	604	**1152**	1.058
41.	**Bernie Federko**, St.L., Det.	14	1000	369	761	**1130**	1.130
42.	**Mike Bossy**, NYI	10	752	573	553	**1126**	1.497
43.	**Darryl Sittler**, Tor., Phi., Det.	15	1096	484	637	**1121**	1.023
44.	**Frank Mahovlich**, Tor., Det., Mtl.	18	1181	533	570	**1103**	.934
* 45.	**Vincent Damphousse**, Tor., Edm., Mtl., S.J.	16	1214	397	706	**1103**	.909
46.	**Glenn Anderson**, Edm., Tor., NYR, St.L.	16	1129	498	601	**1099**	.973
* 47.	**Mark Recchi**, Pit., Phi., Mtl.	14	1012	410	664	**1074**	1.061
48.	**Dave Taylor**, L.A.	17	1111	431	638	**1069**	.962
* 49.	**Pat Verbeek**, N.J., Hfd., NYR, Dal., Det.	20	1424	522	541	**1063**	.746
50.	**Joe Mullen**, St.L., Cgy., Pit., Bos.	17	1062	502	561	**1063**	1.001
* 51.	**Theoren Fleury**, Cgy., Col., NYR	14	1030	443	612	**1055**	1.024
52.	**Denis Potvin**, NYI	15	1060	310	742	**1052**	.992
53.	**Henri Richard**, Mtl.	20	1256	358	688	**1046**	.833
54.	**Bobby Smith**, Min., Mtl.	15	1077	357	679	**1036**	.962
* 55.	**Brendan Shanahan**, N.J., St.L., Hfd., Det.	15	1108	503	527	**1030**	.930
56.	**Brian Bellows**, Min., Mtl., T.B., Ana., Wsh.	17	1188	485	537	**1022**	.860
57.	**Rod Gilbert**, NYR	18	1065	406	615	**1021**	.959
58.	**Dale Hunter**, Que., Wsh., Col.	19	1407	323	697	**1020**	.725
* 59.	**Jeremy Roenick**, Chi., Phx., Phi.	14	983	429	585	**1014**	1.032
60.	**Pat LaFontaine**, NYI, Buf., NYR	15	865	468	545	**1013**	1.171
61.	**Steve Larmer**, Chi., NYR	15	1006	441	571	**1012**	1.006
62.	**Lanny McDonald**, Tor., Col., Cgy.	16	1111	500	506	**1006**	.905
63.	**Brian Propp**, Phi., Bos., Min., Hfd.	15	1016	425	579	**1004**	.988
64.	**Rick Middleton**, NYR, Bos.	14	1005	448	540	**988**	.983
65.	**Dave Keon**, Tor., Hfd.	18	1296	396	590	**986**	.761
* 66.	**Mike Modano**, Min., Dal.	14	946	416	561	**977**	1.033
67.	**Andy Bathgate**, NYR, Tor., Det., Pit.	17	1069	349	624	**973**	.910
* 68.	**Joe Nieuwendyk**, Cgy., Dal., N.J.	16	1033	494	473	**967**	.936
69.	**Maurice Richard**, Mtl.	18	978	544	421	**965**	.987
70.	**Larry Robinson**, Mtl., L.A.	20	1384	208	750	**958**	.692
* 71.	**Kirk Muller**, N.J., Mtl., NYI, Tor., Fla., Dal.	18	1294	356	597	**953**	.736
* 72.	**Rick Tocchet**, Phi., Pit., L.A., Bos., Wsh., Phx.	18	1144	440	512	**952**	.832
* 73.	**Mats Sundin**, Que., Tor.	12	930	397	545	**942**	1.013
74.	**Neal Broten**, Min., Dal., N.J., L.A.	17	1099	289	634	**923**	.840
75.	**Bobby Orr**, Bos., Chi.	12	657	270	645	**915**	1.393
* 76.	**Brian Leetch**, NYR	15	1021	215	700	**915**	.896
* 77.	**Alexander Mogilny**, Buf., Van., N.J., Tor.	13	846	420	478	**898**	1.061
78.	**Ray Ferraro**, Hfd., NYI, NYR, L.A., Atl., St.L.	18	1258	408	490	**898**	.714
79.	**Brad Park**, NYR, Bos., Det.	17	1113	213	683	**896**	.805
80.	**Butch Goring**, L.A., NYI, Bos.	16	1107	375	513	**888**	.802
81.	**Bill Barber**, Phi.	12	903	420	463	**883**	.978
* 82.	**Steve Thomas**, Tor., Chi., NYI, N.J.	18	1110	397	484	**881**	.794
83.	**Dennis Maruk**, Cal., Cle., Min., Wsh.	14	888	356	522	**878**	.989
* 84.	**Scott Stevens**, Wsh., St.L., N.J.	20	1516	189	687	**876**	.578
* 85.	**Chris Chelios**, Mtl., Chi., Det.	19	1260	174	700	**874**	.694
* 86.	**Sergei Fedorov**, Det.	12	828	364	507	**871**	1.052
87.	**Ivan Boldirev**, Bos., Cal., Chi., Atl., Van., Det.	15	1052	361	505	**866**	.823
88.	**Yvan Cournoyer**, Mtl.	16	968	428	435	**863**	.892
89.	**Dean Prentice**, NYR, Bos., Det., Pit., Min.	22	1378	391	469	**860**	.624
90.	**Tomas Sandstrom**, NYR, L.A., Pit., Det., Ana.	15	983	394	462	**856**	.871
* 91.	**Teemu Selanne**, Wpg., Ana., S.J.	10	719	408	447	**855**	1.189
92.	**Ted Lindsay**, Det., Chi.	17	1068	379	472	**851**	.797
* 93.	**Gary Suter**, Cgy., Chi., S.J.	17	1145	203	641	**844**	.737
94.	**Tom Lysiak**, Atl., Chi.	13	919	292	551	**843**	.917
95.	**John MacLean**, N.J., S.J., NYR, Dal.	18	1194	413	429	**842**	.705
96.	**John Tonelli**, NYI, Cgy., L.A., Chi., Que.	14	1028	325	511	**836**	.813
* 97.	**Rod Brind'Amour**, St.L., Phi., Car.	14	983	325	511	**836**	.850
98.	**Jacques Lemaire**, Mtl.	12	853	366	469	**835**	.979
99.	**Brent Sutter**, NYI, Chi.	18	1111	363	466	**829**	.746
100.	**Doug Wilson**, Chi., S.J.	16	1024	237	590	**827**	.808

Top 100 Active Points Leaders

	Player	Seasons	Games	Goals	Assists	Points	Points per game
1.	**Mark Messier**, Edm., NYR, Van.	23	1602	658	1146	**1804**	1.126
2.	**Ron Francis**, Hfd., Pit., Car.	21	1569	514	1187	**1701**	1.084
3.	**Steve Yzerman**, Det.	19	1362	658	1004	**1662**	1.220
4.	**Mario Lemieux**, Pit.	15	812	654	947	**1601**	1.972
5.	**Doug Gilmour**, St.L., Cgy., Tor., N.J., Chi., Buf., Mtl.	19	1412	439	945	**1384**	.980
6.	**Adam Oates**, Det., St.L., Bos., Wsh., Phi.	17	1210	330	1027	**1357**	1.121
7.	**Luc Robitaille**, L.A., Pit., NYR, Det.	16	1205	620	668	**1288**	1.069
8.	**Joe Sakic**, Que., Col.	14	1016	483	774	**1257**	1.237
9.	**Dave Andreychuk**, Buf., Tor., N.J., Bos., Col., T.B.	20	1443	593	654	**1247**	.864
10.	**Brett Hull**, Cgy., St.L., Dal., Det.	17	1101	679	567	**1246**	1.132
11.	**Al MacInnis**, Cgy., St.L.	21	1333	324	880	**1204**	.903
12.	**Phil Housley**, Buf., Wpg., St.L., Cgy., N.J., Wsh., Chi.	20	1437	332	871	**1203**	.837
13.	**Pierre Turgeon**, Buf., NYI, Mtl., St.L., Dal.	15	1074	468	724	**1192**	1.110
14.	**Jaromir Jagr**, Pit., Wsh.	12	875	470	688	**1158**	1.323
15.	**Vincent Damphousse**, Tor., Edm., Mtl., S.J.	16	1214	397	706	**1103**	.909
16.	**Mark Recchi**, Pit., Phi., Mtl.	14	1012	410	664	**1074**	1.061
17.	**Pat Verbeek**, N.J., Hfd., NYR, Dal., Det.	20	1424	522	541	**1063**	.746
18.	**Theoren Fleury**, Cgy., Col., NYR	14	1030	443	612	**1055**	1.024
19.	**Brendan Shanahan**, N.J., St.L., Hfd., Det.	15	1108	503	527	**1030**	.930
20.	**Jeremy Roenick**, Chi., Phx., Phi.	14	983	429	585	**1014**	1.032
21.	**Mike Modano**, Min., Dal.	14	946	416	561	**977**	1.033
22.	**Joe Nieuwendyk**, Cgy., Dal., N.J.	16	1033	494	473	**967**	.936
23.	**Kirk Muller**, N.J., Mtl., NYI, Tor., Fla., Dal.	18	1294	356	597	**953**	.736
24.	**Rick Tocchet**, Phi., Pit., L.A., Bos., Wsh., Phx.	18	1144	440	512	**952**	.832
25.	**Mats Sundin**, Que., Tor.	12	930	397	545	**942**	1.013
26.	**Brian Leetch**, NYR	15	1021	215	700	**915**	.896
27.	**Alexander Mogilny**, Buf., Van., N.J., Tor.	13	846	420	478	**898**	1.061
28.	**Steve Thomas**, Tor., Chi., NYI, N.J.	18	1110	397	484	**881**	.794
29.	**Scott Stevens**, Wsh., St.L., N.J.	20	1516	189	687	**876**	.578
30.	**Chris Chelios**, Mtl., Chi., Det.	19	1260	174	700	**874**	.694
31.	**Sergei Fedorov**, Det.	12	828	364	507	**871**	1.052
32.	**Teemu Selanne**, Wpg., Ana., S.J.	10	719	408	447	**855**	1.189
33.	**Gary Suter**, Cgy., Chi., S.J.	17	1145	203	641	**844**	.737
34.	**Rod Brind'Amour**, St.L., Phi., Car.	14	983	325	511	**836**	.850
35.	**Stephane Richer**, Mtl., N.J., T.B., St.L., Pit.	17	1054	421	398	**819**	.777
36.	**Cliff Ronning**, St.L., Van., Phx., Nsh., L.A.	16	1017	280	517	**797**	.784
37.	**Claude Lemieux**, Mtl., N.J., Col., Phx.	19	1129	371	394	**765**	.678
38.	**Kevin Dineen**, Hfd., Phi., Car., Ott., CBJ	18	1184	355	405	**760**	.642
39.	**Steve Duchesne**, L.A., Phi., Que., St.L., Ott., Det.	16	1113	227	525	**752**	.676
40.	**Gary Roberts**, Cgy., Car., Tor.	15	943	364	386	**750**	.795
41.	**Pavel Bure**, Van., Fla., NYR	11	663	418	331	**749**	1.130
42.	**Trevor Linden**, Van., NYI, Mtl., Wsh.	14	1008	316	421	**737**	.731
43.	**Peter Bondra**, Wsh.	12	831	421	313	**734**	.883
44.	**Eric Lindros**, Phi., NYR	9	558	327	405	**732**	1.312
45.	**Kevin Stevens**, Pit., Bos., L.A., NYR, Phi.	15	874	329	397	**726**	.831
46.	**Tony Amonte**, NYR, Chi.	12	861	352	372	**724**	.841
47.	**Keith Tkachuk**, Wpg., Phx., St.L.	11	725	367	339	**706**	.974
48.	**Doug Weight**, NYR, Edm., St.L.	12	767	195	501	**696**	.907
49.	**Scott Mellanby**, Phi., Edm., Fla., St.L.	17	1143	300	382	**682**	.597
50.	**John LeClair**, Mtl., Phi.	12	763	341	337	**678**	.889
51.	**Shayne Corson**, Mtl., Edm., St.L., Tor.	17	1093	261	407	**668**	.611
52.	**Scott Young**, Hfd., Pit., Que., Col., Ana., St.L.	14	970	293	357	**650**	.670
53.	**Owen Nolan**, Que., Col., S.J.	12	775	301	332	**633**	.817
54.	**Andrew Cassels**, Mtl., Hfd., Cgy., Van.	13	847	174	452	**626**	.739
55.	**Nicklas Lidstrom**, Det.	11	853	145	481	**626**	.734
56.	**Ulf Dahlen**, NYR, Min., Dal., S.J., Chi., Wsh.	13	903	284	334	**618**	.684
57.	**Alexei Zhamnov**, Wpg., Chi.	10	666	222	393	**615**	.923
58.	**James Patrick**, NYR, Hfd., Cgy., Buf.	19	1156	141	471	**612**	.529
59.	**Derek King**, NYI, Hfd., Tor., St.L.	14	830	261	351	**612**	.737
60.	**Adam Graves**, Det., Edm., NYR, S.J.	15	1070	320	278	**598**	.559
61.	**Dmitri Khristich**, Wsh., L.A., Bos., Tor.	12	811	259	337	**596**	.735
62.	**Igor Larionov**, Van., S.J., Det., Fla.	12	798	158	432	**590**	.739
63.	**Paul Kariya**, Ana.	8	524	275	313	**588**	1.122
64.	**Peter Forsberg**, Que., Col.	8	466	169	411	**580**	1.245
65.	**Petr Nedved**, Van., St.L., NYR, Pit.	11	730	255	321	**576**	.789
66.	**Fredrik Olausson**, Wpg., Edm., Ana., Pit., Det.	15	978	145	428	**573**	.586
67.	**Alexei Yashin**, Ott., NYI	8	582	250	316	**566**	.973
68.	**Robert Reichel**, Cgy., NYI, Phx., Tor.	9	680	229	329	**558**	.821
69.	**Alexei Kovalev**, NYR, Pit.	10	693	241	317	**558**	.805
70.	**Bobby Holik**, Hfd., N.J.	12	878	240	311	**551**	.628
71.	**Ziggy Palffy**, NYI, L.A.	9	531	265	280	**545**	1.026
72.	**Keith Primeau**, Det., Hfd., Car., Phi.	12	766	239	305	**544**	.710
73.	**Benoit Hogue**, Buf., NYI, Tor., Dal., T.B., Phx., Bos., Wsh.	15	863	222	321	**543**	.629
74.	**Joe Juneau**, Bos., Wsh., Buf., Ott., Phx., Mtl.	11	686	145	390	**535**	.780
75.	**Mike Ricci**, Phi., Que., Col., S.J.	12	868	215	313	**528**	.608
76.	**Calle Johansson**, Buf., Wsh.	15	1019	116	398	**514**	.504
77.	**Sergei Zubov**, NYR, Pit., Dal.	10	697	105	405	**510**	.732
78.	**Geoff Sanderson**, Hfd., Car., Van., Buf., CBJ	12	766	266	243	**509**	.664
79.	**Eric Desjardins**, Mtl., Phi.	14	971	123	384	**507**	.522
80.	**Bill Guerin**, N.J., Edm., Bos.	11	733	256	248	**504**	.688
81.	**Teppo Numminen**, Wpg., Phx.	14	1020	102	402	**504**	.494
82.	**Valeri Kamensky**, Que., Col., NYR, Dal., N.J.	11	637	200	301	**501**	.786
83.	**Rob Blake**, L.A., Col.	13	750	156	339	**495**	.660
84.	**Shawn McEachern**, Pit., L.A., Bos., Ott.	11	755	227	263	**490**	.649
85.	**Glen Wesley**, Bos., Hfd., Car.	15	1103	123	366	**489**	.443
86.	**Nelson Emerson**, St.L., Wpg., Hfd., Car., Chi., Ott., Atl., L.A.	12	771	195	293	**488**	.633
87.	**Martin Gelinas**, Edm., Que., Van., Car.	14	895	231	237	**468**	.523
88.	**Sandis Ozolinsh**, S.J., Col., Car., Fla.	10	661	141	324	**465**	.703
89.	**Donald Audette**, Buf., L.A., Atl., Dal., Mtl.	13	630	240	225	**465**	.738
90.	**Jason Arnott**, Edm., N.J., Dal.	9	598	200	264	**464**	.776
91.	**Jozef Stumpel**, Bos., L.A.	11	616	129	331	**460**	.747
92.	**Martin Straka**, Pit., Ott., NYI, Fla.	10	616	164	294	**458**	.744
93.	**Mathieu Schneider**, Mtl., NYI, Tor., NYR, L.A.	14	836	138	318	**456**	.545
94.	**Jyrki Lumme**, Mtl., Van., Phx., Dal., Tor.	14	912	108	343	**451**	.495
95.	**Stu Barnes**, Wpg., Fla., Pit., Buf.	11	739	197	248	**445**	.602
96.	**Mike Keane**, Mtl., Col., NYR, Dal., St.L.	14	1032	155	288	**443**	.429
97.	**Bryan Smolinski**, Bos., Pit., NYI, L.A.	10	681	191	251	**442**	.649
98.	**Markus Naslund**, Pit., Van.	9	630	207	234	**441**	.700
99.	**Rob Brown**, Pit., Hfd., Chi., Dal., L.A.	11	543	190	248	**438**	.807
100.	**Vyacheslav Kozlov**, Det., Buf.	11	645	211	226	**437**	.678

In his first season with the Islanders last year, Alexei Yashin led the club in goals (32), assists (43) and points (75). He ranked among the NHL's top 15 scorers and improved to 566 points in his career.

All-Time Games Played Leaders

Regular Season

* active player

	Player	Team	Seasons	GP
1.	**Gordie Howe**	Detroit	25	1687
		Hartford	1	80
		Total	**26**	**1,767**
2.	**Larry Murphy**	Los Angeles	3¼	242
		Washington	5½	453
		Minnesota	1¾	121
		Pittsburgh	4½	336
		Toronto	1¾	151
		Detroit	4¼	312
		Total	**21**	**1,615**
3.	**Raymond Bourque**	Boston	20¾	1,518
		Colorado	1¼	94
		Total	**22**	**1,612**
* 4.	**Mark Messier**	Edmonton	12	851
		NY Rangers	8	544
		Vancouver	3	207
		Total	**23**	**1,602**
* 5.	**Ron Francis**	Hartford	9¾	714
		Pittsburgh	7¼	533
		Carolina	4	322
		Total	**21**	**1,569**
6.	**Alex Delvecchio**	**Detroit**	**24**	**1,549**
7.	**John Bucyk**	Detroit	2	104
		Boston	21	1,436
		Total	**23**	**1,540**
* 8.	**Scott Stevens**	Washington	8	601
		St. Louis	1	78
		New Jersey	11	837
		Total	**20**	**1,516**
9.	**Wayne Gretzky**	Edmonton	9	696
		Los Angeles	7¾	539
		St. Louis	¼	18
		NY Rangers	3	234
		Total	**20**	**1,487**
10.	**Tim Horton**	Toronto	19¾	1,185
		NY Rangers	1¼	93
		Pittsburgh	1	44
		Buffalo	2	124
		Total	**24**	**1,446**
* 11.	**Dave Andreychuk**	Buffalo	11½	837
		Toronto	3¼	223
		New Jersey	3¼	224
		Boston	¾	63
		Colorado	¼	14
		Tampa Bay	1	82
		Total	**20**	**1,443**
* 12.	**Phil Housley**	Buffalo	8	608
		Winnipeg	3	232
		St. Louis	1	26
		Calgary	4¾	328
		New Jersey	¼	22
		Washington	2	141
		Chicago	1	80
		Total	**20**	**1,437**
13.	**Mike Gartner**	Washington	9¾	758
		Minnesota	1	80
		NY Rangers	4	322
		Toronto	2¼	130
		Phoenix	2	142
		Total	**19**	**1,432**
* 14.	**Pat Verbeek**	New Jersey	7	463
		Hartford	5¾	433
		NY Rangers	1¼	88
		Dallas	4	305
		Detroit	2	135
		Total	**20**	**1,424**
* 15.	**Doug Gilmour**	St. Louis	5	384
		Calgary	3½	266
		Toronto	5¼	392
		New Jersey	1¼	83
		Chicago	1¾	135
		Buffalo	1¼	82
		Montreal	1	70
		Total	**19**	**1,412**
16.	**Harry Howell**	NY Rangers	17	1,160
		Oakland	1	55
		California	½	28
		Los Angeles	2½	168
		Total	**21**	**1,411**
17.	**Norm Ullman**	Detroit	12½	875
		Toronto	7½	535
		Total	**20**	**1,410**
18.	**Paul Coffey**	Edmonton	7	532
		Pittsburgh	4¾	331
		Los Angeles	¾	60
		Detroit	3½	231
		Hartford	¼	20
		Philadelphia	1¾	94
		Chicago	¼	10
		Carolina	1¾	113
		Boston	1	18
		Total	**21**	**1,409**
19.	**Dale Hunter**	Quebec	7	523
		Washington	11¾	872
		Colorado	¼	12
		Total	**19**	**1,407**
20.	**Stan Mikita**	**Chicago**	**22**	**1,394**
21.	**Doug Mohns**	Boston	11	710
		Chicago	6½	415
		Minnesota	2½	162
		Atlanta	1	28
		Washington	1	75
		Total	**22**	**1,390**
22.	**Larry Robinson**	Montreal	17	1,202
		Los Angeles	3	182
		Total	**20**	**1,384**
23.	**Dean Prentice**	NY Rangers	10½	666
		Boston	3	170
		Detroit	3½	230
		Pittsburgh	2	144
		Minnesota	3	168
		Total	**22**	**1,378**
* 24.	**Steve Yzerman**	**Detroit**	**19**	**1,362**
25.	**Ron Stewart**	Toronto	13	838
		Boston	2	126
		St. Louis	½	19
		NY Rangers	4	306
		Vancouver	1	42
		NY Islanders	½	22
		Total	**21**	**1,353**
26.	**Marcel Dionne**	Detroit	4	309
		Los Angeles	11¾	921
		NY Rangers	2¼	118
		Total	**18**	**1,348**
* 27.	**Al MacInnis**	Calgary	13	803
		St. Louis	8	530
		Total	**21**	**1,333**
28.	**Guy Carbonneau**	Montreal	13	912
		St. Louis	1	42
		Dallas	5	364
		Total	**19**	**1,318**
29.	**Red Kelly**	Detroit	12½	846
		Toronto	7½	470
		Total	**20**	**1,316**
30.	**Dave Keon**	Toronto	15	1,062
		Hartford	3	234
		Total	**18**	**1,296**
* 31.	**Kirk Muller**	New Jersey	7	556
		Montreal	3¾	267
		NY Islanders	¾	27
		Toronto	1¼	102
		Florida	2¼	162
		Dallas	3	180
		Total	**18**	**1,294**
32.	**Phil Esposito**	Chicago	4	235
		Boston	8¼	625
		NY Rangers	5¾	422
		Total	**18**	**1,282**
33.	**Jean Ratelle**	NY Rangers	15¼	862
		Boston	5¾	419
		Total	**21**	**1,281**
34.	**Bryan Trottier**	NY Islanders	15	1,123
		Pittsburgh	3	156
		Total	**18**	**1,279**
* 35.	**Chris Chelios**	Montreal	7	402
		Chicago	8¾	664
		Detroit	3¼	194
		Total	**19**	**1,260**
36.	**Ray Ferraro**	Hartford	6¼	442
		NY Islanders	4¾	316
		NY Rangers	¾	65
		Los Angeles	3¼	197
		Atlanta	2¾	223
		St. Louis	¼	15
		Total	**18**	**1,258**
37.	**Henri Richard**	**Montreal**	**20**	**1,256**
38.	**Craig Ludwig**	Montreal	8	597
		NY Islanders	1	75
		Minnesota	2	151
		Dallas	6	433
		Total	**17**	**1,256**
39.	**Kevin Lowe**	Edmonton	15	1,037
		NY Rangers	4	217
		Total	**19**	**1,254**
40.	**Jari Kurri**	Edmonton	10	754
		Los Angeles	4¾	331
		NY Rangers	¼	14
		Anaheim	1	82
		Colorado	1	70
		Total	**17**	**1,251**
41.	**Bill Gadsby**	Chicago	8½	468
		NY Rangers	6½	457
		Detroit	5	323
		Total	**20**	**1,248**
42.	**Allan Stanley**	NY Rangers	6¼	307
		Chicago	1¾	111
		Boston	2	129
		Toronto	10	633
		Philadelphia	1	64
		Total	**21**	**1,244**
43.	**Dino Ciccarelli**	Minnesota	8¾	602
		Washington	3¼	223
		Detroit	4	254
		Tampa Bay	1½	111
		Florida	1½	42
		Total	**19**	**1,232**
44.	**Ed Westfall**	Boston	11	734
		NY Islanders	7	493
		Total	**18**	**1,227**
45.	**Brad McCrimmon**	Boston	3	228
		Philadelphia	5	367
		Calgary	3	231
		Detroit	3	203
		Hartford	3	156
		Phoenix	1	37
		Total	**18**	**1,222**
46.	**Eric Nesterenko**	Toronto	5	206
		Chicago	16	1,013
		Total	**21**	**1,219**
* 47.	**Ken Daneyko**	**New Jersey**	**18**	**1,214**
* 48.	**Vincent Damphousse**	Toronto	5	394
		Edmonton	1	80
		Montreal	6¾	519
		San Jose	3¼	221
		Total	**16**	**1,214**
* 49.	**Adam Oates**	Detroit	4	246
		St. Louis	2¾	195
		Boston	5	368
		Washington	5	387
		Philadelphia	¼	14
		Total	**17**	**1,210**
50.	**Marcel Pronovost**	Detroit	16	983
		Toronto	5	223
		Total	**21**	**1,206**
* 51.	**Luc Robitaille**	Los Angeles	12	932
		Pittsburgh	1	46
		NY Rangers	2	146
		Detroit	1	81
		Total	**16**	**1,205**
52.	**Denis Savard**	Chicago	12¼	881
		Montreal	3	210
		Tampa Bay	1¾	105
		Total	**17**	**1,196**
53.	**Dave Babych**	Winnipeg	5¼	390
		Hartford	5¾	349
		Vancouver	6¾	409
		Philadelphia	1	39
		Los Angeles	¼	8
		Total	**19**	**1,195**
54.	**John MacLean**	New Jersey	13¼	934
		San Jose	¾	51
		NY Rangers	2¼	161
		Dallas	1½	48
		Total	**18**	**1,194**
55.	**Gilbert Perrault**	**Buffalo**	**17**	**1,191**
56.	**Dale Hawerchuk**	Winnipeg	9	713
		Buffalo	5	342
		St. Louis	¾	66
		Philadelphia	1¼	67
		Total	**16**	**1,188**
57.	**Brian Bellows**	Minnesota	10	753
		Montreal	3	200
		Tampa Bay	1¼	86
		Anaheim	¾	62
		Washington	2	87
		Total	**17**	**1,188**
58.	**George Armstrong**	**Toronto**	**21**	**1,187**
* 59.	**Kevin Dineen**	Hartford	8¾	587
		Philadelphia	4¼	284
		Carolina	2	121
		Ottawa	1	67
		Columbus	2	125
		Total	**18**	**1,184**
60.	**Frank Mahovlich**	Toronto	11¾	720
		Detroit	2¾	198
		Montreal	3½	263
		Total	**18**	**1,181**
61.	**Bob Carpenter**	Washington	6¼	490
		NY Rangers	½	28
		Los Angeles	1¾	120
		Boston	3½	187
		New Jersey	6	353
		Total	**18**	**1,178**
62.	**Don Marshall**	Montreal	10	585
		NY Rangers	7	479
		Buffalo	1	62
		Toronto	1	50
		Total	**19**	**1,176**

	Player	Team	Seasons	GP
*	63. Sylvain Cote	Hartford	7	382
		Washington	8¾	621
		Toronto	1½	94
		Chicago	½	45
		Dallas	¼	28
		Total	**18**	**1,170**
	64. Bob Gainey	**Montreal**	**16**	**1,160**
	65. Kevin Hatcher	Washington	10	685
		Dallas	2	121
		Pittsburgh	3	220
		NY Rangers	1	74
		Carolina	1	57
		Total	**17**	**1,157**
*	66. James Patrick	NY Rangers	10¼	671
		Hartford	½	47
		Calgary	4¼	217
		Buffalo	4	221
		Total	**19**	**1,156**
	67. Leo Boivin	Toronto	3¼	137
		Boston	11½	717
		Detroit	1¼	85
		Pittsburgh	1½	114
		Minnesota	1½	97
		Total	**19**	**1,150**
	68. Garry Galley	Los Angeles	5½	361
		Washington	1½	76
		Boston	3½	257
		Philadelphia	3¼	236
		Buffalo	2¼	163
		NY Islanders	1	56
		Total	**17**	**1,149**
	69. Borje Salming	Toronto	16	1,099
		Detroit	1	49
		Total	**17**	**1,148**
*	70. Gary Suter	Calgary	8½	617
		Chicago	4½	301
		San Jose	4	227
		Total	**17**	**1,145**
	71. Bobby Clarke	**Philadelphia**	**15**	**1,144**
*	72. Rick Tocchet	Philadelphia	10	621
		Pittsburgh	2¼	150
		Los Angeles	1½	80
		Boston	1¼	67
		Washington	¼	13
		Phoenix	2¾	213
		Total	**18**	**1,144**
*	73. Scott Mellanby	Philadelphia	6	355
		Edmonton	2	149
		Florida	7¾	552
		St. Louis	1¼	87
		Total	**17**	**1,143**
	74. Glenn Anderson	Edmonton	11½	845
		Toronto	2¾	221
		NY Rangers	¼	12
		St. Louis	1½	51
		Total	**16**	**1,129**
	75. Dave Ellett	Winnipeg	6¼	475
		Toronto	6½	446
		New Jersey	¼	20
		Boston	2	136
		St. Louis	1	52
		Total	**16**	**1,129**
*	76. Claude Lemieux	Montreal	7	281
		New Jersey	5¾	423
		Colorado	4¼	297
		Phoenix	2	128
		Total	**19**	**1,129**
	77. Bob Nevin	Toronto	5¾	250
		NY Rangers	7¼	505
		Minnesota	2	138
		Los Angeles	3	235
		Total	**18**	**1,128**
	78. Jamie Macoun	Calgary	8½	586
		Toronto	6½	466
		Detroit	1¼	76
		Total	**16**	**1,128**
	79. Murray Oliver	Detroit	2½	101
		Boston	6½	429
		Toronto	3	226
		Minnesota	5	371
		Total	**17**	**1,127**
	80. Bernie Nicholls	Los Angeles	8½	602
		NY Rangers	1¾	104
		Edmonton	1¼	95
		New Jersey	1½	84
		Chicago	2	107
		San Jose	3	135
		Total	**18**	**1,127**
	81. Guy Lafleur	Montreal	14	961
		NY Rangers	1	67
		Quebec	2	98
		Total	**17**	**1,126**
	82. Jean Beliveau	**Montreal**	**20**	**1,125**
	83. Doug Harvey	Montreal	14	890
		NY Rangers	3	151
		Detroit	1	2
		St. Louis	1	70
		Total	**19**	**1,113**
	84. Brad Park	NY Rangers	7½	465
		Boston	7½	501
		Detroit	2	147
		Total	**17**	**1,113**
*	85. Steve Duchesne	Los Angeles	5¾	442
		Philadelphia	1¼	89
		Quebec	1	82
		St. Louis	3	163
		Ottawa	2	140
		Detroit	3	197
		Total	**16**	**1,113**
	86. Lanny McDonald	Toronto	6½	477
		Colorado	1¾	142
		Calgary	7¾	441
		Total	**16**	**1,111**
	87. Dave Taylor	**Los Angeles**	**17**	**1,111**
	88. Brent Sutter	NY Islanders	11¼	694
		Chicago	6¾	417
		Total	**18**	**1,111**
*	89. Steve Thomas	Toronto	6	377
		Chicago	5¼	265
		NY Islanders	3¾	275
		New Jersey	3	193
		Total	**18**	**1,110**
*	90. Brendan Shanahan	New Jersey	4	281
		St. Louis	4	277
		Hartford	1¼	76
		Detroit	5¾	474
		Total	**15**	**1,108**
	91. Butch Goring	Los Angeles	10¾	736
		NY Islanders	4¾	332
		Boston	½	39
		Total	**16**	**1,107**
	92. Garry Unger	Toronto	½	15
		Detroit	3	216
		St. Louis	8½	662
		Atlanta	1	79
		Los Angeles	¾	58
		Edmonton	2¼	75
		Total	**16**	**1,105**
*	93. Dave Manson	Chicago	6¼	431
		Edmonton	2¾	219
		Winnipeg	2¼	139
		Phoenix	¾	66
		Montreal	1½	101
		Dallas	1¼	60
		Toronto	1¼	87
		Total	**16**	**1,103**
*	94. Glen Wesley	Boston	7	537
		Hartford	3	184
		Carolina	5	382
		Total	**15**	**1,103**
	95. Pit Martin	Detroit	3¼	119
		Boston	1¾	111
		Chicago	10½	740
		Vancouver	1¾	131
		Total	**17**	**1,101**
*	96. Luke Richardson	Toronto	4	278
		Edmonton	6	436
		Philadelphia	5	387
		Total	**15**	**1,101**
*	97. Brett Hull	Calgary	1¾	57
		St. Louis	10¼	744
		Dallas	3	218
		Detroit	1	82
		Total	**16**	**1,101**
	98. Neal Broten	Minnesota	13	876
		Dallas	2	116
		New Jersey	1¾	88
		Los Angeles	¼	19
		Total	**17**	**1,099**
	99. Jay Wells	Los Angeles	9	604
		Philadelphia	1¾	126
		Buffalo	2	85
		NY Rangers	3¼	186
		St. Louis	1	76
		Tampa Bay	1	21
		Total	**18**	**1,098**
	100. Gordie Roberts	Hartford	1½	107
		Minnesota	7	555
		Philadelphia	¼	11
		St. Louis	2½	166
		Pittsburgh	1¾	134
		Boston	2	124
		Total	**15**	**1,097**
	101. Darryl Sittler	Toronto	11½	844
		Philadelphia	2½	191
		Detroit	1	61
		Total	**15**	**1,096**
	102. Craig MacTavish	Boston	5	217
		Edmonton	8¾	701
		NY Rangers	¼	12
		Philadelphia	1¾	100
		St. Louis	1¼	63
		Total	**17**	**1,093**
	103. Ron Sutter	Philadelphia	9	555
		St. Louis	2½	163
		Quebec	½	37
		NY Islanders	1	27
		Boston	1	18
		San Jose	4	272
		Calgary	1	21
		Total	**19**	**1,093**
*	104. Shayne Corson	Montreal	10¾	662
		Edmonton	3	192
		St. Louis	1¼	88
		Toronto	2	151
		Total	**17**	**1,093**
	105. Michel Goulet	Quebec	10¾	813
		Chicago	4¼	276
		Total	**15**	**1,089**
	106. Carol Vadnais	Montreal	2	42
		Oakland	2	152
		California	1¾	94
		Boston	3½	263
		NY Rangers	6¾	485
		New Jersey	1	51
		Total	**17**	**1,087**
	107. Brad Marsh	Atlanta	2	160
		Calgary	1¼	97
		Philadelphia	6¾	514
		Toronto	2¾	181
		Detroit	1¼	75
		Ottawa	1	59
		Total	**15**	**1,086**
	108. Ulf Samuelsson	Hartford	6¾	463
		Pittsburgh	4¼	277
		NY Rangers	3¾	287
		Detroit	¼	4
		Philadelphia	1	49
		Total	**16**	**1,080**
	109. Bob Pulford	Toronto	14	947
		Los Angeles	2	132
		Total	**16**	**1,079**
	110. Bobby Smith	Minnesota	8¼	572
		Montreal	6¾	505
		Total	**15**	**1,077**
*	111. Pierre Turgeon	Buffalo	4¼	322
		NY Islanders	3½	255
		Montreal	1½	104
		St. Louis	4¾	327
		Dallas	1	66
		Total	**15**	**1,074**
	112. Doug Bodger	Pittsburgh	4¼	299
		Buffalo	7	479
		San Jose	2¼	166
		New Jersey	½	49
		Los Angeles	1	65
		Vancouver	1	13
		Total	**16**	**1,071**
	113. Murray Craven	Detroit	2	46
		Philadelphia	7¼	523
		Hartford	1½	128
		Vancouver	1¼	88
		Chicago	3	157
		San Jose	3	129
		Total	**18**	**1,071**
	114. Craig Ramsay	**Buffalo**	**14**	**1,070**
	115. Mike Ramsey	Buffalo	13¾	911
		Pittsburgh	1¼	77
		Detroit	3	82
		Total	**18**	**1,070**
*	116. Adam Graves	Detroit	2¼	78
		Edmonton	1¾	139
		NY Rangers	10	772
		San Jose	1	81
		Total	**18**	**1,070**
	117. Andy Bathgate	NY Rangers	11¾	719
		Toronto	1¼	70
		Detroit	2	130
		Pittsburgh	2	150
		Total	**17**	**1,069**
	118. Ted Lindsay	Detroit	14	862
		Chicago	3	206
		Total	**17**	**1,068**
	119. Terry Harper	Montreal	10	554
		Los Angeles	3	234
		Detroit	4	252
		St. Louis	1	11
		Colorado	1	15
		Total	**19**	**1,066**
	120. Rod Gilbert	**NY Rangers**	**18**	**1,065**
	121. Bobby Hull	Chicago	15	1,036
		Winnipeg	2/3	18
		Hartford	1/3	9
		Total	**16**	**1,063**
	122. Joe Mullen	St. Louis	4½	301
		Calgary	4½	345
		Pittsburgh	6	379
		Boston	1	37
		Total	**16**	**1,062**
	123. Bob Rouse	Minnesota	5¾	351
		Washington	2	130
		Toronto	3¼	237
		Detroit	4	247
		San Jose	2	96
		Total	**17**	**1,061**
	124. Denis Potvin	**NY Islanders**	**15**	**1,060**
*	125. Marc Bergevin	Chicago	4¼	266
		NY Islanders	1¾	76
		Hartford	2	79
		Tampa Bay	3	205
		Detroit	1	70
		St. Louis	5	326
		Pittsburgh	1	38
		Total	**17**	**1,060**
	126. Kelly Miller	NY Rangers	2½	117
		Washington	12½	940
		Total	**15**	**1,057**
	127. Jean Guy Talbot	Montreal	13	791
		Minnesota	¼	4
		Detroit	½	32
		St. Louis	2½	172
		Buffalo	¾	57
		Total	**17**	**1,056**

	Player	Team	Seasons	GP
128.	**Greg Adams**	New Jersey	3	186
		Vancouver	7¾	489
		Dallas	3¼	177
		Phoenix	2	144
		Florida	1	60
		Total	**17**	**1,056**
129.	**Randy Carlyle**	Toronto	2	94
		Pittsburgh	5¾	397
		Winnipeg	9¼	564
		Total	**17**	**1,055**
* 130.	**Stephane Richer**	Montreal	8½	490
		New Jersey	5¼	360
		Tampa Bay	2	110
		St. Louis	½	36
		Pittsburgh	¾	58
		Total	**17**	**1,054**
131.	**Ivan Boldirev**	Boston	1¼	13
		California	2¾	191
		Chicago	4¾	384
		Atlanta	1	65
		Vancouver	2¾	216
		Detroit	2½	183
		Total	**15**	**1,052**
132.	**Geoff Courtnall**	Boston	4¾	259
		Edmonton	¼	12
		Washington	2	159
		St. Louis	5¾	326
		Washington	4¼	292
		Total	**17**	**1,048**
133.	**Eddie Shack**	NY Rangers	2¼	141
		Toronto	8¾	504
		Boston	2	120
		Los Angeles	1¼	84
		Buffalo	1½	111
		Pittsburgh	1½	87
		Total	**17**	**1,047**
134.	**Rob Ramage**	Colorado	3	234
		St. Louis	5¾	441
		Calgary	1¼	80
		Toronto	2	160
		Minnesota	1	34
		Tampa Bay	¾	66
		Montreal	½	14
		Philadelphia	¾	15
		Total	**15**	**1,044**
135.	**Serge Savard**	Montreal	15	917
		Winnipeg	2	123
		Total	**17**	**1,040**
136.	**Ron Ellis**	**Toronto**	**16**	**1,034**
137.	**Harold Snepsts**	Vancouver	11¾	781
		Minnesota	1	71
		Detroit	3	120
		St. Louis	1¼	61
		Total	**17**	**1,033**
* 138.	**Joe Nieuwendyk**	Calgary	9	577
		Dallas	6¾	442
		New Jersey	¼	14
		Total	**16**	**1,033**
139.	**Ralph Backstrom**	Montreal	14½	844
		Los Angeles	2¼	172
		Chicago	¼	16
		Total	**17**	**1,032**
* 140.	**Kelly Buchberger**	Edmonton	13	795
		Atlanta	¾	68
		Los Angeles	2¼	169
		Total	**16**	**1,032**
* 141.	**Mike Keane**	Montreal	7¼	506
		Colorado	2	158
		NY Rangers	¾	70
		Dallas	3¼	242
		St. Louis	¾	56
		Total	**16**	**1,032**
* 142.	**Dave Lowry**	Vancouver	3	165
		St. Louis	5	311
		Florida	4¼	272
		San Jose	2¾	143
		Calgary	2	141
		Total	**17**	**1,032**
143.	**Ed Olczyk**	Chicago	5	322
		Toronto	3¼	257
		Winnipeg	3½	214
		NY Rangers	2¼	103
		Los Angeles	¾	67
		Pittsburgh	1½	68
		Total	**16**	**1,031**
144.	**Dick Duff**	Toronto	9¾	582
		NY Rangers	¾	43
		Montreal	5	305
		Los Angeles	¾	39
		Buffalo	1¾	61
		Total	**18**	**1,030**
* 145.	**Theoren Fleury**	Calgary	10¾	791
		Colorado	¼	15
		NY Rangers	3	224
		Total	**17**	**1,030**
146.	**Russ Courtnall**	Toronto	5¼	309
		Montreal	3¾	250
		Minnesota	1	84
		Dallas	1½	116
		Vancouver	2¼	141
		NY Rangers	¼	14
		Los Angeles	2	115
		Total	**16**	**1,029**
147.	**John Tonelli**	NY Islanders	7¾	584
		Calgary	2¼	161
		Los Angeles	3	231
		Chicago	¾	33
		Quebec	¼	19
		Total	**14**	**1,028**
148.	**Gaetan Duchesne**	Washington	6	451
		Quebec	2	150
		Minnesota	4	297
		San Jose	1¾	117
		Florida	¼	13
		Total	**14**	**1,028**
149.	**Petr Svoboda**	Montreal	7¾	534
		Buffalo	3	139
		Philadelphia	3¾	232
		Tampa Bay	2½	123
		Total	**17**	**1,028**
* 150.	**Grant Ledyard**	NY Rangers	1¼	69
		Los Angeles	2½	142
		Washington	1	82
		Buffalo	4¼	240
		Dallas	4¼	270
		Vancouver	¾	49
		Boston	1¼	69
		Ottawa	1	40
		Tampa Bay	1¾	67
		Total	**18**	**1,028**
151.	**Wayne Cashman**	**Boston**	**17**	**1,027**
152.	**Doug Wilson**	Chicago	14	938
		San Jose	2	86
		Total	**16**	**1,024**
153.	**Jim Neilson**	NY Rangers	12	810
		California	2	98
		Cleveland	2	115
		Total	**16**	**1,023**
154.	**Keith Acton**	Montreal	4¼	228
		Minnesota	4¼	343
		Edmonton	1	72
		Philadelphia	4½	303
		Washington	¼	6
		NY Islanders	¾	71
		Total	**15**	**1,023**
* 155.	**Brian Leetch**	**NY Rangers**	**15**	**1,021**
156.	**Don Lever**	Vancouver	7 2/3	593
		Atlanta	1/3	28
		Calgary	1¼	85
		Colorado	¾	59
		New Jersey	3	216
		Buffalo	2	39
		Total	**15**	**1,020**
* 157.	**Teppo Numinen**	Winnipeg	8	547
		Phoenix	6	473
		Total	**14**	**1,020**
* 158.	**Calle Johansson**	Buffalo	1¾	118
		Washington	13¼	901
		Total	**15**	**1,019**
159.	**Mike Foligno**	Detroit	2½	186
		Buffalo	9	664
		Toronto	2¾	129
		Florida	¾	39
		Total	**15**	**1,018**
160.	**Charlie Huddy**	Edmonton	11	694
		Los Angeles	3¼	226
		Buffalo	2½	85
		St. Louis	¼	12
		Total	**17**	**1,017**
* 161.	**Cliff Ronning**	St. Louis	3¾	180
		Vancouver	5¼	366
		Phoenix	2¼	156
		Nashville	3½	301
		Los Angeles	¼	14
		Total	**17**	**1,017**
162.	**Phil Russell**	Chicago	6¾	504
		Atlanta	1¼	93
		Calgary	3	229
		New Jersey	2¾	172
		Buffalo	1¼	18
		Total	**15**	**1,016**
163.	**Brian Propp**	Philadelphia	10¾	790
		Boston	¼	14
		Minnesota	3	147
		Hartford	1	65
		Total	**15**	**1,016**
* 164.	**Joe Sakic**	Quebec	7	508
		Colorado	7	508
		Total	**14**	**1,016**
* 165.	**Mark Recchi**	Pittsburgh	3¾	225
		Montreal	4¼	346
		Philadelphia	6	441
		Total	**14**	**1,012**
166.	**Laurie Boschman**	Toronto	2¾	187
		Edmonton	1	73
		Winnipeg	7¼	526
		New Jersey	2	153
		Ottawa	1	70
		Total	**14**	**1,009**
167.	**Dave Christian**	Winnipeg	4	230
		Washington	6½	504
		Boston	1½	128
		St. Louis	1	78
		Chicago	2	69
		Total	**15**	**1,009**
168.	**Dave Lewis**	NY Islanders	6¾	514
		Los Angeles	3¼	221
		New Jersey	3	209
		Detroit	2	64
		Total	**15**	**1,008**
169.	**Bob Murray**	**Chicago**	**15**	**1,008**
* 170.	**Trevor Linden**	Vancouver	10½	766
		NY Islanders	1¼	107
		Montreal	1¾	107
		Washington	½	28
		Total	**14**	**1,008**
171.	**Jimmy Roberts**	Montreal	9 2/3	611
		St. Louis	5 1/3	395
		Total	**15**	**1,006**
172.	**Steve Larmer**	Chicago	13	891
		NY Rangers	2	115
		Total	**15**	**1,006**
173.	**Claude Provost**	**Montreal**	**15**	**1,005**
174.	**Rick Middleton**	NY Rangers	2	124
		Boston	12	881
		Total	**14**	**1,005**
175.	**Ryan Walter**	Washington	4	307
		Montreal	9	604
		Vancouver	2	92
		Total	**15**	**1,003**
176.	**Vic Hadfield**	NY Rangers	13	839
		Pittsburgh	3	163
		Total	**16**	**1,002**
* 177.	**Todd Gill**	Toronto	12	639
		Pittsburgh	1¾	143
		St. Louis	¾	39
		Detroit	1¾	104
		Phoenix	¾	41
		Colorado	1	36
		Total	**18**	**1,002**
178.	**Bernie Federko**	St. Louis	14	927
		Detroit	1	73
		Total	**15**	**1,000**

All-Time Penalty-Minute Leaders

* active player

(Regular season. Minimum 2,500 minutes)

	Player	Seas.	GP	Pen. Mins.	Mins./ game
1.	**Tiger Williams**, Tor., Van., Det., L.A., Hfd.	14	962	**3966**	4.12
2.	**Dale Hunter**, Que., Wsh., Col.	19	1407	**3565**	2.53
3.	**Marty McSorley**, Pit., Edm., L.A., NYR, S.J., Bos.	17	961	**3381**	3.52
* 4.	**Bob Probert**, Det., Chi.	16	935	**3300**	3.53
5.	**Tim Hunter**, Cgy., Que., Van., S.J.	16	815	**3146**	3.86
* 6.	**Rob Ray**, Buf.	13	848	**3097**	3.65
* 7.	**Craig Berube**, Phi., Tor., Cgy., Wsh., NYI	16	999	**3049**	3.05
8.	**Chris Nilan**, Mtl., NYR, Bos.	13	688	**3043**	4.42
*9.	**Tie Domi**, Tor., NYR, Wpg.	13	784	**3027**	3.86
*10.	**Rick Tocchet**, Phi., Pit., L.A., Bos., Wsh., Phx.	18	1144	**2972**	2.60
*11.	**Pat Verbeek**, N.J., Hfd., NYR, Dal., Det.	20	1424	**2905**	2.04
*12.	**Dave Manson**, Chi., Edm., Wpg., Phx., Mtl., Dal., Tor.	16	1103	**2792**	2.53
*13.	**Scott Stevens**, Wsh., St.L., N.J.	20	1516	**2722**	1.80
14.	**Willi Plett**, Atl., Cgy., Min., Bos.	13	834	**2572**	3.08
*15.	**Gino Odjick**, Van., NYI, Phi., Mtl.	12	605	**2567**	4.24
*16.	**Chris Chelios**, Mtl., Chi., Det.	19	1260	**2556**	2.03
17.	**Joe Kocur**, Det., NYR, Van.	15	820	**2519**	3.07

Goaltending Records

All-Time Shutout Leaders

Goaltender	Team	Seasons	Games	Shutouts
Terry Sawchuk	Detroit	14	734	85
(1949-1970)	Boston	2	102	11
	Toronto	3	91	4
	Los Angeles	1	36	2
	NY Rangers	1	8	1
	Total	21	971	**103**
George Hainsworth	Montreal	7½	318	75
(1926-1937)	Toronto	3½	147	19
	Total	11	465	**94**
Glenn Hall	Detroit	4	148	17
(1952-1971)	Chicago	10	618	51
	St. Louis	4	140	16
	Total	18	906	**84**
Jacques Plante	Montreal	11	556	58
(1952-1973)	NY Rangers	2	98	5
	St. Louis	2	69	10
	Toronto	2¾	106	7
	Boston	¼	8	2
	Total	18	837	**82**
Tiny Thompson	Boston	10¼	468	74
(1928-1940)	Detroit	1¾	85	7
	Total	12	553	**81**
Alex Connell	Ottawa	8	293	64
(1924-1937)	Detroit	1	48	6
	NY Americans	1	1	0
	Mtl. Maroons	2	75	11
	Total	12	417	**81**
Tony Esposito	Montreal	1	13	2
(1968-1984)	Chicago	15	873	74
	Total	16	886	**76**
Lorne Chabot	NY Rangers	2	80	21
(1926-1937)	Toronto	5	214	33
	Montreal	1	47	8
	Chicago	1	48	8
	Mtl. Maroons	1	16	2
	NY Americans	1	6	1
	Total	11	411	**73**
Harry Lumley	Detroit	6½	324	26
(1943-1960)	NY Rangers	½	1	0
	Chicago	2	134	5
	Toronto	4	267	34
	Boston	3	78	6
	Total	16	804	**71**
Roy Worters	Pittsburgh Pirates	3	123	22
(1925-1937)	NY Americans	9	360	45
	* Montreal		1	0
	Total	12	484	**67**
Turk Broda	Toronto	14	629	62
(1936-1952)				
Dominik Hasek	Chicago	2	25	1
(1990-2002)	Buffalo	9	491	55
	Detroit	1	65	5
	Total	12	581	**61**
Patrick Roy	Montreal	11½	551	29
(1984-2002)	Colorado	6½	415	32
	Total	18	966	**61**
Clint Benedict	Ottawa	7	158	19
(1917-1930)	Mtl. Maroons	6	204	39
	Total	13	362	**58**
John Ross Roach	Toronto	7	222	13
(1921-1935)	NY Rangers	4	89	30
	Detroit	3	180	15
	Total	14	491	**58**
Ed Belfour	Chicago	7⅔	415	30
(1988-2002)	San Jose	⅓	13	1
	Dallas	5	307	27
	Total	13	735	**58**
Martin Brodeur	New Jersey	10	592	**55**
(1991-2002)				
Bernie Parent	Boston	2	57	1
(1965-1979)	Philadelphia	9½	486	50
	Toronto	1½	65	3
	Total	13	608	**54**
Ed Giacomin	NY Rangers	10¼	539	49
(1965-1978)	Detroit	2¾	71	5
	Total	13	610	**54**
Dave Kerr	Mtl. Maroons	3	101	11
(1930-1941)	NY Americans	1	1	0
	NY Rangers	7	324	40
	Total	11	426	**51**
Rogie Vachon	Montreal	5¼	206	13
(1966-1982)	Los Angeles	6¾	389	32
	Detroit	2	109	4
	Boston	2	91	2
	Total	16	795	**51**
Ken Dryden	Montreal	8	397	**46**
(1970-1979)				
Gump Worsley	NY Rangers	10	582	24
(1952-1974)	Montreal	6½	172	16
	Minnesota	4½	107	3
	Total	21	861	**43**
Charlie Gardiner	Chicago	7	316	**42**
(1927-1934)				
Frank Brimsek	Boston	9	444	35
(1938-1950)	Chicago	1	70	5
	Total	10	514	**40**
John Vanbiesbrouck	NY Rangers	11	449	16
(1981-2002)	Florida	5	268	13
	Philadelphia	2	112	9
	NY Islanders	¾	44	1
	New Jersey	1¼	9	1
	Total	20	882	**40**
Johnny Bower	NY Rangers	3	77	5
(1953-1970)	Toronto	12	475	32
	Total	15	552	**37**
Tom Barrasso	Buffalo	5¼	226	13
(1983-2002)	Pittsburgh	11½	460	22
	Ottawa	¼	7	0
	Carolina	¾	34	2
	Toronto	¼	4	0
	Total	18	771	**37**

*Played 1 game for Canadiens in 1929-30.

Ten or More Shutouts, One Season

Number of Shutouts	Goaltender	Team	Season	Length of Schedule
22	George Hainsworth	Montreal	1928-29	44
15	Alex Connell	Ottawa	1925-26	36
	Alex Connell	Ottawa	1927-28	44
	Hal Winkler	Boston	1927-28	44
	Tony Esposito	Chicago	1969-70	76
14	George Hainsworth	Montreal	1926-27	44
13	Clint Benedict	Mtl. Maroons	1926-27	44
	Alex Connell	Ottawa	1926-27	44
	George Hainsworth	Montreal	1927-28	44
	John Ross Roach	NY Rangers	1928-29	44
	Roy Worters	NY Americans	1928-29	44
	Harry Lumley	Toronto	1953-54	70
	Dominik Hasek	Buffalo	1997-98	82
12	Lorne Chabot	Toronto	1928-29	44
	Tiny Thompson	Boston	1928-29	44
	Charlie Gardiner	Chicago	1930-31	44
	Terry Sawchuk	Detroit	1951-52	70
	Terry Sawchuk	Detroit	1953-54	70
	Terry Sawchuk	Detroit	1954-55	70
	Glenn Hall	Detroit	1955-56	70
	Bernie Parent	Philadelphia	1973-74	78
	Bernie Parent	Philadelphia	1974-75	80
11	Lorne Chabot	NY Rangers	1927-28	44
	Hap Holmes	Detroit	1927-28	44
	Roy Worters	Pittsburgh Pirates	1927-28	44
	Clint Benedict	Mtl. Maroons	1928-29	44
	Joe Miller	Pittsburgh Pirates	1928-29	44
	Tiny Thompson	Boston	1932-33	48
	Terry Sawchuk	Detroit	1950-51	70
	Dominik Hasek	Buffalo	2000-01	82
10	Lorne Chabot	NY Rangers	1926-27	44
	Dolly Dolson	Detroit	1928-29	44
	John Ross Roach	Detroit	1932-33	48
	Charlie Gardiner	Chicago	1933-34	48
	Tiny Thompson	Boston	1935-36	48
	Frank Brimsek	Boston	1938-39	48
	Bill Durnan	Montreal	1948-49	60
	Harry Lumley	Toronto	1952-53	70
	Gerry McNeil	Montreal	1952-53	70
	Tony Esposito	Chicago	1973-74	78
	Ken Dryden	Montreal	1976-77	80
	Martin Brodeur	New Jersey	1996-97	82
	Martin Brodeur	New Jersey	1997-98	82
	Byron Dafoe	Boston	1998-99	82
	Roman Cechmanek	Philadelphia	2000-01	82

All-Time Win Leaders

(Minimum 225 Wins)

Wins	Goaltender	GP	Dec.	Losses	Ties
516	* Patrick Roy	966	934	300	118
447	Terry Sawchuk	971	949	330	172
435	Jacques Plante	837	827	247	145
423	Tony Esposito	886	880	306	151
407	Glenn Hall	906	896	326	163
403	Grant Fuhr	868	812	295	114
385	Mike Vernon	781	750	273	92
374	John Vanbiesbrouck	882	839	346	119
372	Andy Moog	713	669	209	88
368	* Tom Barrasso	771	727	273	86
364	* Ed Belfour	735	706	242	100
355	Rogie Vachon	795	773	291	127
346	* Curtis Joseph	706	687	260	81
335	Gump Worsley	861	837	352	150
330	Harry Lumley	803	801	329	142
324	* Martin Brodeur	592	577	168	85
305	Billy Smith	680	643	233	105
302	Turk Broda	629	627	224	101
296	Ron Hextall	608	579	214	69
296	* Mike Richter	653	620	252	72
294	Mike Liut	663	639	271	74
289	Ed Giacomin	610	594	208	97
288	Dominik Hasek	581	557	189	80
286	Dan Bouchard	655	631	232	113
284	Tiny Thompson	553	553	194	75
276	* Sean Burke	693	663	295	92
271	Bernie Parent	608	590	198	121
271	Kelly Hrudey	677	624	265	88
270	Gilles Meloche	788	752	351	131
268	Don Beaupre	667	620	277	75
258	Ken Dryden	397	389	57	74
253	* Chris Osgood	455	440	135	52
252	Frank Brimsek	514	514	182	80
250	Johnny Bower	552	535	195	90
246	George Hainsworth	465	465	145	74
246	Pete Peeters	489	452	155	51
245	Kirk McLean	612	579	262	72
240	Bill Ranford	647	595	279	76
237	* Felix Potvin	565	545	232	76
236	Reggie Lemelin	507	461	162	63
234	Eddie Johnston	592	571	257	80
231	Glenn Resch	571	537	224	82
230	Gerry Cheevers	418	406	102	74
225	Ken Wregget	575	526	248	53

* active player

Active Shutout Leaders

(Minimum 22 Shutouts)

Goaltender	Teams	Seasons	Games	Shutouts
Patrick Roy	Montreal, Colorado	18	966	**61**
Ed Belfour	Chi., S.J., Dal.	13	735	**58**
Martin Brodeur	New Jersey	10	592	**55**
Tom Barrasso	Buf., Pit., Ott., Car., Tor.	18	771	**37**
Curtis Joseph	St.L., Edm., Tor.	13	706	**36**
Chris Osgood	Detroit, NY Islanders	9	455	**34**
Arturs Irbe	S.J., Dal., Van., Car.	11	524	**33**
Sean Burke	N.J., Hfd., Car., Van., Phi., Fla., Phx.	14	693	**31**
Tommy Salo	NY Islanders, Edmonton	8	412	**30**
Nikolai Khabibulin	Wpg., Phx., T.B.	7	356	**28**
Olaf Kolzig	Washington	11	415	**27**
Jocelyn Thibault	Que., Col., Mtl., Chi.	9	460	**27**
Mike Vernon	Cgy., Det., S.J., Fla.	19	781	**27**
Byron Dafoe	Wsh., L.A., Bos.	10	380	**26**
Felix Potvin	Tor., NYI, Van., L.A.	11	565	**25**
Mike Richter	NY Rangers	13	653	**24**
Jeff Hackett	NYI, S.J., Chi., Mtl.	13	437	**22**

Goals Against Average Leaders (Minimum 25 games played)

(Exceptions: Minimum 13 games played, 1994-95; minimum 26 games played, 1992-93 to 1993-94; minimum 15 games played, 1917-18 to 1925-26)

Season	Goaltender and Club	GP	Mins.	GA	SO	AVG.
2001-02	Patrick Roy, Colorado	63	3,773	122	9	1.94
2000-01	Marty Turco, Dallas	26	1,266	40	3	1.90
99-2000	Brian Boucher, Philadelphia	35	2,038	65	4	1.91
1998-99	Ron Tugnutt, Ottawa	43	2,508	75	3	1.79
1997-98	Ed Belfour, Dallas	61	3,581	112	9	1.88
1996-97	Martin Brodeur, New Jersey	67	3,838	120	10	1.88
1995-96	Ron Hextall, Philadelphia	53	3,102	112	4	2.17
1994-95	Dominik Hasek, Buffalo	41	2,416	85	5	2.11
1993-94	Dominik Hasek, Buffalo	58	3,358	109	7	1.95
1992-93	Felix Potvin, Toronto	48	2,781	116	2	2.50
1991-92	Patrick Roy, Montreal	67	3,935	155	5	2.36
1990-91	Ed Belfour, Chicago	74	4,127	170	4	2.47
1989-90	Mike Liut, Hartford, Washington	37	2,161	91	4	2.53
1988-89	Patrick Roy, Montreal	48	2,744	113	4	2.47
1987-88	Pete Peeters, Washington	35	1,896	88	2	2.78
1986-87	Brian Hayward, Montreal	37	2,178	102	1	2.81
1985-86	Bob Froese, Philadelphia	51	2,728	116	5	2.55
1984-85	Tom Barrasso, Buffalo	54	3,248	144	5	2.66
1983-84	Pat Riggin, Washington	41	2,299	102	4	2.66
1982-83	Pete Peeters, Boston	62	3,611	142	8	2.36
1981-82	Denis Herron, Montreal	27	1,547	68	3	2.64
1980-81	Richard Sevigny, Montreal	33	1,777	71	2	2.40
1979-80	Bob Sauve, Buffalo	32	1,880	74	4	2.36
1978-79	Ken Dryden, Montreal	47	2,814	108	5	2.30
1977-78	Ken Dryden, Montreal	52	3,071	105	5	2.05
1976-77	Michel Larocque, Montreal	26	1,525	53	4	2.09
1975-76	Ken Dryden, Montreal	62	3,580	121	8	2.03
1974-75	Bernie Parent, Philadelphia	68	4,041	137	12	2.03
1973-74	Bernie Parent, Philadelphia	73	4,314	136	12	1.89
1972-73	Ken Dryden, Montreal	54	3,165	119	6	2.26
1971-72	Tony Esposito, Chicago	48	2,780	82	9	1.77
1970-71	Jacques Plante, Toronto	40	2,329	73	4	1.88
1969-70	Ernie Wakely, St. Louis	30	1,651	58	4	2.11
1968-69	Jacques Plante, St. Louis	37	2,139	70	5	1.96
1967-68	Gump Worsley, Montreal	40	2,213	73	6	1.98
1966-67	Glenn Hall, Chicago	32	1,664	66	2	2.38
1965-66	Johnny Bower, Toronto	35	1,998	75	3	2.25
1964-65	Johnny Bower, Toronto	34	2,040	81	3	2.38
1963-64	Johnny Bower, Toronto	51	3,009	106	5	2.11
1962-63	Don Simmons, Toronto	28	1,680	69	1	2.46
1961-62	Jacques Plante, Montreal	70	4,200	166	4	2.37
1960-61	Charlie Hodge, Montreal	30	1,800	74	4	2.47
1959-60	Jacques Plante, Montreal	69	4,140	175	3	2.54
1958-59	Jacques Plante, Montreal	67	4,000	144	9	2.16
1957-58	Jacques Plante, Montreal	57	3,386	119	9	2.11
1956-57	Jacques Plante, Montreal	61	3,660	122	9	2.00
1955-56	Jacques Plante, Montreal	64	3,840	119	7	1.86
1954-55	Harry Lumley, Toronto	69	4,140	134	8	1.94
1953-54	Harry Lumley, Toronto	69	4,140	128	13	1.86
1952-53	Terry Sawchuk, Detroit	63	3,780	120	9	1.90
1951-52	Terry Sawchuk, Detroit	70	4,200	133	12	1.90
1950-51	Al Rollins, Toronto	40	2,367	70	5	1.77
1949-50	Bill Durnan, Montreal	64	3,840	141	8	2.20
1948-49	Bill Durnan, Montreal	60	3,600	126	10	2.10
1947-48	Turk Broda, Toronto	60	3,600	143	5	2.38
1946-47	Bill Durnan, Montreal	60	3,600	138	4	2.30
1945-46	Bill Durnan, Montreal	40	2,400	104	4	2.60
1944-45	Bill Durnan, Montreal	50	3,000	121	1	2.42
1943-44	Bill Durnan, Montreal	50	3,000	109	2	2.18
1942-43	Johnny Mowers, Detroit	50	3,010	124	6	2.47
1941-42	Frank Brimsek, Boston	47	2,930	115	3	2.35
1940-41	Turk Broda, Toronto	48	2,970	99	5	2.00
1939-40	Dave Kerr, NY Rangers	48	3,000	77	8	1.54
1938-39	Frank Brimsek, Boston	43	2,610	68	10	1.56
1937-38	Tiny Thompson, Boston	48	2,970	89	7	1.80
1936-37	Normie Smith, Detroit	48	2,980	102	6	2.05
1935-36	Tiny Thompson, Boston	48	2,930	82	10	1.68
1934-35	Lorne Chabot, Chicago	48	2,940	88	8	1.80
1933-34	Wilf Cude, Detroit, Montreal	30	1,920	47	5	1.47
1932-33	Tiny Thompson, Boston	48	3,000	88	11	1.76
1931-32	Charlie Gardiner, Chicago	48	2,989	92	4	1.85
1930-31	Roy Worters, NY Americans	44	2,760	74	8	1.61
1929-30	Tiny Thompson, Boston	44	2,680	98	3	2.19
1928-29	George Hainsworth, Montreal	44	2,800	43	22	0.92
1927-28	George Hainsworth, Montreal	44	2,730	48	13	1.05
1926-27	Clint Benedict, Mtl. Maroons	43	2,748	65	13	1.42
1925-26	Alex Connell, Ottawa	36	2,251	42	15	1.12
1924-25	Georges Vezina, Montreal	30	1,860	56	5	1.81
1923-24	Georges Vezina, Montreal	24	1,459	48	3	1.97
1922-23	Clint Benedict, Ottawa	24	1,478	54	4	2.18
1921-22	Clint Benedict, Ottawa	24	1,508	84	2	3.34
1920-21	Clint Benedict, Ottawa	24	1,457	75	2	3.09
1919-20	Clint Benedict, Ottawa	24	1,444	64	5	2.66
1918-19	Clint Benedict, Ottawa	18	1,113	53	2	2.86
1917-18	Georges Vezina, Montreal	21	1,282	84	1	3.93

All-Time Regular Season NHL Coaching Register

Regular Season, 1917-2002

Coach	Team	Games Coached	Wins	Losses	Ties	Years	Cup Wins	Career
Abel, Sid	Chicago	140	39	79	22	2		
	Detroit	811	340	339	132	12		
	St. Louis	10	3	6	1	1		
	Kansas City	3	0	3	0	1		
	Total	**964**	**382**	**427**	**155**	**16**		**1952-76**
Adams, Jack	**Detroit**	**964**	**413**	**390**	**161**	**20**	**3**	**1927-47**
Allen, Keith	**Philadelphia**	**150**	**51**	**67**	**32**	**2**		**1967-69**
Allison, Dave	**Ottawa**	**25**	**2**	**22**	**1**	**1**		**1995-96**
Anderson, Jim	**Washington**	**54**	**4**	**45**	**5**	**1**		**1974-75**
Angotti, Lou	St. Louis	32	6	20	6	2		
	Pittsburgh	80	16	58	6	1		
	Total	**112**	**22**	**78**	**12**	**3**		**1973-84**
Arbour, Al	St. Louis	107	42	40	25	3		
	NY Islanders	1499	739	537	223	19	4	
	Total	**1606**	**781**	**577**	**248**	**22**	**4**	**1970-94**
Armstrong, George	**Toronto**	**47**	**17**	**26**	**4**	**1**		**1988-89**
Barber, Bill	**Philadelphia**	**136**	**73**	**46**	**17**	**2**		**2000-02**
Barkley, Doug	**Detroit**	**77**	**20**	**46**	**11**	**3**		**1970-76**
Beaulieu, Andre	**Minnesota**	**32**	**6**	**23**	**3**	**1**		**1977-78**
Belisle, Danny	**Washington**	**96**	**28**	**51**	**17**	**2**		**1978-80**
Berenson, Red	**St. Louis**	**204**	**100**	**72**	**32**	**3**		**1979-82**
Bergeron, Michel	Quebec	634	265	283	86	8		
	NY Rangers	158	73	67	18	2		
	Total	**792**	**338**	**350**	**104**	**10**		**1980-90**
Berry, Bob	Los Angeles	240	107	94	39	3		
	Montreal	223	116	71	36	3		
	Pittsburgh	240	88	127	25	3		
	St. Louis	157	73	63	21	2		
	Total	**860**	**384**	**355**	**121**	**11**		**1978-94**
Beverley, Nick	**Toronto**	**17**	**9**	**6**	**2**	**1**		**1995-96**
Blackburn, Don	**Hartford**	**140**	**42**	**63**	**35**	**2**		**1979-81**
Blair, Wren	**Minnesota**	**147**	**48**	**65**	**34**	**3**		**1967-70**
Blake, Toe	**Montreal**	**914**	**500**	**255**	**159**	**13**	**8**	**1955-68**
Boileau, Marc	**Pittsburgh**	**151**	**66**	**61**	**24**	**3**		**1973-76**
Boivin, Leo	**St. Louis**	**97**	**28**	**53**	**16**	**2**		**1975-78**
Boucher, Frank	**NY Rangers**	**527**	**181**	**263**	**83**	**11**	**1**	**1939-54**
Boucher, Georges	Mtl. Maroons	12	6	5	1	1		
	Ottawa	48	13	29	6	1		
	St. Louis	35	9	20	6	1		
	Boston	70	22	32	16	1		
	Total	**165**	**50**	**86**	**29**	**4**		**1930-50**
Bowman, Scotty	St. Louis	238	110	83	45	4		
	Montreal	634	419	110	105	8	5	
	Buffalo	404	210	134	60	7		
	Pittsburgh	164	95	53	16	2	1	
	Detroit	701	410	204	87	9	3	
	Total	**2141**	**1244**	**584**	**313**	**30**	**9**	**1967-02**
Bowness, Rick	Winnipeg	28	8	17	3	1		
	Boston	80	36	32	12	1		
	Ottawa	235	39	178	18	4		
	NY Islanders	100	38	50	12	2		
	Total	**443**	**121**	**277**	**45**	**8**		**1988-98**
Brooks, Herb	NY Rangers	285	131	113	41	4		
	Minnesota	80	19	48	13	1		
	New Jersey	84	40	37	7	1		
	Pittsburgh	58	29	24	5	1		
	Total	**507**	**219**	**222**	**66**	**7**		**1981-00**
Brophy, John	**Toronto**	**193**	**64**	**111**	**18**	**3**		**1986-89**
Burnett, George	**Edmonton**	**35**	**12**	**20**	**3**	**1**		**1994-95**
Burns, Charlie	**Minnesota**	**86**	**22**	**50**	**14**	**2**		**1969-75**
Burns, Pat	Montreal	320	174	104	42	4		
	Toronto	281	133	107	41	4		
	Boston	254	105	103	46	4		
	Total	**855**	**412**	**314**	**129**	**12**		**1988-01**
Bush, Eddie	**Kansas City**	**32**	**1**	**23**	**8**	**1**		**1975-76**
Campbell, Colin	**NY Rangers**	**269**	**118**	**108**	**43**	**4**		**1994-98**
Carpenter, Doug	New Jersey	290	100	166	24	4		
	Toronto	91	39	47	5	2		
	Total	**381**	**139**	**213**	**29**	**6**		**1984-91**
Carroll, Dick	**Toronto**	**40**	**18**	**22**	**0**	**2**	**1**	**1917-19**
Carroll, Frank	**Toronto**	**24**	**15**	**9**	**0**	**1**		**1920-21**
Cashman, Wayne	**Philadelphia**	**61**	**32**	**20**	**9**	**1**		**1997-98**
Chambers, Dave	**Quebec**	**98**	**19**	**64**	**15**	**2**		**1990-92**
Charron, Guy	Calgary	16	6	7	3	1		
	Anaheim	49	14	28	7	1		
	Total	**65**	**20**	**35**	**10**	**2**		**1991-01**
Cheevers, Gerry	**Boston**	**376**	**204**	**126**	**46**	**5**		**1980-85**
Cherry, Don	Boston	400	231	105	64	5		
	Colorado	80	19	48	13	1		
	Total	**480**	**250**	**153**	**77**	**6**		**1974-80**
Clancy, King	Mtl. Maroons	18	6	11	1	1		
	Toronto	210	80	81	49	3		
	Total	**228**	**86**	**92**	**50**	**4**		**1937-56**
Clapper, Dit	**Boston**	**230**	**102**	**88**	**40**	**4**		**1945-49**
Cleghorn, Odie	**Pittsburgh**	**168**	**62**	**86**	**20**	**4**		**1925-29**
Cleghorn, Sprague	**Mtl. Maroons**	**48**	**19**	**22**	**7**	**1**		**1931-32**
Conacher, Charlie	**Chicago**	**162**	**56**	**84**	**22**	**3**		**1947-50**
Conacher, Lionel	**NY Americans**	**44**	**14**	**25**	**5**	**1**		**1929-30**
Constantine, Kevin	San Jose	157	55	78	24	3		
	Pittsburgh	188	86	67	35	3		
	New Jersey	31	20	9	2	1		
	Total	**376**	**161**	**154**	**61**	**7**		**1993-02**
Cook, Bill	**NY Rangers**	**117**	**34**	**59**	**24**	**2**		**1951-53**
Crawford, Marc	Quebec	48	30	13	5	1		
	Colorado	246	135	75	36	3	1	
	Vancouver	283	116	128	39	4		
	Total	**577**	**281**	**216**	**80**	**8**	**1**	**1994-02**
Creamer, Pierre	**Pittsburgh**	**80**	**36**	**35**	**9**	**1**		**1987-88**
Creighton, Fred	Atlanta	348	156	136	56	5		
	Boston	73	40	20	13	1		
	Total	**421**	**196**	**156**	**69**	**6**		**1974-80**
Crisp, Terry	Calgary	240	144	63	33	3	1	
	Tampa Bay	391	142	204	45	6		
	Total	**631**	**286**	**267**	**78**	**9**	**1**	**1987-98**
Crozier, Joe	Buffalo	192	77	80	35	3		
	Toronto	40	13	22	5	1		
	Total	**232**	**90**	**102**	**40**	**4**		**1971-81**
Crozier, Roger	**Washington**	**1**	**0**	**1**	**0**	**1**		**1981-82**
Cunniff, John	Hartford	13	3	9	1	1		
	New Jersey	133	59	56	18	2		
	Total	**146**	**62**	**65**	**19**	**3**		**1982-91**
Curry, Alex	**Ottawa**	**36**	**24**	**8**	**4**	**1**		**1925-26**
Dandurand, Leo	**Montreal**	**163**	**78**	**76**	**9**	**6**	**1**	**1921-35**
Day, Hap	**Toronto**	**546**	**259**	**206**	**81**	**10**	**5**	**1940-50**
Dea, Billy	**Detroit**	**11**	**3**	**8**	**0**	**1**		**1981-82**
Delvecchio, Alex	**Detroit**	**245**	**82**	**131**	**32**	**4**		**1973-77**
Demers, Jacques	Quebec	80	25	44	11	1		
	St. Louis	240	106	106	28	3		
	Detroit	320	137	136	47	4		
	Montreal	221	107	87	27	4	1	
	Tampa Bay	145	34	94	17	2		
	Total	**1006**	**409**	**467**	**130**	**14**	**1**	**1979-99**
Denneny, Cy	Boston	44	26	13	5	1	1	
	Ottawa	48	11	27	10	1		
	Total	**92**	**37**	**40**	**15**	**2**	**1**	**1928-33**
Dineen, Bill	**Philadelphia**	**140**	**60**	**60**	**20**	**2**		**1991-93**
Dudley, Rick	**Buffalo**	**188**	**85**	**72**	**31**	**3**		**1989-92**
Duff, Dick	**Toronto**	**2**	**0**	**2**	**0**	**1**		**1979-80**
Dugal, Jules	**Montreal**	**18**	**9**	**6**	**3**	**1**		**1938-39**
Duncan, Art	Detroit	33	10	21	2	1		
	Toronto	47	21	16	10	2	1	
	Total	**80**	**31**	**37**	**12**	**3**	**1**	**1926-32**
Dutton, Red	NY Americans	288	90	151	47	6		
	Brooklyn	48	16	29	3	1		
	Total	**336**	**106**	**180**	**50**	**7**		**1935-42**
Eddolls, Frank	**Chicago**	**70**	**13**	**40**	**17**	**1**		**1954-55**
Esposito, Phil	**NY Rangers**	**45**	**24**	**21**	**0**	**2**		**1986-89**
Evans, Jack	California	80	27	42	11	1		
	Cleveland	160	47	87	26	2		
	Hartford	374	163	174	37	5		
	Total	**614**	**237**	**303**	**74**	**8**		**1975-88**
Fashoway, Gordie	**Oakland**	**10**	**4**	**5**	**1**	**1**		**1967-68**
Ferguson, John	NY Rangers	121	43	59	19	2		
	Winnipeg	14	7	6	1	1		
	Total	**135**	**50**	**65**	**20**	**3**		**1975-86**
Filion, Maurice	**Quebec**	**6**	**1**	**3**	**2**	**1**		**1980-81**
Francis, Bob	**Phoenix**	**246**	**114**	**98**	**34**	**3**		**1999-02**
Francis, Emile	NY Rangers	654	342	209	103	10		
	St. Louis	124	46	64	14	3		
	Total	**778**	**388**	**273**	**117**	**13**		**1965-83**
Fraser, Curt	**Atlanta**	**246**	**56**	**160**	**30**	**3**		**1999-02**
Fredrickson, Frank	**Pittsburgh**	**44**	**5**	**36**	**3**	**1**		**1929-30**
Ftorek, Robbie	Los Angeles	132	65	56	11	2		
	New Jersey	156	88	49	19	2		
	Boston	82	43	33	6	1		
	Total	**370**	**196**	**138**	**36**	**5**		**1987-02**
Gadsby, Bill	**Detroit**	**78**	**35**	**31**	**12**	**2**		**1968-70**
Gainey, Bob	Minnesota	244	95	119	30	3		
	Dallas	171	70	71	30	3		
	Total	**415**	**165**	**190**	**60**	**6**		**1990-96**
Gardiner, Herb	**Chicago**	**32**	**5**	**23**	**4**	**1**		**1929-30**
Gardner, Jimmy	**Hamilton**	**30**	**19**	**10**	**1**	**1**		**1924-25**
Garvin, Ted	**Detroit**	**11**	**2**	**8**	**1**	**1**		**1973-74**
Geoffrion, Bernie	NY Rangers	43	22	18	3	1		
	Atlanta	208	77	92	39	3		
	Montreal	30	15	9	6	1		
	Total	**281**	**114**	**119**	**48**	**5**		**1968-80**
Gerard, Eddie	Ottawa	22	9	13	0	1		
	Mtl. Maroons	294	129	122	43	7	1	
	NY Americans	92	34	40	18	2		
	St. Louis	13	2	11	0	1		
	Total	**421**	**174**	**186**	**61**	**11**	**1**	**1917-35**
Gilbert, Greg	**Calgary**	**96**	**36**	**46**	**14**	**2**		**2000-02**
Gill, David	**Ottawa**	**132**	**64**	**41**	**27**	**3**	**1**	**1926-29**
Glover, Fred	Oakland	152	51	76	25	2		
	California	204	45	131	28	4		
	Los Angeles	68	18	42	8	1		
	Total	**424**	**114**	**249**	**61**	**7**		**1968-74**
Goodfellow, Ebbie	**Chicago**	**140**	**30**	**91**	**19**	**2**		**1950-52**
Gordon, Jackie	**Minnesota**	**289**	**116**	**123**	**50**	**5**		**1970-75**
Goring, Butch	Boston	93	42	38	13	2		
	NY Islanders	147	41	92	14	2		
	Total	**240**	**83**	**130**	**27**	**4**		**1985-01**

Coach	Team	Games Coached	Wins	Losses	Ties	Years	Cup Wins	Career
Gorman, Tommy	NY Americans	80	31	33	16	2		
	Chicago	73	28	28	17	2	1	
	Mtl. Maroons	174	74	71	29	4	1	
	Total	327	133	132	62	8	2	1925-38
Gottselig, Johnny	Chicago	187	62	105	20	4		1944-48
Goyette, Phil	NY Islanders	48	6	38	4	1		1972-73
Graham, Dirk	Chicago	59	16	35	8	1		1998-99
Green, Gary	Washington	157	50	78	29	3		1979-82
Green, Pete	Ottawa	150	94	52	4	6	3	1919-25
Green, Shorty	NY Americans	44	11	27	6	1		1927-28
Green, Ted	Edmonton	188	65	102	21	3		1991-94
Guidolin, Aldo	Colorado	59	12	39	8	1		1978-79
Guidolin, Bep	Boston	104	72	23	9	2		
	Kansas City	125	26	84	15	2		
	Total	229	98	107	24	4		1972-76
Harkness, Ned	Detroit	38	12	22	4	1		1970-71
Harris, Ted	Minnesota	179	48	104	27	3		1975-78
Hart, Cecil	Montreal	394	196	125	73	9	2	1926-39
Hartley, Bob	Colorado	328	183	106	39	4	1	1998-02
Hartsburg, Craig	Chicago	246	104	102	40	3		
	Anaheim	197	80	88	29	3		
	Total	443	184	190	69	6		1995-01
Harvey, Doug	NY Rangers	70	26	32	12	1		1961-62
Hay, Don	Phoenix	82	38	37	7	1		
	Calgary	68	23	32	13	1		
	Total	150	61	69	20	2		1996-01
Heffernan, Frank	Toronto	12	5	7	0	1		1919-20
Henning, Lorne	Minnesota	158	68	72	18	2		
	NY Islanders	65	19	39	7	2		
	Total	223	87	111	25	4		1985-01
Hitchcock, Ken	Dallas	503	277	166	60	7		1995-02
Hlinka, Ivan	Pittsburgh	86	42	35	9	2		2000-02
Holmgren, Paul	Philadelphia	264	107	126	31	4		
	Hartford	161	54	93	14	4		
	Total	425	161	219	45	8		1988-96
Howell, Harry	Minnesota	11	3	6	2	1		1978-79
Imlach, Punch	Toronto	770	370	275	125	12	4	
	Buffalo	119	32	62	25	2		
	Total	889	402	337	150	14	4	1958-80
Ingarfield, Earl	NY Islanders	30	6	22	2	1		1972-73
Inglis, Bill	Buffalo	56	28	18	10	1		1978-79
Irvin, Dick	Chicago	126	45	62	19	3		
	Toronto	427	216	152	59	9	1	
	Montreal	896	431	313	152	15	3	
	Total	1449	692	527	230	27	4	1928-56
Ivan, Tommy	Detroit	470	262	118	90	7	3	
	Chicago	103	26	56	21	2		
	Total	573	288	174	111	9	3	1947-58
Iverson, Emil	Chicago	21	8	7	6	1		1932-33
Johnson, Bob	Calgary	400	193	155	52	5		
	Pittsburgh	80	41	33	6	1	1	
	Total	480	234	188	58	6	1	1982-91
Johnson, Tom	Boston	208	142	43	23	3	1	1970-73
Johnston, Eddie	Chicago	80	34	27	19	1		
	Pittsburgh	516	232	224	60	7		
	Total	596	266	251	79	8		1979-97
Johnston, Marshall	California	69	13	45	11	2		
	Colorado	56	15	32	9	1		
	Total	125	28	77	20	3		1973-82
Kasper, Steve	Boston	164	66	78	20	2		1995-97
Keats, Duke	Detroit	11	2	7	2	1		1926-27
Keenan, Mike	Philadelphia	320	190	102	28	4		
	Chicago	320	153	126	41	4		
	NY Rangers	84	52	24	8	1	1	
	St. Louis	163	75	66	22	3		
	Vancouver	108	36	54	18	2		
	Boston	74	33	34	7	1		
	Florida	56	16	32	8	1		
	Total	1125	555	438	132	16	1	1984-02
Kehoe, Rick	Pittsburgh	78	28	42	8	1		2001-02
Kelly, Pat	Colorado	101	22	54	25	2		1977-79
Kelly, Red	Los Angeles	150	55	75	20	2		
	Pittsburgh	274	90	132	52	4		
	Toronto	318	133	123	62	4		
	Total	742	278	330	134	10		1967-77
King, Dave	Calgary	216	109	76	31	3		
	Columbus	164	50	97	17	2		
	Total	380	159	173	48	5		1992-02
Kingston, George	San Jose	164	28	129	7	2		1991-93
Kish, Larry	Hartford	49	12	32	5	1		1982-83
Kromm, Bobby	Detroit	231	79	111	41	3		1977-80
Kurtenbach, Orland	Vancouver	125	36	62	27	2		1976-78
LaForge, Bill	Vancouver	20	4	14	2	1		1984-85
Lalonde, Newsy	Montreal	207	96	97	14	8		
	NY Americans	44	17	25	2	1		
	Ottawa	88	31	45	12	2		
	Total	339	144	167	28	11		1917-35
Lapointe, Ron	Quebec	89	33	50	6	2		1987-89
Laviolette, Peter	NY Islanders	82	42	32	8	1		2001-02
Laycoe, Hal	Los Angeles	24	5	18	1	1		
	Vancouver	156	44	96	16	2		
	Total	180	49	114	17	3		1969-72
Lehman, Hugh	Chicago	21	3	17	1	1		1927-28
Lemaire, Jacques	Montreal	97	48	37	12	2		
	New Jersey	378	199	122	57	5	1	
	Minnesota	164	51	88	25	2		
	Total	639	298	247	94	9	1	1983-02
Lepine, Pit	Montreal	48	10	33	5	1		1939-40
LeSueur, Percy	Hamilton	10	3	7	0	1		1923-24
Lewis, Dave*	Detroit	5	4	1	0	1		1998-99
	*Results shared with co-coach Barry Smith							
Ley, Rick	Hartford	160	69	71	20	2		
	Vancouver	124	47	50	27	2		
	Total	284	116	121	47	4		1989-96
Lindsay, Ted	Detroit	29	5	21	3	2		1979-81
Long, Barry	Winnipeg	205	87	93	25	3		1983-86
Loughlin, Clem	Chicago	144	61	63	20	3		1934-37
Lowe, Ron	Edmonton	341	139	162	40	5		
	NY Rangers	164	69	86	9	2		
	Total	505	208	248	49	7		1994-02
Lowe, Kevin	Edmonton	82	32	34	16	1		1999-00
Ludzik, Steve	Tampa Bay	121	31	76	14	2		1999-01
MacDonald, Parker	Minnesota	61	20	30	11	1		
	Los Angeles	42	13	24	5	1		
	Total	103	33	54	16	2		1973-82
MacLean, Doug	Florida	187	83	71	33	3		1995-98
MacMillan, Bill	Colorado	80	22	45	13	1		
	New Jersey	100	19	67	14	2		
	Total	180	41	112	27	3		1980-84
MacNeil, Al	Montreal	55	31	15	9	1	1	
	Atlanta	80	35	32	13	1		
	Calgary	160	68	61	31	2		
	Total	295	134	108	53	4	1	1970-82
MacTavish, Craig	Edmonton	164	77	63	24	2		2000-02
Magnuson, Keith	Chicago	132	49	57	26	2		1980-82
Mahoney, Bill	Minnesota	93	42	39	12	2		1983-85
Maloney, Dan	Toronto	160	45	100	15	2		
	Winnipeg	212	91	93	28	3		
	Total	372	136	193	43	5		1984-89
Maloney, Phil	Vancouver	232	95	105	32	4		1973-77
Mantha, Sylvio	Montreal	48	11	26	11	1		1935-36
Marshall, Bert	Colorado	24	3	17	4	1		1981-82
Martin, Jacques	St. Louis	160	66	71	23	2		
	Ottawa	528	246	204	78	6		
	Total	688	312	275	101	9		1986-02
Matheson, Godfrey	Chicago	2	0	2	0	1		1932-33
Maurice, Paul	Hartford	152	61	72	19	2		
	Carolina	410	177	172	61	5		
	Total	562	238	244	80	7		1995-02
Maxner, Wayne	Detroit	129	34	68	27	2		1980-82
McCammon, Bob	Philadelphia	218	119	68	31	4		
	Vancouver	294	102	156	36	4		
	Total	512	221	224	67	8		1978-91
McCreary, Bill	St. Louis	24	6	14	4	1		
	Vancouver	41	9	25	7	1		
	California	32	8	20	4	1		
	Total	97	23	59	15	3		1971-75
McGuire, Pierre	Hartford	67	23	37	7	1		1993-94
McLellan, John	Toronto	310	126	139	45	4		1969-73
McVie, Tom	Washington	204	49	122	33	3		
	Winnipeg	105	20	67	18	2		
	New Jersey	153	57	74	22	3		
	Total	462	126	263	73	8		1975-92
Meeker, Howie	Toronto	70	21	34	15	1		1956-57
Melrose, Barry	Los Angeles	209	79	101	29	3		1992-95
Milbury, Mike	Boston	160	90	49	21	2		
	NY Islanders	191	56	111	24	4		
	Total	351	146	160	45	6		1989-99
Molleken, Lorne	Chicago	47	18	21	8	2		1998-00
Muckler, John	Minnesota	35	6	23	6	1		
	Edmonton	160	75	65	20	2	1	
	Buffalo	268	125	109	34	4		
	NY Rangers	185	70	91	24	3		
	Total	648	276	288	84	10	1	1968-00
Muldoon, Pete	Chicago	44	19	22	3	1		1926-27
Munro, Dunc	Mtl. Maroons	76	37	29	10	2		1929-31
Murdoch, Bob	Chicago	80	30	41	9	1		
	Winnipeg	160	63	75	22	2		
	Total	240	93	116	31	3		1987-91
Murphy, Mike	Los Angeles	65	20	37	8	2		
	Toronto	164	60	87	17	2		
	Total	229	80	124	25	4		1986-98
Murray, Andy	Los Angeles	246	117	93	36	3		1999-02
Murray, Bryan	Washington	672	343	246	83	9		
	Detroit	244	124	91	29	3		
	Florida	59	17	31	11	1		
	Anaheim	82	29	45	8	1		
	Total	1057	513	413	131	14		1981-02
Murray, Terry	Washington	325	163	134	28	5		
	Philadelphia	212	118	64	30	3		
	Florida	200	79	90	31	3		
	Total	737	360	288	89	11		1989-01
Nanne, Lou	Minnesota	29	7	18	4	1		1977-78
Neale, Harry	Vancouver	407	142	189	76	6		
	Detroit	35	8	23	4	1		
	Total	442	150	212	80	7		1978-86
Neilson, Roger	Toronto	160	75	62	23	2		
	Buffalo	80	39	20	21	1		
	Vancouver	133	51	61	21	3		
	Los Angeles	28	8	17	3	1		
	NY Rangers	280	141	104	35	4		
	Florida	132	53	56	23	2		
	Philadelphia	185	92	60	33	3		
	Ottawa	2	1	1	0	1		
	Total	1000	460	381	159	17		1977-02
Nolan, Ted	Buffalo	164	73	72	19	2		1995-97

Coach	Team	Games Coached	Wins	Losses	Ties	Years	Cup Wins	Career
Nykoluk, Mike	Toronto	280	89	144	47	4		1980-84
O'Donoghue, George	Toronto	29	15	13	1	2	1	1921-23
O'Reilly, Terry	Boston	227	115	86	26	3		1986-89
Oliver, Murray	Minnesota	41	21	12	8	2		1981-83
Olmstead, Bert	Oakland	64	11	37	16	1		1967-68
Paddock, John	Winnipeg	281	106	138	37	4		1991-95
Page, Pierre	Minnesota	160	63	77	20	2		
	Quebec	230	98	103	29	3		
	Calgary	164	66	78	20	2		
	Anaheim	82	26	43	13	1		
	Total	636	253	301	82	8		1988-98
Park, Brad	Detroit	45	9	34	2	1		1985-86
Paterson, Rick	Tampa Bay	8	0	8	0	1		1997-98
Patrick, Craig	NY Rangers	95	37	45	13	2		
	Pittsburgh	74	29	36	9	2		
	Total	169	66	81	22	4		1980-97
Patrick, Frank	Boston	96	48	36	12	2		1934-36
Patrick, Lester	NY Rangers	604	281	216	107	13	2	1926-39
Patrick, Lynn	NY Rangers	107	40	51	16	2		
	Boston	310	117	130	63	5		
	St. Louis	26	8	15	3	3		
	Total	443	165	196	82	10		1948-76
Patrick, Muzz	NY Rangers	136	43	66	27	4		1953-63
Perron, Jean	Montreal	240	126	84	30	3	1	
	Quebec	47	16	26	5	1		
	Total	287	142	110	35	4	1	1985-89
Perry, Don	Los Angeles	168	52	85	31	3		1981-84
Pike, Alf	NY Rangers	123	36	66	21	2		1959-61
Pilous, Rudy	Chicago	387	162	151	74	6	1	1957-63
Plager, Barclay	St. Louis	178	49	96	33	4		1977-83
Plager, Bob	St. Louis	11	4	6	1	1		1992-93
Pleau, Larry	Hartford	224	81	117	26	5		1980-89
Polano, Nick	Detroit	240	79	127	34	3		1982-85
Popein, Larry	NY Rangers	41	18	14	9	1		1973-74
Powers, Eddie	Toronto	66	31	32	3	2		1924-26
Primeau, Joe	Toronto	210	97	71	42	3	1	1950-53
Pronovost, Marcel	Buffalo	104	52	29	23	2		1977-79
Pulford, Bob	Los Angeles	396	178	150	68	5		
	Chicago	433	185	180	68	7		
	Total	829	363	330	136	12		1972-00
Quenneville, Joel	St. Louis	450	237	154	59	6		1996-02
Querrie, Charles	Toronto	72	29	38	5	3		1922-27
Quinn, Mike	Quebec	24	4	20	0	1		1919-20
Quinn, Pat	Philadelphia	262	141	73	48	4		
	Los Angeles	202	75	101	26	3		
	Vancouver	280	141	111	28	5		
	Toronto	328	170	123	35	4		
	Total	1072	527	408	137	16		1978-02
Ramsay, Craig	Buffalo	21	4	15	2	1		
	Philadelphia	28	12	12	4	1		
	Total	49	16	27	6	2		1986-01
Randall, Ken	Hamilton	14	6	8	0	1		1923-24
Reay, Billy	Toronto	90	26	50	14	2		
	Chicago	1012	516	335	161	14		
	Total	1102	542	385	175	16		1957-77
Regan, Larry	Los Angeles	88	27	47	14	2		1970-72
Renney, Tom	Vancouver	101	39	53	9	2		1996-98
Risebrough, Doug	Calgary	144	71	56	17	2		1990-92
Roberts, Jim	Buffalo	45	21	16	8	1		
	Hartford	80	26	41	13	1		
	St. Louis	9	3	3	3	1		
	Total	134	50	60	24	3		1981-97
Robinson, Larry	Los Angeles	328	122	161	45	4		
	New Jersey	141	73	49	19	3	1	
	Total	469	195	210	64	7	1	1995-02
Rodden, Mike	Toronto	2	0	2	0	1		1926-27
Romeril, Alex	Toronto	13	7	5	1	1		1926-27
Ross, Art	Mtl. Wanderers	6	1	5	0	1		
	Hamilton	24	6	18	0	1		
	Boston	728	361	277	90	16	1	
	Total	758	368	300	90	18	1	1917-45
Ruel, Claude	Montreal	305	172	82	51	5	2	1968-81
Ruff, Lindy	Buffalo	410	189	160	61	5		1997-02
Sather, Glen	Edmonton	842	464	268	110	11	4	1979-94
Sator, Ted	NY Rangers	99	41	48	10	2		
	Buffalo	207	96	89	22	3		
	Total	306	137	137	32	5		1985-89
Savard, Andre	Quebec	24	10	13	1	1		1987-88
Schinkel, Ken	Pittsburgh	203	83	92	28	4		1972-77
Schmidt, Milt	Boston	726	245	360	121	11		
	Washington	44	5	34	5	2		
	Total	770	250	394	126	13		1954-76
Schoenfeld, Jim	Buffalo	43	19	19	5	1		
	New Jersey	124	50	59	15	3		
	Washington	249	113	102	34	4		
	Phoenix	164	74	66	24	2		
	Total	580	256	248	78	10		1985-99
Shaughnessy, Tom	Chicago	21	10	8	3	1		1929-30
Shero, Fred	Philadelphia	554	308	151	95	7	2	
	NY Rangers	180	82	74	24	3		
	Total	734	390	225	119	10	2	1971-81
Simpson, Joe	NY Americans	144	42	72	30	3		1932-35
Simpson, Terry	NY Islanders	187	81	82	24	3		
	Philadelphia	84	35	39	10	1		
	Winnipeg	97	43	47	7	2		
	Total	368	159	168	41	6		1986-96
Sims, Al	San Jose	82	27	47	8	1		1996-97
Sinden, Harry	Boston	327	153	116	58	6	1	1966-85
Skinner, Jimmy	Detroit	247	123	78	46	4	1	1954-58
Smeaton, Cooper	Philadelphia	44	4	36	4	1		1930-31
Smith, Alf	Ottawa	18	12	6	0	1		1918-19
Smith, Barry*	Detroit	5	4	1	0	1		1998-99
	*Results shared with co-coach Dave Lewis							
Smith, Floyd	Buffalo	241	143	62	36	4		
	Toronto	68	30	33	5	1		
	Total	309	173	95	41	5		1971-80
Smith, Mike	Winnipeg	23	2	17	4	1		1980-81
Smith, Ron	NY Rangers	44	15	22	7	1		1992-93
Smythe, Conn	Toronto	134	57	57	20	4		1927-31
Sonmor, Glen	Minnesota	417	174	161	82	7		1978-87
Sproule, Harry	Toronto	12	7	5	0	1		1919-20
Stanley, Barney	Chicago	23	4	17	2	1		1927-28
Stasiuk, Vic	Philadelphia	154	45	68	41	2		
	California	75	21	38	16	1		
	Vancouver	78	22	47	9	1		
	Total	307	88	153	66	4		1969-73
Stewart, Bill	Chicago	69	22	35	12	2	1	1937-39
Stewart, Bill	NY Islanders	37	11	19	7	1		1998-99
Stewart, Ron	NY Rangers	39	15	20	4	1		
	Los Angeles	80	31	34	15	1		
	Total	119	46	54	19	2		1975-78
Suhonen, Alpo	Chicago	82	29	45	8	1		2000-01
Sullivan, Red	NY Rangers	196	58	103	35	4		
	Pittsburgh	150	47	79	24	2		
	Washington	18	2	16	0	1		
	Total	364	107	198	59	7		1962-75
Sutherland, Bill	Winnipeg	32	7	22	3	2		1979-81
Sutter, Brian	St. Louis	320	153	124	43	4		
	Boston	216	120	73	23	3		
	Calgary	246	87	122	37	3		
	Chicago	82	41	28	13	1		
	Total	864	401	347	116	11		1988-02
Sutter, Darryl	Chicago	216	110	80	26	3		
	San Jose	410	184	168	58	5		
	Total	626	294	248	84	8		1992-02
Sutter, Duane	Florida	72	22	42	8	2		2000-02
Talbot, Jean-Guy	St. Louis	120	52	53	15	2		
	NY Rangers	80	30	37	13	1		
	Total	200	82	90	28	3		1972-78
Tessier, Orval	Chicago	213	99	93	21	3		1982-85
Therrien, Michel	Montreal	144	59	67	18	2		2000-02
Thompson, Paul	Chicago	272	104	127	41	7		1938-45
Thompson, Percy	Hamilton	48	13	35	0	2		1920-22
Tobin, Bill	Chicago	71	29	29	13	2		1929-32
Tortorella, John	NY Rangers	4	0	3	1	1		
	Tampa Bay	125	39	74	12	2		
	Total	129	39	77	13	3		1999-02
Tremblay, Mario	Montreal	159	71	63	25	2		1995-97
Trotz, Barry	Nashville	328	118	174	36	4		1998-02
Ubriaco, Gene	Pittsburgh	106	50	47	9	2		1988-90
Vachon, Rogie	Los Angeles	10	4	3	3	3		1983-95
Vigneault, Alain	Montreal	266	109	122	35	4		1997-01
Watson, Bryan	Edmonton	18	4	9	5	1		1980-81
Watson, Phil	NY Rangers	295	119	124	52	5		
	Boston	84	16	55	13	2		
	Total	379	135	179	65	7		1955-63
Watt, Tom	Winnipeg	181	72	85	24	3		
	Vancouver	160	52	87	21	2		
	Toronto	149	52	80	17	2		
	Total	490	176	252	62	7		1981-92
Webster, Tom	NY Rangers	18	5	9	4	1		
	Los Angeles	240	115	94	31	3		
	Total	258	120	103	35	4		1986-92
Weiland, Cooney	Boston	96	58	20	18	2	1	1939-41
White, Bill	Chicago	46	16	24	6	1		1976-77
Wiley, Jim	San Jose	57	17	37	3	1		1995-96
Wilson, Johnny	Los Angeles	52	9	34	9	1		
	Detroit	145	67	56	22	2		
	Colorado	80	20	46	14	1		
	Pittsburgh	240	91	105	44	3		
	Total	517	187	241	89	7		1969-80
Wilson, Larry	Detroit	36	3	29	4	1		1976-77
Wilson, Rick	Dallas	32	13	12	7	1		2001-02
Wilson, Ron	Anaheim	296	120	145	31	4		
	Washington	410	192	167	51	5		
	Total	706	312	312	82	9		1993-02
Young, Garry	California	12	2	7	3	1		
	St. Louis	98	41	41	16	2		
	Total	110	43	48	19	3		1972-76

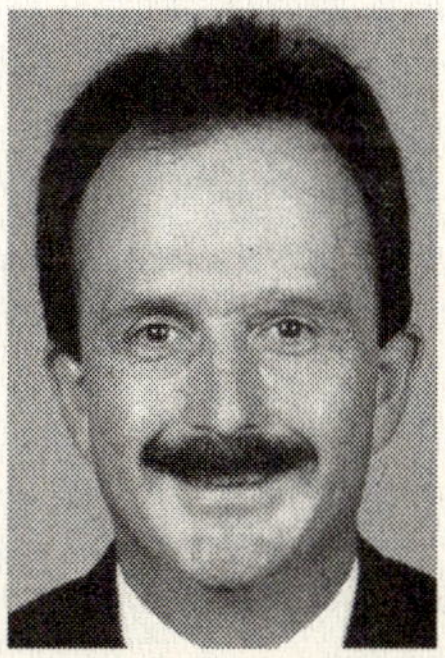

Scotty Bowman (far left) announced his retirement after leading the Red Wings to the Stanley Cup in 2002. It was his record ninth coaching championship. Long-time associate coach Dave Lewis (left) takes over the team this season.

Year-by-Year Individual Regular-Season Leaders

Season	Goals	G	Assists	A	Points	Pts.	Penalty Minutes	PIM
1917-18	Joe Malone	44	Cy Denneny	10	Joe Malone	48	Joe Hall	100
			Reg Noble	10				
			Harry Cameron	10				
1918-19	Newsy Lalonde	22	Newsy Lalonde	10	Newsy Lalonde	32	Joe Hall	135
			Eddie Gerard	10				
1919-20	Joe Malone	39	Frank Nighbor	15	Joe Malone	49	Cully Wilson	86
1920-21	Babe Dye	35	Jack Darragh	15	Newsy Lalonde	43	Bert Corbeau	86
1921-22	Punch Broadbent	32	Harry Cameron	17	Punch Broadbent	46	Sprague Cleghorn	63
1922-23	Babe Dye	26	Edmond Bouchard	12	Babe Dye	37	Georges Boucher	58
1923-24	Cy Denneny	22	Georges Boucher	10	Cy Denneny	24	Bert Corbeau	55
1924-25	Babe Dye	38	Cy Denneny	15	Babe Dye	46	Georges Boucher	95
			Red Green	15				
1925-26	Nels Stewart	34	Frank Nighbor	13	Nels Stewart	42	Bert Corbeau	121
1926-27	Bill Cook	33	Dick Irvin	18	Bill Cook	37	Nels Stewart	133
1927-28	Howie Morenz	33	Howie Morenz	18	Howie Morenz	51	Eddie Shore	165
1928-29	Ace Bailey	22	Frank Boucher	16	Ace Bailey	32	Red Dutton	139
1929-30	Cooney Weiland	43	Frank Boucher	36	Cooney Weiland	73	Joe Lamb	119
1930-31	Charlie Conacher	31	Joe Primeau	32	Howie Morenz	51	Harvey Rockburn	118
1931-32	Charlie Conacher	34	Joe Primeau	37	Busher Jackson	53	Red Dutton	107
	Bill Cook	34						
1932-33	Bill Cook	28	Frank Boucher	28	Bill Cook	50	Red Horner	144
1933-34	Charlie Conacher	32	Joe Primeau	32	Charlie Conacher	52	Red Horner	126 *
1934-35	Charlie Conacher	36	Art Chapman	34	Charlie Conacher	57	Red Horner	125
1935-36	Charlie Conacher	23	Art Chapman	28	Sweeney Schriner	45	Red Horner	167
	Bill Thoms	23						
1936-37	Larry Aurie	23	Syl Apps	29	Sweeney Schriner	46	Red Horner	124
	Nels Stewart	23						
1937-38	Gordie Drillon	26	Syl Apps	29	Gordie Drillon	52	Red Horner	82 *
1938-39	Roy Conacher	26	Bill Cowley	34	Toe Blake	47	Red Horner	85
1939-40	Bryan Hextall	24	Milt Schmidt	30	Milt Schmidt	52	Red Horner	87
1940-41	Bryan Hextall	26	Bill Cowley	45	Bill Cowley	62	Jimmy Orlando	99
1941-42	Lynn Patrick	32	Phil Watson	37	Bryan Hextall	56	Pat Egan	124
1942-43	Doug Bentley	33	Bill Cowley	45	Doug Bentley	73	Jimmy Orlando	89 *
1943-44	Doug Bentley	38	Clint Smith	49	Herb Cain	82	Mike McMahon	98
1944-45	Maurice Richard	50	Elmer Lach	54	Elmer Lach	80	Pat Egan	86
1945-46	Gaye Stewart	37	Elmer Lach	34	Max Bentley	61	Jack Stewart	73
1946-47	Maurice Richard	45	Billy Taylor	46	Max Bentley	72	Gus Mortson	133
1947-48	Ted Lindsay	33	Doug Bentley	37	Elmer Lach	61	Bill Barilko	147
1948-49	Sid Abel	28	Doug Bentley	43	Roy Conacher	68	Bill Ezinicki	145
1949-50	Maurice Richard	43	Ted Lindsay	55	Ted Lindsay	78	Bill Ezinicki	144
1950-51	Gordie Howe	43	Gordie Howe	43	Gordie Howe	86	Gus Mortson	142
			Ted Kennedy	43				
1951-52	Gordie Howe	47	Elmer Lach	50	Gordie Howe	86	Gus Kyle	127
1952-53	Gordie Howe	49	Gordie Howe	46	Gordie Howe	95	Maurice Richard	112
1953-54	Maurice Richard	37	Gordie Howe	48	Gordie Howe	81	Gus Mortson	132
1954-55	Maurice Richard	38	Bert Olmstead	48	Bernie Geoffrion	75	Fern Flaman	150
	Bernie Geoffrion	38						
1955-56	Jean Beliveau	47	Bert Olmstead	56	Jean Beliveau	88	Lou Fontinato	202
1956-57	Gordie Howe	44	Ted Lindsay	55	Gordie Howe	89	Gus Mortson	147
1957-58	Dickie Moore	36	Henri Richard	52	Dickie Moore	84	Lou Fontinato	152
1958-59	Jean Beliveau	45	Dickie Moore	55	Dickie Moore	96	Ted Lindsay	184
1959-60	Bobby Hull	39	Don McKenney	49	Bobby Hull	81	Carl Brewer	150
	Bronco Horvath	39						
1960-61	Bernie Geoffrion	50	Jean Beliveau	58	Bernie Geoffrion	95	Pierre Pilote	165
1961-62	Bobby Hull	50	Andy Bathgate	56	Bobby Hull	84	Lou Fontinato	167
					Andy Bathgate	84		
1962-63	Gordie Howe	38	Henri Richard	50	Gordie Howe	86	Howie Young	273
1963-64	Bobby Hull	43	Andy Bathgate	58	Stan Mikita	89	Vic Hadfield	151
1964-65	Norm Ullman	42	Stan Mikita	59	Stan Mikita	87	Carl Brewer	177
1965-66	Bobby Hull	54	Stan Mikita	48	Bobby Hull	97	Reggie Fleming	166
			Bobby Rousseau	48				
			Jean Beliveau	48				
1966-67	Bobby Hull	52	Stan Mikita	62	Stan Mikita	97	John Ferguson	177
1967-68	Bobby Hull	44	Phil Esposito	49	Stan Mikita	87	Barclay Plager	153
1968-69	Bobby Hull	58	Phil Esposito	77	Phil Esposito	126	Forbes Kennedy	219
1969-70	Phil Esposito	43	Bobby Orr	87	Bobby Orr	120	Keith Magnuson	213
1970-71	Phil Esposito	76	Bobby Orr	102	Phil Esposito	152	Keith Magnuson	291
1971-72	Phil Esposito	66	Bobby Orr	80	Phil Esposito	133	Bryan Watson	212
1972-73	Phil Esposito	55	Phil Esposito	75	Phil Esposito	130	Dave Schultz	259
1973-74	Phil Esposito	68	Bobby Orr	90	Phil Esposito	145	Dave Schultz	348
1974-75	Phil Esposito	61	Bobby Orr	89	Bobby Orr	135	Dave Schultz	472
			Bobby Clarke	89				
1975-76	Reggie Leach	61	Bobby Clarke	89	Guy Lafleur	125	Steve Durbano	370
1976-77	Steve Shutt	60	Guy Lafleur	80	Guy Lafleur	136	Tiger Williams	338
1977-78	Guy Lafleur	60	Bryan Trottier	77	Guy Lafleur	132	Dave Schultz	405
1978-79	Mike Bossy	69	Bryan Trottier	87	Bryan Trottier	134	Tiger Williams	298
1979-80	Charlie Simmer	56	Wayne Gretzky	86	Marcel Dionne	137	Jimmy Mann	287
	Danny Gare	56			Wayne Gretzky	137		
	Blaine Stoughton	56						
1980-81	Mike Bossy	68	Wayne Gretzky	109	Wayne Gretzky	164	Tiger Williams	343
1981-82	Wayne Gretzky	92	Wayne Gretzky	120	Wayne Gretzky	212	Paul Baxter	409
1982-83	Wayne Gretzky	71	Wayne Gretzky	125	Wayne Gretzky	196	Randy Holt	275
1983-84	Wayne Gretzky	87	Wayne Gretzky	118	Wayne Gretzky	205	Chris Nilan	338
1984-85	Wayne Gretzky	73	Wayne Gretzky	135	Wayne Gretzky	208	Chris Nilan	358
1985-86	Jari Kurri	68	Wayne Gretzky	163	Wayne Gretzky	215	Joe Kocur	377
1986-87	Wayne Gretzky	62	Wayne Gretzky	121	Wayne Gretzky	183	Tim Hunter	361
1987-88	Mario Lemieux	70	Wayne Gretzky	109	Mario Lemieux	168	Bob Probert	398
1988-89	Mario Lemieux	85	Mario Lemieux	114	Mario Lemieux	199	Tim Hunter	375
			Wayne Gretzky	114				
1989-90	Brett Hull	72	Wayne Gretzky	102	Wayne Gretzky	142	Basil McRae	351
1990-91	Brett Hull	86	Wayne Gretzky	122	Wayne Gretzky	163	Rob Ray	350
1991-92	Brett Hull	70	Wayne Gretzky	90	Mario Lemieux	131	Mike Peluso	408
1992-93	Teemu Selanne	76	Adam Oates	97	Mario Lemieux	160	Marty McSorley	399
	Alexander Mogilny	76						
1993-94	Pavel Bure	60	Wayne Gretzky	92	Wayne Gretzky	130	Tie Domi	347
1994-95	Peter Bondra	34	Ron Francis	48	Jaromir Jagr	70	Enrico Ciccone	225
					Eric Lindros	70		
1995-96	Mario Lemieux	69	Mario Lemieux	92	Mario Lemieux	161	Matthew Barnaby	335
			Ron Francis	92				
1996-97	Keith Tkachuk	52	Mario Lemieux	72	Mario Lemieux	122	Gino Odjick	371
			Wayne Gretzky	72				
1997-98	Teemu Selanne	52	Jaromir Jagr	67	Jaromir Jagr	102	Donald Brashear	372
	Peter Bondra	52	Wayne Gretzky	67				
1998-99	Teemu Selanne	47	Jaromir Jagr	83	Jaromir Jagr	127	Rob Ray	261
99-2000	Pavel Bure	58	Mark Recchi	63	Jaromir Jagr	96	Denny Lambert	219
2000-01	Pavel Bure	59	Jaromir Jagr	69	Jaromir Jagr	121	Matthew Barnaby	265
			Adam Oates	69				
2001-02	Jarome Iginla	52	Adam Oates	64	Jarome Iginla	96	Peter Worell	354

* Match Misconduct penalty not included in total penalty minutes.
1946-47 was the first season that a Match penalty was automaticaly written into the player's total penalty minutes as 20 minutes.
Beginning in 1947-48 all penalties, Match, Game Misconduct, and Misconduct, are written as 10 minutes.

One Season Scoring Records

Goals-Per-Game Leaders, One Season

(Among players with 20 goals or more in one season)

Player	Team	Season	Games	Goals	Average
Joe Malone	Montreal	1917-18	20	44	2.20
Cy Denneny	Ottawa	1917-18	20	36	1.80
Newsy Lalonde	Montreal	1917-18	14	23	1.64
Joe Malone	Quebec	1919-20	24	39	1.63
Newsy Lalonde	Montreal	1919-20	23	37	1.61
Reg Noble	Toronto	1917-18	20	30	1.50
Babe Dye	Ham., Tor.	1920-21	24	35	1.46
Cy Denneny	Ottawa	1920-21	24	34	1.42
Joe Malone	Hamilton	1920-21	20	28	1.40
Newsy Lalonde	Montreal	1920-21	24	33	1.38
Punch Broadbent	Ottawa	1921-22	24	32	1.33
Babe Dye	Toronto	1924-25	29	38	1.31
Babe Dye	Toronto	1921-22	24	31	1.29
Newsy Lalonde	Montreal	1918-19	17	22	1.29
Odie Cleghorn	Montreal	1918-19	17	22	1.29
Cy Denneny	Ottawa	1921-22	22	27	1.23
Aurel Joliat	Montreal	1924-25	25	30	1.20
Wayne Gretzky	Edmonton	1983-84	74	87	1.18
Babe Dye	Toronto	1922-23	22	26	1.18
Wayne Gretzky	Edmonton	1981-82	80	92	1.15
Mario Lemieux	Pittsburgh	1992-93	60	69	1.15
Frank Nighbor	Ottawa	1919-20	23	26	1.13
Mario Lemieux	Pittsburgh	1988-89	76	85	1.12
Brett Hull	St. Louis	1990-91	78	86	1.10
Cam Neely	Boston	1993-94	49	50	1.02
Maurice Richard	Montreal	1944-45	50	50	1.00
Reg Noble	Toronto	1919-20	24	24	1.00
Corb Denneny	Toronto	1919-20	24	24	1.00
Joe Malone	Hamilton	1921-22	24	24	1.00
Billy Boucher	Montreal	1922-23	24	24	1.00
Cy Denneny	Ottawa	1923-24	22	22	1.00
Alexander Mogilny	Buffalo	1992-93	77	76	0.99
Mario Lemieux	Pittsburgh	1995-96	70	69	0.99
Cooney Weiland	Boston	1929-30	44	43	0.98
Phil Esposito	Boston	1970-71	78	76	0.97
Jari Kurri	Edmonton	1984-85	73	71	0.97

Though better remembered as a brilliant goal scorer (he had six straight 50-goal seasons from 1974-75 to 1979-80), Guy Lafleur was also a great playmaker. He had 80 assists in 80 games in 1976-77 and set up 75 goals in just 74 games in 1979-80.

Assists-Per-Game Leaders, One Season

(Among players with 35 assists or more in one season)

Player	Team	Season	Games	Assists	Average
Wayne Gretzky	Edmonton	1985-86	80	163	2.04
Wayne Gretzky	Edmonton	1987-88	64	109	1.70
Wayne Gretzky	Edmonton	1984-85	80	135	1.69
Wayne Gretzky	Edmonton	1983-84	74	118	1.59
Wayne Gretzky	Edmonton	1982-83	80	125	1.56
Wayne Gretzky	Los Angeles	1990-91	78	122	1.56
Wayne Gretzky	Edmonton	1986-87	79	121	1.53
Mario Lemieux	Pittsburgh	1992-93	60	91	1.52
Wayne Gretzky	Edmonton	1981-82	80	120	1.50
Mario Lemieux	Pittsburgh	1988-89	76	114	1.50
Adam Oates	St. Louis	1990-91	61	90	1.48
Wayne Gretzky	Los Angeles	1988-89	78	114	1.46
Wayne Gretzky	Los Angeles	1989-90	73	102	1.40
Wayne Gretzky	Edmonton	1980-81	80	109	1.36
Mario Lemieux	Pittsburgh	1991-92	64	87	1.36
Mario Lemieux	Pittsburgh	1989-90	59	78	1.32
Bobby Orr	Boston	1970-71	78	102	1.31
Mario Lemieux	Pittsburgh	1995-96	70	92	1.31
Mario Lemieux	Pittsburgh	1987-88	77	98	1.27
Bobby Orr	Boston	1973-74	74	90	1.22
Wayne Gretzky	Los Angeles	1991-92	74	90	1.22
Ron Francis	Pittsburgh	1995-96	77	92	1.19
Mario Lemieux	Pittsburgh	1985-86	79	93	1.18
Bobby Clarke	Philadelphia	1975-76	76	89	1.17
Peter Stastny	Quebec	1981-82	80	93	1.16
Adam Oates	Boston	1992-93	84	97	1.15
Doug Gilmour	Toronto	1992-93	83	95	1.14
Wayne Gretzky	Los Angeles	1993-94	81	92	1.14
Paul Coffey	Edmonton	1985-86	79	90	1.14
Bobby Orr	Boston	1969-70	76	87	1.14
Bryan Trottier	NY Islanders	1978-79	76	87	1.14
Bobby Orr	Boston	1972-73	63	72	1.14
Bill Cowley	Boston	1943-44	36	41	1.14
Pat LaFontaine	Buffalo	1992-93	84	95	1.13
Steve Yzerman	Detroit	1988-89	80	90	1.13
Paul Coffey	Pittsburgh	1987-88	46	52	1.13
Bobby Orr	Boston	1974-75	80	89	1.11
Bobby Clarke	Philadelphia	1974-75	80	89	1.11
Paul Coffey	Pittsburgh	1988-89	75	83	1.11
Wayne Gretzky	Los Angeles	1992-93	45	49	1.11
Denis Savard	Chicago	1982-83	78	86	1.10
Denis Savard	Chicago	1981-82	80	87	1.09
Denis Savard	Chicago	1987-88	80	87	1.09
Wayne Gretzky	Edmonton	1979-80	79	86	1.09
Ron Francis	Pittsburgh	1994-95	44	48	1.09
Paul Coffey	Edmonton	1983-84	80	86	1.08
Elmer Lach	Montreal	1944-45	50	54	1.08
Peter Stastny	Quebec	1985-86	76	81	1.07
Jaromir Jagr	Pittsburgh	1995-96	82	87	1.06
Mark Messier	Edmonton	1989-90	79	84	1.06
Peter Forsberg	Colorado	1995-96	82	86	1.05
Paul Coffey	Edmonton	1984-85	80	84	1.05
Marcel Dionne	Los Angeles	1979-80	80	84	1.05
Bobby Orr	Boston	1971-72	76	80	1.05
Mike Bossy	NY Islanders	1981-82	80	83	1.04
Adam Oates	Boston	1993-94	77	80	1.04
Phil Esposito	Boston	1968-69	74	77	1.04
Bryan Trottier	NY Islanders	1983-84	68	71	1.04
Pete Mahovlich	Montreal	1974-75	80	82	1.03
Kent Nilsson	Calgary	1980-81	80	82	1.03
Peter Stastny	Quebec	1982-83	75	77	1.03
Denis Savard	Chicago	1988-89	58	59	1.02
Jaromir Jagr	Pittsburgh	1998-99	81	83	1.02
Doug Gilmour	Toronto	1993-94	83	84	1.01
Bernie Nicholls	Los Angeles	1988-89	79	80	1.01
Guy Lafleur	Montreal	1979-80	74	75	1.01
Guy Lafleur	Montreal	1976-77	80	80	1.00
Marcel Dionne	Los Angeles	1984-85	80	80	1.00
Brian Leetch	NY Rangers	1991-92	80	80	1.00
Bryan Trottier	NY Islanders	1977-78	77	77	1.00
Mike Bossy	NY Islanders	1983-84	67	67	1.00
Jean Ratelle	NY Rangers	1971-72	63	63	1.00
Steve Yzerman	Detroit	1993-94	58	58	1.00
Ron Francis	Hartford	1985-86	53	53	1.00
Guy Chouinard	Calgary	1980-81	52	52	1.00
Elmer Lach	Montreal	1943-44	48	48	1.00

Points-Per-Game Leaders, One Season

(Among players with 50 points or more in one season)

Player	Team	Season	Games	Points	Average
Wayne Gretzky	Edmonton	1983-84	74	205	2.77
Wayne Gretzky	Edmonton	1985-86	80	215	2.69
Mario Lemieux	Pittsburgh	1992-93	60	160	2.67
Wayne Gretzky	Edmonton	1981-82	80	212	2.65
Mario Lemieux	Pittsburgh	1988-89	76	199	2.62
Wayne Gretzky	Edmonton	1984-85	80	208	2.60
Wayne Gretzky	Edmonton	1982-83	80	196	2.45
Wayne Gretzky	Edmonton	1987-88	64	149	2.33
Wayne Gretzky	Edmonton	1986-87	79	183	2.32
Mario Lemieux	Pittsburgh	1995-96	70	161	2.30
Mario Lemieux	Pittsburgh	1987-88	77	168	2.18
Wayne Gretzky	Los Angeles	1988-89	78	168	2.15
Wayne Gretzky	Los Angeles	1990-91	78	163	2.09
Mario Lemieux	Pittsburgh	1989-90	59	123	2.08
Wayne Gretzky	Edmonton	1980-81	80	164	2.05
Mario Lemieux	Pittsburgh	1991-92	64	131	2.05
Bill Cowley	Boston	1943-44	36	71	1.97
Phil Esposito	Boston	1970-71	78	152	1.95
Wayne Gretzky	Los Angeles	1989-90	73	142	1.95
Steve Yzerman	Detroit	1988-89	80	155	1.94
Bernie Nicholls	Los Angeles	1988-89	79	150	1.90
Adam Oates	St. Louis	1990-91	61	115	1.89
Phil Esposito	Boston	1973-74	78	145	1.86
Jari Kurri	Edmonton	1984-85	73	135	1.85
Mike Bossy	NY Islanders	1981-82	80	147	1.84
Jaromir Jagr	Pittsburgh	1995-96	82	149	1.82
Mario Lemieux	Pittsburgh	1985-86	79	141	1.78
Bobby Orr	Boston	1970-71	78	139	1.78
Jari Kurri	Edmonton	1983-84	64	113	1.77
Mario Lemieux	Pittsburgh	2000-01	43	76	1.77
Pat LaFontaine	Buffalo	1992-93	84	148	1.76
Bryan Trottier	NY Islanders	1978-79	76	134	1.76
Mike Bossy	NY Islanders	1983-84	67	118	1.76
Paul Coffey	Edmonton	1985-86	79	138	1.75
Phil Esposito	Boston	1971-72	76	133	1.75
Peter Stastny	Quebec	1981-82	80	139	1.74
Wayne Gretzky	Edmonton	1979-80	79	137	1.73
Jean Ratelle	NY Rangers	1971-72	63	109	1.73
Marcel Dionne	Los Angeles	1979-80	80	137	1.71
Herb Cain	Boston	1943-44	48	82	1.71
Guy Lafleur	Montreal	1976-77	80	136	1.70
Dennis Maruk	Washington	1981-82	80	136	1.70
Phil Esposito	Boston	1968-69	74	126	1.70
Guy Lafleur	Montreal	1974-75	70	119	1.70
Mario Lemieux	Pittsburgh	1986-87	63	107	1.70
Adam Oates	Boston	1992-93	84	142	1.69
Bobby Orr	Boston	1974-75	80	135	1.69
Marcel Dionne	Los Angeles	1980-81	80	135	1.69
Guy Lafleur	Montreal	1977-78	78	132	1.69
Guy Lafleur	Montreal	1979-80	74	125	1.69
Rob Brown	Pittsburgh	1988-89	68	115	1.69
Jari Kurri	Edmonton	1985-86	78	131	1.68
Brett Hull	St. Louis	1990-91	78	131	1.68
Phil Esposito	Boston	1972-73	78	130	1.67
Cooney Weiland	Boston	1929-30	44	73	1.66
Alexander Mogilny	Buffalo	1992-93	77	127	1.65
Peter Stastny	Quebec	1982-83	75	124	1.65
Bobby Orr	Boston	1973-74	74	122	1.65
Kent Nilsson	Calgary	1980-81	80	131	1.64
Denis Savard	Chicago	1987-88	80	131	1.64
Wayne Gretzky	Los Angeles	1991-92	74	121	1.64
Steve Yzerman	Detroit	1992-93	84	137	1.63
Marcel Dionne	Los Angeles	1978-79	80	130	1.63
Dale Hawerchuk	Winnipeg	1984-85	80	130	1.63
Mark Messier	Edmonton	1989-90	79	129	1.63
Bryan Trottier	NY Islanders	1983-84	68	111	1.63
Pat LaFontaine	Buffalo	1991-92	57	93	1.63
Charlie Simmer	Los Angeles	1980-81	65	105	1.62
Guy Lafleur	Montreal	1978-79	80	129	1.61
Bryan Trottier	NY Islanders	1981-82	80	129	1.61
Phil Esposito	Boston	1974-75	79	127	1.61
Steve Yzerman	Detroit	1989-90	79	127	1.61
Peter Stastny	Quebec	1985-86	76	122	1.61
Mario Lemieux	Pittsburgh	1996-97	76	122	1.61
Michel Goulet	Quebec	1983-84	75	121	1.61
Wayne Gretzky	Los Angeles	1993-94	81	130	1.60
Bryan Trottier	NY Islanders	1977-78	77	123	1.60
Bobby Orr	Boston	1972-73	63	101	1.60
Guy Chouinard	Calgary	1980-81	52	83	1.60
Elmer Lach	Montreal	1944-45	50	80	1.60
Pierre Turgeon	NY Islanders	1992-93	83	132	1.59
Steve Yzerman	Detroit	1987-88	64	102	1.59
Mike Bossy	NY Islanders	1978-79	80	126	1.58
Paul Coffey	Edmonton	1983-84	80	126	1.58
Marcel Dionne	Los Angeles	1984-85	80	126	1.58
Bobby Orr	Boston	1969-70	76	120	1.58
Eric Lindros	Philadelphia	1995-96	73	115	1.58
Charlie Simmer	Los Angeles	1979-80	64	101	1.58
Teemu Selanne	Winnipeg	1992-93	84	132	1.57
Jaromir Jagr	Pittsburgh	1998-99	81	127	1.57
Bobby Clarke	Philadelphia	1975-76	76	119	1.57
Guy Lafleur	Montreal	1975-76	80	125	1.56
Dave Taylor	Los Angeles	1980-81	72	112	1.56
Denis Savard	Chicago	1982-83	78	121	1.55
Ron Francis	Pittsburgh	1995-96	77	119	1.55
Mike Bossy	NY Islanders	1985-86	80	123	1.54
Kevin Stevens	Pittsburgh	1991-92	80	123	1.54
Bobby Orr	Boston	1971-72	76	117	1.54
Mike Bossy	NY Islanders	1984-85	76	117	1.54
Kevin Stevens	Pittsburgh	1992-93	72	111	1.54
Doug Bentley	Chicago	1943-44	50	77	1.54
Doug Gilmour	Toronto	1992-93	83	127	1.53
Marcel Dionne	Los Angeles	1976-77	80	122	1.53
Jaromir Jagr	Pittsburgh	99-2000	63	96	1.52
Eric Lindros	Philadelphia	1996-97	52	79	1.52
Eric Lindros	Philadelphia	1994-95	46	70	1.52
Marcel Dionne	Detroit	1974-75	80	121	1.51
Mike Bossy	NY Islanders	1980-81	79	119	1.51
Paul Coffey	Edmonton	1984-85	80	121	1.51
Dale Hawerchuk	Winnipeg	1987-88	80	121	1.51
Paul Coffey	Pittsburgh	1988-89	75	113	1.51
Jaromir Jagr	Pittsburgh	1996-97	63	95	1.51

One of only five players to win at least five scoring titles, Jaromir Jagr actually had his best offensive production in a year in which he was second in the scoring race. Jagr averaged 1.82 points per game (149 points in 82 games) in 1995-96.

The first draft choice in the history of the Buffalo Sabres, Gilbert Perreault (right) set a rookie record with 38 goals in 1970-71. The record fell just one year later when linemate Rick Martin (left) scored 44 times in 1971-72.

Rookie Scoring Records

All-Time Top 50 Goal-Scoring Rookies

	Rookie	Team	Position	Season	GP	G	A	PTS
1.	* Teemu Selanne	Winnipeg	Right wing	1992-93	84	**76**	56	132
2.	* Mike Bossy	NY Islanders	Right wing	1977-78	73	**53**	38	91
3.	* Joe Nieuwendyk	Calgary	Center	1987-88	75	**51**	41	92
4.	* Dale Hawerchuk	Winnipeg	Center	1981-82	80	**45**	58	103
	* Luc Robitaille	Los Angeles	Left wing	1986-87	79	**45**	39	84
6.	Rick Martin	Buffalo	Left wing	1971-72	73	**44**	30	74
	Barry Pederson	Boston	Center	1981-82	80	**44**	48	92
8.	* Steve Larmer	Chicago	Right wing	1982-83	80	**43**	47	90
	* Mario Lemieux	Pittsburgh	Center	1984-85	73	**43**	57	100
10.	Eric Lindros	Philadelphia	Center	1992-93	61	**41**	34	75
11.	Darryl Sutter	Chicago	Left wing	1980-81	76	**40**	22	62
	Sylvain Turgeon	Hartford	Left wing	1983-84	76	**40**	32	72
	Warren Young	Pittsburgh	Left wing	1984-85	80	**40**	32	72
14.	* Eric Vail	Atlanta	Left wing	1974-75	72	**39**	21	60
	Anton Stastny	Quebec	Left wing	1980-81	80	**39**	46	85
	* Peter Stastny	Quebec	Center	1980-81	77	**39**	70	109
	Steve Yzerman	Detroit	Center	1983-84	80	**39**	48	87
18.	* Gilbert Perreault	Buffalo	Center	1970-71	78	**38**	34	72
	Neal Broten	Minnesota	Center	1981-82	73	**38**	60	98
	Ray Sheppard	Buffalo	Right wing	1987-88	74	**38**	27	65
	Mikael Renberg	Philadelphia	Left wing	1993-94	83	**38**	44	82
22.	Jorgen Pettersson	St. Louis	Left wing	1980-81	62	**37**	36	73
	Jimmy Carson	Los Angeles	Center	1986-87	80	**37**	42	79
24.	Mike Foligno	Detroit	Right wing	1979-80	80	**36**	35	71
	Mike Bullard	Pittsburgh	Center	1981-82	75	**36**	27	63
	Paul MacLean	Winnipeg	Right wing	1981-82	74	**36**	25	61
	Tony Granato	NY Rangers	Right wing	1988-89	78	**36**	27	63
28.	Marian Stastny	Quebec	Right wing	1981-82	74	**35**	54	89
	Brian Bellows	Minnesota	Right wing	1982-83	78	**35**	30	65
	Tony Amonte	NY Rangers	Right wing	1991-92	79	**35**	34	69
31.	Nels Stewart	Mtl. Maroons	Center	1925-26	36	**34**	8	42
	* Danny Grant	Minnesota	Left wing	1968-69	75	**34**	31	65
	Norm Ferguson	Oakland	Right wing	1968-69	76	**34**	20	54
	Brian Propp	Philadelphia	Left wing	1979-80	80	**34**	41	75
	Wendel Clark	Toronto	Left wing	1985-86	66	**34**	11	45
	* Pavel Bure	Vancouver	Right wing	1991-92	65	**34**	26	60
37.	* Willi Plett	Atlanta	Right wing	1976-77	64	**33**	23	56
	Dale McCourt	Detroit	Center	1977-78	76	**33**	39	72
	Mark Pavelich	NY Rangers	Center	1981-82	79	**33**	43	76
	Ron Flockhart	Philadelphia	Center	1981-82	72	**33**	39	72
	Steve Bozek	Los Angeles	Center	1981-82	71	**33**	23	56
	Jason Arnott	Edmonton	Center	1993-94	78	**33**	35	68
43.	Bill Mosienko	Chicago	Right wing	1943-44	50	**32**	38	70
	Michel Bergeron	Detroit	Right wing	1975-76	72	**32**	27	59
	* Bryan Trottier	NY Islanders	Center	1975-76	80	**32**	63	95
	Don Murdoch	NY Rangers	Right wing	1976-77	59	**32**	24	56
	Jari Kurri	Edmonton	Left wing	1980-81	75	**32**	43	75
	Bobby Carpenter	Washington	Center	1981-82	80	**32**	35	67
	Kjell Dahlin	Montreal	Right wing	1985-86	77	**32**	39	71
	Petr Klima	Detroit	Left wing	1985-86	74	**32**	24	56
	Darren Turcotte	NY Rangers	Right wing	1989-90	76	**32**	34	66
	Joe Juneau	Boston	Center	1992-93	84	**32**	70	102

* Calder Trophy Winner

All-Time Top 50 Point-Scoring Rookies

	Rookie	Team	Position	Season	GP	G	A	PTS
1.	* Teemu Selanne	Winnipeg	Right wing	1992-93	84	76	56	**132**
2.	* Peter Stastny	Quebec	Center	1980-81	77	39	70	**109**
3.	* Dale Hawerchuk	Winnipeg	Center	1981-82	80	45	58	**103**
4.	Joe Juneau	Boston	Center	1992-93	84	32	70	**102**
5.	* Mario Lemieux	Pittsburgh	Center	1984-85	73	43	57	**100**
6.	Neal Broten	Minnesota	Center	1981-82	73	38	60	**98**
7.	* Bryan Trottier	NY Islanders	Center	1975-76	80	32	63	**95**
8.	Barry Pederson	Boston	Center	1981-82	80	44	48	**92**
	* Joe Nieuwendyk	Calgary	Center	1987-88	75	51	41	**92**
10.	* Mike Bossy	NY Islanders	Right wing	1977-78	73	53	38	**91**
11.	* Steve Larmer	Chicago	Right wing	1982-83	80	43	47	**90**
12.	Marian Stastny	Quebec	Right wing	1981-82	74	35	54	**89**
13.	Steve Yzerman	Detroit	Center	1983-84	80	39	48	**87**
14.	* Sergei Makarov	Calgary	Right wing	1989-90	80	24	62	**86**
15.	Anton Stastny	Quebec	Left wing	1980-81	80	39	46	**85**
16.	* Luc Robitaille	Los Angeles	Left wing	1986-87	79	45	39	**84**
17.	Mikael Renberg	Philadelphia	Left wing	1993-94	83	38	44	**82**
18.	Jimmy Carson	Los Angeles	Center	1986-87	80	37	42	**79**
	Sergei Fedorov	Detroit	Center	1990-91	77	31	48	**79**
	Alexei Yashin	Ottawa	Center	1993-94	83	30	49	**79**
21.	Marcel Dionne	Detroit	Center	1971-72	78	28	49	**77**
22.	Larry Murphy	Los Angeles	Defense	1980-81	80	16	60	**76**
	Mark Pavelich	NY Rangers	Center	1981-82	79	33	43	**76**
	Dave Poulin	Philadelphia	Center	1983-84	73	31	45	**76**
25.	Brian Propp	Philadelphia	Left wing	1979-80	80	34	41	**75**
	Jari Kurri	Edmonton	Left wing	1980-81	75	32	43	**75**
	Denis Savard	Chicago	Center	1980-81	76	28	47	**75**
	Mike Modano	Minnesota	Center	1989-90	80	29	46	**75**
	Eric Lindros	Philadelphia	Center	1992-93	61	41	34	**75**
30.	Rick Martin	Buffalo	Left wing	1971-72	73	44	30	**74**
	* Bobby Smith	Minnesota	Center	1978-79	80	30	44	**74**
32.	Jorgen Pettersson	St. Louis	Left wing	1980-81	62	37	36	**73**
33.	* Gilbert Perreault	Buffalo	Center	1970-71	78	38	34	**72**
	Dale McCourt	Detroit	Center	1977-78	76	33	39	**72**
	Ron Flockhart	Philadelphia	Center	1981-82	72	33	39	**72**
	Sylvain Turgeon	Hartford	Left wing	1983-84	76	40	32	**72**
	Warren Young	Pittsburgh	Left wing	1984-85	80	40	32	**72**
	Carey Wilson	Calgary	Center	1984-85	74	24	48	**72**
	Alexei Zhamnov	Winnipeg	Center	1992-93	68	25	47	**72**
40.	Mike Foligno	Detroit	Right wing	1979-80	80	36	35	**71**
	Dave Christian	Winnipeg	Center	1980-81	80	28	43	**71**
	Mats Naslund	Montreal	Left wing	1982-83	74	26	45	**71**
	Kjell Dahlin	Montreal	Right wing	1985-86	77	32	39	**71**
	* Brian Leetch	NY Rangers	Defense	1988-89	68	23	48	**71**
45.	Bill Mosienko	Chicago	Right wing	1943-44	50	32	38	**70**
	* Scott Gomez	New Jersey	Center	99-2000	82	19	51	**70**
47.	Roland Eriksson	Minnesota	Center	1976-77	80	25	44	**69**
	Tony Amonte	NY Rangers	Right wing	1991-92	79	35	34	**69**
49.	Jude Drouin	Minnesota	Center	1970-71	75	16	52	**68**
	Pierre Larouche	Pittsburgh	Center	1974-75	79	31	37	**68**
	Ron Francis	Hartford	Center	1981-82	59	25	43	**68**
	* Gary Suter	Calgary	Defense	1985-86	80	18	50	**68**
	Jason Arnott	Edmonton	Center	1993-94	84	33	35	**68**

John Bucyk

Steve Shutt

Rick Kehoe

50-Goal Seasons

Player	Team	Date of 50th Goal	Score	Goaltender	Player's Game No.	Team Game No.	Total Goals	Total Games	Age When First 50th Scored (Yrs. & Mos.)
Maurice Richard	Mtl.	18-3-45	Mtl. 4 at Bos. 2	Harvey Bennett	50	50	50	50	23.7
Bernie Geoffrion	Mtl.	16-3-61	Tor. 2 at Mtl. 5	Cesare Maniago	62	68	50	64	30.1
Bobby Hull	Chi.	25-3-62	Chi. 1 at NYR 4	Gump Worsley	70	70	50	70	23.2
Bobby Hull	Chi.	2-3-66	Det. 4 at Chi. 5	Hank Bassen	52	57	54	65	
Bobby Hull	Chi.	18-3-67	Chi. 5 at Tor. 9	Bruce Gamble	63	66	52	66	
Bobby Hull	Chi.	5-3-69	NYR 4 at Chi. 4	Ed Giacomin	64	66	58	74	
Phil Esposito	Bos.	20-2-71	Bos. 4 at L.A. 5	Denis DeJordy	58	58	76	78	29.0
John Bucyk	Bos.	16-3-71	Bos. 11 at Det. 4	Roy Edwards	69	69	51	78	35.10
Phil Esposito	Bos.	20-2-72	Bos. 3 at Chi. 1	Tony Esposito	60	60	66	76	
Bobby Hull	Chi.	2-4-72	Det. 1 at Chi. 6	Andy Brown	78	78	50	78	
Vic Hadfield	NYR	2-4-72	Mtl. 6 at NYR 5	Denis DeJordy	78	78	50	78	31.6
Phil Esposito	Bos.	25-3-73	Buf. 1 at Bos. 6	Roger Crozier	75	75	55	78	
Mickey Redmond	Det.	27-3-73	Det. 8 at Tor. 1	Ron Low	73	75	52	76	25.3
Rick MacLeish	Phi.	1-4-73	Phi. 4 at Pit. 5	Cam Newton	78	78	50	78	23.2
Phil Esposito	Bos.	20-2-74	Bos. 5 at Min. 5	Cesare Maniago	56	56	68	78	
Mickey Redmond	Det.	23-3-74	NYR 3 at Det. 5	Ed Giacomin	69	71	51	76	
Ken Hodge	Bos.	6-4-74	Bos. 2 at Mtl. 6	Michel Larocque	75	77	50	76	29.10
Rick Martin	Buf.	7-4-74	St.L. 2 at Buf. 5	Wayne Stephenson	78	78	52	78	22.9
Phil Esposito	Bos.	8-2-75	Bos. 8 at Det. 5	Jim Rutherford	54	54	61	79	
Guy Lafleur	Mtl.	29-3-75	K.C. 1 at Mtl. 4	Denis Herron	66	76	53	70	23.6
Danny Grant	Det.	2-4-75	Wsh. 3 at Det. 8	John Adams	78	78	50	80	29.2
Rick Martin	Buf.	3-4-75	Bos. 2 at Buf. 4	Ken Broderick	67	79	52	68	
Reggie Leach	Phi.	14-3-76	Atl. 1 at Phi. 6	Dan Bouchard	69	69	61	80	25.11
Jean Pronovost	Pit.	24-3-76	Bos. 5 at Pit. 5	Gilles Gilbert	74	74	52	80	30.3
Guy Lafleur	Mtl.	27-3-76	K.C. 2 at Mtl. 8	Denis Herron	76	76	56	80	
Bill Barber	Phi.	3-4-76	Buf. 2 at Phi. 5	Al Smith	79	79	50	80	23.9
Pierre Larouche	Pit.	3-4-76	Wsh. 5 at Pit. 4	Ron Low	75	79	53	76	20.5
Danny Gare	Buf.	4-4-76	Tor. 2 at Buf. 5	Gord McRae	79	80	50	79	21.11
Steve Shutt	Mtl.	1-3-77	Mtl. 5 at NYI 4	Glenn Resch	65	65	60	80	24.8
Guy Lafleur	Mtl.	6-3-77	Mtl. 1 at Buf. 4	Don Edwards	68	68	56	80	
Marcel Dionne	L.A.	2-4-77	Min. 2 at L.A. 7	Pete LoPresti	79	79	53	80	25.8
Guy Lafleur	Mtl.	8-3-78	Wsh. 3 at Mtl. 4	Jim Bedard	63	65	60	78	
Mike Bossy	NYI	1-4-78	Wsh. 2 at NYI 3	Bernie Wolfe	69	76	53	73	21.2
Mike Bossy	NYI	24-2-79	Det. 1 at NYI 3	Rogie Vachon	58	58	69	80	
Marcel Dionne	L.A.	11-3-79	L.A. 3 at Phi. 6	Wayne Stephenson	68	68	59	80	
Guy Lafleur	Mtl.	31-3-79	Pit. 3 at Mtl. 5	Denis Herron	76	76	52	80	
Guy Chouinard	Atl.	6-4-79	NYR 2 at Atl. 9	John Davidson	79	79	50	80	22.5
Marcel Dionne	L.A.	12-3-80	L.A. 2 at Pit. 4	Nick Ricci	70	70	53	80	
Mike Bossy	NYI	16-3-80	NYI 6 at Chi. 1	Tony Esposito	68	71	51	75	
Charlie Simmer	L.A.	19-3-80	Det. 3 at L.A. 4	Jim Rutherford	57	73	56	64	26.0
Pierre Larouche	Mtl.	25-3-80	Chi. 4 at Mtl. 8	Tony Esposito	72	75	50	73	
Danny Gare	Buf.	27-3-80	Det. 1 at Buf. 10	Jim Rutherford	71	75	56	76	
Blaine Stoughton	Hfd.	28-3-80	Hfd. 4 at Van. 4	Glen Hanlon	75	75	56	80	27.0
Guy Lafleur	Mtl.	2-4-80	Mtl. 7 at Det. 2	Rogie Vachon	72	78	50	74	
Wayne Gretzky	Edm.	2-4-80	Min. 1 at Edm. 1	Gary Edwards	78	79	51	79	19.2
Reggie Leach	Phi.	3-4-80	Wsh. 2 at Phi. 4	empty net	75	79	50	76	
Mike Bossy	NYI	24-1-81	Que. 3 at NYI 7	Ron Grahame	50	50	68	79	
Charlie Simmer	L.A.	26-1-81	L.A. 7 at Que. 5	Michel Dion	51	51	56	65	
Marcel Dionne	L.A.	8-3-81	L.A. 4 at Wpg. 1	Markus Mattsson	68	68	58	80	
Wayne Babych	St.L.	12-3-81	St.L. 3 at Mtl. 4	Richard Sevigny	70	68	54	78	22.9
Wayne Gretzky	Edm.	15-3-81	Edm. 3 at Cgy. 3	Pat Riggin	69	69	55	80	
Rick Kehoe	Pit.	16-3-81	Pit. 7 at Edm. 6	Eddie Mio	70	70	55	80	29.7
Jacques Richard	Que.	29-3-81	Mtl. 0 at Que. 4	Richard Sevigny	76	75	52	78	28.6
Dennis Maruk	Wsh.	5-4-81	Det. 2 at Wsh. 7	Larry Lozinski	80	80	50	80	25.3
Wayne Gretzky	Edm.	30-12-81	Phi. 5 at Edm. 7	empty net	39	39	92	80	
Dennis Maruk	Wsh.	21-2-82	Wpg. 3 at Wsh. 6	Doug Soetaert	61	61	60	80	
Mike Bossy	NYI	4-3-82	Tor. 1 at NYI 10	Michel Larocque	66	66	64	80	
Dino Ciccarelli	Min.	8-3-82	St.L. 1 at Min. 8	Mike Liut	67	68	55	76	22.1
Rick Vaive	Tor.	24-3-82	St.L. 3 at Tor. 4	Mike Liut	72	75	54	77	22.10
Blaine Stoughton	Hfd.	28-3-82	Min. 5 at Hfd. 2	Gilles Meloche	76	76	52	80	
Rick Middleton	Bos.	28-3-82	Bos. 5 at Buf. 9	Paul Harrison	72	77	51	75	28.11
Marcel Dionne	L.A.	30-3-82	Cgy. 7 at L.A. 5	Pat Riggin	75	77	50	78	
Mark Messier	Edm.	31-3-82	L.A. 3 at Edm. 7	Mario Lessard	78	79	50	78	21.3
Bryan Trottier	NYI	3-4-82	Phi. 3 at NYI 6	Pete Peeters	79	79	50	80	25.9
Lanny McDonald	Cgy.	18-2-83	Cgy. 1 at Buf. 5	Bob Sauve	60	60	66	80	30.0
Wayne Gretzky	Edm.	19-2-83	Edm. 10 at Pit. 7	Nick Ricci	60	60	71	80	
Michel Goulet	Que.	5-3-83	Hfd. 3 at Que. 10	Mike Veisor	67	67	57	80	22.11
Mike Bossy	NYI	12-3-83	Wsh. 2 at NYI 6	Al Jensen	70	71	60	79	
Marcel Dionne	L.A.	17-3-83	Que. 3 at L.A. 4	Dan Bouchard	71	71	56	80	
Al Secord	Chi.	20-3-83	Tor. 3 at Chi. 7	Mike Palmateer	73	73	54	80	25.0
Rick Vaive	Tor.	30-3-83	Tor. 4 at Det. 2	Gilles Gilbert	76	78	51	78	
Wayne Gretzky	Edm.	7-1-84	Hfd. 3 at Edm. 5	Greg Millen	42	42	87	74	
Michel Goulet	Que.	8-3-84	Que. 8 at Pit. 6	Denis Herron	63	69	56	75	
Rick Vaive	Tor.	14-3-84	Min. 3 at Tor. 3	Gilles Meloche	69	72	52	76	
Mike Bullard	Pit.	14-3-84	Pit. 6 at L.A. 7	Markus Mattsson	71	72	51	76	23.0
Jari Kurri	Edm.	15-3-84	Edm. 2 at Mtl. 3	Rick Wamsley	57	73	52	64	23.10
Glenn Anderson	Edm.	21-3-84	Hfd. 3 at Edm. 5	Greg Millen	76	76	54	80	23.6
Tim Kerr	Phi.	22-3-84	Pit. 4 at Phi. 13	Denis Herron	74	75	54	79	24.3
Mike Bossy	NYI	31-3-84	NYI 3 at Wsh. 1	Pat Riggin	67	79	51	67	
Wayne Gretzky	Edm.	26-1-85	Pit. 3 at Edm. 6	Denis Herron	49	49	73	80	
Jari Kurri	Edm.	3-2-85	Hfd. 3 at Edm. 6	Greg Millen	50	53	71	73	
Mike Bossy	NYI	5-3-85	Phi. 5 at NYI 4	Bob Froese	61	65	58	76	
Michel Goulet	Que.	6-3-85	Buf. 3 at Que. 4	Tom Barrasso	62	73	55	69	
Tim Kerr	Phi.	7-3-85	Wsh. 6 at Phi. 9	Pat Riggin	63	65	54	74	
John Ogrodnick	Det.	13-3-85	Det. 6 at Edm. 7	Grant Fuhr	69	69	55	79	25.9
Bob Carpenter	Wsh.	21-3-85	Wsh. 2 at Mtl. 3	Steve Penney	72	72	53	80	21.9

Player	Team	Date of 50th Goal	Score	Goaltender	Player's Game No.	Team Game No.	Total Goals	Total Games	Age When First 50th Scored (Yrs. & Mos.)
Dale Hawerchuk	Wpg.	29-3-85	Chi. 5 at Wpg. 5	W. Skorodenski	77	77	53	80	21.11
Mike Gartner	Wsh.	7-4-85	Pit. 3 at Wsh. 7	Brian Ford	80	80	50	80	25.5
Jari Kurri	Edm.	4-3-86	Edm. 6 at Van. 2	Richard Brodeur	63	65	68	78	
Mike Bossy	NYI	11-3-86	Cgy. 4 at NYI 8	Reggie Lemelin	67	67	61	80	
Glenn Anderson	Edm.	14-3-86	Det. 3 at Edm. 12	Greg Stefan	63	71	54	72	
Michel Goulet	Que.	17-3-86	Que. 8 at Mtl. 6	Patrick Roy	67	72	53	75	
Wayne Gretzky	Edm.	18-3-86	Wpg. 2 at Edm. 6	Brian Hayward	72	72	52	80	
Tim Kerr	Phi.	20-3-86	Pit. 1 at Phi. 5	Roberto Romano	68	72	58	76	
Wayne Gretzky	Edm.	4-2-87	Edm. 6 at Min. 5	Don Beaupre	55	55	62	79	
Dino Ciccarelli	Min.	7-3-87	Pit. 7 at Min. 3	Gilles Meloche	66	66	52	80	
Mario Lemieux	Pit.	12-3-87	Que. 3 at Pit. 6	Mario Gosselin	53	70	54	63	21.5
Tim Kerr	Phi.	17-3-87	NYR 1 at Phi. 4	J. Vanbiesbrouck	67	71	58	75	
Jari Kurri	Edm.	17-3-87	N.J. 4 at Edm. 7	Craig Billington	69	70	54	79	
Mario Lemieux	Pit.	2-2-88	Wsh. 2 at Pit. 3	Pete Peeters	51	54	70	77	
Steve Yzerman	Det.	1-3-88	Buf. 0 at Det. 4	Tom Barrasso	64	64	50	64	22.10
Joe Nieuwendyk	Cgy.	12-3-88	Buf. 4 at Cgy. 10	Tom Barrasso	66	70	51	75	21.5
Craig Simpson	Edm.	15-3-88	Buf. 4 at Edm. 6	Jacques Cloutier	71	71	56	80	21.1
Jimmy Carson	L.A.	26-3-88	Chi. 5 at L.A. 9	Darren Pang	77	77	55	88	19.8
Luc Robitaille	L.A.	1-4-88	L.A. 6 at Cgy. 3	Mike Vernon	79	79	53	80	21.10
Hakan Loob	Cgy.	3-4-88	Min. 1 at Cgy. 4	Don Beaupre	80	80	50	80	27.9
Stephane Richer	Mtl.	3-4-88	Mtl. 4 at Buf. 4	Tom Barrasso	72	80	50	72	21.10
Mario Lemieux	Pit.	20-1-89	Pit. 3 at Wpg. 7	Pokey Reddick	44	46	85	76	
Bernie Nicholls	L.A.	28-1-89	Edm. 7 at L.A. 6	Grant Fuhr	51	51	70	79	27.7
Steve Yzerman	Det.	5-2-89	Det. 6 at Wpg. 2	Pokey Reddick	55	55	65	80	
Wayne Gretzky	L.A.	4-3-89	Phi. 2 at L.A. 6	Ron Hextall	66	67	54	78	
Joe Nieuwendyk	Cgy.	21-3-89	NYI 1 at Cgy. 4	Mark Fitzpatrick	72	74	51	77	
Joe Mullen	Cgy.	31-3-89	Wpg. 1 at Cgy. 4	Bob Essensa	78	79	51	79	32.1
Brett Hull	St.L.	6-2-90	Tor. 4 at St.L. 6	Jeff Reese	54	54	72	80	25.6
Steve Yzerman	Det.	24-2-90	Det. 3 at NYI 3	Glenn Healy	63	63	62	79	
Cam Neely	Bos.	10-3-90	Bos. 3 at NYI 3	Mark Fitzpatrick	69	71	55	76	24.9
Luc Robitaille	L.A.	31-3-90	L.A. 3 at Van. 6	Kirk McLean	79	79	52	80	
Brian Bellows	Min.	22-3-90	Min. 5 at Det. 1	Tim Cheveldae	75	75	55	80	25.6
Pat LaFontaine	NYI	24-3-90	NYI 5 at Edm. 5	Bill Ranford	71	77	54	74	25.1
Stephane Richer	Mtl.	24-3-90	Mtl. 4 at Hfd. 7	Peter Sidorkiewicz	75	77	51	75	
Gary Leeman	Tor.	28-3-90	NYI 6 at Tor. 3	Mark Fitzpatrick	78	78	51	80	26.1
Brett Hull	St.L.	25-1-91	St.L. 9 at Det. 4	David Gagnon	49	49	86	78	
Cam Neely	Bos.	26-3-91	Bos. 7 at Que. 4	empty net	67	78	51	69	
Theoren Fleury	Cgy.	26-3-91	Van. 2 at Cgy. 7	Bob Mason	77	77	51	79	22.9
Steve Yzerman	Det.	30-3-91	NYR 5 at Det. 6	Mike Richter	79	79	51	80	
Brett Hull	St.L.	28-1-92	St.L. 3 at L.A. 3	Kelly Hrudey	50	50	70	73	
Jeremy Roenick	Chi.	7-3-92	Chi. 2 at Bos. 1	Daniel Berthiaume	67	67	53	80	22.2
Kevin Stevens	Pit.	24-3-92	Pit. 3 at Det. 4	Tim Cheveldae	74	74	54	80	26.11
Gary Roberts	Cgy.	31-3-92	Edm. 2 at Cgy. 5	Bill Ranford	73	77	53	76	25.10
Alexander Mogilny	Buf.	3-2-93	Hfd. 2 at Buf. 3	Sean Burke	46	53	76	77	23.11
Teemu Selanne	Wpg.	28-2-93	Min. 6 at Wpg. 7	Darcy Wakaluk	63	63	76	84	22.6
Pavel Bure	Van.	1-3-93	Van. 5 at Buf. 2*	Grant Fuhr	63	63	60	83	21.11
Steve Yzerman	Det.	10-3-93	Det. 6 at Edm. 3	Bill Ranford	70	70	58	84	
Luc Robitaille	L.A.	15-3-93	L.A. 4 at Buf. 2	Grant Fuhr	69	69	63	84	
Brett Hull	St.L.	20-3-93	St.L. 2 at L.A. 3	Robb Stauber	73	73	54	80	
Mario Lemieux	Pit.	21-3-93	Pit. 6 at Edm. 4**	Ron Tugnutt	48	72	69	60	
Kevin Stevens	Pit.	21-3-93	Pit. 6 at Edm. 4**	Ron Tugnutt	62	72	55	72	
Dave Andreychuk	Tor.	23-3-93	Tor. 5 at Wpg. 4	Bob Essensa	72	73	54	83	29.6
Pat LaFontaine	Buf.	28-3-93	Ott. 1 at Buf. 3	Peter Sidorkiewicz	75	75	53	84	
Pierre Turgeon	NYI	2-4-93	NYI 3 at NYR 2	Mike Richter	75	76	58	83	23.8
Mark Recchi	Phi.	3-4-93	T.B. 2 at Phi. 6	J-C Bergeron	77	77	53	84	25.2
Jeremy Roenick	Chi.	15-4-93	Tor. 2 at Chi. 3	Felix Potvin	84	84	50	84	
Brendan Shanahan	St.L.	15-4-93	T.B. 5 at St.L. 6	Pat Jablonski	71	84	51	71	24.3
Cam Neely	Bos.	7-3-94	Wsh. 3 at Bos. 6	Don Beaupre	44	66	50	49	
Sergei Fedorov	Det.	15-3-94	Van. 2 at Det. 5	Kirk McLean	67	69	56	82	24.3
Pavel Bure	Van.	23-3-94	Van. 6 at L.A. 3	empty net	65	73	60	76	
Adam Graves	NYR	23-3-94	NYR 5 at Edm. 3	Bill Ranford	74	74	51	84	25.11
Dave Andreychuk	Tor.	24-3-94	S.J. 2 at Tor. 1	Arturs Irbe	73	74	53	83	
Brett Hull	St.L.	25-3-94	Dal. 3 at St.L. 5	Andy Moog	71	74	52	81	
Ray Sheppard	Det.	29-3-94	Hfd. 2 at Det. 6	Sean Burke	74	76	52	82	27.10
Brendan Shanahan	St.L.	12-4-94	St.L. 5 at Dal. 9	Andy Moog	80	83	52	81	
Mike Modano	Dal.	12-4-94	St.L. 5 at Dal. 9	Curtis Joseph	75	83	50	76	23.11
Mario Lemieux	Pit.	23-2-96	Hfd. 4 at Pit. 5	Sean Burke	50	59	69	70	
Jaromir Jagr	Pit.	23-2-96	Hfd. 4 at Pit. 5	Sean Burke	59	59	62	82	24.0
Alexander Mogilny	Van.	29-2-96	St.L. 2 at Van. 2	Grant Fuhr	60	63	55	79	
Peter Bondra	Wsh.	3-4-96	Wsh. 5 at Buf. 1	Andrei Trefilov	62	77	52	67	28.1
Joe Sakic	Col.	7-4-96	Col. 4 at Dal. 1	empty net	79	79	51	82	26.7
John LeClair	Phi.	10-4-96	Phi. 5 at N.J. 1	Corey Schwab	80	80	51	82	26.7
Keith Tkachuk	Wpg.	12-4-96	L.A. 3 at Wpg. 5	empty net	75	81	50	76	24.0
Paul Kariya	Ana.	14-4-96	Wpg. 2 at Ana. 5	N. Khabibulin	82	82	50	82	21.5
Keith Tkachuk	Phx.	6-4-97	Phx. 1 at Col. 2	Patrick Roy	78	79	52	81	
Teemu Selanne	Ana.	9-4-97	L.A. 1 at Ana. 4	empty net	77	81	51	78	
Mario Lemieux	Pit.	11-4-97	Pit. 2 at Fla. 4	J. Vanbiesbrouck	75	81	50	76	
John LeClair	Phi.	13-4-97	N.J. 4 at Phi. 5	Mike Dunham	82	82	50	82	
Teemu Selanne	Ana.	25-3-98	Ana. 3 at Chi. 2	Jeff Hackett	66	71	52	73	
John LeClair	Phi.	13-4-98	Phi. 1 at Buf. 2	Dominik Hasek	79	79	51	82	
Pavel Bure	Van.	17-4-98	Cgy. 4 at Van. 2	Dwayne Roloson	81	81	51	82	
Peter Bondra	Wsh.	18-4-98	Wsh. 4 at Car. 3	Mike Fountain	75	80	52	76	
Pavel Bure	Fla.	18-3-00	Fla. 4 at NYI 2	empty net	63	71	58	74	
Pavel Bure	Fla.	16-3-01	Pit. 6 at Fla. 3	Johan Hedberg	72	72	59	82	
Joe Sakic	Col.	4-4-01	Ana. 1 at Col. 1	J-S Giguere	80	80	54	82	
Jaromir Jagr	Pit.	4-4-01	T.B. 2 at Pit. 4	Kevin Weekes	80	80	52	81	
Jarome Iginla	Cgy.	7-4-02	Cgy. 2 at Chi. 3	Jocelyn Thibault	79	79	52	82	24.9

* neutral site game played at Hamilton; ** neutral site game played at Cleveland

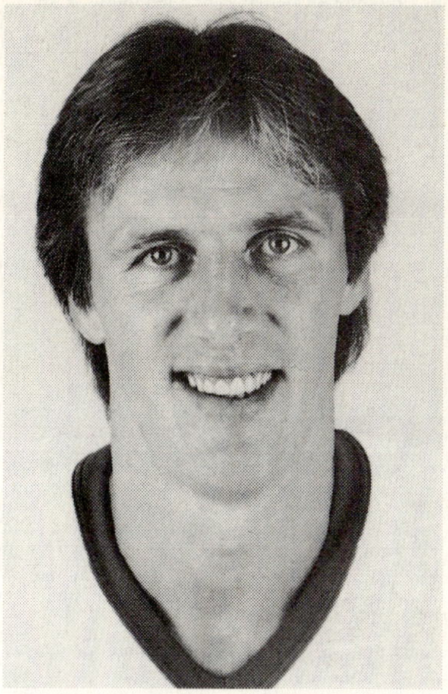

Mike Bossy

John LeClair

Jarome Iginla

Bobby Clarke

Darryl Sittler

Charlie Simmer

100-Point Seasons

Player	Team	Date of 100th Point	G or A	Score	Player's Game No.	Team Game No.	G - A PTS	Total Games	Age when first 100th point scored (Yrs. & Mos.)
Phil Esposito	Bos.	2-3-69	(G)	Pit. 0 at Bos. 4	60	62	49-77 — 126	74	27.1
Bobby Hull	Chi.	20-3-69	(G)	Chi. 5 at Bos. 5	71	71	58-49 — 107	76	30.2
Gordie Howe	Det.	30-3-69	(G)	Det. 5 at Chi. 9	76	76	44-59 — 103	76	41.0
Bobby Orr	Bos.	15-3-70	(G)	Det. 5 at Bos. 5	67	67	33-87 — 120	76	22.11
Phil Esposito	Bos.	6-2-71	(A)	Buf. 3 at Bos. 4	51	51	76-76 — 152	78	
Bobby Orr	Bos.	20-2-71	(A)	Bos. 4 at L.A. 5	58	58	37-102 — 139	78	
John Bucyk	Bos.	13-3-71	(G)	Bos. 6 at Van. 3	68	68	51-65 — 116	78	35.10
Ken Hodge	Bos.	21-3-71	(A)	Buf. 7 at Bos. 5	72	72	43-62 — 105	78	26.9
Jean Ratelle	NYR	18-2-72	(A)	NYR 2 at Cal. 2	58	58	46-63 — 109	63	31.4
Phil Esposito	Bos.	19-2-72	(A)	Bos. 6 at Min. 4	59	59	66-67 — 133	76	
Bobby Orr	Bos.	2-3-72	(A)	Van. 3 at Bos. 7	64	64	37-80 — 117	76	
Vic Hadfield	NYR	25-3-72	(A)	NYR 3 at Mtl. 3	74	74	50-56 — 106	78	31.5
Phil Esposito	Bos.	3-3-73	(A)	Bos. 1 at Mtl. 5	64	64	55-75 — 130	78	
Bobby Clarke	Phi.	29-3-73	(G)	Atl. 2 at Phi. 4	76	76	37-67 — 104	78	23.7
Bobby Orr	Bos.	31-3-73	(G)	Bos. 3 at Tor. 7	62	77	29-72 — 101	63	
Rick MacLeish	Phi.	1-4-73	(G)	Phi. 4 at Pit. 5	78	78	50-50 — 100	78	23.3
Phil Esposito	Bos.	13-2-74	(A)	Bos. 9 at Cal. 6	53	53	68-77 — 145	78	
Bobby Orr	Bos.	12-3-74	(A)	Buf. 0 at Bos. 4	62	66	32-90 — 122	74	
Ken Hodge	Bos.	24-3-74	(A)	Mtl. 3 at Bos. 6	72	72	50-55 — 105	76	
Phil Esposito	Bos.	8-2-75	(A)	Bos. 8 at Det. 5	54	54	61-66 — 127	79	
Bobby Orr	Bos.	13-2-75	(A)	Bos. 1 at Buf. 3	57	57	46-89 — 135	80	
Guy Lafleur	Mtl.	7-3-75	(G)	Wsh. 4 at Mtl. 8	56	66	53-66 — 119	70	24.6
Pete Mahovlich	Mtl.	9-3-75	(G)	Mtl. 5 at NYR 3	67	67	35-82 — 117	80	29.5
Marcel Dionne	Det.	9-3-75	(A)	Det. 5 at Phi. 8	67	67	47-74 — 121	80	23.7
Bobby Clarke	Phi.	22-3-75	(A)	Min. 0 at Phi. 4	72	72	27-89 — 116	80	
Rene Robert	Buf.	5-4-75	(A)	Buf. 4 at Tor. 2	74	80	40-60 — 100	74	26.4
Guy Lafleur	Mtl.	10-3-76	(G)	Mtl. 5 at Chi. 1	69	69	56-69 — 125	80	
Bobby Clarke	Phi.	11-3-76	(A)	Buf. 1 at Phi. 6	64	68	30-89 — 119	76	
Bill Barber	Phi.	18-3-76	(A)	Van. 2 at Phi. 3	71	71	50-62 — 112	80	23.8
Gilbert Perreault	Buf.	21-3-76	(A)	K.C. 1 at Buf. 3	73	73	44-69 — 113	80	25.4
Pierre Larouche	Pit.	24-3-76	(G)	Bos. 5 at Pit. 5	70	74	53-58 — 111	76	20.4
Pete Mahovlich	Mtl.	28-3-76	(A)	Mtl. 2 at Bos. 2	77	77	34-71 — 105	80	
Jean Ratelle	Bos.	30-3-76	(G)	Buf. 4 at Bos. 4	77	77	36-69 — 105	80	
Jean Pronovost	Pit.	3-4-76	(A)	Wsh. 5 at Pit. 4	79	79	52-52 — 104	80	30.4
Darryl Sittler	Tor.	3-4-76	(A)	Bos. 4 at Tor. 2	78	79	41-59 — 100	79	25.7
Guy Lafleur	Mtl.	26-2-77	(A)	Cle. 3 at Mtl. 5	63	63	56-80 — 136	80	
Marcel Dionne	L.A.	5-3-77	(G)	Pit. 3 at L.A. 3	67	67	53-69 — 122	80	
Steve Shutt	Mtl.	27-3-77	(A)	Mtl. 6 at Det. 0	77	77	60-45 — 105	80	24.9
Bryan Trottier	NYI	25-2-78	(A)	Chi. 1 at NYI 7	59	60	46-77 — 123	77	21.7
Guy Lafleur	Mtl.	28-2-78	(G)	Det. 3 at Mtl. 9	69	61	60-72 — 132	78	
Darryl Sittler	Tor.	12-3-78	(A)	Tor. 7 at Pit. 1	67	67	45-72 — 117	80	
Guy Lafleur	Mtl.	27-2-79	(A)	Mtl. 3 at NYI 7	61	61	52-77 — 129	80	
Bryan Trottier	NYI	6-3-79	(A)	Buf. 3 at NYI 2	59	63	47-87 — 134	76	
Marcel Dionne	L.A.	8-3-79	(G)	L.A. 4 at Buf. 6	66	66	59-71 — 130	80	
Mike Bossy	NYI	11-3-79	(G)	NYI 4 at Bos. 4	66	66	69-57 — 126	80	22.2
Bob MacMillan	Atl.	15-3-79	(A)	Atl. 4 at Phi. 5	68	69	37-71 — 108	79	26.6
Guy Chouinard	Atl.	30-3-79	(G)	L.A. 3 at Atl. 5	75	75	50-57 — 107	80	22.5
Denis Potvin	NYI	8-4-79	(A)	NYI 5 at NYR 2	73	80	31-70 — 101	73	25.5
Marcel Dionne	L.A.	6-2-80	(A)	L.A. 3 at Hfd. 7	53	53	53-84 — 137	80	
Guy Lafleur	Mtl.	10-2-80	(A)	Mtl. 3 at Bos. 2	55	55	50-75 — 125	74	
Wayne Gretzky	Edm.	24-2-80	(A)	Bos. 4 at Edm. 2	61	62	51-86 — 137	79	19.2
Bryan Trottier	NYI	30-3-80	(A)	NYI 9 at Que. 6	75	77	42-62 — 104	78	
Gilbert Perreault	Buf.	1-4-80	(A)	Buf. 5 at Atl. 2	77	77	40-66 — 106	80	
Mike Rogers	Hfd.	4-4-80	(A)	Que. 2 at Hfd. 9	79	79	44-61 — 105	80	25.5
Charlie Simmer	L.A.	5-4-80	(G)	Van. 5 at L.A. 3	64	80	56-45 — 101	64	26.0
Blaine Stoughton	Hfd.	6-4-80	(A)	Det. 3 at Hfd. 5	80	80	56-44 — 100	80	27.0
Wayne Gretzky	Edm.	6-2-81	(G)	Wpg. 4 at Edm. 10	53	53	55-109 — 164	80	
Marcel Dionne	L.A.	12-2-81	(A)	L.A. 5 at Chi. 5	58	58	58-77 — 135	80	
Charlie Simmer	L.A.	14-2-81	(A)	Bos. 5 at L.A. 4	59	59	56-49 — 105	65	
Kent Nilsson	Cgy.	27-2-81	(G)	Hfd. 1 at Cgy. 5	64	64	49-82 — 131	80	24.6
Mike Bossy	NYI	3-3-81	(G)	Edm. 8 at NYI 8	65	66	68-51 — 119	79	
Dave Taylor	L.A.	14-3-81	(G)	Min. 4 at L.A. 10	63	70	47-65 — 112	72	25.3
Mike Rogers	Hfd.	22-3-81	(G)	Tor. 3 at Hfd. 3	74	74	40-65 — 105	80	
Bernie Federko	St.L.	28-3-81	(A)	Buf. 4 at St.L. 7	74	76	31-73 — 104	78	24.10
Rick Middleton	Bos.	28-3-81	(A)	Chi. 2 at Bos. 5	76	76	44-59 — 103	80	27.4
Jacques Richard	Que.	29-3-81	(G)	Mtl. 0 at Que. 4	75	76	52-51 — 103	78	28.6
Bryan Trottier	NYI	29-3-81	(G)	NYI 5 at Wsh. 4	69	76	31-72 — 103	73	
Peter Stastny	Que.	29-3-81	(A)	Mtl. 0 at Que. 4	73	76	39-70 — 109	77	24.6
Wayne Gretzky	Edm.	27-12-81	(G)	L.A. 3 at Edm. 10	38	38	92-120 — 212	80	
Mike Bossy	NYI	13-2-82	(A)	Phi. 2 at NYI 8	55	55	64-83 — 147	80	
Peter Stastny	Que.	16-2-82	(A)	Wpg. 3 at Que. 7	60	60	46-93 — 139	80	
Dennis Maruk	Wsh.	20-2-82	(G)	Wsh. 3 at Min. 7	60	60	60-76 — 136	80	26.3
Bryan Trottier	NYI	23-2-82	(G)	Chi. 1 at NYI 5	61	61	50-79 — 129	80	
Denis Savard	Chi.	27-2-82	(A)	Chi. 5 at L.A. 3	64	64	32-87 — 119	80	21.1
Bobby Smith	Min.	3-3-82	(A)	Det. 4 at Min. 6	66	66	43-71 — 114	80	24.1
Marcel Dionne	L.A.	6-3-82	(G)	L.A. 6 at Hfd. 7	64	66	50-67 — 117	78	
Dave Taylor	L.A.	20-3-82	(A)	Pit. 5 at L.A. 7	71	72	39-67 — 106	78	
Dale Hawerchuk	Wpg.	24-3-82	(G)	L.A. 3 at Wpg. 5	74	74	45-58 — 103	80	18.11
Dino Ciccarelli	Min.	27-3-82	(A)	Min. 6 at Bos. 5	72	76	55-52 — 107	76	21.8
Glenn Anderson	Edm.	28-3-82	(G)	Edm. 6 at L.A. 2	78	78	38-67 — 105	80	21.7
Mike Rogers	NYR	2-4-82	(G)	Pit. 7 at NYR 5	79	79	38-65 — 103	80	

Player	Team	Date of 100th Point	G or A	Score	Player's Game No.	Team Game No.	G - A PTS	Total Games	Age when first 100th point scored (Yrs. & Mos.)
Wayne Gretzky	Edm.	5-1-83	(A)	Edm. 8 at Wpg. 3	42	42	71-125 — 196	80	
Mike Bossy	NYI	3-3-83	(A)	Tor. 1 at NYI. 5	66	67	60-58 — 118	79	
Peter Stastny	Que.	5-3-83	(A)	Hfd. 3 at Que. 10	62	67	47-77 — 124	75	
Denis Savard	Chi.	6-3-83	(G)	Mtl. 4 at Chi. 5	65	67	35-86 — 121	78	
Mark Messier	Edm.	23-3-83	(G)	Edm. 4 at Wpg. 7	73	76	48-58 — 106	77	22.2
Barry Pederson	Bos.	26-3-83	(A)	Hfd. 4 at Bos. 7	73	76	46-61 — 107	77	22.0
Marcel Dionne	L.A.	26-3-83	(A)	Edm. 9 at L.A. 3	75	75	56-51 — 107	80	
Michel Goulet	Que.	27-3-83	(A)	Que. 6 at Buf. 6	77	77	57-48 — 105	80	22.11
Glenn Anderson	Edm.	29-3-83	(A)	Edm. 7 at Van. 4	70	78	48-56 — 104	72	
Jari Kurri	Edm.	29-3-83	(A)	Edm. 7 at Van. 4	78	78	45-59 — 104	80	22.10
Kent Nilsson	Cgy.	29-3-83	(G)	L.A. 3 at Cgy. 5	78	78	46-58 — 104	80	
Wayne Gretzky	Edm.	18-12-83	(G)	Edm. 7 at Wpg. 5	34	34	87-118 — 205	74	
Paul Coffey	Edm.	4-3-84	(A)	Mtl. 1 at Edm. 6	68	68	40-86 — 126	80	22.9
Michel Goulet	Que.	4-3-84	(A)	Que. 1 at Buf. 1	62	67	56-65 — 121	75	
Jari Kurri	Edm.	7-3-84	(G)	Chi. 4 at Edm. 7	53	69	52-61 — 113	64	
Peter Stastny	Que.	8-3-84	(A)	Que. 8 at Pit. 6	69	69	46-73 — 119	80	
Mike Bossy	NYI	8-3-84	(G)	Tor. 5 at NYI 9	56	68	51-67 — 118	67	
Barry Pederson	Bos.	14-3-84	(A)	Bos. 4 at Det. 2	71	71	39-77 — 116	80	
Bryan Trottier	NYI	18-3-84	(G)	NYI 4 at Hfd. 5	62	73	40-71 — 111	68	
Bernie Federko	St.L.	20-3-84	(A)	Wpg. 3 at St.L. 9	75	76	41-66 — 107	79	
Rick Middleton	Bos.	27-3-84	(G)	Bos. 6 at Que. 4	77	77	47-58 — 105	80	
Dale Hawerchuk	Wpg.	27-3-84	(G)	Wpg. 3 at L.A. 3	77	77	37-65 — 102	80	
Mark Messier	Edm.	27-3-84	(G)	Edm. 9 at Cgy. 2	72	79	37-64 — 101	73	
Wayne Gretzky	Edm.	29-12-84	(A)	Det. 3 at Edm. 6	35	35	73-135 — 208	80	
Jari Kurri	Edm.	29-1-85	(G)	Edm. 4 at Cgy. 2	48	51	71-64 — 135	73	
Mike Bossy	NYI	23-2-85	(G)	Bos. 1 at NYI 7	56	60	58-59 — 117	76	
Dale Hawerchuk	Wpg.	25-2-85	(A)	Wpg. 12 at NYR 5	64	64	53-77 — 130	80	
Marcel Dionne	L.A.	5-3-85	(A)	Pit. 0 at L.A. 6	66	66	46-80 — 126	80	
Brent Sutter	NYI	12-3-85	(A)	NYI 6 at St.L. 5	68	68	42-60 — 102	72	22.10
John Ogrodnick	Det.	22-3-85	(A)	NYR 3 at Det. 5	73	73	55-50 — 105	79	25.9
Paul Coffey	Edm.	26-3-85	(G)	Edm. 7 at NYI 5	74	74	37-84 — 121	80	
Denis Savard	Chi.	29-3-85	(A)	Chi. 5 at Wpg. 5	75	76	38-67 — 105	79	
Peter Stastny	Que.	2-4-85	(A)	Bos. 4 at Que. 6	74	77	32-68 — 100	75	
Bernie Federko	St.L.	4-4-85	(A)	NYR 5 at St.L. 4	74	78	30-73 — 103	76	
John Tonelli	NYI	6-4-85	(G)	N.J. 5 at NYI 5	80	80	42-58 — 100	80	28.1
Paul MacLean	Wpg.	6-4-85	(A)	Wpg. 6 at Edm. 5	78	79	41-60 — 101	79	27.1
Bernie Nicholls	L.A.	6-4-85	(A)	Van. 4 at L.A. 4	80	80	46-54 — 100	80	22.9
Mike Gartner	Wsh.	7-4-85	(G)	Pit. 3 at Wsh. 7	80	80	50-52 — 102	80	25.6
Mario Lemieux	Pit.	7-4-85	(G)	Pit. 3 at Wsh. 7	73	80	43-57 — 100	73	19.6
Wayne Gretzky	Edm.	4-1-86	(A)	Hfd. 3 at Edm. 4	39	39	52-163 — 215	80	
Mario Lemieux	Pit.	15-2-86	(G)	Van. 4 at Pit. 9	55	56	48-93 — 141	79	
Paul Coffey	Edm.	19-2-86	(A)	Tor. 5 at Edm. 9	59	60	48-90 — 138	79	
Peter Stastny	Que.	1-3-86	(A)	Buf. 8 at Que. 4	66	68	41-81 — 122	76	
Jari Kurri	Edm.	2-3-86	(G)	Phi. 1 at Edm. 2	62	64	68-63 — 131	78	
Mike Bossy	NYI	8-3-86	(G)	Wsh. 6 at NYI 2	65	65	61-62 — 123	80	
Denis Savard	Chi.	12-3-86	(A)	Buf. 7 at Chi. 6	69	69	47-69 — 116	80	
Mats Naslund	Mtl.	13-3-86	(A)	Mtl. 2 at Bos. 3	70	70	43-67 — 110	80	26.4
Michel Goulet	Que.	24-3-86	(A)	Que. 1 at Min. 0	70	75	53-50 — 103	75	
Glenn Anderson	Edm.	25-3-86	(G)	Edm. 7 at Det. 2	66	74	54-48 — 102	72	
Neal Broten	Min.	26-3-86	(A)	Min. 6 at Tor. 1	76	76	29-76 — 105	80	26.4
Dale Hawerchuk	Wpg.	31-3-86	(A)	Wpg. 5 at L.A. 2	78	78	46-59 — 105	80	
Bernie Federko	St.L.	5-4-86	(G)	Chi. 5 at St.L. 7	79	79	34-68 — 102	80	
Wayne Gretzky	Edm.	11-1-87	(A)	Cgy. 3 at Edm. 5	42	42	62-121 — 183	79	
Jari Kurri	Edm.	14-3-87	(A)	Buf. 3 at Edm. 5	67	68	54-54 — 108	79	
Mario Lemieux	Pit.	18-3-87	(A)	St.L. 4 at Pit. 5	55	72	54-53 — 107	63	
Mark Messier	Edm.	19-3-87	(A)	Edm. 4 at Cgy. 5	71	71	37-70 — 107	77	
Dino Ciccarelli	Min.	30-3-87	(A)	NYR 6 at Min. 5	78	78	52-51 — 103	80	
Doug Gilmour	St.L.	2-4-87	(A)	Buf. 3 at St.L. 5	78	78	42-63 — 105	80	23.10
Dale Hawerchuk	Wpg.	5-4-87	(A)	Wpg. 3 at Cgy. 1	80	80	47-53 — 100	80	
Mario Lemieux	Pit.	20-1-88	(G)	Pit. 8 at Chi. 3	45	48	70-98 — 168	77	
Wayne Gretzky	Edm.	11-2-88	(A)	Edm. 7 at Van. 2	43	56	40-109 — 149	64	
Denis Savard	Chi.	12-2-88	(A)	St.L. 3 at Chi. 4	57	57	44-87 — 131	80	
Dale Hawerchuk	Wpg.	23-2-88	(G)	Wpg. 4 at Pit. 3	61	61	44-77 — 121	80	
Steve Yzerman	Det.	27-2-88	(A)	Det. 4 at Que. 5	63	63	50-52 — 102	64	22.10
Peter Stastny	Que.	8-3-88	(A)	Hfd. 4 at Que. 6	63	67	46-65 — 111	76	
Mark Messier	Edm.	15-3-88	(A)	Buf. 4 at Edm. 6	68	71	37-74 — 111	77	
Jimmy Carson	L.A.	26-3-88	(A)	Chi. 5 at L.A. 9	77	77	55-52 — 107	80	19.8
Hakan Loob	Cgy.	26-3-88	(A)	Van. 1 at Cgy. 6	76	76	50-56 — 106	80	27.9
Mike Bullard	Cgy.	26-3-88	(A)	Van. 1 at Cgy. 6	76	76	48-55 — 103	79	27.1
Michel Goulet	Que.	27-3-88	(A)	Pit. 6 at Que. 3	76	76	48-58 — 106	80	
Luc Robitaille	L.A.	30-3-88	(G)	Cgy. 7 at L.A. 9	78	78	53-58 — 111	80	22.1
Mario Lemieux	Pit.	31-12-88	(A)	N.J. 6 at Pit. 8	36	38	85-114 — 199	76	
Wayne Gretzky	L.A.	21-1-89	(A)	L.A. 4 at Hfd. 5	47	48	54-114 — 168	78	
Bernie Nicholls	L.A.	21-1-89	(A)	L.A. 4 at Hfd. 5	48	48	70-80 — 150	79	
Steve Yzerman	Det.	27-1-89	(G)	Tor. 1 at Det. 8	50	50	65-90 — 155	80	
Rob Brown	Pit.	16-3-89	(A)	Pit. 2 at N.J. 1	60	72	49-66 — 115	68	20.11
Paul Coffey	Pit.	20-3-89	(A)	Pit. 2 at Min. 7	69	74	30-83 — 113	75	
Joe Mullen	Cgy.	23-3-89	(A)	L.A. 2 at Cgy. 4	74	75	51-59 — 110	79	32.1
Jari Kurri	Edm.	29-3-89	(A)	Edm. 5 at Van. 2	75	79	44-58 — 102	76	
Jimmy Carson	Edm.	2-4-89	(A)	Edm. 2 at Cgy. 4	80	80	49-51 — 100	80	
Mario Lemieux	Pit.	28-1-90	(G)	Pit. 2 at Buf. 7	50	50	45-78 — 123	59	
Wayne Gretzky	L.A.	30-1-90	(A)	N.J. 2 at L.A. 5	51	51	40-102 — 142	73	
Steve Yzerman	Det.	19-2-90	(A)	Mtl. 5 at Det. 5	61	61	62-65 — 127	79	
Mark Messier	Edm.	20-2-90	(A)	Edm. 4 at Van. 2	62	62	45-84 — 129	79	
Brett Hull	St.L.	3-3-90	(A)	NYI 4 at St.L. 5	67	67	72-41 — 113	80	25.7
Bernie Nicholls	NYR	12-3-90	(A)	L.A. 6 at NYR 2	70	71	39-73 — 112	79	
Pierre Turgeon	Buf.	25-3-90	(G)	N.J. 4 at Buf. 3	76	76	40-66 — 106	80	20.7
Paul Coffey	Pit.	25-3-90	(A)	Pit. 2 at Hfd. 4	77	77	29-74 — 103	80	
Pat LaFontaine	NYI	27-3-90	(G)	Cgy. 4 at NYI 2	72	78	54-51 — 105	74	25.1
Adam Oates	St.L.	29-3-90	(G)	Pit 4 at St.L. 5	79	79	23-79 — 102	80	27.7

Barry Pederson

John Tonelli

Hakan Loob

Steve Larmer

Doug Gilmour

Eric Lindros

Player	Team	Date of 100th Point	G or A	Score	Player's Game No.	Team Game No.	G - A PTS	Total Games	Age when first 100th point scored (Yrs. & Mos.)
Joe Sakic	Que.	31-3-90	(G)	Hfd. 3 at Que. 2	79	79	39-63 — 102	80	20.8
Ron Francis	Hfd.	31-3-90	(G)	Hfd. 3 at Que. 2	79	79	32-69 — 101	80	27.0
Luc Robitaille	L.A.	1-4-90	(A)	L.A. 4 at Cgy. 8	80	80	52-49 — 101	80	
Wayne Gretzky	L.A.	30-1-91	(A)	N.J. 4 at L.A. 2	50	51	41-122 — 163	78	
Brett Hull	St.L.	23-2-91	(G)	Bos. 2 at St.L. 9	60	62	86-45 — 131	78	
Mark Recchi	Pit.	5-3-91	(G)	Van. 1 at Pit. 4	66	67	40-73 — 113	78	23.1
Steve Yzerman	Det.	10-3-91	(G)	Det. 4 at St.L. 1	72	72	51-57 — 108	80	
John Cullen	Hfd.	16-3-91	(G)	N.J. 2 at Hfd. 6	71	71	39-71 — 110	78	26.7
Adam Oates	St.L.	17-3-91	(A)	St.L. 4 at Chi. 6	54	73	25-90 — 115	61	
Joe Sakic	Que.	19-3-91	(G)	Edm. 7 at Que. 6	74	74	48-61 — 109	80	
Steve Larmer	Chi.	24-3-91	(A)	Min. 4 at Chi. 5	76	76	44-57 — 101	80	29.9
Theoren Fleury	Cgy.	26-3-91	(G)	Van. 2 at Cgy. 7	77	77	51-53 — 104	79	22.9
Al MacInnis	Cgy.	28-3-91	(A)	Edm. 4 at Cgy. 4	78	78	28-75 — 103	78	27.8
Brett Hull	St.L.	2-3-92	(G)	St.L. 5 at Van. 3	66	66	70-39 — 109	73	
Wayne Gretzky	L.A.	3-3-92	(A)	Phi. 1 at L.A. 4	60	66	31-90 — 121	74	
Kevin Stevens	Pit.	7-3-92	(A)	Pit. 3 at L.A. 5	66	66	54-69 — 123	80	26.11
Mario Lemieux	Pit.	10-3-92	(A)	Cgy. 2 at Pit. 5	53	67	44-87 — 131	64	
Luc Robitaille	L.A.	17-3-92	(A)	Wpg. 4 at L.A. 5	73	73	44-63 — 107	80	
Mark Messier	NYR	22-3-92	(G)	N.J. 3 at NYR 6	74	75	35-72 — 107	79	
Jeremy Roenick	Chi.	29-3-92	(A)	Tor. 1 at Chi. 5	77	77	53-50 — 103	80	22.2
Steve Yzerman	Det.	14-4-92	(G)	Det. 7 at Min. 4	79	80	45-58 — 103	79	
Brian Leetch	NYR	16-4-92	(G)	Pit. 1 at NYR 7	80	80	22-80 — 102	80	24.1
Mario Lemieux	Pit.	31-12-92	(G)	Tor. 3 at Pit. 3	38	39	69-91 — 160	60	
Pat LaFontaine	Buf.	10-2-93	(A)	Buf. 6 at Wpg. 2	55	55	53-95 — 148	84	
Adam Oates	Bos.	14-2-93	(A)	Bos. 3 at T.B. 3	58	58	45-97 — 142	84	
Steve Yzerman	Det.	24-2-93	(A)	Det. 7 at Buf. 10	64	64	58-79 — 137	84	
Pierre Turgeon	NYI	28-2-93	(G)	NYI 7 at Hfd. 6	62	63	58-74 — 132	83	
Doug Gilmour	Tor.	3-3-93	(A)	Min. 1 at Tor. 3	64	64	32-95 — 127	83	
Alexander Mogilny	Buf.	5-3-93	(A)	Hfd. 4 at Buf. 2	58	65	76-51 — 127	77	24.1
Mark Recchi	Phi.	7-3-93	(G)	Phi. 3 at N.J. 7	66	66	53-70 — 123	84	
Teemu Selanne	Wpg.	9-3-93	(G)	Wpg. 4 at T.B. 2	68	68	76-56 — 132	84	22.7
Luc Robitaille	L.A.	15-3-93	(A)	L.A. 4 at Buf. 2	69	69	63-62 — 125	84	
Kevin Stevens	Pit.	23-3-93	(A)	S.J. 2 at Pit. 7	63	73	55-56 — 111	72	
Mats Sundin	Que.	27-3-93	(G)	Phi. 3 at Que. 8	71	75	47-67 — 114	80	22.1
Pavel Bure	Van.	1-4-93	(G)	Van. 5 at T.B. 3	77	77	60-50 — 110	83	22.0
Jeremy Roenick	Chi.	4-4-93	(G)	St.L. 4 at Chi. 5	79	79	50-57 — 107	84	
Craig Janney	St.L.	4-4-93	(G)	St.L. 4 at Chi. 5	79	79	24-82 — 106	84	25.7
Rick Tocchet	Pit.	7-4-93	(G)	Mtl. 3 at Pit. 4	77	81	48-61 — 109	80	28.11
Joe Sakic	Que.	8-4-93	(A)	Que. 2 at Bos. 6	75	81	48-57 — 105	78	
Ron Francis	Pit.	9-4-93	(A)	Pit. 10 at NYR 4	82	82	24-76 — 100	84	
Brett Hull	St.L.	11-4-93	(G)	Min. 1 at St.L. 5	78	82	54-47 — 101	80	
Theoren Fleury	Cgy.	11-4-93	(G)	Cgy. 3 at Van. 6	82	82	34-66 — 100	83	
Joe Juneau	Bos.	14-4-93	(A)	Bos. 4 at Ott. 2	84	84	32-70 — 102	84	25.3
Wayne Gretzky	L.A.	14-2-94	(A)	Bos. 3 at L.A. 2	56	56	38-92 — 130	81	
Sergei Fedorov	Det.	1-3-94	(A)	Cgy. 2 at Det. 5	63	63	56-64 — 120	82	24.2
Doug Gilmour	Tor.	23-3-94	(G)	Tor. 1 at Fla. 1	74	74	27-84 — 111	83	
Adam Oates	Bos.	26-3-94	(A)	Mtl. 3 at Bos. 6	68	75	32-80 — 112	77	
Mark Recchi	Phi.	27-3-94	(A)	Ana. 3 at Phi. 2	76	76	40-67 — 107	84	
Pavel Bure	Van.	28-3-94	(A)	Tor. 2 at Van. 3	68	76	60-47 — 107	76	
Jeremy Roenick	Chi.	31-3-94	(G)	Chi. 3 at Wsh. 6	78	78	46-61 — 107	84	
Brendan Shanahan	St.L.	12-4-94	(G)	St.L. 5 at Dal. 9	80	83	52-50 — 102	81	25.2
Mario Lemieux	Pit.	16-1-96	(G)	Col. 5 at Pit. 2	38	44	69-92 — 161	70	
Jaromir Jagr	Pit.	6-2-96	(G)	Bos. 5 at Pit. 6	52	52	62-87 — 149	82	23.12
Ron Francis	Pit.	9-3-96	(A)	N.J. 4 at Pit. 3	61	66	27-92 — 119	77	
Peter Forsberg	Col.	9-3-96	(A)	Col. 7 at Van. 5	68	68	30-86 — 116	82	22.7
Joe Sakic	Col.	17-3-96	(A)	Edm. 1 at Col. 8	70	70	51-69 — 120	82	
Teemu Selanne	Ana.	25-3-96	(A)	Ana. 1 at Det. 5	70	73	40-68 — 108	79	
Alexander Mogilny	Van.	25-3-96	(A)	L.A. 1 at Van. 4	72	75	55-52 — 107	79	
Eric Lindros	Phi.	25-3-96	(A)	Hfd. 0 at Phi. 3	65	73	47-68 — 115	73	23.0
Wayne Gretzky	St.L.	28-3-96	(A)	N.J. 4 at St.L. 4	76	75	23-79 — 102	80	
Doug Weight	Edm.	30-3-96	(G)	Tor. 4 at Edm. 3	76	76	25-79 — 104	82	25.3
Sergei Fedorov	Det.	2-4-96	(A)	Det. 3 at S.J. 6	72	76	39-68 — 107	78	
Paul Kariya	Ana.	7-4-96	(G)	Ana. 5 at S.J. 3	78	78	50-58 — 108	82	21.5
Mario Lemieux	Pit.	8-3-97	(A)	Phi. 2 at Pit. 3	61	65	50-72 — 122	76	
Teemu Selanne	Ana.	1-4-97	(A)	Chi. 3 at Ana. 3	74	78	51-58 — 109	78	
Jaromir Jagr	Pit.	15-4-98	(G)	T.B. 1 at Pit. 5	76	80	35-67 — 102	77	
Jaromir Jagr	Pit.	13-3-99	(G)	Phi. 0 at Pit. 4	65	65	44-83 — 127	81	
Teemu Selanne	Ana.	5-4-99	(A)	Ana. 2 at Det. 3	69	76	47-60 — 107	75	
Paul Kariya	Ana.	17-4-99	(G)	Ana. 3 at S.J. 3	82	82	39-62 — 101	82	
Jaromir Jagr	Pit.	10-3-01	(G)	Cgy. 3 at Pit. 6	68	68	52-69 — 121	81	
Joe Sakic	Col.	18-3-01	(G)	Min. 3 at Col. 4	72	72	54-64 — 118	82	

Five-or-more-Goal Games

Player	Team	Date	Score			Opposing Goaltender
SEVEN GOALS						
Joe Malone	Quebec Bulldogs	Jan. 31/20	Tor. 6	at	Que. 10	Ivan Mitchell
SIX GOALS						
Newsy Lalonde	Montreal	Jan. 10/20	Tor. 7	at	Mtl. 14	Ivan Mitchell
Joe Malone	Quebec Bulldogs	Mar. 10/20	Ott. 4	at	Que. 10	Clint Benedict
Corb Denneny	Toronto St. Pats	Jan. 26/21	Ham. 3	at	Tor. 10	Howard Lockhart
Cy Denneny	Ottawa Senators	Mar. 7/21	Ham. 5	at	Ott. 12	Howard Lockhart
Syd Howe	Detroit	Feb. 3/44	NYR 2	at	Det. 12	Ken McAuley
Red Berenson	St. Louis	Nov. 7/68	St.L. 8	at	Phi. 0	Doug Favell
Darryl Sittler	Toronto	Feb. 7/76	Bos. 4	at	Tor. 11	Dave Reece
FIVE GOALS						
Joe Malone	Montreal	Dec. 19/17	Mtl. 7	at	Ott. 4	Clint Benedict
Harry Hyland	Mtl. Wanderers	Dec. 19/17	Tor. 9	at	Mtl. W. 10	Art Brooks
Joe Malone	Montreal	Jan. 12/18	Ott. 4	at	Mtl. 9	Clint Benedict
Joe Malone	Montreal	Feb. 2/18	Tor. 2	at	Mtl. 11	Hap Holmes
Mickey Roach	Toronto St. Pats	Mar. 6/20	Que. 2	at	Tor. 11	Frank Brophy
Newsy Lalonde	Montreal	Feb. 16/21	Ham. 5	at	Mtl. 10	Howard Lockhart
Babe Dye	Toronto St. Pats	Dec. 16/22	Mtl. 2	at	Tor. 7	Georges Vezina
Red Green	Hamilton Tigers	Dec. 5/24	Ham. 10	at	Tor. 3	John Ross Roach
Babe Dye	Toronto St. Pats	Dec. 22/24	Tor. 10	at	Bos. 1	Charles Stewart
Punch Broadbent	Mtl. Maroons	Jan. 7/25	Mtl. 6	at	Ham. 2	Jake Forbes
Pit Lepine	Montreal	Dec. 14/29	Ott. 4	at	Mtl. 6	Alex Connell
Howie Morenz	Montreal	Mar. 18/30	NYA 3	at	Mtl. 8	Roy Worters
Charlie Conacher	Toronto	Jan. 19/32	NYA 3	at	Tor. 11	Roy Worters
Ray Getliffe	Montreal	Feb. 6/43	Bos. 3	at	Mtl. 8	Frank Brimsek
Maurice Richard	Montreal	Dec. 28/44	Det. 1	at	Mtl. 9	Harry Lumley
Howie Meeker	Toronto	Jan. 8/47	Chi. 4	at	Tor. 10	Paul Bibeault
Bernie Geoffrion	Montreal	Feb. 19/55	NYR 2	at	Mtl. 10	Gump Worsley
Bobby Rousseau	Montreal	Feb. 1/64	Det. 3	at	Mtl. 9	Roger Crozier
Yvan Cournoyer	Montreal	Feb. 15/75	Chi. 3	at	Mtl. 12	Mike Veisor
Don Murdoch	NY Rangers	Oct. 12/76	NYR 10	at	Min. 4	Gary Smith
Ian Turnbull	Toronto	Feb. 2/77	Det. 1	at	Tor. 9	Ed Giacomin (2) Jim Rutherford (3)
Bryan Trottier	NY Islanders	Dec. 23/78	NYR 4	at	NYI 9	Wayne Thomas (4) John Davidson (1)
Tim Young	Minnesota	Jan. 15/79	Min. 8	at	NYR 1	Doug Soetaert (3) Wayne Thomas (2)
John Tonelli	NY Islanders	Jan. 6/81	Tor. 3	at	NYI 6	Jiri Crha (4) empty net (1)
Wayne Gretzky	Edmonton	Feb. 18/81	St.L. 2	at	Edm. 9	Mike Liut (3) Ed Staniowski (2)
Wayne Gretzky	Edmonton	Dec. 30/81	Phi. 5	at	Edm. 7	Pete Peeters (4) empty net (1)
Grant Mulvey	Chicago	Feb. 3/82	St.L. 5	at	Chi. 9	Mike Liut (4) Gary Edwards (1)
Bryan Trottier	NY Islanders	Feb. 13/82	Phi. 2	at	NYI 8	Pete Peeters
Willy Lindstrom	Winnipeg	Mar. 2/82	Wpg. 7	at	Phi. 6	Pete Peeters
Mark Pavelich	NY Rangers	Feb. 23/83	Hfd. 3	at	NYR 11	Greg Millen
Jari Kurri	Edmonton	Nov. 19/83	N.J. 4	at	Edm. 13	Glenn Resch (3) Ron Low (2)
Bengt Gustafsson	Washington	Jan. 8/84	Wsh. 7	at	Phi. 1	Pelle Lindbergh
Pat Hughes	Edmonton	Feb. 3/84	Cgy. 5	at	Edm. 10	Don Edwards (3) Reggie Lemelin (2)
Wayne Gretzky	Edmonton	Dec. 15/84	Edm. 8	at	St.L. 2	Rick Wamsley (4) Mike Liut(1)
Dave Andreychuk	Buffalo	Feb. 6/86	Buf. 8	at	Bos. 6	Pat Riggin (1) Doug Keans (4)
Wayne Gretzky	Edmonton	Dec. 6/87	Min. 4	at	Edm. 10	Don Beaupre (4) Kari Takko (1)
Mario Lemieux	Pittsburgh	Dec. 31/88	N.J. 6	at	Pit. 8	Bob Sauve (3) Chris Terreri (2)
Joe Nieuwendyk	Calgary	Jan. 11/89	Wpg. 3	at	Cgy. 8	Daniel Berthiaume
Mats Sundin	Quebec	Mar. 5/92	Que. 10	at	Hfd. 4	Peter Sidorkiewicz (3) Kay Whitmore (2)
Mario Lemieux	Pittsburgh	Apr. 9/93	Pit. 10	at	NYR 4	Corey Hirsch (3) Mike Richter (2)
Peter Bondra	Washington	Feb. 5/94	T.B. 3	at	Wsh. 6	Daren Puppa (4) Pat Jablonski (1)
Mike Ricci	Quebec	Feb. 17/94	Que. 8	at	S.J. 2	Arturs Irbe (3) Jimmy Waite (2)
Alexei Zhamnov	Winnipeg	Apr. 1/95	Wpg. 7	at	L.A. 7	Kelly Hrudey (3) Grant Fuhr (2)
Mario Lemieux	Pittsburgh	Mar. 26/96	St.L. 4	at	Pit. 8	Grant Fuhr (1) Jon Casey (4)
Sergei Fedorov	Detroit	Dec. 26/96	Wsh. 4	at	Det. 5	Jim Carey

Players' 500th Goals

Regular Season

Player	Team	Date	Game No.	Score			Opposing Goaltender	Total Goals	Total Games
Maurice Richard	Montreal	Oct. 19/57	863	Chi. 1	at	Mtl. 3	Glenn Hall	544	978
Gordie Howe	Detroit	Mar. 14/62	1,045	Det. 2	at	NYR 3	Gump Worsley	801	1,767
Bobby Hull	Chicago	Feb. 21/70	861	NYR. 2	at	Chi. 4	Ed Giacomin	610	1,063
Jean Béliveau	Montreal	Feb. 11/71	1,101	Min. 2	at	Mtl. 6	Gilles Gilbert	507	1,125
Frank Mahovlich	Montreal	Mar. 21/73	1,105	Van. 2	at	Mtl. 3	Dunc Wilson	533	1,181
Phil Esposito	Boston	Dec. 22/74	803	Det. 4	at	Bos. 5	Jim Rutherford	717	1,282
John Bucyk	Boston	Oct. 30/75	1,370	St.L. 2	at	Bos. 3	Yves Bélanger	556	1,540
Stan Mikita	Chicago	Feb. 27/77	1,221	Van. 4	at	Chi. 3	Cesare Maniago	541	1,394
Marcel Dionne	Los Angeles	Dec. 14/82	887	L.A. 2	at	Wsh. 7	Al Jensen	731	1,348
Guy Lafleur	Montreal	Dec. 20/83	918	Mtl. 6	at	N.J. 0	Glenn Resch	560	1,126
Mike Bossy	NY Islanders	Jan. 2/86	647	Bos. 5	at	NYI 7	empty net	573	752
Gilbert Perreault	Buffalo	Mar. 9/86	1,159	N.J. 3	at	Buf. 4	Alain Chevrier	512	1,191
Wayne Gretzky	Edmonton	Nov. 22/86	575	Van. 2	at	Edm. 5	empty net	894	1,487
Lanny McDonald	Calgary	Mar. 21/89	1,107	NYI 1	at	Cgy. 4	Mark Fitzpatrick	500	1,111
Bryan Trottier	NY Islanders	Feb. 13/90	1,104	Cgy. 4	at	NYI 2	Rick Wamsley	524	1,279
Mike Gartner	NY Rangers	Oct. 14/91	936	Wsh. 5	at	NYR 3	Mike Liut	708	1,432
Michel Goulet	Chicago	Feb. 16/92	951	Cgy. 5	at	Chi. 5	Jeff Reese	548	1,089
Jari Kurri	Los Angeles	Oct. 17/92	833	Bos. 6	at	L.A. 8	empty net	601	1,251
Dino Ciccarelli	Detroit	Jan. 8/94	946	Det. 6	at	L.A. 3	Kelly Hrudey	608	1,232
*Mario Lemieux	Pittsburgh	Oct. 26/95	605	Pit. 7	at	NYI 5	Tommy Soderstrom	654	812
*Mark Messier	NY Rangers	Nov. 6/95	1,141	Cgy. 2	at	NYR 4	Rick Tabaracci	658	1,602
*Steve Yzerman	Detroit	Jan. 17/96	906	Col. 2	at	Det. 3	Patrick Roy	658	1,362
Dale Hawerchuk	St. Louis	Jan. 31/96	1,103	St.L. 4	at	Tor. 0	Felix Potvin	518	1,188
*Brett Hull	St. Louis	Dec. 22/96	693	L.A. 4	at	St.L. 7	Stephane Fiset	679	1,101
Joe Mullen	Pittsburgh	Mar. 14/97	1,052	Pit. 3	at	Col. 6	Patrick Roy	502	1,062
*Dave Andreychuk	New Jersey	Mar. 15/97	1,070	Wsh. 2	at	N.J. 3	Bill Ranford	593	1,443
*Luc Robitaille	Los Angeles	Jan. 7/99	928	Buf. 2	at	L.A. 4	Dwayne Roloson	620	1,205
*Pat Verbeek	Detroit	Mar. 22/00	1,285	Cgy. 2	at	Det. 2	Fred Brathwaite	522	1,424
*Ron Francis	Carolina	Jan. 2/02	1,533	Bos. 6	at	Car. 3	Byron Dafoe	514	1,569
*Brendan Shanahan	Detroit	Mar. 23/02	1,100	Det. 2	at	Col. 0	Patrick Roy	503	1,108

*Active

Carolina captain Ron Francis reached several milestones during his 21st NHL season, playing in his 1,500th game and collecting his 500th goal and 1,700th point. He also helped the Hurricanes reach the Stanley Cup Finals for the first time.

Players' 1,000th Points

Regular Season

Player	Team	Date	Game No.	G or A	Score	Total Points G A PTS	Total Games
Gordie Howe	Detroit	Nov. 27/60	938	(A)	Tor. 0 at Det. 2	801-1,049–1,850	1,767
Jean Béliveau	Montreal	Mar. 3/68	911	(G)	Mtl. 2 at Det. 5	507-712–1,219	1,125
Alex Delvecchio	Detroit	Feb. 16/69	1,143	(A)	L.A. 3 at Det. 6	456-825–1,281	1,549
Bobby Hull	Chicago	Dec. 13/70	909	(A)	Min. 2 at Chi. 5	610-560–1,170	1,063
Norm Ullman	Toronto	Oct. 16/71	1,113	(A)	NYR 5 at Tor. 3	490-739–1,229	1,410
Stan Mikita	Chicago	Oct. 15/72	924	(A)	St.L. 3 at Chi. 1	541-926–1,467	1,394
John Bucyk	Boston	Nov. 9/72	1,144	(G)	Det. 3 at Bos. 8	556-813–1,369	1,540
Frank Mahovlich	Montreal	Feb. 17/73	1,090	(A)	Phi. 7 at Mtl. 6	533-570–1,103	1,181
Henri Richard	Montreal	Dec. 20/73	1,194	(A)	Mtl. 2 at Buf. 2	358-688–1,046	1,256
Phil Esposito	Boston	Feb. 15/74	745	(A)	Bos. 4 at Van. 2	717-873–1,590	1,282
Rod Gilbert	NY Rangers	Feb. 19/77	1,027	(G)	NYR 2 at NYI 5	406-615–1,021	1,065
Jean Ratelle	Boston	Apr. 3/77	1,007	(A)	Tor. 4 at Bos. 7	491-776–1,267	1,281
Marcel Dionne	Los Angeles	Jan. 7/81	740	(G)	L.A. 5 at Hfd. 3	731-1,040–1,771	1,348
Guy Lafleur	Montreal	Mar. 4/81	720	(G)	Mtl. 9 at Wpg. 3	560-793–1,353	1,126
Bobby Clarke	Philadelphia	Mar. 19/81	922	(G)	Bos. 3 at Phi. 5	358-852–1,210	1,144
Gilbert Perreault	Buffalo	Apr. 3/82	871	(A)	Buf. 5 at Mtl. 4	512-814–1,326	1,191
Darryl Sittler	Philadelphia	Jan. 20/83	927	(G)	Cgy. 2 at Phi. 5	484-637–1,121	1,096
Wayne Gretzky	Edmonton	Dec. 19/84	424	(A)	L.A. 3 at Edm. 7	894-1,963–2,875	1,487
Bryan Trottier	NY Islanders	Jan. 29/85	726	(G)	Min. 4 at NYI 4	524-901–1,425	1,279
Mike Bossy	NY Islanders	Jan. 24/86	656	(A)	NYI 7 at Wsh. 5	573-553–1,126	752
Denis Potvin	NY Islanders	Apr. 4/87	987	(G)	Buf. 6 at NYI 6	310-742–1,052	1,060
Bernie Federko	St. Louis	Mar. 19/88	855	(A)	Hfd. 5 at St.L. 3	369-761–1,130	1,000
Lanny McDonald	Calgary	Mar. 7/89	1,101	(G)	Wpg. 5 at Cgy. 9	500-506–1,006	1,111
Peter Stastny	Quebec	Oct. 19/89	682	(G)	Que. 5 at Chi. 3	450-789–1,239	977
Jari Kurri	Edmonton	Jan. 2/90	716	(A)	Edm. 6 at St.L. 4	601-797–1,398	1,251
Denis Savard	Chicago	Mar. 11/90	727	(A)	St.L. 6 at Chi. 4	473-865–1,338	1,196
Paul Coffey	Pittsburgh	Dec. 22/90	770	(A)	Pit. 4 at NYI 3	396-1,135–1,531	1,409
*Mark Messier	Edmonton	Jan. 13/91	822	(A)	Edm. 5 at Phi. 3	658-1,146–1,804	1,602
Dave Taylor	Los Angeles	Feb. 5/91	930	(A)	L.A. 3 at Phi. 2	431-638–1,069	1,111
Michel Goulet	Chicago	Feb. 23/91	878	(G)	Chi. 3 at Min. 3	548-604–1,152	1,089
Dale Hawerchuk	Buffalo	Mar. 8/91	781	(G)	Chi. 5 at Buf. 3	518-891–1,409	1,188
Bobby Smith	Minnesota	Nov. 30/91	986	(A)	Min. 4 at Tor. 3	357-679–1,036	1,077
Mike Gartner	NY Rangers	Jan. 4/92	971	(G)	NYR 4 at N.J. 6	708-627–1,335	1,432
Raymond Bourque	Boston	Feb. 29/92	933	(A)	Wsh. 5 at Bos. 5	410-1,169–1,579	1,612
*Mario Lemieux	Pittsburgh	Mar. 24/92	513	(A)	Pit. 3 at Det. 4	654-947–1,601	812
Glenn Anderson	Toronto	Feb. 22/93	954	(G)	Tor. 8 at Van. 1	498-601–1,099	1,129
*Steve Yzerman	Detroit	Feb. 24/93	737	(A)	Det. 7 at Buf. 10	658-1,004–1,662	1,362
*Ron Francis	Pittsburgh	Oct. 28/93	893	(G)	Que. 7 at Pit. 3	514-1,187–1,701	1,569
Bernie Nicholls	New Jersey	Feb. 13/94	858	(G)	N.J. 3 at T.B. 3	475-734–1,209	1,127
Dino Ciccarelli	Detroit	Mar. 9/94	957	(G)	Det. 5 at Cgy. 1	608-592–1,200	1,232
Brian Propp	Hartford	Mar. 19/94	1,008	(G)	Hfd. 5 at Phi. 3	425-579–1,004	1,016
Joe Mullen	Pittsburgh	Feb. 7/95	935	(A)	Fla. 3 at Pit. 7	502-561–1,063	1,062
Steve Larmer	NY Rangers	Mar. 8/95	983	(A)	N.J. 4 at NYR 6	441-571–1,012	1,006
*Doug Gilmour	Toronto	Dec. 23/95	935	(A)	Edm. 1 at Tor. 6	439-945–1,384	1,412
Larry Murphy	Toronto	Mar. 27/96	1,228	(G)	Tor. 6 at Van. 2	287-929–1,216	1,615
*Dave Andreychuk	New Jersey	Apr. 7/96	998	(G)	NYR 2 at N.J. 4	593-654–1,247	1,443
*Adam Oates	Washington	Oct. 8/97	830	(G)	Wsh. 6 at NYI 3	330-1,027–1,357	1,210
*Phil Housley	Washington	Nov. 8/97	1,081	(A)	Edm. 1 at Wsh. 2	332-871–1,203	1,437
Dale Hunter	Washington	Jan. 9/98	1,308	(A)	Phi. 1 at Wsh. 4	323-697–1,020	1,407
Pat LaFontaine	NY Rangers	Jan. 22/98	847	(G)	Phi. 4 at NYR 3	468-545–1,013	865
*Luc Robitaille	Los Angeles	Jan. 29/98	882	(A)	Cgy. 3 at L.A. 5	620-668–1,288	1,205
*Al MacInnis	St. Louis	Apr. 7/98	1,056	(A)	St.L. 3 at Det. 5	324-880–1,204	1,333
*Brett Hull	Dallas	Nov. 14/98	815	(A)	Dal. 3 at Bos. 1	679-567–1,246	1,101
Brian Bellows	Washington	Jan. 2/99	1,147	(A)	Tor. 2 at Wsh. 5	485-537–1,022	1,188
*Pierre Turgeon	St. Louis	Oct. 9/99	881	(G)	St.L. 4 at Edm. 3	468-724–1,192	1,074
*Joe Sakic	Colorado	Dec. 27/99	810	(A)	St.L. 1 at Col. 5	483-774–1,257	1,016
*Pat Verbeek	Detroit	Feb. 27/00	1,275	(A)	T.B. 1 at Det. 3	522-541–1,063	1,424
*V. Damphousse	San Jose	Oct. 14/00	1,090	(A)	Bos. 2 at S.J. 5	397-706–1,103	1,214
*Jaromir Jagr	Pittsburgh	Dec. 30/00	763	(G)	Ott. 3 at Pit. 5	470-688-1,158	875
*Mark Recchi	Philadelphia	Mar. 13/01	920	(A)	St.L. 2 at Phi. 5	410-664-1,074	1,012
*Theoren Fleury	NY Rangers	Oct. 29/01	960	(A)	Dal. 2 at NYR 4	443-612-1,055	1,030
*B. Shanahan	Detroit	Jan. 12/02	1,073	(G)	Dal. 2 at Det. 5	503-527-1,030	1,108
*Jeremy Roenick	Philadelphia	Jan. 30/02	961	(G)	Phi. 1 at Ott. 3	429-585-1,014	983

*Active

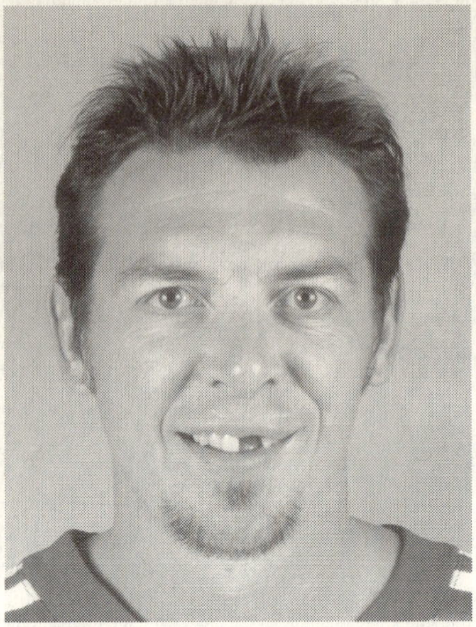

Theoren Fleury (top), Brendan Shanahan (center) and Jeremy Roenick (above) each reached the 1,000-point plateau during the 2001-02 season.

Individual Awards

Hart Memorial Trophy

Art Ross Trophy

Calder Memorial Trophy

James Norris Memorial Trophy

HART MEMORIAL TROPHY

An annual award "to the player adjudged to be the most valuable to his team." Winner selected in a poll by the Professional Hockey Writers' Association in the 30 NHL cities at the end of the regular schedule. The winner receives $10,000 and the runners-up $6,000 and $4,000.

History: The Hart Memorial Trophy was presented by the National Hockey League in 1960 after the original Hart Trophy was retired to the Hockey Hall of Fame. The original Hart Trophy was donated to the NHL in 1923 by Dr. David A. Hart, father of Cecil Hart, former manager-coach of the Montreal Canadiens.

2001-02 Winner: **Jose Theodore, Montreal Canadiens**
Runners-up: **Jarome Iginla, Calgary Flames**
Patrick Roy, Colorado Avalanche

Goaltender Jose Theodore of the Montreal Canadiens captured the Hart Memorial Trophy, edging Calgary Flames right winger Jarome Iginla in the closest NHL awards voting on record. Theodore and Iginla each received 434 points, with Theodore placed first by virtue of the first tie-breaker for NHL trophies – more first-place votes. Theodore was the first selection on 26 ballots, Iginla 23. Patrick Roy had 283 points. Previously, the closest Hart Trophy vote had occurred in 2000, when St. Louis Blues defenseman Chris Pronger edged Pittsburgh Penguins right winger Jaromir Jagr, 396-395. Prior to that, the narrowest margin was two points, when Edmonton's Mark Messier edged Boston's Raymond Bourque 227-225, in 1990.

Theodore led the Canadiens to their first Stanley Cup playoff berth since 1997-98, posting a 30-24-10 record in 67 games - including a seven-game winning streak down the stretch. Theodore led all goaltenders in save percentage, stopping 93.1 percent of the shots he faced, tied for second in shutouts (seven) and placed fourth in goals-against average (2.11). He becomes the sixth goaltender to capture the award, joining Dominik Hasek (1997, 1998), Jacques Plante (1962), Al Rollins (1954), Chuck Rayner (1950) and Roy Worters (1929).

CALDER MEMORIAL TROPHY

An annual award "to the player selected as the most proficient in his first year of competition in the National Hockey League." Winner selected in a poll by the Professional Hockey Writers' Association at the end of the regular schedule. The winner receives $10,000 and the runners-up $6,000 and $4,000.

History: From 1936-37 until his death in 1943, Frank Calder, NHL President, bought a trophy each year to be given permanently to the outstanding rookie. After Calder's death, the NHL presented the Calder Memorial Trophy in his memory and the trophy is to be kept in perpetuity. To be eligible for the award, a player cannot have played more than 25 games in any single preceding season nor in six or more games in each of any two preceding seasons in any major professional league. Beginning in 1990-91, to be eligible for this award a player must not have attained his twenty-sixth birthday by September 15th of the season in which he is eligible.

2001-02 Winner: **Dany Heatley, Atlanta Thrashers**
Runners-up: **Ilya Kovalchuk, Atlanta Thrashers**
Kristian Huselius, Florida Panthers

Right winger Dany Heatley of the Atlanta Thrashers was selected as the winner of the Calder Memorial Trophy. Heatley received 48 of 62 first-place votes and was the second choice on 11 other ballots for 568 points. Thrashers teammate Ilya Kovalchuk finished second in the balloting with 14 first-place votes and 466 points. Kristian Huselius of Florida placed third with 214 points.

The second overall selection in the 2000 Entry Draft, Heatley led all rookies in scoring with 67 points (26 goals, 41 assists), also leading first-year players in assists, power-play points (18) and shots on goal (202). Heatley and Kovalchuk (29 goals, 22 assists) finished the season 1-2 among NHL rookies in scoring, becoming the first teammates to lead the league since Brian Leetch and Tony Granato of the New York Rangers in 1988-89, and the first teammates to finish 1-2 in Calder Trophy balloting since the New York Islanders' Bryan Trottier and Glenn Resch in 1975-76.

ART ROSS TROPHY

An annual award "to the player who leads the league in scoring points at the end of the regular season." The winner receives $10,000 and the runners-up $6,000 and $4,000.

History: Arthur Howie Ross, former manager-coach of the Boston Bruins, presented the trophy to the National Hockey League in 1947. If two players finish the schedule with the same number of points, the trophy is awarded in the following manner: 1. Player with most goals. 2. Player with fewer games played. 3. Player scoring first goal of the season.

2001-02 Winner: **Jarome Iginla, Calgary Flames**
Runners-up: **Markus Naslund, Vancouver Canucks**
Todd Bertuzzi, Vancouver Canucks

Jarome Iginla captured the Art Ross Trophy for the first time in his career, becoming the first player other than Jaromir Jagr, Mario Lemieux or Wayne Gretzky to win the Art Ross Trophy since Marcel Dionne in 1979-80. Also the recipient of the Maurice Richard Trophy as the NHL's top goal scorer, Iginla tallied 96 points (52 goals, 44 assists) in 82 games, highlighted by a 15-game point streak from October 18 through November 22, 2001. Iginla's streak tied with that of Vancouver's Todd Bertuzzi as the NHL's longest in 2001-02. Bertuzzi's teammate Markus Naslund finished second to Iginla in scoring with 90 points (40 goals, 50 assists). Bertuzzi was next with 85 points (36 goals, 49 assists).

JAMES NORRIS MEMORIAL TROPHY

An annual award "to the defense player who demonstrates throughout the season the greatest all-round ability in the position." Winner selected in a poll by the Professional Hockey Writers' Association at the end of the regular schedule. The winner receives $10,000 and the runners-up $6,000 and $4,000.

History: The James Norris Memorial Trophy was presented in 1953 by the four children of the late James Norris in memory of the former owner-president of the Detroit Red Wings.

2001-02 Winner: **Nicklas Lidstrom, Detroit Red Wings**
Runners-up: **Chris Chelios, Detroit Red Wings**
Rob Blake, Colorado Avalanche

Nicklas Lidstrom of the Detroit Red Wings won his second consecutive Norris Trophy after finishing as the runner-up for three straight years. Lidstrom was named on 59 of 62 ballots (29 first-place votes) and received 472 points, edging Red Wings teammate Chris Chelios, who polled 431 points (28 first-place votes). Rob Blake finished third with 321 points. Lidstrom becomes the first winner of the Norris Trophy in consecutive seasons since Boston's Raymond Bourque in 1990 and 1991, while Lidstrom and Chelios become the first teammates to finish 1-2 in Norris voting.

Lidstrom finished second in scoring among defensemen with 59 points (9 goals, 50 assists) in 78 games and led all defensemen in power-play points (30). He logged 28:48 of ice time per game, third in the NHL. The Red Wings won the Presidents' Trophy as the NHL's top club during the regular season (51-17-10-4, 116 points) and allowed 187 goals, tied for the third-lowest total in the league.

Vezina Trophy

Lady Byng Memorial Trophy

Frank J. Selke Trophy

Conn Smythe Trophy

VEZINA TROPHY

An annual award "to the goalkeeper adjudged to be the best at his position" as voted by the general managers of each of the 30 clubs. The winner receives $10,000, and the runners-up $6,000 and $4,000.

History: Leo Dandurand, Louis Letourneau and Joe Cattarinich, former owners of the Montreal Canadiens, presented the trophy to the National Hockey League in 1926-27 in memory of Georges Vezina, outstanding goalkeeper of the Canadiens who collapsed during an NHL game on November 28, 1925, and died of tuberculosis a few months later. Until the 1981-82 season, the goalkeeper(s) of the team allowing the fewest number of goals during the regular season were awarded the Vezina Trophy.

2001-02 Winner: **Jose Theodore, Montreal Canadiens**
Runners-up: **Patrick Roy, Colorado Avalanche**
Sean Burke, Phoenix Coyotes

Jose Theodore of the Montreal Canadiens captured the Vezina Trophy by virtue of receiving more first-place votes than Colorado's Patrick Roy. Theodore and Roy each received 105 points, but Theodore placed first on 15 ballots to Roy's 12. Sean Burke received two first-place votes and was a distant third with 27 points.

Under the NHL's tie-breaking procedure for trophy voting adopted by the league in December, 1954, "in the event of two players receiving the same number of points, the player receiving the greatest number of firsts or other superior choices shall be the winner." The 2001-02 season marked the first time the rule was needed to determine a trophy winner. Theodore also won the Hart Trophy by virtue of receiving more first-place votes.

CONN SMYTHE TROPHY

An annual award "to the most valuable player for his team in the playoffs." Winner selected by the Professional Hockey Writers' Association at the conclusion of the final game in the Stanley Cup Finals. The winner receives $10,000.

History: Presented by Maple Leaf Gardens Limited in 1964 to honor Conn Smythe, the former coach, manager, president and owner-governor of the Toronto Maple Leafs.

2001-02 Winner: **Nicklas Lidstrom, Detroit Red Wings**

Nicklas Lidstrom led all defensemen in postseason scoring with 16 points (five goals, 11 assists) in 23 games, including the game-winning goal in the second game of the Stanley Cup Finals to even the series at 1-1. He logged over 31 minutes of ice time per game, including a league-high 52:03 in the Red Wings' triple-overtime victory against the Carolina Hurricanes in game three of the Finals. Lidstrom is the first European-trained player to win the Conn Smythe Trophy and the seventh defenseman, joining Bobby Orr (twice), Brian Leetch, Al MacInnis, Larry Robinson, Serge Savard and Scott Stevens.

LADY BYNG MEMORIAL TROPHY

An annual award "to the player adjudged to have exhibited the best type of sportsmanship and gentlemanly conduct combined with a high standard of playing ability." Winner selected in a poll by the Professional Hockey Writers' Association at the end of the regular schedule. The winner receives $10,000 and the runners-up $6,000 and $4,000.

History: Lady Byng, wife of Canada's Governor-General at the time, presented the Lady Byng Trophy in the 1924-25 season. After Frank Boucher of the New York Rangers won the award seven times in eight seasons, he was given the trophy to keep and Lady Byng donated another trophy in 1936. After Lady Byng's death in 1949, the National Hockey League presented a new trophy, changing the name to Lady Byng Memorial Trophy.

2001-02 Winner: **Ron Francis, Carolina Hurricanes**
Runners-up: **Joe Sakic, Colorado Avalanche**
Nicklas Lidstrom, Detroit Red Wings

Ron Francis of the Carolina Hurricanes earned the Lady Byng Memorial Trophy for the third time in his career. Francis received 28 first-place votes and was named on 57 of 62 ballots for 427 points, ahead of last year's Lady Byng winner Joe Sakic of the Colorado Avalanche (12 first-place votes, 325 points). Three-time runner-up Nicklas Lidstrom was third in voting with 243 points. Francis led the Hurricanes in scoring with 77 points (27 goals, 50 assists) and was assessed just 18 minutes in penalties in 80 games.

FRANK J. SELKE TROPHY

An annual award "to the forward who best excels in the defensive aspects of the game." Winner selected in a poll by the Professional Hockey Writers' Association at the end of the regular schedule. The winner receives $10,000 and the runners-up $6,000 and $4,000.

History: Presented to the National Hockey League in 1977 by the Board of Governors of the NHL in honor of Frank J. Selke, one of the great architects of NHL championship teams.

2001-02 Winner: **Michael Peca, New York Islanders**
Runners-up: **Craig Conroy, Calgary Flames**
Jere Lehtinen, Dallas Stars

New York Islanders center Michael Peca captured the Frank J. Selke Trophy for the second time in his career. Peca received 25 first-place votes and was named on 55 of 62 ballots for 394 points. Calgary's Craig Conroy was second (seven first-place votes, 178 points), while Jere Lehtinen was third (175 points). Peca becomes the fifth multiple Selke winner, joining Bob Gainey (four), Guy Carbonneau (three), Sergei Fedorov (two) and Jere Lehtinen (two).

Peca was a key to the New York Islanders' resurgence in 2001-02 by posting a +19 rating, playing an average of 20:13 per game and tallying six of the Islanders' league-leading 17 shorthanded goals. The Islanders allowed 48 fewer goals in 2001-02 than the previous season, tied for the top improvement in the NHL.

WILLIAM M. JENNINGS TROPHY

An annual award "to the goalkeeper(s) having played a minimum of 25 games for the team with the fewest goals scored against it." Winners selected on regular-season play. The winner receives $10,000, and the runners-up $6,000 and $4,000.

History: The Jennings Trophy was presented in 1981-82 by the National Hockey League's Board of Governors to honor the late William M. Jennings, longtime governor and president of the New York Rangers and one of the great builders of hockey in the United States.

2001-02 Winner: **Patrick Roy, Colorado Avalanche**
Runners-up: **Tommy Salo, Edmonton Oilers**
Dominik Hasek, Detroit Red Wings
Martin Brodeur, New Jersey Devils

Patrick Roy collected the Jennings Trophy for the fifth time in his career, but the first time since 1992. Also a finalist for the Vezina Trophy and the Hart Trophy, Roy led all goaltenders in goals-against average (1.94) and shutouts (nine) and finished second in save percentage (.925). The Avalanche allowed just 169 goals in 82 games en route to a 45-28-8-1 record for 99 points and their NHL record-tying eighth consecutive division title. Led by Tommy Salo, the Edmonton Oilers surrendered only 182 goals, while the Detroit Red Wings and New Jersey Devils allowed 187.

LESTER B. PEARSON AWARD

An annual award presented to the NHL's outstanding player as selected by the members of the National Hockey League Players' Association. The winner receives $20,000, and the two finalists receive $10,000 each to donate to the grassroots hockey program of their choice, through the NHLPA's Goals & Dreams Fund.

History: The award was first presented in 1970-71 by the NHLPA in honor of the late Lester B. Pearson, former Prime Minister of Canada.

2001-02 Winner: **Jarome Iginla, Calgary Flames**
Runners-up: **Sean Burke, Phoenix Coyotes**
Patrick Roy, Colorado Avalanche

Jarome Iginla, who led the NHL with 52 goals and 96 points, won the Lester B. Pearson Award for the first time in his career. Iginla designated the St. Albert Minor Hockey Association and the Calgary Minor Hockey Association as the beneficiaries of the $20,000 that accompanies the award. Burke allocated $10,000 to the Future Stars Hockey camp in Red Deer, Alberta. Roy also designated his $10,000 to a minor hockey program.

William M. Jennings Trophy

Jack Adams Award

Bill Masterton Trophy

Lester Patrick Trophy

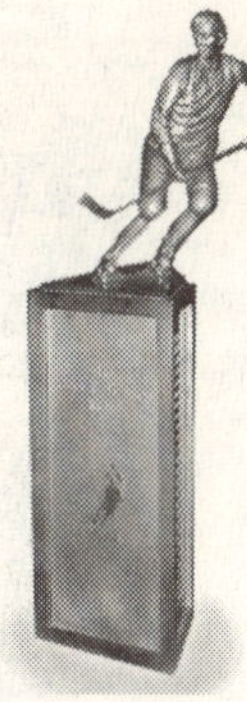

Lester B. Pearson Award

JACK ADAMS AWARD

An annual award presented by the National Hockey League Broadcasters' Association to "the NHL coach adjudged to have contributed the most to his team's success." Winner selected by a poll among members of the NHL Broadcasters' Association at the end of the regular season. The winner receives $1,000 from the NHLBA.

History: The award was presented by the NHL Broadcasters' Association in 1974 to commemorate the late Jack Adams, coach and general manager of the Detroit Red Wings, whose lifetime dedication to hockey serves as an inspiration to all who aspire to further the game.

2001-02 Winner: **Bob Francis, Phoenix Coyotes**
Runners-up: **Brian Sutter, Chicago Blackhawks**
Robbie Ftorek, Boston Bruins

Phoenix Coyotes head coach Bob Francis captured the 2001-02 Jack Adams Award as NHL coach of the year. Francis, the son of Hockey Hall of Famer Emile Francis (builder's category), received 27 first-place votes and 190 points. Chicago's Brian Sutter was second with 15 first-place votes and 151 points. Robbie Ftorek had 10 first-place votes and 108 points.

Despite the departure of key veteran players, Francis guided the Coyotes to a 40-27-9-6 record for 95 points, the second-best record in franchise history, the best record since the club relocated to Phoenix in 1996, and the third 90-or-more point season in as many years behind the Phoenix bench. The Coyotes were the NHL's hottest club following the Olympic break, posting more points (33) than any other NHL club in that span.

BILL MASTERTON MEMORIAL TROPHY

An annual award under the trusteeship of the Professional Hockey Writers' Association to "the National Hockey League player who best exemplifies the qualities of perseverance, sportsmanship and dedication to hockey." Winner selected by a poll among the 30 chapters of the PHWA at the end of the regular season. A $2,500 grant from the PHWA is awarded annually to the Bill Masterton Scholarship Fund, based in Bloomington, MN, in the name of the Masterton Trophy winner.

History: The trophy was presented by the NHL Writers' Association in 1968 to commemorate the late Bill Masterton, a player with the Minnesota North Stars, who exhibited to a high degree the qualities of perseverance, sportsmanship and dedication to hockey, and who died January 15, 1968.

2001-02 Winner: **Saku Koivu, Montreal Canadiens**
Runners-up: **Kevin Dineen, Columbus Blue Jackets**
Ron Francis, Carolina Hurricanes

Diagnosed with non-Hodgkin's lymphoma cancer in early September, 2001, Saku Koivu's courageous battle with the disease was an inspiration to all. Assuming a strong leadership role in the Montreal dressing room and participating in numerous club activities while undergoing treatment and rehabilitation, the Canadiens captain was an uplifting presence. Koivu returned to action on April 9th against the Ottawa Senators and performed well in the playoffs after helping the Canadiens clinch a postseason berth.

LESTER PATRICK TROPHY

An annual award "for outstanding service to hockey in the United States." Eligible recipients are players, officials, coaches, executives and referees. Winners are selected by an award committee consisting of the commissioner of the NHL, an NHL governor, a representative of the New York Rangers, a member of the Hockey Hall of Fame builder's section, a member of the Hockey Hall of Fame player's section, a member of the U.S. Hockey Hall of Fame, a member of the NHL Broadcasters' Association and a member of the Professional Hockey Writers' Association. Each except the League Commissioner is rotated annually. The winner receives a miniature of the trophy.

History: Presented by the New York Rangers in 1966 to honor the late Lester Patrick, longtime general manager and coach of the New York Rangers, whose teams finished out of the playoffs only once in his first 16 years with the club.

2001-02 Winners: **1960 U.S. Olympic Team**
Herb Brooks
Larry Pleau

The 1960 U.S. Olympic Ice Hockey Team, coached by Jack Riley, won America's first Olympic gold medal in hockey by completing upsets of the favored Canadian and Czechoslovakian Olympic teams at Squaw Valley, California. After surprising Canada in the semifinals, the USA posted a six-goal third period to overcome a 4-3 deficit and claim a 9-4 gold medal victory over Czechoslovakia. The 1960 squad is only the third team to be honored in the 35-year history of the Lester Patrick Award, joining the 1980 U.S. Olympic Men's Ice Hockey Team (1980) and the champion 1998 U.S. Olympic Women's Ice Hockey Team (1999). Members of the 1960 team are: Bill Cleary, Bob Cleary, Bill Christian, Roger Christian, Gene Grazia, Paul Johnson, John Mayasich, Jack McCartan, Jack Kirrane, Bob McVey, Dick Meredith, Weldy Olson, Bob Owen, Rod Paavola, Larry Palmer, Dick Rodenhiser and Tommy Williams.

Herb Brooks is best known as the head coach of the 1980 U.S. Olympic "Miracle on Ice" team that posted the legendary upset of the Soviet Union at Lake Placid, New York. He also coached the 2002 U.S. Olympic team and appeared at the Olympics as a player in 1964 and 1968, serving as U.S. team captain in '68. Brooks competed for U.S. national teams in 1961, 1962, 1965, 1967 and 1970. A Minnesota native, Brooks coached the University of Minnesota for seven seasons and led the Golden Gophers to three NCAA Division I championships (1974, 1976 and 1979). Brooks also made an impact in the National Hockey League. He served as coach of the New York Rangers from 1981 to 1985, leading the club to four playoff appearances and reaching the 100-victory mark faster than any other coach in Ranger history. He also coached the Minnesota North Stars during the 1987-88 season and the New Jersey Devils during their 1992-93 campaign. The 2001-02 season marked his eighth season as a scout with the Pittsburgh Penguins. He also coached the team during the 1999-2000 season.

In five seasons as general manager of the St. Louis Blues, Larry Pleau's teams have reached the playoffs for five straight seasons, winning the Presidents' Trophy in 1999-2000 and reaching the Western Conference Finals in 2001. Before joining the Blues, Pleau spent eight seasons as an executive in the New York Rangers organization. He was inducted into the U.S. Hockey Hall of Fame as an administrator in 2000. A veteran of 20 NHL seasons, Pleau spent three years with the Montreal Canadiens (1969-71) then became the first player to sign with the New England Whalers of the World Hockey Association. He would spend 17 seasons with the Whalers organization as a player, assistant coach, head coach, and, later, general manager. Pleau also competed for the 1968 U.S. Olympic team and the 1969 U.S. national team. He went to training camp with Team USA in the 1976 Canada Cup Tournament. Pleau also served as associate general manager of the 2002 U.S. Olympic team.

King Clancy Memorial Trophy

Bud Light Plus-Minus Award

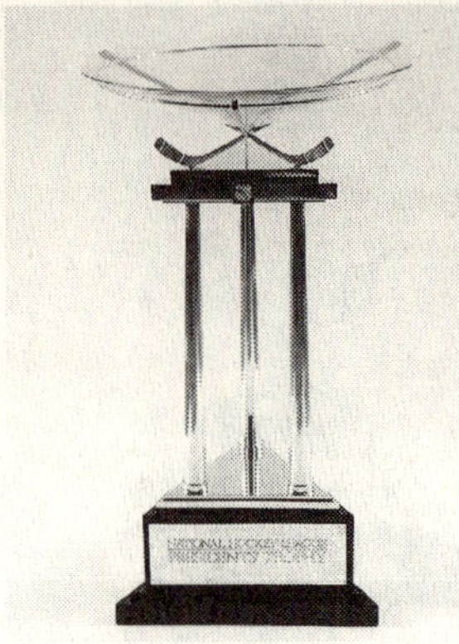

Presidents' Trophy

Maurice "Rocket" Richard Trophy

KING CLANCY MEMORIAL TROPHY

An annual award "to the player who best exemplifies leadership qualities on and off the ice and has made a noteworthy humanitarian contribution in his community."

History: The King Clancy Memorial Trophy was presented to the National Hockey League by the Board of Governors in 1988 to honor the late Frank "King" Clancy.

2001-02 Winner: Ron Francis, Carolina Hurricanes

Carolina Hurricanes captain Ron Francis enjoyed a special year on the ice in his 21st NHL season, reaching the milestones of 500 goals, 1,500 games and 1,700 points. But his contributions extend far beyond the rink and his commitment to community service earned him the 2001-02 King Clancy Memorial Trophy.

Among Francis's most important charitable legacies is the Ron Francis Night Out program, designed to provide respite for children's hospital patients and their families as they deal with life-altering illnesses. The program creates special nights for children who are patients at Duke Children's Hospital in Durham, NC, and their families. It includes a limousine ride, roses for the parents, dinner at the Hurricanes' Arena Club Restaurant, special gifts for the children, and tickets to Francis's luxury suite to watch the Hurricanes play. After the game, the families meet Francis in the locker room area.

Francis also supports Special Olympics, serving as chairman of the Sault Ste. Marie, Ontario, Special Olympics Summer Games, and is on the board of the Carolina Kids 'n' Community Foundation. Francis was instrumental in the creation of Hurricanes Hallway at Duke Children's Hospital, an area that welcomes and entertains pediatric patients. In addition, he has supported charitable efforts related to the September 11th terrorist attacks in New York and Washington.

PRESIDENTS' TROPHY

An annual award to the club finishing the regular-season with the best overall record. The winner receives $350,000, to be split between the team and its players.

History: Presented to the National Hockey League in 1985-86 by the NHL Board of Governors to recognize the team compiling the top regular-season record.

2001-02 Winner: Detroit Red Wings
Runners-up: Boston Bruins
Toronto Maple Leafs

Winners of eight of their first 10 games and with a 22-3-1-1 record after 27, the Detroit Red Wings outdistanced the field early in the 2001-02 regular season and went on to post a mark of 51-17-10-4 for 116 points, the second-highest total in franchise history. Detroit was an NHL best 28-7-5-1 at home and 23-10-5-3 away, highlighted by a 10-game home winning streak (December 19 through January 20) and an eight-game road winning streak (February 4 through March 9). Detroit became just the sixth club in the trophy's history to win the Stanley Cup after leading all clubs in points at the end of the regular season. The Boston Bruins had the NHL's next-best record at 43-24-6-9 for 101 points, while the Toronto Maple Leafs finished third at 43-25-10-4 and 100 points.

MAURICE "ROCKET" RICHARD TROPHY

An annual award "presented to the player finishing the regular season as the League's goal-scoring leader." The winner receives $10,000.

History: A gift to the NHL from the Montreal Canadiens in 1999, the Maurice "Rocket" Richard Trophy honors one of the game's greatest stars. During his 18-year career with the Canadiens from 1942-43 through 1959-60, Richard was the first player in NHL history to score 50 goals in a season and 500 in his career. He played on eight Stanley Cup champions and led the League in goal scoring five times.

2001-02 Winner: Jarome Iginla, Calgary Flames
Runners-up: Bill Guerin, Boston Bruins
Glen Murray, L.A./Boston
Mats Sundin, Toronto Maple Leafs

Jarome Iginla was the NHL's leading goal scorer in 2001-02, tallying 52 times. Iginla scored 18 goals in his first 20 games, helping the Flames jump out to a 13-2-3-2 start, posted a six-game goal-scoring streak following the Olympic break, and finished the season with nine goals in his last 10 games en route to topping the 50-goal plateau for the first time in his career. Iginla accounted for 25.8 percent of Calgary's goals (52 of 201), the highest percentage in the league. The Flames posted a 23-10-5-2 record when he scored.

Finishing behind Iginla in goal scoring was Bill Guerin, Glen Murray and Mats Sundin, each of whom tallied 41 times. The total represented new career highs for Guerin and Murray, while Sundin's output matched his best performance since scoring 47 times in 1992-93.

BUD LIGHT PLUS-MINUS AWARD

An annual award "to the player, having played a minimum of 60 games, who leads the League in plus/minus statistics" at the end of the regular season.
Bud Light will contribute $5,000 on behalf of the winner to the charity of his choice.

History: This award was first presented to the NHL in 1997-98 by Anheuser-Busch Inc. to recognize the League leader in plus-minus statistics. Plus-minus statistics are calculated by giving a player a "plus" when on-ice for an even-strength or short-handed goal scored by his team. He receives a "minus" when on-ice for an even-strength or short-handed goal scored by the opposing team. A plus-minus award has been presented since the 1982-83 season.

2001-02 Winner: Chris Chelios, Detroit Red Wings
Runners-up: Jeremy Roenick, Philadelphia Flyers
Simon Gagne, Philadelphia Flyers
Glen Murray, L.A./Boston

Defenseman Chris Chelios was the NHL's plus-minus leader with a rating of +40. The total was the second-highest career mark for the 40-year-old NHL veteran, who finished second in the NHL behind Chris Pronger with a mark of +48 in 1999-2000. Bud Light donated $5,000 to Cheli's Children's Foundation. Jeremy Roenick of the Philadelphia Flyers finished second to Chelios with a rating of +32. Fellow Flyer Simon Gagne was tied with Boston's Glen Murray at +31.

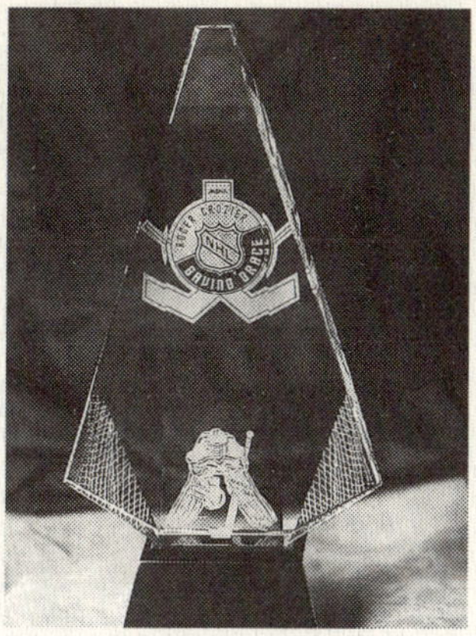

MBNA Roger Crozier Saving Grace Award

Bud Light NHL All-Star Game MVP Award

MBNA ROGER CROZIER SAVING GRACE AWARD

An award "presented to the goaltender having played a minimum of 25 games with the NHL's best save percentage during the regular season." The winner receives $25,000 to be donated to the youth hockey or educational program of his choice.

History: This award was first presented to the league in 1999-2000 by MBNA Corporation. It is named for Roger Crozier, one of the NHL's top goaltenders during his career. Crozier joined MBNA America Bank in 1983. He passed away on Jan. 11, 1996. Save percentage is calculated by dividing total saves by total shots faced.

2001-02 Winner: **Jose Theodore, Montreal Canadiens**
Runners-up: **Patrick Roy, Colorado Avalanche**
Roman Cechmanek, Philadelphia Flyers
Marty Turco, Dallas Stars

Montreal Canadiens goaltender Jose Theodore finished the 2001-02 season with a league-leading save percentage of .931 (1972 shots, 1836 saves) in 67 games. Theodore beat out Patrick Roy of the Colorado Avalanche (.925), as well as Philadelphia's Roman Cechmanek (.921) and last year's winner Marty Turco of the Dallas Stars (.921). Theodore, who also won the Vezina Trophy and the Hart Trophy, enjoyed a breakout season in 2001-02 and helped the Canadiens to their first playoff berth since 1998. His check for $25,000 will be donated to the charity of his choice.

BUD LIGHT NHL ALL-STAR GAME MVP AWARD

1962 Eddie Shack, Tor.
1963 Frank Mahovlich, Tor.
1964 Jean Beliveau, Mtl.
1965 Gordie Howe, Det.
1967 Henri Richard, Mtl.
1968 Bruce Gamble, Tor.
1969 Frank Mahovlich, Det.
1970 Bobby Hull, Chi.
1971 Bobby Hull, Chi.
1972 Bobby Orr, Bos.
1973 Greg Polis, Pit.
1974 Garry Unger, St.L.
1975 Syl Apps Jr., Pit.
1976 Pete Mahovlich, Mtl.
1977 Rick Martin, Buf.
1978 Billy Smith, NYI
1980 Reggie Leach, Phi.
1981 Mike Liut, St.L.
1982 Mike Bossy, NYI
1983 Wayne Gretzky, Edm.
1984 Don Maloney, NYR
1985 Mario Lemieux, Pit.
1986 Grant Fuhr, Edm.
1988 Mario Lemieux, Pit.
1989 Wayne Gretzky, L.A.
1990 Mario Lemieux, Pit.
1991 Vincent Damphousse, Tor.
1992 Brett Hull, St.L.
1993 Mike Gartner, NYR
1994 Mike Richter, NYR
1996 Raymond Bourque, Bos.
1997 Mark Recchi, Mtl.
1998 Teemu Selanne, Ana.
1999 Wayne Gretzky, NYR
2000 Pavel Bure, Fla.
2001 Bill Guerin, Bos.
2002 Eric Daze, Chi.

NHL AWARD MONEY BREAKDOWN — 2001-02

(Players on each club determine how team award money is divided.)

TEAM AWARDS

Stanley Cup Playoffs	Number of Clubs	Share Per Club	Total
Conference Quarter-Final Losers	8	$ 250,121	$2,000,968
Conference Semi-Final Losers	4	435,506	1,742,024
Conference Championship Losers	2	952,835	1,905,670
Stanley Cup Losers	1	1,549,346	1,549,346
Stanley Cup Winners	1	2,261,992	2,261,992
TOTAL PLAYOFF AWARD MONEY			$9,460,000

Final Standings, Regular Season	Number of Clubs	Share Per Club	Total
Presidents' Trophy			
Club's Share	1	$ 100,000	$ 100,000
Players' Share	1	250,000	250,000
Conference First Place*	2	500,000	1,000,000
Conference Second Place*	2	375,000	750,000
Conference Third Place*	2	250,000	500,000
Conference Fourth Place*	2	125,000	250,000
TOTAL REGULAR-SEASON AWARD MONEY			$2,850,000

*based on points.

INDIVIDUAL AWARDS	Winner	First Runner-up	Second Runner-up
Hart, Calder, Norris, Ross, Vezina, Byng, Selke, Jennings, Masterton Trophies	$10,000	$6,000	$4,000
King Clancy Trophy	$ 3,000	$1,000	
Conn Smythe and Maurice Richard Trophies	$10,000		
TOTAL INDIVIDUAL AWARD MONEY			$204,000

ALL-STARS	Number of winners	Per Player	Total
First Team All-Stars	6	$10,000	$ 60,000
Second Team All-Stars	6	5,000	$ 30,000
TOTAL ALL-STAR AWARD MONEY			$ 90,000
TOTAL AWARD MONEY			**$12,604,000**

2001-02 NHL Player of the Week/Month Award Winners

Player of the Week

Week Ending	Player
Oct. 7	**Roman Turek**, Calgary
Oct. 14	**Mark Parrish**, NY Islanders
Oct. 21	**Patrik Elias**, New Jersey
Oct. 28	**Jarome Iginla**, Calgary
Nov. 4	**Brian Boucher**, Philadelphia
Nov. 11	**Jocelyn Thibault**, Chicago
Nov. 18	**Patrick Roy**, Colorado
Nov. 25	**Nikolai Khabibulin**, Tampa Bay
Dec. 2	**Mike Richter**, NY Rangers
Dec. 9	**Joe Sakic**, Colorado
Dec. 16	**Tom Barrasso**, Carolina
Dec. 23	**Dominik Hasek**, Detroit
Dec. 30	**Glen Murray**, Boston
Jan. 6	**Ron Francis**, Carolina
Jan. 13	**Mike Modano**, Dallas
Jan. 20	**Roman Cechmanek**, Philadelphia
Jan. 27	**Brent Sopel**, Vancouver
Feb. 10	**Evgeni Nabokov**, San Jose
Mar. 3	**Marty Turco**, Dallas
Mar. 10	**Marian Gaborik**, Minnesota
Mar. 17	**Dan Cloutier**, Vancouver
Mar. 24	**Tommy Salo**, Edmonton
Mar. 31	**Martin Brodeur**, New Jersey
Apr. 7	**Jose Theodore**, Montreal
Apr. 14	**Darcy Tucker**, Toronto

Player of the Month

Month	Player
October	**Chris Osgood**, NY Islanders
November	**Jarome Iginla**, Calgary
December	**Steve Sullivan**, Chicago
January	**Markus Naslund**, Vancouver
February	**Evgeni Nabokov**, San Jose
March	**Sean Burke**, Phoenix

Rookie of the Month

Month	Player
October	**Kristian Huselius**, Florida
November	**Krys Kolanos**, Phoenix
December	**Dany Heatley**, Atlanta **Ilya Kovalchuk**, Atlanta (co-winners)
January	**Ilya Kovalchuk**, Atlanta
February	**Radim Vrbata**, Colorado
March	**Niklas Hagman**, Florida

NATIONAL HOCKEY LEAGUE INDIVIDUAL AWARD WINNERS

ART ROSS TROPHY

	Winner	Runner-up
2002	Jarome Iginla, Cgy.	Markus Naslund, Van.
2001	Jaromir Jagr, Pit.	Joe Sakic, Col.
2000	Jaromir Jagr, Pit.	Pavel Bure, Fla.
1999	Jaromir Jagr, Pit.	Teemu Selanne, Ana.
1998	Jaromir Jagr, Pit.	Peter Forsberg, Col.
1997	Mario Lemieux, Pit.	Teemu Selanne, Ana.
1996	Mario Lemieux, Pit.	Jaromir Jagr, Pit.
1995	Jaromir Jagr, Pit.	Eric Lindros, Phi.
1994	Wayne Gretzky, L.A.	Sergei Fedorov, Det.
1993	Mario Lemieux, Pit.	Pat LaFontaine, Buf.
1992	Mario Lemieux, Pit.	Kevin Stevens, Pit.
1991	Wayne Gretzky, L.A.	Brett Hull, St.L.
1990	Wayne Gretzky, L.A.	Mark Messier, Edm.
1989	Mario Lemieux, Pit.	Wayne Gretzky, L.A.
1988	Mario Lemieux, Pit.	Wayne Gretzky, Edm.
1987	Wayne Gretzky, Edm.	Jari Kurri, Edm.
1986	Wayne Gretzky, Edm.	Mario Lemieux, Pit.
1985	Wayne Gretzky, Edm.	Jari Kurri, Edm.
1984	Wayne Gretzky, Edm.	Paul Coffey, Edm.
1983	Wayne Gretzky, Edm.	Peter Stastny, Que.
1982	Wayne Gretzky, Edm.	Mike Bossy, NYI
1981	Wayne Gretzky, Edm.	Marcel Dionne, L.A.
1980	Marcel Dionne, L.A.	Wayne Gretzky, Edm.
1979	Bryan Trottier, NYI	Marcel Dionne, L.A.
1978	Guy Lafleur, Mtl.	Bryan Trottier, NYI
1977	Guy Lafleur, Mtl.	Marcel Dionne, L.A.
1976	Guy Lafleur, Mtl.	Bobby Clarke, Phi.
1975	Bobby Orr, Bos.	Phil Esposito, Bos.
1974	Phil Esposito, Bos.	Bobby Orr, Bos.
1973	Phil Esposito, Bos.	Bobby Clarke, Phi.
1972	Phil Esposito, Bos.	Bobby Orr, Bos.
1971	Phil Esposito, Bos.	Bobby Orr, Bos.
1970	Bobby Orr, Bos.	Phil Esposito, Bos.
1969	Phil Esposito, Bos.	Bobby Hull, Chi.
1968	Stan Mikita, Chi.	Phil Esposito, Bos.
1967	Stan Mikita, Chi.	Bobby Hull, Chi.
1966	Bobby Hull, Chi.	Stan Mikita, Chi.
1965	Stan Mikita, Chi.	Norm Ullman, Det.
1964	Stan Mikita, Chi.	Bobby Hull, Chi.
1963	Gordie Howe, Det.	Andy Bathgate, NYR
1962	Bobby Hull, Chi.	Andy Bathgate, NYR
1961	Bernie Geoffrion, Mtl.	Jean Beliveau, Mtl.
1960	Bobby Hull, Chi.	Bronco Horvath, Bos.
1959	Dickie Moore, Mtl.	Jean Beliveau, Mtl.
1958	Dickie Moore, Mtl.	Henri Richard, Mtl.
1957	Gordie Howe, Det.	Ted Lindsay, Det.
1956	Jean Beliveau, Mtl.	Gordie Howe, Det.
1955	Bernie Geoffrion, Mtl.	Maurice Richard, Mtl.
1954	Gordie Howe, Det.	Maurice Richard, Mtl.
1953	Gordie Howe, Det.	Ted Lindsay, Det.
1952	Gordie Howe, Det.	Ted Lindsay, Det.
1951	Gordie Howe, Det.	Maurice Richard, Mtl.
1950	Ted Lindsay, Det.	Sid Abel, Det.
1949	Roy Conacher, Chi.	Doug Bentley, Chi.
1948*	Elmer Lach, Mtl.	Buddy O'Connor, NYR
1947	Max Bentley, Chi.	Maurice Richard, Mtl.
1946	Max Bentley, Chi.	Gaye Stewart, Tor.
1945	Elmer Lach, Mtl.	Maurice Richard, Mtl.
1944	Herb Cain, Bos.	Doug Bentley, Chi.
1943	Doug Bentley, Chi.	Bill Cowley, Bos.
1942	Bryan Hextall, NYR	Lynn Patrick, NYR
1941	Bill Cowley, Bos.	Bryan Hextall, NYR
1940	Milt Schmidt, Bos.	Woody Dumart, Bos.
1939	Toe Blake, Mtl.	Sweeney Schriner, NYA
1938	Gordie Drillon, Tor.	Syl Apps, Tor.
1937	Sweeney Schriner, NYA	Syl Apps, Tor.
1936	Sweeney Schriner, NYA	Marty Barry, Det.
1935	Charlie Conacher, Tor.	Syd Howe, St.L., Det.
1934	Charlie Conacher, Tor.	Joe Primeau, Tor
1933	Bill Cook, NYR	Busher Jackson, Tor.
1932	Busher Jackson, Tor.	Joe Primeau, Tor.
1931	Howie Morenz, Mtl.	Ebbie Goodfellow, Det.
1930	Cooney Weiland, Bos.	Frank Boucher, NYR
1929	Ace Bailey, Tor.	Nels Stewart, Mtl.M
1928	Howie Morenz, Mtl.	Aurel Joliat, Mtl.
1927	Bill Cook, NYR	Dick Irvin, Chi.
1926	Nels Stewart, Mtl.M.	Cy Denneny, Ott.
1925	Babe Dye, Tor.	Cy Denneny, Ott.
1924	Cy Denneny, Ott.	Billy Boucher, Mtl.
1923	Babe Dye, Tor.	Cy Denneny, Ott.
1922	Punch Broadbent, Ott.	Cy Denneny, Ott.
1921	Newsy Lalonde, Mtl.	Babe Dye, Ham., Tor.
1920	Joe Malone, Que.	Newsy Lalonde, Mtl.
1919	Newsy Lalonde, Mtl.	Odie Cleghorn, Mtl.
1918	Joe Malone, Mtl.	Cy Denneny, Ott.

* Trophy first awarded in 1948.
Scoring leaders listed from 1918 to 1947.

HART TROPHY

	Winner	Runner-up
2002	Jose Theodore, Mtl.	Jarome Iginla, Cgy.
2001	Joe Sakic, Col.	Mario Lemieux, Pit.
2000	Chris Pronger, St.L.	Jaromir Jagr, Pit.
1999	Jaromir Jagr, Pit.	Alexei Yashin, Ott.
1998	Dominik Hasek, Buf.	Jaromir Jagr, Pit.
1997	Dominik Hasek, Buf.	Paul Kariya, Ana.
1996	Mario Lemieux, Pit.	Mark Messier, NYR
1995	Eric Lindros, Phi.	Jaromir Jagr, Pit.
1994	Sergei Fedorov, Det.	Dominik Hasek, Buf.
1993	Mario Lemieux, Pit.	Doug Gilmour, Tor.
1992	Mark Messier, NYR	Patrick Roy, Mtl.
1991	Brett Hull, St.L.	Wayne Gretzky, L.A.
1990	Mark Messier, Edm.	Raymond Bourque, Bos.
1989	Wayne Gretzky, L.A.	Mario Lemieux, Pit.
1988	Mario Lemieux, Pit.	Grant Fuhr, Edm.
1987	Wayne Gretzky, Edm.	Raymond Bourque, Bos.
1986	Wayne Gretzky, Edm.	Mario Lemieux, Pit.
1985	Wayne Gretzky, Edm.	Dale Hawerchuk, Wpg.
1984	Wayne Gretzky, Edm.	Rod Langway, Wsh.
1983	Wayne Gretzky, Edm.	Pete Peeters, Bos.
1982	Wayne Gretzky, Edm.	Bryan Trottier, NYI
1981	Wayne Gretzky, Edm.	Mike Liut, St.L.
1980	Wayne Gretzky, Edm.	Marcel Dionne, L.A.
1979	Bryan Trottier, NYI	Guy Lafleur, Mtl
1978	Guy Lafleur, Mtl.	Bryan Trottier, NYI
1977	Guy Lafleur, Mtl.	Bobby Clarke, Phi.
1976	Bobby Clarke, Phi.	Denis Potvin, NYI
1975	Bobby Clarke, Phi.	Rogie Vachon, L.A.
1974	Phil Esposito, Bos.	Bernie Parent, Phi.
1973	Bobby Clarke, Phi.	Phil Esposito, Bos.
1972	Bobby Orr, Bos.	Ken Dryden, Mtl.
1971	Bobby Orr, Bos.	Phil Esposito, Bos.
1970	Bobby Orr, Bos.	Tony Esposito, Chi.
1969	Phil Esposito, Bos.	Jean Beliveau, Mtl.
1968	Stan Mikita, Chi.	Jean Beliveau, Mtl.
1967	Stan Mikita, Chi.	Ed Giacomin, NYR
1966	Bobby Hull, Chi.	Jean Beliveau, Mtl.
1965	Bobby Hull, Chi.	Norm Ullman, Det.
1964	Jean Beliveau, Mtl.	Bobby Hull, Chi.
1963	Gordie Howe, Det.	Stan Mikita, Chi.
1962	Jacques Plante, Mtl.	Doug Harvey, NYR
1961	Bernie Geoffrion, Mtl.	Johnny Bower, Tor.
1960	Gordie Howe, Det.	Bobby Hull, Chi.
1959	Andy Bathgate, NYR	Gordie Howe, Det.
1958	Gordie Howe, Det.	Andy Bathgate, NYR
1957	Gordie Howe, Det.	Jean Beliveau, Mtl.
1956	Jean Beliveau, Mtl.	Tod Sloan, Tor.
1955	Ted Kennedy, Tor.	Harry Lumley, Tor.
1954	Al Rollins, Chi.	Red Kelly, Det.
1953	Gordie Howe, Det.	Al Rollins, Chi.
1952	Gordie Howe, Det.	Elmer Lach, Mtl.
1951	Milt Schmidt, Bos.	Maurice Richard, Mtl.
1950	Chuck Rayner, NYR	Ted Kennedy, Tor.
1949	Sid Abel, Det.	Bill Durnan, Mtl.
1948	Buddy O'Connor, NYR	Frank Brimsek, Bos.
1947	Maurice Richard, Mtl.	Milt Schmidt, Bos.
1946	Max Bentley, Chi.	Gaye Stewart, Tor.
1945	Elmer Lach, Mtl.	Maurice Richard, Mtl.
1944	Babe Pratt, Tor.	Bill Cowley, Bos.
1943	Bill Cowley, Bos.	Doug Bentley, Chi.
1942	Tom Anderson, Bro.	Syl Apps, Tor.
1941	Bill Cowley, Bos.	Dit Clapper, Bos.
1940	Ebbie Goodfellow, Det.	Syl Apps, Tor.
1939	Toe Blake, Mtl.	Syl Apps, Tor.
1938	Eddie Shore, Bos.	Paul Thompson, Chi.
1937	Babe Siebert, Mtl.	Lionel Conacher, Mtl.M
1936	Eddie Shore, Bos.	Hooley Smith, Mtl.M
1935	Eddie Shore, Bos.	Charlie Conacher, Tor.
1934	Aurel Joliat, Mtl.	Lionel Conacher, Chi.
1933	Eddie Shore, Bos.	Bill Cook, NYR
1932	Howie Morenz, Mtl.	Ching Johnson, NYR
1931	Howie Morenz, Mtl.	Eddie Shore, Bos.
1930	Nels Stewart, Mtl.M.	Lionel Hitchman, Bos.
1929	Roy Worters, NYA	Ace Bailey, Tor.
1928	Howie Morenz, Mtl.	Roy Worters, Pit.
1927	Herb Gardiner, Mtl.	Bill Cook, NYR
1926	Nels Stewart, Mtl.M.	Sprague Cleghorn, Bos.
1925	Billy Burch, Ham.	Howie Morenz, Mtl.
1924	Frank Nighbor, Ott.	Sprague Cleghorn, Mtl.

BUD LIGHT PLUS-MINUS AWARD WINNERS

2002	Chris Chelios	Detroit
2001	Patrik Elias	New Jersey
	Joe Sakic	Colorado
2000	Chris Pronger	St. Louis
1999	John LeClair	Philadelphia
1998	Chris Pronger	St. Louis

WILLIAM M. JENNINGS TROPHY WINNERS

	Winner	Runner-up
2002	Patrick Roy, Col.	Tommy Salo, Edm.
2001	Dominik Hasek, Buf.	Ed Belfour, Dal.
		Marty Turco
2000	Roman Turek, St.L.	John Vanbiesbrouck, Phi.
		Brian Boucher
1999	Ed Belfour, Dal.	Dominik Hasek, Buf.
	Roman Turek	
1998	Martin Brodeur, N.J.	Ed Belfour, Dal.
1997	Martin Brodeur, N.J.	Chris Osgood, Det.
	Mike Dunham	Mike Vernon
1996	Chris Osgood, Det.	Martin Brodeur, N.J.
	Mike Vernon	
1995	Ed Belfour, Chi.	Mike Vernon, Det.
		Chris Osgood
1994	Dominik Hasek, Buf.	Martin Brodeur, N.J.
	Grant Fuhr	Chris Terreri
1993	Ed Belfour, Chi.	Felix Potvin, Tor.
		Grant Fuhr
1992	Patrick Roy, Mtl.	Ed Belfour, Chi.
1991	Ed Belfour, Chi.	Patrick Roy, Mtl.
1990	Andy Moog, Bos.	Patrick Roy, Mtl.
	Reggie Lemelin	Brian Hayward
1989	Patrick Roy, Mtl.	Mike Vernon, Cgy.
	Brian Hayward	Rick Wamsley
1988	Patrick Roy, Mtl.	Clint Malarchuk, Wsh.
	Brian Hayward	Pete Peeters
1987	Patrick Roy, Mtl.	Ron Hextall, Phi.
	Brian Hayward	
1986	Bob Froese, Phi.	Al Jensen, Wsh.
	Darren Jensen	Pete Peeters
1985	Tom Barrasso, Buf.	Pat Riggin, Wsh.
	Bob Sauve	
1984	Al Jensen, Wsh.	Tom Barrasso, Buf.
	Pat Riggin	Bob Sauve
1983	Rollie Melanson, NYI	Pete Peeters, Bos.
	Billy Smith	
1982	Rick Wamsley, Mtl.	Billy Smith, NYI
	Denis Herron	Rollie Melanson

BILL MASTERTON TROPHY WINNERS

2002	Saku Koivu	Montreal
2001	Adam Graves	NY Rangers
2000	Ken Daneyko	New Jersey
1999	John Cullen	Tampa Bay
1998	Jamie McLennan	St. Louis
1997	Tony Granato	San Jose
1996	Gary Roberts	Calgary
1995	Pat LaFontaine	Buffalo
1994	Cam Neely	Boston
1993	Mario Lemieux	Pittsburgh
1992	Mark Fitzpatrick	NY Islanders
1991	Dave Taylor	Los Angeles
1990	Gord Kluzak	Boston
1989	Tim Kerr	Philadelphia
1988	Bob Bourne	Los Angeles
1987	Doug Jarvis	Hartford
1986	Charlie Simmer	Boston
1985	Anders Hedberg	NY Rangers
1984	Brad Park	Detroit
1983	Lanny McDonald	Calgary
1982	Glenn Resch	Colorado
1981	Blake Dunlop	St. Louis
1980	Al MacAdam	Minnesota
1979	Serge Savard	Montreal
1978	Butch Goring	Los Angeles
1977	Ed Westfall	NY Islanders
1976	Rod Gilbert	NY Rangers
1975	Don Luce	Buffalo
1974	Henri Richard	Montreal
1973	Lowell MacDonald	Pittsburgh
1972	Bobby Clarke	Philadelphia
1971	Jean Ratelle	NY Rangers
1970	Pit Martin	Chicago
1969	Ted Hampson	Oakland
1968	Claude Provost	Montreal

LADY BYNG TROPHY

	Winner	Runner-up
2002	Ron Francis, Car.	Joe Sakic, Col.
2001	Joe Sakic, Col.	Nicklas Lidstrom, Det.
2000	Pavol Demitra, St.L.	Nicklas Lidstrom, Det.
1999	Wayne Gretzky, NYR.	Nicklas Lidstrom, Det.
1998	Ron Francis, Pit.	Teemu Selanne, Ana.
1997	Paul Kariya, Ana.	Teemu Selanne, Ana.
1996	Paul Kariya, Ana.	Adam Oates, Bos.
1995	Ron Francis, Pit.	Adam Oates, Bos.
1994	Wayne Gretzky, L.A.	Adam Oates, Bos.
1993	Pierre Turgeon, NYI	Adam Oates, Bos.
1992	Wayne Gretzky, L.A.	Joe Sakic, Que.
1991	Wayne Gretzky, L.A.	Brett Hull, St.L.
1990	Brett Hull, St.L.	Wayne Gretzky, L.A.
1989	Joe Mullen, Cgy.	Wayne Gretzky, L.A.
1988	Mats Naslund, Mtl.	Wayne Gretzky, Edm.
1987	Joe Mullen, Cgy.	Wayne Gretzky, Edm.
1986	Mike Bossy, NYI	Jari Kurri, Edm.
1985	Jari Kurri, Edm.	Joe Mullen, St.L.
1984	Mike Bossy, NYI	Rick Middleton, Bos.
1983	Mike Bossy, NYI	Rick Middleton, Bos.
1982	Rick Middleton, Bos.	Mike Bossy, NYI
1981	Rick Kehoe, Pit.	Wayne Gretzky, Edm.
1980	Wayne Gretzky, Edm.	Marcel Dionne, L.A.
1979	Bob MacMillan, Atl.	Marcel Dionne, L.A.
1978	Butch Goring, L.A.	Peter McNab, Bos.
1977	Marcel Dionne, L.A.	Jean Ratelle, Bos.
1976	Jean Ratelle, NYR-Bos.	Jean Pronovost, Pit.
1975	Marcel Dionne, Det.	John Bucyk, Bos.
1974	John Bucyk, Bos.	Lowell MacDonald, Pit.
1973	Gilbert Perreault, Buf.	Jean Ratelle, NYR
1972	Jean Ratelle, NYR	John Bucyk, Bos.
1971	John Bucyk, Bos.	Dave Keon, Tor.
1970	Phil Goyette, St.L.	John Bucyk, Bos.
1969	Alex Delvecchio, Det.	Ted Hampson, Oak.
1968	Stan Mikita, Chi.	John Bucyk, Bos.
1967	Stan Mikita, Chi.	Dave Keon, Tor.
1966	Alex Delvecchio, Det.	Bobby Rousseau, Mtl.
1965	Bobby Hull, Chi.	Alex Delvecchio, Det.
1964	Kenny Wharram, Chi.	Dave Keon, Tor.
1963	Dave Keon, Tor.	Camille Henry, NYR
1962	Dave Keon, Tor.	Claude Provost, Mtl.
1961	Red Kelly, Tor.	Norm Ullman, Det.
1960	Don McKenney, Bos.	Andy Hebenton, NYR
1959	Alex Delvecchio, Det.	Andy Hebenton, NYR
1958	Camille Henry, NYR	Don Marshall, Mtl.
1957	Andy Hebenton, NYR	Dutch Reibel, Det.
1956	Dutch Reibel, Det.	Floyd Curry, Mtl.
1955	Sid Smith, Tor.	Danny Lewicki, NYR
1954	Red Kelly, Det.	Don Raleigh, NYR
1953	Red Kelly, Det.	Wally Hergesheimer, NYR
1952	Sid Smith, Tor.	Red Kelly, Det.
1951	Red Kelly, Det.	Woody Dumart, Bos.
1950	Edgar Laprade, NYR	Red Kelly, Det.
1949	Bill Quackenbush, Det.	Harry Watson, Tor.
1948	Buddy O'Connor, NYR	Syl Apps, Tor.
1947	Bobby Bauer, Bos.	Syl Apps, Tor.
1946	Toe Blake, Mtl.	Clint Smith, Chi.
1945	Bill Mosienko, Chi.	Syd Howe, Det.
1944	Clint Smith, Chi.	Herb Cain, Bos.
1943	Max Bentley, Chi.	Buddy O'Connor, Mtl.
1942	Syl Apps, Tor.	Gordie Drillon, Tor.
1941	Bobby Bauer, Bos.	Gordie Drillon, Tor.
1940	Bobby Bauer, Bos.	Clint Smith, NYR
1939	Clint Smith, NYR	Marty Barry, Det.
1938	Gordie Drillon, Tor.	Clint Smith, NYR
1937	Marty Barry, Det.	Gordie Drillon, Tor.
1936	Doc Romnes, Chi.	Sweeney Schriner, NYA
1935	Frank Boucher, NYR	Russ Blinco, Mtl.M
1934	Frank Boucher, NYR	Joe Primeau, Tor.
1933	Frank Boucher, NYR	Joe Primeau, Tor.
1932	Joe Primeau, Tor.	Frank Boucher, NYR
1931	Frank Boucher, NYR	Normie Himes, NYA
1930	Frank Boucher, NYR	Normie Himes, NYA
1929	Frank Boucher, NYR	Harold Darragh, Pit.
1928	Frank Boucher, NYR	George Hay, Det.
1927	Billy Burch, NYA	Dick Irvin, Chi.
1926	Frank Nighbor, Ott.	Billy Burch, NYA
1925	Frank Nighbor, Ott.	none

KING CLANCY MEMORIAL TROPHY WINNERS

2002	Ron Francis	Carolina
2001	Shjon Podein	Colorado
2000	Curtis Joseph	Toronto
1999	Rob Ray	Buffalo
1998	Kelly Chase	St. Louis
1997	Trevor Linden	Vancouver
1996	Kris King	Winnipeg
1995	Joe Nieuwendyk	Calgary
1994	Adam Graves	NY Rangers
1993	Dave Poulin	Boston
1992	Raymond Bourque	Boston
1991	Dave Taylor	Los Angeles
1990	Kevin Lowe	Edmonton
1989	Bryan Trottier	NY Islanders
1988	Lanny McDonald	Calgary

VEZINA TROPHY

	Winner	Runner-up
2002	Jose Theodore, Mtl.	Patrick Roy, Col.
2001	Dominik Hasek, Buf.	Roman Cechmanek, Phi.
2000	Olaf Kolzig, Wsh.	Roman Turek, St.L.
1999	Dominik Hasek, Buf.	Curtis Joseph, Tor.
1998	Dominik Hasek, Buf.	Martin Brodeur, N.J.
1997	Dominik Hasek, Buf.	Martin Brodeur, N.J.
1996	Jim Carey, Wsh.	Chris Osgood, Det.
1995	Dominik Hasek, Buf.	Ed Belfour, Chi.
1994	Dominik Hasek, Buf.	John Vanbiesbrouck, Fla.
1993	Ed Belfour, Chi.	Tom Barrasso, Pit.
1992	Patrick Roy, Mtl.	Kirk McLean, Van.
1991	Ed Belfour, Chi.	Patrick Roy, Mtl.
1990	Patrick Roy, Mtl.	Daren Puppa, Buf.
1989	Patrick Roy, Mtl.	Mike Vernon, Cgy.
1988	Grant Fuhr, Edm.	Tom Barrasso, Buf.
1987	Ron Hextall, Phi.	Mike Liut, Hfd.
1986	John Vanbiesbrouck, NYR	Bob Froese, Phi.
1985	Pelle Lindbergh, Phi.	Tom Barrasso, Buf.
1984	Tom Barrasso, Buf.	Reggie Lemelin, Cgy.
1983	Pete Peeters, Bos.	Rollie Melanson, NYI
1982	Billy Smith, NYI	Grant Fuhr, Edm.
1981	Richard Sevigny, Mtl.	Pete Peeters, Phi.
	Denis Herron, Mtl.	Rick St. Croix, Phi.
	Michel Larocque, Mtl.	
1980	Bob Sauve, Buf.	Gerry Cheevers, Bos.
	Don Edwards, Buf.	Gilles Gilbert, Bos.
1979	Ken Dryden, Mtl.	Glenn Resch, NYI
	Michel Larocque, Mtl.	Billy Smith, NYI
1978	Ken Dryden, Mtl.	Bernie Parent, Phi.
	Michel Larocque, Mtl.	Wayne Stephenson, Phi.
1977	Ken Dryden, Mtl.	Glenn Resch, NYI
	Michel Larocque, Mtl.	Billy Smith, NYI
1976	Ken Dryden, Mtl.	Glenn Resch, NYI
		Billy Smith, NYI
1975	Bernie Parent, Phi.	Rogie Vachon, L.A.
		Gary Edwards, L.A.
1974	Bernie Parent, Phi. (tie)	Gilles Gilbert, Bos.
	Tony Esposito, Chi. (tie)	
1973	Ken Dryden, Mtl.	Ed Giacomin, NYR
		Gilles Villemure, NYR
1972	Tony Esposito, Chi.	Cesare Maniago, Min.
	Gary Smith, Chi.	Gump Worsley, Min.
1971	Ed Giacomin, NYR	Tony Esposito, Chi.
	Gilles Villemure, NYR	
1970	Tony Esposito, Chi.	Jacques Plante, St.L.
		Ernie Wakely, St.L.
1969	Jacques Plante, St.L.	Ed Giacomin, NYR
	Glenn Hall, St.L.	
1968	Gump Worsley, Mtl.	Johnny Bower, Tor.
	Rogie Vachon, Mtl.	Bruce Gamble, Tor.
1967	Glenn Hall, Chi.	Charlie Hodge, Mtl.
	Denis Dejordy, Chi.	
1966	Gump Worsley, Mtl.	Glenn Hall, Chi.
	Charlie Hodge, Mtl.	
1965	Terry Sawchuk, Tor.	Roger Crozier, Det.
	Johnny Bower, Tor.	
1964	Charlie Hodge, Mtl.	Glenn Hall, Chi.
1963	Glenn Hall, Chi.	Johnny Bower, Tor.
		Don Simmons, Tor.
1962	Jacques Plante, Mtl.	Johnny Bower, Tor.
1961	Johnny Bower, Tor.	Glenn Hall, Chi.
1960	Jacques Plante, Mtl.	Glenn Hall, Chi.
1959	Jacques Plante, Mtl.	Johnny Bower, Tor.
		Ed Chadwick, Tor.
1958	Jacques Plante, Mtl.	Gump Worsley, NYR
		Marcel Paille, NYR
1957	Jacques Plante, Mtl.	Glenn Hall, Det.
1956	Jacques Plante, Mtl.	Glenn Hall, Det.
1955	Terry Sawchuk, Det.	Harry Lumley, Tor.
1954	Harry Lumley, Tor.	Terry Sawchuk, Det.
1953	Terry Sawchuk, Det.	Gerry McNeil, Mtl.
1952	Terry Sawchuk, Det.	Al Rollins, Tor.
1951	Al Rollins, Tor.	Terry Sawchuk, Det.
1950	Bill Durnan, Mtl.	Harry Lumley, Det.
1949	Bill Durnan, Mtl.	Harry Lumley, Det.
1948	Turk Broda, Tor.	Harry Lumley, Det.
1947	Bill Durnan, Mtl.	Turk Broda, Tor.
1946	Bill Durnan, Mtl.	Frank Brimsek, Bos.
1945	Bill Durnan, Mtl.	Frank McCool, Tor. (tie)
		Harry Lumley, Det. (tie)
1944	Bill Durnan, Mtl.	Paul Bibeault, Tor.
1943	Johnny Mowers, Det.	Turk Broda, Tor.
1942	Frank Brimsek, Bos.	Turk Broda, Tor.
1941	Turk Broda, Tor.	Frank Brimsek, Bos. (tie)
		Johnny Mowers, Det. (tie)
1940	Dave Kerr, NYR	Frank Brimsek, Bos.
1939	Frank Brimsek, Bos.	Dave Kerr, NYR
1938	Tiny Thompson, Bos.	Dave Kerr, NYR
1937	Normie Smith, Det.	Dave Kerr, NYR
1936	Tiny Thompson, Bos.	Mike Karakas, Chi.
1935	Lorne Chabot, Chi.	Alex Connell, Mtl.M
1934	Charlie Gardiner, Chi.	Wilf Cude, Det.
1933	Tiny Thompson, Bos.	John Ross Roach, Det.
1932	Charlie Gardiner, Chi.	Alex Connell, Det.
1931	Roy Worters, NYA	Charlie Gardiner, Chi.
1930	Tiny Thompson, Bos.	Charlie Gardiner, Chi.
1929	George Hainsworth, Mtl.	Tiny Thompson, Bos.
1928	George Hainsworth, Mtl.	Alex Connell, Ott.
1927	George Hainsworth, Mtl.	Clint Benedict, Mtl.M

CALDER MEMORIAL TROPHY WINNERS

	Winner	Runner-up
2002	Dany Heatley, Atl.	Ilya Kovalchuk, Atl.
2001	Evgeni Nabokov, S.J.	Brad Richards, T.B.
2000	Scott Gomez, N.J.	Brad Stuart, S.J.
1999	Chris Drury, Col.	Marian Hossa, Ott.
1998	Sergei Samsonov, Bos.	Mattias Ohlund, Van.
1997	Bryan Berard, NYI	Jarome Iginla, Cgy.
1996	Daniel Alfredsson, Ott.	Eric Daze, Chi.
1995	Peter Forsberg, Que.	Jim Carey, Wsh.
1994	Martin Brodeur, N.J.	Jason Arnott, Edm.
1993	Teemu Selanne, Wpg.	Joe Juneau, Bos.
1992	Pavel Bure, Van.	Nicklas Lidstrom, Det
1991	Ed Belfour, Chi.	Sergei Fedorov, Det.
1990	Sergei Makarov, Cgy.	Mike Modano, Min.
1989	Brian Leetch, NYR	Trevor Linden, Van.
1988	Joe Nieuwendyk, Cgy.	Ray Sheppard, Buf.
1987	Luc Robitaille, L.A.	Ron Hextall, Phi.
1986	Gary Suter, Cgy.	Wendel Clark, Tor.
1985	Mario Lemieux, Pit.	Chris Chelios, Mtl.
1984	Tom Barrasso, Buf.	Steve Yzerman, Det.
1983	Steve Larmer, Chi.	Phil Housley, Buf.
1982	Dale Hawerchuk, Wpg.	Barry Pederson, Bos.
1981	Peter Stastny, Que.	Larry Murphy, L.A.
1980	Raymond Bourque, Bos.	Mike Foligno, Det.
1979	Bobby Smith, Min	Ryan Walter, Wsh.
1978	Mike Bossy, NYI	Barry Beck, Col.
1977	Willi Plett, Atl.	Don Murdoch, NYR
1976	Bryan Trottier, NYI	Glenn Resch, NYI
1975	Eric Vail, Atl.	Pierre Larouche, Pit.
1974	Denis Potvin, NYI	Tom Lysiak, Atl.
1973	Steve Vickers, NYR	Bill Barber, Phi.
1972	Ken Dryden, Mtl.	Rick Martin, Buf.
1971	Gilbert Perreault, Buf.	Jude Drouin, Min.
1970	Tony Esposito, Chi.	Bill Fairbairn, NYR
1969	Danny Grant, Min.	Norm Ferguson, Oak.
1968	Derek Sanderson, Bos.	Jacques Lemaire, Mtl.
1967	Bobby Orr, Bos.	Ed Van Impe, Chi.
1966	Brit Selby, Tor.	Bert Marshall, Det.
1965	Roger Crozier, Det.	Ron Ellis, Tor.
1964	Jacques Laperriere, Mtl.	John Ferguson, Mtl.
1963	Kent Douglas, Tor.	Doug Barkley, Det.
1962	Bobby Rousseau, Mtl.	Cliff Pennington, Bos.
1961	Dave Keon, Tor.	Bob Nevin, Tor.
1960	Bill Hay, Chi.	Murray Oliver, Det.
1959	Ralph Backstrom, Mtl.	Carl Brewer, Tor.
1958	Frank Mahovlich, Tor.	Bobby Hull, Chi.
1957	Larry Regan, Bos.	Ed Chadwick, Tor.
1956	Glenn Hall, Det.	Andy Hebenton, NYR
1955	Ed Litzenberger, Chi.	Don McKenney, Bos.
1954	Camille Henry, NYR	Dutch Reibel, Det.
1953	Gump Worsley, NYR	Gord Hannigan, Tor.
1952	Bernie Geoffrion, Mtl.	Hy Buller, NYR
1951	Terry Sawchuk, Det.	Al Rollins, Tor.
1950	Jack Gelineau, Bos.	Phil Maloney, Bos.
1949	Pentti Lund, NYR	Allan Stanley, NYR
1948	Jim McFadden, Det.	Pete Babando, Bos.
1947	Howie Meeker, Tor.	Jim Conacher, Det.
1946	Edgar Laprade, NYR	George Gee, Chi.
1945	Frank McCool, Tor.	Ken Smith, Bos.
1944	Gus Bodnar, Tor.	Bill Durnan, Mtl.
1943	Gaye Stewart, Tor.	Glen Harmon, Mtl.
1942	Grant Warwick, NYR	Buddy O'Connor, Mtl.
1941	John Quilty, Mtl.	Johnny Mowers, Det.
1940	Kilby MacDonald, NYR	Wally Stanowski, Tor.
1939	Frank Brimsek, Bos.	Roy Conacher, Bos.
1938	Cully Dahlstrom, Chi.	Murph Chamberlain, Tor.
1937	Syl Apps, Tor.	Gordie Drillon, Tor.
1936	Mike Karakas, Chi.	Bucko McDonald, Det.
1935	Sweeney Schriner, NYA	Bert Connelly, NYR
1934	Russ Blinco, Mtl.M.	none
1933	Carl Voss, Det.	none

FRANK J. SELKE TROPHY WINNERS

	Winner	Runner-up
2002	Michael Peca, NYI	Craig Conroy, Cgy.
2001	John Madden, N.J.	Joe Sakic, Col.
2000	Steve Yzerman, Det.	Michal Handzus, St.L.
1999	Jere Lehtinen, Dal.	Magnus Arvedson, Ott.
1998	Jere Lehtinen, Dal.	Michael Peca, Buf.
1997	Michael Peca, Buf.	Peter Forsberg, Col.
1996	Sergei Fedorov, Det.	Ron Francis, Pit.
1995	Ron Francis, Pit.	Esa Tikkanen, St.L.
1994	Sergei Fedorov, Det.	Doug Gilmour, Tor.
1993	Doug Gilmour, Tor.	Dave Poulin, Bos.
1992	Guy Carbonneau, Mtl.	Sergei Fedorov, Det.
1991	Dirk Graham, Chi.	Esa Tikkanen, Edm.
1990	Rick Meagher, St.L.	Guy Carbonneau, Mtl.
1989	Guy Carbonneau, Mtl.	Esa Tikkanen, Edm.
1988	Guy Carbonneau, Mtl.	Steve Kasper, Bos.
1987	Dave Poulin, Phi.	Guy Carbonneau, Mtl.
1986	Troy Murray, Chi.	Ron Sutter, Phi.
1985	Craig Ramsay, Buf.	Doug Jarvis, Wsh.
1984	Doug Jarvis, Wsh.	Bryan Trottier, NYI
1983	Bobby Clarke, Phi.	Jari Kurri, Edm.
1982	Steve Kasper, Bos.	Bob Gainey, Mtl.
1981	Bob Gainey, Mtl.	Craig Ramsay, Buf.
1980	Bob Gainey, Mtl.	Craig Ramsay, Buf.
1979	Bob Gainey, Mtl.	Don Marcotte, Bos.
1978	Bob Gainey, Mtl.	Craig Ramsay, Buf.

CONN SMYTHE TROPHY WINNERS

2002	Nicklas Lidstrom	Detroit
2001	Patrick Roy	Colorado
2000	Scott Stevens	New Jersey
1999	Joe Nieuwendyk	Dallas
1998	Steve Yzerman	Detroit
1997	Mike Vernon	Detroit
1996	Joe Sakic	Colorado
1995	Claude Lemieux	New Jersey
1994	Brian Leetch	NY Rangers
1993	Patrick Roy	Montreal
1992	Mario Lemieux	Pittsburgh
1991	Mario Lemieux	Pittsburgh
1990	Bill Ranford	Edmonton
1989	Al MacInnis	Calgary
1988	Wayne Gretzky	Edmonton
1987	Ron Hextall	Philadelphia
1986	Patrick Roy	Montreal
1985	Wayne Gretzky	Edmonton
1984	Mark Messier	Edmonton
1983	Billy Smith	NY Islanders
1982	Mike Bossy	NY Islanders
1981	Butch Goring	NY Islanders
1980	Bryan Trottier	NY Islanders
1979	Bob Gainey	Montreal
1978	Larry Robinson	Montreal
1977	Guy Lafleur	Montreal
1976	Reggie Leach	Philadelphia
1975	Bernie Parent	Philadelphia
1974	Bernie Parent	Philadelphia
1973	Yvan Cournoyer	Montreal
1972	Bobby Orr	Boston
1971	Ken Dryden	Montreal
1970	Bobby Orr	Boston
1969	Serge Savard	Montreal
1968	Glenn Hall	St. Louis
1967	Dave Keon	Toronto
1966	Roger Crozier	Detroit
1965	Jean Beliveau	Montreal

JAMES NORRIS TROPHY WINNERS

	Winner	Runner-up
2002	Nicklas Lidstrom, Det.	Chris Chelios, Det.
2001	Nicklas Lidstrom, Det.	Raymond Bourque, Col.
2000	Chris Pronger, St.L.	Nicklas Lidstrom, Det.
1999	Al MacInnis, St.L.	Nicklas Lidstrom, Det.
1998	Rob Blake, L.A.	Nicklas Lidstrom, Det.
1997	Brian Leetch, NYR	V. Konstantinov, Det.
1996	Chris Chelios, Chi.	Raymond Bourque, Bos.
1995	Paul Coffey, Det.	Chris Chelios, Chi.
1994	Raymond Bourque, Bos.	Scott Stevens, N.J.
1993	Chris Chelios, Chi.	Raymond Bourque, Bos.
1992	Brian Leetch, NYR	Raymond Bourque, Bos.
1991	Raymond Bourque, Bos.	Al MacInnis, Cgy.
1990	Raymond Bourque, Bos.	Al MacInnis, Cgy.
1989	Chris Chelios, Mtl	Paul Coffey, Pit.
1988	Raymond Bourque, Bos.	Scott Stevens, Wsh.
1987	Raymond Bourque, Bos.	Mark Howe, Phi.
1986	Paul Coffey, Edm.	Mark Howe, Phi.
1985	Paul Coffey, Edm.	Raymond Bourque, Bos.
1984	Rod Langway, Wsh.	Paul Coffey, Edm.
1983	Rod Langway, Wsh.	Mark Howe, Phi.
1982	Doug Wilson, Chi.	Raymond Bourque, Bos.
1981	Randy Carlyle, Pit.	Denis Potvin, NYI
1980	Larry Robinson, Mtl.	Borje Salming, Tor.
1979	Denis Potvin, NYI	Larry Robinson, Mtl.
1978	Denis Potvin, NYI	Brad Park, Bos.
1977	Larry Robinson, Mtl.	Borje Salming, Tor.
1976	Denis Potvin, NYI	Brad Park, NYR-Bos.
1975	Bobby Orr, Bos.	Denis Potvin, NYI
1974	Bobby Orr, Bos.	Brad Park, NYR
1973	Bobby Orr, Bos.	Guy Lapointe, Mtl.
1972	Bobby Orr, Bos.	Brad Park, NYR
1971	Bobby Orr, Bos.	Brad Park, NYR
1970	Bobby Orr, Bos.	Brad Park, NYR
1969	Bobby Orr, Bos.	Tim Horton, Tor.
1968	Bobby Orr, Bos.	J.C. Tremblay, Mtl
1967	Harry Howell, NYR	Pierre Pilote, Chi.
1966	Jacques Laperriere, Mtl.	Pierre Pilote, Chi.
1965	Pierre Pilote, Chi.	Jacques Laperriere, Mtl.
1964	Pierre Pilote, Chi.	Tim Horton, Tor.
1963	Pierre Pilote, Chi.	Carl Brewer, Tor.
1962	Doug Harvey, NYR	Pierre Pilote, Chi.
1961	Doug Harvey, Mtl.	Marcel Pronovost, Det.
1960	Doug Harvey, Mtl.	Allan Stanley, Tor.
1959	Tom Johnson, Mtl.	Bill Gadsby, NYR
1958	Doug Harvey, Mtl.	Bill Gadsby, NYR
1957	Doug Harvey, Mtl.	Red Kelly, Det.
1956	Doug Harvey, Mtl.	Bill Gadsby, NYR
1955	Doug Harvey, Mtl.	Red Kelly, Det.
1954	Red Kelly, Det.	Doug Harvey, Mtl.

MAURICE "ROCKET" RICHARD TROPHY WINNER

2002	Jarome Iginla	Calgary
2001	Pavel Bure	Florida
2000	Pavel Bure	Florida
1999	Teemu Selanne	Anaheim

LESTER PATRICK TROPHY WINNERS

2002 1960 U.S. Olympic Team
Herb Brooks
Larry Pleau
2001 Scotty Bowman
David Poile
Gary Bettman
2000 Mario Lemieux
Craig Patrick
Lou Vairo
1999 Harry Sinden
1998 U.S. Olympic Women's Hockey Team
1998 Peter Karmanos
Neal Broten
John Mayasich
Max McNab
1997 Seymour H. Knox III
Bill Cleary
Pat LaFontaine
1996 George Gund
Ken Morrow
Milt Schmidt
1995 Joe Mullen
Brian Mullen
Bob Fleming
1994 Wayne Gretzky
Robert Ridder
1993 *Frank Boucher
*Mervyn "Red" Dutton
Bruce McNall
Gil Stein
1992 Al Arbour
Art Berglund
Lou Lamoriello
1991 Rod Gilbert
Mike Ilitch
1990 Len Ceglarski
1989 Dan Kelly
Lou Nanne
*Lynn Patrick
Bud Poile
1988 Keith Allen
Fred Cusick
Bob Johnson
1987 *Hobey Baker
Frank Mathers
1986 John MacInnes
Jack Riley
1985 Jack Butterfield
Arthur M. Wirtz
1984 John A. Ziegler, Jr.
*Arthur Howie Ross
1983 Bill Torrey
1982 Emile P. Francis
1981 Charles M. Schulz
1980 Bobby Clarke
Edward M. Snider
Frederick A. Shero
1980 U.S. Olympic Hockey Team
1979 Bobby Orr
1978 Phil Esposito
Tom Fitzgerald
William T. Tutt
William W. Wirtz
1977 John P. Bucyk
Murray A. Armstrong
John Mariucci
1976 Stanley Mikita
George A. Leader
Bruce A. Norris
1975 Donald M. Clark
William L. Chadwick
Thomas N. Ivan
1974 Alex Delvecchio
Murray Murdoch
*Weston W. Adams, Sr.
*Charles L. Crovat
1973 Walter L. Bush, Jr.
1972 Clarence S. Campbell
John A. "Snooks" Kelly
Ralph "Cooney" Weiland
*James D. Norris
1971 William M. Jennings
*John B. Sollenberger
*Terrance G. Sawchuk
1970 Edward W. Shore
*James C. V. Hendy
1969 Robert M. Hull
*Edward J. Jeremiah
1968 Thomas F. Lockhart
*Walter A. Brown
*Gen. John R. Kilpatrick
1967 Gordon Howe
*Charles F. Adams
*James Norris, Sr.
1966 J.J. "Jack" Adams

* awarded posthumously

PRESIDENTS' TROPHY

	Winner	Runner-up
2002	Detroit Red Wings	Boston Bruins
2001	Colorado Avalanche	Detroit Red Wings
2000	St. Louis Blues	Detroit Red Wings
1999	Dallas Stars	New Jersey Devils
1998	Dallas Stars	New Jersey Devils
1997	Colorado Avalanche	Dallas Stars
1996	Detroit Red Wings	Colorado Avalanche
1995	Detroit Red Wings	Quebec Nordiques
1994	New York Rangers	New Jersey Devils
1993	Pittsburgh Penguins	Boston Bruins
1992	New York Rangers	Washington Capitals
1991	Chicago Blackhawks	St. Louis Blues
1990	Boston Bruins	Calgary Flames
1989	Calgary Flames	Montreal Canadiens
1988	Calgary Flames	Montreal Canadiens
1987	Edmonton Oilers	Philadelphia Flyers
1986	Edmonton Oilers	Philadelphia Flyers

LESTER B. PEARSON AWARD WINNERS

2002	Jarome Iginla	Calgary
2001	Joe Sakic	Colorado
2000	Jaromir Jagr	Pittsburgh
1999	Jaromir Jagr	Pittsburgh
1998	Dominik Hasek	Buffalo
1997	Dominik Hasek	Buffalo
1996	Mario Lemieux	Pittsburgh
1995	Eric Lindros	Philadelphia
1994	Sergei Fedorov	Detroit
1993	Mario Lemieux	Pittsburgh
1992	Mark Messier	NY Rangers
1991	Brett Hull	St. Louis
1990	Mark Messier	Edmonton
1989	Steve Yzerman	Detroit
1988	Mario Lemieux	Pittsburgh
1987	Wayne Gretzky	Edmonton
1986	Mario Lemieux	Pittsburgh
1985	Wayne Gretzky	Edmonton
1984	Wayne Gretzky	Edmonton
1983	Wayne Gretzky	Edmonton
1982	Wayne Gretzky	Edmonton
1981	Mike Liut	St. Louis
1980	Marcel Dionne	Los Angeles
1979	Marcel Dionne	Los Angeles
1978	Guy Lafleur	Montreal
1977	Guy Lafleur	Montreal
1976	Guy Lafleur	Montreal
1975	Bobby Orr	Boston
1974	Phil Esposito	Boston
1973	Bobby Clarke	Philadelphia
1972	Jean Ratelle	NY Rangers
1971	Phil Esposito	Boston

JACK ADAMS AWARD WINNERS

	Winner	Runner-up
2002	Bob Francis, Phx.	Brian Sutter, Chi.
2001	Bill Barber, Phi.	Scotty Bowman, Det.
2000	Joel Quenneville, St.L.	Alain Vigneault, Mtl.
1999	Jacques Martin, Ott.	Pat Quinn, Tor.
1998	Pat Burns, Bos.	Larry Robinson, L.A.
1997	Ted Nolan, Buf.	Ken Hitchcock, Dal.
1996	Scotty Bowman, Det.	Doug MacLean, Fla.
1995	Marc Crawford, Que.	Scotty Bowman, Det.
1994	Jacques Lemaire, N.J.	Kevin Constantine, S.J.
1993	Pat Burns, Tor.	Brian Sutter, Bos.
1992	Pat Quinn, Van.	Roger Neilson, NYR
1991	Brian Sutter, St.L.	Tom Webster, L.A.
1990	Bob Murdoch, Wpg.	Mike Milbury, Bos.
1989	Pat Burns, Mtl.	Bob McCammon, Van.
1988	Jacques Demers, Det.	Terry Crisp, Cgy.
1987	Jacques Demers, Det.	Jack Evans, Hfd.
1986	Glen Sather, Edm.	Jacques Demers, St.L.
1985	Mike Keenan, Phi.	Barry Long, Wpg.
1984	Bryan Murray, Wsh.	Scotty Bowman, Buf.
1983	Orval Tessier, Chi.	
1982	Tom Watt, Wpg.	
1981	Red Berenson, St.L.	Bob Berry, L.A.
1980	Pat Quinn, Phi.	
1979	Al Arbour, NYI	Fred Shero, NYR
1978	Bobby Kromm, Det.	Don Cherry, Bos.
1977	Scotty Bowman, Mtl.	Tom McVie, Wsh.
1976	Don Cherry, Bos.	
1975	Bob Pulford, L.A.	
1974	Fred Shero, Phi.	

MBNA ROGER CROZIER SAVING GRACE AWARD

	Winner	Runner-up
2002	Jose Theodore, Mtl.	Patrick Roy, Col.
2001	Marty Turco, Dal.	Mike Dunham, N.J.
2000	Ed Belfour, Dal.	Jose Theodore, Mtl.

NHL Amateur and Entry Draft

History

Year	Site	Date	Total Players Drafted
1963	Queen Elizabeth Hotel	June 5	21
1964	Queen Elizabeth Hotel	June 11	24
1965	Queen Elizabeth Hotel	April 27	11
1966	Mount Royal Hotel	April 25	24
1967	Queen Elizabeth Hotel	June 7	18
1968	Queen Elizabeth Hotel	June 13	24
1969	Queen Elizabeth Hotel	June 12	84
1970	Queen Elizabeth Hotel	June 11	115
1971	Queen Elizabeth Hotel	June 10	117
1972	Queen Elizabeth Hotel	June 8	152
1973	Mount Royal Hotel	May 15	168
1974	NHL Montreal Office	May 28	247
1975	NHL Montreal Office	June 3	217
1976	NHL Montreal Office	June 1	135
1977	NHL Montreal Office	June 14	185
1978	Queen Elizabeth Hotel	June 15	234
1979	Queen Elizabeth Hotel	August 9	126
1980	Montreal Forum	June 11	210
1981	Montreal Forum	June 10	211
1982	Montreal Forum	June 9	252
1983	Montreal Forum	June 8	242
1984	Montreal Forum	June 9	250
1985	Toronto Convention Centre	June 15	252
1986	Montreal Forum	June 21	252
1987	Joe Louis Sports Arena	June 13	252
1988	Montreal Forum	June 11	252
1989	Metropolitan Sports Center	June 17	252
1990	B. C. Place	June 16	250
1991	Memorial Auditorium	June 9	264
1992	Montreal Forum	June 20	264
1993	Colisée de Québec	June 26	286
1994	Hartford Civic Center	June 28-29	286
1995	Edmonton Coliseum	July 8	234
1996	Kiel Center	June 22	241
1997	Civic Arena	June 21	246
1998	Marine Midland Arena	June 27	258
1999	FleetCenter	June 26	272
2000	Saddledome	June 24-25	293
2001	National Car Rental Center	June 23-24	289
2002	Air Canada Centre	June 22-23	290

* The NHL Amateur Draft became the NHL Entry Draft in 1979

First Selections

Year	Player	Pos	Drafted By	Drafted From	Age
1969	Rejean Houle	LW	Montreal	Montreal Jr. Canadiens	19.8
1970	Gilbert Perreault	C	Buffalo	Montreal Jr. Canadiens	19.7
1971	Guy Lafleur	RW	Montreal	Quebec Remparts	19.9
1972	Billy Harris	RW	NY Islanders	Toronto Marlboros	20.4
1973	Denis Potvin	D	NY Islanders	Ottawa 67's	19.7
1974	Greg Joly	D	Washington	Regina Pats	20.0
1975	Mel Bridgman	C	Philadelphia	Victoria Cougars	20.1
1976	Rick Green	D	Washington	London Knights	20.3
1977	Dale McCourt	C	Detroit	St. Catharines Fincups	20.4
1978	Bobby Smith	C	Minnesota	Ottawa 67's	20.4
1979	Rob Ramage	D	Colorado	London Knights	20.5
1980	Doug Wickenheiser	C	Montreal	Regina Pats	19.2
1981	Dale Hawerchuk	C	Winnipeg	Cornwall Royals	18.2
1982	Gord Kluzak	D	Boston	Nanaimo Islanders	18.3
1983	Brian Lawton	C	Minnesota	Mount St. Charles High	18.11
1984	Mario Lemieux	C	Pittsburgh	Laval Voisins	18.8
1985	Wendel Clark	LW/D	Toronto	Saskatoon Blades	18.7
1986	Joe Murphy	C	Detroit	Michigan State	18.8
1987	Pierre Turgeon	C	Buffalo	Granby Bisons	17.10
1988	Mike Modano	C	Minnesota	Prince Albert Raiders	18.0
1989	Mats Sundin	RW	Quebec	Nacka IK (Sweden)	18.4
1990	Owen Nolan	RW	Quebec	Cornwall Royals	18.4
1991	Eric Lindros	C	Quebec	Oshawa Generals	18.3
1992	Roman Hamrlik	D	Tampa Bay	ZPS Zlin (Czech)	18.2
1993	Alexandre Daigle	C	Ottawa	Victoriaville Tigres	18.5
1994	Ed Jovanovski	D	Florida	Windsor Spitfires	18.0
1995	Bryan Berard	D	Ottawa	Detroit Jr. Red Wings	18.4
1996	Chris Phillips	D	Ottawa	Prince Albert Raiders	18.3
1997	Joe Thornton	C	Boston	Sault Ste. Marie	17.11
1998	Vincent Lecavalier	C	Tampa Bay	Rimouski Oceanic	18.2
1999	Patrik Stefan	C	Atlanta	Long Beach Ice Dogs	18.9
2000	Rick DiPietro	G	NY Islanders	Boston University	18.9
2001	Ilya Kovalchuk	LW	Atlanta	Krylja Sovetov (Russia)	18.2
2002	Rick Nash	LW	Columbus	London Knights	18.0

Top prospects in Toronto, wearing the jerseys of their 2001-02 teams, as they await the 2002 NHL Entry Draft: Left to right, back row, Jeff Deslauriers (Chicoutimi Sagueneens, G, selected 31st overall by Edmonton); Martin Vagner (Hull Olympiques, D, 26th by Dallas); Eric Nystrom (University of Michigan, LW, 10th by Calgary); Joni Pitkanen (Karpat Oulu, D, 4th by Philadelphia); Christopher Higgins (Yale Bulldogs, C, 14th by Montreal); Steve Eminger (Kitchener Rangers, D, 12th by Washington); Alexander Semin (Chelyabinsk, LW, 13th by Washington). Front row, Kari Lehtonen (Jokerit Helsinki, G, 2nd by Atlanta); Ryan Whitney (Boston University, D, 5th by Pittsburgh); Petr Taticek (Sault Ste. Marie Greyhounds, C, 9th by Florida); Jay Bouwmeester (Medicine Hat Tigers, D, 3rd by Florida); Rick Nash (London Knights, LW, 1st by Columbus); Joffrey Lupul (Medicine Hat Tigers, C, 7th by Anaheim); Scottie Upshall (Kamloops Blazers, RW, 6th by Nashville).

Draft Summary

Following is a summary of the number of players drafted from the Ontario Hockey League (OHL), Western Hockey League (WHL), Quebec Major Junior Hockey League (QMJHL), United States Colleges, United States High Schools, European Leagues and other Leagues throughout North America since 1969:

	OHL	WHL	QMJHL	US Colleges	US HS	International	Other
1969	36	20	11	7	0	1	9
1970	51	22	13	16	0	0	13
1971	41	28	13	22	0	0	13
1972	46	44	30	21	0	0	11
1973	56	49	24	25	0	0	14
1974	69	66	40	41	0	6	25
1975	55	57	28	59	0	6	12
1976	47	33	18	26	0	8	3
1977	42	44	40	49	0	5	5
1978	59	48	22	73	0	16	16
1979	48	37	19	15	0	6	1
1980	73	41	24	42	7	13	10
1981	59	37	28	21	17	32	17
1982	60	55	17	20	47	35	18
1983	57	41	24	14	35	34	37
1984	55	38	16	22	44	40	36
1985	59	47	15	20	48	31	31
1986	66	32	22	22	40	28	42
1987	32	36	17	40	69	38	20
1988	32	30	22	48	56	39	25
1989	39	44	16	48	47	38	20
1990	39	33	14	38	57	53	16
1991	43	40	25	43	37	55	21
1992	57	45	22	9	25	84	22
1993	60	44	23	17	33	78	31
1994	45	66	28	6	28	80	33
1995	54	55	35	5	2	69	14
1996	51	54	31	25	6	58	16
1997	52	63	19	26	4	63	19
1998	50	44	41	27	7	75	14
1999	52	40	20	36	9	94	21
2000	39	41	21	35	7	123	27
2001	41	45	26	24	8	119	26
2002	35	43	23	41	6	110	32
Total	**1700**	**1462**	**787**	**983**	**639**	**1437**	**670**

Total Drafted, 1969-2002: 7,678

Ontario Hockey League

Club	'69	'70	'71	'72	'73	'74	'75	'76	'77	'78	'79	'80	'81	'82	'83	'84	'85	'86	'87	'88	'89	'90	'91	'92	'93	'94	'95	'96	'97	'98	'99	'00	'01	'02	Total
Peterborough	5	5	4	5	9	4	8	1	4	6	9	10	3	5	7	3	9	2	5	2	2	4	3	4	4	2	5	4	5	1	4	1	2	1	148
Oshawa	5	4	3	5	5	7	6	6	1	3	3	2	9	5	5	6	6	6	3	2	4	2	4	4	4	1	10	1	3	4	3	2	1	3	138
Kitchener	1	6	2	8	4	13	3	1	3	4	4	4	5	5	8	4	6	3	2	1	7	5	3	1	4	2	4	2	3	5	–	1	1	4	129
Ottawa	2	4	3	4	6	5	6	5	5	5	3	8	4	9	2	2	3	3	2	1	–	5	5	6	4	1	1	2	5	2	6	2	3	–	124
London	4	9	1	5	6	6	3	5	4	3	6	2	5	5	3	7	1	3	2	6	3	3	1	3	4	1	1	4	1	8	4	1	2	2	124
Sudbury	–	–	–	–	6	6	4	5	4	4	3	7	2	4	–	2	5	3	1	–	1	2	8	2	10	2	2	1	3	5	5	–	2	1	100
S.S. Marie	–	–	–	–	4	5	2	5	1	5	3	3	8	1	6	4	5	7	1	2	3	1	2	7	3	4	3	4	1	4	1	1	1	2	99
Kingston	–	–	–	–	–	4	4	6	4	9	2	8	5	2	1	3	3	4	1	1	–	2	2	3	5	2	3	4	4	1	4	–	2	1	90
Niagara Falls	4	2	1	4	–	–	–	–	2	3	5	8	6	6	–	–	–	–	–	–	4	4	4	4	4	3	2	6	–	–	–	–	–	–	72
Windsor	–	–	–	–	–	–	–	2	1	4	2	3	5	3	2	2	3	7	–	5	2	1	–	3	–	3	4	1	5	1	2	2	2	2	67
Guelph	–	–	–	–	–	–	–	–	–	–	–	–	–	–	1	5	3	8	2	–	4	–	–	2	2	7	5	6	1	5	3	1	4	2	61
North Bay	–	–	–	–	–	–	–	–	–	–	–	–	–	–	4	4	3	3	3	3	1	4	2	5	2	7	2	1	1	2	2	2	3	2	56
Belleville	–	–	–	–	–	–	–	–	–	–	–	–	–	–	3	4	4	5	2	–	4	2	1	4	–	3	3	–	5	2	5	1	3	2	53
Det./Plymouth	–	–	–	–	–	–	–	–	–	–	–	–	–	–	–	–	–	–	–	–	–	–	2	2	7	2	6	3	4	2	2	6	3	3	42
Owen Sound	–	–	–	–	–	–	–	–	–	–	–	–	–	–	–	–	–	–	–	–	–	1	1	2	4	3	2	3	2	1	–	1	–	1	21
Barrie	–	–	–	–	–	–	–	–	–	–	–	–	–	–	–	–	–	–	–	–	–	–	–	–	–	–	–	2	4	3	6	3	1	1	20
Sarnia	–	–	–	–	–	–	–	–	–	–	–	–	–	–	–	–	–	–	–	–	–	–	–	–	–	–	1	7	2	3	1	3	1	2	20
Brampton	–	–	–	–	–	–	–	–	–	–	–	–	–	–	–	–	–	–	–	–	–	–	–	–	–	–	–	–	–	–	2	6	3	3	14
Erie	–	–	–	–	–	–	–	–	–	–	–	–	–	–	–	–	–	–	–	–	–	–	–	–	–	–	–	–	3	1	2	3	2	2	13
St. Michael's	–	–	–	–	–	–	–	–	–	–	–	–	–	–	–	–	–	–	–	–	–	–	–	–	–	–	–	–	–	–	–	1	5	1	7
Mississauga	–	–	–	–	–	–	–	–	–	–	–	–	–	–	–	–	–	–	–	–	–	–	–	–	–	–	–	–	–	–	–	2	–	–	2

Teams no longer operating

Club	'69	'70	'71	'72	'73	'74	'75	'76	'77	'78	'79	'80	'81	'82	'83	'84	'85	'86	'87	'88	'89	'90	'91	'92	'93	'94	'95	'96	'97	'98	'99	'00	'01	'02	Total
Toronto	3	7	6	5	6	8	4	4	7	5	4	10	2	6	4	4	3	4	1	2	2	–	–	–	–	–	–	–	–	–	–	–	–	–	97
Hamilton	2	3	5	4	6	4	7	3	–	8	1	–	–	–	–	–	3	6	4	4	–	–	2	–	–	–	–	–	–	–	–	–	–	–	62
St. Catharines	5	5	8	5	4	7	8	4	6	–	–	–	–	–	–	–	–	–	–	–	–	–	–	–	–	–	–	–	–	–	–	–	–	–	52
Cornwall	–	–	–	–	–	–	–	–	–	–	–	–	–	7	4	3	2	2	3	3	2	3	3	5	–	–	–	–	–	–	–	–	–	–	37
Brantford	–	–	–	–	–	–	–	–	–	–	3	8	5	2	7	2	–	–	–	–	–	–	–	–	–	–	–	–	–	–	–	–	–	–	27
Montreal	5	6	8	1	–	–	–	–	–	–	–	–	–	–	–	–	–	–	–	–	–	–	–	–	–	–	–	–	–	–	–	–	–	–	20
Newmarket	–	–	–	–	–	–	–	–	–	–	–	–	–	–	–	–	–	–	–	–	–	–	–	–	3	2	–	–	–	–	–	–	–	–	5

Year	Total Ontario Drafted	Total Players Drafted	Ontario %
1969	36	84	42.9
1970	51	115	44.3
1971	41	117	35.0
1972	46	152	30.3
1973	56	168	33.3
1974	69	247	27.9
1975	55	217	25.3
1976	47	135	34.8
1977	42	185	22.7
1978	59	234	25.2
1979	48	126	38.1
1980	73	210	34.8
1981	59	211	28.0
1982	60	252	23.8
1983	57	242	23.6
1984	55	250	22.0
1985	59	252	23.4
1986	66	252	26.2
1987	32	252	12.7
1988	32	252	12.7
1989	39	252	15.5
1990	39	250	15.6
1991	43	264	16.3
1992	57	264	21.6
1993	60	286	21.0
1994	45	286	15.7
1995	54	234	23.1
1996	51	241	21.1
1997	52	246	21.1
1998	50	258	19.4
1999	52	272	19.1
2000	39	293	13.3
2001	41	289	14.2
2002	35	290	12.0
Total	**1700**	**7678**	**22.1**

Western Hockey League

Club	'69	'70	'71	'72	'73	'74	'75	'76	'77	'78	'79	'80	'81	'82	'83	'84	'85	'86	'87	'88	'89	'90	'91	'92	'93	'94	'95	'96	'97	'98	'99	'00	'01	'02	Total
Regina	–	–	5	5	1	8	5	3	1	4	1	3	5	6	8	4	4	3	2	–	5	1	–	4	–	3	2	4	3	2	4	2	2	1	101
Saskatoon	1	–	1	3	8	4	5	3	4	1	2	2	3	5	5	3	1	5	4	4	3	2	2	3	2	4	2	2	2	2	4	1	4	–	97
Portland	–	–	–	–	–	–	–	–	4	8	7	8	6	7	7	5	2	4	3	1	4	1	1	4	4	3	2	1	3	3	1	6	–	2	97
Kamloops	–	–	–	–	–	4	4	4	4	–	–	–	–	2	4	4	4	4	3	1	5	4	6	3	2	9	5	4	3	1	4	4	2	5	95
Medicine Hat	–	–	–	4	6	4	5	3	5	4	–	4	2	1	2	1	6	2	5	1	4	1	3	3	1	6	2	7	2	3	1	–	2	3	93
Brandon	–	3	1	5	2	7	4	–	3	1	10	5	2	2	1	3	2	1	3	3	–	1	1	1	2	5	6	2	5	4	–	–	2	4	91
Lethbridge	–	–	–	–	–	–	3	2	3	5	4	1	4	7	2	1	5	1	–	3	3	4	7	3	4	3	3	1	5	1	–	3	1	2	81
Seattle	–	–	–	–	–	–	–	–	–	4	2	3	–	6	–	1	3	1	2	4	2	6	3	2	4	5	5	1	8	2	6	4	5	1	80
Prince Albert	–	–	–	–	–	–	–	–	–	–	–	–	–	4	2	2	6	6	1	3	3	4	6	2	5	3	4	3	5	3	3	2	4	1	72
Spokane	–	–	–	–	–	–	–	–	–	–	–	–	1	–	–	–	–	–	1	3	2	1	5	7	4	4	4	5	4	1	1	2	3	3	51
Swift Current	1	–	1	–	3	6	–	–	–	–	–	–	–	–	–	–	–	–	5	2	2	2	1	1	5	4	4	1	2	2	1	3	1	4	51
Moose Jaw	–	–	–	–	–	–	–	–	–	–	–	–	–	–	–	–	4	1	3	–	3	1	2	3	2	3	4	4	4	2	1	5	3	3	48
Tri-City	–	–	–	–	–	–	–	–	–	–	–	–	–	–	–	–	–	–	–	–	4	3	3	5	2	2	6	6	1	4	1	2	2	3	44
Red Deer	–	–	–	–	–	–	–	–	–	–	–	–	–	–	–	–	–	–	–	–	–	–	–	–	3	5	2	4	3	5	1	1	6	4	34
Calgary	–	–	–	–	–	–	–	–	–	–	–	–	–	–	–	–	–	–	–	–	–	–	–	–	–	–	–	3	–	3	6	4	1	2	19
Kelowna	–	–	–	–	–	–	–	–	–	–	–	–	–	–	–	–	–	–	–	–	–	–	–	–	–	–	–	4	7	2	2	1	1	1	18
Prince George	–	–	–	–	–	–	–	–	–	–	–	–	–	–	–	–	–	–	–	–	–	–	–	–	–	–	2	2	2	4	2	–	4	–	16
Kootenay	–	–	–	–	–	–	–	–	–	–	–	–	–	–	–	–	–	–	–	–	–	–	–	–	–	–	–	–	–	–	2	1	2	3	8
Edmonton	–	–	–	–	–	–	–	–	–	–	–	–	–	–	–	–	–	–	–	–	–	–	–	–	–	–	–	–	4	–	–	–	–	–	4
Vancouver	–	–	–	–	–	–	–	–	–	–	–	–	–	–	–	–	–	–	–	–	–	–	–	–	–	–	–	–	–	–	–	–	–	1	1

Teams no longer operating

Club	'69	'70	'71	'72	'73	'74	'75	'76	'77	'78	'79	'80	'81	'82	'83	'84	'85	'86	'87	'88	'89	'90	'91	'92	'93	'94	'95	'96	'97	'98	'99	'00	'01	'02	Total
Victoria	–	–	–	2	2	5	7	4	3	3	1	8	6	2	3	4	2	1	2	4	4	2	–	1	2	2	–	–	–	–	–	–	–	–	70
Calgary	3	5	2	7	4	8	4	4	4	3	–	2	5	4	3	3	3	2	–	–	–	–	–	–	–	–	–	–	–	–	–	–	–	–	66
New Westm'r	–	–	–	6	8	7	9	5	8	6	5	1	–	–	–	2	1	1	2	1	–	–	–	–	–	–	–	–	–	–	–	–	–	–	62
Flin Flon	4	4	5	2	4	7	4	3	1	5	–	–	–	–	–	–	–	–	–	–	–	–	–	–	–	–	–	–	–	–	–	–	–	–	39
Winnipeg	3	2	4	2	5	4	4	–	4	–	–	–	1	4	1	–	–	–	–	–	–	–	–	–	–	–	–	–	–	–	–	–	–	–	34
Edmonton	4	4	5	6	6	2	3	2	–	–	2	–	–	–	–	–	–	–	–	–	–	–	–	–	–	–	–	–	–	–	–	–	–	–	34
Billings	–	–	–	–	–	–	–	–	–	4	3	4	2	–	–	–	–	–	–	–	–	–	–	–	–	–	–	–	–	–	–	–	–	–	13
Estevan	4	4	4	–	–	–	–	–	–	–	–	–	–	–	–	–	–	–	–	–	–	–	–	–	–	–	–	–	–	–	–	–	–	–	12
Tacoma	–	–	–	–	–	–	–	–	–	–	–	–	–	–	–	–	–	–	–	–	–	–	–	3	2	5	2	–	–	–	–	–	–	–	12
Kelowna	–	–	–	–	–	–	–	–	–	–	–	–	–	–	2	4	5	–	–	–	–	–	–	–	–	–	–	–	–	–	–	–	–	–	11
Nanaimo	–	–	–	–	–	–	–	–	–	–	–	–	–	5	1	–	–	–	–	–	–	–	–	–	–	–	–	–	–	–	–	–	–	–	6
Vancouver	–	–	–	2	–	–	–	–	–	–	–	–	–	–	–	–	–	–	–	–	–	–	–	–	–	–	–	–	–	–	–	–	–	–	2

Year	Total Western Drafted	Total Players Drafted	Western %
1969	20	84	23.8
1970	22	115	19.1
1971	28	117	23.9
1972	44	152	28.9
1973	49	168	29.2
1974	66	247	26.7
1975	57	217	26.3
1976	33	135	24.4
1977	44	185	23.8
1978	48	234	20.5
1979	37	126	29.4
1980	41	210	19.5
1981	37	211	17.5
1982	55	252	21.8
1983	41	242	16.9
1984	37	250	14.8
1985	48	252	19.0
1986	32	252	12.7
1987	36	252	14.3
1988	30	252	11.9
1989	44	252	17.5
1990	33	250	13.2
1991	40	264	15.2
1992	45	264	17.0
1993	44	286	15.4
1994	66	286	23.0
1995	55	234	23.5
1996	54	241	22.4
1997	63	246	25.6
1998	44	258	17.0
1999	40	272	14.7
2000	41	293	14.0
2001	45	289	15.5
2002	43	290	14.8
Total	**1462**	**7678**	**19.0**

Now teammates in Boston, both Brian Rolston (left) and Glen Murray (far left) are products of the 1991 Entry Draft. New Jersey took Rolston with the 11th pick, while Boston selected Murray 18th overall from the Sudbury Wolves.

Quebec Major Junior Hockey League

Club	'69	'70	'71	'72	'73	'74	'75	'76	'77	'78	'79	'80	'81	'82	'83	'84	'85	'86	'87	'88	'89	'90	'91	'92	'93	'94	'95	'96	'97	'98	'99	'00	'01	'02	Total
Shawinigan	3	2	1	6	1	5	3	–	3	–	–	2	2	5	5	2	–	2	1	–	2	–	2	3	1	1	2	4	1	3	1	1	1	2	67
Sherbrooke	–	–	2	2	4	3	7	5	6	3	4	1	5	2	–	–	–	–	–	–	–	–	–	–	3	2	4	–	1	5	–	–	3	–	62
Hull	–	–	–	–	–	–	3	2	2	3	–	3	1	–	3	1	–	4	3	2	2	3	3	3	3	1	3	3	–	3	4	–	2	5	62
Drummondville	2	4	1	4	2	1	–	–	–	–	–	–	–	–	–	1	2	2	2	4	1	–	4	2	2	1	4	3	2	2	–	1	1	–	48
Chicoutimi	–	–	–	–	–	1	–	–	5	1	1	3	6	1	3	–	3	1	2	2	1	1	–	1	1	3	2	–	2	1	–	1	1	3	46
Granby	–	–	–	–	–	–	–	–	–	–	–	–	–	2	1	3	2	2	4	–	2	–	2	–	1	5	2	3	1	–	–	–	–	–	30
Victoriaville	–	–	–	–	–	–	–	–	–	–	–	–	–	–	–	–	–	–	–	4	–	1	–	2	6	1	1	3	2	1	2	3	1	3	30
Beauport	–	–	–	–	–	–	–	–	–	–	–	–	–	–	–	–	–	–	–	–	–	–	1	3	1	3	7	3	3	–	–	–	–	–	21
Val d'Or	–	–	–	–	–	–	–	–	–	–	–	–	–	–	–	–	–	–	–	–	–	–	–	–	–	1	2	4	2	–	3	2	2	1	17
St. Hyacinthe	–	–	–	–	–	–	–	–	–	–	–	–	–	–	–	–	–	–	–	–	–	3	1	2	1	4	–	4	–	–	–	–	–	–	15
Halifax	–	–	–	–	–	–	–	–	–	–	–	–	–	–	–	–	–	–	–	–	–	–	–	–	–	–	3	1	3	3	–	2	3	–	15
Rimouski	–	–	–	–	–	–	–	–	–	–	–	–	–	–	–	–	–	–	–	–	–	–	–	–	–	–	–	–	–	5	2	2	4	–	13
Quebec	–	–	–	–	–	–	–	–	–	–	–	–	–	–	–	–	–	–	–	–	–	–	–	–	–	–	–	–	–	4	3	–	3	1	11
Baie-Comeau	–	–	–	–	–	–	–	–	–	–	–	–	–	–	–	–	–	–	–	–	–	–	–	–	–	–	–	–	–	3	–	2	3	1	9
Rouyn-Noranda	–	–	–	–	–	–	–	–	–	–	–	–	–	–	–	–	–	–	–	–	–	–	–	–	–	–	–	–	–	3	1	4	–	–	8
Moncton	–	–	–	–	–	–	–	–	–	–	–	–	–	–	–	–	–	–	–	–	–	–	–	–	–	–	–	1	1	2	2	–	–	2	8
Cape Breton	–	–	–	–	–	–	–	–	–	–	–	–	–	–	–	–	–	–	–	–	–	–	–	–	–	–	–	–	–	3	–	1	1	2	7
Acadie-Bathurst	–	–	–	–	–	–	–	–	–	–	–	–	–	–	–	–	–	–	–	–	–	–	–	–	–	–	–	–	–	–	2	–	–	2	4
Montreal	–	–	–	–	–	–	–	–	–	–	–	–	–	–	–	–	–	–	–	–	–	–	–	–	–	–	–	–	–	–	–	2	1	1	4

Teams no longer operating

Club	'69	'70	'71	'72	'73	'74	'75	'76	'77	'78	'79	'80	'81	'82	'83	'84	'85	'86	'87	'88	'89	'90	'91	'92	'93	'94	'95	'96	'97	'98	'99	'00	'01	'02	Total
Laval	–	–	–	1	–	2	1	1	4	2	1	–	–	2	1	2	–	5	3	1	3	3	4	1	2	5	4	2	1	3	–	–	–	–	54
Quebec	1	1	2	4	6	6	1	3	7	1	3	2	2	1	2	2	3	–	–	–	–	–	–	–	–	–	–	–	–	–	–	–	–	–	47
Trois Rivieres	–	1	2	2	2	2	3	2	6	3	2	2	2	1	3	–	3	–	1	3	3	1	2	1	–	–	–	–	–	–	–	–	–	–	47
Cornwall	2	1	2	6	4	8	1	3	1	6	1	5	5	–	–	–	–	–	–	–	–	–	–	–	–	–	–	–	–	–	–	–	–	–	45
Montreal	–	–	–	–	4	4	8	1	3	2	4	3	–	3	–	–	–	–	–	–	–	–	–	–	–	–	–	–	–	–	–	–	–	–	32
Sorel	2	3	1	3	1	8	1	1	3	–	–	–	5	–	–	–	–	–	–	–	–	–	–	–	–	–	–	–	–	–	–	–	–	–	28
Verdun	–	1	1	2	–	–	–	–	–	1	3	3	–	–	3	3	–	3	0	3	1	–	–	3	–	–	–	–	–	–	–	–	–	–	27
St. Jean	–	–	–	–	–	–	–	–	–	–	–	–	–	–	2	–	1	1	0	3	1	–	3	1	2	1	1	–	–	–	–	–	–	–	16
Longueuil	–	–	–	–	–	–	–	–	–	–	–	–	–	–	1	2	1	2	1	–	–	2	3	–	–	–	–	–	–	–	–	–	–	–	12
St. Jerome	1	–	1	–	–	–	–	–	–	–	–	–	–	–	–	–	–	–	–	–	–	–	–	–	–	–	–	–	–	–	–	–	–	–	2

Year	Total Quebec Drafted	Total Players Drafted	Quebec %
1969	11	84	13.1
1970	13	115	11.3
1971	13	117	11.1
1972	30	152	19.7
1973	24	168	14.3
1974	40	247	16.2
1975	28	217	12.9
1976	18	135	13.3
1977	40	185	21.6
1978	22	234	9.4
1979	19	126	15.1
1980	24	210	11.4
1981	28	211	13.3
1982	17	252	6.7
1983	24	242	9.9
1984	16	250	6.4
1985	15	252	5.9
1986	22	252	8.7
1987	17	252	6.7
1988	22	252	8.7
1989	16	252	6.3
1990	14	250	5.6
1991	25	264	9.5
1992	22	264	8.3
1993	23	286	8.0
1994	28	286	9.7
1995	35	234	14.9
1996	31	241	12.8
1997	19	246	7.7
1998	41	258	15.9
1999	20	272	7.3
2000	21	293	7.1
2001	26	289	9.0
2002	23	290	7.9
Total	**787**	**7678**	**10.2**

United States Colleges

Club	'69	'70	'71	'72	'73	'74	'75	'76	'77	'78	'79	'80	'81	'82	'83	'84	'85	'86	'87	'88	'89	'90	'91	'92	'93	'94	'95	'96	'97	'98	99	'00	'01	'02	Total
Minnesota	1	3	2	–	–	9	4	4	5	5	2	3	1	1	1	–	–	2	1	1	1	–	–	–	–	–	2	3	2	1	3	3	–	3	63
Michigan	1	–	–	–	2	2	3	3	1	6	–	4	–	–	–	1	1	–	1	2	3	5	4	2	1	1	–	3	1	3	2	1	2	3	58
Boston U.	–	4	–	–	1	1	1	1	4	5	1	–	1	–	–	1	1	2	2	3	1	2	2	1	1	–	1	1	1	2	3	1	2	3	49
Michigan Tech	–	–	3	1	2	5	4	4	1	2	1	4	–	1	–	2	2	2	1	1	2	1	2	–	1	2	–	1	–	–	–	–	1	–	46
Michigan State	–	–	1	–	1	1	1	1	–	–	–	2	–	2	–	2	–	1	1	4	4	5	4	1	1	1	–	1	1	1	2	2	–	4	44
Wisconsin	–	1	2	4	5	4	4	2	3	–	1	–	3	2	–	1	1	–	1	–	1	–	1	–	–	–	–	–	–	–	2	3	–	–	41
Denver	1	3	2	4	2	3	1	2	2	2	2	1	–	1	–	–	1	2	4	1	1	–	–	–	–	–	–	–	3	–	1	–	–	1	40
North Dakota	2	3	3	1	4	2	1	–	1	2	3	3	1	–	–	1	–	–	–	–	2	1	1	–	–	–	–	–	2	–	1	1	–	1	36
Providence	–	–	–	–	–	–	3	2	3	4	–	5	4	1	2	–	1	1	–	–	–	1	–	–	–	–	–	–	1	2	–	2	–	1	33
Harvard	–	–	2	–	–	–	2	–	2	2	–	–	–	1	1	–	2	–	1	1	2	–	–	–	2	1	–	3	1	2	1	2	2	3	33
Boston College	–	1	–	–	–	–	1	1	–	5	–	2	1	1	–	–	–	1	2	–	2	–	–	–	–	–	–	2	3	3	–	3	2	3	33
Clarkson	–	–	2	2	1	–	2	–	2	2	1	1	1	1	1	1	–	–	1	1	1	3	2	1	1	–	–	–	–	3	1	1	–	–	32
New Hampshire	–	–	–	1	1	3	6	–	4	1	1	2	1	1	1	2	–	–	–	1	–	–	–	–	–	–	–	–	1	–	1	2	–	1	30
Cornell	–	–	–	2	1	1	–	1	1	1	–	1	1	1	–	1	2	–	1	2	5	2	–	–	–	–	–	–	1	–	2	2	–	1	29
Colorado	2	1	–	–	–	1	3	1	2	2	–	1	–	–	–	3	–	1	–	1	–	2	–	–	–	–	1	–	–	3	1	2	1	1	29
Notre Dame	–	–	2	3	–	–	7	2	–	3	1	1	–	–	–	–	–	–	–	–	–	–	–	–	–	–	–	1	2	–	2	1	1	2	28
Bowling Green	–	–	–	–	–	1	3	2	1	1	1	1	–	–	1	–	–	–	–	3	2	1	3	1	–	–	–	–	1	1	1	–	1	1	26
RPI	–	–	–	–	1	–	–	–	1	3	–	1	2	1	1	–	1	–	2	2	–	–	3	1	–	–	–	–	1	–	2	2	1	–	25
Lake Superior	–	–	–	–	1	1	1	–	–	3	–	–	–	–	–	1	–	3	–	3	2	3	1	–	1	–	–	1	1	–	1	–	–	1	24
W. Michigan	–	–	–	–	–	–	–	–	2	–	–	2	–	–	2	2	–	2	1	1	1	1	4	–	2	–	–	–	1	1	–	–	–	1	23
St. Lawrence	–	–	–	–	–	–	1	–	1	4	–	–	–	3	–	1	1	1	1	1	1	1	2	–	1	–	–	1	1	–	1	–	–	1	23
Vermont	–	–	–	–	1	–	4	–	1	1	–	1	1	–	1	1	2	–	–	1	–	–	1	–	–	1	–	1	–	–	–	–	2	–	19
Ohio State	–	–	–	–	–	–	–	–	2	1	–	–	–	–	–	1	–	–	2	2	–	1	1	1	1	–	–	1	1	–	1	–	2	2	19
Northern Mich.	–	–	–	–	–	–	–	–	–	4	–	1	2	1	–	–	–	–	4	1	2	–	1	–	–	–	–	1	–	–	–	1	–	–	18
Maine	–	–	–	–	–	–	–	–	–	–	–	–	–	1	1	–	–	1	–	3	2	1	–	–	1	–	–	1	1	1	4	1	–	–	18
Miami of Ohio	–	–	–	–	–	–	–	–	–	–	–	–	–	–	–	–	1	–	2	4	2	–	2	1	1	–	–	–	–	1	–	–	1	1	16
Minn.-Duluth	–	–	2	1	–	–	–	–	1	1	–	–	1	–	–	–	–	–	–	2	1	2	1	–	–	–	–	–	–	1	–	–	–	–	13
Yale	–	–	–	1	–	–	1	–	–	2	–	1	–	–	–	–	–	1	2	–	1	–	–	–	1	–	–	–	–	–	–	–	–	3	13
Brown	–	–	–	–	1	2	1	–	3	2	–	–	–	1	–	–	–	–	–	–	–	–	1	–	–	–	–	–	–	–	–	1	–	–	12
Colgate	–	–	–	–	–	1	–	–	–	2	1	–	–	–	–	–	–	–	–	1	1	2	2	–	–	–	–	–	–	–	–	–	–	1	11
Northeastern	–	–	–	–	–	1	–	–	–	1	–	1	–	–	1	–	1	1	–	–	1	1	–	–	–	–	–	–	–	1	1	–	–	–	10
Princeton	–	–	–	–	–	–	1	–	–	1	–	1	1	–	1	–	–	–	1	–	1	–	–	–	1	–	–	–	–	1	–	–	1	–	10

Colleges with fewer than 10 players drafted:

8 - Ferris State, Merrimack; **6** - Illinois-Chicago, St. Louis, Dartmouth, Lowell; **5** - Pennsylvania, Union College; **4** - Alaska-Anchorage, St. Cloud State; **3** - Babson College; **2** - Alaska-Fairbanks, Nebraska-Omaha, Mass.-Amherst; **1** - Air Force, American International College, Army, Bemidji State, Greenway, Hamilton, Mankato State, St. Anselen College, St. Thomas, Salem State, San Diego U., Wisconsin-River Falls.

Year	Total College Drafted	Total Players Drafted	College %
1969	7	84	8.3
1970	16	115	13.9
1971	22	117	18.8
1972	21	152	13.8
1973	25	168	14.9
1974	41	247	16.6
1975	59	217	27.2
1976	26	135	19.3
1977	49	185	26.5
1978	73	234	31.2
1979	15	126	11.9
1980	42	210	20.0
1981	21	211	10.0
1982	20	252	7.9
1983	14	242	5.8
1984	22	250	8.8
1985	20	252	7.9
1986	22	252	8.7
1987	40	252	15.9
1988	48	252	19.0
1989	48	252	19.0
1990	38	250	15.2
1991	43	264	16.3
1992	9	264	3.4
1993	17	286	5.9
1994	6	286	2.1
1995	5	234	2.1
1996	25	241	10.4
1997	26	246	10.5
1998	27	258	10.4
1999	36	272	13.2
2000	35	293	11.9
2001	24	289	8.3
2002	41	290	14.1
Total	**983**	**7678**	**12.8**

United States High Schools (10 or more players drafted)

Club	'80	'81	'82	'83	'84	'85	'86	'87	'88	'89	'90	'91	'92	'93	'94	'95	'96	'97	'98	'99	'00	'01	'02	Total
Northwood Prep (NY)	–	–	2	1	–	2	2	4	1	1	3	1	–	1	1	–	–	–	–	–	1	–	–	20
Belmont Hill (MA)	–	–	–	1	–	2	1	2	1	3	2	1	2	–	1	–	–	–	–	–	–	–	–	16
Cushing Acad. (MA)	–	–	–	–	1	–	–	–	3	2	3	1	–	2	2	–	1	1	–	–	–	1	–	17
Edina (MN)	–	1	4	2	2	–	–	1	2	2	1	–	1	–	–	–	–	–	–	–	–	–	–	16
Hill-Murray (MN)	–	–	–	–	3	–	3	3	–	2	3	–	–	–	1	–	–	–	–	–	–	–	–	15
Mount St. Charles (RI)	–	1	–	3	1	–	2	1	2	1	1	–	–	–	–	–	–	–	–	–	–	–	–	12
Culver Mil. Acad. (IN)	–	–	–	–	–	–	–	2	1	2	2	1	2	2	–	–	–	–	–	–	–	–	–	12
Catholic Memorial (MA)	–	–	–	–	–	2	–	1	1	2	–	–	2	1	2	–	–	–	1	–	–	–	–	12
Deerfield (IL)	–	–	–	–	1	1	–	1	1	2	–	1	–	–	–	–	–	–	–	1	2	–	1	11
Canterbury (CT)	–	–	–	–	–	–	2	–	3	–	2	–	2	1	–	–	–	–	–	–	–	–	–	10
Matignon (MA)	1	1	1	–	3	–	–	3	–	–	–	1	–	–	–	–	–	–	–	–	–	–	–	10
Roseau (MN)	1	–	1	1	1	–	1	–	–	1	3	1	–	–	–	–	–	–	–	–	–	–	–	10
Choate (CT)	–	–	–	–	–	1	–	2	3	–	1	1	1	–	–	–	–	–	–	–	1	–	–	10
Hotchkiss (CT)	–	–	–	–	–	1	–	–	1	–	1	–	–	–	3	1	2	–	–	–	–	1	–	10

Year	Total USHS Drafted	Total Players Drafted	USHS %
1980	7	210	3.3
1981	17	211	8.1
1982	47	252	18.6
1983	35	242	14.5
1984	44	250	17.6
1985	48	252	19.1
1986	40	252	15.9
1987	69	252	27.4
1988	56	252	22.2
1989	47	252	18.7
1990	57	250	22.8
1991	37	264	14.0
1992	25	264	9.5
1993	33	286	11.5
1994	28	286	9.7
1995	2	234	0.9
1996	6	241	2.4
1997	4	246	1.6
1998	7	258	2.7
1999	9	272	3.3
2000	7	293	2.4
2001	8	289	2.7
2002	6	290	2.0
Total	**639**	**7678**	**8.3**

International

Country	'69	'70	'71	'72	'73	'74	'75	'76	'77	'78	'79	'80	'81	'82	'83	'84	'85	'86	'87	'88	'89	'90	'91	'92	'93	'94	'95	'96	'97	'98	'99	'00	'01	'02	Total
USSR/CIS/Russia	–	–	–	–	–	–	1	–	–	2	–	–	–	3	5	1	2	1	2	11	18	14	25	45	31	35	27	17	16	22	29	44	36	33	420
Sweden	–	–	–	–	–	5	2	5	2	8	5	9	14	14	10	14	16	9	15	14	9	7	11	11	18	17	8	16	14	19	24	24	14	24	358
Czech Republic and Slovakia	–	–	–	–	–	–	–	–	–	–	2	1	–	4	13	13	8	6	11	5	8	21	9	17	15	18	21	14	17	20	20	28	28	21	329
Finland	1	–	–	–	–	1	3	2	3	2	–	4	12	5	9	10	4	10	6	7	3	9	6	8	9	8	12	7	11	12	17	19	29	26	255
Germany	–	–	–	–	–	–	–	–	–	2	–	–	2	–	1	2	1	–	1	2	–	–	1	1	3	1	1	3	1	–	–	1	7	1	32
Switzerland	–	–	–	–	–	–	–	1	–	–	–	–	–	–	–	–	–	–	–	–	–	–	1	–	2	1	–	1	3	2	3	7	5	4	30
Norway	–	–	–	–	–	–	–	–	–	–	–	–	–	–	–	–	–	–	2	–	–	2	1	–	–	–	–	–	1	–	–	–	–	1	7
Denmark	–	–	–	–	–	–	–	–	–	–	–	–	–	–	–	–	–	1	1	–	–	–	–	–	–	–	–	–	–	–	–	–	–	–	2
Scotland	–	–	–	–	–	–	–	–	–	–	–	–	–	–	–	–	–	1	–	–	–	–	–	–	–	–	–	–	–	–	–	–	–	–	1
Poland	–	–	–	–	–	–	–	–	–	–	–	–	–	–	–	–	–	–	–	–	–	–	1	–	–	–	–	–	–	–	–	–	–	–	1
Japan	–	–	–	–	–	–	–	–	–	–	–	–	–	–	–	–	–	–	–	–	–	–	–	1	–	–	–	–	–	–	–	–	–	–	1
Hungary	–	–	–	–	–	–	–	–	–	–	–	–	–	–	–	–	–	–	–	–	–	–	–	–	–	–	–	–	–	–	1	–	–	–	1

Year	Total Int'l Drafted	Total Players Drafted	Int'l %
1969	1	84	1.2
1970	0	115	0
1971	0	117	0
1972	0	152	0
1973	0	168	0
1974	6	247	2.4
1975	6	217	2.8
1976	8	135	5.9
1977	5	185	2.7
1978	16	234	6.8
1979	6	126	4.8
1980	13	210	6.2
1981	32	211	15.2
1982	35	252	13.9
1983	34	242	14.0
1984	40	250	16.0
1985	31	252	12.3
1986	28	252	11.1
1987	38	252	15.1
1988	39	252	15.5
1989	38	252	15.1
1990	53	250	21.2
1991	55	264	20.8
1992	84	264	31.8
1993	78	286	27.3
1994	80	286	27.9
1995	69	234	29.5
1996	58	241	24.0
1997	63	246	25.6
1998	75	258	29.0
1999	94	272	34.5
2000	123	293	42.0
2001	119	289	41.1
2002	110	290	37.9
Total	**1437**	**7678**	**18.7**

USSR/CIS/Russia

Club	'74	'75	'76	'77	'78	'79	'80	'81	'82	'83	'84	'85	'86	'87	'88	'89	'90	'91	'92	'93	'94	'95	'96	'97	'98	99	'00	'01	'02	Total
CSKA Moscow	–	–	–	–	1	–	–	–	1	4	–	1	1	1	5	8	3	4	7	3	5	2	3	–	1	3	–	–	–	53
Dynamo Moscow	–	–	–	–	–	–	–	–	–	–	–	–	–	–	2	3	4	7	10	2	1	7	1	1	1	2	2	–	–	43
Krylja Sovetov Moscow	–	–	–	–	–	–	–	–	–	–	–	–	–	–	1	1	2	4	3	1	5	3	2	1	1	2	1	1	1	29
Spartak Moscow	–	–	–	–	–	–	–	–	1	–	–	1	–	1	–	–	–	1	4	–	6	1	–	–	–	1	–	6	–	22
Lokomotiv-2 Yaroslavl[6]	–	–	–	–	–	–	–	–	–	–	–	–	–	–	–	–	–	–	–	–	–	1	2	2	4	–	9	1	1	20
Lokomotiv Yaroslavl[6]	–	–	–	–	–	–	–	–	–	–	–	–	–	–	–	–	1	2	–	–	1	5	1	1	3	1	1	–	2	18
Traktor Chelyabinsk	–	–	–	–	–	–	–	–	–	–	–	–	–	–	–	2	–	–	2	7	1	1	–	1	1	–	–	–	1	16
Dynamo-2 Moscow	–	–	–	–	–	–	–	–	–	–	–	–	–	–	–	–	–	–	2	1	2	–	–	3	3	–	4	–	–	15
Lada Togliatti	–	–	–	–	–	–	–	–	–	–	–	–	–	–	–	–	–	–	1	2	–	–	1	3	1	2	2	–	2	14
Elemash Elektrostal	–	–	–	–	–	–	–	–	–	–	–	–	–	–	–	–	–	–	3	–	–	–	–	1	–	–	–	1	9	14
Khimik Voskresensk	–	–	–	–	–	–	–	–	–	1	–	–	–	–	–	1	3	1	2	–	1	–	–	–	2	–	1	1	–	13
Sokol Kiev	–	–	–	–	–	–	–	–	–	–	1	–	–	–	1	–	1	2	3	1	–	2	–	–	–	–	–	–	–	11
SKA St. Peterburg[2]	–	–	–	–	1	–	–	–	1	–	–	–	–	–	–	–	–	2	1	–	1	–	–	–	–	2	1	2	–	11
HC CSKA Moscow	–	–	–	–	–	–	–	–	–	–	–	–	–	–	–	–	–	–	–	–	–	–	–	–	–	2	5	–	4	11
Pardaugava Riga[1]	–	1	–	–	–	–	–	–	–	–	–	–	–	–	1	2	–	1	4	1	–	–	–	–	–	–	–	–	–	10
Severstal Cherepovets[5]	–	–	–	–	–	–	–	–	–	–	–	–	–	–	–	–	–	–	1	1	–	1	1	–	–	5	1	–	–	10
Avangard Omsk	–	–	–	–	–	–	–	–	–	–	–	–	–	–	–	–	–	–	–	–	3	–	1	–	–	–	1	3	1	9
Torpedo Ust Kamenogorsk	–	–	–	–	–	–	–	–	–	–	–	–	–	–	–	–	–	–	1	1	2	1	–	–	2	1	–	–	–	8
Salavat Yulayev Ufa	–	–	–	–	–	–	–	–	–	–	–	–	–	–	–	–	–	–	–	2	2	1	1	1	–	–	1	–	–	8
CSKA-2 Moscow	–	–	–	–	–	–	–	–	–	–	–	–	–	–	–	–	–	–	–	–	1	–	2	2	–	–	–	–	–	5
Metallurg Novokuznetsk	–	–	–	–	–	–	–	–	–	–	–	–	–	–	–	–	–	–	–	–	–	–	–	–	–	2	2	1	–	5
Neftekhimik Nizhnekamsk	–	–	–	–	–	–	–	–	–	–	–	–	–	–	–	–	–	–	–	–	–	–	–	1	–	–	2	2	–	5
Lada-2 Togliatti	–	–	–	–	–	–	–	–	–	–	–	–	–	–	–	–	–	–	–	–	–	–	–	–	–	1	2	2	–	5
Molot-Prikamje Perm	–	–	–	–	–	–	–	–	–	–	–	–	–	–	–	–	–	–	–	–	1	1	–	–	–	–	1	1	1	5
Tivali Minsk[3]	–	–	–	–	–	–	–	–	–	–	–	–	–	–	1	–	–	–	–	2	1	–	–	–	–	–	–	–	–	4
Torpedo Nizhny Novgorod[4]	–	–	–	–	–	–	–	–	–	–	–	–	–	–	–	–	1	–	2	–	–	–	–	–	–	–	1	–	–	4
AK Bars Kazan	–	–	–	–	–	–	–	–	–	–	–	–	–	–	–	–	–	–	–	–	1	–	–	–	1	–	1	1	–	4
Krylja Sovetov Moscow 2	–	–	–	–	–	–	–	–	–	–	–	–	–	–	–	–	–	–	–	1	–	–	–	–	–	–	–	–	3	4
Metallurg Magnitogorsk	–	–	–	–	–	–	–	–	–	–	–	–	–	–	–	–	–	–	–	–	–	–	–	–	–	–	–	3	1	4
Avtomobilist Yekaterinburg	–	–	–	–	–	–	–	–	–	–	–	–	–	–	–	–	–	–	–	–	–	–	1	–	1	1	–	–	–	3
CSK VVS Samara	–	–	–	–	–	–	–	–	–	–	–	–	–	–	–	–	–	–	–	–	–	1	–	–	–	1	–	1	–	3
Avangard-2 Omsk	–	–	–	–	–	–	–	–	–	–	–	–	–	–	–	–	–	–	–	–	–	–	–	–	–	–	–	3	–	3
Severstal-2 Cherepovets	–	–	–	–	–	–	–	–	–	–	–	–	–	–	–	–	–	–	–	–	–	–	–	–	–	1	1	–	1	3
Dizelist Penza	–	–	–	–	–	–	–	–	–	–	–	–	–	–	–	–	–	–	1	–	–	–	–	–	–	–	1	–	–	2
Metallurg-2 Novokuznetsk	–	–	–	–	–	–	–	–	–	–	–	–	–	–	–	–	–	–	–	–	–	–	–	–	–	–	2	–	–	2
Kristall Saratov	–	–	–	–	–	–	–	–	–	–	–	–	–	–	–	–	–	–	–	1	–	–	–	–	–	–	–	1	–	2
Ak-Bars-2 Kazan	–	–	–	–	–	–	–	–	–	–	–	–	–	–	–	–	–	–	–	–	–	–	–	–	–	1	–	1	–	2
Torpedo Nizhny Novgorod 2	–	–	–	–	–	–	–	–	–	–	–	–	–	–	–	–	–	–	–	–	–	–	–	–	–	–	–	–	2	2

Former club names: [1]–Dynamo Riga, HC Riga, [2]–SKA Leningrad, [3]–Dynamo Minsk, [4]–Torpedo Gorky, [5]–Metallurg Cherepovets, [6]–Torpedo Yaroslavl

Teams with one player selected:

Amur Khabarovsk, Argus Moscow, Dynamo Khazov, Dynamo-81 Riga, Izohets St. Petersburg, Khimik Novopolotsk, Mechel Chelyabinsk, Neftekhimik Nizhnekamsk, Salavat Novoil Ufa, SKA-2 St. Petersburg, Sibir-2 Novosibirsk-1, THC Tver, Vityaz Podolsk, Vityaz-2 Podolsk, HC CSKA Moscow 2, Khimik Voskresensk 2, Metalurgs Liepaja, Mostovik Kurgan, Torpedo Nizhny Novgorod 2.

Sweden

Club	'74	'75	'76	'77	'78	'79	'80	'81	'82	'83	'84	'85	'86	'87	'88	'89	'90	'91	'92	'93	'94	'95	'96	'97	'98	'99	'00	'01	'02	Total
Djurgarden Stockholm	1	1	1	–	–	1	2	–	1	2	1	–	1	2	–	1	1	2	1	1	–	3	2	2	–	1	4	1	2	34
MoDo Ornskoldsvik	–	–	1	–	–	1	–	–	1	–	2	–	–	1	–	–	–	2	2	5	–	–	3	3	–	7	3	–	3	34
Farjestad Karlstad	–	–	–	2	2	–	1	2	1	1	1	2	–	–	–	1	–	1	2	1	–	2	–	3	6	1	–	1	2	32
Leksand	1	–	–	–	–	1	–	–	1	–	2	2	1	1	2	1	–	2	–	2	2	–	1	–	2	–	5	–	–	26
Vastra Frolunda Goteborg	–	–	–	–	–	–	–	–	–	–	2	1	–	1	1	1	–	–	1	–	3	1	1	–	1	2	4	3	3	25
AIK Solna	–	–	1	–	1	1	–	2	3	1	–	4	–	–	–	–	1	1	1	–	1	–	–	1	1	3	–	1	1	24
Brynas Gavle	1	–	–	–	1	1	1	1	–	1	2	–	–	4	–	–	–	–	–	1	–	–	1	1	2	1	1	2	–	21
HV 71 Jonkoping	–	–	–	–	–	–	–	1	–	–	–	–	–	1	1	–	1	–	–	–	2	–	2	1	4	3	1	–	1	18
Sodertalje	–	–	–	–	1	–	–	1	1	1	2	2	2	–	–	2	–	–	–	1	–	–	–	–	1	–	1	1	–	16
Malmo	–	–	–	–	–	–	–	–	–	–	–	–	–	–	1	–	–	–	1	1	1	–	2	–	–	1	1	–	4	12
Skelleftea	–	1	1	–	–	–	–	1	1	2	1	–	–	–	1	–	–	–	–	1	–	–	–	1	–	–	–	–	–	10
Lulea	–	–	–	–	–	–	1	1	–	–	–	–	1	1	1	–	–	–	1	–	–	–	–	1	1	–	2	–	–	10
Vasteras	–	–	–	–	–	–	–	–	–	–	–	–	–	–	2	2	1	1	–	1	1	–	–	–	–	–	1	–	1	10
Rogle Angelholm	–	–	–	–	–	–	–	–	–	–	–	–	–	1	2	–	–	–	–	2	2	1	–	–	–	1	–	–	–	9
Timra[1]	–	–	–	–	–	–	1	2	–	–	–	1	1	–	–	–	–	–	–	–	1	–	–	–	–	1	–	–	1	8
Hammarby Stockholm	–	–	–	–	–	–	1	1	–	–	–	–	–	–	–	–	–	–	–	1	1	–	–	–	1	–	–	1	–	6
Huddinge	–	–	–	–	–	–	–	–	–	–	–	–	–	–	–	–	–	1	–	–	–	–	2	–	–	1	–	1	1	6
Bjorkloven Umea	–	–	–	–	–	–	2	1	–	–	–	1	–	–	1	–	–	–	–	–	–	–	–	–	–	–	–	–	–	5
Orebro	–	–	1	–	1	–	–	–	–	1	–	1	–	1	–	–	–	–	–	–	–	–	–	–	–	–	–	–	–	5
Mora	–	–	–	–	–	–	–	–	–	1	–	–	1	–	–	–	–	–	–	–	–	–	1	1	–	–	–	1	–	5
Nacka	–	–	–	–	–	–	–	–	–	–	–	–	1	–	–	1	–	–	2	–	–	–	–	–	–	–	–	–	–	4
Falun	–	–	–	–	–	–	–	1	–	–	–	–	–	–	1	–	–	1	–	–	–	–	–	–	–	–	–	–	–	3
Team Kiruna	–	–	–	–	1	–	–	–	1	–	–	–	–	–	–	–	–	–	–	1	–	–	–	–	–	–	–	–	–	3
Boden	1	–	–	–	–	–	–	–	–	–	–	–	1	–	–	–	–	–	–	–	1	–	–	–	–	–	–	–	–	3
Pitea	–	–	–	–	–	–	–	–	1	–	–	–	–	–	–	–	1	–	–	–	1	–	–	–	–	–	–	–	–	3
Troja	–	–	–	–	–	–	–	–	–	–	–	1	–	–	1	–	–	–	–	–	–	–	–	–	–	1	–	–	–	3
Grums	–	–	–	–	–	–	–	–	–	–	–	–	–	–	–	–	–	–	–	–	1	–	–	–	–	1	1	–	–	3
Morrum	–	–	–	–	–	–	–	–	–	–	–	–	–	–	–	–	–	–	–	–	–	–	–	–	–	–	–	1	2	3
Ostersund	–	–	–	–	–	–	–	–	–	–	–	1	–	1	–	–	–	–	–	–	–	–	–	–	–	–	–	–	–	2
Hasten	–	–	–	–	–	–	–	–	–	–	–	–	–	–	–	–	–	–	–	–	–	–	1	–	–	–	–	–	1	2

Teams with one player selected:

Almtuna, Danderyd Hockey, Fagersta, Karskoga, Stocksund, S/G Hockey 83 Gavle, Talje, Tingsryd, Tunabro, Uppsala, Vallentuna, Bofors, Sunne.

Two graduates of the 1991 Entry Draft. Alexei Zhitnik (top) was a product of Sokol Kiev. Michael Nylander (above) was drafted from Huddinge.

Czech Republic and Slovakia

Club	'69	'70	'71	'72	'73	'74	'75	'76	'77	'78	'79	'80	'81	'82	'83	'84	'85	'86	'87	'88	'89	'90	'91	'92	'93	'94	'95	'96	'97	'98	'99	'00	'01	'02	Total
Chemopetrol Litvinov[1]	–	–	–	–	–	–	–	–	–	–	–	–	–	3	1	2	–	–	–	–	2	2	1	3	2	4	2	2	2	1	1	–	1	–	29
Dukla Jihlava	–	–	–	–	–	–	–	–	–	–	–	–	–	2	4	3	1	–	3	1	1	3	2	1	1	1	2	2	–	–	–	1	–	–	28
HC Ceske Budejovice[6]	–	–	–	–	–	–	–	–	–	–	–	–	–	2	1	1	–	–	1	–	1	2	–	–	1	2	3	1	2	1	1	3	2	–	24
Slavia Praha	–	–	–	–	–	–	–	–	–	–	–	–	–	–	–	1	–	–	–	–	–	–	–	–	1	–	–	–	4	5	2	3	5	2	23
Dukla Trencin	–	–	–	–	–	–	–	–	–	–	–	–	–	–	–	1	–	–	–	1	1	2	–	2	2	–	1	2	1	–	2	3	2	–	20
Sparta Praha	–	–	–	–	–	–	–	–	–	–	–	–	–	1	–	2	1	1	1	2	1	2	–	1	1	–	1	–	1	1	–	1	2	1	20
Slovan Bratislava	–	–	–	–	–	–	–	–	–	1	1	–	–	2	–	–	1	1	1	–	–	1	–	–	–	3	–	1	1	1	2	2	–	1	19
ZPS Zlin[2]	–	–	–	–	–	–	–	–	–	–	–	–	–	–	1	–	1	1	1	–	–	–	2	2	1	–	2	–	1	2	2	–	2	–	18
HC Kladno[7]	–	–	–	–	–	–	–	–	–	–	–	–	2	1	–	1	–	1	–	–	–	1	2	–	1	2	–	2	–	–	2	1	1	–	17
HC Vitkovice[8]	–	–	–	–	–	–	–	–	–	1	–	–	1	–	–	–	–	–	–	–	–	1	–	1	3	1	1	1	–	1	1	1	–	2	15
HC Kosice[3]	–	–	–	–	–	–	–	–	–	–	–	–	1	2	–	2	–	1	–	–	–	2	–	–	–	–	–	1	1	1	1	–	1	–	13
Interconex Plzen[9]	–	–	–	–	–	–	–	–	–	–	–	–	–	–	–	–	–	1	–	–	1	1	–	3	–	1	1	–	–	1	–	1	1	2	13
HC Pardubice[4]	–	–	–	–	–	–	–	–	–	–	–	–	–	–	2	–	2	–	1	–	–	–	–	–	–	2	1	–	1	–	–	–	1	3	13
HC Vsetin	–	–	–	–	–	–	–	–	–	–	–	–	–	–	–	–	–	–	–	–	–	–	–	–	–	–	–	–	2	1	–	2	2	3	10
Zetor Brno	–	–	–	–	–	–	–	–	–	–	–	–	–	–	–	–	1	–	3	–	–	2	–	1	–	–	–	–	–	1	–	–	–	–	8
HC Olomouc[5]	–	–	–	–	–	–	–	–	–	–	–	–	–	–	–	–	–	–	–	–	1	–	–	2	–	1	2	–	–	1	–	–	–	–	7
ZTK Zvolen	–	–	–	–	–	–	–	–	–	–	–	–	–	–	–	–	–	–	–	–	–	–	–	1	–	–	1	1	–	–	2	2	–	–	7
AC Nitra	–	–	–	–	–	–	–	–	–	–	–	–	–	–	–	–	–	–	–	–	–	–	2	–	1	–	–	1	–	1	–	–	1	–	6
ZTS Martin	–	–	–	–	–	–	–	–	–	–	–	–	–	–	–	–	–	–	–	1	–	–	–	–	–	–	2	–	–	–	–	1	1	–	5
Zelezarny Trinec	–	–	–	–	–	–	–	–	–	–	–	–	–	–	–	–	–	–	–	–	–	–	–	–	–	–	–	–	–	–	2	1	1	1	5
Havirov	–	–	–	–	–	–	–	–	–	–	–	–	–	–	–	–	–	–	–	–	–	–	–	–	–	–	–	–	–	–	–	1	–	2	3
HC Karlovy Vary	–	–	–	–	–	–	–	–	–	–	–	–	–	–	–	–	–	–	–	–	–	–	–	–	–	–	–	–	–	–	–	1	1	1	3
IS Banska Bystrica	–	–	–	–	–	–	–	–	–	–	–	–	–	–	–	–	–	–	–	–	–	–	–	–	–	1	1	–	–	–	–	–	–	–	2
ZPA Presov	–	–	–	–	–	–	–	–	–	–	–	–	–	–	–	–	–	–	–	–	–	–	–	–	1	–	–	–	1	–	–	–	–	–	2
Partizan Liptovsky Mikulas	–	–	–	–	–	–	–	–	–	–	–	–	–	–	–	–	–	–	–	–	–	1	–	–	–	–	–	–	–	1	–	–	–	–	2
VTJ Pisek	–	–	–	–	–	–	–	–	–	–	–	–	–	–	–	–	–	–	–	–	–	1	–	–	–	–	–	–	–	–	1	–	–	–	2
Michalovce	–	–	–	–	–	–	–	–	–	–	–	–	–	–	–	–	–	–	–	–	–	–	–	–	–	–	–	–	–	–	2	–	–	–	2
Ingstav Brno	–	–	–	–	–	–	–	–	–	–	–	–	–	–	–	–	1	–	–	–	–	–	–	–	–	–	–	–	–	–	–	–	1	–	2
HC Liberec	–	–	–	–	–	–	–	–	–	–	–	–	–	–	–	–	–	–	–	–	–	–	–	–	–	–	–	–	–	–	–	–	2	–	2

Former club names: [1]–CHZ Litvinov, [2]–TJ Gottwaldov, TJ Zlin, [3]–VSZ Kosice, [4]–Tesla Pardubice, [5]–DS Olomouc, [6]–Motor Ceske Budejovice, [7]–Poldi Kladno, [8]–TJ Vitkovice, [9]–Skoda Plzen

Teams with one player selected:

Banik Sokolov, Havlickuv Brod, KLH Chomutov, KC SKP Poprad, KC Skalica, HK Trnava, KHM Zvolen, Spisska Nova Ves.

Finland

Club	'69	'70	'71	'72	'73	'74	'75	'76	'77	'78	'79	'80	'81	'82	'83	'84	'85	'86	'87	'88	'89	'90	'91	'92	'93	'94	'95	'96	'97	'98	'99	'00	'01	'02	Total
TPS Turku	–	–	–	–	–	–	–	–	–	–	–	1	6	–	–	–	1	1	–	–	–	–	–	–	3	2	3	1	3	3	1	3	3	1	32
HIFK Helsinki	1	–	–	–	–	–	1	–	1	–	–	–	1	1	2	2	1	–	–	2	1	–	–	–	2	–	1	–	1	2	4	2	2	5	32
Jokerit	–	–	–	–	–	–	–	–	–	–	–	2	1	–	–	1	–	–	1	1	–	2	–	3	–	1	–	1	1	1	3	3	4	6	31
Ilves	–	–	–	–	–	–	–	–	1	2	–	–	2	–	2	2	–	1	–	1	–	1	1	–	–	–	2	–	–	2	1	3	4	2	27
Tappara	–	–	–	–	–	1	–	–	–	–	–	–	–	2	–	–	–	4	–	1	–	1	–	–	–	–	1	1	2	–	–	1	2	2	18
Karpat	–	–	–	–	–	–	–	–	–	–	–	1	–	1	–	1	–	2	2	–	–	–	1	–	1	–	–	–	1	1	–	–	3	3	17
Lukko	–	–	–	–	–	–	–	2	1	–	–	–	–	–	2	–	1	–	1	–	–	1	–	1	–	–	–	–	–	–	2	1	3	1	16
Assat	–	–	–	–	–	–	2	–	–	–	–	–	1	–	2	2	–	–	–	1	–	1	–	–	1	1	1	–	–	1	–	–	–	1	14
Blues Espoo	–	–	–	–	–	–	–	–	–	–	–	–	–	–	–	–	–	–	–	1	–	1	1	2	–	2	1	–	1	–	2	–	2	1	13
HPK Hameenlinna	–	–	–	–	–	–	–	–	–	–	–	–	–	–	–	–	–	–	–	–	1	–	–	2	–	–	–	–	1	1	1	3	1	1	11
JyP Jyvaskyla	–	–	–	–	–	–	–	–	–	–	–	–	–	–	–	–	–	–	1	–	–	–	–	–	–	–	–	2	1	–	3	–	1	2	10
KalPa Kuopio	–	–	–	–	–	–	–	–	–	–	–	–	–	–	–	–	–	–	1	–	–	–	–	–	1	2	–	–	–	1	–	1	2	–	8
Reipas Lahti	–	–	–	–	–	–	–	–	–	–	–	–	–	1	1	1	–	–	–	–	–	–	2	–	1	–	–	1	–	–	–	–	–	–	7
SaiPa Lappeenranta	–	–	–	–	–	–	–	–	–	–	–	–	1	–	–	–	–	–	–	–	–	1	–	–	–	–	–	–	–	–	–	1	1	–	4
Kiekoo-67 Turku	–	–	–	–	–	–	–	–	–	–	–	–	–	–	–	–	–	–	–	–	–	–	–	–	–	–	3	–	–	–	–	–	–	–	3
Sapko Savonlinna	–	–	–	–	–	–	–	–	–	–	–	–	–	–	–	–	1	1	–	–	–	–	–	–	–	–	–	–	–	–	–	–	–	–	2
Sport Vaasa	–	–	–	–	–	–	–	–	–	–	–	–	–	–	–	–	–	1	–	–	1	–	–	–	–	–	–	–	–	–	–	–	–	–	2

Teams with one player selected:

Ahmat Hyvinkaa, GrIFK Kauniainen, Koo Koo Kouvola, S-Kiekko Seinajoki, Junkkarit Kalajoki, Hermes Kokkola, TuTo.

2002 Entry Draft Analysis

Country of Origin

Country	Players Drafted
Canada	107
USA	60
Russia	33
Czech Republic	26
Finland	25
Sweden	20
Switzerland	5
Slovakia	3
Latvia	3
Norway	2
Hungary	1
Austria	1
Ukraine	1
Belarus	1
Bahamas	1
Denmark	1

Birth Year

Year	Players Drafted
1984	117
1983	108
1982	47
1981	4
1980	1
1979	5
1977	1
1976	1
1975	1
1974	3
1973	1
1970	1

Position

Position	Players Drafted
Defense	92
Center	68
Right Wing	51
Left Wing	46
Goaltender	33

Note: Players drafted in the international category played outside North America in their draft year. European-born players drafted from the OHL, QMJHL, WHL or U.S. Colleges are not counted as International players. See Country of Origin, at left.

Notes on 2002 First Round Selections

1. COLUMBUS • **RICK NASH** • LW • An excellent puck-handler who is strong along the boards and in the corners, Rick Nash is a constant offensive threat and a tough player to defend against. Rookie of the year in the OHL in 2001, Nash has a good wrist shot with a quick release and is very effective at shooting while in full flight. He has solid playmaking and passing skills and is a strong competitor who adapts well to both the finesse and physical aspects of the game. He is used in every game situation.

2. ATLANTA • **KARI LEHTONEN** • G • A tall goaltender (6'3") with impressive quickness and excellent reflexes, Kari Lehtonen was the youngest netminder in the Finnish Elite League in 2001-02. He helped Jokerit win the championship and was named playoff MVP. Lehtonen has great hockey sense and is very mature for his age. He is primarily a standup goaltender, but he can play a strong butterfly style. He anticipates and reads the play very well.

3. FLORIDA • **JAY BOUWMEESTER** • D • The top-rated North American prospect on Central Scouting's final ranking, Jay Bouwmeester is one of the fastest skaters in junior hockey. He has very good acceleration and can control the flow of the game with his speed and change of pace. He has excellent agility and plays an exceptional transition game. At 6'4" and 206 pounds, Bouwmeester has size and strength but he does not play a punishing physical game.

4. PHILADELPHIA • **JONI PITKANEN** • D • Ranked by Central Scouting as the top European prospect, Joni Pitkanen has already spent two seasons in the Finnish Elite League. He is an intelligent player who anticipates the play very well and plays at a very good overall skill level. Pitkanen is a strong skater with good acceleration and excellent lateral movement. He is a tough competitor with a good work ethic and he has very good passing and puck-handling skills.

5. PITTSBURGH • **RYAN WHITNEY** • D • A graduate of USA Hockey's National Team Development Program, Ryan Whitney was named to the Hockey East All-Rookie Team as a freshman at Boston University in 2001-02. He is a very strong skater with a long, smooth stride and he has excellent lateral mobility. Whitney is very confident with the puck and has a low, hard shot from the point. He is very strong along the boards and in front of the net.

6. NASHVILLE • **SCOTTIE UPSHALL** • RW • A tough, gritty competitor, Scottie Upshall has a fierceness to his game that elevates him among the best players in his age group in junior hockey. Upshall is a good skater who is very well balanced and therefore tough to knock off the puck. He is a creative playmaker with keen on-ice perception who uses his teammates well. Upshall is used in all game situations.

7. ANAHEIM • **JOFFREY LUPUL** • C • The top goal-scorer in the Western Hockey League with 56 in 2001-02, Joffrey Lupul has exceptional hockey sense and will sacrifice himself to make the play. A deceptive skater with quick acceleration, Lupul can play all three forward positions and is a constant offensive threat. He is a tough competitor who works well in traffic and has an accurate wrist shot.

8. MINNESOTA • **PIERRE-MARC BOUCHARD** • C • The top scorer in the QMJHL and the CHL player of the year in 2001-02, Pierre-Marc Bouchard is an intelligent player with exceptional hockey sense. Though he stands just 5'10" and weighs only 155 pounds, Bouchard carries the puck with poise and confidence. He is a creative playmaker with strong passing skills and refuses to be intimidated by physical play.

9. FLORIDA • **PETR TATICEK** • C • A creative player and an intelligent playmaker, Petr Taticek adapted well to the North American game as a rookie with Sault Ste. Marie in 2001-02. He has good passing skills and makes excellent decisions around the net. Taticek is a very mature and confident player who is often used to take key face-offs. His father is the coach of the Kladno team in the Czech Republic.

10. CALGARY • **ERIC NYSTROM** • LW • The son of former Islanders star Bob Nystrom, Eric Nystrom plays aggressively along the boards and has a quick and accurate wrist shot. A graduate of USA Hockey's National Team Development Program, Nystrom was named to the CCHA All-Rookie Team with the University of Michigan. He is a good two-way player with good hockey sense who is hard to knock off the puck.

11. BUFFALO • **KEITH BALLARD** • D • A member of the University of Minnesota's 2002 NCAA championship team, Keith Ballard excels at quarterbacking the power-play. He has a good quick wrist shot and a heavy slapshot from the point. Ballard is also an excellent passer who sees the ice well. He is just 5'11" and 196 pounds, but he is an authoritative body checker and is good at blocking shots. Ballard is a graduate of USA Hockey's National Team Development Program.

12. WASHINGTON • **STEVE EMINGER** • D • A solid, agile skater with excellent mobility, Steve Eminger won the puck-control relay at the 2002 Top Prospects Game. Eminger is a standout offensive defenseman with excellent puck-handling and playmaking skills. He is a good decision-maker with impressive hockey sense. Eminger quarterbacked the Kitchener power-play and is used in all game situations. He does not shy away from physical play.

13. WASHINGTON • **ALEXANDER SEMIN** • LW • Thought to be one of the most skilled players available in the draft, Alexander Semin is an extremely fast skater with strong puck-handling abilities. He is a dynamic offensive talent with a good combination of speed and scoring ability, yet he is also a solid two-way player who is aware of his defensive responsibilities. Standing 6' but weighing just 174 pounds, he will have to improve his strength and stamina.

14. MONTREAL • **CHRISTOPHER HIGGINS** • C • The top-scoring American player at the 2002 World Junior Championships (4-2-6), Christopher Higgins is an intelligent player with good hockey sense. He is a good skater with a long, smooth stride and can carry the puck at top speed. He sees the ice well and has an excellent shot with a quick release. Higgins displays good patience with the puck and can deliver the big hit but needs to improve his strength.

15. EDMONTON • **JESSE NIINIMAKI** • C • Though he was ranked 50th among European prospects, the Oilers opted to take Jesse Niinimaki with their first-round selection. Niinimaki stands 6'2" and has good skating speed, good hands and good hockey sense. He split the 2001-02 season between the Ilves Tampere junior team and the Elite League club. Niinimaki patterns his game after Peter Forsberg.

16. OTTAWA • **JAKUB KLEPIS** • C • A fluid skater with impressive agility, Jakub Klepis is a face-off specialist who often plays on special teams. He is very coachable and unselfish and is very effective at reading the play. Klepis has very strong puck-handling skills and works well in heavy traffic. He is primarily a finesse player, but does not shy away from the physical aspects of the game.

17. WASHINGTON • **BOYD GORDON** • RW • A winner of the Memorial Cup with Red Deer in 2001, Boyd Gordon is a fluid skater with tremendous balance and mobility. He is very elusive in one-on-one situations and excels at using his speed to create turnovers. Gordon does not play an overly physical game but he is willing to sacrifice himself to make the play. He has a good shot with a quick release.

18. LOS ANGELES • **DENIS GREBESHKOV** • D • A gold medal winner with Russia at the 2002 World Junior Championships and the 2001 Under-18 Championships, Denis Grebeshkov is an excellent skater with good mobility. He is an intelligent player with a good understanding of the game who has been playing in the Russian senior league since he was 17. Grebeshkov is a stay-at-home defenseman who usually opts for the safe and simple play.

19. PHOENIX • **JAKUB KOREIS** • C • The son of a former Plzen star in the Czech Elite League, Jakub Koreis has a strong hockey sense and has the ability to control the play in the offensive zone. At 6'3" and 205 pounds, he has good size and impressive strength which allows him to work well in the corners and along the boards. Koreis is a creative playmaker with a good shot and good puck-handling skills. He is a mature player who is used in all game situations.

20. BUFFALO • **DAN PAILLE** • LW • An intelligent player with good hockey sense and natural offensive instincts, Dan Paille is a very consistent player who is used in all game situations. He is a tenacious forechecker and a hard worker who plays a solid physical game and is an authoritative body checker. Paille also possesses very good puck-handling and passing skills. He has good speed and agility.

21. CHICAGO • **ANTON BABCHUK** • D • Standing 6'4" and weighing 194 pounds, Anton Babchuk is a good skater who is well-balanced and moves extremely well for his size. He plays an aggressive style and is a tough competitor who is a heavy body checker. Babchuk also possesses very good puck-handling and passing skills and has a good low shot from the point. He is strong in the corners and in front of the net and displays good leadership qualities.

22. NY ISLANDERS • **SEAN BERGENHEIM** • W • A tough competitor and a good team player, Sean Bergenheim was a member of the Finnish team at the 2002 World Junior Championships and helped Jokerit capture the league championship in the Finnish Elite League. Bergenheim is an excellent skater with speed, balance and agility and has a very good shot. He is a hard worker with a winning attitude.

23. PHOENIX • **BEN EAGER** • LW • A very good skater with impressive speed and power, Ben Eager won both the 60' and 150' dash at the 2002 Top Prospects Game. At 6'2" and 210 pounds, he is difficult to knock off the puck and he uses his size and strength to his advantage in the corners and along the boards. Eager is a power forward who drives hard to the net to create scoring chances. He is also a punishing body checker who excels in a physical game.

24. TORONTO • **ALEXANDER STEEN** • C • The son of former Winnipeg Jets captain Thomas Steen, Alexander Steen is a smooth skater with good speed and quickness. He is a very talented all-around player who displays natural leadership qualities and a winning attitude. Steen served as captain of the Swedish team at the 2002 Under-18 Championships. He has strong puck-handling skills and is a creative playmaker with a good shot.

25. CAROLINA • **CAM WARD** • G • A well-balanced and agile goaltender, Cam Ward plays a classic butterfly style and effectively uses the paddle-down style in scramble situations and wraparounds. He plays the angles well, staying square to the shooter, and has very good rebound control. Ward has a good glove hand and quick feet, and he sees the puck well in traffic. He is confident and maintains a cool demeanor.

26. DALLAS • **MARTIN VAGNER** • D • A very good skater both backwards and forwards, Martin Vagner has good mobility and acceleration. He sees the ice well and carries the puck with confidence. Vagner adapted well to the North American game with Hull in 2001-02 and was named to the QMJHL All-Rookie Team. He is a creative offensive playmaker with solid passing skills and an accurate wrist shot. He is an intelligent player with good hockey sense.

27. SAN JOSE • **MIKE MORRIS** • RW • Selected directly out of high school in Massachusetts, Mike Morris served as team captain at St. Sebastien's in 2002 and was named MVP of the Independent School League. He is committed to attend Northeastern University. Morris is a quick and agile skater with a good change of pace. He sees the ice well and is a clever playmaker. Morris has strong puck-handling skills and is very confident with the puck. He is a hard worker.

28. COLORADO • **JONAS JOHANSSON** • W • Rated as an intelligent player with good hockey sense, Jonas Johansson recovered well from a midseason shoulder injury. A creative player who handles the puck well, Johansson spent most of the 2001-02 season in the Swedish junior league but also saw time in the Elite League. He scored five goals in eight games at the World Under-18 Championships. Johansson makes quick decisions and his passing skills are very good.

29. BOSTON • **HANNU TOIVONEN** • G • A big goaltender (6'2" and 191 pounds) with very good mobility and quickness, Hannu Toivonen is a solid competitor with good composure. He prefers to play a butterfly style and cover the top of the net with his body. Toivonen is technically sound with quick reflexes and smooth movement. He is very focused and consistent.

30. ATLANTA • **JIM SLATER** • C • The son of former NFL defensive lineman Bill Slater, Jim Slater brings a high level of intensity to his game. At 6' and 190 pounds, he is thought to be small for an NHL center, but he is a very good skater with tremendous acceleration and impressive breakaway speed. Slater has a quick and accurate shot and anticipates the play well. He is an effective, gritty two-way player.

2002 Entry Draft

Transferred draft choice notation:
Example: Col.-Ana. represents a draft choice transferred **from** Colorado **to** Anaheim.

Players selected first through tenth in the 2002 NHL Entry Draft (All rows left to right):
Top row: 1. Rick Nash, LW, Columbus; 2. Kari Lehtonen, G, Atlanta.
Second row: 3. Jay Bouwmeester, D, Florida; 4. Joni Pitkanen, D, Philadelphia.
Third row: 5. Ryan Whitney, D, Pittsburgh. 6. Scottie Upshall, RW, Nashville.
Fourth row: 7. Joffrey Lupul, C, Anaheim; 8. Pierre-Marc Bouchard, C, Minnesota.
Fifth row: 9. Petr Taticek, C, Florida; 10. Eric Nystrom, LW, Calgary.

Pick	Player	Claimed By	Amateur Club	Position
FIRST ROUND				
1.	NASH, Rick	Fla.-CBJ	London	LW
2.	LEHTONEN, Kari	Atl.	Jokerit	G
3.	BOUWMEESTER, Jay	CBJ-Fla.	Medicine Hat	D
4.	PITKANEN, Joni	T.B.-Phi.	Karpat	D
5.	WHITNEY, Ryan	Pit.	Boston University	D
6.	UPSHALL, Scottie	Nsh.	Kamloops	RW
7.	LUPUL, Joffrey	Ana.	Medicine Hat	C/RW
8.	BOUCHARD, Pierre-Marc	Min.	Chicoutimi	C
9.	TATICEK, Petr	Cgy.-Fla.	Sault Ste. Marie	C
10.	NYSTROM, Eric	NYR-Fla.-Cgy.	U. of Michigan	LW
11.	BALLARD, Keith	Buf.	U. of Minnesota	D
12.	EMINGER, Steve	Wsh.	Kitchener	D
13.	SEMIN, Alexander	Dal.-Wsh.	Chelyabinsk	LW
14.	HIGGINS, Christopher	Edm.-Mtl.	Yale University	C
15.	NIINIMAKI, Jesse	Mtl.-Edm.	Ilves	C
16.	KLEPIS, Jakub	Ott.	Portland	C
17.	GORDON, Boyd	Van.-Wsh.	Red Deer	RW
18.	GREBESHKOV, Denis	L.A	Yaroslavl	D
19.	KOREIS, Jakub	Phx.	Plzen	C
20.	PAILLE, Daniel	N.J.-Dal.-CBJ-Buf.	Guelph	LW
21.	BABCHUK, Anton	Chi.	Elektrostal	D
22.	BERGENHEIM, Sean	NYI	Jokerit	W
23.	EAGER, Ben	St.L.-Phx.	Oshawa	LW
24.	STEEN, Alexander	Tor.	Vastra Frolunda	C
25.	WARD, Cam	Car.	Red Deer	G
26.	VAGNER, Martin	Phi.-Wsh.-Dal.	Hull	D
27.	MORRIS, Mike	S.J.	St. Sebastien's	RW
28.	JOHANSSON, Jonas	Col.	HV 71 Jr.	W
29.	TOIVONEN, Hannu	Bos.	HPK Hameenlinna Jr.	G
30.	SLATER, Jim	Det.-Buf.-CBJ-Atl.	Michigan State	C
SECOND ROUND				
31.	DESLAURIERS, Jeff	Atl.-Buf.-Edm.	Chicoutimi	G
32.	VAS, Janos	CBJ-Dal.	Malmo Jr.	W
33.	FALARDEAU, Lee	Fla.-NYR	Michigan State	C
34.	STEPHAN, Tobias	T.B.-Ott.-Phi.-T.B.-Dal.	Chur	G
35.	NEMEC, Ondrej	Pit.	Vsetin	D
36.	STOLL, Jarret	Nsh.-Buf.-Edm.	Kootenay	C
37.	BRENT, Tim	Ana.	St. Michael's	C
38.	HARDING, Josh	Min.	Regina	G
39.	MCCONNELL, Brian	Cgy.	Boston University	C
40.	GLOBKE, Rob	NYR-Fla.	U. of Notre Dame	C/RW
41.	LINDSTROM, Joakim	Buf.-Atl.-CBJ	MoDo	F
42.	HOLTET, Marius	Wsh.-Dal.	Farjestad Jr.	C/W
43.	DALEY, Trevor	Dal.	Sault Ste. Marie	D
44.	GREENE, Matt	Edm.	Green Bay	D
45.	LINHART, Tomas	Mtl.	Pardubice Jr.	D
46.	LENEVEU, David	Phx.	Cornell University	G
47.	KAIGORODOV, Alexei	Ott.	Magnitogorsk	C
48.	SHKOTOV, Alexei	St.L.	Elektrostal	RW
49.	KOLTSOV, Kiril l	Van.	Omsk	D
50.	ANSHAKOV, Sergei	L.A.	HC CSKA Moscow	LW
51.	KADEIKIN, Anton	N.J.	Elektrostal	D
52.	SPANG, Dan	Phx.-Phi.-T.B.-S.J.	Winchester High	D
53.	TALLACKSON, Barry	N.J.	U. of Minnesota	RW
54.	KEITH, Duncan	Chi.	Michigan State	D
55.	GROT, Denis	NYI-T.B.-Wsh.-Van.	Elektrostal	D
56.	YEVSEYEV, Vladislav	St.L.-Bos.	HC CSKA Moscow	LW
57.	STAJAN, Matthew	Tor.	Belleville	C
58.	HUDLER, Jiri	Car.-Det.	Vsetin	C
59.	DAIGNEAULT, Maxime	Phi.-Wsh.	Val d'Or	G
60.	HENRICH, Adam	S.J.-T.B.	Brampton	LW
61.	BOYCHUK, Johnny	Col.	Calgary	D
62.	MIKHNOV, Andrei	Bos.-St.L.	Sudbury	C/LW
63.	FLEISCHMANN, Tomas	Det.	Vitkovice Jr.	C/W
THIRD ROUND				
64.	RYZNAR, Jason	Atl.-N.J.	U. of Michigan	LW
65.	TOLLEFSEN, Ole Kristian	CBJ	Lillehammer	D
66.	KANKO, Petr	L.A.	Kitchener	RW
67.	CAMPBELL, Gregory	Fla.-NYI-Fla.	Plymouth	C/LW
68.	SKINNER, Brett	T.B.-Phi.-Van.	Des Moines	D
69.	CHRISTENSEN, Erik	Pit.	Kamloops	C
70.	CALLAHAN, Joe	Nsh.-Phi.-Phx.	Yale University	D
71.	LEE, Brian	Ana.	Erie	D
72.	ERICKSON, Michael	Min.	U. of Minnesota	RW
73.	BRUST, Barry	Min.	Spokane	G
74.	FORD, Todd	Cgy.-Tor.	Swift Current	G
75.	LUTTINEN, Arttu	NYR-Ott.	HIFK Helsinki Jr.	F
76.	TESSIER, Michael	Buf.	Acadie-Bathurst	LW
77.	WELLAR, Patrick	Wsh.	Portland	D
78.	WAUGH, Geoff	Dal.	Kindersley	D
79.	RADUNSKE, Brock	Edm.	Michigan State	LW
80.	JONES, Matt	Mtl.-Phx.	U. of North Dakota	D
81.	JONASEN, Marcus	Ott.-NYR	Vasteras Jr.	W
82.	ADAMS, John	Van.-Fla.-Atl.-Buf.	Boston College	D
83.	MENSATOR, Lukas	L.A.-Van.	Karlovy Vary Jr.	G
84.	CHVATAL, Marek	Phx.-N.J.	Trinec Jr.	D
85.	NITTEL, Ahren	N.J.	Windsor	LW
86.	FIDLER, Jonas	Chi.-S.J.	Plymouth	RW
87.	NIELSEN, Frans	NYI	Malmo	C
88.	D'AMOUR, Dominic	Buf.-Nsh.-Tor.	Hull	D
89.	TROLIGA, Tomas	St.L.	Spisska Nova Ves	C/W
90.	LOMBARDI, Matthew	Tor.-Cgy.	Victoriaville	C
91.	LANE, Jesse	Car.-Phi.-Car.	Hull	D
92.	KRESTANOVICH, Derek	Phi.-Wsh.	Moose Jaw	C
93.	KOZHEVNIKOV, Alexander	S.J.-Chi.	Krylja Sovetov 2	LW
94.	LUNDBERG, Eric	Col.	Providence College	D
95.	FILPPULA, Valtteri	Bos.-Ana.-Nsh.-Det.	Jokerit Jr.	C
96.	GENOVY, Jeff	Det.-Atl.-CBJ	Des Moines	C

FOURTH ROUND

Pick	Player	Claimed By	Amateur Club	Position
97.	MONYCH, Lance	Atl.-Phx.	Brandon	RW
98.	TKACHENKO, Ivan	CBJ	Yaroslavl	F
99.	LAMBERT, Michael	Fla.-Cgy.-Mtl.	Montreal	C/LW
100.	KAZIONOV, Dmitri	T.B	HC CSKA Moscow	C
101.	FERNHOLM, Daniel	Pit.	Djurgarden Jr.	D
102.	SEGAL, Brandon	Nsh.	Calgary	RW
103.	VIHKO, Joonas	Ana.	HIFK Helsinki	F
104.	ROME, Aaron	Min.-Dal.-Min.-L.A.	Swift Current	D
105.	RUGGERI, Rosario	Cgy.-Phi.	Chicoutimi	D
106.	KOLTSOV, Ivan	NYR-Edm.	Cherepovets 2	D
107.	KALTEVA, Mikko	Col.	Jokerit Jr.	D
108.	HULVA, Jakub	Buf.	Vitkovice Jr.	C/W
109.	DESAUTELS, Jevon	Wsh.	Spokane	LW
110.	IMMONEN, Jarkko	Dal.	Blues Espoo Jr.	C
111.	ALMTORP, Jonas	Edm.	MoDo Jr.	C
112.	ARTEMENKOV, Yuri	Mtl.-Cgy.	Krylja Sovetov 2	W
113.	DOBBEN, Scott	Ott.	Erie	C/LW
114.	LALIBERTE, John	Van.	N.H. Jr. Monarchs	RW
115.	ROONEEM, Mark	L.A.	Kamloops	LW
116.	DWYER, Patrick	Phx.-Atl.	Western Michigan	RW
117.	JANSSEN, Cam	N.J.	Windsor	RW
118.	DVORAK, Petr	Chi.-Wsh.	Havirov Jr.	C
119.	REDLIHS, Jekabs	NYI.-CBJ	NY Apple Core	D
120.	JONSSON, Robin	St.L.	Bofors	D
121.	MAGERS, Marty	Buf.	Omaha	G
122.	TURON, David	Tor.	Havirov	D
123.	Invalid Pick	Edm.		
124.	MANSON, Lane	Car.-St.L.-Atl.	Moose Jaw	D
125.	BJORK, Johan	Ott.	Malmo Jr.	D
126.	BARANOV, Konstantin	Phi.	Togliatti	F
127.	GUENIN, Nathan	Ott.-NYR	Green Bay	D
128.	ELLISON, Matt	S.J.-Chi.	Cowichan	RW
129.	GILBERT, Tom	Col.	Chicago	D
130.	KUBISTA, Jan	Bos.	Pardubice Jr.	RW
131.	BERGGREN, Johan	Det.	Sunne	D

FIFTH ROUND

Pick	Player	Claimed By	Amateur Club	Position
132.	ZEILER, John	Atl.-Phx.	Sioux City	RW
133.	PIRJETA, Lasse	CBJ	Karpat	C
134.	JAAKOLA, Topi	Fla.	Karpat	D
135.	PEARCE, Joseph	T.B.	N.H. Jr. Monarchs	G
136.	SERTICH, Andrew	Pit.	Greenway High	LW
137.	PADDOCK, Cam	Pit.	Kelowna	C
138.	JARRETT, Patrick	Nsh.	Owen Sound	C
139.	NEWBURY, Kris	S.J.	Sarnia	C/LW
140.	DAVIS, George	Ana.	Cape Breton	RW
141.	CETKOVSKY, Jiri	Min.-Mtl.-Cgy.	Zlin Jr.	C
142.	PETER, Emanuel	Cgy.	Kloten	C
143.	WALSH, Mike	NYR	Compuware	LW
144.	FLACHE, Paul	Buf.-Atl.	Brampton	D
145.	GHERSON, Robert	Wsh.	Sarnia	G
146.	BOBROV, Victor	Cgy.	HC CSKA Moscow 2	F
147.	BARARUK, David	Dal.	Moose Jaw	C/LW
148.	FISHER, Glenn	Edm.	Ft. Saskatchewan	G
149.	PAHLSSON, Markus	Mtl.-NYI	Morrum	W
150.	HOOTON, Brock	Ott.	Quesnel	C
151.	MCVICAR, Rob	Van.	Brandon	G
152.	HOGEBOOM, Greg	L.A.	Miami University	RW
153.	HAMERLIK, Peter	Phx.-Bos.	Kingston	G
154.	REDLIHS , Krisjanis	N.J.	Metalurgs Liepaja	D
155.	BERZINS, Armands	Car.-Min.	Shawinigan	C
156.	WISNIEWSKI, James	Chi.	Plymouth	D
157.	ANDRESEN, Joel	NYI-L.A.	St. Albert	D
158.	BELLISSIMO, Vince	St.L.-Fla.	Topeka	C
159.	PERSSON, Kristofer	Tor.-Cgy.	MoDo Jr.	RW
160.	MANZATO, Daniel	Car.	Victoriaville	G
161.	GRUMET-MORRIS, Dov	Phi.	Harvard	G
162.	DICAIRE, Gerard	S.J.-T.B.	Kootenay	D
163.	WALSH, Tom	S.J.	Deerfield Academy	D
164.	WEIMAN, Tyler	Col.	Tri-City	G
165.	MAISER, Justin	Bos.-St.L.	Boston University	C/LW
166.	KOOPMANS, Logan	Det.	Lethbridge	G

SIXTH ROUND

Pick	Player	Claimed By	Amateur Club	Position
167.	SCHELL, Brad	Atl.	Spokane	C
168.	KONSORADA, Tim	CBJ	Brandon	RW
169.	SWANSON, Jeremy	Fla.	Barrie	D
170.	ATHERTON, P.J.	T.B.	Cedar Rapids	D
171.	GOEPFERT, Robert	Pit.	Cedar Rapids	G
172.	MCKENNA, Mike	Nsh.	St. Lawrence	G
173.	FRITSHAW, Luke	Ana.	Prince Albert	D
174.	AKKANEN, Karri	Car.-T.B.	Ilves Jr.	C/W
175.	FOY, Matt	Min.	Merrimack College	RW
176.	MCELHINNEY, Curtis	Cgy.	Colorado College	G
177.	TAYLOR, Jake	NYR	Green Bay	D
178.	SCHEVJEV , Maxim	Buf.	Elektrostal	C
179.	HAVEL, Marian	Wsh.	Vancouver	C/LW
180.	SIDORENKO, Kirill	Dal.	Kurgan	C
181.	LUOMA, Mikko	Edm.	Tappara	D
182.	DEVEAUX, Andre	Mtl.	Belleville	C
183.	RANGER, Paul	Ott.-T.B.	Oshawa	D
184.	BALASTIK, Jaroslav	Van.-Phi.-CBJ	Zlin	F
185.	MURPHY, Ryan	L.A.	Boston College	RW
186.	PIETRASIAK, Jeff	Phx.	Berkshire	G
187.	JOHANSSON, Eric	N.J.	Tri-City	C
188.	KANTEE, Kevin	Chi.	Jokerit Jr.	D
189.	STONKUS, Alexei	NYI	Elektrostal	D
190.	KING, D.J.	St.L.	Lethbridge	F
191.	WHITE, Ian	Tor.	Swift Current	D
192.	KOROVKIN, Nikita	Car.-Phi.	Kamloops	D
193.	MORMINA, Joey	Phi.	Colgate University	D
194.	HIRSCHOVITS, Kim	S.J.-NYR	HIFK Helsinki	C
195.	CHRISTIE, Taylor	Col.	Bowling Green	D
196.	VUORIO, Mikael	Bos.-Fla.	Lukko Jr.	G
197.	CUDDIHY, James	Det.	Shawinigan	C

SEVENTH ROUND

Pick	Player	Claimed By	Amateur Club	Position
198.	OYSTRICK, Nathan	Atl.	South Surrey	D
199.	MAULDIN, Greg	CBJ	Mass-Amherst	C
200.	YACHMENEV, Denis	Fla.	North Bay	LW
201.	BRUNELLE, Mathieu	T.B.-Phi.	Victoriaville	LW
202.	BARTSCHI, Patrik	Pit.	Kloten	C/W
203.	MORROW, Josh	Nsh.	Tri-City	D
204.	ECKERBLOM, Niklas	Ana.-Min.	Djurgarden Jr.	C
205.	DUFORT, J.F.	Min.-Edm.	Cape Breton	LW
206.	VAN DER GULIK, David	Cgy.	Chilliwack	RW
207.	JOHNSSON, Pierre	NYR-Cgy.	Farjestad Jr.	D
208.	HECL, Radoslav	Buf.	Bratislava	D
209.	LINDLOF, Joni	Wsh.	Tappara Jr.	W
210.	HAMM, Bryan	Dal.	Peterborough	D
211.	MURPHY, Patrick	Edm.	Newmarket	LW
212.	FERLAND, Jonathan	Mtl.	Acadie-Bathurst	RW
213.	NORRENA, Fredrik	Ott.-T.B.	TPS Turku	G
214.	ROY, Marc-Andre	Van.	Baie-Comeau	LW
215.	LYUBUSHIN, Mikhail	L.A	Krylja Sovetov	D
216.	KOUBA, Ladislav	Phx.	Red Deer	LW
217.	CONBOY, Tim	N.J.-Atl.-S.J.	Topeka	D
218.	PIKKARAINEN, Ilkka	N.J.	HIFK Helsinki	RW
219.	KELLERMAN, Tyson	Chi.	North Bay	G
220.	TOPPING, Brad	NYI	Brampton	G
221.	JOHNSON, Jonas	St.L.	Vastra Frolunda	C
222.	MAY, Scott	Tor.	Ohio State	F
223.	KRIKUNOV, IlJa	Van.	Elektrostal	W
224.	TAYLOR, Adam	Car.	Kootenay	C
225.	GOERTZEN, Steven	Phi.-CBJ	Seattle	RW
226.	CRABB, Joseph	S.J.-NYR	Green Bay	RW
227.	STEEVES, Ryan	Col.	Yale University	C/LW
228.	UTKIN, Dmitri	Bos.	Yaroslavl 2	LW
229.	MEECH, Derek	Det.	Red Deer	D

EIGHTH ROUND

Pick	Player	Claimed By	Amateur Club	Position
230.	FRETTER, Colton	Atl.	Chatham	C/RW
231.	KRACIK, Jaroslav	CBJ	Plzen Jr.	C/W
232.	HAFNER, Peter	Fla.	Taft	D
233.	KOSHECHKIN, Vasili	T.B	Togliatti	G
234.	TALBOT, Maxime	Pit.	Hull	C
235.	BETTS, Kaleb	Nsh.	Chilliwack	C
236.	BOLDT, Tyler	Ana.-Atl.	Kamloops	D
237.	BRANDNER, Christoph	Min.	Krefeld	F
238.	MARTTINEN, Jyri	Cgy.	JYP Jyvaskyla	D
239.	LANNON, Ryan	Pit.	Harvard	D
240.	PRUCHA, Petr	NYR	Pardubice	C/W
241.	WIDEMAN, Dennis	Buf.	London	D
242.	IGNATUSHKIN, Igor	Wsh.	Elektrostal	W
243.	MIKKONEN, Tuomas	Dal.	JYP Jyvaskyla	W
244.	HELMINEN, Dwight	Edm.	U. of Michigan	C
245.	MICKA, Tomas	Mtl.-Edm.	Slavia Praha Jr.	W
246.	VAVRA , Josef	Ott.	Vsetin Jr.	LW
247.	VIOLIN, Matt	Van.	Lake Superior	G
248.	PULLIAINEN, Tuukka	L.A	TuTo	C/W
249.	SMITH, Marcus	Phx.	Kitchener	D
250.	GLOVER, Dan	N.J.	Camrose	D
251.	KOSTADINE, Jason	Chi.	Hull	RW
252.	CHABADA, Martin	NYI	Sparta Praha	F
253.	KOIVISTO, Tom	St.L.	Jokerit	D
254.	IMMONEN, Jarkko	Tor.	Assat	C
255.	CRAIG, Ryan	Car.-T.B.	Brandon	C
256.	REID, Darren	Phi.-Car.-T.B.	Medicine Hat	RW
257.	LEVOKARI, Pauli	S.J.-Atl.	HIFK Helsinki	D
258.	SHEMETOV, Sergei	Col.	Elektrostal	W
259.	STASTNY, Yan	Bos.	U. of Notre Dame	C/LW
260.	BEAULIEU, Pierre-Olivier	Det.	Quebec	D

NINTH ROUND

Pick	Player	Claimed By	Amateur Club	Position
261.	CARON, Francois	Atl.-Ana.	Moncton	D
262.	SODERSTROM, Christian	CBJ-Det.	Timra	F
263.	MOZYAKIN, Sergei	Fla.-CBJ	HC CSKA Moscow	F
264.	DAVIS, Matt	T.B.-Col.-Nsh.	Moncton	G
265.	LABROSSE, Dwight	Pit.	Guelph	G
266.	SPENCER, Steven	Nsh.	Swift Current	D
267.	PETROW, Chris	Ana.	Oshawa	D
268.	TYULYAPKIN, Mikhail	Min.	Nizhny Novgorod 2	D
269.	HANNULA, Mika	Cgy.-Min.	Malmo	F
270.	FLYNN, Rob	NYR	Harvard	RW
271.	CIZEK, Martin	Buf.	Slavia Praha Jr.	D
272.	BLOMDAHL, Patric	Wsh.	AIK Solna Jr.	W
273.	HAVERN, Ned	Dal.	Boston College	LW
274.	JOHANSSON, Fredrik	Edm.	Vastra Frolunda Jr.	C
275.	KORNEYEV, Konstantin	Mtl.	Krylja Sovetov 2	D
276.	ATYUSHOV, Vitali	Ott.	Perm	D
277.	NUSSLI, Thomas	Van.	Zug	W
278.	GENS, Matt	Van.	St. Cloud State	D
279.	JAMES, Connor	L.A	U. of Denver	RW
280.	SPENCE, Russell	Phx.	Opaskwayak	C
281.	KINKEL, Bill	N.J.	Kitchener	LW
282.	BURISH, Adam	Chi.	Green Bay	RW
283.	BRAXENHOLM, Per	NYI	Morrum	D
284.	MACMURCHY, Ryan	St.L.	Notre Dame	RW
285.	KRONVALL, Staffan	Tor.	Huddinge	D
286.	GLUKHOV, Alexei	Car.-T.B.	Voskresensk 2	W
287.	TOFFEY, John	Phi.-T.B.	Ohio State	C
288.	HUTCHINS, Michael	S.J.	Des Moines	D
289.	COLLINS, Sean	Col.	U. of New Hampshire	LW
290.	FROLOV, Pavel	Bos.	Nizhny Novgorod 2	F
291.	ERICSSON, Jonathan	Det.	Hasten Jr.	D

First Two Rounds, 2001-1969 Entry/Amateur Drafts

2001

FIRST ROUND

Selection	Claimed By	Amateur Club	
1. KOVALCHUK, Ilya	Atl.	Krylja Sovetov	LW
2. SPEZZA, Jason	NYI-Ott.	Windsor	C
3. SVITOV, Alexander	T.B.	Omsk	C
4. WEISS, Stephen	Fla.	Plymouth	C
5. CHISTOV, Stanislav	Ana.	Omsk	LW
6. KOIVU, Mikko	Min.	TPS Turku	C
7. KOMISAREK, Mike	Mtl.	U. of Michigan	D
8. LECLAIRE, Pascal	CBJ	Halifax	G
9. RUUTU, Tuomo	Chi.	Jokerit	C
10. BLACKBURN, Dan	NYR	Kootenay	G
11. SJOSTROM, Fredrik	Cgy.-Phx.	V. Frolunda Jr.	RW
12. HAMHUIS, Dan	Nsh.	Prince George	D
13. HEMSKY, Ales	Bos.-Edm.	Hull	RW
14. KOBASEW, Chuck	Phx.-Cgy.	Boston College	RW
15. KNYAZEV, Igor	Car.	Spartak	D
16. UMBERGER, R.J.	Van.	Ohio State	C
17. COLAIACOVO, Carlo	Tor.	Erie	D
18. KARLSSON, Jens	L.A.	V. Frolunda Jr.	LW
19. MORRISONN, Shaone	Edm.-Bos.	Kamloops	D
20. GOC, Marcel	S.J.	Schwenningen	C
21. ARMSTRONG, Colby	Pit.	Red Deer	RW
22. NOVOTNY, Jiri	Buf.	Budejovice	C
23. GLEASON, Tim	Phi.-Ott.	Windsor	D
24. KRAJICEK, Lukas	St.L.-N.J.-Fla.	Peterborough	D
25. PEREZHOGIN, Alexander	Wsh.-Mtl.	Omsk 2	RW
26. BACASHIHUA, Jason	Dal.	Chicago Jr.	G
27. WOYWITKA, Jeff	Ott.-Phi.	Red Deer	D
28. FOSTER, Adrian	N.J.	Saskatoon	LW
29. MUNRO, Adam	Det.-Chi.	Erie	G
30. STECKEL, Dave	Col.-L.A.	Ohio State	LW/C

ROUND #2

Selection	Claimed By	Amateur Club	
31. SPILLER, Matthew	NYI-T.B.-Phx.	Seattle	D
32. ROY, Derek	T.B.-Buf.	Kitchener	C
33. SHISHKANOV, Timofei	Atl.-Van.-Nsh.	Spartak	LW
34. WATSON, Greg	Fla.	Prince Albert	C/LW
35. POPOVIC, Mark	Ana.	St. Michael's	D
36. WANVIG, Kyle	Min.	Red Deer	RW
37. MILROY, Duncan	Mtl.	Swift Current	RW
38. JACKMAN, Tim	CBJ	Mankato State	RW
39. PILAR, Karel	Chi.-Tor.	Litvinov	D
40. TUTIN, Fedor	NYR	SKA St.Petersburg	D
41. TARATUKHIN, Andrei	Cgy.-Ana.-Phx.-Cgy.	Omsk 2	C
42. SLOVAK, Tomas	Nsh.	Kosice	D
43. LYNCH, Doug	Bos.-Edm.	Red Deer	D
44. POHANKA, Igor	Phx.-Fla.-N.J.	Prince Albert	C
45. PODLESAK, Martin	Phx.	Lethbridge	LW
46. ZIGOMANIS, Mike	Car.	Kingston	C
47. POLUSHIN, Alexander	Van.-T.B.	Tver	RW
48. PIHLMAN, Tuomas	Van.-Fla.-N.J.	JYP Jyvaskyla	RW
49. CAMMALLERI, Mike	Tor.-L.A.	U. of Michigan	C
50. THORBURN, Chris	Buf.	North Bay	C
51. BEDNAR, Jaroslav	L.A.	HIFK	RW
52. CARON, Ed	Edm.	Phillips-Exeter	LW
53. McLEOD, Kiel	S.J.-Mtl.-CBJ	Kelowna	C
54. WELCH, Noah	Pit.	St. Sebastian's	D
55. POMINVILLE, Jason	Buf.	Shawinigan	RW
56. MEDVEDEV, Andrei	Phi.-Fla.-Cgy.	Spartak Moscow	G
57. McCLEMENT, Jay	St.L.	Brampton	C
58. PAETSCH, Nathan	Wsh.	Moose Jaw	D
59. KEITH, Matt	Dal.-Chi.	Spokane	RW
60. UCHEVATOV, Victor	Ott.-N.J.	Yaroslavl 2	D
61. HOLMQVIST, Andreas	N.J.-Mtl.-Wsh.-T.B.	Hammarby	D
62. GRIGORENKO, Igor	Det.	Samara	RW
63.BUDAJ, Peter	Col.	St. Michael's	G

2000

FIRST ROUND

Selection	Claimed By	Amateur Club	
1. DiPIETRO, Rick	NYI	Boston University	G
2. HEATLEY, Dany	Atl.	U. of Wisconsin	LW
3. GABORIK, Marian	Min.	Dukla Trencin	LW
4. KLESLA, Rostislav	CBJ	Brampton	D
5. TORRES, Raffi	T.B.-NYI	Brampton	LW
6. HARTNELL, Scott	Nsh.	Prince Albert	RW
7. JONSSON, Lars	Bos.	Leksand Jr.	D
8. ALEXEEV, Nikita	NYR-T.B.	Erie	RW
9. KRAHN, Brent	Cgy.	Calgary	G
10. YAKOUBOV, Mikhail	Chi.	Lada Togliatti 2	C
11. VOROBIEV, Pavel	Van.-Chi.	Yaroslavl	RW
12. SMIRNOV, Alexei	Ana.	Tver	LW
13. HAINSEY, Ron	Mtl.	Mass.-Lowell	D
14. NEDOROST, Vaclav	Car.-Col.	Budejovice	C
15. KRYUKOV, Artem	Buf.	Yaroslavl 2	C
16. HOSSA, Marcel	S.J.-Mtl.	Portland	C
17. MIKHNOV, Alexei	Edm.	Yaroslavl 2	W
18. ORPIK, Brooks	Pit.	Boston College	D
19. KOLANOS, Krys	Phx.	Boston College	C
20. FROLOV, Alexander	L.A.	Yaroslavl 2	LW
21. VOLCHENKOV, Anton	Ott.	HC Moscow	D
22. HALE, David	Col.-N.J.	Sioux City	D
23. SMITH, Nathan	Fla.-Van.	Swift Current	C
24. BOYES, Brad	Tor.	Erie	C
25. OTT, Steve	Dal.	Windsor	C
26. SUTHERBY, Brian	Wsh.	Moose Jaw	C
27. SAMUELSSON, Martin	N.J.-Col.-Bos.	MoDo Jr.	W
28. WILLIAMS, Justin	Phi.	Plymouth	RW
29. KRONWALL, Niklas	Det.	Djurgarden	D
30. TAFFE, Jeff	St.L.	U. of Minnesota	C

SECOND ROUND

Selection	Claimed By	Amateur Club	
31. NIKULIN, Ilja	Atl.	Tver	D
32. KURKA, Tomas	CBJ-Col.-Car.	Plymouth	LW
33. SCHULTZ, Nick	Min.	Prince Albert	D
34. ZAINULLIN, Ruslan	T.B.	Kazan	RW
35. WINCHESTER, Brad	NYI-Edm.	U. of Wisconsin	LW
36. WIDING, Daniel	Nsh.	Leksand Jr.	RW
37. HILBERT, Andy	Bos.	U. of Michigan	C
38. KOPECKY, Tomas	NYR-Det.	Dukla Trencin	C
39. LAINE, Teemu	Van.-NYI-N.J.	Jokerit	RW
40. FOSTER, Kurtis	Cgy.	Peterborough	D
41. MAATTA, Tero	Chi-S.J.	Jokerit Jr.	D
42. USTRNUL, Libor	Van.-Atl.	Plymouth	D
43. PETTINGER, Matt	Ana.-Cgy.-Wsh.	Calgary	LW
44. BRYZGALOV, Ilja	Mtl.-Ana.	Lada Togliatti	G
45. CHOUINARD, Mathieu	Ott.	Shawinigan	G
46. STOLL, Jarret	Col.-Cgy.	Kootenay	C
47. AULIN, Jared	Car.-Col.	Kamloops	C
48. DICAIRE, Gerard	Buf.	Seattle	D
49. NORDQVIST, Jonas	S.J.-Chi.	Leksand Jr.	C
50. SOIN, Sergei	Col.	Krylja Sovetov	C
51. VERNARSKY, Kris	Edm.-Tor.	Plymouth	C
52. ENDICOTT, Shane	Pit.	Seattle	C
53. TATARINOV, Alexander	Phx.	Yaroslavl 2	RW
54. LILJA, Andreas	L.A.	Malmo	D
55. VERMETTE, Antoine	Ott.	Victoriaville	C
56. SUGLOBOV, Alexander	N.J.	Yaroslavl 2	C
57. DeMARCHI, Matt	Col.-N.J.	U. of Minnesota	D
58. SAPOZHNIKOV, Vladimir	Fla.	Novokuznetsk 2	D
59. HUML, Ivan	Tor.-Bos.	Langley Jr. A	LW
60. ELLIS, Dan	Dal.	Omaha	G
61. CUTTA, Jakub	Wsh.	Swift Current	D
62. MARTIN, Paul	N.J.	Elk River High	D
63. SAVIELS, Agris	Phi.-Car.-Col.	Owen Sound	D
64. NOVAK, Filip	Det.-NYR	Regina	D
65. MORISSET, David	St.L.	Seattle	RW

1999

FIRST ROUND

Selection	Claimed By	Amateur Club	
1. STEFAN, Patrik	T.B.-Van.-Atl.	Long Beach	C
2. SEDIN, Daniel	Atl.-Van.	MoDo	LW
3. SEDIN, Henrik	Van.	MoDo	C
4. BRENDL, Pavel	Chi.-Van.-T.B.-NYR	Calgary	RW
5. CONNOLLY, Tim	NYI	Erie	C
6. FINLEY, Brian	Nsh.	Barrie	G
7. BEECH, Kris	Wsh.	Calgary	C
8. PYATT, Taylor	L.A.-NYI	Sudbury	LW
9. LUNDMARK, Jamie	Cgy.-NYR	Moose Jaw	C
10. MEZEI, Branislav	Mtl.-NYI	Belleville	D
11. SAPRYKIN, Oleg	NYR-Cgy.	Seattle	C
12. SHVIDKI, Denis	Fla.	Barrie	RW
13. RITA, Jani	Edm.	Jokerit Helsinki	RW
14. JILLSON, Jeff	S.J.	U. of Michigan	D
15. KELMAN, Scott	Ana.-Phx.	Seattle	C
16. TANABE, David	Car.	U. of Wisconsin	D
17. JACKMAN, Barret	St.L.	Regina	D
18. KOLTSOV, Konstantin	Pit.	Cherepovets	LW
19. SAFRONOV, Kirill	Phx.	St. Petersburg	D
20. HEISTEN, Barrett	Buf.	U. of Maine	LW
21. BOYNTON, Nick	Bos.	Ottawa	D
22. OUELLET, Maxime	Phi.	Quebec	G
23. MCCARTHY, Steve	Det.-Chi.	Kootenay	D
24. CEREDA, Luca	Tor.	Ambri	C
25. KULESHOV, Mikhail	Col.	Cherepovets	LW
26. HAVLAT, Martin	Ott.	Trinec	C
27. AHONEN, Ari	N.J.	JyP HT Jr.	G
28. KUDROC, Kristian	Dal.-NYI	Michalovce	D

SECOND ROUND

Selection	Claimed By	Amateur Club	
29. SIVEK, Michal	T.B.-Wsh.	HC Kladno Jr.	C
30. SELLARS, Luke	Atl.	Ottawa	D
31. STEPHENS, Charlie	Van.-Col.-Wsh.	Guelph	F
32. RYAN, Mike	NYI-Dal.	Boston College HS	C
33. ANDERSSON, Jonas	Nsh.	AIK Solna Jr.	RW
34. LUPASCHUK, Ross	Wsh.	Prince Albert	D
35. BARTOVIC, Milan	L.A.-Buf.	Dukla Trencin	RW
36. SEMENOV, Alexei	Edm.	Sudbury	D
37. YONKMAN, Nolan	Wsh.	Kelowna	D
38. CAVANAUGH, Dan	Cgy.	Boston University	F
39. BUTURLIN, Alexander	Mtl.	CSKA Moscow Jr.	LW
40. AULD, Alexander	St.L.-Fla.	North Bay	G
41. SALMELAINEN, Tony	Edm.	HIFK Helsinki	LW
42. COMMODORE, Mike	N.J.	U. of North Dakota	D
43. SHEFER, Andrei	L.A.	Cherepovets	LW
44. LEOPOLD, Jordan	NYR-Ott.-Ana.	U. of Minnesota	D
45. GRENIER, Martin	Fla.-Nsh.-Col.	Quebec	D
46. LEVINSKI, Dmitri	Chi.	Cherepovets	RW
47. KEEFE, Sheldon	S.J.-Det.-T.B.	Barrie	RW
48. LAJEUNESSE, Simon	Ana.-Ott.	Moncton	G
49. LYSAK, Brett	Car.	Regina	C
50. CLOUTHIER, Brett	St.L.-N.J.	Kingston	LW
51. MURLEY, Matt	Pit.	R.P.I.	LW
52. HALL, Adam	Nsh.	Michigan State	RW
53. RALPH, Brad	Phx.	Oshawa	LW
54. HUTCHINSON, Andrew	Col.-Nsh.	Michigan State	D
55. JANIK, Doug	Buf.	U. of Maine	D
56. ZULTEK, Matt	Bos.	Ottawa	LW
57. VAN HOOF, Jeremy	Pit.	Ottawa	D
58. CARKNER, Matt	Phi.-Mtl.	Peterborough	D
59. INMAN, David	Det.-NYR	U. of Notre Dame	C
60. REYNOLDS, Peter	Tor.	London	D
61. HILL, Ed	Col.-Nsh.	Barrie	D
62. SAINOMAA, Teemu	Ott.	Jokerit Helsinki Jr.	LW
63. MOKHOV, Stepan	N.J.-Chi.	Cherepovets	D
64. ZIGOMANIS, Michael	Dal.-Buf.	Kingston	C
65. LASAK, Jan	Nsh.	ZTK Zvolen Jr.	G
66. JANCEVSKI, Dan	St.L.-Dal.	London	D

Selected first overall by Atlanta in 2001, Ilya Kovalchuk (far right) made an immediate impact with the Thrashers last season. Rostislav Klesla (right) was the fourth pick in the 2000 Draft and also spent his first full season in the NHL last year with Columbus.

1998

FIRST ROUND

Selection	Claimed By	Amateur Club	
1. LECAVALIER, Vincent	Fla.-S.J.-T.B.	Rimouski	C
2. LEGWAND, David	T.B.-S.J.-Nsh.	Plymouth	C
3. STUART, Brad	Nsh.-S.J.	Regina	D
4. ALLEN, Bryan	Van.	Oshawa	D
5. VISHNEVSKI, Vitaly	Ana.	Torpedo-2 Yaroslavl	D
6. FATA, Rico	Cgy.	London	C
7. MALHOTRA, Manny	NYR	Guelph	C
8. BELL, Mark	Tor.-Chi.	Ottawa	LW
9. RUPP, Mike	NYI	Erie	LW
10. ANTROPOV, Nik	Chi.-Tor.	Ust-Kamenogorsk	C
11. HEEREMA, Jeff	Car.	Sarnia	RW
12. TANGUAY, Alex	S.J.-Col.	Halifax	C
13. HENRICH, Michael	Edm.	Barrie	RW
14. DESROCHERS, Patrick	Phx.	Sarnia	G
15. CHOUINARD, Mathieu	Ott.	Shawinigan	G
16. CHOUINARD, Eric	Mtl.	Quebec	C
17. SKOULA, Martin	L.A.-Col.	Barrie	D
18. KALININ, Dmitri	Buf.	Traktor Chelyabinsk	D
19. REGEHR, Darcy	Bos.-Col.	Kamloops	D
20. PARKER, Scott	Wsh.-Col.	Kelowna	D
21. BIRON, Mathieu	Col.-L.A.	Shawinigan	D
22. GAGNE, Simon	Phi.-T.B.-Phi.	Quebec	C
23. KRAFT, Milan	T.B.	Keramika Plzen Jr.	C
24. BACKMAN, Christian	St.L.	Vastra Frolunda Jr.	D
25. FISCHER, Jiri	Det.	Hull	D
26. VAN RYN, Mike	N.J.	U. of Michigan	D
27. GOMEZ, Scott	Dal.-N.J.	Tri-City	C

SECOND ROUND

Selection	Claimed By	Amateur Club	
28. ABID, Ramzi	T.B.-Col.	Chicoutimi	LW
29. CHEECHOO, Jonathan	Nsh.-S.J.	Belleville	RW
30. ROSSITER, Kyle	Fla.	Spokane	D
31. CHUBAROV, Artem	Van.	Dynamo Moscow	C
32. PEAT, Stephen	Ana.	Red Deer	D
33. BETTS, Blair	Cgy.	Prince George	C
34. PETERS, Andrew	NYR-Buf.	Oshawa	LW
35. SVOBODA, Petr	Tor.	Havlickuv Brod	D
36. NIELSON, Chris	NYI	Calgary	C
37. BERGLUND, Christian	N.J.	Farjestad Karlstad Jr.	C
38. SAUVE, Philippe	Chi-Col.	Rimouski	G
39. ERSKINE, John	Car.-N.J.-Dal.	London	D
40. COPLEY, Randy	NYR	Cape Breton	RW
41. LINNIK, Maxim	S.J.-Det.-St.L.	St. Thomas Jr. B	D
42. BECKETT, Jason	Edm.-Phi.	Seattle	D
43. VAANANEN, Ossi	Phx.	Jokerit Helsinki Jr.	D
44. FISHER, Mike	Ott.	Sudbury	C
45. RIBEIRO, Mike	Mtl.	Rouyn-Noranda	C
46. PAPINEAU, Justin	L.A.	Belleville	C
47. MILLEY, Norm	Buf.	Sudbury	RW
48. GIRARD, Jonathan	Bos.	Laval	D
49. CRUZ, Jomar	Wsh.	Brandon	G
50. KRISTEK, Jaroslav	Col.-S.J.-Buf.	ZPS Zlin	RW
51. FORBES, Ian	Phi.	Guelph	D
52. ALLEN, Bobby	Bos.	Boston College	D
53. MOORE, Steve	Col.	Harvard	C
54. ZEVAKHIN, Alexander	Pit.	CSKA Moscow	RW
55. BARNES, Ryan	St.L-Det.	Sudbury	LW
56. VALTONEN, Tomek	Det.	Ilves Tampere Jr.	LW
57. BOUCK, Tyler	N.J.-Dal.	Prince George	RW
58. BALA, Chris	Dal.-Phi.-Ott.	Harvard	LW

Ottawa chose Bryan Berard first overall in the 1995 Entry Draft, and later swapped him to the Islanders for #2 pick Wade Redden (above). Both players began their NHL careers in 1996-97.

1997

FIRST ROUND

Selection	Claimed By	Amateur Club	
1. THORNTON, Joe	Bos.	Sault Ste. Marie	C
2. MARLEAU, Patrick	S.J.	Seattle	C
3. JOKINEN, Olli	L.A.	HIFK Helsinki	C
4. LUONGO, Roberto	Tor.-NYI	Val-d'Or	G
5. BREWER, Eric	NYI	Prince George	D
6. TKACZUK, Daniel	Cgy.	Barrie	C
7. MARA, Paul	T.B.	Sudbury	D
8. SAMSONOV, Sergei	Car.-Bos.	Detroit	LW
9. BOYNTON, Nick	Wsh.	Ottawa	D
10. FERENCE, Brad	Van.	Spokane	D
11. WARD, Jason	Mtl.	Erie	C
12. HOSSA, Marian	Ott.	Dukla Trencin	RW
13. CLEARY, Daniel	Chi.	Belleville	LW
14. RIESEN, Michel	Edm.	Biel-Bienne	LW
15. ZULTEK, Matt	St.L-Edm.-St.L.-L.A.	Ottawa	LW
16. JONES, Ty	Pho.-Chi.	Spokane	RW
17. DOME, Robert	Pit.	Las Vegas	RW
18. HOLMQVIST, Mikael	Ana.	Djurgarden	C
19. CHERNESKI, Stefan	NYR	Brandon	RW
20. BROWN, Mike	Fla.	Red Deer	C
21. NORONEN, Mika	Buf.	Tappara Tampere	G
22. TSELIOS, Nikos	Det.-Car.	Belleville	D
23. HANNAN, Scott	Phi.-Car.-S.J.	Kelowna	D
24. DAMPHOUSSE, J-F	N.J.	Moncton	G
25. MORROW, Brenden	Dal.	Portland	LW
26. GRIMES, Kevin	Col.	Kingston	D

SECOND ROUND

Selection	Claimed By	Amateur Club	
27. CLYMER, Ben	Bos.	U. of Minnesota	D
28. DEFAUW, Brad	S.J.-Car.	U. of North Dakota	LW
29. BARNEY, Scott	L.A.	Peterborough	C
30. PELLETIER, Jean-Marc	Tor.-Phi.	Cornell University	G
31. ZEHR, Jeff	NYI	Windsor	LW
32. LINDSAY, Evan	Cgy.	Prince Albert	G
33. KOS, Kyle	T.B.	Red Deer	D
34. BONNI, Ryan	Car.-Van.	Saskatoon	D
35. FORTIN, Jean-Francois	Wsh.	Sherbrooke	D
36. DRUKEN, Harold	Van.	Detroit	LW
37. BAUMGARTNER, Gregor	Mtl.	Laval	C
38. GRON, Stanislav	Ott.-N.J.	Slovan Bratislava Jr.	C
39. REICH, Jeremy	Chi.	Seattle	C
40. RENNETTE, Tyler	St.L.	North Bay	C
41. DOVIGI, Patrick	Edm.	Erie	G
42. TRIPP, John	St.L.-Cgy.	Oshawa	RW
43. GUSTAFSSON, Juha	Pho.	Kiekko-Espoo Jr.	D
44. GAFFANEY, Brian	Pit.	North Iowa Jr. A	D
45. BALMOCHNYKH, Maxim	Ana.	Lada Togliatti	LW
46. JARVIS, Wes	NYR	Kitchener	D
47. HUSELIUS, Kristian	Fla.	Farjestad Karlstad	LW
48. TALLINDER, Henrik	Buf.	AIK Solna	D
49. BUTSAYEV, Yuri	Det.	Lada Togliatti	C
50. KAVANAGH, Pat	Phi.	Peterborough	RW
51. KOKOREV, Dimitri	N.J.-Car.-Cgy.	Dynamo-2 Moscow	D
52. LYASHENKO, Roman	Dal.	Torpedo Yaroslavl	C
53. BELAK, Graham	Col.	Edmonton	D

1995

FIRST ROUND

Selection	Claimed By	Amateur Club	
1. BERARD, Bryan	Ott.	Detroit	D
2. REDDEN, Wade	NYI	Brandon	D
3. BERG, Aki	L.A.	Kiekko-67 Turku	D
4. KILGER, Chad	Ana.	Kingston	C
5. LANGKOW, Daymond	T.B.	Tri-City	C
6. KELLY, Steve	Edm.	Prince Albert	C
7. DOAN, Shane	Wpg.	Kamloops	RW
8. RYAN, Terry	Mtl.	Tri-City	LW
9. McLAREN, Kyle	Hfd.-Bos.	Tacoma	D
10. DVORAK, Radek	Fla.	HC Ceske Budejovice	W
11. IGINLA, Jarome	Dal.	Kamloops	C
12. RIIHIJARVI, Teemu	S.J.	Kiekko-Espoo Jr.	LW
13. GIGUERE, J-Sebastien	NYR-Hfd.	Halifax	G
14. McKEE, Jay	Van.-Buf.	Niagara Falls	D
15. WARE, Jeff	Tor.	Oshawa	D
16. BIRON, Martin	Buf.	Beauport	G
17. CHURCH, Brad	Wsh.	Prince Albert	LW
18. SYKORA, Petr	N.J.	Detroit	C
19. NABOKOV, Dmitri	Chi.	Krylja Sovetov	C
20. GAUTHIER, Denis	Cgy.	Drummondville	D
21. BROWN, Sean	Bos.	Belleville	D
22. BOUCHER, Brian	Phi.	Tri-City	G
23. ELOMO, Miika	St.L.-Wsh.	Kiekko-67 Turku	LW
24. MOROZOV, Aleksey	Pit.	Krylja Sovetov	RW
25. DENIS, Marc	Col.	Chicoutimi	G
26. KUZNETSOV, Maxim	Det.	Dynamo Moscow	D

SECOND ROUND

Selection	Claimed By	Amateur Club	
27. MORO, Marc	Ott.	Kingston	D
28. HLAVAC, Jan	NYI	Sparta Praha	LW
29. WESENBERG, Brian	Ana.	Guelph	RW
30. McBAIN, Mike	T.B.	Red Deer	D
31. LARAQUE, Georges	Edm.	St-Jean	RW
32. CHOUINARD, Marc	Wpg.	Beauport	C
33. MacLEAN, Don	L.A.	Beauport	C
34. DOIG, Jason	Mtl.-Wpg.	Laval	D
35. FEDOTOV, Sergei	Hfd.	Dynamo Moscow	D
36. MacDONALD, Aaron	Fla.	Swift Current	G
37. COTE, Patrick	Dal.	Beauport	LW
38. ROED, Peter	S.J.	White Bear Lake HS	C
39. DUBE, Christian	NYR	Sherbrooke	C
40. McALLISTER, Chris	Van.	Saskatoon	D
41. SMITH, D.J.	Tor.-NYI	Windsor	D
42. DUTIAUME, Mark	Buf.	Brandon	LW
43. HAY, Dwayne	Wsh.	Guelph	LW
44. PERROTT, Nathan	N.J.	Oshawa	RW
45. LAFLAMME, Christian	Chi.	Beauport	D
46. SMIRNOV, Pavel	Cgy.	Molot Perm	RW/C
47. SCHAFER, Paxton	Bos.	Medicine Hat	G
48. KENNY, Shane	Phi.	Owen Sound	D
49. HECHT, Jochen	St.L.	Mannheim	C
50. ROSA, Pavel	Pit.-L.A.	Litvinov Jr.	RW
51. BEAUDOIN, Nic	Col.	Detroit	LW
52. AUDET, Philippe	Det.	Granby	LW

1996

FIRST ROUND

Selection	Claimed By	Amateur Club	
1. PHILLIPS, Chris	Ott.	Prince Albert	D
2. ZYUZIN, Andrei	S.J.	Salavat Yulayev Ufa	D
3. DUMONT, J-P	NYI	Val d'Or	RW
4. VOLCHKOV, Alexander	L.A.-Wsh.	Barrie	C
5. JACKMAN, Richard	Dal.	Sault Ste. Marie	D
6. DEVEREAUX, Boyd	Edm.	Kitchener	C
7. RASMUSSEN, Erik	Buf.	U. of Minnesota	C
8. AITKEN, Johnathan	Hfd.-Bos.	Medicine Hat	D
9. SALEI, Ruslan	Ana.	Las Vegas	D
10. WARD, Lance	N.J.	Red Deer	D
11. FOCHT, Dan	Pho.	Tri-City	D
12. HOLDEN, Josh	Van.	Regina	C
13. MORRIS, Derek	Cgy.	Regina	D
14. REASONER, Marty	St.L.-Edm.-St.L.	Boston College	C
15. ZUBRUS, Dainius	Tor.-Phi.	Pembroke	RW
16. LAROCQUE, Mario	T.B.	Hull	D
17. SVEJKOVSKY, Jaroslav	Wsh.	Tri-City	RW
18. HIGGINS, Matt	Mtl.	Moose Jaw	C
19. DESCOTEAUX, Matthieu	Bos.-Edm.	Shawinigan	D
20. NILSON, Marcus	Fla.	Djurgarden	C
21. STURM, Marco	Chi.-S.J.	Landshut	C
22. BROWN, Jeff	NYR	Sarnia	D
23. HILLIER, Craig	Pit.	Ottawa	G
24. BRIERE, Daniel	Phi.-Pho.	Drummondville	C
25. RATCHUK, Peter	Col.	Shattuck St. Mary's	D
26. WALLIN, Jesse	Det.	Red Deer	D

SECOND ROUND

Selection	Claimed By	Amateur Club	
27. SARICH, Cory	Ott.-St.L.-Buf.	Saskatoon	D
28. SKRBEK, Pavel	S.J.-N.J.-Pit.	HC Kladno	D
29. LACOUTURE, Dan	NYI	Springfield	LW
30. GREEN, Josh	L.A.	Medicine Hat	LW
31. ROYER, Remi	Dal.-Pho.-S.J.-Chi.	St-Hyacinthe	D
32. HAJT, Chris	Edm.	Guelph	D
33. VAN OENE, Darren	Buf.	Brandon	LW
34. WASYLUK, Trevor	Hfd.	Medicine Hat	LW
35. CULLEN, Matt	Ana.	St. Cloud State	C
36. POSMYK, Marek	N.J.-Tor.	Dukla Jihlava	D
37. CISAR, Marian	Pho.-L.A.	Slovan Bratislava	W
38. MASON, Wes	Van.-N.J.	Sarnia	LW
39. BRIGLEY, Travis	Cgy.	Lethbridge	LW
40. BEGIN, Steve	St.L.-Cgy.	Val d'Or	C
41. DEWOLF, Josh	Tor.-Pit.-N.J.	Twin Cities	D
42. PAUL, Jeff	T.B.-Chi.	Niagara Falls	D
43. BULIS, Jan	Wsh.	Barrie	C
44. GARON, Mathieu	Mtl.	Victoriaville	G
45. KUSTER, Henry	Bos.	Medicine Hat	RW
46. PETERS, Geoff	Fla.-S.J.-Chi.	Niagara Falls	C
47. DAGENAIS, Pierre	Chi.-T.B.-N.J.	Moncton	LW
48. GONEAU, Daniel	NYR	Granby	LW
49. WHITE, Colin	Pit.-N.J.	Hull	D
50. LARIVEE, Francis	Phi.-Tor.	Laval	G
51. BABENKO, Yuri	Col.	Krylja Sovetov	C
52. MILLER, Aren	Det.	Spokane	G

1994

FIRST ROUND

Selection	Claimed By	Amateur Club	
1. JOVANOVSKI, Ed	Fla.	Windsor	D
2. TVERDOVSKY, Oleg	Ana.	Soviet Wings	D
3. BONK, Radek	Ott.	Las Vegas	C
4. BONSIGNORE, Jason	Wpg.-Edm.	Niagara Falls	C
5. O'NEILL, Jeff	Hfd.	Guelph	C
6. SMYTH, Ryan	Edm.	Moose Jaw	LW
7. STORR, Jamie	L.A.	Owen Sound	G
8. WIEMER, Jason	T.B.	Portland	LW
9. LINDROS, Brett	Que.-NYI	Kingston	RW
10. BAUMGARTNER, Nolan	Phi.-Que.-Tor.-Wsh.	Kamloops	D
11. FRIESEN, Jeff	S.J.	Regina	LW
12. BELAK, Wade	NYI-Que.	Saskatoon	D
13. OHLUND, Mattias	Van.	Pitea	D
14. MOREAU, Ethan	Chi.	Niagara Falls	LW
15. KHARLAMOV, Alexander	Wsh.	CSKA Moscow	C
16. FICHAUD, Eric	St.L.-Wsh.-Tor.	Chicoutimi	G
17. PRIMEAU, Wayne	Buf.	Owen Sound	C
18. BROWN, Brad	Mtl.	North Bay	C
19. DINGMAN, Chris	Cgy.	Brandon	LW
20. BOTTERILL, Jason	Dal.	U. of Michigan	LW
21. RYABCHIKOV, Evgeni	Bos.	Molot Perm	G
22. KEALTY, Jeffrey	Tor.-Que.	Catholic Memorial	D
23. GOLUBOVSKY, Yan	Det.	CSKA Jr. Moscow	D
24. WELLS, Chris	Pit.	Seattle	G
25. SHARIFIJANOV, Vadim	N.J.	Salavat Yulayev Ufa	RW
26. CLOUTIER, Dan	NYR	Sault Ste. Marie	G

SECOND ROUND

Selection	Claimed By	Amateur Club	
27. WARRENER, Rhett	Fla.	Saskatoon	D
28. DAVIDSSON, Johan	Ana.	HV 71 Jonkoping	C
29. NECKAR, Stan	Ott.	Ceske Budejovice	D
30. QUINT, Deron	Wpg.	Seattle	D
31. PODOLLAN, Jason	Hfd.-Fla.	Spokane	C
32. WATT, Mike	Edm.	Stratford Jr. B	LW
33. JOHNSON, Matt	L.A.	Peterborough	LW
34. CLOUTIER, Colin	T.B.	Brandon	C
35. MARHA, Josef	Que.	Dukla Jihlava	C
36. JOHNSON, Ryan	Phi.-Fla.	Thunder Bay Jr. A	C
37. NIKOLOV, Angel	S.J.	Litvinov	D
38. HOLLAND, Jason	NYI	Kamloops	D
39. GORDON, Robb	Van.	Powell River Jr. A	C
40. LEROUX, Jean-Yves	Chi.	Beauport	LW
41. CHERREY, Scott	Wsh.	North Bay	LW
42. SCATCHARD, Dave	St.L.-Van.	Portland	C
43. BROWN, Curtis	Buf.	Moose Jaw	C
44. THEODORE, Jose	Mtl.	St-Jean	G
45. RYABYKIN, Dmitri	Cgy.	Dynamo-2 Moscow	D
46. JINMAN, Lee	Dal.	North Bay	C
47. GONEAU, Daniel	Bos.	Laval	LW
48. HAGGERTY, Sean	Tor.	Detroit	LW
49. DANDENAULT, Mathieu	Det.	Sherbrooke	RW
50. PARK, Richard	Pit.	Belleville	C
51. ELIAS, Patrik	N.J.	Kladno	LW
52. VERCIK, Rudolf	NYR	Slovan Bratislava	LW

1993

FIRST ROUND

Selection	Claimed By	Amateur Club	
1. DAIGLE, Alexandre	Ott.	Victoriaville	C
2. PRONGER, Chris	S.J.-Hfd.	Peterborough	D
3. GRATTON, Chris	T.B.	Kingston	C
4. KARIYA, Paul	Ana.	University of Maine	LW
5. NIEDERMAYER, Rob	Fla.	Medicine Hat	C
6. KOZLOV, Viktor	Hfd.-S.J.	Dynamo Moscow	LW
7. ARNOTT, Jason	Edm.	Oshawa	C
8. SUNDSTROM, Niklas	NYR	MoDo	LW
9. HARVEY, Todd	Dal.	Detroit	C
10. THIBAULT, Jocelyn	Phi.-Que.	Sherbrooke	G
11. WITT, Brendan	St.L.-Wsh.	Seattle	D
12. JONSSON, Kenny	Buf.-Tor.	Rogle Angelholm	D
13. PEDERSON, Denis	N.J.	Prince Albert	C
14. DEADMARSH, Adam	NYI-Que.	Portland	C
15. LINDGREN, Mats	Wpg.	Skelleftea	C
16. STAJDUHAR, Nick	L.A.-Edm.	London	D
17. ALLISON, Jason	Wsh.	London	C
18. MATTSSON, Jesper	Cgy.	Malmo	C
19. WILSON, Landon	Tor.	Dubuque Jr. A	RW
20. WILSON, Mike	Van.	Sudbury	D
21. KOIVU, Saku	Mtl.	TPS Turku	C
22. ERIKSSON, Anders	Det.	MoDo	D
23. BERTUZZI, Todd	Que.-NYI	Guelph	C
24. LECOMPTE, Eric	Chi.	Hull	LW
25. ADAMS, Kevyn	Bos.	Miami-Ohio	C
26. BERGKVIST, Stefan	Pit.	Leksand	D

SECOND ROUND

Selection	Claimed By	Amateur Club	
27. BICANEK, Radim	Ott.	Dukla Jihlava	D
28. DONOVAN, Shean	S.J.	Ottawa	RW
29. MOSS, Tyler	T.B.	Kingston	G
30. TSULYGIN, Nikolai	Ana.	Salavat Yulayev Ufa	D
31. LANGKOW, Scott	Fla.-Wpg.	Portland	G
32. PANDOLFO, Jay	Hfd.-N.J.	Boston University	LW
33. VYBORNY, David	Edm.	Sparta Praha	C
34. SOROCHAN, Lee	NYR	Lethbridge	D
35. LANGENBRUNNER, Jamie	Dal.	Cloquet	C
36. NIINIMAA, Janne	Phi.	Karpat Oulu	D
37. BETS, Maxim	St.L.	Spokane	LW
38. TSYGUROV, Denis	Buf.	Lada Togliatti	D
39. MORRISON, Brendan	N.J.	Penticton Jr. A	C
40. McCABE, Bryan	NYI	Spokane	D
41. WEEKES, Kevin	Wpg.-Fla.	Owen Sound	G
42. TOPOROWSKI, Shayne	L.A.	Prince Albert	RW
43. BUDAYEV, Alexei	Wsh.-Wpg.	Kristall Elektrostal	C
44. ALLISON, Jamie	Cgy.	Detroit	D
45. KROUPA, Vlastimil	Tor.-Hfd.-S.J.	Chemopetrol Litvinov	D
46. GIRARD, Rick	Van.	Swift Current	C
47. FITZPATRICK, Rory	Mtl.	Sudbury	D
48. COLEMAN, Jon	Det.	Andover Academy	D
49. BUCKBERGER, Ashley	Que.	Swift Current	RW
50. MANLOW, Eric	Chi.	Kitchener	C
51. ALVEY, Matt	Bos.	Springfield Jr. B	RW
52. PITTIS, Domenic	Pit.	Lethbridge	C

1992

FIRST ROUND

Selection	Claimed By	Amateur Club	
1. HAMRLIK, Roman	T.B.	ZPS Zlin	D
2. YASHIN, Alexei	Ott.	Dynamo Moscow	C
3. RATHJE, Mike	S.J.	Medicine Hat	D
4. WARRINER, Todd	Que.	Windsor	LW
5. KASPARAITIS, Darius	Tor.-NYI	Dynamo Moscow	D
6. STILLMAN, Cory	Cgy.	Windsor	C
7. SITTLER, Ryan	Phi.	Nichols High	LW
8. CONVERY, Brandon	NYI-Tor.	Sudbury	C
9. PETROVICKY, Robert	Hfd.	Dukla Trencin	C
10. NAZAROV, Andrei	Min.-S.J.	Dynamo Moscow	LW
11. COOPER, David	Buf.	Medicine Hat	D
12. KRIVOKRASOV, Sergei	Wpg.-Chi.	CSKA Moscow	RW
13. HULBIG, Joe	Edm.	St. Sebastian's	LW
14. GONCHAR, Sergei	St.L.-Wsh.	Chelyabinsk	D
15. BOWEN, Jason	L.A.-Pit.-Phi.	Tri-City	LW
16. KVARTALNOV, Dmitri	Bos.	San Diego	LW
17. BAUTIN, Sergei	Chi.-Wpg.	Dynamo Moscow	D
18. SMITH, Jason	N.J.	Regina	D
19. STRAKA, Martin	Pit.	Skoda Plzen	C
20. WILKIE, David	Mtl.	Kamloops	D
21. POLASEK, Libor	Van.	TJ Vitkovice	C
22. BOWEN, Curtis	Det.	Ottawa	LW
23. MARSHALL, Grant	Wsh.-Tor.	Ottawa	RW
24. FERRARO, Peter	NYR	Waterloo Jr. A	C

SECOND ROUND

Selection	Claimed By	Amateur Club	
25. PENNEY, Chad	Ott.	North Bay	LW
26. BANNISTER, Drew	T.B.	Sault Ste. Marie	D
27. MIRONOV, Boris	S.J.-Chi.-Wpg.	CSKA Moscow	D
28. BROUSSEAU, Paul	Que.	Hull	RW
29. GRONMAN, Tuomas	Tor.-Que.	Tacoma	D
30. O'SULLIVAN, Chris	Cgy.	Catholic Memorial HS	D
31. METLYUK, Denis	Phi.	Lada Togliatti	C
32. CAREY, Jim	NYI-Tor.-Wsh.	Catholic Memorial	G
33. BURE, Valeri	Hfd.-Mtl.	Spokane	LW
34. VARVIO, Jarkko	Min.	HPK	RW
35. CIERNY, Jozef	Buf.	ZTK Zvolen	LW
36. SHANTZ, Jeff	Wpg.-Chi.	Regina	C
37. REICHEL, Martin	Edm.	Freiburg	RW
38. KOROLEV, Igor	St.L.	Dynamo Moscow	RW
39. HOCKING, Justin	L.A.	Spokane	D
40. PECA, Michael	Bos.-Van.	Ottawa	C
41. KLIMOVICH, Sergei	Chi.	Dynamo Moscow	C
42. BRYLIN, Sergei	N.J.	CSKA Moscow	C
43. HUSSEY, Marc	Pit.	Moose Jaw	D
44. CORPSE, Keli	Mtl.	Kingston	C
45. FOUNTAIN, Mike	Van.	Oshawa	G
46. McCARTY, Darren	Det.	Belleville	RW
47. NIKOLISHIN, Andrei	Wsh.-Hfd.	Dynamo Moscow	LW
48. NORSTROM, Mattias	NYR	AIK Solna	D

1991

FIRST ROUND

Selection	Claimed By	Amateur Club	
1. LINDROS, Eric	Que.	Oshawa	C
2. FALLOON, Pat	S.J.	Spokane	RW
3. NIEDERMAYER, Scott	Tor.-N.J.	Kamloops	D
4. LACHANCE, Scott	NYI	Boston University	D
5. WARD, Aaron	Wpg.	U. of Michigan	D
6. FORSBERG, Peter	Phi.	MoDo	C
7. STOJANOV, Alek	Van.	Hamilton	RW
8. MATVICHUK, Richard	Min.	Saskatoon	D
9. POULIN, Patrick	Hfd.	St. Hyacinthe	LW
10. LAPOINTE, Martin	Det.	Laval	RW
11. ROLSTON, Brian	N.J.	Detroit Comp. Jr. A	C
12. WRIGHT, Tyler	Edm.	Swift Current	C
13. BOUCHER, Phillipe	Buf.	Granby	D
14. PEAKE, Pat	Wsh.	Detroit	C
15. KOVALEV, Alexei	NYR	Dynamo Moscow	RW
16. NASLUND, Markus	Pit.	MoDo	RW
17. BILODEAU, Brent	Mtl.	Seattle	D
18. MURRAY, Glen	Bos.	Sudbury	RW
19. SUNDBLAD, Niklas	Cgy.	AIK Solna	RW
20. RUCINSKY, Martin	L.A.-Edm.	CHZ Litvinov	LW
21. HALVERSON, Trevor	St.L.-Wsh.	North Bay	LW
22. McAMMOND, Dean	Chi.	Prince Albert	C

SECOND ROUND

Selection	Claimed By	Amateur Club	
23. WHITNEY, Ray	S.J.	Spokane	C
24. CORBET, Rene	Que.	Drummondville	LW
25. LAVIGNE, Eric	Tor.-Que.-Wsh.	Hull	D
26. PALFFY, Ziggy	NYI	AC Nitra	LW
27. STAIOS, Steve	Wpg.-St.L.	Niagara Falls	D
28. CAMPBELL, Jim	Phi.-Mtl.	Northwood Prep	C
29. CULLIMORE, Jassen	Van.	Peterborough	D
30. OZOLINSH, Sandis	Min.-S.J.	Dynamo Riga	D
31. HAMRLIK, Martin	Hfd.	ZPS Zlin	D
32. PUSHOR, Jamie	Det.	Lethbridge	D
33. HEXTALL, Donevan	N.J.	Prince Albert	LW
34. VERNER, Andrew	Edm.	Peterborough	G
35. DAWE, Jason	Buf.	Peterborough	LW
36. NELSON, Jeff	Wsh.	Prince Albert	C
37. WERENKA, Darcy	NYR	Lethbridge	D
38. FITZGERALD, Rusty	Pit.	East Duluth High	C
39. POMICHTER, Michael	Mtl.-Chi.	Springfield Jr. B	C
40. STUMPEL, Jozef	Bos.	AC Nitra	RW
41. GROLEAU, Francois	Cgy.	Shawinigan	D
42. LEVEQUE, Guy	L.A.	Cornwall	C
43. DARBY, Craig	St.L.-Mtl.	Albany Academy	C
44. MATTHEWS, Jamie	Chi.	Sudbury	C

1990

FIRST ROUND

Selection	Claimed By	Amateur Club	
1. NOLAN, Owen	Que.	Cornwall	RW
2. NEDVED, Petr	Van.	Seattle	C
3. PRIMEAU, Keith	Det.	Niagara Falls	C
4. RICCI, Mike	Phi.	Peterborough	C
5. JAGR, Jaromir	Pit.	Poldi Kladno	LW
6. SCISSONS, Scott	NYI	Saskatoon	C
7. SYDOR, Darryl	L.A.	Kamloops	D
8. HATCHER, Derian	Min.	North Bay	D
9. SLANEY, John	Wsh.	Cornwall	D
10. BEREHOWSKY, Drake	Tor.	Kingston	D
11. KIDD, Trevor	N.J.-Cgy.	Brandon	G
12. STEVENSON, Turner	St.L.-Mtl.	Seattle	RW
13. STEWART, Michael	NYR	Michigan State	D
14. MAY, Brad	Wpg.-Buf.	Niagara Falls	LW
15. GREIG, Mark	Hfd.	Lethbridge	RW
16. DYKHUIS, Karl	Chi.	Hull	D
17. ALLISON, Scott	Edm.	Prince Albert	C
18. ANTOSKI, Shawn	Mtl.-St.L.-Van.	North Bay	LW
19. TKACHUK, Keith	Buf.-Wpg.	Malden Catholic HS	LW
20. BRODEUR, Martin	Cgy.-N.J.	St. Hyacinthe	G
21. SMOLINSKI, Bryan	Bos.	Michigan State	C

SECOND ROUND

Selection	Claimed By	Amateur Club	
22. HUGHES, Ryan	Que.	Cornell	C
23. SLEGR, Jiri	Van.	CHZ Litvinov	D
24. HARLOCK, David	Det.-Cgy.-N.J.	U. of Michigan	D
25. SIMON, Chris	Phi.	Ottawa	LW
26. PERREAULT, Nicolas	Pit.-Cgy.	Hawkesbury Jr. A	D
27. TAYLOR, Chris	NYI	London	C
28. SEMCHUK, Brandy	L.A.	Canadian National	RW
29. GOTZIAMAN, Chris	Min.-Cgy.-N.J.	Roseau High	RW
30. PASMA, Rod	Wsh.	Cornwall	D
31. POTVIN, Felix	Tor.	Chicoutimi	G
32. VIITAKOSKI, Vesa	N.J.-Cgy.	SaiPa	LW
33. JOHNSON, Craig	St.L.	Hill-Murray High	C
34. WEIGHT, Doug	NYR	Lake Superior	C
35. MULLER, Mike	Wpg.	Wayzata High	D
36. SANDERSON, Geoff	Hfd.	Swift Current	C
37. DROPPA, Ivan	Chi.	Kosice Jr.	D
38. LEGAULT, Alexandre	Edm.	Boston University	RW
39. KUWABARA, Ryan	Mtl.	Ottawa	RW
40. RENBERG, Mikael	Buf.-Phi.	Pitea	LW
41. BELZILE, Etienne	Cgy.	Cornell	D
42. SANDWITH, Terran	Bos.-Phi.	Tri-City	D

1989

FIRST ROUND

Selection	Claimed By	Amateur Club	
1. SUNDIN, Mats	Que.	Nacka	RW
2. CHYZOWSKI, Dave	NYI	Kamloops	LW
3. THORNTON, Scott	Tor.	Belleville	C
4. BARNES, Stu	Wpg.	Tri-City	C
5. GUERIN, Bill	N.J.	Springfield Jr. B	RW
6. BENNETT, Adam	Chi.	Sudbury	D
7. ZMOLEK, Doug	Min.	John Marshall High	D
8. HERTER, Jason	Van.	U. of North Dakota	D
9. MARSHALL, Jason	St.L.	Vernon Jr. A	D
10. HOLIK, Robert	Hfd.	Dukla Jihlava	C
11. SILLINGER, Mike	Det.	Regina	C
12. PEARSON, Rob	Phi.-Tor.	Belleville	RW
13. VALLIS, Lindsay	NYR-Mtl.	Seattle	RW
14. HALLER, Kevin	Buf.	Regina	D
15. SOULES, Jason	Edm.	Niagara Falls	D
16. HEWARD, Jamie	Pit.	Regina	RW
17. STEVENSON, Shayne	Bos.	Kitchener	RW
18. MILLER, Jason	L.A.-Edm.-N.J.	Medicine Hat	C
19. KOLZIG, Olaf	Wsh.	Tri-City	G
20. RICE, Steven	Mtl.-NYR	Kitchener	RW
21. BANCROFT, Steve	Cgy.-Tor.	Belleville	D

SECOND ROUND

Selection	Claimed By	Amateur Club	
22. FOOTE, Adam	Que.	Sault Ste. Marie	D
23. GREEN, Travis	NYI	Spokane	C
24. MANDERVILLE, Kent	Tor.-Cgy.	Notre Dame Jr. A	LW
25. RATUSHNY, Dan	Wpg.	Cornell	D
26. SKALDE, Jarrod	N.J.	Oshawa	C
27. SPEER, Michael	Chi.	Guelph	D
28. CRAIG, Mike	Min.	Oshawa	RW
29. WOODWARD, Robert	Van.	Deerfield	LW
30. BRISEBOIS, Patrice	St.L.-Mtl.	Laval	D
31. CORRIVEAU, Rick	Hfd.-St.L.	London	D
32. BOUGHNER, Bob	Det.	Sault Ste. Marie	D
33. JOHNSON, Greg	Phi.	Thunder Bay Jr. A	C
34. JUHLIN, Patrik	NYR-Phi.	Vasteras	LW
35. DAFOE, Byron	Buf.-Wsh.	Portland	G
36. BORGO, Richard	Edm.	Kitchener	G
37. LAUS, Paul	Pit.	Niagara Falls	D
38. PARSON, Mike	Bos.	Guelph	G
39. THOMPSON, Brent	L.A.	Medicine Hat	D
40. PROSOFSKY, Jason	Wsh.-NYR	Medicine Hat	RW
41. LAROUCHE, Steve	Mtl.	Trois-Rivieres	C
42. DRURY, Ted	Cgy.	Fairfield Prep	C

1988

FIRST ROUND

Selection	Claimed By	Amateur Club	
1. MODANO, Mike	Min.	Prince Albert	C
2. LINDEN, Trevor	Van.	Medicine Hat	RW
3. LESCHYSHYN, Curtis	Que.	Saskatoon	D
4. SHANNON, Darrin	Pit.	Windsor	LW
5. DORE, Daniel	NYR-Que.	Drummondville	RW
6. PEARSON, Scott	Tor.	Kingston	LW
7. GELINAS, Martin	L.A.	Hull	LW
8. ROENICK, Jeremy	Chi.	Thayer Academy	C
9. BRIND'AMOUR, Rod	St.L.	Notre Dame Jr. A	C
10. SELANNE, Teemu	Wpg.	Jokerit	RW
11. GOVEDARIS, Chris	Hfd.	Toronto	LW
12. FOSTER, Corey	N.J.	Peterborough	D
13. SAVAGE, Joel	Buf.	Victoria	RW
14. BOIVIN, Claude	Phi.	Drummondville	LW
15. SAVAGE, Reggie	Wsh.	Victoriaville	C
16. CHEVELDAYOFF, Kevin	NYI	Brandon	D
17. KOCUR, Kory	Det.	Saskatoon	RW
18. CIMETTA, Rob	Bos.	Toronto	LW
19. LEROUX, Francois	Edm.	St. Jean	D
20. CHARRON, Eric	Mtl.	Trois-Rivieres	D
21. MUZZATTI, Jason	Cgy.	Michigan State	G

SECOND ROUND

Selection	Claimed By	Amateur Club	
22. MALLETTE, Troy	Min.-NYR	Sault Ste. Marie	C
23. CHRISTIAN, Jeff	Van.-N.J.	London	LW
24. FISET, Stephane	Que.	Victoriaville	G
25. MAJOR, Mark	Pit.	North Bay	D
26. DUVAL, Murray	NYR	Spokane	RW
27. DOMI, Tie	Tor.	Peterborough	RW
28. HOLDEN, Paul	L.A.	London	D
29. DOUCET, Wayne	Chi.-NYI	Hamilton	LW
30. PLAVSIC, Adrien	St.L.	U. of New Hampshire	D
31. ROMANIUK, Russell	Wpg.	St. Boniface Jr. A	LW
32. RICHTER, Barry	Hfd.	Culver Academy	D
33. ROHLIN, Leif	N.J.-Van.	Vasteras	D
34. ST. AMOUR, Martin	Buf.-Mtl.	Verdun	LW
35. MURRAY, Pat	Phi.	Michigan State	LW
36. TAYLOR, Tim	Wsh.	London	C
37. LEBRUN, Sean	NYI	New Westminster	LW
38. ANGLEHART, Serge	Det.	Drummondville	D
39. KOIVUNEN, Petro	Bos.-Edm.	Espoo	C
40. GAETZ, Link	Edm.-Min.	Spokane	D
41. BARTLEY, Wade	Mtl.-St.L.-Wsh.	Dauphin Jr. A	D
42. HARKINS, Todd	Cgy.	Miami-Ohio	RW

1987

FIRST ROUND

Selection	Claimed By	Amateur Club	
1. TURGEON, Pierre	Buf.	Granby	C
2. SHANAHAN, Brendan	N.J.	London	C
3. WESLEY, Glen	Van.-Bos.	Portland	D
4. McBEAN, Wayne	Min.-L.A.	Medicine Hat	D
5. JOSEPH, Chris	Pit.	Seattle	D
6. ARCHIBALD, Dave	L.A.-Min.	Portland	C/LW
7. RICHARDSON, Luke	Tor.	Peterborough	D
8. WAITE, Jimmy	Chi.	Chicoutimi	G
9. FOGARTY, Bryan	Que.	Kingston	D
10. MORE, Jay	NYR	New Westminster	D
11. RACINE, Yves	Det.	Longueuil	D
12. OSBORNE, Keith	St.L.	North Bay	RW
13. CHYNOWETH, Dean	NYI	Medicine Hat	D
14. QUINTAL, Stephane	Bos.	Granby	D
15. SAKIC, Joe	Wsh.-Que.	Swift Current	C
16. MARCHMENT, Bryan	Wpg.	Belleville	D
17. CASSELS, Andrew	Mtl.	Ottawa	C
18. HULL, Jody	Hfd.	Peterborough	RW
19. DEASLEY, Bryan	Cgy.	U. of Michigan	LW
20. RUMBLE, Darren	Phi.	Kitchener	D
21. SOBERLAK, Peter	Edm.	Swift Current	LW

SECOND ROUND

Selection	Claimed By	Amateur Club	
22. MILLER, Brad	Buf.	Regina	D
23. PERSSON, Ricard	N.J.	Ostersund	D
24. MURPHY, Rob	Van.	Laval	C
25. MATTEAU, Stephane	Min.-Cgy.	Hull	LW
26. TABARACCI, Rick	Pit.	Cornwall	G
27. FITZPATRICK, Mark	L.A.	Medicine Hat	G
28. MAROIS, Daniel	Tor.	Chicoutimi	RW
29. McGILL, Ryan	Chi.	Swift Current	D
30. HARDING, Jeff	Que.-Phi.	St. Michael's Jr. B	LW
31. LACROIX, Daniel	NYR	Granby	LW
32. KRUPPKE, Gord	Det.	Prince Albert	D
33. LECLAIR, John	St.L.-Mtl.	Bellows Academy	D
34. HACKETT, Jeff	NYI	Oshawa	G
35. McCRADY, Scott	Bos.-Min.	Medicine Hat	D
36. BALLANTYNE, Jeff	Wsh.	Ottawa	D
37. ERICKSSON, Patrik	Wpg.	Brynas	C
38. DESJARDINS, Eric	Mtl.	Granby	D
39. BURT, Adam	Hfd.	North Bay	D
40. GRANT, Kevin	Cgy.	Kitchener	D
41. WILKIE, Bob	Phi.-Det.	Swift Current	D
42. WERENKA, Brad	Edm.	Northern Michigan	D

1986

FIRST ROUND

Selection	Claimed By	Amateur Club	
1. MURPHY, Joe	Det.	Michigan State	C
2. CARSON, Jimmy	L.A.	Verdun	C
3. BRADY, Neil	N.J.	Medicine Hat	C
4. ZALAPSKI, Zarley	Pit.	Canadian National	D
5. ANDERSON, Shawn	Buf.	Canadian National	D
6. DAMPHOUSSE, Vincent	Tor.	Laval	LW
7. WOODLEY, Dan	Van.	Portland	C
8. ELYNUIK, Pat	Wpg.	Prince Albert	RW
9. LEETCH, Brian	NYR	Avon Old Farms High	D
10. LEMIEUX, Jocelyn	St.L.	Laval	RW
11. YOUNG, Scott	Hfd.	Boston University	RW
12. BABE, Warren	Min.	Lethbridge	LW
13. JANNEY, Craig	Bos.	Boston College	C
14. SANIPASS, Everett	Chi.	Verdun	LW
15. PEDERSON, Mark	Mtl.	Medicine Hat	LW
16. PELAWA, George	Cgy.	Bemidji High	RW
17. FITZGERALD, Tom	NYI	Austin Prep	C
18. McRAE, Ken	Que.	Sudbury	C
19. GREENLAW, Jeff	Wsh.	Canadian National	LW
20. HUFFMAN, Kerry	Phi.	Guelph	D
21. ISSEL, Kim	Edm.	Prince Albert	RW

SECOND ROUND

Selection	Claimed By	Amateur Club	
22. GRAVES, Adam	Det.	Windsor	C
23. SEPPO, Jukka	L.A.-Phi.	Sport Vaasa	LW
24. COPELAND, Todd	N.J.	Belmont Hill High	D
25. CAPUANO, Dave	Pit.	Mt. St. Charles High	C
26. BROWN, Greg	Buf.	St. Mark's	D
27. BRUNET, Benoit	Tor.-Mtl.	Hull	LW
28. HAWLEY, Kent	Van.-Phi.	Ottawa	C
29. NUMMINEN, Teppo	Wpg.	Tappara	D
30. WILKINSON, Neil	NYR-Min.	Selkirk	D
31. POSMA, Mike	St.L.	Buffalo Jr. A	D
32. LaFORGE, Marc	Hfd.	Kingston	D
33. KOLSTAD, Dean	Min.	Prince Albert	D
34. TIRKKONEN, Pekka	Bos.	SaPKo	C
35. KURZAWSKI, Mark	Chi.	Windsor	D
36. SHANNON, Darryl	Mtl.-Tor.	Windsor	D
37. GLYNN, Brian	Cgy.	Saskatoon	D
38. VASKE, Dennis	NYI	Armstrong High	D
39. ROUTHIER, Jean-Marc	Que.	Hull	RW
40. SEFTEL, Steve	Wsh.	Kingston	LW
41. GUERARD, Stephane	Phi.-Que.	Shawinigan	D
42. NICHOLS, Jamie	Edm.	Portland	LW

1985

FIRST ROUND

Selection	Claimed By	Amateur Club	
1. CLARK, Wendel	Tor.	Saskatoon	D
2. SIMPSON, Craig	Pit.	Michigan State	C
3. WOLANIN, Craig	N.J.	Kitchener	D
4. SANDLAK, Jim	Van.	London	RW
5. MURZYN, Dana	Hfd.	Calgary	D
6. DALGARNO, Brad	Min.-NYI	Hamilton	RW
7. DAHLEN, Ulf	NYR	Ostersund	C
8. FEDYK, Brent	Det.	Regina	RW
9. DUNCANSON, Craig	L.A.	Sudbury	LW
10. GRATTON, Dan	Bos.-L.A.	Oshawa	C
11. MANSON, Dave	Chi.	Prince Albert	D
12. CHARBONNEAU, Jose	St.L.-Mtl.	Drummondville	RW
13. KING, Derek	NYI	Sault Ste. Marie	LW
14. JOHANSSON, Calle	Buf.	Vastra Frolunda	D
15. LATTA, Dave	Que.	Kitchener	LW
16. CHORSKE, Tom	Mtl.	Southwest High	LW
17. BIOTTI, Chris	Cgy.	Belmont Hill High	D
18. STEWART, Ryan	Wpg.	Kamloops	C
19. CORRIVEAU, Yvon	Wsh.	Toronto	LW
20. METCALFE, Scott	Edm.	Kingston	LW
21. SEABROOKE, Glen	Phi.	Peterborough	C

SECOND ROUND

Selection	Claimed By	Amateur Club	
22. SPANGLER, Ken	Tor.	Calgary	D
23. GIFFIN, Lee	Pit.	Oshawa	RW
24. BURKE, Sean	N.J.	Toronto	G
25. GAMBLE, Troy	Van.	Medicine Hat	G
26. WHITMORE, Kay	Hfd.	Peterborough	G
27. NIEUWENDYK, Joe	Min.-Cgy.	Cornell	C
28. RICHTER, Mike	NYR	Northwood Prep.	G
29. SHARPLES, Jeff	Det.	Kelowna	D
30. EDLUND, Par	L.A.	Bjorkloven	RW
31. COTE, Alain	Bos.	Quebec	D
32. WEINRICH, Eric	Chi.-N.J.	North Yarmouth	D
33. RICHARDS, Todd	Mtl.	Armstrong High	D
34. LAUER, Brad	NYI	Regina	RW
35. HOGUE, Benoit	Buf.	St-Jean	C
36. LAFRENIERE, Jason	Que.	Hamilton	C
37. RAGLAN, Herb	Mtl.-St.L.	Kingston	RW
38. WENAAS, Jeff	Cgy.	Medicine Hat	C
39. OHMAN, Roger	Wpg.	Leksand	D
40. DRUCE, John	Wsh.	Peterborough	RW
41. CARNELLEY, Todd	Edm.	Kamloops	D
42. RENDALL, Bruce	Phi.	Chatham	LW

1984

FIRST ROUND

Selection	Claimed By	Amateur Club	
1. LEMIEUX, Mario	Pit.	Laval	C
2. MULLER, Kirk	N.J.	Cdn. Nat./Guelph	C
3. OLCZYK, Ed	L.A.-Chi.	U.S. National	RW
4. IAFRATE, Al	Tor.	U.S. National/Belleville	D
5. SVOBODA, Petr	Hfd.-Mtl.	CHZ Litvinov	D
6. REDMOND, Craig	Chi.-L.A.	Canadian National	D
7. BURR, Shawn	Det.	Kitchener	C
8. CORSON, Shayne	St.L.-Mtl.	Brantford	C
9. BODGER, Doug	Wpg.-Pit.	Kamloops Jr. A	D
10. DAIGNEAULT, J.J.	Van.	Cdn. Nat./Longueuil	D
11. COTE, Sylvain	Mtl.-Hfd.	Quebec	D
12. ROBERTS, Gary	Cgy.	Ottawa	LW
13. QUINN, David	Min.	Kent High	D
14. CARKNER, Terry	NYR	Peterborough	D
15. STIENBURG, Trevor	Que.	Guelph	C
16. BELANGER, Roger	Phi.-Pit.	Kingston	C
17. HATCHER, Kevin	Wsh.	North Bay	D
18. ANDERSSON, Mikael	Buf.	Vastra Frolunda	C
19. PASIN, Dave	Bos.	Prince Albert	RW
20. MacPHERSON, Duncan	NYI	Saskatoon	D
21. ODELEIN, Selmar	Edm.	Regina	D

SECOND ROUND

Selection	Claimed By	Amateur Club	
22. SMYTH, Greg	Phi.	London	D
23. BILLINGTON, Craig	N.J.	Belleville	G
24. WILKS, Brian	L.A.	Kitchener	C
25. GILL, Todd	Tor.	Windsor	D
26. BENNING, Brian	Hfd.-St.L.	Portland	D
27. MELLANBY, Scott	Chi.-Phi.	Henry Carr Jr. B	RW
28. HOUDA, Doug	Det.	Calgary	D
29. RICHER, Stephane	St.L.-Mtl.	Granby	C
30. DOURIS, Peter	Wpg.	U. of New Hampshire	C
31. ROHLICEK, Jeff	Van.	Portland	LW
32. HRKAC, Tony	Mtl.-St.L.	Orillia Jr. A	C
33. SABOURIN, Ken	Cgy.	Sault Ste. Marie	D
34. LEACH, Stephen	Min.-Wsh.	Matignon High	RW
35. HELMINEN, Raimo	NYR	Ilves	C
36. BROWN, Jeff	Que.	Sudbury	D
37. CHYCHRUN, Jeff	Phi.	Kingston	D
38. RANHEIM, Paul	Wsh.-Cgy.	Edina High	C
39. TRAPP, Doug	Buf.	Regina	LW
40. PODLOSKI, Ray	Bos.	Portland	C
41. MELANSON, Bruce	NYI	Oshawa	RW
42. REAUGH, Daryl	Edm.	Kamloops Jr. A	G

1983

FIRST ROUND

Selection	Claimed By	Amateur Club	
1. LAWTON, Brian	Pit.-Min.	Mt. St. Charles High	C
2. TURGEON, Sylvain	Hfd.	Hull	C
3. LaFONTAINE, Pat	N.J.-NYI	Verdun	C
4. YZERMAN, Steve	Det.	Peterborough	C
5. BARRASSO, Tom	St.L.-L.A.-Buf.	Acton-Boxboro High	G
6. MacLEAN, John	L.A.-N.J.	Oshawa	RW
7. COURTNALL, Russ	Tor.	Victoria	C
8. McBAIN, Andrew	Wpg.	North Bay	RW
9. NEELY, Cam	Van.	Portland	RW
10. LACOMBE, Normand	Cgy.-Buf.	New Hampshire	RW
11. CREIGHTON, Adam	Que.-Buf.	Ottawa	C
12. GAGNER, Dave	NYR	Brantford	C
13. QUINN, Dan	Buf.-Cgy.	Belleville	C
14. DOLLAS, Bobby	Wsh.-Wpg.	Laval	D
15. ERREY, Bob	Min.-Pit.	Peterborough	LW
16. DIDUCK, Gerald	NYI	Lethbridge	D
17. TURCOTTE, Alfie	Mtl.	Portland	C
18. CASSIDY, Bruce	Chi.	Ottawa	D
19. BEUKEBOOM, Jeff	Edm.	Sault Ste. Marie	D
20. JENSEN, David	Phi.-Hfd.	Lawrence	C
21. MARKWART, Nevin	Bos.	Regina	LW

SECOND ROUND

Selection	Claimed By	Amateur Club	
22. CHARLESWORTH, Todd	Pit.	Oshawa	D
23. SIREN, Ville	Hfd.	Ilves	D
24. EVANS, Shawn	N.J.	Peterborough	D
25. LAMBERT, Lane	Det.	Saskatoon	RW
26. LEMIEUX, Claude	St.L.-Mtl.	Trois-Rivières	RW
27. MOMESSO, Sergio	L.A.-Mtl.	Shawinigan	C
28. JACKSON, Jeff	Tor.	Brantford	LW
29. BERRY, Brad	Wpg.	St. Albert	D
30. BRUCE, David	Van.	Kitchener	RW
31. TUCKER, John	Cgy.-Buf.	Kitchener	C
32. HEROUX, Yves	Que.	Chicoutimi	RW
33. HEATH, Randy	NYR	Portland	LW
34. HAJDU, Richard	Wsh.-Buf.	Kamloops Jr. A	LW
35. FRANCIS, Todd	Mtl.	Brantford	RW
36. PARKS, Malcolm	Min.	St. Albert	C
37. McKECHNEY, Garnet	NYI	Kitchener	RW
38. MUSIL, Frantisek	Mtl.-Min.	Pardubice	D
39. PRESLEY, Wayne	Chi.	Kitchener	RW
40. GOLDEN, Mike	Edm.	Reading High	C
41. ZEZEL, Peter	Phi.	Toronto	C
42. JOHNSTON, Greg	Bos.	Toronto	RW

1982

FIRST ROUND

Selection	Claimed By	Amateur Club	
1. KLUZAK, Gord	Col.-Bos.	Nanaimo	D
2. BELLOWS, Brian	Det.-Min.	Kitchener	RW
3. NYLUND, Gary	Tor.	Portland	D
4. SUTTER, Ron	Hfd.-Phi.	Lethbridge	C
5. STEVENS, Scott	L.A.-Wsh.	Kitchener	D
6. HOUSLEY, Phil	Wsh.-Buf.	S. St. Paul High	D
7. YAREMCHUK, Ken	Chi.	Portland	C
8. TROTTIER, Rocky	St.L.-N.J.	Nanaimo	RW
9. CYR, Paul	Cgy.-Buf.	Victoria	LW
10. SUTTER, Rich	Pit.	Lethbridge	RW
11. PETIT, Michel	Van.	Sherbrooke	D
12. KYTE, Jim	Wpg.	Cornwall	D
13. SHAW, David	Que.	Kitchener	D
14. LAWLESS, Paul	Phi.-Hfd.	Windsor	LW
15. KONTOS, Chris	NYR	Toronto	C
16. ANDREYCHUK, Dave	Buf.	Oshawa	LW
17. CRAVEN, Murray	Min.-Det.	Medicine Hat	C
18. DANEYKO, Ken	Bos.-N.J.	Seattle	D
19. HEROUX, Alain	Mtl.	Chicoutimi	LW
20. PLAYFAIR, Jim	Edm.	Portland	D
21. FLATLEY, Pat	NYI	U. of Wisconsin	RW

SECOND ROUND

Selection	Claimed By	Amateur Club	
22. CURRAN, Brian	Col.-Bos.	Portland	D
23. COURTEAU, Yves	Det.	Laval	RW
24. LEEMAN, Gary	Tor.	Regina	D
25. IHNACAK, Peter	Hfd.-Tor.	Sparta	C
26. ANDERSON, Mike	L.A.-Buf.	N. St. Paul High	D
27. HEIDT, Mike	Wsh.-L.A.	Calgary	D
28. BADEAU, Rene	St.L.-Chi.	Quebec	D
29. REIERSON, Dave	Cgy.	Prince Albert	D
30. JOHANSSON, Jens	Buf.	Pitea	D
31. GAUVREAU, Jocelyn	Pit.-Mtl.	Granby	D
32. CARLSON, Kent	Van.-Mtl.	St. Lawrence University	D
33. MALEY, David	Wpg.-Mtl.	Edina High	C
34. GILLIS, Paul	Que.	Niagara Falls	C
35. PATERSON, Mark	Phi.-Hfd.	Ottawa	D
36. SANDSTROM, Tomas	NYR	Farjestads	RW
37. KROMM, Richard	Buf.-Cgy.	Portland	LW
38. HRYNEWICH, Tim	Min.-Pit.	Sudbury	LW
39. BYERS, Lyndon	Bos.	Regina	RW
40. SANDELIN, Scott	Mtl.	Hibbing High	D
41. GRAVES, Steve	Edm.	Sault Ste. Marie	C
42. SMITH, Vern	NYI	Lethbridge	D

1981

FIRST ROUND

Selection	Claimed By	Amateur Club	
1. HAWERCHUK, Dale	Wpg.	Cornwall	C
2. SMITH, Doug	Det.-L.A.	Ottawa	C
3. CARPENTER, Bob	Col.-Wsh.	St. John's High	C
4. FRANCIS, Ron	Hfd.	Sault Ste. Marie	C
5. CIRELLA, Joe	Wsh.-Col.	Oshawa	D
6. BENNING, Jim	Tor.	Portland	D
7. HUNTER, Mark	Pit.-Mtl.	Brantford	RW
8. FUHR, Grant	Edm.	Victoria	G
9. PATRICK, James	NYR	Prince Albert	D
10. BUTCHER, Garth	Van.	Regina	D
11. MOLLER, Randy	Que.	Lethbridge	D
12. TANTI, Tony	Chi.	Oshawa	RW
13. MEIGHAN, Ron	Min.	Niagara Falls	D
14. LEVEILLE, Normand	Bos.	Chicoutimi	LW
15. MacINNIS, Al	Cgy.	Kitchener	D
16. SMITH, Steve	Phi.	Sault Ste. Marie	D
17. DUDACEK, Jiri	Buf.	Poldi Kladno	RW
18. DELORME, Gilbert	L.A.-Mtl.	Chicoutimi	D
19. INGMAN, Jan	Mtl.	Farjestad	LW
20. RUFF, Marty	St.L.	Lethbridge	D
21. BOUTILIER, Paul	NYI	Sherbrooke	D

SECOND ROUND

Selection	Claimed By	Amateur Club	
22. ARNIEL, Scott	Wpg.	Cornwall	LW
23. LOISELLE, Claude	Det.	Windsor	C
24. YAREMCHUK, Gary	Col.-Tor.	Portland	C
25. GRIFFIN, Kevin	Hfd.-Chi.	Portland	LW
26. CHERNOMAZ, Rich	Wsh.-Col.	Victoria	C
27. DONNELLY, Dave	Tor.-Min.	St. Albert	C
28. GATZOS, Steve	Pit.	Sault Ste. Marie	RW
29. STRUEBY, Todd	Edm.	Regina	LW
30. ERIXON, Jan	NYR	Skelleftea	RW
31. SANDS, Mike	Van.-Min.	Sudbury	G
32. ERIKSSON, Lars	Que.-Mtl.	Brynas	G
33. HIRSCH, Tom	Chi.-Min.	Patrick Henry High	D
34. PREUSS, Dave	Min.	St. Thomas Academy	RW
35. DUFOUR, Luc	Bos.	Chicoutimi	RW
36. NORDIN, Hakan	Cgy.-St.L.	Farjestad	D
37. COSTELLO, Rich	Phi.	Pickering	C
38. VIRTA, Hannu	Buf.	TPS	D
39. KENNEDY, Dean	L.A.	Brandon	D
40. CHELIOS, Chris	Mtl.	Moose Jaw	D
41. WAHLSTEN, Jali	St.L.-Min.	TPS	C
42. DINEEN, Gord	NYI	Sault Ste. Marie	D

1980

FIRST ROUND

Selection	Claimed By	Amateur Club	
1. WICKENHEISER, Doug	Col.-Mtl.	Regina	C
2. BABYCH, Dave	Wpg.	Portland	D
3. SAVARD, Denis	Que.-Chi.	Montreal	C
4. MURPHY, Larry	Det.-L.A.	Peterborough	D
5. VEITCH, Darren	Wsh.	Regina	D
6. COFFEY, Paul	Edm.	Kitchener	D
7. LANZ, Rick	Van.	Oshawa	D
8. ARTHUR, Fred	Hfd.	Cornwall	D
9. BULLARD, Mike	Pit.	Brantford	C
10. FOX, Jim	L.A.	Ottawa	RW
11. BLAISDELL, Mike	Tor.-Det.	Regina	RW
12. WILSON, Rik	St.L.	Kingston	D
13. CYR, Denis	Cgy.	Montreal	RW
14. MALONE, Jim	NYR	Toronto	C
15. DUPONT, Jerome	Chi.	Toronto	D
16. PALMER, Brad	Min.	Victoria	LW
17. SUTTER, Brent	NYI	Red Deer	C
18. PEDERSON, Barry	Bos.	Victoria	C
19. GAGNE, Paul	Mtl.-Col.	Windsor	LW
20. PATRICK, Steve	Buf.	Brandon	RW
21. STOTHERS, Mike	Phi.	Kingston	D

SECOND ROUND

Selection	Claimed By	Amateur Club	
22. WARD, Joe	Col.	Seattle	C
23. MANTHA, Moe	Wpg.	Toronto	D
24. ROCHEFORT, Normand	Que.	Quebec	D
25. MUNI, Craig	Det.-Tor.	Kingston	D
26. McGILL, Bob	Wsh.-Tor.	Victoria	D
27. NATTRESS, Ric	Edm.-Mtl.	Brantford	D
28. LUDZIK, Steve	Van.-Chi.	Niagara Falls	C
29. GALARNEAU, Michel	Hfd.	Hull	C
30. SOLHEIM, Ken	Pit.-Chi.	Medicine Hat	LW
31. CURTALE, Tony	L.A.-Cgy.	Brantford	D
32. LaVALLEE, Kevin	Tor.-Cgy.	Brantford	LW
33. TERRION, Greg	St.L.-L.A.	Brantford	LW
34. MORRISON, Dave	Cgy.-L.A.	Peterborough	RW
35. ALLISON, Mike	NYR	Sudbury	LW
36. DAWES, Len	Chi.	Victoria	D
37. BEAUPRE, Don	Min.	Sudbury	G
38. HRUDEY, Kelly	NYI	Medicine Hat	G
39. KONROYD, Steve	Cgy.	Oshawa	D
40. CHABOT, John	Mtl.	Hull	C
41. MOLLER, Mike	Buf.	Lethbridge	RW
42. FRASER, Jay	Phi.	Ottawa	LW

1979

FIRST ROUND

Selection	Claimed By	Amateur Club	
1. RAMAGE, Rob	Col.	London	D
2. TURNBULL, Perry	St.L.	Portland	C
3. FOLIGNO, Mike	Det.	Sudbury	RW
4. GARTNER, Mike	Wsh.	Niagara Falls	RW
5. VAIVE, Rick	Van.	Sherbrooke	RW
6. HARTSBURG, Craig	Min.	Sault St. Marie	D
7. BROWN, Keith	Chi.	Portland	D
8. BOURQUE, Raymond	L.A.-Bos.	Verdun	D
9. BOSCHMAN, Laurie	Tor.	Brandon	C
10. McCARTHY, Tom	Wsh.-Min.	Oshawa	LW
11. RAMSEY, Mike	Buf.	U. of Minnesota	D
12. REINHART, Paul	Atl.	Kitchener	D
13. SULLIMAN, Doug	NYR	Kitchener	RW
14. PROPP, Brian	Phi.	Brandon	LW
15. McCRIMMON, Brad	Bos.	Brandon	D
16. WELLS, Jay	Mtl.-L.A.	Kingston	D
17. SUTTER, Duane	NYI	Lethbridge	RW
18. ALLISON, Ray	Hfd.	Brandon	RW
19. MANN, Jimmy	Wpg.	Sherbrooke	RW
20. GOULET, Michel	Que.	Quebec	LW
21. LOWE, Kevin	Edm.	Quebec	D

SECOND ROUND

Selection	Claimed By	Amateur Club	
22. WESLEY, Blake	Col.-Phi.	Portland	D
23. PEROVICH, Mike	St.L.-Atl.	Brandon	D
24. RAUSSE, Errol	Det.-Wsh.	Seattle	LW
25. JONSSON, Tomas	Wsh.-NYI	MoDo	D
26. ASHTON, Brent	Van.	Saskatoon	LW
27. GINGRAS, Gaston	Min.-Mtl.	Hamilton	D
28. TRIMPER, Tim	Chi.	Peterborough	LW
29. HOPKINS, Dean	L.A.	London	RW
30. HARDY, Mark	Tor.-L.A.	Montreal	D
31. MARSHALL, Paul	Wsh.-Pit.	Brantford	LW
32. RUFF, Lindy	Buf.	Lethbridge	D
33. RIGGIN, Pat	Atl.	London	G
34. HOSPODAR, Ed	NYR	Ottawa	D
35. LINDBERGH, Pelle	Phi.	AIK Solna	G
36. MORRISON, Doug	Bos.	Lethbridge	RW
37. NASLUND, Mats	Mtl.	Brynas	LW
38. CARROLL, Billy	NYI	London	C
39. SMITH, Stu	Hfd.	Peterborough	D
40. CHRISTIAN, Dave	Wpg.	U. of North Dakota	C
41. HUNTER, Dale	Que.	Sudbury	C
42. BROTEN, Neal	Min.	U. of Minnesota	C

1978

FIRST ROUND

Selection	Claimed By	Amateur Club	
1. SMITH, Bobby	Min.	Ottawa	C
2. WALTER, Ryan	Wsh.	Seattle	LW
3. BABYCH, Wayne	St.L.	Portland	RW
4. DERLAGO, Bill	Van.	Brandon	C
5. GILLIS, Mike	Col.	Kingston	LW
6. WILSON, Behn	Pit.-Phi.	Kingston	D
7. LINSEMAN, Ken	NYR-Phi.	Kingston	C
8. GEOFFRION, Danny	L.A.-Mtl.	Cornwall	RW
9. HUBER, Willie	Det.	Hamilton	D
10. HIGGINS, Tim	Chi.	Ottawa	RW
11. MARSH, Brad	Atl.	London	D
12. PETERSON, Brent	Tor.-Det.	Portland	C
13. PLAYFAIR, Larry	Buf.	Portland	D
14. LUCAS, Danny	Phi.	Sault Ste. Marie	RW
15. TAMBELLINI, Steve	NYI	Lethbridge	C
16. SECORD, Al	Bos.	Hamilton	LW
17. HUNTER, Dave	Mtl.	Sudbury	LW
18. COULIS, Tim	Wsh.	Hamilton	LW

SECOND ROUND

Selection	Claimed By	Amateur Club	
19. PAYNE, Steve	Min.	Ottawa	LW
20. MULVEY, Paul	Wsh.	Portland	RW
21. QUENNEVILLE, Joel	Tor.	Windsor	D
22. FRASER, Curt	Van.	Victoria	LW
23. MacKINNON, Paul	Wsh.	Peterborough	D
24. CHRISTOFF, Steve	Min.	U. of Minnesota	C
25. MEEKER, Mike	Pit.	Peterborough	RW
26. MALONEY, Don	NYR	Kitchener	LW
27. MALINOWSKI, Merlin	Col.	Medicine Hat	C
28. HICKS, Glenn	Det.	Flin Flon	LW
29. LECUYER, Doug	Chi.	Portland	LW
30. YAKIWCHUK, Dale	Mtl.	Portland	C
31. JENSEN, Al	Det.	Hamilton	G
32. McKEGNEY, Tony	Buf.	Kingston	LW
33. SIMURDA, Mike	Phi.	Kingston	RW
34. JOHNSTON, Randy	NYI	Peterborough	D
35. NICOLSON, Graeme	Bos.	Cornwall	D
36. CARTER, Ron	Mtl.	Sherbrooke	RW

1977

FIRST ROUND

Selection	Claimed By	Amateur Club	
1. McCOURT, Dale	Det.	St. Catharines	C
2. BECK, Barry	Col.	New Westminster	D
3. PICARD, Robert	Wsh.	Montreal	D
4. GILLIS, Jere	Van.	Sherbrooke	LW
5. CROMBEEN, Mike	Cle.	Kingston	RW
6. WILSON, Doug	Chi.	Ottawa	D
7. MAXWELL, Brad	Min.	New Westminster	D
8. DEBLOIS, Lucien	NYR	Sorel	C
9. CAMPBELL, Scott	St.L.	London	D
10. NAPIER, Mark	Atl.-Mtl.	Toronto	RW
11. ANDERSON, John	Tor.	Toronto	RW
12. JOHANSEN, Trevor	Pit.-Tor.	Toronto	D
13. DUGUAY, Ron	L.A.-NYR	Sudbury	C
14. SEILING, Ric	Buf.	St. Catharines	RW
15. BOSSY, Mike	NYI	Laval	RW
16. FOSTER, Dwight	Bos.	Kitchener	C/RW
17. McCARTHY, Kevin	Phi.	Winnipeg	D
18. DUPONT, Norm	Mtl.	Montreal	C

SECOND ROUND

Selection	Claimed By	Amateur Club	
19. SAVARD, Jean	Det.-Chi.	Quebec	C
20. ZAHARKO, Miles	Col.-Atl.	New Westminster	D
21. LOFTHOUSE, Mark	Wsh.	New Westminster	RW
22. BANDURA, Jeff	Van.	Portland	D
23. CHICOINE, Dan	Cle.	Sherbrooke	RW
24. GLADNEY, Bob	Chi.-Tor.	Oshawa	D
25. SEMENKO, Dave	Min.	Brandon	LW
26. KEATING, Mike	NYR	St. Catherines	LW
27. LABATTE, Neil	St.L.	Toronto	D
28. LAURENCE, Don	Atl.	Kitchener	C
29. SAGANIUK, Rocky	Tor.	Lethbridge	RW
30. HAMILTON, Jim	Pit.	London	RW
31. HILL, Brian	L.A.-Atl.	Medicine Hat	RW
32. ARESHENKOFF, Ron	Buf.	Medicine Hat	C
33. TONELLI, John	NYI	Toronto	LW
34. PARRO, Dave	Bos.	Saskatoon	G
35. GORENCE, Tom	Phi.	U. of Minnesota	RW
36. LANGWAY, Rod	Mtl.	U. of New Hampshire	D

1976

FIRST ROUND

Selection	Claimed By	Amateur Club	
1. GREEN, Rick	K.C.-Wsh.	London	D
2. CHAPMAN, Blair	Pit.	Saskatoon	RW
3. SHARPLEY, Glen	Min.	Hull	C
4. WILLIAMS, Fred	Det.	Saskatoon	C
5. JOHANSSON, Bjorn	Cal.	Sweden	D
6. MURDOCH, Don	NYR	Medicine Hat	RW
7. FEDERKO, Bernie	St.L.	Saskatoon	C
8. SHAND, Dave	Van.-Atl.	Peterborough	D
9. CLOUTIER, Real	Chi.	Quebec	RW
10. PHILLIPOFF, Harold	Atl.	New Westminster	LW
11. GARDNER, Paul	Pit.-K.C.	Oshawa	C
12. LEE, Peter	Tor.-Mtl.	Ottawa	RW
13. SCHUTT, Rod	L.A.-Mtl.	Sudbury	LW
14. McKENDRY, Alex	NYI	Sudbury	LW
15. CARROLL, Greg	Buf.-Wsh.	Medicine Hat	C
16. PACHAL, Clayton	Bos.	New Westminster	C
17. SUZOR, Mark	Phi.	Kingston	D
18. BAKER, Bruce	Mtl.	Ottawa	RW

SECOND ROUND

Selection	Claimed By	Amateur Club	
19. MALONE, Greg	Wsh.-Pit.	Oshawa	C
20. SUTTER, Brian	K.C.-St.L.	Lethbridge	LW
21. CLIPPINGDALE, Steve	Min.-L.A.	New Westminster	LW
22. LARSON, Reed	Det.	U. of Minnesota	D
23. STENLUND, Vern	Cal.	London	C
24. FARRISH, Dave	NYR	Sudbury	D
25. SMRKE, John	St.L.	Toronto	LW
26. MANNO, Bob	Van.	St. Catharines	D
27. McDILL, Jeff	Chi.	Victoria	RW
28. SIMPSON, Bobby	Atl.	Sherbrooke	LW
29. MARSH, Peter	Pit.	Sherbrooke	RW
30. CARLYLE, Randy	Tor.	Sudbury	D
31. ROBERTS, Jim	L.A.-Min.	Ottawa	LW
32. KASZYCKI, Mike	NYI	Sault Ste. Marie	C
33. KOWAL, Joe	Buf.	Hamilton	LW
34. GLOECKNER, Lorry	Bos.	Victoria	D
35. CALLANDER, Drew	Phi.	Regina	C
36. MELROSE, Barry	Mtl.	Kamloops	D

Allowed to reclaim Wayne Gretzky prior to the 1979 Entry Draft, the Oilers further solidified their future with the selections of Kevin Lowe, Mark Messier and Glenn Anderson with their first three picks.

1975

FIRST ROUND

Selection	Claimed By	Amateur Club	
1. BRIDGMAN, Mel	Wsh.-Phi.	Victoria	C
2. DEAN, Barry	K.C.	Medicine Hat	LW
3. KLASSEN, Ralph	Cal.	Saskatoon	C
4. MAXWELL, Brian	Min.	Medicine Hat	D
5. LAPOINTE, Rick	Det.	Victoria	D
6. ASHBY, Don	Tor.	Calgary	C
7. VAYDIK, Greg	Chi.	Medicine Hat	C
8. MULHERN, Richard	Atl.	Sherbrooke	D
9. SADLER, Robin	St.L.-Mtl.	Edmonton	D
10. BLIGHT, Rick	Van.	Brandon	RW
11. PRICE, Pat	NYI	Saskatoon	D
12. DILLON, Wayne	NYR	Toronto	C
13. LAXTON, Gord	Pit.	New Westminster	G
14. HALWARD, Doug	Bos.	Peterborough	D
15. MONDOU, Pierre	L.A.-Mtl.	Montreal	C
16. YOUNG, Tim	Mtl.-L.A.	Ottawa	C
17. SAUVE, Bob	Buf.	Laval	G
18. FORSYTH, Alex	Phi.-Wsh.	Kingston	C

SECOND ROUND

Selection	Claimed By	Amateur Club	
19. SCAMURRA, Peter	Wsh.	Peterborough	D
20. CAIRNS, Don	K.C.	Victoria	LW
21. MARUK, Dennis	Cal.	London	C
22. ENGBLOM, Brian	Min.-Mtl.	U. of Wisconsin	D
23. ROLLINS, Jerry	Det.	Winnipeg	D
24. JARVIS, Doug	Tor.	Peterborough	C
25. ARNDT, Daniel	Chi.	Saskatoon	LW
26. BOWNESS, Rick	Atl.	Montreal	RW
27. STANIOWSKI, Ed	St.L.	Regina	G
28. GASSOFF, Brad	Van.	Kamloops	D
29. SALVIAN, David	NYI	St. Catharines	RW
30. SOETAERT, Doug	NYR	Edmonton	G
31. ANDERSON, Russ	Pit.	U. of Minnesota	D
32. SMITH, Barry	Bos.	New Westminster	C
33. BUCYK, Terry	L.A.	Lethbridge	RW
34. GREENBANK, Kelvin	Mtl.	Winnipeg	RW
35. BREITENBACH, Ken	Buf.	St. Catharines	D
36. MASTERS, Jamie	Phi.-St.L.	Ottawa	D

1974

FIRST ROUND

Selection	Claimed By	Amateur Club	
1. JOLY, Greg	Wsh.	Regina	D
2. PAIEMENT, Wilfred	K.C.	St. Catharines	RW
3. HAMPTON, Rick	Cal.	St. Catharines	D
4. GILLIES, Clark	NYI	Regina	LW
5. CONNOR, Cam	Van.-Mtl.	Flin Flon	RW
6. HICKS, Doug	Min.	Flin Flon	D
7. RISEBROUGH, Doug	St.L.-Mtl.	Kitchener	C
8. LAROUCHE, Pierre	Pit.	Sorel	C
9. LOCHEAD, Bill	Det.	Oshawa	LW
10. CHARTRAW, Rick	Atl.-Mtl.	Kitchener	D
11. FOGOLIN Jr., Lee	Buf.	Oshawa	D
12. TREMBLAY, Mario	L.A.-Mtl.	Montreal	RW
13. VALIQUETTE, Jack	Tor.	Sault Ste. Marie	C
14. MALONEY, Dave	NYR	Kitchener	D
15. McTAVISH, Gord	Mtl.	Sudbury	C
16. MULVEY, Grant	Chi.	Calgary	RW
17. CHIPPERFIELD, Ron	Phi.-Cal.	Brandon	C
18. LARWAY, Don	Bos.	Swift Current	RW

SECOND ROUND

Selection	Claimed By	Amateur Club	
19. MARSON, Mike	Wsh.	Sudbury	LW
20. BURDON, Glen	K.C.	Regina	C
21. AFFLECK, Bruce	Cal.	U. of Denver	D
22. TROTTIER, Bryan	NYI	Swift Current	C
23. SEDLBAUER, Ron	Van.	Kitchener	LW
24. NANTAIS, Rich	Min.	Quebec	LW
25. HOWE, Mark	St.L.-Bos.	Toronto	D
26. HESS, Bob	Pit.-St.L.	New Westminster	D
27. COSSETTE, Jacques	Det.-Pit.	Sorel	RW
28. CHOUINARD, Guy	Atl.	Quebec	C
29. GARE, Danny	Buf.	Calgary	RW
30. MacGREGOR, Gary	L.A.-Mtl.	Cornwall	C
31. WILLIAMS, Tiger	Tor.	Swift Current	LW
32. GRESCHNER, Ron	NYR	New Westminster	D
33. LUPIEN, Gilles	Mtl.	Montreal	D
34. DAIGLE, Alain	Chi.	Trois-Rivières	RW
35. McLEAN, Don	Phi.	Sudbury	D
36. STURGEON, Peter	Bos.	Kitchener	LW

1973

FIRST ROUND

Selection	Claimed By	Amateur Club	
1. POTVIN, Denis	NYI	Ottawa	D
2. LYSIAK, Tom	Cal.-Mtl.-Atl.	Medicine Hat	C
3. VERVERGAERT, Dennis	Van.	London	RW
4. McDONALD, Lanny	Tor.	Medicine Hat	RW
5. DAVIDSON, John	Atl.-Mtl.-St.L.	Calgary	G
6. SAVARD, Andre	L.A.-Bos.	Quebec	C
7. STOUGHTON, Blaine	Pit.	Flin Flon	RW
8. GAINEY, Bob	St.L.-Mtl.	Peterborough	LW
9. DAILEY, Bob	Min.-Mtl.-Van.	Toronto	D
10. NEELY, Bob	Phi.-Tor.	Peterborough	LW
11. RICHARDSON, Terry	Det.	New Westminster	G
12. TITANIC, Morris	Buf.	Sudbury	LW
13. ROTA, Darcy	Chi.	Edmonton	LW
14. MIDDLETON, Rick	NYR	Oshawa	RW
15. TURNBULL, Ian	Bos.-Tor.	Ottawa	D
16. MERCREDI, Vic	Mtl.-Atl.	New Westminster	C

SECOND ROUND

Selection	Claimed By	Amateur Club	
17. GOLDUP, Glenn	NYI-Mtl.	Toronto	RW
18. DUNLOP, Blake	Cal.-Min.	Ottawa	C
19. BORDELEAU, Paulin	Van.	Toronto	RW
20. GOODENOUGH, Larry	Tor.-Phi.	London	D
21. VAIL, Eric	Atl.	Sudbury	LW
22. MARRIN, Peter	L.A.-Mtl.	Toronto	C
23. BIANCHIN, Wayne	Pit.	Flin Flon	LW
24. PESUT, George	St.L.	Saskatoon	D
25. ROGERS, John	Min.	Edmonton	RW
26. LEAVINS, Brent	Phi.	Swift Current	
27. CAMPBELL, Colin	Det.-Pit.	Peterborough	D
28. LANDRY, Jean	Buf.	Quebec	D
29. THOMAS, Reg	Chi.	London	LW
30. HICKEY, Pat	NYR	Hamilton	LW
31. JONES, Jimmy	Bos.	Peterborough	RW
32. ANDRUFF, Ron	Mtl.	Flin Flon	C

1972

FIRST ROUND

Selection	Claimed By	Amateur Club	
1. HARRIS, Billy	NYI	Toronto	RW
2. RICHARD, Jacques	Atl.	Quebec	LW
3. LEVER, Don	Van.	Niagara Falls	C
4. SHUTT, Steve	L.A.-Mtl.	Toronto	LW
5. SCHOENFELD, Jim	Buf.	Niagara Falls	D
6. LAROCQUE, Michel	Cal.-Mtl.	Ottawa	G
7. BARBER, Bill	Phi.	Kitchener	LW
8. GARDNER, Dave	Pit.-Min.-Mtl.	Toronto	C
9. MERRICK, Wayne	St.L.	Ottawa	C
10. BLANCHARD, Al	Det.-NYR	Kitchener	LW
11. FERGUSON, George	Tor.	Toronto	C
12. BYERS, Jerry	Min.	Kitchener	LW
13. RUSSELL, Phil	Chi.	Edmonton	D
14. VAN BOXMEER, John	Mtl.	Guelph	D
15. MacMILLAN, Bob	NYR	St. Catharines	RW
16. BLOOM, Mike	Bos.	St. Catharines	LW

SECOND ROUND

Selection	Claimed By	Amateur Club	
17. HENNING, Lorne	NYI	New Westminster	C
18. BIALOWAS, Dwight	Atl.	Regina	D
19. McSHEFFREY, Bryan	Van.	Ottawa	RW
20. KOZAK, Don	L.A.	Edmonton	RW
21. SACHARUK, Larry	Buf.-NYR	Saskatoon	D
22. CASSIDY, Tom	Cal.	Kitchener	C
23. BLADON, Tom	Phi.	Edmonton	D
24. LYNCH, Jack	Pit.	Oshawa	D
25. CARRIERE, Larry	St.L.-Buf.	Loyola College	D
26. GUITE, Pierre	Det.	St. Catharines	LW
27. OSBURN, Randy	Tor.	London	LW
28. WEIR, Stan	Min.-Cal.	Medicine Hat	C
29. OGILVIE, Brian	Chi.	Edmonton	C
30. LUKOWICH, Bernie	Mtl.-Pit.	New Westminster	RW
31. VILLEMURE, Rene	NYR	Shawinigan	LW
32. ELDER, Wayne	Bos.	London	D

1971

FIRST ROUND

Selection	Claimed By	Amateur Club	
1. LAFLEUR, Guy	Cal.-Mtl.	Quebec	RW
2. DIONNE, Marcel	Det.	St. Catharines	C
3. GUEVREMONT, Jocelyn	Van.	Montreal	D
4. CARR, Gene	Pit.-St.L.	Flin Flon	C
5. MARTIN, Rick	Buf.	Montreal	LW
6. JONES, Ron	L.A.-Bos.	Edmonton	D
7. ARNASON, Chuck	Min.-Mtl.	Flin Flon	RW
8. WRIGHT, Larry	Phi.	Regina	C
9. PLANTE, Pierre	Tor.-Phi.	Drummondville	RW
10. VICKERS, Steve	St.L.-NYR	Toronto	LW
11. WILSON, Murray	Mtl.	Ottawa	LW
12. SPRING, Dan	Chi.	Edmonton	C
13. DURBANO, Steve	NYR	Toronto	D
14. O'REILLY, Terry	Bos.	Oshawa	RW

SECOND ROUND

Selection	Claimed By	Amateur Club	
15. BAIRD, Ken	Cal.	Flin Flon	D
16. BOUCHA, Henry	Det.	U.S. National	C
17. LALONDE, Bobby	Van.	Montreal	C
18. McKENZIE, Brian	Pit.	St. Catharines	LW
19. RAMSAY, Craig	Buf.	Peterborough	LW
20. ROBINSON, Larry	L.A.-Mtl.	Kitchener	D
21. NORRISH, Rod	Min.	Regina	LW
22. KEHOE, Rick	Phi.-Tor.	Hamilton	RW
23. FORTIER, Dave	Tor.	St. Catharines	D
24. DEGUISE, Michel	St.L.-Mtl.	Sorel	G
25. FRENCH, Terry	Mtl.	Ottawa	C
26. KRYSKOW, Dave	Chi.	Edmonton	LW
27. WILLIAMS, Tom	NYR	Hamilton	LW
28. RIDLEY, Curt	Bos.	Portage	G

1970

FIRST ROUND

Selection	Claimed By	Amateur Club	
1. PERREAULT, Gilbert	Buf.	Montreal	C
2. TALLON, Dale	Van.	Toronto	D
3. LEACH, Reggie	L.A.-Bos.	Flin Flon	LW
4. MacLEISH, Rick	Phi.-Bos.	Peterborough	C
5. MARTYNIUK, Ray	Oak.-Mtl.	Flin Flon	G
6. LEFLEY, Chuck	Min.-Mtl.	Canadian National	C
7. POLIS, Greg	Pit.	Estevan	LW
8. SITTLER, Darryl	Tor.	London	C
9. PLUMB, Ron	Bos.	Peterborough	D
10. ODDLEIFSON, Chris	St.L.-Oak.	Winnipeg	C
11. GRATTON, Norm	Mtl.-NYR	Montreal	LW
12. LAJEUNESSE, Serge	Det.	Montreal	RW
13. STEWART, Bob	Bos.	Oshawa	D
14. MALONEY, Dan	Chi.	London	LW

SECOND ROUND

Selection	Claimed By	Amateur Club	
15. DEADMARSH, Butch	Buf.	Brandon	LW
16. HARGREAVES, Jim	Van.	Winnipeg	D
17. HARVEY, Buster	L.A.-Min.	Hamilton	RW
18. CLEMENT, Bill	Phi.	Ottawa	C
19. LAFRAMBOISE, Pete	Oak.	Ottawa	C
20. BARRETT, Fred	Min.	Toronto	D
21. STEWART, John	Pit.	Flin Flon	LW
22. THOMPSON, Errol	Tor.	Charlottetown	LW
23. KEOGAN, Murray	St.L.	U. of Minnesota	C
24. McDONOUGH, Al	Mtl.-L.A.	St. Catharines	RW
25. MURPHY, Mike	NYR	Toronto	RW
26. GUINDON, Bobby	Det.	Montreal	LW
27. BOUCHARD, Dan	Bos.	London	G
28. ARCHAMBAULT, Michel	Chi.	Drummondville	LW

1969

FIRST ROUND

Selection	Claimed By	Amateur Club	
1. HOULE, Rejean	Mtl.	Montreal	LW
2. TARDIF, Marc	Mtl.	Montreal	LW
3. TANNAHILL, Don	Min.-Bos.	Niagara Falls	LW
4. SPRING, Frank	Pit.-Bos.	Edmonton	RW
5. REDMOND, Dick	L.A.-Mtl.-Min.	St. Catharines	D
6. CURRIER, Bob	Phi.	Cornwall	C
7. FEATHERSTONE, Tony	Oak.	Peterborough	RW
8. DUPONT, André	St.L.-NYR	Montreal	D
9. MOSER, Ernie	Det.-Tor.	Estevan	RW
10. RUTHERFORD, Jim	Det.	Hamilton	G
11. BOLDIREV, Ivan	Bos.	Oshawa	C
12. JARRY, Pierre	NYR	Ottawa	LW
13. BORDELEAU, J.P.	Chi.	Montreal	RW
14. O'BRIEN, Dennis	Min.	St. Catharines	D

SECOND ROUND

Selection	Claimed By	Amateur Club	
15. KESSELL, Rick	Pit.	Oshawa	C
16. HOGANSON, Dale	L.A.	Estevan	D
17. CLARKE, Bobby	Phi.	Flin Flon	C
18. STACKHOUSE, Ron	Oak.	Peterborough	D
19. LOWE, Mike	St.L.	Loyola College	
20. BRINDLEY, Doug	Tor.	Niagara Falls	C
21. GARWASIUK, Ron	Det.	Regina	LW
22. QUOQUOCHI, Art	Bos.	Montreal	
23. WILSON, Bert	NYR	London	LW
24. ROMANCHYCH, Larry	Chi.	Flin Flon	RW
25. GILBERT, Gilles	Min.	London	G
26. BRIERE, Michel	Pit.	Shawinigan	C
27. BODDY, Gregg	L.A.	Edmonton	D
28. BROSSART, Willie	Phi.	Estevan	D

Detroit's Dale McCourt is wrapped up by Barry Beck of the New York Rangers. The Red Wings had made McCourt the #1 pick in the 1977 Amateur Draft. Beck went second overall that year to the Colorado Rockies.

NHL All-Stars

Active Players' All-Star Selection Records

	First Team Selections	Second Team Selections	Total
GOALTENDERS			
Patrick Roy	(4) 1988-89; 1989-90; 1991-92; 2001-02.	(2) 1987-88; 1990-91.	6
Ed Belfour	(2) 1990-91; 1992-93.	(1) 1994-95.	3
Tom Barrasso	(1) 1983-84.	(2) 1984-85, 1992-93.	3
Martin Brodeur	(0)	(2) 1996-97; 1997-98.	2
Olaf Kolzig	(1) 99-2000.	(0)	1
Chris Osgood	(0)	(1) 1995-96.	1
Byron Dafoe	(0)	(1) 1998-99.	1
Roman Turek	(0)	(1) 99-2000.	1
Roman Cechmanek	(0)	(1) 2000-01.	1
Jose Theodore	(0)	(1) 2001-02.	1
DEFENSEMEN			
Chris Chelios	(5) 1988-89; 1992-93; 1994-95; 1995-96; 2001-02.	(2) 1990-91; 1996-97.	7
Al MacInnis	(3) 1989-90; 1990-91; 1998-99.	(3) 1986-87; 1988-89; 1993-94.	6
Nicklas Lidstrom	(5) 1997-98; 1998-99; 99-2000; 2000-01; 2001-02.	(0)	5
Scott Stevens	(2) 1987-88; 1993-94.	(3) 1991-92; 1996-97; 2000-01.	5
Brian Leetch	(2) 1991-92; 1996-97.	(3) 1990-91; 1993-94; 1995-96.	5
Rob Blake	(1) 1997-98.	(3) 99-2000; 2000-01; 2001-02.	4
Chris Pronger	(1) 99-2000.	(1) 1997-98.	2
Eric Desjardins	(0)	(2) 1998-99; 99-2000.	2
Sandis Ozolinsh	(1) 1996-97.	(0)	1
Gary Suter	(0)	(1) 1987-88.	1
Phil Housley	(0)	(1) 1991-92.	1
Scott Niedermayer	(0)	(1) 1997-98.	1
Sergei Gonchar	(0)	(1) 2001-02.	1
CENTERS			
Mario Lemieux	(5) 1987-88; 1988-89; 1992-93; 1995-96; 1996-97.	(4) 1985-86; 1986-87; 1991-92; 2000-01.	9
Mark Messier	(2) 1989-90; 1991-92.	(0)	2
Peter Forsberg	(2) 1997-98; 1998-99.	(0)	2
Joe Sakic	(2) 2000-01; 2001-02.	(0)	2
Eric Lindros	(1) 1994-95.	(1) 1995-96.	2
Sergei Fedorov	(1) 1993-94.	(0)	1
Steve Yzerman	(1) 99-2000.	(0)	1
Adam Oates	(0)	(1) 1990-91.	1
Alexei Zhamnov	(0)	(1) 1994-95.	1
Alexei Yashin	(0)	(1) 1998-99.	1
Mike Modano	(0)	(1) 99-2000.	1
Mats Sundin	(0)	(1) 2001-02.	1
RIGHT WINGERS			
Jaromir Jagr	(6) 1994-95; 1995-96; 1997-98; 1998-99; 99-2000; 2000-01.	(1) 1996-97.	7
Teemu Selanne	(2) 1992-93; 1996-97.	(2) 1997-98; 1998-99.	4
Brett Hull	(3) 1989-90; 1990-91; 1991-92.	(0)	3
Pavel Bure	(1) 1993-94.	(2) 99-2000; 2000-01.	3
Alexander Mogilny	(0)	(2) 1992-93; 1995-96.	2
Jarome Iginla	(1) 2001-02.	(0)	1
Mark Recchi	(0)	(1) 1991-92.	1
Theoren Fleury	(0)	(1) 1994-95.	1
Bill Guerin	(0)	(1) 2001-02.	1
LEFT WINGERS			
Luc Robitaille	(5) 1987-88; 1988-89; 1989-90; 1990-91; 1992-93.	(3) 1986-87; 1991-92; 2000-01.	8
John LeClair	(2) 1994-95, 1997-98.	(3) 1995-96; 1996-97; 1998-99.	5
Paul Kariya	(3) 1995-96; 1996-97; 1998-99.	(1) 99-2000.	4
Mark Messier	(2) 1981-82; 1982-83.	(1) 1983-84.	3
Brendan Shanahan	(2) 1993-94; 99-2000.	(1) 2001-02.	3
Keith Tkachuk	(0)	(2) 1994-95; 1997-98.	2
Patrik Elias	(1) 2000-01.	(0)	1
Markus Naslund	(1) 2001-02.	(0)	1
Adam Graves	(0)	(1) 1993-94.	1

Leading NHL All-Stars 1930-31 to 2001-02

Player	Pos	Team	NHL Seasons	First Team Selections	Second Team Selections	Total Selections
Howe, Gordie	RW	Detroit	26	12	9	21
Bourque, Raymond	D	Bos., Col.	22	13	6	19
Gretzky, Wayne	C	Edm., L.A., NYR	20	8	7	15
Richard, Maurice	RW	Montreal	18	8	6	14
Hull, Bobby	LW	Chicago	16	10	2	12
Harvey, Doug	D	Mtl., NYR	19	10	1	11
Hall, Glenn	G	Det., Chi., St.L.	18	7	4	11
Beliveau, Jean	C	Montreal	20	6	4	10
Seibert, Earl	D	NYR, Chi.	15	4	6	10
Orr, Bobby	D	Boston	12	8	1	9
Lindsay, Ted	LW	Detroit	17	8	1	9
* Lemieux, Mario	C	Pittsburgh	14	5	4	9
Mahovlich, Frank	LW	Tor., Det., Mtl.	18	3	6	9
Shore, Eddie	D	Boston	14	7	1	8
Esposito, Phil	C	Boston	18	6	2	8
Kelly, Red	D	Detroit	20	6	2	8
Mikita, Stan	C	Chicago	22	6	2	8
Bossy, Mike	RW	NY Islanders	10	5	3	8
Pilote, Pierre	D	Chicago	14	5	3	8
* Robitaille, Luc	LW	Los Angeles	16	5	3	8
Coffey, Paul	D	Edm., Pit., Det.	21	4	4	8
Brimsek, Frank	G	Boston	10	2	6	8
* Jagr, Jaromir	RW	Pittsburgh	12	6	1	7
Potvin, Denis	D	NY Islanders	15	5	2	7
Park, Brad	D	NYR, Bos.	17	5	2	7
* Chelios, Chris	D	Mtl., Chi.	19	5	2	7
Plante, Jacques	G	Mtl., Tor.	18	3	4	7
Gadsby, Bill	D	Chi., NYR, Det.	20	3	4	7
Sawchuk, Terry	G	Detroit	21	3	4	7
Durnan, Bill	G	Montreal	7	6	0	6
Hasek, Dominik	G	Buffalo	12	6	0	6
Lafleur, Guy	RW	Montreal	17	6	0	6
Dryden, Ken	G	Montreal	8	5	1	6
* Roy, Patrick	G	Montreal	18	4	2	6
* MacInnis, Al	D	Cgy., St.L.	21	3	3	6
Clapper, Dit	RW/D	Boston	20	3	3	6
Robinson, Larry	D	Montreal	20	3	3	6
Horton, Tim	D	Toronto	24	3	3	6
Salming, Borje	D	Toronto	17	1	5	6
* Lidstrom, Niklas	D	Detroit	11	5	0	5
Cowley, Bill	C	Boston	13	4	1	5
Jackson, Busher	LW	Toronto	15	4	1	5
* Messier, Mark	LW/C	Edm., NYR	23	4	1	5
Conacher, Charlie	RW	Toronto	12	3	2	5
Stewart, Jack	D	Detroit	12	3	2	5
Blake, Toe	LW	Montreal	14	3	2	5
Lach, Elmer	C	Montreal	14	3	2	5
Quackenbush, Bill	D	Det., Bos.	14	3	2	5
Goulet, Michel	LW	Quebec	15	3	2	5
Esposito, Tony	G	Chicago	16	3	2	5
Reardon, Ken	D	Montreal	7	2	3	5
Apps, Syl	C	Toronto	10	2	3	5
* LeClair, John	LW	Mtl., Phi.	12	2	3	5
* Leetch, Brian	D	NY Rangers	14	2	3	5
Giacomin, Ed	G	NY Rangers	13	2	3	5
Kurri, Jari	RW	Edmonton	17	2	3	5
* Stevens, Scott	D	Wsh., N.J.	19	2	3	5

* Active

Position Leaders in All-Star Selections

Position	Player	First Team	Second Team	Total
GOAL	Glenn Hall	7	4	11
	Frank Brimsek	2	6	8
	Jacques Plante	3	4	7
	Terry Sawchuk	3	4	7
	Bill Durnan	6	0	6
	Dominik Hasek	6	0	6
	Ken Dryden	5	1	6
	* Patrick Roy	4	2	6
DEFENSE	Raymond Bourque	13	6	19
	Doug Harvey	10	1	11
	Earl Seibert	4	6	10
	Bobby Orr	8	1	9
	Eddie Shore	7	1	8
	Red Kelly	6	2	8
	Pierre Pilote	5	3	8
	Paul Coffey	4	4	8
LEFT WING	Bobby Hull	10	2	12
	Ted Lindsay	8	1	9
	Frank Mahovlich	3	6	9
	* Luc Robitaille	5	3	8
RIGHT WING	Gordie Howe	12	9	21
	Maurice Richard	8	6	14
	Mike Bossy	5	3	8
	* Jaromir Jagr	6	1	7
	Guy Lafleur	6	0	6
CENTER	Wayne Gretzky	8	7	15
	Jean Beliveau	6	4	10
	* Mario Lemieux	5	4	9
	Phil Esposito	6	2	8
	Stan Mikita	6	2	8

* active player

NHL All★Star Game
SOUTH FLORIDA 03

The 2003 NHL All-Star Weekend will be hosted by the Florida Panthers, February 1st and 2nd, 2003.

All-Star Teams

1930-2002

Voting for the NHL All-Star Team is conducted among the representatives of the Professional Hockey Writers' Association at the end of the season.

Following is a list of the First and Second All-Star Teams since their inception in 1930-31.

2001-02

First Team		Second Team
Roy, Patrick, Col.	G	Theodore, Jose, Mtl.
Lidstrom, Nicklas, Det.	D	Blake, Rob, L.A., Col.
Chelios, Chris, Det.	D	Gonchar, Sergei, Wsh.
Sakic, Joe, Col.	C	Sundin, Mats, Tor.
Iginla, Jarome, Cgy.	RW	Guerin, Bill, Bos.
Naslund, Markus, Van.	LW	Shanahan, Brendan, Det.

2000-01

First Team		Second Team
Hasek, Dominik, Buf.	G	Cechmanek, Roman, Phi.
Lidstrom, Nicklas, Det.	D	Blake, Rob, L.A., Col.
Bourque, Raymond, Col.	D	Stevens, Scott, N.J.
Sakic, Joe, Col.	C	Lemieux, Mario, Pit.
Jagr, Jaromir, Pit.	RW	Bure, Pavel, Fla.
Elias, Patrik, N.J.	LW	Robitaille, Luc, L.A.

1999-2000

First Team		Second Team
Kolzig, Olaf, Wsh.	G	Turek, Roman, St.L.
Pronger, Chris, St.L.	D	Blake, Rob, L.A.
Lidstrom, Nicklas, Det.	D	Desjardins, Eric, Phi.
Yzerman, Steve, Det.	C	Modano, Mike, Dal.
Jagr, Jaromir, Pit.	RW	Bure, Pavel, Fla.
Shanahan, Brendan, Det.	LW	Kariya, Paul, Ana.

1998-99

First Team		Second Team
Hasek, Dominik, Buf.	G	Dafoe, Byron, Bos.
MacInnis, Al, St.L.	D	Bourque, Raymond, Bos.
Lidstrom, Nicklas, Det.	D	Desjardins, Eric, Phi.
Forsberg, Peter, Col.	C	Yashin, Alexei, Ott.
Jagr, Jaromir, Pit.	RW	Selanne, Teemu, Ana.
Kariya, Paul, Ana.	LW	LeClair, John, Phi.

1997-98

First Team		Second Team
Hasek, Dominik, Buf.	G	Brodeur, Martin, N.J.
Lidstrom, Nicklas, Det.	D	Pronger, Chris, St.L.
Blake, Rob, L.A.	D	Niedermayer, Scott, N.J.
Forsberg, Peter, Col.	C	Gretzky, Wayne, NYR
Jagr, Jaromir, Pit.	RW	Selanne, Teemu, Ana.
LeClair, John, Phi.	LW	Tkachuk, Keith, Phx.

1996-97

First Team		Second Team
Hasek, Dominik, Buf.	G	Brodeur, Martin, N.J.
Leetch, Brian, NYR	D	Chelios, Chris, Chi.
Ozolinsh, Sandis, Col.	D	Stevens, Scott, N.J.
Lemieux, Mario, Pit.	C	Gretzky, Wayne, NYR
Selanne, Teemu, Ana.	RW	Jagr, Jaromir, Pit.
Kariya, Paul, Ana.	LW	LeClair, John, Phi.

1995-96

First Team		Second Team
Carey, Jim, Wsh.	G	Osgood, Chris, Det.
Chelios, Chris, Chi.	D	Konstantinov, V., Det.
Bourque, Raymond, Bos.	D	Leetch, Brian, NYR
Lemieux, Mario, Pit.	C	Lindros, Eric, Phi.
Jagr, Jaromir, Pit.	RW	Mogilny, Alexander, Van.
Kariya, Paul, Ana.	LW	LeClair, John, Phi.

1994-95

First Team		Second Team
Hasek, Dominik, Buf.	G	Belfour, Ed, Chi.
Coffey, Paul, Det.	D	Bourque, Raymond, Bos.
Chelios, Chris, Chi.	D	Murphy, Larry, Pit.
Lindros, Eric, Phi.	C	Zhamnov, Alexei, Wpg.
Jagr, Jaromir, Pit.	RW	Fleury, Theoren, Cgy.
LeClair, John, Mtl., Phi.	LW	Tkachuk, Keith, Wpg.

1993-94

First Team		Second Team
Hasek, Dominik, Buf.	G	Vanbiesbrouck, John, Fla.
Bourque, Raymond, Bos.	D	MacInnis, Al, Cgy.
Stevens, Scott, N.J.	D	Leetch, Brian, NYR
Fedorov, Sergei, Det.	C	Gretzky, Wayne, L.A.
Bure, Pavel, Van.	RW	Neely, Cam, Bos.
Shanahan, Brendan, St.L.	LW	Graves, Adam, NYR

1992-93

First Team		Second Team
Belfour, Ed, Chi.	G	Barrasso, Tom, Pit.
Chelios, Chris, Chi.	D	Murphy, Larry, Pit.
Bourque, Raymond, Bos.	D	Iafrate, Al, Wsh.
Lemieux, Mario, Pit.	C	LaFontaine, Pat, Buf.
Selanne, Teemu, Wpg.	RW	Mogilny, Alexander, Buf.
Robitaille, Luc, L.A.	LW	Stevens, Kevin, Pit.

1991-92

First Team		Second Team
Roy, Patrick, Mtl.	G	McLean, Kirk, Van.
Leetch, Brian, NYR	D	Housley, Phil, Wpg.
Bourque, Raymond, Bos.	D	Stevens, Scott, N.J.
Messier, Mark, NYR	C	Lemieux, Mario, Pit.
Hull, Brett, St.L.	RW	Recchi, Mark, Pit., Phi.
Stevens, Kevin, Pit.	LW	Robitaille, Luc, L.A.

1990-91

First Team		Second Team
Belfour, Ed, Chi.	G	Roy, Patrick, Mtl.
Bourque, Raymond, Bos.	D	Chelios, Chris, Chi.
MacInnis, Al, Cgy.	D	Leetch, Brian, NYR
Gretzky, Wayne, L.A.	C	Oates, Adam, St.L.
Hull, Brett, St.L.	RW	Neely, Cam, Bos.
Robitaille, Luc, L.A.	LW	Stevens, Kevin, Pit.

1989-90

First Team		Second Team
Roy, Patrick, Mtl.	G	Puppa, Daren, Buf.
Bourque, Raymond, Bos.	D	Coffey, Paul, Pit.
MacInnis, Al, Cgy.	D	Wilson, Doug, Chi.
Messier, Mark, Edm.	C	Gretzky, Wayne, L.A.
Hull, Brett, St.L.	RW	Neely, Cam, Bos.
Robitaille, Luc, L.A.	LW	Bellows, Brian, Min.

1988-89

First Team		Second Team
Roy, Patrick, Mtl.	G	Vernon, Mike, Cgy.
Chelios, Chris, Mtl.	D	MacInnis, Al, Cgy.
Coffey, Paul, Pit.	D	Bourque, Raymond, Bos.
Lemieux, Mario, Pit.	C	Gretzky, Wayne, L.A.
Mullen, Joe, Cgy.	RW	Kurri, Jari, Edm.
Robitaille, Luc, L.A.	LW	Gallant, Gerard, Det.

1987-88

First Team		Second Team
Fuhr, Grant, Edm.	G	Roy, Patrick, Mtl.
Bourque, Raymond, Bos.	D	Suter, Gary, Cgy.
Stevens, Scott, Wsh.	D	McCrimmon, Brad, Cgy.
Lemieux, Mario, Pit.	C	Gretzky, Wayne, Edm.
Loob, Hakan, Cgy.	RW	Neely, Cam, Bos.
Robitaille, Luc, L.A.	LW	Goulet, Michel, Que.

1986-87

First Team		Second Team
Hextall, Ron, Phi.	G	Liut, Mike, Hfd.
Bourque, Raymond, Bos.	D	Murphy, Larry, Wsh.
Howe, Mark, Phi.	D	MacInnis, Al, Cgy.
Gretzky, Wayne, Edm.	C	Lemieux, Mario, Pit.
Kurri, Jari, Edm.	RW	Kerr, Tim, Phi.
Goulet, Michel, Que.	LW	Robitaille, Luc, L.A.

1985-86

First Team		Second Team
Vanbiesbrouck, John, NYR	G	Froese, Bob, Phi.
Coffey, Paul, Edm.	D	Robinson, Larry, Mtl.
Howe, Mark, Phi.	D	Bourque, Raymond, Bos.
Gretzky, Wayne, Edm.	C	Lemieux, Mario, Pit.
Bossy, Mike, NYI	RW	Kurri, Jari, Edm.
Goulet, Michel, Que.	LW	Naslund, Mats, Mtl.

1984-85

First Team		Second Team
Lindbergh, Pelle, Phi.	G	Barrasso, Tom, Buf.
Coffey, Paul, Edm.	D	Langway, Rod, Wsh.
Bourque, Raymond, Bos.	D	Wilson, Doug, Chi.
Gretzky, Wayne, Edm.	C	Hawerchuk, Dale, Wpg.
Kurri, Jari, Edm.	RW	Bossy, Mike, NYI
Ogrodnick, John, Det.	LW	Tonelli, John, NYI

1983-84

First Team		Second Team
Barrasso, Tom, Buf.	G	Riggin, Pat, Wsh.
Langway, Rod, Wsh.	D	Coffey, Paul, Edm.
Bourque, Raymond, Bos.	D	Potvin, Denis, NYI
Gretzky, Wayne, Edm.	C	Trottier, Bryan, NYI
Bossy, Mike, NYI	RW	Kurri, Jari, Edm.
Goulet, Michel, Que.	LW	Messier, Mark, Edm.

1982-83

First Team		Second Team
Peeters, Pete, Bos.	G	Melanson, Rollie, NYI
Howe, Mark, Phi.	D	Bourque, Raymond, Bos.
Langway, Rod, Wsh.	D	Coffey, Paul, Edm.
Gretzky, Wayne, Edm.	C	Savard, Denis, Chi.
Bossy, Mike, NYI	RW	McDonald, Lanny, Cgy.
Messier, Mark, Edm.	LW	Goulet, Michel, Que.

1981-82

First Team		Second Team
Smith, Billy, NYI	G	Fuhr, Grant, Edm.
Wilson, Doug, Chi.	D	Coffey, Paul, Edm.
Bourque, Raymond, Bos.	D	Engblom, Brian, Mtl.
Gretzky, Wayne, Edm.	C	Trottier, Bryan, NYI
Bossy, Mike, NYI	RW	Middleton, Rick, Bos.
Messier, Mark, Edm.	LW	Tonelli, John, NYI

Though he beat out his childhood hero for both the Vezina and Hart trophies in 2001-02, Montreal's Jose Theodore had to settle for a Second-Team All-Star selection behind First-Team goaltender Patrick Roy of Colorado.

1980-81

First Team		Second Team
Liut, Mike, St.L.	*G*	Lessard, Mario, L.A.
Potvin, Denis, NYI	*D*	Robinson, Larry, Mtl.
Carlyle, Randy, Pit.	*D*	Bourque, Raymond, Bos.
Gretzky, Wayne, Edm.	*C*	Dionne, Marcel, L.A.
Bossy, Mike, NYI	*RW*	Taylor, Dave, L.A.
Simmer, Charlie, L.A.	*LW*	Barber, Bill, Phi.

1979-80

First Team		Second Team
Esposito, Tony, Chi.	*G*	Edwards, Don, Buf.
Robinson, Larry, Mtl.	*D*	Salming, Borje, Tor.
Bourque, Raymond, Bos.	*D*	Schoenfeld, Jim, Buf.
Dionne, Marcel, L.A.	*C*	Gretzky, Wayne, Edm.
Lafleur, Guy, Mtl.	*RW*	Gare, Danny, Buf.
Simmer, Charlie, L.A.	*LW*	Shutt, Steve, Mtl.

1978-79

First Team		Second Team
Dryden, Ken, Mtl.	*G*	Resch, Glenn, NYI
Potvin, Denis, NYI	*D*	Salming, Borje, Tor.
Robinson, Larry, Mtl.	*D*	Savard, Serge, Mtl.
Trottier, Bryan, NYI	*C*	Dionne, Marcel, L.A.
Lafleur, Guy, Mtl.	*RW*	Bossy, Mike, NYI
Gillies, Clark, NYI	*LW*	Barber, Bill, Phi.

1977-78

First Team		Second Team
Dryden, Ken, Mtl.	*G*	Edwards, Don, Buf.
Potvin, Denis, NYI	*D*	Robinson, Larry, Mtl.
Park, Brad, Bos.	*D*	Salming, Borje, Tor.
Trottier, Bryan, NYI	*C*	Sittler, Darryl, Tor.
Lafleur, Guy, Mtl.	*RW*	Bossy, Mike, NYI
Gillies, Clark, NYI	*LW*	Shutt, Steve, Mtl.

1976-77

First Team		Second Team
Dryden, Ken, Mtl.	*G*	Vachon, Rogie, L.A.
Robinson, Larry, Mtl.	*D*	Potvin, Denis, NYI
Salming, Borje, Tor.	*D*	Lapointe, Guy, Mtl.
Dionne, Marcel, L.A.	*C*	Perreault, Gilbert, Buf.
Lafleur, Guy, Mtl.	*RW*	McDonald, Lanny, Tor.
Shutt, Steve, Mtl.	*LW*	Martin, Rick, Buf.

1975-76

First Team		Second Team
Dryden, Ken, Mtl.	*G*	Resch, Glenn, NYI
Potvin, Denis, NYI	*D*	Salming, Borje, Tor.
Park, Brad, Bos.	*D*	Lapointe, Guy, Mtl.
Clarke, Bobby, Phi.	*C*	Perreault, Gilbert, Buf.
Lafleur, Guy, Mtl.	*RW*	Leach, Reggie, Phi.
Barber, Bill, Phi.	*LW*	Martin, Rick, Buf.

1974-75

First Team		Second Team
Parent, Bernie, Phi.	*G*	Vachon, Rogie, L.A.
Orr, Bobby, Bos.	*D*	Lapointe, Guy, Mtl.
Potvin, Denis, NYI	*D*	Salming, Borje, Tor.
Clarke, Bobby, Phi.	*C*	Esposito, Phil, Bos.
Lafleur, Guy, Mtl.	*RW*	Robert, René, Buf.
Martin, Rick, Buf.	*LW*	Vickers, Steve, NYR

1973-74

First Team		Second Team
Parent, Bernie, Phi.	*G*	Esposito, Tony, Chi.
Orr, Bobby, Bos.	*D*	White, Bill, Chi.
Park, Brad, NYR	*D*	Ashbee, Barry, Phi.
Esposito, Phil, Bos.	*C*	Clarke, Bobby, Phi.
Hodge, Ken, Bos.	*RW*	Redmond, Mickey, Det.
Martin, Rick, Buf.	*LW*	Cashman, Wayne, Bos.

1972-73

First Team		Second Team
Dryden, Ken, Mtl.	*G*	Esposito, Tony, Chi.
Orr, Bobby, Bos.	*D*	Park, Brad, NYR
Lapointe, Guy, Mtl.	*D*	White, Bill, Chi.
Esposito, Phil, Bos.	*C*	Clarke, Bobby, Phi.
Redmond, Mickey, Det.	*RW*	Cournoyer, Yvan, Mtl.
Mahovlich, Frank, Mtl.	*LW*	Hull, Dennis, Chi.

1971-72

First Team		Second Team
Esposito, Tony, Chi.	*G*	Dryden, Ken, Mtl.
Orr, Bobby, Bos.	*D*	White, Bill, Chi.
Park, Brad, NYR	*D*	Stapleton, Pat, Chi.
Esposito, Phil, Bos.	*C*	Ratelle, Jean, NYR
Gilbert, Rod, NYR	*RW*	Cournoyer, Yvan, Mtl.
Hull, Bobby, Chi.	*LW*	Hadfield, Vic, NYR

1970-71

First Team		Second Team
Giacomin, Ed, NYR	*G*	Plante, Jacques, Tor.
Orr, Bobby, Bos.	*D*	Park, Brad, NYR
Tremblay, J.C., Mtl.	*D*	Stapleton, Pat, Chi.
Esposito, Phil, Bos.	*C*	Keon, Dave, Tor.
Hodge, Ken, Bos.	*RW*	Cournoyer, Yvan, Mtl.
Bucyk, John, Bos.	*LW*	Hull, Bobby, Chi.

1969-70

First Team		Second Team
Esposito, Tony, Chi.	*G*	Giacomin, Ed, NYR
Orr, Bobby, Bos.	*D*	Brewer, Carl, Det.
Park, Brad, NYR	*D*	Laperriere, Jacques, Mtl.
Esposito, Phil, Bos.	*C*	Mikita, Stan, Chi.
Howe, Gordie, Det.	*RW*	McKenzie, John, Bos.
Hull, Bobby, Chi.	*LW*	Mahovlich, Frank, Det.

1968-69

First Team		Second Team
Hall, Glenn, St.L.	*G*	Giacomin, Ed, NYR
Orr, Bobby, Bos.	*D*	Green, Ted, Bos.
Horton, Tim, Tor.	*D*	Harris, Ted, Mtl.
Esposito, Phil, Bos.	*C*	Béliveau, Jean, Mtl.
Howe, Gordie, Det.	*RW*	Cournoyer, Yvan, Mtl.
Hull, Bobby, Chi.	*LW*	Mahovlich, Frank, Det.

1967-68

First Team		Second Team
Worsley, Gump, Mtl.	*G*	Giacomin, Ed, NYR
Orr, Bobby, Bos.	*D*	Tremblay, J.C., Mtl.
Horton, Tim, Tor.	*D*	Neilson, Jim, NYR
Mikita, Stan, Chi.	*C*	Esposito, Phil, Bos.
Howe, Gordie, Det.	*RW*	Gilbert, Rod, NYR
Hull, Bobby, Chi.	*LW*	Bucyk, John, Bos.

1966-67

First Team		Second Team
Giacomin, Ed, NYR	*G*	Hall, Glenn, Chi.
Pilote, Pierre, Chi.	*D*	Horton, Tim, Tor.
Howell, Harry, NYR	*D*	Orr, Bobby, Bos.
Mikita, Stan, Chi.	*C*	Ullman, Norm, Det.
Wharram, Kenny, Chi.	*RW*	Howe, Gordie, Det.
Hull, Bobby, Chi.	*LW*	Marshall, Don, NYR

1965-66

First Team		Second Team
Hall, Glenn, Chi.	*G*	Worsley, Gump, Mtl.
Laperriere, Jacques, Mtl.	*D*	Stanley, Allan, Tor.
Pilote, Pierre, Chi.	*D*	Stapleton, Pat, Chi.
Mikita, Stan, Chi.	*C*	Béliveau, Jean, Mtl.
Howe, Gordie, Det.	*RW*	Rousseau, Bobby, Mtl.
Hull, Bobby, Chi.	*LW*	Mahovlich, Frank, Tor.

1964-65

First Team		Second Team
Crozier, Roger, Det.	*G*	Hodge, Charlie, Mtl.
Pilote, Pierre, Chi.	*D*	Gadsby, Bill, Det.
Laperriere, Jacques, Mtl.	*D*	Brewer, Carl, Tor.
Ullman, Norm, Det.	*C*	Mikita, Stan, Chi.
Provost, Claude, Mtl.	*RW*	Howe, Gordie, Det.
Hull, Bobby, Chi.	*LW*	Mahovlich, Frank, Tor.

1963-64

First Team		Second Team
Hall, Glenn, Chi.	*G*	Hodge, Charlie, Mtl.
Pilote, Pierre, Chi.	*D*	Vasko, Moose, Chi.
Horton, Tim, Tor.	*D*	Laperriere, Jacques, Mtl.
Mikita, Stan, Chi.	*C*	Béliveau, Jean, Mtl.
Wharram, Kenny, Chi.	*RW*	Howe, Gordie, Det.
Hull, Bobby, Chi.	*LW*	Mahovlich, Frank, Tor.

1962-63

First Team		Second Team
Hall, Glenn, Chi.	*G*	Sawchuk, Terry, Det.
Pilote, Pierre, Chi.	*D*	Horton, Tim, Tor.
Brewer, Carl, Tor.	*D*	Vasko, Moose, Chi.
Mikita, Stan, Chi.	*C*	Richard, Henri, Mtl.
Howe, Gordie, Det.	*RW*	Bathgate, Andy, NYR
Mahovlich, Frank, Tor.	*LW*	Hull, Bobby, Chi.

1961-62

First Team		Second Team
Plante, Jacques, Mtl.	*G*	Hall, Glenn, Chi.
Harvey, Doug, NYR	*D*	Brewer, Carl, Tor.
Talbot, Jean-Guy, Mtl.	*D*	Pilote, Pierre, Chi.
Mikita, Stan, Chi.	*C*	Keon, Dave, Tor.
Bathgate, Andy, NYR	*RW*	Howe, Gordie, Det.
Hull, Bobby, Chi.	*LW*	Mahovlich, Frank, Tor.

1960-61

First Team		Second Team
Bower, Johnny, Tor.	*G*	Hall, Glenn, Chi.
Harvey, Doug, Mtl.	*D*	Stanley, Allan, Tor.
Pronovost, Marcel, Det.	*D*	Pilote, Pierre, Chi.
Béliveau, Jean, Mtl.	*C*	Richard, Henri, Mtl.
Geoffrion, Bernie, Mtl.	*RW*	Howe, Gordie, Det.
Mahovlich, Frank, Tor.	*LW*	Moore, Dickie, Mtl.

1959-60

First Team		Second Team
Hall, Glenn, Chi.	*G*	Plante, Jacques, Mtl.
Harvey, Doug, Mtl.	*D*	Stanley, Allan, Tor.
Pronovost, Marcel, Det.	*D*	Pilote, Pierre, Chi.
Béliveau, Jean, Mtl.	*C*	Horvath, Bronco, Bos.
Howe, Gordie, Det.	*RW*	Geoffrion, Bernie, Mtl.
Hull, Bobby, Chi.	*LW*	Prentice, Dean, NYR

1958-59

First Team		Second Team
Plante, Jacques, Mtl.	*G*	Sawchuk, Terry, Det.
Johnson, Tom, Mtl.	*D*	Pronovost, Marcel, Det.
Gadsby, Bill, NYR	*D*	Harvey, Doug, Mtl.
Béliveau, Jean, Mtl.	*C*	Richard, Henri, Mtl.
Bathgate, Andy, NYR	*RW*	Howe, Gordie, Det.
Moore, Dickie, Mtl.	*LW*	Delvecchio, Alex, Det.

1957-58

First Team		Second Team
Hall, Glenn, Chi.	*G*	Plante, Jacques, Mtl.
Harvey, Doug, Mtl.	*D*	Flaman, Fern, Bos.
Gadsby, Bill, NYR	*D*	Pronovost, Marcel, Det.
Richard, Henri, Mtl.	*C*	Béliveau, Jean, Mtl.
Howe, Gordie, Det.	*RW*	Bathgate, Andy, NYR
Moore, Dickie, Mtl.	*LW*	Henry, Camille, NYR

1956-57

First Team		Second Team
Hall, Glenn, Det.	*G*	Plante, Jacques, Mtl.
Harvey, Doug, Mtl.	*D*	Flaman, Fern, Bos.
Kelly, Red, Det.	*D*	Gadsby, Bill, NYR
Béliveau, Jean, Mtl.	*C*	Litzenberger, Ed, Chi.
Howe, Gordie, Det.	*RW*	Richard, Maurice, Mtl.
Lindsay, Ted, Det.	*LW*	Chevrefils, Real, Bos.

1955-56

First Team		Second Team
Plante, Jacques, Mtl.	*G*	Hall, Glenn, Det.
Harvey, Doug, Mtl.	*D*	Kelly, Red, Det.
Gadsby, Bill, NYR	*D*	Johnson, Tom, Mtl.
Béliveau, Jean, Mtl.	*C*	Sloan, Tod, Tor.
Richard, Maurice, Mtl.	*RW*	Howe, Gordie, Det.
Lindsay, Ted, Det.	*LW*	Olmstead, Bert, Mtl.

1954-55

First Team		Second Team
Lumley, Harry, Tor.	*G*	Sawchuk, Terry, Det.
Harvey, Doug, Mtl.	*D*	Goldham, Bob, Det.
Kelly, Red, Det.	*D*	Flaman, Fern, Bos.
Béliveau, Jean, Mtl.	*C*	Mosdell, Ken, Mtl.
Richard, Maurice, Mtl.	*RW*	Geoffrion, Bernie, Mtl.
Smith, Sid, Tor.	*LW*	Lewicki, Danny, NYR

1953-54

First Team		Second Team
Lumley, Harry, Tor.	G	Sawchuk, Terry, Det.
Kelly, Red, Det.	D	Gadsby, Bill, Chi.
Harvey, Doug, Mtl.	D	Horton, Tim, Tor.
Mosdell, Ken, Mtl.	C	Kennedy, Ted, Tor.
Howe, Gordie, Det.	RW	Richard, Maurice, Mtl.
Lindsay, Ted, Det.	LW	Sandford, Ed, Bos.

1952-53

First Team		Second Team
Sawchuk, Terry, Det.	G	McNeil, Gerry, Mtl.
Kelly, Red, Det.	D	Quackenbush, Bill, Bos.
Harvey, Doug, Mtl.	D	Gadsby, Bill, Chi.
Mackell, Fleming, Bos.	C	Delvecchio, Alex, Det.
Howe, Gordie, Det.	RW	Richard, Maurice, Mtl.
Lindsay, Ted, Det.	LW	Olmstead, Bert, Mtl.

1951-52

First Team		Second Team
Sawchuk, Terry, Det.	G	Henry, Jim, Bos.
Kelly, Red, Det.	D	Buller, Hy, NYR
Harvey, Doug, Mtl.	D	Thomson, Jimmy, Tor.
Lach, Elmer, Mtl.	C	Schmidt, Milt, Bos.
Howe, Gordie, Det.	RW	Richard, Maurice, Mtl.
Lindsay, Ted, Det.	LW	Smith, Sid, Tor.

1950-51

First Team		Second Team
Sawchuk, Terry, Det.	G	Rayner, Chuck, NYR
Kelly, Red, Det.	D	Thomson, Jimmy, Tor.
Quackenbush, Bill, Bos.	D	Reise Jr., Leo, Det.
Schmidt, Milt, Bos.	C	Abel, Sid, Det.
	(tied)	Kennedy, Ted, Tor.
Howe, Gordie, Det.	RW	Richard, Maurice, Mtl.
Lindsay, Ted, Det.	LW	Smith, Sid, Tor.

1949-50

First Team		Second Team
Durnan, Bill, Mtl.	G	Rayner, Chuck, NYR
Mortson, Gus, Tor.	D	Reise Jr., Leo, Det.
Reardon, Ken, Mtl.	D	Kelly, Red, Det.
Abel, Sid, Det.	C	Kennedy, Ted, Tor.
Richard, Maurice, Mtl.	RW	Howe, Gordie, Det.
Lindsay, Ted, Det.	LW	Leswick, Tony, NYR

1948-49

First Team		Second Team
Durnan, Bill, Mtl.	G	Rayner, Chuck, NYR
Quackenbush, Bill, Det.	D	Harmon, Glen, Mtl.
Stewart, Jack, Det.	D	Reardon, Ken, Mtl.
Abel, Sid, Det.	C	Bentley, Doug, Chi.
Richard, Maurice, Mtl.	RW	Howe, Gordie, Det.
Conacher, Roy, Chi.	LW	Lindsay, Ted, Det.

1947-48

First Team		Second Team
Broda, Turk, Tor.	G	Brimsek, Frank, Bos.
Quackenbush, Bill, Det.	D	Reardon, Ken, Mtl.
Stewart, Jack, Det.	D	Colville, Neil, NYR
Lach, Elmer, Mtl.	C	O'Connor, Buddy, NYR
Richard, Maurice, Mtl.	RW	Poile, Bud, Chi.
Lindsay, Ted, Det.	LW	Stewart, Gaye, Chi.

1946-47

First Team		Second Team
Durnan, Bill, Mtl.	G	Brimsek, Frank, Bos.
Reardon, Ken, Mtl.	D	Stewart, Jack, Det.
Bouchard, Butch, Mtl.	D	Quackenbush, Bill, Det.
Schmidt, Milt, Bos.	C	Bentley, Max, Chi.
Richard, Maurice, Mtl.	RW	Bauer, Bobby, Bos.
Bentley, Doug, Chi.	LW	Dumart, Woody, Bos.

1945-46

First Team		Second Team
Durnan, Bill, Mtl.	G	Brimsek, Frank, Bos.
Crawford, Jack, Bos.	D	Reardon, Ken, Mtl.
Bouchard, Butch, Mtl.	D	Stewart, Jack, Det.
Bentley, Max, Chi.	C	Lach, Elmer, Mtl.
Richard, Maurice, Mtl.	RW	Mosienko, Bill, Chi.
Stewart, Gaye, Tor.	LW	Blake, Toe, Mtl.
Irvin, Dick, Mtl.	Coach	Gottselig, Johnny, Chi.

1944-45

First Team		Second Team
Durnan, Bill, Mtl.	G	Karakas, Mike, Chi.
Bouchard, Butch, Mtl.	D	Harmon, Glen, Mtl.
Hollett, Flash, Det.	D	Pratt, Babe, Tor.
Lach, Elmer, Mtl.	C	Cowley, Bill, Bos.
Richard, Maurice, Mtl.	RW	Mosienko, Bill, Chi.
Blake, Toe, Mtl.	LW	Howe, Syd, Det.
Irvin, Dick, Mtl.	Coach	Adams, Jack, Det.

1943-44

First Team		Second Team
Durnan, Bill, Mtl.	G	Bibeault, Paul, Tor.
Seibert, Earl, Chi.	D	Bouchard, Butch, Mtl.
Pratt, Babe, Tor.	D	Clapper, Dit, Bos.
Cowley, Bill, Bos.	C	Lach, Elmer, Mtl.
Carr, Lorne, Tor.	RW	Richard, Maurice, Mtl.
Bentley, Doug, Chi.	LW	Cain, Herb, Bos.
Irvin, Dick, Mtl.	Coach	Day, Hap, Tor.

1942-43

First Team		Second Team
Mowers, Johnny, Det.	G	Brimsek, Frank, Bos.
Seibert, Earl, Chi.	D	Crawford, Jack, Bos.
Stewart, Jack, Det.	D	Hollett, Flash, Bos.
Cowley, Bill, Bos.	C	Apps, Syl, Tor.
Carr, Lorne, Tor.	RW	Hextall, Bryan, NYR
Bentley, Doug, Chi.	LW	Patrick, Lynn, NYR
Adams, Jack, Det.	Coach	Ross, Art, Bos.

1941-42

First Team		Second Team
Brimsek, Frank, Bos.	G	Broda, Turk, Tor.
Seibert, Earl, Chi.	D	Egan, Pat, Bro.
Anderson, Tom, Bro.	D	McDonald, Bucko, Tor.
Apps, Syl, Tor.	C	Watson, Phil, NYR
Hextall, Bryan, NYR	RW	Drillon, Gordie, Tor.
Patrick, Lynn, NYR	LW	Abel, Sid, Det.
Boucher, Frank, NYR	Coach	Thompson, Paul, Chi.

1940-41

First Team		Second Team
Broda, Turk, Tor.	G	Brimsek, Frank, Bos.
Clapper, Dit, Bos.	D	Seibert, Earl, Chi.
Stanowski, Wally, Tor.	D	Heller, Ott, NYR
Cowley, Bill, Bos.	C	Apps, Syl, Tor.
Hextall, Bryan, NYR	RW	Bauer, Bobby, Bos.
Schriner, Sweeney, Tor.	LW	Dumart, Woody, Bos.
Weiland, Cooney, Bos.	Coach	Irvin, Dick, Mtl.

1939-40

First Team		Second Team
Kerr, Dave, NYR	G	Brimsek, Frank, Bos.
Clapper, Dit, Bos.	D	Coulter, Art, NYR
Goodfellow, Ebbie, Det.	D	Seibert, Earl, Chi.
Schmidt, Milt, Bos.	C	Colville, Neil, NYR
Hextall, Bryan, NYR	RW	Bauer, Bobby, Bos.
Blake, Toe, Mtl.	LW	Dumart, Woody, Bos.
Thompson, Paul, Chi.	Coach	Boucher, Frank, NYR

1938-39

First Team		Second Team
Brimsek, Frank, Bos.	G	Robertson, Earl, NYA
Shore, Eddie, Bos.	D	Seibert, Earl, Chi.
Clapper, Dit, Bos.	D	Coulter, Art, NYR
Apps, Syl, Tor.	C	Colville, Neil, NYR
Drillon, Gordie, Tor.	RW	Bauer, Bobby, Bos.
Blake, Toe, Mtl.	LW	Gottselig, Johnny, Chi.
Ross, Art, Bos.	Coach	Dutton, Red, NYA

1937-38

First Team		Second Team
Thompson, Tiny, Bos.	G	Kerr, Dave, NYR
Shore, Eddie, Bos.	D	Coulter, Art, NYR
Siebert, Babe, Mtl.	D	Seibert, Earl, Chi.
Cowley, Bill, Bos.	C	Apps, Syl, Tor.
Dillon, Cecil, NYR	RW	
Drillon, Gordie, Tor.	(tied)	
Thompson, Paul, Chi.	LW	Blake, Toe, Mtl.
Patrick, Lester, NYR	Coach	Ross, Art, Bos.

1936-37

First Team		Second Team
Smith, Normie, Det.	G	Cude, Wilf, Mtl.
Siebert, Babe, Mtl.	D	Seibert, Earl, Chi.
Goodfellow, Ebbie, Det.	D	Conacher, Lionel, Mtl. M.
Barry, Marty, Det.	C	Chapman, Art, NYA
Aurie, Larry, Det.	RW	Dillon, Cecil, NYR
Jackson, Busher, Tor.	LW	Schriner, Sweeney, NYA
Adams, Jack, Det.	Coach	Hart, Cecil, Mtl.

1935-36

First Team		Second Team
Thompson, Tiny, Bos.	G	Cude, Wilf, Mtl.
Shore, Eddie, Bos.	D	Seibert, Earl, Chi.
Siebert, Babe, Bos.	D	Goodfellow, Ebbie, Det.
Smith, Hooley, Mtl. M.	C	Thoms, Bill, Tor.
Conacher, Charlie, Tor.	RW	Dillon, Cecil, NYR
Schriner, Sweeney, NYA	LW	Thompson, Paul, Chi.
Patrick, Lester, NYR	Coach	Gorman, Tommy, Mtl. M.

1934-35

First Team		Second Team
Chabot, Lorne, Chi.	G	Thompson, Tiny, Bos.
Shore, Eddie, Bos.	D	Wentworth, Cy, Mtl. M.
Seibert, Earl, NYR	D	Coulter, Art, Chi.
Boucher, Frank, NYR	C	Weiland, Cooney, Det.
Conacher, Charlie, Tor.	RW	Clapper, Dit, Bos.
Jackson, Busher, Tor.	LW	Joliat, Aurel, Mtl.
Patrick, Lester, NYR	Coach	Irvin, Dick, Tor.

1933-34

First Team		Second Team
Gardiner, Charlie, Chi.	G	Worters, Roy, NYA
Clancy, King, Tor.	D	Shore, Eddie, Bos.
Conacher, Lionel, Chi.	D	Johnson, Ching, NYR
Boucher, Frank, NYR	C	Primeau, Joe, Tor.
Conacher, Charlie, Tor.	RW	Cook, Bill, NYR
Jackson, Busher, Tor.	LW	Joliat, Aurel, Mtl.
Patrick, Lester, NYR	Coach	Irvin, Dick, Tor.

1932-33

First Team		Second Team
Roach, John Ross, Det.	G	Gardiner, Charlie, Chi.
Shore, Eddie, Bos.	D	Clancy, King, Tor.
Johnson, Ching, NYR	D	Conacher, Lionel, Mtl. M.
Boucher, Frank, NYR	C	Morenz, Howie, Mtl.
Cook, Bill, NYR	RW	Conacher, Charlie, Tor.
Northcott, Baldy, Mtl M.	LW	Jackson, Busher, Tor.
Patrick, Lester, NYR	Coach	Irvin, Dick, Tor.

1931-32

First Team		Second Team
Gardiner, Charlie, Chi.	G	Worters, Roy, NYA
Shore, Eddie, Bos.	D	Mantha, Sylvio, Mtl.
Johnson, Ching, NYR	D	Clancy, King, Tor.
Morenz, Howie, Mtl.	C	Smith, Hooley, Mtl. M.
Cook, Bill, NYR	RW	Conacher, Charlie, Tor.
Jackson, Busher, Tor.	LW	Joliat, Aurel, Mtl.
Patrick, Lester, NYR	Coach	Irvin, Dick, Tor.

1930-31

First Team		Second Team
Gardiner, Charlie, Chi.	G	Thompson, Tiny, Bos.
Shore, Eddie, Bos.	D	Mantha, Sylvio, Mtl.
Clancy, King, Tor.	D	Johnson, Ching, NYR
Morenz, Howie, Mtl.	C	Boucher, Frank, NYR
Cook, Bill, NYR	RW	Clapper, Dit, Bos.
Joliat, Aurel, Mtl.	LW	Cook, Bun, NYR
Patrick, Lester, NYR	Coach	Irvin, Dick, Chi.

With his balding head and ever-present smile, Ching Johnson was a fan favorite in the early years of the New York Rangers. He and Eddie Shore were First-Team All-Stars in 1932 and 1933.

All-Star Game Results

Year	Venue	Score	Coaches	Attendance
2002	Los Angeles	World 8, North America 5	Scotty Bowman, Pat Quinn	18,118
2001	Colorado	North America 14, World 12	Joel Quenneville, Jacques Martin	18,646
2000	Toronto	World 9, North America 4	Scotty Bowman, Pat Quinn	19,300
1999	Tampa Bay	North America 8, World 6	Lindy Ruff, Ken Hitchcock	19,758
1998	Vancouver	North America 8, World 7	Jacques Lemaire, Ken Hitchcock	18,422
1997	San Jose	East 11, West 7	Doug MacLean, Ken Hitchcock	17,422
1996	Boston	East 5, West 4	Doug MacLean, Scotty Bowman	17,565
1994	NY Rangers	East 9, West 8	Jacques Demers, Barry Melrose	18,200
1993	Montreal	Wales 16, Campbell 6	Scotty Bowman, Mike Keenan	17,137
1992	Philadelphia	Campbell 10, Wales 6	Bob Gainey, Scotty Bowman	17,380
1991	Chicago	Campbell 11, Wales 5	John Muckler, Mike Milbury	18,472
1990	Pittsburgh	Wales 12, Campbell 7	Pat Burns, Terry Crisp	16,236
1989	Edmonton	Campbell 9, Wales 5	Glen Sather, Terry O'Reilly	17,503
1988	St. Louis	Wales 6, Campbell 5 OT	Mike Keenan, Glen Sather	17,878
1986	Hartford	Wales 4, Campbell 3 OT	Mike Keenan, Glen Sather	15,100
1985	Calgary	Wales 6, Campbell 4	Al Arbour, Glen Sather	16,825
1984	New Jersey	Wales 7, Campbell 6	Al Arbour, Glen Sather	18,939
1983	NY Islanders	Campbell 9, Wales 3	Roger Neilson, Al Arbour	15,230
1982	Washington	Wales 4, Campbell 2	Al Arbour, Glen Sonmor	18,130
1981	Los Angeles	Campbell 4, Wales 1	Pat Quinn, Scotty Bowman	15,761
1980	Detroit	Wales 6, Campbell 3	Scotty Bowman, Al Arbour	21,002
1978	Buffalo	Wales 3, Campbell 2 OT	Scotty Bowman, Fred Shero	16,433
1977	Vancouver	Wales 4, Campbell 3	Scotty Bowman, Fred Shero	15,607
1976	Philadelphia	Wales 7, Campbell 5	Floyd Smith, Fred Shero	16,436
1975	Montreal	Wales 7, Campbell 1	Bep Guidolin, Fred Shero	16,080
1974	Chicago	West 6, East 4	Billy Reay, Scotty Bowman	16,426
1973	NY Rangers	East 5, West 4	Tom Johnson, Billy Reay	16,986
1972	Minnesota	East 3, West 2	Al MacNeil, Billy Reay	15,423
1971	Boston	West 2, East 1	Scotty Bowman, Harry Sinden	14,790
1970	St. Louis	East 4, West 1	Claude Ruel, Scotty Bowman	16,587
1969	Montreal	East 3, West 3	Toe Blake, Scotty Bowman	16,260
1968	Toronto	Toronto 4, All-Stars 3	Punch Imlach, Toe Blake	15,753
1967	Montreal	Montreal 3, All-Stars 0	Toe Blake, Sid Abel	14,284
1965	Montreal	All-Stars 5, Montreal 2	Billy Reay, Toe Blake	13,529
1964	Toronto	All-Stars 3, Toronto 2	Sid Abel, Punch Imlach	14,232
1963	Toronto	All-Stars 3, Toronto 3	Sid Abel, Punch Imlach	14,034
1962	Toronto	Toronto 4, All-Stars 1	Punch Imlach, Rudy Pilous	14,236
1961	Chicago	All-Stars 3, Chicago 1	Sid Abel, Rudy Pilous	14,534
1960	Montreal	All-Stars 2, Montreal 1	Punch Imlach, Toe Blake	13,949
1959	Montreal	Montreal 6, All-Stars 1	Toe Blake, Punch Imlach	13,818
1958	Montreal	Montreal 6, All-Stars 3	Toe Blake, Milt Schmidt	13,989
1957	Montreal	All-Stars 5, Montreal 3	Milt Schmidt, Toe Blake	13,003
1956	Montreal	All-Stars 1, Montreal 1	Jim Skinner, Toe Blake	13,095
1955	Detroit	Detroit 3, All-Stars 1	Jim Skinner, Dick Irvin	10,111
1954	Detroit	All-Stars 2, Detroit 2	King Clancy, Jim Skinner	10,689
1953	Montreal	All-Stars 3, Montreal 1	Lynn Patrick, Dick Irvin	14,153
1952	Detroit	1st Team 1, 2nd Team 1	Tommy Ivan, Dick Irvin	10,680
1951	Toronto	1st Team 2, 2nd Team 2	Joe Primeau, Dick Irvin	11,469
1950	Detroit	Detroit 7, All-Stars 1	Tommy Ivan, Lynn Patrick	9,166
1949	Toronto	All-Stars 3, Toronto 1	Tommy Ivan, Hap Day	13,541
1948	Chicago	All-Stars 3, Toronto 1	Tommy Ivan, Hap Day	12,794
1947	Toronto	All-Stars 4, Toronto 3	Dick Irvin, Hap Day	14,169

There was no All-Star contest during the calendar year of 1966 because the game was moved from the start of season to mid-season. In 1979, the Challenge Cup series between the Soviet Union and Team NHL replaced the All-Star Game. In 1987, Rendez-Vous '87, two games between the Soviet Union and Team NHL replaced the All-Star Game. Rendez-Vous '87 scores: game one, NHL All-Stars 4, Soviet Union 3; game two, Soviet Union 5, NHL All-Stars 3. There was no All-Star Game in 1995 due to a labor disruption.

2001-02 All-Star Game Summary

February 2, 2002 at Los Angeles — World 8, North America 5

PLAYERS ON ICE: **North America** — Burke, Roy, Theodore, R. Blake, Chelios, Jovanovski, Leetch, C. Pronger, Redden, V. Damphousse, Daze, Iginla, P. Kariya, M. Lemieux, O. Nolan, Parrish, Roenick,Sakic, Shanahan, J. Thornton, M. York.

The World — Hasek, Khabibulin, T. Salo, Gonchar, T. Kaberle, Lidstrom, Modry, Ozolinsh, Zhitnik, Demitra, Elias, S. Fedorov, Jagr, S. Kapanen, Knutsen, Naslund, Palffy, Selanne, Sundin, Yashin, Zhamnov.

SUMMARY

First Period

1. North America	Damphousse (1)	(Blake)	0:35
2. North America	Jovanovski (1)	(Damphousse, Daze)	11:22
3. World	Selanne (1)	(Zhamnov)	13:10
4. North America	Daze (1)	(unassisted)	15:05
5. World	Selanne (2)	(Kapanen, Kaberle)	17:26

PENALTIES: None

Second Period

6. North America	Lemieux (1)	(Kariya)	2:02
7. World	Naslund (1)	(Knutsen)	5:26
8. North America	Daze (2)	(Pronger, Damphousse)	11:33

PENALTIES: None

Third Period

9. World	Knutsen (1)	(Sundin, Naslund)	7:52
10. World	Fedorov (1)	(Gonchar)	16:59
11. World	Naslund (2)	(Sundin)	18:17
12. World	Zhamnov (1)	(Yashin)	19:12 (en)
13. World	Kapanen (1)	(Elias, Ozolinsh)	19:56 (en)

PENALTIES: None

SHOTS ON GOAL BY:

World	14	9	16	**39**
North America	13	17	20	**50**

	Goaltenders:	Time	SA	GA	ENG	Dec
World	Hasek	20:00	13	3	0	
World	Salo	20:00	17	2	0	
World	Khabibulin	20:00	20	0	0	W
N. America	Roy	20:00	14	2	0	
N. America	Theodore	20:00	9	1	0	
N. America	Burke	19:06	14	3	2	L

PP Conversions: North America 0/0; World 0/0.

Referees: Dave Jackson, Don Van Massenhoven

Linesmen: Andy McElman, Mark Pare

Attendance: 18,118.

NHL ALL-ROOKIE TEAM

Voting for the NHL All-Rookie Team is conducted among the representatives of the Professional Hockey Writers' Association at the end of the season. The rookie all-star team was first selected for the 1982-83 season.

2001-02

Dan Blackburn, NY Rangers	*Goal*
Nick Boynton, Boston	*Defense*
Rostislav Klesla, Columbus	*Defense*
Dany Heatley, Atlanta	*Forward*
Ilya Kovalchuk, Atlanta	*Forward*
Kristian Huselius, Florida	*Forward*

2000-01

Evgeni Nabokov, San Jose
Lubomir Visnovsky, Los Angeles
Colin White, New Jersey
Martin Havlat, Ottawa
Brad Richards, Tampa Bay
Shane Willis, Carolina

1999-2000

Brian Boucher, Philadelphia	*Goal*
Brian Rafalski, New Jersey	*Defense*
Brad Stuart, San Jose	*Defense*
Simon Gagne, Philadelphia	*Forward*
Scott Gomez, New Jersey	*Forward*
Michael York, NY Rangers	*Forward*

1998-99

Jamie Storr, Los Angeles
Tom Poti, Edmonton
Sami Salo, Ottawa
Chris Drury, Colorado
Milan Hejduk, Colorado
Marian Hossa, Ottawa

1997-98

Jamie Storr, Los Angeles	*Goal*
Mattias Ohlund, Vancouver	*Defense*
Derek Morris, Calgary	*Defense*
Sergei Samsonov, Boston	*Forward*
Patrick Elias, New Jersey	*Forward*
Mike Johnson, Toronto	*Forward*

1996-97

Patrick Lalime, Pittsburgh
Bryan Berard, NY Islanders
Janne Niinimaa, Philadelphia
Jarome Iginla, Calgary
Jim Campbell, St. Louis
Sergei Berezin, Toronto

1995-96

Corey Hirsch, Vancouver	*Goal*
Ed Jovanovski, Florida	*Defense*
Kyle McLaren, Boston	*Defense*
Daniel Alfredsson, Ottawa	*Forward*
Eric Daze, Chicago	*Forward*
Petr Sykora, New Jersey	*Forward*

1994-95

Jim Carey, Washington
Chris Therien, Philadelphia
Kenny Jonsson, Toronto
Peter Forsberg, Quebec
Jeff Friesen, San Jose
Paul Kariya, Anaheim

1993-94

Martin Brodeur, New Jersey	*Goal*
Chris Pronger, Hartford	*Defense*
Boris Mironov, Wpg./Edm.	*Defense*
Jason Arnott, Edmonton	*Center*
Mikael Renberg, Philadelphia	*Wing*
Oleg Petrov, Montreal	*Wing*

1992-93

Felix Potvin, Toronto
Vladimir Malakhov, NY Islanders
Scott Niedermayer, New Jersey
Eric Lindros, Philadelphia
Teemu Selanne, Winnipeg
Joe Juneau, Boston

1991-92

Dominik Hasek, Chicago	*Goal*
Nicklas Lidstrom, Detroit	*Defense*
Vladimir Konstantinov, Detroit	*Defense*
Kevin Todd, New Jersey	*Center*
Tony Amonte, NY Rangers	*Right Wing*
Gilbert Dionne, Montreal	*Left Wing*

1990-91

Ed Belfour, Chicago
Eric Weinrich, New Jersey
Rob Blake, Los Angeles
Sergei Fedorov, Detroit
Ken Hodge, Boston
Jaromir Jagr, Pittsburgh

1989-90

Bob Essensa, Winnipeg	*Goal*
Brad Shaw, Hartford	*Defense*
Geoff Smith, Edmonton	*Defense*
Mike Modano, Minnesota	*Center*
Sergei Makarov, Calgary	*Right Wing*
Rod Brind'Amour, St. Louis	*Left Wing*

1988-89

Peter Sidorkiewicz, Hartford
Brian Leetch, NY Rangers
Zarley Zalapski, Pittsburgh
Trevor Linden, Vancouver
Tony Granato, NY Rangers
David Volek, NY Islanders

1987-88

Darren Pang, Chicago	*Goal*
Glen Wesley, Boston	*Defense*
Calle Johansson, Buffalo	*Defense*
Joe Nieuwendyk, Calgary	*Center*
Ray Sheppard, Buffalo	*Right Wing*
Iain Duncan, Winnipeg	*Left Wing*

1986-87

Ron Hextall, Philadelphia
Steve Duchesne, Los Angeles
Brian Benning, St. Louis
Jimmy Carson, Los Angeles
Jim Sandlak, Vancouver
Luc Robitaille, Los Angeles

1985-86

Patrick Roy, Montreal	*Goal*
Gary Suter, Calgary	*Defense*
Dana Murzyn, Hartford	*Defense*
Mike Ridley, NY Rangers	*Center*
Kjell Dahlin, Montreal	*Right Wing*
Wendel Clark, Toronto	*Left Wing*

1984-85

Steve Penney, Montreal
Chris Chelios, Montreal
Bruce Bell, Quebec
Mario Lemieux, Pittsburgh
Tomas Sandstrom, NY Rangers
Warren Young, Pittsburgh

1983-84

Tom Barrasso, Buffalo	*Goal*
Thomas Eriksson, Philadelphia	*Defense*
Jamie Macoun, Calgary	*Defense*
Steve Yzerman, Detroit	*Center*
Hakan Loob, Calgary	*Right Wing*
Sylvain Turgeon, Hartford	*Left Wing*

1982-83

Pelle Lindbergh, Philadelphia
Scott Stevens, Washington
Phil Housley, Buffalo
Dan Daoust, Mtl./Tor.
Steve Larmer, Chicago
Mats Naslund, Montreal

All-Star Game Records 1947 through 2002

TEAM RECORDS

MOST GOALS, BOTH TEAMS, ONE GAME:
26 — North America 14, World 12, 2001 at Colorado
22 — Wales 16, Campbell 6, 1993 at Montreal
19 — Wales 12, Campbell 7, 1990 at Pittsburgh
18 — East 11, West 7, 1997 at San Jose
17 — East 9, West 8, 1994 at NY Rangers
16 — Campbell 11, Wales 5, 1991 at Chicago
— Campbell 10, Wales 6, 1992 at Philadelphia
15 — North America 8, World 7, 1998 at Vancouver

FEWEST GOALS, BOTH TEAMS, ONE GAME:
2 — NHL All-Stars 1, Montreal Canadiens 1, 1956 at Montreal
— First Team All-Stars 1, Second Team All-Stars 1, 1952 at Detroit
3 — West 2, East 1, 1971 at Boston
— Montreal Canadiens 3, NHL All-Stars 0, 1967 at Montreal
— NHL All-Stars 2, Montreal Canadiens 1, 1960 at Montreal

MOST GOALS, ONE TEAM, ONE GAME:
16 — Wales 16, Campbell 6, 1993 at Montreal
14 — North America 14, World 12, 2001 at Colorado
12 — Wales 12, Campbell 7, 1990 at Pittsburgh
— World 12, North America 14, 2001 at Colorado
11 — Campbell 11, Wales 5, 1991 at Chicago
— East 11, West 7, 1997 at San Jose

FEWEST GOALS, ONE TEAM, ONE GAME:
0 — NHL All-Stars 0, Montreal Canadiens 3, 1967 at Montreal
1 — 17 times (1981, 1975, 1971, 1970, 1962, 1961, 1960, 1959, both teams 1956, 1955, 1953, both teams 1952, 1950, 1949, 1948)

MOST SHOTS, BOTH TEAMS, ONE GAME (SINCE 1955):
102 — 1994 at NY Rangers — East 9 (56 shots), West 8 (46 shots)
98 — 2001 at Colorado — North America 14 (53 shots), World 12 (45 shots)
90 — 1993 at Montreal — Wales 16 (49 shots), Campbell 6 (41 shots)
89 — 2002 at Los Angeles — World 8 (39 shots), North America 5 (50 shots)

FEWEST SHOTS, BOTH TEAMS, ONE GAME (SINCE 1955):
52 — 1978 at Buffalo — Campbell 2 (12 shots), Wales 3 (40 shots)
53 — 1960 at Montreal — NHL All-Stars 2 (27 shots), Montreal Canadiens 1 (26 shots)
55 — 1956 at Montreal — NHL All-Stars 1 (28 shots), Montreal Canadiens 1 (27 shots)
— 1971 at Boston — West 2 (28 shots), East 1 (27 shots)

MOST SHOTS, ONE TEAM, ONE GAME (SINCE 1955):
56 — 1994 at NY Rangers — East (9-8 vs. West)
53 — 2001 at Colorado — North America (14-12 vs. World)
50 — 2002 at Los Angeles — North America (5-8 vs. World)
49 — 1993 at Montreal — Wales (16-6 vs. Campbell)
— 1999 at Tampa Bay — North America (8-6 vs. World)

FEWEST SHOTS, ONE TEAM, ONE GAME (SINCE 1955):
12 — 1978 at Buffalo — Campbell (2-3 vs. Wales)
17 — 1970 at St. Louis — West (1-4 vs. East)
23 — 1961 at Chicago — Chicago Black Hawks (1-3 vs. NHL All-Stars)
24 — 1976 at Philadelphia — Campbell (5-7 vs. Wales)

MOST POWER-PLAY GOALS, BOTH TEAMS, ONE GAME (SINCE 1950):
3 — 1953 at Montreal — NHL All-Stars 3 (2 power-play goals), Montreal Canadiens 1 (1 power-play goal)
— 1954 at Detroit — NHL All-Stars 2 (1 power-play goal) Detroit Red Wings 2 (2 power-play goals)
— 1958 at Montreal — NHL All-Stars 3 (1 power-play goal) Montreal Canadiens 6 (2 power-play goals)

FEWEST POWER-PLAY GOALS, BOTH TEAMS, ONE GAME (SINCE 1950):
0 — 20 times (1952, 1959, 1960, 1967, 1968, 1969, 1972, 1973, 1976, 1980, 1981, 1984, 1985, 1992, 1994, 1996, 1999, 2000, 2001, 2002)

FASTEST TWO GOALS, BOTH TEAMS, FROM START OF GAME:
37 seconds — 1970 at St. Louis — Jacques Laperriere of East scored at 0:20 and Dean Prentice of West scored at 0:37. Final score: East 4, West 1.

2:15 — 1998 at Vancouver — Teemu Selanee scored at 0:53 and Jaromir Jagr scored at 2:15 for World. Final score: North America 8, World 7.

3:37 — 1993 at Montreal — Mike Gartner scored at 3:15 and at 3:37 for Wales. Final score: Wales 16, Campbell 6.

FASTEST TWO GOALS, BOTH TEAMS:
8 seconds — 1997 at San Jose — Owen Nolan scored at 18:54 and 19:02 of second period for West. Final Score: East 11, West 7.

10 seconds — 1976 at Philadelphia — Dennis Ververgaert scored at 4:33 and at 4:43 of third period for Campbell. Final score: Wales 7, Campbell 5.

13 seconds — 1998 at Vancouver — Teemu Selanne scored at 4:00 of first period for World and John LeClair scored at 4:13 for North America. Final score: North America 8, World 7.

FASTEST THREE GOALS, BOTH TEAMS:
1:08 — 1993 at Montreal — all by Wales — Mike Gartner scored at 3:15 and at 3:37 of first period; Peter Bondra scored at 4:23. Final score: Wales 16, Campbell 6.

1:14 — 1994 at NY Rangers — Bob Kudelski scored at 9:46 of first period for East; Sergei Fedorov scored at 10:20 for West; Eric Lindros scored at 11:00 for East. Final score: East 9, West 8.

1:23 — 1999 at Tampa Bay — Mats Sundin scored at 2:57 of third period for World; Darryl Sydor scored at 4:02 for North America; Sergei Zubov scored at 4:20 for World. Final score: North America 8, World 6.

FASTEST FOUR GOALS, BOTH TEAMS:
2:24 — 1997 at San Jose — Brendan Shanahan scored at 16:38 of second period for West; Dale Hawerchuk scored at 17:28 for East; Owen Nolan scored at 18:54 and 19:02 for West. Final score: East 11, West 7.

2:57 — 2002 at Los Angeles — Sergei Fedorov scored at 16:59 of third period for World; Markus Naslund scored at 18:17 for World; Alexei Zhamnov scored at 19:12 for World; Sami Kapanen scored at 19:56 for World. Final score: World 8, North America 5.

3:04 — 1997 at San Jose — Mark Recchi scored at 15:32 of first period for East; Dale Hawerchuk scored at 16:19 for East; Pavel Bure scored at 17:36 for West; Paul Kariya scored at 18:36 for West. Final score: East 11, West 7.

FASTEST TWO GOALS, ONE TEAM, FROM START OF GAME:
2:15 — 1998 at Vancouver — World — Teemu Selanee scored at 0:53 and Jaromir Jagr scored at 2:15. Final score: North America 8, World 7.

3:37 — 1993 at Montreal — Wales — Mike Gartner scored at 3:15 and at 3:37. Final score: Wales 16, Campbell 6.

4:19 — 1980 at Detroit — Wales — Larry Robinson scored at 3:58 and Steve Payne scored at 4:19. Final score: Wales 6, Campbell 3.

FASTEST TWO GOALS, ONE TEAM:
8 seconds — 1997 at San Jose — West — Owen Nolan scored at 18:54 and at 19:02 of second period. Final score: East 11, West 7.

10 seconds — 1976 at Philadelphia — Campbell — Dennis Ververgaert scored at 4:33 and at 4:43 of third period. Final score: Wales 7, Campbell 5.

14 seconds — 1989 at Edmonton — Campbell — Steve Yzerman and Gary Leeman scored at 17:21 and 17:35 of second period. Final score: Campbell 9, Wales 5.

FASTEST THREE GOALS, ONE TEAM:
1:08 — 1993 at Montreal — Wales — Mike Gartner scored at 3:15 and 3:37 of first period; Peter Bondra scored at 4:23. Final score: Wales 16, Campbell 6.

1:32 — 1980 at Detroit — Wales — Ron Stackhouse scored at 11:40 of third period; Craig Hartsburg scored at 12:40; Reed Larson scored at 13:12. Final score: Wales 6, Campbell 3.

1:39 — 2002 at Los Angeles — Markus Naslund scored at 18:17 of third period; Alexei Zhamnov scored at 19:12; Sami Kapanen scored at 19:56. Final score: World 8, North America 5.

FASTEST FOUR GOALS, ONE TEAM:
2:57 — 2002 at Los Angeles — World — Sergei Fedorov scored at 16:59 of third period; Markus Naslund scored at 18:17; Alexei Zhamnov scored at 19:12; Sami Kapanen scored at 19:56. Final score: World 8, North America 5.

4:19 — 1992 at Philadelphia — Campbell — Brian Bellows scored at 7:40 of second period; Jeremy Roenick scored at 8:13; Theoren Fleury scored at 11:06, Brett Hull scored at 11:59. Final score: Campbell 10, Wales 6.

4:26 — 1980 at Detroit — Wales — Ron Stackhouse scored at 11:40 of third period; Craig Hartsburg scored at 12:40; Reed Larson scored at 13:12; Real Cloutier scored at 16:06. Final score: Wales 6, Campbell 3.

MOST GOALS, BOTH TEAMS, ONE PERIOD:
10 — 1997 at San Jose — Second period — East (6), West (4). Final score: East 11, West 7.

— 2001 at Colorado — Second period — North America (6), World (4). Final score: North America 14, World 12.

— 2001 at Colorado — Third period — North America (5), World (5). Final score: North America 14, World 12.

9 — 1990 at Pittsburgh — First period — Wales (7), Campbell (2). Final score: Wales 12, Campbell 7.

MOST GOALS, ONE TEAM, ONE PERIOD:
7 — 1990 at Pittsburgh — First period — Wales. Final score: Wales 12, Campbell 7.
6 — 1983 at NY Islanders — Third period — Campbell. Final score: Campbell 9, Wales 3.
— 1992 at Philadelphia — Second Period — Campbell. Final score: Campbell 10, Wales 6.
— 1993 at Montreal — First period — Wales. Final score: Wales 16, Campbell 6.
— 1993 at Montreal — Second period — Wales. Final score: Wales 16, Campbell 6.
— 1997 at San Jose — Second period — East. Final score: East 11, West 7.
— 2001 at Colorado — Second period — North America. Final score: North America 14, World 12.

MOST SHOTS, BOTH TEAMS, ONE PERIOD:
39 — 1994 at NY Rangers — Second period — West (21), East (18). Final score: East 9, West 8.
— 2001 at Colorado — Third period — World (23), North America (16). Final score: North America 14, World 12.
36 — 1990 at Pittsburgh — Third period — Campbell (22), Wales (14). Final score: Wales 12, Campbell 7.
— 1994 at NY Rangers — First period — East (19), West (17). Final score: East 9, West 8.
— 2002 at Los Angeles — Third period — North America (20), World (16). Final score: World 8, North America 5.

MOST SHOTS, ONE TEAM, ONE PERIOD:
23 — 2001 at Colorado — Third period — World. Final score: North America 14, World 12.
22 — 1990 at Pittsburgh — Third period — Campbell. Final score: Wales 12, Campbell 7.
— 1991 at Chicago — Third Period — Wales. Final score: Campbell 11, Wales 5.
— 1993 at Montreal — First period — Wales. Final score: Wales 16, Campbell 6.

FEWEST SHOTS, BOTH TEAMS, ONE PERIOD:
9 — 1971 at Boston — Third period — East (2), West (7). Final score: West 2, East 1.
— 1980 at Detroit — Second period — Campbell (4), Wales (5). Final score: Wales 6, Campbell 3.
13 — 1982 at Washington — Third period — Campbell (6), Wales (7). Final score: Wales 4, Campbell 2.
14 — 1978 at Buffalo — First period — Campbell (7), Wales (7). Final score: Wales 3, Campbell 2.
— 1986 at Hartford — First period — Campbell (6), Wales (8). Final score: Wales 4, Campbell 3.

FEWEST SHOTS, ONE TEAM, ONE PERIOD:
2 — 1971 at Boston — Third period — East. Final score: West 2, East 1.
— 1978 at Buffalo — Second period — Campbell. Final score: Wales 3, Campbell 2.
3 — 1978 at Buffalo — Third period — Campbell. Final score: Wales 3, Campbell 2.
4 — 1955 at Detroit — First period — NHL All-Stars. Final score: Detroit Red Wings 3, NHL All-Stars 1.
— 1980 at Detroit — Second period — Campbell. Final score: Wales 6, Campbell 3.

INDIVIDUAL RECORDS

Games

MOST GAMES PLAYED:
23 — Gordie Howe from 1948 through 1980
19 — Raymond Bourque from 1981 through 2001
18 — Wayne Gretzky from 1980 through 1999
15 — Frank Mahovlich from 1959 through 1974
14 — Paul Coffey from 1982 through 1997
— Mark Messier from 1982 through 2000

Goals

MOST GOALS (CAREER):
13 — Wayne Gretzky in 18GP
— Mario Lemieux in 10GP
10 — Gordie Howe in 23GP
8 — Frank Mahovlich in 15GP
— Luc Robitaille in 8GP
— Teemu Selanne in 8GP

MOST GOALS, ONE GAME:
4 — Wayne Gretzky, Campbell, 1983
— Mario Lemieux, Wales, 1990
— Vince Damphousse, Campbell, 1991
— Mike Gartner, Wales, 1993
3 — Ted Lindsay, Detroit, 1950
— Mario Lemieux, Wales, 1988
— Pierre Turgeon, Wales, 1993
— Mark Recchi, East, 1997
— Owen Nolan, West, 1997
— Teemu Selanne, World, 1998
— Pavel Bure, World, 2000
— Bill Guerin, North America, 2001

MOST GOALS, ONE PERIOD:
4 — Wayne Gretzky, Campbell, Third period, 1983
3 — Mario Lemieux, Wales, First period, 1990
— Vince Damphousse, Campbell, Third period, 1991
— Mike Gartner, Wales, First period, 1993

Assists

MOST ASSISTS (CAREER):
13 — Mark Messier in 14GP
— Raymond Bourque in 19GP
12 — Adam Oates in 5GP
— Joe Sakic in 8GP
— Wayne Gretzky in 18GP

MOST ASSISTS, ONE GAME:
5 — Mats Naslund, Wales, 1988
4 — Raymond Bourque, Wales, 1985
— Adam Oates, Campbell, 1991
— Adam Oates, Wales, 1993
— Mark Recchi, Wales, 1993
— Pierre Turgeon, East, 1994
— Fredrik Modin, World, 2001

MOST ASSISTS, ONE PERIOD:
4 — Adam Oates, Wales, First period, 1993
3 — Mark Messier, Campbell, Third period, 1983

Fellow Finns Teemu Selanne (left) and Saku Koivu during the first "North America vs. the World" All-Star Game in 1998. Selanne scored three goals in that game. Koivu set up two of them, including the opening goal just 53 seconds into the game.

Points

MOST POINTS, CAREER:
25 — Wayne Gretzky (13G-12A in 18GP)
23 — Mario Lemieux (13G-10A in 10GP)
19 — Gordie Howe (10G-9A in 23GP)
18 — Mark Messier (5G-13A in 14GP)
17 — Raymond Bourque (4G-13A in 19GP)

MOST POINTS, ONE GAME:
6 — Mario Lemieux, Wales, 1988 (3G-3A)
5 — Mats Naslund, Wales, 1988 (5A)
— Adam Oates, Campbell, 1991 (1G-4A)
— Mike Gartner, Wales, 1993 (4G-1A)
— Mark Recchi, Wales, 1993 (1G-4A)
— Pierre Turgeon, Wales, 1993 (3G-2A)
— Bill Guerin, North America, 2001 (3G-2A)

MOST POINTS, ONE PERIOD:
4 — Wayne Gretzky, Campbell, Third period, 1983 (4G)
— Mike Gartner, Wales, First period, 1993 (3G-1A)
— Adam Oates, Wales, First period, 1993 (4A)
3 — Gordie Howe, NHL All-Stars, Second period, 1965 (1G-2A)
— Pete Mahovlich, Wales, First period, 1976 (1G-2A)
— Mark Messier, Campbell, Third period, 1983 (3A)
— Mario Lemieux, Wales, Second period, 1988 (1G-2A)
— Mario Lemieux, Wales, First period, 1990 (3G)
— Vince Damphousse, Campbell, Third period, 1991 (3G)
— Mark Recchi, Wales, Second period, 1993 (1G-2A)
— Tony Amonte, North America, Second period, 2001 (2G-1A)

Power-Play Goals

MOST POWER-PLAY GOALS, CAREER:
6 — Gordie Howe in 23GP
3 — Bobby Hull in 12GP
— Maurice Richard in 13GP

Fastest Goals

FASTEST GOAL FROM START OF GAME:
19 seconds — Ted Lindsay, Detroit, 1950
20 seconds — Jacques Laperriere, East, 1970
21 seconds — Mario Lemieux, Wales, 1990
35 seconds — Vincent Damphousse, North America, 2002
36 seconds — Chico Maki, West, 1971

FASTEST GOAL FROM START OF A PERIOD:
17 seconds — Raymond Bourque, North America, 1999 (second period)
19 seconds — Ted Lindsay, Detroit, 1950 (first period)
— Rick Tocchet, Wales, 1993 (second period)
20 seconds — Jacques Laperriere, East, 1970 (first period)
21 seconds — Mario Lemieux, Wales, 1990 (first period)
26 seconds — Wayne Gretzky, Campbell, 1982 (second period)

FASTEST TWO GOALS (ONE PLAYER) FROM START OF GAME:
3:37 — Mike Gartner, Wales, 1993, at 3:15 and 3:37.
4:00 — Teemu Selanne, World, 1998, at 0:53 and 4:00
5:25 — Wally Hergesheimer, NHL All-Stars, 1953, at 4:06 and 5:25.

FASTEST TWO GOALS (ONE PLAYER) FROM START OF A PERIOD:
3:37 — Mike Gartner, Wales, 1993, at 3:15 and 3:37 of first period.
4:00 — Teemu Selanne, World, 1998, at 0:53 and 4:00 of first period.
4:43 — Dennis Ververgaert, Campbell, 1976, at 4:33 and 4:43 of third period.

FASTEST TWO GOALS (ONE PLAYER):
8 seconds — Owen Nolan, West, 1997. Scored at 18:54 and 19:02 of second period.
10 seconds — Dennis Ververgaert, Campbell, 1976. Scored at 4:33 and 4:43 of third period.
22 seconds — Mike Gartner, Wales, 1993. Scored at 3:15 and 3:37 of first period.

Penalties

MOST PENALTY MINUTES:
25 — Gordie Howe in 23GP
21 — Gus Mortson in 9GP
16 — Harry Howell in 7GP

Goaltenders

MOST GAMES PLAYED:
13 — Glenn Hall from 1955 through 1969
11 — Terry Sawchuk from 1950 through 1968
10 — Patrick Roy from 1988 through 2002
8 — Jacques Plante from 1956 through 1970

MOST MINUTES PLAYED:
540 — Glenn Hall in 13GP
467 — Terry Sawchuk in 11GP
370 — Jacques Plante in 8GP
230 — Patrick Roy in 10GP
209 — Turk Broda in 4GP

MOST GOALS AGAINST:
29 — Patrick Roy in 10GP
22 — Glenn Hall in 13GP
21 — Mike Vernon in 5GP
19 — Terry Sawchuk in 11GP
18 — Jacques Plante in 8GP
— Andy Moog in 4GP

BEST GOALS-AGAINST-AVERAGE AMONG THOSE WITH AT LEAST TWO GAMES PLAYED:
0.68 — Gilles Villemure in 3GP
1.49 — Gerry McNeil in 3GP
1.50 — Johnny Bower in 4GP
1.51 — Frank Brimsek in 2GP
1.64 — Gump Worsley in 4GP

Prince of Wales Conference goaltender Don Beaupre makes a glove save during the 1981 All-Star Game in Los Angeles. Looking on from the edge of the crease is Raymond Bourque in his first of 19 All-Star Game appearances.

Hockey Hall of Fame

(Year of induction is listed after each Honoured Members name)

Location: BCE Place, at the corner of Front and Yonge Streets in the heart of downtown Toronto. Easy access from all major highways running into Toronto. Close to TTC and Union Station.

Telephone: administration (416) 360-7735; information (416) 360-7765.

Summer and Christmas/March break hours: Monday to Saturday 9:30 a.m. to 6 p.m.; Sunday 10:00 a.m. to 6 p.m.

Fall/Winter/Spring hours (except Christmas/March break): Monday to Friday 10 a.m. to 5 p.m.; Saturday 9:30 a.m. to 6 p.m.; Sunday 10:30 a.m. to 5 p.m.

The Hockey Hall of Fame can be booked for private functions after hours.

Website address: www.hhof.com

History: The Hockey Hall of Fame was established in 1943. Members were first honoured in 1945. On August 26, 1961, the Hockey Hall of Fame opened its doors to the public in a building located on the grounds of the Canadian National Exhibition in Toronto. The Hockey Hall of Fame relocated to its new site at BCE Place and welcomed the hockey world on June 18, 1993.

Honour Roll: There are 328 Honoured Members in the Hockey Hall of Fame. 225 have been inducted as players, 89 as builders and 14 as Referees/Linesmen. In addition, there are 68 media honourees.

Founding Sponsors: Special thanks to Blockbuster Video, Ford Motor Company of Canada, Household Financial Services, IBM Canada, Imperial Oil, Kodak Canada, London Life Insurance Company, Molson Canada, Pepsi-Cola Canada, The Sports Network (TSN/RDS), Sun Media (Toronto)/The Toronto Sun, WorldCom Canada.

A defensive star in an era of high scorers, Rod Langway had only three goals in 1982-83 yet won the Norris Trophy anyway. He won it again in 1984. Langway enters the Hall of Fame in 2002 with Clark Gillies, Bernie Federko and Roger Neilson.

PLAYERS

* Abel, Sidney Gerald 1969
* Adams, John James "Jack" 1959
* Apps, Charles Joseph Sylvanus "Syl" 1961
  Armstrong, George Edward 1975
* Bailey, Irvine Wallace "Ace" 1975
* Bain, Donald H. "Dan" 1945
* Baker, Hobart "Hobey" 1945
  Barber, William Charles "Bill" 1990
* Barry, Martin J. "Marty" 1965
  Bathgate, Andrew James "Andy" 1978
* Bauer, Robert Theodore "Bobby" 1996
  Béliveau, Jean Arthur 1972
* Benedict, Clinton S. 1965
* Bentley, Douglas Wagner 1964
* Bentley, Maxwell H. L. 1966
* Blake, Hector Toe 1966
  Boivin, Leo Joseph 1986
* Boon, Richard R. "Dickie" 1952
  Bossy, Michael 1991
  Bouchard, Butch Joseph "Butch" 1966
* Boucher, Frank 1958
* Boucher, George "Buck" 1960
  Bower, John William 1976
* Bowie, Russell 1945
* Brimsek, Francis Charles 1966
* Broadbent, Harry L. "Punch" 1962
* Broda, Walter Edward "Turk" 1967
  Bucyk, John Paul 1981
* Burch, Billy 1974
* Cameron, Harold Hugh "Harry" 1962
  Cheevers, Gerald Michael "Gerry" 1985
* Clancy, Francis Michael "King" 1958
* Clapper, Aubrey "Dit" 1947
  Clarke, Robert "Bobby" 1987
* Cleghorn, Sprague 1958
* Colville, Neil MacNeil 1967
* Conacher, Charles W. 1961
* Conacher, Lionel Pretoria 1994
* Conacher, Roy Gordon 1998
* Connell, Alex 1958
* Cook, Fred "Bun" 1995
* Cook, William Osser 1952
  Coulter, Arthur Edmund 1974
  Cournoyer, Yvan Serge 1982
* Cowley, William Mailes 1968
* Crawford, Samuel Russell "Rusty" 1962
* Darragh, John Proctor "Jack" 1962
* Davidson, Allan M. "Scotty" 1950
* Day, Clarence Henry Hap 1961
  Delvecchio, Alex 1977
* Denneny, Cyril "Cy" 1959
  Dionne, Marcel 1992
* Drillon, Gordon Arthur 1975
* Drinkwater, Charles Graham 1950
  Dryden, Kenneth Wayne 1983
* Dumart, Woodrow "Woody" 1992
* Dunderdale, Thomas 1974
* Durnan, William Ronald 1964
* Dutton, Mervyn A. "Red" 1958
* Dye, Cecil Henry "Babe" 1970
  Esposito, Anthony James "Tony" 1988
  Esposito, Philip Anthony 1984
* Farrell, Arthur F. 1965
  Federko, Bernie 2002
  Fetisov, Viacheslav 2001
  Flaman, Ferdinand Charles "Fern" 1990
* Foyston, Frank 1958
* Fredrickson, Frank 1958
  Gadsby, William Alexander 1970
  Gainey, Bob 1992
* Gardiner, Charles Robert "Chuck" 1945
* Gardiner, Herbert Martin "Herb" 1958
* Gardner, James Henry "Jimmy" 1962
  Gartner, Michael Alfred 2001
  Geoffrion, Jos. A. Bernard "Boom Boom" 1972
* Gerard, Eddie 1945
  Giacomin, Edward "Eddie" 1987
  Gilbert, Rodrigue Gabriel "Rod" 1982
  Gillies, Clark 2002
* Gilmour, Hamilton Livingstone "Billy" 1962
* Goheen, Frank Xavier "Moose" 1952
* Goodfellow, Ebenezer R. "Ebbie" 1963
  Goulet, Michel 1998
* Grant, Michael "Mike" 1950
* Green, Wilfred "Shorty" 1962
  Gretzky, Wayne Douglas 1999
* Griffis, Silas Seth "Si" 1950
* Hainsworth, George 1961
  Hall, Glenn Henry 1975
* Hall, Joseph Henry 1961
* Harvey, Douglas Norman 1973
  Hawerchuk, Dale Martin 2001
* Hay, George 1958
* Hern, William Milton "Riley" 1962
* Hextall, Bryan Aldwyn 1969
* Holmes, Harry Hap 1972
* Hooper, Charles Thomas "Tom" 1962
  Horner, George Reginald "Red" 1965
* Horton, Miles Gilbert "Tim" 1977
  Howe, Gordon 1972
* Howe, Sydney Harris 1965
  Howell, Henry Vernon "Harry" 1979
  Hull, Robert Marvin 1983
* Hutton, John Bower "Bouse" 1962
* Hyland, Harry M. 1962
* Irvin, James Dickenson "Dick" 1958
* Jackson, Harvey "Busher" 1971
* Johnson, Ernest "Moose" 1952
* Johnson, Ivan "Ching" 1958
  Johnson, Thomas Christian 1970
* Joliat, Aurel 1947
* Keats, Gordon "Duke" 1958
  Kelly, Leonard Patrick "Red" 1969
  Kennedy, Theodore Samuel "Teeder" 1966
  Keon, David Michael 1986
  Kurri, Jari 2001
  Lach, Elmer James 1966
  Lafleur, Guy Damien 1988
* Lalonde, Edouard Charles "Newsy" 1950
  Langway, Rod Corry 2002
  Laperriere, Jacques 1987
  Lapointe, Guy 1993
  Laprade, Edgar 1993
* Laviolette, Jean Baptiste "Jack" 1962
* Lehman, Hugh 1958
  Lemaire, Jacques Gerard 1984
  Lemieux, Mario 1997
* LeSueur, Percy 1961
* Lewis, Herbert A. 1989
  Lindsay, Robert Blake Theodore "Ted" 1966
* Lumley, Harry 1980
* MacKay, Duncan "Mickey" 1952
  Mahovlich, Frank William 1981
* Malone, Joseph "Joe" 1950
* Mantha, Sylvio 1960
* Marshall, John "Jack" 1965
* Maxwell, Fred G. "Steamer" 1962
  McDonald, Lanny 1992
* McGee, Frank 1945
* McGimsie, William George "Billy" 1962
* McNamara, George 1958
  Mikita, Stanley 1983
  Moore, Richard Winston "Dickie" 1974
* Moran, Patrick Joseph "Paddy" 1958
* Morenz, Howie 1945
* Mosienko, William "Billy" 1965
  Mullen, Joseph P. 2000
* Nighbor, Frank 1947
* Noble, Edward Reginald "Reg" 1962
* O'Connor, Herbert William "Buddy" 1988
* Oliver, Harry 1967
  Olmstead, Murray Bert "Bert" 1985
  Orr, Robert Gordon 1979
  Parent, Bernard Marcel 1984
  Park, Douglas Bradford "Brad" 1988
* Patrick, Joseph Lynn 1980
* Patrick, Lester 1947
  Perreault, Gilbert 1990
* Phillips, Tommy 1945

Pilote, Joseph Albert Pierre Paul 1975
* Pitre, Didier "Pit" 1962
* Plante, Joseph Jacques Omer 1978
Potvin, Denis 1991
* Pratt, Walter "Babe" 1966
* Primeau, A. Joseph 1963
Pronovost, Joseph René Marcel 1978
Pulford, Bob 1991
* Pulford, Harvey 1945
* Quackenbush, Hubert George "Bill" 1976
* Rankin, Frank 1961
Ratelle, Joseph Gilbert Yvan Jean "Jean" 1985
Rayner, Claude Earl "Chuck" 1973
Reardon, Kenneth Joseph 1966
Richard, Joseph Henri 1979
* Richard, Joseph Henri Maurice "Rocket" 1961
* Richardson, George Taylor 1950
* Roberts, Gordon 1971
Robinson, Larry 1995
* Ross, Arthur Howie 1945
* Russel, Blair 1965
* Russell, Ernest 1965
* Ruttan, J.D. "Jack" 1962
Salming, Borje Anders 1996
Savard, Denis Joseph 2000
Savard, Serge A. 1986
* Sawchuk, Terrance Gordon "Terry" 1971
* Scanlan, Fred 1965
Schmidt, Milton Conrad "Milt" 1961
* Schriner, David "Sweeney" 1962
* Seibert, Earl Walter 1963
* Seibert, Oliver Levi 1961
* Shore, Edward W. "Eddie" 1947
Shutt, Stephen 1993
* Siebert, Albert C. "Babe" 1964
* Simpson, Harold Edward "Bullet Joe" 1962
Sittler, Darryl Glen 1989
* Smith, Alfred E. 1962
Smith, Clint 1991
* Smith, Reginald "Hooley" 1972
* Smith, Thomas James 1973
Smith, William John "Billy" 1993
Stanley, Allan Herbert 1981
* Stanley, Russell "Barney" 1962
Stastny, Peter 1998
* Stewart, John Sherratt "Black Jack" 1964
* Stewart, Nelson "Nels" 1962
* Stuart, Bruce 1961
* Stuart, Hod 1945
* Taylor, Frederic "Cyclone" (O.B.E.) 1947
* Thompson, Cecil R. "Tiny" 1959
Tretiak, Vladislav 1989
* Trihey, Col. Harry J. 1950
Trottier, Bryan 1997
Ullman, Norman V. Alexander "Norm" 1982
* Vezina, Georges 1945
* Walker, John Phillip "Jack" 1960
* Walsh, Martin "Marty" 1962
* Watson, Harry E. 1962
Watson, Harry 1994
* Weiland, Ralph "Cooney" 1971
* Westwick, Harry 1962
* Whitcroft, Fred 1962
* Wilson, Gordon Allan "Phat" 1962
Worsley, Lorne John "Gump" 1980
* Worters, Roy 1969

BUILDERS

* Adams, Charles 1960
* Adams, Weston W. 1972
* Aheam, Thomas Franklin "Frank" 1962
* Ahearne, John Francis "Bunny" 1977
* Allan, Sir Montagu (C.V.O.) 1945
Allen, Keith 1992
Arbour, Alger Joseph "Al" 1996
* Ballard, Harold Edwin 1977
* Bauer, Father David 1989
* Bickell, John Paris 1978
Bowman, Scotty 1991
* Brown, George V. 1961
* Brown, Walter A. 1962
* Buckland, Frank 1975
Bush, Walter Sr. 2000
Butterfield, Jack Arlington 1980
* Calder, Frank 1947
* Campbell, Angus D. 1964
* Campbell, Clarence Sutherland 1966
* Cattarinich, Joseph 1977
* Dandurand, Joseph Viateur "Leo" 1963
* Dilio, Francis Paul 1964
* Dudley, George S. 1958
* Dunn, James A. 1968
Francis, Emile 1982
* Gibson, Dr. John L. "Jack" 1976
* Gorman, Thomas Patrick "Tommy" 1963
* Griffiths, Frank A. 1993
* Hanley, William 1986
* Hay, Charles 1974
* Hendy, James C. 1968
* Hewitt, Foster 1965
* Hewitt, William Abraham 1947
* Hume, Fred J. 1962
* Imlach, George "Punch" 1984
* Ivan, Thomas N. 1974
* Jennings, William M. 1975
* Johnson, Bob 1992
* Juckes, Gordon W. 1979
* Kilpatrick, Gen. John Reed 1960
* Knox, Seymour H. III 1993
* Leader, George Alfred 1969
* LeBel, Robert 1970
* Lockhart, Thomas F. 1965
* Loicq, Paul 1961
* Mariucci, John 1985
Mathers, Frank 1992
* McLaughlin, Major Frederic 1963
* Milford, John "Jake" 1984
Molson, Hon. Hartland de Montarville 1973
Morrison, Ian "Scotty" 1999
* Murray, Monsignor Athol 1998
Neilson, Roger 2002
* Nelson, Francis 1947
* Norris, Bruce A. 1969
* Norris, Sr., James 1958
* Norris, James Dougan 1962
* Northey, William M. 1947
* O'Brien, John Ambrose 1962
O'Neill, Brian 1994
* Page, Fred 1993
Patrick, Craig 2001
* Patrick, Frank 1958
* Pickard, Allan W. 1958
* Pilous, Rudy 1985
Poile, Norman "Bud" 1990
Pollock, Samuel Patterson Smyth 1978
* Raymond, Sen. Donat 1958
* Robertson, John Ross 1947
* Robinson, Claude C. 1947
* Ross, Philip D. 1976
* Sabetzki, Dr. Gunther 1995
Sather, Glen 1997
* Selke, Frank J. 1960
Sinden, Harry James 1983
* Smith, Frank D. 1962
* Smythe, Conn 1958
Snider, Edward M. 1988
* Stanley of Preston, Lord (G.C.B.) 1945
* Sutherland, Cap. James T. 1947
* Tarasov, Anatoli V. 1974
Torrey, Bill 1995
* Turner, Lloyd 1958
* Tutt, William Thayer 1978
* Voss, Carl Potter 1974
* Waghorn, Fred C. 1961
* Wirtz, Arthur Michael 1971
Wirtz, William W. "Bill" 1976
Ziegler, John A. Jr. 1987

REFEREES/LINESMEN

Armstrong, Neil 1991
Ashley, John George 1981
Chadwick, William L. 1964
D'Amico, John 1993
* Elliott, Chaucer 1961
* Hayes, George William 1988
* Hewitson, Robert W. 1963
* Ion, Fred J. "Mickey" 1961
Pavelich, Matt 1987
* Rodden, Michael J. "Mike" 1962
* Smeaton, J. Cooper 1961
Storey, Roy Alvin "Red" 1967
Udvari, Frank Joseph 1973
Van Hellemond, Andy 1999

4th Annual Hockey Hall of Fame Game
Saturday, November 2, 2002
Montreal Canadiens vs.
Toronto Maple Leafs at
Air Canada Centre in Toronto.

Elmer Ferguson Memorial Award Winners

In recognition of distinguished members of the newspaper profession whose words have brought honor to journalism and to hockey. Selected by the Professional Hockey Writers' Association.

* Barton, Charlie, Buffalo-Courier Express 1985
* Beauchamp, Jacques, Montreal Matin/Journal de Montréal 1984
* Brennan, Bill, Detroit News 1987
* Burchard, Jim, New York World Telegram 1984
* Burnett, Red, Toronto Star 1984
* Carroll, Dink, Montreal Gazette 1984
* Coleman, Jim, Southam Newspapers 1984
Conway, Russ, Eagle-Tribune 1999
* Damata, Ted, Chicago Tribune 1984
Delano, Hugh, New York Post 1991
Desjardins, Marcel, Montréal La Presse 1984
Duhatschek, Eric, Calgary Herald/Globe and Mail 2001
* Dulmage, Jack, Windsor Star 1984
Dunnell, Milt, Toronto Star 1984
Dupont, Kevin Paul, Boston Globe 2002
* Ferguson, Elmer, Montreal Herald/Star 1984
Fisher, Red, Montreal Star/Gazette 1985
* Fitzgerald, Tom, Boston Globe 1984
Frayne, Trent, Toronto Telegram/Globe and Mail/Sun 1984
Gatecliff, Jack, St. Catherines Standard 1995
Gross, George, Toronto Telegram/Sun 1985
Johnston, Dick, Buffalo News 1986
* Laney, Al, New York Herald-Tribune 1984
Larochelle, Claude, Le Soleil 1989
L'Esperance, Zotique, Journal de Montréal/le Petit Journal 1985
* MacLeod, Rex, Toronto Globe and Mail/Star 1987
Matheson, Jim, Edmonton Journal 2000
* Mayer, Charles, Journal de Montréal/la Patrie 1985
McKenzie, Ken, The Hockey News 1997
Monahan, Leo, Boston Daily Record/Record-American/Herald American 1986
Moriarty, Tim, UPI/Newsday 1986
* Nichols, Joe, New York Times 1984
* O'Brien, Andy, Weekend Magazine 1985
Orr, Frank, Toronto Star 1989
Olan, Ben, New York Associated Press 1987
* O'Meara, Basil, Montreal Star 1984
Pedneault, Yvon, La Presse/Journal de Montréal 1998
* Proudfoot, Jim, Toronto Star 1988
Raymond, Bertrand, Journal de Montréal 1990
Rosa, Fran, Boston Globe 1987
Strachan, Al, Globe and Mail/Toronto Sun 1993
* Vipond, Jim, Toronto Globe and Mail 1984
Walter, Lewis, Detroit Times 1984
Young, Scott, Toronto Globe and Mail/Telegram 1988

Foster Hewitt Memorial Award Winners

In recognition of members of the radio and television industry who made outstanding contributions to their profession and the game during their career in hockey broadcasting. Selected by the NHL Broadcasters' Association.

Cole, Bob, Hockey Night in Canada 1996
Cusick, Fred, Boston 1984
* Darling, Ted, Buffalo 1994
* Gallivan, Danny, Montreal 1984
Garneau, Richard, Montreal 1999
* Hart, Gene, Philadelphia 1997
* Hewitt, Foster, Toronto 1984
Irvin, Dick, Montreal 1988
* Kelly, Dan, St. Louis 1989
Lange, Mike, Pittsburgh 2001
* Lecavelier, René, Montreal 1984
Lynch, Budd, Detroit 1985
Martyn, Bruce, Detroit 1991
McDonald, Jiggs, Los Angeles, Atlanta, NY Islanders 1990
McFarlane, Brian, Hockey Night in Canada 1995
* McKnight, Wes, Toronto 1986
Meeker, Howie, Hockey Night in Canada 1998
Miller, Bob, Los Angeles 2000
Pettit, Lloyd, Chicago 1986
Robson, Jim, Vancouver 1992
Shaver, Al, Minnesota 1993
* Smith, Doug, Montreal 1985
Tremblay, Gilles, La Soirée du Hockey 2002
Wilson, Bob, Boston 1987

* Deceased

United States Hockey Hall of Fame

The United States Hockey Hall of Fame was opened on June 21, 1973 as the national shrine of American Hockey. It is dedicated to honoring the sport of ice hockey in the United States by preserving those precious memories and legends of the game. It is located in Eveleth, Minnesota, 60 miles north of Duluth on Highway 53. The facility is open Monday to Saturday, 9 a.m. to 5 p.m. and Sundays from 10 a.m. to 3 p.m. Admission is $6.00 for adults, $5.00 for seniors and youths (13-17) and $4.00 for children (6-12). Call for any further information: 1-800-443-7825 or 218-744-5167. Web site address: www.ushockeyhall.com

There are now 110 enshrined members consisting of 66 players, 22 coaches, 19 administrators, one player/administrator, one referee and one team. New members are inducted annually in the fall and must have made a significant contribution towards hockey in the United States during the course of their career. A special Wayne Gretzky Award pays tribute to international individuals who have made major contributions to hockey in the USA. Support for the Hall of Fame comes from sponsorships, admissions, gift store sales, special events and grants from the hockey community and government agencies.

US HOCKEY HALL OF FAME®

EVELETH, MN

PLAYERS

* Abel, Clarence "Taffy" 1973
* Baker, Hobart "Hobey" 1973
* Bartholome, Earl 1977
* Bessone, Peter 1978
Blake, Robert 1985
Boucha, Henry 1995
* Brimsek, Frank 1973
Broten, Neal 2000
Cavanagh, Joe 1994
* Chaisson, Ray 1974
* Chase, John P. 1973
Christian, Dave 2001
Christian, Roger 1989
Christian, William "Bill" 1984
Cleary, Robert 1981
Cleary, William 1976
* Conroy, Anthony 1975
Curran, Mike 1998
* Dahlstrom, Carl "Cully" 1973
* Desjardins, Victor 1974
* Desmond, Richard 1988
* Dill, Robert 1979
* Everett, Doug 1974
Ftorek, Robbie 1991
* Garrison, John B. 1973
Garrity, Jack 1986
* Goheen, Frank "Moose" 1973
Grant, Wally 1994
* Harding, Austin "Austie" 1975
* Iglehart, Stewart 1975
Johnson, Paul 2001
* Johnson, Virgil 1974
* Karakas, Mike 1973
Kirrane, Jack 1987
* Lane, Myles J. 1973
Langevin, David R. 1993
Langway, Rod 1999
Larson, Reed 1996
* Linder, Joseph 1975
* LoPresti, Sam L. 1973
* Mariucci, John 1973
Matchefts, John 1991
* Mather, Bruce 1998
Mayasich, John 1976
McCartan, Jack 1983
* Moe, William 1974
Morrow, Ken 1995
* Moseley, Fred 1975
Mullen, Joe 1998
* Murray, Sr., Hugh "Muzz" 1987
* Nelson, Hubert "Hub" 1978
* Nyrop, William D. 1997
* Olson , Eddie 1977
* Owen, Jr., George 1973
* Palmer, Winthrop 1973
Paradise, Robert 1989
* Purpur, Clifford "Fido" 1974
Ramsey, Mike 2001
Riley, William 1977
Roberts, Gordie 1999
* Romnes, Elwin "Doc" 1973
* Rondeau, Richard 1985
Sheehy, Timothy K. 1997
* Williams, Thomas 1981
* Winters, Frank "Coddy" 1973
* Yackel, Ken 1986

COACHES

* Almquist, Oscar 1983
Bessone, Amo 1992
Brooks, Herbert 1990
Ceglarski, Len 1992
* Fullerton, James 1992
Gambucci, Sergio 1996
* Gordon, Malcolm K. 1973
Harkness, Nevin D. "Ned" 1994
Heyliger, Victor 1974
* Holt, Jr. Charles E. 1997
Ikola, Willard 1990
* Jeremiah, Edward J. 1973
* Johnson, Bob 1991
* Kelley, John "Snooks" 1974
Kelley, John H. "Jack" 1993
Patrick, Craig 1996
* Pleban, Jon "Connie" 1990
Riley, Jack 1979
* Ross, Larry 1988
* Thompson, Clifford, R. 1973
* Stewart, William 1982
* Winsor, Alfred "Ralph" 1973

ADMINISTRATORS

* Brown, George V. 1973
* Brown, Walter A. 1973
Bush, Walter 1980
* Clark, Donald 1978
Claypool, James 1995
* Gibson, J.C. "Doc" 1973
* Jennings, William M. 1981
* Kahler, Nick 1980
* Lockhart, Thomas F. 1973
* Marvin, Cal 1982
Palazzari, Doug 2000
Pleau, Larry 2000
* Ridder, Robert 1976
* Schulz, Charles M. 1993
Trumble, Harold 1985
* Tutt, William Thayer 1973
Watson, Sid 1999
Wirtz, William W. "Bill" 1984
* Wright, Lyle Z.1973

PLAYER/ADMINISTRATOR

Nanne, Lou 1998

REFEREE

Chadwick, William 1974

TEAM

1960 U.S. Olympic Team, 2000

*Deceased

Just 18 years old when he joined the U.S. national team, Mike Ramsey was the youngest player on the "Miracle on Ice" Olympic champions. He joined the Sabres shortly after Lake Placid and spent most of his 18-year NHL career in Buffalo.

NHL League and Team Websites

National Hockey League . . . . . . www.nhl.com
NHL Games on Radio. . . . . www.nhl.com/intheslot/listen/radio/index.html
NHL Site for Kids . . . . www.nhl.com/kids
NHL Merchandise Shop . . . . www.shop.nhl.com
NHL Trivia . . . . nhltrivia.buzztime.com
NHL Job Postings . . . . hockeyjobs.nhl.com
Hockey Fights Cancer . . . . www.nhl.com/nhlhq/hockeyfightscancer/index.html

Official NHL Team Websites:

Anaheim . . . . www.mightyducks.com
Atlanta. . . . . www.atlantathrashers.com
Boston . . . . www.bostonbruins.com
Buffalo . . . . www.sabres.com
Calgary . . . . www.calgaryflames.com
Carolina . . . . www.caneshockey.com
Chicago . . . . www.chicagoblackhawks.com
Colorado . . . . www.coloradoavalanche.com
Columbus . . . . www.bluejackets.com
Dallas . . . . www.dallasstars.com
Detroit . . . . www.detroitredwings.com
Edmonton . . . . www.edmontonoilers.com
Florida . . . . www.floridapanthers.com
Los Angeles . . . . www.lakings.com
Minnesota . . . . www.wild.com
Montreal . . . . www.canadiens.com
Nashville. . . . . www.nashvillepredators.com
New Jersey . . . . www.newjerseydevils.com
NY Islanders . . . . www.newyorkislanders.com
NY Rangers . . . . www.newyorkrangers.com
Ottawa. . . . . www.ottawasenators.com
Philadelphia . . . . www.philadelphiaflyers.com
Phoenix . . . . www.phoenixcoyotes.com
Pittsburgh . . . . www.pittsburghpenguins.com
St. Louis . . . . www.stlouisblues.com
San Jose . . . . www.sjsharks.com
Tampa Bay . . . . www.tampabaylightning.com
Toronto . . . . www.torontomapleleafs.com
Vancouver . . . . www.canucks.com
Washington . . . . www.washingtoncaps.com

2002 Stanley Cup Playoffs

Results

CONFERENCE QUARTER-FINALS
(Best-of-seven series)

Eastern Conference

Series 'A'

Thu. Apr. 18	Montreal 5	at	Boston 2
Sun. Apr. 21	Montreal 4	at	Boston 6
Tue. Apr. 23	Boston 3	at	Montreal 5
Thu. Apr. 25	Boston 5	at	Montreal 2
Sat. Apr. 27	Montreal 2	at	Boston 1
Mon. Apr. 29	Boston 1	at	Montreal 2

(Montreal Won Series 4-2)

Series 'B'

Wed. Apr. 17	Ottawa 0	at	Philadelphia 1*
Sat. Apr. 20	Ottawa 3	at	Philadelphia 0
Mon. Apr. 22	Philadelphia 0	at	Ottawa 3
Wed. Apr. 24	Philadelphia 0	at	Ottawa 3
Fri. Apr. 26	Ottawa 2	at	Philadelphia 1**

*Ruslan Fedotenko Scored at 7:47 of Overtime
**Martin Havlat Scored at 7:33 of Overtime

(Ottawa Won Series 4-1)

Series 'C'

Wed. Apr. 17	New Jersey 1	at	Carolina 2
Fri. Apr. 19	New Jersey 1	at	Carolina 2*
Sun. Apr. 21	Carolina 0	at	New Jersey 4
Tue. Apr. 23	Carolina 1	at	New Jersey 3
Wed. Apr. 24	New Jersey 2	at	Carolina 3**
Sat. Apr. 27	Carolina 1	at	New Jersey 0

*Bates Battaglia Scored at 15:26 of Overtime
**Josef Vasicek Scored at 8:16 of Overtime

(Carolina Won Series 4-2)

Series 'D'

Thu. Apr. 18	NY Islanders 1	at	Toronto 3
Sat. Apr. 20	NY Islanders 0	at	Toronto 2
Tue. Apr. 23	Toronto 1	at	NY Islanders 6
Wed. Apr. 24	Toronto 3	at	NY Islanders 4
Fri. Apr. 26	NY Islanders 3	at	Toronto 6
Sun. Apr. 28	Toronto 3	at	NY Islanders 5
Tue. Apr. 30	NY Islanders 2	at	Toronto 4

(Toronto Won Series 4-3)

Western Conference

Series 'E'

Wed. Apr. 17	Vancouver 4	at	Detroit 3*
Fri. Apr. 19	Vancouver 5	at	Detroit 2
Sun. Apr. 21	Detroit 3	at	Vancouver 1
Tue. Apr. 23	Detroit 4	at	Vancouver 2
Thu. Apr. 25	Vancouver 0	at	Detroit 4
Sat. Apr. 27	Detroit 6	at	Vancouver 4

*Henrik Sedin Scored at 13:59 of Overtime

(Detroit Won Series 4-2)

Series 'F'

Thu. Apr. 18	Los Angeles 3	at	Colorado 4
Sat. Apr. 20	Los Angeles 3	at	Colorado 5
Mon. Apr. 22	Colorado 1	at	Los Angeles 3
Tue. Apr. 23	Colorado 1	at	Los Angeles 0
Thu. Apr. 25	Los Angeles 1	at	Colorado 0*
Sat. Apr. 27	Colorado 1	at	Los Angeles 3
Mon. Apr. 29	Los Angeles 0	at	Colorado 4

*Craig Johnson Scored at 2:19 of Overtime

(Colorado Won Series 4-3)

Series 'G'

Wed. Apr. 17	Phoenix 1	at	San Jose 2
Sat. Apr. 20	Phoenix 3	at	San Jose 1
Mon. Apr. 22	San Jose 4	at	Phoenix 1
Wed. Apr. 24	San Jose 2	at	Phoenix 1
Fri. Apr. 26	Phoenix 1	at	San Jose 4

(San Jose Won Series 4-1)

Series 'H'

Thu. Apr. 18	Chicago 2	at	St. Louis 1
Sat. Apr. 20	Chicago 0	at	St. Louis 2
Sun. Apr. 21	St. Louis 4	at	Chicago 0
Tue. Apr. 23	St. Louis 1	at	Chicago 0
Thu. Apr. 25	Chicago 3	at	St. Louis 5

(St. Louis Won Series 4-1)

CONFERENCE SEMI-FINALS
(Best-of-seven series)

Eastern Conference

Series 'I'

Fri. May 3	Montreal 0	at	Carolina 2
Sun. May 5	Montreal 4	at	Carolina 1
Tue. May 7	Carolina 1	at	Montreal 2*
Thu. May 9	Carolina 4	at	Montreal 3**
Sun. May 12	Montreal 1	at	Carolina 5
Mon. May 13	Carolina 8	at	Montreal 2

*Donald Audette Scored at 2:26 of Overtime
**Niclas Wallin Scored at 3:14 of Overtime

(Carolina Won Series 4-2)

Series 'J'

Thu. May 2	Ottawa 5	at	Toronto 0
Sat. May 4	Ottawa 2	at	Toronto 3*
Mon. May 6	Toronto 2	at	Ottawa 3
Wed. May 8	Toronto 2	at	Ottawa 1
Fri. May 10	Ottawa 4	at	Toronto 2
Sun. May 12	Toronto 4	at	Ottawa 3
Tue. May 14	Ottawa 0	at	Toronto 3

*Gary Roberts Scored at 44:30 of Overtime

(Toronto Won Series 4-3)

Western Conference

Series 'K'

Thu. May 2	St. Louis 0	at	Detroit 2
Sat. May 4	St. Louis 2	at	Detroit 3
Tue. May 7	Detroit 1	at	St. Louis 6
Thu. May 9	Detroit 4	at	St. Louis 3
Sat. May 11	St. Louis 0	at	Detroit 4

(Detroit Won Series 4-1)

Series 'L'

Wed. May 1	San Jose 6	at	Colorado 3
Sat. May 4	San Jose 2	at	Colorado 8
Mon. May 6	Colorado 4	at	San Jose 6
Wed. May 8	Colorado 4	at	San Jose 1
Sat. May 11	San Jose 5	at	Colorado 3
Mon. May 13	Colorado 2	at	San Jose 1*
Wed. May 15	San Jose 0	at	Colorado 1

*Peter Forsberg Scored at 2:47 of Overtime

(Colorado Won Series 4-3)

CONFERENCE FINALS
(Best-of-seven series)

Eastern Conference

Series 'M'

Thu. May 16	Toronto 2	at	Carolina 1
Sun. May 19	Toronto 1	at	Carolina 2*
Tue. May 21	Carolina 2	at	Toronto 1**
Thu. May 23	Carolina 3	at	Toronto 0
Sat. May 25	Toronto 1	at	Carolina 0
Tue. May 28	Carolina 2	at	Toronto 1***

*Niclas Wallin Scored at 13:42 of Overtime
**Jeff O'Neill Scored at 6:01 of Overtime
***Martin Gelinas Scored at 8:05 of Overtime

(Carolina Won Series 4-2)

Western Conference

Series 'N'

Sat. May 18	Colorado 3	at	Detroit 5
Mon. May 20	Colorado 4	at	Detroit 3*
Wed. May 22	Detroit 2	at	Colorado 1**
Sat. May 25	Detroit 2	at	Colorado 3
Mon. May 27	Colorado 2	at	Detroit 1***
Wed. May 29	Detroit 2	at	Colorado 0
Fri. May 31	Colorado 0	at	Detroit 7

*Chris Drury Scored at 2:17 of Overtime
**Fredrik Olausson Scored at 12:44 of Overtime
***Peter Forsberg Scored at 6:24 of Overtime

(Detroit Won Series 4-3)

STANLEY CUP CHAMPIONSHIP
(Best-of-seven series)

Series 'O'

Tue. June 4	Carolina 3	at	Detroit 2*
Thu. June 6	Carolina 1	at	Detroit 3
Sat. June 8	Detroit 3	at	Carolina 2**
Mon. June 10	Detroit 3	at	Carolina 0
Thu. June 13	Carolina 1	at	Detroit 3

*Ron Francis Scored at 0:58 of Overtime
**Igor Larionov Scored at 54:47 of Overtime

(Detroit Won Series 4-1)

Team Playoff Records

	GP	W	L	GF	GA	%
Detroit	23	16	7	72	47	.696
Carolina	23	13	10	47	43	.565
Colorado	21	11	10	54	56	.524
Toronto	20	10	10	44	49	.500
Ottawa	12	7	5	29	18	.583
San Jose	12	7	5	34	32	.583
Montreal	12	6	6	32	39	.500
St. Louis	10	5	5	24	19	.500
NY Islanders	7	3	4	21	22	.429
Los Angeles	7	3	4	13	16	.429
New Jersey	6	2	4	11	9	.333
Boston	6	2	4	18	20	.333
Vancouver	6	2	4	16	22	.333
Phoenix	5	1	4	7	13	.200
Chicago	5	1	4	5	13	.200
Philadelphia	5	1	4	2	11	.200

Individual Leaders

Abbreviations: *– rookie eligible for Calder Trophy; **A** – assists; **G** – goals; **GP** – Games Played; **OT** – overtime goals; **GW** – game-winning goals; **PIM** – penalties in minutes; **PP** – power play goals; **Pts** – points; **S** – shots on goal; **SH** – short-handed goals; **%** – percentage of shots resulting in goals; **+/–** – difference between Goals For (**GF**) scored when a player is on the ice with his team at even strength or short-handed and Goals Against (**GA**) scored when the same player is on the ice with his team at even strength or on a power play.

Playoff Scoring Leaders

Player	Team	GP	G	A	PTS	+/–	PIM	PP	SH	GW	OT	S	%
Peter Forsberg	Colorado	20	9	18	27	8	20	0	0	4	2	35	25.7
Steve Yzerman	Detroit	23	6	17	23	4	10	4	0	2	0	52	11.5
Joe Sakic	Colorado	21	9	10	19	-2	4	4	0	1	0	76	11.8
Brendan Shanahan	Detroit	23	8	11	19	5	20	1	0	2	0	78	10.3
Gary Roberts	Toronto	19	7	12	19	6	56	3	0	1	1	47	14.9
Sergei Fedorov	Detroit	23	5	14	19	4	20	2	1	0	0	88	5.7
Brett Hull	Detroit	23	10	8	18	1	4	3	2	2	0	61	16.4
Ron Francis	Carolina	23	6	10	16	-2	6	4	0	3	1	51	11.8
Nicklas Lidstrom	Detroit	23	5	11	16	6	2	2	1	2	0	41	12.2
Alyn McCauley	Toronto	20	5	10	15	3	4	1	0	2	0	48	10.4
Bates Battaglia	Carolina	23	5	9	14	2	14	1	0	1	1	44	11.4
Chris Chelios	Detroit	23	1	13	14	15	44	1	0	0	0	28	3.6
Jeff O'Neill	Carolina	22	8	5	13	2	27	3	0	1	1	72	11.1
Daniel Alfredsson	Ottawa	12	7	6	13	6	4	3	0	3	0	55	12.7
Alex Tanguay	Colorado	19	5	8	13	-8	0	3	0	0	0	20	25.0
Greg de Vries	Colorado	21	4	9	13	1	2	0	0	1	0	41	9.8
Steve Reinprecht	Colorado	21	7	5	12	7	8	0	0	2	0	29	24.1
Rob Blake	Colorado	20	6	6	12	-1	16	1	0	0	0	67	9.0
Chris Drury	Colorado	21	5	7	12	4	10	1	0	3	1	56	8.9
Rod Brind'Amour	Carolina	23	4	8	12	-3	16	2	1	1	0	48	8.3
Alexander Mogilny	Toronto	20	8	3	11	1	8	2	0	2	0	59	13.6
Tomas Holmstrom	Detroit	23	8	3	11	7	8	3	0	2	0	34	23.5
Patrick Marleau	San Jose	12	6	5	11	3	6	1	0	3	0	21	28.6
Igor Larionov	Detroit	18	5	6	11	5	4	0	0	1	1	24	20.8
Pavol Demitra	St. Louis	10	4	7	11	3	6	2	1	1	0	26	15.4

Playoff Defencemen Scoring Leaders

Player	Team	GP	G	A	PTS	+/–	PIM	PP	SH	GW	OT	S	%
Nicklas Lidstrom	Detroit	23	5	11	16	6	2	2	1	2	0	41	12.2
Chris Chelios	Detroit	23	1	13	14	15	44	1	0	0	0	28	3.6
Greg de Vries	Colorado	21	4	9	13	1	2	0	0	1	0	41	9.8
Rob Blake	Colorado	20	6	6	12	-1	16	1	0	0	0	67	9.0
Bryan McCabe	Toronto	20	5	5	10	4	30	3	0	1	0	64	7.8
Tomas Kaberle	Toronto	20	2	8	10	7	16	0	0	0	0	29	6.9
Sean Hill	Carolina	23	4	4	8	0	20	4	0	1	0	57	7.0
Chris Pronger	St. Louis	9	1	7	8	5	24	0	0	0	0	16	6.3
Adrian Aucoin	NY Islanders	7	2	5	7	-1	4	2	0	0	0	16	12.5
Roman Hamrlik	NY Islanders	7	1	6	7	-6	6	0	0	0	0	21	4.8
Adam Foote	Colorado	21	1	6	7	-2	28	0	0	0	0	26	3.8
Al MacInnis	St. Louis	10	0	7	7	3	4	0	0	0	0	30	.0

GOALTENDING LEADERS

Goals Against Average

Goaltender	Team	GPI	Mins	GA	Avg.
Patrick Lalime	Ottawa	12	778	18	1.39
Kevin Weekes	Carolina	8	408	11	1.62
Arturs Irbe	Carolina	18	1078	30	1.67
Brent Johnson	St. Louis	10	590	18	1.83
Dominik Hasek	Detroit	23	1455	45	1.86

Wins

Goaltender	Team	GPI	Mins	W	L
Dominik Hasek	Detroit	23	1455	16	7
Patrick Roy	Colorado	21	1241	11	10
Arturs Irbe	Carolina	18	1078	10	8
Curtis Joseph	Toronto	20	1253	10	10
Evgeni Nabokov	San Jose	12	712	7	5
Patrick Lalime	Ottawa	12	778	7	5

Save Percentage

Goaltender	Team	GPI	Mins	GA	SA	S%	W	L
Patrick Lalime	Ottawa	12	778	18	332	.946	7	5
Kevin Weekes	Carolina	8	408	11	180	.939	3	2
Arturs Irbe	Carolina	18	1078	30	480	.938	10	8
Brent Johnson	St. Louis	10	590	18	252	.929	5	5
Felix Potvin	Los Angeles	7	417	15	201	.925	3	4
Dominik Hasek	Detroit	23	1455	45	562	.920	16	7

Shutouts

Goaltender	Team	GPI	Mins	SO
Dominik Hasek	Detroit	23	1455	6
Patrick Lalime	Ottawa	12	778	4
Brent Johnson	St. Louis	10	590	3
Patrick Roy	Colorado	21	1241	3
Curtis Joseph	Toronto	20	1253	3
Kevin Weekes	Carolina	8	408	2

Goal Scoring

Name	Team	GP	G
Brett Hull	Detroit	23	10
Peter Forsberg	Colorado	20	9
Joe Sakic	Colorado	21	9
Alexander Mogilny	Toronto	20	8
Jeff O'Neill	Carolina	22	8
Brendan Shanahan	Detroit	23	8
Tomas Holmstrom	Detroit	23	8
Scott Mellanby	St. Louis	10	7
Daniel Alfredsson	Ottawa	12	7
Gary Roberts	Toronto	19	7
Steve Reinprecht	Colorado	21	7
Donald Audette	Montreal	12	6
Patrick Marleau	San Jose	12	6
Rob Blake	Colorado	20	6
Ron Francis	Carolina	23	6
Steve Yzerman	Detroit	23	6
*Erik Cole	Carolina	23	6

Assists

Name	Team	GP	A
Peter Forsberg	Colorado	20	18
Steve Yzerman	Detroit	23	17
Sergei Fedorov	Detroit	23	14
Chris Chelios	Detroit	23	13
Gary Roberts	Toronto	19	12
Brendan Shanahan	Detroit	23	11
Nicklas Lidstrom	Detroit	23	11
Alyn McCauley	Toronto	20	10
Joe Sakic	Colorado	21	10
Ron Francis	Carolina	23	10
Greg de Vries	Colorado	21	9
Bates Battaglia	Carolina	23	9
Alex Tanguay	Colorado	19	8
Tomas Kaberle	Toronto	20	8
Rod Brind'Amour	Carolina	23	8
Brett Hull	Detroit	23	8
Sami Kapanen	Carolina	23	8

Power-play Goals

Name	Team	GP	PP
Scott Mellanby	St. Louis	10	4
Joe Sakic	Colorado	21	4
Ron Francis	Carolina	23	4
Steve Yzerman	Detroit	23	4
Sean Hill	Carolina	23	4

Game-winning Goals

Name	Team	GP	GW
Peter Forsberg	Colorado	20	4
Daniel Alfredsson	Ottawa	12	3
Patrick Marleau	San Jose	12	3
Chris Drury	Colorado	21	3
Ron Francis	Carolina	23	3

Short-handed Goals

Name	Team	GP	SH
Brett Hull	Detroit	23	2
Kirk Maltby	Detroit	23	2

Overtime Goals

Name	Team	GP	OT
Peter Forsberg	Colorado	20	2
Niclas Wallin	Carolina	23	2

Shots

Name	Team	GP	S
Sergei Fedorov	Detroit	23	88
Brendan Shanahan	Detroit	23	78
Joe Sakic	Colorado	21	76
Jeff O'Neill	Carolina	22	72
*Erik Cole	Carolina	23	68

Plus/Minus

Name	Team	GP	+/–
Chris Chelios	Detroit	23	15
Darius Kasparaitis	Colorado	21	10
Benoit Brunet	Ottawa	12	8
Peter Forsberg	Colorado	20	8

TEAMS' PLAYOFF HOME/ROAD RECORD

90 games played.

	HOME						ROAD					
	GP	W	L	GF	GA	%	GP	W	L	GF	GA	%
DET	13	8	5	42	25	.615	10	8	2	30	22	.800
CAR	11	6	5	20	19	.545	12	7	5	27	24	.583
COL	11	6	5	32	26	.545	10	5	5	22	30	.500
TOR	11	6	5	25	24	.545	9	4	5	19	25	.444
OTT	5	3	2	13	8	.600	7	4	3	16	10	.571
S.J.	6	3	3	15	15	.500	6	4	2	19	17	.667
MTL	6	3	3	16	22	.500	6	3	3	16	17	.500
ST.L.	5	3	2	17	10	.600	5	2	3	7	9	.400
NYI	3	3	0	15	7	1.000	4	0	4	6	15	.000
L.A.	3	2	1	6	3	.667	4	1	3	7	13	.250
N.J.	3	2	1	7	2	.667	3	0	3	4	7	.000
BOS	3	1	2	9	11	.333	3	1	2	9	9	.333
VAN	3	0	3	7	13	.000	3	2	1	9	9	.667
PHX	2	0	2	2	6	.000	3	1	2	5	7	.333
CHI	2	0	2	0	5	.000	3	1	2	5	8	.333
PHI	3	1	2	2	5	.333	2	0	2	0	6	.000
Total		**47**	**43**	**228**	**201**	**.522**		**43**	**47**	**201**	**228**	**.478**

TEAM PENALTIES

Abbreviations: GP – games played; **PEN** – total penalty minutes, including bench penalties; **BMI** – total bench minor minutes; **AVG** – average penalty minutes per game. 90 games played.

Team	GP	PEN	BMI	AVG
COL	21	168	6	8.0
N.J.	6	52	2	8.7
OTT	12	115	0	9.6
L.A.	7	72	2	10.3
S.J.	12	124	0	10.3
VAN	6	62	0	10.3
CAR	23	263	4	11.4
DET	23	308	8	13.4
MTL	12	172	2	14.3
CHI	5	73	2	14.6
PHI	5	75	2	15.0
STL	10	150	0	15.0
PHX	5	83	0	16.6
BOS	6	106	0	17.7
TOR	20	389	4	19.5
NYI	7	169	2	24.1
Total		**2381**	**34**	
Two-Team average. PIM/GP				**26.5**

TEAMS' POWER-PLAY RECORD

Abbreviations: ADV-total advantages; **PPGF**-power play goals for; **%** arrived by dividing number of power-play goals by total advantages. 90 games played.

	HOME					ROAD					OVERALL				
	Team	GP	ADV	PPGF	%	Team	GP	ADV	PPGF	%	Team	GP	ADV	PPGF	%
1	NYI	3	20	7	35.0	BOS	3	10	3	30.0	NYI	7	38	11	28.9
2	ST.L.	5	20	7	35.0	NYI	4	18	4	22.2	BOS	6	20	5	25.0
3	N.J.	3	13	4	30.8	OTT	7	32	7	21.9	N.J	6	28	6	21.4
4	MTL	6	27	7	25.9	CAR	12	48	10	20.8	STL	10	45	9	20.0
5	DET	13	60	14	23.3	TOR	9	43	8	18.6	DET	23	99	19	19.2
6	BOS	3	10	2	20.0	CHI	3	7	1	14.3	MTL	12	48	9	18.8
7	VAN	3	17	3	17.6	N.J	3	15	2	13.3	OTT	12	55	9	16.4
8	PHI	3	6	1	16.7	DET	10	39	5	12.8	CAR	23	103	16	15.5
9	S.J.	6	27	4	14.8	COL	10	33	4	12.1	VAN	6	27	4	14.8
10	COL	11	44	6	13.6	VAN	3	10	1	10.0	TOR	20	90	13	14.4
11	CAR	11	55	6	10.9	MTL	6	21	2	9.5	COL	21	77	10	13.0
12	TOR	11	47	5	10.6	L.A	4	12	1	8.3	S.J	12	48	5	10.4
13	OTT	5	23	2	8.7	STL	5	25	2	8.0	PHI	5	16	1	6.3
14	PHX	2	11	0	.0	PHX	3	14	1	7.1	CHI	5	17	1	5.9
15	CHI	2	10	0	.0	S.J	6	21	1	4.8	L.A	7	17	1	5.9
16	L.A.	3	5	0	.0	PHI	2	10	0	.0	PHX	5	25	1	4.0
Total			**395**	**68**	**17.2**			**358**	**52**	**14.5**			**753**	**120**	**15.9**

TEAMS' PENALTY KILLING RECORD

Abbreviations: TSH – Total times short-handed; **PPGA** – power-play goals against; **%** arrived by dividing times short minus power-play goals against by times short. 90 games played.

	HOME					ROAD					OVERALL				
	Team	GP	TSH	PPGA	%	Team	GP	TSH	PPGA	%	Team	GP	TSH	PPGA	%
1	L.A.	3	12	0	100.0	PHX	3	15	1	93.3	PHX	5	27	2	92.6
2	COL	11	31	1	96.8	PHI	2	14	1	92.9	L.A	7	26	2	92.3
3	PHX	2	12	1	91.7	NYI	4	22	2	90.9	S.J	12	48	5	89.6
4	BOS	3	11	1	90.9	S.J	6	28	3	89.3	NYI	7	43	5	88.4
5	S.J.	6	20	2	90.0	STL	5	24	3	87.5	STL	10	42	5	88.1
6	DET	13	56	6	89.3	L.A	4	14	2	85.7	DET	23	104	14	86.5
7	ST.L.	5	18	2	88.9	MTL	6	27	4	85.2	COL	21	62	9	85.5
8	N.J.	3	8	1	87.5	DET	10	48	8	83.3	OTT	12	41	6	85.4
9	OTT	5	24	3	87.5	CAR	12	57	10	82.5	CAR	23	96	15	84.4
10	CAR	11	39	5	87.2	OTT	7	17	3	82.4	PHI	5	24	4	83.3
11	NYI	3	21	3	85.7	TOR	9	47	9	80.9	N.J	6	21	4	81.0
12	CHI	2	11	2	81.8	N.J	3	13	3	76.9	TOR	20	99	20	79.8
13	TOR	11	52	11	78.8	VAN	3	13	3	76.9	BOS	6	27	6	77.8
14	PHI	3	10	3	70.0	COL	10	31	8	74.2	MTL	12	48	11	77.1
15	VAN	3	12	4	66.7	BOS	3	16	5	68.8	CHI	5	20	5	75.0
16	MTL	6	21	7	66.7	CHI	3	9	3	66.7	VAN	6	25	7	72.0
Total			**358**	**52**	**85.5**			**395**	**68**	**82.8**			**753**	**120**	**84.1**

SHORT HAND GOALS

90 games played.

GOALS FOR			GOALS AGAINST		
Team	GP	GF	Team	GP	GA
DET	23	7	TOR	20	0
BOS	6	2	OTT	12	0
ST.L.	10	2	S.J.	12	0
N.J.	6	1	NYI	7	0
CAR	23	1	L.A.	7	0
PHI	5	0	BOS	6	0
PHX	5	0	N.J.	6	0
CHI	5	0	PHI	5	0
VAN	6	0	PHX	5	0
NYI	7	0	ST.L.	10	1
L.A.	7	0	CHI	5	1
MTL	12	0	CAR	23	2
OTT	12	0	DET	23	2
S.J.	12	0	COL	21	2
TOR	20	0	MTL	12	2
COL	21	0	VAN	6	3
Total		**13**	**Total**		**13**

Though most teams prefer to go with one hot goalie in the playoffs, the Carolina Hurricanes used a two-man approach to reach the Stanley Cup Finals in 2002. Both Arturs Irbe (right) and Kevin Weekes ranked among the playoff leaders in goal.

Stanley Cup Record Book

History: The Stanley Cup, the oldest trophy competed for by professional athletes in North America, was donated by Frederick Arthur, Lord Stanley of Preston and son of the Earl of Derby, in 1893. Lord Stanley purchased the trophy for 10 guineas ($50 at that time) for presentation to the amateur hockey champions of Canada. Since 1910, when the National Hockey Association took possession of the Stanley Cup, the trophy has been the symbol of professional hockey supremacy. It has been competed for only by NHL teams since 1926-27 and has been under the exclusive control of the NHL since 1947.

Stanley Cup Standings

1918-2002

(ranked by Cup wins)

Teams	Cup Wins	Yrs.	Series	Wins	Losses	Games	Wins	Losses	Ties	Goals For	Goals Against	Winning %
Montreal	23[1]	73	136[2]	86	49	650	387	255	8	2009	1630	.602
Toronto	13	62	106	57	49	504	242	258	4	1307	1411	.484
Detroit	10	51	96	55	41	478	251	226	1	1355	1248	.526
Boston	5	60	102	47	55	500	238	256	6	1466	1484	.482
Edmonton	5	18	44	31	13	221	135	86	0	857	682	.611
NY Rangers	4	48	86	42	44	386	183	195	8	1091	1114	.484
NY Islanders	4	18	44	30	14	225	131	94	0	769	672	.582
Chicago	3	53	90	40	50	411	188	218	5	1176	1311	.464
Philadelphia	2	28	59	33	26	309	161	148	0	948	913	.521
Pittsburgh	2	21	39	20	19	208	109	99	0	644	641	.524
Colorado[3]	2	16	36	22	14	206	113	93	0	624	580	.549
New Jersey[4]	2	14	28	16	12	164	90	74	0	460	412	.549
Dallas[5]	1	24	48	25	23	260	133	127	0	782	793	.512
Calgary[6]	1	21	32	12	20	156	69	87	0	529	573	.442
St. Louis	0	32	55	23	32	291	134	157	0	827	914	.460
Buffalo	0	25	42	17	25	209	99	110	0	626	639	.474
Los Angeles	0	23	34	11	23	170	65	105	0	511	649	.382
Vancouver	0	18	27	9	18	134	56	78	0	402	460	.418
Washington	0	17	27	10	17	148	67	81	0	452	464	.453
Phoenix[7]	0	16	18	2	16	92	29	63	0	245	343	.315
Carolina[8]	0	11	15	4	11	84	35	49	0	208	256	.417
San Jose	0	7	11	4	7	67	29	38	0	175	232	.433
Ottawa[9]	0	6	8	2	6	44	17	27	0	81	101	.386
Florida	0	3	6	3	3	31	13	18	0	77	82	.419
Anaheim	0	2	3	1	2	15	4	11	0	31	47	.267
Tampa Bay	0	1	1	0	1	6	2	4	0	13	26	.333

1 Montreal also won the Stanley Cup in 1916.
2 1919 final incomplete due to influenza epidemic.
3 Includes totals of Quebec 1979-95.
4 Includes totals of Colorado Rockies 1976-82.
5 Includes totals of Minnesota North Stars 1967-93.
6 Includes totals of Atlanta Flames 1972-80.
7 Includes totals of Winnipeg 1979-96.
8 Includes totals of Hartford 1979-97.
9 Modern Ottawa franchise only 1992 to date.

Stanley Cup Winners Prior to Formation of NHL in 1917

Season	Champions	Manager	Coach
1916-17	Seattle Metropolitans	Pete Muldoon	Pete Muldoon
1915-16	Montreal Canadiens	George Kennedy	George Kennedy
1914-15	Vancouver Millionaires	Frank Patrick	Frank Patrick
1913-14	Toronto Blueshirts	Jack Marshall	Scotty Davidson*
1912-13**	Quebec Bulldogs	M.J. Quinn	Joe Malone*
1911-12	Quebec Bulldogs	M.J. Quinn	C. Nolan
1910-11	Ottawa Senators		Bruce Stuart*
1909-10	Montreal Wanderers	Dickie Boon	Pud Glass*
1908-09	Ottawa Senators		Bruce Stuart*
1907-08	Montreal Wanderers	Dickie Boon	Cecil Blachford
1906-07	Montreal Wanderers (Mar. 1907)	Dickie Boon	Cecil Blachford
1906-07	Kenora Thistles (Jan./Mar. 1907)	F.A. Hudson	Tom Phillips*
1905-06	Montreal Wanderers (Mar. 1906)	Cecil Blachford*	
1905-06	Ottawa Silver Seven (Feb. 1906)		Alf Smith
1904-05	Ottawa Silver Seven		Alf Smith
1903-04	Ottawa Silver Seven		Alf Smith
1902-03	Ottawa Silver Seven (Mar. 1903)		Alf Smith
1902-03	Montreal A.A.A. (Feb. 1903)		C. McKerrow
1901-02	Montreal A.A.A. (Mar. 1902)		C. McKerrow
1901-02	Winnipeg Victorias (Jan. 1902)		
1900-01	Winnipeg Victorias		Dan Bain*
1899-1900	Montreal Shamrocks		Harry Trihey*
1898-99	Montreal Shamrocks (Mar. 1899)		Harry Trihey*
1898-99	Montreal Victorias (Feb. 1899)		Mike Grant*
1897-98	Montreal Victorias		Frank Richardson
1896-97	Montreal Victorias		Mike Grant*
1895-96	Montreal Victorias (Dec. 1896)		Mike Grant*
1895-96	Winnipeg Victorias (Feb. 1896)		Jack Armitage
1894-95	Montreal Victorias		Mike Grant*
1893-94	Montreal A.A.A.		
1892-93	Montreal A.A.A.		

* In the early years the teams were frequently run by the Captain. *Indicates Captain
** Victoria defeated Quebec in challenge series. No official recognition.

Stanley Cup Winners

Year	W-L-T in Finals	Winner	Coach	Finalist	Coach
2002	4-1	Detroit	Scotty Bowman	Carolina	Paul Maurice
2001	4-3	Colorado	Bob Hartley	New Jersey	Larry Robinson
2000	4-2	New Jersey	Larry Robinson	Dallas	Ken Hitchcock
1999	4-2	Dallas	Ken Hitchcock	Buffalo	Lindy Ruff
1998	4-0	Detroit	Scotty Bowman	Washington	Ron Wilson
1997	4-0	Detroit	Scotty Bowman	Philadelphia	Terry Murray
1996	4-0	Colorado	Marc Crawford	Florida	Doug MacLean
1995	4-0	New Jersey	Jacques Lemaire	Detroit	Scotty Bowman
1994	4-3	NY Rangers	Mike Keenan	Vancouver	Pat Quinn
1993	4-1	Montreal	Jacques Demers	Los Angeles	Barry Melrose
1992	4-0	Pittsburgh	Scotty Bowman	Chicago	Mike Keenan
1991	4-2	Pittsburgh	Bob Johnson	Minnesota	Bob Gainey
1990	4-1	Edmonton	John Muckler	Boston	Mike Milbury
1989	4-2	Calgary	Terry Crisp	Montreal	Pat Burns
1988	4-0	Edmonton	Glen Sather	Boston	Terry O'Reilly
1987	4-3	Edmonton	Glen Sather	Philadelphia	Mike Keenan
1986	4-1	Montreal	Jean Perron	Calgary	Bob Johnson
1985	4-1	Edmonton	Glen Sather	Philadelphia	Mike Keenan
1984	4-1	Edmonton	Glen Sather	NY Islanders	Al Arbour
1983	4-0	NY Islanders	Al Arbour	Edmonton	Glen Sather
1982	4-0	NY Islanders	Al Arbour	Vancouver	Roger Neilson
1981	4-1	NY Islanders	Al Arbour	Minnesota	Glen Sonmor
1980	4-2	NY Islanders	Al Arbour	Philadelphia	Pat Quinn
1979	4-1	Montreal	Scotty Bowman	NY Rangers	Fred Shero
1978	4-2	Montreal	Scotty Bowman	Boston	Don Cherry
1977	4-0	Montreal	Scotty Bowman	Boston	Don Cherry
1976	4-0	Montreal	Scotty Bowman	Philadelphia	Fred Shero
1975	4-2	Philadelphia	Fred Shero	Buffalo	Floyd Smith
1974	4-2	Philadelphia	Fred Shero	Boston	Bep Guidolin
1973	4-2	Montreal	Scotty Bowman	Chicago	Billy Reay
1972	4-2	Boston	Tom Johnson	NY Rangers	Emile Francis
1971	4-3	Montreal	Al MacNeil	Chicago	Billy Reay
1970	4-0	Boston	Harry Sinden	St. Louis	Scotty Bowman
1969	4-0	Montreal	Claude Ruel	St. Louis	Scotty Bowman
1968	4-0	Montreal	Toe Blake	St. Louis	Scotty Bowman
1967	4-2	Toronto	Punch Imlach	Montreal	Toe Blake
1966	4-2	Montreal	Toe Blake	Detroit	Sid Abel
1965	4-3	Montreal	Toe Blake	Chicago	Billy Reay
1964	4-3	Toronto	Punch Imlach	Detroit	Sid Abel
1963	4-1	Toronto	Punch Imlach	Detroit	Sid Abel
1962	4-2	Toronto	Punch Imlach	Chicago	Rudy Pilous
1961	4-2	Chicago	Rudy Pilous	Detroit	Sid Abel
1960	4-0	Montreal	Toe Blake	Toronto	Punch Imlach
1959	4-1	Montreal	Toe Blake	Toronto	Punch Imlach
1958	4-2	Montreal	Toe Blake	Boston	Milt Schmidt
1957	4-1	Montreal	Toe Blake	Boston	Milt Schmidt
1956	4-1	Montreal	Toe Blake	Detroit	Jimmy Skinner
1955	4-3	Detroit	Jimmy Skinner	Montreal	Dick Irvin
1954	4-3	Detroit	Tommy Ivan	Montreal	Dick Irvin
1953	4-1	Montreal	Dick Irvin	Boston	Lynn Patrick
1952	4-0	Detroit	Tommy Ivan	Montreal	Dick Irvin
1951	4-1	Toronto	Joe Primeau	Montreal	Dick Irvin
1950	4-3	Detroit	Tommy Ivan	NY Rangers	Lynn Patrick
1949	4-0	Toronto	Hap Day	Detroit	Tommy Ivan
1948	4-0	Toronto	Hap Day	Detroit	Tommy Ivan
1947	4-2	Toronto	Hap Day	Montreal	Dick Irvin
1946	4-1	Montreal	Dick Irvin	Boston	Dit Clapper
1945	4-3	Toronto	Hap Day	Detroit	Jack Adams
1944	4-0	Montreal	Dick Irvin	Chicago	Paul Thompson
1943	4-0	Detroit	Jack Adams	Boston	Art Ross
1942	4-3	Toronto	Hap Day	Detroit	Jack Adams
1941	4-0	Boston	Cooney Weiland	Detroit	Ebbie Goodfellow
1940	4-2	NY Rangers	Frank Boucher	Toronto	Dick Irvin
1939	4-1	Boston	Art Ross	Toronto	Dick Irvin
1938	3-1	Chicago	Bill Stewart	Toronto	Dick Irvin
1937	3-2	Detroit	Jack Adams	NY Rangers	Lester Patrick
1936	3-1	Detroit	Jack Adams	Toronto	Dick Irvin
1935	3-0	Mtl. Maroons	Tommy Gorman	Toronto	Dick Irvin
1934	3-1	Chicago	Tommy Gorman	Detroit	Herbie Lewis
1933	3-1	NY Rangers	Lester Patrick	Toronto	Dick Irvin
1932	3-0	Toronto	Dick Irvin	NY Rangers	Lester Patrick
1931	3-2	Montreal	Cecil Hart	Chicago	Dick Irvin
1930	2-0	Montreal	Cecil Hart	Boston	Art Ross
1929	2-0	Boston	Cy Denneny	NY Rangers	Lester Patrick
1928	3-2	NY Rangers	Lester Patrick	Mtl. Maroons	Eddie Gerard
1927	2-0-2	Ottawa	Dave Gill	Boston	Art Ross
		The National Hockey League assumed control of Stanley Cup competition after 1926			
1926	3-1	Mtl. Maroons	Eddie Gerard	Victoria	Lester Patrick
1925	3-1	Victoria	Lester Patrick	Montreal	Leo Dandurand
1924	2-0	Montreal	Leo Dandurand	Cgy. Tigers	Eddie Oatman
	2-0			Van. Maroons	Art Duncan/Frank Patrick
1923	2-0	Ottawa	Pete Green	Edm. Eskimos	Ken McKenzie
	3-1			Van. Maroons	Lloyd Cook/Frank Patrick
1922	3-2	Tor. St. Pats	George O'Donoghue	Van. Millionaires	Lloyd Cook/Frank Patrick
1921	3-2	Ottawa	Pete Green	Van. Millionaires	Lloyd Cook/Frank Patrick
1920	3-2	Ottawa	Pete Green	Seattle	Pete Muldoon
1919	2-2-1	No decision - series between Montreal and Seattle cancelled due to influenza epidemic			
1918	3-2	Tor. Arenas	Dick Carroll	Van. Millionaires	Frank Patrick

Championship Trophies

PRINCE OF WALES TROPHY

Beginning with the 1993-94 season, the club which advances to the Stanley Cup Finals as the winner of the Eastern Conference Championship is presented with the Prince of Wales Trophy.

History: His Royal Highness, the Prince of Wales, donated the trophy to the National Hockey League in 1924. From 1927-28 through 1937-38, the award was presented to the team finishing first in the American Division of the NHL. (The team finishing first in the Canadian Division received the O'Brien Trophy during these years.) From 1938-39, when the NHL reverted to one section, to 1966-67, it was presented to the team winning the NHL regular-season championship. With expansion in 1967-68, it again became a divisional trophy, awarded to the regular-season champions of the East Division through to the end of the 1973-74 season. Beginning in 1974-75, it was awarded to the regular-season winner of the conference bearing the name of the trophy. From 1981-82 to 1992-93 the trophy was presented to the playoff champion in the Wales Conference. Since 1993-94, the trophy has been presented to the playoff champion in the Eastern Conference.

2001-02 Winner: Carolina Hurricanes

The Carolina Hurricanes won their first Prince of Wales Trophy on May 28, 2002 after defeating the Toronto Maple Leafs 2-1 in overtime in game six of the Eastern Conference Championship series. Before defeating the Leafs, the Hurricanes had series wins over the New Jersey Devils and Montreal Canadiens.

PRINCE OF WALES TROPHY WINNERS

2001-02	Carolina Hurricanes
2000-01	New Jersey Devils
99-2000	New Jersey Devils
1998-99	Buffalo Sabres
1997-98	Washington Capitals
1996-97	Philadelphia Flyers
1995-96	Florida Panthers
1994-95	New Jersey Devils
1993-94	New York Rangers
1992-93	Montreal Canadiens
1991-92	Pittsburgh Penguins
1990-91	Pittsburgh Penguins
1989-90	Boston Bruins
1988-89	Montreal Canadiens
1987-88	Boston Bruins
1986-87	Philadelphia Flyers
1985-86	Montreal Canadiens
1984-85	Philadelphia Flyers
1983-84	New York Islanders
1982-83	New York Islanders
1981-82	New York Islanders
1980-81	Montreal Canadiens
1979-80	Buffalo Sabres
1978-79	Montreal Canadiens
1977-78	Montreal Canadiens
1976-77	Montreal Canadiens
1975-76	Montreal Canadiens
1974-75	Buffalo Sabres
1973-74	Boston Bruins
1972-73	Montreal Canadiens
1971-72	Boston Bruins
1970-71	Boston Bruins
1969-70	Chicago Blackhawks
1968-69	Montreal Canadiens
1967-68	Montreal Canadiens
1966-67	Chicago Blackhawks
1965-66	Montreal Canadiens
1964-65	Detroit Red Wings
1963-64	Montreal Canadiens
1962-63	Toronto Maple Leafs
1961-62	Montreal Canadiens
1960-61	Montreal Canadiens
1959-60	Montreal Canadiens
1958-59	Montreal Canadiens
1957-58	Montreal Canadiens
1956-57	Detroit Red Wings
1955-56	Montreal Canadiens
1954-55	Detroit Red Wings
1953-54	Detroit Red Wings
1952-53	Detroit Red Wings
1951-52	Detroit Red Wings
1950-51	Detroit Red Wings
1949-50	Detroit Red Wings
1948-49	Detroit Red Wings
1947-48	Toronto Maple Leafs
1946-47	Montreal Canadiens
1945-46	Montreal Canadiens
1944-45	Montreal Canadiens
1943-44	Montreal Canadiens
1942-43	Detroit Red Wings
1941-42	New York Rangers
1940-41	Boston Bruins
1939-40	Boston Bruins
1938-39	Boston Bruins
1937-38	Boston Bruins
1936-37	Detroit Red Wings
1935-36	Detroit Red Wings
1934-35	Boston Bruins
1933-34	Detroit Red Wings
1932-33	Boston Bruins
1931-32	New York Rangers
1930-31	Boston Bruins
1929-30	Boston Bruins
1928-29	Boston Bruins
1927-28	Boston Bruins
1926-27	Ottawa Senators
1925-26	Montreal Maroons
1924-25	Montreal Canadiens
1923-24	Montreal Canadiens

CLARENCE S. CAMPBELL BOWL

Beginning with the 1993-94 season, the club which advances to the Stanley Cup Finals as the winner of the Western Conference Championship is presented with the Clarence S. Campbell Bowl.

History: Presented by the member clubs in 1968 for perpetual competition by the National Hockey League in recognition of the services of Clarence S. Campbell, President of the NHL from 1946 to 1977. From 1967-68 through 1973-74, the trophy was awarded to the regular-season champions of the West Division. Beginning in 1974-75, it was awarded to the regular-season winner of the conference bearing the name of the trophy. From 1981-82 to 1992-93 the trophy was presented to the playoff champion in the Campbell Conference. Since 1993-94, the trophy has been presented to the playoff champion in the Western Conference. The trophy itself is a hallmark piece made of sterling silver and was crafted by a British silversmith in 1878.

2001-02 Winner: Detroit Red Wings

The Detroit Red Wings won their first Clarence Campbell Bowl since 1998 on May 31, 2002 after defeating the Colorado Avalanche 7-0 in game seven of the Western Conference Championship series. Before defeating the Avalanche, the Wings had series wins over the Vancouver Canucks and St. Louis Blues.

CLARENCE S. CAMPBELL BOWL WINNERS

2001-02	Detroit Red Wings
2000-01	Colorado Avalanche
99-2000	Dallas Stars
1998-99	Dallas Stars
1997-98	Detroit Red Wings
1996-97	Detroit Red Wings
1995-96	Colorado Avalanche
1994-95	Detroit Red Wings
1993-94	Vancouver Canucks
1992-93	Los Angeles Kings
1991-92	Chicago Blackhawks
1990-91	Minnesota North Stars
1989-90	Edmonton Oilers
1988-89	Calgary Flames
1987-88	Edmonton Oilers
1986-87	Edmonton Oilers
1985-86	Calgary Flames
1984-85	Edmonton Oilers
1983-84	Edmonton Oilers
1982-83	Edmonton Oilers
1981-82	Vancouver Canucks
1980-81	New York Islanders
1979-80	Philadelphia Flyers
1978-79	New York Islanders
1977-78	New York Islanders
1976-77	Philadelphia Flyers
1975-76	Philadelphia Flyers
1974-75	Philadelphia Flyers
1973-74	Philadelphia Flyers
1972-73	Chicago Blackhawks
1971-72	Chicago Blackhawks
1970-71	Chicago Blackhawks
1969-70	St. Louis Blues
1968-69	St. Louis Blues
1967-68	Philadelphia Flyers

Prince of Wales Trophy

Clarence S. Campbell Bowl

Stanley Cup

Stanley Cup Winners

Rosters and Final Series Scores

2001-02 — Detroit Red Wings — Steve Yzerman (Captain), Chris Chelios, Mathieu Dandenault, Pavel Datsyuk, Boyd Devereaux, Kris Draper, Steve Duchesne, Sergei Fedorov, Jiri Fischer, Dominik Hasek, Tomas Holmstrom, Brett Hull, Igor Larionov, Manny Legace, Nicklas Lidstrom, Kirk Maltby, Darren McCarty, Fredrik Olausson, Luc Robitaille, Brendan Shanahan, Jiri Slegr, Jason Williams, Michael Ilitch (Owner/Governor) Marian Ilitch (Owner/Secretary Treasurer), Ronald Ilitch, Michael Ilitch Jr., Lisa Ilitch Murray, Atanas Ilitch, Carole Ilitch Trepeck, Jim Devallano (Senior Vice President), Christopher Ilitch (Vice President), Denise Ilitch (Alternate Governor), Ken Holland (General Manager), Jim Nill (Assistant General Manager), Scotty Bowman (Head Coach), Dave Lewis (Associate Coach), Barry Smith (Associate Coach), Jim Berard (Goaltending Consultant), Joe Kocur (Video Coordinator), John Wharton (Athletic Trainer), Paul Boyer (Equipment Manager), Piet Van Zant (Assistant Athletic Trainer), Tim Abbott (Assistant Equipment Manager), Sergei Tchekmarev (Masseur), Dan Belisle (Pro Scout), Mark Howe (Pro Scout), Bob McCammon (Pro Scout), Hakan Andersson (Director of European Scouting), Mark Leach (Scout), Bruce Haralson (Scout), Joe McDonnell (Scout), Glenn Merkosky (Scout).

Scores: June 4, at Detroit - Carolina 3, Detroit 2; June 6, at Detroit - Detroit 3, Carolina 1; June 8, at Carolina - Detroit 3, Carolina 2; June 10, at Carolina - Detroit 3, Carolina 0; June 13, at Detroit - Detroit 3, Carolina 1.

2000-01 — Colorado Avalanche — Joe Sakic (Captain), David Aebischer, Rob Blake, Raymond Bourque, Greg de Vries, Chris Dingman, Chris Drury, Adam Foote, Peter Forsberg, Milan Hejduk, Dan Hinote, Jon Klemm, Eric Messier, Bryan Muir, Ville Nieminen, Scott Parker, Shjon Podein, Nolan Pratt, Dave Reid, Steve Reinprecht, Patrick Roy, Martin Skoula, Alex Tanguay, Stephane Yelle, E. Stanley Kroenke (Owner/Governor), Pierre Lacroix (President and General Manager), Bob Hartley (Head Coach), Jacques Cloutier (Assistant Coach), Bryan Trottier (Assistant Coach), Paul Fixter (Video Coach), Francois Giguere (Vice President of Hockey Operations), Brian MacDonald (Assistant General Manager), Michel Goulet (Vice President of Player Personnel), Jean Martineau (Vice President of Communications/Team Services), Pat Karns (Head Athletic Trainer), Matthew Sokolowski (Assistant Athletic Trainer), Wayne Flemming (Equipment Manager), Mark Miller (Equipment Manager), Dave Randolph (Assistant Equipment Manager), Paul Goldberg (Strength and Conditioning Coach), Gregorio Pradera (Massage Therapist), Brad Smith (Pro Scout), Jim Hammett (Chief Scout), Garth Joy, Steve Lyons, Joni Lehto, Orval Tessier (Scouts), Charlotte Grahame (Director of Hockey Operations).

Scores: May 26, at Colorado - Colorado 5, New Jersey 0; May 29, at Colorado - New Jersey 2, Colorado 1; May 31, at New Jersey - Colorado 3, New Jersey 1; June 2, at New Jersey - New Jersey 3, Colorado 2; June 4, at Colorado - New Jersey 4, Colorado 1; June 7, at New Jersey - Colorado 4, New Jersey 0; June 9, at Colorado - Colorado 3, New Jersey 1.

1999-2000 — New Jersey Devils — Scott Stevens (Captain), Jason Arnott, Brad Bombardir, Martin Brodeur, Steve Brule, Sergei Brylin, Ken Daneyko, Patrik Elias, Scott Gomez, Bobby Holik, Steve Kelly, Claude Lemieux, John Madden, Vladimir Malakhov, Randy McKay, Alexander Mogilny, Sergei Nemchinov, Scott Niedermayer, Krzysztof Oliwa, Jay Pandolfo, Brian Rafalski, Ken Sutton, Petr Sykora, Chris Terreri, Colin White, Dr. John J. McMullen (Owner/Chairman), Peter S. McMullen (Owner), Lou Lamoriello (President/General Manager), Larry Robinson (Head Coach), Viacheslav Fetisov (Assistant Coach), Bob Carpenter (Assistant Coach), Jacques Caron (Goaltending Coach), John Cunniff (AHL Coach), David Conte (Director of Scouting), Milt Fisher (Scout), Claude Carrier (Assistant Director of Scouting), Dan Labraaten (Scout), Marcel Pronovost (Scout), Bob Hoffmeyer (Pro Scout), Dr. Barry Fisher (Orthopedist), Dennis Gendron (AHL Assistant Coach), Robbie Ftorek (Coach), Vladimir Bure (Consultant), Taran Singleton (Hockey Operations), Marie Carnevale (Hockey Operations), Callie Smith (Hockey Operations), Bill Murray (Medical Trainer), Michael Vasalani (Strength/Conditioning Coordinator), Dana McGuane (Equipment Manager), Juergen Merz (Massage Therapist), Harry Bricker (Assistant Equipment Manager), Lou Centanni (Assistant Equipment Manager).
Scores: May 30, at New Jersey - New Jersey 7, Dallas 3; June 1, at New Jersey - Dallas 2, New Jersey 1; June 3, at Dallas - New Jersey 2, Dallas 1; June 5, at Dallas - New Jersey 3, Dallas 1; June 8, at New Jersey - Dallas 1 - New Jersey 0; at Dallas, New Jersey 2 - Dallas 1.

1998-99 — Dallas Stars — Derian Hatcher (Captain), Ed Belfour, Guy Carbonneau, Shawn Chambers, Benoit Hogue, Tony Hrkac, Brett Hull, Mike Keane, Jamie Langenbrunner, Jere Lehtinen, Craig Ludwig, Grant Marshall, Richard Matvichuk, Mike Modano, Joe Nieuwendyk, Derek Plante, Dave Reid, Jon Sim, Brian Skrudland, Blake Sloan, Darryl Sydor, Roman Turek, Pat Verbeek, Sergei Zubov, Thomas Hicks (Chairman of the Board and Owner), Jim Lites (President), Bob Gainey (Vice President, Hockey Operations and General Manager), Doug Armstrong (Assistant General Manager), Craig Button (Director of Player Personnel), Ken Hitchcock (Head Coach), Doug Jarvis (Assistant Coach), Rick Wilson (Assistant Coach), Rick McLaughlin (Vice President and Chief Financial Officer), Jeff Cogen (Vice President, Marketing and Promotion), Bill Strong (Vice President, Marketing and Broadcasting), Tim Bernhardt (Director of Amateur Scouting), Doug Overton (Director of Pro Scouting), Bob Gernander (Chief Scout), Stu MacGregor (Western Scout), Dave Suprenant (Medical Trainer), Dave Smith (Equipment Manager), Rich Matthews (Equipment Manager), J.J. McQueen (Strength and Conditioning Coach), Rick St. Croix (Goaltending Consultant), Dan Stuchal (Director of Team Services), Larry Kelly (Director of Public Relations).
Scores: June 8, at Dallas - Buffalo 3, Dallas 2; June 10, at Dallas - Dallas 4, Buffalo 2; June 12, at Buffalo - Dallas 2, Buffalo 1; June 15, at Buffalo - Buffalo 2, Dallas 1; June 17, at Dallas - Dallas 2, Buffalo 0; June 19, at Buffalo - Dallas 2, Buffalo 1.

1997-98 — Detroit Red Wings — Steve Yzerman (Captain), Doug Brown, Mathieu Dandenault, Kris Draper, Anders Eriksson, Sergei Fedorov, Viacheslav Fetisov, Brent Gilchrist, Kevin Hodson, Tomas Holmstrom, Mike Knuble, Joe Kocur, Vladimir Konstantinov, Vyacheslav Kozlov, Martin Lapointe, Igor Larionov, Nicklas Lidstrom, Jamie Macoun, Kirk Maltby, Darren McCarty, Dmitri Mironov, Larry Murphy, Chris Osgood, Bob Rouse, Brendan Shanahan, Aaron Ward, Mike Ilitch, (Owner/Chairman), Marian Ilitch (Owner), Atanas Ilitch (Vice President), Christopher Ilitch (Vice President), Denise Ilitch, Ronald Ilitch, Michael Ilitch Jr., Lisa Ilitch Murray, Carole Ilitch Trepeck, Jim Devellano (Senior Vice President), Scotty Bowman (Head Coach), Ken Holland (General Manager), Don Waddell (Assistant General Manager), Barry Smith (Associate Coach), Dave Lewis (Associate Coach), Jim Bedard (Goaltending Consultant), Jim Nill (Director of Player Development), Dan Belisle (Pro Scout), Mark Howe (Pro Scout), Hakan Andersson (Director of European Scouting), Mark Leach (USA Scout), Moe McDonnell (Eastern Scout), Bruce Haralson (Western Scout), John Wharton (Athletic Trainer), Paul Boyer (Equipment Manager) Tim Abbott (Assistant Equipment Manager), Bob Huddleston (Masseur), Sergei Mnatsakanov (Masseur), Wally Crossman (Dressing Room Assistant).
Scores: June 9, at Detroit — Detroit 2, Washington 1; June 11, at Detroit — Detroit 5, Washington 4; June 13, at Washington — Detroit 2, Washington 1; June 16, at Washington — Detroit 4, Washington 1.

1996-97 — Detroit Red Wings — Steve Yzerman (Captain), Doug Brown, Mathieu Dandenault, Kris Draper, Sergei Fedorov, Viacheslav Fetisov, Kevin Hodson, Tomas Holmstrom, Joe Kocur, Vladimir Konstantinov, Vyacheslav Kozlov, Martin Lapointe, Igor Larionov, Nicklas Lidstrom, Kirk Maltby, Darren McCarty, Larry Murphy, Chris Osgood, Jamie Pushor, Bob Rouse, Tomas Sandstrom, Brendan Shanahan, Tim Taylor, Mike Vernon, Aaron Ward, Mike Ilitch (Owner/Chairman), Marian Ilitch (Owner), Atanas Ilitch (Vice President), Christopher Ilitch (Vice President), Denise Ilitch Lites, Ronald Ilitch, Michael Ilitch, Jr., Lisa Ilitch Murray, Carole Ilitch Trepeck, Jim Devellano (Senior Vice President), Scotty Bowman (Head Coach/Director of Player Personnel), Ken Holland (Assistant General Manager), Barry Smith (Associate Coach), Dave Lewis (Associate Coach), Mike Krushelnyski (Assistant Coach). Jim Nill (Director of Player Development), Dan Belisle (Pro Scout), Mark Howe (Pro Scout), Hakan Andersson (Director of European Scouting), John Wharton (Athletic Trainer), Paul Boyer (Equipment Manager) Tim Abbott (Assistant Equipment Manager), Sergei Mnatsakanov (Masseur).
Scores: May 31, at Philadelphia — Detroit 4, Philadelphia 2; June 3, at Philadelphia — Detroit 4, Philadelphia 2; June 5, at Detroit — Detroit 6, Philadelphia 1; June 7, at Detroit — Detroit 2, Philadelphia 1.

1995-96 — Colorado Avalanche — Joe Sakic (Captain), Rene Corbet, Adam Deadmarsh, Stephane Fiset, Adam Foote, Peter Forsberg, Alexei Gusarov, Dave Hannan, Valeri Kamensky, Mike Keane, Jon Klemm, Uwe Krupp, Sylvain Lefebvre, Claude Lemieux, Curtis Leschyshyn, Troy Murray, Sandis Ozolinsh, Mike Ricci, Patrick Roy, Warren Rychel, Chris Simon, Craig Wolanin, Stephane Yelle, Scott Young, Charlie Lyons (Chairman, CEO), Pierre Lacroix (Exec. V.P., G.M.), Marc Crawford (Head Coach), Joel Quenneville (Assistant Coach), Jacques Cloutier (Assistant Coach), Francois Giguere (Assistant General Manager), Michel Goulet (Director of Player Personnel), Dave Draper (Chief Scout), Jean Martineau (Director of Public Relations), Pat Karns (Trainer), Matthew Sokolowski (Assistant Trainer), Rob McLean (Equipment Manager), Mike Kramer (Assistant Equipment Manager), Brock Gibbins (Assistant Equipment Manager), Skip Allen (Strength and Conditioning Coach), Paul Fixter (Video Coordinator), Leo Vyssokov (Massage Therapist).
Scores: June 4, at Colorado — Colorado 3, Florida 1; June 6, at Colorado — Colorado 8, Florida 1; June 8, at Florida — Colorado 3, Florida 2; June 10, at Florida — Colorado 1, Florida 0.

1994-95 — New Jersey Devils — Scott Stevens (Captain), Tommy Albelin, Martin Brodeur, Neal Broten, Sergei Brylin, Bob Carpenter, Shawn Chambers, Tom Chorske, Danton Cole, Ken Daneyko, Kevin Dean, Jim Dowd, Bruce Driver (Alternate Captain), Bill Guerin, Bobby Holik, Claude Lemieux, John MacLean (Alternate Captain), Chris McAlpine, Randy McKay, Scott Niedermayer, Mike Peluso, Stephane Richer, Brian Rolston, Chris Terreri, Valeri Zelepukin, Dr. John J. McMullen (Owner/Chairman), Peter S. McMullen (Owner), Lou Lamoriello (President/General Manager), Jacques Lemaire (Head Coach), Jacques Caron (Goaltender Coach), Dennis Gendron (Assistant Coach), Larry Robinson (Assistant Coach), Robbie Ftorek (AHL Coach), Alex Abasto (Assistant Equipment Manager), Bob Huddleston (Massage Therapist), David Nichols (Equipment Manager), Ted Schuch (Medical Trainer), Mike Vasalani (Strength Coach), David Conte (Director of Scouting) Claude Carrier (Scout), Milt Fisher (Scout), Dan Labraaten (Scout), Marcel Pronovost (Scout).
Scores: June 17, at Detroit — New Jersey 2, Detroit 1; June 20, at Detroit — New Jersey 4, Detroit 2; June 22, at New Jersey — New Jersey 5, Detroit 2; June 24, at New Jersey — New Jersey 5, Detroit 2.

1993-94 — New York Rangers — Mark Messier (Captain), Brian Leetch, Kevin Lowe, Adam Graves, Steve Larmer, Glenn Anderson, Jeff Beukeboom, Greg Gilbert, Mike Hartman, Glenn Healy, Mike Hudson, Alexander Karpovtsev, Joe Kocur, Alexei Kovalev, Nick Kypreos, Doug Lidster, Stephane Matteau, Craig MacTavish, Sergei Nemchinov, Brian Noonan, Ed Olczyk, Mike Richter, Esa Tikkanen, Jay Wells, Sergei Zubov, Neil Smith (President, General Manager and Governor), Robert Gutkowski, Stanley Jaffe, Kenneth Munoz (Governors), Larry Pleau (Assistant General Manager), Mike Keenan (Head Coach), Colin Campbell (Associate Coach), Dick Todd (Assistant Coach), Matthew Loughren (Manager, Team Operations), Barry Watkins (Director, Communications), Christer Rockstrom, Tony Feltrin, Martin Madden, Herb Hammond, Darwin Bennett (Scouts), Dave Smith, Joe Murphy, Mike Folga, Bruce Lifrieri (Trainers).
Scores: May 31, at New York — Vancouver 3, NY Rangers 2; June 2, at New York — NY Rangers 3, Vancouver 1; June 4, at Vancouver — NY Rangers 5, Vancouver 1; June 7, at Vancouver — NY Rangers 4, Vancouver 2; June 9, at New York — Vancouver 6, at NY Rangers 3; June 11, at Vancouver — Vancouver 4, NY Rangers 1; June 14, at New York — NY Rangers 3, Vancouver 2.

1992-93 — Montreal Canadiens — Guy Carbonneau (Captain), Patrick Roy, Mike Keane, Eric Desjardins, Stephan Lebeau, Mathieu Schneider, J-J Daigneault, Denis Savard, Lyle Odelein, Todd Ewen, Kirk Muller, John LeClair, Gilbert Dionne, Benoit Brunet, Patrice Brisebois, Paul Di Pietro, Andre Racicot, Donald Dufresne, Mario Roberge, Sean Hill, Ed Ronan, Kevin Haller, Vincent Damphousse, Brian Bellows, Gary Leeman, Rob Ramage, Ronald Corey (President), Serge Savard (Managing Director & Vice-President Hockey), Jacques Demers (Head Coach), Jacques Laperriere (Assistant Coach), Charles Thiffault (Assistant Coach), Francois Allaire (Goaltending Instructor), Jean Béliveau (Senior Vice-President, Corporate Affairs), Fred Steer (Vice-President, Finance & Adminstration), Aldo Giampaolo (Vice-President, Operations), Bernard Brisset (Vice-President, Marketing & Communications), André Boudrias (Assistant to the Managing Director & Director of Scouting), Jacques Lemaire (Assistant to the Managing Director), Gaeten Lefebvre (Athletic Trainer), John Shipman (Assistant to the Athletic Trainer), Eddy Palchak (Equipment Manager), Pierre Gervais (Assistant to the Equipment Manager), Robert Boulanger (Assistant to the Equipment Manager), Pierre Ouellete (Assistant to the Equipment Manager).
Scores: June 1, at Montreal — Los Angeles 4, Montreal 1; June 2, at Montreal — Montreal 3, Los Angeles 2; June 5, at Los Angeles — Montreal 4, Los Angeles 3; June 7, at Los Angeles — Montreal 3, Los Angeles 2; June 9, at Montreal — Montreal 4, Los Angeles 1.

1991-92 — Pittsburgh Penguins — Mario Lemieux (Captain), Ron Francis, Bryan Trottier, Kevin Stevens, Bob Errey, Phil Bourque, Troy Loney, Rick Tocchet, Joe Mullen, Jaromir Jagr, Jiri Hrdina, Shawn McEachern, Ulf Samuelsson, Kjell Samuelsson, Larry Murphy, Gordie Roberts, Jim Paek, Paul Stanton, Tom Barrasso, Ken Wregget, Jay Caufield, Jamie Leach, Wendell Young, Grant Jennings, Peter Taglianetti, Jock Callander, Dave Michayluk, Mike Needham, Jeff Chychrun, Ken Priestlay, Jeff Daniels, Howard Baldwin (Owner and President), Morris Belzberg (Owner), Thomas Ruta (Owner), Donn Patton (Executive Vice President and Chief Financial Officer), Paul Martha (Executive Vice President and General Counsel), Craig Patrick (Executive Vice President and General Manager), Bob Johnson (Coach), Scotty Bowman (Director of Player Development and Coach), Barry Smith, Rick Kehoe, Pierre McGuire, Gilles Meloche, Rick Paterson (Assistant Coaches), Steve Latin (Equipment Manager), Skip Thayer (Trainer), John Welday (Strength and Conditioning Coach), Greg Malone, Les Binkley, Charlie Hodge, John Gill, Ralph Cox (Scouts).
Scores: May 26, at Pittsburgh — Pittsburgh 5, Chicago 4; May 28, at Pittsburgh — Pittsburgh 3, Chicago 1; May 30, at Chicago — Pittsburgh 1, Chicago 0; June 1, at Chicago — Pittsburgh 6, Chicago 5.

1990-91 — Pittsburgh Penguins — Mario Lemieux (Captain), Paul Coffey, Randy Hillier, Bob Errey, Tom Barrasso, Phil Bourque, Jay Caufield, Ron Francis, Randy Gilhen, Jiri Hrdina, Jaromir Jagr, Grant Jennings, Troy Loney, Joe Mullen, Larry Murphy, Jim Paek, Frank Pietrangelo, Barry Pederson, Mark Recchi, Gordie Roberts, Ulf Samuelsson, Paul Stanton, Kevin Stevens, Peter Taglianetti, Bryan Trottier, Scott Young, Wendell Young, Edward J. DeBartolo, Sr. (Owner), Marie D. DeBartolo York (President), Paul Martha (Vice-President & General Counsel), Craig Patrick (General Manager), Scotty Bowman (Director of Player Development & Recruitment), Bob Johnson (Coach), Rick Kehoe (Assistant Coach), Gilles Meloche (Goaltending Coach & Scout), Rick Paterson (Assistant Coach), Barry Smith (Assistant Coach), Steve Latin (Equipment Manager), Skip Thayer (Trainer), John Welday (Strength & Conditioning Coach), Greg Malone (Scout).
Scores: May 15, at Pittsburgh — Minnesota 5, Pittsburgh 4; May 17, at Pittsburgh — Pittsburgh 4, Minnesota 1; May 19, at Minnesota — Minnesota 3, Pittsburgh 1; May 21, at Minnesota — Pittsburgh 5, Minnesota 3; May 23, at Pittsburgh — Pittsburgh 6, Minnesota 4; May 25, at Minnesota — Pittsburgh 8, Minnesota 0.

1989-90 — Edmonton Oilers — Kevin Lowe, Steve Smith, Jeff Beukeboom, Mark Lamb, Joe Murphy, Glenn Anderson, Mark Messier (Captain), Adam Graves, Craig MacTavish, Kelly Buchberger, Jari Kurri, Craig Simpson, Martin Gelinas, Randy Gregg, Charlie Huddy, Geoff Smith, Reijo Ruotsalainen, Craig Muni, Bill Ranford, Dave Brown, Pokey Reddick, Petr Klima, Esa Tikkanen, Grant Fuhr, Peter Pocklington (Owner), Glen Sather (President/General Manager), John Muckler (Coach), Ted Green (Co-Coach), Ron Low (Ass't Coach), Bruce MacGregor (Ass't General Manager), Barry Fraser (Director of Player Personnel), John Blackwell (Director of Operations, AHL), Ace Bailey, Ed Chadwick, Lorne Davis, Harry Howell, Matti Vaisanen and Albert Reeves (Scouts), Bill Tuele (Director of Public Relations), Werner Baum (Controller), Dr. Gordon Cameron (Medical Chief of Staff), Dr. David Reid (Team Physician), Barrie Stafford (Athletic Trainer), Ken Lowe (Athletic Therapist), Stuart Poirier (Massage Therapist), Lyle Kulchisky (Ass't Trainer).
Scores: May 15, at Boston — Edmonton 3, Boston 2; May 18, at Boston — Edmonton 7, Boston 2; May 20, at Edmonton — Boston 2, Edmonton 1; May 22, at Edmonton — Edmonton 5, Boston 1; May 24, at Boston — Edmonton 4, Boston 1.

1988-89 — Calgary Flames — Mike Vernon, Rick Wamsley, Al MacInnis, Brad McCrimmon, Dana Murzyn, Ric Nattress, Joe Mullen, Lanny McDonald (Co-captain), Gary Roberts, Colin Patterson, Hakan Loob, Theoren Fleury, Jiri Hrdina, Tim Hunter (Ass't. captain), Gary Suter, Mark Hunter, Jim Peplinski (Co-captain), Joe Nieuwendyk, Brian MacLellan, Joel Otto, Jamie Macoun, Doug Gilmour, Rob Ramage. Norman Green, Harley Hotchkiss, Norman Kwong, Sonia Scurfield, B.J. Seaman, D.K. Seaman (Owners), Cliff Fletcher (President and General Manager), Al MacNeil (Ass't General Manager), Al Coates (Ass't to the President), Terry Crisp (Head Coach), Doug Risebrough, Tom Watt (Ass't Coaches), Glenn Hall (Goaltending Consultant), Jim Murray (Trainer), Bob Stewart (Equipment Manager), Al Murray (Ass't Trainer).
Scores: May 14, at Calgary — Calgary 3, Montreal 2; May 17, at Calgary— Montreal 4, Calgary 2; May 19, at Montreal — Montreal 4, Calgary 3; May 21, at Montreal — Calgary 4, Montreal 2; May 23, at Calgary — Calgary 3, Montreal 2; May 25, at Montreal — Calgary 4, Montreal 2.

1987-88 — Edmonton Oilers — Keith Acton, Glenn Anderson, Jeff Beukeboom, Geoff Courtnall, Grant Fuhr, Randy Gregg, Wayne Gretzky (Captain), Dave Hannan, Charlie Huddy, Mike Krushelnyski, Jari Kurri, Normand Lacombe, Kevin Lowe, Craig MacTavish, Kevin McClelland, Marty McSorley, Mark Messier, Craig Muni, Bill Ranford, Craig Simpson, Steve Smith, Esa Tikkanen, Peter Pocklington (Owner), Glen Sather (General Manager/Coach), John Muckler (Co-Coach), Ted Green (Ass't Coach), Bruce MacGregor (Ass't General Manager), Barry Fraser (Director of Player Personnel), Bill Tuele (Director of Public Relations), Dr. Gordon Cameron (Team Physician), Peter Millar (Athletic Therapist), Barrie Stafford (Trainer), Juergen Mers (Massage Therapist), Lyle Kulchisky (Ass't Trainer).
Scores: May 18, at Edmonton — Edmonton 2, Boston 1; May 20, at Edmonton — Edmonton 4, Boston 2; May 22, at Boston — Edmonton 6, Boston 3; May 24, at Boston — Boston 3, Edmonton 3 (suspended due to power failure); May 26, at Edmonton — Edmonton 6, Boston 3.

1986-87 — Edmonton Oilers — Glenn Anderson, Jeff Beukeboom, Kelly Buchberger, Paul Coffey, Grant Fuhr, Randy Gregg, Wayne Gretzky (Captain), Charlie Huddy, Dave Hunter, Mike Krushelnyski, Jari Kurri, Moe Lemay, Kevin Lowe, Craig MacTavish, Kevin McClelland, Marty McSorley, Mark Messier, Andy Moog, Craig Muni, Kent Nilsson, Jaroslav Pouzar, Reijo Ruotsalainen, Steve Smith, Esa Tikkanen, Peter Pocklington (Owner), Glen Sather (General Manager/Coach), John Muckler (Co-Coach), Ted Green (Ass't. Coach), Ron Low (Ass't. Coach), Bruce MacGregor (Ass't. General Manager), Barry Fraser (Director of Player Personnel), Peter Millar (Athletic Therapist), Barrie Stafford (Trainer), Lyle Kulchisky (Ass't Trainer).
Scores: May 17, at Edmonton — Edmonton 4, Philadelphia 2; May 20, at Edmonton — Edmonton 3, Philadelphia 2; May 22, at Philadelphia — Philadelphia 5, Edmonton 3; May 24, at Philadelphia — Edmonton 4, Philadelphia 1; May 26, at Edmonton — Philadelphia 4, Edmonton 3; May 28, at Philadelphia — Philadelphia 3, Edmonton 2; May 31, at Edmonton — Edmonton 3, Philadelphia 1.

1985-86 — Montreal Canadiens — Bob Gainey (Captain), Doug Soetaert, Patrick Roy, Rick Green, David Maley, Ryan Walter, Serge Boisvert, Mario Tremblay, Bobby Smith, Craig Ludwig, Tom Kurvers, Kjell Dahlin, Larry Robinson, Guy Carbonneau, Chris Chelios, Petr Svoboda, Mats Naslund, Lucien DeBlois, Steve Rooney, Gaston Gingras, Mike Lalor, Chris Nilan, John Kordic, Claude Lemieux, Mike McPhee, Brian Skrudland, Stephane Richer, Ronald Corey (President), Serge Savard (General Manager), Jean Perron (Coach), Jacques Laperrière (Ass't. Coach), Jean Béliveau (Vice President), Francois-Xavier Seigneur (Vice President), Fred Steer (Vice President), Jacques Lemaire (Ass't. General Manager), André Boudrias (Ass't. General Manager), Claude Ruel (Scouting), Yves Belanger (Athletic Therapist), Gaetan Lefebvre (Ass't. Athletic Therapist), Eddy Palchak (Trainer), Sylvain Toupin (Ass't. Trainer).
Scores: May 16, at Calgary — Calgary 5, Montreal 2; May 18, at Calgary — Montreal 3, Calgary 2; May 20, at Montreal — Montreal 5, Calgary 3; May 22, at Montreal — Montreal 1, Calgary 0; May 24, at Calgary — Montreal 4, Calgary 3.

1984-85 — Edmonton Oilers — Glenn Anderson, Billy Carroll, Paul Coffey, Lee Fogolin, Grant Fuhr, Randy Gregg, Wayne Gretzky (Captain), Charlie Huddy, Pat Hughes, Dave Hunter, Don Jackson, Mike Krushelnyski, Jari Kurri, Willy Lindstrom, Kevin Lowe, Dave Lumley, Kevin McClelland, Larry Melnyk, Mark Messier, Andy Moog, Mark Napier, Jaroslav Pouzar, Dave Semenko, Esa Tikkanen, Peter Pocklington (Owner), Glen Sather (General Manager/Coach), John Muckler (Ass't. Coach), Ted Green (Ass't. Coach), Bruce MacGregor (Ass't. General Manager), Barry Fraser (Director of Player Personnel/Chief Scout), Peter Millar (Athletic Therapist), Barrie Stafford, Lyle Kulchisky (Trainers)
Scores: May 21, at Philadelphia — Philadelphia 4, Edmonton 1; May 23, at Philadelphia — Edmonton 3, Philadelphia 1; May 25, at Edmonton — Edmonton 4, Philadelphia 3; May 28, at Edmonton — Edmonton 5, Philadelphia 3; May 30, at Edmonton — Edmonton 8, Philadelphia 3.

1983-84 — Edmonton Oilers — Glenn Anderson, Paul Coffey, Pat Conacher, Lee Fogolin, Grant Fuhr, Randy Gregg, Wayne Gretzky (Captain), Charlie Huddy, Pat Hughes, Dave Hunter, Don Jackson, Jari Kurri, Willy Lindstrom, Ken Linseman, Kevin Lowe, Dave Lumley, Kevin McClelland, Mark Messier, Andy Moog, Jaroslav Pouzar, Dave Semenko, Peter Pocklington (Owner), Glen Sather (General Manager/Coach), John Muckler (Ass't. Coach), Ted Green (Ass't. Coach), Bruce MacGregor (Ass't. General Manager), Barry Fraser (Director of Player Personnel/Chief Scout), Peter Millar (Athletic Therapist), Barrie Stafford (Trainer)
Scores: May 10, at New York — Edmonton 1, NY Islanders 0; May 12, at New York — NY Islanders 6, Edmonton 1; May 15, at Edmonton — Edmonton 7, NY Islanders 2; May 17, at Edmonton — Edmonton 7, NY Islanders 2; May 19, at Edmonton — Edmonton 5, NY Islanders 2.

1982-83 — New York Islanders — Mike Bossy, Bob Bourne, Paul Boutilier, Billy Carroll, Greg Gilbert, Clark Gillies, Butch Goring, Mats Hallin, Tomas Jonsson, Anders Kallur, Gord Lane, Dave Langevin, Mike McEwen, Rollie Melanson, Wayne Merrick, Ken Morrow, Bob Nystrom, Stefan Persson, Denis Potvin (Captain), Billy Smith, Brent Sutter, Duane Sutter, John Tonelli, Bryan Trottier, Al Arbour (Coach), Lorne Henning (Ass't. Coach), Bill Torrey (General Manager), Ron Waske, Jim Pickard (Trainers)
Scores: May 10, at Edmonton — NY Islanders 2, Edmonton 0; May 12, at Edmonton — NY Islanders 6, Edmonton 3; May 14, at New York — NY Islanders 5, Edmonton 1; May 17, at New York — NY Islanders 4, Edmonton 2

1981-82 — New York Islanders — Mike Bossy, Bob Bourne, Billy Carroll, Butch Goring, Greg Gilbert, Clark Gillies, Tomas Jonsson, Anders Kallur, Gord Lane, Dave Langevin, Hector Marini, Mike McEwen, Rollie Melanson, Wayne Merrick, Ken Morrow, Bob Nystrom, Stefan Persson, Denis Potvin (Captain), Billy Smith, Brent Sutter, Duane Sutter, John Tonelli, Bryan Trottier, Al Arbour (Coach), Lorne Henning (Ass't. Coach), Bill Torrey (General Manager), Jim Devellano (ass't. general manager/dir. of scouting), Ron Waske, Jim Pickard (Trainers)
Scores: May 8, at New York — NY Islanders 6, Vancouver 5; May 11, at New York — NY Islanders 6, Vancouver 4; May 13, at Vancouver — NY Islanders 3, Vancouver 0; May 16, at Vancouver — NY Islanders 3, Vancouver 1

1980-81 — New York Islanders — Denis Potvin (Captain), Mike McEwen, Ken Morrow, Gord Lane, Bob Lorimer, Stefan Persson, Dave Langevin, Mike Bossy, Bryan Trottier, Butch Goring, Wayne Merrick, Clark Gillies, John Tonelli, Bob Nystrom, Billy Carroll, Bob Bourne, Hector Marini, Anders Kallur, Duane Sutter, Garry Howatt, Lorne Henning, Billy Smith, Rollie Melanson, Al Arbour (Coach), Bill Torrey (General Manager), Jim Devellano (Chief Scout), Ron Waske, Jim Pickard (Trainers).
Scores: May 12, at New York — NY Islanders 6, Minnesota 3; May 14, at New York — NY Islanders 6, Minnesota 3; May 17, at Minnesota — NY Islanders 7, Minnesota 5; May 19, at Minnesota— Minnesota 4, NY Islanders 2; May 21, at New York — NY Islanders 5, Minnesota 1.

1979-80 — New York Islanders — Gord Lane, Jean Potvin, Bob Lorimer, Denis Potvin (Captain), Stefan Persson, Ken Morrow, Dave Langevin, Duane Sutter, Garry Howatt, Clark Gillies, Lorne Henning, Wayne Merrick, Bob Bourne, Steve Tambellini, Bryan Trottier, Mike Bossy, Bob Nystrom, John Tonelli, Anders Kallur, Butch Goring, Alex McKendry, Glenn Resch, Billy Smith, Al Arbour (Coach), Bill Torrey (General Manager), Jim Devellano (Chief Scout), Ron Waske, Jim Pickard (Trainers).
Scores: May 13, at Philadelphia — NY Islanders 4, Philadelphia 3; May 15, at Philadelphia — Philadelphia 8, NY Islanders 3; May 17, at New York — NY Islanders 6, Philadelphia 2; May 19, at New York — NY Islanders 5, Philadelphia 2; May 22 at Philadelphia — Philadelphia 6, NY Islanders 3; May 24, at New York — NY Islanders 5, Philadelphia 4.

1978-79 — Montreal Canadiens — Ken Dryden, Larry Robinson, Serge Savard, Guy Lapointe, Brian Engblom, Gilles Lupien, Rick Chartraw, Guy Lafleur, Steve Shutt, Jacques Lemaire, Yvan Cournoyer (Captain), Réjean Houle, Pierre Mondou, Bob Gainey, Doug Jarvis, Yvon Lambert, Doug Risebrough, Pierre Larouche, Mario Tremblay, Cam Connor, Pat Hughes, Rod Langway, Mark Napier, Michel Larocque, Richard Sévigny, Scotty Bowman (Coach), Irving Grundman (Managing Director), Eddy Palchak, Pierre Meilleur (Trainers).
Scores: May 13, at Montreal — NY Rangers 4, Montreal 1; May 15, at Montreal — Montreal 6, NY Rangers 2; May 17, at New York — Montreal 4, NY Rangers 1; May 19, at New York — Montreal 4, NY Rangers 3; May 21, at Montreal — Montreal 4, NY Rangers 1.

1977-78 — Montreal Canadiens — Ken Dryden, Larry Robinson, Serge Savard, Guy Lapointe, Bill Nyrop, Pierre Bouchard, Brian Engblom, Gilles Lupien, Rick Chartraw, Guy Lafleur, Steve Shutt, Jacques Lemaire, Yvan Cournoyer (Captain), Réjean Houle, Pierre Mondou, Bob Gainey, Doug Jarvis, Yvon Lambert, Doug Risebrough, Pierre Larouche, Mario Tremblay, Michel Larocque, Murray Wilson, Scotty Bowman (Coach), Sam Pollock (General Manager), Eddy Palchak, Pierre Meilleur (Trainers).
Scores: May 13, at Montreal — Montreal 4, Boston 1; May 16, at Montreal — Montreal 3, Boston 2; May 18, at Boston — Boston 4, Montreal 0; May 21, at Boston — Boston 4, Montreal 3; May 23, at Montreal — Montreal 4, Boston 1; May 25, at Boston — Montreal 4, Boston 1.

1976-77 — Montreal Canadiens — Ken Dryden, Guy Lapointe, Larry Robinson, Serge Savard, Jimmy Roberts, Rick Chartraw, Bill Nyrop, Pierre Bouchard, Brian Engblom, Yvan Cournoyer (Captain), Guy Lafleur, Jacques Lemaire, Steve Shutt, Pete Mahovlich, Murray Wilson, Doug Jarvis, Yvon Lambert, Bob Gainey, Doug Risebrough, Mario Tremblay, Rejean Houle, Pierre Mondou, Mike Polich, Michel Larocque, Scotty Bowman (Coach), Sam Pollock (General Manager), Eddy Palchak, Pierre Meilleur (Trainers).
Scores: May 7, at Montreal — Montreal 7, Boston 3; May 10, at Montreal — Montreal 3, Boston 0; May 12, at Boston — Montreal 4, Boston 2; May 14, at Boston — Montreal 2, Boston 1.

1975-76 — Montreal Canadiens — Ken Dryden, Serge Savard, Guy Lapointe, Larry Robinson, Bill Nyrop, Pierre Bouchard, Jimmy Roberts, Guy Lafleur, Steve Shutt, Pete Mahovlich, Yvan Cournoyer (Captain), Jacques Lemaire, Yvon Lambert, Bob Gainey, Doug Jarvis, Doug Risebrough, Murray Wilson, Mario Tremblay, Rick Chartraw, Michel Larocque, Scotty Bowman (Coach), Sam Pollock (General Manager), Eddy Palchak, Pierre Meilleur (Trainers).
Scores: May 9, at Montreal — Montreal 4, Philadelphia 3; May 11, at Montreal — Montreal 2, Philadelphia 1; May 13, at Philadelphia — Montreal 3, Philadelphia 2; May 16, at Philadelphia — Montreal 5, Philadelphia 3.

1974-75 — Philadelphia Flyers — Bernie Parent, Wayne Stephenson, Ed Van Impe, Tom Bladon, André Dupont, Joe Watson, Jimmy Watson, Ted Harris, Larry Goodenough, Rick MacLeish, Bobby Clarke (Captain), Bill Barber, Reggie Leach, Gary Dornhoefer, Ross Lonsberry, Bob Kelly, Terry Crisp, Don Saleski, Dave Schultz, Orest Kindrachuk, Bill Clement, Fred Shero (Coach), Keith Allen (general manager), Frank Lewis, Jim McKenzie (Trainers).
Scores: May 15, at Philadelphia — Philadelphia 4, Buffalo 1; May 18, at Philadelphia — Philadelphia 2, Buffalo 1; May 20, at Buffalo — Buffalo 5, Philadelphia 4; May 22, at Buffalo — Buffalo 4, Philadelphia 2; May 25, at Philadelphia — Philadelphia 5, Buffalo 1; May 27, at Buffalo — Philadelphia 2, Buffalo 0.

1973-74 — Philadelphia Flyers — Bernie Parent, Ed Van Impe, Tom Bladon, André Dupont, Joe Watson, Jimmy Watson, Barry Ashbee, Bill Barber, Dave Schultz, Don Saleski, Gary Dornhoefer, Terry Crisp, Bobby Clarke (Captain), Simon Nolet, Ross Lonsberry, Rick MacLeish, Bill Flett, Orest Kindrachuk, Bill Clement, Bob Kelly, Bruce Cowick, Al MacAdam, Bobby Taylor, Fred Shero (Coach), Keith Allen (General Manager), Frank Lewis, Jim McKenzie (Trainers).
Scores: May 7, at Boston — Boston 3, Philadelphia 2; May 9, at Boston — Philadelphia 3, Boston 2; May 12, at Philadelphia — Philadelphia 4, Boston 1; May 14, at Philadelphia — Philadelphia 4, Boston 2; May 16, at Boston — Boston 5, Philadelphia 1; May 19, at Philadelphia — Philadelphia 1, Boston 0.

1972-73 — Montreal Canadiens — Ken Dryden, Guy Lapointe, Serge Savard, Larry Robinson, Jacques Laperrière, Bob Murdoch, Pierre Bouchard, Jimmy Roberts, Yvan Cournoyer, Frank Mahovlich, Jacques Lemaire, Pete Mahovlich, Marc Tardif, Henri Richard (Captain), Réjean Houle, Guy Lafleur, Chuck Lefley, Claude Larose, Murray Wilson, Steve Shutt, Michel Plasse, Scotty Bowman (Coach), Sam Pollock (General Manager), Eddy Palchak, Bob Williams (Trainers).
Scores: April 29, at Montreal — Montreal 8, Chicago 3; May 1, at Montreal — Montreal 4, Chicago 1; May 3, at Chicago — Chicago 7, Montreal 4; May 6, at Chicago — Montreal 4, Chicago 0; May 8, at Montreal — Chicago 8, Montreal 7; May 10, at Chicago — Montreal 6, Chicago 4.

1971-72 — Boston Bruins — Gerry Cheevers, Eddie Johnston, Bobby Orr, Ted Green, Carol Vadnais, Dallas Smith, Don Awrey, Phil Esposito, Ken Hodge, John Bucyk, Mike Walton, Wayne Cashman, Garnet Bailey, Derek Sanderson, Fred Stanfield, Ed Westfall, John McKenzie, Don Marcotte, Garry Peters, Chris Hayes, Tom Johnson (Coach), Milt Schmidt (General Manager), Dan Canney, John Forristall (Trainers).
Scores: April 30, at Boston — Boston 6, NY Rangers 5; May 2, at Boston — Boston 2, NY Rangers 1; May 4, at New York — NY Rangers 5, Boston 2; May 7, at New York — Boston 3, NY Rangers 2; May 9, at Boston — NY Rangers 3, Boston 2; May 11, at New York — Boston 3, NY Rangers 0.

1970-71 — Montreal Canadiens — Ken Dryden, Rogie Vachon, Jacques Laperrière, J.C. Tremblay, Guy Lapointe, Terry Harper, Pierre Bouchard, Jean Béliveau (Captain), Marc Tardif, Yvan Cournoyer, Réjean Houle, Claude Larose, Henri Richard, Phil Roberto, Pete Mahovlich, Leon Rochefort, John Ferguson, Bobby Sheehan, Jacques Lemaire, Frank Mahovlich, Bob Murdoch, Chuck Lefley, Al MacNeil (Coach), Sam Pollock (General Manager), Yvon Belanger, Eddy Palchak (Trainers).
Scores: May 4, at Chicago — Chicago 2, Montreal 1; May 6, at Chicago — Chicago 5, Montreal 3; May 9, at Montreal — Montreal 4, Chicago 2; May 11, at Montreal — Montreal 5, Chicago 2; May 13, at Chicago — Chicago 2, Montreal 0; May 16, at Montreal — Montreal 4, Chicago 3; May 18, at Chicago — Montreal 3, Chicago 2.

1969-70 — Boston Bruins — Gerry Cheevers, Eddie Johnston, Bobby Orr, Rick Smith, Dallas Smith, Bill Speer, Gary Doak, Don Awrey, Phil Esposito, Ken Hodge, John Bucyk, Wayne Carleton, Wayne Cashman, Derek Sanderson, Fred Stanfield, Ed Westfall, John McKenzie, Jim Lorentz, Don Marcotte, Bill Lesuk, Danny Schock, Harry Sinden (Coach), Milt Schmidt (General Manager), Dan Canney, John Forristall (Trainers).
Scores: May 3, at St. Louis — Boston 6, St. Louis 1; May 5, at St. Louis — Boston 6, St. Louis 2; May 7, at Boston — Boston 4, St. Louis 1; May 10, at Boston — Boston 4, St. Louis 3.

1968-69 — Montreal Canadiens — Gump Worsley, Rogie Vachon, Jacques Laperrière, J.C. Tremblay, Ted Harris, Serge Savard, Terry Harper, Larry Hillman, Jean Béliveau (Captain), Ralph Backstrom, Dick Duff, Yvan Cournoyer, Claude Provost, Bobby Rousseau, Henri Richard, John Ferguson, Christian Bordeleau, Mickey Redmond, Jacques Lemaire, Lucien Grenier, Tony Esposito, Claude Ruel (Coach), Sam Pollock (General Manager), Larry Aubut, Eddy Palchak (Trainers).
Scores: April 27, at Montreal — Montreal 3, St. Louis 1; April 29, at Montreal — Montreal 3, St. Louis 1; May 1 at St. Louis — Montreal 4, St. Louis 0; May 4, at St. Louis — Montreal 2, St. Louis 1.

1967-68 — Montreal Canadiens — Gump Worsley, Rogie Vachon, Jacques Laperrière, J.C. Tremblay, Ted Harris, Serge Savard, Terry Harper, Carol Vadnais, Jean Béliveau (Captain), Gilles Tremblay, Ralph Backstrom, Dick Duff, Claude Larose, Yvan Cournoyer, Claude Provost, Bobby Rousseau, Henri Richard, John Ferguson, Danny Grant, Jacques Lemaire, Mickey Redmond, Toe Blake (Coach), Sam Pollock (General Manager), Larry Aubut, Eddy Palchak (Trainers).
Scores: May 5, at St. Louis — Montreal 3, St. Louis 2; May 7, at St. Louis — Montreal 1, St. Louis 0; May 9, at Montreal — Montreal 4, St. Louis 3; May 11, at Montreal — Montreal 3, St. Louis 2.

1966-67 — Toronto Maple Leafs — Johnny Bower, Terry Sawchuk, Larry Hillman, Marcel Pronovost, Tim Horton, Bob Baun, Aut Erickson, Allan Stanley, Red Kelly, Ron Ellis, George Armstrong (Captain), Pete Stemkowski, Dave Keon, Mike Walton, Jim Pappin, Bob Pulford, Brian Conacher, Eddie Shack, Frank Mahovlich, Milan Marcetta, Larry Jeffrey, Bruce Gamble, Punch Imlach (Manager-Coach), Bob Haggert (Trainer).
Scores: April 20, at Montreal — Toronto 2, Montreal 6; April 22, at Montreal — Toronto 3, Montreal 0; April 25, at Toronto — Toronto 3, Montreal 2; April 27, at Toronto — Toronto 2, Montreal 6; April 29, at Montreal — Toronto 4, Montreal 1; May 2, at Toronto — Toronto 3, Montreal 1.

1965-66 — Montreal Canadiens — Gump Worsley, Charlie Hodge, J.C. Tremblay, Ted Harris, Jean-Guy Talbot, Terry Harper, Jacques Laperrière, Noel Price, Jean Béliveau (Captain), Ralph Backstrom, Dick Duff, Gilles Tremblay, Claude Larose, Yvan Cournoyer, Claude Provost, Bobby Rousseau, Henri Richard, Dave Balon, John Ferguson, Leon Rochefort, Jimmy Roberts, Toe Blake (Coach), Sam Pollock (general manager), Larry Aubut, Andy Galley (Trainers).
Scores: April 24, at Montreal — Detroit 3, Montreal 2; April 26, at Montreal — Detroit 5, Montreal 2; April 28, at Detroit — Montreal 4, Detroit 2; May 1, at Detroit — Montreal 2, Detroit 1; May 3, at Montreal — Montreal 5, Detroit 1; May 5, at Detroit — Montreal 3, Detroit 2.

1964-65 — Montreal Canadiens — Gump Worsley, Charlie Hodge, J.C. Tremblay, Ted Harris, Jean-Guy Talbot, Terry Harper, Jacques Laperrière, Jean Gauthier, Noel Picard, Jean Béliveau (Captain), Ralph Backstrom, Dick Duff, Claude Larose, Yvan Cournoyer, Claude Provost, Bobby Rousseau, Henri Richard, Dave Balon, John Ferguson, Red Berenson, Jimmy Roberts, Toe Blake (Coach), Sam Pollock (general manager), Larry Aubut, Andy Galley (Trainers).
Scores: April 17, at Montreal — Montreal 3, Chicago 2; April 20, at Montreal — Montreal 2, Chicago 0; April 22, at Chicago — Montreal 1, Chicago 3; April 25, at Chicago — Montreal 1, Chicago 5; April 7, at Montreal — Montreal 6, Chicago 0; April 29, at Chicago — Montreal 1, Chicago 2; May 1, at Montreal — Montreal 4, Chicago 0.

1963-64 — Toronto Maple Leafs — Johnny Bower, Don Simmons, Carl Brewer, Tim Horton, Bob Baun, Allan Stanley, Larry Hillman, Al Arbour, Red Kelly, Gerry Ehman, Andy Bathgate, George Armstrong (Captain), Ron Stewart, Dave Keon, Billy Harris, Don McKenney, Jim Pappin, Bob Pulford, Eddie Shack, Frank Mahovlich, Ed Litzenberger, Punch Imlach (Manager-Coach), Bob Haggert (Trainer).
Scores April 11, at Toronto — Toronto 3, Detroit 2; April 14, at Toronto — Toronto 3, Detroit 4; April 16, at Detroit — Toronto 3, Detroit 4; April 18, at Detroit — Toronto 4, Detroit 2; April 21, at Toronto — Toronto 1, Detroit 2; April 23, at Detroit — Toronto 4, Detroit 3; April 25, at Toronto — Toronto 4, Detroit 0.

1962-63 — Toronto Maple Leafs — Johnny Bower, Don Simmons, Carl Brewer, Tim Horton, Kent Douglas, Allan Stanley, Bob Baun, Larry Hillman, Red Kelly, Dick Duff, George Armstrong (Captain), Bob Nevin, Ron Stewart, Dave Keon, Billy Harris, Bob Pulford, Eddie Shack, Ed Litzenberger, Frank Mahovlich, John MacMillan, Punch Imlach (Manager-Coach), Bob Haggert (Trainer).
Scores: April 9, at Toronto — Toronto 4, Detroit 2; April 11, at Toronto — Toronto 4, Detroit 2; April 14, at Detroit — Toronto 2, Detroit 3; April 16, at Detroit — Toronto 4, Detroit 2; April 18, at Toronto — Toronto 3, Detroit 1.

1961-62 — Toronto Maple Leafs — Johnny Bower, Don Simmons, Carl Brewer, Tim Horton, Bob Baun, Allan Stanley, Al Arbour, Larry Hillman, Red Kelly, Dick Duff, George Armstrong (Captain), Frank Mahovlich, Bob Nevin, Ron Stewart, Billy Harris, Bert Olmstead, Bob Pulford, Eddie Shack, Dave Keon, Ed Litzenberger, John MacMillan, Punch Imlach (Manager-Coach), Bob Haggert (Trainer).
Scores: April 10, at Toronto — Toronto 4, Chicago 1; April 12, at Toronto — Toronto 3, Chicago 2; April 15, at Chicago — Toronto 0, Chicago 3; April 17, at Chicago — Toronto 1, Chicago 4; April 19, at Toronto —Toronto 8, Chicago 4; April 22, at Chicago — Toronto 2, Chicago 1.

1960-61 — Chicago Black Hawks — Glenn Hall, Al Arbour, Pierre Pilote, Moose Vasko, Jack Evans, Dollard St. Laurent, Reggie Fleming, Tod Sloan, Ron Murphy, Ed Litzenberger (Captain), Bill Hay, Wayne Hillman, Bobby Hull, Ab McDonald, Eric Nesterenko, Kenny Wharram, Earl Balfour, Stan Mikita, Murray Balfour, Chico Maki, Wayne Hicks, Tommy Ivan (Manager), Rudy Pilous (Coach), Nick Garen (Trainer).
Scores: April 6, at Chicago — Chicago 3, Detroit 2; April 8, at Detroit — Detroit 3, Chicago 1; April 10, at Chicago — Chicago 3, Detroit 1; April 12, at Detroit — Detroit 2, Chicago 1; April 14, at Chicago — Chicago 6, Detroit 3; April 16, at Detroit — Chicago 5, Detroit 1.

1959-60 — Montreal Canadiens — Jacques Plante, Charlie Hodge, Doug Harvey, Tom Johnson, Bob Turner, Jean-Guy Talbot, Albert Langlois, Ralph Backstrom, Jean Béliveau, Marcel Bonin, Bernie Geoffrion, Phil Goyette, Bill Hicke, Don Marshall, Ab McDonald, Dickie Moore, André Pronovost, Claude Provost, Henri Richard, Maurice Richard (Captain), Frank Selke (Manager), Toe Blake (Coach), Hector Dubois, Larry Aubut (Trainers).
Scores: April 7, at Montreal — Montreal 4, Toronto 2; April 9, at Montreal — Montreal 2, Toronto 1; April 12, at Toronto — Montreal 5, Toronto 2; April 14, at Toronto — Montreal 4, Toronto 0.

1958-59 — Montreal Canadiens — Jacques Plante, Charlie Hodge, Doug Harvey, Tom Johnson, Bob Turner, Jean-Guy Talbot, Albert Langlois, Bernie Geoffrion, Ralph Backstrom, Bill Hicke, Maurice Richard (Captain), Dickie Moore, Claude Provost, Ab McDonald, Henri Richard, Marcel Bonin, Phil Goyette, Don Marshall, André Pronovost, Jean Béliveau, Frank Selke (Manager), Toe Blake (Coach), Hector Dubois, Larry Aubut (Trainers).
Scores: April 9, at Montreal — Montreal 5, Toronto 3; April 11, at Montreal — Montreal 3, Toronto 1; April 14, at Toronto — Toronto 3, Montreal 2; April 16, at Toronto — Montreal 3, Toronto 2; April 18, at Montreal — Montreal 5, Toronto 3.

Flyers captain Bobby Clarke and Gary Dornhoefer battle for the puck in front of Bruins netminder Gilles Gilbert during the 1975-76 season. Back in 1974, Philadelphia beat Boston in six games to become the first expansion team to win the Stanley Cup.

1957-58 — Montreal Canadiens — Jacques Plante, Gerry McNeil, Doug Harvey, Tom Johnson, Bob Turner, Dollard St-Laurent, Jean-Guy Talbot, Albert Langlois, Jean Béliveau, Bernie Geoffrion, Maurice Richard (Captain), Dickie Moore, Claude Provost, Floyd Curry, Bert Olmstead, Henri Richard, Marcel Bonin, Phil Goyette, Don Marshall, André Pronovost, Connie Broden, Ab McDonald, Frank Selke (Manager), Toe Blake (Coach), Hector Dubois, Larry Aubut (Trainers).
Scores: April 8, at Montreal —Montreal 2, Boston 1; April 10, at Montreal — Boston 5, Montreal 2; April 13, at Boston — Montreal 3, Boston 0; April 15, at Boston — Boston 3, Montreal 1; April 17, at Montreal — Montreal 3, Boston 2; April 20, at Boston — Montreal 5, Boston 3.

1956-57 — Montreal Canadiens — Jacques Plante, Gerry McNeil, Doug Harvey, Tom Johnson, Bob Turner, Dollard St-Laurent, Jean-Guy Talbot, Jean Béliveau, Bernie Geoffrion, Floyd Curry, Dickie Moore, Maurice Richard (Captain), Claude Provost, Bert Olmstead, Henri Richard, Phil Goyette, Don Marshall, André Pronovost, Connie Broden, Frank Selke (Manager), Toe Blake (Coach), Hector Dubois, Larry Aubut (Trainers).
Scores: April 6, at Montreal — Montreal 5, Boston 1; April 9, at Montreal — Montreal 1, Boston 0; April 11, at Boston — Montreal 4, Boston 2; April 14, at Boston — Boston 2, Montreal 0; April 16, at Montreal — Montreal 5, Boston 1.

1955-56 — Montreal Canadiens — Jacques Plante, Doug Harvey, Butch Bouchard (Captain), Bob Turner, Tom Johnson, Jean-Guy Talbot, Dollard St-Laurent, Jean Béliveau, Bernie Geoffrion, Bert Olmstead, Floyd Curry, Jackie Leclair, Maurice Richard, Dickie Moore, Henri Richard, Ken Mosdell, Don Marshall, Claude Provost, Frank Selke (Manager), Toe Blake (Coach), Hector Dubois (Trainer).
Scores: March 31, at Montreal — Montreal 6, Detroit 4; April 3, at Montreal — Montreal 5, Detroit 1; April 5, at Detroit — Detroit 3, Montreal 1; April 8, at Detroit — Montreal 3, Detroit 0; April 10, at Montreal — Montreal 3, Detroit 1.

1954-55 — Detroit Red Wings — Terry Sawchuk, Red Kelly, Bob Goldham, Marcel Pronovost, Benny Woit, Jim Hay, Larry Hillman, Ted Lindsay (Captain), Tony Leswick, Gordie Howe, Alex Delvecchio, Marty Pavelich, Glen Skov, Earl Reibel, Johnny Wilson, Bill Dineen, Vic Stasiuk, Marcel Bonin, Jack Adams (Manager), Jimmy Skinner (Coach), Carl Mattson (Trainer).
Scores: April 3, at Detroit — Detroit 4, Montreal 2; April 5, at Detroit — Detroit 7, Montreal 1, April 7, at Montreal — Montreal 4, Detroit 2; April 9, at Montreal — Montreal 5, Detroit 3; April 10, at Detroit — Detroit 5, Montreal 1; April 12, at Montreal — Montreal 6, Detroit 3; April 14, at Detroit — Detroit 3, Montreal 1.

1953-54 — Detroit Red Wings — Terry Sawchuk, Red Kelly, Bob Goldham, Benny Woit, Marcel Pronovost, Al Arbour, Keith Allen, Ted Lindsay (Captain), Tony Leswick, Gordie Howe, Marty Pavelich, Alex Delvecchio, Gilles Dube, Metro Prystai, Glen Skov, Johnny Wilson, Bill Dineen, Jimmy Peters, Earl Reibel, Vic Stasiuk, Jack Adams (Manager), Tommy Ivan (Coach), Carl Mattson (Trainer).
Scores: April 4, at Detroit — Detroit 3, Montreal 1; April 6, at Detroit — Montreal 3, Detroit 1; April 8, at Montreal — Detroit 5, Montreal 2; April 10, at Montreal — Detroit 2, Montreal 0; April 11, at Detroit — Montreal 1, Detroit 0; April 13, at Montreal — Montreal 4, Detroit 1; April 16, at Detroit — Detroit 2, Montreal 1.

1952-53 — Montreal Canadiens — Gerry McNeil, Jacques Plante, Doug Harvey, Butch Bouchard (Captain), Tom Johnson, Dollard St. Laurent, Bud MacPherson, Maurice Richard, Elmer Lach, Paul Meger, Bert Olmstead, Bernie Geoffrion, Floyd Curry, Paul Masnick, Billy Reay, Dickie Moore, Ken Mosdell, Dick Gamble, John McCormack, Lorne Davis, Calum MacKay, Eddie Mazur, Frank Selke (Manager), Dick Irvin (Coach), Hector Dubois (Trainer).
Scores: April 9, at Montreal — Montreal 4, Boston 2; April 11, at Montreal — Boston 4, Montreal 1; April 12, at Boston — Montreal 3, Boston 0; April 14, at Boston — Montreal 7, Boston 3; April 16, at Montreal — Montreal 1, Boston 0.

1951-52 — Detroit Red Wings — Terry Sawchuk, Bob Goldham, Benny Woit, Red Kelly, Leo Reise Jr., Marcel Pronovost, Ted Lindsay, Tony Leswick, Gordie Howe, Metro Prystai, Marty Pavelich, Sid Abel (Captain), Glen Skov, Alex Delvecchio, John Wilson, Vic Stasiuk, Larry Zeidel, Jack Adams (Manager) Tommy Ivan (Coach), Carl Mattson (Trainer).
Scores: April 10, at Montreal — Detroit 3, Montreal 1; April 12, at Montreal — Detroit 2, Montreal 1; April 13, at Detroit — Detroit 3, Montreal 0; April 15, at Detroit — Detroit 3, Montreal 0.

1950-51 — Toronto Maple Leafs — Turk Broda, Al Rollins, Jimmy Thomson, Gus Mortson, Bill Barilko, Bill Juzda, Fern Flaman, Hugh Bolton, Ted Kennedy (Captain), Sid Smith, Tod Sloan, Cal Gardner, Howie Meeker, Harry Watson, Max Bentley, Joe Klukay, Danny Lewicki, Ray Timgren, Fleming Mackell, John McCormack, Bob Hassard, Conn Smythe (Manager), Joe Primeau (Coach), Tim Daly (Trainer).
Scores: April 11, at Toronto — Toronto 3, Montreal 2; April 14, at Toronto — Montreal 3, Toronto 2; April 17, at Montreal — Toronto 2, Montreal 1; April 19, at Montreal — Toronto 3, Montreal 2; April 21, at Toronto — Toronto 3, Montreal 2.

1949-50 — Detroit Red Wings — Harry Lumley, Jack Stewart, Leo Reise Jr., Clare Martin, Doug McKay, Al Dewsbury, Lee Fogolin, Marcel Pronovost, Red Kelly, Ted Lindsay, Sid Abel (Captain), Gordie Howe, George Gee, Jimmy Peters, Marty Pavelich, Jim McFadden, Pete Babando, Max McNab, Gerry Couture, Joe Carveth, Steve Black, Johnny Wilson, Larry Wilson, Jack Adams (Manager), Tommy Ivan (Coach), Carl Mattson (Trainer).
Scores: April 11, at Detroit — Detroit 4, NY Rangers 1; April 13, at Toronto* — NY Rangers 3, Detroit 1; April 15, at Toronto — Detroit 4, NY Rangers 0; April 18, at Detroit — NY Rangers 4, Detroit 3; April 20, at Detroit — NY Rangers 2, Detroit 1; April 22, at Detroit — Detroit 5, NY Rangers 4; April 23, at Detroit — Detroit 4, NY Rangers 3.

* Ice was unavailable in Madison Square Garden and Rangers elected to play second and third games on Toronto ice.

1948-49 — Toronto Maple Leafs — Turk Broda, Jimmy Thomson, Gus Mortson, Bill Barilko, Garth Boesch, Bill Juzda, Ted Kennedy (Captain), Howie Meeker, Vic Lynn, Harry Watson, Bill Ezinicki, Cal Gardner, Max Bentley, Joe Klukay, Sid Smith, Don Metz, Ray Timgren, Fleming Mackell, Harry Taylor, Bob Dawes, Tod Sloan, Conn Smythe (Manager), Hap Day (Coach), Tim Daly (Trainer).
Scores: April 8, at Detroit — Toronto 3, Detroit 2; April 10, at Detroit — Toronto 3, Detroit 1; April 13, at Toronto — Toronto 3, Detroit 1; April 16, at Toronto — Toronto 3, Detroit 1.

Ted Lindsay poses with the Cup after Detroit's victory over Montreal in 1954. The Red Wings won the Cup in 1950, 1952, 1954 and 1955. As captain of the team the last two years, Lindsay was the first player to parade the Cup around the ice.

1947-48 — Toronto Maple Leafs — Turk Broda, Jimmy Thomson, Wally Stanowski, Garth Boesch, Bill Barilko, Gus Mortson, Phil Samis, Syl Apps (Captain), Bill Ezinicki, Harry Watson, Ted Kennedy, Howie Meeker, Vic Lynn, Nick Metz, Max Bentley, Joe Klukay, Les Costello, Don Metz, Sid Smith, Conn Smythe (Manager), Hap Day (Coach), Tim Daly (Trainer).
Scores: April 7, at Toronto — Toronto 5, Detroit 3; April 10, at Toronto — Toronto 4, Detroit 2; April 11, at Detroit — Toronto 2, Detroit 0; April 14, at Detroit — Toronto 7, Detroit 2.

1946-47 — Toronto Maple Leafs — Turk Broda, Garth Boesch, Gus Mortson, Jimmy Thomson, Wally Stanowski, Bill Barilko, Harry Watson, Bud Poile, Ted Kennedy, Syl Apps (Captain), Don Metz, Nick Metz, Bill Ezinicki, Vic Lynn, Howie Meeker, Gaye Stewart, Joe Klukay, Gus Bodnar, Bob Goldham, Conn Smythe (Manager), Hap Day (Coach), Tim Daly (Trainer).
Scores: April 8, at Montreal — Montreal 6, Toronto 0; April 10, at Montreal — Toronto 4, Montreal 0; April 12, at Toronto — Toronto 4, Montreal 2; April 15, at Toronto — Toronto 2, Montreal 1; April 17, at Montreal — Montreal 3, Toronto 1; April 19, at Toronto — Toronto 2, Montreal 1.

1945-46 — Montreal Canadiens — Elmer Lach, Toe Blake (Captain), Maurice Richard, Bob Fillion, Dutch Hiller, Murph Chamberlain, Ken Mosdell, Buddy O'Connor, Glen Harmon, Jimmy Peters, Butch Bouchard, Billy Reay, Ken Reardon, Leo Lamoureux, Frank Eddolls, Gerry Plamondon, Bill Durnan, Tommy Gorman (Manager), Dick Irvin (Coach), Ernie Cook (Trainer).
Scores: March 30, at Montreal — Montreal 4, Boston 3; April 2, at Montreal — Montreal 3, Boston 2; April 4, at Boston — Montreal 4, Boston 2; April 7, at Boston — Boston 3, Montreal 2; April 9, at Montreal — Montreal 6, Boston 3.

1944-45 — Toronto Maple Leafs — Don Metz, Frank McCool, Wally Stanowski, Reg Hamilton, Moe Morris, John McCreedy, Tom O'Neill, Ted Kennedy, Babe Pratt, Gus Bodnar, Art Jackson, Jack McLean, Mel Hill, Nick Metz, Bob Davidson (Captain), Sweeney Schriner, Lorne Carr, Conn Smythe (Manager), Frank Selke (Business Manager), Hap Day (Coach), Tim Daly (Trainer).
Scores: April 6, at Detroit — Toronto 1, Detroit 0; April 8, at Detroit — Toronto 2, Detroit 0; April 12, at Toronto — Toronto 1, Detroit 0; April 14, at Toronto — Detroit 5, Toronto 3; April 19, at Detroit — Detroit 2, Toronto 0; April 21, at Toronto — Detroit 1, Toronto 0; April 22, at Detroit — Toronto 2, Detroit 1.

1943-44 — Montreal Canadiens — Toe Blake (Captain), Maurice Richard, Elmer Lach, Ray Getliffe, Murph Chamberlain, Phil Watson, Butch Bouchard, Glen Harmon, Buddy O'Connor, Gerry Heffernan, Mike McMahon, Leo Lamoureux, Fern Majeau, Bob Fillion, Bill Durnan, Tommy Gorman (Manager), Dick Irvin (Coach), Ernie Cook (Trainer).
Scores: April 4, at Montreal — Montreal 5, Chicago 1; April 6, at Chicago — Montreal 3, Chicago 1; April 9, at Chicago — Montreal 3, Chicago 2; April 13, at Montreal — Montreal 5, Chicago 4.

1942-43 — Detroit Red Wings — Jack Stewart, Jimmy Orlando, Sid Abel (Captain), Alex Motter, Harry Watson, Joe Carveth, Mud Bruneteau, Eddie Wares, Johnny Mowers, Cully Simon, Don Grosso, Carl Liscombe, Connie Brown, Syd Howe, Les Douglas, Harold Jackson, Joe Fisher, Jack Adams (Manager), Ebbie Goodfellow (Playing Coach), Honey Walker (Trainer).
Scores: April 1, at Detroit — Detroit 6, Boston 2; April 4, at Detroit — Detroit 4, Boston 3; April 7, at Boston — Detroit 4, Boston 0; April 8, at Boston — Detroit 2, Boston 0.

1941-42 — Toronto Maple Leafs — Wally Stanowski, Syl Apps (Captain), Bob Goldham, Gordie Drillon, Hank Goldup, Ernie Dickens, Sweeney Schriner, Bucko McDonald, Bob Davidson, Nick Metz, Bingo Kampman, Don Metz, Gaye Stewart, Turk Broda, John McCreedy, Lorne Carr, Pete Langelle, Billy Taylor, Conn Smythe (Manager), Hap Day (Coach), Frank Selke (Business Manager), Tim Daly (Trainer).
Scores: April 4, at Toronto — Detroit 3, Toronto 2; April 7, at Toronto — Detroit 4, Toronto 2; April 9, at Detroit — Detroit 5, Toronto 2; April 12, at Detroit — Toronto 4, Detroit 3; April 14, at Toronto — Toronto 9, Detroit 3; April 16, at Detroit — Toronto 3, Detroit 0; April 18, at Toronto — Toronto 3, Detroit 1.

1940-41 — Boston Bruins — Bill Cowley, Des Smith, Dit Clapper (Captain), Frank Brimsek, Flash Hollett, Jack Crawford, Bobby Bauer, Pat McReavy, Herb Cain, Mel Hill, Milt Schmidt, Woody Dumart, Roy Conacher, Terry Reardon, Art Jackson, Eddie Wiseman, Art Ross (Manager), Cooney Weiland (Coach), Win Green (Trainer).
Scores: April 6, at Boston — Detroit 2, Boston 3; April 8, at Boston — Detroit 1, Boston 2; April 10, at Detroit — Boston 4, Detroit 2; April 12, at Detroit — Boston 3, Detroit 1.

1939-40 — New York Rangers — Dave Kerr, Art Coulter (Captain), Ott Heller, Alex Shibicky, Mac Colville, Neil Colville, Phil Watson, Lynn Patrick, Clint Smith, Muzz Patrick, Babe Pratt, Bryan Hextall, Kilby MacDonald, Dutch Hiller, Alf Pike, Stan Smith, Lester Patrick (Manager), Frank Boucher (Coach), Harry Westerby (Trainer).
Scores: April 2, at New York — NY Rangers 2, Toronto 1; April 3, at New York — NY Rangers 6, Toronto 2; April 6, at Toronto — NY Rangers 1, Toronto 2; April 9, at Toronto — NY Rangers 0, Toronto 3; April 11, at Toronto — NY Rangers 2, Toronto 1; April 13, at Toronto — NY Rangers 3, Toronto 2.

1938-39 — Boston Bruins — Bobby Bauer, Mel Hill, Flash Hollett, Roy Conacher, Gord Pettinger, Charlie Sands, Milt Schmidt, Woody Dumart, Jack Crawford, Ray Getliffe, Frank Brimsek, Eddie Shore, Dit Clapper, Bill Cowley, Jack Portland, Red Hamill, Cooney Weiland (Captain), Art Ross (Manager-Coach), Win Green (Trainer).
Scores: April 6, at Boston — Toronto 1, Boston 2; April 9, at Boston — Toronto 3, Boston 2; April 11, at Toronto — Toronto 1, Boston 3; April 13, at Toronto — Toronto 0, Boston 2; April 16, at Boston — Toronto 1, Boston 3.

1937-38 — Chicago Black Hawks — Art Wiebe, Carl Voss, Harold Jackson, Mike Karakas, Mush March, Jack Shill, Earl Seibert, Cully Dahlstrom, Alex Levinsky, Johnny Gottselig (Captain), Lou Trudel, Pete Palangio, Bill MacKenzie, Doc Romnes, Paul Thompson, Roger Jenkins, Alfie Moore, Bert Connelly, Virgil Johnson, Paul Goodman, Bill Stewart (Manager-Coach), Eddie Froelich (Trainer).
Scores: April 5, at Toronto — Chicago 3, Toronto 1; April 7, at Toronto — Chicago 1, Toronto 5; April 10, at Chicago — Chicago 2, Toronto 1; April 12, at Chicago — Chicago 4, Toronto 1.

1936-37 — Detroit Red Wings — Normie Smith, Pete Kelly, Larry Aurie, Herbie Lewis, Hec Kilrea, Mud Bruneteau, Syd Howe, Wally Kilrea, Jimmy Franks, Bucko McDonald, Gord Pettinger, Ebbie Goodfellow, John Gallagher, Ralph Bowman, John Sorrell, Marty Barry, Earl Robertson, John Sherf, Howie Mackie, Rolly Roulston, Doug Young (Captain), Jack Adams (Manager-Coach), Honey Walker (Trainer).
Scores: April 6, at New York — Detroit 1, NY Rangers 5; April 8, at Detroit — Detroit 4, NY Rangers 2; April 11, at Detroit — Detroit 0, NY Rangers 1; April 13, at Detroit — Detroit 1, NY Rangers 0; April 15, at Detroit — Detroit 3, NY Rangers 0.

1935-36 — Detroit Red Wings — John Sorrell, Syd Howe, Marty Barry, Herbie Lewis, Mud Bruneteau, Wally Kilrea, Hec Kilrea, Gord Pettinger, Bucko McDonald, Ralph Bowman, Pete Kelly, Doug Young (Captain), Ebbie Goodfellow, Normie Smith, Larry Aurie, Jack Adams (Manager-Coach), Honey Walker (Trainer).
Scores: April 5, at Detroit — Detroit 3, Toronto 1; April 7, at Detroit — Detroit 9, Toronto 4; April 9, at Toronto — Detroit 3, Toronto 4; April 11, at Toronto — Detroit 3, Toronto 2.

1934-35 — Montreal Maroons — Lionel Conacher, Cy Wentworth, Alex Connell, Toe Blake, Stewart Evans, Earl Robinson, Bill Miller, Dave Trottier, Jimmy Ward, Baldy Northcott, Hooley Smith, Russ Blinco, Al Shields, Sammy McManus, Gus Marker, Bob Gracie, Herb Cain, Tommy Gorman (Manager-Coach), Bill O'Brien (Trainer).
Scores: April 4, at Toronto — Mtl. Maroons 3, Toronto 2; April 6, at Toronto — Mtl. Maroons 3, Toronto 1; April 9, at Montreal — Mtl. Maroons 4, Toronto 1.

1933-34 — Chicago Black Hawks — Clarence Abel, Rosie Couture, Lou Trudel, Lionel Conacher, Paul Thompson, Leroy Goldsworthy, Art Coulter, Roger Jenkins, Don McFadyen, Tom Cook, Doc Romnes, Johnny Gottselig, Mush March, Johnny Sheppard, Charlie Gardiner (Captain), Bill Kendall, Tommy Gorman (Manager-Coach), Eddie Froelich (Trainer).
Scores: April 3, at Detroit — Chicago 2, Detroit 1; April 5, at Detroit — Chicago 4, Detroit 1; April 8, at Chicago — Detroit 5, Chicago 2; April 10, at Chicago — Chicago 1, Detroit 0.

1932-33 — New York Rangers — Ching Johnson, Butch Keeling, Frank Boucher, Art Somers, Babe Siebert, Bun Cook, Andy Aitkenhead, Ott Heller, Oscar Asmundson, Gord Pettinger, Doug Brennan, Cecil Dillon, Bill Cook (Captain), Murray Murdoch, Earl Seibert, Lester Patrick (Manager-Coach), Harry Westerby (Trainer).
Scores: April 4, at New York — NY Rangers 5, Toronto 1; April 8, at Toronto — NY Rangers 3, Toronto 1; April 11, at Toronto — Toronto 3, NY Rangers 2; April 13, at Toronto — NY Rangers 1, Toronto 0.

1931-32 — Toronto Maple Leafs — Charlie Conacher, Busher Jackson, King Clancy, Andy Blair, Red Horner, Lorne Chabot, Alex Levinsky, Joe Primeau, Harold Darragh, Baldy Cotton, Frank Finnigan, Hap Day (Captain), Ace Bailey, Bob Gracie, Fred Robertson, Earl Miller, Conn Smythe (Manager), Dick Irvin (Coach), Tim Daly (Trainer).
Scores: April 5, at New York — Toronto 6, NY Rangers 4; April 7, at Boston* — Toronto 6, NY Rangers 2; April 9, at Toronto — Toronto 6, NY Rangers 4.

* Ice was unavailable in Madison Square Garden and Rangers elected to play the second game on neutral ice.

1930-31 — Montreal Canadiens — George Hainsworth, Wildor Larochelle, Marty Burke, Sylvio Mantha (Captain), Howie Morenz, Johnny Gagnon, Aurel Joliat, Armand Mondou, Pit Lepine, Albert Leduc, Georges Mantha, Art Lesieur, Nick Wasnie, Bert McCaffrey, Gus Rivers, Jean Pusie, Léo Dandurand (Manager), Cecil Hart (Coach), Ed Dufour (Trainer).
Scores: April 3, at Chicago — Montreal 2, Chicago 1; April 5, at Chicago — Chicago 2, Montreal 1; April 9, at Montreal — Chicago 3, Montreal 2; April 11, at Montreal — Montreal 4, Chicago 2; April 14, at Montreal — Montreal 2, Chicago 0.

1929-30 — Montreal Canadiens — George Hainsworth, Marty Burke, Sylvio Mantha (Captain), Howie Morenz, Bert McCaffrey, Aurel Joliat, Albert Leduc, Pit Lepine, Wildor Larochelle, Nick Wasnie, Gerry Carson, Armand Mondou, Georges Mantha, Gus Rivers, Léo Dandurand (Manager), Cecil Hart (Coach), Ed Dufour (Trainer).
Scores: April 1, at Boston — Montreal 3, Boston 0; April 3, at Montreal — Montreal 4, Boston 3.

1928-29 — Boston Bruins — Tiny Thompson, Eddie Shore, Lionel Hitchman (Captain), Percy Galbraith, Eric Pettinger, Frank Fredrickson, Mickey Mackay, Red Green, Dutch Gainor, Harry Oliver, Eddie Rodden, Dit Clapper, Cooney Weiland, Lloyd Klein, Cy Denneny, Bill Carson, George Owen, Myles Lane, Art Ross (Manager-Coach), Win Green (Trainer).
Scores: March 28, at Boston — Boston 2, NY Rangers 0; March 29, at New York — Boston 2, NY Rangers 1.

1927-28 — New York Rangers — Lorne Chabot, Clarence Abel, Leo Bourgeault, Ching Johnson, Bill Cook (Captain), Bun Cook, Frank Boucher, Bill Boyd, Murray Murdoch, Paul Thompson, Alex Gray, Joe Miller, Patsy Callighen, Lester Patrick (Manager-Coach), Harry Westerby (Trainer).
Scores: April 5, at Montreal — Mtl. Maroons 2, NY Rangers 0; April 7, at Montreal — NY Rangers 2, Mtl. Maroons 1; April 10, at Montreal — Mtl. Maroons 2, NY Rangers 0; April 12, at Montreal — NY Rangers 1, Mtl. Maroons 0; April 14, at Montreal — NY Rangers 2, Mtl. Maroons 1.

1926-27 — Ottawa Senators — Alex Connell, King Clancy, Georges Boucher, Ed Gorman, Frank Finnigan, Alex Smith, Hec Kilrea, Hooley Smith, Cy Denneny, Frank Nighbor, Jack Adams, Milt Halliday, Dave Gill (Manager-Coach).
Scores: April 7, at Boston — Ottawa 0, Boston 0; April 9, at Boston — Ottawa 3, Boston 1; April 11, at Ottawa — Boston 1, Ottawa 1; April 13, at Ottawa — Ottawa 3, Boston 1.

1925-26 — Montreal Maroons — Clint Benedict, Reg Noble, Frank Carson, Dunc Munro, Nels Stewart, Punch Broadbent, Babe Siebert, Chuck Dinsmore, Merlyn Phillips, Hobie Kitchen, Sam Rothschild, Albert Holway, George Horne, Bernie Brophy, Eddie Gerard (Manager-Coach), Bill O'Brien (Trainer).
Scores: March 30, at Montreal — Mtl. Maroons 3, Victoria 0; April 1, at Montreal — Mtl. Maroons 3, Victoria 0; April 3, at Montreal — Victoria 3, Mtl. Maroons 2; April 6, at Montreal — Mtl. Maroons 2, Victoria 0.

The series in the spring of 1926 ended the annual playoffs between the champions of the East and the champions of the West. Since 1926-27 the annual playoffs in the National Hockey League have decided the Stanley Cup champions.

1924-25 — Victoria Cougars — Hap Holmes, Clem Loughlin, Gord Fraser, Frank Fredrickson, Jack Walker, Gizzy Hart, Harold Halderson, Frank Foyston, Wally Elmer, Harry Meeking, Jocko Anderson, Lester Patrick (Manager-Coach).
Scores: March 21, at Victoria — Victoria 5, Montreal 2; March 23, at Vancouver — Victoria 3, Montreal 1; March 27, at Victoria — Montreal 4, Victoria 2; March 30, at Victoria — Victoria 6, Montreal 1.

1923-24 — Montreal Canadiens — Georges Vezina, Sprague Cleghorn (Captain), Billy Coutu, Howie Morenz, Aurel Joliat, Billy Boucher, Odie Cleghorn, Sylvio Mantha, Bobby Boucher, Billy Bell, Billy Cameron, Joe Malone, Charles Fortier, Leo Dandurand (Manager-Coach).
Scores: March 18, at Montreal — Montreal 3, Van. Maroons 2; March 20, at Montreal — Montreal 2, Van. Maroons 1. March 22, at Montreal — Montreal 6, Cgy. Tigers 1; March 25, at Ottawa* — Montreal 3, Cgy. Tigers 0.

* Game transferred to Ottawa to benefit from artificial ice surface.

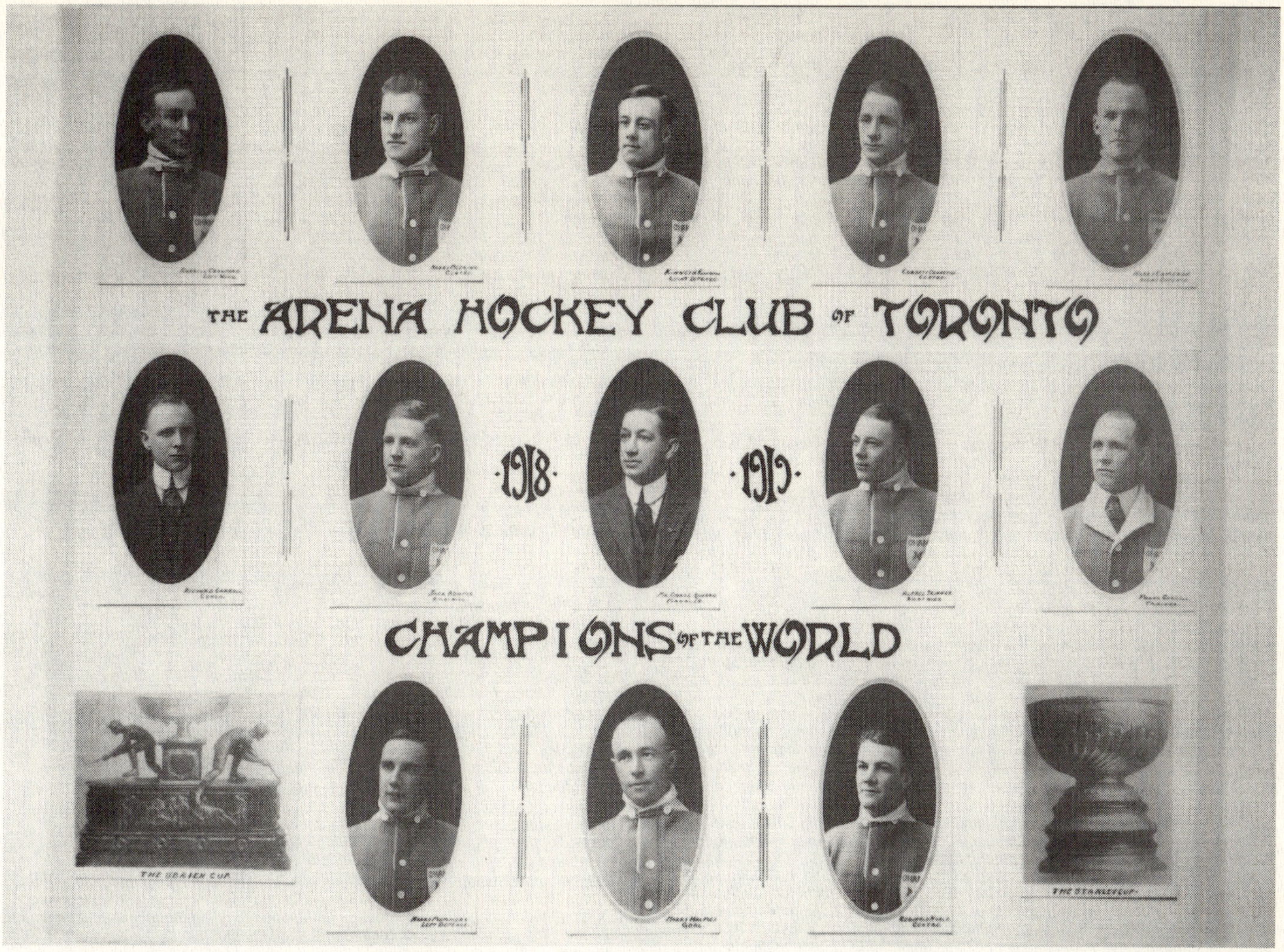

The Toronto Arenas team picture of 1918-19 features both the O'Brien Trophy, emblematic of their NHL championship in the league's inaugural season of 1917-18, and the Stanley Cup, which they claimed by beating the Vancouver Millionaires of the Pacific Coast Hockey Association.

1922-23 — Ottawa Senators — Georges Boucher, Lionel Hitchman, Frank Nighbor, King Clancy, Harry Helman, Clint Benedict, Jack Darragh, Eddie Gerard, Cy Denneny, Punch Broadbent, Tommy Gorman (Manager), Pete Green (Coach), F. Dolan (Trainer).
Scores: March 16, at Vancouver — Ottawa 1, Van. Maroons 0; March 19, at Vancouver — Van. Maroons 4, Ottawa 1; March 23, at Vancouver — Ottawa 3, Van. Maroons 2; March 26, at Vancouver — Ottawa 5, Van. Maroons 1; March 29, at Vancouver — Ottawa 2, Edm. Eskimos 1; March 31, at Vancouver — Ottawa 1, Edm. Eskimos 0.

1921-22 — Toronto St. Pats — Ted Stackhouse, Corb Denneny, Rod Smylie, Lloyd Andrews, John Ross Roach, Harry Cameron, Billy Stuart, Babe Dye, Ken Randall, Reg Noble, Eddie Gerard (borrowed for one game from Ottawa), Stan Jackson, Ivan Mitchell, Charlie Querrie (Manager), George O'Donoghue (Coach).
Scores: March 17, at Toronto — Van. Millionaires 4, Toronto 3; March 20, at Toronto — Toronto 2, Van. Millionaires 1; March 23, at Toronto — Van. Millionaires 3, Toronto 0; March 25, at Toronto — Toronto 6, Van. Millionaires 0; March 28, at Toronto — Toronto 5, Van. Millionaires 1.

1920-21 — Ottawa Senators — Jack MacKell, Jack Darragh, Morley Bruce, Georges Boucher, Eddie Gerard, Clint Benedict, Sprague Cleghorn, Frank Nighbor, Punch Broadbent, Cy Denneny, Leth Graham, Tommy Gorman (Manager),Pete Green (Coach), F. Dolan (Trainer).
Scores: March 21, at Vancouver — Van. Millionaires 2, Ottawa 1; March 24, at Vancouver — Ottawa 4, Van. Millionaires 3; March 28, at Vancouver — Ottawa 3, Van. Millionaires 2; March 31, at Vancouver — Van. Millionaires 3, Ottawa 2; April 4, at Vancouver — Ottawa 2, Van. Millionaires 1

1919-20 — Ottawa Senators — Jack MacKell, Jack Darragh, Morley Bruce, Horrace Merrill, Georges Boucher, Eddie Gerard, Clint Benedict, Sprague Cleghorn, Frank Nighbor, Punch Broadbent, Cy Denneny, Tommy Gorman (Manager), Pete Green (Coach).
Scores: March 22, at Ottawa — Ottawa 3, Seattle 2; March 24, at Ottawa — Ottawa 3, Seattle 0; March 27, at Ottawa — Seattle 3, Ottawa 1; March 30, at Toronto* — Seattle 5, Ottawa 2; April 1, at Toronto* — Ottawa 6, Seattle 1.

* Games transferred to Toronto to benefit from artificial ice surface.

1918-19 — No decision, Series halted by Spanish influenza epidemic, illness of several players and death of Joe Hall of Montreal Canadiens from flu. Five games had been played when the series was halted, each team having won two and tied one. The results are shown:
Scores: March 19, at Seattle — Seattle 7, Montreal 0; March 22, at Seattle — Montreal 4, Seattle 2; March 24, at Seattle — Seattle 7, Montreal 2; March 26, at Seattle — Montreal 0, Seattle 0; March 30, at Seattle — Montreal 4, Seattle 3.

1917-18 — Toronto Arenas — Rusty Crawford, Harry Meeking, Ken Randall, Corb Denneny, Harry Cameron, Jack Adams, Alf Skinner, Harry Mummery, Hap Holmes, Reg Noble, Sammy Hebert, Jack Marks, Jack Coughlin, Charlie Querrie (Manager), Dick Carroll (Coach), Frank Carroll (Trainer).
Scores: March 20, at Toronto — Toronto 5, Van. Millionaires 3; March 23, at Toronto — Van. Millionaires 6, Toronto 4; March 26, at Toronto — Toronto 6, Van. Millionaires 3; March 28, at Toronto — Van. Millionaires 8, Toronto 1; March 30, at Toronto — Toronto 2, Van. Millionaires 1.

1916-17 — Seattle Metropolitans — Hap Holmes, Ed Carpenter, Cully Wilson, Jack Walker, Bernie Morris, Frank Foyston, Roy Rickey, Jim Riley, Bobby Rowe (Captain), Peter Muldoon (Manager).
Scores: March 17, at Seattle — Montreal 8, Seattle 4; March 20, at Seattle — Seattle 6, Montreal 1; March 23, at Seattle — Seattle 4, Montreal 1; March 25, at Seattle — Seattle 9, Montreal 1.

1915-16 — Montreal Canadiens — Georges Vezina, Bert Corbeau, Jack Laviolette, Newsy Lalonde, Louis Berlinquette, Goldie Prodgers, Howard McNamara (Captain), Didier Pitre, Skene Ronan, Amos Arbour, Skinner Poulin, Jack Fournier, George Kennedy (Manager).
Scores: March 20, at Montreal — Portland 2, Montreal 0; March 22, at Montreal — Montreal 2, Portland 1; March 25, at Montreal — Montreal 6, Portland 3; March 28, at Montreal — Portland 6, Montreal 5; March 30, at Montreal — Montreal 2, Portland 1.

1914-15 — Vancouver Millionaires — Ken Mallen, Frank Nighbor, Cyclone Taylor, Hugh Lehman, Lloyd Cook, Mickey Mackay, Barney Stanley, Jim Seaborn, Si Griffis (Captain), Johnny Matz, Frank Patrick (Playing Manager).
Scores: March 22, at Vancouver — Van. Millionaires 6, Ottawa 2; March 24, at Vancouver — Van. Millionaires 8, Ottawa 3; March 26, at Vancouver — Van. Millionaires 12, Ottawa 3.

1913-14 — Toronto Blueshirts — Con Corbeau, Roy McGiffen, Jack Walker, George McNamara, Cully Wilson, Frank Foyston, Harry Cameron, Hap Holmes, Scotty Davidson (Captain), Harriston, Jack Marshall (Playing Manager), Frank and Dick Carroll (Trainers).
Scores: March 14, at Toronto — Toronto 5, Victoria 2; March 17, at Toronto — Toronto 6, Victoria 5; March 19, at Toronto — Toronto 2, Victoria 1.

1912-13 — Quebec Bulldogs — Joe Malone, Joe Hall, Paddy Moran, Harry Mummery, Tommy Smith, Jack Marks, Rusty Crawford, Billy Creighton, Jeff Malone, Rocket Power, M.J. Quinn (Manager), D. Beland (Trainer).
Scores: March 8, at Quebec — Que. Bulldogs 14, Sydney 3; March 10, at Quebec — Que. Bulldogs 6, Sydney 2.

Victoria challenged Quebec but the Bulldogs refused to put the Stanley Cup in competition so the two teams played an exhibition series with Victoria winning two games to one by scores of 7-5, 3-6, 6-1. It was the first meeting between the Eastern champions and the Western champions. The following year, and until the Western Hockey League disbanded after the 1926 playoffs, the Cup went to the winner of the series between East and West.

1911-12 — Quebec Bulldogs — Goldie Prodgers, Joe Hall, Walter Rooney, Paddy Moran, Jack Marks, Jack McDonald, Eddie Oatman, George Leonard, Joe Malone (Captain), C. Nolan (Coach), M.J. Quinn (Manager), D. Beland (Trainer).
Scores: March 11, at Quebec — Que. Bulldogs 9, Moncton 3; March 13, at Quebec — Que. Bulldogs 8, Moncton 0.

Prior to 1912, teams could challenge the Stanley Cup champions for the title, thus there was more than one Championship Series played in most of the seasons between 1894 and 1911.

1910-11 — Ottawa Senators — Hamby Shore, Percy LeSueur, Jack Darragh, Bruce Stuart, Marty Walsh, Bruce Ridpath, Fred Lake, Dubbie Kerr, Alex Currie, Horace Gaul.
Scores: March 13, at Ottawa — Ottawa 7, Galt 4; March 16, at Ottawa — Ottawa 13, Port Arthur 4.

1909-10 (March) **— Montreal Wanderers** — Cecil Blachford, Moose Johnson, Ernie Russell, Riley Hern, Harry Hyland, Jack Marshall, Pud Glass (Captain), Jimmy Gardner, Dickie Boon (Manager).
Scores: March 12, at Montreal — Mtl. Wanderers 7, Berlin (Kitchener) 3.

1909-10 (January) **— Ottawa Senators** — Dubbie Kerr, Fred Lake, Percy LeSueur, Ken Mallen, Bruce Ridpath, Gord Roberts, Hamby Shore, Bruce Stuart, Marty Walsh.
Scores: January 5, at Ottawa — Ottawa 12, Galt 3; January 7, at Ottawa — Ottawa 3, Galt 1; January 18, at Ottawa — Ottawa 8, Edmonton 4; January 20, at Ottawa — Ottawa 13, Edmonton 7.

1908-09 — Ottawa Senators — Fred Lake, Percy LeSueur, Cyclone Taylor, Billy Gilmour, Dubbie Kerr, Edgar Dey, Marty Walsh, Bruce Stuart (Captain).
Scores: Ottawa, as champions of the Eastern Canada Hockey Association took over the Stanley Cup in 1909 and, although a challenge was accepted by the Cup trustees from Winnipeg Shamrocks, games could not be arranged because of the lateness of the season. No other challenges were made in 1909. The following season — 1909-10 — however, the Senators accepted two challenges as defending Cup Champions. The first was against Galt in a two-game, total-goals series, and the second against Edmonton, also a two-game, total-goals series. Results: January 5, at Ottawa —Ottawa 12, Galt 3; January 7, at Ottawa — Ottawa 3, Galt 1. January 18, at Ottawa — Ottawa 8, Edm. Eskimos 4; January 20, at Ottawa — Ottawa 13, Edm. Eskimos 7.

1907-08 — Montreal Wanderers — Riley Hern, Art Ross, Walter Smaill, Pud Glass, Bruce Stuart, Ernie Russell, Moose Johnson, Cecil Blachford (Captain), Tom Hooper, Larry Gilmour, Ernie Liffiton, Dickie Boon (Manager).
Scores: Wanderers accepted four challenges for the Cup: January 9, at Montreal — Mtl. Wanderers 9, Ott. Victorias 3; January 13, at Montreal — Mtl. Wanderers 13, Ott. Victorias 1; March 10, at Montreal — Mtl. Wanderers 11, Wpg. Maple Leafs 5; March 12, at Montreal — Mtl. Wanderers 9, Wpg. Maple Leafs 3; March 14, at Montreal — Mtl. Wanderers 6, Toronto (OPHL) 4. At start of following season, 1908-09, Wanderers were challenged by Edmonton. Results: December 28, at Montreal — Mtl. Wanderers 7, Edm. Eskimos 3; December 30, at Montreal — Edm. Eskimos 7, Mtl. Wanderers 6. Total goals: Mtl. Wanderers 13, Edm. Eskimos 10.

1906-07 — (March 25) **— Montreal Wanderers** — Billy Strachan, Riley Hern, Lester Patrick, Hod Stuart, Pud Glass, Ernie Russell, Cecil Blachford (Captain), Moose Johnson, Rod Kennedy, Jack Marshall, Dickie Boon (Manager).

1906-07 — (March 18) **— Kenora Thistles** — Eddie Geroux, Si Griffis, Tom Hooper, Fred Whitcroft, Alf Smith, Harry Westwick, Roxy Beaudro, Tom Phillips (Captain), Russell Phillips.
Scores: March 16, at Winnipeg — Kenora 8, Brandon 6; March 18, at Winnipeg — Kenora 4, Brandon 1; March 23, at Winnipeg — Mtl. Wanderers 7, Kenora 2; March 25, at Winnipeg — Kenora 6, Mtl. Wanderers 5. Total goals: Mtl. Wanderers 12, Kenora 8.

1906-07 — (January) **— Kenora Thistles** — Eddie Geroux, Art Ross, Si Griffis, Tom Hooper, Billy McGimsie, Roxy Beaudro, Tommy Phillips (Captain), Joe Hall, Russell Phillips.
Scores: January 17, at Montreal — Kenora 4, Mtl. Wanderers 2; Jan. 21, at Montreal — Kenora 8, Mtl. Wanderers 6.

1905-06 — (March) **— Montreal Wanderers** — Henri Menard, Billy Strachan, Rod Kennedy, Lester Patrick, Pud Glass, Ernie Russell, Moose Johnson, Cecil Blachford (Captain), Josh Arnold, Dickie Boon (Manager).
Scores: March 14, at Montreal — Mtl. Wanderers 9, Ottawa 1; March 17, at Ottawa — Ottawa 9, Mtl. Wanderers 3. Total goals: Mtl. Wanderers 12, Ottawa 10. Wanderers accepted a challenge from New Glasgow, N.S., prior to the start of the 1906-07 season. Results: December 27, at Montreal — Mtl. Wanderers 10, New Glasgow 3; December 29, at Montreal — Mtl. Wanderers 7, New Glasgow 2.

1905-06 — (February) **— Ottawa Silver Seven** — Harvey Pulford (Captain), Arthur Moore, Harry Westwick, Frank McGee, Alf Smith (Playing Coach), Billy Gilmour, Billy Hague, Percy LeSueur, Harry Smith, Tommy Smith, Dion, Ebbs.
Scores: February 27, at Ottawa — Ottawa 16, Queen's University 7; February 28, at Ottawa — Ottawa 12, Queen's University 7; March 6, at Ottawa — Ottawa 6, Smiths Falls 5; March 8, at Ottawa — Ottawa 8, Smiths Falls 2.

1904-05 — Ottawa Silver Seven — Dave Finnie, Harvey Pulford (Captain), Arthur Moore, Harry Westwick, Frank McGee, Alf Smith (Playing Coach), Billy Gilmour, Frank White, Horace Gaul, Hamby Shore, Bones Allen.
Scores: January 13, at Ottawa — Ottawa 9, Dawson City 2; January 16, at Ottawa — Ottawa 23, Dawson City 2; March 7, at Ottawa — Rat Portage 9, Ottawa 3; March 9, at Ottawa — Ottawa 4, Rat Portage 2; March 11, at Ottawa — Ottawa 5, Rat Portage 4.

1903-04 — Ottawa Silver Seven — Suddy Gilmour, Arthur Moore, Frank McGee, Bouse Hutton, Billy Gilmour, Jim McGee, Harry Westwick, Harvey Pulford (Captain), Scott, Alf Smith (Playing Coach).
Scores: December 30, at Ottawa — Ottawa 9, Wpg. Rowing Club 1; January 1, at Ottawa — Wpg. Rowing Club 6, Ottawa 2; January 4, at Ottawa — Ottawa 2, Wpg. Rowing Club 0. February 23, at Ottawa — Ottawa 6, Tor. Marlboros 3; February 25, at Ottawa — Ottawa 11, Tor. Marlboros 2; March 2, at Montreal — Ottawa 5, Mtl. Wanderers 5. Following the tie game, a new two-game series was ordered to be played in Ottawa but the Wanderers refused unless the tie game was replayed in Montreal. When no settlement could be reached, the series was abandoned and Ottawa retained the Cup and accepted a two-game challenge from Brandon. Results: (both games at Ottawa), March 9, Ottawa 6, Brandon 3; March 11, Ottawa 9, Brandon 3.

1902-03 — (March) **— Ottawa Silver Seven** — Suddy Gilmour, Percy Sims, Bouse Hutton, Dave Gilmour, Billy Gilmour, Harry Westwick, Frank McGee, F.H. Wood, A.A. Fraser, Charles Spittal, Harvey Pulford (Captain), Arthur Moore, Alf Smith (coach.)
Scores: March 7, at Montreal — Ottawa 1, Mtl. Victorias 1; March 10, at Ottawa — Ottawa 8, Mtl. Victorias 0. Total goals: Ottawa 9, Mtl. Victorias 1; March 12, at Ottawa — Ottawa 6, Rat Portage 2; March 14, at Ottawa — Ottawa 4, Rat Portage 2.

1902-03 — (February) **— Montreal AAA** — Tom Hodge, Dickie Boon, Billy Nicholson, Tommy Phillips, Art Hooper, Billy Bellingham, Charles Liffiton, Jack Marshall, Jimmy Gardner, Cecil Blachford, George Smith.
Scores: January 29, at Montreal — Mtl. AAA 8, Wpg. Victorias 1; January 31, at Montreal — Wpg. Victorias 2, Mtl. AAA 2; February 2, at Montreal — Wpg. Victorias 4, Mtl. AAA 2; February 4, at Montreal — Mtl. AAA 5, Wpg. Victorias 1.

1901-02 — (March) **— Montreal AAA** — Tom Hodge, Dickie Boon, Billy Nicholson, Archie Hooper, Billy Bellingham, Charles Liffiton, Jack Marshall, Roland Elliott, Jimmy Gardner.
Scores: March 13, at Winnipeg — Wpg. Victorias 1, Mtl. AAA 0; March 15, at Winnipeg — Mtl. AAA 5, Wpg. Victorias 0; March 17, at Winnipeg — Mtl. AAA 2, Wpg. Victorias 1.

1901-02 — (January) **— Winnipeg Victorias** — Burke Wood, Tony Gingras, Charles Johnstone, Rod Flett, Magnus Flett, Dan Bain (Captain), Fred Scanlon, F. Cadham, G. Brown.
Scores: January 21, at Winnipeg — Wpg. Victorias 5, Tor Wellingtons 3; January 23, at Winnipeg — Wpg. Victorias 5, Tor. Wellingtons 3.

1900-01 — Winnipeg Victorias — Burke Wood, Jack Marshall, Tony Gingras, Charles Johnstone, Rod Flett, Magnus Flett, Dan Bain (Captain), Art Brown.
Scores: January 29, at Montreal — Wpg. Victorias 4, Mtl. Shamrocks 3; January 31, at Montreal — Wpg. Victorias 2, Mtl. Shamrocks 1.

1899-1900 — Montreal Shamrocks — Joe McKenna, Frank Tansey, Frank Wall, Art Farrell, Fred Scanlon, Harry Trihey (Captain), Jack Brannen.
Scores: February 12, at Montreal — Mtl. Shamrocks 4, Wpg. Victorias 3; February 14, at Montreal — Wpg. Victorias 3, Mtl. Shamrocks 2; February 16, at Montreal — Mtl. Shamrocks 5, Wpg. Victorias 4; March 5, at Montreal — Mtl. Shamrocks 10, Halifax 2; March 7, at Montreal — Mtl. Shamrocks 11, Halifax 0.

1898-99 — (March) **— Montreal Shamrocks** — Joe McKenna, Frank Tansey, Frank Wall, Harry Trihey (Captain), Art Farrell, Fred Scanlon, Jack Brannen, John Dobby, Charles Hoerner.
Scores: March 14, at Montreal — Mtl. Shamrocks 6, Queen's University 2.

1898-99 — (February) **— Montreal Victorias** — Gordon Lewis, Mike Grant, Graham Drinkwater, Cam Davidson, Bob McDougall, Ernie McLea, Frank Richardson, Jack Ewing, Russell Bowie, Douglas Acer, Fred McRobie.
Scores: February 15, at Montreal — Mtl. Victorias 2, Wpg. Victorias 1; February 18, at Montreal — Mtl. Victorias 3, Wpg. Victorias 2.

1897-98 — Montreal Victorias — Gordon Lewis, Hartland McDougall, Mike Grant, Graham Drinkwater, Cam Davidson, Bob McDougall, Ernie McLea, Frank Richardson (Captain), Jack Ewing. The Victorias as champions of the Amateur Hockey Association, retained the Cup and were not called upon to defend it.

1896-97 — Montreal Victorias — Gordon Lewis, Harold Henderson, Mike Grant (Captain), Cam Davidson, Graham Drinkwater, Bob McDougall, Ernie McLea, Shirley Davidson, Hartland McDougall, Jack Ewing, Percy Molson, David Gillilan, McLellan.
Scores: December 27, at Montreal — Mtl. Victorias 15, Ott. Capitals 2.

1895-96 — (December) **— Montreal Victorias** — Harold Henderson, Mike Grant (Captain), Bob McDougall, Graham Drinkwater, Shirley Davidson, Ernie McLea, W. Wallace, Robert Jones, Cam Davidson, David Gillilan, Stanley Willett.
Scores: December 30, at Winnipeg — Mtl. Victorias 6, Wpg. Victorias 5.

1895-96 — (February) **— Winnipeg Victorias** — Whitey Merritt, Rod Flett, Fred Higginbotham, Jack Armitage (Captain), Tote Campbell, Dan Bain, Bobby Benson, Attie Howard.
Scores: February 14, at Montreal — Wpg. Victorias 2, Mtl. Victorias 0.

1894-95 — Montreal Victorias — Robert Jones, Harold Henderson, Mike Grant (Captain), Shirley Davidson, Bob McDougall, Norman Rankin, Graham Drinkwater, Roland Elliot, William Pullan, Hartland McDougall, Art Fenwick, A. McDougall.
Montreal Victorias, as champions of the Amateur Hockey Association, were prepared to defend the Stanley Cup. However, the Stanley Cup trustees had already accepted a challenge match between the 1894 champion Montreal AAA and Queen's University. It was declared that if Montreal AAA defeated Queen's University, Montreal Victorias would be declared Stanley Cup champions. If Queen's University won, the Cup would go to the university club. In a game played March 9, 1895, Montreal AAA defeated Queen's University 5-1. As a result, Montreal Victorias were awarded the Stanley Cup.

1893-94 — Montreal AAA — Herb Collins, Allan Cameron, George James, Billy Barlow, Clare Mussen, Archie Hodgson, Haviland Routh, Alex Irving, James Stewart, E. O'Brien, A.C. (Toad) Wand, A.B. Kingan.
Scores: March 17, at Mtl. Victorias — Mtl. AAA 3, Mtl. Victorias 2; March 22, at Montreal — Mtl. AAA 3, Ott. Capitals 1.

1892-93 — Montreal AAA — Tom Paton, James Stewart, Allan Cameron, Haviland Routh, Archie Hodgson, Billy Barlow, A.B. Kingan, G.S. Lowe.
In accordance with the terms governing the presentation of the Stanley Cup, it was awarded for the first time to the Montreal AAA as champions of the Amateur Hockey Association in 1893. Once Montreal AAA had been declared holders of the Stanley Cup, any Canadian hockey team could challenge for the trophy.

All-Time NHL Playoff Formats

1917-18 — The regular-season was split into two halves. The winners of both halves faced each other in a two-game, total-goals series for the NHL championship and the right to meet the PCHA champion in the best-of-five Stanley Cup Finals.

1918-19 — Same as 1917-18, except that the Stanley Cup Finals was extended to a best-of-seven series.

1919-20 — Same as 1917-1918, except that Ottawa won both halves of the split regular-season schedule to earn an automatic berth into the best-of-five Stanley Cup Finals against the PCHA champions.

1921-22 — The top two teams at the conclusion of the regular-season faced each other in a two-game, total-goals series for the NHL championship. The NHL champion then moved on to play the winner of the PCHA-WCHL playoff series in the best-of-five Stanley Cup Finals.

1922-23 — The top two teams at the conclusion of the regular-season faced each other in a two-game, total-goals series for the NHL championship. The NHL champion then moved on to play the PCHA champion in the best-of-three Stanley Cup Semi-Finals, and the winner of the Semi-Finals played the WCHL champion, which had been given a bye, in the best-of-three Stanley Cup Finals.

1923-24 — The top two teams at the conclusion of the regular-season faced each other in a two-game, total-goals series for the NHL championship. The NHL champion then moved on to play the loser of the PCHA-WCHL playoff (the winner of the PCHA-WCHL playoff earned a bye into the Stanley Cup Finals) in the best-of-three Stanley Cup Semi-Finals. The winner of this series met the PCHA-WCHL playoff winner in the best-of-three Stanley Cup Finals.

1924-25 — The first place team (Hamilton) at the conclusion of the regular-season was supposed to play the winner of a two-game, total-goals series between the second (Toronto) and third (Montreal) place clubs. However, Hamilton refused to abide by this new format, demanding greater compensation than offered by the League. Thus, Toronto and Montreal played their two-game, total-goals series, and the winner (Montreal) earned the NHL title and then played the WCHL champion (Victoria) in the best-of-five Stanley Cup Finals.

1925-26 — The format which was intended for 1924-25 went into effect. The winner of the two-game, total-goals series between the second and third place teams squared off against the first place team in the two-game, total-goals NHL championship series. The NHL champion then moved on to play the WHL champion in the best-of-five Stanley Cup Finals.

After the 1925-26 season, the NHL was the only major professional hockey league still in existence and consequently took over sole control of the Stanley Cup competition.

1926-27 — The 10-team league was divided into two divisions — Canadian and American — of five teams apiece. In each division, the winner of the two-game, total-goals series between the second and third place teams faced the first place team in a two-game, total-goals series for the division title. The two division title winners then met in the best-of-five Stanley Cup Finals.

1928-29 — Both first place teams in the two divisions played each other in a best-of-five series. Both second place teams in the two divisions played each other in a two-game, total-goals series as did the two third place teams. The winners of these latter two series then played each other in a best-of-three series for the right to meet the winner of the series between the two first place clubs. This Stanley Cup Final was a best-of-three.

Series A: First in Canadian Division vs. first in American (best-of-five)
Series B: Second in Canadian Division vs. second in American (two-game, total-goals)
Series C: Third in Canadian Division vs. third in American (two-game, total-goals)
Series D: Winner of Series B vs. winner of Series C (best-of-three)
Series E: Winner of Series A vs. winner of Series D (best-of-three) for Stanley Cup

1931-32 — Same as 1928-29, except that Series D was changed to a two-game, total-goals format and Series E was changed to best-of-five.

1936-37 — Same as 1931-32, except that Series B, C, and D were each best-of-three.

1938-39 — With the NHL reduced to seven teams, the two-division system was replaced by one seven-team league. Based on final regular-season standings, the following playoff format was adopted:

Series A: First vs. Second (best-of-seven)
Series B: Third vs. Fourth (best-of-three)
Series C: Fifth vs. Sixth (best-of-three)
Series D: Winner of Series B vs. winner of Series C (best-of-three)
Series E: Winner of Series A vs. winner of Series D (best-of-seven)

1942-43 — With the NHL reduced to six teams (the "original six"), only the top four finishers qualified for playoff action. The best-of-seven Semi-Finals pitted Team #1 vs. Team #3 and Team #2 vs. Team #4. The winners of each Semi-Final series met in the best-of-seven Stanley Cup Finals.

1967-68 — When it doubled in size from 6 to 12 teams, the NHL once again was divided into two divisions — East and West — of six teams apiece. The top four clubs in each division qualified for the playoffs (all series were best-of-seven):

Series A: Team #1 (East) vs. Team #3 (East)
Series B: Team #2 (East) vs. Team #4 (East)
Series C: Team #1 (West) vs. Team #3 (West)
Series D: Team #2 (West) vs. Team #4 (West)
Series E: Winner of Series A vs. winner of Series B
Series F: Winner of Series C vs. winner of Series D
Series G: Winner of Series E vs. Winner of Series F

1970-71 — Same as 1967-68 except that Series E matched the winners of Series A and D, and Series F matched the winners of Series B and C.

1971-72 — Same as 1970-71, except that Series A and C matched Team #1 vs. Team #4, and Series B and D matched Team #2 vs. Team #3.

1974-75 — With the League now expanded to 18 teams in four divisions, a completely new playoff format was introduced. First, the #2 and #3 teams in each of the four divisions were pooled together in the Preliminary round. These eight (#2 and #3) clubs were ranked #1 to #8 based on regular-season record:

Series A: Team #1 vs. Team #8 (best-of-three)
Series B: Team #2 vs. Team #7 (best-of-three)
Series C: Team #3 vs. Team #6 (best-of-three)
Series D: Team #4 vs. Team #5 (best-of-three)
The winners of this Preliminary round then pooled together with the four division winners, which had received byes into this Quarter-Final round. These eight teams were again ranked #1 to #8 based on regular-season record:
Series E: Team #1 vs. Team #8 (best-of-seven)
Series F: Team #2 vs. Team #7 (best-of-seven)
Series G: Team #3 vs. Team #6 (best-of-seven)
Series H: Team #4 vs. Team #5 (best-of-seven)
The four Quarter-Finals winners, which moved on to the Semi-Finals, were then ranked #1 to #4 based on regular season record:
Series I: Team #1 vs. Team #4 (best-of-seven)
Series J: Team #2 vs. Team #3 (best-of-seven)
Series K: Winner of Series I vs. winner of Series J (best-of-seven)

1977-78 — Same as 1974-75, except that the Preliminary round consisted of the #2 teams in the four divisions and the next four teams based on regular-season record (not their standings within their divisions).

1979-80 — With the addition of four WHA franchises, the League expanded its playoff structure to include 16 of its 21 teams. The four first place teams in the four divisions automatically earned playoff berths. Among the 17 other clubs, the top 12, according to regular-season record, also earned berths. All 16 teams were then pooled together and ranked #1 to #16 based on regular-season record:

Series A: Team #1 vs. Team #16 (best-of-five)
Series B: Team #2 vs. Team #15 (best-of-five)
Series C: Team #3 vs. Team #14 (best-of-five)
Series D: Team #4 vs. Team #13 (best-of-five)
Series E: Team #5 vs. Team #12 (best-of-five)
Series F: Team #6 vs. Team #11 (best-of-five)
Series G: Team #7 vs. Team #10 (best-of-five)
Series H: Team #8 vs. Team # 9 (best-of-five)
The eight Preliminary round winners, ranked #1 to #8 based on regular-season record, moved on to the Quarter-Finals:
Series I: Team #1 vs. Team #8 (best-of-seven)
Series J: Team #2 vs. Team #7 (best-of-seven)
Series K: Team #3 vs. Team #6 (best-of-seven)
Series L: Team #4 vs. Team #5 (best-of-seven)
The four Quarter-Finals winners, ranked #1 to #4 based on regular-season record, moved on to the semi-finals:
Series M: Team #1 vs. Team #4 (best-of-seven)
Series N: Team #2 vs. Team #3 (best-of-seven)
Series O: Winner of Series M vs. winner of Series N (best-of-seven)

1981-82 — The first four teams in each division earned playoff berths. In each division, the first-place team opposed the fourth-place team and the second-place team opposed the third-place team in a best-of-five Division Semi-Final series (DSF). In each division, the two winners of the DSF met in a best-of-seven Division Final series (DF). The two DF winners in each conference met in a best-of-seven Conference Final series (CF). In the Prince of Wales Conference, the Adams Division winner opposed the Patrick Division winner; in the Clarence Campbell Conference, the Smythe Division winner opposed the Norris Division winner. The two CF winners met in a best-of-seven Stanley Cup Final (F) series.

1986-87 — Division Semi-Final series changed from best-of-five to best-of-seven.

1993-94 — The NHL's playoff draw is conference-based rather than division-based. At the conclusion of the regular season, the top eight teams in each of the Eastern and Western Conferences qualify for the playoffs. The teams that finish in first place in each of the League's divisions are seeded first and second in each conference's playoff draw and are assured of home ice advantage in the first two playoff rounds. The remaining teams are seeded based on their regular-season point totals. In each conference, the team seeded #1 plays #8; #2 vs. #7; #3 vs. #6; and #4 vs. #5. All series are best-of-seven with home ice rotating on a 2-2-1-1-1 basis, with the exception of matchups between Central and Pacific Division teams. These matchups will be played on a 2-3-2 basis to reduce travel. In a 2-3-2 series, the team with the most points will have its choice to start the series at home or on the road. The Eastern Conference champion will face the Western Conference champion in the Stanley Cup Final.

1994-95 — Same as 1993-94, except that in first, second or third-round playoff series involving Central and Pacific Division teams, the team with the better record has the choice of using either a 2-3-2 or a 2-2-1-1-1 format. When a 2-3-2 format is selected, the higher-ranked team also has the choice of playing games 1, 2, 6 and 7 at home or playing games 3, 4 and 5 at home. The format for the Stanley Cup Final remains 2-2-1-1-1.

1998-99 — The NHL's clubs are re-aligned into two conferences each consisting of three divisions. The number of teams qualifying for the Stanley Cup Playoffs remains unchanged at 16.

First-round playoff berths will be awarded to the first-place team in each division as well as to the next five best teams based on regular-season point totals in each conference. The three division winners in each conference will be seeded first through third, in order of points, for the playoffs and the next five best teams, in order of points, will be seeded fourth through eighth. In each conference, the team seeded #1 will play #8; #2 vs. #7; #3 vs. #6; and #4 vs. #5 in the quarterfinal round. Home-ice in the Conference Quarter-Finals is granted to those teams seeded first through fourth in each conference.

In the Conference Semi-Finals and Conference Finals, teams will be re-seeded according to the same criteria as the Conference Quarter-Finals. Higher seeded teams will have home-ice advantage.

Home-ice advantage for the Stanley Cup Finals will be determined by points.

All series remain best-of-seven.

Currently, the St. Louis Blues have the NHL's longest streak of consecutive playoff appearances. In this picture, Terry Crisp and Bob Wall (#2) help Ernie Wakely defend against Toronto's Norm Ullman (#9) and Paul Henderson in 1970-71.

Team Records

1918-2002

GAMES PLAYED

MOST GAMES PLAYED BY ALL TEAMS, ONE PLAYOFF YEAR:
92 — 1991. There were 51 DSF, 24 DF, 11 CF and 6 F games.
90 — 1994. There were 48 CQF, 23 CSF, 12 CF and 7 F games.
— 2002. There were 47 CQF, 25 CSF, 13 CF and 5 F games.

MOST GAMES PLAYED, ONE TEAM, ONE PLAYOFF YEAR:
26 — Philadelphia Flyers, 1987. Won DSF 4-2 against NY Rangers, DF 4-3 against NY Islanders, CF 4-2 against Montreal, and lost F 4-3 against Edmonton.
25 — New Jersey Devils, 2001. Won CQF 4-2 against Carolina, CSF 4-3 against Toronto, CF 4-1 against Pittsburgh, and lost F 4-3 against Colorado.
24 — Pittsburgh Penguins,1991. Won DSF 4-3 against New Jersey, DF 4-1 against Washington, CF 4-2 against Boston, and F 4-2 against Minnesota.
— Los Angeles Kings, 1993. Won DSF 4-2 against Calgary, DF 4-2 against Vancouver, CF 4-3 against Toronto, and lost F 4-1 against Montreal.
— Vancouver Canucks, 1994. Won CQF 4-3 against Calgary, CSF 4-1 against Dallas, CF 4-1 against Toronto, and lost F 4-3 against NY Rangers.

PLAYOFF APPEARANCES

MOST STANLEY CUP CHAMPIONSHIPS:
23 — Montreal Canadiens 1924-30-31-44-46-53-56-57-58-59-60-65-66-68-69-71-73-76-77-78-79-86-93
13 — Toronto Maple Leafs 1918-22-32-42-45-47-48-49-51-62-63-64-67
10 — Detroit Red Wings 1936-37-43-50-52-54-55-97-98-02

MOST CONSECUTIVE STANLEY CUP CHAMPIONSHIPS:
5 — Montreal Canadiens (1956-57-58-59-60)
4 — Montreal Canadiens (1976-77-78-79)
— NY Islanders (1980-81-82-83)

MOST FINAL SERIES APPEARANCES:
32 — Montreal Canadiens in 85-year history.
22 — Detroit Red Wings in 75-year history.
21 — Toronto Maple Leafs in 85-year history.

MOST CONSECUTIVE FINAL SERIES APPEARANCES:
10 — Montreal Canadiens, (1951-60, inclusive)
5 — Montreal Canadiens, (1965-69, inclusive)
— NY Islanders, (1980-84, inclusive)

MOST YEARS IN PLAYOFFS:
73 — Montreal Canadiens in 85-year history.
62 — Toronto Maple Leafs in 85-year history.
60 — Boston Bruins in 78-year history.

MOST CONSECUTIVE PLAYOFF APPEARANCES:
29 — Boston Bruins (1968-96, inclusive)
28 — Chicago Blackhawks (1970-97, inclusive)
24 — Montreal Canadiens (1971-94, inclusive)
23 — St. Louis Blues (1980-2002, inclusive)
21 — Montreal Canadiens (1949-69, inclusive)

TEAM WINS

MOST HOME WINS, ONE TEAM, ONE PLAYOFF YEAR:
11 — Edmonton Oilers, 1988 in 11 home games.
10 — Edmonton Oilers, 1985 in 10 home games.
— Montreal Canadiens, 1986 in 11 home games.
— Montreal Canadiens, 1993 in 11 home games.

MOST ROAD WINS, ONE TEAM, ONE PLAYOFF YEAR:
10 — New Jersey Devils, 1995. Won three at Boston in CQF; two at Pittsburgh in CSF; three at Philadelphia in CF; and two at Detroit in F series.
— **New Jersey Devils,** 2000. Won two at Florida in CQF; two at Toronto in CSF; three at Philadelphia in CF; and three at Dallas in F series.
8 — NY Islanders, 1980. Won two at Los Angeles in PR; three at Boston in QF; two at Buffalo in SF; and one at Philadelphia in F series.
— Philadelphia Flyers, 1987. Won two at NY Rangers in DSF; two at NY Islanders in DF; three at Montreal in CF; and one at Edmonton in F series.
— Edmonton Oilers, 1990. Won one at Winnipeg in DSF; two at Los Angeles in DF; two at Chicago in CF and three at Boston in F series.
— Pittsburgh Penguins, 1992. Won two at Washington in DSF; two at NY Rangers in DF; two at Boston in CF; and two at Chicago in F series.
— Vancouver Canucks, 1994. Won three at Calgary in CQF; two at Dallas in CSF; one at Toronto in CF; and two at NY Rangers in F series.
— Colorado Avalanche, 1996. Won two at Vancouver in CQF; two at Chicago in CSF; two at Detroit in CF; and two at Florida in F series.
— Detroit Red Wings, 1998. Won two at Phoenix in CQF; three at St. Louis in CSF; one at Dallas in CF; and two at Washington in F series.
— Colorado Avalanche, 1999. Won three at San Jose in CQF; three at Detroit in CSF; and two at Dallas in CF.
— New Jersey Devils, 2001. Won two at Carolina in CQF; two at Toronto in CSF; two at Pittsburgh in CF; and two at Colorado in F series.
— Detroit Red Wings, 2002. Won three at Vancouver in CQF; one at St. Louis in CSF; two at Colorado in CF; and two at Carolina in F series.

MOST ROAD WINS, ALL TEAMS, ONE PLAYOFF YEAR:
46 — 1987. Of 87 games played, road teams won 46 (22 DSF, 14 DF, 8 CF and 2 in Stanley Cup final).

MOST OVERTIME WINS, ONE TEAM, ONE PLAYOFF YEAR:
10 — Montreal Canadiens, 1993. Two against Quebec in DSF; three against Buffalo in DF; two against NY Islanders in CF; and three against Los Angeles in F. Montreal played 20 games.
7 — Carolina Hurricanes, 2002. Two against New Jersey in CQF; one against Montreal in CSF; three against Toronto in CF; and one against Detroit in F. Carolina played 23 games.

MOST OVERTIME WINS AT HOME, ONE TEAM, ONE PLAYOFF YEAR:
4 — St. Louis Blues, 1968. Won one vs. Philadelphia in QF and three vs. Minnesota in SF.
— **Montreal Canadiens, 1993.** Won one vs. Quebec in DSF, one vs. Buffalo in DF, one vs. NY Islanders in CF and one vs. Los Angeles in F series.

MOST OVERTIME WINS ON THE ROAD, ONE TEAM, ONE PLAYOFF YEAR:
6 — Montreal Canadiens, 1993. Won one vs. Quebec in DSF, two vs. Buffalo in DF, one vs. NY Islanders in CF and two vs. Los Angeles in F series.

TEAM LOSSES

MOST LOSSES, ONE TEAM, ONE PLAYOFF YEAR:

11 — **Philadelphia Flyers, 1987.** Lost two vs. NY Rangers in DSF; three vs. NY Islanders in DF; two vs. Montreal in CF; and four vs. Edmonton in F series.

MOST HOME LOSSES, ONE TEAM, ONE PLAYOFF YEAR:

6 — **Philadelphia Flyers, 1987.** Lost one vs. NY Rangers in DSF; two vs. NY Islanders in DF; two vs. Montreal in CF; and one vs. Edmonton in F series.

— **Washington Capitals, 1998.** Lost two vs. Boston in CQF; two vs. Buffalo in CF; and two vs. Detroit in F series.

— **Colorado Avalanche, 1999.** Lost two vs. San Jose in CQF; two vs. Detroit in CSF; and two vs. Dallas in CF series.

— **New Jersey Devils, 2001.** Lost one vs. Carolina in CQF; two vs. Toronto in CSF; one vs. Pittsburgh in CF series; and two vs Colorado in F series.

MOST ROAD LOSSES, ONE TEAM, ONE PLAYOFF YEAR:

6 — **St. Louis Blues, 1968.** Lost two at Philadelphia in QF; two at Minnesota in SF; and two at Montreal in F series.

— **St. Louis Blues, 1970.** Lost two at Minnesota in QF; two at Pittsburgh in SF; and two at Boston in F series.

— **NY Islanders, 1984.** Lost one at NY Rangers in DSF; two at Montreal in CF; and three at Edmonton in F series.

— **Los Angeles Kings, 1993.** Lost one at Calgary in DSF; one at Vancouver in DF; two at Toronto in CF; and two at Montreal in F series.

MOST OVERTIME LOSSES, ONE TEAM, ONE PLAYOFF YEAR:

4 — **Montreal Canadiens, 1951.** Lost four vs. Toronto in F series.

— **St. Louis Blues, 1968.** Lost one vs. Philadelphia in QF; one vs. Minnesota in SF; and two vs. Montreal in F series.

— **New York Rangers, 1979.** Lost one vs. Philadelphia in QF; two vs. NY Islanders in SF; and one vs. Montreal in F series.

— **Los Angeles Kings, 1991.** Lost one vs. Vancouver in DSF; and three vs. Edmonton in DF series.

— **Los Angeles Kings, 1993.** Lost one vs. Toronto in CF; and three vs. Montreal in F series.

— **New Jersey Devils, 1994.** Lost one vs. Buffalo in CQF; one vs. Boston in CSF; and two vs. NY Rangers in CF.

— **Chicago Blackhawks, 1995.** Lost one vs. Toronto in CQF; and three vs. Detroit in CF.

— **Philadelphia Flyers, 1996.** Lost two vs. Tampa Bay in CQF; and two vs. Florida in CSF series.

— **Dallas Stars, 1999.** Lost two vs. St. Louis in CSF; one vs. Colorado in CF; and one vs. Buffalo in F.

— **Detroit Red Wings, 2002.** Lost one vs. Vancouver in CQF; two vs. Colorado in CF and one vs. Carolina.

MOST OVERTIME LOSSES AT HOME, ONE TEAM, ONE PLAYOFF YEAR:

4 — **Detroit Red Wings, 2002.** Lost one vs. Vancouver in CQF; two vs. Colorado in CF and one vs. Carolina in F series.

MOST OVERTIME LOSSES ON THE ROAD, ONE TEAM, ONE PLAYOFF YEAR:

3 — **Los Angeles Kings, 1991.** Lost one at Vancouver in DSF; and two at Edmonton in DF series.

— **Chicago Blackhawks, 1995.** Lost one at Toronto in CQF; and two at Detroit in CF series.

— **St. Louis Blues, 1996.** Lost two at Toronto in CQF; and one at Detroit in CSF series.

— **Dallas Stars, 1999.** Lost two at St. Louis in CSF; and one at Colorado in CF series.

PLAYOFF WINNING STREAKS

LONGEST PLAYOFF WINNING STREAK:

14 — **Pittsburgh Penguins.** Streak started May 9, 1992, at Pittsburgh with a 5-4 win in fourth game of DF series against NY Rangers, won by Pittsburgh 4-2. Continued with a four-game win over Boston in 1992 CF and a four-game sweep of Chicago in 1992 F. Pittsburgh then won the first three games of 1993 DSF versus New Jersey. New Jersey ended the streak April 25, 1993, at New Jersey with a 4-1 win.

12 — Edmonton Oilers. Streak started May 15, 1984, at Edmonton with a 7-2 win in third game of F series against NY Islanders, won by Edmonton 4-1. Continued with a three-game sweep of Los Angeles in 1985 DSF and a four game sweep of Winnipeg in 1985 DF. Edmonton then won the first two games of 1985 CF versus Chicago. Chicago ended the streak May 9, 1985, at Chicago with a 5-2 win.

MOST CONSECUTIVE WINS, ONE TEAM, ONE PLAYOFF YEAR:

11 — **Chicago Blackhawks** in 1992. Chicago won last three games of DSF against St. Louis to win series 4-2 and then defeated Detroit 4-0 in DF and Edmonton 4-0 in CF.

— **Pittsburgh Penguins** in 1992. Pittsburgh won last three games of DF against NY Rangers to win series 4-2 and then defeated Boston 4-0 in CF and Chicago 4-0 in F.

— **Montreal Canadiens** in 1993. Montreal won last four games of DSF against Quebec to win series 4-2, defeated Buffalo 4-0 in DF and won first three games of CF against NY Islanders.

PLAYOFF LOSING STREAKS

LONGEST PLAYOFF LOSING STREAK:

16 Games — Chicago Blackhawks. Streak started in 1975 QF against Buffalo when Chicago lost last two games. Then Chicago lost four games to Montreal in 1976 QF; two games to NY Islanders in 1977 PR; four games to Boston in 1978 QF and four games to NY Islanders in 1979 QF. Streak ended on April 8, 1980 when Chicago defeated St. Louis 3-2 in the opening game of their 1980 PR series.

14 Games — Los Angeles Kings. Streak started when the Kings lost four consecutive games in the 1993 Stanley Cup Finals against Montreal. The Kings failed to qualify for the playoffs for the next four years. Los Angeles lost four straight games to St. Louis in the 1998 CQF and failed to qualify for the 1999 playoffs. They were defeated in four straight games by Detroit in the 2000 CQF series. Los Angeles lost the first two games of their 2001 CQF series against Detroit before defeating Detroit 2-1 on April 15, 2001 to end the streak.

Scotty Bowman congratulates Nicklas Lidstrom for winning the Conn Smythe Trophy as playoff MVP. Detroit's tenth Stanley Cup victory came in their 22nd appearance in the Finals, ranking the Wing's second behind Montreal's 32 appearances.

MOST GOALS IN A SERIES, ONE TEAM

MOST GOALS, ONE TEAM, ONE PLAYOFF SERIES:
44 — Edmonton Oilers in 1985 CF. Edmonton won best-of-seven series 4-2, outscoring Chicago 44-25.
35 — Edmonton Oilers in 1983 DF. Edmonton won best-of-seven series 4-1, outscoring Calgary 35-13.
— Calgary Flames in 1995 CQF. Calgary lost best-of-seven series 4-3, outscoring San Jose 35-26.

MOST GOALS, ONE TEAM, TWO-GAME SERIES:
11 — Buffalo Sabres in 1977 PR. Buffalo won best-of-three series 2-0, outscoring Minnesota 11-3.
— Toronto Maple Leafs in 1978 PR. Toronto won best-of-three series 2-0, outscoring Los Angeles 11-3.

MOST GOALS, ONE TEAM, THREE-GAME SERIES:
23 — Chicago Blackhawks in 1985 DSF. Chicago won best-of-five series 3-0, outscoring Detroit 23-8.
20 — Minnesota North Stars in 1981 PR. Minnesota won best-of-five series 3-0, outscoring Boston 20-13.
— NY Islanders in 1981 PR. New York won best-of-five series 3-0, outscoring Toronto 20-4.

MOST GOALS, ONE TEAM, FOUR-GAME SERIES:
28 — Boston Bruins in 1972 SF. Boston won best-of-seven series 4-0, outscoring St. Louis 28-8.

MOST GOALS, ONE TEAM, FIVE-GAME SERIES:
35 — Edmonton Oilers in 1983 DF. Edmonton won best-of-seven series 4-1, outscoring Calgary 35-13.
32 — Edmonton Oilers in 1987 DSF. Edmonton won best-of-seven series 4-1, outscoring Los Angeles 32-20.
30 — Calgary Flames in 1988 DSF. Calgary won best-of-seven series 4-1, outscoring Los Angeles 30-18.

MOST GOALS, ONE TEAM, SIX-GAME SERIES:
44 — Edmonton Oilers in 1985 CF. Edmonton won best-of-seven series 4-2, outscoring Chicago 44-25.
33 — Chicago Blackhawks in 1985 DF. Chicago won best-of-seven series 4-2, outscoring Minnesota 33-29.
— Montreal Canadiens in 1973 F. Montreal won best-of-seven series 4-2, outscoring Chicago 33-23.
— Los Angeles Kings in 1993 DSF. Los Angeles won best-of-seven series 4-2, outscoring Calgary 33-28.

MOST GOALS, ONE TEAM, SEVEN-GAME SERIES:
35 — Calgary Flames in 1995 CQF. Calgary lost best-of-seven series 4-3, outscoring San Jose 35-26.
33 — Philadelphia Flyers in 1976 QF. Philadelphia won best-of-seven series 4-3, outscoring Toronto 33-23.
— Boston Bruins in 1983 DF. Boston won best-of-seven series 4-3, outscoring Buffalo 33-23.
— Edmonton Oilers in 1984 DF. Edmonton won best-of-seven series 4-3, outscoring Calgary 33-27.

FEWEST GOALS IN A SERIES, ONE TEAM

FEWEST GOALS, ONE TEAM, TWO-GAME SERIES:
0 — NY Americans in 1929 SF. Lost two-game total-goal series 1-0 against NY Rangers.
— Chicago Blackhawks in 1935 SF. Lost two-game total-goal series 1-0 against Mtl. Maroons.
— Mtl. Maroons in 1937 SF. Lost best-of-three series 2-0 to NY Rangers while being outscored 5-0.
— NY Americans in 1939 QF. Lost best-of-three series 2-0 to Toronto while being outscored 6-0.

FEWEST GOALS, ONE TEAM, THREE-GAME SERIES:
1 — Mtl. Maroons in 1936 SF. Lost best-of-five series 3-0 to Detroit and were outscored 6-1.

FEWEST GOALS, ONE TEAM, FOUR-GAME SERIES:
2 — Boston Bruins in 1935 SF. Lost best-of-five series 3-1 to Toronto and were outscored 7-2.
— Montreal Canadiens in 1952 F. Lost best-of-seven series 4-0 to Detroit and were outscored 11-2.

FEWEST GOALS, ONE TEAM, FIVE-GAME SERIES:
2 — Philadelphia Flyers in 2002 CQF. Ottawa won best-of-seven series 4-1, while outscoring Philadelphia 11-2.

FEWEST GOALS, ONE TEAM, SIX-GAME SERIES:
5 — Boston Bruins in 1951 SF. Toronto won best-of-seven series 4-1 with 1 tie, outscoring Boston 17-5.

FEWEST GOALS, ONE TEAM, SEVEN-GAME SERIES:
9 — Toronto Maple Leafs, in 1945 F. Toronto won best-of- seven series 4-3; teams tied in scoring 9-9.
— Detroit Red Wings, in 1945 F. Toronto won best-of-seven series 4-3; teams tied in scoring 9-9.

After Al Rollins surrendered two goals in the first game, Turk Broda (above) allowed Boston to score just three more in the next four as the Leafs held Boston to a record-low five goals in a six-game series during the 1951 semifinals.

MOST GOALS IN A SERIES, BOTH TEAMS

MOST GOALS, BOTH TEAMS, ONE PLAYOFF SERIES:
69 — Edmonton Oilers, Chicago Blackhawks in 1985 CF. Edmonton won best-of-seven series 4-2, outscoring Chicago 44-25.
62 — Chicago Blackhawks, Minnesota North Stars in 1985 DF. Chicago won best-of-seven series 4-2, outscoring Minnesota 33-29.
61 — Los Angeles Kings, Calgary Flames in 1993 DSF. Los Angeles won best-of-seven series 4-2, outscoring Calgary 33-28.
— San Jose Sharks, Calgary Flames in 1995 CQF. San Jose won best-of-seven series 4-3, while being outscored 35-26.

MOST GOALS, BOTH TEAMS, TWO-GAME SERIES:
17 — Toronto St. Patricks, Montreal Canadiens in 1918 NHL F. Toronto won two-game total goal series 10-7.
15 — Boston Bruins, Chicago Blackhawks in 1927 QF. Boston won two-game total goal series 10-5.
— Pittsburgh Penguins, St. Louis Blues in 1975 PR. Pittsburgh won best-of-three series 2-0, outscoring St. Louis 9-6.

MOST GOALS, BOTH TEAMS, THREE-GAME SERIES:
33 — Minnesota North Stars, Boston Bruins in 1981 PR. Minnesota won best-of-five series 3-0, outscoring Boston 20-13.
31 — Chicago Blackhawks, Detroit Red Wings in 1985 DSF. Chicago won best-of-five series 3-0, outscoring Detroit 23-8.
28 — Toronto Maple Leafs, NY Rangers in 1932 F. Toronto won best-of-five series 3-0, outscoring NY Rangers 18-10.

MOST GOALS, BOTH TEAMS, FOUR-GAME SERIES:
36 — Boston Bruins, St. Louis Blues in 1972 SF. Boston won best-of-seven series 4-0, outscoring St. Louis 28-8.
— Minnesota North Stars, Toronto Maple Leafs in 1983 DSF. Minnesota won best-of-five series 3-1; teams tied in scoring 18-18.
— Edmonton Oilers, Chicago Blackhawks in 1983 CF. Edmonton won best-of-seven series 4-0, outscoring Chicago 25-11.
35 — NY Rangers, Los Angeles Kings in 1981 PR. NY Rangers won best-of-five series 3-1, outscoring Los Angeles 23-12.

MOST GOALS, BOTH TEAMS, FIVE-GAME SERIES:
52 — Edmonton Oilers, Los Angeles Kings in 1987 DSF. Edmonton won best-of-seven series 4-1, outscoring Los Angeles 32-20.
50 — Los Angeles Kings, Edmonton Oilers in 1982 DSF. Los Angeles won best-of-five series 3-2, outscoring Edmonton 27-23.
48 — Edmonton Oilers, Calgary Flames in 1983 DF. Edmonton won best-of-seven series 4-1, outscoring Calgary 35-13.
— Calgary Flames, Los Angeles Kings in 1988 DSF. Calgary won best-of-seven series 4-1, outscoring Los Angeles 30-18.

MOST GOALS, BOTH TEAMS, SIX-GAME SERIES:
69 — Edmonton Oilers, Chicago Blackhawks in 1985 CF. Edmonton won best-of-seven series 4-2, outscoring Chicago 44-25.
62 — Chicago Blackhawks, Minnesota North Stars in 1985 DF. Chicago won best-of-seven series 4-2, outscoring Minnesota 33-29.
61 — Los Angeles Kings, Calgary Flames in 1993 DSF. Los Angeles won best-of-seven series 4-2, outscoring Calgary 33-28.

MOST GOALS, BOTH TEAMS, SEVEN-GAME SERIES:
61 — San Jose Sharks, Calgary Flames in 1995 CQF. San Jose won best-of-seven series 4-3, while being outscored 35-26.
60 — Edmonton Oilers, Calgary Flames in 1984 DF. Edmonton won best-of-seven series 4-3, outscoring Calgary 33-27.

FEWEST GOALS IN A SERIES, BOTH TEAMS

FEWEST GOALS, BOTH TEAMS, TWO-GAME SERIES:
1 — NY Rangers, NY Americans in 1929 SF. NY Rangers defeated NY Americans 1-0 in two-game, total-goal series.
— Mtl. Maroons, Chicago Blackhawks in 1935 SF. Mtl. Maroons defeated Chicago 1-0 in two-game, total-goal series.

FEWEST GOALS, BOTH TEAMS, THREE-GAME SERIES:
7 — Boston Bruins, Montreal Canadiens in 1929 SF. Boston won best-of-five series 3-0, outscoring Montreal 5-2.
— Detroit Red Wings, Montreal Maroons in 1936 SF. Detroit won best-of-five series 3-0, outscoring Mtl. Maroons 6-1.

FEWEST GOALS, BOTH TEAMS, FOUR-GAME SERIES:
9 — Toronto Maple Leafs, Boston Bruins in 1935 SF. Toronto won best-of-five series 3-1, outscoring Boston 7-2.

FEWEST GOALS, BOTH TEAMS, FIVE-GAME SERIES:
11 — NY Rangers, Montreal Maroons in 1928 F. NY Rangers won best-of-five series 3-2, while being outscored by Mtl. Maroons 6-5.

FEWEST GOALS, BOTH TEAMS, SIX-GAME SERIES:
16 — Carolina Hurricanes, Toronto Maple Leafs in 2002 CF. Carolina won best-of-seven series 4-2, while outscoring Toronto 10-6.

FEWEST GOALS, BOTH TEAMS, SEVEN-GAME SERIES:
18 — Toronto Maple Leafs, Detroit Red Wings in 1945 F. Toronto won best-of-seven series 4-3; teams tied in scoring 9-9.

MOST GOALS IN A GAME OR PERIOD

MOST GOALS, ONE TEAM, ONE GAME:
13 — Edmonton Oilers at Edmonton, April 9, 1987. Edmonton 13, Los Angeles 3. Edmonton won best-of-seven DSF 4-1.
12 — Los Angeles Kings at Los Angeles, April 10, 1990. Los Angeles 12, Calgary 4. Los Angeles won best-of-seven DSF 4-2.
11 — Montreal Canadiens at Montreal, March 30, 1944. Montreal 11, Toronto 0. Montreal won best-of-seven SF 4-1.
— Edmonton Oilers at Edmonton, May 4, 1985. Edmonton 11, Chicago 2. Edmonton won best-of-seven CF 4-2.

MOST GOALS, ONE TEAM, ONE PERIOD:
7 — Montreal Canadiens, March 30, 1944, at Montreal in third period, during 11-0 win against Toronto.

MOST GOALS, BOTH TEAMS, ONE GAME:
18 — Los Angeles Kings, Edmonton Oilers at Edmonton, April 7, 1982. Los Angeles 10, Edmonton 8. Los Angeles won best-of-five DSF 3-2.
17 — Pittsburgh Penguins, Philadelphia Flyers at Pittsburgh, April 25, 1989. Pittsburgh 10, Philadelphia 7. Philadelphia won best-of-seven DF 4-3.
16 — Edmonton Oilers, Los Angeles Kings at Edmonton, April 9, 1987. Edmonton 13, Los Angeles 3. Edmonton won best-of-seven DSF 4-1.
— Los Angeles Kings, Calgary Flames at Los Angeles, April 10, 1990. Los Angeles 12, Calgary 4. Los Angeles won best-of-seven DF 4-2.

MOST GOALS, BOTH TEAMS, ONE PERIOD:
9 — NY Rangers, Philadelphia Flyers, at Philadelphia, April 24, 1979, third period. NY Rangers won 8-3, scoring six of nine third-period goals.
— Los Angeles Kings, Calgary Flames, at Los Angeles, April 10, 1990, second period. Los Angeles won 12-4, scoring five of nine second-period goals.
8 — Chicago Blackhawks, Montreal Canadiens, at Montreal, May 8, 1973, second period. Chicago won 8-7, scoring five of eight second-period goals.
— Chicago Blackhawks, Edmonton Oilers, at Chicago, May 12, 1985, first period. Chicago won 8-6, scoring five of eight first-period goals.
— Edmonton Oilers, Winnipeg Jets, at Edmonton, April 6, 1988, third period. Edmonton won 7-4, scoring six of eight third-period goals.
— Hartford Whalers, Montreal Canadiens, at Hartford, April 10, 1988, third period. Hartford won 7-5, scoring five of eight third-period goals.
— Vancouver Canucks, NY Rangers, at NY Rangers, June 9, 1994, third period. Vancouver won 6-3, scoring five of eight third-period goals.

TEAM POWER-PLAY GOALS

MOST POWER-PLAY GOALS BY ALL TEAMS, ONE PLAYOFF YEAR:
199 — 1988 in 83 games.

MOST POWER-PLAY GOALS, ONE TEAM, ONE PLAYOFF YEAR:
35 — Minnesota North Stars, 1991 in 23 games.
32 — Edmonton Oilers, 1988 in 18 games.
31 — NY Islanders, 1981 in 18 games.

MOST POWER-PLAY GOALS, ONE TEAM, ONE SERIES:
15 — NY Islanders in 1980 F against Philadelphia. NY Islanders won series 4-2.
— Minnesota North Stars in 1991 DSF against Chicago. Minnesota won series 4-2.
13 — NY Islanders in 1981 QF against Edmonton. NY Islanders won series 4-2.
— Calgary Flames in 1986 CF against St. Louis. Calgary won series 4-3.
12 — Toronto Maple Leafs in 1976 QF against Philadelphia. Philadelphia won series 4-3.

MOST POWER-PLAY GOALS, BOTH TEAMS, ONE SERIES:
21 — NY Islanders, Philadelphia Flyers in 1980 F, won by NY Islanders 4-2. NY Islanders had 15 and Philadelphia 6.
— NY Islanders, Edmonton Oilers in 1981 QF, won by NY Islanders 4-2. NY Islanders had 13 and Edmonton 8.
— Philadelphia Flyers, Pittsburgh Penguins in 1989 DF, won by Philadelphia 4-3. Philadelphia had 11 and Pittsburgh 10.
— Minnesota North Stars, Chicago Blackhawks in 1991 DSF, won by Minnesota 4-2. Minnesota had 15 and Chicago 6.
20 — Toronto Maple Leafs, Philadelphia Flyers in 1976 QF, won by Philadelphia 4-3. Toronto had 12 and Philadelphia 8.

MOST POWER-PLAY GOALS, ONE TEAM, ONE GAME:
6 — Boston Bruins, April 2, 1969, at Boston against Toronto. Boston won 10-0.

MOST POWER-PLAY GOALS, BOTH TEAMS, ONE GAME:
8 — Minnesota North Stars, St. Louis Blues, April 24, 1991 at Minnesota. Minnesota had 4, St. Louis 4. Minnesota won 8-4.
7 — Minnesota North Stars, Edmonton Oilers, April 28, 1984 at Minnesota. Minnesota had 4, Edmonton 3. Edmonton won 8-5.
— Philadelphia Flyers, NY Rangers, April 13, 1985 at NY Rangers. Philadelphia had 4, NY Rangers 3. Philadelphia won 6-5.
— Edmonton Oilers, Chicago Blackhawks, May 14, 1985 at Edmonton. Chicago had 5, Edmonton 2. Edmonton won 10-5.
— Edmonton Oilers, Los Angeles Kings, April 9, 1987 at Edmonton. Edmonton had 5, Los Angeles 2. Edmonton won 13-3.
— Vancouver Canucks, Calgary Flames, April 9, 1989 at Vancouver. Vancouver had 4, Calgary 3. Vancouver won 5-3.

MOST POWER-PLAY GOALS, ONE TEAM, ONE PERIOD:
4 — Toronto Maple Leafs, March 26, 1936, second period against Boston at Toronto. Toronto won 8-3.
— Minnesota North Stars, April 28, 1984, second period against Edmonton at Minnesota. Edmonton won 8-5.
— Boston Bruins, April 11, 1991, third period against Hartford at Boston. Boston won 6-1.
— Minnesota North Stars, April 24, 1991, second period against St. Louis at Minnesota. Minnesota won 8-4.
— St. Louis Blues, April 27, 1998, third period at Los Angeles. St. Louis won 4-3.

MOST POWER-PLAY GOALS, BOTH TEAMS, ONE PERIOD:
5 — Minnesota North Stars, Edmonton Oilers, April 28, 1984, second period, at Minnesota. Minnesota had 4 and Edmonton 1. Edmonton won 8-5.
— Vancouver Canucks, Calgary Flames, April 9, 1989, third period, at Vancouver. Vancouver had 3 and Calgary 2. Vancouver won 5-3.
— Minnesota North Stars, St. Louis Blues, April 24, 1991, second period, at Minnesota. Minnesota had 4 and St. Louis 1. Minnesota won 8-4.

TEAM SHORTHAND GOALS

MOST SHORTHAND GOALS BY ALL TEAMS, ONE PLAYOFF YEAR:
33 — 1988, in 83 games.

MOST SHORTHAND GOALS, ONE TEAM, ONE PLAYOFF YEAR:
10 — Edmonton Oilers, 1983, in 16 games.
9 — NY Islanders, 1981, in 19 games.
8 — Philadelphia Flyers, 1989, in 19 games.

MOST SHORTHAND GOALS, ONE TEAM, ONE SERIES:
6 — Calgary Flames in 1995 against San Jose in best-of-seven CQF won by San Jose 4-3.
— Vancouver Canucks in 1995 against St. Louis in best-of-seven CQF won by Vancouver 4-3.
5 — NY Rangers in 1979 against Philadelphia in best-of-seven QF won by NY Rangers 4-1.
— Edmonton Oilers in 1983 against Calgary in best-of-seven DF won by Edmonton 4-1.

MOST SHORTHAND GOALS, BOTH TEAMS, ONE SERIES:
7 — Boston Bruins (4), NY Rangers (3), in 1958 SF won by Boston 4-2.
— Edmonton Oilers (5), Calgary Flames (2), in 1983 DF won by Edmonton 4-1.
— Vancouver Canucks (6), St. Louis Blues (1), in 1995 CQF won by Vancouver 4-3.

MOST SHORTHAND GOALS, ONE TEAM, ONE GAME:
3 — Boston Bruins, April 11, 1981, at Minnesota. Minnesota won 6-3.
— NY Islanders, April 17, 1983, at NY Rangers. NY Rangers won 7-6.
— Toronto Maple Leafs, May 8, 1994, at San Jose. Toronto won 8-3.

MOST SHORTHAND GOALS, BOTH TEAMS, ONE GAME:
4 — Boston Bruins, Minnesota North Stars, April 11, 1981, at Minnesota. Boston had 3 shorthand goals, Minnesota 1. Minnesota won 6-3.
— NY Islanders, NY Rangers, April 17, 1983, at NY Rangers. NY Islanders had 3 shorthand goals, NY Rangers 1. NY Rangers won 7-6.
— San Jose Sharks, Toronto Maple Leafs, May 8, 1994, at San Jose. Toronto had 3 shorthand goals, San Jose 1. Toronto won 8-3.
3 — Toronto Maple Leafs, Detroit Red Wings, April 5, 1947, at Toronto. Toronto had 2 shorthand goals, Detroit 1. Toronto won 6-1.
— NY Rangers, Boston Bruins, April 1, 1958, at Boston. NY Rangers had 2 shorthand goals, Boston 1. NY Rangers won 5-2.
— Minnesota North Stars, Philadelphia Flyers, May 4, 1980, at Minnesota. Minnesota had 2 shorthand goals, Philadelphia 1. Philadelphia won 5-3.
— Edmonton Oilers, Winnipeg Jets, April 9, 1988, at Winnipeg. Winnipeg had 2 shorthand goals, Edmonton 1. Winnipeg won 6-4.
— New Jersey Devils, NY Islanders, April 14, 1988, at New Jersey. NY Islanders had 2 shorthand goals, New Jersey 1. New Jersey won 6-5.
— Montreal Canadiens, New Jersey Devils, April 17, 1997, at New Jersey. Montreal had 2 shorthand goals, New Jersey 1. New Jersey won 5-2.
— Dallas Stars, San Jose Sharks, May 5, 2000, at San Jose. Dallas had 2 shorthand goals, San Jose 1. Dallas won 5-4.

MOST SHORTHAND GOALS, ONE TEAM, ONE PERIOD:
2 — Toronto Maple Leafs, April 5, 1947, at Toronto against Detroit, first period. Toronto won 6-1.
— Toronto Maple Leafs, April 13, 1965, at Toronto against Montreal, first period. Montreal won 4-3.
— Boston Bruins, April 20, 1969, at Boston against Montreal, first period. Boston won 3-2.
— Boston Bruins, April 8, 1970, at Boston against NY Rangers, second period. Boston won 8-2.
— Boston Bruins, April 30, 1972, at Boston against NY Rangers, first period. Boston won 6-5.
— Chicago Blackhawks, May 3, 1973, at Chicago against Montreal, first period. Chicago won 7-4.
— Montreal Canadiens, April 23, 1978, at Detroit, first period. Montreal won 8-0.
— NY Islanders, April 8, 1980, at NY Islanders against Los Angeles, second period. NY Islanders won 8-1.
— Los Angeles Kings, April 9, 1980, at NY Islanders, first period. Los Angeles won 6-3.
— Boston Bruins, April 13, 1980, at Pittsburgh, second period. Boston won 8-3.
— Minnesota North Stars, May 4, 1980, at Minnesota against Philadelphia, second period. Philadelphia won 5-3.
— Boston Bruins, April 11, 1981, at Minnesota, third period. Minnesota won 6-3.
— NY Islanders, May 12, 1981, at NY Islanders against Minnesota, first period. NY Islanders won 6-3.
— Montreal Canadiens, April 7, 1982, at Montreal against Quebec, third period. Montreal won 5-1.
— Edmonton Oilers, April 24, 1983, at Edmonton against Chicago, third period. Edmonton won 8-4.
— Winnipeg Jets, April 14, 1985, at Calgary, second period. Winnipeg won 5-3.
— Boston Bruins, April 6, 1988, at Boston against Buffalo, first period. Boston won 7-3.
— NY Islanders, April 14, 1988, at New Jersey, third period. New Jersey won 6-5.
— Detroit Red Wings, April 29, 1993, at Toronto, second period. Detroit won 7-3.
— Toronto Maple Leafs, May 8, 1994, at San Jose, third period. Toronto won 8-3.
— Calgary Flames, May 11, 1995, at San Jose, first period. Calgary won 9-2.
— Vancouver Canucks, May 15, 1995, at St. Louis, second period. Vancouver won 6-5.
— Montreal Canadiens, April 17, 1997, at New Jersey, second period. New Jersey won 5-2.
— Philadelphia Flyers, April 26, 1997, at Philadelphia against Pittsburgh, first period. Philadelphia won 6-3.
— Phoenix Coyotes, April 24, 1998, at Detroit, second period. Phoenix won 7-4.
— Buffalo Sabres, April 27, 1998, at Buffalo against Philadelphia, second period. Buffalo won 6-1.
— San Jose Sharks, April 30, 1999, at Colorado, third period. San Jose won 7-3.
— Detroit Red Wings, April 27, 2002, at Vancouver, second period. Detroit won 6-4.

MOST SHORTHAND GOALS, BOTH TEAMS, ONE PERIOD:
3 — Toronto Maple Leafs, Detroit Red Wings, April 5, 1947, at Toronto, first period. Toronto had 2 shorthand goals, Detroit 1. Toronto won 6-1.
— Toronto Maple Leafs, San Jose Sharks, May 8, 1994, at San Jose, third period. Toronto had 2 shorthand goals, San Jose 1. Toronto won 8-3.

FASTEST GOALS

FASTEST FIVE GOALS, BOTH TEAMS:
3 Minutes, 6 Seconds — Chicago Blackhawks, Minnesota North Stars at Chicago, April 21, 1985. Keith Brown scored for Chicago at 1:12 of second period; Ken Yaremchuk, Chicago, 1:27; Dino Ciccarelli, Minnesota, 2:48; Tony McKegney, Minnesota, 4:07; and Curt Fraser, Chicago, 4:18. Chicago won 6-2 and best-of-seven DF 4-2.
3 Minutes, 20 Seconds — Minnesota North Stars, Philadelphia Flyers at Philadelphia, April 29, 1980. Paul Shmyr scored for Minnesota at 13:20 of first period; Steve Christoff, Minnesota, 13:59; Ken Linseman, Philadelphia, 14:54; Tom Gorence, Philadelphia, 15:36; and Linseman, 16:40. Minnesota won 6-5. Philadelphia won best-of-seven SF 4-1.
4 Minutes — Detroit Red Wings, Los Angeles Kings at Detroit, April 15, 2000. Brendan Shanahan scored for Detroit at 0:55 of first period; Martin Lapointe, Detroit, 1:33; Luc Robitaille, Los Angeles, 2:04; Kris Draper, Detroit, 3:32; and Ziggy Palffy, Los Angeles, 4:55. Detroit won 8-5 and best-of-seven CQF 4-0.

FASTEST FIVE GOALS, ONE TEAM:
3 Minutes, 36 Seconds — Montreal Canadiens at Montreal, March 30, 1944, against Toronto. Toe Blake scored at 7:58 and 8:37 of third period; Maurice Richard, 9:17; Ray Getliffe, 10:33; and Buddy O'Connor, 11:34. Montreal won 11-0 and best-of-seven SF 4-1.

FASTEST FOUR GOALS, BOTH TEAMS:
1 Minute, 33 Seconds — Philadelphia Flyers, Toronto Maple Leafs at Philadelphia, April 20, 1976. Don Saleski scored for Philadelphia at 10:04 of second period; Bob Neely, Toronto, 10:42; Gary Dornhoefer, Philadelphia, 11:24; and Don Saleski, 11:37. Philadelphia won 7-1 and best-of-seven QF 4-3.
1 Minute, 34 seconds — Montreal Canadiens, Calgary Flames at Montreal, May 20, 1986. Joel Otto scored for Calgary at 17:59 of first period; Bobby Smith, Montreal, 18:25; Mats Naslund, Montreal, 19:17; and Bob Gainey, Montreal, 19:33. Montreal won 5-3 and best-of-seven F 4-1.
1 Minute, 38 Seconds — Boston Bruins, Philadelphia Flyers at Philadelphia, April 26, 1977. Gregg Sheppard scored for Boston at 14:01 of second period; Mike Milbury, Boston, 15:01; Gary Dornhoefer, Philadelphia, 15:16; and Jean Ratelle, Boston, 15:39. Boston won 5-4 and best-of-seven SF 4-0.

FASTEST FOUR GOALS, ONE TEAM:
2 Minutes, 35 Seconds — Montreal Canadiens at Montreal, March 30, 1944, against Toronto. Toe Blake scored at 7:58 and 8:37 of third period; Maurice Richard, 9:17; and Ray Getliffe, 10:33. Montreal won 11-0 and best-of-seven SF 4-1.

FASTEST THREE GOALS, BOTH TEAMS:
21 Seconds — Edmonton Oilers, Chicago Blackhawks at Edmonton, May 7, 1985. Behn Wilson scored for Chicago at 19:22 of third period, Jari Kurri at 19:36 and Glenn Anderson at 19:43 for Edmonton. Edmonton won 7-3 and best-of-seven CF 4-2.
27 Seconds — Phoenix Coyotes, Detroit Red Wings at Detroit, April 24, 1998. Jeremy Roenick scored for Phoenix at 13:24 of the second period. Mathieu Dandenault scored for Detroit at 13:32, and Keith Tkachuk scored for Phoenix at 13:51. Phoenix won 7-4, Detroit won the best-of-seven CQF 4-2.
30 Seconds — Chicago Blackhawks, Pittsburgh Penguins at Chicago, June 1, 1992. Dirk Graham scored for Chicago at 6:21 of first period, Kevin Stevens for Pittsburgh at 6:33 and Dirk Graham at 6:51. Pittsburgh won 6-5 and best-of-seven F 4-0.

FASTEST THREE GOALS, ONE TEAM:
23 Seconds — Toronto Maple Leafs at Toronto, April 12, 1979, against Atlanta. Darryl Sittler scored at 4:04 and 4:16 of first period and Ron Ellis at 4:27. Leafs won 7-4 and best-of-three PR 2-0.
38 Seconds — NY Rangers at NY Rangers, April 12, 1986 against Philadelphia. Jim Wiemer scored at 12:29 of third period, Bob Brooke at 12:43 and Ron Greschner at 13:07. NY Rangers won 5-2 and best-of-five DSF 3-2.
— Colorado Avalanche at Vancouver, April 18, 2001. Peter Forsberg scored at 9:11 of third period, Joe Sakic at 9:28 and Eric Messier at 9:49. Colorado won 5-1 and best-of-seven CQF 4-0.

FASTEST TWO GOALS, BOTH TEAMS:
5 Seconds — Pittsburgh Penguins, Buffalo Sabres at Buffalo, April 14, 1979. Gilbert Perreault scored for Buffalo at 12:59 and Jim Hamilton for Pittsburgh at 13:04 of first period. Pittsburgh won 4-3 and best-of-three PR 2-1.
8 Seconds — Minnesota North Stars, St. Louis Blues at Minnesota, April 9, 1989. Bernie Federko scored for St. Louis at 2:28 and Perry Berezan for Minnesota at 2:36 of third period. Minnesota won 5-4. St. Louis won best-of-seven DSF 4-1.
— Phoenix Coyotes, Detroit Red Wings at Detroit, April 24, 1998. Jeremy Roenick scored for Phoenix at 13:24 and Mathieu Dandenault for Detroit at 13:32 of second period. Phoenix won 7-4. Detroit won best-of-seven CQF 4-2.
9 Seconds — NY Islanders, Washington Capitals at Washington, April 10, 1986. Bryan Trottier scored for NY Islanders at 18:26 and Scott Stevens for Washington at 18:35 of second period. Washington won 5-2, and best-of-five DSF 3-0.
— Buffalo Sabres, Toronto Maple Leafs at Toronto, May 23, 1999. Vaclav Varada scored for Buffalo at 4:23 and Mats Sundin for Toronto at 4:32 of first period. Buffalo won 5-4, and best-of-seven CF 4-1.

FASTEST TWO GOALS, ONE TEAM:
5 Seconds — Detroit Red Wings at Detroit, April 11, 1965, against Chicago. Norm Ullman scored at 17:35 and 17:40, second period. Detroit won 4-2. Chicago won best-of-seven SF 4-3.

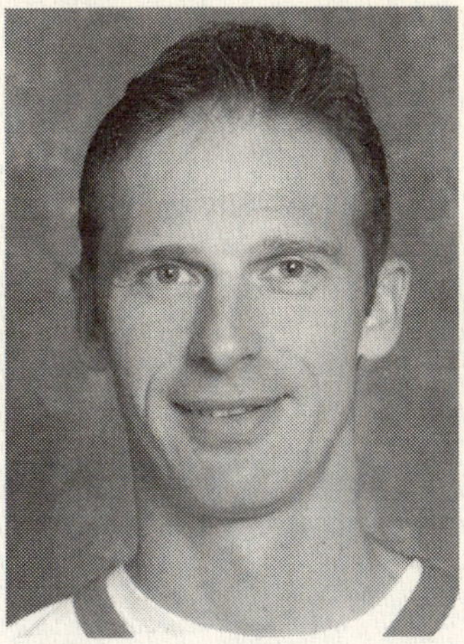

Ottawa's Patrick Lalime (above) had three straight shutouts and four overall during two rounds of the 2002 playoffs. Detroit's Dominik Hasek (top) would set a new playoff record with six shutouts en route to the Stanley Cup.

OVERTIME

SHORTEST OVERTIME:
9 Seconds — Montreal Canadiens, Calgary Flames at Calgary, May 18, 1986. Montreal won 3-2 on Brian Skrudland's goal and captured the best-of-seven F 4-1.
11 Seconds — NY Islanders, NY Rangers at NY Rangers, April 11, 1975. NY Islanders won 4-3 on Jean-Paul Parise's goal and captured the best-of-three PR 2-1.

LONGEST OVERTIME:
116 Minutes, 30 Seconds — Detroit Red Wings, Mtl. Maroons at Montreal, March 24, 25, 1936. Detroit 1, Mtl. Maroons 0. Mud Bruneteau scored, assisted by Hec Kilrea, at 16:30 of sixth overtime period, or after 176 minutes, 30 seconds from start of game, which ended at 2:25 a.m. Detroit won best-of-five SF 3-0.

MOST OVERTIME GAMES, ONE PLAYOFF YEAR:
28 — 1993. Of 85 games played, 28 went into overtime.
26 — 2001. Of 86 games played, 26 went into overtime.
21 — 1999. Of 86 games played, 21 went into overtime.
19 — 1996. Of 86 games played, 19 went into overtime.
— 1998. Of 82 games played, 19 went into overtime.

FEWEST OVERTIME GAMES, ONE PLAYOFF YEAR:
0 — 1963. None of the 16 games went into overtime, the only year since 1926 that no overtime was required in any playoff series.

MOST OVERTIME GAMES, ONE SERIES:
5 — Toronto Maple Leafs, Montreal Canadiens in 1951. Toronto won best-of-seven F 4-1.
4 — Toronto Maple Leafs, Boston Bruins in 1933. Toronto won best-of-five SF 3-2.
— Boston Bruins, NY Rangers in 1939. Boston won best-of-seven SF 4-3.
— St. Louis Blues, Minnesota North Stars in 1968. St. Louis won best-of-seven SF 4-3.
— Dallas Stars, St. Louis Blues in 1999. Dallas won best-of-seven CSF 4-2.
— Dallas Stars, Edmonton Oilers in 2001. Dallas won best-of-seven CQF 4-2.

THREE-OR-MORE GOAL GAMES

MOST THREE-OR-MORE GOAL GAMES BY ALL TEAMS, ONE PLAYOFF YEAR:
12 — 1983 in 66 games.
— **1988** in 83 games.
11 — 1985 in 70 games.
— 1992 in 86 games.

MOST THREE-OR-MORE GOAL GAMES, ONE TEAM, ONE PLAYOFF YEAR:
6 — Edmonton Oilers in 16 games, 1983.
— **Edmonton Oilers** in 18 games, 1985.

SHUTOUTS

MOST SHUTOUTS, ONE PLAYOFF YEAR, ALL TEAMS:
25 — 2002. Of 90 games played, Detroit had 6; Ottawa had 4; Carolina, Colorado, St. Louis and Toronto had 3 each; while Los Angeles, New Jersey and Philadelphia had 1 each.
19 — 2001. Of 86 games played, Colorado and New Jersey had 4 each, Toronto had 3, Pittsburgh and Los Angeles had 2 each, while Buffalo, Washington, Detroit and San Jose had 1 each.
18 — 1997. Of 82 games played, Colorado and NY Rangers had 3 each, Edmonton, St. Louis and New Jersey had 2 each, while Anaheim, Buffalo, Detroit, Florida, Ottawa and Phoenix had 1 each.

FEWEST SHUTOUTS, ONE PLAYOFF YEAR, ALL TEAMS:
0 — 1959. 18 games played.

MOST SHUTOUTS, BOTH TEAMS, ONE SERIES:
5 — 1945 F, Toronto Maple Leafs, Detroit Red Wings. Toronto had 3 shutouts, Detroit 2. Toronto won best-of-seven series 4-3.
— **1950 SF, Toronto Maple Leafs, Detroit Red Wings.** Toronto had 3 shutouts, Detroit 2. Detroit won best-of-seven series 4-3.

TEAM PENALTIES

FEWEST PENALTIES, BOTH TEAMS, BEST-OF-SEVEN SERIES:
19 — Detroit Red Wings, Toronto Maple Leafs in 1945 F, won by Toronto 4-3. Detroit received 10 minors, Toronto had 9 minors.

FEWEST PENALTIES, ONE TEAM, BEST-OF-SEVEN SERIES:
9 — Toronto Maple Leafs in 1945 F, won by Toronto 4-3 against Detroit.

MOST PENALTIES, BOTH TEAMS, ONE SERIES:
218 — New Jersey Devils, Washington Capitals in 1988 DF won by New Jersey 4-3. New Jersey received 98 minors, 11 majors, 9 misconducts and 1 match penalty. Washington received 80 minors, 11 majors, 8 misconducts and 1 match penalty.

MOST PENALTY MINUTES, BOTH TEAMS, ONE SERIES:
654 — New Jersey Devils, Washington Capitals in 1988 DF won by New Jersey 4-3. New Jersey had 351 minutes; Washington 305.

MOST PENALTIES, ONE TEAM, ONE SERIES:
118 — New Jersey Devils in 1988 DF won by New Jersey against Washington. New Jersey received 98 minors, 11 majors, 9 misconducts and 1 match penalty.

MOST PENALTY MINUTES, ONE TEAM, ONE SERIES:
349 — New Jersey Devils in 1988 DF won by New Jersey 4-3 against Washington.

MOST PENALTIES, BOTH TEAMS, ONE GAME:
66 — Detroit Red Wings, St. Louis Blues at St. Louis, April 12, 1991. Detroit received 33 penalties; St. Louis 33. St. Louis won 6-1.
63 — Minnesota North Stars, Chicago Blackhawks at Chicago, April 6, 1990. Minnesota received 34 penalties; Chicago 29. Chicago won 5-3.
62 — New Jersey Devils, Washington Capitals at New Jersey, April 22, 1988. New Jersey received 32 penalties; Washington 30. New Jersey won 10-4.

MOST PENALTY MINUTES, BOTH TEAMS, ONE GAME:
298 Minutes — Detroit Red Wings, St. Louis Blues at St. Louis, April 12, 1991. Detroit received 33 penalties for 152 minutes; St. Louis 33 penalties for 146 minutes. St. Louis won 6-1.
267 Minutes — NY Rangers, Los Angeles Kings at Los Angeles, April 9, 1981. NY Rangers received 31 penalties for 142 minutes; Los Angeles 28 penalties for 125 minutes. Los Angeles won 5-4.

MOST PENALTIES, ONE TEAM, ONE GAME:
34 — Minnesota North Stars, at Chicago, April 6, 1990. Chicago won 5-3.
33 — Detroit Red Wings, at St. Louis, April 12,1991. St. Louis won 6-1.
— St. Louis Blues, at St. Louis, April 12, 1991. St. Louis won 6-1.
32 — New Jersey Devils, at Washington, April 22,1988. New Jersey won 10-4.
31 — NY Rangers, at Los Angeles, April 9, 1981. Los Angeles won 5-4.

MOST PENALTY MINUTES, ONE TEAM, ONE GAME:
152 — Detroit Red Wings, at St. Louis, April 12, 1991. St. Louis won 6-1.
146 — St. Louis Blues, at St. Louis, April 12, 1991. St. Louis won 6-1.
142 — NY Rangers, at Los Angeles, April 9, 1981. Los Angeles won 5-4.

MOST PENALTIES, BOTH TEAMS, ONE PERIOD:
43 — NY Rangers, Los Angeles Kings at Los Angeles, April 9, 1981, first period. NY Rangers had 24 penalties; Los Angeles 19. Los Angeles won 5-4.

MOST PENALTY MINUTES, BOTH TEAMS, ONE PERIOD:
248 — NY Islanders, Boston Bruins at Boston, April 17, 1980, first period. Each team received 124 minutes. NY Islanders won 5-4.

MOST PENALTIES, ONE TEAM, ONE PERIOD:
24 — NY Rangers, at Los Angeles, April 9, 1981, first period. Los Angeles won 5-4.

Individual Records

GAMES PLAYED

MOST YEARS IN PLAYOFFS:
21 — Raymond Bourque, Boston, Colorado (1980-96 incl.; 98-01 incl.)
20 — Gordie Howe, Detroit, Hartford
— Larry Robinson, Montreal, Los Angeles
— Larry Murphy, Los Angeles, Washington, Minnesota, Pittsburgh, Toronto, Detroit
19 — Red Kelly, Detroit, Toronto
— Scott Stevens, Washington, St. Louis, New Jersey

MOST CONSECUTIVE YEARS IN PLAYOFFS:
20 — Larry Robinson, Montreal, Los Angeles (1973-92, inclusive).
18 — Larry Murphy, Los Angeles, Washington, Minnesota, Pittsburgh, Toronto, Detroit (1984-2001, inclusive).
17 — Brad Park, NY Rangers, Boston, Detroit (1969-85, inclusive).
— Raymond Bourque, Boston (1980-96, inclusive).
— Brett Hull, Calgary, St. Louis, Dallas, Detroit (1986-02, inclusive).

MOST PLAYOFF GAMES:
236 — Mark Messier, Edmonton, NY Rangers
231 — Guy Carbonneau, Montreal, St. Louis, Dallas
227 — Larry Robinson, Montreal, Los Angeles
226 — Claude Lemieux, Montreal, New Jersey, Colorado, Phoenix
225 — Glenn Anderson, Edmonton, Toronto, NY Rangers, St. Louis

GOALS

MOST GOALS IN PLAYOFFS (CAREER):
122 — Wayne Gretzky, Edmonton, Los Angeles, St. Louis, NY Rangers
109 — Mark Messier, Edmonton, NY Rangers
106 — Jari Kurri, Edmonton, Los Angeles, NY Rangers, Anaheim
100 — Brett Hull, Calgary, St. Louis, Dallas, Detroit
93 — Glenn Anderson, Edmonton, Toronto, NY Rangers, St. Louis

MOST GOALS, ONE PLAYOFF YEAR:
19 — Reggie Leach, Philadelphia, 1976. 16 games.
— Jari Kurri, Edmonton, 1985. 18 games.
18 — Joe Sakic, Colorado, 1996. 22 games.
17 — Newsy Lalonde, Montreal, 1919. 10 games.
— Mike Bossy, NY Islanders, 1981. 18 games.
— Steve Payne, Minnesota, 1981. 19 games.
— Mike Bossy, NY Islanders, 1982. 19 games.
— Mike Bossy, NY Islanders, 1983. 19 games
— Wayne Gretzky, Edmonton, 1985. 18 games.
— Kevin Stevens, Pittsburgh, 1991. 24 games.

MOST GOALS IN ONE SERIES (OTHER THAN FINAL):
12 — Jari Kurri, Edmonton, in 1985 CF, 6 games vs. Chicago.
11 — Newsy Lalonde, Montreal, in 1919 NHL F, 5 games vs. Ottawa.
10 — Tim Kerr, Philadelphia, in 1989 DF, 7 games vs. Pittsburgh.
9 — Reggie Leach, Philadelphia, in 1976 SF, 5 games vs. Boston.
— Bill Barber, Philadelphia, in 1980 SF, 5 games vs. Minnesota.
— Mike Bossy, NY Islanders, in 1983 CF, 6 games vs. Boston.
— Mario Lemieux, Pittsburgh, in 1989 DF, 7 games vs. Philadelphia.

MOST GOALS IN FINAL SERIES (NHL PLAYERS ONLY):
9 — Babe Dye, Toronto, in 1922, 5 games vs. Van. Millionaires.
8 — Alf Skinner, Toronto, in 1918, 5 games vs. Van. Millionaires.
7 — Jean Beliveau, Montreal, in 1956, 5 games vs. Detroit.
— Mike Bossy, NY Islanders, in 1982, 4 games vs. Vancouver.
— Wayne Gretzky, Edmonton, in 1985, 5 games vs. Philadelphia.

MOST GOALS, ONE GAME:
5 — Newsy Lalonde, Montreal, March 1, 1919, at Montreal. Final score: Montreal 6, Ottawa 3.
— Maurice Richard, Montreal, March 23, 1944, at Montreal. Final score: Montreal 5, Toronto 1.
— Darryl Sittler, Toronto, April 22, 1976, at Toronto. Final score: Toronto 8, Philadelphia 5.
— Reggie Leach, Philadelphia, May 6, 1976, at Philadelphia. Final score: Philadelphia 6, Boston 3.
— Mario Lemieux, Pittsburgh, April 25, 1989, at Pittsburgh. Final score: Pittsburgh 10, Philadelphia 7.

MOST GOALS, ONE PERIOD:
4 — Tim Kerr, Philadelphia, April 13, 1985, at NY Rangers, second period. Final score: Philadelphia 6, NY Rangers 5.
— Mario Lemieux, Pittsburgh, April 25, 1989, at Pittsburgh vs. Philadelphia, first period. Final score: Pittsburgh 10, Philadelphia 7.

ASSISTS

MOST ASSISTS IN PLAYOFFS (CAREER):
260 — Wayne Gretzky, Edmonton, Los Angeles, St. Louis, NY Rangers
186 — Mark Messier, Edmonton, NY Rangers
139 — Raymond Bourque, Boston, Colorado
137 — Paul Coffey, Edmonton, Pittsburgh, Los Angeles, Detroit, Philadelphia, Carolina
128 — Doug Gilmour, St. Louis, Calgary, Toronto, New Jersey, Buffalo, Montreal

MOST ASSISTS, ONE PLAYOFF YEAR:
31 — Wayne Gretzky, Edmonton, 1988. 19 games.
30 — Wayne Gretzky, Edmonton, 1985. 18 games.
29 — Wayne Gretzky, Edmonton, 1987. 21 games.
28 — Mario Lemieux, Pittsburgh, 1991. 23 games.
26 — Wayne Gretzky, Edmonton, 1983. 16 games.

MOST ASSISTS IN ONE SERIES (OTHER THAN FINAL):
14 — Rick Middleton, Boston, in 1983 DF, 7 games vs. Buffalo.
— Wayne Gretzky, Edmonton, in 1985 CF, 6 games vs. Chicago.
13 — Wayne Gretzky, Edmonton, in 1987 DSF, 5 games vs. Los Angeles.
— Doug Gilmour, Toronto, in 1994 CSF, 7 games vs. San Jose.
11 — Al MacInnis, Calgary, in 1984 DF, 7 games vs. Edmonton.
— Mark Messier, Edmonton, in 1989 DSF, 7 games vs. Los Angeles.
— Mike Ridley, Washington, in 1992 DSF, 7 games vs. Pittsburgh.
— Ron Francis, Pittsburgh, in 1995 CQF, 7 games vs. Washington.
10 — Fleming Mackell, Boston, in 1958 SF, 6 games vs. NY Rangers.
— Stan Mikita, Chicago, in 1962 SF, 6 games vs. Montreal.
— Bob Bourne, NY Islanders, in 1983 DF, 6 games vs. NY Rangers.
— Wayne Gretzky, Edmonton, in 1988 DSF, 5 games vs. Winnipeg.
— Mario Lemieux, Pittsburgh, in 1992 DSF, 6 games vs. Washington.

MOST ASSISTS IN FINAL SERIES:
10 — Wayne Gretzky, Edmonton, in 1988, 4 games plus suspended game vs. Boston.
9 — Jacques Lemaire, Montreal, in 1973, 6 games vs. Chicago.
— Wayne Gretzky, Edmonton, in 1987, 7 games vs. Philadelphia.
— Larry Murphy, Pittsburgh, in 1991, 6 games vs. Minnesota.

MOST ASSISTS, ONE GAME:
6 — Mikko Leinonen, NY Rangers, April 8, 1982, at NY Rangers. Final score: NY Rangers 7, Philadelphia 3.
— Wayne Gretzky, Edmonton, April 9, 1987, at Edmonton. Final score: Edmonton 13, Los Angeles 3.
5 — Toe Blake, Montreal, March 23, 1944, at Montreal. Final score: Montreal 5, Toronto 1.
— Maurice Richard, Montreal, March 27, 1956, at Montreal. Final score: Montreal 7, NY Rangers 0.
— Bert Olmstead, Montreal, March 30, 1957, at Montreal. Final score: Montreal 8, NY Rangers 3.
— Don McKenney, Boston, April 5, 1958, at Boston. Final score: Boston 8, NY Rangers 2.
— Stan Mikita, Chicago, April 4, 1973, at Chicago. Final score: Chicago 7, St. Louis 1.
— Wayne Gretzky, Edmonton, April 8, 1981, at Montreal. Final score: Edmonton 6, Montreal 3.
— Paul Coffey, Edmonton, May 14, 1985, at Edmonton. Final score: Edmonton 10, Chicago 5.
— Doug Gilmour, St. Louis, April 15, 1986, at Minnesota. Final score: St. Louis 6, Minnesota 3.
— Risto Siltanen, Quebec, April 14, 1987, at Hartford. Final score: Quebec 7, Hartford 5.
— Patrik Sundstrom, New Jersey, April 22, 1988, at New Jersey. Final score: New Jersey 10, Washington 4.
— Geoff Courtnall, St. Louis, April 23, 1998, at St. Louis. Final score: St. Louis 8, Los Angeles 3.

MOST ASSISTS, ONE PERIOD:
3 — Three assists by one player in one period of a playoff game has been recorded on 74 occasions. Chris Chelios of the Detroit Red Wings is the most recent to equal this mark with 3 assists in the second period against Vancouver, April 27, 2002. Final score: Detroit 6, Vancouver 4.
— Wayne Gretzky has had 3 assists in one period 5 times; Raymond Bourque, 3 times; Toe Blake, Jean Beliveau, Doug Harvey and Bobby Orr, twice. Nick Metz of Toronto was the first player to be credited with 3 assists in one period of a playoff game Mar. 21, 1941 at Toronto vs. Boston.

POINTS

MOST POINTS IN PLAYOFFS (CAREER):
382 — Wayne Gretzky, Edmonton, Los Angeles, St. Louis, NY Rangers, 122G, 260A
295 — Mark Messier, Edmonton, NY Rangers, 109G, 186A
233 — Jari Kurri, Edmonton, Los Angeles, NY Rangers, Anaheim, 106G, 127A
214 — Glenn Anderson, Edmonton, Toronto, NY Rangers, St. Louis, 93G, 121A
196 — Paul Coffey, Edmonton, Pittsburgh, Los Angeles, Detroit, Philadelphia, Carolina, 59G, 137A

MOST POINTS, ONE PLAYOFF YEAR:
47 — Wayne Gretzky, Edmonton, in 1985. 17 goals, 30 assists in 18 games.
44 — Mario Lemieux, Pittsburgh, in 1991. 16 goals, 28 assists in 23 games.
43 — Wayne Gretzky, Edmonton, in 1988. 12 goals, 31 assists in 19 games.
40 — Wayne Gretzky, Los Angeles, in 1993. 15 goals, 25 assists in 24 games.
38 — Wayne Gretzky, Edmonton, in 1983. 12 goals, 26 assists in 16 games.

MOST POINTS IN ONE SERIES (OTHER THAN FINAL):

19 — Rick Middleton, Boston, in 1983 DF, 7 games vs. Buffalo. 5 goals, 14 assists.
18 — Wayne Gretzky, Edmonton, in 1985 CF, 6 games vs. Chicago. 4 goals, 14 assists.
17 — Mario Lemieux, Pittsburgh, in 1992 DSF, 6 games vs. Washington. 7 goals, 10 assists.
16 — Barry Pederson, Boston, in 1983 DF, 7 games vs. Buffalo. 7 goals, 9 assists.
— Doug Gilmour, Toronto, in 1994 CSF, 7 games vs. San Jose. 3 goals, 13 assists.
15 — Jari Kurri, Edmonton, in 1985 CF, 6 games vs. Chicago. 12 goals, 3 assists.
— Wayne Gretzky, Edmonton, in 1987 DSF, 5 games vs. Los Angeles. 2 goals, 13 assists.
— Tim Kerr, Philadelphia, in 1989 DF, 7 games vs. Pittsburgh. 10 goals, 5 assists.
— Mario Lemieux, Pittsburgh, in 1991 CF, 6 games vs. Boston. 6 goals, 9 assists.

MOST POINTS IN FINAL SERIES:

13 — Wayne Gretzky, Edmonton, in 1988, 4 games plus suspended game vs. Boston. 3 goals, 10 assists.
12 — Gordie Howe, Detroit, in 1955, 7 games vs. Montreal. 5 goals, 7 assists.
— Yvan Cournoyer, Montreal, in 1973, 6 games vs. Chicago. 6 goals, 6 assists.
— Jacques Lemaire, Montreal, in 1973, 6 games vs. Chicago. 3 goals, 9 assists.
— Mario Lemieux, Pittsburgh, in 1991, 5 games vs. Minnesota. 5 goals, 7 assists.

MOST POINTS, ONE GAME:

8 — Patrik Sundstrom, New Jersey, April 22, 1988, at New Jersey during 10-4 win over Washington. Sundstrom had 3 goals, 5 assists.
— Mario Lemieux, Pittsburgh, April 25, 1989, at Pittsburgh during 10-7 win over Philadelphia. Lemieux had 5 goals, 3 assists.
7 — Wayne Gretzky, Edmonton, April 17, 1983, at Calgary during 10-2 win. Gretzky had 4 goals, 3 assists.
— Wayne Gretzky, Edmonton, April 25,1985, at Winnipeg during 8-3 win. Gretzky had 3 goals, 4 assists.
— Wayne Gretzky, Edmonton, April 9, 1987, at Edmonton during 13-3 win over Los Angeles. Gretzky had 1 goal, 6 assists.
6 — Dickie Moore, Montreal, March 25, 1954, at Montreal during 8-1 win over Boston. Moore had 2 goals, 4 assists.
— Phil Esposito, Boston, April 2, 1969, at Boston during 10-0 win over Toronto. Esposito had 4 goals, 2 assists.
— Darryl Sittler, Toronto, April 22, 1976, at Toronto during 8-5 win over Philadelphia. Sittler had 5 goals, 1 assist.
— Guy Lafleur, Montreal, April 11, 1977, at Montreal during 7-2 win over St. Louis. Lafleur had 3 goals, 3 assists.
— Mikko Leinonen, NY Rangers, April 8, 1982, at NY Rangers during 7-3 win over Philadelphia. Leinonen had 6 assists.
— Paul Coffey, Edmonton, May 14, 1985, at Edmonton during 10-5 win over Chicago. Coffey had 1 goal, 5 assists.
— John Anderson, Hartford, April 12, 1986, at Hartford during 9-4 win over Quebec. Anderson had 2 goals, 4 assists.
— Mario Lemieux, Pittsburgh, April 23, 1992, at Pittsburgh during 6-4 win over Washington. Lemieux had 3 goals, 3 assists.
— Geoff Courtnall, St. Louis Blues, April 23, 1998, at St. Louis during 8-3 win over Los Angeles. Courtnall had 1 goal, 5 assists.

MOST POINTS, ONE PERIOD:

4 — Maurice Richard, Montreal, March 29, 1945, at Montreal vs. Toronto. Third period, 3 goals, 1 assist. Final score: Montreal 10, Toronto 3.
— Dickie Moore, Montreal, March 25, 1954, at Montreal vs. Boston. First period, 2 goals, 2 assists. Final score: Montreal 8, Boston 1.
— Barry Pederson, Boston, April 8, 1982, at Boston vs. Buffalo. Second period, 3 goals, 1 assist. Final score: Boston 7, Buffalo 3.
— Peter McNab, Boston, April 11, 1982, at Buffalo. Second period, 1 goal, 3 assists. Final score: Boston 5, Buffalo 2.
— Tim Kerr, Philadelphia, April 13, 1985, at NY Rangers. Second period, 4 goals. Final score: Philadelphia 6, NY Rangers 5.
— Ken Linseman, Boston, April 14, 1985, at Boston vs. Montreal. Second period, 2 goals, 2 assists. Final score: Boston 7, Montreal 6.
— Wayne Gretzky, Edmonton, April 12, 1987, at Los Angeles. Third period, 1 goal, 3 assists. Final score: Edmonton 6, Los Angeles 3.
— Glenn Anderson, Edmonton, April 6, 1988, at Edmonton vs. Winnipeg. Third period, 3 goals, 1 assist. Final score: Edmonton 7, Winnipeg 4.
— Mario Lemieux, Pittsburgh, April 25, 1989, at Pittsburgh vs. Philadelphia. First period, 4 goals. Final score: Pittsburgh 10, Philadelphia 7.
— Dave Gagner, Minnesota, April 8, 1991, at Minnesota vs. Chicago. First period, 2 goals, 2 assists. Final score: Chicago 6, Minnesota 5.
— Mario Lemieux, Pittsburgh, April 23, 1992, at Pittsburgh vs. Washington. Second period, 2 goals, 2 assists. Final score: Pittsburgh 6, Washington 4.
— Alexander Mogilny, New Jersey, April 28, 2001, at New Jersey vs. Toronto. Second period, 1 goal, 3 assists. Final score: New Jersey 6, Toronto 5.

POWER-PLAY GOALS

MOST POWER-PLAY GOALS IN PLAYOFFS (CAREER):

37 — Brett Hull, St. Louis, Dallas, Detroit
35 — Mike Bossy, NY Islanders
34 — Dino Ciccarelli, Minnesota, Washington, Detroit
— Wayne Gretzky, Edmonton, Los Angeles, St. Louis, NY Rangers
29 — Mario Lemieux, Pittsburgh

MOST POWER-PLAY GOALS, ONE PLAYOFF YEAR:

9 — Mike Bossy, NY Islanders, 1981. 18 games against Toronto, Edmonton, NY Rangers and Minnesota.
— Cam Neely, Boston, 1991. 19 games against Hartford, Montreal and Pittsburgh.
8 — Tim Kerr, Philadelphia, 1989. 19 games.
— John Druce, Washington, 1990. 15 games.
— Brian Propp, Minnesota, 1991. 23 games.
— Mario Lemieux, Pittsburgh, 1992. 15 games.

MOST POWER-PLAY GOALS, ONE PLAYOFF SERIES:

6 — Chris Kontos, Los Angeles, 1989, DSF vs. Edmonton, won by Los Angeles 4-3.
5 — Andy Bathgate, Detroit, 1966, SF vs. Chicago, won by Detroit 4-2.
— Denis Potvin, NY Islanders, 1981, QF vs. Edmonton, won by NY Islanders 4-2.
— Ken Houston, Calgary, 1981, QF vs. Philadelphia, won by Calgary 4-3.
— Rick Vaive, Chicago, 1988, DSF vs. St. Louis, won by St. Louis 4-1.
— Tim Kerr, Philadelphia, 1989, DF vs. Pittsburgh, won by Philadelphia 4-3.
— Mario Lemieux, Pittsburgh, 1989, DF vs. Philadelphia, won by Philadelphia 4-3.
— John Druce, Washington, 1990, DF vs. NY Rangers, won by Washington 4-1.
— Pat LaFontaine, Buffalo, 1992, DSF vs. Boston, won by Boston 4-3.
— Adam Graves, NY Rangers, 1996, CQF vs Montreal, won by NY Rangers 4-2.

MOST POWER-PLAY GOALS, ONE GAME:

3 — Syd Howe, Detroit, March 23, 1939, at Detroit vs. Montreal. Detroit won 7-3.
— Sid Smith, Toronto, April 10, 1949, at Detroit. Toronto won 3-1.
— Phil Esposito, Boston, April 2, 1969, at Boston vs. Toronto. Boston won 10-0.
— John Bucyk, Boston, April 21, 1974, at Boston vs. Chicago. Boston won 8-6.
— Denis Potvin, NY Islanders, April 17, 1981, at NY Islanders vs. Edmonton. NY Islanders won 6-3.
— Tim Kerr, Philadelphia, April 13, 1985, at NY Rangers. Philadelphia won 6-5.
— Jari Kurri, Edmonton, April 9, 1987, at Edmonton vs. Los Angeles. Edmonton won 13-3.
— Mark Johnson, New Jersey, April 22, 1988, at New Jersey vs. Washington. New Jersey won 10-4.
— Dino Ciccarelli, Detroit, April 29, 1993, at Toronto. Detroit won 7-3.
— Dino Ciccarelli, Detroit, May 11, 1995, at Dallas. Detroit won 5-1.
— Valeri Kamensky, Colorado, April 24, 1997, at Colorado vs. Chicago. Colorado won 7-0.

MOST POWER-PLAY GOALS, ONE PERIOD:

3 — Tim Kerr, Philadelphia, April 13, 1985, at NY Rangers, second period in 6-5 win.
2 — Two power-play goals have been scored by one player in one period on 53 occasions. Charlie Conacher of Toronto was the first to score two power-play goals in one period, setting the mark on March 26, 1936. Brendan Shanahan of the Detroit Red Wings is the most recent to equal this mark with two power-play goals May 3, 1998, at Phoenix, first period in 5-2 win.

SHORTHAND GOALS

MOST SHORTHAND GOALS IN PLAYOFFS (CAREER):

14 — Mark Messier, Edmonton, NY Rangers
11 — Wayne Gretzky, Edmonton, Los Angeles, St. Louis
10 — Jari Kurri, Edmonton, Los Angeles, NY Rangers
8 — Ed Westfall, Boston, NY Islanders
— Hakan Loob, Calgary

MOST SHORTHAND GOALS, ONE PLAYOFF YEAR:

3 — Derek Sanderson, Boston, 1969. 1 against Toronto in QF, won by Boston 4-0; 2 against Montreal in SF, won by Montreal, 4-2.
— Bill Barber, Philadelphia, 1980. All against Minnesota in SF, won by Philadelphia 4-1.
— Lorne Henning, NY Islanders, 1980. 1 against Boston in QF, won by NY Islanders 4-1; 1 against Buffalo in SF, won by NY Islanders 4-2, 1 against Philadelphia in F, won by NY Islanders 4-2.
— Wayne Gretzky, Edmonton, 1983. 2 against Winnipeg in DSF, won by Edmonton 3-0; 1 against Calgary in DF, won by Edmonton 4-1.
— Wayne Presley, Chicago, 1989. All against Detroit in DSF, won by Chicago 4-2.
— Todd Marchant, Edmonton, 1997. 1 against Dallas in CQF, won by Edmonton 4-3; 2 against Colorado in CSF, won by Colorado 4-1.

Derek Sanderson of the Boston Bruins was the first player in NHL history to score three shorthand goals in one playoff year back in 1969. It's a feat that has been matched just five times.

MOST SHORTHAND GOALS, ONE PLAYOFF SERIES:
3 — Bill Barber, Philadelphia, 1980, SF vs. Minnesota, won by Philadelphia 4-1.
— Wayne Presley, Chicago, 1989, DSF vs. Detroit, won by Chicago 4-2.
2 — Mac Colville, NY Rangers, 1940, SF vs. Boston, won by NY Rangers 4-2.
— Jerry Toppazzini, Boston, 1958, SF vs. NY Rangers, won by Boston 4-2.
— Dave Keon, Toronto, 1963, F vs. Detroit, won by Toronto 4-1.
— Bob Pulford, Toronto, 1964, F vs. Detroit, won by Toronto 4-3.
— Serge Savard, Montreal, 1968, F vs. St. Louis, won by Montreal 4-0.
— Derek Sanderson, Boston, 1969, SF vs. Montreal, won by Montreal 4-2.
— Bryan Trottier, NY Islanders, 1980, PR vs. Los Angeles, won by NY Islanders 3-1.
— Bobby Lalonde, Boston, 1981, PR vs. Minnesota, won by Minnesota 3-0.
— Butch Goring, NY Islanders, 1981, SF vs. NY Rangers, won by NY Islanders 4-0.
— Wayne Gretzky, Edmonton, 1983, DSF vs. Winnipeg, won by Edmonton 3-0.
— Mark Messier, Edmonton, 1983, DF vs. Calgary, won by Edmonton 4-1.
— Jari Kurri, Edmonton, 1983, CF vs. Chicago, won by Edmonton 4-0.
— Wayne Gretzky, Edmonton, 1985, DF vs. Winnipeg, won by Edmonton 4-0.
— Kevin Lowe, Edmonton, 1987, F vs. Philadelphia, won by Edmonton 4-3.
— Bob Gould, Washington, 1988, DSF vs. Philadelphia, won by Washington 4-3.
— Dave Poulin, Philadelphia, 1989, DF vs. Pittsburgh, won by Philadelphia 4-3.
— Russ Courtnall, Montreal, 1991, DF vs. Boston, won by Boston 4-3.
— Sergei Fedorov, Detroit, 1992, DSF vs. Minnesota, won by Detroit 4-3.
— Mark Messier, NY Rangers, 1992, DSF vs. New Jersey, won by NY Rangers 4-3.
— Tom Fitzgerald, NY Islanders, 1993, DF vs. Pittsburgh, won by NY Islanders 4-3.
— Mark Osborne, Toronto, 1994, CSF vs. San Jose, won by Toronto 4-3.
— Tony Amonte, Chicago, 1997, CQF vs. Colorado, won by Colorado 4-2.
— Brian Rolston, New Jersey, 1997, CQF vs. Montreal, won by New Jersey 4-1.
— Rod Brind'Amour, Philadelphia, 1997, CQF vs. Pittsburgh, won by Philadelphia 4-1.
— Todd Marchant, Edmonton, 1997, CSF vs. Colorado, won by Colorado 4-1.
— Jeremy Roenick, Phoenix, 1998, CQF vs. Detroit, won by Detroit 4-2.
— Vincent Damphousse, San Jose, 1999, CQF vs. Colorado, won by Colorado 4-2.
— Dixon Ward, Buffalo, 1999, CF vs. Toronto, won by Buffalo 4-1.
— Curtis Brown, Buffalo, 2001, CSF vs. Pittsburgh, won by Pittsburgh 4-3.

MOST SHORTHAND GOALS, ONE GAME:
2 — Dave Keon, Toronto, April 18, 1963, at Toronto, in 3-1 win vs. Detroit.
— Bryan Trottier, NY Islanders, April 8, 1980, at NY Islanders, in 8-1 win vs. Los Angeles.
— Bobby Lalonde, Boston, April 11, 1981, at Minnesota, in 6-3 loss vs. Minnesota.
— Wayne Gretzky, Edmonton, April 6, 1983, at Edmonton, in 6-3 win vs. Winnipeg.
— Jari Kurri, Edmonton, April 24, 1983, at Edmonton, in 8-3 win vs. Chicago.
— Mark Messier, NY Rangers, April 21, 1992, at NY Rangers, in 7-3 loss vs. New Jersey.
— Tom Fitzgerald, NY Islanders, May 8, 1993, at NY Islanders, in 6-5 win vs. Pittsburgh.
— Rod Brind'Amour, Philadelphia, April 26, 1997, at Philadelphia, in 6-3 win vs. Pittsburgh.
— Jeremy Roenick, Phoenix, April 24, 1998, at Detroit, in 7-4 win by Phoenix.
— Vincent Damphousse, San Jose, April 30, 1999, at Colorado, in 7-3 win by San Jose.

MOST SHORTHAND GOALS, ONE PERIOD:
2 — Bryan Trottier, NY Islanders, April 8, 1980, second period at NY Islanders in 8-1 win vs. Los Angeles.
— Bobby Lalonde, Boston, April 11, 1981, third period at Minnesota in 6-3 loss vs. Minnesota.
— Jari Kurri, Edmonton, April 24, 1983, third period at Edmonton in 8-4 win vs. Chicago.
— Rod Brind'Amour, Philadelphia, April 26, 1997, first period at Philadelphia in 6-3 win vs. Pittsburgh.
— Jeremy Roenick, Phoenix, April 24, 1998, second period at Detroit in 7-4 win by Phoenix.
— Vincent Damphousse, San Jose, April 30, 1999, third period at Colorado in 7-3 win by San Jose.

GAME-WINNING GOALS

MOST GAME-WINNING GOALS IN PLAYOFFS (CAREER):
24 — Wayne Gretzky, Edmonton, Los Angeles, St. Louis, NY Rangers
23 — Brett Hull, St. Louis, Dallas, Detroit
19 — Claude Lemieux, Montreal, New Jersey, Colorado
18 — Maurice Richard, Montreal
17 — Mike Bossy, NY Islanders
— Glenn Anderson, Edmonton, Toronto, NY Rangers, St. Louis

MOST GAME-WINNING GOALS, ONE PLAYOFF YEAR:
6 — Joe Sakic, Colorado, 1996. 22 games.
— Joe Nieuwendyk, Dallas, 1999. 23 games.
5 — Mike Bossy, NY Islanders, 1983. 19 games.
— Jari Kurri, Edmonton, 1987. 21 games.
— Bobby Smith, Minnesota, 1991. 23 games.
— Mario Lemieux, Pittsburgh, 1992. 15 games.

MOST GAME-WINNING GOALS, ONE PLAYOFF SERIES:
4 — Mike Bossy, NY Islanders, 1983, CF vs. Boston, won by NY Islanders 4-2.

OVERTIME GOALS

MOST OVERTIME GOALS IN PLAYOFFS (CAREER):
6 — Maurice Richard, Montreal (1 in 1946; 3 in 1951; 1 in 1957; 1 in 1958.)
5 — Glenn Anderson, Edmonton, Toronto, St. Louis
4 — Bob Nystrom, NY Islanders
— Dale Hunter, Quebec, Washington
— Wayne Gretzky, Edmonton, Los Angeles
— Stephane Richer, Montreal, New Jersey
— Joe Murphy, Edmonton, Chicago
— Esa Tikkanen, Edmonton, NY Rangers
— Jaromir Jagr, Pittsburgh
— Kirk Muller, Montreal, Dallas
— Joe Sakic, Colorado

MOST OVERTIME GOALS, ONE PLAYOFF YEAR:
3 — Mel Hill, Boston, 1939. All against NY Rangers in best-of-seven SF, won by Boston 4-3.
— Maurice Richard, Montreal, 1951. 2 against Detroit in best-of-seven SF, won by Montreal 4-2; 1 against Toronto best-of-seven F, won by Toronto 4-1.

MOST OVERTIME GOALS, ONE PLAYOFF SERIES:
3 — Mel Hill, Boston, 1939, SF vs. NY Rangers, won by Boston 4-3. Hill scored at 59:25 of overtime March 21 for a 2-1 win; at 8:24, March 23 for a 3-2 win; and at 48:00, April 2 for a 2-1 win.

Joe Sakic scored the game-winning goal in six of Colorado's 16 victories when they won their first Stanley Cup title in 1996. After winning the Cup again in 2001, the Avalanche pushed Detroit to seven games in the 2002 Western Conference Final.

SCORING BY A DEFENSEMAN

MOST GOALS BY A DEFENSEMAN, ONE PLAYOFF YEAR:
12 — Paul Coffey, Edmonton, 1985. 18 games.
11 — Brian Leetch, NY Rangers, 1994. 23 games.
9 — Bobby Orr, Boston, 1970. 14 games.
— Brad Park, Boston, 1978. 15 games.
8 — Denis Potvin, NY Islanders, 1981. 18 games.
— Raymond Bourque, Boston, 1983. 17 games.
— Denis Potvin, NY Islanders, 1983. 20 games.
— Paul Coffey, Edmonton, 1984. 19 games.

MOST GOALS BY A DEFENSEMAN, ONE GAME:
3 — Bobby Orr, Boston, April 11, 1971, at Montreal. Final score: Boston 5, Montreal 2.
— Dick Redmond, Chicago, April 4, 1973, at Chicago. Final score: Chicago 7, St. Louis 1.
— Denis Potvin, NY Islanders, April 17, 1981, at NY Islanders. Final score: NY Islanders 6, Edmonton 3.
— Paul Reinhart, Calgary, April 14, 1983, at Edmonton. Final score: Edmonton 6, Calgary 3.
— Doug Halward, Vancouver, April 7, 1984, at Vancouver. Final score: Vancouver 7, Calgary 0.
— Paul Reinhart, Calgary, April 8, 1984, at Vancouver. Final score: Calgary 5, Vancouver 1.
— Al Iafrate, Washington, April 26, 1993, at Washington. Final score: Washington 6, NY Islanders 4.
— Eric Desjardins, Montreal, June 3, 1993, at Montreal. Final score: Montreal 3, Los Angeles 2.
— Gary Suter, Chicago, April 24, 1994, at Chicago. Final score: Chicago 4, Toronto 3.
— Brian Leetch, NY Rangers, May 22, 1995, at Philadelphia. Final score: Philadelphia 4, NY Rangers 3.
— Andy Delmore, Philadelphia, May 7, 2000, at Philadelphia. Final score: Philadelphia 6, Pittsburgh 3.

MOST ASSISTS BY A DEFENSEMAN, ONE PLAYOFF YEAR:
25 — Paul Coffey, Edmonton, 1985. 18 games.
24 — Al MacInnis, Calgary, 1989. 22 games.
23 — Brian Leetch, NY Rangers, 1994. 23 games.
19 — Bobby Orr, Boston, 1972. 15 games.
18 — Raymond Bourque, Boston, 1988. 23 games.
— Raymond Bourque, Boston, 1991. 19 games.
— Larry Murphy, Pittsburgh, 1991. 23 games.

MOST ASSISTS BY A DEFENSEMAN, ONE GAME:
5 — Paul Coffey, Edmonton, May 14, 1985 at Edmonton vs. Chicago. Edmonton won 10-5.
— Risto Siltanen, Quebec, April 14, 1987 at Hartford. Quebec won 7-5.

MOST POINTS BY A DEFENSEMAN, ONE PLAYOFF YEAR:
37 — Paul Coffey, Edmonton, in 1985. 12 goals, 25 assists in 18 games.
34 — Brian Leetch, NY Rangers, in 1994. 11 goals, 23 assists in 23 games.
31 — Al MacInnis, Calgary, in 1989. 7 goals, 24 assists in 22 games.
25 — Denis Potvin, NY Islanders, in 1981. 8 goals, 17 assists in 18 games.
— Raymond Bourque, Boston, in 1991. 7 goals, 18 assists in 19 games.

MOST POINTS BY A DEFENSEMAN, ONE GAME:
6 — Paul Coffey, Edmonton, May 14, 1985 at Edmonton vs. Chicago. 1 goal, 5 assists. Edmonton won 10-5.
5 — Eddie Bush, Detroit, April 9, 1942, at Detroit vs. Toronto. 1 goal, 4 assists. Detroit won 5-2.
— Bob Dailey, Philadelphia, May 1, 1980, at Philadelphia vs. Minnesota. 1 goal, 4 assists. Philadelphia won 7-0.
— Denis Potvin, NY Islanders, April 17, 1981, at NY Islanders vs. Edmonton. 3 goals, 2 assists. NY Islanders won 6-3.
— Risto Siltanen, Quebec, April 14, 1987, at Hartford. 5 assists. Quebec won 7-5.

SCORING BY A ROOKIE

MOST GOALS BY A ROOKIE, ONE PLAYOFF YEAR:
14 — Dino Ciccarelli, Minnesota, 1981. 19 games.
11 — Jeremy Roenick, Chicago, 1990. 20 games.
10 — Claude Lemieux, Montreal, 1986. 20 games.
9 — Pat Flatley, NY Islanders, 1984. 21 games
8 — Steve Christoff, Minnesota, 1980. 14 games.
— Brad Palmer, Minnesota, 1981. 19 games.
— Mike Krushelnyski, Boston, 1983. 17 games.
— Bob Joyce, Boston, 1988. 23 games.

MOST POINTS BY A ROOKIE, ONE PLAYOFF YEAR:
21 — Dino Ciccarelli, Minnesota, in 1981. 14 goals, 7 assists in 19 games.
20 — Don Maloney, NY Rangers, in 1979. 7 goals, 13 assists in 18 games.

THREE-OR-MORE-GOAL GAMES

MOST THREE-OR-MORE-GOAL GAMES IN PLAYOFFS (CAREER):
10 — Wayne Gretzky, Edmonton, Los Angeles, NY Rangers. Eight three-goal games; two four-goal games.
7 — Maurice Richard, Montreal. Four three-goal games; two four-goal games; one five-goal game.
— Jari Kurri, Edmonton. Six three-goal games; one four-goal game.
6 — Dino Ciccarelli, Minnesota, Washington, Detroit. Five three-goal games; one four-goal game.
5 — Mike Bossy, NY Islanders. Four three-goal games; one four-goal game.

MOST THREE-OR-MORE-GOAL GAMES, ONE PLAYOFF YEAR:
4 — Jari Kurri, Edmonton, 1985. 1 four-goal game, 3 three-goal games.
3 — Mark Messier, Edmonton, 1983. 3 three-goal games.
— Mike Bossy, NY Islanders, 1983. 1 four-goal game, 2 three-goal games
2 — Newsy Lalonde, Montreal, 1919. 1 five-goal game, 1 four-goal game.
— Maurice Richard, Montreal, 1944. 1 five-goal game; 1 three-goal game.
— Doug Bentley, Chicago, 1944. 2 three-goal games.
— Norm Ullman, Detroit, 1964. 2 three-goal games.
— Phil Esposito, Boston, 1970. 2 three-goal games.
— Pit Martin, Chicago, 1973. 2 three-goal games.
— Rick MacLeish, Philadelphia, 1975. 2 three-goal games.
— Lanny McDonald, Toronto, 1977. 1 four-goal game; 1 three-goal game.
— Wayne Gretzky, Edmonton, 1981. 2 three-goal games.
— Wayne Gretzky, Edmonton, 1983. 2 four-goal games.
— Wayne Gretzky, Edmonton, 1985. 2 three-goal games.
— Petr Klima, Detroit, 1988. 2 three-goal games.
— Cam Neely, Boston, 1991. 2 three-goal games.
— Wayne Gretzky, NY Rangers, 1997. 2 three-goal games.
— Daniel Alfredsson, Ottawa, 1998. 2 three-goal games.

MOST THREE-OR-MORE-GOAL GAMES, ONE PLAYOFF SERIES:
3 — Jari Kurri, Edmonton, 1985, CF vs. Chicago, won by Edmonton 4-2. Kurri scored 3 G May 7 at Edmonton in 7-3 win, 3 G May 14 at Edmonton in 10-5 win and 4 G May 16 at Chicago in 8-2 win.
2 — Doug Bentley, Chicago, 1944, SF vs. Detroit, won by Chicago 4-1. Bentley scored 3 G Mar. 28 at Chicago in 7-1 win and 3 G Mar. 30 at Detroit in 5-2 win.
— Norm Ullman, Detroit, 1964, SF vs. Chicago, won by Detroit 4-3. Ullman scored 3 G Mar. 29 at Chicago in 5-4 win and 3 G April 7 at Detroit in 7-2 win.
— Mark Messier, Edmonton, 1983, DF vs. Calgary, won by Edmonton 4-1. Messier scored 4 G April 14 at Edmonton in 6-3 win and 3 G April 17 at Calgary in 10-2 win.
— Mike Bossy, NY Islanders, 1983, CF vs. Boston, won by NY Islanders 4-2. Bossy scored 3 G May 3 at NY Islanders in 8-3 win and 4 G May 7 at New York in 8-4 win.

SCORING STREAKS

LONGEST CONSECUTIVE GOAL-SCORING STREAK, ONE PLAYOFF YEAR:
10 Games — Reggie Leach, Philadelphia, 1976. Streak started April 17 at Toronto and ended May 9 at Montreal. He scored one goal in each of eight games; two in one game; and five in another; a total of 15 goals.

LONGEST CONSECUTIVE POINT-SCORING STREAK, ONE PLAYOFF YEAR:
18 games — Bryan Trottier, NY Islanders, 1981. 11 goals, 18 assists, 29 points.
17 games — Wayne Gretzky, Edmonton, 1988. 12 goals, 29 assists, 41 points.
— Al MacInnis, Calgary, 1989. 7 goals, 19 assists, 26 points.

LONGEST CONSECUTIVE POINT-SCORING STREAK, MORE THAN ONE PLAYOFF YEAR:
27 games — Bryan Trottier, NY Islanders, 1980, 1981 and 1982. 7 games in 1980 (3 G, 5 A, 8 PTS), 18 games in 1981 (11 G, 18 A, 29 PTS), and two games in 1982 (2 G, 3 A, 5 PTS). Total points, 42.
19 games — Wayne Gretzky, Edmonton, Los Angeles, 1988 and 1989. 17 games in 1988 (12 G, 29 A, 41 PTS with Edmonton), 2 games in 1989 (1 G, 2 A, 3 PTS with Los Angeles). Total points, 44.
— Al MacInnis, Calgary, 1989 and 1990. 17 games in 1989 (7 G, 19 A, 26 PTS), and two games in 1990 (2G, 1A, 3 PTS). Total points, 29.

FASTEST GOALS

FASTEST GOAL FROM START OF GAME:
6 Seconds — Don Kozak, Los Angeles, April 17, 1977, at Los Angeles vs. Boston and goaltender Gerry Cheevers. Los Angeles won 7-4.
7 Seconds — Bob Gainey, Montreal, May 5, 1977, at NY Islanders vs. goaltender Chico Resch. Montreal won 2-1.
— Terry Murray, Philadelphia, April 12, 1981, at Quebec vs. goaltender Dan Bouchard. Quebec won 4-3 in overtime.

FASTEST GOAL FROM START OF PERIOD (OTHER THAN FIRST):
6 Seconds — Pelle Eklund, Philadelphia, April 25, 1989, at Pittsburgh vs. goaltender Tom Barrasso, second period. Pittsburgh won 10-7.
9 Seconds — Bill Collins, Minnesota, April 9, 1968, at Minnesota vs. Los Angeles and goaltender Wayne Rutledge, third period. Minnesota won 7-5.
— Dave Balon, Minnesota, April 25, 1968, at St. Louis vs. goaltender Glenn Hall, third period. Minnesota won 5-1.
— Murray Oliver, Minnesota, April 8, 1971, at St. Louis vs. goaltender Ernie Wakely, third period. St. Louis won 4-2.
— Clark Gillies, NY Islanders, April 15, 1977, at Buffalo vs. goaltender Don Edwards, third period. NY Islanders won 4-3.
— Eric Vail, Atlanta, April 11, 1978, at Atlanta vs. Detroit and goaltender Ron Low, third period. Detroit won 5-3.
— Stan Smyl, Vancouver, April 10, 1979, at Philadelphia vs. goaltender Wayne Stephenson, third period. Vancouver won 3-2.
— Wayne Gretzky, Edmonton, April 6, 1983, at Edmonton vs. Winnipeg and goaltender Brian Hayward, second period. Edmonton won 6-3.
— Mark Messier, Edmonton, April 16, 1984, at Calgary vs. goaltender Don Edwards, third period. Edmonton won 5-3.
— Brian Skrudland, Montreal, May 18, 1986, at Calgary vs. goaltender Mike Vernon, first overtime period. Montreal won 3-2.

FASTEST TWO GOALS:
5 Seconds — Norm Ullman, Detroit, April 11, 1965, at Detroit vs. Chicago and goaltender Glenn Hall. Ullman scored at 17:35 and 17:40 of second period. Detroit won 4-2.

FASTEST TWO GOALS FROM START OF A GAME:
1 Minute, 8 Seconds — Dick Duff, Toronto, April 9, 1963, at Toronto vs. Detroit and goaltender Terry Sawchuk. Duff scored at 0:49 and 1:08. Final score: Toronto 4, Detroit 2.

FASTEST TWO GOALS FROM START OF A PERIOD:
35 Seconds — Pat LaFontaine, NY Islanders, May 19, 1984, at Edmonton vs. goaltender Andy Moog. LaFontaine scored at 0:13 and 0:35 of third period. Final score: Edmonton 5, NY Islanders 2.

PENALTIES

MOST PENALTY MINUTES IN PLAYOFFS (CAREER):
729 — Dale Hunter, Quebec, Washington, Colorado
541 — Chris Nilan, Montreal, NY Rangers, Boston
519 — Claude Lemieux, Montreal, New Jersey, Colorado, Phoenix
471 — Rick Tocchet, Philadelphia, Pittsburgh, Boston, Phoenix
466 — Willi Plett, Atlanta, Calgary, Minnesota, Boston

MOST PENALTIES, ONE GAME:
8 — Forbes Kennedy, Toronto, April 2, 1969, at Boston. Four minors, 2 majors, 1 10-minute misconduct, 1 game misconduct. Final score: Boston 10, Toronto 0.
— Kim Clackson, Pittsburgh, April 14, 1980, at Boston. Five minors, 2 majors, 1 10-minute misconduct. Final score: Boston 6, Pittsburgh 2

MOST PENALTY MINUTES, ONE GAME:
42 — Dave Schultz, Philadelphia, April 22, 1976, at Toronto. One minor, 2 majors, 1 10-minute misconduct and 2 game-misconducts. Final score: Toronto 8, Philadelphia 5.

MOST PENALTIES, ONE PERIOD AND MOST PENALTY MINUTES, ONE PERIOD:
6 Penalties; 39 Minutes — Ed Hospodar, NY Rangers, April 9, 1981, at Los Angeles, first period. Two minors, 1 major, 1 10-minute misconduct, 2 game misconducts. Final score: Los Angeles 5, NY Rangers 4.

GOALTENDING

MOST PLAYOFF GAMES APPEARED IN BY A GOALTENDER (CAREER):
240 — Patrick Roy, Montreal, Colorado
150 — Grant Fuhr, Edmonton, Buffalo, St. Louis
141 — Ed Belfour, Chicago, Dallas
138 — Mike Vernon, Calgary, Detroit, San Jose, Florida
132 — Billy Smith, NY Islanders
— Andy Moog, Edmonton, Boston, Dallas, Montreal

MOST MINUTES PLAYED BY A GOALTENDER (CAREER):
14,786 — Patrick Roy, Montreal, Colorado
8,834 — Grant Fuhr, Edmonton, Buffalo, St. Louis
8,639 — Ed Belfour, Chicago, Dallas
8,214 — Mike Vernon, Calgary, Detroit, San Jose, Florida
7,645 — Billy Smith, NY Islanders

MOST MINUTES PLAYED BY A GOALTENDER, ONE PLAYOFF YEAR:
1,544 — Kirk McLean, Vancouver, 1994. 24 games.
— Ed Belfour, Dallas, 1999. 23 games.
1,540 — Ron Hextall, Philadelphia, 1987. 26 games.
1,505 — Martin Brodeur, New Jersey, 2001. 25 games.

MOST SHUTOUTS IN PLAYOFFS (CAREER):
22 — Patrick Roy, Montreal, Colorado
15 — Clint Benedict, Ottawa, Mtl. Maroons
— Curtis Joseph, St. Louis, Edmonton, Toronto
14 — Jacques Plante, Montreal, St. Louis
13 — Turk Broda, Toronto
— Martin Brodeur, New Jersey
12 — Terry Sawchuk, Detroit, Los Angeles
— Dominik Hasek, Chicago, Buffalo, Detroit

MOST SHUTOUTS, ONE PLAYOFF YEAR:
6 — Dominik Hasek, Detroit, 2002. 23 games.
4 — Four shutouts by a goaltender in one playoff year has been recorded 14 times: Clint Benedict (1928, 1929); Dave Kerr (1937); Frank McCool (1945); Terry Sawchuk (1952); Bernie Parent (1975); Ken Dryden (1977); Mike Richter (1994); Kirk McLean (1994); Olaf Kolzig (1998); Ed Belfour (2000); Patrick Roy (2001); Martin Brodeur (2001); Patrick Lalime (2002).

MOST SHUTOUTS, ONE PLAYOFF SERIES:
3 — Clint Benedict, Mtl. Maroons, in 1926 F, 4 games vs. Victoria.
— Dave Kerr, NY Rangers, in 1940 SF, 6 games vs. Boston.
— Frank McCool, Toronto, in 1945 F, 7 games vs. Detroit.
— Turk Broda, Toronto, in 1950 SF, 7 games vs. Detroit.
— Felix Potvin, Toronto, in 1994 CQF, 6 games vs. Chicago.
— Martin Brodeur, New Jersey, in 1995 CQF, 5 games vs. Boston.
— Brent Johnson, St. Louis, in 2002 CQF, 5 games vs. Chicago.
— Patrick Lalime, Ottawa, in 2002 CQF, 5 games vs. Philadelphia.

MOST WINS BY A GOALTENDER, (CAREER):
148 — Patrick Roy, Montreal, Colorado
92 — Grant Fuhr, Edmonton, Buffalo, St. Louis
88 — Billy Smith, NY Islanders
80 — Ken Dryden, Montreal
79 — Ed Belfour, Chicago, Dallas

MOST WINS BY A GOALTENDER, ONE PLAYOFF YEAR:
16 — Sixteen wins by a goaltender in one playoff year has been recorded on 14 occasions. Dominik Hasek of the Detroit Red Wings is the most recent to equal this mark, posting a record of 16 wins and 7 losses in 23 games in 2002. It was first accomplished by Grant Fuhr in 1988.

MOST CONSECUTIVE WINS BY A GOALTENDER, MORE THAN ONE PLAYOFF YEAR:
14 — Tom Barrasso, Pittsburgh, 1992, 1993; 3 wins against NY Rangers in 1992 DF, won by Pittsburgh 4-2; 4 wins against Boston in 1992 CF, won by Pittsburgh 4-0; 4 wins against Chicago in 1992 F, won by Pittsburgh 4-0; 3 wins against New Jersey in 1993 DSF, won by Pittsburgh 4-1.

MOST CONSECUTIVE WINS BY A GOALTENDER, ONE PLAYOFF YEAR:
11 — Ed Belfour, Chicago, 1992. 3 wins against St. Louis in DSF, won by Chicago 4-2; 4 wins against Detroit in DF, won by Chicago 4-0; and 4 wins against Edmonton in CF, won by Chicago 4-0.
— Tom Barrasso, Pittsburgh, 1992. 3 wins against NY Rangers in DF, won by Pittsburgh 4-2; 4 wins against Boston in CF, won by Pittsburgh 4-0; and 4 wins against Chicago in F, won by Pittsburgh 4-0.
— Patrick Roy, Montreal, 1993. 4 wins against Quebec in DSF, won by Montreal 4-2; 4 wins against Buffalo in DF, won by Montreal 4-0; and 3 wins against NY Islanders in CF, won by Montreal 4-1.

LONGEST SHUTOUT SEQUENCE:
248 Minutes, 32 Seconds — Normie Smith, Detroit, 1936. In best-of-five SF, Smith shut out Mtl. Maroons twice, 1-0, March 24, in 116:30 overtime; shut out Maroons 3-0 in second game, March 26; and was scored against at 12:02 of first period, March 29, by Gus Marker. Detroit won SF 3-0.

MOST CONSECUTIVE SHUTOUTS:
3 — Clint Benedict, Mtl. Maroons, 1926. Benedict shut out Ottawa 1-0, Mar. 27; he then shut out Victoria twice, 3-0, Mar. 30; 3-0, Apr. 1. Mtl. Maroons won NHL F vs. Ottawa 2 goals to 1 and won the best-of-five F vs. Victoria 3-1.
— John Ross Roach, NY Rangers, 1929. Roach shut out NY Americans twice, 0-0, Mar. 19; 1-0, Mar. 21; he then shut out Toronto 1-0, Mar. 24. NY Rangers won QF vs. NY Americans 1 goal to 0 and won the best-of-three SF vs. Toronto 2-0.
— Frank McCool, Toronto, 1945. McCool shut out Detroit 1-0, April 6; 2-0, April 8; 1-0, April 12. Toronto won the best-of-seven F 4-3.
— Brent Johnson, St. Louis, 2002. Johnson shut out Chicago 2-0, April 20; 4-0, April 21; 1-0, April 23. St. Louis won the best-of-seven CQF 4-1.
— Patrick Lalime, Ottawa, 2002. Lalime shut out Philadelphia 3-0, April 20; 3-0, April 22; 3-0, April 24. Ottawa won the best-of-seven CQF 4-1.

Early Playoff Records

1893-1918

Team Records

MOST GOALS, BOTH TEAMS, ONE GAME:
25 — Ottawa Silver Seven, Dawson City at Ottawa, Jan. 16, 1905. Ottawa 23, Dawson City 2. Ottawa won best-of-three series 2-0.

MOST GOALS, ONE TEAM, ONE GAME:
23 — Ottawa Silver Seven at Ottawa, Jan. 16, 1905. Ottawa defeated Dawson City 23-2.

MOST GOALS, BOTH TEAMS, BEST-OF-THREE SERIES:
42 — Ottawa Silver Seven, Queen's University at Ottawa, 1906. Ottawa defeated Queen's 16-7, Feb. 27, and 12-7, Feb. 28.

MOST GOALS, ONE TEAM, BEST-OF-THREE SERIES:
32 — Ottawa Silver Seven in 1905 at Ottawa. Defeated Dawson City 9-2, Jan. 13, and 23-2, Jan. 16.

MOST GOALS, BOTH TEAMS, BEST-OF-FIVE SERIES:
39 — Toronto Arenas, Vancouver Millionaires at Toronto, 1918. Toronto won 5-3, Mar. 20; 6-3, Mar. 26; 2-1, Mar. 30. Vancouver won 6-4, Mar. 23, and 8-1, Mar. 28. Toronto scored 18 goals; Vancouver 21.

MOST GOALS, ONE TEAM, BEST-OF-FIVE SERIES:
26 — Vancouver Millionaires in 1915 at Vancouver. Defeated Ottawa Senators 6-2, Mar. 22; 8-3, Mar. 24; and 12-3, Mar. 26.

Individual Records

MOST GOALS IN PLAYOFFS:
63 — Frank McGee, Ottawa Silver Seven, in 22 playoff games. Seven goals in four games, 1903; 21 goals in eight games, 1904; 18 goals in four games, 1905; 17 goals in six games, 1906.

MOST GOALS, ONE PLAYOFF SERIES:
15 — Frank McGee, Ottawa Silver Seven, in two games in 1905 at Ottawa. Scored one goal, Jan. 13, in 9-2 victory over Dawson City and 14 goals, Jan. 16, in 23-2 victory.

MOST GOALS, ONE PLAYOFF GAME:
14 — Frank McGee, Ottawa Silver Seven, at Ottawa, Jan. 16, 1905, in 23-2 victory over Dawson City.

FASTEST THREE GOALS:
40 Seconds — Marty Walsh, Ottawa Senators, at Ottawa, March 16, 1911, at 3:00, 3:10, and 3:40 of third period. Ottawa defeated Port Arthur 13-4.

All-Time Playoff Goal Leaders since 1918

(40 or more goals)

Player	Teams	Yrs.	GP	G
Wayne Gretzky	Edm., L.A., St.L., NYR	16	208	122
* Mark Messier	Edm., NYR	17	236	109
Jari Kurri	Edm., L.A., NYR, Ana., Col.	15	200	106
* Brett Hull	Cgy., St.L., Dal., Det.	17	186	100
Glenn Anderson	Edm., Tor., NYR, St.L.	15	225	93
Mike Bossy	NYI	10	129	85
Maurice Richard	Mtl.	15	133	82
* Claude Lemieux	Mtl., N.J., Col., Phx.	16	226	80
Jean Beliveau	Mtl.	17	162	79
* Mario Lemieux	Pit.	8	107	76
Dino Ciccarelli	Min., Wsh., Det.	14	141	73
Esa Tikkanen	Edm., NYR, St.L., Van., Wsh.	13	186	72
Bryan Trottier	NYI, Pit.	17	221	71
Gordie Howe	Det., Hfd.	20	157	68
* Steve Yzerman	Det.	17	177	67
Denis Savard	Chi., Mtl.	16	169	66
* Joe Sakic	Que., Col.	9	135	65
* Jaromir Jagr	Pit.	11	140	65
Yvan Cournoyer	Mtl.	12	147	64
Brian Propp	Phi., Bos., Min.	13	160	64
Bobby Smith	Min., Mtl.	13	184	64
Bobby Hull	Chi., Hfd.	14	119	62
Phil Esposito	Chi., Bos., NYR	15	130	61
Jacques Lemaire	Mtl.	11	145	61
Joe Mullen	St.L., Cgy., Pit.	15	143	60
* Doug Gilmour	St.L., Cgy., Tor., N.J., Buf., Mtl.	17	182	60
Stan Mikita	Chi.	18	155	59
Paul Coffey	Edm., Pit., L.A., Det., Phi., Car.	16	194	59
Guy Lafleur	Mtl., NYR	14	128	58
Bernie Geoffrion	Mtl., NYR	16	132	58
Cam Neely	Van., Bos.	9	93	57
* Joe Nieuwendyk	Cgy., Dal., N.J.	14	132	57
* Luc Robitaille	L.A., Pit., NYR, Det.	14	155	57
Steve Larmer	Chi., NYR	13	140	56
Denis Potvin	NYI	14	185	56
Rick MacLeish	Phi., Pit., Det.	11	114	54
Bill Barber	Phi.	11	129	53
* Stephane Richer	Mtl., N.J., St.L.	12	131	53
* Rick Tocchet	Phi., Pit., Bos., Phx.	13	145	52
* Peter Forsberg	Que., Col.	8	115	51
Frank Mahovlich	Tor., Det., Mtl.	14	137	51
Brian Bellows	Min., Mtl., T.B., Ana., Wsh.	13	143	51
Steve Shutt	Mtl., L.A.	12	99	50
* Brendan Shanahan	N.J., St.L., Det.	13	135	50
* Steve Thomas	Tor., Chi., NYI, N.J.	14	147	50
* Sergei Fedorov	Det.	12	158	49
Henri Richard	Mtl.	18	180	49
Reggie Leach	Bos., Phi.	8	94	47
Ted Lindsay	Det., Chi.	16	133	47
Clark Gillies	NYI, Buf.	13	164	47
* Kevin Stevens	Pit.	7	103	46
Dickie Moore	Mtl., Tor., St.L.	14	135	46
* Ron Francis	Hfd., Pit., Car.	16	159	46
Rick Middleton	NYR, Bos.	12	114	45
* Mike Modano	Min., Dal.	10	127	45
* Jeremy Roenick	Chi., Phx., Phi.	13	105	44
Lanny McDonald	Tor., Cgy.	13	117	44
Ken Linseman	Phi., Edm., Bos.	11	113	43
Mike Gartner	Wsh., Min., NYR, Tor., Phx.	15	122	43
* Vyacheslav Kozlov	Det.	9	114	42
Bernie Nicholls	L.A., NYR, Edm., N.J., Chi., S.J.	13	118	42
Bobby Clarke	Phi.	13	136	42
Dale Hunter	Que., Wsh., Col.	18	186	42
John Bucyk	Det., Bos.	14	124	41
Raymond Bourque	Bos., Col.	21	214	41
Tim Kerr	Phi., NYR	10	81	40
Peter McNab	Buf., Bos., Van.	10	107	40
Bob Bourne	NYI, L.A.	13	139	40
John Tonelli	NYI, Cgy., L.A.	13	172	40

* Active

All-Time Playoff Assist Leaders since 1918

(60 or more assists)

Player	Teams	Yrs.	GP	A
Wayne Gretzky	Edm., L.A., St.L., NYR	16	208	260
* Mark Messier	Edm., NYR	17	236	186
Raymond Bourque	Bos., Col.	21	214	139
Paul Coffey	Edm., Pit., L.A., Det., Phi., Car.	16	194	137
* Doug Gilmour	St.L., Cgy., Tor., N.J., Buf., Mtl.	17	182	128
Jari Kurri	Edm., L.A., NYR, Ana., Col.	15	200	127
Glenn Anderson	Edm., Tor., NYR, St.L.	15	225	121
* Al MacInnis	Cgy., St.L.	18	174	120
Larry Robinson	Mtl., L.A.	20	227	116
Larry Murphy	L.A., Wsh., Min., Pit., Tor., Det.	20	215	115
Bryan Trottier	NYI, Pit.	17	221	113
* Sergei Fedorov	Det.	12	158	111
Denis Savard	Chi., Mtl.	16	169	109
* Steve Yzerman	Det.	17	177	108
Denis Potvin	NYI	14	185	108
* Chris Chelios	Mtl., Chi., Det.	18	210	106
* Adam Oates	Det., St.L., Bos., Wsh., Phi.	14	142	105
Jean Beliveau	Mtl.	17	162	97
* Mario Lemieux	Pit.	8	107	96
Bobby Smith	Min., Mtl.	13	184	96
* Ron Francis	Hfd., Pit., Car.	16	159	93
Gordie Howe	Det., Hfd.	20	157	92
Stan Mikita	Chi.	18	155	91
Brad Park	NYR, Bos., Det.	17	161	90
Craig Janney	Bos., St.L., S.J., Wpg., Phx.	11	120	86
* Scott Stevens	Wsh., St.L., N.J.	19	209	86
* Peter Forsberg	Que., Col.	8	115	84
Brian Propp	Phi., Bos., Min.	13	160	84
* Brett Hull	Cgy., St.L., Dal., Det.	17	186	84
* Joe Sakic	Que., Col.	9	135	83
* Jaromir Jagr	Pit.	11	140	82
Henri Richard	Mtl.	18	180	80
Jacques Lemaire	Mtl.	11	145	78
Ken Linseman	Phi., Edm., Bos.	11	113	77
Bobby Clarke	Phi.	13	136	77
* Claude Lemieux	Mtl., N.J., Col., Phx.	16	226	77
Guy Lafleur	Mtl., NYR	14	128	76
Phil Esposito	Chi., Bos., NYR	15	130	76
Dale Hunter	Que., Wsh., Col.	18	186	76
Mike Bossy	NYI	10	129	75
Steve Larmer	Chi., NYR	13	140	75
* Nicklas Lidstrom	Det.	11	152	75
John Tonelli	NYI, Cgy., L.A.	13	172	75
Peter Stastny	Que., N.J., St.L.	12	93	72
Bernie Nicholls	L.A., NYR, Edm., N.J., Chi., S.J.	13	118	72
Brian Bellows	Min., Mtl., T.B., Ana., Wsh.	13	143	71
Gilbert Perreault	Buf.	11	90	70
Geoff Courtnall	Bos., Edm., Wsh., St.L., Van.	15	156	70
Dale Hawerchuk	Wpg., Buf., Phi.	15	97	69
Alex Delvecchio	Det.	14	121	69
* Luc Robitaille	L.A., Pit., NYR, Det.	14	155	69
* Sergei Zubov	NYR, Pit., Dal.	8	125	68
Bobby Hull	Chi., Hfd.	14	119	67
Frank Mahovlich	Tor., Det., Mtl.	14	137	67
Bobby Orr	Bos.	8	74	66
Bernie Federko	St.L.	11	91	66
Jean Ratelle	NYR, Bos.	15	123	66
* Igor Larionov	Van., S.J., Det.	11	145	66
Charlie Huddy	Edm., L.A., Buf., St.L.	14	183	66
* Mike Modano	Min., Dal.	10	127	64
Dickie Moore	Mtl., Tor., St.L.	14	135	64
Doug Harvey	Mtl., NYR, St.L.	15	137	64
Neal Broten	Min., Dal., N.J.	13	135	63
Yvan Cournoyer	Mtl.	12	147	63
John Bucyk	Det., Bos.	14	124	62
* Brian Leetch	NYR	7	82	61
Doug Wilson	Chi.	12	95	61
* Sandis Ozolinsh	S.J., Col., Car.	8	113	61
* Steve Duchesne	L.A., Que., St.L., Ott., Phi., Det.	14	121	61
* Kevin Stevens	Pit.	7	103	60
Bernie Geoffrion	Mtl., NYR	16	132	60
* Rick Tocchet	Phi., Pit., Bos., Phx.	13	145	60
Esa Tikkanen	Edm., NYR, St.L., Van., Wsh.	13	186	60

All-Time Playoff Point Leaders since 1918

(100 or more points)

Player	Teams	Yrs.	GP	G	A	Pts.
Wayne Gretzky	Edm., L.A., St.L., NYR	16	208	122	260	382
* Mark Messier	Edm., NYR	17	236	109	186	295
Jari Kurri	Edm., L.A., NYR, Ana., Col.	15	200	106	127	233
Glenn Anderson	Edm., Tor., NYR, St.L.	15	225	93	121	214
Paul Coffey	Edm., Pit., L.A., Det., Phi., Car.	16	194	59	137	196
* Doug Gilmour	St.L., Cgy., Tor., N.J., Buf., Mtl.	17	182	60	128	188
* Brett Hull	Cgy., St.L., Dal., Det.	17	186	100	84	184
Bryan Trottier	NYI, Pit.	17	221	71	113	184
Raymond Bourque	Bos., Col.	21	214	41	139	180
Jean Beliveau	Mtl.	17	162	79	97	176
Denis Savard	Chi., Mtl.	16	169	66	109	175
* Steve Yzerman	Det.	17	177	67	108	175
* Mario Lemieux	Pit.	8	107	76	96	172
Denis Potvin	NYI	14	185	56	108	164
Mike Bossy	NYI	10	129	85	75	160
Gordie Howe	Det., Hfd.	20	157	68	92	160
* Sergei Fedorov	Det.	12	158	49	111	160
Bobby Smith	Min., Mtl.	13	184	64	96	160
* Al MacInnis	Cgy., St.L.	18	174	39	120	159
* Claude Lemieux	Mtl., N.J., Col., Phx.	16	226	80	77	157
Larry Murphy	L.A., Wsh., Min., Pit., Tor., Det.	20	215	37	115	152
Stan Mikita	Chi.	18	155	59	91	150
* Joe Sakic	Que., Col.	9	135	65	83	148
Brian Propp	Phi., Bos., Min.	13	160	64	84	148
* Jaromir Jagr	Pit.	11	140	65	82	147
Larry Robinson	Mtl., L.A.	20	227	28	116	144
* Adam Oates	Det., St.L., Bos., Wsh., Phi.	14	142	38	105	143
Jacques Lemaire	Mtl.	11	145	61	78	139
* Ron Francis	Hfd., Pit., Car.	16	159	46	93	139
Phil Esposito	Chi., Bos., NYR	15	130	61	76	137
* Chris Chelios	Mtl., Chi., Det.	18	210	30	106	136
* Peter Forsberg	Que., Col.	8	115	51	84	135
Guy Lafleur	Mtl., NYR	14	128	58	76	134
Esa Tikkanen	Edm., NYR, St.L., Van., Wsh.	13	186	72	60	132
Steve Larmer	Chi., NYR	13	140	56	75	131
Bobby Hull	Chi., Hfd.	14	119	62	67	129
Henri Richard	Mtl.	18	180	49	80	129
Yvan Cournoyer	Mtl.	12	147	64	63	127
Maurice Richard	Mtl.	15	133	82	44	126
* Luc Robitaille	L.A., Pit., NYR, Det.	14	155	57	69	126
Brad Park	NYR, Bos., Det.	17	161	35	90	125
Brian Bellows	Min., Mtl., T.B., Ana., Wsh.	13	143	51	71	122
Ken Linseman	Phi., Edm., Bos.	11	113	43	77	120
Bobby Clarke	Phi.	13	136	42	77	119
Bernie Geoffrion	Mtl., NYR	16	132	58	60	118
Frank Mahovlich	Tor., Det., Mtl.	14	137	51	67	118
Dino Ciccarelli	Min., Wsh., Det.	14	141	73	45	118
Dale Hunter	Que., Wsh., Col.	18	186	42	76	118
John Tonelli	NYI, Cgy., L.A.	13	172	40	75	115
Bernie Nicholls	L.A., NYR, Edm., N.J., Chi., S.J.	13	118	42	72	114
* Rick Tocchet	Phi., Pit., Bos., Phx.	13	145	52	60	112
Craig Janney	Bos., St.L., S.J., Wpg., Phx.	11	120	24	86	110
Dickie Moore	Mtl., Tor., St.L.	14	135	46	64	110
* Mike Modano	Min., Dal.	10	127	45	64	109
* Brendan Shanahan	N.J., St.L., Det.	13	135	50	59	109
Geoff Courtnall	Bos., Edm., Wsh., St.L., Van.	15	156	39	70	109
* Scott Stevens	Wsh., St.L., N.J.	19	209	23	86	109
Bill Barber	Phi.	11	129	53	55	108
Rick MacLeish	Phi., Pit., Det.	11	114	54	53	107
* Nicklas Lidstrom	Det.	11	152	32	75	107
* Kevin Stevens	Pit.	7	103	46	60	106
Joe Mullen	St.L., Cgy., Pit.	15	143	60	46	106
Peter Stastny	Que., N.J., St.L.	12	93	33	72	105
Alex Delvecchio	Det.	14	121	35	69	104
Gilbert Perreault	Buf.	11	90	33	70	103
John Bucyk	Det., Bos.	14	124	41	62	103
Bernie Federko	St.L.	11	91	35	66	101
* Joe Nieuwendyk	Cgy., Dal., N.J.	14	132	57	44	101
Rick Middleton	NYR, Bos.	12	114	45	55	100

Three-or-more-Goal Games, Playoffs 1918–2002

Player	Team	Date	City	Total Goals	Opposing Goaltender	Score	
Wayne Gretzky (10)	Edm.	Apr. 11/81	Edm.	3	Richard Sevigny	Edm. 6	Mtl. 2
		Apr. 19/81	Edm.	3	Billy Smith	Edm. 5	NYI 2
		Apr. 6/83	Edm.	4	Brian Hayward	Edm. 6	Wpg. 3
		Apr. 17/83	Cgy.	4	Reggie Lemelin	Edm. 10	Cgy. 2
		Apr. 25/85	Wpg.	3	Brian Hayward (2) Marc Behrend (1)	Edm. 8	Wpg. 3
		May 25/85	Edm.	3	Pelle Lindbergh	Edm. 4	Phi. 3
		Apr. 24/86	Cgy.	3	Mike Vernon	Edm. 7	Cgy. 4
	L.A.	May 29/93	Tor.	3	Felix Potvin	L.A. 5	Tor. 4
	NYR	Apr. 23/97	NYR	3	John Vanbiesbrouck	NYR 3	Fla. 2
		May 18/97	Phi.	3	Garth Snow	NYR 5	Phi. 4
Maurice Richard (7)	Mtl.	Mar. 23/44	Mtl.	5	Paul Bibeault	Mtl. 5	Tor. 1
		Apr. 7/44	Chi.	3	Mike Karakas	Mtl. 3	Chi. 1
		Mar. 29/45	Mtl.	4	Frank McCool	Mtl. 10	Tor. 3
		Apr. 14/53	Bos.	3	Gord Henry	Mtl. 7	Bos. 3
		Mar. 20/56	Mtl.	3	Gump Worsley	Mtl. 7	NYR 1
		Apr. 6/57	Mtl.	4	Don Simmons	Mtl. 5	Bos. 1
		Apr. 1/58	Det.	3	Terry Sawchuk	Mtl. 4	Det. 3
Jari Kurri (7)	Edm.	Apr. 4/84	Edm.	3	Doug Soetaert (1) Mike Veisor (2)	Edm. 9	Wpg. 2
		Apr. 25/85	Wpg.	3	Brian Hayward (2) Marc Behrend (1)	Edm. 8	Wpg. 3
		May 7/85	Edm.	3	Murray Bannerman	Edm. 7	Chi. 3
		May 14/85	Edm.	3	Murray Bannerman	Edm. 10	Chi. 5
		May 16/85	Chi.	4	Murray Bannerman	Edm. 8	Chi. 2
		Apr. 9/87	Edm.	4	Rollie Melanson (2) Darren Eliot (2)	Edm. 13	L.A. 3
		May 18/90	Bos.	3	Andy Moog (2) Reggie Lemelin (1)	Edm. 7	Bos. 2
Dino Ciccarelli (6)	Min.	May 5/81	Min.	3	Pat Riggin	Min. 7	Cgy. 4
		Apr. 10/82	Min.	3	Murray Bannerman	Min. 7	Chi. 1
	Wsh.	Apr. 5/90	N.J.	3	Sean Burke	Wsh. 5	N.J. 4
		Apr. 25/92	Pit.	4	Tom Barrasso (1) Ken Wregget (3)	Wsh. 7	Pit. 2
	Det.	Apr. 29/93	Tor.	3	Felix Potvin (2) Daren Puppa (1)	Det. 7	Tor. 3
		May 11/95	Dal.	3	Andy Moog (2) Darcy Wakaluk (1)	Det. 5	Dal. 1
Mike Bossy (5)	NYI	Apr. 16/79	NYI	3	Tony Esposito	NYI 6	Chi. 2
		May 8/82	NYI	3	Richard Brodeur	NYI 6	Van. 5
		Apr. 10/83	Wsh.	3	Al Jensen	NYI 6	Wsh. 3
		May 3/83	NYI	3	Pete Peeters	NYI 8	Bos. 3
		May 7/83	NYI	4	Pete Peeters	NYI 8	Bos. 4
Phil Esposito (4)	Bos.	Apr. 2/69	Bos.	4	Bruce Gamble	Bos. 10	Tor. 0
		Apr. 8/70	Bos.	3	Ed Giacomin	Bos. 8	NYR 2
		Apr. 19/70	Chi.	3	Tony Esposito	Bos. 6	Chi. 3
		Apr. 8/75	Bos.	3	Tony Esposito (2) Michel Dumas (1)	Bos. 8	Chi. 2
Mark Messier (4)	Edm.	Apr. 14/83	Edm.	4	Reggie Lemelin	Edm. 6	Cgy. 3
		Apr. 17/83	Cgy.	3	Reggie Lemelin (1) Don Edwards (2)	Edm. 10	Cgy. 2
		Apr. 26/83	Edm.	3	Murray Bannerman	Edm. 8	Chi. 2
	NYR	May 25/94	N.J.	3	Martin Brodeur (2) ENG (1)	NYR 4	N.J. 2
Steve Yzerman (4)	Det.	Apr. 6/89	Det.	3	Alain Chevrier	Chi. 5	Det. 4
		Apr. 4/91	St.L.	3	Vincent Riendeau (2) Pat Jablonski (1)	Det. 6	St.L. 3
		May 8/96	St.L.	3	Jon Casey	St.L. 5	Det. 4
		Apr. 21/99	Det.	3	Guy Hebert (2) Pat Jablonski (1)	Det. 5	Ana. 3
Bernie Geoffrion (3)	Mtl.	Mar. 27/52	Mtl.	3	Jim Henry	Mtl. 4	Bos. 0
		Apr. 7/55	Mtl.	3	Terry Sawchuk	Mtl. 4	Det. 2
		Mar. 30/57	Mtl.	3	Gump Worsley	Mtl. 8	NYR 3
Norm Ullman (3)	Det.	Mar. 29/64	Chi.	3	Glenn Hall	Det. 5	Chi. 4
		Apr. 7/64	Det.	3	Glenn Hall (2) Denis DeJordy (1)	Det. 7	Chi. 2
		Apr. 11/65	Det.	3	Glenn Hall	Det. 4	Chi. 2
John Bucyk (3)	Bos.	May 3/70	St.L.	3	Jacques Plante (1) Ernie Wakely (2)	Bos. 6	St.L. 1
		Apr. 20/72	Bos.	3	Jacques Caron (1) Ernie Wakely (2)	Bos. 10	St.L. 2
		Apr. 21/74	Bos.	3	Tony Esposito	Bos. 8	Chi. 6
Rick MacLeish (3)	Phi.	Apr. 11/74	Phi.	3	Phil Myre	Phi. 5	Atl. 1
		Apr. 13/75	Phi.	3	Gord McRae	Phi. 6	Tor. 3
		May 13/75	Phi.	3	Glenn Resch	Phi. 4	NYI 1
Denis Savard (3)	Chi.	Apr. 19/82	Chi.	3	Mike Liut	Chi. 7	StL. 4
		Apr. 10/86	Chi.	4	Ken Wregget	Tor. 6	Chi. 4
		Apr. 9/88	St.L.	3	Greg Millen	Chi. 6	St.L. 3
Tim Kerr (3)	Phi.	Apr. 13/85	NYR	4	Glen Hanlon	Phi. 6	NYR 5
		Apr. 20/87	Phi.	3	Kelly Hrudey	Phi. 4	NYI 2
		Apr. 19/89	Pit.	3	Tom Barrasso	Phi. 4	Pit. 2
Cam Neely (3)	Bos.	Apr. 9/87	Mtl.	3	Patrick Roy	Mtl. 4	Bos. 3
		Apr. 5/91	Bos.	3	Peter Sidorkiewicz	Bos. 4	Hfd. 3
		Apr. 25/91	Bos.	3	Patrick Roy	Bos. 4	Mtl. 1
Petr Klima (3)	Det.	Apr. 7/88	Tor.	3	Alan Bester (2) Ken Wregett (1)	Det. 6	Tor. 2
		Apr. 21/88	St.L.	3	Greg Millen	Det. 6	St.L. 0
	Edm.	May 4/91	Edm.	3	Jon Casey	Edm. 7	Min. 2
Esa Tikkanen (3)	Edm.	May 22/88	Edm.	3	Reggie Lemelin	Edm. 6	Bos. 3
		Apr. 16/91	Cgy.	3	Mike Vernon	Edm. 5	Cgy. 4
		Apr. 26/92	L.A.	3	Kelly Hrudey (2) Tom Askey (1)	Edm. 5	L.A. 2
Mike Gartner (3)	NYR	Apr. 13/90	NYR	3	Mark Fitzpatrick (2) Glenn Healy (1)	NYR 6	NYI 5
		Apr. 27/92	NYR	3	Chris Terreri	NYR 8	N.J. 5
	Tor.	Apr. 25/96	Tor.	3	Jon Casey	Tor. 5	St.L. 4
Mario Lemieux (3)	Pit.	Apr. 25/89	Pit.	5	Ron Hextall	Pit. 10	Phi. 7
		Apr. 23/92	Pit.	3	Don Beaupre	Pit. 6	Wsh. 4
		May 11/96	Pit.	3	Mike Richter	Pit. 7	NYR 3

Player	Team	Date	City	Total Goals	Opposing Goaltender	Score	
Newsy Lalonde (2)	Mtl.	Mar. 1/19	Mtl.	5	Clint Benedict	Mtl. 6	Ott. 3
		Mar. 22/19	Sea.	4	Hap Holmes	Mtl. 4	Sea. 2
Howie Morenz (2)	Mtl.	Mar. 22/24	Mtl.	3	Charles Reid	Mtl. 6	Cgy.T. 1
		Mar. 27/25	Mtl.	3	Hap Holmes	Mtl. 4	Vic. 2
Doug Bentley (2)	Chi.	Mar. 28/44	Chi.	3	Connie Dion	Chi. 7	Det. 1
		Mar. 30/44	Det.	3	Connie Dion	Chi. 5	Det. 2
Toe Blake (2)	Mtl.	Mar. 22/38	Mtl.	3	Mike Karakas	Mtl. 6	Chi. 4
		Mar. 26/46	Chi.	3	Mike Karakas	Mtl. 7	Chi. 2
Ted Kennedy (2)	Tor.	Apr. 14/45	Tor.	3	Harry Lumley	Det. 5	Tor. 3
		Mar. 27/48	Tor.	4	Frank Brimsek	Tor. 5	Bos. 3
F. St. Marseille (2)	St.L.	Apr. 28/70	St.L.	3	Al Smith	St.L. 5	Pit. 0
		Apr. 6/72	Min.	3	Cesare Maniago	Min. 6	St.L. 5
Bobby Hull (2)	Chi.	Apr. 7/63	Det.	3	Terry Sawchuk	Det. 7	Chi. 4
		Apr. 9/72	Pit.	3	Jim Rutherford	Chi. 6	Pit. 5
Pit Martin (2)	Chi.	Apr. 4/73	Chi.	3	Wayne Stephenson	Chi. 7	St.L. 1
		May 10/73	Chi.	3	Ken Dryden	Mtl. 6	Chi. 4
Yvan Cournoyer (2)	Mtl.	Apr. 5/73	Mtl.	3	Dave Dryden	Mtl. 7	Buf. 3
		Apr. 11/74	Mtl.	3	Ed Giacomin	Mtl. 4	NYR 1
Guy Lafleur (2)	Mtl.	May 1/75	Mtl.	3	Roger Crozier (1) Gerry Desjardins (2)	Mtl. 7	Buf. 0
		Apr. 11/77	Mtl.	3	Ed Staniowski	Mtl. 7	St.L. 2
Lanny McDonald (2)	Tor.	Apr. 9/77	Pit.	3	Denis Herron	Tor. 5	Pit. 2
		Apr. 17/77	Tor.	4	Wayne Stephenson	Phi. 6	Tor. 5
Bill Barber (2)	Phi.	May 4/80	Min.	4	Gilles Meloche	Phi. 5	Min. 3
		Apr. 9/81	Phi.	3	Dan Bouchard	Phi. 8	Que. 5
Bryan Trottier (2)	NYI	Apr. 8/80	NYI	3	Doug Keans	NYI 8	L.A. 1
		Apr. 9/81	NYI	3	Michel Larocque	NYI 5	Tor. 1
Butch Goring (2)	L.A.	Apr. 9/77	L.A.	3	Phil Myre	L.A. 4	Atl. 2
	NYI	May 17/81	Min.	3	Gilles Meloche	NYI 7	Min. 5
Paul Reinhart (2)	Cgy.	Apr. 14/83	Edm.	3	Andy Moog	Edm. 6	Cgy. 3
		Apr. 8/84	Van	3	Richard Brodeur	Cgy. 5	Van. 1
Brian Propp (2)	Phi.	Apr. 22/81	Phi.	3	Pat Riggin	Phi. 9	Cgy. 4
		Apr. 21/85	Phi.	3	Billy Smith	Phi. 5	NYI 2
Peter Stastny (2)	Que.	Apr. 5/83	Bos.	3	Pete Peeters	Bos. 4	Que. 3
		Apr. 11/87	Que.	3	Mike Liut (2) Steve Weeks (1)	Que. 5	Hfd. 1
Michel Goulet (2)	Que.	Apr. 23/85	Que.	3	Steve Penney	Que. 7	Mtl. 6
		Apr. 12/87	Que.	3	Mike Liut	Que. 4	Hfd. 1
Glenn Anderson (2)	Edm.	Apr. 26/83	Edm.	4	Murray Bannerman	Edm. 8	Chi. 2
		Apr. 6/88	Wpg.	3	Daniel Berthiaume	Edm. 7	Wpg. 4
Peter Zezel (2)	Phi.	Apr. 13/86	NYR	3	John Vanbiesbrouck	Phi. 7	NYR 1
	St.L.	Apr. 11/89	St.L.	3	Jon Casey (2) Kari Takko (1)	St.L. 6	Min. 1
Geoff Courtnall (2)	Van.	Apr. 4/91	L.A.	3	Kelly Hrudey	Van. 6	L.A. 5
		Apr. 30/92	Van.	3	Rick Tabaracci	Van. 5	Win. 0
Joe Sakic (2)	Que.	May 6/95	Que.	3	Mike Richter	Que. 5	NYR 4
	Col.	Apr. 25/96	Col.	3	Corey Hirsch	Col. 5	Van. 4
Daniel Alfredsson (2)	Ott.	Apr. 28/98	Ott.	3	Martin Brodeur	Ott. 4	N.J. 3
		May 11/98	Ott.	3	Olaf Kolzig	Ott. 4	Wsh. 3
Harry Meeking	Tor.	Mar. 11/18	Tor.	3	Georges Vezina	Tor. 7	Mtl. 3
Alf Skinner	Tor.	Mar. 23/18	Tor.	3	Hugh Lehman	Van.M. 6	Tor. 4
Joe Malone	Mtl.	Feb. 23/19	Mtl.	3	Clint Benedict	Mtl. 8	Ott. 4
Odie Cleghorn	Mtl.	Feb. 27/19	Ott.	3	Clint Benedict	Mtl. 5	Ott. 3
Jack Darragh	Ott.	Apr. 1/20	Tor.	3	Hap Holmes	Ott. 6	Sea. 1
George Boucher	Ott.	Mar. 10/21	Ott.	3	Jake Forbes	Ott. 5	Tor. 0
Babe Dye	Tor.	Mar. 28/22	Tor.	4	Hugh Lehman	Tor. 5	Van.M. 1
Percy Galbraith	Bos.	Mar. 31/27	Bos.	3	Hugh Lehman	Bos. 4	Chi. 4
Busher Jackson	Tor.	Apr. 5/32	NYR	3	John Ross Roach	Tor. 6	NYR 4
Frank Boucher	NYR	Apr. 9/32	Tor.	3	Lorne Chabot	Tor. 6	NYR 4
Charlie Conacher	Tor.	Mar. 26/36	Tor.	3	Tiny Thompson	Tor. 8	Bos. 3
Syd Howe	Det.	Mar. 23/39	Det.	3	Claude Bourque	Det. 7	Mtl. 3
Bryan Hextall	NYR	Apr. 3/40	NYR	3	Turk Broda	NYR 6	Tor. 2
Joe Benoit	Mtl.	Mar. 22/41	Mtl.	3	Sam LoPresti	Mtl. 4	Chi. 3
Syl Apps	Tor.	Mar. 25/41	Tor.	3	Frank Brimsek	Tor. 7	Bos. 2
Jack McGill	Bos.	Mar. 29/42	Bos.	3	Johnny Mowers	Det. 6	Bos. 4
Don Metz	Tor.	Apr. 14/42	Tor.	3	Johnny Mowers	Tor. 9	Det. 3
Mud Bruneteau	Det.	Apr. 1/43	Det.	3	Frank Brimsek	Det. 6	Bos. 2
Don Grosso	Det.	Apr. 7/43	Bos.	3	Frank Brimsek	Det. 4	Bos. 0
Carl Liscombe	Det.	Apr. 3/45	Det.	4	Paul Bibeault	Det. 5	Bos. 3
Billy Reay	Mtl.	Apr. 1/47	Bos.	4	Frank Brimsek	Mtl. 5	Bos. 1
Gerry Plamondon	Mtl.	Mar. 24/49	Det.	3	Harry Lumley	Mtl. 4	Det. 3
Sid Smith	Tor.	Apr. 10/49	Det.	3	Harry Lumley	Tor. 3	Det. 1
Pentti Lund	NYR	Apr. 2/50	NYR	3	Bill Durnan	NYR 4	Mtl. 1
Ted Lindsay	Det.	Apr. 5/55	Det.	4	Charlie Hodge (1) Jacques Plante (3)	Det. 7	Mtl. 1
Gordie Howe	Det.	Apr. 10/55	Det.	3	Jacques Plante	Det. 5	Mtl. 1
Phil Goyette	Mtl.	Mar. 25/58	Mtl.	3	Terry Sawchuk	Mtl. 8	Det. 1
Jerry Toppazzini	Bos.	Apr. 5/58	Bos.	3	Gump Worsley	Bos. 8	NYR 2
Bob Pulford	Tor.	Apr. 19/62	Tor.	3	Glenn Hall	Tor. 8	Chi. 4
Dave Keon	Tor.	Apr. 9/64	Mtl.	3	Charlie Hodge	Tor. 3	Mtl. 1
Henri Richard	Mtl.	Apr. 20/67	Mtl.	3	Terry Sawchuk (2) Johnny Bower (1)	Mtl. 6	Tor. 2
Rosaire Paiement	Phi.	Apr. 13/68	Phi.	3	Glenn Hall (1) Seth Martin (2)	Phi. 6	St.L. 1
Jean Beliveau	Mtl.	Apr. 20/68	Mtl.	3	Denis DeJordy	Mtl. 4	Chi. 1
Red Berenson	St.L.	Apr. 15/69	St.L.	3	Gerry Desjardins	St.L. 4	L.A. 0
Ken Schinkel	Pit.	Apr. 11/70	Oak.	3	Gary Smith	Pit. 5	Oak. 2
Jim Pappin	Chi.	Apr. 11/71	Phi.	3	Bruce Gamble	Chi. 6	Phi. 2
Bobby Orr	Bos.	Apr. 11/71	Mtl.	3	Ken Dryden	Bos. 5	Mtl. 2
Jacques Lemaire	Mtl.	Apr. 20/71	Mtl.	3	Gump Worsley	Mtl. 7	Min. 2
Vic Hadfield	NYR	Apr. 22/71	NYR	3	Tony Esposito	NYR 4	Chi. 1
Fred Stanfield	Bos.	Apr. 18/72	Bos.	3	Jacques Caron	Bos. 6	St.L. 1
Ken Hodge	Bos.	Apr. 30/72	Bos.	3	Ed Giacomin	Bos. 6	NYR 5
Dick Redmond	Chi.	Apr. 4/73	Chi.	3	Wayne Stephenson	Chi. 7	St.L. 1
Steve Vickers	NYR	Apr. 10/73	Bos.	3	Ross Brooks (2) Eddie Johnston (1)	NYR 6	Bos. 3
Tom Williams	L.A.	Apr. 14/74	L.A.	3	Mike Veisor	L.A. 5	Chi. 1
Marcel Dionne	L.A.	Apr. 15/76	L.A.	3	Gilles Gilbert	L.A. 6	Bos. 4
Don Saleski	Phi.	Apr. 20/76	Phi.	3	Wayne Thomas	Phi. 7	Tor. 1

Player	Team	Date	City	Total Goals	Opposing Goaltender	Score	
Darryl Sittler	Tor.	Apr. 22/76	Tor.	5	Bernie Parent	Tor. 8	Phi. 5
Reggie Leach	Phi.	May 6/76	Phi.	5	Gilles Gilbert	Phi. 6	Bos. 3
Jim Lorentz	Buf.	Apr. 7/77	Min.	3	Pete LoPresti (2) Gary Smith (1)	Buf. 7	Min. 1
Bobby Schmautz	Bos.	Apr. 11/77	Bos.	3	Rogie Vachon	Bos. 8	L.A. 3
Billy Harris	NYI	Apr. 23/77	Mtl.	3	Ken Dryden	Mtl. 4	NYI 3
George Ferguson	Tor.	Apr. 11/78	Tor.	3	Rogie Vachon	Tor. 7	L.A. 3
Jean Ratelle	Bos.	May 3/79	Bos.	3	Ken Dryden	Bos. 4	Mtl. 3
Stan Jonathan	Bos.	May 8/79	Bos.	3	Ken Dryden	Bos. 5	Mtl. 2
Ron Duguay	NYR	Apr. 20/80	NYR	3	Pete Peeters	NYR 4	Phi. 2
Steve Shutt	Mtl.	Apr. 22/80	Mtl.	3	Gilles Meloche	Mtl. 6	Min. 2
Gilbert Perreault	Buf.	May 6/80	NYI	3	Billy Smith (2) ENG (1)	Buf. 7	NYI 4
Paul Holmgren	Phi.	May 15/80	Phi.	3	Billy Smith	Phi. 8	NYI 3
Steve Payne	Min.	Apr. 8/81	Bos.	3	Rogie Vachon	Min. 5	Bos. 4
Denis Potvin	NYI	Apr. 17/81	NYI	3	Andy Moog	NYI 6	Edm. 3
Barry Pederson	Bos.	Apr. 8/82	Bos.	3	Don Edwards	Bos. 7	Buf. 3
Duane Sutter	NYI	Apr. 15/83	NYI	3	Glen Hanlon	NYI 5	NYR 0
Doug Halward	Van.	Apr. 7/84	Van.	3	Reggie Lemelin (2) Don Edwards (1)	Van. 7	Cgy. 0
Jorgen Pettersson	St.L.	Apr. 8/84	Det.	3	Eddie Mio	St.L. 3	Det. 2
Clark Gillies	NYI	May 12/84	NYI	3	Grant Fuhr	NYI 6	Edm. 1
Ken Linseman	Bos.	Apr. 14/85	Bos.	3	Steve Penney	Bos. 7	Mtl. 6
Dave Andreychuk	Buf.	Apr. 14/85	Buf.	3	Dan Bouchard	Buf. 7	Que. 4
Greg Paslawski	St.L.	Apr. 15/86	Min.	3	Don Beaupre	St.L. 6	Min. 3
Doug Risebrough	Cgy.	May 4/86	Cgy.	3	Rick Wamsley	Cgy. 8	St.L. 2
Mike McPhee	Mtl.	Apr. 11/87	Bos.	3	Doug Keans	Mtl. 5	Bos. 4
John Ogrodnick	Que.	Apr. 14/87	Hfd.	3	Mike Liut	Que. 7	Hfd. 5
Pelle Eklund	Phi.	May 10/87	Mtl.	3	Patrick Roy (1) Brian Hayward (2)	Phi. 6	Mtl. 3
John Tucker	Buf.	Apr. 9/88	Bos.	4	Andy Moog	Buf. 6	Bos. 2
Tony Hrkac	St.L.	Apr. 10/88	St.L.	4	Darren Pang	St.L. 6	Chi. 5
Hakan Loob	Cgy.	Apr. 10/88	Cgy.	3	Glenn Healy	Cgy. 7	L.A. 3
Ed Olczyk	Tor.	Apr. 12/88	Tor.	3	Greg Stefan (2) Glen Hanlon (1)	Tor. 6	Det. 5
Aaron Broten	N.J.	Apr. 20/88	N.J.	3	Pete Peeters	N.J. 5	Wsh. 2
Mark Johnson	N.J.	Apr. 22/88	Wsh.	4	Pete Peeters	N.J. 10	Wsh. 4
Patrik Sundstrom	N.J.	Apr. 22/88	Wsh.	3	Pete Peeters (2) Clint Malarchuk (1)	N.J. 10	Wsh. 4
Bob Brooke	Min.	Apr. 5/89	St.L.	3	Greg Millen	St.L. 4	Min. 3
Chris Kontos	L.A.	Apr. 6/89	L.A.	3	Grant Fuhr	L.A. 5	Edm. 2
Wayne Presley	Chi.	Apr. 13/89	Chi.	3	Greg Stefan (1) Glen Hanlon (2)	Chi. 7	Det. 1
Tony Granato	L.A.	Apr. 10/90	L.A.	3	Mike Vernon (1) Rick Wamsley (2)	L.A. 12	Cgy. 4
Tomas Sandstrom	L.A.	Apr. 10/90	L.A.	3	Mike Vernon (1) Rick Wamsley (2)	L.A. 12	Cgy. 4
Dave Taylor	L.A.	Apr. 10/90	L.A.	3	Mike Vernon (1) Rick Wamsley (2)	L.A. 12	Cgy. 4
Bernie Nicholls	NYR	Apr. 19/90	NYR	3	Mike Liut	NYR 7	Wsh. 3
John Druce	Wsh.	Apr. 21/90	NYR	3	John Vanbiesbrouck	Wsh. 6	NYR 3
Adam Oates	St.L.	Apr. 12/91	St.L.	3	Tim Chevaldae	St.L. 6	Det. 1
Luc Robitaille	L.A.	Apr. 26/91	L.A.	3	Grant Fuhr	L.A. 5	Edm. 2
Ray Sheppard	Det.	Apr. 24/92	Min.	3	Jon Casey	Min. 5	Det. 2
Pavel Bure	Van.	Apr. 28/92	Wpg.	3	Rick Tabaracci	Van. 8	Wpg. 3
Joe Murphy	Edm.	May 6/92	Edm.	3	Kirk McLean	Edm. 5	Van. 2
Ron Francis	Pit.	May 9/92	Pit.	3	Mike Richter (2) John V'brouck (1)	Pit. 5	NYR. 4
Kevin Stevens	Pit.	May 21/92	Bos.	4	Andy Moog	Pit. 5	Bos. 2
Dirk Graham	Chi.	June 1/92	Chi.	3	Tom Barrasso	Pit. 5	Chi. 2
Brian Noonan	Chi.	Apr. 18/93	Chi.	3	Curtis Joseph	St.L. 4	Chi. 3
Dale Hunter	Wsh.	Apr. 20/93	Wsh.	3	Glenn Healy	NYI 5	Wsh. 4
Teemu Selanne	Wpg.	Apr. 23/93	Wpg.	3	Kirk McLean	Wpg. 5	Van. 4
Ray Ferraro	NYI	Apr. 26/93	Wsh.	4	Don Beaupre	Wsh. 6	NYI 4
Al Iafrate	Wsh.	Apr. 26/93	Wsh.	3	Glenn Healy (2) Mark Fitzpatrick (1)	Wsh. 6	NYI 4
Paul Di Pietro	Mtl.	Apr. 28/93	Mtl.	3	Ron Hextall	Mtl. 6	Que. 2
Wendel Clark	Tor.	May 27/93	L.A.	3	Kelly Hrudey	L.A. 5	Tor. 4
Eric Desjardins	Mtl.	Jun. 3/93	Mtl.	3	Kelly Hrudey	Mtl. 3	L.A. 2
Tony Amonte	Chi.	Apr. 23/94	Chi.	4	Felix Potvin	Chi. 5	Tor. 4
Gary Suter	Chi.	Apr. 24/94	Chi.	3	Felix Potvin	Chi. 4	Tor. 3
Ulf Dahlen	S.J.	May 6/94	S.J.	3	Felix Potvin	S.J. 5	Tor. 2
Mike Sullivan	Cgy.	May 11/95	S.J.	3	Arturs Irbe (2) Wade Flaherty (1)	Cgy. 9	S.J. 2
Theoren Fleury	Cgy.	May 13/95	S.J.	4	Arturs Irbe (3) ENG (1)	Cgy. 6	S.J. 4
Brendan Shanahan	St.L.	May 13/95	Van.	3	Kirk McLean	St.L. 5	Van. 2
John LeClair	Phi.	May 21/95	Phi.	3	Mike Richter	Phi. 5	NYR 4
Brian Leetch	NYR	May 22/95	Phi.	3	Ron Hextall	Phi. 4	NYR 3
Trevor Linden	Van.	Apr. 25/96	Col.	3	Patrick Roy	Col. 5	Van. 4
Jaromir Jagr	Pit.	May 11/96	Pit.	3	Mike Richter	Pit. 7	NYR 3
Peter Forsberg	Col.	Jun. 6/96	Col.	3	John Vanbiesbrouck	Col. 8	Fla. 1
Valeri Zelepukin	N.J.	Apr. 22/97	Mtl.	3	Jocelyn Thibault	N.J. 6	Mtl. 4
Valeri Kamensky	Col.	Apr. 24/97	Col.	3	Jeff Hackett (2) Chris Terreri (1)	Col. 7	Chi. 0
Eric Lindros	Phi.	May 20/97	NYR	3	Mike Richter	Phi. 6	NYR 3
Matthew Barnaby	Buf.	May 10/98	Buf.	3	Andy Moog (2) ENG (1)	Buf. 6	Mtl. 3
Martin Straka	Pit.	Apr. 25/99	Pit.	3	Martin Brodeur	Pit. 4	N.J. 2
Martin Lapointe	Det.	Apr. 15/00	Det.	3	Stephane Fiset (2) Jamie Storr (1)	Det. 8	L.A. 5
Doug Weight	Edm.	Apr. 16/00	Edm.	3	Ed Belfour	Edm. 5	Dal. 2
Bill Guerin	Edm.	Apr. 18/00	Edm.	3	Ed Belfour	Dal. 4	Edm. 3
Scott Young	St.L.	Apr. 23/00	S.J.	3	Steve Shields	St.L. 6	S.J. 2
Andy Delmore	Phi.	May 7/00	Phi.	3	Ron Tugnutt (2) Peter Skudra (1)	Phi. 6	Pit. 3
Brett Hull	Det.	Apr. 27/02	Van.	3	Peter Skudra	Det. 6	Van. 4
Keith Tkachuk	St.L.	May 7/02	St.L.	3	Dominik Hasek	St.L. 6	Det. 1
Darren McCarty	Det.	May 18/02	Det.	3	Patrick Roy	Det. 5	Col. 3

Leading Playoff Scorers, 1918–2002

Season	Player and Club	Games Played	Goals	Assists	Points
2001-02	Peter Forsberg, Colorado	20	9	18	27
2000-01	Joe Sakic, Colorado	21	13	13	26
99-2000	Brett Hull, Dallas	23	11	13	24
1998-99	Peter Forsberg, Colorado	19	8	16	24
1997-98	Steve Yzerman, Detroit	22	6	18	24
1996-97	Eric Lindros, Philadelphia	19	12	14	26
1995-96	Joe Sakic, Colorado	22	18	16	34
1994-95	Sergei Fedorov, Detroit	17	7	17	24
1993-94	Brian Leetch, NY Rangers	23	11	23	34
1992-93	Wayne Gretzky, Los Angeles	24	15	25	40
1991-92	Mario Lemieux, Pittsburgh	15	16	18	34
1990-91	Mario Lemieux, Pittsburgh	23	16	28	44
1989-90	Craig Simpson, Edmonton	22	16	15	31
	Mark Messier, Edmonton	22	9	22	31
1988-89	Al MacInnis, Calgary	22	7	24	31
1987-88	Wayne Gretzky, Edmonton	19	12	31	43
1986-87	Wayne Gretzky, Edmonton	21	5	29	34
1985-86	Doug Gilmour, St. Louis	19	9	12	21
	Bernie Federko, St. Louis	19	7	14	21
1984-85	Wayne Gretzky, Edmonton	18	17	30	47
1983-84	Wayne Gretzky, Edmonton	19	13	22	35
1982-83	Wayne Gretzky, Edmonton	16	12	26	38
1981-82	Bryan Trottier, NY Islanders	19	6	23	29
1980-81	Mike Bossy, NY Islanders	18	17	18	35
1979-80	Bryan Trottier, NY Islanders	21	12	17	29
1978-79	Jacques Lemaire, Montreal	16	11	12	23
	Guy Lafleur, Montreal	16	10	13	23
1977-78	Guy Lafleur, Montreal	15	10	11	21
	Larry Robinson, Montreal	15	4	17	21
1976-77	Guy Lafleur, Montreal	14	9	17	26
1975-76	Reggie Leach, Philadelphia	16	19	5	24
1974-75	Rick MacLeish, Philadelphia	17	11	9	20
1973-74	Rick MacLeish, Philadelphia	17	13	9	22
1972-73	Yvan Cournoyer, Montreal	17	15	10	25
1971-72	Phil Esposito, Boston	15	9	15	24
	Bobby Orr, Boston	15	5	19	24
1970-71	Frank Mahovlich, Montreal	20	14	13	27
1969-70	Phil Esposito, Boston	14	13	14	27
1968-69	Phil Esposito, Boston	10	8	10	18
1967-68	Bill Goldsworthy, Minnesota	14	8	7	15
1966-67	Jim Pappin, Toronto	12	7	8	15
1965-66	Norm Ullman, Detroit	12	6	9	15
1964-65	Bobby Hull, Chicago	14	10	7	17
1963-64	Gordie Howe, Detroit	14	9	10	19
1962-63	Gordie Howe, Detroit	11	7	9	16
	Norm Ullman, Detroit	11	4	12	16
1961-62	Stan Mikita, Chicago	12	6	15	21
1960-61	Gordie Howe, Detroit	11	4	11	15
	Pierre Pilote, Chicago	12	3	12	15
1959-60	Henri Richard, Montreal	8	3	9	12
	Bernie Geoffrion, Montreal	8	2	10	12
1958-59	Dickie Moore, Montreal	11	5	12	17
1957-58	Fleming Mackell, Boston	12	5	14	19
1956-57	Bernie Geoffrion, Montreal	11	11	7	18
1955-56	Jean Béliveau, Montreal	10	12	7	19
1954-55	Gordie Howe, Detroit	11	9	11	20
1953-54	Dickie Moore, Montreal	11	5	8	13
1952-53	Ed Sandford, Boston	11	8	3	11
1951-52	Ted Lindsay, Detroit	8	5	2	7
	Floyd Curry, Montreal	11	4	3	7
	Metro Prystai, Detroit	8	2	5	7
	Gordie Howe, Detroit	8	2	5	7
1950-51	Maurice Richard, Montreal	11	9	4	13
	Max Bentley, Toronto	11	2	11	13
1949-50	Pentti Lund, NY Rangers	12	6	5	11
1948-49	Gordie Howe, Detroit	11	8	3	11
1947-48	Ted Kennedy, Toronto	9	8	6	14
1946-47	Maurice Richard, Montreal	10	6	5	11
1945-46	Elmer Lach, Montreal	9	5	12	17
1944-45	Joe Carveth, Detroit	14	5	6	11
1943-44	Toe Blake, Montreal	9	7	11	18
1942-43	Carl Liscombe, Detroit	10	6	8	14
1941-42	Don Grosso, Detroit	12	8	6	14
	Syl Apps, Toronto	13	5	9	14
1940-41	Milt Schmidt, Boston	11	5	6	11
1939-40	Phil Watson, NY Rangers	12	3	6	9
	Neil Colville, NY Rangers	12	2	7	9
1938-39	Bill Cowley, Boston	12	3	11	14
1937-38	Johnny Gottselig, Chicago	10	5	3	8
	Gordie Drillon, Toronto	7	7	1	8
1936-37	Marty Barry, Detroit	10	4	7	11
1935-36	Frank Boll, Toronto	9	7	3	10
1934-35	Baldy Northcott, Mtl. Maroons	7	4	1	5
	Busher Jackson, Toronto	7	3	2	5
	Cy Wentworth, Mtl. Maroons	7	3	2	5
	Charlie Conacher, Toronto	7	1	4	5
1933-34	Larry Aurie, Detroit	9	3	7	10
1932-33	Cecil Dillon, NY Rangers	8	8	2	10
1931-32	Frank Boucher, NY Rangers	7	3	6	9
1930-31	Cooney Weiland, Boston	5	6	3	9
1929-30	Marty Barry, Boston	6	3	3	6
	Cooney Weiland, Boston	6	1	5	6
1928-29	Andy Blair, Toronto	4	3	0	3
	Butch Keeling, NY Rangers	6	3	0	3
	Ace Bailey, Toronto	4	1	2	3
1927-28	Frank Boucher, NY Rangers	9	7	3	10
1926-27	Harry Oliver, Boston	8	4	2	6
	Percy Galbraith, Boston	8	3	3	6
1925-26	Nels Stewart, Mtl. Maroons	8	6	3	9
1924-25	Howie Morenz, Montreal	6	7	1	8
1923-24	Howie Morenz, Montreal	6	7	2	9
1922-23	Punch Broadbent, Ottawa	8	6	1	7
1921-22	Babe Dye, Toronto	7	11	1	12
1920-21	Cy Denneny, Ottawa	7	4	2	6
1919-20	Frank Nighbor, Ottawa	5	6	1	7
	Jack Darragh, Ottawa	5	5	2	7
1918-19	Newsy Lalonde, Montreal	10	17	2	19
1917-18	Alf Skinner, Toronto	7	8	3	11

Overtime Games since 1918

Abbreviations: Teams/Cities: — **Ana.** - Anaheim; **Atl.** - Atlanta; **Bos.** - Boston; **Buf.** - Buffalo; **Cgy.** - Calgary; **Cgy. T.** - Calgary Tigers (Western Canada Hockey League); **Chi.** - Chicago; **Col.** - Colorado; **Dal.** - Dallas; **Det.** - Detroit; **Edm.** - Edmonton; **Edm. E.** - Edmonton Eskimos (WCHL); **Fla.** - Florida; **Hfd.** - Hartford; **K.C.** - Kansas City; **L.A.** - Los Angeles; **Min.** - Minnesota; **Mtl.** - Montreal; **Mtl. M.** - Montreal Maroons; **N.J.** - New Jersey; **NYA** - NY Americans; **NYI** - New York Islanders; **NYR** - New York Rangers; **Oak.** - Oakland; **Ott.** - Ottawa; **Phi.** - Philadelphia; **Phx.** - Phoenix; **Pit.** - Pittsburgh; **Que.** - Quebec; **St.L.** - St. Louis; **Sea.** - Seattle Metropolitans (Pacific Coast Hockey Association); **S.J.** - San Jose; **T.B.** - Tampa Bay; **Tor.** - Toronto; **Van.** - Vancouver; **Van. M.** - Vancouver Millionaires (PCHA); **Vic.** - Victoria Cougars (WCHL); **Wpg.** - Winnipeg; **Wsh.** - Washington.

SERIES — **CF** - conference final; **CSF** - conference semi-final; **CQF** - conference quarter-final; **DF** - division final; **DSF** - division semi-final; **F** - final; **PRE** - preliminary round; **QF** - quarter final; **SF** - semi-final.

Date	City	Series	Score		Scorer	Overtime	Series Winner
Mar. 26/19	Sea.	F	Mtl. 0	Sea. 0	no scorer	20:00	....
Mar. 30/19	Sea.	F	Mtl. 4	Sea. 3	Odie Cleghorn	15:57	....
Mar. 20/22	Tor.	F	Tor. 2	Van. M. 1	Babe Dye	4:50	Tor.
Mar. 29/23	Van.	F	Ott. 2	Edm. E. 1	Cy Denneny	2:08	Ott.
Mar. 31/27	Mtl.	QF	Mtl. 1	Mtl. M. 0	Howie Morenz	12:05	Mtl.
Apr. 7/27	Bos.	F	Ott. 0	Bos. 0	no scorer	20:00	Ott.
Apr. 11/27	Ott.	F	Bos. 1	Ott. 1	no scorer	20:00	Ott.
Apr. 3/28	Mtl.	QF	Mtl. M. 1	Mtl. 0	Russell Oatman	8:20	Mtl. M.
Apr. 7/28	Mtl.	F	NYR 2	Mtl. M. 1	Frank Boucher	7:05	NYR
Mar. 21/29	NYR	QF	NYR 1	NYA 0	Butch Keeling	29:50	NYR
Mar. 26/29	Tor.	SF	NYR 2	Tor. 1	Frank Boucher	2:03	NYR
Mar. 20/30	Mtl.	SF	Bos. 2	Mtl. M. 1	Harry Oliver	45:35	Bos.
Mar. 25/30	Bos.	SF	Mtl. M. 1	Bos. 0	Archie Wilcox	26:27	Bos.
Mar. 26/30	Mtl.	QF	Chi. 2	Mtl. 2	Howie Morenz (Mtl.)	51:43	Mtl.
Mar. 28/30	Mtl.	SF	Mtl. 2	NYR 1	Gus Rivers	68:52	Mtl.
Mar. 24/31	Bos.	SF	Bos. 5	Mtl. 4	Cooney Weiland	18:56	Mtl.
Mar. 26/31	Chi.	QF	Chi. 2	Tor. 1	Stew Adams	19:20	Chi.
Mar. 28/31	Mtl.	SF	Mtl. 4	Bos. 3	Georges Mantha	5:10	Mtl.
Apr. 1/31	Mtl.	SF	Mtl. 3	Bos. 2	Wildor Larochelle	19:00	Mtl.
Apr. 5/31	Chi.	F	Chi. 2	Mtl. 1	Johnny Gottselig	24:50	Mtl.
Apr. 9/31	Mtl.	F	Chi. 3	Mtl. 2	Cy Wentworth	53:50	Mtl.
Mar. 26/32	Mtl.	SF	NYR 4	Mtl. 3	Fred Cook	59:32	NYR
Apr. 2/32	Tor.	SF	Tor. 3	Mtl. M. 2	Bob Gracie	17:59	Tor.
Mar. 25/33	Bos.	SF	Bos. 2	Tor. 1	Marty Barry	14:14	Tor.
Mar. 28/33	Bos.	SF	Tor. 1	Bos. 0	Busher Jackson	15:03	Tor.
Mar. 30/33	Tor.	SF	Bos. 2	Tor. 1	Eddie Shore	4:23	Tor.
Apr. 3/33	Tor.	SF	Tor. 1	Bos. 0	Ken Doraty	104:46	Tor.
Apr. 13/33	Tor.	F	NYR 1	Tor. 0	Bill Cook	7:33	NYR
Mar. 22/34	Tor.	SF	Det. 2	Tor. 1	Herbie Lewis	1:33	Det.
Mar. 25/34	Chi.	QF	Chi. 1	Mtl. 1	Mush March (Chi)	11:05	Chi.
Apr. 3/34	Det.	F	Chi. 2	Det. 1	Paul Thompson	21:10	Chi.
Apr. 10/34	Chi.	F	Chi. 1	Det. 0	Mush March	30:05	Chi.
Mar. 23/35	Bos.	SF	Bos. 1	Tor. 0	Dit Clapper	33:26	Tor.
Mar. 26/35	Chi.	QF	Mtl. M. 1	Chi. 0	Baldy Northcott	4:02	Mtl. M.
Mar. 30/35	Tor.	SF	Tor. 2	Bos. 1	Pep Kelly	1:36	Tor.
Apr. 4/35	Tor.	F	Mtl. M. 3	Tor. 2	Dave Trottier	5:28	Mtl. M.
Mar. 24/36	Mtl.	SF	Det. 1	Mtl. M. 0	Mud Bruneteau	116:30	Det.
Apr. 9/36	Tor.	F	Tor. 4	Det. 3	Buzz Boll	0:31	Det.
Mar. 25/37	NYR	QF	NYR 2	Tor. 1	Babe Pratt	13:05	NYR
Apr. 1/37	Mtl.	SF	Det. 2	Mtl. 1	Hec Kilrea	51:49	Det.
Mar. 22/38	NYR	QF	NYA 2	NYR 1	John Sorrell	21:25	NYA
Mar. 24/38	Tor.	SF	Tor. 1	Bos. 0	George Parsons	21:31	Tor.
Mar. 26/38	Mtl.	QF	Chi. 3	Mtl. 2	Paul Thompson	11:49	Chi.
Mar. 27/38	NYR	QF	NYA 3	NYR 2	Lorne Carr	60:40	NYA
Mar. 29/38	Bos.	SF	Tor. 3	Bos. 2	Gordie Drillon	10:04	Tor.
Mar. 31/38	Chi.	SF	Chi. 1	NYA 0	Cully Dahlstrom	33:01	Chi.
Mar. 21/39	NYR	SF	Bos. 2	NYR 1	Mel Hill	59:25	Bos.
Mar. 23/39	Bos.	SF	Bos. 3	NYR 2	Mel Hill	8:24	Bos.
Mar. 26/39	Det.	QF	Det. 1	Mtl. 0	Marty Barry	7:47	Det.
Mar. 30/39	Bos.	SF	NYR 2	Bos. 1	Clint Smith	17:19	Bos.
Apr. 1/39	Tor.	SF	Tor. 5	Det. 4	Gordie Drillon	5:42	Tor.
Apr. 2/39	Bos.	SF	Bos. 2	NYR 1	Mel Hill	48:00	Bos.
Apr. 9/39	Bos.	F	Tor. 3	Bos. 2	Doc Romnes	10:38	Bos.
Mar. 19/40	Det.	QF	Det. 2	NYA 1	Syd Howe	0:25	Det.
Mar. 19/40	Tor.	QF	Tor. 3	Chi. 2	Syl Apps	6:35	Tor.
Apr. 2/40	NYR	F	NYR 2	Tor. 1	Alf Pike	15:30	NYR
Apr. 11/40	Tor.	F	NYR 2	Tor. 1	Muzz Patrick	31:43	NYR
Apr. 13/40	Tor.	F	NYR 3	Tor. 2	Bryan Hextall	2:07	NYR
Mar. 20/41	Det.	QF	Det. 2	NYR 1	Gus Giesebrecht	12:01	Det.
Mar. 22/41	Mtl.	QF	Mtl. 4	Chi. 3	Charlie Sands	34:04	Chi.
Mar. 29/41	Bos.	SF	Tor. 2	Bos. 1	Pete Langelle	17:31	Bos.
Mar. 30/41	Chi.	SF	Det. 2	Chi. 1	Gus Giesebrecht	9:15	Det.
Mar. 22/42	Chi.	QF	Bos. 2	Chi. 1	Des Smith	6:51	Bos.
Mar. 21/43	Bos.	SF	Bos. 5	Mtl. 4	Don Gallinger	12:30	Bos.
Mar. 23/43	Det.	SF	Tor. 3	Det. 2	Jack McLean	70:18	Det.
Mar. 25/43	Mtl.	SF	Bos. 3	Mtl. 2	Busher Jackson	3:20	Bos.
Mar. 30/43	Tor.	SF	Det. 3	Tor. 2	Adam Brown	9:21	Det.
Mar. 30/43	Bos.	SF	Bos. 5	Mtl. 4	Ab DeMarco	3:41	Bos.
Apr. 13/44	Mtl.	F	Mtl. 5	Chi. 4	Toe Blake	9:12	Mtl.
Mar. 27/45	Tor.	SF	Tor. 4	Mtl. 3	Gus Bodnar	12:36	Tor.
Mar. 29/45	Det.	SF	Det. 3	Bos. 2	Mud Bruneteau	17:12	Det.
Apr. 21/45	Tor.	F	Det. 1	Tor. 0	Ed Bruneteau	14:16	Tor.
Mar. 28/46	Bos.	SF	Bos. 4	Det. 3	Don Gallinger	9:51	Bos.
Mar. 30/46	Mtl.	F	Mtl. 4	Bos. 3	Maurice Richard	9:08	Mtl.
Apr. 2/46	Mtl.	F	Mtl. 3	Bos. 2	Jimmy Peters	16:55	Mtl.
Apr. 7/46	Bos.	F	Bos. 3	Mtl. 2	Terry Reardon	15:13	Mtl.
Mar. 26/47	Tor.	SF	Tor. 3	Det. 2	Howie Meeker	3:05	Tor.
Mar. 27/47	Mtl.	SF	Mtl. 2	Bos. 1	Ken Mosdell	5:38	Mtl.
Apr. 3/47	Mtl.	SF	Mtl. 4	Bos. 3	John Quilty	36:40	Mtl.
Apr. 15/47	Tor.	F	Tor. 2	Mtl. 1	Syl Apps	16:36	Tor.
Mar. 24/48	Tor.	SF	Tor. 5	Bos. 4	Nick Metz	17:03	Tor.
Mar. 22/49	Det.	SF	Det. 2	Mtl. 1	Max McNab	44:52	Det.
Mar. 24/49	Det.	SF	Mtl. 4	Det. 3	Gerry Plamondon	2:59	Det.
Mar. 26/49	Tor.	SF	Bos. 5	Tor. 4	Woody Dumart	16:14	Tor.
Apr. 8/49	Det.	F	Tor. 3	Det. 2	Joe Klukay	17:31	Tor.
Apr. 4/50	Tor.	SF	Det. 2	Tor. 1	Leo Reise	20:38	Det.
Apr. 4/50	Mtl.	SF	Mtl. 3	NYR 2	Elmer Lach	15:19	NYR
Apr. 9/50	Det.	SF	Det. 1	Tor. 0	Leo Reise Jr.	8:39	Det.
Apr. 18/50	Det.	F	NYR 4	Det. 3	Don Raleigh	8:34	Det.
Apr. 20/50	Det.	F	NYR 2	Det. 1	Don Raleigh	1:38	Det.
Apr. 23/50	Det.	F	Det. 4	NYR 3	Pete Babando	28:31	Det.
Mar. 27/51	Det.	SF	Mtl. 3	Det. 2	Maurice Richard	61:09	Mtl.
Mar. 29/51	Det.	SF	Mtl. 1	Det. 0	Maurice Richard	42:20	Mtl.
Mar. 31/51	Tor.	SF	Bos. 1	Tor. 1	no scorer	20:00	Tor.
Apr. 11/51	Tor.	F	Tor. 3	Mtl. 2	Sid Smith	5:51	Tor.
Apr. 14/51	Tor.	F	Mtl. 3	Tor. 2	Maurice Richard	2:55	Tor.
Apr. 17/51	Mtl.	F	Tor. 2	Mtl. 1	Ted Kennedy	4:47	Tor.
Apr. 19/51	Mtl.	F	Tor. 3	Mtl. 2	Harry Watson	5:15	Tor.
Apr. 21/51	Tor.	F	Tor. 3	Mtl. 2	Bill Barilko	2:53	Tor.
Apr. 6/52	Bos.	SF	Mtl. 3	Bos. 2	Paul Masnick	27:49	Mtl.
Mar. 29/53	Bos.	SF	Bos. 2	Det. 1	Jack McIntyre	12:29	Bos.
Mar. 29/53	Chi.	SF	Chi. 2	Mtl. 1	Al Dewsbury	5:18	Mtl.
Apr. 16/53	Mtl.	F	Mtl. 1	Bos. 0	Elmer Lach	1:22	Mtl.
Apr. 1/54	Det.	SF	Det. 4	Tor. 3	Ted Lindsay	21:01	Det.
Apr. 11/54	Det.	F	Mtl. 1	Det. 0	Ken Mosdell	5:45	Det.
Apr. 16/54	Det.	F	Det. 2	Mtl. 1	Tony Leswick	4:29	Det.
Mar. 29/55	Bos.	SF	Mtl. 4	Bos. 3	Don Marshall	3:05	Mtl.
Mar. 24/56	Tor.	SF	Det. 5	Tor. 4	Ted Lindsay	4:22	Det.
Mar. 28/57	NYR	SF	NYR 4	Mtl. 3	Andy Hebenton	13:38	Mtl.
Apr. 4/57	Mtl.	SF	Mtl. 4	NYR 3	Maurice Richard	1:11	Mtl.
Mar. 27/58	NYR	SF	Bos. 4	NYR 3	Jerry Toppazzini	4:46	Bos.
Mar. 30/58	Det.	SF	Mtl. 2	Det. 1	André Pronovost	11:52	Mtl.
Apr. 17/58	Mtl.	F	Mtl. 3	Bos. 2	Maurice Richard	5:45	Mtl.
Mar. 28/59	Tor.	SF	Tor. 3	Bos. 2	Gerry Ehman	5:02	Tor.
Mar. 31/59	Tor.	SF	Tor. 3	Bos. 2	Frank Mahovlich	11:21	Tor.
Apr. 14/59	Tor.	F	Tor. 3	Mtl. 2	Dick Duff	10:06	Mtl.
Mar. 26/60	Mtl.	SF	Mtl. 4	Chi. 3	Doug Harvey	8:38	Mtl.
Mar. 27/60	Det.	SF	Tor. 5	Det. 4	Frank Mahovlich	43:00	Tor.
Mar. 29/60	Det.	SF	Det. 2	Tor. 1	Gerry Melnyk	1:54	Tor.
Mar. 22/61	Tor.	SF	Tor. 3	Det. 2	George Armstrong	24:51	Det.
Mar. 26/61	Chi.	SF	Chi. 2	Mtl. 1	Murray Balfour	52:12	Chi.
Apr. 5/62	Tor.	SF	Tor. 3	NYR 2	Red Kelly	24:23	Tor.
Apr. 2/64	Det.	SF	Chi. 3	Det. 2	Murray Balfour	8:21	Det.
Apr. 14/64	Tor.	F	Det. 4	Tor. 3	Larry Jeffrey	7:52	Tor.
Apr. 23/64	Det.	F	Tor. 4	Det. 3	Bob Baun	1:43	Tor.
Apr. 6/65	Tor.	SF	Tor. 3	Mtl. 2	Dave Keon	4:17	Mtl.
Apr. 13/65	Tor.	SF	Mtl. 4	Tor. 3	Claude Provost	16:33	Mtl.
May 5/66	Det.	F	Mtl. 3	Det. 2	Henri Richard	2:20	Mtl.
Apr. 13/67	NYR	SF	Mtl. 2	NYR 1	John Ferguson	6:28	Mtl.
Apr. 25/67	Tor.	F	Tor. 3	Mtl. 2	Bob Pulford	28:26	Tor.
Apr. 10/68	St.L.	QF	St.L. 3	Phi. 2	Larry Keenan	24:10	St.L.
Apr. 16/68	St.L.	QF	Phi. 2	St.L. 1	Don Blackburn	31:18	St.L.
Apr. 16/68	Min.	QF	Min. 4	L.A. 3	Milan Marcetta	9:11	Min.
Apr. 22/68	Min.	SF	Min. 3	St.L. 2	Parker MacDonald	3:41	St.L.
Apr. 27/68	St.L.	SF	St.L. 4	Min. 3	Gary Sabourin	1:32	St.L.
Apr. 28/68	Mtl.	SF	Mtl. 4	Chi. 3	Jacques Lemaire	2:14	Mtl.
Apr. 29/68	St.L.	SF	St.L. 3	Min. 2	Bill McCreary	17:27	St.L.
May 3/68	St.L.	SF	St.L. 2	Min. 1	Ron Schock	22:50	St.L.
May 5/68	St.L.	F	Mtl. 3	St.L. 2	Jacques Lemaire	1:41	Mtl.
May 9/68	Mtl.	F	Mtl. 4	St.L. 3	Bobby Rousseau	1:13	Mtl.
Apr. 2/69	Oak.	QF	L.A. 5	Oak. 4	Ted Irvine	0:19	L.A.
Apr. 10/69	Mtl.	SF	Mtl. 3	Bos. 2	Ralph Backstrom	0:42	Mtl.
Apr. 13/69	Mtl.	SF	Mtl. 4	Bos. 3	Mickey Redmond	4:55	Mtl.
Apr. 24/69	Bos.	SF	Mtl. 2	Bos. 1	Jean Béliveau	31:28	Mtl.
Apr. 12/70	Oak.	QF	Pit. 3	Oak. 2	Michel Briere	8:28	Pit.
May 10/70	Bos.	F	Bos. 4	St.L. 3	Bobby Orr	0:40	Bos.
Apr. 15/71	Tor.	QF	NYR 2	Tor. 1	Bob Nevin	9:07	NYR
Apr. 18/71	Chi.	SF	NYR 2	Chi. 1	Pete Stemkowski	1:37	Chi.
Apr. 27/71	Chi.	SF	Chi. 3	NYR 2	Bobby Hull	6:35	Chi.
Apr. 29/71	NYR	SF	NYR 3	Chi. 2	Pete Stemkowski	41:29	Chi.
May 4/71	Chi.	F	Chi. 2	Mtl. 1	Jim Pappin	21:11	Mtl.
Apr. 6/72	Bos.	QF	Tor. 4	Bos. 3	Jim Harrison	2:58	Bos.
Apr. 6/72	Min.	QF	Min. 6	St.L. 5	Bill Goldsworthy	1:36	St.L.
Apr. 9/72	Pit.	QF	Chi. 6	Pit. 5	Pit Martin	0:12	Chi.
Apr. 16/72	Min.	QF	St.L. 2	Min. 1	Kevin O'Shea	10:07	St.L.
Apr. 1/73	Mtl.	QF	Buf. 3	Mtl. 2	René Robert	9:18	Mtl.
Apr. 10/73	Phi.	QF	Phi. 3	Min. 2	Gary Dornhoefer	8:35	Phi.
Apr. 14/73	Mtl.	SF	Phi. 5	Mtl. 4	Rick MacLeish	2:56	Mtl.
Apr. 17/73	Mtl.	SF	Mtl. 4	Phi. 3	Larry Robinson	6:45	Mtl.
Apr. 14/74	Tor.	QF	Bos. 4	Tor. 3	Ken Hodge	1:27	Bos.
Apr. 14/74	Atl.	QF	Phi. 4	Atl. 3	Dave Schultz	5:40	Phi.
Apr. 16/74	Mtl.	QF	NYR 3	Mtl. 2	Ron Harris	4:07	NYR
Apr. 23/74	Chi.	SF	Chi. 4	Bos. 3	Jim Pappin	3:48	Bos.
Apr. 28/74	NYR	SF	NYR 2	Phi. 1	Rod Gilbert	4:20	Phi.
May 9/74	Bos.	F	Phi. 3	Bos. 2	Bobby Clarke	12:01	Phi.
Apr. 8/75	L.A.	PRE	L.A. 3	Tor. 2	Mike Murphy	8:53	Tor.
Apr. 10/75	Tor.	PRE	Tor. 3	L.A. 2	Blaine Stoughton	10:19	Tor.
Apr. 10/75	Chi.	PRE	Chi. 4	Bos. 3	Ivan Boldirev	7:33	Chi.
Apr. 11/75	NYR	PRE	NYI 4	NYR 3	Jean-Paul Parise	0:11	NYI
Apr. 17/75	Chi.	QF	Chi. 5	Buf. 4	Stan Mikita	2:31	Buf.
Apr. 19/75	Tor.	QF	Phi. 4	Tor. 3	André Dupont	1:45	Phi.
Apr. 22/75	Mtl.	QF	Mtl. 5	Van. 4	Guy Lafleur	17:06	Mtl.
Apr. 27/75	Buf.	SF	Buf. 6	Mtl. 5	Danny Gare	4:42	Buf.
May 1/75	Phi.	SF	Phi. 5	NYI 4	Bobby Clarke	2:56	Phi.
May 6/75	Buf.	SF	Buf. 5	Mtl. 4	René Robert	5:56	Buf.
May 7/75	NYI	SF	NYI 4	Phi. 3	Jude Drouin	1:53	Phi.
May 20/75	Buf.	F	Buf. 5	Phi. 4	René Robert	18:29	Phi.
Apr. 8/76	Buf.	PRE	Buf. 3	St.L. 2	Danny Gare	11:43	Buf.
Apr. 9/76	Buf.	PRE	Buf. 2	St.L. 1	Don Luce	14:27	Buf.
Apr. 13/76	Bos.	QF	L.A. 3	Bos. 2	Butch Goring	0:27	Bos.
Apr. 13/76	Buf.	QF	Buf. 3	NYI 2	Danny Gare	14:04	NYI
Apr. 22/76	L.A.	QF	L.A. 4	Bos. 3	Butch Goring	18:28	Bos.
Apr. 29/76	Phi.	SF	Phi. 2	Bos. 1	Reggie Leach	13:38	Phi.
Apr. 15/77	Tor.	QF	Phi. 4	Tor. 3	Rick MacLeish	2:55	Phi.
Apr. 17/77	Tor.	QF	Phi. 6	Tor. 5	Reggie Leach	19:10	Phi.
Apr. 24/77	Phi.	SF	Bos. 4	Phi. 3	Rick Middleton	2:57	Bos.
Apr. 26/77	Phi.	SF	Bos. 5	Phi. 4	Terry O'Reilly	30:07	Bos.
May 3/77	Mtl.	SF	NYI 4	Mtl. 3	Billy Harris	3:58	Mtl.

Date	City	Series	Score		Scorer	Overtime	Series Winner
May 14/77	Bos.	F	Mtl. 2	Bos. 1	Jacques Lemaire	4:32	Mtl.
Apr. 11/78	Phi.	PRE	Phi. 3	Col. 2	Mel Bridgman	0:23	Phi.
Apr. 13/78	NYR	PRE	NYR 4	Buf. 3	Don Murdoch	1:37	Buf.
Apr. 19/78	Bos.	QF	Bos. 4	Chi. 3	Terry O'Reilly	1:50	Bos.
Apr. 19/78	NYI	QF	NYI 3	Tor. 2	Mike Bossy	2:50	Tor.
Apr. 21/78	Chi.	QF	Bos. 4	Chi. 3	Peter McNab	10:17	Bos.
Apr. 25/78	NYI	QF	NYI 2	Tor. 1	Bob Nystrom	8:02	Tor.
Apr. 29/78	NYI	QF	Tor. 2	NYI 1	Lanny McDonald	4:13	Tor.
May 2/78	Bos.	SF	Bos. 3	Phi. 2	Rick Middleton	1:43	Bos.
May 16/78	Mtl.	F	Mtl. 3	Bos. 2	Guy Lafleur	13:09	Mtl.
May 21/78	Bos.	F	Bos. 4	Mtl. 3	Bobby Schmautz	6:22	Mtl.
Apr. 12/79	L.A.	PRE	NYR 2	L.A. 1	Phil Esposito	6:11	NYR
Apr. 14/79	Buf.	PRE	Pit. 4	Buf. 3	George Ferguson	0:47	Pit.
Apr. 16/79	Phi.	QF	Phi. 3	NYR 2	Ken Linseman	0:44	NYR
Apr. 18/79	NYI	QF	NYI 1	Chi. 0	Mike Bossy	2:31	NYI
Apr. 21/79	Tor.	QF	Mtl. 4	Tor. 3	Cam Connor	25:25	Mtl.
Apr. 22/79	Tor.	QF	Mtl. 5	Tor. 4	Larry Robinson	4:14	Mtl.
Apr. 28/79	NYI	SF	NYI 4	NYR 3	Denis Potvin	8:02	NYR
May 3/79	NYR	SF	NYI 3	NYR 2	Bob Nystrom	3:40	NYR
May 3/79	Bos.	SF	Bos. 4	Mtl. 3	Jean Ratelle	3:46	Mtl.
May 10/79	Mtl.	SF	Mtl. 5	Bos. 4	Yvon Lambert	9:33	Mtl.
May 19/79	NYR	F	Mtl. 4	NYR 3	Serge Savard	7:25	Mtl.
Apr. 8/80	NYR	PRE	NYR 2	Atl. 1	Steve Vickers	0:33	NYR
Apr. 8/80	Phi.	PRE	Phi. 4	Edm. 3	Bobby Clarke	8:06	Phi.
Apr. 8/80	Chi.	PRE	Chi. 3	St.L. 2	Doug Lecuyer	12:34	Chi.
Apr. 11/80	Hfd.	PRE	Mtl. 4	Hfd. 3	Yvon Lambert	0:29	Mtl.
Apr. 11/80	Tor.	PRE	Min. 4	Tor. 3	Al MacAdam	0:32	Min.
Apr. 11/80	L.A.	PRE	NYI 4	L.A. 3	Ken Morrow	6:55	NYI
Apr. 11/80	Edm.	PRE	Phi. 3	Edm. 2	Ken Linseman	23:56	Phi.
Apr. 16/80	Bos.	QF	NYI 2	Bos. 1	Clark Gillies	1:02	NYI
Apr. 17/80	Bos.	QF	NYI 5	Bos. 4	Bob Bourne	1:24	NYI
Apr. 21/80	NYI	QF	Bos. 4	NYI 3	Terry O'Reilly	17:13	NYI
May 1/80	Buf.	SF	NYI 2	Buf. 1	Bob Nystrom	21:20	NYI
May 13/80	Phi.	F	NYI 4	Phi. 3	Denis Potvin	4:07	NYI
May 24/80	NYI	F	NYI 5	Phi. 4	Bob Nystrom	7:11	NYI
Apr. 8/81	Buf.	PRE	Buf. 3	Van. 2	Alan Haworth	5:00	Buf.
Apr. 8/81	Bos.	PRE	Min. 5	Bos. 4	Steve Payne	3:34	Min.
Apr. 11/81	Chi.	PRE	Cgy. 5	Chi. 4	Willi Plett	35:17	Cgy.
Apr. 12/81	Que.	PRE	Que. 4	Phi. 3	Dale Hunter	0:37	Phi.
Apr. 14/81	St.L.	PRE	St.L. 4	Pit. 3	Mike Crombeen	25:16	St.L.
Apr. 16/81	Buf.	QF	Min. 4	Buf. 3	Steve Payne	0:22	Min.
Apr. 20/81	Min.	QF	Buf. 5	Min. 4	Craig Ramsay	16:32	Min.
Apr. 20/81	Edm.	QF	NYI 5	Edm. 4	Ken Morrow	5:41	NYI
Apr. 7/82	Min.	DSF	Chi. 3	Min. 2	Greg Fox	3:34	Chi.
Apr. 8/82	Edm.	DSF	Edm. 3	L.A. 2	Wayne Gretzky	6:20	L.A.
Apr. 8/82	Van.	DSF	Van. 2	Cgy. 1	Dave Williams	14:20	Van.
Apr. 10/82	Pit.	DSF	Pit. 2	NYI 1	Rick Kehoe	4:14	NYI
Apr. 10/82	L.A.	DSF	L.A. 6	Edm. 5	Daryl Evans	2:35	L.A.
Apr. 13/82	Mtl.	DSF	Que. 3	Mtl. 2	Dale Hunter	0:22	Que.
Apr. 13/82	NYI	DSF	NYI 4	Pit. 3	John Tonelli	6:19	NYI
Apr. 16/82	Van.	DF	L.A. 3	Van. 2	Steve Bozek	4:33	Van.
Apr. 18/82	Que.	DF	Que. 3	Bos. 2	Wilf Paiement	11:44	Que.
Apr. 18/82	NYR	DF	NYI 4	NYR 3	Bryan Trottier	3:00	NYI
Apr. 18/82	L.A.	DF	Van. 4	L.A. 3	Colin Campbell	1:23	Van.
Apr. 21/82	St.L.	DF	St.L. 3	Chi. 2	Bernie Federko	3:28	Chi.
Apr. 23/82	Que.	DF	Bos. 6	Que. 5	Peter McNab	10:54	Que.
Apr. 27/82	Chi.	CF	Van. 2	Chi. 1	Jim Nill	28:58	Van.
May 1/82	Que.	CF	NYI 5	Que. 4	Wayne Merrick	16:52	NYI
May 8/82	NYI	F	NYI 6	Van. 5	Mike Bossy	19:58	NYI
Apr. 5/83	Bos.	DSF	Bos. 4	Que. 3	Barry Pederson	1:46	Bos.
Apr. 6/83	Cgy.	DSF	Cgy. 4	Van. 3	Eddy Beers	12:27	Cgy.
Apr. 7/83	Min.	DSF	Min. 5	Tor. 4	Bobby Smith	5:03	Min.
Apr. 10/83	Tor.	DSF	Min. 5	Tor. 4	Dino Ciccarelli	8:05	Min.
Apr. 10/83	Van.	DSF	Cgy. 4	Van. 3	Greg Meredith	1:06	Cgy.
Apr. 18/83	Min.	DF	Chi. 4	Min. 3	Rich Preston	10:34	Chi.
Apr. 24/83	Bos.	DF	Bos. 3	Buf. 2	Brad Park	1:52	Bos.
Apr. 5/84	Edm.	DSF	Edm. 5	Wpg. 4	Randy Gregg	0:21	Edm.
Apr. 7/84	Det.	DSF	St.L. 4	Det. 3	Mark Reeds	37:07	St.L.
Apr. 8/84	Det.	DSF	St.L. 3	Det. 2	Jorgen Pettersson	2:42	St.L.
Apr. 10/84	NYI	DSF	NYI 3	NYR 2	Ken Morrow	8:56	NYI
Apr. 13/84	Min.	DF	St.L. 4	Min. 3	Doug Gilmour	16:16	Min.
Apr. 13/84	Edm.	DF	Cgy. 6	Edm. 5	Carey Wilson	3:42	Edm.
Apr. 13/84	NYI	DF	NYI 5	Wsh. 4	Anders Kallur	7:35	NYI
Apr. 16/84	Mtl.	DF	Que. 4	Mtl. 3	Bo Berglund	3:00	Mtl.
Apr. 20/84	Cgy.	DF	Cgy. 5	Edm. 4	Lanny McDonald	1:04	Edm.
Apr. 22/84	Min.	DF	Min. 4	St.L. 3	Steve Payne	6:00	Min.
Apr. 10/85	Phi.	DSF	Phi. 5	NYR 4	Mark Howe	8:01	Phi.
Apr. 10/85	Wsh.	DSF	Wsh. 4	NYI 3	Alan Haworth	2:28	NYI
Apr. 10/85	Edm.	DSF	Edm. 3	L.A. 2	Lee Fogolin	3:01	Edm.
Apr. 10/85	Wpg.	DSF	Wpg. 5	Cgy. 4	Brian Mullen	7:56	Wpg.
Apr. 11/85	Wsh.	DSF	Wsh. 2	NYI 1	Mike Gartner	21:23	NYI
Apr. 13/85	L.A.	DSF	Edm. 4	L.A. 3	Glenn Anderson	0:46	Edm.
Apr. 18/85	Mtl.	DF	Que. 2	Mtl. 1	Mark Kumpel	12:23	Que.
Apr. 23/85	Que.	DF	Que. 7	Mtl. 6	Dale Hunter	18:36	Que.
Apr. 25/85	Min.	DF	Chi. 7	Min. 6	Darryl Sutter	21:57	Chi.
Apr. 28/85	Chi.	DF	Min. 5	Chi. 4	Dennis Maruk	1:14	Chi.
Apr. 30/85	Min.	DF	Chi. 6	Min. 5	Darryl Sutter	15:41	Chi.
May 2/85	Mtl.	DF	Que. 3	Mtl. 2	Peter Stastny	2:22	Que.
May 5/85	Que.	CF	Que. 2	Phi. 1	Peter Stastny	6:20	Phi.
Apr. 9/86	Que.	DSF	Hfd. 3	Que. 2	Sylvain Turgeon	2:36	Hfd.
Apr. 12/86	Wpg.	DSF	Cgy. 4	Wpg. 3	Lanny McDonald	8:25	Cgy.
Apr. 17/86	Wsh.	DF	NYR 4	Wsh. 3	Brian MacLellan	1:16	NYR
Apr. 20/86	Edm.	DF	Edm. 6	Cgy. 5	Glenn Anderson	1:04	Cgy.
Apr. 23/86	Hfd.	DF	Hfd. 2	Mtl. 1	Kevin Dineen	1:07	Mtl.
Apr. 23/86	NYR	DF	NYR 6	Wsh. 5	Bob Brooke	2:40	NYR
Apr. 26/86	St L.	DF	St L. 4	Tor. 3	Mark Reeds	7:11	St L.
Apr. 29/86	Mtl.	DF	Mtl. 2	Hfd. 1	Claude Lemieux	5:55	Mtl.
May 5/86	NYR	CF	Mtl. 4	NYR 3	Claude Lemieux	9:41	Mtl.
May 12/86	St L.	CF	St L. 6	Cgy. 5	Doug Wickenheiser	7:30	Cgy.
May 18/86	Cgy.	F	Mtl. 3	Cgy. 2	Brian Skrudland	0:09	Mtl.
Apr. 8/87	Hfd.	DSF	Hfd. 3	Que. 2	Paul MacDermid	2:20	Que.
Apr. 9/87	Mtl.	DSF	Mtl. 4	Bos. 3	Mats Naslund	2:38	Mtl.

Date	City	Series	Score		Scorer	Overtime	Series Winner
Apr. 9/87	St.L.	DSF	Tor. 3	St.L. 2	Rick Lanz	10:17	Tor.
Apr. 11/87	Wpg.	DSF	Cgy. 3	Wpg. 2	Mike Bullard	3:53	Wpg.
Apr. 11/87	Chi.	DSF	Det. 4	Chi. 3	Shawn Burr	4:51	Det.
Apr. 16/87	Que.	DSF	Que. 5	Hfd. 4	Peter Stastny	6:05	Que.
Apr. 18/87	Wsh.	DSF	NYI 3	Wsh. 2	Pat LaFontaine	68:47	NYI
Apr. 21/87	Edm.	DF	Edm. 3	Wpg. 2	Glenn Anderson	0:36	Edm.
Apr. 26/87	Que.	DF	Mtl. 3	Que. 2	Mats Naslund	5:30	Mtl.
Apr. 27/87	Tor.	DF	Tor. 3	Det. 2	Mike Allison	9:31	Det.
May 4/87	Phi.	CF	Phi. 4	Mtl. 3	Ilkka Sinislao	9:11	Phi.
May 20/87	Edm.	F	Edm. 3	Phi. 2	Jari Kurri	6:50	Edm.
Apr. 6/88	NYI	DSF	NYI 4	N.J. 3	Pat LaFontaine	6:11	N.J.
Apr. 10/88	Phi.	DSF	Phi. 5	Wsh. 4	Murray Craven	1:18	Wsh.
Apr. 10/88	N.J.	DSF	NYI 5	N.J. 4	Brent Sutter	15:07	N.J.
Apr. 10/88	Buf.	DSF	Buf. 6	Bos. 5	John Tucker	5:32	Bos.
Apr. 12/88	Det.	DSF	Tor. 6	Det. 5	Ed Olczyk	0:34	Det.
Apr. 16/88	Wsh.	DSF	Wsh. 5	Phi. 4	Dale Hunter	5:57	Wsh.
Apr. 21/88	Cgy.	DF	Edm. 5	Cgy. 4	Wayne Gretzky	7:54	Edm.
May 4/88	Bos.	CF	N.J. 3	Bos. 2	Doug Brown	17:46	Bos.
May 9/88	Det.	CF	Edm. 4	Det. 3	Jari Kurri	11:02	Edm.
Apr. 5/89	St.L.	DSF	St.L. 4	Min. 3	Brett Hull	11:55	St.L.
Apr. 5/89	Cgy.	DSF	Van. 4	Cgy. 3	Paul Reinhart	2:47	Cgy.
Apr. 6/89	St.L.	DSF	St.L. 4	Min. 3	Rick Meagher	5:30	St.L.
Apr. 6/89	Det.	DSF	Chi. 5	Det. 4	Duane Sutter	14:36	Chi.
Apr. 8/89	Hfd.	DSF	Mtl. 5	Hfd. 4	Stephane Richer	5:01	Mtl.
Apr. 8/89	Phi.	DSF	Wsh. 4	Phi. 3	Kelly Miller	0:51	Phi.
Apr. 9/89	Hfd.	DSF	Mtl. 4	Hfd. 3	Russ Courtnall	15:12	Mtl.
Apr. 15/89	Cgy.	DSF	Cgy. 4	Van. 3	Joel Otto	19:21	Cgy.
Apr. 18/89	Cgy.	DF	Cgy. 4	L.A. 3	Doug Gilmour	7:47	Cgy.
Apr. 19/89	Mtl.	DF	Mtl. 3	Bos. 2	Bobby Smith	12:24	Mtl.
Apr. 20/89	St.L.	DF	St.L. 5	Chi. 4	Tony Hrkac	33:49	Chi.
Apr. 21/89	Phi.	DF	Pit. 4	Phi. 3	Phil Bourque	12:08	Phi.
May 8/89	Chi.	CF	Cgy. 2	Chi. 1	Al MacInnis	15:05	Cgy.
May 9/89	Mtl.	CF	Phi. 2	Mtl. 1	Dave Poulin	5:02	Mtl.
May 19/89	Mtl.	F	Mtl. 4	Cgy. 3	Ryan Walter	38:08	Cgy.
Apr. 5/90	N.J.	DSF	Wsh. 5	N.J. 4	Dino Ciccarelli	5:34	Wsh.
Apr. 6/90	Edm.	DSF	Edm. 3	Wpg. 2	Mark Lamb	4:21	Edm.
Apr. 8/90	Tor.	DSF	St.L. 6	Tor. 5	Sergio Momesso	6:04	St.L.
Apr. 8/90	L.A.	DSF	L.A. 2	Cgy. 1	Tony Granato	8:37	L.A.
Apr. 9/90	Mtl.	DSF	Mtl. 2	Buf. 1	Brian Skrudland	12:35	Mtl.
Apr. 9/90	NYI	DSF	NYI 4	NYR 3	Brent Sutter	20:59	NYR
Apr. 10/90	Wpg.	DSF	Wpg. 4	Edm. 3	Dave Ellett	21:08	Edm.
Apr. 14/90	L.A.	DSF	L.A. 4	Cgy. 3	Mike Krushelnyski	23:14	L.A.
Apr. 15/90	Hfd.	DSF	Hfd. 3	Bos. 2	Kevin Dineen	12:30	Bos.
Apr. 21/90	Bos.	DF	Bos. 5	Mtl. 4	Garry Galley	3:42	Bos.
Apr. 24/90	L.A.	DF	Edm. 6	L.A. 5	Joe Murphy	4:42	Edm.
Apr. 25/90	Wsh.	DF	Wsh. 4	NYR 3	Rod Langway	0:34	Wsh.
Apr. 27/90	NYR	DF	Wsh. 2	NYR 1	John Druce	6:48	Wsh.
May 15/90	Bos.	F	Edm. 3	Bos. 2	Petr Klima	55:13	Edm.
Apr. 4/91	Chi.	DSF	Min. 4	Chi. 3	Brian Propp	4:14	Min.
Apr. 5/91	Pit.	DSF	Pit. 5	N.J. 4	Jaromir Jagr	8:52	Pit.
Apr. 6/91	L.A.	DSF	L.A. 3	Van. 2	Wayne Gretzky	11:08	L.A.
Apr. 8/91	Van.	DSF	Van. 2	L.A. 1	Cliff Ronning	3:12	L.A.
Apr. 11/91	NYR	DSF	Wsh. 5	NYR 4	Dino Ciccarelli	6:44	Wsh.
Apr. 11/91	Mtl.	DSF	Mtl. 4	Buf. 3	Russ Courtnall	5:56	Mtl.
Apr. 14/91	Edm.	DSF	Cgy. 2	Edm. 1	Theoren Fleury	4:40	Edm.
Apr. 16/91	Cgy.	DSF	Edm. 5	Cgy. 4	Esa Tikkanen	6:58	Edm.
Apr. 18/91	L.A.	DF	L.A. 4	Edm. 3	Luc Robitaille	2:13	Edm.
Apr. 19/91	Bos.	DF	Mtl. 4	Bos. 3	Stephane Richer	0:27	Bos.
Apr. 19/91	Pit.	DF	Pit. 7	Wsh. 6	Kevin Stevens	8:10	Pit.
Apr. 20/91	L.A.	DF	Edm. 4	L.A. 3	Petr Klima	24:48	Edm.
Apr. 22/91	Edm.	DF	Edm. 4	L.A. 3	Esa Tikkanen	20:48	Edm.
Apr. 27/91	Mtl.	DF	Mtl. 3	Bos. 2	Shayne Corson	17:47	Bos.
Apr. 28/91	Edm.	DF	Edm. 4	L.A. 3	Craig MacTavish	16:57	Edm.
May 3/91	Bos.	CF	Bos. 5	Pit. 4	Vladimir Ruzicka	8:14	Pit.
Apr. 21/92	Bos.	DSF	Bos. 3	Buf. 2	Adam Oates	11:14	Bos.
Apr. 22/92	Min.	DSF	Det. 5	Min. 4	Yves Racine	1:15	Det.
Apr. 22/92	St.L.	DSF	St.L. 5	Chi. 4	Brett Hull	23:33	Chi.
Apr. 25/92	Buf.	DSF	Bos. 5	Buf. 4	Ted Donato	2:08	Bos.
Apr. 28/92	Min.	DSF	Det. 1	Min. 0	Sergei Fedorov	16:13	Det.
Apr. 29/92	Hfd.	DSF	Hfd. 2	Mtl. 1	Yvon Corriveau	0:24	Mtl.
May 1/92	Mtl.	DSF	Mtl. 3	Hfd. 2	Russ Courtnall	25:26	Mtl.
May 3/92	Van.	DF	Edm. 4	Van. 3	Joe Murphy	8:36	Edm.
May 5/92	Mtl.	DF	Bos. 3	Mtl. 2	Peter Douris	3:12	Bos.
May 7/92	Pit.	DF	NYR 6	Pit. 5	Kris King	1:29	Pit.
May 9/92	Pit.	DF	Pit. 5	NYR 4	Ron Francis	2:47	Pit.
May 17/92	Pit.	CF	Pit. 4	Bos. 3	Jaromir Jagr	9:44	Pit.
May 20/92	Edm.	CF	Chi. 4	Edm. 3	Jeremy Roenick	2:45	Chi.
Apr. 18/93	Bos.	DSF	Buf. 5	Bos. 4	Bob Sweeney	11:03	Buf.
Apr. 18/93	Que.	DSF	Que. 3	Mtl. 2	Scott Young	16:49	Mtl.
Apr. 20/93	Wsh.	DSF	NYI 5	Wsh. 4	Brian Mullen	34:50	NYI
Apr. 22/93	Mtl.	DSF	Mtl. 2	Que. 1	Vincent Damphousse	10:30	Mtl.
Apr. 22/93	Buf.	DSF	Buf. 4	Bos. 3	Yuri Khmylev	1:05	Buf.
Apr. 22/93	NYI	DSF	NYI 4	Wsh. 3	Ray Ferraro	4:46	NYI
Apr. 24/93	Buf.	DSF	Buf. 6	Bos. 5	Brad May	4:48	Buf.
Apr. 24/93	NYI	DSF	NYI 4	Wsh. 3	Ray Ferraro	25:40	NYI
Apr. 25/93	St.L.	DSF	St.L. 4	Chi. 3	Craig Janney	10:43	St.L.
Apr. 26/93	Que.	DSF	Mtl. 5	Que. 4	Kirk Muller	8:17	Mtl.
Apr. 27/93	Det.	DSF	Tor. 5	Det. 4	Mike Foligno	2:05	Tor.
Apr. 27/93	Van.	DSF	Wpg. 4	Van. 3	Teemu Selanne	6:18	Van.
Apr. 29/93	Wpg.	DSF	Van. 4	Wpg. 3	Greg Adams	4:30	Van.
May 1/93	Det.	DSF	Tor. 4	Det. 3	Nikolai Borschevsky	2:35	Tor.
May 3/93	Tor.	DF	Tor. 2	St.L. 1	Doug Gilmour	23:16	Tor.
May 4/93	Mtl.	DF	Mtl. 4	Buf. 3	Guy Carbonneau	2:50	Mtl.
May 5/93	Tor.	DF	St.L. 2	Tor. 1	Jeff Brown	23:03	Tor.
May 6/93	Buf.	DF	Mtl. 4	Buf. 3	Gilbert Dionne	8:28	Mtl.
May 8/93	Buf.	DF	Mtl. 4	Buf. 3	Kirk Muller	11:37	Mtl.
May 11/93	Van.	DF	L.A. 4	Van. 3	Gary Shuchuk	26:31	L.A.
May 14/93	Pit.	DF	NYI 4	Pit. 3	Dave Volek	5:16	NYI
May 18/93	Mtl.	CF	Mtl. 4	NYI 3	Stephan Lebeau	26:21	Mtl.
May 20/93	NYI	CF	Mtl. 2	NYI 1	Guy Carbonneau	12:34	Mtl.
May 25/93	Tor.	CF	Tor. 3	L.A. 2	Glenn Anderson	19:20	L.A.
May 27/93	L.A.	CF	L.A. 5	Tor. 4	Wayne Gretzky	1:41	L.A.

Date	City	Series	Score		Scorer	Overtime	Series Winner
Jun. 3/93	Mtl.	F	Mtl. 3	L.A. 2	Eric Desjardins	0:51	Mtl.
Jun. 5/93	L.A.	F	Mtl. 4	L.A. 3	John LeClair	0:34	Mtl.
Jun. 7/93	L.A.	F	Mtl. 3	L.A. 2	John LeClair	14:37	Mtl.
Apr. 20/94	Tor.	CQF	Tor. 1	Chi. 0	Todd Gill	2:15	Tor.
Apr. 22/94	St.L.	CQF	Dal. 5	St.L. 4	Paul Cavallini	8:34	Dal.
Apr. 24/94	Chi.	CQF	Chi. 4	Tor. 3	Jeremy Roenick	1:23	Tor.
Apr. 25/94	Bos.	CQF	Mtl. 2	Bos. 1	Kirk Muller	17:18	Bos.
Apr. 26/94	Cgy.	CQF	Van. 2	Cgy. 1	Geoff Courtnall	7:15	Van.
Apr. 27/94	Buf.	CQF	Buf. 1	N.J. 0	Dave Hannan	65:43	N.J.
Apr. 28/94	Van.	CQF	Van. 3	Cgy. 2	Trevor Linden	16:43	Van.
Apr. 30/94	Cgy.	CQF	Van. 4	Cgy. 3	Pavel Bure	22:20	Van.
May 3/94	N.J.	CSF	Bos. 6	N.J. 5	Don Sweeney	9:08	N.J.
May 7/94	Bos.	CSF	N.J. 5	Bos. 4	Stephane Richer	14:19	N.J.
May 8/94	Van.	CSF	Van. 2	Dal. 1	Sergio Momesso	11:01	Van.
May 12/94	Tor.	CSF	Tor. 3	S.J. 2	Mike Gartner	8:53	Tor.
May 15/94	NYR	CF	N.J. 4	NYR 3	Stephane Richer	35:23	NYR
May 16/94	Tor.	CF	Tor. 3	Van. 2	Peter Zezel	16:55	Van.
May 19/94	N.J.	CF	NYR 3	N.J. 2	Stephane Matteau	26:13	NYR
May 24/94	Van.	CF	Van. 4	Tor. 3	Greg Adams	20:14	Van.
May 27/94	NYR	CF	NYR 2	N.J. 1	Stephane Matteau	24:24	NYR
May 31/94	NYR	F	Van. 3	NYR 2	Greg Adams	19:26	NYR
May 7/95	Phi.	CQF	Phi. 4	Buf. 3	Karl Dykhuis	10:06	Phi.
May 9/95	Cgy.	CQF	S.J. 5	Cgy. 4	Ulf Dahlen	12:21	S.J.
May 12/95	NYR	CQF	NYR 3	Que. 2	Steve Larmer	8:09	NYR
May 12/95	N.J.	CQF	N.J. 1	Bos. 0	Randy McKay	8:51	N.J.
May 14/95	Pit.	CQF	Pit. 6	Wsh. 5	Luc Robitaille	4:30	Pit.
May 15/95	St.L.	CQF	Van. 6	St.L. 5	Cliff Ronning	1:48	Van.
May 17/95	Tor.	CQF	Tor. 5	Chi. 4	Randy Wood	10:00	Chi.
May 19/95	Cgy.	CQF	S.J. 5	Cgy. 4	Ray Whitney	21:54	S.J.
May 21/95	Phi.	CSF	Phi. 5	NYR 4	Eric Desjardins	7:03	Phi.
May 21/95	Chi.	CSF	Chi. 2	Van. 1	Joe Murphy	9:04	Chi.
May 22/95	Phi.	CSF	Phi. 4	NYR 3	Kevin Haller	0:25	Phi.
May 25/95	Van.	CSF	Chi. 3	Van. 2	Chris Chelios	6:22	Chi.
May 26/95	N.J.	CSF	N.J. 2	Pit. 1	Neal Broten	18:36	N.J.
May 27/95	Van.	CSF	Chi. 4	Van. 3	Chris Chelios	5:35	Chi.
Jun. 1/95	Det.	CF	Det. 2	Chi. 1	Nicklas Lidstrom	1:01	Det.
Jun. 6/95	Chi.	CF	Det. 4	Chi. 3	Vladimir Konstantinov	29:25	Det.
Jun. 7/95	N.J.	CF	Phi. 3	N.J. 2	Eric Lindros	4:19	N.J.
Jun. 11/95	Det.	CF	Det. 2	Chi. 1	Vyacheslav Kozlov	22:25	Det.
Apr. 16/96	NYR	CQF	Mtl. 3	NYR 2	Vincent Damphousse	5:04	NYR
Apr. 18/96	Tor.	CQF	Tor. 5	St.L. 4	Mats Sundin	4:02	St.L.
Apr. 18/96	Phi.	CQF	T.B. 2	Phi. 1	Brian Bellows	9:05	Phi.
Apr. 21/96	St.L.	CQF	St.L. 3	Tor. 2	Glenn Anderson	1:24	St.L.
Apr. 21/96	T.B.	CQF	T.B. 5	Phi. 4	Alexander Selivanov	2:04	Phi.
Apr. 23/96	Cgy.	CQF	Chi. 2	Cgy. 1	Joe Murphy	50:02	Chi.
Apr. 24/96	Wsh.	CQF	Pit. 3	Wsh. 2	Petr Nedved	79:15	Pit.
Apr. 25/96	Col.	CQF	Col. 5	Van. 4	Joe Sakic	0:51	Col.
Apr. 25/96	Tor.	CQF	Tor. 5	St.L. 4	Mike Gartner	7:31	St.L.
May 2/96	Col.	CSF	Chi. 3	Col. 2	Jeremy Roenick	6:29	Col.
May 6/96	Chi.	CSF	Chi. 4	Col. 3	Sergei Krivokrasov	0:46	Col.
May 8/96	St.L.	CSF	St.L. 5	Det. 4	Igor Kravchuk	3:23	Det.
May 8/96	Chi.	CSF	Col. 3	Chi. 2	Joe Sakic	44:33	Col.
May 9/96	Fla.	CSF	Fla. 4	Phi. 3	Dave Lowry	4:06	Fla.
May 12/96	Phi.	CSF	Fla. 2	Phi. 1	Mike Hough	28:05	Fla.
May 13/96	Chi.	CSF	Col. 4	Chi. 3	Sandis Ozolinsh	25:18	Col.
May 16/96	Det.	CSF	Det. 1	St.L. 0	Steve Yzerman	21:15	Det.
May 19/96	Det.	CF	Col. 3	Det. 2	Mike Keane	17:31	Col.
Jun. 10/96	Fla.	F	Col. 1	Fla. 0	Uwe Krupp	44:31	Col.
Apr. 20/97	Chi.	CQF	Chi. 4	Col. 3	Sergei Krivokrasov	31:03	Col.
Apr. 20/97	Edm.	CQF	Edm. 4	Dal. 3	Kelly Buchberger	9:15	Edm.
Apr. 22/97	NYR	CQF	NYR 4	Fla. 3	Esa Tikkanen	16:29	NYR
Apr. 23/97	Ott.	CQF	Ott. 1	Buf. 0	Daniel Alfredsson	2:34	Buf.
Apr. 24/97	Mtl.	CQF	Mtl. 4	N.J. 3	Patrice Brisebois	47:37	N.J.
Apr. 25/97	Fla.	CQF	NYR 3	Fla. 2	Esa Tikkanen	12:02	NYR
Apr. 25/97	Dal.	CQF	Edm. 1	Dal. 0	Ryan Smyth	20:22	Edm.
Apr. 27/97	Phx.	CQF	Ana. 3	Phx. 2	Paul Kariya	7:29	Ana.
Apr. 29/97	Buf.	CQF	Buf. 3	Ott. 2	Derek Plante	5:24	Buf.
Apr. 29/97	Dal.	CQF	Edm. 4	Dal. 3	Todd Marchant	12:26	Edm.
May 2/97	Det.	CSF	Det. 2	Ana. 1	Martin Lapointe	0:59	Det.
May 4/97	Det.	CSF	Det. 3	Ana. 2	Vyacheslav Kozlov	41:31	Det.
May 8/97	Ana.	CSF	Det. 3	Ana. 2	Brendan Shanahan	37:03	Det.
May 9/97	Phi.	CSF	Buf. 5	Phi. 4	Ed Ronan	6:24	Phi.
May 9/97	Edm.	CSF	Col. 3	Edm. 2	Claude Lemieux	8:35	Col.
May 11/97	N.J.	CSF	NYR 2	N.J. 1	Adam Graves	14:08	NYR
Apr. 22/98	N.J.	CQF	Ott. 2	N.J. 1	Bruce Gardiner	5:58	Ott.
Apr. 23/98	Pit.	CQF	Mtl. 3	Pit. 2	Benoit Brunet	18:43	Mtl.
Apr. 24/98	Wsh.	CQF	Bos. 4	Wsh. 3	Darren Van Impe	20:54	Wsh.
Apr. 26/98	Ott.	CQF	Ott. 2	N.J. 1	Alexei Yashin	2:47	Ott.
Apr. 26/98	Bos.	CQF	Wsh. 3	Bos. 2	Joe Juneau	26:31	Wsh.
Apr. 26/98	Edm.	CQF	Col. 5	Edm. 4	Joe Sakic	15:25	Edm.
Apr. 28/98	S.J.	CQF	S.J. 1	Dal. 0	Andrei Zyuzin	6:31	Dal.
May 1/98	Phi.	CQF	Buf. 3	Phi. 2	Michal Grosek	5:40	Buf.
May 2/98	S.J.	CQF	Dal. 3	S.J. 2	Mike Keane	3:43	Dal.
May 3/98	Bos.	CQF	Wsh. 3	Bos. 2	Brian Bellows	15:24	Wsh.
May 3/98	Buf.	CSF	Buf. 3	Mtl. 2	Geoff Sanderson	2:37	Buf.
May 11/98	Edm.	CSF	Dal. 1	Edm. 0	Benoit Hogue	13:07	Dal.
May 12/98	Mtl.	CSF	Buf. 5	Mtl. 4	Michael Peca	21:24	Buf.
May 12/98	St.L.	CSF	Det. 3	St.L. 2	Brendan Shanahan	31:12	Det.
May 25/98	Wsh.	CF	Wsh. 3	Buf. 2	Todd Krygier	3:01	Wsh.
May 28/98	Buf.	CF	Wsh. 4	Buf. 3	Peter Bondra	9:37	Wsh.
Jun. 3/98	Dal.	CF	Dal. 3	Det. 2	Jamie Langenbrunner	0:46	Det.
Jun. 4/98	Buf.	CF	Wsh. 3	Buf. 2	Joe Juneau	6:24	Wsh.
Jun. 11/98	Det.	F	Det. 5	Wsh. 4	Kris Draper	15:24	Det.
Apr. 23/99	Ott.	CQF	Buf. 3	Ott. 2	Miroslav Satan	30:35	Buf.
Apr. 24/99	Car.	CQF	Car. 3	Bos. 2	Ray Sheppard	17:05	Bos.
Apr. 24/99	Phx.	CQF	Phx. 4	St.L. 3	Shane Doan	8:58	St.L.
Apr. 26/99	S.J.	CQF	Col. 2	S.J. 1	Milan Hejduk	7:53	Col.
Apr. 27/99	Edm.	CQF	Dal. 3	Edm. 2	Joe Nieuwendyk	57:34	Dal.
Apr. 30/99	Tor.	CQF	Tor. 2	Phi. 1	Yanic Perreault	11:51	Tor.
Apr. 30/99	Car.	CQF	Bos. 4	Car. 3	Anson Carter	34:45	Bos.
Apr. 30/99	Phx.	CQF	St.L. 2	Phx. 1	Scott Young	5:43	St.L.
May 2/99	Pit.	CQF	Pit. 3	N.J. 2	Jaromir Jagr	8:59	Pit.
May 3/99	S.J.	CQF	Col. 3	S.J. 2	Milan Hejduk	13:12	Col.
May 4/99	Phx.	CQF	St.L. 1	Phx. 0	Pierre Turgeon	17:59	St.L.
May 7/99	Col.	CSF	Det. 3	Col. 2	Kirk Maltby	4:18	Col.
May 8/99	Dal.	CSF	Dal. 5	St.L. 4	Joe Nieuwendyk	8:22	Dal.
May 10/99	St.L.	CSF	St.L. 3	Dal. 2	Pavol Demitra	2:43	Dal.
May 12/99	St.L.	CSF	St.L. 3	Dal. 2	Pierre Turgeon	5:52	Dal.
May 13/99	Pit.	CSF	Tor. 3	Pit. 2	Sergei Berezin	2:18	Tor.
May 17/99	Pit.	CSF	Tor. 4	Pit. 3	Garry Valk	1:57	Tor.
May 17/99	St.L.	CSF	Dal. 2	St.L. 1	Mike Modano	2:21	Dal.
May 28/99	Col.	CF	Col. 3	Dal. 2	Chris Drury	19:29	Dal.
Jun. 8/99	Dal.	F	Buf. 3	Dal. 2	Jason Woolley	15:30	Dal.
Jun. 19/99	Buf.	F	Dal. 2	Buf. 1	Brett Hull	54:51	Dal.
Apr. 15/00	Pit.	CQF	Pit. 2	Wsh. 1	Jaromir Jagr	5:49	Pit.
Apr. 18/00	Buf.	CQF	Buf. 3	Phi. 2	Stu Barnes	4:42	Phi.
Apr. 22/00	Tor.	CQF	Tor. 2	Ott. 1	Steve Thomas	14:47	Tor.
May 2/00	Pit.	CSF	Phi. 4	Pit. 3	Andy Delmore	11:01	Phi.
May 3/00	Det.	CSF	Col. 3	Det. 2	Chris Drury	10:21	Col.
May 4/00	Pit.	CSF	Phi. 2	Pit. 1	Keith Primeau	92:01	Phi.
May 23/00	Dal.	CF	Dal. 3	Col. 2	Joe Nieuwendyk	12:10	Dal.
Jun. 8/00	N.J.	F	Dal. 1	N.J. 0	Mike Modano	46:21	N.J.
Jun. 10/00	Dal.	F	N.J. 2	Dal. 1	Jason Arnott	28:20	N.J.
Apr. 11/01	Dal.	CQF	Dal. 2	Edm. 1	Jamie Langenbrunner	2:08	Dal.
Apr. 13/01	Ott.	CQF	Tor. 1	Ott. 0	Mats Sundin	10:49	Tor.
Apr. 14/01	Phi.	CQF	Buf. 4	Phi. 3	Jay McKee	18:02	Buf.
Apr. 15/01	Edm.	CQF	Dal. 3	Edm. 2	Benoit Hogue	19:48	Dal.
Apr. 16/01	Tor.	CQF	Tor. 3	Ott. 2	Cory Cross	2:16	Tor.
Apr. 16/01	Van.	CQF	Col. 4	Van. 3	Peter Forsberg	2:50	Col.
Apr. 17/01	Buf.	CQF	Buf. 4	Phi. 3	Curtis Brown	6:13	Buf.
Apr. 17/01	Edm.	CQF	Edm. 2	Dal. 1	Mike Comrie	17:19	Dal.
Apr. 18/01	Car.	CQF	Car. 3	N.J. 2	Rod Brind'Amour	:46	N.J.
Apr. 18/01	Pit.	CQF	Wsh. 4	Pit. 3	Jeff Halpern	4:01	Pit.
Apr. 18/01	L.A.	CQF	L.A. 4	Det. 3	Eric Belanger	2:36	L.A.
Apr. 19/01	Dal.	CQF	Dal. 4	Edm. 3	Kirk Muller	8:01	Dal.
Apr. 19/01	St.L.	CQF	St.L. 3	S.J. 2	Bryce Salvador	9:54	St.L.
Apr. 23/01	Pit.	CQF	Pit. 4	Wsh. 3	Martin Straka	13:04	Pit.
Apr. 23/01	L.A.	CQF	L.A. 3	Det. 2	Adam Deadmarsh	4:48	L.A.
Apr. 26/01	Col.	CSF	L.A. 4	Col. 3	Jaroslav Modry	14:23	Col.
Apr. 28/01	N.J.	CSF	N.J. 6	Tor. 5	Randy McKay	5:31	N.J.
May 1/01	Tor.	CSF	N.J. 3	Tor. 2	Brian Rafalski	7:00	N.J.
May 1/01	St.L.	CSF	St.L. 3	Dal. 2	Cory Stillman	29:26	St.L.
May 5/01	Buf.	CSF	Buf. 3	Pit. 2	Stu Barnes	8:34	Pit.
May 6/01	L.A.	CSF	L.A. 1	Col. 0	Glen Murray	22:41	Col.
May 8/01	Pit.	CSF	Pit. 3	Buf. 2	Martin Straka	11:29	Pit.
May 10/01	Buf.	CSF	Pit. 3	Buf. 2	Darius Kasparaitis	13:01	Pit.
May 16/01	St.L.	CF	St.L. 4	Col. 3	Scott Young	30:27	Col.
May 18/01	St.L.	CF	Col. 4	St.L. 3	Stephane Yelle	4:23	Col.
May 21/01	Col.	CF	Col. 2	St.L. 1	Joe Sakic	:24	Col.
Apr. 17/02	Phi.	CQF	Phi. 1	Ott. 0	Ruslan Fedotenko	7:47	Ott.
Apr. 17/02	Det.	CQF	Van. 4	Det. 3	Henrik Sedin	13:59	Det.
Apr. 19/02	Car.	CQF	Car. 2	N.J. 1	Bates Battaglia	15:26	Car.
Apr. 24/02	Car.	CQF	Car. 3	N.J. 2	Josef Vasicek	8:16	Car.
Apr. 25/02	Col.	CQF	L.A. 1	Col. 0	Craig Johnson	2:19	Col.
Apr. 26/02	Phi.	CQF	Ott. 2	Phi. 1	Martin Havlat	7:33	Ott.
May 4/02	Tor.	CSF	Tor. 3	Ott. 2	Gary Roberts	44:30	Tor.
May 7/02	Mtl.	CSF	Mtl. 2	Car. 1	Donald Audette	2:26	Car.
May 9/02	Mtl.	CSF	Car. 4	Mtl. 3	Niclas Wallin	3:14	Car.
May 13/02	S.J.	CSF	Col. 2	S.J. 1	Peter Forsberg	2:47	Col.
May 19/02	Car.	CF	Car. 2	Tor. 1	Niclas Wallin	13:42	Car.
May 20/02	Det.	CF	Col. 4	Det. 3	Chris Drury	2:17	Det.
May 21/02	Tor.	CF	Car. 2	Tor. 1	Jeff O'Neill	6:01	Car.
May 22/02	Col.	CF	Det. 2	Col. 1	Fredrik Olausson	12:44	Det.
May 27/02	Det.	CF	Col. 2	Det. 1	Peter Forsberg	6:24	Det.
May 28/02	Tor.	CF	Car. 2	Tor. 1	Martin Gelinas	8:05	Car.
Jun. 4/02	Det.	F	Car. 3	Det. 2	Ron Francis	:58	Det.
Jun. 8/02	Car.	F	Det. 3	Car. 2	Igor Larionov	54:47	Det.

An overtime goal by Jeff O'Neill in game three against Toronto helped turn the Eastern Conference Final in Carolina's favor. The Hurricanes won seven games and lost only two in overtime during their surprising run to the Stanley Cup Finals.

NHL Playoff Coaching Records

Coach	Team	Games Coached	Wins	Losses	Ties	Playoff Years	Cup Wins	Career
Abel, Sid	Chicago	7	3	4	0	1		
	Detroit	69	29	40	0	8		
	Total	76	32	44	0	9		1952-76
Adams, Jack	Detroit	105	52	52	1	15	3	1927-47
Allen, Keith	Philadelphia	11	3	8	0	2		1967-69
Arbour, Al	St. Louis	11	4	7	0	1		
	NY Islanders	198	119	79	0	15	4	
	Total	209	123	86	0	16	4	1970-94
Barber, Bill	Philadelphia	11	3	8	0	2		2000-02
Berenson, Red	St. Louis	14	5	9	0	2		1979-82
Bergeron, Michel	Quebec	68	31	37	0	7		1980-90
Berry, Bob	Los Angeles	10	2	8	0	3		
	Montreal	8	2	6	0	2		
	St. Louis	15	7	8	0	2		
	Total	33	11	22	0	7		1978-94
Beverley, Nick	Toronto	6	2	4	0	1		1995-96
Blackburn, Don	Hartford	3	0	3	0	1		1979-81
Blair, Wren	Minnesota	14	7	7	0	1		1967-70
Blake, Toe	Montreal	119	82	37	0	13	8	1955-68
Boileau, Marc	Pittsburgh	9	5	4	0	1		1973-76
Boivin, Leo	St. Louis	3	1	2	0	1		1975-78
Boucher, Frank	NY Rangers	27	13	14	0	4	1	1939-54
Boucher, Georges	Mtl. Maroons	2	0	2	0	1		1930-50
Bowman, Scotty	St. Louis	52	26	26	0	4		
	Montreal	98	70	28	0	8	5	
	Buffalo	36	18	18	0	5		
	Pittsburgh	33	23	10	0	2	1	
	Detroit	134	86	48	0	9	3	
	Total	353	223	130	0	28	9	1967-02
Bowness, Rick	Boston	15	8	7	0	1		1988-98
Brooks, Herb	NY Rangers	24	12	12	0	3		
	New Jersey	5	1	4	0	1		
	Pittsburgh	11	6	5	0	1		
	Total	40	19	21	0	5		1981-00
Brophy, John	Toronto	19	9	10	0	2		1986-89
Burns, Charlie	Minnesota	6	2	4	0	1		1969-75
Burns, Pat	Montreal	56	30	26	0	4		
	Toronto	46	23	23	0	3		
	Boston	18	8	10	0	3		
	Total	120	61	59	0	9		1988-01
Campbell, Colin	NY Rangers	36	18	18	0	3		1994-98
Carpenter, Doug	Toronto	5	1	4	0	1		1984-91
Carroll, Dick	Toronto	9	4	5	0	2	1	1917-19
Cheevers, Gerry	Boston	34	15	19	0	4		1980-85
Cherry, Don	Boston	55	31	24	0	5		1974-80
Clancy, King	Toronto	14	2	12	0	3		1937-56
Clapper, Dit	Boston	25	8	17	0	4		1945-49
Cleghorn, Odie	Pittsburgh	4	1	2	1	2		1925-29
Cleghorn, Sprague	Mtl. Maroons	4	1	1	2	1		1931-32
Constantine, Kevin	San Jose	25	11	14	0	2		
	Pittsburgh	19	8	11	0	2		
	New Jersey	6	2	4	0	1		
	Total	50	21	29	0	5		1993-02
Crawford, Marc	Quebec	6	2	4	0	1		
	Colorado	46	29	17	0	3	1	
	Vancouver	10	2	8	0	2		
	Total	62	33	29	0	6	1	1994-02
Creighton, Fred	Atlanta	9	2	7	0	4		1974-80
Crisp, Terry	Calgary	37	22	15	0	3	1	
	Tampa Bay	6	2	4	0	1		
	Total	43	24	19	0	4	1	1987-98
Crozier, Joe	Buffalo	6	2	4	0	1		1971-81
Cunniff, John	New Jersey	6	2	4	0	1		1982-91
Curry, Alex	Ottawa	2	0	1	1	1		1925-26
Dandurand, Leo	Montreal	16	10	6	0	4	1	1921-35
Day, Hap	Toronto	80	49	31	0	9	5	1940-50
Demers, Jacques	St. Louis	33	16	17	0	3		
	Detroit	38	20	18	0	3		
	Montreal	27	19	8	0	2	1	
	Total	98	55	43	0	8	1	1979-99
Denneny, Cy	Boston	5	5	0	0	1	1	1928-33
Dudley, Rick	Buffalo	12	4	8	0	2		1989-92
Dugal, Jules	Montreal	3	1	2	0	1		1938-39
Duncan, Art	Toronto	2	0	1	1	1	1	1926-32
Dutton, Red	NY Americans	16	6	10	0	4		1935-42
Esposito, Phil	NY Rangers	10	2	8	0	2		1986-89
Evans, Jack	Hartford	16	8	8	0	2		1975-88
Ferguson, John	Winnipeg	3	0	3	0	1		1975-86
Francis, Bob	Phoenix	10	2	8	0	2		1999-02
Francis, Emile	NY Rangers	75	34	41	0	9		
	St. Louis	14	5	9	0	2		
	Total	89	39	50	0	11		1965-83
Ftorek, Robbie	Los Angeles	16	5	11	0	2		
	New Jersey	7	3	4	0	1		
	Boston	6	2	4	0	1		
	Total	29	10	19	0	4		1987-02
Gainey, Bob	Minnesota	30	17	13	0	2		
	Dallas	14	6	8	0	2		
	Total	44	23	21	0	4		1990-96
Geoffrion, Bernie	Atlanta	4	0	4	0	1		1968-80
Gerard, Eddie	Mtl. Maroons	25	11	9	5	5	1	1917-35
Gill, David	Ottawa	8	3	2	3	2	1	1926-29
Glover, Fred	Oakland	11	3	8	0	2		1968-74
Gordon, Jackie	Minnesota	25	11	14	0	3		1970-75
Goring, Butch	Boston	3	0	3	0	1		1985-01
Gorman, Tommy	NY Americans	2	0	1	1	1		
	Chicago	8	6	1	1	1	1	
	Mtl. Maroons	15	7	6	2	3	1	
	Total	25	13	8	4	5	2	1925-38
Gottselig, Johnny	Chicago	4	0	4	0	1		1944-48
Green, Pete	Ottawa	26	14	9	3	6	3	1919-25
Green, Ted	Edmonton	16	8	8	0	1		1991-94
Guidolin, Bep	Boston	21	11	10	0	2		1972-76
Harris, Ted	Minnesota	2	0	2	0	1		1975-78
Hart, Cecil	Montreal	37	16	17	4	8	2	1926-39
Hartley, Bob	Colorado	80	49	31	0	4	1	1998-02
Hartsburg, Craig	Chicago	16	8	8	0	2		
	Anaheim	4	0	4	0	1		
	Total	20	8	12	0	3		1995-01
Harvey, Doug	NY Rangers	6	2	4	0	1		1961-62
Hay, Don	Phoenix	7	3	4	0	1		1996-01
Henning, Lorne	Minnesota	5	2	3	0	1		1985-01
Hitchcock, Ken	Dallas	80	47	33	0	5	1	1995-02
Hlinka, Ivan	Pittsburgh	18	9	9	0	1		2000-02
Holmgren, Paul	Philadelphia	19	10	9	0	1		1988-96
Imlach, Punch	Toronto	92	44	48	0	11	4	1958-80
Inglis, Bill	Buffalo	3	1	2	0	1		1978-79
Irvin, Dick	Chicago	9	5	3	1	1		
	Toronto	66	33	32	1	9	1	
	Montreal	115	62	53	0	14	3	
	Total	190	100	88	2	24	4	1928-56
Ivan, Tommy	Detroit	67	36	31	0	7	3	1947-58
Johnson, Bob	Calgary	52	25	27	0	5		
	Pittsburgh	24	16	8	0	1	1	
	Total	76	41	35	0	6	1	1982-91
Johnson, Tom	Boston	22	15	7	0	2	1	1970-73
Johnston, Eddie	Chicago	7	3	4	0	1		
	Pittsburgh	46	22	24	0	5		
	Total	53	25	28	0	6		1979-97
Kasper, Steve	Boston	5	1	4	0	1		1995-97
Keenan, Mike	Philadelphia	57	32	25	0	4		
	Chicago	60	33	27	0	4		
	NY Rangers	23	16	7	0	1	1	
	St. Louis	20	10	10	0	2		
	Total	160	91	69	0	11	1	1984-02
Kelly, Pat	Colorado	2	0	2	0	1		1977-79
Kelly, Red	Los Angeles	18	7	11	0	2		
	Pittsburgh	14	6	8	0	2		
	Toronto	30	11	19	0	4		
	Total	62	24	38	0	8		1967-77
King, Dave	Calgary	20	8	12	0	3		1992-02
Kromm, Bobby	Detroit	7	3	4	0	1		1977-80
Lalonde, Newsy	Montreal	16	7	6	3	4		
	Ottawa	2	0	1	1	1		
	Total	18	7	7	4	5		1917-35
Laviolette, Peter	NY Islanders	7	3	4	0	1		2001-02
Lemaire, Jacques	Montreal	27	15	12	0	2		
	New Jersey	56	34	22	0	4	1	
	Total	83	49	34	0	6	1	1983-02
Ley, Rick	Hartford	13	5	8	0	2		
	Vancouver	11	4	7	0	1		
	Total	24	9	15	0	3		1989-96
Long, Barry	Winnipeg	11	3	8	0	2		1983-86
Loughlin, Clem	Chicago	4	1	2	1	2		1934-37
Low, Ron	Edmonton	28	10	18	0	3		1994-02
Lowe, Kevin	Edmonton	5	1	4	0	1		1999-00
MacLean, Doug	Florida	27	13	14	0	2		1995-98
MacNeil, Al	Montreal	20	12	8	0	1	1	
	Atlanta	4	1	3	0	1		
	Calgary	19	9	10	0	2		
	Total	43	22	21	0	4	1	1970-82
MacTavish, Craig	Edmonton	6	2	4	0	1		2000-02
Magnuson, Keith	Chicago	3	0	3	0	1		1980-82
Mahoney, Bill	Minnesota	16	7	9	0	1		1983-85
Maloney, Dan	Toronto	10	6	4	0	1		
	Winnipeg	15	5	10	0	2		
	Total	25	11	14	0	3		1984-89
Maloney, Phil	Vancouver	7	1	6	0	2		1973-77
Martin, Jacques	St. Louis	16	7	9	0	2		
	Ottawa	44	17	27	0	5		
	Total	60	24	36	0	8		1986-02
Maurice, Paul	Carolina	35	17	18	0	3		1995-02
McCammon, Bob	Philadelphia	10	1	9	0	3		
	Vancouver	7	3	4	0	1		
	Total	17	4	13	0	4		1978-91
McLellan, John	Toronto	11	3	8	0	2		1969-73
McVie, Tom	New Jersey	14	6	8	0	2		1975-92
Melrose, Barry	Los Angeles	24	13	11	0	1		1992-95
Milbury, Mike	Boston	40	23	17	0	2		1989-98
Muckler, John	Edmonton	40	25	15	0	2	1	
	Buffalo	27	11	16	0	4		
	Total	67	36	31	0	6	1	1968-00
Muldoon, Pete	Chicago	2	0	1	1	1		1926-27
Munro, Dunc	Mtl. Maroons	4	1	3	0	1		1929-31

Coach	Team	Games Coached	Wins	Losses	Ties	Playoff Year	Cup Wins	Career
Murdoch, Bob	Chicago	5	1	4	0	1		
	Winnipeg	7	3	4	0	1		
	Total	**12**	**4**	**8**	**0**	**2**		**1987-91**
Murphy, Mike	**Los Angeles**	**5**	**1**	**4**	**0**	**1**		**1986-98**
Murray, Andy	**Los Angeles**	**24**	**10**	**14**	**0**	**3**		**1999-02**
Murray, Bryan	Washington	53	24	29	0	7		
	Detroit	25	10	15	0	3		
	Total	**78**	**34**	**44**	**0**	**10**		**1981-02**
Murray, Terry	Washington	39	18	21	0	4		
	Philadelphia	46	28	18	0	3		
	Florida	4	0	4	0	1		
	Total	**89**	**46**	**43**	**0**	**8**		**1989-01**
Neale, Harry	**Vancouver**	**14**	**3**	**11**	**0**	**4**		**1978-86**
Neilson, Roger	Toronto	19	8	11	0	2		
	Buffalo	8	4	4	0	1		
	Vancouver	21	12	9	0	2		
	NY Rangers	29	13	16	0	3		
	Philadelphia	29	14	15	0	3		
	Total	**106**	**51**	**55**	**0**	**11**		**1977-02**
Nolan, Ted	**Buffalo**	**12**	**5**	**7**	**0**	**1**		**1995-97**
Nykoluk, Mike	**Toronto**	**7**	**1**	**6**	**0**	**2**		**1980-84**
O'Donoghue, George	**Toronto**	**7**	**4**	**2**	**1**	**1**		**1921-23**
O'Reilly, Terry	**Boston**	**37**	**17**	**19**	**1**	**3**		**1986-89**
Oliver, Murray	**Minnesota**	**13**	**5**	**8**	**0**	**2**		**1981-83**
Paddock, John	**Winnipeg**	**13**	**5**	**8**	**0**	**2**		**1991-95**
Page, Pierre	Minnesota	12	4	8	0	2		
	Quebec	6	2	4	0	1		
	Calgary	4	0	4	0	1		
	Total	**22**	**6**	**16**	**0**	**4**		**1988-98**
Patrick, Craig	NY Rangers	17	7	10	0	2		
	Pittsburgh	5	1	4	0	1		
	Total	**22**	**8**	**14**	**0**	**3**		**1980-97**
Patrick, Frank	**Boston**	**6**	**2**	**4**	**0**	**2**		**1934-36**
Patrick, Lester	**NY Rangers**	**65**	**32**	**26**	**7**	**12**	**2**	**1926-39**
Patrick, Lynn	NY Rangers	12	7	5	0	1		
	Boston	28	9	18	1	4		
	Total	**40**	**16**	**23**	**1**	**5**		**1948-76**
Perron, Jean	**Montreal**	**48**	**30**	**18**	**0**	**3**	**1**	**1985-89**
Perry, Don	**Los Angeles**	**10**	**4**	**6**	**0**	**1**		**1981-84**
Pilous, Rudy	**Chicago**	**41**	**19**	**22**	**0**	**5**	**1**	**1957-63**
Plager, Barclay	**St. Louis**	**4**	**1**	**3**	**0**	**1**		**1977-83**
Pleau, Larry	**Hartford**	**10**	**2**	**8**	**0**	**2**		**1980-89**
Polano, Nick	**Detroit**	**7**	**1**	**6**	**0**	**2**		**1982-85**
Powers, Eddie	**Toronto**	**2**	**0**	**2**	**0**	**1**		**1924-26**
Primeau, Joe	**Toronto**	**15**	**8**	**6**	**1**	**2**	**1**	**1950-53**
Pronovost, Marcel	**Buffalo**	**8**	**3**	**5**	**0**	**1**		**1977-79**
Pulford, Bob	Los Angeles	26	10	16	0	4		
	Chicago	45	17	28	0	6		
	Total	**71**	**27**	**44**	**0**	**10**		**1972-00**
Quenneville, Joel	**St. Louis**	**61**	**31**	**30**	**0**	**6**		**1996-02**
Quinn, Pat	Philadelphia	39	22	17	0	3		
	Los Angeles	3	0	3	0	1		
	Vancouver	61	31	30	0	5		
	Toronto	60	32	28	0	4		
	Total	**163**	**85**	**78**	**0**	**13**		**1978-02**
Reay, Billy	**Chicago**	**116**	**56**	**60**	**0**	**12**		**1957-77**
Risebrough, Doug	**Calgary**	**7**	**3**	**4**	**0**	**1**		**1990-92**
Roberts, Jim	**Hartford**	**7**	**3**	**4**	**0**	**1**		**1981-97**
Robinson, Larry	Los Angeles	4	0	4	0	1		
	New Jersey	48	31	17	0	2	1	
	Total	**52**	**31**	**21**	**0**	**3**	**1**	**1995-02**

Coach	Team	Games Coached	Wins	Losses	Ties	Playoff Years	Cup Wins	Career
Ross, Art	**Boston**	**65**	**27**	**33**	**5**	**11**	**1**	**1917-45**
Ruel, Claude	**Montreal**	**27**	**18**	**9**	**0**	**3**	**2**	**1968-81**
Ruff, Lindy	**Buffalo**	**54**	**32**	**22**	**0**	**4**		**1997-02**
Sather, Glen	**Edmonton**	**127**	**89**	**37**	**1**	**10**	**4**	**1979-94**
Sator, Ted	NY Rangers	16	8	8	0	1		
	Buffalo	11	3	8	0	2		
	Total	**27**	**11**	**16**	**0**	**3**		**1985-89**
Schinkel, Ken	**Pittsburgh**	**6**	**2**	**4**	**0**	**2**		**1972-77**
Schmidt, Milt	**Boston**	**34**	**15**	**19**	**0**	**4**		**1954-76**
Schoenfeld, Jim	New Jersey	20	11	9	0	1		
	Washington	24	10	14	0	3		
	Phoenix	13	5	8	0	2		
	Total	**57**	**26**	**31**	**0**	**6**		**1985-99**
Shero, Fred	Philadelphia	83	48	35	0	6	2	
	NY Rangers	27	15	12	0	2		
	Total	**110**	**63**	**47**	**0**	**8**	**2**	**1971-81**
Simpson, Terry	NY Islanders	20	9	11	0	2		
	Winnipeg	6	2	4	0	1		
	Total	**26**	**11**	**15**	**0**	**3**		**1986-96**
Sinden, Harry	**Boston**	**43**	**24**	**19**	**0**	**5**	**1**	**1966-85**
Skinner, Jimmy	**Detroit**	**26**	**14**	**12**	**0**	**3**	**1**	**1954-58**
Smith, Alf	**Ottawa**	**5**	**1**	**4**	**0**	**1**		**1918-19**
Smith, Floyd	**Buffalo**	**32**	**16**	**16**	**0**	**3**		**1971-80**
Smythe, Conn	**Toronto**	**4**	**2**	**2**	**0**	**1**		**1927-31**
Sonmor, Glen	**Minnesota**	**43**	**25**	**18**	**0**	**3**		**1978-87**
Stasiuk, Vic	**Philadelphia**	**4**	**0**	**4**	**0**	**1**		**1969-73**
Stewart, Bill	**Chicago**	**10**	**7**	**3**	**0**	**1**	**1**	**1937-39**
Stewart, Ron	**Los Angeles**	**2**	**0**	**2**	**0**	**1**		**1975-78**
Sutter, Brian	St. Louis	41	20	21	0	4		
	Boston	22	7	15	0	3		
	Chicago	5	1	4	0	1		
	Total	**68**	**28**	**40**	**0**	**8**		**1988-02**
Sutter, Darryl	Chicago	26	11	15	0	3		
	San Jose	42	18	24	0	5		
	Total	**68**	**29**	**39**	**0**	**8**		**1992-02**
Talbot, Jean-Guy	St. Louis	5	1	4	0	1		
	NY Rangers	3	1	2	0	1		
	Total	**8**	**2**	**6**	**0**	**2**		**1972-78**
Tessier, Orval	**Chicago**	**18**	**9**	**9**	**0**	**2**		**1982-85**
Therrien, Michel	**Montreal**	**12**	**6**	**6**	**0**	**1**		**2000-02**
Thompson, Paul	**Chicago**	**19**	**7**	**12**	**0**	**4**		**1938-45**
Tobin, Bill	**Chicago**	**4**	**1**	**2**	**1**	**2**		**1929-32**
Tremblay, Mario	**Montreal**	**11**	**3**	**8**	**0**	**2**		**1995-97**
Ubriaco, Gene	**Pittsburgh**	**11**	**7**	**4**	**0**	**1**		**1988-90**
Vigneault, Alain	**Montreal**	**10**	**4**	**6**	**0**	**1**		**1997-01**
Watson, Phil	**NY Rangers**	**16**	**4**	**12**	**0**	**3**		**1955-63**
Watt, Tom	Winnipeg	7	1	6	0	2		
	Vancouver	3	0	3	0	1		
	Total	**10**	**1**	**9**	**0**	**3**		**1981-92**
Webster, Tom	**Los Angeles**	**28**	**12**	**16**	**0**	**3**		**1986-92**
Weiland, Cooney	**Boston**	**17**	**10**	**7**	**0**	**2**	**1**	**1939-41**
White, Bill	**Chicago**	**2**	**0**	**2**	**0**	**1**		**1976-77**
Wilson, Johnny	**Pittsburgh**	**12**	**4**	**8**	**0**	**2**		**1969-80**
Wilson, Ron	Anaheim	11	4	7	0	1		
	Washington	32	15	17	0	3		
	Total	**43**	**19**	**24**	**0**	**4**		**1993-02**
Young, Garry	**St. Louis**	**2**	**0**	**2**	**0**	**1**		**1972-76**

Shawn Bates beat Curtis Joseph cleanly on a penalty shot with just 2:30 left in the game to give the Islanders a 4-3 victory and even their 2002 opening-round matchup at two wins apiece. Toronto bounced back to win the series in seven games.

Penalty Shots in Stanley Cup Playoff Games

Date	Player	Goaltender	Scored	Final Score					Series
Mar. 25/37	Lionel Conacher, Mtl. Maroons	Tiny Thompson, Boston	No	Mtl. M.	0	at	Bos.	4	QF
Apr. 15/37	Alex Shibicky, NY Rangers	Earl Robertson, Detroit	No	NYR	0	at	Det.	3	F
Apr. 13/44	Virgil Johnson, Chicago	Bill Durnan, Montreal	No	Chi.	4	at	Mtl.	5*	F
Apr. 9/68	Wayne Connelly, Minnesota	Terry Sawchuk, Los Angeles	Yes	L.A.	5	at	Min.	7	QF
Apr. 27/68	Jim Roberts, St. Louis	Cesare Maniago, Minnesota	No	St.L.	4	at	Min.	3	SF
May 16/71	Frank Mahovlich, Montreal	Tony Esposito, Chicago	No	Chi.	3	at	Mtl.	4	F
May 7/75	Bill Barber, Philadelphia	Chico Resch, NY Islanders	No	Phi.	3	at	NYI	4*	SF
Apr. 20/79	Mike Walton, Chicago	Chico Resch, NY Islanders	No	NYI	4	at	Chi.	0	QF
Apr. 9/81	Peter McNab, Boston	Don Beaupre, Minnesota	No	Min.	5	at	Bos.	4*	PR
Apr. 17/81	Anders Hedberg, NY Rangers	Mike Liut, St. Louis	Yes	NYR	6	at	St.L.	4	QF
Apr. 9/83	Denis Potvin, NY Islanders	Pat Riggin, Washington	No	NYI	6	at	Wsh.	2	DSF
Apr. 28/84	Wayne Gretzky, Edmonton	Don Beaupre, Minnesota	Yes	Edm.	8	at	Min.	5	CF
May 1/84	Mats Naslund, Montreal	Billy Smith, NY Islanders	No	Mtl.	1	at	NYI	3	CF
Apr. 14/85	Bob Carpenter, Washington	Billy Smith, NY Islanders	No	Wsh.	4	at	NYI.	6	DF
May 28/85	Ron Sutter, Philadelphia	Grant Fuhr, Edmonton	No	Phi.	3	at	Edm.	5	F
May 30/85	Dave Poulin, Philadelphia	Grant Fuhr, Edmonton	No	Phi.	3	at	Edm.	8	F
Apr. 9/88	John Tucker, Buffalo	Andy Moog, Boston	Yes	Bos.	2	at	Buf.	6	DSF
Apr. 9/88	Petr Klima, Detroit	Allan Bester, Toronto	Yes	Det.	6	at	Tor.	3	DSF
Apr. 8/89	Neal Broten, Minnesota	Greg Millen, St. Louis	Yes	St.L.	5	at	Min.	3	DSF
Apr. 4/90	Al MacInnis, Calgary	Kelly Hrudey, Los Angeles	Yes	L.A.	5	at	Cgy.	3	DSF
Apr. 5/90	Randy Wood, NY Islanders	Mike Richter, NY Rangers	No	NYI	1	at	NYR	2	DSF
May 3/90	Kelly Miller, Washington	Andy Moog, Boston	No	Wsh.	3	at	Bos.	5	CF
May 18/90	Petr Klima, Edmonton	Reggie Lemelin, Boston	No	Edm.	7	at	Bos.	2	F
Apr. 6/91	Basil McRae, Minnesota	Ed Belfour, Chicago	Yes	Min.	2	at	Chi.	5	DSF
Apr. 10/91	Steve Duchesne, Los Angeles	Kirk McLean, Vancouver	Yes	L.A.	6	at	Van.	1	DSF
May 11/92	Jaromir Jagr, Pittsburgh	John Vanbiesbrouck, NYR	Yes	Pit.	3	at	NYR	2	DF
May 13/92	Shawn McEachern, Pittsburgh	John Vanbiesbrouck, NYR	No	NYR	1	at	Pit.	5	DF
June 7/94	Pavel Bure, Vancouver	Mike Richter, NYR	No	NYR	4	at	Van.	2	F
May 9/95	Patrick Poulin, Chicago	Felix Potvin, Toronto	No	Tor.	3	at	Chi.	0	CQF
May 10/95	Michal Pivonka, Washington	Tom Barrasso, Pittsburgh	No	Pit.	2	at	Wsh.	6	CQF
Apr. 24/96	Joe Juneau, Washington	Ken Wregget, Pittsburgh	No	Pit.	3	at	Wsh.	2**	CQF
May 11/97	Eric Lindros, Philadelphia	Steve Shields, Buffalo	Yes	Phi.	6	at	Buf.	3	CSF
Apr. 23/98	Alexei Morozov, Pittsburgh	Andy Moog, Montreal	No	Mtl.	3	at	Pit.	2**	CQF
Apr. 22/99	Mats Sundin, Toronto	John Vanbiesbrouck, Phi.	No	Phi.	3	at	Tor.	0	CQF
May 29/99	Mats Sundin, Toronto	Dominik Hasek, Buffalo	Yes	Tor.	2	at	Buf.	5	CF
Apr. 16/00	Eric Desjardins, Philadelphia	Dominik Hasek, Buffalo	No	Phi.	2	at	Buf.	0	CQF
Apr. 11/01	Mark Recchi, Philadelphia	Dominik Hasek, Buffalo	No	Buf.	2	at	Phi.	1	CQF
May 2/01	Martin Straka, Pittsburgh	Dominik Hasek, Buffalo	No	Buf.	5	at	Pit.	2	CSF
May 12/01	Joe Sakic, Colorado	Roman Turek, St. Louis	Yes	St.L.	1	at	Col.	4	CF
Apr. 21/02	Todd Bertuzzi, Vancouver	Dominik Hasek, Detroit	No	Det.	3	at	Van.	1	CQF
Apr. 24/02	Shawn Bates, NY Islanders	Curtis Joseph, Toronto	Yes	Tor.	3	at	NYI	4	CQF
Apr. 26/02	Mike Johnson, Phoenix	Evgeni Nabokov, San Jose	Yes	Phx.	1	at	S.J.	4	CQF

* Game was decided in overtime, but shot taken during regulation time.
** Shot taken in overtime.

Peter Forsberg scored two overtime goals for the Avalanche in last year's playoffs. After missing the entire regular season due to injuries, Forsberg led the playoffs in scoring with 27 points (9 goals, 18 assists) in just 20 games.

Overtime Record of Current Teams

(Listed by number of OT games played)

	Overall				Home					Road				
Team	GP	W	L	T	GP	W	L	T	Last OT Game	GP	W	L	T	Last OT Game
Montreal	122	70	50	2	57	37	19	1	May 9/02	65	33	31	1	May 8/98
Toronto	101	52	48	1	64	34	29	1	May 28/02	37	18	19	0	May 19/02
Boston	98	38	57	3	45	20	24	1	May 3/98	53	18	33	2	Apr. 30/99
Detroit	72	33	39	0	43	16	27	0	Jun. 4/02	29	17	12	0	Jun. 8/02
NY Rangers	63	30	33	0	27	12	15	0	Apr. 22/97	36	18	18	0	May 11/97
Chicago	62	30	30	2	30	16	13	1	Apr. 20/97	32	14	17	1	May 2/96
Philadelphia	51	24	27	0	23	11	12	0	Apr. 26/02	28	13	15	0	May 4/00
Dallas[1]	50	23	27	0	23	10	13	0	Apr. 19/01	27	13	14	0	May 1/01
St. Louis	49	27	22	0	26	20	6	0	May 18/01	23	7	16	0	May 21/01
Buffalo	46	25	21	0	26	16	10	0	May 10/01	20	9	11	0	May 8/01
Colorado[2]	45	27	18	0	18	9	9	0	May 22/02	27	18	9	0	May 27/02
NY Islanders	38	29	9	0	17	14	3	0	May 20/93	21	15	6	0	May 18/93
Edmonton	38	21	17	0	21	11	10	0	Apr. 17/01	17	10	7	0	Apr. 19/01
Los Angeles	35	17	18	0	19	11	8	0	May 6/01	16	6	10	0	Apr. 25/02
Vancouver	31	14	17	0	13	5	8	0	Apr. 16/01	18	9	9	0	Apr. 17/02
Calgary[3]	30	11	19	0	14	4	10	0	Apr. 23/96	16	7	9	0	Apr. 28/94
Washington	29	14	15	0	10	5	5	0	May 25/98	19	9	10	0	Apr. 23/01
Pittsburgh	28	15	13	0	18	10	8	0	May 8/01	10	5	5	0	May 10/01
New Jersey[4]	27	8	19	0	11	3	8	0	Apr. 28/01	16	5	11	0	Apr. 24/02
Carolina[5]	23	14	9	0	14	9	5	0	Jun. 8/02	9	5	4	0	Jun. 4/02
Phoenix[6]	12	5	7	0	8	3	5	0	May 4/99	4	2	2	0	Apr. 27/93
Ottawa	11	4	7	0	4	2	2	0	Apr. 13/01	7	2	5	0	May 4/02
San Jose	9	3	6	0	5	1	4	0	May 13/02	4	2	2	0	Apr. 19/01
Florida	5	2	3	0	3	1	2	0	Apr. 25/97	2	1	1	0	Apr. 22/97
Anaheim	4	1	3	0	1	0	1	0	May 8/97	3	1	2	0	May 4/97
Tampa Bay	2	2	0	0	1	1	0	0	Apr. 21/96	1	1	0	0	Apr. 18/96

[1] Totals include those of Minnesota North Stars 1967-93.
[2] Totals include those of Quebec 1979-95.
[3] Totals include those of Atlanta Flames 1972-80.
[4] Totals include those of Kansas City and Colorado Rockies 1974-82.
[5] Totals include those of Hartford 1979-97.
[6] Totals include those of Winnipeg 1979-96.

Igor Larionov's goal at 14:47 of triple overtime gave Detroit a 3-2 win over Carolina in game three of last year's Stanley Cup Finals. Larionov's goal turned the series in Detroit's favor after the Hurricanes surprised the Red Wings in game one.

Ten Longest Overtime Games

Date	City	Series	Score		Scorer	Overtime	Series Winner
Mar. 24/36	Mtl.	SF	Det. 1	Mtl. M. 0	Mud Bruneteau	116:30	Det.
Apr. 3/33	Tor.	SF	Tor. 1	Bos. 0	Ken Doraty	104:46	Tor.
May 4/00	Pit.	CSF	Phi. 2	Pit. 1	Keith Primeau	92:01	Phi.
Apr. 24/96	Wsh.	CQF	Pit. 3	Wsh. 2	Petr Nedved	79:15	Pit.
Mar. 23/43	Det.	SF	Tor. 3	Det. 2	Jack McLean	70:18	Det.
Mar. 28/30	Mtl.	SF	Mtl. 2	NYR 1	Gus Rivers	68:52	Mtl.
Apr. 18/87	Wsh.	DSF	NYI 3	Wsh. 2	Pat LaFontaine	68:47	NYI
Apr. 27/94	Buf.	CQF	Buf. 1	N.J. 0	Dave Hannan	65:43	N.J.
Mar. 27/51	Det.	SF	Mtl. 3	Det. 2	Maurice Richard	61:09	Mtl.
Mar. 27/38	NYR	QF	NYA 3	NYR 2	Lorne Carr	60:40	NYA

Key to Prospect, NHL Player and Goaltender Registers

Demographics: Position, shooting side (catching hand for goaltenders), height, weight, place and date of birth as well as draft information, if any, is located on this line.

Major Junior, NCAA, minor pro, senior European and NHL clubs form a permanent part of each player's data panel. If a player sees action with more than one club in any of the above categories, a separate line is included for each one.

Olympic Team statistics are also listed.

Player's NHL organization as of August 23, 2002. This includes players under contract, unsigned draft choices and other players on reserve lists. Free agents as of August 23, 2002 show a blank here.

The complete career data panels of players with NHL experience who announced their retirement before the start of the 2002-03 season are included in the 2002-03 Player Register. These newly-retired players also show a blank here.

Each NHL club's minor-pro affiliates are listed on page 14.

YZERMAN, Steve (IGH-zuhr-muhn, STEEV) **DET.**

Center. Shoots right. 5'11", 185 lbs. Born, Cranbrook, B.C., May 9, 1965. Detroit's 1st choice, 4th overall, in 1983 Entry Draft.

			Regular Season																	Playoffs							
Season	Club	League	GP	G	A	Pts	PIM	PP	SH	GW	S	%	+/-	TF	F%	H	SB	Min	GP	G	A	Pts	PIM	PP	S[illegible]	GW
1980-81	Nepean Raiders	OCJHL	50	38	*54	92	44	....	....	....	....	....	....	....	....	....	....	....	....	....	....	....	....	....	....	....
1981-82	Peterborough	OHL	58	21	43	64	65	....	....	....	....	....	....	....	....	....	....	....	6	0	1	1	16	....	....	....
1982-83	Peterborough	OHL	56	42	49	91	33	....	....	....	....	....	....	....	....	....	....	....	4	1	4	5	0	....	....	....
1983-84	**Detroit**	**NHL**	**80**	**39**	**48**	**87**	**33**	**13**	**0**	**2**	**177**	**22.0**	**−17**	....	....	....	....	....	**4**	**3**	**3**	**6**	**0**	**1**	**0**	**1**
1984-85	**Detroit**	**NHL**	**80**	**30**	**59**	**89**	**58**	**9**	**0**	**3**	**231**	**13.0**	**−17**	....	....	....	....	....	**3**	**2**	**1**	**3**	**2**	**0**	**0**	**0**
1985-86	**Detroit**	**NHL**	**51**	**14**	**28**	**42**	**16**	**3**	**0**	**3**	**132**	**10.6**	**−24**	....	....	....	....	....	....	....	....	....	....	....	....	....
1986-87	**Detroit**	**NHL**	**80**	**31**	**59**	**90**	**43**	**9**	**1**	**2**	**217**	**14.3**	**−1**	....	....	....	....	....	**16**	**5**	**13**	[illegible]	**8**	**1**	**0**	**0**
1987-88	**Detroit**	**NHL**	**64**	**50**	**52**	**102**	**44**	**10**	**6**	**6**	**242**	**20.7**	**30**	....	....	....	....	....	**3**	**1**	**3**	**4**	**6**	**0**	**0**	**0**
1988-89	**Detroit**	**NHL**	**80**	**65**	**90**	**155**	**61**	**17**	**3**	**7**	**388**	**16.8**	**17**	....	....	....	....	....	**6**	**5**	**5**	**10**	**2**	**2**	**0**	**0**
1989-90	**Detroit**	**NHL**	**79**	**62**	**65**	**127**	**79**	**16**	**7**	**8**	**332**	**18.7**	**−6**	....	....	....	....	....	....	....	....	....	....	....	....	....
1990-91	**Detroit**	**NHL**	**80**	**51**	**57**	**108**	**34**	**12**	**6**	**4**	**326**	**15.6**	**−2**	....	....	....	....	....	**7**	**3**	**3**	**6**	**4**	**1**	**0**	**0**
1991-92	**Detroit**	**NHL**	**79**	**45**	**58**	**103**	**64**	**9**	**8**	**9**	**295**	**15.3**	**26**	....	....	....	....	....	**11**	**3**	**5**	**8**	**12**	**0**	**1**	**1**
1992-93	**Detroit**	**NHL**	**84**	**58**	**79**	**137**	**44**	**13**	**7**	**6**	**307**	**18.9**	**33**	....	....	....	....	....	**7**	**4**	**3**	**7**	**4**	**1**	**1**	**1**
1993-94	**Detroit**	**NHL**	**58**	**24**	**58**	**82**	**36**	**7**	**3**	**3**	**217**	**11.1**	**11**	....	....	....	....	....	**3**	**1**	**3**	**4**	[illegible]	**0**	**0**	**0**
1994-95	**Detroit**	**NHL**	**47**	**12**	**26**	**38**	**40**	**4**	**0**	**1**	**134**	**9.0**	**6**	....	....	....	....	....	**15**	**4**	**8**	**12**	[illegible]	**2**	**0**	**1**
1995-96	**Detroit**	**NHL**	**80**	**36**	**59**	**95**	**64**	**16**	**2**	**8**	**220**	**16.4**	**29**	....	....	....	....	....	**18**	**8**	**12**	**20**	**4**	**4**	**0**	**1**
1996-97 ♦	**Detroit**	**NHL**	**81**	**22**	**63**	**85**	**78**	**8**	**0**	**3**	**232**	**9.5**	**22**	....	....	....	....	....	**20**	[illegible]	**6**	**13**	**4**	**3**	**0**	**2**
19[illegible]-98 ♦	**Detroit**	**NHL**	**75**	**24**	**45**	**69**	**46**	**6**	**2**	**0**	**188**	**12.8**	**3**	....	....	....	....	....	**22**	**6**	***18**	***24**	**22**	**3**	**1**	**0**
	Canada	Olympics	6	1	1	2	10	....	....	....	....	....	....	....	....	....	....	....	....	....	....	....	....	....	....	....
1998-99	**Detroit**	**NHL**	**80**	**29**	**45**	**74**	**42**	**13**	**2**	**4**	**231**	**12.6**	**8**	**1600**	**56.9**	**46**	**58**	**21:35**	**10**	**9**	**4**	**13**	**0**	**4**	**0**	**2**
99-2000	**Detroit**	**NHL**	**78**	**35**	**44**	**79**	**34**	**15**	**2**	**6**	**234**	**15.0**	**28**	**1868**	**56.8**	**67**	**56**	**21:07**	**8**	**0**	**4**	**4**	**0**	**0**	**0**	**0**
[illegible]-01	**Detroit**	**NHL**	**54**	**18**	**34**	**52**	**18**	**5**	**0**	**7**	**155**	**11.6**	**4**	**1197**	**59.6**	**38**	**51**	**22:14**	**1**	**0**	**0**	**0**	**0**	[illegible]	**0**	**0**
2001-02 ♦	**Detroit**	**NHL**	**52**	**13**	**35**	**48**	**18**	**5**	**1**	**5**	**104**	**12.5**	**11**	**1182**	**58.4**	**30**	**43**	**20:35**	**23**	**6**	**17**	**23**	**10**	[illegible]	**0**	**2**
	Canada	Olympics	6	2	4	6	2	....	....	....	....	....	....	....	....	....	....	....	....	....	....	....	....	....	....	....
	NHL Totals		**1362**	**658**	**1004**	**1662**	**852**	**190**	**50**	**87**	**4362**	**15.1**		**5847**	**57.7**	**181**	**208**	**21:23**	**177**	**67**	**108**	**175**	**78**	**26**	**3**	**11**

NHL All-Rookie Team (1984) • Won Lester B. Pearson Award (1989) • Won Conn Smythe Trophy (1998) • NHL First All-Star Team (2000) • Won Frank J. Selke Trophy (2000) • Played in NHL All-Star Game (1984, 1988, 1989, 1990, 1991, 1992, 1993, 1997, 2000)

Diamond (♦) indicates member of Stanley Cup-winning team.

Asterisk (*) indicates league leader in this statistical category.

Trade and free agent signing dates are based on when the player's contract is filed with NHL Central Registry. This date often differs from the date when the club announces that it has made a trade or come to terms with a free agent.

All-Star Team selections and awards are listed below player's year-by-year data.

NHL All-Star Game appearances are listed above trade notes.

All trades, free agent signings and other transactions involving NHL clubs are listed in chronological order. First draft selection for players who re-enter the NHL Entry Draft is noted here. Other special notes are also listed here. These are highlighted with a bullet (•).

THIS 71ST EDITION OF THE *NHL Official Guide & Record Book* is the fourth to include additional statistical categories for forwards and defensemen in the National Hockey League. These new categories are, from left to right in the sample panel above, power-play goals (PP), shorthand goals (SH), game-winning goals (GW), shots on goal (S), percentage of shots that score (%), plus-minus rating (+/–), total faceoffs taken (TF), faceoff winning percentage (F%), hits (H), shots blocked (SB) and average time-on-ice per game played (Min).

To integrate this data, the Player Register has been is split into two sections. The Prospect Register presents data on players who have yet to play in the NHL. The NHL Player Register, containing more information and a photo of each player, lists all active players who have appeared in an NHL regular-season or playoff game at any time.

Goaltenders, whether prospects or active NHLers, are included in one register.

Registers (with their starting page) are presented in the following order: Prospects (269), NHL Players (338), Goaltenders (576), Retired Players (600) and Retired Goaltenders (632).

Late additions to the Registers and a list of league abbreviations are found on page 337.

Some information is unavailable at press time. Readers are encouraged to contribute. See page 5 for contact names and addresses.

Pronunciation of Player Names

United Press International phonetic style.

AY	long A as in mate
A	short A as in cat
AI	nasal A as on air
AH	short A as in father
AW	broad A as in talk
EE	long E as in meat
EH	short E as in get
UH	hollow E as in the
AY	French long E with acute accent as in Pathe
IH	middle E as in pretty
EW	EW dipthong as in few
IGH	long I as in time
EE	French long I as in machine
IH	short I as in pity
OH	long O as in note
AH	short O as in hot
AW	broad O as in fought
OI	OI dipthong as in noise
OO	long double OO as in fool
U	short double O as in foot
OW	OW dipthong as in how
EW	long U as in mule
OO	long U as in rule
U	middle U as in put
UH	short U as in shut or hurt
K	hard C as in cat
S	soft C as in cease
SH	soft CH as in machine
CH	hard CH or TCH as in catch
Z	hard S as in bells
S	soft S as in sun
G	hard G as in gang
J	soft G as in general
ZH	soft J as in French version of Joliet
KH	gutteral CH as in Scottish version of Loch

2002-03 Prospect Register

Note: The 2002-03 Prospect Register lists forwards and defensemen only. Goaltenders are listed separately. The Prospect Register lists every player drafted in the first five rounds of the 2002 Entry Draft, players on NHL Reserve Lists and other players who have not yet played in the NHL. Trades and roster changes are current as of August 23, 2002.

Abbreviations: A – assists; **G** – goals; **GP** – games played; **Lea** – league; **PIM** – penalties in minutes; **TP** – total points; * – league-leading total.

NHL Player Register begins on page 338.

Goaltender Register begins on page 576.

League Abbreviations are listed on page 268.

ABBOTT, Jim (A-buht, JIHM) **PIT.**

Left wing. Shoots left. 6'1", 185 lbs. Born, New York, NY, May 3, 1980.
(Pittsburgh's 7th choice, 216th overall, in 2000 Entry Draft).

			Regular Season					Playoffs				
Season	**Club**	**Lea**	**GP**	**G**	**A**	**TP**	**PIM**	**GP**	**G**	**A**	**TP**	**PIM**
1997-98	Pittsburgh	MTJHL	50	30	23	53	128	5	5	3	8	0
1998-99	St. Louis Sting	NAJHL	56	*45	33	78	35	9	7	2	9	12
99-2000	New Hampshire	H-East	28	7	6	13	22	....	....	....	....	....
2000-01	New Hampshire	H-East	33	10	12	22	52	....	....	....	....	....
2001-02	New Hampshire	H-East	38	9	20	29	48	....	....	....	....	....

MTJHL Rookie of the Year (1998) • NAJHL First All-Star Team (1999)

ABID, Ramzi (a-BIHD, RAM-zee) **PHX.**

Left wing. Shoots left. 6'2", 210 lbs. Born, Montreal, Que., March 24, 1980.
(Phoenix's 3rd choice, 85th overall, in 2000 Entry Draft).

			Regular Season					Playoffs				
Season	**Club**	**Lea**	**GP**	**G**	**A**	**TP**	**PIM**	**GP**	**G**	**A**	**TP**	**PIM**
1995-96	Richelieu	QAAA	42	10	14	24	18	4	1	2	3	2
1996-97	Chicoutimi	QMJHL	65	13	24	37	141	21	2	12	14	28
1997-98	Chicoutimi	QMJHL	68	50	*85	*135	266	6	3	4	7	10
1998-99	Chicoutimi	QMJHL	21	11	15	26	97	....	....	....	....	....
	Acadie-Bathurst	QMJHL	24	14	22	36	102	23	14	20	34	*84
99-2000	Acadie-Bathurst	QMJHL	13	10	11	21	61	....	....	....	....	....
	Halifax	QMJHL	59	57	80	137	148	10	10	13	23	18
2000-01	Springfield	AHL	17	6	4	10	38	....	....	....	....	....
2001-02	Springfield	AHL	66	18	25	43	214	....	....	....	....	....

• Re-entered NHL Entry Draft. Originally Colorado's 5th choice, 28th overall, in 1998 Entry Draft.

Won Michel Briere Trophy (QMJHL MVP) (1998) •. Won Jean Beliveau Trophy (QMJHL Leading Scorer) (1998) • QMJHL First All-Star Team (1998, 2000) • Canadian Major Junior First All-Star Team (2000) • Won Ed Chynoweth Trophy (Memorial Cup Tournament Leading Scorer) (2000)

Traded to **Halifax** (QMJHL) by **Acadie-Bathurst** (QMJHL) with future considerations for Samuel Seguin, David Hemsworth and future considerations, October 10, 1999. • Missed majority of 2000-01 season recovering from wrist injury originally suffered in game vs. Louisville (AHL), October 27, 2000.

ADAMS, John (A-duhms, JAWN) **BUF.**

Defense. Shoots left. 6'2", 188 lbs. Born, Orono, ME, December 21, 1982.
(Buffalo's 4th choice, 82nd overall, in 2002 Entry Draft).

			Regular Season					Playoffs				
Season	**Club**	**Lea**	**GP**	**G**	**A**	**TP**	**PIM**	**GP**	**G**	**A**	**TP**	**PIM**
99-2000	Breck Mustangs	Hi-School	25	6	30	36	....	....	....	....	....	....
2000-01	Breck Mustangs	Hi-School	25	13	29	42	....	....	....	....	....	....
2001-02	Boston College	H-East	25	0	5	5	20	....	....	....	....	....

All-Metro Second All-Star Team (2000) • All-Metro First All-Star Team (2001)

ADDUONO, Jeremy (uh-DOO-noh, JAIR-eh-mee) **BUF.**

Right wing. Shoots right. 6', 182 lbs. Born, Thunder Bay, Ont., August 4, 1978.
(Buffalo's 8th choice, 184th overall, in 1997 Entry Draft).

			Regular Season					Playoffs				
Season	**Club**	**Lea**	**GP**	**G**	**A**	**TP**	**PIM**	**GP**	**G**	**A**	**TP**	**PIM**
1994-95	Thunder Bay	USHL	40	11	10	21	8	....	....	....	....	....
1995-96	Sudbury Wolves	OHL	66	15	22	37	14	....	....	....	....	....
1996-97	Sudbury Wolves	OHL	66	29	40	69	24	....	....	....	....	....
1997-98	Sudbury Wolves	OHL	66	37	69	106	40	10	5	5	10	10
1998-99	Team Canada	Nat-Tm	44	10	18	28	10	....	....	....	....	....
99-2000	Rochester	AHL	51	23	22	45	20	21	6	11	17	2
2000-01	Rochester	AHL	76	24	30	54	53	4	1	0	1	4
2001-02	Rochester	AHL	79	15	20	35	38	1	1	0	1	0

AHMAOJA, Timo (ahkh-mah-OH-yah, TIH-moo) **ANA.**

Defense. Shoots right. 6'1", 180 lbs. Born, Jyvaskyla, Finland, August 8, 1978.
(Anaheim's 5th choice, 172nd overall, in 1996 Entry Draft).

			Regular Season					Playoffs				
Season	**Club**	**Lea**	**GP**	**G**	**A**	**TP**	**PIM**	**GP**	**G**	**A**	**TP**	**PIM**
1993-94	JYP Jyvaskyla-C	Finn-Jr.	32	6	9	15	34	6	2	0	2	4
1994-95	JYP Jyvaskyla-B	Finn-Jr.	6	1	3	4	8	....	....	....	....	....
	JYP Jyvaskyla Jr.	Finn-Jr.	30	1	1	2	2	7	0	0	0	0
1995-96	JYP Jyvaskyla Jr.	Finn-Jr.	28	0	7	7	16	6	1	0	1	2
	JyP HT Jyvaskyla	Finland	4	0	0	0	4	....	....	....	....	....
1996-97	JYP Jyvaskyla Jr.	Finn-Jr.	23	4	6	10	26	7	2	2	4	12
	JyP HT Jyvaskyla	Finland	35	0	4	4	8	....	....	....	....	....
1997-98	JYP Jyvaskyla	Finland	10	0	0	0	4	....	....	....	....	....
	Diskos Jyvaskyla	Finland-2	23	1	2	3	14	....	....	....	....	....
	Lukko Rauma	Finland	10	0	0	0	0	....	....	....	....	....
1998-99	KalPa Kuopio Jr.	Finn-Jr.	3	0	1	1	10	4	1	1	2	6
	KalPa Kuopio	Finland	52	0	1	1	16	....	....	....	....	....
99-2000	SaPKo Savonlinna	Finland-2	14	4	3	7	38	....	....	....	....	....
	SaiPa	Finland	8	0	0	0	0	....	....	....	....	....
	HIFK Helsinki	Finland	5	0	0	0	2	....	....	....	....	....
	Pelicans Lahti	Finland	29	0	0	0	4	....	....	....	....	....
2000-01	HPK Hameenlinna	Finland	21	1	0	1	12	....	....	....	....	....
	Assat Pori	Finland	32	1	3	4	10	....	....	....	....	....
2001-02	Assat Pori	Finland	54	3	4	7	18	....	....	....	....	....

AHOSILTA, Marko (ah-hoh-SIHL-tuh, mahr-KOH) **N.J.**

Center. Shoots left. 5'8", 165 lbs. Born, Kuopio, Finland, January 24, 1980.
(New Jersey's 11th choice, 227th overall, in 1998 Entry Draft).

			Regular Season					Playoffs				
Season	**Club**	**Lea**	**GP**	**G**	**A**	**TP**	**PIM**	**GP**	**G**	**A**	**TP**	**PIM**
1994-95	KalPa Kuopio-C	Finn-Jr.	13	4	6	10	10	....	....	....	....	....
1995-96	KalPa Kuopio Jr.	Finn-Jr.	12	4	8	12	10	....	....	....	....	....
1996-97	KalPa Kuopio Jr.	Finn-Jr.	35	15	21	36	36	5	2	0	2	2
1997-98	KalPa Kuopio Jr.	Finn-Jr.	14	14	13	27	10	....	....	....	....	....
	KalPa Kuopio	Finland	2	0	0	0	0	....	....	....	....	....
1998-99	KalPa Kuopio	Finland	1	0	0	0	0	....	....	....	....	....
	KalPa Kuopio Jr.	Finn-Jr.	24	7	7	14	10	....	....	....	....	....
99-2000	KalPa Kuopio Jr.	Finn-Jr.	10	6	4	10	8	....	....	....	....	....
	KJT Jarvenpaa	Finland-2	30	19	5	24	14	....	....	....	....	....
2000-01			DID NOT PLAY									
2001-02	KalPa Kuopio	Finland-2	44	17	20	37	24	3	0	1	1	2

AIKINS, Justin (AY-kihns, JUHS-tihn) **CBJ**

Center. Shoots left. 6', 176 lbs. Born, Surrey, B.C., January 12, 1982.
(Columbus' 7th choice, 173rd overall, in 2001 Entry Draft).

			Regular Season					Playoffs				
Season	**Club**	**Lea**	**GP**	**G**	**A**	**TP**	**PIM**	**GP**	**G**	**A**	**TP**	**PIM**
1998-99	Langley Hornets	BCHL	52	6	21	27	16	....	....	....	....	....
99-2000	Langley Hornets	BCHL	59	29	38	67	46	....	....	....	....	....
2000-01	Langley Hornets	BCHL	59	30	61	91	47	....	....	....	....	....
2001-02	New Hampshire	H-East	29	4	4	8	6	....	....	....	....	....

ALBERTS, Andrew (AL-buhrts, AN-droo) **BOS.**

Defense. Shoots left. 6'4", 218 lbs. Born, Minneapolis, MN, June 30, 1981.
(Boston's 5th choice, 179th overall, in 2001 Entry Draft).

			Regular Season					Playoffs				
Season	**Club**	**Lea**	**GP**	**G**	**A**	**TP**	**PIM**	**GP**	**G**	**A**	**TP**	**PIM**
1998-99	Benilde High	Hi-School	26	10	25	35	....	....	....	....	....	....
99-2000	Waterloo	USHL	49	2	2	4	55	4	0	0	0	12
2000-01	Waterloo	USHL	54	4	10	14	128	....	....	....	....	....
2001-02	Boston College	H-East	38	2	10	12	52	....	....	....	....	....

ALINC, Jan (AHL-lihnch, YAHN) PIT.

Center. Shoots left. 6'2", 190 lbs. Born, Louny, Czech., May 27, 1972.
(Pittsburgh's 7th choice, 163rd overall, in 1992 Entry Draft).

			Regular Season					Playoffs				
Season	Club	Lea	GP	G	A	TP	PIM	GP	G	A	TP	PIM
1990-91	HC CHZ Litvinov	Czech	7	1	1	2	....	....	....	....	....	....
1991-92	Litvinov	Czech	45	21	16	37	24	....	....	....	....	....
1992-93	Litvinov	Czech	36	16	13	29	....	....	....	....	....	....
1993-94	Litvinov	Czech	36	16	25	41	....	4	1	4	5	....
	Czech Republic	Olympics	6	2	0	2	4	....	....	....	....	....
1994-95	Litvinov	Czech	42	16	32	48	50	4	3	2	5	2
1995-96	Litvinov	Czech	38	15	29	44	....	16	2	5	7	....
1996-97	Assat Pori	Finland	47	9	16	25	16	4	0	4	4	2
1997-98	Assat Pori	Finland	15	2	8	10	10	....	....	....	....	....
	Litvinov	Czech	33	12	32	44	14	4	1	2	3	12
1998-99	MoDo	Sweden	48	7	11	18	22	9	1	0	1	0
99-2000	Litvinov	Czech	48	19	15	34	72	6	0	1	1	25
2000-01	HC Slavia Praha	Czech	44	12	35	47	18	11	1	6	7	0
2001-02	HC Slavia Praha	Czech	12	1	6	7	45	....	....	....	....	....
	HC Karlovy Vary	Czech	33	9	18	27	18	....	....	....	....	....

ALLEN, Bobby (AHL-lehn, BAW-bee) EDM.

Defense. Shoots left. 6'1", 205 lbs. Born, Braintree, MA, November 14, 1978.
(Boston's 2nd choice, 52nd overall, in 1998 Entry Draft).

			Regular Season					Playoffs				
Season	Club	Lea	GP	G	A	TP	PIM	GP	G	A	TP	PIM
1996-97	Cushing Academy	Hi-School	36	11	33	44	28	....	....	....	....	....
1997-98	Boston College	H-East	40	7	21	28	49	....	....	....	....	....
1998-99	Boston College	H-East	43	9	23	32	34	....	....	....	....	....
99-2000	Boston College	H-East	42	4	23	27	40	....	....	....	....	....
2000-01	Boston College	H-East	42	5	18	23	28	....	....	....	....	....
2001-02	Providence	AHL	49	5	10	15	18	....	....	....	....	....
	Hamilton	AHL	10	1	6	7	0	14	0	3	3	6

Hockey East Second All-Star Team (2000) • Hockey East First All-Star Team (2001) • NCAA East First All-American Team (2001)

Traded to **Edmonton** by **Boston** for Sean Brown, March 19, 2002.

ALMTORP, Jonas (AHLM-tohrp, YOH-nuhs) EDM.

Center. Shoots left. 6'1", 189 lbs. Born, Uppsala, Sweden, November 17, 1983.
(Edmonton's 7th choice, 111th overall, in 2002 Entry Draft).

			Regular Season					Playoffs				
Season	Club	Lea	GP	G	A	TP	PIM	GP	G	A	TP	PIM
99-2000	MoDo-18	Swede-Jr.	22	*19	12	31	*55	....	....	....	....	....
	MoDo Jr.	Swede-Jr.	7	1	0	1	0	....	....	....	....	....
2000-01	MoDo Jr.	Swede-Jr.	27	19	7	26	38	7	6	1	7	10
2001-02	MoDo Jr.	Swede-Jr.	37	26	18	44	102	2	1	1	2	4
	MoDo	Sweden	3	0	0	0	0	....	....	....	....	....

ALTAREV, Dmitri (al-ta-REHV, dih-MEE-tree) NYI

Left wing. Shoots left. 6'3", 191 lbs. Born, Penza, USSR, August 12, 1980.
(NY Islanders' 8th choice, 264th overall, in 2000 Entry Draft).

			Regular Season					Playoffs				
Season	Club	Lea	GP	G	A	TP	PIM	GP	G	A	TP	PIM
1997-98	Dizelist Penza 2	Russia-3	57	15	8	23	83	....	....	....	....	....
1998-99	Dizelist Penza 2	Russia-4	35	4	3	7	30	....	....	....	....	....
	Dizelist Penza	Russia-2	6	2	0	2	8	....	....	....	....	....
99-2000	Dizelist Penza 2	Russia-3	44	10	8	18	25	....	....	....	....	....
2000-01	Nizhny Novgorod	Russia	36	1	2	3	26	....	....	....	....	....
2001-02	Niz. Novgorod 2	Russia-3	10	5	6	11	12	....	....	....	....	....
	Nizhny Novgorod	Russia	32	2	3	5	42	....	....	....	....	....

ANDERSON, Erik (AN-duhr-suhn, AIR-ihk) NSH.

Center. Shoots left. 5'9", 190 lbs. Born, Plymouth, MI, March 6, 1978.

			Regular Season					Playoffs				
Season	Club	Lea	GP	G	A	TP	PIM	GP	G	A	TP	PIM
1995-96	Stratford	OJHL-B	38	29	43	72	10	....	....	....	....	....
1996-97	Stratford	OJHL-B	48	54	*91	*145	40	....	....	....	....	....
1997-98	St. Lawrence	ECAC	33	5	13	18	12	....	....	....	....	....
1998-99	St. Lawrence	ECAC	39	10	30	40	18	....	....	....	....	....
99-2000	St. Lawrence	ECAC	36	14	25	39	20	....	....	....	....	....
2000-01	St. Lawrence	ECAC	32	17	34	51	4	....	....	....	....	....
2001-02	Milwaukee	AHL	49	4	6	10	10	....	....	....	....	....
	Cincinnati	ECHL	17	6	13	19	19	....	....	....	....	....

ECAC First All-Star Team (2001) • ECAC Player of the Year (2001) • NCAA East First All-American Team (2001)

Signed as a free agent by **Nashville**, July 5, 2001.

ANDRESEN, Joel (AN-druh-suhn, JOHL) L.A.

Defense. Shoots left. 6'3", 200 lbs. Born, St. Albert, Alta., April 11, 1983.
(Los Angeles' 7th choice, 157th overall, in 2002 Entry Draft).

			Regular Season					Playoffs				
Season	Club	Lea	GP	G	A	TP	PIM	GP	G	A	TP	PIM
2000-01	St. Albert	AJHL	45	2	9	11	47	7	1	1	2	6
2001-02	St. Albert	AJHL	53	9	19	28	62	6	0	2	2	12

AJHL Top Defenseman (2002)

• Signed Letter of Intent to attend **Nebraska-Omaha** (CCHA), October 30, 2001.

ANDREWS, Bobby (AN-drooz, BAW-bee) NYR

Center. Shoots left. 6'1", 200 lbs. Born, Birtle, Man., January 5, 1978.

			Regular Season					Playoffs				
Season	Club	Lea	GP	G	A	TP	PIM	GP	G	A	TP	PIM
1996-97	Cleveland	NAJHL	45	21	22	43	259	....	....	....	....	....
1997-98	Langley Thunder	BCHL	59	26	48	74	143	....	....	....	....	....
1998-99	Alaska-Fairbanks	CCHA	31	5	11	16	38	....	....	....	....	....
99-2000	Alaska-Fairbanks	CCHA	34	13	12	25	72	....	....	....	....	....
2000-01	Alaska-Fairbanks	CCHA	36	9	16	25	46	....	....	....	....	....
2001-02	Alaska-Fairbanks	CCHA	37	14	23	37	46	....	....	....	....	....
	Hartford	AHL	1	0	0	0	0	7	0	0	0	2

BCHL Central First All-Star Team (1998) • CCHA Second All-Star Team (2002)

Signed as a free agent by **NY Rangers**, March 19, 2002.

ANDREWS, Daryl (AN-drews, DAI-rihl) N.J.

Defense. Shoots left. 6'3", 215 lbs. Born, Campbell River, B.C., April 27, 1977.
(New Jersey's 11th choice, 173rd overall, in 1996 Entry Draft).

			Regular Season					Playoffs				
Season	Club	Lea	GP	G	A	TP	PIM	GP	G	A	TP	PIM
1995-96	Melfort Mustangs	SJHL	55	2	12	14	51	....	....	....	....	....
1996-97	West-Michigan	CCHA	37	6	20	26	86	....	....	....	....	....
1997-98	West-Michigan	CCHA	36	3	0	3	81	....	....	....	....	....
1998-99	West-Michigan	CCHA	33	3	11	14	42	....	....	....	....	....
99-2000	West-Michigan	CCHA	36	3	16	19	52	....	....	....	....	....
	Albany	AHL	9	0	2	2	9	5	0	0	0	0
2000-01	Albany	AHL	80	2	8	10	49	....	....	....	....	....
2001-02	Albany	AHL	69	3	10	13	60	....	....	....	....	....

CCHA Rookie of the Year (1997)

ANGELSTAD, Mel (AN-gehl-stahd, MEHL)

Left wing. Shoots left. 6'2", 214 lbs. Born, Saskatoon, Sask., October 31, 1972.

			Regular Season					Playoffs				
Season	Club	Lea	GP	G	A	TP	PIM	GP	G	A	TP	PIM
1988-89	Allan	MAHA	35	15	23	38	256	....	....	....	....	....
1989-90	Warman Valley	MJHL	38	1	5	6	411	....	....	....	....	....
1990-91	Flin Flon	MJHL	62	6	11	17	463	....	....	....	....	....
1991-92	Dauphin Kings	MJHL	44	8	29	37	*296	....	....	....	....	....
1992-93	Thunder Bay	ColHL	45	2	5	7	256	5	0	0	0	10
	Nashville	ECHL	1	0	0	0	14	....	....	....	....	....
1993-94	Thunder Bay	ColHL	58	1	20	21	374	9	1	2	3	65
	P.E.I. Senators	AHL	1	0	0	0	5	....	....	....	....	....
1994-95	Thunder Bay	ColHL	46	0	8	8	317	7	0	3	3	62
	P.E.I. Senators	AHL	3	0	0	0	16	....	....	....	....	....
1995-96	Thunder Bay	ColHL	51	3	3	6	335	16	0	6	6	94
	Phoenix	IHL	5	0	0	0	43	....	....	....	....	....
1996-97	Thunder Bay	ColHL	66	10	21	31	422	7	0	1	1	21
1997-98	Fort Worth	WPHL	19	1	6	7	102	....	....	....	....	....
	Las Vegas	IHL	3	0	0	0	5	....	....	....	....	....
	Orlando	IHL	63	1	3	4	321	8	0	0	0	29
1998-99	Michigan K-Wings	IHL	78	3	5	8	421	5	1	0	1	16
99-2000	Michigan K-Wings	IHL	33	3	4	7	144	....	....	....	....	....
2000-01	Manitoba Moose	IHL	67	1	5	6	232	8	0	0	0	26
2001-02	Portland Pirates	AHL	53	1	7	8	212	....	....	....	....	....

Signed as a free agent by **Dallas**, July 29, 1998.

ANGER, Niklas (AN-guhr, NIHK-lahs) MTL.

Right wing. Shoots left. 6'1", 185 lbs. Born, Gavle, Sweden, July 31, 1977.
(Montreal's 5th choice, 112th overall, in 1995 Entry Draft).

			Regular Season					Playoffs				
Season	Club	Lea	GP	G	A	TP	PIM	GP	G	A	TP	PIM
1994-95	Djurgarden Jr.	Swede-Jr.	30	14	12	26	26	....	....	....	....	....
	Djurgarden	Sweden	1	0	0	0	0	....	....	....	....	....
1995-96	Djurgarden	Sweden	10	0	0	0	2	....	....	....	....	....
1996-97	Djurgarden Jr.	Swede-Jr.	2	1	2	3	0	....	....	....	....	....
	Arlanda Mastra	Swede-2	16	5	9	14	6	....	....	....	....	....
	Linkopings HC	Swede-2	10	2	2	4	10	14	3	7	10	2
	Djurgarden	Sweden	4	0	0	0	0	....	....	....	....	....
1997-98	Djurgarden	Sweden	45	2	5	7	37	12	0	1	1	2
1998-99	AIK Solna	Sweden	47	6	6	12	16	....	....	....	....	....
99-2000	AIK Solna	Sweden	50	11	13	24	14	....	....	....	....	....
2000-01	AIK Solna	Sweden	50	5	10	15	22	5	0	1	1	2
2001-02	AIK Solna	Sweden	50	12	20	32	16	....	....	....	....	....

ANISIMOV, Artem (ah-NIH-sih-mohv, AHR-tehm) MIN.

Defense. Shoots left. 6'1", 187 lbs. Born, Kazan, USSR, July 27, 1976.
(Philadelphia's 1st choice, 62nd overall, in 1994 Entry Draft).

			Regular Season					Playoffs				
Season	Club	Lea	GP	G	A	TP	PIM	GP	G	A	TP	PIM
1993-94	Itil Kazan	CIS	38	0	1	1	12	5	0	0	0	0
1994-95	Itil Kazan	CIS	46	3	2	5	55	1	0	0	0	0
1995-96	Ak Bars Kazan	CIS	30	0	2	2	8	....	....	....	....	....
1996-97	Ak Bars Kazan	Russia	5	0	1	1	2	....	....	....	....	....
1997-98	Ak Bars Kazan	Russia	44	0	0	0	26	....	....	....	....	....
1998-99	Ak Bars Kazan	Russia	39	0	4	4	10	9	0	0	0	2
99-2000	Ak Bars Kazan	Russia	11	0	1	1	2	5	1	1	2	4
2000-01	Ak Bars Kazan	Russia	4	1	0	1	6	....	....	....	....	....
	Amur Khabarovsk	Russia	16	1	0	1	12	....	....	....	....	....
2001-02	Ufa	Russia	44	1	2	3	24	....	....	....	....	....

Selected by **Minnesota** from **Philadelphia** in Expansion Draft, June 23, 2000.

ANSHAKOV, Sergei (an-sha-KAHV, SAIR-gay) L.A.

Left wing. Shoots left. 6'3", 179 lbs. Born, Moscow, USSR, January 13, 1984.
(Los Angeles' 2nd choice, 50th overall, in 2002 Entry Draft).

			Regular Season					Playoffs				
Season	Club	Lea	GP	G	A	TP	PIM	GP	G	A	TP	PIM
2001-02	H.C. CSKA 2	Russia-3	3	3	1	4	0	....	....	....	....	....
	H.C. CSKA	Russia-2	46	20	12	22	10	....	....	....	....	....

ANTIPOV, Vladimir (an-TIH-pahv, vla-DIH-meer) TOR.

Right wing. Shoots left. 5'11", 180 lbs. Born, Appatity, USSR, January 17, 1978.
(Toronto's 6th choice, 103rd overall, in 1996 Entry Draft).

			Regular Season					Playoffs				
Season	Club	Lea	GP	G	A	TP	PIM	GP	G	A	TP	PIM
1995-96	Yaroslavl	CIS	39	14	11	25	....	....	....	....	....	....
1996-97	Yaroslavl 2	Russia-3	14	3	4	7	26	....	....	....	....	....
	Yaroslavl	Russia	27	6	4	10	22	2	0	0	0	0
1997-98	Yaroslavl	Russia	41	9	3	12	57	....	....	....	....	....
	Yaroslavl	EuroHL	7	0	2	2	2	....	....	....	....	....
1998-99	Yaroslavl	Russia	42	7	13	20	30	10	2	2	4	10
99-2000	St. John's	AHL	45	6	7	13	14	....	....	....	....	....
	South Carolina	ECHL	4	0	4	4	2	....	....	....	....	....
	Long Beach	IHL	16	3	3	6	18	1	0	0	0	0
2000-01	St. John's	AHL	9	1	1	2	2	....	....	....	....	....
	Yaroslavl	Russia	22	5	5	10	56	11	2	1	3	8
2001-02	Yaroslavl	Russia	50	8	12	20	48	9	3	2	5	14

Signed as a free agent by **Yaroslavl** (Russia) with **Toronto** retaining NHL rights, November 14, 2000.

APPS, Syl (APPS, sihl) TOR.

Center. Shoots right. 6', 195 lbs. Born, Pittsburgh, PA, June 2, 1976.

			Regular Season					Playoffs				
Season	Club	Lea	GP	G	A	TP	PIM	GP	G	A	TP	PIM
1994-95	St. Michael's B	OJHL-B	6	3	1	4	2					
1995-96	Princeton	ECAC	26	4	6	10	30					
1996-97	Princeton	ECAC	27	3	6	9	40					
1997-98	Princeton	ECAC	35	10	8	18	65					
1998-99	Princeton	ECAC	34	13	21	34	45					
99-2000	St. John's	AHL	58	5	7	12	87					
2000-01	St. John's	AHL	69	6	8	14	73	4	0	0	0	0
2001-02	Norfolk Admirals	AHL	1	0	0	0	0					
	Jackson Bandits	ECHL	12	3	2	5	19					
	Springfield	AHL	6	1	0	1	0					
	Trenton Titans	ECHL	42	8	15	23	56	7	1	1	2	16

Signed as a free agent by **Toronto**, July 22, 1999.

AQUINO, Anthony (a-KEE-noh, AN-thuh-nee) DAL.

Right wing. Shoots right. 5'10", 180 lbs. Born, Toronto, Ont., August 1, 1982.
(Dallas' 3rd choice, 92nd overall, in 2001 Entry Draft).

			Regular Season					Playoffs				
Season	Club	Lea	GP	G	A	TP	PIM	GP	G	A	TP	PIM
1997-98	Mississauga	OPJHL	50	10	13	23	14					
1998-99	Bramalea Blues	OPJHL	47	31	44	75	31					
99-2000	Merrimack	H-East	36	15	14	29	12					
2000-01	Merrimack	H-East	38	17	25	42	22					
2001-02	Merrimack	H-East	36	24	20	44	20					

Hockey East All-Rookie Team (2000) • Hockey East Second All-Star Team (2001)

ARCHER, Andrew (AHR-chuhr, AN-droo) MTL.

Defense. Shoots right. 6'4", 194 lbs. Born, Calgary, Alta., May 15, 1983.
(Montreal's 7th choice, 203rd overall, in 2001 Entry Draft Draft).

			Regular Season					Playoffs				
Season	Club	Lea	GP	G	A	TP	PIM	GP	G	A	TP	PIM
1998-99	Richmond Hill	OMHA	STATISTICS NOT AVAILABLE									
99-2000	Oshawa Generals	OHL	47	0	1	1	24	3	0	1	1	2
2000-01	Oshawa Generals	OHL	2	0	0	0	4					
	Guelph Storm	OHL	50	0	2	2	59	4	0	0	0	4
2001-02	Guelph Storm	OHL	58	3	10	13	76	9	0	2	2	16

Traded to **Guelph** (OHL) by **Oshawa** (OHL) with Nick Lees and Oshawa's 5th choice (Tyler Haskins) in 2002 Midget Draft for Jon Hedberg, Chris Whitley and Kevin Mitchell, September 29, 2000.

ARMSTRONG, Colby (AHRM-stawng, KOHL-bee) PIT.

Right wing. Shoots right. 5'11", 185 lbs. Born, Lloydminster, Sask., November 23, 1982.
(Pittsburgh's 1st choice, 21st overall, in 2001 Entry Draft).

			Regular Season					Playoffs				
Season	Club	Lea	GP	G	A	TP	PIM	GP	G	A	TP	PIM
1998-99	Sask. Contacts	SMHL	33	21	19	40	103					
	Red Deer Rebels	WHL	1	0	1	1	0					
99-2000	Red Deer Rebels	WHL	68	13	25	38	122	2	0	1	1	11
2000-01	Red Deer Rebels	WHL	72	36	42	78	156	21	6	6	12	29
2001-02	Red Deer Rebels	WHL	64	27	41	68	115	23	6	10	16	32

ARTEMENKOV, Yuri (ahr-TUH-mehn-kahv, YOO-ree) CGY.

Right wing. Shoots left. 6'1", 174 lbs. Born, Moscow, USSR, February 3, 1984.
(Calgary's 4th choice, 112th overall, in 2002 Entry Draft).

			Regular Season					Playoffs				
Season	Club	Lea	GP	G	A	TP	PIM	GP	G	A	TP	PIM
99-2000	Team Russia	Nat-Tm	5	1	1	2	0					
2000-01	Krylja Sovetov 2	Russia-3	2	0	0	0	4					
2001-02	Krylja Sovetov 2	Russia-3	32	24	15	42	10					

ARTUKHIN, Evgeni (ahr-TYEW-khin, yehv-GEH-nee) T.B.

Right wing. Shoots left. 6'4", 215 lbs. Born, Moscow, USSR, April 4, 1983.
(Tampa Bay's 4th choice, 94th overall, in 2001 Entry Draft).

			Regular Season					Playoffs				
Season	Club	Lea	GP	G	A	TP	PIM	GP	G	A	TP	PIM
99-2000	Vityaz Podolsk 2	Russia-3	26	9	8	17	46					
	Vityaz Podolsk	Russia-2	3	0	0	0	2					
2000-01	Vityaz Podolsk	Russia	24	0	1	1	14					
2001-02	Vityaz Podolsk 2	Russia-3	4	3	1	4	6					
	Vityaz Podolsk	Russia-2	49	15	7	22	94	12	0	1	1	18

ASLUND, Calle (AZ-luhnd, KAL-ee) BUF.

Defense. Shoots left. 6'2", 198 lbs. Born, Haninge, Sweden, March 29, 1983.
(Buffalo's 6th choice, 234th overall, in 2001 Entry Draft).

			Regular Season					Playoffs				
Season	Club	Lea	GP	G	A	TP	PIM	GP	G	A	TP	PIM
99-2000	Huddinge IK-18	Swede-Jr.	17	1	5	6	48					
2000-01	Huddinge IK-18	Swede-Jr.	8	1	3	4	30					
	Huddinge IK Jr.	Swede-Jr.	7	0	1	1	14					
2001-02	Huddinge IK Jr.	Swede-Jr.	27	1	6	7	66					
	Huddinge IK	Swede-2	26	0	0	0	47					

AUFIERO, Patrick (ow-fee-AIR-oh, PAT-rihk) NYR

Defense. Shoots right. 6'2", 186 lbs. Born, Winchester, MA, July 1, 1980.
(NY Rangers' 5th choice, 90th overall, in 1999 Entry Draft).

			Regular Season					Playoffs				
Season	Club	Lea	GP	G	A	TP	PIM	GP	G	A	TP	PIM
1995-96	Winchester High	Hi-School	25	13	18	31	....					
1996-97	Winchester High	Hi-School	27	27	21	48	....					
1997-98	Team USA	USDP-18	56	10	11	21	111					
1998-99	Boston University	H-East	22	3	4	7	14					
99-2000	Boston University	H-East	38	3	20	23	37					
2000-01	Boston University	H-East	34	5	8	13	30					
2001-02	Boston University	H-East	26	2	10	12	16					

Hockey East Second All-Star Team (2000)

AULIN, Jared (AW-lihn, JAIR-ehd) L.A.

Center. Shoots right. 6', 180 lbs. Born, Calgary, Alta., March 15, 1982.
(Colorado's 2nd choice, 47th overall, in 2000 Entry Draft).

			Regular Season					Playoffs				
Season	Club	Lea	GP	G	A	TP	PIM	GP	G	A	TP	PIM
1997-98	Airdrie Xtreme	AAHA	55	42	61	103	60					
	Kamloops Blazers	WHL	2	0	0	0	0					
1998-99	Kamloops Blazers	WHL	55	7	19	26	23	13	1	3	4	2
99-2000	Kamloops Blazers	WHL	57	17	38	55	70	4	0	1	1	6
2000-01	Kamloops Blazers	WHL	70	31	*77	108	62	4	0	2	2	0
2001-02	Kamloops Blazers	WHL	46	33	34	67	80	4	1	2	3	2

WHL West First All-Star Team (2001, 2002)

Traded to **LA Kings** by **Colorado** to complete transaction that sent Rob Blake and Steve Reinprecht to Colorado (February 21, 2001), March 22, 2001.

BABCHUK, Anton (bab-CHUHK, an-TAWN) CHI.

Defense. Shoots right. 6'4", 199 lbs. Born, Kiev, USSR, May 6, 1984.
(Chicago's 1st choice, 21st overall, in 2002 Entry Draft).

			Regular Season					Playoffs				
Season	Club	Lea	GP	G	A	TP	PIM	GP	G	A	TP	PIM
99-2000	Elektrostal-18	Russia-Jr.	5	0	0	0	18					
	Elektrostal Jr.	Russia-Jr.	6	0	0	0	8					
	Elektrostal 2	Russia-3	18	0	1	1	18					
2000-01	Elektrostal	Russia-2	7	0	0	0	12					
2001-02	Elektrostal	Russia-2	40	7	8	15	90					
	Elektrostal 2	Russia-3	3	0	0	0	8					

BABY, Stephan (BAY-bee, STEE-vehn) ATL.

Right wing. Shoots right. 6'5", 235 lbs. Born, Chicago, IL, January 31, 1980.
(Atlanta's 8th choice, 188th overall, in 1999 Entry Draft).

			Regular Season					Playoffs				
Season	Club	Lea	GP	G	A	TP	PIM	GP	G	A	TP	PIM
1997-98	Green Bay	USHL	56	17	17	34	85	4	1	3	4	8
1998-99	Green Bay	USHL	55	23	24	47	83	6	1	1	2	4
99-2000	Cornell Big Red	ECAC	31	4	10	14	52					
2000-01	Cornell Big Red	ECAC	32	8	20	28	47					
2001-02	Cornell Big Red	ECAC	35	9	23	32	42					

ECAC Second All-Star Team (2002)

BACKER, Per (BAK-uhr, PAIR) DET.

Right wing. Shoots left. 6'1", 161 lbs. Born, Grums, Sweden, January 4, 1982.
(Detroit's 7th choice, 187th overall, in 2000 Entry Draft).

			Regular Season					Playoffs				
Season	Club	Lea	GP	G	A	TP	PIM	GP	G	A	TP	PIM
1998-99	Grums IK	Swede-2	17	1	1	2	4					
99-2000	Grums IK	Swede-2	46	12	10	22	24					
2000-01	Bofors IK	Swede-2	41	20	15	35	47					
2001-02	Farjestad	Sweden	47	4	8	12	12	10	5	1	6	10

BACKMAN, Christian (BAK-man, KRIH-stan) ST.L.

Defense. Shoots left. 6'4", 198 lbs. Born, Alingsas, Sweden, April 28, 1980.
(St. Louis' 1st choice, 24th overall, in 1998 Entry Draft).

			Regular Season					Playoffs				
Season	Club	Lea	GP	G	A	TP	PIM	GP	G	A	TP	PIM
1996-97	V. Frolunda Jr.	Swede-Jr.	26	2	5	7	16					
1997-98	V. Frolunda-18	Swede-Jr.	4	4	1	5	2	5	2	2	4	2
	V. Frolunda Jr.	Swede-Jr.	28	5	14	19	12	2	0	1	1	4
1998-99	V. Frolunda Jr.	Swede-Jr.	4	0	2	2	4					
	Vastra Frolunda	Sweden	49	0	4	4	4	4	0	0	0	0
99-2000	V. Frolunda Jr.	Swede-Jr.	5	1	1	2	0	3	1	1	2	0
	Gislaveds IK	Swede-2	21	5	2	7	8					
	Vastra Frolunda	Sweden	27	1	0	1	14	5	0	0	0	0
2000-01	Vastra Frolunda	Sweden	50	1	10	11	32	3	0	2	2	2
2001-02	Vastra Frolunda	Sweden	44	7	4	11	38	10	0	0	0	8

BAHEN, Chris (BAY-hehn, KRIHS) COL.

Defense. Shoots left. 6', 180 lbs. Born, Montreal, Que., November 16, 1980.
(Colorado's 10th choice, 189th overall, in 2000 Entry Draft).

			Regular Season					Playoffs				
Season	Club	Lea	GP	G	A	TP	PIM	GP	G	A	TP	PIM
1994-95	Thornhill	OMHA	40	12	38	50	82					
1995-96	Thornhill	OMHA	40	18	51	69	96					
1996-97	Thornhill	OMHA	40	23	42	65	108					
1997-98	Thornhill	MTJHL	45	5	9	14	131					
1998-99	Milton Merchants	OPJHL	32	3	14	17	20					
99-2000	Clarkson Knights	ECAC	34	8	10	18	54					
2000-01	Clarkson Knights	ECAC	34	3	7	10	45					
2001-02	Clarkson Knights	ECAC	37	2	6	8	36					

ECAC All-Academic Team (2002)

BAINES, Ajay (BAYNZ, AY-JAY) CHI.

Center. Shoots left. 5'10", 178 lbs. Born, Kamloops, B.C., March 25, 1978.

			Regular Season					Playoffs				
Season	Club	Lea	GP	G	A	TP	PIM	GP	G	A	TP	PIM
1994-95	Kamloops	BCAHA	52	45	79	124	139					
1995-96	Kamloops Blazers	WHL	68	14	29	43	43					
1996-97	Kamloops Blazers	WHL	70	32	43	75	106	5	4	1	5	6
1997-98	Kamloops Blazers	WHL	72	34	25	59	88					
1998-99	Kamloops Blazers	WHL	72	33	32	65	145	15	7	6	13	20
99-2000	Greenville	ECHL	67	24	31	55	102	15	2	5	7	13
2000-01	Norfolk Admirals	AHL	73	18	18	36	92	9	0	1	1	2
2001-02	Norfolk Admirals	AHL	80	16	28	44	70	4	0	1	1	0

Signed as a free agent by **Chicago**, August 1, 2001.

BALAN, Scott (BAY-luhn, SKAWT) CHI.

Defense. Shoots right. 6'3", 191 lbs. Born, Medicine Hat, Alta., May 29, 1982.
(Chicago's 5th choice, 106th overall, in 2000 Entry Draft).

			Regular Season					Playoffs				
Season	Club	Lea	GP	G	A	TP	PIM	GP	G	A	TP	PIM
1997-98	Regina Chiefs	SMHL	56	6	25	31	139					
	Regina Pats	WHL	1	0	1	1	0					
1998-99	Regina Pats	WHL	63	1	8	9	42					
99-2000	Regina Pats	WHL	67	3	11	14	157	7	0	1	1	17
2000-01	Regina Pats	WHL	42	2	5	7	69					
	Saskatoon Blades	WHL	28	0	5	5	49					
2001-02	Saskatoon Blades	WHL	27	1	10	11	40					

Traded to **Saskatoon** (WHL) by **Regina** (WHL) with future considerations for Garnet Exelby, January 15, 2001. • Missed majority of 2001-02 season recovering from knee injury suffered in game vs. Saskatoon (WHL), December 18, 2001.

BALEJ, Josef (BAH-lay, YOH-zehf) MTL.

Right wing. Shoots right. 6'1", 187 lbs. Born, Myjava, Czech., February 22, 1982.
(Montreal's 3rd choice, 78th overall, in 2000 Entry Draft).

			Regular Season					Playoffs				
Season	Club	Lea	GP	G	A	TP	PIM	GP	G	A	TP	PIM
1997-98	Dukla Trencin Jr.	Slovak-Jr.	52	57	40	97	60	....	....	....	....	....
1998-99	Thunder Bay	USHL	38	8	7	15	9	....	....	....	....	....
	Rochester	USHL	17	0	1	1	2	....	....	....	....	....
99-2000	Portland	WHL	65	22	23	45	33	....	....	....	....	....
2000-01	Portland	WHL	46	32	21	53	18	16	9	6	15	6
2001-02	Portland	WHL	65	51	41	92	52	7	0	2	2	6

WHL West First All-Star Team (2002)

BALLANTYNE, Paul (BAL-uhn-tughn, PAWL) DET.

Defense. Shoots right. 6'3", 200 lbs. Born, Waterloo, Ont., July 16, 1982.
(Detroit's 8th choice, 196th overall, in 2000 Entry Draft).

			Regular Season					Playoffs				
Season	Club	Lea	GP	G	A	TP	PIM	GP	G	A	TP	PIM
1997-98	Waterloo Lions	OMHA	30	4	16	20	55	....	....	....	....	....
	Waterloo Siskins	OPJHL	1	0	0	0	0	....	....	....	....	....
1998-99	Sault Ste. Marie	OHL	53	0	6	6	33	5	1	0	1	4
99-2000	Sault Ste. Marie	OHL	58	4	15	19	60	17	2	3	5	17
2000-01	Sault Ste. Marie	OHL	63	12	28	40	60	....	....	....	....	....
2001-02	Sault Ste. Marie	OHL	68	4	24	28	40	6	1	1	2	8

BALLARD, Keith (BAL-uhrd, KEETH) BUF.

Defense. Shoots left. 5'11", 202 lbs. Born, Baudette, MN, November 26, 1982.
(Buffalo's 1st choice, 11th overall, in 2002 Entry Draft).

			Regular Season					Playoffs				
Season	Club	Lea	GP	G	A	TP	PIM	GP	G	A	TP	PIM
99-2000	Team USA	USDP-18	58	12	21	33	....	....	....	....	....	....
2000-01	Omaha Lancers	USHL	56	22	29	51	168	10	1	6	7	8
2001-02	U. of Minnesota	WCHA	41	10	13	23	42	....	....	....	....	....

USHL First All-Star Team (2000) • WCHA All-Rookie Team (2002)

BARANOV, Konstantin (buh-RA-nawf, kawn-stuhn-TEEN) PHI.

Right wing. Shoots left. 6'2", 185 lbs. Born, Omsk, USSR, January 11, 1982.
(Philadelphia's 3rd choice, 126th overall, in 2002 Entry Draft).

			Regular Season					Playoffs				
Season	Club	Lea	GP	G	A	TP	PIM	GP	G	A	TP	PIM
1998-99	Omsk 2	Russia-4	23	18	8	26	40	....	....	....	....	....
	Avangard Omsk	Russia	1	0	0	0	0	2	0	0	0	0
99-2000	Omsk 2	Russia-3	33	15	8	23	46	....	....	....	....	....
	Avangard Omsk	Russia	1	0	0	0	2	....	....	....	....	....
2000-01	Kristall Saratov	Russia-2	26	6	9	15	26	....	....	....	....	....
	Ufa	Russia	8	1	0	1	4	....	....	....	....	....
2001-02	Avangard Omsk	Russia	5	0	0	0	6	....	....	....	....	....
	Mechel	Russia	6	1	2	3	2	....	....	....	....	....
	Lada Togliatti	Russia	20	2	4	6	18	3	0	2	2	0

BARARUK, David (BAIR-a-ruhk, DAY-vihd) DAL.

Center. Shoots left. 6', 175 lbs. Born, Moose Jaw, Sask., May 26, 1983.
(Dallas' 8th choice, 147th overall, in 2002 Entry Draft).

			Regular Season					Playoffs				
Season	Club	Lea	GP	G	A	TP	PIM	GP	G	A	TP	PIM
99-2000	Moose Jaw	WHL	21	0	2	2	0	2	0	0	0	0
2000-01	Moose Jaw	WHL	53	6	9	15	9	3	0	0	0	0
2001-02	Moose Jaw	WHL	72	33	29	62	31	12	3	2	5	0

BARBER, Greg (BAHR-buhr, GREHG) BOS.

Right wing. Shoots right. 6', 185 lbs. Born, Dawson Creek, B.C., May 26, 1980.
(Boston's 7th choice, 207th overall, in 1999 Entry Draft).

			Regular Season					Playoffs				
Season	Club	Lea	GP	G	A	TP	PIM	GP	G	A	TP	PIM
1996-97	Kelowna Spartans	BCAHA	54	36	48	84	90	....	....	....	....	....
1997-98	Victoria Salsa	BCHL	60	15	22	37	24	7	2	2	4	4
1998-99	Victoria Salsa	BCHL	60	41	41	82	95	....	....	....	....	....
99-2000	U. of Denver	WCHA	40	7	8	15	24	....	....	....	....	....
2000-01	U. of Denver	WCHA	35	7	8	15	22	....	....	....	....	....
2001-02	U. of Denver	WCHA	38	18	17	35	28	....	....	....	....	....

BARCH, Krys (BAHR-ch, KRIHS) WSH.

Left wing. Shoots left. 6'2", 200 lbs. Born, Guelph, Ont., March 26, 1980.
(Washington's 3rd choice, 106th overall, in 1998 Entry Draft).

			Regular Season					Playoffs				
Season	Club	Lea	GP	G	A	TP	PIM	GP	G	A	TP	PIM
1995-96	Georgetown	OPJHL	41	6	8	14	10	....	....	....	....	....
1996-97	Georgetown	OPJHL	51	18	26	44	58	....	....	....	....	....
1997-98	London Knights	OHL	65	9	27	36	62	16	4	3	7	16
1998-99	London Knights	OHL	66	18	20	38	66	25	9	17	26	15
99-2000	London Knights	OHL	56	23	26	49	78	....	....	....	....	....
	Portland Pirates	AHL	....	....	....	....	....	4	0	2	2	2
2000-01	Portland Pirates	AHL	76	10	15	25	91	2	0	0	0	0
2001-02	Portland Pirates	AHL	29	3	8	11	28	....	....	....	....	....
	Richmond	ECHL	25	6	4	10	43	....	....	....	....	....

BARKUNOV, Alexander (bahr-koo-NAHF, al-ehx-AN-duhr) CHI.

Defense. Shoots right. 6'1", 199 lbs. Born, Novosibirsk, USSR, May 13, 1981.
(Chicago's 7th choice, 151st overall, in 2000 Entry Draft).

			Regular Season					Playoffs				
Season	Club	Lea	GP	G	A	TP	PIM	GP	G	A	TP	PIM
1996-97	Yaroslavl 2	Russia-3	1	0	0	0	0	....	....	....	....	....
1997-98	Yaroslavl	Russia	20	1	1	2	6	....	....	....	....	....
1998-99	Yaroslavl 2	Russia-3	19	0	3	3	8	....	....	....	....	....
99-2000	Yaroslavl	Russia	38	5	9	14	16	....	....	....	....	....
2000-01	Yaroslavl	Russia	34	5	1	6	10	2	0	0	0	2
2001-02	Yaroslavl 2	Russia-3	10	1	4	5	2	....	....	....	....	....
	Yaroslavl	Russia	15	0	0	0	2	....	....	....	....	....
	Amur Khabarovsk	Russia	7	0	0	0	0	....	....	....	....	....

BARNES, Ryan (BAHR-nz, RIGH-uhn) DET.

Left wing. Shoots left. 6'1", 201 lbs. Born, Dunnville, Ont., January 30, 1980.
(Detroit's 2nd choice, 55th overall, in 1998 Entry Draft).

			Regular Season					Playoffs				
Season	Club	Lea	GP	G	A	TP	PIM	GP	G	A	TP	PIM
1996-97	Quinte Hawks	MTJHL	46	15	19	34	245	....	....	....	....	....
1997-98	Sudbury Wolves	OHL	46	13	18	31	111	10	0	2	2	24
1998-99	Sudbury Wolves	OHL	8	2	0	2	23	....	....	....	....	....
	St. Michael's	OHL	31	11	14	25	*215	....	....	....	....	....
	Barrie Colts	OHL	24	16	14	30	*161	12	2	4	6	40
99-2000	Barrie Colts	OHL	31	17	12	29	98	25	7	7	14	49
2000-01	Cincinnati	AHL	1	0	0	0	7	....	....	....	....	....
	Toledo Storm	ECHL	16	2	4	6	31	....	....	....	....	....
2001-02	Cincinnati	AHL	46	2	3	5	152	....	....	....	....	....
	Toledo Storm	ECHL	1	1	0	1	0	....	....	....	....	....

Traded to **Guelph** (OHL) by **Sudbury** (OHL) with Sudbury's 3rd round choice in 1999 OHL Priority Draft for Brian McGratton and Guelph's 4th round choice (Steve Ellis) in 1999 OHL Priority Draft, October 14, 1998. Traded to **St. Michael's** (OHL) by **Guelph** (OHL) with Sudbury's 3rd round choice (previously acquired, St. Michael's selected Steve Farquharson) in 1999 OHL Priority Draft for Charlie Stephens and St. Michael's 5th round choice (Derek Hennessey) in 1999 OHL Priority Draft, October 14, 1998. Traded to **Barrie** (OHL) by **St. Michael's** (OHL) with Mike Jefferson, Sheldon Keefe and Shawn Cation for Keith Delaney, Darryl Bootland, Adam DeLeew and Brad Pierce, January 11, 1999. • Suspended for 25 games for stick-swining incident in game vs. Oshawa (OHL), October 3, 1999. • Missed majority of 2000-01 season recovering from head injury suffered in training camp, September, 2000.

BARRETT, Nathan (BAIR-uht, NAY-thun) TOR.

Center. Shoots left. 5'11", 192 lbs. Born, Vancouver, B.C., August 3, 1981.
(Vancouver's 6th choice, 241st overall, in 2000 Entry Draft).

			Regular Season					Playoffs				
Season	Club	Lea	GP	G	A	TP	PIM	GP	G	A	TP	PIM
1996-97	Langley Bantams	BCAHA	90	107	109	216	96	....	....	....	....	....
1997-98	Tri-City	WHL	47	1	1	2	23	....	....	....	....	....
1998-99	Tri-City	WHL	33	9	9	18	19	....	....	....	....	....
	Lethbridge	WHL	22	12	9	21	19	4	1	0	1	0
99-2000	Lethbridge	WHL	72	44	38	82	38	....	....	....	....	....
2000-01	Lethbridge	WHL	70	46	53	99	66	5	1	1	2	6
2001-02	Lethbridge	WHL	72	45	*62	*107	100	4	0	1	1	6

WHL East Second All-Star Team (2001) • WHL East First All-Star Team (2002)

Traded to **Lethbridge** (WHL) by **Tri-City** (WHL) with Ryan Jorde for Chris Huppe and Andrew Guindon, February 2, 1999. Signed as a free agent by **Toronto**, July 31, 2002.

BARTEK, Martin (BAHR-tehk, MAHR-tehn) NSH.

Center. Shoots left. 6'1", 210 lbs. Born, Kindgseed Jill, Czech., July 17, 1980.
(Nashville's 7th choice, 202nd overall, in 1998 Entry Draft).

			Regular Season					Playoffs				
Season	Club	Lea	GP	G	A	TP	PIM	GP	G	A	TP	PIM
1995-96	HKm Zvolen Jr.	Slovak-Jr.	46	75	43	118	....	....	....	....	....	....
1996-97	King's Edge Hill	Hi-School	35	45	40	85	32	....	....	....	....	....
1997-98	Rouyn-Noranda	QMJHL	28	9	19	28	12	....	....	....	....	....
	Rimouski Oceanic	QMJHL	13	3	4	7	6	....	....	....	....	....
	Sherbrooke	QMJHL	25	11	12	23	38	....	....	....	....	....
1998-99	HKm Zvolen Jr.	Slovak-Jr.	17	14	17	31	89	....	....	....	....	....
	HKm Zvolen	Slovakia	28	10	8	18	18	2	1	0	1	0
99-2000	Moncton Wildcats	QMJHL	69	32	44	76	36	16	10	13	23	24
2000-01	Milwaukee	IHL	14	1	0	1	2	....	....	....	....	....
	New Orleans	ECHL	51	30	33	63	16	8	5	4	9	4
2001-02	Milwaukee	AHL	62	14	8	22	25	....	....	....	....	....
	Cincinnati	ECHL	2	2	2	4	0	....	....	....	....	....

BARTLETT, Russ (BAHRT-leht, RUHS) TOR.

Center. Shoots left. 6'2", 185 lbs. Born, Windham, NH, February 10, 1978.
(Toronto's 7th choice, 194th overall, in 1997 Entry Draft).

			Regular Season					Playoffs				
Season	Club	Lea	GP	G	A	TP	PIM	GP	G	A	TP	PIM
1996-97	Phillips Exeter	Hi-School	31	27	63	90	24	....	....	....	....	....
1997-98	Boston University	H-East	38	8	11	19	44	....	....	....	....	....
1998-99	Boston University	H-East	35	12	20	32	18	....	....	....	....	....
99-2000	St. Lawrence	ECAC	DID NOT PLAY – TRANSFERRED COLLEGES									
2000-01	St. Lawrence	ECAC	37	18	25	43	20	....	....	....	....	....
2001-02	St. Lawrence	ECAC	34	4	18	22	22	....	....	....	....	....

BARTOVIC, Milan (BAHR-tuh-vihch, MIH-lan) BUF.

Right wing. Shoots left. 5'11", 192 lbs. Born, Trencin, Czech., April 9, 1981.
(Buffalo's 2nd choice, 35th overall, in 1999 Entry Draft).

			Regular Season					Playoffs				
Season	Club	Lea	GP	G	A	TP	PIM	GP	G	A	TP	PIM
1997-98	Dukla Trencin Jr.	Slovak-Jr.	26	2	6	8	27	....	....	....	....	....
1998-99	Dukla Trencin Jr.	Slovak-Jr.	46	36	35	71	62	6	9	3	12	10
99-2000	Tri-City	WHL	18	8	9	17	12	....	....	....	....	....
	Brandon	WHL	38	18	22	40	28	....	....	....	....	....
2000-01	Brandon	WHL	34	15	25	40	40	6	1	2	3	8
	Rochester	AHL	2	1	1	2	0	4	0	1	1	2
2001-02	Rochester	AHL	73	15	11	26	56	2	0	0	0	0

Traded to **Brandon** (WHL) by **Tri-City** (WHL) for Jomar Cruz, November 23, 1999. • Missed majority of 2000-01 season recovering from shoulder injury originally suffered in game vs. Red Deer (WHL), October 10, 2000.

BATEMAN, Jeff (BAYT-mahn, JEHF) DAL.

Center. Shoots left. 5'11", 184 lbs. Born, Belleville, Ont., August 29, 1981.
(Dallas' 4th choice, 126th overall, in 1999 Entry Draft).

			Regular Season					Playoffs				
Season	Club	Lea	GP	G	A	TP	PIM	GP	G	A	TP	PIM
1996-97	Belleville	OMHA	51	102	64	166	....	....	....	....	....	....
1997-98	Wellington Dukes	MTJHL	50	26	35	61	68	....	....	....	....	....
1998-99	Brampton	OHL	68	23	35	58	27	....	....	....	....	....
99-2000	Brampton	OHL	64	23	41	64	62	6	2	1	3	4
2000-01	Brampton	OHL	53	15	44	59	81	9	5	7	12	18
2001-02	Fort Worth	CHL	27	7	11	18	32	1	1	0	1	0
	Utah Grizzlies	AHL	35	1	6	7	26	4	0	2	2	2

BAUM, Dan (BAWM, DAN) EDM.

Center. Shoots left. 6', 170 lbs. Born, Biggar, Sask., June 4, 1983.
(Edmonton's 8th choice, 215th overall, in 2001 Entry Draft).

			Regular Season					Playoffs				
Season	Club	Lea	GP	G	A	TP	PIM	GP	G	A	TP	PIM
99-2000	Prince George	WHL	55	6	10	16	82	12	1	0	1	19
2000-01	Prince George	WHL	59	9	14	23	169	6	1	2	3	27
2001-02	Prince George	WHL	72	32	35	67	197	6	5	3	8	25

BAYDA, Ryan (BAY-duh, RIGH-uhn) CAR.

Left wing. Shoots left. 5'11", 185 lbs. Born, Saskatoon, Sask., December 9, 1980.
(Carolina's 2nd choice, 80th overall, in 2000 Entry Draft).

			Regular Season					Playoffs				
Season	Club	Lea	GP	G	A	TP	PIM	GP	G	A	TP	PIM
1995-96	Saskatoon Flyers	SMHL	60	85	74	159	85	....	....	....	....	....
1996-97	Sask. Contacts	SMHL	44	22	23	45	18	....	....	....	....	....
1997-98	Sask. Contacts	SMHL	41	29	49	78	103	....	....	....	....	....
1998-99	Vernon Vipers	BCHL	45	24	58	82	15	....	....	....	....	....
99-2000	North Dakota	WCHA	44	17	23	40	30	....	....	....	....	....
2000-01	North Dakota	WCHA	46	25	34	59	48	....	....	....	....	....
2001-02	North Dakota	WCHA	37	19	28	47	52	....	....	....	....	....
	Lowell	AHL	3	1	1	2	0	5	3	0	3	0

BCHL Rookie of the Year (1999) • WCHA All-Rookie Team (2000) • WCHA Second All-Star Team (2001, 2002)

BEAUCHEMIN, Francois (boh-sheh-MEH, frahn-SWUH) MTL.

Defense. Shoots left. 6', 206 lbs. Born, Sorel, Que., June 4, 1980.
(Montreal's 3rd choice, 75th overall, in 1998 Entry Draft).

			Regular Season					Playoffs				
Season	Club	Lea	GP	G	A	TP	PIM	GP	G	A	TP	PIM
1995-96	Richelieu	QAAA	40	9	23	32	59	....	....	....	....	....
1996-97	Laval Titan	QMJHL	66	7	21	28	132	3	0	0	0	2
1997-98	Laval Titan	QMJHL	70	12	35	47	132	16	1	3	4	23
1998-99	Acadie-Bathurst	QMJHL	31	4	17	21	53	23	2	16	18	55
99-2000	Acadie-Bathurst	QMJHL	38	11	36	47	64	....	....	....	....	....
	Moncton Wildcats	QMJHL	33	8	31	39	35	16	2	11	13	14
2000-01	Quebec	AHL	56	3	6	9	44	....	....	....	....	....
2001-02	Quebec	AHL	56	8	11	19	88	3	0	1	1	0
	Mississippi	ECHL	7	1	3	4	2	....	....	....	....	....

QMJHL All-Rookie Team (1997) • QMJHL Second All-Star Team (2000)

Traded to **Moncton** (QMJHL) by **Acadie-Bathurst** (QMJHL) with Mathieu Benoit and Martin Lavergne for Tyson Maloney, Daniel MacLeod and future considerations, December 27, 1999.

BECKETT, Jason (Beh-keht, JAY-suhn) NSH.

Defense. Shoots right. 6'3", 208 lbs. Born, Lethbridge, Alta., July 23, 1980.
(Philadelphia's 2nd choice, 42nd overall, in 1998 Entry Draft).

			Regular Season					Playoffs				
Season	Club	Lea	GP	G	A	TP	PIM	GP	G	A	TP	PIM
1996-97	Lethbridge	AMHL	34	7	10	17	118	....	....	....	....	....
1997-98	Seattle	WHL	71	1	11	12	241	5	0	0	0	16
1998-99	Seattle	WHL	70	4	26	30	195	11	0	1	1	40
99-2000	Seattle	WHL	70	3	15	18	183	7	1	1	2	12
2000-01	Trenton Titans	ECHL	17	2	2	4	24	15	0	1	1	26
	Philadelphia	AHL	56	2	10	12	107	....	....	....	....	....
2001-02	Philadelphia	AHL	9	0	0	0	11	....	....	....	....	....
	Milwaukee	AHL	28	2	4	6	56	....	....	....	....	....
	Trenton Titans	ECHL	14	1	1	2	49	3	0	0	0	4

Traded to **Nashville** by **Philadelphia** with Petr Hubacek for Yves Sarault and a conditional choice in 2003 Entry Draft, January 11, 2002.

BELL, Brendan (BEHL, BREHN-duhn) TOR.

Defense. Shoots left. 6'1", 198 lbs. Born, Ottawa, Ont., March 31, 1983.
(Toronto's 3rd choice, 65th overall, in 2001 Entry Draft).

			Regular Season					Playoffs				
Season	Club	Lea	GP	G	A	TP	PIM	GP	G	A	TP	PIM
1998-99	Ottawa Jr. Sens	OCJHL	54	7	20	27	46	....	....	....	....	....
99-2000	Ottawa 67's	OHL	48	1	32	33	34	5	0	1	1	4
2000-01	Ottawa 67's	OHL	68	7	32	39	59	20	1	11	12	22
2001-02	Ottawa 67's	OHL	67	10	36	46	56	13	2	5	7	25

OCJHL All-Rookie Team (1999) • OCJHL Rookie of the Year (1999)

BELL, Thatcher (BEHL, THA-tchuhr)

Center. Shoots left. 6', 172 lbs. Born, Charlottetown, P.E.I., February 1, 1982.
(Vancouver's 2nd choice, 71st overall, in 2000 Entry Draft).

			Regular Season					Playoffs				
Season	Club	Lea	GP	G	A	TP	PIM	GP	G	A	TP	PIM
1997-98	Upper Canada	Hi-School	44	35	50	85	....	....	....	....	....	....
1998-99	Rimouski Oceanic	QMJHL	64	16	38	54	67	11	3	1	4	0
99-2000	Rimouski Oceanic	QMJHL	53	26	43	69	61	5	0	0	0	15
2000-01	Rimouski Oceanic	QMJHL	46	27	32	59	77	11	6	7	13	14
2001-02	Rimouski Oceanic	QMJHL	29	12	27	39	50	7	8	7	15	6

BELLISSIMO, Vince (behl-IHS-ih-moh, VIHNTS) FLA.

Center. Shoots right. 6', 199 lbs. Born, Toronto, Ont., December 14, 1982.
(Florida's 6th choice, 158th overall, in 2002 Entry Draft).

			Regular Season					Playoffs				
Season	Club	Lea	GP	G	A	TP	PIM	GP	G	A	TP	PIM
99-2000	St. Michael's B	OPJHL	47	30	29	59	31	....	....	....	....	....
2000-01	St. Michael's B	OPJHL	47	32	64	96	28	6	6	8	14	....
2001-02	Topeka	USHL	61	37	39	76	33	....	....	....	....	....

USHL First All-Star Team (2002) • USHL Top Forward (2002)

• Signed Letter of Intent to attend **Western Michigan** (CCHA), September 20, 2001.

BEMBRIDGE, Garrett (bem-BRIHDJ, GAHR-reht) CGY.

Right wing. Shoots right. 6', 180 lbs. Born, Melfort, Sask., July 6, 1981.
(Calgary's 8th choice, 207th overall, in 2001 Entry Draft).

			Regular Season					Playoffs				
Season	Club	Lea	GP	G	A	TP	PIM	GP	G	A	TP	PIM
1997-98	Sask. Blazers	SMHL	44	29	45	74	74	....	....	....	....	....
	Saskatoon Blades	WHL	6	1	1	2	0	1	0	0	0	0
1998-99	Saskatoon Blades	WHL	68	23	27	50	30	....	....	....	....	....
99-2000	Saskatoon Blades	WHL	72	27	31	58	41	11	5	5	10	2
2000-01	Saskatoon Blades	WHL	72	38	40	78	40	....	....	....	....	....
2001-02	Saint John	AHL	66	9	12	21	22	....	....	....	....	....

• Re-entered NHL Entry Draft. Originally NY Rangers' 6th choice, 137th overall, in 1999 Entry Draft.

BERG, Reggie (BUHRG, REH-jee) CAR.

Center. Shoots left. 5'10", 180 lbs. Born, Coon Rapids, MN, September 18, 1976.
(Toronto's 12th choice, 178th overall, in 1996 Entry Draft).

			Regular Season					Playoffs				
Season	Club	Lea	GP	G	A	TP	PIM	GP	G	A	TP	PIM
1993-94	Des Moines	USHL	27	13	24	37	48	....	....	....	....	....
1994-95	Des Moines	USHL	35	30	35	65	75	12	11	8	19	12
1995-96	U. of Minnesota	WCHA	40	23	11	34	69	....	....	....	....	....
1996-97	U. of Minnesota	WCHA	38	11	26	37	48	....	....	....	....	....
1997-98	U. of Minnesota	WCHA	39	20	19	39	53	....	....	....	....	....
1998-99	U. of Minnesota	WCHA	43	20	28	48	64	....	....	....	....	....
99-2000	Florida	ECHL	52	27	25	52	64	5	2	1	3	6
	Orlando	IHL	2	0	0	0	0	....	....	....	....	....
2000-01	Cincinnati	IHL	3	0	0	0	2	....	....	....	....	....
	Lowell	AHL	30	6	7	13	22	....	....	....	....	....
	Florida	ECHL	33	19	29	48	39	5	1	3	4	6
2001-02	Florida	ECHL	29	14	11	25	41	....	....	....	....	....
	Lowell	AHL	29	4	8	12	15	5	1	0	1	4

WCHA Second All-Star Team (1998)

Signed as a free agent by **Carolina**, August 21, 2000.

BERGENHEIM, Sean (BUHR-gehn-highm, SHAWN) NYI

Left wing. Shoots left. 5'11", 194 lbs. Born, Helsinki, Finland, February 8, 1984.
(NY Islanders' 1st choice, 22nd overall, in 2002 Entry Draft).

			Regular Season					Playoffs				
Season	Club	Lea	GP	G	A	TP	PIM	GP	G	A	TP	PIM
2000-01	Jokerit Jr.	Finn-Jr.	30	22	11	33	34	3	1	0	1	0
	Jokerit-B	Finn-Jr.	1	1	0	1	4	2	0	0	0	2
2001-02	Jokerit Jr.	Finn-Jr.	23	11	19	30	36	6	6	2	8	20
	Kiekko Vantaa	Finland-2	4	0	0	0	52	....	....	....	....	....
	Jokerit Helsinki	Finland	28	2	2	4	4	....	....	....	....	....

BERGERON, Antoine (BAIR-zhuhr-uhn, ahn-TWAHN) ST.L.

Defense. Shoots left. 6'2", 212 lbs. Born, Valcourt, Que., December 14, 1981.
(St. Louis' 4th choice, 96th overall, in 2000 Entry Draft).

			Regular Season					Playoffs				
Season	Club	Lea	GP	G	A	TP	PIM	GP	G	A	TP	PIM
1996-97	Magog	QAAA	34	4	11	15	20	....	....	....	....	....
1997-98	Magog	QAAA	17	7	10	17	39	10	3	4	7	12
1998-99	Rimouski Oceanic	QMJHL	20	0	4	4	28	....	....	....	....	....
	Victoriaville	QMJHL	30	3	6	9	73	2	0	0	0	0
99-2000	Victoriaville	QMJHL	24	2	8	10	30	....	....	....	....	....
	Val-d'Or Foreurs	QMJHL	8	3	1	4	10	....	....	....	....	....
2000-01	Val-d'Or Foreurs	QMJHL	36	16	26	42	95	....	....	....	....	....
	Acadie-Bathurst	QMJHL	26	8	19	27	72	19	6	10	16	46
2001-02	Acadie-Bathurst	QMJHL	51	23	22	45	126	16	6	16	22	18

• Missed majority of 1999-2000 season recovering from leg injury suffered during game vs. Halifax (QMJHL), November 24, 1999. Traded to **Acadie-Bathurst** (QMJHL) by **Val-d'Or** (QMJHL) with Jean-Francois Laniel, Eric Labelle and future considerations for Simon Lajeunesse and the return of Val-d'Or's 4th choice (previously acquired, Val d'Or selected Mathieu Curadeau) in 2001 QMJHL Priority Draft, January 7, 2001.

BERGERON, Marc-Andre (BAIR-zhur-uhn, MAHRK-AWN-dray) EDM.

Defense. Shoots left. 5'9", 185 lbs. Born, St-Louis-de-France, Que., October 13, 1980.

			Regular Season					Playoffs				
Season	Club	Lea	GP	G	A	TP	PIM	GP	G	A	TP	PIM
1996-97	Cap-de-Madelaine	QAAA	4	0	1	1	0	2	0	0	0	0
1997-98	Baie-Comeau	QMJHL	40	6	14	20	48	....	....	....	....	....
1998-99	Baie-Comeau	QMJHL	46	8	14	22	57	....	....	....	....	....
	Shawinigan	QMJHL	24	6	7	13	66	5	2	2	4	24
99-2000	Shawinigan	QMJHL	70	24	50	74	173	13	4	7	11	45
2000-01	Shawinigan	QMJHL	69	42	59	101	185	10	4	11	15	24
2001-02	Hamilton	AHL	50	2	13	15	61	9	1	4	5	8

QMJHL First All-Star Team (2001) • Canadian Major Junior First All-Star Team (2001) • Canadian Major Junior Defenseman of the Year (2001)

Signed as a free agent by **Edmonton**, July 20, 2001.

BERGFORS, Henrik (BAIRG-fohrz, HEHN-rihk) T.B.

Defense. Shoots right. 6' 4", 225 lbs. Born, Stockholm, Sweden, May 15, 1982.
(Tampa Bay's 14th choice, 289th overall, in 2001 Entry Draft).

			Regular Season					Playoffs				
Season	Club	Lea	GP	G	A	TP	PIM	GP	G	A	TP	PIM
99-2000	AIK Solna-18	Swede-Jr.	19	1	1	2	55	....	....	....	....	....
	AIK Solna Jr.	Swede-Jr.	12	0	1	1	6	....	....	....	....	....
2000-01	Sodertalje Jr.	Swede-Jr.	14	1	0	1	16	....	....	....	....	....
2001-02	Sodertalje Jr.	Swede-Jr.	40	4	4	8	62	2	1	0	1	2

BERGGREN, Johan (BUHR-gruhn, YOH-han) DET.

Defense. Shoots left. 6'3", 176 lbs. Born, Vastra Amtevik, Sweden, May 18, 1984.
(Detroit's 4th choice, 131st overall, in 2002 Entry Draft).

			Regular Season					Playoffs				
Season	Club	Lea	GP	G	A	TP	PIM	GP	G	A	TP	PIM
2001-02	HC Sunne	Swede-3	25	1	7	8	22	....	....	....	....	....

BERNIKOV, Ruslan (BAIR-nih-kahf, roos-LAHN) DAL.

Right wing. Shoots left. 6'3", 198 lbs. Born, Vidnoye, USSR, December 4, 1977.
(Dallas' 6th choice, 139th overall, in 2000 Entry Draft).

			Regular Season					Playoffs				
Season	Club	Lea	GP	G	A	TP	PIM	GP	G	A	TP	PIM
1996-97	DynamoMoscow2	Russia-3	32	11	4	15	20	....	....	....	....	....
	Dynamo Moscow	Russia	2	0	0	0	0	....	....	....	....	....
1997-98	Yekaterinburg 2	Russia-3	2	1	1	2	0	....	....	....	....	....
	Yekaterinburg	Russia	43	7	7	14	55	....	....	....	....	....
1998-99	Dynamo Moscow	Russia	6	0	1	1	2	....	....	....	....	....
	Krylja Sovetov	Russia	20	3	1	4	24	....	....	....	....	....
	CSKA Moscow	Russia	1	0	0	0	0	....	....	....	....	....
	Cherepovets	Russia	5	0	0	0	0	1	0	0	0	0
99-2000	Dynamo Moscow	Russia	6	2	1	3	2	....	....	....	....	....
	Amur Khabarovsk	Russia	14	3	6	9	10	5	3	1	4	2
2000-01	Amur Khabarovsk	Russia	33	1	4	5	40	....	....	....	....	....
2001-02	Amur Khabarovsk	Russia	38	7	10	17	20	....	....	....	....	....

BERZINS, Armands (BUHR-zihnsh, AHR-muhnds) MIN.

Center. Shoots left. 6'3", 227 lbs. Born, Riga, Latvia, December 27, 1983.
(Minnesota's 5th choice, 155th overall, in 2002 Entry Draft).

			Regular Season					Playoffs				
Season	Club	Lea	GP	G	A	TP	PIM	GP	G	A	TP	PIM
99-2000	HC Essamika-B	EEHL	9	2	1	3	0	....	....	....	....	....
	Prizma Riga-18	Latvia-Jr.	STATISTICS NOT AVAILABLE									
2000-01	Prizma Riga-18	Latvia-Jr.	21	9	11	20	....	....	....	....	....	....
	HC Riga	EEHL	12	1	2	3	....	....	....	....	....	....
	HC Riga	Latvia	4	0	0	0	....	....	....	....	....	....
2001-02	Shawinigan	QMJHL	63	17	18	35	59	12	2	2	4	6

BEZINA, Goran (BEH-zee-nuh, GOH-ran) PHX.

Defense. Shoots left. 6'2", 220 lbs. Born, Split, Yugoslavia, March 21, 1980.
(Phoenix's 8th choice, 234th overall, in 1999 Entry Draft).

			Regular Season					Playoffs				
Season	Club	Lea	GP	G	A	TP	PIM	GP	G	A	TP	PIM
1998-99	Fribourg Jr.	Swiss-Jr.	22	11	6	17	64	....	....	....	....	....
	Fribourg	Swiss	38	0	0	0	14	4	0	0	0	2
	Fribourg	EuroHL	6	0	0	0	0	....	....	....	....	....
99-2000	Fribourg Jr.	Swiss-Jr.	2	0	1	1	16	2	1	1	2	8
	Fribourg	Swiss	44	3	6	9	10	4	0	0	0	6
	EHC Visp	Swiss-2	2	0	0	0	2	....	....	....	....	....
2000-01	Fribourg	Swiss	44	10	10	20	44	5	1	1	2	12
2001-02	Springfield	AHL	66	2	11	13	50	....	....	....	....	....

BEZRUKOV, Dmitri (behz-ROO-kahv, dih-MEE-tree) T.B.

Left wing. Shoots left. 6'3", 187 lbs. Born, Kazan, USSR, November 9, 1977.
(Tampa Bay's 11th choice, 259th overall, in 2001 Entry Draft).

			Regular Season					Playoffs				
Season	Club	Lea	GP	G	A	TP	PIM	GP	G	A	TP	PIM
1997-98	Nizhnekamsk 2	Russia-3	8	0	1	1	6	....	....	....	....	....
	Nizhnekamsk	Russia	14	5	3	8	4	....	....	....	....	....
1998-99	Nizhnekamsk 2	Russia-4	1	3	0	3	0	....	....	....	....	....
	Nizhnekamsk	Russia	39	4	6	10	18	3	1	0	1	2
99-2000	Nizhnekamsk 2	Russia-3	4	0	0	0	6	....	....	....	....	....
	Leninogorsk	Russia-2	8	2	1	3	8	....	....	....	....	....
	Nizhnekamsk	Russia	28	5	6	11	45	3	0	1	1	2
2000-01	Nizhnekamsk	Russia	35	7	10	17	54	4	0	2	2	2
2001-02	Nizhnekamsk	Russia	38	5	6	11	45	....	....	....	....	....

BIEKSA, Kevin (BEEKS-ah, KEH-vihn) VAN.

Defense. Shoots right. 6'1", 180 lbs. Born, Grimsby, Ont., June 16, 1981.
(Vancouver's 4th choice, 151st overall, in 2001 Entry Draft).

			Regular Season					Playoffs				
Season	Club	Lea	GP	G	A	TP	PIM	GP	G	A	TP	PIM
1997-98	Burlington	OPJHL	27	0	3	3	10	....	....	....	....	....
1998-99	Burlington	OPJHL	49	8	29	37	83	....	....	....	....	....
99-2000	Burlington	OPJHL	49	6	27	33	139	....	....	....	....	....
2000-01	Bowling Green	CCHA	35	4	9	13	90	....	....	....	....	....
2001-02	Bowling Green	CCHA	40	5	10	15	68	....	....	....	....	....

BIRBRAER, Max (beer-BRIEGH-uhr, max) N.J.

Right wing. Shoots left. 6'2", 195 lbs. Born, Ust-Kamenogorsk, USSR, December 15, 1980.
(New Jersey's 6th choice, 67th overall, in 2000 Entry Draft).

			Regular Season					Playoffs				
Season	Club	Lea	GP	G	A	TP	PIM	GP	G	A	TP	PIM
1997-98	Shelburne Wolves	MTJHL	12	7	11	18	8	....	....	....	....	....
1998-99	Shelburne Wolves	OPJHL	35	20	22	42	25	....	....	....	....	....
99-2000	Newmarket	OPJHL	47	50	32	82	52	....	....	....	....	....
2000-01	Albany	AHL	50	7	6	13	24	....	....	....	....	....
2001-02	Albany	AHL	40	6	7	13	22	....	....	....	....	....

BISHAI, Mike (BIHSH-igh, MIGHK) EDM.

Left wing. Shoots left. 5'11", 185 lbs. Born, Edmonton, Alta., May 30, 1979.

			Regular Season					Playoffs				
Season	Club	Lea	GP	G	A	TP	PIM	GP	G	A	TP	PIM
1996-97	South Surrey	BCHL	38	6	13	19	10	....	....	....	....	....
1997-98	South Surrey	BCHL	47	48	52	100	36	....	....	....	....	....
1998-99	West-Michigan	CCHA	26	0	3	3	20	....	....	....	....	....
99-2000	West-Michigan	CCHA	35	18	19	37	52	....	....	....	....	....
2000-01	West-Michigan	CCHA	37	23	45	68	37	....	....	....	....	....
2001-02	West-Michigan	CCHA	34	10	27	37	28	....	....	....	....	....
	Hamilton	AHL	3	0	0	0	0	....	....	....	....	....

CCHA Second All-Star Team (2001) • NCAA West Second All-American Team (2001)

Signed as a free agent by **Edmonton**, April 4, 2002.

BIZYAYEV, Vasili (bihz-AY-yehv, va-SEE-lee) BUF.

Right wing. Shoots left. 6'1", 185 lbs. Born, Moscow, USSR, June 6, 1982.
(Buffalo's 5th choice, 213th overall, in 2000 Entry Draft).

			Regular Season					Playoffs				
Season	Club	Lea	GP	G	A	TP	PIM	GP	G	A	TP	PIM
99-2000	CSKA Moscow Jr.	Russia-Jr.	20	28	14	42	22	....	....	....	....	....
	H.C. CSKA	Russia-2	STATISTICS NOT AVAILABLE									
2000-01	Kitchener	OHL	53	10	14	24	10	....	....	....	....	....
2001-02	CSKA Moscow	Russia	3	0	1	1	0	....	....	....	....	....
	H.C. CSKA	Russia-2	4	2	0	2	0	....	....	....	....	....

BJORK, Johan (b'YAWRK, YOH-han) OTT.

Defense. Shoots left. 6'1", 176 lbs. Born, Malmo, Sweden, August 28, 1984.
(Ottawa's 5th choice, 125th overall, in 2002 Entry Draft).

			Regular Season					Playoffs				
Season	Club	Lea	GP	G	A	TP	PIM	GP	G	A	TP	PIM
99-2000	Malmo IF Jr.	Swede-Jr.	8	0	1	1	8	....	....	....	....	....
2000-01	Malmo IF Jr.	Swede-Jr.	25	0	2	2	20	....	....	....	....	....
2001-02	Malmo IF Jr.	Swede-Jr.	39	1	4	5	72	7	0	0	0	8

BLAIS, Ben (BLAYS, BEHN) NYI

Defense. Shoots left. 6'4", 195 lbs. Born, Berlin, NH, February 16, 1978.
(NY Islanders' 8th choice, 237th overall, in 1998 Entry Draft).

			Regular Season					Playoffs				
Season	Club	Lea	GP	G	A	TP	PIM	GP	G	A	TP	PIM
1997-98	Salisbury High	Hi-School	20	5	21	26	....	....	....	....	....	....
	Walpole Jr.	EJHL	34	7	18	25	75	....	....	....	....	....
1998-99	St. Lawrence	ECAC	3	0	1	1	6	....	....	....	....	....
99-2000	Quinnipiac	MAAC	DID NOT PLAY – TRANSFERRED COLLEGES									
2000-01	Quinnipiac	MAAC	33	3	11	14	54	....	....	....	....	....
2001-02	Quinnipiac	MAAC	36	0	6	6	76	....	....	....	....	....

EJHL Defenseman of the Year (1998) • MAAC All-Tournament Team (2001)

BLANAR, Jan (BLAH-nuhr, YAN) FLA.

Defense. Shoots left. 6'3", 185 lbs. Born, Trencin, Czech., June 6, 1983.
(Florida's 11th choice, 263rd overall, in 2001 Entry Draft).

			Regular Season					Playoffs				
Season	Club	Lea	GP	G	A	TP	PIM	GP	G	A	TP	PIM
2000-01	Dukla Trencin Jr.	Slovak-Jr.	35	3	6	9	16	....	....	....	....	....
2001-02	Dukla Trencin Jr.	Slovak-Jr.	42	3	12	15	....	....	....	....	....	....
	Dukla Trencin	Slovakia	11	0	0	0	2	....	....	....	....	....

BLATAK, Miroslav (BLAT-ak, MEER-oh-slahv) DET.

Defense. Shoots left. 5'11", 172 lbs. Born, Gottwaldov, Czech., May 25, 1982.
(Detroit's 3rd choice, 129th overall, in 2001 Entry Draft).

			Regular Season					Playoffs				
Season	Club	Lea	GP	G	A	TP	PIM	GP	G	A	TP	PIM
99-2000	Vsetin-18	Czech-Jr.	30	0	0	0	12	....	....	....	....	....
	Vsetin Jr.	Czech-Jr.	12	0	2	2	10	....	....	....	....	....
2000-01	Vsetin-18	Czech-Jr.	33	7	8	15	56	....	....	....	....	....
	Vsetin Jr.	Czech-Jr.	12	2	4	6	54	7	0	6	6	6
	Zlin	Czech	8	0	2	2	0	6	0	0	0	0
2001-02	Dukla Jihlava Jr.	Czech-Jr.	3	0	0	0	0	....	....	....	....	....
	Zlin Jr.	Czech-Jr.	3	0	0	0	8	....	....	....	....	....
	HC Dukla Jihlava	Czech-2	1	0	0	0	0	2	0	0	0	0
	Zlin	Czech	39	4	7	11	18	11	1	2	3	8

BLATNY, Zdenek (BLAT-nee, z-DEHN-ehk) ATL.

Left wing/Center. Shoots left. 6'1", 190 lbs. Born, Brno, Czech., January 14, 1981.
(Atlanta's 3rd choice, 68th overall, in 1999 Entry Draft).

			Regular Season					Playoffs				
Season	Club	Lea	GP	G	A	TP	PIM	GP	G	A	TP	PIM
1997-98	Kometa Brno Jr.	Czech-Jr.	42	22	21	43	40	....	....	....	....	....
1998-99	Seattle	WHL	44	18	15	33	25	11	4	0	4	24
99-2000	Seattle	WHL	7	4	5	9	12	....	....	....	....	....
	Kootenay Ice	WHL	61	43	39	82	119	21	10	*17	27	46
2000-01	Kootenay Ice	WHL	58	37	48	85	120	11	8	10	18	24
2001-02	Greenville	ECHL	12	5	5	10	17	9	2	8	10	14
	Chicago Wolves	AHL	41	4	3	7	30	3	2	0	2	0

WHL East Second All-Star Team (2000)

Traded to **Kootenay** (WHL) by **Seattle** (WHL) for future considerations, October 22, 1999.

BLAZEK, Michal (BLA-zhehk, MEE-kuhl) DAL.

Defense. Shoots left. 6'2", 187 lbs. Born, Vsetin, Czech., April 2, 1982.
(Dallas' 6th choice, 167th overall, in 2001 Entry Draft).

			Regular Season					Playoffs				
Season	Club	Lea	GP	G	A	TP	PIM	GP	G	A	TP	PIM
99-2000	HC Vsetin Jr.	Czech-Jr.	49	0	3	3	67	....	....	....	....	....
2000-01	HC Vsetin Jr.	Czech-Jr.	52	9	18	27	116	....	....	....	....	....
2001-02	HC Vsetin Jr.	Czech-Jr.	40	6	12	18	160	11	0	3	3	22

BLINDENBACHER, Severin (blihn-duhn-BAH-khur, SEH-vuhr-ihn) PHX.

Defense. Shoots right. 5'11", 189 lbs. Born, Bulach, Switz., March 15, 1983.
(Phoenix's 9th choice, 273rd overall, in 2001 Entry Draft).

			Regular Season					Playoffs				
Season	Club	Lea	GP	G	A	TP	PIM	GP	G	A	TP	PIM
1998-99	EHC Kloten Jr.	Swiss-Jr.	5	0	2	2	4	....	....	....	....	....
99-2000	EHC Kloten Jr.	Swiss-Jr.	30	8	10	18	18	4	1	1	2	8
2000-01	EHC Kloten	Swiss	27	0	2	2	17	9	0	0	0	10
2001-02	Kloten Flyers	Swiss	38	1	3	4	22	10	0	2	2	10
	Kloten Flyers	Swiss-Jr.	....	....	....	....	....	1	0	1	1	2

BOBROV, Viktor (bawb-RAWV, VIHK-tohr) CGY.

Center. Shoots left. 6'1", 176 lbs. Born, Novocheboksarsk, USSR, January 1, 1984.
(Calgary's 7th choice, 146th overall, in 2002 Entry Draft).

			Regular Season					Playoffs				
Season	Club	Lea	GP	G	A	TP	PIM	GP	G	A	TP	PIM
2001-02	CSKA Moscow 2	Russia-3	36	11	16	27	20	....	....	....	....	....

BOCHENSKI, Brandon (boh-CHEHN-skee, BRAHN-duhn) OTT.

Right wing. Shoots right. 6', 180 lbs. Born, Blaine, MN, April 4, 1982.
(Ottawa's 9th choice, 223rd overall, in 2001 Entry Draft).

			Regular Season					Playoffs				
Season	Club	Lea	GP	G	A	TP	PIM	GP	G	A	TP	PIM
99-2000	Blaine Bengals	Hi-School	28	32	30	62	....	....	....	....	....	....
2000-01	Lincoln Stars	USHL	55	*47	33	80	22	11	5	7	12	4
2001-02	North Dakota	WCHA	36	17	15	32	34	....	....	....	....	....

USHL First All-Star Team (2001) • USHL Rookie of the Year (2001) • WCHA All-Rookie Team (2002) • WCHA Rookie of the Year (2002)

BOHAC, Jan (BOH-hach, YAHN) OTT.

Center. Shoots left. 6'4", 201 lbs. Born, Tabor, Czech., February 3, 1982.
(Ottawa's 4th choice, 87th overall, in 2000 Entry Draft).

			Regular Season					Playoffs				
Season	Club	Lea	GP	G	A	TP	PIM	GP	G	A	TP	PIM
1997-98	Slavia Praha Jr.	Czech-Jr.	39	8	12	20	12	....	....	....	....	....
1998-99	Slavia Praha Jr.	Czech-Jr.	35	6	6	12	10	....	....	....	....	....
	HC Slavia Praha	Czech	2	0	0	0	0	....	....	....	....	....
99-2000	Slavia Praha Jr.	Czech-Jr.	22	7	8	15	6	7	0	1	1	4
	HC Slavia Praha	Czech	25	1	2	3	4	....	....	....	....	....
2000-01	HC Slavia Praha	Czech	19	2	0	2	0	2	0	0	0	0
	HC Banik Most	Czech-2	5	0	2	2	0	....	....	....	....	....
2001-02	Trinec Jr.	Czech-Jr.	2	0	1	1	0	....	....	....	....	....
	Trinec	Czech	16	0	1	1	4	....	....	....	....	....
	HC Banik Most	Czech-3	5	0	2	2	0	....	....	....	....	....
	Liberec	Czech-2	10	0	2	2	0	2	0	0	0	0

BOIS, Danny (BOIZ, DA-nee) COL.

Right wing. Shoots right. 6', 190 lbs. Born, Thunder Bay, Ont., June 1, 1983.
(Colorado's 2nd choice, 97th overall, in 2001 Entry Draft).

			Regular Season					Playoffs				
Season	Club	Lea	GP	G	A	TP	PIM	GP	G	A	TP	PIM
1998-99	T. Bay Kings	TBMHL	15	7	12	19	28	....	....	....	....	....
99-2000	Wellington Dukes	MTJHL	37	15	20	35	115	....	....	....	....	....
2000-01	London Knights	OHL	66	21	16	37	218	5	2	1	3	19
2001-02	London Knights	OHL	62	16	14	30	256	12	2	2	4	47

BOISVERT, Hugo (bwuh-VAIR, HEW-goh)

Center. Shoots left. 6', 200 lbs. Born, St-Eustache, Que., February 11, 1976.

			Regular Season					Playoffs				
Season	Club	Lea	GP	G	A	TP	PIM	GP	G	A	TP	PIM
1994-95	Cornwall Colts	OCJHL	27	13	19	32	26	....	....	....	....	....
1995-96	Cornwall Colts	OCJHL	54	40	90	130	102	15	15	20	35	44
1996-97	Ohio State	CCHA	38	11	27	38	44	....	....	....	....	....
1997-98	Ohio State	CCHA	42	23	*35	58	70	....	....	....	....	....
1998-99	Ohio State	CCHA	41	24	27	51	54	....	....	....	....	....
99-2000	Team Canada	Nat-Tm	39	10	14	24	12	....	....	....	....	....
2000-01	Orlando	IHL	68	6	12	18	41	16	4	5	9	23
2001-02	Grand Rapids	AHL	74	11	18	29	48	5	1	3	4	4

CCHA First All-Star Team (1998, 1999) • NCAA West First All-American Team (1998) • NCAA West Second All-American Team (1999)

Signed as a free agent by **Atlanta**, June 25, 1999.

BOLIBRUCK, Kevin (BOH-lee-bruhk, KEH-vihn)

Defense. Shoots left. 6'1", 200 lbs. Born, Peterborough, Ont., February 8, 1977.
(Edmonton's 7th choice, 176th overall, in 1997 Entry Draft).

			Regular Season					Playoffs				
Season	Club	Lea	GP	G	A	TP	PIM	GP	G	A	TP	PIM
1993-94	Thorold	OJHL-B	38	6	18	24	78	....	....	....	....	....
1994-95	Peterborough	OHL	66	2	16	18	88	11	1	1	2	14
1995-96	Peterborough	OHL	57	6	21	27	105	24	3	6	9	46
1996-97	Peterborough	OHL	46	4	26	30	63	11	3	3	6	14
1997-98	Team Canada	Nat-Tm	49	2	5	7	65	....	....	....	....	....
1998-99	Hamilton	AHL	64	1	6	7	42	11	0	1	1	4
99-2000	Hamilton	AHL	54	1	4	5	67	10	0	1	1	4
2000-01	Rochester	AHL	76	2	5	7	52	3	0	0	0	0
2001-02	Houston Aeros	AHL	60	3	15	18	59	11	1	2	3	10

• Re-entered NHL Entry Draft. Originally Ottawa's 4th choice, 89th overall, in 1995 Entry Draft.

OHL First All-Star Team (1996)

Rights traded to **Chicago** by **Ottawa** with Denis Chasse and Ottawa's 6th round choice (later traded back to Ottawa - Ottawa selected Christopher Neil) in 1998 Entry Draft for Mike Prokopec, March 18, 1997. Signed as a free agent by **Rochester** (AHL), September 20, 2000.

BOOGAARD, Derek (BOO-gard, DAIR-ihk) MIN.

Left wing. Shoots right. 6'7", 249 lbs. Born, Saskatoon, Sask., June 23, 1982.
(Minnesota's 6th choice, 202nd overall, in 2001 Entry Draft).

			Regular Season					Playoffs				
Season	Club	Lea	GP	G	A	TP	PIM	GP	G	A	TP	PIM
1998-99	Regina Caps	SJHL	35	2	3	5	166	....	....	....	....	....
99-2000	Regina Pats	WHL	5	0	0	0	17	....	....	....	....	....
	Prince George	WHL	33	0	0	0	149	....	....	....	....	....
2000-01	Prince George	WHL	61	1	8	9	245	6	1	0	1	31
2001-02	Prince George	WHL	2	0	0	0	16	....	....	....	....	....
	Medicine Hat	WHL	46	1	8	9	178	....	....	....	....	....

Traded to **Prince George** (WHL) by **Regina** (WHL) for Jonathan Parker, October 26, 1999. Traded to **Medicine Hat** (WHL) by **Prince George** (WHL) for Danny Johnston, October 26, 2001.

BOOTLAND, Darryl (BOOT-land, DAIR-ihl)

Right wing. Shoots right. 6'1", 200 lbs. Born, Toronto, Ont., November 2, 1981.
(Colorado's 12th choice, 252nd overall, in 2000 Entry Draft).

			Regular Season					Playoffs				
Season	Club	Lea	GP	G	A	TP	PIM	GP	G	A	TP	PIM
1997-98	Orangeville	OJHL-B	44	22	26	48	177	....	....	....	....	....
1998-99	Barrie Colts	OHL	38	18	11	29	89	....	....	....	....	....
	St. Michael's	OHL	28	12	6	18	80	....	....	....	....	....
99-2000	St. Michael's	OHL	65	24	30	54	166	....	....	....	....	....
2000-01	St. Michael's	OHL	56	32	33	65	136	11	3	1	4	20
2001-02	St. Michael's	OHL	61	41	56	97	137	15	8	10	18	50

Traded to **St. Michael's** (OHL) by **Barrie** (OHL) with Keith Delaney, Adam DeLeew and Brad Pierce for Sheldon Keefe, Mike Jefferson, Ryan Barnes and Shawn Cation, January 11, 1999.

BOOTLAND, Nick (BOOT-land, NIHK)

Left wing. Shoots left. 6'3", 215 lbs. Born, Shelbourne, Ont., July 31, 1978.
(Dallas' 8th choice, 220th overall, in 1996 Entry Draft).

			Regular Season					Playoffs				
Season	Club	Lea	GP	G	A	TP	PIM	GP	G	A	TP	PIM
1994-95	Orangeville	OJHL-B	47	11	10	21	57	....	....	....	....	....
1995-96	Guelph Storm	OHL	64	8	7	15	90	16	1	0	1	21
1996-97	Guelph Storm	OHL	64	35	23	58	117	18	11	7	18	36
1997-98	Guelph Storm	OHL	64	23	37	60	128	12	7	6	13	22
1998-99	Hershey Bears	AHL	62	3	6	9	122	....	....	....	....	....
99-2000	Hershey Bears	AHL	59	5	13	18	108	14	2	2	4	26
2000-01	Hershey Bears	AHL	71	6	8	14	110	12	1	3	4	2
2001-02	Cleveland Barons	AHL	19	2	3	5	15	....	....	....	....	....
	Cincinnati	ECHL	60	25	30	55	102	3	0	1	1	11

Signed as a free agent by **Colorado**, August 6, 1998.

BORNHAMMAR, David WSH.

Defense. Shoots left. 6', 176 lbs. Born, Lidingo, Sweden, June 15, 1981.
(Washington's 8th choice, 192nd overall, in 1999 Entry Draft).

			Regular Season					Playoffs				
Season	Club	Lea	GP	G	A	TP	PIM	GP	G	A	TP	PIM
1997-98	AIK Solna Jr.	Swede-Jr.	13	0	1	1	10	....	....	....	....	....
1998-99	AIK Solna Jr.	Swede-Jr.	33	5	5	10	30	....	....	....	....	....
99-2000	Kelowna Rockets	WHL	62	3	24	27	36	2	0	0	0	0
2000-01	AIK Solna Jr.	Swede-Jr.	10	0	3	3	37	....	....	....	....	....
	AIK Solna	Sweden	37	4	2	6	57	4	0	1	1	0
2001-02	AIK Solna	Sweden	48	1	6	7	34	....	....	....	....	....

• Name when drafted was David Johannson.

BOUCHARD, Francois (BOO-shahrd, fran-SWUH) TOR.

Defense. Shoots right. 6', 189 lbs. Born, Brossard, Que., August 8, 1973.
(Tampa Bay's 1st choice, 8th overall, in 1994 Supplemental Draft).

			Regular Season					Playoffs				
Season	Club	Lea	GP	G	A	TP	PIM	GP	G	A	TP	PIM
1989-90	Richelieu	QAAA	39	4	22	26	....	4	1	1	2	....
1990-91			STATISTICS NOT AVAILABLE									
1991-92	Northeastern	H-East	34	4	9	13	28	....	....	....	....	....
1992-93	Northeastern	H-East	26	4	6	10	20	....	....	....	....	....
1993-94	Northeastern	H-East	39	15	15	30	34	....	....	....	....	....
1994-95	Northeastern	H-East	31	7	16	23	39	....	....	....	....	....
1995-96	Charlotte	ECHL	66	9	18	27	64	16	4	4	8	18
1996-97	Karpat Oulu	Finland-2	50	22	20	42	6	....	....	....	....	....
1997-98	HPK Hameenlinna	Finland	10	1	0	1	20	....	....	....	....	....
	Augsburg	Germany	36	7	9	16	36	....	....	....	....	....
1998-99	Karpat Oulu	Finland-2	42	6	19	25	60	5	1	0	1	2
99-2000	MoDo	Sweden	41	2	6	8	38	13	0	0	0	16
2000-01	Djurgarden	Sweden	48	10	11	21	58	16	2	3	5	14
2001-02	St. John's	AHL	51	3	5	8	40	11	0	1	1	4

Hockey East First All-Star Team (1992) • NCAA East First All-American Team (1992)

Signed as a free agent by **Toronto**, July 16, 2001. Signed as a free agent by **Ingolstadt** (Germany) with **Toronto** retaining NHL rights, May 28, 2002.

BOUCHARD, Pierre-Marc (BOO-shahrd, PEE-air- MAHRK) MIN.

Center. Shoots left. 5'10", 165 lbs. Born, Sherbrooke, Que., April 27, 1984.
(Minnesota's 1st choice, 8th overall, in 2002 Entry Draft).

			Regular Season					Playoffs				
Season	Club	Lea	GP	G	A	TP	PIM	GP	G	A	TP	PIM
99-2000	Charles-Lemoyne	QAAA	42	28	*45	*74	20	9	4	8	12	6
2000-01	Chicoutimi	QMJHL	67	38	57	95	20	6	5	8	13	0
2001-02	Chicoutimi	QMJHL	69	46	*94	*140	54	4	2	3	5	4

QAAA First All-Star Team (2000) • QMJHL Rookie of the Year (2001) • QMJHL First All-Star Team (2002) • Canadian Major Junior First All-Star Team (2002) • Canadian Major Junior Player of the Year (2002)

BOUWMEESTER, Jay (BOW-mee-stuhr, JAY) FLA.

Defense. Shoots left. 6'4", 210 lbs. Born, Edmonton, Alta., September 27, 1983.
(Florida's 1st choice, 3rd overall, in 2002 Entry Draft).

			Regular Season					Playoffs				
Season	Club	Lea	GP	G	A	TP	PIM	GP	G	A	TP	PIM
1998-99	Edmonton SSAC	AMHL	32	14	29	43	36	....	....	....	....	....
	Medicine Hat	WHL	8	2	1	3	2	....	....	....	....	....
99-2000	Medicine Hat	WHL	64	13	21	34	26	....	....	....	....	....
2000-01	Medicine Hat	WHL	61	14	39	53	44	....	....	....	....	....
2001-02	Medicine Hat	WHL	61	11	50	61	42	....	....	....	....	....

WHL East First All-Star Team (2002)

BOWNESS, Ryan (BOH-nehs, RIGH-uhn) CBJ

Right wing. Shoots right. 6'2", 198 lbs. Born, Halifax, N.S., August 5, 1983.
(Columbus' 10th choice, 236th overall, in 2001 Entry Draft).

			Regular Season					Playoffs				
Season	Club	Lea	GP	G	A	TP	PIM	GP	G	A	TP	PIM
1998-99	Ottawa West	OMHA	28	10	20	30	60	....	....	....	....	....
99-2000	Kanata Valley	OCJHL	51	10	25	35	98	....	....	....	....	....
2000-01	Brampton	OHL	52	5	10	15	51	9	0	2	2	9
2001-02	Brampton	OHL	66	7	10	17	104	....	....	....	....	....

BOYCHUK, Johnny (BOY-chuhk, JAW-nee) COL.

Defense. Shoots right. 6'2", 205 lbs. Born, Edmonton, Alta., January 19, 1984.
(Colorado's 2nd choice, 61st overall, in 2002 Entry Draft).

			Regular Season					Playoffs				
Season	Club	Lea	GP	G	A	TP	PIM	GP	G	A	TP	PIM
1998-99	Edm. Cycle	AMBHL	36	8	20	28	59	....	....	....	....	....
99-2000	Edm. Cycle	AMHL	35	6	17	23	59	....	....	....	....	....
2000-01	Calgary Hitmen	WHL	66	4	8	12	61	12	1	1	2	17
2001-02	Calgary Hitmen	WHL	70	8	32	40	85	7	1	1	2	6

BOYES, Brad (BOIZ, BRAD) TOR.

Center. Shoots right. 6', 181 lbs. Born, Mississauga, Ont., April 17, 1982.
(Toronto's 1st choice, 24th overall, in 2000 Entry Draft).

			Regular Season					Playoffs				
Season	Club	Lea	GP	G	A	TP	PIM	GP	G	A	TP	PIM
1997-98	Mississauga Reps	MTHL	44	27	50	77	....	....	....	....	....	....
1998-99	Erie Otters	OHL	59	24	36	60	30	5	1	2	3	10
99-2000	Erie Otters	OHL	68	36	46	82	38	13	6	8	14	10
2000-01	Erie Otters	OHL	59	45	45	90	42	15	10	13	23	8
2001-02	Erie Otters	OHL	47	36	41	77	42	21	22	*19	41	27

Canadian Major Junior Scholastic Player of the Year (2000) • OHL Second All-Star Team (2001) • OHL First All-Star Team (2002)

BRAY, Mike (BRAY, MIGHK) NYI

Right wing. Shoots right. 6'4", 208 lbs. Born, Halifax, N.S., December 17, 1982.
(NY Islanders' 5th choice, 228th overall, in 2001 Entry Draft).

			Regular Season					Playoffs				
Season	Club	Lea	GP	G	A	TP	PIM	GP	G	A	TP	PIM
1998-99	Hfx. McDonald's	NSMHL	STATISTICS NOT AVAILABLE									
	Halifax	QMJHL	28	0	0	0	4	....	....	....	....	....
2000-01	Quebec Remparts	QMJHL	41	10	10	20	317	4	1	1	2	35
2001-02	Quebec Remparts	QMJHL	28	2	4	6	146	....	....	....	....	....
	Montreal Rocket	QMJHL	30	3	6	9	166	7	2	1	3	36

Traded to **Quebec** (QMJHL) by **Halifax** (QMJHL) for Quebec's 7th round choice (Brian McConnell) in 2000 QMJHL Priority Draft, June 9, 2000. Traded to **Montreal** (QMJHL) by **Quebec** (QMJHL) with Jeff MacAuley, Cory Urquhart and Yan Turcotte for Jeff Montgomery, Karl St-Pierre and Pierre-Olivier Beaulieu, December 12, 2001.

BRENK, Jake (BREHNK, JAYK) EDM.

Center. Shoots right. 6'2", 187 lbs. Born, Detroit Lakes, MN, April 16, 1982.
(Edmonton's 6th choice, 154th overall, in 2001 Entry Draft).

			Regular Season					Playoffs				
Season	Club	Lea	GP	G	A	TP	PIM	GP	G	A	TP	PIM
99-2000	Breck Mustangs	Hi-School	28	18	23	41	....	....	....	....	....	....
2000-01	Breck Mustangs	Hi-School	22	28	30	58	22	....	....	....	....	....
2001-02	Minnesota State	WCHA	21	3	3	6	6	....	....	....	....	....

BRENT, Tim (BREHNT, TIHM) **ANA.**

Center. Shoots right. 6', 175 lbs. Born, Cambridge, Ont., March 10, 1984.
(Anaheim's 2nd choice, 37th overall, in 2002 Entry Draft).

			Regular Season					Playoffs				
Season	Club	Lea	GP	G	A	TP	PIM	GP	G	A	TP	PIM
99-2000	Cambridge	OJHL-B	40	19	16	35	42	....	....	....	....	....
2000-01	St. Michael's	OHL	64	9	19	28	31	18	2	8	10	6
2001-02	St. Michael's	OHL	61	19	40	59	52	14	7	12	19	20

BROOKBANK, Wade (BRUK-bank, WAYD) **OTT.**

Defense. Shoots left. 6'4", 219 lbs. Born, Lanigan, Sask., September 29, 1977.

			Regular Season					Playoffs				
Season	Club	Lea	GP	G	A	TP	PIM	GP	G	A	TP	PIM
1997-98	Melville	SJHL	58	8	21	29	330	....	....	....	....	....
	Anchorage Aces	WCHL	7	0	0	0	46	4	0	0	0	20
1998-99	Anchorage Aces	WCHL	56	0	4	4	337	....	....	....	....	....
99-2000	Oklahoma City	CHL	68	3	9	12	354	7	1	1	2	29
2000-01	Orlando	IHL	29	0	1	1	122	4	0	0	0	6
	Oklahoma City	CHL	46	1	13	14	267	5	0	0	0	24
2001-02	Grand Rapids	AHL	73	1	6	7	337	3	0	1	1	14

Signed as a free agent by **Orlando** (IHL), September 1, 2000. Signed as a free agent by **Ottawa**, July 27, 2001.

BROOKS, Alex (BROOKS, AL-ehx) **N.J.**

Defense. Shoots right. 6'2", 200 lbs. Born, Madison, WI, August 21, 1976.

			Regular Season					Playoffs				
Season	Club	Lea	GP	G	A	TP	PIM	GP	G	A	TP	PIM
1993-94	Madison Capitols	USHL	13	3	11	14	....	....	....	....	....	....
1994-95	Madison West	Hi-School	24	13	28	41	....	....	....	....	....	....
1995-96	Green Bay	USHL	46	3	22	25	....	....	....	....	....	....
1996-97	U. of Wisconsin	WCHA	DID NOT PLAY – INJURED									
1997-98	U. of Wisconsin	WCHA	40	1	4	5	72	....	....	....	....	....
1998-99	U. of Wisconsin	WCHA	37	0	3	3	73	....	....	....	....	....
99-2000	U. of Wisconsin	WCHA	41	4	10	14	78	....	....	....	....	....
2000-01	U. of Wisconsin	WCHA	41	3	16	19	76	....	....	....	....	....
2001-02	Jokerit Helsinki	Finland	53	1	3	4	109	12	0	0	0	11

• Missed entire 1996-97 season recovering from back injury suffered during off-season training, August, 1996. • Signed as a free agent by **New Jersey**, July 12, 2002.

BROS, Michal (BROHSH, MEE-khahl) **MIN.**

Center. Shoots right. 6'1", 195 lbs. Born, Olomouc, Czech., January 25, 1976.
(San Jose's 6th choice, 130th overall, in 1995 Entry Draft).

			Regular Season					Playoffs				
Season	Club	Lea	GP	G	A	TP	PIM	GP	G	A	TP	PIM
1994-95	HC Olomouc Jr.	Czech-Jr.	34	29	32	61	....	....	....	....	....	....
1995-96	HC Olomouc	Czech	35	8	11	19	....	4	2	0	2	....
1996-97	HC Olomouc	Czech	50	13	14	27	28	....	....	....	....	....
1997-98	HC Petra Vsetin	Czech	47	14	18	32	28	10	3	1	4	2
	HC Petra Vsetin	EuroHL	9	3	0	3	2	....	....	....	....	....
1998-99	Vsetin	Czech	42	10	18	28	18	12	1	3	4	....
99-2000	HC Sparta Praha	Czech	49	6	30	36	49	9	1	3	4	4
2000-01	HC Sparta Praha	Czech	27	3	13	16	22	13	4	6	10	8
2001-02	HC Sparta Praha	Czech	48	19	29	48	71	13	7	*11	*18	8

Selected by **Minnesota** from **San Jose** in Expansion Draft, June 23, 2000.

BROWN, Marc (BROWN, MAHRK) **ST.L.**

Left wing. Shoots left. 6'1", 196 lbs. Born, Surrey, B.C., March 10, 1979.

			Regular Season					Playoffs				
Season	Club	Lea	GP	G	A	TP	PIM	GP	G	A	TP	PIM
1995-96	Abbotsford	PIJHL	35	20	15	35	15	....	....	....	....	....
1996-97	Spokane Chiefs	WHL	52	4	8	12	37	2	0	0	0	0
1997-98	Spokane Chiefs	WHL	41	10	15	25	35	....	....	....	....	....
	Prince Albert	WHL	23	6	8	14	21	....	....	....	....	....
1998-99	Prince Albert	WHL	72	35	45	80	47	14	12	6	18	6
99-2000	Worcester	AHL	72	13	11	24	24	17	3	1	4	0
2000-01	Worcester	AHL	34	9	10	19	27	10	0	1	1	6
2001-02	Worcester	AHL	74	30	25	55	42	3	2	2	4	0

Signed as a free agent by **St. Louis**, September 24, 1999. • Missed majority of 2000-01 season recovering from abdominal injury suffered in training camp, September 27, 2000.

BRUCE, Kyle (BROOS, KIGHL) **FLA.**

Right wing. Shoots right. 6', 181 lbs. Born, Sechelt, B.C., March 30, 1983.
(Florida's 10th choice, 231st overall, in 2001 Entry Draft).

			Regular Season					Playoffs				
Season	Club	Lea	GP	G	A	TP	PIM	GP	G	A	TP	PIM
1998-99	CAT Systems	BCAHA	STATISTICS NOT AVAILABLE									
99-2000	Port Coquitlam	PIJHL	30	12	14	26	206	....	....	....	....	....
	Kamloops Blazers	WHL	1	0	0	0	2	....	....	....	....	....
	Prince Albert	WHL	8	2	1	3	13	2	0	0	0	2
2000-01	Prince Albert	WHL	68	8	8	16	193	....	....	....	....	....
2001-02	Prince Albert	WHL	70	7	8	15	189	....	....	....	....	....

BRUNEL, Craig (broo-NEHL, KRAYG) **PHI.**

Right wing. Shoots right. 6', 200 lbs. Born, Winnipeg, Man., November 12, 1979.
(Buffalo's 12th choice, 263rd overall, in 1999 Entry Draft).

			Regular Season					Playoffs				
Season	Club	Lea	GP	G	A	TP	PIM	GP	G	A	TP	PIM
1995-96	Notre Dame Argos	SMHL	36	4	9	13	79	....	....	....	....	....
1996-97	Prince Albert	WHL	57	5	2	7	208	4	0	0	0	13
1997-98	Prince Albert	WHL	58	6	12	18	247	....	....	....	....	....
1998-99	Prince Albert	WHL	50	10	8	18	173	14	4	2	6	48
99-2000	Prince Albert	WHL	17	0	2	2	59	....	....	....	....	....
	Red Deer Rebels	WHL	35	3	2	5	140	4	0	1	1	22
2000-01	Rochester	AHL	37	3	2	5	96	....	....	....	....	....
	South Carolina	ECHL	8	0	3	3	34	....	....	....	....	....
	Long Beach	WCHL	6	2	0	2	57	....	....	....	....	....
2001-02	Philadelphia	AHL	43	1	2	3	213	....	....	....	....	....
	Trenton Titans	ECHL	5	1	3	4	10	7	0	0	0	42

• Re-entered NHL Entry Draft. Originally Nashville's 6th choice, 147th overall, in 1998 Entry Draft.
Traded to **Red Deer** (WHL) by **Prince Albert** (WHL) with Russ Lupaschuk for Regan Darby, Steven MacIntyre and Scott McQueen, November 19, 1999. • Signed as a free agent by **Philadelphia**, September, 2001.

BUCKLEY, Brendan (BUHK-lee, BREHN-duhn) **PIT.**

Defense. Shoots right. 6'1", 200 lbs. Born, Boston, MA, February 26, 1977.
(Anaheim's 3rd choice, 117th overall, in 1996 Entry Draft).

			Regular Season					Playoffs				
Season	Club	Lea	GP	G	A	TP	PIM	GP	G	A	TP	PIM
1994-95	Boston Jr. Bruins	Exhib.	48	22	43	65	164	....	....	....	....	....
1995-96	Boston College	H-East	34	0	4	4	72	....	....	....	....	....
1996-97	Boston College	H-East	38	2	6	8	90	....	....	....	....	....
1997-98	Boston College	H-East	41	1	12	13	69	....	....	....	....	....
1998-99	Boston College	H-East	43	1	13	14	75	....	....	....	....	....
99-2000	Cincinnati	AHL	4	0	0	0	6	....	....	....	....	....
	Quad City	UHL	61	1	10	11	73	9	1	0	1	10
2000-01	Wilkes-Barre	AHL	63	2	8	10	62	21	0	2	2	33
2001-02	Wilkes-Barre	AHL	80	1	19	20	116	....	....	....	....	....

Signed as a free agent by **Pittsburgh**, September 28, 2000.

BULATOV, Alexei (boo-LA-tahf, al-EHX-ay) **NYR**

Right wing. Shoots left. 6'1", 185 lbs. Born, Sverdlovsk, USSR, January 24, 1978.
(NY Rangers' 11th choice, 254th overall, in 1999 Entry Draft).

			Regular Season					Playoffs				
Season	Club	Lea	GP	G	A	TP	PIM	GP	G	A	TP	PIM
1996-97	Yekaterinburg	Russia	20	2	4	6	2	....	....	....	....	....
	Yekaterinburg	Russia-Q	22	6	6	12	10	....	....	....	....	....
1997-98	Yekaterinburg	Russia	15	5	3	8	10	....	....	....	....	....
	Yekaterinburg	Russia-Q	22	7	6	13	14	....	....	....	....	....
1998-99	Yekaterinburg	Russia-2	47	21	16	37	24	....	....	....	....	....
99-2000	Cherepovets	Russia	9	1	0	1	0	....	....	....	....	....
	Ufa	Russia	6	0	3	3	2	....	....	....	....	....
	CSK VVS Samara	Russia	6	0	0	0	2	....	....	....	....	....
2000-01	Magnitogorsk	Russia	32	2	3	5	4	....	....	....	....	....
2001-02	Novokuznetsk	Russia	25	2	6	8	6	....	....	....	....	....

BUMAGIN, Yevgeny (boo-MA-gihn, yehv-GEH-nee) **DET.**

Center. Shoots left. 6', 170 lbs. Born, Belgorod, USSR, April 7, 1982.
(Detroit's 11th choice, 260th overall, in 2000 Entry Draft).

			Regular Season					Playoffs				
Season	Club	Lea	GP	G	A	TP	PIM	GP	G	A	TP	PIM
1997-98	Lada Togliatti 2	Russia-3	7	0	0	0	2	....	....	....	....	....
1998-99	Lada Togliatti 2	Russia-4	40	7	8	15	22	....	....	....	....	....
99-2000	Lada Togliatti 2	Russia-3	36	23	8	31	....	....	....	....	....	....
2000-01	CSK VVS Samara	Russia-2	13	1	0	1	10	....	....	....	....	....
2001-02	Dizelist Penza	Russia-2	40	2	0	2	0	....	....	....	....	....

BURNETT, Garrett (buhr-NEHT, GAIR-eht)

Left wing. Shoots left. 6'3", 230 lbs. Born, Coquitlam, B.C., September 23, 1975.

			Regular Season					Playoffs				
Season	Club	Lea	GP	G	A	TP	PIM	GP	G	A	TP	PIM
1993-94	Trail	RIJHL	26	2	1	3	248	....	....	....	....	....
1994-95	Sault Ste. Marie	OHL	14	0	1	1	78	....	....	....	....	....
	Kitchener	OHL	22	0	1	1	74	3	0	1	1	23
1995-96	Utica Blizzard	ColHL	15	0	1	1	78	....	....	....	....	....
	Oklahoma City	CHL	3	0	0	0	20	....	....	....	....	....
	Tulsa Oilers	CHL	6	1	0	1	94	....	....	....	....	....
	Nashville	ECHL	3	0	0	0	22	....	....	....	....	....
	Jacksonville	ECHL	8	0	1	1	38	1	0	0	0	0
1996-97	Knoxville	ECHL	50	5	11	16	321	....	....	....	....	....
1997-98	Johnstown Chiefs	ECHL	34	1	1	2	331	....	....	....	....	....
	Philadelphia	AHL	14	1	2	3	129	....	....	....	....	....
1998-99	Kentucky	AHL	31	1	0	1	186	....	....	....	....	....
99-2000	Kentucky	AHL	58	3	3	6	*506	4	0	0	0	31
2000-01	Cleveland	IHL	54	2	4	6	250	....	....	....	....	....
2001-02	New Haven	UHL	4	1	0	1	40	....	....	....	....	....
	Cincinnati	AHL	32	1	0	1	175	....	....	....	....	....

Signed as a free agent by **San Jose**, July 2, 1998. • Missed majority of 2001-02 season recovering from knee injury suffered in game vs. New Haven (AHL), January 15, 2002.

BUT, Anton (BOOT, AN-tawn) **T.B.**

Left wing. Shoots left. 6'1", 190 lbs. Born, Kharkov, USSR, July 3, 1980.
(New Jersey's 7th choice, 119th overall, in 1998 Entry Draft).

			Regular Season					Playoffs				
Season	Club	Lea	GP	G	A	TP	PIM	GP	G	A	TP	PIM
1995-96	Yaroslavl 2	CIS-2	60	30	12	42	10	....	....	....	....	....
1996-97	Yaroslavl 2	Russia-3	70	30	20	50	20	....	....	....	....	....
1997-98	Yaroslavl 2	Russia-2	48	12	5	17	28	....	....	....	....	....
1998-99	Yaroslavl 2	Russia-3	22	12	8	20	59	....	....	....	....	....
	Yaroslavl	Russia	5	0	0	0	0	....	....	....	....	....
99-2000	Yaroslavl 2	Russia-3	1	0	0	0	2	....	....	....	....	....
	Yaroslavl	Russia	26	2	5	7	16	8	2	1	3	0
2000-01	Yaroslavl	Russia	42	14	6	20	14	11	1	3	4	8
2001-02	Yaroslavl	Russia	48	14	11	25	14	6	0	1	1	2

Rights traded to **Tampa Bay** by **New Jersey** with Josef Boumedienne and Sascha Goc for Andrei Zyuzin, November 9, 2001.

BUTURLIN, Alexander (boo-tuhr-LIHN, AL-ehx-an-DEHR) **MTL.**

Right wing. Shoots left. 5'11", 182 lbs. Born, Moscow, USSR, September 3, 1981.
(Montreal's 1st choice, 39th overall, in 1999 Entry Draft).

			Regular Season					Playoffs				
Season	Club	Lea	GP	G	A	TP	PIM	GP	G	A	TP	PIM
1997-98	CSKA Moscow 2	Russia-3	50	12	15	27	46	....	....	....	....	....
	H.C. CSKA	Russia	2	0	0	0	0	....	....	....	....	....
1998-99	CSKA Moscow	Russia	16	1	0	1	6	3	1	0	1	2
99-2000	Sarnia Sting	OHL	57	20	27	47	46	7	4	2	6	12
2000-01	Sarnia Sting	OHL	57	28	37	65	27	4	3	1	4	0
2001-02	Ufa	Russia	32	3	3	6	42	....	....	....	....	....

BYFUGLIEN, Derrick (bigh-FEWG-lehn, DEHR-ihk) **OTT.**

Defense. Shoots left. 6'1", 212 lbs. Born, Roseau, MN, December 23, 1980.
(Ottawa's 5th choice, 122nd overall, in 2000 Entry Draft).

			Regular Season					Playoffs				
Season	Club	Lea	GP	G	A	TP	PIM	GP	G	A	TP	PIM
1998-99	Roseau Rams	Hi-School	28	15	19	34	....	....	....	....	....	....
	Fargo-Moorhead	USHL	12	2	3	5	71	....	....	....	....	....
99-2000	Fargo-Moorhead	USHL	50	5	11	16	106	....	....	....	....	....
2000-01	North Dakota	WCHA	6	0	1	1	2	....	....	....	....	....
	Erie Otters	OHL	31	1	4	5	37	15	3	1	4	29
2001-02	Florida	ECHL	43	2	4	6	81	....	....	....	....	....
	Roanoke Express	ECHL	21	3	2	5	39	4	0	1	1	6

• Left **North Dakota** (WCHA) and signed as a free agent by **Erie** (OHL), December 27, 2000.

BYKOV, Dmitri (BEE-kawv, dih-MEE-tree) **DET.**

Defense. Shoots left. 5'10", 169 lbs. Born, Izhevsk, USSR, May 5, 1977.
(Detroit's 6th choice, 258th overall, in 2001 Entry Draft).

			Regular Season					Playoffs				
Season	Club	Lea	GP	G	A	TP	PIM	GP	G	A	TP	PIM
1995-96	CSK VVS Samara	CIS	50	1	2	3	39	....	....	....	....	....
1996-97	CSK VVS Samara	Russia	44	1	6	7	20	2	0	0	0	2
1997-98	Lada Togliatti	Russia	9	0	1	1	4	....	....	....	....	....
	CSK VVS Samara	Russia	27	0	5	5	14	....	....	....	....	....
	Yaroslavl	Russia	10	1	2	3	6	7	0	1	1	10
1998-99	Lada Togliatti	Russia	39	0	6	6	24	7	1	0	1	8
	CSK VVS Samara	Russia	2	0	0	0	0	....	....	....	....	....
99-2000	Ak Bars Kazan	Russia	35	3	8	11	18	18	0	2	2	8
	Ak Bars Kazan 2	Russia-3	3	0	1	1	4	....	....	....	....	....
2000-01	Ak Bars Kazan	Russia	39	3	8	11	28	4	0	1	1	4
2001-02	Ak Bars Kazan	Russia	44	1	1	2	38	11	0	0	0	4

BYRNE, Trevor (BUHR-ne, TREH-vohr) **ST.L.**

Defense. Shoots left. 6'3", 205 lbs. Born, Hingham, MA, May 7, 1980.
(St. Louis' 4th choice, 143rd overall, in 1999 Entry Draft).

			Regular Season					Playoffs				
Season	Club	Lea	GP	G	A	TP	PIM	GP	G	A	TP	PIM
1997-98	Deerfield	Hi-School	25	5	14	19	16	....	....	....	....	....
1998-99	Deerfield	Hi-School	25	9	19	28	22	....	....	....	....	....
99-2000	Dartmouth	ECAC	30	3	9	12	40	....	....	....	....	....
2000-01	Dartmouth	ECAC	34	5	21	26	52	....	....	....	....	....
2001-02	Dartmouth	ECAC	32	5	16	21	38	....	....	....	....	....

ECAC Second All-Star Team (2001, 2002)

CABANA, Paul (CA-ba-NA, PAWL) **VAN.**

Right wing. Shoots right. 6'1", 185 lbs. Born, Calgary, Alta., September 28, 1978.
(Vancouver's 8th choice, 149th overall, in 1998 Entry Draft).

			Regular Season					Playoffs				
Season	Club	Lea	GP	G	A	TP	PIM	GP	G	A	TP	PIM
1996-97	Fort McMurray	AJHL	58	24	22	46	....	3	1	3	4	2
1997-98	Fort McMurray	AJHL	52	48	32	80	111	....	....	....	....	....
1998-99	Michigan Tech	WCHA	38	12	9	21	50	....	....	....	....	....
99-2000	Michigan Tech	WCHA	36	10	5	15	94	....	....	....	....	....
2000-01	Michigan Tech	WCHA	35	15	6	21	41	....	....	....	....	....
2001-02	Michigan Tech	WCHA	37	8	11	19	29	....	....	....	....	....
	Columbia Inferno	ECHL	3	1	1	2	2	....	....	....	....	....
	Manitoba Moose	AHL	4	0	1	1	4	6	0	0	0	4

AJHL All-Rookie Team (1997) • AJHL First All-Star Team (1998)

CAJANEK, Petr (chuh-YA-nihk, PEE-tuhr) **ST.L.**

Center. Shoots left. 5'11", 176 lbs. Born, Gottwaldov, Czech., August 18, 1975.
(St. Louis' 6th choice, 253rd overall, in 2001 Entry Draft).

			Regular Season					Playoffs				
Season	Club	Lea	GP	G	A	TP	PIM	GP	G	A	TP	PIM
1993-94	AC ZPS Zlin	Czech	34	5	4	9	....	3	0	0	0	....
1994-95	AC ZPS Zlin	Czech	35	7	9	16	8	12	2	6	8	4
1995-96	AC ZPS Zlin	Czech	36	8	11	19	32	8	2	6	8	8
1996-97	AC ZPS Zlin	Czech	50	9	30	39	46	....	....	....	....	....
1997-98	Zlin	Czech	46	19	27	46	117	....	....	....	....	....
1998-99	Zlin	Czech	49	15	33	48	123	11	5	7	12	12
99-2000	Zlin	Czech	50	23	34	57	66	4	1	0	1	0
2000-01	Zlin	Czech	52	18	31	49	105	6	0	4	4	22
2001-02	Zlin	Czech	49	20	44	64	64	11	5	7	12	10
	Czech Republic	Olympics	4	0	0	0	0	....	....	....	....	....

CALDWELL, Ryan (KAWLD-wehl, RIGH-uhn) **NYI**

Defense. Shoots left. 6'2", 174 lbs. Born, Deloraine, Man., June 15, 1981.
(NY Islanders' 7th choice, 202nd overall, in 2000 Entry Draft).

			Regular Season					Playoffs				
Season	Club	Lea	GP	G	A	TP	PIM	GP	G	A	TP	PIM
1998-99	Shat.-St. Mary's	Hi-School	29	24	55	79	22	....	....	....	....	....
99-2000	Thunder Bay	USHL	46	3	20	23	152	....	....	....	....	....
2000-01	U. of Denver	WCHA	36	3	20	23	76	....	....	....	....	....
2001-02	U. of Denver	WCHA	40	3	16	19	76	....	....	....	....	....

WCHA All-Rookie Team (2001)

CALLAHAN, Joe (kal-AH-han, JOH) **PHX.**

Defense. Shoots right. 6'3", 219 lbs. Born, Brockton, MA, December 20, 1982.
(Phoenix's 4th choice, 70th overall, in 2002 Entry Draft).

			Regular Season					Playoffs				
Season	Club	Lea	GP	G	A	TP	PIM	GP	G	A	TP	PIM
2000-01	B.C. High Irish	Hi-School	STATISTICS NOT AVAILABLE									
2001-02	Yale	ECAC	31	3	8	11	20	....	....	....	....	....

CAMERON, Scott (KAM-erh-RAWN, SCAWT) **N.J.**

Center. Shoots left. 6', 190 lbs. Born, Sudbury, Ont., April 11, 1981.
(New Jersey's 6th choice, 185th overall, in 1999 Entry Draft).

			Regular Season					Playoffs				
Season	Club	Lea	GP	G	A	TP	PIM	GP	G	A	TP	PIM
1997-98	Port Colborne	OJHL-B	40	16	35	51	73	....	....	....	....	....
1998-99	Barrie Colts	OHL	66	10	32	42	14	12	2	2	4	2
99-2000	Barrie Colts	OHL	21	2	4	6	19	....	....	....	....	....
	North Bay	OHL	49	26	28	54	13	6	3	0	3	4
	Albany	AHL	3	1	0	1	0	1	0	0	0	0
2000-01	North Bay	OHL	68	37	43	80	45	4	0	1	1	2
2001-02	Albany	AHL	54	3	6	9	12	....	....	....	....	....

OJHL-B Rookie of the Year (1998)

CAMMALLERI, Mike (kam-UH-LAIR-ee, MIGHK) **L.A.**

Center. Shoots left. 5'9", 180 lbs. Born, Richmond Hill, Ont., June 8, 1982.
(Los Angeles' 3rd choice, 49th overall, in 2001 Entry Draft).

			Regular Season					Playoffs				
Season	Club	Lea	GP	G	A	TP	PIM	GP	G	A	TP	PIM
1997-98	Bramalea Blues	OPJHL	46	36	52	88	30	....	....	....	....	....
1998-99	Bramalea Blues	OPJHL	41	31	72	103	51	....	....	....	....	....
99-2000	U. of Michigan	CCHA	39	13	13	26	32	....	....	....	....	....
2000-01	U. of Michigan	CCHA	42	*29	32	61	24	....	....	....	....	....
2001-02	U. of Michigan	CCHA	29	23	21	44	28	....	....	....	....	....

OPJHL Rookie of the Year (1998) • CCHA First All-Star Team (2001) • NCAA West Second All-American Team (2001) • CCHA Second All-Star Team (2002) • NCAA West First All-American Team (2002)

CAMPBELL, Eddy (KAM-behl, EH-dee) **DET.**

Defense. Shoots left. 6'2", 212 lbs. Born, Worcester, MA, November 26, 1974.
(NY Rangers' 9th choice, 190th overall, in 1993 Entry Draft).

			Regular Season					Playoffs				
Season	Club	Lea	GP	G	A	TP	PIM	GP	G	A	TP	PIM
1992-93	Omaha Lancers	USHL	42	9	19	28	160	....	....	....	....	....
1993-94	U. Mass-Lowell	H-East	40	8	16	24	114	....	....	....	....	....
1994-95	U. Mass-Lowell	H-East	34	6	24	30	105	....	....	....	....	....
1995-96	U. Mass-Lowell	H-East	39	6	33	39	*107	....	....	....	....	....
1996-97	Binghamton	AHL	74	5	17	22	108	4	0	0	0	2
1997-98	Hartford	AHL	9	0	1	1	9	14	0	2	2	33
	Fort Wayne	IHL	50	10	5	15	147	....	....	....	....	....
1998-99	Hartford	AHL	18	0	3	3	24	7	0	3	3	14
	Fort Wayne	IHL	46	1	16	17	137	....	....	....	....	....
99-2000	Orlando	IHL	81	2	7	9	217	3	0	0	0	6
2000-01	Worcester	AHL	78	5	27	32	207	10	1	0	1	10
2001-02	Worcester	AHL	70	3	15	18	172	3	0	0	0	5

Signed as a free agent by **St. Louis**, July 1, 2001. Signed as a free agent by **Detroit**, August 5, 2002.

CAMPBELL, Gregory (KAM-behl, GREH-goh-ree) **FLA.**

Center. Shoots left. 6', 191 lbs. Born, London, Ont., December 17, 1983.
(Florida's 4th choice, 67th overall, in 2002 Entry Draft).

			Regular Season					Playoffs				
Season	Club	Lea	GP	G	A	TP	PIM	GP	G	A	TP	PIM
1998-99	Aylmer Aces	OJHL-B	49	5	9	14	44	....	....	....	....	....
99-2000	St. Thomas Stars	OJHL-B	51	12	8	20	51	....	....	....	....	....
2000-01	Plymouth Whalers	OHL	65	2	12	14	40	10	0	0	0	7
2001-02	Plymouth Whalers	OHL	65	17	36	53	105	6	0	2	2	13

CAMPBELL, Joe (KAM-behl, JOH) **CGY.**

Defense. Shoots left. 6'4", 172 lbs. Born, Duluth, MN, June 26, 1982.
(Calgary's 10th choice, 233rd overall, in 2001 Entry Draft).

			Regular Season					Playoffs				
Season	Club	Lea	GP	G	A	TP	PIM	GP	G	A	TP	PIM
2000-01	Des Moines	USHL	56	7	11	18	53	3	0	1	1	4
2001-02	U. of Wisconsin	WCHA	7	0	0	0	0	....	....	....	....	....

CARKNER, Matt (KARK-nehr, MAT) **S.J.**

Defense. Shoots right. 6'4", 229 lbs. Born, Winchester, Ont., November 3, 1980.
(Montreal's 2nd choice, 58th overall, in 1999 Entry Draft).

			Regular Season					Playoffs				
Season	Club	Lea	GP	G	A	TP	PIM	GP	G	A	TP	PIM
1996-97	Winchester Hawks	OJHL-B	29	1	18	19	....	....	....	....	....	....
1997-98	Peterborough	OHL	57	0	6	6	121	4	0	0	0	2
1998-99	Peterborough	OHL	60	2	16	18	173	5	0	0	0	20
99-2000	Peterborough	OHL	62	3	13	16	177	5	0	1	1	6
2000-01	Peterborough	OHL	53	8	8	16	128	7	0	3	3	25
2001-02	Cleveland Barons	AHL	74	0	3	3	335	....	....	....	....	....

Signed as a free agent by **San Jose**, June 6, 2001.

CARON, Ed (kahr-OHN, EHD) **EDM.**

Left wing. Shoots left. 6'2", 214 lbs. Born, Nashua, NH, April 30, 1982.
(Edmonton's 3rd choice, 52nd overall, in 2001 Entry Draft).

			Regular Season					Playoffs				
Season	Club	Lea	GP	G	A	TP	PIM	GP	G	A	TP	PIM
1998-99	Phillips Exeter	Hi-School	31	39	30	69	28	....	....	....	....	....
99-2000	Phillips Exeter	Hi-School	26	22	26	48	24	....	....	....	....	....
2000-01	Phillips Exeter	Hi-School	17	30	20	50	42	....	....	....	....	....
2001-02	New Hampshire	H-East	34	6	7	13	51	....	....	....	....	....

CARTELLI, Mario (car-TEHL-lee, MAHR-ee-oh) **ATL.**

Defense. Shoots right. 6'1", 196 lbs. Born, Trinec, Czech., November 16, 1979.
(Atlanta's 9th choice, 262nd overall, in 2001 Entry Draft).

			Regular Season					Playoffs				
Season	Club	Lea	GP	G	A	TP	PIM	GP	G	A	TP	PIM
1998-99	Trinec	Czech	34	1	1	2	14	8	2	1	3	4
99-2000	Trinec Jr.	Czech-Jr.	5	3	0	3	0	....	....	....	....	....
	Trinec	Czech	42	7	4	11	14	4	0	0	0	0
2000-01	Trinec	Czech	46	11	16	27	24	....	....	....	....	....
2001-02	Trinec	Czech	20	0	1	1	12	....	....	....	....	....
	Kladno	Czech	24	3	8	11	14	....	....	....	....	....

CASS, Bill (KAS, BIHL) **ANA.**

Defense. Shoots left. 6', 208 lbs. Born, Hingham, MA, September 30, 1980.
(Anaheim's 5th choice, 153rd overall, in 2000 Entry Draft).

			Regular Season					Playoffs				
Season	Club	Lea	GP	G	A	TP	PIM	GP	G	A	TP	PIM
1997-98	Team USA	USDP-18	38	4	9	13	75	2	0	0	0	0
1998-99	Team USA	USDP-18	41	0	6	6	22	....	....	....	....	....
99-2000	Boston College	H-East	41	1	8	9	26	....	....	....	....	....
2000-01	Boston College	H-East	40	0	6	6	52	....	....	....	....	....
2001-02	Boston College	H-East	36	3	5	8	40	....	....	....	....	....

CAVANAGH, Tom (KAV-a-NAW, TAWM) **S.J.**

Right wing. Shoots left. 5'10", 178 lbs. Born, Warwick, RI, March 24, 1982.
(San Jose's 6th choice, 182nd overall, in 2001 Entry Draft).

			Regular Season					Playoffs				
Season	Club	Lea	GP	G	A	TP	PIM	GP	G	A	TP	PIM
1997-98	Toll Gate Titans	Hi-School	15	5	17	22	6	4	2	8	10	4
1998-99	Toll Gate Titans	Hi-School	15	9	20	29	26	5	5	4	9	6
99-2000	Toll Gate Titans	Hi-School	18	25	29	*54	28	5	0	12	12	9
2000-01	Phillips Exeter	Hi-School	31	*42	40	82	34	....	....	....	....	....
2001-02	Harvard Crimson	ECAC	34	8	17	25	4	....	....	....	....	....

CAVANAUGH, Dan (KAV-a-naw, DAN) MIN.

Center. Shoots right. 6'1", 190 lbs. Born, Springfield, MA, March 3, 1980.
(Calgary's 2nd choice, 38th overall, in 1999 Entry Draft).

			Regular Season					Playoffs				
Season	**Club**	**Lea**	**GP**	**G**	**A**	**TP**	**PIM**	**GP**	**G**	**A**	**TP**	**PIM**
1995-96	New England	EJHL	43	8	7	15	....	....	....	....	....	....
1996-97	New England	EJHL	56	23	46	69	....	....	....	....	....	....
1997-98	New England	EJHL	38	31	*47	*78	58	13	8	12	30	....
1998-99	Boston University	H-East	36	6	8	14	60	....	....	....	....	....
99-2000	Boston University	H-East	40	9	25	34	62	....	....	....	....	....
2000-01	Boston University	H-East	35	7	21	28	43	....	....	....	....	....
2001-02	Houston Aeros	AHL	70	3	16	19	41	5	0	0	0	0

EJHL Most Valuable Player (1998)

Rights traded to **Minnesota** by **Calgary** with Calgary's 8th round choice (Jake Riddle) in 2001 Entry Draft for Mike Vernon, June 23, 2000.

CAVOSIE, Marc (kuh-VOI-see, MAHRK) MIN.

Left wing. Shoots left. 6', 173 lbs. Born, Albany, NY, August 6, 1981.
(Minnesota's 3rd choice, 99th overall, in 2000 Entry Draft).

			Regular Season					Playoffs				
Season	**Club**	**Lea**	**GP**	**G**	**A**	**TP**	**PIM**	**GP**	**G**	**A**	**TP**	**PIM**
1995-96	Albany	Hi-School	22	8	27	31	....	....	....	....	....	....
1996-97	Albany	Hi-School	28	26	45	71	....	....	....	....	....	....
1997-98	Albany	Hi-School	28	38	33	71	....	....	....	....	....	....
1998-99	Albany	Hi-School	28	23	20	43	32	....	....	....	....	....
99-2000	RPI Engineers	ECAC	29	11	17	28	10	....	....	....	....	....
2000-01	RPI Engineers	ECAC	28	13	16	29	47	....	....	....	....	....
2001-02	RPI Engineers	ECAC	36	23	*27	*50	44	....	....	....	....	....

ECAC First All-Star Team (2002) • ECAC Player of the Year (2002)

CEREDA, Luca (suh-REH-duh, LOO-ka) TOR.

Center. Shoots left. 6'2", 210 lbs. Born, Lugano, Switz., September 7, 1981.
(Toronto's 1st choice, 24th overall, in 1999 Entry Draft).

			Regular Season					Playoffs				
Season	**Club**	**Lea**	**GP**	**G**	**A**	**TP**	**PIM**	**GP**	**G**	**A**	**TP**	**PIM**
1996-97	HC Ambri-Piotta	Swiss	35	13	8	21	....	....	....	....	....	....
1997-98	HC Ambri-Piotta	Swiss	28	17	27	44	24	....	....	....	....	....
1998-99	Ambri Jr.	Swiss-Jr.	3	4	3	7	20	....	....	....	....	....
	HC Ambri-Piotta	Swiss	38	6	10	16	8	15	0	6	6	4
99-2000	HC Ambri-Piotta	Swiss	44	1	5	6	14	9	0	1	1	2
2000-01	Ottawa 67's	OHL			DID NOT PLAY							
2001-02	St. John's	AHL	71	5	8	13	23	11	2	1	3	10

• Missed entire 2000-01 season recovering from heart surgery, September, 2000.

CETKOVSKY, Jiri (tseht-KAWF-skee, YIH-ree) CGY.

Center. Shoots left. 6'4", 209 lbs. Born, Prostejov, Czech., November 4, 1983.
(Calgary's 5th choice, 141st overall, in 2002 Entry Draft).

			Regular Season					Playoffs				
Season	**Club**	**Lea**	**GP**	**G**	**A**	**TP**	**PIM**	**GP**	**G**	**A**	**TP**	**PIM**
99-2000	HC Olomouc-18	Czech-Jr.	25	7	3	10	18	....	....	....	....	....
	Prostejov Jr.	Czech-Jr.	5	3	2	5	0	....	....	....	....	....
2000-01	Prostejov Jr.	Czech-Jr.	28	7	7	14	75	....	....	....	....	....
2001-02	Zlin Jr.	Czech-Jr.	30	9	6	15	44	....	....	....	....	....

CHABADA, Martin NYI

Right wing. Shoots right. 6'1", 203 lbs. Born, Prague, Czech., June 14, 1977.
(NY Islanders' 6th choice, 252nd overall, in 2002 Entry Draft).

			Regular Season					Playoffs				
Season	**Club**	**Lea**	**GP**	**G**	**A**	**TP**	**PIM**	**GP**	**G**	**A**	**TP**	**PIM**
1996-97	HC Sparta Praha	EuroHL	1	0	0	0	0	4	0	0	0	2
	HC Sparta Praha	Czech	16	1	2	3	4	4	0	1	1	2
1997-98	HC Sparta Praha	EuroHL	4	0	1	1	4	....	....	....	....	....
	HC Sparta Praha	Czech	40	7	12	19	57	11	1	0	1	4
1998-99	HC Sparta Praha	EuroHL	6	1	1	2	10	2	2	1	3	6
	HC Sparta Praha	Czech	39	3	9	12	14	1	0	0	0	0
99-2000	HC Sparta Praha	Czech	35	7	12	19	12	9	2	2	4	4
2000-01	HC Sparta Praha	Czech	43	9	10	19	16	12	0	0	0	0
2001-02	HC Sparta Praha	Czech	51	19	21	40	113	13	4	7	11	4

CHAGODAYEV, Alexandr (cheh-goh-digh-ehv, al-ehx-AN-duhr) ANA.

Center. Shoots left. 6'1", 185 lbs. Born, Perm, USSR, January 15, 1981.
(Anaheim's 3rd choice, 105th overall, in 1999 Entry Draft).

			Regular Season					Playoffs				
Season	**Club**	**Lea**	**GP**	**G**	**A**	**TP**	**PIM**	**GP**	**G**	**A**	**TP**	**PIM**
1997-98	H.C. CSKA	Russia-2	5	1	0	1	0	....	....	....	....	....
	H.C. CSKA	Russia	1	0	0	0	0	....	....	....	....	....
1998-99	H.C. CSKA	Russia-2	35	9	9	18	16	....	....	....	....	....
99-2000	H.C. CSKA	Russia-2	40	5	7	12	....	....	....	....	....	....
2000-01	Krylja Sovetov	Russia-2	7	0	1	1	0	....	....	....	....	....
	St. Petersburg	Russia	8	0	1	1	0	....	....	....	....	....
	Yaroslavl	Russia	6	2	0	2	2	....	....	....	....	....
	Ufa	Russia	1	0	0	0	0	....	....	....	....	....
2001-02	Lada Togliatti	Russia	15	1	1	2	6	....	....	....	....	....

CHALMERS, James (CHAWL-muhrz, JAYMS)

Center. Shoots left. 6'2", 200 lbs. Born, Mississauga, Ont., August 19, 1977.

			Regular Season					Playoffs				
Season	**Club**	**Lea**	**GP**	**G**	**A**	**TP**	**PIM**	**GP**	**G**	**A**	**TP**	**PIM**
1995-96	Shelburne Wolves	MTJHL	14	5	5	10	17	....	....	....	....	....
1996-97	Shelburne Wolves	MTJHL	40	35	24	59	58	....	....	....	....	....
	Summerside	MJrHL	32	24	21	45	56	....	....	....	....	....
1997-98	Nebraska-Omaha	NCAA-2	30	8	12	20	20	....	....	....	....	....
1998-99	Nebraska-Omaha	NCAA-2	28	6	11	17	53	....	....	....	....	....
99-2000	Nebraska-Omaha	CCHA	37	5	8	13	72	....	....	....	....	....
2000-01	Nebraska-Omaha	CCHA	37	6	15	21	96	....	....	....	....	....
2001-02	Philadelphia	AHL	61	3	8	11	67	....	....	....	....	....

Signed as a free agent by **Philadelphia**, June 14, 2001.

CHARPENTIER, Marco (shar-PUHNT-yay, MAHR-koh) NYI

Center. Shoots left. 6', 200 lbs. Born, Montreal, Que., January 23, 1980.

			Regular Season					Playoffs				
Season	**Club**	**Lea**	**GP**	**G**	**A**	**TP**	**PIM**	**GP**	**G**	**A**	**TP**	**PIM**
1996-97	Mtl-Bourassa	QAAA	42	15	23	38	28	....	....	....	....	....
1997-98	Quebec Remparts	QMJHL	55	4	7	11	12	3	0	1	1	0
1998-99	Quebec Remparts	QMJHL	12	4	7	11	4	....	....	....	....	....
	Baie-Comeau	QMJHL	52	16	23	39	34	....	....	....	....	....
99-2000	Baie-Comeau	QMJHL	72	51	62	113	39	6	3	2	5	16
2000-01	Baie-Comeau	QMJHL	71	57	55	112	88	11	9	10	19	10
2001-02	Bridgeport	AHL	16	1	1	2	2	....	....	....	....	....
	Trenton Titans	ECHL	54	20	21	41	31	7	5	2	7	4

Traded to **Baie-Comeau** (QMJHL) by **Quebec** (QMJHL) for Joey Fetta, October 10, 1998. Signed as a free agent by **NY Islanders**, December 12, 2000.

CHARRON, Craig (shah-ROHN, KRAYG)

Center. Shoots right. 5'10", 175 lbs. Born, North Easton, MA, November 15, 1967.
(Montreal's 1st choice, 25th overall, in 1989 Supplemental Draft).

			Regular Season					Playoffs				
Season	**Club**	**Lea**	**GP**	**G**	**A**	**TP**	**PIM**	**GP**	**G**	**A**	**TP**	**PIM**
1986-87	U. Mass-Lowell	H-East	36	11	16	27	48	....	....	....	....	....
1987-88	U. Mass-Lowell	H-East	39	22	18	40	32	....	....	....	....	....
1988-89	U. Mass-Lowell	H-East	32	14	21	35	32	....	....	....	....	....
1989-90	U. Mass-Lowell	H-East	35	17	29	46	10	....	....	....	....	....
1990-91	Winston-Salem	ECHL	30	11	16	27	10	....	....	....	....	....
	Albany Choppers	IHL	5	0	2	2	0	....	....	....	....	....
	Fredericton	AHL	24	2	5	7	4	5	0	3	3	0
1991-92	Cincinnati	ECHL	64	41	55	96	97	9	5	5	10	10
1992-93	Birmingham Bulls	ECHL	23	9	17	26	18	....	....	....	....	....
	Cincinnati	IHL	27	6	8	14	8	....	....	....	....	....
1993-94	Olofstroms IK	Swede-2	32	38	34	72	66	....	....	....	....	....
1994-95	Dayton Bombers	ECHL	48	35	47	82	82	9	9	13	22	10
	Kalamazoo Wings	IHL	2	0	0	0	0	....	....	....	....	....
	Fort Wayne	IHL	2	1	0	1	4	....	....	....	....	....
	Cornwall Aces	AHL	6	5	0	5	0	2	0	0	0	0
1995-96	Rochester	AHL	72	43	52	95	79	19	7	10	17	12
1996-97	Rochester	AHL	72	24	41	65	42	10	8	8	16	2
1997-98	Rochester	AHL	75	25	53	78	51	4	1	1	2	0
1998-99	Lowell	AHL	71	22	39	61	41	3	1	2	3	8
99-2000	St. John's	AHL	32	11	18	29	14	....	....	....	....	....
	Lowell	AHL	22	8	13	21	14	7	2	3	5	4
2000-01	Rochester	AHL	73	18	32	50	53	4	0	1	1	2
2001-02	Rochester	AHL	43	12	12	24	24	2	0	0	0	0

Signed as a free agent by **NY Islanders**, September 3, 1998. Traded to **Toronto** by **NY Islanders** for Niklas Andersson, August 17, 1999. Traded to **LA Kings** by **Toronto** for Donald MacLean, February 23, 2000. Signed as a free agent by **Rochester** (AHL), July 20, 2000.

CHARTIER, Chris (SHAR-tee-yay, KRIHS-t'yehn) TOR.

Defense. Shoots left. 6', 216 lbs. Born, Russell, Man., December 29, 1980.
(Edmonton's 8th choice, 199th overall, in 1999 Entry Draft).

			Regular Season					Playoffs				
Season	**Club**	**Lea**	**GP**	**G**	**A**	**TP**	**PIM**	**GP**	**G**	**A**	**TP**	**PIM**
1995-96	Yellowhead Pass	MMHL	36	8	23	31	68	....	....	....	....	....
1996-97	Saskatoon Blades	WHL	64	2	23	25	32	....	....	....	....	....
1997-98	Saskatoon Blades	WHL	68	8	33	41	43	6	0	3	3	12
1998-99	Saskatoon Blades	WHL	62	2	14	16	71	....	....	....	....	....
99-2000	Saskatoon Blades	WHL	11	2	2	4	4	....	....	....	....	....
	Prince George	WHL	57	16	36	52	60	13	4	9	13	12
2000-01	Prince George	WHL	63	12	56	68	99	6	1	4	5	6
2001-02	St. John's	AHL	65	5	10	15	18	11	0	4	4	4

WHL West Second All-Star Team (2000) • WHL West First All-Star Team (2001)

Traded to **Prince George** (WHL) by **Saskatoon** (WHL) with Saskatoon's 5th round choice (Clayton Bartnel) in 2001 WHL Bantam Draft for future considerations, January 10, 2000. Signed as a free agent by **Toronto**, June 25, 2001.

CHEECHOO, Jonathan (CHEE-choo, JAWN-ah-thuhn) S.J.

Right wing. Shoots right. 6', 205 lbs. Born, Moose Factory, Ont., July 15, 1980.
(San Jose's 2nd choice, 29th overall, in 1998 Entry Draft).

			Regular Season					Playoffs				
Season	**Club**	**Lea**	**GP**	**G**	**A**	**TP**	**PIM**	**GP**	**G**	**A**	**TP**	**PIM**
1996-97	Kitchener	OJHL-B	43	35	41	76	33	....	....	....	....	....
1997-98	Belleville Bulls	OHL	64	31	45	76	62	10	4	2	6	10
1998-99	Belleville Bulls	OHL	63	35	47	82	74	21	15	15	30	27
99-2000	Belleville Bulls	OHL	66	45	46	91	102	16	5	12	17	16
2000-01	Kentucky	AHL	75	32	34	66	63	3	0	0	0	0
2001-02	Cleveland Barons	AHL	53	21	25	46	54	....	....	....	....	....

OHL All-Rookie Team (1998)

CHERNOV, Artem (chair-NAHF, AR-tehm) DAL.

Center. Shoots left. 5'10", 176 lbs. Born, Novokuznetsk, USSR, April 28, 1982.
(Dallas' 7th choice, 162nd overall, in 2000 Entry Draft).

			Regular Season					Playoffs				
Season	**Club**	**Lea**	**GP**	**G**	**A**	**TP**	**PIM**	**GP**	**G**	**A**	**TP**	**PIM**
1997-98	Novokuznetsk 2	Russia-3	4	0	0	0	0	....	....	....	....	....
1998-99	Novokuznetsk 2	Russia-4	32	9	7	16	14	....	....	....	....	....
99-2000	Magnitogorsk	Russia	10	2	3	5	0	5	0	0	0	0
2000-01	Magnitogorsk	Russia	44	15	17	32	30	....	....	....	....	....
2001-02	Avangard Omsk	Russia	43	6	5	11	4	....	....	....	....	....

CHISTOV, Stanislav (chihs-TAHV, STAHN-his-LAHV) ANA.

Left wing. Shoots right. 5'10", 178 lbs. Born, Chelyabinsk, USSR, April 17, 1983.
(Anaheim's 1st choice, 5th overall, in 2001 Entry Draft).

			Regular Season					Playoffs				
Season	**Club**	**Lea**	**GP**	**G**	**A**	**TP**	**PIM**	**GP**	**G**	**A**	**TP**	**PIM**
1998-99	Chelyabinsk 2	Russia-4	1	0	0	0	0	....	....	....	....	....
	Georgetown	OPJHL	14	10	7	17	21	....	....	....	....	....
99-2000	Omsk Jr.	Russia-Jr.	5	4	3	7	8	....	....	....	....	....
	Omsk 2	Russia-3	18	12	4	16	24	....	....	....	....	....
	Novokuznetsk	Russia	9	7	4	11	18	....	....	....	....	....
	Avangard Omsk	Russia	3	1	0	1	2	....	....	....	....	....
2000-01	Omsk 2	Russia-3	8	5	4	9	2	....	....	....	....	....
	Avangard Omsk	Russia	24	4	8	12	12	5	0	0	0	2
2001-02	Avangard Omsk	Russia	9	0	0	0	4	....	....	....	....	....
	CSKA Moscow 2	Russia-3	1	1	2	3	0	....	....	....	....	....

CHRISTEEN, Mats (KRIHS-teen, MATS) NSH.

Defense. Shoots left. 6'1", 181 lbs. Born, Sodertalje, Sweden, February 13, 1982.
(Nashville's 11th choice, 236th overall, in 2000 Entry Draft).

			Regular Season					Playoffs				
Season	Club	Lea	GP	G	A	TP	PIM	GP	G	A	TP	PIM
99-2000	Sodertalje SK-18	Swede-Jr.	5	0	1	1	8					
	Sodertalje Jr.	Swede-Jr.	30	1	0	1	30	4	1	0	1	6
2000-01	Sodertalje Jr.	Swede-Jr.	19	2	8	10	16					
	Sodertalje SK	Swede-2	15	0	3	3	4					
2001-02	Sodertalje Jr.	Swede-Jr.	19	3	7	10	47	2	0	0	0	4
	Tierp HK	Swede-2	15	3	1	4	14					
	Sodertalje SK	Sweden	6	0	0	0	0					

CHRISTENSEN, Erik (KRIHS-tehn-suhn, AIR-ihk) PIT.

Center. Shoots left. 6'1", 178 lbs. Born, Edmonton, Alta., December 17, 1983.
(Pittsburgh's 3rd choice, 69th overall, in 2002 Entry Draft).

			Regular Season					Playoffs				
Season	Club	Lea	GP	G	A	TP	PIM	GP	G	A	TP	PIM
1998-99	Leduc Oil Kings	AMBHL	36	34	42	76	70					
99-2000	Kamloops Blazers	WHL	66	9	5	14	41	4	0	0	0	2
2000-01	Kamloops Blazers	WHL	72	21	23	44	36	4	1	1	2	0
2001-02	Kamloops Blazers	WHL	70	22	36	58	68	4	0	0	0	4

CHVATAL, Marek (KHVA-tuhl, MAIR-ehk) N.J.

Defense. Shoots left. 6', 183 lbs. Born, Pardubice, Czech., January 27, 1984.
(New Jersey's 4th choice, 84th overall, in 2002 Entry Draft).

			Regular Season					Playoffs				
Season	Club	Lea	GP	G	A	TP	PIM	GP	G	A	TP	PIM
99-2000	Dukla Jihlava-18	Czech-Jr.	36	7	8	15	110					
	Dukla Jihlava Jr.	Czech-Jr.	1	0	0	0	2					
2000-01	Trinec-18	Czech-Jr.	9	3	4	7	10					
	Trinec Jr.	Czech-Jr.	44	2	2	4	22					
2001-02	Trinec Jr.	Czech-Jr.	40	10	13	23	91					
	Trinec	Czech	3	0	0	0	2	1	0	0	0	0

CLARK, Kyle (KLAHRK, KIGHLE) WSH.

Right wing. Shoots right. 6'6", 224 lbs. Born, Burlington, VT, February 14, 1980.
(Washington's 7th choice, 175th overall, in 1999 Entry Draft).

			Regular Season					Playoffs				
Season	Club	Lea	GP	G	A	TP	PIM	GP	G	A	TP	PIM
1997-98	Team USA	USDP-18	65	14	11	25	287					
1998-99	Harvard Crimson	ECAC	20	0	2	2	30					
99-2000	Harvard Crimson	ECAC	22	0	3	3	30					
2000-01	Harvard Crimson	ECAC	18	0	2	2	20					
	Portland Pirates	AHL	2	0	0	0	2					
	Richmond	ECHL	18	1	2	3	110	3	0	0	0	5
2001-02	Portland Pirates	AHL	57	2	4	6	208					

CLARKE, Noah (KLAHRK, NOH-uh) L.A.

Left wing. Shoots left. 5'9", 175 lbs. Born, LaVerne, CA, June 11, 1979.
(Los Angeles' 10th choice, 250th overall, in 1999 Entry Draft).

			Regular Season					Playoffs				
Season	Club	Lea	GP	G	A	TP	PIM	GP	G	A	TP	PIM
1996-97	Shat.-St. Mary's	Hi-School	30	33	44	77	....					
1997-98	Des Moines	USHL	54	19	30	49	29	12	2	9	11	23
1998-99	Des Moines	USHL	52	31	32	63	47	13	8	2	10	16
99-2000	Colorado College	WCHA	39	17	20	37	30					
2000-01	Colorado College	WCHA	41	12	20	32	22					
2001-02	Colorado College	WCHA	42	13	24	37	32					

USHL All-Rookie Team (1998) • USHL First All-Star Team (1999) • Won Curt Hammer Award (Most Gentlemanly Player - USHL) (1999) • WCHA All-Rookie Team (2000)

CLOUTHIER, Brett (KLOO-tyay, BREHT) N.J.

Left wing. Shoots left. 6'5", 245 lbs. Born, Ottawa, Ont., June 9, 1981.
(New Jersey's 3rd choice, 50th overall, in 1999 Entry Draft).

			Regular Season					Playoffs				
Season	Club	Lea	GP	G	A	TP	PIM	GP	G	A	TP	PIM
1997-98	Kanata Valley	OCJHL	50	12	10	22	135					
1998-99	Kingston	OHL	64	8	14	22	227	5	1	1	2	4
99-2000	Kingston	OHL	65	13	26	39	*266	5	2	0	2	17
2000-01	Kingston	OHL	68	28	29	57	165	4	1	0	1	10
2001-02	Albany	AHL	62	4	0	4	109					

CLOUTIER, David (KLOO-tyay, DAY-vihd) S.J.

Defense. Shoots right. 6'2", 200 lbs. Born, Quebec City, Que., December 17, 1981.

			Regular Season					Playoffs				
Season	Club	Lea	GP	G	A	TP	PIM	GP	G	A	TP	PIM
1998-99	Sherbrooke	QMJHL	46	2	2	4	39	4	0	0	0	2
99-2000	Montreal Rocket	QMJHL	33	2	7	9	36					
	Val-d'Or Foreurs	QMJHL	30	5	8	13	24					
2000-01	Val-d'Or Foreurs	QMJHL	72	14	20	34	200	21	3	9	12	36
2001-02	Val-d'Or Foreurs	QMJHL	38	16	16	32	123					
	Cape Breton	QMJHL	22	9	17	26	60	16	7	15	22	24

Traded to **Cape Breton** (QMJHL) **by Val d'Or** (QMJHL) for Cape Breton's 4th round choice in 2003 Priority Draft and future considerations, January 5, 2002. Signed as a free agent by **San Jose**, July 10, 2002.

CLOWE, Ryan (KLOH, RIGH-uhn) S.J.

Right wing. Shoots right. 6'2", 190 lbs. Born, St. John's, Nfld., September 30, 1982.
(San Jose's 5th choice, 175th overall, in 2001 Entry Draft).

			Regular Season					Playoffs				
Season	Club	Lea	GP	G	A	TP	PIM	GP	G	A	TP	PIM
99-2000	St. John's	NFAHA		STATISTICS NOT AVAILABLE								
2000-01	Moncton	MJrHL		STATISTICS NOT AVAILABLE								
	Rimouski Oceanic	QMJHL	32	15	10	25	43	11	8	1	9	12
2001-02	Rimouski Oceanic	QMJHL	53	28	45	73	120	7	1	6	7	2

COLAGIACOMO, Adam (coh-lah-JAH-coh-moh, A-dam)

Right wing. Shoots right. 6'2", 200 lbs. Born, Rexdale, Ont., March 17, 1979.
(San Jose's 3rd choice, 82nd overall, in 1997 Entry Draft).

			Regular Season					Playoffs				
Season	Club	Lea	GP	G	A	TP	PIM	GP	G	A	TP	PIM
1994-95	North York	MTJHL	33	39	20	59	48					
1995-96	London Knights	OHL	66	28	38	66	88					
1996-97	London Knights	OHL	26	11	11	22	37					
	Oshawa Generals	OHL	23	14	10	24	32	13	1	5	6	4
1997-98	Oshawa Generals	OHL	58	25	31	56	80	7	1	0	1	2
1998-99	Plymouth Whalers	OHL	67	40	68	108	89	10	6	9	15	14
99-2000	Kentucky	AHL	42	5	8	13	27	1	1	0	1	0
	New Orleans	ECHL	18	7	7	14	12	3	0	2	2	2
2000-01	Kentucky	AHL	67	10	11	21	48	2	0	0	0	0
2001-02	Cleveland Barons	AHL	59	11	14	25	32					

OHL All-Rookie Team (1996)

Traded to **Oshawa** (OHL) by **London** (OHL) for Jason Metcalfe, January 4, 1997. Traded to **Plymouth** (OHL) by **Oshawa** (OHL) for Brian Passmore, Plymouth's 4th round choice (Brandon Cullen) in 1998 OHL Priority Draft and future considerations, June 11, 1998.

COLAIACOVO, Carlo (koh-lee-A-KOH-voh, KAR-loh) TOR.

Defense. Shoots left. 6'1", 184 lbs. Born, Toronto, Ont., January 27, 1983.
(Toronto's 1st choice, 17th overall, in 2001 Entry Draft).

			Regular Season					Playoffs				
Season	Club	Lea	GP	G	A	TP	PIM	GP	G	A	TP	PIM
1998-99	Mississauga Reps	GTHL	44	10	12	23	28					
99-2000	Erie Otters	OHL	52	4	18	22	12	13	2	4	6	9
2000-01	Erie Otters	OHL	62	12	27	39	59	14	4	7	11	16
2001-02	Erie Otters	OHL	60	13	27	40	49	21	7	10	17	20

OHL Second All-Star Team (2002)

COLE, Phil (KOHL, FIHL) N.J.

Defense. Shoots left. 6'4", 205 lbs. Born, Winnipeg, Man., September 6, 1982.
(New Jersey's 8th choice, 125th overall, in 2000 Entry Draft).

			Regular Season					Playoffs				
Season	Club	Lea	GP	G	A	TP	PIM	GP	G	A	TP	PIM
1997-98	Winnipeg Sharks	MMHL	45	0	18	18	68	5	0	4	4	2
1998-99	Lethbridge	WHL	45	2	1	3	64	4	0	0	0	0
99-2000	Lethbridge	WHL	51	1	6	7	112					
2000-01	Lethbridge	WHL	63	6	15	21	129	1	0	0	0	2
2001-02	Lethbridge	WHL	33	3	13	16	87					
	Vancouver Giants	WHL	6	0	1	1	18					
	Medicine Hat	WHL	15	1	5	6	49					

Traded to **Vancouver** (WHL) by **Lethbridge** (WHL) for Jeremy Jackson, December 28, 2001. Traded to **Medicine Hat** (WHL) by **Vancouver** (WHL) with Warrem McCutcheon and Andrew Davidson for Rory Rawlyk, Eric Clark, Adam Courchaine, Mitch Bartley, Mitch McGillvary and Medicine Hat's 5th round choice (Brandon Brown) in 2002 WHL Bantam Draft, January 11, 2002

COLEMAN, Jon (KOHL-man, JAWN)

Defense. Shoots right. 6'1", 205 lbs. Born, Boston, MA, March 9, 1975.
(Detroit's 2nd choice, 48th overall, in 1993 Entry Draft).

			Regular Season					Playoffs				
Season	Club	Lea	GP	G	A	TP	PIM	GP	G	A	TP	PIM
1992-93	Phillips Andover	Hi-School	24	14	33	47	40					
1993-94	Boston University	H-East	29	1	14	15	26					
1994-95	Boston University	H-East	40	5	23	28	42					
1995-96	Boston University	H-East	40	7	31	38	58					
1996-97	Boston University	H-East	39	5	27	32	20					
1997-98	Detroit Vipers	IHL	1	0	0	0	0					
	Adirondack	AHL	54	2	29	31	23	2	0	0	0	0
1998-99	Adirondack	AHL	72	12	26	38	32	3	0	0	0	0
99-2000	Kentucky	AHL	66	1	14	15	43	9	2	4	6	2
2000-01	Providence	AHL	49	4	15	19	14					
	Orlando	IHL	19	1	7	8	6	12	0	1	1	2
2001-02	Amur Khabarovsk	Russia	44	0	5	5	40					

Hockey East Second All-Star Team (1996, 1997) • NCAA East Second All-American Team (1996) • NCAA East First All-American Team (1997)

Signed as a free agent by **San Jose**, August 26, 1999. Signed as a free agent by **Providence** (AHL), December 12, 2000. Loaned to **Orlando** (IHL) by **Providence** (AHL) for future considerations, March 9. 2001.

COLLINS, Brian (kAW-lihns, BRIGH-uhn) NYI

Center. Shoots left. 6'1", 190 lbs. Born, Worcester, MA, September 13, 1980.
(NY Islanders' 6th choice, 87th overall, in 1999 Entry Draft).

			Regular Season					Playoffs				
Season	Club	Lea	GP	G	A	TP	PIM	GP	G	A	TP	PIM
1998-99	St. John's Prep	Hi-School	28	38	35	73	20					
99-2000	Boston University	H-East	42	13	11	24	61					
2000-01	Boston University	H-East	37	14	16	30	16					
2001-02	Boston University	H-East	38	12	8	20	22					

COLLYMORE, Shawn (KAW-lee-mohr, SHAWN) NYR

Right wing. Shoots right. 5'11", 180 lbs. Born, Ville de LaSalle, Que., May 2, 1983.
(NY Rangers' 5th choice, 139th overall, in 2001 Entry Draft).

			Regular Season					Playoffs				
Season	Club	Lea	GP	G	A	TP	PIM	GP	G	A	TP	PIM
1998-99	Lac St-Louis	QAAA	41	12	20	32	36					
99-2000	Quebec Remparts	QMJHL	64	8	16	24	22	11	0	2	2	0
2000-01	Quebec Remparts	QMJHL	71	24	43	67	32	4	0	3	3	0
2001-02	Quebec Remparts	QMJHL	52	23	34	57	37	9	2	4	6	0

CONCANNON, Mark (KAHN-kan-nuhn, MAHRK) S.J.

Left wing. Shoots left. 6', 200 lbs. Born, Boston, MA, June 12, 1980.
(San Jose's 2nd choice, 82nd overall, in 1999 Entry Draft).

			Regular Season					Playoffs				
Season	Club	Lea	GP	G	A	TP	PIM	GP	G	A	TP	PIM
1997-98	Hull High	Hi-School	20	38	35	73	....					
1998-99	Winchendon High	Hi-School	26	23	38	61	11					
99-2000	U. Mass-Lowell	H-East	23	4	3	7	8					
2000-01	U. Mass-Lowell	H-East	19	2	7	9	6					
2001-02	U. Mass-Lowell	H-East	38	8	13	21	14					

CONNE, Flavien (KAW-neh, FLA-vee-ehn) L.A.

Center. Shoots left. 5'9", 176 lbs. Born, Geneva, Switz., April 1, 1980.
(Los Angeles' 10th choice, 250th overall, in 2000 Entry Draft).

			Regular Season					Playoffs				
Season	Club	Lea	GP	G	A	TP	PIM	GP	G	A	TP	PIM
1995-96	Geneva Jr.	Swiss-Jr.	22	39	27	56	32	....	....	....	....	....
1996-97	Geneva	Swiss-2	30	9	8	17	8	5	1	1	2	4
1997-98	Geneva Jr.	Swiss-Jr.	37	15	12	27	57	3	0	2	2	2
	Geneva	Swiss-2	9	11	6	17	12	....	....	....	....	....
	HC Ambri-Piotta	Swiss	1	0	0	0	0	....	....	....	....	....
1998-99	Fribourg Jr.	Swiss-Jr.	1	1	1	2	2	....	....	....	....	....
	Fribourg	Swiss	37	14	14	28	59	4	4	1	5	6
	Fribourg	EuroHL	3	0	0	0	0	....	....	....	....	....
99-2000	Fribourg	Swiss	44	19	22	41	38	4	0	1	1	0
2000-01	HC Lugano	Swiss	42	9	14	23	8	15	2	6	8	37
2001-02	HC Lugano	Swiss	44	12	13	25	20	5	1	1	2	0
	Switzerland	Olympics	1	0	0	0	0	....	....	....	....	....

CONNOLLY, Sean (KAW-nuhl-lee, SHAWN) OTT.

Defense. Shoots right. 6'1", 187 lbs. Born, Dearborne, MI, October 8, 1980.
(Ottawa's 8th choice, 158th overall, in 2000 Entry Draft).

			Regular Season					Playoffs				
Season	Club	Lea	GP	G	A	TP	PIM	GP	G	A	TP	PIM
1997-98	Markham Waxers	OJHL	45	10	28	38	181	....	....	....	....	....
1998-99	North-Michigan	CCHA	33	4	18	22	62	....	....	....	....	....
99-2000	North-Michigan	CCHA	35	2	14	16	64	....	....	....	....	....
2000-01	North-Michigan	CCHA	38	5	14	19	80	....	....	....	....	....
2001-02	North-Michigan	CCHA	40	5	10	15	73	....	....	....	....	....

COOK, Jesse (KUK, JEH-see) CGY.

Defense. Shoots right. 6'6", 210 lbs. Born, Denver, CO, October 11, 1979.
(Calgary's 6th choice, 153rd overall, in 1999 Entry Draft).

			Regular Season					Playoffs				
Season	Club	Lea	GP	G	A	TP	PIM	GP	G	A	TP	PIM
1997-98	Calgary Royals	AJHL	34	5	24	29	35	3	1	2	3	4
1998-99	U. of Denver	WCHA	33	0	10	10	22	....	....	....	....	....
99-2000	U. of Denver	WCHA	41	2	12	14	40	....	....	....	....	....
2000-01	U. of Denver	WCHA	37	2	15	17	22	....	....	....	....	....
2001-02	U. of Denver	WCHA	41	2	16	18	30	....	....	....	....	....

CORAZZINI, Carl (koh-ra-ZEE-nee, KAHRL) BOS.

Left wing. Shoots right. 5'10", 170 lbs. Born, Framingham, MA, April 21, 1979.

			Regular Season					Playoffs				
Season	Club	Lea	GP	G	A	TP	PIM	GP	G	A	TP	PIM
1996-97	St. Sebastian's	Hi-School	25	29	31	60	....	....	....	....	....	....
1997-98	Boston University	H-East	36	9	6	15	4	....	....	....	....	....
1998-99	Boston University	H-East	37	15	9	24	12	....	....	....	....	....
99-2000	Boston University	H-East	42	22	20	42	44	....	....	....	....	....
2000-01	Boston University	H-East	35	16	20	36	48	....	....	....	....	....
2001-02	Providence	AHL	61	7	8	15	10	....	....	....	....	....

Hockey East All-Rookie Team (1998) • Hockey East First All-Star Team (2001) • NCAA East Second All-American Team (2001)

Signed as a free agent by **Boston**, August 8, 2001.

CORBEIL, Nicolas (kohr-BAY, NIH-coh-las), TOR.

Center. Shoots right. 5'10", 177 lbs. Born, Laval, Que., March 30, 1983.
(Toronto's 5th choice, 88th overall, in 2001 Entry Draft).

			Regular Season					Playoffs				
Season	Club	Lea	GP	G	A	TP	PIM	GP	G	A	TP	PIM
1998-99	Gatineau	QAAA	40	15	22	37	104	....	....	....	....	....
99-2000	Sherbrooke	QMJHL	64	9	12	21	28	5	1	0	1	0
2000-01	Sherbrooke	QMJHL	68	33	51	84	159	3	1	3	4	4
2001-02	Sherbrooke	QMJHL	62	41	52	93	99	....	....	....	....	....

CORVO, Joe (KOHR-voh, JOH-sehf) L.A.

Defense. Shoots right. 6', 205 lbs. Born, Oak Park, IL, June 20, 1977.
(Los Angeles' 4th choice, 83rd overall, in 1997 Entry Draft).

			Regular Season					Playoffs				
Season	Club	Lea	GP	G	A	TP	PIM	GP	G	A	TP	PIM
1995-96	West-Michigan	CCHA	41	5	25	30	38	....	....	....	....	....
1996-97	West-Michigan	CCHA	32	12	21	33	85	....	....	....	....	....
1997-98	West-Michigan	CCHA	32	5	12	17	93	....	....	....	....	....
1998-99	Springfield	AHL	50	5	15	20	32	....	....	....	....	....
	Hampton Roads	ECHL	5	0	0	0	15	4	0	1	1	0
99-2000					DID NOT PLAY							
2000-01	Lowell	AHL	77	10	23	33	31	4	3	1	4	0
2001-02	Manchester	AHL	80	13	37	50	30	5	0	5	5	0

CCHA All-Rookie Team (1996) • CCHA Second All-Star Team (1997)
• Missed entire 1999-2000 season after failing to come to contract terms with **LA Kings**

COTE, Jean-Philippe (KOH-tay, zhawn-fihl-EEP) TOR.

Defense. Shoots left. 6'1", 195 lbs. Born, Charlesbourg, Que., April 22, 1982.
(Toronto's 10th choice, 265th overall, in 2000 Entry Draft).

			Regular Season					Playoffs				
Season	Club	Lea	GP	G	A	TP	PIM	GP	G	A	TP	PIM
1998-99	Ste-Foy	QAAA	38	10	24	34	34	17	1	8	9	17
	Quebec Remparts	QMJHL	8	0	0	0	2	....	....	....	....	....
99-2000	Quebec Remparts	QMJHL	34	0	10	10	15	....	....	....	....	....
	Cape Breton	QMJHL	28	0	4	4	21	4	0	1	1	4
2000-01	Cape Breton	QMJHL	71	6	29	35	90	12	0	0	0	18
2001-02	Cape Breton	QMJHL	61	4	20	24	72	16	1	6	7	38

Traded to **Cape Breton** (QMJHL) by **Quebec** (QMJHL) with Stuart MacRae for Chris Lyness, January 10, 2000.

COX, Justin (KAWKS, JUH-stihn) DAL.

Right wing. Shoots right. 6', 173 lbs. Born, Merritt, B.C., March 13, 1981.
(Dallas' 6th choice, 184th overall, in 1999 Entry Draft).

			Regular Season					Playoffs				
Season	Club	Lea	GP	G	A	TP	PIM	GP	G	A	TP	PIM
1996-97	Spruce Grove	AMHL	78	57	92	149	86	....	....	....	....	....
1997-98	Prince George	WHL	40	1	4	5	15	2	0	0	0	0
1998-99	Prince George	WHL	72	9	13	22	51	7	1	0	1	13
99-2000	Prince George	WHL	71	33	38	71	74	13	2	4	6	16
2000-01	Prince George	WHL	71	30	38	68	91	6	3	4	7	20
2001-02	Fort Worth	CHL	2	2	0	2	0	....	....	....	....	....
	Utah Grizzlies	AHL	74	10	7	17	53	5	0	1	1	0

CRAIN, Jason (KRAYN, JAY-suhn) L.A.

Defense. Shoots left. 6'3", 190 lbs. Born, Pittsburgh, PA, January 3, 1980.
(Los Angeles' 2nd choice, 74th overall, in 1999 Entry Draft).

			Regular Season					Playoffs				
Season	Club	Lea	GP	G	A	TP	PIM	GP	G	A	TP	PIM
1996-97	St. Thomas Stars	OJHL-B	50	9	18	27	67	....	....	....	....	....
1997-98	St. Thomas Stars	OJHL-B	43	6	33	39	49	....	....	....	....	....
1998-99	Ohio State	CCHA	41	3	14	17	18	....	....	....	....	....
99-2000	Ohio State	CCHA	35	2	9	11	32	....	....	....	....	....
2000-01	Ohio State	CCHA	36	2	4	6	36	....	....	....	....	....
2001-02	Ohio State	CCHA	40	5	7	12	48	....	....	....	....	....

CRAMPTON, Steven (KRAMP-tuhn, STEE-vehn) PIT.

Right wing. Shoots right. 6'2", 200 lbs. Born, Winnipeg, Man., April 12, 1982.
(Pittsburgh's 8th choice, 248th overall, in 2000 Entry Draft).

			Regular Season					Playoffs				
Season	Club	Lea	GP	G	A	TP	PIM	GP	G	A	TP	PIM
1997-98	Winnipeg Sharks	MMHL	29	33	38	71	76	....	....	....	....	....
1998-99	Moose Jaw	WHL	52	7	5	12	31	9	1	1	2	4
99-2000	Moose Jaw	WHL	69	22	20	42	91	4	0	3	3	9
2000-01	Moose Jaw	WHL	72	26	33	59	153	4	1	2	3	6
2001-02	Moose Jaw	WHL	37	19	24	43	71	12	4	10	14	20

CRONIN, John (KROH-nihhn, JAWN) BOS.

Defense. Shoots right. 6'2", 200 lbs. Born, Duxbury, MA, May 1, 1980.
(Boston's 8th choice, 236th overall, in 1999 Entry Draft).

			Regular Season					Playoffs				
Season	Club	Lea	GP	G	A	TP	PIM	GP	G	A	TP	PIM
1997-98	Noble-Greenough	Hi-School	30	8	22	30	14	....	....	....	....	....
1998-99	Noble-Greenough	Hi-School	30	8	26	34	24	....	....	....	....	....
99-2000	Boston University	H-East	28	3	5	8	26	....	....	....	....	....
2000-01	Boston University	H-East	37	2	11	13	36	....	....	....	....	....
2001-02	Boston University	H-East	32	1	8	9	20	....	....	....	....	....

CULL, Trent (KUHL, TREHNT)

Defense. Shoots left. 6'2", 215 lbs. Born, Brampton, Ont., September 27, 1973.

			Regular Season					Playoffs				
Season	Club	Lea	GP	G	A	TP	PIM	GP	G	A	TP	PIM
1988-89	Georgetown	OJHL-B	36	1	5	6	51	....	....	....	....	....
1989-90	Owen Sound	OHL	57	0	5	5	53	12	0	2	2	11
1990-91	Owen Sound	OHL	24	1	2	3	19	....	....	....	....	....
	Windsor	OHL	33	1	6	7	34	11	0	0	0	8
1991-92	Windsor	OHL	32	0	6	6	66	....	....	....	....	....
	Kingston	OHL	18	0	0	0	31	....	....	....	....	....
1992-93	Kingston	OHL	60	11	28	39	144	16	2	8	10	37
1993-94	Kingston	OHL	50	2	30	32	147	6	0	1	1	6
1994-95	St. John's	AHL	43	0	1	1	53	....	....	....	....	....
	Brantford Smoke	ColHL	4	0	0	0	14	....	....	....	....	....
1995-96	St. John's	AHL	46	2	1	3	118	4	0	0	0	6
1996-97	St. John's	AHL	75	4	5	9	219	8	0	1	1	18
1997-98	Houston Aeros	IHL	72	4	8	12	201	4	0	0	0	4
1998-99	Houston Aeros	IHL	72	2	14	16	232	19	0	2	2	34
99-2000	Springfield	AHL	28	0	2	2	74	....	....	....	....	....
	Houston Aeros	IHL	35	2	7	9	133	5	0	0	0	24
2000-01	Wilkes-Barre	AHL	71	11	15	26	166	21	3	2	5	28
2001-02	Houston Aeros	AHL	74	1	11	12	158	12	0	2	2	8

Signed as a free agent by **Toronto**, June 4, 1994. Signed as a free agent by **Phoenix**, August 26, 1999. Signed as a free agent by **Pittsburgh**, August 28, 2000. Signed as a free agent by **Minnesota**, July 13, 2001.

CULLEN, Joe (KUH-lehn, JOH) EDM.

Center. Shoots left. 6'1", 190 lbs. Born, Virginia, MN, February 14, 1981.
(Edmonton's 7th choice, 211th overall, in 2000 Entry Draft).

			Regular Season					Playoffs				
Season	Club	Lea	GP	G	A	TP	PIM	GP	G	A	TP	PIM
1997-98	Moorhead Spuds	Hi-School	23	18	18	36	....	....	....	....	....	....
1998-99	Team USA	USDP-17	52	11	15	26	33	....	....	....	....	....
99-2000	Colorado College	WCHA	29	4	6	10	30	....	....	....	....	....
2000-01	Colorado College	WCHA	34	8	12	20	38	....	....	....	....	....
2001-02	Colorado College	WCHA	43	9	12	21	42	....	....	....	....	....

CURRY, Sean (KUH-ree, SHAWN) CAR.

Defense. Shoots right. 6'4", 230 lbs. Born, Burnsville, MN, April 29, 1982.
(Carolina's 6th choice, 211th overall, in 2001 Entry Draft).

			Regular Season					Playoffs				
Season	Club	Lea	GP	G	A	TP	PIM	GP	G	A	TP	PIM
99-2000	Burnsville	Hi-School	23	8	18	26	....	....	....	....	....	....
2000-01	Tri-City	WHL	72	5	12	17	113	....	....	....	....	....
2001-02	Tri-City	WHL	36	6	6	12	84	....	....	....	....	....
	Medicine Hat	WHL	24	4	13	17	43	....	....	....	....	....

Traded to **Medicine Hat** (WHL) by **Tri-City** (WHL) for Josh Morrow and Medicine Hat's 2nd round choice (Dave Falk) in 2002 WHL Bantam Draft, January 10, 2002.

DALEY, Trevor (DAY-lee, TREH-vuhr) DAL.

Defense. Shoots left. 5'9", 197 lbs. Born, Toronto, Ont., October 9, 1983.
(Dallas' 5th choice, 43rd overall, in 2002 Entry Draft).

			Regular Season					Playoffs				
Season	Club	Lea	GP	G	A	TP	PIM	GP	G	A	TP	PIM
1998-99	Vaughan Vipers	OPJHL	44	10	36	46	79	....	....	....	....	....
99-2000	Sault Ste. Marie	OHL	54	16	30	46	77	15	3	7	10	12
2000-01	Sault Ste. Marie	OHL	58	14	27	41	105	....	....	....	....	....
2001-02	Sault Ste. Marie	OHL	47	9	39	48	38	6	2	2	4	4

DALLMAN, Kevin BOS.

Defense. Shoots right. 5'11", 195 lbs. Born, Niagara Falls, Ont., February 26, 1981.

			Regular Season					Playoffs				
Season	Club	Lea	GP	G	A	TP	PIM	GP	G	A	TP	PIM
1996-97	Niagrara Falls	OJHL-B	3	0	1	1	2	....	....	....	....	....
1997-98	Niagrara Falls	OJHL-B	47	13	25	38	42	....	....	....	....	....
1998-99	Guelph Storm	OHL	68	8	30	38	52	11	1	4	5	2
99-2000	Guelph Storm	OHL	67	13	46	59	38	6	0	2	2	11
2000-01	Guelph Storm	OHL	66	25	52	77	88	1	0	0	0	0
2001-02	Guelph Storm	OHL	67	23	63	86	68	9	8	8	16	22

Signed as a free agent by **Boston**, July 18, 2002.

D'AMOUR, Dominic (da+F276h-MOOR, DOHM-ihn-ihk) TOR.

Defense. Shoots left. 6'2", 203 lbs. Born, La Salle, Que., January 28, 1984.
(Toronto's 4th choice, 88th overall, in 2002 Entry Draft).

			Regular Season					Playoffs				
Season	Club	Lea	GP	G	A	TP	PIM	GP	G	A	TP	PIM
99-2000	Charles-Lemoyne	QAAA	35	3	8	11	47	16	1	1	2	14
2000-01	Charles-Lemoyne	QAAA	11	1	4	5	36	....	....	....	....	....
	Rouyn-Noranda	QMJHL	18	0	0	0	10	....	....	....	....	....
2001-02	Hull Olympiques	QMJHL	68	5	5	10	225	12	0	3	3	32

Traded to **Hull** (QMJHL) by **Rouyn-Noranda** (QMJHL) with Maxime Talbot and Rouyn-Noranda's 1st round choice (Charles Fontaine) in 2001 QMJHL Priority Draft for Alexandre Giroux, January 8, 2001.

DARBY, Regan (DAHR-bee, REE-gan) VAN.

Defense. Shoots left. 6'2", 200 lbs. Born, Estevan, Sask., July 17, 1980.
(Vancouver's 5th choice, 90th overall, in 1998 Entry Draft).

			Regular Season					Playoffs				
Season	Club	Lea	GP	G	A	TP	PIM	GP	G	A	TP	PIM
1996-97	Swift Current	SMHL	36	10	20	30	210	....	....	....	....	....
1997-98	Spokane Chiefs	WHL	7	0	1	1	28	....	....	....	....	....
	Tri-City	WHL	32	1	2	3	125	....	....	....	....	....
1998-99	Tri-City	WHL	38	2	4	6	152	....	....	....	....	....
	Red Deer Rebels	WHL	19	1	6	7	90	9	0	1	1	18
99-2000	Red Deer Rebels	WHL	18	3	6	9	79	....	....	....	....	....
	Prince Albert	WHL	44	1	9	10	143	6	0	1	1	23
2000-01	Kansas City	IHL	55	1	5	6	164	....	....	....	....	....
2001-02	Manitoba Moose	AHL	39	0	1	1	196	....	....	....	....	....
	Columbia Inferno	ECHL	7	0	2	2	14	3	0	0	0	0

Traded to **Tri-City** (WHL) by **Spokane** (WHL) with Blake Evans for Zenith Komarniski, October 6, 1997. Traded to **Red Deer** (WHL) by **Tri-City** (WHL) with Jarrett Thompson for Stephen Peat, January 4, 1999. Traded to **Prince Albert** (WHL) by **Red Deer** (WHL) with Brent Hobday, Steven MacIntyre and Scott McQueen for Craig Brunel and Russ Lupaschuk, November 19, 1999.

DAVIS, George (DAY-vihs, JOHRJ) ANA.

Right wing. Shoots right. 6'2", 225 lbs. Born, North Sydney, N.S., July 28, 1983.
(Anaheim's 5th choice, 140th overall, in 2002 Entry Draft).

			Regular Season					Playoffs				
Season	Club	Lea	GP	G	A	TP	PIM	GP	G	A	TP	PIM
2000-01	Cape Breton	QMJHL	44	0	1	1	196	5	0	0	0	14
2001-02	Cape Breton	QMJHL	46	4	1	5	274	16	1	2	3	18

DAVIS, Greg (DAY-vihs, GREHG) ST.L.

Left wing. Shoots right. 6'4", 195 lbs. Born, Calgary, Alta., July 26, 1979.

			Regular Season					Playoffs				
Season	Club	Lea	GP	G	A	TP	PIM	GP	G	A	TP	PIM
1997-98	Olds Grizzlies	AJHL	56	12	21	33	37	8	0	2	2	10
1998-99	Olds Grizzlies	AJHL	61	19	28	47	97	....	....	....	....	....
99-2000	McGill Redmen	OUAA	38	23	16	39	22	....	....	....	....	....
2000-01	McGill Redmen	OUAA	33	30	35	65	38	....	....	....	....	....
2001-02	Worcester	AHL	57	12	6	18	10	....	....	....	....	....
	Peoria Rivermen	ECHL	3	0	1	1	0	....	....	....	....	....

Signed as a free agent by **St. Louis**, May 5, 2001.

DAVIS, Ken (DAY-vihs, KEHN) DET.

Right wing. Shoots right. 6'4", 210 lbs. Born, Calgary, Alta., March 20, 1981.
(Detroit's 6th choice, 266th overall, in 1999 Entry Draft).

			Regular Season					Playoffs				
Season	Club	Lea	GP	G	A	TP	PIM	GP	G	A	TP	PIM
1996-97	Calgary Royals	AMHL	54	24	27	51	84	....	....	....	....	....
1997-98	Portland	WHL	66	8	11	19	27	14	1	0	1	7
1998-99	Portland	WHL	72	13	14	27	76	4	0	1	1	11
99-2000	Portland	WHL	6	1	2	3	18	....	....	....	....	....
	Medicine Hat	WHL	61	23	10	33	86	....	....	....	....	....
2000-01	Medicine Hat	WHL	69	19	15	34	75	....	....	....	....	....
2001-02	Prince George	WHL	64	19	9	28	83	7	2	2	4	6

Traded to **Medicine Hat** (WHL) by **Portland** (WHL) for Kevin Young, October 6, 1999. Traded to **Prince George** (WHL) by **Medicine Hat** (WHL) for Prince George's 4th round choice (Gord Baldwin) in 2002 WHL Bantam Draft, June 29, 2001.

DAVIS, Wade (DAY-vihs, WAYD) CGY.

Defense. Shoots right. 6'4", 185 lbs. Born, Kamloops, B.C., April 13, 1982.
(Calgary's 5th choice, 141st overall, in 2000 Entry Draft).

			Regular Season					Playoffs				
Season	Club	Lea	GP	G	A	TP	PIM	GP	G	A	TP	PIM
1997-98	Fernie	RMJHL	42	8	17	25	19	....	....	....	....	....
	Calgary Hitmen	WHL	2	0	0	0	0	1	0	0	0	0
1998-99	Calgary Hitmen	WHL	38	0	2	2	21	2	0	0	0	0
99-2000	Calgary Hitmen	WHL	61	3	15	18	59	13	0	2	2	15
2000-01	Calgary Hitmen	WHL	67	11	21	32	77	12	0	2	2	16
2001-02	Calgary Hitmen	WHL	68	21	34	55	87	7	3	2	5	12

DAVISON, Rob (DAY-vihs-ohn, RAWB) S.J.

Defense. Shoots left. 6'2", 220 lbs. Born, St. Catharines, Ont., May 1, 1980.
(San Jose's 4th choice, 98th overall, in 1998 Entry Draft).

			Regular Season					Playoffs				
Season	Club	Lea	GP	G	A	TP	PIM	GP	G	A	TP	PIM
1995-96	St. Michael's B	OJHL-B	21	0	0	0	21	....	....	....	....	....
1996-97	St. Michael's B	OJHL-B	45	2	6	8	93	....	....	....	....	....
1997-98	North Bay	OHL	59	0	11	11	200	....	....	....	....	....
1998-99	North Bay	OHL	59	2	17	19	150	4	0	1	1	12
99-2000	North Bay	OHL	67	4	6	10	194	6	0	1	1	8
2000-01	Kentucky	AHL	72	0	4	4	230	3	0	0	0	0
2001-02	Cleveland Barons	AHL	70	1	3	4	206	....	....	....	....	....

DEFAUW, Brad (duh-FOU, BRAD) CAR.

Left wing. Shoots left. 6'2", 210 lbs. Born, Edina, MN, November 10, 1977.
(Carolina's 2nd choice, 28th overall, in 1997 Entry Draft).

			Regular Season					Playoffs				
Season	Club	Lea	GP	G	A	TP	PIM	GP	G	A	TP	PIM
1995-96	Apple Valley	Hi-School	28	21	34	55	14	....	....	....	....	....
1996-97	North Dakota	WCHA	37	7	6	13	39	....	....	....	....	....
1997-98	North Dakota	WCHA	36	9	11	20	34	....	....	....	....	....
1998-99	North Dakota	WCHA	34	11	12	23	64	....	....	....	....	....
99-2000	North Dakota	WCHA	43	13	9	22	52	....	....	....	....	....
2000-01	Cincinnati	IHL	82	20	31	51	39	4	2	0	2	8
2001-02	Lowell	AHL	63	17	21	38	29	5	2	1	3	6

DELEEUW, Adam (DEH-lee-EW, A-dam) DET.

Left wing. Shoots left. 6', 206 lbs. Born, Brampton, Ont., February 29, 1980.
(Detroit's 7th choice, 151st overall, in 1998 Entry Draft).

			Regular Season					Playoffs				
Season	Club	Lea	GP	G	A	TP	PIM	GP	G	A	TP	PIM
1995-96	Brampton	OPJHL	1	0	0	0	0	....	....	....	....	....
1996-97	Brampton	OPJHL	45	11	17	28	97	....	....	....	....	....
1997-98	Barrie Colts	OHL	56	10	6	16	224	....	....	....	....	....
1998-99	Barrie Colts	OHL	39	15	16	31	146	....	....	....	....	....
	St. Michael's	OHL	29	10	5	15	55	....	....	....	....	....
99-2000	St. Michael's	OHL	45	11	19	30	107	....	....	....	....	....
	Dayton Bombers	ECHL	2	0	0	0	2	3	0	0	0	2
2000-01	St. Michael's	OHL	54	11	14	25	122	18	1	1	2	21
2001-02			DID NOT PLAY – INJURED									

Traded to **St. Michael's** (OHL) by **Barrie** (OHL) with Keith Delaney, Darryl Bootland and Brad Pierce for Sheldon Keefe, Mike Jefferson, Ryan Barnes and Shawn Cation, January 11, 1999.
• Missed entire 2001-02 season recovering from shoulder injury originally suffered during 2000-01 season, October 3, 2001.

DELISLE, Miguel (duh-LIGHL, mih-GEHL) TOR.

Right wing. Shoots right. 6'2", 202 lbs. Born, Cornwall, Ont., April 6, 1982.
(Toronto's 5th choice, 100th overall, in 2000 Entry Draft).

			Regular Season					Playoffs				
Season	Club	Lea	GP	G	A	TP	PIM	GP	G	A	TP	PIM
1997-98	Caledon	MTJHL	46	20	24	44	142	....	....	....	....	....
1998-99	Ottawa 67's	OHL	57	16	17	33	34	9	1	0	1	4
99-2000	Ottawa 67's	OHL	54	20	29	49	73	11	4	1	5	28
2000-01	Ottawa 67's	OHL	61	34	38	72	89	20	8	17	25	30
2001-02	Ottawa 67's	OHL	67	55	40	95	73	13	6	6	12	20

DeMARCHI, Matt (dih-MAHR-shee, MAT) N.J.

Defense. Shoots left. 6'3", 180 lbs. Born, Bemidji, MN, May 4, 1981.
(New Jersey's 4th choice, 57th overall, in 2000 Entry Draft).

			Regular Season					Playoffs				
Season	Club	Lea	GP	G	A	TP	PIM	GP	G	A	TP	PIM
1997-98	North Iowa	USHL	34	1	2	3	66	10	0	1	1	19
1998-99	North Iowa	USHL	53	4	14	18	131	....	....	....	....	....
99-2000	U. of Minnesota	WCHA	39	1	6	7	82	....	....	....	....	....
2000-01	U. of Minnesota	WCHA	39	4	9	13	*149	....	....	....	....	....
2001-02	U. of Minnesota	WCHA	36	3	8	11	112	....	....	....	....	....

DEMIDOV, Ilja (deh-MEE-dahf, ihl-YA) OTT.

Defense. Shoots left. 6'3", 185 lbs. Born, Moscow, USSR, April 14, 1979.
(Calgary's 10th choice, 140th overall, in 1997 Entry Draft).

			Regular Season					Playoffs				
Season	Club	Lea	GP	G	A	TP	PIM	GP	G	A	TP	PIM
1995-96	Dynamo Moscow	CIS	10	0	14	14	....	....	....	....	....	....
1996-97	DynamoMoscow2	Russia-3	32	1	0	1	60	....	....	....	....	....
1997-98	Oshawa Generals	OHL	61	4	16	20	67	7	0	1	1	2
1998-99	Oshawa Generals	OHL	62	4	23	27	72	15	2	5	7	24
99-2000	Oshawa Generals	OHL	62	11	37	48	105	5	0	2	2	24
2000-01	Grand Rapids	IHL	54	1	4	5	65	2	0	1	1	21
2001-02	Grand Rapids	AHL	7	0	1	1	8	....	....	....	....	....
	Mobile Mysticks	ECHL	57	2	7	9	93	....	....	....	....	....

Signed as a free agent by **Ottawa**, February 25, 2000.

DENISOV, Denis (den-NEES-ahf, deh-NEES) BUF.

Left wing. Shoots left. 6', 183 lbs. Born, Kalinin, USSR, December 31, 1981.
(Buffalo's 4th choice, 149th overall, in 2000 Entry Draft).

			Regular Season					Playoffs				
Season	Club	Lea	GP	G	A	TP	PIM	GP	G	A	TP	PIM
1997-98	H.C. CSKA	Russia	7	0	0	0	4	....	....	....	....	....
1998-99	H.C. CSKA	Russia-2	42	1	6	7	16	....	....	....	....	....
99-2000	H.C. CSKA	Russia-2	39	1	8	9	16	....	....	....	....	....
2000-01	H.C. CSKA	Russia-2	41	0	3	3	6	....	....	....	....	....
2001-02	Krylja Sovetov	Russia	47	3	4	7	37	....	....	....	....	....
	Krylja Sovetov 2	Russia-3	3	0	1	1	18	....	....	....	....	....

DESAUTELS, Jevon (DEH-soh-tehl, jeh-VAWN) WSH.

Left wing. Shoots left. 6'3", 215 lbs. Born, Redvers, Sask., March 15, 1984.
(Washington's 7th choice, 109th overall, in 2002 Entry Draft).

			Regular Season					Playoffs				
Season	Club	Lea	GP	G	A	TP	PIM	GP	G	A	TP	PIM
2000-01	Spokane Chiefs	WHL	49	1	2	3	77	....	....	....	....	....
2001-02	Spokane Chiefs	WHL	60	6	11	17	162	11	1	0	1	20

DESSNER, Jeff (DEHS-nehr, JEHF)

Defense. Shoots left. 6'2", 195 lbs. Born, Skokie, IL, April 16, 1977.
(NY Rangers' 6th choice, 185th overall, in 1996 Entry Draft).

			Regular Season					Playoffs				
Season	Club	Lea	GP	G	A	TP	PIM	GP	G	A	TP	PIM
1995-96	Taft Eagles	Hi-School	25	12	18	30	....	....	....	....	....	....
1996-97	U. of Wisconsin	WCHA		DID NOT PLAY – INJURED								
1997-98	U. of Wisconsin	WCHA	19	1	3	4	43	....	....	....	....	....
1998-99	U. of Wisconsin	WCHA	37	7	14	21	46	....	....	....	....	....
99-2000	U. of Wisconsin	WCHA	40	11	15	26	59	....	....	....	....	....
2000-01	U. of Wisconsin	WCHA	39	7	12	19	58	....	....	....	....	....
2001-02	Chicago Wolves	AHL	40	1	11	12	80	....	....	....	....	....
	Greenville	ECHL	5	1	2	3	8	13	3	5	8	10

WCHA First All-Star Team (2000) • NCAA West First All-American Team (2000)

• Missed entire 1996-97 season recovering from back surgery, June, 1996. Traded to **Atlanta** by **NY Rangers** for Atlanta's 8th round choice (Leonid Zhvachkin) in 2001 Entry Draft, June 23, 2001.

DEZAINDE, Joel (duh-ZAYND, JOHL) N.J.

Defense. Shoots left. 6', 200 lbs. Born, Simcoe, Ont., November 2, 1978.

			Regular Season					Playoffs				
Season	Club	Lea	GP	G	A	TP	PIM	GP	G	A	TP	PIM
1993-94	Oshweken	OJHL-B	47	4	2	6	111	....	....	....	....	....
1994-95	London Knights	OHL	53	0	4	4	60	4	0	3	3	0
1995-96	London Knights	OHL	64	9	33	42	99	....	....	....	....	....
1996-97	Owen Sound	OHL	34	6	8	14	39	....	....	....	....	....
	Belleville Bulls	OHL	29	5	12	17	35	6	1	5	6	8
1997-98	Belleville Bulls	OHL	65	13	45	58	81	10	3	5	8	16
1998-99	Mississauga	OHL	24	4	20	24	35	....	....	....	....	....
	Barrie Colts	OHL	36	15	33	48	52	12	2	9	11	....
99-2000	Team Canada	Nat-Tm	5	1	0	1	4	....	....	....	....	....
	Arkansas	WPHL	51	15	25	40	57	....	....	....	....	....
	Detroit Vipers	IHL	14	1	2	3	20	....	....	....	....	....
2000-01	Mississippi	ECHL	58	10	38	48	77	....	....	....	....	....
2001-02	Albany	AHL	66	5	10	15	30	....	....	....	....	....

Signed tryout contract and invited to training camp by **New Jersey**, May 2, 2001. Signed as a free agent by **New Jersey**, October 3, 2001.

DICAIRE, Gerard (dih-KAIR, zhehr-AHR) T.B.

Defense. Shoots left. 6'2", 198 lbs. Born, Faro, Yukon, September 14, 1982.
(Tampa Bay's 4th choice, 162nd overall, in 2002 Entry Draft).

			Regular Season					Playoffs				
Season	Club	Lea	GP	G	A	TP	PIM	GP	G	A	TP	PIM
1997-98	Tumber Ridge	NWJHL	35	15	28	43	63	....	....	....	....	....
1998-99	Prince George	BCHL	51	6	22	28	28	....	....	....	....	....
99-2000	Seattle	WHL	68	11	25	36	38	7	0	1	1	6
2000-01	Seattle	WHL	69	15	36	51	33	9	0	2	2	2
2001-02	Seattle	WHL	41	4	25	29	25	....	....	....	....	....
	Kootenay Ice	WHL	25	2	21	23	9	22	1	14	15	24

• Re-entered NHL Entry Draft. Originally Buffalo's 2nd choice, 48th overall, in 2000 Entry Draft.

WHL West Second All-Star Team (2001)

Traded to **Kootenay** (WHL) by **Seattle** (WHL) for Trevor Johnson, January 14, 2002.

DiCASMIRRO, Nate (dee-CAZ-MIHR-oh, NAYT) EDM.

Left wing. Shoots left. 5'11", 205 lbs. Born, Burnsville, MN, September 27, 1978.

			Regular Season					Playoffs				
Season	Club	Lea	GP	G	A	TP	PIM	GP	G	A	TP	PIM
1996-97	North Iowa	USHL	51	18	22	40	86	12	0	6	6	22
1997-98	North Iowa	USHL	52	29	45	74	118	11	5	5	10	34
1998-99	St. Cloud State	WCHA	34	6	8	14	46	....	....	....	....	....
99-2000	St. Cloud State	WCHA	40	19	24	43	26	....	....	....	....	....
2000-01	St. Cloud State	WCHA	32	9	20	29	26	....	....	....	....	....
2001-02	St. Cloud State	WCHA	41	17	33	50	58	....	....	....	....	....
	Hamilton	AHL	1	0	0	0	0	10	0	5	5	6

USHL First All-Star Team (1998) • USHL MVP (1998)

Signed as a free agent by **Edmonton**, April 8, 2002.

DIMITRAKOS, Nico (DIH-mih-tra-kohs, NIK-oh) S.J.

Right wing. Shoots right. 5'11", 190 lbs. Born, Boston, MA, May 21, 1979.
(San Jose's 4th choice, 155th overall, in 1999 Entry Draft).

			Regular Season					Playoffs				
Season	Club	Lea	GP	G	A	TP	PIM	GP	G	A	TP	PIM
1994-95	Matignon	Hi-School	23	10	12	22	....	....	....	....	....	....
1995-96	Matignon	Hi-School	25	12	28	40	....	....	....	....	....	....
1996-97	Matignon	Hi-School	25	23	32	55	....	....	....	....	....	....
1997-98	Avon Old Farms	Hi-School	26	27	28	55	....	....	....	....	....	....
1998-99	U. of Maine	H-East	35	8	19	27	33	....	....	....	....	....
99-2000	U. of Maine	H-East	32	11	16	27	16	....	....	....	....	....
2000-01	U. of Maine	H-East	29	11	14	25	43	....	....	....	....	....
2001-02	U. of Maine	H-East	43	20	31	51	44	....	....	....	....	....

NCAA Championship All-Tournament Team (1999) • Hockey East Second All-Star Team (2002)

DiPENTA, Joe (DIH-pehn-tah, JOH) ATL.

Defense. Shoots left. 6'2", 235 lbs. Born, Barrie, Ont., February 25, 1979.
(Florida's 2nd choice, 61st overall, in 1998 Entry Draft).

			Regular Season					Playoffs				
Season	Club	Lea	GP	G	A	TP	PIM	GP	G	A	TP	PIM
1996-97	Smiths Falls	OCJHL	54	13	22	35	92	....	....	....	....	....
1997-98	Boston University	H-East	38	2	16	18	50	....	....	....	....	....
1998-99	Boston University	H-East	36	2	15	17	72	....	....	....	....	....
99-2000	Halifax	QMJHL	63	13	43	56	83	10	3	4	7	26
2000-01	Philadelphia	AHL	71	3	5	8	65	10	1	2	3	15
2001-02	Philadelphia	AHL	61	2	4	6	71	....	....	....	....	....
	Chicago Wolves	AHL	15	0	2	2	15	25	1	3	4	22

Selected by **Halifax** (QMJHL) 146th overall in 1996 QMJHL Midget Draft, June 6, 1996. • Left **Boston U.** (H-East) and signed with **Halifax** (QMJHL), May 2, 1999. Signed as a free agent by **Philadelphia**, July 12, 2000. Traded to **Atlanta** by **Philadelphia** for Jarrod Skalde, March 5, 2002.

DISALVATORE, Jon (dih-sal-vuh-TOH-ray, JAWN) S.J.

Right wing. Shoots right. 6'1", 180 lbs. Born, Bangor, ME, March 30, 1981.
(San Jose's 2nd choice, 104th overall, in 2000 Entry Draft).

			Regular Season					Playoffs				
Season	Club	Lea	GP	G	A	TP	PIM	GP	G	A	TP	PIM
1997-98	New England	EJHL	38	24	41	65	....	....	....	....	....	....
1998-99	New England	EJHL	48	44	76	*120	38	....	....	....	....	....
99-2000	Providence	H-East	38	15	12	27	12	....	....	....	....	....
2000-01	Providence	H-East	36	9	16	25	29	....	....	....	....	....
2001-02	Providence	H-East	38	16	26	42	6	....	....	....	....	....

EJHL First All-Star Team (1999) • EJHL MVP (1999)

DOBBEN, Scott (DAW-behn, SKAWT) OTT.

Left wing. Shoots left. 6'1", 193 lbs. Born, Palmerston, Ont., April 10, 1983.
(Ottawa's 4th choice, 113th overall, in 2002 Entry Draft).

			Regular Season					Playoffs				
Season	Club	Lea	GP	G	A	TP	PIM	GP	G	A	TP	PIM
99-2000	Elmira	OJHL-B	48	6	15	21	24	....	....	....	....	....
2000-01	Erie Otters	OHL	57	8	9	17	18	15	0	1	1	10
2001-02	Erie Otters	OHL	68	31	32	63	72	21	4	11	15	24

DOBRYSHKIN, Yuri (doh-BRIHSH-kihn, yew-REE) ATL.

Left wing. Shoots right. 6', 190 lbs. Born, Penza, USSR, July 19, 1979.
(Atlanta's 7th choice, 159th overall, in 1999 Entry Draft).

			Regular Season					Playoffs				
Season	Club	Lea	GP	G	A	TP	PIM	GP	G	A	TP	PIM
1996-97	Krylja Sovetov 2	Russia-3	35	13	5	18	42	....	....	....	....	....
	Krylja Sovetov	Russia	2	0	0	0	0	2	0	0	0	0
1997-98	Krylja Sovetov 2	Russia-3	26	12	5	17	68	....	....	....	....	....
	Krylja Sovetov	Russia	22	4	0	4	12	....	....	....	....	....
1998-99	Krylja Sovetov	Russia	50	11	5	16	86	....	....	....	....	....
99-2000	Ak Bars Kazan	Russia	27	6	9	15	24	17	2	0	2	10
2000-01	Ak Bars Kazan	Russia	40	10	5	15	32	4	2	0	2	2
2001-02	Ak Bars Kazan	Russia	38	9	8	17	22	11	0	2	2	6

DOMAN, Matt (DOH-man, MAT) CGY.

Right wing. Shoots right. 6'1", 218 lbs. Born, St. Cloud, MN, February 10, 1980.
(Calgary's 5th choice, 135th overall, in 1999 Entry Draft).

			Regular Season					Playoffs				
Season	Club	Lea	GP	G	A	TP	PIM	GP	G	A	TP	PIM
1997-98	Team USA	USDP-18	55	24	22	46	208	....	....	....	....	....
1998-99	U. of Wisconsin	WCHA	34	5	5	10	52	....	....	....	....	....
99-2000	U. of Wisconsin	WCHA	22	1	12	13	53	....	....	....	....	....
2000-01	U. of Wisconsin	WCHA	41	9	10	19	68	....	....	....	....	....
2001-02	U. of Wisconsin	WCHA	38	11	12	23	68	....	....	....	....	....
	Saint John	AHL	6	0	1	1	2	....	....	....	....	....

DONIKA, Mikhail (DAW-nih-ka, mihk-high-EHL) DAL.

Defense. Shoots left. 6', 185 lbs. Born, Yaroslavl, USSR, May 15, 1979.
(Dallas' 11th choice, 272nd overall, in 1999 Entry Draft).

			Regular Season					Playoffs				
Season	Club	Lea	GP	G	A	TP	PIM	GP	G	A	TP	PIM
1996-97	Yaroslavl 2	Russia-3	15	3	5	8	6	....	....	....	....	....
	Yaroslavl	Russia	22	1	0	1	6	2	0	0	0	0
1997-98	Yaroslavl 2	Russia-2	19	1	2	3	32	....	....	....	....	....
	Yaroslavl	Russia	30	0	2	2	14	....	....	....	....	....
1998-99	Yaroslavl 2	Russia-3	6	2	1	3	4	....	....	....	....	....
	Yaroslavl	Russia	37	0	1	1	10	....	....	....	....	....
99-2000	Yaroslavl	Russia	35	0	1	1	22	10	0	0	0	4
2000-01	Dynamo Moscow	Russia	43	1	3	4	12	....	....	....	....	....
2001-02	Amur Khabarovsk	Russia	51	1	3	4	66	....	....	....	....	....

DOULL, Doug (DOOL, DUHG) TOR.

Left wing. Shoots left. 6'2", 216 lbs. Born, Glace Bay, N.S., May 31, 1974.

			Regular Season					Playoffs				
Season	Club	Lea	GP	G	A	TP	PIM	GP	G	A	TP	PIM
1990-91	Wexford Hawks	MTHL	39	22	36	58	141	....	....	....	....	....
1991-92	Belleville Bulls	OHL	62	6	11	17	123	....	....	....	....	....
1992-93	Belleville Bulls	OHL	65	19	37	56	143	....	....	....	....	....
1993-94	Belleville Bulls	OHL	62	13	24	37	143	....	....	....	....	....
1994-95	Belleville Bulls	OHL	29	7	12	19	71	16	2	13	15	39
1995-96	St. Mary's U.	AUAA	11	4	4	8	54	....	....	....	....	....
1996-97	St. Mary's U.	AUAA	18	3	10	13	138	....	....	....	....	....
1997-98	St. Mary's U.	AUAA	25	4	11	15	227	....	....	....	....	....
1998-99	Michigan K-Wings	IHL	55	4	11	15	227	3	1	1	2	4
99-2000	Detroit Vipers	IHL	17	0	2	2	69	....	....	....	....	....
	Manitoba Moose	IHL	45	4	4	8	184	2	0	0	0	2
2000-01	Manchester Storm	Britain	15	1	6	7	51	....	....	....	....	....
	Saint John	AHL	49	3	10	13	167	16	0	1	1	32
2001-02	St. John's	AHL	36	5	8	13	166	9	0	1	1	17

Signed as a free agent by **Manchester** (Britain), August 15, 2000. Signed to a 25-game tryout contract by **Saint John** (AHL) after securing release from **Manchester** (Britain), December 19, 2000. Signed as a free agent by **Saint John** (AHL), February 18, 2001. Signed as a free agent by **Toronto**, July 25, 2001. • Missed majority of 2001-02 season recovering from ankle injury suffered in game vs. Manitoba (AHL), October 19, 2001.

DOUVILLE, Thierry (doo-VEEL, tee-AIR-ee) PHI.

Defense. Shoots left. 6'4", 212 lbs. Born, Laval, Que., April 18, 1983.
(Philadelphia's 8th choice, 208th overall, in 2001 Entry Draft).

			Regular Season					Playoffs				
Season	Club	Lea	GP	G	A	TP	PIM	GP	G	A	TP	PIM
99-2000	Laval Laurentide	QAAA	40	2	5	7	118	9	0	3	3	42
2000-01	Baie-Comeau	QMJHL	65	0	4	4	408	11	0	0	0	16
2001-02	Baie-Comeau	QMJHL	67	0	7	7	380	5	0	1	1	17

DOWN, Blaine (DOWN, BLAYN) NYI

Left wing. Shoots left. 5'11", 170 lbs. Born, Whitby, Ont., July 16, 1982.

			Regular Season					Playoffs				
Season	Club	Lea	GP	G	A	TP	PIM	GP	G	A	TP	PIM
1998-99	Oshawa	OPJHL	36	18	22	40	75	....	....	....	....	....
99-2000	Barrie Colts	OHL	43	17	22	39	49	22	10	6	16	18
2000-01	Barrie Colts	OHL	62	35	38	73	80	5	2	0	2	10
2001-02	Barrie Colts	OHL	63	25	36	61	92	20	15	10	25	34

Signed as a free agent by **NY Islanders**, August 13, 2002.

DRANEY, Brett (DRAY-nee, BREHT) DAL.

Left wing. Shoots left. 6'1", 195 lbs. Born, Merritt, B.C., March 12, 1981.
(Dallas' 7th choice, 186th overall, in 1999 Entry Draft).

			Regular Season					Playoffs				
Season	Club	Lea	GP	G	A	TP	PIM	GP	G	A	TP	PIM
1996-97	Kamloops	BCAHA	43	69	73	132	54	....	....	....	....	....
1997-98	Kamloops Blazers	WHL	42	2	2	4	16	7	0	0	0	0
1998-99	Kamloops Blazers	WHL	58	7	10	17	48	15	1	1	2	8
99-2000	Kamloops Blazers	WHL	62	18	27	45	63	4	1	1	2	11
2000-01	Medicine Hat	WHL	57	20	28	48	84	....	....	....	....	....
2001-02	Medicine Hat	WHL	51	18	37	55	78	....	....	....	....	....
	Fort Worth	CHL	....	....	....	....	....	4	0	1	1	4
	Utah Grizzlies	AHL	6	1	2	3	6	5	0	0	0	4

Traded to **Medicine Hat** (WHL) by **Kamloops** (WHL) for Paul Elliot and Kevin Labbe, September 7, 2000.

DROZDETSKY, Alexander (drawz-DEHT-skee, al-ehx-AN-duhr) PHI.

Right wing. Shoots left. 6', 180 lbs. Born, Moscow, USSR, November 10, 1981.
(Philadelphia's 2nd choice, 94th overall, in 2000 Entry Draft).

			Regular Season					Playoffs				
Season	Club	Lea	GP	G	A	TP	PIM	GP	G	A	TP	PIM
1997-98	St. Petersburg 2	Russia-3	19	0	1	1	0	....	....	....	....	....
1998-99	St. Petersburg 2	Russia-4	24	5	3	8	12	....	....	....	....	....
99-2000	St. Petersburg 2	Russia-3	4	4	1	5	2	....	....	....	....	....
	St. Petersburg	Russia	32	2	0	2	10	4	0	0	0	0
2000-01	St. Petersburg	Russia	42	6	7	13	74	....	....	....	....	....
2001-02	CSKA Moscow	Russia	49	11	6	17	26	....	....	....	....	....

DUBEC, Marek (DOO-behts, MAIR-ehk) BUF.

Left wing. Shoots left. 6', 179 lbs. Born, Bratislava, Czech., February 26, 1982.
(Buffalo's 7th choice, 247th overall, in 2001 Entry Draft).

			Regular Season					Playoffs				
Season	Club	Lea	GP	G	A	TP	PIM	GP	G	A	TP	PIM
2000-01	HC Vsetin Jr.	Czech-Jr.	45	27	15	42	52	8	7	2	9	20
2001-02	HC Vsetin Jr.	Czech-Jr.	18	12	9	21	32	....	....	....	....	....
	HC Vsetin	Czech	5	1	0	1	0	4	2	0	2	4

DUBEN, Premysl (DUH-behn, PREHM-uh-suhl) NYR

Defense. Shoots left. 6'3", 220 lbs. Born, Jihlava, Czech., October 5, 1981.
(NY Rangers' 3rd choice, 112th overall, in 2000 Entry Draft).

			Regular Season					Playoffs				
Season	Club	Lea	GP	G	A	TP	PIM	GP	G	A	TP	PIM
1997-98	Dukla Jihlava Jr.	Czech-Jr.	25	1	6	7	34	....	....	....	....	....
1998-99	Dukla Jihlava Jr.	Czech-Jr.	41	1	5	6	18	....	....	....	....	....
99-2000	Dukla Jihlava Jr.	Czech-Jr.	27	4	2	6	36	....	....	....	....	....
	HC Dukla Jihlava	Czech-2	19	0	1	1	10	14	0	1	1	4
2000-01	HC Dukla Jihlava	Czech-2	5	0	0	0	10	....	....	....	....	....
	Baie-Comeau	QMJHL	32	0	8	8	36	9	1	0	1	12
2001-02	Dukla Jihlava Jr.	Czech-Jr.	13	0	5	5	20	....	....	....	....	....
	HC Dukla Jihlava	Czech-2	33	0	1	1	36	6	0	0	0	2

DUDA, Radek (DOO-duh, RA-dehk) CGY.

Right wing. Shoots left. 6'1", 193 lbs. Born, Skolov, Czech., January 28, 1979.
(Calgary's 7th choice, 192nd overall, in 1998 Entry Draft).

			Regular Season					Playoffs				
Season	Club	Lea	GP	G	A	TP	PIM	GP	G	A	TP	PIM
1994-95	Sokolov Jr.	Czech-Jr.	36	67	37	104	....	....	....	....	....	....
1995-96	Sparta Praha Jr.	Czech-Jr.	39	15	10	25	....	....	....	....	....	....
1996-97	Sparta Praha Jr.	Czech-Jr.	21	9	14	23	....	....	....	....	....	....
	Sokolov Jr.	Czech	1	0	0	0	....	....	....	....	....	....
	HC Sparta Praha	Czech	....	....	....	....	....	1	0	0	0	0
1997-98	HC Sparta Praha	Czech	39	3	3	6	41	10	0	2	2	6
1998-99	Regina Pats	WHL	65	24	31	55	139	....	....	....	....	....
99-2000	Lethbridge	WHL	69	42	64	106	193	....	....	....	....	....
2000-01	Plzen	Czech	24	5	6	11	49	....	....	....	....	....
	HC Sparta Praha	Czech	18	2	1	3	66	....	....	....	....	....
2001-02	Plzen	Czech	49	17	18	35	156	6	1	5	6	18

DUFORT, J.F. (doo-FOHR, JAY-EHF) EDM.

Left wing. Shoots left. 6'2", 202 lbs. Born, Drummondville, Que., March 9, 1982.
(Edmonton's choice, 205th overall, in 2002 Entry Draft).

			Regular Season					Playoffs				
Season	Club	Lea	GP	G	A	TP	PIM	GP	G	A	TP	PIM
1997-98	Antoine-Girouard	QAAA	38	13	16	29	....	....	....	....	....	....
1998-99	Magog	QAAA	40	23	30	53	107	....	....	....	....	....
	Shawinigan	QMJHL	3	0	0	0	0	....	....	....	....	....
99-2000	Shawinigan	QMJHL	70	14	34	48	218	6	2	4	6	2
2000-01	Shawinigan	QMJHL	46	21	32	53	194	10	3	3	6	18
2001-02	Shawinigan	QMJHL	42	11	32	43	147	....	....	....	....	....
	Cape Breton	QMJHL	26	11	9	20	120	16	4	9	13	24

DUMA, Pavel (DOO-muh, PAH-vehl) VAN.

Center. Shoots left. 6'1", 183 lbs. Born, Karaganda, USSR, June 20, 1981.
(Vancouver's 4th choice, 144th overall, in 2000 Entry Draft).

			Regular Season					Playoffs				
Season	Club	Lea	GP	G	A	TP	PIM	GP	G	A	TP	PIM
1997-98	Nizhnekamsk 2	Russia-3	36	6	2	8	18	....	....	....	....	....
1998-99	Nizhnekamsk 2	Russia-4	34	13	12	25	20	....	....	....	....	....
	Nizhnekamsk	Russia	4	0	0	0	2	3	0	2	2	2
99-2000	Nizhnekamsk	Russia	29	2	5	7	14	....	....	....	....	....
	Ak Bars Kazan	Russia	8	0	1	1	2	7	0	0	0	2
2000-01	Nizhnekamsk	Russia	31	3	1	4	22	....	....	....	....	....
2001-02	Perm	Russia	42	5	4	9	12	....	....	....	....	....

DURAK, Miroslav (DOO-rak, MEE-roh-slav) NSH.

Defense. Shoots right. 6'4", 212 lbs. Born, Topolcany, Czech., June 9, 1981.
(Nashville's 14th choice, 220th overall, in 1999 Entry Draft).

			Regular Season					Playoffs				
Season	Club	Lea	GP	G	A	TP	PIM	GP	G	A	TP	PIM
1997-98	S. Bratislava Jr.	Slovak-Jr.	36	4	10	14	34	....	....	....	....	....
1998-99	S. Bratislava Jr.	Slovak-Jr.	38	1	7	8	48	....	....	....	....	....
99-2000	Des Moines	USHL	55	5	6	11	93	9	2	1	3	12
2000-01	Sherbrooke	QMJHL	34	0	17	17	24	....	....	....	....	....
	Acadie-Bathurst	QMJHL	27	5	13	18	95	18	4	3	7	45
2001-02	Acadie-Bathurst	QMJHL	67	11	36	47	131	16	2	13	15	42

Traded to **Acadie-Bathurst** (QMJHL) by **Sherbrooke** (QMJHL) for future considerations, January 5, 2001.

DUSABLON, Benoit NYR

Center. Shoots left. 6'1", 207 lbs. Born, Ste Anne de la Perad, Que., August 1, 1979.

			Regular Season					Playoffs				
Season	Club	Lea	GP	G	A	TP	PIM	GP	G	A	TP	PIM
1996-97	Halifax	QMJHL	61	7	7	14	181	....	....	....	....	....
1997-98	Halifax	QMJHL	7	1	0	1	7	....	....	....	....	....
	Val d'Or Foreurs	QMJHL	57	14	11	25	56	19	2	9	11	33
1998-99	Val d'Or Foreurs	QMJHL	67	42	74	116	63	6	2	6	8	4
99-2000	Val d'Or Foreurs	QMJHL	41	29	53	82	45	....	....	....	....	....
	Halifax	QMJHL	31	18	35	53	18	10	6	7	13	12
2000-01	Johnstown Chiefs	ECHL	11	2	3	5	4	....	....	....	....	....
	Tallahassee	ECHL	49	21	29	50	33	....	....	....	....	....
2001-02	Charlotte	ECHL	19	12	13	25	2	....	....	....	....	....
	Hartford	AHL	38	8	15	23	16	9	1	2	3	4

DVORAK, Petr (duv-VOHR-ak, PEE-tuhr) WSH.

Center. Shoots right. 6', 194 lbs. Born, Roznov, Czech., October 11, 1983.
(Washington's 8th choice, 118th overall, in 2002 Entry Draft).

			Regular Season					Playoffs				
Season	Club	Lea	GP	G	A	TP	PIM	GP	G	A	TP	PIM
99-2000	Havirov Jr.	Czech-Jr.	47	28	28	56	79	....	....	....	....	....
2000-01	Havirov Jr.	Czech-Jr.	36	12	15	27	18	....	....	....	....	....
	HC Femax Havirov	Czech	1	0	0	0	0	....	....	....	....	....
2001-02	Havirov Jr.	Czech-Jr.	40	24	25	49	90	....	....	....	....	....
	Sumperk	Czech-2	1	0	1	1	0	....	....	....	....	....
	HC Femax Havirov	Czech	7	0	0	0	0	....	....	....	....	....

DWYER, Jeff (DWIGH-uhr, JEHF) ATL.

Defense. Shoots left. 6'2", 205 lbs. Born, Greenwich, CT, November 22, 1980.
(Atlanta's 8th choice, 178th overall, in 2000 Entry Draft).

			Regular Season					Playoffs				
Season	Club	Lea	GP	G	A	TP	PIM	GP	G	A	TP	PIM
1996-97	Choate-Rosemary	Hi-School	28	5	11	16	....	....	....	....	....	....
1997-98	Choate-Rosemary	Hi-School	28	9	14	23	....	....	....	....	....	....
1998-99	Choate-Rosemary	Hi-School	27	8	22	30	....	....	....	....	....	....
99-2000	Choate-Rosemary	Hi-School	25	11	30	41	25	....	....	....	....	....
2000-01	Yale	ECAC	31	3	18	21	16	....	....	....	....	....
2001-02	Yale	ECAC	31	6	9	15	16	....	....	....	....	....

DWYER, Patrick (DWIGH-uhr, PAT-rihk) ATL.

Right wing. Shoots right. 5'10", 170 lbs. Born, Spokane, WA, June 22, 1983.
(Atlanta's 3rd choice, 116th overall, in 2002 Entry Draft).

			Regular Season					Playoffs				
Season	Club	Lea	GP	G	A	TP	PIM	GP	G	A	TP	PIM
2000-01	Great Falls	NWJHL	40	33	57	90	106	12	10	12	22	....
2001-02	West-Michigan	CCHA	38	17	17	34	26	....	....	....	....	....

NWJHL First All-Star Team (2001) • NWJHL MVP (2001) • CCHA All-Rookie Team (2002) • CCHA Rookie of the Year (2002)

DYMENT, Chris (DIGH-mehnt, KRIHS) MIN.

Defense. Shoots right. 6'3", 210 lbs. Born, Stoneham, MA, October 24, 1979.
(Montreal's 3rd choice, 97th overall, in 1999 Entry Draft).

			Regular Season					Playoffs				
Season	Club	Lea	GP	G	A	TP	PIM	GP	G	A	TP	PIM
1997-98	Reading High	Hi-School	22	22	22	44	15	....	....	....	....	....
1998-99	Boston University	H-East	25	1	5	6	16	....	....	....	....	....
99-2000	Boston University	H-East	42	11	20	31	42	....	....	....	....	....
2000-01	Boston University	H-East	37	1	10	11	38	....	....	....	....	....
2001-02	Boston University	H-East	38	7	18	25	24	....	....	....	....	....

Hockey East First All-Star Team (2000) • NCAA East Second All-American Team (2000) • Hockey East Second All-Star Team (2002)

Traded to **Minnesota** by **Montreal** for Minnesota's 5th round choice (later traded to Calgary - Calgary selected Jiri Cetkovsky) in 2002 Entry Draft, May 25, 2002.

EAGER, Ben (EE-guhr, BEHN) PHX.

Left wing. Shoots left. 6'2", 215 lbs. Born, Ottawa, Ont., January 22, 1984.
(Phoenix's 2nd choice, 23rd overall, in 2002 Entry Draft).

			Regular Season					Playoffs				
Season	Club	Lea	GP	G	A	TP	PIM	GP	G	A	TP	PIM
99-2000	Ottawa Jr. Sens	OCJHL	50	8	11	19	119	....	....	....	....	....
2000-01	Oshawa Generals	OHL	61	4	6	10	120	....	....	....	....	....
2001-02	Oshawa Generals	OHL	63	14	23	37	255	5	0	1	1	13

EAVES, Ben (EEVZ, BEHN) PIT.

Center. Shoots right. 5'8", 174 lbs. Born, Minneapolis, MN, March 27, 1982.
(Pittsburgh's 6th choice, 131st overall, in 2001 Entry Draft).

			Regular Season					Playoffs				
Season	Club	Lea	GP	G	A	TP	PIM	GP	G	A	TP	PIM
1998-99	Minnesota	USAHA	71	68	88	156	12	....	....	....	....	....
99-2000	Shat.-St. Mary's	Hi-School	57	47	71	118	16	....	....	....	....	....
2000-01	Boston College	H-East	40	13	26	39	12	....	....	....	....	....
2001-02	Boston College	H-East	23	13	26	39	12	....	....	....	....	....

Hockey East Second All-Star Team (2002)

EBERLY, Dan (EH-bur-lee, DAN) NYR

Defense. Shoots left. 6', 180 lbs. Born, Newton, MA, October 28, 1980.
(NY Rangers' 8th choice, 238th overall, in 2000 Entry Draft).

			Regular Season					Playoffs				
Season	Club	Lea	GP	G	A	TP	PIM	GP	G	A	TP	PIM
1998-99	Catholic Memorial	Hi-School	STATISTICS NOT AVAILABLE									
99-2000	RPI Engineers	ECAC	21	2	5	7	8	....	....	....	....	....
2000-01	RPI Engineers	ECAC	28	3	8	11	20	....	....	....	....	....
2001-02	RPI Engineers	ECAC	35	3	14	17	26	....	....	....	....	....

EHRHOFF, Christian (AIR-hawf, KRIHS-tyan) S.J.

Defense. Shoots left. 6'2", 187 lbs. Born, Moers, West Germany, July 6, 1982.
(San Jose's 2nd choice, 106th overall, in 2001 Entry Draft).

			Regular Season					Playoffs				
Season	Club	Lea	GP	G	A	TP	PIM	GP	G	A	TP	PIM
1998-99	Krefeld Jr.	Ger.-Jr.	22	10	14	24	46	....	....	....	....	....
99-2000	Duisburger SC	German-3	41	3	12	15	50	....	....	....	....	....
	Krefeld Pinguine	Germany	9	1	0	1	6	3	0	0	0	0
2000-01	Duisburger SC	German-3	6	1	2	3	12	....	....	....	....	....
	Krefeld Pinguine	Germany	58	3	11	14	73	....	....	....	....	....
2001-02	Krefeld Pinguine	Germany	46	7	17	24	81	3	0	0	0	2
	Germany	Olympics	7	0	0	0	8	....	....	....	....	....

EICHELBERGER, John (IGH-kehl-buhr-guhr, JAWN) PHI.

Center. Shoots left. 6'2", 185 lbs. Born, Atlanta, GA, February 23, 1981.
(Philadelphia's 5th choice, 210th overall, in 2000 Entry Draft).

			Regular Season					Playoffs				
Season	Club	Lea	GP	G	A	TP	PIM	GP	G	A	TP	PIM
1997-98	Team USA	USDP-18	65	11	28	34	54	....	....	....	....	....
1998-99	Team USA	USDP-18	31	5	11	16	24	....	....	....	....	....
99-2000	Green Bay	USHL	49	19	50	69	58	....	....	....	....	....
2000-01	U. of Wisconsin	WCHA	25	0	5	5	2	....	....	....	....	....
2001-02	U. of Wisconsin	WCHA	36	2	7	9	29	....	....	....	....	....

ELLIOTT, Paul (EHL-lee-awt, PAWL) FLA.

Defense. Shoots left. 6'1", 216 lbs. Born, White Rock, B.C., June 2, 1980.
(Edmonton's 5th choice, 128th overall, in 1998 Entry Draft).

			Regular Season					Playoffs				
Season	Club	Lea	GP	G	A	TP	PIM	GP	G	A	TP	PIM
1995-96	Surrey Eagles	BCAHA	45	23	60	83	77	....	....	....	....	....
	Lethbridge	WHL	2	0	0	0	0	....	....	....	....	....
1996-97	Lethbridge	WHL	46	0	8	8	17	1	0	0	0	0
1997-98	Lethbridge	WHL	48	4	18	22	35	....	....	....	....	....
	Medicine Hat	WHL	24	7	9	16	12	....	....	....	....	....
1998-99	Medicine Hat	WHL	71	11	36	47	80	....	....	....	....	....
99-2000	Medicine Hat	WHL	66	15	28	43	91	....	....	....	....	....
2000-01	Kamloops Blazers	WHL	40	10	33	43	37	....	....	....	....	....
	Regina Pats	WHL	27	8	17	25	51	6	0	4	4	2
2001-02	Utah Grizzlies	AHL	35	2	2	4	14	....	....	....	....	....
	Florida	ECHL	5	2	2	4	0	6	2	3	5	10

Memorial Cup All-Star Team (2001)

Traded to **Kamloops** (WHL) by **Medicine Hat** (WHL) with Kevin Labbe for Brett Draney, September 7, 2000. Traded to **Regina** (WHL) by **Kamloops** (WHL) for Shawn Belle, January 11, 2001. Signed as a free agent by **Florida**, July 3, 2001.

ELLISON, Matt (EHL-ih-suhn, MAT) CHI.

Right wing. Shoots right. 5'11", 185 lbs. Born, Duncan, B.C., December 8, 1983.
(Chicago's 4th choice, 128th overall, in 2002 Entry Draft).

			Regular Season					Playoffs				
Season	Club	Lea	GP	G	A	TP	PIM	GP	G	A	TP	PIM
1997-98	Cowichan Valley	BCAHA	24	27	31	58	10	....	....	....	....	....
1998-99	Kerry Park	VIJHL	38	40	47	87	110	....	....	....	....	....
99-2000	Cowichan	BCHL	60	11	23	34	95	....	....	....	....	....
2000-01	Cowichan	BCHL	60	22	44	66	102	....	....	....	....	....
2001-02	Cowichan	BCHL	60	42	*75	*117	76	....	....	....	....	....

BCHL Coastal Division First All-Star Team (2002) • BCHL Coastal Division MVP (2002)

• Signed Letter of Intent to attend **Nebraska-Omaha** (CCHA), November 20, 2001.

ELOFSSON, Jonas (EHL-uhf-suhn, YOHuhs) CHI.

Defense. Shoots left. 6'1", 180 lbs. Born, Ulricehamn, Sweden, January 31, 1979.
(Edmonton's 4th choice, 94th overall, in 1997 Entry Draft).

			Regular Season					Playoffs				
Season	Club	Lea	GP	G	A	TP	PIM	GP	G	A	TP	PIM
1995-96	Farjestad Jr.	Swede-Jr.	26	6	11	17	18	....	....	....	....	....
1996-97	Farjestad	Sweden	3	0	0	0	0	5	0	1	1	0
	Farjestad	EuroHL	2	1	1	2	0	....	....	....	....	....
	Sweden	EJC-A	6	3	0	3	6	....	....	....	....	....
1997-98	Farjestad	Sweden	29	3	2	5	14	12	0	1	1	6
	Farjestad	EuroHL	7	1	1	2	4	....	....	....	....	....
1998-99	Farjestad	Sweden	40	2	7	9	18	4	0	0	0	0
	Farjestad	EuroHL	5	1	0	1	4	....	....	....	....	....
99-2000	Farjestad	Sweden	47	3	6	9	44	4	0	0	0	6
2000-01	HV 71 Jonkoping	Sweden	30	1	3	4	12	....	....	....	....	....
	TPS Turku	Finland	10	0	0	0	0	....	....	....	....	....
2001-02	Leksands IF	Sweden	30	8	20	28	12	....	....	....	....	....
	Leksands IF	Swede-Q	10	1	4	5	4	10	1	1	2	10

Traded to **Chicago** by **Edmonton** with Boris Mironov and Dean McAmmond for Chad Kilger, Daniel Cleary, Ethan Moreau and Christian Laflamme, March 20, 1999.

ELOMO, Teemu (eh-LOH-moh, TEE-moo) DAL.

Left wing. Shoots left. 5'11", 176 lbs. Born, Turku, Finland, January 13, 1979.
(Dallas' 5th choice, 132nd overall, in 1997 Entry Draft).

			Regular Season					Playoffs				
Season	Club	Lea	GP	G	A	TP	PIM	GP	G	A	TP	PIM
1993-94	TPS Turku-C	Finn-Jr.	34	7	25	32	56	....	....	....	....	....
1994-95	TPS Turku-C	Finn-Jr.	24	19	22	41	82	....	....	....	....	....
	TPS Turku-B	Finn-Jr.	8	5	2	7	12	2	1	0	1	0
1995-96	TPS Turku-B	Finn-Jr.	17	9	8	17	28	....	....	....	....	....
	TPS Turku Jr.	Finn-Jr.	2	0	0	0	0	....	....	....	....	....
	Kiekko-67 Turku	Finland-2	11	1	0	1	14	6	2	0	2	8
1996-97	TPS Turku Jr.	Finn-Jr.	9	6	2	8	16	....	....	....	....	....
	Kiekko-67 Turku	Finland-2	15	4	3	7	24	....	....	....	....	....
	TPS Turku	Finland	6	0	1	1	0	3	0	0	0	2
1997-98	TPS Turku	Finland	26	3	3	6	14	3	1	0	1	2
	TPS Turku	EuroHL	3	0	0	0	2	....	....	....	....	....
1998-99	TPS Turku	Finland	34	4	8	12	16	5	0	0	0	2
99-2000	TPS Turku	Finland	52	9	7	16	28	11	3	2	5	0
2000-01	TPS Turku	Finland	56	2	10	12	44	10	1	3	4	0
2001-02	Blues Espoo	Finland	49	12	14	26	66	3	0	0	0	14

EMINGER, Steve (EH-mihn-juhr, STEEV) WSH.

Defense. Shoots right. 6'1", 196 lbs. Born, Woodbridge, Ont., October 31, 1983.
(Washington's 1st choice, 12th overall, in 2002 Entry Draft).

			Regular Season					Playoffs				
Season	Club	Lea	GP	G	A	TP	PIM	GP	G	A	TP	PIM
1997-98	Vaughan Rangers	OMHA		STATISTICS NOT AVAILABLE								
1998-99	Bramalea Blues	OPJHL	47	6	9	15	81	....	....	....	....	....
99-2000	Kitchener	OHL	50	2	14	16	74	5	0	0	0	0
2000-01	Kitchener	OHL	54	6	26	32	66	....	....	....	....	....
2001-02	Kitchener	OHL	64	19	39	58	93	4	0	2	2	10

OHL Second All-Star Team (2002)

EMOND, Pierre-Luc (ee-MOHN, pee-AIR-LOOK) COL.

Center. Shoots left. 6', 195 lbs. Born, Valleyfield, Que., October 10, 1982.
(Colorado's 7th choice, 165th overall, in 2001 Entry Draft).

			Regular Season					Playoffs				
Season	Club	Lea	GP	G	A	TP	PIM	GP	G	A	TP	PIM
1998-99	Gatineau	QAAA	42	7	17	24	24	....	....	....	....	....
99-2000	Drummondville	QMJHL	58	8	11	19	74	16	6	3	9	10
2000-01	Drummondville	QMJHL	62	10	36	46	110	6	2	0	2	10
2001-02	Cape Breton	QMJHL	72	19	29	48	63	16	3	2	5	4

Traded to **Cape Breton** (QMJHL) by **Drummondville** (QMJHL) with Steve Villeneuve for Cape Breton's 3rd round choice (Patrick Tessier) in QMJHL 2001 Priority Draft, June 16, 2001.

ENEQVIST, Johan (EHN-uh-kvist, YOH-han) MTL.

Left wing. Shoots left. 6', 183 lbs. Born, Nacka, Sweden, January 21, 1982.
(Montreal's 5th choice, 109th overall, in 2000 Entry Draft).

			Regular Season					Playoffs				
Season	Club	Lea	GP	G	A	TP	PIM	GP	G	A	TP	PIM
99-2000	Leksands IF-18	Swede-Jr.	5	1	1	2	28	....	....	....	....	....
	Leksands IF Jr.	Swede-Jr.	36	10	13	23	36	2	0	0	0	0
2000-01	Leksands IF Jr.	Swede-Jr.	21	19	15	34	2	2	1	1	2	4
	Leksands IF	Sweden	2	0	0	0	0	....	....	....	....	....
2001-02	Leksands IF Jr.	Swede-Jr.	7	3	6	9	18	....	....	....	....	....
	Leksands IF	Sweden	45	3	7	10	24	....	....	....	....	....

ERICKSON, Mike (AIR-ihk-suhn, MIGHK) MIN.

Right wing. Shoots right. 6'2", 186 lbs. Born, Minneapolis, MN, April 12, 1983.
(Minnesota's 3rd choice, 72nd overall, in 2002 Entry Draft).

			Regular Season					Playoffs				
Season	Club	Lea	GP	G	A	TP	PIM	GP	G	A	TP	PIM
1998-99	Eden Prairie	Hi-School	23	24	18	42	....	....	....	....	....	....
99-2000	Eden Prairie	Hi-School	25	38	25	63	....	....	....	....	....	....
2000-01	Eden Prairie	Hi-School	22	24	26	50	....	....	....	....	....	....
2001-02	U. of Minnesota	WCHA	9	1	2	3	2	....	....	....	....	....

All-Conference All-Star Team (1999, 2000, 2001)

ERIKSSON, Tim (AIR-ihk-suhn, TIHM) L.A.

Center. Shoots left. 5'9", 161 lbs. Born, Sodertalje, Sweden, February 5, 1982.
(Los Angeles' 7th choice, 206th overall, in 2000 Entry Draft).

			Regular Season					Playoffs				
Season	Club	Lea	GP	G	A	TP	PIM	GP	G	A	TP	PIM
1997-98	Sodertalje Jr.	Swede-Jr.	11	7	12	19	10	....	....	....	....	....
1998-99	V. Frolunda Jr.	Swede-Jr.	29	7	8	15	6	....	....	....	....	....
99-2000	V. Frolunda Jr.	Swede-Jr.	36	16	25	41	82	....	....	....	....	....
2000-01	Hammarby	Swede-2	1	2	1	3	0	....	....	....	....	....
	Hammarby	Swede-2	38	9	22	31	10	14	2	4	6	8
2001-02	Hammarby	Swede-2	44	10	30	40	34	2	0	1	1	0

ESTRADA, Kevin (eh-STRA-duh, KEH-vihn) CAR.

Right wing. Shoots left. 5'11", 185 lbs. Born, Surrey, B.C., May 28, 1982.
(Carolina's 3rd choice, 91st overall, in 2001 Entry Draft).

			Regular Season					Playoffs				
Season	Club	Lea	GP	G	A	TP	PIM	GP	G	A	TP	PIM
1997-98	Chilliwack	BCHL	35	1	5	6	17	....	....	....	....	....
1998-99	Chilliwack	BCHL	58	13	29	42	58	....	....	....	....	....
99-2000	Chilliwack	BCHL	45	9	20	29	29	30	6	27	33	14
2000-01	Chilliwack	BCHL	59	34	*84	*118	65	....	....	....	....	....
2001-02	Michigan State	CCHA	40	4	7	11	24	....	....	....	....	....

BCHL Coastal Conference First All-Star Team (2001)

EVANS, Blake (EH-vans, BLAYK) ST.L.

Center. Shoots right. 6'1", 221 lbs. Born, Kindersley, Sask., July 2, 1980.
(Washington's 10th choice, 251st overall, in 1998 Entry Draft).

			Regular Season					Playoffs				
Season	Club	Lea	GP	G	A	TP	PIM	GP	G	A	TP	PIM
1995-96	Sask. Contacts	SMHL	41	15	23	38	84	....	....	....	....	....
1996-97	Spokane Chiefs	WHL	53	4	7	11	19	7	0	0	0	0
1997-98	Spokane Chiefs	WHL	16	6	5	11	29	....	....	....	....	....
	Tri-City	WHL	57	13	29	42	102	....	....	....	....	....
1998-99	Tri-City	WHL	72	18	29	47	131	12	0	4	4	16
99-2000	Tri-City	WHL	72	27	43	70	110	4	1	0	1	6
2000-01	Tri-City	WHL	40	28	31	59	70	....	....	....	....	....
	Regina Pats	WHL	27	24	19	43	50	6	6	2	8	8
2001-02	Worcester	AHL	28	4	5	9	18	3	0	0	0	2
	Peoria Rivermen	ECHL	43	17	20	37	26	....	....	....	....	....

WHL East Second All-Star Team (2001)

Traded to **Tri-City** (WHL) by **Spokane** (WHL) with Regan Darby for Zenith Komarniski, October 6, 1997. Traded to **Regina** (WHL) by **Tri-City** (WHL) with Jeff Feniak for Shawn Bell, Joey Bastien, Justin Lucyshyn and future considerations, January 8, 2001. Signed as a free agent by **St. Louis**, April 11, 2001.

EVANS, David (EHV-vuhns, DAY-vihd) CAR.

Center. Shoots right. 6'3", 185 lbs. Born, Albany, NY, February 17, 1980.
(Carolina's 7th choice, 231st overall, in 1999 Entry Draft).

			Regular Season					Playoffs				
Season	Club	Lea	GP	G	A	TP	PIM	GP	G	A	TP	PIM
1997-98	Capital District	Exhib.	51	53	60	113	....	....	....	....	....	....
1998-99	Clarkson Knights	ECAC	33	6	10	16	6	....	....	....	....	....
99-2000	Clarkson Knights	ECAC	34	11	17	28	18	....	....	....	....	....
2000-01	Clarkson Knights	ECAC	33	12	19	31	8	....	....	....	....	....
2001-02	Clarkson Knights	ECAC	36	11	12	23	16	....	....	....	....	....

ECAC All-Academic Team (2002)

EXELBY, Garnet (EHX-uhl-bee, GAHR-neht) ATL.

Defense. Shoots left. 6'1", 210 lbs. Born, Craik, Sask., August 16, 1981.
(Atlanta's 9th choice, 217th overall, in 1999 Entry Draft).

			Regular Season					Playoffs				
Season	Club	Lea	GP	G	A	TP	PIM	GP	G	A	TP	PIM
1997-98	Winnipeg South	MJHL	46	5	11	16	110	....	....	....	....	....
1998-99	Saskatoon Blades	WHL	61	5	3	8	91	....	....	....	....	....
99-2000	Saskatoon Blades	WHL	63	1	8	9	79	11	0	2	2	21
2000-01	Saskatoon Blades	WHL	43	5	10	15	110	....	....	....	....	....
	Regina Pats	WHL	22	2	8	10	51	6	0	2	2	2
2001-02	Chicago Wolves	AHL	75	3	4	7	257	25	0	4	4	49

Traded to **Regina** (WHL) by **Saskatoon** (WHL) for Scott Balan and future considerations, January 15, 2001.

FABUS, Peter (fah-BUSH, PEE-tuhr) PHX.

Center. Shoots left. 6'1", 191 lbs. Born, Ilava, Czech., July 15, 1979.
(Phoenix's 8th choice, 281st overall, in 2000 Entry Draft).

			Regular Season					Playoffs				
Season	Club	Lea	GP	G	A	TP	PIM	GP	G	A	TP	PIM
1997-98	Dubnica Jr.	Slovak-Jr.	28	9	14	23	96	....	....	....	....	....
	Dubnica	Slovak-2	34	4	4	8	24	....	....	....	....	....
1998-99	HK SKP Zilina	Slovak-2	26	7	8	15	24	....	....	....	....	....
99-2000	Dukla Trencin	Slovakia	54	20	11	31	34	5	1	1	2	0
2000-01	Dukla Trencin	Slovakia	52	31	22	53	94	14	4	6	10	*60
2001-02	Springfield	AHL	6	0	0	0	2	....	....	....	....	....
	Dukla Trencin	Slovakia	28	14	11	25	65	5	0	1	1	2

FADRNY, Jan — (FAHD-uhr-nee, YAN) — PIT.

Center. Shoots right. 6', 195 lbs. Born, Brno, Czech., June 14, 1980.
(Pittsburgh's 6th choice, 169th overall, in 1998 Entry Draft).

			Regular Season					Playoffs				
Season	Club	Lea	GP	G	A	TP	PIM	GP	G	A	TP	PIM
1995-96	Kometa Brno Jr.	Czech-Jr.	36	22	15	37	26					
1996-97	HC Olomouc Jr.	Czech-Jr.	38	16	24	40	32					
1997-98	Slavia Praha Jr.	Czech-Jr.	14	7	4	11	12					
	HC Slavia Praha	Czech	18	1	1	2	2	3	0	0	0	4
1998-99	Brandon	WHL	45	4	17	21	36	5	1	2	3	4
99-2000	Brandon	WHL	55	26	25	51	56					
2000-01	Brandon	WHL	2	1	1	2	6					
	Kelowna Rockets	WHL	56	32	45	77	58	6	3	4	7	8
2001-02	Wilkes-Barre	AHL	60	8	14	22	54					

Traded to **Kelowna** (WHL) by **Brandon** (WHL) with Bart Rushmer for Nolan Yonkman and Kelowna's 6th round choice (Jeff Wollin) in 2001 WHL Bantam Draft, October 12, 2000.

FAHEY, Brian — (FAY-hee, BRIGH-uhn) — COL.

Defense. Shoots right. 6', 200 lbs. Born, Des Plaines, IL, March 2, 1981.
(Colorado's 7th choice, 119th overall, in 2000 Entry Draft).

			Regular Season					Playoffs				
Season	Club	Lea	GP	G	A	TP	PIM	GP	G	A	TP	PIM
1997-98	Team USA	USDP-18	68	7	21	28	57					
1998-99	Team USA	USDP-18	52	9	9	18	34					
99-2000	U. of Wisconsin	WCHA	41	6	11	17	42					
2000-01	U. of Wisconsin	WCHA	38	1	5	6	16					
2001-02	U. of Wisconsin	WCHA	38	2	8	10	55					

WCHA All-Rookie Team (2000)

FAHEY, Jim — (FA-hee, JIHM) — S.J.

Defense. Shoots right. 6', 215 lbs. Born, Boston, MA, May 11, 1979.
(San Jose's 9th choice, 212th overall, in 1998 Entry Draft).

			Regular Season					Playoffs				
Season	Club	Lea	GP	G	A	TP	PIM	GP	G	A	TP	PIM
1997-98	Catholic Memorial	Hi-School	24	12	32	44	28					
1998-99	Northeastern	H-East	32	5	13	18	34					
99-2000	Northeastern	H-East	36	3	17	20	62					
2000-01	Northeastern	H-East	36	4	23	27	48					
2001-02	Northeastern	H-East	39	14	32	46	50					

Hockey East Second All-Star Team (2001) • Hockey East First All-Star Team (2002)

FALARDEAU, Lee — (FAL-ahr-doh, LEE) — NYR

Center. Shoots left. 6'4", 203 lbs. Born, Midland, MI, July 22, 1983.
(NY Rangers' 1st choice, 33rd overall, in 2002 Entry Draft).

			Regular Season					Playoffs				
Season	Club	Lea	GP	G	A	TP	PIM	GP	G	A	TP	PIM
1998-99	Det. Honeybaked	MMHL	STATISTICS NOT AVAILABLE									
99-2000	Team USA	USDP-17	61	9	14	23	....					
2000-01	Team USA	USDP-18	61	10	21	31	26					
2001-02	Michigan State	CCHA	34	4	10	14	24					

FAST, Brad — (FAST, BRAD) — CAR.

Defense. Shoots left. 6', 185 lbs. Born, Fort St. John, B.C., February 21, 1980.
(Carolina's 2nd choice, 84th overall, in 1999 Entry Draft).

			Regular Season					Playoffs				
Season	Club	Lea	GP	G	A	TP	PIM	GP	G	A	TP	PIM
1994-95	Fort St. John	BCAHA	40	9	26	35	40					
1995-96	Fort St. John	BCAHA	60	53	52	105	70					
1996-97	Prince George	BCHL	49	3	7	10	19					
1997-98	Prince George	BCHL	59	10	33	43	22					
1998-99	Prince George	BCHL	59	27	46	73	....					
99-2000	Michigan State	CCHA	42	5	9	14	20					
2000-01	Michigan State	CCHA	42	4	24	28	16					
2001-02	Michigan State	CCHA	41	10	16	26	26					

FATA, Drew — (FA-tuh, DROO) — PIT.

Defense. Shoots left. 6'1", 209 lbs. Born, Sault Ste. Marie, Ont., July 28, 1983.
(Pittsburgh's 3rd choice, 86th overall, in 2001 Entry Draft).

			Regular Season					Playoffs				
Season	Club	Lea	GP	G	A	TP	PIM	GP	G	A	TP	PIM
99-2000	St. Michael's B	OJHL-B	49	9	18	27	144					
2000-01	St. Michael's	OHL	58	5	15	20	134	18	1	3	4	26
2001-02	St. Michael's	OHL	67	7	21	28	175	15	1	9	10	38

FEDOROV, Fedor — (FEH-duh-rahf, feh-DUHR) — VAN.

Left wing. Shoots left. 6'3", 202 lbs. Born, Appatity, USSR, June 11, 1981.
(Vancouver's 2nd choice, 66th overall, in 2001 Entry Draft).

			Regular Season					Playoffs				
Season	Club	Lea	GP	G	A	TP	PIM	GP	G	A	TP	PIM
1997-98	Det. Caesars	MNHL	13	3	7	10	18					
1998-99	Port Huron	UHL	42	2	5	7	20					
99-2000	Windsor	OHL	60	7	10	17	115	12	1	0	1	4
2000-01	Sudbury Wolves	OHL	67	33	45	78	88	12	4	6	10	36
2001-02	Manitoba Moose	AHL	8	2	1	3	6					
	Columbia Inferno	ECHL	2	0	2	2	0					

• Re-entered NHL Entry Draft. Originally Tampa Bay's 7th choice, 182nd overall, in 1999 Entry Draft.

Signed as an underage free agent by **Detroit** (IHL), August 5, 1998. Released by **Detroit** (IHL), September 30, 1998. Signed as an underage free agent by **Port Huron** (UHL), October 1, 1998. Traded to **Sudbury** (OHL) by **Windsor** (OHL) for future considerations, September 16, 2001. • Missed majority of 2001-02 season recovering from eye injury suffered in game vs. Macon (ECHL), November 17, 2001.

FEDOROV, Yevgeny — (FEH-duh-rahf, yehv-GEH-nee) — L.A.

Center. Shoots left. 5'10", 187 lbs. Born, Sverdlovsk, USSR, November 11, 1980.
(Los Angeles' 6th choice, 201st overall, in 2000 Entry Draft).

			Regular Season					Playoffs				
Season	Club	Lea	GP	G	A	TP	PIM	GP	G	A	TP	PIM
1997-98	Krylja Sovetov 2	Russia-3	20	1	6	7	48					
	Krylja Sovetov	Russia	32	1	0	1	12					
1998-99	Krylja Sovetov	Russia	52	5	3	8	61					
99-2000	Perm	Russia	37	5	5	10	20	3	0	1	1	4
2000-01	Perm	Russia	43	9	9	18	18					
2001-02	Ak Bars Kazan	Russia	45	10	12	22	12	11	0	0	0	2

FEMENELLA, Arthur — (feh-meh-NEHL-uh, AHR-thuhr) — T.B.

Defense. Shoots right. 6'7", 235 lbs. Born, New York, NY, June 6, 1982.
(Tampa Bay's 7th choice, 188th overall, in 2001 Entry Draft).

			Regular Season					Playoffs				
Season	Club	Lea	GP	G	A	TP	PIM	GP	G	A	TP	PIM
99-2000	Team USA	USDP-17	54	0	8	8	156					
2000-01	Sioux City	USHL	52	1	1	2	*252	3	0	0	0	2
2001-02	Sioux City	USHL	56	1	10	11	215	12	0	1	1	32

FERGUSON, Troy — (fuhr-GUH-suhn, TROI) — CAR.

Right wing. Shoots right. 5'9", 165 lbs. Born, Calgary, Alta., September 30, 1980.
(Carolina's 7th choice, 276th overall, in 2000 Entry Draft).

			Regular Season					Playoffs				
Season	Club	Lea	GP	G	A	TP	PIM	GP	G	A	TP	PIM
1996-97	Kitchener	OJHL-B	47	14	15	29	16					
1997-98	Team USA	USDP-18	54	11	9	20	14					
1998-99	Team USA	USDP-17	24	3	10	13	14					
	Team USA	USDP-18	29	4	4	8	20					
99-2000	Michigan State	CCHA	42	5	7	12	10					
2000-01	Michigan State	CCHA	41	4	10	14	12					
2001-02	Michigan State	CCHA	39	1	5	6	14					

FERNHOLM, Daniel — (FUHRN-hohlm, DAN-yehl) — PIT.

Defense. Shoots left. 6'4", 218 lbs. Born, Stockholm, Sweden, December 20, 1983.
(Pittsburgh's 4th choice, 101st overall, in 2002 Entry Draft).

			Regular Season					Playoffs				
Season	Club	Lea	GP	G	A	TP	PIM	GP	G	A	TP	PIM
99-2000	Mora IK Jr.	Swede-Jr.	33	3	3	6	8	1	0	0	0	0
2000-01	Mora IK Jr.	Swede-Jr.	3	0	1	1	2					
	Mora IK	Swede-2	2	0	0	0	0					
2001-02	Djurgarden Jr.	Swede-Jr.	8	6	13	19	12	3	0	0	0	0

FIBIGER, Jesse — (feh-BEH-gehr, JEH-see) — S.J.

Defense. Shoots left. 6'3", 210 lbs. Born, Victoria, B.C., April 4, 1978.
(Anaheim's 5th choice, 178th overall, in 1998 Entry Draft).

			Regular Season					Playoffs				
Season	Club	Lea	GP	G	A	TP	PIM	GP	G	A	TP	PIM
1996-97	Victoria Salsa	BCHL	53	6	18	24	88					
1997-98	U. Minn-Duluth	WCHA	40	3	6	9	82					
1998-99	U. Minn-Duluth	WCHA	36	4	16	20	61					
99-2000	U. Minn-Duluth	WCHA	37	4	6	10	83					
2000-01	U. Minn-Duluth	WCHA	37	0	8	8	56					
2001-02	Cleveland Barons	AHL	79	6	12	18	94					

Signed as a free agent by **San Jose**, August 15, 2001.

FIDDLER, Vernon — (FIHD-luhr, VUHR-nuhn) — NSH.

Center. Shoots left. 5'11", 195 lbs. Born, Edmonton, Alta., May 9, 1980.

			Regular Season					Playoffs				
Season	Club	Lea	GP	G	A	TP	PIM	GP	G	A	TP	PIM
1997-98	Kelowna Rockets	WHL	65	10	11	21	31	7	0	1	1	4
1998-99	Kelowna Rockets	WHL	68	22	21	43	82	6	2	0	2	8
99-2000	Kelowna Rockets	WHL	64	20	28	48	60	5	1	3	4	4
2000-01	Kelowna Rockets	WHL	3	0	2	2	0					
	Medicine Hat	WHL	67	33	38	71	100					
	Arkansas	ECHL	3	0	1	1	2	5	3	0	3	5
2001-02	Roanoke Express	ECHL	44	27	28	55	71					
	Norfolk Admirals	AHL	38	8	5	13	28	4	1	3	4	2

Traded to **Medicine Hat** (WHL) by **Kelowna** (WHL) for future considerations, October 3, 2000. Signed as a free agent by **Arkansas** (ECHL), March 31, 2001. Traded to **Roanoke** (ECHL) by **Arkansas** (ECHL) for Calvin Elfring, August 11, 2001. Signed as a free agent by **Nashville**, May 6, 2002.

ECHL All-Rookie Team (2002)

FIDLER, Jonas — (FIHD-luhr, YOH-nahsh) — S.J.

Right wing. Shoots right. 6'2", 173 lbs. Born, Jihlava, Czech., May 29, 1984.
(San Jose's 3rd choice, 86th overall, in 2002 Entry Draft).

			Regular Season					Playoffs				
Season	Club	Lea	GP	G	A	TP	PIM	GP	G	A	TP	PIM
99-2000	Dukla Jihlava Jr.	Czech-Jr.	48	11	11	22	48					
2000-01	Dukla Jihlava Jr.	Czech-Jr.	44	29	33	62	167					
2001-02	Plymouth Whalers	OHL	68	8	12	20	27	6	0	1	1	4

FILIPOWICZ, Jayme — (fihl-ih-POW-its, JAY-mee)

Defense. Shoots left. 6'2", 215 lbs. Born, Arlington Heights, IL, June 15, 1976.

			Regular Season					Playoffs				
Season	Club	Lea	GP	G	A	TP	PIM	GP	G	A	TP	PIM
1994/96	Dubuque	USHL	127	18	55	73	....					
1996-97	New Hampshire	H-East	35	3	16	19	43					
1997-98	New Hampshire	H-East	38	3	28	31	47					
1998-99	New Hampshire	H-East	41	8	30	38	56					
99-2000	Milwaukee	IHL	76	9	23	32	118	3	0	1	1	0
2000-01	Milwaukee	IHL	68	0	13	13	101	2	0	0	0	2
2001-02	Quebec	AHL	63	0	7	7	107	1	0	0	0	2

• Statistics for **Dubuque** (USHL) are career totals for 1994-1996 seasons. • Hockey East First All-Star Team (1999) • NCAA East Second All-American Team (1999) • NCAA Championship All-Tournament Team (1999)

Signed as a free agent by **Nashville**, June 17, 1999.

FILPPULA, Valtteri — (FIHL-poo-luh, VAL-tuhr-ee) — DET.

Center. Shoots left. 5'11", 172 lbs. Born, Vantaa, Finland, March 20, 1984.
(Detroit's 3rd choice, 95th overall, in 2002 Entry Draft).

			Regular Season					Playoffs				
Season	Club	Lea	GP	G	A	TP	PIM	GP	G	A	TP	PIM
2000-01	Jokerit-B	Finn-Jr.	31	18	29	47	4					
	Jokerit Jr.	Finn-Jr.	1	0	1	1	0					
2001-02	Jokerit-B	Finn-Jr.	1	0	1	1	0					
	Jokerit Jr.	Finn-Jr.	40	8	15	23	14	9	4	9	13	2

FINGER, Jeff — (FIHN-guhr, JEHF) — COL.

Defense. Shoots left. 6'1", 195 lbs. Born, Hancock, MI, December 18, 1979.
(Colorado's 11th choice, 240th overall, in 1999 Entry Draft).

			Regular Season					Playoffs				
Season	Club	Lea	GP	G	A	TP	PIM	GP	G	A	TP	PIM
1997-98	Green Bay	USHL	51	5	9	14	208	4	0	0	0	18
1998-99	Green Bay	USHL	54	11	28	39	199	6	0	3	3	14
99-2000	Green Bay	USHL	55	13	35	48	15	5	1	6	7	22
2000-01	St. Cloud State	WCHA	41	4	5	9	84					
2001-02	St. Cloud State	WCHA	42	6	20	26	105					

FINNSTROM, Johan (FIHN-struhm, YOH-hahn) CGY.

Defense. Shoots left. 6'3", 205 lbs. Born, Broby, Sweden, March 27, 1976.
(Calgary's 5th choice, 97th overall, in 1994 Entry Draft).

			Regular Season					Playoffs				
Season	Club	Lea	GP	G	A	TP	PIM	GP	G	A	TP	PIM
1992-93	Rogle-18	Swede-Jr.	16	1	4	5	18	....	....	....	....	....
1993-94	Rogle	Sweden	7	1	1	2	4	1	0	0	0	0
1994-95	Rogle Jr.	Swede-Jr.	4	0	0	0	12	....	....	....	....	....
	Rogle	Sweden	19	0	0	0	10	....	....	....	....	....
	Rogle	Swede-Q	17	1	2	3	8	11	2	1	3	12
1995-96	Rogle	Sweden	18	0	0	0	10	....	....	....	....	....
	Rogle	Swede-Q	15	1	3	4	12	12	1	1	2	12
1996-97	Rogle	Swede-2	31	1	5	6	59	....	....	....	....	....
1997-98	Lulea HF	Sweden	45	0	1	1	17	3	0	0	0	0
	Lulea HF	EuroHL	6	0	0	0	4	....	....	....	....	....
1998-99	Lulea HF	Sweden	49	1	8	9	63	9	0	4	4	12
99-2000	Lulea HF	Sweden	46	1	4	5	71	9	0	0	0	4
2000-01	Lulea HF	Sweden	46	1	6	7	40	12	0	2	2	12
2001-02	Lulea HF	Sweden	46	1	7	8	46	4	0	1	1	2

FITZRANDOLPH, Colin (fihts-RAN-dawlf, KAW-lihn) ATL.

Center. Shoots left. 6'3", 195 lbs. Born, Potsdam, NY, April 23, 1982.
(Atlanta's 8th choice, 201st overall, in 2001 Entry Draft).

			Regular Season					Playoffs				
Season	Club	Lea	GP	G	A	TP	PIM	GP	G	A	TP	PIM
2000-01	Phillips Exeter	Hi-School	23	7	33	40	....	....	....	....	....	....
2001-02	St. Lawrence	ECAC	13	0	3	3	21	....	....	....	....	....

FLACHE, Paul (FLAK, PAWL) ATL.

Defense. Shoots right. 6'5", 215 lbs. Born, Toronto, Ont., March 4, 1982.
(Atlanta's 5th choice, 144th overall, in 2002 Entry Draft).

			Regular Season					Playoffs				
Season	Club	Lea	GP	G	A	TP	PIM	GP	G	A	TP	PIM
1998-99	Cobourg Cougars	OPJHL	41	1	6	7	50	....	....	....	....	....
99-2000	Brampton	OHL	54	1	0	1	59	6	0	0	0	8
2000-01	Brampton	OHL	68	8	16	24	100	9	1	1	2	18
2001-02	Brampton	OHL	68	9	35	44	148	....	....	....	....	....

• Re-entered NHL Entry Draft. Originally Edmonton's 5th choice, 152nd overall, in 2000 Entry Draft.

FLEISCHMANN, Tomas (FLIGHSH-muhn, TAW-mash) DET.

Left wing. Shoots left. 6', 165 lbs. Born, Koprivnice, Czech., May 16, 1984.
(Detroit's 2nd choice, 63rd overall, in 2002 Entry Draft).

			Regular Season					Playoffs				
Season	Club	Lea	GP	G	A	TP	PIM	GP	G	A	TP	PIM
99-2000	HC Vitkovice Jr.	Czech-Jr.	46	9	13	22	6	....	....	....	....	....
2000-01	HC Vitkovice-18	Czech-Jr.	30	28	34	62	8	....	....	....	....	....
	HC Vitkovice Jr.	Czech-Jr.	21	4	9	13	8	....	....	....	....	....
2001-02	HC Vitkovice Jr.	Czech-Jr.	46	26	35	51	16	....	....	....	....	....
	TJ Novy Jicin	Czech-3	8	3	2	5	8	7	3	4	7	35

FOLEY, Patrick (FOH-lee, PAT-rihk) PIT.

Left wing. Shoots left. 6', 216 lbs. Born, Boston, MA, January 24, 1981.
(Pittsburgh's 6th choice, 185th overall, in 2000 Entry Draft).

			Regular Season					Playoffs				
Season	Club	Lea	GP	G	A	TP	PIM	GP	G	A	TP	PIM
1996-97	St. Sebastian's	Hi-School	23	11	12	23	....	....	....	....	....	....
1997-98	St. Sebastian's	Hi-School	25	17	24	41	....	....	....	....	....	....
	Team USA	USDP-18	8	3	3	6	8	....	....	....	....	....
1998-99	Team USA	USDP-17	52	7	9	16	146	....	....	....	....	....
99-2000	New Hampshire	H-East	30	3	7	10	61	....	....	....	....	....
2000-01	New Hampshire	H-East		DID NOT PLAY – INJURED								
2001-02	New Hampshire	H-East	35	8	4	12	58	....	....	....	....	....

• Missed entire 2000-01 season recovering from head injury originally suffered in game vs. U. Mass-Lowell (H-East), February 4, 2000.

FORBES, Ian (FOHRBZ, EE-an) PHI.

Defense. Shoots left. 6'6", 215 lbs. Born, Brampton, Ont., August 2, 1980.
(Philadelphia's 3rd choice, 51st overall, in 1998 Entry Draft).

			Regular Season					Playoffs				
Season	Club	Lea	GP	G	A	TP	PIM	GP	G	A	TP	PIM
1996-97	Mississauga Reps	MTHL	39	10	32	42	178	....	....	....	....	....
1997-98	Guelph Fire	OJHL-B	3	0	1	1	19	....	....	....	....	....
	Guelph Storm	OHL	61	2	3	5	164	12	0	0	0	16
1998-99	Guelph Storm	OHL	60	1	8	9	182	5	0	1	1	8
99-2000	Guelph Storm	OHL	62	2	7	9	143	6	0	0	0	11
2000-01	Trenton Titans	ECHL	33	0	2	2	133	....	....	....	....	....
2001-02	Trenton Titans	ECHL	42	0	5	5	204	....	....	....	....	....
	Philadelphia	AHL	18	1	0	1	50	2	0	0	0	0

FORREST, J.D. (FOH-rehst, JAY-DEE) CAR.

Defense. Shoots left. 5'8", 167 lbs. Born, Westchester, NY, April 15, 1981.
(Carolina's 5th choice, 181st overall, in 2000 Entry Draft).

			Regular Season					Playoffs				
Season	Club	Lea	GP	G	A	TP	PIM	GP	G	A	TP	PIM
1997-98	Team USA	USDP-18	74	7	26	33	41	....	....	....	....	....
1998-99	Team USA	USDP-17	2	1	0	1	4	....	....	....	....	....
	Team USA	USDP-18	48	5	21	26	34	....	....	....	....	....
99-2000	Team USA	USDP-17	49	6	28	34	36	....	....	....	....	....
	Team USA	USDP-18	8	0	0	0	2	....	....	....	....	....
2000-01	Boston College	H-East	38	6	17	23	40	....	....	....	....	....
2001-02	Boston College	H-East	35	8	19	27	28	....	....	....	....	....

NAJHL All-League First All-Star Team (2000)

FORSANDER, Johan (fohr-SAHN-duhr, YOH-hahn) DET.

Left wing. Shoots left. 6'1", 174 lbs. Born, Jonkoping, Sweden, April 28, 1978.
(Detroit's 3rd choice, 108th overall, in 1996 Entry Draft).

			Regular Season					Playoffs				
Season	Club	Lea	GP	G	A	TP	PIM	GP	G	A	TP	PIM
1994-95	HV 71 Jr.	Swede-Jr.	25	2	2	4	6	....	....	....	....	....
1995-96	HV 71 Jr.	Swede-Jr.	27	15	8	23	12	....	....	....	....	....
	HV 71 Jonkoping	Sweden	6	0	0	0	0	3	0	0	0	2
1996-97	HV 71 Jonkoping	Sweden	44	3	2	5	6	5	0	0	0	0
1997-98	HV 71 Jonkoping	Sweden	46	3	2	5	12	5	0	0	0	0
1998-99	HV 71 Jonkoping	Sweden	48	5	4	9	6	....	....	....	....	....
99-2000	HV 71 Jonkoping	Sweden	48	9	9	18	6	6	0	0	0	4
2000-01	HV 71 Jonkoping	Sweden		DID NOT PLAY – INJURED								
2001-02	Djurgarden	Sweden	48	5	6	11	16	5	0	0	0	4

• Missed entire 2000-01 season recovering from foot injury suffered in training camp, September 2, 2000.

FORSTER, Beat (FOHRS-tuhr, BEE-at) PHX.

Defense. Shoots left. 6'1", 209 lbs. Born, Herisau, Switz., February 2, 1983.
(Phoenix's 4th choice, 78th overall, in 2001 Entry Draft).

			Regular Season					Playoffs				
Season	Club	Lea	GP	G	A	TP	PIM	GP	G	A	TP	PIM
99-2000	HC Davos Jr.	Swiss-Jr.	34	4	15	19	40	6	1	0	1	6
2000-01	HC Davos Jr.	Swiss-Jr.	27	6	7	13	44	....	....	....	....	....
	SC Hersiau	Swiss-2	3	0	0	0	16	....	....	....	....	....
	HC Davos	Swiss	7	0	0	0	6	3	0	0	0	2
2001-02	HC Davos	Swiss-Jr.	4	2	2	4	12	....	....	....	....	....
	HC Davos	Swiss	33	1	3	4	51	16	0	1	1	2

FORSTER, Nate (FOHRS-tuhr, NAY-than) WSH.

Defense. Shoots right. 6'1", 195 lbs. Born, Vancouver, B.C., July 3, 1980.
(Washington's 7th choice, 179th overall, in 1998 Entry Draft).

			Regular Season					Playoffs				
Season	Club	Lea	GP	G	A	TP	PIM	GP	G	A	TP	PIM
1995-96	Victoria Lions	BCAHA	49	25	48	73	122	....	....	....	....	....
1996-97	Seattle	WHL	22	1	2	3	39	4	0	0	0	0
1997-98	Seattle	WHL	68	1	12	13	153	5	0	1	1	8
1998-99	Seattle	WHL	65	5	17	22	153	10	0	2	2	26
99-2000	Seattle	WHL	51	9	28	37	104	7	1	4	5	30
2000-01	Richmond	ECHL	30	2	7	9	55	....	....	....	....	....
	Portland Pirates	AHL	27	0	2	2	25	3	0	0	0	0
2001-02	Portland Pirates	AHL	27	1	1	2	26	....	....	....	....	....
	Richmond	ECHL	24	3	6	9	60	....	....	....	....	....

FOSTER, Adrian (FAW-stuhr, AY-dree-uhn) N.J.

Center. Shoots left. 6'1", 200 lbs. Born, Lethbridge, Alta., January 15, 1982.
(New Jersey's 1st choice, 28th overall, in 2001 Entry Draft).

			Regular Season					Playoffs				
Season	Club	Lea	GP	G	A	TP	PIM	GP	G	A	TP	PIM
1997-98	Cgy. Buffaloes	AMHL	36	26	54	80	50	9	3	14	17	18
1998-99	Calgary Canucks	AJHL	18	15	17	32	18	....	....	....	....	....
99-2000	Saskatoon Blades	WHL	7	1	2	3	6	....	....	....	....	....
2000-01	Saskatoon Blades	WHL	5	0	5	5	4	....	....	....	....	....
2001-02	Saskatoon Blades	WHL	13	9	3	12	18	....	....	....	....	....
	Brandon	WHL	14	5	10	15	23	15	4	11	15	14

• Missed majority of 1998-99 season recovering from ankle injury. • Missed majority of 1999-2000, 2000-01 and 2001-02 seasons recovering from abdominal injury, October, 1999. Traded to **Brandon** (WHL) by **Saskatoon** (WHL) for Richard Meuller, January 14, 2002.

FOSTER, Kurtis (FAW-stuhr, KUHR-this) ATL.

Defense. Shoots right. 6'5", 230 lbs. Born, Carp, Ont., November 24, 1981.
(Calgary's 2nd choice, 40th overall, in 2000 Entry Draft).

			Regular Season					Playoffs				
Season	Club	Lea	GP	G	A	TP	PIM	GP	G	A	TP	PIM
1997-98	Ottawa Valley	OMHA	36	7	18	25	88	....	....	....	....	....
	Peterborough	OHL	39	1	1	2	45	4	0	0	0	2
1998-99	Peterborough	OHL	54	2	13	15	59	5	0	0	0	6
99-2000	Peterborough	OHL	68	6	18	24	116	5	1	2	3	4
2000-01	Peterborough	OHL	62	17	24	41	78	7	1	1	2	10
2001-02	Peterborough	OHL	33	10	4	14	58	....	....	....	....	....
	Chicago Wolves	AHL	39	6	9	15	59	14	1	1	2	21

Rights traded to **Atlanta** by **Calgary** with Jeff Cowan for Petr Buzek, December 18, 2001.

FRIED, Robert (FREED, RAW-buhrt) FLA.

Right wing. Shoots right. 6'4", 210 lbs. Born, Philadelphia, PA, March 8, 1981.
(Florida's 2nd choice, 77th overall, in 2000 Entry Draft).

			Regular Season					Playoffs				
Season	Club	Lea	GP	G	A	TP	PIM	GP	G	A	TP	PIM
1998-99	Deerfield	Hi-School	24	13	20	33	28	....	....	....	....	....
99-2000	Deerfield	Hi-School	26	25	20	45	35	....	....	....	....	....
2000-01	Harvard Crimson	ECAC	30	4	1	5	22	....	....	....	....	....
2001-02	Harvard Crimson	ECAC	31	7	5	12	16	....	....	....	....	....

FROGREN, Jonas (FREW-grehn, YOH-nuhs) CGY.

Defense. Shoots left. 6'1", 190 lbs. Born, Falun, Sweden, August 28, 1980.
(Calgary's 8th choice, 206th overall, in 1998 Entry Draft).

			Regular Season					Playoffs				
Season	Club	Lea	GP	G	A	TP	PIM	GP	G	A	TP	PIM
1996-97	Farjestad Jr.	Swede-Jr.	20	2	7	9	4	....	....	....	....	....
1997-98	Farjestad Jr.	Swede-Jr.	28	5	6	11	12	2	1	0	1	0
1998-99	Farjestad Jr.	Swede-Jr.	14	1	5	6	12	....	....	....	....	....
	Farjestad	Sweden	22	0	0	0	2	....	....	....	....	....
99-2000	Bofors IK	Swede-2	43	2	7	9	40	....	....	....	....	....
2000-01	Farjestad	Sweden	49	3	0	3	10	16	0	0	0	4
2001-02	Farjestad	Sweden	50	3	6	9	18	10	0	0	0	4

FROLOV, Alexander (froh-LAHF, al-ehx-AN-duhr) L.A.

Left wing. Shoots right. 6'4", 191 lbs. Born, Moscow, USSR, June 19, 1982.
(Los Angeles' 1st choice, 20th overall, in 2000 Entry Draft).

			Regular Season					Playoffs				
Season	Club	Lea	GP	G	A	TP	PIM	GP	G	A	TP	PIM
1998-99	Spartak Moscow	Russia	1	0	0	0	0	....	....	....	....	....
99-2000	Yaroslavl 2	Russia-3	36	27	13	40	30	....	....	....	....	....
2000-01	Spartak Moscow	Russia-2	44	20	19	39	8	....	....	....	....	....
2001-02	Krylja Sovetov 2	Russia-3	2	0	0	0	4	....	....	....	....	....
	Krylja Sovetov	Russia	43	18	12	30	16	3	1	0	1	0

FUSSEY, Owen (FOO-see, OH-when) WSH.

Right wing. Shoots left. 6', 185 lbs. Born, Winnipeg, Man., April 2, 1983.
(Washington's 2nd choice, 90th overall, in 2001 Entry Draft).

			Regular Season					Playoffs				
Season	Club	Lea	GP	G	A	TP	PIM	GP	G	A	TP	PIM
1998-99	Wpg. Warriors	MMHL	40	38	33	71	24	....	....	....	....	....
99-2000	Calgary Hitmen	WHL	51	7	6	13	35	12	3	4	7	2
2000-01	Calgary Hitmen	WHL	48	15	10	25	33	12	2	1	3	6
2001-02	Calgary Hitmen	WHL	72	43	27	70	61	7	3	1	4	4

GABINET, Mike (GA-bihn-AY, MIGHK) L.A.

Defense. Shoots left. 6'3", 180 lbs. Born, Edmonton, Alta., September 26, 1981.
(Los Angeles' 10th choice, 237th overall, in 2001 Entry Draft).

			Regular Season					Playoffs				
Season	Club	Lea	GP	G	A	TP	PIM	GP	G	A	TP	PIM
1998-99	Edmonton Leafs	AMHL	33	1	15	16	34	....	....	....	....	....
99-2000	Lloydminster	AJHL	56	4	25	29	30	9	2	4	6	4
2000-01	Nebraska-Omaha	CCHA	30	2	13	15	14	....	....	....	....	....
2001-02	Nebraska-Omaha	CCHA	21	0	4	4	8	....	....	....	....	....

AJHL North First All-Star Team (2000)

GAGNON, Jonathan (GAN-YAW, JAWN-ah-thuhn) **TOR.**

Center. Shoots left. 6'1", 190 lbs. Born, Montreal, Que., May 20, 1980.
(Toronto's 7th choice, 181st overall, in 1998 Entry Draft).

			Regular Season					Playoffs				
Season	**Club**	**Lea**	**GP**	**G**	**A**	**TP**	**PIM**	**GP**	**G**	**A**	**TP**	**PIM**
1996-97	Val-d'Or Foreurs	QMJHL	65	5	15	20	35	13	1	0	1	2
1997-98	Val-d'Or Foreurs	QMJHL	40	9	19	28	54	....	....	....	....	....
	Cape Breton	QMJHL	29	6	7	13	25	4	2	3	5	12
1998-99	Cape Breton	QMJHL	68	27	37	64	39	5	2	4	6	2
99-2000	Cape Breton	QMJHL	26	8	12	20	25	....	....	....	....	....
	Halifax	QMJHL	19	11	4	15	37	....	....	....	....	....
	Drummondville	QMJHL	27	16	21	37	38	14	4	12	16	8
2000-01	Pensacola	ECHL	70	24	15	39	52	....	....	....	....	....
2001-02	Memphis	CHL	54	34	26	60	46	....	....	....	....	....
	St. John's	AHL	12	0	3	3	4	8	0	0	0	0

Traded to **Halifax** (QMJHL) by **Cape Breton** (QMJHL) with Robbie Sutherland for Frederic Belanger and Jonathan Andrews, November 12, 1999. Traded to **Drummondville** (QMJHL) by **Halifax** (QMJHL) with Eric Dubois for Jonathan St-Louis, January 6, 2000.

GAJIC, Milan (GAY-jihk, MEE-lan) **ATL.**

Center. Shoots right. 5'11", 182 lbs. Born, Vancouver, B.C., June 1, 1981.
(Atlanta's 4th choice, 112th overall, in 2001 Entry Draft).

			Regular Season					Playoffs				
Season	**Club**	**Lea**	**GP**	**G**	**A**	**TP**	**PIM**	**GP**	**G**	**A**	**TP**	**PIM**
1997-98	Merritt	BCHL	51	6	14	20	....	....	....	....	....	....
1998-99	Burnaby Bulldogs	BCHL	56	29	30	59	41	....	....	....	....	....
99-2000	Burnaby Bulldogs	BCHL	56	34	47	81	42	....	....	....	....	....
2000-01	Burnaby Bulldogs	BCHL	50	46	52	98	84	....	....	....	....	....
2001-02	U. of Michigan	CCHA	39	9	13	22	22	....	....	....	....	....

GAMACHE, Simon (ga-MOHSH, see-MOHN) **ATL.**

Center. Shoots left. 5'9", 185 lbs. Born, Montreal, Que., January 3, 1981.
(Atlanta's 14th choice, 290th overall, in 2000 Entry Draft).

			Regular Season					Playoffs				
Season	**Club**	**Lea**	**GP**	**G**	**A**	**TP**	**PIM**	**GP**	**G**	**A**	**TP**	**PIM**
1997-98	Levis-Lauzon	QAAA	42	28	26	54	....	4	1	1	2	....
1998-99	Val-d'Or Foreurs	QMJHL	70	19	43	62	54	6	1	2	3	4
99-2000	Val-d'Or Foreurs	QMJHL	72	64	79	143	74	....	....	....	....	....
2000-01	Val-d'Or Foreurs	QMJHL	72	*74	*110	*184	70	21	*22	*35	*57	18
2001-02	Chicago Wolves	AHL	26	2	4	6	11	....	....	....	....	....
	Greenville	ECHL	31	19	19	38	35	17	*15	9	*24	22

Canadian Major Junior Second All-Star Team (2000) • QMJHL First All-Star Team (2001) • Won Michel Briere Trophy (MVP - QMJHL) (2001) • Canadian Major Junior First All-Star Team (2001) • Canadian Major Junior Player of the Year (2001) • Won Sheetrock CHL Top Scorer Award (2001) • Memorial Cup All-Star Team (2001) • Won Ed Chynoweth Trophy (Memorial Cup Tournament Leading Scorer) (2001) • ECHL All-Rookie Team (2002) • ECHL Playoff MVP (2002) - tied with Tyrone Garner

GAMALEI, Yevgeny (ga-mah-LAY, yehv-GEH-nee) **N.J.**

Defense. Shoots left. 6'2", 191 lbs. Born, Surgut, USSR, August 27, 1982.
(New Jersey's 11th choice, 257th overall, in 2001 Entry Draft).

			Regular Season					Playoffs				
Season	**Club**	**Lea**	**GP**	**G**	**A**	**TP**	**PIM**	**GP**	**G**	**A**	**TP**	**PIM**
99-2000	Omsk 2	Russia-3	13	1	1	2	12	....	....	....	....	....
2000-01	Voskresensk	Russia-2	8	0	0	0	4	....	....	....	....	....
2001-02	Voskresensk 2	Russia-3	33	3	7	10	72	....	....	....	....	....
	Voskresensk	Russia-2	11	0	0	0	6	....	....	....	....	....

GAUSTAD, Paul (GAW-stad, PAWL) **BUF.**

Center/Left wing. Shoots left. 6'4", 217 lbs. Born, Fargo, ND, February 3, 1982.
(Buffalo's 6th choice, 220th overall, in 2000 Entry Draft).

			Regular Season					Playoffs				
Season	**Club**	**Lea**	**GP**	**G**	**A**	**TP**	**PIM**	**GP**	**G**	**A**	**TP**	**PIM**
1998-99	Portland Hawks	USAHA	45	47	53	100	81	....	....	....	....	....
99-2000	Portland	WHL	56	6	8	14	110	....	....	....	....	....
2000-01	Portland	WHL	70	11	30	41	168	16	10	6	16	59
2001-02	Portland	WHL	72	36	44	80	202	6	3	1	4	16

GAUVREAU, Brent (GAWV-roh, BREHNT) **PHX.**

Right wing. Shoots right. 6'3", 196 lbs. Born, Sudbury, Ont., June 29, 1980.
(Phoenix's 5th choice, 186th overall, in 2000 Entry Draft).

			Regular Season					Playoffs				
Season	**Club**	**Lea**	**GP**	**G**	**A**	**TP**	**PIM**	**GP**	**G**	**A**	**TP**	**PIM**
1995-96	Sudbury Cubs	NOHA	61	43	49	92	68	....	....	....	....	....
1996-97	Oshawa Generals	OHL	59	8	13	21	13	18	1	5	6	2
1997-98	Oshawa Generals	OHL	66	25	42	67	39	7	3	1	4	2
1998-99	Oshawa Generals	OHL	68	33	62	95	57	15	9	11	20	15
99-2000	Oshawa Generals	OHL	59	34	53	87	74	5	1	2	3	4
	Saint John	AHL	2	0	0	0	0	....	....	....	....	....
2000-01	Mississippi	ECHL	67	11	21	32	63	....	....	....	....	....
	Springfield	AHL	1	0	0	0	0	....	....	....	....	....
2001-02	Mississippi	ECHL	66	29	43	72	89	10	6	9	15	6
	Springfield	AHL	3	0	0	0	0	....	....	....	....	....

• Re-entered NHL Entry Draft. Originally Calgary's 6th choice, 120th overall, in 1998 Entry Draft.

GELLARD, Mike (GEHL-ahrd, MIGHK) **BOS.**

Left wing. Shoots left. 6'1", 193 lbs. Born, Markham, Ont., October 10, 1978.

			Regular Season					Playoffs				
Season	**Club**	**Lea**	**GP**	**G**	**A**	**TP**	**PIM**	**GP**	**G**	**A**	**TP**	**PIM**
1995-96	Thornhill	MTJHL	49	15	24	39	2	16	6	9	15	0
1996-97	Thornhill	MTJHL	43	29	37	66	4	9	3	10	13	0
1997-98	St. Lawrence	ECAC	31	4	6	10	18	....	....	....	....	....
1998-99	St. Lawrence	ECAC	39	10	11	21	22	....	....	....	....	....
99-2000	St. Lawrence	ECAC	36	14	22	36	36	....	....	....	....	....
2000-01	St. Lawrence	ECAC	37	19	*38	*57	14	....	....	....	....	....
2001-02			DID NOT PLAY									

MTJHL Metro All-Star Team (1997) • MTJHL Central Sportsmanlike Player of the Year (1997) • ECAC First All-Star Team (2001)

Signed as a free agent by **Boston**, August 2, 2001. • Missed entire 2001-02 season recovering from viral infection diagnosed during training camp, September 20, 2001.

GENOVY, Jeff (jeh-NOH-vee, JEHF) **CBJ**

Center. Shoots left. 6'3", 191 lbs. Born, Kalamazoo, MI, December 4, 1982.
(Columbus' 4th choice, 96th overall, in 2002 Entry Draft).

			Regular Season					Playoffs				
Season	**Club**	**Lea**	**GP**	**G**	**A**	**TP**	**PIM**	**GP**	**G**	**A**	**TP**	**PIM**
99-2000	West-Michigan	MMHL		STATISTICS NOT AVAILABLE								
	Det. Compuware	NAJHL	9	0	3	3	2	2	0	0	0	0
2000-01	Sault Ste. Marie	NAJHL	54	12	14	26	26	8	0	1	1	4
2001-02	Des Moines	USHL	52	23	23	46	86	3	1	4	5	2

• Signed Letter of Intent to attend **Clarkson University** (ECAC), January 29, 2002.

GILBERT, Tom (GIHL-buhrt, TAWM) **COL.**

Defense. Shoots right. 6'2", 190 lbs. Born, Minneapolis, MN, January 10, 1983.
(Colorado's 5th choice, 129th overall, in 2002 Entry Draft).

			Regular Season					Playoffs				
Season	**Club**	**Lea**	**GP**	**G**	**A**	**TP**	**PIM**	**GP**	**G**	**A**	**TP**	**PIM**
2000-01	Bloomington-Jeff.	Hi-School	23	20	18	38	....	....	....	....	....	....
2001-02	Chicago Steel	USHL	57	13	15	28	62	4	0	0	0	4

• Signed Letter of Intent to attend **U. of Wisconsin** (WCHA), November 15, 2001.

GIROUX, Alexandre (ZHIH-roo, al-ehx-AN-dreh) **OTT.**

Center. Shoots left. 6'3", 182 lbs. Born, Quebec, Que., June 16, 1981.
(Ottawa's 9th choice, 213th overall, in 1999 Entry Draft).

			Regular Season					Playoffs				
Season	**Club**	**Lea**	**GP**	**G**	**A**	**TP**	**PIM**	**GP**	**G**	**A**	**TP**	**PIM**
1997-98	Ste-Foy	QAAA	42	28	30	58	96	....	....	....	....	....
1998-99	Hull Olympiques	QMJHL	67	15	22	37	124	22	2	2	4	8
99-2000	Hull Olympiques	QMJHL	72	52	47	99	117	15	12	6	18	30
2000-01	Hull Olympiques	QMJHL	38	31	32	63	62	....	....	....	....	....
	Rouyn-Noranda	QMJHL	25	13	14	27	56	9	2	6	8	22
2001-02	Grand Rapids	AHL	70	11	16	27	74	....	....	....	....	....

Traded to **Rouyn-Noranda** (QMJHL) by **Hull** (QMJHL) for Maxime Talbot, Dominic D'amour and Rouyn-Noranda's 1st choice (Charles Fontaine) in 2001 QMJHL Priority Draft, January 8, 2001.

GLADSKIKH, Evgeny (glad-SKEEKH, ehv-GEH-nee) **VAN.**

Right wing. Shoots left. 6', 176 lbs. Born, Magnitogorsk, USSR, April 24, 1982.
(Vancouver's 3rd choice, 114th overall, in 2001 Entry Draft).

			Regular Season					Playoffs				
Season	**Club**	**Lea**	**GP**	**G**	**A**	**TP**	**PIM**	**GP**	**G**	**A**	**TP**	**PIM**
1998-99	Magnitogorsk 2	Russia-4	16	3	3	6	6	....	....	....	....	....
99-2000	Magnitogorsk 2	Russia-3	39	17	2	19	24	....	....	....	....	....
	Magnitogorsk	Russia	1	0	0	0	0	....	....	....	....	....
2000-01	Magnitogorsk 2	Russia-3	11	10	7	17	6	....	....	....	....	....
	Magnitogorsk	Russia	31	3	5	8	10	12	0	2	2	2
2001-02	Magnitogorsk	Russia	32	5	6	11	6	4	0	0	0	4

GLEASON, Tim (GLEE-suhn, TIHM) **OTT.**

Defense. Shoots left. 6'1", 202 lbs. Born, Southfield, MI, January 29, 1983.
(Ottawa's 2nd choice, 23rd overall, in 2001 Entry Draft).

			Regular Season					Playoffs				
Season	**Club**	**Lea**	**GP**	**G**	**A**	**TP**	**PIM**	**GP**	**G**	**A**	**TP**	**PIM**
1998-99	Leamington	OJHL-B	52	5	26	31	76	....	....	....	....	....
99-2000	Windsor	OHL	55	5	13	18	101	12	2	4	6	14
2000-01	Windsor	OHL	47	8	28	36	124	9	1	2	3	23
2001-02	Windsor	OHL	67	17	42	59	109	16	7	13	20	40

GLENN, Ryan (GLEHN, RIGH-unh) **MTL.**

Defense. Shoots left. 6'3", 210 lbs. Born, New Westminster, B.C., June 7, 1980.
(Montreal's 7th choice, 145th overall, in 2000 Entry Draft).

			Regular Season					Playoffs				
Season	**Club**	**Lea**	**GP**	**G**	**A**	**TP**	**PIM**	**GP**	**G**	**A**	**TP**	**PIM**
99-2000	Walpole Jr.	EJHL	42	19	40	59	54	8	4	13	17	....
2000-01	St. Lawrence	ECAC	37	3	1	4	34	....	....	....	....	....
2001-02	St. Lawrence	ECAC	32	3	6	9	27	....	....	....	....	....

EJHL First All-Star Team (2000) • EJHL Defencemen of the Year (2000)

GLOBKE, Rob (GLAWB-kee, RAWB) **FLA.**

Center. Shoots right. 6'2", 200 lbs. Born, Farmington, MI, October 24, 1982.
(Florida's 3rd choice, 40th overall, in 2002 Entry Draft).

			Regular Season					Playoffs				
Season	**Club**	**Lea**	**GP**	**G**	**A**	**TP**	**PIM**	**GP**	**G**	**A**	**TP**	**PIM**
1998-99	Det. Compuware	NAJHL	55	8	14	22	111	7	1	2	3	2
99-2000	Team USA	USDP-18	54	15	21	36	68	....	....	....	....	....
2000-01	U. of Notre Dame	CCHA	33	17	9	26	74	....	....	....	....	....
2001-02	U. of Notre Dame	CCHA	33	11	11	22	79	....	....	....	....	....

GOC, Marcel (GAWCH, mahr-SEHL) **S.J.**

Center. Shoots left. 6', 189 lbs. Born, Calw, West Germany, August 24, 1983.
(San Jose's 1st choice, 20th overall, in 2001 Entry Draft).

			Regular Season					Playoffs				
Season	**Club**	**Lea**	**GP**	**G**	**A**	**TP**	**PIM**	**GP**	**G**	**A**	**TP**	**PIM**
1998-99	Schwenningen Jr.	Ger.-Jr.	12	23	10	33	12	....	....	....	....	....
99-2000	Schwenningen	Germany	51	0	3	3	4	11	1	1	2	2
2000-01	Schwenningen	Germany	58	13	28	41	12	....	....	....	....	....
2001-02	Schwenningen	Germany	45	8	9	17	24	....	....	....	....	....
	Adler Mannheim	Germany	8	0	2	2	0	....	....	....	....	....

GODARD, Eric (GAW-duhrd, AIR-ihk) **NYI**

Right wing. Shoots right. 6'4", 227 lbs. Born, Vernon, B.C., March 7, 1980.

			Regular Season					Playoffs				
Season	**Club**	**Lea**	**GP**	**G**	**A**	**TP**	**PIM**	**GP**	**G**	**A**	**TP**	**PIM**
1997-98	Lethbridge	WHL	7	0	0	0	26	2	0	0	0	0
1998-99	Lethbridge	WHL	66	2	5	7	213	4	0	0	0	14
99-2000	Lethbridge	WHL	60	3	5	8	*310	....	....	....	....	....
	Louisville	AHL	4	0	1	1	16	....	....	....	....	....
2000-01	Louisville	AHL	45	0	0	0	132	....	....	....	....	....
2001-02	Bridgeport	AHL	67	1	4	5	198	20	0	4	4	30

Signed as a free agent by **Florida**, September 24, 1999. Traded to **NY Islanders** by **Florida** for Florida's 3rd round choice (previously acquired, Florida selected Gregory Campbell) in 2002 Entry Draft, June 22, 2002.

GOLOVIN, Alexander (goh-loh-VEEN, al-ehx-AN-duhr) **CHI.**

Left wing. Shoots right. 5'11", 194 lbs. Born, Ust-Kamenogorsk, USSR, March 26, 1983.
(Chicago's 9th choice, 174th overall, in 2001 Entry Draft).

			Regular Season					Playoffs				
Season	**Club**	**Lea**	**GP**	**G**	**A**	**TP**	**PIM**	**GP**	**G**	**A**	**TP**	**PIM**
1998-99	Omsk 2	Russia-4	6	2	4	6	4	....	....	....	....	....
99-2000	Omsk 2	Russia-3	24	12	14	26	10	....	....	....	....	....
2000-01	Omsk 2	Russia-3	40	22	34	56	12	....	....	....	....	....
2001-02	Mostovik Kurgan	Russia-2	56	16	23	39	20	....	....	....	....	....

GORBUNOV, Vladimir (gohr-buh-NAHF, vla-DIH-meer) **NYI**

Center. Shoots left. 6', 174 lbs. Born, Moscow, USSR, April 22, 1982.
(NY Islanders' 4th choice, 105th overall, in 2000 Entry Draft).

			Regular Season					Playoffs				
Season	**Club**	**Lea**	**GP**	**G**	**A**	**TP**	**PIM**	**GP**	**G**	**A**	**TP**	**PIM**
99-2000	H.C. CSKA	Russia-2	22	11	7	18	32	....	....	....	....	....
2000-01	H.C. CSKA	Russia-2	43	10	14	24	63	....	....	....	....	....
2001-02	H.C. CSKA	Russia-2	46	16	18	34	22	....	....	....	....	....
	CSKA Moscow 2	Russia-3	3	1	1	2	0	....	....	....	....	....

GORDON, Boyd (GOHR-duhn, BOYD) **WSH.**

Right wing. Shoots right. 6', 192 lbs. Born, Unity, Sask., October 19, 1983.
(Washington's 3rd choice, 17th overall, in 2002 Entry Draft).

			Regular Season					Playoffs				
Season	**Club**	**Lea**	**GP**	**G**	**A**	**TP**	**PIM**	**GP**	**G**	**A**	**TP**	**PIM**
1998-99	Regina Rangers	SMBHL	60	70	102	172	53	....	....	....	....	....
99-2000	Red Deer Rebels	WHL	66	10	26	36	24	4	0	1	1	16
2000-01	Red Deer Rebels	WHL	72	12	27	39	39	22	3	6	9	2
2001-02	Red Deer Rebels	WHL	66	22	29	51	19	23	10	12	22	8

GORNICK, Brian (GOHR-nihk, BRIGH-uhn) **ANA.**

Center. Shoots left. 6'4", 200 lbs. Born, St. Paul, MN, March 17, 1980.
(Anaheim's 7th choice, 258th overall, in 1999 Entry Draft).

			Regular Season					Playoffs				
Season	**Club**	**Lea**	**GP**	**G**	**A**	**TP**	**PIM**	**GP**	**G**	**A**	**TP**	**PIM**
1998-99	Air Force	CHA	34	10	11	21	20	....	....	....	....	....
99-2000	Air Force	CHA	39	13	25	38	26	....	....	....	....	....
2000-01	Air Force	CHA	36	16	17	33	18	....	....	....	....	....
2001-02	Air Force	CHA	21	6	5	11	14	....	....	....	....	....

CHA First All-Star Team (2001)

GOROVIKOV, Konstantin (goh-roh-vih-KAHF, kawn-stehn-TEEN) **OTT.**

Center. Shoots left. 5'11", 172 lbs. Born, Novosibirsk, USSR, August 31, 1977.
(Ottawa's 10th choice, 269th overall, in 1999 Entry Draft).

			Regular Season					Playoffs				
Season	**Club**	**Lea**	**GP**	**G**	**A**	**TP**	**PIM**	**GP**	**G**	**A**	**TP**	**PIM**
1994-95	St. Petersburg 2	CIS-2	38	6	5	11	22	....	....	....	....	....
	St. Petersburg	CIS	13	1	0	1	4	2	0	0	0	0
1995-96	St. Petersburg 2	CIS-2	3	2	0	2	0	....	....	....	....	....
	St. Petersburg	CIS	45	2	4	6	18	2	0	0	0	0
1996-97	St. Petersburg	Russia	37	4	2	6	20	....	....	....	....	....
1997-98	St. Petersburg	Russia	44	6	12	18	22	....	....	....	....	....
1998-99	St. Petersburg	Russia	42	12	7	19	14	....	....	....	....	....
99-2000	Grand Rapids	IHL	57	9	14	23	30	8	1	0	1	4
2000-01	Grand Rapids	IHL	68	7	19	26	48	....	....	....	....	....
2001-02	Ufa	Russia	51	13	18	31	32	....	....	....	....	....

GOSSELIN, Christian (gawz-LEH, KRIHST-an)

Defense. Shoots right. 6'5", 235 lbs. Born, Laval, Que., August 21, 1976.
(New Jersey's 5th choice, 129th overall, in 1994 Entry Draft).

			Regular Season					Playoffs				
Season	**Club**	**Lea**	**GP**	**G**	**A**	**TP**	**PIM**	**GP**	**G**	**A**	**TP**	**PIM**
1992-93	Hull Olympiques	QMJHL	49	1	3	4	24	....	....	....	....	....
1993-94	St-Hyacinthe	QMJHL	12	3	2	5	16	....	....	....	....	....
1994-95	St-Hyacinthe	QMJHL	60	5	10	15	202	5	0	0	0	11
1995-96	Laval Titan	QMJHL	21	1	8	9	69	....	....	....	....	....
1996-97	Macon Whoopee	CHL	63	8	10	18	229	5	0	0	0	29
1997-98	Pensacola	ECHL	42	6	5	11	181	18	0	1	1	52
	Fredericton	AHL	6	0	0	0	17	....	....	....	....	....
1998-99	Kentucky	AHL	31	1	1	2	107	....	....	....	....	....
99-2000	Kentucky	AHL	68	0	4	4	266	9	0	0	0	34
2000-01	Kentucky	AHL	42	2	3	5	145	3	0	1	1	6
2001-02	Charlotte	ECHL	16	0	1	1	49	....	....	....	....	....
	Hartford	AHL	39	0	0	0	164	4	0	0	0	17

Signed as a free agent by **San Jose**, July 15, 1998. Traded to **NY Rangers** by **San Jose** with Mikael Samuelsson for Adam Graves and future considerations, June 24, 2001. Signed as a free agent by **Bracknell** (Britain), August 19, 2002.

GRASBERG, Gustav (GRAHS-buhrg, GOO-stahv) **NSH.**

Center. Shoots left. 6', 193 lbs. Born, Furudal, Sweden, April 6, 1983.
(Nashville's 8th choice, 240th overall, in 2001 Entry Draft).

			Regular Season					Playoffs				
Season	**Club**	**Lea**	**GP**	**G**	**A**	**TP**	**PIM**	**GP**	**G**	**A**	**TP**	**PIM**
99-2000	Mora IK-18	Swede-Jr.	1	0	0	0	6	....	....	....	....	....
	Mora IK Jr.	Swede-Jr.	37	10	15	25	44	....	....	....	....	....
2000-01	Mora IK Jr.	Swede-Jr.	15	6	4	10	40	2	1	2	3	4
	Mora IK	Swede-2	12	1	0	1	4	....	....	....	....	....
2001-02	Mora IK Jr.	Swede-Jr.	1	0	0	0	8	....	....	....	....	....
	Mora IK	Swede-2	46	13	7	20	72	3	0	0	0	2

GREBESHKOV, Denis (greh-behsh-KAHV, DEH-nihs) **L.A.**

Defense. Shoots left. 6', 190 lbs. Born, Yaroslavl, USSR, October 11, 1983.
(Los Angeles' 1st choice, 18th overall, in 2002 Entry Draft).

			Regular Season					Playoffs				
Season	**Club**	**Lea**	**GP**	**G**	**A**	**TP**	**PIM**	**GP**	**G**	**A**	**TP**	**PIM**
99-2000	Yaroslavl 2	Russia-3	42	2	1	3	12	....	....	....	....	....
2000-01	Yaroslavl 2	Russia-3	34	7	2	9	20	....	....	....	....	....
2001-02	Yaroslavl 2	Russia-3	7	1	1	2	2	....	....	....	....	....
	Yaroslavl	Russia	26	1	2	3	10	....	....	....	....	....

GREEN, Mike (GREEN, MIGHK) **FLA.**

Center/Right wing. Shoots right. 5'11", 192 lbs. Born, Calgary, Alta., August 23, 1979.

			Regular Season					Playoffs				
Season	**Club**	**Lea**	**GP**	**G**	**A**	**TP**	**PIM**	**GP**	**G**	**A**	**TP**	**PIM**
1996-97	Cgy. North Stars	AMHL	35	34	27	61	78	....	....	....	....	....
	Edmonton Ice	WHL	7	0	2	2	0	....	....	....	....	....
1997-98	Edmonton Ice	WHL	71	15	26	41	16	....	....	....	....	....
1998-99	Kootenay Ice	WHL	71	35	45	80	37	7	2	2	4	4
99-2000	Kootenay Ice	WHL	69	43	49	92	63	21	9	16	25	20
2000-01	Port Huron	UHL	11	1	5	6	6	....	....	....	....	....
	Louisville	AHL	24	2	1	3	4	....	....	....	....	....
	Knoxville Speed	UHL	48	18	24	42	35	1	0	0	0	0
2001-02	Macon Whoopee	ECHL	54	27	35	62	18	....	....	....	....	....
	Cincinnati	AHL	22	2	9	11	4	3	0	0	0	0

WHL East Second All-Star Team (2000)

Signed as a free agent by **Florida**, April 7, 2000.

GREENE, Matt (GREEN, MAT) **EDM.**

Defense. Shoots right. 6'2", 210 lbs. Born, Grand Ledge, MI, May 13, 1983.
(Edmonton's 4th choice, 44th overall, in 2002 Entry Draft).

			Regular Season					Playoffs				
Season	**Club**	**Lea**	**GP**	**G**	**A**	**TP**	**PIM**	**GP**	**G**	**A**	**TP**	**PIM**
2000-01	Team USA	USDP-18	54	0	10	10	59	....	....	....	....	....
2001-02	Green Bay	USHL	55	4	20	24	150	7	0	1	1	31

USHL Second All-Star Team (2002)

• Signed Letter of Intent to attend **North Dakota** (WCHA), November 13, 2001.

GRIGORENKO, Igor (grih-goh-REHN-koh, EE-gohr) **DET.**

Right wing. Shoots right. 5'10", 178 lbs. Born, Togliatti, USSR, April 9, 1983.
(Detroit's 1st choice, 62nd overall, in 2001 Entry Draft).

			Regular Season					Playoffs				
Season	**Club**	**Lea**	**GP**	**G**	**A**	**TP**	**PIM**	**GP**	**G**	**A**	**TP**	**PIM**
1998-99	Lada Togliatti 2	Russia-4	19	3	3	6	2	....	....	....	....	....
99-2000	Lada Togliatti 2	Russia-3	38	17	18	35	36	....	....	....	....	....
2000-01	Lada Togliatti 2	Russia-3	6	5	4	9	....	....	....	....	....	....
	CSK VVS Samara	Russia-2	39	10	10	20	....	....	....	....	....	....
	Lada Togliatti	Russia	....	....	....	....	....	5	1	0	1	4
2001-02	Lada Togliatti	Russia	41	8	9	17	58	4	1	0	1	2

GROSCHL, Tamas (GROH-shuhl, TAW-mahsh) **EDM.**

Right wing. Shoots left. 6'2", 183 lbs. Born, Budapest, Hungary, August 21, 1980.
(Edmonton's 9th choice, 256th overall, in 1999 Entry Draft).

			Regular Season					Playoffs				
Season	**Club**	**Lea**	**GP**	**G**	**A**	**TP**	**PIM**	**GP**	**G**	**A**	**TP**	**PIM**
1995-96	Ujpesti	Hungary	20	9	3	12	0	....	....	....	....	....
1996-97	SE Dunaujvaros	Hungary	20	0	1	1	2	....	....	....	....	....
1997-98	Ujpesti	Hungary	20	3	3	6	29	....	....	....	....	....
1998-99	Ujpesti	Hungary	22	9	10	19	8	....	....	....	....	....
99-2000	Leksands IF Jr.	Swede-Jr.	33	15	14	29	46	....	....	....	....	....
	Leksands IF	Sweden	2	0	0	0	0	....	....	....	....	....
2000-01	Erfurt	German-2	28	0	1	1	52	....	....	....	....	....
	Augusta Lynx	ECHL	59	5	11	16	52	3	0	0	0	0
2001-02	Szekesfehervar	Hungary	12	9	8	17	4	....	....	....	....	....

GROT, Denis (GROHT, DEH-nihs) **VAN.**

Defense. Shoots left. 6', 185 lbs. Born, Minsk, USSR, June 1, 1984.
(Vancouver's 2nd choice, 55th overall, in 2002 Entry Draft).

			Regular Season					Playoffs				
Season	**Club**	**Lea**	**GP**	**G**	**A**	**TP**	**PIM**	**GP**	**G**	**A**	**TP**	**PIM**
2000-01	Yaroslavl 2	Russia-3	34	5	1	6	10	....	....	....	....	....
	Team Russia	Nat-Tm	5	0	2	2	8	....	....	....	....	....
2001-02	Yaroslavl 2	Russia-3	14	1	0	1	10	....	....	....	....	....
	Elektrostal 2	Russia-3	3	0	1	1	2	....	....	....	....	....
	Elektrostal	Russia-2	33	1	1	2	42	....	....	....	....	....

GROULX, Danny (GROO, DA-nee) **DET.**

Defense. Shoots left. 6', 205 lbs. Born, LaSalle, Que., June 23, 1981.

			Regular Season					Playoffs				
Season	**Club**	**Lea**	**GP**	**G**	**A**	**TP**	**PIM**	**GP**	**G**	**A**	**TP**	**PIM**
1996-97	Charles-Lemoyne	QAAA	40	2	26	28	....	15	3	15	18	....
1997-98	Val d'Or Foreurs	QMJHL	63	4	16	20	61	19	1	4	5	18
1998-99	Val d'Or Foreurs	QMJHL	36	3	26	29	55	....	....	....	....	....
	Acadie-Bathurst	QMJHL	36	2	15	17	51	18	0	2	2	6
99-2000	Victoriaville	QMJHL	66	12	55	67	131	6	0	4	4	14
2000-01	Victoriaville	QMJHL	72	16	71	87	164	13	2	19	21	46
2001-02	Victoriaville	QMJHL	68	29	83	112	165	14	7	20	27	52

QMJHL First All-Star Team (2001, 2002) • Canadian Major Junior First All-Star Team (2002) • Memorial Cup MVP (2002) • Memorial Cup All-Star Team (2002)

Signed as a free agent by **Detroit**, August 12, 2002.

GUENIN, Nate (GEH-nihn, NAYT) **NYR**

Defense. Shoots right. 6'2", 191 lbs. Born, Sewickley, PA, December 10, 1982.
(NY Rangers' 3rd choice, 127th overall, in 2002 Entry Draft).

			Regular Season					Playoffs				
Season	**Club**	**Lea**	**GP**	**G**	**A**	**TP**	**PIM**	**GP**	**G**	**A**	**TP**	**PIM**
99-2000	Pittsburgh	AAHA	40	3	10	13	122	....	....	....	....	....
2000-01	Green Bay	USHL	54	2	11	13	70	4	1	1	2	6
2001-02	Green Bay	USHL	56	4	11	15	150	7	3	3	6	10

USHL All-Rookie Team (2001)

• Signed Letter of Intent to attend **Ohio State** (CCHA), November 20, 2001.

GUITE, Ben (GEE-tay, BEHN) **ANA.**

Right wing. Shoots right. 6'1", 205 lbs. Born, Montreal, Que., July 17, 1978.
(Montreal's 8th choice, 172nd overall, in 1997 Entry Draft).

			Regular Season					Playoffs				
Season	**Club**	**Lea**	**GP**	**G**	**A**	**TP**	**PIM**	**GP**	**G**	**A**	**TP**	**PIM**
1994-95	Lac St-Louis	QAAA	40	9	12	21	....	4	0	0	0	0
1995-96	Capital District	Exhib.	STATISTICS NOT AVAILABLE									
1996-97	U. of Maine	H-East	34	7	7	14	21	....	....	....	....	....
1997-98	U. of Maine	H-East	32	6	12	18	20	....	....	....	....	....
1998-99	U. of Maine	H-East	40	12	16	28	30	....	....	....	....	....
99-2000	U. of Maine	H-East	40	22	14	36	36	....	....	....	....	....
2000-01	Tallahassee	ECHL	68	11	18	29	34	....	....	....	....	....
2001-02	Bridgeport	AHL	68	12	18	30	39	....	....	....	....	....
	Cincinnati	AHL	10	2	5	7	4	3	0	0	0	2

Signed as a free agent by **NY Islanders**, August, 2001. Traded to **Anaheim** by **NY Islanders** with the rights to Bjorn Mellin for Dave Roche, March 19, 2002.

GUSAKOV, Yevgeny (goo-sawk-KAHF, yehv-GEH-nee) NYR

Right wing. Shoots left. 6'6", 225 lbs. Born, Togliatti, USSR, March 6, 1981.
(NY Rangers' 9th choice, 226th overall, in 1999 Entry Draft).

			Regular Season					Playoffs				
Season	**Club**	**Lea**	**GP**	**G**	**A**	**TP**	**PIM**	**GP**	**G**	**A**	**TP**	**PIM**
1997-98	Lada Togliatti 2	Russia-3	24	4	2	6	10	....	....	....	....	....
1998-99	Lada Togliatti 2	Russia-4	42	12	3	15	28	....	....	....	....	....
99-2000	Baie-Comeau	QMJHL	61	25	22	47	135	6	5	1	6	10
	Hartford	AHL	1	0	0	0	0	....	....	....	....	....
2000-01	Baie-Comeau	QMJHL	32	11	20	31	68	....	....	....	....	....
	Lada Togliatti	Russia	1	0	0	0	0	....	....	....	....	....

GUSEV, Vladimir (GOO-sehv, vla-DIH-meer) CHI.

Defense. Shoots left. 6'1", 205 lbs. Born, Novosibirsk, USSR, November 24, 1982.
(Chicago's 6th choice, 115th overall, in 2001 Entry Draft).

			Regular Season					Playoffs				
Season	**Club**	**Lea**	**GP**	**G**	**A**	**TP**	**PIM**	**GP**	**G**	**A**	**TP**	**PIM**
99-2000	Novokuznetsk 2	Russia-3	21	1	0	1	52	....	....	....	....	....
	Magnitogorsk	Russia	....	....	....	....	....	3	0	0	0	0
2000-01	Amur Khabarovsk	Russia	1	0	0	0	0	....	....	....	....	....
	Novosibirsk 2	Russia-3	3	0	0	0	8	....	....	....	....	....
	Novosibirsk	Russia-2	1	0	0	0	0	....	....	....	....	....
	Omsk 2	Russia-3	4	1	0	1	4	....	....	....	....	....
2001-02	Novosibirsk	Russia-2	42	1	3	4	82	....	....	....	....	....

GUSTAFSSON, Juha (GOOS-tahf-suhn, YOO-huh) PHX.

Defense. Shoots left. 6'3", 200 lbs. Born, Helsinki, Finland, April 26, 1979.
(Phoenix's 1st choice, 43rd overall, in 1997 Entry Draft).

			Regular Season					Playoffs				
Season	**Club**	**Lea**	**GP**	**G**	**A**	**TP**	**PIM**	**GP**	**G**	**A**	**TP**	**PIM**
1994-95	Kiekko Espoo-B	Finn-Jr.	24	1	4	5	26	....	....	....	....	....
	Kiekko Espoo Jr.	Finn-Jr.	2	0	0	0	0	4	0	0	0	4
1995-96	Kiekko Espoo Jr.	Finn-Jr.	33	1	5	6	28	4	0	0	0	2
	Kiekko Espoo	Finland	1	0	0	0	0	....	....	....	....	....
1996-97	Kiekko Espoo Jr.	Finn-Jr.	32	1	3	4	30	....	....	....	....	....
	Kiekko Espoo	Finland	3	0	0	0	0	3	0	0	0	0
1997-98	Kiekko Espoo Jr.	Finn-Jr.	33	3	3	6	18	....	....	....	....	....
	Kiekko Espoo	Finland	2	0	0	0	0	....	....	....	....	....
1998-99	Ahmat Hyvinkaa	Finland-2	37	3	7	10	36	....	....	....	....	....
99-2000	Blues Espoo	Finland	33	1	4	5	26	....	....	....	....	....
2000-01	KJT Jarvenpaa	Finland-2	6	2	1	3	4	....	....	....	....	....
	Blues Espoo	Finland	46	1	2	3	34	....	....	....	....	....
2001-02	Blues Espoo	Finland	53	1	3	4	46	3	1	0	1	6

HAAKANA, Kari (HA-kuh-nuh, KAH-ree) EDM.

Defense. Shoots left. 6'1", 222 lbs. Born, Outokumpu, Finland, November 8, 1973.
(Edmonton's 9th choice, 248th overall, in 2001 Entry Draft).

			Regular Season					Playoffs				
Season	**Club**	**Lea**	**GP**	**G**	**A**	**TP**	**PIM**	**GP**	**G**	**A**	**TP**	**PIM**
1990-91	Kiekko Espoo Jr.	Finn-Jr.	36	3	3	6	34	....	....	....	....	....
	Kiekko Espoo	Finland-2	4	0	1	1	0	....	....	....	....	....
1991-92	Kiekko Espoo Jr.	Finn-Jr.	26	0	4	4	34	....	....	....	....	....
1992-93	Lukko Rauma Jr.	Finn-Jr.	36	4	16	20	60	....	....	....	....	....
	Lukko Rauma	Finland	4	0	0	0	0	....	....	....	....	....
1993-94	Kiekko Espoo Jr.	Finn-Jr.	5	0	2	2	2	....	....	....	....	....
	Kiekko Espoo	Finland	47	3	2	5	40	....	....	....	....	....
1994-95	Kiekko Espoo	Finland	48	4	3	7	54	4	0	0	0	0
1995-96	Kiekko Espoo	Finland	45	1	7	8	48	....	....	....	....	....
1996-97	Kiekko Espoo	Finland	48	0	12	12	69	....	....	....	....	....
1997-98	Kiekko Espoo	Finland	47	4	1	5	59	8	0	1	1	6
1998-99	Rosenheim	Germany	51	1	9	10	58	....	....	....	....	....
99-2000	Rosenheim	Germany	51	3	5	8	46	10	1	4	5	28
2000-01	Jokerit Helsinki	Finland	52	2	8	10	98	5	0	0	0	2
2001-02	Hamilton	AHL	6	0	2	2	25	....	....	....	....	....
	Jokerit Helsinki	Finland	36	0	3	3	40	12	2	2	4	2

HAGGLUND, Johan (HAG-luhnd, YOH-hahn) T.B.

Center. Shoots left. 6'2", 197 lbs. Born, Ornskoldsvik, Sweden, June 9, 1982.
(Tampa Bay's 4th choice, 126th overall, in 2000 Entry Draft).

			Regular Season					Playoffs				
Season	**Club**	**Lea**	**GP**	**G**	**A**	**TP**	**PIM**	**GP**	**G**	**A**	**TP**	**PIM**
1998-99	MoDo Jr.	Swede-Jr.	28	16	22	38	52	....	....	....	....	....
99-2000	MoDo-18	Swede-Jr.	7	1	2	3	8	....	....	....	....	....
	MoDo Jr.	Swede-Jr.	35	7	10	17	75	2	1	0	1	0
2000-01	MoDo Jr.	Swede-Jr.	21	9	9	18	66	....	....	....	....	....
2001-02	Orebro IK	Swede-2	36	6	4	10	50	....	....	....	....	....

HAGOS, Yared (HA-gohs, YAIR-ehd) DAL.

Center. Shoots left. 6'1", 202 lbs. Born, Stockholm, Sweden, March 27, 1983.
(Dallas' 2nd choice, 70th overall, in 2001 Entry Draft).

			Regular Season					Playoffs				
Season	**Club**	**Lea**	**GP**	**G**	**A**	**TP**	**PIM**	**GP**	**G**	**A**	**TP**	**PIM**
1998-99	AIK Solna Jr.	Swede-Jr.	32	8	12	20	22	....	....	....	....	....
99-2000	AIK Solna-18	Swede-Jr.	13	4	6	10	6	....	....	....	....	....
	AIK Solna Jr.	Swede-Jr.	17	6	4	10	10	....	....	....	....	....
2000-01	AIK Solna Jr.	Swede-Jr.	24	8	13	21	46	2	2	1	3	2
	AIK Solna	Sweden	....	....	....	....	....	5	0	0	0	0
2001-02	AIK Solna Jr.	Swede-Jr.	1	0	3	3	2	1	0	2	2	0
	AIK Solna	Sweden	45	4	6	10	36	....	....	....	....	....
	AIK Solna	Swede-Q	9	0	0	0	12	....	....	....	....	....

HAINSEY, Ron (HAYN-zee, RAWN) MTL.

Defense. Shoots left. 6'3", 200 lbs. Born, Bolton, CT, March 24, 1981.
(Montreal's 1st choice, 13th overall, in 2000 Entry Draft).

			Regular Season					Playoffs				
Season	**Club**	**Lea**	**GP**	**G**	**A**	**TP**	**PIM**	**GP**	**G**	**A**	**TP**	**PIM**
1997-98	Team USA	USDP-18	66	6	15	21	44	....	....	....	....	....
1998-99	Team USA	USDP-18	48	5	12	17	45	....	....	....	....	....
99-2000	U. Mass-Lowell	H-East	30	3	8	11	20	....	....	....	....	....
2000-01	U. Mass-Lowell	H-East	33	10	26	36	51	....	....	....	....	....
	Quebec	AHL	4	1	0	1	0	1	0	0	0	0
2001-02	Quebec	AHL	63	7	24	31	26	3	0	0	0	0

Hockey East First All-Star Team (2001) • NCAA East Second All-American Team (2001) • AHL All-Rookie Team (2002)

HAJEK, David (HIGH-ehk, DAV-vihd) CGY.

Defense. Shoots left. 5'11", 165 lbs. Born, Chomutov, Czech., June 13, 1980.
(Calgary's 8th choice, 239th overall, in 2000 Entry Draft).

			Regular Season					Playoffs				
Season	**Club**	**Lea**	**GP**	**G**	**A**	**TP**	**PIM**	**GP**	**G**	**A**	**TP**	**PIM**
1996-97	KLH Chomutov Jr.	Czech-Jr.	36	8	14	22	30	4	0	0	0	2
1997-98	KLH Chomutov Jr.	Czech-Jr.	36	3	9	12	18	....	....	....	....	....
1998-99	Melville	SJHL	25	10	19	29	....	....	....	....	....	....
	Spokane Chiefs	WHL	27	0	3	3	10	....	....	....	....	....
99-2000	KLH Chomutov Jr.	Czech-Jr.	7	1	5	6	14	....	....	....	....	....
	KLH Chomutov	Czech-2	28	1	5	6	14	12	1	3	4	10
2000-01	Kladno	Czech	40	1	1	2	66	....	....	....	....	....
2001-02	Kladno	Czech	48	0	7	7	16	5	0	1	1	4

HAKANSSON, Mikael (HAK-ahn-suhn, mihk-AIL) TOR.

Center. Shoots left. 6'2", 204 lbs. Born, Stockholm, Sweden, May 31, 1974.
(Toronto's 7th choice, 125th overall, in 1992 Entry Draft).

			Regular Season					Playoffs				
Season	**Club**	**Lea**	**GP**	**G**	**A**	**TP**	**PIM**	**GP**	**G**	**A**	**TP**	**PIM**
1990-91	Nacka HK	Swede-2	27	2	5	7	6	....	....	....	....	....
1991-92	Nacka HK	Swede-2	29	3	15	18	24	....	....	....	....	....
1992-93	Djurgarden	Sweden	40	0	1	1	6	3	0	0	0	0
1993-94	Djurgarden	Sweden	37	3	3	6	12	4	0	0	0	0
1994-95	MoDo	Sweden	37	3	7	10	16	....	....	....	....	....
1995-96	MoDo	Sweden	40	8	4	12	18	8	2	0	2	0
1996-97	Djurgarden	Sweden	48	8	12	20	12	4	0	0	0	0
1997-98	Djurgarden	Sweden	43	9	2	11	8	15	3	0	3	14
1998-99	Djurgarden	Sweden	48	13	12	25	14	4	1	0	1	0
	Djurgarden	EuroHL	6	0	1	1	20	....	....	....	....	....
99-2000	Djurgarden	Sweden	48	17	17	34	26	13	3	8	11	12
2000-01	St. John's	AHL	64	10	40	50	46	4	0	0	0	0
2001-02	Djurgarden	Sweden	49	13	24	37	86	2	0	0	0	4

Signed as a free agent by **Djurgarden** (Sweden) with **Toronto** retaining NHL rights, August 2, 2001.

HAKEWILL, James (HAYK-wihl, JAYMZ) CGY.

Defense. Shoots left. 6'3", 205 lbs. Born, Wilmette, IL, June 7, 1982.
(Calgary's 6th choice, 145th overall, in 2001 Entry Draft).

			Regular Season					Playoffs				
Season	**Club**	**Lea**	**GP**	**G**	**A**	**TP**	**PIM**	**GP**	**G**	**A**	**TP**	**PIM**
99-2000	Westminster High	Hi-School	23	3	13	16	22	....	....	....	....	....
2000-01	Westminster High	Hi-School	23	4	15	19	30	....	....	....	....	....
2001-02	Northeastern	ECAC	30	2	2	4	18	....	....	....	....	....

HALE, David (HAYL, DAY-vihd) N.J.

Defense. Shoots left. 6'2", 204 lbs. Born, Colorado Springs, CO, June 18, 1981.
(New Jersey's 1st choice, 22nd overall, in 2000 Entry Draft).

			Regular Season					Playoffs				
Season	**Club**	**Lea**	**GP**	**G**	**A**	**TP**	**PIM**	**GP**	**G**	**A**	**TP**	**PIM**
1997-98	Colorado North	Hi-School	25	11	33	44	154	....	....	....	....	....
1998-99	Sioux City	USHL	56	3	15	18	127	5	0	0	0	18
99-2000	Sioux City	USHL	54	6	18	24	187	5	0	2	2	6
2000-01	North Dakota	WCHA	44	4	5	9	79	....	....	....	....	....
2001-02	North Dakota	WCHA	34	4	5	9	63	....	....	....	....	....

USHL First All-Star Team (2000)

HALVARDSSON, Johan (HAL-vahrds-sohn, YOH-hahn) NYI

Defense. Shoots left. 6'3", 198 lbs. Born, Jonkoping, Sweden, December 26, 1979.
(NY Islanders' 8th choice, 102nd overall, in 1999 Entry Draft).

			Regular Season					Playoffs				
Season	**Club**	**Lea**	**GP**	**G**	**A**	**TP**	**PIM**	**GP**	**G**	**A**	**TP**	**PIM**
1997-98	HV 71 Jr.	Swede-Jr.	28	5	5	10	65	....	....	....	....	....
1998-99	HV 71 Jonkoping	Sweden	17	1	2	3	33	....	....	....	....	....
99-2000	HV 71 Jonkoping	Sweden	46	0	3	3	75	5	0	0	0	8
2000-01	HV 71 Jonkoping	Sweden	33	0	0	0	24	....	....	....	....	....
2001-02	HV 71 Jonkoping	Sweden	3	0	0	0	0	....	....	....	....	....

• Missed majority of 2001-02 season recovering from knee injury suffered in game vs. Sodertalje (Sweden), September 23, 2001.

HAMALAINEN, Ville (ha-muh-LAY-nuhn, VIHL-ee) CGY.

Left wing. Shoots left. 5'11", 178 lbs. Born, Lappeenranta, Finland, July 6, 1981.
(Calgary's 11th choice, 251st overall, in 2001 Entry Draft).

			Regular Season					Playoffs				
Season	**Club**	**Lea**	**GP**	**G**	**A**	**TP**	**PIM**	**GP**	**G**	**A**	**TP**	**PIM**
1997-98	SaiPa Jr.	Finn-Jr.	31	8	27	35	24	....	....	....	....	....
1998-99	SaiPa Jr.	Finn-Jr.	12	6	26	32	26	14	10	12	22	22
	SaiPa	Finland	12	1	0	1	8	....	....	....	....	....
99-2000	SaiPa	Finland	47	6	4	10	20	....	....	....	....	....
	KooKoo Kouvola	Finland-2	3	0	1	1	2	....	....	....	....	....
	SaiPa Jr.	Finn-Jr.	1	2	2	4	2	2	1	1	2	2
2000-01	SaiPa	Finland	42	0	4	4	10	....	....	....	....	....
	SaiPa Jr.	Finn-Jr.	1	1	1	2	0	....	....	....	....	....
2001-02	SaiPa	Finland	51	4	9	13	24	....	....	....	....	....

HAMHUIS, Dan (HAM-yoos, DAN) NSH.

Defense. Shoots left. 6', 208 lbs. Born, Smithers, B.C., December 13, 1982.
(Nashville's 1st choice, 12th overall, in 2001 Entry Draft).

			Regular Season					Playoffs				
Season	**Club**	**Lea**	**GP**	**G**	**A**	**TP**	**PIM**	**GP**	**G**	**A**	**TP**	**PIM**
1997-98	Smithers A's	BCAHA	59	59	72	131	59	....	....	....	....	....
1998-99	Prince George	WHL	56	1	3	4	45	7	1	2	3	8
99-2000	Prince George	WHL	70	10	23	33	140	13	2	3	5	35
2000-01	Prince George	WHL	62	13	47	60	125	6	2	3	5	15
2001-02	Prince George	WHL	59	10	50	60	135	7	0	5	5	16

WHL West First All-Star Team (2001, 2002) • WHL Player of the Year (2002) • Canadian Major Junior First All-Star Team (2002) • Canadian Major Junior Defenseman of the Year (2002)

HAMILTON, Jeff (HAM-ihl-tuhn, JEHF) NYI

Center. Shoots right. 5'10", 180 lbs. Born, Englewood, OH, September 4, 1977.

			Regular Season					Playoffs				
Season	Club	Lea	GP	G	A	TP	PIM	GP	G	A	TP	PIM
1996-97	Yale	ECAC	31	10	13	23	26	....	....	....	....	....
1997-98	Yale	ECAC	33	27	20	47	28	....	....	....	....	....
1998-99	Yale	ECAC	30	20	28	48	51	....	....	....	....	....
99-2000	Yale	ECAC	2	0	1	1	0	....	....	....	....	....
2000-01	Yale	ECAC	31	23	32	55	39	....	....	....	....	....
2001-02	Karpat Oulu	Finland	39	18	15	33	16	3	0	0	0	0

ECAC All-Rookie Team (1997) • ECAC First All-Star Team (1998, 1999, 2001) • Ivy League First All-Star Team (1998, 1999, 2001) • NCAA East Second All-American Team (1998, 1999) • NCAA East First All-American Team (2001) • Ivy League Player of the Year (2001)

• Missed majority of 1999-2000 season recovering from abdominal injury originally suffered in game vs. U. of Michigan (CCHA), October 30, 1999. • Announced official withdrawl from **Yale** (ECAC) for 1999-2000 semester, December 3, 1999. Signed as a free agent by **Karpat** (Finland), October 4, 2001. Signed as a free agent by **NY Islanders**, August 6, 2002.

HANNUS, Tommi (HA-nuhs, TAW-mee) L.A.

Center. Shoots right. 6', 180 lbs. Born, Vantaa, Finland, June 27, 1980.
(Los Angeles' 7th choice, 190th overall, in 1998 Entry Draft).

			Regular Season					Playoffs				
Season	Club	Lea	GP	G	A	TP	PIM	GP	G	A	TP	PIM
1994-95	TPS Turku-C	Finn-Jr.	27	24	13	37	43	....	....	....	....	....
1995-96	TPS Turku-C	Finn-Jr.	29	34	22	56	67	....	....	....	....	....
	TPS Turku-B	Finn-Jr.	1	1	0	1	2	....	....	....	....	....
1996-97	TPS Turku Jr.	Finn-Jr.	27	7	7	14	6	....	....	....	....	....
	TPS Turku-B	Finn-Jr.	18	7	7	14	6	6	4	2	6	10
1997-98	TPS Turku Jr.	Finn-Jr.	19	6	3	9	4	....	....	....	....	....
	TPS Turku-B	Finn-Jr.	9	5	4	9	8	....	....	....	....	....
1998-99	TPS Turku Jr.	Finn-Jr.	8	5	4	9	22	....	....	....	....	....
	TuTo Turku	Finland-2	18	6	4	10	16	8	0	2	2	8
99-2000	TPS Turku Jr.	Finn-Jr.	3	4	3	7	6	....	....	....	....	....
	TuTo Turku	Finland-2	27	11	6	17	20	....	....	....	....	....
	Assat Pori	Finland	13	1	0	1	4	....	....	....	....	....
2000-01	TuTo Turku	Finland-2	29	9	8	17	16	11	*7	7	14	*24
2001-02	TPS Turku	Finland	47	4	2	6	6	3	1	0	1	0

HARANT, Tomas (HAH-rant, TAW-mahsh) NSH.

Defense. Shoots left. 6'3", 201 lbs. Born, Zilina, Czech., April 28, 1980.
(Nashville's 8th choice, 173rd overall, in 2000 Entry Draft).

			Regular Season					Playoffs				
Season	Club	Lea	GP	G	A	TP	PIM	GP	G	A	TP	PIM
1997-98	Zilina Jr.	Slovak-Jr.	41	5	7	12	72	....	....	....	....	....
	HK SKP Zilina	Slovak-2	5	0	0	0	0	....	....	....	....	....
1998-99	Zilina Jr.	Slovak-Jr.	33	1	6	7	108	....	....	....	....	....
99-2000	SKP Zilina	Slovak-2	26	0	3	3	34	....	....	....	....	....
2000-01	Trinec Jr.	Czech-Jr.	5	1	2	3	8	1	0	0	0	4
	Trinec	Czech	15	1	2	3	14	....	....	....	....	....
2001-02	MsHK SKP Zilina	Slovakia	51	2	3	5	46	4	0	0	0	4

HARIKKALA, Jaakko (HAHR-ee-kuh-lah, YAH-koh) BOS.

Defense. Shoots left. 6'2", 215 lbs. Born, Kalanti, Finland, March 30, 1981.
(Boston's 4th choice, 118th overall, in 1999 Entry Draft).

			Regular Season					Playoffs				
Season	Club	Lea	GP	G	A	TP	PIM	GP	G	A	TP	PIM
1997-98	Jaa-Kotkat	Finland-3	5	0	2	2	8	....	....	....	....	....
	Jaa-Kotkat	Finland-2	45	2	6	8	65	....	....	....	....	....
1998-99	Lukko Rauma Jr.	Finn-Jr.	11	1	3	4	22	....	....	....	....	....
	Lukko Rauma	Finland	35	0	0	0	10	....	....	....	....	....
99-2000	Lukko Rauma	Finland	24	0	0	0	2	....	....	....	....	....
2000-01	Lukko Rauma Jr.	Finn-Jr.	2	2	0	2	2	....	....	....	....	....
	Jaa-Kotkat	Finland-2	10	0	2	2	16	....	....	....	....	....
	Lukko Rauma	Finland	10	0	0	0	0	....	....	....	....	....
2001-02	Lukko Rauma	Finland	47	4	8	12	40	....	....	....	....	....

HARLTON, Tyler (HAHRL-tawn, TIGH-luhr) TOR.

Defense. Shoots left. 6'2", 212 lbs. Born, Pense, Sask., January 11, 1976.
(St. Louis' 2nd choice, 94th overall, in 1994 Entry Draft).

			Regular Season					Playoffs				
Season	Club	Lea	GP	G	A	TP	PIM	GP	G	A	TP	PIM
1993-94	Vernon Vipers	BCJHL	60	3	18	21	102	....	....	....	....	....
1994-95	Michigan State	CCHA	39	1	3	4	55	....	....	....	....	....
1995-96	Michigan State	CCHA	39	1	6	7	51	....	....	....	....	....
1996-97	Michigan State	CCHA	39	2	9	11	75	....	....	....	....	....
1997-98	Michigan State	CCHA	44	1	12	13	68	....	....	....	....	....
1998-99	Worcester	AHL	58	2	5	7	94	....	....	....	....	....
	Peoria Rivermen	ECHL	6	0	2	2	40	....	....	....	....	....
99-2000	Worcester	AHL	3	0	0	0	4	....	....	....	....	....
	St. John's	AHL	56	2	6	8	62	....	....	....	....	....
2000-01	St. John's	AHL	80	5	18	23	68	4	0	0	0	0
2001-02	Austin Ice Bats	CHL	1	0	0	0	2	....	....	....	....	....
	Houston Aeros	AHL	16	0	0	0	2	....	....	....	....	....

CCHA First All-Star Team (1998) • NCAA West Second All-American Team (1998)

Traded to **Toronto** by **St. Louis** with future considerations for Derek King, October 20, 1999. Signed to 25 game try-out contract by **Houston** (AHL), November 5, 2001. Released by **Houston** (AHL), January 7, 2002.

HARRISON, Jay (HAIR-ih-suhn, JAY) TOR.

Defense. Shoots left. 6'3", 200 lbs. Born, Oshawa, Ont., November 3, 1982.
(Toronto's 4th choice, 82nd overall, in 2001 Entry Draft).

			Regular Season					Playoffs				
Season	Club	Lea	GP	G	A	TP	PIM	GP	G	A	TP	PIM
1997-98	Oshawa	OJHL-B	42	1	11	12	143	....	....	....	....	....
1998-99	Brampton	OHL	63	1	14	15	108	....	....	....	....	....
99-2000	Brampton	OHL	68	2	18	20	139	6	0	2	2	15
2000-01	Brampton	OHL	53	4	15	19	112	9	1	1	2	17
2001-02	Brampton	OHL	61	12	31	43	116	....	....	....	....	....
	Memphis	CHL	....	....	....	....	....	1	0	0	0	2
	St. John's	AHL	7	0	1	1	2	10	0	0	0	4

OHL All-Rookie Team (1999)

HARTSBURG, Chris (HAHRTZ-buhrg, KRIHS) N.J.

Left wing. Shoots right. 6', 190 lbs. Born, Edina, MN, May 30, 1980.
(New Jersey's 7th choice, 214th overall, in 1999 Entry Draft).

			Regular Season					Playoffs				
Season	Club	Lea	GP	G	A	TP	PIM	GP	G	A	TP	PIM
1995-96	Cambridge	OJHL-B	46	12	15	27	10	....	....	....	....	....
1996-97	Cambridge	OJHL-B	47	14	19	33	29	....	....	....	....	....
1997-98	Omaha Lancers	USHL	54	16	19	35	58	12	2	2	4	20
1998-99	Colorado College	WCHA	34	6	4	10	60	....	....	....	....	....
99-2000	Colorado College	WCHA	33	3	2	5	50	....	....	....	....	....
2000-01	Colorado College	WCHA	41	8	7	15	38	....	....	....	....	....
2001-02	Colorado College	WCHA	40	14	8	22	50	....	....	....	....	....

HAVELKA, Petr (huh-VEHL-kah, PEE-tuhr) PIT.

Left wing. Shoots left. 6'2", 187 lbs. Born, Most, Czech., March 4, 1979.
(Pittsburgh's 6th choice, 152nd overall, in 1997 Entry Draft).

			Regular Season					Playoffs				
Season	Club	Lea	GP	G	A	TP	PIM	GP	G	A	TP	PIM
1995-96	Sparta Praha Jr.	Czech-Jr.	40	15	10	25	....	....	....	....	....	....
1996-97	Sparta Praha Jr.	Czech-Jr.	22	14	13	27	....	....	....	....	....	....
	HC Sparta Praha	Czech	....	....	....	....	....	1	0	0	0	0
1997-98	Sparta Praha Jr.	Czech-Jr.	DID NOT PLAY – INJURED									
1998-99	Sparta Praha Jr.	Czech-Jr.	5	7	3	10	....	4	1	1	2	....
	Kladno	Czech	5	0	0	0	0	....	....	....	....	....
99-2000	Sparta Praha Jr.	Czech-Jr.	2	1	0	1	0	....	....	....	....	....
	Beroun	Czech-2	5	2	4	6	29	....	....	....	....	....
	Kladno	Czech	6	1	2	3	2	....	....	....	....	....
	HC Sparta Praha	Czech	10	0	1	1	0	3	0	0	0	0
2000-01	Beroun	Czech-2	3	0	0	0	0	....	....	....	....	....
	HC Sparta Praha	Czech	....	....	....	....	....	7	1	0	1	2
2001-02	Sparta Praha Jr.	Czech-Jr.	2	1	0	1	0	....	....	....	....	....
	Usti nad Labem	Czech-3	2	0	0	0	2	....	....	....	....	....
	HC Sparta Praha	Czech	35	3	4	7	12	6	0	0	0	2

HAY, Darrell (HAY, DAIR-ehl) VAN.

Defense. Shoots right. 6', 190 lbs. Born, Kamloops, B.C, April 2, 1980.
(Vancouver's 8th choice, 271st overall, in 1999 Entry Draft).

			Regular Season					Playoffs				
Season	Club	Lea	GP	G	A	TP	PIM	GP	G	A	TP	PIM
1995-96	Kamloops Lions	BCAHA	65	34	57	91	155	....	....	....	....	....
1996-97	Tri-City	WHL	61	0	10	10	41	....	....	....	....	....
1997-98	Tri-City	WHL	71	5	33	38	90	....	....	....	....	....
1998-99	Tri-City	WHL	72	13	49	62	87	12	2	10	12	22
99-2000	Tri-City	WHL	64	15	36	51	85	4	0	1	1	8
2000-01	Florida	ECHL	40	5	4	9	26	5	2	1	3	0
	Kansas City	IHL	9	0	0	0	19	....	....	....	....	....
2001-02	Manitoba Moose	AHL	53	5	8	13	29	7	1	1	2	2
	Columbia Inferno	ECHL	15	0	13	13	15	....	....	....	....	....

WHL West Second All-Star Team (2000)

HAYDAR, Darren (HAY-duhr, DAIR-ehn) NSH.

Right wing. Shoots left. 5'9", 170 lbs. Born, Toronto, Ont., October 22, 1979.
(Nashville's 14th choice, 248th overall, in 1999 Entry Draft).

			Regular Season					Playoffs				
Season	Club	Lea	GP	G	A	TP	PIM	GP	G	A	TP	PIM
1995-96	Milton Merchants	OPJHL	6	1	2	3	4	....	....	....	....	....
1996-97	Milton Merchants	OPJHL	51	32	68	100	68	....	....	....	....	....
1997-98	Milton Merchants	OPJHL	51	*71	*69	*140	65	....	....	....	....	....
1998-99	New Hampshire	H-East	41	31	30	61	34	....	....	....	....	....
99-2000	New Hampshire	H-East	38	22	19	41	42	....	....	....	....	....
2000-01	New Hampshire	H-East	39	18	23	41	38	....	....	....	....	....
2001-02	New Hampshire	H-East	40	31	*45	*76	28	....	....	....	....	....

OPJHL First All-Star Team (1998) • OPJHL Player of the Year (1998) • Hockey East Second All-Star Team (1999, 2000) • Hockey East Rookie of the Year (1999) • Hockey East First All-Star Team (2002)

HEALEY, Eric (HEE-lee, AIR-ihk)

Left wing. Shoots left. 5'11", 196 lbs. Born, Hull, MA, January 20, 1975.

			Regular Season					Playoffs				
Season	Club	Lea	GP	G	A	TP	PIM	GP	G	A	TP	PIM
1993-94	New England	NEJHL	37	61	76	137	....	....	....	....	....	....
1994-95	RPI Engineers	ECAC	37	13	11	24	35	....	....	....	....	....
1995-96	RPI Engineers	ECAC	35	18	22	40	57	....	....	....	....	....
1996-97	RPI Engineers	ECAC	36	30	26	56	63	....	....	....	....	....
1997-98	RPI Engineers	ECAC	35	21	27	48	42	....	....	....	....	....
1998-99	Saint John	AHL	64	14	24	38	77	....	....	....	....	....
	Orlando	IHL	13	5	4	9	13	8	1	0	1	12
99-2000	Springfield	AHL	32	14	15	29	51	1	0	0	0	2
2000-01	Springfield	AHL	66	16	17	33	53	....	....	....	....	....
2001-02	Manchester	AHL	65	24	34	58	45	5	2	2	4	8
	Jackson Bandits	ECHL	2	1	1	2	0	....	....	....	....	....

ECAC Second All-Star Team (1997) • NCAA East Second All-American Team (1997, 1998) • ECAC First All-Star Team (1998)

Signed as a free agent by **Calgary**, September 22, 1998. Signed as a free agent by **Phoenix**, July 26, 1999. Signed to try-out contract by **Manchester** (AHL), September 30, 2001.

HECL, Radoslav BUF.

Defense. Shoots left. 6'1", 196 lbs. Born, Partizanske, Czech., October 11, 1974.
(Buffalo's 8th choice, 208th overall, in 2002 Entry Draft).

			Regular Season					Playoffs				
Season	Club	Lea	GP	G	A	TP	PIM	GP	G	A	TP	PIM
1996-97	Slovan Bratislava	EuroHL	4	0	0	0	0	2	0	0	0	0
	Slovan Bratislava	Slovakia	36	1	5	6	6	....	....	....	....	....
1997-98	Slovan Bratislava	EuroHL	6	2	1	3	6	2	0	0	0	0
	Slovan Bratislava	Slovakia	35	7	7	14	41	10	0	4	4	6
1998-99	Slovan Bratislava	EuroHL	6	0	0	0	14	....	....	....	....	....
	Slovan Bratislava	Slovakia	40	5	10	15	61	10	2	4	6	31
99-2000	Slovan Bratislava	Slovakia	41	2	7	9	61	....	....	....	....	....
2000-01	Slovan Bratislava	Slovakia	25	2	5	7	6	6	1	1	2	18
2001-02	Slovan Bratislava	Slovakia	50	7	13	20	22	19	3	5	8	10

HEDIN, Pierre (heh-DEEN, PEE-air) TOR.

Defense. Shoots left. 6'1", 198 lbs. Born, Ornskoldsvik, Sweden, February 19, 1978.
(Toronto's 8th choice, 239th overall, in 1999 Entry Draft).

			Regular Season					Playoffs				
Season	Club	Lea	GP	G	A	TP	PIM	GP	G	A	TP	PIM
1994-95	MoDo Jr.	Swede-Jr.	21	0	3	3	20	....	....	....	....	....
1996-97	MoDo	Sweden	19	1	2	3	6	....	....	....	....	....
1997-98	MoDo Jr.	Swede-Jr.	7	1	6	7	10	....	....	....	....	....
	MoDo	Sweden	29	2	1	3	26	9	1	1	2	4
1998-99	MoDo	Sweden	41	6	5	11	28	13	1	1	2	12
99-2000	MoDo	Sweden	48	9	5	14	36	13	0	2	2	8
2000-01	MoDo	Sweden	46	5	8	13	59	7	3	0	3	4
2001-02	MoDo	Sweden	39	7	9	16	20	14	*8	2	10	10

HEDSTROM, Jonathan (HEHD-struhm, JAWN-ah-thuhn) ANA.

Right wing. Shoots left. 6', 200 lbs. Born, Skelleftea, Sweden, December 27, 1977.
(Toronto's 8th choice, 221st overall, in 1997 Entry Draft).

			Regular Season					Playoffs				
Season	Club	Lea	GP	G	A	TP	PIM	GP	G	A	TP	PIM
1995-96	Skelleftea AIK	Swede-2	7	0	0	0	0	....	....	....	....	....
1996-97	Skelleftea Jr.	Swede-Jr.	9	4	4	8	....	....	....	....	....	....
	Skelleftea AIK	Swede-2	12	1	1	2	10	6	0	0	0	2
1997-98	Skelleftea Jr.	Swede-Jr.	1	0	0	0	2	....	....	....	....	....
	Skelleftea AIK	Swede-2	16	2	3	5	....	....	....	....	....	....
1998-99	Skelleftea AIK	Swede-2	36	15	28	43	74	....	....	....	....	....
99-2000	Lulea HF	Sweden	48	9	17	26	46	9	2	1	3	12
2000-01	Lulea HF	Sweden	46	9	19	28	68	12	1	6	7	16
2001-02	Lulea HF	Sweden	47	11	7	18	38	4	2	1	3	6

Rights traded to **Anaheim** by **Toronto** for Anaheim's 6th (Vadim Sozinov) and 7th (Markus Seikola) round choices in 2000 Entry Draft, June 25, 2000.

HEEREMA, Jeff (HEER-eh-muh, JEHF) CAR.

Right wing. Shoots right. 6'1", 190 lbs. Born, Thunder Bay, Ont., January 17, 1980.
(Carolina's 1st choice, 11th overall, in 1998 Entry Draft).

			Regular Season					Playoffs				
Season	Club	Lea	GP	G	A	TP	PIM	GP	G	A	TP	PIM
1996-97	T. Bay Kings	TBMHL	54	42	29	71	112	....	....	....	....	....
1997-98	Sarnia Sting	OHL	63	32	40	72	88	5	4	1	5	10
1998-99	Sarnia Sting	OHL	62	31	39	70	113	6	5	1	6	0
99-2000	Sarnia Sting	OHL	67	36	41	77	62	7	4	2	6	10
2000-01	Cincinnati	IHL	73	17	16	33	42	4	0	0	0	0
2001-02	Lowell	AHL	76	33	37	70	90	5	2	3	5	2

HEFFERNAN, Scott (HEH-fuhr-nuhn, SKAWT) CBJ

Defense. Shoots left. 6'5", 200 lbs. Born, Montreal, Que., March 9, 1982.
(Columbus' 4th choice, 138th overall, in 2000 Entry Draft).

			Regular Season					Playoffs				
Season	Club	Lea	GP	G	A	TP	PIM	GP	G	A	TP	PIM
1998-99	Pembroke	OCJHL	44	3	5	8	51	....	....	....	....	....
99-2000	Sarnia Sting	OHL	55	5	10	15	24	7	0	1	1	4
2000-01	Sarnia Sting	OHL	51	2	20	22	30	4	0	1	1	0
2001-02	Sarnia Sting	OHL	10	0	9	9	16	....	....	....	....	....
	St. Michael's	OHL	58	4	17	21	77	9	1	2	3	4

Traded to **St. Michael's** (OHL) by **Sarnia** (OHL) with future considerations for Matt Hannan, October 13, 2001.

HEID, Chris (HIGHD, KRIHS) MIN.

Defense. Shoots left. 6'2", 205 lbs. Born, Langley, B.C., March 14, 1983.
(Minnesota's 3rd choice, 74th overall, in 2001 Entry Draft).

			Regular Season					Playoffs				
Season	Club	Lea	GP	G	A	TP	PIM	GP	G	A	TP	PIM
1998-99	Kamloops	BCAHA	58	26	34	60	65	....	....	....	....	....
	Spokane Chiefs	WHL	1	0	0	0	0	....	....	....	....	....
99-2000	Spokane Chiefs	WHL	44	1	7	8	25	6	0	0	0	4
2000-01	Spokane Chiefs	WHL	51	2	15	17	76	12	0	4	4	12
2001-02	Spokane Chiefs	WHL	69	7	28	35	56	11	1	4	5	8

HELBLING, Timo (HEHL-blihng, TEE-moh) NSH.

Defense. Shoots right. 6'3", 209 lbs. Born, Basel, Switz., July 21, 1981.
(Nashville's 11th choice, 162nd overall, in 1999 Entry Draft).

			Regular Season					Playoffs				
Season	Club	Lea	GP	G	A	TP	PIM	GP	G	A	TP	PIM
1997-98	HC Davos Jr.	Swiss-Jr.	34	6	6	12	38	....	....	....	....	....
1998-99	HC Davos Jr.	Swiss-Jr.	28	5	10	15	116	2	1	3	4	35
	HC Davos	Swiss	44	0	0	0	8	4	0	0	0	0
99-2000	HC Davos	Swiss	44	0	0	0	49	5	0	0	0	0
2000-01	Windsor	OHL	54	7	14	21	90	7	0	2	2	11
	Milwaukee	IHL	....	....	....	....	....	1	0	0	0	0
2001-02	Milwaukee	AHL	67	2	6	8	59	....	....	....	....	....

HELFENSTEIN, Sven (hehl-fehn-SHTIGHN, SVEHN) NYR

Left wing. Shoots right. 5'10", 176 lbs. Born, Winterthur, Switz., July 30, 1982.
(NY Rangers' 6th choice, 175th overall, in 2000 Entry Draft).

			Regular Season					Playoffs				
Season	Club	Lea	GP	G	A	TP	PIM	GP	G	A	TP	PIM
1997-98	EHC Kloten Jr.	Swiss-Jr.	31	6	8	14	14	....	....	....	....	....
1998-99	EHC Kloten Jr.	Swiss-Jr.	33	25	18	43	14	7	5	3	8	2
	EHC Kloten	Swiss	2	0	0	0	0	....	....	....	....	....
99-2000	EHC Kloten	Swiss	40	6	3	9	28	6	0	1	1	0
2000-01	Kloten Jr.	Swiss-Jr.	2	3	3	6	0	....	....	....	....	....
	EHC Kloten	Swiss	8	1	1	2	0	....	....	....	....	....
	Chaux-de-Fonds	Swiss	23	2	8	10	6	12	1	3	4	6
	HC Thurgau	Swiss-2	4	2	1	3	8	....	....	....	....	....
2001-02	SC Bern	Swiss	35	2	10	12	39	6	0	0	0	4
	SC Bern Jr.	Swiss-Jr.	....	....	....	....	....	4	4	2	6	2

HEMINGWAY, Colin (HEH-mihng-way, CAW-lihn) ST.L.

Right wing. Shoots right. 6', 170 lbs. Born, Regina, Sask., August 12, 1980.
(St. Louis' 7th choice, 221st overall, in 1999 Entry Draft).

			Regular Season					Playoffs				
Season	Club	Lea	GP	G	A	TP	PIM	GP	G	A	TP	PIM
1996-97	Port Coquitlam	PIJHL	34	23	24	47	52	....	....	....	....	....
1997-98	South Surrey	BCHL	58	12	16	28	46	....	....	....	....	....
1998-99	South Surrey	BCHL	59	40	64	104	52	....	....	....	....	....
99-2000	New Hampshire	H-East	22	3	5	8	6	....	....	....	....	....
2000-01	New Hampshire	H-East	37	9	18	27	16	....	....	....	....	....
2001-02	New Hampshire	H-East	40	*33	33	66	30	....	....	....	....	....

Hockey East First All-Star Team (2002)

HEMSKY, Ales (HEHM-skee, ahl-EHSH) EDM.

Right wing. Shoots right. 6', 191 lbs. Born, Pardubice, Czech., August 13, 1983.
(Edmonton's 1st choice, 13th overall, in 2001 Entry Draft).

			Regular Season					Playoffs				
Season	Club	Lea	GP	G	A	TP	PIM	GP	G	A	TP	PIM
99-2000	Pardubice Jr.	Czech-Jr.	45	20	36	56	54	7	4	14	18	36
	Pardubice	Czech	4	0	1	1	0	....	....	....	....	....
2000-01	Hull Olympiques	QMJHL	68	36	64	100	67	5	2	3	5	2
2001-02	Hull Olympiques	QMJHL	53	27	70	97	86	10	6	10	16	6

QMJHL Second All-Star Team (2002)

HENDRICKS, Matt (HEHN-drihks, MAT) NSH.

Center. Shoots left. 6', 215 lbs. Born, Blaine, MN, June 17, 1981.
(Nashville's 5th choice, 131st overall, in 2000 Entry Draft).

			Regular Season					Playoffs				
Season	Club	Lea	GP	G	A	TP	PIM	GP	G	A	TP	PIM
1998-99	Blaine Bengals	Hi-School	22	23	34	57	42	....	....	....	....	....
99-2000	Blaine Bengals	Hi-School	21	23	30	53	28	....	....	....	....	....
2000-01	St. Cloud State	WCHA	37	3	9	12	23	....	....	....	....	....
2001-02	St. Cloud State	WCHA	42	19	20	39	74	....	....	....	....	....

HENKEL, Jim (HEHN-kehl, JIHM) L.A.

Center. Shoots left. 6'2", 180 lbs. Born, Red Bank, NJ, May 25, 1979.
(Los Angeles' 8th choice, 217th overall, in 1998 Entry Draft).

			Regular Season					Playoffs				
Season	Club	Lea	GP	G	A	TP	PIM	GP	G	A	TP	PIM
1997-98	New England	EJHL	37	34	37	71	....	11	7	17	24	....
1998-99	RPI Engineers	ECAC	20	0	4	4	14	....	....	....	....	....
99-2000	RPI Engineers	ECAC	34	2	5	7	28	....	....	....	....	....
2000-01	RPI Engineers	ECAC	34	11	19	30	44	....	....	....	....	....
2001-02	RPI Engineers	ECAC	35	9	15	24	22	....	....	....	....	....

HENNING, Petter (HEH-nihng, PEH-tehr) NYR

Right wing. Shoots left. 6', 209 lbs. Born, Ornskoldsvik, Sweden, September 15, 1980.
(NY Rangers' 10th choice, 251st overall, in 1999 Entry Draft).

			Regular Season					Playoffs				
Season	Club	Lea	GP	G	A	TP	PIM	GP	G	A	TP	PIM
1997-98	MoDo Jr.	Swede-Jr.	27	7	6	13	12	....	....	....	....	....
1998-99	MoDo Jr.	Swede-Jr.	38	10	10	20	74	....	....	....	....	....
	MoDo	Sweden	1	0	0	0	0	....	....	....	....	....
99-2000	Sodertalje Jr.	Swede-Jr.	3	1	0	1	0	....	....	....	....	....
	Skelleftea AIK	Swede-2	8	0	0	0	2	....	....	....	....	....
2000-01	Tingsryds AIF	Swede-2	40	3	5	8	18	3	0	0	0	6
2001-02	Tingsryds AIF	Swede-2	32	9	4	13	18	....	....	....	....	....
	Tingsryds AIF	Swede-Q	13	2	0	2	12	2	0	0	0	0

HENRICH, Adam (HEHN-rihch, A-duhm) T.B.

Left wing. Shoots left. 6'4", 219 lbs. Born, Toronto, Ont., January 19, 1984.
(Tampa Bay's 1st choice, 60th overall, in 2002 Entry Draft).

			Regular Season					Playoffs				
Season	Club	Lea	GP	G	A	TP	PIM	GP	G	A	TP	PIM
99-2000	Don Mills Flyers	GTHL	54	30	52	82	86	....	....	....	....	....
2000-01	Brampton	OHL	48	5	4	9	27	9	0	0	0	6
2001-02	Brampton	OHL	66	33	30	63	92	....	....	....	....	....

HENRICH, Michael (HEHN-rihch, MIGH-kuhl) EDM.

Right wing. Shoots right. 6'2", 206 lbs. Born, Thornhill, Ont., March 3, 1980.
(Edmonton's 1st choice, 13th overall, in 1998 Entry Draft).

			Regular Season					Playoffs				
Season	Club	Lea	GP	G	A	TP	PIM	GP	G	A	TP	PIM
1995-96	Wexford Raiders	MTJHL	4	1	0	1	0	....	....	....	....	....
1996-97	Barrie Colts	OHL	52	9	15	24	19	9	0	5	5	0
1997-98	Barrie Colts	OHL	66	41	22	63	75	5	1	3	4	0
1998-99	Barrie Colts	OHL	62	38	33	71	42	12	0	2	2	4
99-2000	Barrie Colts	OHL	66	38	48	86	69	25	10	18	28	30
2000-01	Tallahassee	ECHL	6	1	1	2	0	....	....	....	....	....
	Hamilton	AHL	73	5	10	15	36	....	....	....	....	....
2001-02	Hamilton	AHL	67	14	24	38	24	9	2	2	4	2

HENRY, Alex (HEHN-ree, AL-ehx) EDM.

Defense. Shoots left. 6'5", 220 lbs. Born, Elliot Lake, Ont., October 18, 1979.
(Edmonton's 2nd choice, 67th overall, in 1998 Entry Draft).

			Regular Season					Playoffs				
Season	Club	Lea	GP	G	A	TP	PIM	GP	G	A	TP	PIM
1995-96	Timmins Titans	NOHA	30	2	4	11	15	....	....	....	....	....
	Timmins	NOJHA	2	0	0	0	0	....	....	....	....	....
1996-97	London Knights	OHL	61	1	10	11	65	....	....	....	....	....
1997-98	London Knights	OHL	62	5	9	14	97	16	0	3	3	14
1998-99	London Knights	OHL	68	5	23	28	105	25	3	10	13	22
99-2000	Hamilton	AHL	60	1	0	1	69	....	....	....	....	....
2000-01	Hamilton	AHL	56	2	3	5	87	....	....	....	....	....
2001-02	Hamilton	AHL	69	4	8	12	143	15	1	2	3	16

HENRY, Burke (HEHN-ree, BUHRK) CGY.

Defense. Shoots left. 6'3", 190 lbs. Born, Ste. Rose, Man., January 21, 1979.
(NY Rangers' 3rd choice, 73rd overall, in 1997 Entry Draft).

			Regular Season					Playoffs				
Season	Club	Lea	GP	G	A	TP	PIM	GP	G	A	TP	PIM
1995-96	Brandon	WHL	50	6	11	17	58	19	0	4	4	19
1996-97	Brandon	WHL	55	6	25	31	81	6	1	3	4	4
1997-98	Brandon	WHL	72	18	65	83	153	18	3	16	19	37
1998-99	Brandon	WHL	68	18	58	76	151	5	1	6	7	9
99-2000	Hartford	AHL	64	3	12	15	47	5	0	0	0	2
2000-01	Hartford	AHL	80	8	30	38	133	5	0	0	0	2
2001-02	Saint John	AHL	58	0	17	17	92	....	....	....	....	....

WHL East First All-Star Team (1998) • WHL East Second All-Star Team (1999)

Traded to **Calgary** by **NY Rangers** for Chris St. Croix, June 23, 2001.

HIGGINS, Christopher (HIH-gihns, KRIHS-toh-fuhr) MTL.

Center. Shoots left. 5'11", 192 lbs. Born, Smithtown, NY, June 2, 1983.
(Montreal's 1st choice, 14th overall, in 2002 Entry Draft).

			Regular Season					Playoffs				
Season	Club	Lea	GP	G	A	TP	PIM	GP	G	A	TP	PIM
99-2000	Avon Old Farms	Hi-School	27	19	20	39	10	....	....	....	....	....
2000-01	Avon Old Farms	Hi-School	24	22	14	36	29	....	....	....	....	....
2001-02	Yale	ECAC	27	14	17	31	32	....	....	....	....	....

ECAC All-Rookie Team (2002) • ECAC Second All-Star Team (2002) • ECAC Rookie of the Year (2002)

HILL, Ed (HIHL, EHD) **CAR.**

Defense. Shoots left. 6'3", 215 lbs. Born, Newburyport, MA, October 24, 1980.
(Nashville's 5th choice, 61st overall, in 1999 Entry Draft).

			Regular Season					Playoffs				
Season	Club	Lea	GP	G	A	TP	PIM	GP	G	A	TP	PIM
1996-97	Green Bay	USHL	61	4	11	15	36	17	0	1	1	14
1997-98	Green Bay	USHL	51	1	16	17	76	3	0	0	0	0
1998-99	Barrie Colts	OHL	53	7	17	24	42	12	0	2	2	8
99-2000	Barrie Colts	OHL	66	1	18	19	63	25	1	3	4	14
2000-01	Barrie Colts	OHL	60	3	19	22	104	5	0	1	1	2
2001-02	Florida	ECHL	34	2	12	14	22	....	....	....	....	....
	Lowell	AHL	37	0	0	0	37	5	0	1	1	12

Signed as a free agent by **Carolina**, July 16, 2002.

HIMELFARB, Eric (HIH-muhl-FAHRB, AIR-ihk) **MTL.**

Center. Shoots right. 5'9", 161 lbs. Born, Thornhill, Ont., January 1, 1983.
(Montreal's 6th choice, 171st overall, in 2001 Entry Draft).

			Regular Season					Playoffs				
Season	Club	Lea	GP	G	A	TP	PIM	GP	G	A	TP	PIM
1998-99	Don Mills Flyers	GTHL	40	40	31	71	42	....	....	....	....	....
99-2000	Sarnia Sting	OHL	62	14	33	47	26	7	1	4	5	4
2000-01	Sarnia Sting	OHL	49	31	44	75	48	4	1	7	8	4
2001-02	Sarnia Sting	OHL	67	35	48	83	67	5	1	4	5	11

HINZ, Chad (HIHNZ, CHAD) **EDM.**

Right wing. Shoots right. 5'10", 190 lbs. Born, Saskatoon, Sask., March 21, 1979.
(Edmonton's 8th choice, 187th overall, in 1997 Entry Draft).

			Regular Season					Playoffs				
Season	Club	Lea	GP	G	A	TP	PIM	GP	G	A	TP	PIM
1994-95	Sask. Contacts	SMHL	29	25	21	46	41	....	....	....	....	....
1995-96	Moose Jaw	WHL	70	22	32	54	65	....	....	....	....	....
1996-97	Moose Jaw	WHL	72	37	47	84	47	12	4	1	5	11
1997-98	Moose Jaw	WHL	72	20	57	77	45	4	1	2	3	2
1998-99	Moose Jaw	WHL	71	42	*75	117	40	11	4	12	16	12
	Hamilton	AHL	3	0	0	0	2	....	....	....	....	....
99-2000	Hamilton	AHL	18	1	4	5	2	5	1	0	1	0
	Tallahassee	ECHL	49	15	25	40	35	....	....	....	....	....
2000-01	Hamilton	AHL	78	13	22	35	30	....	....	....	....	....
2001-02	Hamilton	AHL	71	6	13	19	27	15	2	8	10	8

WHL East First All-Star Team (1999)

HIRVONEN, Tomi (HIHR-voh-nehn, TAW-mee) **COL.**

Center. Shoots left. 5'11", 185 lbs. Born, Tampere, Finland, January 11, 1977.
(Colorado's 8th choice, 207th overall, in 1995 Entry Draft).

			Regular Season					Playoffs				
Season	Club	Lea	GP	G	A	TP	PIM	GP	G	A	TP	PIM
1992-93	Ilves Tampere-C	Finn-Jr.	34	32	20	52	71	....	....	....	....	....
1993-94	Ilves Tampere-B	Finn-Jr.	28	13	14	27	96	....	....	....	....	....
	Ilves Jr.	Finn-Jr.	1	0	0	0	0	....	....	....	....	....
1994-95	Ilves Tampere-B	Finn-Jr.	6	1	5	6	14	....	....	....	....	....
	Ilves Jr.	Finn-Jr.	28	9	13	22	30	8	4	2	6	14
1995-96	Ilves Jr.	Finn-Jr.	5	2	2	4	37	7	5	10	15	8
	KooVee Tampere	Finland-2	7	4	1	5	26	....	....	....	....	....
	Ilves Tampere	Finland	28	1	0	1	24	....	....	....	....	....
1996-97	Ilves Jr.	Finn-Jr.	5	2	2	4	37	4	1	3	4	8
	Ilves Tampere	Finland	40	0	7	7	22	6	0	0	0	4
1997-98	Ilves Tampere	Finland	48	10	12	22	54	9	0	0	0	2
	Ilves Jr.	Finn-Jr.	....	....	....	....	....	2	5	0	5	4
1998-99	Ilves Tampere	Finland	50	5	15	20	94	4	0	0	0	8
	Ilves Tampere	EuroHL	6	1	3	4	2	....	....	....	....	....
99-2000	Ilves Tampere	Finland	45	4	7	11	62	3	0	1	1	6
2000-01	JYP Jyvaskyla	Finland	52	5	8	13	76	....	....	....	....	....
2001-02	JYP Jyvaskyla	Finland	55	9	10	19	53	....	....	....	....	....

HLINKA, Martin (huh-LIHN-kuh, MAHR-tihn)

Left wing. Shoots left. 6'1", 200 lbs. Born, Bratislava, Czech., September 25, 1976.

			Regular Season					Playoffs				
Season	Club	Lea	GP	G	A	TP	PIM	GP	G	A	TP	PIM
1995-96	Augsburg Auggies	MIAC	15	6	5	11	....	....	....	....	....	....
1996-97	Augsburg Auggies	MIAC	24	14	23	37	....	....	....	....	....	....
1997-98	Augsburg Auggies	MIAC	24	14	30	44	....	....	....	....	....	....
1998-99	Augsburg Auggies	MIAC	22	6	27	33	24	....	....	....	....	....
	Quad City	UHL	2	1	1	2	0	1	0	0	0	0
99-2000	Quad City	UHL	71	21	46	67	74	14	0	7	7	24
	Chicago Wolves	IHL	1	0	0	0	0	....	....	....	....	....
2000-01	Quad City	UHL	11	4	10	14	0	5	1	1	2	....
	Portland Pirates	AHL	60	13	19	32	50	3	1	2	3	2
2001-02	Portland Pirates	AHL	50	5	18	23	34	....	....	....	....	....

MIAC First All-Star Team (1997, 1998, 1999) • MIAC All-Conference All-Star Team (1999) • Also played football at Augsburg College and set season and career records for field goals attempted, field goals made and total points scored. • UHL All-Rookie Team (2000)

Selected by **Quad City** (UHL) in UHL Over-age Priority Draft, March 12, 1999. Signed as a free agent by **Portland** (AHL), November 10, 2000. Signed as a free agent by **Washington**, May 3, 2001.

HOGEBOOM, Greg (HOH-guh-BOOM, GREHG) **L.A.**

Right wing. Shoots right. 6', 190 lbs. Born, Toronto, Ont., September 26, 1982.
(Los Angeles' 6th choice, 152nd overall, in 2002 Entry Draft).

			Regular Season					Playoffs				
Season	Club	Lea	GP	G	A	TP	PIM	GP	G	A	TP	PIM
1998-99	North York	GTHL	STATISTICS NOT AVAILABLE									
99-2000	Wexford Raiders	OPJHL	48	32	47	79	44	....	....	....	....	....
2000-01	Miami-Ohio	CCHA	38	8	5	13	20	....	....	....	....	....
2001-02	Miami-Ohio	CCHA	36	14	9	23	22	....	....	....	....	....

HOHENER, Martin (HOH-ehn-uhr, MAHR-tihn) **NSH.**

Defense. Shoots left. 6'1", 192 lbs. Born, Zurich, Switz., June 23, 1980.
(Nashville's 12th choice, 284th overall, in 2000 Entry Draft).

			Regular Season					Playoffs				
Season	Club	Lea	GP	G	A	TP	PIM	GP	G	A	TP	PIM
1996-97	EHC Kloten Jr.	Swiss-Jr.	37	4	7	11	....	....	....	....	....	....
1997-98	EHC Kloten Jr.	Swiss-Jr.	25	2	9	11	31	....	....	....	....	....
	EHC Bulach	Swiss-2	4	0	0	0	0	....	....	....	....	....
1998-99	EHC Kloten Jr.	Swiss-Jr.	21	5	8	13	22	2	1	0	1	2
	EHC Kloten	Swiss	20	0	1	1	2	9	0	1	1	0
99-2000	EHC Kloten	Swiss	44	4	2	6	20	5	0	1	1	2
2000-01	EHC Kloten	Swiss	35	4	5	9	32	9	0	1	1	6
2001-02	Kloten Flyers	Swiss	44	4	12	16	12	9	0	1	1	2
	Switzerland	Olympics	4	0	0	0	0					

HOLLIS, Scott (HAWL-lihs, SKAWT)

Right wing. Shoots right. 6', 185 lbs. Born, Kingston, Ont., September 18, 1972.
(Vancouver's 9th choice, 165th overall, in 1992 Entry Draft).

			Regular Season					Playoffs				
Season	Club	Lea	GP	G	A	TP	PIM	GP	G	A	TP	PIM
1988-89	Kingston Lions	OMHA	14	14	14	28	15	....	....	....	....	....
1989-90	Oshawa Generals	OHL	50	4	6	10	33	9	0	1	1	2
1990-91	Oshawa Generals	OHL	66	24	33	57	91	16	5	3	8	20
1991-92	Oshawa Generals	OHL	66	47	54	101	183	7	7	3	10	8
1992-93	Oshawa Generals	OHL	62	49	53	102	148	13	8	15	23	22
1993-94	Las Vegas	IHL	23	3	1	4	65	....	....	....	....	....
	Knoxville	ECHL	29	20	16	36	99	3	3	1	4	8
1994-95	Adirondack	AHL	48	12	15	27	118	....	....	....	....	....
1995-96	Toledo Storm	ECHL	7	7	11	18	5	....	....	....	....	....
	Adirondack	AHL	55	18	17	37	111	3	0	1	1	4
1996-97	San Antonio	IHL	73	17	17	34	187	9	1	1	2	6
1997-98	San Antonio	IHL	19	15	6	21	21	....	....	....	....	....
	Orlando	IHL	48	16	23	39	68	17	5	4	9	30
1998-99	Orlando	IHL	3	1	1	2	6	....	....	....	....	....
	Long Beach	IHL	13	2	5	7	21	....	....	....	....	....
	Las Vegas	IHL	53	20	25	45	67	....	....	....	....	....
99-2000	Rosenheim	Germany	63	21	16	37	80	....	....	....	....	....
2000-01	Syracuse Crunch	AHL	38	10	10	20	92	....	....	....	....	....
	Houston Aeros	IHL	22	7	6	13	22	7	1	0	1	10
2001-02	Flint Generals	UHL	69	35	39	74	163	5	1	2	3	18

Signed as a free agent by **Columbus**, August 7, 2000.

HOLLWEG, Ryan (HOHL-wehg, RIGH-uhn) **NYR**

Center. Shoots left. 5' 9", 201 lbs. Born, Downey, CA, April 23, 1983.
(NY Rangers' 10th choice, 238th overall, in 2001 Entry Draft).

			Regular Season					Playoffs				
Season	Club	Lea	GP	G	A	TP	PIM	GP	G	A	TP	PIM
1998-99	Langley Hornets	BCHL	58	14	40	54	187	....	....	....	....	....
	Grandview	PIJHL	41	23	27	50	135	....	....	....	....	....
99-2000	Medicine Hat	WHL	54	19	27	46	107	....	....	....	....	....
2000-01	Medicine Hat	WHL	65	19	39	58	125	....	....	....	....	....
2001-02	Medicine Hat	WHL	58	30	40	70	121	....	....	....	....	....
	Hartford	AHL	8	1	1	2	2	9	0	2	2	19

HOLMQVIST, Andreas (HOHLM-kvihst, ahn-DRAY-uhs) **T.B.**

Defense. Shoots right. 6'4", 190 lbs. Born, Stockholm, Sweden, July 23, 1981.
(Tampa Bay's 3rd choice, 61st overall, in 2001 Entry Draft).

			Regular Season					Playoffs				
Season	Club	Lea	GP	G	A	TP	PIM	GP	G	A	TP	PIM
99-2000	Hammarby Jr.	Swede-Jr.	33	8	12	20	16	6	1	2	3	4
2000-01	Hammarby Jr.	Swede-Jr.	47	6	15	21	40	....	....	....	....	....
2001-02	Hammarby	Swede-2	42	11	13	24	97	....	....	....	....	....

HOLMQVIST, Mikael (HOHLM-kvihst, MIHK-al) **ANA.**

Center. Shoots left. 6'3", 189 lbs. Born, Stockholm, Sweden, June 8, 1979.
(Anaheim's 1st choice, 18th overall, in 1997 Entry Draft).

			Regular Season					Playoffs				
Season	Club	Lea	GP	G	A	TP	PIM	GP	G	A	TP	PIM
1995-96	Djurgarden Jr.	Swede-Jr.	24	7	2	9	4	....	....	....	....	....
1996-97	Djurgarden Jr.	Swede-Jr.	39	29	35	64	110	....	....	....	....	....
	Djurgarden	Sweden	9	0	0	0	0	....	....	....	....	....
1997-98	Farjestad	Sweden	41	2	3	5	6	7	0	0	0	0
	Farjestad	EuroHL	5	2	2	4	2	....	....	....	....	....
1998-99	Farjestad Jr.	Swede-Jr.	2	2	2	4	2	....	....	....	....	....
	Farjestad	EuroHL	3	0	0	0	0	1	0	0	0	0
	Farjestad	Sweden	15	0	0	0	6	....	....	....	....	....
	Hammarby	Swede-2	3	2	0	2	0	....	....	....	....	....
99-2000	TPS Turku	Finland	54	12	3	15	14	11	2	3	5	4
2000-01	TPS Turku	Finland	46	4	5	9	8	10	1	3	4	2
2001-02	TPS Turku	Finland	56	9	13	22	16	8	1	0	1	12

HOLTET, Marius (HOHL-teht, MAIR-ee-uhs) **DAL.**

Center. Shoots right. 6', 183 lbs. Born, Hamar, Norway, August 31, 1984.
(Dallas' 4th choice, 42nd overall, in 2002 Entry Draft).

			Regular Season					Playoffs				
Season	Club	Lea	GP	G	A	TP	PIM	GP	G	A	TP	PIM
2000-01	Farjestad Jr.	Swede-Jr.	26	5	4	9	34	....	....	....	....	....
2001-02	Farjestad Jr.	Swede-Jr.	37	12	7	19	70	....	....	....	....	....

HOLUB, Jan (HOH-luhb, YAN) **NYI**

Defense. Shoots left. 6'3", 185 lbs. Born, Liberec, Czech., May 3, 1983.
(NY Islanders' 4th choice, 197th overall, in 2001 Entry Draft).

			Regular Season					Playoffs				
Season	Club	Lea	GP	G	A	TP	PIM	GP	G	A	TP	PIM
99-2000	HC Liberec Jr.	Czech-Jr.	2	0	0	0	0	....	....	....	....	....
	HC Liberec-18	Czech-Jr.	33	2	5	7	65	....	....	....	....	....
2000-01	HC Liberec Jr.	Czech-Jr.	43	0	6	6	42	....	....	....	....	....
2001-02	HC Liberec Jr.	Czech-Jr.	11	2	0	2	24	....	....	....	....	....
	Jablonec n. Nisou	Czech-3	3	0	0	0	2	....	....	....	....	....
	HC Tygri Liberec	Czech-3	30	1	3	4	22	12	0	0	0	2

HOOTON, Brock (HOO-tuhn, BRAWK) **OTT.**

Center. Shoots right. 6'1", 185 lbs. Born, Smithers, B.C., March 20, 1983.
(Ottawa's 6th choice, 150th overall, in 2002 Entry Draft).

			Regular Season					Playoffs				
Season	Club	Lea	GP	G	A	TP	PIM	GP	G	A	TP	PIM
1998-99	Smithers Selects	BCAHA	40	30	55	85	30	....	....	....	....	....
99-2000	Campbell River	VIJHL	40	8	17	25	8	....	....	....	....	....
2000-01	Quesnel	BCHL	60	11	26	37	....	....	....	....	....	....
2001-02	Quesnel	BCHL	60	34	50	84	33	....	....	....	....	....

BCHL Interior First All-Star Team (2002)

• Signed Letter of Intent to attend **St. Cloud State** (WCHA), May 2, 2002.

HORACEK, Jan (HOHR-uh-chehk, YAN) EDM.

Defense. Shoots right. 6'4", 206 lbs. Born, Benesov, Czech., May 22, 1979.
(St. Louis' 3rd choice, 98th overall, in 1997 Entry Draft).

			Regular Season					Playoffs				
Season	**Club**	**Lea**	**GP**	**G**	**A**	**TP**	**PIM**	**GP**	**G**	**A**	**TP**	**PIM**
1995-96	Slavia Praha Jr.	Czech-Jr.	18	1	5	6	0	....	....	....	....	....
	Karlovy Vary	Czech-2	11	0	0	0	0	....	....	....	....	....
	HC Slavia Praha	Czech	8	0	1	1	4	....	....	....	....	....
1996-97	Slavia Praha Jr.	Czech-Jr.	25	4	14	18	....	....	....	....	....	....
	H+S Beroun	Czech-2	2	0	0	0	....	....	....	....	....	....
	HC Slavia Praha	Czech	9	0	0	0	6	3	0	0	0	0
1997-98	Moncton Wildcats	QMJHL	54	3	18	21	146	10	1	5	6	20
1998-99	HC Slavia Praha	Czech	1	0	0	0	2	....	....	....	....	....
	Worcester	AHL	53	1	13	14	119	4	0	0	0	6
99-2000	Worcester	AHL	68	1	8	9	145	9	0	0	0	2
2000-01	Peoria Rivermen	ECHL	6	0	3	3	8	....	....	....	....	....
	Worcester	AHL	11	0	1	1	20	2	0	0	0	2
2001-02	Hamilton	AHL	44	0	5	5	99	13	0	1	1	10

• Missed majority of 2000-01 season recovering from wrist surgery, October 23, 2000. Traded to **Edmonton** by **St. Louis** with Marty Reasoner and Jochen Hecht for Doug Weight and Michel Riesen, July 1, 2001.

HOUSE, Bobby (HOWSE, BAW-bee) TOR.

Right wing. Shoots right. 6'1", 205 lbs. Born, Whitehorse, Yukon, January 7, 1973.
(Chicago's 4th choice, 66th overall, in 1991 Entry Draft).

			Regular Season					Playoffs				
Season	**Club**	**Lea**	**GP**	**G**	**A**	**TP**	**PIM**	**GP**	**G**	**A**	**TP**	**PIM**
1988-89	Whitehorse Elks	AAHL	28	36	27	63	28	....	....	....	....	....
1989-90	Spokane Chiefs	WHL	64	18	16	34	74	5	0	0	0	6
1990-91	Spokane Chiefs	WHL	38	11	19	30	63	....	....	....	....	....
	Brandon	WHL	23	18	7	25	14	....	....	....	....	....
1991-92	Brandon	WHL	71	35	42	77	133	....	....	....	....	....
1992-93	Brandon	WHL	61	57	39	96	87	4	2	2	4	0
1993-94	Indianapolis Ice	IHL	42	10	8	18	51	....	....	....	....	....
	Flint Generals	ColHL	4	3	3	6	0	....	....	....	....	....
1994-95	Columbus Chill	ECHL	9	11	6	17	2	....	....	....	....	....
	Indianapolis Ice	IHL	26	2	3	5	26	....	....	....	....	....
	Albany	AHL	26	4	7	11	12	8	1	1	2	0
1995-96	Albany	AHL	77	37	49	86	57	4	0	0	0	4
1996-97	Albany	AHL	68	18	16	34	65	16	3	2	5	23
1997-98	Albany	AHL	19	10	10	20	10	....	....	....	....	....
	Hershey Bears	AHL	20	2	6	8	8	....	....	....	....	....
	Quebec Rafales	IHL	24	5	7	12	12	....	....	....	....	....
	Syracuse Crunch	AHL	9	5	6	11	6	5	2	0	2	4
1998-99	Augusta Lynx	ECHL	5	1	0	1	15	....	....	....	....	....
	Albany	AHL	1	0	0	0	0	....	....	....	....	....
	Springfield	AHL	56	11	18	29	27	3	1	0	1	2
99-2000	St. John's	AHL	68	24	29	53	46	....	....	....	....	....
2000-01	St. John's	AHL	65	36	33	69	49	....	....	....	....	....
2001-02	St. John's	AHL	71	36	20	56	40	11	6	2	8	6

WHL East Second All-Star Team (1993) • AHL Second All-Star Team (2001)

Traded to **New Jersey** by **Chicago** for cash, May 21, 1996. Signed as a free agent by **Toronto**, August 20, 1999.

HUBL, Viktor (HEW-buhl, VIHK-tohr) WSH.

Left wing. Shoots left. 6', 183 lbs. Born, Chomutov, Czech., August 13, 1978.
(Washington's 10th choice, 284th overall, in 2001 Entry Draft).

			Regular Season					Playoffs				
Season	**Club**	**Lea**	**GP**	**G**	**A**	**TP**	**PIM**	**GP**	**G**	**A**	**TP**	**PIM**
1997-98	KLH Chomutov	Czech-2	50	18	17	35	....	....	....	....	....	....
1998-99	KLH Chomutov	Czech-2	5	2	1	3	....	....	....	....	....	....
	Litvinov	Czech	27	2	6	8	10	....	....	....	....	....
99-2000	KLH Chomutov	Czech-2	8	2	4	6	4	....	....	....	....	....
	Litvinov	Czech	36	6	9	15	16	....	....	....	....	....
2000-01	HC Slavia Praha	Czech	50	16	24	40	24	11	0	2	2	6
2001-02	HC Slavia Praha	Czech	48	11	14	25	34	8	0	1	1	6

HUDLER, Jiri (HUHD-luhr, YIH-ree) DET.

Center. Shoots left. 5'9", 178 lbs. Born, Olomouc, Czech., January 4, 1984.
(Detroit's 1st choice, 58th overall, in 2002 Entry Draft).

			Regular Season					Playoffs				
Season	**Club**	**Lea**	**GP**	**G**	**A**	**TP**	**PIM**	**GP**	**G**	**A**	**TP**	**PIM**
99-2000	Vsetin Jr.	Czech-Jr.	53	29	31	60	75	....	....	....	....	....
	Vsetin	Czech	2	0	1	1	4	....	....	....	....	....
2000-01	Vsetin Jr.	Czech-Jr.	16	8	14	22	16	....	....	....	....	....
	Vsetin	Czech	22	1	4	5	10	....	....	....	....	....
	HC Femax Havirov	Czech	15	5	1	6	12	....	....	....	....	....
2001-02	HC Vsetin	Czech	46	15	31	46	54	....	....	....	....	....
	HC Olomouc	Czech-3	1	0	2	2	4	....	....	....	....	....
	HC Liberec	Czech-2	13	9	7	16	10	....	....	....	....	....

HULVA, Jakub (HUHL-vuh, YA-kuhb) BUF.

Right wing. Shoots left. 6', 172 lbs. Born, Opava, Czech., May 6, 1984.
(Buffalo's 5th choice, 108th overall, in 2002 Entry Draft).

			Regular Season					Playoffs				
Season	**Club**	**Lea**	**GP**	**G**	**A**	**TP**	**PIM**	**GP**	**G**	**A**	**TP**	**PIM**
99-2000	HC Vitkovice Jr.	Czech-Jr.	49	20	34	54	38	....	....	....	....	....
2000-01	HC Vitkovice-18	Czech-Jr.	30	28	40	68	22	....	....	....	....	....
	HC Vitkovice Jr.	Czech-Jr.	20	5	6	11	6	....	....	....	....	....
	HC Vitkovice	Czech	1	0	0	0	0	....	....	....	....	....
2001-02	HC Vitkovice Jr.	Czech-Jr.	45	25	24	49	36	....	....	....	....	....
	HC Opava	Czech-2	4	0	0	0	2	....	....	....	....	....
	HC Vitkovice	Czech	1	0	0	0	0	2	0	0	0	0

HUSKINS, Kent (HUHS-kihns, KEHNT) CHI.

Defense. Shoots left. 6'2", 190 lbs. Born, Ottawa, Ont., May 4, 1979.
(Chicago's 3rd choice, 156th overall, in 1998 Entry Draft).

			Regular Season					Playoffs				
Season	**Club**	**Lea**	**GP**	**G**	**A**	**TP**	**PIM**	**GP**	**G**	**A**	**TP**	**PIM**
1995-96	Kanata Valley	OCJHL	49	6	21	27	18	....	....	....	....	....
1996-97	Kanata Valley	OCJHL	53	11	36	47	89	....	....	....	....	....
1997-98	Clarkson Knights	ECAC	35	2	8	10	46	....	....	....	....	....
1998-99	Clarkson Knights	ECAC	37	5	11	16	28	....	....	....	....	....
99-2000	Clarkson Knights	ECAC	28	2	16	18	30	....	....	....	....	....
2000-01	Clarkson Knights	ECAC	35	6	28	34	22	....	....	....	....	....
2001-02	Norfolk Admirals	AHL	65	4	11	15	44	4	0	1	1	0

ECAC First All-Star Team (2000, 2001) • NCAA East First All-American Team (2001)

HUSSEY, Matt (HUH-see, MAT) PIT.

Center. Shoots left. 6'2", 195 lbs. Born, New Haven, CT, May 28, 1979.
(Pittsburgh's 10th choice, 254th overall, in 1998 Entry Draft).

			Regular Season					Playoffs				
Season	**Club**	**Lea**	**GP**	**G**	**A**	**TP**	**PIM**	**GP**	**G**	**A**	**TP**	**PIM**
1996-97	Wayzata High	Hi-School	48	34	31	65	....	....	....	....	....	....
1997-98	Avon Old Farms	Hi-School	26	26	23	49	20	....	....	....	....	....
1998-99	U. of Wisconsin	WCHA	37	10	5	15	18	....	....	....	....	....
99-2000	U. of Wisconsin	WCHA	35	5	11	16	8	....	....	....	....	....
2000-01	U. of Wisconsin	WCHA	40	9	11	20	24	....	....	....	....	....
2001-02	U. of Wisconsin	WCHA	39	18	15	33	16	....	....	....	....	....

HUTCHINSON, Andrew (HUHT-chihn-suhn, AN-droo) NSH.

Defense. Shoots right. 6'2", 198 lbs. Born, Evanston, IL, March 24, 1980.
(Nashville's 4th choice, 54th overall, in 1999 Entry Draft).

			Regular Season					Playoffs				
Season	**Club**	**Lea**	**GP**	**G**	**A**	**TP**	**PIM**	**GP**	**G**	**A**	**TP**	**PIM**
1996-97	Det. Caesars	MNHL	82	15	41	56	....	....	....	....	....	....
1997-98	Team USA	USDP-18	59	7	21	28	53	....	....	....	....	....
1998-99	Michigan State	CCHA	37	3	12	15	26	....	....	....	....	....
99-2000	Michigan State	CCHA	42	5	12	17	64	....	....	....	....	....
2000-01	Michigan State	CCHA	42	5	19	24	46	....	....	....	....	....
2001-02	Michigan State	CCHA	39	6	16	22	24	....	....	....	....	....
	Milwaukee	AHL	5	0	1	1	0	....	....	....	....	....

CCHA Second All-Star Team (2001, 2002) • NCAA West Second All-American Team (2002)

HYMOVITZ, David (HIH-moh-vihtz, DAY-vihd) OTT.

Left wing. Shoots left. 5'11", 170 lbs. Born, Randolph, MA, May 30, 1974.
(Chicago's 9th choice, 209th overall, in 1992 Entry Draft).

			Regular Season					Playoffs				
Season	**Club**	**Lea**	**GP**	**G**	**A**	**TP**	**PIM**	**GP**	**G**	**A**	**TP**	**PIM**
1991-92	Thayer Academy	Hi-School	26	28	21	49	22	....	....	....	....	....
1992-93	Boston College	H-East	37	7	6	13	6	....	....	....	....	....
1993-94	Boston College	H-East	36	18	14	32	18	....	....	....	....	....
1994-95	Boston College	H-East	35	21	19	40	22	....	....	....	....	....
1995-96	Boston College	H-East	36	26	18	44	32	....	....	....	....	....
1996-97	Columbus Chill	ECHL	58	39	32	71	29	5	4	1	5	2
	Indianapolis Ice	IHL	6	0	1	1	0	1	0	0	0	0
1997-98	Indianapolis Ice	IHL	63	11	15	26	20	5	1	1	2	6
1998-99	Indianapolis Ice	IHL	78	46	30	76	42	5	2	3	5	2
99-2000	Lowell	AHL	67	19	17	36	30	....	....	....	....	....
	Houston Aeros	IHL	18	10	3	13	16	11	3	4	7	8
2000-01	Lowell	AHL	60	14	15	29	52	4	2	1	3	17
2001-02	Grand Rapids	AHL	80	15	33	48	22	5	1	2	3	0

IHL Second All-Star Team (1999)

Signed as a free agent by **LA Kings**, June 10, 1999. Traded to **Houston** (IHL) by **Lowell** (AHL) with LA Kings retaining NHL rights for Jeff Daw, March 17, 2000. Signed as a free agent by **Ottawa**, July 13, 2001.

INMAN, David (IHN-man, DAY-vihd)

Center. Shoots left. 6'1", 180 lbs. Born, New York, NY, June 13, 1980.
(NY Rangers' 3rd choice, 59th overall, in 1999 Entry Draft).

			Regular Season					Playoffs				
Season	**Club**	**Lea**	**GP**	**G**	**A**	**TP**	**PIM**	**GP**	**G**	**A**	**TP**	**PIM**
1995-96	Wexford Raiders	MTJHL	4	2	1	3	0	....	....	....	....	....
1996-97	Wexford Raiders	MTJHL	43	32	56	88	59	....	....	....	....	....
1997-98	Wexford Raiders	MTJHL	37	36	44	80	82	....	....	....	....	....
1998-99	U. of Notre Dame	CCHA	38	10	10	20	74	....	....	....	....	....
99-2000	U. of Notre Dame	CCHA	32	13	7	20	12	....	....	....	....	....
2000-01	U. of Notre Dame	CCHA	37	11	6	17	10	....	....	....	....	....
2001-02	U. of Notre Dame	CCHA	38	19	18	37	24	....	....	....	....	....

IRGL, Zbynek (UHR-guhl, ZBIH-nehk) NSH.

Center. Shoots right. 5'11", 183 lbs. Born, Vitkovice, Czech., November 29, 1980.
(Nashville's 9th choice, 197th overall, in 2000 Entry Draft).

			Regular Season					Playoffs				
Season	**Club**	**Lea**	**GP**	**G**	**A**	**TP**	**PIM**	**GP**	**G**	**A**	**TP**	**PIM**
1996-97	HC Vitkovice Jr.	Czech-Jr.	43	44	22	66	....	....	....	....	....	....
1997-98	HC Vitkovice Jr.	Czech-Jr.	37	17	10	27	....	....	....	....	....	....
1998-99	HC Vitkovice Jr.	Czech-Jr.	18	9	8	17	....	....	....	....	....	....
	HC Vitkovice	Czech	33	2	2	4	6	4	0	0	0	....
99-2000	HC Dukla Jihlava	Czech-2	1	0	0	0	0	....	....	....	....	....
	HC Vitkovice	Czech	47	7	5	12	10	....	....	....	....	....
2000-01	HC Vitkovice	Czech	37	0	1	1	8	4	0	0	0	0
2001-02	HC Vitkovice	Czech	39	2	8	10	8	13	4	0	4	6

ISOSALO, Samu (ee-soh-SA-low, SA-moo) ATL.

Right wing. Shoots left. 6'3", 205 lbs. Born, Rauma, Finland, October 10, 1981.
(Atlanta's 10th choice, 230th overall, in 2000 Entry Draft).

			Regular Season					Playoffs				
Season	**Club**	**Lea**	**GP**	**G**	**A**	**TP**	**PIM**	**GP**	**G**	**A**	**TP**	**PIM**
1996-97	Lukko Rauma-B	Finn-Jr.	21	5	6	11	45	....	....	....	....	....
1997-98	Lukko Rauma Jr.	Finn-Jr.	34	21	19	40	48	....	....	....	....	....
1998-99	North Bay	OHL	59	13	12	25	19	4	0	0	0	4
99-2000	North Bay	OHL	48	17	25	42	26	3	0	0	0	0
2000-01	Lukko Rauma Jr.	Finn-Jr.	14	11	9	20	42	3	1	2	3	0
	Jaa-Kotkat	Finland-2	3	2	1	3	2	....	....	....	....	....
	Lukko Rauma	Finland	31	1	1	2	33	1	0	0	0	0
2001-02	Lukko Rauma Jr.	Finn-Jr.	2	0	3	3	2	....	....	....	....	....
	Lukko Rauma	Finland	4	0	0	0	2	....	....	....	....	....

JAAKOLA, Topi (YAH-koh-luh, TOH-pee) FLA.

Defense. Shoots left. 6'1", 187 lbs. Born, Oulu, Finland, November 15, 1983.
(Florida's 5th choice, 134th overall, in 2002 Entry Draft).

			Regular Season					Playoffs				
Season	**Club**	**Lea**	**GP**	**G**	**A**	**TP**	**PIM**	**GP**	**G**	**A**	**TP**	**PIM**
99-2000	Karpat Oulu Jr.	Finn-Jr.	36	5	12	17	71	5	1	1	2	4
2000-01	Karpat Oulu Jr.	Finn-Jr.	35	4	10	14	47	6	0	2	2	6
	Karpat Oulu	Finland	4	0	0	0	2	....	....	....	....	....
2001-02	Karpat Oulu Jr.	Finn-Jr.	3	0	0	0	2	2	1	0	1	4
	Karpat Oulu	Finland	44	0	4	4	18	4	0	0	0	4

JAASKELAINEN, Teemu (yas-keh-LIGH-nuhn, TEE-moo) CHI.

Defense. Shoots left. 6'1", 202 lbs. Born, Tampere, Finland, June 6, 1983.
(Chicago's 11th choice, 205th overall, in 2001 Entry Draft).

			Regular Season					Playoffs				
Season	Club	Lea	GP	G	A	TP	PIM	GP	G	A	TP	PIM
1998-99	Ilves Tampere-C	Finn-Jr.	22	1	4	5	8	3	0	0	0	2
99-2000	Ilves Tampere-B	Finn-Jr.	36	5	1	6	20	....	....	....	....	....
2000-01	Ilves Tampere-B	Finn-Jr.	5	1	1	2	16	....	....	....	....	....
	Ilves Jr.	Finn-Jr.	41	4	1	5	30	....	....	....	....	....
2001-02	Ilves Jr.	Finn-Jr.	13	1	6	7	8	....	....	....	....	....
	Ilves Tampere	Finland	38	0	0	0	24	3	0	0	0	0

JACKMAN, Tim (JAK-man, TIHM) CBJ

Right wing. Shoots right. 6'3", 190 lbs. Born, Minot, ND, November 14, 1981.
(Columbus' 2nd choice, 38th overall, in 2001 Entry Draft).

			Regular Season					Playoffs				
Season	Club	Lea	GP	G	A	TP	PIM	GP	G	A	TP	PIM
1998-99	Park Center High	Hi-School	22	22	22	44	....	....	....	....	....	....
99-2000	Park Center High	Hi-School	19	34	22	56	....	....	....	....	....	....
	Twin Cities	USHL	25	11	9	20	58	13	8	5	13	12
2000-01	Minnesota State	WCHA	37	11	14	25	92	....	....	....	....	....
2001-02	Minnesota State	WCHA	36	14	14	28	86	....	....	....	....	....

Minnesota High School All-Conference Team (1999, 2000) • Minnesota All-State Team (2000)

JACKSON, Todd (JAK-suhn, TAWD) DET.

Right wing. Shoots right. 5'11", 170 lbs. Born, Syracuse, NY, April 10, 1981.
(Detroit's 10th choice, 251st overall, in 2000 Entry Draft).

			Regular Season					Playoffs				
Season	Club	Lea	GP	G	A	TP	PIM	GP	G	A	TP	PIM
1998-99	Team USA	USDP-17	53	11	9	20	56	....	....	....	....	....
99-2000	Team USA	USDP-17	29	8	10	18	25	....	....	....	....	....
	Team USA	USDP-18	23	8	6	14	12	....	....	....	....	....
2000-01	U. of Maine	H-East	39	4	8	12	8	....	....	....	....	....
2001-02	U. of Maine	H-East	39	7	21	28	10	....	....	....	....	....

JACOBSEN, Grant (JAY-kawb-suhn, GRANT) ST.L.

Center. Shoots left. 6' 2", 201 lbs. Born, Souris, Man., March 4, 1983.
(St. Louis' 7th choice, 270th overall, in 2001 Entry Draft).

			Regular Season					Playoffs				
Season	Club	Lea	GP	G	A	TP	PIM	GP	G	A	TP	PIM
99-2000	Regina Pats	WHL	56	7	7	14	20	7	0	2	2	0
2000-01	Regina Pats	WHL	64	10	25	35	40	6	0	1	1	4
2001-02	Regina Pats	WHL	58	16	28	44	48	6	1	1	2	4

JAKES, Jiri (YA-kesh, YOO-ree) BOS.

Right wing. Shoots left. 6'4", 210 lbs. Born, Prague, Czech., October 4, 1982.
(Boston's 4th choice, 147th overall, in 2001 Entry Draft).

			Regular Season					Playoffs				
Season	Club	Lea	GP	G	A	TP	PIM	GP	G	A	TP	PIM
99-2000	Sparta Praha Jr.	Czech-Jr.	27	5	7	12	....	....	....	....	....	....
2000-01	Brandon	WHL	64	22	16	38	73	6	1	1	2	4
2001-02	Brandon	WHL	56	16	18	34	73	19	3	3	6	30

JAMIESON, Dusty (JAY-mih-suhn, DUHS-tee)

Left wing. Shoots left. 6'3", 176 lbs. Born, Sarnia, Ont., May 26, 1981.
(Montreal's 5th choice, 136th overall, in 1999 Entry Draft).

			Regular Season					Playoffs				
Season	Club	Lea	GP	G	A	TP	PIM	GP	G	A	TP	PIM
1996-97	St. Thomas Stars	OJHL-B	50	15	16	31	19	....	....	....	....	....
1997-98	Guelph Storm	OHL	48	3	4	7	0	12	0	1	1	0
1998-99	Guelph Storm	OHL	12	2	7	9	0	....	....	....	....	....
	Sarnia Sting	OHL	54	14	21	35	10	6	1	0	1	2
99-2000	Sarnia Sting	OHL	50	15	17	32	15	7	3	2	5	0
2000-01	Sarnia Sting	OHL	66	14	35	49	22	4	4	4	8	0
2001-02	Sarnia Sting	OHL	68	44	53	97	33	5	5	1	6	4

Traded to **Sarnia** (OHL) by **Guelph** (OHL) for Darryl Knight, November 12, 1998.

JAMINKI, Tommi (yah-MIHN-kee,TAW-mee) CHI.

Left wing. Shoots right. 6', 180 lbs. Born, Turku, Finland, February 11, 1983.
(Chicago's 8th choice, 142nd overall, in 2001 Entry Draft).

			Regular Season					Playoffs				
Season	Club	Lea	GP	G	A	TP	PIM	GP	G	A	TP	PIM
99-2000	KJT Kerawa Jr.	Finn-Jr.	35	9	10	19	106	....	....	....	....	....
2000-01	Blues Espoo Jr.	Finn-Jr.	36	6	6	12	16	....	....	....	....	....
2001-02	Blues Espoo Jr.	Finn-Jr.	31	9	7	16	66	2	0	0	0	2
	Blues Espoo	Finland	4	0	0	0	0	....	....	....	....	....

JAMTIN, Andreas (yahm-TEEN, ahn-DRAY-uhs) DET.

Right wing. Shoots left. 5'11", 185 lbs. Born, Stockholm, Sweden, May 4, 1983.
(Detroit's 4th choice, 157th overall, in 2001 Entry Draft).

			Regular Season					Playoffs				
Season	Club	Lea	GP	G	A	TP	PIM	GP	G	A	TP	PIM
1998-99	AIK Solna Jr.	Swede-Jr.	44	33	29	62	105	....	....	....	....	....
99-2000	Farjestad Jr.	Swede-Jr.	28	6	6	12	36	....	....	....	....	....
2000-01	Farjestad-18	Swede-Jr.	1	1	0	1	2	....	....	....	....	....
	Farjestad Jr.	Swede-Jr.	13	5	8	13	83	....	....	....	....	....
	Farjestad	Sweden	1	0	0	0	0	....	....	....	....	....
2001-02	AIK Solna	Sweden	42	2	3	5	55	....	....	....	....	....
	AIK Stockholm	Swede-2	12	12	15	27	61	1	2	2	4	2
	AIK Stockholm	Swede-Q	10	1	2	3	4	....	....	....	....	....

JANCEVSKI, Dan (jan-SEHV-skee, DAN) DAL.

Defense. Shoots left. 6'3", 212 lbs. Born, Windsor, Ont., June 15, 1981.
(Dallas' 2nd choice, 66th overall, in 1999 Entry Draft).

			Regular Season					Playoffs				
Season	Club	Lea	GP	G	A	TP	PIM	GP	G	A	TP	PIM
1995-96	Riverside	OMHA	59	9	22	31	67	....	....	....	....	....
1996-97	Windsor Lions	OMHA	47	6	20	26	99	....	....	....	....	....
1997-98	Tecumseh	OJHL-B	49	3	11	14	145	....	....	....	....	....
1998-99	London Knights	OHL	68	2	12	14	115	25	1	7	8	24
99-2000	London Knights	OHL	59	8	15	23	138	....	....	....	....	....
2000-01	London Knights	OHL	39	4	23	27	95	....	....	....	....	....
	Sudbury Wolves	OHL	31	3	14	17	42	12	0	9	9	17
2001-02	Utah Grizzlies	AHL	77	0	13	13	147	5	0	0	0	4

Traded to **Sudbury** (OHL) by **London** (OHL) with Chris Kelly for Dennis Wideman and future considerations, January 10, 2001.

JANIK, Doug (JAN-nihk, DUHG) BUF.

Defense. Shoots left. 6'2", 209 lbs. Born, Agawam, MA, March 26, 1980.
(Buffalo's 3rd choice, 55th overall, in 1999 Entry Draft).

			Regular Season					Playoffs				
Season	Club	Lea	GP	G	A	TP	PIM	GP	G	A	TP	PIM
1995-96	Springfield	NEJHL	48	16	38	54	....	....	....	....	....	....
1996-97	Springfield	NEJHL	39	12	24	36	22	11	5	9	14	10
1997-98	Team USA	USDP-18	65	8	26	34	105	....	....	....	....	....
1998-99	U. of Maine	H-East	35	3	13	16	44	....	....	....	....	....
99-2000	U. of Maine	H-East	36	6	14	20	54	....	....	....	....	....
2000-01	U. of Maine	H-East	39	3	15	18	52	....	....	....	....	....
2001-02	Rochester	AHL	80	6	17	23	100	2	0	0	0	0

JANSSEN, Cam (JAN-suhn, KAM) N.J.

Right wing. Shoots right. 5'11", 200 lbs. Born, St. Louis, MO, April 15, 1984.
(New Jersey's 6th choice, 117th overall, in 2002 Entry Draft).

			Regular Season					Playoffs				
Season	Club	Lea	GP	G	A	TP	PIM	GP	G	A	TP	PIM
2000-01	St. Louis Jr.	NAJHL	45	1	2	3	244	....	....	....	....	....
2001-02	Windsor	OHL	64	5	17	22	268	10	0	0	0	13

JARRETT, Cole (JAIR-reht, KOHL) CBJ

Defense. Shoots left. 6', 195 lbs. Born, Sault Ste. Marie, Ont., January 4, 1983.
(Columbus' 6th choice, 141st overall, in 2001 Entry Draft).

			Regular Season					Playoffs				
Season	Club	Lea	GP	G	A	TP	PIM	GP	G	A	TP	PIM
1998-99	Waterloo Siskins	OPJHL	44	6	10	16	43	....	....	....	....	....
99-2000	Plymouth Whalers	OHL	57	3	7	10	47	23	3	7	10	19
2000-01	Plymouth Whalers	OHL	60	12	36	48	98	19	6	12	18	29
2001-02	Plymouth Whalers	OHL	51	14	24	38	92	6	1	1	2	18

JARRETT, Patrick (JEHR-eht, PAT-rihk) NSH.

Center. Shoots left. 5'11", 181 lbs. Born, Sault Ste. Marie, Ont., February 6, 1984.
(Nashville's 3rd choice, 138th overall, in 2002 Entry Draft).

			Regular Season					Playoffs				
Season	Club	Lea	GP	G	A	TP	PIM	GP	G	A	TP	PIM
1998-99	Soo Thunder	NOHA	35	32	39	*71	....	....	....	....	....	....
99-2000	Soo Thunder	NOBHL	39	22	32	54	57	....	....	....	....	....
2000-01	Mississauga	OHL	60	15	38	53	22	....	....	....	....	....
2001-02	Mississauga	OHL	20	3	10	13	14	....	....	....	....	....
	Owen Sound	OHL	27	3	14	17	10	....	....	....	....	....

NOHA Rookie of the Year (1999)

Traded to **Owen Sound** (OHL) by **Mississauga** (OHL) with Matt Passfield for Daniel Sisca and Greg Jacina, January 5, 2002.

JENSEN, Erik (JEHN-sehn, AIR-ihk) N.J.

Right wing. Shoots right. 6'1", 195 lbs. Born, Madison, WI, September 4, 1979.
(New Jersey's 10th choice, 199th overall, in 1998 Entry Draft).

			Regular Season					Playoffs				
Season	Club	Lea	GP	G	A	TP	PIM	GP	G	A	TP	PIM
1997-98	Des Moines	USHL	41	12	14	26	90	....	....	....	....	....
1998-99	Des Moines	USHL	26	5	11	16	62	14	4	3	7	35
99-2000	U. of Wisconsin	WCHA	22	3	3	6	30	....	....	....	....	....
2000-01	U. of Wisconsin	WCHA	39	6	7	13	57	....	....	....	....	....
2001-02	U. of Wisconsin	WCHA	36	1	5	6	94	....	....	....	....	....

JINDRICH, Robert (IHN-drihkh, RAW-buhrt) S.J.

Defense. Shoots left. 5'11", 195 lbs. Born, Plzen, Czech., October 14, 1976.
(San Jose's 10th choice, 168th overall, in 1995 Entry Draft).

			Regular Season					Playoffs				
Season	Club	Lea	GP	G	A	TP	PIM	GP	G	A	TP	PIM
1993-94	HC Skoda Plzen	Czech	18	0	2	2	....	....	....	....	....	....
1994-95	Plzen	Czech	11	1	0	1	4	....	....	....	....	....
1995-96	HC ZKZ Plzen	Czech	37	1	3	4	....	3	0	0	0	....
1996-97	HC ZKZ Plzen	Czech	49	7	9	16	44	....	....	....	....	....
1997-98	H+S Beroun	Czech-2	13	4	2	6	0	....	....	....	....	....
	Plzen	Czech	39	1	6	7	18	4	0	0	0	0
1998-99	Plzen	Czech	52	6	12	18	24	5	1	0	1	....
99-2000	Kentucky	AHL	78	2	21	23	51	9	0	4	4	6
2000-01	Kentucky	AHL	62	4	16	20	36	2	0	0	0	0
2001-02	Timra IK	Sweden	43	3	4	7	46	....	....	....	....	....

JOHANSSON, Daniel (yoh-HAN-suhn, DAN-yehl) L.A.

Center. Shoots left. 5'11", 176 lbs. Born, Ornskoldsvik, Sweden, July 5, 1981.
(Los Angeles' 6th choice, 125th overall, in 1999 Entry Draft).

			Regular Season					Playoffs				
Season	Club	Lea	GP	G	A	TP	PIM	GP	G	A	TP	PIM
1997-98	MoDo Jr.	Swede-Jr.	6	0	0	0	0	....	....	....	....	....
1998-99	MoDo Jr.	Swede-Jr.	43	10	19	29	....	....	....	....	....	....
99-2000	MoDo Jr.	Swede-Jr.	35	11	21	32	34	....	....	....	....	....
	MoDo	EuroHL	2	0	0	0	0	....	....	....	....	....
2000-01	Bodens IK	Swede-2	36	4	3	7	10	4	0	0	0	0
2001-02	Bodens IK	Swede-2	37	2	3	5	10	....	....	....	....	....
	Vaxjo HC	Swede-3	8	2	5	7	6	....	....	....	....	....

JOHANSSON, Eric (joh-HAHN-suhn, AIR-ihk) N.J.

Center. Shoots left. 6', 190 lbs. Born, Edmonton, Alta., January 7, 1982.
(New Jersey's choice, 187th overall, in 2002 Entry Draft).

			Regular Season					Playoffs				
Season	Club	Lea	GP	G	A	TP	PIM	GP	G	A	TP	PIM
1997-98	Edmonton CAC	AMHA	22	13	10	23	19	....	....	....	....	....
1998-99	Tri-City	WHL	48	8	14	22	20	6	1	1	2	2
99-2000	Tri-City	WHL	72	24	36	60	38	4	0	0	0	2
2000-01	Tri-City	WHL	72	36	44	80	72	....	....	....	....	....
2001-02	Tri-City	WHL	69	44	59	103	73	5	1	2	3	5

• Re-entered NHL Entry Draft. Originally Minnesota's 9th choice, 255th overall, in 2000 Entry Draft.

WHL West Second All-Star Team (2002)

JOHANSSON, Jonas (yoh-HAHN-suhn, YOH-nuhs) COL.

Right wing. Shoots right. 6'1", 180 lbs. Born, Jonkoping, Sweden, March 18, 1984.
(Colorado's 1st choice, 28th overall, in 2002 Entry Draft).

			Regular Season					Playoffs				
Season	Club	Lea	GP	G	A	TP	PIM	GP	G	A	TP	PIM
99-2000	HV 71 Jr.	Swede-Jr.	9	6	3	9	2	2	0	0	0	4
2000-01	HV 71 Jr.	Swede-Jr.	27	13	8	21	14	2	1	0	1	0
2001-02	HV 71 Jr.	Swede-Jr.	26	15	19	34	20	....	....	....	....	....
	HV 71 Jonkoping	Sweden	5	0	0	0	0	2	0	0	0	0

JOHANSSON, Mathias (yoh-HAHN-suhn) CGY.

Center. Shoots left. 6'2", 185 lbs. Born, Oskarshamn, Sweden, February 22, 1974.
(Calgary's 3rd choice, 54th overall, in 1992 Entry Draft).

			Regular Season					Playoffs				
Season	Club	Lea	GP	G	A	TP	PIM	GP	G	A	TP	PIM
1990-91	Farjestad	Sweden	3	0	0	0	0	....	....	....	....	....
1991-92	Farjestad	Sweden	16	0	0	0	2	1	0	0	0	0
1992-93	Grums IK	Swede-2	25	8	6	14	12	....	....	....	....	....
	Farjestad	Sweden	12	2	3	5	4	3	0	0	0	0
1993-94	Farjestad	Sweden	16	2	1	3	4	....	....	....	....	....
1994-95	Farjestad	Sweden	40	7	8	15	30	4	4	3	7	2
1995-96	Farjestad	Sweden	40	8	21	29	10	8	2	1	3	4
1996-97	Farjestad	Sweden	48	12	15	27	14	14	4	4	8	12
	Farjestad	EuroHL	5	0	1	1	0	....	....	....	....	....
1997-98	Farjestad	Sweden	46	8	21	29	36	12	2	1	3	10
	Farjestad	EuroHL	8	3	4	7	2	....	....	....	....	....
1998-99	Farjestad	Sweden	50	9	15	24	14	3	0	1	1	4
	Farjestad	EuroHL	5	1	3	4	6	2	0	0	0	4
99-2000	Farjestad	Sweden	49	20	19	39	40	7	2	1	3	4
2000-01	Farjestad	Sweden	49	15	20	35	42	16	4	9	13	18
2001-02	Farjestad	Sweden	50	4	11	15	22	10	6	1	7	8
	Sweden	Olympics	4	1	0	1	0	....	....	....	....	....

JOHANSSON, Tobias (yoh-HAHN-suhn, TOH-bigh-as) ANA.

Left wing. Shoots left. 5'11", 180 lbs. Born, Malmo, Sweden, October 22, 1977.
(Anaheim's 7th choice, 224th overall, in 1996 Entry Draft).

			Regular Season					Playoffs				
Season	Club	Lea	GP	G	A	TP	PIM	GP	G	A	TP	PIM
1995-96	Malmo IF Jr.	Swede-Jr.	30	7	13	20	38	....	....	....	....	....
1996-97	Malmo IF Jr.	Swede-Jr.	15	6	8	14	63	....	....	....	....	....
1997-98	Tranas AIF	Swede-2	31	7	2	9	18	....	....	....	....	....
1998-99	Tranas AIF	Swede-2	32	11	5	16	28	....	....	....	....	....
99-2000	Tranas AIF	Swede-2	44	10	8	18	83	....	....	....	....	....
2000-01	Tranas AIF	Swede-2	42	23	11	34	50	3	0	0	0	0
2001-02	Tranas IK	Swede-2	30	11	8	19	36	....	....	....	....	....

JOHNER, Dustin (JAW-nuhr, DUHS-tihn) FLA.

Center. Shoots right. 5'11", 170 lbs. Born, Estevan, Sask., March 6, 1983.
(Florida's 8th choice, 169th overall, in 2001 Entry Draft).

			Regular Season					Playoffs				
Season	Club	Lea	GP	G	A	TP	PIM	GP	G	A	TP	PIM
1998-99	Red Deer Rebels	AMBHL	36	35	29	64	42	....	....	....	....	....
99-2000	Red Deer Chiefs	AMHL	36	24	31	55	80	....	....	....	....	....
	Seattle	WHL	6	0	1	1	0	....	....	....	....	....
2000-01	Seattle	WHL	72	25	31	56	45	9	1	5	6	6
2001-02	Seattle	WHL	71	33	48	81	71	8	4	3	7	8

JOHNSON, Aaron (JAWN-suhn, AIR-ruhn) CBJ

Defense. Shoots left. 6', 186 lbs. Born, Port Hawkesbury, N.S., April 30, 1983.
(Columbus' 4th choice, 85th overall, in 2001 Entry Draft).

			Regular Season					Playoffs				
Season	Club	Lea	GP	G	A	TP	PIM	GP	G	A	TP	PIM
99-2000	Rimouski Oceanic	QMJHL	63	1	14	15	57	8	0	0	0	0
2000-01	Rimouski Oceanic	QMJHL	64	12	41	53	128	11	2	4	6	35
2001-02	Rimouski Oceanic	QMJHL	68	17	49	66	172	7	1	2	3	12

JOHNSON, Gregg (JAWN-suhn, GREHG) OTT.

Center. Shoots left. 5'11", 183 lbs. Born, Windsor, CT, June 18, 1982.
(Ottawa's 11th choice, 256th overall, in 2001 Entry Draft).

			Regular Season					Playoffs				
Season	Club	Lea	GP	G	A	TP	PIM	GP	G	A	TP	PIM
1997-98	New England	EJHL	40	13	24	37	....	....	....	....	....	....
1998-99	New England	EJHL	40	29	27	56	....	....	....	....	....	....
99-2000	New England	EJHL	40	40	69	109	....	....	....	....	....	....
2000-01	Boston University	H-East	35	5	5	10	20	....	....	....	....	....
2001-02	Boston University	H-East	33	5	18	23	34	....	....	....	....	....

EJHL Rookie of the Year (1998) • EJHL MVP (2000)

JOKELA, Mikko (YOH-kih-lah, MIH-koh) N.J.

Defense. Shoots right. 6'1", 210 lbs. Born, Lappeenranta, Finland, March 4, 1980.
(New Jersey's 5th choice, 96th overall, in 1998 Entry Draft).

			Regular Season					Playoffs				
Season	Club	Lea	GP	G	A	TP	PIM	GP	G	A	TP	PIM
1994-95	KalPa Kuopio-C	Finn-Jr.	29	7	12	19	36	....	....	....	....	....
1995-96	KalPa Kuopio-C	Finn-Jr.	23	10	19	29	103	6	3	5	8	4
	KalPa Kuopio-B	Finn-Jr.	9	2	1	3	20	....	....	....	....	....
	KalPa Kuopio Jr.	Finn-Jr.	11	2	1	3	20	....	....	....	....	....
1996-97	KalPa Kuopio-B	Finn-Jr.	11	3	2	5	8	5	1	1	2	4
	KalPa Kuopio Jr.	Finn-Jr.	22	2	4	6	4	....	....	....	....	....
	KalPa Kuopio-B	Finn-Jr.	....	....	....	....	....	12	0	1	1	14
1997-98	HIFK Jr.	Finn-Jr.	22	2	5	7	14	....	....	....	....	....
	Hermes Kokkola	Finland-2	6	0	1	1	2	....	....	....	....	....
	HIFK Helsinki	Finland	16	0	0	0	0	....	....	....	....	....
1998-99	KalPa Kuopio Jr.	Finn-Jr.	1	0	1	1	2	....	....	....	....	....
	KalPa Kuopio	Finland	42	1	2	3	18	....	....	....	....	....
	HIFK Helsinki	Finland	3	0	0	0	2	....	....	....	....	....
	KalPa Kuopio	Finland-2	....	....	....	....	....	6	0	0	0	2
99-2000	SaiPa	Finland	48	0	5	5	50	....	....	....	....	....
	SaiPa Jr.	Finn-Jr.	....	....	....	....	....	1	0	0	0	0
2000-01	SaiPa Jr.	Finn-Jr.	4	2	2	4	2	3	0	0	0	2
	KooKoo Kouvola	Finland-2	5	3	0	3	0	....	....	....	....	....
	SaiPa	Finland	50	1	0	1	24	....	....	....	....	....
2001-02	Albany	AHL	56	5	13	18	28	....	....	....	....	....

JOKILA, Janne (YOHK-ih-luh, YAHN-ee) CBJ

Left wing. Shoots left. 5'9", 174 lbs. Born, Turku, Finland, April 22, 1982.
(Columbus' 7th choice, 200th overall, in 2000 Entry Draft).

			Regular Season					Playoffs				
Season	Club	Lea	GP	G	A	TP	PIM	GP	G	A	TP	PIM
1996-97	TPS Turku-C	Finn-Jr.	30	25	17	42	24	6	*5	*5	*10	2
1997-98	TPS Turku-B	Finn-Jr.	30	11	10	21	28	6	3	1	4	2
1998-99	TPS Turku Jr.	Finn-Jr.	36	17	15	32	77	....	....	....	....	....
99-2000	TPS Turku Jr.	Finn-Jr.	35	10	7	17	32	13	3	3	6	4
2000-01	TPS Turku Jr.	Finn-Jr.	22	15	15	30	30	2	0	1	1	4
	TPS Turku	Finland	2	0	0	0	0	....	....	....	....	....
	SaiPa	Finland	8	1	1	2	0	....	....	....	....	....
2001-02	TPS Turku	Finland	14	1	1	2	4	....	....	....	....	....
	SaiPa	Finland	12	0	0	0	2	....	....	....	....	....
	Lukko Rauma	Finland	14	0	1	1	18	....	....	....	....	....

JOKINEN, Jussi (YOH-kih-nihn, YOO-see) DAL.

Center. Shoots left. 5'11", 183 lbs. Born, Kalajoki, Finland, April 1, 1983.
(Dallas' 7th choice, 192nd overall, in 2001 Entry Draft).

			Regular Season					Playoffs				
Season	Club	Lea	GP	G	A	TP	PIM	GP	G	A	TP	PIM
1998-99	Karpat Oulu-C	Finn-Jr.	27	29	34	63	12	....	....	....	....	....
99-2000	Karpat Oulu Jr.	Finn-Jr.	28	4	6	10	14	....	....	....	....	....
	Karpat Oulu-B	Finn-Jr.	15	6	25	31	14	6	2	3	5	0
2000-01	Karpat Oulu-B	Finn-Jr.	1	2	1	3	0	....	....	....	....	....
	Karpat Oulu Jr.	Finn-Jr.	41	18	31	49	69	6	2	2	4	0
2001-02	Karpat Oulu	Finland	54	10	6	16	34	4	1	0	1	0
	Karpat Oulu Jr.	Finn-Jr.	2	4	1	5	2	1	1	1	2	0

JONASEN, Marcus (YOH-nuh-suhn, MAHR-kuhs) NYR

Left wing. Shoots right. 6'4", 220 lbs. Born, Vasteras, Sweden, January 12, 1984.
(NY Rangers' 2nd choice, 81st overall, in 2002 Entry Draft).

			Regular Season					Playoffs				
Season	Club	Lea	GP	G	A	TP	PIM	GP	G	A	TP	PIM
2000-01	Vasteras IK Jr.	Swede-Jr.	3	2	1	3	0	....	....	....	....	....
	Vasteras IK	Sweden	13	3	5	8	6	....	....	....	....	....
2001-02	Vasteras IK Jr.	Swede-Jr.	3	0	0	0	4	....	....	....	....	....
	Vasteras IK	Swede-2	16	2	1	3	6	....	....	....	....	....
	Vasteras IK	Swede-Q	2	0	0	0	0	....	....	....	....	....

JONES, Matt (JOHNZ, MAT) PHX.

Defense. Shoots left. 6', 214 lbs. Born, Downers Grove, IL, August 8, 1983.
(Phoenix's 5th choice, 80th overall, in 2002 Entry Draft).

			Regular Season					Playoffs				
Season	Club	Lea	GP	G	A	TP	PIM	GP	G	A	TP	PIM
99-2000	Green Bay	USHL	54	1	4	5	59	13	0	0	0	2
2000-01	Green Bay	USHL	52	3	10	13	58	4	0	0	0	2
2001-02	North Dakota	WCHA	37	2	5	7	20	....	....	....	....	....

JONES, Mike (JOHNZ, MIGHK) T.B.

Defense. Shoots left. 6'3", 190 lbs. Born, Toledo, OH, May 18, 1976.

			Regular Season					Playoffs				
Season	Club	Lea	GP	G	A	TP	PIM	GP	G	A	TP	PIM
1994-95	Cleveland Barons	NAJHL	45	15	33	48	....	....	....	....	....	....
1995-96	Cleveland Barons	NAJHL	46	25	*45	70	....	....	....	....	....	....
1996-97	Bowling Green	CCHA	27	1	6	7	47	....	....	....	....	....
1997-98	Bowling Green	CCHA	28	3	12	15	69	....	....	....	....	....
1998-99	Bowling Green	CCHA	38	8	21	29	80	....	....	....	....	....
99-2000	Bowling Green	CCHA	34	6	13	19	71	....	....	....	....	....
2000-01	Detroit Vipers	IHL	71	9	17	26	41	....	....	....	....	....
2001-02	Pensacola	ECHL	6	2	1	3	2	3	1	1	2	4

NAJHL Defenseman of the Year (1996) • CCHA Second All-Star Team (1999)

Signed as a free agent by **Tampa Bay**, April 13, 2000. • Missed majority of 2001-02 season recovering from heel injury suffered during off-season training, August, 2001.

JONSSON, Lars (YAWN-suhn, LARZ) BOS.

Defense. Shoots left. 6'1", 198 lbs. Born, Borlange, Sweden, January 2, 1982.
(Boston's 1st choice, 7th overall, in 2000 Entry Draft).

			Regular Season					Playoffs				
Season	Club	Lea	GP	G	A	TP	PIM	GP	G	A	TP	PIM
1998-99	Leksands IF Jr.	Swede-Jr.	40	4	8	12	42	....	....	....	....	....
99-2000	Leksands IF Jr.	Swede-Jr.	34	16	22	38	50	2	0	0	0	0
	Leksands IF	Sweden	5	0	0	0	4	....	....	....	....	....
2000-01	Leksands IF Jr.	Swede-Jr.	7	1	3	4	6	....	....	....	....	....
	Leksands IF	Sweden	31	2	1	3	12	....	....	....	....	....
2001-02	Leksands IF Jr.	Swede-Jr.	3	2	1	3	4	1	0	0	0	0
	Leksands IF	Swede-2	28	1	7	8	59	....	....	....	....	....
	Leksands IF	Swede-Q	12	1	7	8	35	8	0	2	2	6

JONSSON, Robin (YAWN-suhn, RAW-bihn) ST.L.

Defense. Shoots right. 6'2", 194 lbs. Born, Upplands Vasby, Sweden, December 10, 1983.
(St. Louis' 4th choice, 120th overall, in 2002 Entry Draft).

			Regular Season					Playoffs				
Season	Club	Lea	GP	G	A	TP	PIM	GP	G	A	TP	PIM
99-2000	Farjestad-18	Swede-Jr.	8	0	0	0	24	....	....	....	....	....
	Farjestad Jr.	Swede-Jr.	18	2	2	4	12	....	....	....	....	....
2000-01	Farjestad Jr.	Swede-Jr.	24	3	8	11	34	....	....	....	....	....
	Farjestad	Sweden	....	....	....	....	....	1	0	0	0	0
2001-02	Bofors IK	Swede-2	55	3	4	7	36	....	....	....	....	....
	Farjestad	Sweden	1	0	0	0	0	....	....	....	....	....

JORDE, Ryan (JOHR-dee, RIGH-uhn) BUF.

Defense. Shoots right. 6'3", 223 lbs. Born, Kelowna, B.C., March 23, 1982.
(Buffalo's 8th choice, 279th overall, in 2001 Entry Draft).

			Regular Season					Playoffs				
Season	Club	Lea	GP	G	A	TP	PIM	GP	G	A	TP	PIM
1997-98	Tri-City	WHL	3	0	1	1	2	....	....	....	....	....
1998-99	Tri-City	WHL	19	0	1	1	7	....	....	....	....	....
	Lethbridge	WHL	21	2	5	7	22	4	0	0	0	0
99-2000	Lethbridge	WHL	54	1	6	7	112	....	....	....	....	....
2000-01	Lethbridge	WHL	11	0	0	0	49	....	....	....	....	....
	Tri-City	WHL	56	1	6	7	170	....	....	....	....	....
2001-02	Tri-City	WHL	29	0	6	6	68	....	....	....	....	....
	Moose Jaw	WHL	23	0	4	4	62	8	0	1	1	2

Traded to **Lethbridge** (WHL) by **Tri-City** (WHL) with Nathan Barrett for Chris Huppe and Andrew Guindon, February 2, 1999.Traded to **Tri-City** (WHL) by **Lethbridge** (WHL) with Colin Johnson for Adam Johnson and Ryley Layden, October 25, 2000. Traded to **Moose Jaw** (WHL) by **Tri-City** (WHL) for Moose Jaw's 3rd round choice (later traded to Swift Current, Swift Current selected Mark Dafoe) in 2002 WHL Bantam Draft, January 10, 2002.

JUNTUNEN, Henrik (YUN-tuh-nehn, HEHN-rihk) L.A.

Right wing. Shoots right. 6'2", 185 lbs. Born, Goteborg, Sweden, April 24, 1983.
(Los Angeles' 5th choice, 83rd overall, in 2001 Entry Draft).

			Regular Season					Playoffs				
Season	Club	Lea	GP	G	A	TP	PIM	GP	G	A	TP	PIM
99-2000	Karpat Oulu Jr.	Finn-Jr.	34	16	7	23	18	5	0	0	0	2
2000-01	Karpat Oulu Jr.	Finn-Jr.	17	4	4	8	12	....	....	....	....	....
	Karpat Oulu	Finland	....	....	....	....	....	2	0	0	0	0
2001-02	Karpat Oulu Jr.	Finn-Jr.	34	19	11	30	42	3	1	1	2	0
	Karpat Oulu	Finland	13	0	0	0	2	....	....	....	....	....

JURCINA, Milan — (YEWR-chee-nah, MEE-lan) — BOS.

Defense. Shoots right. 6' 4", 198 lbs. Born, Liptovsky Mikulas, Czech., June 7, 1983.
(Boston's 7th choice, 241st overall, in 2001 Entry Draft).

			Regular Season					Playoffs				
Season	Club	Lea	GP	G	A	TP	PIM	GP	G	A	TP	PIM
99-2000	L. Mikulas Jr.	Slovak-Jr.		STATISTICS NOT AVAILABLE								
2000-01	Halifax	QMJHL	68	0	5	5	56	6	0	2	2	12
2001-02	Halifax	QMJHL	61	4	16	20	58	13	5	3	8	10

KACZOWKA, David — (kuh-ZOW-kuh, DAY-vihd) — ATL.

Left wing. Shoots left. 6'3", 220 lbs. Born, Regina, Sask., July 5, 1981.
(Atlanta's 4th choice, 98th overall, in 1999 Entry Draft).

			Regular Season					Playoffs				
Season	Club	Lea	GP	G	A	TP	PIM	GP	G	A	TP	PIM
1997-98	Prince Albert	SMHL	56	5	13	18	334	....	....	....	....	....
1998-99	Seattle	WHL	60	3	2	5	247	9	0	0	0	24
99-2000	Seattle	WHL	63	3	3	6	211	1	0	0	0	0
2000-01	Regina Pats	WHL	63	4	6	10	*414	6	0	0	0	6
2001-02	Chicago Wolves	AHL	1	0	0	0	0	....	....	....	....	....
	Greenville	ECHL	32	1	1	2	182	....	....	....	....	....

Traded to **Regina** (WHL) by **Seattle** (WHL) for Regina's 3rd round choice (later traded to Kamloops, Kamloops chose Scott Wagner) in 2002 WHL Bantam Draft, September 3, 2000.
• Missed majority of 2001-02 season recovering from head injury suffered in game vs. Florida (ECHL), February 12, 2002.

KADEIKIN, Anton — (ka-DAY-kihn, an-TAWN) — N.J.

Defense. Shoots left. 6'2", 180 lbs. Born, Elektrostal, USSR, May 17, 1984.
(New Jersey's 1st choice, 51st overall, in 2002 Entry Draft).

			Regular Season					Playoffs				
Season	Club	Lea	GP	G	A	TP	PIM	GP	G	A	TP	PIM
99-2000	Elektrostal 2	Russia-3	5	0	0	0	0	....	....	....	....	....
2000-01	Elektrostal 2	Russia-3		STATISTICS NOT AVAILABLE								
2001-02	Elektrostal 2	Russia-3	3	0	1	1	2	....	....	....	....	....
	Elektrostal	Russia-2	20	0	0	0	16	....	....	....	....	....

KAHNBERG, Magnus — (KAHN-buhrg, MAHG-nus) — CAR.

Left wing. Shoots right. 6'1", 185 lbs. Born, Goteborg, Sweden, February 25, 1980.
(Carolina's 6th choice, 212th overall, in 2000 Entry Draft).

			Regular Season					Playoffs				
Season	Club	Lea	GP	G	A	TP	PIM	GP	G	A	TP	PIM
1997-98	V. Frolunda-18	Swede-Jr.	11	15	6	21	6	8	5	8	13	6
	V. Frolunda Jr.	Swede-Jr.	28	6	7	13	8	2	0	0	0	0
1998-99	V. Frolunda Jr.	Swede-Jr.	34	23	18	41	4	4	1	1	2	0
99-2000	V. Frolunda Jr.	Swede-Jr.	35	45	21	66	30	6	7	4	11	4
	Vastra Frolunda	Sweden	4	0	0	0	0	....	....	....	....	....
2000-01	V. Frolunda-18	Swede-Jr.	1	8	1	9	0	....	....	....	....	....
	V. Frolunda Jr.	Swede-Jr.	2	*2	1	3	2	....	....	....	....	....
	Vastra Frolunda	Sweden	50	8	6	14	6	5	0	0	0	2
2001-02	Vastra Frolunda	Sweden	50	14	11	25	24	10	5	0	5	2

KAIGORODOV, Alexei — (kay-goh-ROH-dahv, al-EHX-ay) — OTT.

Center. Shoots left. 6'1", 183 lbs. Born, Chelyabinsk, USSR, July 29, 1983.
(Ottawa's 2nd choice, 47th overall, in 2002 Entry Draft).

			Regular Season					Playoffs				
Season	Club	Lea	GP	G	A	TP	PIM	GP	G	A	TP	PIM
1998-99	Magnitogorsk 2	Russia-4	10	6	4	10	2	....	....	....	....	....
99-2000	Magnitogorsk 2	Russia-3	19	2	3	5	8	....	....	....	....	....
2000-01	Magnitogorsk 2	Russia-3	45	12	30	42	26	....	....	....	....	....
2001-02	Magnitogorsk	Russia	46	4	12	16	20	9	0	3	3	2

KALLARSSON, Tomi — (KAL-ahr-suhn, TAW-mee) — NYR

Defense. Shoots left. 6'3", 194 lbs. Born, Lempaala, Finland, March 15, 1979.
(NY Rangers' 4th choice, 93rd overall, in 1997 Entry Draft).

			Regular Season					Playoffs				
Season	Club	Lea	GP	G	A	TP	PIM	GP	G	A	TP	PIM
1994-95	Tappara Jr.	Finn-Jr.	22	9	4	13	26	....	....	....	....	....
1995-96	Tappara Jr.	Finn-Jr.	31	3	5	8	24	6	0	0	0	0
1996-97	HPK Jr.	Finn-Jr.	31	1	3	4	26	6	0	0	0	2
1997-98	HPK Jr.	Finn-Jr.	32	5	9	14	71	....	....	....	....	....
	Pelicans Lahti	Finland-2	3	0	0	0	0	....	....	....	....	....
	HPK Hameenlinna	Finland	12	0	0	0	2	....	....	....	....	....
1998-99	HPK Jr.	Finn-Jr.	2	0	1	1	6	....	....	....	....	....
	Ahmat Hyvinkaa	Finland-2	21	2	7	9	52	....	....	....	....	....
	HPK Hameenlinna	Finland	25	0	0	0	22	8	0	0	0	8
99-2000	HPK Jr.	Finn-Jr.	3	2	2	4	4	....	....	....	....	....
	HPK Hameenlinna	Finland	50	0	5	5	58	8	0	0	0	4
2000-01	Timra IK	Sweden	37	0	0	0	54	....	....	....	....	....
2001-02	Timra IK	Sweden	43	4	4	8	76	....	....	....	....	....
	Ilves Tampere	Finland	13	1	1	2	35	3	0	0	0	4

KALTEVA, Mikko — (KAL-tuh-vah, MEE-koh) — COL.

Defense. Shoots left. 6'3", 190 lbs. Born, Hyvinkaa, Finland, May 25, 1984.
(Colorado's 4th choice, 107th overall, in 2002 Entry Draft).

			Regular Season					Playoffs				
Season	Club	Lea	GP	G	A	TP	PIM	GP	G	A	TP	PIM
2000-01	Jokerit-B	Finn-Jr.	11	1	4	5	6	....	....	....	....	....
	Jokerit Jr.	Finn-Jr.	34	0	3	3	12	....	....	....	....	....
2001-02	Jokerit Jr.	Finn-Jr.	29	5	3	8	10	....	....	....	....	....
	Jokerit-B	Finn-Jr.	....	....	....	....	....	8	2	2	4	0

KANE, Boyd — (KAYN, BOIYD) — NYR

Left wing. Shoots left. 6'2", 218 lbs. Born, Swift Current, Sask., April 18, 1978.
(NY Rangers' 4th choice, 114th overall, in 1998 Entry Draft).

			Regular Season					Playoffs				
Season	Club	Lea	GP	G	A	TP	PIM	GP	G	A	TP	PIM
1994-95	Regina Pats	WHL	25	6	5	11	6	4	0	0	0	0
1995-96	Regina Pats	WHL	72	21	42	63	155	11	5	7	12	12
1996-97	Regina Pats	WHL	66	25	50	75	154	5	1	1	2	15
1997-98	Regina Pats	WHL	68	48	45	93	133	9	5	7	12	29
1998-99	Hartford	AHL	56	3	5	8	23	....	....	....	....	....
	Charlotte	ECHL	12	5	6	11	14	....	....	....	....	....
99-2000	Charlotte	ECHL	47	10	19	29	110	....	....	....	....	....
	Hartford	AHL	8	0	0	0	9	....	....	....	....	....
	Binghamton	UHL	3	0	2	2	4	1	0	0	0	0
2000-01	Charlotte	ECHL	12	9	8	17	6	....	....	....	....	....
	Hartford	AHL	56	11	17	28	81	5	2	0	2	2
2001-02	Hartford	AHL	78	17	22	39	193	10	1	2	3	50

• Re-entered NHL Entry Draft. Originally Pittsburgh's 3rd choice, 72nd overall, in 1996 Entry Draft.

KANKAANPERA, Markus — (kan-kahn-PEHR-a, MAHR-kus) — VAN.

Defense. Shoots left. 6'1", 191 lbs. Born, Skelleftea, Sweden, April 27, 1980.
(Vancouver's 7th choice, 218th overall, in 1999 Entry Draft).

			Regular Season					Playoffs				
Season	Club	Lea	GP	G	A	TP	PIM	GP	G	A	TP	PIM
1995-96	JyP HT-B	Finn-Jr.	9	1	1	2	4	....	....	....	....	....
1996-97	JyP HT Jr.	Finn-Jr.	33	3	5	8	83	....	....	....	....	....
1997-98	JYP Jyvaskyla-B	Finn-Jr.	13	4	10	14	18	5	3	2	5	6
	JYP Jr.	Finn-Jr.	32	0	0	0	2	....	....	....	....	....
1998-99	JYP Jr.	Finn-Jr.	1	0	0	0	4	....	....	....	....	....
	JYP Jyvaskyla	Finland	50	0	2	2	85	3	0	0	0	0
99-2000	JYP Jr.	Finn-Jr.	3	1	1	2	2	3	0	1	1	6
	JYP Jyvaskyla	Finland	47	0	5	5	87	....	....	....	....	....
2000-01	JYP Jyvaskyla	Finland	53	5	4	9	60	....	....	....	....	....
2001-02	HPK Hameenlinna	Finland	51	4	3	7	80	8	0	0	0	10

KANKO, Petr — (KAN-koh, PEE-tuhr) — L.A.

Right wing. Shoots left. 5'9", 195 lbs. Born, Pribram, Czech., February 7, 1984.
(Los Angeles' 3rd choice, 66th overall, in 2002 Entry Draft).

			Regular Season					Playoffs				
Season	Club	Lea	GP	G	A	TP	PIM	GP	G	A	TP	PIM
2000-01	Sparta Praha Jr.	Czech-Jr.	43	27	10	37	80	....	....	....	....	....
	HC Sparta Praha	Czech	6	1	0	1	0	....	....	....	....	....
2001-02	Kitchener	OHL	61	28	32	60	54	4	0	2	2	0

KARLIN, Mattias — (KAR-lihn, MAT-teeuhs) — BOS.

Center/Right wing. Shoots left. 5'11", 183 lbs. Born, Domsjo, Sweden, July 4, 1979.
(Boston's 4th choice, 54th overall, in 1997 Entry Draft).

			Regular Season					Playoffs				
Season	Club	Lea	GP	G	A	TP	PIM	GP	G	A	TP	PIM
1995-96	MoDo Jr.	Swede-Jr.	30	12	23	35	16	....	....	....	....	....
1996-97	MoDo	Sweden	6	0	0	0	0	....	....	....	....	....
1997-98	MoDo	Sweden	32	0	2	2	8	1	0	0	0	0
1998-99	MoDo	Sweden	50	2	5	7	14	13	1	1	2	4
99-2000	MoDo	Sweden	42	0	5	5	10	12	0	0	0	2
2000-01	Providence	AHL	68	3	12	15	28	16	1	2	3	2
2001-02	Providence	AHL	28	2	3	5	12	....	....	....	....	....

Signed as a free agent by **MoDo** (Sweden) with Boston retaining NHL rights, August 15, 2002.

KARLSSON, Gabriel — (KARLS-suhn, ga-BREE-ehl) — DAL.

Center. Shoots left. 6'1", 189 lbs. Born, Borlange, Sweden, January 22, 1980.
(Dallas' 3rd choice, 86th overall, in 1998 Entry Draft).

			Regular Season					Playoffs				
Season	Club	Lea	GP	G	A	TP	PIM	GP	G	A	TP	PIM
1996-97	HV 71 Jr.	Swede-Jr.	25	7	9	16	....	....	....	....	....	....
1997-98	HV 71 Jr.	Swede-Jr.	27	11	15	26	32	....	....	....	....	....
	HV 71 Jonkoping	Sweden	1	0	0	0	0	....	....	....	....	....
1998-99	HV 71 Jr.	Swede-Jr.	12	4	9	13	4	....	....	....	....	....
	HV 71 Jonkoping	Sweden	33	2	1	3	2	....	....	....	....	....
99-2000	HV 71 Jonkoping	Sweden	50	5	3	8	12	6	0	0	0	2
2000-01	Assat Pori	Finland	17	2	2	4	6	....	....	....	....	....
	Leksands IF	Sweden	35	9	8	17	10	....	....	....	....	....
2001-02	Sodertalje SK	Sweden	47	8	7	15	18	....	....	....	....	....

KARLSSON, Jens — (KARLS-suhn, YEHNZ) — L.A.

Left wing. Shoots right. 6'3", 205 lbs. Born, Goteborg, Sweden, November 7, 1982.
(Los Angeles' 1st choice, 18th overall, in 2001 Entry Draft).

			Regular Season					Playoffs				
Season	Club	Lea	GP	G	A	TP	PIM	GP	G	A	TP	PIM
1997-98	V. Frolunda-16	Swede-Jr.	8	9	3	12	32	....	....	....	....	....
	V. Frolunda Jr.	Swede-Jr.	17	2	3	5	4	....	....	....	....	....
1998-99	V. Frolunda-18	Swede-Jr.	32	27	17	44	110	4	2	2	4	0
99-2000	V. Frolunda-18	Swede-Jr.	3	6	3	9	6	....	....	....	....	....
	V. Frolunda Jr.	Swede-Jr.	32	24	13	37	82	6	3	0	3	42
2000-01	V. Frolunda Jr.	Swede-Jr.	25	20	15	35	154	1	0	1	1	2
	Molndals IF	Swede-2	5	1	1	2	35	....	....	....	....	....
	Vastra Frolunda	Sweden	19	2	0	2	4	5	1	3	4	50
2001-02	Vastra Frolunda	Sweden	46	6	9	15	44	10	1	0	1	18
	V. Frolunda Jr.	Swede-Jr.	....	....	....	....	....	2	0	1	1	12

KASPARIK, Pavel — (kas-PAHR-ihk, PAH-vehl) — PHI.

Center. Shoots left. 6'2", 198 lbs. Born, Pisek, Czech., November 11, 1979.
(Philadelphia's 4th choice, 200th overall, in 1999 Entry Draft).

			Regular Season					Playoffs				
Season	Club	Lea	GP	G	A	TP	PIM	GP	G	A	TP	PIM
1996-97	IHC Pisek Jr.	Czech-Jr.	36	12	5	17	....	....	....	....	....	....
1997-98	IHC Pisek Jr.	Czech-Jr.	39	21	19	40	....	....	....	....	....	....
	IHC Pisek	Czech-2	15	3	3	6	....	....	....	....	....	....
1998-99	IHC Pisek Jr.	Czech-Jr.	7	2	3	5	....	....	....	....	....	....
	IHC Pisek	Czech-2	51	20	23	43	....	....	....	....	....	....
99-2000	IHC Pisek	Czech-2	24	6	9	15	....	....	....	....	....	....
	HC Femax Havirov	Czech	1	0	0	0	0	....	....	....	....	....
	HC Sparta Praha	Czech	22	1	1	2	0	....	....	....	....	....
2000-01	HC Karlovy Vary	Czech	29	1	2	3	20	....	....	....	....	....
	HC Sparta Praha	Czech	19	5	1	6	18	13	2	0	2	12
2001-02	HC Sparta Praha	Czech	50	14	11	25	10	13	2	0	2	4

KAUPPINEN, Marko — (KOW-pih-nehn, MAHR-koh) — PHI.

Defense. Shoots left. 6', 178 lbs. Born, Mikkeli, Finland, March 23, 1979.
(Philadelphia's 7th choice, 214th overall, in 1997 Entry Draft).

			Regular Season					Playoffs				
Season	Club	Lea	GP	G	A	TP	PIM	GP	G	A	TP	PIM
1994-95	Jukurit Mikkeli-C	Finn-Jr.	31	12	11	23	48	....	....	....	....	....
1995-96	Jukurit Mikkeli	Finland-3	19	1	5	6	10	3	0	0	0	0
1996-97	JyP HT Jr.	Finn-Jr.	29	2	3	5	14	7	0	0	0	29
1997-98	JYP Jr.	Finn-Jr.	16	2	4	6	16	....	....	....	....	....
	Diskos Jyvaskyla	Finland-2	2	2	1	3	0	....	....	....	....	....
	JYP Jyvaskyla	Finland	33	2	6	8	26	....	....	....	....	....
1998-99	JYP Jr.	Finn-Jr.	3	2	1	3	6	....	....	....	....	....
	JYP Jyvaskyla	Finland	49	5	7	12	56	3	0	0	0	6
99-2000	Jokerit Jr.	Finn-Jr.	5	0	3	3	8	1	1	0	1	4
	Jokerit Helsinki	Finland	47	4	9	13	16	10	1	3	4	2
2000-01	AIK Solna	Sweden	4	0	0	0	4	....	....	....	....	....
	Jokerit Helsinki	Finland	48	5	7	12	41	5	0	0	0	8
2001-02	TPS Turku	Finland	55	4	9	13	55	5	1	0	1	2

KAZIONOV, Dmitri (ka-zee-OH-nahv, dih-MEE-tree) **T.B.**

Center. Shoots left. 6'3", 185 lbs. Born, Moscow, USSR, May 13, 1984.
(Tamba Bay's 2nd choice, 100th overall, in 2002 Entry Draft).

			Regular Season					Playoffs				
Season	Club	Lea	GP	G	A	TP	PIM	GP	G	A	TP	PIM
99-2000	Dynamo Moscow 2	Russia-3	2	1	0	1	0	....	....	....	....	....
2000-01	THC Tver	Russia-2	33	1	1	2	6	....	....	....	....	....
2001-02	H.C. CSKA	Russia-2	2	0	1	1	0	....	....	....	....	....
	H.C. CSKA 2	Russia-3	10	1	0	1	4	....	....	....	....	....
	Lada Togliatti	Russia	3	0	0	0	0	....	....	....	....	....

KEITH, Duncan (KEETH, DUHN-kuhn) **CHI.**

Defense. Shoots left. 6', 168 lbs. Born, Winnipeg, Man., July 16, 1983.
(Chicago's 2nd choice, 54th overall, in 2002 Entry Draft).

			Regular Season					Playoffs				
Season	Club	Lea	GP	G	A	TP	PIM	GP	G	A	TP	PIM
1998-99	Penticton	BCAHA	44	51	57	108	45	....	....	....	....	....
99-2000	Penticton	BCHL	59	9	27	36	37	....	....	....	....	....
2000-01	Penticton	BCHL	60	18	64	82	61	9	4	6	10	18
2001-02	Michigan State	CCHA	41	3	12	15	18	....	....	....	....	....

BCHL First All-Star Team (2001) • BCHL Top Defenseman (2001)

KEITH, Matt (KEETH, MAT) **CHI.**

Right wing. Shoots right. 6'2", 194 lbs. Born, Edmonton, Alta., April 11, 1983.
(Chicago's 3rd choice, 59th overall, in 2001 Entry Draft).

			Regular Season					Playoffs				
Season	Club	Lea	GP	G	A	TP	PIM	GP	G	A	TP	PIM
1998-99	Banff Icemen	HJHL	STATISTICS NOT AVAILABLE									
	Spokane Chiefs	WHL	7	1	0	1	4	....	....	....	....	....
99-2000	Spokane Chiefs	WHL	39	1	3	4	37	15	1	2	3	11
2000-01	Spokane Chiefs	WHL	33	13	14	27	63	12	1	3	4	14
2001-02	Spokane Chiefs	WHL	68	34	33	67	71	11	5	5	10	16

• Missed majority of 2000-01 season recovering from shoulder injury originally suffered in game vs. Tri-City (WHL), September 22, 2000.

KELLY, Chris (KEHL-lee, KRIHS) **OTT.**

Center/Left wing. Shoots left. 6', 190 lbs. Born, Toronto, Ont., November 11, 1980.
(Ottawa's 4th choice, 94th overall, in 1999 Entry Draft).

			Regular Season					Playoffs				
Season	Club	Lea	GP	G	A	TP	PIM	GP	G	A	TP	PIM
1995-96	Toronto Marlies	MTHL	42	25	45	70	25	....	....	....	....	....
1996-97	Aurora Tigers	OPJHL	49	14	20	34	11	....	....	....	....	....
1997-98	London Knights	OHL	54	15	14	29	4	16	4	5	9	12
1998-99	London Knights	OHL	68	36	41	77	60	25	9	17	26	22
99-2000	London Knights	OHL	63	29	43	72	57	....	....	....	....	....
2000-01	London Knights	OHL	31	21	34	55	46	....	....	....	....	....
	Sudbury Wolves	OHL	19	5	16	21	17	12	11	5	16	14
2001-02	Muskegon Fury	UHL	4	1	2	3	0	....	....	....	....	....
	Grand Rapids	AHL	31	3	3	6	20	5	1	1	2	5

Traded to **Sudbury** (OHL) by **London** (OHL) with Dan Jancevski for Dennis Wideman and future considerations, January 10, 2001.

KELLY, Regan (KEHL-lee, REE-guhn) **TOR.**

Defense. Shoots left. 6'2", 185 lbs. Born, Watrous, Sask., March 9, 1981.
(Philadelphia's 7th choice, 259th overall, in 2000 Entry Draft).

			Regular Season					Playoffs				
Season	Club	Lea	GP	G	A	TP	PIM	GP	G	A	TP	PIM
1997-98	Tisdale Trojans	SMHL	41	2	13	15	24	....	....	....	....	....
1998-99	Nipawin Hawks	SJHL	52	4	14	18	30	....	....	....	....	....
99-2000	Nipawin Hawks	SJHL	46	8	21	29	20	....	....	....	....	....
2000-01	Providence	H-East	36	4	21	25	58	....	....	....	....	....
2001-02	Providence	H-East	38	6	10	16	48	....	....	....	....	....

SJHL All-Rookie Team (1999) • Hockey East All-Rookie Team (2001) • Hockey East All-Tournament Team (2001)

Rights traded to **Toronto** by **Philadelphia** for Chris McAllister, September 26, 2000.

KESA, Teemu (KEH-sah, TEE-moo) **N.J.**

Defense. Shoots right. 6', 185 lbs. Born, Helsinki, Finland, June 7, 1981.
(New Jersey's 5th choice, 100th overall, in 1999 Entry Draft).

			Regular Season					Playoffs				
Season	Club	Lea	GP	G	A	TP	PIM	GP	G	A	TP	PIM
1996-97	Tappara Jr.	Finn-Jr.	32	1	5	6	58	4	1	0	4	29
1997-98	Ilves Tampere-B	Finn-Jr.	33	8	1	9	78	....	....	....	....	....
1998-99	Ilves Tampere-B	Finn-Jr.	24	4	5	9	146	....	....	....	....	....
	Ilves Jr.	Finn-Jr.	6	0	1	1	10	....	....	....	....	....
99-2000	Ilves Jr.	Finn-Jr.	31	1	7	8	92	....	....	....	....	....
	Ilves Tampere	Finland	5	0	0	0	8	....	....	....	....	....
2000-01	Ilves Jr.	Finn-Jr.	4	1	0	1	4	....	....	....	....	....
	Sport Vassa	Finland-2	1	0	1	1	0	....	....	....	....	....
2001-02	Lukko Rauma	Finland	34	2	0	2	32	....	....	....	....	....

KETOLA, Juha-Pekka (KEH-toh-luh, YOO-ha-PEH-kuh) **NYI**

Right wing. Shoots left. 6'1", 175 lbs. Born, Rauma, Finland, January 21, 1983.
(NY Islanders' 8th choice, 287th overall, in 2001 Entry Draft).

			Regular Season					Playoffs				
Season	Club	Lea	GP	G	A	TP	PIM	GP	G	A	TP	PIM
1998-99	Lukko Rauma Jr.	Finn-Jr.	15	1	5	6	14	....	....	....	....	....
99-2000	Lukko Rauma Jr.	Finn-Jr.	37	6	16	22	38	....	....	....	....	....
2000-01	Lukko Rauma Jr.	Finn-Jr.	44	6	10	16	34	3	0	0	0	2
2001-02	Sherbrooke	QMJHL	67	8	27	35	58	....	....	....	....	....

Signed as a free agent by **Lukka Rauma** (Finland) with NY Islanders retaining NHL rights, July 29, 2002.

KHOMITSKY, Vadim (khoh-MIHT-skee, va-DEEM) **DAL.**

Defense. Shoots left. 6'1", 185 lbs. Born, Voskresensk, USSR, July 21, 1982.
(Dallas' 5th choice, 123rd overall, in 2000 Entry Draft).

			Regular Season					Playoffs				
Season	Club	Lea	GP	G	A	TP	PIM	GP	G	A	TP	PIM
1998-99	Voskresensk	Russia	9	0	0	0	10	....	....	....	....	....
99-2000	Voskresensk	Russia-2	17	0	0	0	31	....	....	....	....	....
	H.C. CSKA	Russia-2	11	0	1	1	10	....	....	....	....	....
2000-01	H.C. CSKA	Russia-2	44	2	7	9	89	....	....	....	....	....
2001-02	H.C. CSKA	Russia-2	68	2	18	20	63	....	....	....	....	....

KINCH, Matt (KIHNCH, MATT) **NYR**

Defense. Shoots left. 6', 195 lbs. Born, Red Deer, Alta., February 17, 1980.
(Buffalo's 8th choice, 146th overall, in 1999 Entry Draft).

			Regular Season					Playoffs				
Season	Club	Lea	GP	G	A	TP	PIM	GP	G	A	TP	PIM
1995-96	Red Deer	AMHL	35	6	17	23	31	....	....	....	....	....
	Calgary Hitmen	WHL	1	0	1	1	2	....	....	....	....	....
1996-97	Calgary Hitmen	WHL	64	10	22	32	31	....	....	....	....	....
1997-98	Calgary Hitmen	WHL	55	7	24	31	13	18	3	2	5	4
1998-99	Calgary Hitmen	WHL	68	14	69	83	16	21	8	15	23	59
99-2000	Calgary Hitmen	WHL	62	14	61	75	24	13	2	12	14	8
2000-01	Calgary Hitmen	WHL	70	18	66	84	52	12	3	6	9	6
2001-02	Hartford	AHL	40	1	7	8	4	....	....	....	....	....
	Charlotte	ECHL	26	3	12	15	13	5	3	2	5	0

WHL East First All-Star Team (1999, 2001) • Memorial Cup All-Star Team (1999) • WHL East Second All-Star Team (2000) • Canadian Major Junior First All-Star Team (2001)

Signed as a free agent by **NY Rangers**, June 26, 2001.

KING, Colt (KIHNG, KOHLT) **COL.**

Left wing. Shoots left. 6'2", 220 lbs. Born, Thunder Bay, Ont., March 4, 1983.
(Colorado's 3rd choice, 130th overall, in 2001 Entry Draft).

			Regular Season					Playoffs				
Season	Club	Lea	GP	G	A	TP	PIM	GP	G	A	TP	PIM
1998-99	St. Thomas Stars	OJHL-B	41	11	13	24	89	....	....	....	....	....
99-2000	Guelph Storm	OHL	53	2	2	4	41	6	0	0	0	5
2000-01	Guelph Storm	OHL	65	25	27	52	129	4	0	1	1	8
2001-02	Guelph Storm	OHL	20	4	5	9	49	....	....	....	....	....
	North Bay	OHL	39	11	7	18	92	5	4	1	5	20

Traded to **North Bay** (OHL) by **Guelph** (OHL) with Jeremy Day for Andrew Penner and North Bay's 4th round choice (Mark Verstegg-Lytwyn) in 2002 OHL Priority Draft, November 22, 2001.

KING, Jason (KIHNG, JAY-suhn) **VAN.**

Right wing. Shoots left. 6'1", 195 lbs. Born, Corner Brook, Nfld., September 14, 1981.
(Vancouver's 5th choice, 212th overall, in 2001 Entry Draft).

			Regular Season					Playoffs				
Season	Club	Lea	GP	G	A	TP	PIM	GP	G	A	TP	PIM
99-2000	Halifax	QMJHL	53	3	7	10	8	10	0	0	0	2
2000-01	Halifax	QMJHL	72	48	41	89	78	6	3	2	5	16
2001-02	Halifax	QMJHL	61	*63	36	99	39	13	9	8	17	13

QMJHL Second All-Star Team (2001)

KINOS, Lauri (KEE-nohs, LOH-ree) **ST.L.**

Defense. Shoots left. 6'2", 195 lbs. Born, Jyvaskyla, Finland, June 29, 1980.
(St. Louis' 9th choice, 293rd overall, in 2000 Entry Draft).

			Regular Season					Playoffs				
Season	Club	Lea	GP	G	A	TP	PIM	GP	G	A	TP	PIM
1996-97	Diskos Jr.	Finn-Jr.	24	2	0	2	32	....	....	....	....	....
1997-98	JYP Jr.	Finn-Jr.	29	8	6	14	30	2	0	0	0	0
1998-99	JYP Jr.	Finn-Jr.	27	3	1	4	30	....	....	....	....	....
99-2000	Montreal Rocket	QMJHL	69	12	17	29	80	5	1	1	2	6
2000-01	Worcester	AHL	11	0	0	0	8	....	....	....	....	....
	Peoria Rivermen	ECHL	55	4	12	16	56	9	0	1	1	6
2001-02	Peoria Rivermen	ECHL	46	2	9	11	30	5	0	1	1	4
	Worcester	AHL	10	0	1	1	8	....	....	....	....	....

KLEMA, David (KLEE-mah, DAY-vihd) **PHX.**

Center. Shoots left. 6', 178 lbs. Born, Roseau, MN, April 3, 1982.
(Phoenix's 5th choice, 148th overall, in 2001 Entry Draft).

			Regular Season					Playoffs				
Season	Club	Lea	GP	G	A	TP	PIM	GP	G	A	TP	PIM
1998-99	Roseau Rams	Hi-School	28	25	35	63	....	....	....	....	....	....
99-2000	Roseau Rams	Hi-School	28	16	31	47	8	....	....	....	....	....
	Fargo-Moorhead	USHL	6	0	2	2	4	....	....	....	....	....
	Des Moines	USHL	4	1	1	2	0	6	1	1	2	0
2000-01	Des Moines	USHL	56	13	40	53	58	3	1	0	1	0
2001-02	Boston University	H-East	38	6	11	17	4	....	....	....	....	....

KLEPIS, Jakub (KLEH-pihsh, YA-kuhb) **OTT.**

Center. Shoots right. 6', 200 lbs. Born, Prague, Czech., June 5, 1984.
(Ottawa's 1st choice, 16th overall, in 2002 Entry Draft).

			Regular Season					Playoffs				
Season	Club	Lea	GP	G	A	TP	PIM	GP	G	A	TP	PIM
99-2000	Slavia Praha Jr.	Czech-Jr.	48	14	26	40	30	....	....	....	....	....
2000-01	Slavia Praha Jr.	Czech-Jr.	52	21	25	46	82	....	....	....	....	....
2001-02	Portland	WHL	70	14	50	64	111	7	0	3	3	22

KLYAZMIN, Sergei (klee-YAZ-mihn, SAIR-gay) **COL.**

Left wing. Shoots left. 6'3", 190 lbs. Born, Moscow, USSR, March 1, 1982.
(Colorado's 6th choice, 92nd overall, in 2000 Entry Draft).

			Regular Season					Playoffs				
Season	Club	Lea	GP	G	A	TP	PIM	GP	G	A	TP	PIM
1998-99	Krylja Sovetov 2	Russia-4	21	1	4	5	14	....	....	....	....	....
99-2000	DynamoMoscow2	Russia-3	16	3	2	5	18	....	....	....	....	....
2000-01	Halifax	QMJHL	65	33	28	61	76	6	1	2	3	16
2001-02	Halifax	QMJHL	46	24	37	61	34	....	....	....	....	....

KNOEPFLI, Mike (NAWF-lee, MIGHK) **TOR.**

Left wing. Shoots left. 6' 1", 185 lbs. Born, Georgetown, Ont., April 9, 1982.
(Toronto's 12th choice, 276th overall, in 2001 Entry Draft).

			Regular Season					Playoffs				
Season	Club	Lea	GP	G	A	TP	PIM	GP	G	A	TP	PIM
99-2000	Georgetown	OPJHL	48	37	47	84	75	....	....	....	....	....
2000-01	Georgetown	OPJHL	47	45	38	83	....	....	....	....	....	....
2001-02	Cornell Big Red	ECAC	34	4	11	15	14	....	....	....	....	....

KNOPP, Ben (KUH-nawp, BEHN) **CBJ**

Right wing. Shoots right. 6'1", 190 lbs. Born, Calgary, Alta., April 8, 1982.
(Columbus' 2nd choice, 69th overall, in 2000 Entry Draft).

			Regular Season					Playoffs				
Season	Club	Lea	GP	G	A	TP	PIM	GP	G	A	TP	PIM
1997-98	Cgy. Buffaloes	AMHL	35	28	48	76	34	10	10	6	16	30
1998-99	Calgary Canucks	AJHL	55	37	45	82	161	6	6	6	12	6
99-2000	Moose Jaw	WHL	72	30	30	60	101	4	2	2	4	4
2000-01	Moose Jaw	WHL	58	22	34	56	105	4	0	0	0	11
2001-02	Moose Jaw	WHL	37	18	14	32	56	....	....	....	....	....
	Kamloops Blazers	WHL	41	26	25	51	63	4	0	1	1	4

Traded to **Kamloops** (WHL) by **Moose Jaw** (WHL) for Derek Krestanovich and Kamloops' 3rd choice in 2003 WHL Bantam Draft, December 12, 2001.

KNYAZEV, Igor (kuh-NYA-zhev, EE-gohr) CAR.

Defense. Shoots left. 6', 191 lbs. Born, Elektrostal, USSR, January 27, 1983.
(Carolina's 1st choice, 15th overall, in 2001 Entry Draft).

			Regular Season					Playoffs				
Season	Club	Lea	GP	G	A	TP	PIM	GP	G	A	TP	PIM
99-2000	Spartak Moscow 2	Russia-3	13	2	4	6	74	....	....	....	....	....
	Spartak Moscow	Russia-2	26	1	1	2	6	....	....	....	....	....
2000-01	Spartak Moscow	Russia-2	53	6	5	11	101	....	....	....	....	....
2001-02	Spartak Moscow	Russia	3	0	0	0	8	....	....	....	....	....
	Spartak Moscow 2	Russia-3	2	0	1	1	0	....	....	....	....	....
	Ak Bars Kazan	Russia	14	0	1	1	4	3	0	0	0	0

KOALSKA, Matt (KOHL-skuh, MAT) NSH.

Center. Shoots left. 6'1", 196 lbs. Born, St. Paul, MN, May 16, 1980.
(Nashville's 7th choice, 154th overall, in 2000 Entry Draft).

			Regular Season					Playoffs				
Season	Club	Lea	GP	G	A	TP	PIM	GP	G	A	TP	PIM
1998-99	Hill-Murray	Hi-School	26	20	50	70	18	....	....	....	....	....
99-2000	Twin Cities	USHL	57	24	34	58	19	13	5	5	10	4
2000-01	U. of Minnesota	WCHA	42	10	24	34	36	....	....	....	....	....
2001-02	U. of Minnesota	WCHA	44	10	23	33	34	....	....	....	....	....

First All-Conference Schoolboy Team (1999) • First All-State Schoolboy Team (1999)

KOBASEW, Chuck (KOH-buh-soo, CHUK) CGY.

Right wing. Shoots right. 5'11", 195 lbs. Born, Osoyoos, B.C., April 17, 1982.
(Calgary's 1st choice, 14th overall, in 2001 Entry Draft).

			Regular Season					Playoffs				
Season	Club	Lea	GP	G	A	TP	PIM	GP	G	A	TP	PIM
1997-98	Osoyoos Heat	KIJHL	6	2	2	4	2	....	....	....	....	....
1998-99	Osoyoos Heat	KIJHL	23	25	24	49	....	....	....	....	....	....
	Penticton	BCHL	30	11	17	28	18	....	....	....	....	....
99-2000	Penticton	BCHL	58	*54	52	106	83	....	....	....	....	....
2000-01	Boston College	H-East	43	27	22	49	38	....	....	....	....	....
2001-02	Kelowna Rockets	WHL	55	41	21	62	114	15	10	5	15	22

BCHL First All-Star Team (2000) • BCHL Interior Division MVP (2000) • Hockey East Second All-Star Team (2001) • Hockey East Rookie of the Year (2001) • NCAA Championship All-Tournament Team (2001) • NCAA Championship Tournament MVP (2001)

Rights traded to **Kelowna** (WHL) by **Prince George** (WHL) for Chuck Di Ubaldo and future considerations, August 1, 2001. • Left **Boston College** (H-East) and signed with **Kelowna** (WHL), August 13, 2001.

KOCI, David (KOH-chee, DAY-vihd) PIT.

Defense. Shoots left. 6'6", 225 lbs. Born, Prague, Czech., May 12, 1981.
(Pittsburgh's 5th choice, 146th overall, in 2000 Entry Draft).

			Regular Season					Playoffs				
Season	Club	Lea	GP	G	A	TP	PIM	GP	G	A	TP	PIM
1997-98	Sparta Praha Jr.	Czech-Jr.	41	2	9	11	105	....	....	....	....	....
1998-99	Hvezda Praha Jr.	Czech-Jr.	22	1	3	4	36	....	....	....	....	....
	Sparta Praha Jr.	Czech-Jr.	7	0	0	0	4	....	....	....	....	....
99-2000	Sparta Praha Jr.	Czech-Jr.	47	0	6	6	124	....	....	....	....	....
2000-01	Prince George	WHL	70	2	7	9	155	6	0	0	0	20
2001-02	Wheeling Nailers	ECHL	33	2	4	6	105	....	....	....	....	....
	Wilkes-Barre	AHL	26	1	3	4	98	....	....	....	....	....

KOIVISTO, Tom (KOI-vihs-toh, TAWM) ST.L.

Defense. Shoots right. 5'10", 194 lbs. Born, Turku, Finland, June 4, 1974.
(St. Louis' 8th choice, 253rd overall, in 2002 Entry Draft).

			Regular Season					Playoffs				
Season	Club	Lea	GP	G	A	TP	PIM	GP	G	A	TP	PIM
1991-92	TPS Turku Jr.	Finn-Jr.	36	4	9	13	40	8	1	3	4	12
1992-93	TPS Turku Jr.	Finn-Jr.	21	10	8	18	30	5	1	1	2	6
	TPS Turku	Fiinland	1	0	0	0	0	1	0	0	0	0
	Kiekoo-67 Turku	Finland-2	16	0	4	4	4	....	....	....	....	....
1993-94	TPS Turku Jr.	Finn-Jr.	10	4	3	7	4	7	0	5	5	8
	TPS Turku	Finland	18	2	5	7	4	1	0	1	1	2
	Kiekko-67 Turku	Finland-2	16	4	12	16	2	....	....	....	....	....
1994-95	TPS Turku Jr.	Finn-Jr.	5	0	2	2	8	....	....	....	....	....
	TPS Turku	Finland	4	0	0	0	4	....	....	....	....	....
	Kiekko-67 Turku	Finland-2	14	7	3	10	6	....	....	....	....	....
	HPK Hameenlina	Finland	25	3	3	6	16	....	....	....	....	....
1995-96	HPK Hameenlina	Finland	50	8	11	19	52	9	1	2	3	6
1996-97	HPK Hameenlina	Finland	46	18	17	35	50	10	4	2	6	6
1997-98	HPK Hameenlina	EuroHL	3	1	0	1	0	....	....	....	....	....
	HPK Hameenlina	Finland	23	6	6	12	28	....	....	....	....	....
1998-99	HPK Hameenlina	Finland	52	13	26	39	91	8	5	1	6	14
99-2000	Jokerit Helsinki	Finland	43	8	20	28	58	11	2	1	3	2
2000-01	Jokerit Helsinki	Finland	47	8	15	23	36	5	0	0	0	2
2001-02	Jokerit Helsinki	Finland	43	8	14	22	30	12	3	8	11	2

Finnish Elite Leagye Best Defenseman (2002) • Finnish Elite League All-Star Team (2002)

KOIVISTO, Toni (KOI-vihs-toh, TOH-nee) FLA.

Left wing. Shoots left. 6', 189 lbs. Born, Ylitornio, Finland, November 5, 1982.
(Florida's 9th choice, 200th overall, in 2001 Entry Draft).

			Regular Season					Playoffs				
Season	Club	Lea	GP	G	A	TP	PIM	GP	G	A	TP	PIM
1997-98	Lukko Rauma-B	Finn-Jr.	10	10	4	14	0	....	....	....	....	....
1998-99	Lukko Rauma-B	Finn-Jr.	8	6	5	11	2	....	....	....	....	....
	Lukko Rauma Jr.	Finn-Jr.	24	5	2	7	8	....	....	....	....	....
99-2000	Lukko Rauma Jr.	Finn-Jr.	32	21	12	33	8	8	2	1	3	0
	Lukko Rauma	Finland	11	1	0	1	2	....	....	....	....	....
2000-01	Lukko Rauma	Finland	47	5	1	6	6	2	0	0	0	0
	Lukko Rauma Jr.	Finn-Jr.	16	9	15	24	6	1	0	0	0	0
	Jaa-Kotkat	Finland-2	2	0	0	0	0	....	....	....	....	....
2001-02	Lukko Rauma Jr.	Finn-Jr.	6	7	3	10	2	....	....	....	....	....
	Lukko Rauma	Finland	52	4	10	14	8	....	....	....	....	....

KOIVU, Mikko (KOI-voo, MEE-koh) MIN.

Center. Shoots left. 6'2", 183 lbs. Born, Turku, Finland, March 12, 1983.
(Minnesota's 1st choice, 6th overall, in 2001 Entry Draft).

			Regular Season					Playoffs				
Season	Club	Lea	GP	G	A	TP	PIM	GP	G	A	TP	PIM
1997-98	TPS Turku-C	Finn-Jr.	32	6	12	18	34	....	....	....	....	....
1998-99	TPS Turku-C	Finn-Jr.	30	17	42	59	26	5	2	*9	*11	25
99-2000	TPS Turku Jr.	Finn-Jr.	41	8	17	25	40	13	1	4	5	8
2000-01	TPS Turku Jr.	Finn-Jr.	26	9	36	45	26	4	2	2	4	8
	TPS Turku	Finland	21	0	1	1	2	....	....	....	....	....
2001-02	TPS Turku	Finland	48	4	3	7	34	8	0	3	3	4
	TPS Turku Jr.	Finn-Jr.	2	0	1	1	12	....	....	....	....	....

KOKOREV, Dimitri (KOH-koh-rehf, DEH-mee-tree) CGY.

Defense. Shoots left. 6'3", 198 lbs. Born, Moscow, USSR, January 9, 1979.
(Calgary's 4th choice, 51st overall, in 1997 Entry Draft).

			Regular Season					Playoffs				
Season	Club	Lea	GP	G	A	TP	PIM	GP	G	A	TP	PIM
1996-97	DynamoMoscow2	Russia-3	27	2	4	6	24	....	....	....	....	....
	Dynamo Moscow	Russia	1	0	0	0	0	....	....	....	....	....
1997-98	Dynamo Moscow	Russia	24	1	2	3	20	....	....	....	....	....
1998-99	Dynamo Moscow	Russia	26	0	1	1	20	8	1	0	1	0
	DynamoMoscow2	Russia-3	14	0	2	2	20	....	....	....	....	....
99-2000	THC Tver	Russia-2	20	6	3	9	32	....	....	....	....	....
	Dynamo Moscow	Russia	22	0	1	1	14	1	0	0	0	0
2000-01	Dynamo Moscow	Russia	5	0	1	1	4	....	....	....	....	....
2001-02	CSKA Moscow 2	Russia-3	11	2	2	4	22	....	....	....	....	....
	CSKA Moscow	Russia	32	1	3	4	32	....	....	....	....	....

KOLARIK, Tyler (koh-LAHR-ihk, TIGH-luhr) CBJ

Center. Shoots right. 5'10", 185 lbs. Born, Philadelphia, PA, January 26, 1981.
(Columbus' 5th choice, 150th overall, in 2000 Entry Draft).

			Regular Season					Playoffs				
Season	Club	Lea	GP	G	A	TP	PIM	GP	G	A	TP	PIM
99-2000	Deerfield	Hi-School	26	31	22	53	8	....	....	....	....	....
	NY/Mid-Atlantic	MBHL	3	4	3	7	0	....	....	....	....	....
2000-01	Harvard Crimson	ECAC	32	13	15	28	36	....	....	....	....	....
2001-02	Harvard Crimson	ECAC	32	9	21	30	32	....	....	....	....	....

KOLOZVARY, Ivan (KOH-lohzh-vah-ree, EE-vahn) TOR.

Center. Shoots left. 6', 163 lbs. Born, Ilava, Czech., February 16, 1983.
(Toronto's 9th choice, 198th overall, in 2001 Entry Draft).

			Regular Season					Playoffs				
Season	Club	Lea	GP	G	A	TP	PIM	GP	G	A	TP	PIM
1998-99	Dukla Trencin Jr.	Slovak-Jr.	46	13	24	37	12	....	....	....	....	....
99-2000	Dukla Trencin Jr.	Slovak-Jr.	51	10	27	37	14	....	....	....	....	....
2000-01	Dukla Trencin Jr.	Slovak-Jr.	39	13	10	23	14	....	....	....	....	....
	Dukla Trencin	Slovakia	14	0	0	0	2	7	0	2	2	2
2001-02	Dukla Trencin	Slovakia	31	1	5	6	2	5	1	1	2	0

KOLTSOV, Ivan (kohlt-SAHV, ee-VAHN) EDM.

Defense. Shoots left. 6'2", 182 lbs. Born, Cherepovets, USSR, March 7, 1984.
(Edmonton's 6th choice, 106th overall, in 2002 Entry Draft).

			Regular Season					Playoffs				
Season	Club	Lea	GP	G	A	TP	PIM	GP	G	A	TP	PIM
2000-01	Cherepovets 2	Russia-3	5	0	1	1	2	....	....	....	....	....
	Team Russia-18	Nat-Tm	5	1	1	2	2	....	....	....	....	....
2001-02	Cherepovets 2	Russia-3	27	2	2	4	24	....	....	....	....	....

KOLTSOV, Kirill (kohlt-SAHV, kih-RIHL) VAN.

Defense. Shoots left. 5'11", 183 lbs. Born, Chelyabinsk, USSR, February 1, 1983.
(Vancouver's 1st choice, 49th overall, in 2002 Entry Draft).

			Regular Season					Playoffs				
Season	Club	Lea	GP	G	A	TP	PIM	GP	G	A	TP	PIM
1997-98	California	CBHL	STATISTICS NOT AVAILABLE									
1998-99	Streetsville	OPJHL	20	5	7	12	4	....	....	....	....	....
99-2000	Omsk 2	Russia-3	27	0	7	7	30	....	....	....	....	....
2000-01	Avangard Omsk	Russia	65	1	4	5	34	....	....	....	....	....
2001-02	Avangard Omsk	Russia	41	1	5	6	34	11	1	0	1	8

KOLTSOV, Konstantin (kohlt-SAHV, kawn-stuhn-TEEN) PIT.

Right wing. Shoots left. 6', 190 lbs. Born, Minsk, USSR, April 17, 1981.
(Pittsburgh's 1st choice, 18th overall, in 1999 Entry Draft).

			Regular Season					Playoffs				
Season	Club	Lea	GP	G	A	TP	PIM	GP	G	A	TP	PIM
1997-98	Cherepovets 2	Russia-3	44	11	12	23	16	....	....	....	....	....
	Cherepovets	Russia	2	0	0	0	2	....	....	....	....	....
1998-99	Cherepovets 3	Russia-4	2	0	1	1	2	....	....	....	....	....
	Cherepovets 2	Russia-3	11	1	4	5	18	....	....	....	....	....
	Cherepovets	Russia	33	3	0	3	8	1	0	0	0	2
99-2000	Magnitogorsk	Russia	30	3	4	7	12	11	1	1	2	6
2000-01	Ak Bars Kazan	Russia	24	7	8	15	10	2	0	0	0	4
2001-02	Ak Bars Kazan	Russia	10	1	2	3	2	....	....	....	....	....
	Spartak Moscow 2	Russia-3	2	0	1	1	0	....	....	....	....	....
	Spartak Moscow	Russia	23	1	0	1	12	....	....	....	....	....

KOMADOSKI, Neil (koh-mah-DAW-skee, NEEL) OTT.

Defense. Shoots left. 6'4", 215 lbs. Born, Chesterfield, MO, February 10, 1982.
(Ottawa's 3rd choice, 81st overall, in 2001 Entry Draft).

			Regular Season					Playoffs				
Season	Club	Lea	GP	G	A	TP	PIM	GP	G	A	TP	PIM
1997-98	Aurora Tigers	OPJHL	1	0	0	0	0	....	....	....	....	....
1998-99	Team USA	USDP-18	50	8	7	15	202	....	....	....	....	....
99-2000	Team USA	USDP-18	49	3	11	14	222	....	....	....	....	....
2000-01	U. of Notre Dame	CCHA	30	2	5	7	106	....	....	....	....	....
2001-02	U. of Notre Dame	CCHA	37	2	9	11	100	....	....	....	....	....

KOMAROV, Alexei (KOH-muh-rahf, al-EHX-ay) DAL.

Defense. Shoots left. 6'4", 194 lbs. Born, Moscow, USSR, June 11, 1978.
(Dallas' 8th choice, 216th overall, in 1997 Entry Draft).

			Regular Season					Playoffs				
Season	Club	Lea	GP	G	A	TP	PIM	GP	G	A	TP	PIM
1996-97	DynamoMoscow2	Russia-3	32	2	3	5	12	....	....	....	....	....
1997-98	Yekaterinburg	Russia	19	0	0	0	6	....	....	....	....	....
	Yekaterinburg	Russia-Q	22	0	1	1	14	....	....	....	....	....
1998-99	Spartak Moscow	Russia	21	1	0	1	6	....	....	....	....	....
99-2000	Spartak Moscow	Russia-2	32	0	5	5	22	....	....	....	....	....
2000-01	Spartak Moscow	Russia-2	33	2	5	7	28	12	0	3	3	8
2001-02	Spartak Moscow 2	Russia-3	2	1	2	3	0	....	....	....	....	....
	Spartak Moscow	Russia	36	3	5	8	43	....	....	....	....	....

KOMISAREK, Mike (koh-mih-SAIR-ehk, MIGHK) MTL.

Defense. Shoots right. 6'4", 240 lbs. Born, Islip Terrace, NY, January 19, 1982.
(Montreal's 1st choice, 7th overall, in 2001 Entry Draft).

			Regular Season					Playoffs				
Season	Club	Lea	GP	G	A	TP	PIM	GP	G	A	TP	PIM
1998-99	New England	EJHL	53	17	24	51	....	....	....	....	....	....
99-2000	Team USA	USDP-18	51	5	8	13	124	....	....	....	....	....
2000-01	U. of Michigan	CCHA	41	4	12	16	77	....	....	....	....	....
2001-02	U. of Michigan	CCHA	40	11	19	30	70	....	....	....	....	....

CCHA First All-Star Team (2002) • NCAA West First All-American Team (2002)

KONDRATJEV, Maxim (kohn-DRAT-yehv, mahx-EEM) **TOR.**

Defense. Shoots left. 6'1", 176 lbs. Born, Togliatti, USSR, January 20, 1983.
(Toronto's 7th choice, 168th overall, in 2001 Entry Draft).

			Regular Season					Playoffs				
Season	Club	Lea	GP	G	A	TP	PIM	GP	G	A	TP	PIM
99-2000	Lada Togliatti 2	Russia-3	16	0	2	2	6	....	....	....	....	....
	Lada Togliatti	Russia-2	20	1	1	2	....	....	....	....	....	....
2000-01	Lada Togliatti 2	Russia-3		STATISTICS NOT AVAILABLE								
	CSK VVS Samara	Russia-2	18	2	1	3	24	....	....	....	....	....
2001-02	Lada Togliatti	Russia	43	3	3	6	32	4	0	0	0	0

KOPECKY, Milan (koh-PEHTS-kee, MEE-lan) **PHI.**

Left wing. Shoots left. 6', 180 lbs. Born, Kolin, Czech., May 11, 1981.
(Philadelphia's 8th choice, 287th overall, in 2000 Entry Draft).

			Regular Season					Playoffs				
Season	Club	Lea	GP	G	A	TP	PIM	GP	G	A	TP	PIM
1998-99	Sparta Praha Jr.	Czech-Jr.	49	14	21	35	....	....	....	....	....	....
99-2000	Slavia Praha Jr.	Czech-Jr.	36	17	7	24	24	7	5	3	8	0
	HC Slavia Praha	Czech	2	0	0	0	0	....	....	....	....	....
2000-01	Slavia Praha Jr.	Czech-Jr.	38	19	17	36	36	....	....	....	....	....
	SC Kalin	Czech-2	8	2	4	6	2	....	....	....	....	....
2001-02	Slavia Praha Jr.	Czech-Jr.	3	2	0	2	4	....	....	....	....	....
	Beroun	Czech-2	36	10	9	19	16	....	....	....	....	....
	HC Slavia Praha	Czech	....	....	....	....	....	1	0	0	0	0

KOPECKY, Tomas (koh-PEHTS-kee, TAW-mahsh) **DET.**

Center. Shoots left. 6'3", 187 lbs. Born, Ilava, Czech., February 5, 1982.
(Detroit's 2nd choice, 38th overall, in 2000 Entry Draft).

			Regular Season					Playoffs				
Season	Club	Lea	GP	G	A	TP	PIM	GP	G	A	TP	PIM
1997-98	Dukla Trencin Jr.	Slovak-Jr.	41	19	22	41	....	....	....	....	....	....
1998-99	Dukla Trencin Jr.	Slovak-Jr.	44	13	16	29	18	....	....	....	....	....
99-2000	Dukla Trencin Jr.	Slovak-Jr.	14	8	9	17	36	....	....	....	....	....
	Dukla Trencin	Slovakia	52	3	4	7	24	5	0	0	0	0
2000-01	Lethbridge	WHL	49	22	28	50	52	5	1	1	2	6
	Cincinnati	AHL	1	0	0	0	0	....	....	....	....	....
2001-02	Lethbridge	WHL	60	34	42	76	94	4	2	1	3	15
	Cincinnati	AHL	2	1	1	2	6	2	0	0	0	0

KOREIS, Jakub (KOHR-ays, YA-kuhb) **PHX.**

Center. Shoots left. 6'2", 212 lbs. Born, Plzen, Czech, June 26, 1984.
(Phoenix's 1st choice, 19th overall, in 2002 Entry Draft).

			Regular Season					Playoffs				
Season	Club	Lea	GP	G	A	TP	PIM	GP	G	A	TP	PIM
99-2000	Plzen Jr.	Czech-Jr.	44	23	23	46	46	....	....	....	....	....
2000-01	Plzen Jr.	Czech-Jr.	34	7	2	9	20	....	....	....	....	....
2001-02	Plzen Jr.	Czech-Jr.	23	14	14	28	38	....	....	....	....	....
	Plzen	Czech	20	3	0	3	10	....	....	....	....	....

KORSUNOV, Vladimir KOHR-suhn-ahv, vla-DIH-meer) **ANA.**

Defense. Shoots left. 6'2", 202 lbs. Born, Moscow, USSR, March 16, 1983.
(Anaheim's 5th choice, 105th overall, in 2001 Entry Draft).

			Regular Season					Playoffs				
Season	Club	Lea	GP	G	A	TP	PIM	GP	G	A	TP	PIM
99-2000	Spartak Moscow 2	Russia-3	22	1	11	12	60	....	....	....	....	....
2000-01	Spartak Moscow	Russia-2	10	0	0	0	2	....	....	....	....	....
2001-02	Spartak Moscow 2	Russia-3	2	1	0	1	8	....	....	....	....	....
	Spartak Moscow	Russia	40	0	3	3	28	....	....	....	....	....

KOTARY, Sean (koh-TAH-ree, SHAWN) **COL.**

Center. Shoots right. 6', 180 lbs. Born, New Hartford, NY, April 28, 1981.
(Colorado's 13th choice, 266th overall, in 2000 Entry Draft).

			Regular Season					Playoffs				
Season	Club	Lea	GP	G	A	TP	PIM	GP	G	A	TP	PIM
1995/98	New Hartford	Hi-School	74	54	55	109	....	....	....	....	....	....
1998-99	Loomis-Chaffee	Hi-School	25	32	26	58	....	....	....	....	....	....
99-2000	Northfield Prep	Hi-School	41	49	57	106	21	....	....	....	....	....
2000-01	Bowling Green	CCHA	5	0	0	0	4	....	....	....	....	....
2001-02	Des Moines	USHL	46	12	14	26	39	3	0	0	0	0

• Statistics for **New Hartford** (Hi-School) are career totals from the 1995-1998 seasons

KOVAC, Kristian (KOH-vach, KRIHST-yan) **COL.**

Right wing. Shoots right. 6'2", 205 lbs. Born, Kosice, Czech., January 1, 1981.
(Colorado's 5th choice, 122nd overall, in 1999 Entry Draft).

			Regular Season					Playoffs				
Season	Club	Lea	GP	G	A	TP	PIM	GP	G	A	TP	PIM
1997-98	HC Kosice Jr.	Slovak-Jr.	47	22	11	33	103	....	....	....	....	....
1998-99	HC Kosice Jr.	Slovak-Jr.	39	30	20	50	73	2	1	0	1	2
	HC Kosice	Slovakia	6	0	0	0	2	....	....	....	....	....
99-2000	Victoriaville	QMJHL	65	11	18	29	50	5	0	0	0	4
2000-01	Victoriaville	QMJHL	51	10	20	30	38	13	2	3	5	4
2001-02	HC Kosice	Slovakia	26	5	9	14	10	11	0	0	0	4

KOZHEVNIKOV, Alexander (kuh-ZHEHV-nih-kahv, al-ehx-AN-duhr **CHI.**

Left wing. Shoots left. 6', 185 lbs. Born, Moscow, USSR, April 12, 1984.
(Chicago's 3rd choice, 93rd overall, in 2002 Entry Draft).

			Regular Season					Playoffs				
Season	Club	Lea	GP	G	A	TP	PIM	GP	G	A	TP	PIM
2001-02	Krylja Sovetov 2	Russia-3	32	12	15	27	59	....	....	....	....	....

KRAFT, Ryan (KRAFT, RIGH-uhn) **S.J.**

Center. Shoots left. 5'9", 190 lbs. Born, Bottineau, ND, November 7, 1975.
(San Jose's 11th choice, 194th overall, in 1995 Entry Draft).

			Regular Season					Playoffs				
Season	Club	Lea	GP	G	A	TP	PIM	GP	G	A	TP	PIM
1993-94	Moorhead Spuds	Hi-School	25	40	45	85	....	....	....	....	....	....
1994-95	U. of Minnesota	WCHA	44	13	33	46	44	....	....	....	....	....
1995-96	U. of Minnesota	WCHA	41	13	24	37	24	....	....	....	....	....
1996-97	U. of Minnesota	WCHA	42	25	21	46	37	....	....	....	....	....
1997-98	U. of Minnesota	WCHA	32	11	26	37	16	....	....	....	....	....
1998-99	Richmond	ECHL	63	28	36	64	35	18	10	10	20	4
99-2000	Richmond	ECHL	44	32	35	67	32	....	....	....	....	....
	Cleveland	IHL	1	0	1	1	0	....	....	....	....	....
	Kentucky	AHL	15	7	6	13	2	5	3	1	4	0
2000-01	Kentucky	AHL	77	38	50	88	36	3	2	0	2	0
2001-02	Cleveland Barons	AHL	63	19	41	60	42	....	....	....	....	....

WCHA All-Rookie Team (1995) • WCHA All-Academic Team (1996) • AHL Second All-Star Team (2001) • Won Dudley "Red" Garrett Memorial Trophy (Top Rookie - AHL) (2001)

KRESTANOVICH, Derek (krehs-TAN-oh-vihch, DAIR-ihk) **WSH.**

Center. Shoots left. 6'1", 175 lbs. Born, Surrey, B.C., April 29, 1983.
(Washington's 6th choice, 92nd overall, in 2002 Entry Draft).

			Regular Season					Playoffs				
Season	Club	Lea	GP	G	A	TP	PIM	GP	G	A	TP	PIM
2000-01	Kamloops Blazers	WHL	67	4	13	17	72	3	0	0	0	6
2001-02	Kamloops Blazers	WHL	30	6	13	19	41	....	....	....	....	....
	Moose Jaw	WHL	33	9	20	29	52	12	7	4	11	16

Traded to **Moose Jaw** (WHL) by **Kamloops** (WHL) for Ben Knopp and Moose Jaw's 3rd round choice in 2003 WHL Bantam Draft, December 14, 2001.

KRISTEK, Jaroslav (KRIHSH-tehk, YAH-roh-slahv) **BUF.**

Right wing. Shoots left. 6'1", 188 lbs. Born, Zlin, Czech., March 16, 1980.
(Buffalo's 4th choice, 50th overall, in 1998 Entry Draft).

			Regular Season					Playoffs				
Season	Club	Lea	GP	G	A	TP	PIM	GP	G	A	TP	PIM
1995-96	AC ZPS Zlin Jr.	Czech-Jr.	34	33	20	53	....	....	....	....	....	....
1996-97	AC ZPS Zlin Jr.	Czech-Jr.	44	28	27	55	....	....	....	....	....	....
1997-98	Zlin Jr.	Czech-Jr.	7	8	5	13	....	....	....	....	....	....
	HC Prostejov	Czech-2	4	0	0	0	....	....	....	....	....	....
	Zlin	Czech	37	2	8	10	20	....	....	....	....	....
1998-99	Tri-City	WHL	70	38	48	86	55	12	4	3	7	2
99-2000	Tri-City	WHL	45	26	25	51	16	2	0	0	0	0
2000-01	Rochester	AHL	35	5	3	8	20	....	....	....	....	....
2001-02	Rochester	AHL	43	3	6	9	20	1	0	0	0	0

KRISTOFFERSSON, Marcus (KRIHST-aw-fuhr-SOHN, MAHRK) **DAL.**

Right wing. Shoots left. 6'3", 217 lbs. Born, Ostersund, Sweden, January 22, 1979.
(Dallas' 4th choice, 105th overall, in 1997 Entry Draft).

			Regular Season					Playoffs				
Season	Club	Lea	GP	G	A	TP	PIM	GP	G	A	TP	PIM
1995-96	Mora IK Jr.	Swede-Jr.	16	2	2	4	28	....	....	....	....	....
	Mora IK	Swede-2	26	1	0	1	20	5	0	0	0	2
1996-97	Mora IK	Swede-2	33	1	5	6	26	....	....	....	....	....
1997-98	Mora IK	Swede-2	27	7	6	13	40	....	....	....	....	....
1998-99	HV 71 Jr.	Swede-Jr.	3	0	1	1	27	....	....	....	....	....
	HV 71 Jonkoping	Sweden	34	0	1	1	65	....	....	....	....	....
99-2000	HV 71 Jr.	Swede-Jr.	7	5	9	14	41	....	....	....	....	....
	HV 71 Jonkoping	Sweden	10	0	0	0	0	....	....	....	....	....
	Blues Espoo	Finland	29	7	4	11	42	1	1	0	1	0
2000-01	Assat Pori	Finland	9	1	0	1	16	....	....	....	....	....
	Djurgarden	Sweden	29	4	2	6	72	12	2	2	4	45
2001-02	Utah Grizzlies	AHL	50	6	10	16	42	....	....	....	....	....

KRONWALL, Niklas (KRAHN-wuhl, NIHK-las) **DET.**

Defense. Shoots left. 5'11", 165 lbs. Born, Stockholm, Sweden, January 12, 1981.
(Detroit's 1st choice, 29th overall, in 2000 Entry Draft).

			Regular Season					Playoffs				
Season	Club	Lea	GP	G	A	TP	PIM	GP	G	A	TP	PIM
1996-97	Djurgarden Jr.	Swede-Jr.	1	0	0	0	0	....	....	....	....	....
1997-98	Djurgarden Jr.	Swede-Jr.	27	4	3	7	71	2	0	0	0	2
1998-99	Huddinge IK	Swede-2	24	1	1	2	24	....	....	....	....	....
99-2000	Djurgarden	Sweden	37	1	4	5	16	8	0	0	0	8
2000-01	Djurgarden	Sweden	31	1	9	10	32	15	0	1	1	8
2001-02	Djurgarden	Sweden	48	5	7	12	34	5	0	0	0	0

KRUCHININ, Andrei (kroo-CHIHN-ihn, AWN-dray) **MTL.**

Defense. Shoots left. 5'11", 187 lbs. Born, Karaganda, USSR, May 18, 1978.
(Montreal's 7th choice, 189th overall, in 1998 Entry Draft).

			Regular Season					Playoffs				
Season	Club	Lea	GP	G	A	TP	PIM	GP	G	A	TP	PIM
1996-97	Lada Togliatti	Russia	19	0	1	1	8	11	0	0	0	0
1997-98	Lada Togliatti	Russia	43	0	4	4	73	....	....	....	....	....
1998-99	Lada Togliatti	Russia	41	1	4	5	56	6	0	1	1	2
99-2000	Lada Togliatti	Russia	25	1	2	3	24	6	1	0	1	4
	CSK VVS Samara	Russia	6	1	0	1	0	....	....	....	....	....
2000-01	Perm	Russia	14	1	3	4	10	....	....	....	....	....
	Lada Togliatti	Russia	14	0	2	2	12	3	0	0	0	0
2001-02	Avangard Omsk	Russia	21	0	0	0	6	....	....	....	....	....
	Nizhnekamsk	Russia	17	1	3	4	8	....	....	....	....	....

KRYUKOV, Artem (KREE-oo-kahf, AHR-tehm) **BUF.**

Center. Shoots left. 6'3", 180 lbs. Born, Novosibirsk, USSR, March 5, 1982.
(Buffalo's 1st choice, 15th overall, in 2000 Entry Draft).

			Regular Season					Playoffs				
Season	Club	Lea	GP	G	A	TP	PIM	GP	G	A	TP	PIM
1997-98	Yaroslavl	Russia	7	0	0	0	2	....	....	....	....	....
1998-99	Yaroslavl 2	Russia-3	20	2	2	4	6	....	....	....	....	....
99-2000	Yaroslavl 2	Russia-3	14	1	1	2	12	....	....	....	....	....
	Yaroslavl	Russia	3	0	0	0	4	....	....	....	....	....
2000-01	Yaroslavl 2	Russia-3	6	0	0	0	2	11	0	0	0	8
	St. Petersburg	Russia	14	0	2	2	14	....	....	....	....	....
2001-02	Yaroslavl	Russia	15	1	3	4	10	6	1	0	1	8

KUBISTA, Jan (KOO-bihsh-tuh, YAHN) **BOS.**

Right wing. Shoots left. 5'11", 184 lbs. Born, Kolin, Czech., April 12, 1984.
(Boston's 3rd choice, 130th overall, in 2002 Entry Draft).

			Regular Season					Playoffs				
Season	Club	Lea	GP	G	A	TP	PIM	GP	G	A	TP	PIM
99-2000	Pardubice-18	Czech-Jr.	44	16	18	34	10	....	....	....	....	....
	SC Kolin Jr.	Czech-Jr.	5	4	2	6	0	....	....	....	....	....
2000-01	Pardubice-18	Czech-Jr.	53	33	24	57	61	8	6	7	13	16
	Pardubice Jr.	Czech-Jr.	1	0	1	1	0	....	....	....	....	....
2001-02	Pardubice Jr.	Czech-Jr.	48	12	12	24	36	....	....	....	....	....

KUKHTINOV, Roman (kukh-TEEN-nawv, ROH-muhn) **NYI**

Defense. Shoots right. 6'1", 207 lbs. Born, Belgorod, USSR, December 1, 1975.
(NY Islanders' 7th choice, 280th overall, in 2001 Entry Draft).

			Regular Season					Playoffs				
Season	Club	Lea	GP	G	A	TP	PIM	GP	G	A	TP	PIM
1993-94	Krylja Sovetov	CIS	4	0	0	0	0	....	....	....	....	....
1994-95	DynamoMoscow2	CIS-2		STATISTICS NOT AVAILABLE								
	St. Petersburg	CIS	5	1	0	1	2	....	....	....	....	....
1995-96	Raichikhinsk	CIS-2		STATISTICS NOT AVAILABLE								
	Nizhnekamsk	CIS	7	0	0	0	0	....	....	....	....	....
1996-97	Raichikhinsk	Russia-3		STATISTICS NOT AVAILABLE								
1997-98	Raichikhinsk	Russia-3	36	18	8	26	46	....	....	....	....	....
1998-99	Novokuznetsk	Russia	42	2	9	11	26	6	0	0	0	2
99-2000	Novokuznetsk	Russia	37	2	6	8	28	14	1	1	2	8
2000-01	Novokuznetsk	Russia	44	7	10	17	36	....	....	....	....	....
2001-02	Ufa	Russia	51	11	10	21	74	....	....	....	....	....

KULESHOV, Mikhail (koo-leh-SHAWV, mihk-AIL) COL.

Left wing. Shoots right. 6'2", 205 lbs. Born, Perm, USSR, January 7, 1981.
(Colorado's 1st choice, 25th overall, in 1999 Entry Draft).

			Regular Season					Playoffs				
Season	**Club**	**Lea**	**GP**	**G**	**A**	**TP**	**PIM**	**GP**	**G**	**A**	**TP**	**PIM**
1997-98	Omsk 2	Russia-3	12	12	3	15	12	....	....	....	....	....
	Avangard Omsk	Russia	4	1	0	1	4	....	....	....	....	....
1998-99	Cherepovets 3	Russia-4	3	2	1	3	32	....	....	....	....	....
	Cherepovets 2	Russia-3	25	7	5	12	12	....	....	....	....	....
	Cherepovets	Russia	15	2	0	2	8	3	0	0	0	4
99-2000	Cherepovets	Russia	8	0	0	0	4	3	0	0	0	2
2000-01	St. Petersburg	Russia	7	0	0	0	8	....	....	....	....	....
	Hershey Bears	AHL	3	0	0	0	4	11	1	0	1	0
2001-02	Hershey Bears	AHL	60	8	11	19	43	7	0	1	1	4

KURKA, Tomas (KUHR-kuh, TAW-mahsh) CAR.

Left wing. Shoots left. 5'11", 190 lbs. Born, Most, Czech., December 14, 1981.
(Carolina's 1st choice, 32nd overall, in 2000 Entry Draft).

			Regular Season					Playoffs				
Season	**Club**	**Lea**	**GP**	**G**	**A**	**TP**	**PIM**	**GP**	**G**	**A**	**TP**	**PIM**
1996-97	Litvinov Jr.	Czech-Jr.	38	25	20	45	20	....	....	....	....	....
1997-98	Litvinov Jr.	Czech-Jr.	44	38	23	61	90	....	....	....	....	....
1998-99	Litvinov Jr.	Czech-Jr.	42	23	16	39	47	....	....	....	....	....
	Litvinov	Czech	6	0	0	0	0	....	....	....	....	....
99-2000	Plymouth Whalers	OHL	64	36	28	64	37	17	7	6	13	6
2000-01	Plymouth Whalers	OHL	47	15	29	44	20	16	8	13	21	13
2001-02	Lowell	AHL	71	13	15	28	24	5	1	1	2	2

KUZNETSOV, Sergei (kooz-NEHT-zahv, SAIR-gay) PHX.

Center. Shoots left. 6'1", 195 lbs. Born, Yaroslavl, USSR, January 29, 1980.
(Tampa Bay's 6th choice, 146th overall, in 1998 Entry Draft).

			Regular Season					Playoffs				
Season	**Club**	**Lea**	**GP**	**G**	**A**	**TP**	**PIM**	**GP**	**G**	**A**	**TP**	**PIM**
1995-96	Yaroslavl Jr.	CIS-Jr.	28	14	14	28	20	....	....	....	....	....
1996-97	Yaroslavl 2	Russia-3	62	16	15	31	35	....	....	....	....	....
1997-98	Yaroslavl	Russia	42	10	13	23	30	....	....	....	....	....
1998-99	Peterborough	OHL	65	8	17	25	39	5	2	1	3	2
99-2000	Peterborough	OHL	68	33	32	65	54	5	0	4	4	2
2000-01	Mississippi	ECHL	55	11	17	28	34	....	....	....	....	....
	Springfield	AHL	22	0	4	4	10	....	....	....	....	....
2001-02	Magnitogorsk	Russia	2	0	0	0	0	....	....	....	....	....
	Mississippi	ECHL	66	12	25	37	48	10	0	3	3	14

Signed as a free agent by **Phoenix**, August 15, 2000.

LAATIKAINEN, Arto (lah-tee-KIGH-nuhn, AHR-toh) NYR

Defense. Shoots left. 6', 187 lbs. Born, Espoo, Finland, May 24, 1980.
(NY Rangers' 8th choice, 197th overall, in 1999 Entry Draft).

			Regular Season					Playoffs				
Season	**Club**	**Lea**	**GP**	**G**	**A**	**TP**	**PIM**	**GP**	**G**	**A**	**TP**	**PIM**
1996-97	Kiekko Espoo Jr.	Finn-Jr.	34	2	9	11	34	....	....	....	....	....
1997-98	Kiekko Espoo B	Finn-Jr.	7	2	1	3	6	....	....	....	....	....
	Kiekko Espoo Jr.	Finn-Jr.	35	7	9	16	24	5	2	0	2	2
1998-99	Blues Espoo Jr.	Finn-Jr.	3	0	1	1	4	1	1	1	2	0
	Blues Espoo	Finland	48	0	6	6	14	4	0	2	2	2
99-2000	Blues Espoo Jr.	Finn-Jr.	1	0	0	0	2	....	....	....	....	....
	Blues Espoo	Finland	51	6	5	11	12	4	1	0	1	4
2000-01	Blues Espoo	Finland	54	5	9	14	38	....	....	....	....	....
2001-02	Blues Espoo	Finland	56	5	6	11	32	3	0	0	0	0

LAICH, Brooks (LAYCH, BROOKS) OTT.

Center. Shoots left. 6' 2", 194 lbs. Born, Medicine Hat, Alta., June 23, 1983.
(Ottawa's 7th choice, 193rd overall, in 2001 Entry Draft).

			Regular Season					Playoffs				
Season	**Club**	**Lea**	**GP**	**G**	**A**	**TP**	**PIM**	**GP**	**G**	**A**	**TP**	**PIM**
99-2000	Tisdale Trojans	SMHL		STATISTICS NOT AVAILABLE								
2000-01	Moose Jaw	WHL	71	9	21	30	28	4	0	0	0	5
2001-02	Moose Jaw	WHL	28	6	14	20	12	....	....	....	....	....
	Seattle	WHL	47	22	36	58	42	11	5	3	8	11

Traded to **Seattle** (WHL) by **Moose Jaw** (WHL) with Tomas Mojzis for Craig Olynick, Stanislav Avksentiev and Seattle's 3rd round choice (Tanner Gillies) in 2002 WHL Bantam Draft, November 20, 2001.

LAINE, Teemu (LIGH-neh, TEE-moo) N.J.

Right wing. Shoots left. 6', 200 lbs. Born, Helsinki, Finland, August 9, 1982.
(New Jersey's 2nd choice, 39th overall, in 2000 Entry Draft).

			Regular Season					Playoffs				
Season	**Club**	**Lea**	**GP**	**G**	**A**	**TP**	**PIM**	**GP**	**G**	**A**	**TP**	**PIM**
1997-98	Jokerit-B	Finn-Jr.	20	20	24	44	54	5	1	3	4	4
1998-99	Jokerit Jr.	Finn-Jr.	29	20	17	37	83	6	0	2	2	6
99-2000	Jokerit Jr.	Finn-Jr.	23	5	9	14	42	....	....	....	....	....
	Jokerit Helsinki	Finland	14	1	1	2	8	....	....	....	....	....
2000-01	Jokerit Jr.	Finn-Jr.	4	1	2	3	18	1	0	0	0	0
	Kiekko Vantaa	Finland-2	18	2	4	6	30	....	....	....	....	....
	Jokerit Helsinki	Finland	25	3	2	5	10	5	1	0	1	2
2001-02	Jokerit Jr.	Finn-Jr.	8	9	5	14	50	1	0	0	0	2
	Kiekko Vantaa	Finland-2	9	6	4	10	4	....	....	....	....	....
	Jokerit Helsinki	Finland	38	0	1	1	45	7	0	0	0	2

LALIBERTE, John (lal-IH-buhr-tee, JAWN) VAN.

Right wing. Shoots left. 6'1", 185 lbs. Born, Portland, ME, August 5, 1983.
(Vancouver's 5th choice, 114th overall, in 2002 Entry Draft).

			Regular Season					Playoffs				
Season	**Club**	**Lea**	**GP**	**G**	**A**	**TP**	**PIM**	**GP**	**G**	**A**	**TP**	**PIM**
99-2000	Exeter Eagles	Hi-School	32	37	40	77	60	....	....	....	....	....
2000-01	N.H. Jr. Monarchs	EJHL	53	33	42	75	42	....	....	....	....	....
2001-02	N.H. Jr. Monarchs	EJHL	35	39	44	*83	60	....	....	....	....	....

EJHL First All-Star Team (2002) • EJHL MVP (2002)

• Signed Letter of Intent to attend **Boston University** (H-East), January 31, 2002.

LAMBERT, Michael (lam-BAIR, MIGH-kuhl) MTL.

Left wing. Shoots left. 6'2", 180 lbs. Born, Trois-Rivieres, Que., March 10, 1984.
(Montreal's 3rd choice, 99th overall, in 2002 Entry Draft).

			Regular Season					Playoffs				
Season	**Club**	**Lea**	**GP**	**G**	**A**	**TP**	**PIM**	**GP**	**G**	**A**	**TP**	**PIM**
1998-99	Cap-d-Madeleine	QAAA	3	0	0	0	0	....	....	....	....	....
99-2000	Cap-d-Madeleine	QAAA	42	20	16	36	38	1	0	2	2	0
2000-01	Acadie-Bathurst	QMJHL	23	2	5	7	15	....	....	....	....	....
	Montreal Rocket	QMJHL	33	6	12	18	14	....	....	....	....	....
2001-02	Montreal Rocket	QMJHL	71	29	24	53	111	7	1	6	7	15

Traded to **Montreal** (QMJHL) by **Acadie-Bathurst** (QMJHL) with Kevin Lavallee for Yann Joseph, Michael Lanthier and Montreal's 3rd (Josh Rathbone) and 6th round (Shawn Doucet) choices in 2001 QMJHL Priority Draft, December 5, 2000.

LAMPMAN, Bryce (LAMP-man, BRIGHS) NYR

Defense. Shoots left. 6'1", 193 lbs. Born, Rochester, MN, August 31, 1982.
(NY Rangers' 4th choice, 113th overall, in 2001 Entry Draft).

			Regular Season					Playoffs				
Season	**Club**	**Lea**	**GP**	**G**	**A**	**TP**	**PIM**	**GP**	**G**	**A**	**TP**	**PIM**
1998-99	Rochester	USHL	53	3	8	11	33	....	....	....	....	....
99-2000	Rochester	USHL	10	0	0	0	14	....	....	....	....	....
	Omaha Lancers	USHL	11	1	2	3	38	4	0	0	0	0
2000-01	Omaha Lancers	USHL	55	10	11	21	77	12	1	4	5	12
2001-02	Nebraska-Omaha	CCHA	26	0	4	4	28	....	....	....	....	....

LANE, Jesse (LAYN, JEH-see) CAR.

Defense. Shoots left. 6'1", 210 lbs. Born, Boston, MA, February 3, 1983.
(Carolina's 2nd choice, 91st overall, in 2002 Entry Draft).

			Regular Season					Playoffs				
Season	**Club**	**Lea**	**GP**	**G**	**A**	**TP**	**PIM**	**GP**	**G**	**A**	**TP**	**PIM**
1998-99	Walpole Jr.	EJHL		STATISTICS NOT AVAILABLE								
99-2000	Des Moines	USHL	54	3	15	18	89	9	1	1	2	6
2000-01	Team USA	USDP-18	62	7	20	27	84	....	....	....	....	....
2001-02	Harvard Crimson	ECAC	3	0	1	1	....	....	....	....	....	....
	Hull Olympiques	QMJHL	48	16	24	40	83	11	1	9	10	18

• Left **Harvard** (ECAC) and signed as a free agent by **Hull** (QMJHL), November 10, 2001.

LAPLANTE, Eric (LA-plawnt, AIR-ihk) S.J.

Left wing. Shoots left. 6', 185 lbs. Born, St-Maurice, Que., December 1, 1979.
(San Jose's 3rd choice, 65th overall, in 1998 Entry Draft).

			Regular Season					Playoffs				
Season	**Club**	**Lea**	**GP**	**G**	**A**	**TP**	**PIM**	**GP**	**G**	**A**	**TP**	**PIM**
1995-96	Cap-d-Madeleine	QAAA	41	13	18	31	138	....	....	....	....	....
1996-97	Halifax	QMJHL	68	20	30	50	245	18	3	11	14	28
1997-98	Halifax	QMJHL	40	19	22	41	193	....	....	....	....	....
1998-99	Drummondville	QMJHL	42	14	25	39	258	....	....	....	....	....
	Quebec Remparts	QMJHL	23	4	17	21	58	13	8	7	15	45
99-2000	Quebec Remparts	QMJHL	47	24	35	59	234	10	3	5	8	*83
2000-01	Kentucky	AHL	62	5	8	13	181	3	0	0	0	6
2001-02	Cleveland Barons	AHL	78	7	14	21	326	....	....	....	....	....

Traded to **Quebec** (QMJHL) by **Drummondville** (QMJHL) for Eric Jean, January 11, 1999.

LAROSE, Cory (la-ROHZ, KOH-ree) MIN.

Left wing. Shoots left. 6', 188 lbs. Born, Campbellton, N.B., May 14, 1975.

			Regular Season					Playoffs				
Season	**Club**	**Lea**	**GP**	**G**	**A**	**TP**	**PIM**	**GP**	**G**	**A**	**TP**	**PIM**
1993-94	Kimball Union	Hi-School	21	18	11	29	14	....	....	....	....	....
1994-95	Langley Thunder	BCJHL		STATISTICS NOT AVAILABLE								
1995-96	Langley Thunder	BCJHL	54	28	46	74	61	....	....	....	....	....
1996-97	U. of Maine	H-East	35	10	27	37	32	....	....	....	....	....
1997-98	U. of Maine	H-East	34	15	25	40	22	....	....	....	....	....
1998-99	U. of Maine	H-East	38	21	31	52	34	....	....	....	....	....
99-2000	U. of Maine	H-East	39	15	*36	51	45	....	....	....	....	....
2000-01	Cleveland	IHL	4	1	1	2	6	....	....	....	....	....
	Jackson Bandits	ECHL	63	21	32	53	73	5	2	2	4	12
2001-02	Houston Aeros	AHL	78	32	32	64	73	14	6	8	14	15

BCJHL Playoff MVP (1996) • Hockey East First All-Star Team (2000) • NCAA East Second All-American Team (2000) • AHL All-Rookie Team (2002)

Signed as a free agent by **Minnesota**, May 10, 2000.

LARRIVEE, Christian (la-ree-VAY, krihs-TYEH) MTL.

Center. Shoots left. 6'3", 192 lbs. Born, Gaspe, Que., August 25, 1982.
(Montreal's 6th choice, 114th overall, in 2000 Entry Draft).

			Regular Season					Playoffs				
Season	**Club**	**Lea**	**GP**	**G**	**A**	**TP**	**PIM**	**GP**	**G**	**A**	**TP**	**PIM**
1998-99	Jonquiere Elites	QAAA	42	26	36	62	10	....	....	....	....	....
99-2000	Chicoutimi	QMJHL	69	8	15	23	18	....	....	....	....	....
2000-01	Chicoutimi	QMJHL	72	32	48	80	46	7	3	1	4	4
2001-02	Chicoutimi	QMJHL	72	48	52	100	60	4	2	5	7	0

LAUZON, Ryan (LOH-zohn, RIGH-yan) PHX.

Center. Shoots left. 5'9", 193 lbs. Born, Halifax, N.S., October 8, 1980.
(Phoenix's 5th choice, 116th overall, in 1999 Entry Draft).

			Regular Season					Playoffs				
Season	**Club**	**Lea**	**GP**	**G**	**A**	**TP**	**PIM**	**GP**	**G**	**A**	**TP**	**PIM**
1995-96	Halifax Hawks	NSMHL	89	96	128	224	....	....	....	....	....	....
1996-97	Hull Olympiques	QMJHL	65	8	10	18	16	14	1	0	1	0
1997-98	Hull Olympiques	QMJHL	64	33	56	89	37	11	8	13	21	2
1998-99	Hull Olympiques	QMJHL	57	21	47	68	36	23	3	20	23	10
99-2000	Hull Olympiques	QMJHL	49	24	35	59	37	15	6	8	14	8
2000-01	Springfield	AHL	37	5	10	15	6	....	....	....	....	....
2001-02	Springfield	AHL	2	0	0	0	0	....	....	....	....	....
	Mississippi	ECHL	58	7	12	19	12	2	0	1	1	4

LAVRENTIEV, Anton (lahv-REHN-tee-yehv, an-TAWN) NSH.

Defense. Shoots right. 6'4", 196 lbs. Born, Kazan, USSR, August 25, 1983.
(Nashville's 7th choice, 178th overall, in 2001 Entry Draft).

			Regular Season					Playoffs				
Season	**Club**	**Lea**	**GP**	**G**	**A**	**TP**	**PIM**	**GP**	**G**	**A**	**TP**	**PIM**
2000-01	Ak Bars Kazan 2	Russia-3		STATISTICS NOT AVAILABLE								
2001-02	Sudbury Wolves	OHL	10	0	0	0	17	....	....	....	....	....
	Ak Bars Kazan 2	Russia-3		STATISTICS NOT AVAILABLE								

LAZAREV, Yevgeny (LA-zahr-ehv, YEHV-geh-nee) COL.

Left wing. Shoots left. 6'2", 205 lbs. Born, Kharkov, USSR, April 25, 1980.
(Colorado's 8th choice, 79th overall, in 1998 Entry Draft).

			Regular Season					Playoffs				
Season	Club	Lea	GP	G	A	TP	PIM	GP	G	A	TP	PIM
1995-96	Yaroslavl	CIS	60	32	30	62	45	....	....	....	....	....
1996-97	Yaroslavl 2	Russia-3	44	18	15	33	38	16	23	21	44	22
	Yaroslavl	Russia	1	0	0	0	0	....	....	....	....	....
1997-98	Kitchener	OJHL-B	11	9	13	22	19	5	5	2	7	17
1998-99	Hershey Bears	AHL	53	6	15	21	18	....	....	....	....	....
99-2000	Hershey Bears	AHL	46	2	11	13	44	8	0	1	1	2
	Pensacola	ECHL	11	3	5	8	23	....	....	....	....	....
2000-01	Hershey Bears	AHL	80	17	21	38	50	12	3	8	11	10
2001-02	Hershey Bears	AHL	50	11	11	22	56	7	2	2	4	27

LEACH, Jay (LEECH, JAY) PHX.

Defense. Shoots left. 6'4", 232 lbs. Born, Syracuse, NY, September 2, 1979.
(Phoenix's 5th choice, 115th overall, in 1998 Entry Draft).

			Regular Season					Playoffs				
Season	Club	Lea	GP	G	A	TP	PIM	GP	G	A	TP	PIM
1994-95	John Marshall	Hi-School	10	0	0	0	14	....	....	....	....	....
1995-96	John Marshall	Hi-School	11	1	2	3	8	4	0	0	0	0
	Capital District	Exhib.	53	3	8	11	33	....	....	....	....	....
1996-97	Capital District	Exhib.	57	8	50	58	140	....	....	....	....	....
1997-98	Providence	H-East	32	0	8	8	29	....	....	....	....	....
1998-99	Providence	H-East	33	1	8	9	42	....	....	....	....	....
99-2000	Providence	H-East	37	1	9	10	101	....	....	....	....	....
2000-01	Providence	H-East	40	4	21	25	104	....	....	....	....	....
2001-02	Mississippi	ECHL	70	3	13	16	116	10	1	1	2	8

Hockey East All-Academic Team (2000)

LEAHY, Patrick (LEH-hey, PAT-rihk)

Right wing. Shoots right. 6'3", 190 lbs. Born, Brighton, MA, June 9, 1979.
(NY Rangers' 5th choice, 122nd overall, in 1998 Entry Draft).

			Regular Season					Playoffs				
Season	Club	Lea	GP	G	A	TP	PIM	GP	G	A	TP	PIM
1996-97	B.C. High Irish	Hi-School	25	24	24	48	....	....	....	....	....	....
1997-98	Miami-Ohio	CCHA	28	0	1	1	24	....	....	....	....	....
1998-99	Miami-Ohio	CCHA	34	10	20	30	40	....	....	....	....	....
99-2000	Miami-Ohio	CCHA	36	16	22	38	89	....	....	....	....	....
2000-01	Miami-Ohio	CCHA	37	13	19	32	14	....	....	....	....	....
2001-02	Trenton Titans	ECHL	41	20	21	41	64	....	....	....	....	....
	Hershey Bears	AHL	9	1	2	3	8	....	....	....	....	....
	Portland Pirates	AHL	9	1	1	2	8	....	....	....	....	....
	Bridgeport	AHL	14	2	2	4	2	20	3	4	7	4

LEBLANC, Robin (luh-BLAWNK, RAW-bihn) N.J.

Right wing. Shoots right. 6'1", 175 lbs. Born, Chur, Switz., January 11, 1983.
(New Jersey's 5th choice, 67th overall, in 2001 Entry Draft).

			Regular Season					Playoffs				
Season	Club	Lea	GP	G	A	TP	PIM	GP	G	A	TP	PIM
99-2000	Baie-Comeau	QMJHL	51	6	11	17	40	5	2	1	3	8
2000-01	Baie-Comeau	QMJHL	61	24	38	62	33	11	7	8	15	8
2001-02	Baie-Comeau	QMJHL	55	17	21	38	103	5	1	2	3	6

LEE, Brian (LEE, BRIGH-uhn) ANA.

Defense. Shoots left. 6'2", 187 lbs. Born, Berrien Springs, MI, July 5, 1984.
(Anaheim's 3rd choice, 71st overall, in 2002 Entry Draft).

			Regular Season					Playoffs				
Season	Club	Lea	GP	G	A	TP	PIM	GP	G	A	TP	PIM
99-2000	Det. Honeybaked	MMHL	65	15	35	50	....	....	....	....	....	....
2000-01	Erie Otters	OHL	50	0	3	3	35	9	0	0	0	4
2001-02	Erie Otters	OHL	66	5	14	19	115	21	1	6	7	41

LEGAULT, Jay (LEH-goh, JAY)

Left wing. Shoots left. 6'4", 214 lbs. Born, Peterborough, Ont., May 15, 1979.
(Anaheim's 3rd choice, 72nd overall, in 1997 Entry Draft).

			Regular Season					Playoffs				
Season	Club	Lea	GP	G	A	TP	PIM	GP	G	A	TP	PIM
1994-95	Peterboro AA	OMHA	34	41	70	111	68	....	....	....	....	....
1995-96	Oshawa Generals	OHL	61	2	11	13	37	5	0	1	1	8
1996-97	Oshawa Generals	OHL	39	13	26	39	50	....	....	....	....	....
	London Knights	OHL	28	6	13	19	37	....	....	....	....	....
1997-98	London Knights	OHL	61	39	56	95	87	16	1	8	9	34
1998-99	London Knights	OHL	65	43	51	94	99	25	8	18	26	40
99-2000	Cincinnati	AHL	70	15	19	34	75	....	....	....	....	....
	Dayton Bombers	ECHL	2	0	0	0	2	....	....	....	....	....
2000-01	Baton Rouge	ECHL	8	3	5	8	6	....	....	....	....	....
	Cincinnati	AHL	57	18	19	37	44	4	0	3	3	0
2001-02	Cincinnati	AHL	21	4	6	10	46	....	....	....	....	....
	Milwaukee	AHL	43	11	5	16	18	....	....	....	....	....

Traded to **Nashville** by **Anaheim** for Bert Robertsson, December 4, 2001.

LEGG, Chris (LEHG, KRIHS) EDM.

Center. Shoots left. 5'11", 177 lbs. Born, London, Ont., February 19, 1980.
(Edmonton's 7th choice, 171st overall, in 1999 Entry Draft).

			Regular Season					Playoffs				
Season	Club	Lea	GP	G	A	TP	PIM	GP	G	A	TP	PIM
1996-97	London Nationals	OJHL-B	43	6	16	22	31	....	....	....	....	....
1997-98	London Nationals	OJHL-B	50	36	32	68	45	....	....	....	....	....
1998-99	London Nationals	OJHL-B	52	38	40	78	28	....	....	....	....	....
99-2000	Brown U.	ECAC	23	2	3	5	4	....	....	....	....	....
2000-01	Brown U.	ECAC	26	3	5	8	6	....	....	....	....	....
2001-02	Brown U.	ECAC	29	4	6	10	4	....	....	....	....	....

LEHOUX, Jason (luh-HOO, JAY-suhn) N.J.

Left wing. Shoots left. 6'2", 220 lbs. Born, Ste-Marie-Beauce, Que., July 21, 1979.

			Regular Season					Playoffs				
Season	Club	Lea	GP	G	A	TP	PIM	GP	G	A	TP	PIM
1995-96	Cap-d-Madeleine	QAAA	42	22	23	45	....	5	2	1	3	16
1996-97	Rimouski Oceanic	QMJHL	16	1	2	3	111	....	....	....	....	....
1997-98	Rouyn-Noranda	QMJHL	28	6	1	7	95	6	3	1	4	6
1998-99	Rouyn-Noranda	QMJHL	64	13	20	33	288	6	1	2	3	49
99-2000	Rouyn-Noranda	QMJHL	14	4	7	11	54	....	....	....	....	....
	Hull Olympiques	QMJHL	29	11	7	18	109	15	6	4	10	14
2000-01	Albany	AHL	52	8	7	15	101	....	....	....	....	....
2001-02	Albany	AHL	67	4	6	10	135	....	....	....	....	....

Signed as a free agent by **New Jersey**, June 27, 2000.

LEHOUX, Yanick (luh-HOO, YAH-nihk) L.A.

Center. Shoots right. 6', 170 lbs. Born, Montreal, Que., April 8, 1982.
(Los Angeles' 3rd choice, 86th overall, in 2000 Entry Draft).

			Regular Season					Playoffs				
Season	Club	Lea	GP	G	A	TP	PIM	GP	G	A	TP	PIM
1997-98	Cap-d-Madeleine	QAAA	42	29	50	79	26	....	....	....	....	....
1998-99	Baie-Comeau	QMJHL	63	10	20	30	31	....	....	....	....	....
99-2000	Baie-Comeau	QMJHL	67	31	61	92	14	6	1	2	3	2
2000-01	Baie-Comeau	QMJHL	70	67	68	135	62	11	8	16	24	0
2001-02	Baie-Comeau	QMJHL	66	56	69	125	63	5	5	4	9	0
	Manchester	AHL	....	....	....	....	....	1	0	0	0	0

QMJHL Second All-Star Team (2002)

LEHTONEN, Mikko (LEHT-oh-nehn, MEE-koh) NSH.

Defense. Shoots left. 6'1", 194 lbs. Born, Oulu, Finland, June 12, 1979.
(Nashville's 9th choice, 271st overall, in 2001 Entry Draft).

			Regular Season					Playoffs				
Season	Club	Lea	GP	G	A	TP	PIM	GP	G	A	TP	PIM
1995-96	Karpat Oulu-B	Finn-Jr.	15	4	5	9	30	....	....	....	....	....
1996-97	Karpat Oulu Jr.	Finn-Jr.	35	6	19	25	82	....	....	....	....	....
1997-98	Karpat Oulu Jr.	Finn-Jr.	20	3	4	7	40	....	....	....	....	....
	Karpat Oulu-B	Finn-Jr.	11	5	7	12	31	....	....	....	....	....
1998-99	Karpat Oulu Jr.	Finn-Jr.	22	7	8	15	51	....	....	....	....	....
	Karpat Oulu	Finland-2	2	0	0	0	0	....	....	....	....	....
	Karpat Oulu-B	Finn-Jr.	13	6	14	20	14	....	....	....	....	....
99-2000	Karpat Oulu	Finland-2	45	5	10	15	26	6	0	0	0	4
2000-01	Karpat Oulu	Finland	54	6	9	15	58	9	0	3	3	4
2001-02	Karpat Oulu	Finland	55	8	11	19	32	4	1	1	2	4

LEOPOLD, Jordan (LEE-oh-pohld, JOHR-dan) CGY.

Defense. Shoots left. 6', 193 lbs. Born, Golden Valley, MN, August 3, 1980.
(Anaheim's 1st choice, 44th overall, in 1999 Entry Draft).

			Regular Season					Playoffs				
Season	Club	Lea	GP	G	A	TP	PIM	GP	G	A	TP	PIM
1995-96	Armstrong	Hi-School	19	11	14	25	30	....	....	....	....	....
1996-97	Armstrong	Hi-School	30	24	36	60	....	....	....	....	....	....
1997-98	Team USA	USDP-18	60	11	12	23	16	....	....	....	....	....
1998-99	U. of Minnesota	WCHA	39	7	16	23	20	....	....	....	....	....
99-2000	U. of Minnesota	WCHA	39	6	18	24	20	....	....	....	....	....
2000-01	U. of Minnesota	WCHA	42	12	37	49	38	....	....	....	....	....
2001-02	U. of Minnesota	WCHA	44	20	28	48	28	....	....	....	....	....

WCHA All-Rookie Team (1999) • WCHA Second All-Star Team (2000) • WCHA First All-Star Team (2001, 2002) • NCAA West First All-American Team (2001) • Won Hobey Baker Memorial Award (Top U.S. Collegiate Player) (2002)

Traded to **Calgary** by **Anaheim** for Andrei Nazarov and Calgary's 2nd round choice (later traded to Phoenix - later traded back to Calgary - Calgary selected Andrei Taratukhin) in 2001 Entry Draft, September 26, 2000.

LEPHART, Mike (LEHP-huhrt, MIGHK) PHI.

Left wing. Shoots right. 5'11", 194 lbs. Born, Niskayuna, NY, April 3, 1977.

			Regular Season					Playoffs				
Season	Club	Lea	GP	G	A	TP	PIM	GP	G	A	TP	PIM
1994-95	Springfield	NEJHL	45	20	30	50	....	....	....	....	....	....
1995-96	Omaha Lancers	USHL	45	12	12	24	40	....	....	....	....	....
1996-97	Omaha Lancers	USHL	54	40	50	*90	76	10	4	5	9	16
1997-98	Boston College	H-East	40	15	12	27	24	....	....	....	....	....
1998-99	Boston College	H-East	36	11	16	27	28	....	....	....	....	....
99-2000	Boston College	H-East	42	14	19	33	66	....	....	....	....	....
2000-01	Boston College	H-East	43	15	19	34	46	....	....	....	....	....
2001-02	Philadelphia	AHL	43	10	8	18	18	5	0	1	1	0

USHL First All-Star Team (1997) • Hockey East All-Academic Team (2000, 2001) • Hockey East Defensive Player of the Year (2001)

Signed as a free agent by **Philadelphia**, June 11, 2001.

LEVESQUE, Willie (luh-VEHK, WIHL-lee) S.J.

Right wing. Shoots right. 6', 195 lbs. Born, Oak Bluffs, MA, January 22, 1980.
(San Jose's 3rd choice, 111th overall, in 1999 Entry Draft).

			Regular Season					Playoffs				
Season	Club	Lea	GP	G	A	TP	PIM	GP	G	A	TP	PIM
1997-98	Team USA	USDP-18	60	12	24	36	118	....	....	....	....	....
1998-99	Northeastern	H-East	34	12	10	22	38	....	....	....	....	....
99-2000	Northeastern	H-East	33	9	13	22	45	....	....	....	....	....
2000-01	Northeastern	H-East	35	13	16	29	62	....	....	....	....	....
2001-02	Northeastern	H-East	30	6	10	16	42	....	....	....	....	....

Hockey East All-Rookie Team (1999)

LEVINSKI, Dimitri (leh-VIHN-skee, DEH-mih-TREE) CHI.

Right wing. Shoots left. 6'1", 183 lbs. Born, Ust-Kamenogorsk, USSR, June 23, 1981.
(Chicago's 2nd choice, 46th overall, in 1999 Entry Draft).

			Regular Season					Playoffs				
Season	Club	Lea	GP	G	A	TP	PIM	GP	G	A	TP	PIM
1996-97	Avangard Omsk 2	Russia-3	15	6	2	8	8	....	....	....	....	....
1997-98	Avangard Omsk 2	Russia-3	18	5	2	7	8	....	....	....	....	....
1998-99	Cherepovets 3	Russia-4	3	2	0	2	2	....	....	....	....	....
	Cherepovets 2	Russia-3	26	4	2	6	39	....	....	....	....	....
	Cherepovets	Russia	1	0	0	0	0	....	....	....	....	....
99-2000	St. Petersburg	Russia	25	0	2	2	4	4	0	0	0	0
2000-01	Khabarovsk 2	Russia-3	22	0	2	2	2	....	....	....	....	....
	Amur Khabarovsk	Russia	19	0	2	2	2	....	....	....	....	....
2001-02	H.C. CSKA	Russia-2	49	6	2	8	28	....	....	....	....	....

LEVOKARI, Pauli (leh-voh-KAHR-ee, PAWL-ee) ATL.

Defense. Shoots left. 6'7", 260 lbs. Born, Luvia, Finland, April 7, 1979.
(Atlanta's 10th choice, 257th overall, in 2002 Entry Draft).

			Regular Season					Playoffs				
Season	**Club**	**Lea**	**GP**	**G**	**A**	**TP**	**PIM**	**GP**	**G**	**A**	**TP**	**PIM**
1993-94	Assat Pori-C	Finn-Jr.	32	1	2	3	50	....	....	....	....	....
1994-95	Assat Pori-C	Finn-Jr.	STATISTICS NOT AVAILABLE									
1995-96	Assat Pori-B	Finn-Jr.	28	5	6	11	42	....	....	....	....	....
	Assat Pori Jr.	Finn-Jr.	....	....	....	....	....	10	0	1	1	22
1996-97	Assat Pori Jr.	Finn-Jr.	35	3	8	11	78	4	0	0	0	10
	Assat Pori	Finland	1	0	0	0	0	....	....	....	....	....
1997-98	Assat Pori Jr.	Finn-Jr.	27	0	2	2	79	....	....	....	....	....
	Assat Pori	Finland	16	0	0	0	0	....	....	....	....	....
1998-99	Assat Pori Jr.	Finn-Jr.	19	4	3	7	68	....	....	....	....	....
	Assat Pori	Finland	27	0	2	2	8	....	....	....	....	....
99-2000	Assat Pori	Finland	51	3	2	5	80	....	....	....	....	....
2000-01	Assat Pori	Finland	12	1	0	1	12	....	....	....	....	....
	Jokerit Helsinki	Finland	10	0	0	0	0	....	....	....	....	....
	Kiekko-Vantaa	Finland-2	22	3	6	9	68	....	....	....	....	....
2001-02	HIFK Helsinki	Finland	29	3	5	8	86	....	....	....	....	....

LEWERSTROM, Erik (LEH-vuhr-struhm, AIR-ihk) PHX.

Defense. Shoots left. 6'2", 198 lbs. Born, Grums, Sweden, May 28, 1980.
(Phoenix's 7th choice, 168th overall, in 1999 Entry Draft).

			Regular Season					Playoffs				
Season	**Club**	**Lea**	**GP**	**G**	**A**	**TP**	**PIM**	**GP**	**G**	**A**	**TP**	**PIM**
1996-97	Grums IK	Swede-2	1	0	0	0	0	....	....	....	....	....
1997-98	Grums IK	Swede-2	4	0	0	0	2	....	....	....	....	....
1998-99	Grums IK	Swede-2	29	3	9	12	42	....	....	....	....	....
99-2000	Grums IK	Swede-2	37	5	7	12	88	....	....	....	....	....
2000-01	Farjestad Jr.	Swede-Jr.	1	0	0	0	0	....	....	....	....	....
	Farjestad	Sweden	45	4	0	4	45	14	0	0	0	27
2001-02	Farjestad Jr.	Swede-Jr.	5	2	0	2	6	....	....	....	....	....
	Malmo IF	Sweden	2	0	0	0	2	....	....	....	....	....
	Farjestad	Sweden	39	0	0	0	51	6	0	0	0	0

LEWIS, Carlyle (LOO-ihs, KAHR-lighl)

Right wing. Shoots right. 6'3", 230 lbs. Born, Middleton, N.S., March 1, 1978.

			Regular Season					Playoffs				
Season	**Club**	**Lea**	**GP**	**G**	**A**	**TP**	**PIM**	**GP**	**G**	**A**	**TP**	**PIM**
1994-95	Summerside	MJrHL	20	3	5	8	63	....	....	....	....	....
1995-96	Beauport	QMJHL	30	0	2	2	90	1	0	0	0	0
1996-97	Beauport	QMJHl	66	7	8	15	353	4	1	1	2	7
1997-98	Laval Titan	QMJHL	59	9	13	22	310	16	1	3	4	53
1998-99	Halifax	QMJHL	65	20	27	47	425	5	1	1	2	18
99-2000	Albany	AHL	69	1	2	3	181	4	0	0	0	0
2000-01	Albany	AHL	68	1	7	8	175	....	....	....	....	....
2001-02	Albany	AHL	16	0	0	0	10	....	....	....	....	....
	Columbus	ECHL	51	13	12	25	224	....	....	....	....	....

Signed as a free agent by **New Jersey**, January 26, 1999.

LILES, John-Michael (LIGH-uhls, JAWN-MIGHK-uhl) COL.

Defense. Shoots left. 5'10", 185 lbs. Born, Zionsville, IN, November 25, 1980.
(Colorado's 8th choice, 159th overall, in 2000 Entry Draft).

			Regular Season					Playoffs				
Season	**Club**	**Lea**	**GP**	**G**	**A**	**TP**	**PIM**	**GP**	**G**	**A**	**TP**	**PIM**
1997-98	Team USA	USDP-17	67	6	14	20	44	....	....	....	....	....
1998-99	Team USA	USDP-17	13	2	5	7	6	....	....	....	....	....
	Team USA	USDP-18	46	4	14	18	47	....	....	....	....	....
99-2000	Michigan State	CCHA	40	8	20	28	26	....	....	....	....	....
2000-01	Michigan State	CCHA	42	7	18	25	28	....	....	....	....	....
2001-02	Michigan State	CCHA	41	13	22	35	18	....	....	....	....	....

CCHA Second All-Star Team (2001) • NCAA West Second All-American Team (2002) • CCHA First All-Star Team (2002)

LINDSTROM, Andreas (LIHND-struhm, an-DRAY-uhs) BOS.

Right wing. Shoots left. 6'5", 210 lbs. Born, Lulea, Sweden, September 1, 1982.
(Boston's 12th choice, 279th overall, in 2000 Entry Draft).

			Regular Season					Playoffs				
Season	**Club**	**Lea**	**GP**	**G**	**A**	**TP**	**PIM**	**GP**	**G**	**A**	**TP**	**PIM**
99-2000	Lulea HF Jr.	Swede-Jr.	9	2	2	4	14	....	....	....	....	....
2000-01	Lulea HF Jr.	Swede-Jr.	21	8	6	14	18	....	....	....	....	....
	Lulea HF	Sweden	3	0	0	0	0	8	1	0	1	0
2001-02	Lulea HF Jr.	Swede-Jr.	17	5	4	9	40	2	1	0	1	6
	Lulea HF	Sweden	....	....	....	....	....	1	0	0	0	4

LINDSTROM, Joakim (LIHND-struhm, YOH-ah-kihm) CBJ

Left wing. Shoots left. 6', 187 lbs. Born, Skelleftea, Sweden, December 5, 1983.
(Columbus' 2nd choice, 41st overall, in 2002 Entry Draft).

			Regular Season					Playoffs				
Season	**Club**	**Lea**	**GP**	**G**	**A**	**TP**	**PIM**	**GP**	**G**	**A**	**TP**	**PIM**
99-2000	MoDo-18	Swede-Jr.	17	6	*14	20	32	....	....	....	....	....
	MoDo Jr.	Swede-Jr.	10	4	4	8	2	....	....	....	....	....
2000-01	MoDo Jr.	Swede-Jr.	12	7	14	21	46	....	....	....	....	....
	MoDo	Sweden	10	2	3	5	2	....	....	....	....	....
2001-02	MoDo Jr.	Swede-Jr.	10	9	6	15	67	....	....	....	....	....
	IF Troja-Ljungby	Swede-2	3	0	0	0	12	....	....	....	....	....
	MoDo	Sweden	42	4	3	7	20	14	3	5	8	8

LINDSTROM, Sanny (LIHND-struhm, SAN-nee) COL.

Defense. Shoots left. 6'2", 205 lbs. Born, Stockholm, Sweden, December 24, 1979.
(Colorado's 4th choice, 112th overall, in 1999 Entry Draft).

			Regular Season					Playoffs				
Season	**Club**	**Lea**	**GP**	**G**	**A**	**TP**	**PIM**	**GP**	**G**	**A**	**TP**	**PIM**
1997-98	Huddinge IK	Swede-2	32	6	6	12	46	....	....	....	....	....
1998-99	Huddinge IK	Swede-2	37	4	4	8	65	....	....	....	....	....
99-2000	Hershey Bears	AHL	42	1	2	3	57	....	....	....	....	....
	Baton Rouge	ECHL	11	1	2	3	16	....	....	....	....	....
2000-01	Hershey Bears	AHL	24	0	0	0	61	....	....	....	....	....
	Quad City	UHL	5	1	1	2	10	....	....	....	....	....
2001-02	Quad City	UHL	38	4	23	27	71	12	0	3	3	20
	Hershey Bears	AHL	2	0	0	0	0	....	....	....	....	....

• Missed majority of 2000-01 season recovering from knee injury originally suffered in practice, March 5, 2000.

LINHART, Tomas (LIHN-hart, TAW-mash) MTL.

Defense. Shoots left. 6'2", 209 lbs. Born, Pardubice, Czech., February 16, 1984.
(Montreal's 2nd choice, 45th overall, in 2002 Entry Draft).

			Regular Season					Playoffs				
Season	**Club**	**Lea**	**GP**	**G**	**A**	**TP**	**PIM**	**GP**	**G**	**A**	**TP**	**PIM**
99-2000	Pardubice-18	Czech-Jr.	45	2	8	10	83	....	....	....	....	....
2000-01	Pardubice-18	Czech-Jr.	23	5	7	12	82	....	....	....	....	....
	Pardubice Jr.	Czech-Jr.	29	1	6	7	12	4	0	0	0	0
2001-02	Pardubice Jr.	Czech-Jr.	38	2	4	6	28	....	....	....	....	....
	Sumperk	Czech-2	1	0	0	0	2	....	....	....	....	....

LITVINENKO, Alexei (liht-vihn-EHN-koh, al-EHX-ay) PHX.

Defense. Shoots left. 6'4", 180 lbs. Born, Ust-Kamenogorsk, USSR, March 7, 1980.
(Phoenix's 9th choice, 262nd overall, in 1999 Entry Draft).

			Regular Season					Playoffs				
Season	**Club**	**Lea**	**GP**	**G**	**A**	**TP**	**PIM**	**GP**	**G**	**A**	**TP**	**PIM**
1997-98	Ust-Kamenog. 2	Russia-3	12	0	0	0	8	....	....	....	....	....
	Ust-Kamenogorsk	Russia-2	2	0	0	0	0	....	....	....	....	....
1998-99	Ust-Kamenog. 2	Russia-4	31	3	4	7	52	....	....	....	....	....
	Ust-Kamenog. 2	Russia-4	16	0	4	4	14	....	....	....	....	....
99-2000	Dynamo Moscow	Russia	7	0	0	0	4	....	....	....	....	....
2000-01	Dynamo Moscow	Russia	6	0	0	0	0	....	....	....	....	....
	Yekaterinburg	Russia	26	0	0	0	42	....	....	....	....	....
2001-02	Magnitogorsk	Russia	20	0	3	3	29	9	1	1	2	20

LIUBIMOV, Alexander (loo-BEE-mahf, al-ehx-AN-duhr) EDM.

Defense. Shoots left. 6'3", 196 lbs. Born, Ust-Kamenogorsk, USSR, February 15, 1980.
(Edmonton's 3rd choice, 83rd overall, in 2000 Entry Draft).

			Regular Season					Playoffs				
Season	**Club**	**Lea**	**GP**	**G**	**A**	**TP**	**PIM**	**GP**	**G**	**A**	**TP**	**PIM**
1995-96	Ust-Kamenog. Jr.	CIS-Jr.	30	9	5	14	20	....	....	....	....	....
	Kamenogorsk 2	CIS-2	2	0	0	0	0	....	....	....	....	....
1996-97	Togliatti Jr.	Russia-Jr.	STATISTICS NOT AVAILABLE									
	Lada Togliatti 2	Russia-3	STATISTICS NOT AVAILABLE									
	Lada Togliatti	Russia	2	0	0	0	0	....	....	....	....	....
1997-98	Lada Togliatti 2	Russia-3	38	7	6	13	63	....	....	....	....	....
1998-99	Lada Togliatti 2	Russia-4	44	5	4	9	36	....	....	....	....	....
99-2000	CSK VVS Samara	Russia	17	0	1	1	14	....	....	....	....	....
	Lada Togliatti	Russia	8	0	0	0	6	7	0	1	1	2
2000-01	Lada Togliatti	Russia	23	1	2	3	8	....	....	....	....	....
2001-02	Odessa	CHL	56	4	20	24	112	5	0	1	1	0

LOBB, Aaron (LAWB, AIR-ruhn) T.B.

Right wing. Shoots right. 6'4", 195 lbs. Born, Brucefield, Ont., June 10, 1983.
(Tampa Bay's 5th choice, 123rd overall, in 2001 Entry Draft).

			Regular Season					Playoffs				
Season	**Club**	**Lea**	**GP**	**G**	**A**	**TP**	**PIM**	**GP**	**G**	**A**	**TP**	**PIM**
1998-99	Strathroy	OJHL-B	39	6	14	20	30	....	....	....	....	....
99-2000	London Knights	OHL	58	2	9	11	23	....	....	....	....	....
2000-01	London Knights	OHL	67	23	25	48	93	5	0	2	2	12
2001-02	London Knights	OHL	10	1	2	3	14	....	....	....	....	....
	Guelph Storm	OHL	7	1	6	7	4	....	....	....	....	....

Traded to **Guelph** (OHL) by **London** (OHL) for Charlie Stephens, Guelph's 2nd round choice (Alan Nemeth) in 2002 OHL Priority Draft and Guelph's 1st round choice (Alexander Skorohod) in 2002 CHL Import Draft, October 19, 2001. • Missed majority of 2001-02 season recovering from knee injury suffered in game vs. London (OHL), November 4, 2001.

LOMBARDI, Matthew (lawm-BAHR-dee, MA-thew) CGY.

Center. Shoots left. 5'11", 191 lbs. Born, Montreal, Que., March 18, 1982.
(Calgary's 3rd choice, 90th overall, in 2002 Entry Draft).

			Regular Season					Playoffs				
Season	**Club**	**Lea**	**GP**	**G**	**A**	**TP**	**PIM**	**GP**	**G**	**A**	**TP**	**PIM**
1997-98	Gatineau	QAAA	42	10	13	23	....	13	4	7	11	....
1998-99	Victoriaville	QMJHL	47	6	10	16	8	5	0	0	0	0
99-2000	Victoriaville	QMJHL	65	18	26	44	28	6	0	0	0	6
2000-01	Victoriaville	QMJHL	72	28	39	67	66	13	12	6	18	10
2001-02	Victoriaville	QMJHL	66	57	73	130	70	22	*17	18	35	18

• Re-entered NHL Entry Draft. Originally Edmonton's 7th choice, 215th overall, in 2000 Entry Draft.

Memorial Cup All-Star Team (2002)

LOVDAHL, Anders (LUHV-duhl, AN-duhrs) COL.

Center. Shoots left. 6'4", 190 lbs. Born, Borlange, Sweden, February 4, 1981.
(Colorado's 8th choice, 158th overall, in 1999 Entry Draft).

			Regular Season					Playoffs				
Season	**Club**	**Lea**	**GP**	**G**	**A**	**TP**	**PIM**	**GP**	**G**	**A**	**TP**	**PIM**
1997-98	HV 71 Jr.	Swede-Jr.	26	5	5	10	18	....	....	....	....	....
1998-99	HV 71 Jr.	Swede-Jr.	DID NOT PLAY – INJURED									
99-2000	Calgary Hitmen	WHL	36	5	9	14	28	....	....	....	....	....
	Moose Jaw	WHL	31	3	0	3	8	4	0	0	0	2
2000-01	Tranas AIF	Swede-2	16	3	4	7	4	....	....	....	....	....
2001-02	Tranas IK	Swede-2	30	2	4	6	12	8	2	2	4	2

Traded to **Moose Jaw** (WHL) by **Calgary** (WHL) with Sean Connors for Radislav Stana and Cory Hintz , January 10, 2000.

LOVEN, Fredrik (LUH-vehn, FREHD-rihk) PHX.

Center. Shoots left. 6'2", 183 lbs. Born, Stockholm, Sweden, March 14, 1977.
(Winnipeg's 10th choice, 189th overall, in 1995 Entry Draft).

			Regular Season					Playoffs				
Season	**Club**	**Lea**	**GP**	**G**	**A**	**TP**	**PIM**	**GP**	**G**	**A**	**TP**	**PIM**
1994-95	Djurgarden Jr.	Swede-Jr.	29	6	10	16	14	....	....	....	....	....
1995-96	Djurgarden	Sweden	4	0	0	0	0	4	0	0	0	0
1996-97	Djurgarden Jr.	Swede-Jr.	4	2	2	4	8	....	....	....	....	....
	Arlanda Mastra	Swede-2	5	0	3	3	4	....	....	....	....	....
	Djurgarden	Sweden	7	0	0	0	0	....	....	....	....	....
1997-98	Bjorkloven	Swede-2	32	5	6	11	8	14	0	2	2	10
1998-99	Hammarby	Swede-2	37	8	17	25	47	5	2	0	2	4
99-2000	Hammarby	Swede-2	45	2	8	10	34	2	0	0	0	0
2000-01	Hammarby	Swede-2	39	3	4	7	12	5	0	3	3	4
2001-02	Tingsryds AIF	Swede-2	46	7	12	19	34	2	0	0	0	0

Signed as a free agent by **Tingsryds AIF** (Sweden-2), August 8, 2001.

LOYA, Cliff (LOI-uh, KLIHF) CHI.

Defense. Shoots left. 6'2", 200 lbs. Born, Pittsburgh, PA, May 8, 1981.
(Chicago's 10th choice, 207th overall, in 2000 Entry Draft).

			Regular Season					Playoffs				
Season	Club	Lea	GP	G	A	TP	PIM	GP	G	A	TP	PIM
1998-99	Shat.-St. Mary's	Hi-School	52	7	29	36	34	....	....	....	....	....
99-2000	U. of Maine	H-East	31	0	5	5	22	....	....	....	....	....
2000-01	U. of Maine	H-East	36	1	2	3	28	....	....	....	....	....
2001-02	U. of Maine	H-East	43	0	5	5	44	....	....	....	....	....

LUCHINKIN, Sergei (loo-CHIHN-kihn, SAIR-gay) CBJ

Left wing. Shoots left. 5'11", 172 lbs. Born, Dmitrov, USSR, October 16, 1976.
(Dallas' 9th choice, 202nd overall, in 1995 Entry Draft).

			Regular Season					Playoffs				
Season	Club	Lea	GP	G	A	TP	PIM	GP	G	A	TP	PIM
1994-95	Dynamo Moscow	CIS	6	1	0	1	4	....	....	....	....	....
1995-96	Dynamo Moscow	CIS	21	6	2	8	14	10	0	1	1	6
1996-97	Dynamo Moscow	Russia	18	1	5	6	4	....	....	....	....	....
1997-98	Dynamo Moscow	EuroHL	1	0	0	0	0	....	....	....	....	....
	Dynamo Moscow	Russia	6	0	1	1	0	....	....	....	....	....
	Spartak Moscow	Russia	10	0	1	1	4	....	....	....	....	....
1998-99	Spartak Moscow	Russia	33	4	5	9	18	....	....	....	....	....
99-2000	Spartak Moscow 2	Russia-3	1	2	0	2	2	....	....	....	....	....
	Spartak Moscow	Russia-2	58	19	17	36	54	....	....	....	....	....
2000-01	Spartak Moscow	Russia-2	57	23	18	41	46	....	....	....	....	....
2001-02	H.C. CSKA	Russia-2	60	14	20	34	32	....	....	....	....	....

Selected by **Columbus** from **Dallas** in Expansion Draft, June 23, 2000.

LUCHKIN, Vladislav (LOOCH-kihn, VLA-dihs-lav) CHI.

Center. Shoots left. 6'1", 185 lbs. Born, Cherepovets, USSR, February 3, 1982.
(Chicago's 11th choice, 225th overall, in 2000 Entry Draft).

			Regular Season					Playoffs				
Season	Club	Lea	GP	G	A	TP	PIM	GP	G	A	TP	PIM
1997-98	Cherepovets 2	Russia-3	23	2	5	7	12	....	....	....	....	....
1998-99	Cherepovets 2	Russia-3	25	6	4	10	8	....	....	....	....	....
	Cherepovets 3	Russia-4	8	1	3	4	10	....	....	....	....	....
99-2000	Cherepovets 2	Russia-3	30	23	10	33	36	....	....	....	....	....
2000-01	Cherepovets	Russia	28	2	3	5	10	6	2	0	2	4
2001-02	Cherepovets 2	Russia-3	6	5	6	11	0	....	....	....	....	....
	Cherepovets	Russia	27	4	6	10	12	....	....	....	....	....

LUCKY, Jeff (LUH-kee, JEHF) WSH.

Right wing. Shoots right. 6'1", 193 lbs. Born, Regina, Sask., March 17, 1983.
(Washington's 3rd choice, 125th overall, in 2001 Entry Draft).

			Regular Season					Playoffs				
Season	Club	Lea	GP	G	A	TP	PIM	GP	G	A	TP	PIM
1998-99	Yorkton Mallers	SMHL	42	28	40	68	4	8	6	7	13	0
	Spokane Chiefs	WHL	1	0	0	0	0	....	....	....	....	....
99-2000	Spokane Chiefs	WHL	57	8	10	18	12	13	1	0	1	2
2000-01	Spokane Chiefs	WHL	53	20	21	41	26	....	....	....	....	....
2001-02	Spokane Chiefs	WHL	32	9	11	20	10	6	1	2	3	4

LUKES, Frantisek (LOO-kehsh, FRAHN-tih-sehk) PHX.

Left wing. Shoots right. 5'9", 162 lbs. Born, Kadan, Czech., September 25, 1982.
(Phoenix's 8th choice, 243rd overall, in 2001 Entry Draft).

			Regular Season					Playoffs				
Season	Club	Lea	GP	G	A	TP	PIM	GP	G	A	TP	PIM
99-2000	Litvinov Jr.	Czech-Jr.	36	15	13	28	....	....	....	....	....	....
2000-01	St. Michael's	OHL	61	23	33	56	37	18	4	9	13	12
2001-02	St. Michael's	OHL	63	27	37	64	50	15	7	11	18	16

LUNDBERG, Eric (LUHND-buhrg, AIR-ihk) COL.

Defense. Shoots right. 6'3", 200 lbs. Born, Vernon, CT, April 13, 1983.
(Colorado's 3rd choice, 94th overall, in 2002 Entry Draft).

			Regular Season					Playoffs				
Season	Club	Lea	GP	G	A	TP	PIM	GP	G	A	TP	PIM
99-2000	New England	EJHL	36	8	36	42	106	....	....	....	....	....
2000-01	New England	EJHL	36	6	28	34	92	10	2	10	12	14
2001-02	Providence	H-East	36	0	9	9	28	....	....	....	....	....

LUNDBOHM, Andy (LUHND-bawm, AN-dee) FLA.

Center. Shoots left. 6'3", 225 lbs. Born, Roseau, MN, March 24, 1977.

			Regular Season					Playoffs				
Season	Club	Lea	GP	G	A	TP	PIM	GP	G	A	TP	PIM
1995-96	Army	NCAA	37	21	25	46	42	....	....	....	....	....
1996-97	Army	NCAA	29	19	27	46	16	....	....	....	....	....
1997-98	Army	NCAA	31	19	25	44	42	....	....	....	....	....
1998-99	Army	NCAA	26	17	15	32	30	....	....	....	....	....
99-2000	New Orleans	ECHL	12	2	2	4	4	....	....	....	....	....
	Kentucky	AHL	22	2	1	3	8	2	0	0	0	5
2000-01	Kentucky	AHL	63	7	15	22	41	....	....	....	....	....
2001-02	Cleveland Barons	AHL	51	4	4	8	38	....	....	....	....	....

Signed as a free agent by **San Jose**, June 11, 1999. Signed as a free agent by **Florida**, July 16, 2002.

LUNDBOHM, Bryan (LUHND-bawm, BRIGH-uhn) NSH.

Center. Shoots left. 5'10", 184 lbs. Born, Roseau, MN, August 24, 1977.

			Regular Season					Playoffs				
Season	Club	Lea	GP	G	A	TP	PIM	GP	G	A	TP	PIM
1996-97	Lincoln Stars	USHL	52	13	33	45	33	14	8	4	12	33
1997-98	Lincoln Stars	USHL	55	26	38	64	10	9	2	7	9	0
1998-99	North Dakota	WCHA	32	2	9	11	4	....	....	....	....	....
99-2000	North Dakota	WCHA	44	22	22	44	14	....	....	....	....	....
2000-01	North Dakota	WCHA	46	*32	27	69	38	....	....	....	....	....
2001-02	Milwaukee	AHL	79	11	23	34	63	....	....	....	....	....

USHL First All-Star Team (1998) • WCHA First All-Star Team (2001) • NCAA West Second All-American Team (2001) • NCAA Championship All-Tournament Team (2001)

Signed as a free agent by **Nashville**, May 1, 2001.

LUNDMARK, Jamie (LUHND-mahrk, JAY-mee) NYR

Center. Shoots right. 6', 174 lbs. Born, Edmonton, Alta., January 16, 1981.
(NY Rangers' 2nd choice, 9th overall, in 1999 Entry Draft).

			Regular Season					Playoffs				
Season	Club	Lea	GP	G	A	TP	PIM	GP	G	A	TP	PIM
1996-97	St. Albert	AJHL	35	10	9	19	8	....	....	....	....	....
1997-98	St. Albert	AJHL	57	33	58	91	171	19	13	18	31	5
1998-99	Moose Jaw	WHL	70	40	51	91	121	11	5	4	9	24
99-2000	Moose Jaw	WHL	37	21	27	48	33	....	....	....	....	....
2000-01	Seattle	WHL	52	35	42	77	49	9	4	4	8	16
2001-02	Hartford	AHL	79	27	32	59	56	10	3	4	7	16

WHL All-Rookie Team (1999) • WHL East Second All-Star Team (1999) • WHL West First All-Star Team (2001)

Traded to **Seattle** (WHL) by **Moose Jaw** (WHL) for Scott Kelman, October 27, 2000.

LUNDQVIST, Joel (LOOND-kvihst, JOHL) DAL.

Center. Shoots left. 6', 185 lbs. Born, Are, Sweden, March 2, 1982.
(Dallas' 3rd choice, 68th overall, in 2000 Entry Draft).

			Regular Season					Playoffs				
Season	Club	Lea	GP	G	A	TP	PIM	GP	G	A	TP	PIM
1997-98	Rogle Jr.	Swede-Jr.	59	36	40	76	....	....	....	....	....	....
1998-99	V. Frolunda-18	Swede-Jr.	32	26	38	64	37	4	3	1	4	2
99-2000	V. Frolunda-18	Swede-Jr.	4	2	4	6	4	....	....	....	....	....
	V. Frolunda Jr.	Swede-Jr.	25	7	12	19	2	6	2	3	5	2
2000-01	V. Frolunda Jr.	Swede-Jr.	18	14	27	41	12	....	....	....	....	....
	Molndals HS	Swede-2	26	18	13	31	22	....	....	....	....	....
	Vastra Frolunda	Sweden	9	0	0	0	0	....	....	....	....	....
2001-02	Vastra Frolunda	Sweden	46	12	14	26	28	10	1	3	4	8
	V. Frolunda Jr.	Swede-Jr.	....	....	....	....	....	1	0	0	0	0

LUNDQVIST, Stefan (LUHND-kvihst, STEH-fan) NYR

Right wing. Shoots left. 6'3", 209 lbs. Born, Gavle, Sweden, February 18, 1978.
(NY Rangers' 7th choice, 180th overall, in 1998 Entry Draft).

			Regular Season					Playoffs				
Season	Club	Lea	GP	G	A	TP	PIM	GP	G	A	TP	PIM
1994-95	Avesta BK	Swede-2	3	0	1	1	0	....	....	....	....	....
1995-96	Avesta BK	Swede-3	27	24	13	37	10	....	....	....	....	....
1996-97	Avesta BK	Swede-3	31	37	29	66	....	....	....	....	....	....
1997-98	Brynas Jr.	Swede-Jr.	21	23	15	38	2	....	....	....	....	....
	Brynas IF Gavle	Sweden	27	2	2	4	0	1	0	0	0	0
1998-99	Brynas IF Gavle	Sweden	13	0	0	0	0	....	....	....	....	....
	Uppsala	Swede-2	15	7	8	15	0	....	....	....	....	....
	Mora IK	Swede-2	23	10	7	17	22	4	2	2	4	2
99-2000	Brynas IF Gavle	Sweden	48	6	4	10	12	11	1	0	1	0
	Brynas IF Gavle	EuroHL	6	1	1	2	2	....	....	....	....	....
2000-01	Skelleftea AIK	Swede-2	35	21	11	32	14	1	0	0	0	0
2001-02	Skelleftea AIK	Swede-2	40	28	14	42	20	5	1	0	1	4

LUNDSTROM, Per-Anton (LUHND-struhm, PAIR-AN-tawn) PHX.

Defense. Shoots left. 6'2", 185 lbs. Born, Umea, Sweden, September 29, 1977.
(Phoenix's 3rd choice, 62nd overall, in 1996 Entry Draft).

			Regular Season					Playoffs				
Season	Club	Lea	GP	G	A	TP	PIM	GP	G	A	TP	PIM
1993-94	AIK Tegs	Swede-3	13	4	4	8	8	....	....	....	....	....
1994-95	MoDo Jr.	Swede-Jr.	20	3	4	7	18	....	....	....	....	....
1995-96	MoDo Jr.	Swede-Jr.	25	3	3	6	28	2	1	0	1	4
	MoDo	Sweden	19	1	1	2	29	4	0	0	0	2
1996-97	MoDo	Sweden	35	0	0	0	42	....	....	....	....	....
1997-98	Bjorkloven	Swede-2	31	6	13	19	71	....	....	....	....	....
1998-99	Bjorkloven	Sweden	39	1	2	3	36	....	....	....	....	....
99-2000	AIK Solna	Sweden	50	4	5	9	48	....	....	....	....	....
2000-01	AIK Solna	Sweden	50	2	4	6	63	3	1	0	1	0
2001-02	AIK Solna	Sweden	49	6	2	8	104	....	....	....	....	....

LUPASCHUK, Ross (LOO-puhs-chuhk, RAWS) PIT.

Defense. Shoots right. 6'1", 217 lbs. Born, Edmonton, Alta., January 19, 1981.
(Washington's 4th choice, 34th overall, in 1999 Entry Draft).

			Regular Season					Playoffs				
Season	Club	Lea	GP	G	A	TP	PIM	GP	G	A	TP	PIM
1996-97	Edmonton Mets	AJHL	65	5	22	27	87	....	....	....	....	....
1997-98	Prince Albert	WHL	67	6	12	18	170	....	....	....	....	....
1998-99	Prince Albert	WHL	67	8	20	28	127	14	4	9	13	16
99-2000	Prince Albert	WHL	22	8	8	16	42	....	....	....	....	....
	Red Deer Rebels	WHL	46	13	27	40	116	4	0	1	1	10
2000-01	Red Deer Rebels	WHL	65	28	37	65	135	22	5	10	15	54
2001-02	Wilkes-Barre	AHL	72	9	20	29	91	....	....	....	....	....

WHL East Second All-Star Team (2001) • Memorial Cup All-Star Team (2001)

Traded to **Red Deer** (WHL) by **Prince Albert** (WHL) with Craig Brunel for Brent Hobday, Steven MacIntyre, Regan Darby and Scott McQueen, November 19, 1999. Traded to **Pittsburgh** by **Washington** with Kris Beech, Michal Sivek and future considerations for Jaromir Jagr and Frantisek Kucera, July 11, 2001.

LUPUL, Joffrey (LOO-puhl, JAWF-ree) ANA.

Center. Shoots right. 6'1", 194 lbs. Born, Edmonton, Alta., September 23, 1983.
(Anaheim's 1st choice, 7th overall, in 2002 Entry Draft).

			Regular Season					Playoffs				
Season	Club	Lea	GP	G	A	TP	PIM	GP	G	A	TP	PIM
1998-99	Ft. Saskatchewan	ABHL	36	40	50	90	40	....	....	....	....	....
99-2000	Ft. Saskatchewan	AMHL	34	43	30	*73	47	4	0	1	1	2
2000-01	Medicine Hat	WHL	69	30	26	56	39	22	3	6	9	2
2001-02	Medicine Hat	WHL	72	*56	50	106	95	....	....	....	....	....

WHL East First All-Star Team (2002) • Canadian Major Junior First All-Star Team (2002)

LUTES, Brett (LOOTZ, BREHT)

Left wing. Shoots left. 6', 182 lbs. Born, Moncton, N.B., February 2, 1982.
(St. Louis' 7th choice, 229th overall, in 2000 Entry Draft).

			Regular Season					Playoffs				
Season	Club	Lea	GP	G	A	TP	PIM	GP	G	A	TP	PIM
99-2000	Montreal Rocket	QMJHL	4	2	1	3	0	....	....	....	....	....
2000-01	Montreal Rocket	QMJHL	72	29	38	67	34	....	....	....	....	....
2001-02	Montreal Rocket	QMJHL	63	35	44	79	107	7	4	3	7	33

LUTTINEN, Arttu (LOO-tuh-nehn, AHR-too) **OTT.**

Center. Shoots left. 5'10", 205 lbs. Born, Helsinki, Finland, September 3, 1983.
(Ottawa's 3rd choice, 75th overall, in 2002 Entry Draft).

			Regular Season					Playoffs				
Season	Club	Lea	GP	G	A	TP	PIM	GP	G	A	TP	PIM
99-2000	HIFK Jr.	Finn-Jr.	17	5	9	14	10	2	0	0	0	2
2000-01	HIFK Helsinki-B	Finn-Jr.	20	14	20	34	141	....	....	....	....	....
	HIFK Jr.	Finn-Jr.	8	4	2	6	4	8	0	1	1	2
2001-02	HIFK Jr.	Finn-Jr.	24	16	17	23	60	1	0	0	0	0

LYNCH, Doug (LIHNCH, DUHG) **EDM.**

Defense. Shoots left. 6'3", 205 lbs. Born, North Vancouver, B.C., April 4, 1983.
(Edmonton's 2nd choice, 43rd overall, in 2001 Entry Draft).

			Regular Season					Playoffs				
Season	Club	Lea	GP	G	A	TP	PIM	GP	G	A	TP	PIM
1998-99	Port Coquitlam	BCAHA	45	47	48	95	120	....	....	....	....	....
	Red Deer Rebels	WHL	2	0	1	1	2	....	....	....	....	....
99-2000	Red Deer Rebels	WHL	65	9	5	14	57	4	0	0	0	5
2000-01	Red Deer Rebels	WHL	72	12	37	49	181	21	1	9	10	30
2001-02	Red Deer Rebels	WHL	71	21	27	48	202	22	5	4	9	12

LYNCH, Paul (LIHNCH, PAWL) **T.B.**

Defense. Shoots left. 6'4", 195 lbs. Born, Salem, MA, April 23, 1982.
(Tampa Bay's 6th choice, 138th overall, in 2001 Entry Draft).

			Regular Season					Playoffs				
Season	Club	Lea	GP	G	A	TP	PIM	GP	G	A	TP	PIM
99-2000	Brooks High	Hi-School	23	21	25	46	34	....	....	....	....	....
2000-01	Valley Juniors	EJHL	21	3	4	7	143	....	....	....	....	....
2001-02	U. of Maine	H-East	22	2	4	6	24	....	....	....	....	....

LYSAK, Brett (LIGH-sak, BREHT) **CAR.**

Center. Shoots left. 6', 190 lbs. Born, Edmonton, Alta., December 30, 1980.
(Carolina's 2nd choice, 49th overall, in 1999 Entry Draft).

			Regular Season					Playoffs				
Season	Club	Lea	GP	G	A	TP	PIM	GP	G	A	TP	PIM
1995-96	St. Albert	AJHL	35	20	23	43	68	....	....	....	....	....
1996-97	Regina Pats	WHL	66	11	14	25	41	5	0	1	1	5
1997-98	Regina Pats	WHL	70	22	38	60	82	9	6	2	8	8
1998-99	Regina Pats	WHL	61	39	49	88	84	....	....	....	....	....
99-2000	Regina Pats	WHL	70	38	40	78	24	7	5	4	9	2
2000-01	Regina Pats	WHL	64	35	48	83	44	6	5	1	6	4
2001-02	Lowell	AHL	53	6	8	14	26	3	0	0	0	0
	Florida	ECHL	16	2	7	9	14	6	3	1	4	6

WHL East Second All-Star Team (1999) • Memorial Cup All-Star Team (2001)

MAATTA, Tero (MAH-tuh, TEH-roh) **S.J.**

Defense. Shoots left. 6'1", 205 lbs. Born, Vantaa, Finland, January 2, 1982.
(San Jose's 1st choice, 41st overall, in 2000 Entry Draft).

			Regular Season					Playoffs				
Season	Club	Lea	GP	G	A	TP	PIM	GP	G	A	TP	PIM
1996-97	Kiekko Vantaa-C	Finn-Jr.	20	2	6	8	6	....	....	....	....	....
	Haukat-C	Finn-Jr.	8	3	1	4	10	....	....	....	....	....
1997-98	Jokerit-B	Finn-Jr.	28	4	7	11	10	2	0	0	0	2
1998-99	Jokerit Jr.	Finn-Jr.	38	4	8	12	75	8	1	3	4	6
99-2000	Jokerit-B	Finn-Jr.	13	4	10	14	24	1	0	0	0	25
	Jokerit Jr.	Finn-Jr.	31	4	4	8	53	....	....	....	....	....
2000-01	Blues Espoo Jr.	Finn-Jr.	6	0	1	1	6	....	....	....	....	....
	KJT Jarvenpaa	Finland-2	6	0	3	3	31	....	....	....	....	....
	Blues Espoo	Finland	44	4	4	8	24	....	....	....	....	....
2001-02	Blues Espoo Jr.	Finn-Jr.	8	2	6	8	2	....	....	....	....	....
	Blues Espoo	Finland	51	4	6	10	65	3	0	0	0	0

MacDONALD, Jason (MAK-DAWN-uhld, JAY-suhn) **PIT.**

Right wing. Shoots right. 5'11", 205 lbs. Born, Charlottetown, P.E.I., April 1, 1974.
(Detroit's 5th choice, 142nd overall, in 1992 Entry Draft).

			Regular Season					Playoffs				
Season	Club	Lea	GP	G	A	TP	PIM	GP	G	A	TP	PIM
1989-90	Charlottetown	MJrHL	29	11	29	40	206	....	....	....	....	....
1990-91	North Bay	OHL	57	12	15	27	126	10	3	3	6	15
1991-92	North Bay	OHL	17	5	8	13	50	....	....	....	....	....
	Owen Sound	OHL	42	17	19	36	129	5	0	3	3	16
1992-93	Owen Sound	OHL	56	46	43	89	197	8	6	5	11	28
1993-94	Owen Sound	OHL	66	55	61	116	177	9	7	11	18	36
	Adirondack	AHL	....	....	....	....	....	1	0	0	0	0
1994-95	Adirondack	AHL	68	14	21	35	238	4	0	0	0	2
1995-96	Adirondack	AHL	43	9	13	22	99	....	....	....	....	....
	Toledo Storm	ECHL	9	5	5	10	26	9	3	1	4	39
1996-97	Adirondack	AHL	1	0	0	0	2	....	....	....	....	....
	Fredericton	AHL	63	22	25	47	189	....	....	....	....	....
1997-98	Team Canada	Nat-Tm	51	15	20	35	133	....	....	....	....	....
	Saint John	AHL	6	2	0	2	27	11	1	3	4	17
1998-99	Manitoba Moose	IHL	82	25	27	52	283	5	2	2	4	13
99-2000	Manitoba Moose	IHL	30	5	10	15	77	....	....	....	....	....
	Orlando	IHL	29	7	7	14	113	4	0	0	0	19
2000-01	Wilkes-Barre	AHL	74	17	16	33	290	17	1	3	4	*66
2001-02	Wilkes-Barre	AHL	57	8	13	21	330	....	....	....	....	....

OHL Second All-Star Team (1994)

Traded to **Montreal** by **Detroit** for cash, November 8, 1996. Signed as a free agent by **Pittsburgh**, July 18, 2001.

MACHO, Michal (MA-khoh, MEE-khuhl) **S.J.**

Center. Shoots right. 6'1", 169 lbs. Born, Martin, Czech., January 17, 1982.
(San Jose's 5th choice, 183rd overall, in 2000 Entry Draft).

			Regular Season					Playoffs				
Season	Club	Lea	GP	G	A	TP	PIM	GP	G	A	TP	PIM
1997-98	MHC Martin Jr.	Slovak-Jr.	55	44	58	102	....	....	....	....	....	....
1998-99	King's Edge Hill	Hi-School	50	45	55	100	....	....	....	....	....	....
99-2000	MHC Martin Jr.	Slovak-Jr.	30	38	44	82	....	....	....	....	....	....
	MHC Martin	Slovak-2	8	1	5	6	4	....	....	....	....	....
2000-01	MHC Martin	Slovakia	37	5	10	15	12	3	1	1	2	2
2001-02	MHC Martin	Slovakia	40	12	8	20	20	....	....	....	....	....

MacISAAC, Dave (muh-KIGH-zuhk, DAYV)

Defense. Shoots left. 6'2", 225 lbs. Born, Cambridge, MA, April 23, 1972.

			Regular Season					Playoffs				
Season	Club	Lea	GP	G	A	TP	PIM	GP	G	A	TP	PIM
1992-93	U. of Maine	H-East	35	5	32	37	14	....	....	....	....	....
1993-94	U. of Maine	H-East	31	4	20	24	22	....	....	....	....	....
1994-95	U. of Maine	H-East	44	5	13	18	44	....	....	....	....	....
	Milwaukee	IHL	2	0	0	0	5	9	0	2	2	2
1995-96	Milwaukee	IHL	71	7	16	23	165	....	....	....	....	....
1996-97	Philadelphia	AHL	61	3	15	18	187	10	0	1	1	31
1997-98	Philadelphia	AHL	80	7	21	28	241	18	5	13	18	20
1998-99	Philadelphia	AHL	47	6	15	21	98	16	2	5	7	50
99-2000	Lowell	AHL	77	1	25	26	179	7	0	1	1	4
2000-01	Kentucky	AHL	73	9	24	33	178	3	0	0	0	6
2001-02	Hershey Bears	AHL	32	1	5	6	42	....	....	....	....	....
	Hartford	AHL	24	1	4	5	21	10	1	1	2	*56

Signed as a free agent by **Philadelphia**, July 30, 1996. Signed as a free agent by **LA Kings**, August 25, 1999. Signed as a free agent by **San Jose**, August 10, 2000.

MacLELLAN, Brent (mak-LEHL-uhn, BREHNT) **CHI.**

Defense. Shoots right. 6'4", 218 lbs. Born, Halifax, N.S., March 23, 1983.
(Chicago's 5th choice, 104th overall, in 2001 Entry Draft).

			Regular Season					Playoffs				
Season	Club	Lea	GP	G	A	TP	PIM	GP	G	A	TP	PIM
99-2000	Rimouski Oceanic	QMJHL	69	2	13	15	98	14	2	1	3	4
2000-01	Rimouski Oceanic	QMJHL	62	7	19	26	155	11	0	0	0	41
2001-02	Rimouski Oceanic	QMJHL	67	7	23	30	165	7	0	0	0	4

MacMILLAN, Jeff (muhk-MIHL-uhn, JEHF) **DAL.**

Defense. Shoots left. 6'3", 206 lbs. Born, Durham, Ont., March 30, 1979.
(Dallas' 8th choice, 215th overall, in 1999 Entry Draft).

			Regular Season					Playoffs				
Season	Club	Lea	GP	G	A	TP	PIM	GP	G	A	TP	PIM
1995-96	Hanover Barons	OJHL-C	29	7	13	20	26	....	....	....	....	....
1996-97	Oshawa Generals	OHL	39	0	4	4	15	15	0	0	0	4
1997-98	Oshawa Generals	OHL	64	3	12	15	72	7	0	3	3	11
1998-99	Oshawa Generals	OHL	65	3	18	21	109	15	3	6	9	28
99-2000	Michigan K-Wings	IHL	53	0	3	3	54	....	....	....	....	....
	Fort Wayne	UHL	7	1	1	2	25	9	0	2	2	10
2000-01	Utah Grizzlies	IHL	81	5	15	20	105	....	....	....	....	....
2001-02	Utah Grizzlies	AHL	77	6	9	15	146	5	1	0	1	17

MacNEIL, Ian (muhk-NEEL, EE-an) **PHI.**

Center. Shoots left. 6'2", 190 lbs. Born, Halifax, N.S., April 27, 1977.
(Hartford's 3rd choice, 85th overall, in 1995 Entry Draft).

			Regular Season					Playoffs				
Season	Club	Lea	GP	G	A	TP	PIM	GP	G	A	TP	PIM
1993-94	Whitby Lions	OMHA	50	30	22	52	102	....	....	....	....	....
1994-95	Oshawa Generals	OHL	60	7	21	28	62	7	0	2	2	0
1995-96	Oshawa Generals	OHL	49	15	17	32	54	5	1	2	3	8
1996-97	Oshawa Generals	OHL	64	23	20	43	96	18	2	3	5	37
1997-98	New Haven	AHL	68	12	21	33	67	3	1	0	1	10
1998-99	New Haven	AHL	47	6	4	10	62	....	....	....	....	....
99-2000	Cincinnati	IHL	81	19	18	37	100	11	3	2	5	25
2000-01	Cincinnati	IHL	82	17	22	39	139	5	0	1	1	4
2001-02	Lowell	AHL	79	14	20	34	128	2	0	1	1	4

Rights transferred to **Carolina** after **Hartford** franchise relocated, June 25, 1997. Signed as a free agent by **Philadelphia**, July 2, 2002.

MAGLIONE, Matt (US, MAT) **WSH.**

Defense. Shoots left. 6'1", 185 lbs. Born, Syracuse, NY, April 20, 1982.
(Washington's 7th choice, 249th overall, in 2001 Entry Draft).

			Regular Season					Playoffs				
Season	Club	Lea	GP	G	A	TP	PIM	GP	G	A	TP	PIM
1996-97	Syracuse	MTJHL	40	0	4	4	20	....	....	....	....	....
1997-98	Syracuse	MTJHL	40	3	13	16	86	....	....	....	....	....
1998-99	Auburn	OPJHL	44	12	24	36	44	....	....	....	....	....
99-2000	Team USA	USDP-18	48	4	6	10	23	....	....	....	....	....
2000-01	Princeton	ECAC	30	4	5	9	12	....	....	....	....	....
2001-02	Princeton	ECAC	27	2	6	8	12	....	....	....	....	....

MAGNUSON, Will (MAG-nuh-sohn, Wihl) **COL.**

Defense. Shoots right. 6'5", 235 lbs. Born, Anchorage, AK, February 19, 1980.
(Colorado's 6th choice, 142nd overall, in 1999 Entry Draft).

			Regular Season					Playoffs				
Season	Club	Lea	GP	G	A	TP	PIM	GP	G	A	TP	PIM
1997-98	Team USA	USDP-18	69	2	15	17	62	....	....	....	....	....
1998-99	Lake Superior	CCHA	32	0	1	1	44	....	....	....	....	....
99-2000	Lake Superior	CCHA	27	0	2	2	38	....	....	....	....	....
2000-01	Lake Superior	CCHA	33	0	3	3	54	....	....	....	....	....
2001-02	Lake Superior	CCHA	33	4	4	8	31	....	....	....	....	....

MAGOWAN, Ken (muh-GOW-uhn, KEHN) **N.J.**

Left wing. Shoots left. 6'2", 207 lbs. Born, Kelowna, B.C., July 22, 1981.
(New Jersey's 11th choice, 198th overall, in 2000 Entry Draft).

			Regular Season					Playoffs				
Season	Club	Lea	GP	G	A	TP	PIM	GP	G	A	TP	PIM
1996-97	Kelowna Rockets	BCAHA	57	68	66	134	78	....	....	....	....	....
1997-98	Kelowna Vikings	BCAHA	47	45	45	90	80	....	....	....	....	....
1998-99	Vernon Vipers	BCHL	60	15	25	40	40	....	....	....	....	....
99-2000	Vernon Vipers	BCHL	58	31	36	68	....	....	....	....	....	....
2000-01	Boston University	H-East	34	5	1	6	22	....	....	....	....	....
2001-02	Boston University	H-East	37	6	15	21	28	....	....	....	....	....

BCHL Interior First All-Star Team (1999)

MAISER, Justin (MAY-zuhr, JUHS-tihn) **ST.L.**

Center. Shoots left. 6'1", 191 lbs. Born, Milwaukee, WI, June 29, 1983.
(St. Louis' 5th choice, 165th overall, in 2002 Entry Draft).

			Regular Season					Playoffs				
Season	Club	Lea	GP	G	A	TP	PIM	GP	G	A	TP	PIM
99-2000	Team USA	USDP-17	48	11	20	31	114	....	....	....	....	....
2000-01	Team USA	USDP-18	58	20	18	38	127	....	....	....	....	....
2001-02	Boston University	H-East	36	8	14	22	65	....	....	....	....	....

MAJESKY, Ivan (migh-EHV-skee, EE-vahn) FLA.

Defense. Shoots right. 6'5", 224 lbs. Born, Banska Bystrica, Czech., September 2, 1976.
(Florida's 12th choice, 267th overall, in 2001 Entry Draft).

			Regular Season					Playoffs				
Season	**Club**	**Lea**	**GP**	**G**	**A**	**TP**	**PIM**	**GP**	**G**	**A**	**TP**	**PIM**
1995-96	Banska Bystrica	Slovakia	17	0	0	0	18	....	....	....	....	....
1996-97	Banska Bystrica	Slovakia	49	2	4	6	....	....	....	....	....	....
1997-98	Banska Bystrica	Slovak-2	43	6	7	13	50	....	....	....	....	....
1998-99	Banska Bystrica	Slovak-2	48	7	7	14	68	....	....	....	....	....
	HKm Zvolen	Slovakia	....	....	....	....	....	6	0	2	2	2
99-2000	HKm Zvolen	Slovakia	51	7	9	16	68	10	0	4	4	2
2000-01	Ilves Tampere	Finland	54	2	14	16	99	9	0	1	1	6
2001-02	Ilves Tampere	Finland	44	6	6	12	84	....	....	....	....	....
	Slovakia	Olympics	4	0	1	1	4	....	....	....	....	....

MAKELA, Tuukka (MA-kuh-luh TUH-kuh) BOS.

Defense. Shoots left. 6'2", 202 lbs. Born, Helsinki, Finland, May 24, 1982.
(Boston's 5th choice, 66th overall, in 2000 Entry Draft).

			Regular Season					Playoffs				
Season	**Club**	**Lea**	**GP**	**G**	**A**	**TP**	**PIM**	**GP**	**G**	**A**	**TP**	**PIM**
1997-98	HIFK Jr.	Finn-Jr.	5	0	0	0	4	....	....	....	....	....
1998-99	HIFK Jr.	Finn-Jr.	32	1	1	2	20	3	0	1	1	0
99-2000	HIFK Jr.	Finn-Jr.	36	2	5	7	22	....	....	....	....	....
2000-01	Montreal Rocket	QMJHL	9	2	1	3	14	....	....	....	....	....
2001-02	HPK-Jr.	Finn-Jr.	12	3	0	3	26	7	2	2	4	18
	HPK Hameenlinna	Finland	49	2	3	5	44	8	0	0	0	10

• Missed majority of 2000-01 season recovering from head injury suffered in game vs. Rouyn-Noranda (QMJHL), September 20, 2000.

MAKI, Tomi (MA-kee, TAW-mee) CGY.

Right wing. Shoots left. 5'11", 172 lbs. Born, Helsinki, Finland, August 19, 1983.
(Calgary's 4th choice, 108th overall, in 2001 Entry Draft).

			Regular Season					Playoffs				
Season	**Club**	**Lea**	**GP**	**G**	**A**	**TP**	**PIM**	**GP**	**G**	**A**	**TP**	**PIM**
1997-98	Jokerit-C	Finn-Jr.	4	0	1	1	0	2	0	0	0	0
1998-99	Jokerit-C	Finn-Jr.	32	20	22	42	40	....	....	....	....	....
99-2000	Jokerit Jr.	Finn-Jr.	33	6	1	7	12	3	0	0	0	0
2000-01	Jokerit-B	Finn-Jr.	10	4	9	13	4	6	4	3	7	0
	Jokerit Jr.	Finn-Jr.	39	7	8	15	10	2	0	0	0	2
2001-02	Jokerit Jr.	Finn-Jr.	29	12	13	25	12	1	0	0	0	2
	Kiekko Vantaa	Finland-2	5	0	0	0	0	....	....	....	....	....
	Jokerit Helsinki	Finland	8	0	1	1	2	....	....	....	....	....

MALEC, Tomas (MA-lehts, TAW-mahsh) CAR.

Defense. Shoots left. 6'2", 193 lbs. Born, Skalica, Czech., May 13, 1982.
(Florida's 4th choice, 64th overall, in 2001 Entry Draft).

			Regular Season					Playoffs				
Season	**Club**	**Lea**	**GP**	**G**	**A**	**TP**	**PIM**	**GP**	**G**	**A**	**TP**	**PIM**
99-2000	Skalica Jr.	Slovak-Jr.	46	6	5	11	150	....	....	....	....	....
2000-01	Rimouski Oceanic	QMJHL	64	13	50	63	198	11	0	11	11	26
2001-02	Rimouski Oceanic	QMJHL	51	14	32	46	164	7	3	1	4	10
	Lowell	AHL	....	....	....	....	....	4	0	0	0	4

Traded to **Carolina** by **Florida** with Bret Hedican, Kevyn Adams and a conditional 3rd round choice in 2003 Entry Draft for Sandis Ozolinsh and Byron Ritchie, January 16, 2002.

MALENKIKH, Vladimir (MAH-lihn-keh, vla-DIH-meer) PIT.

Defense. Shoots left. 6'1", 187 lbs. Born, Togliatti, USSR, October 1, 1980.
(Pittsburgh's 7th choice, 157th overall, in 1999 Entry Draft).

			Regular Season					Playoffs				
Season	**Club**	**Lea**	**GP**	**G**	**A**	**TP**	**PIM**	**GP**	**G**	**A**	**TP**	**PIM**
1997-98	Lada Togliatti 2	Russia-3	39	6	4	10	112	....	....	....	....	....
1998-99	Lada Togliatti 2	Russia-4	38	6	3	9	68	....	....	....	....	....
	Lada Togliatti	Russia	9	0	0	0	2	....	....	....	....	....
99-2000	CSK VVS Samara	Russia	7	0	1	1	14	....	....	....	....	....
2000-01	Lada Togliatti	Russia	25	1	1	2	14	5	0	0	0	26
2001-02	Lada Togliatti	Russia	47	5	4	9	88	4	0	0	0	2

MALLETTE, Carl (muh-LEHT, KAHRL)

Center. Shoots right. 6'1", 188 lbs. Born, Pointe Claire, Que., November 17, 1981.
(Atlanta's 4th choice, 107th overall, in 2000 Entry Draft).

			Regular Season					Playoffs				
Season	**Club**	**Lea**	**GP**	**G**	**A**	**TP**	**PIM**	**GP**	**G**	**A**	**TP**	**PIM**
1996-97	Lac St-Louis	QAAA	32	14	22	36	35	7	2	6	8	16
1997-98	Victoriaville	QMJHL	55	8	7	15	30	6	1	1	2	4
1998-99	Victoriaville	QMJHL	62	27	46	73	51	6	1	2	3	2
99-2000	Victoriaville	QMJHL	69	49	76	125	97	6	6	3	9	28
2000-01	Victoriaville	QMJHL	61	28	55	83	99	13	10	8	18	42
2001-02	Victoriaville	QMJHL	71	39	83	122	191	22	12	21	33	73

MALMIVAARA, Olli (mal-MIH-vah-ruh, OH-lee) CHI.

Defense. Shoots left. 6'5", 213 lbs. Born, Kajaani, Finland, March 13, 1982.
(Chicago's 6th choice, 117th overall, in 2000 Entry Draft).

			Regular Season					Playoffs				
Season	**Club**	**Lea**	**GP**	**G**	**A**	**TP**	**PIM**	**GP**	**G**	**A**	**TP**	**PIM**
1998-99	Jokerit-B	Finn-Jr.	35	1	8	9	10	7	0	0	0	2
99-2000	Jokerit-B	Finn-Jr.	8	3	4	7	8	1	0	0	0	2
	Jokerit Jr.	Finn-Jr.	27	3	3	6	12	12	0	2	2	2
2000-01	Jokerit Jr.	Finn-Jr.	33	10	13	23	24	2	0	0	0	0
	Kiekko Vantaa	Finland-2	4	1	0	1	2	....	....	....	....	....
	Jokerit Helsinki	Finland	5	0	0	0	2	....	....	....	....	....
2001-02	Jokerit Jr.	Finn-Jr.	2	0	1	1	0	....	....	....	....	....
	Jokerit Helsinki	Finland	53	0	6	6	16	11	0	0	0	2

MALONE, Ryan (MA-lohn, RIGH-yan) PIT.

Left wing. Shoots left. 6'3", 190 lbs. Born, Pittsburgh, PA, December 1, 1979.
(Pittsburgh's 5th choice, 115th overall, in 1999 Entry Draft).

			Regular Season					Playoffs				
Season	**Club**	**Lea**	**GP**	**G**	**A**	**TP**	**PIM**	**GP**	**G**	**A**	**TP**	**PIM**
1997-98	Shat.-St. Mary's	Hi-School	50	41	44	85	69	....	....	....	....	....
1998-99	Omaha Lancers	USHL	51	14	22	36	81	....	....	....	....	....
99-2000	St. Cloud State	WCHA	38	9	21	30	68	....	....	....	....	....
2000-01	St. Cloud State	WCHA	36	7	18	25	52	....	....	....	....	....
2001-02	St. Cloud State	WCHA	41	24	25	49	76	....	....	....	....	....

MAMANE, Shawn (muh-MA-nee, SHAWN) ST.L.

Left wing. Shoots left. 6', 195 lbs. Born, Montreal, Que., February 26, 1979.

			Regular Season					Playoffs				
Season	**Club**	**Lea**	**GP**	**G**	**A**	**TP**	**PIM**	**GP**	**G**	**A**	**TP**	**PIM**
1997-98	Laval Titan	QMJHL	15	2	1	3	8	....	....	....	....	....
	Nipawin Hawks	SJHL	36	3	6	9	111	....	....	....	....	....
1998-99	Nipawin Hawks	SJHL	STATISTICS NOT AVAILABLE									
99-2000	Nipawin Hawks	SJHL	35	35	25	58	79	....	....	....	....	....
	Wayne State	CHA	DID NOT PLAY									
	Worcester	AHL	1	0	0	0	0	....	....	....	....	....
2000-01	Peoria Rivermen	ECHL	19	9	8	17	47	....	....	....	....	....
	Worcester	AHL	51	4	6	10	68	4	0	0	0	2
2001-02	Peoria Rivermen	ECHL	8	0	0	0	12	....	....	....	....	....
	Worcester	AHL	3	0	1	1	7	....	....	....	....	....
	Arkansas	ECHL	48	14	17	31	119	....	....	....	....	....

• Signed Letter of Intent to attend **Wayne State** (CHA), November 4, 1999. • Ruled ineligible to play first semester of 1999-2000 season due to NCAA Transfer Regulations, January 5, 2000. Signed as a free agent by **St. Louis**, April 18, 2000.

MANNING, Paul (MAN-nihng, PAWL) CBJ

Defense. Shoots left. 6'4", 205 lbs. Born, Red Deer, Alta., April 15, 1979.
(Calgary's 3rd choice, 62nd overall, in 1998 Entry Draft).

			Regular Season					Playoffs				
Season	**Club**	**Lea**	**GP**	**G**	**A**	**TP**	**PIM**	**GP**	**G**	**A**	**TP**	**PIM**
1995-96	Red Deer	AMHL	32	8	32	40	....	....	....	....	....	....
1996-97	Red Deer Vipers	HJHL	36	9	33	42	....	....	....	....	....	....
1997-98	Colorado College	WCHA	30	1	5	6	16	....	....	....	....	....
1998-99	Colorado College	WCHA	41	3	10	13	75	....	....	....	....	....
99-2000	Colorado College	WCHA	39	6	17	23	26	....	....	....	....	....
2000-01	Colorado College	WCHA	34	2	28	30	48	....	....	....	....	....
2001-02	Syracuse Crunch	AHL	35	1	4	5	16	....	....	....	....	....
	Elmira Jackals	UHL	1	1	0	1	0	....	....	....	....	....

WCHA Second All-Star Team (2001)

Rights traded to **Columbus** by **Calgary** for Buffalo's 5th round choice (previously acquired, later traded to Detroit - Detroit selected Andreas Jamtin) in 2001 Entry Draft, June 24, 2001.

MANSON, Lane (MAN-suhn, LAYN) ATL.

Defense. Shoots left. 6'8", 222 lbs. Born, Watrous, Sask., February 14, 1984.
(Atlanta's 4th choice, 124th overall, in 2002 Entry Draft).

			Regular Season					Playoffs				
Season	**Club**	**Lea**	**GP**	**G**	**A**	**TP**	**PIM**	**GP**	**G**	**A**	**TP**	**PIM**
99-2000	North Battleford	SMBHL	41	7	12	19	110	....	....	....	....	....
2000-01	North Battleford	MMHL	40	14	12	26	180	....	....	....	....	....
2001-02	Moose Jaw	WHL	67	4	3	7	88	12	0	1	1	6

MANTYLA, Tuukka (man-TYEW-la, TOO-OO-kuh) L.A.

Defense. Shoots left. 5'9", 172 lbs. Born, Tampere, Finland, May 25, 1981.
(Los Angeles' 8th choice, 153rd overall, in 2001 Entry Draft).

			Regular Season					Playoffs				
Season	**Club**	**Lea**	**GP**	**G**	**A**	**TP**	**PIM**	**GP**	**G**	**A**	**TP**	**PIM**
1995-96	Tappara-C	Finn-Jr.	32	2	2	4	28	....	....	....	....	....
1996-97	Tappara-C	Finn-Jr.	32	12	25	37	52	4	1	2	3	4
	Tappara-B	Finn-Jr.	2	0	0	0	2	....	....	....	....	....
1997-98	Tappara-B	Finn-Jr.	31	2	23	25	49	....	....	....	....	....
	Tappara Jr.	Finn-Jr.	2	0	0	0	0	6	0	0	0	4
1998-99	Tappara-B	Finn-Jr.	10	4	4	8	42	....	....	....	....	....
	Tappara Jr.	Finn-Jr.	34	5	13	18	42	....	....	....	....	....
99-2000	Tappara Jr.	Finn-Jr.	7	2	5	7	22	5	2	5	7	4
	Tappara Tampere	Finland	43	2	8	10	16	4	0	0	0	0
2000-01	Tappara Tampere	Finland	53	6	14	20	32	10	2	2	4	10
2001-02	Tappara Tampere	Finland	56	5	10	15	70	10	2	4	6	8

MAPLETOFT, Justin (MAPLE-tawft, JUH-stihn) NYI

Center. Shoots left. 6'1", 180 lbs. Born, Lloydminster, Sask., January 11, 1981.
(NY Islanders' 9th choice, 130th overall, in 1999 Entry Draft).

			Regular Season					Playoffs				
Season	**Club**	**Lea**	**GP**	**G**	**A**	**TP**	**PIM**	**GP**	**G**	**A**	**TP**	**PIM**
1996-97	Calgary Royals	AMHL	36	25	36	51	....	....	....	....	....	....
	Red Deer Rebels	WHL	2	0	0	0	0	....	....	....	....	....
1997-98	Red Deer Rebels	WHL	65	9	4	13	41	....	....	....	....	....
1998-99	Red Deer Rebels	WHL	72	24	22	46	81	....	....	....	....	....
99-2000	Red Deer Rebels	WHL	72	39	57	96	135	4	2	1	3	28
2000-01	Red Deer Rebels	WHL	70	43	*77	*120	111	22	13	*21	34	59
2001-02	Bridgeport	AHL	80	13	20	33	60	20	7	10	17	23

WHL East First All-Star Team (2000, 2001) • Canadian Major Junior First All-Star Team (2001)

MAROIS, Jerome (MAIR-wuh, jair-OHM) MTL.

Left wing. Shoots left. 6'1", 199 lbs. Born, Quebec, Que., January 27, 1981.
(Montreal's 11th choice, 253rd overall, in 1999 Entry Draft).

			Regular Season					Playoffs				
Season	**Club**	**Lea**	**GP**	**G**	**A**	**TP**	**PIM**	**GP**	**G**	**A**	**TP**	**PIM**
1996-97	Ste-Foy	QAAA	44	29	24	53	....	10	8	8	16	....
1997-98	Quebec Remparts	QMJHL	55	5	12	17	12	12	2	0	2	2
1998-99	Quebec Remparts	QMJHL	52	8	15	23	48	12	0	4	4	13
99-2000	Cape Breton	QMJHL	66	28	34	62	95	4	2	1	3	14
2000-01	Rouyn-Noranda	QMJHL	68	36	47	83	119	6	1	0	1	17
2001-02	Quebec	AHL	10	0	0	0	2	....	....	....	....	....
	Mississippi	ECHL	44	12	10	22	47	10	5	2	7	4

Traded to **Rouyn-Noranda** (QMJHL) by **Cape Breton** (QMJHL) with Cape Breton's 8th round choice (Etienne Gagner) in 2001 QMJHL Priority Draft for Kevin Cloutier and Hull's 4th round choice (previously acquired, Cape Breton selected Marc-Olivier Vary) in 2000 QMJHL Priority Draft, September 10, 2000.

MARS, Per (MAHRZ, PAIR) CBJ

Center. Shoots left. 6'3", 203 lbs. Born, Ostersund, Sweden, October 23, 1982.
(Columbus' 5th choice, 87th overall, in 2001 Entry Draft).

			Regular Season					Playoffs				
Season	**Club**	**Lea**	**GP**	**G**	**A**	**TP**	**PIM**	**GP**	**G**	**A**	**TP**	**PIM**
2000-01	Brynas Jr.	Swede-Jr.	23	7	7	14	62	....	....	....	....	....
	Brynas IF Gavle	Sweden	6	0	0	0	0	2	0	0	0	0
2001-02	Brynas IF Gavle	Sweden	12	0	0	0	14	....	....	....	....	....
	Tierp HK	Swede-2	29	1	4	5	22	11	1	0	1	35

MARTENSSON, Tony (MOHR-tehn-suhn, TOH-nee) ANA.

Right wing. Shoots left. 6', 189 lbs. Born, Upplands Vasby, Sweden, June 23, 1980.
(Anaheim's 9th choice, 224th overall, in 2001 Entry Draft).

			Regular Season					Playoffs				
Season	Club	Lea	GP	G	A	TP	PIM	GP	G	A	TP	PIM
1997-98	Arlanda Mastra	Swede-2	12	2	4	6	0	2	0	0	0	0
1998-99	Arlanda Mastra	Swede-2	37	8	22	30	8	2	0	0	0	0
99-2000	Arlanda Mastra	Swede-2	44	21	28	49	14	....	....	....	....	....
2000-01	Brynas IF Gavle	Sweden	50	15	11	26	20	4	0	1	1	2
	Brynas Jr.	Swede-Jr.	1	1	0	1	0	....	....	....	....	....
2001-02	Brynas IF Gavle	Sweden	50	9	17	26	14	4	1	3	4	0

MARTIN, Joey (MAHR-tihn, JOH-ee) CHI.

Defense. Shoots left. 6'3", 198 lbs. Born, Fridley, MN, July 17, 1981.
(Chicago's 9th choice, 193rd overall, in 2000 Entry Draft).

			Regular Season					Playoffs				
Season	Club	Lea	GP	G	A	TP	PIM	GP	G	A	TP	PIM
1998-99	Buffalo MN High	Hi-School	23	10	9	19	19	....	....	....	....	....
99-2000	Omaha Lancers	USHL	56	1	4	5	41	4	0	0	0	0
2000-01	U. of Minnesota	WCHA	18	0	2	2	2	....	....	....	....	....
2001-02	U. of Minnesota	WCHA	11	0	4	4	14	....	....	....	....	....

MARTIN, Mike (MAHR-tihn, MIGHK) CGY.

Defense. Shoots right. 6'2", 205 lbs. Born, Stratford, Ont., October 27, 1976.
(NY Rangers' 2nd choice, 65th overall, in 1995 Entry Draft).

			Regular Season					Playoffs				
Season	Club	Lea	GP	G	A	TP	PIM	GP	G	A	TP	PIM
1991-92	Stratford	OJHL-B	16	2	3	5	14	....	....	....	....	....
1992-93	Windsor	OHL	61	2	7	9	80	....	....	....	....	....
1993-94	Windsor	OHL	64	2	29	31	94	4	1	2	3	4
1994-95	Windsor	OHL	53	9	28	37	79	10	1	3	4	21
1995-96	Windsor	OHL	65	19	48	67	128	7	0	6	6	14
1996-97	Binghamton	AHL	62	2	7	9	45	3	0	1	1	2
1997-98	Hartford	AHL	60	4	11	15	70	4	0	0	0	2
1998-99	Fort Wayne	IHL	75	6	20	26	89	2	0	0	0	4
99-2000	Michigan K-Wings	IHL	74	8	15	23	99	....	....	....	....	....
2000-01	Saint John	AHL	60	7	16	23	69	16	1	2	3	14
2001-02	Saint John	AHL	13	1	8	9	14	....	....	....	....	....

Signed as a free agent by **Calgary**, August 18, 2000.

MARTIN, Paul (MAHR-tihn, PAWL) N.J.

Defense. Shoots left. 6'1", 170 lbs. Born, Minneapolis, MN, March 5, 1981.
(New Jersey's 5th choice, 62nd overall, in 2000 Entry Draft).

			Regular Season					Playoffs				
Season	Club	Lea	GP	G	A	TP	PIM	GP	G	A	TP	PIM
1998-99	Elk River Elks	Hi-School	24	9	11	20	....	....	....	....	....	....
99-2000	Elk River Elks	Hi-School	24	15	35	50	26	....	....	....	....	....
2000-01	U. of Minnesota	WCHA	38	3	17	20	8	....	....	....	....	....
2001-02	U. of Minnesota	WCHA	44	8	30	38	22	....	....	....	....	....

Minnesota High School Player of the Year (1999) • WCHA Second All-Star Team (2002)

MARTYNYUK, Denis (mahr-tih-nyook, DEH-nihs) VAN.

Left wing. Shoots left. 6'3", 190 lbs. Born, Kapfenberg, Austria, July 26, 1979.
(Vancouver's 11th choice, 201st overall, in 1997 Entry Draft).

			Regular Season					Playoffs				
Season	Club	Lea	GP	G	A	TP	PIM	GP	G	A	TP	PIM
1994-95	CSKA Moscow Jr.	CIS-Jr.	34	25	25	50	20	....	....	....	....	....
1995-96	CSKA Moscow Jr.	CIS-Jr.	36	10	15	25	20	....	....	....	....	....
	H.C. CSKA	Russia-2	25	2	5	7	20	....	....	....	....	....
1996-97	CSKA Moscow 2	Russia-3	41	7	4	11	12	....	....	....	....	....
	H.C. CSKA	Russia	3	1	0	1	0	....	....	....	....	....
1997-98	Spartak Moscow 2	Russia-3	45	9	5	14	34	....	....	....	....	....
1998-99	Spartak Moscow	Russia	17	0	0	0	8	....	....	....	....	....
99-2000	Spartak Moscow	Russia-2		STATISTICS NOT AVAILABLE								
2000-01	Spartak Moscow 2	Russia-3	6	1	3	4	0	....	....	....	....	....
	Amur Khabarovsk	Russia	1	0	0	0	0	....	....	....	....	....
2001-02	Columbia Inferno	ECHL	52	14	8	22	42	....	....	....	....	....
	Manitoba Moose	AHL	15	1	2	3	0	5	0	0	0	0

MARTZ, Nathan (MAHRTZ, NAY-thun) NYR

Center. Shoots left. 6'3", 169 lbs. Born, Chilliwack, B.C., March 4, 1981.
(NY Rangers' 4th choice, 140th overall, in 2000 Entry Draft).

			Regular Season					Playoffs				
Season	Club	Lea	GP	G	A	TP	PIM	GP	G	A	TP	PIM
1997-98	Chilliwack	BCHL	59	8	13	21	86	....	....	....	....	....
1998-99	Chilliwack	BCHL	59	20	38	58	....	....	....	....	....	....
99-2000	Chilliwack	BCHL	59	35	75	110	97	....	....	....	....	....
2000-01	New Hampshire	H-East	37	5	14	19	20	....	....	....	....	....
2001-02	New Hampshire	H-East	28	3	7	10	12	....	....	....	....	....

MASSEN, James (MA-suhn, JAYMS) N.J.

Right wing. Shoots right. 6'1", 228 lbs. Born, Bismarck, ND, January 13, 1982.
(New Jersey's 9th choice, 194th overall, in 2001 Entry Draft).

			Regular Season					Playoffs				
Season	Club	Lea	GP	G	A	TP	PIM	GP	G	A	TP	PIM
99-2000	Sioux Falls	USHL	53	16	17	33	25	3	0	0	0	2
2000-01	Sioux Falls	USHL	56	37	38	75	56	8	6	2	8	4
2001-02	North Dakota	WCHA	34	5	8	13	18	....	....	....	....	....

USHL First All-Star Team (2001)

MATEJOVSKY, Radek (ma-teh-YAHV-skee, ra-DEHK) NYI

Right wing. Shoots right. 6'1", 187 lbs. Born, Praha, Czech., November 17, 1977.
(NY Islanders' 9th choice, 250th overall, in 1998 Entry Draft).

			Regular Season					Playoffs				
Season	Club	Lea	GP	G	A	TP	PIM	GP	G	A	TP	PIM
1992-93	C. Budejovice Jr.	Czech-Jr.	25	38	24	62	....	....	....	....	....	....
1993-94	Slavia Praha Jr.	Czech-Jr.	45	30	26	56	....	....	....	....	....	....
1994-95	Slavia Praha Jr.	Czech-Jr.	28	7	8	15	12	....	....	....	....	....
1995-96	Slavia Praha Jr.	Czech-Jr.	47	37	21	58	24	....	....	....	....	....
1996-97	Slavia Praha Jr.	Czech-Jr.	4	1	1	2	....	....	....	....	....	....
	H+S Beroun	Czech-2	12	3	1	4	18	....	....	....	....	....
	HC Slavia Praha	Czech	41	3	4	7	10	3	0	0	0	0
1997-98	HC Slavia Praha	Czech	52	9	4	13	24	3	0	0	0	0
1998-99	HC Dukla Jihlava	Czech	52	12	10	22	57	....	....	....	....	....
99-2000	HC Slavia Praha	Czech	25	4	3	7	22	....	....	....	....	....
	Pardubice	Czech	25	2	4	6	18	3	0	0	0	0
2000-01	HC Slavia Praha	Czech	39	6	8	14	63	11	1	2	3	18
2001-02	HC Slavia Praha	Czech	35	3	1	4	42	9	0	0	0	6

MAXIMENKO, Andrei (max-EE-mehn-koh, AWN-dray) DET.

Left wing. Shoots right. 5'11", 172 lbs. Born, Moscow, USSR, January 10, 1981.
(Detroit's 2nd choice, 149th overall, in 1999 Entry Draft).

			Regular Season					Playoffs				
Season	Club	Lea	GP	G	A	TP	PIM	GP	G	A	TP	PIM
1997-98	Krylja Sovetov 2	Russia-3	42	2	4	6	12	....	....	....	....	....
1998-99	Krylja Sovetov	Russia	28	1	2	3	24	....	....	....	....	....
99-2000	Krylja Sovetov 2	Russia-3	6	4	2	6	26	....	....	....	....	....
	Krylja Sovetov	Russia-2	39	6	7	13	41	....	....	....	....	....
2000-01	Krylja Sovetov	Russia-2	28	2	5	7	8	....	....	....	....	....
2001-02	THC Tver	Russia-2	20	3	7	10	6	....	....	....	....	....
	Krylja Sovetov 2	Russia-3	7	3	5	8	2	....	....	....	....	....
	Perm	Russia	9	0	1	1	2	....	....	....	....	....

McASLAN, Sean (mihk-AZ-luhn, SHAWN) EDM.

Right wing. Shoots right. 6'1", 190 lbs. Born, Okotoks, Alta., January 12, 1980.

			Regular Season					Playoffs				
Season	Club	Lea	GP	G	A	TP	PIM	GP	G	A	TP	PIM
1996-97	Calgary Hitmen	WHL	26	2	3	5	22	....	....	....	....	....
1997-98	Calgary Hitmen	WHL	69	8	16	24	83	....	....	....	....	....
1998-99	Calgary Hitmen	WHL	71	7	16	23	110	21	2	1	3	18
99-2000	Calgary Hitmen	WHL	72	18	16	34	117	13	4	2	6	49
2000-01	Calgary Hitmen	WHL	51	21	32	53	137	12	3	5	8	29
2001-02	Columbus	ECHL	72	16	21	37	139	....	....	....	....	....

Signed as a free agent by **Edmonton**, March 14, 2001.

McCAMBRIDGE, Keith (muh-KAYM-brihdj, KEETH)

Defense. Shoots left. 6'2", 205 lbs. Born, Thompson, Man., February 1, 1974.
(Calgary's 10th choice, 201st overall, in 1994 Entry Draft).

			Regular Season					Playoffs				
Season	Club	Lea	GP	G	A	TP	PIM	GP	G	A	TP	PIM
1991-92	Swift Current	WHL	72	1	4	5	84	8	0	0	0	2
1992-93	Swift Current	WHL	70	0	6	6	87	17	0	1	1	27
1993-94	Swift Current	WHL	71	0	10	10	179	7	0	0	0	4
1994-95	Swift Current	WHL	48	5	7	12	120	....	....	....	....	....
	Kamloops Blazers	WHL	21	0	6	6	90	21	0	5	5	49
1995-96	Saint John	AHL	48	1	3	4	89	16	0	0	0	6
1996-97	Saint John	AHL	56	2	1	3	109	....	....	....	....	....
1997-98	Saint John	AHL	56	4	4	8	118	....	....	....	....	....
	Las Vegas	IHL	10	0	1	1	16	4	0	0	0	9
1998-99	Las Vegas	IHL	18	1	2	3	56	....	....	....	....	....
	Long Beach	IHL	52	2	5	7	200	8	0	0	0	20
99-2000	Providence	AHL	47	0	2	2	135	....	....	....	....	....
	Manitoba Moose	IHL	3	0	1	1	4	....	....	....	....	....
2000-01	Providence	AHL	63	1	5	6	215	10	0	0	0	18
2001-02	Providence	AHL	59	1	3	4	263	2	0	0	0	5

Signed as free agent by **Boston**, August 20, 1999.

McCARTHY, Jeremiah (mih-KAHR-thee, jeh-rih-MIGH-uh)

Defense. Shoots left. 6', 210 lbs. Born, Boston, MA, March 1, 1976.

			Regular Season					Playoffs				
Season	Club	Lea	GP	G	A	TP	PIM	GP	G	A	TP	PIM
1994-95	Harvard Crimson	ECAC	25	3	5	8	4	....	....	....	....	....
1995-96	Harvard Crimson	ECAC	32	4	12	16	20	....	....	....	....	....
1996-97	Harvard Crimson	ECAC	32	4	9	13	22	....	....	....	....	....
1997-98	Harvard Crimson	ECAC	28	11	10	21	38	....	....	....	....	....
1998-99	Peoria Rivermen	ECHL	6	1	2	3	6	....	....	....	....	....
	Worcester	AHL	59	5	10	15	37	4	0	2	2	0
99-2000	Missouri	UHL	33	10	25	35	45	....	....	....	....	....
	Springfield	AHL	43	5	9	14	16	5	1	1	2	0
2000-01	Cincinnati	IHL	69	6	12	18	34	3	0	0	0	0
2001-02	Lowell	AHL	74	7	28	35	43	5	0	2	2	0

Signed as a free agent by **Carolina**, August 21, 2000. Signed as a free agent by **Amur** (Russia), July 19, 2002.

McCLEMENT, Jay (muh-KLEHM-ehnt, JAY) ST.L.

Center. Shoots left. 6'1", 193 lbs. Born, Kingston, Ont., March 2, 1983.
(St. Louis' 1st choice, 57th overall, in 2001 Entry Draft).

			Regular Season					Playoffs				
Season	Club	Lea	GP	G	A	TP	PIM	GP	G	A	TP	PIM
1997-98	Kingston	OPJHL	48	3	8	11	15	....	....	....	....	....
1998-99	Kingston	OPJHL	51	25	28	53	34	....	....	....	....	....
99-2000	Brampton	OHL	63	13	16	29	34	6	0	4	4	8
2000-01	Brampton	OHL	66	30	19	49	61	9	4	2	6	10
2001-02	Brampton	OHL	61	26	29	55	43	....	....	....	....	....

McCONNELL, Brian (mih-CAW-nuhl, BRIGH-uhn) CGY.

Center. Shoots left. 6'2", 190 lbs. Born, Boston, MA, February 1, 1983.
(Calgary's 2nd choice, 39th overall, in 2002 Entry Draft).

			Regular Season					Playoffs				
Season	Club	Lea	GP	G	A	TP	PIM	GP	G	A	TP	PIM
1998-99	Thayer Academy	Hi-School	23	12	27	39	....	....	....	....	....	....
99-2000	Team USA	USDP-17	48	8	11	19	76	....	....	....	....	....
2000-01	Team USA	USDP-18	62	19	25	44	143	....	....	....	....	....
2001-02	Boston University	H-East	38	11	15	26	68	....	....	....	....	....

McCORMICK, Cody (muh-KOHR-mihk, KOH-dee) COL.

Center. Shoots right. 6'2", 200 lbs. Born, London, Ont., April 18, 1983.
(Colorado's 5th choice, 144th overall, in 2001 Entry Draft).

			Regular Season					Playoffs				
Season	Club	Lea	GP	G	A	TP	PIM	GP	G	A	TP	PIM
1998-99	Elgin-Middlesex	OMHA	58	22	40	62	81	....	....	....	....	....
99-2000	Belleville Bulls	OHL	45	3	4	7	42	9	1	0	1	10
2000-01	Belleville Bulls	OHL	66	7	16	23	135	10	1	1	2	23
2001-02	Belleville Bulls	OHL	63	10	17	27	118	11	2	4	6	24

McDONALD, Brent (muhk-DAW-nuhld, BREHNT) CAR.

Center. Shoots right. 5'11", 180 lbs. Born, Olds, Alta., October 7, 1979.
(Carolina's 10th choice, 239th overall, in 1998 Entry Draft).

			Regular Season					Playoffs				
Season	Club	Lea	GP	G	A	TP	PIM	GP	G	A	TP	PIM
1994-95	Red Deer Vipers	AMHL	26	16	24	40	55	....	....	....	....	....
1995-96	Red Deer Rebels	WHL	68	1	7	8	55	6	0	1	1	2
1996-97	Red Deer Rebels	WHL	69	11	17	28	94	16	4	3	7	38
1997-98	Red Deer Rebels	WHL	69	18	27	45	93	5	0	2	2	4
1998-99	Red Deer Rebels	WHL	38	17	18	35	64	....	....	....	....	....
	Prince George	WHL	34	13	13	26	40	7	1	1	2	18
99-2000	Prince George	WHL	7	1	3	4	6	....	....	....	....	....
	Spokane Chiefs	WHL	61	28	30	58	77	15	5	8	13	42
2000-01	Florida	ECHL	67	11	11	22	55	5	1	0	1	4
2001-02	Florida	ECHL	69	12	20	32	92	6	2	2	4	4
	Lowell	AHL	2	0	0	0	0	2	0	0	0	0

Traded to **Prince George** (WHL) by **Red Deer** (WHL) for Jordan Walker, January 4, 1999. Traded to **Spokane** (WHL) by **Prince George** (WHL) for future considerations, November 2, 1999.

McDONELL, Kent (MAHK-dah-NEHL, KEHNT) CBJ

Right wing. Shoots right. 6'2", 205 lbs. Born, Williamstown, Ont., March 1, 1979.
(Detroit's 3rd choice, 181st overall, in 1999 Entry Draft).

			Regular Season					Playoffs				
Season	Club	Lea	GP	G	A	TP	PIM	GP	G	A	TP	PIM
1995-96	Cornwall Colts	OCJHL	33	21	14	35	64	....	....	....	....	....
1996-97	Guelph Storm	OHL	56	7	5	12	57	16	0	2	2	4
1997-98	Guelph Storm	OHL	64	28	23	51	76	12	7	4	11	18
1998-99	Guelph Storm	OHL	60	31	38	69	110	11	4	3	7	36
99-2000	Guelph Storm	OHL	56	35	35	70	100	6	1	4	5	6
2000-01	Dayton Bombers	ECHL	28	16	9	25	94	3	0	0	0	4
	Syracuse Crunch	AHL	32	3	3	6	36	3	1	0	1	0
2001-02	Syracuse Crunch	AHL	72	18	13	31	122	3	0	2	2	0

• Re-entered NHL Entry Draft. Originally Carolina's 9th choice, 225th overall, in 1997 Entry Draft.
Traded to **Columbus** by **Detroit** for future considerations, August 14, 2000.

McLACHLAN, Darren (muhk-LAWK-luhn, DAIR-rehn) BOS.

Left wing. Shoots left. 6'1", 230 lbs. Born, Penticton, B.C., February 16, 1983.
(Boston's 2nd choice, 77th overall, in 2001 Entry Draft).

			Regular Season					Playoffs				
Season	Club	Lea	GP	G	A	TP	PIM	GP	G	A	TP	PIM
1998-99	Campbell River	VIJHL	31	15	25	40	212	....	....	....	....	....
	Seattle	WHL	2	0	1	1	7	....	....	....	....	....
99-2000	Seattle	WHL	54	5	1	6	175	7	0	0	0	9
2000-01	Seattle	WHL	42	10	9	19	161	9	1	3	4	18
2001-02	Seattle	WHL	51	15	16	31	153	10	1	0	1	20

McLAREN, Steve (muh-KLAIR-uhn, STEEV) ST.L.

Left wing. Shoots left. 6', 200 lbs. Born, Owen Sound, Ont., February 3, 1975.
(Chicago's 3rd choice, 85th overall, in 1994 Entry Draft).

			Regular Season					Playoffs				
Season	Club	Lea	GP	G	A	TP	PIM	GP	G	A	TP	PIM
1992-93	N. Bay Trappers	NOJHA	30	15	18	33	110	....	....	....	....	....
1993-94	North Bay	OHL	55	2	15	17	130	18	0	3	3	50
1994-95	North Bay	OHL	27	3	10	13	119	6	2	1	3	23
1995-96	Indianapolis Ice	IHL	54	1	2	3	170	3	0	0	0	2
1996-97	Indianapolis Ice	IHL	63	2	5	7	309	4	0	0	0	10
1997-98	Indianapolis Ice	IHL	61	3	5	8	208	5	0	0	0	24
1998-99	Philadelphia	AHL	52	4	3	7	216	7	0	0	0	2
99-2000	Philadelphia	AHL	64	1	2	3	247	....	....	....	....	....
2000-01	Philadelphia	AHL	48	3	1	4	177	8	0	0	0	38
2001-02	Worcester	AHL	58	0	4	4	251	1	0	0	0	2

Signed as a free agent by **Philadelphia**, August 24, 1998. Signed as a free agent by **St. Louis**, July 16, 2001.

McLEAN, Brett (muh-CLAYN, BREHT) CHI.

Center. Shoots left. 5'11", 194 lbs. Born, Comox, B.C., August 14, 1978.
(Dallas' 9th choice, 242nd overall, in 1997 Entry Draft).

			Regular Season					Playoffs				
Season	Club	Lea	GP	G	A	TP	PIM	GP	G	A	TP	PIM
1993-94	Notre Dame	SMBHL	71	109	124	233	70	....	....	....	....	....
1994-95	Tacoma Rockets	WHL	67	11	23	34	33	4	0	1	1	0
1995-96	Kelowna Rockets	WHL	71	37	42	79	60	6	2	2	4	6
1996-97	Kelowna Rockets	WHL	72	44	60	104	98	6	4	2	6	12
1997-98	Kelowna Rockets	WHL	54	42	45	87	91	7	4	5	9	17
1998-99	Kelowna Rockets	WHL	44	32	38	70	46	....	....	....	....	....
	Brandon	WHL	21	15	16	31	20	5	1	6	7	8
	Cincinnati	AHL	7	0	3	3	6	....	....	....	....	....
99-2000	Johnstown Chiefs	ECHL	8	4	7	11	6	....	....	....	....	....
	Saint John	AHL	72	15	23	38	115	3	0	1	1	2
2000-01	Cleveland	IHL	74	20	24	44	54	4	0	0	0	18
2001-02	Houston Aeros	AHL	78	24	21	45	71	14	1	6	7	12

WHL West Second All-Star Team (1998)

Traded to **Brandon** (WHL) by **Kelowna** (WHL) for Ryan Johnson, Andrew Kaminsky and future considerations, February 2, 1999. Signed as a free agent by **Calgary**, September, 1999. Signed as a free agent by **Minnesota**, July 13, 2000. Signed as a free agent by **Chicago**, July 23, 2002.

McLEOD, Kiel (muk-KLOWD, KIGHL) CBJ

Center. Shoots right. 6'6", 229 lbs. Born, Ft. Saskatchewan, Alta., December 30, 1982.
(Columbus' 3rd choice, 53rd overall, in 2001 Entry Draft).

			Regular Season					Playoffs				
Season	Club	Lea	GP	G	A	TP	PIM	GP	G	A	TP	PIM
1997-98	North Delta	BCAHA	55	57	55	112	202	....	....	....	....	....
1998-99	Kelowna Rockets	WHL	55	12	15	27	48	6	0	1	1	2
99-2000	Kelowna Rockets	WHL	59	17	13	30	100	5	2	1	3	2
2000-01	Kelowna Rockets	WHL	65	38	28	66	94	4	4	1	5	8
2001-02	Kelowna Rockets	WHL	41	17	31	48	62	15	3	10	13	14

McMEEKIN, Brian (muhk-MEE-khin, BRIGH-uhn) ST.L.

Defense. Shoots right. 6'4", 195 lbs. Born, Trail, B.C., June 20, 1979.
(St. Louis' 9th choice, 260th overall, in 1999 Entry Draft).

			Regular Season					Playoffs				
Season	Club	Lea	GP	G	A	TP	PIM	GP	G	A	TP	PIM
1994-95	Rossland/Trail	BCAHA	70	15	35	50	12	....	....	....	....	....
1995-96	Rossland/Trail	BCAHA	60	15	40	55	10	....	....	....	....	....
1996-97	Trail	BCHL	37	1	10	11	20	....	....	....	....	....
1997-98	Trail	BCHL	49	4	7	11	66	10	0	1	1	10
1998-99	Cornell Big Red	ECAC	26	0	1	1	12	....	....	....	....	....
99-2000	Cornell Big Red	ECAC	11	0	1	1	8	....	....	....	....	....
2000-01	Cornell Big Red	ECAC	31	2	1	3	24	....	....	....	....	....
2001-02	Cornell Big Red	ECAC	35	1	6	7	40	....	....	....	....	....

McMORROW, Sean (muhk-MOHR-roh, SHAWN) BUF.

Defense. Shoots right. 6'4", 214 lbs. Born, Vancouver, B.C., January 19, 1982.
(Buffalo's 7th choice, 258th overall, in 2000 Entry Draft).

			Regular Season					Playoffs				
Season	Club	Lea	GP	G	A	TP	PIM	GP	G	A	TP	PIM
1998-99	Pickering	OPJHL	35	2	10	12	175	....	....	....	....	....
99-2000	Sarnia Sting	OHL	31	0	1	1	75	....	....	....	....	....
	Kitchener	OHL	31	0	1	1	67	4	0	0	0	12
2000-01	Mississauga	OHL	13	0	0	0	34	....	....	....	....	....
	Kingston	OHL	7	0	1	1	22	....	....	....	....	....
	London Knights	OHL	29	0	3	3	75	5	0	0	0	0
2001-02	London Knights	OHL	38	0	1	1	107	....	....	....	....	....
	Oshawa Generals	OHL	27	6	1	7	63	5	1	0	1	12

Traded to **Kitchener** (OHL) by **Sarnia** (OHL) for future considerations, January 9, 2000. Traded to **Mississauga** (OHL) by **Kitchener** (OHL) with Michael Wehrstedt and Brent Labre for Marcus Smith and Kitchener's 1st round choice (later traded back to Kitchener - Kitchener selected Vasily Bizyayev) in 2000 CHL Import Draft, July 2, 2000. Traded to **Kingston** (OHL) by **Mississauga** (OHL) for Matt Timmons, December 13, 2000. Traded to **London** (OHL) by **Kingston** (OHL) with Kingston's 2nd round choice (Gerald Coleman) in 2001 OHL Midget Draft for Lou Dickenson and London's 10th round choice (Dayne Davis) in 2001 OHL Midget Draft, January 8, 2001. Traded to **Oshawa** (OHL) by **London** (OHL) for Oshawa's 4th and 6th round choices in OHL 2003 Priority Draft, January 10, 2002.

McNEILL, Grant (muhk-NEEL, GRANT) FLA.

Defense. Shoots left. 6'2", 210 lbs. Born, Vermillion, Alta., June 8, 1983.
(Florida's 5th choice, 68th overall, in 2001 Entry Draft).

			Regular Season					Playoffs				
Season	Club	Lea	GP	G	A	TP	PIM	GP	G	A	TP	PIM
99-2000	Prince Albert	WHL	58	1	1	2	43	6	0	1	1	0
2000-01	Prince Albert	WHL	61	2	6	8	280	....	....	....	....	....
2001-02	Prince Albert	WHL	70	7	6	13	326	....	....	....	....	....

McPHERSON, Andrew (muhk-FUHR-suhn, AN-droo) PIT.

Left wing. Shoots left. 6'2", 175 lbs. Born, Ottawa, Ont., April 28, 1979.
(Pittsburgh's 11th choice, 261st overall, in 1999 Entry Draft).

			Regular Season					Playoffs				
Season	Club	Lea	GP	G	A	TP	PIM	GP	G	A	TP	PIM
1997-98	Dauphin Kings	MJHL	61	24	37	61	132	....	....	....	....	....
1998-99	RPI Engineers	ECAC	29	5	4	9	12	....	....	....	....	....
99-2000	RPI Engineers	ECAC	37	10	6	16	22	....	....	....	....	....
2000-01	RPI Engineers	ECAC	31	5	6	11	34	....	....	....	....	....
2001-02	RPI Engineers	ECAC	35	6	14	20	44	....	....	....	....	....

McRAE, Mark (muh-KRAY, MAHRK) ATL.

Defense. Shoots right. 6', 175 lbs. Born, Toronto, Ont., April 29, 1981.
(Atlanta's 13th choice, 288th overall, in 2000 Entry Draft).

			Regular Season					Playoffs				
Season	Club	Lea	GP	G	A	TP	PIM	GP	G	A	TP	PIM
1997-98	Brampton	OPJHL	46	23	41	64	25	....	....	....	....	....
1998-99	Brampton	OPJHL	50	23	41	64	43	....	....	....	....	....
99-2000	Cornell Big Red	ECAC	27	5	16	21	10	....	....	....	....	....
2000-01	Cornell Big Red	ECAC	33	8	11	19	14	....	....	....	....	....
2001-02	Cornell Big Red	ECAC	35	8	22	30	22	....	....	....	....	....

ECAC Second All-Star Team (2002)

McRAE, Matt (muh-KRAY, MAT) ATL.

Center. Shoots right. 6', 180 lbs. Born, Toronto, Ont., April 29, 1981.
(Atlanta's 6th choice, 147th overall, in 2000 Entry Draft).

			Regular Season					Playoffs				
Season	Club	Lea	GP	G	A	TP	PIM	GP	G	A	TP	PIM
1997-98	Brampton	OPJHL	50	26	26	52	22	....	....	....	....	....
1998-99	Brampton	OPJHL	50	34	40	74	29	....	....	....	....	....
99-2000	Cornell Big Red	ECAC	31	8	16	24	22	....	....	....	....	....
2000-01	Cornell Big Red	ECAC	27	3	6	9	10	....	....	....	....	....
2001-02	Cornell Big Red	ECAC	34	11	6	17	26	....	....	....	....	....

MEHALKO, Brad (muh-HAL-koh, BOYD)

Right wing. Shoots right. 6'1", 190 lbs. Born, Enchant, Alta., November 4, 1977.
(San Jose's 9th choice, 167th overall, in 1995 Entry Draft).

			Regular Season					Playoffs				
Season	Club	Lea	GP	G	A	TP	PIM	GP	G	A	TP	PIM
1992-93	Lethbridge Elite	AMHL	54	41	54	95	146	....	....	....	....	....
1993-94	Lethbridge	WHL	62	9	9	18	48	....	....	....	....	....
1994-95	Lethbridge	WHL	52	11	15	26	83	....	....	....	....	....
1995-96	Lethbridge	WHL	48	15	37	52	97	....	....	....	....	....
	Prince George	WHL	23	6	15	21	30	....	....	....	....	....
1996-97	Prince George	WHL	58	16	29	45	104	15	1	8	9	29
1997-98	Calgary Hitmen	WHL	55	26	49	75	107	....	....	....	....	....
1998-99	Tacoma Sabercats	WCHL	49	14	22	36	75	....	....	....	....	....
	Las Vegas	IHL	8	1	0	1	7	....	....	....	....	....
99-2000	Team Canada	Nat-Tm	56	10	21	31	86	....	....	....	....	....
	Kansas City	IHL	11	1	3	4	2	....	....	....	....	....
2000-01	Charlotte	ECHL	26	8	11	19	47	....	....	....	....	....
	Hartford	AHL	27	4	7	11	60	3	0	0	0	10
2001-02	Hartford	AHL	41	6	11	17	72	3	0	0	0	4

Signed as a free agent by **Hartford** (AHL), May 7, 2001.

MELENOVSKY, Marek (meh-leh-NAHF-skee, MAIR-ehk) TOR.

Center. Shoots left. 5'10", 180 lbs. Born, Humpolec, Czech., March 30, 1977.
(Toronto's 5th choice, 171st overall, in 1995 Entry Draft).

			Regular Season					Playoffs				
Season	Club	Lea	GP	G	A	TP	PIM	GP	G	A	TP	PIM
1994-95	Dukla Jihlava Jr.	Czech-Jr.	28	23	11	34	....	....	....	....	....	....
	HC Dukla Jihlava	Czech	3	0	0	0	0	5	1	3	4	0
1995-96	HC Dukla Jihlava	Czech	33	3	3	6	....	5	1	2	3	....
1996-97	HC Dukla Jihlava	Czech	46	5	13	18	22	....	....	....	....	....
	St. John's	AHL	2	1	2	3	0	2	0	0	0	0
1997-98	HC Dukla Jihlava	Czech	18	10	14	24	30	....	....	....	....	....
1998-99	HC Dukla Jihlava	Czech	52	7	10	17	26	....	....	....	....	....
	Dukla Jihlava	EuroHL	6	0	2	2	2	....	....	....	....	....
99-2000	HC Femax Havirov	Czech	49	12	12	24	32	....	....	....	....	....
2000-01	HC Femax Havirov	Czech	48	22	20	42	43	....	....	....	....	....
2001-02	HC Vitkovice	Czech	50	14	22	36	36	14	1	*11	12	31

MELIN, Bjorn (MEH-lihn, b-YUHRN) ANA.

Right wing. Shoots right. 6'1", 178 lbs. Born, Jonkoping, Sweden, July 4, 1981.
(NY Islanders' 11th choice, 163rd overall, in 1999 Entry Draft).

			Regular Season					Playoffs				
Season	**Club**	**Lea**	**GP**	**G**	**A**	**TP**	**PIM**	**GP**	**G**	**A**	**TP**	**PIM**
1997-98	HV 71 Jr.	Swede-Jr.	8	0	3	3	2	....	....	....	....	....
1998-99	HV 71 Jr.	Swede-Jr.	30	12	7	19	50	....	....	....	....	....
99-2000	HV 71 Jr.	Swede-Jr.	24	19	16	35	70	....	....	....	....	....
	HV 71 Jonkoping	Sweden	23	3	0	3	2	5	0	0	0	0
2000-01	HV 71 Jr.	Swede-Jr.	10	6	5	11	66	....	....	....	....	....
	HV 71 Jonkoping	Sweden	43	2	1	3	26	....	....	....	....	....
2001-02	HV 71 Jonkoping	Sweden	50	7	9	16	40	8	0	0	0	6

Rights traded to **Anaheim** by **NY Islanders** with Ben Guite for Dave Roche, March 19, 2002.

METCALF, Peter (MEHT-kaf, PEE-tuhr) BOS.

Defense. Shoots left. 6', 200 lbs. Born, Colorado Springs, CO, February 25, 1979.
(Toronto's 9th choice, 267th overall, in 1999 Entry Draft).

			Regular Season					Playoffs				
Season	**Club**	**Lea**	**GP**	**G**	**A**	**TP**	**PIM**	**GP**	**G**	**A**	**TP**	**PIM**
1997-98	Cushing Academy	Hi-School	25	18	48	66	....	....	....	....	....	....
1998-99	U. of Maine	H-East	33	6	17	23	34	....	....	....	....	....
99-2000	U. of Maine	H-East	40	4	17	21	56	....	....	....	....	....
2000-01	U. of Maine	H-East	31	5	9	14	44	....	....	....	....	....
2001-02	U. of Maine	H-East	44	9	41	50	66	....	....	....	....	....

Hockey East First All-Star Team (2002) • NCAA Championship All-Tournament Team (2002)

Signed as a free agent by **Boston**, August 8, 2002.

METHOT, Francois (meh-THOH, FRAN-swaw) BUF.

Center. Shoots right. 6', 203 lbs. Born, Montreal, Que., April 26, 1978.
(Buffalo's 4th choice, 54th overall, in 1996 Entry Draft).

			Regular Season					Playoffs				
Season	**Club**	**Lea**	**GP**	**G**	**A**	**TP**	**PIM**	**GP**	**G**	**A**	**TP**	**PIM**
1993-94	Mtl-Bourassa	QAAA	44	17	38	55	....	4	3	1	4	4
1994-95	St-Hyacinthe	QMJHL	60	14	38	52	22	5	0	1	1	0
1995-96	St-Hyacinthe	QMJHL	68	32	62	94	22	12	6	6	12	4
1996-97	Rouyn-Noranda	QMJHL	47	21	30	51	22	....	....	....	....	....
	Shawinigan	QMJHL	18	8	17	25	2	7	2	6	8	2
1997-98	Shawinigan	QMJHL	36	23	42	65	10	6	1	3	4	5
1998-99	Rochester	AHL	58	5	8	13	8	9	0	1	1	0
99-2000	Rochester	AHL	80	14	18	32	20	21	2	4	6	16
2000-01	Rochester	AHL	79	22	33	55	35	4	1	3	4	0
2001-02	Rochester	AHL	59	17	17	34	28	2	1	0	1	0

MEYER, Doug (MIGH-yuhr, DUHG) PIT.

Left wing. Shoots left. 6'1", 197 lbs. Born, Bloomington, MN, February 21, 1980.
(Pittsburgh's 8th choice, 176th overall, in 1999 Entry Draft).

			Regular Season					Playoffs				
Season	**Club**	**Lea**	**GP**	**G**	**A**	**TP**	**PIM**	**GP**	**G**	**A**	**TP**	**PIM**
1997-98	Team USA	USDP-18	68	21	16	37	30	....	....	....	....	....
1998-99	U. of Minnesota	WCHA	36	4	4	8	18	....	....	....	....	....
99-2000	U. of Minnesota	WCHA	26	1	3	4	18	....	....	....	....	....
2000-01	Des Moines	USHL	50	29	27	56	36	3	0	0	0	15
2001-02	St. Cloud State	WCHA	20	0	4	4	23	....	....	....	....	....

• Ruled academically ineligible to play 2000-01 season by **U. of Minnesota** (WCHA), May 1, 2000.

MIETTINEN, Antti (mih-EHT-tih-nehn, AN-tee) DAL.

Center. Shoots right. 5'11", 180 lbs. Born, Hameenlinna, Finland, July 3, 1980.
(Dallas' 10th choice, 224th overall, in 2000 Entry Draft).

			Regular Season					Playoffs				
Season	**Club**	**Lea**	**GP**	**G**	**A**	**TP**	**PIM**	**GP**	**G**	**A**	**TP**	**PIM**
1996-97	HPK-B	Finn-Jr.	36	24	29	53	34	....	....	....	....	....
1997-98	HPK-B	Finn-Jr.	34	13	28	41	63	8	1	0	1	2
	HPK Jr.	Finn-Jr.	8	1	0	1	2	....	....	....	....	....
1998-99	HPK Jr.	Finn-Jr.	35	17	22	39	28	....	....	....	....	....
	FoPS Forssa	Finland-2	4	3	1	4	6	....	....	....	....	....
	HPK Hameenlinna	Finland	13	0	0	0	2	4	0	0	0	0
99-2000	HPK Jr.	Finn-Jr.	16	11	13	24	16	....	....	....	....	....
	HPK Hameenlinna	Finland	39	2	1	3	8	7	1	0	1	0
2000-01	HPK Jr.	Finn-Jr.	4	3	10	13	2	....	....	....	....	....
	HPK Hameenlinna	Finland	55	13	11	24	20	....	....	....	....	....
2001-02	HPK Hameenlinna	Finland	56	19	37	56	50	8	2	4	6	8

MIETTINEN, Tommi (mih-EHT-tih-nehn,TAW-mee) ANA.

Center. Shoots left. 5'10", 165 lbs. Born, Kuopio, Finland, December 3, 1975.
(Anaheim's 9th choice, 236th overall, in 1994 Entry Draft).

			Regular Season					Playoffs				
Season	**Club**	**Lea**	**GP**	**G**	**A**	**TP**	**PIM**	**GP**	**G**	**A**	**TP**	**PIM**
1991-92	KalPa Kuopio Jr.	Finn-Jr.	37	9	16	25	12	....	....	....	....	....
1992-93	KalPa Kuopio-B	Finn-Jr.	7	3	8	11	2	....	....	....	....	....
	KalPa Kuopio Jr.	Finn-Jr.	26	16	27	43	14	....	....	....	....	....
	KalPa Kuopio	Finland	14	0	0	0	0	....	....	....	....	....
1993-94	KalPa Kuopio Jr.	Finn-Jr.	9	5	9	14	10	....	....	....	....	....
	KalPa Kuopio	Finland	47	5	7	12	14	....	....	....	....	....
1994-95	KalPa Kuopio Jr.	Finn-Jr.	2	1	3	4	2	....	....	....	....	....
	KalPa Kuopio	Finland	48	13	16	29	26	3	1	1	2	2
1995-96	TPS Turku	Finland	36	3	10	13	10	10	2	1	3	29
1996-97	TPS Turku	Finland	41	6	15	21	6	12	3	4	7	8
1997-98	TPS Turku	Finland	42	8	6	14	26	4	0	0	0	0
	TPS Turku	EuroHL	3	0	0	0	2	....	....	....	....	....
1998-99	TPS Turku	Finland	54	10	17	27	26	10	4	4	8	0
99-2000	Ilves Tampere	Finland	54	13	20	33	24	....	....	....	....	....
2000-01	Ilves Tampere	Finland	55	10	20	30	38	9	0	1	1	4
2001-02	Ilves Tampere	Finland	55	11	31	42	36	3	0	0	0	2

MIKHAILOV, Konstantin (mih-KHIGH-lawv, kawn-stuhn-TEEN) VAN.

Center. Shoots left. 5'11", 174 lbs. Born, Moscow, USSR, February 12, 1983.
(Vancouver's 6th choice, 245th overall, in 2001 Entry Draft).

			Regular Season					Playoffs				
Season	**Club**	**Lea**	**GP**	**G**	**A**	**TP**	**PIM**	**GP**	**G**	**A**	**TP**	**PIM**
99-2000	DynamoMoscow2	Russia-3	14	3	2	5	14	....	....	....	....	....
2000-01	Nizhnekamsk	Russia	24	0	2	2	12	....	....	....	....	....
2001-02	Nizhnekamsk	Russia	38	3	6	9	10	....	....	....	....	....

MIKHNOV, Alexei (MIHKH-nahf, al-EHX-ay) EDM.

Left wing. Shoots left. 6'5", 198 lbs. Born, Kiev, USSR, August 31, 1982.
(Edmonton's 1st choice, 17th overall, in 2000 Entry Draft).

			Regular Season					Playoffs				
Season	**Club**	**Lea**	**GP**	**G**	**A**	**TP**	**PIM**	**GP**	**G**	**A**	**TP**	**PIM**
1997-98	Yaroslavl	Russia	6	0	0	0	0	....	....	....	....	....
1998-99	Yaroslavl 2	Russia-3	14	2	2	4	4	....	....	....	....	....
99-2000	Yaroslavl 2	Russia-3	53	24	17	41	10	....	....	....	....	....
2000-01	H.C. CSKA	Russia-2	4	0	0	0	2	....	....	....	....	....
	THC Tver	Russia-2	22	5	11	16	6	....	....	....	....	....
2001-02	DynamoMoscow2	Russia-3	8	8	6	14	0	....	....	....	....	....
	Dynamo Moscow	Russia	35	2	1	3	4	3	0	0	0	2

MIKHNOV, Andrei (mihkh-NAWV, AWN-dray) ST.L.

Center. Shoots left. 6'5", 192 lbs. Born, Kiev, USSR, November 26, 1983.
(St. Louis' 2nd choice, 62nd overall, in 2002 Entry Draft).

			Regular Season					Playoffs				
Season	**Club**	**Lea**	**GP**	**G**	**A**	**TP**	**PIM**	**GP**	**G**	**A**	**TP**	**PIM**
2001-02	Sudbury Wolves	OHL	67	14	18	32	43	....	....	....	....	....

MIKKOLA, Ilkka (mih-KOHLA, IHL-ka) MTL.

Defense. Shoots left. 6', 189 lbs. Born, Oulu, Finland, January 18, 1979.
(Montreal's 3rd choice, 65th overall, in 1997 Entry Draft).

			Regular Season					Playoffs				
Season	**Club**	**Lea**	**GP**	**G**	**A**	**TP**	**PIM**	**GP**	**G**	**A**	**TP**	**PIM**
1993-94	Karpat Oulu-C	Finn-Jr.	22	3	6	9	4	4	0	1	1	4
1994-95	Karpat Oulu-C	Finn-Jr.	31	17	27	44	24	3	0	0	0	8
1995-96	Karpat Oulu-B	Finn-Jr.	4	2	0	2	0	....	....	....	....	....
	Karpat Oulu Jr.	Finn-Jr.	21	2	3	5	20	....	....	....	....	....
	Karpat Oulu	Finland-2	10	0	4	4	29	2	0	0	0	2
1996-97	Karpat Oulu Jr.	Finn-Jr.	40	7	12	19	32	6	0	0	0	4
1997-98	Karpat Oulu Jr.	Finn-Jr.	8	4	2	6	10	....	....	....	....	....
	Karpat Oulu	Finland-2	27	7	2	9	34	....	....	....	....	....
1998-99	TPS Turku	Finland	42	1	3	4	41	10	1	0	1	6
99-2000	TPS Turku	Finland	54	2	6	8	48	11	0	0	0	6
2000-01	TPS Turku	Finland	37	3	6	9	22	10	0	1	1	4
2001-02	Jokerit Helsinki	Finland	55	3	2	5	14	12	1	3	4	6

MILES, Jeff (MIGHLZ, JEHF) CHI.

Center. Shoots right. 5'11", 188 lbs. Born, Pickering, Ont., July 12, 1981.
(Chicago's 13th choice, 268th overall, in 2001 Entry Draft).

			Regular Season					Playoffs				
Season	**Club**	**Lea**	**GP**	**G**	**A**	**TP**	**PIM**	**GP**	**G**	**A**	**TP**	**PIM**
1998-99	Thunder Bay	USHL	56	11	19	30	57	3	2	0	2	2
99-2000	Thornhill	OPJHL	31	40	38	78	32	20	17	16	33	38
2000-01	U. of Vermont	ECAC	34	8	23	31	14	....	....	....	....	....
2001-02	U. of Vermont	ECAC	30	5	18	23	18	....	....	....	....	....

MILLER, Nate (MIHL-luhr, NAYT)

Right wing. Shoots left. 6'3", 192 lbs. Born, Anoka, MN, June 3, 1976.

			Regular Season					Playoffs				
Season	**Club**	**Lea**	**GP**	**G**	**A**	**TP**	**PIM**	**GP**	**G**	**A**	**TP**	**PIM**
1995-96	Twin Cities	USHL	53	25	34	59	91	....	....	....	....	....
1996-97	U. of Minnesota	WCHA	40	5	9	14	36	....	....	....	....	....
1997-98	U. of Minnesota	WCHA	39	8	6	14	42	....	....	....	....	....
1998-99	U. of Minnesota	WCHA	43	6	8	14	70	....	....	....	....	....
99-2000	U. of Minnesota	WCHA	41	16	19	35	38	....	....	....	....	....
2000-01	Lowell	AHL	80	15	10	25	61	4	0	0	0	2
2001-02	Reading Royals	ECHL	8	0	3	3	2	....	....	....	....	....
	Manchester	AHL	58	2	8	10	34	1	0	0	0	0

Signed as a free agent by **LA Kings**, August 11, 2000.

MILROY, Duncan (MIHL-roi, DUHN-can) MTL.

Right wing. Shoots right. 6', 197 lbs. Born, Edmonton, Alta., February 8, 1983.
(Montreal's 3rd choice, 37th overall, in 2001 Entry Draft).

			Regular Season					Playoffs				
Season	**Club**	**Lea**	**GP**	**G**	**A**	**TP**	**PIM**	**GP**	**G**	**A**	**TP**	**PIM**
1998-99	Edm. Leafs	AMHL	34	34	36	70	73	....	....	....	....	....
	Swift Current	WHL	3	0	0	0	0	....	....	....	....	....
99-2000	Swift Current	WHL	68	15	15	30	20	12	3	5	8	12
2000-01	Swift Current	WHL	68	38	54	92	51	19	9	12	21	6
2001-02	Swift Current	WHL	26	20	11	31	20	....	....	....	....	....
	Kootenay Ice	WHL	38	25	31	56	24	22	*17	*20	*37	26

Traded to **Kootenay** (WHL) by **Swift Current** (WHL) with B.J. Boxma and future considerations for Jason Jaffray, Aaron Rome, Jeff Harvey and future considerations, January 4, 2002.

MINAKOV, Oleg (mih-nah-KAHV, OH-lehg) CHI.

Right wing. Shoots left. 6'1", 198 lbs. Born, Elektrostal, USSR, February 18, 1983.
(Chicago's 12th choice, 216th overall, in 2001 Entry Draft).

			Regular Season					Playoffs				
Season	**Club**	**Lea**	**GP**	**G**	**A**	**TP**	**PIM**	**GP**	**G**	**A**	**TP**	**PIM**
99-2000	Elektrostal 2	Russia-3	19	10	2	12	16	....	....	....	....	....
	Elektrostal	Russia	4	0	0	0	0	....	....	....	....	....
2000-01	Elektrostal	Russia-2	25	0	3	3	6	....	....	....	....	....
2001-02	Elektrostal 2	Russia-3	26	9	7	16	14	....	....	....	....	....
	Elektrostal	Russia-2	16	0	1	1	14	....	....	....	....	....

MINK, Graham (MIHNK, GRAY-uhm) WSH.

Left wing. Shoots right. 6'3", 210 lbs. Born, Stowe, VT, May 12, 1979.

			Regular Season					Playoffs				
Season	**Club**	**Lea**	**GP**	**G**	**A**	**TP**	**PIM**	**GP**	**G**	**A**	**TP**	**PIM**
1997-98	Mount Hermon	Hi-School	25	17	25	42	....	....	....	....	....	....
1998-99	U. of Vermont	ECAC	27	4	2	6	34	....	....	....	....	....
99-2000	U. of Vermont	ECAC	17	7	4	11	14	....	....	....	....	....
2000-01	U. of Vermont	ECAC	32	17	12	29	52	....	....	....	....	....
2001-02	Richmond	ECHL	29	8	9	17	78	....	....	....	....	....
	Portland Pirates	AHL	56	17	17	34	50	....	....	....	....	....

Signed as a free agent by **Portland** (AHL), September 30, 2001. Signed as a free agent by **Washington**, April 9, 2002.

MISCHLER, Greg (MIH-schluhr, GREHG) S.J.

Center. Shoots left. 6'3", 174 lbs. Born, Holbrook, NY, September 15, 1978.
(Vancouver's 10th choice, 204th overall, in 1998 Entry Draft).

			Regular Season					Playoffs				
Season	**Club**	**Lea**	**GP**	**G**	**A**	**TP**	**PIM**	**GP**	**G**	**A**	**TP**	**PIM**
1997-98	Northeastern	H-East	39	7	13	20	22	....	....	....	....	....
1998-99	Northeastern	H-East	33	8	15	23	36	....	....	....	....	....
99-2000	Northeastern	H-East	34	9	14	23	22	....	....	....	....	....
2000-01	Northeastern	H-East	36	10	*32	42	34	....	....	....	....	....
2001-02	Cleveland Barons	AHL	65	10	14	24	71	....	....	....	....	....

Signed as a free agent by **San Jose**, June 13, 2001.

MOEN, Travis (MOH-ehn, TRA-vihs) CGY.

Left wing. Shoots left. 6'2", 198 lbs. Born, Swift Current, Sask., April 6, 1982.
(Calgary's 6th choice, 155th overall, in 2000 Entry Draft).

			Regular Season					Playoffs				
Season	**Club**	**Lea**	**GP**	**G**	**A**	**TP**	**PIM**	**GP**	**G**	**A**	**TP**	**PIM**
1998-99	Swift Current	SMHL	STATISTICS NOT AVAILABLE									
	Kelowna Rockets	WHL	4	0	0	0	0	....	....	....	....	....
99-2000	Kelowna Rockets	WHL	66	9	6	15	96	5	1	1	2	2
2000-01	Kelowna Rockets	WHL	40	8	8	16	106	....	....	....	....	....
2001-02	Kelowna Rockets	WHL	71	10	17	27	197	13	1	0	1	28

MOJZIS, Tomas (MOI-shihsh, TAW-mash) TOR.

Defense. Shoots left. 6'1", 186 lbs. Born, Kolin, Czech., May 2, 1982.
(Toronto's 11th choice, 246th overall, in 2001 Entry Draft).

			Regular Season					Playoffs				
Season	**Club**	**Lea**	**GP**	**G**	**A**	**TP**	**PIM**	**GP**	**G**	**A**	**TP**	**PIM**
99-2000	Pardubice Jr.	Czech-Jr.	40	7	1	8	....	....	....	....	....	....
2000-01	Moose Jaw	WHL	72	11	25	36	115	4	0	1	1	8
2001-02	Moose Jaw	WHL	28	2	11	13	43	....	....	....	....	....
	Seattle	WHL	36	8	15	23	66	11	1	3	4	20

Traded to **Seattle** (WHL) by **Moose Jaw** (WHL) with Brooks Laich for Craig Olynick, Stanislav Avksentiev and Seattle's 3rd round choice (Tanner Gillies) in 2002 WHL Bantam Draft, November 20, 2001.

MOKHOV, Stepan (MOH-khohv, STEH-pan) CHI.

Defense. Shoots left. 6'2", 190 lbs. Born, Ust-Kamenogorsk, USSR, January 22, 1981.
(Chicago's 3rd choice, 63rd overall, in 1999 Entry Draft).

			Regular Season					Playoffs				
Season	**Club**	**Lea**	**GP**	**G**	**A**	**TP**	**PIM**	**GP**	**G**	**A**	**TP**	**PIM**
1997-98	Omsk 2	Russia-3	18	0	1	1	8	....	....	....	....	....
1998-99	Cherepovets 2	Russia-3	25	2	0	2	34	....	....	....	....	....
	Cherepovets 3	Russia-4	3	0	0	0	6	....	....	....	....	....
	Cherepovets	Russia	1	0	0	0	0	....	....	....	....	....
99-2000	Magnitogorsk	Russia	18	0	1	1	6	14	1	0	1	4
2000-01	Krylja Sovetov	Russia-2	30	0	5	5	4	....	....	....	....	....
2001-02	Spartak Moscow 2	Russia-3	16	2	9	11	16	....	....	....	....	....
	Spartak Moscow	Russia	11	1	0	1	2	....	....	....	....	....

MONYCH, Lance (MOH-nihch, LANTS) PHX.

Right wing. Shoots right. 6'2", 199 lbs. Born, Red Deer, Alta., July 25, 1984.
(Phoenix's 6th choice, 97th overall, in 2002 Entry Draft).

			Regular Season					Playoffs				
Season	**Club**	**Lea**	**GP**	**G**	**A**	**TP**	**PIM**	**GP**	**G**	**A**	**TP**	**PIM**
99-2000	Brandon Hawks	MBHL	30	32	34	66	98	....	....	....	....	....
	Brandon	WHL	3	0	0	0	0	....	....	....	....	....
2000-01	Brandon	WHL	53	14	8	22	34	6	1	0	1	0
2001-02	Brandon	WHL	71	18	30	48	96	19	4	3	7	20

MOORE, Dominic (MOOR, DOHM-ih-nihk) NYR

Center. Shoots left. 6', 180 lbs. Born, Thornhill, Ont., August 3, 1980.
(NY Rangers' 2nd choice, 95th overall, in 2000 Entry Draft).

			Regular Season					Playoffs				
Season	**Club**	**Lea**	**GP**	**G**	**A**	**TP**	**PIM**	**GP**	**G**	**A**	**TP**	**PIM**
1996-97	Thornhill	MTJHL	29	4	6	10	48	1	0	1	1	0
1997-98	Aurora Tigers	OPJHL	51	10	15	25	16	....	....	....	....	....
1998-99	Aurora Tigers	OPJHL	51	34	53	87	70	....	....	....	....	....
99-2000	Harvard Crimson	ECAC	30	12	24	28	16	....	....	....	....	....
2000-01	Harvard Crimson	ECAC	32	15	28	43	40	....	....	....	....	....
2001-02	Harvard Crimson	ECAC	32	13	16	29	37	....	....	....	....	....

ECAC All-Rookie Team (2000) • ECAC Second All-Star Team (2001)

MORGAN, Gavin (MOHR-guhn, GA-vign) DAL.

Center. Shoots right. 5'11", 191 lbs. Born, Scarborough, Ont., July 9, 1976.

			Regular Season					Playoffs				
Season	**Club**	**Lea**	**GP**	**G**	**A**	**TP**	**PIM**	**GP**	**G**	**A**	**TP**	**PIM**
1992-93	Wexford Raiders	MTJHL	3	0	1	1	0	....	....	....	....	....
1993-94	Wexford Raiders	MTJHL	49	18	32	50	91	....	....	....	....	....
1994-95	Wexford Raiders	MTJHL	49	26	39	65	170	....	....	....	....	....
1995-96	U. of Denver	WCHA	28	2	9	11	47	....	....	....	....	....
1996-97	U. of Denver	WCHA	41	8	15	23	46	....	....	....	....	....
1997-98	U. of Denver	WCHA	37	9	8	17	42	....	....	....	....	....
1998-99	U. of Denver	WCHA	40	13	16	29	85	....	....	....	....	....
99-2000	Idaho Steelheads	WCHL	54	17	33	50	150	3	0	3	3	4
	Long Beach	IHL	7	0	1	1	10	....	....	....	....	....
	Utah Grizzlies	IHL	10	0	2	2	4	2	1	0	1	2
2000-01	Utah Grizzlies	IHL	79	7	14	21	187	....	....	....	....	....
2001-02	Utah Grizzlies	AHL	76	8	24	32	249	5	0	1	1	2

Signed as a free agent by **Idaho** (WCHL), August 25, 1999. Signed as a free agent by **Utah** (IHL), June 26, 2000. Signed as a free agent by **Dallas**, July 17, 2001.

MORRIS, Mike (MOHR-his, MIGHK) S.J.

Right wing. Shoots right. 6', 182 lbs. Born, Dorchester, MA, July 14, 1983.
(San Jose's 1st choice, 27th overall, in 2002 Entry Draft).

			Regular Season					Playoffs				
Season	**Club**	**Lea**	**GP**	**G**	**A**	**TP**	**PIM**	**GP**	**G**	**A**	**TP**	**PIM**
2000-01	St. Sebastian's	Hi-School	28	20	28	48	18	....	....	....	....	....
2001-02	St. Sebastian's	Hi-School	31	29	29	58	26	....	....	....	....	....

• Signed Letter of Intent to attend **Northeastern University** (H-East), May 7, 2001 • Massachusetts Independent League MVP (2002)

MORRISON, Justin (MOHR-rihs-OHN, JUHS-tihn) VAN.

Right wing. Shoots right. 6'3", 205 lbs. Born, Los Angeles, CA, September 10, 1979.
(Vancouver's 4th choice, 81st overall, in 1998 Entry Draft).

			Regular Season					Playoffs				
Season	**Club**	**Lea**	**GP**	**G**	**A**	**TP**	**PIM**	**GP**	**G**	**A**	**TP**	**PIM**
1996-97	Omaha Lancers	USHL	62	12	24	36	44	10	2	4	6	8
1997-98	Colorado College	WCHA	42	4	9	13	8	....	....	....	....	....
1998-99	Colorado College	WCHA	38	23	15	38	33	....	....	....	....	....
99-2000	Colorado College	WCHA	38	7	19	26	28	....	....	....	....	....
2000-01	Colorado College	WCHA	41	21	14	35	42	....	....	....	....	....
2001-02	Manitoba Moose	AHL	64	10	9	19	37	7	1	0	1	4

MORRISONN, Shaone (MOHR-rih-sohn, SHAWN) BOS.

Defense. Shoots left. 6'3", 205 lbs. Born, Vancouver, B.C., December 23, 1982.
(Boston's 1st choice, 19th overall, in 2001 Entry Draft).

			Regular Season					Playoffs				
Season	**Club**	**Lea**	**GP**	**G**	**A**	**TP**	**PIM**	**GP**	**G**	**A**	**TP**	**PIM**
1997-98	Vancouver T-Birds	BCAHA	45	16	44	60	75	....	....	....	....	....
1998-99	South Surrey	BCHL	19	0	2	2	13	....	....	....	....	....
99-2000	Kamloops Blazers	WHL	57	1	6	7	80	4	0	0	0	6
2000-01	Kamloops Blazers	WHL	61	13	25	38	132	4	0	0	0	6
2001-02	Kamloops Blazers	WHL	61	11	26	37	106	4	0	2	2	2

MORROW, Josh (MOHR-oh, JAWSH) NSH.

Defense. Shoots left. 6'1", 200 lbs. Born, Edmonton, Alta., June 12, 1983.
(Nashville's choice, 203rd overall, in 2002 Entry Draft).

			Regular Season					Playoffs				
Season	**Club**	**Lea**	**GP**	**G**	**A**	**TP**	**PIM**	**GP**	**G**	**A**	**TP**	**PIM**
1998-99	Ft-Saskatchewan	AJHL	STATISTICS NOT AVAILABLE									
	Medicine Hat	WHL	10	0	1	1	2	....	....	....	....	....
99-2000	Medicine Hat	WHL	49	1	8	9	57	....	....	....	....	....
2000-01	Medicine Hat	WHL	46	0	7	7	111	....	....	....	....	....
2001-02	Medicine Hat	WHL	46	4	15	19	137	....	....	....	....	....
	Tri-City	WHL	30	6	12	18	62	....	....	....	....	....

Traded to **Tri-City** (WHL) by **Medicine Hat** (WHL) with Medicine Hat's 2nd round choice (Dave Falk) in 2002 WHL Bantam Draft for Sean Curry, January 10, 2002.

MOSOVSKY, Karel (moh-SAWV-skee, KA-rehl) BUF.

Left wing. Shoots right. 6'2", 198 lbs. Born, Piesk, Czech., August 22, 1981.
(Buffalo's 6th choice, 117th overall, in 1999 Entry Draft).

			Regular Season					Playoffs				
Season	**Club**	**Lea**	**GP**	**G**	**A**	**TP**	**PIM**	**GP**	**G**	**A**	**TP**	**PIM**
1997-98	C. Budejovice Jr.	Czech-Jr.	36	15	17	32	52	....	....	....	....	....
1998-99	Regina Pats	WHL	68	26	25	51	58	....	....	....	....	....
99-2000	Regina Pats	WHL	56	24	34	58	80	7	3	1	4	12
2000-01	Regina Pats	WHL	61	25	26	51	59	6	1	3	4	8
2001-02	Rochester	AHL	5	1	1	2	0	....	....	....	....	....

• Missed majority of 2001-02 season recovering from shoulder injury suffered in training camp, September 9, 2001.

MOSS, David (MAWS, DAY-vihd) CGY.

Left wing. Shoots left. 6'3", 185 lbs. Born, Dearborn, MI, December 28, 1981.
(Calgary's 9th choice, 220th overall, in 2001 Entry Draft).

			Regular Season					Playoffs				
Season	**Club**	**Lea**	**GP**	**G**	**A**	**TP**	**PIM**	**GP**	**G**	**A**	**TP**	**PIM**
99-2000	Catholic Central	Hi-School	28	18	20	28	20	....	....	....	....	....
2000-01	St. Louis Sting	NAJHL	9	2	2	4	2	....	....	....	....	....
	Cedar Rapids	USHL	51	20	18	38	14	4	0	1	1	2
2001-02	U. of Michigan	CCHA	43	4	9	13	10	....	....	....	....	....

MRAZEK, Frantisek (muh-RA-zehk, FRAN-tih-SEHK) TOR.

Left wing. Shoots left. 6'4", 220 lbs. Born, Ceske Budejovice, Czech., May 16, 1979.
(Toronto's 3rd choice, 111th overall, in 1997 Entry Draft).

			Regular Season					Playoffs				
Season	**Club**	**Lea**	**GP**	**G**	**A**	**TP**	**PIM**	**GP**	**G**	**A**	**TP**	**PIM**
1995-96	C. Budejovice Jr.	Czech-Jr.	19	8	3	11	....	....	....	....	....	....
1996-97	C. Budejovice Jr.	Czech-Jr.	40	18	15	33	....	....	....	....	....	....
1997-98	Red Deer Rebels	WHL	65	30	24	54	71	5	1	0	1	2
1998-99	Red Deer Rebels	WHL	60	34	42	76	79	9	6	4	10	16
99-2000	St. John's	AHL	1	0	0	0	0	....	....	....	....	....
2000-01	St. John's	AHL	51	4	9	13	24	....	....	....	....	....
2001-02	Ceske Budejovice	Czech	17	2	3	5	24	....	....	....	....	....
	St. John's	AHL	16	0	4	4	12	....	....	....	....	....

• Missed majority of 1999-2000 season recovering from knee injury suffered in training camp, October 1, 1999.

MROZIK, Rick (muh-ROH-zihk, RIHK)

Defense. Shoots left. 6'2", 185 lbs. Born, Duluth, MN, January 2, 1975.
(Dallas' 4th choice, 136th overall, in 1993 Entry Draft).

			Regular Season					Playoffs				
Season	**Club**	**Lea**	**GP**	**G**	**A**	**TP**	**PIM**	**GP**	**G**	**A**	**TP**	**PIM**
1992-93	Cloquet High	Hi-School	28	9	38	47	12	....	....	....	....	....
1993-94	U. Minn-Duluth	WCHA	38	2	9	11	38	....	....	....	....	....
1994-95	U. Minn-Duluth	WCHA	3	0	0	0	2	....	....	....	....	....
1995-96	U. Minn-Duluth	WCHA	35	3	19	22	63	....	....	....	....	....
1996-97	U. Minn-Duluth	WCHA	38	11	23	34	56	....	....	....	....	....
1997-98	Portland Pirates	AHL	75	2	15	17	52	10	1	3	4	2
1998-99	Portland Pirates	AHL	70	4	8	12	63	....	....	....	....	....
99-2000	Worcester	AHL	3	0	0	0	0	....	....	....	....	....
	Pee Dee Pride	ECHL	60	9	19	28	44	5	2	0	2	6
	Syracuse Crunch	AHL	1	0	0	0	0	....	....	....	....	....
2000-01	Saint John	AHL	76	5	11	16	26	19	1	1	2	6
2001-02	Saint John	AHL	55	2	5	7	27	....	....	....	....	....

WCHA Second All-Star Team (1997)

Traded to **Washington** by **Dallas** with Mark Tinordi for Kevin Hatcher, January 18, 1995. Signed as a free agent by **Calgary**, August 6, 2001.

MULICK, Robert (muhl-LIHK, RAW-buhrt) S.J.

Defense. Shoots right. 6'2", 210 lbs. Born, Toronto, Ont., October 23, 1979.
(San Jose's 8th choice, 185th overall, in 1998 Entry Draft).

			Regular Season					Playoffs				
Season	**Club**	**Lea**	**GP**	**G**	**A**	**TP**	**PIM**	**GP**	**G**	**A**	**TP**	**PIM**
1994-95	Mississauga Reps	MTHL	52	8	23	31	80	....	....	....	....	....
1995-96	Sault Ste. Marie	OHL	54	0	3	3	58	3	0	0	0	0
1996-97	Sault Ste. Marie	OHL	60	2	8	10	49	11	0	1	1	12
1997-98	Sault Ste. Marie	OHL	61	0	10	10	109	....	....	....	....	....
1998-99	Sault Ste. Marie	OHL	66	1	12	13	83	5	0	1	1	10
99-2000	Kentucky	AHL	52	0	0	0	52	9	0	0	0	10
2000-01	Kentucky	AHL	71	0	6	6	55	2	0	0	0	0
2001-02	Cleveland Barons	AHL	60	0	2	2	77	....	....	....	....	....

MURATOV, Yevgeny (muhr-A-tahf, ehv-GEH-nee) EDM.

Left wing. Shoots right. 5'10", 178 lbs. Born, Nizhny Tagil, USSR, January 28, 1981.
(Edmonton's 10th choice, 274th overall, in 2000 Entry Draft).

			Regular Season					Playoffs				
Season	Club	Lea	GP	G	A	TP	PIM	GP	G	A	TP	PIM
1997-98	Nizhnekamsk 2	Russia-3	39	7	7	14	2	....	....	....	....	....
1998-99	Nizhnekamsk 2	Russia-4	37	26	9	35	32	....	....	....	....	....
	Nizhnekamsk	Russia	4	0	0	0	0	3	1	0	1	2
99-2000	Nizhnekamsk	Russia	29	9	7	16	2	....	....	....	....	....
	Ak Bars Kazan	Russia	8	2	2	4	2	9	0	0	0	2
2000-01	Nizhnekamsk	Russia	42	9	8	17	14	4	0	0	0	0
2001-02	Nizhnekamsk	Russia	45	5	13	18	4	....	....	....	....	....

MURLEY, Matt (MUHR-lee, MAT) PIT.

Left wing. Shoots left. 6'1", 192 lbs. Born, Troy, NY, December 17, 1979.
(Pittsburgh's 2nd choice, 51st overall, in 1999 Entry Draft).

			Regular Season					Playoffs				
Season	Club	Lea	GP	G	A	TP	PIM	GP	G	A	TP	PIM
1996-97	Syracuse	MTJHL	48	52	58	110	111	....	....	....	....	....
1997-98	Syracuse	MTJHL	49	56	70	126	103	....	....	....	....	....
1998-99	RPI Engineers	ECAC	36	17	32	49	32	....	....	....	....	....
99-2000	RPI Engineers	ECAC	35	9	29	38	42	....	....	....	....	....
2000-01	RPI Engineers	ECAC	34	*24	18	42	34	....	....	....	....	....
2001-02	RPI Engineers	ECAC	32	*24	22	46	26	....	....	....	....	....

ECAC First All-Star Team (2002)

MURPHY, Curtis (MUHR-fee, KUHR-this) MIN.

Defense. Shoots right. 5'8", 185 lbs. Born, Kerrobert, Sask., December 3, 1975.

			Regular Season					Playoffs				
Season	Club	Lea	GP	G	A	TP	PIM	GP	G	A	TP	PIM
1993-94	Nipawin Hawks	SJHL	60	21	33	54	....	....	....	....	....	....
1994-95	North Dakota	WCHA	33	6	10	16	28	....	....	....	....	....
1995-96	North Dakota	WCHA	38	6	12	18	58	....	....	....	....	....
1996-97	North Dakota	WCHA	43	12	30	42	36	....	....	....	....	....
1997-98	North Dakota	WCHA	39	8	34	42	78	....	....	....	....	....
1998-99	Orlando	IHL	80	22	35	57	60	17	4	5	9	16
99-2000	Orlando	IHL	81	8	43	51	59	6	0	2	2	6
2000-01	Orlando	IHL	51	19	30	49	55	10	2	9	11	12
2001-02	Houston Aeros	AHL	80	12	35	47	75	14	2	4	6	10

WCHA First All-Star Team (1997, 1998) • NCAA West Second All-American Team (1997) • WCHA Player of the Year (1998) • WCHA All-Tournament Team (1998) • NCAA West First All-American Team (1998) • IHL First All-Star Team (2001)

Signed as a free agent by **Minnesota**, June 18, 2001.

MURPHY, Joe (MUHR-fee, JOH) OTT.

Right wing. Shoots right. 6', 200 lbs. Born, Didsbury, Alta., January 21, 1975.

			Regular Season					Playoffs				
Season	Club	Lea	GP	G	A	TP	PIM	GP	G	A	TP	PIM
1993-94	Olds Grizzlies	AJHL	52	34	38	72	121	....	....	....	....	....
1994-95	Olds Grizzlies	AJHL	52	30	29	59	65	....	....	....	....	....
1995-96	U. of Denver	WCHA	29	1	6	7	16	....	....	....	....	....
1996-97	U. of Denver	WCHA	36	8	14	22	40	....	....	....	....	....
1997-98	U. of Denver	WCHA	38	7	14	21	60	....	....	....	....	....
1998-99	U. of Denver	WCHA	33	3	13	16	25	....	....	....	....	....
	Huntsville	CHL	....	....	....	....	....	13	1	5	6	20
99-2000	Roanoke Express	ECHL	34	12	19	31	34	....	....	....	....	....
	Rochester	AHL	32	0	6	6	8	21	4	2	6	27
2000-01	Rochester	AHL	74	20	15	35	43	4	0	1	1	4
2001-02	Grand Rapids	AHL	57	8	14	22	24	5	1	0	1	0

Signed as a free agent by **Buffalo**, July 28, 1998. Signed as a free agent by **Ottawa**, August 1, 2001.

MURPHY, Mark (MUHR-fee, MAHRK) WSH.

Left wing. Shoots left. 5'11", 200 lbs. Born, Stoughton, MA, August 6, 1976.
(Toronto's 6th choice, 197th overall, in 1995 Entry Draft).

			Regular Season					Playoffs				
Season	Club	Lea	GP	G	A	TP	PIM	GP	G	A	TP	PIM
1994-95	Stratford	OJHL-B	47	52	56	108	64	....	....	....	....	....
1995-96	Stratford	OJHL-B	1	0	0	0	0	....	....	....	....	....
	RPI Engineers	ECAC	32	1	1	2	50	....	....	....	....	....
1996-97	RPI Engineers	ECAC	34	9	18	27	56	....	....	....	....	....
1997-98	RPI Engineers	ECAC	35	8	27	35	63	....	....	....	....	....
1998-99	RPI Engineers	ECAC	37	11	30	41	76	....	....	....	....	....
99-2000	Wilkes-Barre	AHL	38	11	22	33	35	....	....	....	....	....
	Trenton Titans	ECHL	37	21	18	39	60	12	2	8	10	17
	Philadelphia	AHL	....	....	....	....	....	2	0	0	0	0
2000-01	Portland Pirates	AHL	76	29	41	70	92	3	2	0	2	2
2001-02	Portland Pirates	AHL	77	20	37	57	56	....	....	....	....	....

Signed as a free agent by **Washington**, July 13, 2000.

MURPHY, Ryan (MUHR-fee, RIGH-yan) CAR.

Left wing. Shoots left. 6'1", 192 lbs. Born, Van Nuys, CA, March 21, 1979.
(Carolina's 4th choice, 113th overall, in 1999 Entry Draft).

			Regular Season					Playoffs				
Season	Club	Lea	GP	G	A	TP	PIM	GP	G	A	TP	PIM
1995-96	Thornhill	MTJHL	32	13	16	29	49	1	0	0	0	0
1996-97	Thornhill	MTJHL	41	22	32	54	36	12	7	8	15	....
1997-98	Bowling Green	CCHA	36	3	9	12	27	....	....	....	....	....
1998-99	Bowling Green	CCHA	34	10	23	33	38	....	....	....	....	....
99-2000	Bowling Green	CCHA	36	9	10	19	63	....	....	....	....	....
2000-01	Bowling Green	CCHA	38	23	15	38	22	....	....	....	....	....
2001-02	Florida	ECHL	66	13	18	31	38	6	1	2	3	4

MURRAY, Andrew (MUH-ree, AN-droo) CBJ

Center. Shoots left. 6'2", 210 lbs. Born, Selkirk, Man., November 6, 1981.
(Columbus' 11th choice, 242nd overall, in 2001 Entry Draft).

			Regular Season					Playoffs				
Season	Club	Lea	GP	G	A	TP	PIM	GP	G	A	TP	PIM
2000-01	Selkirk Steelers	MJHL	64	46	56	102	72	5	3	0	3	6
2001-02	Bemidji State	CHA	35	15	15	30	22	....	....	....	....	....

MJHL First All-Star Team (2001) • CHA All-Rookie Team (2002)

MURRAY, Craig (MUH-ree, KRAYG) MTL.

Center. Shoots left. 6', 175 lbs. Born, Souris, Man., February 22, 1979.
(Montreal's 8th choice, 201st overall, in 1998 Entry Draft).

			Regular Season					Playoffs				
Season	Club	Lea	GP	G	A	TP	PIM	GP	G	A	TP	PIM
1994-95	Penticton	BCAHA	60	74	76	150	42	....	....	....	....	....
1995-96	Penticton	BCJHL	55	14	6	20	54	....	....	....	....	....
1996-97	Penticton	BCHL	45	25	31	56	39	....	....	....	....	....
1997-98	Penticton	BCHL	57	45	52	97	50	7	6	8	14	14
1998-99	U. of Michigan	CCHA	16	0	1	1	6	....	....	....	....	....
99-2000	U. of Michigan	CCHA	23	2	1	3	12	....	....	....	....	....
2000-01	U. of Michigan	CCHA	42	10	7	17	32	....	....	....	....	....
2001-02	U. of Michigan	CCHA	39	7	7	14	10	....	....	....	....	....

MURRAY, Doug (MUH-ree, DUHG) S.J.

Defense. Shoots left. 6'3", 220 lbs. Born, Bromma, Sweden, March 12, 1980.
(San Jose's 6th choice, 241st overall, in 1999 Entry Draft).

			Regular Season					Playoffs				
Season	Club	Lea	GP	G	A	TP	PIM	GP	G	A	TP	PIM
1998-99	NY Apple Core	MJBHL	60	17	47	64	62	....	....	....	....	....
99-2000	Cornell Big Red	ECAC	32	3	6	9	38	....	....	....	....	....
2000-01	Cornell Big Red	ECAC	25	5	13	18	39	....	....	....	....	....
2001-02	Cornell Big Red	ECAC	35	11	21	32	67	....	....	....	....	....

ECAC First All-Star Team (2002)

MURRAY, Garth (MUH-ree, GARTH) NYR

Center. Shoots left. 6'1", 205 lbs. Born, Regina, Sask., September 17, 1982.
(NY Rangers' 3rd choice, 79th overall, in 2001 Entry Draft).

			Regular Season					Playoffs				
Season	Club	Lea	GP	G	A	TP	PIM	GP	G	A	TP	PIM
1997-98	Cgy. Buffaloes	AMHL	56	26	34	60	110	....	....	....	....	....
	Regina Pats	WHL	4	0	0	0	2	2	0	0	0	0
1998-99	Regina Pats	WHL	60	3	5	8	101	....	....	....	....	....
99-2000	Regina Pats	WHL	68	14	26	40	155	7	1	1	2	7
2000-01	Regina Pats	WHL	72	28	16	44	183	6	1	1	2	10
2001-02	Regina Pats	WHL	62	33	30	63	154	6	2	3	5	9
	Hartford	AHL	4	0	0	0	0	9	1	3	4	6

NASH, Rick (NASH, RIHK) CBJ

Left wing. Shoots left. 6'3", 188 lbs. Born, Brampton, Ont., June 16, 1984.
(Columbus' 1st choice, 1st overall, in 2002 Entry Draft).

			Regular Season					Playoffs				
Season	Club	Lea	GP	G	A	TP	PIM	GP	G	A	TP	PIM
99-2000	Tor. Marlboros	GTHL	34	61	54	115	34	....	....	....	....	....
2000-01	London Knights	OHL	58	31	35	66	56	4	3	3	6	8
2001-02	London Knights	OHL	54	32	40	72	88	12	10	9	19	21

OHL All-Rookie Team (2001) • OHL Rookie of the Year (2001) • CHL All-Rookie Team (2001)

NAUMENKO, Nick (NAH-mehn-koh, NIHK)

Defense. Shoots right. 5'11", 185 lbs. Born, Chicago, IL, July 7, 1974.
(St. Louis' 9th choice, 182nd overall, in 1992 Entry Draft).

			Regular Season					Playoffs				
Season	Club	Lea	GP	G	A	TP	PIM	GP	G	A	TP	PIM
1991-92	Dubuque	USHL	24	6	19	25	4	....	....	....	....	....
1992-93	North Dakota	WCHA	38	10	24	34	26	....	....	....	....	....
1993-94	North Dakota	WCHA	32	4	22	26	22	....	....	....	....	....
1994-95	North Dakota	WCHA	39	13	26	39	78	....	....	....	....	....
1995-96	North Dakota	WCHA	37	11	30	41	32	....	....	....	....	....
1996-97	Worcester	AHL	54	6	22	28	72	1	0	0	0	0
1997-98	Worcester	AHL	71	12	34	46	63	11	1	7	8	8
1998-99	Utah Grizzlies	IHL	20	4	3	7	20	....	....	....	....	....
	Las Vegas	IHL	34	5	16	21	37	....	....	....	....	....
	Kansas City	IHL	21	3	8	11	4	3	1	2	3	4
99-2000	Kansas City	IHL	54	9	27	36	79	....	....	....	....	....
2000-01	Cleveland	IHL	77	5	45	50	60	4	0	0	0	4
2001-02	Portland Pirates	AHL	75	15	35	50	40	....	....	....	....	....

WCHA First All-Star Team (1995, 1996)

Signed as a free agent by **Minnesota**, September 6, 2000. Signed as a free agent by **Mannheim** (Germany), April 23, 2002.

NEHRLING, Lucas (NEHR-lihng, LEW-cas) N.J.

Defense. Shoots right. 6'5", 225 lbs. Born, Peterborough, Ont., August 14, 1979.
(New Jersey's 3rd choice, 104th overall, in 1997 Entry Draft).

			Regular Season					Playoffs				
Season	Club	Lea	GP	G	A	TP	PIM	GP	G	A	TP	PIM
1995-96	Quinte Hawks	MTJHL	25	0	9	9	59	....	....	....	....	....
1996-97	Sarnia Sting	OHL	63	3	12	15	74	12	0	2	2	23
1997-98	Sarnia Sting	OHL	22	0	2	2	46	....	....	....	....	....
	Kingston	OHL	39	1	8	9	83	12	0	1	1	19
1998-99	Kingston	OHL	2	0	1	1	13	....	....	....	....	....
	Guelph Storm	OHL	60	5	14	19	131	11	0	1	1	35
99-2000	Augusta Lynx	ECHL	9	1	0	1	12	....	....	....	....	....
	Arkansas	ECHL	16	0	1	1	51	....	....	....	....	....
	Muskegon Fury	UHL	22	1	2	3	86	2	0	0	0	2
	Albany	AHL	18	0	0	0	27	4	0	0	0	0
2000-01	Albany	AHL	33	1	1	2	92	....	....	....	....	....
	Adirondack	UHL	23	1	4	5	99	5	0	1	1	19
2001-02	Albany	AHL	29	0	0	0	43	....	....	....	....	....
	Adirondack	UHL	24	1	2	3	189	1	0	0	0	0

Traded to **Kingston** (OHL) by **Sarnia** (OHL) with Lucas Nehrling for Matt Price and Curtis Cruickshank, November 26, 1998.

NEMEC, Ondrej (NEH-mehts, AWN-dray) PIT.

Defense. Shoots right. 5'11", 194 lbs. Born, Trebic, Czech., April 18, 1984.
(Pittsburgh's 2nd choice, 35th overall, in 2002 Entry Draft).

			Regular Season					Playoffs				
Season	Club	Lea	GP	G	A	TP	PIM	GP	G	A	TP	PIM
99-2000	Vsetin Jr.	Czech-Jr.	5	0	2	2	4	....	....	....	....	....
	Vsetin-18	Czech-Jr.	48	6	18	24	62	....	....	....	....	....
2000-01	Vsetin Jr.	Czech-Jr.	42	10	12	22	77	....	....	....	....	....
	Vsetin-18	Czech-Jr.	8	3	7	10	20	....	....	....	....	....
2001-02	HC Vsetin Jr.	Czech-Jr.	8	4	7	11	47	....	....	....	....	....
	Trebic	Czech-2	9	1	0	1	18	....	....	....	....	....
	HC Vsetin	Czech	44	5	3	8	79	....	....	....	....	....

NEPRYAYEV, Ivan (neh-pree-YIGH-ehv, IGH-van) WSH.

Center. Shoots left. 6'1", 178 lbs. Born, Yaroslavl, USSR, February 4, 1982.
(Washington's 5th choice, 163rd overall, in 2000 Entry Draft).

			Regular Season					Playoffs				
Season	Club	Lea	GP	G	A	TP	PIM	GP	G	A	TP	PIM
1997-98	Yaroslavl	Russia	6	0	0	0	0					
1998-99	Yaroslavl 2	Russia-3	15	1	0	1	0					
99-2000	Yaroslavl 2	Russia-3	40	8	14	22	....					
2000-01	Yaroslavl	Russia	10	0	0	0	2					
2001-02	Yaroslavl 2	Russia-3	2	1	0	1	18					
	Yaroslavl	Russia	36	3	8	11	28					

NEWBURY, Kris (new-BUHR-ee, KRIHS) S.J.

Left wing. Shoots left. 5'10", 197 lbs. Born, Brampton, Ont., February 19, 1982.
(San Jose's 4th choice, 139th overall, in 2002 Entry Draft).

			Regular Season					Playoffs				
Season	Club	Lea	GP	G	A	TP	PIM	GP	G	A	TP	PIM
1996-97	Brampton	OPJHL	28	9	4	13	36					
1997-98	Brampton	OPJHL	46	11	21	32	161					
1998-99	Belleville Bulls	OHL	51	6	8	14	89					
99-2000	Belleville Bulls	OHL	34	6	18	24	72					
	Sarnia Sting	OHL	27	6	8	14	44	7	0	3	3	16
2000-01	Sarnia Sting	OHL	64	28	30	58	126	4	1	3	4	20
2001-02	Sarnia Sting	OHL	66	42	62	104	141	5	1	3	4	15

Traded to **Sarnia** (OHL) by **Belleville** (OHL) for David Cornacchia, January 2, 2000.

NEWMAN, Jared (NOO-muhn, JAH-rehd) CAR.

Defense. Shoots right. 6'2", 201 lbs. Born, Detroit, MI, March 7, 1982.
(Carolina's 4th choice, 110th overall, in 2000 Entry Draft).

			Regular Season					Playoffs				
Season	Club	Lea	GP	G	A	TP	PIM	GP	G	A	TP	PIM
1997-98	Plymouth	NAJHL	49	1	4	5	69					
1998-99	Plymouth Whalers	OHL	66	2	15	17	57	11	1	2	3	9
99-2000	Plymouth Whalers	OHL	50	1	15	16	123	23	0	5	5	34
2000-01	Plymouth Whalers	OHL	34	0	4	4	114	6	0	3	3	26
2001-02	Plymouth Whalers	OHL	60	2	14	16	106	6	0	1	1	8

NIELSEN, Evan (NEEL-suhn, EH-vuhn) ATL.

Defense. Shoots right. 6'2", 195 lbs. Born, Evanston, IL, May 28, 1981.
(Atlanta's 11th choice, 242nd overall, in 2000 Entry Draft).

			Regular Season					Playoffs				
Season	Club	Lea	GP	G	A	TP	PIM	GP	G	A	TP	PIM
1996-97	Evanston High	Hi-School	23	14	17	31	....					
1997-98	Taft Eagles	Hi-School	20	5	12	17	....					
1998-99	Taft Eagles	Hi-School	20	7	5	12	12					
99-2000	U. of Notre Dame	CCHA	48	4	10	14	63					
2000-01	U. of Notre Dame	CCHA	37	2	10	12	54					
2001-02	U. of Notre Dame	CCHA	38	7	13	20	44					

NIELSEN, Frans (NEEL-sehn, FRAHNS) NYI

Center. Shoots left. 5'11", 172 lbs. Born, Herning, Denmark, April 24, 1984.
(NY Islanders' 2nd choice, 87th overall, in 2002 Entry Draft).

			Regular Season					Playoffs				
Season	Club	Lea	GP	G	A	TP	PIM	GP	G	A	TP	PIM
99-2000	Herning IK Jr.	Denmark	STATISTICS NOT AVAILABLE									
2000-01	Herning IK	Denmark	38	18	19	37	6					
2001-02	Malmo IF	Sweden	20	0	1	1	0					

NIINIMAKI, Jesse (NIH-nee-ma-kee, JEH-see) EDM.

Center. Shoots left. 6'2", 183 lbs. Born, Tampere, Finland, August 19, 1983.
(Edmonton's 1st choice, 15th overall, in 2002 Entry Draft).

			Regular Season					Playoffs				
Season	Club	Lea	GP	G	A	TP	PIM	GP	G	A	TP	PIM
1998-99	Tappara-C	Finn-Jr.	22	7	22	29	18	4	0	2	2	2
99-2000	Tappara Jr.	Finn-Jr.	14	0	5	5	6					
2000-01	Ilves Tampere-18	Finn-Jr.	16	3	5	8	40					
	Ilves Jr.	Finn-Jr.	18	2	4	6	6					
2001-02	Ilves Jr.	Finn-Jr.	27	9	23	32	54					
	Ilves Tampere	Finland	16	2	4	6	4	3	0	0	0	0

NIKOLOV, Angel (NIH-koh-lohv, AYN-jehl) S.J.

Defense. Shoots left. 6'2", 205 lbs. Born, Most, Czech., November 18, 1975.
(San Jose's 2nd choice, 37th overall, in 1994 Entry Draft).

			Regular Season					Playoffs				
Season	Club	Lea	GP	G	A	TP	PIM	GP	G	A	TP	PIM
1993-94	Litvinov	Czech	10	2	2	4	....	3	0	0	0	....
1994-95	Litvinov	Czech	41	1	4	5	18	4	0	0	0	27
1995-96	Litvinov	Czech	40	1	7	8	....	10	0	1	1	....
1996-97	Litvinov	Czech	47	0	9	9	44					
1997-98	Litvinov	Czech	51	1	4	5	53	4	0	3	3	27
1998-99	Litvinov	Czech	51	5	12	17	54					
99-2000	Litvinov	Czech	51	6	15	21	32	7	1	2	3	2
2000-01	Litvinov	Czech	48	5	11	16	85	6	0	1	1	6
2001-02	JYP Jyvaskyla	Finland	53	4	16	20	56					

NIKULIN, Ilja (nij-KOO-lihn, ihl-YUH) ATL.

Defense. Shoots left. 6'3", 210 lbs. Born, Moscow, USSR, March 12, 1982.
(Atlanta's 2nd choice, 31st overall, in 2000 Entry Draft).

			Regular Season					Playoffs				
Season	Club	Lea	GP	G	A	TP	PIM	GP	G	A	TP	PIM
1998-99	DynamoMoscow2	Russia-3	23	0	2	2	18					
99-2000	DynamoMoscow2	Russia-3	4	2	1	3	10					
	THC Tver	Russia-2	39	3	6	9	84					
2000-01	Dynamo Moscow	Russia	44	0	4	4	61					
2001-02	DynamoMoscow2	Russia-3	2	0	1	1	2					
	Dynamo Moscow	Russia	47	2	1	3	44	3	0	0	0	0

NILSSON, Magnus (NIHL-suhn, MAG-nuhs) DET.

Right wing. Shoots left. 6'1", 187 lbs. Born, Finspang, Sweden, February 1, 1978.
(Detroit's 5th choice, 144th overall, in 1996 Entry Draft).

			Regular Season					Playoffs				
Season	Club	Lea	GP	G	A	TP	PIM	GP	G	A	TP	PIM
1995-96	Vita Hasten	Swede-2	28	3	3	6	16					
1996-97	Malmo IF Jr.	Swede-Jr.	14	10	9	19	45					
	Malmo IF	Sweden	12	0	0	0	0					
1997-98	Malmo IF	Sweden	45	6	1	7	6					
1998-99	Malmo IF	Sweden	42	0	0	0	10	4	0	0	0	0
99-2000	Malmo IF	Sweden	44	5	5	10	63	6	0	0	0	25
2000-01	Louisiana	ECHL	68	13	11	24	66	11	3	2	5	8
2001-02	Toledo Storm	ECHL	57	15	20	35	63					

NILSSON, Mattias (NIHL-suhn, MA-tee-uhs) NSH.

Defense. Shoots left. 6'3", 195 lbs. Born, Ornskoldsvik, Sweden, February 16, 1982.
(Nashville's 3rd choice, 72nd overall, in 2000 Entry Draft).

			Regular Season					Playoffs				
Season	Club	Lea	GP	G	A	TP	PIM	GP	G	A	TP	PIM
1997-98	MoDo Jr.	Swede-Jr.	40	20	14	34	34					
1998-99	MoDo Jr.	Swede-Jr.	30	7	7	14	26					
99-2000	MoDo Jr.	Swede-Jr.	33	5	5	10	56	2	0	0	0	0
	MoDo-18	Swede-Jr.	....	....	....	....	....					
2000-01	MoDo Jr.	Swede-Jr.	18	2	3	5	62					
2001-02	Hammarby Jr.	Swede-Jr.	25	3	8	11	40					
	Hammarby	Swede-2	20	0	1	1	18	2	1	0	1	0

NITTEL, Ahren (NIH-tuhl, AH-rehn) N.J.

Left wing. Shoots left. 6'3", 215 lbs. Born, Waterloo, Ont., December 6, 1984.
(New Jersey's 5th choice, 85th overall, in 2002 Entry Draft).

			Regular Season					Playoffs				
Season	Club	Lea	GP	G	A	TP	PIM	GP	G	A	TP	PIM
99-2000	Streetsville	OPJHL	11	1	5	6	10					
2000-01	Windsor	OHL	46	6	4	10	56	7	3	1	4	16
2001-02	Windsor	OHL	52	19	11	30	100	9	4	1	5	23

NOLAN, Brandon (NOH-lan, BRAN-duhn) N.J.

Center. Shoots left. 6', 180 lbs. Born, Sault Ste. Marie, Ont., July 18, 1983.
(New Jersey's 6th choice, 72nd overall, in 2001 Entry Draft).

			Regular Season					Playoffs				
Season	Club	Lea	GP	G	A	TP	PIM	GP	G	A	TP	PIM
99-2000	St. Catharines	OJHL-B	47	18	13	31	10					
2000-01	Oshawa Generals	OHL	52	15	23	38	21					
2001-02	Oshawa Generals	OHL	57	30	28	58	78	5	2	4	6	4

NORDGREN, Niklas (NORHD-grehn, NIHK-lahs) CAR.

Left wing. Shoots right. 5'11", 190 lbs. Born, Ornskoldsvik, Sweden, June 28, 1979.
(Carolina's 7th choice, 195th overall, in 1997 Entry Draft).

			Regular Season					Playoffs				
Season	Club	Lea	GP	G	A	TP	PIM	GP	G	A	TP	PIM
1995-96	MoDo-18	Swede-Jr.	30	37	27	64	....					
1996-97	MoDo Jr.	Swede-Jr.	22	14	6	20	....					
	MoDo	Sweden	5	0	0	0	0					
1997-98	MoDo Jr.	Swede-Jr.	28	15	15	30	52					
1998-99	MoDo	Sweden	7	0	0	0	2					
	MoDo Jr.	Swede-Jr.	22	7	4	11	22	3	2	1	3	0
99-2000	IF Sundsvall	Swede-2	27	21	11	32	58					
	MoDo	EuroHL	1	0	0	0	0	1	0	0	0	0
2000-01	IF Sundsvall	Swede-2	35	22	19	41	45					
2001-02	Timra IK Jr.	Swede-Jr.	1	1	1	2	0					
	Timra IK	Sweden	49	8	6	14	16					
	Timra IK	Swede-Q	10	3	3	5	4					

NORDQVIST, Jonas (NAWRD-kvihst, YOH-nuhs) CHI.

Center. Shoots left. 6'2", 198 lbs. Born, Leksand, Sweden, April 26, 1982.
(Chicago's 3rd choice, 49th overall, in 2000 Entry Draft).

			Regular Season					Playoffs				
Season	Club	Lea	GP	G	A	TP	PIM	GP	G	A	TP	PIM
1997-98	Leksands IF Jr.	Swede-Jr.	42	26	35	61	....					
1998-99	Leksands IF Jr.	Swede-Jr.	32	14	25	39	....					
99-2000	Leksands IF Jr.	Swede-Jr.	34	15	24	39	32	2	0	0	0	2
	Leksands IF	Sweden	3	0	0	0	0					
	Leksands IF-18	Swede-Jr.	2	0	2	2	0	4	3	5	8	0
2000-01	Leksands IF Jr.	Swede-Jr.	10	6	13	19	6	5	1	6	7	2
	Leksands IF	Sweden	42	3	4	7	4					
2001-02	Leksands IF Jr.	Swede-Jr.	8	14	7	21	6	1	0	1	1	0
	Leksands IF	Swede-2	40	8	7	15	16					

NOVAK, Filip (NOH-vak, FIH-lihp) FLA.

Defense. Shoots left. 6'1", 185 lbs. Born, Ceske Budejovice, Czech., May 7, 1982.
(NY Rangers' 1st choice, 64th overall, in 2000 Entry Draft).

			Regular Season					Playoffs				
Season	Club	Lea	GP	G	A	TP	PIM	GP	G	A	TP	PIM
1998-99	C. Budejovice Jr.	Czech-Jr.	68	8	10	18	34					
99-2000	Regina Pats	WHL	47	7	32	39	70	7	1	4	5	5
2000-01	Regina Pats	WHL	64	17	50	67	75	6	1	4	5	6
2001-02	Regina Pats	WHL	60	12	46	58	125	6	2	2	4	19

WHL East Second All-Star Team (2001) • WHL East First All-Star Team (2002)

Traded to **Florida** by **NY Rangers** with Igor Ulanov, NY Rangers' 1st (later traded to Calgary - Calgary selected Eric Nystrom) and 2nd (Rob Globke) round choices in 2002 Entry Draft and NY Rangers' 4th round choice in 2003 Entry Draft for Pavel Bure and Florida's 2nd round choice (Lee Falardeau) in 2002 Entry Draft, March 18, 2002.

NOVAK, Zbynek (NOH-vahk, z'BIHN-nehk) WSH.

Left wing. Shoots left. 6'2", 194 lbs. Born, Kutna Hora, Czech., July 23, 1983.
(Washington's 5th choice, 191st overall, in 2001 Entry Draft).

			Regular Season					Playoffs				
Season	Club	Lea	GP	G	A	TP	PIM	GP	G	A	TP	PIM
99-2000	Slavia Praha Jr.	Czech-Jr.	15	1	2	3	8	2	1	0	1	0
	Slavia Praha-18	Czech-Jr.	31	14	12	26	12					
2000-01	Slavia Praha Jr.	Czech-Jr.	44	10	8	18	24	3	1	0	1	2
2001-02	Slavia Praha-18	Czech-Jr.	3	1	0	1	2					
	Slavia Praha Jr.	Czech-Jr.	23	11	12	23	8					
	HC Slavia Praha	Czech	27	0	1	1	2					
	HC Brod	Czech-3	5	0	2	2	0	2	1	0	1	2

NOVOTNY, Jiri (NOH-vaht-nee, YOO-ree) BUF.

Center. Shoots right. 6'2", 194 lbs. Born, Pelhrimov, Czech., August 12, 1983.
(Buffalo's 1st choice, 22nd overall, in 2001 Entry Draft).

			Regular Season					Playoffs				
Season	Club	Lea	GP	G	A	TP	PIM	GP	G	A	TP	PIM
99-2000	C. Budejovice Jr.	Czech-Jr.	39	11	12	23	14					
2000-01	C. Budejovice Jr.	Czech-Jr.	33	10	10	20	....					
2001-02	C. Budejovice Jr.	Czech-Jr.	7	4	4	8	4					
	Jindrichuv Hradec	Czech-3	3	1	3	4	0					
	Ceske Budejovice	Czech	41	8	6	14	6					

NOWAK, Brett (NOH-wak, BREHT) BOS.

Center. Shoots left. 6'2", 192 lbs. Born, New Haven, CT, May 20, 1981.
(Boston's 7th choice, 102nd overall, in 2000 Entry Draft).

			Regular Season					Playoffs				
Season	Club	Lea	GP	G	A	TP	PIM	GP	G	A	TP	PIM
1997-98	Hotchkiss High	Hi-School	21	24	42	66	42	....	....	....	....	....
1998-99	Hotchkiss High	Hi-School	20	21	36	57	6	....	....	....	....	....
99-2000	Harvard Crimson	ECAC	26	6	11	17	20	....	....	....	....	....
2000-01	Harvard Crimson	ECAC	24	7	9	16	26	....	....	....	....	....
2001-02	Harvard Crimson	ECAC	33	14	17	31	50	....	....	....	....	....

ECAC Second All-Star Team (2002)

NUUTINEN, Sami (NOO-tih-nehn, SA-mee) EDM.

Defense. Shoots left. 6'1", 189 lbs. Born, Espoo, Finland, June 11, 1971.
(Edmonton's 12th choice, 248th overall, in 1990 Entry Draft).

			Regular Season					Playoffs				
Season	Club	Lea	GP	G	A	TP	PIM	GP	G	A	TP	PIM
1988-89	Kiekko Espoo Jr.	Finn-Jr.	6	1	7	8	10	4	2	4	6	2
	Kiekko Espoo	Finland-2	39	18	10	28	46	....	....	....	....	....
1989-90	Kiekko Espoo Jr.	Finn-Jr.	8	3	5	8	6	5	3	1	4	8
	Kiekko Espoo	Finland-2	40	8	15	23	....	....	....	....	....	....
1990-91	Karhu-Kissat	Finland-2	3	1	0	1	0	....	....	....	....	....
	HIFK Helsinki	Finland	27	1	3	4	6	3	0	0	0	0
1991-92	HIFK Helsinki	Finland	44	5	6	11	10	9	0	1	1	4
1992-93	Kiekko Espoo	Finland	48	7	11	18	59	....	....	....	....	....
1993-94	Kiekko Espoo	Finland	46	9	15	24	36	....	....	....	....	....
1994-95	Kiekko Espoo	Finland	50	8	24	32	38	4	0	1	1	0
1995-96	Kiekko Espoo	Finland	49	7	7	14	54	....	....	....	....	....
1996-97	Vasteras IK	Sweden	50	7	7	14	22	....	....	....	....	....
1997-98	Kiekko Espoo	Finland	48	4	17	21	51	8	0	2	2	6
1998-99	Jokerit Helsinki	Finland	54	10	18	28	22	3	0	2	2	0
99-2000	Rosenheim	Germany	52	2	11	13	16	....	....	....	....	....
2000-01	HPK Hameenlinna	Finland	48	3	6	9	52	....	....	....	....	....
2001-02	IF Frisk Asker	Norway	42	14	28	42	68	8	2	3	5	16

NYCHOLAT, Lawrence (NIH-coh-lat, LAW-rehnts) MIN.

Defense. Shoots left. 6', 192 lbs. Born, Calgary, Alta., May 7, 1979.

			Regular Season					Playoffs				
Season	Club	Lea	GP	G	A	TP	PIM	GP	G	A	TP	PIM
1995-96	Notre Dame Argos	SMHL	42	10	36	46	66	....	....	....	....	....
1996-97	Swift Current	WHL	67	8	13	21	82	10	0	0	0	24
1997-98	Swift Current	WHL	71	13	35	48	108	1	0	0	0	0
1998-99	Swift Current	WHL	72	16	44	60	125	6	2	2	4	12
99-2000	Swift Current	WHL	70	22	58	80	92	2	0	0	0	0
2000-01	Jackson Bandits	ECHL	5	1	2	3	5	....	....	....	....	....
	Cleveland	IHL	42	3	7	10	69	4	0	0	0	2
2001-02	Houston Aeros	AHL	72	3	11	14	92	14	1	0	1	23

Signed as a free agent by **Minnesota**, August 31, 2000.

NYSTROM, David (NEW-strawm, DAY-vihd) PHI.

Left wing. Shoots right. 6', 174 lbs. Born, Hagersten, Sweden, February 21, 1980.
(Philadelphia's 6th choice, 224th overall, in 1999 Entry Draft).

			Regular Season					Playoffs				
Season	Club	Lea	GP	G	A	TP	PIM	GP	G	A	TP	PIM
1996-97	V. Frolunda Jr.	Swede-Jr.	23	6	2	8	....	....	....	....	....	....
1997-98	V. Frolunda Jr.	Swede-Jr.	29	17	18	35	26	2	0	0	0	0
1998-99	V. Frolunda Jr.	Swede-Jr.	28	15	16	31	20	....	....	....	....	....
99-2000	IF Troja-Ljungby	Swede-2	45	20	15	35	34	2	0	0	0	0
2000-01	IF Troja-Ljungby	Swede-2	38	20	11	31	30	4	0	0	0	0
2001-02	Skelleftea AIK	Swede-2	45	4	5	9	16	6	0	0	0	4

NYSTROM, Eric (NIGH-stuhm, AIR-ihk) CGY.

Left wing. Shoots left. 6'1", 195 lbs. Born, Syosset, NY, February 14, 1983.
(Calgary's 1st choice, 10th overall, in 2002 Entry Draft).

			Regular Season					Playoffs				
Season	Club	Lea	GP	G	A	TP	PIM	GP	G	A	TP	PIM
1998-99	Nassau Lions	NYAHA		STATISTICS NOT AVAILABLE								
99-2000	Team USA	USDP-17	55	7	16	23	57	....	....	....	....	....
2000-01	Team USA	USDP-18	66	15	17	32	102	....	....	....	....	....
2001-02	U. of Michigan	CCHA	40	18	13	31	42	....	....	....	....	....

CCHA All-Rookie Team (2002)

O'CONNOR, Sean (oh-KAW-nuhr, SHAWN) FLA.

Right wing. Shoots right. 6'3", 220 lbs. Born, Victoria, B.C., October 19, 1981.
(Florida's 3rd choice, 82nd overall, in 2000 Entry Draft).

			Regular Season					Playoffs				
Season	Club	Lea	GP	G	A	TP	PIM	GP	G	A	TP	PIM
1995-96	Victoria Racquet	BCAHA	60	23	45	68	70	....	....	....	....	....
1996-97	Victoria Racquet	BCAHA	50	59	73	132	190	....	....	....	....	....
1997-98	Victoria Salsa	BCHL	50	9	17	26	145	....	....	....	....	....
1998-99	Victoria Salsa	BCHL	53	17	18	35	197	....	....	....	....	....
99-2000	Moose Jaw	WHL	51	5	8	13	166	2	0	0	0	2
2000-01	Moose Jaw	WHL	71	34	15	49	192	4	0	1	1	15
2001-02	Moose Jaw	WHL	62	15	19	34	135	12	0	3	3	4

O'DETTE, Matt (oh-DEHT, MAT) MTL.

Left wing. Shoots right. 6'5", 228 lbs. Born, Oshawa, Ont., November 9, 1975.
(Florida's 7th choice, 157th overall, in 1994 Entry Draft).

			Regular Season					Playoffs				
Season	Club	Lea	GP	G	A	TP	PIM	GP	G	A	TP	PIM
1990-91	Oshawa Elites	OMHA	75	4	24	28	84	....	....	....	....	....
1991-92	Milton Merchants	OPJHL	20	2	9	11	13	....	....	....	....	....
1992-93	Kitchener	OHL	28	0	0	0	6	....	....	....	....	....
1993-94	Kitchener	OHL	46	1	3	4	107	....	....	....	....	....
1994-95	Kitchener	OHL	10	1	2	3	29	....	....	....	....	....
	Sault Ste. Marie	OHL	42	3	12	15	94	....	....	....	....	....
1995-96	Kitchener	OHL	57	6	7	13	146	4	0	0	0	2
1996-97	Roanoke Express	ECHL	69	6	16	22	139	4	0	0	0	10
1997-98	Saint John	AHL	58	0	5	5	92	15	0	1	1	10
	Roanoke Express	ECHL	5	0	0	0	29	....	....	....	....	....
1998-99	Saint John	AHL	42	0	1	1	82	2	0	0	0	14
99-2000	Saint John	AHL	69	1	4	5	177	3	0	0	0	0
2000-01	Quebec	AHL	38	1	3	4	124	7	0	0	0	16
2001-02	Quebec	AHL	48	7	1	8	136	3	0	0	0	2

Signed as a free agent by **Calgary**, September, 1999. Signed as a free agent by **Montreal**, July 4, 2002.

ODUYA, Johnny (oh-DOO-yuh, JAW-nee) WSH.

Defense. Shoots left. 6', 200 lbs. Born, Stockholm, Sweden, October 1, 1981.
(Washington's 6th choice, 221st overall, in 2001 Entry Draft).

			Regular Season					Playoffs				
Season	Club	Lea	GP	G	A	TP	PIM	GP	G	A	TP	PIM
1997-98	Hammarby Jr.	Swede-Jr.	26	3	11	14	70	....	....	....	....	....
1998-99	Hammarby Jr.	Swede-Jr.	38	14	31	45	45	....	....	....	....	....
99-2000	Hammarby Jr.	Swede-Jr.	32	3	18	21	48	6	1	2	3	4
	Hammarby	Swede-2	1	0	0	0	0	1	0	0	0	0
2000-01	Moncton Wildcats	QMJHL	44	11	38	49	147	....	....	....	....	....
	Victoriaville	QMJHL	24	3	16	19	112	13	4	9	13	10
2001-02	Hammarby	Swede-2	46	11	14	25	66	2	1	0	1	4

Traded to **Victoriaville** (QMJHL) by **Moncton** (QMJHL) for Teddy Kyres and Simon St-Pierre, January 10, 2002.

O'LEARY, Pat (OH-leer-ree, PAT) PHX.

Center. Shoots left. 6'2", 190 lbs. Born, Minneapolis, MN, September 2, 1979.
(Phoenix's 3rd choice, 73rd overall, in 1998 Entry Draft).

			Regular Season					Playoffs				
Season	Club	Lea	GP	G	A	TP	PIM	GP	G	A	TP	PIM
1996-97	Robbinsdale	Hi-School	22	28	27	55	42	....	....	....	....	....
1997-98	Robbinsdale	Hi-School	24	22	27	49	28	....	....	....	....	....
1998-99	U. of Minnesota	WCHA	17	0	2	2	8	....	....	....	....	....
99-2000	U. of Minnesota	WCHA	25	6	1	7	8	....	....	....	....	....
2000-01	U. of Minnesota	WCHA	38	5	4	9	44	....	....	....	....	....
2001-02	U. of Minnesota	WCHA	40	4	2	6	18	....	....	....	....	....

OLSON, Josh (OHL-suhn, JAWSH) FLA.

Left wing. Shoots left. 6'5", 225 lbs. Born, Grand Forks, ND, July 13, 1981.
(Florida's 6th choice, 190th overall, in 2000 Entry Draft).

			Regular Season					Playoffs				
Season	Club	Lea	GP	G	A	TP	PIM	GP	G	A	TP	PIM
99-2000	Fargo-Moorhead	USHL	18	2	5	7	37	....	....	....	....	....
	Omaha Lancers	USHL	43	6	7	13	44	4	0	0	0	4
2000-01	Portland	WHL	72	22	38	60	86	16	5	4	9	17
2001-02	Utah Grizzlies	AHL	1	0	0	0	0	1	0	0	0	0
	Portland	WHL	72	40	48	88	85	7	4	3	7	8

OLYNICK, Craig (oh-LIHN-ihk, KRAYG)

Defense. Shoots right. 6'1", 185 lbs. Born, Saskatoon, Sask., August 21, 1982.
(Los Angeles' 8th choice, 218th overall, in 2000 Entry Draft).

			Regular Season					Playoffs				
Season	Club	Lea	GP	G	A	TP	PIM	GP	G	A	TP	PIM
1997-98	Humboldt	SAHA	57	37	72	109	230	....	....	....	....	....
1998-99	Seattle	WHL	51	1	9	10	89	11	0	0	0	14
99-2000	Seattle	WHL	65	1	9	10	120	7	1	0	1	17
2000-01	Seattle	WHL	62	5	20	25	105	9	0	2	2	14
2001-02	Seattle	WHL	21	5	8	13	44	....	....	....	....	....
	Moose Jaw	WHL	39	3	15	18	92	12	2	3	5	32

Traded to **Moose Jaw** (WHL) by **Seattle** (WHL) with Stas Avksentiev and future considerations for Tomas Mojzis and Brooks Laich, November 18, 2001.

OREKHOVSKY, Oleg (oh-reh-KHOHV-skee, OH-lehg) MIN.

Defense. Shoots right. 6', 183 lbs. Born, Krasnoyarsk, USSR, November 3, 1977.
(Washington's 11th choice, 206th overall, in 1996 Entry Draft).

			Regular Season					Playoffs				
Season	Club	Lea	GP	G	A	TP	PIM	GP	G	A	TP	PIM
1994-95	Dynamo Moscow	CIS	30	0	1	1	18	....	....	....	....	....
1995-96	Dynamo Moscow	CIS	22	1	2	3	14	8	0	0	0	6
1996-97	Dynamo Moscow	Russia	32	4	2	6	16	4	2	1	3	2
1997-98	Dynamo Moscow	EuroHL	7	2	1	3	12	....	....	....	....	....
	Dynamo Moscow	Russia	40	4	5	9	34	....	....	....	....	....
1998-99	Dynamo Moscow	EuroHL	3	1	1	2	2	6	0	0	0	6
	Dynamo Moscow	Russia	42	1	1	2	22	16	1	2	3	6
99-2000	Dynamo Moscow	EuroHL	6	1	0	1	2	....	....	....	....	....
	Dynamo Moscow	Russia	37	3	6	9	38	15	1	0	1	4
2000-01	Dynamo Moscow	Russia	40	2	4	6	44	....	....	....	....	....
2001-02	Dynamo Moscow	Russia	50	2	15	17	50	3	1	0	1	2

Selected by **Minnesota** from **Washington** in Expansion Draft, June 23, 2000.

ORLOV, Maxim (ohr-LAHF, max-EEM) WSH.

Center. Shoots left. 6', 176 lbs. Born, Moscow, USSR, March 31, 1981.
(Washington's 9th choice, 219th overall, in 1999 Entry Draft).

			Regular Season					Playoffs				
Season	Club	Lea	GP	G	A	TP	PIM	GP	G	A	TP	PIM
1998-99	CSKA Moscow	Russia	2	0	0	0	2	1	0	0	0	0
99-2000	CSKA Moscow	Russia	25	0	0	0	2	2	0	0	0	2
2000-01	CSKA Moscow	Russia	41	5	4	9	14	....	....	....	....	....
2001-02	CSKA Moscow 2	Russia-3	7	7	4	11	4	....	....	....	....	....
	CSKA Moscow	Russia	35	3	5	8	14	....	....	....	....	....

ORPIK, Brooks (OHR-pihk, BRUKS) PIT.

Defense. Shoots left. 6'2", 222 lbs. Born, San Francisco, CA, September 26, 1980.
(Pittsburgh's 1st choice, 18th overall, in 2000 Entry Draft).

			Regular Season					Playoffs				
Season	Club	Lea	GP	G	A	TP	PIM	GP	G	A	TP	PIM
1996-97	Thayer Academy	Hi-School	20	4	1	5	....	....	....	....	....	....
1997-98	Thayer Academy	Hi-School	22	0	7	7	....	....	....	....	....	....
1998-99	Boston College	H-East	41	1	10	11	*96	....	....	....	....	....
99-2000	Boston College	H-East	38	1	9	10	102	....	....	....	....	....
2000-01	Boston College	H-East	40	0	20	20	*124	....	....	....	....	....
2001-02	Wilkes-Barre	AHL	78	2	18	20	99	....	....	....	....	....

ORR, Colton (OHR, KOHL-tuhn) BOS.

Right wing. Shoots right. 6'2", 210 lbs. Born, Winnipeg, Man., March 3, 1982.

			Regular Season					Playoffs				
Season	Club	Lea	GP	G	A	TP	PIM	GP	G	A	TP	PIM
1998-99	Swift Current	WHL	2	0	0	0	0	....	....	....	....	....
99-2000	Swift Current	WHL	61	3	2	5	130	12	1	0	1	25
2000-01	Swift Current	WHL	19	0	4	4	67	....	....	....	....	....
	Kamloops Blazers	WHL	41	8	1	9	179	3	0	0	0	20
2001-02	Kamloops Blazers	WHL	1	0	0	0	7	2	0	0	0	2

Traded to **Kamloops** (WHL) by **Swift Current** (WHL) for Paul Diniset and Kamloops' 3rd round choice (Kyle Moir) in 2001 WHL Bantam Draft, November 24, 2000. Signed as a free agent by **Boston**, September 19, 2001. • Missed majority of 2001-02 season recovering from wrist injury suffered in game vs. Red Deer (WHL), October 20, 2001.

OTT, Steve (AWT, STEEV) DAL.

Center. Shoots left. 6', 160 lbs. Born, Summerside, P.E.I., August 19, 1982.
(Dallas' 1st choice, 25th overall, in 2000 Entry Draft).

			Regular Season					Playoffs				
Season	**Club**	**Lea**	**GP**	**G**	**A**	**TP**	**PIM**	**GP**	**G**	**A**	**TP**	**PIM**
1998-99	Leamington	OJHL-B	48	14	30	44	110	....	....	....	....	....
99-2000	Windsor	OHL	66	23	39	62	131	12	3	5	8	21
2000-01	Windsor	OHL	55	50	37	87	164	9	3	8	11	27
2001-02	Windsor	OHL	53	43	45	88	178	14	6	10	16	49

OHL Second All-Star Team (2002)

OTTOSSON, Kristofer (AW-toh-suhn, KRIHS-tuh-fuhr) NYI

Right wing. Shoots left. 5'10", 187 lbs. Born, Stockholm, Sweden, January 9, 1976.
(NY Islanders' 6th choice, 148th overall, in 2000 Entry Draft).

			Regular Season					Playoffs				
Season	**Club**	**Lea**	**GP**	**G**	**A**	**TP**	**PIM**	**GP**	**G**	**A**	**TP**	**PIM**
1993-94	Djurgarden Jr.	Swede-Jr.	13	3	6	9	4	....	....	....	....	....
1994-95	Djurgarden Jr.	Swede-Jr.	15	8	23	31	2	....	....	....	....	....
	Djurgarden	Sweden	30	0	0	0	2	3	0	0	0	0
1995-96	Djurgarden Jr.	Swede-Jr.	15	5	12	17	4	....	....	....	....	....
	Djurgarden	Sweden	32	1	0	1	2	2	0	0	0	0
1996-97	Arlanda Mastra	Swede-2	6	6	0	6	0	....	....	....	....	....
	Djurgarden	Sweden	20	0	0	0	2	....	....	....	....	....
	Huddinge IK	Swede-2	13	1	3	4	2	2	0	1	1	0
1997-98	Huddinge IK	Swede-2	30	17	20	37	14	....	....	....	....	....
1998-99	Huddinge IK	Swede-2	41	15	19	34	16	....	....	....	....	....
99-2000	Djurgarden	Sweden	47	25	15	40	12	13	7	2	9	2
2000-01	Djurgarden	Sweden	46	17	24	41	14	14	*7	4	11	4
2001-02	Djurgarden	Sweden	41	13	8	21	12	4	0	0	0	0

OUELLET, Michel (oo-LEHT, mih-SHEHL) PIT.

Right wing. Shoots right. 6', 190 lbs. Born, Rimouski, Que., March 5, 1982.
(Pittsburgh's 4th choice, 124th overall, in 2000 Entry Draft).

			Regular Season					Playoffs				
Season	**Club**	**Lea**	**GP**	**G**	**A**	**TP**	**PIM**	**GP**	**G**	**A**	**TP**	**PIM**
1997-98	Jonquiere Elites	QAAA	33	20	32	52	52	....	....	....	....	....
1998-99	Rimouski Oceanic	QMJHL	28	7	13	20	10	11	0	1	1	6
99-2000	Rimouski Oceanic	QMJHL	72	36	53	89	38	14	4	5	9	14
2000-01	Rimouski Oceanic	QMJHL	63	42	50	92	50	11	6	7	13	8
2001-02	Rimouski Oceanic	QMJHL	61	40	58	98	66	7	3	6	9	4

PACKARD, Dennis (PA-kuhrd, DEH-nihs) T.B.

Left wing. Shoots left. 6'4", 195 lbs. Born, St. Catherines, Ont., February 9, 1982.
(Tampa Bay's 8th choice, 219th overall, in 2001 Entry Draft).

			Regular Season					Playoffs				
Season	**Club**	**Lea**	**GP**	**G**	**A**	**TP**	**PIM**	**GP**	**G**	**A**	**TP**	**PIM**
99-2000	Team USA	USDP-18	55	11	14	25	85	....	....	....	....	....
2000-01	Harvard Crimson	ECAC	33	4	4	8	28	....	....	....	....	....
2001-02	Harvard Crimson	ECAC	32	9	10	19	34	....	....	....	....	....

PADDOCK, Cam (PA-dawk, KAM) PIT.

Center. Shoots right. 6', 178 lbs. Born, Vancouver, B.C., March 22, 1983.
(Pittsburgh's 6th choice, 137th overall, in 2002 Entry Draft).

			Regular Season					Playoffs				
Season	**Club**	**Lea**	**GP**	**G**	**A**	**TP**	**PIM**	**GP**	**G**	**A**	**TP**	**PIM**
99-2000	Kelowna Rockets	WHL	46	5	5	10	42	5	0	0	0	0
2000-01	Kelowna Rockets	WHL	72	14	10	24	110	6	0	0	0	4
2001-02	Kelowna Rockets	WHL	72	38	35	73	122	15	8	6	14	35

PAETSCH, Nathan (PASH, NAY-thuhn) WSH.

Defense. Shoots left. 6', 195 lbs. Born, Humboldt, Sask., March 30, 1983.
(Washington's 1st choice, 58th overall, in 2001 Entry Draft).

			Regular Season					Playoffs				
Season	**Club**	**Lea**	**GP**	**G**	**A**	**TP**	**PIM**	**GP**	**G**	**A**	**TP**	**PIM**
1998-99	Tisdale Trojans	SMHL	74	20	55	75	120	....	....	....	....	....
	Moose Jaw	WHL	2	0	0	0	0	....	....	....	....	....
99-2000	Moose Jaw	WHL	68	9	35	44	49	4	0	1	1	0
2000-01	Moose Jaw	WHL	70	8	54	62	118	4	1	2	3	6
2001-02	Moose Jaw	WHL	59	16	36	52	86	12	0	4	4	16

PAILLE, Dan (PIGH-yay, DAN) BUF.

Left wing. Shoots left. 6', 200 lbs. Born, Welland, Ont., April 15, 1984.
(Buffalo's 2nd choice, 20th overall, in 2002 Entry Draft).

			Regular Season					Playoffs				
Season	**Club**	**Lea**	**GP**	**G**	**A**	**TP**	**PIM**	**GP**	**G**	**A**	**TP**	**PIM**
99-2000	Welland Cougars	OJHL-B	42	14	17	31	19	16	16	16	32	....
2000-01	Guelph Storm	OHL	64	22	31	53	57	4	2	0	2	2
2001-02	Guelph Storm	OHL	62	27	30	57	54	9	5	2	7	9

PANDOLFO, Mike (pan-DAHL-foh, MIGHK) CBJ

Left wing. Shoots left. 6'3", 221 lbs. Born, Winchester, MA, September 15, 1979.
(Buffalo's 5th choice, 77th overall, in 1998 Entry Draft).

			Regular Season					Playoffs				
Season	**Club**	**Lea**	**GP**	**G**	**A**	**TP**	**PIM**	**GP**	**G**	**A**	**TP**	**PIM**
1996-97	St. Sebastian's	Hi-School	32	27	28	55	30	....	....	....	....	....
1997-98	St. Sebastian's	Hi-School	28	29	23	52	18	....	....	....	....	....
1998-99	Boston University	H-East	34	13	4	17	26	....	....	....	....	....
99-2000	Boston University	H-East	41	13	10	23	37	....	....	....	....	....
2000-01	Boston University	H-East	37	16	13	29	30	....	....	....	....	....
2001-02	Boston University	H-East	38	22	18	40	22	....	....	....	....	....

Rights traded to **Columbus** by **Buffalo** with Detroit's 1st round choice (previously acquired, later traded to Atlanta - Atlanta selected Jim Slater) in 2002 Entry Draft for New Jersey's 1st round choice (previously acquired, Buffalo selected Dan Paille) in 2002 Entry Draft, June 22, 2002.

PANOV, Konstantin (PAN-ahv, KAWN-stan-tihn) NSH.

Right wing. Shoots left. 6', 195 lbs. Born, Chelyabinsk, USSR, June 29, 1980.
(Nashville's 10th choice, 131st overall, in 1999 Entry Draft).

			Regular Season					Playoffs				
Season	**Club**	**Lea**	**GP**	**G**	**A**	**TP**	**PIM**	**GP**	**G**	**A**	**TP**	**PIM**
1996-97	Chelyabinsk 2	Russia-3	25	18	30	48	22	....	....	....	....	....
1997-98	Yunior-T Kurgan	Russia-3	20	7	3	10	6	....	....	....	....	....
	Chelyabinsk	Russia	6	2	0	2	4	2	0	0	0	0
1998-99	Kamloops Blazers	WHL	62	33	30	63	62	13	5	3	8	10
99-2000	Kamloops Blazers	WHL	64	43	30	73	47	....	....	....	....	....
2000-01	Kamloops Blazers	WHL	69	44	56	100	54	4	1	0	1	2
2001-02	Milwaukee	AHL	15	1	5	6	2	....	....	....	....	....

WHL West Second All-Star Team (2000) • WHL West First All-Star Team (2001)

PANZER, Jeff (PAN-zuhr, JEHF) ST.L.

Center. Shoots left. 5'10", 160 lbs. Born, Grand Forks, ND, April 7, 1978.

			Regular Season					Playoffs				
Season	**Club**	**Lea**	**GP**	**G**	**A**	**TP**	**PIM**	**GP**	**G**	**A**	**TP**	**PIM**
1996-97	Fargo-Moorhead	USHL	49	30	40	70	52	6	3	7	10	0
1997-98	North Dakota	WCHA	37	14	23	37	18	....	....	....	....	....
1998-99	North Dakota	WCHA	37	21	26	47	14	....	....	....	....	....
99-2000	North Dakota	WCHA	44	19	*44	63	16	....	....	....	....	....
2000-01	North Dakota	WCHA	46	26	*55	*81	28	....	....	....	....	....
2001-02	Worcester	AHL	70	26	27	53	29	3	0	2	2	0

WCHA Second All-Star Team (1999) • WCHA First All-Star Team (2000, 2001) • NCAA West First All-American Team (2000, 2001)

Signed as a free agent by **St. Louis**, April 30, 2001.

PARADISE, Chris BOS.

Center. Shoots right. 6'2", 200 lbs. Born, St. Paul, MN, August 6, 1977.

			Regular Season					Playoffs				
Season	**Club**	**Lea**	**GP**	**G**	**A**	**TP**	**PIM**	**GP**	**G**	**A**	**TP**	**PIM**
1996-97	Twin Cities	USHL	53	16	21	37	92	5	1	0	1	28
1997-98	Omaha Lancers	USHL	55	23	22	45	65	14	*11	5	16	6
1998-99	U. of Denver	WCHA	32	5	4	9	31	....	....	....	....	....
99-2000	U. of Denver	WCHA	40	4	19	23	46	....	....	....	....	....
2000-01	U. of Denver	WCHA	38	17	16	33	50	....	....	....	....	....
2001-02	U. of Denver	WCHA	40	22	19	41	40	....	....	....	....	....

Signed as a free agent by **Boston**, August 6, 2002.

PARENTEAU, Pierre (pair-ehn-TOH, PEE-air) ANA.

Center. Shoots right. 5'11", 156 lbs. Born, Hull, Que., March 24, 1983.
(Anaheim's 11th choice, 264th overall, in 2001 Entry Draft).

			Regular Season					Playoffs				
Season	**Club**	**Lea**	**GP**	**G**	**A**	**TP**	**PIM**	**GP**	**G**	**A**	**TP**	**PIM**
99-2000	Charles-Lemoyne	QAAA	40	25	40	65	18	16	4	9	13	8
2000-01	Moncton Wildcats	QMJHL	45	10	19	29	38	....	....	....	....	....
	Chicoutimi	QMJHL	28	10	13	23	14	7	4	7	11	2
2001-02	Chicoutimi	QMJHL	68	51	67	118	120	4	3	1	4	10

Traded to **Chicoutimi** (QMJHL) by **Moncton** (QMJHL) for Francois Caron, January 10, 2001.

PAROULEK, Martin (PAHR-oh-lehk, MAHR-tihn) CBJ

Right wing. Shoots right. 6', 193 lbs. Born, Uherske Hradiste, Czech., November 4, 1979.
(Columbus' 9th choice, 278th overall, in 2000 Entry Draft).

			Regular Season					Playoffs				
Season	**Club**	**Lea**	**GP**	**G**	**A**	**TP**	**PIM**	**GP**	**G**	**A**	**TP**	**PIM**
1998-99	Vsetin Jr.	Czech-Jr.	45	25	19	44	....	....	....	....	....	....
	Vsetin	Czech	11	1	1	2	....	7	0	1	1	0
99-2000	Vsetin	Czech	48	11	14	25	24	8	1	1	2	4
2000-01	Sumperk	Czech-2	14	0	1	1	27	....	....	....	....	....
	Vsetin	Czech	28	10	4	14	16	14	3	3	6	10
2001-02	Syracuse Crunch	AHL	59	11	14	25	31	9	1	1	2	4

PARROS, George (PAIR-ohs, JOHRJ) L.A.

Right wing. Shoots right. 6'4", 210 lbs. Born, Washington, PA, December 29, 1979.
(Los Angeles' 9th choice, 222nd overall, in 1999 Entry Draft).

			Regular Season					Playoffs				
Season	**Club**	**Lea**	**GP**	**G**	**A**	**TP**	**PIM**	**GP**	**G**	**A**	**TP**	**PIM**
1997-98	Delbarton Wave	Hi-School		STATISTICS NOT AVAILABLE								
1998-99	Chicago Freeze	NAJHL	54	30	20	50	126	....	....	....	....	....
99-2000	Princeton	ECAC	27	4	2	6	14	....	....	....	....	....
2000-01	Princeton	ECAC	31	7	10	17	38	....	....	....	....	....
2001-02	Princeton	ECAC	31	9	13	22	36	....	....	....	....	....

NAJHL All-Rookie Team (1999) • NAJHL Rookie of the Year (1999)

PARSONS, Steve (PAHR-suhnz, STEEV)

Left wing. Shoots left. 6'4", 235 lbs. Born, Vancouver, B.C., March 12, 1975.

			Regular Season					Playoffs				
Season	**Club**	**Lea**	**GP**	**G**	**A**	**TP**	**PIM**	**GP**	**G**	**A**	**TP**	**PIM**
1992-93	Kamloops Blazers	WHL	1	0	0	0	0	....	....	....	....	....
1993-94				STATISTICS NOT AVAILABLE								
1994-95	Laval Titan	QMJHL	27	0	1	1	176	14	0	0	0	52
1995-96	Nanaimo Clippers	BCJHL	48	18	36	54	275	....	....	....	....	....
1996-97	Reno Renegades	WCHL	7	0	1	1	27	....	....	....	....	....
	Bakersfield Fog	WCHL	5	0	1	1	38	....	....	....	....	....
1997-98	Concordia	ACAC		STATISTICS NOT AVAILABLE								
1998-99	Concordia	ACAC	25	6	20	26	178	....	....	....	....	....
99-2000	Madison Monsters	UHL	39	3	6	9	241	....	....	....	....	....
	Milwaukee	IHL	2	0	1	1	2	....	....	....	....	....
	Fort Wayne	UHL	17	0	3	3	108	11	0	0	0	29
2000-01	Wilkes-Barre	AHL	4	0	0	0	12	....	....	....	....	....
	Wheeling Nailers	ECHL	19	5	4	9	178	....	....	....	....	....
	Hershey Bears	AHL	30	1	0	1	98	1	0	0	0	0
2001-02	Wilkes-Barre	AHL	52	1	1	2	366	....	....	....	....	....

Signed to 25-game contract by **Wilkes-Barre** (AHL), October 4, 2000. Released by **Wilkes-Barre** (AHL) and assigned to **Wheeling** (ECHL), November 6, 2000. Signed to 25-game tryout contract by **Hershey** (AHL), January 9, 2001. Signed as a free agent by **Pittsburgh**, July 30, 2001. Traded to **Nashville** by **Pittsburgh** for future considerations, May 15, 2002.

PAUL, Jeff (PAWL, JEHF) COL.

Defense. Shoots right. 6'3", 200 lbs. Born, London, Ont., March 1, 1978.
(Chicago's 2nd choice, 42nd overall, in 1996 Entry Draft).

			Regular Season					Playoffs				
Season	**Club**	**Lea**	**GP**	**G**	**A**	**TP**	**PIM**	**GP**	**G**	**A**	**TP**	**PIM**
1993-94	Woodstock Vets	OJHL-C	36	1	6	7	73	....	....	....	....	....
1994-95	Niagara Falls	OHL	57	3	10	13	64	6	0	2	2	0
1995-96	Niagara Falls	OHL	48	1	7	8	81	10	0	4	4	37
1996-97	Erie Otters	OHL	60	4	23	27	152	5	2	0	2	12
1997-98	Erie Otters	OHL	48	3	17	20	108	7	0	2	2	13
1998-99	Portland Pirates	AHL	6	0	0	0	4	....	....	....	....	....
	Indianapolis Ice	IHL	55	0	7	7	120	7	0	2	2	12
99-2000	Cleveland	IHL	69	6	6	12	210	9	1	0	1	12
2000-01	Norfolk Admirals	AHL	59	5	6	11	171	9	0	2	2	12
2001-02	Hershey Bears	AHL	58	1	13	14	201	7	0	1	1	6

Signed as a free agent by **Colorado**, August 8, 2001.

PAULSSON, Marcus (POWL-suhn, MAHR-kuhs) NYI

Left wing. Shoots left. 6'1", 185 lbs. Born, Karlskrona, Sweden, January 10, 1984.
(NY Islanders' 3rd choice, 149th overall, in 2002 Entry Draft).

			Regular Season					Playoffs				
Season	**Club**	**Lea**	**GP**	**G**	**A**	**TP**	**PIM**	**GP**	**G**	**A**	**TP**	**PIM**
2000-01	Morrums GoIS IK	Swede-2	12	0	1	1	4	....	....	....	....	....
2001-02	Morrums GoIS IK	Swede-2	41	0	1	1	4	....	....	....	....	....

PAVLIKOVSKY, Rastislav (pahv-lih-KAWV-skee, RA-this-LAHV) MIN.

Center. Shoots left. 6'1", 180 lbs. Born, Dubnica, Czech., March 22, 1977.
(Ottawa's 10th choice, 246th overall, in 1998 Entry Draft).

			Regular Season					Playoffs				
Season	Club	Lea	GP	G	A	TP	PIM	GP	G	A	TP	PIM
1993-94	Dukla Trencin	Slovakia	3	0	1	1	0	....	....	....	....	....
1994-95	Dukla Trencin	Slovakia	14	0	7	7	4	6	1	1	2	0
1995-96	Sault Ste. Marie	OHL	13	0	3	3	6	....	....	....	....	....
	Dukla Trencin	Slovakia	10	1	1	2	6	12	4	1	5	....
1996-97	Dukla Trencin	Slovakia	35	9	11	20	....	7	1	2	3	....
1997-98	Dukla Trencin	Slovakia	3	2	1	3	0	....	....	....	....	....
	Las Vegas	IHL	1	0	0	0	0	....	....	....	....	....
	Utah Grizzlies	IHL	74	17	29	46	54	2	0	0	0	6
1998-99	Cincinnati	IHL	31	4	12	16	28	....	....	....	....	....
	Cincinnati	AHL	36	12	23	35	59	2	0	1	1	4
99-2000	Grand Rapids	IHL	15	0	5	5	24	....	....	....	....	....
	Philadelphia	AHL	12	4	7	11	8	....	....	....	....	....
	Cincinnati	AHL	17	3	5	8	10	....	....	....	....	....
2000-01	Grand Rapids	IHL	1	0	0	0	0	....	....	....	....	....
	Jokerit Helsinki	Finland	7	0	1	1	0	....	....	....	....	....
	HV 71 Jonkoping	Sweden	23	6	8	14	93	....	....	....	....	....
2001-02	HV 71 Jonkoping	Sweden	46	14	16	30	102	8	2	2	4	6
	Slovakia	Olympics	4	2	3	5	6	....	....	....	....	....

Signed as a free agent by **Minnesota**, May 31, 2002.

PAVLOV, Yevgeny (pahv-lohv, YEHV-jeh-nee) NSH.

Center. Shoots right. 6'1", 205 lbs. Born, Togliatti, USSR, January 10, 1981.
(Nashville's 8th choice, 121st overall, in 1999 Entry Draft).

			Regular Season					Playoffs				
Season	Club	Lea	GP	G	A	TP	PIM	GP	G	A	TP	PIM
1997-98	Lada Togliatti 2	Russia-3	27	8	1	9	4	....	....	....	....	....
1998-99	Lada Togliatti 2	Russia-4	33	17	3	20	8	....	....	....	....	....
	Lada Togliatti	Russia	9	0	1	1	2	....	....	....	....	....
99-2000	Lada Togliatti 2	Russia-3	55	30	17	47	18	....	....	....	....	....
	Ufa	Russia	5	1	0	1	10	....	....	....	....	....
	Lada Togliatti	Russia	2	0	0	0	0	1	0	0	0	0
2000-01	Lada Togliatti	Russia	28	2	1	3	4	....	....	....	....	....
2001-02	Cincinnati	ECHL	54	17	14	31	14	3	0	0	0	0
	Milwaukee	AHL	7	1	2	3	0	....	....	....	....	....

PECKER, Cory (PEH-kuhr, KOH-ree) ANA.

Center. Shoots right. 6', 195 lbs. Born, Montreal, Que., March 20, 1981.
(Calgary's 7th choice, 166th overall, in 1999 Entry Draft).

			Regular Season					Playoffs				
Season	Club	Lea	GP	G	A	TP	PIM	GP	G	A	TP	PIM
1995-96	Lac St-Louis	QAAA	5	0	0	0	0	....	....	....	....	....
1996-97	Lac St-Louis	QAAA	40	30	40	70	....	7	4	2	6	....
1997-98	Sault Ste. Marie	OHL	29	3	4	7	15	....	....	....	....	....
1998-99	Sault Ste. Marie	OHL	68	25	34	59	24	5	1	2	3	2
99-2000	Sault Ste. Marie	OHL	65	33	36	69	38	12	6	8	14	8
2000-01	Sault Ste. Marie	OHL	31	24	16	40	37	....	....	....	....	....
	Erie Otters	OHL	30	17	22	39	32	15	14	9	23	16
2001-02	Erie Otters	OHL	56	*53	46	99	108	21	*25	17	*42	36

OHL Second All-Star Team (2001) • OHL First All-Star Team (2002) • Memorial Cup All-Star Team (2002)

• Missed majority of 1997-98 season after being diagnosed with Chron's Disease. Traded to **Erie** (OHL) by **Sault Ste. Marie** (OHL) for Troy Ilijow, Patrick Lamesse and Erie's 4th round choice (Jordan Owens) in 2002 OHL Priority Draft, January 3, 2001. Signed as a free agent by **Anaheim**, July 8, 2002.

PEREZ, Bryan (PAIR-ehz, BRIGH-uhn) NYI

Left wing. Shoots left. 6'1", 198 lbs. Born, Blaine, MN, February 5, 1982.
(NY Islanders' 6th choice, 260th overall, in 2001 Entry Draft).

			Regular Season					Playoffs				
Season	Club	Lea	GP	G	A	TP	PIM	GP	G	A	TP	PIM
1998-99	Team USA	USDP-18	51	5	7	12	103	....	....	....	....	....
99-2000	Team USA	USDP-18	51	7	14	21	119	....	....	....	....	....
2000-01	Michigan Tech	WCHA	DID NOT PLAY – FRESHMAN									
2001-02	Michigan Tech	WCHA	32	6	10	16	20	....	....	....	....	....

PEREZHOGIN, Alexander (pehr-eh-ZHO-ghin, al-ehx-AN-duhr) MTL.

Right wing. Shoots left. 6', 185 lbs. Born, Ust-Kamenogorsk, USSR, August 10, 1983.
(Montreal's 2nd choice, 25th overall, in 2001 Entry Draft).

			Regular Season					Playoffs				
Season	Club	Lea	GP	G	A	TP	PIM	GP	G	A	TP	PIM
1998-99	Avangard Omsk 2	Russia-4	4	0	1	1	0	....	....	....	....	....
	Avangard Omsk	Russia	22	12	11	23	12	....	....	....	....	....
99-2000	Avangard Omsk 2	Russia-3	22	12	11	23	12	....	....	....	....	....
	Avangard Omsk	Russia	1	0	0	0	0	....	....	....	....	....
2000-01	Omsk Jr.	Russia-Jr.	6	1	5	6	4	....	....	....	....	....
	Avangard Omsk 2	Russia-3	41	47	24	71	40	....	....	....	....	....
	Avangard Omsk	Russia	....	....	....	....	....	1	0	0	0	0
2001-02	Avangard Omsk	Russia	4	1	0	1	4	....	....	....	....	....
	Mostovik Kurgan	Russia-2	19	14	10	24	10	....	....	....	....	....

PERIARD, Michel (pair-EE-ahr, mee-SHEHL) FLA.

Defense. Shoots left. 5'11", 183 lbs. Born, Montreal, Que., November 10, 1979.
(Ottawa's 8th choice, 188th overall, in 1998 Entry Draft).

			Regular Season					Playoffs				
Season	Club	Lea	GP	G	A	TP	PIM	GP	G	A	TP	PIM
1996-97	Charles-Lemoyne	QAAA	40	8	15	23	64	15	7	20	27	....
1997-98	Shawinigan	QMJHL	68	14	30	44	64	5	0	0	0	18
1998-99	Shawinigan	QMJHL	64	14	40	54	90	6	1	2	3	4
99-2000	Rimouski Oceanic	QMJHL	70	25	75	100	58	14	5	17	22	16
2000-01	Port Huron	UHL	23	1	8	9	30	....	....	....	....	....
	Rockford IceHogs	UHL	31	3	14	17	20	....	....	....	....	....
	Louisville	AHL	7	0	1	1	0	....	....	....	....	....
2001-02	Macon Whoopee	ECHL	72	4	19	23	26	....	....	....	....	....

QMJHL First All-Star Team (2000) • Canadian Major Junior First All-Star Team (2000) • Memorial Cup All-Star Team (2000)

Signed as a free agent by **Florida**, August 1, 2000.

PERREAULT, Joel (PAIR-oh, JOHL) ANA.

Right wing. Shoots right. 6'1", 163 lbs. Born, Montreal, Que., April 6, 1983.
(Anaheim's 7th choice, 137th overall, in 2001 Entry Draft).

			Regular Season					Playoffs				
Season	Club	Lea	GP	G	A	TP	PIM	GP	G	A	TP	PIM
99-2000	Antoine-Girouard	QAAA	19	4	7	11	6	....	....	....	....	....
2000-01	Baie-Comeau	QMJHL	68	10	14	24	46	11	1	1	2	10
2001-02	Baie-Comeau	QMJHL	57	18	44	62	96	5	2	0	2	6

PERSSON, Kristofer (PAIR-suhn, KRIHS-toh-fuhr) CGY.

Right wing. Shoots left. 6'3", 194 lbs. Born, Umea, Sweden, January 14, 1984.
(Calgary's 8th choice, 159th overall, in 2002 Entry Draft).

			Regular Season					Playoffs				
Season	Club	Lea	GP	G	A	TP	PIM	GP	G	A	TP	PIM
99-2000	Bjorkloven Jr.	Swede-Jr.	8	4	1	5	2	....	....	....	....	....
	Team Sweden-16	Nat-Tm	3	1	0	1	0	....	....	....	....	....
2000-01	MoDo Jr.	Swede-Jr.	16	8	3	11	4	....	....	....	....	....
2001-02	MoDo Jr.	Swede-Jr.	26	9	7	16	2	3	2	0	2	0

PETER, Emanuel (PEE-tuhr, ih-MAN-yew-ehl) CGY.

Center. Shoots left. 6', 198 lbs. Born, Nieder Uzwil, Switz., June 9, 1984.
(Calgary's 6th choice, 142nd overall, in 2002 Entry Draft).

			Regular Season					Playoffs				
Season	Club	Lea	GP	G	A	TP	PIM	GP	G	A	TP	PIM
99-2000	SC Herisau-Jr.	Swiss-Jr.	18	2	12	14	....	....	....	....	....	....
2000-01	EHC Uzwil Hawks	Swiss-Jr.	26	6	20	26	14	1	0	0	0	0
2001-02	Kloten Flyers	Swiss	39	1	7	8	14	3	0	0	0	0
	EHC Kloten	Swiss-Jr.	4	1	2	3	4	3	1	0	1	4

PETERS, Andrew (PEE-tuhrs, AN-droo) BUF.

Left wing. Shoots left. 6'4", 223 lbs. Born, St. Catharines, Ont., May 5, 1980.
(Buffalo's 2nd choice, 34th overall, in 1998 Entry Draft).

			Regular Season					Playoffs				
Season	Club	Lea	GP	G	A	TP	PIM	GP	G	A	TP	PIM
1996-97	Georgetown	OPJHL	46	11	16	27	65	....	....	....	....	....
1997-98	Oshawa Generals	OHL	60	11	7	18	220	7	2	0	2	19
1998-99	Oshawa Generals	OHL	54	14	10	24	137	15	2	7	9	36
99-2000	Kitchener	OHL	42	6	13	19	95	4	0	1	1	14
2000-01	Rochester	AHL	49	0	4	4	118	....	....	....	....	....
2001-02	Rochester	AHL	67	4	1	5	*388	....	....	....	....	....

Traded to **Kitchener** (OHL) by **Oshawa** (OHL) for Barry Graham, October 11, 1999.

PETERS, Dan (PEE-tuhrs, DAN) PHI.

Defense. Shoots left. 5'10", 183 lbs. Born, Cottage Grove, MN, November 24, 1977.

			Regular Season					Playoffs				
Season	Club	Lea	GP	G	A	TP	PIM	GP	G	A	TP	PIM
1995-96	Omaha Lancers	USHL	54	11	36	47	....	....	....	....	....	....
1996-97	Colorado College	WCHA	36	4	12	16	62	....	....	....	....	....
1997-98	Colorado College	WCHA	37	5	16	21	108	....	....	....	....	....
1998-99	Colorado College	WCHA	36	8	21	29	82	....	....	....	....	....
99-2000	Colorado College	WCHA	27	2	8	10	58	....	....	....	....	....
2000-01	Philadelphia	AHL	73	2	14	16	71	6	1	4	5	8
2001-02	Philadelphia	AHL	66	6	12	18	91	5	0	1	1	2

WCHA Second All-Star Team (1999)

Signed as a free agent by **Philadelphia**, May 5, 2000.

PETERS, Geoff (PEE-tuhrs, JEHF)

Center. Shoots left. 6'1", 185 lbs. Born, Hamilton, Ont., April 30, 1978.
(Chicago's 3rd choice, 46th overall, in 1996 Entry Draft).

			Regular Season					Playoffs				
Season	Club	Lea	GP	G	A	TP	PIM	GP	G	A	TP	PIM
1993-94	Wexford Hawks	MTHL	34	39	26	65	26	....	....	....	....	....
1994-95	Niagara Falls	OHL	57	11	9	20	37	6	2	0	2	4
1995-96	Niagara Falls	OHL	64	25	34	59	51	10	4	4	8	8
1996-97	Erie Otters	OHL	28	12	10	22	39	5	1	3	4	7
1997-98	Erie Otters	OHL	31	15	11	26	36	....	....	....	....	....
	North Bay	OHL	20	11	14	25	22	....	....	....	....	....
	Indianapolis Ice	IHL	2	0	0	0	10	....	....	....	....	....
1998-99	Team Canada	Nat-Tm	38	9	4	13	50	....	....	....	....	....
	Portland Pirates	AHL	4	1	1	2	9	....	....	....	....	....
99-2000	Cleveland	IHL	68	10	4	14	87	7	0	3	3	4
2000-01	Norfolk Admirals	AHL	73	11	10	21	48	6	0	1	1	6
2001-02	Trenton Titans	ECHL	43	13	21	34	55	....	....	....	....	....
	Columbus	ECHL	22	5	13	18	7	....	....	....	....	....
	Rochester	AHL	10	1	4	5	18	2	0	1	1	0

Traded to **North Bay** (OHL) by **Erie** (OHL) with Brett Gibson for Steve Montador, January 16, 1998.

PETIOT, Richard (PEH-tee-awt, RIH-chuhrd) L.A.

Defense. Shoots left. 6'2", 190 lbs. Born, Daysland, Alta., August 20, 1982.
(Los Angeles' 6th choice, 116th overall, in 2001 Entry Draft).

			Regular Season					Playoffs				
Season	Club	Lea	GP	G	A	TP	PIM	GP	G	A	TP	PIM
99-2000	Camrose Nordics	AAHA	STATISTICS NOT AVAILABLE									
2000-01	Camrose Kodiacs	AJHL	55	8	16	24	81	8	2	1	3	8
2001-02	Colorado College	WCHA	39	4	6	10	35	....	....	....	....	....

AJHL All-Rookie Team (2001) • AJHL South Second All-Star Team (2001)

PETRASEK, David (PEH-truh-sehk, DAY-vihd) DET.

Defense. Shoots right. 6', 187 lbs. Born, Jonkoping, Sweden, February 1, 1976.
(Detroit's 10th choice, 226th overall, in 1998 Entry Draft).

			Regular Season					Playoffs				
Season	Club	Lea	GP	G	A	TP	PIM	GP	G	A	TP	PIM
1993-94	HV 71 Jr.	Swede-Jr.	14	3	3	6	26	....	....	....	....	....
1994-95	HV 71 Jr.	Swede-Jr.	19	8	9	17	55	....	....	....	....	....
	HV 71 Jonkoping	Sweden	30	0	1	1	6	11	0	0	0	0
1995-96	HV 71 Jr.	Swede-Jr.	12	1	5	6	16	....	....	....	....	....
	HV 71 Jonkoping	Sweden	36	0	1	1	14	1	0	0	0	0
1996-97	HV 71 Jr.	Swede-Jr.	3	0	0	0	0	....	....	....	....	....
	HV 71 Jonkoping	Sweden	49	2	4	6	14	5	0	0	0	4
1997-98	HV 71 Jonkoping	Sweden	43	6	7	13	80	5	2	2	4	14
1998-99	HV 71 Jonkoping	Sweden	45	3	4	7	48	....	....	....	....	....
99-2000	HV 71 Jonkoping	Sweden	46	4	6	10	54	5	1	1	2	41
2000-01	Malmo IF	Sweden	47	7	7	14	74	9	1	1	2	8
2001-02	Malmo IF	Sweden	47	2	9	11	48	2	1	0	1	2

PETRE, Henrik (PEH-truh, HEHN-rihk) WSH.

Defense. Shoots left. 6'1", 187 lbs. Born, Stockholm, Sweden, April 9, 1979.
(Washington's 5th choice, 143rd overall, in 1997 Entry Draft).

			Regular Season					Playoffs				
Season	Club	Lea	GP	G	A	TP	PIM	GP	G	A	TP	PIM
1995-96	Djurgarden Jr.	Swede-Jr.	21	6	4	10	8	....	....	....	....	....
1996-97	Djurgarden Jr.	Swede-Jr.	20	7	6	13	....	....	....	....	....	....
1997-98	Huddinge IK	Swede-2	30	4	4	8	30	....	....	....	....	....
	Djurgarden	Sweden	3	0	0	0	0	....	....	....	....	....
1998-99	Huddinge IK	Swede-2	14	0	1	1	20	....	....	....	....	....
	Djurgarden	Sweden	9	0	0	0	10	....	....	....	....	....
99-2000	Brynas IF Gavle	Sweden	47	3	3	6	73	11	1	0	1	12
	Brynas IF Gavle	EuroHL	5	0	2	2	4	....	....	....	....	....
2000-01	Brynas IF Gavle	Sweden	27	2	3	5	20	4	0	1	1	27
2001-02	Brynas IF Gavle	Sweden	24	2	1	3	49	4	0	0	0	4

PETRILAINEN, Pasi (peh-trih-LAI-nehn, PAH-see) N.J.

Defense. Shoots left. 5'10", 185 lbs. Born, Tampere, Finland, May 5, 1978.
(New Jersey's 14th choice, 225th overall, in 1996 Entry Draft).

			Regular Season					Playoffs				
Season	Club	Lea	GP	G	A	TP	PIM	GP	G	A	TP	PIM
1992-93	Tappara-C	Finn-Jr.	1	0	0	0	2	....	....	....	....	....
1993-94	Tappara-C	Finn-Jr.	30	6	20	26	36	7	2	4	6	4
	Tappara Jr.	Finn-Jr.	2	0	0	0	0	....	....	....	....	....
1994-95	Tappara-B	Finn-Jr.	8	1	5	6	4	7	1	3	4	2
	Tappara Jr.	Finn-Jr.	14	3	4	7	6	....	....	....	....	....
	Tappara Tampere	Finland	25	3	0	3	14	....	....	....	....	....
1995-96	Tappara Jr.	Finn-Jr.	5	0	2	2	4	....	....	....	....	....
	Tappara Tampere	Finland	40	0	4	4	18	4	0	0	0	2
	Tappara-B	Finn-Jr.	....	....	....	....	....	1	1	1	2	2
1996-97	Tappara Tampere	Finland	43	2	9	11	46	3	0	0	0	2
1997-98	Tappara Tampere	Finland	48	4	7	11	34	4	0	0	0	4
1998-99	Tappara Tampere	Finland	35	3	3	6	26	....	....	....	....	....
99-2000	Tappara Tampere	Finland	39	1	9	10	22	4	0	1	1	0
2000-01	Timra IK	Sweden	49	3	10	13	22	....	....	....	....	....
2001-02	TPS Turku	Finland	56	1	6	7	32	8	0	0	0	4

PETROCHININ, Evgeny (peht-roh-CHIH-nihn, ehv-GEH-nee) CBJ

Defense. Shoots left. 6'2", 190 lbs. Born, Murmansk, USSR, February 7, 1976.
(Dallas' 5th choice, 150th overall, in 1994 Entry Draft).

			Regular Season					Playoffs				
Season	Club	Lea	GP	G	A	TP	PIM	GP	G	A	TP	PIM
1993-94	Spartak Moscow	CIS	2	0	0	0	0	....	....	....	....	....
1994-95	Spartak Moscow	CIS	45	0	2	2	14	....	....	....	....	....
1995-96	Spartak Moscow	CIS	50	5	17	22	18	5	3	0	3	0
1996-97	Spartak Moscow	Russia	32	5	6	11	52	....	....	....	....	....
1997-98	Spartak Moscow	Russia	46	12	6	18	100	....	....	....	....	....
1998-99	Spartak Moscow	Russia	21	4	6	10	14	....	....	....	....	....
	Ak Bars Kazan	Russia	6	0	2	2	2	9	1	1	2	24
99-2000	Magnitogorsk	Russia	33	7	10	17	38	14	2	1	3	26
2000-01	Cherepovets	Russia	40	8	7	15	38	9	2	0	2	40
2001-02	Cherepovets	Russia	35	35	2	8	10	1	0	0	0	0

Rights traded to **Columbus** by **Dallas** for Kirk Muller, September 28, 2001.

PETRUIC, Neil (peh-TROO-ihk, NEEL) OTT.

Defense. Shoots left. 6'1", 183 lbs. Born, Regina, Sask., July 30, 1982.
(Ottawa's 10th choice, 235th overall, in 2001 Entry Draft).

			Regular Season					Playoffs				
Season	Club	Lea	GP	G	A	TP	PIM	GP	G	A	TP	PIM
99-2000	Kindersley	SJHL	68	5	25	30	....	....	....	....	....	....
2000-01	Kindersley	SJHL	68	18	24	42	123	....	....	....	....	....
2001-02	U. Minn-Duluth	WCHA	40	3	6	9	54	....	....	....	....	....

SJHL First All-Star Team (2001)

PETTERSTROM, Pontus (PEH-tuhr-stawm, PAWN-tuhs) NYR

Left wing. Shoots left. 6', 174 lbs. Born, Nybro, Sweden, April 21, 1982.
(NY Rangers' 8th choice, 226th overall, in 2001 Entry Draft).

			Regular Season					Playoffs				
Season	Club	Lea	GP	G	A	TP	PIM	GP	G	A	TP	PIM
99-2000	Leksands IF-18	Swede-Jr.	7	1	2	3	4	....	....	....	....	....
	Leksands IF Jr.	Swede-Jr.	37	13	10	23	26	2	0	0	0	4
2000-01	Tingsryds AIF	Swede-2	24	6	5	11	24	....	....	....	....	....
	Tingsryds AIF	Swede-Q	14	2	0	2	0	3	0	2	2	2
2001-02	Tingsryds AIF	Swede-2	41	8	7	15	0	....	....	....	....	....

PETTINEN, Tomi (peh-TIHN-ehn, TAW-mee) NYI

Defense. Shoots left. 6'3", 220 lbs. Born, Ylojarvi, Finland, June 17, 1977.
(NY Islanders' 9th choice, 267th overall, in 2000 Entry Draft).

			Regular Season					Playoffs				
Season	Club	Lea	GP	G	A	TP	PIM	GP	G	A	TP	PIM
1993-94	Ilves Tampere-C	Finn-Jr.	31	2	2	4	4	....	....	....	....	....
1994-95	Ilves Tampere-B	Finn-Jr.	31	1	6	7	46	4	0	0	0	2
	Ilves Jr.	Finn-Jr.	1	0	0	0	0	....	....	....	....	....
1995-96	KooVee Jr.	Finn-Jr.	21	0	0	0	64	....	....	....	....	....
	Ilves Jr.	Finn-Jr.	12	1	1	2	18	....	....	....	....	....
1996-97	Ilves Jr.	Finn-Jr.	26	3	8	11	44	....	....	....	....	....
	Ilves Tampere	Finland	16	1	0	1	12	....	....	....	....	....
1997-98	Ilves Jr.	Finn-Jr.	14	1	4	5	24	....	....	....	....	....
	Ilves Tampere	Finland	3	0	0	0	0	....	....	....	....	....
	Lukko Rauma Jr.	Finn-Jr.	11	2	6	8	14	....	....	....	....	....
	Lukko Rauma	Finland-2	4	0	0	0	2	....	....	....	....	....
	Lukko Rauma	Finland	27	0	2	2	16	....	....	....	....	....
1998-99	Hermes Kokkola	Finland-2	42	8	6	14	76	3	0	0	0	6
	HIFK Helsinki	Finland	4	0	0	0	2	....	....	....	....	....
99-2000	Ilves Tampere	Finland	51	1	6	7	78	3	1	2	3	2
2000-01	Ilves Tampere	Finland	56	2	2	4	86	9	0	0	0	4
2001-02	Ilves Tampere	Finland	48	5	4	9	51	3	0	0	0	4
	Bridgeport	AHL	....	....	....	....	....	9	0	1	1	0

PIHLMAN, Thomas (PIHL-mahn, TAWH-muhs) N.J.

Right wing. Shoots left. 6'2", 205 lbs. Born, Espoo, Finland, November 13, 1982.
(New Jersey's 3rd choice, 48th overall, in 2001 Entry Draft).

			Regular Season					Playoffs				
Season	Club	Lea	GP	G	A	TP	PIM	GP	G	A	TP	PIM
1996-97	JYP Jyvaskyla-C	Finn-Jr.	25	9	10	19	28	....	....	....	....	....
1997-98	JYP Jyvaskyla-C	Finn-Jr.	1	0	1	1	0	....	....	....	....	....
	JYP Jyvaskyla-B	Finn-Jr.	30	2	5	7	18	4	1	3	4	6
1998-99	JYP Jyvaskyla-B	Finn-Jr.	35	21	20	41	64	6	1	1	2	12
99-2000	JYP Jr.	Finn-Jr.	20	4	4	8	54	4	0	0	0	8
	JYP Jyvaskyla	Finland	17	0	0	0	18	....	....	....	....	....
2000-01	JYP Jr.	Finn-Jr.	1	1	0	1	0	....	....	....	....	....
	JYP Jyvaskyla	Finland	47	3	6	9	59	....	....	....	....	....
2001-02	JYP Jr.	Finn-Jr.	3	1	1	2	4	....	....	....	....	....
	JYP Jyvaskyla	Finland	44	9	2	11	93	....	....	....	....	....

PINC, Michal (PIHNTS, MIH-khuhl) S.J.

Center. Shoots left. 5'11", 180 lbs. Born, Litvinov, Czech., December 2, 1981.
(San Jose's 3rd choice, 142nd overall, in 2000 Entry Draft).

			Regular Season					Playoffs				
Season	Club	Lea	GP	G	A	TP	PIM	GP	G	A	TP	PIM
1997-98	Litvinov Jr.	Czech-Jr.	47	18	36	54	....	....	....	....	....	....
1998-99	Litvinov Jr.	Czech-Jr.	41	24	17	41	....	....	....	....	....	....
	Litvinov	Czech	7	0	0	0	4	....	....	....	....	....
99-2000	Hull Olympiques	QMJHL	31	11	27	38	67	....	....	....	....	....
	Rouyn-Noranda	QMJHL	27	8	16	24	64	11	0	5	5	18
2000-01	Rouyn-Noranda	QMJHL	64	16	46	62	220	9	0	3	3	12
2001-02	KLH Chomutov Jr.	Czech-Jr.	3	2	2	4	18	....	....	....	....	....
	KLH Chomutov	Czech-2	25	3	4	7	36	....	....	....	....	....
	HC Banik Most	Czech-3	7	3	1	4	14	....	....	....	....	....
	Litvinov Jr.	Czech-Jr.	14	6	7	13	58	....	....	....	....	....
	Litvinov	Czech	1	0	0	0	0	....	....	....	....	....

PIRJETA, Lasse (PEER-yeh-tuh, LAH-see) CBJ

Center. Shoots left. 6'3", 222 lbs. Born, Oulu, Finland, April 4, 1974.
(Columbus' 7th choice, 133rd overall, in 2002 Entry Draft).

			Regular Season					Playoffs				
Season	Club	Lea	GP	G	A	TP	PIM	GP	G	A	TP	PIM
1991-92	Tacoma Rockets	WHL	16	5	2	7	4	....	....	....	....	....
	Karpat Oulu Jr.	Swede-Jr.	7	1	4	5	8	....	....	....	....	....
	Karpat Oulu	Finland-2	2	0	0	0	0	....	....	....	....	....
1992-93	Karpat Oulu Jr.	Finn-Jr.	24	13	19	32	34	....	....	....	....	....
	Karpat Oulu	Finland-2	20	4	3	7	6	....	....	....	....	....
1993-94	TPS Turku Jr.	Finn-Jr.	6	2	2	4	2	4	6	2	8	2
	Kiekko-67 Turku	Finland-2	2	2	1	3	0	....	....	....	....	....
	TPS Turku	Finland	43	9	9	18	14	11	4	0	4	2
1994-95	TPS Turku Jr.	Finn-Jr.	1	0	0	0	0	....	....	....	....	....
	TPS Turku	Finland	49	7	13	20	64	8	0	1	1	29
1995-96	TPS Turku	Finland	45	13	14	27	34	11	6	3	9	4
1996-97	Vastra Frolunda	Sweden	50	14	8	22	36	3	0	1	1	4
	Vastra Frolunda	EuroHL	5	2	6	8	4	....	....	....	....	....
1997-98	Tappara Tampere	Finland	48	24	22	46	20	4	1	1	2	2
1998-99	Tappara Tampere	Finland	54	22	19	41	32	....	....	....	....	....
99-2000	HIFK Helsinki	Finland	54	18	25	43	24	9	2	3	5	10
	HIFK Helsinki	EuroHL	6	1	3	4	6	2	1	1	2	0
2000-01	HIFK Helsinki	Finland	55	15	18	33	24	....	....	....	....	....
2001-02	Karpat Oulu	Finland	55	15	26	41	24	4	2	2	4	2

PISANI, Fernando (pih-ZAN-ee, FUHR-nan-DOH) EDM.

Center/Left wing. Shoots left. 6'1", 185 lbs. Born, Edmonton, Alta., December 27, 1976.
(Edmonton's 9th choice, 195th overall, in 1996 Entry Draft).

			Regular Season					Playoffs				
Season	Club	Lea	GP	G	A	TP	PIM	GP	G	A	TP	PIM
1994-95	Bonnyville	AJHL	56	30	55	85	105	....	....	....	....	....
1995-96	St. Albert	AJHL	58	40	63	103	134	18	7	22	29	28
1996-97	Providence	H-East	35	12	18	30	36	....	....	....	....	....
1997-98	Providence	H-East	36	16	18	34	20	....	....	....	....	....
1998-99	Providence	H-East	38	14	37	51	42	....	....	....	....	....
99-2000	Providence	H-East	38	14	24	38	56	....	....	....	....	....
2000-01	Hamilton	AHL	52	12	13	25	28	....	....	....	....	....
2001-02	Hamilton	AHL	79	26	34	60	60	15	4	6	10	4

PITKANEN, Joni (PIHT-ka-nuhn, YOH-nee) PHI.

Defense. Shoots left. 6'3", 200 lbs. Born, Oulu, Finland, September 19, 1983.
(Philadelphia's 1st choice, 4th overall, in 2002 Entry Draft).

			Regular Season					Playoffs				
Season	Club	Lea	GP	G	A	TP	PIM	GP	G	A	TP	PIM
1998-99	Karpat Oulu Jr.	Finn-Jr.	30	1	5	6	12	....	....	....	....	....
99-2000	Karpat Oulu Jr.	Finn-Jr.	38	12	14	26	26	6	1	4	5	2
2000-01	Karpat Oulu Jr.	Finn-Jr.	24	6	11	17	77	....	....	....	....	....
	Karpat Oulu	Finland	21	0	0	0	10	2	0	0	0	2
2001-02	Karpat Oulu	Finland	49	4	15	19	65	4	0	0	0	12
	Karpat Oulu Jr.	Finn-Jr.	....	....	....	....	....	1	0	0	0	0

PIVKO, Libor (PIHV-koh, LEE-bohr) NSH.

Left wing. Shoots left. 6'2", 195 lbs. Born, Novy Vicin, Czech., March 29, 1980.
(Nashville's 4th choice, 89th overall, in 2000 Entry Draft).

			Regular Season					Playoffs				
Season	Club	Lea	GP	G	A	TP	PIM	GP	G	A	TP	PIM
1995-96	Slezan Opava Jr.	Czech-Jr.	37	19	14	33	30	....	....	....	....	....
1996-97	Opava Jr.	Czech-Jr.	16	12	9	21	22	....	....	....	....	....
1997-98	HC Opava Jr.	Czech-Jr.	37	15	11	26	36	....	....	....	....	....
1998-99	HC Opava Jr.	Czech-Jr.	38	21	14	35	....	....	....	....	....	....
	HC Opava	Czech	5	0	1	1	0	....	....	....	....	....
99-2000	Havirov Jr.	Czech-Jr.	5	1	3	4	4	....	....	....	....	....
	HC Femax Havirov	Czech	40	11	11	22	41	....	....	....	....	....
	HC Ytong Brno	Czech-3	....	....	....	....	....	4	3	4	7	0
2000-01	HC Femax Havirov	Czech	45	7	12	19	58	....	....	....	....	....
2001-02	Zlin	Czech	46	8	20	28	36	9	5	3	8	8

PLATIL, Jan (PLA-tihl, YAN) OTT.

Defense. Shoots left. 6'1", 193 lbs. Born, Kladno, Czech., February 9, 1983.
(Ottawa's 8th choice, 218th overall, in 2001 Entry Draft).

			Regular Season					Playoffs				
Season	Club	Lea	GP	G	A	TP	PIM	GP	G	A	TP	PIM
99-2000	Kladno Jr.	Czech-Jr.	39	5	6	11	....	....	....	....	....	....
2000-01	Barrie Colts	OHL	60	6	18	24	114	5	0	0	0	12
2001-02	Barrie Colts	OHL	68	13	34	47	136	20	1	5	6	51

PLATONOV, Denis (PLAH-tah-nahv, DIHN-ihs) **NSH.**

Center. Shoots left. 6'1", 194 lbs. Born, Saratov, USSR, November 6, 1981.
(Nashville's 4th choice, 75th overall, in 2001 Entry Draft).

			Regular Season					Playoffs				
Season	**Club**	**Lea**	**GP**	**G**	**A**	**TP**	**PIM**	**GP**	**G**	**A**	**TP**	**PIM**
1997-98	Saratov 2	Russia-3	20	4	2	6	34	....	....	....	....	....
1998-99	Kristall Saratov	Russia-2	14	1	0	1	61	....	....	....	....	....
99-2000	Saratov 2	Russia-3	5	0	0	0	37	....	....	....	....	....
	Kristall Saratov	Russia-2	32	9	4	13	60	....	....	....	....	....
2000-01	Kristall Saratov	Russia-2	51	14	6	20	75	....	....	....	....	....
2001-02	Kristall Saratov	Russia-2	50	18	14	32	96	....	....	....	....	....

PLATT, Jason (PLAT, JAY-suhn) **EDM.**

Defense. Shoots left. 6'1", 210 lbs. Born, San Francisco, CA, April 29, 1981.
(Edmonton's 9th choice, 247th overall, in 2000 Entry Draft).

			Regular Season					Playoffs				
Season	**Club**	**Lea**	**GP**	**G**	**A**	**TP**	**PIM**	**GP**	**G**	**A**	**TP**	**PIM**
1998-99	Omaha Lancers	USHL	56	2	9	11	65	11	0	0	0	8
99-2000	Omaha Lancers	USHL	49	1	6	7	65	4	0	0	0	9
2000-01	Providence	H-East	26	0	2	2	12	....	....	....	....	....
2001-02	Providence	H-East	36	2	5	7	60	....	....	....	....	....

PLEKANEC, Tomas (pleh-KA-nyehts, TAW-mahsh) **MTL.**

Left wing. Shoots left. 5'10", 189 lbs. Born, Kladno, Czech., October 31, 1982.
(Montreal's 4th choice, 71st overall, in 2001 Entry Draft).

			Regular Season					Playoffs				
Season	**Club**	**Lea**	**GP**	**G**	**A**	**TP**	**PIM**	**GP**	**G**	**A**	**TP**	**PIM**
1998-99	Kladno	Czech	3	0	0	0	0	....	....	....	....	....
99-2000	Kladno Jr.	Czech-Jr.	43	14	16	30	....	....	....	....	....	....
2000-01	Kladno	Czech	47	9	9	18	24	....	....	....	....	....
2001-02	Kladno	Czech	48	7	16	23	28	....	....	....	....	....
	Kladno	Czech-Q	5	0	1	1	0	....	....	....	....	....

PLIHAL, Tomas (PLEE-hahl, TAW-mahsh) **S.J.**

Center. Shoots left. 6'1", 176 lbs. Born, Frydlant v Cechach, Czech., March 28, 1983.
(San Jose's 4th choice, 140th overall, in 2001 Entry Draft).

			Regular Season					Playoffs				
Season	**Club**	**Lea**	**GP**	**G**	**A**	**TP**	**PIM**	**GP**	**G**	**A**	**TP**	**PIM**
2000-01	HC Liberec Jr.	Czech-Jr.	33	16	12	28	....	....	....	....	....	....
2001-02	Kootenay Ice	WHL	72	32	54	86	28	22	4	10	14	14

PODHRADSKY, Peter (pohd-RAD-skee, PEE-tuhr) **ANA.**

Defense. Shoots right. 6'2", 204 lbs. Born, Bratislava, Czech., December 10, 1979.
(Anaheim's 4th choice, 134th overall, in 2000 Entry Draft).

			Regular Season					Playoffs				
Season	**Club**	**Lea**	**GP**	**G**	**A**	**TP**	**PIM**	**GP**	**G**	**A**	**TP**	**PIM**
1995-96	S. Bratislava Jr.	Slovak-Jr.	50	14	17	31	....	....	....	....	....	....
1996-97	S. Bratislava Jr.	Slovak-Jr.	41	2	6	8	28	....	....	....	....	....
1997-98	S. Bratislava Jr.	Slovak-Jr.	48	11	13	24	58	....	....	....	....	....
1998-99	S. Bratislava Jr.	Slovak-Jr.	25	9	14	23	57	2	0	1	1	0
	Slov. Bratislava	Slovakia	23	1	4	5	37	4	0	0	0	0
99-2000	HK Kabat Trnava	Slovak-2	1	0	0	0	0	....	....	....	....	....
	Slov. Bratislava	Slovakia	40	4	11	15	63	8	1	0	1	2
2000-01	Cincinnati	AHL	59	4	5	9	27	2	0	0	0	0
2001-02	Cincinnati	AHL	49	0	8	8	19	....	....	....	....	....

PODLESAK, Martin (PAWD-leh-shahk, MAHR-tihn) **PHX.**

Left wing. Shoots left. 6'6", 218 lbs. Born, Melnik, Czech., September 26, 1982.
(Phoenix's 3rd choice, 45th overall, in 2001 Entry Draft).

			Regular Season					Playoffs				
Season	**Club**	**Lea**	**GP**	**G**	**A**	**TP**	**PIM**	**GP**	**G**	**A**	**TP**	**PIM**
99-2000	Sparta Praha Jr.	Czech-Jr.	24	6	5	11	....	11	6	2	8	....
2000-01	Tri-City	WHL	39	13	13	26	36	....	....	....	....	....
	Lethbridge	WHL	21	8	6	14	23	3	1	1	2	2
2001-02	Lethbridge	WHL	34	14	20	34	33	....	....	....	....	....

Traded to **Lethbridge** (WHL) by **Tri-City** (WHL) for Derrick Atkinson, Lethbridge's 1st choice (Logan Stephenson) in 2001 WHL Bantam Draft and Lethbridge's 1st choice (Peter Lorentzen) in 2001 CHL Import Draft, January 15, 2001.

POHANKA, Igor (poh-HAHN-kah, EE-gohr) **ANA.**

Center. Shoots left. 6'3", 185 lbs. Born, Piestany, Czech., July 5, 1983.
(New Jersey's 2nd choice, 44th overall, in 2001 Entry Draft).

			Regular Season					Playoffs				
Season	**Club**	**Lea**	**GP**	**G**	**A**	**TP**	**PIM**	**GP**	**G**	**A**	**TP**	**PIM**
99-2000	S. Bratislava Jr.	Slovak-Jr.	57	36	41	77	62	....	....	....	....	....
2000-01	Prince Albert	WHL	70	16	33	49	24	....	....	....	....	....
2001-02	Prince Albert	WHL	58	25	43	68	18	....	....	....	....	....

Traded to **Anaheim** by **New Jersey** with Petr Sykora, Mike Commodore and Jean-Francois Damphousse for Jeff Friesen, Oleg Tverdovsky and Maxim Balmochnykh, July 6, 2002.

POHL, John (PAWL, JAWN) **ST.L.**

Center. Shoots right. 6', 186 lbs. Born, Rochester, MN, June 29, 1979.
(St. Louis' 8th choice, 255th overall, in 1998 Entry Draft).

			Regular Season					Playoffs				
Season	**Club**	**Lea**	**GP**	**G**	**A**	**TP**	**PIM**	**GP**	**G**	**A**	**TP**	**PIM**
1997-98	Red Wing Wingers	Hi-School	28	30	77	107	18	....	....	....	....	....
	Twin Cities	USHL	10	5	3	8	10	....	....	....	....	....
1998-99	U. of Minnesota	WCHA	42	7	10	17	18	....	....	....	....	....
99-2000	U. of Minnesota	WCHA	41	18	41	59	26	....	....	....	....	....
2000-01	U. of Minnesota	WCHA	38	19	26	45	24	....	....	....	....	....
2001-02	U. of Minnesota	WCHA	44	27	*52	*79	26	....	....	....	....	....

Minnesota High School Player of the Year (1998) • WCHA Second All-Star Team (2000) • WCHA First All-Star Team (2002) • NCAA Championship All-Tournament Team (2002)

POLASKI, Scott (poh-LAHZ-kee, SKAWT) **PHX.**

Right wing. Shoots right. 6'2", 182 lbs. Born, Colorado Springs, CO, August 4, 1982.
(Phoenix's 6th choice, 180th overall, in 2001 Entry Draft).

			Regular Season					Playoffs				
Season	**Club**	**Lea**	**GP**	**G**	**A**	**TP**	**PIM**	**GP**	**G**	**A**	**TP**	**PIM**
1998-99	Pikes Point	AAHA	60	40	46	86	....	....	....	....	....	....
99-2000	Sioux City	USHL	58	15	22	37	46	5	3	6	9	0
2000-01	Sioux City	USHL	51	18	26	44	65	3	0	1	1	0
2001-02	Colorado College	WCHA	38	4	12	16	28	....	....	....	....	....

POLCIK, Peter (POHL-chihk, PEE-tuhr) **WSH.**

Right wing. Shoots left. 6'4", 187 lbs. Born, Nitra, Czech., July 23, 1983.
(Washington's 8th choice, 254th overall, in 2001 Entry Draft).

			Regular Season					Playoffs				
Season	**Club**	**Lea**	**GP**	**G**	**A**	**TP**	**PIM**	**GP**	**G**	**A**	**TP**	**PIM**
1998-99	MHC Nitra Jr.	Slovak-Jr.	35	13	16	29	14	....	....	....	....	....
99-2000	MHC Nitra Jr.	Slovak-Jr.	40	26	20	46	82	....	....	....	....	....
2000-01	MHC Nitra Jr.	Slovak-Jr.	42	8	10	18	32	....	....	....	....	....
2001-02	Montreal Rocket	QMJHL	70	9	12	21	35	2	0	0	0	0

POLLOCK, Jame (PAWL-lawk, JAYM)

Defense. Shoots right. 6'1", 210 lbs. Born, Quebec, Que., June 16, 1979.
(St. Louis' 4th choice, 106th overall, in 1997 Entry Draft).

			Regular Season					Playoffs				
Season	**Club**	**Lea**	**GP**	**G**	**A**	**TP**	**PIM**	**GP**	**G**	**A**	**TP**	**PIM**
1994-95	Victoria Legion	BCAHA	43	22	56	78	96	....	....	....	....	....
1995-96	Seattle	WHL	32	0	1	1	15	....	....	....	....	....
1996-97	Seattle	WHL	66	15	19	34	94	15	3	5	8	16
1997-98	Seattle	WHL	66	11	36	47	78	5	0	1	1	17
1998-99	Seattle	WHL	59	10	32	42	78	11	3	4	7	8
99-2000	Worcester	AHL	56	12	12	24	50	9	5	3	8	6
2000-01	Worcester	AHL	55	15	8	23	36	11	1	7	8	10
2001-02	Worcester	AHL	71	23	43	66	89	3	1	0	1	2

POLUSHIN, Alexander (puh-LOOSH-ihn, al-ehx-AN-duhr) **T.B.**

Right wing. Shoots left. 6'3", 200 lbs. Born, Kirovo-Chepetsk, USSR, May 8, 1983.
(Tampa Bay's 2nd choice, 47th overall, in 2001 Entry Draft).

			Regular Season					Playoffs				
Season	**Club**	**Lea**	**GP**	**G**	**A**	**TP**	**PIM**	**GP**	**G**	**A**	**TP**	**PIM**
99-2000	DynamoMoscow2	Russia-3	18	4	3	7	14	....	....	....	....	....
	Spartak Moscow	Russia-2	14	1	0	1	2	....	....	....	....	....
2000-01	THC Tver	Russia-2	38	10	5	15	10	....	....	....	....	....
2001-02	H.C. CSKA	Russia-2	55	28	21	49	18	....	....	....	....	....

POMINVILLE, Jason (paw-MIHN-vihl, JAY-suhn) **BUF.**

Right wing. Shoots right. 6', 178 lbs. Born, Repentigny, Que., November 30, 1982.
(Buffalo's 4th choice, 55th overall, in 2001 Entry Draft).

			Regular Season					Playoffs				
Season	**Club**	**Lea**	**GP**	**G**	**A**	**TP**	**PIM**	**GP**	**G**	**A**	**TP**	**PIM**
1998-99	Rive-Nord Elites	QAHA	STATISTICS NOT AVAILABLE									
	Shawinigan	QMJHL	2	0	0	0	0	....	....	....	....	....
99-2000	Shawinigan	QMJHL	60	4	17	21	12	13	2	3	5	0
2000-01	Shawinigan	QMJHL	71	46	67	113	24	10	6	6	12	0
2001-02	Shawinigan	QMJHL	66	57	64	121	32	2	0	0	0	0

QMJHL First All-Star Team (2002)

POPOVIC, Mark (poh-PUH-vihk, MAHRK) **ANA.**

Defense. Shoots left. 6'1", 191 lbs. Born, Stoney Creek, Ont., October 11, 1982.
(Anaheim's 2nd choice, 35th overall, in 2001 Entry Draft).

			Regular Season					Playoffs				
Season	**Club**	**Lea**	**GP**	**G**	**A**	**TP**	**PIM**	**GP**	**G**	**A**	**TP**	**PIM**
1997-98	Mississauga	OPJHL	51	10	16	26	32	....	....	....	....	....
1998-99	St. Michael's	OHL	60	6	26	32	46	....	....	....	....	....
99-2000	St. Michael's	OHL	68	11	29	40	68	....	....	....	....	....
2000-01	St. Michael's	OHL	61	7	35	42	54	18	3	5	8	22
2001-02	St. Michael's	OHL	58	12	29	41	42	15	1	11	12	10

OHL First All-Star Team (2002)

POSNOV, Andrei (pawz-NAWF, AN-dray) **N.J.**

Left wing. Shoots right. 6', 185 lbs. Born, Vorkuta, USSR, November 19, 1981.
(New Jersey's 7th choice, 128th overall, in 2001 Entry Draft).

			Regular Season					Playoffs				
Season	**Club**	**Lea**	**GP**	**G**	**A**	**TP**	**PIM**	**GP**	**G**	**A**	**TP**	**PIM**
1998-99	Spartak Mos. 2	Russia-4	8	3	1	4	8	....	....	....	....	....
	Spartak Moscow	Russia-Q	1	0	0	0	0	....	....	....	....	....
99-2000	Krylja Sovetov 2	Russia-3	26	7	8	15	52	....	....	....	....	....
2000-01	Krylja Sovetov	Russia-2	15	0	5	5	4	....	....	....	....	....
2001-02	Krylja Sovetov 2	Russia-3	7	4	5	9	12	....	....	....	....	....
	Krylja Sovetov	Russia	49	12	6	18	99	3	0	0	0	22

POTULNY, Grant (puh-TUHL-nee, GRANT) **OTT.**

Center. Shoots left. 6'2", 194 lbs. Born, Grand Forks, ND, March 4, 1980.
(Ottawa's 7th choice, 157th overall, in 2000 Entry Draft).

			Regular Season					Playoffs				
Season	**Club**	**Lea**	**GP**	**G**	**A**	**TP**	**PIM**	**GP**	**G**	**A**	**TP**	**PIM**
1998-99	Lincoln Stars	USHL	46	7	11	18	76	10	2	1	3	7
99-2000	Lincoln Stars	USHL	56	25	30	55	85	10	3	4	7	4
2000-01	U. of Minnesota	WCHA	42	22	11	33	38	....	....	....	....	....
2001-02	U. of Minnesota	WCHA	43	15	19	34	38	....	....	....	....	....

NCAA Championship All-Tournament Team (2002) • NCAA Championship Tournament MVP (2002)

PRATT, Harlan (PRAT, HAR-lahn) **T.B.**

Defense. Shoots left. 6'1", 195 lbs. Born, Fort McMurray, Alta., December 10, 1978.
(Pittsburgh's 5th choice, 124th overall, in 1997 Entry Draft).

			Regular Season					Playoffs				
Season	**Club**	**Lea**	**GP**	**G**	**A**	**TP**	**PIM**	**GP**	**G**	**A**	**TP**	**PIM**
1994-95	Seattle	WHL	33	1	0	1	17	1	0	0	0	0
1995-96	Red Deer Rebels	WHL	60	2	3	5	22	10	0	0	0	4
1996-97	Red Deer Rebels	WHL	2	0	0	0	2	....	....	....	....	....
	Prince Albert	WHL	65	7	26	33	49	4	1	1	2	4
1997-98	Prince Albert	WHL	37	6	14	20	12	....	....	....	....	....
	Regina Pats	WHL	24	2	6	8	23	9	2	2	4	2
1998-99	Portland	WHL	10	1	3	4	10	....	....	....	....	....
	Toledo Storm	ECHL	61	4	35	39	32	3	0	0	0	0
99-2000	Florida	ECHL	68	4	29	33	38	5	0	1	1	2
2000-01	Cincinnati	IHL	73	6	23	29	45	2	0	1	1	2
2001-02	Lowell	AHL	17	1	6	7	6	....	....	....	....	....
	Florida	ECHL	13	0	7	7	4	....	....	....	....	....
	Springfield	AHL	19	0	4	4	17	....	....	....	....	....
	Pensacola	ECHL	17	2	8	10	14	3	0	4	4	2

Traded to **Red Deer** (WHL) by **Seattle** for Dan Tompkins, September 26, 1995. Traded to **Prince Albert** (WHL) by **Red Deer** (WHL) for future considerations, October 2, 1996. Traded to **Portland** (WHL) by **Regina** (WHL) for future considerations, September 28, 1999. Signed as a free agent by **Carolina**, August 21, 2000. Traded to **Tampa Bay** by **Carolina** for Kaspars Astashenko, December 28, 2001.

PRESTBERG, Pelle (PREHST-buhrg, PEHL-lee) ANA.

Left wing. Shoots left. 5'10", 170 lbs. Born, Jonkoping, Sweden, February 5, 1975.
(Anaheim's 7th choice, 233rd overall, in 1998 Entry Draft).

			Regular Season					Playoffs				
Season	Club	Lea	GP	G	A	TP	PIM	GP	G	A	TP	PIM
1990-91	IFK Munkfors	Swede-3	3	0	3	3	....	....	....	....	....	....
1991-92	IFK Munkfors	Swede-3	26	6	10	16	18	....	....	....	....	....
1992-93	IFK Munkfors	Swede-3	36	8	8	16	20	....	....	....	....	....
1993-94	Sunne IK	Swede-3	32	8	6	14	16	....	....	....	....	....
1994-95	IFK Munkfors	Swede-3	27	13	9	22	44	....	....	....	....	....
1995-96	IFK Munkfors	Swede-3	30	20	11	31	32	....	....	....	....	....
1996-97	IFK Munkfors	Swede-3	32	28	10	38	50	....	....	....	....	....
1997-98	Farjestad	Sweden	45	29	15	44	22	12	*9	2	11	8
1998-99	Farjestad	Sweden	48	18	15	33	28	4	0	1	1	4
99-2000	Farjestad	Sweden	48	13	9	22	26	7	1	1	2	18
2000-01	Vastra Frolunda	Sweden	50	14	9	23	18	5	0	0	0	8
2001-02	Vastra Frolunda	Sweden	50	14	11	25	28	10	5	0	5	12

PREUCIL, Petr (PREE-oo-chihl, PEE-tuhr) NYR

Center. Shoots left. 6'1", 168 lbs. Born, Most, Czech., January 21, 1983.
(NY Rangers' 7th choice, 206th overall, in 2001 Entry Draft).

			Regular Season					Playoffs				
Season	Club	Lea	GP	G	A	TP	PIM	GP	G	A	TP	PIM
99-2000	Litvinov Jr.	Czech-Jr.	23	7	4	11	....	....	....	....	....	....
2000-01	Quebec Remparts	QMJHL	70	12	35	47	121	4	1	0	1	11
2001-02	Quebec Remparts	QMJHL	57	14	20	34	116	9	2	3	5	24

PRIECHODSKY, Marek (pree-HOHD-skee, MA-rehk) T.B.

Defense. Shoots left. 6'2", 194 lbs. Born, Bratislava, Czech., October 24, 1979.
(Tampa Bay's 7th choice, 222nd overall, in 2000 Entry Draft).

			Regular Season					Playoffs				
Season	Club	Lea	GP	G	A	TP	PIM	GP	G	A	TP	PIM
1997-98	S. Bratislava Jr.	Slovak-Jr.	33	1	3	4	24	....	....	....	....	....
1998-99	HC Dukla Senica	Slovak-2	STATISTICS NOT AVAILABLE									
99-2000	HK Kabat Trnava	Slovak-2	28	2	8	10	54	....	....	....	....	....
	Slov. Bratislava	Slovakia	6	0	0	0	2	....	....	....	....	....
2000-01	Slov. Bratislava	Slovakia	44	0	5	5	6	....	....	....	....	....
2001-02	Pensacola	ECHL	52	3	11	14	33	3	0	0	0	0

PRINTZ, David (PRIHNTS, DAY-vihd) PHI.

Defense. Shoots left. 6'5", 220 lbs. Born, Stockholm, Sweden, July 24, 1980.
(Philadelphia's 9th choice, 225th overall, in 2001 Entry Draft).

			Regular Season					Playoffs				
Season	Club	Lea	GP	G	A	TP	PIM	GP	G	A	TP	PIM
1996-97	AIK Solna Jr.	Swede-Jr.	1	0	0	0	0	....	....	....	....	....
1997-98	AIK Solna Jr.	Swede-Jr.	8	0	0	0	6	....	....	....	....	....
1998-99	AIK Solna Jr.	Swede-Jr.	23	1	0	1	14	....	....	....	....	....
99-2000	AIK Solna Jr.	Swede-Jr.	36	8	4	12	53	....	....	....	....	....
2000-01	Great Falls	AWHL	54	13	23	36	93	13	3	5	8	16
2001-02	AIK Solna Jr.	Swede-Jr.	8	2	3	5	20	....	....	....	....	....
	AIK Solna	Sweden	37	3	2	5	59	....	....	....	....	....
	AIK Solna	Swede-Q	10	0	0	0	12	....	....	....	....	....

PUDLICK, Michael (PUHD-lihk, MIGHK-uhl) L.A.

Defense. Shoots left. 6'3", 190 lbs. Born, Blaine, MN, February 24, 1978.

			Regular Season					Playoffs				
Season	Club	Lea	GP	G	A	TP	PIM	GP	G	A	TP	PIM
1995-96	Blaine Bengals	Hi-School	25	9	30	39	....	....	....	....	....	....
1996-97	Twin Cities	USHL	49	10	19	29	93	5	0	2	2	4
1997-98	Twin Cities	USHL	50	3	14	17	138	....	....	....	....	....
1998-99	St. Cloud State	WCHA	37	13	12	25	74	....	....	....	....	....
99-2000	St. Cloud State	WCHA	40	8	22	30	65	....	....	....	....	....
2000-01	Lowell	AHL	57	7	13	20	39	4	0	1	1	2
2001-02	Manchester	AHL	64	9	6	15	42	3	0	0	0	6

WCHA First All-Star Team (2000) • NCAA West Second All-American Team (2000)

Signed as a free agent by **LA Kings** April 5, 2000.

PUNCOCHAR, Petr (POON-choh-hahr, PEE-tuhr) CHI.

Defense. Shoots right. 6'1", 202 lbs. Born, Tabor, Czech., June 8, 1983.
(Chicago's 10th choice, 186th overall, in 2001 Entry Draft).

			Regular Season					Playoffs				
Season	Club	Lea	GP	G	A	TP	PIM	GP	G	A	TP	PIM
1998-99	C. Budejovice Jr.	Czech-Jr.	45	6	12	18	20	....	....	....	....	....
99-2000	C. Budejovice Jr.	Czech-Jr.	18	1	3	4	4	....	....	....	....	....
	Karlovy Vary Jr.	Czech-Jr.	23	1	1	2	4	2	0	0	0	2
	HC Karlovy Vary	Czech	1	0	0	0	0	....	....	....	....	....
2000-01	Karlovy Vary Jr.	Czech-Jr.	35	8	5	13	14	....	....	....	....	....
	HC Karlovy Vary	Czech	8	1	0	1	4	....	....	....	....	....
	HC Banik Most	Czech-3	1	0	0	0	0	....	....	....	....	....
2001-02	Karlovy Vary Jr.	Czech-Jr.	5	0	0	0	6	....	....	....	....	....
	HC Banik	Czech-3	7	2	1	3	0	....	....	....	....	....
	HC Karlovy Vary	Czech	32	0	2	2	36	....	....	....	....	....

RACHUNEK, Ivan (ra-KHOO-nuhk, EE-vahn) T.B.

Right wing. Shoots left. 5'9", 180 lbs. Born, Gottwaldov, Czech., July 6, 1981.
(Tampa Bay's 8th choice, 187th overall, in 1999 Entry Draft).

			Regular Season					Playoffs				
Season	Club	Lea	GP	G	A	TP	PIM	GP	G	A	TP	PIM
1997-98	Zlin Jr.	Czech-Jr.	48	15	25	40	172	....	....	....	....	....
1998-99	Zlin Jr.	Czech-Jr.	40	37	22	59	70	....	....	....	....	....
	Zlin	Czech	5	0	0	0	0	....	....	....	....	....
99-2000	Zlin	Czech	5	0	1	1	2	....	....	....	....	....
	Windsor	OHL	15	2	2	4	21	....	....	....	....	....
2000-01	Zlin	Czech	50	8	9	17	95	6	0	1	1	8
2001-02	Zlin	Czech	48	9	11	20	143	11	3	4	7	6

RADULOV, Igor (rah-DOO-lahf, EE-gohr) CHI.

Left wing. Shoots left. 6', 194 lbs. Born, Nizhny Tagil, USSR, August 23, 1982.
(Chicago's 4th choice, 74th overall, in 2000 Entry Draft).

			Regular Season					Playoffs				
Season	Club	Lea	GP	G	A	TP	PIM	GP	G	A	TP	PIM
1997-98	Yaroslavl	Russia	5	0	2	2	4	....	....	....	....	....
1998-99	Yaroslavl 2	Russia-3	21	2	3	5	4	....	....	....	....	....
99-2000	Yaroslavl 2	Russia-3	31	17	16	33	....	....	....	....	....	....
2000-01	Kristall Saratov	Russia-2	4	0	2	2	2	....	....	....	....	....
	St. Petersburg	Russia	8	1	0	1	6	....	....	....	....	....
2001-02	Mississauga	OHL	62	33	30	63	30	....	....	....	....	....

RADUNSKE, Brock (ra-DOON-skee, BRAWK) EDM.

Left wing. Shoots left. 6'4", 187 lbs. Born, Kitchener, Ont., April 5, 1983.
(Edmonton's 5th choice, 79th overall, in 2002 Entry Draft).

			Regular Season					Playoffs				
Season	Club	Lea	GP	G	A	TP	PIM	GP	G	A	TP	PIM
99-2000	Aurora Tigers	OPJHL	42	6	14	20	23	4	4	8	12	2
2000-01	Newmarket	OPJHL	48	30	39	69	65	....	....	....	....	....
2001-02	Michigan State	CCHA	41	4	9	13	28	....	....	....	....	....

RAJAMAKI, Erkki (righ-ya-MA-kee, UHR-kee) T.B.

Left wing. Shoots left. 6'2", 205 lbs. Born, Vantaa, Finland, October 30, 1978.
(Tampa Bay's 9th choice, 216th overall, in 1999 Entry Draft).

			Regular Season					Playoffs				
Season	Club	Lea	GP	G	A	TP	PIM	GP	G	A	TP	PIM
1993-94	Vantaa HT-C	Finn-Jr.	10	0	1	1	0	....	....	....	....	....
1994-95	K. Vantaa Jr.	Finn-Jr.	2	0	0	0	0	....	....	....	....	....
1995-96	K. Vantaa Jr.	Finn-Jr.	DID NOT PLAY – INJURED									
1996-97	Kiekko Vantaa-B	Finn-Jr.	33	14	19	33	32	....	....	....	....	....
1997-98	HIFK Jr.	Finn-Jr.	14	1	2	3	2	....	....	....	....	....
	Kiekko Vantaa-B	Finn-Jr.	....	....	....	....	....	10	3	0	3	2
1998-99	HIFK Helsinki-B	Finn-Jr.	14	2	2	4	8	....	....	....	....	....
	HIFK Helsinki	Finland	14	0	0	0	2	....	....	....	....	....
	HIFK Jr.	Finn-Jr.	....	....	....	....	....	13	7	3	10	45
99-2000	Colgate	ECAC	31	1	6	7	20	....	....	....	....	....
2000-01	Newcastle	Britain	11	1	0	1	0	....	....	....	....	....
	FoPS Forssa	Finland-2	4	1	3	4	0	....	....	....	....	....
	HIFK Helsinki	Finland	50	1	2	3	10	5	0	0	0	2
2001-02	HPK Hameenlinna	Finland	56	12	7	19	75	8	1	2	3	2

RAJAMAKI, Tommi (righ-YAH-ma-kee, TAW-mee) CBJ

Defense. Shoots left. 6'3", 206 lbs. Born, Pori, Finland, February 29, 1976.
(Toronto's 6th choice, 178th overall, in 1994 Entry Draft).

			Regular Season					Playoffs				
Season	Club	Lea	GP	G	A	TP	PIM	GP	G	A	TP	PIM
1992-93	Assat Pori Jr.	Finn-Jr.	35	3	7	10	16	4	0	1	1	0
1993-94	Assat Pori Jr.	Finn-Jr.	27	2	8	10	34	....	....	....	....	....
1994-95	Assat Pori Jr.	Finn-Jr.	29	11	17	28	30	....	....	....	....	....
	Assat Pori	Finland	12	4	1	5	8	7	0	1	1	2
1995-96	Assat Pori	Finland	45	5	2	7	26	3	0	0	0	6
	Assat Pori Jr.	Finn-Jr.	....	....	....	....	....	6	1	3	4	2
1996-97	Assat Pori	Finland	46	0	1	1	16	4	0	0	0	2
1997-98	TPS Turku	Finland	46	1	1	2	24	4	0	0	0	0
1998-99	TPS Turku	Finland	53	3	3	6	24	10	0	0	0	6
99-2000	TPS Turku	Finland	53	2	6	8	51	11	0	1	1	2
	TPS Turku	EuroHL	6	0	3	3	0	5	0	0	0	0
2000-01	TPS Turku	Finland	56	6	3	9	20	10	1	1	2	2
2001-02	Timra IK	Sweden	50	1	5	6	57	....	....	....	....	....
	Timra IK	Swede-Q	10	1	0	1	4	....	....	....	....	....

Selected by **Columbus** from **Toronto** in Expansion Draft, June 23, 2000.

RAKHMATULLIN, Ashkat (rahkh-ma-TOO-lihn, ahs-KHAHT) MIN.

Left wing. Shoots left. 5'11", 165 lbs. Born, Ufa, USSR, May 31, 1978.
(Hartford's 10th choice, 231st overall, in 1996 Entry Draft).

			Regular Season					Playoffs				
Season	Club	Lea	GP	G	A	TP	PIM	GP	G	A	TP	PIM
1996-97	Ufa	Russia	28	1	3	4	8	3	0	0	0	0
1997-98	Ufa	Russia	14	0	1	1	6	....	....	....	....	....
1998-99	Asheville Smoke	UHL	31	6	10	16	23	4	1	0	1	0
	Fayetteville	CHL	4	0	0	0	4	....	....	....	....	....
	Florida	ECHL	6	0	1	1	2	....	....	....	....	....
99-2000	Ufa	Russia	34	5	8	13	14	....	....	....	....	....
2000-01	Ufa	Russia	44	7	15	22	22	....	....	....	....	....
2001-02	St. Petersburg	Russia	30	7	7	14	14	....	....	....	....	....

Rights transferred to **Carolina** after **Hartford** franchise relocated, June 25, 1997. Traded to **Minnesota** by **Carolina** with Carolina's 3rd round choice (later traded to NY Rangers - NY Rangers selected Garth Murray) in 2001 Entry Draft and Carolina's compensatory 5th round choice (Armands Berzins) in 2002 Entry Draft for Scott Pellerin, March 1, 2001.

RAMHOLT, Arne (RAM-hohlt, AHR-neh)

Defense. Shoots right. 6'3", 215 lbs. Born, Zurich, Switz., May 20, 1976.
(Chicago's 15th choice, 291st overall, in 2000 Entry Draft).

			Regular Season					Playoffs				
Season	Club	Lea	GP	G	A	TP	PIM	GP	G	A	TP	PIM
1993-94	Zurcher SC Jr.	Swiss-Jr.	15	0	0	0	2	....	....	....	....	....
1994-95	Zurcher SC Jr.	Swiss-Jr.	25	0	2	2	4	11	0	0	0	0
1995-96	Zurcher SC Jr.	Swiss-Jr.	5	0	0	0	0	....	....	....	....	....
1996-97	Zurcher SC	Swiss	5	0	0	0	0	....	....	....	....	....
1997-98	St. Lawrence	ECAC	14	0	0	0	2	....	....	....	....	....
1998-99	ZSC Lions Zurich	Swiss-2	29	5	10	15	24	....	....	....	....	....
	ZSC Lions Zurich	Swiss	5	0	0	0	2	....	....	....	....	....
	EHC Kloten	Swiss	8	0	0	0	2	12	0	1	1	8
99-2000	EHC Kloten	Swiss	34	0	4	4	12	7	0	2	2	16
2000-01	Norfolk Admirals	AHL	62	3	8	11	25	....	....	....	....	....
2001-02	EV Zug	Swiss	42	3	3	6	28	6	1	2	3	2

RAWLYK, Rory (RAW-lihk, ROHR-ee) NYR

Defense. Shoots right. 6'3", 175 lbs. Born, Edmonton, Alta., September 9, 1983.

			Regular Season					Playoffs				
Season	Club	Lea	GP	G	A	TP	PIM	GP	G	A	TP	PIM
1998-99	Edm. Maple Leafs	AMBHL	36	6	18	24	58	....	....	....	....	....
99-2000	Edm. United Cycle	AMHL	28	3	14	17	34	....	....	....	....	....
2000-01	Camrose Kodiaks	AJHL	24	3	6	9	16	16	1	5	6	32
2001-02	Medicine Hat	WHL	40	2	9	11	59	....	....	....	....	....
	Vancouver Giants	WHL	28	3	7	10	21	....	....	....	....	....

Signed as a free agent by **NY Rangers**, September 15, 2001. Traded to **Vancouver** (WHL) by **Medicine Hat** (WHL) with Eric Clark, Adam Courchaine, Mitch Bartley and Medicine Hat's 1st choice (later traded to Prince Albert - Prince Albert selected Michael Gauhier) in 2002 WHL Bantam Draft for Warren McCutcheon, Andrew Davidson and Phil Cole, January 11, 2002.

RAZIN, Andrei (RAH-zihn, AN-dray) **PHI.**

Center. Shoots left. 5'11", 180 lbs. Born, Togliatti, Russia, October 23, 1973.
(Philadelphia's 7th choice, 177th overall, in 2001 Entry Draft).

			Regular Season					Playoffs				
Season	Club	Lea	GP	G	A	TP	PIM	GP	G	A	TP	PIM
1990-91	Mayak Samara	USSR-3	2	0	0	0	2	....	....	....	....	....
1991-92	Mayak Samara	CIS-3	41	15	16	31	26	....	....	....	....	....
	Lada Togliatti	CIS	7	0	0	0	2	....	....	....	....	....
1992-93	Mayak Samara	CIS-2	29	8	7	15	12	....	....	....	....	....
	Lada Togliatti	CIS	12	1	2	3	0	1	0	0	0	0
1993-94	Lada Togliatti 2	CIS-3	4	4	0	4	0	....	....	....	....	....
	Lada Togliatti	CIS	15	1	1	2	0	....	....	....	....	....
1994-95	Magnitogorsk	CIS	49	11	14	25	12	7	3	2	5	16
1995-96	Magnitogorsk	CIS	40	6	11	17	28	4	0	0	0	2
	Magnitogorsk 2	CIS-2	5	3	0	3	0	....	....	....	....	....
1996-97	Magnitogorsk 2	Russia-3	4	5	3	8	4	....	....	....	....	....
	CSK VVS Samara	Russia	32	7	8	15	12	2	0	1	1	2
1997-98	Magnitogorsk 2	Russia-3	1	1	2	3	0	....	....	....	....	....
	Magnitogorsk	Russia	46	6	32	38	12	....	....	....	....	....
1998-99	Magnitogorsk	EuroHL	6	2	4	6	2	6	2	5	7	4
	Magnitogorsk 2	Russia-4	1	2	2	4	0	....	....	....	....	....
	Magnitogorsk	Russia	39	7	25	32	14	16	4	3	7	6
99-2000	Magnitogorsk	EuroHL	2	0	2	2	4	5	2	0	2	20
	Magnitogorsk 2	Russia-3	4	3	5	8	0	....	....	....	....	....
	Magnitogorsk	Russia	29	11	10	21	8	12	3	2	5	4
2000-01	Magnitogorsk	Russia	44	16	*31	*47	78	12	7	6	13	20
2001-02	Dynamo Moscow	Russia	51	11	*32	43	96	3	2	2	4	0

RAZIN, Gennady (RAH-zihn, gen-AH-dee)

Defense. Shoots left. 6'4", 207 lbs. Born, Kharkov, USSR, February 3, 1978.
(Montreal's 6th choice, 122nd overall, in 1997 Entry Draft).

			Regular Season					Playoffs				
Season	Club	Lea	GP	G	A	TP	PIM	GP	G	A	TP	PIM
1995-96	St. Albert	AJHL	52	3	16	19	113	18	1	10	11	8
1996-97	Kamloops Blazers	WHL	63	7	19	26	56	3	0	0	0	4
1997-98	Kamloops Blazers	WHL	70	2	11	13	64	7	0	0	0	4
1998-99	Fredericton	AHL	48	0	3	3	16	4	0	0	0	2
99-2000	Quebec	AHL	66	2	9	11	29	3	0	0	0	0
2000-01	Quebec	AHL	69	3	19	22	25	9	0	0	0	4
2001-02	Quebec	AHL	75	2	7	9	14	3	1	1	2	2

READY, Ryan (REH-dee, RIGH-yan) **VAN.**

Left wing. Shoots left. 6'2", 195 lbs. Born, Peterborough, Ont., November 7, 1978.
(Calgary's 8th choice, 100th overall, in 1997 Entry Draft).

			Regular Season					Playoffs				
Season	Club	Lea	GP	G	A	TP	PIM	GP	G	A	TP	PIM
1994-95	Peterborough	OPJHL	48	20	33	53	65	....	....	....	....	....
1995-96	Belleville Bulls	OHL	63	5	13	18	54	10	0	2	2	2
1996-97	Belleville Bulls	OHL	66	23	24	47	102	6	1	3	4	4
1997-98	Belleville Bulls	OHL	66	33	39	72	80	10	5	2	7	12
1998-99	Belleville Bulls	OHL	63	33	59	92	73	21	10	28	38	22
99-2000	Syracuse Crunch	AHL	70	4	12	16	59	2	0	0	0	0
2000-01	Kansas City	IHL	67	10	15	25	75	....	....	....	....	....
2001-02	Manitoba Moose	AHL	72	23	32	55	73	7	5	1	6	4

OHL First All-Star Team (1999)

Signed as a free agent by **Vancouver**, June 16, 1999.

REDLIHS, Jekabs (REHD-lihs, YEH-kabs) **CBJ**

Defense. Shoots left. 6'2", 185 lbs. Born, Riga, Latvia, March 29, 1982.
(Columbus' 6th choice, 119th overall, in 2002 Entry Draft).

			Regular Season					Playoffs				
Season	Club	Lea	GP	G	A	TP	PIM	GP	G	A	TP	PIM
1998-99	Dynamo Riga-18	Latvia-Jr.	STATISTICS NOT AVAILABLE									
99-2000	HC Essamika-Jr.	EEHL	16	1	4	5	6	....	....	....	....	....
	Liepaja	Latvia	1	0	0	0	0	....	....	....	....	....
	Liepaja	EEHL	11	0	0	0	2	....	....	....	....	....
2000-01	Liepaja	EEHL	31	1	3	4	....	....	....	....	....	....
	Liepaja	Latvia	23	4	5	9	....	....	....	....	....	....
2001-02	NY Apple Core	EJHL	38	3	16	19	24	....	....	....	....	....

EJHL First All-Star Team (2002) • EJHL Defensive Player of the Year (2002)

Signed Letter of Intent to attend **Boston University** (H-East), February 14, 2002.

REDLIHS, Krisjanis (REHD-lihs, krihs-JA-nihs) **N.J.**

Defense. Shoots left. 6'2", 185 lbs. Born, Riga, Latvia, January 15, 1981.
(New Jersey's 7th choice, 154th overall, in 2002 Entry Draft).

			Regular Season					Playoffs				
Season	Club	Lea	GP	G	A	TP	PIM	GP	G	A	TP	PIM
1998-99	Dynamo Riga-18	Latvia-Jr.	STATISTICS NOT AVAILABLE									
99-2000	Liepaja	EEHL	12	1	3	4	0	....	....	....	....	....
2000-01	Liepaja	EEHL	27	2	2	4	....	....	....	....	....	....
	Liepaja	Latvia	22	1	6	7	....	....	....	....	....	....
2001-02	Liepaja	EEHL	32	0	2	2	....	11	1	1	2	....
	Liepaja	Latvia	13	0	6	6	4	3	2	2	4	0

REED, Josh (REED, JAWSH) **VAN.**

Defense. Shoots right. 6'2", 204 lbs. Born, Vernon, B.C., May 21, 1979.
(Vancouver's 5th choice, 172nd overall, in 1999 Entry Draft).

			Regular Season					Playoffs				
Season	Club	Lea	GP	G	A	TP	PIM	GP	G	A	TP	PIM
1994-95	Vernon	BCAHA	56	13	41	54	136	....	....	....	....	....
1995-96	Vernon Vikings	BCAHA	50	9	43	52	150	....	....	....	....	....
1996-97	Cowichan	BCHL	42	3	3	6	61	....	....	....	....	....
1997-98	Cowichan	BCHL	50	5	20	25	115	....	....	....	....	....
1998-99	Vernon Vipers	BCHL	54	16	38	54	110	....	....	....	....	....
99-2000	U. Mass-Lowell	H-East	30	2	10	12	24	....	....	....	....	....
2000-01	U. Mass-Lowell	H-East	17	1	6	7	22	....	....	....	....	....
2001-02	U. Mass-Lowell	H-East	37	4	9	13	38	....	....	....	....	....

REHNBERG, Henrik (REHN-buhrg, HEHN-rihk) **N.J.**

Defense. Shoots left. 6'2", 195 lbs. Born, Grava, Sweden, July 20, 1977.
(New Jersey's 6th choice, 96th overall, in 1995 Entry Draft).

			Regular Season					Playoffs				
Season	Club	Lea	GP	G	A	TP	PIM	GP	G	A	TP	PIM
1994-95	Farjestad Jr.	Swede-Jr.	24	1	2	3	62	....	....	....	....	....
1995-96	Farjestad Jr.	Swede-Jr.	21	1	4	5	38	....	....	....	....	....
	Farjestad	Sweden	4	0	0	0	0	....	....	....	....	....
1996-97	Farjestad	Sweden	42	2	3	5	38	14	1	1	2	16
1997-98	Farjestad	Sweden	32	0	1	1	24	10	0	0	0	12
	Farjestad	EuroHL	6	0	0	0	39	....	....	....	....	....
1998-99	Albany	AHL	55	1	4	5	49	2	0	0	0	0
99-2000	Farjestad	Sweden	46	1	3	4	66	7	0	1	1	18
2000-01	Albany	AHL	80	1	8	9	78	....	....	....	....	....
2001-02	Brynas IF Gavle	Sweden	27	1	1	2	47	4	1	0	1	6

REICH, Jeremy (REECH, JAIR-eh-MEE) **CBJ**

Center. Shoots left. 6'1", 204 lbs. Born, Craik, Sask., February 11, 1979.
(Chicago's 3rd choice, 39th overall, in 1997 Entry Draft).

			Regular Season					Playoffs				
Season	Club	Lea	GP	G	A	TP	PIM	GP	G	A	TP	PIM
1993-94	Pilote Butte	SAHA	80	70	65	135	120	....	....	....	....	....
1994-95	Sask. Contacts	SMHL	35	13	20	33	81	....	....	....	....	....
1995-96	Seattle	WHL	65	11	11	22	88	5	0	1	1	10
1996-97	Seattle	WHL	62	19	31	50	134	15	2	5	7	36
1997-98	Seattle	WHL	43	24	23	47	121	....	....	....	....	....
	Swift Current	WHL	22	8	8	16	47	12	5	6	11	37
1998-99	Swift Current	WHL	67	21	28	49	220	6	0	3	3	26
99-2000	Swift Current	WHL	72	33	58	91	167	12	2	10	12	19
2000-01	Syracuse Crunch	AHL	56	6	9	15	108	5	0	0	0	6
2001-02	Syracuse Crunch	AHL	59	9	7	16	178	10	4	0	4	16

Traded to **Swift Current** (WHL) by **Seattle** (WHL) for Jeffrey Beatch, December 20, 1997. Signed as a free agent by **Columbus**, May 17, 2000.

REID, Brandon (REED, BRAN-duhn) **VAN.**

Center. Shoots right. 5'8", 165 lbs. Born, Kirkland, Que., March 9, 1981.
(Vancouver's 5th choice, 208th overall, in 2000 Entry Draft).

			Regular Season					Playoffs				
Season	Club	Lea	GP	G	A	TP	PIM	GP	G	A	TP	PIM
1996-97	Lac-St-Louis	QAAA	44	17	34	51	....	7	2	3	5	....
1997-98	Halifax	QMJHL	67	13	21	36	6	5	1	0	1	15
1998-99	Halifax	QMJHL	70	32	25	57	33	5	2	2	4	0
99-2000	Halifax	QMJHL	62	44	80	124	10	10	7	11	18	4
2000-01	Val-d'Or Foreurs	QMJHL	57	45	81	126	18	21	13	29	42	14
2001-02	Manitoba Moose	AHL	60	18	19	37	6	7	0	3	3	0

QMJHL Second All-Star Team (2000) • Won George Parsons Trophy (Memorial Cup Tournament Most Sportsmanlike Player) (2000, 2001) • QMJHL First All-Star Team (2001)

Traded to **Val d'Or** (QMJHL) by **Halifax** (QMJHL) with Jonathan Jolette to complete earlier transaction that sent Benoit Dusablon and Nick Greenough to Halifax (January 7, 2000), June 6, 2000.

REITZ, Erik (RIGHTZ, AIR-ihk) **MIN.**

Defense. Shoots right. 6', 192 lbs. Born, Detroit, MI, July 29, 1982.
(Minnesota's 5th choice, 170th overall, in 2000 Entry Draft).

			Regular Season					Playoffs				
Season	Club	Lea	GP	G	A	TP	PIM	GP	G	A	TP	PIM
1998-99	Leamington	OJHL-B	50	5	10	15	80	....	....	....	....	....
99-2000	Barrie Colts	OHL	63	2	10	12	85	25	0	5	5	44
2000-01	Barrie Colts	OHL	68	5	21	26	178	5	1	0	1	21
2001-02	Barrie Colts	OHL	61	13	27	40	153	20	4	16	20	40

Memorial Cup All-Star Team (2000) • OHL First All-Star Team (2002)

RENNETTE, Tyler (REHN-neht, TIGH-luhr) **ST.L.**

Center. Shoots right. 6'1", 179 lbs. Born, North Bay, Ont., April 16, 1979.
(St. Louis' 1st choice, 40th overall, in 1997 Entry Draft).

			Regular Season					Playoffs				
Season	Club	Lea	GP	G	A	TP	PIM	GP	G	A	TP	PIM
1994-95	N. Bay Athletics	NOHA	52	24	42	66	72	....	....	....	....	....
1995-96	Waterloo	USHL	45	27	47	74	64	....	....	....	....	....
1996-97	North Bay	OHL	63	24	34	58	42	....	....	....	....	....
1997-98	North Bay	OHL	31	17	14	31	37	....	....	....	....	....
	Erie Otters	OHL	24	16	17	33	20	6	3	3	6	2
1998-99	Erie Otters	OHL	61	30	37	67	40	5	6	1	7	8
99-2000	Worcester	AHL	55	8	17	25	16	7	2	3	5	4
2000-01	Worcester	AHL	29	3	8	11	24	....	....	....	....	....
	Peoria Rivermen	ECHL	7	5	1	6	6	14	9	2	11	14
2001-02	Worcester	AHL	4	0	1	1	8	....	....	....	....	....
	Peoria Rivermen	ECHL	62	23	22	45	24	5	1	0	1	8

Traded to **Erie** (OHL) by **North Bay** (OHL) with Steve Montador for Brett Gibson and Geoff Peters, January 16, 1998.

REYNOLDS, Peter (REH-nolds, PEE-tuhr) **CAR.**

Defense. Shoots right. 6'3", 195 lbs. Born, Waterloo, Ont., April 27, 1981.
(Carolina's 8th choice, 274th overall, in 2001 Entry Draft).

			Regular Season					Playoffs				
Season	Club	Lea	GP	G	A	TP	PIM	GP	G	A	TP	PIM
1996-97	Caledon	MTJHL	45	1	10	11	69	....	....	....	....	....
1997-98	London Knights	OHL	55	0	8	8	30	16	0	0	0	10
1998-99	London Knights	OHL	59	2	25	27	55	23	2	3	5	24
99-2000	North Bay	OHL	61	3	29	32	53	6	1	3	4	10
2000-01	North Bay	OHL	58	2	27	29	85	4	0	1	1	7
	St. John's	AHL	2	0	0	0	0	1	0	0	0	0
2001-02	Lowell	AHL	45	1	2	3	38	....	....	....	....	....
	Florida	ECHL	7	0	0	0	10	....	....	....	....	....

• Re-entered NHL Entry Draft. Originally Toronto's 2nd choice, 60th overall, in 1999 Entry Draft.

Traded to **North Bay** (OHL) by **London** (OHL) for Brett Gibson and Brett Angel, Sdeptember 24, 1999.

RIAZANTSEV, Alexander (ree-ZAHNT-sehv, al-ehx-AN-duhr) COL.

Defense. Shoots right. 6', 210 lbs. Born, Moscow, USSR, March 15, 1980.
(Colorado's 10th choice, 167th overall, in 1998 Entry Draft).

			Regular Season					Playoffs				
Season	Club	Lea	GP	G	A	TP	PIM	GP	G	A	TP	PIM
1996-97	Spartak Moscow 2	Russia-3	18	0	0	0	8	....	....	....	....	....
	Spartak Moscow	Russia	20	1	2	3	4	....	....	....	....	....
1997-98	Spartak Moscow 2	Russia-3	31	3	8	11	26	....	....	....	....	....
	Victoriaville	QMJHL	22	6	9	15	14	4	0	0	0	0
1998-99	Victoriaville	QMJHL	64	17	40	57	57	6	0	3	3	10
	Hershey Bears	AHL	2	0	0	0	0	....	....	....	....	....
99-2000	Victoriaville	QMJHL	48	17	45	62	45	6	2	5	7	20
	Hershey Bears	AHL	2	0	1	1	2	6	1	1	2	0
2000-01	Hershey Bears	AHL	66	5	18	23	26	11	0	0	0	2
2001-02	Hershey Bears	AHL	76	5	19	24	28	5	0	1	1	4

RICHTER, Martin (RIHKH-tuhr, MAHR-tihn) NYR

Defense. Shoots right. 6'1", 196 lbs. Born, Prostejov, Czech., June 2, 1977.
(NY Rangers' 9th choice, 269th overall, in 2000 Entry Draft).

			Regular Season					Playoffs				
Season	Club	Lea	GP	G	A	TP	PIM	GP	G	A	TP	PIM
1995-96	HC Olomouc	Czech	3	0	0	0	0	1	0	0	0	0
1996-97	HC Olomouc	Czech	27	1	0	1	26	....	....	....	....	....
1997-98	HC Karlovy Vary	Czech	42	1	2	3	32	....	....	....	....	....
1998-99	HC Karlovy Vary	Czech	51	3	6	9	44	....	....	....	....	....
99-2000	HC Karlovy Vary	Czech	24	0	5	5	18	....	....	....	....	....
	SaiPa	Finland	26	1	3	4	54	....	....	....	....	....
2000-01	SaiPa	Finland	41	4	5	9	80	....	....	....	....	....
	Hartford	AHL	1	0	0	0	0	....	....	....	....	....
2001-02	Hartford	AHL	29	1	1	2	36	....	....	....	....	....
	HC Sparta Praha	Czech	8	0	0	0	14	13	0	0	0	10

RIDDLE, Jake (RIH-duhl, JAYK) MIN.

Left wing. Shoots left. 6'1", 205 lbs. Born, Fridley, MN, April 22, 1983.
(Minnesota's 7th choice, 239th overall, in 2001 Entry Draft).

			Regular Season					Playoffs				
Season	Club	Lea	GP	G	A	TP	PIM	GP	G	A	TP	PIM
1998-99	St. Margaret's	Hi-School		STATISTICS NOT AVAILABLE								
99-2000	Team USA	USDP-17	33	7	6	13	93	....	....	....	....	....
2000-01	Seattle	WHL	67	13	20	33	109	6	0	2	2	2
2001-02	Seattle	WHL	70	16	15	31	109	11	1	2	3	18

RIDDLE, Troy (RIH-duhl, TROI) ST.L.

Center. Shoots right. 6', 172 lbs. Born, Minneapolis, MN, August 24, 1981.
(St. Louis' 5th choice, 129th overall, in 2000 Entry Draft).

			Regular Season					Playoffs				
Season	Club	Lea	GP	G	A	TP	PIM	GP	G	A	TP	PIM
1997-98	St. Margaret's	Hi-School	29	33	35	68	....	....	....	....	....	....
1998-99	St. Margaret's	Hi-School	29	54	45	99	....	....	....	....	....	....
99-2000	Des Moines	USHL	53	36	30	66	95	8	2	2	4	31
2000-01	U. of Minnesota	WCHA	38	16	14	30	49	....	....	....	....	....
2001-02	U. of Minnesota	WCHA	44	16	31	47	46	....	....	....	....	....

USHL Second All-Star Team (2000) • USHL Rookie of the Year (2000)

RIVA, Danny (REE-vuh, DAN-nee)

Center. Shoots right. 6', 190 lbs. Born, Framingham, MA, September 17, 1975.

			Regular Season					Playoffs				
Season	Club	Lea	GP	G	A	TP	PIM	GP	G	A	TP	PIM
1995-96	RPI Engineers	ECAC	35	3	7	10	30	....	....	....	....	....
1996-97	RPI Engineers	ECAC	36	12	14	26	30	....	....	....	....	....
1997-98	RPI Engineers	ECAC	35	10	18	28	16	....	....	....	....	....
1998-99	RPI Engineers	ECAC	36	*22	*35	*57	35	....	....	....	....	....
	Milwaukee	IHL	8	0	2	2	4	1	0	1	1	0
99-2000	Milwaukee	IHL	67	8	12	20	18	3	0	0	0	2
2000-01	Milwaukee	IHL	75	12	9	21	23	5	0	1	1	2
2001-02	Reading Royals	ECHL	37	10	17	27	38	....	....	....	....	....
	Manchester	AHL	29	4	1	5	8	3	0	0	0	12

ECAC First All-Star Team (1999)

Signed as a free agent by **Nashville**, May 4, 1999.

ROBINSON, Darcy (RAW-bihn-suhn, DAHR-see) PIT.

Defense. Shoots right. 6'3", 221 lbs. Born, Kamloops, B.C., May 3, 1981.
(Pittsburgh's 10th choice, 233rd overall, in 1999 Entry Draft).

			Regular Season					Playoffs				
Season	Club	Lea	GP	G	A	TP	PIM	GP	G	A	TP	PIM
1996-97	Kamloops	BCAHA	59	18	42	60	188	....	....	....	....	....
1997-98	Saskatoon Blades	WHL	62	1	2	3	84	4	0	0	0	2
1998-99	Saskatoon Blades	WHL	48	3	6	9	86	....	....	....	....	....
99-2000	Saskatoon Blades	WHL	59	5	9	14	91	10	1	3	4	13
2000-01	Saskatoon Blades	WHL	41	2	6	8	80	....	....	....	....	....
	Red Deer Rebels	WHL	30	1	5	6	70	20	1	1	2	20
2001-02	Wheeling Nailers	ECHL	10	2	3	5	43	....	....	....	....	....
	Wilkes-Barre	AHL	40	0	5	5	35	....	....	....	....	....

Traded to **Red Deer** (WHL) by **Saskatoon** (WHL) with Martin Erat and Cam Ondrik for Michael Garnett, Justin Wallin, Martin Vymazzal and future considerations, January 11, 2001.

ROCHEFORT, Richard (ROHSH-fohr, RIH-chahrd)

Center. Shoots right. 5'10", 195 lbs. Born, North Bay, Ont., January 7, 1977.
(New Jersey's 9th choice, 174th overall, in 1995 Entry Draft).

			Regular Season					Playoffs				
Season	Club	Lea	GP	G	A	TP	PIM	GP	G	A	TP	PIM
1993-94	Waterloo	USHL	45	21	32	53	41	....	....	....	....	....
1994-95	Sudbury Wolves	OHL	57	21	44	65	26	13	3	7	10	6
1995-96	Sudbury Wolves	OHL	56	25	40	65	38	....	....	....	....	....
1996-97	Sudbury Wolves	OHL	28	18	24	42	40	....	....	....	....	....
	Sarnia Sting	OHL	18	5	23	28	23	12	3	9	12	8
1997-98	Albany	AHL	59	7	14	21	16	13	1	0	1	4
1998-99	Albany	AHL	70	16	10	26	26	5	1	0	1	0
99-2000	Albany	AHL	55	12	12	24	22	5	0	0	0	0
2000-01	Albany	AHL	75	16	24	40	18	....	....	....	....	....
2001-02	Albany	AHL	59	15	13	28	12	....	....	....	....	....

OHL All-Rookie Team (1995)

Traded to **Sarnia** (OHL) by **Sudbury** with Louis Blackbird for Brandon Sugden, Ryan McKie and future considerations (Brad Simms, June 8, 1997), January 8, 1997

RODMAN, Marcel (RAWD-muhn, mahr-SEHL) BOS.

Right wing. Shoots right. 6'1", 183 lbs. Born, Jesenice, Yugoslavia, September 25, 1981.
(Boston's 8th choice, 282nd overall, in 2001 Entry Draft).

			Regular Season					Playoffs				
Season	Club	Lea	GP	G	A	TP	PIM	GP	G	A	TP	PIM
1997-98	Jesenice Jr.	Sloven.-Jr.	44	29	44	73	14	....	....	....	....	....
1998-99	Pickering	OPJHL	37	30	21	51	8	....	....	....	....	....
99-2000	Peterborough	OHL	61	17	20	37	16	5	1	2	3	0
2000-01	Peterborough	OHL	61	36	35	71	14	7	4	2	6	2
2001-02	Acroni Jesenice	EEHL	7	4	2	6	4	....	....	....	....	....
	Acroni Jesenice	Slovenia	9	12	6	18	4	....	....	....	....	....

Signed as a free agent by **EC Graz** (Austria) with Boston retaining NHL rights, July 20, 2002.

ROGERS, Brandon (RAW-juhrs, BRAN-duhn) ANA.

Defense. Shoots right. 6'1", 190 lbs. Born, Rochester, NH, February 27, 1982.
(Anaheim's 6th choice, 118th overall, in 2001 Entry Draft).

			Regular Season					Playoffs				
Season	Club	Lea	GP	G	A	TP	PIM	GP	G	A	TP	PIM
99-2000	Hotchkiss High	Hi-School	25	9	12	21	35	....	....	....	....	....
2000-01	Hotchkiss High	Hi-School	22	10	13	23	45	....	....	....	....	....
2001-02	U. of Michigan	CCHA	32	2	1	3	30	....	....	....	....	....

ROME, Aaron (ROHM, AIR-uhn) L.A.

Defense. Shoots left. 6'1", 197 lbs. Born, Nesbitt, Man., September 27, 1983.
(Los Angeles' 4th choice, 104th overall, in 2002 Entry Draft).

			Regular Season					Playoffs				
Season	Club	Lea	GP	G	A	TP	PIM	GP	G	A	TP	PIM
1998-99	Sask. Contacts	SMHL		STATISTICS NOT AVAILABLE								
	Saskatoon Blades	WHL	1	0	0	0	0	....	....	....	....	....
99-2000	Saskatoon Blades	WHL	47	0	6	6	22	1	0	0	0	0
2000-01	Saskatoon Blades	WHL	3	0	0	0	2	....	....	....	....	....
	Kootenay Ice	WHL	53	2	8	10	43	11	1	3	4	6
2001-02	Kootenay Ice	WHL	33	4	13	17	55	....	....	....	....	....
	Swift Current	WHL	37	3	11	14	113	10	1	4	5	23

Traded to **Kootenay** (WHL) by **Saskatoon** (WHL) with Jeff Harvey, Jason Jaffray and future considerations for B.J. Boxma, Duncan Milroy and Kootenay's 5th round choice in 2004 WHL Bantam Draft, October 4, 2000. Traded to **Swift Current** (WHL) by **Kootenay** (WHL) with Jeff Harvey, Jason Jaffray and Kootenay's 1st round choice (Ryan Sawka) in 2002 WHL Bantam Draft for B.J. Boxma, Duncan Milroy and Swift Current's 5th round choice in 2005 WHL Bantam Draft, December 4, 2001.

ROONEEM, Mark (ROO-neem, MAHRK) L.A.

Left wing. Shoots left. 6'1", 185 lbs. Born, Hinton, Alta., January 9, 1983.
(Los Angeles' 5th choice, 115th overall, in 2002 Entry Draft).

			Regular Season					Playoffs				
Season	Club	Lea	GP	G	A	TP	PIM	GP	G	A	TP	PIM
1998-99	Spruce Grove	AMBHL	36	32	30	62	183	....	....	....	....	....
99-2000	Kamloops Blazers	WHL	50	3	8	11	39	4	0	0	0	4
2000-01	Kamloops Blazers	WHL	62	8	9	17	77	4	1	0	1	8
2001-02	Kamloops Blazers	WHL	69	18	23	41	77	4	0	0	0	10

ROSA, Marco (ROH-zuh, MAHR-koh) DAL.

Center. Shoots left. 6', 170 lbs. Born, Scarborough, Ont., January 15, 1982.
(Dallas' 8th choice, 255th overall, in 2001 Entry Draft).

			Regular Season					Playoffs				
Season	Club	Lea	GP	G	A	TP	PIM	GP	G	A	TP	PIM
1998-99	Wexford Raiders	OPJHL	42	12	13	25	35	....	....	....	....	....
99-2000	Wexford Raiders	OPJHL	49	29	52	81	23	6	1	7	9	0
2000-01	Merrimack	H-East	33	6	18	24	22	....	....	....	....	....
2001-02	Merrimack	H-East	36	5	21	26	22	....	....	....	....	....

ROULEAU, Alexandre (ROO-loh, al-ehx-AHN-druh) PIT.

Defense. Shoots left. 6'1", 184 lbs. Born, Mont-Laurier, Que., July 29, 1983.
(Pittsburgh's 4th choice, 96th overall, in 2001 Entry Draft).

			Regular Season					Playoffs				
Season	Club	Lea	GP	G	A	TP	PIM	GP	G	A	TP	PIM
1998-99	Amos Forestiers	QAAA	41	7	6	13	144	....	....	....	....	....
99-2000	Val-d'Or Foreurs	QMJHL	41	3	3	6	39	....	....	....	....	....
2000-01	Val-d'Or Foreurs	QMJHL	70	8	17	25	124	21	1	0	1	46
2001-02	Val-d'Or Foreurs	QMJHL	69	14	25	39	174	7	0	2	2	16

ROURKE, Allan (RAWRK, AL-lan) TOR.

Defense. Shoots left. 6'1", 214 lbs. Born, Mississauga, Ont., March 6, 1980.
(Toronto's 6th choice, 154th overall, in 1998 Entry Draft).

			Regular Season					Playoffs				
Season	Club	Lea	GP	G	A	TP	PIM	GP	G	A	TP	PIM
1995-96	Mississauga Reps	MTHL	38	15	25	40	173	....	....	....	....	....
1996-97	Kitchener	OHL	25	1	1	2	12	6	0	0	0	0
1997-98	Kitchener	OHL	48	5	17	22	59	6	1	1	2	6
1998-99	Kitchener	OHL	66	11	28	39	79	1	0	0	0	2
99-2000	Kitchener	OHL	67	31	43	74	57	5	0	6	6	13
2000-01	St. John's	AHL	64	9	19	28	36	....	....	....	....	....
2001-02	St. John's	AHL	62	2	9	11	48	10	0	2	2	6

OHL Second All-Star Team (2000)

ROY, Derek (ROI, DEHR-ihk) BUF.

Center. Shoots left. 5'9", 186 lbs. Born, Ottawa, Ont., May 4, 1983.
(Buffalo's 2nd choice, 32nd overall, in 2001 Entry Draft).

			Regular Season					Playoffs				
Season	Club	Lea	GP	G	A	TP	PIM	GP	G	A	TP	PIM
1998-99	Ontario East	OMHA	34	61	31	92	42	....	....	....	....	....
99-2000	Kitchener	OHL	66	34	53	87	44	5	4	1	5	6
2000-01	Kitchener	OHL	65	42	39	81	114	....	....	....	....	....
2001-02	Kitchener	OHL	62	43	46	89	92	4	1	2	3	2

OHL All-Rookie Team (2000) • OHL Rookie of the Year (2000) • CHL All-Rookie Team (2000) • Won CHL Plus/Minus Award (2000) • CHL Most Sportsmanlike Playerr (2000)

ROY, Jimmy (ROI, JIHM-mee)

Center. Shoots right. 5'11", 170 lbs. Born, Sioux Lookout, Ont., September 22, 1975.
(Dallas' 7th choice, 254th overall, in 1994 Entry Draft).

			Regular Season					Playoffs				
Season	Club	Lea	GP	G	A	TP	PIM	GP	G	A	TP	PIM
1993-94	Thunder Bay	USHL	46	21	33	54	101	....	....	....	....	....
1994-95	Michigan Tech	WCHA	38	5	11	16	62	....	....	....	....	....
1995-96	Michigan Tech	WCHA	42	17	17	34	84	....	....	....	....	....
1996-97	Team Canada	Nat-Tm	55	10	17	27	82	....	....	....	....	....
1997-98	Manitoba Moose	IHL	61	8	10	18	133	3	0	0	0	6
1998-99	Manitoba Moose	IHL	78	10	16	26	185	5	0	1	1	6
99-2000	Manitoba Moose	IHL	74	12	9	21	187	1	0	0	0	16
2000-01	Manitoba Moose	IHL	77	18	13	31	150	12	1	1	2	22
2001-02	Manitoba Moose	AHL	73	16	22	38	167	7	2	0	2	28

ROZAKOV, Rail (roh-zah-KAWF, righ-EEL) CGY.

Defense. Shoots left. 6'1", 198 lbs. Born, Murmansk, USSR, March 29, 1981.
(Calgary's 4th choice, 106th overall, in 1999 Entry Draft).

			Regular Season					Playoffs				
Season	Club	Lea	GP	G	A	TP	PIM	GP	G	A	TP	PIM
1997-98	Lada Togliatti 2	Russia-3	36	0	2	2	43	....	....	....	....	....
1998-99	Lada Togliatti 2	Russia-4	30	0	0	0	14	....	....	....	....	....
99-2000	Krylja Sovetov	Russia-2	23	0	1	1	41	....	....	....	....	....
2000-01	Magnitogorsk	Russia	21	0	1	1	10	....	....	....	....	....
2001-02	CSK VVS Samara	Russia-2	5	0	0	0	0	....	....	....	....	....
	CSKA Moscow	Russia	16	1	1	2	8	....	....	....	....	....

RUDENKO, Konstantin (roo-DEHN-koh, KOHN-stan-tihn) PHI.

Left wing. Shoots right. 5'11", 180 lbs. Born, Ust-Kamenogorsk, USSR, July 23, 1981.
(Philadelphia's 3rd choice, 160th overall, in 1999 Entry Draft).

			Regular Season					Playoffs				
Season	Club	Lea	GP	G	A	TP	PIM	GP	G	A	TP	PIM
1997-98	Omsk 2	Russia-3	22	7	8	15	4	....	....	....	....	....
1998-99	Cherepovets	Russia	28	15	9	24	67	....	....	....	....	....
	Cherepovets 2	Russia-3	3	0	1	1	4	....	....	....	....	....
99-2000	St. Petersburg	Russia	7	2	4	6	2	....	....	....	....	....
	St. Petersburg	Russia	19	1	1	2	10	1	0	0	0	0
2000-01	Yaroslavl	Russia	18	2	3	5	28	9	2	1	3	8
2001-02	Yaroslavl 2	Russia-3	2	1	1	2	2	....	....	....	....	....
	Yaroslavl	Russia	8	0	2	2	12	1	0	0	0	0

RUGGERI, Rosario (ROO-gee-AIR-ee, roh-ZAHR-ee-oh) PHI.

Defense. Shoots left. 6'1", 202 lbs. Born, Montreal, Que., June 8, 1984.
(Philadelphia's 2nd choice, 105th overall, in 2002 Entry Draft).

			Regular Season					Playoffs				
Season	Club	Lea	GP	G	A	TP	PIM	GP	G	A	TP	PIM
99-2000	Lac St-Louis	QAAA	40	0	7	7	70	....	....	....	....	....
2000-01	Lac St-Louis	QAAA	24	6	11	17	117	5	1	3	4	4
	Montreal Rocket	QMJHL	9	0	0	0	8	....	....	....	....	....
2001-02	Chicoutimi	QMJHL	60	2	15	17	131	4	1	1	2	10

Traded to **Chicoutimi** (QMJHL) by **Montreal** (QMJHL) for Karl St-Pierre, June 15, 2001.

RULLIER, Joe (ROO-yay, JOH) L.A.

Defense. Shoots right. 6'3", 200 lbs. Born, Montreal, Que., January 28, 1980.
(Los Angeles' 5th choice, 133rd overall, in 1998 Entry Draft).

			Regular Season					Playoffs				
Season	Club	Lea	GP	G	A	TP	PIM	GP	G	A	TP	PIM
1996-97	Mtl-Bourassa	QAAA	24	5	10	15	....	....	....	....	....	....
	Rimouski Oceanic	QMJHL	23	0	3	3	87	4	0	0	0	11
1997-98	Rimouski Oceanic	QMJHL	55	1	10	11	176	16	1	4	5	34
1998-99	Rimouski Oceanic	QMJHL	54	7	32	39	202	11	2	3	5	26
99-2000	Rimouski Oceanic	QMJHL	49	3	32	35	161	14	1	8	9	34
2000-01	Lowell	AHL	63	1	1	2	162	4	0	1	1	2
2001-02	Manchester	AHL	62	2	2	4	133	3	0	0	0	5

RUPP, Mike (RUHP, MIGHK) N.J.

Right wing. Shoots left. 6'5", 235 lbs. Born, Cleveland, OH, January 13, 1980.
(New Jersey's 7th choice, 76th overall, in 2000 Entry Draft).

			Regular Season					Playoffs				
Season	Club	Lea	GP	G	A	TP	PIM	GP	G	A	TP	PIM
1996-97	St. Edward's	Hi-School	20	26	24	50	....	....	....	....	....	....
1997-98	Windsor	OHL	38	9	8	17	60	....	....	....	....	....
	Erie Otters	OHL	26	7	3	10	57	7	3	1	4	6
1998-99	Erie Otters	OHL	63	22	25	47	102	5	0	2	2	25
99-2000	Erie Otters	OHL	58	32	21	53	134	13	5	5	10	22
2000-01	Albany	AHL	71	10	10	20	63	....	....	....	....	....
2001-02	Albany	AHL	78	13	17	30	90	....	....	....	....	....

• Re-entered NHL Entry Draft. Originally NY Islanders' 1st choice, 9th overall, in 1998 Entry Draft.
Traded to **Erie** (OHL) by **Windsor** (OHL) for Jason Ward, January 6, 1998.

RUUTU, Mikko (ROO-too, MIH-koh) OTT.

Left wing. Shoots left. 6'4", 183 lbs. Born, Vantaa, Finland, September 10, 1978.
(Ottawa's 7th choice, 201st overall, in 1999 Entry Draft).

			Regular Season					Playoffs				
Season	Club	Lea	GP	G	A	TP	PIM	GP	G	A	TP	PIM
1997-98	HIFK Jr.	Finn-Jr.	24	4	2	6	37	....	....	....	....	....
1998-99	HIFK Jr.	Finn-Jr.	23	13	8	21	30	....	....	....	....	....
	HIFK Helsinki	Finland	31	3	1	4	12	4	0	0	0	2
99-2000	Clarkson Knights	ECAC	33	5	6	11	26	....	....	....	....	....
2000-01	Jokerit Helsinki	Finland	56	5	6	11	38	5	0	1	1	2
2001-02	Jokerit Helsinki	Finland	47	3	3	6	65	5	1	0	1	27

RUUTU, Tuomo (ROO-too, TOO-oh-moh) CHI.

Center. Shoots left. 6', 201 lbs. Born, Vantaa, Finland, February 16, 1983.
(Chicago's 1st choice, 9th overall, in 2001 Entry Draft).

			Regular Season					Playoffs				
Season	Club	Lea	GP	G	A	TP	PIM	GP	G	A	TP	PIM
1997-98	HIFK Helsinki-C	Finn-Jr.	22	4	11	15	10	3	0	1	1	4
1998-99	HIFK Helsinki-C	Finn-Jr.	3	6	3	9	25	5	*4	2	6	8
	HIFK Helsinki-B	Finn-Jr.	25	9	11	20	88	2	1	1	2	2
99-2000	HIFK Jr.	Finn-Jr.	35	11	16	27	32	3	0	1	1	4
	HIFK Helsinki	Finland	1	0	0	0	2	....	....	....	....	....
2000-01	HIFK Jr.	Finn-Jr.	2	1	0	1	0	....	....	....	....	....
	Jokerit Helsinki	Finland	47	11	11	22	86	5	0	0	0	4
2001-02	Jokerit Helsinki	Finland	51	7	16	23	69	10	0	6	6	29

RYAN, Michael (RIGH-yan, MIGHK-uhl)) DAL.

Center. Shoots left. 6'1", 180 lbs. Born, Boston, MA, May 16, 1980.
(Dallas' 1st choice, 32nd overall, in 1999 Entry Draft).

			Regular Season					Playoffs				
Season	Club	Lea	GP	G	A	TP	PIM	GP	G	A	TP	PIM
1997-98	B.C. High Irish	Hi-School	23	22	14	36	28	....	....	....	....	....
1998-99	B.C. High Irish	Hi-School	21	20	24	44	22	....	....	....	....	....
99-2000	Northeastern	H-East	32	4	9	13	47	....	....	....	....	....
2000-01	Northeastern	H-East	33	17	12	29	52	....	....	....	....	....
2001-02	Northeastern	H-East	36	24	15	39	54	....	....	....	....	....

RYBIN, Maxim (ray-bihn, max-EEM) ANA.

Left wing. Shoots right. 5'8", 182 lbs. Born, Zhukovsky, USSR, June 15, 1981.
(Anaheim's 4th choice, 141st overall, in 1999 Entry Draft).

			Regular Season					Playoffs				
Season	Club	Lea	GP	G	A	TP	PIM	GP	G	A	TP	PIM
1996-97	Spartak Moscow 2	Russia-3	5	0	0	0	4	....	....	....	....	....
	Spartak Moscow	Russia	6	0	0	0	0	....	....	....	....	....
1997-98	Spartak Moscow 2	Russia-3	25	13	5	18	26	....	....	....	....	....
	Spartak Moscow	Russia	5	0	0	0	2	....	....	....	....	....
1998-99	Spartak Moscow	Russia	53	15	12	27	83	....	....	....	....	....
99-2000	Sarnia Sting	OHL	66	29	27	56	47	7	4	1	5	2
2000-01	Sarnia Sting	OHL	67	34	36	70	60	4	0	3	3	2
2001-02	Ufa	Russia	41	6	4	10	30	....	....	....	....	....

RYDER, Michael (RIGH-duhr, MIGH-kuhl) MTL.

Center. Shoots right. 6'1", 195 lbs. Born, St. John's, Nfld., March 31, 1980.
(Montreal's 9th choice, 216th overall, in 1998 Entry Draft).

			Regular Season					Playoffs				
Season	Club	Lea	GP	G	A	TP	PIM	GP	G	A	TP	PIM
1996-97	Bonavista Saints	NFAHA	23	31	17	48	....	....	....	....	....	....
1997-98	Hull Olympiques	QMJHL	69	34	28	62	41	10	4	2	6	4
1998-99	Hull Olympiques	QMJHL	69	44	43	87	65	23	*20	16	36	39
99-2000	Hull Olympiques	QMJHL	63	50	58	108	50	15	11	17	28	28
2000-01	Tallahassee	ECHL	5	4	5	9	6	....	....	....	....	....
	Quebec	AHL	61	6	9	15	14	....	....	....	....	....
2001-02	Mississippi	ECHL	20	14	13	27	2	....	....	....	....	....
	Quebec	AHL	50	11	17	28	9	3	0	1	1	2

RYZNAR, Jason (RIHZ-nuhr, JAY-suhn) N.J.

Left wing. Shoots left. 6'3", 205 lbs. Born, Anchorage, AK, February 19, 1983.
(New Jersey's 3rd choice, 64th overall, in 2002 Entry Draft).

			Regular Season					Playoffs				
Season	Club	Lea	GP	G	A	TP	PIM	GP	G	A	TP	PIM
1998-99	Alaska All-Stars	AAHA		STATISTICS NOT AVAILABLE								
99-2000	Team USA	USDP-17	52	5	10	15	22	....	....	....	....	....
2000-01	Team USA	USDP-18	66	15	17	32	102	....	....	....	....	....
2001-02	U. of Michigan	CCHA	40	9	7	16	22	....	....	....	....	....

SAARINEN, Pasi (SAH-rih-nehn, PAH-see) S.J.

Defense. Shoots right. 5'11", 194 lbs. Born, Hyvinkaa, Finland, April 17, 1977.
(San Jose's 7th choice, 256th overall, in 2000 Entry Draft).

			Regular Season					Playoffs				
Season	Club	Lea	GP	G	A	TP	PIM	GP	G	A	TP	PIM
1993-94	Ilves Jr.	Finn-Jr.	34	5	4	9	24	6	1	2	3	4
1994-95	Ilves Jr.	Finn-Jr.	24	4	5	9	55	....	....	....	....	....
	Ilves Tampere	Finland	1	0	0	0	0	....	....	....	....	....
1995-96	Ilves Tampere	Finn-Jr.	8	2	2	4	24	....	....	....	....	....
	KooVee Tampere	Finland-2	7	1	1	2	22	....	....	....	....	....
	Ilves Tampere	Finland	7	0	1	1	10	....	....	....	....	....
1996-97	Ilves Jr.	Finn-Jr.	9	3	3	6	22	....	....	....	....	....
	Ilves Tampere	Finland	44	4	3	7	83	6	1	0	1	10
1997-98	Ilves Tampere	Finland	36	11	8	19	73	....	....	....	....	....
1998-99	Ilves Tampere	Finland	45	2	8	10	56	4	1	0	1	8
99-2000	Ilves Tampere	Finland	50	9	19	28	79	3	0	0	0	26
2000-01	Jokerit Helsinki	Finland	52	6	9	15	79	5	2	0	2	4
2001-02	Jokerit Helsinki	Finland	32	0	0	0	48	10	0	2	2	0

SACHL, Petr (SAH-khuhl, PEE-tuhr)

Left wing. Shoots right. 6'2", 205 lbs. Born, Jindrivichuk Hradec, Czech., December 2, 1977.
(NY Islanders' 6th choice, 128th overall, in 1996 Entry Draft).

			Regular Season					Playoffs				
Season	Club	Lea	GP	G	A	TP	PIM	GP	G	A	TP	PIM
1994-95	C. Budejovice Jr.	Czech-Jr.	40	6	8	14	....	....	....	....	....	....
1995-96	C. Budejovice Jr.	Czech-Jr.	39	19	17	36	....	....	....	....	....	....
	Ceske Budejovice	Czech	....	....	....	....	....	2	0	0	0	....
1996-97	Ceske Budejovice	Czech	2	0	0	0	0	....	....	....	....	....
	Tri-City	WHL	63	13	24	37	32	....	....	....	....	....
1997-98	Ceske Budejovice	Czech	20	1	3	4	4	....	....	....	....	....
1998-99	Ceske Budejovice	Czech	4	0	1	1	2	....	....	....	....	....
99-2000	Tacoma Sabercats	WCHL	1	0	0	0	0	....	....	....	....	....
	Asheville Smoke	UHL	3	0	1	1	4	....	....	....	....	....
	Fort Wayne	UHL	55	30	24	54	28	10	4	7	11	8
2000-01	Milwaukee	IHL	76	12	17	29	33	5	0	0	0	4
2001-02	Milwaukee	AHL	79	14	29	43	52	....	....	....	....	....

Traded to **Nashville** by **NY Islanders** for Nashville's 9th round choice (Tomi Pettinen) in 2000 Entry Draft, March 14, 2000. Signed as a free agent by **Assat** (Finland), July 19, 2002.

SAINOMAA, Teemu (SIGH-noh-muh, TEE-moo) OTT.

Left wing. Shoots left. 6'3", 202 lbs. Born, Helsinki, Finland, May 15, 1981.
(Ottawa's 3rd choice, 62nd overall, in 1999 Entry Draft).

			Regular Season					Playoffs				
Season	Club	Lea	GP	G	A	TP	PIM	GP	G	A	TP	PIM
1997-98	Jokerit-B	Finn-Jr.	12	3	5	8	8	3	1	1	2	6
1998-99	Jokerit Jr.	Finn-Jr.	11	4	5	9	0	....	....	....	....	....
99-2000	Jokerit Jr.	Finn-Jr.	30	6	7	13	59	11	6	2	8	20
	Jokerit Helsinki	Finland	6	0	0	0	0	....	....	....	....	....
2000-01	Jokerit Jr.	Finn-Jr.	24	17	8	25	37	1	0	1	1	0
	Jokerit Helsinki	Finland	28	1	3	4	2	4	0	1	1	0
2001-02	Jokerit Jr.	Finn-Jr.	6	2	2	4	27	....	....	....	....	....
	Kiekko Vantaa	Finland-2	1	0	0	0	0	....	....	....	....	....
	Jokerit Helsinki	Finland	44	1	4	5	4	1	0	0	0	0

ST. CROIX, Chris (SAINT KWAH, KRIHS)

Defense. Shoots right. 6'1", 199 lbs. Born, Voorhees, NJ, May 2, 1979.
(Calgary's 7th choice, 92nd overall, in 1997 Entry Draft).

			Regular Season					Playoffs				
Season	Club	Lea	GP	G	A	TP	PIM	GP	G	A	TP	PIM
1993-94	Winnipeg Blues	MMHL	35	2	40	42	30	....	....	....	....	....
1994-95	Winnipeg Blues	MMHL	40	9	32	41	20	....	....	....	....	....
1995-96	Kamloops Blazers	WHL	61	4	5	9	29	13	0	2	2	4
1996-97	Kamloops Blazers	WHL	67	11	39	50	67	5	0	1	1	2
1997-98	Kamloops Blazers	WHL	46	3	13	16	51	7	1	1	2	6
1998-99	Kamloops Blazers	WHL	64	8	27	35	123	14	0	4	4	16
99-2000	Saint John	AHL	75	5	16	21	51	3	0	1	1	2
2000-01	Saint John	AHL	69	0	4	4	66	5	0	1	1	4
2001-02	Hartford	AHL	73	0	10	10	67	10	0	1	1	10

Traded to **NY Rangers** by **Calgary** for Burke Henry, June 23, 2001.

SALMELAINEN, Tony (sal-meh-LIGH-nehn, TOH-nee) **EDM.**

Left wing. Shoots right. 5'9", 176 lbs. Born, Espoo, Finland, August 8, 1981.
(Edmonton's 3rd choice, 41st overall, in 1999 Entry Draft).

			Regular Season					Playoffs				
Season	Club	Lea	GP	G	A	TP	PIM	GP	G	A	TP	PIM
1996-97	Kiekko Espoo Jr.	Finn-Jr.	30	8	5	13	38	....	....	....	....	....
1997-98	Kiekko Espoo-B	Finn-Jr.	5	2	2	4	10	....	....	....	....	....
	HIFK Helsinki-B	Finn-Jr.	28	23	16	39	30	....	....	....	....	....
	HIFK Jr.	Finn-Jr.	5	0	0	0	0	....	....	....	....	....
1998-99	HIFK Helsinki-B	Finn-Jr.	21	13	10	23	45	....	....	....	....	....
	HIFK Jr.	Finn-Jr.	....	....	....	....	....	10	10	8	18	10
99-2000	HIFK Jr.	Finn-Jr.	1	0	1	1	0	....	....	....	....	....
	HIFK Helsinki	Finland	1	1	0	1	0	....	....	....	....	....
2000-01	HIFK Jr.	Finn-Jr.	3	3	3	6	0	....	....	....	....	....
	HIFK Helsinki	Finland	19	1	0	1	6	....	....	....	....	....
	Ilves Jr.	Finn-Jr.	3	1	2	3	2	....	....	....	....	....
	Ilves Tampere	Finland	26	3	10	13	4	3	0	0	0	0
2001-02	Ilves Jr.	Finn-Jr.	6	2	2	4	27	....	....	....	....	....
	Ilves Tampere	Finland	49	10	9	19	30	3	0	0	0	2

SAMOILOV, Igor (sam-OI-lawf, EE-gohr) **PHX.**

Defense. Shoots left. 5'11", 195 lbs. Born, Moscow, USSR, January 23, 1982.
(Phoenix's 6th choice, 217th overall, in 2000 Entry Draft).

			Regular Season					Playoffs				
Season	Club	Lea	GP	G	A	TP	PIM	GP	G	A	TP	PIM
1998-99	Yaroslavl 2	Russia-3	16	0	1	1	6	....	....	....	....	....
99-2000	Yaroslavl 2	Russia-3	41	1	3	4	50	....	....	....	....	....
2000-01	St. Petersburg	Russia	38	0	2	2	22	....	....	....	....	....
2001-02	Cherepovets	Russia	35	1	2	3	14	4	0	0	0	2

SAMUELSSON, Martin (SAM-yuhl-suhn, MAHR-tihn) **BOS.**

Right wing. Shoots left. 6'2", 194 lbs. Born, Upplands Vasby, Sweden, January 25, 1982.
(Boston's 2nd choice, 27th overall, in 2000 Entry Draft).

			Regular Season					Playoffs				
Season	Club	Lea	GP	G	A	TP	PIM	GP	G	A	TP	PIM
1996-97	Hammarby Jr.	Swede-Jr.	6	1	1	2	0	....	....	....	....	....
1997-98	Hammarby Jr.	Swede-Jr.	20	13	12	25	....	....	....	....	....	....
	Hammarby	Swede-2	2	0	0	0	0	....	....	....	....	....
1998-99	MoDo Jr.	Swede-Jr.	31	18	13	31	10	....	....	....	....	....
99-2000	MoDo-18	Swede-Jr.	6	3	0	3	4	....	....	....	....	....
	MoDo Jr.	Swede-Jr.	19	9	8	17	18	2	1	0	1	2
2000-01	Hammarby	Swede-Q	14	2	2	4	10	....	....	....	....	....
	Hammarby	Swede-2	24	13	4	17	8	5	0	0	0	2
	Hammarby Jr.	Swede-Jr.	1	0	1	1	0	....	....	....	....	....
2001-02	Hammarby Jr.	Swede-Jr.	2	5	2	7	2	....	....	....	....	....
	Hammarby	Swede-2	44	13	10	23	45	2	0	0	0	2
	Hammarby	Swede-Q	10	3	1	4	10	....	....	....	....	....

SANDSTROM, Jan (SAND-struhm, YAN) **ANA.**

Defense. Shoots left. 6', 190 lbs. Born, Pitea, Sweden, January 24, 1978.
(Anaheim's 5th choice, 173rd overall, in 1999 Entry Draft).

			Regular Season					Playoffs				
Season	Club	Lea	GP	G	A	TP	PIM	GP	G	A	TP	PIM
1994-95	Pitea HC	Swede-2	12	1	0	1	4	....	....	....	....	....
1995-96	Pitea HC	Swede-2	29	1	11	12	18	....	....	....	....	....
1996-97	Pitea HC	Swede-2	28	3	4	7	28	....	....	....	....	....
1997-98	AIK Solna	Sweden	38	0	2	2	16	....	....	....	....	....
1998-99	AIK Solna	Sweden	47	2	7	9	18	....	....	....	....	....
99-2000	AIK Solna	Sweden	49	2	6	8	22	....	....	....	....	....
2000-01	Skelleftea AIK	Swede-2	22	0	5	5	14	....	....	....	....	....
	AIK Solna	Sweden	18	1	3	4	10	....	....	....	....	....
2001-02	Lulea HF	Sweden	41	2	4	6	12	5	0	2	2	2

SANNITZ, Raffaele (ZAH-nihts, ra-FIGH-ehl-lay) **CBJ**

Center. Shoots left. 6'1", 187 lbs. Born, Mendrisio, Switz., May 18, 1983.
(Columbus' 9th choice, 204th overall, in 2001 Entry Draft).

			Regular Season					Playoffs				
Season	Club	Lea	GP	G	A	TP	PIM	GP	G	A	TP	PIM
1997-98	HC Lugano Jr.	Swiss-Jr.	33	7	12	19	54	....	....	....	....	....
1998-99	HC Lugano Jr.	Swiss-Jr.	38	5	12	17	62	....	....	....	....	....
	HC Lugano	Swiss	8	0	1	1	0	....	....	....	....	....
99-2000	HC Lugano Jr.	Swiss-Jr.	33	13	16	29	47	....	....	....	....	....
	HC Lugano	Swiss	1	0	0	0	2	....	....	....	....	....
2000-01	HC Sierre	Swiss-2	2	0	0	0	0	....	....	....	....	....
	HC Lugano Jr.	Swiss-Jr.	35	22	30	52	152	2	0	0	0	0
	HC Lugano	Swiss	13	1	0	1	0	2	0	0	0	0
2001-02	HC Lugano Jr.	Swiss-Jr.	14	14	13	27	18	3	3	2	5	4
	HC Lugano	Swiss	38	3	4	7	37	12	1	1	2	2

SANTALA, Tommi (SAHN-tah-luh, TAW-mee) **ATL.**

Center. Shoots right. 6'2", 198 lbs. Born, Helsinki, Finland, June 27, 1979.
(Atlanta's 10th choice, 245th overall, in 1999 Entry Draft).

			Regular Season					Playoffs				
Season	Club	Lea	GP	G	A	TP	PIM	GP	G	A	TP	PIM
1994-95	Jokerit-C	Finn-Jr.	27	8	13	21	40	6	0	0	0	0
1995-96	Jokerit-B	Finn-Jr.	25	8	4	12	12	6	1	2	3	4
1996-97	Jokerit Jr.	Finn-Jr.	20	0	2	2	10	5	0	0	0	0
1997-98	Jokerit Jr.	Finn-Jr.	36	10	28	38	48	8	0	1	1	4
1998-99	Jokerit Jr.	Finn-Jr.	30	20	24	44	20	8	1	3	4	22
	Jokerit Helsinki	Finland	30	0	0	0	14	3	0	0	0	0
99-2000	Jokerit Jr.	Finn-Jr.	5	6	4	10	4	....	....	....	....	....
	Jokerit Helsinki	Finland	14	0	1	1	6	....	....	....	....	....
	HPK Jr.	Finn-Jr.	5	6	4	10	4	....	....	....	....	....
	HPK Hameenlinna	Finland	38	8	19	27	65	8	3	4	7	10
2000-01	HPK Hameenlinna	Finland	56	16	24	40	90	....	....	....	....	....
2001-02	HPK Hameenlinna	Finland	17	6	16	22	14	....	....	....	....	....

SAPOZHNIKOV, Vladimir (suh-POHZH-nih-kahf, vla-DIH-meer) **FLA.**

Defense. Shoots left. 6'3", 205 lbs. Born, Seversk, USSR, August 2, 1982.
(Florida's 1st choice, 58th overall, in 2000 Entry Draft).

			Regular Season					Playoffs				
Season	Club	Lea	GP	G	A	TP	PIM	GP	G	A	TP	PIM
1997-98	Novosibirsk 2	Russia-3	15	0	0	0	2	....	....	....	....	....
1998-99	Novokuznetsk Jr.	Russia-Jr.	STATISTICS NOT AVAILABLE									
99-2000	Novokuznetsk 2	Russia-3	STATISTICS NOT AVAILABLE									
	Saratov Jr.	Russia-Jr.	4	0	0	0	6	....	....	....	....	....
2000-01	North Bay	OHL	53	0	6	6	70	4	0	0	0	2
2001-02	North Bay	OHL	29	0	4	4	28	5	0	0	0	0
	Utah Grizzlies	AHL	4	0	0	0	4	....	....	....	....	....

SARNO, Peter (SAHR-noh, PEE-tuhr) **EDM.**

Center. Shoots left. 5'11", 185 lbs. Born, Toronto, Ont., July 26, 1979.
(Edmonton's 6th choice, 141st overall, in 1997 Entry Draft).

			Regular Season					Playoffs				
Season	Club	Lea	GP	G	A	TP	PIM	GP	G	A	TP	PIM
1995-96	North York	MTJHL	52	39	57	96	27	....	....	....	....	....
1996-97	Windsor	OHL	66	20	63	83	59	5	0	3	3	6
1997-98	Windsor	OHL	64	33	*88	*121	18	....	....	....	....	....
	Hamilton	AHL	8	1	1	2	2	....	....	....	....	....
1998-99	Sarnia Sting	OHL	68	37	*93	*130	49	6	1	7	8	2
99-2000	Hamilton	AHL	67	10	36	46	31	....	....	....	....	....
2000-01	Hamilton	AHL	79	19	46	65	64	....	....	....	....	....
2001-02	Hamilton	AHL	76	12	40	52	38	15	6	7	13	4

OHL All-Rookie Team (1997) • OHL Rookie of the Year (1997)

Traded to **Sarnia** (OHL) by **Windsor** (OHL) for future considerations, June 11, 1998.

SAUER, Kurt (SAW-uhr, KUHRT) **ANA.**

Defense. Shoots left. 6'4", 225 lbs. Born, St. Cloud, MN, January 16, 1981.
(Colorado's 5th choice, 88th overall, in 2000 Entry Draft).

			Regular Season					Playoffs				
Season	Club	Lea	GP	G	A	TP	PIM	GP	G	A	TP	PIM
1998-99	North Iowa	USHL	52	1	4	5	67	....	....	....	....	....
99-2000	Spokane Chiefs	WHL	71	3	12	15	48	15	2	1	3	8
2000-01	Spokane Chiefs	WHL	48	5	10	15	85	3	1	0	1	2
2001-02	Spokane Chiefs	WHL	61	4	20	24	73	11	0	3	3	12

WHL West First All-Star Team (2002)

Signed as a free agent by **Anaheim**, June 6, 2002.

SAVIELS, Agris (sah-VEE-ehls, AG-rihs) **COL.**

Defense. Shoots left. 6'2", 200 lbs. Born, Riga, Latvia, January 15, 1982.
(Colorado's 4th choice, 63rd overall, in 2000 Entry Draft).

			Regular Season					Playoffs				
Season	Club	Lea	GP	G	A	TP	PIM	GP	G	A	TP	PIM
1996-97	Dynamo Riga	Lat.-Jr.	15	1	2	3	4	....	....	....	....	....
	HK Lido-Nafta	Latvia	40	4	15	19	40	....	....	....	....	....
1997-98	Dynamo Riga	Lat.-Jr.	15	1	2	3	4	....	....	....	....	....
	HK Lido-Nafta	Latvia	40	7	21	28	30	....	....	....	....	....
1998-99	Notre Dame	SMBHL	18	6	9	15	25	....	....	....	....	....
	Notre Dame	SJHL	30	6	13	19	....	....	....	....	....	....
99-2000	Owen Sound	OHL	65	7	25	32	56	....	....	....	....	....
2000-01	Owen Sound	OHL	68	14	37	51	46	5	0	1	1	2
2001-02	Owen Sound	OHL	60	5	27	32	37	....	....	....	....	....

SCHADILOV, Igor (sha-DEE-lahf, EE-gor) **WSH.**

Defense. Shoots left. 6'2", 189 lbs. Born, Moscow, USSR, June 7, 1980.
(Washington's 10th choice, 249th overall, in 1999 Entry Draft).

			Regular Season					Playoffs				
Season	Club	Lea	GP	G	A	TP	PIM	GP	G	A	TP	PIM
1996-97	Dyno. Moscow Jr.	Russia-Jr.	30	3	7	10	30	....	....	....	....	....
1997-98	Dynamo Moscow	Russia	38	1	0	1	6	....	....	....	....	....
1998-99	Dynamo Moscow	Russia	28	2	9	11	15	....	....	....	....	....
	Dynamo Moscow	Russia	2	0	0	0	0	....	....	....	....	....
	Krylja Sovetov	Russia	9	0	0	0	0	....	....	....	....	....
99-2000	THC Tver	Russia-2	14	0	3	3	6	....	....	....	....	....
	Dynamo Moscow	Russia	26	0	2	2	8	16	0	0	0	2
2000-01	Dynamo Moscow	Russia	34	1	5	6	12	....	....	....	....	....
2001-02	Cherepovets	Russia	33	7	3	10	10	4	0	0	0	2

SCHAUER, Stefan (SHOW-uhr, SHTEH-fuhn) **OTT.**

Defense. Shoots left. 6'1", 180 lbs. Born, Schongau, West Germany, January 12, 1983.
(Ottawa's 6th choice, 162nd overall, in 2001 Entry Draft).

			Regular Season					Playoffs				
Season	Club	Lea	GP	G	A	TP	PIM	GP	G	A	TP	PIM
99-2000	Riessersee-16	Ger.-Jr.	14	4	8	12	85	....	....	....	....	....
	Riessersee Jr.	Ger.-Jr.	30	7	14	21	82	....	....	....	....	....
	Riessersee	German-3	3	0	0	0	0	....	....	....	....	....
2000-01	Riessersee Jr.	Ger.-Jr.	11	1	7	8	24	....	....	....	....	....
	Riessersee	German-3	39	1	2	3	12	5	0	1	1	0
2001-02	SC Riessersee	German-2	46	2	18	20	46	8	0	2	2	18

SCHEFFELMAIER, Brett (sch-EHFEHL-mai-uhr, BREHT) **ST.L.**

Defense. Shoots right. 6'5", 200 lbs. Born, Coronation, Alta., March 31, 1981.
(St. Louis' 5th choice, 190th overall, in 2001 Entry Draft).

			Regular Season					Playoffs				
Season	Club	Lea	GP	G	A	TP	PIM	GP	G	A	TP	PIM
1997-98	Red Deer	AMHL	13	0	4	4	36	....	....	....	....	....
	Medicine Hat	WHL	25	0	1	1	69	....	....	....	....	....
1998-99	Medicine Hat	WHL	69	3	10	13	252	....	....	....	....	....
99-2000	Medicine Hat	WHL	71	1	9	10	281	....	....	....	....	....
2000-01	Medicine Hat	WHL	62	3	10	13	279	....	....	....	....	....
2001-02	Medicine Hat	WHL	45	4	6	10	180	....	....	....	....	....

• Re-entered NHL Entry Draft. Originally Tampa Bay's 3rd choice, 75th overall, in 1999 Entry Draft.

SCHILL, Jonathan (SHIHL, JAWN-ah-thuhn) CBJ

Left wing. Shoots left. 6'2", 201 lbs. Born, Kitchener, Ont., June 28, 1979.

			Regular Season					Playoffs				
Season	Club	Lea	GP	G	A	TP	PIM	GP	G	A	TP	PIM
1995-96	Kitchener Lions	OMHA	30	17	12	29	100	....	....	....	....	....
	Kitchener	OJHL-B	1	0	0	0	0	....	....	....	....	....
1996-97	Kingston	OHL	52	3	7	10	16	4	0	0	0	0
1997-98	Kingston	OHL	64	14	17	31	34	12	1	1	2	4
1998-99	Kingston	OHL	68	32	27	59	93	5	2	1	3	6
99-2000	Kingston	OHL	65	39	48	87	79	3	1	1	2	6
2000-01	Syracuse Crunch	AHL	4	0	1	1	8	....	....	....	....	....
	Dayton Bombers	ECHL	58	18	6	24	153	6	2	3	5	22
2001-02	Dayton Bombers	ECHL	5	0	1	1	15	....	....	....	....	....
	Syracuse Crunch	AHL	44	3	4	7	61	9	0	0	0	4

Signed as a free agent by **Columbus**, May 8, 2000.

SCHMIDT, Chris L.A.

Center. Shoots left. 6'3", 212 lbs. Born, Beaverlodge, Alta., March 1, 1976.
(Los Angeles' 4th choice, 111th overall, in 1994 Entry Draft).

			Regular Season					Playoffs				
Season	Club	Lea	GP	G	A	TP	PIM	GP	G	A	TP	PIM
1992-93	Seattle	WHL	61	6	7	13	17	5	0	1	1	0
1993-94	Seattle	WHL	68	7	17	24	26	9	3	1	4	2
1994-95	Seattle	WHL	61	21	11	32	31	3	0	0	0	0
1995-96	Seattle	WHL	61	39	23	62	135	5	1	5	6	9
1996-97	Mississippi	ECHL	18	7	7	14	35	....	....	....	....	....
	Phoenix	IHL	37	3	6	9	60	....	....	....	....	....
1997-98	Fredericton	AHL	69	8	5	13	67	4	0	0	0	2
1998-99	Springfield	AHL	17	3	2	5	19	1	0	0	0	0
	Mississippi	ECHL	6	1	0	1	2	18	6	8	14	10
99-2000	Team Canada	Nat-Tm	33	1	9	10	28	....	....	....	....	....
	Lowell	AHL	38	8	10	18	38	7	2	1	3	8
2000-01	Lowell	AHL	79	21	32	53	84	4	2	2	4	2
2001-02	Manchester	AHL	62	9	12	21	43	5	0	2	2	0

SCHNEIDER, Andrew (SHNIGH-duhr, AN-droo) PIT.

Defense. Shoots left. 6', 220 lbs. Born, Grand Forks, ND, July 31, 1981.
(Pittsburgh's 7th choice, 156th overall, in 2001 Entry Draft).

			Regular Season					Playoffs				
Season	Club	Lea	GP	G	A	TP	PIM	GP	G	A	TP	PIM
1998-99	Lincoln Stars	USHL	9	0	4	4	8	4	0	0	0	2
99-2000	Lincoln Stars	USHL	46	7	10	17	102	10	6	4	10	27
2000-01	Lincoln Stars	USHL	54	12	24	36	134	....	....	....	....	....
2001-02	Munchen Barons	Germany	53	11	29	40	101	....	....	....	....	....

SCHUBERT, Christoph (SHOO-buhrt, KRIHS-tawf) OTT.

Defense. Shoots left. 6'3", 198 lbs. Born, Munich, West Germany, February 5, 1982.
(Ottawa's 5th choice, 127th overall, in 2001 Entry Draft).

			Regular Season					Playoffs				
Season	Club	Lea	GP	G	A	TP	PIM	GP	G	A	TP	PIM
1998-99	EV Landshut Jr.	Ger.-Jr.	28	15	20	35	77	....	....	....	....	....
99-2000	EV Landshut Jr.	Ger.-Jr.	11	14	11	25	51	....	....	....	....	....
	EV Landshut	German-3	55	7	5	12	68	....	....	....	....	....
2000-01	Munchen Barons	Germany	55	6	3	9	80	10	0	2	2	27
2001-02	Munchen Barons	Germany	50	5	11	16	125	....	....	....	....	....

SCHUELLER, Doug (SHOO-luhr, DUHG)

Defense. Shoots right. 6'1", 210 lbs. Born, Inver Grove Heights, MN, March 30, 1977.
(Florida's 9th choice, 211th overall, in 1997 Entry Draft).

			Regular Season					Playoffs				
Season	Club	Lea	GP	G	A	TP	PIM	GP	G	A	TP	PIM
1994-95	White Bear Lake	Hi-School	25	7	17	24	32	....	....	....	....	....
1995-96	White Bear Lake	Hi-School	5	0	2	2	6	....	....	....	....	....
	Twin Cities	USHL	39	5	13	18	186	....	....	....	....	....
1996-97	Twin Cities	USHL	54	7	29	36	83	5	0	2	2	29
1997-98	Bowling Green	CCHA	36	5	14	19	62	....	....	....	....	....
1998-99	Bowling Green	CCHA	37	7	5	12	54	....	....	....	....	....
99-2000	Bowling Green	CCHA	32	2	5	7	85	....	....	....	....	....
2000-01	Bowling Green	CCHA	33	2	6	8	50	....	....	....	....	....
	Louisville	AHL	11	1	4	5	11	....	....	....	....	....
2001-02	Macon Whoopee	ECHL	55	3	9	12	47	....	....	....	....	....
	Utah Grizzlies	AHL	2	0	1	1	2	....	....	....	....	....
	Jackson Bandits	ECHL	15	0	4	4	16	9	0	1	1	15

SCHUTTE, Michael (SHOOT, MIGH-kuhl) PHX.

Defense. Shoots left. 6'2", 194 lbs. Born, Burlington, Ont., July 28, 1979.

			Regular Season					Playoffs				
Season	Club	Lea	GP	G	A	TP	PIM	GP	G	A	TP	PIM
1998-99	Burlington	OPJHL	47	26	44	70	....	....	....	....	....	....
99-2000	U. of Maine	H-East	23	2	7	9	14	....	....	....	....	....
2000-01	U. of Maine	H-East	38	15	10	25	20	....	....	....	....	....
2001-02	U. of Maine	H-East	39	13	18	31	31	....	....	....	....	....

OPJHL Defenseman of the Year (1999) • NCAA Championship All-Tournament Team (2002)

Signed as a free agent by **Phoenix**, May 30, 2002.

SCISSONS, Jeff (SKIH-zuhns, JEHF) VAN.

Center. Shoots left. 6'1", 190 lbs. Born, Saskatoon, Sask., November 24, 1976.
(Vancouver's 7th choice, 201st overall, in 1996 Entry Draft).

			Regular Season					Playoffs				
Season	Club	Lea	GP	G	A	TP	PIM	GP	G	A	TP	PIM
1992-93	Sask. Blazers	SMHL	35	8	11	19	32	....	....	....	....	....
1993-94	Sask. Contacts	SMHL	35	20	37	57	42	....	....	....	....	....
1994-95	Vernon Vipers	BCJHL	50	10	20	30	....	....	....	....	....	....
1995-96	Vernon Vipers	BCJHL	60	26	48	74	28	30	14	18	32	....
1996-97	U. Minn-Duluth	WCHA	38	3	14	17	30	....	....	....	....	....
1997-98	U. Minn-Duluth	WCHA	40	17	24	41	50	....	....	....	....	....
1998-99	U. Minn-Duluth	WCHA	38	18	19	37	42	....	....	....	....	....
99-2000	U. Minn-Duluth	WCHA	37	14	19	33	32	....	....	....	....	....
2000-01	Kansas City	IHL	68	5	19	24	24	....	....	....	....	....
2001-02	HIFK Helsinki	Finland	11	2	0	2	4	....	....	....	....	....
	Manitoba Moose	AHL	7	0	0	0	2	....	....	....	....	....

WCHA Student Athlete of the Year (2000) • Established **Minnesota-Duluth** (WCHA) record by appearing in 153 consecutive games, March 23, 2000.

Signed as a free agent by **HIFK Helsinki** (Finland), December 4, 2001. Released by **HIFK Helsinki** (Finland), January 15, 2002.

SCUDERI, Rob (SKUD-uhree, RAWB) PIT.

Defense. Shoots left. 6', 208 lbs. Born, Syosset, NY, December 30, 1978.
(Pittsburgh's 5th choice, 134th overall, in 1998 Entry Draft).

			Regular Season					Playoffs				
Season	Club	Lea	GP	G	A	TP	PIM	GP	G	A	TP	PIM
1995-96	NY Apple Core	MJBHL	76	18	60	78	....	....	....	....	....	....
1996-97	NY Apple Core	MJBHL	82	42	70	112	64	....	....	....	....	....
1997-98	Boston College	H-East	42	0	24	24	12	....	....	....	....	....
1998-99	Boston College	H-East	41	2	8	10	20	....	....	....	....	....
99-2000	Boston College	H-East	42	1	12	13	22	....	....	....	....	....
2000-01	Boston College	H-East	43	4	19	23	42	....	....	....	....	....
2001-02	Wilkes-Barre	AHL	75	1	22	23	66	....	....	....	....	....

NCAA Championship All-Tournament Team (2001)

SEDOV, Pavel (se-DAHF), PAH-vehl) T.B.

Left wing. Shoots left. 6'3", 200 lbs. Born, Voskresensk, USSR, January 12, 1982.
(Tampa Bay's 5th choice, 161st overall, in 2000 Entry Draft).

			Regular Season					Playoffs				
Season	Club	Lea	GP	G	A	TP	PIM	GP	G	A	TP	PIM
99-2000	Voskresensk	Russia-2	10	0	0	0	2	....	....	....	....	....
2000-01	Voskresensk	Russia-2	38	2	1	3	10	....	....	....	....	....
2001-02	Voskresensk 2	Russia-3	12	4	1	5	0	....	....	....	....	....
	Voskresensk	Russia-2	18	3	1	4	0	....	....	....	....	....

SEELEY, Richard (SEE-lee, RIH-chahrd) L.A.

Defense. Shoots left. 6'2", 205 lbs. Born, Powell River, B.C., April 30, 1979.
(Los Angeles' 6th choice, 137th overall, in 1997 Entry Draft).

			Regular Season					Playoffs				
Season	Club	Lea	GP	G	A	TP	PIM	GP	G	A	TP	PIM
1995-96	Powell River	BCJHL	44	1	8	9	42	....	....	....	....	....
1996-97	Lethbridge	WHL	3	0	0	0	11	....	....	....	....	....
	Prince Albert	WHL	18	0	1	1	9	4	0	0	0	2
1997-98	Prince Albert	WHL	65	8	21	29	114	....	....	....	....	....
1998-99	Prince Albert	WHL	61	10	48	58	110	14	1	11	12	14
99-2000	Lowell	AHL	36	5	1	6	37	....	....	....	....	....
2000-01	Lowell	AHL	55	2	8	10	102	....	....	....	....	....
	Trenton Titans	ECHL	9	0	2	2	18	....	....	....	....	....
2001-02	Manchester	AHL	61	2	10	12	78	5	0	0	0	6

SEGAL, Brandon (SEE-guhl, BRAN-duhn) NSH.

Right wing. Shoots right. 6'3", 206 lbs. Born, Richmond, B.C., July 12, 1983.
(Nashville's 2nd choice, 102nd overall, in 2002 Entry Draft).

			Regular Season					Playoffs				
Season	Club	Lea	GP	G	A	TP	PIM	GP	G	A	TP	PIM
99-2000	Calgary Hitmen	WHL	44	2	6	8	76	13	1	1	2	13
	Delta Ice Hawks	PIJHL	....	....	....	....	....	3	0	1	1	2
2000-01	Calgary Hitmen	WHL	72	16	11	27	103	12	1	1	2	17
2001-02	Calgary Hitmen	WHL	71	43	40	83	122	7	1	4	5	16

SEIDENBERG, Denis (ZIGH-dehn-buhrg, DEH-nihs) PHI.

Defense. Shoots left. 6', 180 lbs. Born, Schwenningen, West Germany, July 18, 1981.
(Philadelphia's 6th choice, 172nd overall, in 2001 Entry Draft).

			Regular Season					Playoffs				
Season	Club	Lea	GP	G	A	TP	PIM	GP	G	A	TP	PIM
99-2000	Mannheim Jr.	Ger.-Jr.	52	12	28	40	28	....	....	....	....	....
	Adler Mannheim	Germany	3	0	0	0	0	....	....	....	....	....
2000-01	Mannheim Jr.	Ger.-Jr.	9	3	8	11	20	....	....	....	....	....
	Adler Mannheim	Germany	55	2	5	7	6	12	0	1	1	10
2001-02	Adler Mannheim	Germany	55	7	13	20	56	8	0	0	0	2

SEIKKULA, Timo (SAY-koo-lah, TEE-moh) PIT.

Center. Shoots left. 6'2", 183 lbs. Born, Kalajoki, Finland, May 27, 1978.
(Pittsburgh's 8th choice, 238th overall, in 1996 Entry Draft).

			Regular Season					Playoffs				
Season	Club	Lea	GP	G	A	TP	PIM	GP	G	A	TP	PIM
1994-95	Junkkarit	Finland-2	44	1	2	3	8	....	....	....	....	....
1995-96	Junkkarit	Finland-2	41	9	9	18	58	....	....	....	....	....
	Junkkarit	Finland-3	....	....	....	....	....	5	1	2	3	22
1996-97	TPS Turku Jr.	Finn-Jr.	4	1	0	1	4	....	....	....	....	....
	Kiekko-67 Turku	Finland-2	43	5	5	10	38	....	....	....	....	....
	Kiekko-67 Turku	Finland-3	....	....	....	....	....	3	1	0	1	4
1997-98	TPS Turku Jr.	Finn-Jr.	12	6	2	8	10	7	3	4	7	12
	TuTo Turku	Finland-2	17	4	4	8	29	....	....	....	....	....
	TPS Turku	Finland	1	0	1	1	0	....	....	....	....	....
1998-99	KalPa Kuopio Jr.	Finn-Jr.	1	1	0	1	4	....	....	....	....	....
	KalPa Kuopio	Finland	52	3	8	11	40	....	....	....	....	....
	KalPa Kuopio-B	Finn-Jr.	....	....	....	....	....	1	0	0	0	0
	KalPa Kuopio	Finland-2	....	....	....	....	....	6	0	0	0	0
99-2000	Hermes Kokkola	Finland-2	48	3	11	14	34	3	1	0	1	2
2000-01	FoPS Forssa	Finland-2	22	5	8	13	28	....	....	....	....	....
2001-02	IK Oskarshamn	Swede-2	46	6	5	11	52	....	....	....	....	....

SEIKOLA, Markus (SAY-koh-la, MAHR-kuhs) TOR.

Defense. Shoots right. 6'1", 194 lbs. Born, Laitila, Finland, June 5, 1982.
(Toronto's 7th choice, 209th overall, in 2000 Entry Draft).

			Regular Season					Playoffs				
Season	Club	Lea	GP	G	A	TP	PIM	GP	G	A	TP	PIM
1996-97	TPS Turku-C	Finn-Jr.	4	1	0	1	4	1	0	0	0	0
1997-98	TPS Turku Jr.	Finn-Jr.	2	0	0	0	0	1	0	0	0	2
1998-99	TPS Turku Jr.	Finn-Jr.	36	2	10	12	24	....	....	....	....	....
99-2000	TPS Turku-B	Finn-Jr.	9	2	1	3	6	....	....	....	....	....
2000-01	TPS Turku Jr.	Finn-Jr.	26	13	7	20	6	3	1	0	1	0
	TPS Turku	Finland	23	1	0	1	16	....	....	....	....	....
2001-02	TPS Turku	Finland	51	4	4	8	20	7	1	1	2	12
	TPS Turku Jr.	Finn-Jr.	1	0	1	1	2	2	1	0	1	2

SELIG, Scott (SEH-lihg, SKAWT) MTL.

Center. Shoots right. 6'3", 178 lbs. Born, Philadelphia, PA, March 2, 1981.
(Montreal's 8th choice, 172nd overall, in 2000 Entry Draft).

			Regular Season					Playoffs				
Season	Club	Lea	GP	G	A	TP	PIM	GP	G	A	TP	PIM
99-2000	Thayer Academy	Hi-School	28	32	25	57	25	....	....	....	....	....
2000-01	Northeastern	H-East	35	7	8	15	34	....	....	....	....	....
2001-02	Northeastern	H-East	26	3	1	4	28	....	....	....	....	....

SELUYANOV, Alexander (sehl-oo-YA-nahf, al-ehx-AN-duhr) DET.

Defense. Shoots right. 5'11", 172 lbs. Born, Ufa, USSR, March 24, 1982.
(Detroit's 5th choice, 128th overall, in 2000 Entry Draft).

			Regular Season					Playoffs				
Season	Club	Lea	GP	G	A	TP	PIM	GP	G	A	TP	PIM
1997-98	Novoil Ufa	Russia-3	19	0	1	1	8					
1998-99	Novoil Ufa	Russia-4	20	3	3	6	8					
99-2000	Ufa 2	Russia-3	18	3	4	7	10					
	Ufa	Russia	13	1	2	3	4					
2000-01	Ufa	Russia	30	0	3	3	10					
2001-02	CSK VVS Samara	Russia-2	30	2	7	9	58					
	Lada Togliatti	Russia	6	0	0	0	0					

SEMENOV, Alexei (seh-MEH-nahv, al-EHX-ay) EDM.

Defense. Shoots left. 6'6", 210 lbs. Born, Murmansk, USSR, April 10, 1981.
(Edmonton's 2nd choice, 36th overall, in 1999 Entry Draft).

			Regular Season					Playoffs				
Season	Club	Lea	GP	G	A	TP	PIM	GP	G	A	TP	PIM
1997-98	Krylja Sovetov 2	Russia-3	52	1	2	3	48					
1998-99	St. Petersburg 2	Russia-4	19	0	1	1	20					
	Sudbury Wolves	OHL	28	0	3	3	28	2	0	0	0	4
99-2000	Sudbury Wolves	OHL	65	9	35	44	135	12	1	3	4	23
	Hamilton	AHL	....	....	....	....	....	3	0	0	0	0
2000-01	Sudbury Wolves	OHL	65	21	42	63	106	12	4	13	17	17
2001-02	Hamilton	AHL	78	5	11	16	67					

OHL First All-Star Team (2001)

SEMENOV, Dmitri (seh-MEH-nahv, dih-MEE-tree) DET.

Right wing. Shoots left. 5'10", 178 lbs. Born, Moscow, USSR, April 19, 1982.
(Detroit's 4th choice, 127th overall, in 2000 Entry Draft).

			Regular Season					Playoffs				
Season	Club	Lea	GP	G	A	TP	PIM	GP	G	A	TP	PIM
1997-98	Dynamo Moscow	Russia	13	2	1	3	2					
1998-99	DynamoMoscow2	Russia-3	26	14	4	18	16					
99-2000	THC Tver	Russia-2	16	4	2	6	63					
2000-01	Dynamo Moscow	Russia	12	0	0	0	8					
	Yekaterinburg	Russia	24	0	0	0	18					
2001-02	DynamoMoscow2	Russia-3	6	5	2	7	6					
	Yekaterinburg	Russia-2	6	1	2	3	2					
	Dynamo Moscow	Russia	25	2	1	3	8					

SEMIN, Alexander (SEH-min, al-ehx-AN-duhr) WSH.

Left wing. Shoots left. 6', 174 lbs. Born, Krasjonarsk, USSR, March 3, 1984.
(Washington's 2nd choice, 13th overall, in 2002 Entry Draft).

			Regular Season					Playoffs				
Season	Club	Lea	GP	G	A	TP	PIM	GP	G	A	TP	PIM
99-2000	Chelyabinsk-18	Russia-Jr.	6	2	2	4	14					
2000-01	Team Russia	Nat-Tm	9	4	1	5	30					
2001-02	Chelyabinsk-18	Russia-Jr.	4	6	0	6	24					
	Chelyabinsk	Russia-2	48	15	8	23	52					

SEMIN, Dmitri (SEH-min, dih-MEE-tree) ST.L.

Center. Shoots left. 5'10", 165 lbs. Born, Moscow, USSR, August 14, 1983.
(St. Louis' 4th choice, 159th overall, in 2001 Entry Draft).

			Regular Season					Playoffs				
Season	Club	Lea	GP	G	A	TP	PIM	GP	G	A	TP	PIM
99-2000	Spartak Moscow 2	Russia-3	27	9	10	19	10					
	Spartak Moscow	Russia-2	1	0	0	0	0					
2000-01	Spartak Moscow	Russia-2	32	8	6	14	8					
2001-02	Spartak Moscow 2	Russia-3	4	5	0	5	4					
	Spartak Moscow	Russia	44	2	6	8	14					

SENEZ, Francois (seh-NAY, fran-SWUH) DET.

Defense. Shoots left. 6'2", 215 lbs. Born, Montreal, Que., March 20, 1982.
(Detroit's 7th choice, 288th overall, in 2001 Entry Draft).

			Regular Season					Playoffs				
Season	Club	Lea	GP	G	A	TP	PIM	GP	G	A	TP	PIM
1997-98	Laval-Laurentide	QAAA	13	0	1	1	0					
1998-99	Laval-Laurentide	QAAA	42	1	12	13	58					
2000-01	RPI Engineers	ECAC	21	0	3	3	14					
2001-02	RPI Engineers	ECAC	7	0	0	0	6					
	Drummondville	QMJHL	19	0	0	0	51	12	1	2	3	25

• Left **RPI** (ECAC) and signed as a free agent with **Drummondville** (QMJHL), January 16, 2002.

SERTICH, Andrew (SUHR-tihch, AN-droo) PIT.

Left wing. Shoots left. 5'11", 161 lbs. Born, Coleraine, MN, May 6, 1983.
(Pittsburgh's 5th choice, 136th overall, in 2002 Entry Draft).

			Regular Season					Playoffs				
Season	Club	Lea	GP	G	A	TP	PIM	GP	G	A	TP	PIM
2000-01	Greenway Raiders	Hi-School	31	35	45	80	14					
2001-02	Greenway Raiders	Hi-School	26	24	48	72	35					
	Sioux Falls	USHL	13	2	4	6	0	2	0	0	0	2

All-State First All-Star Team (2002)

• Signed Letter of Intent to attend **U. of Minnesota** (WCHA), July 2, 2001.

SESSA, Jason (SEH-sa, JAY-suhn) TOR.

Right wing. Shoots right. 6'1", 190 lbs. Born, Long Island, NY, July 17, 1977.
(Toronto's 5th choice, 86th overall, in 1996 Entry Draft).

			Regular Season					Playoffs				
Season	Club	Lea	GP	G	A	TP	PIM	GP	G	A	TP	PIM
1994-95	Rochester	USHL	47	45	22	67	81					
1995-96	Lake Superior	CCHA	30	9	5	14	12					
1996-97	Lake Superior	CCHA	34	22	22	44	91					
1997-98	Lake Superior	CCHA	32	16	13	29	55					
	St. John's	AHL	5	0	0	0	6					
1998-99	St. John's	AHL	56	9	4	13	25					
99-2000	St. John's	AHL	30	7	14	21	54					
	Louisiana	ECHL	17	1	4	5	14					
	South Carolina	ECHL	15	10	8	18	30	7	4	1	5	2
2000-01	South Carolina	ECHL	58	34	30	64	113					
	St. John's	AHL	9	4	4	8	4	3	1	0	1	6
2001-02	South Carolina	ECHL	40	21	17	38	40					
	Cincinnati	AHL	1	0	0	0	0					
	Springfield	AHL	3	0	0	0	0					
	Dayton Bombers	ECHL	8	9	3	12	25	14	7	3	10	42

CCHA Second All-Star Team (1997)

Signed as a free agent by **Sheffield** (Britain), July 30, 2002.

SETZINGER, Oliver (SEHT-zihn-guhr, AW-lih-vuhr) NSH.

Center. Shoots left. 6', 215 lbs. Born, Horn, Austria, July 11, 1983.
(Nashville's 5th choice, 76th overall, in 2001 Entry Draft).

			Regular Season					Playoffs				
Season	Club	Lea	GP	G	A	TP	PIM	GP	G	A	TP	PIM
1998-99	Wiener EV Jr.	Austria-Jr.	30	25	27	52	30					
99-2000	Ilves Jr.	Finn-Jr.	35	6	4	10	65					
	Ilves Tampere-B	Finn-Jr.	18	16	9	25	38					
	Ilves Tampere	Finland	1	0	0	0	2					
	Ilves Tampere	Finland-2	18	16	9	25	38					
2000-01	Ilves Jr.	Finn-Jr.	31	8	12	20	74					
	Ilves Tampere	Finland	14	0	1	1	10					
2001-02	Ilves Jr.	Finn-Jr.	1	0	0	0	2					
	Ilves Tampere	Finland	10	1	0	1	4					
	Sport Vaasa	Finland-2	8	5	2	7	6					
	Austria	Olympics	4	1	0	1	2					
	EHC Linz	Austria	8	6	7	13	4	13	4	14	18	14

SHARP, Patrick (SHAHRP, PAT-rihk) PHI.

Center. Shoots right. 6', 188 lbs. Born, Thunder Bay, Ont., December 27, 1981.
(Philadelphia's 2nd choice, 95th overall, in 2001 Entry Draft).

			Regular Season					Playoffs				
Season	Club	Lea	GP	G	A	TP	PIM	GP	G	A	TP	PIM
1998-99	Thunder Bay	USHL	55	19	24	43	48	3	1	1	2	0
99-2000	Thunder Bay	USHL	56	20	35	55	41					
2000-01	U. of Vermont	ECAC	34	12	15	27	36					
2001-02	U. of Vermont	ECAC	31	13	13	26	50					

SHASBY, Matt (SHAS-bee, MAT) MTL.

Defense. Shoots left. 6'3", 188 lbs. Born, Sioux Falls, SD, July 2, 1980.
(Montreal's 7th choice, 150th overall, in 1999 Entry Draft).

			Regular Season					Playoffs				
Season	Club	Lea	GP	G	A	TP	PIM	GP	G	A	TP	PIM
1997-98	Lincoln Stars	USHL	43	1	15	16	30	8	0	0	0	2
1998-99	Des Moines	USHL	49	4	22	26	34	11	0	1	1	12
99-2000	Alaska-Anchorage	WCHA	32	1	8	9	36					
2000-01	Alaska-Anchorage	WCHA	35	4	14	18	32					
2001-02	Alaska-Anchorage	WCHA	35	7	20	27	72					

WCHA Second All-Star Team (2002)

SHASTIN, Yegor (SHAS-tihn, yeh-GOHR) CGY.

Left wing. Shoots left. 5'9", 172 lbs. Born, Kiev, USSR, September 10, 1982.
(Calgary's 5th choice, 124th overall, in 2001 Entry Draft).

			Regular Season					Playoffs				
Season	Club	Lea	GP	G	A	TP	PIM	GP	G	A	TP	PIM
1997-98	Omsk 2	Russia-3	4	0	1	1	0					
1998-99	Omsk 2	Russia-4	19	11	17	28	30					
	Avangard Omsk	Russia	4	0	0	0	0	4	0	1	1	0
99-2000	Omsk 2	Russia-3	11	6	5	11	20					
	Avangard Omsk	Russia	26	2	4	6	20	7	3	1	4	16
2000-01	Omsk 2	Russia-3	14	13	9	22	62					
	Avangard Omsk	Russia	35	3	11	14	59	9	1	0	1	18
2001-02	Avangard Omsk	Russia	26	2	5	7	10	11	1	0	1	8

SHEFER, Andrei (SHEH-fuhr, AN-dray) L.A.

Right wing. Shoots left. 6'1", 194 lbs. Born, Sverdlovsk, USSR, July 26, 1981.
(Los Angeles' 1st choice, 43rd overall, in 1999 Entry Draft).

			Regular Season					Playoffs				
Season	Club	Lea	GP	G	A	TP	PIM	GP	G	A	TP	PIM
1997-98	Yekaterinburg 2	Russia-3	16	3	3	6	18					
1998-99	Cherepovets 3	Russia-4	6	2	2	4	18					
	Cherepovets 2	Russia-3	21	6	5	11	20					
	Cherepovets	Russia	8	1	0	1	4					
99-2000	Halifax	QMJHL	72	34	42	76	30	10	0	5	5	4
2000-01	St. Petersburg	Russia	11	6	1	7	4					
	Cherepovets	Russia	20	1	1	2	10	6	1	0	1	0
2001-02	Cherepovets 2	Russia-3	3	1	2	3	2					
	Cherepovets	Russia	8	0	0	0	6					
	St. Petersburg	Russia	28	4	4	8	10					

SHIELDS, Colin (SHEELDZ, KAW-lihn) PHI.

Right wing. Shoots right. 6', 175 lbs. Born, Glasgow, Scotland, January 27, 1980.
(Philadelphia's 4th choice, 195th overall, in 2000 Entry Draft).

			Regular Season					Playoffs				
Season	Club	Lea	GP	G	A	TP	PIM	GP	G	A	TP	PIM
1998-99	Cleveland Barons	NAJHL	55	30	30	60	131					
99-2000	Cleveland Barons	NAJHL	55	46	*49	*95	40	3	1	3	4	2
2000-01	U. of Maine	H-East	DID NOT PLAY – ACADEMICALLY INELIGIBLE									
2001-02	U. of Maine	H-East	42	29	17	46	39					

NAJHL First All-Star Team (2000) • Hockey East All-Rookie Team (2002)

• Ruled ineligible to play 2000-01 season by Hockey East officials due to academic violations while playing U.S. junior hockey, October 10, 2000.

SHIKHANOV, Sergei (shih-KHAHN-ohf, SAIR-gay) CHI.

Right wing. Shoots left. 6'2", 190 lbs. Born, Togliatti, USSR, April 8, 1978.
(Chicago's 10th choice, 204th overall, in 1997 Entry Draft).

			Regular Season					Playoffs				
Season	Club	Lea	GP	G	A	TP	PIM	GP	G	A	TP	PIM
1996-97	Lada Togliatti	Russia	19	4	4	8	20	8	1	0	1	10
	Nizhnekamsk	Russia	5	1	1	2	2					
1997-98	Nizhnekamsk 2	Russia-3	32	9	7	16	14					
	Lada Togliatti	Russia	19	3	0	3	4					
1998-99	Lada Togliatti	Russia	29	4	5	9	65	3	0	0	0	2
	CSK VVS Samara	Russia	11	3	4	7	10	3	1	0	1	2
99-2000	Yaroslavl	Russia	31	2	3	5	24	7	0	2	2	4
2000-01	Magnitogorsk	Russia	24	4	4	8	39	10	1	0	1	4
2001-02	Ufa	Russia	37	5	5	10	57					

SHINKAR, Alexander (shihn-KAHR, al-ehx-AN-duhr) TOR.

Right wing. Shoots left. 6', 176 lbs. Born, Ufa, USSR, July 3, 1981.
(Toronto's 9th choice, 254th overall, in 2000 Entry Draft).

			Regular Season					Playoffs				
Season	Club	Lea	GP	G	A	TP	PIM	GP	G	A	TP	PIM
1997-98	Novoil Ufa	Russia-3	22	6	2	8	4					
1998-99	Cherepovets 2	Russia-3	25	7	2	9	8					
	Cherepovets 3	Russia-4	8	0	4	4	4					
99-2000	Cherepovets	Russia	18	1	1	2	2	8	0	0	0	0
2000-01	St. Petersburg	Russia	43	7	4	11	50					
2001-02	Ufa	Russia	27	3	3	6	8					
	St. Petersburg	Russia	18	3	4	7	10					

SHISHKANOV, Timofei (SHIHSH-kuh-nahv, tee-moh-FAY) **NSH.**

Left wing. Shoots right. 6'1", 213 lbs. Born, Moscow, USSR, June 10, 1983.
(Nashville's 2nd choice, 33rd overall, in 2001 Entry Draft).

Season	Club	Lea	Regular Season GP	G	A	TP	PIM	Playoffs GP	G	A	TP	PIM
99-2000	Spartak Moscow 2	Russia-3	14	6	5	11	10					
	Spartak Moscow	Russia-2	14	1	0	1	2					
2000-01	Spartak Moscow 2	Russia-3	STATISTICS NOT AVAILABLE									
	Spartak Moscow	Russia-2	12	0	0	0	2					
2001-02	H.C. CSKA	Russia-2	23	7	6	13	8					
	H.C. CSKA 2	Russia-3	13	7	9	16	14					

SHKOTOV, Alexei (SHKOH-tahv, al-EHX-ay) **ST.L.**

Right wing. Shoots left. 5'11", 161 lbs. Born, Elektrostal, USSR, June 22, 1984.
(St. Louis' 1st choice, 48th overall, in 2002 Entry Draft).

Season	Club	Lea	Regular Season GP	G	A	TP	PIM	Playoffs GP	G	A	TP	PIM
99-2000	Elektrostal 2	Russia-3	2	0	0	0	0					
2000-01	Elektrostal 2	Russia-3	STATISTICS NOT AVAILABLE									
2001-02	Elektrostal 2	Russia-3	4	4	5	9	2					
	Elektrostal	Russia-2	52	17	9	26	40					

SHMYR, Jason (SHMEER, JAY-suhn)

Left wing. Shoots left. 6'4", 220 lbs. Born, Fairview, Alta., July 27, 1975.

Season	Club	Lea	Regular Season GP	G	A	TP	PIM	Playoffs GP	G	A	TP	PIM
1995-96	Bonnyville	AJHL	42	10	22	32	270					
1996-97	Anchorage Aces	WCHL	51	8	12	20	388	9	1	1	2	50
	Pensacola	ECHL	1	0	0	0	2					
1997-98	Anchorage Aces	WCHL	31	4	6	10	177					
	Utah Grizzlies	IHL	3	0	0	0	7					
	San Diego Gulls	WCHL	14	0	3	3	50	11	3	3	6	78
1998-99	Long Beach	IHL	8	0	0	0	35					
	San Diego Gulls	WCHL	2	0	0	0	7					
	Manitoba Moose	IHL	57	1	1	2	227	3	0	0	0	0
99-2000	Portland Pirates	AHL	53	3	4	7	170	2	0	0	0	2
2000-01	Portland Pirates	AHL	32	0	0	0	141					
	Utah Grizzlies	IHL	25	0	1	1	167					
2001-02	Houston Aeros	AHL	74	3	3	6	169	3	1	0	1	0

Signed as a free agent by **Washington**, April 27, 1999.

SIDYAKIN, Andrei (sihd-YA-kihn, AN-dray) **MTL.**

Right wing. Shoots left. 5'11", 169 lbs. Born, Ufa, USSR, January 20, 1979.
(Montreal's 10th choice, 202nd overall, in 1997 Entry Draft).

Season	Club	Lea	Regular Season GP	G	A	TP	PIM	Playoffs GP	G	A	TP	PIM
1994-95	Ufa	CIS	7	0	1	1	0					
1995-96	Ufa	CIS	25	1	0	1	4	3	0	0	0	2
1996-97	Ufa	Russia	29	3	5	8	4					
1997-98	Ufa	Russia	42	5	4	9	32					
1998-99	Ufa	Russia	36	6	4	10	14	2	0	0	0	2
99-2000	Ufa	Russia	38	7	2	9	32					
2000-01	Ufa	Russia	44	10	13	23	42					
2001-02	Ufa	Russia	43	9	7	16	20					

SIMONS, Mikael (SIH-mawns, mih-KIGHL) **L.A.**

Center. Shoots left. 6'2", 194 lbs. Born, Falun, Sweden, January 15, 1978.
(Los Angeles' 4th choice, 84th overall, in 1996 Entry Draft).

Season	Club	Lea	Regular Season GP	G	A	TP	PIM	Playoffs GP	G	A	TP	PIM
1994-95	Mora IK Jr.	Swede-Jr.	26	5	3	8	57					
1995-96	Mora IK Jr.	Swede-Jr.	10	4	4	8	12					
	Mora IK	Swede-2	33	6	3	9	22	6	0	2	2	2
1996-97	Mora IK Jr.	Swede-Jr.	1	0	0	0	0					
	Mora IK	Swede-2	33	6	3	9	22	6	0	2	2	2
1997-98	Mora IK	Swede-2	30	16	9	25	88	4	3	0	3	2
1998-99	Mora IK	Swede-2	41	14	9	23	36	14	4	3	7	14
99-2000	Mora IK	Swede-2	36	13	10	23	56	5	0	1	1	4
2000-01	Mora IK	Swede-2	34	9	15	24	55					
2001-02	Timra IK	Sweden	46	1	2	3	14					

SIPOTZ, Brian (SIHP-awtz, BRIGH-uhn) **ATL.**

Defense. Shoots right. 6'6", 225 lbs. Born, South Bend, IN, September 16, 1981.
(Atlanta's 3rd choice, 100th overall, in 2001 Entry Draft).

Season	Club	Lea	Regular Season GP	G	A	TP	PIM	Playoffs GP	G	A	TP	PIM
99-2000	Culver Academy	Hi-School	45	14	22	36	56					
2000-01	Miami Redhawks	CCHA	32	0	1	1	48					
2001-02	Miami Redhawks	CCHA	25	0	1	1	28					

SIVEK, Michal (sih-VIHK, mee-KHAHL) **PIT.**

Center. Shoots left. 6'3", 209 lbs. Born, Nachod, Czech., January 21, 1981.
(Washington's 2nd choice, 29th overall, in 1999 Entry Draft).

Season	Club	Lea	Regular Season GP	G	A	TP	PIM	Playoffs GP	G	A	TP	PIM
1997-98	Sparta Praha Jr.	Czech-Jr.	31	13	8	21	....					
	HC Sparta Praha	Czech	25	1	1	2	10	5	1	0	1	0
1998-99	HC Sparta Praha	Czech	1	1	0	1	....					
	Kladno	Czech	34	3	8	11	24					
99-2000	Prince Albert	WHL	53	23	37	60	65	6	1	4	5	10
2000-01	HC Sparta Praha	Czech	32	6	7	13	28	13	4	2	6	8
2001-02	Wilkes-Barre	AHL	25	4	8	12	30					
	HC Sparta Praha	Czech	17	5	3	8	20	12	0	1	1	10

Traded to **Pittsburgh** by **Washington** with Kris Beech, Ross Lupaschuk and future considerations for Jaromir Jagr and Frantisek Kucera, July 11, 2001.

SJOSTROM, Fredrik (SHAW-strahm, FREHD-rihk) **PHX.**

Right wing. Shoots left. 6', 210 lbs. Born, Fargelanda, Sweden, May 6, 1983.
(Phoenix's 1st choice, 11th overall, in 2001 Entry Draft).

Season	Club	Lea	Regular Season GP	G	A	TP	PIM	Playoffs GP	G	A	TP	PIM
99-2000	MoDo-18	Swede-Jr.	4	0	2	2	6					
	MoDo Jr.	Swede-Jr.	18	4	6	10	8					
2000-01	V. Frolunda Jr.	Swede-Jr.	11	3	7	10	12	4	1	2	3	6
	Vastra Frolunda	Sweden	31	3	2	5	6	5	0	0	0	2
2001-02	Calgary Hitmen	WHL	58	19	31	50	51	4	1	1	2	8

SKINNER, Brett (SKIH-nuhr, BREHT) **VAN.**

Defense. Shoots left. 6'1", 170 lbs. Born, Brandon, Man., June 28, 1983.
(Vancouver's 3rd choice, 68th overall, in 2002 Entry Draft).

Season	Club	Lea	Regular Season GP	G	A	TP	PIM	Playoffs GP	G	A	TP	PIM
1998-99	Brandon Kings	MMBHL	29	3	18	21	20					
99-2000	Brandon Kings	MMHL	40	8	27	35	48					
2000-01	Trail	BCHL	59	11	24	35	43					
2001-02	Des Moines	USHL	44	9	38	47	25	3	0	1	1	0

MMBHL First All-Star Team (1999) • MMHL First All-Star Team (2000) • USHL First All-Star Team (2002) • USHL Defenseman of the Year (2002)

• Signed Letter of Intent to attend **University of Denver**, November 14, 2001.

SKLADANY, Frantisek (sklah-DAH-nee, FRAN-tih-shehk) **COL.**

Left wing. Shoots left. 6', 185 lbs. Born, Martin, Czech., April 22, 1982.
(Colorado's 4th choice, 143rd overall, in 2001 Entry Draft).

Season	Club	Lea	Regular Season GP	G	A	TP	PIM	Playoffs GP	G	A	TP	PIM
1998-99	Martin	Slovakia	1	0	0	0	0					
99-2000	Martin Jr.	Slovak-Jr.	STATISTICS NOT AVAILABLE									
	Martin	Slovak-2	13	1	4	5	2					
2000-01	Boston University	H-East	35	4	5	9	4					
2001-02	Boston University	H-East	33	13	13	26	23					

SKLENAR, Jaroslav (SKLEH-nahr, YAHR-oh-slav) **TOR.**

Right wing. Shoots right. 6', 167 lbs. Born, Ivancice, Czech., November 22, 1982.
(Toronto's 8th choice, 183rd overall, in 2001 Entry Draft).

Season	Club	Lea	Regular Season GP	G	A	TP	PIM	Playoffs GP	G	A	TP	PIM
99-2000	Trinec Jr.	Czech-Jr.	32	5	6	11	2					
2000-01	Ytong Brno Jr.	Czech-Jr.	26	10	11	21	22					
	HC Ytong Brno	Czech-3	21	4	4	8	6					
2001-02	Znojmo	Czech	4	0	0	0	2					
	HC Ytong Brno	Czech-2	5	0	0	0	4					
	Rosice	Czech-2	8	3	1	4	0					

SKOOG, Simon (SKOOG, SEE-muhn) **ST.L.**

Defense. Shoots left. 6'2", 218 lbs. Born, Solvesborg, Sweden, February 17, 1983.
(St. Louis' 8th choice, 283rd overall, in 2001 Entry Draft).

Season	Club	Lea	Regular Season GP	G	A	TP	PIM	Playoffs GP	G	A	TP	PIM
99-2000	Malmo IF Jr.	Swede-Jr.	4	0	0	0	0					
2000-01	Morrums GoIS IK	Swede-2	27	0	1	1	12					
2001-02	Morrums GoIS IK	Swede-2	53	3	8	11	46					

SKRLAC, Rob (SKUHR-lak, RAWB) **N.J.**

Left wing. Shoots left. 6'5", 245 lbs. Born, Port McNeill, B.C., June 10, 1976.
(Buffalo's 11th choice, 224th overall, in 1995 Entry Draft).

Season	Club	Lea	Regular Season GP	G	A	TP	PIM	Playoffs GP	G	A	TP	PIM
1993-94	Richmond	BCAHA	49	44	55	99	56					
1994-95	Kamloops Blazers	WHL	23	0	1	1	177					
1995-96	Kamloops Blazers	WHL	63	1	4	5	216	13	0	0	0	52
1996-97	Kamloops Blazers	WHL	61	8	10	18	278	5	0	0	0	35
1997-98	Albany	AHL	53	0	2	2	256					
1998-99	Albany	AHL	61	1	1	2	213	1	0	0	0	0
99-2000	Albany	AHL	37	0	0	0	115					
2000-01	Albany	AHL	38	0	0	0	105					
2001-02	Albany	AHL	2	0	0	0	22					
	Mississippi	ECHL	29	1	3	4	161					
	Portland Pirates	AHL	33	0	3	3	87					

Signed as a free agent by **New Jersey**, June 17, 1997.

SKVARIDLO, Tomas (SHKVAHR-ihd-loh, TOH-mas) **PIT.**

Left wing. Shoots left. 6'1", 180 lbs. Born, Zvolen, Czech., June 19, 1981.
(Pittsburgh's 6th choice, 144th overall, in 1999 Entry Draft).

Season	Club	Lea	Regular Season GP	G	A	TP	PIM	Playoffs GP	G	A	TP	PIM
1997-98	HKm Zvolen Jr.	Slovak-Jr.	51	11	11	22	22					
1998-99	HKm Zvolen Jr.	Slovak-Jr.	35	21	11	32	18	6	0	4	4	2
	HKm Zvolen	Slovakia	9	1	1	2	2					
	HKm Zvolen	Slovak-Q	1	0	0	0	0					
99-2000	Kingston	OHL	66	19	25	44	14	5	0	0	0	2
2000-01	Kingston	OHL	58	10	19	29	33	4	2	0	2	6
2001-02	MsHK SKP Zilina	Slovakia	35	2	1	3	8	4	0	0	0	0

SLATER, Jim (SLAY-tuhr, JIHM) **ATL.**

Center. Shoots left. 6', 182 lbs. Born, Petoskey, MI, December 9, 1982.
(Atlanta's 2nd choice, 30th overall, in 2002 Entry Draft).

Season	Club	Lea	Regular Season GP	G	A	TP	PIM	Playoffs GP	G	A	TP	PIM
1998-99	Cleveland Barons	NAJHL	50	13	20	33	58	2	0	0	0	2
99-2000	Cleveland Barons	NAJHL	56	35	50	85	129	3	1	3	4	4
2000-01	Cleveland Barons	NAJHL	48	27	37	64	122	6	6	6	12	6
2001-02	Michigan State	CCHA	37	11	21	32	50					

NAJHL First All-Star Team (2000, 2001) • CCHA All-Rookie Team (2002)

SLOAN, Tyler **CBJ**

Defense. Shoots L. 6'4", 190 lbs. Born, Calgary, Alta., March 15, 1981.

Season	Club	Lea	Regular Season GP	G	A	TP	PIM	Playoffs GP	G	A	TP	PIM
1997-98	Cgy. Buffaloes	AMHL	36	2	11	13	24	10	0	4	4	2
1998-99	Calgary Royals	AJHL	STATISTICS NOT AVAILABLE									
99-2000	Calgary Royals	AJHL	45	5	26	31	80					
2000-01	Kamloops Blazers	WHL	70	5	28	33	146	4	0	0	0	4
2001-02	Kamloops Blazers	WHL	70	3	29	32	89	4	0	0	0	15
	Syracuse Crunch	AHL	2	0	0	0	5					

AJHL First All-Star Team (1999)

Signed as a free agent by **Columbus**, September 24, 2000.

SLOVAK, Tomas (SLOHW-vahk, TAW-mawsh) **NSH.**

Defense. Shoots right. 6'1", 203 lbs. Born, Kosice, Czech., April 5, 1983.
(Nashville's 3rd choice, 42nd overall, in 2001 Entry Draft).

Season	Club	Lea	Regular Season GP	G	A	TP	PIM	Playoffs GP	G	A	TP	PIM
99-2000	HC Kosice	Slovakia	2	0	0	0	0					
2000-01	HC Kosice	Slovakia	43	5	5	10	28	3	1	0	1	2
2001-02	Kelowna Rockets	WHL	53	2	24	26	41	15	0	0	0	8

SMIRNOV, Alexei (smihr-NAHV, al-EHX-ay) ANA.

Left wing. Shoots left. 6'3", 211 lbs. Born, Tver, USSR, January 28, 1982.
(Anaheim's 1st choice, 12th overall, in 2000 Entry Draft).

			Regular Season					Playoffs				
Season	Club	Lea	GP	G	A	TP	PIM	GP	G	A	TP	PIM
1997-98	DynamoMoscow2	Russia-2	11	1	1	2	4	....	....	....	....	....
1998-99	DynamoMoscow2	Russia-3	27	9	3	12	24	....	....	....	....	....
99-2000	DynamoMoscow2	Russia-3	12	5	3	8	34	....	....	....	....	....
	THC Tver	Russia-2	35	3	5	8	24	....	....	....	....	....
	Dynamo Moscow	Russia	1	0	0	0	0	....	....	....	....	....
2000-01	Dynamo Moscow	Russia	29	2	0	2	16	....	....	....	....	....
2001-02	CSKA Moscow 2	Russia-3	2	1	0	1	0	....	....	....	....	....
	CSKA Moscow	Russia	51	5	11	16	42	....	....	....	....	....

SMIRNOV, Oleg (smihr-NOHF, OH-lehg) EDM.

Left wing. Shoots right. 5'11", 176 lbs. Born, Elektrostal, USSR, April 8, 1980.
(Edmonton's 6th choice, 144th overall, in 1998 Entry Draft).

			Regular Season					Playoffs				
Season	Club	Lea	GP	G	A	TP	PIM	GP	G	A	TP	PIM
1996-97	Elektrostal 2	Russia-3	38	2	2	4	8	....	....	....	....	....
1997-98	Elektrostal 2	Russia-3	10	0	0	0	2	....	....	....	....	....
	Elektrostal	Russia	6	0	2	2	0	....	....	....	....	....
1998-99	Chelyabinsk	Russia	27	0	3	3	6	....	....	....	....	....
	HC Lipetsk	Russia	3	0	0	0	0	....	....	....	....	....
	Spartak Moscow	Russia	14	4	0	4	4	....	....	....	....	....
99-2000	Spartak Moscow 2	Russia-3		STATISTICS NOT AVAILABLE								
2000-01	Dynamo Moscow	Russia	18	0	0	0	6	....	....	....	....	....
2001-02	CSKA Moscow 2	Russia-3	1	0	2	2	4	....	....	....	....	....
	CSKA Moscow	Russia	23	6	4	10	4	....	....	....	....	....

SMITH, Don (SMIHTH, DAWN) CAR.

Center. Shoots left. 6'3", 195 lbs. Born, Buffalo, NY, March 17, 1979.
(Carolina's 7th choice, 184th overall, in 1998 Entry Draft).

			Regular Season					Playoffs				
Season	Club	Lea	GP	G	A	TP	PIM	GP	G	A	TP	PIM
1996-97	Nichols High	Hi-School	31	25	29	54	....	....	....	....	....	....
1997-98	Clarkson Knights	ECAC	30	4	6	10	8	....	....	....	....	....
1998-99	Clarkson Knights	ECAC	37	9	12	21	18	....	....	....	....	....
99-2000	Clarkson Knights	ECAC	35	7	9	16	20	....	....	....	....	....
2000-01	Clarkson Knights	ECAC	31	12	13	25	18	....	....	....	....	....
2001-02	Florida	ECHL	71	18	15	33	35	6	0	2	2	2

SMITH, Jarrett (SMIHTH, JAHR-reht) ANA.

Center. Shoots left. 6'2", 204 lbs. Born, Edmonton, Alta., June 15, 1979.
(NY Islanders' 4th choice, 59th overall, in 1997 Entry Draft).

			Regular Season					Playoffs				
Season	Club	Lea	GP	G	A	TP	PIM	GP	G	A	TP	PIM
1994-95	Sherwood Park	AMHL	34	21	26	47	45	....	....	....	....	....
1995-96	Prince George	WHL	18	2	0	2	6	....	....	....	....	....
1996-97	Prince George	WHL	67	20	22	42	58	15	2	2	4	5
1997-98	Prince George	WHL	42	12	22	34	21	11	3	1	4	8
1998-99	Prince George	WHL	49	20	37	57	54	3	0	2	2	2
99-2000	Prince Albert	WHL	72	27	32	59	53	6	2	4	6	20
2000-01	Cincinnati	AHL	46	5	9	14	28	....	....	....	....	....
	Baton Rouge	ECHL	18	4	7	11	16	....	....	....	....	....
2001-02	Cincinnati	AHL	64	13	15	28	38	3	0	3	3	0

Traded to **Prince Albert** (WHL) by **Prince George** (WHL) for Prince Albert's 3rd round choice (Justin Craig) in 2000 WHL Bantam Draft, September 20, 1999. Signed as a free agent by **Anaheim**, June 13, 2000.

SMITH, Kenny (SMIHTH, KEHN-nee) EDM.

Defense. Shoots right. 6'2", 209 lbs. Born, Stoneham, MA, December 31, 1981.
(Edmonton's 4th choice, 84th overall, in 2001 Entry Draft).

			Regular Season					Playoffs				
Season	Club	Lea	GP	G	A	TP	PIM	GP	G	A	TP	PIM
1998-99	Team USA	USDP-17	29	2	5	7	32	....	....	....	....	....
99-2000	Team USA	USDP-18	27	4	6	10	77	....	....	....	....	....
2000-01	Harvard Crimson	ECAC	21	0	2	2	37	....	....	....	....	....
2001-02	Harvard Crimson	ECAC	33	3	10	13	48	....	....	....	....	....

SMITH, Kenton (SMIHTH, KEHN-tuhn) T.B.

Forward/Defense. Shoots left. 5'11", 177 lbs. Born, Edmonton, Alta., September 10, 1979.

			Regular Season					Playoffs				
Season	Club	Lea	GP	G	A	TP	PIM	GP	G	A	TP	PIM
1994-95	Edmonton SSAC	AMHL	33	20	25	45	30	....	....	....	....	....
1995-96	Calgary Hitmen	WHL	53	3	17	20	32	....	....	....	....	....
1996-97	Calgary Hitmen	WHL	72	7	26	33	63	....	....	....	....	....
1997-98	Calgary Hitmen	WHL	69	7	26	33	81	18	1	6	7	26
1998-99	Calgary Hitmen	WHL	69	19	35	54	138	21	1	14	15	34
99-2000	Calgary Hitmen	WHL	71	7	46	53	128	13	3	8	11	25
2000-01	Detroit Vipers	IHL	59	1	3	4	24	....	....	....	....	....
	Johnstown Chiefs	ECHL	13	0	3	3	12	4	0	1	1	2
2001-02	Pensacola	ECHL	62	3	15	18	28	3	0	1	1	0
	Springfield	AHL	9	0	1	1	2	....	....	....	....	....

Signed as a free agent by **Tampa Bay**, March 30, 2000.

SMITH, Nathan (SMIHTH, NAY-thun) VAN.

Center. Shoots left. 6'2", 192 lbs. Born, Edmonton, Alta., February 9, 1982.
(Vancouver's 1st choice, 23rd overall, in 2000 Entry Draft).

			Regular Season					Playoffs				
Season	Club	Lea	GP	G	A	TP	PIM	GP	G	A	TP	PIM
1997-98	Sherwood Park	AMHL	35	15	13	28	24	....	....	....	....	....
1998-99	Swift Current	WHL	47	5	8	13	26	....	....	....	....	....
99-2000	Swift Current	WHL	70	21	28	49	72	12	1	6	7	4
2000-01	Swift Current	WHL	67	28	62	90	78	19	4	3	7	20
2001-02	Swift Current	WHL	47	22	38	60	52	12	3	6	9	18

SMITH, Tim (SMIHTH, TIHM) VAN.

Center. Shoots left. 5'9", 160 lbs. Born, Whitecourt, Alta., July 21, 1981.
(Vancouver's 7th choice, 272nd overall, in 2000 Entry Draft).

			Regular Season					Playoffs				
Season	Club	Lea	GP	G	A	TP	PIM	GP	G	A	TP	PIM
1997-98	Lebret Eagles	SJHL	36	6	7	13	12	....	....	....	....	....
1998-99	Spokane Chiefs	WHL	57	5	20	25	21	....	....	....	....	....
99-2000	Spokane Chiefs	WHL	71	26	70	96	65	15	7	7	14	32
2000-01	Spokane Chiefs	WHL	38	19	37	56	65	....	....	....	....	....
	Swift Current	WHL	30	12	22	34	38	19	10	14	24	38
2001-02	Swift Current	WHL	67	28	47	75	123	12	7	6	13	20

Traded to **Swift Current** (WHL) by **Spokane** (WHL) for Scott Henkleman, Scott Scherger and future considerations, January 15, 2001.

SMITHSON, Jerred (SMIHTH-suhn, JEHR-rehd) L.A.

Right wing. Shoots right. 6'2", 190 lbs. Born, Vernon, B.C., February 4, 1979.

			Regular Season					Playoffs				
Season	Club	Lea	GP	G	A	TP	PIM	GP	G	A	TP	PIM
1994-95	Vernon	BCAHA	64	39	46	85	120	....	....	....	....	....
1995-96	Calgary Hitmen	WHL	60	4	2	6	16	....	....	....	....	....
1996-97	Calgary Hitmen	WHL	65	3	6	9	49	....	....	....	....	....
1997-98	Calgary Hitmen	WHL	65	12	9	21	65	18	0	2	2	25
1998-99	Calgary Hitmen	WHL	63	14	22	36	108	21	3	7	10	17
99-2000	Calgary Hitmen	WHL	66	14	25	39	111	10	1	1	2	16
2000-01	Lowell	AHL	24	1	1	2	10	4	0	0	0	2
2001-02	Manchester	AHL	78	5	13	18	45	5	0	1	1	4

Signed as a free agent by **LA Kings**, February 18, 2000.

SOCHOR, Jan (soh-KHAWR, YAN) TOR.

Left wing. Shoots right. 6', 198 lbs. Born, Usti nad Labem, Czech., January 17, 1980.
(Toronto's 6th choice, 161st overall, in 1999 Entry Draft).

			Regular Season					Playoffs				
Season	Club	Lea	GP	G	A	TP	PIM	GP	G	A	TP	PIM
1996-97	Slavia Praha Jr.	Czech-Jr.	26	11	12	23	....	....	....	....	....	....
1997-98	Slavia Praha Jr.	Czech-Jr.	33	26	12	38	....	....	....	....	....	....
	HC Slavia Praha	Czech	14	1	1	2	2	1	0	0	0	0
1998-99	Slavia Praha Jr.	Czech-Jr.	7	2	1	3	2	....	....	....	....	....
	HC Slavia Praha	Czech	47	10	10	20	14	....	....	....	....	....
99-2000	Slavia Praha Jr.	Czech-Jr.	12	5	5	10	6	....	....	....	....	....
	HC Slavia Praha	Czech	39	7	11	18	37	....	....	....	....	....
	Slovan Labem	Czech-2	6	2	3	5	8	....	....	....	....	....
2000-01	Vsetin Jr.	Czech-Jr.	8	5	3	8	0	....	....	....	....	....
	Vsetin	Czech	41	5	4	9	28	3	0	0	0	0
2001-02	HC Vsetin	Czech	27	4	0	4	8	....	....	....	....	....

SODERBERG, Anders (SOH-dehr-buhrg, AN-duhrs) BOS.

Right wing. Shoots right. 5'6", 161 lbs. Born, Ornskoldsvik, Sweden, October 7, 1975.
(Boston's 10th choice, 234th overall, in 1996 Entry Draft).

			Regular Season					Playoffs				
Season	Club	Lea	GP	G	A	TP	PIM	GP	G	A	TP	PIM
1992-93	MoDo Jr.	Swede-Jr.	13	6	12	18	2	....	....	....	....	....
	MoDo	Sweden	1	0	0	0	0	....	....	....	....	....
1993-94	MoDo Jr.	Swede-Jr.	9	8	5	13	10	....	....	....	....	....
	MoDo	Sweden	19	0	0	0	2	9	0	0	0	0
1994-95	MoDo	Sweden	38	9	14	23	2	....	....	....	....	....
1995-96	MoDo	Sweden	40	10	18	28	10	8	3	3	6	0
1996-97	MoDo	Sweden	39	9	13	22	16	....	....	....	....	....
1997-98	MoDo	Sweden	44	15	10	25	4	9	5	1	6	2
1998-99	MoDo	Sweden	49	6	15	21	18	13	3	6	9	4
99-2000	MoDo	Sweden	43	15	10	25	18	9	1	2	3	0
2000-01	MoDo	Sweden	42	11	4	15	12	6	1	6	7	0
2001-02	MoDo	Sweden	30	6	9	15	6	11	0	0	0	0

SOIN, Sergei (SOY-ihn, SAIR-gay) COL.

Center. Shoots left. 6', 175 lbs. Born, Moscow, USSR, March 31, 1982.
(Colorado's 3rd choice, 50th overall, in 2000 Entry Draft).

			Regular Season					Playoffs				
Season	Club	Lea	GP	G	A	TP	PIM	GP	G	A	TP	PIM
1997-98	Krylja Sovetov 2	Russia-3	2	0	0	0	0	....	....	....	....	....
1998-99	Krylja Sovetov	Russia	34	1	4	5	12	....	....	....	....	....
99-2000	Krylja Sovetov	Russia-2	33	8	7	5	24	....	....	....	....	....
2000-01	Krylja Sovetov 2	Russia-3	8	2	3	5	12	....	....	....	....	....
	Krylja Sovetov	Russia-2	30	8	5	13	10	....	....	....	....	....
2001-02	Krylja Sovetov 2	Russia-3	5	2	6	8	20	....	....	....	....	....
	Krylja Sovetov	Russia	41	5	7	12	8	....	....	....	....	....

SOLAREV, Ilja (SOH-luh-rehv, IHL-yuh) T.B.

Left wing. Shoots left. 6'3", 176 lbs. Born, Perm, USSR, August 2, 1982.
(Tampa Bay's 13th choice, 281st overall, in 2001 Entry Draft).

			Regular Season					Playoffs				
Season	Club	Lea	GP	G	A	TP	PIM	GP	G	A	TP	PIM
1997-98	Perm 2	Russia-3	4	1	0	1	0	....	....	....	....	....
1998-99	Perm 2	Russia-4	20	2	6	8	10	....	....	....	....	....
99-2000	Perm 2	Russia-3	35	3	2	5	24	....	....	....	....	....
2000-01	Perm 2	Russia-3		STATISTICS NOT AVAILABLE								
	Perm	Russia	5	0	1	1	0	....	....	....	....	....
2001-02	Leninogorsk	Russia-2	31	3	5	8	20	....	....	....	....	....
	HC Tambov	Russia-3	2	0	0	0	0	....	....	....	....	....

SOMERVUORI, Eero (soh-muhr-VOH-ree, AIR-oh) T.B.

Right wing. Shoots right. 5'10", 167 lbs. Born, Jarvenpaa, Finland, February 7, 1979.
(Tampa Bay's 9th choice, 170th overall, in 1997 Entry Draft).

			Regular Season					Playoffs				
Season	Club	Lea	GP	G	A	TP	PIM	GP	G	A	TP	PIM
1993-94	Jokerit-C	Finn-Jr.	29	23	31	54	16	....	....	....	....	....
1994-95	Jokerit-C	Finn-Jr.	17	23	16	39	8	6	7	3	10	2
	Jokerit-B	Finn-Jr.	15	8	10	18	4	....	....	....	....	....
	Jokerit Jr.	Finn-Jr.	10	1	1	2	2	....	....	....	....	....
1995-96	Jokerit Jr.	Finn-Jr.	28	14	12	26	10	9	4	1	5	4
	Jokerit-B	Finn-Jr.	13	10	13	23	6	....	....	....	....	....
	Haukat Jarvenpaa	Finland-2	1	0	0	0	0	....	....	....	....	....
	Jokerit Helsinki	Finland	6	1	2	3	0	....	....	....	....	....
1996-97	Jokerit Jr.	Finn-Jr.	28	20	19	39	30	5	3	0	3	4
	Jokerit Helsinki	EuroHL	3	0	0	0	0	2	2	0	0	0
	Jokerit Helsinki	Finland	35	1	1	2	2	5	0	0	0	0
1997-98	Jokerit Jr.	Finn-Jr.	14	4	8	12	2	....	....	....	....	....
	Jokerit Helsinki	Finland	42	3	7	10	12	8	2	1	3	6
	Jokerit Helsinki	EuroHL	5	0	0	0	0	....	....	....	....	....
1998-99	Jokerit Jr.	Finn-Jr.	4	1	1	2	2	4	3	0	3	2
	Jokerit Helsinki	Finland	50	7	8	15	24	3	1	0	1	6
	Jokerit Helsinki	EuroHL	6	0	0	0	0	1	0	0	0	0
99-2000	Jokerit Helsinki	Finland	54	6	6	12	10	11	1	0	1	0
2000-01	HPK Hameenlinna	Finland	56	14	6	20	35	....	....	....	....	....
2001-02	HPK Hameenlinna	Finland	56	25	23	48	34	8	2	2	4	6

SOMIK, Radovan (SAW-mihk, RAH-doh-vahn) **PHI.**

Right wing. Shoots right. 6'2", 194 lbs. Born, Martin, Czech., May 5, 1977.
(Philadelphia's 3rd choice, 100th overall, in 1995 Entry Draft).

			Regular Season					Playoffs				
Season	Club	Lea	GP	G	A	TP	PIM	GP	G	A	TP	PIM
1993-94	Martin	Slovakia	1	0	0	0	0	....	....	....	....	....
1994-95	No ZTS??	Slovakia	25	3	0	3	39	3	1	0	1	2
1995-96	Martin	Slovakia	25	3	6	9	8	9	1	0	1	....
1996-97	Martin	Slovakia	35	3	5	8	....	3	0	0	0	....
1997-98	Martin	Slovakia	26	6	9	15	10	3	0	0	0	0
1998-99	Dukla Trencin	Slovakia	26	1	4	5	6	....	....	....	....	....
99-2000	Martin	Slovak-2	40	38	28	66	32	....	....	....	....	....
2000-01	Zlin	Czech	46	15	10	25	22	6	1	0	1	0
2001-02	Zlin	Czech	37	14	14	28	22	11	4	3	7	37

SOUCY, J-F (SOO-cee, JAY-EHF) **T.B.**

Center. Shoots left. 6'3", 180 lbs. Born, Riviere Du Loup, Que., March 25, 1983.
(Tampa Bay's 10th choice, 252nd overall, in 2001 Entry Draft).

			Regular Season					Playoffs				
Season	Club	Lea	GP	G	A	TP	PIM	GP	G	A	TP	PIM
1998-99	Levis	QAAA	42	6	17	23	82	....	....	....	....	....
99-2000	Val-d'Or Foreurs	QMJHL	55	1	4	5	9	....	....	....	....	....
2000-01	Val-d'Or Foreurs	QMJHL	38	3	4	7	57	....	....	....	....	....
	Montreal Rocket	QMJHL	27	3	8	11	37	....	....	....	....	....
2001-02	Montreal Rocket	QMJHL	49	8	13	21	94	7	0	2	2	8

SOUZA, Mike (SOO-zah, MIGHK) **CHI.**

Left wing. Shoots left. 6'1", 210 lbs. Born, Melrose, MA, January 28, 1978.
(Chicago's 4th choice, 67th overall, in 1997 Entry Draft).

			Regular Season					Playoffs				
Season	Club	Lea	GP	G	A	TP	PIM	GP	G	A	TP	PIM
1992-93	Wakefield High	Hi-School	20	22	13	35	....	....	....	....	....	....
1993-94	Wakefield High	Hi-School	22	28	34	62	....	....	....	....	....	....
1994-95	Wakefield High	Hi-School	21	22	31	53	....	....	....	....	....	....
1995-96	Wakefield High	Hi-School	21	25	31	56	22	....	....	....	....	....
1996-97	New Hampshire	H-East	39	15	11	26	20	....	....	....	....	....
1997-98	New Hampshire	H-East	38	13	12	25	36	....	....	....	....	....
1998-99	New Hampshire	H-East	41	23	42	65	38	....	....	....	....	....
99-2000	New Hampshire	H-East	38	15	25	40	58	....	....	....	....	....
2000-01	Norfolk Admirals	AHL	75	14	17	31	44	3	0	0	0	2
2001-02	Norfolk Admirals	AHL	66	20	11	31	58	....	....	....	....	....

NCAA Championship All-Tournament Team (1999) • Hockey East Second All-Star Team (2000)

SOZINOV, Vadim (SOH-zih-nahf, va-DEEM) **TOR.**

Center. Shoots left. 6'1", 185 lbs. Born, Ust-Kamenogorsk, USSR, June 17, 1981.
(Toronto's 6th choice, 179th overall, in 2000 Entry Draft).

			Regular Season					Playoffs				
Season	Club	Lea	GP	G	A	TP	PIM	GP	G	A	TP	PIM
1997-98	Novokuznetsk 2	Russia-3	24	3	0	3	0	....	....	....	....	....
1998-99	Novokuznetsk 2	Russia-4	43	7	10	17	16	....	....	....	....	....
99-2000	Kristall Saratov	Russia-2	2	1	0	1	0	....	....	....	....	....
2000-01	Ottawa 67's	OHL	57	21	18	39	57	20	7	8	15	4
2001-02	Prokopievsk	Russia-2	19	1	0	1	14	....	....	....	....	....

SPANG, Dan (SPANG, DAN) **S.J.**

Defense. Shoots left. 6', 200 lbs. Born, Winchester, MA, August 18, 1983.
(San Jose's 2nd choice, 52nd overall, in 2002 Entry Draft).

			Regular Season					Playoffs				
Season	Club	Lea	GP	G	A	TP	PIM	GP	G	A	TP	PIM
2000-01	Winchester High	Hi-School	24	8	37	45	14	....	....	....	....	....
2001-02	Winchester High	Hi-School	6	9	8	17	14	....	....	....	....	....

• Signed Letter of Intent to attend **Boston University** (H-East), March 24, 2001. • Missed majority of 2001-02 season recovering from head injuries suffered in automobile accident, October 2001.

SPEZZA, Jason (SPEHT-zah, JAY-suhn) **OTT.**

Center. Shoots right. 6'2", 214 lbs. Born, Mississauga, Ont., June 13, 1983.
(Ottawa's 1st choice, 2nd overall, in 2001 Entry Draft).

			Regular Season					Playoffs				
Season	Club	Lea	GP	G	A	TP	PIM	GP	G	A	TP	PIM
1997-98	Tor. Red Wings	MTHL	54	53	61	114	42	....	....	....	....	....
1998-99	Brampton	OHL	67	22	49	71	18	....	....	....	....	....
99-2000	Mississauga	OHL	52	24	37	61	33	....	....	....	....	....
2000-01	Mississauga	OHL	15	7	23	30	11	....	....	....	....	....
	Windsor	OHL	41	36	50	86	32	9	4	5	9	10
2001-02	Windsor	OHL	27	19	26	45	16	....	....	....	....	....
	Belleville Bulls	OHL	26	23	37	60	26	11	5	6	11	18
	Grand Rapids	AHL	....	....	....	....	....	3	1	0	1	2

OHL All-Rookie Team (1999)

Traded to **Windsor** (OHL) by **Mississauga** (OHL) with Mark Rideout and Brett Angel for Ryan Courtney, Mike James, Steve Rawski and Tyler Eady, November 15, 2000. Traded to **Belleville** (OHL) by **Windsor** (OHL) for Kyle Wellwood and future considerations, January 10, 2002.

SPILLER, Matthew (SPIHL-uhr, MA-thew) **PHX.**

Defense. Shoots left. 6'5", 225 lbs. Born, Daysland, Alta., February 7, 1983.
(Phoenix's 2nd choice, 31st overall, in 2001 Entry Draft).

			Regular Season					Playoffs				
Season	Club	Lea	GP	G	A	TP	PIM	GP	G	A	TP	PIM
1998-99	East Central	AMHL	36	8	19	27	140	....	....	....	....	....
99-2000	Seattle	WHL	60	1	10	11	108	7	0	0	0	25
2000-01	Seattle	WHL	71	4	7	11	174	....	....	....	....	....
2001-02	Seattle	WHL	72	8	23	31	168	1	0	0	0	4

SPRUKTS, Janis (SPRUKTS, YAN-ish) **FLA.**

Center. Shoots left. 6'3", 224 lbs. Born, Riga, Latvia, January 31, 1982.
(Florida's 7th choice, 234th overall, in 2000 Entry Draft).

			Regular Season					Playoffs				
Season	Club	Lea	GP	G	A	TP	PIM	GP	G	A	TP	PIM
99-2000	Lukko Rauma Jr.	Finn-Jr.	26	2	5	7	6	3	0	0	0	0
2000-01	Lukko Rauma Jr.	Finn-Jr.	36	15	22	37	24	3	0	0	0	0
	Lukko Rauma	Finland	9	0	0	0	2	....	....	....	....	....
2001-02	Acadie-Bathurst	QMJHL	63	35	44	79	46	16	14	8	22	12

SRDINKO, Jan (suhr-DIHN-koh, YAN) **N.J.**

Defense. Shoots left. 5'11", 195 lbs. Born, Vsetin, Czech., February 22, 1974.
(New Jersey's 8th choice, 241st overall, in 1997 Entry Draft).

			Regular Season					Playoffs				
Season	Club	Lea	GP	G	A	TP	PIM	GP	G	A	TP	PIM
1995-96	HC Petra Vsetin	Czech	31	0	3	3	....	9	0	0	0	....
1996-97	HC Petra Vsetin	Czech	49	2	8	10	71	10	0	3	3	29
1997-98	HC Petra Vsetin	Czech	47	1	4	5	95	10	0	3	3	4
	HC Petra Vsetin	EuroHL	9	0	1	1	4	....	....	....	....	....
1998-99	Vsetin	Czech	50	2	7	9	58	12	0	1	1	....
99-2000	Vsetin	Czech	48	3	10	13	50	9	2	0	2	29
2000-01	Vsetin	Czech	47	8	9	17	81	14	1	1	2	24
2001-02	HC Sparta Praha	Czech	50	6	5	11	100	2	0	0	0	0

SRYUBKO, Andrei (SHROOB-koh, AN-dray)

Defense. Shoots left. 6'3", 205 lbs. Born, Kiev, USSR, October 21, 1975.

			Regular Season					Playoffs				
Season	Club	Lea	GP	G	A	TP	PIM	GP	G	A	TP	PIM
1996-97	Toledo Storm	ECHL	62	0	8	8	238	5	0	0	0	4
1997-98	Toledo Storm	ECHL	50	1	12	13	165	7	0	0	0	8
	Las Vegas	IHL	13	0	0	0	57	....	....	....	....	....
1998-99	Port Huron	UHL	22	2	2	4	78	2	0	0	0	4
	Fort Wayne	IHL	1	0	0	0	0	....	....	....	....	....
	Las Vegas	IHL	51	0	8	8	164	....	....	....	....	....
99-2000	Port Huron	UHL	2	0	0	0	7	....	....	....	....	....
	Utah Grizzlies	IHL	5	1	1	2	32	....	....	....	....	....
	Grand Rapids	IHL	28	0	1	1	109	2	0	0	0	9
2000-01	Syracuse Crunch	AHL	70	1	6	7	169	5	0	0	0	0
2001-02	Syracuse Crunch	AHL	58	0	10	10	181	9	0	0	0	18
	Ukraine	Olympics	4	0	2	2	0	....	....	....	....	....

Signed as a free agent by **Columbus**, August 3, 2000. Signed as a free agent by **Molot Perm** (Russia), July 20, 2002.

STAAL, Kim (STOHL, KIHM) **MTL.**

Center. Shoots right. 6', 185 lbs. Born, Herlev, Denmark, March 10, 1978.
(Montreal's 4th choice, 92nd overall, in 1996 Entry Draft).

			Regular Season					Playoffs				
Season	Club	Lea	GP	G	A	TP	PIM	GP	G	A	TP	PIM
1994-95	Malmo IF Jr.	Swede-Jr.	17	4	2	6	4	....	....	....	....	....
1995-96	Malmo IF Jr.	Swede-Jr.	30	24	20	44	14	....	....	....	....	....
1996-97	Malmo IF Jr.	Swede-Jr.	3	6	4	10	2	....	....	....	....	....
	Malmo IF	Sweden	4	0	1	1	2	....	....	....	....	....
1997-98	Malmo IF Jr.	Swede-Jr.	20	13	11	24	36	....	....	....	....	....
	Malmo IF	Sweden	13	0	1	1	1	....	....	....	....	....
1998-99	Malmo IF	Sweden	48	1	5	6	14	4	0	0	0	0
99-2000	Malmo IF	Sweden	50	14	10	24	24	6	1	1	2	4
2000-01	Malmo IF	Sweden	48	16	15	31	32	9	4	2	6	4
2001-02	MoDo	Sweden	49	14	23	37	16	12	3	7	10	2

STAJAN, Matthew (STAY-juhn, MAT-thew) **TOR.**

Center. Shoots left. 6'1", 178 lbs. Born, Mississauga, Ont., December 19, 1983.
(Toronto's 2nd choice, 57th overall, in 2002 Entry Draft).

			Regular Season					Playoffs				
Season	Club	Lea	GP	G	A	TP	PIM	GP	G	A	TP	PIM
99-2000	Mississauga Reps	GTHL		STATISTICS NOT AVAILABLE								
2000-01	Belleville Bulls	OHL	57	9	18	27	27	7	1	6	7	5
2001-02	Belleville Bulls	OHL	68	33	52	85	50	11	3	8	11	14

STALS, Juris (STAHLS, YOO-rihs) **NYR**

Left wing. Shoots left. 6'3", 187 lbs. Born, Riga, Latvia, August 4, 1982.
(NY Rangers' 11th choice, 269th overall, in 2001 Entry Draft).

			Regular Season					Playoffs				
Season	Club	Lea	GP	G	A	TP	PIM	GP	G	A	TP	PIM
99-2000	Lukko Rauma Jr.	Finn-Jr.	2	1	0	1	2	....	....	....	....	....
2000-01	Lukko Rauma Jr.	Finn-Jr.	45	23	22	45	26	3	1	0	1	0
2001-02	Sarnia Sting	OHL	60	23	22	45	12	5	0	1	1	2

STATE, Jeff (STAYT, JEHF) **NYR**

Defense. Shoots right. 6'6", 235 lbs. Born, Tonowanda, NY, September 17, 1979.

			Regular Season					Playoffs				
Season	Club	Lea	GP	G	A	TP	PIM	GP	G	A	TP	PIM
99-2000	Burlington	OPJHL	44	12	18	30	50	....	....	....	....	....
2000-01	Merrimack	H-East	35	1	2	3	58	....	....	....	....	....
2001-02	Merrimack	H-East	35	0	9	9	84	....	....	....	....	....

Signed as a free agent by **NY Rangers**, July 11, 2002.

STAYZER, Blair (STAY-zuhr, BLAIR) **CGY.**

Left wing. Shoots left. 6'3", 205 lbs. Born, Dunnville, Ont., October 4, 1980.
(Calgary's 9th choice, 190th overall, in 1999 Entry Draft).

			Regular Season					Playoffs				
Season	Club	Lea	GP	G	A	TP	PIM	GP	G	A	TP	PIM
1995-96	Welland Cougars	OJHL-B	41	4	7	11	79	....	....	....	....	....
1996-97	Windsor	OHL	58	3	5	8	54	3	0	0	0	2
1997-98	Windsor	OHL	57	4	8	12	132	....	....	....	....	....
1998-99	Windsor	OHL	62	12	19	31	140	5	2	0	2	14
99-2000	Windsor	OHL	44	14	5	19	100	11	1	3	4	14
2000-01	Johnstown Chiefs	ECHL	37	1	3	4	104	....	....	....	....	....
2001-02	Johnstown Chiefs	ECHL	41	1	8	9	100	5	1	0	1	9
	Saint John	AHL	4	0	0	0	4	....	....	....	....	....

STECKEL, Dave (STEH-kuhl, DAYV) **L.A.**

Center. Shoots left. 6'5", 200 lbs. Born, Milwaukee, WI, March 15, 1982.
(Los Angeles' 2nd choice, 30th overall, in 2001 Entry Draft).

			Regular Season					Playoffs				
Season	Club	Lea	GP	G	A	TP	PIM	GP	G	A	TP	PIM
1998-99	Team USA	USDP-17	51	3	14	17	18	....	....	....	....	....
	Team USA	USDP-18	2	0	0	0	2	....	....	....	....	....
99-2000	Team USA	USDP-18	52	13	13	26	94	....	....	....	....	....
2000-01	Ohio State	CCHA	33	17	18	35	80	....	....	....	....	....
2001-02	Ohio State	CCHA	36	6	16	22	75	....	....	....	....	....

CCHA All-Rookie Team (2001)

STEEN, Alexander (STEEN, al-ehx-AN-duhr) TOR.

Center. Shoots left. 5'11", 183 lbs. Born, Winnipeg, Man., March 1, 1984.
(Toronto's 1st choice, 24th overall, in 2002 Entry Draft).

			Regular Season					Playoffs				
Season	Club	Lea	GP	G	A	TP	PIM	GP	G	A	TP	PIM
99-2000	V. Frolunda-18	Swede-Jr.	14	3	5	8	16	....	....	....	....	....
2000-01	V. Frolunda Jr.	Swede-Jr.	23	11	12	23	15	3	1	0	1	2
2001-02	V. Frolunda Jr.	Swede-Jr.	23	21	17	38	47	2	1	1	2	2
	Vastra Frolunda	Sweden	26	0	3	3	14	10	1	2	3	0

STEEN, Calle (STEEN, CAL-lee) DET.

Right wing. Shoots left. 5'11", 198 lbs. Born, Stockholm, Sweden, May 16, 1980.
(Detroit's 6th choice, 142nd overall, in 1998 Entry Draft).

			Regular Season					Playoffs				
Season	Club	Lea	GP	G	A	TP	PIM	GP	G	A	TP	PIM
1995-96	Hammarby Jr.	Swede-Jr.	5	0	0	0	0	....	....	....	....	....
1996-97	Hammarby Jr.	Swede-Jr.	24	4	9	13	....	....	....	....	....	....
1997-98	Hammarby	Swede-2	21	1	3	4	22	....	....	....	....	....
1998-99	Hammarby	Swede-2	33	4	16	20	28	5	0	2	2	6
99-2000	Mora IK	Swede-2	32	4	4	8	48	9	0	2	2	10
2000-01	Bofors IK	Swede-2	31	4	7	11	69	....	....	....	....	....
	JYP Jyvaskyla	Finland	5	0	0	0	0	....	....	....	....	....
2001-02	Bofors IK	Swede-2	53	12	22	34	78	6	2	2	4	12

STEPHENS, Charlie (STEE-vuhns, CHAHR-lee) COL.

Center/Right wing. Shoots right. 6'3", 225 lbs. Born, London, Ont., April 5, 1981.
(Colorado's 9th choice, 196th overall, in 2001 Entry Draft).

			Regular Season					Playoffs				
Season	Club	Lea	GP	G	A	TP	PIM	GP	G	A	TP	PIM
1995-96	Elgin-Middlesex	OMHA	60	25	29	54	60	....	....	....	....	....
1996-97	Leamington	OJHL-B	50	26	36	62	103	....	....	....	....	....
1997-98	St. Michael's	OHL	58	9	21	30	38	....	....	....	....	....
1998-99	St. Michael's	OHL	7	2	4	6	8	....	....	....	....	....
	Guelph Storm	OHL	61	24	28	52	72	11	3	5	8	19
99-2000	Guelph Storm	OHL	56	16	34	50	87	6	1	3	4	15
2000-01	Guelph Storm	OHL	67	38	38	76	53	4	0	2	2	2
2001-02	Guelph Storm	OHL	4	1	2	3	2	....	....	....	....	....
	London Knights	OHL	56	23	33	56	55	12	6	10	16	18
	Hershey Bears	AHL	....	....	....	....	....	1	0	0	0	0

• Re-entered NHL Entry Draft. Originally Washington's 3rd choice, 31st overall, in 1999 Entry Draft.

Traded to **Guelph** (OHL) by **St. Michael's** (OHL) with St. Michael's 5th round choice (Derek Hennessey) in 1999 OHL Priority Draft for Ryan Barnes and Sudbury's 3rd round choice (previously acquired, Toronto selected Steve Farquharson) in 1999 OHL Priority Draft, October 14, 1998. Traded to **London** (OHL) by **Guelph** (OHL) with Guelph's 2nd round choice (Adam Nemeth) in 2002 OHL Priority Draft and Guelph's 1st round choice (Alexander Skorohod) in 2002 CHL Import Draft for Aaron Lobb, October 19, 2001.

STEPHENSON, Shay (STEE-vuhn-suhn, SHAY) EDM.

Left wing. Shoots left. 6'2", 180 lbs. Born, Outlook, Sask., September 13, 1983.
(Edmonton's 11th choice, 278th overall, in 2001 Entry Draft).

			Regular Season					Playoffs				
Season	Club	Lea	GP	G	A	TP	PIM	GP	G	A	TP	PIM
99-2000	Notre Dame	SMHL	42	23	7	30	46	....	....	....	....	....
2000-01	Red Deer Rebels	WHL	44	1	4	5	30	22	0	0	0	15
2001-02	Red Deer Rebels	WHL	59	9	10	19	55	23	0	3	3	14

STEPP, Joel (STEHP, JOHL) ANA.

Center. Shoots left. 6', 185 lbs. Born, Estevan, Sask., February 11, 1983.
(Anaheim's 3rd choice, 69th overall, in 2001 Entry Draft).

			Regular Season					Playoffs				
Season	Club	Lea	GP	G	A	TP	PIM	GP	G	A	TP	PIM
1998-99	Estevan	SMBHL	60	65	70	135	120	....	....	....	....	....
	Red Deer Rebels	WHL	2	0	0	0	0	....	....	....	....	....
99-2000	Red Deer Rebels	WHL	65	11	13	24	59	4	1	0	1	8
2000-01	Red Deer Rebels	WHL	70	24	13	37	89	22	6	3	9	24
2001-02	Red Deer Rebels	WHL	70	27	26	53	59	23	11	11	22	24

STILLMAN, Cory (STIHL-mahn, KOHR-ee) NYI

Center. Shoots left. 6'2", 204 lbs. Born, Lindsay, Ont., March 2, 1983.
(NY Islanders' 1st choice, 101st overall, in 2001 Entry Draft).

			Regular Season					Playoffs				
Season	Club	Lea	GP	G	A	TP	PIM	GP	G	A	TP	PIM
1998-99	Lindsay Muskies	OPJHL	49	16	15	31	12	....	....	....	....	....
99-2000	Kingston	OHL	61	13	10	23	27	5	0	0	0	0
2000-01	Kingston	OHL	68	29	27	56	39	4	2	0	2	0
2001-02	Kingston	OHL	48	14	27	41	58	1	0	0	0	0

STOLL, Jarret (STOHL, JEHR-eht) EDM.

Center. Shoots right. 6'1", 199 lbs. Born, Melville, Sask., June 25, 1982.
(Edmonton's 3rd choice, 36th overall, in 2002 Entry Draft).

			Regular Season					Playoffs				
Season	Club	Lea	GP	G	A	TP	PIM	GP	G	A	TP	PIM
1997-98	Sask. Blazers	SMHL	44	45	44	*89	78	....	....	....	....	....
	Edmonton Ice	WHL	8	2	3	5	4	....	....	....	....	....
1998-99	Kootenay Ice	WHL	57	13	21	34	38	4	0	0	0	2
99-2000	Kootenay Ice	WHL	71	37	38	75	64	20	7	9	16	24
2000-01	Kootenay Ice	WHL	62	40	66	106	105	11	5	9	14	22
2001-02	Kootenay Ice	WHL	47	32	34	66	64	22	6	14	20	35

• Re-entered NHL Entry Draft. Originally Calgary's 3rd choice, 46th overall, in 2000 Entry Draft.

WHL East First All-Star Team (2001) • Canadian Major Junior First All-Star Team (2001) • WHL West First All-Star Team (2002)

STREIT, Martin (STRIGHT, MAHR-tihn) CBJ

Left wing. Shoots right. 6'2", 202 lbs. Born, Vyskov, Czech., February 2, 1977.
(Philadelphia's 7th choice, 178th overall, in 1995 Entry Draft).

			Regular Season					Playoffs				
Season	Club	Lea	GP	G	A	TP	PIM	GP	G	A	TP	PIM
1995-96	HC Olomouc Jr.	Czech-Jr.	19	10	6	16	....	....	....	....	....	....
	HC Olomouc	Czech	10	0	0	0	....	....	....	....	....	....
1996-97	HC Olomouc	Czech	18	1	2	3	14	....	....	....	....	....
1997-98	HC Karlovy Vary	Czech	48	5	13	18	24	....	....	....	....	....
1998-99	HC Karlovy Vary	Czech	48	9	4	13	34	....	....	....	....	....
99-2000	HC Karlovy Vary	Czech	18	0	2	2	20	....	....	....	....	....
	HC Vitkovice	Czech	31	3	6	9	28	....	....	....	....	....
2000-01	HC Karlovy Vary	Czech	14	0	1	1	12	....	....	....	....	....
	HC Femax Havirov	Czech	11	0	0	0	6	....	....	....	....	....
2001-02	HC Vsetin	Czech	35	3	4	7	65	....	....	....	....	....

Selected by **Columbus** from **Philadelphia** in Expansion Draft, June 23, 2000.

STROM, Peter (STRUHM, PEE-tuhr) MTL.

Left wing. Shoots right. 6', 178 lbs. Born, Snotorp, Sweden, January 14, 1975.
(Montreal's 10th choice, 200th overall, in 1994 Entry Draft).

			Regular Season					Playoffs				
Season	Club	Lea	GP	G	A	TP	PIM	GP	G	A	TP	PIM
1992-93	V. Frolunda-18	Swede-Jr.	16	12	6	18	8	....	....	....	....	....
	Vastra Frolunda	Sweden	3	0	0	0	0	....	....	....	....	....
1993-94	Vastra Frolunda	Sweden	29	0	0	0	8	....	....	....	....	....
1994-95	V. Frolunda Jr.	Swede-Jr.	9	12	6	18	6	....	....	....	....	....
	Vastra Frolunda	Sweden	14	0	3	3	10	....	....	....	....	....
	Vastra Frolunda	Swede-Q	16	13	11	24	12	5	2	3	5	0
1995-96	V. Frolunda Jr.	Swede-Jr.	6	2	3	5	0	....	....	....	....	....
	Vastra Frolunda	Sweden	35	7	8	15	10	13	0	3	3	0
1996-97	Vastra Frolunda	Sweden	49	7	16	23	24	3	0	0	0	2
	Vastra Frolunda	EuroHL	4	3	2	5	0	4	0	3	3	0
1997-98	Vastra Frolunda	Sweden	46	6	15	21	22	7	0	4	4	0
1998-99	Vastra Frolunda	Sweden	49	17	15	32	18	4	0	2	2	0
99-2000	Vastra Frolunda	Sweden	50	12	26	38	20	5	0	1	1	2
2000-01	Vastra Frolunda	Sweden	50	15	18	33	16	5	0	3	3	6
2001-02	Vastra Frolunda	Sweden	50	12	11	23	12	10	2	2	4	2

STUART, Colin (STOO-uhrt, CAW-lihn) ATL.

Center. Shoots left. 6'1", 195 lbs. Born, Rochester, MN, July 8, 1982.
(Atlanta's 5th choice, 135th overall, in 2001 Entry Draft).

			Regular Season					Playoffs				
Season	Club	Lea	GP	G	A	TP	PIM	GP	G	A	TP	PIM
1998-99	Lourdes High	Hi-School	23	22	32	54	....	....	....	....	....	....
99-2000	Lincoln Stars	USHL	53	18	19	37	38	....	....	....	....	....
2000-01	Colorado College	WCHA	41	2	7	9	26	....	....	....	....	....
2001-02	Colorado College	WCHA	43	13	9	22	34	....	....	....	....	....

STUSSI, Rene (SHTOO-see, REH-nay) ANA.

Center. Shoots right. 5'11", 183 lbs. Born, Muri, Switz., December 13, 1978.
(Anaheim's 7th choice, 209th overall, in 1997 Entry Draft).

			Regular Season					Playoffs				
Season	Club	Lea	GP	G	A	TP	PIM	GP	G	A	TP	PIM
1995-96	HC Thurgau	Swiss-2	34	2	4	6	10	7	3	0	3	2
1996-97	HC Thurgau	Swiss-2	42	20	31	51	24	8	5	4	9	4
1997-98	EHC Kloten Jr.	Swiss-Jr.	1	5	0	5	0	....	....	....	....	....
	EHC Bulach	Swiss-2	4	3	5	8	20	....	....	....	....	....
	EHC Kloten	Swiss	38	9	8	17	10	7	1	0	1	4
1998-99	EHC Kloten	Swiss	36	5	5	10	14	....	....	....	....	....
	ZSC Lions Zurich	Swiss	7	1	1	2	6	7	1	1	2	4
99-2000	EV Zug	Swiss	45	7	5	12	13	9	0	1	1	0
2000-01	EHC Chur	Swiss	36	7	7	14	26	11	1	1	2	26
2001-02	EHC Chur	Swiss	21	3	2	5	2	....	....	....	....	....
	EHC Basel	Swiss-2	16	11	11	22	2	....	....	....	....	....
	HC Ajoie	Swiss-2	....	....	....	....	....	8	1	7	8	0

SUBBOTIN, Dmitri (soo-BOH-tihn, DIH-mih-TREE) CBJ

Left wing. Shoots left. 6'1", 183 lbs. Born, Tomsk, USSR, October 20, 1977.
(NY Rangers' 3rd choice, 76th overall, in 1996 Entry Draft).

			Regular Season					Playoffs				
Season	Club	Lea	GP	G	A	TP	PIM	GP	G	A	TP	PIM
1993-94	Yekaterinburg	CIS	12	0	3	3	4	....	....	....	....	....
1994-95	Yekaterinburg	CIS	52	9	6	15	75	2	0	0	0	2
1995-96	CSKA Moscow	CIS	41	6	5	11	62	3	0	0	0	0
1996-97	H.C. CSKA	Russia-2	8	1	0	1	8	....	....	....	....	....
	H.C. CSKA	Russia	17	5	3	8	22	2	0	0	0	2
1997-98	H.C. CSKA	Russia	16	1	1	2	47	....	....	....	....	....
1998-99	Dynamo Moscow	Russia	1	0	1	1	0	....	....	....	....	....
	Lada Togliatti	Russia	31	8	3	11	47	7	0	0	0	4
99-2000	Lada Togliatti	Russia	2	0	1	1	0	....	....	....	....	....
	Lada Togliatti	Russia	27	10	4	14	26	7	1	1	2	4
2000-01	Dynamo Moscow	Russia	39	11	15	26	48	....	....	....	....	....
2001-02	Magnitogorsk	Russia	38	8	3	11	18	9	0	2	2	10

Selected by **Columbus** from **NY Rangers** in Expansion Draft, June 23, 2000.

SUDERMAN, Matt (SOO-duhr-man, MAT) ATL.

Defense. Shoots left. 6'3", 228 lbs. Born, Winkler, Man., January 27, 1983.
(Atlanta's 7th choice, 199th overall, in 2001 Entry Draft).

			Regular Season					Playoffs				
Season	Club	Lea	GP	G	A	TP	PIM	GP	G	A	TP	PIM
2000-01	Saskatoon Blades	WHL	70	1	8	9	116	....	....	....	....	....
2001-02	Saskatoon Blades	WHL	65	1	7	8	128	6	0	1	1	9

SUGLOBOV, Alexander (suh-GLOH-bahf, al-ehx-AN-duhr) N.J.

Right wing. Shoots left. 6', 176 lbs. Born, Elektrostal, USSR, January 15, 1982.
(New Jersey's 3rd choice, 56th overall, in 2000 Entry Draft).

			Regular Season					Playoffs				
Season	Club	Lea	GP	G	A	TP	PIM	GP	G	A	TP	PIM
1998-99	Spartak Mos. 2	Russia-4	1	0	0	0	0	....	....	....	....	....
	Spartak Moscow	Russia	1	0	0	0	0	....	....	....	....	....
99-2000	Yaroslavl 2	Russia-3	38	23	10	33	....	....	....	....	....	....
2000-01	St. Petersburg	Russia	8	1	0	1	6	....	....	....	....	....
	Ufa	Russia	6	0	0	0	4	....	....	....	....	....
	Yaroslavl	Russia	4	0	0	0	2	11	1	2	3	6
2001-02	Yaroslavl 2	Russia-3	6	5	2	7	20	....	....	....	....	....
	Yaroslavl	Russia	25	4	2	6	26	5	1	1	2	18

SULLIVAN, Brian (suh-LIH-vuhn, BRIGH-uhn) DAL.

Defense. Shoots left. 6'3", 185 lbs. Born, Marshfield, MA, June 27, 1980.
(Dallas' 9th choice, 243rd overall, in 1999 Entry Draft).

			Regular Season					Playoffs				
Season	Club	Lea	GP	G	A	TP	PIM	GP	G	A	TP	PIM
1997-98	Thayer Academy	Hi-School	26	0	5	5	17	....	....	....	....	....
1998-99	Thayer Academy	Hi-School	21	0	7	7	10	....	....	....	....	....
99-2000	Northeastern	H-East	23	0	1	1	8	....	....	....	....	....
2000-01	Northeastern	H-East	34	1	2	3	37	....	....	....	....	....
2001-02	Northeastern	H-East	6	0	0	0	2	....	....	....	....	....

SULLIVAN, Dale (SUHL-lih-vahn, DAYL) DAL.

Right wing. Shoots right. 6', 180 lbs. Born, St. John's, Nfld., July 30, 1983.
(Dallas' 9th choice, 265th overall, in 2001 Entry Draft).

			Regular Season					Playoffs				
Season	Club	Lea	GP	G	A	TP	PIM	GP	G	A	TP	PIM
2000-01	Hull Olympiques	QMJHL	72	8	14	22	31	5	2	1	3	0
2001-02	Hull Olympiques	QMJHL	61	18	20	38	39	12	6	8	14	6

SULLIVAN, Jeff (SUHL-lih-vahn, JEHF)

Defense. Shoots left. 6'1", 185 lbs. Born, St. John's, Nfld., September 18, 1978.
(Ottawa's 5th choice, 146th overall, in 1997 Entry Draft).

			Regular Season					Playoffs				
Season	Club	Lea	GP	G	A	TP	PIM	GP	G	A	TP	PIM
1994-95	St. John's	NFAHA	40	10	15	25	120	....	....	....	....	....
1995-96	East Hants	MJrHL	52	15	20	35	270	....	....	....	....	....
1996-97	Granby	QMJHL	25	4	8	12	47	....	....	....	....	....
	Halifax	QMJHL	45	4	23	27	200	18	0	5	5	96
1997-98	Halifax	QMJHL	69	9	27	36	377	5	0	1	1	21
1998-99	Halifax	QMJHL	69	7	30	37	320	5	1	1	2	14
99-2000	Saint John	AHL	9	0	3	3	16	....	....	....	....	....
	Johnstown Chiefs	ECHL	58	2	8	10	181	7	0	2	2	32
2000-01	Johnstown Chiefs	ECHL	69	2	9	11	302	4	0	0	0	9
	Kentucky	AHL	1	0	0	0	4	....	....	....	....	....
	Saint John	AHL	2	0	0	0	0	....	....	....	....	....
2001-02	Saint John	AHL	41	1	1	2	96	....	....	....	....	....
	Johnstown Chiefs	ECHL	36	3	5	8	155	5	0	1	1	20

QMJHL All-Rookie Team (1997)

SURMA, Damian (SUHR-ma, DAY-mee-an) CAR.

Left wing. Shoots left. 5'9", 200 lbs. Born, Lincoln Park, MI, June 22, 1981.
(Carolina's 5th choice, 174th overall, in 1999 Entry Draft).

			Regular Season					Playoffs				
Season	Club	Lea	GP	G	A	TP	PIM	GP	G	A	TP	PIM
1997-98	Det. Compuware	NAJHL	50	12	17	29	50	6	3	1	4	4
1998-99	Plymouth Whalers	OHL	65	17	15	32	62	11	3	6	9	15
99-2000	Plymouth Whalers	OHL	66	34	44	78	114	20	9	8	17	10
2000-01	Plymouth Whalers	OHL	55	26	34	60	62	19	8	9	17	25
2001-02	Plymouth Whalers	OHL	55	28	27	55	68	6	3	0	3	10
	Lowell	AHL	1	0	0	0	0	4	0	0	0	0

SUROVY, Tomas (suh-ROH-vee, TAW-mahsh) PIT.

Center. Shoots left. 6'1", 187 lbs. Born, Banska Bystrica, Czech., September 24, 1981.
(Pittsburgh's 5th choice, 120th overall, in 2001 Entry Draft).

			Regular Season					Playoffs				
Season	Club	Lea	GP	G	A	TP	PIM	GP	G	A	TP	PIM
99-2000	Banska Bystrica	Slovak-2	39	25	29	54	4	....	....	....	....	....
2000-01	HC SKP Poprad	Slovakia	53	22	28	50	30	6	2	1	3	14
2001-02	Wilkes-Barre	AHL	65	23	10	33	37	....	....	....	....	....

SUTTER, Shaun (SUH-tuhr, SHAWN) CGY.

Center. Shoots right. 6'1", 175 lbs. Born, Red Deer, Alta., June 2, 1980.
(Calgary's 4th choice, 102nd overall, in 1998 Entry Draft).

			Regular Season					Playoffs				
Season	Club	Lea	GP	G	A	TP	PIM	GP	G	A	TP	PIM
1995-96	Red Deer	AMHL	23	4	6	10	62	....	....	....	....	....
1996-97	Red Deer	AMHL	33	15	24	39	143	....	....	....	....	....
	Lethbridge	WHL	1	0	0	0	0	....	....	....	....	....
1997-98	Lethbridge	WHL	69	11	9	20	146	4	0	0	0	4
1998-99	Lethbridge	WHL	35	8	4	12	43	....	....	....	....	....
	Medicine Hat	WHL	23	9	5	14	38	....	....	....	....	....
99-2000	Medicine Hat	WHL	29	1	7	8	43	....	....	....	....	....
	Calgary Hitmen	WHL	6	0	1	1	8	....	....	....	....	....
2000-01	Calgary Hitmen	WHL	63	29	35	64	102	12	1	2	3	12
	Saint John	AHL	1	0	0	0	0	....	....	....	....	....
2001-02	Saint John	AHL	11	0	2	2	10	....	....	....	....	....
	Johnstown Chiefs	ECHL	34	13	7	20	34	8	0	1	1	4

Traded to **Medicine Hat** (WHL) by **Lethbridge** (WHL) for Blair Simpson, January 6, 1999. Traded to **Calgary** (WHL) by **Medicine Hat** (WHL) with Shaune Draper for futire considerations, January 10, 2000.

SVATOS, Marek (SVA-tohsh, MAIR-ehk) COL.

Right wing. Shoots right. 5'9", 170 lbs. Born, Kosice, Czech., July 17, 1982.
(Colorado's 10th choice, 227th overall, in 2001 Entry Draft).

			Regular Season					Playoffs				
Season	Club	Lea	GP	G	A	TP	PIM	GP	G	A	TP	PIM
99-2000	HC Kosice Jr.	Slovak-Jr.	39	43	30	73	28	....	....	....	....	....
	HC Kosice	Slovakia	19	2	2	4	0	....	....	....	....	....
2000-01	Kootenay Ice	WHL	39	23	18	41	47	11	7	2	9	26
2001-02	Kootenay Ice	WHL	53	38	39	77	58	21	12	6	18	40

WHL West Second All-Star Team (2002)

SVENSK, Mikael (SVEHNSK, mih-KIGH-ehl) EDM.

Defense. Shoots right. 6'2", 191 lbs. Born, Gällstad, Sweden, February 28, 1983.
(Edmonton's 7th choice, 185th overall, in 2001 Entry Draft).

			Regular Season					Playoffs				
Season	Club	Lea	GP	G	A	TP	PIM	GP	G	A	TP	PIM
99-2000	V. Frolunda-18	Swede-Jr.	18	2	6	8	6	2	0	0	0	2
	V. Frolunda Jr.	Swede-Jr.	10	1	1	2	12	2	0	0	0	0
2000-01	V. Frolunda-18	Swede-Jr.	6	0	1	1	4	3	0	1	1	2
	V. Frolunda Jr.	Swede-Jr.	15	1	0	1	4	....	....	....	....	....
2001-02	V. Frolunda Jr.	Swede-Jr.	34	4	1	5	16	5	1	0	1	2

SVENSSON, Jimmie (SVEHN-sohn, JIH-mee) DET.

Center. Shoots left. 6'1", 183 lbs. Born, Vasteras, Sweden, February 25, 1982.
(Detroit's 9th choice, 228th overall, in 2000 Entry Draft).

			Regular Season					Playoffs				
Season	Club	Lea	GP	G	A	TP	PIM	GP	G	A	TP	PIM
99-2000	Vasteras IK-18	Swede-Jr.	5	1	3	4	20	....	....	....	....	....
	Vasteras IK Jr.	Swede-Jr.	29	10	2	12	121	....	....	....	....	....
2000-01	Malmo IF Jr.	Swede-Jr.	23	3	1	4	74	....	....	....	....	....
2001-02	Malmo IF Jr.	Swede-Jr.	38	18	10	28	105	5	0	1	1	10

SVITOV, Alexander (SVEE-tawf, al-ehx-AN-duhr) T.B.

Center. Shoots left. 6'3", 198 lbs. Born, Omsk, USSR, November 3, 1982.
(Tampa Bay's 1st choice, 3rd overall, in 2001 Entry Draft).

			Regular Season					Playoffs				
Season	Club	Lea	GP	G	A	TP	PIM	GP	G	A	TP	PIM
1997-98	Novokuznetsk 2	Russia-3	4	0	0	0	0	....	....	....	....	....
1998-99	Omsk 2	Russia-4	27	15	8	23	20	....	....	....	....	....
	Avangard Omsk	Russia	....	....	....	....	....	1	0	0	0	0
99-2000	Omsk 2	Russia-3	14	13	9	22	62	....	....	....	....	....
	Avangard Omsk	Russia	18	3	3	6	45	6	1	0	1	16
2000-01	Avangard Omsk	Russia	39	8	6	14	115	14	2	1	3	34
2001-02	CSKA Moscow 2	Russia-3	2	1	0	1	2	....	....	....	....	....
	Avangard Omsk	Russia	2	0	1	1	2	....	....	....	....	....

SZYSKY, Chris (SHIHS-kee, KRIHS)

Right wing. Shoots right. 6', 208 lbs. Born, White City, Sask., June 8, 1976.
(Dallas' 8th choice, 280th overall, in 1994 Entry Draft).

			Regular Season					Playoffs				
Season	Club	Lea	GP	G	A	TP	PIM	GP	G	A	TP	PIM
1992-93	Swift Current	SMHL	35	23	24	47	150	....	....	....	....	....
	Swift Current	WHL	....	....	....	....	....	1	0	0	0	0
1993-94	Swift Current	WHL	60	6	10	16	82	7	0	1	1	12
1994-95	Swift Current	WHL	61	6	6	12	105	6	2	0	2	10
1995-96	Swift Current	WHL	63	19	16	35	115	6	3	2	5	21
1996-97	Swift Current	WHL	66	28	30	58	181	10	5	10	15	18
1997-98	Team Canada	Nat-Tm	50	9	20	29	111	....	....	....	....	....
1998-99	Team Canada	Nat-Tm	41	9	13	22	56	....	....	....	....	....
	Grand Rapids	IHL	6	1	1	2	10	....	....	....	....	....
99-2000	Grand Rapids	IHL	32	5	4	9	44	15	2	3	5	*45
2000-01	Grand Rapids	IHL	70	15	13	28	108	7	0	2	2	13
2001-02	Grand Rapids	AHL	42	4	6	10	80	5	0	1	1	14

Signed as a free agent by **Ottawa**, June 20, 1999.

TABACEK, Jan (tah-BA-chehk, YAN) ANA.

Defense. Shoots left. 5'11", 169 lbs. Born, Martin, Czech., April 7, 1980.
(Anaheim's 8th choice, 170th overall, in 2001 Entry Draft).

			Regular Season					Playoffs				
Season	Club	Lea	GP	G	A	TP	PIM	GP	G	A	TP	PIM
1997-98	MHC Martin Jr.	Slovak-Jr.	25	1	4	5	70	....	....	....	....	....
1998-99	MHC Martin	Slovakia	2	1	0	1	0	....	....	....	....	....
99-2000	MHC Martin	Slovak-2	35	1	7	8	22	....	....	....	....	....
2000-01	MHC Martin	Slovakia	44	8	5	13	62	3	1	2	3	0
2001-02	Slov. Bratislava	Slovakia	45	3	7	10	30	19	1	0	1	16

TAFFE, Jeff (TAYF, JEHF) PHX.

Center. Shoots left. 6'3", 195 lbs. Born, Hastings, MN, February 19, 1981.
(St. Louis' 1st choice, 30th overall, in 2000 Entry Draft).

			Regular Season					Playoffs				
Season	Club	Lea	GP	G	A	TP	PIM	GP	G	A	TP	PIM
1996-97	Hastings Huskies	Hi-School	25	21	37	58	....	....	....	....	....	....
1997-98	Hastings Huskies	Hi-School	28	37	29	66	....	....	....	....	....	....
1998-99	Hastings Huskies	Hi-School	28	39	51	90	....	....	....	....	....	....
99-2000	U. of Minnesota	WCHA	39	10	10	20	22	....	....	....	....	....
2000-01	U. of Minnesota	WCHA	38	12	23	35	56	....	....	....	....	....
2001-02	U. of Minnesota	WCHA	43	34	24	58	86	....	....	....	....	....

Minnesota High School Player of the Year (1999)

Rights traded to **Phoenix** by **St. Louis** with Michal Handzus, Ladislav Nagy and St. Louis' 1st round choice (Ben Eager) in 2002 Entry Draft for Keith Tkachuk, March 13, 2001.

TALLACKSON, Barry (TAL-ak-suhn, BAIR-ee) N.J.

Right wing. Shoots right. 6'4", 196 lbs. Born, Grafton, ND, April 14, 1983.
(New Jersey's 2nd choice, 53rd overall, in 2002 Entry Draft).

			Regular Season					Playoffs				
Season	Club	Lea	GP	G	A	TP	PIM	GP	G	A	TP	PIM
99-2000	Team USA	USDP-17	53	14	6	20	90	....	....	....	....	....
2000-01	Team USA	USDP-18	63	23	24	47	77	....	....	....	....	....
2001-02	U. of Minnesota	WCHA	44	13	10	23	44	....	....	....	....	....

TARATUKHIN, Andrei (tahr-a-TOO-khin, AN-dray) CGY.

Center. Shoots left. 6', 198 lbs. Born, Omsk, USSR, February 22, 1983.
(Calgary's 2nd choice, 41st overall, in 2001 Entry Draft).

			Regular Season					Playoffs				
Season	Club	Lea	GP	G	A	TP	PIM	GP	G	A	TP	PIM
1998-99	Omsk 2	Russia-4	1	0	0	0	0	....	....	....	....	....
99-2000	Omsk 2	Russia-3	27	10	6	16	16	....	....	....	....	....
	Avangard Omsk	Russia	....	....	....	....	....	1	1	0	1	0
2000-01	Omsk 2	Russia-3	41	19	28	47	69	....	....	....	....	....
2001-02	Mostovik Kurgan	Russia-2	44	13	22	35	30	....	....	....	....	....
	Yaroslavl 2	Russia-3	5	5	2	7	12	....	....	....	....	....

TARVAINEN, Jussi (tahr-VIGH-nehn, YU-see) EDM.

Right wing. Shoots right. 6'3", 215 lbs. Born, Lahti, Finland, May 31, 1976.
(Edmonton's 7th choice, 95th overall, in 1994 Entry Draft).

			Regular Season					Playoffs				
Season	Club	Lea	GP	G	A	TP	PIM	GP	G	A	TP	PIM
1991-92	KalPa Kuopio-C	Finn-Jr.	3	0	3	3	4	....	....	....	....	....
1992-93	KalPa Kuopio-B	Finn-Jr.	18	13	9	22	38	....	....	....	....	....
	KalPa Kuopio Jr.	Finn-Jr.	17	3	6	9	35	....	....	....	....	....
1993-94	KalPa Kuopio Jr.	Finn-Jr.	16	9	14	23	12	....	....	....	....	....
	Junkkarit	Finland-2	1	0	0	0	0	....	....	....	....	....
	KalPa Kuopio	Finland	42	3	4	7	20	....	....	....	....	....
1994-95	KalPa Kuopio Jr.	Finn-Jr.	3	4	0	4	2	....	....	....	....	....
	KalPa Kuopio	Finland	45	10	7	17	34	3	0	0	0	2
1995-96	KalPa Kuopio Jr.	Finn-Jr.	3	2	3	5	10	7	3	4	7	12
	KalPa Kuopio	Finland	47	8	11	19	50	....	....	....	....	....
1996-97	KalPa Kuopio	Finland	49	14	26	40	62	....	....	....	....	....
	KalPa Kuopio Jr.	Finn-Jr.	....	....	....	....	....	1	0	0	0	2
	KalPa Kuopio	Finland-2	....	....	....	....	....	6	4	3	7	4
1997-98	JYP Jyvaskyla	Finland	43	12	26	38	59	....	....	....	....	....
1998-99	JYP Jyvaskyla	Finland	54	17	24	41	84	3	0	0	0	8
99-2000	Tappara Tampere	Finland	52	20	27	47	91	4	1	0	1	2
2000-01	Tappara Tampere	Finland	56	23	32	55	36	10	*8	2	10	2
2001-02	Tappara Tampere	Finland	56	24	26	50	42	10	4	3	7	0

TATARINOV, Alexander (ta-TAHR-ee-nahf, al-ehx-AN-duhr) PHX.

Right wing. Shoots left. 5'11", 176 lbs. Born, Sverdlovsk, USSR, April 14, 1982.
(Phoenix's 2nd choice, 53rd overall, in 2000 Entry Draft).

			Regular Season					Playoffs				
Season	Club	Lea	GP	G	A	TP	PIM	GP	G	A	TP	PIM
1998-99	Spartak Moscow	Russia	3	0	0	0	0	....	....	....	....	....
99-2000	Yaroslavl 2	Russia-3	35	12	12	24	36	....	....	....	....	....
2000-01	Kristall Saratov	Russia-2	24	3	3	6	8	....	....	....	....	....
	Yaroslavl	Russia	2	0	1	1	0	1	0	0	0	0
2001-02	Yaroslavl	Russia	21	4	3	7	8	....	....	....	....	....
	Amur Khabarovsk	Russia	11	0	0	0	0	....	....	....	....	....

TATICEK, Petr (TA-tih-chehk, PEE-tuhr) FLA.

Center. Shoots left. 6'2", 188 lbs. Born, Rakovnik, Czech., September 22, 1983.
(Florida's 2nd choice, 9th overall, in 2002 Entry Draft).

			Regular Season					Playoffs				
Season	Club	Lea	GP	G	A	TP	PIM	GP	G	A	TP	PIM
99-2000	Kladno Jr.	Czech-Jr.	48	11	16	27	26	....	....	....	....	....
	Kladno	Czech	4	0	0	0	2	....	....	....	....	....
2000-01	Kladno Jr.	Czech-Jr.	30	7	12	19	54	....	....	....	....	....
	Kladno	Czech	3	0	0	0	0	....	....	....	....	....
2001-02	Sault Ste. Marie	OHL	60	21	42	63	32	6	3	3	6	4

TERESCHENKO, Alexei (teh-REH-shehn-koh, al-EHX-ay) DAL.

Center. Shoots left. 5'11", 176 lbs. Born, Mozhaisk, USSR, December 16, 1980.
(Dallas' 4th choice, 91st overall, in 2000 Entry Draft).

			Regular Season					Playoffs				
Season	Club	Lea	GP	G	A	TP	PIM	GP	G	A	TP	PIM
1996-97	DynamoMoscow2	Russia-3	9	0	0	0	2	....	....	....	....	....
1997-98	Dynamo Moscow	Russia	26	6	7	13	30	....	....	....	....	....
1998-99	DynamoMoscow2	Russia-3	28	4	17	21	20	....	....	....	....	....
	THC Tver	Russia-2	12	3	4	7	4	....	....	....	....	....
	Dynamo Moscow	Russia	1	0	1	1	0	2	0	0	0	0
99-2000	Dynamo Moscow	Russia	27	1	1	2	10	17	1	1	2	8
2000-01	Dynamo Moscow	Russia	39	3	2	5	18	....	....	....	....	....
2001-02	Yaroslavl 2	Russia-3	1	0	0	0	0	....	....	....	....	....
	Dynamo Moscow	Russia	40	3	6	9	20	3	0	0	0	0

TERNAVSKY, Artem (tuhr-NAV-skee, ahr-TEHM) WSH.

Defense. Shoots left. 6'3", 213 lbs. Born, Magnitogorsk, USSR, June 2, 1983.
(Washington's 4th choice, 160th overall, in 2001 Entry Draft).

			Regular Season					Playoffs				
Season	Club	Lea	GP	G	A	TP	PIM	GP	G	A	TP	PIM
99-2000	CSKA Moscow Jr.	Russia-Jr.	2	0	1	1	0	....	....	....	....	....
	H.C. CSKA 2	Russia-3	25	0	4	4	42	....	....	....	....	....
2000-01	Sherbrooke	QMJHL	65	3	15	18	143	....	....	....	....	....
2001-02	Mostovik Kurgan	Russia-2	25	0	0	0	46	....	....	....	....	....

TESSIER, Michael (teh-SEE-ay, MIGH-kuhl) BUF.

Left wing. Shoots left. 6'2", 180 lbs. Born, Granby, Que., August 14, 1984.
(Buffalo's 3rd choice, 76th overall, in 2002 Entry Draft).

			Regular Season					Playoffs				
Season	Club	Lea	GP	G	A	TP	PIM	GP	G	A	TP	PIM
99-2000	Antoine-Girouard	QAAA	38	18	22	40	22	7	5	3	8	6
2000-01	Acadie-Bathurst	QMJHL	55	4	8	12	110	13	1	0	1	2
2001-02	Acadie-Bathurst	QMJHL	66	22	48	70	83	14	5	7	12	12

THINEL, Marc-Andre (tih-nehl, MAHRK-AWN-dray) MTL.

Right wing. Shoots left. 6', 178 lbs. Born, St-Jerome, Que., March 24, 1981.
(Montreal's 6th choice, 145th overall, in 1999 Entry Draft).

			Regular Season					Playoffs				
Season	Club	Lea	GP	G	A	TP	PIM	GP	G	A	TP	PIM
1996-97	Laval Laurentide	QAAA	40	12	10	22	....	13	0	5	5	....
1997-98	Victoriaville	QMJHL	58	7	10	17	20	6	0	3	3	4
1998-99	Victoriaville	QMJHL	66	45	58	103	16	6	5	3	8	4
99-2000	Victoriaville	QMJHL	71	59	73	132	55	6	5	6	11	18
2000-01	Victoriaville	QMJHL	70	62	88	150	101	13	12	13	25	18
2001-02	Quebec	AHL	73	6	4	10	8	2	0	0	0	0

QMJHL First All-Star Team (2000) • QMJHL Second All-Star Team (2001)

THORBURN, Chris (THOHR-buhrn, KRIHS) BUF.

Center. Shoots right. 6'3", 207 lbs. Born, Sault Ste. Marie, Ont., June 3, 1983.
(Buffalo's 3rd choice, 50th overall, in 2001 Entry Draft).

			Regular Season					Playoffs				
Season	Club	Lea	GP	G	A	TP	PIM	GP	G	A	TP	PIM
1998-99	Elliot Lake	NOJHA	40	21	12	33	28	....	....	....	....	....
99-2000	North Bay	OHL	56	12	8	20	33	6	0	2	2	0
2000-01	North Bay	OHL	66	22	32	54	64	4	0	1	1	9
2001-02	North Bay	OHL	67	15	43	58	112	5	1	2	3	8

THORNTON, Shawn (THOHRN-tohn, SHAWN) CHI.

Right wing. Shoots right. 6'1", 196 lbs. Born, Oshawa, Ont., July 23, 1977.
(Toronto's 6th choice, 190th overall, in 1997 Entry Draft).

			Regular Season					Playoffs				
Season	Club	Lea	GP	G	A	TP	PIM	GP	G	A	TP	PIM
1995-96	Peterborough	OHL	63	4	10	14	192	24	3	0	3	25
1996-97	Peterborough	OHL	61	19	10	29	204	11	2	4	6	20
1997-98	St. John's	AHL	59	0	3	3	225	....	....	....	....	....
1998-99	St. John's	AHL	78	8	11	19	354	5	0	0	0	9
99-2000	St. John's	AHL	60	4	12	16	316	....	....	....	....	....
2000-01	St. John's	AHL	79	5	12	17	320	3	1	2	3	2
2001-02	Norfolk Admirals	AHL	70	8	14	22	281	4	0	0	0	4

Traded to **Chicago** by **Toronto** for Marty Wilford, September 30, 2001.

TIILIKAINEN, Jukka (TEE-ee-lee-kigh-nehn, yoo-KUH) L.A.

Left wing. Shoots left. 6', 190 lbs. Born, Espoo, Finland, April 4, 1974.
(Los Angeles' 8th choice, 255th overall, in 1992 Entry Draft).

			Regular Season					Playoffs				
Season	Club	Lea	GP	G	A	TP	PIM	GP	G	A	TP	PIM
1992-93	Vantaa HT	Finland-2	18	7	3	10	10	....	....	....	....	....
	Kiekko Espoo	Finland	5	0	0	0	4	....	....	....	....	....
1993-94	Kiekko Espoo	Finland	33	2	4	6	12	....	....	....	....	....
1994-95	TPS Turku	Finland	38	5	4	9	8	11	1	0	1	8
1995-96	TPS Turku	Finland	38	6	13	19	28	10	0	2	2	2
1996-97	Lukko Rauma	Finland	49	15	12	27	42	....	....	....	....	....
1997-98	Assat Pori	Finland	48	13	13	26	18	3	0	2	2	0
1998-99	Jokerit Helsinki	Finland	51	7	8	15	52	2	0	0	0	4
99-2000	Jokerit Helsinki	Finland	19	1	2	3	31	....	....	....	....	....
	AIK Solna	Sweden	29	8	6	14	22	....	....	....	....	....
2000-01	Blues Espoo	Finland	53	2	12	14	50	....	....	....	....	....
2001-02	Sodertalje SK	Sweden	50	10	6	16	45	....	....	....	....	....

TIMMONS, K.C. (TIHM-mohns, KAY-SEE) COL.

Left wing. Shoots left. 6'4", 215 lbs. Born, Victoria, B.C., April 6, 1980.
(Colorado's 9th choice, 141st overall, in 1998 Entry Draft).

			Regular Season					Playoffs				
Season	Club	Lea	GP	G	A	TP	PIM	GP	G	A	TP	PIM
1995-96	Victoria Lions	BCAHA	68	82	101	183	190	....	....	....	....	....
1996-97	Tri-City	WHL	52	0	5	5	27	....	....	....	....	....
1997-98	Tri-City	WHL	72	11	7	18	139	....	....	....	....	....
1998-99	Tri-City	WHL	69	13	11	24	113	12	1	1	2	36
99-2000	Tri-City	WHL	69	24	24	48	193	4	0	0	0	4
	Hershey Bears	AHL	....	....	....	....	....	4	0	0	0	15
2000-01	Hershey Bears	AHL	39	3	9	12	69	....	....	....	....	....
2001-02	Hershey Bears	AHL	63	7	10	17	104	7	1	1	2	0

• Missed majority of 2000-01 season recovering from shoulder injury suffered in game vs. Rochester (AHL), January 13, 2001.

TIMOFEEV, Denis (teh-moh-FAY-ehf, DEH-nihs) BOS.

Defense. Shoots left. 6'6", 210 lbs. Born, Moscow, USSR, January 14, 1979.
(Boston's 7th choice, 135th overall, in 1997 Entry Draft).

			Regular Season					Playoffs				
Season	Club	Lea	GP	G	A	TP	PIM	GP	G	A	TP	PIM
1996-97	CSKA Moscow 2	Russia-3	11	0	0	0	2	....	....	....	....	....
	H.C. CSKA	Russia	41	6	8	14	....	....	....	....	....	....
1997-98	CSKA Moscow Jr.	Russia-Jr.	STATISTICS NOT AVAILABLE									
1998-99	CSKA Moscow Jr.	Russia-Jr.	31	4	12	16	16	....	....	....	....	....
99-2000	Providence	AHL	16	0	1	1	14	....	....	....	....	....
	Greenville	ECHL	24	2	3	5	30	10	0	1	1	24
2000-01	Greenville	ECHL	3	0	0	0	8	....	....	....	....	....
	Pensacola	ECHL	20	0	4	4	28	....	....	....	....	....
	New Orleans	ECHL	38	2	6	8	49	3	0	0	0	0
2001-02	Nizhnekamsk	Russia	9	0	0	0	6	....	....	....	....	....

TIMONEN, Jussi (TEEM-oh-nehn, YU-see) PHI.

Defense. Shoots left. 6', 200 lbs. Born, Kuopio, Finland, June 29, 1983.
(Philadelphia's 3rd choice, 146th overall, in 2001 Entry Draft).

			Regular Season					Playoffs				
Season	Club	Lea	GP	G	A	TP	PIM	GP	G	A	TP	PIM
99-2000	KalPa Kuopio Jr.	Finn-Jr.	33	4	2	6	16	4	0	0	0	4
2000-01	Kalpa Kuopio-B	Finn-Jr.	38	6	7	13	22	....	....	....	....	....
	KalPa Kuopio Jr.	Finn-Jr.	1	0	1	1	0	....	....	....	....	....
2001-02	KalPa Kuopio Jr.	Finn-Jr.	10	1	1	2	10	....	....	....	....	....
	Kalpa Kuopio	Finland-2	41	3	8	11	10	8	0	2	2	0

TJARNQVIST, Mathias (TUH-yahrn-kvihst, MAT-ee-uhs) DAL.

Right wing. Shoots left. 6'1", 183 lbs. Born, Umea, Sweden, April 15, 1979.
(Dallas' 3rd choice, 96th overall, in 1999 Entry Draft).

			Regular Season					Playoffs				
Season	Club	Lea	GP	G	A	TP	PIM	GP	G	A	TP	PIM
1995-96	Rogle Jr.	Swede-Jr.	4	2	0	2	0	....	....	....	....	....
1996-97	Rogle Jr.	Swede-Jr.	18	5	8	13	....	....	....	....	....	....
	Rogle	Swede-2	15	1	4	5	4	....	....	....	....	....
1997-98	Rogle	Swede-2	31	12	11	23	30	....	....	....	....	....
1998-99	Rogle	Swede-2	34	18	16	34	44	5	4	1	5	4
99-2000	Djurgarden	Sweden	50	12	12	24	20	13	3	2	5	16
2000-01	Djurgarden	Sweden	47	11	8	19	53	16	1	2	3	6
2001-02	Djurgarden	Sweden	6	0	1	1	4	2	0	0	0	2

TKACHENKO, Ivan (t'kuh-CHEHN-koh, ee-VAHN) CBJ

Left wing. Shoots left. 5'10", 183 lbs. Born, Yaroslavl, USSR, November 9, 1979.
(Columbus' 5th choice, 98th overall, in 2002 Entry Draft).

			Regular Season					Playoffs				
Season	Club	Lea	GP	G	A	TP	PIM	GP	G	A	TP	PIM
1997-98	Yaroslavl 2	Russia-2	STATISTICS NOT AVAILABLE									
	Yaroslavl	Russia	....	....	....	....	....	1	0	0	0	0
1998-99	Yaroslavl 2	Russia-3	28	15	13	28	26	....	....	....	....	....
99-2000	Yaroslavl 2	Russia-3	1	1	0	1	0	....	....	....	....	....
	Motor Zavolzhje	Russia-2	43	15	14	29	22	....	....	....	....	....
	Nizhnekamsk 2	Russia-3	8	6	3	9	24	....	....	....	....	....
	Nizhnekamsk	Russia	5	1	0	1	0	4	0	1	1	0
2000-01	Nizhnekamsk	Russia	28	2	2	4	14	4	0	1	1	0
2001-02	Yaroslavl 2	Russia-3	1	0	1	1	2	....	....	....	....	....
	Yaroslavl	Russia	44	13	20	33	57	9	5	2	7	4

TOLKUNOV, Dmitri (tohl-ku-NAWF, di-MEE-tree) CHI.

Defense. Shoots right. 6'2", 200 lbs. Born, Kiev, USSR, May 5, 1979.

			Regular Season					Playoffs				
Season	Club	Lea	GP	G	A	TP	PIM	GP	G	A	TP	PIM
1996-97	Hull Olympiques	QMJHL	34	3	8	11	99	....	....	....	....	....
	Beauport	QMJHL	27	3	7	10	18	4	0	1	1	4
1997-98	Quebec Remparts	QMJHL	66	10	25	35	81	14	3	9	12	22
1998-99	Quebec Rafales	QMJHL	69	11	57	68	110	13	2	7	9	22
99-2000	Cleveland	IHL	65	3	12	15	54	8	0	0	0	2
2000-01	Norfolk Admirals	AHL	78	5	18	23	93	9	0	1	1	4
2001-02	Norfolk Admirals	AHL	51	1	18	19	20	4	0	0	0	2

QMJHL Second All-Star Team (1999)

Signed as a free agent by **Chicago**, October 8, 1998.

TOLLEFSEN, Ole-Kristian (TOHL-uhf-suhn, OH-lay-KRIHS-tyahn) CBJ

Defense. Shoots left. 6'2", 200 lbs. Born, Oslo, Norway, March 29, 1984.
(Columbus' 3rd choice, 65th overall, in 2002 Entry Draft).

			Regular Season					Playoffs				
Season	Club	Lea	GP	G	A	TP	PIM	GP	G	A	TP	PIM
2000-01	Lillehammer IHK	Norway	4	0	0	0	2	....	....	....	....	....
2001-02	Lillehammer IHK	Norway	37	1	5	6	63	6	1	1	2	10
	Lillehammer IHK	Norge-Jr.	....	....	....	....	....	1	0	2	2	4

TOLSA, Jari (TOHL-suh, YA-ree) DET.

Center. Shoots left. 6', 172 lbs. Born, Goteborg, Sweden, April 20, 1981.
(Detroit's 1st choice, 120th overall, in 1999 Entry Draft).

			Regular Season					Playoffs				
Season	Club	Lea	GP	G	A	TP	PIM	GP	G	A	TP	PIM
1997-98	V. Frolunda Jr.	Swede-Jr.	26	18	25	43	30	....	....	....	....	....
1998-99	V. Frolunda Jr.	Swede-Jr.	35	16	21	37	51	4	1	3	4	2
99-2000	V. Frolunda Jr.	Swede-Jr.	33	12	47	59	43	6	3	8	11	4
	Vastra Frolunda	Sweden	10	0	0	0	0	....	....	....	....	....
2000-01	V. Frolunda Jr.	Swede-Jr.	11	4	6	10	8	2	2	4	6	4
	Molndals HK	Swede-2	1	0	2	2	0	....	....	....	....	....
	Vastra Frolunda	Sweden	42	2	5	7	18	5	0	0	0	0
2001-02	Vastra Frolunda	Sweden	48	8	16	24	18	10	0	0	0	4

TOMICA, Marek (TAW-miht-suh, MAIR-ehk) DAL.

Left wing. Shoots left. 6', 178 lbs. Born, Prague, Czech., January 1, 1981.
(Dallas' 10th choice, 285th overall, in 2001 Entry Draft).

			Regular Season					Playoffs				
Season	Club	Lea	GP	G	A	TP	PIM	GP	G	A	TP	PIM
99-2000	Slavia Praha Jr.	Czech-Jr.	41	13	8	21	16	7	2	3	5	0
	HC Slavia Praha	Czech	13	0	0	0	2	....	....	....	....	....
2000-01	Slavia Praha Jr.	Czech-Jr.	5	2	2	4	4	....	....	....	....	....
	Beroun	Czech-2	8	0	2	2	2	....	....	....	....	....
	Mlada Boleslav	Czech-3	2	0	0	0	6	....	....	....	....	....
	HC Slavia Praha	Czech	37	3	9	12	8	11	0	0	0	0
2001-02	HC Slavia Praha	Czech	49	6	8	14	14	8	0	0	0	4

TOOTOO, Jordin (TOO-TOO, JOHR-dihn) NSH.

Right wing. Shoots right. 5'9", 198 lbs. Born, Churchill, Man., February 2, 1983.
(Nashville's 6th choice, 98th overall, in 2001 Entry Draft).

			Regular Season					Playoffs				
Season	Club	Lea	GP	G	A	TP	PIM	GP	G	A	TP	PIM
1998-99	OCN Blizzard	MJHL	47	16	21	37	251	....	....	....	....	....
99-2000	Brandon	WHL	45	6	10	16	214	....	....	....	....	....
2000-01	Brandon	WHL	60	20	28	48	172	6	2	4	6	18
2001-02	Brandon	WHL	64	32	39	71	272	16	4	3	7	*58

TRAVNICEK, Michal (TRAV-nih-chehk, MEE-khuhl) TOR.

Right wing. Shoots left. 6'1", 198 lbs. Born, Decin, Czech., March 14, 1980.
(Toronto's 9th choice, 228th overall, in 1998 Entry Draft).

			Regular Season					Playoffs				
Season	Club	Lea	GP	G	A	TP	PIM	GP	G	A	TP	PIM
1995-96	Litvinov-18	Czech-Jr.	45	26	22	48	....	....	....	....	....	....
1996-97	Litvinov Jr.	Czech-Jr.	45	35	22	57	....	....	....	....	....	....
1997-98	Litvinov Jr.	Czech-Jr.	43	18	20	38	....	....	....	....	....	....
1998-99	Litvinov	Czech	49	7	7	14	65	....	....	....	....	....
	Litvinov	EuroHL	5	0	1	1	8	....	....	....	....	....
99-2000	Litvinov	Czech	51	3	6	9	28	7	0	0	0	0
2000-01	St. John's	AHL	80	3	21	24	70	4	0	1	1	4
2001-02	St. John's	AHL	23	3	2	5	17	....	....	....	....	....
	Litvinov	Czech	10	1	2	3	4	....	....	....	....	....

TREILLE, Yorick (TRAYL, YOH-rihk) CHI.

Right wing. Shoots right. 6'3", 205 lbs. Born, Cannes, France, July 15, 1980.
(Chicago's 7th choice, 195th overall, in 1999 Entry Draft).

			Regular Season					Playoffs				
Season	Club	Lea	GP	G	A	TP	PIM	GP	G	A	TP	PIM
1997-98	Notre Dame	SJHL	54	18	28	46	42	....	....	....	....	....
1998-99	U. Mass-Lowell	H-East	30	6	5	11	24	....	....	....	....	....
99-2000	U. Mass-Lowell	H-East	33	10	12	22	34	....	....	....	....	....
2000-01	U. Mass-Lowell	H-East	31	10	14	24	35	....	....	....	....	....
2001-02	U. Mass-Lowell	H-East	30	10	16	26	24	....	....	....	....	....

TREVISANI, Carter (treh-vih-SAN-ee, KAHR-tuhr) CAR.

Defense. Shoots left. 6'1", 185 lbs. Born, Carlisle, Ont., June 15, 1982.
(Carolina's 7th choice, 244th overall, in 2001 Entry Draft).

			Regular Season					Playoffs				
Season	Club	Lea	GP	G	A	TP	PIM	GP	G	A	TP	PIM
1997-98	Kitchener	OJHL-B	39	4	10	14	27	....	....	....	....	....
1998-99	Kitchener	OJHL-B	12	4	5	9	20	....	....	....	....	....
	Milton Merchants	OJHL	34	7	11	18	38	....	....	....	....	....
99-2000	Kitchener	OJHL-B	44	10	26	36	65	....	....	....	....	....
2000-01	Ohio State	CCHA	10	0	1	1	2	....	....	....	....	....
	Ottawa 67's	OHL	35	9	10	19	22	20	1	4	5	26
2001-02	Ottawa 67's	OHL	67	4	24	28	96	13	0	7	7	8

• Left **Ohio State** (CCHA) and signed as a free agent with **Ottawa** (OHL), December 14, 2000.

TROLIGA, Tomas (TROH-lih-guh, TAW-mash) ST.L.

Center. Shoots right. 6'4", 207 lbs. Born, Presov, Czech., April 24, 1984.
(St. Louis' 3rd choice, 89th overall, in 2002 Entry Draft).

			Regular Season					Playoffs				
Season	Club	Lea	GP	G	A	TP	PIM	GP	G	A	TP	PIM
99-2000	Presov Jr.	Slovak-Jr.	39	6	1	7	56	....	....	....	....	....
2000-01	Presov Jr.	Slovak-Jr.	30	14	16	30	108	....	....	....	....	....
	HK VTJ Presov	Slovak-2	7	1	0	1	8	....	....	....	....	....
2001-02	Nova Ves Jr.	Slovak-Jr.		STATISTICS NOT AVAILABLE								
	Spisska Nova Ves	Slovak-2	11	2	5	7	0	....	....	....	....	....

TROSCHINSKY, Andrei (troh-SCHIHN-skee, AN-dray) ST.L.

Center. Shoots left. 6'5", 187 lbs. Born, Ust-Kamenogorsk, USSR, February 14, 1978.
(St. Louis' 5th choice, 170th overall, in 1998 Entry Draft).

			Regular Season					Playoffs				
Season	Club	Lea	GP	G	A	TP	PIM	GP	G	A	TP	PIM
1996-97	Ust-Kamenogorsk	Russia-2	9	1	1	2	8	....	....	....	....	....
1997-98	Ust-Kamenogorsk	Russia-2	47	10	16	26	34	....	....	....	....	....
1998-99	Ust-Kamenog. 2	Russia-4	4	4	4	8	6	....	....	....	....	....
	Ust-Kamenog. 2	Russia-4	42	11	21	32	62	....	....	....	....	....
99-2000	Ust-Kamenog. 2	Russia-3	50	20	46	66	....	....	....	....	....	....
2000-01	Worcester	AHL	78	17	28	45	32	11	2	2	4	2
2001-02	Worcester	AHL	70	13	10	23	32	2	0	0	0	0

TRUBACHEV, Yuri (troo-bah-CHEHV, YOO-ree) CGY.

Center. Shoots left. 5'9", 187 lbs. Born, Cherepovets, USSR, March 9, 1983.
(Calgary's 7th choice, 164th overall, in 2001 Entry Draft).

			Regular Season					Playoffs				
Season	Club	Lea	GP	G	A	TP	PIM	GP	G	A	TP	PIM
1997-98	Cherepovets 2	Russia-3	1	0	0	0	0	....	....	....	....	....
1998-99	Cherepovets 3	Russia-4	9	5	1	6	0	....	....	....	....	....
	Cherepovets 2	Russia-3	2	0	0	0	0	....	....	....	....	....
99-2000	Cherepovets 2	Russia-3	42	13	19	32	76	....	....	....	....	....
2000-01	St. Petersburg	Russia	34	6	5	11	24	....	....	....	....	....
2001-02	Cherepovets 2	Russia-3	5	3	3	6	2	....	....	....	....	....
	Cherepovets	Russia	32	2	1	3	6	4	0	2	2	0

TUKIO, Arto (TOO-kee-oh, AHR-toh) NYI

Defense. Shoots left. 5'10", 176 lbs. Born, Tampere, Finland, April 4, 1981.
(NY Islanders' 3rd choice, 101st overall, in 2000 Entry Draft).

			Regular Season					Playoffs				
Season	Club	Lea	GP	G	A	TP	PIM	GP	G	A	TP	PIM
1995-96	Ilves Tampere-C	Finn-Jr.	28	3	5	8	20	....	....	....	....	....
1996-97	Ilves Tampere-C	Finn-Jr.	6	1	3	4	4	....	....	....	....	....
	Ilves Tampere-B	Finn-Jr.	20	0	2	2	6	1	1	0	1	0
1997-98	Ilves Tampere-B	Finn-Jr.	39	7	7	14	64	....	....	....	....	....
	Ilves Jr.	Finn-Jr.	....	....	....	....	....	4	1	0	1	4
1998-99	Ilves Tampere-B	Finn-Jr.	10	1	5	6	18	....	....	....	....	....
	Ilves Jr.	Finn-Jr.	18	1	4	5	12	10	0	0	0	2
99-2000	Ilves Jr.	Finn-Jr.	11	2	1	3	24	....	....	....	....	....
	Hermes Kokkola	Finland-2	1	0	0	0	0	....	....	....	....	....
	Ilves Tampere	Finland	42	2	1	3	20	3	0	0	0	0
2000-01	Ilves Jr.	Finn-Jr.	3	1	1	2	0	....	....	....	....	....
	Ilves Tampere	Finland	43	5	10	15	26	7	2	1	3	4
2001-02	Ilves Tampere	Finland	47	8	9	17	28	3	0	1	1	0

TUOKKO, Marco (too-OH-koh, MAHR-koh) DAL.

Center. Shoots left. 6', 185 lbs. Born, Raisio, Finland, March 27, 1979.
(Dallas' 9th choice, 219th overall, in 2000 Entry Draft).

			Regular Season					Playoffs				
Season	Club	Lea	GP	G	A	TP	PIM	GP	G	A	TP	PIM
1995-96	TPS Turku-B	Finn-Jr.	31	9	13	22	49	3	1	0	1	0
	Kiekko-67 Turku	Finland-2	1	0	0	0	0	....	....	....	....	....
1996-97	TPS Turku-B	Finn-Jr.	5	1	3	4	0	6	1	6	7	4
	TPS Turku Jr.	Finn-Jr.	30	8	7	15	32	....	....	....	....	....
	Kiekko-67 Turku	Finland-2	6	3	1	4	31	....	....	....	....	....
1997-98	TPS Turku Jr.	Finn-Jr.	26	5	13	18	77	7	1	4	5	14
1998-99	TPS Turku Jr.	Finn-Jr.	1	0	0	0	0	....	....	....	....	....
	TPS Turku	Finland	48	5	4	9	24	10	0	1	1	10
99-2000	TPS Turku	Finland	54	10	10	20	73	11	2	2	4	6
2000-01	TPS Turku	Finland	53	5	11	16	54	10	2	2	4	16
2001-02	TPS Turku	Finland	42	2	9	11	69	8	1	1	2	4

TURON, David (TUHR-awn, DAY-vihd) TOR.

Defense. Shoots right. 6'2", 200 lbs. Born, Havirov, Czech., October 4, 1983.
(Toronto's 5th choice, 122nd overall, in 2002 Entry Draft).

			Regular Season					Playoffs				
Season	Club	Lea	GP	G	A	TP	PIM	GP	G	A	TP	PIM
99-2000	SK Karvina Jr.	Czech-Jr.	2	0	0	0	0	....	....	....	....	....
	Havirov Jr.	Czech-Jr.	41	14	11	25	54	....	....	....	....	....
2000-01	Havirov Jr.	Czech-Jr.	43	12	7	19	26	....	....	....	....	....
	HC Femax Havirov	Czech	4	0	0	0	4	....	....	....	....	....
2001-02	Havirov Jr.	Czech-Jr.	41	5	11	16	75	....	....	....	....	....
	HC Femax Havirov	Czech	14	0	1	1	10	....	....	....	....	....

TUTIN, Fedor (TYOO-tihn, feh-DUHR) NYR

Defense. Shoots left. 6'2", 196 lbs. Born, Izhevsk, USSR, July 19, 1983.
(NY Rangers' 2nd choice, 40th overall, in 2001 Entry Draft).

			Regular Season					Playoffs				
Season	Club	Lea	GP	G	A	TP	PIM	GP	G	A	TP	PIM
1998-99	Magnitogorsk 2	Russia-4	7	0	1	1	2	....	....	....	....	....
99-2000	Izhevsk 2	Russia-3	38	11	8	19	68	....	....	....	....	....
	Izhstal Izhevsk	Russia-2	10	0	1	1	12	....	....	....	....	....
2000-01	St. Petersburg	Russia	34	2	4	6	20	....	....	....	....	....
2001-02	Guelph Storm	OHL	53	19	40	59	54	9	2	8	10	8

TVRDON, Roman (t-vahr-DAWN, ROH-muhn) WSH.

Center. Shoots left. 6'1", 189 lbs. Born, Trencin, Czech., January 29, 1981.
(Washington's 6th choice, 132nd overall, in 1999 Entry Draft).

			Regular Season					Playoffs				
Season	Club	Lea	GP	G	A	TP	PIM	GP	G	A	TP	PIM
1997-98	Dukla Trencin Jr.	Slovak-Jr.	48	4	12	16	39	....	....	....	....	....
1998-99	Dukla Trencin Jr.	Slovak-Jr.	49	23	23	46	20	6	4	4	8	4
99-2000	Spokane Chiefs	WHL	69	26	44	70	40	15	4	7	11	16
2000-01	Spokane Chiefs	WHL	62	28	34	62	55	12	5	11	16	0
2001-02	Portland Pirates	AHL	49	5	9	14	22	....	....	....	....	....

UCHEVATOV, Victor (oo-cheh-VA-tawf, VIHK-tohr) N.J.

Defense. Shoots left. 6'4", 215 lbs. Born, Angarsk, USSR, February 10, 1983.
(New Jersey's 4th choice, 60th overall, in 2001 Entry Draft).

			Regular Season					Playoffs				
Season	Club	Lea	GP	G	A	TP	PIM	GP	G	A	TP	PIM
2000-01	Yaroslavl 2	Russia-3	28	1	1	2	74	....	....	....	....	....
2001-02	Albany	AHL	64	0	2	2	50	....	....	....	....	....

UJCIK, Viktor (OOY-chehk, VIHK-tohr) MTL.

Right wing. Shoots left. 5'10", 194 lbs. Born, Jihlava, Czech., May 24, 1972.
(Montreal's 8th choice, 266th overall, in 2001 Entry Draft).

			Regular Season					Playoffs				
Season	Club	Lea	GP	G	A	TP	PIM	GP	G	A	TP	PIM
1990-91	Dukla Jihlava	Czech	2	0	0	0	0	....	....	....	....	....
1991-92	Dukla Jihlava	Czech	35	10	9	19	32	8	3	4	7	0
1992-93	Dukla Jihlava	Czech	30	16	16	32	....	....	....	....	....	....
1993-94	HC Dukla Jihlava	Czech	44	17	30	47	....	4	3	3	6	....
1994-95	HC Dukla Jihlava	Czech	42	20	16	36	65	2	1	2	3	2
1995-96	HC Slavia Praha	Czech	39	*37	19	56	59	7	8	4	12	6
1996-97	HC Slavia Praha	Czech	40	26	21	47	41	3	1	1	2	2
1997-98	HC Slavia Praha	Czech	17	13	8	21	12	....	....	....	....	....
	Trinec	Czech	31	21	22	43	63	13	8	10	18	4
1998-99	Trinec	Czech	44	20	23	43	55	10	4	4	8	18
99-2000	Trinec	Czech	42	14	20	34	32	4	1	0	1	28
2000-01	Trinec	Czech	31	8	12	20	20	....	....	....	....	....
	HC Slavia Praha	Czech	19	9	9	18	8	11	8	8	16	10
2001-02	HC Slavia Praha	Czech	52	25	23	48	51	9	3	0	3	6

ULMER, Layne (UHL-muhr, LAYN) NYR

Center. Shoots left. 6'1", 205 lbs. Born, North Battleford, Sask., September 14, 1980.
(Ottawa's 8th choice, 209th overall, in 1999 Entry Draft).

			Regular Season					Playoffs				
Season	Club	Lea	GP	G	A	TP	PIM	GP	G	A	TP	PIM
1996-97	Swift Current	SMHL	43	35	49	84	31	....	....	....	....	....
1997-98	Swift Current	WHL	50	8	9	17	23	12	3	1	4	0
1998-99	Swift Current	WHL	72	40	35	75	34	6	2	1	3	4
99-2000	Swift Current	WHL	71	50	54	104	66	12	12	6	18	16
2000-01	Swift Current	WHL	68	*63	56	119	75	19	7	3	10	20
2001-02	Hartford	AHL	22	0	5	5	17	....	....	....	....	....
	Charlotte	ECHL	38	18	17	35	12	5	2	2	4	4

WHL East First All-Star Team (2000, 2001)

Signed as a free agent by **NY Rangers**, June 13, 2001.

UMBERGER, R.J. (UHM-buhr-guhr, AHR-JAY) VAN.

Center. Shoots left. 6'2", 200 lbs. Born, Pittsburgh, PA, May 3, 1982.
(Vancouver's 1st choice, 16th overall, in 2001 Entry Draft).

			Regular Season					Playoffs				
Season	Club	Lea	GP	G	A	TP	PIM	GP	G	A	TP	PIM
1997-98	Plum Mustangs	Hi-School	26	*60	*56	*116	....	....	....	....	....	....
1998-99	Team USA	USDP-17	50	29	29	58	....	....	....	....	....	....
99-2000	Team USA	USDP-18	57	33	35	68	20	....	....	....	....	....
2000-01	Ohio State	CCHA	32	14	23	37	18	....	....	....	....	....
2001-02	Ohio State	CCHA	37	18	21	39	31	....	....	....	....	....

CCHA All-Rookie Team (2001) • CCHA Rookie of the Year (2001)

UPPER, Dmitri (OO-puhr, dih-MEE-tree) NYI

Center. Shoots right. 6'1", 185 lbs. Born, Ust-Kamenogorsk, USSR, July 27, 1978.
(NY Islanders' 5th choice, 136th overall, in 2000 Entry Draft).

			Regular Season					Playoffs				
Season	Club	Lea	GP	G	A	TP	PIM	GP	G	A	TP	PIM
1997-98	Ust-Kamenogorsk	Russia-2	47	16	12	28	44	....	....	....	....	....
1998-99	Ust-Kamenog. 2	Russia-4	29	10	11	21	44	....	....	....	....	....
	Nizhny Novgorod	Russia-2	28	10	16	26	65	....	....	....	....	....
99-2000	Nizhny Novgorod	Russia	36	14	6	20	50	5	1	1	2	4
2000-01	Nizhny Novgorod	Russia	6	0	2	2	4	....	....	....	....	....
	Ak Bars Kazan	Russia	31	7	4	11	6	1	0	0	0	0
2001-02	Spartak Moscow	Russia	51	16	9	25	74	....	....	....	....	....

UPSHALL, Scottie (UHP-shuhl, SKAW-tee) NSH.

Right wing. Shoots left. 6', 184 lbs. Born, Fort McMurray, Alta., October 7, 1983.
(Nashville's 1st choice, 6th overall, in 2002 Entry Draft).

			Regular Season					Playoffs				
Season	Club	Lea	GP	G	A	TP	PIM	GP	G	A	TP	PIM
99-2000	Fort McMurray	AJHL	52	26	26	52	65	....	....	....	....	....
2000-01	Kamloops Blazers	WHL	70	42	45	87	111	4	0	2	2	10
2001-02	Kamloops Blazers	WHL	61	32	51	83	139	4	1	2	3	21

WHL All-Rookie Team (2001) • WHL Rookie of the Year (2001) • CHL All-Rookie Team (2001) • Canadian Major Junior Rookie of the Year (2001) • WHL West Second All-Star Team (2002)

USTRNUL, Libor (OOS-tuhr-nuhl, LEE-bohr) ATL.

Defense. Shoots left. 6'5", 230 lbs. Born, Steruberk, Czech., February 20, 1982.
(Atlanta's 3rd choice, 42nd overall, in 2000 Entry Draft).

			Regular Season					Playoffs				
Season	Club	Lea	GP	G	A	TP	PIM	GP	G	A	TP	PIM
1997-98	HC Olomouc Jr.	Czech-Jr.	45	2	11	13	54	....	....	....	....	....
1998-99	Thunder Bay	USHL	52	2	5	7	65	18	1	4	5	95
99-2000	Plymouth Whalers	OHL	68	0	15	15	208	23	0	3	3	29
2000-01	Plymouth Whalers	OHL	35	3	13	16	66	19	1	4	5	19
2001-02	Plymouth Whalers	OHL	43	1	8	9	84	2	0	0	0	6
	Chicago Wolves	AHL	1	0	0	0	0	1	0	0	0	5

VAGNER, Martin (VAHG-nuhr, MAHR-tihn) DAL.

Defense. Shoots left. 6'1", 214 lbs. Born, Jaromer, Czech., March 16, 1984.
(Dallas' 1st choice, 26th overall, in 2002 Entry Draft).

			Regular Season					Playoffs				
Season	Club	Lea	GP	G	A	TP	PIM	GP	G	A	TP	PIM
99-2000	Sparta Praha Jr.	Czech-Jr.	46	2	8	10	34	....	....	....	....	....
2000-01	Pardubice Jr.	Czech-Jr.	53	2	12	14	46	....	....	....	....	....
2001-02	Hull Olympiques	QMJHL	64	6	28	34	81	8	0	1	1	10

QMJHL All-Rookie Team (2002)

VALENTINE, Curtis (VAL-lehn-tighn, KUHR-tihs) VAN.

Left wing. Shoots left. 6'5", 195 lbs. Born, Haileybury, Ont., July 22, 1979.
(Vancouver's 11th choice, 219th overall, in 1998 Entry Draft).

			Regular Season					Playoffs				
Season	Club	Lea	GP	G	A	TP	PIM	GP	G	A	TP	PIM
1996-97	Capital District	Exhib.	56	53	60	113	28	....	....	....	....	....
1997-98	Bowling Green	CCHA	38	7	8	15	34	....	....	....	....	....
1998-99	Bowling Green	CCHA	38	4	8	12	40	....	....	....	....	....
99-2000	Bowling Green	CCHA	37	5	7	12	24	....	....	....	....	....
2000-01	Bowling Green	CCHA	40	9	7	16	8	....	....	....	....	....
	Pensacola	ECHL	3	1	0	1	0	....	....	....	....	....
2001-02	Pee Dee Pride	ECHL	40	4	9	13	6	....	....	....	....	....
	New Orleans	ECHL	16	4	5	9	4	1	0	0	0	0

VALEYEV, Igor (val-AY-ehv, EE-gohr) ST.L.

Left wing. Shoots left. 5'11", 203 lbs. Born, Snezhinsk, USSR, January 9, 1981.
(St. Louis' 3rd choice, 122nd overall, in 2001 Entry Draft).

			Regular Season					Playoffs				
Season	Club	Lea	GP	G	A	TP	PIM	GP	G	A	TP	PIM
1998-99	Lethbridge	WHL	8	2	3	5	13	....	....	....	....	....
	Saskatoon Blades	WHL	23	2	2	4	36	....	....	....	....	....
99-2000	Swift Current	WHL	36	7	5	12	78	4	0	1	1	22
2000-01	North Bay	OHL	62	17	61	78	175	4	0	1	1	22
	Muskegon Fury	UHL	....	....	....	....	....	3	0	1	1	2
2001-02	Worcester	AHL	29	3	6	9	72	....	....	....	....	....

Selected by **Lethbridge** (WHL) 64th overall in 1998 CHL Import Draft, July 2, 1998. Traded to **Saskatoon** (WHL) by **Lethbridge** (WHL) for future considerations, November 3, 1998. Signed as a free agent by **Lethbridge** (WHL), September 1, 1999. Selected by **North Bay** (OHL) 52nd overall in 2000 CHL Import Draft, June 1, 1999. • Missed majority of 2001-02 season recovering from head injury suffered in game vs. Springfield (AHL), December 14, 2001.

VALTONEN, Tomek (VAL-tuh-nehn, Toh-MEHK) DET.

Left wing. Shoots left. 6'1", 198 lbs. Born, Piotrkow Trybunalski, Poland, January 8, 1980.
(Detroit's 3rd choice, 56th overall, in 1998 Entry Draft).

			Regular Season					Playoffs				
Season	Club	Lea	GP	G	A	TP	PIM	GP	G	A	TP	PIM
1995-96	Ilves Tampere-C	Finn-Jr.	9	3	3	6	24	....	....	....	....	....
	Ilves Tampere-B	Finn-Jr.	11	7	7	14	28	....	....	....	....	....
	Ilves Jr.	Finn-Jr.	....	....	....	....	....	1	0	0	0	0
1996-97	Ilves Tampere-B	Finn-Jr.	26	10	9	19	82	3	0	1	1	6
	Ilves Jr.	Finn-Jr.	1	0	0	0	0	....	....	....	....	....
1997-98	JoKP Joensuu Jr.	Finn-Jr.	3	0	0	0	12	....	....	....	....	....
	JoKP Joensuu	Finland-2	6	1	2	3	39	....	....	....	....	....
	Ilves Jr.	Finn-Jr.	13	3	2	5	36	....	....	....	....	....
	Ilves Tampere	Finland	19	1	0	1	14	3	0	0	0	0
	Ilves Tampere-B	Finn-Jr.	....	....	....	....	....	7	0	2	2	16
1998-99	Plymouth Whalers	OHL	43	8	16	24	53	7	1	0	1	0
99-2000	Jokerit Helsinki	Finland	41	0	3	3	63	9	1	0	1	8
2000-01	Jokerit Helsinki	Finland	45	3	2	5	138	3	0	0	0	0
2001-02	Jokerit Helsinki	Finland	55	4	4	8	65	11	2	1	3	2

VAN HOOF, Jeremy (van-HOOF, JAIR-reh-mee) T.B.

Defense. Shoots left. 6'2", 208 lbs. Born, Lindsay, Ont., August 12, 1981.
(Tampa Bay's 9th choice, 222nd overall, in 2001 Entry Draft).

			Regular Season					Playoffs				
Season	Club	Lea	GP	G	A	TP	PIM	GP	G	A	TP	PIM
1997-98	Lindsay Muskies	OPJHL	50	2	8	10	40	....	....	....	....	....
1998-99	Ottawa 67's	OHL	54	0	13	13	46	5	1	0	1	2
99-2000	Ottawa 67's	OHL	66	4	14	18	71	11	1	0	1	12
2000-01	Ottawa 67's	OHL	65	1	14	15	49	20	3	4	7	27
2001-02	Pensacola	ECHL	72	2	14	16	86	3	0	0	0	9

• Re-entered NHL Entry Draft. Originally Pittsburgh's 3rd choice, 57th overall, in 1999 Entry Draft.

VanBUSKIRK, Ryan (van-BUHS-kuhrk, RIGH-uhn) WSH.

Defense. Shoots left. 6'1", 190 lbs. Born, Sault Ste. Marie, MI, January 12, 1980.
(Washington's 4th choice, 121st overall, in 2000 Entry Draft).

			Regular Season					Playoffs				
Season	Club	Lea	GP	G	A	TP	PIM	GP	G	A	TP	PIM
1995-96	Petrolia Jets	OJHL-B	48	6	16	22	124	....	....	....	....	....
1996-97	Petrolia Jets	OJHL-B	43	7	28	35	133	....	....	....	....	....
1997-98	Sarnia Sting	OHL	61	8	17	25	84	5	1	2	3	4
1998-99	Sarnia Sting	OHL	66	15	33	48	85	6	1	2	3	4
99-2000	Sarnia Sting	OHL	45	8	20	28	62	7	1	2	3	16
	Springfield	AHL	1	0	0	0	0	....	....	....	....	....
2000-01	Portland Pirates	AHL	18	0	0	0	16	....	....	....	....	....
	Richmond	ECHL	15	0	2	2	16	3	0	2	2	6
2001-02	Portland Pirates	AHL	42	2	3	5	62	....	....	....	....	....

• Re-entered NHL Entry Draft. Originally Phoenix's 4th choice, 100th overall, in 1998 Entry Draft.

• Missed majority of 2000-01 season recovering from shoulder injury suffered in game vs. St. John's (AHL), January 6, 2001.

VANDERMEER, Jim (VAN-duhr-meer, JIHM) PHI.

Defense. Shoots left. 6'1", 208 lbs. Born, Caroline, Alta., February 21, 1980.

			Regular Season					Playoffs				
Season	Club	Lea	GP	G	A	TP	PIM	GP	G	A	TP	PIM
1997-98	Red Deer	AMHL	26	4	8	12	51	....	....	....	....	....
	Red Deer Rebels	WHL	35	0	3	3	55	2	0	0	0	0
1998-99	Red Deer Rebels	WHL	70	5	23	28	258	9	0	1	1	24
99-2000	Red Deer Rebels	WHL	71	8	30	38	221	4	0	1	1	16
2000-01	Red Deer Rebels	WHL	65	28	37	65	180	22	3	13	16	43
2001-02	Philadelphia	AHL	74	1	13	14	88	5	0	2	2	14

WHL East First All-Star Team (2001) • Canadian Major Junior Humanitarian Player of the Year (2001)

Signed as a free agent by **Philadelphia**, December 21, 2000.

VANDERMEER, Peter (VAN-duhr-meer, PEE-tuhr) PHI.

Left wing. Shoots left. 6', 195 lbs. Born, Caroline, Alta., October 14, 1975.

			Regular Season					Playoffs				
Season	Club	Lea	GP	G	A	TP	PIM	GP	G	A	TP	PIM
1992-93	Red Deer	AMHL	34	26	30	56	172	....	....	....	....	....
	Red Deer Rebels	WHL	2	0	0	0	2	....	....	....	....	....
1993-94	Red Deer Rebels	WHL	54	4	9	13	170	....	....	....	....	....
1994-95	Red Deer Rebels	WHL	61	16	16	32	218	....	....	....	....	....
1995-96	Red Deer Rebels	WHL	63	21	40	61	207	....	....	....	....	....
1996-97	Columbus Chill	ECHL	30	6	11	17	195	7	2	1	3	26
1997-98	Columbus Chill	ECHL	20	4	7	11	78	....	....	....	....	....
	Richmond	ECHL	18	2	5	7	165	....	....	....	....	....
	Rochester	AHL	30	4	2	6	140	4	1	0	1	13
1998-99	Binghamton	UHL	62	15	21	36	*390	5	2	2	4	0
	Rochester	AHL	2	1	0	1	16	16	1	0	1	38
99-2000	Richmond	ECHL	58	31	25	56	*457	3	0	1	1	20
	Wilkes-Barre	AHL	4	0	0	0	7	....	....	....	....	....
	Providence	AHL	....	....	....	....	....	9	0	3	3	2
2000-01	Providence	AHL	62	19	18	37	240	4	0	0	0	16
2001-02	Philadelphia	AHL	61	5	1	6	313	5	0	0	0	8
	Trenton Titans	ECHL	2	0	1	1	2	....	....	....	....	....

Signed as a free agent by **Philadelphia**, July 6, 2001.

VAN OENE, Darren (van OH-uhn, DAIR-rehn) BOS.

Left wing. Shoots left. 6'4", 216 lbs. Born, Edmonton, Alta., January 18, 1978.
(Buffalo's 3rd choice, 33rd overall, in 1996 Entry Draft).

			Regular Season					Playoffs				
Season	Club	Lea	GP	G	A	TP	PIM	GP	G	A	TP	PIM
1993-94	Edmonton SSAC	AMHL	34	15	16	31	121	....	....	....	....	....
1994-95	Brandon	WHL	58	5	13	18	106	18	1	1	2	34
1995-96	Brandon	WHL	47	10	18	28	126	18	1	6	7	*78
1996-97	Brandon	WHL	56	21	27	48	139	6	2	3	5	19
1997-98	Brandon	WHL	51	23	24	47	161	18	6	8	14	51
1998-99	Rochester	AHL	73	11	20	31	143	12	2	4	6	8
99-2000	Rochester	AHL	80	20	18	38	153	21	1	3	4	24
2000-01	Rochester	AHL	64	10	12	22	147	4	1	0	1	4
2001-02	Rochester	AHL	52	8	6	14	73	2	0	0	0	4

Signed as a free agent by **Boston**, July 29, 2002.

VAS, Janos (VAHSH, YAH-nohsh) DAL.

Left wing. Shoots left. 6'1", 183 lbs. Born, Dunaferr, Hungary, January 29, 1984.
(Dallas' 2nd choice, 32nd overall, in 2002 Entry Draft).

			Regular Season					Playoffs				
Season	Club	Lea	GP	G	A	TP	PIM	GP	G	A	TP	PIM
99-2000	Dunaferr SE	Hungary	2	2	0	0	0	....	....	....	....	....
2000-01	Malmo IF Jr.	Swede-Jr.	26	6	4	10	16	....	....	....	....	....
2001-02	Malmo IF Jr.	Swede-Jr.	36	15	19	34	52	7	8	2	10	4

VAUCLAIR, Julien (voh-KLAIR, JEW-lee-ehn) OTT.

Defense. Shoots left. 6', 198 lbs. Born, Delemont, Switz., October 2, 1979.
(Ottawa's 4th choice, 74th overall, in 1998 Entry Draft).

			Regular Season					Playoffs				
Season	Club	Lea	GP	G	A	TP	PIM	GP	G	A	TP	PIM
1995-96	HC Ajoie	Swiss-3	20	4	10	14	....	....	....	....	....	....
1996-97	HC Ajoie	Swiss-2	40	0	6	6	24	9	0	2	2	8
1997-98	HC Lugano	Swiss	36	1	2	3	12	7	0	0	0	25
1998-99	HC Lugano	Swiss	38	0	3	3	8	....	....	....	....	....
99-2000	HC Lugano	Swiss	45	3	3	6	16	14	0	0	0	0
2000-01	HC Lugano	Swiss	42	3	4	7	57	18	0	1	1	4
2001-02	Grand Rapids	AHL	71	5	14	19	18	4	0	1	1	4
	Switzerland	Olympics	4	1	0	1	2	....	....	....	....	....

VEILLEUX, Stephane (VAY-oo, STEH-fan) MIN.

Right wing. Shoots left. 6'1", 187 lbs. Born, Beaureville, Que., November 16, 1981.
(Minnesota's 4th choice, 93rd overall, in 2001 Entry Draft).

			Regular Season					Playoffs				
Season	Club	Lea	GP	G	A	TP	PIM	GP	G	A	TP	PIM
1997-98	Beauce-Amiante	QAAA	21	20	17	37	....	....	....	....	....	....
	Levis-Lauzon	QAAA	14	3	5	8	....	1	0	0	0	0
1998-99	Victoriaville	QMJHL	65	6	13	19	35	6	1	3	4	2
99-2000	Victoriaville	QMJHL	22	1	4	5	17	....	....	....	....	....
	Val-d'Or Foreurs	QMJHL	50	14	28	42	100	....	....	....	....	....
2000-01	Val-d'Or Foreurs	QMJHL	68	48	67	115	90	21	15	18	33	42
2001-02	Houston Aeros	AHL	77	13	22	35	113	14	2	4	6	20

• Traded to **Victoriaville** (QMJHL) by **Val D'or** (QMJHL) for Jonathan Fauteux, November 11, 1999.

VELEBNY, Lubos (vehl-EHB-nee, LOO-bohsh) TOR.

Defense. Shoots left. 6'1", 189 lbs. Born, Zvolen, Czech., February 9, 1982.
(Toronto's 8th choice, 223rd overall, in 2000 Entry Draft).

			Regular Season					Playoffs				
Season	Club	Lea	GP	G	A	TP	PIM	GP	G	A	TP	PIM
1997-98	HKm Zvolen Jr.	Slovak-Jr.	45	23	19	42	99	....	....	....	....	....
1998-99	HKm Zvolen Jr.	Slovak-Jr.	37	12	17	29	91	....	....	....	....	....
99-2000	HKm Zvolen Jr.	Slovak-Jr.	41	6	8	14	11	....	....	....	....	....
	HKm Zvolen	Slovakia	7	0	0	0	0	....	....	....	....	....
2000-01	Waterloo	USHL	45	11	25	36	179	....	....	....	....	....
2001-02	London Knights	OHL	29	3	9	12	50	....	....	....	....	....
	Belleville Bulls	OHL	24	6	18	24	31	10	1	1	2	30

Traded to **Belleville** (OHL) by **London** (OHL) with Glen Ridler for Jan Chovan and Alex White, December 5, 2001.

VENALAINEN, Sami (veh-na-LIGH-nehn, SA-mee) PHX.

Right wing. Shoots right. 5'11", 183 lbs. Born, Kangasala, Finland, October 14, 1981.
(Phoenix's 7th choice, 249th overall, in 2000 Entry Draft).

			Regular Season					Playoffs				
Season	Club	Lea	GP	G	A	TP	PIM	GP	G	A	TP	PIM
1996-97	Tappara-C	Finn-Jr.	32	21	17	38	31	4	2	1	3	0
1997-98	Tappara-C	Finn-Jr.	2	2	0	2	6	6	4	5	9	6
	Tappara-B	Finn-Jr.	33	18	7	25	12	....	....	....	....	....
1998-99	Tappara-B	Finn-Jr.	31	27	17	44	45	....	....	....	....	....
	Tappara Jr.	Finn-Jr.	10	3	2	5	29	....	....	....	....	....
99-2000	Tappara Jr.	Finn-Jr.	37	8	9	17	18	....	....	....	....	....
2000-01	Tappara Jr.	Finn-Jr.	21	6	13	19	12	9	3	1	4	0
	Tappara Tampere	Finland	36	0	1	1	2	1	0	0	0	0
2001-02	Tappara Tampere	Finland	56	5	8	13	24	10	0	1	1	8

VERENIKIN, Sergei (veh-rih-NEE-kihn, SAIR-gay) OTT.

Right wing. Shoots left. 5'11", 187 lbs. Born, Yaroslavl, USSR, September 8, 1979.
(Ottawa's 9th choice, 223rd overall, in 1998 Entry Draft).

			Regular Season					Playoffs				
Season	Club	Lea	GP	G	A	TP	PIM	GP	G	A	TP	PIM
1997-98	Yaroslavl 2	Russia-2	44	11	4	15	100	....	....	....	....	....
	Yaroslavl	Russia	3	0	0	0	0	....	....	....	....	....
1998-99	Yaroslavl	Russia	37	2	5	7	16	8	0	0	0	18
99-2000	Yaroslavl	Russia	25	4	1	5	18	4	0	0	0	2
2000-01	Magnitogorsk	Russia	31	1	0	1	14	....	....	....	....	....
2001-02	Perm	Russia	47	6	4	10	26	....	....	....	....	....

VERMETTE, Antoine (vuhr-MEHT, AN-twuhn) OTT.

Center. Shoots left. 6', 184 lbs. Born, St-Agapit, Que., July 20, 1982.
(Ottawa's 3rd choice, 55th overall, in 2000 Entry Draft).

			Regular Season					Playoffs				
Season	Club	Lea	GP	G	A	TP	PIM	GP	G	A	TP	PIM
1997-98	Quebec Select	QAHA	19	11	20	31	36	....	....	....	....	....
	Levis-Lauzon	QAAA	8	1	1	2	4	1	0	0	0	0
1998-99	Quebec Remparts	QMJHL	57	9	17	26	32	13	0	0	0	2
99-2000	Victoriaville	QMJHL	71	30	41	71	87	6	0	1	1	6
2000-01	Victoriaville	QMJHL	71	57	62	119	102	9	4	6	10	14
2001-02	Victoriaville	QMJHL	4	0	2	2	6	22	10	16	26	10

• Missed majority of 2001-02 season recovering from neck injury suffered at Team Canada Jr. Selection Camp, June 3, 2001.

VERNARSKY, Kris (veh-NAHR-skee, KRIHS) BOS.

Center. Shoots left. 6'3", 201 lbs. Born, Detroit, MI, April 5, 1982.
(Toronto's 2nd choice, 51st overall, in 2000 Entry Draft).

			Regular Season					Playoffs				
Season	Club	Lea	GP	G	A	TP	PIM	GP	G	A	TP	PIM
1997-98	Team USA	USDP-18	69	11	18	29	97	....	....	....	....	....
1998-99	Plymouth Whalers	OHL	45	3	14	17	30	11	0	0	0	2
99-2000	Plymouth Whalers	OHL	64	16	22	38	63	19	3	6	9	24
2000-01	Plymouth Whalers	OHL	60	14	21	35	35	19	7	10	17	19
2001-02	Plymouth Whalers	OHL	59	19	36	55	98	6	1	2	3	15

Rights traded to **Boston** by **Toronto** for Richard Jackman, May 13, 2002.

VEROT, Darcy (vuhr-AWT, DAHR-see) CGY.

Left wing. Shoots left. 6', 190 lbs. Born, Radville, Sask., July 13, 1976.

			Regular Season					Playoffs				
Season	Club	Lea	GP	G	A	TP	PIM	GP	G	A	TP	PIM
1994-95	Weyburn	SJHL	57	8	18	26	240	16	5	2	7	50
1995-96	Weyburn	SJHL	64	15	30	45	191	3	1	0	1	20
1996-97	Weyburn	SJHL	61	26	51	77	218	13	3	8	11	24
1997-98	Lake Charles	WPHL	68	11	26	37	269	4	0	1	1	25
1998-99	Lake Charles	WPHL	68	17	23	40	236	9	2	4	6	53
99-2000	Wheeling Nailers	ECHL	44	7	12	19	240	....	....	....	....	....
	Wilkes-Barre	AHL	23	5	5	10	96	....	....	....	....	....
2000-01	Wilkes-Barre	AHL	78	10	15	25	347	21	2	3	5	40
2001-02	Wilkes-Barre	AHL	71	6	10	16	387	....	....	....	....	....

Signed as a free agent by **Wilkes-Barre** (AHL), February 25, 2000. Signed as a free agent by **Pittsburgh**, July 28, 2000. Signed as a free agent by **Calgary**, July 9, 2002.

VIHKO, Joonas (VIH-koh, YOO-nuhs) ANA.

Center. Shoots right. 5'9", 176 lbs. Born, Helsinki, Finland, April 6, 1981.
(Anaheim's 4th choice, 103rd overall, in 2002 Entry Draft).

			Regular Season					Playoffs				
Season	Club	Lea	GP	G	A	TP	PIM	GP	G	A	TP	PIM
1998-99	HIFK Jr.	Finn-Jr.	32	15	14	29	10	3	1	1	2	0
99-2000	HIFK Jr.	Finn-Jr.	38	12	15	27	36	3	0	0	0	0
2000-01	HIFK Jr.	Finn-Jr.	27	23	17	40	85	7	5	2	7	18
	HIFK Helsinki	Finland	3	0	0	0	2	....	....	....	....	....
2001-02	HIFK Helsinki	Finland	48	13	11	24	89	....	....	....	....	....
	HIFK Jr.	Finn-Jr.	....	....	....	....	....	2	0	2	2	4

VIITANEN, Mikko (vee-EE-tan-ehn, MEE-koh) COL.

Defense. Shoots left. 6'3", 220 lbs. Born, Rajamaki, Finland, February 18, 1982.
(Colorado's 6th choice, 149th overall, in 2001 Entry Draft).

			Regular Season					Playoffs				
Season	Club	Lea	GP	G	A	TP	PIM	GP	G	A	TP	PIM
1998-99	HPK-B	Finn-Jr.	36	3	6	9	40	....	....	....	....	....
	HPK Jr.	Finn-Jr.	1	0	0	0	0	....	....	....	....	....
99-2000	Chicago Freeze	NAJHL	53	4	6	10	126	....	....	....	....	....
2000-01	Ahmat Jr.	Finn-Jr.	9	3	4	7	41	....	....	....	....	....
	Ahmat Hyvinkaa	Finland-2	41	3	9	12	66	3	0	0	0	0
2001-02	Blues Espoo Jr.	Finn-Jr.	10	1	3	4	16	....	....	....	....	....
	Blues Espoo	Finland	3	0	0	0	6	....	....	....	....	....
	Jukurit Mikkeli	Finland-2	21	0	3	3	26	....	....	....	....	....

VIKINGSTAD, Tore (VIH-kihng-stahd, TOO-reh) ST.L.

Center. Shoots left. 6'4", 200 lbs. Born, Stavenger, Norway, October 8, 1975.
(St. Louis' 5th choice, 180th overall, in 1999 Entry Draft).

			Regular Season					Playoffs				
Season	Club	Lea	GP	G	A	TP	PIM	GP	G	A	TP	PIM
1994-95	Viking	Norway	28	5	3	8	8	....	....	....	....	....
1995-96	Viking	Norway	27	12	11	23	....	....	....	....	....	....
1996-97	Stjernen	Norway	42	23	35	58	20	....	....	....	....	....
1997-98	Stjernen	Norway	42	26	31	57	18	....	....	....	....	....
1998-99	Farjestad	Sweden	49	9	11	20	18	4	2	3	5	0
99-2000	Farjestad	Sweden	47	8	19	27	26	7	3	0	3	6
2000-01	Leksands IF	Sweden	41	10	15	25	24	....	....	....	....	....
2001-02	DEG Metro Stars	Germany	58	18	30	48	6	....	....	....	....	....

VLCEK, Ladislav (vuhl-CHEHK, LA-dih-dlav) DAL.

Right wing. Shoots left. 5'11", 184 lbs. Born, Kladno, Czech., September 26, 1981.
(Dallas' 8th choice, 192nd overall, in 2000 Entry Draft).

			Regular Season					Playoffs				
Season	Club	Lea	GP	G	A	TP	PIM	GP	G	A	TP	PIM
1998-99	Kladno Jr.	Czech-Jr.	46	13	27	40	....	....	....	....	....	....
	Kladno	Czech	4	0	1	1	0	....	....	....	....	....
99-2000	Kladno Jr.	Czech-Jr.	34	16	15	31	16	....	....	....	....	....
	HC CKD Slany	Czech-3	4	3	0	3	0	....	....	....	....	....
	Kralupy	Czech-3	1	0	1	1	0	5	2	2	4	4
	Kladno	Czech	21	3	2	5	4	....	....	....	....	....
2000-01	Kladno	Czech	45	6	10	16	22	....	....	....	....	....
2001-02	Kladno	Czech	20	3	1	4	14	....	....	....	....	....
	Trinec	Czech	28	3	6	9	18	6	1	0	1	2

VOLCHENKOV, Anton (vohl-chen-KAHF, an-TUHN) OTT.

Defense. Shoots left. 6', 209 lbs. Born, Moscow, USSR, February 25, 1982.
(Ottawa's 1st choice, 21st overall, in 2000 Entry Draft).

			Regular Season					Playoffs				
Season	Club	Lea	GP	G	A	TP	PIM	GP	G	A	TP	PIM
99-2000	H.C. CSKA 2	Russia-3	6	0	1	1	10	....	....	....	....	....
	H.C. CSKA	Russia-2	30	2	9	11	36	....	....	....	....	....
2000-01	Krylja Sovetov	Russia-2	34	3	4	7	56	....	....	....	....	....
2001-02	Krylja Sovetov 2	Russia-3	1	0	0	0	0	....	....	....	....	....
	Krylja Sovetov	Russia	47	4	16	20	50	3	0	0	0	29

VOLRAB, Daniel (VOHL-rab, DAN-yehl) DAL.

Center. Shoots left. 6', 182 lbs. Born, Decin, Czech., March 11, 1983.
(Dallas' 4th choice, 126th overall, in 2001 Entry Draft).

			Regular Season					Playoffs				
Season	Club	Lea	GP	G	A	TP	PIM	GP	G	A	TP	PIM
1994-95	HC Decin-14	Czech-Jr.	28	16	15	31	....	....	....	....	....	....
1995-96	HC Decin-14	Czech-Jr.	28	32	15	47	....	....	....	....	....	....
1996-97	Litvinov-14	Czech-Jr.	3	10	1	11	....	....	....	....	....	....
	Litvinov-18	Czech-Jr.	32	24	24	48	....	....	....	....	....	....
1997-98	Litvinov-18	Czech-Jr.	10	8	7	15	....	....	....	....	....	....
	Litvinov-Jr.	Czech-Jr.	35	4	3	7	....	....	....	....	....	....
1998-99	Litvinov Jr.	Czech-Jr.	7	1	1	2	....	....	....	....	....	....
	Sparta Praha Jr.	Czech-Jr.	37	23	20	43	....	....	....	....	....	....
99-2000	Sparta Praha Jr.	Czech-Jr.	39	33	25	58	18	7	4	6	10	2
2000-01	Sparta Praha Jr.	Czech-Jr.	27	9	10	19	30	2	1	0	1	0
	HC Sparta Praha	Czech	1	0	0	0	0	....	....	....	....	....
2001-02	Saskatoon Blades	WHL	72	13	28	41	61	7	1	1	2	6

VONDRKA, Michal (VOHND-rah-ka, MEE-khahl) **BUF.**

Left wing. Shoots right. 6', 178 lbs. Born, Ceske Budejovice, Czech., May 17, 1983.
(Buffalo's 5th choice, 155th overall, in 2001 Entry Draft).

			Regular Season					Playoffs				
Season	Club	Lea	GP	G	A	TP	PIM	GP	G	A	TP	PIM
1998-99	C. Budejovice-18	Czech-Jr.	50	28	15	43	....	....	....	....	....	....
99-2000	C. Budejovice Jr.	Czech-Jr.	31	12	7	19	16	....	....	....	....	....
2000-01	C. Budejovice Jr.	Czech-Jr.	37	9	14	23	37	....	....	....	....	....
	Ceske Budejovice	Czech	8	1	0	1	2	....	....	....	....	....
	Hradec Kralove	Czech-2	1	0	1	1	2	....	....	....	....	....
2001-02	C. Budejovice Jr.	Czech-Jr.	14	3	9	12	8	....	....	....	....	....
	IHC Pisek	Czech-2	2	1	0	1	0	....	....	....	....	....
	Ceske Budejovice	Czech	33	1	1	2	2	....	....	....	....	....

VOROBIEV, Pavel (voh-roh-BEE-ehf, PAH-vehl) **CHI.**

Right wing. Shoots left. 6', 183 lbs. Born, Karaganda, USSR, May 5, 1982.
(Chicago's 2nd choice, 11th overall, in 2000 Entry Draft).

			Regular Season					Playoffs				
Season	Club	Lea	GP	G	A	TP	PIM	GP	G	A	TP	PIM
1996-97	Molot-Perm 2	Russia-3	2	0	0	0	0	....	....	....	....	....
1997-98	Yaroslavl 2	Russia-2	16	2	0	2	6	....	....	....	....	....
1998-99	Yaroslavl 2	Russia-3	17	0	1	1	0	....	....	....	....	....
99-2000	Yaroslavl 2	Russia-3	40	19	15	34	20	....	....	....	....	....
	Yaroslavl	Russia	8	2	0	2	4	10	2	2	4	0
2000-01	Yaroslavl	Russia	36	8	8	16	28	10	4	1	5	8
2001-02	Yaroslavl	Russia	9	3	2	5	6	7	0	0	0	4

VOROS, Aaron (VOH-ruhs, AIR-uhn) **N.J.**

Center. Shoots left. 6'3", 178 lbs. Born, Vancouver, B.C., July 2, 1981.
(New Jersey's 10th choice, 229th overall, in 2001 Entry Draft).

			Regular Season					Playoffs				
Season	Club	Lea	GP	G	A	TP	PIM	GP	G	A	TP	PIM
1998-99	Nor-West Caps	PIJHL	STATISTICS NOT AVAILABLE									
99-2000	Victoria Salsa	BCHL	58	14	21	35	285	....	....	....	....	....
2000-01	Victoria Salsa	BCHL	57	34	34	68	....	30	16	15	31	....
2001-02	Alaska-Fairbanks	CCHA	37	18	12	30	*101	....	....	....	....	....

BCHL Coastal Conference Second All-Star Team (2001) • CCHA All-Rookie Team (2002)

VOSTRIKOV, Artem (VAWS-trih-kawv, ahr-TEHM) **CBJ**

Center. Shoots left. 6'1", 175 lbs. Born, Togliatti, Russia, March 23, 1983.
(Columbus' 8th choice, 187th overall, in 2001 Entry Draft).

			Regular Season					Playoffs				
Season	Club	Lea	GP	G	A	TP	PIM	GP	G	A	TP	PIM
99-2000	Lada Togliatti 2	Russia-3	STATISTICS NOT AVAILABLE									
2000-01	Lada Togliatti 2	Russia-3	STATISTICS NOT AVAILABLE									
2001-02	Lada Togliatti	Russia	12	1	1	2	0	....	....	....	....	....
	CSK VVS Samara	Russia-2	23	1	4	5	40	....	....	....	....	....

VYDARENY, Rene (vih-DAH-reh-nay, REH-nay) **VAN.**

Defense. Shoots left. 6'1", 198 lbs. Born, Bratislava, Czech., May 6, 1981.
(Vancouver's 3rd choice, 69th overall, in 1999 Entry Draft).

			Regular Season					Playoffs				
Season	Club	Lea	GP	G	A	TP	PIM	GP	G	A	TP	PIM
1997-98	S. Bratislava Jr.	Slovak-Jr.	50	5	14	19	26	....	....	....	....	....
1998-99	S. Bratislava Jr.	Slovak-Jr.	42	4	7	11	65	2	0	0	0	2
	HK Kabat Trnava	Slovak-2	20	1	6	7	6	....	....	....	....	....
99-2000	Rimouski Oceanic	QMJHL	51	7	23	30	41	14	2	2	4	20
2000-01	Kansas City	IHL	39	0	1	1	25	....	....	....	....	....
2001-02	Manitoba Moose	AHL	61	3	11	14	15	7	0	2	2	4
	Columbia Inferno	ECHL	10	2	1	3	9	....	....	....	....	....

• Missed majority of 2000-01 season due to dispute over ownership of playing rights between **Vancouver** and **HC Bratislava** (Slovakia), November 28, 2000.

WALBY, Steffon (WAHL-bee, STEH-fohn)

Right wing. Shoots right. 6'1", 198 lbs. Born, Madison, WI, November 22, 1972.

			Regular Season					Playoffs				
Season	Club	Lea	GP	G	A	TP	PIM	GP	G	A	TP	PIM
1990-91	Madison Capitols	USHL	45	23	16	39	89	....	....	....	....	....
1991-92	Kelowna Spartans	BCJHL	24	18	13	31	10	....	....	....	....	....
1992-93	Kelowna Spartans	BCJHL	59	53	68	121	76	....	....	....	....	....
1993-94	St. John's	AHL	63	15	22	37	79	2	0	0	0	2
1994-95	St. John's	AHL	70	23	23	46	30	5	1	1	2	4
1995-96	St. John's	AHL	57	23	31	54	61	4	2	2	4	17
1996-97	Hershey Bears	AHL	74	24	23	47	61	23	9	5	14	38
1997-98	Fort Wayne	IHL	77	28	26	54	53	4	1	1	2	6
1998-99	Rochester	AHL	48	15	13	28	52	....	....	....	....	....
	Kentucky	AHL	11	8	4	12	6	12	3	2	5	14
99-2000	Hershey Bears	AHL	49	19	13	32	50	14	3	9	12	11
2000-01	Hershey Bears	AHL	69	12	21	33	48	12	2	6	8	12
2001-02	Mississippi	ECHL	63	42	51	93	52	10	5	8	13	12

ECHL First All-Star Team (2002)

Signed as a free agent by **Toronto**, August 20, 1993. Signed as a free agent by **Colorado**, September, 1996. Signed as a free agent by **Buffalo**, August 31, 1998. Signed as a free agent by **Missisipi** (ECHL), September 18, 2001.

WALKER, Matt (WAHL-kuhr, MAT) **ST.L.**

Defense. Shoots right. 6'2", 222 lbs. Born, Beaverlodge, Alta., April 7, 1980.
(St. Louis' 3rd choice, 83rd overall, in 1998 Entry Draft).

			Regular Season					Playoffs				
Season	Club	Lea	GP	G	A	TP	PIM	GP	G	A	TP	PIM
1996-97	Grand Prairie	AAHA	68	22	62	74	186	....	....	....	....	....
1997-98	Portland	WHL	64	2	13	15	124	16	0	0	0	21
1998-99	Portland	WHL	64	1	10	11	151	4	0	1	1	6
99-2000	Portland	WHL	38	2	7	9	97	....	....	....	....	....
	Kootenay Ice	WHL	31	4	19	23	53	21	5	13	18	24
2000-01	Peoria Rivermen	ECHL	8	1	0	1	70	....	....	....	....	....
	Worcester	AHL	61	4	8	12	131	11	0	0	0	6
2001-02	Worcester	AHL	49	2	11	13	164	3	0	0	0	8

Traded to **Kootenay** (WHL) by **Portland** (WHL) for Jesse Ferguson, January 10, 2000.

WALLIN, Rickard (WAHL-in, RIH-kahrd) **MIN.**

Center. Shoots left. 6'2", 185 lbs. Born, Stockholm, Sweden, April 19, 1980.
(Phoenix's 8th choice, 160th overall, in 1998 Entry Draft).

			Regular Season					Playoffs				
Season	Club	Lea	GP	G	A	TP	PIM	GP	G	A	TP	PIM
1996-97	Vasteras IK Jr.	Swede-Jr.	26	3	3	6	....	....	....	....	....	....
1997-98	Farjestad Jr.	Swede-Jr.	29	20	30	50	32	2	1	1	2	2
1998-99	Farjestad Jr.	Swede-Jr.	21	11	15	26	30	....	....	....	....	....
	Farjestad	Sweden	5	0	0	0	0	....	....	....	....	....
99-2000	IF Troja-Ljungby	Swede-2	46	15	22	37	54	....	....	....	....	....
2000-01	Farjestad	Sweden	47	9	22	31	24	16	11	3	14	4
2001-02	Farjestad	Sweden	50	12	31	43	56	10	4	9	13	8

Rights traded to **Minnesota** by **Phoenix** for Joe Juneau, June 23, 2000.

WALLIN, Viktor (WAHL-in, VIHK-tohr) **ANA.**

Defense. Shoots left. 6'3", 200 lbs. Born, Jonkoping, Sweden, January 17, 1980.
(Anaheim's 3rd choice, 112th overall, in 1998 Entry Draft).

			Regular Season					Playoffs				
Season	Club	Lea	GP	G	A	TP	PIM	GP	G	A	TP	PIM
1996-97	HV 71 Jr.	Swede-Jr.	16	1	2	3	....	....	....	....	....	....
1997-98	HV 71 Jr.	Swede-Jr.	28	9	15	24	42	....	....	....	....	....
1998-99	HV 71 Jonkoping	Sweden	23	0	0	0	4	....	....	....	....	....
99-2000	HV 71 Jr.	Swede-Jr.	6	3	1	4	2	....	....	....	....	....
	HV 71 Jonkoping	Sweden	43	2	2	4	16	6	1	1	2	4
2000-01	HV 71 Jonkoping	Sweden	5	0	1	1	2	....	....	....	....	....
2001-02	Timra IK	Sweden	35	0	3	3	18	....	....	....	....	....

WALSH, Brendan (WAHLSH, BREHN-duhn)

Right wing. Shoots right. 5'9", 181 lbs. Born, Dorchester, MA, October 22, 1974.

			Regular Season					Playoffs				
Season	Club	Lea	GP	G	A	TP	PIM	GP	G	A	TP	PIM
1995-96	Boston University	H-East	38	8	16	24	90	....	....	....	....	....
1996-97	Boston University	H-East	27	5	8	13	83	....	....	....	....	....
1997-98	U. of Maine	H-East	DID NOT PLAY – TRANSFERRED COLLEGES									
1998-99	U. of Maine	H-East	30	7	13	20	58	....	....	....	....	....
99-2000	U. of Maine	H-East	39	9	21	30	*106	....	....	....	....	....
2000-01	Jackson Bandits	ECHL	25	3	6	9	179	....	....	....	....	....
	Cleveland	IHL	10	1	1	2	45	4	0	0	0	11
2001-02	Wheeling Nailers	ECHL	29	7	15	22	168	....	....	....	....	....
	Wilkes-Barre	AHL	29	2	2	4	184	....	....	....	....	....

Signed as a free agent by **Minnesota**, May 18, 2000.

WALSH, Mike (WAWLSH, MIGHK) **NYR**

Left wing. Shoots left. 6'2", 194 lbs. Born, Royal Oak, MI, March 4, 1983.
(NY Rangers' 4th choice, 143rd overall, in 2002 Entry Draft).

			Regular Season					Playoffs				
Season	Club	Lea	GP	G	A	TP	PIM	GP	G	A	TP	PIM
2000-01	Det. Compuware	NAJHL	50	10	12	22	59	3	1	0	1	4
2001-02	Det. Compuware	NAJHL	53	25	25	49	69	6	4	2	6	8

• Signed Letter of Intent to attend **Notre Dame** (CCHA), January 2, 2002.

WALSH, Tom (WAWLSH, TAWM) **S.J.**

Defense. Shoots left. 6', 190 lbs. Born, Arlington, MA, April 22, 1983.
(San Jose's 5th choice, 163rd overall, in 2002 Entry Draft).

			Regular Season					Playoffs				
Season	Club	Lea	GP	G	A	TP	PIM	GP	G	A	TP	PIM
2001-02	Deerfield	Hi-School	21	3	18	21	8	....	....	....	....	....

• Signed Letter of Intent to attend **Harvard** (ECAC), December 8, 2000.

WANVIG, Kyle (WEHN-vihg, KIGHL) **MIN.**

Right wing. Shoots right. 6'2", 219 lbs. Born, Calgary, Alta., January 29, 1981.
(Minnesota's 2nd choice, 36th overall, in 2001 Entry Draft).

			Regular Season					Playoffs				
Season	Club	Lea	GP	G	A	TP	PIM	GP	G	A	TP	PIM
1996-97	Calgary Blazers	AMHL	26	31	48	79	85	....	....	....	....	....
1997-98	Edmonton Ice	WHL	62	17	12	29	69	....	....	....	....	....
1998-99	Kootenay Ice	WHL	71	12	20	32	119	7	1	3	4	18
99-2000	Kootenay Ice	WHL	6	2	2	4	12	....	....	....	....	....
	Red Deer Rebels	WHL	58	21	18	39	123	4	1	0	1	4
2000-01	Red Deer Rebels	WHL	69	55	46	101	202	22	10	12	22	47
2001-02	Houston Aeros	AHL	34	6	7	13	43	9	0	1	1	23

• Re-entered NHL Entry Draft. Originally Boston's 3rd choice, 89th overall, in 1999 Entry Draft.

WHL East Second All-Star Team (2001) • Memorial Cup All-Star Team (2001) • Won Stafford Smythe Memorial Trophy (Memorial Cup Tournament MVP) (2001)

Traded to **Seattle** (WHL) by **Kootenay** (WHL) for future considerations, October 21, 1999. Traded to **Red Deer** (WHL) by **Kootenay** (WHL) for future considerations, October 21, 1999.
• Missed majority of 2001-02 season recovering from ankle injury suffered in game vs. Grand Rapids (AHL), December 30, 2001.

WARREN, Morgan (WAWR-ihn, MOHR-gan) **TOR.**

Right wing. Shoots right. 6'2", 193 lbs. Born, Summerside, P.E.I., March 6, 1980.
(Toronto's 5th choice, 126th overall, in 1998 Entry Draft).

			Regular Season					Playoffs				
Season	Club	Lea	GP	G	A	TP	PIM	GP	G	A	TP	PIM
1996-97	Quinte Hawks	MTJHL	49	32	38	70	65	....	....	....	....	....
1997-98	Moncton Wildcats	QMJHL	58	11	10	21	80	10	2	2	4	2
1998-99	Moncton Wildcats	QMJHL	48	20	16	36	68	1	0	0	0	2
99-2000	Moncton Wildcats	QMJHL	65	29	36	65	53	16	7	5	12	4
2000-01	St. John's	AHL	57	2	10	12	16	4	0	0	0	0
2001-02	St. John's	AHL	80	11	14	25	46	11	0	0	0	8

WATSON, Dan (WAWT-suhn, DAN) **CBJ**

Defense. Shoots right. 6'2", 220 lbs. Born, Glencoe, Ont., October 5, 1979.

			Regular Season					Playoffs				
Season	Club	Lea	GP	G	A	TP	PIM	GP	G	A	TP	PIM
1995-96	Strathroy	OJHL-B	46	6	13	19	12	....	....	....	....	....
1996-97	Strathroy	OJHL-B	49	5	25	30	33	....	....	....	....	....
	Sarnia Sting	OHL	10	0	2	2	7	....	....	....	....	....
1997-98	Sarnia Sting	OHL	66	6	15	21	19	5	0	1	1	4
1998-99	Sarnia Sting	OHL	68	2	18	20	27	6	0	0	0	4
99-2000	Sarnia Sting	OHL	68	1	15	16	40	7	0	0	0	4
2000-01	Elmira Jackals	UHL	1	0	0	0	0	....	....	....	....	....
	Syracuse Crunch	AHL	59	3	4	7	12	4	0	0	0	6
2001-02	Dayton Bombers	ECHL	4	0	0	0	6	....	....	....	....	....
	Syracuse Crunch	AHL	53	1	3	4	25	....	....	....	....	....

Signed as a free agent by **Columbus**, May 29, 2000.

WATSON, Greg (WAWT-suhn, GREHG) FLA.

Center. Shoots left. 6'2", 198 lbs. Born, Eastend, Sask., March 2, 1983.
(Florida's 3rd choice, 34th overall, in 2001 Entry Draft).

			Regular Season					Playoffs				
Season	Club	Lea	GP	G	A	TP	PIM	GP	G	A	TP	PIM
1998-99	Cgy. Buffaloes	AMHL	71	23	23	46	120	....	....	....	....	....
	Prince Albert	WHL	2	0	0	0	5	....	....	....	....	....
99-2000	Prince Albert	WHL	67	10	5	15	63	6	0	2	2	2
2000-01	Prince Albert	WHL	71	22	28	50	72	....	....	....	....	....
2001-02	Prince Albert	WHL	51	22	30	52	88	....	....	....	....	....

WAUGH, Geoff (WAW, JEHF) DAL.

Defense. Shoots right. 6'3", 210 lbs. Born, Winnipeg, Man., August 25, 1983.
(Dallas' 6th choice, 78th overall, in 2002 Entry Draft).

			Regular Season					Playoffs				
Season	Club	Lea	GP	G	A	TP	PIM	GP	G	A	TP	PIM
2000-01	Kindersley	SJHL	57	2	5	7	74	....	....	....	....	....
2001-02	Kindersley	SJHL	59	4	21	25	125	18	0	8	8	59

SJHL West First All-Star Team (2002)

• Signed Letter of Intent to attend **Northeastern** (H-East), May 13, 2002.

WEINHANDL, Mattias (vayn-hanh-duhl, mah-TEE-uhs) NYI

Right wing. Shoots right. 6', 183 lbs. Born, Ljungby, Sweden, June 1, 1980.
(NY Islanders' 5th choice, 78th overall, in 1999 Entry Draft).

			Regular Season					Playoffs				
Season	Club	Lea	GP	G	A	TP	PIM	GP	G	A	TP	PIM
1995-96	Troja-Ljungby Jr.	Swede-Jr.	28	38	40	78	....	....	....	....	....	....
1996-97	Troja-Ljungby Jr.	Swede-Jr.	48	61	69	130	46	....	....	....	....	....
1997-98	IF Troja-Ljungby	Swede-2	28	3	2	5	2	5	0	0	0	2
1998-99	IF Troja-Ljungby	Swede-2	38	20	20	40	30	5	4	3	7	4
99-2000	MoDo Jr.	Swede-Jr.	1	2	2	4	2	....	....	....	....	....
	MoDo	Sweden	32	15	9	24	6	13	5	3	8	8
2000-01	MoDo	Sweden	48	16	16	32	14	6	1	3	4	6
2001-02	MoDo	Sweden	50	18	16	34	10	14	4	*11	*15	4

WELCH, Dan (WEHLCH, DAN) L.A.

Right wing. Shoots right. 5'10", 199 lbs. Born, Lansing, MI, February 23, 1981.
(Los Angeles' 9th choice, 245th overall, in 2000 Entry Draft).

			Regular Season					Playoffs				
Season	Club	Lea	GP	G	A	TP	PIM	GP	G	A	TP	PIM
1996/99	Hastings Huskies	Hi-School	90	76	123	199	....	....	....	....	....	....
99-2000	U. of Minnesota	WCHA	36	6	8	14	31	....	....	....	....	....
2000-01	Omaha Lancers	USHL	52	30	27	57	103	12	9	13	22	20
2001-02	U. of Minnesota	WCHA	19	4	7	11	12	....	....	....	....	....

• Statistics for **Hastings** (Hi-School) are career totals for 1996-1999 seasons. • Ruled academically ineligible to play 2000-01 WCHA season by University of Minnesota (WCHA).

WELCH, Noah (WEHLCH, NOH-ah) PIT.

Defense. Shoots left. 6'3", 212 lbs. Born, Brighton, MA, August 26, 1982.
(Pittsburgh's 2nd choice, 54th overall, in 2001 Entry Draft).

			Regular Season					Playoffs				
Season	Club	Lea	GP	G	A	TP	PIM	GP	G	A	TP	PIM
99-2000	St. Sebastian's	Hi-School	26	4	11	15	35	....	....	....	....	....
	Eastern-Mass	MBAHL	4	0	3	3	6	....	....	....	....	....
2000-01	St. Sebastian's	Hi-School	30	11	20	31	37	....	....	....	....	....
2001-02	Harvard Crimson	ECAC	27	5	6	11	56	....	....	....	....	....

ECAC All-Rookie Team (2002)

WELLAR, Patrick (WEHL-uhr, PAT-rihk) WSH.

Defense. Shoots left. 6'3", 210 lbs. Born, Carrot River, Sask., April 12, 1983.
(Washington's 5th choice, 77th overall, in 2002 Entry Draft).

			Regular Season					Playoffs				
Season	Club	Lea	GP	G	A	TP	PIM	GP	G	A	TP	PIM
99-2000	Sask. Contacts	SMHL	44	5	15	20	120	....	....	....	....	....
	Portland	WHL	1	0	0	0	0	....	....	....	....	....
2000-01	Portland	WHL	57	2	7	9	65	10	0	1	1	13
2001-02	Portland	WHL	61	3	10	13	125	7	0	2	2	4

WELLER, Craig (WEH-luhr) NYR

Defense. Shoots right. 6'3", 195 lbs. Born, Calgary, Alta., January 17, 1981.
(St. Louis' 6th choice, 167th overall, in 2000 Entry Draft).

			Regular Season					Playoffs				
Season	Club	Lea	GP	G	A	TP	PIM	GP	G	A	TP	PIM
1997-98	Calgary Flames	AMHL	33	2	10	12	65	3	0	1	1	2
1998-99	Calgary Canucks	AJHL	49	4	14	18	80	13	0	1	1	10
99-2000	Calgary Canucks	AJHL	53	3	14	17	100	4	0	0	0	4
2000-01	U. Minn-Duluth	WCHA	6	0	1	1	0	....	....	....	....	....
	Kootenay Ice	WHL	30	1	5	6	40	11	0	2	2	26
2001-02	Kootenay Ice	WHL	69	5	13	18	127	22	3	7	10	27

WHL West Second All-Star Team (2002)

• Left **Minnesota-Duluth** (WCHA) and signed as a free agent by **Kootenay** (WHL), January 7, 2001. Signed as a free agent by **NY Rangers**, July 11, 2002.

WELLWOOD, Kyle (WEHL-wud, KIGHL) TOR.

Center. Shoots right. 5'9", 190 lbs. Born, Windsor, Ont., May 16, 1983.
(Toronto's 6th choice, 134th overall, in 2001 Entry Draft).

			Regular Season					Playoffs				
Season	Club	Lea	GP	G	A	TP	PIM	GP	G	A	TP	PIM
1998-99	Tecumseh	OJHL-B	51	22	41	63	12	....	....	....	....	....
99-2000	Belleville Bulls	OHL	65	14	37	51	14	16	3	7	10	6
2000-01	Belleville Bulls	OHL	68	35	*83	*118	24	10	3	16	19	4
2001-02	Belleville Bulls	OHL	28	16	24	40	4	....	....	....	....	....
	Windsor	OHL	26	14	21	35	0	16	12	12	24	0

OHL First All-Star Team (2001)

Traded to **Windsor** (OHL) by **Belleville** (OHL) with future considerations for Jason Spezza, January 10, 2002.

WENDELL, Erik (WEHN-dehl, AIR-ihk) WSH.

Center. Shoots left. 6'1", 197 lbs. Born, Minneapolis, MN, August 23, 1979.
(Washington's 6th choice, 125th overall, in 1998 Entry Draft).

			Regular Season					Playoffs				
Season	Club	Lea	GP	G	A	TP	PIM	GP	G	A	TP	PIM
1997-98	Maple Grove	Hi-School	24	24	23	47	38	....	....	....	....	....
	Twin Cities	USHL	17	7	2	9	64	....	....	....	....	....
1998-99	U. of Minnesota	WCHA	41	7	7	14	46	....	....	....	....	....
99-2000	U. of Minnesota	WCHA	32	4	2	6	26	....	....	....	....	....
2000-01	U. of Minnesota	WCHA	33	5	2	7	38	....	....	....	....	....
2001-02	U. of Minnesota	WCHA	44	8	9	17	42	....	....	....	....	....

WENNERBERG, Mattias (VEH-nuhr-buhrg, MA-tee-uhs) CHI.

Center. Shoots left. 5'11", 191 lbs. Born, Uma, Sweden, August 6, 1981.
(Chicago's 6th choice, 194th overall, in 1999 Entry Draft).

			Regular Season					Playoffs				
Season	Club	Lea	GP	G	A	TP	PIM	GP	G	A	TP	PIM
1996-97	Vilhelmina HC	Swede-4	20	7	12	19	14	....	....	....	....	....
1997-98	MoDo Jr.	Swede-Jr.	30	10	17	27	....	....	....	....	....	....
1998-99	MoDo Jr.	Swede-Jr.	43	13	12	25	....	....	....	....	....	....
99-2000	MoDo Jr.	Swede-Jr.	32	14	6	20	102	....	....	....	....	....
2000-01	Bodens IK	Swede-2	34	9	4	13	36	....	....	....	....	....
2001-02	Bjorkloven	Swede-2	21	8	10	18	43	....	....	....	....	....
	MoDo	Sweden	24	2	2	4	16	13	4	4	8	*39

WESTRUM, Erik (WEHST-ruhm, AIR-ihk) PHX.

Center. Shoots left. 6', 204 lbs. Born, Minneapolis, MN, July 26, 1979.
(Phoenix's 9th choice, 187th overall, in 1998 Entry Draft).

			Regular Season					Playoffs				
Season	Club	Lea	GP	G	A	TP	PIM	GP	G	A	TP	PIM
1995/97	Apple Valley	Hi-School	78	56	84	140	....	....	....	....	....	....
1997-98	U. of Minnesota	WCHA	39	6	12	18	43	....	....	....	....	....
1998-99	U. of Minnesota	WCHA	41	10	26	36	81	....	....	....	....	....
99-2000	U. of Minnesota	WCHA	39	27	26	53	99	....	....	....	....	....
2000-01	U. of Minnesota	WCHA	42	26	35	61	84	....	....	....	....	....
2001-02	Springfield	AHL	73	13	29	42	116	....	....	....	....	....

• Statistics for **Apple Valley** (Hi-School) are career totals for 1995-1997 seasons. • WCHA Second All-Star Team (2001)

WHITNEY, Ryan (WIHT-nee, RIGH-uhn) PIT.

Defense. Shoots left. 6'3", 200 lbs. Born, Boston, MA, February 19, 1983.
(Pittsburgh's 1st choice, 5th overall, in 2002 Entry Draft).

			Regular Season					Playoffs				
Season	Club	Lea	GP	G	A	TP	PIM	GP	G	A	TP	PIM
99-2000	Thayer Academy	Hi-School	22	5	33	38	....	....	....	....	....	....
2000-01	Team USA	USDP-17	60	9	31	40	86	....	....	....	....	....
2001-02	Boston University	H-East	35	4	17	21	46	....	....	....	....	....

Hockey East All-Rookie Team (2002)

WICHSER, Adrian (WIH-shuhr, A-dree-uhn) FLA.

Center. Shoots left. 6', 180 lbs. Born, Winterthur, Switz., March 18, 1980.
(Florida's 9th choice, 231st overall, in 1998 Entry Draft).

			Regular Season					Playoffs				
Season	Club	Lea	GP	G	A	TP	PIM	GP	G	A	TP	PIM
1997-98	EHC Kloten	Swiss	35	6	5	11	31	7	0	1	1	8
1998-99	EHC Kloten	Swiss	40	11	14	25	14	9	7	0	7	8
	EHC Kloten Jr.	Swiss-Jr.	....	....	....	....	....	1	1	2	3	2
99-2000	EHC Kloten	Swiss	33	8	15	23	12	6	2	1	3	0
2000-01	EHC Kloten	Swiss	31	9	9	18	12	9	2	3	5	0
2001-02	Kloten Flyers	Swiss	41	18	27	45	16	11	4	4	8	2

WIDING, Daniel (VEE-dihng, DAN-yehl) NSH.

Right wing. Shoots right. 6'1", 197 lbs. Born, Gavle, Sweden, April 13, 1982.
(Nashville's 2nd choice, 36th overall, in 2000 Entry Draft).

			Regular Season					Playoffs				
Season	Club	Lea	GP	G	A	TP	PIM	GP	G	A	TP	PIM
99-2000	Leksands IF-18	Swede-Jr.	6	2	1	3	20	....	....	....	....	....
	Leksands IF Jr.	Swede-Jr.	34	15	12	27	65	2	1	0	1	4
	Leksands IF	Sweden	3	0	0	0	2	....	....	....	....	....
2000-01	Leksands IF Jr.	Swede-Jr.	6	2	3	5	31	....	....	....	....	....
	Leksands IF	Sweden	40	6	5	11	18	....	....	....	....	....
2001-02	Leksands IF Jr.	Swede-Jr.	3	2	6	8	2	....	....	....	....	....
	Leksands IF	Swede-2	55	12	12	24	92	....	....	....	....	....

WIKSTROM, John (WIHK-strohm, JAWN) DET.

Defense. Shoots left. 6'3", 200 lbs. Born, Lulea, Sweden, January 30, 1979.
(Detroit's 4th choice, 129th overall, in 1997 Entry Draft).

			Regular Season					Playoffs				
Season	Club	Lea	GP	G	A	TP	PIM	GP	G	A	TP	PIM
1995-96	Lulea HF	Sweden	9	0	0	0	2	....	....	....	....	....
1996-97	Lulea HF	Sweden	9	0	0	0	0	3	0	0	0	0
1997-98	Lulea HF	Sweden	1	0	0	0	0	....	....	....	....	....
	Lulea HF	EuroHL	1	0	0	0	0	....	....	....	....	....
	Pitea HC	Swede-2	4	0	0	0	0	....	....	....	....	....
1998-99	Morrums GoIS IK	Swede-2	23	0	1	1	28	....	....	....	....	....
99-2000	Louisiana	ECHL	10	0	0	0	4	....	....	....	....	....
	Wheeling Nailers	ECHL	48	4	4	8	23	....	....	....	....	....
2000-01	Cincinnati	AHL	43	3	1	4	35	2	0	0	0	2
2001-02	Cincinnati	AHL	23	1	3	4	15	....	....	....	....	....
	Toledo Storm	ECHL	17	1	1	2	20	....	....	....	....	....

WILDE, Martin (WIGHLD, MAHR-tihn) BOS.

Defense. Shoots left. 6'4", 215 lbs. Born, Linkoping, Sweden, June 28, 215.

			Regular Season					Playoffs				
Season	Club	Lea	GP	G	A	TP	PIM	GP	G	A	TP	PIM
1994-95	Richland High	Hi-School		STATISTICS NOT AVAILABLE								
1995-96	Linkopings HC	Swede-2	12	0	0	0	0	2	0	0	0	0
1996-97	Linkopings HC	Swede-2	24	0	1	1	0	14	0	0	0	18
1997-98	U. of Vermont	ECAC	34	1	16	17	34	....	....	....	....	....
1998-99	U. of Vermont	ECAC	33	1	13	14	22	....	....	....	....	....
99-2000	U. of Vermont	ECAC	11	1	3	4	12	....	....	....	....	....
2000-01	U. of Vermont	ECAC	33	3	13	16	32	....	....	....	....	....
2001-02	Providence	AHL	71	2	9	11	46	2	0	0	0	0

Signed as a free agent by **Providence** (AHL), September 17, 2001. Signed as a free agent by **Boston**, July 18, 2002.

WILFORD, Marty (WIHL-fohrd, MAHR-tee)

Defense. Shoots left. 6'1", 216 lbs. Born, Cobourg, Ont., April 17, 1977.
(Chicago's 7th choice, 149th overall, in 1995 Entry Draft).

			Regular Season					Playoffs				
Season	Club	Lea	GP	G	A	TP	PIM	GP	G	A	TP	PIM
1993-94	Peterborough	OPJHL	40	3	19	22	*107	....	....	....	....	....
1994-95	Oshawa Generals	OHL	63	1	6	7	95	7	1	1	2	4
1995-96	Oshawa Generals	OHL	65	3	24	27	107	5	0	1	1	4
1996-97	Oshawa Generals	OHL	62	19	43	62	126	16	2	18	20	28
1997-98	Columbus Chill	ECHL	46	8	27	35	123	....	....	....	....	....
	Indianapolis Ice	IHL	26	0	4	4	16	....	....	....	....	....
1998-99	Indianapolis Ice	IHL	80	3	13	16	116	7	0	1	1	16
99-2000	Cleveland	IHL	7	0	3	3	24	....	....	....	....	....
	Houston Aeros	IHL	45	0	9	9	30	11	2	2	4	18
2000-01	Norfolk Admirals	AHL	80	7	41	48	102	9	1	5	6	8
2001-02	Milwaukee	AHL	8	1	3	4	12	....	....	....	....	....
	St. John's	AHL	60	4	21	25	70	....	....	....	....	....
	Hartford	AHL	9	0	2	2	2	10	3	3	6	4

OHL Second All-Star Team (1997)

Traded to **Toronto** by **Chicago** for Shawn Thornton, September 30, 2001. Traded to **Nashville** by **Toronto** with D.J. Smith for Marc Moro, March 1, 2002.

WILLIS, Tyler (WHIL-lihs, TIGH-luhr)

Right wing. Shoots right. 5'9", 171 lbs. Born, Princeton, B.C., April 8, 1977.
(Vancouver's 8th choice, 196th overall, in 1995 Entry Draft).

			Regular Season					Playoffs				
Season	Club	Lea	GP	G	A	TP	PIM	GP	G	A	TP	PIM
1992-93	Merritt	BCJHL	54	11	22	33	129	....	....	....	....	....
1993-94	Swift Current	WHL	71	19	26	45	263	....	....	....	....	....
1994-95	Swift Current	WHL	71	21	29	50	284	6	0	0	0	20
1995-96	Swift Current	WHL	40	9	38	47	196	....	....	....	....	....
	Seattle	WHL	15	1	3	4	71	5	1	5	6	13
1996-97	Seattle	WHL	72	12	40	52	302	15	1	7	8	68
1997-98	Worcester	AHL	24	2	1	3	140	....	....	....	....	....
	Baton Rouge	ECHL	21	4	10	14	112	....	....	....	....	....
1998-99	Worcester	AHL	55	8	10	18	227	....	....	....	....	....
99-2000	Worcester	AHL	32	3	10	13	98	9	1	2	3	8
	Peoria Rivermen	ECHL	19	5	6	11	89	....	....	....	....	....
2000-01	Peoria Rivermen	ECHL	58	14	18	32	251	14	1	2	3	45
	Worcester	AHL	16	2	1	3	50	....	....	....	....	....
2001-02	Augusta Lynx	ECHL	66	9	23	32	*320	....	....	....	....	....

Signed as a free agent by **St. Louis**, October 3, 1997.

WINCHESTER, Brad (WIHN-chehst-uhr, BRAD) **EDM.**

Left wing. Shoots left. 6'5", 208 lbs. Born, Madison, WI, March 1, 1981.
(Edmonton's 2nd choice, 35th overall, in 2000 Entry Draft).

			Regular Season					Playoffs				
Season	Club	Lea	GP	G	A	TP	PIM	GP	G	A	TP	PIM
1997-98	Team USA	USDP-18	74	22	23	45	162	....	....	....	....	....
1998-99	Team USA	USDP-18	65	21	23	44	103	....	....	....	....	....
99-2000	U. of Wisconsin	WCHA	33	9	9	18	48	....	....	....	....	....
2000-01	U. of Wisconsin	WCHA	41	7	9	16	71	....	....	....	....	....
2001-02	U. of Wisconsin	WCHA	38	14	20	34	38	....	....	....	....	....

WISEMAN, Chad (WIGHZ-man, CHAD) **S.J.**

Left wing. Shoots left. 6', 190 lbs. Born, Burlington, Ont., March 25, 1981.
(San Jose's 8th choice, 246th overall, in 2000 Entry Draft).

			Regular Season					Playoffs				
Season	Club	Lea	GP	G	A	TP	PIM	GP	G	A	TP	PIM
1997-98	Burlington	OPJHL	50	28	36	64	31	....	....	....	....	....
1998-99	Mississauga	OHL	64	11	25	36	29	....	....	....	....	....
99-2000	Mississauga	OHL	68	23	45	68	53	....	....	....	....	....
2000-01	Mississauga	OHL	30	15	29	44	22	....	....	....	....	....
	Plymouth Whalers	OHL	32	11	16	27	12	19	12	8	20	22
2001-02	Cleveland Barons	AHL	76	21	29	50	61	....	....	....	....	....

Traded to **Plymouth** (OHL) by **Mississauga** (OHL) for Nathan O'Nabigon and Plymouth's 3rd round choice (later traded to North Bay/Saginaw - Saginaw selected Kevin Porter) in 2002 OHL Priority Draft, December 28, 2000.

WISNIEWSKI, James (wihs-NEHV-skee, JAYMS) **CHI.**

Defense. Shoots right. 5'11", 197 lbs. Born, Canton, MI, February 21, 1984.
(Chicago's 5th choice, 156th overall, in 2002 Entry Draft).

			Regular Season					Playoffs				
Season	Club	Lea	GP	G	A	TP	PIM	GP	G	A	TP	PIM
99-2000	Det. Compuware	NAJHL	50	5	11	16	67	5	0	3	3	4
2000-01	Plymouth Whalers	OHL	53	6	23	29	72	19	3	10	13	34
2001-02	Plymouth Whalers	OHL	62	11	25	36	100	6	1	2	3	6

WOODFORD, Mike (WUD-fohrd, MIGHK) **FLA.**

Right wing. Shoots right. 5'11", 183 lbs. Born, Boston, MA, October 4, 1981.
(Florida's 6th choice, 117th overall, in 2001 Entry Draft).

			Regular Season					Playoffs				
Season	Club	Lea	GP	G	A	TP	PIM	GP	G	A	TP	PIM
99-2000	Cushing Academy	Hi-School	31	35	35	70	60	....	....	....	....	....
2000-01	Cushing Academy	Hi-School	36	34	39	73	68	....	....	....	....	....
2001-02	U. of Michigan	CCHA	43	8	11	19	46	....	....	....	....	....

WOYWITKA, Jeff (WOI-wiht-ka, JEHF) **PHI.**

Defense. Shoots left. 6'2", 209 lbs. Born, Vermilion, Alta., September 1, 1983.
(Philadelphia's 1st choice, 27th overall, in 2001 Entry Draft).

			Regular Season					Playoffs				
Season	Club	Lea	GP	G	A	TP	PIM	GP	G	A	TP	PIM
1998-99	Wainwright	AAHA	26	7	15	22	60	....	....	....	....	....
99-2000	Red Deer Rebels	WHL	67	4	12	16	40	4	0	3	3	2
2000-01	Red Deer Rebels	WHL	72	7	28	35	113	22	2	8	10	25
2001-02	Red Deer Rebels	WHL	72	14	23	37	109	23	2	10	12	22

WHL East Second All-Star Team (2002)

YAKUBOV, Mikhail (yuh-KOO-bahf, mih-KIGH-eel) **CHI.**

Center. Shoots left. 6'3", 208 lbs. Born, Barnaul, USSR, February 16, 1982.
(Chicago's 1st choice, 10th overall, in 2000 Entry Draft).

			Regular Season					Playoffs				
Season	Club	Lea	GP	G	A	TP	PIM	GP	G	A	TP	PIM
1997-98	Lada Togliatti 2	Russia-3	7	0	0	0	0	....	....	....	....	....
1998-99	Lada Togliatti 2	Russia-4	38	11	4	15	32	....	....	....	....	....
99-2000	Lada Togliatti 2	Russia-3	26	12	19	31	14	....	....	....	....	....
2000-01	Lada Togliatti	Russia	25	0	0	0	4	4	0	0	0	0
2001-02	Red Deer Rebels	WHL	71	32	57	89	54	23	14	9	23	28

YERSHOV, Andrei (yuhr-SHAWF, AWN-dray) **CHI.**

Defense. Shoots left. 6', 216 lbs. Born, Voskresensk, USSR, August 22, 1976.
(Chicago's 9th choice, 240th overall, in 1998 Entry Draft).

			Regular Season					Playoffs				
Season	Club	Lea	GP	G	A	TP	PIM	GP	G	A	TP	PIM
1994-95	Voskresensk	CIS	16	0	0	0	6	....	....	....	....	....
1995-96	Voskresensk	CIS	18	1	0	1	28	....	....	....	....	....
1996-97	Voskresensk	Russia	23	3	1	4	32	2	0	0	0	2
1997-98	Voskresensk	Russia	45	5	8	13	60	....	....	....	....	....
1998-99	Voskresensk	Russia	33	6	6	12	88	....	....	....	....	....
99-2000	Lada Togliatti	Russia	10	0	1	1	12	4	0	0	0	6
2000-01	Vityaz Podolsk	Russia	30	1	5	6	32	....	....	....	....	....
2001-02	Voskresensk 2	Russia-3	1	0	1	1	4	....	....	....	....	....
	Voskresensk	Russia-2	62	18	15	33	101	....	....	....	....	....

YEVSEYEV, Vladislav (yehv-SAY-ehv, VLAD-ih-slav) **BOS.**

Left wing. Shoots left. 6'2", 200 lbs. Born, Moscow, USSR, September 10, 1984.
(Boston's 2nd choice, 56th overall, in 2002 Entry Draft).

			Regular Season					Playoffs				
Season	Club	Lea	GP	G	A	TP	PIM	GP	G	A	TP	PIM
2000-01	Dyno. Moscow Jr.	Russia-Jr.	6	5	2	7	2	....	....	....	....	....
2001-02	CSKA Moscow 2	Russia-3	8	2	1	3	2	....	....	....	....	....
	H.C. CSKA	Russia-2	15	2	5	7	10	....	....	....	....	....

YTFELDT, David (YOOT-fehld, DAY-vihd) **VAN.**

Defense. Shoots left. 6'1", 187 lbs. Born, Ornskoldsvik, Sweden, September 29, 1979.
(Vancouver's 6th choice, 136th overall, in 1998 Entry Draft).

			Regular Season					Playoffs				
Season	Club	Lea	GP	G	A	TP	PIM	GP	G	A	TP	PIM
1996-97	Leksands IF Jr.	Swede-Jr.	25	3	5	8	....	....	....	....	....	....
1997-98	Leksands IF Jr.	Swede-Jr.	23	13	10	23	101	....	....	....	....	....
	Leksands IF	Sweden	10	0	0	0	2	....	....	....	....	....
1998-99	Leksands IF	Sweden	39	0	4	4	65	4	0	1	1	4
99-2000	Leksands IF	Sweden	50	3	9	12	72	....	....	....	....	....
2000-01	V. Frolunda Jr.	Swede-Jr.	2	2	1	3	0	....	....	....	....	....
	JYP Jyvaskyla	Finland	11	0	4	4	26	....	....	....	....	....
	Vastra Frolunda	Sweden	9	0	1	1	8	5	0	1	1	4
2001-02	Linkopings HC	Sweden	8	0	0	0	4	....	....	....	....	....

• Name when drafted was David Jonsson. His last name was legally changed to Ytfeldt.

ZAINULLIN, Ruslan (zihj-NOO-luhn, roos-LAHN) **ATL.**

Right wing. Shoots left. 6'2", 202 lbs. Born, Kazan, USSR, February 14, 1982.
(Tampa Bay's 2nd choice, 34th overall, in 2000 Entry Draft).

			Regular Season					Playoffs				
Season	Club	Lea	GP	G	A	TP	PIM	GP	G	A	TP	PIM
1997-98	Ak Bars Kazan 2	Russia-3	27	0	1	1	2	....	....	....	....	....
1998-99	Ak Bars Kazan 2	Russia-4	36	13	8	21	22	....	....	....	....	....
99-2000	Ak Bars Kazan 2	Russia-3	12	13	6	19	....	....	....	....	....	....
	Ak Bars Kazan	Russia	14	1	1	2	4	....	....	....	....	....
2000-01	Ak Bars Kazan	Russia	29	1	3	4	14	1	0	0	0	0
2001-02	Ak Bars Kazan	Russia	21	0	2	2	8	3	0	0	0	2

Traded to **Phoenix** by **Tampa Bay** with Mike Johnson, Paul Mara and NY Islanders' 2nd round choice (previously acquired, Phoenix selected Matthew Spiller) in 2001 Entry Draft for Nikolai Khabibulin and Stan Neckar, March 5, 2001. Rights traded to **Atlanta** by **Phoenix** with Kirill Safronov and Phoenix's 4th round choice (Patrick Dwyer) in 2002 Entry Draft for Darcy Hordichuk and Atlanta's 4th (Lance Monych) and 5th (John Zeiler) round choices in 2002 Entry Draft, March 19, 2002.

ZALESAK, Miroslav (zah-LIH-sahk, MEER-oh-slav) **S.J.**

Right wing. Shoots left. 6', 185 lbs. Born, Skalica, Czech., January 2, 1980.
(San Jose's 5th choice, 104th overall, in 1998 Entry Draft).

			Regular Season					Playoffs				
Season	Club	Lea	GP	G	A	TP	PIM	GP	G	A	TP	PIM
1995-96	HC Nitra Jr.	Slovak-Jr.	49	53	29	82	....	....	....	....	....	....
1996-97	HC Nitra Jr.	Slovak-Jr.	58	51	31	82	....	....	....	....	....	....
1997-98	Nitra Jr.	Slovak-Jr.	27	32	29	61	30	....	....	....	....	....
	Nitra	Slovakia	30	8	6	14	0	....	....	....	....	....
1998-99	Nitra	Slovakia	15	4	3	7	10	....	....	....	....	....
	Drummondville	QMJHL	45	24	27	51	18	....	....	....	....	....
99-2000	Drummondville	QMJHL	60	50	61	111	40	16	7	11	18	4
2000-01	Kentucky	AHL	60	14	11	25	26	3	0	1	1	4
2001-02	Cleveland Barons	AHL	74	22	20	42	44	....	....	....	....	....

ZANON, Greg (ZA-nuhn, GREHG) **OTT.**

Defense. Shoots left. 5'11", 200 lbs. Born, Burnaby, B.C., June 5, 1980.
(Ottawa's 6th choice, 156th overall, in 2000 Entry Draft).

			Regular Season					Playoffs				
Season	Club	Lea	GP	G	A	TP	PIM	GP	G	A	TP	PIM
1995-96	Burnaby Beavers	BCAHA	49	16	27	43	142	....	....	....	....	....
1996-97	Victoria Salsa	BCHL	53	4	13	17	124	....	....	....	....	....
1997-98	Victoria Salsa	BCHL	59	11	21	32	108	7	0	2	2	10
1998-99	South Surrey	BCHL	59	17	54	71	154	....	....	....	....	....
99-2000	Nebraska-Omaha	CCHA	42	3	26	29	56	....	....	....	....	....
2000-01	Nebraska-Omaha	CCHA	39	12	16	28	64	....	....	....	....	....
2001-02	Nebraska-Omaha	CCHA	41	9	16	25	54	....	....	....	....	....

CCHA First All-Star Team (2001) • NCAA West Second All-American Team (2001, 2002) • CCHA Second All-Star Team (2002)

ZAVORAL, Vaclav (ZA-vohr-uhl, VATS-lahf) **TOR.**

Defense. Shoots left. 6'3", 207 lbs. Born, Teplice, Czech., May 22, 1981.
(Toronto's 5th choice, 151st overall, in 1999 Entry Draft).

			Regular Season					Playoffs				
Season	Club	Lea	GP	G	A	TP	PIM	GP	G	A	TP	PIM
1997-98	Litvinov Jr.	Czech-Jr.	46	0	5	5	....	....	....	....	....	....
1998-99	Litvinov Jr.	Czech-Jr.	43	2	10	12	....	....	....	....	....	....
	Litvinov	Czech	1	0	1	1	2	....	....	....	....	....
99-2000	Sault Ste. Marie	OHL	57	3	11	14	89	14	0	2	2	28
2000-01	Sault Ste. Marie	OHL	55	4	6	10	116	....	....	....	....	....
2001-02	Flint Generals	UHL	65	2	13	15	123	5	0	0	0	12

ZEILER, John (ZIGH-luhr, JAWN) **PHX.**

Right wing. Shoots right. 6', 193 lbs. Born, Pittsburgh, PA, November 21, 1982.
(Phoenix's 7th choice, 132nd overall, in 2002 Entry Draft).

			Regular Season					Playoffs				
Season	Club	Lea	GP	G	A	TP	PIM	GP	G	A	TP	PIM
99-2000	Pittsburgh	PAHA	27	17	15	32	94	....	....	....	....	....
2000-01	Sioux City	USHL	56	8	20	28	45	....	....	....	....	....
2001-02	Sioux City	USHL	60	23	27	50	116	12	2	3	5	25

• Signed Letter of Intent to attend **St. Lawrence** (ECAC), September 1, 2000.

ZETTERBERG, Henrik (ZEH-tuhr-buhrg, HEHN-rihk) DET.

Left wing. Shoots left. 5'11", 176 lbs. Born, Njurunda, Sweden, October 9, 1980.
(Detroit's 4th choice, 210th overall, in 1999 Entry Draft).

			Regular Season					Playoffs				
Season	Club	Lea	GP	G	A	TP	PIM	GP	G	A	TP	PIM
1997-98	Timra IK Jr.	Swede-Jr.	18	9	5	14	4	....	....	....	....	....
	Timra IK	Swede-2	16	1	2	3	4	4	0	1	1	0
1998-99	Timra IK	Swede-2	37	15	13	28	2	4	2	1	3	2
99-2000	Timra IK	Swede-2	32	20	14	34	20	10	10	4	14	4
2000-01	Timra IK	Sweden	47	15	31	46	24	....	....	....	....	....
2001-02	Timra IK	Sweden	48	10	22	32	20	....	....	....	....	....
	Sweden	Olympics	4	0	1	1	0	....	....	....	....	....

Swedish Elite League Rookie of the Year (2001)

ZEVAKHIN, Alexander (zeh-VAH-khin, al-ehx-AN-duhr) PIT.

Right wing. Shoots left. 5'11", 208 lbs. Born, Perm, USSR, June 4, 1980.
(Pittsburgh's 2nd choice, 54th overall, in 1998 Entry Draft).

			Regular Season					Playoffs				
Season	Club	Lea	GP	G	A	TP	PIM	GP	G	A	TP	PIM
1995-96	CSKA Moscow Jr.	CIS-Jr.	65	52	30	82	30	....	....	....	....	....
1996-97	CSKA Moscow 2	Russia-3	30	15	18	33	10	....	....	....	....	....
	H.C. CSKA	Russia	29	7	3	10	10	....	....	....	....	....
1997-98	CSKA Moscow 2	Russia-3	32	13	14	27	20	....	....	....	....	....
	H.C. CSKA	Russia	10	1	0	1	0	....	....	....	....	....
1998-99	CSKA Moscow	Russia	42	7	4	11	16	3	0	0	0	0
99-2000	CSKA Moscow	Russia	15	1	0	1	6	....	....	....	....	....
2000-01	Wilkes-Barre	AHL	77	14	11	25	16	21	2	4	6	0
2001-02	Wilkes-Barre	AHL	74	7	16	23	29	....	....	....	....	....

ZHVACHKIN, Leonid (ZNVAHCH-kihn, lay-oh-NEED) NYR

Defense. Shoots left. 6'3", 189 lbs. Born, Tula, USSR, February 24, 1983.
(NY Rangers' 9th choice, 230th overall, in 2001 Entry Draft).

			Regular Season					Playoffs				
Season	Club	Lea	GP	G	A	TP	PIM	GP	G	A	TP	PIM
99-2000	H.C. CSKA 2	Russia-3	9	0	1	1	12	....	....	....	....	....
2000-01	Vityaz Podolsk 2	Russia-3		STATISTICS NOT AVAILABLE								
2001-02	Guelph Storm	OHL	62	2	2	4	58	9	0	0	0	5

ZIB, Lukas (ZIHB, LOO-kahsh) EDM.

Defense. Shoots right. 6'1", 200 lbs. Born, Ceske Budejovice, Czech., February 24, 1977.
(Edmonton's 3rd choice, 57th overall, in 1995 Entry Draft).

			Regular Season					Playoffs				
Season	Club	Lea	GP	G	A	TP	PIM	GP	G	A	TP	PIM
1994-95	Ceske Budejovice	Czech	13	2	0	2	16	9	1	0	1	6
1995-96	C. Budejovice Jr.	Czech-Jr.	11	5	1	6	....	....	....	....	....	....
	Ceske Budejovice	Czech	10	1	0	1	....	2	0	0	0	....
1996-97	Ceske Budejovice	Czech	13	0	0	0	4	2	0	0	0	0
1997-98	Ceske Budejovice	Czech	47	5	6	11	22	....	....	....	....	....
1998-99	Ceske Budejovice	Czech	24	1	4	5	18	....	....	....	....	....
99-2000	Ceske Budejovice	Czech	38	3	6	9	10	1	0	0	0	0
2000-01	Ceske Budejovice	Czech	22	2	3	5	16	....	....	....	....	....
	Zlin	Czech	19	4	3	7	8	....	....	....	....	....
2001-02	Blues Espoo	Finland	5	0	0	0	2	....	....	....	....	....
	HC Karlovy Vary	Czech	36	4	10	14	20	....	....	....	....	....

ZIDLICKY, Marek (zhihd-LIHTS-kee, MAIR-ehk) NYR

Defense. Shoots right. 5'11", 187 lbs. Born, Most, Czech., February 3, 1977.
(NY Rangers' 6th choice, 176th overall, in 2001 Entry Draft).

			Regular Season					Playoffs				
Season	Club	Lea	GP	G	A	TP	PIM	GP	G	A	TP	PIM
1994-95	HC Kladno	Czech	30	2	2	4	38	11	1	1	2	10
1995-96	HC Poldi Kladno	Czech	37	4	5	9	74	7	1	1	2	8
1996-97	HC Poldi Kladno	Czech	49	5	16	21	60	2	0	0	0	0
1997-98	Kladno	Czech	51	2	13	15	121	....	....	....	....	....
1998-99	Kladno	Czech	50	10	12	22	94	....	....	....	....	....
99-2000	HIFK Helsinki	EuroHL	4	2	2	4	10	1	0	0	0	0
	HIFK Helsinki	Finland	47	4	16	20	66	9	3	2	5	24
2000-01	HIFK Helsinki	Finland	51	12	25	37	146	5	0	1	1	6
2001-02	HIFK Helsinki	Finland	56	11	29	40	107	....	....	....	....	....

ZIGOMANIS, Mike (zih-goh-MAN-his, MIGHK) CAR.

Center. Shoots right. 6'1", 189 lbs. Born, North York, Ont., January 17, 1981.
(Carolina's 2nd choice, 46th overall, in 2001 Entry Draft).

			Regular Season					Playoffs				
Season	Club	Lea	GP	G	A	TP	PIM	GP	G	A	TP	PIM
1996-97	Wexford Hawks	MTHL	40	37	48	85	23	....	....	....	....	....
	Wexford Raiders	MTJHL	8	2	5	7	2	....	....	....	....	....
1997-98	Kingston	OHL	62	23	51	74	30	12	1	6	7	2
1998-99	Kingston	OHL	67	29	56	85	36	5	1	7	8	2
99-2000	Kingston	OHL	59	40	54	94	49	5	0	4	4	0
2000-01	Kingston	OHL	52	40	37	77	44	....	....	....	....	....
2001-02	Lowell	AHL	79	18	30	48	24	5	1	1	2	2

• Re-entered NHL Entry Draft. Originally Buffalo's 4th choice, 64th overall, in 1999 Entry Draft.

ZIMAKOV, Sergei (zih-MAH-kahv, SAIR-gay) WSH.

Defense. Shoots left. 6'1", 194 lbs. Born, Moscow, USSR, January 15, 1978.
(Washington's 4th choice, 58th overall, in 1996 Entry Draft).

			Regular Season					Playoffs				
Season	Club	Lea	GP	G	A	TP	PIM	GP	G	A	TP	PIM
1994-95	Omaha Lancers	USHL	48	14	46	60	22	....	....	....	....	....
1995-96	Krylja Sovetov	CIS	49	2	7	9	36	....	....	....	....	....
1996-97	Krylja Sovetov	Russia	39	4	3	7	57	2	0	0	0	0
1997-98	Krylja Sovetov	Russia	42	4	1	5	48	....	....	....	....	....
1998-99	Ak Bars Kazan	Russia	28	1	0	1	6	8	0	1	1	6
99-2000	Perm	Russia	31	1	2	3	34	3	0	1	1	0
2000-01	CSKA Moscow 2	Russia-3	3	2	2	4	2	....	....	....	....	....
	CSKA Moscow	Russia	26	1	5	6	28	....	....	....	....	....
2001-02	CSKA Moscow	Russia	42	3	10	13	74	....	....	....	....	....

ZINGER, Dwayne (ZIHN-guhr, DWAYN WSH.

Defense. Shoots left. 6'4", 225 lbs. Born, Coronation, Alta., July 5, 1976.

			Regular Season					Playoffs				
Season	Club	Lea	GP	G	A	TP	PIM	GP	G	A	TP	PIM
1995-96	Melville	SJHL	64	7	17	24	....	....	....	....	....	....
1996-97	Alaska-Fairbanks	CCHA	32	1	5	6	45	....	....	....	....	....
1997-98	Alaska-Fairbanks	CCHA	32	1	3	4	91	....	....	....	....	....
1998-99	Alaska-Fairbanks	CCHA	33	4	14	18	42	....	....	....	....	....
99-2000	Alaska-Fairbanks	CCHA	34	10	4	14	34	....	....	....	....	....
	Cincinnati	AHL	13	0	2	2	33	....	....	....	....	....
2000-01	Cincinnati	AHL	68	6	9	15	120	4	1	1	2	2
2001-02	Cincinnati	AHL	67	6	13	19	156	3	0	1	1	2

SJHL First All-Star Team (1996)

Signed as a free agent by **Detroit**, March 13, 2000. Signed as a free agent by **Washington**, July 9, 2002.

ZINGONI, Peter (zihn-GOH-nee, PEE-tuhr) CBJ

Center. Shoots left. 6', 180 lbs. Born, Bridgeport, CT, April 28, 1981.
(Columbus' 8th choice, 231st overall, in 2000 Entry Draft).

			Regular Season					Playoffs				
Season	Club	Lea	GP	G	A	TP	PIM	GP	G	A	TP	PIM
1998-99	New England	EJHL	40	26	22	48	....	....	....	....	....	....
99-2000	New England	EJHL	40	39	38	77	85	3	1	1	2	0
2000-01	Providence	H-East	28	2	6	8	38	....	....	....	....	....
2001-02	Providence	H-East	31	7	10	17	27	....	....	....	....	....

ZINOVJEV, Sergei (zih-NOH-vee-ehv, SAIR-gay) BOS.

Center/Left wing. Shoots left. 5'11", 176 lbs. Born, Novokuznetsk, USSR, March 4, 1980.
(Boston's 6th choice, 73rd overall, in 2000 Entry Draft).

			Regular Season					Playoffs				
Season	Club	Lea	GP	G	A	TP	PIM	GP	G	A	TP	PIM
1997-98	Novokuznetsk 2	Russia-3	40	7	7	14	36	....	....	....	....	....
1998-99	Novokuznetsk 2	Russia-4	4	0	1	1	8	....	....	....	....	....
	Magnitogorsk	Russia	31	2	4	6	14	3	0	0	0	0
99-2000	Magnitogorsk	Russia	28	0	2	2	16	....	....	....	....	....
2000-01	Yaroslavl	Russia	27	2	10	12	36	....	....	....	....	....
	Ufa	Russia	8	4	5	9	6	....	....	....	....	....
2001-02	Spartak Moscow	Russia	51	12	18	30	43	....	....	....	....	....

ZION, Jon (ZIGH-awn, JAWN) TOR.

Defense. Shoots left. 6', 200 lbs. Born, Nepean, Ont., May 21, 1981.
(Toronto's 4th choice, 110th overall, in 1999 Entry Draft).

			Regular Season					Playoffs				
Season	Club	Lea	GP	G	A	TP	PIM	GP	G	A	TP	PIM
1996-97	Nepean Raiders	OCJHL	47	8	26	34	14	....	....	....	....	....
1997-98	Ottawa 67's	OHL	53	4	19	23	20	13	3	12	15	2
1998-99	Ottawa 67's	OHL	60	8	33	41	10	9	2	3	5	8
99-2000	Ottawa 67's	OHL	66	7	52	59	16	11	3	10	13	8
2000-01	Ottawa 67's	OHL	59	22	51	73	38	20	3	*19	22	18
2001-02	Ottawa 67's	OHL	52	16	31	47	60	13	0	7	7	5

OHL Second All-Star Team (2001)

ZIZKA, Tomas (ZHIHZH-kuh, TAW-mahsh) L.A.

Defense. Shoots left. 6'1", 198 lbs. Born, Sternberk, Czech., October 10, 1979.
(Los Angeles' 6th choice, 163rd overall, in 1998 Entry Draft).

			Regular Season					Playoffs				
Season	Club	Lea	GP	G	A	TP	PIM	GP	G	A	TP	PIM
1994-95	AC ZPS Zlin Jr.	Czech-Jr.	39	1	10	11	....	....	....	....	....	....
1995-96	AC ZPS Zlin Jr.	Czech-Jr.	47	2	8	10	....	....	....	....	....	....
1996-97	AC ZPS Zlin Jr.	Czech-Jr.	14	1	0	1	....	....	....	....	....	....
1997-98	Zlin Jr.	Czech-Jr.	11	3	4	7	....	....	....	....	....	....
	Zlin	Czech	33	0	3	3	2	....	....	....	....	....
1998-99	Zlin	Czech	44	3	7	10	14	11	1	2	3	....
99-2000	Zlin	Czech	46	4	6	10	30	4	1	0	1	4
2000-01	Zlin	Czech	43	2	11	13	16	6	0	0	0	6
2001-02	Manchester	AHL	58	4	17	21	22	4	1	0	1	14

ZOTKIN, Alexei (ZOHT-kihn, al-EHX-ay) CHI.

Left wing. Shoots left. 6', 200 lbs. Born, Magnitogorsk, USSR, February 5, 1982.
(Chicago's 7th choice, 119th overall, in 2001 Entry Draft).

			Regular Season					Playoffs				
Season	Club	Lea	GP	G	A	TP	PIM	GP	G	A	TP	PIM
1997-98	Magnitogorsk 2	Russia-3	2	0	1	1	2	....	....	....	....	....
1998-99	Magnitogorsk 2	Russia-4	23	5	1	6	20	....	....	....	....	....
99-2000	Magnitogorsk 2	Russia-3	38	22	27	49	86	....	....	....	....	....
	Magnitogorsk	Russia	1	0	0	0	0	....	....	....	....	....
2000-01	Magnitogorsk 2	Russia-3	5	6	2	8	26	....	....	....	....	....
	Magnitogorsk	Russia	40	2	3	5	34	12	1	1	2	20
2001-02	Magnitogorsk	Russia	25	3	0	3	10	8	0	1	1	2

ZULTEK, Matt (ZUHL-tehk, MAT) PHI.

Left wing. Shoots left. 6'4", 222 lbs. Born, Windsor, Ont., March 12, 1979.
(Boston's 2nd choice, 56th overall, in 1999 Entry Draft).

			Regular Season					Playoffs				
Season	Club	Lea	GP	G	A	TP	PIM	GP	G	A	TP	PIM
1994-95	Toronto Marlies	MTHL	94	45	36	81	110	....	....	....	....	....
1995-96	Caledon	MTJHL	50	19	14	33	40	....	....	....	....	....
1996-97	Ottawa 67's	OHL	63	27	13	40	76	21	7	6	13	27
1997-98	Ottawa 67's	OHL	62	28	28	56	156	13	6	12	18	20
1998-99	Ottawa 67's	OHL	56	33	33	66	71	9	6	2	8	4
99-2000	Ottawa 67's	OHL	28	9	6	15	34	11	3	5	8	12
2000-01	St. Thomas	AUAA	17	12	7	19	68	....	....	....	....	....
	Philadelphia	AHL	16	1	4	5	6	9	0	1	1	8
2001-02	Trenton Titans	ECHL	48	15	13	28	148	7	4	1	5	12
	Philadelphia	AHL	....	....	....	....	....	1	0	0	0	0

• Re-entered NHL Entry Draft. Originally Los Angeles' 2nd choice, 15th overall, in 1997 Entry Draft.

• Missed majority of 1999-2000 season recovering from off-season knee injury that required surgery, November 12, 1999. Rights traded to **Philadelphia** by **Boston** for Philadelphia's 9th round choice (Michael Rodman) in 2001 Entry Draft, February 13, 2001.

Late Additions to Player Register

CULLEN, Mark (KUH-lehn, MAHRK) MIN.

Center. Shoots left. 5 11", 175 lbs. Born, Moorhead, MN, October 28, 1978.

			Regular Season					Playoffs				
Season	Club	League	GP	G	A	TP	PIM	GP	G	A	TP	PIM
1998-99	Colorado College	WCHA	42	8	25	33	22	....	....	....	....	....
99-2000	Colorado College	WCHA	37	11	20	31	22	....	....	....	....	....
2000-01	Colorado College	WCHA	31	20	33	53	26	....	....	....	....	....
2001-02	Colorado College	WCHA	43	14	36	50	14	....	....	....	....	....

WCHA First All-Star Team (2001) • NCAA West Second All-American Team (2001)

Signed as a free agent by **Minnesota**, April 8, 2002.

DeWOLF, Josh (duh-WOOLF, JAWSH) ANA.

Defense. Shoots left. 6' 2", 203 lbs. Born, Bloomington, MN, July 25, 1977.
(New Jersey's 3rd choice, 41st overall, in 1996 Entry Draft)

			Regular Season					Playoffs				
Season	Club	League	GP	G	A	TP	PIM	GP	G	A	TP	PIM
1996-97	St. Cloud State	WCHA	31	3	11	14	62	....	....	....	....	....
1997-98	St. Cloud State	WCHA	37	9	9	18	78	....	....	....	....	....
	Albany	AHL	2	0	0	0	0	....	....	....	....	....
1998-99	Albany	AHL	75	1	17	18	111	5	0	0	0	2
99-2000	Albany	AHL	58	3	11	14	38	....	....	....	....	....
	Quebec	AHL	15	1	0	1	17	3	0	1	1	0
2000-01	Quebec	AHL	58	3	8	11	65	7	0	1	1	16
2001-02	Cincinnati	AHL	76	4	11	15	88	3	0	0	0	2

Traded to **Montreal** by **New Jersey** with Sheldon Souray and New Jersey's 2nd round choice (later traded to Washington – later traded to Tampa Bay – Tampa Bay selected Andreas Holmqvist) in 2001 Entry Draft for Vladimir Malakhov, March 1, 2000. Signed as a free agent by **Cincinnati** (AHL), September 25, 2001. Signed as a free agent by **Anaheim**, August 22, 2002.

HOGGAN, Jeff MIN.

Right wing. Shoots right. 6', 200 lbs. Born, Hope B.C., February 1, 1978.

			Regular Season					Playoffs				
Season	Club	League	GP	G	A	TP	PIM	GP	G	A	TP	PIM
99-2000	Nebraska-Omaha	CCHA	34	16	9	25	82	....	....	....	....	....
2000-01	Nebraska-Omaha	CCHA	42	12	17	29	78	....	....	....	....	....
2001-02	Nebraska-Omaha	CCHA	41	24	21	45	92	....	....	....	....	....
	Houston	AHL	....	....	....	....	....	4	0	0	0	2

CCHA First All-Star Team (2002) • NCAA West Second All-American Team (2002)

Signed to a try-out contract by **Houston** (AHL), April 4, 2002. Signed as a free agent by **Minnesota**, August 20, 2002.

HUNTER, J.J. (HUHN-tuhr, JAY-JAY) EDM.

Center. Shoots left. 6' 1", 185 lbs. Born, Shaunavon, Sask., July 6, 1980.

			Regular Season					Playoffs				
Season	Club	League	GP	G	A	TP	PIM	GP	G	A	TP	PIM
1989-90	Kelowna	WHL	66	18	32	50	61	6	1	3	3	2
1999-00	Kelowna	WHL	66	22	26	47	61	5	1	0	1	2
2000-01	Kelowna	WHL	12	1	5	6	4	....	....	....	....	....
	Prince Albert	WHL	58	28	17	45	40	....	....	....	....	....
2001-02	Columbus	ECHL	60	23	22	45	59	....	....	....	....	....
	Hamilton	AHL	1	0	0	0	0	1	0	0	0	0

Signed as a free agent by **Edmonton**, August 19, 2002.

SEVERSON, Cam (SEH-vuhr-SOHN, KAM) ANA.

Left wing. Shoots left. 6' 1", 215 lbs. Born, Canora, Sask., January 15, 1978.
(San Jose's 6th choice, 192nd overall, in 1997 Entry Draft)

			Regular Season					Playoffs				
Season	Club	League	GP	G	A	TP	PIM	GP	G	A	TP	PIM
1996-97	Lethbridge	WHL	45	12	13	25	169	....	....	....	....	....
	Prince Albert	WHL	16	5	13	18	54	4	4	0	4	8
1997-98	Prince Albert	WHL	41	23	25	48	129	....	....	....	....	....
	Spokane	WHL	23	9	11	20	88	18	11	4	15	51
1998-99	Spokane	WHL	46	16	17	33	190	....	....	....	....	....
	Oklahoma City	CHL	5	6	3	9	4	10	4	0	4	26
99-2000	Louisiana	ECHL	7	0	2	2	22	....	....	....	....	....
	Peoria	ECHL	56	19	8	27	138	18	3	4	7	41
2000-01	Quad City	UHL	46	22	26	48	129	....	....	....	....	....
	Portland	AHL	8	0	0	0	11	....	....	....	....	....
	Cincinnati	AHL	20	4	7	11	60	3	1	1	2	0
2001-02	Hartford	AHL	65	11	10	21	116	5	0	0	0	7

Signed as a free agent by **Hartford** (AHL), September 24, 2001. Signed as a free agent by **Anaheim**, August 22, 2002.

SKOLNEY, Wade (SKOHL-nee, WAYD) PHI.

Defense. Shoots right. 6', 185 lbs. Born, Wynyard, Sask., June 24, 1981.

			Regular Season					Playoffs				
Season	Club	League	GP	G	A	TP	PIM	GP	G	A	TP	PIM
1997-98	Brandon	WHL	42	1	11	12	49	3	0	0	0	0
1998-99	Brandon	WHL	39	3	10	13	60	5	0	1	1	16
99-2000	Brandon	WHL	13	0	2	2	23	....	....	....	....	....
2000-01	Brandon	WHL	28	2	9	11	37	....	....	....	....	....
2001-02	Brandon	WHL	50	4	12	16	179	19	2	7	9	56

Signed as a free agent by **Philadelphia**, May 20, 2002.

SMITH, Jeff (SMIHTH, JEHF) PHI.

Left wing. Shoots left. 6' 6", 212 lbs. Born, Regina, Sask., January 2, 1981.

			Regular Season					Playoffs				
Season	Club	League	GP	G	A	TP	PIM	GP	G	A	TP	PIM
1998-99	Red Deer	WHL	25	0	0	0	0	....	....	....	....	....
	Reg. Pat Cdns.	SMHL	35	28	19	47	71	....	....	....	....	....
99-2000	Red Deer	WHL	63	9	6	15	74	....	....	....	....	....
2000-01	Red Deer	WHL	72	22	11	33	187	23	6	6	12	30

Returned to **Regina** (SMHL) by **Regina** (WHL), January 11, 1999. Signed as a free agent by **Philadelphia**, August 20, 2002.

FREE AGENT SIGNINGS

BELANGER, Francis – Signed as a free agent by **Anaheim**, August 22, 2002.

SMITH, Nick – Signed as a free agent by **Anaheim**, August 22, 2002.

NOTE:

Heights and weights for some players have been adjusted on their club's player personnel table only, as the player register had already gone to press. In the event of a discrepancy between club player personnel tables and player register panels, the data in the club player personnel table is most current.

League Abbreviations

AAHA	Alberta Amateur Hockey Association
AAHL	Alaska Amateur Hockey League
ACHL	Atlantic Coast Hockey League
AFHL	American Frontier Hockey League
AHL	American Hockey League
AJHL	Alberta Junior Hockey Leagues
Alpenliga	Alpenliga (Austria, Italy, Slovenia 1994-1999)
AMBHL	Alberta Major Bantam Hockey League
AMHL	Alberta Midget AAA Hockey League
AUAA	Atlantic University Athletic Association
BCAHA	British Columbia Amateur Hockey Association
BCHL	British Columbia (Junior) Hockey League (also BCJHL)
CCHA	Central Collegiate Hockey Association
CEGEP	Quebec College Prep
CHA	College Hockey America
CHL	Central Hockey League
CIS	Commonwealth of Independent States
ColHL	Colonial Hockey League
CWUAA	Canadian Western University Athletic Association
ECAC	Eastern College Athletic Conference
ECHL	East Coast Hockey League
EEHL	Eastern European Hockey League
EJHL	Eastern Junior Hockey League
EuroHL	European Hockey League
Exhib.	Exhibition Games, Series or Season
G.N.	Great Northern
GPAC	Great Plains Athletic Conference
GTHL	Greater Toronto Hockey League
H-East	Hockey East
HJHL	Heritage Junior Hockey League
Hi-School	High School (also H.S.)
IEL	Internationale Eishockey Liga
IHL	International Hockey League
IJHL	Interstate Junior Hockey League
KIDHL	Kootenay International Junior B Hockey League
MAAC	Metro Atlantic Athletic Conference
MAHA	Manitoba Amateur Hockey Association
MBHL	Metropolitan Boston Hockey League
MEHL	Midwest Elite Hockey League
MIAC	Minnesota Intercollegiate Athletic Conference
MJHL	Manitoba Junior Hockey League
MJrHL	Maritime Junior Major Hockey League
MMHL	Manitoba Midget AAA Hockey League
MNHL	Michigan National Hockey League
MTJHL	Metropolitan Toronto Junior Hockey League
MTHL	Metro Toronto Hockey League
NAJHL	North American Junior Hockey League
Nat-Team	National Team (also Nt.-Team)
NBAHA	New Brunswick Amateur Hockey Association
NCAA	National Collegiate Athletic Association
NEJHL	New England Junior Hockey League
NFAHA	Newfoundland Amateur Hockey Association
NHL	National Hockey League
NOHA	Northern Ontario Hockey Association
NOJHL	Northern Ontario Junior Hockey League
NSMHL	Nova Scotia Midget AAA Hockey League
OCJHL	Ontario Central Junior A Hockey League
OHL	Ontario Hockey League
OJHL-B	Ontario Junior B Hockey Leagues
OMHA	Ontario Minor Hockey Association
OMJHL	Ontario Major Junior Hockey League
OPJHL	Ontario Provincial Junior A Hockey League
OUAA	Ontario Universities Athletic Association
QAAA	Quebec Amateur Athletic Association
QAHA	Quebec Amateur Hockey Association
QJHL	Quebec Junior Hockey League
QMJHL	Quebec Major Junior Hockey League
PCJHL	Pacific Coast Junior Hockey League
PIJHL	Pacific International Junior Hockey League
RAMHL	Rural Alberta Midget Hockey League
RMJHL	Rocky Mountain Junior Hockey League
SAHA	Saskatchewan Amateur Hockey Association
SBHL	Saskatchewan Bantam Hockey League
SJHL	Saskatchewan Junior Hockey League
SMHL	Saskatchewan Midget AAA Hockey League
SSJHL	Southern Saskatchewan Junior B Hockey League
SunHL	Sunshine Hockey League
TBAHA	Thunder Bay Amateur Hockey Association
TBJHL	Thunder Bay Junior Hockey League
TBMHL	Thunder Bay Midget Hockey League
UHL	United Hockey League
USAHA	United States Amateur Hockey Association
USDP-17	United States National Development Under-17 Program
USDP-18	United States National Development Under-18 Program
USHL	United States (Junior A) Hockey League
VIJHL	Vancouver Island Junior Hockey League
WCHA	Western Collegiate Hockey Association
WCHL	West Coast Hockey League
WHA	World Hockey Association
WHL	Western Hockey League
WNYHA	Western New York Hockey Association
WPHL	Western Professional Hockey League
WSJHL	Western States Junior Hockey League

2002-03 NHL Player Register

Note: The 2002-03 NHL Player Register lists forwards and defensemen only. Goaltenders are listed separately. The NHL Player Register lists every skater who has played in the NHL. Trades and roster changes are current as of August 23, 2002.

Abbreviations: A – assists; **F%** – faceoff winning percentage; **G** – goals; **GP** – games played; **GT** – game-tying goals scored; **GW** – game-winning goals scored; **H** – HITS: any legal contact by one player on an opposing player that impedes the opposing players' progress; **Lea** – league; **Min** – average time on ice; **PIM** – penalties in minutes; **+/–** – plus/minus rating; **PP** – powerplay goals scored; **Pts** – points; **S** – shots on goal; **S%** – shooting percentage; **SB** – shots blocked; **SH** – shorthand goal scored; **TF** – Total faceoffs taken; ***** – league-leading total; ♦ – member of Stanley Cup-winning team.

Prospect Register begins on page 269.

Goaltender Register begins on page 576.

League abbreviations are listed on page 268.

AALTO, Antti

(AL-toh, AN-tee)

Center. Shoots left. 6'2", 210 lbs. Born, Lappeenranta, Finland, March 4, 1975. Anaheim's 6th choice, 134th overall, in 1993 Entry Draft.

			Regular Season																Playoffs							
Season	Club	League	GP	G	A	Pts	PIM	PP	SH	GW	S	%	+/-	TF	F%	H	SB	Min	GP	G	A	Pts	PIM	PP	SH	GW
1991-92	SaiPa-B	Finn-Jr.	13	7	9	16	38	....	....	....	....	....	....	....	....	....	....	....	6	3	1	4	6	....	....	....
	SaiPa	Finland-2	20	6	6	12	20	....	....	....	....	....	....	....	....	....	....	....	....	....	....	....	....	....	....	....
1992-93	SaiPa	Finn-Jr.	3	0	1	1	2	....	....	....	....	....	....	....	....	....	....	....	....	....	....	....	....	....	....	....
	SaiPa-B	Finn-Jr.	3	4	2	6	2	....	....	....	....	....	....	....	....	....	....	....	....	....	....	....	....	....	....	....
	TPS Turku Jr.	Finn-Jr.	14	6	8	14	18	....	....	....	....	....	....	....	....	....	....	....	6	2	2	4	8	....	....	....
	TPS Turku	Finland	1	0	0	0	0	....	....	....	....	....	....	....	....	....	....	....	....	....	....	....	....	....	....	....
	SaiPa	Finland-2	23	6	8	14	14	....	....	....	....	....	....	....	....	....	....	....	....	....	....	....	....	....	....	....
1993-94	TPS Turku Jr.	Finn-Jr.	10	3	8	11	14	....	....	....	....	....	....	....	....	....	....	....	5	1	4	5	12	....	....	....
	Kiekko-67 Turku	Finland-2	4	2	2	4	27	....	....	....	....	....	....	....	....	....	....	....	....	....	....	....	....	....	....	....
	TPS Turku	Finland	33	5	9	14	16	....	....	....	....	....	....	....	....	....	....	....	10	1	1	2	4	....	....	....
1994-95	Kiekko-67 Jr.	Finn-Jr.	2	1	2	3	2	....	....	....	....	....	....	....	....	....	....	....	....	....	....	....	....	....	....	....
	Kiekko-67 Turku	Finland-2	1	1	0	1	29	....	....	....	....	....	....	....	....	....	....	....	....	....	....	....	....	....	....	....
	TPS Turku	Finland	44	11	7	18	18	....	....	....	....	....	....	....	....	....	....	....	5	0	1	1	2	....	....	....
1995-96	Kiekko-67 Turku	Finland-2	2	0	2	2	2	....	....	....	....	....	....	....	....	....	....	....	....	....	....	....	....	....	....	....
	TPS Turku	Finland	40	15	16	31	22	....	....	....	....	....	....	....	....	....	....	....	11	3	5	8	14	....	....	....
1996-97	TPS Turku	Finland	44	15	19	34	60	....	....	....	....	....	....	....	....	....	....	....	11	5	6	11	31	....	....	....
	TPS Turku	EuroHL	5	3	3	6	2	....	....	....	....	....	....	....	....	....	....	....	2	1	1	2	0	....	....	....
1997-98	**Anaheim**	**NHL**	**3**	**0**	**0**	**0**	**0**	**0**	**0**	**0**	**1**	**0.0**	**–1**	....	....	....	....	....	....	....	....	....	....	....	....	....
	Cincinnati	AHL	29	4	9	13	30	....	....	....	....	....	....	....	....	....	....	....	....	....	....	....	....	....	....	....
1998-99	**Anaheim**	**NHL**	**73**	**3**	**5**	**8**	**24**	**2**	**0**	**0**	**61**	**4.9**	**–12**	**22**	**22.7**	**64**	**12**	**9:23**	**4**	**0**	**0**	**0**	**2**	**0**	**0**	**0**
99-2000	**Anaheim**	**NHL**	**63**	**7**	**11**	**18**	**26**	**1**	**0**	**1**	**102**	**6.9**	**–13**	**827**	**51.0**	**83**	**12**	**13:11**	....	....	....	....	....	....	....	....
2000-01	**Anaheim**	**NHL**	**12**	**1**	**1**	**2**	**2**	**0**	**0**	**0**	**18**	**5.6**	**1**	**57**	**40.4**	**19**	**1**	**10:37**	....	....	....	....	....	....	....	....
	Cincinnati	AHL	40	14	26	40	39	....	....	....	....	....	....	....	....	....	....	....	3	2	1	3	2	....	....	....
2001-02	Jokerit Helsinki	Finland	48	8	21	29	87	....	....	....	....	....	....	....	....	....	....	....	12	1	*8	9	12	....	....	....
	Finland	Olympics	4	0	0	0	4	....	....	....	....	....	....	....	....	....	....	....	....	....	....	....	....	....	....	....
	NHL Totals		**151**	**11**	**17**	**28**	**52**	**3**	**0**	**1**	**182**	**6.0**		**906**	**49.7**	**166**	**25**	**11:06**	**4**	**0**	**0**	**0**	**2**	**0**	**0**	**0**

ADAMS, Bryan

(A-duhms, BRIGH-uhn) **DET.**

Left wing. Shoots left. 6', 185 lbs. Born, Fort St. James, B.C., March 20, 1977.

Season	Club	League	GP	G	A	Pts	PIM	PP	SH	GW	S	%	+/-	TF	F%	H	SB	Min	GP	G	A	Pts	PIM	PP	SH	GW
1994-95	Prince George	RMJHL	48	37	53	90	....	....	....	....	....	....	....	....	....	....	....	....	....	....	....	....	....	....	....	....
1995-96	Michigan State	CCHA	42	3	8	11	12	....	....	....	....	....	....	....	....	....	....	....	....	....	....	....	....	....	....	....
1996-97	Michigan State	CCHA	29	7	7	14	51	....	....	....	....	....	....	....	....	....	....	....	....	....	....	....	....	....	....	....
1997-98	Michigan State	CCHA	31	9	21	30	39	....	....	....	....	....	....	....	....	....	....	....	....	....	....	....	....	....	....	....
1998-99	Michigan State	CCHA	42	21	16	37	56	....	....	....	....	....	....	....	....	....	....	....	....	....	....	....	....	....	....	....
99-2000	**Atlanta**	**NHL**	**2**	**0**	**0**	**0**	**0**	**0**	**0**	**0**	**1**	**0.0**	**–1**	**0**	**0.0**	**5**	**0**	**10:57**	....	....	....	....	....	....	....	....
	Orlando	IHL	64	16	18	34	27	....	....	....	....	....	....	....	....	....	....	....	4	0	1	1	6	....	....	....
2000-01	**Atlanta**	**NHL**	**9**	**0**	**1**	**1**	**2**	**0**	**0**	**0**	**3**	**0.0**	**–4**	**1**	**0.0**	**8**	**3**	**9:37**	....	....	....	....	....	....	....	....
	Orlando	IHL	61	18	28	46	43	....	....	....	....	....	....	....	....	....	....	....	16	3	4	7	20	....	....	....
2001-02	Chicago Wolves	AHL	64	10	17	27	25	....	....	....	....	....	....	....	....	....	....	....	5	0	0	0	2	....	....	....
	NHL Totals		**11**	**0**	**1**	**1**	**2**	**0**	**0**	**0**	**4**	**0.0**		**1**	**0.0**	**13**	**3**	**9:51**	....	....	....	....	....	....	....	....

Signed as a free agent by **Atlanta**, July 6, 1999. Signed as a free agent by **Detroit**, August 5, 2002.

ADAMS, Craig

(A-duhms, KRAYG) **CAR.**

Right wing. Shoots right. 6', 200 lbs. Born, Seria, Brunei, April 26, 1977. Hartford's 9th choice, 223rd overall, in 1996 Entry Draft.

Season	Club	League	GP	G	A	Pts	PIM	PP	SH	GW	S	%	+/-	TF	F%	H	SB	Min	GP	G	A	Pts	PIM	PP	SH	GW
1994-95	Calgary Canucks	AJHL	STATISTICS NOT AVAILABLE																							
1995-96	Harvard Crimson	ECAC	34	8	9	17	56	....	....	....	....	....	....	....	....	....	....	....	....	....	....	....	....	....	....	....
1996-97	Harvard Crimson	ECAC	32	6	4	10	36	....	....	....	....	....	....	....	....	....	....	....	....	....	....	....	....	....	....	....
1997-98	Harvard Crimson	ECAC	12	6	6	12	12	....	....	....	....	....	....	....	....	....	....	....	....	....	....	....	....	....	....	....
1998-99	Harvard Crimson	ECAC	31	9	14	23	53	....	....	....	....	....	....	....	....	....	....	....	....	....	....	....	....	....	....	....
99-2000	Cincinnati	IHL	73	12	12	24	124	....	....	....	....	....	....	....	....	....	....	....	8	0	1	1	14	....	....	....
2000-01	**Carolina**	**NHL**	**44**	**1**	**0**	**1**	**20**	**0**	**0**	**0**	**15**	**6.7**	**–7**	**4**	**25.0**	**61**	**5**	**4:30**	**3**	**0**	**0**	**0**	**0**	**0**	**0**	**0**
	Cincinnati	IHL	4	0	1	1	9	....	....	....	....	....	....	....	....	....	....	....	1	0	0	0	2	....	....	....
2001-02	**Carolina**	**NHL**	**33**	**0**	**1**	**1**	**38**	**0**	**0**	**0**	**17**	**0.0**	**2**	**9**	**33.3**	**59**	**9**	**5:54**	**1**	**0**	**0**	**0**	**0**	**0**	**0**	**0**
	Lowell	AHL	22	5	4	9	51	....	....	....	....	....	....	....	....	....	....	....	....	....	....	....	....	....	....	....
	NHL Totals		**77**	**1**	**1**	**2**	**58**	**0**	**0**	**0**	**32**	**3.1**		**13**	**30.8**	**120**	**14**	**5:06**	**4**	**0**	**0**	**0**	**0**	**0**	**0**	**0**

Rights transferred to **Carolina** after **Hartford** franchise relocated, June 25, 1997. • Missed majority of 1997-98 season recovering from shoulder injury suffered in game vs. University of Wisconsin (WCHA), December 27, 1997.

ADAMS, Kevyn

(A-duhms, KEH-vihn) **CAR.**

Center. Shoots right. 6'1", 195 lbs. Born, Washington, DC, October 8, 1974. Boston's 1st choice, 25th overall, in 1993 Entry Draft.

Season	Club	League	GP	G	A	Pts	PIM	PP	SH	GW	S	%	+/-	TF	F%	H	SB	Min	Playoffs GP	G	A	Pts	PIM	PP	SH	GW
1990-91	Niagara Scenics	NAJHL	55	17	20	37	24																			
1991-92	Niagara Scenics	NAJHL	40	25	33	58	51																			
1992-93	Miami-Ohio	CCHA	40	17	15	32	18																			
1993-94	Miami-Ohio	CCHA	36	15	28	43	24																			
1994-95	Miami-Ohio	CCHA	38	20	29	49	30																			
1995-96	Miami-Ohio	CCHA	36	17	30	47	30																			
1996-97	Grand Rapids	IHL	82	22	25	47	47												5	1	1	2	4			
1997-98	**Toronto**	**NHL**	**5**	**0**	**0**	**0**	**7**	**0**	**0**	**0**	**3**	**0.0**	**0**													
	St. John's	AHL	59	17	20	37	99												4	0	0	0	4			
1998-99	**Toronto**	**NHL**	**1**	**0**	**0**	**0**	**0**	**0**	**0**	**0**	**1**	**0.0**	**0**	**9**	**44.4**	**2**	**0**	**7:56**	**7**	**0**	**2**	**2**	**14**	**0**	**0**	**0**
	St. John's	AHL	80	15	35	50	85												5	2	0	2	4			
99-2000	**Toronto**	**NHL**	**52**	**5**	**8**	**13**	**39**	**0**	**0**	**0**	**70**	**7.1**	**−7**	**604**	**56.5**	**78**	**13**	**12:23**	**12**	**1**	**0**	**1**	**7**	**0**	**1**	**0**
	St. John's	AHL	23	6	11	17	24																			
2000-01	**Columbus**	**NHL**	**66**	**8**	**12**	**20**	**52**	**0**	**0**	**1**	**84**	**9.5**	**−4**	**1152**	**57.4**	**104**	**41**	**15:18**								
	Florida	**NHL**	**12**	**3**	**6**	**9**	**2**	**0**	**0**	**2**	**21**	**14.3**	**7**	**198**	**47.5**	**16**	**5**	**17:25**								
2001-02	**Florida**	**NHL**	**44**	**4**	**8**	**12**	**28**	**0**	**0**	**1**	**71**	**5.6**	**−3**	**572**	**57.9**	**34**	**24**	**13:21**								
	Carolina	**NHL**	**33**	**2**	**3**	**5**	**15**	**0**	**0**	**1**	**37**	**5.4**	**−2**	**187**	**58.8**	**32**	**10**	**9:05**	**23**	**1**	**0**	**1**	**4**	**0**	**0**	**0**
	NHL Totals		**213**	**22**	**37**	**59**	**143**	**0**	**0**	**5**	**287**	**7.7**		**2722**	**56.6**	**266**	**93**	**13:16**	**42**	**2**	**2**	**4**	**25**	**0**	**1**	**0**

CCHA Second All-Star Team (1995)

Signed as a free agent by **Toronto**, August 7, 1997. Selected by **Columbus** from **Toronto** in Expansion Draft, June 23, 2000. Traded to **Florida** by **Columbus** with Columbus's 4th round choice (Michael Woodford) in 2001 Entry Draft for Ray Whitney and future considerations, March 13, 2001. Traded to **Carolina** by **Florida** with Bret Hedican, Tomas Malec and a conditional 3rd round choice in 2003 Entry Draft for Sandis Ozolinsh and Byron Ritchie, January 16, 2002.

AFANASENKOV, Dmitry

(a-fahn-A-sehn-kahv, dih-MEE-tree) **T.B.**

Left wing. Shoots right. 6'2", 200 lbs. Born, Arkhangelsk, USSR, May 12, 1980. Tampa Bay's 3rd choice, 72nd overall, in 1998 Entry Draft.

Season	Club	League	GP	G	A	Pts	PIM	PP	SH	GW	S	%	+/-	TF	F%	H	SB	Min	Playoffs GP	G	A	Pts	PIM	PP	SH	GW
1995-96	Yaroslavl Jr.	CIS-Jr.	35	28	16	44	8																			
	Yaroslavl	CIS	25	10	5	15	10																			
1996-97	Yaroslavl 2	Russia-3	45	20	15	35	14																			
1997-98	Yaroslavl	Russia	48	14	7	21	20																			
1998-99	Moncton Wildcats	QMJHL	15	5	5	10	12																			
	Sherbrooke	QMJHL	51	23	30	53	22												13	10	6	16	6			
99-2000	Sherbrooke	QMJHL	60	56	43	99	70												5	3	2	5	4			
2000-01	**Tampa Bay**	**NHL**	**9**	**1**	**1**	**2**	**4**	**0**	**0**	**0**	**8**	**12.5**	**1**	**7**	**28.6**	**4**	**2**	**11:24**								
	Detroit Vipers	IHL	65	15	22	37	26																			
2001-02	**Tampa Bay**	**NHL**	**5**	**0**	**0**	**0**	**0**	**0**	**0**	**0**	**1**	**0.0**	**−1**	**0**	**0.0**	**3**	**0**	**4:54**								
	Springfield	AHL	28	4	5	9	4																			
	Grand Rapids	AHL	18	1	2	3	2																			
	NHL Totals		**14**	**1**	**1**	**2**	**4**	**0**	**0**	**0**	**9**	**11.1**		**7**	**28.6**	**7**	**2**	**9:05**								

Traded to **Sherbrooke** (QMJHL) by **Moncton** (QMJHL) for Sherbrooke's 6th round choice (later traded back to Sherbrooke - Sherbrooke selected Pascal Vandal) in 1999 QMJHL Priority Draft and 1st choice (later traded to Baie-Comeau - Baie-Comeau selected Evgeny Gusakov) in 1999 CHL Import Draft, October 30, 1998.

AFINOGENOV, Maxim

(ah-fihn-ah-GEHN-ahf, mahx-EEM) **BUF.**

Right wing. Shoots left. 6', 190 lbs. Born, Moscow, USSR, September 4, 1979. Buffalo's 3rd choice, 69th overall, in 1997 Entry Draft.

Season	Club	League	GP	G	A	Pts	PIM	PP	SH	GW	S	%	+/-	TF	F%	H	SB	Min	Playoffs GP	G	A	Pts	PIM	PP	SH	GW
1996-97	Dynamo Moscow	Russia	29	6	5	11	10												4	0	2	2	0			
	Dynamo Moscow	EuroHL	3	0	0	0	0												3	1	0	1	4			
1997-98	Dynamo Moscow	Russia	35	10	5	15	53																			
	Dynamo Moscow	EuroHL	6	3	1	4	27																			
1998-99	Dynamo Moscow	Russia	38	8	13	21	24												16	*10	6	*16	14			
	Dynamo Moscow	EuroHL	5	3	5	8	29												4	2	1	3	27			
99-2000	**Buffalo**	**NHL**	**65**	**16**	**18**	**34**	**41**	**2**	**0**	**2**	**128**	**12.5**	**−4**	**0**	**0.0**	**32**	**10**	**13:09**	**5**	**0**	**1**	**1**	**2**	**0**	**0**	**0**
	Rochester	AHL	15	6	12	18	8												8	3	1	4	4			
2000-01	**Buffalo**	**NHL**	**78**	**14**	**22**	**36**	**40**	**3**	**0**	**5**	**190**	**7.4**	**1**	**2**	**0.0**	**33**	**17**	**14:32**	**11**	**2**	**3**	**5**	**4**	**0**	**0**	**0**
2001-02	**Buffalo**	**NHL**	**81**	**21**	**19**	**40**	**69**	**3**	**1**	**0**	**234**	**9.0**	**−9**	**1**	**100.0**	**47**	**11**	**15:22**								
	Russia	Olympics	6	2	2	4	4																			
	NHL Totals		**224**	**51**	**59**	**110**	**150**	**8**	**1**	**7**	**552**	**9.2**		**3**	**33.3**	**112**	**38**	**14:26**	**16**	**2**	**4**	**6**	**6**	**0**	**0**	**0**

AITKEN, Johnathan

(ATE-kin, JAWN-uh-thuhn)

Defense. Shoots left. 6'4", 215 lbs. Born, Edmonton, Alta., May 24, 1978. Boston's 1st choice, 8th overall, in 1996 Entry Draft.

Season	Club	League	GP	G	A	Pts	PIM	PP	SH	GW	S	%	+/-	TF	F%	H	SB	Min	Playoffs GP	G	A	Pts	PIM	PP	SH	GW
1993-94	Sherwood Park	AMHL	31	4	9	13	54																			
1994-95	Medicine Hat	WHL	53	0	5	5	71												5	0	0	0	0			
1995-96	Medicine Hat	WHL	71	6	14	20	131												5	1	0	1	6			
1996-97	Brandon	WHL	65	4	18	22	211												6	0	0	0	4			
1997-98	Brandon	WHL	69	9	25	34	183												18	0	8	8	67			
1998-99	Providence	AHL	65	2	9	11	92												13	0	0	0	17			
99-2000	**Boston**	**NHL**	**3**	**0**	**0**	**0**	**0**	**0**	**0**	**0**	**2**	**0.0**	**−3**	**0**	**0.0**	**5**	**3**	**18:57**								
	Providence	AHL	70	2	12	14	121												11	1	0	1	26			
2000-01	HC Sparta Praha	Czech	24	0	3	3	62																			
2001-02	Norfolk Admirals	AHL	28	0	1	1	43												4	0	0	0	2			
	Jackson Bandits	ECHL	43	1	9	10	141																			
	NHL Totals		**3**	**0**	**0**	**0**	**0**	**0**	**0**	**0**	**2**	**0.0**		**0**	**0.0**	**5**	**3**	**18:57**								

WHL East Second All-Star Team (1998)

Signed as a free agent by **Norfolk** (AHL) with **Boston** retaining NHL rights, September 6, 2001.

ALATALO, Mika

(a-luh-TAH-loh, MEE-kuh)

Left wing. Shoots left. 6', 202 lbs. Born, Oulu, Finland, June 11, 1971. Winnipeg's 11th choice, 203rd overall, in 1990 Entry Draft.

Season	Club	League	GP	G	A	Pts	PIM	PP	SH	GW	S	%	+/-	TF	F%	H	SB	Min	Playoffs GP	G	A	Pts	PIM	PP	SH	GW
1988-89	KooKoo Kouvola	Finland	34	8	6	14	10																			
1989-90	KooKoo Kouvola	Finland	41	3	5	8	22																			
1990-91	Lukko Rauma	Finland	39	10	1	11	10																			
1991-92	Lukko Rauma	Finland	43	20	17	37	32												2	0	0	0	0			
1992-93	Lukko Rauma	Finland	48	16	19	35	38												3	0	0	0	0			
1993-94	Lukko Rauma	Finland	45	19	15	34	77												9	2	2	4	4			
	Finland	Olympics	7	2	1	3	2																			
1994-95	TPS Turku	Finland	44	23	13	36	79												13	2	5	7	8			
1995-96	TPS Turku	Finland	49	19	18	37	44												11	3	4	7	8			
1996-97	Lulea HF	Sweden	50	19	18	37	54												10	2	3	5	22			
	Lulea HF	EuroHL	6	4	3	7	4																			
1997-98	Lulea HF	Sweden	45	14	10	24	22												2	0	0	0	0			
	Lulea HF	EuroHL	6	2	1	3	6																			
1998-99	TPS Turku	Finland	53	14	23	37	44												10	6	3	9	6			
99-2000	**Phoenix**	**NHL**	**82**	**10**	**17**	**27**	**36**	**1**	**0**	**1**	**107**	**9.3**	**−3**	**3**	**0.0**	**92**	**17**	**12:37**	**5**	**0**	**0**	**0**	**2**	**0**	**0**	**0**
2000-01	**Phoenix**	**NHL**	**70**	**7**	**12**	**19**	**22**	**0**	**0**	**1**	**64**	**10.9**	**1**	**2**	**0.0**	**78**	**13**	**10:59**								
2001-02	TPS Turku	Finland	56	20	22	42	95												8	1	1	2	12			
	NHL Totals		**152**	**17**	**29**	**46**	**58**	**1**	**0**	**2**	**171**	**9.9**		**5**	**0.0**	**170**	**30**	**11:52**	**5**	**0**	**0**	**0**	**2**	**0**	**0**	**0**

Rights transferred to **Phoenix** after **Winnipeg** franchise relocated, July 1, 1996.

ALBELIN, Tommy

(AHL-buh-LEEN, TAW-mee) **N.J.**

Defense. Shoots left. 6'2", 195 lbs. Born, Stockholm, Sweden, May 21, 1964. Quebec's 7th choice, 158th overall, in 1983 Entry Draft.

			Regular Season																Playoffs							
Season	Club	League	GP	G	A	Pts	PIM	PP	SH	GW	S	%	+/-	TF	F%	H	SB	Min	GP	G	A	Pts	PIM	PP	SH	GW
1980-81	Stocksunds IF	Swede-3	18	6	1	7	....	....	....	....	....	....	....	....	....	....	....	....	....	....	....	....	....	....	....	....
1981-82	Stocksunds IF	Swede-3	22	6	2	8	....	....	....	....	....	....	....	....	....	....	....	....	....	....	....	....	....	....	....	....
1982-83	Djurgarden	Sweden	19	2	5	7	4	....	....	....	....	....	....	....	....	....	....	....	6	1	0	1	2	....	....	....
1983-84	Djurgarden	Sweden	30	9	5	14	26	....	....	....	....	....	....	....	....	....	....	....	4	0	1	1	2	....	....	....
1984-85	Djurgarden	Sweden	32	9	8	17	22	....	....	....	....	....	....	....	....	....	....	....	8	2	1	3	4	....	....	....
1985-86	Djurgarden	Sweden	35	4	8	12	26	....	....	....	....	....	....	....	....	....	....	....	....	....	....	....	....	....	....	....
1986-87	Djurgarden	Sweden	33	7	5	12	49	....	....	....	....	....	....	....	....	....	....	....	2	0	0	0	0	....	....	....
1987-88	**Quebec**	**NHL**	**60**	**3**	**23**	**26**	**47**	**0**	**0**	**0**	**98**	**3.1**	**-7**	....	....	....	....	....	....	....	....	....	....	....	....	....
1988-89	**Quebec**	**NHL**	**14**	**2**	**4**	**6**	**27**	**1**	**0**	**1**	**16**	**12.5**	**-6**	....	....	....	....	....	....	....	....	....	....	....	....	....
	Halifax Citadels	AHL	8	2	5	7	4	....	....	....	....	....	....	....	....	....	....	....	....	....	....	....	....	....	....	....
	New Jersey	**NHL**	**46**	**7**	**24**	**31**	**40**	**1**	**1**	**1**	**82**	**8.5**	**18**	....	....	....	....	....	....	....	....	....	....	....	....	....
1989-90	**New Jersey**	**NHL**	**68**	**6**	**23**	**29**	**63**	**4**	**0**	**0**	**125**	**4.8**	**-1**	....	....	....	....	....	....	....	....	....	....	....	....	....
1990-91	**New Jersey**	**NHL**	**47**	**2**	**12**	**14**	**44**	**1**	**0**	**0**	**66**	**3.0**	**1**	....	....	....	....	....	**3**	**0**	**1**	**1**	**2**	**0**	**0**	**0**
	Utica Devils	AHL	14	4	2	6	10	....	....	....	....	....	....	....	....	....	....	....	....	....	....	....	....	....	....	....
1991-92	**New Jersey**	**NHL**	**19**	**0**	**4**	**4**	**4**	**0**	**0**	**0**	**18**	**0.0**	**7**	....	....	....	....	....	**1**	**1**	**1**	**2**	**0**	**0**	**0**	**0**
	Utica Devils	AHL	11	4	6	10	4	....	....	....	....	....	....	....	....	....	....	....	....	....	....	....	....	....	....	....
1992-93	**New Jersey**	**NHL**	**36**	**1**	**5**	**6**	**14**	**1**	**0**	**1**	**33**	**3.0**	**0**	....	....	....	....	....	**5**	**2**	**0**	**2**	**0**	**1**	**0**	**1**
1993-94	**New Jersey**	**NHL**	**62**	**2**	**17**	**19**	**36**	**1**	**0**	**1**	**62**	**3.2**	**20**	....	....	....	....	....	**20**	**2**	**5**	**7**	**14**	**1**	**0**	**1**
	Albany	AHL	4	0	2	2	17	....	....	....	....	....	....	....	....	....	....	....	....	....	....	....	....	....	....	....
1994-95♦	**New Jersey**	**NHL**	**48**	**5**	**10**	**15**	**20**	**2**	**0**	**0**	**60**	**8.3**	**9**	....	....	....	....	....	**20**	**1**	**7**	**8**	**2**	**0**	**0**	**0**
1995-96	**New Jersey**	**NHL**	**53**	**1**	**12**	**13**	**14**	**0**	**0**	**0**	**90**	**1.1**	**0**	....	....	....	....	....	....	....	....	....	....	....	....	....
	Calgary	**NHL**	**20**	**0**	**1**	**1**	**4**	**0**	**0**	**0**	**31**	**0.0**	**1**	....	....	....	....	....	**4**	**0**	**0**	**0**	**0**	**0**	**0**	**0**
1996-97	**Calgary**	**NHL**	**72**	**4**	**11**	**15**	**14**	**2**	**0**	**0**	**103**	**3.9**	**-8**	....	....	....	....	....	....	....	....	....	....	....	....	....
1997-98	**Calgary**	**NHL**	**69**	**2**	**17**	**19**	**32**	**1**	**0**	**2**	**88**	**2.3**	**9**	....	....	....	....	....	....	....	....	....	....	....	....	....
	Sweden	Olympics	3	0	0	0	4	....	....	....	....	....	....	....	....	....	....	....	....	....	....	....	....	....	....	....
1998-99	**Calgary**	**NHL**	**60**	**1**	**5**	**6**	**8**	**0**	**0**	**0**	**54**	**1.9**	**-11**	**1**	**0.0**	**34**	**55**	**19:08**	....	....	....	....	....	....	....	....
99-2000	**Calgary**	**NHL**	**41**	**4**	**6**	**10**	**12**	**1**	**1**	**1**	**37**	**10.8**	**-3**	**0**	**0.0**	**21**	**60**	**21:35**	....	....	....	....	....	....	....	....
2000-01	**Calgary**	**NHL**	**77**	**1**	**19**	**20**	**22**	**1**	**0**	**0**	**69**	**1.4**	**2**	**0**	**0.0**	**27**	**69**	**20:53**	....	....	....	....	....	....	....	....
2001-02	**New Jersey**	**NHL**	**42**	**1**	**3**	**4**	**4**	**0**	**0**	**1**	**33**	**3.0**	**0**	**0**	**0.0**	**21**	**28**	**13:20**	**6**	**0**	**0**	**0**	**0**	**0**	**0**	**0**
	NHL Totals		**834**	**42**	**196**	**238**	**405**	**16**	**2**	**8**	**1065**	**3.9**		**1**	**0.0**	**103**	**212**	**19:06**	**59**	**6**	**14**	**20**	**18**	**2**	**0**	**2**

Traded to **New Jersey** by **Quebec** for New Jersey's 4th round choice (Niklas Andersson) in 1989 Entry Draft, December 12, 1988. Traded to **Calgary** by **New Jersey** with Cale Hulse and Jocelyn Lemieux for Phil Housley and Dan Keczmer, February 26, 1996. Signed as a free agent by **New Jersey**, July 9, 2001.

ALEXEEV, Nikita

(uh-LEHX-ee-ehv, nih-KEE-tuh) **T.B.**

Right wing. Shoots left. 6'5", 210 lbs. Born, Murmansk, USSR, December 27, 1981. Tampa Bay's 1st choice, 8th overall, in 2000 Entry Draft.

Season	Club	League	GP	G	A	Pts	PIM	PP	SH	GW	S	%	+/-	TF	F%	H	SB	Min	GP	G	A	Pts	PIM	PP	SH	GW
1996-97	Krylja Sovetov Jr.	Russia-Jr.	45	4	6	10	8	....	....	....	....	....	....	....	....	....	....	....	....	....	....	....	....	....	....	....
1997-98	Krylja Sovetov 2	Russia-3	61	11	4	15	36	....	....	....	....	....	....	....	....	....	....	....	....	....	....	....	....	....	....	....
1998-99	Erie Otters	OHL	61	17	18	35	14	....	....	....	....	....	....	....	....	....	....	....	5	1	1	2	4	....	....	....
99-2000	Erie Otters	OHL	64	24	29	53	42	....	....	....	....	....	....	....	....	....	....	....	13	4	3	7	6	....	....	....
2000-01	Erie Otters	OHL	64	31	41	72	45	....	....	....	....	....	....	....	....	....	....	....	12	7	7	14	12	....	....	....
2001-02	**Tampa Bay**	**NHL**	**44**	**4**	**4**	**8**	**8**	**1**	**0**	**1**	**47**	**8.5**	**-9**	**0**	**0.0**	**32**	**6**	**11:26**	....	....	....	....	....	....	....	....
	Springfield	AHL	35	5	9	14	16	....	....	....	....	....	....	....	....	....	....	....	....	....	....	....	....	....	....	....
	NHL Totals		**44**	**4**	**4**	**8**	**8**	**1**	**0**	**1**	**47**	**8.5**		**0**	**0.0**	**32**	**6**	**11:26**	....	....	....	....	....	....	....	....

ALFREDSSON, Daniel

(AHL-frehd-suhn, DAN-yehl) **OTT.**

Right wing. Shoots right. 5'11", 195 lbs. Born, Goteborg, Sweden, December 11, 1972. Ottawa's 5th choice, 133rd overall, in 1994 Entry Draft.

Season	Club	League	GP	G	A	Pts	PIM	PP	SH	GW	S	%	+/-	TF	F%	H	SB	Min	GP	G	A	Pts	PIM	PP	SH	GW
1990-91	IF Molndal	Swede-2	3	0	0	0	2	....	....	....	....	....	....	....	....	....	....	....	8	4	4	8	4	....	....	....
1991-92	IF Molndal	Swede-2	32	12	8	20	43	....	....	....	....	....	....	....	....	....	....	....	....	....	....	....	....	....	....	....
1992-93	Vastra Frolunda	Sweden	20	1	5	6	8	....	....	....	....	....	....	....	....	....	....	....	....	....	....	....	....	....	....	....
1993-94	Vastra Frolunda	Sweden	39	20	10	30	18	....	....	....	....	....	....	....	....	....	....	....	4	1	1	2	....	....	....	....
1994-95	Vastra Frolunda	Sweden	22	7	11	18	22	....	....	....	....	....	....	....	....	....	....	....	....	....	....	....	....	....	....	....
1995-96	**Ottawa**	**NHL**	**82**	**26**	**35**	**61**	**28**	**8**	**2**	**3**	**212**	**12.3**	**-18**	....	....	....	....	....	....	....	....	....	....	....	....	....
1996-97	**Ottawa**	**NHL**	**76**	**24**	**47**	**71**	**30**	**11**	**1**	**1**	**247**	**9.7**	**5**	....	....	....	....	....	**7**	**5**	**2**	**7**	**6**	**3**	**0**	**2**
1997-98	**Ottawa**	**NHL**	**55**	**17**	**28**	**45**	**18**	**7**	**0**	**7**	**149**	**11.4**	**7**	....	....	....	....	....	**11**	**7**	**2**	**9**	**20**	**2**	**1**	**1**
	Sweden	Olympics	4	2	3	5	2	....	....	....	....	....	....	....	....	....	....	....	....	....	....	....	....	....	....	....
1998-99	**Ottawa**	**NHL**	**58**	**11**	**22**	**33**	**14**	**3**	**0**	**5**	**163**	**6.7**	**8**	**7**	**57.1**	**67**	**21**	**17:22**	**4**	**1**	**2**	**3**	**4**	**1**	**0**	**0**
99-2000	**Ottawa**	**NHL**	**57**	**21**	**38**	**59**	**28**	**4**	**2**	**0**	**164**	**12.8**	**11**	**3**	**66.7**	**58**	**29**	**18:45**	**6**	**1**	**3**	**4**	**2**	**1**	**0**	**0**
2000-01	**Ottawa**	**NHL**	**68**	**24**	**46**	**70**	**30**	**10**	**0**	**3**	**206**	**11.7**	**11**	**8**	**50.0**	**85**	**26**	**18:47**	**4**	**1**	**0**	**1**	**2**	**0**	**0**	**0**
2001-02	**Ottawa**	**NHL**	**78**	**37**	**34**	**71**	**45**	**9**	**1**	**4**	**243**	**15.2**	**3**	**30**	**30.0**	**96**	**29**	**20:19**	**12**	**7**	**6**	**13**	**4**	**3**	**0**	**3**
	Sweden	Olympics	4	1	4	5	2	....	....	....	....	....	....	....	....	....	....	....	....	....	....	....	....	....	....	....
	NHL Totals		**474**	**160**	**250**	**410**	**193**	**52**	**6**	**23**	**1384**	**11.6**		**48**	**39.6**	**306**	**105**	**18:55**	**44**	**22**	**15**	**37**	**38**	**10**	**1**	**6**

NHL All-Rookie Team (1996) • Won Calder Memorial Trophy (1996) • Played in NHL All-Star Game (1996, 1997, 1998)

ALLEN, Bryan

(AHL-lehn, BRIGH-uhn) **VAN.**

Defense. Shoots left. 6'4", 215 lbs. Born, Kingston, Ont., August 21, 1980. Vancouver's 1st choice, 4th overall, in 1998 Entry Draft.

Season	Club	League	GP	G	A	Pts	PIM	PP	SH	GW	S	%	+/-	TF	F%	H	SB	Min	GP	G	A	Pts	PIM	PP	SH	GW
1995-96	Ernestown Jets	OJHL-C	36	1	16	17	71	....	....	....	....	....	....	....	....	....	....	....	....	....	....	....	....	....	....	....
1996-97	Oshawa Generals	OHL	60	2	4	6	76	....	....	....	....	....	....	....	....	....	....	....	18	1	3	4	26	....	....	....
1997-98	Oshawa Generals	OHL	48	6	13	19	126	....	....	....	....	....	....	....	....	....	....	....	5	0	5	5	18	....	....	....
1998-99	Oshawa Generals	OHL	37	7	15	22	77	....	....	....	....	....	....	....	....	....	....	....	15	0	3	3	26	....	....	....
99-2000	Oshawa Generals	OHL	3	0	2	2	12	....	....	....	....	....	....	....	....	....	....	....	3	0	0	0	13	....	....	....
	Syracuse Crunch	AHL	9	1	1	2	11	....	....	....	....	....	....	....	....	....	....	....	2	0	0	0	2	....	....	....
2000-01	**Vancouver**	**NHL**	**6**	**0**	**0**	**0**	**0**	**0**	**0**	**0**	**2**	**0.0**	**0**	**0**	**0.0**	**9**	**6**	**9:20**	**2**	**0**	**0**	**0**	**2**	**0**	**0**	**0**
	Kansas City	IHL	75	5	20	25	99	....	....	....	....	....	....	....	....	....	....	....	....	....	....	....	....	....	....	....
2001-02	**Vancouver**	**NHL**	**11**	**0**	**0**	**0**	**6**	**0**	**0**	**0**	**4**	**0.0**	**1**	**0**	**0.0**	**20**	**2**	**10:47**	....	....	....	....	....	....	....	....
	Manitoba Moose	AHL	68	7	18	25	121	....	....	....	....	....	....	....	....	....	....	....	5	0	1	1	8	....	....	....
	NHL Totals		**17**	**0**	**0**	**0**	**6**	**0**	**0**	**0**	**6**	**0.0**		**0**	**0.0**	**29**	**8**	**10:16**	**2**	**0**	**0**	**0**	**2**	**0**	**0**	**0**

OHL First All-Star Team (1999) • Missed majority of 1999-2000 season recovering from knee injury suffered in training camp, September 21, 1999.

ALLISON, Jamie

(AHL-lih-sohn, JAY-mee) **CBJ**

Defense. Shoots left. 6'1", 200 lbs. Born, Lindsay, Ont., May 13, 1975. Calgary's 2nd choice, 44th overall, in 1993 Entry Draft.

Season	Club	League	GP	G	A	Pts	PIM	PP	SH	GW	S	%	+/-	TF	F%	H	SB	Min	GP	G	A	Pts	PIM	PP	SH	GW
1990-91	Waterloo Siskins	OJHL-B	38	3	8	11	91	....	....	....	....	....	....	....	....	....	....	....	....	....	....	....	....	....	....	....
1991-92	Windsor	OHL	59	4	8	12	70	....	....	....	....	....	....	....	....	....	....	....	4	1	1	2	2	....	....	....
1992-93	Detroit	OHL	61	0	13	13	64	....	....	....	....	....	....	....	....	....	....	....	15	2	5	7	23	....	....	....
1993-94	Detroit	OHL	40	2	22	24	69	....	....	....	....	....	....	....	....	....	....	....	17	2	9	11	35	....	....	....
1994-95	Detroit	OHL	50	1	14	15	119	....	....	....	....	....	....	....	....	....	....	....	18	2	7	9	35	....	....	....
	Calgary	**NHL**	**1**	**0**	**0**	**0**	**0**	**0**	**0**	**0**	**0**	**0.0**	**0**	....	....	....	....	....	....	....	....	....	....	....	....	....
1995-96	Saint John	AHL	71	3	16	19	223	....	....	....	....	....	....	....	....	....	....	....	14	0	2	2	16	....	....	....
1996-97	**Calgary**	**NHL**	**20**	**0**	**0**	**0**	**35**	**0**	**0**	**0**	**8**	**0.0**	**-4**	....	....	....	....	....	....	....	....	....	....	....	....	....
	Saint John	AHL	46	3	6	9	139	....	....	....	....	....	....	....	....	....	....	....	5	0	1	1	4	....	....	....
1997-98	**Calgary**	**NHL**	**43**	**3**	**8**	**11**	**104**	**0**	**0**	**1**	**27**	**11.1**	**3**	....	....	....	....	....	....	....	....	....	....	....	....	....
	Saint John	AHL	16	0	5	5	49	....	....	....	....	....	....	....	....	....	....	....	....	....	....	....	....	....	....	....
1998-99	Saint John	AHL	5	0	0	0	23	....	....	....	....	....	....	....	....	....	....	....	....	....	....	....	....	....	....	....
	Chicago	**NHL**	**39**	**2**	**2**	**4**	**62**	**0**	**0**	**0**	**24**	**8.3**	**0**	**0**	**0.0**	**50**	**12**	**14:01**	....	....	....	....	....	....	....	....
	Indianapolis Ice	IHL	3	1	0	1	10	....	....	....	....	....	....	....	....	....	....	....	....	....	....	....	....	....	....	....
99-2000	**Chicago**	**NHL**	**59**	**1**	**3**	**4**	**102**	**0**	**0**	**0**	**24**	**4.2**	**-5**	**0**	**0.0**	**87**	**29**	**14:14**	....	....	....	....	....	....	....	....
2000-01	**Chicago**	**NHL**	**44**	**1**	**3**	**4**	**53**	**0**	**0**	**0**	**16**	**6.3**	**7**	**1**	**100.0**	**57**	**33**	**14:33**	....	....	....	....	....	....	....	....
2001-02	**Calgary**	**NHL**	**37**	**0**	**2**	**2**	**24**	**0**	**0**	**0**	**14**	**0.0**	**-3**	**2**	**0.0**	**33**	**5**	**7:44**	....	....	....	....	....	....	....	....
	Columbus	**NHL**	**7**	**0**	**0**	**0**	**28**	**0**	**0**	**0**	**2**	**0.0**	**-4**	**0**	**0.0**	**12**	**2**	**11:23**	....	....	....	....	....	....	....	....
	NHL Totals		**250**	**7**	**18**	**25**	**408**	**0**	**0**	**1**	**115**	**6.1**		**3**	**33.3**	**239**	**81**	**12:52**	....	....	....	....	....	....	....	....

Traded to **Chicago** by **Calgary** with Marty McInnis and Erik Andersson for Jeff Shantz and Steve Dubinsky, October 27, 1998. Claimed by **Calgary** from **Chicago** in Waiver Draft, September 28, 2001. Traded to **Columbus** by **Calgary** for Blake Sloan, March 19, 2002.

ALLISON, Jason

(AHL-lih-sohn, JAY-suhn) **L.A.**

Center. Shoots right. 6'3", 215 lbs. Born, North York, Ont., May 29, 1975. Washington's 2nd choice, 17th overall, in 1993 Entry Draft.

Season	Club	League	GP	G	A	Pts	PIM	PP	SH	GW	S	%	+/-	TF	F%	H	SB	Min	Playoffs GP	G	A	Pts	PIM	PP	SH	GW
1990-91	North York	MTJHL	63	53	41	94																				
1991-92	London Knights	OHL	65	11	19	30	15												7	0	0	0	0			
1992-93	London Knights	OHL	66	42	76	118	50												12	7	13	20	8			
1993-94	London Knights	OHL	56	55	87	*142	68												5	2	13	15	13			
	Washington	NHL	2	0	1	1	0	0	0	0	5	0.0	1													
	Portland Pirates	AHL																	6	2	1	3	0			
1994-95	London Knights	OHL	15	15	21	36	43																			
	Washington	NHL	12	2	1	3	6	2	0	0	9	22.2	-3													
	Portland Pirates	AHL	8	5	4	9	2												7	3	8	11	2			
1995-96	Washington	NHL	19	0	3	3	2	0	0	0	18	0.0	-3													
	Portland Pirates	AHL	57	28	41	69	42												6	1	6	7	9			
1996-97	Washington	NHL	53	5	17	22	25	1	0	1	71	7.0	-3													
	Boston	NHL	19	3	9	12	9	1	0	0	28	10.7	-3													
1997-98	Boston	NHL	81	33	50	83	60	5	0	8	158	20.9	33						6	2	6	8	4	1	0	0
1998-99	Boston	NHL	82	23	53	76	68	5	1	3	158	14.6	5	1760	52.2	91	28	22:23	12	2	9	11	6	1	0	0
99-2000	Boston	NHL	37	10	18	28	20	3	0	1	66	15.2	5	100	60.0	61	20	21:33								
2000-01	Boston	NHL	82	36	59	95	85	11	3	6	185	19.5	-8	1897	51.9	108	48	23:21								
2001-02	Los Angeles	NHL	73	19	55	74	68	5	0	2	139	13.7	2	1698	54.5	96	29	21:47	7	3	3	6	4	0	0	1
	NHL Totals		460	131	266	397	343	33	4	21	837	15.7		5455	53.0	356	125	22:24	25	7	18	25	14	2	0	1

OHL First All-Star Team (1994) • OHL MVP (1994) • Canadian Major Junior First All-Star Team (1994) • Canadian Major Junior Player of the Year (1994) • Played in NHL All-Star Game (2001)

Traded to **Boston** by **Washington** with Jim Carey, Anson Carter and Washington's 3rd round choice (Lee Goren) in 1997 Entry Draft for Bill Ranford, Adam Oates and Rick Tocchet, March 1, 1997. • Missed majority of 1999-2000 season recovering from thumb injury suffered in game vs. NY Islanders, January 8, 2000. Traded to **LA Kings** by **Boston** with Mikko Eloranta for Jozef Stumpel and Glen Murray, October 24, 2001.

AMONTE, Tony

(uh-MAHN-tee, TOH-nee) **PHX.**

Right wing. Shoots left. 6', 200 lbs. Born, Hingham, MA, August 2, 1970. NY Rangers' 3rd choice, 68th overall, in 1988 Entry Draft.

Season	Club	League	GP	G	A	Pts	PIM	PP	SH	GW	S	%	+/-	TF	F%	H	SB	Min	Playoffs GP	G	A	Pts	PIM	PP	SH	GW
1985-86	Thayer Academy	Hi-School	2	0	0	0	0																			
1986-87	Thayer Academy	Hi-School	25	25	32	57																				
1987-88	Thayer Academy	Hi-School	28	30	38	68																				
1988-89	Thayer Academy	Hi-School	25	35	38	73																				
1989-90	Boston University	H-East	41	25	33	58	52																			
1990-91	Boston University	H-East	38	31	37	68	82																			
	NY Rangers	NHL																	2	0	2	2	2	0	0	0
1991-92	NY Rangers	NHL	79	35	34	69	55	9	0	4	234	15.0	12						13	3	6	9	2	2	0	0
1992-93	NY Rangers	NHL	83	33	43	76	49	13	0	4	270	12.2	0													
1993-94	NY Rangers	NHL	72	16	22	38	31	3	0	4	179	8.9	5													
	Chicago	NHL	7	1	3	4	6	1	0	0	16	6.3	-5						6	4	2	6	4	1	0	1
1994-95	HC Fassa	Euroliga	14	22	16	38	10																			
	HC Fassa	EuroHL	2	5	1	6	0																			
	Chicago	NHL	48	15	20	35	41	6	1	3	105	14.3	7						16	3	3	6	10	0	0	0
1995-96	Chicago	NHL	81	31	32	63	62	5	4	5	216	14.4	10						7	2	4	6	6	1	0	0
1996-97	Chicago	NHL	81	41	36	77	64	9	2	4	266	15.4	35						6	4	2	6	8	0	0	0
1997-98	Chicago	NHL	82	31	42	73	66	7	3	5	296	10.5	21													
	United States	Olympics	4	0	1	1	4																			
1998-99	Chicago	NHL	82	44	31	75	60	14	3	8	256	17.2	0	8	12.5	55	40	22:12								
99-2000	Chicago	NHL	82	43	41	84	48	11	5	2	260	16.5	10	22	22.7	34	44	21:54								
2000-01	Chicago	NHL	82	35	29	64	54	9	1	3	256	13.7	-22	27	40.7	52	39	22:09								
2001-02	Chicago	NHL	82	27	39	66	67	6	1	4	232	11.6	11	30	43.3	26	71	21:18	5	0	1	1	4	0	0	0
	United States	Olympics	6	2	2	4	0																			
	NHL Totals		861	352	372	724	603	93	20	46	2586	13.6		87	34.5	167	194	21:53	55	16	20	36	36	4	0	1

Hockey East Second All-Star Team (1991) • NCAA Championship All-Tournament Team (1991) • NHL All-Rookie Team (1992) • Played in NHL All-Star Game (1997, 1998, 1999, 2000, 2001)

• Missed majority of 1985-86 season recovering from knee injury, October, 1985. Traded to **Chicago** by **NY Rangers** with the rights to Matt Oates for Stephane Matteau and Brian Noonan, March 21, 1994. Signed as a free agent by **Phoenix**, July 12, 2002.

ANDERSSON, Jonas

(AN-duhr-suhn, JOH-nas) **NSH.**

Right wing. Shoots left. 6'3", 202 lbs. Born, Stockholm, Sweden, February 24, 1981. Nashville's 2nd choice, 33rd overall, in 1999 Entry Draft.

Season	Club	League	GP	G	A	Pts	PIM	PP	SH	GW	S	%	+/-	TF	F%	H	SB	Min	Playoffs GP	G	A	Pts	PIM	PP	SH	GW
1997-98	AIK Solna Jr.	Swede-Jr.	33	14	16	30	32																			
1998-99	AIK Solna Jr.	Swede-Jr.	16	3	7	10	18																			
	London Knights	Britain	12	2	3	5	0																			
99-2000	North Bay	OHL	67	31	36	67	27												6	2	2	4	2			
	Milwaukee	IHL	2	1	0	1	0												2	0	0	0	2			
2000-01	Milwaukee	IHL	52	6	7	13	44												5	0	0	0	2			
2001-02	Nashville	NHL	5	0	0	0	2	0	0	0	4	0.0	-2	0	0.0	5	0	9:06								
	Milwaukee	AHL	71	13	17	30	19																			
	NHL Totals		5	0	0	0	2	0	0	0	4	0.0		0	0.0	5	0	9:06								

ANDERSSON, Niklas

(AN-duhr-suhn, NIHK-las)

Left wing. Shoots left. 5'9", 180 lbs. Born, Kungalv, Sweden, May 20, 1971. Quebec's 5th choice, 68th overall, in 1989 Entry Draft.

Season	Club	League	GP	G	A	Pts	PIM	PP	SH	GW	S	%	+/-	TF	F%	H	SB	Min	Playoffs GP	G	A	Pts	PIM	PP	SH	GW
1987-88	Vastra Frolunda	Swede-2	15	5	4	9	6												8	6	4	10	4			
1988-89	Vastra Frolunda	Swede-2	30	12	24	36	24												10	4	6	10	4			
1989-90	Vastra Frolunda	Sweden	38	10	21	31	14																			
1990-91	Vastra Frolunda	Sweden	39	14	25	39	38												10	6	3	9	24			
1991-92	Halifax Citadels	AHL	57	8	26	34	41																			
1992-93	Quebec	NHL	3	0	1	1	2	0	0	0	4	0.0	0													
	Halifax Citadels	AHL	76	32	50	82	42																			
1993-94	Cornwall Aces	AHL	42	18	34	52	8																			
1994-95	Denver Grizzlies	IHL	66	22	39	61	28												15	8	13	21	10			
1995-96	NY Islanders	NHL	47	14	12	26	12	3	2	1	89	15.7	-3													
	Utah Grizzlies	IHL	30	13	22	35	25																			
1996-97	NY Islanders	NHL	74	12	31	43	57	1	1	1	122	9.8	4													
1997-98	San Jose	NHL	5	0	0	0	2	0	0	0	6	0.0	-1													
	Kentucky	AHL	37	10	28	38	54																			
	Utah Grizzlies	IHL	21	6	20	26	24												4	3	1	4	4			
1998-99	Chicago Wolves	IHL	65	17	47	64	49												10	2	2	4	10			
99-2000	NY Islanders	NHL	17	3	7	10	8	1	0	0	24	12.5	-3	0	0.0	10	4	13:19								
	Chicago Wolves	IHL	52	20	21	41	59												9	6	1	7	4			
	Nashville	NHL	7	0	1	1	0	0	0	0	7	0.0	0	0	0.0	1	0	12:50								
2000-01	Calgary	NHL	11	0	1	1	4	0	0	0	8	0.0	0	4	25.0	5	2	11:10								
	Chicago Wolves	IHL	66	33	39	72	81												16	1	*14	15	14			
2001-02	Vastra Frolunda	Sweden	41	14	32	46	64												7	0	2	2	6			
	NHL Totals		164	29	53	82	85	5	3	2	260	11.2		4	25.0	16	6	12:33								

IHL Second All-Star Team (2000) • IHL First All-Star Team (2001)

Signed as a free agent by **NY Islanders**, July 15, 1994. Signed as a free agent by **San Jose**, September 17, 1997. Signed as a free agent by **Toronto**, September 4, 1998. Traded to **NY Islanders** by **Toronto** for Craig Charron, August 17, 1999. Claimed on waivers by **Nashville** from **NY Islanders**, January 20, 2000. Claimed on waivers by **NY Islanders** from **Nashville**, February 19, 2000. Signed as a free agent by **Calgary**, August 29, 2000.

ANDREYCHUK, Dave (AN-druh-chuhk, DAYV) T.B.

Left wing. Shoots right. 6'4", 220 lbs. Born, Hamilton, Ont., September 29, 1963. Buffalo's 3rd choice, 16th overall, in 1982 Entry Draft.

Season	Club	League	GP	G	A	Pts	PIM	PP	SH	GW	S	%	+/-	TF	F%	H	SB	Min	Playoffs GP	G	A	Pts	PIM	PP	SH	GW
1979-80	Hamilton Hawks	OMHA	21	25	24	49	....	....	....	....	....	....	....	....	....	....	....	....	....	....	....	....	....	....	....	....
1980-81	Oshawa Generals	OMJHL	67	22	22	44	80	....	....	....	....	....	....	....	....	....	....	....	10	3	2	5	20	....	....	....
1981-82	Oshawa Generals	OHL	67	57	43	100	71	....	....	....	....	....	....	....	....	....	....	....	3	1	4	5	16	....	....	....
1982-83	Oshawa Generals	OHL	14	8	24	32	6	....	....	....	....	....	....	....	....	....	....	....	....	....	....	....	....	....	....	....
	Buffalo	**NHL**	**43**	**14**	**23**	**37**	**16**	**3**	**0**	**1**	**66**	**21.2**	**6**	....	....	....	....	....	**4**	**1**	**0**	**1**	**4**	**0**	**0**	**0**
1983-84	**Buffalo**	**NHL**	**78**	**38**	**42**	**80**	**42**	**10**	**0**	**7**	**178**	**21.3**	**20**	....	....	....	....	....	**2**	**0**	**1**	**1**	**2**	**0**	**0**	**0**
1984-85	**Buffalo**	**NHL**	**64**	**31**	**30**	**61**	**54**	**14**	**0**	**2**	**153**	**20.3**	**–4**	....	....	....	....	....	**5**	**4**	**2**	**6**	**4**	**0**	**0**	**2**
1985-86	**Buffalo**	**NHL**	**80**	**36**	**51**	**87**	**61**	**12**	**0**	**3**	**225**	**16.0**	**3**	....	....	....	....	....	....	....	....	....	....	....	....	....
1986-87	**Buffalo**	**NHL**	**77**	**25**	**48**	**73**	**46**	**13**	**0**	**2**	**255**	**9.8**	**2**	....	....	....	....	....	....	....	....	....	....	....	....	....
1987-88	**Buffalo**	**NHL**	**80**	**30**	**48**	**78**	**112**	**15**	**0**	**5**	**253**	**11.9**	**1**	....	....	....	....	....	**6**	**2**	**4**	**6**	**0**	**1**	**0**	**0**
1988-89	**Buffalo**	**NHL**	**56**	**28**	**24**	**52**	**40**	**7**	**0**	**3**	**145**	**19.3**	**0**	....	....	....	....	....	**5**	**0**	**3**	**3**	**0**	**0**	**0**	**0**
1989-90	**Buffalo**	**NHL**	**73**	**40**	**42**	**82**	**42**	**18**	**0**	**3**	**206**	**19.4**	**6**	....	....	....	....	....	**6**	**2**	**5**	**7**	**2**	**1**	**0**	**0**
1990-91	**Buffalo**	**NHL**	**80**	**36**	**33**	**69**	**32**	**13**	**0**	**4**	**234**	**15.4**	**11**	....	....	....	....	....	**6**	**2**	**2**	**4**	**8**	**1**	**0**	**0**
1991-92	**Buffalo**	**NHL**	**80**	**41**	**50**	**91**	**71**	**28**	**0**	**2**	**337**	**12.2**	**–9**	....	....	....	....	....	**7**	**1**	**3**	**4**	**12**	**0**	**0**	**0**
1992-93	**Buffalo**	**NHL**	**52**	**29**	**32**	**61**	**48**	**20**	**0**	**2**	**171**	**17.0**	**–8**	....	....	....	....	....	....	....	....	....	....	....	....	....
	Toronto	**NHL**	**31**	**25**	**13**	**38**	**8**	**12**	**0**	**2**	**139**	**18.0**	**12**	....	....	....	....	....	**21**	**12**	**7**	**19**	**35**	**4**	**0**	**3**
1993-94	**Toronto**	**NHL**	**83**	**53**	**46**	**99**	**98**	**21**	**5**	**8**	**333**	**15.9**	**22**	....	....	....	....	....	**18**	**5**	**5**	**10**	**16**	**3**	**1**	**0**
1994-95	**Toronto**	**NHL**	**48**	**22**	**16**	**38**	**34**	**8**	**0**	**2**	**168**	**13.1**	**–7**	....	....	....	....	....	**7**	**3**	**2**	**5**	**25**	**2**	**0**	**0**
1995-96	**Toronto**	**NHL**	**61**	**20**	**24**	**44**	**54**	**12**	**2**	**3**	**200**	**10.0**	**–11**	....	....	....	....	....	....	....	....	....	....	....	....	....
	New Jersey	**NHL**	**15**	**8**	**5**	**13**	**10**	**2**	**0**	**0**	**41**	**19.5**	**2**	....	....	....	....	....	....	....	....	....	....	....	....	....
1996-97	**New Jersey**	**NHL**	**82**	**27**	**34**	**61**	**48**	**4**	**1**	**2**	**233**	**11.6**	**38**	....	....	....	....	....	**1**	**0**	**0**	**0**	**0**	**0**	**0**	**0**
1997-98	**New Jersey**	**NHL**	**75**	**14**	**34**	**48**	**26**	**4**	**0**	**2**	**180**	**7.8**	**19**	....	....	....	....	....	**6**	**1**	**0**	**1**	**4**	**1**	**0**	**0**
1998-99	**New Jersey**	**NHL**	**52**	**15**	**13**	**28**	**20**	**4**	**0**	**3**	**110**	**13.6**	**1**	**9**	**44.4**	**36**	**17**	**15:32**	**4**	**2**	**0**	**2**	**4**	**0**	**0**	**0**
99-2000	**Boston**	**NHL**	**63**	**19**	**14**	**33**	**28**	**7**	**0**	**2**	**192**	**9.9**	**–11**	**446**	**52.0**	**68**	**66**	**19:50**	....	....	....	....	....	....	....	....
	Colorado	**NHL**	**14**	**1**	**2**	**3**	**2**	**1**	**0**	**1**	**41**	**2.4**	**–9**	**15**	**60.0**	**10**	**9**	**17:16**	**17**	**3**	**2**	**5**	**18**	**2**	**0**	**0**
2000-01	**Buffalo**	**NHL**	**74**	**20**	**13**	**33**	**32**	**8**	**0**	**4**	**119**	**16.8**	**0**	**187**	**49.7**	**36**	**25**	**11:60**	**13**	**1**	**2**	**3**	**4**	**1**	**0**	**0**
2001-02	**Tampa Bay**	**NHL**	**82**	**21**	**17**	**38**	**109**	**9**	**1**	**5**	**161**	**13.0**	**–12**	**1393**	**53.0**	**41**	**74**	**16:26**	....	....	....	....	....	....	....	....
	NHL Totals		**1443**	**593**	**654**	**1247**	**1033**	**245**	**9**	**68**	**4140**	**14.3**		**2050**	**52.5**	**191**	**191**	**15:54**	**128**	**39**	**38**	**77**	**138**	**16**	**1**	**5**

Played in NHL All-Star Game (1990, 1994)

Traded to **Toronto** by **Buffalo** with Daren Puppa and Buffalo's 1st round choice (Kenny Jonsson) in 1993 Entry Draft for Grant Fuhr and Toronto's 5th round choice (Kevin Popp) in 1995 Entry Draft, February 2, 1993. Traded to **New Jersey** by **Toronto** for New Jersey's 2nd round choice (Marek Posmyk) in 1996 Entry Draft and New Jersey's 3rd round choice (later traded back to New Jersey - New Jersey selected Andre Lakos) in 1999 Entry Draft, March 13, 1996. Signed as a free agent by Boston, July 29, 1999. Traded to **Colorado** by **Boston** with Raymond Bourque for Brian Rolston, Martin Grenier, Sami Pahlsson and New Jersey's 1st round choice (previously acquired, Boston selected Martin Samuelsson) in 2000 Entry Draft, March 6, 2000. Signed as a free agent by **Buffalo**, July 13, 2000. Signed as a free agent by **Tampa Bay**, July 13, 2001.

ANTROPOV, Nik (an-TROH-pahv, NIHK) TOR.

Center. Shoots left. 6'5", 203 lbs. Born, Vost, USSR, February 18, 1980. Toronto's 1st choice, 10th overall, in 1998 Entry Draft.

Season	Club	League	GP	G	A	Pts	PIM	PP	SH	GW	S	%	+/-	TF	F%	H	SB	Min	Playoffs GP	G	A	Pts	PIM	PP	SH	GW
1995-96	Ust-Kamenog. Jr.	CIS-Jr.	20	18	20	38	30	....	....	....	....	....	....	....	....	....	....	....	....	....	....	....	....	....	....	....
1996-97	Ust-Kamenogorsk	Russia-2	8	2	1	3	6	....	....	....	....	....	....	....	....	....	....	....	....	....	....	....	....	....	....	....
1997-98	Ust-Kamenogorsk	Russia-2	42	15	24	39	62	....	....	....	....	....	....	....	....	....	....	....	....	....	....	....	....	....	....	....
1998-99	Dynamo Moscow	Russia	30	5	9	14	30	....	....	....	....	....	....	....	....	....	....	....	11	0	1	1	4	....	....	....
99-2000	**Toronto**	**NHL**	**66**	**12**	**18**	**30**	**41**	**0**	**0**	**2**	**89**	**13.5**	**14**	**501**	**46.3**	**79**	**22**	**12:48**	**3**	**0**	**0**	**0**	**4**	**0**	**0**	**0**
	St. John's	AHL	2	0	0	0	4	....	....	....	....	....	....	....	....	....	....	....	....	....	....	....	....	....	....	....
2000-01	**Toronto**	**NHL**	**52**	**6**	**11**	**17**	**30**	**0**	**0**	**1**	**71**	**8.5**	**5**	**431**	**44.3**	**53**	**8**	**10:02**	**9**	**2**	**1**	**3**	**12**	**1**	**0**	**1**
2001-02	**Toronto**	**NHL**	**11**	**1**	**1**	**2**	**4**	**0**	**0**	**0**	**12**	**8.3**	**–1**	**31**	**38.7**	**10**	**1**	**8:57**	....	....	....	....	....	....	....	....
	St. John's	AHL	34	11	24	35	47	....	....	....	....	....	....	....	....	....	....	....	....	....	....	....	....	....	....	....
	NHL Totals		**129**	**19**	**30**	**49**	**75**	**0**	**0**	**3**	**172**	**11.0**		**963**	**45.2**	**142**	**31**	**11:21**	**12**	**2**	**1**	**3**	**16**	**1**	**0**	**1**

ARKHIPOV, Denis (AHR-kih-pahv, DIHN-ihs) NSH.

Right wing. Shoots left. 6'3", 214 lbs. Born, Kazan, USSR, May 19, 1979. Nashville's 2nd choice, 60th overall, in 1998 Entry Draft.

Season	Club	League	GP	G	A	Pts	PIM	PP	SH	GW	S	%	+/-	TF	F%	H	SB	Min	Playoffs GP	G	A	Pts	PIM	PP	SH	GW
1994-95	Kazan Jr.	CIS-Jr.	40	20	12	32	10	....	....	....	....	....	....	....	....	....	....	....	....	....	....	....	....	....	....	....
1995-96	Kazan Jr.	CIS-Jr.	40	15	8	23	30	....	....	....	....	....	....	....	....	....	....	....	....	....	....	....	....	....	....	....
	Ak Bars Kazan	CIS	15	10	8	18	10	....	....	....	....	....	....	....	....	....	....	....	....	....	....	....	....	....	....	....
1996-97	Ak Bars Kazan 2	Russia-3	50	17	23	40	20	....	....	....	....	....	....	....	....	....	....	....	....	....	....	....	....	....	....	....
	Ak Bars Kazan	Russia	1	1	0	1	0	....	....	....	....	....	....	....	....	....	....	....	....	....	....	....	....	....	....	....
1997-98	Ak Bars Kazan	Russia	29	2	2	4	2	....	....	....	....	....	....	....	....	....	....	....	....	....	....	....	....	....	....	....
1998-99	Ak Bars Kazan	Russia	34	12	1	13	22	....	....	....	....	....	....	....	....	....	....	....	9	2	3	5	6	....	....	....
	Ak Bars Kazan	EuroHL	4	0	0	0	0	....	....	....	....	....	....	....	....	....	....	....	1	0	0	0	0	....	....	....
99-2000	Ak Bars Kazan	Russia	32	7	9	16	14	....	....	....	....	....	....	....	....	....	....	....	18	5	5	10	6	....	....	....
2000-01	**Nashville**	**NHL**	**40**	**6**	**7**	**13**	**4**	**0**	**0**	**0**	**42**	**14.3**	**0**	**299**	**43.8**	**16**	**10**	**9:56**	....	....	....	....	....	....	....	....
	Milwaukee	IHL	40	9	8	17	11	....	....	....	....	....	....	....	....	....	....	....	....	....	....	....	....	....	....	....
2001-02	**Nashville**	**NHL**	**82**	**20**	**22**	**42**	**16**	**7**	**0**	**6**	**118**	**16.9**	**–18**	**1108**	**44.8**	**53**	**19**	**15:43**	....	....	....	....	....	....	....	....
	NHL Totals		**122**	**26**	**29**	**55**	**20**	**7**	**0**	**6**	**160**	**16.3**		**1407**	**44.6**	**69**	**29**	**13:49**	....	....	....	....	....	....	....	....

ARMSTRONG, Chris (ahrm-STRAWNG, KRIHS)

Defense. Shoots left. 6', 205 lbs. Born, Regina, Sask., June 26, 1975. Florida's 3rd choice, 57th overall, in 1993 Entry Draft.

Season	Club	League	GP	G	A	Pts	PIM	PP	SH	GW	S	%	+/-	TF	F%	H	SB	Min	Playoffs GP	G	A	Pts	PIM	PP	SH	GW
1990-91	Whitewood	SMHL	40	25	30	55	40	....	....	....	....	....	....	....	....	....	....	....	....	....	....	....	....	....	....	....
1991-92	Moose Jaw	WHL	43	2	7	9	19	....	....	....	....	....	....	....	....	....	....	....	4	0	0	0	0	....	....	....
1992-93	Moose Jaw	WHL	67	9	35	44	104	....	....	....	....	....	....	....	....	....	....	....	....	....	....	....	....	....	....	....
1993-94	Moose Jaw	WHL	64	13	55	68	54	....	....	....	....	....	....	....	....	....	....	....	....	....	....	....	....	....	....	....
	Cincinnati	IHL	1	0	0	0	0	....	....	....	....	....	....	....	....	....	....	....	10	1	3	4	2	....	....	....
1994-95	Moose Jaw	WHL	66	17	54	71	61	....	....	....	....	....	....	....	....	....	....	....	10	2	12	14	22	....	....	....
	Cincinnati	IHL	....	....	....	....	....	....	....	....	....	....	....	....	....	....	....	....	9	1	3	4	10	....	....	....
1995-96	Carolina	AHL	78	9	33	42	65	....	....	....	....	....	....	....	....	....	....	....	....	....	....	....	....	....	....	....
1996-97	Carolina	AHL	66	9	23	32	38	....	....	....	....	....	....	....	....	....	....	....	....	....	....	....	....	....	....	....
1997-98	Fort Wayne	IHL	79	8	36	44	66	....	....	....	....	....	....	....	....	....	....	....	4	0	2	2	4	....	....	....
1998-99	Milwaukee	IHL	5	0	3	3	4	....	....	....	....	....	....	....	....	....	....	....	....	....	....	....	....	....	....	....
	Hershey Bears	AHL	65	12	32	44	30	....	....	....	....	....	....	....	....	....	....	....	5	0	1	1	0	....	....	....
99-2000	Kentucky	AHL	78	9	48	57	77	....	....	....	....	....	....	....	....	....	....	....	9	1	5	6	4	....	....	....
2000-01	**Minnesota**	**NHL**	**3**	**0**	**0**	**0**	**0**	**0**	**0**	**0**	**4**	**0.0**	**–3**	**0**	**0.0**	**4**	**2**	**18:06**	....	....	....	....	....	....	....	....
	Cleveland	IHL	77	9	32	41	42	....	....	....	....	....	....	....	....	....	....	....	4	0	2	2	2	....	....	....
2001-02	Bridgeport	AHL	80	10	38	48	49	....	....	....	....	....	....	....	....	....	....	....	20	3	8	11	4	....	....	....
	NHL Totals		**3**	**0**	**0**	**0**	**0**	**0**	**0**	**0**	**4**	**0.0**		**0**	**0.0**	**4**	**2**	**18:06**	....	....	....	....	....	....	....	....

WHL East First All-Star Team (1994) • Canadian Major Junior Second All-Star Team (1994) • WHL East Second All-Star Team (1995)

Claimed by **Nashville** from **Florida** in Expansion Draft, June 26, 1998. Signed as a free agent by **San Jose**, September 2, 1999. Selected by **Minnesota** from **San Jose** in Expansion Draft, June 23, 2000. Signed as a free agent by **NY Islanders**, August 8, 2001. Signed as a free agent by **EV Zug** (Swiss), June 14, 2002.

ARMSTRONG, Derek (ahrm-STRAWNG, DEHR-ehk) L.A.

Center. Shoots right. 5'11", 188 lbs. Born, Ottawa, Ont., April 23, 1973. NY Islanders' 5th choice, 128th overall, in 1992 Entry Draft.

Season	Club	League	GP	G	A	Pts	PIM	PP	SH	GW	S	%	+/-	TF	F%	H	SB	Min	Playoffs GP	G	A	Pts	PIM	PP	SH	GW
1989-90	Hawkesbury	OCJHL	48	8	10	18	30	....	....	....	....	....	....	....	....	....	....	....	....	....	....	....	....	....	....	....
1990-91	Hawkesbury	OCJHL	54	27	45	72	49	....	....	....	....	....	....	....	....	....	....	....	....	....	....	....	....	....	....	....
	Sudbury Wolves	OHL	2	0	2	2	0	....	....	....	....	....	....	....	....	....	....	....	....	....	....	....	....	....	....	....
1991-92	Sudbury Wolves	OHL	66	31	54	85	22	....	....	....	....	....	....	....	....	....	....	....	9	2	2	4	2	....	....	....
1992-93	Sudbury Wolves	OHL	66	44	62	106	56	....	....	....	....	....	....	....	....	....	....	....	14	9	10	19	26	....	....	....
1993-94	**NY Islanders**	**NHL**	**1**	**0**	**0**	**0**	**0**	**0**	**0**	**0**	**2**	**0.0**	**0**	....	....	....	....	....	....	....	....	....	....	....	....	....
	Salt Lake	IHL	76	23	35	58	61	....	....	....	....	....	....	....	....	....	....	....	....	....	....	....	....	....	....	....
1994-95	Denver Grizzlies	IHL	59	13	18	31	65	....	....	....	....	....	....	....	....	....	....	....	6	0	2	2	0	....	....	....
1995-96	**NY Islanders**	**NHL**	**19**	**1**	**3**	**4**	**14**	**0**	**0**	**0**	**23**	**4.3**	**–6**	....	....	....	....	....	....	....	....	....	....	....	....	....
	Worcester	AHL	51	11	15	26	33	....	....	....	....	....	....	....	....	....	....	....	4	2	1	3	0	....	....	....
1996-97	**NY Islanders**	**NHL**	**50**	**6**	**7**	**13**	**33**	**0**	**0**	**2**	**36**	**16.7**	**–8**	....	....	....	....	....	....	....	....	....	....	....	....	....
	Utah Grizzlies	IHL	17	4	8	12	10	....	....	....	....	....	....	....	....	....	....	....	6	0	4	4	4	....	....	....

Season	Club	League	GP	G	A	Pts	PIM	PP	SH	GW	S	%	+/-	TF	F%	H	SB	Min	Playoffs GP	G	A	Pts	PIM	PP	SH	GW
1997-98	**Ottawa**	**NHL**	**9**	**2**	**0**	**2**	**9**	**0**	**0**	**1**	**8**	**25.0**	**1**													
	Detroit Vipers	IHL	10	0	1	1	2																			
	Hartford	AHL	54	16	30	46	40												15	2	6	8	22			
1998-99	**NY Rangers**	**NHL**	**3**	**0**	**0**	**0**	**0**	**0**	**0**	**0**	**1**	**0.0**	**0**	**0**	**0.0**	**0**	**0**	**2:50**								
	Hartford	AHL	59	29	51	80	73												7	5	4	9	10			
99-2000	**NY Rangers**	**NHL**	**1**	**0**	**0**	**0**	**0**	**0**	**0**	**0**	**1**	**0.0**	**0**	**3**	**33.3**	**0**	**0**	**3:10**								
	Hartford	AHL	77	28	54	82	101												23	7	16	23	24			
2000-01	**NY Rangers**	**NHL**	**3**	**0**	**0**	**0**	**0**	**0**	**0**	**0**	**6**	**0.0**	**0**	**30**	**50.0**	**4**	**0**	**11:22**								
	Hartford	AHL	75	32	*69	*101	73												5	0	6	6	6			
2001-02	SC Bern	Swiss	44	17	36	53	62												6	3	5	8	8			
	NHL Totals		**86**	**9**	**10**	**19**	**56**	**0**	**0**	**3**	**77**	**11.7**		**33**	**48.5**	**4**	**0**	**6:32**								

AHL Second All-Star Team (2000) • AHL First All-Star Team (2001) • Won John P. Sollenberger Trophy (Top Scorer - AHL) (2001) • Won Les Cunningham Award (MVP - AHL) (2001)

Signed as a free agent by **Ottawa**, July 28, 1997. Loaned to **Hartford** (AHL) by **Ottawa**, October 28, 1997. Signed as a free agent by **NY Rangers**, August 10, 1998. Signed as a free agent by **SC Bern** (Swiss) with NY Rangers retaining NHL rights, July 18, 2001. Traded to **LA Kings** by **NY Rangers** for a conditional choice in 2003 Entry Draft, July 16, 2002.

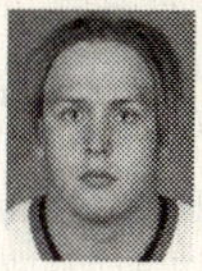

ARNASON, Tyler

(AHR-na-suhn, TIGH-luhr) **CHI.**

Center. Shoots left. 5'11", 207 lbs. Born, Oklahoma City, OK, March 16, 1979. Chicago's 6th choice, 183rd overall, in 1998 Entry Draft.

Season	Club	League	GP	G	A	Pts	PIM	PP	SH	GW	S	%	+/-	TF	F%	H	SB	Min	Playoffs GP	G	A	Pts	PIM	PP	SH	GW
1996-97	Winnipeg South	MJHL	50	35	50	85	15												6	3	3	6	18			
1997-98	Fargo-Moorhead	USHL	52	37	45	82	16												4	1	1	2	2			
1998-99	St. Cloud State	WCHA	38	14	17	31	16																			
99-2000	St. Cloud State	WCHA	39	19	30	49	18																			
2000-01	St. Cloud State	WCHA	41	28	28	56	14																			
2001-02	**Chicago**	**NHL**	**21**	**3**	**1**	**4**	**4**	**0**	**0**	**0**	**19**	**15.8**	**−3**	**112**	**41.1**	**1**	**1**	**9:28**	**3**	**0**	**0**	**0**	**0**	**0**	**0**	**0**
	Norfolk Admirals	AHL	60	26	30	56	42																			
	NHL Totals		**21**	**3**	**1**	**4**	**4**	**0**	**0**	**0**	**19**	**15.8**		**112**	**41.1**	**1**	**1**	**9:28**	**3**	**0**	**0**	**0**	**0**	**0**	**0**	**0**

MJHL Rookie of the Year (1997) • USHL First All-Star Team (1998) • WCHA All-Rookie Team (1999) • WCHA Second All-Star Team (2000) • AHL All-Rookie Team (2002) • Won Dudley "Red" Garrett Memorial Trophy (Top Rookie - AHL) (2002)

ARNOTT, Jason

(AHR-niht, JAY-suhn) **DAL.**

Center. Shoots right. 6'4", 225 lbs. Born, Collingwood, Ont., October 11, 1974. Edmonton's 1st choice, 7th overall, in 1993 Entry Draft.

Season	Club	League	GP	G	A	Pts	PIM	PP	SH	GW	S	%	+/-	TF	F%	H	SB	Min	Playoffs GP	G	A	Pts	PIM	PP	SH	GW
1989-90	Stayner Siskins	OJHL-C	34	21	31	52	12																			
1990-91	Lindsay Bears	OJHL-B	42	17	44	61	10												8	9	8	17	6			
1991-92	Oshawa Generals	OHL	57	9	15	24	12																			
1992-93	Oshawa Generals	OHL	56	41	57	98	74												13	9	9	18	20			
1993-94	**Edmonton**	**NHL**	**78**	**33**	**35**	**68**	**104**	**10**	**0**	**4**	**194**	**17.0**	**1**													
1994-95	**Edmonton**	**NHL**	**42**	**15**	**22**	**37**	**128**	**7**	**0**	**1**	**156**	**9.6**	**−14**													
1995-96	**Edmonton**	**NHL**	**64**	**28**	**31**	**59**	**87**	**8**	**0**	**5**	**244**	**11.5**	**−6**													
1996-97	**Edmonton**	**NHL**	**67**	**19**	**38**	**57**	**92**	**10**	**1**	**2**	**248**	**7.7**	**−21**						**12**	**3**	**6**	**9**	**18**	**1**	**0**	**0**
1997-98	**Edmonton**	**NHL**	**35**	**5**	**13**	**18**	**78**	**1**	**0**	**0**	**100**	**5.0**	**−16**													
	New Jersey	**NHL**	**35**	**5**	**10**	**15**	**21**	**3**	**0**	**2**	**99**	**5.1**	**−8**						**5**	**0**	**2**	**2**	**0**	**0**	**0**	**0**
1998-99	**New Jersey**	**NHL**	**74**	**27**	**27**	**54**	**79**	**8**	**0**	**3**	**200**	**13.5**	**10**	**872**	**49.3**	**196**	**16**	**15:24**	**7**	**2**	**2**	**4**	**4**	**1**	**0**	**0**
99-2000♦	**New Jersey**	**NHL**	**76**	**22**	**34**	**56**	**51**	**7**	**0**	**4**	**244**	**9.0**	**22**	**1172**	**46.9**	**194**	**18**	**17:05**	**23**	**8**	**12**	**20**	**18**	**3**	**0**	**1**
2000-01	**New Jersey**	**NHL**	**54**	**21**	**34**	**55**	**75**	**8**	**0**	**3**	**138**	**15.2**	**23**	**760**	**49.6**	**108**	**14**	**16:12**	**23**	**8**	**7**	**15**	**16**	**5**	**0**	**0**
2001-02	**New Jersey**	**NHL**	**63**	**22**	**19**	**41**	**59**	**8**	**0**	**1**	**169**	**13.0**	**3**	**934**	**47.8**	**97**	**19**	**17:13**								
	Dallas	**NHL**	**10**	**3**	**1**	**4**	**6**	**2**	**0**	**2**	**28**	**10.7**	**−1**	**77**	**52.0**	**23**	**2**	**18:13**								
	NHL Totals		**598**	**200**	**264**	**464**	**780**	**72**	**1**	**27**	**1820**	**11.0**		**3815**	**48.3**	**618**	**69**	**16:32**	**70**	**21**	**29**	**50**	**56**	**10**	**0**	**1**

NHL All-Rookie Team (1994) • Played in NHL All-Star Game (1997)

Traded to **New Jersey** by **Edmonton** with Bryan Muir for Valeri Zelepukin and Bill Guerin, January 4, 1998. Traded to **Dallas** by **New Jersey** with Randy McKay and New Jersey's 1st round choice (later traded to Columbus - later traded to Buffalo - Buffalo selected Dan Paille) in 2002 Entry Draft for Joe Nieuwendyk and Jamie Langenbrunner, March 19, 2002.

ARVEDSON, Magnus

(AHR-vehd-suhn, MAGH-nuhs) **OTT.**

Center. Shoots left. 6'2", 198 lbs. Born, Karlstad, Sweden, November 25, 1971. Ottawa's 4th choice, 119th overall, in 1997 Entry Draft.

Season	Club	League	GP	G	A	Pts	PIM	PP	SH	GW	S	%	+/-	TF	F%	H	SB	Min	Playoffs GP	G	A	Pts	PIM	PP	SH	GW
1990-91	Orebro IK	Swede-2	29	7	11	18	12												2	0	1	1	2			
1991-92	Orebro IK	Swede-2	32	12	21	33	30												7	4	4	8	4			
1992-93	Orebro IK	Swede-2	36	11	18	29	34												6	2	1	3	0			
1993-94	Farjestad	Sweden	16	1	7	8	10																			
1994-95	Farjestad Jr.	Swede-Jr.	1	0	0	0	0																			
	Farjestad	Sweden	36	1	6	7	45												4	0	0	0	6			
1995-96	Farjestad	Sweden	40	10	14	24	40												8	0	3	3	10			
1996-97	Farjestad	Sweden	48	13	11	24	36												14	4	7	11	8			
	Farjestad	EuroHL	5	1	0	1	2												2	0	1	1	2			
1997-98	**Ottawa**	**NHL**	**61**	**11**	**15**	**26**	**36**	**0**	**1**	**0**	**90**	**12.2**	**2**						**11**	**0**	**1**	**1**	**6**	**0**	**0**	**0**
1998-99	**Ottawa**	**NHL**	**80**	**21**	**26**	**47**	**50**	**0**	**4**	**6**	**136**	**15.4**	**33**	**25**	**20.0**	**48**	**42**	**17:08**	**3**	**0**	**1**	**1**	**2**	**0**	**0**	**0**
99-2000	**Ottawa**	**NHL**	**47**	**15**	**13**	**28**	**36**	**1**	**1**	**4**	**91**	**16.5**	**4**	**11**	**45.5**	**33**	**43**	**18:04**	**6**	**0**	**0**	**0**	**6**	**0**	**0**	**0**
2000-01	**Ottawa**	**NHL**	**51**	**17**	**16**	**33**	**24**	**1**	**2**	**4**	**79**	**21.5**	**23**	**7**	**28.6**	**36**	**30**	**16:01**	**2**	**0**	**0**	**0**	**0**	**0**	**0**	**0**
2001-02	**Ottawa**	**NHL**	**74**	**12**	**27**	**39**	**35**	**0**	**0**	**1**	**121**	**9.9**	**27**	**8**	**37.5**	**62**	**33**	**17:44**	**12**	**2**	**1**	**3**	**4**	**0**	**0**	**0**
	Sweden	Olympics	4	0	0	0	0																			
	NHL Totals		**313**	**76**	**97**	**173**	**181**	**2**	**8**	**15**	**517**	**14.7**		**51**	**29.4**	**179**	**148**	**17:15**	**34**	**2**	**3**	**5**	**18**	**0**	**0**	**0**

ASHAM, Arron

(ASH-uhm, AIR-ruhn) **NYI**

Right wing. Shoots right. 5'11", 209 lbs. Born, Portage La Prairie, Man., April 13, 1978. Montreal's 3rd choice, 71st overall, in 1996 Entry Draft.

Season	Club	League	GP	G	A	Pts	PIM	PP	SH	GW	S	%	+/-	TF	F%	H	SB	Min	Playoffs GP	G	A	Pts	PIM	PP	SH	GW
1993-94	Portage	MAHA	21	18	19	37	82																			
1994-95	Red Deer Rebels	WHL	62	11	16	27	126																			
1995-96	Red Deer Rebels	WHL	70	32	45	77	174												10	6	3	9	20			
1996-97	Red Deer Rebels	WHL	67	45	51	96	149												16	12	14	26	36			
1997-98	Red Deer Rebels	WHL	67	43	49	92	153												5	0	2	2	8			
	Fredericton	AHL	2	1	1	2	0												2	0	1	1	0			
1998-99	**Montreal**	**NHL**	**7**	**0**	**0**	**0**	**0**	**0**	**0**	**0**	**5**	**0.0**	**−4**	**0**	**0.0**	**8**	**2**	**7:27**								
	Fredericton	AHL	60	16	18	34	118												13	8	6	14	11			
99-2000	**Montreal**	**NHL**	**33**	**4**	**2**	**6**	**24**	**0**	**1**	**1**	**29**	**13.8**	**−7**	**1**	**0.0**	**47**	**10**	**10:14**								
	Quebec	AHL	13	4	5	9	32												2	0	0	0	2			
2000-01	**Montreal**	**NHL**	**46**	**2**	**3**	**5**	**59**	**0**	**0**	**0**	**32**	**6.3**	**−9**	**3**	**100.0**	**93**	**10**	**8:28**								
	Quebec	AHL	15	7	9	16	51												7	1	2	3	2			
2001-02	**Montreal**	**NHL**	**35**	**5**	**4**	**9**	**55**	**0**	**0**	**0**	**30**	**16.7**	**7**	**4**	**25.0**	**50**	**5**	**8:13**	**3**	**0**	**1**	**1**	**0**	**0**	**0**	**0**
	Quebec	AHL	24	9	14	23	35																			
	NHL Totals		**121**	**11**	**9**	**20**	**138**	**0**	**1**	**1**	**96**	**11.5**		**8**	**50.0**	**198**	**27**	**8:49**	**3**	**0**	**1**	**1**	**0**	**0**	**0**	**0**

Traded to **NY Islanders** by **Montreal** with Montreal's 5th round choice (Markus Pahlsson) in 2002 Entry Draft for Mariusz Czerkawski, June 22, 2002.

ASTASHENKO, Kaspars

(ahs-tuh-SHEHN-koh, KAHS-pars) **CAR.**

Wing. Shoots left. 6'2", 183 lbs. Born, Riga, Latvia, February 17, 1975. Tampa Bay's 5th choice, 127th overall, in 1999 Entry Draft.

Season	Club	League	GP	G	A	Pts	PIM	PP	SH	GW	S	%	+/-	TF	F%	H	SB	Min	Playoffs GP	G	A	Pts	PIM	PP	SH	GW
1993-94	Pardaugava Riga	CIS	4	0	0	0	10																			
1994-95	Pardaugava Riga	CIS	25	0	0	0	24																			
1995-96	CSKA Moscow	CIS	26	0	1	1	10																			
1996-97	H.C. CSKA	Russia	41	0	0	0	48												2	0	1	1	4			
1997-98	H.C. CSKA	Russia	25	1	3	4	6																			
1998-99	Cincinnati	IHL	74	3	11	14	166												3	0	2	2	6			
	Dayton Bombers	ECHL	2	0	1	1	4																			
99-2000	**Tampa Bay**	**NHL**	**8**	**0**	**1**	**1**	**4**	**0**	**0**	**0**	**3**	**0.0**	**−2**	**0**	**0.0**	**11**	**6**	**16:48**								
	Detroit Vipers	IHL	51	1	10	11	86																			
	Long Beach	IHL	14	0	3	3	10																			
2000-01	**Tampa Bay**	**NHL**	**15**	**1**	**1**	**2**	**4**	**0**	**0**	**0**	**4**	**25.0**	**−4**	**0**	**0.0**	**16**	**1**	**6:50**								
	Detroit Vipers	IHL	51	6	10	16	58																			

Season	Club	League	GP	G	A	Pts	PIM	PP	SH	GW	S	%	+/-	TF	F%	H	SB	Min	GP	G	A	Pts	PIM	PP	SH	GW
					Regular Season														Playoffs							
2001-02	Springfield	AHL	11	0	2	2	15																			
	Lowell	AHL	37	2	8	10	39												5	1	1	2	2			
	Latvia	Olympics	1	0	0	0	0																			
	NHL Totals		**23**	**1**	**2**	**3**	**8**	**0**	**0**	**0**	**7**	**14.3**		**0**	**0.0**	**27**	**7**	**10:18**								

Traded to **Carolina** by **Tampa Bay** for Harlan Pratt, December 28, 2001.

AUBIN, Serge

(oh-BEHN, SAIRZH)

Center. Shoots left. 6'1", 194 lbs. Born, Val d'Or, Que., February 15, 1975. Pittsburgh's 9th choice, 161st overall, in 1994 Entry Draft.

Season	Club	League	GP	G	A	Pts	PIM	PP	SH	GW	S	%	+/-	TF	F%	H	SB	Min	GP	G	A	Pts	PIM	PP	SH	GW
1990-91	Abitibi	QAAA	27	2	4	6	10																			
1991-92	Abitibi	QAAA	42	28	32	60	36												1	0	1	1	0			
1992-93	Drummondville	QMJHL	65	16	34	50	30												8	0	1	1	16			
1993-94	Granby Bisons	QMJHL	63	42	32	74	80												7	2	3	5	8			
1994-95	Granby Bisons	QMJHL	60	37	73	110	55												11	8	15	23	4			
1995-96	Hampton Roads	ECHL	62	24	62	86	74												3	1	4	5	10			
	Cleveland	IHL	2	0	0	0	0												2	0	0	0	0			
1996-97	Cleveland	IHL	57	9	16	25	38												2	0	0	0	0			
1997-98	Syracuse Crunch	AHL	55	6	14	20	57																			
	Hershey Bears	AHL	5	2	1	3	0												7	1	3	4	6			
1998-99	Hershey Bears	AHL	64	30	39	69	58												3	0	1	1	2			
	Colorado	**NHL**	**1**	**0**	**0**	**0**	**0**	**0**	**0**	**0**	**1**	**0.0**	**0**	**1**	**0.0**	**0**	**0**	**4:16**								
99-2000	**Colorado**	**NHL**	**15**	**2**	**1**	**3**	**6**	**0**	**0**	**1**	**14**	**14.3**	**1**	**79**	**50.6**	**15**	**3**	**6:37**	**17**	**0**	**1**	**1**	**6**	**0**	**0**	**0**
	Hershey Bears	AHL	58	42	38	80	56																			
2000-01	**Columbus**	**NHL**	**81**	**13**	**17**	**30**	**107**	**0**	**0**	**2**	**110**	**11.8**	**–20**	**1346**	**51.3**	**144**	**73**	**16:20**								
2001-02	**Columbus**	**NHL**	**71**	**8**	**8**	**16**	**32**	**1**	**0**	**1**	**86**	**9.3**	**–20**	**780**	**50.5**	**129**	**80**	**15:30**								
	NHL Totals		**168**	**23**	**26**	**49**	**145**	**1**	**0**	**4**	**211**	**10.9**		**2206**	**51.0**	**288**	**156**	**15:03**	**17**	**0**	**1**	**1**	**6**	**0**	**0**	**0**

AHL First All-Star Team (2000)

Signed as a free agent by **Hershey** (AHL), July 24, 1998. Signed as a free agent by **Colorado**, December 22, 1998. Signed as a free agent by **Columbus**, July 11, 2000.

AUCOIN, Adrian

(oh-KOIN, AY-dree-an) **NYI**

Defense. Shoots right. 6'2", 214 lbs. Born, Ottawa, Ont., July 3, 1973. Vancouver's 7th choice, 117th overall, in 1992 Entry Draft.

Season	Club	League	GP	G	A	Pts	PIM	PP	SH	GW	S	%	+/-	TF	F%	H	SB	Min	GP	G	A	Pts	PIM	PP	SH	GW
1989-90	Nepean Raiders	OCJHL	54	2	14	16	95												4	0	1	1				
1990-91	Nepean Raiders	OCJHL	56	17	33	50	125																			
1991-92	Boston University	H-East	32	2	10	12	60																			
1992-93	Team Canada	Nat-Tm	42	8	10	18	71																			
1993-94	Team Canada	Nat-Tm	59	5	12	17	80																			
	Canada	Olympics	4	0	0	0	2																			
	Hamilton Canucks	AHL	13	1	2	3	19												4	0	2	2	6			
1994-95	Syracuse Crunch	AHL	71	13	18	31	52																			
	Vancouver	**NHL**	**1**	**1**	**0**	**1**	**0**	**0**	**0**	**0**	**2**	**50.0**	**1**						**4**	**1**	**0**	**1**	**0**	**1**	**0**	**0**
1995-96	**Vancouver**	**NHL**	**49**	**4**	**14**	**18**	**34**	**2**	**0**	**0**	**85**	**4.7**	**8**						**6**	**0**	**0**	**0**	**2**	**0**	**0**	**0**
	Syracuse Crunch	AHL	29	5	13	18	47																			
1996-97	**Vancouver**	**NHL**	**70**	**5**	**16**	**21**	**63**	**1**	**0**	**0**	**116**	**4.3**	**0**													
1997-98	**Vancouver**	**NHL**	**35**	**3**	**3**	**6**	**21**	**1**	**0**	**1**	**44**	**6.8**	**–4**													
1998-99	**Vancouver**	**NHL**	**82**	**23**	**11**	**34**	**77**	**18**	**2**	**3**	**174**	**13.2**	**–14**	**1**	**100.0**	**208**	**50**	**23:52**								
99-2000	**Vancouver**	**NHL**	**57**	**10**	**14**	**24**	**30**	**4**	**0**	**1**	**126**	**7.9**	**7**	**0**	**0.0**	**123**	**36**	**23:06**								
2000-01	**Vancouver**	**NHL**	**47**	**3**	**13**	**16**	**20**	**1**	**0**	**0**	**99**	**3.0**	**13**	**0**	**0.0**	**62**	**14**	**18:21**								
	Tampa Bay	**NHL**	**26**	**1**	**11**	**12**	**25**	**1**	**0**	**0**	**60**	**1.7**	**–8**	**0**	**0.0**	**43**	**25**	**23:34**								
2001-02	**NY Islanders**	**NHL**	**81**	**12**	**22**	**34**	**62**	**7**	**0**	**1**	**232**	**5.2**	**23**	**0**	**0.0**	**189**	**79**	**28:54**	**7**	**2**	**5**	**7**	**4**	**2**	**0**	**0**
	NHL Totals		**448**	**62**	**104**	**166**	**332**	**35**	**2**	**6**	**938**	**6.6**		**1**	**100.0**	**625**	**204**	**24:12**	**17**	**3**	**5**	**8**	**6**	**3**	**0**	**0**

• Missed majority of 1997-98 season recovering from ankle injury suffered in game vs. Anaheim (October 4, 1997) and groin injury suffered in game vs. Pittsburgh (November 1, 1997). Traded to **Tampa Bay** by **Vancouver** with Vancouver's 2nd round choice (Alexander Polushin) in 2001 Entry Draft for Dan Cloutier, February 7, 2001. Traded to **NY Islanders** by **Tampa Bay** with Alexander Kharitonov for Mathieu Biron and NY Islanders' 2nd round choice (later traded to Washington - later traded to Vancouver - Vancouver selected Denis Grot) in 2002 Entry Draft, June 22, 2001.

AUDETTE, Donald

(aw-DEHT, DAW-nohld) **MTL.**

Right wing. Shoots right. 5'8", 190 lbs. Born, Laval, Que., September 23, 1969. Buffalo's 8th choice, 183rd overall, in 1989 Entry Draft.

Season	Club	League	GP	G	A	Pts	PIM	PP	SH	GW	S	%	+/-	TF	F%	H	SB	Min	GP	G	A	Pts	PIM	PP	SH	GW
1985-86	Laval Laurentide	QAAA	41	32	38	70	51												8	5	9	14	10			
1986-87	Laval Titan	QMJHL	66	17	22	39	36												14	2	6	8	10			
1987-88	Laval Titan	QMJHL	63	48	61	109	56												14	7	12	19	20			
1988-89	Laval Titan	QMJHL	70	76	85	161	123												17	17	12	29	43			
1989-90	Rochester	AHL	70	42	46	88	78												15	9	8	17	29			
	Buffalo	**NHL**																	**2**	**0**	**0**	**0**	**0**	**0**	**0**	**0**
1990-91	**Buffalo**	**NHL**	**8**	**4**	**3**	**7**	**4**	**2**	**0**	**1**	**17**	**23.5**	**–1**													
	Rochester	AHL	5	4	0	4	2																			
1991-92	**Buffalo**	**NHL**	**63**	**31**	**17**	**48**	**75**	**5**	**0**	**6**	**153**	**20.3**	**–1**													
1992-93	**Buffalo**	**NHL**	**44**	**12**	**7**	**19**	**51**	**2**	**0**	**0**	**92**	**13.0**	**–8**						**8**	**2**	**2**	**4**	**6**	**0**	**0**	**0**
	Rochester	AHL	6	8	4	12	10																			
1993-94	**Buffalo**	**NHL**	**77**	**29**	**30**	**59**	**41**	**16**	**1**	**4**	**207**	**14.0**	**2**						**7**	**0**	**1**	**1**	**6**	**0**	**0**	**0**
1994-95	**Buffalo**	**NHL**	**46**	**24**	**13**	**37**	**27**	**13**	**0**	**7**	**124**	**19.4**	**–3**						**5**	**1**	**1**	**2**	**4**	**1**	**0**	**0**
1995-96	**Buffalo**	**NHL**	**23**	**12**	**13**	**25**	**18**	**8**	**0**	**1**	**92**	**13.0**	**0**													
1996-97	**Buffalo**	**NHL**	**73**	**28**	**22**	**50**	**48**	**8**	**0**	**5**	**182**	**15.4**	**–6**						**11**	**4**	**5**	**9**	**6**	**3**	**0**	**0**
1997-98	**Buffalo**	**NHL**	**75**	**24**	**20**	**44**	**59**	**10**	**0**	**5**	**198**	**12.1**	**10**						**15**	**5**	**8**	**13**	**10**	**3**	**0**	**2**
1998-99	**Los Angeles**	**NHL**	**49**	**18**	**18**	**36**	**51**	**6**	**0**	**2**	**152**	**11.8**	**7**	**4**	**50.0**	**28**	**9**	**16:50**								
99-2000	**Los Angeles**	**NHL**	**49**	**12**	**20**	**32**	**45**	**1**	**0**	**3**	**112**	**10.7**	**6**	**4**	**50.0**	**12**	**8**	**14:56**								
	Atlanta	**NHL**	**14**	**7**	**4**	**11**	**12**	**0**	**1**	**1**	**50**	**14.0**	**–4**	**0**	**0.0**	**8**	**4**	**21:35**								
2000-01	**Atlanta**	**NHL**	**64**	**32**	**39**	**71**	**64**	**13**	**1**	**2**	**187**	**17.1**	**–3**	**7**	**57.1**	**18**	**23**	**20:18**								
	Buffalo	**NHL**	**12**	**2**	**6**	**8**	**12**	**1**	**0**	**1**	**38**	**5.3**	**1**	**1**	**0.0**	**1**	**1**	**17:25**	**13**	**3**	**6**	**9**	**4**	**0**	**0**	**0**
2001-02	**Dallas**	**NHL**	**20**	**4**	**8**	**12**	**12**	**3**	**0**	**2**	**49**	**8.2**	**2**	**2**	**50.0**	**4**	**8**	**12:26**								
	Montreal	**NHL**	**13**	**1**	**5**	**6**	**8**	**0**	**0**	**1**	**33**	**3.0**	**1**	**1**	**100.0**	**7**	**5**	**17:42**	**12**	**6**	**4**	**10**	**10**	**2**	**0**	**2**
	NHL Totals		**630**	**240**	**225**	**465**	**527**	**88**	**3**	**41**	**1686**	**14.2**		**19**	**52.6**	**78**	**58**	**17:24**	**73**	**21**	**27**	**48**	**46**	**9**	**0**	**4**

QMJHL First All-Star Team (1989) • AHL First All-Star Team (1990) • Won Dudley "Red" Garret Memorial Trophy (Top Rookie - AHL) (1990) • Played in NHL ALL-Star Game (2001)

• Missed majority of 1990-91 season recovering from knee injury originally suffered in game vs. Edmonton, November 16, 1990. • Missed majority of 1995-96 season recovering from knee injury suffered in training camp, September 23, 1995. Traded to **LA Kings** by **Buffalo** for LA Kings' 2nd round choice (Milan Bartovic) in 1999 Entry Draft, December 18, 1998. Traded to **Atlanta** by **LA Kings** with Frantisek Kaberle for Kelly Buchberger and Nelson Emerson, March 13, 2000. Traded to **Buffalo** by **Atlanta** for the rights to Kamil Piros and Buffalo's 4th round choice (later traded to St. Louis - St. Louis selected Igor Valeyev) in 2001 Entry Draft, March 13, 2001. Signed as a free agent by **Dallas**, July 2, 2001. Traded to **Montreal** by **Dallas** with Shaun Van Allen for Martin Rucinsky and Benoit Brunet, November 21, 2001. • Missed majority of 2001-02 season recovering from wrist injury suffered in game vs. NY Rangers, December 1, 2001.

AVERY, Sean

(AY-vuhr-ee, SHAWN) **DET.**

Center. Shoots left. 5'10", 185 lbs. Born, North York, Ont., April 10, 1980.

Season	Club	League	GP	G	A	Pts	PIM	PP	SH	GW	S	%	+/-	TF	F%	H	SB	Min	GP	G	A	Pts	PIM	PP	SH	GW
1995-96	Markham	OMHA	70	34	81	115	180																			
	Markham Waxers	OJHL	1	0	0	0	4																			
1996-97	Owen Sound	OHL	58	10	21	31	86												4	1	0	1	4			
1997-98	Owen Sound	OHL	47	13	41	54	105																			
1998-99	Owen Sound	OHL	28	22	23	45	70																			
	Kingston	OHL	33	14	25	39	88												5	1	3	4	13			
99-2000	Kingston	OHL	55	28	56	84	215												5	2	2	4	26			
2000-01	Cincinnati	AHL	58	8	15	23	304												4	1	0	1	19			
2001-02	**Detroit**	**NHL**	**36**	**2**	**2**	**4**	**68**	**0**	**0**	**1**	**30**	**6.7**	**1**	**299**	**51.8**	**71**	**8**	**7:51**								
	Cincinnati	AHL	36	14	7	21	106																			
	NHL Totals		**36**	**2**	**2**	**4**	**68**	**0**	**0**	**1**	**30**	**6.7**		**299**	**51.8**	**71**	**8**	**7:51**								

Traded to **Kingston** (OHL) by **Owen Sound** (OHL) with Steve Lafleur for Aaron Fransen and D.J. Maracle, January 11, 1999. Signed as a free agent by **Detroit**, September 21, 1999.

AXELSSON, P.J.
(AHX-ehl-suhn, PEE-jay) **BOS.**

Left wing. Shoots left. 6'1", 175 lbs. Born, Kungalv, Sweden, February 26, 1975. Boston's 7th choice, 177th overall, in 1995 Entry Draft.

Season	Club	League	GP	G	A	Pts	PIM	PP	SH	GW	S	%	+/-	TF	F%	H	SB	Min	Playoffs GP	G	A	Pts	PIM	PP	SH	GW
1992-93	V. Frolunda Jr.	Swede-Jr.	16	9	5	14	12																			
	Vastra Frolunda	Sweden	1	0	0	0	0																			
1993-94	Vastra Frolunda	Sweden	11	0	0	0	4												4	0	0	0	0			
1994-95	V. Frolunda Jr.	Swede-Jr.	19	16	9	25	22																			
	Vastra Frolunda	Sweden	11	2	1	3	6												5	0	0	0	0			
1995-96	Vastra Frolunda	Sweden	36	15	5	20	10												13	3	0	3	10			
1996-97	Vastra Frolunda	Sweden	50	19	15	34	34												3	0	2	2	0			
	Vastra Frolunda	EuroHL	3	1	1	2	0												3	0	0	0	2			
1997-98	**Boston**	**NHL**	**82**	**8**	**19**	**27**	**38**	**2**	**0**	**1**	**144**	**5.6**	**–14**						**6**	**1**	**0**	**1**	**0**	**0**	**0**	**0**
1998-99	**Boston**	**NHL**	**77**	**7**	**10**	**17**	**18**	**0**	**0**	**2**	**146**	**4.8**	**–14**	**8**	**75.0**	**66**	**22**	**16:38**	**12**	**1**	**1**	**2**	**4**	**0**	**0**	**0**
99-2000	**Boston**	**NHL**	**81**	**10**	**16**	**26**	**24**	**0**	**0**	**4**	**186**	**5.4**	**1**	**22**	**27.3**	**84**	**22**	**16:43**								
2000-01	**Boston**	**NHL**	**81**	**8**	**15**	**23**	**27**	**0**	**0**	**2**	**146**	**5.5**	**–12**	**41**	**36.6**	**94**	**14**	**12:30**								
2001-02	**Boston**	**NHL**	**78**	**7**	**17**	**24**	**16**	**0**	**2**	**0**	**127**	**5.5**	**6**	**17**	**35.3**	**71**	**16**	**14:42**	**6**	**2**	**1**	**3**	**6**	**0**	**1**	**1**
	Sweden	Olympics	4	0	0	0	2																			
	NHL Totals		**399**	**40**	**77**	**117**	**123**	**2**	**2**	**9**	**749**	**5.3**		**88**	**37.5**	**315**	**74**	**15:07**	**24**	**4**	**2**	**6**	**10**	**0**	**1**	**1**

BABENKO, Yuri
(bah-BEHN-koh, EW-ree)

Center. Shoots left. 6'1", 200 lbs. Born, Penza, USSR, January 2, 1978. Colorado's 2nd choice, 51st overall, in 1996 Entry Draft.

Season	Club	League	GP	G	A	Pts	PIM	PP	SH	GW	S	%	+/-	TF	F%	H	SB	Min	Playoffs GP	G	A	Pts	PIM	PP	SH	GW
1995-96	Krylja Sovetov	CIS	21	0	0	0	16																			
1996-97	Krylja Sovetov 2	Russia-3	26	8	10	18	24																			
	H.C. CSKA	Russia-2	24	3	3	6	12																			
	Krylja Sovetov	Russia	4	1	0	1	4																			
1997-98	Plymouth Whalers	OHL	59	22	34	56	22												15	3	7	10	24			
1998-99	Hershey Bears	AHL	74	11	15	26	47												2	0	1	1	0			
99-2000	Hershey Bears	AHL	75	20	25	45	53												14	4	3	7	37			
2000-01	**Colorado**	**NHL**	**3**	**0**	**0**	**0**	**0**	**0**	**0**	**0**	**2**	**0.0**	**0**	**26**	**23.1**	**4**	**0**	**10:34**								
	Hershey Bears	AHL	71	17	18	35	80												12	2	1	3	6			
2001-02	Hershey Bears	AHL	67	7	23	30	95												8	2	2	4	6			
	NHL Totals		**3**	**0**	**0**	**0**	**0**	**0**	**0**	**0**	**2**	**0.0**		**26**	**23.1**	**4**	**0**	**10:34**								

BALA, Chris
(BA-la, KRIHS) **OTT.**

Left wing. Shoots left. 6'1", 180 lbs. Born, Alexandria, VA, September 24, 1978. Ottawa's 3rd choice, 58th overall, in 1998 Entry Draft.

Season	Club	League	GP	G	A	Pts	PIM	PP	SH	GW	S	%	+/-	TF	F%	H	SB	Min	Playoffs GP	G	A	Pts	PIM	PP	SH	GW
1996-97	Hill-Murray	Hi-School	23	28	33	61	36																			
1997-98	Harvard Crimson	ECAC	33	16	14	30	23																			
1998-99	Harvard Crimson	ECAC	28	5	10	15	16																			
99-2000	Harvard Crimson	ECAC	30	10	14	24	18																			
2000-01	Harvard Crimson	ECAC	32	14	16	30	24																			
2001-02	**Ottawa**	**NHL**	**6**	**0**	**1**	**1**	**0**	**0**	**0**	**0**	**2**	**0.0**	**1**	**0**	**0.0**	**0**	**2**	**5:43**								
	Grand Rapids	AHL	70	21	16	37	9												4	0	1	1	0			
	NHL Totals		**6**	**0**	**1**	**1**	**0**	**0**	**0**	**0**	**2**	**0.0**		**0**	**0.0**	**0**	**2**	**5:43**								

BALMOCHNYKH, Maxim
(bahl-MAWCH-nihky, mahx-EEM) **N.J.**

Left wing. Shoots left. 6'1", 180 lbs. Born, Lipetsk, USSR, March 7, 1979. Anaheim's 2nd choice, 45th overall, in 1997 Entry Draft.

Season	Club	League	GP	G	A	Pts	PIM	PP	SH	GW	S	%	+/-	TF	F%	H	SB	Min	Playoffs GP	G	A	Pts	PIM	PP	SH	GW
1994-95	HC Lipetsk	CIS-2	3	0	1	1	4																			
1995-96	HC Lipetsk	CIS-2	40	15	5	20	60																			
1996-97	Lada Togliatti	Russia	18	6	1	7	22																			
1997-98	Lada Togliatti	Russia	37	10	4	14	46																			
	Chelyabinsk	Russia	2	0	0	0	2																			
1998-99	Lada Togliatti	Russia	15	2	1	3	10												4	0	1	1	8			
	Quebec Remparts	QMJHL	21	9	22	31	38																			
99-2000	**Anaheim**	**NHL**	**6**	**0**	**1**	**1**	**2**	**0**	**0**	**0**	**6**	**0.0**	**2**	**0**	**0.0**	**5**	**0**	**6:44**								
	Cincinnati	AHL	40	9	12	21	82																			
2000-01	Cincinnati	AHL	65	6	9	15	45																			
2001-02	Cincinnati	AHL	23	6	4	10	33																			
	NHL Totals		**6**	**0**	**1**	**1**	**2**	**0**	**0**	**0**	**6**	**0.0**		**0**	**0.0**	**5**	**0**	**6:44**								

• Traded to **New Jersey** by **Anaheim** with Jeff Friesen and Oleg Tverdovsky for Petr Sykora, Mike Commodore, Jean-Francois Damphousse and Igor Pohanka, July 6, 2002.

BANCROFT, Steve
(BAN-crawft, STEEV) **ST.L.**

Defense. Shoots left. 6'1", 214 lbs. Born, Toronto, Ont., October 6, 1970. Toronto's 3rd choice, 21st overall, in 1989 Entry Draft.

Season	Club	League	GP	G	A	Pts	PIM	PP	SH	GW	S	%	+/-	TF	F%	H	SB	Min	Playoffs GP	G	A	Pts	PIM	PP	SH	GW
1985-86	Madoc	OJHL-C	7	1	0	1	21																			
	Trenton Bobcats	OJHL-B	16	1	5	6	16																			
1986-87	St. Catharines	OJHL-B	11	5	8	13	20																			
	Trenton Bobcats	OJHL-B	13	2	3	5	45																			
1987-88	Belleville Bulls	OHL	56	1	8	9	42																			
1988-89	Belleville Bulls	OHL	66	7	30	37	99												5	0	2	2	10			
1989-90	Belleville Bulls	OHL	53	10	33	43	135												11	3	9	12	38			
1990-91	Newmarket Saints	AHL	9	0	3	3	22																			
	Maine Mariners	AHL	53	2	12	14	46												2	0	0	0	2			
1991-92	Maine Mariners	AHL	26	1	3	4	45																			
	Indianapolis Ice	IHL	36	8	23	31	49																			
1992-93	**Chicago**	**NHL**	**1**	**0**	**0**	**0**	**0**	**0**	**0**	**0**	**0**	**0.0**	**0**													
	Indianapolis Ice	IHL	53	10	35	45	138																			
	Moncton Hawks	AHL	21	3	13	16	16												5	0	0	0	16			
1993-94	Cleveland	IHL	33	2	12	14	58																			
1994-95	Detroit Vipers	IHL	6	1	3	4	0																			
	Fort Wayne	IHL	50	7	17	24	100																			
	St. John's	AHL	4	2	0	2	2												5	0	3	3	8			
1995-96	Los Angeles	IHL	15	3	10	13	22																			
	Chicago Wolves	IHL	64	9	41	50	91												9	1	7	8	22			
1996-97	Chicago Wolves	IHL	39	6	10	16	66																			
	Las Vegas	IHL	36	9	28	37	64												3	0	0	0	2			
1997-98	Las Vegas	IHL	70	15	44	59	148																			
	Saint John	AHL	9	0	4	4	12												19	2	11	13	30			
1998-99	Saint John	AHL	8	1	4	5	22																			
	Providence	AHL	62	7	34	41	78												15	0	6	6	28			
99-2000	Cincinnati	IHL	39	6	14	20	37																			
	Houston Aeros	IHL	37	2	18	20	47												10	2	6	8	40			
2000-01	Kentucky	AHL	80	23	50	73	162												3	0	2	2	8			
2001-02	**San Jose**	**NHL**	**5**	**0**	**1**	**1**	**2**	**0**	**0**	**0**	**5**	**0.0**	**–2**	**0**	**0.0**	**9**	**3**	**10:24**								
	Cleveland Barons	AHL	72	6	38	44	226																			
	NHL Totals		**6**	**0**	**1**	**1**	**2**	**0**	**0**	**0**	**5**	**0.0**		**0**	**0.0**	**9**	**3**	**10:24**								

AHL First All-Star Team (2001)

Traded to **Boston** by **Toronto** for Rob Cimetta, November 9, 1990. Traded to **Chicago** by **Boston** with Boston's 11th round choice (later traded to Winnipeg - Winnipeg selected Russ Hewson) in 1993 Entry Draft for Chicago's 11th round choice (Evgeny Pavlov) in 1992 Entry Draft, January 8, 1992. Traded to **Winnipeg** by **Chicago** with future considerations for Troy Murray, February 21, 1993. Claimed by **Florida** from **Winnipeg** in Expansion Draft, June 24, 1993. Signed as a free agent by **Pittsburgh**, August 2, 1993. Signed as a free agent by **Los Angeles** (IHL), August 30, 1995. Signed as a free agent by **Carolina**, August 4, 1999. Traded to **Houston** (IHL) by **Cincinnati** (IHL) for Brian Felsner with Carolina retaining his NHL rights, January 19, 2000. Signed as a free agent by **San Jose**, August 10, 2000. Signed as a free agent by **St. Louis**, July 16, 2002.

BANNISTER, Drew
(BAN-nihs-stuhr, DREW)

Defense. Shoots right. 6'2", 200 lbs. Born, Belleville, Ont., September 4, 1974. Tampa Bay's 2nd choice, 26th overall, in 1992 Entry Draft.

			Regular Season																Playoffs							
Season	Club	League	GP	G	A	Pts	PIM	PP	SH	GW	S	%	+/-	TF	F%	H	SB	Min	GP	G	A	Pts	PIM	PP	SH	GW
1989-90	Sudbury Legion	NOHA	26	13	14	27	98																			
1990-91	Sault Ste. Marie	OHL	41	2	8	10	51												4	0	0	0	0			
1991-92	Sault Ste. Marie	OHL	64	4	21	25	122												16	3	10	13	36			
1992-93	Sault Ste. Marie	OHL	59	5	28	33	114												18	2	7	9	12			
1993-94	Sault Ste. Marie	OHL	58	7	43	50	108												14	6	9	15	20			
1994-95	Atlanta Knights	IHL	72	5	7	12	74												5	0	2	2	22			
1995-96	Tampa Bay	NHL	13	0	1	1	4	0	0	0	10	0.0	-1													
	Atlanta Knights	IHL	61	3	13	16	105												3	0	0	0	4			
1996-97	Tampa Bay	NHL	64	4	13	17	44	1	0	0	57	7.0	-21													
	Edmonton	NHL	1	0	1	1	0	0	0	0	2	0.0	-2						12	0	0	0	30	0	0	0
1997-98	Edmonton	NHL	34	0	2	2	42	0	0	0	27	0.0	-7													
	Anaheim	NHL	27	0	6	6	47	0	0	0	23	0.0	-2													
1998-99	Las Vegas	IHL	16	2	1	3	73																			
	Tampa Bay	NHL	21	1	2	3	24	0	0	0	29	3.4	-4	0	0.0	25	11	15:49								
99-2000	Hartford	AHL	44	6	14	20	121												18	2	9	11	53			
2000-01	NY Rangers	NHL	3	0	0	0	0	0	0	0	3	0.0	-1	0	0.0	3	3	10:47								
	Hartford	AHL	73	9	30	39	143												5	0	2	2	6			
2001-02	Anaheim	NHL	1	0	0	0	0	0	0	0	1	0.0	0	0	0.0	3	1	13:19								
	Cincinnati	AHL	30	1	10	11	57												3	0	1	1	6			
	NHL Totals		164	5	25	30	161	1	0	0	152	3.3		0	0.0	31	15	15:07	12	0	0	0	30	0	0	0

Memorial Cup All-Star Team (1993) • OHL Second All-Star Team (1994)

Traded to **Edmonton** by **Tampa Bay** with Tampa Bay's 6th round choice (Peter Sarno) in 1997 Entry Draft for Jeff Norton, March 18, 1997. Traded to **Anaheim** by **Edmonton** for Bobby Dollas, January 9, 1998. Traded to **Tampa Bay** by **Anaheim** for Tampa Bay's 5th round choice (Peter Podhradsky) in 2000 Entry Draft, December 10, 1998. Signed as a free agent by **NY Rangers**, October 3, 1999. Signed as a free agent by **Anaheim**, July 27, 2001. • Missed majority of 2001-02 season recovering from shoulder injury suffered in game vs. Utah (AHL), November 30, 2001.

BARNABY, Matthew
(BAHR-na-BEE, MA-thew) **NYR**

Right wing. Shoots left. 6', 189 lbs. Born, Ottawa, Ont., May 4, 1973. Buffalo's 5th choice, 83rd overall, in 1992 Entry Draft.

			Regular Season																Playoffs							
Season	Club	League	GP	G	A	Pts	PIM	PP	SH	GW	S	%	+/-	TF	F%	H	SB	Min	GP	G	A	Pts	PIM	PP	SH	GW
1989-90	Hull Frontaliers	QAHA	50	43	50	93	149																			
	L'Outaouais	QAAA	2	0	0	0	0																			
1990-91	Beauport	QMJHL	52	9	5	14	262																			
1991-92	Beauport	QMJHL	63	29	37	66	*476																			
1992-93	Victoriaville	QMJHL	65	44	67	111	*448												6	2	4	6	44			
	Buffalo	NHL	2	1	0	1	10	1	0	0	8	12.5	0						1	0	1	1	4	0	0	0
1993-94	Buffalo	NHL	35	2	4	6	106	1	0	0	13	15.4	-7						3	0	0	0	17	0	0	0
	Rochester	AHL	42	10	32	42	153																			
1994-95	Rochester	AHL	56	21	29	50	274																			
	Buffalo	NHL	23	1	1	2	116	0	0	0	27	3.7	-2													
1995-96	Buffalo	NHL	73	15	16	31	*335	0	0	0	131	11.5	-2													
1996-97	Buffalo	NHL	68	19	24	43	249	2	0	1	121	15.7	16						8	0	4	4	36	0	0	0
1997-98	Buffalo	NHL	72	5	20	25	289	0	0	2	96	5.2	8						15	7	6	13	22	3	0	1
1998-99	Buffalo	NHL	44	4	14	18	143	0	0	3	52	7.7	-2	6	16.7	45	8	13:56								
	Pittsburgh	NHL	18	2	2	4	34	1	0	0	27	7.4	-10	3	66.7	40	6	13:33	13	0	0	0	35	0	0	0
99-2000	Pittsburgh	NHL	64	12	12	24	197	0	0	3	80	15.0	3	75	44.0	99	10	12:38	11	0	2	2	29	0	0	0
2000-01	Pittsburgh	NHL	47	1	4	5	*168	0	0	0	38	2.6	-7	15	33.3	55	5	7:49								
	Tampa Bay	NHL	29	4	4	8	*97	1	0	0	29	13.8	-3	1	100.0	32	7	12:34								
2001-02	Tampa Bay	NHL	29	0	0	0	70	0	0	0	13	0.0	-7	1	0.0	22	2	7:54								
	NY Rangers	NHL	48	8	13	21	144	0	0	1	56	14.3	-3	12	33.3	59	13	11:24								
	NHL Totals		552	74	114	188	1958	6	0	10	691	10.7		113	40.7	352	51	11:23	51	7	13	20	143	3	0	1

Traded to **Pittsburgh** by **Buffalo** for Stu Barnes, March 11, 1999. Traded to **Tampa Bay** by **Pittsburgh** for Wayne Primeau, February 1, 2001. Traded to **NY Rangers** by **Tampa Bay** for Zdeno Ciger, December 12, 2001.

BARNES, Stu
(BAHRNZ, STEW) **BUF.**

Center. Shoots right. 5'11", 180 lbs. Born, Spruce Grove, Alta., December 25, 1970. Winnipeg's 1st choice, 4th overall, in 1989 Entry Draft.

			Regular Season																Playoffs							
Season	Club	League	GP	G	A	Pts	PIM	PP	SH	GW	S	%	+/-	TF	F%	H	SB	Min	GP	G	A	Pts	PIM	PP	SH	GW
1986-87	St. Albert	AJHL	53	41	34	*75	103												19	7	15	22				
1987-88	New Westminster	WHL	71	37	64	101	88												5	2	3	5	6			
1988-89	Tri-City	WHL	70	59	82	141	117												7	6	5	11	10			
1989-90	Tri-City	WHL	63	52	92	144	165												7	1	5	6	26			
1990-91	Team Canada	Nat-Tm	53	22	27	49	68																			
1991-92	Winnipeg	NHL	46	8	9	17	26	4	0	0	75	10.7	-2													
	Moncton Hawks	AHL	30	13	19	32	10												11	3	9	12	6			
1992-93	Winnipeg	NHL	38	12	10	22	10	3	0	3	73	16.4	-3						6	1	3	4	2	0	0	0
	Moncton Hawks	AHL	42	23	31	54	58																			
1993-94	Winnipeg	NHL	18	5	4	9	8	2	0	0	24	20.8	-1													
	Florida	NHL	59	18	20	38	30	6	1	3	148	12.2	5													
1994-95	Florida	NHL	41	10	19	29	8	1	0	2	93	10.8	7													
1995-96	Florida	NHL	72	19	25	44	46	8	0	5	158	12.0	-12						22	6	10	16	4	2	0	2
1996-97	Florida	NHL	19	2	8	10	10	1	0	0	44	4.5	-3													
	Pittsburgh	NHL	62	17	22	39	16	4	0	3	132	12.9	-20						5	0	1	1	0	0	0	0
1997-98	Pittsburgh	NHL	78	30	35	65	30	15	1	5	196	15.3	15						6	3	3	6	2	0	0	1
1998-99	Pittsburgh	NHL	64	20	12	32	20	13	0	3	155	12.9	-12	720	51.9	57	13	17:52								
	Buffalo	NHL	17	0	4	4	10	0	0	0	25	0.0	1	236	51.3	15	4	18:20	21	7	3	10	6	4	0	1
99-2000	Buffalo	NHL	82	20	25	45	16	8	2	2	137	14.6	-3	778	48.5	22	32	17:23	5	3	0	3	2	2	0	1
2000-01	Buffalo	NHL	75	19	24	43	26	3	2	5	160	11.9	-2	1470	48.3	19	35	19:06	13	4	4	8	2	2	0	2
2001-02	Buffalo	NHL	68	17	31	48	26	5	0	4	127	13.4	6	984	47.2	24	33	18:35								
	NHL Totals		739	197	248	445	282	73	6	35	1547	12.7		4188	48.9	137	117	18:13	78	24	24	48	18	10	0	7

WHL West Second All-Star Team (1988, 1989) • WHL Rookie of the Year (1988) • WHL MVP (1989)

Traded to **Florida** by **Winnipeg** with St. Louis' 6th round choice (previously acquired, later traded to Edmonton - later traded back to Winnipeg - Winnipeg selected Chris Kibermanis) in 1994 Entry Draft for Randy Gilhen, November 25, 1993. Traded to **Pittsburgh** by **Florida** with Jason Woolley for Chris Wells, November 19, 1996. Traded to **Buffalo** by **Pittsburgh** for Matthew Barnaby, March 11, 1999.

BARON, Murray
(BAIR-uhn, MUHR-ray) **VAN.**

Defense. Shoots left. 6'3", 215 lbs. Born, Prince George, B.C., June 1, 1967. Philadelphia's 7th choice, 167th overall, in 1986 Entry Draft.

			Regular Season																Playoffs							
Season	Club	League	GP	G	A	Pts	PIM	PP	SH	GW	S	%	+/-	TF	F%	H	SB	Min	GP	G	A	Pts	PIM	PP	SH	GW
1984-85	Vernon Lakers	BCJHL	37	5	9	14	93												13	5	6	11	107			
1985-86	Vernon Lakers	BCJHL	46	12	32	44	179												7	1	2	3	13			
1986-87	North Dakota	WCHA	41	4	10	14	62																			
1987-88	North Dakota	WCHA	41	1	10	11	95																			
1988-89	North Dakota	WCHA	40	2	6	8	92																			
	Hershey Bears	AHL	9	0	3	3	8																			
1989-90	Philadelphia	NHL	16	2	2	4	12	0	0	0	18	11.1	-1													
	Hershey Bears	AHL	50	0	10	10	101																			
1990-91	Philadelphia	NHL	67	8	8	16	74	3	0	1	86	9.3	-3													
	Hershey Bears	AHL	6	2	3	5	0																			
1991-92	St. Louis	NHL	67	3	8	11	94	0	0	0	55	5.5	-3						2	0	0	0	2	0	0	0
1992-93	St. Louis	NHL	53	2	2	4	59	0	0	1	42	4.8	-5						11	0	0	0	12	0	0	0
1993-94	St. Louis	NHL	77	5	9	14	123	0	0	0	73	6.8	-14						4	0	0	0	10	0	0	0
1994-95	St. Louis	NHL	39	0	5	5	93	0	0	0	28	0.0	9						7	1	1	2	2	0	0	0
1995-96	St. Louis	NHL	82	2	9	11	190	0	0	0	86	2.3	3						13	1	0	1	20	0	1	0
1996-97	St. Louis	NHL	11	0	2	2	11	0	0	0	7	0.0	-4													
	Montreal	NHL	60	1	5	6	107	0	0	0	52	1.9	-16													
	Phoenix	NHL	8	0	0	0	4	0	0	0	5	0.0	0						1	0	0	0	0	0	0	0
1997-98	Phoenix	NHL	45	1	5	6	106	0	0	0	23	4.3	-10						6	0	2	2	6	0	0	0
1998-99	Vancouver	NHL	81	2	6	8	115	0	0	0	53	3.8	-23	0	0.0	192	100	18:14								
99-2000	Vancouver	NHL	81	2	10	12	67	0	0	0	48	4.2	8	2	50.0	187	185	21:36								

			Regular Season																Playoffs							
Season	Club	League	GP	G	A	Pts	PIM	PP	SH	GW	S	%	+/-	TF	F%	H	SB	Min	GP	G	A	Pts	PIM	PP	SH	GW
2000-01	**Vancouver**	**NHL**	**82**	**3**	**8**	**11**	**63**	**0**	**0**	**1**	**56**	**5.4**	**–13**	**3**	**66.7**	**168**	**170**	**19:24**	**4**	**0**	**0**	**0**	**0**	**0**	**0**	**0**
2001-02	**Vancouver**	**NHL**	**61**	**1**	**6**	**7**	**68**	**0**	**0**	**0**	**38**	**2.6**	**8**	**2**	**50.0**	**98**	**91**	**16:57**	**6**	**0**	**1**	**1**	**10**	**0**	**0**	**0**
	NHL Totals		**830**	**32**	**85**	**117**	**1186**	**3**	**0**	**3**	**670**	**4.8**		**7**	**57.1**	**645**	**546**	**19:11**	**54**	**2**	**4**	**6**	**62**	**0**	**1**	**0**

Traded to **St. Louis** by **Philadelphia** with Ron Sutter for Dan Quinn and Rod Brind'Amour, September 22, 1991. Traded to **Montreal** by **St. Louis** with Shayne Corson and St. Louis' 5th round choice (Gennady Razin) in 1997 Entry Draft for Pierre Turgeon, Rory Fitzpatrick and Craig Conroy, October 29, 1996. Traded to **Phoenix** by **Montreal** with Chris Murray for Dave Manson, March 18, 1997. Signed as a free agent by **Vancouver**, July 14, 1998.

BARTECKO, Lubos

(bahr-TESHK-oh, LOO-bohsh) **ATL.**

Left wing. Shoots left. 5'11", 200 lbs. Born, Kezmarok, Czech., July 14, 1976.

Season	Club	League	GP	G	A	Pts	PIM	PP	SH	GW	S	%	+/-	TF	F%	H	SB	Min	GP	G	A	Pts	PIM	PP	SH	GW
1994-95	Poprad	Slovakia	3	1	0	1	0	....	....	....	....	....	....	....	....	....	....	....	....	....	....	....	....	....	....	....
1995-96	Chicoutimi	QMJHL	70	32	41	73	50	....	....	....	....	....	....	....	....	....	....	....	17	8	15	23	10	....	....	....
1996-97	Drummondville	QMJHL	58	40	51	91	49	....	....	....	....	....	....	....	....	....	....	....	8	1	8	9	4	....	....	....
1997-98	Worcester	AHL	34	10	12	22	24	....	....	....	....	....	....	....	....	....	....	....	10	4	2	6	2	....	....	....
1998-99	HC SKP Poprad	Slovakia	1	1	0	1	0	....	....	....	....	....	....	....	....	....	....	....	....	....	....	....	....	....	....	....
	St. Louis	**NHL**	**32**	**5**	**11**	**16**	**6**	**0**	**0**	**1**	**37**	**13.5**	**4**	**0**	**0.0**	**34**	**5**	**13:13**	**5**	**0**	**0**	**0**	**2**	**0**	**0**	**0**
	Worcester	AHL	49	14	24	38	22	....	....	....	....	....	....	....	....	....	....	....	....	....	....	....	....	....	....	....
99-2000	**St. Louis**	**NHL**	**67**	**16**	**23**	**39**	**51**	**3**	**0**	**3**	**75**	**21.3**	**24**	**10**	**50.0**	**46**	**12**	**13:33**	**7**	**1**	**1**	**2**	**0**	**0**	**0**	**0**
	Worcester	AHL	12	4	7	11	4	....	....	....	....	....	....	....	....	....	....	....	....	....	....	....	....	....	....	....
2000-01	**St. Louis**	**NHL**	**50**	**5**	**8**	**13**	**12**	**0**	**0**	**3**	**51**	**9.8**	**–1**	**2**	**50.0**	**64**	**5**	**10:25**	....	....	....	....	....	....	....	....
2001-02	**Atlanta**	**NHL**	**71**	**13**	**14**	**27**	**30**	**1**	**0**	**0**	**96**	**13.5**	**–15**	**4**	**25.0**	**91**	**17**	**14:28**	....	....	....	....	....	....	....	....
	Slovakia	Olympics	4	0	1	1	0	....	....	....	....	....	....	....	....	....	....	....	....	....	....	....	....	....	....	....
	NHL Totals		**220**	**39**	**56**	**95**	**99**	**4**	**0**	**7**	**259**	**15.1**		**16**	**43.8**	**235**	**39**	**13:05**	**12**	**1**	**1**	**2**	**2**	**0**	**0**	**0**

Signed as a free agent by **St. Louis**, October 3, 1997. Traded to **Atlanta** by **St. Louis** for Buffalo's 4th round choice (previously acquired, St. Louis selected Igor Valeyev) in 2001 Entry Draft, June 23, 2001.

BARTOS, Peter

(bahr-TAWSH, PEE-tuhr)

Left wing. Shoots right. 6', 185 lbs. Born, Martin, Czech., September 5, 1973. Minnesota's 7th choice, 214th overall, in 2000 Entry Draft.

Season	Club	League	GP	G	A	Pts	PIM	PP	SH	GW	S	%	+/-	TF	F%	H	SB	Min	GP	G	A	Pts	PIM	PP	SH	GW
1991-92	Hutnik Martin	Czech-2	33	13	8	21	16	....	....	....	....	....	....	....	....	....	....	....	....	....	....	....	....	....	....	....
1992-93	Hutnik Martin	Czech-2	22	6	5	11	4	....	....	....	....	....	....	....	....	....	....	....	....	....	....	....	....	....	....	....
	Dukla Trencin	Czech	28	1	2	3	....	....	....	....	....	....	....	....	....	....	....	....	10	1	1	2	....	....	....	....
1993-94	Martin	Slovakia	36	12	9	21	10	....	....	....	....	....	....	....	....	....	....	....	6	2	1	3	8	....	....	....
1994-95	Martin	Slovakia	34	14	20	34	20	....	....	....	....	....	....	....	....	....	....	....	3	0	0	0	0	....	....	....
1995-96	Martin	Slovakia	36	23	16	39	8	....	....	....	....	....	....	....	....	....	....	....	13	4	4	8	4	....	....	....
1996-97	Martin	Slovakia	46	22	15	37	....	....	....	....	....	....	....	....	....	....	....	....	5	1	5	6	....	....	....	....
1997-98	Martin	Slovakia	36	20	26	46	20	....	....	....	....	....	....	....	....	....	....	....	3	0	2	2	0	....	....	....
1998-99	Ceske Budejovice	Czech	52	22	26	48	24	....	....	....	....	....	....	....	....	....	....	....	3	3	0	3	....	....	....	....
99-2000	Ceske Budejovice	Czech	52	23	25	48	24	....	....	....	....	....	....	....	....	....	....	....	3	0	1	1	6	....	....	....
2000-01	**Minnesota**	**NHL**	**13**	**4**	**2**	**6**	**6**	**1**	**0**	**1**	**18**	**22.2**	**2**	**3**	**33.3**	**5**	**4**	**13:50**	....	....	....	....	....	....	....	....
	Cleveland	IHL	60	18	28	46	18	....	....	....	....	....	....	....	....	....	....	....	4	0	1	1	2	....	....	....
2001-02	Ceske Budejovice	Czech	43	18	10	28	22	....	....	....	....	....	....	....	....	....	....	....	....	....	....	....	....	....	....	....
	NHL Totals		**13**	**4**	**2**	**6**	**6**	**1**	**0**	**1**	**18**	**22.2**		**3**	**33.3**	**5**	**4**	**13:50**	....	....	....	....	....	....	....	....

BASHKIROV, Andrei

(bahsh-KIHR-ahf, AWN-dray)

Left wing. Shoots left. 6', 215 lbs. Born, Shelekhov, USSR, June 22, 1970. Montreal's 4th choice, 132nd overall, in 1998 Entry Draft.

Season	Club	League	GP	G	A	Pts	PIM	PP	SH	GW	S	%	+/-	TF	F%	H	SB	Min	GP	G	A	Pts	PIM	PP	SH	GW
1991-92	Voskresensk	CIS	11	2	0	2	4	....	....	....	....	....	....	....	....	....	....	....	....	....	....	....	....	....	....	....
1992-93	Yermak Angarsk	CIS-3	STATISTICS NOT AVAILABLE																							
1993-94	Charlotte	ECHL	62	28	42	70	25	....	....	....	....	....	....	....	....	....	....	....	3	1	0	1	0	....	....	....
	Providence	AHL	1	0	0	0	2	....	....	....	....	....	....	....	....	....	....	....	....	....	....	....	....	....	....	....
1994-95	Charlotte	ECHL	61	19	27	46	20	....	....	....	....	....	....	....	....	....	....	....	3	0	0	0	0	....	....	....
1995-96	Huntington	ECHL	55	19	39	58	35	....	....	....	....	....	....	....	....	....	....	....	....	....	....	....	....	....	....	....
1996-97	Huntington	ECHL	47	29	41	70	12	....	....	....	....	....	....	....	....	....	....	....	....	....	....	....	....	....	....	....
	Detroit Vipers	IHL	2	0	0	0	0	....	....	....	....	....	....	....	....	....	....	....	....	....	....	....	....	....	....	....
	Las Vegas	IHL	27	10	12	22	0	....	....	....	....	....	....	....	....	....	....	....	2	0	0	0	0	....	....	....
1997-98	Las Vegas	IHL	15	2	3	5	5	....	....	....	....	....	....	....	....	....	....	....	....	....	....	....	....	....	....	....
	Port Huron	UHL	3	1	3	4	0	....	....	....	....	....	....	....	....	....	....	....	....	....	....	....	....	....	....	....
	Fort Wayne	IHL	65	28	48	76	16	....	....	....	....	....	....	....	....	....	....	....	4	2	2	4	2	....	....	....
1998-99	**Montreal**	**NHL**	**10**	**0**	**0**	**0**	**0**	**0**	**0**	**0**	**4**	**0.0**	**–3**	**0**	**0.0**	**3**	**3**	**6:57**	....	....	....	....	....	....	....	....
	Fredericton	AHL	13	7	5	12	4	....	....	....	....	....	....	....	....	....	....	....	....	....	....	....	....	....	....	....
	Fort Wayne	IHL	34	11	25	36	10	....	....	....	....	....	....	....	....	....	....	....	....	....	....	....	....	....	....	....
99-2000	**Montreal**	**NHL**	**2**	**0**	**0**	**0**	**0**	**0**	**0**	**0**	**0**	**0.0**	**0**	**0**	**0.0**	**0**	**0**	**5:09**	....	....	....	....	....	....	....	....
	Quebec	AHL	78	28	33	61	17	....	....	....	....	....	....	....	....	....	....	....	3	0	3	3	0	....	....	....
2000-01	**Montreal**	**NHL**	**18**	**0**	**3**	**3**	**0**	**0**	**0**	**0**	**22**	**0.0**	**–2**	**3**	**100.0**	**6**	**8**	**11:41**	....	....	....	....	....	....	....	....
	Quebec	AHL	53	17	25	42	6	....	....	....	....	....	....	....	....	....	....	....	6	1	1	2	0	....	....	....
2001-02	HC Lausanne	Swiss	33	7	19	26	8	....	....	....	....	....	....	....	....	....	....	....	5	1	6	7	2	....	....	....
	NHL Totals		**30**	**0**	**3**	**3**	**0**	**0**	**0**	**0**	**26**	**0.0**		**3**	**100.0**	**9**	**11**	**9:40**	....	....	....	....	....	....	....	....

Signed as a free agent by **HC Lausanne** (Swiss), August 7, 2001.

BAST, Ryan

(BAST, RIGH-yuhn) **CAR.**

Defense. Shoots left. 6'2", 190 lbs. Born, Spruce Grove, Alta., August 27, 1975.

Season	Club	League	GP	G	A	Pts	PIM	PP	SH	GW	S	%	+/-	TF	F%	H	SB	Min	GP	G	A	Pts	PIM	PP	SH	GW
1992-93	St. Albert	AMHL	35	1	18	19	51	....	....	....	....	....	....	....	....	....	....	....	....	....	....	....	....	....	....	....
1993-94	Portland	WHL	6	0	0	0	4	....	....	....	....	....	....	....	....	....	....	....	....	....	....	....	....	....	....	....
	Prince Albert	WHL	47	2	8	10	139	....	....	....	....	....	....	....	....	....	....	....	....	....	....	....	....	....	....	....
1994-95	Prince Albert	WHL	42	1	10	11	149	....	....	....	....	....	....	....	....	....	....	....	14	0	3	3	13	....	....	....
1995-96	Prince Albert	WHL	44	7	15	22	129	....	....	....	....	....	....	....	....	....	....	....	....	....	....	....	....	....	....	....
	Calgary Hitmen	WHL	3	0	0	0	24	....	....	....	....	....	....	....	....	....	....	....	....	....	....	....	....	....	....	....
	Swift Current	WHL	25	2	3	5	50	....	....	....	....	....	....	....	....	....	....	....	6	1	0	1	21	....	....	....
1996-97	Las Vegas	IHL	49	2	3	5	266	....	....	....	....	....	....	....	....	....	....	....	....	....	....	....	....	....	....	....
	Toledo Storm	ECHL	12	2	2	4	75	....	....	....	....	....	....	....	....	....	....	....	....	....	....	....	....	....	....	....
	Saint John	AHL	12	0	0	0	21	....	....	....	....	....	....	....	....	....	....	....	5	0	0	0	4	....	....	....
1997-98	Saint John	AHL	77	3	8	11	187	....	....	....	....	....	....	....	....	....	....	....	21	0	1	1	55	....	....	....
1998-99	Saint John	AHL	2	0	0	0	5	....	....	....	....	....	....	....	....	....	....	....	....	....	....	....	....	....	....	....
	Philadelphia	**NHL**	**2**	**0**	**1**	**1**	**0**	**0**	**0**	**0**	**1**	**0.0**	**0**	**0**	**0.0**	**0**	**0**	**13:24**	....	....	....	....	....	....	....	....
	Philadelphia	AHL	69	0	11	11	160	....	....	....	....	....	....	....	....	....	....	....	16	0	0	0	30	....	....	....
99-2000	Philadelphia	AHL	71	1	9	10	198	....	....	....	....	....	....	....	....	....	....	....	5	0	0	0	6	....	....	....
2000-01	Hartford	AHL	50	1	1	2	146	....	....	....	....	....	....	....	....	....	....	....	....	....	....	....	....	....	....	....
2001-02	Pee Dee Pride	ECHL	16	0	4	4	32	....	....	....	....	....	....	....	....	....	....	....	....	....	....	....	....	....	....	....
	Lowell	AHL	59	1	7	8	88	....	....	....	....	....	....	....	....	....	....	....	5	0	1	1	4	....	....	....
	NHL Totals		**2**	**0**	**1**	**1**	**0**	**0**	**0**	**0**	**1**	**0.0**		**0**	**0.0**	**0**	**0**	**13:24**	....	....	....	....	....	....	....	....

AHL Second All-Star Team (1998)

Signed as a free agent by **Las Vegas** (IHL), September 30, 1996. Traded to **Saint John** (AHL) by **Las Vegas** (IHL) for loan of Sasha Lakovic, March 20, 1997. Signed as a free agent by **Philadelphia**, May 18, 1998. • Calgary Flames filed official protest contesting Philadelphia's signing of Bast under the contention that he was property of AHL's Saint John Flames, May 20, 1998. • NHL ruled that Bast was not under contract to Calgary since he was never drafted and had no NHL clause in contract, May 22, 1998. NHL also ruled that Bast was not property of Philadelphia because Flyers' contract offer exceeded NHL rookie salary cap, May 22, 1998. A compromise was reached that traded Bast to **Philadelphia** by **Calgary** with Calgary's 8th round choice (David Nystrom) in 1999 Entry Draft for Philadelphia's 3rd round choice (later traded to NY Rangers - NY Rangers selected Patrick Aufiero) in 1999 Entry Draft, October 13, 1998. Signed as a free agent by **Hartford** (AHL), September 18, 2000. Signed as a free agent by **Lowell** (AHL), November 24, 2001. Signed as a free agent by **Carolina**, July 16, 2002.

BATES, Shawn

(BAYTS, SHAWN) **NYI**

Center. Shoots right. 6', 205 lbs. Born, Melrose, MA, April 3, 1975. Boston's 4th choice, 103rd overall, in 1993 Entry Draft.

Season	Club	League	GP	G	A	Pts	PIM	PP	SH	GW	S	%	+/-	TF	F%	H	SB	Min	GP	G	A	Pts	PIM	PP	SH	GW
1990-91	Medford	Hi-School	22	18	43	61	6	....	....	....	....	....	....	....	....	....	....	....	....	....	....	....	....	....	....	....
1991-92	Medford	Hi-School	22	38	41	79	10	....	....	....	....	....	....	....	....	....	....	....	....	....	....	....	....	....	....	....
1992-93	Medford	Hi-School	25	49	46	95	20	....	....	....	....	....	....	....	....	....	....	....	....	....	....	....	....	....	....	....
1993-94	Boston University	H-East	41	10	19	29	24	....	....	....	....	....	....	....	....	....	....	....	....	....	....	....	....	....	....	....
1994-95	Boston University	H-East	38	18	12	30	48	....	....	....	....	....	....	....	....	....	....	....	....	....	....	....	....	....	....	....
1995-96	Boston University	H-East	40	28	22	50	54	....	....	....	....	....	....	....	....	....	....	....	....	....	....	....	....	....	....	....
1996-97	Boston University	H-East	41	17	18	35	64	....	....	....	....	....	....	....	....	....	....	....	....	....	....	....	....	....	....	....

			Regular Season																Playoffs							
Season	Club	League	GP	G	A	Pts	PIM	PP	SH	GW	S	%	+/-	TF	F%	H	SB	Min	GP	G	A	Pts	PIM	PP	SH	GW
1997-98	Boston	NHL	13	2	0	2	2	0	0	0	12	16.7	−3													
	Providence	AHL	50	15	19	34	22																			
1998-99	Boston	NHL	33	5	4	9	2	0	0	0	30	16.7	3	178	51.1	47	5	8:35	12	0	0	0	4	0	0	0
	Providence	AHL	37	25	21	46	39																			
99-2000	Boston	NHL	44	5	7	12	14	0	0	1	65	7.7	−17	460	47.0	76	8	10:52								
2000-01	Boston	NHL	45	2	3	5	26	0	0	0	59	3.4	−12	413	50.6	51	8	9:19								
	Providence	AHL	11	5	8	13	12												8	2	6	8	8			
2001-02	NY Islanders	NHL	71	17	35	52	30	1	4	4	150	11.3	18	306	49.4	135	29	18:45	7	2	4	6	11	1	0	1
	NHL Totals		206	31	49	80	74	1	4	5	316	9.8		1357	49.2	309	50	13:01	19	2	4	6	15	1	0	1

NCAA Championship All-Tournament Team (1995)
Signed as a free agent by **NY Islanders**, July 8, 2001.

BATTAGLIA, Bates

(buh-TAG-lee-ah, BAYTS) **CAR.**

Left wing. Shoots left. 6'2", 205 lbs. Born, Chicago, IL, December 13, 1975. Anaheim's 6th choice, 132nd overall, in 1994 Entry Draft.

Season	Club	League	GP	G	A	Pts	PIM	PP	SH	GW	S	%	+/-	TF	F%	H	SB	Min	GP	G	A	Pts	PIM	PP	SH	GW
1992-93	Team Illinois	MEHL	60	42	42	84	68																			
1993-94	Caledon	MTJHL	44	15	33	48	104																			
1994-95	Lake Superior	CCHA	38	6	14	20	34																			
1995-96	Lake Superior	CCHA	40	13	22	35	48																			
1996-97	Lake Superior	CCHA	38	12	27	39	80																			
1997-98	Carolina	NHL	33	2	4	6	10	0	0	1	21	9.5	−1													
	New Haven	AHL	48	15	21	36	48												1	0	0	0	0			
1998-99	Carolina	NHL	60	7	11	18	97	0	0	0	52	13.5	7	144	39.6	67	9	9:53	6	0	3	3	8	0	0	0
99-2000	Carolina	NHL	77	16	18	34	39	3	0	3	86	18.6	20	23	26.1	128	19	15:12								
2000-01	Carolina	NHL	80	12	15	27	76	2	0	3	133	9.0	−14	5	60.0	155	10	14:28	6	0	2	2	2	0	0	0
2001-02	Carolina	NHL	82	21	25	46	44	5	1	2	167	12.6	−6	12	33.3	87	34	19:05	23	5	9	14	14	1	0	1
	NHL Totals		332	58	73	131	266	10	1	9	459	12.6		184	38.0	437	72	15:00	35	5	14	19	24	1	0	1

Traded to **Hartford** by **Anaheim** with Anaheim's 4th round choice (Josef Vasicek) in 1998 Entry Draft for Mark Janssens, March 18, 1997. Rights transferred to **Carolina** after **Hartford** franchise relocated, June 25, 1997.

BAUMGARTNER, Nolan

(BAWM-gahrt-nuhr, NOH-lan) **VAN.**

Defense. Shoots right. 6'2", 205 lbs. Born, Calgary, Alta., March 23, 1976. Washington's 1st choice, 10th overall, in 1994 Entry Draft.

Season	Club	League	GP	G	A	Pts	PIM	PP	SH	GW	S	%	+/-	TF	F%	H	SB	Min	GP	G	A	Pts	PIM	PP	SH	GW
1991-92	Calgary Flames	AMHL	39	11	29	40	40																			
1992-93	Kamloops Blazers	WHL	43	0	5	5	30												11	1	1	2	0			
1993-94	Kamloops Blazers	WHL	69	13	42	55	109												19	3	14	17	33			
1994-95	Kamloops Blazers	WHL	62	8	36	44	71												21	4	13	17	16			
1995-96	Kamloops Blazers	WHL	28	13	15	28	45												16	1	9	10	26			
	Washington	NHL	1	0	0	0	0	0	0	0	0	0.0	−1						1	0	0	0	10	0	0	0
1996-97	Portland Pirates	AHL	8	2	2	4	4																			
1997-98	Washington	NHL	4	0	1	1	0	0	0	0	4	0.0	0													
	Portland Pirates	AHL	70	2	24	26	70												10	1	4	5	10			
1998-99	Washington	NHL	5	0	0	0	0	0	0	0	1	0.0	−3	0	0.0	1	0	8:41								
	Portland Pirates	AHL	38	5	14	19	62																			
99-2000	Washington	NHL	8	0	1	1	2	0	0	0	6	0.0	1	0	0.0	8	2	10:31								
	Portland Pirates	AHL	71	5	18	23	56												4	1	2	3	10			
2000-01	Chicago	NHL	8	0	0	0	6	0	0	0	7	0.0	−4	2	50.0	6	4	12:40								
	Norfolk Admirals	AHL	63	5	28	33	75												9	2	3	5	11			
2001-02	Norfolk Admirals	AHL	76	10	24	34	72												4	0	1	1	2			
	NHL Totals		26	0	2	2	8	0	0	0	18	0.0		2	50.0	15	6	10:54	1	0	0	0	10	0	0	0

Memorial Cup All-Star Team (1994, 1995) • WHL West First All-Star Team (1995, 1996) • Canadian Major Junior First All-Star Team (1995) • Canadian Major Junior Defenseman of the Year (1995)
Traded to **Chicago** by **Washington** for Remi Royer, July 20, 2000. Signed as a free agent by **Vancouver**, July 11, 2002.

BEAUDOIN, Eric

(boh-DWEH, AIR-ihk) **FLA.**

Left wing. Shoots left. 6'5", 204 lbs. Born, Ottawa, Ont., May 3, 1980. Tampa Bay's 4th choice, 92nd overall, in 1998 Entry Draft.

Season	Club	League	GP	G	A	Pts	PIM	PP	SH	GW	S	%	+/-	TF	F%	H	SB	Min	GP	G	A	Pts	PIM	PP	SH	GW
1996-97	Ottawa Jr. Sens	OCJHL	54	12	19	31	55																			
1997-98	Guelph Storm	OHL	62	9	13	22	43												12	3	2	5	4			
1998-99	Guelph Storm	OHL	66	28	43	71	79												11	5	3	8	12			
99-2000	Guelph Storm	OHL	68	38	34	72	126												6	3	0	3	2			
2000-01	Louisville	AHL	71	15	10	25	78																			
2001-02	Florida	NHL	8	1	3	4	4	0	0	1	11	9.1	−2	1	0.0	19	12	17:27								
	Utah Grizzlies	AHL	44	5	16	21	83																			
	NHL Totals		8	1	3	4	4	0	0	1	11	9.1		1	0.0	19	12	17:27								

Traded to **Florida** by **Tampa Bay** for Florida's 7th round choice (Marek Priechodsky) in 2000 Entry Draft, June 1, 2000.

BEAUFAIT, Mark

(BOH-fayt, MAHRK)

Center. Shoots right. 5'9", 170 lbs. Born, Livonia, MI, May 13, 1970. San Jose's 2nd choice, 7th overall, in 1991 Supplemental Draft.

Season	Club	League	GP	G	A	Pts	PIM	PP	SH	GW	S	%	+/-	TF	F%	H	SB	Min	GP	G	A	Pts	PIM	PP	SH	GW
1987-88	Redford Royals	NAJHL	STATISTICS NOT AVAILABLE																							
1988-89	North-Michigan	WCHA	11	2	1	3	2																			
1989-90	North-Michigan	WCHA	34	10	14	24	12																			
1990-91	North-Michigan	WCHA	47	19	30	49	18																			
1991-92	North-Michigan	WCHA	39	31	44	75	43																			
1992-93	San Jose	NHL	5	1	0	1	0	0	0	0	3	33.3	−1													
	Kansas City	IHL	66	19	40	59	22												9	1	1	2	8			
1993-94	Team USA	Nat-Tm	51	22	29	51	36																			
	United States	Olympics	8	1	4	5	2																			
	Kansas City	IHL	21	12	9	21	18																			
1994-95	San Diego Gulls	IHL	68	24	39	63	22												5	2	2	4	2			
1995-96	Orlando	IHL	77	30	79	109	87												22	9	*19	*28	22			
1996-97	Orlando	IHL	80	26	65	91	63												10	5	8	13	18			
1997-98	Orlando	IHL	76	24	61	85	56												17	6	16	22	10			
1998-99	Orlando	IHL	71	28	43	71	38												15	2	12	14	14			
99-2000	Orlando	IHL	78	28	49	77	87												6	2	0	2	4			
2000-01	Orlando	IHL	54	23	42	65	34												9	1	9	10	2			
2001-02	Houston Aeros	AHL	63	14	37	51	36												12	5	4	9	4			
	NHL Totals		5	1	0	1	0	0	0	0	3	33.3														

Won Ken McKenzie Trophy (U.S.- Born Rookie of the Year - IHL) (1993) • IHL Second All-Star Team (1997)
Signed as a free agent by **San Diego** (IHL), August 30, 1994. Selected by **Orlando** (IHL) from **San Diego** (IHL) in IHL Expansion Draft, July 13, 1995. Signed as a free agent by **Minnesota**, July 16, 2001. Signed as a free agent by **Eisbaren Berlin**, (Germany), May 3, 2002.

BEDNAR, Jaroslav

(BEHD-nahr, YA-roh-slahv) **L.A.**

Right wing. Shoots right. 5'11", 198 lbs. Born, Prague, Czech., November 8, 1976. Los Angeles' 4th choice, 51st overall, in 2001 Entry Draft.

Season	Club	League	GP	G	A	Pts	PIM	PP	SH	GW	S	%	+/-	TF	F%	H	SB	Min	GP	G	A	Pts	PIM	PP	SH	GW
1994-95	HC Slavia Praha	Czech	20	6	7	13	4												3	0	0	0	0			
1995-96	HC Slavia Praha	Czech	20	3	1	4	6												3	0	0	0	0			
1996-97	HC Slavia Praha	Czech	45	18	12	30	18																			
1997-98	HC Slavia Praha	Czech	14	2	5	7	6																			
	Plzen	Czech	34	26	15	41	16												5	2	4	6	4			
	Plzen	EuroHL																	5	4	2	6	4			
1998-99	HC Sparta Praha	Czech	52	23	14	37	30												8	5	2	7	0			
99-2000	JYP Jyvaskyla	Finland	53	34	28	62	56																			
2000-01	HIFK Helsinki	Finland	56	*32	28	60	51												5	3	1	4	0			
2001-02	Los Angeles	NHL	22	4	2	6	8	1	0	2	20	20.0	−4	2	50.0	33	2	10:42	3	0	0	0	0	0	0	0
	Manchester	AHL	48	16	21	37	16																			
	NHL Totals		22	4	2	6	8	1	0	2	20	20.0		2	50.0	33	2	10:42	3	0	0	0	0	0	0	0

BEECH, Kris (BEECH, KRIHS) PIT.

Center. Shoots left. 6'3", 199 lbs. Born, Salmon Arm, B.C., February 5, 1981. Washington's 1st choice, 7th overall, in 1999 Entry Draft.

			Regular Season																Playoffs							
Season	Club	League	GP	G	A	Pts	PIM	PP	SH	GW	S	%	+/-	TF	F%	H	SB	Min	GP	G	A	Pts	PIM	PP	SH	GW
1996-97	Sicamous Eagles	KIJHL	49	34	36	70	80																			
	Calgary Hitmen	WHL	8	1	1	2	0																			
1997-98	Calgary Hitmen	WHL	58	10	25	35	24												12	4	5	9	32			
1998-99	Calgary Hitmen	WHL	68	26	41	67	103												6	1	4	5	8			
99-2000	Calgary Hitmen	WHL	66	32	54	86	99												5	3	5	8	16			
2000-01	**Washington**	**NHL**	**4**	**0**	**0**	**0**	**2**	**0**	**0**	**0**	**0**	**0.0**	**–2**	**25**	**36.0**	**1**	**0**	**7:29**								
	Calgary Hitmen	WHL	40	22	44	66	103												10	2	8	10	26			
2001-02	**Pittsburgh**	**NHL**	**79**	**10**	**15**	**25**	**45**	**2**	**0**	**0**	**126**	**7.9**	**–25**	**604**	**45.2**	**71**	**20**	**13:33**								
	NHL Totals		**83**	**10**	**15**	**25**	**47**	**2**	**0**	**0**	**126**	**7.9**		**629**	**44.8**	**72**	**20**	**13:16**								

Returned to **Calgary** (WHL) by **Washington**, October 24, 2000. Traded to **Pittsburgh** by **Washington** with Michal Sivek, Ross Lupaschuk and future considerations for Jaromir Jagr and Frantisek Kucera, July 11, 2001.

BEGIN, Steve (bay-ZHIN, STEEV) CGY.

Center. Shoots left. 5'11", 190 lbs. Born, Trois-Rivieres, Que., June 14, 1978. Calgary's 3rd choice, 40th overall, in 1996 Entry Draft.

			Regular Season																Playoffs							
Season	Club	League	GP	G	A	Pts	PIM	PP	SH	GW	S	%	+/-	TF	F%	H	SB	Min	GP	G	A	Pts	PIM	PP	SH	GW
1993-94	Cap-d-Madeleine	QAAA	8	0	1	1	6												2	0	0	0	0			
1994-95	Cap-d-Madeleine	QAAA	35	9	15	24	48												3	0	0	0	2			
1995-96	Val-d'Or Foreurs	QMJHL	64	13	23	36	218												13	1	3	4	33			
1996-97	Val-d'Or Foreurs	QMJHL	58	13	33	46	229												10	0	3	3	8			
	Saint John	AHL																	4	0	2	2	6			
1997-98	Val-d'Or Foreurs	QMJHL	35	18	17	35	73												15	2	12	14	34			
	Calgary	**NHL**	**5**	**0**	**0**	**0**	**23**	**0**	**0**	**0**	**2**	**0.0**	**0**													
1998-99	Saint John	AHL	73	11	9	20	156												7	2	0	2	18			
99-2000	**Calgary**	**NHL**	**13**	**1**	**1**	**2**	**18**	**0**	**0**	**0**	**3**	**33.3**	**–3**	**19**	**47.4**	**23**	**2**	**7:13**								
	Saint John	AHL	47	13	12	25	99																			
2000-01	**Calgary**	**NHL**	**4**	**0**	**0**	**0**	**21**	**0**	**0**	**0**	**3**	**0.0**	**0**	**0**	**0.0**	**7**	**1**	**6:04**								
	Saint John	AHL	58	14	14	28	109												19	10	7	17	18			
2001-02	**Calgary**	**NHL**	**51**	**7**	**5**	**12**	**79**	**1**	**0**	**0**	**65**	**10.8**	**–3**	**129**	**53.5**	**92**	**18**	**9:25**								
	NHL Totals		**73**	**8**	**6**	**14**	**141**	**1**	**0**	**0**	**73**	**11.0**		**148**	**52.7**	**122**	**21**	**8:48**								

Won Jack A. Butterfield Trophy (Playoff MVP - AHL) (2001)

BEKAR, Derek (BEH-kahr, DAIR-ehk) L.A.

Left wing. Shoots left. 6'2", 205 lbs. Born, Burnaby, B.C., September 15, 1975. St. Louis' 7th choice, 205th overall, in 1995 Entry Draft.

			Regular Season																Playoffs							
Season	Club	League	GP	G	A	Pts	PIM	PP	SH	GW	S	%	+/-	TF	F%	H	SB	Min	GP	G	A	Pts	PIM	PP	SH	GW
1992-93	Notre Dame	SMHL	29	25	24	49	68																			
1993-94	Notre Dame	SJHL	62	20	31	51	77																			
1994-95	Powell River	BCJHL	46	33	29	62	35																			
1995-96	New Hampshire	H-East	34	15	18	33	4																			
1996-97	New Hampshire	H-East	39	18	21	39	34																			
1997-98	New Hampshire	H-East	35	32	28	60	46																			
1998-99	Worcester	AHL	51	16	20	36	6												4	0	0	0	0			
99-2000	**St. Louis**	**NHL**	**1**	**0**	**0**	**0**	**0**	**0**	**0**	**0**	**0**	**0.0**	**0**	**0**	**0.0**	**1**	**0**	**5:14**								
	Worcester	AHL	71	21	19	40	26												7	0	3	3	2			
2000-01	Worcester	AHL	18	5	2	7	10																			
	Portland Pirates	AHL	58	19	16	35	49												3	0	0	0	0			
2001-02	Manchester	AHL	74	27	20	47	42												5	1	4	5	2			
	NHL Totals		**1**	**0**	**0**	**0**	**0**	**0**	**0**	**0**	**0**	**0.0**		**0**	**0.0**	**1**	**0**	**5:14**								

Hockey East Second All-Star Team (1998)

Traded to **Washington** by **St. Louis** for Mike Peluso, November 29, 2000. Signed as a free agent by **LA Kings**, September 25, 2001.

BELAK, Wade (BEE-lak, WAYD) TOR.

Defense. Shoots right. 6'4", 225 lbs. Born, Saskatoon, Sask., July 3, 1976. Quebec's 1st choice, 12th overall, in 1994 Entry Draft.

			Regular Season																Playoffs							
Season	Club	League	GP	G	A	Pts	PIM	PP	SH	GW	S	%	+/-	TF	F%	H	SB	Min	GP	G	A	Pts	PIM	PP	SH	GW
1991-92	North Battleford	SMBHL	57	6	20	26	186																			
1992-93	North Battleford	SJHL	50	5	15	20	146																			
	Saskatoon Blades	WHL	7	0	0	0	23												7	0	0	0	0			
1993-94	Saskatoon Blades	WHL	69	4	13	17	226												16	2	2	4	43			
1994-95	Saskatoon Blades	WHL	72	4	14	18	290												9	0	0	0	36			
	Cornwall Aces	AHL																	11	1	2	3	40			
1995-96	Saskatoon Blades	WHL	63	3	15	18	207												4	0	0	0	9			
	Cornwall Aces	AHL	5	0	0	0	18												2	0	0	0	2			
1996-97	**Colorado**	**NHL**	**5**	**0**	**0**	**0**	**11**	**0**	**0**	**0**	**1**	**0.0**	**–1**													
	Hershey Bears	AHL	65	1	7	8	320												16	0	1	1	61			
1997-98	**Colorado**	**NHL**	**8**	**1**	**1**	**2**	**27**	**0**	**0**	**1**	**2**	**50.0**	**–3**													
	Hershey Bears	AHL	11	0	0	0	30																			
1998-99	**Colorado**	**NHL**	**22**	**0**	**0**	**0**	**71**	**0**	**0**	**0**	**5**	**0.0**	**–2**	**0**	**0.0**	**18**	**10**	**6:48**								
	Hershey Bears	AHL	17	0	1	1	49																			
	Calgary	**NHL**	**9**	**0**	**1**	**1**	**23**	**0**	**0**	**0**	**2**	**0.0**	**3**	**0**	**0.0**	**9**	**7**	**10:46**								
	Saint John	AHL	12	0	2	2	43												6	0	1	1	23			
99-2000	**Calgary**	**NHL**	**40**	**0**	**2**	**2**	**122**	**0**	**0**	**0**	**11**	**0.0**	**–4**	**1**	**0.0**	**41**	**23**	**7:33**								
2000-01	**Calgary**	**NHL**	**23**	**0**	**0**	**0**	**79**	**0**	**0**	**0**	**8**	**0.0**	**–2**	**0**	**0.0**	**26**	**4**	**6:54**								
	Toronto	**NHL**	**16**	**1**	**1**	**2**	**31**	**0**	**0**	**0**	**8**	**12.5**	**–4**	**0**	**0.0**	**32**	**16**	**13:38**								
2001-02	**Toronto**	**NHL**	**63**	**1**	**3**	**4**	**142**	**0**	**0**	**0**	**47**	**2.1**	**2**	**0**	**0.0**	**102**	**20**	**9:14**	**16**	**1**	**0**	**1**	**18**	**0**	**0**	**0**
	NHL Totals		**186**	**3**	**8**	**11**	**506**	**0**	**0**	**1**	**84**	**3.6**		**1**	**0.0**	**228**	**80**	**8:43**	**16**	**1**	**0**	**1**	**18**	**0**	**0**	**0**

Rights transferred to **Colorado** after **Quebec** franchise relocated, June 21, 1995. Traded to **Calgary** by **Colorado** with Rene Corbet, Robyn Regehr and Colorado's 2nd round compensatory choice (Jarret Stoll) in 2000 Entry Draft for Theoren Fleury and Chris Dingman, February 28, 1999. • Missed majority of 1999-2000 and 2000-01 seasons recovering from shoulder injury suffered in game vs. Colorado, February 10, 2000. Claimed on waivers by **Toronto** from **Calgary**, February 16, 2001.

BELANGER, Eric (buh-LAWN-zhay, AIR-ihk) L.A.

Center. Shoots left. 6', 185 lbs. Born, Sherbrooke, Que., December 16, 1977. Los Angeles' 5th choice, 96th overall, in 1996 Entry Draft.

			Regular Season																Playoffs							
Season	Club	League	GP	G	A	Pts	PIM	PP	SH	GW	S	%	+/-	TF	F%	H	SB	Min	GP	G	A	Pts	PIM	PP	SH	GW
1993-94	Magog	QAAA	32	19	24	43	24												13	5	6	11	36			
1994-95	Beauport	QMJHL	71	12	28	40	24												18	5	9	14	25			
1995-96	Beauport	QMJHL	59	35	48	83	18												20	13	14	27	6			
1996-97	Beauport	QMJHL	31	13	37	50	30																			
	Rimouski Oceanic	QMJHL	31	26	41	67	36												4	2	3	5	10			
1997-98	Fredericton	AHL	56	17	34	51	28												4	2	1	3	2			
1998-99	Springfield	AHL	33	8	18	26	10												3	0	1	1	2			
	Long Beach	IHL	1	0	0	0	0																			
99-2000	Lowell	AHL	65	15	25	40	20												7	3	3	6	2			
	Mohawk Valley	UHL	3	0	0	0	0																			
2000-01	**Los Angeles**	**NHL**	**62**	**9**	**12**	**21**	**16**	**1**	**2**	**1**	**80**	**11.3**	**14**	**849**	**56.4**	**136**	**28**	**13:25**	**13**	**1**	**4**	**5**	**2**	**0**	**0**	**1**
	Lowell	AHL	13	8	10	18	4																			
2001-02	**Los Angeles**	**NHL**	**53**	**8**	**16**	**24**	**21**	**2**	**1**	**1**	**67**	**11.9**	**2**	**882**	**57.7**	**75**	**22**	**14:33**	**7**	**0**	**0**	**0**	**4**	**0**	**0**	**0**
	NHL Totals		**115**	**17**	**28**	**45**	**37**	**3**	**3**	**2**	**147**	**11.6**		**1731**	**57.1**	**211**	**50**	**13:56**	**20**	**1**	**4**	**5**	**6**	**0**	**0**	**1**

BELANGER, Francis (buh-LAWN-zhay, FRAN-sihs)

Left wing. Shoots left. 6'3", 228 lbs. Born, Bellefeuille, Que., January 15, 1978. Philadelphia's 5th choice, 124th overall, in 1998 Entry Draft.

			Regular Season																Playoffs							
Season	Club	League	GP	G	A	Pts	PIM	PP	SH	GW	S	%	+/-	TF	F%	H	SB	Min	GP	G	A	Pts	PIM	PP	SH	GW
1994-95	Laval Laurentide	QAAA	25	11	8	19	78																			
1995-96	Hull Olympiques	QMJHL	1	0	0	0	0																			
1996-97	Hull Olympiques	QMJHL	53	13	13	26	134												8	2	2	4	57			
1997-98	Hull Olympiques	QMJHL	33	22	23	45	133																			
	Rimouski Oceanic	QMJHL	30	18	10	28	248												17	14	8	22	61			
1998-99	Philadelphia	AHL	58	13	13	26	242												16	4	3	7	16			

			Regular Season																Playoffs							
Season	Club	League	GP	G	A	Pts	PIM	PP	SH	GW	S	%	+/-	TF	F%	H	SB	Min	GP	G	A	Pts	PIM	PP	SH	GW
99-2000	Philadelphia	AHL	35	5	6	11	112																			
	Trenton Titans	ECHL	9	1	1	2	29																			
2000-01	Philadelphia	AHL	13	1	3	4	32																			
	Montreal	NHL	10	0	0	0	29	0	0	0	2	0.0	–3	0	0.0	13	0	4:30								
	Quebec	AHL	22	15	4	19	101												9	2	5	7	20			
2001-02	Quebec	AHL	69	15	26	41	165												3	1	1	2	0			
	NHL Totals		10	0	0	0	29	0	0	0	2	0.0		0	0.0	13	0	4:30								

Traded to **Rimouski** (QMJHL) by **Hull** (QMJHL) for Rimouski's 1st (Andrew Carver) and 3rd (Marco Joly) round choices in 1998 QMJHL Priority Draft, December 26, 1997. Signed as a free agent by **Montreal**, February 15, 2001.

BELANGER, Ken (buh-LAWN-zhay, KEHN) L.A.

Left wing. Shoots left. 6'4", 225 lbs. Born, Sault Ste. Marie, Ont., May 14, 1974. Hartford's 7th choice, 153rd overall, in 1992 Entry Draft.

Season	Club	League	GP	G	A	Pts	PIM	PP	SH	GW	S	%	+/-	TF	F%	H	SB	Min	GP	G	A	Pts	PIM	PP	SH	GW
1990-91	Soo Legion	NOHA	43	24	29	53	169																			
1991-92	Ottawa 67's	OHL	51	4	4	8	174												11	0	0	0	24			
1992-93	Ottawa 67's	OHL	34	6	12	18	139																			
	Guelph Storm	OHL	29	10	14	24	86												5	2	1	3	14			
1993-94	Guelph Storm	OHL	55	11	22	33	185												9	2	3	5	30			
1994-95	St. John's	AHL	47	5	5	10	246												4	0	0	0	30			
	Toronto	NHL	3	0	0	0	9	0	0	0	1	0.0	0													
1995-96	St. John's	AHL	40	16	14	30	222																			
	NY Islanders	NHL	7	0	0	0	27	0	0	0	0	0.0	–2													
1996-97	NY Islanders	NHL	18	0	2	2	102	0	0	0	5	0.0	–1													
	Kentucky	AHL	38	10	12	22	164												4	0	1	1	27			
1997-98	NY Islanders	NHL	37	3	1	4	101	0	0	1	10	30.0	1													
1998-99	NY Islanders	NHL	9	1	1	2	30	0	0	0	3	33.3	1	0	0.0	15	1	5:05								
	Boston	NHL	45	1	4	5	152	0	0	0	16	6.3	–2	1	0.0	66	4	4:38	12	1	0	1	16	0	0	0
99-2000	Boston	NHL	37	2	2	4	44	0	0	0	20	10.0	–4	1	0.0	79	4	5:17								
2000-01	Boston	NHL	40	2	2	4	121	0	0	1	35	5.7	–6	1	100.0	74	1	7:06								
	Providence	AHL	10	1	4	5	47												2	0	0	0	4			
2001-02	Los Angeles	NHL	43	2	0	2	85	0	0	0	22	9.1	–5	0	0.0	65	3	4:22								
	NHL Totals		239	11	12	23	671	0	0	2	112	9.8		3	33.3	299	13	5:18	12	1	0	1	16	0	0	0

Traded to **Toronto** by **Hartford** for Toronto's 9th round choice (Matt Ball) in 1994 Entry Draft, March 18, 1994. Traded to **NY Islanders** by **Toronto** with Damian Rhodes for future considerations (Kirk Muller and Don Beaupre, January 23, 1996), January 23, 1996. Traded to **Boston** by **NY Islanders** for Ted Donato, November 7, 1998. • Missed majority of 1999-2000 season recovering from head injury suffered in game vs. Toronto, November 11, 1999. Signed as a free agent by **LA Kings**, July 2, 2001.

BELL, Mark (BEHL, MAWRK) CHI.

Center. Shoots left. 6'3", 198 lbs. Born, St. Paul's, Ont., August 5, 1980. Chicago's 1st choice, 8th overall, in 1998 Entry Draft.

Season	Club	League	GP	G	A	Pts	PIM	PP	SH	GW	S	%	+/-	TF	F%	H	SB	Min	GP	G	A	Pts	PIM	PP	SH	GW
1995-96	Stratford	OJHL-B	47	8	15	23	32																			
1996-97	Ottawa 67's	OHL	65	8	12	20	40												24	4	7	11	13			
1997-98	Ottawa 67's	OHL	55	34	26	60	87												13	6	5	11	14			
1998-99	Ottawa 67's	OHL	44	29	26	55	69												9	6	5	11	8			
99-2000	Ottawa 67's	OHL	48	34	38	72	95												2	0	1	1	0			
2000-01	Chicago	NHL	13	0	1	1	4	0	0	0	14	0.0	0	141	48.9	16	5	12:00								
	Norfolk Admirals	AHL	61	15	27	42	126												9	4	3	7	10			
2001-02	Chicago	NHL	80	12	16	28	124	1	0	1	120	10.0	–6	47	42.6	168	17	12:39	5	0	0	0	8	0	0	0
	NHL Totals		93	12	17	29	128	1	0	1	134	9.0		188	47.3	184	22	12:33	5	0	0	0	8	0	0	0

BELLEFEUILLE, Blake (BEHL-fay, BLAYK) CBJ

Right wing. Shoots right. 5'10", 208 lbs. Born, Framingham, MA, December 27, 1977.

Season	Club	League	GP	G	A	Pts	PIM	PP	SH	GW	S	%	+/-	TF	F%	H	SB	Min	GP	G	A	Pts	PIM	PP	SH	GW
1994-95	Framingham	Hi-School	30	42	50	92																				
1995-96	Framingham	Hi-School	30	31	60	91																				
1996-97	Boston College	H-East	34	16	19	35	20																			
1997-98	Boston College	H-East	41	19	20	39	35																			
1998-99	Boston College	H-East	43	24	25	49	80																			
99-2000	Boston College	H-East	39	18	31	49	28																			
2000-01	Syracuse Crunch	AHL	50	5	5	10	18												5	0	0	0	0			
2001-02	Columbus	NHL	2	0	1	1	0	0	0	0	2	0.0	1	15	60.0	2	0	8:21								
	Syracuse Crunch	AHL	75	11	19	30	33												4	2	0	2	0			
	NHL Totals		2	0	1	1	0	0	0	0	2	0.0		15	60.0	2	0	8:21								

• All-time leading scorer in Massachusetts High School history with career totals of 120-182-302. • Hockey East Second All-Star Team (2000)

Signed as a free agent by **Columbus**, May 26, 2000.

BENYSEK, Ladislav (BEHN-ih-sihk, LAD-ihs-SLAHV) MIN.

Defense. Shoots left. 6'2", 190 lbs. Born, Olomouc, Czech., March 24, 1975. Edmonton's 16th choice, 266th overall, in 1994 Entry Draft.

Season	Club	League	GP	G	A	Pts	PIM	PP	SH	GW	S	%	+/-	TF	F%	H	SB	Min	GP	G	A	Pts	PIM	PP	SH	GW
1992-93	HC Olomouc	Czech	3	0	0	0	0																			
1993-94	HC Olomouc Jr.	Czech-Jr.	STATISTICS NOT AVAILABLE																							
1994-95	Cape Breton	AHL	58	2	7	9	54																			
1995-96	HC Olomouc	Czech	33	1	4	5													4	0	0	0				
1996-97	HC Olomouc	Czech	14	0	1	1	8																			
	HC Sparta Praha	Czech	36	5	5	10	28												5	0	1	1	2			
	HC Sparta Praha	EuroHL	3	0	0	0	4												4	0	0	0	0			
1997-98	HC Sparta Praha	Czech	1	0	0	0	0																			
	Edmonton	NHL	2	0	0	0	0																			
	Hamilton	AHL	53	2	14	16	29												9	1	1	2	2			
1998-99	HC Sparta Praha	Czech	52	8	11	19	47												8	0	1	1				
	HC Sparta Praha	EuroHL	7	0	0	0	2												2	0	0	0	0			
99-2000	HC Sparta Praha	Czech	51	1	5	6	45												9	0	0	0	4			
	HC Sparta Praha	EuroHL	5	1	1	2	6												4	0	0	0	4			
2000-01	Minnesota	NHL	71	2	5	7	38	1	0	0	48	4.2	–11	0	0.0	122	109	18:25								
2001-02	Minnesota	NHL	74	1	7	8	28	0	0	0	44	2.3	–12	1	100.0	118	157	19:09								
	NHL Totals		147	3	12	15	66	1	0	0	92	3.3		1	100.0	240	266	18:47								

Claimed by **Anaheim** from **Edmonton** in Waiver Draft, September 27, 1999. Selected by **Minnesota** from **Anaheim** in Expansion Draft, June 23, 2000.

BERANEK, Josef (buh-RAH-nehk, JOH-sehf)

Left wing/Center. Shoots left. 6'2", 195 lbs. Born, Litvinov, Czech., October 25, 1969. Edmonton's 3rd choice, 78th overall, in 1989 Entry Draft.

Season	Club	League	GP	G	A	Pts	PIM	PP	SH	GW	S	%	+/-	TF	F%	H	SB	Min	GP	G	A	Pts	PIM	PP	SH	GW
1987-88	CHZ Litvinov	Czech	14	7	4	11	12																			
1988-89	CHZ Litvinov	Czech	32	18	10	28	47																			
1989-90	Dukla Trencin	Czech	40	16	21	37													9	3	2	5				
1990-91	HC CHZ Litvinov	Czech	58	29	31	60	98																			
1991-92	Edmonton	NHL	58	12	16	28	18	0	0	1	79	15.2	–2						12	2	1	3	0	1	0	1
1992-93	Edmonton	NHL	26	2	6	8	28	0	0	0	44	4.5	–7													
	Cape Breton	AHL	6	1	2	3	8																			
	Philadelphia	NHL	40	13	12	25	50	1	0	0	86	15.1	–1													
1993-94	Philadelphia	NHL	80	28	21	49	85	6	0	2	182	15.4	–2													
1994-95	HC Dadak Vsetin	Czech	16	7	7	14	26																			
	Philadelphia	NHL	14	5	5	10	2	1	0	0	39	12.8	3													
	Vancouver	NHL	37	8	13	21	28	2	0	0	95	8.4	–10						11	1	1	2	12	0	0	0
1995-96	Vancouver	NHL	61	6	14	20	60	0	0	1	131	4.6	–11						3	2	1	3	0	0	0	0
1996-97	HC Petra Vsetin	Czech	39	19	24	43	115												3	3	2	5	4			
	Pittsburgh	NHL	8	3	1	4	4	1	0	0	15	20.0	–1						5	0	0	0	2	0	0	0
1997-98	HC Petra Vsetin	Czech	45	24	27	51	92												10	2	8	10	14			
	HC Petra Vsetin	EuroHL	8	5	4	9	10																			
	Czech Republic	Olympics	6	1	0	1	4																			
1998-99	Edmonton	NHL	66	19	30	49	23	7	0	2	160	11.9	6	1261	50.2	47	21	16:25	2	0	0	0	4	0	0	0

			Regular Season																Playoffs							
Season	Club	League	GP	G	A	Pts	PIM	PP	SH	GW	S	%	+/-	TF	F%	H	SB	Min	GP	G	A	Pts	PIM	PP	SH	GW
99-2000	**Edmonton**	**NHL**	**58**	**9**	**8**	**17**	**39**	**3**	**0**	**1**	**107**	**8.4**	**–6**	**558**	**53.4**	**42**	**6**	**13:19**	....	....	....	....	....	....	....	....
	Pittsburgh	**NHL**	**13**	**4**	**4**	**8**	**18**	**1**	**0**	**0**	**32**	**12.5**	**–6**	**73**	**42.5**	**13**	**4**	**19:46**	**11**	**0**	**3**	**3**	**4**	**0**	**0**	**0**
2000-01	**Pittsburgh**	**NHL**	**70**	**9**	**14**	**23**	**43**	**2**	**0**	**2**	**152**	**5.9**	**–7**	**172**	**44.8**	**56**	**19**	**14:27**	**13**	**0**	**2**	**2**	**2**	**0**	**0**	**0**
2001-02	HC Slavia Praha	Czech	36	13	19	32	71	....	....	....	....	....	....	....	....	....	....	....	9	1	2	3	6	....	....	....
	NHL Totals		**531**	**118**	**144**	**262**	**398**	**24**	**0**	**9**	**1122**	**10.5**		**2064**	**50.3**	**158**	**50**	**15:05**	**57**	**5**	**8**	**13**	**24**	**1**	**0**	**1**

Traded to **Philadelphia** by **Edmonton** with Greg Hawgood for Brian Benning, January 16, 1993. Traded to **Vancouver** by **Philadelphia** for Shawn Antoski, February 15, 1995. Traded to **Pittsburgh** by **Vancouver** for future considerations, March 18, 1997. Traded to **Edmonton** by **Pittsburgh** for Bobby Dollas and Tony Hrkac, June 16, 1998. Traded to **Pittsburgh** by **Edmonton** for German Titov, March 14, 2000.

BERARD, Bryan

(buh-RAHRD, BRIGH-uhn) **BOS.**

Defense. Shoots left. 6'2", 195 lbs. Born, Woonsocket, RI, March 5, 1977. Ottawa's 1st choice, 1st overall, in 1995 Entry Draft.

Season	Club	League	GP	G	A	Pts	PIM	PP	SH	GW	S	%	+/-	TF	F%	H	SB	Min	GP	G	A	Pts	PIM	PP	SH	GW
1991-92	Mount St. Charles	Hi-School	15	3	15	18	4	....	....	....	....	....	....	....	....	....	....	....	....	....	....	....	....	....	....	....
1992-93	Mount St. Charles	Hi-School	15	8	12	20	18	....	....	....	....	....	....	....	....	....	....	....	....	....	....	....	....	....	....	....
1993-94	Mount St. Charles	Hi-School	15	11	26	37	4.5	....	....	....	....	....	....	....	....	....	....	....	4	3	3	6	6	....	....	....
1994-95	Detroit	OHL	58	20	55	75	97	....	....	....	....	....	....	....	....	....	....	....	21	4	20	24	38	....	....	....
1995-96	Detroit	OHL	56	31	58	89	116	....	....	....	....	....	....	....	....	....	....	....	17	7	18	25	41	....	....	....
1996-97	**NY Islanders**	**NHL**	**82**	**8**	**40**	**48**	**86**	**3**	**0**	**1**	**172**	**4.7**	**1**	....	....	....	....	....	....	....	....	....	....	....	....	....
1997-98	**NY Islanders**	**NHL**	**75**	**14**	**32**	**46**	**59**	**8**	**1**	**2**	**192**	**7.3**	**–32**	....	....	....	....	....	....	....	....	....	....	....	....	....
	United States	Olympics	2	0	0	0	0	....	....	....	....	....	....	....	....	....	....	....	....	....	....	....	....	....	....	....
1998-99	**NY Islanders**	**NHL**	**31**	**4**	**11**	**15**	**26**	**2**	**0**	**3**	**72**	**5.6**	**–6**	**0**	**0.0**	**16**	**32**	**24:45**	....	....	....	....	....	....	....	....
	Toronto	**NHL**	**38**	**5**	**14**	**19**	**22**	**2**	**0**	**2**	**63**	**7.9**	**7**	**0**	**0.0**	**34**	**42**	**22:38**	**17**	**1**	**8**	**9**	**8**	**1**	**0**	**0**
99-2000	**Toronto**	**NHL**	**64**	**3**	**27**	**30**	**42**	**1**	**0**	**0**	**98**	**3.1**	**11**	**0**	**0.0**	**64**	**58**	**19:34**	....	....	....	....	....	....	....	....
2000-01	**Toronto**	**NHL**	DID NOT PLAY – INJURED																....	....	....	....	....	....	....	....
2001-02	**NY Rangers**	**NHL**	**82**	**2**	**21**	**23**	**60**	**0**	**0**	**0**	**132**	**1.5**	**–1**	**0**	**0.0**	**78**	**111**	**19:38**	....	....	....	....	....	....	....	....
	NHL Totals		**372**	**36**	**145**	**181**	**295**	**16**	**1**	**8**	**729**	**4.9**		**0**	**0.0**	**192**	**243**	**20:53**	**17**	**1**	**8**	**9**	**8**	**1**	**0**	**0**

OHL All-Rookie Team (1995) • OHL First All-Star Team (1995, 1996) • OHL Rookie of the Year (1995) • Canadian Major Junior First All-Star Team (1995, 1996) • Canadian Major Junior Rookie of the Year (1995) • Canadian Major Junior Defenseman of the Year (1996) • NHL All-Rookie Team (1997) • Won Calder Memorial Trophy (1997)

Traded to **NY Islanders** by **Ottawa** with Don Beaupre and Martin Straka for Damian Rhodes and Wade Redden, January 23, 1996. Traded to **Toronto** by **NY Islanders** with NY Islanders' 6th round choice (Jan Sochor) in 1999 Entry Draft for Felix Potvin and Toronto's 6th round choice (later traded to Tampa Bay - Tampa Bay selected Fedor Fedorov) in 1999 Entry Draft, January 9, 1999. • Missed remainder of 1999-2000 and entire 2000-01 seasons recovering from eye injury suffered in game vs. Ottawa, March 11, 2000. Signed as a free agent by **NY Rangers**, October 5, 2001. Signed as a free agent by **Boston**, August 13, 2002.

BEREHOWSKY, Drake

(beh-reh-HOW-skee, DRAYK) **PHX.**

Defense. Shoots right. 6'2", 225 lbs. Born, Toronto, Ont., January 3, 1972. Toronto's 1st choice, 10th overall, in 1990 Entry Draft.

Season	Club	League	GP	G	A	Pts	PIM	PP	SH	GW	S	%	+/-	TF	F%	H	SB	Min	GP	G	A	Pts	PIM	PP	SH	GW
1987-88	Barrie Colts	OJHL-B	40	10	36	46	81	....	....	....	....	....	....	....	....	....	....	....	....	....	....	....	....	....	....	....
1988-89	Kingston Raiders	OHL	63	7	39	46	85	....	....	....	....	....	....	....	....	....	....	....	....	....	....	....	....	....	....	....
1989-90	Kingston	OHL	9	3	11	14	28	....	....	....	....	....	....	....	....	....	....	....	....	....	....	....	....	....	....	....
1990-91	**Toronto**	**NHL**	**8**	**0**	**1**	**1**	**25**	**0**	**0**	**0**	**4**	**0.0**	**–6**	....	....	....	....	....	....	....	....	....	....	....	....	....
	Kingston	OHL	13	5	13	18	38	....	....	....	....	....	....	....	....	....	....	....	....	....	....	....	....	....	....	....
	North Bay	OHL	26	7	23	30	51	....	....	....	....	....	....	....	....	....	....	....	10	2	7	9	21	....	....	....
1991-92	North Bay	OHL	62	19	63	82	147	....	....	....	....	....	....	....	....	....	....	....	21	7	24	31	22	....	....	....
	Toronto	**NHL**	**1**	**0**	**0**	**0**	**0**	**0**	**0**	**0**	**0**	**0.0**	**0**	....	....	....	....	....	....	....	....	....	....	....	....	....
	St. John's	AHL	....	....	....	....	....	....	....	....	....	....	....	....	....	....	....	....	6	0	5	5	21	....	....	....
1992-93	**Toronto**	**NHL**	**41**	**4**	**15**	**19**	**61**	**1**	**0**	**1**	**41**	**9.8**	**1**	....	....	....	....	....	....	....	....	....	....	....	....	....
	St. John's	AHL	28	10	17	27	38	....	....	....	....	....	....	....	....	....	....	....	....	....	....	....	....	....	....	....
1993-94	**Toronto**	**NHL**	**49**	**2**	**8**	**10**	**63**	**2**	**0**	**2**	**29**	**6.9**	**–3**	....	....	....	....	....	....	....	....	....	....	....	....	....
	St. John's	AHL	18	3	12	15	40	....	....	....	....	....	....	....	....	....	....	....	....	....	....	....	....	....	....	....
1994-95	**Toronto**	**NHL**	**25**	**0**	**2**	**2**	**15**	**0**	**0**	**0**	**12**	**0.0**	**–10**	....	....	....	....	....	....	....	....	....	....	....	....	....
	Pittsburgh	**NHL**	**4**	**0**	**0**	**0**	**13**	**0**	**0**	**0**	**2**	**0.0**	**1**	....	....	....	....	....	**1**	**0**	**0**	**0**	**0**	**0**	**0**	**0**
1995-96	**Pittsburgh**	**NHL**	**1**	**0**	**0**	**0**	**0**	**0**	**0**	**0**	**0**	**0.0**	**1**	....	....	....	....	....	....	....	....	....	....	....	....	....
	Cleveland	IHL	74	6	28	34	141	....	....	....	....	....	....	....	....	....	....	....	3	0	3	3	6	....	....	....
1996-97	Carolina	AHL	49	2	15	17	55	....	....	....	....	....	....	....	....	....	....	....	....	....	....	....	....	....	....	....
	San Antonio	IHL	16	3	4	7	36	....	....	....	....	....	....	....	....	....	....	....	....	....	....	....	....	....	....	....
1997-98	**Edmonton**	**NHL**	**67**	**1**	**6**	**7**	**169**	**1**	**0**	**1**	**58**	**1.7**	**1**	....	....	....	....	....	**12**	**1**	**2**	**3**	**14**	**0**	**0**	**1**
	Hamilton	AHL	8	2	0	2	21	....	....	....	....	....	....	....	....	....	....	....	....	....	....	....	....	....	....	....
1998-99	**Nashville**	**NHL**	**74**	**2**	**15**	**17**	**140**	**0**	**0**	**0**	**79**	**2.5**	**–9**	**1**	**100.0**	**140**	**109**	**21:43**	....	....	....	....	....	....	....	....
99-2000	**Nashville**	**NHL**	**79**	**12**	**20**	**32**	**87**	**5**	**0**	**1**	**102**	**11.8**	**–4**	**0**	**0.0**	**140**	**110**	**22:39**	....	....	....	....	....	....	....	....
2000-01	**Nashville**	**NHL**	**66**	**6**	**18**	**24**	**100**	**3**	**0**	**1**	**94**	**6.4**	**–9**	**1**	**0.0**	**115**	**56**	**21:38**	....	....	....	....	....	....	....	....
	Vancouver	**NHL**	**14**	**1**	**1**	**2**	**21**	**1**	**0**	**0**	**13**	**7.7**	**0**	**0**	**0.0**	**13**	**16**	**17:10**	**4**	**0**	**0**	**0**	**12**	**0**	**0**	**0**
2001-02	**Vancouver**	**NHL**	**25**	**1**	**2**	**3**	**18**	**0**	**0**	**1**	**15**	**6.7**	**–5**	**1**	**0.0**	**46**	**18**	**13:28**	....	....	....	....	....	....	....	....
	Phoenix	**NHL**	**32**	**1**	**4**	**5**	**42**	**0**	**0**	**0**	**23**	**4.3**	**5**	**0**	**0.0**	**65**	**26**	**12:20**	**5**	**0**	**1**	**1**	**4**	**0**	**0**	**0**
	NHL Totals		**486**	**30**	**92**	**122**	**754**	**13**	**0**	**7**	**472**	**6.4**		**3**	**33.3**	**519**	**335**	**19:59**	**22**	**1**	**3**	**4**	**30**	**0**	**0**	**1**

OHL First All-Star Team (1992) • Canadian Major Junior Defenseman of the Year (1992)

Traded to **Pittsburgh** by **Toronto** for Grant Jennings, April 7, 1995. Signed as a free agent by **Edmonton**, September 30, 1997. Traded to **Nashville** by **Edmonton** with Eric Fichaud and Greg de Vries for Mikhail Shtalenkov and Jim Dowd, October 1, 1998. Traded to **Vancouver** by **Nashville** for Atlanta's 2nd round choice (previously acquired, Nashville selected Timofei Shishkanov) in 2001 Entry Draft, March 9, 2001. Traded to **Phoenix** by **Vancouver** with Denis Pederson for Todd Warriner, Trevor Letowski, Tyler Bouck and Phoenix's 3rd round choice in 2003 Entry Draft, December 28, 2001.

BERENZWEIG, Bubba

(BAIR-ehn-zwighg, BUH-buh) **NSH.**

Defense. Shoots left. 6'1", 217 lbs. Born, Arlington Heights, IL, August 8, 1977. NY Islanders' 5th choice, 109th overall, in 1996 Entry Draft.

Season	Club	League	GP	G	A	Pts	PIM	PP	SH	GW	S	%	+/-	TF	F%	H	SB	Min	GP	G	A	Pts	PIM	PP	SH	GW
1992-93	Loomis-Chaffee	Hi-School	22	5	13	18	....	....	....	....	....	....	....	....	....	....	....	....	....	....	....	....	....	....	....	....
1993-94	Loomis-Chaffee	Hi-School	22	12	27	39	....	....	....	....	....	....	....	....	....	....	....	....	....	....	....	....	....	....	....	....
1994-95	Loomis-Chaffee	Hi-School	23	19	23	42	10	....	....	....	....	....	....	....	....	....	....	....	....	....	....	....	....	....	....	....
1995-96	U. of Michigan	CCHA	42	4	8	12	4	....	....	....	....	....	....	....	....	....	....	....	....	....	....	....	....	....	....	....
1996-97	U. of Michigan	CCHA	38	7	12	19	49	....	....	....	....	....	....	....	....	....	....	....	....	....	....	....	....	....	....	....
1997-98	U. of Michigan	CCHA	45	8	11	19	32	....	....	....	....	....	....	....	....	....	....	....	....	....	....	....	....	....	....	....
1998-99	U. of Michigan	CCHA	42	7	24	31	38	....	....	....	....	....	....	....	....	....	....	....	....	....	....	....	....	....	....	....
99-2000	**Nashville**	**NHL**	**2**	**0**	**0**	**0**	**0**	**0**	**0**	**0**	**3**	**0.0**	**–1**	**0**	**0.0**	**5**	**4**	**17:31**	....	....	....	....	....	....	....	....
	Milwaukee	IHL	79	4	23	27	48	....	....	....	....	....	....	....	....	....	....	....	3	1	2	3	0	....	....	....
2000-01	**Nashville**	**NHL**	**5**	**0**	**0**	**0**	**0**	**0**	**0**	**0**	**0**	**0.0**	**0**	**0**	**0.0**	**1**	**0**	**11:57**	....	....	....	....	....	....	....	....
	Milwaukee	IHL	72	10	26	36	38	....	....	....	....	....	....	....	....	....	....	....	5	0	4	4	4	....	....	....
2001-02	**Nashville**	**NHL**	**26**	**3**	**7**	**10**	**14**	**0**	**0**	**1**	**27**	**11.1**	**–3**	**1**	**0.0**	**21**	**17**	**13:45**	....	....	....	....	....	....	....	....
	Milwaukee	AHL	23	2	5	7	23	....	....	....	....	....	....	....	....	....	....	....	....	....	....	....	....	....	....	....
	NHL Totals		**33**	**3**	**7**	**10**	**14**	**0**	**0**	**1**	**30**	**10.0**		**1**	**0.0**	**27**	**21**	**13:42**	....	....	....	....	....	....	....	....

CCHA Second All-Star Team (1998) • NCAA Championship All-Tournament Team (1998) • Won Ken McKenzie Trophy (Outstanding U.S.- Born Player - IHL) (2000) • IHL Second All-Star Team (2001)

Traded to **Nashville** by **NY Islanders** for Nashville's 4th round choice (Johan Halvardsson) in 1999 Entry Draft, April 14, 1999.

BEREZIN, Sergei

(BEH-reh-zihn, SAIR-gay) **CHI.**

Left wing. Shoots right. 5'10", 200 lbs. Born, Voskresensk, USSR, November 5, 1971. Toronto's 8th choice, 256th overall, in 1994 Entry Draft.

Season	Club	League	GP	G	A	Pts	PIM	PP	SH	GW	S	%	+/-	TF	F%	H	SB	Min	GP	G	A	Pts	PIM	PP	SH	GW
1990-91	Voskresensk	USSR	30	6	2	8	4	....	....	....	....	....	....	....	....	....	....	....	....	....	....	....	....	....	....	....
1991-92	Voskresensk	CIS	36	7	5	12	10	....	....	....	....	....	....	....	....	....	....	....	....	....	....	....	....	....	....	....
1992-93	Voskresensk	CIS	38	9	3	12	12	....	....	....	....	....	....	....	....	....	....	....	2	1	0	1	0	....	....	....
1993-94	Voskresensk	CIS	40	31	10	41	16	....	....	....	....	....	....	....	....	....	....	....	3	2	0	2	2	....	....	....
	Russia	Olympics	8	3	2	5	2	....	....	....	....	....	....	....	....	....	....	....	....	....	....	....	....	....	....	....
1994-95	Kolner Haie	Germany	43	*38	19	57	8	....	....	....	....	....	....	....	....	....	....	....	18	*17	8	25	14	....	....	....
1995-96	Kolner Haie	Germany	45	*49	31	80	8	....	....	....	....	....	....	....	....	....	....	....	14	*13	9	22	10	....	....	....
1996-97	**Toronto**	**NHL**	**73**	**25**	**16**	**41**	**2**	**7**	**0**	**2**	**177**	**14.1**	**–3**	....	....	....	....	....	....	....	....	....	....	....	....	....
1997-98	**Toronto**	**NHL**	**68**	**16**	**15**	**31**	**10**	**3**	**0**	**3**	**167**	**9.6**	**–3**	....	....	....	....	....	....	....	....	....	....	....	....	....
1998-99	**Toronto**	**NHL**	**76**	**37**	**22**	**59**	**12**	**9**	**1**	**4**	**263**	**14.1**	**16**	**26**	**57.7**	**24**	**11**	**15:32**	**17**	**6**	**6**	**12**	**4**	**2**	**0**	**2**
99-2000	**Toronto**	**NHL**	**61**	**26**	**13**	**39**	**2**	**5**	**0**	**4**	**241**	**10.8**	**8**	**11**	**45.5**	**18**	**9**	**16:52**	**12**	**4**	**4**	**8**	**0**	**0**	**0**	**1**
2000-01	**Toronto**	**NHL**	**79**	**22**	**28**	**50**	**8**	**10**	**0**	**3**	**256**	**8.6**	**2**	**4**	**75.0**	**33**	**10**	**15:41**	**11**	**2**	**5**	**7**	**2**	**0**	**0**	**2**

Season	Club	League	GP	G	A	Pts	PIM	PP	SH	GW	S	%	+/-	TF	F%	H	SB	Min	GP	G	A	Pts	PIM	PP	SH	GW
			Regular Season																Playoffs							
2001-02	Phoenix	NHL	41	7	9	16	4	1	0	4	120	5.8	-1	2	50.0	17	9	15:33	....	....	....	....	....	....	....	....
	Montreal	NHL	29	4	6	10	4	3	0	1	80	5.0	3	2	0.0	7	3	14:01	6	1	1	2	0	1	0	0
	NHL Totals		427	137	109	246	42	38	1	21	1304	10.5		45	53.3	99	42	15:42	46	13	16	29	6	3	0	5

NHL All-Rookie Team (1997)

Traded to **Phoenix** by **Toronto** for Mikael Renberg, June 23, 2001. Traded to **Montreal** by **Phoenix** for Brian Savage, Montreal's 3rd round choice (Matt Jones) in 2002 Entry Draft and future considerations, January 25, 2002. Traded to **Chicago** by **Montreal** for Chicago's 4th round choice in 2004 Entry Draft, June 30, 2002.

BERG, Aki

(BUHRG, AH-kee) **TOR.**

Defense. Shoots left. 6'3", 220 lbs. Born, Turku, Finland, July 28, 1977. Los Angeles' 1st choice, 3rd overall, in 1995 Entry Draft.

Season	Club	League	GP	G	A	Pts	PIM	PP	SH	GW	S	%	+/-	TF	F%	H	SB	Min	GP	G	A	Pts	PIM	PP	SH	GW
1992-93	TPS Turku Jr.	Finn-Jr.	39	18	24	42	24	....	....	....	....	....	....	....	....	....	....	....	....	....	....	....	....	....	....	....
1993-94	TPS Turku Jr.	Finn-Jr.	21	3	11	14	24	....	....	....	....	....	....	....	....	....	....	....	7	0	0	0	10	....	....	....
	Kiekko-67 Turku	Finland-2	12	1	1	2	16	....	....	....	....	....	....	....	....	....	....	....	....	....	....	....	....	....	....	....
	TPS Turku	Finland	6	0	3	3	4	....	....	....	....	....	....	....	....	....	....	....	....	....	....	....	....	....	....	....
1994-95	TPS Turku Jr.	Finn-Jr.	8	1	0	1	30	....	....	....	....	....	....	....	....	....	....	....	....	....	....	....	....	....	....	....
	Kiekko-67 Turku	Finland-2	21	3	9	12	24	....	....	....	....	....	....	....	....	....	....	....	7	0	0	0	10	....	....	....
	TPS Turku	Finland	5	0	0	0	4	....	....	....	....	....	....	....	....	....	....	....	....	....	....	....	....	....	....	....
1995-96	Los Angeles	NHL	51	0	7	7	29	0	0	0	56	0.0	-13	....	....	....	....	....	....	....	....	....	....	....	....	....
	Phoenix	IHL	20	0	3	3	18	....	....	....	....	....	....	....	....	....	....	....	2	0	0	0	4	....	....	....
1996-97	Los Angeles	NHL	41	2	6	8	24	2	0	0	65	3.1	-9	....	....	....	....	....	....	....	....	....	....	....	....	....
	Phoenix	IHL	23	1	3	4	21	....	....	....	....	....	....	....	....	....	....	....	....	....	....	....	....	....	....	....
1997-98	Los Angeles	NHL	72	0	8	8	61	0	0	0	58	0.0	3	....	....	....	....	....	4	0	3	3	0	0	0	0
	Finland	Olympics	6	0	0	0	6	....	....	....	....	....	....	....	....	....	....	....	....	....	....	....	....	....	....	....
1998-99	TPS Turku	Finland	48	8	7	15	137	....	....	....	....	....	....	....	....	....	....	....	9	1	1	2	45	....	....	....
99-2000	Los Angeles	NHL	70	3	13	16	45	0	0	0	70	4.3	-1	0	0.0	197	83	16:39	2	0	0	0	2	0	0	0
2000-01	Los Angeles	NHL	47	0	4	4	43	0	0	0	31	0.0	3	0	0.0	130	73	14:54	....	....	....	....	....	....	....	....
	Toronto	NHL	12	3	0	3	2	3	0	1	12	25.0	-6	0	0.0	22	13	18:13	11	0	2	2	4	0	0	0
2001-02	Toronto	NHL	81	1	10	11	46	0	0	0	66	1.5	14	1	100.0	169	94	18:43	20	0	1	1	37	0	0	0
	Finland	Olympics	4	1	0	1	2	....	....	....	....	....	....	....	....	....	....	....	....	....	....	....	....	....	....	....
	NHL Totals		374	9	48	57	250	5	0	1	358	2.5		1	100.0	518	263	17:09	37	0	6	6	43	0	0	0

Traded to **Toronto** by **LA Kings** for Adam Mair and Toronto's 2nd round choice (Mike Cammalleri) in 2001 Entry Draft, March 13, 2001.

BERGEVIN, Marc

(BUHR-zheh-vihn, MAHRK) **PIT.**

Defense. Shoots left. 6'1", 214 lbs. Born, Montreal, Que., August 11, 1965. Chicago's 3rd choice, 60th overall, in 1983 Entry Draft.

Season	Club	League	GP	G	A	Pts	PIM	PP	SH	GW	S	%	+/-	TF	F%	H	SB	Min	GP	G	A	Pts	PIM	PP	SH	GW
1981-82	Mtl-Concordia	QAAA	44	10	20	30	54	....	....	....	....	....	....	....	....	....	....	....	5	0	2	2	4	....	....	....
1982-83	Chicoutimi	QMJHL	64	3	27	30	113	....	....	....	....	....	....	....	....	....	....	....	....	....	....	....	....	....	....	....
1983-84	Chicoutimi	QMJHL	70	10	35	45	125	....	....	....	....	....	....	....	....	....	....	....	....	....	....	....	....	....	....	....
	Springfield	AHL	7	0	1	1	2	....	....	....	....	....	....	....	....	....	....	....	....	....	....	....	....	....	....	....
1984-85	Chicago	NHL	60	0	6	6	54	0	0	0	41	0.0	-9	....	....	....	....	....	6	0	3	3	2	0	0	0
	Springfield	AHL	....	....	....	....	....	....	....	....	....	....	....	....	....	....	....	....	4	0	0	0	0	....	....	....
1985-86	Chicago	NHL	71	7	7	14	60	0	0	1	50	14.0	0	....	....	....	....	....	3	0	0	0	0	0	0	0
1986-87	Chicago	NHL	66	4	10	14	66	0	0	0	56	7.1	4	....	....	....	....	....	3	1	0	1	2	0	0	0
1987-88	Chicago	NHL	58	1	6	7	85	0	0	0	51	2.0	-19	....	....	....	....	....	....	....	....	....	....	....	....	....
	Saginaw Hawks	IHL	10	2	7	9	20	....	....	....	....	....	....	....	....	....	....	....	....	....	....	....	....	....	....	....
1988-89	Chicago	NHL	11	0	0	0	18	0	0	0	9	0.0	-3	....	....	....	....	....	....	....	....	....	....	....	....	....
	NY Islanders	NHL	58	2	13	15	62	1	0	0	56	3.6	2	....	....	....	....	....	....	....	....	....	....	....	....	....
1989-90	NY Islanders	NHL	18	0	4	4	30	0	0	0	12	0.0	-8	....	....	....	....	....	....	....	....	....	....	....	....	....
	Springfield	AHL	47	7	16	23	66	....	....	....	....	....	....	....	....	....	....	....	17	2	11	13	16	....	....	....
1990-91	Capital District	AHL	7	0	5	5	6	....	....	....	....	....	....	....	....	....	....	....	....	....	....	....	....	....	....	....
	Hartford	NHL	4	0	0	0	4	0	0	0	2	0.0	-3	....	....	....	....	....	....	....	....	....	....	....	....	....
	Springfield	AHL	58	4	23	27	85	....	....	....	....	....	....	....	....	....	....	....	18	0	7	7	26	....	....	....
1991-92	Hartford	NHL	75	7	17	24	64	4	1	1	96	7.3	-13	....	....	....	....	....	5	0	0	0	2	0	0	0
1992-93	Tampa Bay	NHL	78	2	12	14	66	0	0	0	69	2.9	-16	....	....	....	....	....	....	....	....	....	....	....	....	....
1993-94	Tampa Bay	NHL	83	1	15	16	87	0	0	1	76	1.3	-5	....	....	....	....	....	....	....	....	....	....	....	....	....
1994-95	Tampa Bay	NHL	44	2	4	6	51	0	1	0	32	6.3	-6	....	....	....	....	....	....	....	....	....	....	....	....	....
1995-96	Detroit	NHL	70	1	9	10	33	0	0	0	26	3.8	7	....	....	....	....	....	17	1	0	1	14	1	0	0
1996-97	St. Louis	NHL	82	0	4	4	53	0	0	0	30	0.0	-9	....	....	....	....	....	6	1	0	1	8	0	0	0
1997-98	St. Louis	NHL	81	3	7	10	90	0	0	0	40	7.5	-2	....	....	....	....	....	10	0	1	1	8	0	0	0
1998-99	St. Louis	NHL	52	1	1	2	99	0	0	0	40	2.5	-14	0	0.0	90	40	16:09	....	....	....	....	....	....	....	....
99-2000	St. Louis	NHL	81	1	8	9	75	0	0	0	54	1.9	27	0	0.0	100	98	21:16	7	0	1	1	6	0	0	0
2000-01	St. Louis	NHL	2	0	0	0	0	0	0	0	1	0.0	1	0	0.0	2	2	14:51	....	....	....	....	....	....	....	....
	Pittsburgh	NHL	36	1	4	5	26	0	0	0	11	9.1	5	0	0.0	56	48	16:57	12	0	1	1	2	0	0	0
2001-02	St. Louis	NHL	30	0	3	3	2	0	0	0	13	0.0	6	0	0.0	22	25	12:15	7	0	0	0	4	0	0	0
	Worcester	AHL	2	0	0	0	0	....	....	....	....	....	....	....	....	....	....	....	....	....	....	....	....	....	....	....
	NHL Totals		1060	33	130	163	1025	5	2	3	765	4.3		0	0.0	270	213	17:46	76	3	6	9	48	1	0	0

Traded to **NY Islanders** by **Chicago** with Gary Nylund for Steve Konroyd and Bob Bassen, November 25, 1988. Traded to **Hartford** by **NY Islanders** for Hartford's 5th round choice (Ryan Duthie) in 1992 Entry Draft, October 30, 1990. Signed as a free agent by **Tampa Bay**, July 9, 1992. Traded to **Detroit** by **Tampa Bay** with Ben Hankinson for Shawn Burr and Detroit's 3rd round choice (later traded to Boston - Boston selected Jason Doyle) in 1996 Entry Draft, August 17, 1995. Signed as a free agent by **St. Louis**, July 31, 1996. Traded to **Pittsburgh** by **St. Louis** for Dan Trebil, December 28, 2000. • Missed majority of 2000-01 season recovering from thumb injury suffered in game vs. Phoenix (October 5, 2000) and knee injury suffered in game vs. Detroit (February 23, 2001). Signed as a free agent by **St. Louis**, November 6, 2001. Signed as a free agent by **Pittsburgh**, July 18, 2002.

BERGLUND, Christian

(BUHRG-luhnd, KRIH-stan) **N.J.**

Left wing. Shoots left. 5'11", 195 lbs. Born, Orebro, Sweden, March 12, 1980. New Jersey's 3rd choice, 37th overall, in 1998 Entry Draft.

Season	Club	League	GP	G	A	Pts	PIM	PP	SH	GW	S	%	+/-	TF	F%	H	SB	Min	GP	G	A	Pts	PIM	PP	SH	GW
1994-95	Kariskoga IK	Swede-4	20	14	13	27	....	....	....	....	....	....	....	....	....	....	....	....	....	....	....	....	....	....	....	....
1995-96	Kristinehamn SK	Swede-3	23	8	8	16	12	....	....	....	....	....	....	....	....	....	....	....	....	....	....	....	....	....	....	....
1996-97	Farjestad Jr.	Swede-Jr.	21	2	3	5	24	....	....	....	....	....	....	....	....	....	....	....	....	....	....	....	....	....	....	....
1997-98	Farjestad Jr.	Swede-Jr.	29	23	19	42	88	....	....	....	....	....	....	....	....	....	....	....	2	0	0	0	0	....	....	....
	Farjestad	Sweden	1	0	0	0	0	....	....	....	....	....	....	....	....	....	....	....	....	....	....	....	....	....	....	....
1998-99	Farjestad Jr.	Swede-Jr.	5	3	4	7	22	....	....	....	....	....	....	....	....	....	....	....	....	....	....	....	....	....	....	....
	Farjestad	Sweden	37	2	4	6	37	....	....	....	....	....	....	....	....	....	....	....	4	1	0	1	4	....	....	....
99-2000	Farjestad Jr.	Swede-Jr.	5	3	5	8	8	....	....	....	....	....	....	....	....	....	....	....	....	....	....	....	....	....	....	....
	Bofors IK	Swede-2	6	2	0	2	12	....	....	....	....	....	....	....	....	....	....	....	....	....	....	....	....	....	....	....
	Farjestad	Sweden	43	8	6	14	44	....	....	....	....	....	....	....	....	....	....	....	7	2	1	3	10	....	....	....
2000-01	Farjestad	Sweden	49	17	20	37	*142	....	....	....	....	....	....	....	....	....	....	....	16	7	7	14	22	....	....	....
2001-02	New Jersey	NHL	15	2	7	9	8	0	0	0	22	9.1	-3	2	50.0	20	0	12:26	3	0	0	0	2	0	0	0
	Albany	AHL	60	21	26	47	69	....	....	....	....	....	....	....	....	....	....	....	....	....	....	....	....	....	....	....
	NHL Totals		15	2	7	9	8	0	0	0	22	9.1		2	50.0	20	0	12:26	3	0	0	0	2	0	0	0

BERRY, Rick

(BAIR-ree, RIHK) **PIT.**

Defense. Shoots left. 6'2", 210 lbs. Born, Birtle, Man., November 4, 1978. Colorado's 3rd choice, 55th overall, in 1997 Entry Draft.

Season	Club	League	GP	G	A	Pts	PIM	PP	SH	GW	S	%	+/-	TF	F%	H	SB	Min	GP	G	A	Pts	PIM	PP	SH	GW
1994-95	Yellowhead Pass	MMHL	33	12	19	31	90	....	....	....	....	....	....	....	....	....	....	....	....	....	....	....	....	....	....	....
1995-96	Seattle	WHL	59	4	9	13	103	....	....	....	....	....	....	....	....	....	....	....	1	0	0	0	0	....	....	....
1996-97	Seattle	WHL	72	12	21	33	125	....	....	....	....	....	....	....	....	....	....	....	15	3	7	10	23	....	....	....
1997-98	Seattle	WHL	37	5	12	17	100	....	....	....	....	....	....	....	....	....	....	....	....	....	....	....	....	....	....	....
	Spokane Chiefs	WHL	22	4	9	13	31	....	....	....	....	....	....	....	....	....	....	....	17	1	4	5	26	....	....	....
1998-99	Hershey Bears	AHL	62	2	6	8	153	....	....	....	....	....	....	....	....	....	....	....	....	....	....	....	....	....	....	....
99-2000	Hershey Bears	AHL	64	9	16	25	148	....	....	....	....	....	....	....	....	....	....	....	13	2	3	5	24	....	....	....
2000-01	Colorado	NHL	19	0	4	4	38	0	0	0	10	0.0	5	0	0.0	29	15	12:08	....	....	....	....	....	....	....	....
	Hershey Bears	AHL	48	6	17	23	87	....	....	....	....	....	....	....	....	....	....	....	12	2	2	4	18	....	....	....
2001-02	Colorado	NHL	57	0	0	0	60	0	0	0	29	0.0	1	0	0.0	72	50	9:29	....	....	....	....	....	....	....	....
	Pittsburgh	NHL	13	0	2	2	21	0	0	0	20	0.0	-4	0	0.0	29	15	19:39	....	....	....	....	....	....	....	....
	NHL Totals		89	0	6	6	119	0	0	0	59	0.0		0	0.0	130	80	11:32	....	....	....	....	....	....	....	....

Traded to **Spokane** (WHL) by **Seattle** (WHL) for Justin Ossachuk, December 11, 1997. Traded to **Pittsburgh** by **Colorado** with Ville Nieminen for Darius Kasparaitis, March 19, 2002.

BERTUZZI, Todd

(buhr-TOO-zee, TAWD) **VAN.**

Center. Shoots left. 6'3", 235 lbs. Born, Sudbury, Ont., February 2, 1975. NY Islanders' 1st choice, 23rd overall, in 1993 Entry Draft.

			Regular Season																Playoffs							
Season	Club	League	GP	G	A	Pts	PIM	PP	SH	GW	S	%	+/-	TF	F%	H	SB	Min	GP	G	A	Pts	PIM	PP	SH	GW
1990-91	Sudbury Legion	NOHA	48	25	46	71	247																			
	Sud. Cub Wolves	NOJHA	3	3	2	5	10																			
1991-92	Guelph Storm	OHL	47	7	14	21	145																			
1992-93	Guelph Storm	OHL	59	27	32	59	164												5	2	2	4	6			
1993-94	Guelph Storm	OHL	61	28	54	82	165												9	2	6	8	30			
1994-95	Guelph Storm	OHL	62	54	65	119	58												14	*15	18	33	41			
1995-96	**NY Islanders**	**NHL**	**76**	**18**	**21**	**39**	**83**	**4**	**0**	**2**	**127**	**14.2**	**-14**													
1996-97	**NY Islanders**	**NHL**	**64**	**10**	**13**	**23**	**68**	**3**	**0**	**1**	**79**	**12.7**	**-3**													
	Utah Grizzlies	IHL	13	5	5	10	16																			
1997-98	**NY Islanders**	**NHL**	**52**	**7**	**11**	**18**	**58**	**1**	**0**	**1**	**63**	**11.1**	**-19**													
	Vancouver	**NHL**	**22**	**6**	**9**	**15**	**63**	**1**	**1**	**1**	**39**	**15.4**	**2**													
1998-99	**Vancouver**	**NHL**	**32**	**8**	**8**	**16**	**44**	**1**	**0**	**3**	**72**	**11.1**	**-6**	**191**	**43.5**	**53**	**12**	**18:28**								
99-2000	**Vancouver**	**NHL**	**80**	**25**	**25**	**50**	**126**	**4**	**0**	**2**	**173**	**14.5**	**-2**	**476**	**46.6**	**187**	**16**	**15:24**								
2000-01	**Vancouver**	**NHL**	**79**	**25**	**30**	**55**	**93**	**14**	**0**	**3**	**203**	**12.3**	**-18**	**84**	**45.2**	**132**	**25**	**17:13**	**4**	**2**	**2**	**4**	**8**	**0**	**0**	**0**
2001-02	**Vancouver**	**NHL**	**72**	**36**	**49**	**85**	**110**	**14**	**0**	**3**	**203**	**17.7**	**21**	**151**	**49.0**	**123**	**14**	**19:40**	**6**	**2**	**2**	**4**	**14**	**1**	**0**	**0**
	NHL Totals		**477**	**135**	**166**	**301**	**645**	**42**	**1**	**16**	**959**	**14.1**		**902**	**46.2**	**495**	**67**	**17:29**	**10**	**4**	**4**	**8**	**22**	**1**	**0**	**0**

OHL Second All-Star team (1995)

Traded to **Vancouver** by **NY Islanders** with Bryan McCabe and NY Islanders' 3rd round choice (Jarkko Ruutu) in 1998 Entry Draft for Trevor Linden, February 6, 1998. • Missed majority of 1998-99 season recovering from leg injury suffered in game vs. Washington, November 1, 1998.

BERUBE, Craig

(buh-ROO-bee, KRAYG) **CGY.**

Left wing. Shoots left. 6'1", 205 lbs. Born, Calahoo, Alta., December 17, 1965.

			Regular Season																Playoffs							
Season	Club	League	GP	G	A	Pts	PIM	PP	SH	GW	S	%	+/-	TF	F%	H	SB	Min	GP	G	A	Pts	PIM	PP	SH	GW
1982-83	Williams Lake	PCJHL	33	9	24	33	99																			
	Kamloops	WHL	4	0	0	0	0																			
1983-84	New Westminster	WHL	70	11	20	31	104												8	1	2	3	5			
1984-85	New Westminster	WHL	70	25	44	69	191												10	3	2	5	4			
1985-86	Kamloops Blazers	WHL	32	17	14	31	119																			
	Medicine Hat	WHL	34	14	16	30	95												25	7	8	15	102			
1986-87	**Philadelphia**	**NHL**	**7**	**0**	**0**	**0**	**57**	**0**	**0**	**0**	**4**	**0.0**	**2**						**5**	**0**	**0**	**0**	**17**	**0**	**0**	**0**
	Hershey Bears	AHL	63	7	17	24	325																			
1987-88	**Philadelphia**	**NHL**	**27**	**3**	**2**	**5**	**108**	**0**	**0**	**2**	**13**	**23.1**	**1**													
	Hershey Bears	AHL	31	5	9	14	119																			
1988-89	**Philadelphia**	**NHL**	**53**	**1**	**1**	**2**	**199**	**0**	**0**	**0**	**31**	**3.2**	**-15**						**16**	**0**	**0**	**0**	**56**	**0**	**0**	**0**
	Hershey Bears	AHL	7	0	2	2	19																			
1989-90	**Philadelphia**	**NHL**	**74**	**4**	**14**	**18**	**291**	**0**	**0**	**0**	**52**	**7.7**	**-7**													
1990-91	**Philadelphia**	**NHL**	**74**	**8**	**9**	**17**	**293**	**0**	**0**	**0**	**46**	**17.4**	**-6**													
1991-92	**Toronto**	**NHL**	**40**	**5**	**7**	**12**	**109**	**1**	**0**	**1**	**42**	**11.9**	**-2**													
	Calgary	**NHL**	**36**	**1**	**4**	**5**	**155**	**0**	**0**	**0**	**27**	**3.7**	**-3**													
1992-93	**Calgary**	**NHL**	**77**	**4**	**8**	**12**	**209**	**0**	**0**	**2**	**58**	**6.9**	**-6**						**6**	**0**	**1**	**1**	**21**	**0**	**0**	**0**
1993-94	**Washington**	**NHL**	**84**	**7**	**7**	**14**	**305**	**0**	**0**	**0**	**48**	**14.6**	**-4**						**8**	**0**	**0**	**0**	**21**	**0**	**0**	**0**
1994-95	**Washington**	**NHL**	**43**	**2**	**4**	**6**	**173**	**0**	**0**	**0**	**22**	**9.1**	**-5**						**7**	**0**	**0**	**0**	**29**	**0**	**0**	**0**
1995-96	**Washington**	**NHL**	**50**	**2**	**10**	**12**	**151**	**1**	**0**	**1**	**28**	**7.1**	**1**						**2**	**0**	**0**	**0**	**19**	**0**	**0**	**0**
1996-97	**Washington**	**NHL**	**80**	**4**	**3**	**7**	**218**	**0**	**0**	**1**	**55**	**7.3**	**-11**													
1997-98	**Washington**	**NHL**	**74**	**6**	**9**	**15**	**189**	**0**	**0**	**0**	**68**	**8.8**	**-3**						**21**	**1**	**0**	**1**	**21**	**0**	**0**	**1**
1998-99	**Washington**	**NHL**	**66**	**5**	**4**	**9**	**166**	**0**	**0**	**0**	**45**	**11.1**	**-7**	**14**	**42.9**	**85**	**10**	**6:47**								
	Philadelphia	**NHL**	**11**	**0**	**0**	**0**	**28**	**0**	**0**	**0**	**7**	**0.0**	**-3**	**0**	**0.0**	**8**	**1**	**8:15**	**6**	**1**	**0**	**1**	**4**	**0**	**0**	**0**
99-2000	**Philadelphia**	**NHL**	**77**	**4**	**8**	**12**	**162**	**0**	**0**	**0**	**63**	**6.3**	**3**	**4**	**25.0**	**111**	**7**	**8:00**	**18**	**1**	**0**	**1**	**23**	**0**	**0**	**1**
2000-01	**Washington**	**NHL**	**22**	**0**	**1**	**1**	**18**	**0**	**0**	**0**	**8**	**0.0**	**-3**	**0**	**0.0**	**11**	**3**	**5:34**								
	NY Islanders	**NHL**	**38**	**0**	**2**	**2**	**54**	**0**	**0**	**0**	**27**	**0.0**	**-5**	**11**	**45.5**	**46**	**4**	**6:02**								
2001-02	**Calgary**	**NHL**	**66**	**3**	**1**	**4**	**164**	**1**	**0**	**0**	**34**	**8.8**	**-2**	**28**	**53.6**	**55**	**9**	**6:31**								
	NHL Totals		**999**	**59**	**94**	**153**	**3049**	**3**	**0**	**7**	**678**	**8.7**		**57**	**47.4**	**316**	**34**	**6:55**	**89**	**3**	**1**	**4**	**211**	**0**	**0**	**2**

Signed as a free agent by **Philadelphia**, March 19, 1986. Traded to **Edmonton** by **Philadelphia** with Craig Fisher and Scott Mellanby for Dave Brown, Corey Foster and Jari Kurri, May 30, 1991. Traded to **Toronto** by **Edmonton** with Grant Fuhr and Glenn Anderson for Vincent Damphousse, Peter Ing, Scott Thornton and Luke Richardson, September 19, 1991. Traded to **Calgary** by **Toronto** with Alexander Godynyuk, Gary Leeman, Michel Petit and Jeff Reese for Doug Gilmour, Jamie Macoun, Ric Nattress, Rick Wamsley and Kent Manderville, January 2, 1992. Traded to **Washington** by **Calgary** for Washington's 5th round choice (Darryl Lafrance) in 1993 Entry Draft, June 26, 1993. Traded to **Philadelphia** by **Washington** for cash, March 23, 1999. Signed as a free agent by **Washington**, July 7, 2000. Traded to **NY Islanders** by **Washington** for Vancouver's 9th round choice (previously acquired, Washington selected Robert Muller) in 2001 Entry Draft, January 11, 2001. Signed as a free agent by **Calgary**, September 18, 2001.

BETTS, Blair

(BEHTS, BLAIR) **CGY.**

Center. Shoots left. 6'1", 200 lbs. Born, Edmonton, Alta., February 16, 1980. Calgary's 2nd choice, 33rd overall, in 1998 Entry Draft.

			Regular Season																Playoffs							
Season	Club	League	GP	G	A	Pts	PIM	PP	SH	GW	S	%	+/-	TF	F%	H	SB	Min	GP	G	A	Pts	PIM	PP	SH	GW
1995-96	Sherwood Park	AMHL	34	22	19	41	69																			
1996-97	Prince George	WHL	58	12	18	30	19												15	2	2	4	6			
1997-98	Prince George	WHL	71	35	41	76	38												11	4	6	10	8			
1998-99	Prince George	WHL	42	20	22	42	39												7	3	2	5	8			
99-2000	Prince George	WHL	44	24	35	59	38												13	11	11	22	6			
2000-01	Saint John	AHL	75	13	15	28	28												19	2	3	5	4			
2001-02	**Calgary**	**NHL**	**6**	**1**	**0**	**1**	**2**	**0**	**0**	**1**	**4**	**25.0**	**-1**	**39**	**48.7**	**5**	**0**	**7:05**								
	Saint John	AHL	67	20	29	49	10																			
	NHL Totals		**6**	**1**	**0**	**1**	**2**	**0**	**0**	**1**	**4**	**25.0**		**39**	**48.7**	**5**	**0**	**7:05**								

BICANEK, Radim

(BEE-chah-nehk, RA-dihm) **CBJ**

Defense. Shoots left. 6'1", 209 lbs. Born, Uherske Hradiste, Czech., January 18, 1975. Ottawa's 2nd choice, 27th overall, in 1993 Entry Draft.

			Regular Season																Playoffs							
Season	Club	League	GP	G	A	Pts	PIM	PP	SH	GW	S	%	+/-	TF	F%	H	SB	Min	GP	G	A	Pts	PIM	PP	SH	GW
1992-93	Dukla Jihlava	Czech	43	2	3	5																				
1993-94	Belleville Bulls	OHL	63	16	27	43	49												12	2	8	10	21			
1994-95	Belleville Bulls	OHL	49	13	26	39	61												16	6	5	11	30			
	Ottawa	**NHL**	**6**	**0**	**0**	**0**	**0**	**0**	**0**	**0**	**6**	**0.0**	**3**													
	P.E.I. Senators	AHL																	3	0	1	1	0			
1995-96	P.E.I. Senators	AHL	74	7	19	26	87												5	0	2	2	6			
1996-97	**Ottawa**	**NHL**	**21**	**0**	**1**	**1**	**8**	**0**	**0**	**0**	**27**	**0.0**	**-4**						**7**	**0**	**0**	**0**	**8**	**0**	**0**	**0**
	Worcester	AHL	44	1	15	16	22																			
1997-98	**Ottawa**	**NHL**	**1**	**0**	**0**	**0**	**0**	**0**	**0**	**0**	**0**	**0.0**	**0**													
	Detroit Vipers	IHL	9	1	3	4	16																			
	Manitoba Moose	IHL	42	1	7	8	52																			
1998-99	**Ottawa**	**NHL**	**7**	**0**	**0**	**0**	**4**	**0**	**0**	**0**	**6**	**0.0**	**-1**	**0**	**0.0**	**11**	**4**	**10:51**								
	Grand Rapids	IHL	46	8	17	25	48																			
	Chicago	**NHL**	**7**	**0**	**0**	**0**	**6**	**0**	**0**	**0**	**7**	**0.0**	**-3**	**0**	**0.0**	**7**	**8**	**15:57**								
99-2000	**Chicago**	**NHL**	**11**	**0**	**3**	**3**	**4**	**0**	**0**	**0**	**8**	**0.0**	**7**	**0**	**0.0**	**22**	**7**	**18:27**								
	Cleveland	IHL	70	5	27	32	125												9	2	2	4	8			
2000-01	**Columbus**	**NHL**	**9**	**0**	**2**	**2**	**6**	**0**	**0**	**0**	**11**	**0.0**	**1**	**1**	**100.0**	**24**	**6**	**17:08**								
	Syracuse Crunch	AHL	68	22	43	65	124												5	4	2	6	2			
2001-02	**Columbus**	**NHL**	**60**	**1**	**5**	**6**	**34**	**0**	**0**	**0**	**43**	**2.3**	**-15**	**0**	**0.0**	**63**	**45**	**13:32**								
	NHL Totals		**122**	**1**	**11**	**12**	**62**	**0**	**0**	**0**	**108**	**0.9**		**1**	**100.0**	**127**	**70**	**14:26**	**7**	**0**	**0**	**0**	**8**	**0**	**0**	**0**

AHL Second All-Star Team (2001)

Traded to **Chicago** by **Ottawa** for LA Kings' 6th round choice (previously acquired, Ottawa selected Martin Prusek) in 1999 Entry Draft, March 12, 1999. Selected by **Columbus** from **Chicago** in Expansion Draft, June 23, 2000.

Season	Club	League	Regular Season																Playoffs							
Season	Club	League	GP	G	A	Pts	PIM	PP	SH	GW	S	%	+/-	TF	F%	H	SB	Min	GP	G	A	Pts	PIM	PP	SH	GW

BICEK, Jiri (bee-CHEHK, YEH-ree) N.J.

Left wing. Shoots left. 5'10", 190 lbs. Born, Kosice, Czech., December 3, 1978. New Jersey's 4th choice, 131st overall, in 1997 Entry Draft.

Season	Club	League	GP	G	A	Pts	PIM	PP	SH	GW	S	%	+/-	TF	F%	H	SB	Min	GP	G	A	Pts	PIM	PP	SH	GW
1994-95	HC Kosice Jr.	Slovak-Jr.	42	38	36	74	18																			
1995-96	HC Kosice	Slovakia	30	10	15	25	16												9	2	4	6	0			
1996-97	HC Kosice	Slovakia	44	11	14	25	20												7	1	3	4				
1997-98	Albany	AHL	50	10	10	20	22												13	1	6	7	4			
1998-99	Albany	AHL	79	15	45	60	102												5	2	2	4	2			
99-2000	Albany	AHL	80	7	36	43	51												4	0	2	2	0			
2000-01	**New Jersey**	**NHL**	**5**	**1**	**0**	**1**	**4**	**0**	**0**	**0**	**10**	**10.0**	**0**	**0**	**0.0**	**3**	**1**	**13:04**								
	Albany	AHL	73	12	29	41	73																			
2001-02	**New Jersey**	**NHL**	**1**	**0**	**0**	**0**	**0**	**0**	**0**	**0**	**2**	**0.0**	**–1**	**0**	**0.0**	**1**	**0**	**13:10**								
	Albany	AHL	62	15	19	34	45																			
	NHL Totals		**6**	**1**	**0**	**1**	**4**	**0**	**0**	**0**	**12**	**8.3**		**0**	**0.0**	**4**	**1**	**13:05**								

BIRON, Mathieu (BEE-rawn, mat-yoo) T.B.

Defense. Shoots right. 6'6", 226 lbs. Born, Lac-St-Charles, Que., April 29, 1980. Los Angeles' 1st choice, 21st overall, in 1998 Entry Draft.

Season	Club	League	GP	G	A	Pts	PIM	PP	SH	GW	S	%	+/-	TF	F%	H	SB	Min	GP	G	A	Pts	PIM	PP	SH	GW
1996-97	Ste-Foy	QAAA	40	4	22	26	49												10	3	4	7				
1997-98	Shawinigan	QMJHL	59	8	28	36	60												6	0	1	1	10			
1998-99	Shawinigan	QMJHL	69	13	32	45	116												6	0	2	2	6			
99-2000	**NY Islanders**	**NHL**	**60**	**4**	**4**	**8**	**38**	**2**	**0**	**2**	**70**	**5.7**	**–13**	**2**	**0.0**	**120**	**65**	**15:02**								
2000-01	**NY Islanders**	**NHL**	**14**	**0**	**1**	**1**	**12**	**0**	**0**	**0**	**10**	**0.0**	**2**	**0**	**0.0**	**26**	**14**	**12:21**								
	Lowell	AHL	22	1	3	4	17																			
	Springfield	AHL	34	0	6	6	18																			
2001-02	**Tampa Bay**	**NHL**	**36**	**0**	**0**	**0**	**12**	**0**	**0**	**0**	**35**	**0.0**	**–16**	**0**	**0.0**	**53**	**38**	**14:47**								
	Springfield	AHL	35	4	9	13	16																			
	NHL Totals		**110**	**4**	**5**	**9**	**62**	**2**	**0**	**2**	**115**	**3.5**		**2**	**0.0**	**199**	**117**	**14:36**								

Traded to **NY Islanders** by **LA Kings** with Olli Jokinen, Josh Green and LA Kings' 1st round choice (Taylor Pyatt) in 1999 Entry Draft for Ziggy Palffy, Brian Smolinski, Marcel Cousineau and New Jersey's 4th round choice (previously acquired, LA Kings selected Daniel Johansson) in 1999 Entry Draft, June 20, 1999. Traded to **Tampa Bay** by **NY Islanders** with NY Islanders' 2nd round choice (later traded to Washington - later traded to Vancouver - Vancouver selected Denis Grot) in 2002 Entry Draft for Adrian Aucoin and Alexander Kharitonov, June 22, 2001.

BLACK, James (BLAK, JAYMS)

Left wing. Shoots left. 6', 202 lbs. Born, Regina, Sask., August 15, 1969. Hartford's 4th choice, 94th overall, in 1989 Entry Draft.

Season	Club	League	GP	G	A	Pts	PIM	PP	SH	GW	S	%	+/-	TF	F%	H	SB	Min	GP	G	A	Pts	PIM	PP	SH	GW
1986-87	Edmonton Mets	AJHL	41	36	48	84	58																			
1987-88	Portland	WHL	72	30	50	80	50																			
1988-89	Portland	WHL	71	45	51	96	57												19	13	6	19	28			
1989-90	**Hartford**	**NHL**	**1**	**0**	**0**	**0**	**0**	**0**	**0**	**0**	**0**	**0.0**	**0**													
	Binghamton	AHL	80	37	35	72	34																			
1990-91	**Hartford**	**NHL**	**1**	**0**	**0**	**0**	**0**	**0**	**0**	**0**	**0**	**0.0**	**0**													
	Springfield	AHL	79	35	61	96	34												18	9	9	18	6			
1991-92	**Hartford**	**NHL**	**30**	**4**	**6**	**10**	**10**	**1**	**0**	**1**	**54**	**7.4**	**–4**													
	Springfield	AHL	47	15	25	40	33												10	3	2	5	18			
1992-93	**Minnesota**	**NHL**	**10**	**2**	**1**	**3**	**4**	**0**	**0**	**0**	**10**	**20.0**	**0**													
	Kalamazoo Wings	IHL	63	25	45	70	40																			
1993-94	**Dallas**	**NHL**	**13**	**2**	**3**	**5**	**2**	**2**	**0**	**0**	**16**	**12.5**	**–4**													
	Buffalo	**NHL**	**2**	**0**	**0**	**0**	**0**	**2**	**0**	**0**	**2**	**0.0**	**0**													
	Rochester	AHL	45	19	32	51	28												4	2	3	5	0			
1994-95	Las Vegas	IHL	78	29	44	73	54												10	1	6	7	4			
1995-96	**Chicago**	**NHL**	**13**	**3**	**3**	**6**	**16**	**0**	**0**	**1**	**23**	**13.0**	**1**						**8**	**1**	**0**	**1**	**2**	**0**	**0**	**0**
	Indianapolis Ice	IHL	67	32	50	82	56																			
1996-97	**Chicago**	**NHL**	**64**	**12**	**11**	**23**	**20**	**0**	**0**	**3**	**122**	**9.8**	**6**						**5**	**1**	**1**	**2**	**2**	**0**	**0**	**0**
1997-98	**Chicago**	**NHL**	**52**	**10**	**5**	**15**	**8**	**2**	**1**	**3**	**90**	**11.1**	**–8**													
1998-99	Chicago Wolves	IHL	5	6	0	6	0																			
	Washington	**NHL**	**75**	**16**	**14**	**30**	**14**	**1**	**1**	**3**	**135**	**11.9**	**5**	**10**	**50.0**	**42**	**30**	**15:04**								
99-2000	**Washington**	**NHL**	**49**	**8**	**9**	**17**	**6**	**1**	**0**	**1**	**71**	**11.3**	**–1**	**18**	**38.9**	**27**	**31**	**12:03**								
2000-01	**Washington**	**NHL**	**42**	**1**	**5**	**6**	**4**	**0**	**0**	**0**	**34**	**2.9**	**–3**	**7**	**57.1**	**23**	**13**	**9:04**								
	Portland Pirates	AHL	5	2	3	5	0																			
2001-02	Grand Rapids	AHL	29	6	10	16	10																			
	NHL Totals		**352**	**58**	**57**	**115**	**84**	**9**	**2**	**12**	**557**	**10.4**		**35**	**45.7**	**92**	**74**	**12:40**	**13**	**2**	**1**	**3**	**4**	**0**	**0**	**0**

Traded to **Minnesota** by **Hartford** for Mark Janssens, September 3, 1992. Transferred to **Dallas** after **Minnesota** franchise relocated, June 9, 1993. Traded to **Buffalo** by **Dallas** with Dallas' 7th round choice (Steve Webb) in 1994 Entry Draft for Gord Donnelly, December 15, 1993. Signed as a free agent by **Chicago**, September 18, 1995. Traded to **Washington** by **Chicago** for Washington's 9th round choice (later traded back to Washington - Washington selected Igor Schadilov) in 1999 Entry Draft, October 15, 1998. Signed as a free agent **Grand Rapids** (AHL), September 24, 2001.

BLAKE, Jason (BLAYK, JAY-suhn) NYI

Center. Shoots left. 5'10", 180 lbs. Born, Moorhead, MN, September 2, 1973.

Season	Club	League	GP	G	A	Pts	PIM	PP	SH	GW	S	%	+/-	TF	F%	H	SB	Min	GP	G	A	Pts	PIM	PP	SH	GW
1991-92	Moorhead Spuds	Hi-School	25	30	30	60																				
1992-93	Waterloo	USHL	45	24	27	51	107																			
1993-94	Waterloo	USHL	47	50	50	100	76																			
1994-95	Ferris State	CCHA	36	16	16	32	46																			
1995-96	North Dakota	CCHA	DID NOT PLAY – TRANSFERRED COLLEGES																							
1996-97	North Dakota	WCHA	43	19	32	51	44																			
1997-98	North Dakota	WCHA	38	24	27	51	62																			
1998-99	North Dakota	WCHA	38	*28	*41	*69	49																			
	Los Angeles	**NHL**	**1**	**1**	**0**	**1**	**0**	**0**	**0**	**0**	**5**	**20.0**	**1**	**14**	**35.7**	**1**	**0**	**17:13**								
	Orlando	IHL	5	3	5	8	6												13	3	4	7	20			
99-2000	**Los Angeles**	**NHL**	**64**	**5**	**18**	**23**	**26**	**0**	**0**	**1**	**131**	**3.8**	**4**	**269**	**43.9**	**68**	**14**	**11:17**	**3**	**0**	**0**	**0**	**0**	**0**	**0**	**0**
	Long Beach	IHL	7	3	6	9	2																			
2000-01	**Los Angeles**	**NHL**	**17**	**1**	**3**	**4**	**10**	**0**	**0**	**0**	**27**	**3.7**	**–8**	**13**	**61.5**	**18**	**4**	**10:03**								
	Lowell	AHL	2	0	1	1	2																			
	NY Islanders	**NHL**	**30**	**4**	**8**	**12**	**24**	**1**	**1**	**0**	**73**	**5.5**	**–12**	**118**	**44.1**	**80**	**17**	**15:43**								
2001-02	**NY Islanders**	**NHL**	**82**	**8**	**10**	**18**	**36**	**0**	**0**	**1**	**136**	**5.9**	**–11**	**23**	**43.5**	**99**	**46**	**12:54**	**7**	**0**	**1**	**1**	**13**	**0**	**0**	**0**
	NHL Totals		**194**	**19**	**39**	**58**	**96**	**1**	**1**	**2**	**372**	**5.1**		**437**	**44.2**	**266**	**81**	**12:34**	**10**	**0**	**1**	**1**	**13**	**0**	**0**	**0**

WCHA First All-Star Team (1997, 1998, 1999) • NCAA West Second All-American Team (1998) • WCHA Player of the Year (1999) • NCAA West First All-American Team (1999)

Signed as a free agent by **LA Kings**, April 20, 1999. Traded to **NY Islanders** by **LA Kings** for NY Islanders' 5th round choice (Joel Andresen) in 2002 Entry Draft, January 3, 2001.

BLAKE, Rob (BLAYK, RAWB) COL.

Defense. Shoots right. 6'4", 225 lbs. Born, Simcoe, Ont., December 10, 1969. Los Angeles' 4th choice, 70th overall, in 1988 Entry Draft.

Season	Club	League	GP	G	A	Pts	PIM	PP	SH	GW	S	%	+/-	TF	F%	H	SB	Min	GP	G	A	Pts	PIM	PP	SH	GW
1985-86	Brantford	OJHL-B	39	3	13	16	43																			
1986-87	Stratford	OJHL-B	31	11	20	31	115																			
1987-88	Bowling Green	CCHA	43	5	8	13	88																			
1988-89	Bowling Green	CCHA	46	11	21	32	140																			
1989-90	Bowling Green	CCHA	42	23	36	59	140																			
	Los Angeles	**NHL**	**4**	**0**	**0**	**0**	**4**	**0**	**0**	**0**	**3**	**0.0**	**0**						**8**	**1**	**3**	**4**	**4**	**1**	**0**	**0**
1990-91	**Los Angeles**	**NHL**	**75**	**12**	**34**	**46**	**125**	**9**	**0**	**2**	**150**	**8.0**	**3**						**12**	**1**	**4**	**5**	**26**	**1**	**0**	**0**
1991-92	**Los Angeles**	**NHL**	**57**	**7**	**13**	**20**	**102**	**5**	**0**	**0**	**131**	**5.3**	**–5**						**6**	**2**	**1**	**3**	**12**	**0**	**0**	**0**
1992-93	**Los Angeles**	**NHL**	**76**	**16**	**43**	**59**	**152**	**10**	**0**	**4**	**243**	**6.6**	**18**						**23**	**4**	**6**	**10**	**46**	**1**	**1**	**0**
1993-94	**Los Angeles**	**NHL**	**84**	**20**	**48**	**68**	**137**	**7**	**0**	**6**	**304**	**6.6**	**–7**													
1994-95	**Los Angeles**	**NHL**	**24**	**4**	**7**	**11**	**38**	**4**	**0**	**1**	**76**	**5.3**	**–16**													
1995-96	**Los Angeles**	**NHL**	**6**	**1**	**2**	**3**	**8**	**0**	**0**	**0**	**13**	**7.7**	**0**													
1996-97	**Los Angeles**	**NHL**	**62**	**8**	**23**	**31**	**82**	**4**	**0**	**1**	**169**	**4.7**	**–28**													
1997-98	**Los Angeles**	**NHL**	**81**	**23**	**27**	**50**	**94**	**11**	**0**	**4**	**261**	**8.8**	**–3**						**4**	**0**	**0**	**0**	**6**	**0**	**0**	**0**
	Canada	Olympics	6	1	1	2	2																			
1998-99	**Los Angeles**	**NHL**	**62**	**12**	**23**	**35**	**128**	**5**	**1**	**2**	**216**	**5.6**	**–7**	**0**	**0.0**	**132**	**139**	**24:52**								
99-2000	**Los Angeles**	**NHL**	**77**	**18**	**39**	**57**	**112**	**12**	**0**	**5**	**327**	**5.5**	**10**	**0**	**0.0**	**202**	**181**	**28:30**	**4**	**0**	**2**	**2**	**4**	**0**	**0**	**0**
2000-01	**Los Angeles**	**NHL**	**54**	**17**	**32**	**49**	**69**	**9**	**0**	**1**	**223**	**7.6**	**–8**	**0**	**0.0**	**144**	**132**	**28:11**								
	♦ Colorado	**NHL**	**13**	**2**	**8**	**10**	**8**	**1**	**0**	**1**	**44**	**4.5**	**11**	**0**	**0.0**	**31**	**23**	**26:03**	**23**	**6**	**13**	**19**	**16**	**3**	**0**	**0**

Season	Club	League	GP	G	A	Pts	PIM	PP	SH	GW	S	%	+/-	TF	F%	H	SB	Min	GP	G	A	Pts	PIM	PP	SH	GW
			Regular Season																Playoffs							
2001-02	**Colorado**	**NHL**	**75**	**16**	**40**	**56**	**58**	**10**	**0**	**2**	**229**	**7.0**	**16**	**0**	**0.0**	**193**	**190**	**27:35**	**20**	**6**	**6**	**12**	**16**	**1**	**0**	**0**
	Canada	Olympics	6	1	2	3	2																			
	NHL Totals		**750**	**156**	**339**	**495**	**1117**	**87**	**1**	**29**	**2389**	**6.5**		**0**	**0.0**	**702**	**665**	**27:16**	**100**	**20**	**35**	**55**	**130**	**7**	**1**	**0**

CCHA Second All-Star Team (1989) • CCHA First All-Star Team (1990) • NCAA West First All-American Team (1990) • NHL All-Rookie Team (1991) • NHL First All-Star Team (1998) • Won James Norris Memorial Trophy (1998) • NHL Second All-Star Team (2000, 2001, 2002) • Played in NHL All-Star Game (1994, 1999, 2000, 2001, 2002)

• Missed majority of 1995-96 season recovering from knee injury suffered in game vs. Washington, October 20, 1995. Traded to **Colorado** by **LA Kings** with Steve Reinprecht for Adam Deadmarsh, Aaron Miller, a player to be named later (Jared Aulin, March 22, 2001), Colorado's 1st round choice (Dave Steckel) in 2001 Entry Draft and future considerations, February 21, 2001.

BLOUIN, Sylvain
(bluh-WHEN, SIHL-veh) **MIN.**

Left wing. Shoots left. 6'2", 207 lbs. Born, Montreal, Que., May 21, 1974. NY Rangers' 5th choice, 104th overall, in 1994 Entry Draft.

Season	Club	League	GP	G	A	Pts	PIM	PP	SH	GW	S	%	+/-	TF	F%	H	SB	Min	GP	G	A	Pts	PIM	PP	SH	GW
1991-92	Laval Titan	QMJHL	28	0	0	0	23												9	0	0	0	35			
1992-93	Laval Titan	QMJHL	68	0	10	10	373												13	1	0	1	*66			
1993-94	Laval Titan	QMJHL	62	18	22	40	*492												21	4	13	17	*177			
1994-95	Chicago Wolves	IHL	1	0	0	0	2																			
	Charlotte	ECHL	50	5	7	12	280												3	0	0	0	6			
	Binghamton	AHL	10	1	0	1	46												2	0	0	0	24			
1995-96	Binghamton	AHL	71	5	8	13	*352												4	0	3	3	4			
1996-97	**NY Rangers**	**NHL**	**6**	**0**	**0**	**0**	**18**	**0**	**0**	**0**	**1**	**0.0**	**−1**													
	Binghamton	AHL	62	13	17	30	301												4	2	1	3	16			
1997-98	**NY Rangers**	**NHL**	**1**	**0**	**0**	**0**	**5**	**0**	**0**	**0**	**0**	**0.0**	**0**													
	Hartford	AHL	53	8	9	17	286												9	0	1	1	63			
1998-99	**Montreal**	**NHL**	**5**	**0**	**0**	**0**	**19**	**0**	**0**	**0**	**1**	**0.0**	**0**	**0**	**0.0**	**2**	**0**	**3:37**								
	Fredericton	AHL	67	6	10	16	333												15	2	0	2	*87			
99-2000	Worcester	AHL	70	16	18	34	337												8	3	5	8	30			
2000-01	**Minnesota**	**NHL**	**41**	**3**	**2**	**5**	**117**	**0**	**0**	**0**	**37**	**8.1**	**−5**	**2**	**100.0**	**118**	**10**	**9:54**								
2001-02	**Minnesota**	**NHL**	**43**	**0**	**2**	**2**	**130**	**0**	**0**	**0**	**28**	**0.0**	**−11**	**0**	**0.0**	**104**	**17**	**9:51**								
	NHL Totals		**96**	**3**	**4**	**7**	**289**	**0**	**0**	**0**	**67**	**4.5**		**2**	**100.0**	**224**	**27**	**9:32**								

Traded to **Montreal** by **NY Rangers** with NY Rangers' 6th round choice (later traded to Phoenix - Phoenix selected Erik Lewerstrom) in 1999 Entry Draft for Peter Popovic, June 30, 1998. Signed as a free agent by **St. Louis**, August 25, 1999. Signed as a free agent by **Montreal**, July 7, 2000. Claimed by **Minnesota** from **Montreal** in Waiver Draft, September 29, 2000. • Missed majority of 2000-01 season recovering from shoulder injury suffered in game vs. Chicago, December 7, 2000.

BOGUNIECKI, Eric
(BOH-guhn-ih-kee, AIR-ihk) **ST.L.**

Center. Shoots right. 5'8", 192 lbs. Born, New Haven, CT, May 6, 1975. St. Louis' 6th choice, 193rd overall, in 1993 Entry Draft.

Season	Club	League	GP	G	A	Pts	PIM	PP	SH	GW	S	%	+/-	TF	F%	H	SB	Min	GP	G	A	Pts	PIM	PP	SH	GW
1992-93	Westminster High	Hi-School	24	30	24	54	55																			
1993-94	New Hampshire	H-East	40	17	16	33	66																			
1994-95	New Hampshire	H-East	34	12	16	28	62																			
1995-96	New Hampshire	H-East	32	23	28	51	46																			
1996-97	New Hampshire	H-East	36	26	31	57	58																			
1997-98	Dayton Bombers	ECHL	26	19	18	37	36																			
	Fort Wayne	IHL	35	4	8	12	29												4	1	2	3	10			
1998-99	Fort Wayne	IHL	72	32	34	66	100												2	0	1	1	2			
99-2000	**Florida**	**NHL**	**4**	**0**	**0**	**0**	**2**	**0**	**0**	**0**	**5**	**0.0**	**−1**	**25**	**36.0**	**8**	**0**	**8:35**								
	Louisville	AHL	57	33	42	75	148												4	3	2	5	20			
2000-01	Louisville	AHL	28	13	12	25	56																			
	St. Louis	**NHL**	**1**	**0**	**0**	**0**	**0**	**0**	**0**	**0**	**1**	**0.0**	**−1**	**0**	**0.0**	**1**	**1**	**13:44**								
	Worcester	AHL	45	17	28	45	100												9	3	2	5	10			
2001-02	**St. Louis**	**NHL**	**8**	**0**	**1**	**1**	**4**	**0**	**0**	**0**	**10**	**0.0**	**−2**	**21**	**38.1**	**17**	**0**	**11:42**	**1**	**0**	**1**	**1**	**0**	**0**	**0**	**0**
	Worcester	AHL	63	*38	46	84	181												3	2	0	2	4			
	NHL Totals		**13**	**0**	**1**	**1**	**6**	**0**	**0**	**0**	**16**	**0.0**		**46**	**37.0**	**26**	**1**	**10:54**	**1**	**0**	**1**	**1**	**0**	**0**	**0**	**0**

Hockey East Second All-Star Team (1997) • AHL First All-Star Team (2002) • Won Les Cunningham Plaque (MVP - AHL) (2002)

Signed as a free agent by **Florida**, July 7, 1999. Traded to **St. Louis** by **Florida** for Andrei Podkonicky, December 17, 2000.

BOHONOS, Lonny
(boh-HOH-nohz, LAW-nee) **TOR.**

Right wing. Shoots right. 5'11", 190 lbs. Born, Winnipeg, Man., May 20, 1973.

Season	Club	League	GP	G	A	Pts	PIM	PP	SH	GW	S	%	+/-	TF	F%	H	SB	Min	GP	G	A	Pts	PIM	PP	SH	GW
1990-91	Winnipeg South	MJHL	46	33	22	55	70																			
1991-92	Winnipeg South	MJHL	40	53	36	89	42																			
	Moose Jaw	WHL	8	1	1	2	0																			
1992-93	Seattle	WHL	46	13	13	26	27																			
	Portland	WHL	27	20	17	37	16												15	8	13	21	19			
1993-94	Portland	WHL	70	*62	*90	*152	80												10	8	11	19	13			
1994-95	Syracuse Crunch	AHL	67	30	45	75	71																			
1995-96	**Vancouver**	**NHL**	**3**	**0**	**1**	**1**	**0**	**0**	**0**	**0**	**3**	**0.0**	**1**													
	Syracuse Crunch	AHL	74	40	39	79	82												16	14	8	22	16			
1996-97	**Vancouver**	**NHL**	**36**	**11**	**11**	**22**	**10**	**2**	**0**	**1**	**67**	**16.4**	**−3**													
	Syracuse Crunch	AHL	41	22	30	52	28												3	2	2	4	4			
1997-98	**Vancouver**	**NHL**	**31**	**2**	**1**	**3**	**4**	**0**	**0**	**0**	**37**	**5.4**	**−9**													
	Syracuse Crunch	AHL	17	12	12	24	8																			
	Toronto	**NHL**	**6**	**3**	**3**	**6**	**4**	**0**	**0**	**0**	**13**	**23.1**	**1**													
	St. John's	AHL	11	7	9	16	10												2	1	1	2	2			
1998-99	**Toronto**	**NHL**	**7**	**3**	**0**	**3**	**4**	**0**	**0**	**0**	**13**	**23.1**	**3**	**2**	**50.0**	**3**	**2**	**13:47**	**9**	**3**	**6**	**9**	**2**	**0**	**0**	**0**
	St. John's	AHL	70	34	48	82	40												5	2	4	6	2			
99-2000	Manitoba Moose	IHL	63	18	33	51	45												2	0	0	0	2			
2000-01	HC Davos	Swiss	43	28	32	*60	42												4	0	1	1	2			
2001-02	HC Davos	Swiss	43	19	26	45	22												16	*13	8	*21	4			
	NHL Totals		**83**	**19**	**16**	**35**	**22**	**2**	**0**	**1**	**133**	**14.3**		**2**	**50.0**	**3**	**2**	**13:47**	**9**	**3**	**6**	**9**	**2**	**0**	**0**	**0**

WHL West First All-Star Team (1994) • Canadian Major Junior First All-Star Team (1994)

Signed as a free agent by **Vancouver**, May 31, 1994. Traded to **Toronto** by **Vancouver** for Brandon Convery, March 7, 1998. Signed as a free agent by **HC Davos** (Swiss), August 1, 2000.

BOIKOV, Alexandre
(bohy-KAHV, al-ehx-AN-duhr)

Defense. Shoots left. 6', 200 lbs. Born, Chelyabinsk, USSR, February 7, 1975.

Season	Club	League	GP	G	A	Pts	PIM	PP	SH	GW	S	%	+/-	TF	F%	H	SB	Min	GP	G	A	Pts	PIM	PP	SH	GW
1993-94	Victoria Cougars	WHL	70	4	31	35	250																			
1994-95	Prince George	WHL	46	5	23	28	115																			
	Tri-City	WHL	24	3	13	16	63												17	1	7	8	30			
1995-96	Tri-City	WHL	71	3	49	52	230												11	2	4	6	28			
1996-97	Kentucky	AHL	61	1	19	20	182												4	0	1	1	4			
1997-98	Kentucky	AHL	69	5	14	19	153												3	0	1	1	8			
1998-99	Kentucky	AHL	55	5	13	18	116																			
	Rochester	AHL	13	0	1	1	15												17	1	3	4	24			
99-2000	**Nashville**	**NHL**	**2**	**0**	**0**	**0**	**2**	**0**	**0**	**0**	**1**	**0.0**	**0**	**0**	**0.0**	**5**	**0**	**8:31**								
	Milwaukee	IHL	58	1	6	7	120																			
2000-01	**Nashville**	**NHL**	**8**	**0**	**0**	**0**	**13**	**0**	**0**	**0**	**3**	**0.0**	**−1**	**0**	**0.0**	**19**	**3**	**6:18**								
	Milwaukee	IHL	56	2	11	13	147												5	2	1	3	0			
2001-02	Milwaukee	AHL	56	4	6	10	102																			
	NHL Totals		**10**	**0**	**0**	**0**	**15**	**0**	**0**	**0**	**4**	**0.0**		**0**	**0.0**	**24**	**3**	**6:45**								

Signed as a free agent by **San Jose**, April 22, 1996. Signed as a free agent by **Nashville**, July 26, 1999.

BOILEAU, Patrick
(BWOI-loh, PA-trihk) **DET.**

Defense. Shoots right. 6', 202 lbs. Born, Montreal, Que., February 22, 1975. Washington's 3rd choice, 69th overall, in 1993 Entry Draft.

Season	Club	League	GP	G	A	Pts	PIM	PP	SH	GW	S	%	+/-	TF	F%	H	SB	Min	GP	G	A	Pts	PIM	PP	SH	GW
1990-91	Laval Laurentide	QAAA	3	0	1	1	0																			
1991-92	Laval Laurentide	QAAA	42	9	36	45	94												12	3	5	8	10			
1992-93	Laval Titan	QMJHL	69	4	19	23	73												13	1	2	3	10			
1993-94	Laval Titan	QMJHL	64	13	57	70	56												21	1	7	8	24			
1994-95	Laval Titan	QMJHL	38	8	25	33	46												20	4	16	20	24			
1995-96	Portland Pirates	AHL	78	10	28	38	41												19	1	3	4	12			

Season	Club	League	GP	G	A	Pts	PIM	PP	SH	GW	S	%	+/-	TF	F%	H	SB	Min	Playoffs GP	G	A	Pts	PIM	PP	SH	GW
1996-97	**Washington**	**NHL**	**1**	**0**	**0**	**0**	**0**	**0**	**0**	**0**	**0**	**0.0**	**0**													
	Portland Pirates	AHL	67	16	28	44	63												5	1	1	2	4			
1997-98	Portland Pirates	AHL	47	6	21	27	53												10	0	1	1	8			
1998-99	**Washington**	**NHL**	**4**	**0**	**1**	**1**	**2**	**0**	**0**	**0**	**7**	**0.0**	**–4**	**0**	**0.0**	**8**	**2**	**15:56**								
	Portland Pirates	AHL	52	6	18	24	52																			
	Indianapolis Ice	IHL	29	8	13	21	27												4	0	1	1	2			
99-2000	Portland Pirates	AHL	63	2	15	17	61												4	0	0	0	4			
2000-01	Portland Pirates	AHL	77	6	14	20	50												3	0	0	0	8			
2001-02	**Washington**	**NHL**	**2**	**0**	**0**	**0**	**2**	**0**	**0**	**0**	**0**	**0.0**	**–1**	**0**	**0.0**	**0**	**2**	**11:52**								
	Portland Pirates	AHL	75	17	19	36	43																			
	NHL Totals		**7**	**0**	**1**	**1**	**4**	**0**	**0**	**0**	**7**	**0.0**		**0**	**0.0**	**8**	**4**	**14:35**								

Canadian Major Junior Scholastic Player of the Year (1994)

Loaned to **Indianapolis** (IHL) by **Washington** (Portland-AHL), February 4, 1999. Signed as a free agent by **Detroit**, August 5, 2002.

BOMBARDIR, Brad (bawm-bahr-DEER, BRAD) MIN.

Defense. Shoots left. 6'1", 205 lbs. Born, Powell River, B.C., May 5, 1972. New Jersey's 5th choice, 56th overall, in 1990 Entry Draft.

Season	Club	League	GP	G	A	Pts	PIM	PP	SH	GW	S	%	+/-	TF	F%	H	SB	Min	Playoffs GP	G	A	Pts	PIM	PP	SH	GW
1988-89	Powell River	BCJHL	30	6	5	11	24												6	0	0	0	0			
1989-90	Powell River	BCJHL	60	10	35	45	93												8	2	3	5	4			
1990-91	North Dakota	WCHA	33	3	6	9	18																			
1991-92	North Dakota	WCHA	35	3	14	17	54																			
1992-93	North Dakota	WCHA	38	8	15	23	34																			
1993-94	North Dakota	WCHA	38	5	17	22	38																			
1994-95	Albany	AHL	77	5	22	27	22												14	0	3	3	6			
1995-96	Albany	AHL	80	6	25	31	63												3	0	1	1	4			
1996-97	Albany	AHL	32	0	8	8	6												16	1	3	4	8			
1997-98	**New Jersey**	**NHL**	**43**	**1**	**5**	**6**	**8**	**0**	**0**	**0**	**16**	**6.3**	**11**													
	Albany	AHL	5	0	0	0	0																			
1998-99	**New Jersey**	**NHL**	**56**	**1**	**7**	**8**	**16**	**0**	**0**	**0**	**47**	**2.1**	**–4**	**1**	**0.0**	**48**	**57**	**15:03**	**5**	**0**	**0**	**0**	**0**	**0**	**0**	**0**
99-2000♦	**New Jersey**	**NHL**	**32**	**3**	**1**	**4**	**6**	**0**	**0**	**0**	**24**	**12.5**	**–6**	**0**	**0.0**	**31**	**37**	**15:54**	**1**	**0**	**0**	**0**	**0**	**0**	**0**	**0**
2000-01	**Minnesota**	**NHL**	**70**	**0**	**15**	**15**	**42**	**0**	**0**	**0**	**81**	**0.0**	**–6**	**1**	**0.0**	**61**	**164**	**20:50**								
2001-02	**Minnesota**	**NHL**	**28**	**1**	**2**	**3**	**14**	**1**	**0**	**0**	**24**	**4.2**	**–6**	**0**	**0.0**	**11**	**67**	**20:33**								
	NHL Totals		**229**	**6**	**30**	**36**	**86**	**1**	**0**	**0**	**192**	**3.1**		**2**	**0.0**	**151**	**325**	**18:12**	**6**	**0**	**0**	**0**	**0**	**0**	**0**	**0**

AHL Second All-Star Team (1996)

• Missed majority of 1999-2000 season recovering from esophagus injury suffered in game vs. Philadelphia, October 30, 1999. Traded to **Minnesota** by **New Jersey** for Chris Terreri and Minnesota's 9th round choice (later traded to Tampa Bay - Tampa Bay selected Thomas Ziegler) in 2000 Entry Draft, June 23, 2000. • Missed majority of 2001-02 season recovering from ankle injury suffered in game vs. San Jose, October 16, 2001.

BONDRA, Peter (BAWN-druh, PEE-tuhr) WSH.

Right wing. Shoots left. 6', 200 lbs. Born, Luck, USSR, February 7, 1968. Washington's 9th choice, 156th overall, in 1990 Entry Draft.

Season	Club	League	GP	G	A	Pts	PIM	PP	SH	GW	S	%	+/-	TF	F%	H	SB	Min	Playoffs GP	G	A	Pts	PIM	PP	SH	GW
1986-87	VSZ Kosice	Czech	32	4	5	9	24																			
1987-88	VSZ Kosice	Czech	45	27	11	38	20																			
1988-89	VSZ Kosice	Czech	40	30	10	40	20																			
1989-90	VSZ Kosice	Czech	44	29	17	46	....												5	7	2	9	....			
1990-91	**Washington**	**NHL**	**54**	**12**	**16**	**28**	**47**	**4**	**0**	**1**	**95**	**12.6**	**–10**						**4**	**0**	**1**	**1**	**2**	**0**	**0**	**0**
1991-92	**Washington**	**NHL**	**71**	**28**	**28**	**56**	**42**	**4**	**0**	**3**	**158**	**17.7**	**16**						**7**	**6**	**2**	**8**	**4**	**1**	**0**	**0**
1992-93	**Washington**	**NHL**	**83**	**37**	**48**	**85**	**70**	**10**	**0**	**7**	**239**	**15.5**	**8**						**6**	**0**	**6**	**6**	**0**	**0**	**0**	**0**
1993-94	**Washington**	**NHL**	**69**	**24**	**19**	**43**	**40**	**4**	**0**	**2**	**200**	**12.0**	**22**						**9**	**2**	**4**	**6**	**4**	**0**	**0**	**1**
1994-95	HC Kosice	Slovakia	2	1	0	1	0																			
	Washington	**NHL**	**47**	***34**	**9**	**43**	**24**	**12**	**6**	**3**	**177**	**19.2**	**9**						**7**	**5**	**3**	**8**	**10**	**2**	**0**	**1**
1995-96	Detroit Vipers	IHL	7	8	1	9	0																			
	Washington	**NHL**	**67**	**52**	**28**	**80**	**40**	**11**	**4**	**7**	**322**	**16.1**	**18**						**6**	**3**	**2**	**5**	**8**	**2**	**0**	**1**
1996-97	**Washington**	**NHL**	**77**	**46**	**31**	**77**	**72**	**10**	**4**	**3**	**314**	**14.6**	**7**													
1997-98	**Washington**	**NHL**	**76**	***52**	**26**	**78**	**44**	**11**	**5**	**13**	**284**	**18.3**	**14**						**17**	**7**	**5**	**12**	**12**	**3**	**0**	**2**
	Slovakia	Olympics	2	1	0	1	25																			
1998-99	**Washington**	**NHL**	**66**	**31**	**24**	**55**	**56**	**6**	**3**	**5**	**284**	**10.9**	**–1**	**1**	**0.0**	**87**	**22**	**20:35**								
99-2000	**Washington**	**NHL**	**62**	**21**	**17**	**38**	**30**	**5**	**3**	**5**	**187**	**11.2**	**5**	**2**	**50.0**	**81**	**22**	**18:48**	**5**	**1**	**1**	**2**	**4**	**1**	**0**	**0**
2000-01	**Washington**	**NHL**	**82**	**45**	**36**	**81**	**60**	**22**	**4**	**8**	**305**	**14.8**	**8**	**2**	**50.0**	**131**	**29**	**20:48**	**6**	**2**	**0**	**2**	**2**	**2**	**0**	**1**
2001-02	**Washington**	**NHL**	**77**	**39**	**31**	**70**	**80**	**17**	**1**	**8**	**333**	**11.7**	**–2**	**2**	**50.0**	**76**	**25**	**21:43**								
	NHL Totals		**831**	**421**	**313**	**734**	**605**	**116**	**30**	**65**	**2898**	**14.5**		**7**	**42.9**	**375**	**98**	**20:34**	**67**	**26**	**24**	**50**	**46**	**11**	**0**	**6**

Played in NHL All-Star Game (1993, 1996, 1997, 1998, 1999)

BONIN, Brian (BAWN-ihn, BRIGH-uhn)

Center. Shoots left. 5'10", 186 lbs. Born, St. Paul, MN, November 28, 1973. Pittsburgh's 9th choice, 211th overall, in 1992 Entry Draft.

Season	Club	League	GP	G	A	Pts	PIM	PP	SH	GW	S	%	+/-	TF	F%	H	SB	Min	Playoffs GP	G	A	Pts	PIM	PP	SH	GW
1991-92	White Bear Lake	Hi-School	23	22	35	57	8																			
1992-93	U. of Minnesota	WCHA	38	10	18	28	10																			
1993-94	U. of Minnesota	WCHA	42	24	20	44	14																			
1994-95	U. of Minnesota	WCHA	44	32	31	*63	28																			
1995-96	U. of Minnesota	WCHA	42	34	*47	*81	30																			
1996-97	Cleveland	IHL	60	13	26	39	18												1	1	0	1	0			
1997-98	Syracuse Crunch	AHL	67	31	38	69	46												5	1	3	4	6			
1998-99	**Pittsburgh**	**NHL**	**5**	**0**	**0**	**0**	**0**	**0**	**0**	**0**	**2**	**0.0**	**–2**	**28**	**39.3**	**4**	**2**	**12:19**	**3**	**0**	**0**	**0**	**0**	**0**	**0**	**0**
	Kansas City	IHL	19	2	5	7	10																			
	Adirondack	AHL	54	19	16	35	31												2	0	0	0	0			
99-2000	Syracuse Crunch	AHL	67	19	28	47	20												4	0	1	1	0			
2000-01	**Minnesota**	**NHL**	**7**	**0**	**0**	**0**	**0**	**0**	**0**	**0**	**7**	**0.0**	**–3**	**20**	**50.0**	**5**	**2**	**9:46**								
	Cleveland	IHL	72	35	42	77	45												4	2	0	2	0			
2001-02	Langnau	Swiss	39	20	17	37	28												8	4	11	15	0			
	NHL Totals		**12**	**0**	**0**	**0**	**0**	**0**	**0**	**0**	**9**	**0.0**		**48**	**43.8**	**9**	**4**	**10:50**	**3**	**0**	**0**	**0**	**0**	**0**	**0**	**0**

Minnesota High School Player of the Year (1992) • WCHA First All-Star Team (1995, 1996) • WCHA Player of the Year (1995, 1996) • NCAA West First All-American Team (1995, 1996) • Won Hobey Baker Memorial Award (Top U.S. Collegiate Player) (1996) • IHL Second All-Star Team (2001)

Signed as a free agent by **Vancouver**, September 9, 1999. Signed as a free agent by **Minnesota**, July 6, 2000.

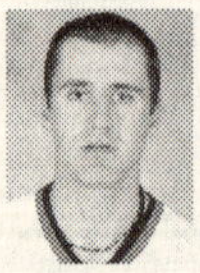

BONK, Radek (BOHNK, RA-dehk) OTT.

Center. Shoots left. 6'3", 210 lbs. Born, Krnov, Czech., January 9, 1976. Ottawa's 1st choice, 3rd overall, in 1994 Entry Draft.

Season	Club	League	GP	G	A	Pts	PIM	PP	SH	GW	S	%	+/-	TF	F%	H	SB	Min	Playoffs GP	G	A	Pts	PIM	PP	SH	GW
1990-91	Slezan Opava Jr.	Czech-Jr.	35	47	42	89	25																			
1991-92	AC ZPS Zlin Jr.	Czech-Jr.	45	47	36	83	30																			
1992-93	AC ZPS Zlin	Czech	30	5	5	10	10																			
1993-94	Las Vegas	IHL	76	42	45	87	208												5	1	2	3	10			
1994-95	Las Vegas	IHL	33	7	13	20	62																			
	Ottawa	**NHL**	**42**	**3**	**8**	**11**	**28**	**1**	**0**	**0**	**40**	**7.5**	**–5**													
	P.E.I. Senators	AHL																	1	0	0	0	0			
1995-96	**Ottawa**	**NHL**	**76**	**16**	**19**	**35**	**36**	**5**	**0**	**1**	**161**	**9.9**	**–5**													
1996-97	**Ottawa**	**NHL**	**53**	**5**	**13**	**18**	**14**	**0**	**1**	**0**	**82**	**6.1**	**–4**						**7**	**0**	**1**	**1**	**4**	**0**	**0**	**0**
1997-98	**Ottawa**	**NHL**	**65**	**7**	**9**	**16**	**16**	**1**	**0**	**0**	**93**	**7.5**	**–13**						**5**	**0**	**0**	**0**	**2**	**0**	**0**	**0**
1998-99	**Ottawa**	**NHL**	**81**	**16**	**16**	**32**	**48**	**0**	**1**	**6**	**110**	**14.5**	**15**	**1184**	**50.1**	**225**	**30**	**13:44**	**4**	**0**	**0**	**0**	**6**	**0**	**0**	**0**
99-2000	Pardubice	Czech	3	1	0	1	4																			
	Ottawa	**NHL**	**80**	**23**	**37**	**60**	**53**	**10**	**0**	**5**	**167**	**13.8**	**–2**	**1654**	**52.0**	**211**	**45**	**18:14**	**6**	**0**	**0**	**0**	**8**	**0**	**0**	**0**
2000-01	**Ottawa**	**NHL**	**74**	**23**	**36**	**59**	**52**	**5**	**2**	**3**	**139**	**16.5**	**27**	**1506**	**51.2**	**136**	**52**	**18:16**	**2**	**0**	**0**	**0**	**2**	**0**	**0**	**0**
2001-02	**Ottawa**	**NHL**	**82**	**25**	**45**	**70**	**52**	**6**	**2**	**5**	**170**	**14.7**	**3**	**1530**	**51.0**	**130**	**52**	**17:57**	**12**	**3**	**7**	**10**	**6**	**2**	**0**	**1**
	NHL Totals		**553**	**118**	**183**	**301**	**299**	**28**	**6**	**20**	**962**	**12.3**		**5874**	**51.1**	**702**	**179**	**17:01**	**36**	**3**	**8**	**11**	**28**	**2**	**0**	**1**

Won Garry F. Longman Memorial Trophy (Top Rookie - IHL) (1994) • Played in NHL All-Star Game (2000, 2001)

BONNI, Ryan

(baw-NEE, RIGH-uhn) **TOR.**

Defense. Shoots left. 6'4", 190 lbs. Born, Winnipeg, Man., February 18, 1979. Vancouver's 2nd choice, 34th overall, in 1997 Entry Draft.

			Regular Season																Playoffs							
Season	Club	League	GP	G	A	Pts	PIM	PP	SH	GW	S	%	+/-	TF	F%	H	SB	Min	GP	G	A	Pts	PIM	PP	SH	GW
1994-95	Winnipeg Sharks	MMHL	24	3	19	22	59	....	....	....	....	....	....	....	....	....	....	....	....	....	....	....	....	....	....	....
1995-96	Saskatoon Blades	WHL	63	1	7	8	78	....	....	....	....	....	....	....	....	....	....	....	3	0	0	0	0	....	....	....
1996-97	Saskatoon Blades	WHL	69	11	19	30	219	....	....	....	....	....	....	....	....	....	....	....	....	....	....	....	....	....	....	....
1997-98	Saskatoon Blades	WHL	42	5	14	19	100	....	....	....	....	....	....	....	....	....	....	....	0	0	0	0	0	....	....	....
1998-99	Saskatoon Blades	WHL	51	6	26	32	211	....	....	....	....	....	....	....	....	....	....	....	....	....	....	....	....	....	....	....
	Red Deer Rebels	WHL	20	3	10	13	41	....	....	....	....	....	....	....	....	....	....	....	9	0	4	4	25	....	....	....
99-2000	**Vancouver**	**NHL**	**3**	**0**	**0**	**0**	**0**	**0**	**0**	**0**	**1**	**0.0**	**−1**	**0**	**0.0**	**1**	**2**	**9:37**	....	....	....	....	....	....	....	....
	Syracuse Crunch	AHL	71	5	13	18	125	....	....	....	....	....	....	....	....	....	....	....	2	0	1	1	2	....	....	....
2000-01	Kansas City	IHL	80	2	9	11	127	....	....	....	....	....	....	....	....	....	....	....	....	....	....	....	....	....	....	....
2001-02	Manitoba Moose	AHL	11	0	1	1	33	....	....	....	....	....	....	....	....	....	....	....	2	0	0	0	2	....	....	....
	Columbia Inferno	ECHL	46	3	18	21	128	....	....	....	....	....	....	....	....	....	....	....	4	0	4	4	10	....	....	....
	NHL Totals		**3**	**0**	**0**	**0**	**0**	**0**	**0**	**0**	**1**	**0.0**		**0**	**0.0**	**1**	**2**	**9:37**	....	....	....	....	....	....	....	....

Traded to **Red Deer** (WHL) by **Saskatoon** (WHL) for Chris Ovington, February 1, 1999. Traded to **Toronto** by **Vancouver** for future considerations, June 25, 2002.

BONVIE, Dennis

(BOHN-vee, DEHN-his)

Right wing/Defense. Shoots right. 5'11", 205 lbs. Born, Antigonish, N.S., July 23, 1973.

			Regular Season																Playoffs							
Season	Club	League	GP	G	A	Pts	PIM	PP	SH	GW	S	%	+/-	TF	F%	H	SB	Min	GP	G	A	Pts	PIM	PP	SH	GW
1989-90	Antigonish	NSMHL	50	15	30	45	52	....	....	....	....	....	....	....	....	....	....	....	....	....	....	....	....	....	....	....
1990-91	Antigonish	MJrHL	40	1	8	9	347	....	....	....	....	....	....	....	....	....	....	....	....	....	....	....	....	....	....	....
1991-92	Kitchener	OHL	7	1	1	2	23	....	....	....	....	....	....	....	....	....	....	....	....	....	....	....	....	....	....	....
	North Bay	OHL	49	0	12	12	261	....	....	....	....	....	....	....	....	....	....	....	21	0	1	1	91	....	....	....
1992-93	North Bay	OHL	64	3	21	24	*316	....	....	....	....	....	....	....	....	....	....	....	5	0	0	0	34	....	....	....
1993-94	Cape Breton	AHL	63	1	10	11	278	....	....	....	....	....	....	....	....	....	....	....	4	0	0	0	11	....	....	....
1994-95	Cape Breton	AHL	74	5	15	20	422	....	....	....	....	....	....	....	....	....	....	....	....	....	....	....	....	....	....	....
	Edmonton	**NHL**	**2**	**0**	**0**	**0**	**0**	**0**	**0**	**0**	**0**	**0.0**	**0**	....	....	....	....	....	....	....	....	....	....	....	....	....
1995-96	**Edmonton**	**NHL**	**8**	**0**	**0**	**0**	**47**	**0**	**0**	**0**	**0**	**0.0**	**−3**	....	....	....	....	....	....	....	....	....	....	....	....	....
	Cape Breton	AHL	38	13	14	27	269	....	....	....	....	....	....	....	....	....	....	....	....	....	....	....	....	....	....	....
1996-97	Hamilton	AHL	73	9	20	29	*522	....	....	....	....	....	....	....	....	....	....	....	22	3	11	14	*91	....	....	....
1997-98	**Edmonton**	**NHL**	**4**	**0**	**0**	**0**	**27**	**0**	**0**	**0**	**0**	**0.0**	**0**	....	....	....	....	....	....	....	....	....	....	....	....	....
	Hamilton	AHL	57	11	19	30	295	....	....	....	....	....	....	....	....	....	....	....	9	0	5	5	18	....	....	....
1998-99	**Chicago**	**NHL**	**11**	**0**	**0**	**0**	**44**	**0**	**0**	**0**	**1**	**0.0**	**−4**	**0**	**0.0**	**10**	**0**	**3:59**	....	....	....	....	....	....	....	....
	Portland Pirates	AHL	3	1	0	1	16	....	....	....	....	....	....	....	....	....	....	....	....	....	....	....	....	....	....	....
	Philadelphia	AHL	37	4	10	14	158	....	....	....	....	....	....	....	....	....	....	....	14	3	3	6	26	....	....	....
99-2000	**Pittsburgh**	**NHL**	**28**	**0**	**0**	**0**	**80**	**0**	**0**	**0**	**6**	**0.0**	**−2**	**0**	**0.0**	**28**	**4**	**3:14**	....	....	....	....	....	....	....	....
	Wilkes-Barre	AHL	42	5	26	31	243	....	....	....	....	....	....	....	....	....	....	....	....	....	....	....	....	....	....	....
2000-01	**Pittsburgh**	**NHL**	**3**	**0**	**0**	**0**	**0**	**0**	**0**	**0**	**1**	**0.0**	**−1**	**0**	**0.0**	**3**	**1**	**3:30**	....	....	....	....	....	....	....	....
	Wilkes-Barre	AHL	65	5	18	23	221	....	....	....	....	....	....	....	....	....	....	....	21	0	4	4	35	....	....	....
2001-02	**Boston**	**NHL**	**23**	**1**	**2**	**3**	**84**	**0**	**0**	**0**	**5**	**20.0**	**3**	**0**	**0.0**	**34**	**3**	**5:05**	**1**	**0**	**0**	**0**	**0**	**0**	**0**	**0**
	Providence	AHL	55	8	8	16	290	....	....	....	....	....	....	....	....	....	....	....	....	....	....	....	....	....	....	....
	NHL Totals		**79**	**1**	**2**	**3**	**282**	**0**	**0**	**0**	**13**	**7.7**		**0**	**0.0**	**75**	**8**	**4:02**	**1**	**0**	**0**	**0**	**0**	**0**	**0**	**0**

Signed as a free agent by **Edmonton**, August 25, 1994. Claimed by **Chicago** from **Edmonton** in Waiver Draft, October 5, 1998. Traded to **Philadelphia** by **Chicago** for Frank Bialowas, January 8, 1999. Signed as a free agent by **Pittsburgh**, September 20, 1999. Signed as a free agent by **Boston**, October 5, 2001.

BORDELEAU, Sebastien

(BOHR-duh-loh, SEH-bas-tyehn)

Center. Shoots right. 5'11", 185 lbs. Born, Vancouver, B.C., February 15, 1975. Montreal's 3rd choice, 73rd overall, in 1993 Entry Draft.

			Regular Season																Playoffs							
Season	Club	League	GP	G	A	Pts	PIM	PP	SH	GW	S	%	+/-	TF	F%	H	SB	Min	GP	G	A	Pts	PIM	PP	SH	GW
1990-91	Laval Laurentide	QAAA	39	27	36	63	....	....	....	....	....	....	....	....	....	....	....	....	....	....	....	....	....	....	....	....
1991-92	Hull Olympiques	QMJHL	62	26	32	58	91	....	....	....	....	....	....	....	....	....	....	....	5	0	3	3	23	....	....	....
1992-93	Hull Olympiques	QMJHL	60	18	39	57	95	....	....	....	....	....	....	....	....	....	....	....	10	3	8	11	20	....	....	....
1993-94	Hull Olympiques	QMJHL	60	26	57	83	147	....	....	....	....	....	....	....	....	....	....	....	17	6	14	20	26	....	....	....
1994-95	Hull Olympiques	QMJHL	68	52	76	128	142	....	....	....	....	....	....	....	....	....	....	....	18	*13	19	*32	25	....	....	....
	Fredericton	AHL	....	....	....	....	....	....	....	....	....	....	....	....	....	....	....	....	1	0	0	0	0	....	....	....
1995-96	**Montreal**	**NHL**	**4**	**0**	**0**	**0**	**0**	**0**	**0**	**0**	**0**	**0.0**	**−1**	....	....	....	....	....	....	....	....	....	....	....	....	....
	Fredericton	AHL	43	17	29	46	68	....	....	....	....	....	....	....	....	....	....	....	7	0	2	2	8	....	....	....
1996-97	**Montreal**	**NHL**	**28**	**2**	**9**	**11**	**2**	**0**	**0**	**0**	**27**	**7.4**	**−3**	....	....	....	....	....	....	....	....	....	....	....	....	....
	Fredericton	AHL	34	18	22	40	50	....	....	....	....	....	....	....	....	....	....	....	....	....	....	....	....	....	....	....
1997-98	**Montreal**	**NHL**	**53**	**6**	**8**	**14**	**36**	**2**	**1**	**0**	**55**	**10.9**	**5**	....	....	....	....	....	**5**	**0**	**0**	**0**	**2**	**0**	**0**	**0**
1998-99	**Nashville**	**NHL**	**72**	**16**	**24**	**40**	**26**	**1**	**2**	**3**	**168**	**9.5**	**−14**	**1368**	**57.1**	**76**	**22**	**15:18**	....	....	....	....	....	....	....	....
99-2000	**Nashville**	**NHL**	**60**	**10**	**13**	**23**	**30**	**0**	**2**	**1**	**127**	**7.9**	**−12**	**942**	**54.4**	**43**	**22**	**13:56**	....	....	....	....	....	....	....	....
2000-01	**Nashville**	**NHL**	**14**	**2**	**3**	**5**	**14**	**0**	**0**	**0**	**20**	**10.0**	**−4**	**192**	**59.4**	**11**	**4**	**12:40**	....	....	....	....	....	....	....	....
	Worcester	AHL	2	0	2	2	9	....	....	....	....	....	....	....	....	....	....	....	11	1	7	8	23	....	....	....
2001-02	**Minnesota**	**NHL**	**14**	**1**	**4**	**5**	**8**	**0**	**0**	**0**	**25**	**4.0**	**−1**	**115**	**59.1**	**9**	**6**	**13:35**	....	....	....	....	....	....	....	....
	Houston Aeros	AHL	16	4	7	11	23	....	....	....	....	....	....	....	....	....	....	....	....	....	....	....	....	....	....	....
	Phoenix	**NHL**	**6**	**0**	**0**	**0**	**2**	**0**	**0**	**0**	**3**	**0.0**	**−1**	**35**	**51.4**	**3**	**1**	**7:46**	....	....	....	....	....	....	....	....
	Springfield	AHL	34	9	10	19	54	....	....	....	....	....	....	....	....	....	....	....	....	....	....	....	....	....	....	....
	NHL Totals		**251**	**37**	**61**	**98**	**118**	**3**	**5**	**4**	**425**	**8.7**		**2652**	**56.3**	**142**	**55**	**14:10**	**5**	**0**	**0**	**0**	**2**	**0**	**0**	**0**

QMJHL All-Rookie Team (1992) • QMJHL First All-Star Team (1995)

Traded to **Nashville** by **Montreal** for future considerations, June 26, 1998. • Missed majority of 2000-01 season recovering from abdominal injury suffered in game vs. Detroit, November 18, 2000. Claimed on waivers by **St. Louis** from **Nashville**, March 13, 2001. Claimed by **Minnesota** from **St. Louis** in Waiver Draft, September 28, 2001. Traded to **Phoenix** by **Minnesota** for David Cullen, January 4, 2002.

BOTTERILL, Jason

(BOH-tuhr-ihl, JAY-suhn) **BUF.**

Left wing. Shoots left. 6'4", 220 lbs. Born, Edmonton, Alta., May 19, 1976. Dallas' 1st choice, 20th overall, in 1994 Entry Draft.

			Regular Season																Playoffs							
Season	Club	League	GP	G	A	Pts	PIM	PP	SH	GW	S	%	+/-	TF	F%	H	SB	Min	GP	G	A	Pts	PIM	PP	SH	GW
1992-93	St. Paul's Prep	Hi-School	22	22	26	48	....	....	....	....	....	....	....	....	....	....	....	....	....	....	....	....	....	....	....	....
1993-94	U. of Michigan	CCHA	36	20	19	39	94	....	....	....	....	....	....	....	....	....	....	....	....	....	....	....	....	....	....	....
1994-95	U. of Michigan	CCHA	34	14	14	28	117	....	....	....	....	....	....	....	....	....	....	....	....	....	....	....	....	....	....	....
1995-96	U. of Michigan	CCHA	37	*32	25	57	*143	....	....	....	....	....	....	....	....	....	....	....	....	....	....	....	....	....	....	....
1996-97	U. of Michigan	CCHA	42	*37	24	61	129	....	....	....	....	....	....	....	....	....	....	....	....	....	....	....	....	....	....	....
1997-98	**Dallas**	**NHL**	**4**	**0**	**0**	**0**	**19**	**0**	**0**	**0**	**2**	**0.0**	**−1**	....	....	....	....	....	....	....	....	....	....	....	....	....
	Michigan K-Wings	IHL	50	11	11	22	82	....	....	....	....	....	....	....	....	....	....	....	4	0	0	0	5	....	....	....
1998-99	**Dallas**	**NHL**	**17**	**0**	**0**	**0**	**23**	**0**	**0**	**0**	**8**	**0.0**	**−2**	**0**	**0.0**	**27**	**0**	**8:19**	....	....	....	....	....	....	....	....
	Michigan K-Wings	IHL	56	13	25	38	106	....	....	....	....	....	....	....	....	....	....	....	5	2	1	3	4	....	....	....
99-2000	**Atlanta**	**NHL**	**25**	**1**	**4**	**5**	**17**	**0**	**0**	**1**	**17**	**5.9**	**−7**	**2**	**50.0**	**51**	**3**	**11:16**	....	....	....	....	....	....	....	....
	Orlando	IHL	17	7	8	15	27	....	....	....	....	....	....	....	....	....	....	....	....	....	....	....	....	....	....	....
	Calgary	**NHL**	**2**	**0**	**0**	**0**	**0**	**0**	**0**	**0**	**2**	**0.0**	**−4**	**0**	**0.0**	**1**	**0**	**8:00**	....	....	....	....	....	....	....	....
	Saint John	AHL	21	3	4	7	39	....	....	....	....	....	....	....	....	....	....	....	3	0	0	0	19	....	....	....
2000-01	Saint John	AHL	60	13	20	33	101	....	....	....	....	....	....	....	....	....	....	....	19	2	7	9	30	....	....	....
2001-02	**Calgary**	**NHL**	**4**	**1**	**0**	**1**	**2**	**1**	**0**	**1**	**4**	**25.0**	**−3**	**0**	**0.0**	**4**	**2**	**9:26**	....	....	....	....	....	....	....	....
	Saint John	AHL	71	21	21	42	121	....	....	....	....	....	....	....	....	....	....	....	....	....	....	....	....	....	....	....
	NHL Totals		**52**	**2**	**4**	**6**	**61**	**1**	**0**	**2**	**33**	**6.1**		**2**	**50.0**	**83**	**5**	**9:56**	....	....	....	....	....	....	....	....

CCHA Second All-Star Team (1996) • NCAA West Second All-American Team (1997)

Traded to **Atlanta** by **Dallas** for Jamie Pushor, July 15, 1999. Traded to **Calgary** by **Atlanta** with Darryl Shannon for Hnat Domenichelli and Dmitri Vlasenkov, February 11, 2000. Signed as a free agent by **Buffalo**, August 12, 2002.

BOUCHARD, Joel

(BOO-shahrd, JOHL) **NYR**

Defense. Shoots left. 6'1", 209 lbs. Born, Montreal, Que., January 23, 1974. Calgary's 7th choice, 129th overall, in 1992 Entry Draft.

			Regular Season																Playoffs							
Season	Club	League	GP	G	A	Pts	PIM	PP	SH	GW	S	%	+/-	TF	F%	H	SB	Min	GP	G	A	Pts	PIM	PP	SH	GW
1989-90	Mtl-Bourassa	QAAA	41	7	17	24	10	....	....	....	....	....	....	....	....	....	....	....	1	1	0	1	0	....	....	....
1990-91	Longueuil	QMJHL	53	3	19	22	34	....	....	....	....	....	....	....	....	....	....	....	8	1	0	1	11	....	....	....
1991-92	Verdun	QMJHL	70	9	20	29	55	....	....	....	....	....	....	....	....	....	....	....	19	1	7	8	20	....	....	....
1992-93	Verdun	QMJHL	60	10	49	59	126	....	....	....	....	....	....	....	....	....	....	....	4	0	2	2	4	....	....	....
1993-94	Verdun	QMJHL	60	15	55	70	62	....	....	....	....	....	....	....	....	....	....	....	4	1	0	1	6	....	....	....
	Saint John	AHL	1	0	0	0	0	....	....	....	....	....	....	....	....	....	....	....	2	0	0	0	0	....	....	....
1994-95	Saint John	AHL	77	6	25	31	63	....	....	....	....	....	....	....	....	....	....	....	5	1	0	1	4	....	....	....
	Calgary	**NHL**	**2**	**0**	**0**	**0**	**0**	**0**	**0**	**0**	**0**	**0.0**	**0**	....	....	....	....	....	....	....	....	....	....	....	....	....

			Regular Season																Playoffs							
Season	Club	League	GP	G	A	Pts	PIM	PP	SH	GW	S	%	+/-	TF	F%	H	SB	Min	GP	G	A	Pts	PIM	PP	SH	GW
1995-96	Calgary	NHL	4	0	0	0	4	0	0	0	0	0.0	0													
	Saint John	AHL	74	8	25	33	104												16	1	4	5	10			
1996-97	Calgary	NHL	76	4	5	9	49	0	1	0	61	6.6	–23													
1997-98	Calgary	NHL	44	5	7	12	57	0	1	1	51	9.8	0													
	Saint John	AHL	3	2	1	3	6																			
1998-99	Nashville	NHL	64	4	11	15	60	0	0	0	78	5.1	–10	0	0.0	109	67	22:34								
99-2000	Nashville	NHL	52	1	4	5	23	0	0	0	60	1.7	–11	0	0.0	88	48	18:41								
	Dallas	NHL	2	0	0	0	2	0	0	0	1	0.0	1	0	0.0	2	2	9:45								
2000-01	Phoenix	NHL	32	1	2	3	22	0	0	0	26	3.8	–8	0	0.0	32	27	14:28								
	Grand Rapids	IHL	19	3	9	12	8																			
2001-02	New Jersey	NHL	1	0	1	1	0	0	0	0	0	0.0	1	0	0.0	0	2	19:26								
	Albany	AHL	70	9	22	31	28																			
	NHL Totals		277	15	30	45	217	0	2	1	277	5.4		0	0.0	231	146	19:19								

QMJHL First All-Star Team (1994)

Claimed by **Nashville** from **Calgary** in Expansion Draft, June 26, 1998. Claimed on waivers by **Dallas** from **Nashville**, March 14, 2000. Signed as a free agent by **Phoenix**, August 31, 2000. Signed as a free agent by **New Jersey**, October 25, 2001. Signed as a free agent by **NY Rangers**, August 5, 2002.

BOUCHER, Philippe

(boo-SHAY, fihl-EEP) **DAL.**

Defense. Shoots right. 6'2", 221 lbs. Born, Ste-Apollinaire, Que., March 24, 1973. Buffalo's 1st choice, 13th overall, in 1991 Entry Draft.

Season	Club	League	GP	G	A	Pts	PIM	PP	SH	GW	S	%	+/-	TF	F%	H	SB	Min	GP	G	A	Pts	PIM	PP	SH	GW
1988-89	Ste-Foy	QAAA	5	0	0	0	2																			
1989-90	Ste-Foy	QAAA	42	26	60	86	76												12	6	*19	25	16			
1990-91	Granby Bisons	QMJHL	69	21	46	67	92																			
1991-92	Granby Bisons	QMJHL	49	22	37	59	47																			
	Laval Titan	QMJHL	16	7	11	18	36												10	5	6	11	8			
1992-93	Laval Titan	QMJHL	16	12	15	27	37												13	6	15	21	12			
	Buffalo	NHL	18	0	4	4	14	0	0	0	28	0.0	1													
	Rochester	AHL	5	4	3	7	8												3	0	1	1	2			
1993-94	Buffalo	NHL	38	6	8	14	29	4	0	1	67	9.0	–1						7	1	1	2	2	1	0	0
	Rochester	AHL	31	10	22	32	51																			
1994-95	Rochester	AHL	43	14	27	41	26																			
	Buffalo	NHL	9	1	4	5	0	0	0	0	15	6.7	6													
	Los Angeles	NHL	6	1	0	1	4	0	0	0	15	6.7	–3													
1995-96	Los Angeles	NHL	53	7	16	23	31	5	0	1	145	4.8	–26													
	Phoenix	IHL	10	4	3	7	4																			
1996-97	Los Angeles	NHL	60	7	18	25	25	2	0	1	159	4.4	0													
1997-98	Los Angeles	NHL	45	6	10	16	49	1	0	0	80	7.5	6													
	Long Beach	IHL	2	0	1	1	4																			
1998-99	Los Angeles	NHL	45	2	6	8	32	1	0	0	87	2.3	–12	0	0.0	58	63	17:51								
99-2000	Los Angeles	NHL	1	0	0	0	0	0	0	0	3	0.0	0	0	0.0	4	3	17:04								
	Long Beach	IHL	14	4	11	15	8												6	0	9	9	8			
2000-01	Los Angeles	NHL	22	2	4	6	20	2	0	0	40	5.0	4	0	0.0	42	24	18:25	13	0	1	1	2	0	0	0
	Manitoba Moose	IHL	45	10	22	32	39																			
2001-02	Los Angeles	NHL	80	7	23	30	94	4	0	2	198	3.5	0	0	0.0	148	141	21:36	5	0	1	1	2	0	0	0
	NHL Totals		377	39	93	132	298	19	0	5	837	4.7		0	0.0	252	231	19:57	25	1	3	4	6	1	0	0

QMJHL Second All-Star Team (1991, 1992) • QMJHL Defensive Rookie of the Year (1991) • Canadian Major Junior Rookie of the Year (1991)

Traded to **LA Kings** by **Buffalo** with Denis Tsygurov and Grant Fuhr for Alexei Zhitnik, Robb Stauber, Charlie Huddy and LA Kings' 5th round choice (Marian Menhart) in 1995 Entry Draft, February 14, 1995. • Missed majority of 1999-2000 season recovering from foot injury suffered in training camp, September, 1999. Signed as a free agent by **Dallas**, July 2, 2002.

BOUCK, Tyler

(BOWK, TIGH-luhr) **VAN.**

Right wing. Shoots left. 6', 196 lbs. Born, Camrose, Alta., January 13, 1980. Dallas' 2nd choice, 57th overall, in 1998 Entry Draft.

Season	Club	League	GP	G	A	Pts	PIM	PP	SH	GW	S	%	+/-	TF	F%	H	SB	Min	GP	G	A	Pts	PIM	PP	SH	GW
1995-96	Sherwood Park	AMHL	22	10	21	31	58																			
1996-97	Prince George	WHL	12	0	2	2	11																			
1997-98	Prince George	WHL	65	11	26	37	90												11	1	0	1	21			
1998-99	Prince George	WHL	56	22	25	47	178												2	0	2	2	10			
99-2000	Prince George	WHL	57	30	33	63	183												13	6	13	19	36			
2000-01	Dallas	NHL	48	2	5	7	29	0	0	1	41	4.9	–3	1	0.0	81	2	8:59	1	0	0	0	0	0	0	0
	Utah Grizzlies	IHL	24	2	6	8	39																			
2001-02	Phoenix	NHL	7	0	0	0	4	0	0	0	3	0.0	–1	0	0.0	9	0	6:54								
	Springfield	AHL	21	1	2	3	33																			
	Manitoba Moose	AHL	20	4	4	8	25																			
	NHL Totals		55	2	5	7	33	0	0	1	44	4.5		1	0.0	90	2	8:43	1	0	0	0	0	0	0	0

WHL West First All-Star Team (2000)

Traded to **Phoenix** by **Dallas** for Jyrki Lumme, June 23, 2001. Traded to **Vancouver** by **Phoenix** with Todd Warriner, Trevor Letowski and Phoenix's 3rd round choice in 2003 Entry Draft for Drake Berehowsky and Denis Pederson, December 28, 2001.

BOUGHNER, Bob

(BOOG-nuhr, BAWB) **CGY.**

Defense. Shoots right. 6', 203 lbs. Born, Windsor, Ont., March 8, 1971. Detroit's 2nd choice, 32nd overall, in 1989 Entry Draft.

Season	Club	League	GP	G	A	Pts	PIM	PP	SH	GW	S	%	+/-	TF	F%	H	SB	Min	GP	G	A	Pts	PIM	PP	SH	GW
1986-87	Belle River	OJHL-C	37	3	11	14	88																			
1987-88	St. Mary's	OJHL-B	36	4	18	22	177																			
1988-89	Sault Ste. Marie	OHL	64	6	15	21	182																			
1989-90	Sault Ste. Marie	OHL	49	7	23	30	122																			
1990-91	Sault Ste. Marie	OHL	64	13	33	46	156												14	2	9	11	35			
1991-92	Toledo Storm	ECHL	28	3	10	13	79												5	2	0	2	15			
	Adirondack	AHL	1	0	0	0	7																			
1992-93	Adirondack	AHL	69	1	16	17	190																			
1993-94	Adirondack	AHL	72	8	14	22	292												10	1	1	2	18			
1994-95	Cincinnati	IHL	81	2	14	16	192												10	0	0	0	18			
1995-96	Carolina	AHL	46	2	15	17	127																			
	Buffalo	NHL	31	0	1	1	104	0	0	0	14	0.0	3													
1996-97	Buffalo	NHL	77	1	7	8	225	0	0	0	34	2.9	12						11	0	1	1	9	0	0	0
1997-98	Buffalo	NHL	69	1	3	4	165	0	0	0	26	3.8	5						14	0	4	4	15	0	0	0
1998-99	Nashville	NHL	79	3	10	13	137	0	0	1	59	5.1	–6	0	0.0	233	91	18:31								
99-2000	Nashville	NHL	62	2	4	6	97	0	0	0	32	6.3	–13	0	0.0	207	67	17:20								
	Pittsburgh	NHL	11	1	0	1	69	1	0	1	8	12.5	2	0	0.0	26	14	17:05	11	0	2	2	15	0	0	0
2000-01	Pittsburgh	NHL	58	1	3	4	147	0	0	0	46	2.2	18	0	0.0	171	65	16:30	18	0	1	1	22	0	0	0
2001-02	Calgary	NHL	79	2	4	6	170	0	0	0	58	3.4	9	0	0.0	239	99	18:43								
	NHL Totals		466	11	32	43	1114	1	0	2	277	4.0		0	0.0	876	336	17:51	54	0	8	8	61	0	0	0

Signed as a free agent by **Florida**, July 25, 1994. Traded to **Buffalo** by **Florida** for Buffalo's 3rd round choice (Chris Allen) in 1996 Entry Draft, February 1, 1996. Claimed by **Nashville** from **Buffalo** in Expansion Draft, June 26, 1998. Traded to **Pittsburgh** by **Nashville** for Pavel Skrbek, March 13, 2000. Signed as a free agent by **Calgary**, July 2, 2001.

BOUILLON, Francis

(BOO-liawn, FRAN-sihs) **MTL.**

Defense. Shoots left. 5'8", 194 lbs. Born, New York, NY, October 17, 1975.

Season	Club	League	GP	G	A	Pts	PIM	PP	SH	GW	S	%	+/-	TF	F%	H	SB	Min	GP	G	A	Pts	PIM	PP	SH	GW
1991-92	Mtl-Bourassa	QAAA	42	2	5	7	28												9	1	0	1	6			
1992-93	Laval Titan	QMJHL	46	0	7	7	45																			
1993-94	Laval Titan	QMJHL	68	3	15	18	129												19	2	9	11	48			
1994-95	Laval Titan	QMJHL	72	8	25	33	115												20	3	11	14	21			
1995-96	Granby	QMJHL	68	11	35	46	156												21	2	12	14	30			
1996-97	Wheeling Nailers	ECHL	69	10	32	42	77												3	0	2	2	10			
1997-98	Quebec Rafales	IHL	71	8	27	35	76																			
1998-99	Fredericton	AHL	79	19	36	55	174												5	2	1	3	0			
99-2000	Montreal	NHL	74	3	13	16	38	2	0	1	76	3.9	–7	1	0.0	68	52	15:52								
2000-01	Montreal	NHL	29	0	6	6	26	0	0	0	24	0.0	3	0	0.0	42	16	13:24								
	Quebec	AHL	4	0	0	0	0																			

Season	Club	League	GP	G	A	Pts	PIM	PP	SH	GW	S	%	+/-	TF	F%	H	SB	Min	GP	G	A	Pts	PIM	PP	SH	GW
			Regular Season																Playoffs							
2001-02	Montreal	NHL	28	0	5	5	33	0	0	0	24	0.0	-5	0	0.0	52	42	18:47								
	Quebec	AHL	38	8	14	22	30																			
	NHL Totals		131	3	24	27	97	2	0	1	124	2.4		1	0.0	162	110	15:56								

Signed as a free agent by **Montreal**, August 18, 1998 • Missed majority of 2000-01 season recovering from ankle injury suffered in game vs. Calgary, December 31, 2000..

BOULERICE, Jesse
(BOO-luhr-ighs, JEHS-see) **CAR.**

Right wing. Shoots right. 6'1", 215 lbs. Born, Plattsburgh, NY, August 10, 1978. Philadelphia's 4th choice, 133rd overall, in 1996 Entry Draft.

Season	Club	League	GP	G	A	Pts	PIM	PP	SH	GW	S	%	+/-	TF	F%	H	SB	Min	GP	G	A	Pts	PIM	PP	SH	GW
1994-95	Hawkesbury	OCJHL	46	1	8	9	160																			
1995-96	Detroit	OHL	64	2	5	7	150												16	0	0	0	12			
1996-97	Detroit	OHL	33	10	14	24	209																			
1997-98	Plymouth Whalers	OHL	53	20	23	43	170												13	2	4	6	35			
1998-99	Philadelphia	AHL	24	1	2	3	82																			
	New Orleans	ECHL	12	0	1	1	38																			
99-2000	Philadelphia	AHL	40	3	4	7	85												4	0	2	2	4			
	Trenton Titans	ECHL	25	8	8	16	90																			
2000-01	Philadelphia	AHL	60	3	4	7	256												10	1	1	2	28			
2001-02	Philadelphia	NHL	3	0	0	0	5	0	0	0	1	0.0	-1	0	0.0	2	1	4:18								
	Philadelphia	AHL	41	2	5	7	204																			
	Lowell	AHL	15	2	4	6	80												5	0	2	2	6			
	NHL Totals		3	0	0	0	5	0	0	0	1	0.0		0	0.0	2	1	4:18								

Traded to **Carolina** by **Philadelphia** for Greg Koehler, February 13, 2002.

BOULTON, Eric
(BOHL-tuhn, AIR-ihk) **BUF.**

Left wing. Shoots left. 6', 222 lbs. Born, Halifax, N.S., August 17, 1976. NY Rangers' 12th choice, 234th overall, in 1994 Entry Draft.

Season	Club	League	GP	G	A	Pts	PIM	PP	SH	GW	S	%	+/-	TF	F%	H	SB	Min	GP	G	A	Pts	PIM	PP	SH	GW
1992-93	Cole Harbour	MJrHL	44	12	15	27	212																			
1993-94	Oshawa Generals	OHL	45	4	3	7	149												5	0	0	0	16			
1994-95	Oshawa Generals	OHL	27	7	5	12	125																			
	Sarnia Sting	OHL	24	3	7	10	134												4	0	1	1	10			
1995-96	Sarnia Sting	OHL	66	14	29	43	243												9	0	3	3	29			
1996-97	Binghamton	AHL	23	2	3	5	67												3	0	0	0	4			
	Charlotte	ECHL	44	14	11	25	325												3	0	1	1	6			
1997-98	Charlotte	ECHL	53	11	16	27	202												4	1	0	1	0			
	Fort Wayne	IHL	8	0	2	2	42																			
1998-99	Kentucky	AHL	34	3	3	6	154												10	0	1	1	36			
	Florida	ECHL	26	9	13	22	143																			
	Houston Aeros	IHL	7	1	0	1	41																			
99-2000	Rochester	AHL	76	2	2	4	276												18	2	1	3	53			
2000-01	Buffalo	NHL	35	1	2	3	94	0	0	0	20	5.0	-1	2	0.0	44	6	5:42								
2001-02	Buffalo	NHL	35	2	3	5	129	0	0	1	21	9.5	-1	0	0.0	48	7	6:08								
	NHL Totals		70	3	5	8	223	0	0	1	41	7.3		2	0.0	92	13	5:55								

Signed as a free agent by **Buffalo**, September 14, 1999.

BOUMEDIENNE, Josef
(BOO-mih-dyehn, JOH-sehf) **OTT.**

Defense. Shoots left. 6'1", 200 lbs. Born, Stockholm, Sweden, January 12, 1978. New Jersey's 7th choice, 91st overall, in 1996 Entry Draft.

Season	Club	League	GP	G	A	Pts	PIM	PP	SH	GW	S	%	+/-	TF	F%	H	SB	Min	GP	G	A	Pts	PIM	PP	SH	GW
1994-95	Huddinge IK Jr.	Swede-Jr.	10	0	2	2	57																			
1995-96	Huddinge IK Jr.	Swede-Jr.	25	2	4	6	66																			
	Huddinge IK	Swede-2	7	0	0	0	14																			
1996-97	Sodertalje SK	Sweden	32	1	1	2	32																			
1997-98	Sodertalje SK	Sweden	26	3	3	6	28																			
1998-99	Tappara Tampere	Finland	51	6	8	14	119																			
99-2000	Tappara Tampere	Finland	50	8	24	32	160												4	1	2	3	10			
2000-01	Albany	AHL	79	8	28	36	117																			
2001-02	New Jersey	NHL	1	1	0	1	2	0	0	0	1	100.0	-1	0	0.0	0	0	20:23								
	Albany	AHL	9	0	3	3	10																			
	Tampa Bay	NHL	3	0	0	0	4	0	0	0	0	0.0	-1	0	0.0	2	1	10:58								
	Springfield	AHL	53	7	25	32	57																			
	NHL Totals		4	1	0	1	6	0	0	0	1	100.0		0	0.0	2	1	13:19								

Traded to **Tampa Bay** by **New Jersey** with Sascha Goc and the rights to Anton But for Andrei Zyuzin, November 9, 2001. Traded to **Ottawa** by **Tampa Bay** for Ottawa's 7th round choice (Fredrik Norrena) in 2002 Entry Draft, June 23, 2002.

BOWLER, Bill
(BOH-luhr, BIHL) **BOS.**

Center. Shoots left. 5'9", 180 lbs. Born, Toronto, Ont., September 25, 1974.

Season	Club	League	GP	G	A	Pts	PIM	PP	SH	GW	S	%	+/-	TF	F%	H	SB	Min	GP	G	A	Pts	PIM	PP	SH	GW
1990-91	Tor. Red Wings	MTHL	69	58	99	157																				
1991-92	Windsor	OHL	66	25	63	88	28												7	2	3	5	13			
1992-93	Windsor	OHL	57	44	77	121	41																			
1993-94	Windsor	OHL	66	47	76	123	39																			
1994-95	Windsor	OHL	61	33	102	135	63												10	7	15	22	13			
	Las Vegas	IHL																	1	0	0	0	0			
1995-96	Las Vegas	IHL	75	31	55	86	26												14	3	5	8	22			
1996-97	Houston Aeros	IHL	78	22	43	65	79												13	2	5	7	6			
1997-98	Hamilton	AHL	46	7	24	31	22																			
	Manitoba Moose	IHL	30	9	26	34	30												3	0	2	2	4			
1998-99	Manitoba Moose	IHL	82	26	67	93	59												5	6	5	11	6			
99-2000	Manitoba Moose	IHL	75	20	42	62	59												2	1	2	3	6			
2000-01	Columbus	NHL	9	0	2	2	8	0	0	0	3	0.0	-3	48	47.9	1	0	9:44								
	Syracuse Crunch	AHL	72	21	58	79	50												4	1	3	4	2			
2001-02	Milwaukee	AHL	17	2	5	7	6																			
	Norfolk Admirals	AHL	28	3	16	19	35																			
	NHL Totals		9	0	2	2	8	0	0	0	3	0.0		48	47.9	1	0	9:44								

IHL Second All-Star Team (1999)

Signed as a free agent by **Manitoba** (IHL), August 28, 1998. Signed as a free agent by **Columbus**, August 3, 2000. Claimed on waivers by **Nashville** from **Columbus**, June 1, 2001. Signed as a free agent by **Boston**, July 18, 2002.

BOYLE, Dan
(BOIL, DAN) **T.B.**

Defense. Shoots right. 5'11", 190 lbs. Born, Ottawa, Ont., July 12, 1976.

Season	Club	League	GP	G	A	Pts	PIM	PP	SH	GW	S	%	+/-	TF	F%	H	SB	Min	GP	G	A	Pts	PIM	PP	SH	GW
1992-93	Gloucester	OCJHL	55	22	51	73	60																			
1993-94	Gloucester	OCJHL	53	27	54	81	155																			
1994-95	Miami-Ohio	CCHA	35	8	18	26	24																			
1995-96	Miami-Ohio	CCHA	36	7	20	27	70																			
1996-97	Miami-Ohio	CCHA	40	11	43	54	52																			
1997-98	Miami-Ohio	CCHA	37	14	26	40	58																			
1998-99	Florida	NHL	22	3	5	8	6	1	0	1	31	9.7	0	1	100.0	27	14	18:50								
	Kentucky	AHL	53	8	34	42	87												12	3	5	8	16			
99-2000	Florida	NHL	13	0	3	3	4	0	0	0	9	0.0	-2	0	0.0	9	12	16:57								
	Louisville	AHL	58	14	38	52	75												4	0	2	2	8			
2000-01	Florida	NHL	69	4	18	22	28	1	0	0	83	4.8	-14	0	0.0	95	30	16:56								
	Louisville	AHL	6	0	5	5	12																			
2001-02	Florida	NHL	25	3	3	6	12	1	0	0	31	9.7	-1	2	50.0	15	19	15:40								
	Tampa Bay	NHL	41	5	15	20	27	2	0	1	68	7.4	-15	0	0.0	42	31	22:28								
	NHL Totals		170	15	44	59	77	5	0	2	222	6.8		3	66.7	188	106	18:20								

CCHA First All-Star Team (1997, 1998) • NCAA West First All-American Team (1997, 1998) • AHL Second All-Star Team (1999, 2000)

Signed as a free agent by **Florida**, March 30, 1998. Traded to **Tampa Bay** by **Florida** for Tampa Bay's 5th round choice in 2003 Entry Draft, January 7, 2002.

BOYNTON, Nick

(BOIN-tuhn, NIHK) **BOS.**

Defense. Shoots right. 6'2", 210 lbs. Born, Nobleton, Ont., January 14, 1979. Boston's 1st choice, 21st overall, in 1999 Entry Draft.

			Regular Season																Playoffs							
Season	Club	League	GP	G	A	Pts	PIM	PP	SH	GW	S	%	+/-	TF	F%	H	SB	Min	GP	G	A	Pts	PIM	PP	SH	GW
1993-94	Caledon	MTJHL	4	0	1	1	0	....	....	....	....	....	....	....	....	....	....	....	....	....	....	....	....	....	....	....
1994-95	Caledon	MTJHL	44	10	35	45	139	....	....	....	....	....	....	....	....	....	....	....	....	....	....	....	....	....	....	....
1995-96	Ottawa 67's	OHL	64	10	14	24	90	....	....	....	....	....	....	....	....	....	....	....	4	0	3	3	10	....	....	....
1996-97	Ottawa 67's	OHL	63	13	51	64	143	....	....	....	....	....	....	....	....	....	....	....	24	4	*24	28	38	....	....	....
1997-98	Ottawa 67's	OHL	40	7	31	38	94	....	....	....	....	....	....	....	....	....	....	....	13	0	4	4	24	....	....	....
1998-99	Ottawa 67's	OHL	51	11	48	59	83	....	....	....	....	....	....	....	....	....	....	....	9	1	9	10	18	....	....	....
99-2000	**Boston**	**NHL**	**5**	**0**	**0**	**0**	**0**	**0**	**0**	**0**	**6**	**0.0**	**−5**	**0**	**0.0**	**6**	**8**	**21:21**	....	....	....	....	....	....	....	....
	Providence	AHL	53	5	14	19	66	....	....	....	....	....	....	....	....	....	....	....	12	1	0	1	6	....	....	....
2000-01	**Boston**	**NHL**	**1**	**0**	**0**	**0**	**0**	**0**	**0**	**0**	**1**	**0.0**	**−1**	**0**	**0.0**	**2**	**3**	**14:27**	....	....	....	....	....	....	....	....
	Providence	AHL	78	6	27	33	105	....	....	....	....	....	....	....	....	....	....	....	17	0	2	2	35	....	....	....
2001-02	**Boston**	**NHL**	**80**	**4**	**14**	**18**	**107**	**0**	**0**	**1**	**136**	**2.9**	**18**	**0**	**0.0**	**119**	**76**	**18:30**	**6**	**1**	**2**	**3**	**8**	**0**	**0**	**0**
	NHL Totals		**86**	**4**	**14**	**18**	**107**	**0**	**0**	**1**	**143**	**2.8**		**0**	**0.0**	**127**	**87**	**18:37**	**6**	**1**	**2**	**3**	**8**	**0**	**0**	**0**

• Re-entered NHL Entry Draft. Originally Washington's 1st choice, 9th overall, in 1997 Entry Draft.

OHL All-Rookie Team (1996) • Memorial Cup All-Star Team (1999) • Won Stafford Smythe Memorial Trophy (Memorial Cup Tournament MVP) (1999) • NHL All-Rookie Team (2002)

BRADLEY, Matt

(BRAD-lee, MAT) **S.J.**

Right wing. Shoots right. 6'2", 195 lbs. Born, Stittsville, Ont., June 13, 1978. San Jose's 4th choice, 102nd overall, in 1996 Entry Draft.

Season	Club	League	GP	G	A	Pts	PIM	PP	SH	GW	S	%	+/-	TF	F%	H	SB	Min	GP	G	A	Pts	PIM	PP	SH	GW
1994-95	Cumberland	OCJHL	49	13	20	33	18	....	....	....	....	....	....	....	....	....	....	....	....	....	....	....	....	....	....	....
1995-96	Kingston	OHL	55	10	14	24	17	....	....	....	....	....	....	....	....	....	....	....	6	0	1	1	6	....	....	....
1996-97	Kingston	OHL	65	24	24	48	41	....	....	....	....	....	....	....	....	....	....	....	5	0	4	4	2	....	....	....
	Kentucky	AHL	1	0	1	1	0	....	....	....	....	....	....	....	....	....	....	....	....	....	....	....	....	....	....	....
1997-98	Kingston	OHL	55	33	50	83	24	....	....	....	....	....	....	....	....	....	....	....	8	3	4	7	7	....	....	....
1998-99	Kentucky	AHL	79	23	20	43	57	....	....	....	....	....	....	....	....	....	....	....	10	1	4	5	4	....	....	....
99-2000	Kentucky	AHL	80	22	19	41	81	....	....	....	....	....	....	....	....	....	....	....	9	6	3	9	9	....	....	....
2000-01	**San Jose**	**NHL**	**21**	**1**	**1**	**2**	**19**	**0**	**0**	**0**	**16**	**6.3**	**0**	**0**	**0.0**	**49**	**1**	**6:58**	....	....	....	....	....	....	....	....
	Kentucky	AHL	22	5	8	13	16	....	....	....	....	....	....	....	....	....	....	....	1	1	0	1	5	....	....	....
2001-02	**San Jose**	**NHL**	**54**	**9**	**13**	**22**	**43**	**0**	**0**	**2**	**63**	**14.3**	**22**	**2**	**0.0**	**113**	**8**	**8:27**	**10**	**0**	**0**	**0**	**0**	**0**	**0**	**0**
	NHL Totals		**75**	**10**	**14**	**24**	**62**	**0**	**0**	**2**	**79**	**12.7**		**2**	**0.0**	**162**	**9**	**8:02**	**10**	**0**	**0**	**0**	**0**	**0**	**0**	**0**

Won William Hanley Award (Most Gentlemanly Player - OHL) (1998)

BRASHEAR, Donald

(bra-SHEER, DAWN-ohld) **PHI.**

Left wing. Shoots left. 6'2", 225 lbs. Born, Bedford, IN, January 7, 1972.

Season	Club	League	GP	G	A	Pts	PIM	PP	SH	GW	S	%	+/-	TF	F%	H	SB	Min	GP	G	A	Pts	PIM	PP	SH	GW
1988-89	Ste-Foy	QAAA	10	1	2	3	10	....	....	....	....	....	....	....	....	....	....	....	....	....	....	....	....	....	....	....
1989-90	Longueuil	QMJHL	64	12	14	26	169	....	....	....	....	....	....	....	....	....	....	....	7	0	0	0	11	....	....	....
1990-91	Longueuil	QMJHL	68	12	26	38	195	....	....	....	....	....	....	....	....	....	....	....	8	0	3	3	33	....	....	....
1991-92	Verdun	QMJHL	65	18	24	42	283	....	....	....	....	....	....	....	....	....	....	....	18	4	2	6	98	....	....	....
1992-93	Fredericton	AHL	76	11	3	14	261	....	....	....	....	....	....	....	....	....	....	....	5	0	0	0	8	....	....	....
1993-94	**Montreal**	**NHL**	**14**	**2**	**2**	**4**	**34**	**0**	**0**	**0**	**15**	**13.3**	**0**	....	....	....	....	....	**2**	**0**	**0**	**0**	**0**	**0**	**0**	**0**
	Fredericton	AHL	62	38	28	66	250	....	....	....	....	....	....	....	....	....	....	....	....	....	....	....	....	....	....	....
1994-95	Fredericton	AHL	29	10	9	19	182	....	....	....	....	....	....	....	....	....	....	....	17	7	5	12	77	....	....	....
	Montreal	**NHL**	**20**	**1**	**1**	**2**	**63**	**0**	**0**	**1**	**10**	**10.0**	**−5**	....	....	....	....	....	....	....	....	....	....	....	....	....
1995-96	**Montreal**	**NHL**	**67**	**0**	**4**	**4**	**223**	**0**	**0**	**0**	**25**	**0.0**	**−10**	....	....	....	....	....	**6**	**0**	**0**	**0**	**2**	**0**	**0**	**0**
1996-97	**Montreal**	**NHL**	**10**	**0**	**0**	**0**	**38**	**0**	**0**	**0**	**6**	**0.0**	**−2**	....	....	....	....	....	....	....	....	....	....	....	....	....
	Vancouver	**NHL**	**59**	**8**	**5**	**13**	**207**	**0**	**0**	**2**	**55**	**14.5**	**−6**	....	....	....	....	....	....	....	....	....	....	....	....	....
1997-98	**Vancouver**	**NHL**	**77**	**9**	**9**	**18**	***372**	**0**	**0**	**1**	**64**	**14.1**	**−9**	....	....	....	....	....	....	....	....	....	....	....	....	....
1998-99	**Vancouver**	**NHL**	**82**	**8**	**10**	**18**	**209**	**2**	**0**	**1**	**112**	**7.1**	**−25**	**6**	**16.7**	**126**	**20**	**13:25**	....	....	....	....	....	....	....	....
99-2000	**Vancouver**	**NHL**	**60**	**11**	**2**	**13**	**136**	**1**	**0**	**3**	**83**	**13.3**	**−9**	**11**	**36.4**	**125**	**18**	**13:07**	....	....	....	....	....	....	....	....
2000-01	**Vancouver**	**NHL**	**79**	**9**	**19**	**28**	**145**	**0**	**0**	**1**	**127**	**7.1**	**0**	**6**	**16.7**	**183**	**24**	**13:27**	**4**	**0**	**0**	**0**	**0**	**0**	**0**	**0**
2001-02	**Vancouver**	**NHL**	**31**	**5**	**8**	**13**	**90**	**1**	**0**	**0**	**45**	**11.1**	**−8**	**4**	**25.0**	**68**	**8**	**13:58**	....	....	....	....	....	....	....	....
	Philadelphia	**NHL**	**50**	**4**	**15**	**19**	**109**	**0**	**0**	**2**	**62**	**6.5**	**0**	**1**	**0.0**	**131**	**18**	**13:00**	**5**	**0**	**0**	**0**	**19**	**0**	**0**	**0**
	NHL Totals		**549**	**57**	**75**	**132**	**1626**	**4**	**0**	**11**	**604**	**9.4**		**28**	**25.0**	**633**	**88**	**13:21**	**17**	**0**	**0**	**0**	**21**	**0**	**0**	**0**

Signed as a free agent by **Montreal**, July 28, 1992. Traded to **Vancouver** by **Montreal** for Jassen Cullimore, November 13, 1996. Traded to **Philadelphia** by **Vancouver** with Vancouver's 6th round choice (later traded to Columbus - Columbus selected Jaroslav Balastik) in 2002 Entry Draft for Jan Hlavac and Tampa Bay's 3rd round choice (previously acquired, Vancouver selected Brett Skinner) in 2002 Entry Draft, December 17, 2001.

BRENDL, Pavel

(BREHN-duhl, PAH-vehl) **PHI.**

Right wing. Shoots right. 6'1", 204 lbs. Born, Opocno, Czech., March 23, 1981. NY Rangers' 1st choice, 4th overall, in 1999 Entry Draft.

Season	Club	League	GP	G	A	Pts	PIM	PP	SH	GW	S	%	+/-	TF	F%	H	SB	Min	GP	G	A	Pts	PIM	PP	SH	GW
1996-97	HC Olomouc Jr.	Czech-Jr.	40	35	17	52	....	....	....	....	....	....	....	....	....	....	....	....	....	....	....	....	....	....	....	....
1997-98	HC Olomouc Jr.	Czech-Jr.	38	29	23	52	....	....	....	....	....	....	....	....	....	....	....	....	....	....	....	....	....	....	....	....
	HC Olomouc	Czech-2	12	1	1	2	....	....	....	....	....	....	....	....	....	....	....	....	....	....	....	....	....	....	....	....
1998-99	Calgary Hitmen	WHL	68	*73	61	*134	40	....	....	....	....	....	....	....	....	....	....	....	20	*21	*25	*46	18	....	....	....
99-2000	Calgary Hitmen	WHL	61	*59	52	111	94	....	....	....	....	....	....	....	....	....	....	....	10	7	12	19	8	....	....	....
	Hartford	AHL	....	....	....	....	....	....	....	....	....	....	....	....	....	....	....	....	2	0	0	0	0	....	....	....
2000-01	Calgary Hitmen	WHL	49	40	35	75	66	....	....	....	....	....	....	....	....	....	....	....	10	7	6	13	6	....	....	....
2001-02	Philadelphia	AHL	64	15	22	37	22	....	....	....	....	....	....	....	....	....	....	....	5	4	1	5	0	....	....	....
	Philadelphia	**NHL**	**8**	**1**	**0**	**1**	**2**	**0**	**0**	**0**	**6**	**16.7**	**−1**	**21**	**19.1**	**1**	**2**	**8:59**	**2**	**0**	**0**	**0**	**0**	**0**	**0**	**0**
	NHL Totals		**8**	**1**	**0**	**1**	**2**	**0**	**0**	**0**	**6**	**16.7**		**21**	**19.0**	**1**	**2**	**8:59**	**2**	**0**	**0**	**0**	**0**	**0**	**0**	**0**

WHL East First All-Star Team (1999) • Canadian Major Junior First All-Star Team (1999) • Canadian Major Junior Rookie of the Year (1999) • Memorial Cup All-Star Team (1999) • WHL East Second All-Star Team (2000)

Traded to **Philadelphia** by **NY Rangers** with Jan Hlavac, Kim Johnsson and NY Rangers' 3rd round choice in 2003 Entry Draft for Eric Lindros, August 20, 2001.

BRENNAN, Kip

(BREHN-nan, KIHP) **L.A.**

Left wing. Shoots left. 6'4", 210 lbs. Born, Kingston, Ont., August 27, 1980. Los Angeles' 4th choice, 103rd overall, in 1998 Entry Draft.

Season	Club	League	GP	G	A	Pts	PIM	PP	SH	GW	S	%	+/-	TF	F%	H	SB	Min	GP	G	A	Pts	PIM	PP	SH	GW
1995-96	St. Michael's B	OJHL-B	40	0	11	11	155	....	....	....	....	....	....	....	....	....	....	....	....	....	....	....	....	....	....	....
1996-97	Windsor	OHL	42	0	10	10	156	....	....	....	....	....	....	....	....	....	....	....	5	0	1	1	16	....	....	....
1997-98	Windsor	OHL	24	0	7	7	103	....	....	....	....	....	....	....	....	....	....	....	....	....	....	....	....	....	....	....
	Sudbury Wolves	OHL	24	0	3	3	85	....	....	....	....	....	....	....	....	....	....	....	....	....	....	....	....	....	....	....
1998-99	Sudbury Wolves	OHL	38	9	12	21	160	....	....	....	....	....	....	....	....	....	....	....	....	....	....	....	....	....	....	....
99-2000	Sudbury Wolves	OHL	55	16	16	32	228	....	....	....	....	....	....	....	....	....	....	....	12	3	3	6	67	....	....	....
2000-01	Lowell	AHL	23	2	3	5	117	....	....	....	....	....	....	....	....	....	....	....	....	....	....	....	....	....	....	....
	Sudbury Wolves	OHL	27	7	14	21	94	....	....	....	....	....	....	....	....	....	....	....	12	5	6	11	*92	....	....	....
2001-02	**Los Angeles**	**NHL**	**4**	**0**	**0**	**0**	**22**	**0**	**0**	**0**	**0**	**0.0**	**1**	**0**	**0.0**	**4**	**0**	**4:40**	....	....	....	....	....	....	....	....
	Manchester	AHL	44	4	1	5	269	....	....	....	....	....	....	....	....	....	....	....	4	0	1	1	26	....	....	....
	NHL Totals		**4**	**0**	**0**	**0**	**22**	**0**	**0**	**0**	**0**	**0.0**		**0**	**0.0**	**4**	**0**	**4:40**								

Traded to **Sudbury** (OHL) by **Windsor** (OHL) with Glenn Crawford and future considerations for Steve Valiquette and Paul Mara, December 16, 1997. • Returned to **Sudbury** (OHL) by **LA Kings** (Lowell-AHL), January 10, 2001.

BRENNAN, Rich

(BREHN-nan, RIHCH) **BOS.**

Defense. Shoots right. 6'2", 200 lbs. Born, Schenectady, NY, November 26, 1972. Quebec's 3rd choice, 46th overall, in 1991 Entry Draft.

Season	Club	League	GP	G	A	Pts	PIM	PP	SH	GW	S	%	+/-	TF	F%	H	SB	Min	GP	G	A	Pts	PIM	PP	SH	GW
1988-89	Albany	Hi-School	25	17	30	47	57	....	....	....	....	....	....	....	....	....	....	....	....	....	....	....	....	....	....	....
1989-90	Tabor Academy	Hi-School	33	12	14	26	68	....	....	....	....	....	....	....	....	....	....	....	....	....	....	....	....	....	....	....
1990-91	Tabor Academy	Hi-School	34	13	37	50	91	....	....	....	....	....	....	....	....	....	....	....	....	....	....	....	....	....	....	....
1991-92	Boston University	H-East	30	4	13	17	50	....	....	....	....	....	....	....	....	....	....	....	....	....	....	....	....	....	....	....
1992-93	Boston University	H-East	40	9	11	20	68	....	....	....	....	....	....	....	....	....	....	....	....	....	....	....	....	....	....	....
1993-94	Boston University	H-East	41	8	27	35	82	....	....	....	....	....	....	....	....	....	....	....	....	....	....	....	....	....	....	....
1994-95	Boston University	H-East	31	5	22	27	56	....	....	....	....	....	....	....	....	....	....	....	....	....	....	....	....	....	....	....
1995-96	Brantford Smoke	ColHL	5	1	2	3	2	....	....	....	....	....	....	....	....	....	....	....	....	....	....	....	....	....	....	....
	Cornwall Aces	AHL	36	4	8	12	61	....	....	....	....	....	....	....	....	....	....	....	7	0	0	0	6	....	....	....
1996-97	**Colorado**	**NHL**	**2**	**0**	**0**	**0**	**0**	**0**	**0**	**0**	**0**	**0.0**	**0**	....	....	....	....	....	....	....	....	....	....	....	....	....
	Hershey Bears	AHL	74	11	45	56	88	....	....	....	....	....	....	....	....	....	....	....	23	2	*16	18	22	....	....	....

			Regular Season																Playoffs							
Season	Club	League	GP	G	A	Pts	PIM	PP	SH	GW	S	%	+/-	TF	F%	H	SB	Min	GP	G	A	Pts	PIM	PP	SH	GW
1997-98	San Jose	NHL	11	1	2	3	2	1	0	0	24	4.2	–4													
	Kentucky	AHL	42	11	17	28	71																			
	Hartford	AHL	9	2	4	6	12												15	4	5	9	14			
1998-99	NY Rangers	NHL	24	1	3	4	23	0	0	0	36	2.8	–4	0	0.0	40	22	13:02								
	Hartford	AHL	47	4	24	28	42																			
99-2000	Lowell	AHL	67	15	30	45	110												7	1	5	6	0			
2000-01	Los Angeles	NHL	2	0	0	0	0	0	0	0	1	0.0	–3	0	0.0	4	2	14:44								
	Lowell	AHL	69	10	31	41	146																			
2001-02	Nashville	NHL	4	0	0	0	2	0	0	0	0	0.0	0	0	0.0	8	3	14:31								
	Milwaukee	AHL	23	4	8	12	27																			
	Manchester	AHL	16	2	5	7	6												5	1	1	2	16			
	NHL Totals		43	2	5	7	27	1	0	0	61	3.3		0	0.0	52	27	13:21								

Hockey East First All-Star Team (1994) • NCAA East Second All-American Team (1994)

Rights transferred to **Colorado** after **Quebec** franchise relocated, June 21, 1995. Signed as a free agent by **San Jose**, July 9, 1997. Traded to **NY Rangers** by **San Jose** for Jason Muzzatti, March 24, 1998. Signed as a free agent by **Nashville**, September 23, 1999. Claimed by **LA Kings** from **Nashville** in Waiver Draft, September 27, 1999. Signed as a free agent by **Nashville**, August 8, 2001. Traded to **LA Kings** by **Nashville** for Brett Hauer, December 19, 2001. Signed as a free agent by **Boston**, July 18, 2002.

BREWER, Eric

(BREW-uhr, AIR-ihk) **EDM.**

Defense. Shoots left. 6'3", 220 lbs. Born, Vernon, B.C., April 17, 1979. NY Islanders' 2nd choice, 5th overall, in 1997 Entry Draft.

Season	Club	League	GP	G	A	Pts	PIM	PP	SH	GW	S	%	+/-	TF	F%	H	SB	Min	GP	G	A	Pts	PIM	PP	SH	GW
1994-95	Kamloops	BCAHA	40	19	19	38	62																			
1995-96	Prince George	WHL	63	4	10	14	25																			
1996-97	Prince George	WHL	71	5	24	29	81												15	2	4	6	16			
1997-98	Prince George	WHL	34	5	28	33	45												11	4	2	6	19			
1998-99	NY Islanders	NHL	63	5	6	11	32	2	0	0	63	7.9	–14	0	0.0	89	32	15:28								
99-2000	NY Islanders	NHL	26	0	2	2	20	0	0	0	30	0.0	–11	0	0.0	55	32	18:33								
	Lowell	AHL	25	2	2	4	26												7	0	0	0	0			
2000-01	Edmonton	NHL	77	7	14	21	53	2	0	2	91	7.7	15	0	0.0	162	91	18:31	6	1	5	6	2	1	0	0
2001-02	Edmonton	NHL	81	7	18	25	45	6	0	2	165	4.2	–5	0	0.0	195	122	23:56								
	Canada	Olympics	6	2	0	2	0																			
	NHL Totals		247	19	40	59	150	10	0	4	349	5.4		0	0.0	501	277	19:31	6	1	5	6	2	1	0	0

WHL West Second All-Star Team (1998)

Traded to **Edmonton** by **NY Islanders** with Josh Green and NY Islanders' 2nd round choice (Brad Winchester) in 2000 Entry Draft for Roman Hamrlik, June 24, 2000.

BRIERE, Daniel

(bree-AIR, DAN-yehl) **PHX.**

Center. Shoots right. 5'10", 181 lbs. Born, Gatineau, Que., October 6, 1977. Phoenix's 2nd choice, 24th overall, in 1996 Entry Draft.

Season	Club	League	GP	G	A	Pts	PIM	PP	SH	GW	S	%	+/-	TF	F%	H	SB	Min	GP	G	A	Pts	PIM	PP	SH	GW
1992-93	Abitibi Regents	QAAA	42	24	30	54	28												3	0	3	3	8			
1993-94	Gatineau	QAAA	44	56	47	103	56																			
1994-95	Drummondville	QMJHL	72	51	72	123	54												4	2	3	5	2			
1995-96	Drummondville	QMJHL	67	*67	*96	*163	84												6	6	12	18	8			
1996-97	Drummondville	QMJHL	59	52	78	130	94												8	7	7	14	14			
1997-98	Phoenix	NHL	5	1	0	1	2	0	0	0	4	25.0	1													
	Springfield	AHL	68	36	56	92	42												4	1	2	3	4			
1998-99	Phoenix	NHL	64	8	14	22	30	2	0	2	90	8.9	–3	484	47.5	15	8	11:13								
	Las Vegas	IHL	1	1	1	2	0																			
	Springfield	AHL	13	2	6	8	20												3	0	1	1	2			
99-2000	Phoenix	NHL	13	1	1	2	0	0	0	0	9	11.1	0	65	49.2	4	1	7:41	1	0	0	0	0	0	0	0
	Springfield	AHL	58	29	42	71	56																			
2000-01	Phoenix	NHL	30	11	4	15	12	9	0	1	43	25.6	–2	210	50.0	7	2	10:50								
	Springfield	AHL	30	21	25	46	30																			
2001-02	Phoenix	NHL	78	32	28	60	52	12	0	5	149	21.5	6	951	51.8	59	18	15:44	5	2	1	3	2	1	0	1
	NHL Totals		190	53	47	100	96	23	0	8	295	18.0		1710	50.3	85	29	12:49	6	2	1	3	2	1	0	1

QMJHL All-Rookie Team (1995) • QMJHL Offensive Rookie of the Year (1995) • QMJHL Second All-Star Team (1996, 1997) • AHL First All-Star Team (1998) • Won Dudley "Red" Garrett Memorial Trophy (Top Rookie - AHL) (1998)

BRIGLEY, Travis

(BRIH-glee), TRA-vihs) **ANA.**

Left wing. Shoots left. 6'1", 200 lbs. Born, Coronation, Alta., June 16, 1977. Calgary's 2nd choice, 39th overall, in 1996 Entry Draft.

Season	Club	League	GP	G	A	Pts	PIM	PP	SH	GW	S	%	+/-	TF	F%	H	SB	Min	GP	G	A	Pts	PIM	PP	SH	GW
1992-93	Leduc Oil Barons	AMHL	32	36	24	60	56																			
1993-94	Leduc Oil Barons	AMHL	34	29	44	73	141																			
	Lethbridge	WHL	1	0	0	0	0																			
1994-95	Lethbridge	WHL	64	14	18	32	14																			
1995-96	Lethbridge	WHL	69	34	43	77	94												4	2	3	5	8			
1996-97	Lethbridge	WHL	71	43	47	90	56												19	9	9	18	31			
1997-98	Calgary	NHL	2	0	0	0	2	0	0	0	1	0.0	0													
	Saint John	AHL	79	17	15	32	28												8	0	0	0	0			
1998-99	Saint John	AHL	74	15	35	50	48												7	3	1	4	2			
99-2000	Calgary	NHL	17	0	2	2	4	0	0	0	17	0.0	–6	2	0.0	16	8	14:14								
	Saint John	AHL	9	3	1	4	4																			
	Detroit Vipers	IHL	29	6	10	16	24																			
	Philadelphia	AHL	15	2	2	4	15												5	1	0	1	4			
2000-01	Knoxville Speed	UHL	4	2	4	6	4																			
	Cardiff Devils	Britain	12	5	9	14	6																			
	Louisville	AHL	49	14	21	35	34																			
2001-02	Macon Whoopee	ECHL	8	4	3	7	2																			
	Cincinnati	AHL	70	22	21	43	40												3	2	0	2	0			
	NHL Totals		19	0	2	2	6	0	0	0	18	0.0		2	0.0	16	8	14:14								

Traded to **Philadelphia** by **Calgary** with Calgary's 6th round choice (Andrei Razin) in 2001 Entry Draft for Marc Bureau, March 6, 2000. Signed as a free agent by **Cardiff** (Britain), November 3, 2000. Signed as a free agent by **Florida**, December 16, 2000. Signed as a free agent by **Anaheim**, January 22, 2002.

BRIMANIS, Aris

(brih-MAN-ihs, AR-ihs) **ST.L.**

Defense. Shoots right. 6'3", 210 lbs. Born, Cleveland, OH, March 14, 1972. Philadelphia's 3rd choice, 86th overall, in 1991 Entry Draft.

Season	Club	League	GP	G	A	Pts	PIM	PP	SH	GW	S	%	+/-	TF	F%	H	SB	Min	GP	G	A	Pts	PIM	PP	SH	GW
1988-89	Culver Eagles	Hi-School	38	10	13	23	24																			
1989-90	Culver Eagles	Hi-School	37	15	10	25	52																			
1990-91	Bowling Green	CCHA	38	3	6	9	42																			
1991-92	Bowling Green	CCHA	32	2	9	11	38																			
1992-93	Brandon	WHL	71	8	50	58	110												4	2	1	3	7			
1993-94	Philadelphia	NHL	1	0	0	0	0	0	0	0	1	0.0	–1													
	Hershey Bears	AHL	75	8	15	23	65												11	2	3	5	12			
1994-95	Hershey Bears	AHL	76	8	17	25	68												6	1	1	2	14			
1995-96	Philadelphia	NHL	17	0	2	2	12	0	0	0	11	0.0	–1													
	Hershey Bears	AHL	54	9	22	31	64												5	1	2	3	4			
1996-97	Philadelphia	NHL	3	0	1	1	0	0	0	0	1	0.0	0													
	Philadelphia	AHL	65	14	18	32	69												10	2	2	4	13			
1997-98	Philadelphia	AHL	30	1	11	12	26																			
	Michigan K-Wings	IHL	35	3	9	12	24												4	1	0	1	4			
1998-99	Grand Rapids	IHL	66	16	21	37	70																			
	Fredericton	AHL	8	2	4	6	6												15	3	10	13	18			
99-2000	NY Islanders	NHL	18	2	1	3	6	2	0	0	16	12.5	–5	1	0.0	32	28	20:00								
	Kansas City	IHL	46	5	17	22	28																			
	Providence	AHL	7	0	2	2	2												14	3	4	7	10			
2000-01	NY Islanders	NHL	56	0	8	8	26	0	0	0	66	0.0	–12	0	0.0	137	52	15:36								
	Chicago Wolves	IHL	20	2	2	4	14												16	3	1	4	8			

			Regular Season																Playoffs							
Season	Club	League	GP	G	A	Pts	PIM	PP	SH	GW	S	%	+/-	TF	F%	H	SB	Min	GP	G	A	Pts	PIM	PP	SH	GW
2001-02	**Anaheim**	**NHL**	**5**	**0**	**0**	**0**	**9**	**0**	**0**	**0**	**2**	**0.0**	**–1**	**0**	**0.0**	**11**	**2**	**9:31**	....	....	....	....	....	....	....	....
	Cincinnati	AHL	72	2	9	11	44	....	....	....	....	....	....	....	....	....	....	....	3	1	0	1	0	....	....	....
	NHL Totals		**100**	**2**	**12**	**14**	**53**	**2**	**0**	**0**	**97**	**2.1**		**1**	**0.0**	**180**	**82**	**16:13**	....	....	....	....	....	....	....	....

Signed as a free agent by **NY Islanders**, August 16, 1999. Loaned to **Providence** (AHL) by **NY Islanders**, March 14, 2000. Signed as a free agent by **Anaheim**, August 1, 2001. Signed as a free agent by **St. Louis**, August 15, 2002.

BRIND'AMOUR, Rod

(BRIHND-uh-MOHR, RAWD) **CAR.**

Center. Shoots left. 6'1", 202 lbs. Born, Ottawa, Ont., August 9, 1970. St. Louis' 1st choice, 9th overall, in 1988 Entry Draft.

Season	Club	League	GP	G	A	Pts	PIM	PP	SH	GW	S	%	+/-	TF	F%	H	SB	Min	GP	G	A	Pts	PIM	PP	SH	GW
1986-87	Notre Dame	SMHL	33	38	50	88	66	....	....	....	....	....	....	....	....	....	....	....	....	....	....	....	....	....	....	....
1987-88	Notre Dame	SJHL	56	46	61	107	136	....	....	....	....	....	....	....	....	....	....	....	....	....	....	....	....	....	....	....
1988-89	Michigan State	CCHA	42	27	32	59	63	....	....	....	....	....	....	....	....	....	....	....	....	....	....	....	....	....	....	....
	St. Louis	**NHL**	....	....	....	....	....	....	....	....	....	....	....	....	....	....	....	....	**5**	**2**	**0**	**2**	**4**	**0**	**0**	**0**
1989-90	**St. Louis**	**NHL**	**79**	**26**	**35**	**61**	**46**	**10**	**0**	**1**	**160**	**16.3**	**23**	....	....	....	....	....	**12**	**5**	**8**	**13**	**6**	**1**	**0**	**0**
1990-91	**St. Louis**	**NHL**	**78**	**17**	**32**	**49**	**93**	**4**	**0**	**3**	**169**	**10.1**	**2**	....	....	....	....	....	**13**	**2**	**5**	**7**	**10**	**1**	**0**	**0**
1991-92	**Philadelphia**	**NHL**	**80**	**33**	**44**	**77**	**100**	**8**	**4**	**5**	**202**	**16.3**	**–3**	....	....	....	....	....	....	....	....	....	....	....	....	....
1992-93	**Philadelphia**	**NHL**	**81**	**37**	**49**	**86**	**89**	**13**	**4**	**4**	**206**	**18.0**	**–8**	....	....	....	....	....	....	....	....	....	....	....	....	....
1993-94	**Philadelphia**	**NHL**	**84**	**35**	**62**	**97**	**85**	**14**	**1**	**4**	**230**	**15.2**	**–9**	....	....	....	....	....	....	....	....	....	....	....	....	....
1994-95	**Philadelphia**	**NHL**	**48**	**12**	**27**	**39**	**33**	**4**	**1**	**2**	**86**	**14.0**	**–4**	....	....	....	....	....	**15**	**6**	**9**	**15**	**8**	**2**	**1**	**1**
1995-96	**Philadelphia**	**NHL**	**82**	**26**	**61**	**87**	**110**	**4**	**4**	**5**	**213**	**12.2**	**20**	....	....	....	....	....	**12**	**2**	**5**	**7**	**6**	**1**	**0**	**0**
1996-97	**Philadelphia**	**NHL**	**82**	**27**	**32**	**59**	**41**	**8**	**2**	**3**	**205**	**13.2**	**2**	....	....	....	....	....	**19**	***13**	**8**	**21**	**10**	**4**	**2**	**1**
1997-98	**Philadelphia**	**NHL**	**82**	**36**	**38**	**74**	**54**	**10**	**2**	**8**	**205**	**17.6**	**–2**	....	....	....	....	....	**5**	**2**	**2**	**4**	**7**	**0**	**0**	**0**
	Canada	Olympics	6	1	2	3	0	....	....	....	....	....	....	....	....	....	....	....	....	....	....	....	....	....	....	....
1998-99	**Philadelphia**	**NHL**	**82**	**24**	**50**	**74**	**47**	**10**	**0**	**3**	**191**	**12.6**	**3**	**1773**	**56.5**	**90**	**31**	**21:29**	**6**	**1**	**3**	**4**	**0**	**0**	**0**	**0**
99-2000	**Philadelphia**	**NHL**	**12**	**5**	**3**	**8**	**4**	**4**	**0**	**0**	**26**	**19.2**	**–1**	**291**	**60.5**	**17**	**11**	**20:50**	....	....	....	....	....	....	....	....
	Carolina	**NHL**	**33**	**4**	**10**	**14**	**22**	**0**	**1**	**1**	**61**	**6.6**	**–12**	**704**	**55.5**	**62**	**13**	**20:35**	....	....	....	....	....	....	....	....
2000-01	**Carolina**	**NHL**	**79**	**20**	**36**	**56**	**47**	**5**	**1**	**5**	**163**	**12.3**	**–7**	**1907**	**60.4**	**117**	**51**	**22:07**	**6**	**1**	**3**	**4**	**6**	**0**	**0**	**1**
2001-02	**Carolina**	**NHL**	**81**	**23**	**32**	**55**	**40**	**5**	**2**	**5**	**162**	**14.2**	**3**	**2058**	**59.2**	**109**	**66**	**22:07**	**23**	**4**	**8**	**12**	**16**	**2**	**1**	**1**
	NHL Totals		**983**	**325**	**511**	**836**	**811**	**99**	**22**	**49**	**2279**	**14.3**		**6733**	**58.5**	**395**	**172**	**21:42**	**116**	**38**	**51**	**89**	**73**	**11**	**4**	**4**

CCHA Rookie of the Year (1989) • NHL All-Rookie Team (1990) • Played in NHL All-Star Game (1992)

Traded to **Philadelphia** by **St. Louis** with Dan Quinn for Ron Sutter and Murray Baron, September 22, 1991. Traded to **Carolina** by **Philadelphia** with Jean-Marc Pelletier and Philadelphia's 2nd round choice (later traded to Colorado - Colorado selected Agris Saviels) in 2000 Entry Draft for Keith Primeau and Carolina's 5th round choice (later traded to NY Islanders - NY Islanders selected Kristofer Ottosson) in 2000 Entry Draft, January 23, 2000.

BRISEBOIS, Patrice

(BREES-bwah, pa-TREEZ) **MTL.**

Defense. Shoots right. 6'2", 203 lbs. Born, Montreal, Que., January 27, 1971. Montreal's 2nd choice, 30th overall, in 1989 Entry Draft.

Season	Club	League	GP	G	A	Pts	PIM	PP	SH	GW	S	%	+/-	TF	F%	H	SB	Min	GP	G	A	Pts	PIM	PP	SH	GW
1986-87	Mtl-Bourassa	QAAA	39	15	19	34	66	....	....	....	....	....	....	....	....	....	....	....	....	....	....	....	....	....	....	....
1987-88	Laval Titan	QMJHL	48	10	34	44	95	....	....	....	....	....	....	....	....	....	....	....	6	0	2	2	2	....	....	....
1988-89	Laval Titan	QMJHL	50	20	45	65	95	....	....	....	....	....	....	....	....	....	....	....	17	8	14	22	45	....	....	....
1989-90	Laval Titan	QMJHL	56	18	70	88	108	....	....	....	....	....	....	....	....	....	....	....	13	7	9	16	26	....	....	....
1990-91	Drummondville	QMJHL	54	17	44	61	72	....	....	....	....	....	....	....	....	....	....	....	14	6	18	24	49	....	....	....
	Montreal	**NHL**	**10**	**0**	**2**	**2**	**4**	**0**	**0**	**0**	**11**	**0.0**	**1**	....	....	....	....	....	....	....	....	....	....	....	....	....
1991-92	**Montreal**	**NHL**	**26**	**2**	**8**	**10**	**20**	**0**	**0**	**1**	**37**	**5.4**	**9**	....	....	....	....	....	**11**	**2**	**4**	**6**	**6**	**1**	**0**	**1**
	Fredericton	AHL	53	12	27	39	51	....	....	....	....	....	....	....	....	....	....	....	....	....	....	....	....	....	....	....
1992-93♦	**Montreal**	**NHL**	**70**	**10**	**21**	**31**	**79**	**4**	**0**	**2**	**123**	**8.1**	**6**	....	....	....	....	....	**20**	**0**	**4**	**4**	**18**	**0**	**0**	**0**
1993-94	**Montreal**	**NHL**	**53**	**2**	**21**	**23**	**63**	**1**	**0**	**0**	**71**	**2.8**	**5**	....	....	....	....	....	**7**	**0**	**4**	**4**	**6**	**0**	**0**	**0**
1994-95	**Montreal**	**NHL**	**35**	**4**	**8**	**12**	**26**	**0**	**0**	**2**	**67**	**6.0**	**–2**	....	....	....	....	....	....	....	....	....	....	....	....	....
1995-96	**Montreal**	**NHL**	**69**	**9**	**27**	**36**	**65**	**3**	**0**	**1**	**127**	**7.1**	**10**	....	....	....	....	....	**6**	**1**	**2**	**3**	**6**	**0**	**0**	**0**
1996-97	**Montreal**	**NHL**	**49**	**2**	**13**	**15**	**24**	**0**	**0**	**1**	**72**	**2.8**	**–7**	....	....	....	....	....	**3**	**1**	**1**	**2**	**24**	**0**	**0**	**1**
1997-98	**Montreal**	**NHL**	**79**	**10**	**27**	**37**	**67**	**5**	**0**	**1**	**125**	**8.0**	**16**	....	....	....	....	....	**10**	**1**	**0**	**1**	**0**	**0**	**0**	**0**
1998-99	**Montreal**	**NHL**	**54**	**3**	**9**	**12**	**28**	**1**	**0**	**1**	**90**	**3.3**	**–8**	**0**	**0.0**	**62**	**80**	**22:26**	....	....	....	....	....	....	....	....
99-2000	**Montreal**	**NHL**	**54**	**10**	**25**	**35**	**18**	**5**	**0**	**2**	**88**	**11.4**	**–1**	**0**	**0.0**	**82**	**79**	**23:14**	....	....	....	....	....	....	....	....
2000-01	**Montreal**	**NHL**	**77**	**15**	**21**	**36**	**28**	**11**	**0**	**4**	**178**	**8.4**	**–31**	**1**	**100.0**	**100**	**122**	**24:43**	....	....	....	....	....	....	....	....
2001-02	**Montreal**	**NHL**	**71**	**4**	**29**	**33**	**25**	**2**	**1**	**1**	**95**	**4.2**	**9**	**0**	**0.0**	**91**	**97**	**23:53**	**10**	**1**	**1**	**2**	**2**	**0**	**0**	**0**
	NHL Totals		**647**	**71**	**211**	**282**	**447**	**32**	**1**	**16**	**1084**	**6.5**		**1**	**100.0**	**335**	**378**	**23:42**	**67**	**6**	**16**	**22**	**62**	**1**	**0**	**2**

QMJHL Second All-Star Team (1990) • QMJHL First All-Star Team (1991) • Canadian Major Junior Defenseman of the Year (1991) • Memorial Cup All-Star Team (1991)

BROUSSEAU, Paul

(BROO-soh, PAWL)

Right wing. Shoots right. 6'2", 203 lbs. Born, Pierrefonds, Que., September 18, 1973. Quebec's 2nd choice, 28th overall, in 1992 Entry Draft.

Season	Club	League	GP	G	A	Pts	PIM	PP	SH	GW	S	%	+/-	TF	F%	H	SB	Min	GP	G	A	Pts	PIM	PP	SH	GW
1988-89	Lac St-Louis	QAAA	37	6	17	23	28	....	....	....	....	....	....	....	....	....	....	....	3	3	1	4	2	....	....	....
1989-90	Chicoutimi	QMJHL	57	17	24	41	32	....	....	....	....	....	....	....	....	....	....	....	7	0	3	3	0	....	....	....
1990-91	Trois-Rivieres	QMJHL	67	30	66	96	48	....	....	....	....	....	....	....	....	....	....	....	6	3	2	5	2	....	....	....
1991-92	Hull Olympiques	QMJHL	57	35	61	96	54	....	....	....	....	....	....	....	....	....	....	....	6	3	5	8	10	....	....	....
1992-93	Hull Olympiques	QMJHL	59	27	48	75	49	....	....	....	....	....	....	....	....	....	....	....	10	7	8	15	6	....	....	....
1993-94	Cornwall Aces	AHL	69	18	26	44	35	....	....	....	....	....	....	....	....	....	....	....	1	0	0	0	0	....	....	....
1994-95	Cornwall Aces	AHL	57	19	17	36	29	....	....	....	....	....	....	....	....	....	....	....	7	2	1	3	10	....	....	....
1995-96	**Colorado**	**NHL**	**8**	**1**	**1**	**2**	**2**	**0**	**0**	**0**	**10**	**10.0**	**1**	....	....	....	....	....	....	....	....	....	....	....	....	....
	Cornwall Aces	AHL	63	21	22	43	60	....	....	....	....	....	....	....	....	....	....	....	8	4	0	4	2	....	....	....
1996-97	**Tampa Bay**	**NHL**	**6**	**0**	**0**	**0**	**0**	**0**	**0**	**0**	**3**	**0.0**	**–4**	....	....	....	....	....	....	....	....	....	....	....	....	....
	Adirondack	AHL	66	35	31	66	25	....	....	....	....	....	....	....	....	....	....	....	4	1	2	3	0	....	....	....
1997-98	**Tampa Bay**	**NHL**	**11**	**0**	**2**	**2**	**27**	**0**	**0**	**0**	**6**	**0.0**	**0**	....	....	....	....	....	....	....	....	....	....	....	....	....
	Adirondack	AHL	67	45	20	65	18	....	....	....	....	....	....	....	....	....	....	....	3	1	1	2	0	....	....	....
1998-99	Milwaukee	IHL	5	1	1	2	2	....	....	....	....	....	....	....	....	....	....	....	....	....	....	....	....	....	....	....
	Hershey Bears	AHL	39	11	21	32	15	....	....	....	....	....	....	....	....	....	....	....	5	1	1	2	0	....	....	....
99-2000	Louisville	AHL	36	19	24	43	10	....	....	....	....	....	....	....	....	....	....	....	4	1	1	2	12	....	....	....
2000-01	**Florida**	**NHL**	**1**	**0**	**0**	**0**	**0**	**0**	**0**	**0**	**0**	**0.0**	**0**	**0**	**0.0**	**0**	**0**	**1:39**	....	....	....	....	....	....	....	....
	Louisville	AHL	73	29	39	68	21	....	....	....	....	....	....	....	....	....	....	....	....	....	....	....	....	....	....	....
2001-02	SaiPa	Finland	39	15	11	26	24	....	....	....	....	....	....	....	....	....	....	....	....	....	....	....	....	....	....	....
	AIK Solna	Sweden	13	1	3	4	6	....	....	....	....	....	....	....	....	....	....	....	....	....	....	....	....	....	....	....
	NHL Totals		**26**	**1**	**3**	**4**	**29**	**0**	**0**	**0**	**19**	**5.3**		**0**	**0.0**	**0**	**0**	**1:39**	....	....	....	....	....	....	....	....

AHL Second All-Star Team (1998)

Rights transferred to **Colorado** after **Quebec** franchise relocated, June 21, 1995. Signed as a free agent by **Tampa Bay**, September 10, 1996. Claimed by **Nashville** from **Tampa Bay** in Expansion Draft, June 26, 1998. Signed as a free agent by **Florida**, September 20, 1999. • Missed majority of 1999-2000 season recovering from knee injury suffered in game vs. Rochester (AHL), January 8, 2000.

BROWN, Brad

(BROWN, BRAD) **MIN.**

Defense. Shoots right. 6'4", 220 lbs. Born, Baie Verte, Nfld., December 27, 1975. Montreal's 1st choice, 18th overall, in 1994 Entry Draft.

Season	Club	League	GP	G	A	Pts	PIM	PP	SH	GW	S	%	+/-	TF	F%	H	SB	Min	GP	G	A	Pts	PIM	PP	SH	GW
1990-91	Tor. Red Wings	MTHL	80	15	45	60	105	....	....	....	....	....	....	....	....	....	....	....	....	....	....	....	....	....	....	....
	St. Michael's B	OJHL-B	2	0	0	0	0	....	....	....	....	....	....	....	....	....	....	....	....	....	....	....	....	....	....	....
1991-92	North Bay	OHL	49	2	9	11	170	....	....	....	....	....	....	....	....	....	....	....	18	0	6	6	43	....	....	....
1992-93	North Bay	OHL	61	4	9	13	228	....	....	....	....	....	....	....	....	....	....	....	2	0	2	2	13	....	....	....
1993-94	North Bay	OHL	66	8	24	32	196	....	....	....	....	....	....	....	....	....	....	....	18	3	12	15	33	....	....	....
1994-95	North Bay	OHL	64	8	38	46	172	....	....	....	....	....	....	....	....	....	....	....	6	1	4	5	8	....	....	....
1995-96	Barrie Colts	OHL	27	3	13	16	82	....	....	....	....	....	....	....	....	....	....	....	....	....	....	....	....	....	....	....
	Fredericton	AHL	38	0	3	3	148	....	....	....	....	....	....	....	....	....	....	....	10	2	1	3	6	....	....	....
1996-97	**Montreal**	**NHL**	**8**	**0**	**0**	**0**	**22**	**0**	**0**	**0**	**0**	**0.0**	**–1**	....	....	....	....	....	....	....	....	....	....	....	....	....
	Fredericton	AHL	64	3	7	10	368	....	....	....	....	....	....	....	....	....	....	....	....	....	....	....	....	....	....	....
1997-98	Fredericton	AHL	64	1	8	9	297	....	....	....	....	....	....	....	....	....	....	....	4	0	0	0	29	....	....	....
1998-99	**Montreal**	**NHL**	**5**	**0**	**0**	**0**	**21**	**0**	**0**	**0**	**0**	**0.0**	**0**	**0**	**0.0**	**1**	**5**	**6:02**	....	....	....	....	....	....	....	....
	Chicago	**NHL**	**61**	**1**	**7**	**8**	**184**	**0**	**0**	**0**	**26**	**3.8**	**–4**	**0**	**0.0**	**145**	**63**	**15:08**	....	....	....	....	....	....	....	....
99-2000	**Chicago**	**NHL**	**57**	**0**	**9**	**9**	**134**	**0**	**0**	**0**	**15**	**0.0**	**–1**	**0**	**0.0**	**93**	**64**	**14:12**	....	....	....	....	....	....	....	....
2000-01	**NY Rangers**	**NHL**	**48**	**1**	**3**	**4**	**107**	**0**	**0**	**0**	**14**	**7.1**	**0**	**0**	**0.0**	**98**	**71**	**14:31**	....	....	....	....	....	....	....	....
2001-02	**Minnesota**	**NHL**	**51**	**0**	**4**	**4**	**123**	**0**	**0**	**0**	**23**	**0.0**	**–11**	**0**	**0.0**	**120**	**64**	**15:54**	....	....	....	....	....	....	....	....
	NHL Totals		**230**	**2**	**23**	**25**	**591**	**0**	**0**	**0**	**78**	**2.6**		**0**	**0.0**	**457**	**267**	**14:44**	....	....	....	....	....	....	....	....

OHL All-Rookie Team (1992)

Traded to **Chicago** by **Montreal** with Jocelyn Thibault and Dave Manson for Jeff Hackett, Eric Weinrich, Alain Nasreddine and Tampa Bay's 4th round choice (previously acquired, Montreal selected Chris Dyment) in 1999 Entry Draft, November 16, 1998. Traded to **NY Rangers** by **Chicago** with Michal Grosek for future considerations, October 5, 2000. Signed as a free agent by **Minnesota**, July 31, 2001.

BROWN, Curtis

(BROWN, KUHR-tihs) **BUF.**

Center/Left wing. Shoots left. 6', 197 lbs. Born, Unity, Sask., February 12, 1976. Buffalo's 2nd choice, 43rd overall, in 1994 Entry Draft.

Season	Club	League	Regular Season GP	G	A	Pts	PIM	PP	SH	GW	S	%	+/-	TF	F%	H	SB	Min	Playoffs GP	G	A	Pts	PIM	PP	SH	GW
1990-91	Unity Bantams	SMHL	60	93	104	197	55																			
1991-92	Moose Jaw	SMHL	36	35	30	65	44																			
1992-93	Moose Jaw	WHL	71	13	16	29	30																			
1993-94	Moose Jaw	WHL	72	27	38	65	82																			
1994-95	Moose Jaw	WHL	70	51	53	104	63												10	8	7	15	20			
	Buffalo	**NHL**	**1**	**1**	**1**	**2**	**2**	**0**	**0**	**0**	**4**	**25.0**	**2**													
1995-96	Moose Jaw	WHL	25	20	18	38	30																			
	Prince Albert	WHL	19	12	21	33	8												18	10	15	25	18			
	Buffalo	**NHL**	**4**	**0**	**0**	**0**	**0**	**0**	**0**	**0**	**1**	**0.0**	**0**													
	Rochester	AHL																	12	0	1	1	2			
1996-97	**Buffalo**	**NHL**	**28**	**4**	**3**	**7**	**18**	**0**	**0**	**1**	**31**	**12.9**	**4**													
	Rochester	AHL	51	22	21	43	30												10	4	6	10	4			
1997-98	**Buffalo**	**NHL**	**63**	**12**	**12**	**24**	**34**	**1**	**1**	**2**	**91**	**13.2**	**11**						**13**	**1**	**2**	**3**	**10**	**1**	**0**	**0**
1998-99	**Buffalo**	**NHL**	**78**	**16**	**31**	**47**	**56**	**5**	**1**	**3**	**128**	**12.5**	**23**	**1198**	**45.0**	**83**	**64**	**17:30**	**21**	**7**	**6**	**13**	**10**	**3**	**0**	**3**
99-2000	**Buffalo**	**NHL**	**74**	**22**	**29**	**51**	**42**	**5**	**0**	**4**	**149**	**14.8**	**19**	**1318**	**48.6**	**48**	**65**	**18:11**	**5**	**1**	**3**	**4**	**6**	**1**	**0**	**0**
2000-01	**Buffalo**	**NHL**	**70**	**10**	**22**	**32**	**34**	**2**	**1**	**0**	**105**	**9.5**	**15**	**1159**	**50.4**	**45**	**62**	**16:34**	**13**	**5**	**0**	**5**	**8**	**0**	**2**	**1**
2001-02	**Buffalo**	**NHL**	**82**	**20**	**17**	**37**	**32**	**4**	**1**	**5**	**171**	**11.7**	**-4**	**1608**	**49.0**	**71**	**94**	**17:48**								
	NHL Totals		**400**	**85**	**115**	**200**	**218**	**17**	**4**	**15**	**680**	**12.5**		**5283**	**48.3**	**247**	**285**	**17:32**	**52**	**14**	**11**	**25**	**34**	**5**	**2**	**4**

WHL East First All-Star Team (1995) • WHL East Second All-Star Team (1996)

BROWN, Kevin

(BROWN, KEH-vihn)

Right wing. Shoots right. 6'1", 212 lbs. Born, Birmingham, England, May 11, 1974. Los Angeles' 3rd choice, 87th overall, in 1992 Entry Draft.

Season	Club	League	Regular Season GP	G	A	Pts	PIM	PP	SH	GW	S	%	+/-	TF	F%	H	SB	Min	Playoffs GP	G	A	Pts	PIM	PP	SH	GW
1989-90	Georgetown	OJHL-B	31	3	8	11	59																			
1990-91	Waterloo	USHL	46	25	33	58	116																			
1991-92	Belleville Bulls	OHL	66	24	24	48	52												5	1	4	5	8			
1992-93	Belleville Bulls	OHL	6	2	5	7	4																			
	Detroit	OHL	56	48	86	134	76												15	10	18	28	18			
1993-94	Detroit	OHL	57	54	81	135	85												17	14	*26	*40	28			
1994-95	Phoenix	IHL	48	19	31	50	64																			
	Los Angeles	**NHL**	**23**	**2**	**3**	**5**	**18**	**0**	**0**	**0**	**25**	**8.0**	**-7**													
1995-96	**Los Angeles**	**NHL**	**7**	**1**	**0**	**1**	**4**	**0**	**0**	**0**	**9**	**11.1**	**-2**													
	Phoenix	IHL	45	10	16	26	39																			
	P.E.I. Senators	AHL	8	3	6	9	2												3	1	3	4	0			
1996-97	**Hartford**	**NHL**	**11**	**0**	**4**	**4**	**6**	**0**	**0**	**0**	**12**	**0.0**	**-6**													
	Springfield	AHL	48	32	16	48	45												17	*11	6	17	24			
1997-98	**Carolina**	**NHL**	**4**	**0**	**0**	**0**	**0**	**0**	**0**	**0**	**0**	**0.0**	**-2**													
	New Haven	AHL	67	28	44	72	65												3	0	2	2	0			
1998-99	**Edmonton**	**NHL**	**12**	**4**	**2**	**6**	**0**	**2**	**0**	**0**	**13**	**30.8**	**-2**	**1**	**0.0**	**19**	**3**	**9:11**								
	Hamilton	AHL	32	9	14	23	47																			
	Hartford	AHL	9	3	2	5	14												5	1	3	4	4			
99-2000	Hamilton	AHL	54	21	38	59	53												4	2	2	4	8			
	Edmonton	**NHL**	**7**	**0**	**0**	**0**	**0**	**0**	**0**	**0**	**5**	**0.0**	**0**	**1**	**0.0**	**6**	**1**	**8:21**	**1**	**0**	**0**	**0**	**0**	**0**	**0**	**0**
2000-01	Manchester Storm	Britain	36	17	32	49	118																			
	Phoenix Mustangs	WCHL	13	7	7	14	32																			
	Anchorage Aces	WCHL	10	10	10	20	8												3	2	3	5	20			
2001-02	Hamilton	AHL	70	28	36	64	154												10	2	9	11	6			
	NHL Totals		**64**	**7**	**9**	**16**	**28**	**2**	**0**	**0**	**64**	**10.9**		**2**	**0.0**	**25**	**4**	**8:53**	**1**	**0**	**0**	**0**	**0**	**0**	**0**	**0**

OHL Second All-Star Team (1993) • OHL First All-Star Team (1994) • Canadian Major Junior Second All-Star Team (1994)

Traded to **Ottawa** by **LA Kings** for Jaroslav Modry and Ottawa's 8th round choice (Stephen Valiquette) in 1996 Entry Draft, March 20, 1996. Traded to **Anaheim** by **Ottawa** for Mike Maneluk, July 1, 1996. Traded to **Hartford** by **Anaheim** for the rights to Espen Knutsen, October 1, 1996. Transferred to **Carolina** after **Hartford** franchise relocated, June 25, 1997. Signed as a free agent by **Edmonton**, August 14, 1998. Traded to **NY Rangers** by **Edmonton** for Vladimir Vorobiev, March 23, 1999. Signed as a free agent by **Edmonton**, March 7, 2000. Signed as a free agent by **Manchester** (Britain), September 10, 2000. Signed as a free agent by **Phoenix** (WCHL) after securing release from Manchester, February 16, 2001. Traded to **Anchorage** (WCHL) by **Phoenix** (WCHL) for Derry Minard, March 23, 2001. Signed as a free agent by **Hamilton** (AHL), October 3, 2001.

BROWN, Mike

(BROWN, MIGHK) **VAN.**

Left wing. Shoots left. 6'5", 185 lbs. Born, Surrey, B.C., April 27, 1979. Florida's 1st choice, 20th overall, in 1997 Entry Draft.

Season	Club	League	Regular Season GP	G	A	Pts	PIM	PP	SH	GW	S	%	+/-	TF	F%	H	SB	Min	Playoffs GP	G	A	Pts	PIM	PP	SH	GW
1993-94	Penticton	BCJHL	50	52	48	100	100																			
1994-95	Merritt	BCJHL	45	3	4	7	145																			
1995-96	Red Deer Rebels	WHL	62	4	5	9	125												10	0	0	0	18			
1996-97	Red Deer Rebels	WHL	70	19	13	32	243												16	1	2	3	47			
1997-98	Kamloops Blazers	WHL	72	23	33	56	305												7	2	1	3	22			
1998-99	Kamloops Blazers	WHL	69	28	16	44	*285												15	3	7	10	*68			
99-2000	Syracuse Crunch	AHL	71	13	18	31	284												4	0	0	0	0			
2000-01	**Vancouver**	**NHL**	**1**	**0**	**0**	**0**	**5**	**0**	**0**	**0**	**1**	**0.0**	**0**	**0**	**0.0**	**1**	**0**	**4:48**								
	Kansas City	IHL	78	14	13	27	214																			
2001-02	**Vancouver**	**NHL**	**15**	**0**	**0**	**0**	**72**	**0**	**0**	**0**	**2**	**0.0**	**1**	**0**	**0.0**	**7**	**0**	**3:29**								
	Manitoba Moose	AHL	31	7	9	16	155												6	0	1	1	16			
	NHL Totals		**16**	**0**	**0**	**0**	**77**	**0**	**0**	**0**	**3**	**0.0**		**0**	**0.0**	**8**	**0**	**3:34**								

Traded to **Kamloops** (WHL) by **Red Deer** (WHL) for Shawn McNeil, June 25, 1997. Traded to **Vancouver** by **Florida** with Ed Jovanovski, Dave Gagner, Kevin Weekes and Florida's 1st round choice (Nathan Smith) in 2000 Entry Draft for Pavel Bure, Bret Hedican, Brad Ference and Vancouver's 3rd round choice (Robert Fried) in 2000 Entry Draft, January 17, 1999.

BROWN, Rob

(BROWN, RAWB)

Right wing. Shoots left. 5'10", 177 lbs. Born, Kingston, Ont., April 10, 1968. Pittsburgh's 4th choice, 67th overall, in 1986 Entry Draft.

Season	Club	League	Regular Season GP	G	A	Pts	PIM	PP	SH	GW	S	%	+/-	TF	F%	H	SB	Min	Playoffs GP	G	A	Pts	PIM	PP	SH	GW
1982-83	St. Albert	AMHL	61	137	122	259	200																			
1983-84	St. Albert	AJHL	1	0	0	0	0																			
	Kamloops	WHL	50	16	42	58	80												15	1	2	3	17			
1984-85	Kamloops Blazers	WHL	60	29	50	79	95												15	8	8	26	28			
1985-86	Kamloops Blazers	WHL	69	58	*115	*173	171												16	*18	*28	*46	14			
1986-87	Kamloops Blazers	WHL	63	*76	*136	*212	101												5	6	5	11	6			
1987-88	**Pittsburgh**	**NHL**	**51**	**24**	**20**	**44**	**56**	**13**	**0**	**1**	**80**	**30.0**	**8**													
1988-89	**Pittsburgh**	**NHL**	**68**	**49**	**66**	**115**	**118**	**24**	**0**	**6**	**169**	**29.0**	**27**						**11**	**5**	**3**	**8**	**22**	**1**	**0**	**3**
1989-90	**Pittsburgh**	**NHL**	**80**	**33**	**47**	**80**	**102**	**12**	**0**	**3**	**157**	**21.0**	**-10**													
1990-91	**Pittsburgh**	**NHL**	**25**	**6**	**10**	**16**	**31**	**2**	**0**	**0**	**32**	**18.8**	**0**													
	Hartford	**NHL**	**44**	**18**	**24**	**42**	**101**	**10**	**0**	**2**	**94**	**19.1**	**-7**						**5**	**1**	**0**	**1**	**7**	**1**	**0**	**1**
1991-92	**Hartford**	**NHL**	**42**	**16**	**15**	**31**	**39**	**13**	**0**	**2**	**65**	**24.6**	**-14**													
	Chicago	**NHL**	**25**	**5**	**11**	**16**	**34**	**3**	**0**	**1**	**41**	**12.2**	**-1**						**8**	**2**	**4**	**6**	**4**	**1**	**0**	**0**
1992-93	**Chicago**	**NHL**	**15**	**1**	**6**	**7**	**33**	**0**	**0**	**0**	**16**	**6.3**	**6**													
	Indianapolis Ice	IHL	19	14	19	33	32												2	0	1	1	2			
1993-94	**Dallas**	**NHL**	**1**	**0**	**0**	**0**	**0**	**0**	**0**	**0**	**1**	**0.0**	**-1**													
	Kalamazoo Wings	IHL	79	42	*113	*155	188												5	1	3	4	6			
1994-95	Phoenix	IHL	69	34	73	107	135												9	4	12	16	0			
	Los Angeles	**NHL**	**2**	**0**	**0**	**0**	**0**	**0**	**0**	**0**	**1**	**0.0**	**-2**													
1995-96	Chicago Wolves	IHL	79	52	*91	*143	100												9	4	11	15	6			
1996-97	Chicago Wolves	IHL	76	37	*80	*117	98												4	2	4	6	16			
1997-98	**Pittsburgh**	**NHL**	**82**	**15**	**25**	**40**	**59**	**4**	**0**	**4**	**172**	**8.7**	**-1**						**6**	**1**	**0**	**1**	**4**	**1**	**0**	**0**
1998-99	**Pittsburgh**	**NHL**	**58**	**13**	**11**	**24**	**16**	**9**	**0**	**1**	**78**	**16.7**	**-15**	**18**	**38.9**	**115**	**18**	**12:35**	**13**	**2**	**5**	**7**	**8**	**2**	**0**	**0**
99-2000	**Pittsburgh**	**NHL**	**50**	**10**	**13**	**23**	**10**	**4**	**0**	**3**	**73**	**13.7**	**-13**	**14**	**35.7**	**70**	**6**	**10:37**	**11**	**1**	**2**	**3**	**0**	**0**	**0**	**0**
2000-01	Chicago Wolves	IHL	75	24	53	77	99												16	4	13	17	26			
2001-02	Chicago Wolves	AHL	80	29	*54	83	103												25	7	*26	*33	34			
	NHL Totals		**543**	**190**	**248**	**438**	**599**	**94**	**0**	**23**	**979**	**19.4**		**32**	**21.9**	**185**	**24**	**12:35**	**54**	**12**	**14**	**26**	**45**	**6**	**0**	**4**

WHL West First All-Star Team (1986, 1987) • WHL West MVP (1986, 1987) • Canadian Major Junior Player of the Year (1987) • IHL First All-Star Team (1994, 1996, 1997) • Won Leo P. Lamoureux Memorial Trophy (Top Scorer - IHL) (1994, 1996, 1997) • Won James Gatschene Memorial Trophy (MVP - IHL) (1994) • IHL Second All-Star Team (1995) • Played in NHL All-Star Game (1989)

Traded to **Hartford** by **Pittsburgh** for Scott Young, December 21, 1990. Traded to **Chicago** by **Hartford** for Steve Konroyd, January 24, 1992. Signed as a free agent by **Dallas**, August 12, 1993. Signed as a free agent by **LA Kings**, June 14, 1994. Signed as a free agent by **Pittsburgh**, October 1, 1997.

BROWN, Sean — (BROWN, SHAWN) — BOS.

Defense. Shoots left. 6'3", 205 lbs. Born, Oshawa, Ont., November 5, 1976. Boston's 2nd choice, 21st overall, in 1995 Entry Draft.

			Regular Season																Playoffs							
Season	Club	League	GP	G	A	Pts	PIM	PP	SH	GW	S	%	+/-	TF	F%	H	SB	Min	GP	G	A	Pts	PIM	PP	SH	GW
1992-93	Oshawa	OJHL-B	15	0	1	1	9																			
1993-94	Wellington Dukes	MTJHL	32	5	14	19	165																			
	Belleville Bulls	OHL	28	1	2	3	53												8	0	0	0	17			
1994-95	Belleville Bulls	OHL	58	2	16	18	200												16	4	2	6	*67			
1995-96	Belleville Bulls	OHL	37	10	23	33	150																			
	Sarnia Sting	OHL	26	8	17	25	112												10	1	0	1	38			
1996-97	**Edmonton**	**NHL**	**5**	**0**	**0**	**0**	**4**	**0**	**0**	**0**	**2**	**0.0**	**–1**													
	Hamilton	AHL	61	1	7	8	238												19	1	0	1	47			
1997-98	**Edmonton**	**NHL**	**18**	**0**	**1**	**1**	**43**	**0**	**0**	**0**	**9**	**0.0**	**–1**													
	Hamilton	AHL	43	4	6	10	166												6	0	2	2	38			
1998-99	**Edmonton**	**NHL**	**51**	**0**	**7**	**7**	**188**	**0**	**0**	**0**	**27**	**0.0**	**1**	**0**	**0.0**	**104**	**29**	**12:14**	**1**	**0**	**0**	**0**	**10**	**0**	**0**	**0**
99-2000	**Edmonton**	**NHL**	**72**	**4**	**8**	**12**	**192**	**0**	**0**	**2**	**36**	**11.1**	**1**	**0**	**0.0**	**146**	**46**	**12:41**	**3**	**0**	**0**	**0**	**23**	**0**	**0**	**0**
2000-01	**Edmonton**	**NHL**	**62**	**2**	**3**	**5**	**110**	**0**	**0**	**0**	**30**	**6.7**	**2**	**0**	**0.0**	**106**	**32**	**11:07**								
2001-02	**Edmonton**	**NHL**	**61**	**6**	**4**	**10**	**127**	**3**	**0**	**1**	**58**	**10.3**	**8**	**0**	**0.0**	**97**	**37**	**12:36**								
	Boston	**NHL**	**12**	**0**	**1**	**1**	**47**	**0**	**0**	**0**	**6**	**0.0**	**–1**	**0**	**0.0**	**25**	**7**	**16:22**	**4**	**0**	**0**	**0**	**2**	**0**	**0**	**0**
	NHL Totals		**281**	**12**	**24**	**36**	**711**	**3**	**0**	**3**	**168**	**7.1**		**0**	**0.0**	**478**	**151**	**12:22**	**8**	**0**	**0**	**0**	**35**	**0**	**0**	**0**

OHL Second All-Star Team (1996)

Rights traded to **Edmonton** by **Boston** with Mariusz Czerkawski and Boston's 1st round choice (Matthieu Descoteaux) in 1996 Entry Draft for Bill Ranford, January 11, 1996. Traded to **Boston** by **Edmonton** for Bobby Allen, March 19, 2002.

BRULE, Steve — (broo-LAY, STEEV) — COL.

Right wing. Shoots right. 6', 200 lbs. Born, Montreal, Que., January 15, 1975. New Jersey's 6th choice, 143rd overall, in 1993 Entry Draft.

Season	Club	League	GP	G	A	Pts	PIM	PP	SH	GW	S	%	+/-	TF	F%	H	SB	Min	GP	G	A	Pts	PIM	PP	SH	GW
1990-91	L'est Cantonniers	QAHA	32	25	30	55	20																			
1991-92	Mtl-Bourassa	QAAA	40	33	37	70	46												9	9	7	16	10			
1992-93	St-Jean Lynx	QMJHL	70	33	47	80	46												4	0	0	0	9			
1993-94	St-Jean Lynx	QMJHL	66	41	64	105	46												5	2	1	3	0			
1994-95	St-Jean Lynx	QMJHL	69	44	64	108	42												7	3	4	7	8			
	Albany	AHL	3	1	4	5	0												14	9	5	14	4			
1995-96	Albany	AHL	80	30	21	51	37												4	0	0	0	17			
1996-97	Albany	AHL	79	28	48	76	27												16	7	7	14	12			
1997-98	Albany	AHL	80	34	43	77	34												13	8	3	11	4			
1998-99	Albany	AHL	78	32	52	84	35												5	3	1	4	4			
99-2000	Albany	AHL	75	30	46	76	18												5	1	2	3	0			
	♦ **New Jersey**	**NHL**																	**1**	**0**	**0**	**0**	**0**	**0**	**0**	**0**
2000-01	Manitoba Moose	IHL	78	21	48	69	22												13	3	10	13	12			
2001-02	Cincinnati	AHL	77	21	42	63	50												3	0	1	1	0			
	NHL Totals																		**1**	**0**	**0**	**0**	**0**	**0**	**0**	**0**

QMJHL All-Rookie Team (1993) • QMJHL Offensive Rookie of the Year (1993) • QMJHL Second All-Star Team (1995)

Signed as a free agent by **Detroit**, July 20, 2000. Signed as a free agent by **Colorado**, July 22, 2002.

BRUNET, Benoit — (broo-NAY, BEHN-wah)

Left wing. Shoots left. 6', 203 lbs. Born, Ste-Anne-de-Bellevue, Que., August 24, 1968. Montreal's 2nd choice, 27th overall, in 1986 Entry Draft.

Season	Club	League	GP	G	A	Pts	PIM	PP	SH	GW	S	%	+/-	TF	F%	H	SB	Min	GP	G	A	Pts	PIM	PP	SH	GW
1985-86	Hull Olympiques	QMJHL	71	33	37	70	81																			
1986-87	Hull Olympiques	QMJHL	60	43	67	110	105												6	7	5	12	8			
1987-88	Hull Olympiques	QMJHL	62	54	89	143	131												10	3	10	13	11			
1988-89	**Montreal**	**NHL**	**2**	**0**	**1**	**1**	**0**	**0**	**0**	**0**	**1**	**0.0**	**0**													
	Sherbrooke	AHL	73	41	*76	117	95												6	2	0	2	4			
1989-90	Sherbrooke	AHL	72	32	35	67	82												12	8	7	15	20			
1990-91	**Montreal**	**NHL**	**17**	**1**	**3**	**4**	**0**	**0**	**0**	**0**	**12**	**8.3**	**–1**													
	Fredericton	AHL	24	13	18	31	16												6	5	6	11	2			
1991-92	**Montreal**	**NHL**	**18**	**4**	**6**	**10**	**14**	**0**	**0**	**0**	**37**	**10.8**	**4**													
	Fredericton	AHL	6	7	9	16	27																			
1992-93♦	**Montreal**	**NHL**	**47**	**10**	**15**	**25**	**19**	**0**	**0**	**1**	**71**	**14.1**	**13**						**20**	**2**	**8**	**10**	**8**	**1**	**0**	**1**
1993-94	**Montreal**	**NHL**	**71**	**10**	**20**	**30**	**20**	**0**	**3**	**1**	**92**	**10.9**	**14**						**7**	**1**	**4**	**5**	**16**	**0**	**0**	**0**
1994-95	**Montreal**	**NHL**	**45**	**7**	**18**	**25**	**16**	**1**	**1**	**2**	**80**	**8.8**	**7**													
1995-96	**Montreal**	**NHL**	**26**	**7**	**8**	**15**	**17**	**3**	**1**	**4**	**48**	**14.6**	**–4**						**3**	**0**	**2**	**2**	**0**	**0**	**0**	**0**
	Fredericton	AHL	3	2	1	3	6																			
1996-97	**Montreal**	**NHL**	**39**	**10**	**13**	**23**	**14**	**2**	**0**	**2**	**63**	**15.9**	**6**						**4**	**1**	**3**	**4**	**4**	**0**	**1**	**0**
1997-98	**Montreal**	**NHL**	**68**	**12**	**20**	**32**	**61**	**1**	**2**	**2**	**87**	**13.8**	**11**						**8**	**1**	**0**	**1**	**4**	**0**	**0**	**1**
1998-99	**Montreal**	**NHL**	**60**	**14**	**17**	**31**	**31**	**4**	**2**	**0**	**115**	**12.2**	**–1**	**375**	**41.6**	**27**	**37**	**17:47**								
99-2000	**Montreal**	**NHL**	**50**	**14**	**15**	**29**	**13**	**6**	**1**	**2**	**103**	**13.6**	**3**	**293**	**42.7**	**27**	**22**	**17:59**								
2000-01	**Montreal**	**NHL**	**35**	**3**	**11**	**14**	**12**	**0**	**0**	**0**	**61**	**4.9**	**–4**	**24**	**45.8**	**20**	**21**	**16:40**								
2001-02	**Montreal**	**NHL**	**16**	**0**	**2**	**2**	**4**	**0**	**0**	**0**	**19**	**0.0**	**–4**	**3**	**0.0**	**1**	**11**	**13:33**								
	Dallas	**NHL**	**32**	**4**	**9**	**13**	**8**	**0**	**1**	**1**	**33**	**12.1**	**5**	**25**	**36.0**	**16**	**14**	**12:34**								
	Utah Grizzlies	AHL	5	3	1	4	6																			
	Ottawa	**NHL**	**13**	**5**	**3**	**8**	**0**	**1**	**0**	**2**	**20**	**25.0**	**–4**	**36**	**47.2**	**11**	**9**	**17:31**	**12**	**0**	**3**	**3**	**0**	**0**	**0**	**0**
	NHL Totals		**539**	**101**	**161**	**262**	**229**	**18**	**11**	**17**	**842**	**12.0**		**756**	**42.1**	**102**	**114**	**16:29**	**54**	**5**	**20**	**25**	**32**	**1**	**1**	**2**

QMJHL Second All-Star Team (1987) • AHL First All-Star Team (1989)

• Missed majority of 2000-01 season recovering from knee injury suffered in game vs. Pittsburgh, December 16, 2000. Traded to **Dallas** by **Montreal** with Martin Rucinsky for Donald Audette and Shaun Van Allen, November 21, 2001. Traded to **Ottawa** by **Dallas** for a conditional choice in 2003 Entry Draft, March 16, 2002.

BRUNETTE, Andrew — (broo-NEHT, AN-droo) — MIN.

Left wing. Shoots left. 6'1", 210 lbs. Born, Sudbury, Ont., August 24, 1973. Washington's 6th choice, 174th overall, in 1993 Entry Draft.

Season	Club	League	GP	G	A	Pts	PIM	PP	SH	GW	S	%	+/-	TF	F%	H	SB	Min	GP	G	A	Pts	PIM	PP	SH	GW
1989-90	Rayside-Balfour	NOJHA	32	38	*65	*103																				
	Rayside-Balfour	NOJHA	4	1	1	2	0																			
1990-91	Owen Sound	OHL	63	15	20	35	15																			
1991-92	Owen Sound	OHL	66	51	47	98	42												5	5	0	5	8			
1992-93	Owen Sound	OHL	66	*62	*100	*162	91												8	8	6	14	16			
1993-94	Portland Pirates	AHL	23	9	11	20	10												2	0	1	1	0			
	Providence	AHL	3	0	0	0	0																			
	Hampton Roads	ECHL	20	12	18	30	32												7	7	6	13	18			
1994-95	Portland Pirates	AHL	79	30	50	80	53												7	3	3	6	10			
1995-96	**Washington**	**NHL**	**11**	**3**	**3**	**6**	**0**	**0**	**0**	**1**	**16**	**18.8**	**5**						**6**	**1**	**3**	**4**	**0**	**0**	**0**	**0**
	Portland Pirates	AHL	69	28	66	94	125												20	11	18	29	15			
1996-97	**Washington**	**NHL**	**23**	**4**	**7**	**11**	**12**	**2**	**0**	**0**	**23**	**17.4**	**–3**													
	Portland Pirates	AHL	50	22	51	73	48												5	1	2	3	0			
1997-98	**Washington**	**NHL**	**28**	**11**	**12**	**23**	**12**	**4**	**0**	**2**	**42**	**26.2**	**2**													
	Portland Pirates	AHL	43	21	46	67	64												10	1	11	12	12			
1998-99	**Nashville**	**NHL**	**77**	**11**	**20**	**31**	**26**	**7**	**0**	**1**	**65**	**16.9**	**–10**	**8**	**50.0**	**13**	**10**	**13:13**								
99-2000	**Atlanta**	**NHL**	**81**	**23**	**27**	**50**	**30**	**9**	**0**	**2**	**107**	**21.5**	**–32**	**8**	**25.0**	**45**	**17**	**15:42**								
2000-01	**Atlanta**	**NHL**	**77**	**15**	**44**	**59**	**26**	**6**	**0**	**4**	**104**	**14.4**	**–5**	**11**	**54.6**	**38**	**12**	**16:58**								
2001-02	**Minnesota**	**NHL**	**81**	**21**	**48**	**69**	**18**	**10**	**0**	**2**	**106**	**19.8**	**–4**	**111**	**58.6**	**25**	**16**	**16:02**								
	NHL Totals		**378**	**88**	**161**	**249**	**124**	**38**	**0**	**12**	**463**	**19.0**		**138**	**55.8**	**121**	**55**	**15:29**	**6**	**1**	**3**	**4**	**0**	**0**	**0**	**0**

OHL First All-Star Team (1993) • Canadian Major Junior Second All-Star Team (1993) • AHL Second All-Star Team (1995)

Claimed by **Nashville** from **Washington** in Expansion Draft, June 26, 1998. Traded to **Atlanta** by **Nashville** for Atlanta's 5th round choice (Matt Hendricks) in 2000 Entry Draft, June 21, 1999. Signed as a free agent by **Minnesota**, July 17, 2001.

BRYLIN, Sergei

(BRIH-lin, SAIR-gay) **N.J.**

Center. Shoots left. 5'10", 190 lbs. Born, Moscow, USSR, January 13, 1974. New Jersey's 2nd choice, 42nd overall, in 1992 Entry Draft.

Season	Club	League	Regular Season GP	G	A	Pts	PIM	PP	SH	GW	S	%	+/-	TF	F%	H	SB	Min	Playoffs GP	G	A	Pts	PIM	PP	SH	GW
1991-92	CSKA Moscow	CIS	44	1	6	7	4																			
1992-93	CSKA Moscow	CIS	42	5	4	9	36																			
1993-94	CSKA Moscow	CIS	39	4	6	10	36												3	1	0	1	2			
	Russian Penguins	IHL	13	4	5	9	18																			
1994-95	Albany	AHL	63	19	35	54	78																			
	◆ New Jersey	**NHL**	**26**	**6**	**8**	**14**	**8**	**0**	**0**	**0**	**41**	**14.6**	**12**						**12**	**1**	**2**	**3**	**4**	**0**	**0**	**0**
1995-96	**New Jersey**	**NHL**	**50**	**4**	**5**	**9**	**26**	**0**	**0**	**1**	**51**	**7.8**	**−2**													
1996-97	**New Jersey**	**NHL**	**29**	**2**	**2**	**4**	**20**	**0**	**0**	**0**	**34**	**5.9**	**−13**													
	Albany	AHL	43	17	24	41	38												16	4	8	12	12			
1997-98	**New Jersey**	**NHL**	**18**	**2**	**3**	**5**	**0**	**0**	**0**	**0**	**20**	**10.0**	**4**													
	Albany	AHL	44	21	22	43	60																			
1998-99	**New Jersey**	**NHL**	**47**	**5**	**10**	**15**	**28**	**3**	**0**	**1**	**51**	**9.8**	**8**	**184**	**50.5**	**61**	**7**	**12:55**	**5**	**3**	**1**	**4**	**4**	**1**	**0**	**1**
99-2000◆	**New Jersey**	**NHL**	**64**	**9**	**11**	**20**	**20**	**1**	**0**	**1**	**84**	**10.7**	**0**	**72**	**41.7**	**107**	**21**	**13:23**	**17**	**3**	**5**	**8**	**0**	**0**	**0**	**0**
2000-01	**New Jersey**	**NHL**	**75**	**23**	**29**	**52**	**24**	**3**	**1**	**0**	**130**	**17.7**	**25**	**43**	**44.2**	**108**	**16**	**15:31**	**20**	**3**	**4**	**7**	**6**	**1**	**0**	**1**
2001-02	**New Jersey**	**NHL**	**76**	**16**	**28**	**44**	**10**	**5**	**0**	**3**	**133**	**12.0**	**21**	**17**	**47.1**	**93**	**25**	**17:11**	**6**	**0**	**2**	**2**	**2**	**0**	**0**	**0**
	NHL Totals		**385**	**67**	**96**	**163**	**136**	**12**	**1**	**6**	**544**	**12.3**		**316**	**47.5**	**369**	**69**	**15:01**	**60**	**10**	**14**	**24**	**16**	**2**	**0**	**2**

BUCHBERGER, Kelly

(BUK-buhr-guhr, KEHL-lee) **PHX.**

Right wing. Shoots left. 6'2", 210 lbs. Born, Langenburg, Sask., December 2, 1966. Edmonton's 8th choice, 188th overall, in 1985 Entry Draft.

Season	Club	League	Regular Season GP	G	A	Pts	PIM	PP	SH	GW	S	%	+/-	TF	F%	H	SB	Min	Playoffs GP	G	A	Pts	PIM	PP	SH	GW
1983-84	Melville	SJHL	60	14	11	25	139																			
1984-85	Moose Jaw	WHL	51	12	17	29	114																			
1985-86	Moose Jaw	WHL	72	14	22	36	206												13	11	4	15	37			
1986-87	Nova Scotia	AHL	70	12	20	32	257												5	0	1	1	23			
	◆ Edmonton	**NHL**																	**3**	**0**	**1**	**1**	**5**	**0**	**0**	**0**
1987-88	**Edmonton**	**NHL**	**19**	**1**	**0**	**1**	**81**	**0**	**0**	**0**	**10**	**10.0**	**−1**													
	Nova Scotia	AHL	49	21	23	44	206												2	0	0	0	11			
1988-89	**Edmonton**	**NHL**	**66**	**5**	**9**	**14**	**234**	**1**	**0**	**1**	**57**	**8.8**	**−14**													
1989-90◆	**Edmonton**	**NHL**	**55**	**2**	**6**	**8**	**168**	**0**	**0**	**2**	**35**	**5.7**	**−8**						**19**	**0**	**5**	**5**	**13**	**0**	**0**	**0**
1990-91	**Edmonton**	**NHL**	**64**	**3**	**1**	**4**	**160**	**0**	**0**	**2**	**54**	**5.6**	**−6**						**12**	**2**	**1**	**3**	**25**	**0**	**0**	**0**
1991-92	**Edmonton**	**NHL**	**79**	**20**	**24**	**44**	**157**	**0**	**4**	**3**	**90**	**22.2**	**9**						**16**	**1**	**4**	**5**	**32**	**0**	**0**	**0**
1992-93	**Edmonton**	**NHL**	**83**	**12**	**18**	**30**	**133**	**1**	**2**	**3**	**92**	**13.0**	**−27**													
1993-94	**Edmonton**	**NHL**	**84**	**3**	**18**	**21**	**199**	**0**	**0**	**0**	**93**	**3.2**	**−20**													
1994-95	**Edmonton**	**NHL**	**48**	**7**	**17**	**24**	**82**	**2**	**1**	**5**	**73**	**9.6**	**0**													
1995-96	**Edmonton**	**NHL**	**82**	**11**	**14**	**25**	**184**	**0**	**2**	**3**	**119**	**9.2**	**−20**													
1996-97	**Edmonton**	**NHL**	**81**	**8**	**30**	**38**	**159**	**0**	**0**	**3**	**78**	**10.3**	**4**						**12**	**5**	**2**	**7**	**16**	**0**	**0**	**1**
1997-98	**Edmonton**	**NHL**	**82**	**6**	**17**	**23**	**122**	**1**	**1**	**1**	**86**	**7.0**	**−10**						**12**	**1**	**2**	**3**	**25**	**0**	**0**	**0**
1998-99	**Edmonton**	**NHL**	**52**	**4**	**4**	**8**	**68**	**0**	**2**	**1**	**29**	**13.8**	**−6**	**23**	**26.1**	**41**	**27**	**11:49**	**4**	**0**	**0**	**0**	**0**	**0**	**0**	**0**
99-2000	**Atlanta**	**NHL**	**68**	**5**	**12**	**17**	**139**	**0**	**0**	**0**	**56**	**8.9**	**−34**	**577**	**45.6**	**137**	**47**	**16:21**								
	Los Angeles	**NHL**	**13**	**2**	**1**	**3**	**13**	**0**	**0**	**0**	**20**	**10.0**	**−2**	**6**	**16.7**	**42**	**5**	**14:43**	**4**	**0**	**0**	**0**	**4**	**0**	**0**	**0**
2000-01	**Los Angeles**	**NHL**	**82**	**6**	**14**	**20**	**75**	**0**	**0**	**1**	**66**	**9.1**	**−10**	**155**	**40.7**	**175**	**52**	**14:26**	**8**	**1**	**0**	**1**	**2**	**0**	**0**	**0**
2001-02	**Los Angeles**	**NHL**	**74**	**6**	**7**	**13**	**105**	**0**	**0**	**1**	**39**	**15.4**	**−13**	**126**	**41.3**	**90**	**31**	**10:21**	**7**	**0**	**0**	**0**	**7**	**0**	**0**	**0**
	NHL Totals		**1032**	**101**	**192**	**293**	**2079**	**5**	**12**	**26**	**997**	**10.1**		**887**	**43.4**	**485**	**162**	**13:23**	**97**	**10**	**15**	**25**	**129**	**0**	**0**	**1**

Claimed by **Atlanta** from **Edmonton** in Expansion Draft, June 25, 1999. Traded to **LA Kings** by **Atlanta** with Nelson Emerson for Donald Audette and Frantisek Kaberle, March 13, 2000. Signed as a free agent by **Phoenix**, July 7, 2002.

BULIS, Jan

(BOO-lihs, YAHN) **MTL.**

Center. Shoots left. 6'2", 201 lbs. Born, Pardubice, Czech., March 18, 1978. Washington's 3rd choice, 43rd overall, in 1996 Entry Draft.

Season	Club	League	Regular Season GP	G	A	Pts	PIM	PP	SH	GW	S	%	+/-	TF	F%	H	SB	Min	Playoffs GP	G	A	Pts	PIM	PP	SH	GW
1993-94	HC Pardubice Jr.	Czech-Jr.	25	16	11	27																				
1994-95	Kelowna Spartans	BCJHL	51	23	25	48	36												17	7	9	16	0			
1995-96	Barrie Colts	OHL	59	29	30	59	22												7	2	3	5	2			
1996-97	Barrie Colts	OHL	64	42	61	103	42												9	3	7	10	10			
1997-98	Kingston	OHL	2	0	1	1	0												12	8	10	18	12			
	Washington	**NHL**	**48**	**5**	**11**	**16**	**18**	**0**	**0**	**0**	**37**	**13.5**	**−5**													
	Portland Pirates	AHL	3	1	4	5	12																			
1998-99	**Washington**	**NHL**	**38**	**7**	**16**	**23**	**6**	**3**	**0**	**3**	**57**	**12.3**	**3**	**599**	**48.9**	**48**	**3**	**14:27**								
	Cincinnati	IHL	10	2	2	4	14																			
99-2000	**Washington**	**NHL**	**56**	**9**	**22**	**31**	**30**	**0**	**0**	**1**	**92**	**9.8**	**7**	**609**	**45.5**	**66**	**14**	**13:55**								
2000-01	**Washington**	**NHL**	**39**	**5**	**13**	**18**	**26**	**1**	**0**	**0**	**41**	**12.2**	**0**	**224**	**46.9**	**34**	**11**	**11:53**								
	Portland Pirates	AHL	4	0	2	2	0																			
	Montreal	**NHL**	**12**	**0**	**5**	**5**	**0**	**0**	**0**	**0**	**20**	**0.0**	**−1**	**230**	**48.3**	**8**	**6**	**18:25**								
2001-02	**Montreal**	**NHL**	**53**	**9**	**10**	**19**	**8**	**1**	**0**	**3**	**87**	**10.3**	**−2**	**156**	**43.0**	**47**	**12**	**13:34**	**6**	**0**	**0**	**0**	**6**	**0**	**0**	**0**
	NHL Totals		**246**	**35**	**77**	**112**	**88**	**5**	**0**	**7**	**334**	**10.5**		**1818**	**46.9**	**203**	**46**	**13:48**	**6**	**0**	**0**	**0**	**6**	**0**	**0**	**0**

Traded to **Montreal** by **Washington** with Richard Zednik and Washington's 1st round choice (Alexander Perezhogin) in 2001 Entry Draft for Trevor Linden, Dainius Zubrus and New Jersey's 2nd round choice (previously acquired by Montreal - later traded to Tampa Bay - Tampa Bay selected Andreas Holmqvist) in 2001 Entry Draft, March 13, 2001.

BURE, Pavel

(boo-RAY, PAH-vehl) **NYR**

Right wing. Shoots left. 5'10", 189 lbs. Born, Moscow, USSR, March 31, 1971. Vancouver's 4th choice, 113th overall, in 1989 Entry Draft.

Season	Club	League	Regular Season GP	G	A	Pts	PIM	PP	SH	GW	S	%	+/-	TF	F%	H	SB	Min	Playoffs GP	G	A	Pts	PIM	PP	SH	GW
1987-88	CSKA Moscow	USSR	5	1	1	2	0																			
1988-89	CSKA Moscow	USSR	32	17	9	26	8																			
1989-90	CSKA Moscow	USSR	46	14	10	24	20																			
1990-91	CSKA Moscow	USSR	44	35	11	46	24																			
1991-92	**Vancouver**	**NHL**	**65**	**34**	**26**	**60**	**30**	**7**	**3**	**6**	**268**	**12.7**	**0**						**13**	**6**	**4**	**10**	**14**	**0**	**0**	**0**
1992-93	**Vancouver**	**NHL**	**83**	**60**	**50**	**110**	**69**	**13**	**7**	**9**	**407**	**14.7**	**35**						**12**	**5**	**7**	**12**	**8**	**0**	**0**	**1**
1993-94	**Vancouver**	**NHL**	**76**	***60**	**47**	**107**	**86**	**25**	**4**	**9**	**374**	**16.0**	**1**						**24**	***16**	**15**	**31**	**40**	**3**	**0**	**2**
1994-95	EV Landshut	Germany	1	3	0	3	2																			
	Spartak Moscow	CIS	1	2	0	2	2																			
	Vancouver	**NHL**	**44**	**20**	**23**	**43**	**47**	**6**	**2**	**2**	**198**	**10.1**	**−8**						**11**	**7**	**6**	**13**	**10**	**2**	**2**	**0**
1995-96	**Vancouver**	**NHL**	**15**	**6**	**7**	**13**	**8**	**1**	**1**	**0**	**78**	**7.7**	**−2**													
1996-97	**Vancouver**	**NHL**	**63**	**23**	**32**	**55**	**40**	**4**	**1**	**2**	**265**	**8.7**	**−14**													
1997-98	**Vancouver**	**NHL**	**82**	**51**	**39**	**90**	**48**	**13**	**6**	**4**	**329**	**15.5**	**5**													
	Russia	Olympics	6	*9	0	9	2																			
1998-99	**Florida**	**NHL**	**11**	**13**	**3**	**16**	**4**	**5**	**1**	**0**	**44**	**29.5**	**3**	**1**	**0.0**	**2**	**5**	**21:41**								
99-2000	**Florida**	**NHL**	**74**	***58**	**36**	**94**	**16**	**11**	**2**	**14**	**360**	**16.1**	**25**	**1**	**0.0**	**25**	**12**	**24:23**	**4**	**1**	**3**	**4**	**2**	**1**	**0**	**0**
2000-01	**Florida**	**NHL**	**82**	***59**	**33**	**92**	**58**	**19**	**5**	**8**	**384**	**15.4**	**−2**	**5**	**20.0**	**33**	**22**	**26:52**								
2001-02	**Florida**	**NHL**	**56**	**22**	**27**	**49**	**56**	**9**	**1**	**1**	**238**	**9.2**	**−14**	**7**	**28.6**	**14**	**22**	**25:18**								
	Russia	Olympics	6	2	1	3	8																			
	NY Rangers	**NHL**	**12**	**12**	**8**	**20**	**6**	**3**	**0**	**1**	**49**	**24.5**	**9**	**1**	**0.0**	**2**	**10**	**23:43**								
	NHL Totals		**663**	**418**	**331**	**749**	**468**	**116**	**33**	**56**	**2994**	**14.0**		**15**	**20.0**	**76**	**71**	**25:18**	**64**	**35**	**35**	**70**	**74**	**6**	**2**	**3**

Won Calder Memorial Trophy (1992) • NHL First All-Star Team (1994) • Best Forward at Olympic Games (1998) • NHL Second All-Star Team (2000, 2001) • Won Maurice "Rocket" Richard Trophy (2000, 2001) • Played in NHL All-Star Game (1993, 1994, 1997, 1998, 2000, 2001)

Traded to **Florida** by **Vancouver** with Bret Hedican, Brad Ference and Vancouver's 3rd round choice (Robert Fried) in 2000 Entry Draft for Ed Jovanovski, Dave Gagner, Mike Brown, Kevin Weekes and Florida's 1st round choice (Nathan Smith) in 2000 Entry Draft, January 17, 1999. • Missed majority of 1998-99 season after demanding trade (August 10, 1998) and recovering from knee injury suffered in game vs. Pittsburgh, February 5, 1999. Traded to **NY Rangers** by **Florida** with Florida's 2nd round choice (Lee Falardeau) in 2002 Entry Draft for Igor Ulanov, Filip Novak, NY Rangers' 1st (later traded to Calgary - Calgary selected Eric Nystrom) and 2nd (Rob Globke) round choices in 2002 Entry Draft and NY Rangers' 4th round choice in 2003 Entry Draft, March 18, 2002.

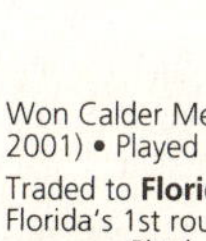

BURE, Valeri

(boo-RAY, VAL-uhr-ee) **FLA.**

Right wing. Shoots right. 5'10", 185 lbs. Born, Moscow, USSR, June 13, 1974. Montreal's 2nd choice, 33rd overall, in 1992 Entry Draft.

Season	Club	League	Regular Season GP	G	A	Pts	PIM	PP	SH	GW	S	%	+/-	TF	F%	H	SB	Min	Playoffs GP	G	A	Pts	PIM	PP	SH	GW
1990-91	CSKA Moscow	USSR	3	0	0	0	0																			
1991-92	Spokane Chiefs	WHL	53	27	22	49	78												10	11	6	17	10			
1992-93	Spokane Chiefs	WHL	66	68	79	147	49												9	6	11	17	14			
1993-94	Spokane Chiefs	WHL	59	40	62	102	48												3	5	3	8	2			
1994-95	Fredericton	AHL	45	23	25	48	32																			
	Montreal	**NHL**	**24**	**3**	**1**	**4**	**6**	**0**	**0**	**1**	**39**	**7.7**	**−1**													

Season	Club	League	GP	G	A	Pts	PIM	PP	SH	GW	S	%	+/-	TF	F%	H	SB	Min	GP	G	A	Pts	PIM	PP	SH	GW
			Regular Season																Playoffs							
1995-96	**Montreal**	**NHL**	**77**	**22**	**20**	**42**	**28**	**5**	**0**	**1**	**143**	**15.4**	**10**						**6**	**0**	**1**	**1**	**6**	**0**	**0**	**0**
1996-97	**Montreal**	**NHL**	**64**	**14**	**21**	**35**	**6**	**4**	**0**	**2**	**131**	**10.7**	**4**						**5**	**0**	**1**	**1**	**2**	**0**	**0**	**0**
1997-98	**Montreal**	**NHL**	**50**	**7**	**22**	**29**	**33**	**2**	**0**	**1**	**134**	**5.2**	**–5**													
	Calgary	**NHL**	**16**	**5**	**4**	**9**	**2**	**0**	**0**	**1**	**45**	**11.1**	**0**													
	Russia	Olympics	6	1	0	1	0																			
1998-99	**Calgary**	**NHL**	**80**	**26**	**27**	**53**	**22**	**7**	**0**	**4**	**260**	**10.0**	**0**	**15**	**40.0**	**25**	**8**	**16:11**								
99-2000	**Calgary**	**NHL**	**82**	**35**	**40**	**75**	**50**	**13**	**0**	**6**	**308**	**11.4**	**–7**	**8**	**25.0**	**28**	**23**	**20:58**								
2000-01	**Calgary**	**NHL**	**78**	**27**	**28**	**55**	**26**	**16**	**0**	**2**	**276**	**9.8**	**–21**	**9**	**11.1**	**21**	**16**	**19:01**								
2001-02	**Florida**	**NHL**	**31**	**8**	**10**	**18**	**12**	**2**	**0**	**1**	**100**	**8.0**	**–3**	**30**	**33.3**	**4**	**10**	**18:36**								
	Russia	Olympics	6	1	0	1	2																			
	NHL Totals		**502**	**147**	**173**	**320**	**185**	**49**	**0**	**19**	**1436**	**10.2**		**62**	**30.6**	**78**	**57**	**18:43**	**11**	**0**	**2**	**2**	**8**	**0**	**0**	**0**

WHL West First All-Star Team (1993) • WHL West Second All-Star Team (1994) • Played in NHL All-Star Game (2000)

Traded to **Calgary** by **Montreal** with Montreal's 4th round choice (Shaun Sutter) in 1998 Entry Draft for Jonas Hoglund and Zarley Zalapski, February 1, 1998. Traded to **Florida** by **Calgary** with Jason Wiemer for Rob Niedermayer and Philadelphia's 2nd round choice (previously acquired, Calgary selected Andrei Medvedev) in 2001 Entry Draft, June 24, 2001. • Missed majority of 2001-02 season recovering from knee injury suffered in game vs. Vancouver, October 16, 2001.

BUTENSCHON, Sven

(BUH-tehn-shohn, SVEHN) **FLA.**

Defense. Shoots left. 6'4", 215 lbs. Born, Itzehoe, West Germany, March 22, 1976. Pittsburgh's 3rd choice, 57th overall, in 1994 Entry Draft.

Season	Club	League	GP	G	A	Pts	PIM	PP	SH	GW	S	%	+/-	TF	F%	H	SB	Min	GP	G	A	Pts	PIM	PP	SH	GW
1991-92	Eastman Selects	MMHL	36	2	10	12	110																			
1992-93	Eastman Selects	MMHL	35	14	22	36	101																			
1993-94	Brandon	WHL	70	3	19	22	51												4	0	0	0	6			
1994-95	Brandon	WHL	21	1	5	6	44												18	1	2	3	11			
1995-96	Brandon	WHL	70	4	37	41	99												19	1	12	13	18			
1996-97	Cleveland	IHL	75	3	12	15	68												10	0	1	1	4			
1997-98	**Pittsburgh**	**NHL**	**8**	**0**	**0**	**0**	**6**	**0**	**0**	**0**	**4**	**0.0**	**–1**													
	Syracuse Crunch	AHL	65	14	23	37	66												5	1	2	3	0			
1998-99	**Pittsburgh**	**NHL**	**17**	**0**	**0**	**0**	**6**	**0**	**0**	**0**	**8**	**0.0**	**–7**	**0**	**0.0**	**13**	**12**	**13:08**								
	Houston Aeros	IHL	57	1	4	5	81																			
99-2000	**Pittsburgh**	**NHL**	**3**	**0**	**0**	**0**	**0**	**0**	**0**	**0**	**2**	**0.0**	**3**	**0**	**0.0**	**1**	**2**	**16:25**								
	Wilkes-Barre	AHL	75	19	21	40	101																			
2000-01	**Pittsburgh**	**NHL**	**5**	**0**	**1**	**1**	**2**	**0**	**0**	**0**	**6**	**0.0**	**1**	**0**	**0.0**	**5**	**6**	**17:51**								
	Wilkes-Barre	AHL	55	7	28	35	85																			
	Edmonton	**NHL**	**7**	**1**	**1**	**2**	**2**	**0**	**0**	**0**	**3**	**33.3**	**2**	**0**	**0.0**	**5**	**3**	**11:07**								
2001-02	**Edmonton**	**NHL**	**14**	**0**	**0**	**0**	**4**	**0**	**0**	**0**	**8**	**0.0**	**0**	**0**	**0.0**	**6**	**8**	**9:39**								
	Hamilton	AHL	61	9	35	44	88																			
	NHL Totals		**54**	**1**	**2**	**3**	**20**	**0**	**0**	**0**	**31**	**3.2**		**0**	**0.0**	**30**	**31**	**12:30**								

Traded to **Edmonton** by **Pittsburgh** for Dan LaCouture, March 13, 2001. Signed as a free agent by **Florida**, July 9, 2002.

BUTSAYEV, Yuri

(buht-SIGH-ehv, YOO-ree) **ATL.**

Center. Shoots left. 6', 195 lbs. Born, Togliatti, USSR, October 11, 1978. Detroit's 1st choice, 49th overall, in 1997 Entry Draft.

Season	Club	League	GP	G	A	Pts	PIM	PP	SH	GW	S	%	+/-	TF	F%	H	SB	Min	GP	G	A	Pts	PIM	PP	SH	GW
1995-96	Lada Togliatti 2	CIS-2	35	19	7	26																				
	Lada Togliatti	CIS	1	0	0	0	0																			
1996-97	Lada Togliatti	Russia	42	13	11	24	38												11	2	2	4	8			
1997-98	Lada Togliatti	Russia	44	8	9	17	63																			
	Lada Togliatti	EuroHL	6	2	0	2	8																			
1998-99	Dynamo Moscow	Russia	1	0	1	1	0																			
	Lada Togliatti	Russia	39	10	7	17	55												7	1	2	3	14			
99-2000	**Detroit**	**NHL**	**57**	**5**	**3**	**8**	**12**	**0**	**0**	**0**	**46**	**10.9**	**–6**	**22**	**40.9**	**29**	**11**	**9:36**								
	Cincinnati	AHL	9	0	1	1	0																			
2000-01	**Detroit**	**NHL**	**15**	**1**	**1**	**2**	**4**	**0**	**0**	**0**	**18**	**5.6**	**–2**	**6**	**33.3**	**4**	**1**	**9:09**								
	Cincinnati	AHL	54	29	17	46	26												4	0	2	2	2			
2001-02	**Detroit**	**NHL**	**3**	**0**	**0**	**0**	**0**	**0**	**0**	**0**	**4**	**0.0**	**–1**	**0**	**0.0**	**0**	**1**	**9:44**								
	Cincinnati	AHL	61	21	23	44	44																			
	Atlanta	**NHL**	**8**	**2**	**0**	**2**	**4**	**0**	**0**	**0**	**6**	**33.3**	**1**	**1**	**100.0**	**3**	**4**	**13:28**								
	Chicago Wolves	AHL	4	1	1	2	0												22	7	4	11	20			
	NHL Totals		**83**	**8**	**4**	**12**	**20**	**0**	**0**	**0**	**74**	**10.8**		**29**	**41.4**	**36**	**17**	**9:54**								

Traded to **Atlanta** by **Detroit** with Detroit's 3rd round choice (later traded to Columbus - Columbus selected Jeff Genovy) in 2002 Entry Draft for Jiri Slegr, March 19, 2002.

BUZEK, Petr

(BOO-zehk, PEE-tuhr)

Defense. Shoots left. 6'1", 220 lbs. Born, Jihlava, Czech., April 26, 1977. Dallas' 3rd choice, 63rd overall, in 1995 Entry Draft.

Season	Club	League	GP	G	A	Pts	PIM	PP	SH	GW	S	%	+/-	TF	F%	H	SB	Min	GP	G	A	Pts	PIM	PP	SH	GW
1993-94	Dukla Jihlava Jr.	Czech-Jr.	3	0	0	0																				
1994-95	HC Dukla Jihlava	Czech	43	2	5	7	47												2	0	0	0	2			
1995-96	Michigan K-Wings	IHL	DID NOT PLAY – INJURED																							
1996-97	Michigan K-Wings	IHL	67	4	6	10	48																			
1997-98	**Dallas**	**NHL**	**2**	**0**	**0**	**0**	**2**	**0**	**0**	**0**	**0**	**0.0**	**1**													
	Michigan K-Wings	IHL	60	10	15	25	58												2	0	1	1	17			
1998-99	**Dallas**	**NHL**	**2**	**0**	**0**	**0**	**2**	**0**	**0**	**0**	**0**	**0.0**	**0**	**0**	**0.0**	**3**	**0**	**13:50**								
	Michigan K-Wings	IHL	74	5	14	19	68												5	0	0	0	10			
99-2000	**Atlanta**	**NHL**	**63**	**5**	**14**	**19**	**41**	**3**	**0**	**0**	**90**	**5.6**	**–22**	**0**	**0.0**	**139**	**76**	**18:24**								
2000-01	**Atlanta**	**NHL**	**5**	**0**	**0**	**0**	**8**	**0**	**0**	**0**	**11**	**0.0**	**2**	**0**	**0.0**	**14**	**2**	**17:26**								
2001-02	**Atlanta**	**NHL**	**9**	**0**	**0**	**0**	**13**	**0**	**0**	**0**	**2**	**0.0**	**–4**	**0**	**0.0**	**18**	**11**	**15:58**								
	Chicago Wolves	AHL	4	0	1	1	2																			
	Calgary	**NHL**	**32**	**1**	**3**	**4**	**14**	**0**	**0**	**0**	**34**	**2.9**	**4**	**0**	**0.0**	**40**	**37**	**17:12**								
	NHL Totals		**113**	**6**	**17**	**23**	**80**	**3**	**0**	**0**	**137**	**4.4**		**0**	**0.0**	**214**	**126**	**17:44**								

Played in NHL All-Star Game (2000)

• Missed entire 1995-96 season recovering from injuries suffered in automobile accident, July, 1995. Claimed by **Atlanta** from **Dallas** in Expansion Draft, June 25, 1999. • Missed majority of 2000-01 season recovering from neck injury suffered in game vs. Anaheim, October 17, 2000. Traded to **Calgary** by **Atlanta** for Jeff Cowan and the rights to Kurtis Foster, December 18, 2001.

BYLSMA, Dan

(BEEL-smah, DAN) **ANA.**

Right wing. Shoots left. 6'2", 212 lbs. Born, Grand Haven, MI, September 19, 1970. Winnipeg's 7th choice, 109th overall, in 1989 Entry Draft.

Season	Club	League	GP	G	A	Pts	PIM	PP	SH	GW	S	%	+/-	TF	F%	H	SB	Min	GP	G	A	Pts	PIM	PP	SH	GW
1986-87	Oakville Blades	OJHL-B	10	4	9	13	21																			
	St. Mary's	OJHL-B	27	14	28	42	21																			
1987-88	St. Mary's	OJHL-B	40	30	39	69	33												8	8	18	26				
1988-89	Bowling Green	CCHA	32	3	7	10	10																			
1989-90	Bowling Green	CCHA	44	13	17	30	30																			
1990-91	Bowling Green	CCHA	40	9	12	21	48																			
1991-92	Bowling Green	CCHA	34	11	14	25	24																			
1992-93	Greensboro	ECHL	60	25	35	60	66												1	0	1	1	10			
	Rochester	AHL	2	0	1	1	0																			
1993-94	Greensboro	ECHL	25	14	16	30	52																			
	Albany	AHL	3	0	1	1	2																			
	Moncton Hawks	AHL	50	12	16	28	25												21	3	4	7	31			
1994-95	Phoenix	IHL	81	19	23	42	41												9	4	4	8	4			
1995-96	**Los Angeles**	**NHL**	**4**	**0**	**0**	**0**	**0**	**0**	**0**	**0**	**6**	**0.0**	**0**													
	Phoenix	IHL	78	22	20	42	48												4	1	0	1	2			
1996-97	**Los Angeles**	**NHL**	**79**	**3**	**6**	**9**	**32**	**0**	**0**	**0**	**86**	**3.5**	**–15**													
1997-98	**Los Angeles**	**NHL**	**65**	**3**	**9**	**12**	**33**	**0**	**0**	**0**	**57**	**5.3**	**9**						**2**	**0**	**0**	**0**	**0**	**0**	**0**	**0**
	Long Beach	IHL	8	2	3	5	0																			
1998-99	**Los Angeles**	**NHL**	**8**	**0**	**0**	**0**	**2**	**0**	**0**	**0**	**3**	**0.0**	**–1**	**0**	**0.0**	**15**	**4**	**9:51**								
	Springfield	AHL	2	0	2	2	2																			
	Long Beach	IHL	58	10	8	18	53												4	0	0	0	8			
99-2000	**Los Angeles**	**NHL**	**64**	**3**	**6**	**9**	**55**	**0**	**1**	**0**	**43**	**7.0**	**–2**	**62**	**43.6**	**168**	**33**	**10:23**	**3**	**0**	**0**	**0**	**0**	**0**	**0**	**0**
	Long Beach	IHL	6	0	3	3	2																			
	Lowell	AHL	2	1	1	2	2																			

			Regular Season																Playoffs							
Season	Club	League	GP	G	A	Pts	PIM	PP	SH	GW	S	%	+/-	TF	F%	H	SB	Min	GP	G	A	Pts	PIM	PP	SH	GW
2000-01	**Anaheim**	**NHL**	**82**	**1**	**9**	**10**	**22**	**0**	**0**	**0**	**50**	**2.0**	**–12**	**10**	**50.0**	**109**	**88**	**11:45**	....	....	....	....	....	....	....	....
2001-02	**Anaheim**	**NHL**	**77**	**8**	**9**	**17**	**28**	**0**	**1**	**2**	**72**	**11.1**	**5**	**257**	**42.4**	**88**	**76**	**11:36**	....	....	....	....	....	....	....	....
	NHL Totals		**379**	**18**	**39**	**57**	**172**	**0**	**2**	**2**	**317**	**5.7**		**329**	**42.9**	**380**	**201**	**11:15**	**5**	**0**	**0**	**0**	**0**	**0**	**0**	**0**

CCHA All-Academic Team (1991, 1992)

Signed as a free agent by **LA Kings**, July 7, 1994. Signed as a free agent by **Anaheim**, July 13, 2000.

CAIRNS, Eric
(KAIRNZ, AIR-ihk) **NYI**

Defense. Shoots left. 6'6", 230 lbs. Born, Oakville, Ont., June 27, 1974. NY Rangers' 3rd choice, 72nd overall, in 1992 Entry Draft.

Season	Club	League	GP	G	A	Pts	PIM	PP	SH	GW	S	%	+/-	TF	F%	H	SB	Min	GP	G	A	Pts	PIM	PP	SH	GW
1990-91	Burlington	OJHL-B	37	5	16	21	120	....	....	....	....	....	....	....	....	....	....	....	....	....	....	....	....	....	....	....
1991-92	Detroit	OHL	64	1	11	12	237	....	....	....	....	....	....	....	....	....	....	....	7	0	0	0	31	....	....	....
1992-93	Detroit	OHL	64	3	13	16	194	....	....	....	....	....	....	....	....	....	....	....	15	0	3	3	24	....	....	....
1993-94	Detroit	OHL	59	7	35	42	204	....	....	....	....	....	....	....	....	....	....	....	17	0	4	4	46	....	....	....
1994-95	Birmingham Bulls	ECHL	11	1	3	4	49	....	....	....	....	....	....	....	....	....	....	....	....	....	....	....	....	....	....	....
	Binghamton	AHL	27	0	3	3	134	....	....	....	....	....	....	....	....	....	....	....	9	1	1	2	28	....	....	....
1995-96	Binghamton	AHL	46	1	13	14	192	....	....	....	....	....	....	....	....	....	....	....	4	0	0	0	37	....	....	....
	Charlotte	ECHL	6	0	1	1	34	....	....	....	....	....	....	....	....	....	....	....	....	....	....	....	....	....	....	....
1996-97	**NY Rangers**	**NHL**	**40**	**0**	**1**	**1**	**147**	**0**	**0**	**0**	**17**	**0.0**	**–7**	....	....	....	....	....	**3**	**0**	**0**	**0**	**0**	**0**	**0**	**0**
	Binghamton	AHL	10	1	1	2	96	....	....	....	....	....	....	....	....	....	....	....	....	....	....	....	....	....	....	....
1997-98	**NY Rangers**	**NHL**	**39**	**0**	**3**	**3**	**92**	**0**	**0**	**0**	**17**	**0.0**	**–3**	....	....	....	....	....	....	....	....	....	....	....	....	....
	Hartford	AHL	7	1	2	3	43	....	....	....	....	....	....	....	....	....	....	....	....	....	....	....	....	....	....	....
1998-99	Hartford	AHL	11	0	2	2	49	....	....	....	....	....	....	....	....	....	....	....	....	....	....	....	....	....	....	....
	NY Islanders	**NHL**	**9**	**0**	**3**	**3**	**23**	**0**	**0**	**0**	**2**	**0.0**	**1**	**0**	**0.0**	**13**	**5**	**10:15**	....	....	....	....	....	....	....	....
	Lowell	AHL	24	0	0	0	91	....	....	....	....	....	....	....	....	....	....	....	3	1	0	1	32	....	....	....
99-2000	**NY Islanders**	**NHL**	**67**	**2**	**7**	**9**	**196**	**0**	**0**	**0**	**55**	**3.6**	**–5**	**0**	**0.0**	**182**	**73**	**17:43**	....	....	....	....	....	....	....	....
	Providence	AHL	4	1	1	2	14	....	....	....	....	....	....	....	....	....	....	....	....	....	....	....	....	....	....	....
2000-01	**NY Islanders**	**NHL**	**45**	**2**	**2**	**4**	**106**	**0**	**0**	**0**	**21**	**9.5**	**–18**	**1**	**0.0**	**71**	**58**	**16:24**	....	....	....	....	....	....	....	....
2001-02	**NY Islanders**	**NHL**	**74**	**2**	**5**	**7**	**176**	**0**	**0**	**1**	**34**	**5.9**	**–2**	**0**	**0.0**	**124**	**29**	**11:09**	**7**	**0**	**0**	**0**	**15**	**0**	**0**	**0**
	NHL Totals		**274**	**6**	**21**	**27**	**740**	**0**	**0**	**1**	**146**	**4.1**		**1**	**0.0**	**390**	**165**	**14:34**	**10**	**0**	**0**	**0**	**15**	**0**	**0**	**0**

Claimed on waivers by **NY Islanders** from **NY Rangers**, December 22, 1998. Loaned to **Providence** (AHL) by **NY Islanders**, October 6, 1999 and recalled October 13, 1999. • Missed most of 2000-01 season recovering from hand injury suffered in game vs. Tampa Bay, October 6, 2000.

CALDER, Kyle
(KAWL-dehr, KIGHL) **CHI.**

Center. Shoots left. 5'11", 180 lbs. Born, Mannville, Alta., January 5, 1979. Chicago's 7th choice, 130th overall, in 1997 Entry Draft.

Season	Club	League	GP	G	A	Pts	PIM	PP	SH	GW	S	%	+/-	TF	F%	H	SB	Min	GP	G	A	Pts	PIM	PP	SH	GW
1994-95	Leduc Oil Barons	AMHL	27	25	32	57	22	....	....	....	....	....	....	....	....	....	....	....	....	....	....	....	....	....	....	....
1995-96	Regina Pats	WHL	27	1	7	8	10	....	....	....	....	....	....	....	....	....	....	....	11	0	0	0	0	....	....	....
1996-97	Regina Pats	WHL	62	25	34	59	17	....	....	....	....	....	....	....	....	....	....	....	5	3	0	3	6	....	....	....
1997-98	Regina Pats	WHL	62	27	50	77	58	....	....	....	....	....	....	....	....	....	....	....	2	0	1	1	0	....	....	....
1998-99	Regina Pats	WHL	34	23	28	51	29	....	....	....	....	....	....	....	....	....	....	....	....	....	....	....	....	....	....	....
	Kamloops Blazers	WHL	27	19	18	37	30	....	....	....	....	....	....	....	....	....	....	....	15	6	10	16	6	....	....	....
99-2000	**Chicago**	**NHL**	**8**	**1**	**1**	**2**	**2**	**0**	**0**	**0**	**5**	**20.0**	**–3**	**2**	**0.0**	**5**	**1**	**9:59**	....	....	....	....	....	....	....	....
	Cleveland	IHL	74	14	22	36	43	....	....	....	....	....	....	....	....	....	....	....	9	2	2	4	14	....	....	....
2000-01	**Chicago**	**NHL**	**43**	**5**	**10**	**15**	**14**	**0**	**0**	**1**	**63**	**7.9**	**–4**	**2**	**0.0**	**31**	**8**	**12:43**	....	....	....	....	....	....	....	....
	Norfolk Admirals	AHL	37	12	15	27	21	....	....	....	....	....	....	....	....	....	....	....	9	2	6	8	2	....	....	....
2001-02	**Chicago**	**NHL**	**81**	**17**	**36**	**53**	**47**	**6**	**0**	**3**	**133**	**12.8**	**8**	**0**	**0.0**	**79**	**21**	**16:33**	**5**	**2**	**0**	**2**	**2**	**1**	**0**	**0**
	NHL Totals		**132**	**23**	**47**	**70**	**63**	**6**	**0**	**4**	**201**	**11.4**		**4**	**0.0**	**115**	**30**	**14:54**	**5**	**2**	**0**	**2**	**2**	**1**	**0**	**0**

Traded to **Kamloops** (WHL) by **Regina** (WHL) for Alan Manness, January 13, 1999.

CALOUN, Jan
(CHAH-loon, YAHN)

Right wing. Shoots right. 5'10", 190 lbs. Born, Usti-Nad-Labem, Czech., December 20, 1972. San Jose's 4th choice, 75th overall, in 1992 Entry Draft.

Season	Club	League	GP	G	A	Pts	PIM	PP	SH	GW	S	%	+/-	TF	F%	H	SB	Min	GP	G	A	Pts	PIM	PP	SH	GW
1990-91	HC CHZ Litvinov	Czech	50	28	19	47	12	....	....	....	....	....	....	....	....	....	....	....	....	....	....	....	....	....	....	....
1991-92	Litvinov	Czech	46	39	13	52	24	....	....	....	....	....	....	....	....	....	....	....	....	....	....	....	....	....	....	....
1992-93	Litvinov	Czech	36	37	16	53	....	....	....	....	....	....	....	....	....	....	....	....	11	8	6	14	....	....	....	....
1993-94	Litvinov	Czech	38	25	17	42	....	....	....	....	....	....	....	....	....	....	....	....	4	2	2	4	....	....	....	....
1994-95	Kansas City	IHL	76	34	39	73	50	....	....	....	....	....	....	....	....	....	....	....	21	13	10	23	18	....	....	....
1995-96	**San Jose**	**NHL**	**11**	**8**	**3**	**11**	**0**	**2**	**0**	**0**	**20**	**40.0**	**4**	....	....	....	....	....	....	....	....	....	....	....	....	....
	Kansas City	IHL	61	38	30	68	58	....	....	....	....	....	....	....	....	....	....	....	5	0	1	1	6	....	....	....
1996-97	**San Jose**	**NHL**	**2**	**0**	**0**	**0**	**0**	**0**	**0**	**0**	**3**	**0.0**	**–2**	....	....	....	....	....	....	....	....	....	....	....	....	....
	Kentucky	AHL	66	43	43	86	68	....	....	....	....	....	....	....	....	....	....	....	4	0	1	1	4	....	....	....
1997-98	HIFK Helsinki	Finland	41	22	26	48	73	....	....	....	....	....	....	....	....	....	....	....	9	6	*11	*17	6	....	....	....
	Czech Republic	Olympics	3	0	0	0	6	....	....	....	....	....	....	....	....	....	....	....	....	....	....	....	....	....	....	....
1998-99	HIFK Helsinki	Finland	51	24	*57	*81	95	....	....	....	....	....	....	....	....	....	....	....	8	*8	6	*14	31	....	....	....
	HIFK Helsinki	EuroHL	5	4	2	6	26	....	....	....	....	....	....	....	....	....	....	....	3	1	1	2	30	....	....	....
99-2000	HIFK Helsinki	Finland	44	38	34	72	94	....	....	....	....	....	....	....	....	....	....	....	9	3	6	9	10	....	....	....
	HIFK Helsinki	EuroHL	4	1	3	4	6	....	....	....	....	....	....	....	....	....	....	....	1	0	1	1	0	....	....	....
2000-01	**Columbus**	**NHL**	**11**	**0**	**3**	**3**	**2**	**0**	**0**	**0**	**14**	**0.0**	**–8**	**0**	**0.0**	**2**	**3**	**12:40**	....	....	....	....	....	....	....	....
	HIFK Helsinki	Finland	24	8	14	22	42	....	....	....	....	....	....	....	....	....	....	....	....	....	....	....	....	....	....	....
2001-02	Blues Espoo	Finland	48	15	43	58	49	....	....	....	....	....	....	....	....	....	....	....	3	1	1	2	25	....	....	....
	NHL Totals		**24**	**8**	**6**	**14**	**2**	**2**	**0**	**0**	**37**	**21.6**		**0**	**0.0**	**2**	**3**	**12:40**	....	....	....	....	....	....	....	....

AHL Second All-Star Team (1997)

Traded to **Columbus** by **San Jose** with San Jose's 9th round choice (Martin Paroulek) in 2000 Entry Draft for future considerations, June 12, 2000. Signed as a free agent by **HIFK Helsinki** (Finland) after securing release from **Columbus**, November 28, 2000.

CAMPBELL, Brian
(KAM-behl, BRIGH-uhn) **BUF.**

Defense. Shoots left. 6', 190 lbs. Born, Strathroy, Ont., May 23, 1979. Buffalo's 7th choice, 156th overall, in 1997 Entry Draft.

Season	Club	League	GP	G	A	Pts	PIM	PP	SH	GW	S	%	+/-	TF	F%	H	SB	Min	GP	G	A	Pts	PIM	PP	SH	GW
1994-95	Petrolia	OJHL-B	49	11	27	38	43	....	....	....	....	....	....	....	....	....	....	....	....	....	....	....	....	....	....	....
1995-96	Ottawa 67's	OHL	66	5	22	27	23	....	....	....	....	....	....	....	....	....	....	....	4	0	1	1	2	....	....	....
1996-97	Ottawa 67's	OHL	66	7	36	43	12	....	....	....	....	....	....	....	....	....	....	....	24	2	11	13	8	....	....	....
1997-98	Ottawa 67's	OHL	66	14	39	53	31	....	....	....	....	....	....	....	....	....	....	....	13	1	14	15	0	....	....	....
1998-99	Ottawa 67's	OHL	62	12	75	87	27	....	....	....	....	....	....	....	....	....	....	....	9	2	10	12	6	....	....	....
	Rochester	AHL	....	....	....	....	....	....	....	....	....	....	....	....	....	....	....	....	2	0	0	0	0	....	....	....
99-2000	**Buffalo**	**NHL**	**12**	**1**	**4**	**5**	**4**	**0**	**0**	**0**	**10**	**10.0**	**–2**	**0**	**0.0**	**8**	**3**	**15:48**	....	....	....	....	....	....	....	....
	Rochester	AHL	67	2	24	26	22	....	....	....	....	....	....	....	....	....	....	....	21	0	3	3	0	....	....	....
2000-01	**Buffalo**	**NHL**	**8**	**0**	**0**	**0**	**2**	**0**	**0**	**0**	**7**	**0.0**	**–2**	**0**	**0.0**	**8**	**8**	**15:40**	....	....	....	....	....	....	....	....
	Rochester	AHL	65	7	25	32	24	....	....	....	....	....	....	....	....	....	....	....	4	0	1	1	0	....	....	....
2001-02	**Buffalo**	**NHL**	**29**	**3**	**3**	**6**	**12**	**0**	**0**	**0**	**30**	**10.0**	**0**	**1**	**0.0**	**21**	**13**	**15:18**	....	....	....	....	....	....	....	....
	Rochester	AHL	45	2	35	37	13	....	....	....	....	....	....	....	....	....	....	....	....	....	....	....	....	....	....	....
	NHL Totals		**49**	**4**	**7**	**11**	**18**	**0**	**0**	**0**	**47**	**8.5**		**1**	**0.0**	**37**	**24**	**15:29**	....	....	....	....	....	....	....	....

OHL First All-Star Team (1999) • OHL MVP (1999) • Canadian Major Junior First All-Star Team (1999) • Canadian Major Junior Player of the Year (1999) • Won George Parsons Trophy (Memorial Cup Tournament Most Sportsmanlike Player) (1999)

CAMPBELL, Jim
(KAM-behl, JIHM) **FLA.**

Right wing. Shoots right. 6'2", 205 lbs. Born, Worcester, MA, April 3, 1973. Montreal's 2nd choice, 28th overall, in 1991 Entry Draft.

Season	Club	League	GP	G	A	Pts	PIM	PP	SH	GW	S	%	+/-	TF	F%	H	SB	Min	GP	G	A	Pts	PIM	PP	SH	GW
1988-89	Northfield Prep	Hi-School	12	12	8	20	6	....	....	....	....	....	....	....	....	....	....	....	....	....	....	....	....	....	....	....
1989-90	Northfield Prep	Hi-School	8	14	7	21	8	....	....	....	....	....	....	....	....	....	....	....	....	....	....	....	....	....	....	....
1990-91	Lawrence School	Hi-School	26	36	47	83	26	....	....	....	....	....	....	....	....	....	....	....	....	....	....	....	....	....	....	....
1991-92	Hull Olympiques	QMJHL	64	41	44	85	51	....	....	....	....	....	....	....	....	....	....	....	6	7	3	10	8	....	....	....
1992-93	Hull Olympiques	QMJHL	50	42	29	71	66	....	....	....	....	....	....	....	....	....	....	....	8	11	4	15	43	....	....	....
1993-94	Team USA	Nat-Tm	56	24	33	57	59	....	....	....	....	....	....	....	....	....	....	....	....	....	....	....	....	....	....	....
	United States	Olympics	8	0	0	0	6	....	....	....	....	....	....	....	....	....	....	....	....	....	....	....	....	....	....	....
	Fredericton	AHL	19	6	17	23	6	....	....	....	....	....	....	....	....	....	....	....	....	....	....	....	....	....	....	....
1994-95	Fredericton	AHL	77	27	24	51	103	....	....	....	....	....	....	....	....	....	....	....	12	0	7	7	8	....	....	....
1995-96	Fredericton	AHL	44	28	23	51	24	....	....	....	....	....	....	....	....	....	....	....	....	....	....	....	....	....	....	....
	Anaheim	**NHL**	**16**	**2**	**3**	**5**	**36**	**1**	**0**	**0**	**25**	**8.0**	**0**	....	....	....	....	....	....	....	....	....	....	....	....	....
	Baltimore	AHL	16	13	7	20	8	....	....	....	....	....	....	....	....	....	....	....	12	7	5	12	10	....	....	....

			Regular Season																Playoffs							
Season	Club	League	GP	G	A	Pts	PIM	PP	SH	GW	S	%	+/-	TF	F%	H	SB	Min	GP	G	A	Pts	PIM	PP	SH	GW
1996-97	St. Louis	NHL	68	23	20	43	68	5	0	6	169	13.6	3	....	....	....	....	....	4	1	0	1	6	1	0	0
1997-98	St. Louis	NHL	76	22	19	41	55	7	0	6	147	15.0	0	....	....	....	....	....	10	7	3	10	12	4	0	2
1998-99	St. Louis	NHL	55	4	21	25	41	1	0	0	99	4.0	-8	7	42.9	65	7	13:34	....	....	....	....	....	....	....	....
99-2000	Manitoba Moose	IHL	10	1	3	4	10	....	....	....	....	....	....	....	....	....	....	....	....	....	....	....	....	....	....	....
	St. Louis	NHL	2	0	0	0	9	0	0	0	6	0.0	0	0	0.0	1	0	15:17	....	....	....	....	....	....	....	....
	Worcester	AHL	66	31	34	65	88	....	....	....	....	....	....	....	....	....	....	....	9	1	2	3	6	....	....	....
2000-01	Montreal	NHL	57	9	11	20	53	6	0	1	81	11.1	-3	14	42.9	23	10	10:19	....	....	....	....	....	....	....	....
	Quebec	AHL	3	5	0	5	6	....	....	....	....	....	....	....	....	....	....	....	....	....	....	....	....	....	....	....
2001-02	Chicago	NHL	9	1	1	2	4	0	0	0	12	8.3	-3	1	0.0	4	1	13:21	....	....	....	....	....	....	....	....
	Norfolk Admirals	AHL	44	11	14	25	26	....	....	....	....	....	....	....	....	....	....	....	4	3	1	4	2	....	....	....
	NHL Totals		283	61	75	136	266	20	0	13	539	11.3		22	40.9	93	18	12:04	14	8	3	11	18	5	0	2

NHL All-Rookie Team (1997)

Traded to **Anaheim** by **Montreal** for Robert Dirk, January 21, 1996. Signed as a free agent by **St. Louis**, July 11, 1996. Loaned to **Manitoba** (IHL) by **St. Louis**, October 4, 1999 and recalled November 1, 1999. Signed as a free agent by **Montreal**, August 21, 2000. Signed as a free agent by **Chicago**, November 19, 2001. Signed as a free agent by **Florida**, July 19, 2002.

CARNEY, Keith (KAHRN-nee, KEETH) ANA.

Defense. Shoots left. 6'2", 211 lbs. Born, Providence, RI, February 3, 1970. Buffalo's 3rd choice, 76th overall, in 1988 Entry Draft.

Season	Club	League	GP	G	A	Pts	PIM	PP	SH	GW	S	%	+/-	TF	F%	H	SB	Min	GP	G	A	Pts	PIM	PP	SH	GW
1987-88	Mount St. Charles	Hi-School	23	12	43	55	....	....	....	....	....	....	....	....	....	....	....	....	....	....	....	....	....	....	....	....
1988-89	U. of Maine	H-East	40	4	22	26	24	....	....	....	....	....	....	....	....	....	....	....	....	....	....	....	....	....	....	....
1989-90	U. of Maine	H-East	41	3	41	44	43	....	....	....	....	....	....	....	....	....	....	....	....	....	....	....	....	....	....	....
1990-91	U. of Maine	H-East	40	7	49	56	38	....	....	....	....	....	....	....	....	....	....	....	....	....	....	....	....	....	....	....
1991-92	Team USA	Nat-Tm	49	2	17	19	16	....	....	....	....	....	....	....	....	....	....	....	....	....	....	....	....	....	....	....
	Buffalo	NHL	14	1	2	3	18	1	0	0	17	5.9	-3	....	....	....	....	....	7	0	3	3	0	0	0	0
	Rochester	AHL	24	1	10	11	2	....	....	....	....	....	....	....	....	....	....	....	2	0	2	2	0	....	....	....
1992-93	Buffalo	NHL	30	2	4	6	55	0	0	1	26	7.7	3	....	....	....	....	....	8	0	3	3	6	0	0	0
	Rochester	AHL	41	5	21	26	32	....	....	....	....	....	....	....	....	....	....	....	....	....	....	....	....	....	....	....
1993-94	Buffalo	NHL	7	1	3	4	4	0	0	0	6	16.7	-1	....	....	....	....	....	....	....	....	....	....	....	....	....
	Chicago	NHL	30	3	5	8	35	0	0	0	31	9.7	15	....	....	....	....	....	6	0	1	1	4	0	0	0
	Indianapolis Ice	IHL	28	0	14	14	20	....	....	....	....	....	....	....	....	....	....	....	....	....	....	....	....	....	....	....
1994-95	Chicago	NHL	18	1	0	1	11	0	0	1	14	7.1	-1	....	....	....	....	....	4	0	1	1	0	0	0	0
1995-96	Chicago	NHL	82	5	14	19	94	1	0	1	69	7.2	31	....	....	....	....	....	10	0	3	3	4	0	0	0
1996-97	Chicago	NHL	81	3	15	18	62	0	0	1	77	3.9	26	....	....	....	....	....	6	1	1	2	2	0	0	0
1997-98	Chicago	NHL	60	2	13	15	73	0	1	0	53	3.8	-7	....	....	....	....	....	....	....	....	....	....	....	....	....
	United States	Olympics	4	0	0	0	2	....	....	....	....	....	....	....	....	....	....	....	....	....	....	....	....	....	....	....
	Phoenix	NHL	20	1	6	7	18	1	0	0	18	5.6	5	....	....	....	....	....	6	0	0	0	4	0	0	0
1998-99	Phoenix	NHL	82	2	14	16	62	0	2	0	62	3.2	15	0	0.0	133	87	22:46	7	1	2	3	10	0	0	0
99-2000	Phoenix	NHL	82	4	20	24	87	0	0	1	73	5.5	11	0	0.0	189	98	21:12	5	0	0	0	17	0	0	0
2000-01	Phoenix	NHL	82	2	14	16	86	0	0	0	65	3.1	15	0	0.0	194	94	20:53	....	....	....	....	....	....	....	....
2001-02	Anaheim	NHL	60	5	9	14	30	0	0	1	66	7.6	14	0	0.0	83	64	20:47	....	....	....	....	....	....	....	....
	NHL Totals		648	32	119	151	635	3	3	6	577	5.5		0	0.0	599	343	21:33	59	2	14	16	47	0	0	0

Hockey East Second All-Star Team (1990) • NCAA East Second All-American Team (1990) • Hockey East First All-Star Team (1991) • NCAA East First All-American Team (1991)

Traded to **Chicago** by **Buffalo** with Buffalo's 6th round choice (Marc Magliarditi) in 1995 Entry Draft for Craig Muni and Chicago's 5th round choice (Daniel Bienvenue) in 1995 Entry Draft, October 26, 1993. Traded to **Phoenix** by **Chicago** with Jim Cummins for Chad Kilger and Jayson More, March 4, 1998. Traded to **Anaheim** by **Phoenix** for Calgary's 2nd round choice (previously acquired, later traded back to Calgary - Calgary selected Andrei Taratukhin) in 2001 Entry Draft, June 19, 2001.

CARTER, Anson (KAHR-tuhr, AN-sohn) EDM.

Right wing. Shoots right. 6'1", 200 lbs. Born, Toronto, Ont., June 6, 1974. Quebec's 11th choice, 220th overall, in 1992 Entry Draft.

Season	Club	League	GP	G	A	Pts	PIM	PP	SH	GW	S	%	+/-	TF	F%	H	SB	Min	GP	G	A	Pts	PIM	PP	SH	GW
1989-90	Don Mills	MTHL	40	15	47	62	105	....	....	....	....	....	....	....	....	....	....	....	....	....	....	....	....	....	....	....
1990-91	Don Mills	MTHL	67	69	73	142	43	....	....	....	....	....	....	....	....	....	....	....	....	....	....	....	....	....	....	....
1991-92	Wexford Raiders	MTJHL	42	18	22	40	24	....	....	....	....	....	....	....	....	....	....	....	....	....	....	....	....	....	....	....
1992-93	Michigan State	CCHA	34	15	7	22	20	....	....	....	....	....	....	....	....	....	....	....	....	....	....	....	....	....	....	....
1993-94	Michigan State	CCHA	39	30	24	54	36	....	....	....	....	....	....	....	....	....	....	....	....	....	....	....	....	....	....	....
1994-95	Michigan State	CCHA	39	34	17	51	40	....	....	....	....	....	....	....	....	....	....	....	....	....	....	....	....	....	....	....
1995-96	Michigan State	CCHA	42	23	20	43	36	....	....	....	....	....	....	....	....	....	....	....	....	....	....	....	....	....	....	....
1996-97	Washington	NHL	19	3	2	5	7	1	0	1	28	10.7	0	....	....	....	....	....	....	....	....	....	....	....	....	....
	Portland Pirates	AHL	27	19	19	38	11	....	....	....	....	....	....	....	....	....	....	....	....	....	....	....	....	....	....	....
	Boston	NHL	19	8	5	13	2	1	1	1	51	15.7	-7	....	....	....	....	....	....	....	....	....	....	....	....	....
1997-98	Boston	NHL	78	16	27	43	31	6	0	4	179	8.9	7	....	....	....	....	....	6	1	1	2	0	0	0	0
1998-99	Utah Grizzlies	IHL	6	1	1	2	0	....	....	....	....	....	....	....	....	....	....	....	....	....	....	....	....	....	....	....
	Boston	NHL	55	24	16	40	22	6	0	6	123	19.5	7	172	43.0	62	5	18:44	12	4	3	7	0	1	0	1
99-2000	Boston	NHL	59	22	25	47	14	4	0	1	144	15.3	8	793	48.2	87	9	20:31	....	....	....	....	....	....	....	....
2000-01	Edmonton	NHL	61	16	26	42	23	7	1	4	102	15.7	1	80	47.5	66	11	18:13	6	3	1	4	4	1	0	1
2001-02	Edmonton	NHL	82	28	32	60	25	12	0	6	181	15.5	3	316	46.5	59	27	19:18	....	....	....	....	....	....	....	....
	NHL Totals		373	117	133	250	124	37	2	23	808	14.5		1361	47.1	274	52	19:12	24	8	5	13	4	2	0	2

CCHA First All-Star Team (1994, 1995) • NCAA West Second All-American Team (1995) • CCHA Second All-Star Team (1996)

Rights transferred to **Colorado** after **Quebec** franchise relocated, June 21, 1995. Traded to **Washington** by **Colorado** for Washington's 4th round choice (Ben Storey) in 1996 Entry Draft, April 3, 1996. Traded to **Boston** by **Washington** with Jim Carey, Jason Allison and Washington's 3rd round choice (Lee Goren) in 1997 Entry Draft for Bill Ranford, Adam Oates and Rick Tocchet, March 1, 1997. Signed as a free agent by **Utah** (IHL) with Boston retaining NHL rights, October 20, 1998. Traded to **Edmonton** by **Boston** with Boston's 1st (Ales Hemsky) and 2nd (Doug Lynch) round choices in 2001 Entry Draft for Bill Guerin and future considerations, November 15, 2000.

CASSELS, Andrew (KAS-uhls, AN-droo) CBJ

Center. Shoots left. 6'1", 185 lbs. Born, Bramalea, Ont., July 23, 1969. Montreal's 1st choice, 17th overall, in 1987 Entry Draft.

Season	Club	League	GP	G	A	Pts	PIM	PP	SH	GW	S	%	+/-	TF	F%	H	SB	Min	GP	G	A	Pts	PIM	PP	SH	GW
1985-86	Bramalea Blues	OPJHL	33	18	25	43	26	....	....	....	....	....	....	....	....	....	....	....	....	....	....	....	....	....	....	....
1986-87	Ottawa 67's	OHL	66	26	66	92	28	....	....	....	....	....	....	....	....	....	....	....	11	5	9	14	7	....	....	....
1987-88	Ottawa 67's	OHL	61	48	*103	*151	39	....	....	....	....	....	....	....	....	....	....	....	16	8	*24	*32	13	....	....	....
1988-89	Ottawa 67's	OHL	56	37	97	134	66	....	....	....	....	....	....	....	....	....	....	....	12	5	10	15	10	....	....	....
1989-90	Montreal	NHL	6	2	0	2	2	0	0	1	5	40.0	1	....	....	....	....	....	....	....	....	....	....	....	....	....
	Sherbrooke	AHL	55	22	45	67	25	....	....	....	....	....	....	....	....	....	....	....	12	2	11	13	6	....	....	....
1990-91	Montreal	NHL	54	6	19	25	20	1	0	3	55	10.9	2	....	....	....	....	....	8	0	2	2	2	0	0	0
1991-92	Hartford	NHL	67	11	30	41	18	2	2	3	99	11.1	3	....	....	....	....	....	7	2	4	6	6	1	0	0
1992-93	Hartford	NHL	84	21	64	85	62	8	3	1	134	15.7	-11	....	....	....	....	....	....	....	....	....	....	....	....	....
1993-94	Hartford	NHL	79	16	42	58	37	8	1	3	126	12.7	-21	....	....	....	....	....	....	....	....	....	....	....	....	....
1994-95	Hartford	NHL	46	7	30	37	18	1	0	1	74	9.5	-3	....	....	....	....	....	....	....	....	....	....	....	....	....
1995-96	Hartford	NHL	81	20	43	63	39	6	0	1	135	14.8	8	....	....	....	....	....	....	....	....	....	....	....	....	....
1996-97	Hartford	NHL	81	22	44	66	46	8	0	2	142	15.5	-16	....	....	....	....	....	....	....	....	....	....	....	....	....
1997-98	Calgary	NHL	81	17	27	44	32	6	1	2	138	12.3	-7	....	....	....	....	....	....	....	....	....	....	....	....	....
1998-99	Calgary	NHL	70	12	25	37	18	4	1	3	97	12.4	-12	1322	51.1	25	40	18:58	....	....	....	....	....	....	....	....
99-2000	Vancouver	NHL	79	17	45	62	16	6	0	1	109	15.6	8	1127	48.3	45	36	19:19	....	....	....	....	....	....	....	....
2000-01	Vancouver	NHL	66	12	44	56	10	2	0	1	104	11.5	1	1164	49.9	41	31	19:22	....	....	....	....	....	....	....	....
2001-02	Vancouver	NHL	53	11	39	50	22	7	0	1	64	17.2	5	866	50.4	42	14	17:27	6	2	1	3	0	1	0	0
	NHL Totals		847	174	452	626	340	59	8	23	1282	13.6		4479	49.9	153	121	18:52	21	4	7	11	8	2	0	0

OHL Rookie of the Year (1987) • OHL First All-Star Team (1988,1989) • OHL MVP (1988)

Traded to **Hartford** by **Montreal** for Hartford's 2nd round choice (Valeri Bure) in 1992 Entry Draft, September 17, 1991. Transferred to **Carolina** after **Hartford** franchise relocated, June 25, 1997. Traded to **Calgary** by **Carolina** with Jean-Sebastien Giguere for Gary Roberts and Trevor Kidd, August 25, 1997. Signed as a free agent by **Vancouver**, August 19, 1999. Signed as a free agent by **Columbus**, August 15, 2002.

CHARA, Zdeno (KHAH-rah, ZDEH-noh) OTT.

Defense. Shoots left. 6'9", 255 lbs. Born, Trencin, Czech., March 18, 1977. NY Islanders' 3rd choice, 56th overall, in 1996 Entry Draft.

Season	Club	League	GP	G	A	Pts	PIM	PP	SH	GW	S	%	+/-	TF	F%	H	SB	Min	GP	G	A	Pts	PIM	PP	SH	GW
1994-95	Dukla Trencin-B	Slovak-Jr.	30	22	22	44	113	....	....	....	....	....	....	....	....	....	....	....	....	....	....	....	....	....	....	....
	Dukla Trencin Jr.	Slovak-Jr.	2	0	0	0	0	....	....	....	....	....	....	....	....	....	....	....	....	....	....	....	....	....	....	....
1995-96	Dukla Trencin Jr.	Slovak-Jr.	22	1	13	14	80	....	....	....	....	....	....	....	....	....	....	....	....	....	....	....	....	....	....	....
	HK VTJ Piestany	Slovak-2	10	1	3	4	10	....	....	....	....	....	....	....	....	....	....	....	....	....	....	....	....	....	....	....
	Sparta Praha Jr.	Czech-Jr.	15	1	2	3	42	....	....	....	....	....	....	....	....	....	....	....	....	....	....	....	....	....	....	....
	HC Sparta Praha	Czech	1	0	0	0	0	....	....	....	....	....	....	....	....	....	....	....	....	....	....	....	....	....	....	....
1996-97	Prince George	WHL	49	3	19	22	120	....	....	....	....	....	....	....	....	....	....	....	15	1	7	8	45	....	....	....

Season	Club	League	GP	G	A	Pts	PIM	PP	SH	GW	S	%	+/-	TF	F%	H	SB	Min	GP	G	A	Pts	PIM	PP	SH	GW
			Regular Season																Playoffs							
1997-98	NY Islanders	NHL	25	0	1	1	50	0	0	0	10	0.0	1													
	Kentucky	AHL	48	4	9	13	125												1	0	0	0	4			
1998-99	NY Islanders	NHL	59	2	6	8	83	0	1	0	56	3.6	−8	0	0.0	214	55	18:54								
	Lowell	AHL	23	2	2	4	47																			
99-2000	NY Islanders	NHL	65	2	9	11	57	0	0	1	47	4.3	−27	0	0.0	309	100	22:52								
2000-01	NY Islanders	NHL	82	2	7	9	157	0	1	0	83	2.4	−27	0	0.0	373	126	22:20								
2001-02	Dukla Trencin	Slovakia	8	2	2	4	32																			
	Ottawa	NHL	75	10	13	23	156	4	1	2	105	9.5	30	0	0.0	299	73	22:16	10	0	1	1	12	0	0	0
	NHL Totals		306	16	36	52	503	4	3	3	301	5.3		0	0.0	1195	354	21:43	10	0	1	1	12	0	0	0

Traded to **Ottawa** by **NY Islanders** with Bill Muckalt and NY Islanders' 1st round choice (Jason Spezza) in 2001 Entry Draft for Alexei Yashin, June 23, 2001.

CHARTRAND, Brad

(SHAR-trand, BRAD) **L.A.**

Right wing. Shoots left. 5'11", 191 lbs. Born, Winnipeg, Man., December 14, 1974.

Season	Club	League	GP	G	A	Pts	PIM	PP	SH	GW	S	%	+/-	TF	F%	H	SB	Min	GP	G	A	Pts	PIM	PP	SH	GW
1988-89	Winnipeg Hawks	MMHL	24	30	50	80	40																			
1989-90	Winnipeg Hawks	MMHL	24	26	55	81	40																			
1990-91	Winnipeg Hawks	MMHL	34	26	45	71	40																			
1991-92	St. James	MJHL	45	24	25	49	32																			
1992-93	Cornell Big Red	ECAC	26	10	6	16	16																			
1993-94	Cornell Big Red	ECAC	30	4	14	18	48																			
1994-95	Cornell Big Red	ECAC	28	9	9	18	10																			
1995-96	Cornell Big Red	ECAC	34	24	19	43	16																			
1996-97	Team Canada	Nat-Tm	54	10	14	24	42																			
1997-98	Team Canada	Nat-Tm	60	24	30	54	47																			
	Rapperswil	Swiss	8	2	3	5	4																			
1998-99	St. John's	AHL	64	16	14	30	48												5	0	2	2	2			
99-2000	Los Angeles	NHL	50	6	6	12	17	0	1	3	51	11.8	4	62	53.2	70	18	11:03	4	0	0	0	6	0	0	0
	Lowell	AHL	16	5	10	15	8																			
	Long Beach	IHL	1	0	0	0	0												3	0	0	0	0			
2000-01	Los Angeles	NHL	4	1	0	1	2	0	0	1	6	16.7	−2	0	0.0	7	0	11:37								
	Lowell	AHL	72	17	34	51	44												4	0	1	1	8			
2001-02	Los Angeles	NHL	46	7	9	16	40	0	0	1	49	14.3	5	481	53.2	72	17	12:05	7	1	1	2	2	0	0	1
	Manchester	AHL	22	10	12	22	31																			
	NHL Totals		100	14	15	29	59	0	1	5	106	13.2		543	53.2	149	35	11:33	11	1	1	2	8	0	0	1

Signed as a free agent by **LA Kings**, July 15, 1999. Loaned to **Lowell** (AHL) by **LA Kings**, January 26, 2000.

CHEBATURKIN, Vladimir

(cheh-bah-TOOR-kihn) **NYR**

Defense. Shoots left. 6'2", 226 lbs. Born, Tyumen, USSR, April 23, 1975. NY Islanders' 3rd choice, 66th overall, in 1993 Entry Draft.

Season	Club	League	GP	G	A	Pts	PIM	PP	SH	GW	S	%	+/-	TF	F%	H	SB	Min	GP	G	A	Pts	PIM	PP	SH	GW
1993-94	Elektrostal	CIS-2	42	4	4	8	38																			
1994-95	Elektrostal	CIS	52	2	6	8	90																			
1995-96	Elektrostal	CIS	44	1	6	7	30																			
1996-97	Utah Grizzlies	IHL	68	0	4	4	34												1	0	0	0	0			
1997-98	NY Islanders	NHL	2	0	2	2	0	0	0	0	0	0.0	−1													
	Kentucky	AHL	54	6	8	14	52												2	0	0	0	4			
1998-99	NY Islanders	NHL	8	0	0	0	12	0	0	0	4	0.0	6	0	0.0	27	8	17:12								
	Lowell	AHL	69	2	12	14	85												3	0	0	0	0			
99-2000	NY Islanders	NHL	17	1	1	2	8	0	0	0	9	11.1	−3	0	0.0	58	25	16:57								
	Lowell	AHL	63	1	8	9	118												7	0	4	4	11			
2000-01	Worcester	AHL	33	0	7	7	73												10	1	0	1	10			
	St. Louis	NHL	22	1	2	3	26	0	0	0	5	20.0	5	0	0.0	54	15	13:09								
2001-02	Chicago	NHL	13	0	2	2	6	0	0	0	8	0.0	0	0	0.0	40	3	12:42	3	0	0	0	2	0	0	0
	Norfolk Admirals	AHL	57	2	8	10	78																			
	NHL Totals		62	2	7	9	52	0	0	0	26	7.7		0	0.0	179	51	14:40	3	0	0	0	2	0	0	0

Signed as a free agent by **St. Louis**, June 9, 2000. Signed as a free agent by **Chicago**, September 5, 2001. Signed as a free agent by **NY Rangers**, July 18, 2002.

CHELIOS, Chris

(CHELL-EE-ohs, KRIHS) **DET.**

Defense. Shoots right. 6'1", 190 lbs. Born, Chicago, IL, January 25, 1962. Montreal's 5th choice, 40th overall, in 1981 Entry Draft.

Season	Club	League	GP	G	A	Pts	PIM	PP	SH	GW	S	%	+/-	TF	F%	H	SB	Min	GP	G	A	Pts	PIM	PP	SH	GW
1979-80	Moose Jaw	SJHL	53	12	31	43	118																			
1980-81	Moose Jaw	SJHL	54	23	64	87	175																			
1981-82	U. of Wisconsin	WCHA	43	6	43	49	50																			
1982-83	U. of Wisconsin	WCHA	26	9	17	26	50																			
1983-84	Team USA	Nat-Tm	60	14	35	49	58																			
	United States	Olympics	6	0	4	4	8																			
	Montreal	NHL	12	0	2	2	12	0	0	0	23	0.0	−5						15	1	9	10	17	1	0	0
1984-85	Montreal	NHL	74	9	55	64	87	2	1	0	199	4.5	11						9	2	8	10	17	2	0	0
1985-86♦	Montreal	NHL	41	8	26	34	67	2	0	0	101	7.9	4						20	2	9	11	49	1	0	0
1986-87	Montreal	NHL	71	11	33	44	124	6	0	2	141	7.8	−5						17	4	9	13	38	2	1	0
1987-88	Montreal	NHL	71	20	41	61	172	10	1	5	199	10.1	14						11	3	1	4	29	1	0	0
1988-89	Montreal	NHL	80	15	58	73	185	8	0	6	206	7.3	35						21	4	15	19	28	1	0	2
1989-90	Montreal	NHL	53	9	22	31	136	1	2	1	123	7.3	20						5	0	1	1	8	0	0	0
1990-91	Chicago	NHL	77	12	52	64	192	5	2	2	187	6.4	23						6	1	7	8	46	1	0	0
1991-92	Chicago	NHL	80	9	47	56	245	2	2	2	239	3.8	24						18	6	15	21	37	3	0	1
1992-93	Chicago	NHL	84	15	58	73	282	8	0	2	290	5.2	14						4	0	2	2	14	0	0	0
1993-94	Chicago	NHL	76	16	44	60	212	7	1	2	219	7.3	12						6	1	1	2	8	1	0	0
1994-95	EHC Biel-Bienne	Swiss	3	0	3	3	4																			
	Chicago	NHL	48	5	33	38	72	3	1	0	166	3.0	17						16	4	7	11	12	0	1	3
1995-96	Chicago	NHL	81	14	58	72	140	7	0	3	219	6.4	25						9	0	3	3	8	0	0	0
1996-97	Chicago	NHL	72	10	38	48	112	2	0	2	194	5.2	16						6	0	1	1	8	0	0	0
1997-98	Chicago	NHL	81	3	39	42	151	1	0	0	205	1.5	−7													
	United States	Olympics	4	2	0	2	2																			
1998-99	Chicago	NHL	65	8	26	34	89	2	1	0	172	4.7	−4	4	25.0	72	109	27:19								
	Detroit	NHL	10	1	1	2	4	1	0	1	15	6.7	5	0	0.0	11	7	22:21	10	0	4	4	14	0	0	0
99-2000	Detroit	NHL	81	3	31	34	103	0	0	0	135	2.2	48	0	0.0	120	86	25:16	9	0	1	1	8	0	0	0
2000-01	Detroit	NHL	24	0	3	3	45	0	0	0	26	0.0	4	0	0.0	41	32	22:51	5	1	0	1	2	0	0	0
2001-02♦	Detroit	NHL	79	6	33	39	126	1	0	1	128	4.7	40	0	0.0	106	110	25:18	23	1	13	14	44	1	0	0
	United States	Olympics	6	1	0	1	4																			
	NHL Totals		1260	174	700	874	2556	68	11	29	3187	5.5		4	25.0	350	344	25:27	210	30	106	136	387	14	2	6

WCHA Second All-Star Team (1983) • NCAA Championship All-Tournament Team (1983) • NHL All-Rookie Team (1985) • NHL First All-Star Team (1989, 1993, 1995, 1996, 2002) • Won James Norris Memorial Trophy (1989, 1993, 1996) • NHL Second All-Star Team (1991, 1997) • Played in NHL All-Star Game (1985, 1990, 1991, 1992, 1993, 1994, 1996, 1997, 1998, 2000, 2002)

Traded to **Chicago** by **Montreal** with Montreal's 2nd round choice (Michael Pomichter) in 1991 Entry Draft for Denis Savard, June 29, 1990. Traded to **Detroit** by **Chicago** for Anders Eriksson and Detroit's 1st round choices in 1999 (Steve McCarthy) and 2001 (Adam Munro) Entry Drafts, March 23, 1999. • Missed majority of 2000-01 season recovering from knee injury suffered in game vs. Dallas, November 17, 2000.

CHIMERA, Jason

(chihm-AIR-a, JAY-suhn) **EDM.**

Center. Shoots left. 6', 215 lbs. Born, Edmonton, Alta., May 2, 1979. Edmonton's 5th choice, 121st overall, in 1997 Entry Draft.

Season	Club	League	GP	G	A	Pts	PIM	PP	SH	GW	S	%	+/-	TF	F%	H	SB	Min	GP	G	A	Pts	PIM	PP	SH	GW
1994-95	Edmonton Pats	AMHL	33	27	31	58	42																			
1995-96	Edmonton Pats	AMHL	34	23	24	47	44																			
1996-97	Medicine Hat	WHL	71	16	23	39	64												4	0	1	1	4			
1997-98	Medicine Hat	WHL	72	34	32	66	93																			
	Hamilton	AHL	4	0	0	0	8																			
1998-99	Medicine Hat	WHL	37	18	22	40	84																			
	Brandon	WHL	21	14	12	26	32												5	4	1	5	8			
99-2000	Hamilton	AHL	78	15	13	28	77												10	0	2	2	12			
2000-01	Edmonton	NHL	1	0	0	0	0	0	0	0	0	0.0	0	0	0.0	0	0	6:58								
	Hamilton	AHL	78	29	25	54	93																			

Season	Club	League	GP	G	A	Pts	PIM	PP	SH	GW	S	%	+/-	TF	F%	H	SB	Min	GP	G	A	Pts	PIM	PP	SH	GW
			Regular Season																Playoffs							
2001-02	**Edmonton**	**NHL**	**3**	**1**	**0**	**1**	**0**	**0**	**0**	**0**	**3**	**33.3**	**–3**	**0**	**0.0**	**8**	**0**	**12:44**	....	....	....	....	....	....	....	....
	Hamilton	AHL	77	26	51	77	158	....	....	....	....	....	....	....	....	....	....	....	15	4	6	10	10	....	....	....
	NHL Totals		**4**	**1**	**0**	**1**	**0**	**0**	**0**	**0**	**3**	**33.3**		**0**	**0.0**	**8**	**0**	**11:18**	....	....	....	....	....	....	....	....

AHL First All-Star Team (2002)
Traded to **Brandon** (WHL) by **Medicine Hat** (WHL) for Justin Yeomans and future considerations, February 2, 1999.

CHOUINARD, Eric

(shwee-NAHR, AIR-ihk) **MTL.**

Center. Shoots left. 6'3", 205 lbs. Born, Atlanta, GA, July 8, 1980. Montreal's 1st choice, 16th overall, in 1998 Entry Draft.

Season	Club	League	GP	G	A	Pts	PIM	PP	SH	GW	S	%	+/-	TF	F%	H	SB	Min	GP	G	A	Pts	PIM	PP	SH	GW
1995-96	Magog	QAHA	22	12	14	26	12	....	....	....	....	....	....	....	....	....	....	....	....	....	....	....	....	....	....	....
	Ste-Foy	QAAA	17	2	5	7	....	....	....	....	....	....	....	....	....	....	....	....	15	7	12	19	12	....	....	....
1996-97	Ste-Foy	QAAA	40	29	41	70	40	....	....	....	....	....	....	....	....	....	....	....	10	14	9	23	....	....	....	....
1997-98	Quebec Remparts	QMJHL	68	41	42	83	18	....	....	....	....	....	....	....	....	....	....	....	14	7	10	17	6	....	....	....
1998-99	Quebec Remparts	QMJHL	62	50	59	109	56	....	....	....	....	....	....	....	....	....	....	....	13	8	10	18	8	....	....	....
	Fredericton	AHL	....	....	....	....	....	....	....	....	....	....	....	....	....	....	....	....	6	3	2	5	0	....	....	....
99-2000	Quebec Remparts	QMJHL	50	57	47	104	105	....	....	....	....	....	....	....	....	....	....	....	11	14	4	18	8	....	....	....
2000-01	**Montreal**	**NHL**	**13**	**1**	**3**	**4**	**0**	**1**	**0**	**0**	**11**	**9.1**	**0**	**31**	**54.8**	**5**	**1**	**10:59**	....	....	....	....	....	....	....	....
	Quebec	AHL	48	12	21	33	6	....	....	....	....	....	....	....	....	....	....	....	9	2	0	2	2	....	....	....
2001-02	Quebec	AHL	65	19	23	42	18	....	....	....	....	....	....	....	....	....	....	....	2	0	0	0	0	....	....	....
	NHL Totals		**13**	**1**	**3**	**4**	**0**	**1**	**0**	**0**	**11**	**9.1**		**31**	**54.8**	**5**	**1**	**10:59**	....	....	....	....	....	....	....	....

CHOUINARD, Marc

(shwee-NAHR, MAHRK) **ANA.**

Center. Shoots right. 6'5", 210 lbs. Born, Charlesbourg, Que., May 6, 1977. Winnipeg's 2nd choice, 32nd overall, in 1995 Entry Draft.

Season	Club	League	GP	G	A	Pts	PIM	PP	SH	GW	S	%	+/-	TF	F%	H	SB	Min	GP	G	A	Pts	PIM	PP	SH	GW
1992-93	Beauboury	QAHA	28	26	45	71	42	....	....	....	....	....	....	....	....	....	....	....	....	....	....	....	....	....	....	....
1993-94	Beauport	QMJHL	62	11	19	30	23	....	....	....	....	....	....	....	....	....	....	....	13	2	5	7	2	....	....	....
1994-95	Beauport	QMJHL	68	24	40	64	32	....	....	....	....	....	....	....	....	....	....	....	18	1	6	7	4	....	....	....
1995-96	Beauport	QMJHL	30	14	21	35	19	....	....	....	....	....	....	....	....	....	....	....	....	....	....	....	....	....	....	....
	Halifax	QMJHL	24	6	12	18	17	....	....	....	....	....	....	....	....	....	....	....	6	2	1	3	2	....	....	....
1996-97	Halifax	QMJHL	63	24	49	73	74	....	....	....	....	....	....	....	....	....	....	....	18	9	16	25	12	....	....	....
1997-98	Cincinnati	AHL	8	1	2	3	4	....	....	....	....	....	....	....	....	....	....	....	....	....	....	....	....	....	....	....
1998-99	Cincinnati	AHL	69	7	8	15	20	....	....	....	....	....	....	....	....	....	....	....	3	0	0	0	4	....	....	....
99-2000	Cincinnati	AHL	70	17	16	33	29	....	....	....	....	....	....	....	....	....	....	....	....	....	....	....	....	....	....	....
2000-01	**Anaheim**	**NHL**	**44**	**3**	**4**	**7**	**12**	**0**	**0**	**1**	**26**	**11.5**	**–5**	**414**	**60.9**	**55**	**11**	**7:50**	....	....	....	....	....	....	....	....
	Cincinnati	AHL	32	10	9	19	4	....	....	....	....	....	....	....	....	....	....	....	....	....	....	....	....	....	....	....
2001-02	**Anaheim**	**NHL**	**45**	**4**	**5**	**9**	**10**	**0**	**0**	**0**	**40**	**10.0**	**2**	**581**	**54.9**	**52**	**13**	**10:36**	....	....	....	....	....	....	....	....
	NHL Totals		**89**	**7**	**9**	**16**	**22**	**0**	**0**	**1**	**66**	**10.6**		**995**	**57.4**	**107**	**24**	**9:14**	....	....	....	....	....	....	....	....

Traded to **Anaheim** by **Winnipeg** with Teemu Selanne and Winnipeg's 4th round choice (later traded to Toronto - later traded to Montreal - Montreal selected Kim Staal) in 1996 Entry Draft for Chad Kilger, Oleg Tverdovsky and Anaheim's 3rd round choice (Per-Anton Lundstrom) in 1996 Entry Draft, February 7, 1996.

CHRISTIAN, Jeff

(KRIHS-tyan, JEHF)

Left wing. Shoots left. 6'2", 210 lbs. Born, Burlington, Ont., July 30, 1970. New Jersey's 2nd choice, 23rd overall, in 1988 Entry Draft.

Season	Club	League	GP	G	A	Pts	PIM	PP	SH	GW	S	%	+/-	TF	F%	H	SB	Min	GP	G	A	Pts	PIM	PP	SH	GW
1986-87	Dundas Blues	OJHL-C	29	20	34	54	42	....	....	....	....	....	....	....	....	....	....	....	....	....	....	....	....	....	....	....
1987-88	London Knights	OHL	64	15	29	44	154	....	....	....	....	....	....	....	....	....	....	....	9	1	5	6	27	....	....	....
1988-89	London Knights	OHL	60	27	30	57	221	....	....	....	....	....	....	....	....	....	....	....	20	3	4	7	56	....	....	....
1989-90	London Knights	OHL	18	14	7	21	64	....	....	....	....	....	....	....	....	....	....	....	....	....	....	....	....	....	....	....
	Owen Sound	OHL	37	19	26	45	145	....	....	....	....	....	....	....	....	....	....	....	10	6	7	13	43	....	....	....
1990-91	Utica Devils	AHL	80	24	42	66	165	....	....	....	....	....	....	....	....	....	....	....	....	....	....	....	....	....	....	....
1991-92	**New Jersey**	**NHL**	**2**	**0**	**0**	**0**	**2**	**0**	**0**	**0**	**1**	**0.0**	**0**	....	....	....	....	....	....	....	....	....	....	....	....	....
	Utica Devils	AHL	76	27	24	51	198	....	....	....	....	....	....	....	....	....	....	....	4	0	0	0	16	....	....	....
1992-93	Utica Devils	AHL	22	4	6	10	39	....	....	....	....	....	....	....	....	....	....	....	....	....	....	....	....	....	....	....
	Hamilton Canucks	AHL	11	2	5	7	35	....	....	....	....	....	....	....	....	....	....	....	....	....	....	....	....	....	....	....
	Cincinnati	IHL	36	5	12	17	113	....	....	....	....	....	....	....	....	....	....	....	....	....	....	....	....	....	....	....
1993-94	Albany	AHL	76	34	43	77	227	....	....	....	....	....	....	....	....	....	....	....	5	1	2	3	19	....	....	....
1994-95	Cleveland	IHL	56	13	24	37	126	....	....	....	....	....	....	....	....	....	....	....	2	0	1	1	8	....	....	....
	Pittsburgh	**NHL**	**1**	**0**	**0**	**0**	**0**	**0**	**0**	**0**	**2**	**0.0**	**0**	....	....	....	....	....	....	....	....	....	....	....	....	....
1995-96	**Pittsburgh**	**NHL**	**3**	**0**	**0**	**0**	**2**	**0**	**0**	**0**	**0**	**0.0**	**0**	....	....	....	....	....	....	....	....	....	....	....	....	....
	Cleveland	IHL	66	23	32	55	131	....	....	....	....	....	....	....	....	....	....	....	3	0	1	1	8	....	....	....
1996-97	**Pittsburgh**	**NHL**	**11**	**2**	**2**	**4**	**13**	**0**	**0**	**0**	**18**	**11.1**	**–3**	....	....	....	....	....	....	....	....	....	....	....	....	....
	Cleveland	IHL	69	40	40	80	262	....	....	....	....	....	....	....	....	....	....	....	12	6	8	14	44	....	....	....
1997-98	**Phoenix**	**NHL**	**1**	**0**	**0**	**0**	**0**	**0**	**0**	**0**	**0**	**0.0**	**–1**	....	....	....	....	....	....	....	....	....	....	....	....	....
	Las Vegas	IHL	30	12	15	27	90	....	....	....	....	....	....	....	....	....	....	....	4	2	2	4	20	....	....	....
1998-99	Houston Aeros	IHL	80	45	41	86	252	....	....	....	....	....	....	....	....	....	....	....	18	4	12	16	32	....	....	....
99-2000	Cleveland	IHL	77	29	35	64	202	....	....	....	....	....	....	....	....	....	....	....	9	1	4	5	20	....	....	....
2000-01	Krefeld Pinguine	Germany	51	17	22	39	205	....	....	....	....	....	....	....	....	....	....	....	....	....	....	....	....	....	....	....
2001-02	Krefeld Pinguine	Germany	53	31	18	49	116	....	....	....	....	....	....	....	....	....	....	....	3	2	0	2	12	....	....	....
	NHL Totals		**18**	**2**	**2**	**4**	**17**	**0**	**0**	**0**	**21**	**9.5**		....	....	....	....	....	....	....	....	....	....	....	....	....

Signed as a free agent by **Pittsburgh**, August 2, 1994. Signed as a free agent by **Phoenix**, July 28, 1997. Signed as a free agent by **Chicago**, August 25, 1999.

CHRISTIE, Ryan

(KRIHS-tee, RIGH-yuhn) **CGY.**

Left wing. Shoots left. 6'3", 200 lbs. Born, Beamsville, Ont., July 3, 1978. Dallas' 4th choice, 112th overall, in 1996 Entry Draft.

Season	Club	League	GP	G	A	Pts	PIM	PP	SH	GW	S	%	+/-	TF	F%	H	SB	Min	GP	G	A	Pts	PIM	PP	SH	GW
1994-95	St. Catharines	OJHL-B	40	10	11	21	96	....	....	....	....	....	....	....	....	....	....	....	....	....	....	....	....	....	....	....
1995-96	Owen Sound	OHL	66	29	17	46	93	....	....	....	....	....	....	....	....	....	....	....	6	1	1	2	0	....	....	....
1996-97	Owen Sound	OHL	66	23	29	52	136	....	....	....	....	....	....	....	....	....	....	....	4	1	1	2	8	....	....	....
1997-98	Owen Sound	OHL	66	39	41	80	208	....	....	....	....	....	....	....	....	....	....	....	11	3	5	8	13	....	....	....
1998-99	Michigan K-Wings	IHL	48	4	5	9	74	....	....	....	....	....	....	....	....	....	....	....	3	1	1	2	2	....	....	....
99-2000	**Dallas**	**NHL**	**5**	**0**	**0**	**0**	**0**	**0**	**0**	**0**	**1**	**0.0**	**–1**	**0**	**0.0**	**3**	**0**	**2:29**	....	....	....	....	....	....	....	....
	Michigan K-Wings	IHL	76	24	25	49	140	....	....	....	....	....	....	....	....	....	....	....	....	....	....	....	....	....	....	....
2000-01	Utah Grizzlies	IHL	69	22	16	38	88	....	....	....	....	....	....	....	....	....	....	....	....	....	....	....	....	....	....	....
2001-02	**Calgary**	**NHL**	**2**	**0**	**0**	**0**	**0**	**0**	**0**	**0**	**0**	**0.0**	**–1**	**0**	**0.0**	**3**	**1**	**6:10**	....	....	....	....	....	....	....	....
	Saint John	AHL	77	21	18	39	61	....	....	....	....	....	....	....	....	....	....	....	....	....	....	....	....	....	....	....
	NHL Totals		**7**	**0**	**0**	**0**	**0**	**0**	**0**	**0**	**1**	**0.0**		**0**	**0.0**	**6**	**1**	**3:32**	....	....	....	....	....	....	....	....

Signed as a free agent by **Calgary**, July 1, 2001.

CHUBAROV, Artem

(choo-BAH-rahf, AHR-tehm) **VAN.**

Center. Shoots left. 6'1", 189 lbs. Born, Gorky, USSR, December 12, 1979. Vancouver's 2nd choice, 31st overall, in 1998 Entry Draft.

Season	Club	League	GP	G	A	Pts	PIM	PP	SH	GW	S	%	+/-	TF	F%	H	SB	Min	GP	G	A	Pts	PIM	PP	SH	GW
1994-95	Niz. Novgorod Jr.	CIS-Jr.	60	20	30	50	20	....	....	....	....	....	....	....	....	....	....	....	....	....	....	....	....	....	....	....
1995-96	Niz. Novgorod Jr.	CIS-Jr.	60	22	25	47	20	....	....	....	....	....	....	....	....	....	....	....	....	....	....	....	....	....	....	....
1996-97	Niz. Novgorod 2	Russia-3	40	24	5	29	16	....	....	....	....	....	....	....	....	....	....	....	....	....	....	....	....	....	....	....
	Nizhny Novgorod	Russia	15	1	1	2	8	....	....	....	....	....	....	....	....	....	....	....	....	....	....	....	....	....	....	....
1997-98	Dynamo Moscow	Russia	30	1	4	5	4	....	....	....	....	....	....	....	....	....	....	....	....	....	....	....	....	....	....	....
1998-99	Dynamo Moscow	Russia	34	8	2	10	10	....	....	....	....	....	....	....	....	....	....	....	12	0	0	0	4	....	....	....
99-2000	**Vancouver**	**NHL**	**49**	**1**	**8**	**9**	**10**	**0**	**0**	**1**	**53**	**1.9**	**–4**	**488**	**48.0**	**38**	**17**	**11:43**	....	....	....	....	....	....	....	....
	Syracuse Crunch	AHL	14	7	6	13	4	....	....	....	....	....	....	....	....	....	....	....	1	0	0	0	0	....	....	....
2000-01	**Vancouver**	**NHL**	**1**	**0**	**0**	**0**	**0**	**0**	**0**	**0**	**0**	**0.0**	**–1**	**17**	**52.9**	**1**	**0**	**15:08**	....	....	....	....	....	....	....	....
	Kansas City	IHL	10	7	4	11	12	....	....	....	....	....	....	....	....	....	....	....	....	....	....	....	....	....	....	....
2001-02	**Vancouver**	**NHL**	**51**	**5**	**5**	**10**	**10**	**0**	**0**	**3**	**73**	**6.8**	**–3**	**517**	**53.6**	**38**	**24**	**12:37**	**6**	**0**	**1**	**1**	**0**	**0**	**0**	**0**
	Manitoba Moose	AHL	19	7	12	19	4	....	....	....	....	....	....	....	....	....	....	....	....	....	....	....	....	....	....	....
	NHL Totals		**101**	**6**	**13**	**19**	**20**	**0**	**0**	**4**	**126**	**4.8**		**1022**	**50.9**	**77**	**41**	**12:12**	**6**	**0**	**1**	**1**	**0**	**0**	**0**	**0**

• Missed majority of 2000-01 season recovering from shoulder injury suffered in game vs. Manitoba (IHL), November 15, 2000.

CHURCH, Brad

(CHUHRCH, BRAD)

Left wing. Shoots left. 6'1", 210 lbs. Born, Dauphin, Man., November 14, 1976. Washington's 1st choice, 17th overall, in 1995 Entry Draft.

			Regular Season																Playoffs							
Season	Club	League	GP	G	A	Pts	PIM	PP	SH	GW	S	%	+/-	TF	F%	H	SB	Min	GP	G	A	Pts	PIM	PP	SH	GW
1991-92	Parkland Rangers	MMHL	26	15	17	32	62																			
1992-93	Dauphin Kings	MJHL	45	15	23	38	80																			
1993-94	Prince Albert	WHL	71	33	20	53	197																			
1994-95	Prince Albert	WHL	62	26	24	50	184												15	6	9	15	32			
1995-96	Prince Albert	WHL	69	42	46	88	123												18	15	*20	*35	74			
1996-97	Portland Pirates	AHL	50	4	8	12	92												1	0	0	0	0			
1997-98	**Washington**	**NHL**	**2**	**0**	**0**	**0**	**0**	**0**	**0**	**0**	**4**	**0.0**	**0**													
	Portland Pirates	AHL	59	6	5	11	98												9	2	4	6	14			
1998-99	Portland Pirates	AHL	10	1	3	4	18																			
	Hampton Roads	ECHL	24	10	9	19	129																			
	Hamilton	AHL	9	0	2	2	4																			
	New Orleans	ECHL	5	3	4	7	4												11	1	1	2	22			
99-2000	Hampton Roads	ECHL	11	4	3	7	31																			
	Portland Pirates	AHL	56	9	17	26	52												4	1	1	2	4			
2000-01	Portland Pirates	AHL	61	14	18	32	90												3	1	1	2	18			
2001-02	Portland Pirates	AHL	45	6	13	19	52																			
	Richmond	ECHL	8	6	5	11	13																			
	Lowell	AHL	9	2	2	4	2												4	0	0	0	4			
	NHL Totals		**2**	**0**	**0**	**0**	**0**	**0**	**0**	**0**	**4**	**0.0**														

Traded to **Edmonton** by **Washington** for the rights to Barrie Moore, February 3, 1999. Signed as a free agent by **Portland** (AHL), June 15, 2000.

CIBAK, Martin

(TSEE-bak, MAHR-tihn) **T.B.**

Center. Shoots left. 6'1", 195 lbs. Born, Liptovmikulas, Czech., May 17, 1980. Tampa Bay's 11th choice, 252nd overall, in 1998 Entry Draft.

			Regular Season																Playoffs							
Season	Club	League	GP	G	A	Pts	PIM	PP	SH	GW	S	%	+/-	TF	F%	H	SB	Min	GP	G	A	Pts	PIM	PP	SH	GW
1995-96	L. Mikulas Jr.	Slovak-Jr.	48	38	35	73																				
1996-97	L. Mikulas Jr.	Slovak-Jr.	45	22	18	40																				
1997-98	L. Mikulas Jr.	Slovak-Jr.	42	31	21	52																				
	Liptov. Mikulas	Slovakia	28	1	3	4	10																			
1998-99	Medicine Hat	WHL	66	21	26	47	72																			
99-2000	Medicine Hat	WHL	58	16	29	45	77																			
2000-01	Detroit Vipers	IHL	79	10	28	38	88																			
2001-02	**Tampa Bay**	**NHL**	**26**	**1**	**5**	**6**	**8**	**0**	**0**	**0**	**22**	**4.5**	**−6**	**85**	**34.1**	**37**	**4**	**11:08**								
	Springfield	AHL	52	5	9	14	44																			
	NHL Totals		**26**	**1**	**5**	**6**	**8**	**0**	**0**	**0**	**22**	**4.5**		**85**	**34.1**	**37**	**4**	**11:08**								

CIERNIK, Ivan

(CHAIR-nihk, ee-VAHN) **WSH.**

Left wing. Shoots left. 6'1", 234 lbs. Born, Levice, Czech., October 30, 1977. Ottawa's 6th choice, 216th overall, in 1996 Entry Draft.

			Regular Season																Playoffs							
Season	Club	League	GP	G	A	Pts	PIM	PP	SH	GW	S	%	+/-	TF	F%	H	SB	Min	GP	G	A	Pts	PIM	PP	SH	GW
1994-95	HC Nitra Jr.	Slovak-Jr.	30	22	15	37	36																			
	HC Nitra	Slovakia	7	1	0	1	2																			
1995-96	HC Nitra	Slovakia	35	9	7	16	36												8	3	3	6				
1996-97	Nitra	Slovakia	41	11	19	30																				
1997-98	**Ottawa**	**NHL**	**2**	**0**	**0**	**0**	**0**	**0**	**0**	**0**	**0**	**0.0**	**0**													
	Worcester	AHL	53	9	12	21	38												1	0	0	0	2			
1998-99	Adirondack	AHL	21	1	4	5	4																			
	Cincinnati	AHL	32	10	3	13	10												2	0	0	0	2			
99-2000	Grand Rapids	IHL	66	13	12	25	64												6	0	6	6	2			
2000-01	**Ottawa**	**NHL**	**4**	**2**	**0**	**2**	**2**	**0**	**0**	**0**	**7**	**28.6**	**2**	**0**	**0.0**	**0**	**0**	**7:41**								
	Grand Rapids	IHL	66	27	38	65	53												10	5	6	11	26			
2001-02	**Ottawa**	**NHL**	**23**	**1**	**2**	**3**	**4**	**0**	**0**	**0**	**18**	**5.6**	**0**	**7**	**71.4**	**16**	**3**	**7:51**								
	Grand Rapids	AHL	2	2	1	3	0																			
	Washington	**NHL**	**6**	**0**	**1**	**1**	**2**	**0**	**0**	**0**	**5**	**0.0**	**0**	**0**	**0.0**	**1**	**4**	**9:15**								
	Portland Pirates	AHL	26	10	5	15	28																			
	NHL Totals		**35**	**3**	**3**	**6**	**8**	**0**	**0**	**0**	**30**	**10.0**		**7**	**71.4**	**17**	**7**	**8:05**								

Loaned to **Cincinnati** (AHL) by **Ottawa** with Ratislav Pavlikovsky and Erich Goldmann, January 12, 1999. Claimed on waivers by **Washington** from **Ottawa**, January 19, 2002.

CIGER, Zdeno

(SEE-guhr, ZDEH-noh)

Left wing. Shoots left. 6'1", 190 lbs. Born, Martin, Czech., October 19, 1969. New Jersey's 3rd choice, 54th overall, in 1988 Entry Draft.

			Regular Season																Playoffs							
Season	Club	League	GP	G	A	Pts	PIM	PP	SH	GW	S	%	+/-	TF	F%	H	SB	Min	GP	G	A	Pts	PIM	PP	SH	GW
1987-88	Dukla Trencin	Czech	8	3	4	7	2																			
1988-89	Dukla Trencin	Czech	43	18	13	31	18																			
1989-90	Dukla Trencin	Czech	44	17	24	41													9	1	4	5				
1990-91	**New Jersey**	**NHL**	**45**	**8**	**17**	**25**	**8**	**2**	**0**	**1**	**82**	**9.8**	**3**						**6**	**0**	**2**	**2**	**4**	**0**	**0**	**0**
	Utica Devils	AHL	8	5	4	9	2																			
1991-92	**New Jersey**	**NHL**	**20**	**6**	**5**	**11**	**10**	**1**	**0**	**0**	**33**	**18.2**	**−2**						**7**	**2**	**4**	**6**	**0**	**0**	**0**	**1**
1992-93	**New Jersey**	**NHL**	**27**	**4**	**8**	**12**	**2**	**2**	**0**	**1**	**39**	**10.3**	**−8**													
	Edmonton	**NHL**	**37**	**9**	**15**	**24**	**6**	**0**	**0**	**1**	**67**	**13.4**	**−5**													
1993-94	**Edmonton**	**NHL**	**84**	**22**	**35**	**57**	**8**	**8**	**0**	**1**	**158**	**13.9**	**−11**													
1994-95	Dukla Trencin	Slovakia	34	23	25	48	8												9	2	9	11	2			
	Edmonton	**NHL**	**5**	**2**	**2**	**4**	**0**	**1**	**0**	**1**	**10**	**20.0**	**−1**													
1995-96	**Edmonton**	**NHL**	**78**	**31**	**39**	**70**	**41**	**12**	**0**	**3**	**184**	**16.8**	**−15**													
1996-97	Slov. Bratislava	Slovakia	44	26	27	53													2	1	3	4				
	Slov. Bratislava	EuroHL	6	4	5	9	2												2	1	1	2	2			
1997-98	Slov. Bratislava	Slovakia	36	14	*31	45	2												11	6	*10	*16	4			
	Slov. Bratislava	EuroHL	8	1	5	6	6																			
	Slovakia	Olympics	4	1	1	2	4																			
1998-99	Slov. Bratislava	Slovakia	40	26	32	*58	8												9	3	*10	13	2			
	Slov. Bratislava	EuroHL	6	4	5	9	8																			
99-2000	Slov. Bratislava	Slovakia	51	23	39	62	48												8	1	*8	*9	0			
	Slov. Bratislava	EuroHL	6	3	4	7	2												2	0	2	2	4			
2000-01	Slov. Bratislava	Slovakia	53	17	32	49	22												8	6	3	9	16			
2001-02	**NY Rangers**	**NHL**	**29**	**6**	**7**	**13**	**16**	**1**	**0**	**1**	**48**	**12.5**	**−3**	**0**	**0.0**	**19**	**13**	**14:30**								
	Tampa Bay	**NHL**	**27**	**6**	**6**	**12**	**10**	**0**	**0**	**2**	**44**	**13.6**	**−12**	**8**	**62.5**	**17**	**7**	**16:08**								
	NHL Totals		**352**	**94**	**134**	**228**	**101**	**27**	**0**	**11**	**665**	**14.1**		**8**	**62.5**	**36**	**20**	**15:17**	**13**	**2**	**6**	**8**	**4**	**0**	**0**	**1**

Traded to **Edmonton** by **New Jersey** with Kevin Todd for Bernie Nicholls, January 13, 1993. Claimed by **Nashville** from **Edmonton** in Waiver Draft, October 5, 1998. Claimed by **Minnesota** from **Nashville** in Waiver Draft, September 29, 2000. Signed as a free agent by **NY Rangers**, July 17, 2001. Traded to **Tampa Bay** by **NY Rangers** for Matthew Barnaby, December 12, 2001.

CISAR, Marian

(SIH-sahr, MAIR-eean) **NSH.**

Right wing. Shoots right. 6', 197 lbs. Born, Bratislava, Czech., February 25, 1978. Los Angeles' 2nd choice, 37th overall, in 1996 Entry Draft.

			Regular Season																Playoffs							
Season	Club	League	GP	G	A	Pts	PIM	PP	SH	GW	S	%	+/-	TF	F%	H	SB	Min	GP	G	A	Pts	PIM	PP	SH	GW
1994-95	S. Bratislava Jr.	Slovak-Jr.	38	42	28	70	16																			
1995-96	S. Bratislava Jr.	Slovak-Jr.	16	26	17	43	2																			
	Slov. Bratislava	Slovakia	13	3	3	6	0												6	3	0	3	0			
1996-97	Spokane Chiefs	WHL	70	31	35	66	52												9	6	2	8	4			
1997-98	Spokane Chiefs	WHL	52	33	40	73	34												18	8	5	13	8			
1998-99	Milwaukee	IHL	51	11	17	28	31												2	0	0	0	12			
99-2000	**Nashville**	**NHL**	**3**	**0**	**0**	**0**	**4**	**0**	**0**	**0**	**2**	**0.0**	**−2**	**0**	**0.0**	**1**	**1**	**8:18**								
	Milwaukee	IHL	78	20	32	52	82												1	0	0	0	0			
2000-01	**Nashville**	**NHL**	**60**	**12**	**15**	**27**	**45**	**5**	**0**	**1**	**97**	**12.4**	**−7**	**0**	**0.0**	**50**	**13**	**13:10**								
	Milwaukee	IHL	15	4	7	11	4																			
2001-02	**Nashville**	**NHL**	**10**	**1**	**2**	**3**	**8**	**1**	**0**	**0**	**16**	**6.3**	**−3**	**1**	**100.0**	**14**	**4**	**12:55**								
	Milwaukee	AHL	2	1	0	1	2																			
	NHL Totals		**73**	**13**	**17**	**30**	**57**	**6**	**0**	**1**	**115**	**11.3**		**1**	**100.0**	**65**	**18**	**12:56**								

Traded to **Nashville** by **LA Kings** for future considerations, June 1, 1998. • Missed majority of 2001-02 season recovering from hip injury suffered in training camp, September 29, 2001. Signed as a free agent by **HC Znojemsti** (Czech) with Nashville retaining NHL rights, July 22, 2002.

CLARK, Brett

(KLAHRK, BREHT) **COL.**

Defense. Shoots left. 6'1", 195 lbs. Born, Wapella, Sask., December 23, 1976. Montreal's 7th choice, 154th overall, in 1996 Entry Draft.

			Regular Season																Playoffs							
Season	Club	League	GP	G	A	Pts	PIM	PP	SH	GW	S	%	+/-	TF	F%	H	SB	Min	GP	G	A	Pts	PIM	PP	SH	GW
1994-95	Melville	SJHL	62	19	32	51	77	....	....	....	....	....	....	....	....	....	....	....	....	....	....	....	....	....	....	....
1995-96	U. of Maine	H-East	39	7	31	38	22	....	....	....	....	....	....	....	....	....	....	....	....	....	....	....	....	....	....	....
1996-97	Team Canada	Nat-Tm	57	6	21	27	52	....	....	....	....	....	....	....	....	....	....	....	....	....	....	....	....	....	....	....
1997-98	**Montreal**	**NHL**	**41**	**1**	**0**	**1**	**20**	**0**	**0**	**0**	**26**	**3.8**	**–3**	....	....	....	....	....	....	....	....	....	....	....	....	....
	Fredericton	AHL	20	0	6	6	6	....	....	....	....	....	....	....	....	....	....	....	4	0	1	1	17	....	....	....
1998-99	**Montreal**	**NHL**	**61**	**2**	**2**	**4**	**16**	**0**	**0**	**0**	**36**	**5.6**	**–3**	**0**	**0.0**	**62**	**43**	**13:11**	....	....	....	....	....	....	....	....
	Fredericton	AHL	3	1	0	1	0	....	....	....	....	....	....	....	....	....	....	....	....	....	....	....	....	....	....	....
99-2000	**Atlanta**	**NHL**	**14**	**0**	**1**	**1**	**4**	**0**	**0**	**0**	**13**	**0.0**	**–12**	**0**	**0.0**	**27**	**19**	**16:51**	....	....	....	....	....	....	....	....
	Orlando	IHL	63	9	17	26	31	....	....	....	....	....	....	....	....	....	....	....	6	0	1	1	0	....	....	....
2000-01	**Atlanta**	**NHL**	**28**	**1**	**2**	**3**	**14**	**0**	**0**	**0**	**35**	**2.9**	**–12**	**0**	**0.0**	**42**	**21**	**18:02**	....	....	....	....	....	....	....	....
	Orlando	IHL	43	2	9	11	32	....	....	....	....	....	....	....	....	....	....	....	15	1	6	7	2	....	....	....
2001-02	**Atlanta**	**NHL**	**2**	**0**	**0**	**0**	**0**	**0**	**0**	**0**	**0**	**0.0**	**–3**	**1**	**100.0**	**1**	**2**	**15:32**	....	....	....	....	....	....	....	....
	Chicago Wolves	AHL	42	3	17	20	18	....	....	....	....	....	....	....	....	....	....	....	....	....	....	....	....	....	....	....
	Hershey Bears	AHL	32	7	9	16	12	....	....	....	....	....	....	....	....	....	....	....	8	0	2	2	6	....	....	....
	NHL Totals		**146**	**4**	**5**	**9**	**54**	**0**	**0**	**0**	**110**	**3.6**		**1**	**100.0**	**132**	**85**	**15:01**	....	....	....	....	....	....	....	....

Claimed by **Atlanta** from **Montreal** in Expansion Draft, June 25, 1999. Traded to **Colorado** by **Atlanta** for Frederic Cassivi, January 24, 2002.

CLARK, Chris

(KLAHRK, KRIHS) **CGY.**

Right wing. Shoots right. 6', 200 lbs. Born, South Windsor, CT, March 8, 1976. Calgary's 3rd choice, 77th overall, in 1994 Entry Draft.

			Regular Season																Playoffs							
Season	Club	League	GP	G	A	Pts	PIM	PP	SH	GW	S	%	+/-	TF	F%	H	SB	Min	GP	G	A	Pts	PIM	PP	SH	GW
1990-91	South Windsor	Hi-School	23	16	15	31	24	....	....	....	....	....	....	....	....	....	....	....	....	....	....	....	....	....	....	....
1991-92	Springfield	NEJHL	49	21	29	50	56	....	....	....	....	....	....	....	....	....	....	....	....	....	....	....	....	....	....	....
1992-93	Springfield	NEJHL	43	17	60	77	120	....	....	....	....	....	....	....	....	....	....	....	....	....	....	....	....	....	....	....
1993-94	Springfield	NEJHL	35	31	26	57	185	....	....	....	....	....	....	....	....	....	....	....	....	....	....	....	....	....	....	....
1994-95	Clarkson Knights	ECAC	32	12	11	23	92	....	....	....	....	....	....	....	....	....	....	....	....	....	....	....	....	....	....	....
1995-96	Clarkson Knights	ECAC	38	10	8	18	108	....	....	....	....	....	....	....	....	....	....	....	....	....	....	....	....	....	....	....
1996-97	Clarkson Knights	ECAC	37	23	25	48	*86	....	....	....	....	....	....	....	....	....	....	....	....	....	....	....	....	....	....	....
1997-98	Clarkson Knights	ECAC	35	18	21	39	*106	....	....	....	....	....	....	....	....	....	....	....	....	....	....	....	....	....	....	....
1998-99	Saint John	AHL	73	13	27	40	123	....	....	....	....	....	....	....	....	....	....	....	7	2	4	6	15	....	....	....
99-2000	**Calgary**	**NHL**	**22**	**0**	**1**	**1**	**14**	**0**	**0**	**0**	**17**	**0.0**	**–3**	**0**	**0.0**	**22**	**3**	**9:02**	....	....	....	....	....	....	....	....
	Saint John	AHL	48	16	17	33	134	....	....	....	....	....	....	....	....	....	....	....	....	....	....	....	....	....	....	....
2000-01	**Calgary**	**NHL**	**29**	**5**	**1**	**6**	**38**	**1**	**0**	**0**	**43**	**11.6**	**0**	**3**	**33.3**	**40**	**10**	**11:56**	....	....	....	....	....	....	....	....
	Saint John	AHL	48	18	17	35	131	....	....	....	....	....	....	....	....	....	....	....	18	4	10	14	49	....	....	....
2001-02	**Calgary**	**NHL**	**64**	**10**	**7**	**17**	**79**	**2**	**1**	**4**	**109**	**9.2**	**–12**	**21**	**33.3**	**85**	**29**	**13:57**	....	....	....	....	....	....	....	....
	NHL Totals		**115**	**15**	**9**	**24**	**131**	**3**	**1**	**4**	**169**	**8.9**		**24**	**33.3**	**147**	**42**	**12:30**	....	....	....	....	....	....	....	....

ECAC Second All-Star Team (1998)

CLARKE, Dale

(KLAHRK, DAIL) **ST.L.**

Defense. Shoots right. 6'2", 193 lbs. Born, Belleville, Ont., March 23, 1978.

			Regular Season																Playoffs							
Season	Club	League	GP	G	A	Pts	PIM	PP	SH	GW	S	%	+/-	TF	F%	H	SB	Min	GP	G	A	Pts	PIM	PP	SH	GW
1994-95	Wellington Dukes	MTJHL	48	2	13	15	18	....	....	....	....	....	....	....	....	....	....	....	....	....	....	....	....	....	....	....
1995-96	Wellington Dukes	MTJHL	51	6	24	30	78	....	....	....	....	....	....	....	....	....	....	....	....	....	....	....	....	....	....	....
1996-97	St. Lawrence	ECAC	34	1	6	7	20	....	....	....	....	....	....	....	....	....	....	....	....	....	....	....	....	....	....	....
1997-98	St. Lawrence	ECAC	33	1	6	7	66	....	....	....	....	....	....	....	....	....	....	....	....	....	....	....	....	....	....	....
1998-99	St. Lawrence	ECAC	39	3	13	16	44	....	....	....	....	....	....	....	....	....	....	....	....	....	....	....	....	....	....	....
99-2000	St. Lawrence	ECAC	36	6	17	23	24	....	....	....	....	....	....	....	....	....	....	....	....	....	....	....	....	....	....	....
	Worcester	AHL	....	....	....	....	....	....	....	....	....	....	....	....	....	....	....	....	2	0	0	0	0	....	....	....
2000-01	**St. Louis**	**NHL**	**3**	**0**	**0**	**0**	**0**	**0**	**0**	**0**	**5**	**0.0**	**1**	**0**	**0.0**	**3**	**1**	**13:38**	....	....	....	....	....	....	....	....
	Peoria Rivermen	ECHL	2	1	0	1	0	....	....	....	....	....	....	....	....	....	....	....	....	....	....	....	....	....	....	....
	Worcester	AHL	67	7	25	32	26	....	....	....	....	....	....	....	....	....	....	....	1	0	0	0	0	....	....	....
2001-02	Worcester	AHL	72	2	10	12	32	....	....	....	....	....	....	....	....	....	....	....	....	....	....	....	....	....	....	....
	NHL Totals		**3**	**0**	**0**	**0**	**0**	**0**	**0**	**0**	**5**	**0.0**		**0**	**0.0**	**3**	**1**	**13:38**	....	....	....	....	....	....	....	....

Signed as a free agent by **St. Louis**, July 24, 1999.

CLASSEN, Greg

(KLAW-sihn, GREHG) **NSH.**

Center. Shoots left. 6'1", 200 lbs. Born, Aylsham, Sask., August 24, 1977.

			Regular Season																Playoffs							
Season	Club	League	GP	G	A	Pts	PIM	PP	SH	GW	S	%	+/-	TF	F%	H	SB	Min	GP	G	A	Pts	PIM	PP	SH	GW
1997-98	Nipawin Hawks	SJHL	59	32	50	82	50	....	....	....	....	....	....	....	....	....	....	....	14	8	13	21	6	....	....	....
1998-99	Merrimack	H-East	36	14	11	25	28	....	....	....	....	....	....	....	....	....	....	....	....	....	....	....	....	....	....	....
99-2000	Merrimack	H-East	36	14	16	30	16	....	....	....	....	....	....	....	....	....	....	....	....	....	....	....	....	....	....	....
	Milwaukee	IHL	11	1	0	1	2	....	....	....	....	....	....	....	....	....	....	....	2	0	0	0	2	....	....	....
2000-01	**Nashville**	**NHL**	**27**	**2**	**4**	**6**	**14**	**1**	**0**	**0**	**18**	**11.1**	**–4**	**195**	**42.1**	**21**	**4**	**10:16**	....	....	....	....	....	....	....	....
	Milwaukee	IHL	23	5	10	15	31	....	....	....	....	....	....	....	....	....	....	....	5	0	0	0	0	....	....	....
2001-02	**Nashville**	**NHL**	**55**	**5**	**6**	**11**	**30**	**0**	**1**	**0**	**32**	**15.6**	**1**	**389**	**43.2**	**72**	**15**	**10:09**	....	....	....	....	....	....	....	....
	Milwaukee	AHL	8	2	4	6	12	....	....	....	....	....	....	....	....	....	....	....	....	....	....	....	....	....	....	....
	NHL Totals		**82**	**7**	**10**	**17**	**44**	**1**	**1**	**0**	**50**	**14.0**		**584**	**42.8**	**93**	**19**	**10:11**	....	....	....	....	....	....	....	....

Hockey East All-Rookie Team (1999)

Signed as a free agent by **Nashville**, March 27, 2000.

CLEARY, Daniel

(KLIH-ree, DAN-yehl) **EDM.**

Left wing. Shoots left. 6', 203 lbs. Born, Carbonear, Nfld., December 18, 1978. Chicago's 1st choice, 13th overall, in 1997 Entry Draft.

			Regular Season																Playoffs							
Season	Club	League	GP	G	A	Pts	PIM	PP	SH	GW	S	%	+/-	TF	F%	H	SB	Min	GP	G	A	Pts	PIM	PP	SH	GW
1993-94	Kingston	MTJHL	41	18	28	46	33	....	....	....	....	....	....	....	....	....	....	....	2	0	1	1	0	....	....	....
1994-95	Belleville Bulls	OHL	62	26	55	81	62	....	....	....	....	....	....	....	....	....	....	....	16	7	10	17	23	....	....	....
1995-96	Belleville Bulls	OHL	64	53	62	115	74	....	....	....	....	....	....	....	....	....	....	....	14	10	17	27	40	....	....	....
1996-97	Belleville Bulls	OHL	64	32	48	80	88	....	....	....	....	....	....	....	....	....	....	....	6	3	4	7	6	....	....	....
1997-98	Belleville Bulls	OHL	30	16	31	47	14	....	....	....	....	....	....	....	....	....	....	....	10	6	*17	*23	10	....	....	....
	Chicago	**NHL**	**6**	**0**	**0**	**0**	**0**	**0**	**0**	**0**	**4**	**0.0**	**–2**	....	....	....	....	....	....	....	....	....	....	....	....	....
	Indianapolis Ice	IHL	4	2	1	3	6	....	....	....	....	....	....	....	....	....	....	....	....	....	....	....	....	....	....	....
1998-99	**Chicago**	**NHL**	**35**	**4**	**5**	**9**	**24**	**0**	**0**	**0**	**49**	**8.2**	**–1**	**13**	**46.2**	**28**	**9**	**14:21**	....	....	....	....	....	....	....	....
	Portland Pirates	AHL	30	9	17	26	74	....	....	....	....	....	....	....	....	....	....	....	....	....	....	....	....	....	....	....
	Hamilton	AHL	9	0	1	1	7	....	....	....	....	....	....	....	....	....	....	....	3	0	0	0	0	....	....	....
99-2000	**Edmonton**	**NHL**	**17**	**3**	**2**	**5**	**8**	**0**	**0**	**1**	**18**	**16.7**	**–1**	**1**	**100.0**	**16**	**2**	**9:44**	**4**	**0**	**1**	**1**	**2**	**0**	**0**	**0**
	Hamilton	AHL	58	22	52	74	108	....	....	....	....	....	....	....	....	....	....	....	5	2	3	5	18	....	....	....
2000-01	**Edmonton**	**NHL**	**81**	**14**	**21**	**35**	**37**	**2**	**0**	**2**	**107**	**13.1**	**5**	**13**	**23.1**	**73**	**26**	**12:58**	**6**	**1**	**1**	**2**	**8**	**1**	**0**	**0**
2001-02	**Edmonton**	**NHL**	**65**	**10**	**19**	**29**	**51**	**2**	**1**	**1**	**75**	**13.3**	**–1**	**5**	**60.0**	**49**	**21**	**12:43**	....	....	....	....	....	....	....	....
	NHL Totals		**204**	**31**	**47**	**78**	**120**	**4**	**1**	**4**	**253**	**12.3**		**32**	**40.6**	**166**	**58**	**12:51**	**10**	**1**	**2**	**3**	**10**	**1**	**0**	**0**

OHL All-Rookie Team (1995) • OHL First All-Star Team (1996, 1997) • AHL Second All-Star Team (2000)

Traded to **Edmonton** by **Chicago** with Chad Kilger, Ethan Moreau and Christian Laflamme for Boris Mironov, Dean McAmmond and Jonas Elofsson, March 20, 1999.

CLYMER, Ben

(KLIH-mehr, BEHN) **T.B.**

Left wing. Shoots right. 6'1", 199 lbs. Born, Edina, MN, April 11, 1978. Boston's 3rd choice, 27th overall, in 1997 Entry Draft.

			Regular Season																Playoffs							
Season	Club	League	GP	G	A	Pts	PIM	PP	SH	GW	S	%	+/-	TF	F%	H	SB	Min	GP	G	A	Pts	PIM	PP	SH	GW
1993-94	Jefferson High	Hi-School	23	3	7	10	20	....	....	....	....	....	....	....	....	....	....	....	....	....	....	....	....	....	....	....
1994-95	Jefferson High	Hi-School	28	11	22	33	36	....	....	....	....	....	....	....	....	....	....	....	....	....	....	....	....	....	....	....
1995-96	Jefferson High	Hi-School	18	12	34	46	34	....	....	....	....	....	....	....	....	....	....	....	5	0	6	6	6	....	....	....
1996-97	U. of Minnesota	WCHA	29	7	13	20	64	....	....	....	....	....	....	....	....	....	....	....	....	....	....	....	....	....	....	....
1997-98	U. of Minnesota	WCHA	1	0	0	0	2	....	....	....	....	....	....	....	....	....	....	....	....	....	....	....	....	....	....	....
1998-99	Seattle	WHL	70	12	44	56	93	....	....	....	....	....	....	....	....	....	....	....	11	1	5	6	12	....	....	....
99-2000	**Tampa Bay**	**NHL**	**60**	**2**	**6**	**8**	**87**	**2**	**0**	**0**	**98**	**2.0**	**–26**	**3**	**66.7**	**121**	**41**	**19:37**	....	....	....	....	....	....	....	....
	Detroit Vipers	IHL	19	1	9	10	30	....	....	....	....	....	....	....	....	....	....	....	....	....	....	....	....	....	....	....

Season	Club	League	GP	G	A	Pts	PIM	PP	SH	GW	S	%	+/-	TF	F%	H	SB	Min	GP	G	A	Pts	PIM	PP	SH	GW
			Regular Season																Playoffs							
2000-01	Tampa Bay	NHL	23	5	1	6	21	3	0	0	25	20.0	–7	8	25.0	33	5	13:03								
	Detroit Vipers	IHL	53	5	8	13	88																			
2001-02	Tampa Bay	NHL	81	14	20	34	36	4	0	2	151	9.3	–10	14	28.6	123	33	17:26								
	NHL Totals		164	21	27	48	144	9	0	2	274	7.7		25	32.0	277	79	17:37								

• Missed majority of 1997-98 season recovering from shoulder injury suffered in game vs. U. of Michigan (CCHA), October 10, 1997. Signed as a free agent by **Tampa Bay**, October 2, 1999.

COLE, Erik

(KOHL, AIR-ihk) **CAR.**

Left wing. Shoots left. 6'1", 200 lbs. Born, Oswego, NY, November 6, 1978. Carolina's 3rd choice, 71st overall, in 1998 Entry Draft.

Season	Club	League	GP	G	A	Pts	PIM	PP	SH	GW	S	%	+/-	TF	F%	H	SB	Min	GP	G	A	Pts	PIM	PP	SH	GW
1995-96	Oswego	Hi-School	40	49	41	90																				
1996-97	Des Moines	USHL	48	30	34	64	140												5	2	0	2	6			
1997-98	Clarkson Knights	ECAC	34	11	20	31	55																			
1998-99	Clarkson Knights	ECAC	36	*22	20	42	50																			
99-2000	Clarkson Knights	ECAC	33	19	11	30	46																			
	Cincinnati	IHL	9	4	3	7	2												7	1	1	2	2			
2000-01	Cincinnati	IHL	69	23	20	43	28												5	1	0	1	2			
2001-02	Carolina	NHL	81	16	24	40	35	3	0	2	159	10.1	–10	17	47.1	257	25	16:04	23	6	3	9	30	1	0	1
	NHL Totals		81	16	24	40	35	3	0	2	159	10.1		17	47.1	257	25	16:04	23	6	3	9	30	1	0	1

ECAC Rookie of the Year (Shared with Willie Mitchell) (1998) • ECAC First All-Star Team (1999) • NCAA East Second All-American Team (1999) • ECAC Second All-Star Team (2000)

COMMODORE, Mike

(KAWM-uh-dohr, MIGHK) **ANA.**

Defense. Shoots right. 6'4", 230 lbs. Born, Fort Saskatchewan, Alta., November 7, 1979. New Jersey's 2nd choice, 42nd overall, in 1999 Entry Draft.

Season	Club	League	GP	G	A	Pts	PIM	PP	SH	GW	S	%	+/-	TF	F%	H	SB	Min	GP	G	A	Pts	PIM	PP	SH	GW
1996-97	Ft. Saskatchewan	AJHL	51	3	8	11	244																			
1997-98	North Dakota	WCHA	29	0	5	5	74																			
1998-99	North Dakota	WCHA	39	5	8	13	154																			
99-2000	North Dakota	WCHA	38	5	7	12	*154																			
2000-01	New Jersey	NHL	20	1	4	5	14	0	0	0	11	9.1	5	0	0.0	43	17	12:46								
	Albany	AHL	41	2	5	7	59																			
2001-02	New Jersey	NHL	37	0	1	1	30	0	0	0	22	0.0	–12	0	0.0	82	35	12:37								
	Albany	AHL	14	0	3	3	31																			
	NHL Totals		57	1	5	6	44	0	0	0	33	3.0		0	0.0	125	52	12:40								

NCAA Championship All-Tournament Team (2000)

Traded to **Anaheim** by **New Jersey** with Petr Sykora, Jean-Francois Damphousse and Igor Pohanka for Jeff Friesen, Oleg Tverdovsky and Maxim Balmochnykh, July 6, 2002.

COMRIE, Mike

(KAWM-ree, MIGHK) **EDM.**

Center. Shoots left. 5'9", 175 lbs. Born, Edmonton, Alta., September 11, 1980. Edmonton's 5th choice, 91st overall, in 1999 Entry Draft.

Season	Club	League	GP	G	A	Pts	PIM	PP	SH	GW	S	%	+/-	TF	F%	H	SB	Min	GP	G	A	Pts	PIM	PP	SH	GW
1995-96	Edmonton SSAC	AMHL	33	51	52	103																				
1996-97	St. Albert	AJHL	63	37	41	78	44																			
1997-98	St. Albert	AJHL	58	*60	*78	*138	134												19	*24	*24	*48	51			
1998-99	U. of Michigan	CCHA	42	19	25	44	38																			
99-2000	U. of Michigan	CCHA	40	24	35	59	95																			
2000-01	Kootenay Ice	WHL	37	39	40	79	79																			
	Edmonton	NHL	41	8	14	22	14	3	0	1	62	12.9	6	372	43.3	12	13	11:23	6	1	2	3	0	1	0	1
2001-02	Edmonton	NHL	82	33	27	60	45	8	0	5	170	19.4	16	1198	47.3	27	42	17:32								
	NHL Totals		123	41	41	82	59	11	0	6	232	17.7		1570	46.4	39	55	15:29	6	1	2	3	0	1	0	1

AJHL Rookie of the Year (1997) • AJHL MVP (1998) • Canadian Junior "A" Player of the Year (1998) • CCHA All-Rookie Team (1999) • CCHA First All-Star Team (1999) • CCHA Rookie of the Year (1999) • CCHA First All-Star Team (2000) • NCAA West Second All-American Team (2000)

• Left **University of Michigan** (CCHA) and signed as a free agent by **Kootenay** (WHL), August 23, 2000. • Left **Kootenay** (WHL) and signed with **Edmonton**, December 30, 2000.

CONNOLLY, Tim

(KAHN-noh-lee, TIHM) **BUF.**

Center. Shoots right. 6'1", 182 lbs. Born, Syracuse, NY, May 7, 1981. NY Islanders' 1st choice, 5th overall, in 1999 Entry Draft.

Season	Club	League	GP	G	A	Pts	PIM	PP	SH	GW	S	%	+/-	TF	F%	H	SB	Min	GP	G	A	Pts	PIM	PP	SH	GW
1996-97	Syracuse	MTJHL	50	42	62	104	34																			
1997-98	Erie Otters	OHL	59	30	32	62	32												7	1	6	7	6			
1998-99	Erie Otters	OHL	46	34	34	68	50																			
99-2000	NY Islanders	NHL	81	14	20	34	44	2	1	1	114	12.3	–25	786	36.3	51	29	16:18								
2000-01	NY Islanders	NHL	82	10	31	41	42	5	0	0	171	5.8	–14	989	41.7	37	47	20:02								
2001-02	Buffalo	NHL	82	10	35	45	34	3	0	3	126	7.9	4	1074	39.6	45	37	16:58								
	NHL Totals		245	34	86	120	120	10	1	4	411	8.3		2849	39.4	133	113	17:46								

Traded to **Buffalo** by **NY Islanders** with Taylor Pyatt for Michael Peca, June 24, 2001.

CONROY, Craig

(KAWN-roi, KRAYG) **CGY.**

Center. Shoots right. 6'2", 197 lbs. Born, Potsdam, NY, September 4, 1971. Montreal's 7th choice, 123rd overall, in 1990 Entry Draft.

Season	Club	League	GP	G	A	Pts	PIM	PP	SH	GW	S	%	+/-	TF	F%	H	SB	Min	GP	G	A	Pts	PIM	PP	SH	GW
1989-90	Northfield Prep	Hi-School	31	33	43	76																				
1990-91	Clarkson Knights	ECAC	40	8	21	29	24																			
1991-92	Clarkson Knights	ECAC	31	19	17	36	36																			
1992-93	Clarkson Knights	ECAC	35	10	23	33	26																			
1993-94	Clarkson Knights	ECAC	34	26	*40	*66	46																			
1994-95	Fredericton	AHL	55	26	18	44	29												11	7	3	10	6			
	Montreal	NHL	6	1	0	1	0	0	0	0	4	25.0	–1													
1995-96	Montreal	NHL	7	0	0	0	2	0	0	0	1	0.0	–4													
	Fredericton	AHL	67	31	38	69	65												10	5	7	12	6			
1996-97	Fredericton	AHL	9	10	6	16	10																			
	St. Louis	NHL	61	6	11	17	43	0	0	1	74	8.1	0						6	0	0	0	8	0	0	0
	Worcester	AHL	5	5	6	11	2																			
1997-98	St. Louis	NHL	81	14	29	43	46	0	3	1	118	11.9	20						10	1	2	3	8	0	0	1
1998-99	St. Louis	NHL	69	14	25	39	38	0	1	1	134	10.4	14	1190	54.6	77	35	16:39	13	2	1	3	6	0	0	0
99-2000	St. Louis	NHL	79	12	15	27	36	1	2	3	98	12.2	5	1339	53.6	105	29	14:48	7	0	2	2	2	0	0	0
2000-01	St. Louis	NHL	69	11	14	25	46	0	3	2	101	10.9	2	729	55.1	101	30	14:01								
	Calgary	NHL	14	3	4	7	14	0	1	0	32	9.4	0	264	52.7	16	11	18:08								
2001-02	Calgary	NHL	81	27	48	75	32	7	2	4	146	18.5	24	1654	54.3	99	37	20:56								
	NHL Totals		467	88	146	234	257	8	12	12	708	12.4		5176	54.2	398	142	16:47	36	3	5	8	24	0	0	1

ECAC First All-Star Team (1994) • NCAA East First All-American Team (1994) • NCAA Final Four All-Tournament Team (1994)

Traded to **St. Louis** by **Montreal** with Pierre Turgeon and Rory Fitzpatrick for Murray Baron, Shayne Corson and St. Louis' 5th round choice (Gennady Razin) in 1997 Entry Draft, October 29, 1996. Traded to **Calgary** by **St. Louis** with St. Louis' 7th round choice (David Moss) in 2001 Entry Draft for Cory Stillman, March 13, 2001.

COOKE, Matt

(KUK, MAT) **VAN.**

Left wing. Shoots left. 5'11", 205 lbs. Born, Belleville, Ont., September 7, 1978. Vancouver's 8th choice, 144th overall, in 1997 Entry Draft.

Season	Club	League	GP	G	A	Pts	PIM	PP	SH	GW	S	%	+/-	TF	F%	H	SB	Min	GP	G	A	Pts	PIM	PP	SH	GW
1994-95	Wellington Dukes	MTJHL	46	9	23	32	62																			
1995-96	Windsor	OHL	61	8	11	19	102												7	1	3	4	6			
1996-97	Windsor	OHL	65	45	50	95	146												5	5	5	10	10			
1997-98	Windsor	OHL	23	14	19	33	50																			
	Kingston	OHL	25	8	13	21	49												12	8	8	16	20			
1998-99	Vancouver	NHL	30	0	2	2	27	0	0	0	22	0.0	–12	189	40.2	43	7	8:07								
	Syracuse Crunch	AHL	37	15	18	33	119																			
99-2000	Vancouver	NHL	51	5	7	12	39	0	1	1	58	8.6	3	71	39.4	124	14	11:48								
	Syracuse Crunch	AHL	18	5	8	13	27																			
2000-01	Vancouver	NHL	81	14	13	27	94	0	2	0	121	11.6	5	321	43.0	198	40	14:35	4	0	0	0	4	0	0	0
2001-02	Vancouver	NHL	82	13	20	33	111	1	0	2	103	12.6	4	28	32.1	238	21	14:03	6	3	2	5	0	1	0	0
	NHL Totals		244	32	42	74	271	1	3	3	304	10.5		609	41.2	603	82	13:02	10	3	2	5	4	1	0	0

Traded to **Kingston** (OHL) by **Windsor** (OHL) for Brent L'Hereux, December 17, 1997.

COOPER, David

(KOO-puhr, DAY-vihd)

Defense. Shoots left. 6'2", 204 lbs. Born, Ottawa, Ont., November 2, 1973. Buffalo's 1st choice, 11th overall, in 1992 Entry Draft.

			Regular Season																Playoffs							
Season	Club	League	GP	G	A	Pts	PIM	PP	SH	GW	S	%	+/-	TF	F%	H	SB	Min	GP	G	A	Pts	PIM	PP	SH	GW
1988-89	Edmonton Mets	AJHL	32	24	22	46	151																			
1989-90	Medicine Hat	WHL	61	4	11	15	65												3	0	2	2	2			
1990-91	Medicine Hat	WHL	64	12	31	43	66												11	1	3	4	23			
1991-92	Medicine Hat	WHL	72	17	47	64	176												4	1	4	5	8			
1992-93	Medicine Hat	WHL	63	15	50	65	88												10	2	2	4	32			
	Rochester	AHL																	2	0	0	0	2			
1993-94	Rochester	AHL	68	10	25	35	82												4	1	1	2	2			
1994-95	Rochester	AHL	21	2	4	6	48																			
	South Carolina	ECHL	39	9	19	28	90												9	3	8	11	24			
1995-96	Rochester	AHL	67	9	18	27	79												8	0	1	1	12			
1996-97	**Toronto**	**NHL**	**19**	**3**	**3**	**6**	**16**	**2**	**0**	**0**	**23**	**13.0**	**-3**													
	St. John's	AHL	44	16	19	35	65																			
1997-98	**Toronto**	**NHL**	**9**	**0**	**4**	**4**	**8**	**0**	**0**	**0**	**13**	**0.0**	**2**													
	St. John's	AHL	60	19	23	42	117												4	0	1	1	6			
1998-99	Saint John	AHL	65	18	24	42	121												7	1	4	5	10			
99-2000	Kassel Huskies	Germany	55	11	13	24	82												6	2	1	3	38			
2000-01	**Toronto**	**NHL**	**2**	**0**	**0**	**0**	**0**	**0**	**0**	**0**	**3**	**0.0**	**-1**	**0**	**0.0**	**2**	**0**	**11:33**								
	St. John's	AHL	71	16	26	42	117												4	1	1	2	10			
2001-02	Eisbaren Berlin	Germany	54	15	13	28	155												4	0	1	1	6			
	NHL Totals		**30**	**3**	**7**	**10**	**24**	**2**	**0**	**0**	**39**	**7.7**		**0**	**0.0**	**2**	**0**	**11:33**								

WHL East First All-Star Team (1992) • AHL Second All-Star Team (1998)

Signed as a free agent by **Toronto**, September 26, 1996. Traded to **Calgary** by **Toronto** for Ladislav Kohn, July 2, 1998. Signed as a free agent by **Toronto**, October 16, 2000.

CORKUM, Bob

(KOHR-kuhm, BAWB)

Center. Shoots right. 6'2", 225 lbs. Born, Salisbury, MA, December 18, 1967. Buffalo's 3rd choice, 47th overall, in 1986 Entry Draft.

			Regular Season																Playoffs							
Season	Club	League	GP	G	A	Pts	PIM	PP	SH	GW	S	%	+/-	TF	F%	H	SB	Min	GP	G	A	Pts	PIM	PP	SH	GW
1984-85	Triton High	Hi-School	18	35	36	71																				
1985-86	U. of Maine	H-East	39	7	26	33	53																			
1986-87	U. of Maine	H-East	35	18	11	29	24																			
1987-88	U. of Maine	H-East	40	14	18	32	64																			
1988-89	U. of Maine	H-East	45	17	31	48	64																			
1989-90	**Buffalo**	**NHL**	**8**	**2**	**0**	**2**	**4**	**0**	**0**	**1**	**6**	**33.3**	**2**						**5**	**1**	**0**	**1**	**4**	**0**	**0**	**0**
	Rochester	AHL	43	8	11	19	45												12	2	5	7	16			
1990-91	Rochester	AHL	69	13	21	34	77												15	4	4	8	4			
1991-92	**Buffalo**	**NHL**	**20**	**2**	**4**	**6**	**21**	**0**	**0**	**0**	**23**	**8.7**	**-9**						**4**	**1**	**0**	**1**	**0**	**1**	**0**	**0**
	Rochester	AHL	52	16	12	28	47												8	0	6	6	8			
1992-93	**Buffalo**	**NHL**	**68**	**6**	**4**	**10**	**38**	**0**	**1**	**1**	**69**	**8.7**	**-3**						**5**	**0**	**0**	**0**	**2**	**0**	**0**	**0**
1993-94	**Anaheim**	**NHL**	**76**	**23**	**28**	**51**	**18**	**3**	**3**	**0**	**180**	**12.8**	**4**													
1994-95	**Anaheim**	**NHL**	**44**	**10**	**9**	**19**	**25**	**0**	**0**	**1**	**100**	**10.0**	**-7**													
1995-96	**Anaheim**	**NHL**	**48**	**5**	**7**	**12**	**26**	**0**	**0**	**1**	**88**	**5.7**	**0**													
	Philadelphia	**NHL**	**28**	**4**	**3**	**7**	**8**	**0**	**0**	**2**	**38**	**10.5**	**3**						**12**	**1**	**2**	**3**	**6**	**0**	**0**	**0**
1996-97	**Phoenix**	**NHL**	**80**	**9**	**11**	**20**	**40**	**0**	**1**	**3**	**119**	**7.6**	**-7**						**7**	**2**	**2**	**4**	**4**	**0**	**0**	**1**
1997-98	**Phoenix**	**NHL**	**76**	**12**	**9**	**21**	**28**	**0**	**5**	**0**	**105**	**11.4**	**-7**						**6**	**1**	**0**	**1**	**4**	**0**	**0**	**0**
1998-99	**Phoenix**	**NHL**	**77**	**9**	**10**	**19**	**17**	**0**	**0**	**0**	**146**	**6.2**	**-9**	**1644**	**51.6**	**116**	**28**	**17:15**	**7**	**0**	**1**	**1**	**4**	**0**	**0**	**0**
99-2000	**Los Angeles**	**NHL**	**45**	**5**	**6**	**11**	**14**	**0**	**0**	**0**	**45**	**11.1**	**0**	**910**	**55.3**	**90**	**22**	**14:26**	**4**	**0**	**0**	**0**	**0**	**0**	**0**	**0**
2000-01	**Los Angeles**	**NHL**	**58**	**4**	**6**	**10**	**18**	**1**	**0**	**0**	**47**	**8.5**	**-12**	**1020**	**53.2**	**96**	**40**	**12:59**								
	New Jersey	**NHL**	**17**	**3**	**1**	**4**	**4**	**0**	**0**	**0**	**19**	**15.8**	**4**	**150**	**58.7**	**31**	**3**	**11:19**	**12**	**1**	**2**	**3**	**0**	**0**	**0**	**0**
2001-02	**Atlanta**	**NHL**	**65**	**3**	**4**	**7**	**16**	**0**	**0**	**0**	**70**	**4.3**	**-30**	**1249**	**51.5**	**112**	**30**	**15:47**								
	Buffalo	**NHL**	**10**	**0**	**1**	**1**	**4**	**0**	**0**	**0**	**10**	**0.0**	**-2**	**126**	**53.2**	**16**	**2**	**10:44**								
	NHL Totals		**720**	**97**	**103**	**200**	**281**	**4**	**10**	**9**	**1065**	**9.1**		**5099**	**52.8**	**461**	**125**	**14:55**	**62**	**7**	**7**	**14**	**24**	**1**	**0**	**1**

Claimed by **Anaheim** from **Buffalo** in Expansion Draft, June 24, 1993. Traded to **Philadelphia** by **Anaheim** for Chris Herperger and Winnipeg's 7th round choice (previously acquired, Anaheim selected Tony Mohagen) in 1997 Entry Draft, February 6, 1996. Claimed by **Phoenix** from **Philadelphia** in Waiver Draft, September 30, 1996. Signed as a free agent by **LA Kings**, December 28, 1999. Traded to **New Jersey** by **LA Kings** for future considerations (Steve Kelly, February 27, 2001), February 23, 2001. Signed as a free agent by **Atlanta**, July 16, 2001. Traded to **Buffalo** by **Atlanta** for Buffalo's 5th round choice (Paul Flache) in 2002 Entry Draft, March 19, 2002.

CORRINET, Chris

(KOHR-rih-neht, KRIHS) **WSH.**

Right wing. Shoots right. 6'3", 220 lbs. Born, Derby, CT, October 29, 1978. Washington's 4th choice, 107th overall, in 1998 Entry Draft.

			Regular Season																Playoffs							
Season	Club	League	GP	G	A	Pts	PIM	PP	SH	GW	S	%	+/-	TF	F%	H	SB	Min	GP	G	A	Pts	PIM	PP	SH	GW
1996-97	Deerfield	Hi-School	16	6	15	21	10																			
1997-98	Princeton	ECAC	31	3	6	9	22																			
1998-99	Princeton	ECAC	32	10	6	16	38																			
99-2000	Princeton	ECAC	30	10	14	24	41																			
2000-01	Princeton	ECAC	31	13	12	25	30																			
	Portland Pirates	AHL	6	0	1	1	4												2	1	0	1	0			
2001-02	**Washington**	**NHL**	**8**	**0**	**1**	**1**	**6**	**0**	**0**	**0**	**8**	**0.0**	**-4**	**0**	**0.0**	**17**	**2**	**10:04**								
	Portland Pirates	AHL	51	15	18	33	64																			
	NHL Totals		**8**	**0**	**1**	**1**	**6**	**0**	**0**	**0**	**8**	**0.0**		**0**	**0.0**	**17**	**2**	**10:04**								

CORSO, Daniel

(KOHR-soh, DAN-yehl) **ST.L.**

Center. Shoots left. 5'10", 187 lbs. Born, Montreal, Que., April 3, 1978. St. Louis' 6th choice, 169th overall, in 1996 Entry Draft.

			Regular Season																Playoffs							
Season	Club	League	GP	G	A	Pts	PIM	PP	SH	GW	S	%	+/-	TF	F%	H	SB	Min	GP	G	A	Pts	PIM	PP	SH	GW
1993-94	Magog	QAAA	36	17	22	39													12	10	12	22				
1994-95	Victoriaville	QMJHL	65	27	26	53	6												4	2	5	7	2			
1995-96	Victoriaville	QMJHL	65	49	65	114	77												12	6	7	13	4			
1996-97	Victoriaville	QMJHL	54	51	68	119	50																			
1997-98	Victoriaville	QMJHL	35	24	51	75	20												3	1	1	2	2			
1998-99	Worcester	AHL	63	14	14	28	26																			
99-2000	Worcester	AHL	71	21	34	55	19												9	2	3	5	10			
2000-01	**St. Louis**	**NHL**	**28**	**10**	**3**	**13**	**14**	**5**	**0**	**4**	**42**	**23.8**	**0**	**296**	**56.1**	**21**	**3**	**13:56**	**12**	**0**	**1**	**1**	**0**	**0**	**0**	**0**
	Worcester	AHL	52	19	37	56	47																			
2001-02	**St. Louis**	**NHL**	**41**	**4**	**7**	**11**	**6**	**1**	**0**	**2**	**25**	**16.0**	**3**	**423**	**54.9**	**32**	**15**	**11:13**	**2**	**0**	**0**	**0**	**0**	**0**	**0**	**0**
	NHL Totals		**69**	**14**	**10**	**24**	**20**	**6**	**0**	**6**	**67**	**20.9**		**719**	**55.4**	**53**	**18**	**12:19**	**14**	**0**	**1**	**1**	**0**	**0**	**0**	**0**

QMJHL All-Rookie Team (1995) • QMJHL First All-Star Team (1997) • QMJHL MVP (1997)

• Spent majority of 2001-02 season on practice roster, October 22, 2001.

CORSON, Shayne

(KOHR-sohn, SHAYN) **TOR.**

Left wing. Shoots left. 6'1", 202 lbs. Born, Barrie, Ont., August 13, 1966. Montreal's 2nd choice, 8th overall, in 1984 Entry Draft.

			Regular Season																Playoffs							
Season	Club	League	GP	G	A	Pts	PIM	PP	SH	GW	S	%	+/-	TF	F%	H	SB	Min	GP	G	A	Pts	PIM	PP	SH	GW
1982-83	Barrie Colts	OJHL-B	23	13	29	42	87																			
1983-84	Brantford	OHL	66	25	46	71	165												6	4	1	5	26			
1984-85	Hamilton	OHL	54	27	63	90	154												11	3	7	10	19			
1985-86	Hamilton	OHL	47	41	57	98	153																			
	Montreal	**NHL**	**3**	**0**	**0**	**0**	**2**	**0**	**0**	**0**	**1**	**0.0**	**-3**													
1986-87	**Montreal**	**NHL**	**55**	**12**	**11**	**23**	**144**	**0**	**1**	**3**	**69**	**17.4**	**10**						**17**	**6**	**5**	**11**	**30**	**1**	**1**	**1**
1987-88	**Montreal**	**NHL**	**71**	**12**	**27**	**39**	**152**	**2**	**0**	**2**	**90**	**13.3**	**22**						**3**	**1**	**0**	**1**	**12**	**0**	**0**	**0**
1988-89	**Montreal**	**NHL**	**80**	**26**	**24**	**50**	**193**	**10**	**0**	**3**	**133**	**19.5**	**-1**						**21**	**4**	**5**	**9**	**65**	**2**	**0**	**2**
1989-90	**Montreal**	**NHL**	**76**	**31**	**44**	**75**	**144**	**7**	**0**	**6**	**192**	**16.1**	**33**						**11**	**2**	**8**	**10**	**20**	**0**	**0**	**0**
1990-91	**Montreal**	**NHL**	**71**	**23**	**24**	**47**	**138**	**7**	**0**	**2**	**164**	**14.0**	**9**						**13**	**9**	**6**	**15**	**36**	**4**	**1**	**3**
1991-92	**Montreal**	**NHL**	**64**	**17**	**36**	**53**	**118**	**3**	**0**	**2**	**165**	**10.3**	**15**						**10**	**2**	**5**	**7**	**15**	**0**	**0**	**0**
1992-93	**Edmonton**	**NHL**	**80**	**16**	**31**	**47**	**209**	**9**	**2**	**1**	**164**	**9.8**	**-19**													
1993-94	**Edmonton**	**NHL**	**64**	**25**	**29**	**54**	**118**	**11**	**0**	**3**	**171**	**14.6**	**-8**													
1994-95	**Edmonton**	**NHL**	**48**	**12**	**24**	**36**	**86**	**2**	**0**	**1**	**131**	**9.2**	**-17**													
1995-96	**St. Louis**	**NHL**	**77**	**18**	**28**	**46**	**192**	**13**	**0**	**0**	**150**	**12.0**	**3**						**13**	**8**	**6**	**14**	**22**	**6**	**1**	**1**
1996-97	**St. Louis**	**NHL**	**11**	**2**	**1**	**3**	**24**	**1**	**0**	**0**	**19**	**10.5**	**-4**													
	Montreal	**NHL**	**47**	**6**	**15**	**21**	**80**	**2**	**0**	**2**	**96**	**6.3**	**-5**						**5**	**1**	**0**	**1**	**4**	**0**	**1**	**0**
1997-98	**Montreal**	**NHL**	**62**	**21**	**34**	**55**	**108**	**14**	**1**	**1**	**142**	**14.8**	**2**						**10**	**3**	**6**	**9**	**26**	**1**	**0**	**1**
	Canada	Olympics	6	1	1	2	2																			

			Regular Season																Playoffs							
Season	Club	League	GP	G	A	Pts	PIM	PP	SH	GW	S	%	+/-	TF	F%	H	SB	Min	GP	G	A	Pts	PIM	PP	SH	GW
1998-99	**Montreal**	**NHL**	**63**	**12**	**20**	**32**	**147**	**7**	**0**	**4**	**142**	**8.5**	**–10**	**184**	**45.1**	**71**	**38**	**20:42**	....	....	....	....	....	....	....	....
99-2000	**Montreal**	**NHL**	**70**	**8**	**20**	**28**	**115**	**2**	**0**	**1**	**121**	**6.6**	**–2**	**445**	**43.4**	**121**	**48**	**19:05**	....	....	....	....	....	....	....	....
2000-01	**Toronto**	**NHL**	**77**	**8**	**18**	**26**	**189**	**0**	**0**	**2**	**102**	**7.8**	**1**	**602**	**48.8**	**151**	**42**	**15:50**	**11**	**1**	**1**	**2**	**14**	**0**	**0**	**0**
2001-02	**Toronto**	**NHL**	**74**	**12**	**21**	**33**	**120**	**0**	**1**	**1**	**111**	**10.8**	**11**	**598**	**45.5**	**87**	**35**	**17:04**	**19**	**1**	**6**	**7**	**33**	**0**	**0**	**0**
	NHL Totals		**1093**	**261**	**407**	**668**	**2279**	**90**	**5**	**34**	**2163**	**12.1**		**1829**	**46.0**	**430**	**163**	**18:02**	**133**	**38**	**48**	**86**	**277**	**14**	**4**	**8**

Played in NHL All-Star Game (1990, 1994, 1998)

Traded to **Edmonton** by **Montreal** with Brent Gilchrist and Vladimir Vujtek for Vincent Damphousse and Edmonton's 4th round choice (Adam Wiesel) in 1993 Entry Draft, August 27, 1992. Signed as a free agent by **St. Louis**, July 28, 1995. Traded to **Montreal** by **St. Louis** with Murray Baron and St. Louis' 5th round choice (Gennady Razin) in 1997 Entry Draft for Pierre Turgeon, Rory Fitzpatrick and Craig Conroy, October 29, 1996. Signed as a free agent by **Toronto**, July 4, 2000.

COTE, Sylvain

(KOH-tay, SIHL-vayn) **WSH.**

Defense. Shoots right. 5'11", 201 lbs. Born, Quebec City, Que., January 19, 1966. Hartford's 1st choice, 11th overall, in 1984 Entry Draft.

Season	Club	League	GP	G	A	Pts	PIM	PP	SH	GW	S	%	+/-	TF	F%	H	SB	Min	GP	G	A	Pts	PIM	PP	SH	GW
1981-82	Ste-Foy	QAAA	46	18	29	47	117	....	....	....	....	....	....	....	....	....	....	....	5	0	3	3	8	....	....	....
1982-83	Quebec Remparts	QMJHL	66	10	24	34	50	....	....	....	....	....	....	....	....	....	....	....	....	....	....	....	....	....	....	....
1983-84	Quebec Remparts	QMJHL	66	15	50	65	89	....	....	....	....	....	....	....	....	....	....	....	5	1	1	2	0	....	....	....
1984-85	**Hartford**	**NHL**	**67**	**3**	**9**	**12**	**17**	**1**	**0**	**1**	**90**	**3.3**	**–30**	....	....	....	....	....	....	....	....	....	....	....	....	....
1985-86	Hull Olympiques	QMJHL	26	10	33	43	14	....	....	....	....	....	....	....	....	....	....	....	13	6	*28	34	22	....	....	....
	Hartford	**NHL**	**2**	**0**	**0**	**0**	**0**	**0**	**0**	**0**	**0**	**0.0**	**1**	....	....	....	....	....	....	....	....	....	....	....	....	....
	Binghamton	AHL	12	2	4	6	0	....	....	....	....	....	....	....	....	....	....	....	....	....	....	....	....	....	....	....
1986-87	**Hartford**	**NHL**	**67**	**2**	**8**	**10**	**20**	**0**	**0**	**0**	**100**	**2.0**	**11**	....	....	....	....	....	**2**	**0**	**2**	**2**	**2**	**0**	**0**	**0**
1987-88	**Hartford**	**NHL**	**67**	**7**	**21**	**28**	**30**	**0**	**1**	**0**	**142**	**4.9**	**–8**	....	....	....	....	....	**6**	**1**	**1**	**2**	**4**	**1**	**0**	**0**
1988-89	**Hartford**	**NHL**	**78**	**8**	**9**	**17**	**49**	**1**	**0**	**0**	**130**	**6.2**	**–7**	....	....	....	....	....	**3**	**0**	**1**	**1**	**4**	**0**	**0**	**0**
1989-90	**Hartford**	**NHL**	**28**	**4**	**2**	**6**	**14**	**1**	**0**	**1**	**50**	**8.0**	**2**	....	....	....	....	....	**5**	**0**	**0**	**0**	**2**	**0**	**0**	**0**
1990-91	**Hartford**	**NHL**	**73**	**7**	**12**	**19**	**17**	**1**	**0**	**0**	**154**	**4.5**	**–17**	....	....	....	....	....	**6**	**0**	**2**	**2**	**2**	**0**	**0**	**0**
1991-92	**Washington**	**NHL**	**78**	**11**	**29**	**40**	**31**	**6**	**0**	**2**	**151**	**7.3**	**7**	....	....	....	....	....	**7**	**1**	**2**	**3**	**4**	**0**	**0**	**0**
1992-93	**Washington**	**NHL**	**77**	**21**	**29**	**50**	**34**	**8**	**2**	**3**	**206**	**10.2**	**28**	....	....	....	....	....	**6**	**1**	**1**	**2**	**4**	**0**	**0**	**0**
1993-94	**Washington**	**NHL**	**84**	**16**	**35**	**51**	**66**	**3**	**2**	**2**	**212**	**7.5**	**30**	....	....	....	....	....	**9**	**1**	**8**	**9**	**6**	**0**	**0**	**0**
1994-95	**Washington**	**NHL**	**47**	**5**	**14**	**19**	**53**	**1**	**0**	**2**	**124**	**4.0**	**2**	....	....	....	....	....	**7**	**1**	**3**	**4**	**2**	**0**	**0**	**0**
1995-96	**Washington**	**NHL**	**81**	**5**	**33**	**38**	**40**	**3**	**0**	**2**	**212**	**2.4**	**5**	....	....	....	....	....	**6**	**2**	**0**	**2**	**12**	**1**	**0**	**0**
1996-97	**Washington**	**NHL**	**57**	**6**	**18**	**24**	**28**	**2**	**0**	**0**	**131**	**4.6**	**11**	....	....	....	....	....	....	....	....	....	....	....	....	....
1997-98	**Washington**	**NHL**	**59**	**1**	**15**	**16**	**36**	**0**	**0**	**0**	**83**	**1.2**	**–5**	....	....	....	....	....	....	....	....	....	....	....	....	....
	Toronto	**NHL**	**12**	**3**	**6**	**9**	**6**	**1**	**0**	**1**	**20**	**15.0**	**2**	....	....	....	....	....	....	....	....	....	....	....	....	....
1998-99	**Toronto**	**NHL**	**79**	**5**	**24**	**29**	**28**	**0**	**0**	**1**	**119**	**4.2**	**22**	**1**	**0.0**	**76**	**95**	**21:04**	**17**	**2**	**1**	**3**	**10**	**0**	**0**	**0**
99-2000	**Toronto**	**NHL**	**3**	**0**	**1**	**1**	**0**	**0**	**0**	**0**	**3**	**0.0**	**1**	**0**	**0.0**	**6**	**3**	**21:40**	....	....	....	....	....	....	....	....
	Chicago	**NHL**	**45**	**6**	**18**	**24**	**14**	**5**	**0**	**2**	**78**	**7.7**	**–4**	**1**	**100.0**	**49**	**62**	**23:22**	....	....	....	....	....	....	....	....
	Dallas	**NHL**	**28**	**2**	**8**	**10**	**14**	**0**	**0**	**0**	**47**	**4.3**	**6**	**0**	**0.0**	**33**	**23**	**18:10**	**23**	**2**	**1**	**3**	**8**	**2**	**0**	**0**
2000-01	**Washington**	**NHL**	**68**	**7**	**11**	**18**	**18**	**1**	**1**	**1**	**86**	**8.1**	**–3**	**0**	**0.0**	**125**	**72**	**17:47**	**5**	**0**	**0**	**0**	**2**	**0**	**0**	**0**
2001-02	**Washington**	**NHL**	**70**	**3**	**11**	**14**	**26**	**1**	**0**	**2**	**101**	**3.0**	**–15**	**0**	**0.0**	**84**	**90**	**19:44**	....	....	....	....	....	....	....	....
	NHL Totals		**1170**	**122**	**313**	**435**	**541**	**35**	**6**	**20**	**2239**	**5.4**		**2**	**50.0**	**373**	**345**	**20:04**	**102**	**11**	**22**	**33**	**62**	**4**	**0**	**0**

QMJHL Second All-Star Team (1984) • QMJHL First All-Star Team (1986)

Traded to **Washington** by **Hartford** for Washington's 2nd round choice (Andrei Nikolishin) in 1992 Entry Draft, September 8, 1991. Traded to **Toronto** by **Washington** for Jeff Brown, March 24, 1998. Traded to **Chicago** by **Toronto** for Chicago's 2nd round choice (Karel Pilar) in 2001 Entry Draft, October 8, 1999. Traded to **Dallas** by **Chicago** with Dave Manson for Kevin Dean, Derek Plante and Dallas' 2nd round choice (Matt Keith) in 2001 Entry Draft, February 8, 2000. Signed as a free agent by **Washington**, July 7, 2000.

COWAN, Jeff

(KOW-an, JEHF) **ATL.**

Left wing. Shoots left. 6'2", 215 lbs. Born, Scarborough, Ont., September 27, 1976.

Season	Club	League	GP	G	A	Pts	PIM	PP	SH	GW	S	%	+/-	TF	F%	H	SB	Min	GP	G	A	Pts	PIM	PP	SH	GW
1992-93	Guelph Platers	OJHL-B	45	8	8	16	22	....	....	....	....	....	....	....	....	....	....	....	....	....	....	....	....	....	....	....
1993-94	Guelph Platers	OJHL-B	43	30	26	56	96	....	....	....	....	....	....	....	....	....	....	....	....	....	....	....	....	....	....	....
	Guelph Storm	OHL	17	1	0	1	5	....	....	....	....	....	....	....	....	....	....	....	....	....	....	....	....	....	....	....
1994-95	Guelph Storm	OHL	51	10	7	17	14	....	....	....	....	....	....	....	....	....	....	....	14	1	1	2	0	....	....	....
1995-96	Barrie Colts	OHL	66	38	14	52	29	....	....	....	....	....	....	....	....	....	....	....	5	1	2	3	6	....	....	....
1996-97	Saint John	AHL	22	5	5	10	8	....	....	....	....	....	....	....	....	....	....	....	....	....	....	....	....	....	....	....
	Roanoke Express	ECHL	47	21	13	34	42	....	....	....	....	....	....	....	....	....	....	....	....	....	....	....	....	....	....	....
1997-98	Saint John	AHL	69	15	13	28	23	....	....	....	....	....	....	....	....	....	....	....	13	4	1	5	14	....	....	....
1998-99	Saint John	AHL	71	7	12	19	117	....	....	....	....	....	....	....	....	....	....	....	4	0	1	1	10	....	....	....
99-2000	**Calgary**	**NHL**	**13**	**4**	**1**	**5**	**16**	**0**	**0**	**0**	**26**	**15.4**	**2**	**0**	**0.0**	**22**	**5**	**10:22**	....	....	....	....	....	....	....	....
	Saint John	AHL	47	15	10	25	77	....	....	....	....	....	....	....	....	....	....	....	....	....	....	....	....	....	....	....
2000-01	**Calgary**	**NHL**	**51**	**9**	**4**	**13**	**74**	**2**	**0**	**1**	**48**	**18.8**	**–8**	**5**	**20.0**	**51**	**8**	**9:06**	....	....	....	....	....	....	....	....
2001-02	**Calgary**	**NHL**	**19**	**1**	**0**	**1**	**40**	**0**	**0**	**1**	**13**	**7.7**	**–3**	**2**	**50.0**	**19**	**7**	**7:44**	....	....	....	....	....	....	....	....
	Atlanta	**NHL**	**38**	**4**	**1**	**5**	**50**	**0**	**0**	**1**	**51**	**7.8**	**–11**	**5**	**20.0**	**50**	**15**	**12:27**	....	....	....	....	....	....	....	....
	NHL Totals		**121**	**18**	**6**	**24**	**180**	**2**	**0**	**3**	**138**	**13.0**		**12**	**25.0**	**142**	**35**	**10:05**	....	....	....	....	....	....	....	....

Signed as a free agent by **Calgary**, October 2, 1995. Traded to **Atlanta** by **Calgary** with the rights to Kurtis Foster for Petr Buzek, December 18, 2001.

CRAIG, Mike

(KRAYG, MIGHK) **S.J.**

Right wing. Shoots right. 6'1", 185 lbs. Born, London, Ont., June 6, 1971. Minnesota's 2nd choice, 28th overall, in 1989 Entry Draft.

Season	Club	League	GP	G	A	Pts	PIM	PP	SH	GW	S	%	+/-	TF	F%	H	SB	Min	GP	G	A	Pts	PIM	PP	SH	GW
1986-87	Woodstock	OJHL-C	32	29	19	48	64	....	....	....	....	....	....	....	....	....	....	....	....	....	....	....	....	....	....	....
1987-88	Oshawa Generals	OHL	61	6	10	16	39	....	....	....	....	....	....	....	....	....	....	....	7	7	0	1	11	....	....	....
1988-89	Oshawa Generals	OHL	63	36	36	72	34	....	....	....	....	....	....	....	....	....	....	....	6	3	1	4	6	....	....	....
1989-90	Oshawa Generals	OHL	43	36	40	76	85	....	....	....	....	....	....	....	....	....	....	....	17	10	16	26	46	....	....	....
1990-91	**Minnesota**	**NHL**	**39**	**8**	**4**	**12**	**32**	**1**	**0**	**2**	**59**	**13.6**	**–11**	....	....	....	....	....	**10**	**1**	**1**	**2**	**20**	**1**	**0**	**1**
1991-92	**Minnesota**	**NHL**	**67**	**15**	**16**	**31**	**155**	**4**	**0**	**4**	**136**	**11.0**	**–12**	....	....	....	....	....	**4**	**1**	**0**	**1**	**7**	**0**	**0**	**0**
1992-93	**Minnesota**	**NHL**	**70**	**15**	**23**	**38**	**106**	**7**	**0**	**0**	**131**	**11.5**	**–11**	....	....	....	....	....	....	....	....	....	....	....	....	....
1993-94	**Dallas**	**NHL**	**72**	**13**	**24**	**37**	**139**	**3**	**0**	**2**	**150**	**8.7**	**–14**	....	....	....	....	....	**4**	**0**	**0**	**0**	**2**	**0**	**0**	**0**
1994-95	**Toronto**	**NHL**	**37**	**5**	**5**	**10**	**12**	**1**	**0**	**1**	**61**	**8.2**	**–21**	....	....	....	....	....	**2**	**0**	**1**	**1**	**2**	**0**	**0**	**0**
1995-96	**Toronto**	**NHL**	**70**	**8**	**12**	**20**	**42**	**1**	**0**	**1**	**108**	**7.4**	**–8**	....	....	....	....	....	**6**	**0**	**0**	**0**	**18**	**0**	**0**	**0**
1996-97	**Toronto**	**NHL**	**65**	**7**	**13**	**20**	**62**	**1**	**0**	**0**	**128**	**5.5**	**–20**	....	....	....	....	....	....	....	....	....	....	....	....	....
1997-98	San Antonio	IHL	12	4	1	5	18	....	....	....	....	....	....	....	....	....	....	....	....	....	....	....	....	....	....	....
	Kansas City	IHL	59	14	33	47	68	....	....	....	....	....	....	....	....	....	....	....	11	5	5	10	28	....	....	....
1998-99	**San Jose**	**NHL**	**1**	**0**	**0**	**0**	**0**	**0**	**0**	**0**	**1**	**0.0**	**–1**	**0**	**0.0**	**1**	**1**	**11:25**	....	....	....	....	....	....	....	....
	Kentucky	AHL	52	27	17	44	72	....	....	....	....	....	....	....	....	....	....	....	12	5	4	9	18	....	....	....
99-2000	Kentucky	AHL	76	39	39	78	116	....	....	....	....	....	....	....	....	....	....	....	9	5	5	10	14	....	....	....
2000-01	Hershey Bears	AHL	57	21	22	43	73	....	....	....	....	....	....	....	....	....	....	....	12	3	2	5	20	....	....	....
2001-02	**San Jose**	**NHL**	**2**	**0**	**0**	**0**	**2**	**0**	**0**	**0**	**2**	**0.0**	**0**	**0**	**0.0**	**0**	**0**	**7:57**	....	....	....	....	....	....	....	....
	Cleveland Barons	AHL	69	35	25	60	87	....	....	....	....	....	....	....	....	....	....	....	....	....	....	....	....	....	....	....
	NHL Totals		**423**	**71**	**97**	**168**	**550**	**18**	**0**	**10**	**776**	**9.1**		**0**	**0.0**	**1**	**1**	**9:06**	**26**	**2**	**2**	**4**	**49**	**1**	**0**	**1**

Transferred to **Dallas** after **Minnesota** franchise relocated, June 9, 1993. Signed as a free agent by **Toronto**, July 29, 1994. Signed as a free agent by **San Jose**, July 13, 1998. Signed as a free agent by **Colorado**, August 2, 2000. Signed as a free agent by **San Jose**, September 6, 2001. Signed as a free agent by **SC Langnau** (Swiss), May 1, 2002.

CROSS, Cory

(KRAWS, KOHR-ee) **TOR.**

Defense. Shoots left. 6'5", 220 lbs. Born, Lloydminster, Alta., January 3, 1971. Tampa Bay's 1st choice, 1st overall, in 1992 Supplemental Draft.

Season	Club	League	GP	G	A	Pts	PIM	PP	SH	GW	S	%	+/-	TF	F%	H	SB	Min	GP	G	A	Pts	PIM	PP	SH	GW
1990-91	U. of Alberta	CWUAA	20	2	5	7	16	....	....	....	....	....	....	....	....	....	....	....	....	....	....	....	....	....	....	....
1991-92	U. of Alberta	CWUAA	41	4	11	15	82	....	....	....	....	....	....	....	....	....	....	....	....	....	....	....	....	....	....	....
1992-93	U. of Alberta	CWUAA	43	11	28	39	105	....	....	....	....	....	....	....	....	....	....	....	....	....	....	....	....	....	....	....
	Atlanta Knights	IHL	7	0	1	1	2	....	....	....	....	....	....	....	....	....	....	....	4	0	0	0	6	....	....	....
1993-94	**Tampa Bay**	**NHL**	**5**	**0**	**0**	**0**	**6**	**0**	**0**	**0**	**5**	**0.0**	**–3**	....	....	....	....	....	....	....	....	....	....	....	....	....
	Atlanta Knights	IHL	70	4	14	18	72	....	....	....	....	....	....	....	....	....	....	....	9	1	2	3	14	....	....	....
1994-95	Atlanta Knights	IHL	41	5	10	15	67	....	....	....	....	....	....	....	....	....	....	....	....	....	....	....	....	....	....	....
	Tampa Bay	**NHL**	**43**	**1**	**5**	**6**	**41**	**0**	**0**	**1**	**35**	**2.9**	**–6**	....	....	....	....	....	....	....	....	....	....	....	....	....
1995-96	**Tampa Bay**	**NHL**	**75**	**2**	**14**	**16**	**66**	**0**	**0**	**0**	**57**	**3.5**	**4**	....	....	....	....	....	**6**	**0**	**0**	**0**	**22**	**0**	**0**	**0**
1996-97	**Tampa Bay**	**NHL**	**72**	**4**	**5**	**9**	**95**	**0**	**0**	**2**	**75**	**5.3**	**6**	....	....	....	....	....	....	....	....	....	....	....	....	....
1997-98	**Tampa Bay**	**NHL**	**74**	**3**	**6**	**9**	**77**	**0**	**1**	**0**	**72**	**4.2**	**–24**	....	....	....	....	....	....	....	....	....	....	....	....	....
1998-99	**Tampa Bay**	**NHL**	**67**	**2**	**16**	**18**	**92**	**0**	**0**	**0**	**96**	**2.1**	**–25**	**0**	**0.0**	**127**	**82**	**22:38**	....	....	....	....	....	....	....	....
99-2000	**Toronto**	**NHL**	**71**	**4**	**11**	**15**	**64**	**0**	**0**	**1**	**60**	**6.7**	**13**	**0**	**0.0**	**154**	**53**	**15:59**	**12**	**0**	**2**	**2**	**2**	**0**	**0**	**0**

			Regular Season																Playoffs							
Season	Club	League	GP	G	A	Pts	PIM	PP	SH	GW	S	%	+/-	TF	F%	H	SB	Min	GP	G	A	Pts	PIM	PP	SH	GW
2000-01	Toronto	NHL	41	3	5	8	50	1	0	1	34	8.8	7	0	0.0	109	55	18:00	11	2	1	3	10	0	0	1
2001-02	Toronto	NHL	50	3	9	12	54	0	0	1	39	7.7	11	0	0.0	132	35	15:18	12	0	0	0	8	0	0	0
	NHL Totals		498	22	71	93	545	1	1	6	473	4.7		0	0.0	522	225	18:08	41	2	3	5	42	0	0	1

Traded to **Toronto** by **Tampa Bay** with Tampa Bay's 7th round choice (Ivan Kolozvary) in 2001 Entry Draft for Fredrik Modin, October 1, 1999.

CROWLEY, Mike

(KROH-lee, MIGHK) **MIN.**

Defense. Shoots left. 5'11", 190 lbs. Born, Bloomington, MN, July 4, 1975. Philadelphia's 5th choice, 140th overall, in 1993 Entry Draft.

Season	Club	League	GP	G	A	Pts	PIM	PP	SH	GW	S	%	+/-	TF	F%	H	SB	Min	GP	G	A	Pts	PIM	PP	SH	GW
1990-91	Jefferson High	Hi-School	20	3	9	12	2																			
1991-92	Jefferson High	Hi-School	28	5	18	23	8																			
1992-93	Jefferson High	Hi-School	22	10	32	42	18																			
1993-94	Jefferson High	Hi-School	28	23	54	77	26																			
1994-95	U. of Minnesota	WCHA	41	11	27	38	60																			
1995-96	U. of Minnesota	WCHA	42	17	46	63	28																			
1996-97	U. of Minnesota	WCHA	42	9	*47	*56	24																			
1997-98	Anaheim	NHL	8	2	2	4	8	0	0	1	17	11.8	0													
	Cincinnati	AHL	76	12	26	38	91																			
1998-99	Anaheim	NHL	20	2	3	5	16	1	0	1	41	4.9	−10	0	0.0	8	20	16:36								
	Cincinnati	AHL	44	5	23	28	42												3	0	3	3	2			
99-2000	Long Beach	IHL	67	9	39	48	35												4	2	1	3	6			
2000-01	Anaheim	NHL	39	1	10	11	20	0	0	1	45	2.2	−16	1	0.0	30	34	17:21								
	Grand Rapids	IHL	22	4	12	16	10																			
2001-02	Houston Aeros	AHL	11	3	3	6	4												11	1	2	3	0			
	NHL Totals		67	5	15	20	44	1	0	3	103	4.9		1	0.0	38	54	17:06								

Minnesota High School Player of the Year (1994) • WCHA Rookie of the Year (1995) • WCHA First All-Star Team (1996, 1997) • NCAA West First All-American Team (1996, 1997) • WCHA Player of the Year (1997) • IHL First All-Star Team (2000)

Traded to **Anaheim** by **Philadelphia** with Anatoli Semenov for Brian Wesenberg, March 19, 1996. Signed as a free agent by **Long Beach** (IHL), August 24, 1999. Signed as a free agent by **Anaheim**, December 8, 2000. Signed as a free agent by **Minnesota**, July 25, 2001. • Missed majority of 2001-02 season recovering from Achilles tendon injury suffered in training camp, October 2, 2001.

CROZIER, Greg

(KROH-zhuhr, GREHG) **MIN.**

Left wing. Shoots left. 6'3", 200 lbs. Born, Calgary, Alta., July 6, 1976. Pittsburgh's 4th choice, 73rd overall, in 1994 Entry Draft.

Season	Club	League	GP	G	A	Pts	PIM	PP	SH	GW	S	%	+/-	TF	F%	H	SB	Min	GP	G	A	Pts	PIM	PP	SH	GW
1991-92	Amherst Broncos	Hi-School	46	61	47	108	47																			
1992-93	Lawrence School	Hi-School	22	22	14	36	....																			
1993-94	Lawrence School	Hi-School	18	22	26	48	12																			
1994-95	Lawrence School	Hi-School	31	45	32	77	22																			
1995-96	U. of Michigan	CCHA	42	14	10	24	46																			
1996-97	U. of Michigan	CCHA	31	5	15	20	45																			
1997-98	U. of Michigan	CCHA	45	12	10	22	26																			
1998-99	U. of Michigan	CCHA	39	7	6	13	63																			
99-2000	Wilkes-Barre	AHL	71	22	22	44	33																			
2000-01	Pittsburgh	NHL	1	0	0	0	0	0	0	0	0	0.0	0	0	0.0	0	0	4:10								
	Wilkes-Barre	AHL	77	24	36	60	81												21	6	5	11	16			
2001-02	Providence	AHL	54	5	6	11	62																			
	Houston Aeros	AHL	10	1	5	6	10												14	2	1	3	26			
	NHL Totals		1	0	0	0	0	0	0	0	0	0.0		0	0.0	0	0	4:10								

Signed as a free agent by **Boston**, August 8, 2001. Traded to **Minnesota** by **Boston** for Darryl Laplante, March 19, 2002.

CULLEN, David

(KUH-lehn, DAY-vihd) **MIN.**

Defense. Shoots right. 6'2", 209 lbs. Born, St. Catharines, Ont., December 30, 1976.

Season	Club	League	GP	G	A	Pts	PIM	PP	SH	GW	S	%	+/-	TF	F%	H	SB	Min	GP	G	A	Pts	PIM	PP	SH	GW
1992-93	Thorold	OJHL-B	34	4	6	10	28																			
1993-94	Thorold	OJHL-B	40	10	35	45	26																			
1994-95	Thorold	OJHL-B	36	16	30	46	12																			
1995-96	U. of Maine	H-East	34	2	4	6	22																			
1996-97	U. of Maine	H-East	35	5	25	30	8																			
1997-98	U. of Maine	H-East	36	10	27	37	24																			
1998-99	U. of Maine	H-East	41	11	33	44	24																			
99-2000	Springfield	AHL	78	10	21	31	57												2	0	0	0	2			
2000-01	Springfield	AHL	69	13	29	42	40																			
	Phoenix	NHL	2	0	0	0	0	0	0	0	0	0.0	1	0	0.0	1	0	12:25								
2001-02	Phoenix	NHL	14	0	0	0	6	0	0	0	3	0.0	−5	0	0.0	7	6	12:21								
	Springfield	AHL	15	1	4	5	4																			
	Minnesota	NHL	3	0	0	0	0	0	0	0	0	0.0	−3	0	0.0	2	3	14:02								
	Houston Aeros	AHL	38	5	15	20	4												13	0	6	6	6			
	NHL Totals		19	0	0	0	6	0	0	0	3	0.0		0	0.0	10	9	12:37								

Hockey East First All-Star Team (1999) • NCAA East First All-American Team (1999) • NCAA Championship All-Tournament Team (1999)

Signed as a free agent by **Phoenix**, April 16, 1999. Traded to **Minnesota** by **Phoenix** for Sebastien Bordeleau, January 4, 2002.

CULLEN, Matt

(KUH-lehn, MAT) **ANA.**

Center. Shoots left. 6', 205 lbs. Born, Virginia, MN, November 2, 1976. Anaheim's 2nd choice, 35th overall, in 1996 Entry Draft.

Season	Club	League	GP	G	A	Pts	PIM	PP	SH	GW	S	%	+/-	TF	F%	H	SB	Min	GP	G	A	Pts	PIM	PP	SH	GW
1994-95	Moorhead Spuds	Hi-School	28	47	42	89	78																			
1995-96	St. Cloud State	WCHA	39	12	29	41	28																			
1996-97	St. Cloud State	WCHA	36	15	30	45	70																			
	Baltimore	AHL	6	3	3	6	7												3	0	2	2	0			
1997-98	Anaheim	NHL	61	6	21	27	23	2	0	0	75	8.0	−4													
	Cincinnati	AHL	18	15	12	27	2																			
1998-99	Anaheim	NHL	75	11	14	25	47	5	1	1	112	9.8	−12	1047	47.7	41	22	15:31	4	0	0	0	0	0	0	0
	Cincinnati	AHL	3	1	2	3	8																			
99-2000	Anaheim	NHL	80	13	26	39	24	1	0	1	137	9.5	5	1247	44.6	74	39	16:54								
2000-01	Anaheim	NHL	82	10	30	40	38	4	0	1	159	6.3	−23	1478	48.0	49	37	18:15								
2001-02	Anaheim	NHL	79	18	30	48	24	3	1	4	164	11.0	−1	1283	51.4	57	18	17:01								
	NHL Totals		377	58	121	179	156	15	2	7	647	9.0		5055	48.0	221	116	16:57	4	0	0	0	0	0	0	0

WCHA Second All-Star Team (1997)

CULLIMORE, Jassen

(KUHL-ih-mohr, JAY-sehn) **T.B.**

Defense. Shoots left. 6'5", 244 lbs. Born, Simcoe, Ont., December 4, 1972. Vancouver's 2nd choice, 29th overall, in 1991 Entry Draft.

Season	Club	League	GP	G	A	Pts	PIM	PP	SH	GW	S	%	+/-	TF	F%	H	SB	Min	GP	G	A	Pts	PIM	PP	SH	GW
1986-87	Caledonia	OJHL-C	18	2	0	2	9																			
1987-88	Simcoe Rams	OJHL-C	35	11	14	25	92																			
1988-89	Peterboro B's	OJHL-B	29	11	17	28	88																			
	Peterborough	OHL	20	2	1	3	6																			
1989-90	Peterborough	OHL	59	2	6	8	61												11	0	2	2	8			
1990-91	Peterborough	OHL	62	8	16	24	74												4	1	0	1	7			
1991-92	Peterborough	OHL	54	9	37	46	65												10	3	6	9	8			
1992-93	Hamilton Canucks	AHL	56	5	7	12	60																			
1993-94	Hamilton Canucks	AHL	71	8	20	28	86												3	0	1	1	2			
1994-95	Syracuse Crunch	AHL	33	2	7	9	66																			
	Vancouver	NHL	34	1	2	3	39	0	0	0	30	3.3	−2						11	0	0	0	12	0	0	0
1995-96	Vancouver	NHL	27	1	1	2	21	0	0	1	12	8.3	4													
1996-97	Vancouver	NHL	3	0	0	0	2	0	0	0	2	0.0	−2													
	Montreal	NHL	49	2	6	8	42	0	1	1	52	3.8	4						2	0	0	0	2	0	0	0
1997-98	Montreal	NHL	3	0	0	0	4	0	0	0	1	0.0	0													
	Fredericton	AHL	5	1	0	1	8																			
	Tampa Bay	NHL	25	1	2	3	22	1	0	0	17	5.9	−4													
1998-99	Tampa Bay	NHL	78	5	12	17	81	1	1	1	73	6.8	−22	0	0.0	161	67	20:14								
99-2000	Providence	AHL	16	5	10	15	31																			
	Tampa Bay	NHL	46	1	1	2	66	0	0	0	23	4.3	−12	2	0.0	92	45	15:38								

			Regular Season																Playoffs							
Season	Club	League	GP	G	A	Pts	PIM	PP	SH	GW	S	%	+/-	TF	F%	H	SB	Min	GP	G	A	Pts	PIM	PP	SH	GW
2000-01	Tampa Bay	NHL	74	1	6	7	80	0	0	0	56	1.8	–6	0	0.0	186	114	19:43								
2001-02	Tampa Bay	NHL	78	4	9	13	58	0	0	1	84	4.8	–1	0	0.0	171	104	20:07								
	NHL Totals		417	16	39	55	415	2	2	4	350	4.6		2	0.0	610	330	19:17	13	0	0	0	14	0	0	0

OHL Second All-Star Team (1992)

Traded to **Montreal** by **Vancouver** for Donald Brashear, November 13, 1996. Claimed on waivers by **Tampa Bay** from **Montreal**, January 22, 1998. Loaned to **Providence** (AHL) by **Tampa Bay**, October 1, 1999.

CUMMINS, Jim

(KUH-mihns, JIHM)

Right wing. Shoots right. 6'2", 212 lbs. Born, Dearborn, MI, May 17, 1970. NY Rangers' 5th choice, 67th overall, in 1989 Entry Draft.

			Regular Season																Playoffs							
Season	Club	League	GP	G	A	Pts	PIM	PP	SH	GW	S	%	+/-	TF	F%	H	SB	Min	GP	G	A	Pts	PIM	PP	SH	GW
1987-88	Det. Compuware	NAJHL	31	11	15	26	146																			
1988-89	Michigan State	CCHA	30	3	8	11	98																			
1989-90	Michigan State	CCHA	41	8	7	15	94																			
1990-91	Michigan State	CCHA	34	9	6	15	110																			
1991-92	Detroit	NHL	1	0	0	0	7	0	0	0	0	0.0	0													
	Adirondack	AHL	65	7	13	20	338												5	0	0	0	19			
1992-93	Detroit	NHL	7	1	1	2	58	0	0	0	5	20.0	0													
	Adirondack	AHL	43	16	4	20	179												9	3	1	4	4			
1993-94	Philadelphia	NHL	22	1	2	3	71	0	0	0	17	5.9	0													
	Hershey Bears	AHL	17	6	6	12	70																			
	Tampa Bay	NHL	4	0	0	0	13	0	0	0	3	0.0	–1													
	Atlanta Knights	IHL	7	4	5	9	14												13	1	2	3	90			
1994-95	Tampa Bay	NHL	10	1	0	1	41	0	0	1	3	33.3	–3													
	Chicago	NHL	27	3	1	4	117	0	0	0	20	15.0	–3						14	1	1	2	4	0	0	1
1995-96	Chicago	NHL	52	2	4	6	180	0	0	2	34	5.9	–1						10	0	0	0	2	0	0	0
1996-97	Chicago	NHL	65	6	6	12	199	0	0	0	61	9.8	4						6	0	0	0	24	0	0	0
1997-98	Chicago	NHL	55	0	2	2	178	0	0	0	33	0.0	–9													
	Phoenix	NHL	20	0	0	0	47	0	0	0	10	0.0	–7						3	0	0	0	4	0	0	0
1998-99	Phoenix	NHL	55	1	7	8	190	0	0	0	26	3.8	3	0	0.0	74	7	7:07	3	0	1	1	0	0	0	0
99-2000	Montreal	NHL	47	3	5	8	92	0	0	0	33	9.1	–5	4	25.0	59	12	8:58								
2000-01	Anaheim	NHL	79	5	6	11	167	0	0	1	45	11.1	–11	7	28.6	70	12	7:14								
2001-02	Anaheim	NHL	2	0	0	0	0	0	0	0	0	0.0	–1	0	0.0	6	1	7:27								
	Cincinnati	AHL	11	1	4	5	39																			
	NY Islanders	NHL	10	0	0	0	31	0	0	0	3	0.0	–5	1	100.0	12	0	4:26	1	0	0	0	9	0	0	0
	NHL Totals		456	23	34	57	1391	0	0	4	293	7.8		12	33.3	221	32	7:29	37	1	2	3	43	0	0	1

Traded to **Detroit** by **NY Rangers** with Kevin Miller and Dennis Vial for Joe Kocur and Per Djoos, March 5, 1991. Traded to **Philadelphia** by **Detroit** with Philadelphia's 4th round choice (previously acquired by Detroit - later traded to Boston - Boston selected Charles Paquette) in 1993 Entry Draft for Greg Johnson and Philadelphia's 5th round choice (Frederic Deschenes) in 1994 Entry Draft, June 20, 1993. Traded to **Tampa Bay** by **Philadelphia** with Philadelphia's 4th round choice (later traded back to Philadelphia - Philadelphia selected Radovan Somik) in 1995 Entry Draft for Rob DiMaio, March 18, 1994. Traded to **Chicago** by **Tampa Bay** with Tom Tilley and Jeff Buchanan for Paul Ysebaert and Rich Sutter, February 22, 1995. Traded to **Phoenix** by **Chicago** with Keith Carney for Chad Kilger and Jayson More, March 4, 1998. Traded to **Montreal** by **Phoenix** for NY Rangers' 6th round choice (previously acquired, Phoenix selected Erik Lewerstrom) in 1999 Entry Draft, June 26, 1999. Signed as a free agent by **Anaheim**, July 5, 2000. Traded to **NY Islanders** by **Anaheim** for Dave Roche, January 14, 2002.

CUTTA, Jakub

(KOO-tuh, YA-kuhb) **WSH.**

Defense. Shoots left. 6'3", 217 lbs. Born, Jablonec nad Nisou, Czech., December 29, 1981. Washington's 3rd choice, 61st overall, in 2000 Entry Draft.

			Regular Season																Playoffs							
Season	Club	League	GP	G	A	Pts	PIM	PP	SH	GW	S	%	+/-	TF	F%	H	SB	Min	GP	G	A	Pts	PIM	PP	SH	GW
1997-98	HC Liberec Jr.	Czech-Jr.	29	3	13	16	70																			
1998-99	Swift Current	WHL	59	3	3	6	63																			
99-2000	Swift Current	WHL	71	2	12	14	114												12	0	2	2	24			
2000-01	Washington	NHL	3	0	0	0	0	0	0	0	1	0.0	–1	0	0.0	2	0	11:33								
	Swift Current	WHL	47	5	8	13	102												16	1	3	4	32			
2001-02	Washington	NHL	2	0	0	0	0	0	0	0	2	0.0	–3	0	0.0	2	2	16:02								
	Portland Pirates	AHL	56	1	3	4	69																			
	NHL Totals		5	0	0	0	0	0	0	0	3	0.0		0	0.0	4	2	13:20								

Returned to **Swift Current** (WHL) by **Washington**, October 16, 2000.

CZERKAWSKI, Mariusz

(chehr-KAWV-skee, MAIR-ee-UHZ) **MTL.**

Right wing. Shoots left. 6', 200 lbs. Born, Radomsko, Poland, April 13, 1972. Boston's 5th choice, 106th overall, in 1991 Entry Draft.

			Regular Season																Playoffs							
Season	Club	League	GP	G	A	Pts	PIM	PP	SH	GW	S	%	+/-	TF	F%	H	SB	Min	GP	G	A	Pts	PIM	PP	SH	GW
1990-91	GKS Tychy	Poland	24	25	15	40																				
1991-92	Djurgarden	Sweden	39	8	5	13	4												3	0	0	0	2			
	Poland	Olympics	5	0	1	1	4																			
1992-93	Hammarby	Swede-2	32	*39	30	*69	74												13	*16	7	*23	34			
1993-94	Djurgarden	Sweden	39	13	21	34	20												6	3	1	4	2			
	Boston	NHL	4	2	1	3	0	1	0	0	11	18.2	–2						13	3	3	6	4	1	0	0
1994-95	Kiekko Espoo	Finland	7	9	3	12	10																			
	Boston	NHL	47	12	14	26	31	1	0	2	126	9.5	4						5	1	0	1	0	0	0	0
1995-96	Boston	NHL	33	5	6	11	10	1	0	0	63	7.9	–11													
	Edmonton	NHL	37	12	17	29	8	2	0	1	79	15.2	7													
1996-97	Edmonton	NHL	76	26	21	47	16	4	0	3	182	14.3	0						12	2	1	3	10	0	0	0
1997-98	NY Islanders	NHL	68	12	13	25	23	2	0	1	136	8.8	11													
1998-99	NY Islanders	NHL	78	21	17	38	14	4	0	1	205	10.2	–10	2	0.0	47	14	14:18								
99-2000	NY Islanders	NHL	79	35	35	70	34	16	0	4	276	12.7	–16	4	25.0	77	25	17:45								
2000-01	NY Islanders	NHL	82	30	32	62	48	10	1	0	287	10.5	–24	8	50.0	86	31	18:44								
2001-02	NY Islanders	NHL	82	22	29	51	48	6	0	6	169	13.0	–8	10	10.0	52	13	15:58	7	2	2	4	4	1	0	0
	NHL Totals		586	177	185	362	232	47	1	18	1534	11.5		24	25.0	262	83	16:42	37	8	6	14	18	2	0	0

Played in NHL All-Star Game (2000)

Traded to **Edmonton** by **Boston** with Sean Brown and Boston's 1st round choice (Matthieu Descoteaux) in 1996 Entry Draft for Bill Ranford, January 11, 1996. Traded to **NY Islanders** by **Edmonton** for Dan LaCouture, August 25, 1997. Traded to **Montreal** by **NY Islanders** for Arron Asham and Montreal's 5th round choice (Markus Pahlsson) in 2002 Entry Draft, June 22, 2002.

DACKELL, Andreas

(DA-kuhl, an-DRAY-uhs) **MTL.**

Right wing. Shoots right. 5'11", 194 lbs. Born, Gavle, Sweden, December 29, 1972. Ottawa's 3rd choice, 136th overall, in 1996 Entry Draft.

			Regular Season																Playoffs							
Season	Club	League	GP	G	A	Pts	PIM	PP	SH	GW	S	%	+/-	TF	F%	H	SB	Min	GP	G	A	Pts	PIM	PP	SH	GW
1991-92	Brynas Jr.	Swede-Jr.	26	17	24	41	42												2	3	1	4	2			
	Brynas IF Gavle	Sweden	4	0	0	0	2												2	0	1	1	4			
1992-93	Brynas IF Gavle	Sweden	40	12	15	27	12												10	4	5	9	2			
1993-94	Brynas IF Gavle	Sweden	38	12	17	29	47												7	2	2	4	8			
	Sweden	Olympics	4	0	0	0	0																			
1994-95	Brynas IF Gavle	Sweden	39	17	16	33	34												14	3	3	6	14			
1995-96	Brynas IF Gavle	Sweden	40	25	22	47	79												10	9	6	15	12			
1996-97	Ottawa	NHL	79	12	19	31	8	2	0	3	79	15.2	–6						7	1	0	1	0	0	0	0
1997-98	Ottawa	NHL	82	15	18	33	24	3	2	2	130	11.5	–11						11	1	1	2	2	1	0	0
1998-99	Ottawa	NHL	77	15	35	50	30	6	0	3	107	14.0	9	5	40.0	34	29	17:18	4	0	1	1	0	0	0	0
99-2000	Ottawa	NHL	82	10	25	35	18	0	0	1	99	10.1	5	1	100.0	34	35	16:17	6	2	1	3	2	0	0	1
2000-01	Ottawa	NHL	81	13	18	31	24	1	0	3	72	18.1	7	13	15.4	29	39	14:02	4	0	0	0	0	0	0	0
2001-02	Montreal	NHL	79	15	18	33	24	2	3	2	83	18.1	–3	16	31.3	34	55	17:14	12	1	2	3	6	0	0	0
	NHL Totals		480	80	133	213	128	14	5	14	570	14.0		35	28.6	131	158	16:11	44	5	5	10	10	1	0	1

Traded to **Montreal** by **Ottawa** for Montreal's 8th round choice (Neil Petruic) in 2001 Entry Draft, June 24, 2001.

DAGENAIS, Pierre

(da-ZHUH-nay, PEE-air) **FLA.**

Right wing. Shoots left. 6'5", 215 lbs. Born, Blainville, Que., March 4, 1978. New Jersey's 6th choice, 105th overall, in 1998 Entry Draft.

			Regular Season																Playoffs							
Season	Club	League	GP	G	A	Pts	PIM	PP	SH	GW	S	%	+/-	TF	F%	H	SB	Min	GP	G	A	Pts	PIM	PP	SH	GW
1994-95	Laval Laurentide	QAAA	34	28	14	42	68												13	10	9	19	32			
1995-96	Moncton Alpines	QMJHL	67	43	25	68	59																			
1996-97	Moncton Wildcats	QMJHL	6	4	2	6	0																			
	Laval Titan	QMJHL	37	16	14	30	40																			
	Rouyn-Noranda	QMJHL	27	21	8	29	22																			
1997-98	Rouyn-Noranda	QMJHL	60	*66	67	133	50												6	6	2	8	2			
1998-99	Albany	AHL	69	17	13	30	37												4	0	0	0	0			
99-2000	Albany	AHL	80	35	30	65	47												5	1	0	1	14			

			Regular Season																Playoffs							
Season	Club	League	GP	G	A	Pts	PIM	PP	SH	GW	S	%	+/-	TF	F%	H	SB	Min	GP	G	A	Pts	PIM	PP	SH	GW
2000-01	**New Jersey**	**NHL**	9	3	2	5	6	1	0	1	20	15.0	1	8	37.5	6	2	12:22								
	Albany	AHL	69	34	28	62	52																			
2001-02	**New Jersey**	**NHL**	16	3	3	6	4	1	0	1	30	10.0	–5	5	40.0	12	2	10:54								
	Albany	AHL	6	0	2	2	2																			
	Florida	**NHL**	26	7	1	8	4	2	0	0	47	14.9	–5	4	75.0	18	3	11:04								
	Utah Grizzlies	AHL	4	1	1	2	2																			
	NHL Totals		51	13	6	19	14	4	0	2	97	13.4		17	47.1	36	7	11:15								

• Re-entered NHL Entry Draft. Originally New Jersey's 4th choice, 47th overall, in 1996 Entry Draft.

QMJHL All-Rookie Team (1996) • QMJHL Second All-Star Team (1998) • AHL Second All-Star Team (2001)

Claimed on waivers by **Florida** from **New Jersey**, January 12, 2002.

DAHL, Kevin

(DAHL, KEH-vihn)

Defense. Shoots right. 5'11", 190 lbs. Born, Regina, Sask., December 30, 1968. Montreal's 12th choice, 230th overall, in 1988 Entry Draft.

Season	Club	League	GP	G	A	Pts	PIM	PP	SH	GW	S	%	+/-	TF	F%	H	SB	Min	GP	G	A	Pts	PIM	PP	SH	GW
1985-86	Stratford	OJHL-B	29	8	15	23	99																			
1986-87	Bowling Green	CCHA	32	2	6	8	54																			
1987-88	Bowling Green	CCHA	44	2	23	25	78																			
1988-89	Bowling Green	CCHA	46	9	26	35	51																			
1989-90	Bowling Green	CCHA	43	8	22	30	74																			
1990-91	Fredericton	AHL	32	1	15	16	45												9	0	1	1	11			
	Winston-Salem	ECHL	36	7	17	24	58																			
1991-92	Team Canada	Nat-Tm	45	2	15	17	44																			
	Canada	Olympics	8	2	0	2	6																			
	Salt Lake	IHL	13	0	2	2	12												5	0	0	0	13			
1992-93	**Calgary**	**NHL**	61	2	9	11	56	1	0	0	40	5.0	9						6	0	2	2	8	0	0	0
1993-94	**Calgary**	**NHL**	33	0	3	3	23	0	0	0	20	0.0	–2						6	0	0	0	4	0	0	0
	Saint John	AHL	2	0	0	0	0																			
1994-95	**Calgary**	**NHL**	34	4	8	12	38	0	0	0	30	13.3	8						3	0	0	0	0	0	0	0
1995-96	**Calgary**	**NHL**	32	1	1	2	26	0	0	1	17	5.9	–2						1	0	0	0	0	0	0	0
	Saint John	AHL	23	4	11	15	37																			
1996-97	**Phoenix**	**NHL**	2	0	0	0	0	0	0	0	2	0.0	0													
	Las Vegas	IHL	73	10	21	31	101												3	0	0	0	2			
1997-98	**Calgary**	**NHL**	19	0	1	1	6	0	0	0	17	0.0	–3													
	Chicago Wolves	IHL	45	8	9	17	61												20	1	8	9	32			
1998-99	**Toronto**	**NHL**	3	0	0	0	2	0	0	0	0	0.0	0	0	0.0	6	1	15:19								
	Chicago Wolves	IHL	34	3	6	9	61												10	2	3	5	8			
99-2000	Chicago Wolves	IHL	27	1	2	3	44												3	0	1	1	2			
2000-01	**Columbus**	**NHL**	4	0	0	0	2	0	0	0	3	0.0	1	0	0.0	3	3	10:34								
	Chicago Wolves	IHL	72	2	6	8	63												16	2	2	4	16			
2001-02	Nurnberg	Germany	48	2	12	14	95												4	0	0	0	6			
	NHL Totals		188	7	22	29	153	1	0	1	129	5.4		0	0.0	9	4	12:36	16	0	2	2	12	0	0	0

Signed as a free agent by **Calgary**, July 27, 1991. Signed as a free agent by **Phoenix**, September 4, 1996. Signed as a free agent by **Calgary**, September 8, 1997. Signed as a free agent by **St. Louis**, September 4, 1998. Claimed by **Toronto** from **St. Louis** in NHL Waiver Draft, October 5, 1998. Signed as a free agent by **NY Islanders**, August 12, 1999. Signed as a free agent by **Columbus**, August 24, 2000. Signed as a free agent by **Nurnberg** (Germany), June 1, 2001.

DAHLEN, Ulf

(DAH-lehn, UHLF) **DAL.**

Right wing. Shoots left. 6'2", 199 lbs. Born, Ostersund, Sweden, January 12, 1967. NY Rangers' 1st choice, 7th overall, in 1985 Entry Draft.

Season	Club	League	GP	G	A	Pts	PIM	PP	SH	GW	S	%	+/-	TF	F%	H	SB	Min	GP	G	A	Pts	PIM	PP	SH	GW
1983-84	Ostersunds IK	Swede-2	36	15	11	26	10																			
1984-85	Ostersunds IK	Swede-2	31	27	*26	*53	20												5	6	0	6	4			
1985-86	Bjorkloven	Sweden	22	4	3	7	8																			
1986-87	Bjorkloven	Sweden	31	9	12	21	20												6	6	2	8	4			
1987-88	**NY Rangers**	**NHL**	70	29	23	52	26	11	0	4	159	18.2	5													
	Colorado Rangers	IHL	2	2	2	4	0																			
1988-89	**NY Rangers**	**NHL**	56	24	19	43	50	8	0	1	147	16.3	–6						4	0	0	0	0	0	0	0
1989-90	**NY Rangers**	**NHL**	63	18	18	36	30	13	0	4	111	16.2	–4													
	Minnesota	**NHL**	13	2	4	6	0	0	0	0	24	8.3	1						7	1	4	5	2	0	0	0
1990-91	**Minnesota**	**NHL**	66	21	18	39	6	4	0	3	133	15.8	7						15	2	6	8	4	0	0	0
1991-92	**Minnesota**	**NHL**	79	36	30	66	10	16	1	5	216	16.7	–5						7	0	3	3	2	0	0	0
1992-93	**Minnesota**	**NHL**	83	35	39	74	6	13	0	6	223	15.7	–20													
1993-94	**Dallas**	**NHL**	65	19	38	57	10	12	0	3	147	12.9	–1													
	San Jose	**NHL**	13	6	6	12	0	3	0	2	43	14.0	0						14	6	2	8	0	3	0	1
1994-95	**San Jose**	**NHL**	46	11	23	34	11	4	1	4	85	12.9	–2						11	5	4	9	0	3	0	1
1995-96	**San Jose**	**NHL**	59	16	12	28	27	5	0	2	103	15.5	–21													
1996-97	**San Jose**	**NHL**	43	8	11	19	8	3	0	1	78	10.3	–11													
	Chicago	**NHL**	30	6	8	14	10	1	0	3	53	11.3	9						5	0	1	1	0	0	0	0
1997-98	HV 71 Jonkoping	Sweden	29	9	22	31	16												5	1	3	4	12			
	Sweden	Olympics	4	1	0	1	2																			
1998-99	HV 71 Jonkoping	Sweden	25	14	15	29	4																			
99-2000	**Washington**	**NHL**	75	15	23	38	8	5	0	4	106	14.2	11	115	48.7	59	11	12:40	5	0	1	1	2	0	0	0
2000-01	**Washington**	**NHL**	73	15	33	48	6	6	0	2	145	10.3	11	6	66.7	35	13	14:48	6	0	1	1	2	0	0	0
2001-02	**Washington**	**NHL**	69	23	29	52	8	7	0	4	141	16.3	–5	54	61.1	18	14	16:09								
	Sweden	Olympics	4	1	2	3	0																			
	NHL Totals		903	284	334	618	216	111	2	48	1914	14.8		175	53.1	112	38	14:29	74	14	22	36	12	6	0	2

Traded to **Minnesota** by **NY Rangers** with LA Kings' 4th round choice (previously acquired, Minnesota selected Cal McGowan) in 1990 Entry Draft for Mike Gartner, March 6, 1990. Transferred to **Dallas** after **Minnesota** franchise relocated, June 9, 1993. Traded to **San Jose** by **Dallas** with Dallas' 7th round choice (Brad Mehalko) in 1995 Entry Draft for Doug Zmolek and Mike Lalor, March 19, 1994. Traded to **Chicago** by **San Jose** with Chris Terreri and Michal Sykora for Ed Belfour, January 25, 1997. Signed as a free agent by **Washington**, August 16, 1999. Signed as a free agent by **Dallas**, August 13, 2002.

DAHLMAN, Toni

(DAHL-muhn, TOH-nee) **OTT.**

Right wing. Shoots right. 6', 193 lbs. Born, Helsinki, Finland, September 3, 1979. Ottawa's 12th choice, 286th overall, in 2001 Entry Draft.

Season	Club	League	GP	G	A	Pts	PIM	PP	SH	GW	S	%	+/-	TF	F%	H	SB	Min	GP	G	A	Pts	PIM	PP	SH	GW
1996-97	Karhu-Kissat Jr.	Finn-Jr.	24	22	18	30	6																			
1997-98	Jokerit Jr.	Finn-Jr.	21	13	11	24	14												7	2	2	4	0			
1998-99	Jokerit Jr.	Finn-Jr.	33	10	22	32	6												9	4	3	7	2			
	Jokerit Helsinki	Finland	5	0	0	0	0												3	0	1	1	0			
99-2000	Jokerit Jr.	Finn-Jr.	12	6	8	14	4												12	7	7	14	4			
	Hermes Kokkola	Finland-2	23	6	3	9	4																			
	Jokerit Helsinki	Finland	1	0	0	0	0																			
2000-01	Ilves Tampere	Finland	56	10	18	28	16												9	3	2	5	2			
2001-02	**Ottawa**	**NHL**	10	0	1	1	0	0	0	0	5	0.0	–1	0	0.0	10	4	7:04								
	Grand Rapids	AHL	50	6	8	14	25												4	0	0	0	0			
	NHL Totals		10	0	1	1	0	0	0	0	5	0.0		0	0.0	10	4	7:04								

DAIGLE, Alexandre

(DAYG, al-EHX-an-dreh) **PIT.**

Center. Shoots left. 6', 195 lbs. Born, Montreal, Que., February 7, 1975. Ottawa's 1st choice, 1st overall, in 1993 Entry Draft.

Season	Club	League	GP	G	A	Pts	PIM	PP	SH	GW	S	%	+/-	TF	F%	H	SB	Min	GP	G	A	Pts	PIM	PP	SH	GW
1990-91	Laval Laurentide	QAAA	42	*50	*60	*110	98												13	5	9	14	23			
1991-92	Victoriaville	QMJHL	66	35	75	110	63																			
1992-93	Victoriaville	QMJHL	53	45	92	137	85												6	5	6	11	4			
1993-94	**Ottawa**	**NHL**	84	20	31	51	40	4	0	2	168	11.9	–45													
1994-95	Victoriaville	QMJHL	18	14	20	34	16																			
	Ottawa	**NHL**	47	16	21	37	14	4	1	2	105	15.2	–22													
1995-96	**Ottawa**	**NHL**	50	5	12	17	24	1	0	0	77	6.5	–30													
1996-97	**Ottawa**	**NHL**	82	26	25	51	33	4	0	5	203	12.8	–33						7	0	0	0	2	0	0	0
1997-98	**Ottawa**	**NHL**	38	7	9	16	8	4	0	2	68	10.3	–7													
	Philadelphia	**NHL**	37	9	17	26	6	4	0	3	78	11.5	–1						5	0	2	2	0	0	0	0
1998-99	**Philadelphia**	**NHL**	31	3	2	5	2	1	0	1	26	11.5	–1	53	39.6	5	3	7:59								
	Tampa Bay	**NHL**	32	6	6	12	2	3	0	0	56	10.7	–12	4	50.0	6	8	13:59								
99-2000	**NY Rangers**	**NHL**	58	8	18	26	23	1	0	1	52	15.4	–5	339	53.1	22	7	10:59								
	Hartford	AHL	16	6	13	19	4																			

Season	Club	League	GP	G	A	Pts	PIM	PP	SH	GW	S	%	+/-	TF	F%	H	SB	Min	GP	G	A	Pts	PIM	PP	SH	GW
			Regular Season																Playoffs							
2000-01			OUT OF HOCKEY – RETIRED																							
2001-02			OUT OF HOCKEY – RETIRED																							
	NHL Totals		**459**	**100**	**141**	**241**	**152**	**26**	**1**	**16**	**833**	**12.0**		**396**	**51.3**	**33**	**18**	**11:00**	**12**	**0**	**2**	**2**	**2**	**0**	**0**	**0**

QMJHL Second All-Star Team (1992) • QMJHL Offensive Rookie of the Year (1992) • Canadian Major Junior Rookie of the Year (1992) • QMJHL First All-Star Team (1993)

Traded to **Philadelphia** by **Ottawa** for Vaclav Prospal, Pat Falloon and Dallas' 2nd round choice (previously acquired, Ottawa selected Chris Bala) in 1998 Entry Draft, January 17, 1998. Traded to **Edmonton** by **Philadelphia** for Andrei Kovalenko, January 29, 1999. Traded to **Tampa Bay** by **Edmonton** for Alexander Selivanov, January 29, 1999. Traded to **NY Rangers** by **Tampa Bay** for cash, October 3, 1999. Signed to a free agent tryout contract by **Pittsburgh**, August 13, 2002.

DAMPHOUSSE, Vincent (DAHM-fooz, VIHN-seht) S.J.

Center. Shoots left. 6'1", 200 lbs. Born, Montreal, Que., December 17, 1967. Toronto's 1st choice, 6th overall, in 1986 Entry Draft.

Season	Club	League	GP	G	A	Pts	PIM	PP	SH	GW	S	%	+/-	TF	F%	H	SB	Min	GP	G	A	Pts	PIM	PP	SH	GW
1982-83	Mtl-Bourassa	QAAA	48	33	45	78	22												10	4	4	8	12			
1983-84	Laval Voisins	QMJHL	66	29	36	65	25																			
1984-85	Laval Voisins	QMJHL	68	35	68	103	62																			
1985-86	Laval Titan	QMJHL	69	45	110	155	70												14	9	27	36	12			
1986-87	**Toronto**	**NHL**	**80**	**21**	**25**	**46**	**26**	**4**	**0**	**1**	**142**	**14.8**	**–6**						**12**	**1**	**5**	**6**	**8**	**1**	**0**	**0**
1987-88	**Toronto**	**NHL**	**75**	**12**	**36**	**48**	**40**	**1**	**0**	**2**	**111**	**10.8**	**2**						**6**	**0**	**1**	**1**	**10**	**0**	**0**	**0**
1988-89	**Toronto**	**NHL**	**80**	**26**	**42**	**68**	**75**	**6**	**0**	**4**	**190**	**13.7**	**–8**													
1989-90	**Toronto**	**NHL**	**80**	**33**	**61**	**94**	**56**	**9**	**0**	**5**	**229**	**14.4**	**2**						**5**	**0**	**2**	**2**	**2**	**0**	**0**	**0**
1990-91	**Toronto**	**NHL**	**79**	**26**	**47**	**73**	**65**	**10**	**1**	**4**	**247**	**10.5**	**–31**													
1991-92	**Edmonton**	**NHL**	**80**	**38**	**51**	**89**	**53**	**12**	**1**	**8**	**247**	**15.4**	**10**						**16**	**6**	**8**	**14**	**8**	**1**	**0**	**0**
1992-93♦	**Montreal**	**NHL**	**84**	**39**	**58**	**97**	**98**	**9**	**3**	**8**	**287**	**13.6**	**5**						**20**	**11**	**12**	**23**	**16**	**5**	**0**	**3**
1993-94	**Montreal**	**NHL**	**84**	**40**	**51**	**91**	**75**	**13**	**0**	**10**	**274**	**14.6**	**0**						**7**	**1**	**2**	**3**	**8**	**0**	**0**	**0**
1994-95	Ratingen	Germany	11	5	7	12	24																			
	Montreal	**NHL**	**48**	**10**	**30**	**40**	**42**	**4**	**0**	**4**	**123**	**8.1**	**15**													
1995-96	**Montreal**	**NHL**	**80**	**38**	**56**	**94**	**158**	**11**	**4**	**3**	**254**	**15.0**	**5**						**6**	**4**	**4**	**8**	**0**	**0**	**1**	**2**
1996-97	**Montreal**	**NHL**	**82**	**27**	**54**	**81**	**82**	**7**	**2**	**3**	**244**	**11.1**	**–6**						**5**	**0**	**0**	**0**	**2**	**0**	**0**	**0**
1997-98	**Montreal**	**NHL**	**76**	**18**	**41**	**59**	**58**	**2**	**1**	**5**	**164**	**11.0**	**14**						**10**	**3**	**6**	**9**	**22**	**1**	**0**	**0**
1998-99	**Montreal**	**NHL**	**65**	**12**	**24**	**36**	**46**	**3**	**2**	**2**	**147**	**8.2**	**–7**	**1425**	**48.4**	**41**	**37**	**20:27**								
	San Jose	**NHL**	**12**	**7**	**6**	**13**	**4**	**3**	**0**	**1**	**43**	**16.3**	**3**	**230**	**51.3**	**9**	**1**	**19:21**	**6**	**3**	**2**	**5**	**6**	**0**	**2**	**0**
99-2000	**San Jose**	**NHL**	**82**	**21**	**49**	**70**	**58**	**3**	**1**	**1**	**204**	**10.3**	**4**	**1642**	**49.0**	**54**	**38**	**20:26**	**12**	**1**	**7**	**8**	**16**	**1**	**0**	**0**
2000-01	**San Jose**	**NHL**	**45**	**9**	**37**	**46**	**62**	**4**	**0**	**3**	**101**	**8.9**	**17**	**1027**	**52.7**	**30**	**17**	**20:49**	**6**	**2**	**1**	**3**	**14**	**0**	**1**	**0**
2001-02	**San Jose**	**NHL**	**82**	**20**	**38**	**58**	**60**	**7**	**2**	**4**	**172**	**11.6**	**8**	**1690**	**49.9**	**45**	**37**	**19:38**	**12**	**2**	**6**	**8**	**12**	**1**	**0**	**0**
	NHL Totals		**1214**	**397**	**706**	**1103**	**1058**	**108**	**17**	**68**	**3179**	**12.5**		**6014**	**49.8**	**179**	**130**	**20:13**	**123**	**34**	**56**	**90**	**124**	**10**	**4**	**5**

QMJHL Second All-Star Team (1986) • Played in NHL All-Star Game (1991, 1992, 2002)

Traded to **Edmonton** by **Toronto** with Peter Ing, Scott Thornton and Luke Richardson for Grant Fuhr, Glenn Anderson and Craig Berube, September 19, 1991. Traded to **Montreal** by **Edmonton** with Edmonton's 4th round choice (Adam Wiesel) in 1993 Entry Draft for Shayne Corson, Brent Gilchrist and Vladimir Vujtek, August 27, 1992. Traded to **San Jose** by **Montreal** for Phoenix's 5th round choice (previously acquired, Montreal selected Marc-Andre Thinel) in 1999 Entry Draft, San Jose's 1st round choice (Marcel Hossa) in 2000 Entry Draft and 2nd round choice (later traded to Columbus - Columbus selected Kiel McLeod) in 2001 Entry Draft, March 23, 1999.

DANDENAULT, Mathieu (DAHN-deh-noh, MAT-yoo) DET.

Right wing/Defense. Shoots right. 6', 200 lbs. Born, Sherbrooke, Que., February 3, 1976. Detroit's 2nd choice, 49th overall, in 1994 Entry Draft.

Season	Club	League	GP	G	A	Pts	PIM	PP	SH	GW	S	%	+/-	TF	F%	H	SB	Min	GP	G	A	Pts	PIM	PP	SH	GW
1990-91	Gloucester	OMHA	44	52	50	102	30																			
1991-92	Vanier Voyageurs	OCJHL	33	27	31	58	20																			
	Gloucester	OCJHL	6	3	4	7	0																			
1992-93	Gloucester	OCJHL	55	11	26	37	64																			
1993-94	Sherbrooke	QMJHL	67	17	36	53	67												12	4	10	14	12			
1994-95	Sherbrooke	QMJHL	67	37	70	107	76												7	1	7	8	10			
1995-96	**Detroit**	**NHL**	**34**	**5**	**7**	**12**	**6**	**1**	**0**	**0**	**32**	**15.6**	**6**													
	Adirondack	AHL	4	0	0	0	0																			
1996-97♦	**Detroit**	**NHL**	**65**	**3**	**9**	**12**	**28**	**0**	**0**	**0**	**81**	**3.7**	**–10**													
1997-98♦	**Detroit**	**NHL**	**68**	**5**	**12**	**17**	**43**	**0**	**0**	**0**	**75**	**6.7**	**5**						**3**	**1**	**0**	**1**	**0**	**1**	**0**	**0**
1998-99	**Detroit**	**NHL**	**75**	**4**	**10**	**14**	**59**	**0**	**0**	**0**	**94**	**4.3**	**17**	**3**	**0.0**	**109**	**37**	**15:10**	**10**	**0**	**1**	**1**	**0**	**0**	**0**	**0**
99-2000	**Detroit**	**NHL**	**81**	**6**	**12**	**18**	**20**	**0**	**0**	**0**	**98**	**6.1**	**–12**	**1**	**100.0**	**108**	**28**	**12:10**	**6**	**0**	**0**	**0**	**2**	**0**	**0**	**0**
2000-01	**Detroit**	**NHL**	**73**	**10**	**15**	**25**	**38**	**2**	**0**	**2**	**95**	**10.5**	**11**	**0**	**0.0**	**90**	**54**	**16:06**	**6**	**0**	**1**	**1**	**0**	**0**	**0**	**0**
2001-02♦	**Detroit**	**NHL**	**81**	**8**	**12**	**20**	**44**	**2**	**0**	**3**	**97**	**8.2**	**–5**	**1**	**0.0**	**90**	**87**	**16:43**	**23**	**1**	**2**	**3**	**8**	**0**	**1**	**0**
	NHL Totals		**477**	**41**	**77**	**118**	**238**	**5**	**0**	**5**	**572**	**7.2**		**5**	**20.0**	**397**	**206**	**15:01**	**48**	**2**	**4**	**6**	**10**	**1**	**1**	**0**

DANEYKO, Ken (DAN-ee-KOH, KEHN) N.J.

Defense. Shoots left. 6'1", 215 lbs. Born, Windsor, Ont., April 17, 1964. New Jersey's 2nd choice, 18th overall, in 1982 Entry Draft.

Season	Club	League	GP	G	A	Pts	PIM	PP	SH	GW	S	%	+/-	TF	F%	H	SB	Min	GP	G	A	Pts	PIM	PP	SH	GW
1980-81	St. Albert	AJHL	1	0	0	0	4																			
	Spokane Flyers	WHL	62	6	13	19	140												4	0	0	0	6			
1981-82	Spokane Flyers	WHL	26	1	11	12	147																			
	Seattle Breakers	WHL	38	1	22	23	151												14	1	9	10	49			
1982-83	Seattle Breakers	WHL	69	17	43	60	150												4	1	3	4	14			
1983-84	Kamloops	WHL	19	6	28	34	52												17	4	9	13	28			
	New Jersey	**NHL**	**11**	**1**	**4**	**5**	**17**	**0**	**0**	**0**	**17**	**5.9**	**–1**													
1984-85	**New Jersey**	**NHL**	**1**	**0**	**0**	**0**	**10**	**0**	**0**	**0**	**1**	**0.0**	**–1**													
	Maine Mariners	AHL	80	4	9	13	206												11	1	3	4	36			
1985-86	**New Jersey**	**NHL**	**44**	**0**	**10**	**10**	**100**	**0**	**0**	**0**	**48**	**0.0**	**0**													
	Maine Mariners	AHL	21	3	2	5	75																			
1986-87	**New Jersey**	**NHL**	**79**	**2**	**12**	**14**	**183**	**0**	**0**	**0**	**113**	**1.8**	**–13**													
1987-88	**New Jersey**	**NHL**	**80**	**5**	**7**	**12**	**239**	**1**	**0**	**0**	**82**	**6.1**	**–3**						**20**	**1**	**6**	**7**	**83**	**0**	**0**	**1**
1988-89	**New Jersey**	**NHL**	**80**	**5**	**5**	**10**	**283**	**1**	**0**	**0**	**108**	**4.6**	**–22**													
1989-90	**New Jersey**	**NHL**	**74**	**6**	**15**	**21**	**219**	**0**	**1**	**1**	**64**	**9.4**	**15**						**6**	**2**	**0**	**2**	**21**	**0**	**0**	**0**
1990-91	**New Jersey**	**NHL**	**80**	**4**	**16**	**20**	**249**	**1**	**2**	**1**	**106**	**3.8**	**–10**						**7**	**0**	**1**	**1**	**10**	**0**	**0**	**0**
1991-92	**New Jersey**	**NHL**	**80**	**1**	**7**	**8**	**170**	**0**	**0**	**0**	**57**	**1.8**	**7**						**7**	**0**	**3**	**3**	**16**	**0**	**0**	**0**
1992-93	**New Jersey**	**NHL**	**84**	**2**	**11**	**13**	**236**	**0**	**0**	**0**	**71**	**2.8**	**4**						**5**	**0**	**0**	**0**	**8**	**0**	**0**	**0**
1993-94	**New Jersey**	**NHL**	**78**	**1**	**9**	**10**	**176**	**0**	**0**	**1**	**60**	**1.7**	**27**						**20**	**0**	**1**	**1**	**45**	**0**	**0**	**0**
1994-95♦	**New Jersey**	**NHL**	**25**	**1**	**2**	**3**	**54**	**0**	**0**	**0**	**27**	**3.7**	**4**						**20**	**1**	**0**	**1**	**22**	**0**	**0**	**0**
1995-96	**New Jersey**	**NHL**	**80**	**2**	**4**	**6**	**115**	**0**	**0**	**0**	**67**	**3.0**	**–10**													
1996-97	**New Jersey**	**NHL**	**77**	**2**	**7**	**9**	**70**	**0**	**0**	**0**	**63**	**3.2**	**24**						**10**	**0**	**0**	**0**	**28**	**0**	**0**	**0**
1997-98	**New Jersey**	**NHL**	**37**	**0**	**1**	**1**	**57**	**0**	**0**	**0**	**18**	**0.0**	**3**						**6**	**0**	**1**	**1**	**10**	**0**	**0**	**0**
1998-99	**New Jersey**	**NHL**	**82**	**2**	**9**	**11**	**63**	**0**	**0**	**0**	**63**	**3.2**	**27**	**1**	**0.0**	**182**	**159**	**20:03**	**7**	**0**	**0**	**0**	**8**	**0**	**0**	**0**
99-2000♦	**New Jersey**	**NHL**	**78**	**0**	**6**	**6**	**98**	**0**	**0**	**0**	**74**	**0.0**	**13**	**0**	**0.0**	**183**	**161**	**18:06**	**23**	**1**	**2**	**3**	**14**	**0**	**0**	**0**
2000-01	**New Jersey**	**NHL**	**77**	**0**	**4**	**4**	**87**	**0**	**0**	**0**	**50**	**0.0**	**8**	**0**	**0.0**	**173**	**122**	**17:17**	**25**	**0**	**3**	**3**	**21**	**0**	**0**	**0**
2001-02	**New Jersey**	**NHL**	**67**	**0**	**6**	**6**	**60**	**0**	**0**	**0**	**44**	**0.0**	**2**	**0**	**0.0**	**103**	**113**	**15:31**	**6**	**0**	**0**	**0**	**8**	**0**	**0**	**0**
	NHL Totals		**1214**	**34**	**135**	**169**	**2486**	**3**	**3**	**3**	**1133**	**3.0**		**1**	**0.0**	**641**	**555**	**17:51**	**162**	**5**	**17**	**22**	**294**	**0**	**0**	**1**

• Missed majority of 1997-98 season after voluntarily entering NHL/NHLPA substance abuse program, November 6, 1997. • Won Bill Masterton Memorial Trophy (2000)

DANIELS, Jeff (DAN-yehls, JEHF) CAR.

Left wing. Shoots left. 6'1", 200 lbs. Born, Oshawa, Ont., June 24, 1968. Pittsburgh's 6th choice, 109th overall, in 1986 Entry Draft.

Season	Club	League	GP	G	A	Pts	PIM	PP	SH	GW	S	%	+/-	TF	F%	H	SB	Min	GP	G	A	Pts	PIM	PP	SH	GW
1983-84	Oshawa	OJHL-B	57	59	72	131	22																			
1984-85	Oshawa	OJHL-B	7	7	2	9	11																			
	Oshawa Generals	OHL	59	7	11	18	16																			
1985-86	Oshawa Generals	OHL	62	13	19	32	23												6	0	1	1	0			
1986-87	Oshawa Generals	OHL	54	14	9	23	22												15	3	2	5	5			
1987-88	Oshawa Generals	OHL	64	29	39	68	59												4	2	3	5	0			
1988-89	Muskegon	IHL	58	21	21	42	58												11	3	5	8	11			
1989-90	Muskegon	IHL	80	30	47	77	39												6	1	1	2	7			
1990-91	**Pittsburgh**	**NHL**	**11**	**0**	**2**	**2**	**2**	**0**	**0**	**0**	**6**	**0.0**	**0**													
	Muskegon	IHL	62	23	29	52	18												5	1	3	4	2			
1991-92	**Pittsburgh**	**NHL**	**2**	**0**	**0**	**0**	**0**	**0**	**0**	**0**	**0**	**0.0**	**0**													
	Muskegon	IHL	44	19	16	35	38												10	5	4	9	9			
1992-93	**Pittsburgh**	**NHL**	**58**	**5**	**4**	**9**	**14**	**0**	**0**	**1**	**30**	**16.7**	**–5**						**12**	**3**	**2**	**5**	**0**	**0**	**0**	**1**
	Cleveland	IHL	3	2	1	3	0																			

			Regular Season																Playoffs							
Season	Club	League	GP	G	A	Pts	PIM	PP	SH	GW	S	%	+/-	TF	F%	H	SB	Min	GP	G	A	Pts	PIM	PP	SH	GW
1993-94	Pittsburgh	NHL	63	3	5	8	20	0	0	1	46	6.5	–1													
	Florida	NHL	7	0	0	0	0	0	0	0	6	0.0	0													
1994-95	Florida	NHL	3	0	0	0	0	0	0	0	0	0.0	0													
	Detroit Vipers	IHL	25	8	12	20	6												5	1	0	1	0			
1995-96	Springfield	AHL	72	22	20	42	32												10	3	0	3	2			
1996-97	Hartford	NHL	10	0	2	2	0	0	0	0	6	0.0	2													
	Springfield	AHL	38	18	14	32	19												16	7	3	10	4			
1997-98	Carolina	NHL	2	0	0	0	0	0	0	0	1	0.0	0													
	New Haven	AHL	71	24	27	51	34												3	0	1	1	0			
1998-99	Nashville	NHL	9	1	3	4	2	0	0	0	8	12.5	–1	1	0.0	9	1	10:56								
	Milwaukee	IHL	62	12	31	43	19												2	1	1	2	0			
99-2000	Carolina	NHL	69	3	4	7	10	0	0	0	28	10.7	–8	47	46.8	54	37	7:42								
2000-01	Carolina	NHL	67	1	1	2	15	0	0	0	42	2.4	–3	125	53.6	51	27	7:30	6	0	2	2	2	0	0	0
2001-02	Carolina	NHL	65	4	1	5	12	0	1	0	40	10.0	–6	178	46.6	48	23	8:07	23	0	1	1	0	0	0	0
	NHL Totals		366	17	22	39	75	0	1	2	213	8.0		351	49.0	162	88	7:54	41	3	5	8	2	0	0	1

Traded to **Florida** by **Pittsburgh** for Greg Hawgood, March 19, 1994. Signed as a free agent by **Hartford**, August 18, 1995. Transferred to **Carolina** after **Hartford** franchise relocated, June 25, 1997. Claimed by **Nashville** from **Carolina** in Expansion Draft, June 26, 1998. Signed as a free agent by **Carolina**, August 31, 1999.

DANTON, Mike
(DAHN-tuhn, MIGHK) **N.J.**

Center. Shoots right. 5'9", 190 lbs. Born, Brampton, Ont., October 21, 1980. New Jersey's 8th choice, 135th overall, in 2000 Entry Draft.

Season	Club	League	GP	G	A	Pts	PIM	PP	SH	GW	S	%	+/-	TF	F%	H	SB	Min	GP	G	A	Pts	PIM	PP	SH	GW
1996-97	Quinte Hawks	MTJHL	35	10	18	28	281																			
1997-98	Sarnia Sting	OHL	12	6	1	7	37																			
	St. Michael's	OHL	18	4	6	10	77																			
1998-99	St. Michael's	OHL	27	18	22	40	116																			
	Barrie Colts	OHL	26	15	20	35	62												9	6	5	11	38			
99-2000	Barrie Colts	OHL	58	34	53	87	203												25	7	16	23	*107			
2000-01	New Jersey	NHL	2	0	0	0	6	0	0	0	3	0.0	0	6	50.0	8	0	7:52								
	Albany	AHL	69	19	15	34	195																			
2001-02	Albany	AHL	DID NOT PLAY – SUSPENDED																							
	NHL Totals		2	0	0	0	6	0	0	0	3	0.0		6	50.0	8	0	7:52								

• Legally changed last name from **Jefferson** to **Danton**, July 25, 2002.

Traded to **Barrie** (OHL) by **St. Michael's** (OHL) with Sheldon Keefe, Ryan Barnes and Shawn Cation for Keith Delaney, Darryl Bootland, Adam DeLeew and Brad Pierce, January 11, 1999. • Suspended for 2001-02 season by New Jersey for refusing to report to Albany (AHL), October 2, 2001.

DARBY, Craig
(DAHR-bee, KRAYG) **N.J.**

Center. Shoots right. 6'3", 200 lbs. Born, Oneida, NY, September 26, 1972. Montreal's 3rd choice, 43rd overall, in 1991 Entry Draft.

Season	Club	League	GP	G	A	Pts	PIM	PP	SH	GW	S	%	+/-	TF	F%	H	SB	Min	GP	G	A	Pts	PIM	PP	SH	GW
1987-88	Albany	Hi-School	29	11	27	38																				
1988-89	Albany	Hi-School	29	36	40	*76																				
1989-90	Albany	Hi-School	29	32	53	85																				
1990-91	Albany	Hi-School	29	33	61	*94													4	8	1	9				
1991-92	Providence	H-East	35	17	24	41	47																			
1992-93	Providence	H-East	35	11	21	32	62																			
1993-94	Fredericton	AHL	66	23	33	56	51																			
1994-95	Fredericton	AHL	64	21	47	68	82																			
	Montreal	NHL	10	0	2	2	0	0	0	0	4	0.0	–5													
	NY Islanders	NHL	3	0	0	0	0	0	0	0	1	0.0	–1													
1995-96	NY Islanders	NHL	10	0	2	2	0	0	0	0	1	0.0	–1													
	Worcester	AHL	68	22	28	50	47												4	1	1	2	2			
1996-97	Philadelphia	NHL	9	1	4	5	2	0	1	0	13	7.7	2													
	Philadelphia	AHL	59	26	33	59	24												10	3	6	9	0			
1997-98	Philadelphia	NHL	3	1	0	1	0	0	0	0	3	33.3	0													
	Philadelphia	AHL	77	*42	45	87	34												20	5	9	14	4			
1998-99	Milwaukee	IHL	81	32	22	54	33												2	3	0	3	0			
99-2000	Montreal	NHL	76	7	10	17	14	0	1	2	90	7.8	–14	1068	48.3	46	18	13:45								
2000-01	Montreal	NHL	78	12	16	28	16	0	1	0	97	12.4	–17	1214	46.9	64	33	15:53								
2001-02	Montreal	NHL	2	0	0	0	0	0	0	0	0	0.0	0	10	30.0	0	1	5:20								
	Quebec	AHL	66	16	55	71	18												3	2	1	3	0			
	NHL Totals		191	21	34	55	32	0	3	2	209	10.0		2292	47.5	110	52	14:42								

Hockey East Rookie of the Year (Shared with Ian Moran) (1992) • AHL First All-Star Team (1998)

Traded to **NY Islanders** by **Montreal** with Kirk Muller and Mathieu Schneider for Pierre Turgeon and Vladimir Malakhov, April 5, 1995. Claimed on waivers by **Philadelphia** from **NY Islanders**, June 4, 1996. Claimed by **Nashville** from **Philadelphia** in Expansion Draft, June 26, 1998. Signed as a free agent by **Montreal**, August 4, 1999. Signed as a free agent by **New Jersey**, July 12, 2002.

DARCHE, Mathieu
(DAHRSH, MATH-you) **CBJ**

Left wing. Shoots left. 6'1", 210 lbs. Born, St-Laurent, Que., November 26, 1976.

Season	Club	League	GP	G	A	Pts	PIM	PP	SH	GW	S	%	+/-	TF	F%	H	SB	Min	GP	G	A	Pts	PIM	PP	SH	GW
1995-96	Choate-Rosemary	Hi-School	STATISTICS NOT AVAILABLE																							
1996-97	McGill Redmen	OUAA	23	1	2	3	27																			
1997-98	McGill Redmen	OUAA	40	28	17	45	69																			
1998-99	McGill Redmen	OUAA	32	16	24	40	60																			
99-2000	McGill Redmen	OUAA	33	31	41	*72	38												5	2	8	10	16			
2000-01	Columbus	NHL	9	0	0	0	0	0	0	0	9	0.0	–4	1	0.0	12	0	10:07								
	Syracuse Crunch	AHL	66	16	24	40	21												5	0	1	1	4			
2001-02	Columbus	NHL	14	1	1	2	6	0	0	0	15	6.7	–5	3	33.3	22	2	9:49								
	Syracuse Crunch	AHL	63	22	23	45	26												10	2	5	7	2			
	NHL Totals		23	1	1	2	6	0	0	0	24	4.2		4	25.0	34	2	9:56								

• Played CIAU Football (1996-97) • OUAA East Second All-Star Team (1998) • OUAA East First All-Star Team (1999) • OUAA First All-Star Team (2000) • CIAU All-Canadian Team (2000) • Won Randy Gregg Trophy (Athletics and Academics) (2000)

Signed as a free agent by **Columbus**, May 16, 2000.

DATSYUK, Pavel
(daht-SOOK, PAH-vehl) **DET.**

Center. Shoots left. 5'11", 180 lbs. Born, Sverdlovsk, USSR, July 20, 1978. Detroit's 8th choice, 171st overall, in 1998 Entry Draft.

Season	Club	League	GP	G	A	Pts	PIM	PP	SH	GW	S	%	+/-	TF	F%	H	SB	Min	GP	G	A	Pts	PIM	PP	SH	GW
1996-97	Yekaterinburg	Russia-Q	18	2	2	4	4																			
	Yekaterinburg	Russia	36	12	10	22	12																			
1997-98	Yekaterinburg	Russia	24	3	5	8	4																			
	Yekaterinburg	Russia	22	7	8	15	4																			
1998-99	Yekaterinburg 2	Russia-4	10	14	14	28	4																			
	Yekaterinburg	Russia-2	35	21	23	44	14												9	3	7	10	10			
99-2000	Yekaterinburg	Russia	15	1	3	4	4																			
2000-01	Ak Bars Kazan	Russia	42	9	18	27	10												4	0	1	1	2			
2001-02♦	Detroit	NHL	70	11	24	35	4	2	0	1	79	13.9	4	794	47.7	27	23	13:39	21	3	3	6	2	1	0	1
	Russia	Olympics	6	1	2	3	0																			
	NHL Totals		70	11	24	35	4	2	0	1	79	13.9		794	47.7	27	23	13:39	21	3	3	6	2	1	0	1

• Spent majority of 1999-2000 season on **Ak Bars Kazan** (Russia) reserve squad.

DAVIDSON, Matt
(DAY-vihd-SOHN, MAT) **CBJ**

Right wing. Shoots right. 6'3", 196 lbs. Born, Flin Flon, Man., August 9, 1977. Buffalo's 5th choice, 94th overall, in 1995 Entry Draft.

Season	Club	League	GP	G	A	Pts	PIM	PP	SH	GW	S	%	+/-	TF	F%	H	SB	Min	GP	G	A	Pts	PIM	PP	SH	GW
1992-93	Sask. Contacts	SMHL	36	14	18	32	36																			
1993-94	Portland	WHL	59	4	12	16	18												10	0	0	0	4			
1994-95	Portland	WHL	72	17	20	37	51												9	1	3	4	0			
1995-96	Portland	WHL	70	24	26	50	96												7	2	2	4	2			
1996-97	Portland	WHL	72	44	27	71	47												6	0	1	1	2			
1997-98	Rochester	AHL	72	15	12	27	12												3	1	0	1	2			
1998-99	Rochester	AHL	80	26	15	41	44												18	2	1	3	6			
99-2000	Rochester	AHL	80	12	20	32	30												19	4	2	6	8			

Season	Club	League	GP	G	A	Pts	PIM	PP	SH	GW	S	%	+/-	TF	F%	H	SB	Min	GP	G	A	Pts	PIM	PP	SH	GW
			Regular Season																Playoffs							
2000-01	**Columbus**	**NHL**	**5**	**0**	**0**	**0**	**0**	**0**	**0**	**0**	**2**	**0.0**	**2**	**0**	**0.0**	**4**	**2**	**7:14**								
	Syracuse Crunch	AHL	72	14	11	25	24												5	1	2	3	2			
2001-02	**Columbus**	**NHL**	**17**	**1**	**2**	**3**	**10**	**0**	**0**	**0**	**18**	**5.6**	**−11**	**7**	**0.0**	**19**	**11**	**14:34**								
	Syracuse Crunch	AHL	47	9	11	20	64												8	1	3	4	4			
	NHL Totals		**22**	**1**	**2**	**3**	**10**	**0**	**0**	**0**	**20**	**5.0**		**7**	**0.0**	**23**	**13**	**12:54**								

Traded to **Columbus** by **Buffalo** with Jean-Luc Grand-Pierre, San Jose's 5th round choice (previously acquired, Columbus selected Tyler Kolarik) in 2000 Entry Draft and Buffalo's 5th round choice (later traded to Calgary - later traded to Detroit - Detroit selected Andreas Jamtin) in 2001 Entry Draft to complete Expansion Draft agreement which had Columbus select Geoff Sanderson and Dwayne Roloson from Buffalo, June 23, 2000.

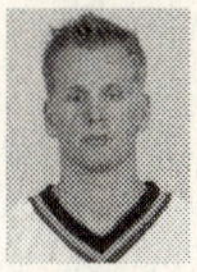

DAVIDSSON, Johan

(DAH-vihd-suhn, YOH-hahn) **VAN.**

Center. Shoots right. 6'1", 190 lbs. Born, Jonkoping, Sweden, January 6, 1976. Anaheim's 2nd choice, 28th overall, in 1994 Entry Draft.

Season	Club	League	GP	G	A	Pts	PIM	PP	SH	GW	S	%	+/-	TF	F%	H	SB	Min	GP	G	A	Pts	PIM	PP	SH	GW
1992-93	HV 71 Jonkoping	Sweden	8	1	0	1	0																			
1993-94	HV 71 Jr.	Swede-Jr.	5	2	3	5	0																			
	HV 71 Jonkoping	Sweden	38	2	5	7	4																			
1994-95	HV 71 Jr.	Swede-Jr.	3	4	1	5	0																			
	HV 71 Jonkoping	Sweden	37	4	7	11	20												13	3	2	5	0			
1995-96	HV 71 Jonkoping	Sweden	39	7	11	18	20												4	0	2	2	0			
1996-97	HV 71 Jonkoping	Sweden	50	18	21	39	18												5	0	3	3	2			
1997-98	HIFK Helsinki	Finland	43	10	30	40	8												9	3	10	13	0			
1998-99	**Anaheim**	**NHL**	**64**	**3**	**5**	**8**	**14**	**1**	**0**	**1**	**48**	**6.3**	**−9**	**516**	**37.0**	**34**	**7**	**10:37**	**1**	**0**	**0**	**0**	**0**	**0**	**0**	**0**
	Cincinnati	AHL	9	1	6	7	2																			
99-2000	**Anaheim**	**NHL**	**5**	**1**	**0**	**1**	**2**	**0**	**0**	**1**	**8**	**12.5**	**0**	**38**	**42.1**	**8**	**0**	**10:42**								
	Cincinnati	AHL	56	9	31	40	24																			
	NY Islanders	**NHL**	**14**	**2**	**4**	**6**	**0**	**0**	**0**	**0**	**21**	**9.5**	**0**	**130**	**40.0**	**9**	**9**	**11:32**								
2000-01	Blues Espoo	Finland	35	12	17	29	34																			
2001-02	HV 71 Jonkoping	Sweden	50	13	27	40	24												8	2	3	5	2			
	NHL Totals		**83**	**6**	**9**	**15**	**16**	**1**	**0**	**2**	**77**	**7.8**		**684**	**37.9**	**51**	**16**	**10:47**	**1**	**0**	**0**	**0**	**0**	**0**	**0**	**0**

Traded to **NY Islanders** by **Anaheim** with future considerations for Jorgen Jonsson, March 11, 2000. Signed as a free agent by **Vancouver**, September 6, 2000.

DAW, Jeff

(DAW, JEHF)

Center. Shoots right. 6'3", 190 lbs. Born, Carlisle, Ont., February 28, 1972.

Season	Club	League	GP	G	A	Pts	PIM	PP	SH	GW	S	%	+/-	TF	F%	H	SB	Min	GP	G	A	Pts	PIM	PP	SH	GW
1989-90	Milton Merchants	OPJHL	42	19	29	48	2																			
1990-91	Milton Merchants	OPJHL	34	21	41	62	22																			
1991-92	Milton Merchants	OPJHL	41	33	33	66	20																			
1992-93	U. Mass-Lowell	H-East	37	12	18	30	14																			
1993-94	U. Mass-Lowell	H-East	40	6	12	18	12																			
1994-95	U. Mass-Lowell	H-East	40	27	15	42	24																			
1995-96	U. Mass-Lowell	H-East	40	23	28	51	10																			
1996-97	Wheeling Nailers	ECHL	13	3	8	11	26																			
	Hamilton	AHL	56	11	8	19	39												19	4	5	9	0			
1997-98	Hamilton	AHL	79	28	35	63	20												9	6	3	9	0			
1998-99	Hamilton	AHL	66	18	29	47	10												11	0	3	3	4			
99-2000	Cleveland	IHL	9	4	1	5	2																			
	Houston Aeros	IHL	44	9	8	17	12																			
	Lowell	AHL	10	0	5	5	4												7	1	2	3	6			
2000-01	Lowell	AHL	65	28	28	56	33												3	0	1	1	2			
	Cleveland	IHL	8	2	3	5	2																			
2001-02	**Colorado**	**NHL**	**1**	**0**	**1**	**1**	**0**	**0**	**0**	**0**	**2**	**0.0**	**0**	**0**	**0.0**	**1**	**1**	**12:34**								
	Hershey Bears	AHL	79	26	25	51	22												8	1	1	2	4			
	NHL Totals		**1**	**0**	**1**	**1**	**0**	**0**	**0**	**0**	**2**	**0.0**		**0**	**0.0**	**1**	**1**	**12:34**								

Signed as a free agent by **Edmonton**, August 1, 1996. Signed as a free agent by **Chicago**, July 22, 1999. Traded to **Lowell** (AHL) by **Houston** (IHL) with Chicago retaining NHL rights for Dave Hymovitz, March 17, 2000. Selected by **Minnesota** from **Chicago** in Expansion Draft, June 23, 2000. Signed as a free agent by **Colorado**, July 23, 2001.

DAWE, Jason

(DAW, JAY-suhn) **ST.L.**

Right wing. Shoots left. 5'10", 189 lbs. Born, North York, Ont., May 29, 1973. Buffalo's 2nd choice, 35th overall, in 1991 Entry Draft.

Season	Club	League	GP	G	A	Pts	PIM	PP	SH	GW	S	%	+/-	TF	F%	H	SB	Min	GP	G	A	Pts	PIM	PP	SH	GW
1988-89	Don Mills	MTHL	44	35	28	63	103																			
1989-90	Peterborough	OHL	50	15	18	33	19												12	4	7	11	4			
1990-91	Peterborough	OHL	66	43	27	70	43												4	3	1	4	0			
1991-92	Peterborough	OHL	66	53	55	108	55												4	5	0	5	0			
1992-93	Peterborough	OHL	59	58	68	126	80												21	18	33	51	18			
	Rochester	AHL																	3	1	0	1	0			
1993-94	**Buffalo**	**NHL**	**32**	**6**	**7**	**13**	**12**	**3**	**0**	**1**	**35**	**17.1**	**1**						**6**	**0**	**1**	**1**	**6**	**0**	**0**	**0**
	Rochester	AHL	48	22	14	36	44																			
1994-95	Rochester	AHL	44	27	19	46	24																			
	Buffalo	**NHL**	**42**	**7**	**4**	**11**	**19**	**0**	**1**	**2**	**51**	**13.7**	**−6**						**5**	**2**	**1**	**3**	**6**	**0**	**0**	**0**
1995-96	**Buffalo**	**NHL**	**67**	**25**	**25**	**50**	**33**	**8**	**1**	**0**	**130**	**19.2**	**−8**													
	Rochester	AHL	7	5	4	9	2																			
1996-97	**Buffalo**	**NHL**	**81**	**22**	**26**	**48**	**32**	**4**	**1**	**3**	**136**	**16.2**	**14**						**11**	**2**	**1**	**3**	**6**	**0**	**0**	**0**
1997-98	**Buffalo**	**NHL**	**68**	**19**	**17**	**36**	**36**	**4**	**1**	**3**	**115**	**16.5**	**10**													
	NY Islanders	**NHL**	**13**	**1**	**2**	**3**	**6**	**0**	**0**	**0**	**19**	**5.3**	**−2**													
1998-99	**NY Islanders**	**NHL**	**22**	**2**	**3**	**5**	**8**	**0**	**0**	**0**	**29**	**6.9**	**0**	**4**	**25.0**	**27**	**5**	**11:55**								
	Montreal	**NHL**	**37**	**4**	**5**	**9**	**14**	**1**	**0**	**1**	**52**	**7.7**	**0**	**4**	**0.0**	**44**	**4**	**11:00**								
99-2000	Milwaukee	IHL	41	11	13	24	24																			
	NY Rangers	**NHL**	**3**	**0**	**1**	**1**	**2**	**0**	**0**	**0**	**8**	**0.0**	**0**	**1**	**100.0**	**3**	**0**	**13:35**								
	Hartford	AHL	27	9	9	18	24												21	10	7	17	37			
2000-01	Hartford	AHL	4	2	0	2	2																			
2001-02	**NY Rangers**	**NHL**	**1**	**0**	**0**	**0**	**0**	**0**	**0**	**0**	**1**	**0.0**	**−1**	**0**	**0.0**	**2**	**0**	**9:21**								
	Hartford	AHL	79	28	37	65	46												9	4	0	4	15			
	NHL Totals		**366**	**86**	**90**	**176**	**162**	**20**	**4**	**10**	**576**	**14.9**		**9**	**22.2**	**76**	**9**	**11:25**	**22**	**4**	**3**	**7**	**18**	**0**	**0**	**0**

OHL First All-Star Team (1993) • Canadian Major Junior Second All-Star Team (1993) • Won George Parsons Trophy (Memorial Cup Tournament Most Sportsmanlike Player) (1993)

Traded to **NY Islanders** by **Buffalo** for Jason Holland and Paul Kruse, March 24, 1998. Claimed on waivers by **Montreal** from **NY Islanders**, December 15, 1998. Signed as a free agent by **Nashville**, October 2, 1999. Traded to **NY Rangers** by **Nashville** for John Namestnikov, February 3, 2000. • Missed majority of 2000-01 season recovering from ankle injury originally suffered in game vs. Springfield (AHL), October 6, 2000. Signed as a free agent by **St. Louis**, July 23, 2002.

DAZE, Eric

(dah-ZAY, AIR-ihk) **CHI.**

Left wing. Shoots left. 6'6", 234 lbs. Born, Montreal, Que., July 2, 1975. Chicago's 5th choice, 90th overall, in 1993 Entry Draft.

Season	Club	League	GP	G	A	Pts	PIM	PP	SH	GW	S	%	+/-	TF	F%	H	SB	Min	GP	G	A	Pts	PIM	PP	SH	GW
1990-91	Laval Laurentide	QAHA	30	25	20	45	30																			
1991-92	Laval Laurentide	QAAA	35	30	29	59	40												12	8	10	18	8			
1992-93	Beauport	QMJHL	68	19	36	55	24																			
1993-94	Beauport	QMJHL	66	59	48	107	31												15	16	8	24	2			
1994-95	Beauport	QMJHL	57	54	45	99	20												16	9	12	21	23			
	Chicago	**NHL**	**4**	**1**	**1**	**2**	**2**	**0**	**0**	**0**	**1**	**100.0**	**2**						**16**	**0**	**1**	**1**	**4**	**0**	**0**	**0**
1995-96	**Chicago**	**NHL**	**80**	**30**	**23**	**53**	**18**	**2**	**0**	**2**	**167**	**18.0**	**16**						**10**	**3**	**5**	**8**	**0**	**0**	**0**	**1**
1996-97	**Chicago**	**NHL**	**71**	**22**	**19**	**41**	**16**	**11**	**0**	**4**	**176**	**12.5**	**−4**						**6**	**2**	**1**	**3**	**2**	**0**	**0**	**0**
1997-98	**Chicago**	**NHL**	**80**	**31**	**11**	**42**	**22**	**10**	**0**	**7**	**216**	**14.4**	**4**													
1998-99	**Chicago**	**NHL**	**72**	**22**	**20**	**42**	**22**	**8**	**0**	**2**	**189**	**11.6**	**−13**	**4**	**0.0**	**92**	**22**	**16:16**								
99-2000	**Chicago**	**NHL**	**59**	**23**	**13**	**36**	**28**	**6**	**0**	**1**	**143**	**16.1**	**−16**	**9**	**22.2**	**91**	**29**	**16:15**								
2000-01	**Chicago**	**NHL**	**79**	**33**	**24**	**57**	**16**	**9**	**1**	**8**	**205**	**16.1**	**1**	**3**	**33.3**	**77**	**35**	**17:45**								
2001-02	**Chicago**	**NHL**	**82**	**38**	**32**	**70**	**36**	**12**	**0**	**5**	**264**	**14.4**	**17**	**5**	**0.0**	**159**	**42**	**17:07**	**5**	**0**	**0**	**0**	**2**	**0**	**0**	**0**
	NHL Totals		**527**	**200**	**143**	**343**	**160**	**58**	**1**	**29**	**1361**	**14.7**		**21**	**14.3**	**419**	**128**	**16:54**	**37**	**5**	**7**	**12**	**8**	**0**	**0**	**1**

QMJHL First All-Star Team (1994, 1995) • Canadian Major Junior Most Sportsmanlike Player of the Year (1995) • NHL All-Rookie Team (1996) • Played in NHL All-Star Game (2002)

DEADMARSH, Adam
(DEHD-mahrsh, A-duhm) **L.A.**

Left wing/Center. Shoots right. 6', 195 lbs. Born, Trail, B.C., May 10, 1975. Quebec's 2nd choice, 14th overall, in 1993 Entry Draft.

			Regular Season																Playoffs							
Season	Club	League	GP	G	A	Pts	PIM	PP	SH	GW	S	%	+/-	TF	F%	H	SB	Min	GP	G	A	Pts	PIM	PP	SH	GW
1990-91	Beaver Valley	KIJHL	35	28	44	72	95	....	....	....	....	....	....	....	....	....	....	....	....	....	....	....	....	....	....	....
1991-92	Portland	WHL	68	30	30	60	81	....	....	....	....	....	....	....	....	....	....	....	6	3	3	6	13	....	....	....
1992-93	Portland	WHL	58	33	36	69	126	....	....	....	....	....	....	....	....	....	....	....	16	7	8	15	29	....	....	....
1993-94	Portland	WHL	65	43	56	99	212	....	....	....	....	....	....	....	....	....	....	....	10	9	8	17	33	....	....	....
1994-95	Portland	WHL	29	28	20	48	129	....	....	....	....	....	....	....	....	....	....	....	....	....	....	....	....	....	....	....
	Quebec	**NHL**	**48**	**9**	**8**	**17**	**56**	**0**	**0**	**0**	**48**	**18.8**	**16**	....	....	....	....	....	**6**	**0**	**1**	**1**	**0**	**0**	**0**	**0**
1995-96♦	**Colorado**	**NHL**	**78**	**21**	**27**	**48**	**142**	**3**	**0**	**2**	**151**	**13.9**	**20**	....	....	....	....	....	**22**	**5**	**12**	**17**	**25**	**1**	**0**	**0**
1996-97	**Colorado**	**NHL**	**78**	**33**	**27**	**60**	**136**	**10**	**3**	**4**	**198**	**16.7**	**8**	....	....	....	....	....	**17**	**3**	**6**	**9**	**24**	**1**	**0**	**1**
1997-98	**Colorado**	**NHL**	**73**	**22**	**21**	**43**	**125**	**10**	**0**	**6**	**187**	**11.8**	**0**	....	....	....	....	....	**7**	**2**	**0**	**2**	**4**	**1**	**0**	**0**
	United States	Olympics	4	1	0	1	2	....	....	....	....	....	....	....	....	....	....	....	....	....	....	....	....	....	....	....
1998-99	**Colorado**	**NHL**	**66**	**22**	**27**	**49**	**99**	**10**	**0**	**3**	**152**	**14.5**	**−2**	**621**	**45.9**	**121**	**51**	**20:46**	**19**	**8**	**4**	**12**	**20**	**3**	**0**	**0**
99-2000	**Colorado**	**NHL**	**71**	**18**	**27**	**45**	**106**	**5**	**0**	**4**	**153**	**11.8**	**−10**	**430**	**46.5**	**133**	**45**	**20:27**	**17**	**4**	**11**	**15**	**21**	**1**	**0**	**1**
2000-01	**Colorado**	**NHL**	**39**	**13**	**13**	**26**	**59**	**7**	**0**	**2**	**86**	**15.1**	**−2**	**56**	**55.4**	**81**	**10**	**17:38**	....	....	....	....	....	....	....	....
	Los Angeles	**NHL**	**18**	**4**	**2**	**6**	**4**	**0**	**0**	**0**	**40**	**10.0**	**3**	**21**	**57.1**	**49**	**7**	**18:47**	**13**	**3**	**3**	**6**	**4**	**0**	**0**	**2**
2001-02	**Los Angeles**	**NHL**	**76**	**29**	**33**	**62**	**71**	**12**	**0**	**5**	**139**	**20.9**	**8**	**60**	**38.3**	**126**	**34**	**19:17**	**4**	**1**	**3**	**4**	**2**	**0**	**0**	**0**
	United States	Olympics	6	1	1	2	2	....	....	....	....	....	....	....	....	....	....	....	....	....	....	....	....	....	....	....
	NHL Totals		**547**	**171**	**185**	**356**	**798**	**57**	**3**	**26**	**1154**	**14.8**		**1188**	**46.4**	**510**	**147**	**19:41**	**105**	**26**	**40**	**66**	**100**	**7**	**0**	**4**

Transferred to **Colorado** after **Quebec** franchise relocated, June 21, 1995. Traded to **LA Kings** by **Colorado** with Aaron Miller, a player to be named later (Jared Aulin, March 22, 2001), Colorado's 1st round choice (Dave Steckel) in 2001 Entry Draft and future considerations for Rob Blake and Steve Reinprecht, February 21, 2001.

DEAN, Kevin
(DEEN, KEH-vihn)

Defense. Shoots left. 6'3", 210 lbs. Born, Madison, WI, April 1, 1969. New Jersey's 4th choice, 86th overall, in 1987 Entry Draft.

			Regular Season																Playoffs							
Season	Club	League	GP	G	A	Pts	PIM	PP	SH	GW	S	%	+/-	TF	F%	H	SB	Min	GP	G	A	Pts	PIM	PP	SH	GW
1985-86	Culver Eagles	Hi-School	35	28	44	72	48	....	....	....	....	....	....	....	....	....	....	....	....	....	....	....	....	....	....	....
1986-87	Culver Eagles	Hi-School	25	19	25	44	30	....	....	....	....	....	....	....	....	....	....	....	....	....	....	....	....	....	....	....
1987-88	New Hampshire	H-East	27	1	6	7	34	....	....	....	....	....	....	....	....	....	....	....	....	....	....	....	....	....	....	....
1988-89	New Hampshire	H-East	34	1	12	13	28	....	....	....	....	....	....	....	....	....	....	....	....	....	....	....	....	....	....	....
1989-90	New Hampshire	H-East	39	2	6	8	42	....	....	....	....	....	....	....	....	....	....	....	....	....	....	....	....	....	....	....
1990-91	New Hampshire	H-East	31	10	12	22	22	....	....	....	....	....	....	....	....	....	....	....	....	....	....	....	....	....	....	....
	Utica Devils	AHL	7	0	1	1	2	....	....	....	....	....	....	....	....	....	....	....	....	....	....	....	....	....	....	....
1991-92	Utica Devils	AHL	23	0	3	3	6	....	....	....	....	....	....	....	....	....	....	....	....	....	....	....	....	....	....	....
	Cincinnati	ECHL	30	3	22	25	43	....	....	....	....	....	....	....	....	....	....	....	9	1	6	7	8	....	....	....
1992-93	Cincinnati	IHL	13	2	1	3	15	....	....	....	....	....	....	....	....	....	....	....	....	....	....	....	....	....	....	....
	Utica Devils	AHL	57	2	16	18	76	....	....	....	....	....	....	....	....	....	....	....	5	1	0	1	8	....	....	....
1993-94	Albany	AHL	70	9	33	42	92	....	....	....	....	....	....	....	....	....	....	....	5	0	2	2	7	....	....	....
1994-95	Albany	AHL	68	5	37	42	66	....	....	....	....	....	....	....	....	....	....	....	8	0	4	4	4	....	....	....
	♦ New Jersey	**NHL**	**17**	**0**	**1**	**1**	**4**	**0**	**0**	**0**	**11**	**0.0**	**6**	....	....	....	....	....	**3**	**0**	**2**	**2**	**0**	**0**	**0**	**0**
1995-96	**New Jersey**	**NHL**	**41**	**0**	**6**	**6**	**28**	**0**	**0**	**0**	**29**	**0.0**	**4**	....	....	....	....	....	....	....	....	....	....	....	....	....
	Albany	AHL	1	1	0	1	2	....	....	....	....	....	....	....	....	....	....	....	....	....	....	....	....	....	....	....
1996-97	**New Jersey**	**NHL**	**28**	**2**	**4**	**6**	**6**	**0**	**0**	**0**	**21**	**9.5**	**2**	....	....	....	....	....	**1**	**1**	**0**	**1**	**0**	**0**	**0**	**1**
	Albany	AHL	2	0	1	1	4	....	....	....	....	....	....	....	....	....	....	....	....	....	....	....	....	....	....	....
1997-98	**New Jersey**	**NHL**	**50**	**1**	**8**	**9**	**12**	**1**	**0**	**0**	**28**	**3.6**	**12**	....	....	....	....	....	**5**	**1**	**0**	**1**	**2**	**0**	**0**	**0**
	Albany	AHL	2	0	1	1	2	....	....	....	....	....	....	....	....	....	....	....	....	....	....	....	....	....	....	....
1998-99	**New Jersey**	**NHL**	**62**	**1**	**10**	**11**	**22**	**1**	**0**	**0**	**51**	**2.0**	**4**	**0**	**0.0**	**77**	**59**	**15:42**	**7**	**0**	**0**	**0**	**0**	**0**	**0**	**0**
99-2000	**Atlanta**	**NHL**	**23**	**1**	**0**	**1**	**14**	**0**	**1**	**0**	**9**	**11.1**	**−5**	**1**	**0.0**	**48**	**39**	**16:53**	....	....	....	....	....	....	....	....
	Dallas	**NHL**	**14**	**0**	**0**	**0**	**10**	**0**	**0**	**0**	**6**	**0.0**	**−1**	**0**	**0.0**	**11**	**15**	**11:19**	....	....	....	....	....	....	....	....
	Chicago	**NHL**	**27**	**2**	**8**	**10**	**12**	**0**	**0**	**0**	**32**	**6.3**	**9**	**0**	**0.0**	**34**	**42**	**18:36**	....	....	....	....	....	....	....	....
2000-01	**Chicago**	**NHL**	**69**	**0**	**11**	**11**	**30**	**0**	**0**	**0**	**75**	**0.0**	**−16**	**0**	**0.0**	**76**	**103**	**19:27**	....	....	....	....	....	....	....	....
2001-02	Milwaukee	AHL	76	5	14	19	33	....	....	....	....	....	....	....	....	....	....	....	....	....	....	....	....	....	....	....
	NHL Totals		**331**	**7**	**48**	**55**	**138**	**2**	**1**	**0**	**262**	**2.7**		**1**	**0.0**	**246**	**258**	**17:15**	**16**	**2**	**2**	**4**	**2**	**0**	**0**	**1**

AHL First All-Star Team (1995)

Claimed by **Atlanta** from **New Jersey** in Expansion Draft, June 25, 1999. Traded to **Dallas** by **Atlanta** for Dallas' 9th round choice (Mark McRae) in 2000 Entry Draft, December 15, 1999. Traded to **Chicago** by **Dallas** with Derek Plante and Dallas' 2nd round choice (Matt Keith) in 2001 Entry Draft for Sylvain Cote and Dave Manson, February 8, 2000.

DeBRUSK, Louie
(duh-BRUHSK, LEW-ee)

Left wing. Shoots left. 6'2", 238 lbs. Born, Cambridge, Ont., March 19, 1971. NY Rangers' 4th choice, 49th overall, in 1989 Entry Draft.

			Regular Season																Playoffs							
Season	Club	League	GP	G	A	Pts	PIM	PP	SH	GW	S	%	+/-	TF	F%	H	SB	Min	GP	G	A	Pts	PIM	PP	SH	GW
1986-87	Port Elgin	OJHL-C	10	2	1	3	4	....	....	....	....	....	....	....	....	....	....	....	....	....	....	....	....	....	....	....
1987-88	Stratford	OJHL-B	45	13	14	27	205	....	....	....	....	....	....	....	....	....	....	....	....	....	....	....	....	....	....	....
1988-89	London Knights	OHL	59	11	11	22	149	....	....	....	....	....	....	....	....	....	....	....	19	1	1	2	43	....	....	....
1989-90	London Knights	OHL	61	21	19	40	198	....	....	....	....	....	....	....	....	....	....	....	6	2	2	4	24	....	....	....
1990-91	London Knights	OHL	61	31	33	64	*223	....	....	....	....	....	....	....	....	....	....	....	7	2	2	4	14	....	....	....
	Binghamton	AHL	2	0	0	0	7	....	....	....	....	....	....	....	....	....	....	....	2	0	0	0	9	....	....	....
1991-92	**Edmonton**	**NHL**	**25**	**2**	**1**	**3**	**124**	**0**	**0**	**1**	**7**	**28.6**	**4**	....	....	....	....	....	....	....	....	....	....	....	....	....
	Cape Breton	AHL	28	2	2	4	73	....	....	....	....	....	....	....	....	....	....	....	....	....	....	....	....	....	....	....
1992-93	**Edmonton**	**NHL**	**51**	**8**	**2**	**10**	**205**	**0**	**0**	**1**	**33**	**24.2**	**−16**	....	....	....	....	....	....	....	....	....	....	....	....	....
1993-94	**Edmonton**	**NHL**	**48**	**4**	**6**	**10**	**185**	**0**	**0**	**0**	**27**	**14.8**	**−9**	....	....	....	....	....	....	....	....	....	....	....	....	....
	Cape Breton	AHL	5	3	1	4	58	....	....	....	....	....	....	....	....	....	....	....	....	....	....	....	....	....	....	....
1994-95	**Edmonton**	**NHL**	**34**	**2**	**0**	**2**	**93**	**0**	**0**	**0**	**14**	**14.3**	**−4**	....	....	....	....	....	....	....	....	....	....	....	....	....
1995-96	**Edmonton**	**NHL**	**38**	**1**	**3**	**4**	**96**	**0**	**0**	**0**	**17**	**5.9**	**−7**	....	....	....	....	....	....	....	....	....	....	....	....	....
1996-97	**Edmonton**	**NHL**	**32**	**2**	**0**	**2**	**94**	**0**	**0**	**0**	**10**	**20.0**	**−6**	....	....	....	....	....	**6**	**0**	**0**	**0**	**4**	**0**	**0**	**0**
1997-98	**Tampa Bay**	**NHL**	**54**	**1**	**2**	**3**	**166**	**0**	**0**	**0**	**14**	**7.1**	**−2**	....	....	....	....	....	....	....	....	....	....	....	....	....
	San Antonio	IHL	17	7	4	11	130	....	....	....	....	....	....	....	....	....	....	....	....	....	....	....	....	....	....	....
1998-99	**Phoenix**	**NHL**	**15**	**0**	**0**	**0**	**34**	**0**	**0**	**0**	**6**	**0.0**	**−2**	**0**	**0.0**	**16**	**0**	**5:57**	**6**	**2**	**0**	**2**	**6**	**0**	**0**	**0**
	Las Vegas	IHL	26	3	6	9	160	....	....	....	....	....	....	....	....	....	....	....	....	....	....	....	....	....	....	....
	Springfield	AHL	3	1	0	1	0	....	....	....	....	....	....	....	....	....	....	....	....	....	....	....	....	....	....	....
	Long Beach	IHL	24	5	5	10	134	....	....	....	....	....	....	....	....	....	....	....	....	....	....	....	....	....	....	....
99-2000	**Phoenix**	**NHL**	**61**	**4**	**3**	**7**	**78**	**0**	**0**	**0**	**24**	**16.7**	**1**	**0**	**0.0**	**55**	**4**	**5:21**	**3**	**0**	**0**	**0**	**0**	**0**	**0**	**0**
2000-01	**Phoenix**	**NHL**	**39**	**0**	**0**	**0**	**79**	**0**	**0**	**0**	**12**	**0.0**	**−5**	**0**	**0.0**	**18**	**5**	**4:49**	....	....	....	....	....	....	....	....
2001-02	Quebec	AHL	9	0	0	0	44	....	....	....	....	....	....	....	....	....	....	....	....	....	....	....	....	....	....	....
	Hamilton	AHL	20	3	5	8	42	....	....	....	....	....	....	....	....	....	....	....	13	1	0	1	30	....	....	....
	NHL Totals		**397**	**24**	**17**	**41**	**1154**	**0**	**0**	**2**	**164**	**14.6**		**0**	**0.0**	**89**	**9**	**5:15**	**15**	**2**	**0**	**2**	**10**	**0**	**0**	**0**

Traded to **Edmonton** by **NY Rangers** with Bernie Nicholls and Steven Rice for Mark Messier and future considerations (Jeff Beukeboom for David Shaw, November 12, 1991), October 4, 1991. Signed as a free agent by **Tampa Bay**, September 23, 1997. Traded to **Phoenix** by **Tampa Bay** with Tampa Bay's 5th round choice (Jay Leach) in 1998 Entry Draft for Craig Janney, June 11, 1998. Signed to 25-game try-out contract by **Quebec** (AHL), November 25, 2001. Released by **Quebec** (AHL) and signed as a free agent by **Hamilton** (AHL), December 21, 2001.

DELISLE, Jonathan
(duh-LIGHL, JAWN-ah-thuhn)

Right wing. Shoots right. 5'10", 180 lbs. Born, Ste-Anne-des-Plaines, Que., June 30, 1977. Montreal's 4th choice, 86th overall, in 1995 Entry Draft.

			Regular Season																Playoffs							
Season	Club	League	GP	G	A	Pts	PIM	PP	SH	GW	S	%	+/-	TF	F%	H	SB	Min	GP	G	A	Pts	PIM	PP	SH	GW
1992-93	Laval Laurentide	QAAA	14	3	3	6	12	....	....	....	....	....	....	....	....	....	....	....	13	2	5	7	24	....	....	....
1993-94	Verdun	QMJHL	61	16	17	33	130	....	....	....	....	....	....	....	....	....	....	....	4	0	1	1	14	....	....	....
1994-95	Hull Olympiques	QMJHL	60	21	38	59	218	....	....	....	....	....	....	....	....	....	....	....	19	11	8	19	43	....	....	....
1995-96	Hull Olympiques	QMJHL	62	31	57	88	193	....	....	....	....	....	....	....	....	....	....	....	18	6	13	19	64	....	....	....
1996-97	Hull Olympiques	QMJHL	61	35	54	89	228	....	....	....	....	....	....	....	....	....	....	....	14	11	13	24	46	....	....	....
1997-98	Fredericton	AHL	78	15	21	36	138	....	....	....	....	....	....	....	....	....	....	....	4	0	1	1	7	....	....	....
1998-99	**Montreal**	**NHL**	**1**	**0**	**0**	**0**	**0**	**0**	**0**	**0**	**0**	**0.0**	**0**	**0**	**0.0**	**1**	**0**	**4:32**	....	....	....	....	....	....	....	....
	Fredericton	AHL	78	7	29	36	118	....	....	....	....	....	....	....	....	....	....	....	15	3	6	9	39	....	....	....
99-2000	Quebec	AHL	62	7	19	26	142	....	....	....	....	....	....	....	....	....	....	....	3	0	0	0	4	....	....	....
2000-01	Quebec	AHL	71	6	18	24	201	....	....	....	....	....	....	....	....	....	....	....	6	1	0	1	53	....	....	....
2001-02	Quebec	AHL	24	0	3	3	37	....	....	....	....	....	....	....	....	....	....	....	....	....	....	....	....	....	....	....
	New Mexico	CHL	32	13	17	30	102	....	....	....	....	....	....	....	....	....	....	....	....	....	....	....	....	....	....	....
	NHL Totals		**1**	**0**	**0**	**0**	**0**	**0**	**0**	**0**	**0**	**0.0**		**0**	**0.0**	**1**	**0**	**4:32**	....	....	....	....	....	....	....	....

DELISLE, Xavier

(duh-LIGHL, ehx-AY-vee-uhr)

Center. Shoots right. 5'11", 193 lbs. Born, Quebec City, Que., May 24, 1977. Tampa Bay's 5th choice, 157th overall, in 1996 Entry Draft.

			Regular Season																Playoffs							
Season	Club	League	GP	G	A	Pts	PIM	PP	SH	GW	S	%	+/-	TF	F%	H	SB	Min	GP	G	A	Pts	PIM	PP	SH	GW
1992-93	Ste-Foy	QAAA	41	20	23	43	10												12	8	10	18	2			
1993-94	Granby Bisons	QMJHL	46	11	22	33	25												7	2	0	2	0			
1994-95	Granby Bisons	QMJHL	72	18	36	54	48												13	2	6	8	4			
1995-96	Granby	QMJHL	67	45	75	120	45												20	13	*27	*40	12			
1996-97	Granby	QMJHL	59	36	56	92	20												5	1	4	5	6			
1997-98	Adirondack	AHL	76	10	19	29	47												3	0	0	0	0			
1998-99	**Tampa Bay**	**NHL**	**2**	**0**	**0**	**0**	**0**	**0**	**0**	**0**	**1**	**0.0**	**0**	**11**	**45.5**	**1**	**0**	**5:51**								
	Cleveland	IHL	77	15	29	44	36																			
99-2000	Detroit Vipers	IHL	20	2	6	8	18																			
	Toledo Storm	ECHL	2	0	1	1	0																			
	Quebec	AHL	42	17	28	45	8												3	1	2	3	0			
2000-01	**Montreal**	**NHL**	**14**	**3**	**2**	**5**	**6**	**1**	**0**	**0**	**15**	**20.0**	**–5**	**3**	**33.3**	**5**	**1**	**10:14**								
	Quebec	AHL	62	18	29	47	34												9	1	5	6	2			
2001-02	Quebec	AHL	50	8	17	25	19																			
	NHL Totals		**16**	**3**	**2**	**5**	**6**	**1**	**0**	**0**	**16**	**18.8**		**14**	**42.9**	**6**	**1**	**9:41**								

QMJHL Second All-Star Team (1996) • Memorial Cup All-Star Team (1996)

Signed as a free agent by **Montreal**, August 8, 2000. Signed as a free agent by **Ausburg** (Germany), May 8, 2002.

DELMORE, Andy

(DEHL-mohr, AN-dee) **NSH.**

Defense. Shoots right. 6'1", 200 lbs. Born, LaSalle, Ont., December 26, 1976.

Season	Club	League	GP	G	A	Pts	PIM	PP	SH	GW	S	%	+/-	TF	F%	H	SB	Min	GP	G	A	Pts	PIM	PP	SH	GW
1992-93	Chatham	OJHL-B	47	4	21	25	38																			
1993-94	North Bay	OHL	45	2	7	9	33												17	0	0	0	2			
1994-95	North Bay	OHL	40	2	14	16	21																			
	Sarnia Sting	OHL	27	5	13	18	27												3	0	0	0	2			
1995-96	Sarnia Sting	OHL	64	21	38	59	45												10	3	7	10	2			
1996-97	Sarnia Sting	OHL	64	18	60	78	39												12	2	10	12	10			
	Fredericton	AHL	4	0	1	1	0																			
1997-98	Philadelphia	AHL	73	9	30	39	46												18	4	4	8	21			
1998-99	**Philadelphia**	**NHL**	**2**	**0**	**1**	**1**	**0**	**0**	**0**	**0**	**2**	**0.0**	**–1**	**0**	**0.0**	**1**	**1**	**20:42**								
	Philadelphia	AHL	70	5	18	23	51												15	1	4	5	6			
99-2000	**Philadelphia**	**NHL**	**27**	**2**	**5**	**7**	**8**	**0**	**0**	**1**	**55**	**3.6**	**–1**	**0**	**0.0**	**24**	**31**	**17:17**	**18**	**5**	**2**	**7**	**14**	**1**	**0**	**1**
	Philadelphia	AHL	39	12	14	26	31																			
2000-01	**Philadelphia**	**NHL**	**66**	**5**	**9**	**14**	**16**	**2**	**0**	**0**	**119**	**4.2**	**2**	**0**	**0.0**	**70**	**65**	**17:39**	**2**	**1**	**0**	**1**	**2**	**0**	**0**	**1**
2001-02	**Nashville**	**NHL**	**73**	**16**	**22**	**38**	**22**	**11**	**0**	**3**	**175**	**9.1**	**–13**	**0**	**0.0**	**82**	**36**	**19:40**								
	NHL Totals		**168**	**23**	**37**	**60**	**46**	**13**	**0**	**4**	**351**	**6.6**		**0**	**0.0**	**177**	**133**	**18:30**	**20**	**6**	**2**	**8**	**16**	**1**	**0**	**2**

OHL First All-Star Team (1997)

Signed as a free agent by **Philadelphia**, June 9, 1997. Traded to **Nashville** by **Philadelphia** for Nashville's 3rd round choice (later traded to Phoenix - Phoenix selected Joe Callahan) in 2002 Entry Draft, July 31, 2001.

DEMITRA, Pavol

(deh-MEET-rah, PAH-vohl) **ST.L.**

Left wing. Shoots left. 5'11", 203 lbs. Born, Dubnica, Czech., November 29, 1974. Ottawa's 9th choice, 227th overall, in 1993 Entry Draft.

Season	Club	League	GP	G	A	Pts	PIM	PP	SH	GW	S	%	+/-	TF	F%	H	SB	Min	GP	G	A	Pts	PIM	PP	SH	GW
1991-92	Dubnica	Czech-2	28	13	10	23	12																			
1992-93	CAPEH Dubnica	Czech-2	4	3	0	3	....																			
	Dukla Trencin	Czech	46	11	17	28	0																			
1993-94	**Ottawa**	**NHL**	**12**	**1**	**1**	**2**	**4**	**1**	**0**	**0**	**10**	**10.0**	**–7**													
	P.E.I. Senators	AHL	41	18	23	41	8																			
1994-95	P.E.I. Senators	AHL	61	26	48	74	23												5	0	7	7	0			
	Ottawa	**NHL**	**16**	**4**	**3**	**7**	**0**	**1**	**0**	**0**	**21**	**19.0**	**–4**													
1995-96	**Ottawa**	**NHL**	**31**	**7**	**10**	**17**	**6**	**2**	**0**	**1**	**66**	**10.6**	**–3**													
	P.E.I. Senators	AHL	48	28	53	81	44																			
1996-97	Dukla Trencin	Slovakia	1	1	1	2	....																			
	Las Vegas	IHL	22	8	13	21	10																			
	St. Louis	**NHL**	**8**	**3**	**0**	**3**	**2**	**2**	**0**	**1**	**15**	**20.0**	**0**						**6**	**1**	**3**	**4**	**6**	**0**	**0**	**0**
	Grand Rapids	IHL	42	20	30	50	24																			
1997-98	**St. Louis**	**NHL**	**61**	**22**	**30**	**52**	**22**	**4**	**4**	**6**	**147**	**15.0**	**11**						**10**	**3**	**3**	**6**	**2**	**0**	**0**	**0**
1998-99	**St. Louis**	**NHL**	**82**	**37**	**52**	**89**	**16**	**14**	**0**	**10**	**259**	**14.3**	**13**	**250**	**44.0**	**31**	**15**	**20:10**	**13**	**5**	**4**	**9**	**4**	**3**	**0**	**1**
99-2000	**St. Louis**	**NHL**	**71**	**28**	**47**	**75**	**8**	**8**	**0**	**4**	**241**	**11.6**	**34**	**41**	**39.0**	**15**	**15**	**19:13**								
2000-01	**St. Louis**	**NHL**	**44**	**20**	**25**	**45**	**16**	**5**	**0**	**5**	**124**	**16.1**	**27**	**8**	**37.5**	**9**	**5**	**18:03**	**15**	**2**	**4**	**6**	**2**	**0**	**0**	**1**
2001-02	**St. Louis**	**NHL**	**82**	**35**	**43**	**78**	**46**	**11**	**0**	**10**	**212**	**16.5**	**13**	**1224**	**48.1**	**34**	**33**	**19:11**	**10**	**4**	**7**	**11**	**6**	**2**	**1**	**1**
	Slovakia	Olympics	2	1	2	3	2																			
	NHL Totals		**407**	**157**	**211**	**368**	**120**	**48**	**4**	**37**	**1095**	**14.3**		**1523**	**47.1**	**89**	**68**	**19:18**	**54**	**15**	**21**	**36**	**20**	**5**	**1**	**3**

Won Lady Byng Trophy (2000) • Played in NHL All-Star Game (1999, 2000, 2002)

Traded to **St. Louis** by **Ottawa** for Christer Olsson, November 27, 1996. • Missed most of 2000-01 season recovering from eye (January 1, 2001 vs. Edmonton) and leg (February 1, 2001 vs. Columbus) injuries.

DEMPSEY, Nathan

(DEHMP-see, NAY-thun) **CHI.**

Defense. Shoots right. 6', 190 lbs. Born, Spruce Grove, Alta., July 14, 1974. Toronto's 12th choice, 245th overall, in 1992 Entry Draft.

Season	Club	League	GP	G	A	Pts	PIM	PP	SH	GW	S	%	+/-	TF	F%	H	SB	Min	GP	G	A	Pts	PIM	PP	SH	GW
1990-91	St. Albert	AJHL	34	11	20	31	73																			
1991-92	Regina Pats	WHL	70	4	22	26	72																			
1992-93	Regina Pats	WHL	72	12	29	41	95												13	3	8	11	14			
	St. John's	AHL	....	....	....	....	....												2	0	0	0	0			
1993-94	Regina Pats	WHL	56	14	36	50	100												4	0	0	0	4			
1994-95	St. John's	AHL	74	7	30	37	91												5	1	0	1	11			
1995-96	St. John's	AHL	73	5	15	20	103												4	1	0	1	9			
1996-97	**Toronto**	**NHL**	**14**	**1**	**1**	**2**	**2**	**0**	**0**	**0**	**11**	**9.1**	**–2**													
	St. John's	AHL	52	8	18	26	108												6	1	0	1	4			
1997-98	St. John's	AHL	68	12	16	28	85												4	0	0	0	0			
1998-99	St. John's	AHL	67	2	29	31	70												5	0	1	1	2			
99-2000	**Toronto**	**NHL**	**6**	**0**	**2**	**2**	**2**	**0**	**0**	**0**	**3**	**0.0**	**2**	**1**	**0.0**	**4**	**2**	**13:40**								
	St. John's	AHL	44	15	12	27	40																			
2000-01	**Toronto**	**NHL**	**25**	**1**	**9**	**10**	**4**	**1**	**0**	**0**	**31**	**3.2**	**13**	**0**	**0.0**	**37**	**18**	**15:53**								
	St. John's	AHL	55	11	28	39	60												4	0	4	4	8			
2001-02	**Toronto**	**NHL**	**3**	**0**	**0**	**0**	**0**	**0**	**0**	**0**	**3**	**0.0**	**1**	**0**	**0.0**	**3**	**0**	**14:00**	**6**	**0**	**2**	**2**	**0**	**0**	**0**	**0**
	St. John's	AHL	75	13	48	61	66												11	1	5	6	8			
	NHL Totals		**48**	**2**	**12**	**14**	**8**	**1**	**0**	**0**	**48**	**4.2**		**1**	**0.0**	**44**	**20**	**15:20**	**6**	**0**	**2**	**2**	**0**	**0**	**0**	**0**

WHL East Second All-Star Team (1994) • AHL Second All-Star Team (2002) • Won Fred Hunt Memorial Trophy (Sportsmanship - AHL) (2002)

Signed as a free agent by **Chicago**, July 13, 2002.

DESCOTEAUX, Matthieu

(DAY-koh-toh, MAT-yoo) **MTL.**

Defense. Shoots left. 6'3", 216 lbs. Born, Pierreville, Que., September 23, 1977. Edmonton's 2nd choice, 19th overall, in 1996 Entry Draft.

Season	Club	League	GP	G	A	Pts	PIM	PP	SH	GW	S	%	+/-	TF	F%	H	SB	Min	GP	G	A	Pts	PIM	PP	SH	GW
1993-94	Cap-d-Madeleine	QAAA	43	2	5	7	26																			
1994-95	Shawinigan	QMJHL	50	3	2	5	28												15	1	1	2	19			
1995-96	Shawinigan	QMJHL	69	2	13	15	129												6	0	0	0	6			
1996-97	Shawinigan	QMJHL	38	6	18	24	121																			
	Hull Olympiques	QMJHL	32	6	19	25	34												14	1	8	9	29			
1997-98	Hamilton	AHL	67	2	8	10	70												2	0	0	0	0			
1998-99	Hamilton	AHL	74	6	12	18	49												4	0	0	0	0			
99-2000	Hamilton	AHL	49	5	7	12	29																			
	Quebec	AHL	12	0	6	6	6												2	0	1	1	0			

			Regular Season																Playoffs							
Season	Club	League	GP	G	A	Pts	PIM	PP	SH	GW	S	%	+/-	TF	F%	H	SB	Min	GP	G	A	Pts	PIM	PP	SH	GW
2000-01	**Montreal**	**NHL**	**5**	**1**	**1**	**2**	**4**	**1**	**0**	**0**	**6**	**16.7**	**−2**	**0**	**0.0**	**5**	**3**	**15:10**	....	....	....	....	....	....	....	....
	Quebec	AHL	73	16	27	43	38	....	....	....	....	....	....	....	....	....	....	....	7	0	3	3	4	....	....	....
2001-02	Quebec	AHL	65	6	14	20	34	....	....	....	....	....	....	....	....	....	....	....	3	1	2	3	0	....	....	....
	NHL Totals		**5**	**1**	**1**	**2**	**4**	**1**	**0**	**0**	**6**	**16.7**		**0**	**0.0**	**5**	**3**	**15:10**	....	....	....	....	....	....	....	....

Traded to **Montreal** by **Edmonton** with Christian Laflamme for Igor Ulanov and Alain Nasreddine, March 9, 2000.

DESJARDINS, Eric
(deh-ZHAHR-dai, AIR-ihk) **PHI.**

Defense. Shoots right. 6'1", 205 lbs. Born, Rouyn, Que., June 14, 1969. Montreal's 3rd choice, 38th overall, in 1987 Entry Draft.

Season	Club	League	GP	G	A	Pts	PIM	PP	SH	GW	S	%	+/-	TF	F%	H	SB	Min	GP	G	A	Pts	PIM	PP	SH	GW
1985-86	Laval Laurentide	QAAA	42	6	30	36	54	....	....	....	....	....	....	....	....	....	....	....	8	2	10	12	14	....	....	....
1986-87	Granby Bisons	QMJHL	66	14	24	38	178	....	....	....	....	....	....	....	....	....	....	....	8	3	2	5	10	....	....	....
1987-88	Granby Bisons	QMJHL	62	18	49	67	138	....	....	....	....	....	....	....	....	....	....	....	5	0	3	3	10	....	....	....
	Sherbrooke	AHL	3	0	0	0	6	....	....	....	....	....	....	....	....	....	....	....	4	0	2	2	2	....	....	....
1988-89	**Montreal**	**NHL**	**36**	**2**	**12**	**14**	**26**	**1**	**0**	**0**	**39**	**5.1**	**9**	....	....	....	....	....	**14**	**1**	**1**	**2**	**6**	**1**	**0**	**0**
1989-90	**Montreal**	**NHL**	**55**	**3**	**13**	**16**	**51**	**1**	**0**	**0**	**48**	**6.3**	**1**	....	....	....	....	....	**6**	**0**	**0**	**0**	**10**	**0**	**0**	**0**
1990-91	**Montreal**	**NHL**	**62**	**7**	**18**	**25**	**27**	**0**	**0**	**1**	**114**	**6.1**	**7**	....	....	....	....	....	**13**	**1**	**4**	**5**	**8**	**1**	**0**	**0**
1991-92	**Montreal**	**NHL**	**77**	**6**	**32**	**38**	**50**	**4**	**0**	**2**	**141**	**4.3**	**17**	....	....	....	....	....	**11**	**3**	**3**	**6**	**4**	**1**	**0**	**0**
1992-93♦	**Montreal**	**NHL**	**82**	**13**	**32**	**45**	**98**	**7**	**0**	**1**	**163**	**8.0**	**20**	....	....	....	....	....	**20**	**4**	**10**	**14**	**23**	**1**	**0**	**1**
1993-94	**Montreal**	**NHL**	**84**	**12**	**23**	**35**	**97**	**6**	**1**	**3**	**193**	**6.2**	**−1**	....	....	....	....	....	**7**	**0**	**2**	**2**	**4**	**0**	**0**	**0**
1994-95	**Montreal**	**NHL**	**9**	**0**	**6**	**6**	**2**	**0**	**0**	**0**	**14**	**0.0**	**2**	....	....	....	....	....	....	....	....	....	....	....	....	....
	Philadelphia	**NHL**	**34**	**5**	**18**	**23**	**12**	**1**	**0**	**1**	**79**	**6.3**	**10**	....	....	....	....	....	**15**	**4**	**4**	**8**	**10**	**1**	**0**	**2**
1995-96	**Philadelphia**	**NHL**	**80**	**7**	**40**	**47**	**45**	**5**	**0**	**2**	**184**	**3.8**	**19**	....	....	....	....	....	**12**	**0**	**6**	**6**	**2**	**0**	**0**	**0**
1996-97	**Philadelphia**	**NHL**	**82**	**12**	**34**	**46**	**50**	**5**	**1**	**1**	**183**	**6.6**	**25**	....	....	....	....	....	**19**	**2**	**8**	**10**	**12**	**0**	**0**	**0**
1997-98	**Philadelphia**	**NHL**	**77**	**6**	**27**	**33**	**36**	**2**	**1**	**0**	**150**	**4.0**	**11**	....	....	....	....	....	**5**	**0**	**1**	**1**	**0**	**0**	**0**	**0**
	Canada	Olympics	6	0	0	0	2	....	....	....	....	....	....	....	....	....	....	....	....	....	....	....	....	....	....	....
1998-99	**Philadelphia**	**NHL**	**68**	**15**	**36**	**51**	**38**	**6**	**0**	**2**	**190**	**7.9**	**18**	**0**	**0.0**	**36**	**108**	**25:48**	**6**	**2**	**2**	**4**	**4**	**1**	**0**	**1**
99-2000	**Philadelphia**	**NHL**	**81**	**14**	**41**	**55**	**32**	**8**	**0**	**4**	**207**	**6.8**	**20**	**1**	**0.0**	**28**	**141**	**27:01**	**18**	**2**	**10**	**12**	**2**	**1**	**0**	**1**
2000-01	**Philadelphia**	**NHL**	**79**	**15**	**33**	**48**	**50**	**6**	**1**	**4**	**187**	**8.0**	**−3**	**3**	**100.0**	**38**	**122**	**26:27**	**6**	**1**	**1**	**2**	**0**	**0**	**0**	**0**
2001-02	**Philadelphia**	**NHL**	**65**	**6**	**19**	**25**	**24**	**2**	**1**	**0**	**117**	**5.1**	**−1**	**2**	**0.0**	**21**	**88**	**22:12**	**5**	**0**	**1**	**1**	**2**	**0**	**0**	**0**
	NHL Totals		**971**	**123**	**384**	**507**	**638**	**54**	**5**	**21**	**2009**	**6.1**		**6**	**50.0**	**123**	**459**	**25:31**	**157**	**20**	**53**	**73**	**87**	**7**	**0**	**5**

QMJHL Second All-Star Team (1987) • QMJHL First All-Star Team (1988) • NHL Second All-Star Team (1999, 2000) • Played in NHL All-Star Game (1992, 1996, 2000)
Traded to **Philadelphia** by **Montreal** with Gilbert Dionne and John LeClair for Mark Recchi and Philadelphia's 3rd round choice (Martin Hohenberger) in 1995 Entry Draft, February 9, 1995.

DEULING, Jarrett
(DEW-lihng, JAIR-uht)

Left wing. Shoots left. 6', 205 lbs. Born, Vernon, B.C., March 4, 1974. NY Islanders' 2nd choice, 56th overall, in 1992 Entry Draft.

Season	Club	League	GP	G	A	Pts	PIM	PP	SH	GW	S	%	+/-	TF	F%	H	SB	Min	GP	G	A	Pts	PIM	PP	SH	GW
1989-90	Whitehorse Bears	AAHL	28	34	48	72	84	....	....	....	....	....	....	....	....	....	....	....	....	....	....	....	....	....	....	....
1990-91	Kamloops Blazers	WHL	48	4	12	16	43	....	....	....	....	....	....	....	....	....	....	....	12	5	2	7	7	....	....	....
1991-92	Kamloops Blazers	WHL	68	28	26	54	79	....	....	....	....	....	....	....	....	....	....	....	17	10	6	16	18	....	....	....
1992-93	Kamloops Blazers	WHL	68	31	32	63	93	....	....	....	....	....	....	....	....	....	....	....	13	6	7	13	14	....	....	....
1993-94	Kamloops Blazers	WHL	70	44	59	103	171	....	....	....	....	....	....	....	....	....	....	....	18	*13	8	21	43	....	....	....
1994-95	Worcester	AHL	63	11	8	19	37	....	....	....	....	....	....	....	....	....	....	....	....	....	....	....	....	....	....	....
1995-96	**NY Islanders**	**NHL**	**14**	**0**	**1**	**1**	**11**	**0**	**0**	**0**	**11**	**0.0**	**−1**	....	....	....	....	....	....	....	....	....	....	....	....	....
	Worcester	AHL	57	16	7	23	57	....	....	....	....	....	....	....	....	....	....	....	4	1	2	3	2	....	....	....
1996-97	**NY Islanders**	**NHL**	**1**	**0**	**0**	**0**	**0**	**0**	**0**	**0**	**0**	**0.0**	**0**	....	....	....	....	....	....	....	....	....	....	....	....	....
	Kentucky	AHL	58	15	31	46	57	....	....	....	....	....	....	....	....	....	....	....	4	3	0	3	8	....	....	....
1997-98	Milwaukee	IHL	64	18	18	36	84	....	....	....	....	....	....	....	....	....	....	....	10	4	3	7	36	....	....	....
1998-99	Kentucky	AHL	60	22	31	53	68	....	....	....	....	....	....	....	....	....	....	....	12	3	6	9	8	....	....	....
99-2000	Kentucky	AHL	75	17	25	42	83	....	....	....	....	....	....	....	....	....	....	....	8	1	1	2	6	....	....	....
2000-01	Kentucky	AHL	54	10	29	39	61	....	....	....	....	....	....	....	....	....	....	....	3	0	0	0	0	....	....	....
	NHL Totals		**15**	**0**	**1**	**1**	**11**	**0**	**0**	**0**	**11**	**0.0**		....	....	....	....	....	....	....	....	....	....	....	....	....

Signed as a free agent by **San Jose**, August 27, 1998.

DEVEREAUX, Boyd
(DEH-vuhr-oh, BOID) **DET.**

Center. Shoots left. 6'2", 195 lbs. Born, Seaforth, Ont., April 16, 1978. Edmonton's 1st choice, 6th overall, in 1996 Entry Draft.

Season	Club	League	GP	G	A	Pts	PIM	PP	SH	GW	S	%	+/-	TF	F%	H	SB	Min	GP	G	A	Pts	PIM	PP	SH	GW
1992-93	Seaforth Sailors	OJHL-D	34	7	20	27	13	....	....	....	....	....	....	....	....	....	....	....	....	....	....	....	....	....	....	....
1993-94	Stratford	OJHL-B	46	12	27	39	8	....	....	....	....	....	....	....	....	....	....	....	....	....	....	....	....	....	....	....
1994-95	Stratford	OJHL-B	45	31	74	105	21	....	....	....	....	....	....	....	....	....	....	....	....	....	....	....	....	....	....	....
1995-96	Kitchener	OHL	66	20	38	58	35	....	....	....	....	....	....	....	....	....	....	....	12	3	7	10	4	....	....	....
1996-97	Kitchener	OHL	54	28	41	69	37	....	....	....	....	....	....	....	....	....	....	....	13	4	11	15	8	....	....	....
	Hamilton	AHL	....	....	....	....	....	....	....	....	....	....	....	....	....	....	....	....	1	0	1	1	0	....	....	....
1997-98	**Edmonton**	**NHL**	**38**	**1**	**4**	**5**	**6**	**0**	**0**	**0**	**27**	**3.7**	**−5**	....	....	....	....	....	....	....	....	....	....	....	....	....
	Hamilton	AHL	14	5	6	11	6	....	....	....	....	....	....	....	....	....	....	....	9	1	1	2	8	....	....	....
1998-99	**Edmonton**	**NHL**	**61**	**6**	**8**	**14**	**23**	**0**	**1**	**4**	**39**	**15.4**	**2**	**409**	**42.8**	**32**	**32**	**10:09**	**1**	**0**	**0**	**0**	**0**	**0**	**0**	**0**
	Hamilton	AHL	7	4	6	10	2	....	....	....	....	....	....	....	....	....	....	....	8	0	3	3	4	....	....	....
99-2000	**Edmonton**	**NHL**	**76**	**8**	**19**	**27**	**20**	**0**	**1**	**2**	**108**	**7.4**	**7**	**241**	**34.9**	**54**	**26**	**12:36**	....	....	....	....	....	....	....	....
2000-01	**Detroit**	**NHL**	**55**	**5**	**6**	**11**	**14**	**0**	**0**	**0**	**66**	**7.6**	**1**	**124**	**37.1**	**45**	**18**	**10:08**	**2**	**0**	**0**	**0**	**0**	**0**	**0**	**0**
2001-02♦	**Detroit**	**NHL**	**79**	**9**	**16**	**25**	**24**	**0**	**0**	**2**	**116**	**7.8**	**9**	**12**	**33.3**	**70**	**13**	**11:30**	**21**	**2**	**4**	**6**	**4**	**0**	**0**	**0**
	NHL Totals		**309**	**29**	**53**	**82**	**87**	**0**	**2**	**8**	**356**	**8.1**		**786**	**39.3**	**201**	**89**	**11:14**	**24**	**2**	**4**	**6**	**4**	**0**	**0**	**0**

Canadian Major Junior Scholastic Player of the Year (1996)
Signed as a free agent by **Detroit**, August 23, 2000.

de VRIES, Greg
(deh-VREES, GREHG) **COL.**

Defense. Shoots left. 6'3", 215 lbs. Born, Sundridge, Ont., January 4, 1973.

Season	Club	League	GP	G	A	Pts	PIM	PP	SH	GW	S	%	+/-	TF	F%	H	SB	Min	GP	G	A	Pts	PIM	PP	SH	GW
1988-89	Cortina Astros	OMHA	35	28	40	68	....	....	....	....	....	....	....	....	....	....	....	....	....	....	....	....	....	....	....	....
1989-90	Aurora Eagles	OJHL	42	1	16	17	32	....	....	....	....	....	....	....	....	....	....	....	....	....	....	....	....	....	....	....
1990-91	Stratford	OJHL-B	40	8	32	40	120	....	....	....	....	....	....	....	....	....	....	....	3	2	1	3	20	....	....	....
1991-92	Thorold Eagles	OJHL-B	3	0	0	0	0	....	....	....	....	....	....	....	....	....	....	....	....	....	....	....	....	....	....	....
	Bowling Green	CCHA	24	0	3	3	20	....	....	....	....	....	....	....	....	....	....	....	....	....	....	....	....	....	....	....
1992-93	Niagara Falls	OHL	62	3	23	26	86	....	....	....	....	....	....	....	....	....	....	....	4	0	1	1	6	....	....	....
1993-94	Niagara Falls	OHL	64	5	40	45	135	....	....	....	....	....	....	....	....	....	....	....	....	....	....	....	....	....	....	....
	Cape Breton	AHL	9	0	0	0	11	....	....	....	....	....	....	....	....	....	....	....	1	0	0	0	0	....	....	....
1994-95	Cape Breton	AHL	77	5	19	24	68	....	....	....	....	....	....	....	....	....	....	....	....	....	....	....	....	....	....	....
1995-96	**Edmonton**	**NHL**	**13**	**1**	**1**	**2**	**12**	**0**	**0**	**0**	**8**	**12.5**	**−2**	....	....	....	....	....	....	....	....	....	....	....	....	....
	Cape Breton	AHL	58	9	30	39	174	....	....	....	....	....	....	....	....	....	....	....	....	....	....	....	....	....	....	....
1996-97	**Edmonton**	**NHL**	**37**	**0**	**4**	**4**	**52**	**0**	**0**	**0**	**31**	**0.0**	**−2**	....	....	....	....	....	**12**	**0**	**1**	**1**	**8**	**0**	**0**	**0**
	Hamilton	AHL	34	4	14	18	26	....	....	....	....	....	....	....	....	....	....	....	....	....	....	....	....	....	....	....
1997-98	**Edmonton**	**NHL**	**65**	**7**	**4**	**11**	**80**	**1**	**0**	**0**	**53**	**13.2**	**−17**	....	....	....	....	....	**7**	**0**	**0**	**0**	**21**	**0**	**0**	**0**
1998-99	**Nashville**	**NHL**	**6**	**0**	**0**	**0**	**4**	**0**	**0**	**0**	**1**	**1.0**	**−4**	**0**	**0.0**	**12**	**7**	**18:11**	....	....	....	....	....	....	....	....
	Colorado	**NHL**	**67**	**1**	**3**	**4**	**60**	**0**	**0**	**0**	**56**	**56.0**	**−3**	**1**	**100.0**	**85**	**69**	**16:23**	**19**	**0**	**2**	**2**	**22**	**0**	**0**	**0**
99-2000	**Colorado**	**NHL**	**69**	**2**	**7**	**9**	**73**	**0**	**0**	**0**	**40**	**5.0**	**−7**	**0**	**0.0**	**87**	**61**	**14:59**	**5**	**0**	**0**	**0**	**4**	**0**	**0**	**0**
2000-01♦	**Colorado**	**NHL**	**79**	**5**	**12**	**17**	**51**	**0**	**0**	**0**	**76**	**6.6**	**23**	**0**	**0.0**	**142**	**93**	**17:06**	**23**	**0**	**1**	**1**	**20**	**0**	**0**	**0**
2001-02	**Colorado**	**NHL**	**82**	**8**	**12**	**20**	**57**	**1**	**1**	**3**	**148**	**5.4**	**18**	**1**	**0.0**	**135**	**180**	**23:03**	**21**	**4**	**9**	**13**	**2**	**0**	**0**	**1**
	NHL Totals		**418**	**24**	**43**	**67**	**389**	**2**	**1**	**3**	**413**	**5.8**		**2**	**50.0**	**461**	**410**	**18:05**	**87**	**4**	**13**	**17**	**77**	**0**	**0**	**1**

Signed as a free agent by **Edmonton**, March 20, 1994. Traded to **Nashville** by **Edmonton** with Eric Fichaud and Drake Berehowsky for Mikhail Shtalenkov and Jim Dowd, October 1, 1998. Traded to **Colorado** by **Nashville** for Colorado's 2nd round choice (Ed Hill) in 1999 Entry Draft, October 24, 1998.

DiMAIO, Rob
(duh-MIGH-oh, RAWB) **DAL.**

Center. Shoots right. 5'10", 190 lbs. Born, Calgary, Alta., February 19, 1968. NY Islanders' 6th choice, 118th overall, in 1987 Entry Draft.

Season	Club	League	GP	G	A	Pts	PIM	PP	SH	GW	S	%	+/-	TF	F%	H	SB	Min	GP	G	A	Pts	PIM	PP	SH	GW
1984-85	Kamloops Blazers	WHL	55	9	18	27	29	....	....	....	....	....	....	....	....	....	....	....	7	1	3	4	2	....	....	....
1985-86	Kamloops Blazers	WHL	6	1	0	1	0	....	....	....	....	....	....	....	....	....	....	....	....	....	....	....	....	....	....	....
	Medicine Hat	WHL	55	20	30	50	82	....	....	....	....	....	....	....	....	....	....	....	22	6	6	12	39	....	....	....
1986-87	Medicine Hat	WHL	70	27	43	70	130	....	....	....	....	....	....	....	....	....	....	....	20	7	11	18	46	....	....	....
1987-88	Medicine Hat	WHL	54	47	43	90	120	....	....	....	....	....	....	....	....	....	....	....	14	12	19	*31	59	....	....	....
1988-89	**NY Islanders**	**NHL**	**16**	**1**	**0**	**1**	**30**	**0**	**0**	**1**	**16**	**6.3**	**−6**	....	....	....	....	....	....	....	....	....	....	....	....	....
	Springfield	AHL	40	13	18	31	67	....	....	....	....	....	....	....	....	....	....	....	....	....	....	....	....	....	....	....

			Regular Season																Playoffs							
Season	Club	League	GP	G	A	Pts	PIM	PP	SH	GW	S	%	+/-	TF	F%	H	SB	Min	GP	G	A	Pts	PIM	PP	SH	GW
1989-90	**NY Islanders**	**NHL**	**7**	**0**	**0**	**0**	**2**	**0**	**0**	**0**	**2**	**0.0**	**0**	....	....	....	....	....	**1**	**1**	**0**	**1**	**4**	**0**	**0**	**0**
	Springfield	AHL	54	25	27	52	69	....	....	....	....	....	....	....	....	....	....	....	16	4	7	11	45	....	....	....
1990-91	**NY Islanders**	**NHL**	**1**	**0**	**0**	**0**	**0**	**0**	**0**	**0**	**0**	**0.0**	**0**	....	....	....	....	....	....	....	....	....	....	....	....	....
	Capital District	AHL	12	3	4	7	22	....	....	....	....	....	....	....	....	....	....	....	....	....	....	....	....	....	....	....
1991-92	**NY Islanders**	**NHL**	**50**	**5**	**2**	**7**	**43**	**0**	**2**	**0**	**43**	**11.6**	**–23**	....	....	....	....	....	....	....	....	....	....	....	....	....
1992-93	**Tampa Bay**	**NHL**	**54**	**9**	**15**	**24**	**62**	**2**	**0**	**0**	**75**	**12.0**	**0**	....	....	....	....	....	....	....	....	....	....	....	....	....
1993-94	**Tampa Bay**	**NHL**	**39**	**8**	**7**	**15**	**40**	**2**	**0**	**1**	**51**	**15.7**	**–5**	....	....	....	....	....	....	....	....	....	....	....	....	....
	Philadelphia	**NHL**	**14**	**3**	**5**	**8**	**6**	**0**	**0**	**1**	**30**	**10.0**	**1**	....	....	....	....	....	....	....	....	....	....	....	....	....
1994-95	**Philadelphia**	**NHL**	**36**	**3**	**1**	**4**	**53**	**0**	**0**	**0**	**34**	**8.8**	**8**	....	....	....	....	....	**15**	**2**	**4**	**6**	**4**	**0**	**1**	**1**
1995-96	**Philadelphia**	**NHL**	**59**	**6**	**15**	**21**	**58**	**1**	**1**	**0**	**49**	**12.2**	**0**	....	....	....	....	....	**3**	**0**	**0**	**0**	**0**	**0**	**0**	**0**
1996-97	**Boston**	**NHL**	**72**	**13**	**15**	**28**	**82**	**0**	**3**	**2**	**152**	**8.6**	**–21**	....	....	....	....	....	....	....	....	....	....	....	....	....
1997-98	**Boston**	**NHL**	**79**	**10**	**17**	**27**	**82**	**0**	**0**	**4**	**112**	**8.9**	**–13**	....	....	....	....	....	**6**	**1**	**0**	**1**	**8**	**0**	**0**	**0**
1998-99	**Boston**	**NHL**	**71**	**7**	**14**	**21**	**95**	**1**	**0**	**0**	**121**	**5.8**	**–14**	**83**	**45.8**	**106**	**26**	**16:41**	**12**	**2**	**0**	**2**	**8**	**0**	**0**	**1**
99-2000	**Boston**	**NHL**	**50**	**5**	**16**	**21**	**42**	**0**	**0**	**0**	**93**	**5.4**	**–1**	**278**	**44.2**	**103**	**20**	**16:47**	....	....	....	....	....	....	....	....
	NY Rangers	**NHL**	**12**	**1**	**3**	**4**	**8**	**0**	**0**	**0**	**18**	**5.6**	**–8**	**1**	**0.0**	**16**	**6**	**15:30**	....	....	....	....	....	....	....	....
2000-01	**Carolina**	**NHL**	**74**	**6**	**18**	**24**	**54**	**0**	**2**	**1**	**99**	**6.1**	**–14**	**46**	**43.5**	**107**	**31**	**14:58**	**6**	**0**	**0**	**0**	**4**	**0**	**0**	**0**
2001-02	Utah Grizzlies	AHL	3	1	1	2	0	....	....	....	....	....	....	....	....	....	....	....	....	....	....	....	....	....	....	....
	Dallas	**NHL**	**61**	**6**	**6**	**12**	**25**	**0**	**2**	**2**	**63**	**9.5**	**–2**	**76**	**44.7**	**71**	**23**	**10:52**	....	....	....	....	....	....	....	....
	NHL Totals		**695**	**83**	**134**	**217**	**682**	**6**	**10**	**12**	**958**	**8.7**		**484**	**44.4**	**403**	**106**	**14:51**	**43**	**6**	**4**	**10**	**28**	**0**	**1**	**2**

Won Stafford Smythe Memorial Trophy (Memorial Cup Tournament MVP) (1988)

Claimed by **Tampa Bay** from **NY Islanders** in Expansion Draft, June 18, 1992. Traded to **Philadelphia** by **Tampa Bay** for Jim Cummins and Philadelphia's 4th round choice (later traded back to Philadelphia - Philadelphia selected Radovan Somik) in 1995 Entry Draft, March 18, 1994. Claimed by **San Jose** from **Philadelphia** in NHL Waiver Draft, September 30, 1996. Traded to **Boston** by **San Jose** for Boston's 5th round choice (Adam Nittel) in 1997 Entry Draft, September 30, 1996. Traded to **NY Rangers** by **Boston** for Mike Knuble, March 10, 2000. Traded to **Carolina** by **NY Rangers** with Darren Langdon for Sandy McCarthy and Carolina's 4th round choice (Bryce Lampman) in 2001 Entry Draft, August 4, 2000. Signed as a free agent by **Dallas**, July 1, 2001.

DINEEN, Kevin (DIH-neen, KEH-vihn) CBJ

Right wing. Shoots right. 5'11", 198 lbs. Born, Quebec City, Que., October 28, 1963. Hartford's 3rd choice, 56th overall, in 1982 Entry Draft.

Season	Club	League	GP	G	A	Pts	PIM	PP	SH	GW	S	%	+/-	TF	F%	H	SB	Min	GP	G	A	Pts	PIM	PP	SH	GW
1980-81	St. Michael's B	OJHL-B	40	15	28	43	167	....	....	....	....	....	....	....	....	....	....	....	....	....	....	....	....	....	....	....
1981-82	U. of Denver	WCHA	26	10	10	20	70	....	....	....	....	....	....	....	....	....	....	....	....	....	....	....	....	....	....	....
1982-83	U. of Denver	WCHA	36	16	13	29	108	....	....	....	....	....	....	....	....	....	....	....	....	....	....	....	....	....	....	....
1983-84	Team Canada	Nat-Tm	52	5	11	16	2	....	....	....	....	....	....	....	....	....	....	....	....	....	....	....	....	....	....	....
	Canada	Olympics	7	0	0	0	8	....	....	....	....	....	....	....	....	....	....	....	....	....	....	....	....	....	....	....
1984-85	**Hartford**	**NHL**	**57**	**25**	**16**	**41**	**120**	**8**	**4**	**2**	**141**	**17.7**	**–6**	....	....	....	....	....	....	....	....	....	....	....	....	....
	Binghamton	AHL	25	15	8	23	41	....	....	....	....	....	....	....	....	....	....	....	....	....	....	....	....	....	....	....
1985-86	**Hartford**	**NHL**	**57**	**33**	**35**	**68**	**124**	**6**	**0**	**8**	**167**	**19.8**	**16**	....	....	....	....	....	**10**	**6**	**7**	**13**	**18**	**1**	**0**	**2**
1986-87	**Hartford**	**NHL**	**78**	**40**	**39**	**79**	**110**	**11**	**0**	**7**	**234**	**17.1**	**7**	....	....	....	....	....	**6**	**2**	**1**	**3**	**31**	**1**	**0**	**0**
1987-88	**Hartford**	**NHL**	**74**	**25**	**25**	**50**	**217**	**5**	**0**	**4**	**223**	**11.2**	**–14**	....	....	....	....	....	**6**	**4**	**4**	**8**	**8**	**1**	**0**	**1**
1988-89	**Hartford**	**NHL**	**79**	**45**	**44**	**89**	**167**	**20**	**1**	**4**	**294**	**15.3**	**–6**	....	....	....	....	....	**4**	**1**	**0**	**1**	**10**	**0**	**0**	**0**
1989-90	**Hartford**	**NHL**	**67**	**25**	**41**	**66**	**164**	**8**	**2**	**2**	**214**	**11.7**	**7**	....	....	....	....	....	**6**	**3**	**2**	**5**	**18**	**0**	**0**	**1**
1990-91	**Hartford**	**NHL**	**61**	**17**	**30**	**47**	**104**	**4**	**0**	**2**	**161**	**10.6**	**–15**	....	....	....	....	....	**6**	**1**	**0**	**1**	**16**	**0**	**0**	**0**
1991-92	**Hartford**	**NHL**	**16**	**4**	**2**	**6**	**23**	**1**	**0**	**1**	**28**	**14.3**	**–6**	....	....	....	....	....	....	....	....	....	....	....	....	....
	Philadelphia	**NHL**	**64**	**26**	**30**	**56**	**130**	**5**	**3**	**4**	**197**	**13.2**	**1**	....	....	....	....	....	....	....	....	....	....	....	....	....
1992-93	**Philadelphia**	**NHL**	**83**	**35**	**28**	**63**	**201**	**6**	**3**	**7**	**241**	**14.5**	**14**	....	....	....	....	....	....	....	....	....	....	....	....	....
1993-94	**Philadelphia**	**NHL**	**71**	**19**	**23**	**42**	**113**	**5**	**1**	**2**	**156**	**12.2**	**–9**	....	....	....	....	....	....	....	....	....	....	....	....	....
1994-95	Houston Aeros	IHL	17	6	4	10	42	....	....	....	....	....	....	....	....	....	....	....	....	....	....	....	....	....	....	....
	Philadelphia	**NHL**	**40**	**8**	**5**	**13**	**39**	**4**	**0**	**2**	**55**	**14.5**	**–1**	....	....	....	....	....	**15**	**6**	**4**	**10**	**18**	**1**	**0**	**1**
1995-96	**Philadelphia**	**NHL**	**26**	**0**	**2**	**2**	**50**	**0**	**0**	**0**	**31**	**0.0**	**–8**	....	....	....	....	....	....	....	....	....	....	....	....	....
	Hartford	**NHL**	**20**	**2**	**7**	**9**	**67**	**0**	**0**	**0**	**35**	**5.7**	**7**	....	....	....	....	....	....	....	....	....	....	....	....	....
1996-97	**Hartford**	**NHL**	**78**	**19**	**29**	**48**	**141**	**8**	**0**	**5**	**185**	**10.3**	**–6**	....	....	....	....	....	....	....	....	....	....	....	....	....
1997-98	**Carolina**	**NHL**	**54**	**7**	**16**	**23**	**105**	**0**	**0**	**1**	**96**	**7.3**	**–7**	....	....	....	....	....	....	....	....	....	....	....	....	....
1998-99	**Carolina**	**NHL**	**67**	**8**	**10**	**18**	**97**	**0**	**0**	**1**	**86**	**9.3**	**5**	**6**	**16.7**	**77**	**6**	**9:58**	**6**	**0**	**0**	**0**	**8**	**0**	**0**	**0**
99-2000	**Ottawa**	**NHL**	**67**	**4**	**8**	**12**	**57**	**0**	**0**	**1**	**71**	**5.6**	**2**	**10**	**50.0**	**67**	**14**	**9:31**	....	....	....	....	....	....	....	....
2000-01	**Columbus**	**NHL**	**66**	**8**	**7**	**15**	**126**	**0**	**0**	**3**	**74**	**10.8**	**2**	**13**	**30.8**	**101**	**18**	**10:37**	....	....	....	....	....	....	....	....
2001-02	**Columbus**	**NHL**	**59**	**5**	**8**	**13**	**62**	**0**	**0**	**0**	**73**	**6.8**	**–6**	**9**	**11.1**	**68**	**11**	**10:07**	....	....	....	....	....	....	....	....
	NHL Totals		**1184**	**355**	**405**	**760**	**2217**	**91**	**14**	**56**	**2762**	**12.9**		**38**	**28.9**	**313**	**49**	**10:03**	**59**	**23**	**18**	**41**	**127**	**4**	**0**	**5**

Won Bud Light/NHL Man of the Year Award (1991) • Played in NHL All-Star Game (1988, 1989)

Traded to **Philadelphia** by **Hartford** for Murray Craven and Philadelphia's 4th round choice (Kevin Smyth) in 1992 Entry Draft, November 13, 1991. Traded to **Hartford** by **Philadelphia** for Hartford/Carolina's 3rd (Kris Mallette) and 7th (later traded back to Hartford/Carolina - Carolina selected Andrew Merrick) round choices in 1997 Entry Draft, December 28, 1995. Transferred to **Carolina** after **Hartford** franchise relocated, June 25, 1997. Signed as a free agent by **Ottawa**, September 1, 1999. Selected by **Columbus** from **Ottawa** in Expansion Draft, June 23, 2000.

DINGMAN, Chris (DIHNG-man, KRIHS) T.B.

Left wing. Shoots left. 6'4", 225 lbs. Born, Edmonton, Alta., July 6, 1976. Calgary's 1st choice, 19th overall, in 1994 Entry Draft.

Season	Club	League	GP	G	A	Pts	PIM	PP	SH	GW	S	%	+/-	TF	F%	H	SB	Min	GP	G	A	Pts	PIM	PP	SH	GW
1991-92	Edm. Mercurys	AMHL	36	23	18	41	72	....	....	....	....	....	....	....	....	....	....	....	....	....	....	....	....	....	....	....
1992-93	Brandon	WHL	50	10	17	27	64	....	....	....	....	....	....	....	....	....	....	....	4	0	0	0	0	....	....	....
1993-94	Brandon	WHL	45	21	20	41	77	....	....	....	....	....	....	....	....	....	....	....	13	1	7	8	39	....	....	....
1994-95	Brandon	WHL	66	40	43	83	201	....	....	....	....	....	....	....	....	....	....	....	3	1	0	1	9	....	....	....
1995-96	Brandon	WHL	40	16	29	45	109	....	....	....	....	....	....	....	....	....	....	....	19	12	11	23	60	....	....	....
	Saint John	AHL	....	....	....	....	....	....	....	....	....	....	....	....	....	....	....	....	1	0	0	0	0	....	....	....
1996-97	Saint John	AHL	71	5	6	11	195	....	....	....	....	....	....	....	....	....	....	....	....	....	....	....	....	....	....	....
1997-98	**Calgary**	**NHL**	**70**	**3**	**3**	**6**	**149**	**1**	**0**	**0**	**47**	**6.4**	**–11**	....	....	....	....	....	....	....	....	....	....	....	....	....
1998-99	**Calgary**	**NHL**	**2**	**0**	**0**	**0**	**17**	**0**	**0**	**0**	**1**	**0.0**	**–2**	**0**	**0.0**	**3**	**1**	**8:11**	....	....	....	....	....	....	....	....
	Saint John	AHL	50	5	7	12	140	....	....	....	....	....	....	....	....	....	....	....	....	....	....	....	....	....	....	....
	Colorado	**NHL**	**1**	**0**	**0**	**0**	**7**	**0**	**0**	**0**	**0**	**0.0**	**0**	**0**	**0.0**	**0**	**0**	**0:30**	....	....	....	....	....	....	....	....
	Hershey Bears	AHL	17	1	3	4	102	....	....	....	....	....	....	....	....	....	....	....	5	0	2	2	6	....	....	....
99-2000	**Colorado**	**NHL**	**68**	**8**	**3**	**11**	**132**	**2**	**0**	**1**	**54**	**14.8**	**–2**	**2**	**0.0**	**73**	**18**	**6:29**	....	....	....	....	....	....	....	....
2000-01♦	**Colorado**	**NHL**	**41**	**1**	**1**	**2**	**108**	**0**	**0**	**0**	**33**	**3.0**	**–3**	**0**	**0.0**	**74**	**9**	**6:26**	**16**	**0**	**4**	**4**	**14**	**0**	**0**	**0**
2001-02	**Carolina**	**NHL**	**30**	**0**	**1**	**1**	**77**	**0**	**0**	**0**	**17**	**0.0**	**–2**	**2**	**100.0**	**35**	**6**	**6:54**	....	....	....	....	....	....	....	....
	Tampa Bay	**NHL**	**14**	**0**	**4**	**4**	**26**	**0**	**0**	**0**	**24**	**0.0**	**–8**	**0**	**0.0**	**19**	**5**	**10:43**	....	....	....	....	....	....	....	....
	NHL Totals		**226**	**12**	**12**	**24**	**516**	**3**	**0**	**1**	**176**	**6.8**		**4**	**50.0**	**204**	**39**	**6:55**	**16**	**0**	**4**	**4**	**14**	**0**	**0**	**0**

Traded to **Colorado** by **Calgary** with Theoren Fleury for Rene Corbet, Wade Belak, Robyn Regehr and Colorado's 2nd round compensatory choice (Jarret Stoll) in 2000 Entry Draft, February 28, 1999. • Missed majority of 2000-01 season recovering from knee injury suffered in game vs. Ottawa, November 15, 2000. Traded to **Carolina** by **Colorado** for Carolina's 5th round choice (Mikko Viitanen) in 2001 Entry Draft, June 24, 2001. Traded to **Tampa Bay** by **Carolina** with Shane Willis for Kevin Weekes, March 5, 2002.

DIVISEK, Tomas (DIH-vih-sehk, TOH-mahs) PHI.

Center. Shoots left. 6'2", 204 lbs. Born, Most, Czech., July 19, 1979. Philadelphia's 9th choice, 195th overall, in 1998 Entry Draft.

Season	Club	League	GP	G	A	Pts	PIM	PP	SH	GW	S	%	+/-	TF	F%	H	SB	Min	GP	G	A	Pts	PIM	PP	SH	GW
1995-96	Slavia Praha Jr.	Czech-Jr.	36	20	27	47	12	....	....	....	....	....	....	....	....	....	....	....	....	....	....	....	....	....	....	....
1996-97	Slavia Praha Jr.	Czech-Jr.	41	17	25	42	18	....	....	....	....	....	....	....	....	....	....	....	....	....	....	....	....	....	....	....
	HC Slavia Praha	Czech	1	0	0	0	0	....	....	....	....	....	....	....	....	....	....	....	....	....	....	....	....	....	....	....
1997-98	Slavia Praha Jr.	Czech-Jr.	27	20	16	36	12	....	....	....	....	....	....	....	....	....	....	....	....	....	....	....	....	....	....	....
	HC Slavia Praha	Czech	22	2	0	2	8	....	....	....	....	....	....	....	....	....	....	....	....	....	....	....	....	....	....	....
1998-99	HC Slavia Praha	Czech	45	8	4	12	26	....	....	....	....	....	....	....	....	....	....	....	....	....	....	....	....	....	....	....
99-2000	Philadelphia	AHL	59	18	31	49	30	....	....	....	....	....	....	....	....	....	....	....	5	0	3	3	2	....	....	....
2000-01	**Philadelphia**	**NHL**	**2**	**0**	**0**	**0**	**0**	**0**	**0**	**0**	**2**	**0.0**	**–1**	**0**	**0.0**	**3**	**0**	**11:30**	....	....	....	....	....	....	....	....
	Philadelphia	AHL	45	10	22	32	33	....	....	....	....	....	....	....	....	....	....	....	10	4	9	13	4	....	....	....
2001-02	**Philadelphia**	**NHL**	**3**	**1**	**0**	**1**	**0**	**0**	**0**	**1**	**3**	**33.3**	**1**	**28**	**50.0**	**2**	**0**	**8:12**	....	....	....	....	....	....	....	....
	Philadelphia	AHL	65	13	18	31	54	....	....	....	....	....	....	....	....	....	....	....	....	....	....	....	....	....	....	....
	Springfield	AHL	9	2	1	3	8	....	....	....	....	....	....	....	....	....	....	....	....	....	....	....	....	....	....	....
	NHL Totals		**5**	**1**	**0**	**1**	**0**	**0**	**0**	**1**	**5**	**20.0**		**28**	**50.0**	**5**	**0**	**9:31**	....	....	....	....	....	....	....	....

Signed as a free agent by **HC Pardubice** (Czech) with Philadelphia retaining NHL rights, July 23, 2002.

			Regular Season																Playoffs							
Season	Club	League	GP	G	A	Pts	PIM	PP	SH	GW	S	%	+/-	TF	F%	H	SB	Min	GP	G	A	Pts	PIM	PP	SH	GW

DOAN, Shane

(DOHN, SHAYN) **PHX.**

Right wing. Shoots right. 6'2", 223 lbs. Born, Halkirk, Alta., October 10, 1976. Winnipeg's 1st choice, 7th overall, in 1995 Entry Draft.

Season	Club	League	GP	G	A	Pts	PIM	PP	SH	GW	S	%	+/-	TF	F%	H	SB	Min	GP	G	A	Pts	PIM	PP	SH	GW
1991-92	Killam Selects	AAHA	56	80	84	164	74																			
1992-93	Kamloops Blazers	WHL	51	7	12	19	65												13	0	1	1	8			
1993-94	Kamloops Blazers	WHL	52	24	24	48	88																			
1994-95	Kamloops Blazers	WHL	71	37	57	94	106												21	6	10	16	16			
1995-96	**Winnipeg**	**NHL**	**74**	**7**	**10**	**17**	**101**	**1**	**0**	**3**	**106**	**6.6**	**-9**						**6**	**0**	**0**	**0**	**6**	**0**	**0**	**0**
1996-97	**Phoenix**	**NHL**	**63**	**4**	**8**	**12**	**49**	**0**	**0**	**0**	**100**	**4.0**	**-3**						**4**	**0**	**0**	**0**	**2**	**0**	**0**	**0**
1997-98	**Phoenix**	**NHL**	**33**	**5**	**6**	**11**	**35**	**0**	**0**	**3**	**42**	**11.9**	**-3**						**6**	**1**	**0**	**1**	**6**	**0**	**0**	**0**
	Springfield	AHL	39	21	21	42	64																			
1998-99	**Phoenix**	**NHL**	**79**	**6**	**16**	**22**	**54**	**0**	**0**	**0**	**156**	**3.8**	**-5**	**6**	**16.7**	**161**	**15**	**12:42**	**7**	**2**	**2**	**4**	**6**	**0**	**0**	**2**
99-2000	**Phoenix**	**NHL**	**81**	**26**	**25**	**51**	**66**	**1**	**1**	**4**	**221**	**11.8**	**6**	**25**	**36.0**	**225**	**11**	**16:51**	**4**	**1**	**2**	**3**	**8**	**1**	**0**	**0**
2000-01	**Phoenix**	**NHL**	**76**	**26**	**37**	**63**	**89**	**6**	**1**	**6**	**220**	**11.8**	**0**	**15**	**40.0**	**206**	**19**	**19:32**								
2001-02	**Phoenix**	**NHL**	**81**	**20**	**29**	**49**	**61**	**6**	**0**	**2**	**205**	**9.8**	**11**	**52**	**44.2**	**254**	**14**	**18:10**	**5**	**2**	**2**	**4**	**6**	**0**	**0**	**0**
	NHL Totals		**487**	**94**	**131**	**225**	**455**	**14**	**2**	**18**	**1050**	**9.0**		**98**	**39.8**	**846**	**59**	**16:48**	**32**	**6**	**6**	**12**	**34**	**1**	**0**	**2**

Memorial Cup All-Star Team (1995) • Won Stafford Smythe Memorial Trophy (Memorial Cup Tournament MVP) (1995)

Transferred to **Phoenix** after **Winnipeg** franchise relocated, July 1, 1996.

DOIG, Jason

(DOIG, JAY-suhn)

Defense. Shoots right. 6'3", 228 lbs. Born, Montreal, Que., January 29, 1977. Winnipeg's 3rd choice, 34th overall, in 1995 Entry Draft.

Season	Club	League	GP	G	A	Pts	PIM	PP	SH	GW	S	%	+/-	TF	F%	H	SB	Min	GP	G	A	Pts	PIM	PP	SH	GW
1990-91	North Shore	QAHA	31	30	33	63	53																			
1991-92	North Shore	QAHA	29	11	11	22	20																			
1992-93	Lac St-Louis	QAAA	35	11	16	27	40												7	5	5	10	16			
1993-94	St-Jean Lynx	QMJHL	63	8	17	25	65												5	0	2	2	2			
1994-95	Laval Titan	QMJHL	55	13	42	55	259												20	4	13	17	39			
1995-96	Laval Titan	QMJHL	5	3	6	9	20																			
	Granby	QMJHL	24	4	30	34	91												20	10	22	32	*110			
	Winnipeg	**NHL**	**15**	**1**	**1**	**2**	**28**	**0**	**0**	**0**	**7**	**14.3**	**-2**													
	Springfield	AHL	5	0	0	0	28																			
1996-97	Granby	QMJHL	39	14	33	47	211												5	0	4	4	27			
	Las Vegas	IHL	6	0	1	1	19																			
	Springfield	AHL	5	0	3	3	2												17	1	4	5	37			
1997-98	**Phoenix**	**NHL**	**4**	**0**	**1**	**1**	**12**	**0**	**0**	**0**	**1**	**0.0**	**-4**													
	Springfield	AHL	46	2	25	27	153												3	0	0	0	2			
1998-99	**Phoenix**	**NHL**	**9**	**0**	**1**	**1**	**10**	**0**	**0**	**0**	**0**	**0.0**	**2**	**0**	**0.0**	**1**	**2**	**5:08**								
	Springfield	AHL	32	3	5	8	67																			
	Hartford	AHL	8	1	4	5	40												7	1	1	2	39			
99-2000	**NY Rangers**	**NHL**	**7**	**0**	**1**	**1**	**22**	**0**	**0**	**0**	**3**	**0.0**	**-2**	**0**	**0.0**	**9**	**6**	**8:50**								
	Hartford	AHL	27	3	11	14	70												21	1	5	6	20			
2000-01	**NY Rangers**	**NHL**	**3**	**0**	**0**	**0**	**0**	**0**	**0**	**0**	**1**	**0.0**	**0**	**0**	**0.0**	**1**	**2**	**6:35**								
	Hartford	AHL	52	4	20	24	178												5	0	1	1	4			
2001-02	Grand Rapids	AHL	57	1	17	18	103												5	0	0	0	18			
	NHL Totals		**38**	**1**	**4**	**5**	**72**	**0**	**0**	**0**	**12**	**8.3**		**0**	**0.0**	**11**	**10**	**6:43**								

QMJHL All-Rookie Team (1994) • Memorial Cup All-Star Team (1996)

Transferred to **Phoenix** after **Winnipeg** franchise relocated, July 1, 1996. Traded to **NY Rangers** by **Phoenix** with Phoenix's 6th round choice (Jay Dardis) in 1999 Entry Draft for Stan Neckar, March 23, 1999. Traded to **Ottawa** by **NY Rangers** with Jeff Ulmer for Sean Gagnon, June 29, 2001.

DOME, Robert

(doh-MAY, RAW-buhrt) **CGY.**

Right wing. Shoots left. 6', 210 lbs. Born, Skalica, Czech., January 29, 1979. Pittsburgh's 1st choice, 17th overall, in 1997 Entry Draft.

Season	Club	League	GP	G	A	Pts	PIM	PP	SH	GW	S	%	+/-	TF	F%	H	SB	Min	GP	G	A	Pts	PIM	PP	SH	GW
1994-95	Dukla Trencin Jr.	Slovak-Jr.	36	36	43	79	39																			
1995-96	Utah Grizzlies	IHL	56	10	9	19	28																			
1996-97	Long Beach	IHL	13	4	6	10	14																			
	Las Vegas	IHL	43	10	7	17	22																			
1997-98	**Pittsburgh**	**NHL**	**30**	**5**	**2**	**7**	**12**	**1**	**0**	**0**	**29**	**17.2**	**-1**													
	Syracuse Crunch	AHL	36	21	25	46	77																			
1998-99	Syracuse Crunch	AHL	48	18	17	35	70																			
	Houston Aeros	IHL	20	2	4	6	24																			
99-2000	**Pittsburgh**	**NHL**	**22**	**2**	**5**	**7**	**0**	**0**	**0**	**0**	**27**	**7.4**	**1**	**5**	**40.0**	**12**	**6**	**9:51**								
	Wilkes-Barre	AHL	51	12	26	38	83																			
2000-01	Kladno	Czech	29	9	12	21	57																			
	Trinec	Czech	5	0	3	3	4																			
2001-02	Wilkes-Barre	AHL	39	9	8	17	53																			
	NHL Totals		**52**	**7**	**7**	**14**	**12**	**1**	**0**	**0**	**56**	**12.5**		**5**	**40.0**	**12**	**6**	**9:51**								

• Missed majority of 2001-02 season recovering from heel injury suffered during off-season training, July 10, 2001. Signed as a free agent by **Calgary**, July 17, 2002.

DOMENICHELLI, Hnat

(daw-meh-CHEHL-ee, NAT) **MIN.**

Center. Shoots left. 6', 195 lbs. Born, Edmonton, Alta., February 17, 1976. Hartford's 2nd choice, 83rd overall, in 1994 Entry Draft.

Season	Club	League	GP	G	A	Pts	PIM	PP	SH	GW	S	%	+/-	TF	F%	H	SB	Min	GP	G	A	Pts	PIM	PP	SH	GW
1991-92	Edmonton Freeze	AMHL	34	34	49	83	101																			
1992-93	Kamloops Blazers	WHL	45	12	8	20	15												11	1	1	2	2			
1993-94	Kamloops Blazers	WHL	69	27	40	67	31												19	10	12	22	0			
1994-95	Kamloops Blazers	WHL	72	52	62	114	34												19	9	9	18	9			
1995-96	Kamloops Blazers	WHL	62	59	89	148	37												16	7	9	16	29			
1996-97	**Hartford**	**NHL**	**13**	**2**	**1**	**3**	**7**	**1**	**0**	**0**	**14**	**14.3**	**-4**													
	Springfield	AHL	39	24	24	48	12																			
	Calgary	**NHL**	**10**	**1**	**2**	**3**	**2**	**1**	**0**	**0**	**16**	**6.3**	**1**													
	Saint John	AHL	1	1	1	2	0												5	5	0	5	2			
1997-98	**Calgary**	**NHL**	**31**	**9**	**7**	**16**	**6**	**1**	**0**	**1**	**70**	**12.9**	**4**													
	Saint John	AHL	48	33	13	46	24												19	7	8	15	14			
1998-99	**Calgary**	**NHL**	**23**	**5**	**5**	**10**	**11**	**3**	**0**	**0**	**45**	**11.1**	**-4**	**3**	**0.0**	**27**	**0**	**12:59**								
	Saint John	AHL	51	25	21	46	26												7	4	4	8	2			
99-2000	**Calgary**	**NHL**	**32**	**5**	**9**	**14**	**12**	**1**	**0**	**1**	**57**	**8.8**	**0**	**78**	**46.2**	**37**	**7**	**12:39**								
	Saint John	AHL	12	6	7	13	8																			
	Atlanta	**NHL**	**27**	**6**	**9**	**15**	**4**	**0**	**0**	**0**	**68**	**8.8**	**-21**	**9**	**55.6**	**36**	**3**	**16:55**								
2000-01	**Atlanta**	**NHL**	**63**	**15**	**12**	**27**	**18**	**4**	**0**	**1**	**150**	**10.0**	**-9**	**24**	**37.5**	**70**	**12**	**14:21**								
2001-02	**Atlanta**	**NHL**	**40**	**8**	**11**	**19**	**34**	**1**	**0**	**1**	**87**	**9.2**	**-18**	**9**	**33.3**	**57**	**9**	**14:51**								
	Minnesota	**NHL**	**27**	**1**	**5**	**6**	**10**	**0**	**0**	**0**	**57**	**1.8**	**-5**	**10**	**30.0**	**24**	**4**	**12:40**								
	NHL Totals		**266**	**52**	**61**	**113**	**104**	**12**	**0**	**4**	**564**	**9.2**		**133**	**42.1**	**251**	**35**	**14:09**								

WHL West Second All-Star Team (1995) • WHL West First All-Star Team (1996) • Canadian Major Junior First All-Star Team (1996) • Canadian Major Junior Most Sportsmanlike Player of the Year (1996)

Traded to **Calgary** by **Hartford** with Glen Featherstone, New Jersey's 2nd round choice (previously acquired, Calgary selected Dimitri Kokorev) in 1997 Entry Draft and Vancouver's 3rd round choice (previously acquired, Calgary selected Paul Manning) in 1998 Entry Draft for Steve Chiasson and Colorado's 3rd round choice (previously acquired, Carolina selected Francis Lessard) in 1997 Entry Draft, March 5, 1997. Traded to **Atlanta** by **Calgary** with Dmitri Vlasenkov for Darryl Shannon and Jason Botterill, February 11, 2000. Traded to **Minnesota** by **Atlanta** for Andy Sutton, January 22, 2002.

DOMI, Tie

(DOH-mee, TIGH) **TOR.**

Right wing. Shoots right. 5'10", 200 lbs. Born, Windsor, Ont., November 1, 1969. Toronto's 2nd choice, 27th overall, in 1988 Entry Draft.

Season	Club	League	GP	G	A	Pts	PIM	PP	SH	GW	S	%	+/-	TF	F%	H	SB	Min	GP	G	A	Pts	PIM	PP	SH	GW
1984-85	Belle River	OJHL-C	28	7	5	12	98																			
1985-86	Windsor Bulldogs	OJHL-B	42	8	17	25	*346																			
1986-87	Peterboro B's	OJHL-B	2	0	0	0	10																			
	Peterborough	OHL	18	1	1	2	79																			
1987-88	Peterborough	OHL	60	22	21	43	*292												12	3	9	12	24			
1988-89	Peterborough	OHL	43	14	16	30	175												17	10	9	19	*70			
1989-90	**Toronto**	**NHL**	**2**	**0**	**0**	**0**	**42**	**0**	**0**	**0**	**0**	**0.0**	**0**													
	Newmarket Saints	AHL	57	14	11	25	285																			

Season	Club	League	GP	G	A	Pts	PIM	PP	SH	GW	S	%	+/-	TF	F%	H	SB	Min	GP	G	A	Pts	PIM	PP	SH	GW
			Regular Season																Playoffs							
1990-91	**NY Rangers**	**NHL**	**28**	**1**	**0**	**1**	**185**	**0**	**0**	**0**	**5**	**20.0**	**–5**													
	Binghamton	AHL	25	11	6	17	219												7	3	2	5	16			
1991-92	**NY Rangers**	**NHL**	**42**	**2**	**4**	**6**	**246**	**0**	**0**	**1**	**20**	**10.0**	**–4**						**6**	**1**	**1**	**2**	**32**	**0**	**0**	**0**
1992-93	**NY Rangers**	**NHL**	**12**	**2**	**0**	**2**	**95**	**0**	**0**	**0**	**11**	**18.2**	**–1**													
	Winnipeg	**NHL**	**49**	**3**	**10**	**13**	**249**	**0**	**0**	**0**	**29**	**10.3**	**2**						**6**	**1**	**0**	**1**	**23**	**0**	**0**	**0**
1993-94	**Winnipeg**	**NHL**	**81**	**8**	**11**	**19**	***347**	**0**	**0**	**1**	**98**	**8.2**	**–8**													
1994-95	**Winnipeg**	**NHL**	**31**	**4**	**4**	**8**	**128**	**0**	**0**	**0**	**34**	**11.8**	**–6**													
	Toronto	**NHL**	**9**	**0**	**1**	**1**	**31**	**0**	**0**	**0**	**12**	**0.0**	**1**						**7**	**1**	**0**	**1**	**0**	**0**	**0**	**0**
1995-96	**Toronto**	**NHL**	**72**	**7**	**6**	**13**	**297**	**0**	**0**	**1**	**61**	**11.5**	**–3**						**6**	**0**	**2**	**2**	**4**	**0**	**0**	**0**
1996-97	**Toronto**	**NHL**	**80**	**11**	**17**	**28**	**275**	**2**	**0**	**1**	**98**	**11.2**	**–17**													
1997-98	**Toronto**	**NHL**	**80**	**4**	**10**	**14**	**365**	**0**	**0**	**0**	**72**	**5.6**	**–5**													
1998-99	**Toronto**	**NHL**	**72**	**8**	**14**	**22**	**198**	**0**	**0**	**1**	**65**	**12.3**	**5**	**9**	**44.4**	**100**	**3**	**9:42**	**14**	**0**	**2**	**2**	**24**	**0**	**0**	**0**
99-2000	**Toronto**	**NHL**	**70**	**5**	**9**	**14**	**198**	**0**	**0**	**2**	**64**	**7.8**	**–5**	**4**	**25.0**	**85**	**10**	**9:58**	**12**	**0**	**1**	**1**	**20**	**0**	**0**	**0**
2000-01	**Toronto**	**NHL**	**82**	**13**	**7**	**20**	**214**	**1**	**0**	**1**	**60**	**21.7**	**2**	**5**	**80.0**	**129**	**9**	**8:23**	**8**	**0**	**1**	**1**	**20**	**0**	**0**	**0**
2001-02	**Toronto**	**NHL**	**74**	**9**	**10**	**19**	**157**	**0**	**0**	**2**	**93**	**9.7**	**3**	**22**	**45.5**	**102**	**7**	**10:20**	**19**	**1**	**3**	**4**	***61**	**0**	**0**	**1**
	NHL Totals		**784**	**77**	**103**	**180**	**3027**	**3**	**0**	**10**	**722**	**10.7**		**40**	**47.5**	**416**	**29**	**9:33**	**78**	**4**	**10**	**14**	**184**	**0**	**0**	**1**

Traded to **NY Rangers** by **Toronto** with Mark LaForest for Greg Johnston, June 28, 1990. Traded to **Winnipeg** by **NY Rangers** with Kris King for Ed Olczyk, December 28, 1992. Traded to **Toronto** by **Winnipeg** for Mike Eastwood and Toronto's 3rd round choice (Brad Isbister) in 1995 Entry Draft, April 7, 1995. Traded to **Nashville** by **Toronto** for Nashville's 8th round choice in 2003 Entry Draft, June 30, 2002. Signed as a free agent by **Toronto**, July 14, 2002.

DONATO, Ted

(duh-NAH-toh, TEHD) **NYR**

Left wing. Shoots left. 5'10", 178 lbs. Born, Boston, MA, April 28, 1969. Boston's 6th choice, 98th overall, in 1987 Entry Draft.

Season	Club	League	GP	G	A	Pts	PIM	PP	SH	GW	S	%	+/-	TF	F%	H	SB	Min	GP	G	A	Pts	PIM	PP	SH	GW
1986-87	Catholic Memorial	Hi-School	22	29	34	63	30																			
1987-88	Harvard Crimson	ECAC	28	12	14	26	24																			
1988-89	Harvard Crimson	ECAC	34	14	37	51	30																			
1989-90	Harvard Crimson	ECAC	16	5	6	11	34																			
1990-91	Harvard Crimson	ECAC	27	19	*37	56	26																			
1991-92	Team USA	Nat-Tm	52	11	22	33	24																			
	United States	Olympics	8	4	3	7	8																			
	Boston	**NHL**	**10**	**1**	**2**	**3**	**8**	**0**	**0**	**0**	**13**	**7.7**	**–1**						**15**	**3**	**4**	**7**	**4**	**0**	**0**	**1**
1992-93	**Boston**	**NHL**	**82**	**15**	**20**	**35**	**61**	**3**	**2**	**5**	**118**	**12.7**	**2**						**4**	**0**	**1**	**1**	**0**	**0**	**0**	**0**
1993-94	**Boston**	**NHL**	**84**	**22**	**32**	**54**	**59**	**9**	**2**	**1**	**158**	**13.9**	**0**						**13**	**4**	**2**	**6**	**10**	**2**	**0**	**1**
1994-95	TuTo Turku	Finland	14	5	5	10	47																			
	Boston	**NHL**	**47**	**10**	**10**	**20**	**10**	**1**	**0**	**1**	**71**	**14.1**	**3**						**5**	**0**	**0**	**0**	**4**	**0**	**0**	**0**
1995-96	**Boston**	**NHL**	**82**	**23**	**26**	**49**	**46**	**7**	**0**	**1**	**152**	**15.1**	**6**						**5**	**1**	**2**	**3**	**2**	**1**	**0**	**0**
1996-97	**Boston**	**NHL**	**67**	**25**	**26**	**51**	**37**	**6**	**2**	**2**	**172**	**14.5**	**–9**													
1997-98	**Boston**	**NHL**	**79**	**16**	**23**	**39**	**54**	**3**	**0**	**5**	**129**	**12.4**	**6**						**5**	**0**	**0**	**0**	**2**	**0**	**0**	**0**
1998-99	**Boston**	**NHL**	**14**	**1**	**3**	**4**	**4**	**0**	**0**	**0**	**22**	**4.5**	**0**	**18**	**44.4**	**6**	**1**	**15:21**								
	NY Islanders	**NHL**	**55**	**7**	**11**	**18**	**27**	**2**	**0**	**0**	**68**	**10.3**	**–10**	**142**	**45.8**	**23**	**2**	**12:09**								
	Ottawa	**NHL**	**13**	**3**	**2**	**5**	**10**	**1**	**0**	**0**	**16**	**18.8**	**2**	**4**	**25.0**	**7**	**5**	**11:10**	**1**	**0**	**0**	**0**	**0**	**0**	**0**	**0**
99-2000	**Anaheim**	**NHL**	**81**	**11**	**19**	**30**	**26**	**2**	**0**	**3**	**138**	**8.0**	**–3**	**212**	**41.5**	**72**	**23**	**14:35**								
2000-01	**Dallas**	**NHL**	**65**	**8**	**17**	**25**	**26**	**1**	**0**	**3**	**71**	**11.3**	**6**	**16**	**37.5**	**62**	**8**	**10:13**	**8**	**0**	**1**	**1**	**0**	**0**	**0**	**0**
2001-02	**NY Islanders**	**NHL**	**1**	**0**	**0**	**0**	**0**	**0**	**0**	**0**	**1**	**0.0**	**–1**	**0**	**0.0**	**1**	**0**	**7:55**								
	Bridgeport	AHL	1	0	0	0	0																			
	St. Louis	**NHL**	**2**	**0**	**0**	**0**	**2**	**0**	**0**	**0**	**0**	**0.0**	**–2**	**15**	**53.3**	**0**	**0**	**9:17**								
	Los Angeles	**NHL**	**2**	**0**	**0**	**0**	**2**	**0**	**0**	**0**	**1**	**0.0**	**–2**	**10**	**20.0**	**1**	**1**	**7:20**								
	Manchester	AHL	36	18	25	43	19												5	1	3	4	0			
	NHL Totals		**684**	**142**	**191**	**333**	**372**	**35**	**6**	**21**	**1130**	**12.6**		**417**	**42.7**	**172**	**40**	**12:31**	**56**	**8**	**10**	**18**	**22**	**3**	**0**	**2**

NCAA Championship All-Tournament Team (1989) • NCAA Championship Tournament MVP (1989) • ECAC First All-Star Team (1991)

Traded to **NY Islanders** by **Boston** for Ken Belanger, November 7, 1998. Traded to **Ottawa** by **NY Islanders** for Ottawa's 4th round choice (later traded to Phoenix - Phoenix selected Preston Mizzi) in 1999 Entry Draft, March 20, 1999. Traded to **Anaheim** by **Ottawa** with the rights to Antti-Jussi Niemi for Patrick Lalime, June 18, 1999. Signed as a free agent agent by **Dallas**, August 17, 2000. Signed as a free agent by **NY Islanders**, January 16, 2002. Claimed on waivers by **LA Kings** from **NY Islanders**, January 28, 2002. Claimed on waivers by **St. Louis** from **LA Kings**, March 6, 2002. Claimed on waivers by **LA Kings** from **St. Louis**, March 19, 2002. • Missed majority of 2001-02 season recovering from shoulder injury suffered in game vs. St. John's (AHL), January 9, 2002. Signed as a free agent by **NY Rangers**, July 8, 2002.

DONOVAN, Shean

(DAW-nuh-vuhn, SHAWN) **PIT.**

Right wing. Shoots right. 6'2", 200 lbs. Born, Timmins, Ont., January 22, 1975. San Jose's 2nd choice, 28th overall, in 1993 Entry Draft.

Season	Club	League	GP	G	A	Pts	PIM	PP	SH	GW	S	%	+/-	TF	F%	H	SB	Min	GP	G	A	Pts	PIM	PP	SH	GW
1990-91	Kanata Valley	OCJHL	44	8	5	13	8																			
1991-92	Ottawa 67's	OHL	58	11	8	19	14												11	1	0	1	5			
1992-93	Ottawa 67's	OHL	66	29	23	52	33																			
1993-94	Ottawa 67's	OHL	62	35	49	84	63												17	10	11	21	14			
1994-95	Ottawa 67's	OHL	29	22	19	41	41																			
	San Jose	**NHL**	**14**	**0**	**0**	**0**	**6**	**0**	**0**	**0**	**13**	**0.0**	**–6**						**7**	**0**	**1**	**1**	**6**	**0**	**0**	**0**
	Kansas City	IHL	5	0	2	2	7												14	5	3	8	23			
1995-96	**San Jose**	**NHL**	**74**	**13**	**8**	**21**	**39**	**0**	**1**	**2**	**73**	**17.8**	**–17**													
	Kansas City	IHL	4	0	0	0	8												5	0	0	0	8			
1996-97	**San Jose**	**NHL**	**73**	**9**	**6**	**15**	**42**	**0**	**1**	**0**	**115**	**7.8**	**–18**													
	Kentucky	AHL	3	1	3	4	18																			
1997-98	**San Jose**	**NHL**	**20**	**3**	**3**	**6**	**22**	**0**	**0**	**0**	**24**	**12.5**	**3**													
	Colorado	**NHL**	**47**	**5**	**7**	**12**	**48**	**0**	**0**	**0**	**57**	**8.8**	**3**													
1998-99	**Colorado**	**NHL**	**68**	**7**	**12**	**19**	**37**	**1**	**0**	**1**	**81**	**8.6**	**4**	**9**	**22.2**	**35**	**8**	**8:46**	**5**	**0**	**0**	**0**	**2**	**0**	**0**	**0**
99-2000	**Colorado**	**NHL**	**18**	**1**	**0**	**1**	**8**	**0**	**0**	**0**	**13**	**7.7**	**–4**	**1**	**0.0**	**7**	**1**	**5:20**								
	Atlanta	**NHL**	**33**	**4**	**7**	**11**	**18**	**1**	**0**	**1**	**53**	**7.5**	**–13**	**22**	**31.8**	**36**	**11**	**14:19**								
2000-01	**Atlanta**	**NHL**	**63**	**12**	**11**	**23**	**47**	**1**	**3**	**1**	**93**	**12.9**	**–14**	**218**	**45.9**	**42**	**20**	**14:03**								
2001-02	**Atlanta**	**NHL**	**48**	**6**	**6**	**12**	**40**	**1**	**0**	**2**	**64**	**9.4**	**–16**	**12**	**50.0**	**51**	**16**	**13:30**								
	Pittsburgh	**NHL**	**13**	**2**	**1**	**3**	**4**	**0**	**0**	**0**	**18**	**11.1**	**–5**	**4**	**0.0**	**21**	**3**	**14:34**								
	NHL Totals		**471**	**62**	**61**	**123**	**311**	**4**	**5**	**7**	**604**	**10.3**		**266**	**43.2**	**192**	**59**	**11:53**	**12**	**0**	**1**	**1**	**8**	**0**	**0**	**0**

Traded to **Colorado** by **San Jose** with San Jose's 1st round choice (Alex Tanguay) in 1998 Entry Draft for Mike Ricci and Colorado's 2nd round choice (later traded to Buffalo - Buffalo selected Jaroslav Kristek), in 1998 Entry Draft, November 21, 1997. Traded to **Atlanta** by **Colorado** for Rick Tabaracci, December 8, 1999. Claimed on waivers by **Pittsburgh** from **Atlanta**, March 15, 2002.

DOPITA, Jiri

(doh-PEE-tuh, YIH-ree) **EDM.**

Center. Shoots left. 6'4", 210 lbs. Born, Sumperk, Czech., December 2, 1968. NY Islanders' 4th choice, 123rd overall, in 1998 Entry Draft.

Season	Club	League	GP	G	A	Pts	PIM	PP	SH	GW	S	%	+/-	TF	F%	H	SB	Min	GP	G	A	Pts	PIM	PP	SH	GW
1989-90	Dukla Jihlava	Czech	5	1	2	3	0																			
1990-91	TJ DS Olomouc	Czech	42	11	13	24	26																			
1991-92	TJ DS Olomouc	Czech	38	24	20	44	28												3	1	4	5	0			
1992-93	HC Olomouc	Czech	28	12	17	29	16																			
	Eisbaren Berlin	Germany	11	7	8	15	49												4	3	5	8	5			
1993-94	Eisbaren Berlin	Germany	42	23	21	44	52																			
	HC Olomouc	Czech																	12	4	7	11				
1994-95	Eisbaren Berlin	Germany	42	28	40	68	55																			
1995-96	HC Petra Vsetin	Czech	38	19	20	39	20												13	9	11	20	10			
1996-97	HC Petra Vsetin	Czech	52	*30	31	61	55												10	7	4	11	22			
1997-98	HC Petra Vsetin	Czech	52	21	34	55	64												10	*12	6	18	4			
	HC Petra Vsetin	EuroHL	6	2	4	6	4																			
1998-99	Vsetin	Czech	50	19	32	51	43												12	1	6	7				
99-2000	Vsetin	Czech	49	*30	29	59	83												9	0	4	4	6			
	Vsetin	EuroHL	3	0	2	2	2																			
2000-01	Vsetin	Czech	46	19	31	50	53												14	8	*13	*21	18			
2001-02	**Philadelphia**	**NHL**	**52**	**11**	**16**	**27**	**8**	**3**	**0**	**2**	**79**	**13.9**	**9**	**705**	**48.7**	**31**	**28**	**15:14**								
	Czech Republic	Olympics	4	2	2	4	2																			
	NHL Totals		**52**	**11**	**16**	**27**	**8**	**3**	**0**	**2**	**79**	**13.9**		**705**	**48.7**	**31**	**28**	**15:14**								

• Re-entered NHL Entry Draft. Originally Boston's 4th choice, 133rd overall, in 1992 Entry Draft.

Rights traded to **Florida** by **NY Islanders** for San Jose's 5th round choice (previously acquired, NY Islanders selected Adam Johnson) in 1999 Entry Draft, June 26, 1999. Rights traded to **Philadelphia** by **Florida** for Philadelphia's 2nd round choice (later traded to Calgary - Calgary selected Andrei Medvedev) in 2001 Entry Draft, June 23, 2001. Traded to **Edmonton** by **Philadelphia** for Edmonton's 3rd round choice in 2003 Entry Draft and a conditional 5th round choice in 2004 Entry Draft, June 19, 2002.

			Regular Season																Playoffs							
Season	Club	League	GP	G	A	Pts	PIM	PP	SH	GW	S	%	+/-	TF	F%	H	SB	Min	GP	G	A	Pts	PIM	PP	SH	GW

DOWD, Jim

(DOWD, JIHM) **MIN.**

Center. Shoots right. 6'1", 190 lbs. Born, Brick, NJ, December 25, 1968. New Jersey's 7th choice, 149th overall, in 1987 Entry Draft.

Season	Club	League	GP	G	A	Pts	PIM	PP	SH	GW	S	%	+/-	TF	F%	H	SB	Min	GP	G	A	Pts	PIM	PP	SH	GW
1983-84	Brick High	Hi-School	20	19	30	49																				
1984-85	Brick High	Hi-School	24	58	55	113																				
1985-86	Brick High	Hi-School	24	47	51	98																				
1986-87	Brick High	Hi-School	24	22	33	55																				
1987-88	Lake Superior	CCHA	45	18	27	45	16																			
1988-89	Lake Superior	CCHA	46	24	35	59	40																			
1989-90	Lake Superior	CCHA	46	25	*67	92	30																			
1990-91	Lake Superior	CCHA	44	24	*54	*78	53																			
1991-92	**New Jersey**	**NHL**	**1**	**0**	**0**	**0**	**0**	**0**	**0**	**0**	**0**	**0.0**	**0**													
	Utica Devils	AHL	78	17	42	59	47												4	2	2	4	4			
1992-93	**New Jersey**	**NHL**	**1**	**0**	**0**	**0**	**0**	**0**	**0**	**0**	**1**	**0.0**	**−1**													
	Utica Devils	AHL	78	27	45	72	62												5	1	7	8	10			
1993-94	**New Jersey**	**NHL**	**15**	**5**	**10**	**15**	**0**	**2**	**0**	**0**	**26**	**19.2**	**8**						**19**	**2**	**6**	**8**	**8**	**0**	**0**	**0**
	Albany	AHL	58	26	37	63	76																			
1994-95♦	**New Jersey**	**NHL**	**10**	**1**	**4**	**5**	**0**	**1**	**0**	**0**	**14**	**7.1**	**−5**						**11**	**2**	**1**	**3**	**8**	**0**	**0**	**1**
1995-96	**New Jersey**	**NHL**	**28**	**4**	**9**	**13**	**17**	**0**	**0**	**0**	**41**	**9.8**	**−1**													
	Vancouver	**NHL**	**38**	**1**	**6**	**7**	**6**	**0**	**0**	**0**	**35**	**2.9**	**−8**						**1**	**0**	**0**	**0**	**0**	**0**	**0**	**0**
1996-97	**NY Islanders**	**NHL**	**3**	**0**	**0**	**0**	**0**	**0**	**0**	**0**	**0**	**0.0**	**−1**													
	Utah Grizzlies	IHL	48	10	21	31	27																			
	Saint John	AHL	24	5	11	16	18												5	1	2	3	0			
1997-98	**Calgary**	**NHL**	**48**	**6**	**8**	**14**	**12**	**0**	**1**	**0**	**58**	**10.3**	**10**													
	Saint John	AHL	35	8	30	38	20												19	3	13	16	10			
1998-99	**Edmonton**	**NHL**	**1**	**0**	**0**	**0**	**0**	**0**	**0**	**0**	**1**	**0.0**	**0**	7	14.3	1	0	9:47								
	Hamilton	AHL	51	15	29	44	82												11	3	6	9	8			
99-2000	**Edmonton**	**NHL**	**69**	**5**	**18**	**23**	**45**	**2**	**0**	**1**	**103**	**4.9**	**10**	720	54.0	51	25	13:08	**5**	**2**	**1**	**3**	**4**	**0**	**0**	**0**
2000-01	**Minnesota**	**NHL**	**68**	**7**	**22**	**29**	**80**	**0**	**0**	**0**	**92**	**7.6**	**−6**	1154	50.7	41	39	17:50								
2001-02	**Minnesota**	**NHL**	**82**	**13**	**30**	**43**	**54**	**5**	**0**	**1**	**111**	**11.7**	**−14**	1243	52.9	59	40	15:34								
	NHL Totals		**364**	**42**	**107**	**149**	**214**	**10**	**1**	**2**	**482**	**8.7**		**3124**	**52.3**	**152**	**104**	**15:29**	**36**	**6**	**8**	**14**	**20**	**0**	**0**	**1**

CCHA Second All-Star Team (1990) • NCAA West Second All-American Team (1990) • CCHA First All-Star Team (1991) • CCHA Player of the Year (1991) • NCAA West First All-American Team (1991)

• Missed majority of 1994-95 season recovering from shoulder injury suffered in game vs. Quebec, February 2, 1995. Traded to **Hartford** by **New Jersey** with New Jersey's 2nd round choice (later traded to Calgary - Calgary selected Dmitri Kokorev) in 1997 Entry Draft for Jocelyn Lemieux and Hartford's 2nd round choice (later traded to Dallas - Dallas selected John Erskine) in 1998 Entry Draft, December 19, 1995. Traded to **Vancouver** by **Hartford** with Frantisek Kucera and Hartford's 2nd round choice (Ryan Bonni) in 1997 Entry Draft for Jeff Brown and Vancouver's 3rd round choice (later traded to Calgary - Calgary selected Paul Manning) in 1998 Entry Draft, December 19, 1995. Claimed by **NY Islanders** from **Vancouver** in NHL Waiver Draft, September 30, 1996. Signed as a free agent by **Calgary**, August, 1997. Traded to **Nashville** by **Calgary** for future considerations, June 26, 1998. Traded to **Edmonton** by **Nashville** with Mikhail Shtalenkov for Eric Fichaud, Drake Berehowsky and Greg de Vries, October 1, 1998. Selected by **Minnesota** from **Edmonton** in Expansion Draft, June 23, 2000.

DOWNEY, Aaron

(DOW-nee, AIR-ruhn) **DAL.**

Right wing. Shoots right. 6'1", 216 lbs. Born, Shelburne, Ont., August 27, 1974.

Season	Club	League	GP	G	A	Pts	PIM	PP	SH	GW	S	%	+/-	TF	F%	H	SB	Min	GP	G	A	Pts	PIM	PP	SH	GW
1990-91	Grand Valley	OJHL-C	27	6	8	14	57																			
1991-92	Collingwood	OJHL-B	40	9	8	17	111																			
1992-93	Guelph Storm	OHL	53	3	3	6	88												5	1	0	1	0			
1993-94	Cole Harbour	NSMHL	35	8	20	28	210																			
1994-95	Cole Harbour	NSMHL	40	10	31	41	320																			
1995-96	Hampton Roads	ECHL	65	12	11	23	354																			
1996-97	Manitoba Moose	IHL	2	0	0	0	17																			
	Portland Pirates	AHL	3	0	0	0	19																			
	Hampton Roads	ECHL	64	8	8	16	338												9	0	3	3	26			
1997-98	Providence	AHL	78	5	10	15	*407																			
1998-99	Providence	AHL	75	10	12	22	*401												19	1	1	2	46			
99-2000	**Boston**	**NHL**	**1**	**0**	**0**	**0**	**0**	**0**	**0**	**0**	**0**	**0.0**	**0**	0	0.0	1	0	8:31								
	Providence	AHL	47	6	4	10	221												14	1	0	1	24			
2000-01	**Chicago**	**NHL**	**3**	**0**	**0**	**0**	**6**	**0**	**0**	**0**	**2**	**0.0**	**−1**	0	0.0	3	1	5:30								
	Norfolk Admirals	AHL	67	6	15	21	234												9	0	0	0	4			
2001-02	**Chicago**	**NHL**	**36**	**1**	**0**	**1**	**76**	**0**	**0**	**1**	**10**	**10.0**	**−2**	0	0.0	57	7	5:06	**4**	**0**	**0**	**0**	**8**	**0**	**0**	**0**
	Norfolk Admirals	AHL	12	0	2	2	21																			
	NHL Totals		**40**	**1**	**0**	**1**	**82**	**0**	**0**	**1**	**12**	**8.3**		**0**	**0.0**	**61**	**8**	**5:13**	**4**	**0**	**0**	**0**	**8**	**0**	**0**	**0**

Signed as a free agent by **Boston**, January 20, 1998. Signed as a free agent by **Chicago**, August 13, 2000. Signed as a free agent by **Dallas**, July 3, 2002.

DRAKE, Dallas

(DRAYK, DAL-uhs) **ST.L.**

Right wing. Shoots left. 6'1", 187 lbs. Born, Trail, B.C., February 4, 1969. Detroit's 6th choice, 116th overall, in 1989 Entry Draft.

Season	Club	League	GP	G	A	Pts	PIM	PP	SH	GW	S	%	+/-	TF	F%	H	SB	Min	GP	G	A	Pts	PIM	PP	SH	GW
1984-85	Rossland	KIJHL	30	13	37	50																				
1985-86	Rossland	KIJHL	41	53	73	126																				
1986-87	Rossland	KIJHL	40	55	80	135																				
1987-88	Vernon Lakers	BCJHL	47	39	85	124	50												11	9	17	26	30			
1988-89	North-Michigan	WCHA	38	17	22	39	22												7	1	2	3	4			
1989-90	North-Michigan	WCHA	36	13	24	37	42																			
1990-91	North-Michigan	WCHA	44	22	36	58	89																			
1991-92	North-Michigan	WCHA	38	*39	41	*80	46																			
1992-93	**Detroit**	**NHL**	**72**	**18**	**26**	**44**	**93**	**3**	**2**	**5**	**89**	**20.2**	**15**						**7**	**3**	**3**	**6**	**6**	**1**	**0**	**0**
1993-94	**Detroit**	**NHL**	**47**	**10**	**22**	**32**	**37**	**0**	**1**	**2**	**78**	**12.8**	**5**													
	Adirondack	AHL	1	2	0	2	0																			
	Winnipeg	**NHL**	**15**	**3**	**5**	**8**	**12**	**1**	**1**	**1**	**34**	**8.8**	**−6**													
1994-95	**Winnipeg**	**NHL**	**43**	**8**	**18**	**26**	**30**	**0**	**0**	**1**	**66**	**12.1**	**−6**													
1995-96	**Winnipeg**	**NHL**	**69**	**19**	**20**	**39**	**36**	**4**	**4**	**2**	**121**	**15.7**	**−7**						**3**	**0**	**0**	**0**	**0**	**0**	**0**	**0**
1996-97	**Phoenix**	**NHL**	**63**	**17**	**19**	**36**	**52**	**5**	**1**	**1**	**113**	**15.0**	**−11**						**7**	**0**	**1**	**1**	**2**	**0**	**0**	**0**
1997-98	**Phoenix**	**NHL**	**60**	**11**	**29**	**40**	**71**	**3**	**0**	**2**	**112**	**9.8**	**17**						**4**	**0**	**1**	**1**	**2**	**0**	**0**	**0**
1998-99	**Phoenix**	**NHL**	**53**	**9**	**22**	**31**	**65**	**0**	**0**	**3**	**105**	**8.6**	**17**	5	60.0	105	17	15:38	**7**	**4**	**3**	**7**	**4**	**2**	**0**	**1**
99-2000	**Phoenix**	**NHL**	**79**	**15**	**30**	**45**	**62**	**0**	**2**	**5**	**127**	**11.8**	**11**	4	25.0	176	46	15:48	**5**	**0**	**1**	**1**	**4**	**0**	**0**	**0**
2000-01	**St. Louis**	**NHL**	**82**	**12**	**29**	**41**	**71**	**2**	**0**	**3**	**142**	**8.5**	**18**	11	45.5	163	33	14:44	**15**	**4**	**2**	**6**	**16**	**0**	**1**	**1**
2001-02	**St. Louis**	**NHL**	**80**	**11**	**15**	**26**	**87**	**1**	**3**	**2**	**116**	**9.5**	**8**	92	32.6	165	37	13:26	**8**	**0**	**0**	**0**	**8**	**0**	**0**	**0**
	NHL Totals		**663**	**133**	**235**	**368**	**616**	**19**	**14**	**27**	**1103**	**12.1**		**112**	**34.8**	**609**	**133**	**14:50**	**56**	**11**	**11**	**22**	**42**	**3**	**1**	**2**

WCHA First All-Star Team (1992) • NCAA West First All-American Team (1992)

Traded to **Winnipeg** by **Detroit** with Tim Cheveldae for Bob Essensa and Sergei Bautin, March 8, 1994. Transferred to **Phoenix** after **Winnipeg** franchise relocated, July 1, 1996. Selected by **Minnesota** from **Phoenix** in Expansion Draft, June 23, 2000. Signed as a free agent by **St. Louis**, July 1, 2000.

DRAPER, Kris

(DRAY-puhr, KRIHS) **DET.**

Center. Shoots left. 5'11", 190 lbs. Born, Toronto, Ont., May 24, 1971. Winnipeg's 4th choice, 62nd overall, in 1989 Entry Draft.

Season	Club	League	GP	G	A	Pts	PIM	PP	SH	GW	S	%	+/-	TF	F%	H	SB	Min	GP	G	A	Pts	PIM	PP	SH	GW
1987-88	Don Mills	MTHL	40	35	32	67	46																			
1988-89	Team Canada	Nat-Tm	60	11	15	26	16																			
1989-90	Team Canada	Nat-Tm	61	12	22	34	44																			
1990-91	Ottawa 67's	OHL	39	19	42	61	35												17	8	11	19	20			
	Winnipeg	**NHL**	**3**	**1**	**0**	**1**	**5**	**0**	**0**	**0**	**1**	**100.0**	**0**													
	Moncton Hawks	AHL	7	2	1	3	2																			
1991-92	**Winnipeg**	**NHL**	**10**	**2**	**0**	**2**	**2**	**0**	**0**	**0**	**19**	**10.5**	**0**						**2**	**0**	**0**	**0**	**0**	**0**	**0**	**0**
	Moncton Hawks	AHL	61	11	18	29	113												4	0	1	1	6			
1992-93	**Winnipeg**	**NHL**	**7**	**0**	**0**	**0**	**2**	**0**	**0**	**0**	**5**	**0.0**	**−6**													
	Moncton Hawks	AHL	67	12	23	35	40												5	2	2	4	18			
1993-94	**Detroit**	**NHL**	**39**	**5**	**8**	**13**	**31**	**0**	**1**	**0**	**55**	**9.1**	**11**						**7**	**2**	**2**	**4**	**4**	**0**	**1**	**0**
	Adirondack	AHL	46	20	23	43	49																			
1994-95	**Detroit**	**NHL**	**36**	**2**	**6**	**8**	**22**	**0**	**0**	**0**	**44**	**4.5**	**1**						**18**	**4**	**1**	**5**	**12**	**0**	**1**	**1**
1995-96	**Detroit**	**NHL**	**52**	**7**	**9**	**16**	**32**	**0**	**1**	**0**	**51**	**13.7**	**2**						**18**	**4**	**2**	**6**	**18**	**0**	**1**	**0**
1996-97♦	**Detroit**	**NHL**	**76**	**8**	**5**	**13**	**73**	**1**	**0**	**1**	**85**	**9.4**	**−11**						**20**	**2**	**4**	**6**	**12**	**0**	**1**	**0**
1997-98♦	**Detroit**	**NHL**	**64**	**13**	**10**	**23**	**45**	**1**	**0**	**4**	**96**	**13.5**	**5**						**19**	**1**	**3**	**4**	**12**	**0**	**0**	**1**
1998-99	**Detroit**	**NHL**	**80**	**4**	**14**	**18**	**79**	**0**	**1**	**1**	**78**	**5.1**	**2**	887	54.6	82	18	12:43	**10**	**0**	**1**	**1**	**6**	**0**	**0**	**0**
99-2000	**Detroit**	**NHL**	**51**	**5**	**7**	**12**	**28**	**0**	**0**	**3**	**76**	**6.6**	**3**	380	57.6	69	13	13:33	**9**	**2**	**0**	**2**	**6**	**0**	**0**	**0**

Season	Club	League	GP	G	A	Pts	PIM	PP	SH	GW	S	%	+/-	TF	F%	H	SB	Min	GP	G	A	Pts	PIM	PP	SH	GW
			Regular Season																Playoffs							
2000-01	**Detroit**	**NHL**	**75**	**8**	**17**	**25**	**38**	**0**	**1**	**1**	**123**	**6.5**	**17**	**997**	**56.5**	**109**	**23**	**13:26**	**6**	**0**	**1**	**1**	**2**	**0**	**0**	**0**
2001-02♦	**Detroit**	**NHL**	**82**	**15**	**15**	**30**	**56**	**0**	**2**	**3**	**137**	**10.9**	**26**	**756**	**53.2**	**111**	**28**	**15:35**	**23**	**2**	**3**	**5**	**20**	**0**	**0**	**0**
	NHL Totals		**575**	**70**	**91**	**161**	**413**	**2**	**6**	**13**	**770**	**9.1**		**3020**	**55.2**	**371**	**82**	**13:52**	**132**	**17**	**17**	**34**	**92**	**0**	**4**	**2**

Traded to **Detroit** by **Winnipeg** for future considerations, June 30, 1993.

DRUKEN, Harold

(DROO-kehn, HAIR-ohld) **VAN.**

Center. Shoots left. 6', 205 lbs. Born, St. John's, Nfld., January 26, 1979. Vancouver's 3rd choice, 36th overall, in 1997 Entry Draft.

Season	Club	League	GP	G	A	Pts	PIM	PP	SH	GW	S	%	+/-	TF	F%	H	SB	Min	GP	G	A	Pts	PIM	PP	SH	GW
1995-96	Noble-Greenough	Hi-School	30	37	28	65	28																			
1996-97	Detroit	OHL	63	27	31	58	14												5	3	2	5	0			
1997-98	Plymouth Whalers	OHL	64	38	44	82	12												15	9	11	20	4			
1998-99	Plymouth Whalers	OHL	60	*58	45	103	34												11	9	12	21	14			
99-2000	**Vancouver**	**NHL**	**33**	**7**	**9**	**16**	**10**	**2**	**0**	**0**	**69**	**10.1**	**14**	**307**	**47.9**	**13**	**10**	**13:01**								
	Syracuse Crunch	AHL	47	20	25	45	32												4	1	2	3	6			
2000-01	**Vancouver**	**NHL**	**55**	**15**	**15**	**30**	**14**	**6**	**0**	**3**	**82**	**18.3**	**2**	**598**	**43.8**	**16**	**17**	**11:59**	**4**	**0**	**1**	**1**	**0**	**0**	**0**	**0**
	Kansas City	IHL	15	5	9	14	20																			
2001-02	**Vancouver**	**NHL**	**27**	**4**	**4**	**8**	**6**	**1**	**0**	**2**	**33**	**12.1**	**–1**	**269**	**54.7**	**12**	**8**	**11:09**								
	Manitoba Moose	AHL	11	2	9	11	4																			
	NHL Totals		**115**	**26**	**28**	**54**	**30**	**9**	**0**	**5**	**184**	**14.1**		**1174**	**47.4**	**41**	**35**	**12:05**	**4**	**0**	**1**	**1**	**0**	**0**	**0**	**0**

OHL All-Rookie Team (1997) • OHL Second All-Star Team (1999) • Missed majority of 2001-02 season recovering from ankle injury suffered in game vs. Dallas, December 2, 2001.

DRURY, Chris

(DROO-ree, KRIHS) **COL.**

Center. Shoots right. 5'10", 180 lbs. Born, Trumbull, CT, August 20, 1976. Quebec's 5th choice, 72nd overall, in 1994 Entry Draft.

Season	Club	League	GP	G	A	Pts	PIM	PP	SH	GW	S	%	+/-	TF	F%	H	SB	Min	GP	G	A	Pts	PIM	PP	SH	GW
1991-92	Fairfield Prep	Hi-School	25	22	27	49																				
1992-93	Fairfield Prep	Hi-School	24	25	32	57	15																			
1993-94	Fairfield Prep	Hi-School	24	37	18	55																				
1994-95	Boston University	H-East	39	12	15	27	38																			
1995-96	Boston University	H-East	37	35	33	*68	46																			
1996-97	Boston University	H-East	41	*38	24	62	64																			
1997-98	Boston University	H-East	38	28	29	57	88																			
1998-99	**Colorado**	**NHL**	**79**	**20**	**24**	**44**	**62**	**6**	**0**	**3**	**138**	**14.5**	**9**	**418**	**46.9**	**88**	**39**	**13:15**	**19**	**6**	**2**	**8**	**4**	**0**	**0**	**4**
99-2000	**Colorado**	**NHL**	**82**	**20**	**47**	**67**	**42**	**7**	**0**	**2**	**213**	**9.4**	**8**	**1321**	**53.1**	**68**	**50**	**18:33**	**17**	**4**	**10**	**14**	**4**	**1**	**0**	**2**
2000-01♦	**Colorado**	**NHL**	**71**	**24**	**41**	**65**	**47**	**11**	**0**	**5**	**204**	**11.8**	**6**	**552**	**55.1**	**65**	**32**	**18:03**	**23**	**11**	**5**	**16**	**4**	**2**	**0**	**2**
2001-02	**Colorado**	**NHL**	**82**	**21**	**25**	**46**	**38**	**5**	**0**	**6**	**236**	**8.9**	**1**	**1139**	**53.2**	**107**	**55**	**17:57**	**21**	**5**	**7**	**12**	**10**	**1**	**0**	**3**
	United States	Olympics	6	0	0	0	0																			
	NHL Totals		**314**	**85**	**137**	**222**	**189**	**29**	**0**	**16**	**791**	**10.7**		**3430**	**52.7**	**328**	**176**	**16:57**	**80**	**26**	**24**	**50**	**22**	**4**	**0**	**11**

Hockey East Second All-Star Team (1996, 1997) • Hockey East Player of the Year (1997, 1998) • NCAA East Second All-American Team (1996) • NCAA East First All-American Team (1997, 1998) • NCAA Championship All-Tournament Team (1997) • Hockey East First All-Star Team (1998) • Won Hobey Baker Memorial Award (Top U.S. Collegiate Player) (1998) • NHL All-Rookie Team (1999) • Won Calder Memorial Trophy (1999)

Rights transferred to **Colorado** after **Quebec** franchise relocated, June 21, 1995.

DRURY, Ted

(DROO-ree, TEHD)

Center. Shoots left. 6'2", 210 lbs. Born, Boston, MA, September 13, 1971. Calgary's 2nd choice, 42nd overall, in 1989 Entry Draft.

Season	Club	League	GP	G	A	Pts	PIM	PP	SH	GW	S	%	+/-	TF	F%	H	SB	Min	GP	G	A	Pts	PIM	PP	SH	GW
1987-88	Fairfield Prep	Hi-School	24	21	28	49																				
1988-89	Fairfield Prep	Hi-School	25	35	31	66																				
1989-90	Harvard Crimson	ECAC	17	9	13	22	10																			
1990-91	Harvard Crimson	ECAC	25	18	18	36	22																			
1991-92	Team USA	Nat-Tm	53	11	23	34	30																			
	United States	Olympics	7	1	1	2	0																			
1992-93	Harvard Crimson	ECAC	31	22	*41	*63	28																			
1993-94	**Calgary**	**NHL**	**34**	**5**	**7**	**12**	**26**	**0**	**1**	**1**	**43**	**11.6**	**–5**													
	Team USA	Nat-Tm	11	1	4	5	11																			
	United States	Olympics	7	1	2	3	2																			
	Hartford	**NHL**	**16**	**1**	**5**	**6**	**10**	**0**	**0**	**0**	**37**	**2.7**	**–10**													
1994-95	**Hartford**	**NHL**	**34**	**3**	**6**	**9**	**21**	**0**	**0**	**0**	**31**	**9.7**	**–3**													
	Springfield	AHL	2	0	1	1	0																			
1995-96	**Ottawa**	**NHL**	**42**	**9**	**7**	**16**	**54**	**1**	**0**	**1**	**80**	**11.3**	**–19**													
1996-97	**Anaheim**	**NHL**	**73**	**9**	**9**	**18**	**54**	**1**	**0**	**2**	**114**	**7.9**	**–9**						**10**	**1**	**0**	**1**	**4**	**0**	**0**	**0**
1997-98	**Anaheim**	**NHL**	**73**	**6**	**10**	**16**	**82**	**0**	**1**	**0**	**110**	**5.5**	**–10**													
1998-99	**Anaheim**	**NHL**	**75**	**5**	**6**	**11**	**83**	**0**	**0**	**0**	**79**	**6.3**	**2**	**449**	**47.9**	**77**	**13**	**8:15**	**4**	**0**	**0**	**0**	**0**	**0**	**0**	**0**
99-2000	**Anaheim**	**NHL**	**11**	**1**	**1**	**2**	**6**	**0**	**0**	**0**	**9**	**11.1**	**–1**	**82**	**41.5**	**12**	**2**	**7:06**								
	NY Islanders	**NHL**	**55**	**2**	**1**	**3**	**31**	**1**	**0**	**0**	**48**	**4.2**	**–8**	**194**	**47.9**	**46**	**15**	**7:27**								
2000-01	**Columbus**	**NHL**	**1**	**0**	**0**	**0**	**0**	**0**	**0**	**0**	**3**	**0.0**	**–3**	**10**	**80.0**	**0**	**1**	**14:44**								
	Chicago Wolves	IHL	68	21	21	42	53												14	5	4	9	4			
2001-02	Albany	AHL	51	8	10	18	23																			
	Lowell	AHL	16	6	5	11	10												5	0	5	5	6			
	NHL Totals		**414**	**41**	**52**	**93**	**367**	**3**	**2**	**4**	**554**	**7.4**		**735**	**47.6**	**135**	**31**	**7:54**	**14**	**1**	**0**	**1**	**4**	**0**	**0**	**0**

ECAC First All-Star Team (1993) • ECAC Player of the Year (1993) • NCAA East First All-America Team (1993)

Traded to **Hartford** by **Calgary** with Gary Suter and Paul Ranheim for James Patrick, Zarley Zalapski and Michael Nylander, March 10, 1994. Claimed by **Ottawa** from **Hartford** in NHL Waiver Draft, October 2, 1995. Traded to **Anaheim** by **Ottawa** with the rights to Marc Moro for Jason York and Shaun Van Allen, October 1, 1996. Traded to **NY Islanders** by **Anaheim** for Tony Hrkac and Dean Malkoc, October 29, 1999. Selected by **Columbus** from **NY Islanders** in Expansion Draft, June 23, 2000. Signed as a free agent by **New Jersey**, August 21, 2001. Traded to **Carolina** by **New Jersey** for Mike Rucinski, March 4, 2002.

DUBINSKY, Steve

(doo-BIHN-skee, STEEV) **ST.L.**

Center. Shoots left. 6', 190 lbs. Born, Montreal, Que., July 9, 1970. Chicago's 9th choice, 226th overall, in 1990 Entry Draft.

Season	Club	League	GP	G	A	Pts	PIM	PP	SH	GW	S	%	+/-	TF	F%	H	SB	Min	GP	G	A	Pts	PIM	PP	SH	GW
1989-90	Clarkson Knights	ECAC	35	7	10	17	24																			
1990-91	Clarkson Knights	ECAC	39	13	23	36	26																			
1991-92	Clarkson Knights	ECAC	32	20	31	51	40																			
1992-93	Clarkson Knights	ECAC	35	18	26	44	58																			
1993-94	**Chicago**	**NHL**	**27**	**2**	**6**	**8**	**16**	**0**	**0**	**0**	**20**	**10.0**	**1**						**6**	**0**	**0**	**0**	**10**	**0**	**0**	**0**
	Indianapolis Ice	IHL	54	15	25	40	63																			
1994-95	**Chicago**	**NHL**	**16**	**0**	**0**	**0**	**8**	**0**	**0**	**0**	**16**	**0.0**	**–5**													
	Indianapolis Ice	IHL	62	16	11	27	29																			
1995-96	**Chicago**	**NHL**	**43**	**2**	**3**	**5**	**14**	**0**	**0**	**0**	**33**	**6.1**	**3**													
	Indianapolis Ice	IHL	16	8	8	16	10																			
1996-97	**Chicago**	**NHL**	**5**	**0**	**0**	**0**	**0**	**0**	**0**	**0**	**4**	**0.0**	**2**						**4**	**1**	**0**	**1**	**4**	**0**	**0**	**0**
	Indianapolis Ice	IHL	77	32	40	72	53												1	3	1	4	0			
1997-98	**Chicago**	**NHL**	**82**	**5**	**13**	**18**	**57**	**0**	**1**	**0**	**112**	**4.5**	**–6**													
1998-99	**Chicago**	**NHL**	**1**	**0**	**0**	**0**	**0**	**0**	**0**	**0**	**1**	**0.0**	**0**	**5**	**60.0**	**1**	**0**	**5:11**								
	Calgary	**NHL**	**61**	**4**	**10**	**14**	**14**	**0**	**2**	**0**	**69**	**5.8**	**–7**	**223**	**46.6**	**161**	**75**	**14:38**								
99-2000	**Calgary**	**NHL**	**23**	**0**	**1**	**1**	**4**	**0**	**0**	**0**	**29**	**0.0**	**–12**	**207**	**49.3**	**58**	**26**	**12:05**								
2000-01	**Chicago**	**NHL**	**60**	**6**	**4**	**10**	**33**	**0**	**1**	**0**	**70**	**8.6**	**–4**	**714**	**56.7**	**105**	**36**	**10:60**								
	Norfolk Admirals	AHL	14	6	5	11	4																			
2001-02	**Chicago**	**NHL**	**3**	**1**	**0**	**1**	**4**	**0**	**0**	**1**	**6**	**16.7**	**1**	**1**	**100.0**	**6**	**5**	**11:50**								
	Norfolk Admirals	AHL	16	7	3	10	6																			
	Nashville	**NHL**	**26**	**5**	**2**	**7**	**10**	**0**	**0**	**0**	**42**	**11.9**	**–2**	**270**	**56.3**	**53**	**22**	**15:54**								
	Milwaukee	AHL	36	13	13	26	16																			
	NHL Totals		**347**	**25**	**39**	**64**	**160**	**0**	**4**	**1**	**402**	**6.2**		**1420**	**54.0**	**384**	**164**	**13:08**	**10**	**1**	**0**	**1**	**14**	**0**	**0**	**0**

Traded to **Calgary** by **Chicago** with Jeff Shantz for Marty McInnis, Jamie Allison and Eric Andersson, October 27, 1998. • Missed remainder of 1999-2000 season recovering from knee injury suffered in game vs. Chicago, December 12, 1999. Signed as a free agent by **Chicago**, August 25, 2000. Traded to **Nashville** by **Chicago** for future considerations, February 6, 2002. Signed as a free agent by **St. Louis**, July 16, 2002.

DUCHESNE, Steve

(doo-SHAYN, STEEV)

Defense. Shoots left. 5'11", 195 lbs. Born, Sept-Iles, Que., June 30, 1965.

Season	Club	League	GP	G	A	Pts	PIM	PP	SH	GW	S	%	+/-	TF	F%	H	SB	Min	GP	G	A	Pts	PIM	PP	SH	GW
			Regular Season																Playoffs							
1983-84	Wawa Travellers	NOJHA	10	9	23	32	9	....	....	....	....	....	....	....	....	....	....	....	....	....	....	....	....	....	....	....
	Drummondville	QMJHL	67	1	34	35	79	....	....	....	....	....	....	....	....	....	....	....	....	....	....	....	....	....	....	....
1984-85	Drummondville	QMJHL	65	22	54	76	94	....	....	....	....	....	....	....	....	....	....	....	5	4	7	11	8	....	....	....
1985-86	New Haven	AHL	75	14	35	49	76	....	....	....	....	....	....	....	....	....	....	....	5	0	2	2	9	....	....	....
1986-87	**Los Angeles**	**NHL**	**75**	**13**	**25**	**38**	**74**	**5**	**0**	**2**	**113**	**11.5**	**8**	....	....	....	....	....	**5**	**2**	**2**	**4**	**4**	**1**	**0**	**0**
1987-88	**Los Angeles**	**NHL**	**71**	**16**	**39**	**55**	**109**	**5**	**0**	**4**	**190**	**8.4**	**0**	....	....	....	....	....	**5**	**1**	**3**	**4**	**14**	**1**	**0**	**0**
1988-89	**Los Angeles**	**NHL**	**79**	**25**	**50**	**75**	**92**	**8**	**5**	**2**	**215**	**11.6**	**31**	....	....	....	....	....	**11**	**4**	**4**	**8**	**12**	**2**	**0**	**0**
1989-90	**Los Angeles**	**NHL**	**79**	**20**	**42**	**62**	**36**	**6**	**0**	**1**	**224**	**8.9**	**–3**	....	....	....	....	....	**10**	**2**	**9**	**11**	**6**	**1**	**0**	**0**
1990-91	**Los Angeles**	**NHL**	**78**	**21**	**41**	**62**	**66**	**8**	**0**	**3**	**171**	**12.3**	**19**	....	....	....	....	....	**12**	**4**	**8**	**12**	**8**	**1**	**0**	**0**
1991-92	**Philadelphia**	**NHL**	**78**	**18**	**38**	**56**	**86**	**7**	**2**	**3**	**229**	**7.9**	**–7**	....	....	....	....	....	....	....	....	....	....	....	....	....
1992-93	**Quebec**	**NHL**	**82**	**20**	**62**	**82**	**57**	**8**	**0**	**2**	**227**	**8.8**	**15**	....	....	....	....	....	**6**	**0**	**5**	**5**	**6**	**0**	**0**	**0**
1993-94	**St. Louis**	**NHL**	**36**	**12**	**19**	**31**	**14**	**8**	**0**	**1**	**115**	**10.4**	**1**	....	....	....	....	....	**4**	**0**	**2**	**2**	**2**	**0**	**0**	**0**
1994-95	**St. Louis**	**NHL**	**47**	**12**	**26**	**38**	**36**	**1**	**0**	**1**	**116**	**10.3**	**29**	....	....	....	....	....	**7**	**0**	**4**	**4**	**2**	**0**	**0**	**0**
1995-96	**Ottawa**	**NHL**	**62**	**12**	**24**	**36**	**42**	**7**	**0**	**2**	**163**	**7.4**	**–23**	....	....	....	....	....	....	....	....	....	....	....	....	....
1996-97	**Ottawa**	**NHL**	**78**	**19**	**28**	**47**	**38**	**10**	**2**	**3**	**208**	**9.1**	**–9**	....	....	....	....	....	**7**	**1**	**4**	**5**	**0**	**1**	**0**	**1**
1997-98	**St. Louis**	**NHL**	**80**	**14**	**42**	**56**	**32**	**5**	**1**	**1**	**153**	**9.2**	**9**	....	....	....	....	....	**10**	**0**	**4**	**4**	**6**	**0**	**0**	**0**
1998-99	**Los Angeles**	**NHL**	**60**	**4**	**19**	**23**	**22**	**1**	**0**	**1**	**99**	**4.0**	**–6**	**2**	**50.0**	**38**	**95**	**21:11**	....	....	....	....	....	....	....	....
	Philadelphia	**NHL**	**11**	**2**	**5**	**7**	**2**	**1**	**0**	**1**	**19**	**10.5**	**0**	**0**	**0.0**	**5**	**14**	**22:28**	**6**	**0**	**2**	**2**	**2**	**0**	**0**	**0**
99-2000	**Detroit**	**NHL**	**79**	**10**	**31**	**41**	**42**	**1**	**0**	**1**	**154**	**6.5**	**12**	**1**	**0.0**	**41**	**104**	**21:14**	**9**	**0**	**4**	**4**	**10**	**0**	**0**	**0**
2000-01	**Detroit**	**NHL**	**54**	**6**	**19**	**25**	**48**	**2**	**0**	**0**	**76**	**7.9**	**9**	**0**	**0.0**	**40**	**75**	**18:20**	**6**	**2**	**4**	**6**	**0**	**2**	**0**	**0**
2001-02♦	**Detroit**	**NHL**	**64**	**3**	**15**	**18**	**28**	**1**	**0**	**2**	**70**	**4.3**	**3**	**1**	**100.0**	**32**	**62**	**15:53**	**23**	**0**	**6**	**6**	**24**	**0**	**0**	**0**
	NHL Totals		**1113**	**227**	**525**	**752**	**824**	**84**	**10**	**30**	**2542**	**8.9**		**4**	**50.0**	**156**	**350**	**19:25**	**121**	**16**	**61**	**77**	**96**	**9**	**0**	**1**

QMJHL First All-Star Team (1985) • NHL All-Rookie Team (1987) • Played in NHL All-Star Game (1989, 1990, 1993)

Signed as a free agent by **LA Kings,** October 1, 1984. Traded to **Philadelphia** by **LA Kings** with Steve Kasper and LA Kings' 4th round choice (Aris Brimanis) in 1991 Entry Draft for Jari Kurri and Jeff Chychrun, May 30, 1991. Traded to **Quebec** by **Philadelphia** with Peter Forsberg, Kerry Huffman, Mike Ricci, Ron Hextall, Philadelphia's 1st round choice (Jocelyn Thibault) in 1993 Entry Draft, $15,000,000 and future considerations (Chris Simon and Philadelphia's 1st round choice (later traded to Toronto - later traded to Washington - Washington selected Nolan Baumgartner) in 1994 Entry Draft, July 21, 1992) for Eric Lindros, June 30, 1992. Traded to **St. Louis** by **Quebec** with Denis Chasse for Garth Butcher, Ron Sutter and Bob Bassen, January 23, 1994. Traded to **Ottawa** by **St. Louis** for Ottawa's 2nd round choice (later traded to Buffalo - Buffalo selected Cory Sarich) in 1996 Entry Draft, August 4, 1995. Traded to **St. Louis** by **Ottawa** for Igor Kravchuk, August 25, 1997. Signed as a free agent by **LA Kings**, July 2, 1998. Traded to **Philadelphia** by **LA Kings** for Dave Babych and Philadelphia's 5th round choice (Nathan Marsters) in 2000 Entry Draft, March 23, 1999. Signed as a free agent by **Detroit**, September 3, 1999.

DUMONT, J-P

(DOO-mawnt, JAY-pee) **BUF.**

Right wing. Shoots left. 6'1", 205 lbs. Born, Montreal, Que., April 1, 1978. NY Islanders' 1st choice, 3rd overall, in 1996 Entry Draft.

Season	Club	League	GP	G	A	Pts	PIM	PP	SH	GW	S	%	+/-	TF	F%	H	SB	Min	GP	G	A	Pts	PIM	PP	SH	GW
1993-94	Mtl-Bourassa	QAAA	44	27	20	47	44	....	....	....	....	....	....	....	....	....	....	....	4	2	3	5	4	....	....	....
1994-95	Val-d'Or Foreurs	QMJHL	48	5	14	19	24	....	....	....	....	....	....	....	....	....	....	....	....	....	....	....	....	....	....	....
1995-96	Val-d'Or Foreurs	QMJHL	66	48	57	105	109	....	....	....	....	....	....	....	....	....	....	....	13	12	8	20	22	....	....	....
1996-97	Val-d'Or Foreurs	QMJHL	62	44	64	108	86	....	....	....	....	....	....	....	....	....	....	....	13	9	7	16	12	....	....	....
1997-98	Val-d'Or Foreurs	QMJHL	55	57	42	99	63	....	....	....	....	....	....	....	....	....	....	....	19	31	15	46	18	....	....	....
1998-99	**Chicago**	**NHL**	**25**	**9**	**6**	**15**	**10**	**0**	**0**	**2**	**42**	**21.4**	**7**	**10**	**50.0**	**22**	**8**	**14:14**	....	....	....	....	....	....	....	....
	Portland Pirates	AHL	50	32	14	46	39	....	....	....	....	....	....	....	....	....	....	....	....	....	....	....	....	....	....	....
	Chicago Wolves	IHL	....	....	....	....	....	....	....	....	....	....	....	....	....	....	....	....	10	4	1	5	6	....	....	....
99-2000	**Chicago**	**NHL**	**47**	**10**	**8**	**18**	**18**	**0**	**0**	**1**	**86**	**11.6**	**–6**	**12**	**33.3**	**49**	**7**	**12:54**	....	....	....	....	....	....	....	....
	Cleveland	IHL	7	5	2	7	8	....	....	....	....	....	....	....	....	....	....	....	....	....	....	....	....	....	....	....
	Rochester	AHL	13	7	10	17	18	....	....	....	....	....	....	....	....	....	....	....	21	14	7	21	32	....	....	....
2000-01	**Buffalo**	**NHL**	**79**	**23**	**28**	**51**	**54**	**9**	**0**	**5**	**156**	**14.7**	**1**	**3**	**33.3**	**132**	**20**	**15:01**	**13**	**4**	**3**	**7**	**8**	**0**	**0**	**0**
2001-02	**Buffalo**	**NHL**	**76**	**23**	**21**	**44**	**42**	**7**	**0**	**3**	**154**	**14.9**	**–10**	**4**	**50.0**	**128**	**17**	**15:14**	....	....	....	....	....	....	....	....
	NHL Totals		**227**	**65**	**63**	**128**	**124**	**16**	**0**	**11**	**438**	**14.8**		**29**	**41.4**	**331**	**52**	**14:34**	**13**	**4**	**3**	**7**	**8**	**0**	**0**	**0**

QMJHL Second All-Star Team (1997)

Rights traded to **Chicago** by **NY Islanders** with Chicago's 5th round choice (later traded to Philadelphia - Philadelphia selected Francis Belanger) in 1998 Entry Draft for Dmitri Nabokov, May 30, 1998. Traded to **Buffalo** by **Chicago** with Doug Gilmour for Michal Grosek, March 10, 2000.

DuPONT, Micki

(DOO-pawnt, MIH-kee) **CGY.**

Defense. Shoots right. 5'9", 180 lbs. Born, Calgary, Alta., April 15, 1980. Calgary's 9th choice, 270th overall, in 2000 Entry Draft.

Season	Club	League	GP	G	A	Pts	PIM	PP	SH	GW	S	%	+/-	TF	F%	H	SB	Min	GP	G	A	Pts	PIM	PP	SH	GW
1995-96	Calgary Blazers	AMHL	35	10	35	45	68	....	....	....	....	....	....	....	....	....	....	....	....	....	....	....	....	....	....	....
1996-97	Kamloops Blazers	WHL	59	8	27	35	39	....	....	....	....	....	....	....	....	....	....	....	5	0	4	4	8	....	....	....
1997-98	Kamloops Blazers	WHL	71	13	41	54	91	....	....	....	....	....	....	....	....	....	....	....	7	0	1	1	10	....	....	....
1998-99	Kamloops Blazers	WHL	59	8	27	35	110	....	....	....	....	....	....	....	....	....	....	....	15	2	8	10	22	....	....	....
99-2000	Kamloops Blazers	WHL	70	26	62	88	156	....	....	....	....	....	....	....	....	....	....	....	4	0	2	2	17	....	....	....
	Long Beach	IHL	1	0	0	0	0	....	....	....	....	....	....	....	....	....	....	....	....	....	....	....	....	....	....	....
	San Diego Gulls	WCHL	....	....	....	....	....	....	....	....	....	....	....	....	....	....	....	....	7	2	2	4	0	....	....	....
2000-01	Saint John	AHL	67	8	21	29	28	....	....	....	....	....	....	....	....	....	....	....	19	1	9	10	14	....	....	....
2001-02	**Calgary**	**NHL**	**2**	**0**	**0**	**0**	**2**	**0**	**0**	**0**	**2**	**0.0**	**0**	**0**	**0.0**	**1**	**1**	**14:44**	....	....	....	....	....	....	....	....
	Saint John	AHL	77	7	33	40	77	....	....	....	....	....	....	....	....	....	....	....	....	....	....	....	....	....	....	....
	NHL Totals		**2**	**0**	**0**	**0**	**2**	**0**	**0**	**0**	**2**	**0.0**		**0**	**0.0**	**1**	**1**	**14:44**	....	....	....	....	....	....	....	....

DUPUIS, Pascal

(doo-PWEE, pas-KAL) **MIN.**

Left wing. Shoots Right. 6', 195 lbs. Born, Laval, Que., April 7, 1979.

Season	Club	League	GP	G	A	Pts	PIM	PP	SH	GW	S	%	+/-	TF	F%	H	SB	Min	GP	G	A	Pts	PIM	PP	SH	GW
1995-96	Laval Laurentide	QAAA	41	10	15	25	....	....	....	....	....	....	....	....	....	....	....	....	14	11	11	22	....	....	....	....
1996-97	Rouyn-Noranda	QMJHL	44	9	15	24	20	....	....	....	....	....	....	....	....	....	....	....	....	....	....	....	....	....	....	....
1997-98	Rouyn-Noranda	QMJHL	39	9	17	26	36	....	....	....	....	....	....	....	....	....	....	....	....	....	....	....	....	....	....	....
	Shawinigan	QMJHL	28	7	13	20	10	....	....	....	....	....	....	....	....	....	....	....	6	2	0	2	4	....	....	....
1998-99	Shawinigan	QMJHL	57	30	42	72	118	....	....	....	....	....	....	....	....	....	....	....	6	1	8	9	18	....	....	....
99-2000	Shawinigan	QMJHL	61	50	55	105	99	....	....	....	....	....	....	....	....	....	....	....	13	*15	7	22	4	....	....	....
2000-01	**Minnesota**	**NHL**	**4**	**1**	**0**	**1**	**4**	**1**	**0**	**0**	**8**	**12.5**	**0**	**0**	**0.0**	**12**	**1**	**15:36**	....	....	....	....	....	....	....	....
	Cleveland	IHL	70	19	24	43	37	....	....	....	....	....	....	....	....	....	....	....	4	0	0	0	0	....	....	....
2001-02	**Minnesota**	**NHL**	**76**	**15**	**12**	**27**	**16**	**3**	**2**	**0**	**154**	**9.7**	**–10**	**40**	**32.5**	**61**	**26**	**15:08**	....	....	....	....	....	....	....	....
	NHL Totals		**80**	**16**	**12**	**28**	**20**	**4**	**2**	**0**	**162**	**9.9**		**40**	**32.5**	**73**	**27**	**15:09**	....	....	....	....	....	....	....	....

Signed as a free agent by **Minnesota**, August 18, 2000.

DVORAK, Radek

(duh-VOHR-ak, RA-dehk) **NYR**

Right wing. Shoots right. 6'1", 194 lbs. Born, Tabor, Czech., March 9, 1977. Florida's 1st choice, 10th overall, in 1995 Entry Draft.

Season	Club	League	GP	G	A	Pts	PIM	PP	SH	GW	S	%	+/-	TF	F%	H	SB	Min	GP	G	A	Pts	PIM	PP	SH	GW
1992-93	C. Budejovice Jr.	Czech-Jr.	35	44	46	90	....	....	....	....	....	....	....	....	....	....	....	....	....	....	....	....	....	....	....	....
1993-94	C. Budejovice Jr.	Czech-Jr.	20	17	18	35	....	....	....	....	....	....	....	....	....	....	....	....	....	....	....	....	....	....	....	....
	Ceske Budejovice	Czech	8	0	0	0	0	....	....	....	....	....	....	....	....	....	....	....	....	....	....	....	....	....	....	....
1994-95	Ceske Budejovice	Czech	10	3	5	8	2	....	....	....	....	....	....	....	....	....	....	....	9	5	1	6	....	....	....	....
1995-96	**Florida**	**NHL**	**77**	**13**	**14**	**27**	**20**	**0**	**0**	**4**	**126**	**10.3**	**5**	....	....	....	....	....	**16**	**1**	**3**	**4**	**0**	**0**	**0**	**0**
1996-97	**Florida**	**NHL**	**78**	**18**	**21**	**39**	**30**	**2**	**0**	**1**	**139**	**12.9**	**–2**	....	....	....	....	....	**3**	**0**	**0**	**0**	**0**	**0**	**0**	**0**
1997-98	**Florida**	**NHL**	**64**	**12**	**24**	**36**	**33**	**2**	**3**	**0**	**112**	**10.7**	**–1**	....	....	....	....	....	....	....	....	....	....	....	....	....
1998-99	**Florida**	**NHL**	**82**	**19**	**24**	**43**	**29**	**0**	**4**	**0**	**182**	**10.4**	**7**	**98**	**46.9**	**30**	**33**	**16:13**	....	....	....	....	....	....	....	....
99-2000	**Florida**	**NHL**	**35**	**7**	**10**	**17**	**6**	**0**	**0**	**1**	**67**	**10.4**	**5**	**16**	**37.5**	**5**	**11**	**15:25**	....	....	....	....	....	....	....	....
	NY Rangers	**NHL**	**46**	**11**	**22**	**33**	**10**	**2**	**1**	**0**	**90**	**12.2**	**0**	**34**	**35.3**	**22**	**16**	**18:24**	....	....	....	....	....	....	....	....
2000-01	**NY Rangers**	**NHL**	**82**	**31**	**36**	**67**	**20**	**5**	**2**	**3**	**230**	**13.5**	**9**	**20**	**30.0**	**29**	**41**	**19:04**	....	....	....	....	....	....	....	....
2001-02	**NY Rangers**	**NHL**	**65**	**17**	**20**	**37**	**14**	**3**	**3**	**1**	**210**	**8.1**	**–20**	**5**	**0.0**	**54**	**43**	**19:44**	....	....	....	....	....	....	....	....
	Czech Republic	Olympics	4	0	0	0	0	....	....	....	....	....	....	....	....	....	....	....	....	....	....	....	....	....	....	....
	NHL Totals		**529**	**128**	**171**	**299**	**162**	**14**	**13**	**10**	**1156**	**11.1**		**173**	**40.5**	**140**	**144**	**17:57**	**19**	**1**	**3**	**4**	**0**	**0**	**0**	**0**

Traded to **San Jose** by **Florida** for Mike Vernon and San Jose's 3rd round choice (Sean O'Connor) in 2000 Entry Draft, December 30, 1999. Traded to **NY Rangers** by **San Jose** for Todd Harvey and NY Rangers' 4th round choice (Dimitri Patzold) in 2001 Entry Draft, December 30, 1999.

			Regular Season																Playoffs							
Season	Club	League	GP	G	A	Pts	PIM	PP	SH	GW	S	%	+/-	TF	F%	H	SB	Min	GP	G	A	Pts	PIM	PP	SH	GW

DWYER, Gordie

(DWIGH-uhr, GOHR-dee) **T.B.**

Left wing. Shoots left. 6'3", 215 lbs. Born, Dalhousie, N.B., January 25, 1978. Montreal's 5th choice, 152nd overall, in 1998 Entry Draft.

Season	Club	League	GP	G	A	Pts	PIM	PP	SH	GW	S	%	+/-	TF	F%	H	SB	Min	GP	G	A	Pts	PIM	PP	SH	GW
1993-94	Magog	QAAA	42	7	15	22	62	....	....	....	....	....	....	....	....	....	....	....	4	2	1	3	0	....	....	....
1994-95	Hull Olympiques	QMJHL	57	3	7	10	204	....	....	....	....	....	....	....	....	....	....	....	17	1	3	4	54	....	....	....
1995-96	Hull Olympiques	QMJHL	25	5	9	14	199	....	....	....	....	....	....	....	....	....	....	....	....	....	....	....	....	....	....	....
	Laval Titan	QMJHL	22	5	17	22	72	....	....	....	....	....	....	....	....	....	....	....	....	....	....	....	....	....	....	....
	Beauport	QMJHL	22	4	9	13	87	....	....	....	....	....	....	....	....	....	....	....	20	3	5	8	104	....	....	....
1996-97	Drummondville	QMJHL	66	21	48	69	393	....	....	....	....	....	....	....	....	....	....	....	8	6	1	7	39	....	....	....
1997-98	Quebec Remparts	QMJHL	59	18	27	45	365	....	....	....	....	....	....	....	....	....	....	....	14	4	9	13	67	....	....	....
1998-99	Fredericton	AHL	14	0	0	0	46	....	....	....	....	....	....	....	....	....	....	....	....	....	....	....	....	....	....	....
	New Orleans	ECHL	36	1	3	4	163	....	....	....	....	....	....	....	....	....	....	....	11	0	0	0	27	....	....	....
99-2000	Quebec	AHL	7	0	0	0	37	....	....	....	....	....	....	....	....	....	....	....	....	....	....	....	....	....	....	....
	Tampa Bay	**NHL**	**24**	**0**	**1**	**1**	**135**	**0**	**0**	**0**	**7**	**0.0**	**–6**	**0**	**0.0**	**53**	**4**	**4:57**	....	....	....	....	....	....	....	....
	Detroit Vipers	IHL	27	0	2	2	147	....	....	....	....	....	....	....	....	....	....	....	....	....	....	....	....	....	....	....
2000-01	**Tampa Bay**	**NHL**	**28**	**0**	**1**	**1**	**96**	**0**	**0**	**0**	**12**	**0.0**	**–7**	**2**	**50.0**	**48**	**2**	**5:47**	....	....	....	....	....	....	....	....
	Detroit Vipers	IHL	24	2	3	5	169	....	....	....	....	....	....	....	....	....	....	....	....	....	....	....	....	....	....	....
2001-02	**Tampa Bay**	**NHL**	**26**	**0**	**2**	**2**	**60**	**0**	**0**	**0**	**6**	**0.0**	**–4**	**0**	**0.0**	**40**	**4**	**5:05**	....	....	....	....	....	....	....	....
	Springfield	AHL	17	1	3	4	80	....	....	....	....	....	....	....	....	....	....	....	....	....	....	....	....	....	....	....
	NHL Totals		**78**	**0**	**4**	**4**	**291**	**0**	**0**	**0**	**25**	**0.0**		**2**	**50.0**	**141**	**10**	**5:17**	....	....	....	....	....	....	....	....

• Re-entered NHL Entry Draft. Originally St. Louis' 2nd choice, 67th overall, in 1996 Entry Draft.

Traded to **Tampa Bay** by **Montreal** for Mike McBain, November 26, 1999.

DYKHUIS, Karl

(DIGH-kowz, KAHRL) **MTL.**

Defense. Shoots left. 6'3", 214 lbs. Born, Sept-Iles, Que., July 8, 1972. Chicago's 1st choice, 16th overall, in 1990 Entry Draft.

Season	Club	League	GP	G	A	Pts	PIM	PP	SH	GW	S	%	+/-	TF	F%	H	SB	Min	GP	G	A	Pts	PIM	PP	SH	GW
1987-88	Lac St-Jean	QAAA	37	2	12	14	....	....	....	....	....	....	....	....	....	....	....	....	2	0	1	1	2	....	....	....
1988-89	Hull Olympiques	QMJHL	63	2	29	31	59	....	....	....	....	....	....	....	....	....	....	....	9	1	9	10	6	....	....	....
1989-90	Hull Olympiques	QMJHL	69	10	46	56	119	....	....	....	....	....	....	....	....	....	....	....	11	2	5	7	2	....	....	....
1990-91	Team Canada	Nat-Tm	37	2	9	11	16	....	....	....	....	....	....	....	....	....	....	....	....	....	....	....	....	....	....	....
	Longueuil	QMJHL	3	1	4	5	6	....	....	....	....	....	....	....	....	....	....	....	8	2	5	7	6	....	....	....
1991-92	Team Canada	Nat-Tm	19	1	2	3	16	....	....	....	....	....	....	....	....	....	....	....	....	....	....	....	....	....	....	....
	Verdun	QMJHL	29	5	19	24	55	....	....	....	....	....	....	....	....	....	....	....	17	0	12	12	14	....	....	....
	Chicago	**NHL**	**6**	**1**	**3**	**4**	**4**	**1**	**0**	**0**	**12**	**8.3**	**–1**	....	....	....	....	....	....	....	....	....	....	....	....	....
1992-93	**Chicago**	**NHL**	**12**	**0**	**5**	**5**	**0**	**0**	**0**	**0**	**10**	**0.0**	**2**	....	....	....	....	....	....	....	....	....	....	....	....	....
	Indianapolis Ice	IHL	59	5	18	23	76	....	....	....	....	....	....	....	....	....	....	....	5	1	1	2	8	....	....	....
1993-94	Indianapolis Ice	IHL	73	7	25	32	132	....	....	....	....	....	....	....	....	....	....	....	....	....	....	....	....	....	....	....
1994-95	Indianapolis Ice	IHL	52	2	21	23	63	....	....	....	....	....	....	....	....	....	....	....	....	....	....	....	....	....	....	....
	Philadelphia	**NHL**	**33**	**2**	**6**	**8**	**37**	**1**	**0**	**1**	**46**	**4.3**	**7**	....	....	....	....	....	**15**	**4**	**4**	**8**	**14**	**2**	**0**	**2**
	Hershey Bears	AHL	1	0	0	0	0	....	....	....	....	....	....	....	....	....	....	....	....	....	....	....	....	....	....	....
1995-96	**Philadelphia**	**NHL**	**82**	**5**	**15**	**20**	**101**	**1**	**0**	**0**	**104**	**4.8**	**12**	....	....	....	....	....	**12**	**2**	**2**	**4**	**22**	**1**	**0**	**0**
1996-97	**Philadelphia**	**NHL**	**62**	**4**	**15**	**19**	**35**	**2**	**0**	**1**	**101**	**4.0**	**6**	....	....	....	....	....	**18**	**0**	**3**	**3**	**2**	**0**	**0**	**0**
1997-98	**Tampa Bay**	**NHL**	**78**	**5**	**9**	**14**	**110**	**0**	**1**	**0**	**91**	**5.5**	**–8**	....	....	....	....	....	....	....	....	....	....	....	....	....
1998-99	**Tampa Bay**	**NHL**	**33**	**2**	**1**	**3**	**18**	**0**	**0**	**0**	**27**	**7.4**	**–21**	**0**	**0.0**	**44**	**39**	**20:14**	....	....	....	....	....	....	....	....
	Philadelphia	**NHL**	**45**	**2**	**4**	**6**	**32**	**1**	**0**	**0**	**61**	**3.3**	**–2**	**0**	**0.0**	**38**	**47**	**18:15**	**5**	**1**	**0**	**1**	**4**	**0**	**0**	**0**
99-2000	**Philadelphia**	**NHL**	**5**	**0**	**1**	**1**	**6**	**0**	**0**	**0**	**5**	**0.0**	**–2**	**0**	**0.0**	**9**	**3**	**14:54**	....	....	....	....	....	....	....	....
	Montreal	**NHL**	**67**	**7**	**12**	**19**	**40**	**3**	**1**	**0**	**64**	**10.9**	**–3**	**0**	**0.0**	**97**	**98**	**19:53**	....	....	....	....	....	....	....	....
2000-01	**Montreal**	**NHL**	**67**	**8**	**9**	**17**	**44**	**2**	**0**	**1**	**66**	**12.1**	**9**	**4**	**100.0**	**89**	**63**	**15:40**	....	....	....	....	....	....	....	....
2001-02	**Montreal**	**NHL**	**80**	**5**	**7**	**12**	**32**	**0**	**0**	**1**	**85**	**5.9**	**16**	**0**	**0.0**	**96**	**111**	**19:54**	**12**	**1**	**1**	**2**	**8**	**0**	**0**	**0**
	NHL Totals		**570**	**41**	**87**	**128**	**459**	**11**	**2**	**4**	**672**	**6.1**		**4**	**100.0**	**373**	**361**	**18:39**	**62**	**8**	**10**	**18**	**50**	**3**	**0**	**2**

QMJHL All-Rookie Team (1989) • QMJHL Defensive Rookie of the Year (1989) • QMJHL First All-Star Team (1990)

Traded to **Philadelphia** by **Chicago** for Bob Wilkie and Philadelphia's 5th round choice (Kyle Calder) in 1997 Entry Draft, February 16, 1995. Traded to **Tampa Bay** by **Philadelphia** with Mikael Renberg for Philadelphia's 1st round choices (previously acquired by Tampa Bay) in 1998 (Simon Gagne), 1999 (Maxime Ouellet), 2000 (Justin Williams) and 2001 (later traded to Ottawa - Ottawa selected Tim Gleason) Entry Drafts, August 20, 1997. Traded to **Philadelphia** by **Tampa Bay** for Petr Svoboda, December 28, 1998. Traded to **Montreal** by **Philadelphia** for cash, October 20, 1999.

EAKINS, Dallas

(EE-kins, DAL-las) **ATL.**

Defense. Shoots left. 6'2", 195 lbs. Born, Dade City, FL, February 27, 1967. Washington's 11th choice, 208th overall, in 1985 Entry Draft.

Season	Club	League	GP	G	A	Pts	PIM	PP	SH	GW	S	%	+/-	TF	F%	H	SB	Min	GP	G	A	Pts	PIM	PP	SH	GW
1983-84	Peterboro AA	OMHA	29	7	20	27	67	....	....	....	....	....	....	....	....	....	....	....	....	....	....	....	....	....	....	....
	Peterborough	OJHL-B	5	0	3	3	4	....	....	....	....	....	....	....	....	....	....	....	....	....	....	....	....	....	....	....
1984-85	Peterborough	OHL	48	0	8	8	96	....	....	....	....	....	....	....	....	....	....	....	7	0	0	0	18	....	....	....
1985-86	Peterborough	OHL	60	6	16	22	134	....	....	....	....	....	....	....	....	....	....	....	16	0	1	1	30	....	....	....
1986-87	Peterborough	OHL	54	3	11	14	145	....	....	....	....	....	....	....	....	....	....	....	12	1	4	5	37	....	....	....
1987-88	Peterborough	OHL	64	11	27	38	129	....	....	....	....	....	....	....	....	....	....	....	12	3	12	15	16	....	....	....
1988-89	Baltimore	AHL	62	0	10	10	139	....	....	....	....	....	....	....	....	....	....	....	....	....	....	....	....	....	....	....
1989-90	Moncton Hawks	AHL	75	2	11	13	189	....	....	....	....	....	....	....	....	....	....	....	....	....	....	....	....	....	....	....
1990-91	Moncton Hawks	AHL	75	1	12	13	132	....	....	....	....	....	....	....	....	....	....	....	9	0	1	1	44	....	....	....
1991-92	Moncton Hawks	AHL	67	3	13	16	136	....	....	....	....	....	....	....	....	....	....	....	11	2	1	3	16	....	....	....
1992-93	**Winnipeg**	**NHL**	**14**	**0**	**2**	**2**	**38**	**0**	**0**	**0**	**9**	**0.0**	**2**	....	....	....	....	....	....	....	....	....	....	....	....	....
	Moncton Hawks	AHL	55	4	6	10	132	....	....	....	....	....	....	....	....	....	....	....	....	....	....	....	....	....	....	....
1993-94	**Florida**	**NHL**	**1**	**0**	**0**	**0**	**0**	**0**	**0**	**0**	**2**	**0.0**	**0**	....	....	....	....	....	....	....	....	....	....	....	....	....
	Cincinnati	IHL	80	1	18	19	143	....	....	....	....	....	....	....	....	....	....	....	8	0	1	1	41	....	....	....
1994-95	Cincinnati	IHL	59	6	12	18	69	....	....	....	....	....	....	....	....	....	....	....	....	....	....	....	....	....	....	....
	Florida	**NHL**	**17**	**0**	**1**	**1**	**35**	**0**	**0**	**0**	**3**	**0.0**	**2**	....	....	....	....	....	....	....	....	....	....	....	....	....
1995-96	**St. Louis**	**NHL**	**16**	**0**	**1**	**1**	**34**	**0**	**0**	**0**	**6**	**0.0**	**–2**	....	....	....	....	....	....	....	....	....	....	....	....	....
	Worcester	AHL	4	0	0	0	12	....	....	....	....	....	....	....	....	....	....	....	....	....	....	....	....	....	....	....
	Winnipeg	**NHL**	**2**	**0**	**0**	**0**	**0**	**0**	**0**	**0**	**0**	**0.0**	**1**	....	....	....	....	....	....	....	....	....	....	....	....	....
1996-97	**Phoenix**	**NHL**	**4**	**0**	**0**	**0**	**10**	**0**	**0**	**0**	**2**	**0.0**	**–3**	....	....	....	....	....	....	....	....	....	....	....	....	....
	Springfield	AHL	38	6	7	13	63	....	....	....	....	....	....	....	....	....	....	....	....	....	....	....	....	....	....	....
	NY Rangers	**NHL**	**3**	**0**	**0**	**0**	**6**	**0**	**0**	**0**	**2**	**0.0**	**–1**	....	....	....	....	....	**4**	**0**	**0**	**0**	**4**	**0**	**0**	**0**
	Binghamton	AHL	19	1	7	8	15	....	....	....	....	....	....	....	....	....	....	....	....	....	....	....	....	....	....	....
1997-98	**Florida**	**NHL**	**23**	**0**	**1**	**1**	**44**	**0**	**0**	**0**	**16**	**0.0**	**1**	....	....	....	....	....	....	....	....	....	....	....	....	....
	New Haven	AHL	4	0	1	1	7	....	....	....	....	....	....	....	....	....	....	....	....	....	....	....	....	....	....	....
1998-99	**Toronto**	**NHL**	**18**	**0**	**2**	**2**	**24**	**0**	**0**	**0**	**11**	**0.0**	**3**	**0**	**0.0**	**20**	**10**	**16:28**	**1**	**0**	**0**	**0**	**0**	**0**	**0**	**0**
	Chicago Wolves	IHL	2	0	0	0	0	....	....	....	....	....	....	....	....	....	....	....	....	....	....	....	....	....	....	....
	St. John's	AHL	20	3	7	10	16	....	....	....	....	....	....	....	....	....	....	....	5	0	1	1	6	....	....	....
99-2000	**NY Islanders**	**NHL**	**2**	**0**	**1**	**1**	**2**	**0**	**0**	**0**	**4**	**0.0**	**3**	**0**	**0.0**	**2**	**1**	**21:28**	....	....	....	....	....	....	....	....
	Chicago Wolves	IHL	68	5	26	31	99	....	....	....	....	....	....	....	....	....	....	....	16	1	4	5	16	....	....	....
2000-01	**Calgary**	**NHL**	**17**	**0**	**1**	**1**	**11**	**0**	**0**	**0**	**4**	**0.0**	**–1**	**0**	**0.0**	**3**	**11**	**12:18**	....	....	....	....	....	....	....	....
	Chicago Wolves	IHL	64	3	16	19	49	....	....	....	....	....	....	....	....	....	....	....	14	0	0	0	24	....	....	....
2001-02	**Calgary**	**NHL**	**3**	**0**	**0**	**0**	**4**	**0**	**0**	**0**	**0**	**0.0**	**1**	**0**	**0.0**	**3**	**5**	**14:21**	....	....	....	....	....	....	....	....
	Chicago Wolves	AHL	54	2	15	17	58	....	....	....	....	....	....	....	....	....	....	....	25	0	6	6	53	....	....	....
	NHL Totals		**120**	**0**	**9**	**9**	**208**	**0**	**0**	**0**	**59**	**0.0**		**0**	**0.0**	**28**	**27**	**14:47**	**5**	**0**	**0**	**0**	**4**	**0**	**0**	**0**

IHL Second All-Star Team (2000)

Signed as a free agent by **Winnipeg**, October 17, 1989. Signed as a free agent by **Florida**, July 8, 1993. Traded to **St. Louis** by **Florida** for St. Louis' 4th round choice (Ivan Novoseltsev) in 1997 Entry Draft, September 28, 1995. Claimed on waivers by **Winnipeg** from **St. Louis**, March 20, 1996. Transferred to **Phoenix** after **Winnipeg** franchise relocated, July 1, 1996. Traded to **NY Rangers** by **Phoenix** with Mike Eastwood for Jayson More, February 6, 1997. Signed as a free agent by **Florida**, July 30, 1997. Signed as a free agent by **Toronto**, July 28, 1998. Signed as a free agent by **NY Islanders**, August 12, 1999. Traded to **Chicago** by **NY Islanders** for future considerations, March 3, 2000. Signed as a free agent by **Calgary**, July 27, 2000. Signed as a free agent by **Atlanta**, July 23, 2002.

EASTWOOD, Mike

(EEST-wuhd, MIGHK) **ST.L.**

Center. Shoots right. 6'3", 213 lbs. Born, Ottawa, Ont., July 1, 1967. Toronto's 5th choice, 91st overall, in 1987 Entry Draft.

Season	Club	League	GP	G	A	Pts	PIM	PP	SH	GW	S	%	+/-	TF	F%	H	SB	Min	GP	G	A	Pts	PIM	PP	SH	GW
1984-85	Nepean Raiders	OCJHL	46	10	13	23	18	....	....	....	....	....	....	....	....	....	....	....	....	....	....	....	....	....	....	....
1985-86	Nepean Raiders	OCJHL	7	4	2	6	6	....	....	....	....	....	....	....	....	....	....	....	....	....	....	....	....	....	....	....
1986-87	Pembroke	OCJHL	54	58	45	103	62	....	....	....	....	....	....	....	....	....	....	....	23	36	11	47	32	....	....	....
1987-88	West-Michigan	CCHA	42	5	8	13	14	....	....	....	....	....	....	....	....	....	....	....	....	....	....	....	....	....	....	....
1988-89	West-Michigan	CCHA	40	10	13	23	87	....	....	....	....	....	....	....	....	....	....	....	....	....	....	....	....	....	....	....
1989-90	West-Michigan	CCHA	40	25	27	52	36	....	....	....	....	....	....	....	....	....	....	....	....	....	....	....	....	....	....	....
1990-91	West-Michigan	CCHA	42	29	32	61	84	....	....	....	....	....	....	....	....	....	....	....	....	....	....	....	....	....	....	....

			Regular Season																Playoffs							
Season	Club	League	GP	G	A	Pts	PIM	PP	SH	GW	S	%	+/-	TF	F%	H	SB	Min	GP	G	A	Pts	PIM	PP	SH	GW
1991-92	**Toronto**	**NHL**	**9**	**0**	**2**	**2**	**4**	**0**	**0**	**0**	**6**	**0.0**	**–4**	....	....	....	....	....	....	....	....	....	....	....	....	....
	St. John's	AHL	61	18	25	43	28	....	....	....	....	....	....	....	....	....	....	....	16	9	10	19	16	....	....	....
1992-93	**Toronto**	**NHL**	**12**	**1**	**6**	**7**	**21**	**0**	**0**	**0**	**11**	**9.1**	**–2**	....	....	....	....	....	**10**	**1**	**2**	**3**	**8**	**0**	**0**	**0**
	St. John's	AHL	60	24	35	59	32	....	....	....	....	....	....	....	....	....	....	....	....	....	....	....	....	....	....	....
1993-94	**Toronto**	**NHL**	**54**	**8**	**10**	**18**	**28**	**1**	**0**	**2**	**41**	**19.5**	**2**	....	....	....	....	....	**18**	**3**	**2**	**5**	**12**	**1**	**0**	**1**
1994-95	**Toronto**	**NHL**	**36**	**5**	**5**	**10**	**32**	**0**	**0**	**0**	**38**	**13.2**	**–12**	....	....	....	....	....	....	....	....	....	....	....	....	....
	Winnipeg	**NHL**	**13**	**3**	**6**	**9**	**4**	**0**	**0**	**0**	**17**	**17.6**	**3**	....	....	....	....	....	....	....	....	....	....	....	....	....
1995-96	**Winnipeg**	**NHL**	**80**	**14**	**14**	**28**	**20**	**2**	**0**	**3**	**94**	**14.9**	**–14**	....	....	....	....	....	**6**	**0**	**1**	**1**	**2**	**0**	**0**	**0**
1996-97	**Phoenix**	**NHL**	**33**	**1**	**3**	**4**	**4**	**0**	**0**	**0**	**22**	**4.5**	**–3**	....	....	....	....	....	....	....	....	....	....	....	....	....
	NY Rangers	**NHL**	**27**	**1**	**7**	**8**	**10**	**0**	**0**	**0**	**22**	**4.5**	**2**	....	....	....	....	....	**15**	**1**	**2**	**3**	**22**	**0**	**0**	**0**
1997-98	**NY Rangers**	**NHL**	**48**	**5**	**5**	**10**	**16**	**0**	**0**	**0**	**34**	**14.7**	**–2**	....	....	....	....	....	....	....	....	....	....	....	....	....
	St. Louis	**NHL**	**10**	**1**	**0**	**1**	**6**	**0**	**0**	**1**	**4**	**25.0**	**0**	....	....	....	....	....	**3**	**1**	**0**	**1**	**0**	**0**	**0**	**1**
1998-99	**St. Louis**	**NHL**	**82**	**9**	**21**	**30**	**36**	**0**	**0**	**0**	**76**	**11.8**	**6**	**1235**	**56.6**	**40**	**35**	**14:59**	**13**	**1**	**1**	**2**	**6**	**0**	**0**	**0**
99-2000	**St. Louis**	**NHL**	**79**	**19**	**15**	**34**	**32**	**1**	**3**	**3**	**83**	**22.9**	**5**	**872**	**52.2**	**43**	**37**	**15:08**	**7**	**1**	**1**	**2**	**6**	**0**	**0**	**0**
2000-01	**St. Louis**	**NHL**	**77**	**6**	**17**	**23**	**28**	**0**	**2**	**1**	**51**	**11.8**	**4**	**1230**	**53.4**	**41**	**35**	**14:08**	**15**	**0**	**2**	**2**	**2**	**0**	**0**	**0**
2001-02	**St. Louis**	**NHL**	**71**	**7**	**10**	**17**	**41**	**0**	**0**	**2**	**60**	**11.7**	**–2**	**1110**	**53.8**	**44**	**45**	**12:55**	**10**	**0**	**0**	**0**	**6**	**0**	**0**	**0**
	NHL Totals		**631**	**80**	**121**	**201**	**282**	**4**	**5**	**12**	**559**	**14.3**		**4447**	**54.1**	**168**	**152**	**14:20**	**97**	**8**	**11**	**19**	**64**	**1**	**0**	**2**

CCHA Second All-Star Team (1991)

Traded to **Winnipeg** by **Toronto** with Toronto's 3rd round choice (Brad Isbister) in 1995 Entry Draft for Tie Domi, April 7, 1995. Transferred to **Phoenix** after **Winnipeg** franchise relocated, July 1, 1996. Traded to **NY Rangers** by **Phoenix** with Dallas Eakins for Jayson More, February 6, 1997. Traded to **St. Louis** by **NY Rangers** for Harry York, March 24, 1998.

EATON, Mark

(EE-tohn, MAHRK) **NSH.**

Defense. Shoots left. 6'2", 205 lbs. Born, Wilmington, DE, May 6, 1977.

Season	Club	League	GP	G	A	Pts	PIM	PP	SH	GW	S	%	+/-	TF	F%	H	SB	Min	GP	G	A	Pts	PIM	PP	SH	GW
1995-96	Waterloo	USHL	50	4	21	25	....	....	....	....	....	....	....	....	....	....	....	....	....	....	....	....	....	....	....	....
1996-97	Waterloo	USHL	50	6	32	38	62	....	....	....	....	....	....	....	....	....	....	....	....	....	....	....	....	....	....	....
1997-98	U. of Notre Dame	CCHA	41	12	17	29	32	....	....	....	....	....	....	....	....	....	....	....	....	....	....	....	....	....	....	....
1998-99	Philadelphia	AHL	74	9	27	36	38	....	....	....	....	....	....	....	....	....	....	....	16	4	8	12	0	....	....	....
99-2000	**Philadelphia**	**NHL**	**27**	**1**	**1**	**2**	**8**	**0**	**0**	**1**	**25**	**4.0**	**1**	**0**	**0.0**	**21**	**35**	**18:17**	**7**	**0**	**0**	**0**	**0**	**0**	**0**	**0**
	Philadelphia	AHL	47	9	17	26	6	....	....	....	....	....	....	....	....	....	....	....	....	....	....	....	....	....	....	....
2000-01	**Nashville**	**NHL**	**34**	**3**	**8**	**11**	**14**	**1**	**0**	**1**	**32**	**9.4**	**7**	**0**	**0.0**	**29**	**23**	**17:13**	....	....	....	....	....	....	....	....
	Milwaukee	IHL	34	3	12	15	27	....	....	....	....	....	....	....	....	....	....	....	....	....	....	....	....	....	....	....
2001-02	**Nashville**	**NHL**	**58**	**3**	**5**	**8**	**24**	**0**	**0**	**0**	**52**	**5.8**	**–12**	**0**	**0.0**	**59**	**59**	**17:12**	....	....	....	....	....	....	....	....
	NHL Totals		**119**	**7**	**14**	**21**	**46**	**1**	**0**	**2**	**109**	**6.4**		**0**	**0.0**	**109**	**117**	**17:27**	**7**	**0**	**0**	**0**	**0**	**0**	**0**	**0**

Won Curt Hammer Award (Most Gentlemanly Player - USHL) (1997) • USHL Second All-Star Team (1997) • CCHA Rookie of the Year (1998)

Signed as a free agent by **Philadelphia**, August 4, 1998. Traded to **Nashville** by **Philadelphia** for Detroit's 3rd round choice (previously acquired, Philadelphia selected Patrick Sharp) in 2001 Entry Draft, September 29, 2000.

EKMAN, Nils

(EHK-mahn, NIHLS) **NYR**

Left wing. Shoots left. 5'11", 185 lbs. Born, Stockholm, Sweden, March 11, 1976. Calgary's 6th choice, 107th overall, in 1994 Entry Draft.

Season	Club	League	GP	G	A	Pts	PIM	PP	SH	GW	S	%	+/-	TF	F%	H	SB	Min	GP	G	A	Pts	PIM	PP	SH	GW
1993-94	Hammarby Jr.	Swede-Jr.	11	4	5	9	14	....	....	....	....	....	....	....	....	....	....	....	....	....	....	....	....	....	....	....
	Hammarby	Swede-2	18	7	2	9	4	....	....	....	....	....	....	....	....	....	....	....	....	....	....	....	....	....	....	....
1994-95	Hammarby Jr.	Swede-Jr.	2	2	1	3	0	....	....	....	....	....	....	....	....	....	....	....	....	....	....	....	....	....	....	....
	Hammarby	Swede-2	32	10	8	18	18	....	....	....	....	....	....	....	....	....	....	....	....	....	....	....	....	....	....	....
1995-96	Hammarby	Swede-2	26	9	7	16	53	....	....	....	....	....	....	....	....	....	....	....	1	0	0	0	0	....	....	....
1996-97	Kiekko Espoo	Finland	50	24	19	43	60	....	....	....	....	....	....	....	....	....	....	....	4	2	0	2	4	....	....	....
1997-98	Kiekko Espoo	Finland	43	14	14	28	86	....	....	....	....	....	....	....	....	....	....	....	7	2	2	4	27	....	....	....
	Saint John	AHL	....	....	....	....	....	....	....	....	....	....	....	....	....	....	....	....	1	0	0	0	2	....	....	....
1998-99	Blues Espoo	Finland	52	20	14	34	96	....	....	....	....	....	....	....	....	....	....	....	3	1	1	2	6	....	....	....
99-2000	Detroit Vipers	IHL	10	7	2	9	8	....	....	....	....	....	....	....	....	....	....	....	....	....	....	....	....	....	....	....
	Tampa Bay	**NHL**	**28**	**2**	**2**	**4**	**36**	**1**	**0**	**0**	**42**	**4.8**	**–8**	**3**	**0.0**	**23**	**8**	**11:12**	....	....	....	....	....	....	....	....
	Long Beach	IHL	27	11	12	23	26	....	....	....	....	....	....	....	....	....	....	....	5	3	3	6	4	....	....	....
2000-01	**Tampa Bay**	**NHL**	**43**	**9**	**11**	**20**	**40**	**2**	**1**	**1**	**72**	**12.5**	**–15**	**16**	**37.5**	**21**	**13**	**15:45**	....	....	....	....	....	....	....	....
	Detroit Vipers	IHL	33	22	14	36	63	....	....	....	....	....	....	....	....	....	....	....	....	....	....	....	....	....	....	....
2001-02	Djurgarden	Sweden	38	16	15	31	57	....	....	....	....	....	....	....	....	....	....	....	4	1	0	1	32	....	....	....
	NHL Totals		**71**	**11**	**13**	**24**	**76**	**3**	**1**	**1**	**114**	**9.6**		**19**	**31.6**	**44**	**21**	**13:57**	....	....	....	....	....	....	....	....

Won Garry F. Longman Memorial Trophy (Top Rookie - IHL) (2000)

Traded to **Tampa Bay** by **Calgary** with Calgary's 4th round choice (later traded to NY Islanders - NY Islanders selected Vladimir Gorbunov) in 2000 Entry Draft for Andreas Johansson, November 20, 1999. Traded to **NY Rangers** by **Tampa Bay** with Kyle Freadrich for Tim Taylor, June 30, 2001.

ELIAS, Patrik

(ehl-EE-ahsh, PA-trihk) **N.J.**

Center. Shoots left. 6'1", 195 lbs. Born, Trebic, Czech., April 13, 1976. New Jersey's 2nd choice, 51st overall, in 1994 Entry Draft.

Season	Club	League	GP	G	A	Pts	PIM	PP	SH	GW	S	%	+/-	TF	F%	H	SB	Min	GP	G	A	Pts	PIM	PP	SH	GW
1992-93	Poldi Kladno	Czech	2	0	0	0	....	....	....	....	....	....	....	....	....	....	....	....	....	....	....	....	....	....	....	....
1993-94	HC Kladno	Czech	15	1	2	3	....	....	....	....	....	....	....	....	....	....	....	....	11	2	2	4	....	....	....	....
1994-95	HC Kladno	Czech	28	4	3	7	37	....	....	....	....	....	....	....	....	....	....	....	7	1	2	3	12	....	....	....
1995-96	**New Jersey**	**NHL**	**1**	**0**	**0**	**0**	**0**	**0**	**0**	**0**	**2**	**0.0**	**–1**	....	....	....	....	....	....	....	....	....	....	....	....	....
	Albany	AHL	74	27	36	63	83	....	....	....	....	....	....	....	....	....	....	....	4	1	1	2	2	....	....	....
1996-97	**New Jersey**	**NHL**	**17**	**2**	**3**	**5**	**2**	**0**	**0**	**0**	**23**	**8.7**	**–4**	....	....	....	....	....	**8**	**2**	**3**	**5**	**4**	**1**	**0**	**0**
	Albany	AHL	57	24	43	67	76	....	....	....	....	....	....	....	....	....	....	....	6	1	2	3	8	....	....	....
1997-98	**New Jersey**	**NHL**	**74**	**18**	**19**	**37**	**28**	**5**	**0**	**6**	**147**	**12.2**	**18**	....	....	....	....	....	**4**	**0**	**1**	**1**	**0**	**0**	**0**	**0**
	Albany	AHL	3	3	0	3	2	....	....	....	....	....	....	....	....	....	....	....	....	....	....	....	....	....	....	....
1998-99	**New Jersey**	**NHL**	**74**	**17**	**33**	**50**	**34**	**3**	**0**	**2**	**157**	**10.8**	**19**	**99**	**38.4**	**86**	**13**	**15:50**	**7**	**0**	**5**	**5**	**6**	**0**	**0**	**0**
99-2000	Trebic	Czech-2	2	2	1	3	2	....	....	....	....	....	....	....	....	....	....	....	....	....	....	....	....	....	....	....
	Pardubice	Czech	5	1	4	5	31	....	....	....	....	....	....	....	....	....	....	....	....	....	....	....	....	....	....	....
	♦ **New Jersey**	**NHL**	**72**	**35**	**37**	**72**	**58**	**9**	**0**	**9**	**183**	**19.1**	**16**	**134**	**45.5**	**112**	**12**	**17:28**	**23**	**7**	***13**	**20**	**9**	**2**	**1**	**1**
2000-01	**New Jersey**	**NHL**	**82**	**40**	**56**	**96**	**51**	**8**	**3**	**6**	**220**	**18.2**	**45**	**155**	**41.3**	**85**	**20**	**18:44**	**25**	**9**	**14**	**23**	**10**	**3**	**1**	**2**
2001-02	**New Jersey**	**NHL**	**75**	**29**	**32**	**61**	**36**	**8**	**1**	**8**	**199**	**14.6**	**4**	**128**	**45.3**	**70**	**17**	**18:57**	**6**	**2**	**4**	**6**	**6**	**2**	**0**	**0**
	Czech Republic	Olympics	4	1	1	2	0	....	....	....	....	....	....	....	....	....	....	....	....	....	....	....	....	....	....	....
	NHL Totals		**395**	**141**	**180**	**321**	**209**	**33**	**4**	**31**	**931**	**15.1**		**516**	**42.8**	**353**	**62**	**17:46**	**73**	**20**	**40**	**60**	**35**	**8**	**2**	**3**

NHL All-Rookie Team (1998) • NHL First All-Star Team (2001) • Played in NHL All-Star Game (2000, 2002)

ELICH, Matt

(EHL-ihch, MAT) **T.B.**

Right wing. Shoots right. 6'3", 196 lbs. Born, Detroit, MI, September 22, 1979. Tampa Bay's 3rd choice, 61st overall, in 1997 Entry Draft.

Season	Club	League	GP	G	A	Pts	PIM	PP	SH	GW	S	%	+/-	TF	F%	H	SB	Min	GP	G	A	Pts	PIM	PP	SH	GW
1993-94	Det. Caesars	MNHL	40	20	20	40	110	....	....	....	....	....	....	....	....	....	....	....	....	....	....	....	....	....	....	....
1994-95	Det. Caesars	MNHL	45	31	22	53	170	....	....	....	....	....	....	....	....	....	....	....	....	....	....	....	....	....	....	....
1995-96	Windsor	OHL	52	10	2	12	17	....	....	....	....	....	....	....	....	....	....	....	5	1	0	1	2	....	....	....
1996-97	Windsor	OHL	58	15	13	28	19	....	....	....	....	....	....	....	....	....	....	....	5	0	1	1	6	....	....	....
1997-98	Windsor	OHL	20	9	12	21	8	....	....	....	....	....	....	....	....	....	....	....	....	....	....	....	....	....	....	....
	Kingston	OHL	34	14	4	18	2	....	....	....	....	....	....	....	....	....	....	....	12	2	4	6	2	....	....	....
1998-99	Kingston	OHL	67	44	30	74	32	....	....	....	....	....	....	....	....	....	....	....	5	3	5	8	0	....	....	....
99-2000	**Tampa Bay**	**NHL**	**8**	**1**	**1**	**2**	**0**	**0**	**0**	**0**	**5**	**20.0**	**–1**	**0**	**0.0**	**7**	**0**	**6:07**	....	....	....	....	....	....	....	....
	Detroit Vipers	IHL	48	12	4	16	12	....	....	....	....	....	....	....	....	....	....	....	....	....	....	....	....	....	....	....
2000-01	**Tampa Bay**	**NHL**	**8**	**0**	**0**	**0**	**0**	**0**	**0**	**0**	**7**	**0.0**	**–5**	**0**	**0.0**	**1**	**2**	**8:59**	....	....	....	....	....	....	....	....
	Detroit Vipers	IHL	60	12	16	28	12	....	....	....	....	....	....	....	....	....	....	....	....	....	....	....	....	....	....	....
2001-02	Springfield	AHL	22	2	2	4	4	....	....	....	....	....	....	....	....	....	....	....	....	....	....	....	....	....	....	....
	NHL Totals		**16**	**1**	**1**	**2**	**0**	**0**	**0**	**0**	**12**	**8.3**		**0**	**0.0**	**8**	**2**	**7:33**	....	....	....	....	....	....	....	....

ELOMO, Miika

(eh-LOH-moh, MEE-ka)

Left wing. Shoots left. 6', 200 lbs. Born, Turku, Finland, April 21, 1977. Washington's 2nd choice, 23rd overall, in 1995 Entry Draft.

Season	Club	League	GP	G	A	Pts	PIM	PP	SH	GW	S	%	+/-	TF	F%	H	SB	Min	GP	G	A	Pts	PIM	PP	SH	GW
1993-94	TPS Turku Jr.	Finn-Jr.	30	8	5	13	24	....	....	....	....	....	....	....	....	....	....	....	5	1	1	2	2	....	....	....
1994-95	TPS Turku Jr.	Finn-Jr.	14	3	8	11	24	....	....	....	....	....	....	....	....	....	....	....	....	....	....	....	....	....	....	....
	Kiekko-67 Turku	Finland-2	14	9	2	11	39	....	....	....	....	....	....	....	....	....	....	....	....	....	....	....	....	....	....	....
1995-96	TPS Turku Jr.	Finn-Jr.	6	0	2	2	18	....	....	....	....	....	....	....	....	....	....	....	....	....	....	....	....	....	....	....
	Kiekko-67 Turku	Finland-2	21	9	6	15	100	....	....	....	....	....	....	....	....	....	....	....	....	....	....	....	....	....	....	....
	TPS Turku	Finland	10	1	1	2	8	....	....	....	....	....	....	....	....	....	....	....	3	0	0	0	2	....	....	....

			Regular Season																Playoffs							
Season	Club	League	GP	G	A	Pts	PIM	PP	SH	GW	S	%	+/-	TF	F%	H	SB	Min	GP	G	A	Pts	PIM	PP	SH	GW
1996-97	Portland Pirates	AHL	52	8	9	17	37																			
1997-98	Portland Pirates	AHL	33	1	1	2	54																			
	HIFK Helsinki	Finland	16	4	1	5	6												9	4	3	7	6			
1998-99	TPS Turku	Finland	36	5	10	15	76												10	3	5	8	6			
99-2000	**Washington**	**NHL**	**2**	**0**	**1**	**1**	**2**	**0**	**0**	**0**	**3**	**0.0**	**1**	**4**	**100.0**	**5**	**0**	**11:12**								
	Portland Pirates	AHL	59	21	14	35	50																			
2000-01	Saint John	AHL	72	10	21	31	109												6	0	2	2	12			
2001-02	TPS Turku	Finland	24	3	4	7	102																			
	Blues Espoo	Finland	28	2	2	4	44												3	0	0	0	4			
	NHL Totals		**2**	**0**	**1**	**1**	**2**	**0**	**0**	**0**	**3**	**0.0**		**4**	**100.0**	**5**	**0**	**11:12**								

Traded to **Calgary** by **Washington** with Buffalo's compensatory 4th round choice (previously acquired, Calgary selected Levente Szuper) in 2000 Entry Draft for Anaheim's 2nd round choice (previously acquired, Washington selected Matt Pettinger) in 2000 Entry Draft, June 24, 2000.

ELORANTA, Mikko

(ehl-oh-RAN-tuh, MEE-koh) **L.A.**

Left wing. Shoots left. 6', 190 lbs. Born, Turku, Finland, August 24, 1972. Boston's 9th choice, 247th overall, in 1999 Entry Draft.

Season	Club	League	GP	G	A	Pts	PIM	PP	SH	GW	S	%	+/-	TF	F%	H	SB	Min	GP	G	A	Pts	PIM	PP	SH	GW
1989-90	TPS Turku Jr.	Finn-Jr.	2	0	0	0	0																			
1990-91	TPS Turku Jr.	Finn-Jr.	35	8	8	16	18																			
1991-92	TPS Turku Jr.	Finn-Jr.	19	3	1	4	8												8	0	0	0	0			
1992-93	TPS Turku Jr.	Finn-Jr.	31	11	6	17	20												6	0	4	4	6			
1993-94	Kiekko-67 Turku	Finland-2	45	3	4	7	24																			
1994-95	Kiekko-67 Turku	Finland-2	47	18	14	32	52												3	3	0	3	4			
1995-96	Kiekko-67 Turku	Finland-2	8	6	7	13	2																			
	Ilves Tampere	Finland	43	18	15	33	86												3	0	2	2	2			
1996-97	TPS Turku	EuroHL	6	3	1	4	6												1	0	0	0	0			
	TPS Turku	Finland	31	6	15	21	52												10	5	2	7	6			
1997-98	TPS Turku	EuroHL	3	1	0	1	12																			
	TPS Turku	Finland	46	23	14	37	82												2	0	0	0	8			
1998-99	TPS Turku	Finland	52	19	21	40	103												10	1	6	7	26			
99-2000	**Boston**	**NHL**	**50**	**6**	**12**	**18**	**36**	**1**	**0**	**0**	**59**	**10.2**	**−10**	**77**	**35.1**	**74**	**10**	**12:18**								
2000-01	**Boston**	**NHL**	**62**	**12**	**11**	**23**	**38**	**1**	**1**	**2**	**89**	**13.5**	**2**	**82**	**23.2**	**75**	**14**	**10:26**								
2001-02	**Boston**	**NHL**	**6**	**0**	**0**	**0**	**2**	**0**	**0**	**0**	**15**	**0.0**	**−1**	**8**	**25.0**	**8**	**3**	**16:12**								
	Los Angeles	**NHL**	**71**	**9**	**9**	**18**	**54**	**1**	**0**	**2**	**121**	**7.4**	**0**	**1**	**100.0**	**83**	**13**	**11:27**	**7**	**1**	**1**	**2**	**2**	**0**	**0**	**0**
	Finland	Olympics	4	2	0	2	2																			
	NHL Totals		**189**	**27**	**32**	**59**	**130**	**3**	**1**	**4**	**284**	**9.5**		**168**	**29.2**	**240**	**40**	**11:30**	**7**	**1**	**1**	**2**	**2**	**0**	**0**	**0**

Traded to **LA Kings** by **Boston** with Jason Allison for Jozef Stumpel and Glen Murray, October 24, 2001.

EMERSON, Nelson

(EH-muhr-SOHN, NEHL-sohn)

Right wing. Shoots right. 5'11", 180 lbs. Born, Hamilton, Ont., August 17, 1967. St. Louis' 2nd choice, 44th overall, in 1985 Entry Draft.

Season	Club	League	GP	G	A	Pts	PIM	PP	SH	GW	S	%	+/-	TF	F%	H	SB	Min	GP	G	A	Pts	PIM	PP	SH	GW
1984-85	Stratford	OJHL-B	40	23	38	61	70																			
1985-86	Stratford	OJHL-B	39	*54	58	*112	91																			
1986-87	Bowling Green	CCHA	45	26	35	61	28																			
1987-88	Bowling Green	CCHA	45	34	49	83	54																			
1988-89	Bowling Green	CCHA	44	22	46	68	46																			
1989-90	Bowling Green	CCHA	44	30	52	82	42																			
	Peoria Rivermen	IHL	3	1	1	2	0																			
1990-91	**St. Louis**	**NHL**	**4**	**0**	**3**	**3**	**2**	**0**	**0**	**0**	**3**	**0.0**	**−2**													
	Peoria Rivermen	IHL	73	36	79	115	91												17	9	12	21	16			
1991-92	**St. Louis**	**NHL**	**79**	**23**	**36**	**59**	**66**	**3**	**0**	**2**	**143**	**16.1**	**−5**						**6**	**3**	**3**	**6**	**21**	**2**	**0**	**0**
1992-93	**St. Louis**	**NHL**	**82**	**22**	**51**	**73**	**62**	**5**	**2**	**4**	**196**	**11.2**	**2**						**11**	**1**	**6**	**7**	**6**	**0**	**0**	**0**
1993-94	**Winnipeg**	**NHL**	**83**	**33**	**41**	**74**	**80**	**4**	**5**	**6**	**282**	**11.7**	**−38**													
1994-95	**Winnipeg**	**NHL**	**48**	**14**	**23**	**37**	**26**	**4**	**1**	**1**	**122**	**11.5**	**−12**													
1995-96	**Hartford**	**NHL**	**81**	**29**	**29**	**58**	**78**	**12**	**2**	**5**	**247**	**11.7**	**−7**													
1996-97	**Hartford**	**NHL**	**66**	**9**	**29**	**38**	**34**	**2**	**1**	**2**	**194**	**4.6**	**−21**													
1997-98	**Carolina**	**NHL**	**81**	**21**	**24**	**45**	**50**	**6**	**0**	**4**	**203**	**10.3**	**−17**													
1998-99	**Carolina**	**NHL**	**35**	**8**	**13**	**21**	**36**	**3**	**0**	**0**	**84**	**9.5**	**1**	**7**	**42.9**	**6**	**7**	**14:30**								
	Chicago	**NHL**	**27**	**4**	**10**	**14**	**13**	**0**	**0**	**1**	**94**	**4.3**	**8**	**169**	**41.4**	**8**	**9**	**19:37**								
	Ottawa	**NHL**	**3**	**1**	**1**	**2**	**2**	**0**	**0**	**0**	**10**	**10.0**	**−1**	**5**	**60.0**	**1**	**0**	**17:05**	**4**	**1**	**3**	**4**	**0**	**0**	**0**	**0**
99-2000	**Atlanta**	**NHL**	**58**	**14**	**19**	**33**	**47**	**4**	**0**	**0**	**183**	**7.7**	**−24**	**124**	**39.5**	**27**	**33**	**19:10**								
	Los Angeles	**NHL**	**5**	**1**	**1**	**2**	**0**	**0**	**0**	**0**	**13**	**7.7**	**1**	**0**	**0.0**	**2**	**0**	**14:39**	**1**	**0**	**0**	**0**	**0**	**0**	**0**	**0**
2000-01	**Los Angeles**	**NHL**	**78**	**11**	**11**	**22**	**54**	**0**	**0**	**1**	**157**	**7.0**	**−13**	**28**	**32.1**	**37**	**30**	**13:51**	**13**	**2**	**2**	**4**	**4**	**0**	**0**	**0**
2001-02	**Los Angeles**	**NHL**	**41**	**5**	**2**	**7**	**25**	**0**	**0**	**1**	**40**	**12.5**	**−8**	**24**	**29.2**	**10**	**9**	**9:12**	**5**	**0**	**1**	**1**	**2**	**0**	**0**	**0**
	NHL Totals		**771**	**195**	**293**	**488**	**575**	**43**	**11**	**27**	**1971**	**9.9**		**357**	**39.5**	**91**	**88**	**15:06**	**40**	**7**	**15**	**22**	**33**	**2**	**0**	**0**

CCHA Rookie of the Year (1987) • CCHA First All-Star Team (1988, 1990) • NCAA West Second All-American Team (1988) • CCHA Second All-Star Team (1989) • NCAA West First All-American Team (1990) • IHL First All-Star Team (1991) • Won Garry F. Longman Memorial Trophy (Top Rookie - IHL) (1991)

Traded to **Winnipeg** by **St. Louis** with Stephane Quintal for Phil Housley, September 24, 1993. Traded to **Hartford** by **Winnipeg** for Darren Turcotte, October 6, 1995. Transferred to **Carolina** after **Hartford** franchise relocated, June 25, 1997. Traded to **Chicago** by **Carolina** for Paul Coffey, December 29, 1998. Traded to **Ottawa** by **Chicago** for Chris Murray, March 23, 1999. Signed as a free agent by **Atlanta**, August 3, 1999. Traded to **Los Angeles** by **Atlanta** with Kelly Buchberger for Donald Audette and Frantisek Kaberle, March 13, 2000.

EMMA, David

(EH-muh, DAY-vihd)

Center. Shoots left. 5'10", 185 lbs. Born, Cranston, RI, January 14, 1969. New Jersey's 6th choice, 110th overall, in 1989 Entry Draft.

Season	Club	League	GP	G	A	Pts	PIM	PP	SH	GW	S	%	+/-	TF	F%	H	SB	Min	GP	G	A	Pts	PIM	PP	SH	GW
1987-88	Boston College	H-East	30	19	16	35	30																			
1988-89	Boston College	H-East	36	20	31	51	36																			
1989-90	Boston College	H-East	42	38	34	*72	46																			
1990-91	Boston College	H-East	39	*35	46	*81	44																			
1991-92	Team USA	Nat-Tm	55	15	16	31	32																			
	United States	Olympics	6	0	1	1	6																			
	Utica Devils	AHL	15	4	7	11	12												4	1	1	2	2			
1992-93	**New Jersey**	**NHL**	**2**	**0**	**0**	**0**	**0**	**0**	**0**	**0**	**2**	**0.0**	**0**													
	Utica Devils	AHL	61	21	40	61	47												5	2	1	3	6			
1993-94	**New Jersey**	**NHL**	**15**	**5**	**5**	**10**	**2**	**1**	**0**	**2**	**24**	**20.8**	**0**													
	Albany	AHL	56	26	29	55	53												5	1	2	3	8			
1994-95	**New Jersey**	**NHL**	**6**	**0**	**1**	**1**	**0**	**0**	**0**	**0**	**4**	**0.0**	**−2**													
	Albany	AHL	1	0	0	0	0																			
1995-96	Detroit Vipers	IHL	79	30	32	62	75												11	5	2	7	2			
1996-97	**Boston**	**NHL**	**5**	**0**	**0**	**0**	**0**	**0**	**0**	**0**	**3**	**0.0**	**−1**													
	Providence	AHL	53	10	18	28	24																			
	Phoenix	IHL	8	0	4	4	4																			
1997-98	Klagenfurter AC	Alpenliga	16	6	17	23	....																			
	Klagenfurter AC	Austria	33	22	22	44	48																			
1998-99	Klagenfurter AC	Alpenliga	26	15	32	47	49																			
	Klagenfurter AC	Austria	15	8	7	15	16																			
99-2000	Klagenfurter AC	IEL	32	26	28	54	28																			
	Klagenfurter AC	Austria	15	9	6	15	18																			
2000-01	**Florida**	**NHL**	**6**	**0**	**0**	**0**	**0**	**0**	**0**	**0**	**6**	**0.0**	**−1**	**1**	**0.0**	**2**	**4**	**7:23**								
	Louisville	AHL	55	22	28	50	63																			
	Portland Pirates	AHL	16	2	8	10	6												2	0	0	0	0			
2001-02	Nurnberg	Germany	DID NOT PLAY – INJURED																							
	NHL Totals		**34**	**5**	**6**	**11**	**2**	**1**	**0**	**2**	**39**	**12.8**		**1**	**0.0**	**2**	**4**	**7:23**								

Hockey East Second All-Star Team (1989) • Hockey East First All-Star Team (1990, 1991) • NCAA East First All-American Team (1990, 1991) • Hockey East Player of the Year (1991) • Won Hobey Baker Memorial Award (Top U.S. Collegiate Player) (1991)

Signed as a free agent by **Boston**, August 27, 1996. Signed as a free agent by **Florida**, August 1, 2000. Traded to **Washington** by **Florida** for Remi Royer, March 3, 2001. Signed as a free agent by **Nurnberg** (Germany), May 22, 2001. • Missed entire 2001-02 season recovering from back injury originally suffered during off-season training, August 9, 2001.

EMMONS, John

(eh-mohns, JAWN) **BOS.**

Center. Shoots left. 6'1", 203 lbs. Born, San Jose, CA, August 17, 1974. Calgary's 7th choice, 122nd overall, in 1993 Entry Draft.

			Regular Season															Playoffs							
Season	Club	League	GP	G	A	Pts	PIM	PP	SH	GW	S	%	+/-	TF	F%	H	SB	Min	GP	G	A	Pts	PIM	PP	SH	GW
1990-91	New Canaan	Hi-School	20	19	37	56	20	....	....	....	....	....	....	....	....	....	....	....	....	....	....	....	....	....	....	....
1991-92	New Canaan	Hi-School	22	24	49	73	24	....	....	....	....	....	....	....	....	....	....	....	....	....	....	....	....	....	....	....
1992-93	Yale	ECAC	28	3	5	8	66	....	....	....	....	....	....	....	....	....	....	....	....	....	....	....	....	....	....	....
1993-94	Yale	ECAC	25	5	12	17	66	....	....	....	....	....	....	....	....	....	....	....	....	....	....	....	....	....	....	....
1994-95	Yale	ECAC	28	4	16	20	57	....	....	....	....	....	....	....	....	....	....	....	....	....	....	....	....	....	....	....
1995-96	Yale	ECAC	31	8	20	28	124	....	....	....	....	....	....	....	....	....	....	....	....	....	....	....	....	....	....	....
1996-97	Dayton Bombers	ECHL	69	20	37	57	62	....	....	....	....	....	....	....	....	....	....	....	4	0	1	1	2	....	....	....
	Fort Wayne	IHL	1	0	0	0	0	....	....	....	....	....	....	....	....	....	....	....	....	....	....	....	....	....	....	....
1997-98	Michigan K-Wings	IHL	81	9	25	34	85	....	....	....	....	....	....	....	....	....	....	....	4	1	1	2	10	....	....	....
1998-99	Detroit Vipers	IHL	75	13	22	35	172	....	....	....	....	....	....	....	....	....	....	....	11	4	5	9	22	....	....	....
99-2000	**Ottawa**	**NHL**	**10**	**0**	**0**	**0**	**6**	**0**	**0**	**0**	**3**	**0.0**	**-2**	**62**	**58.1**	**11**	**2**	**7:27**	....	....	....	....	....	....	....	....
	Grand Rapids	IHL	64	10	16	26	78	....	....	....	....	....	....	....	....	....	....	....	16	1	4	5	28	....	....	....
2000-01	**Ottawa**	**NHL**	**41**	**1**	**1**	**2**	**20**	**0**	**0**	**0**	**28**	**3.6**	**-5**	**315**	**51.4**	**29**	**21**	**7:46**	....	....	....	....	....	....	....	....
	Grand Rapids	IHL	9	1	0	1	4	....	....	....	....	....	....	....	....	....	....	....	....	....	....	....	....	....	....	....
	Tampa Bay	**NHL**	**12**	**1**	**1**	**2**	**22**	**0**	**0**	**0**	**9**	**11.1**	**0**	**160**	**52.5**	**7**	**6**	**13:07**	....	....	....	....	....	....	....	....
2001-02	**Boston**	**NHL**	**22**	**0**	**2**	**2**	**16**	**0**	**0**	**0**	**20**	**0.0**	**-4**	**153**	**56.9**	**12**	**4**	**8:04**	....	....	....	....	....	....	....	....
	Providence	AHL	46	4	8	12	32	....	....	....	....	....	....	....	....	....	....	....	2	1	1	2	0	....	....	....
	NHL Totals		**85**	**2**	**4**	**6**	**64**	**0**	**0**	**0**	**60**	**3.3**		**690**	**53.5**	**59**	**33**	**8:34**	....	....	....	....	....	....	....	....

Signed as a free agent by **Ottawa**, August 7, 1998. Traded to **Tampa Bay** by **Ottawa** for Craig Millar, March 13, 2001. Signed as a free agent by **Boston**, August 8, 2001.

ENDICOTT, Shane

(ehn-DIH-kawt, SHAYN) **PIT.**

Center. Shoots left. 6'4", 200 lbs. Born, Saskatoon, Sask., December 21, 1981. Pittsburgh's 2nd choice, 52nd overall, in 2000 Entry Draft.

			Regular Season															Playoffs							
Season	Club	League	GP	G	A	Pts	PIM	PP	SH	GW	S	%	+/-	TF	F%	H	SB	Min	GP	G	A	Pts	PIM	PP	SH	GW
1997-98	Sask. Contacts	SMHL	43	31	32	63	42	....	....	....	....	....	....	....	....	....	....	....	....	....	....	....	....	....	....	....
	Seattle	WHL	....	....	....	....	....	....	....	....	....	....	....	....	....	....	....	....	5	0	0	0	0	....	....	....
1998-99	Seattle	WHL	72	13	26	39	27	....	....	....	....	....	....	....	....	....	....	....	11	0	1	1	0	....	....	....
99-2000	Seattle	WHL	70	23	32	55	62	....	....	....	....	....	....	....	....	....	....	....	7	1	6	7	6	....	....	....
2000-01	Seattle	WHL	72	36	43	79	86	....	....	....	....	....	....	....	....	....	....	....	9	4	5	9	12	....	....	....
2001-02	**Pittsburgh**	**NHL**	**4**	**0**	**1**	**1**	**4**	**0**	**0**	**0**	**2**	**0.0**	**-1**	**18**	**33.3**	**1**	**1**	**8:28**	....	....	....	....	....	....	....	....
	Wilkes-Barre	AHL	63	19	20	39	46	....	....	....	....	....	....	....	....	....	....	....	....	....	....	....	....	....	....	....
	NHL Totals		**4**	**0**	**1**	**1**	**4**	**0**	**0**	**0**	**2**	**0.0**		**18**	**33.3**	**1**	**1**	**8:28**	....	....	....	....	....	....	....	....

ERAT, Martin

(EE-rat, mahr-TIHN) **NSH.**

Left wing. Shoots left. 6', 195 lbs. Born, Trebic, Czech., August 28, 1981. Nashville's 12th choice, 191st overall, in 1999 Entry Draft.

			Regular Season															Playoffs							
Season	Club	League	GP	G	A	Pts	PIM	PP	SH	GW	S	%	+/-	TF	F%	H	SB	Min	GP	G	A	Pts	PIM	PP	SH	GW
1997-98	Zlin Jr.	Czech-Jr.	46	35	30	65	....	....	....	....	....	....	....	....	....	....	....	....	....	....	....	....	....	....	....	....
1998-99	Zlin Jr.	Czech-Jr.	35	21	23	44	....	....	....	....	....	....	....	....	....	....	....	....	....	....	....	....	....	....	....	....
	Zlin	Czech	5	0	0	0	2	....	....	....	....	....	....	....	....	....	....	....	....	....	....	....	....	....	....	....
99-2000	Saskatoon Blades	WHL	66	27	26	53	82	....	....	....	....	....	....	....	....	....	....	....	11	4	8	12	16	....	....	....
2000-01	Saskatoon Blades	WHL	31	19	35	54	48	....	....	....	....	....	....	....	....	....	....	....	....	....	....	....	....	....	....	....
	Red Deer Rebels	WHL	17	4	24	28	24	....	....	....	....	....	....	....	....	....	....	....	22	*15	*21	*36	32	....	....	....
2001-02	**Nashville**	**NHL**	**80**	**9**	**24**	**33**	**32**	**2**	**0**	**2**	**84**	**10.7**	**-11**	**3**	**66.7**	**61**	**11**	**13:10**	....	....	....	....	....	....	....	....
	NHL Totals		**80**	**9**	**24**	**33**	**32**	**2**	**0**	**2**	**84**	**10.7**		**3**	**66.7**	**61**	**11**	**13:10**	....	....	....	....	....	....	....	....

Traded to **Red Deer** (WHL) by **Saskatoon** (WHL) with Darcy Robinson and Cam Ondik for Michael Garnett, Justin Wallin, Martin Vymazzi and future considerations, January 11, 2001.

ERIKSSON, Anders

(AIR-ihk-suhn, AND-uhrs) **TOR.**

Defense. Shoots left. 6'2", 220 lbs. Born, Bollnas, Sweden, January 9, 1975. Detroit's 1st choice, 22nd overall, in 1993 Entry Draft.

			Regular Season															Playoffs							
Season	Club	League	GP	G	A	Pts	PIM	PP	SH	GW	S	%	+/-	TF	F%	H	SB	Min	GP	G	A	Pts	PIM	PP	SH	GW
1992-93	MoDo Jr.	Swede-Jr.	10	5	3	8	14	....	....	....	....	....	....	....	....	....	....	....	....	....	....	....	....	....	....	....
	MoDo	Sweden	20	0	2	2	2	....	....	....	....	....	....	....	....	....	....	....	1	0	0	0	0	....	....	....
1993-94	MoDo Jr.	Swede-Jr.	3	1	2	3	34	....	....	....	....	....	....	....	....	....	....	....	....	....	....	....	....	....	....	....
	MoDo	Sweden	38	2	8	10	42	....	....	....	....	....	....	....	....	....	....	....	11	0	0	0	8	....	....	....
1994-95	MoDo	Sweden	39	3	6	9	54	....	....	....	....	....	....	....	....	....	....	....	....	....	....	....	....	....	....	....
1995-96	**Detroit**	**NHL**	**1**	**0**	**0**	**0**	**2**	**0**	**0**	**0**	**0**	**0.0**	**1**	....	....	....	....	....	**3**	**0**	**0**	**0**	**0**	**0**	**0**	**0**
	Adirondack	AHL	75	6	36	42	64	....	....	....	....	....	....	....	....	....	....	....	3	0	0	0	0	....	....	....
1996-97	**Detroit**	**NHL**	**23**	**0**	**6**	**6**	**10**	**0**	**0**	**0**	**27**	**0.0**	**5**	....	....	....	....	....	....	....	....	....	....	....	....	....
	Adirondack	AHL	44	3	25	28	36	....	....	....	....	....	....	....	....	....	....	....	4	0	1	1	4	....	....	....
1997-98♦	**Detroit**	**NHL**	**66**	**7**	**14**	**21**	**32**	**1**	**0**	**2**	**91**	**7.7**	**21**	....	....	....	....	....	**18**	**0**	**5**	**5**	**16**	**0**	**0**	**0**
1998-99	**Detroit**	**NHL**	**61**	**2**	**10**	**12**	**34**	**0**	**0**	**1**	**67**	**3.0**	**5**	**0**	**0.0**	**72**	**60**	**15:54**	....	....	....	....	....	....	....	....
	Chicago	**NHL**	**11**	**0**	**8**	**8**	**0**	**0**	**0**	**0**	**12**	**0.0**	**6**	**0**	**0.0**	**15**	**20**	**22:51**	....	....	....	....	....	....	....	....
99-2000	**Chicago**	**NHL**	**73**	**3**	**25**	**28**	**20**	**0**	**0**	**1**	**86**	**3.5**	**4**	**1**	**100.0**	**86**	**80**	**21:03**	....	....	....	....	....	....	....	....
2000-01	**Chicago**	**NHL**	**13**	**2**	**3**	**5**	**2**	**1**	**0**	**0**	**19**	**10.5**	**-4**	**0**	**0.0**	**9**	**17**	**21:20**	....	....	....	....	....	....	....	....
	Florida	**NHL**	**60**	**0**	**21**	**21**	**28**	**0**	**0**	**0**	**80**	**0.0**	**2**	**1**	**0.0**	**79**	**60**	**21:02**	....	....	....	....	....	....	....	....
2001-02	**Toronto**	**NHL**	**34**	**0**	**2**	**2**	**12**	**0**	**0**	**0**	**31**	**0.0**	**-1**	**0**	**0.0**	**37**	**23**	**15:55**	**10**	**0**	**0**	**0**	**0**	**0**	**0**	**0**
	St. John's	AHL	25	4	6	10	14	....	....	....	....	....	....	....	....	....	....	....	11	0	5	5	6	....	....	....
	NHL Totals		**342**	**14**	**89**	**103**	**140**	**2**	**0**	**4**	**413**	**3.4**		**2**	**50.0**	**298**	**260**	**19:12**	**31**	**0**	**5**	**5**	**16**	**0**	**0**	**0**

Traded to **Chicago** by **Detroit** with Detroit's 1st round choices in 1999 (Steve McCarthy) and 2001 (Adam Munro) Entry Drafts for Chris Chelios, March 23, 1999. Traded to **Florida** by **Chicago** for Jaroslav Spacek, November 6, 2000. Signed as a free agent by **Toronto**, July 9, 2001.

ERSKINE, John

(AIR-skign, JAWN) **DAL.**

Defense. Shoots left. 6'4", 215 lbs. Born, Kingston, Ont., June 26, 1980. Dallas' 1st choice, 39th overall, in 1998 Entry Draft.

			Regular Season															Playoffs							
Season	Club	League	GP	G	A	Pts	PIM	PP	SH	GW	S	%	+/-	TF	F%	H	SB	Min	GP	G	A	Pts	PIM	PP	SH	GW
1996-97	Quinte Hawks	MTJHL	48	4	16	20	241	....	....	....	....	....	....	....	....	....	....	....	....	....	....	....	....	....	....	....
1997-98	London Knights	OHL	55	0	9	9	205	....	....	....	....	....	....	....	....	....	....	....	16	0	5	5	25	....	....	....
1998-99	London Knights	OHL	57	8	12	20	208	....	....	....	....	....	....	....	....	....	....	....	25	5	10	15	38	....	....	....
99-2000	London Knights	OHL	58	12	31	43	177	....	....	....	....	....	....	....	....	....	....	....	....	....	....	....	....	....	....	....
2000-01	Utah Grizzlies	IHL	77	1	8	9	284	....	....	....	....	....	....	....	....	....	....	....	....	....	....	....	....	....	....	....
2001-02	**Dallas**	**NHL**	**33**	**0**	**1**	**1**	**62**	**0**	**0**	**0**	**16**	**0.0**	**-8**	**0**	**0.0**	**63**	**19**	**10:44**	....	....	....	....	....	....	....	....
	Utah Grizzlies	AHL	39	2	6	8	118	....	....	....	....	....	....	....	....	....	....	....	3	0	0	0	10	....	....	....
	NHL Totals		**33**	**0**	**1**	**1**	**62**	**0**	**0**	**0**	**16**	**0.0**		**0**	**0.0**	**63**	**19**	**10:44**	....	....	....	....	....	....	....	....

OHL First All-Star Team (2000)

FAIRCHILD, Kelly

(FAIR-chighld, KEHL-lee)

Center. Shoots left. 5'11", 180 lbs. Born, Hibbing, MN, April 9, 1973. Los Angeles' 6th choice, 152nd overall, in 1991 Entry Draft.

			Regular Season															Playoffs							
Season	Club	League	GP	G	A	Pts	PIM	PP	SH	GW	S	%	+/-	TF	F%	H	SB	Min	GP	G	A	Pts	PIM	PP	SH	GW
1988-89	Hibbing High	Hi-School	22	9	8	17	24	....	....	....	....	....	....	....	....	....	....	....	....	....	....	....	....	....	....	....
1989-90	Grand Rapids	Hi-School	28	12	17	29	73	....	....	....	....	....	....	....	....	....	....	....	....	....	....	....	....	....	....	....
1990-91	Grand Rapids	Hi-School	28	28	45	73	25	....	....	....	....	....	....	....	....	....	....	....	....	....	....	....	....	....	....	....
1991-92	U. of Wisconsin	WCHA	37	11	10	21	45	....	....	....	....	....	....	....	....	....	....	....	....	....	....	....	....	....	....	....
1992-93	U. of Wisconsin	WCHA	42	25	29	54	54	....	....	....	....	....	....	....	....	....	....	....	....	....	....	....	....	....	....	....
1993-94	U. of Wisconsin	WCHA	42	20	44	*64	81	....	....	....	....	....	....	....	....	....	....	....	....	....	....	....	....	....	....	....
1994-95	St. John's	AHL	53	27	23	50	51	....	....	....	....	....	....	....	....	....	....	....	4	0	2	2	4	....	....	....
1995-96	**Toronto**	**NHL**	**1**	**0**	**1**	**1**	**2**	**0**	**0**	**0**	**1**	**0.0**	**1**	....	....	....	....	....	....	....	....	....	....	....	....	....
	St. John's	AHL	78	29	49	78	85	....	....	....	....	....	....	....	....	....	....	....	2	0	1	1	4	....	....	....
1996-97	**Toronto**	**NHL**	**22**	**0**	**2**	**2**	**2**	**0**	**0**	**0**	**14**	**0.0**	**-5**	....	....	....	....	....	....	....	....	....	....	....	....	....
	St. John's	AHL	29	9	22	31	36	....	....	....	....	....	....	....	....	....	....	....	....	....	....	....	....	....	....	....
	Orlando	IHL	25	9	6	15	20	....	....	....	....	....	....	....	....	....	....	....	9	6	5	11	16	....	....	....
1997-98	St. John's	AHL	17	5	2	7	24	....	....	....	....	....	....	....	....	....	....	....	....	....	....	....	....	....	....	....
	Orlando	IHL	22	2	6	8	20	....	....	....	....	....	....	....	....	....	....	....	....	....	....	....	....	....	....	....
	Milwaukee	IHL	40	20	24	44	32	....	....	....	....	....	....	....	....	....	....	....	10	5	2	7	4	....	....	....
1998-99	**Dallas**	**NHL**	**1**	**0**	**0**	**0**	**0**	**0**	**0**	**0**	**4**	**0.0**	**0**	**12**	**25.0**	**1**	**1**	**12:37**	....	....	....	....	....	....	....	....
	Michigan K-Wings	IHL	74	17	33	50	88	....	....	....	....	....	....	....	....	....	....	....	5	2	2	4	16	....	....	....
99-2000	Michigan K-Wings	IHL	78	21	41	62	89	....	....	....	....	....	....	....	....	....	....	....	....	....	....	....	....	....	....	....
2000-01	Hershey Bears	AHL	70	23	40	63	68	....	....	....	....	....	....	....	....	....	....	....	12	2	9	11	10	....	....	....

			Regular Season																Playoffs							
Season	Club	League	GP	G	A	Pts	PIM	PP	SH	GW	S	%	+/-	TF	F%	H	SB	Min	GP	G	A	Pts	PIM	PP	SH	GW
2001-02	**Colorado**	**NHL**	**10**	**2**	**0**	**2**	**2**	**0**	**0**	**0**	**6**	**33.3**	**1**	**32**	**43.8**	**13**	**3**	**6:35**								
	Hershey Bears	AHL	63	22	25	47	92												8	2	1	3	6			
	NHL Totals		**34**	**2**	**3**	**5**	**6**	**0**	**0**	**0**	**25**	**8.0**		**44**	**38.6**	**14**	**4**	**7:08**								

WCHA First All-Star Team (1994)

Traded to **Toronto** by **LA Kings** with Dixon Ward, Guy Leveque and Shayne Toporowski for Eric Lacroix, Chris Snell and Toronto's 4th round choice (Eric Belanger) in 1996 Entry Draft, October 3, 1994. Traded to **Milwaukee** (IHL) by **Orlando** (IHL) with Dave McIntyre for Sean McCann and Dave Mackey, January 11, 1998. Signed as a free agent by **Dallas**, July 2, 1998. Signed as a free agent by **Colorado**, August 29, 2000. Signed as a free agent by **Eisbaren Berlin** (Germany) with **Colorado** retaining NHL rights, July 28, 2002.

FARKAS, Jeff

(FAHR-kuhs, JEHF) **VAN.**

Right wing. Shoots left. 6', 185 lbs. Born, Amherst, MA, January 24, 1978. Toronto's 1st choice, 57th overall, in 1997 Entry Draft.

Season	Club	League	GP	G	A	Pts	PIM	PP	SH	GW	S	%	+/-	TF	F%	H	SB	Min	GP	G	A	Pts	PIM	PP	SH	GW
1993-94	Nichols High	Hi-School	28	27	57	84	25																			
1994-95	Niagara Scenics	EJHL	47	54	55	99	70																			
1995-96	Niagara Scenics	MTJHL	47	42	70	112	75																			
1996-97	Boston College	H-East	35	13	23	36	34																			
1997-98	Boston College	H-East	40	11	28	39	42																			
1998-99	Boston College	H-East	43	32	25	57	56																			
99-2000	Boston College	H-East	41	32	26	*58	61																			
	Toronto	**NHL**																	**3**	**1**	**0**	**1**	**0**	**0**	**0**	**0**
2000-01	**Toronto**	**NHL**	**2**	**0**	**0**	**0**	**2**	**0**	**0**	**0**	**1**	**0.0**	**−1**	**0**	**0.0**	**1**	**2**	**14:09**								
	St. John's	AHL	77	28	40	68	62												4	1	2	3	4			
2001-02	**Toronto**	**NHL**	**6**	**0**	**2**	**2**	**4**	**0**	**0**	**0**	**3**	**0.0**	**1**	**1**	**0.0**	**3**	**1**	**9:37**	**2**	**0**	**0**	**0**	**0**	**0**	**0**	**0**
	St. John's	AHL	71	16	34	50	49												4	0	0	0	0			
	NHL Totals		**8**	**0**	**2**	**2**	**6**	**0**	**0**	**0**	**4**	**0.0**		**1**	**0.0**	**4**	**3**	**10:45**	**5**	**1**	**0**	**1**	**0**	**0**	**0**	**0**

Hockey East First All-Star Team (2000) • NCAA East First All-American Team (2000) • NCAA Championship All-Tournament Team (2000)

Traded to **Vancouver** by **Toronto** for Josh Holden, June 23, 2002.

FARRELL, Michael

(FAHR-ehl, MIHK-ehl) **WSH.**

Right wing. Shoots right. 6', 222 lbs. Born, Edina, MN, October 20, 1978. Washington's 9th choice, 220th overall, in 1998 Entry Draft.

Season	Club	League	GP	G	A	Pts	PIM	PP	SH	GW	S	%	+/-	TF	F%	H	SB	Min	GP	G	A	Pts	PIM	PP	SH	GW
1996-97	Culver Eagles	Hi-School	STATISTICS NOT AVAILABLE																							
1997-98	Providence	H-East	33	5	8	13	32																			
1998-99	Providence	H-East	29	3	12	15	51																			
99-2000	Providence	H-East	36	3	6	9	71																			
	Portland Pirates	AHL	7	2	0	2	0												4	0	1	1	0			
2000-01	Portland Pirates	AHL	79	6	18	24	61												3	0	2	2	2			
2001-02	**Washington**	**NHL**	**8**	**0**	**0**	**0**	**0**	**0**	**0**	**0**	**1**	**0.0**	**−1**	**0**	**0.0**	**13**	**2**	**5:35**								
	Portland Pirates	AHL	61	12	15	27	62																			
	NHL Totals		**8**	**0**	**0**	**0**	**0**	**0**	**0**	**0**	**1**	**0.0**		**0**	**0.0**	**13**	**2**	**5:35**								

FATA, Rico

(FA-tuh, REE-koh) **NYR**

Center. Shoots left. 5'11", 200 lbs. Born, Sault Ste. Marie, Ont., February 12, 1980. Calgary's 1st choice, 6th overall, in 1998 Entry Draft.

Season	Club	League	GP	G	A	Pts	PIM	PP	SH	GW	S	%	+/-	TF	F%	H	SB	Min	GP	G	A	Pts	PIM	PP	SH	GW
1994-95	Soo Legion	NOHA	51	52	51	103																				
1995-96	Sault Ste. Marie	OHL	62	11	15	26	52												4	0	0	0	0			
1996-97	London Knights	OHL	59	19	34	53	76																			
1997-98	London Knights	OHL	64	43	33	76	110												16	9	5	14	*49			
1998-99	**Calgary**	**NHL**	**20**	**0**	**1**	**1**	**4**	**0**	**0**	**0**	**13**	**0.0**	**0**	**2**	**50.0**	**10**	**5**	**7:36**								
	London Knights	OHL	23	15	18	33	41												25	10	12	22	42			
99-2000	**Calgary**	**NHL**	**2**	**0**	**0**	**0**	**0**	**0**	**0**	**0**	**0**	**0.0**	**−1**	**0**	**0.0**	**3**	**0**	**10:06**								
	Saint John	AHL	76	29	29	58	65												3	0	0	0	4			
2000-01	**Calgary**	**NHL**	**5**	**0**	**0**	**0**	**6**	**0**	**0**	**0**	**6**	**0.0**	**−3**	**0**	**0.0**	**3**	**0**	**9:25**								
	Saint John	AHL	70	23	29	52	129												19	2	3	5	22			
2001-02	**NY Rangers**	**NHL**	**10**	**0**	**0**	**0**	**0**	**0**	**0**	**0**	**8**	**0.0**	**−2**	**55**	**47.3**	**5**	**5**	**8:31**								
	Hartford	AHL	61	35	36	71	36												10	2	5	7	4			
	NHL Totals		**37**	**0**	**1**	**1**	**10**	**0**	**0**	**0**	**27**	**0.0**		**57**	**47.4**	**21**	**10**	**8:14**								

AHL Second All-Star Team (2002)

• Returned to **London** (OHL) by **Calgary** following WJC-A tournament, January 10, 1999. Claimed on waivers by **NY Rangers** from **Calgary**, October 3, 2001.

FEDOROV, Sergei

(FEH-duh-rahf, SAIR-gay) **DET.**

Center. Shoots left. 6'1", 200 lbs. Born, Pskov, USSR, December 13, 1969. Detroit's 4th choice, 74th overall, in 1989 Entry Draft.

Season	Club	League	GP	G	A	Pts	PIM	PP	SH	GW	S	%	+/-	TF	F%	H	SB	Min	GP	G	A	Pts	PIM	PP	SH	GW
1985-86	Dynamo Minsk	USSR-2	15	6	1	7	10																			
1986-87	CSKA Moscow	USSR	29	6	6	12	12																			
1987-88	CSKA Moscow	USSR	48	7	9	16	20																			
1988-89	CSKA Moscow	USSR	44	9	8	17	35																			
1989-90	CSKA Moscow	USSR	48	19	10	29	22																			
1990-91	**Detroit**	**NHL**	**77**	**31**	**48**	**79**	**66**	**11**	**3**	**5**	**259**	**12.0**	**11**						**7**	**1**	**5**	**6**	**4**	**0**	**0**	**1**
1991-92	**Detroit**	**NHL**	**80**	**32**	**54**	**86**	**72**	**7**	**2**	**5**	**249**	**12.9**	**26**						**11**	**5**	**5**	**10**	**8**	**1**	**2**	**1**
1992-93	**Detroit**	**NHL**	**73**	**34**	**53**	**87**	**72**	**13**	**4**	**3**	**217**	**15.7**	**33**						**7**	**3**	**6**	**9**	**23**	**1**	**1**	**0**
1993-94	**Detroit**	**NHL**	**82**	**56**	**64**	**120**	**34**	**13**	**4**	**10**	**337**	**16.6**	**48**						**7**	**1**	**7**	**8**	**6**	**0**	**0**	**0**
1994-95	**Detroit**	**NHL**	**42**	**20**	**30**	**50**	**24**	**7**	**3**	**5**	**147**	**13.6**	**6**						**17**	**7**	***17**	***24**	**6**	**3**	**0**	**0**
1995-96	**Detroit**	**NHL**	**78**	**39**	**68**	**107**	**48**	**11**	**3**	**11**	**306**	**12.7**	**49**						**19**	**2**	***18**	**20**	**10**	**0**	**0**	**2**
1996-97♦	**Detroit**	**NHL**	**74**	**30**	**33**	**63**	**30**	**9**	**2**	**4**	**273**	**11.0**	**29**						**20**	**8**	**12**	**20**	**12**	**3**	**0**	**4**
1997-98	Russia	Olympics	6	1	5	6	8																			
	♦ Detroit	**NHL**	**21**	**6**	**11**	**17**	**25**	**2**	**0**	**2**	**68**	**8.8**	**10**						**22**	***10**	**10**	**20**	**12**	**2**	**1**	**1**
1998-99	**Detroit**	**NHL**	**77**	**26**	**37**	**63**	**66**	**6**	**2**	**3**	**224**	**11.6**	**9**	**1414**	**51.7**	**77**	**24**	**19:21**	**10**	**1**	**8**	**9**	**8**	**0**	**0**	**0**
99-2000	**Detroit**	**NHL**	**68**	**27**	**35**	**62**	**22**	**4**	**4**	**7**	**263**	**10.3**	**8**	**1274**	**53.8**	**62**	**23**	**20:05**	**9**	**4**	**4**	**8**	**4**	**2**	**0**	**1**
2000-01	**Detroit**	**NHL**	**75**	**32**	**37**	**69**	**40**	**14**	**2**	**7**	**268**	**11.9**	**12**	**1601**	**55.8**	**75**	**34**	**21:05**	**6**	**2**	**5**	**7**	**0**	**1**	**0**	**1**
2001-02♦	**Detroit**	**NHL**	**81**	**31**	**37**	**68**	**36**	**10**	**0**	**6**	**256**	**12.1**	**20**	**1160**	**51.7**	**53**	**34**	**19:33**	**23**	**5**	**14**	**19**	**20**	**2**	**1**	**0**
	Russia	Olympics	6	2	2	4	4																			
	NHL Totals		**828**	**364**	**507**	**871**	**535**	**107**	**29**	**68**	**2867**	**12.7**		**5449**	**53.4**	**267**	**115**	**20:00**	**158**	**49**	**111**	**160**	**113**	**15**	**5**	**11**

NHL All-Rookie Team (1991) • NHL First All-Star Team (1994) • Won Frank J. Selke Trophy (1994, 1996) • Won Lester B. Pearson Award (1994) • Won Hart Trophy (1994) • Played in NHL All-Star Game (1992, 1994, 1996, 2001, 2002)

• Missed majority of 1997-98 season after failing to come to contract terms with **Detroit**.

FEDORUK, Todd

(FEH-duh-ruhk, TAWD) **PHI.**

Left wing. Shoots left. 6'2", 235 lbs. Born, Redwater, Alta., February 13, 1979. Philadelphia's 6th choice, 164th overall, in 1997 Entry Draft.

Season	Club	League	GP	G	A	Pts	PIM	PP	SH	GW	S	%	+/-	TF	F%	H	SB	Min	GP	G	A	Pts	PIM	PP	SH	GW
1994-95	Ft. Saskatchewan	AMHL	STATISTICS NOT AVAILABLE																							
1995-96	Kelowna Rockets	WHL	44	1	1	2	83												4	0	0	0	6			
1996-97	Kelowna Rockets	WHL	31	1	5	6	87												6	0	0	0	13			
1997-98	Kelowna Rockets	WHL	31	3	5	8	120																			
	Regina Pats	WHL	21	4	3	7	80												9	1	2	3	23			
1998-99	Regina Pats	WHL	39	12	12	24	107																			
	Prince Albert	WHL	28	6	4	10	75												13	1	6	7	49			
99-2000	Trenton Titans	ECHL	18	2	5	7	118																			
	Philadelphia	AHL	19	1	2	3	40												5	0	1	1	2			
2000-01	**Philadelphia**	**NHL**	**53**	**5**	**5**	**10**	**109**	**0**	**0**	**0**	**28**	**17.9**	**0**	**0**	**0.0**	**65**	**2**	**7:02**	**2**	**0**	**0**	**0**	**20**	**0**	**0**	**0**
	Philadelphia	AHL	14	0	1	1	49																			
2001-02	**Philadelphia**	**NHL**	**55**	**3**	**4**	**7**	**141**	**0**	**0**	**0**	**21**	**14.3**	**−2**	**5**	**0.0**	**69**	**1**	**6:21**	**3**	**0**	**0**	**0**	**0**	**0**	**0**	**0**
	Philadelphia	AHL	7	0	1	1	54																			
	NHL Totals		**108**	**8**	**9**	**17**	**250**	**0**	**0**	**0**	**49**	**16.3**		**5**	**0.0**	**134**	**3**	**6:41**	**5**	**0**	**0**	**0**	**20**	**0**	**0**	**0**

Traded to **Prince Albert** (WHL) by **Regina** (WHL) for future considerations, January 13, 1999.

FEDOTENKO, Ruslan
(feh-doh-TEHN-koh, roos-LAHN) **T.B.**

Left wing. Shoots left. 6'2", 195 lbs. Born, Kiev, Ukraine, January 18, 1979.

			Regular Season																Playoffs							
Season	Club	League	GP	G	A	Pts	PIM	PP	SH	GW	S	%	+/-	TF	F%	H	SB	Min	GP	G	A	Pts	PIM	PP	SH	GW
1997-98	Melfort Mustangs	SJHL	68	35	31	66	55																			
1998-99	Sioux City	USHL	55	43	34	77	139												5	5	1	6	9			
99-2000	Trenton Titans	ECHL	8	5	3	8	9																			
	Philadelphia	AHL	67	16	34	50	42												2	0	0	0	0			
2000-01	**Philadelphia**	**NHL**	**74**	**16**	**20**	**36**	**72**	**3**	**0**	**4**	**119**	**13.4**	**8**	**7**	**71.4**	**94**	**28**	**14:38**	**6**	**0**	**1**	**1**	**4**	**0**	**0**	**0**
	Philadelphia	AHL	8	1	0	1	8																			
2001-02	**Philadelphia**	**NHL**	**78**	**17**	**9**	**26**	**43**	**0**	**1**	**3**	**121**	**14.0**	**15**	**41**	**43.9**	**89**	**25**	**13:56**	**5**	**1**	**0**	**1**	**2**	**0**	**0**	**1**
	Ukraine	Olympics	1	1	0	1	4																			
	NHL Totals		**152**	**33**	**29**	**62**	**115**	**3**	**1**	**7**	**240**	**13.8**		**48**	**47.9**	**183**	**53**	**14:16**	**11**	**1**	**1**	**2**	**6**	**0**	**0**	**1**

Signed as a free agent by **Philadelphia**, August 3, 1999. Traded to **Tampa Bay** by **Philadelphia** with Tampa Bay's 2nd round choice (previously acquired, later traded to Dallas - Dallas selected Tobias Stephan) in 2002 Entry Draft and Phoenix's 2nd round choice (previously acquired, later traded to San Jose - San Jose selected Dan Spang) in 2002 Entry Draft for Tampa Bay's 1st round choice (Joni Pitkanen) in 2002 Entry Draft, June 21, 2002.

FERENCE, Andrew
(fuhr-EHNS, AN-droo) **PIT.**

Defense. Shoots left. 5'10", 196 lbs. Born, Edmonton, Alta., March 17, 1979. Pittsburgh's 8th choice, 208th overall, in 1997 Entry Draft.

			Regular Season																Playoffs							
Season	Club	League	GP	G	A	Pts	PIM	PP	SH	GW	S	%	+/-	TF	F%	H	SB	Min	GP	G	A	Pts	PIM	PP	SH	GW
1994-95	Sherwood Park	AMHL	31	4	14	18	74																			
	Portland	WHL	2	0	0	0	4																			
1995-96	Portland	WHL	72	9	31	40	159												7	1	3	4	12			
1996-97	Portland	WHL	72	12	32	44	163												6	1	2	3	12			
1997-98	Portland	WHL	72	11	57	68	142												16	2	18	20	28			
1998-99	Portland	WHL	40	11	21	32	104												4	1	4	5	10			
	Kansas City	IHL	5	1	2	3	4												3	0	0	0	9			
99-2000	**Pittsburgh**	**NHL**	**30**	**2**	**4**	**6**	**20**	**0**	**0**	**1**	**26**	**7.7**	**3**	**0**	**0.0**	**61**	**26**	**16:19**								
	Wilkes-Barre	AHL	44	8	20	28	58																			
2000-01	Wilkes-Barre	AHL	43	6	18	24	95												3	1	0	1	12			
	Pittsburgh	**NHL**	**36**	**4**	**11**	**15**	**28**	**1**	**0**	**1**	**47**	**8.5**	**6**	**0**	**0.0**	**82**	**38**	**18:51**	**18**	**3**	**7**	**10**	**16**	**1**	**0**	**1**
2001-02	**Pittsburgh**	**NHL**	**75**	**4**	**7**	**11**	**73**	**1**	**0**	**0**	**82**	**4.9**	**–12**	**2**	**0.0**	**149**	**60**	**18:34**								
	NHL Totals		**141**	**10**	**22**	**32**	**121**	**2**	**0**	**2**	**155**	**6.5**		**2**	**0.0**	**292**	**124**	**18:10**	**18**	**3**	**7**	**10**	**16**	**1**	**0**	**1**

WHL West First All-Star Team (1998) • WHL West Second All-Star Team (1999)

FERENCE, Brad
(FAIR-ehns, BRAD) **FLA.**

Defense. Shoots right. 6'3", 210 lbs. Born, Calgary, Alta., April 2, 1979. Vancouver's 1st choice, 10th overall, in 1997 Entry Draft.

			Regular Season																Playoffs							
Season	Club	League	GP	G	A	Pts	PIM	PP	SH	GW	S	%	+/-	TF	F%	H	SB	Min	GP	G	A	Pts	PIM	PP	SH	GW
1994-95	Calgary Royals	ABHL	60	19	47	66	220																			
1995-96	Calgary Royals	ABHL	22	7	21	28	140																			
	Spokane Chiefs	WHL	5	0	2	2	18																			
1996-97	Spokane Chiefs	WHL	67	6	20	26	324												9	0	4	4	21			
1997-98	Spokane Chiefs	WHL	54	9	30	39	213												18	0	7	7	59			
1998-99	Spokane Chiefs	WHL	31	3	22	25	125																			
	Tri-City	WHL	20	6	15	21	116												12	1	9	10	63			
99-2000	**Florida**	**NHL**	**13**	**0**	**2**	**2**	**46**	**0**	**0**	**0**	**10**	**0.0**	**2**	**0**	**0.0**	**13**	**17**	**13:40**								
	Louisville	AHL	58	2	7	9	231												2	0	0	0	2			
2000-01	**Florida**	**NHL**	**14**	**0**	**1**	**1**	**14**	**0**	**0**	**0**	**5**	**0.0**	**–10**	**0**	**0.0**	**17**	**10**	**13:03**								
	Louisville	AHL	52	3	21	24	200																			
2001-02	**Florida**	**NHL**	**80**	**2**	**15**	**17**	**254**	**0**	**0**	**0**	**65**	**3.1**	**–13**	**1**	**0.0**	**92**	**80**	**19:44**								
	NHL Totals		**107**	**2**	**18**	**20**	**314**	**0**	**0**	**0**	**80**	**2.5**		**1**	**0.0**	**122**	**107**	**18:07**								

Memorial Cup All-Star Team (1998)

Traded to **Tri-City** (WHL) by **Spokane** (WHL) for David Boychuk, February 2, 1999. Traded to **Florida** by **Vancouver** with Pavel Bure, Bret Hedican and Vancouver's 3rd round choice (Robert Fried) in 2000 Entry Draft for Ed Jovanovski, Dave Gagner, Mike Brown, Kevin Weekes and Florida's 1st round choice (Nathan Smith) in 2000 Entry Draft, January 17, 1999.

FERGUSON, Scott
(fuhr-GUH-sohn, SKAWT) **EDM.**

Defense. Shoots left. 6'1", 195 lbs. Born, Camrose, Alta., January 6, 1973.

			Regular Season																Playoffs							
Season	Club	League	GP	G	A	Pts	PIM	PP	SH	GW	S	%	+/-	TF	F%	H	SB	Min	GP	G	A	Pts	PIM	PP	SH	GW
1990-91	Sherwood Park	AJHL	32	2	9	11	91																			
	Kamloops Blazers	WHL	4	0	0	0	0																			
1991-92	Kamloops Blazers	WHL	62	4	10	14	138												12	0	2	2	21			
1992-93	Kamloops Blazers	WHL	71	4	19	23	206												13	0	2	2	24			
1993-94	Kamloops Blazers	WHL	68	5	49	54	180												19	5	11	16	48			
1994-95	Cape Breton	AHL	58	4	6	10	103																			
	Wheeling	ECHL	5	1	5	6	16																			
1995-96	Cape Breton	AHL	80	5	16	21	196																			
1996-97	Hamilton	AHL	74	6	14	20	115												21	5	7	12	59			
1997-98	**Edmonton**	**NHL**	**1**	**0**	**0**	**0**	**0**	**0**	**0**	**0**	**0**	**0.0**	**1**													
	Hamilton	AHL	77	7	17	24	150												9	0	3	3	16			
1998-99	**Anaheim**	**NHL**	**2**	**0**	**1**	**1**	**0**	**0**	**0**	**0**	**1**	**0.0**	**0**	**0**	**0.0**	**1**	**4**	**15:09**								
	Cincinnati	AHL	78	4	31	35	59												3	0	0	0	4			
99-2000	Cincinnati	AHL	77	7	25	32	166																			
2000-01	**Edmonton**	**NHL**	**20**	**0**	**1**	**1**	**13**	**0**	**0**	**0**	**8**	**0.0**	**2**	**0**	**0.0**	**27**	**11**	**10:55**	**6**	**0**	**0**	**0**	**0**	**0**	**0**	**0**
	Hamilton	AHL	42	3	18	21	79																			
2001-02	**Edmonton**	**NHL**	**50**	**3**	**2**	**5**	**75**	**0**	**0**	**0**	**27**	**11.1**	**11**	**0**	**0.0**	**61**	**39**	**13:40**								
	NHL Totals		**73**	**3**	**4**	**7**	**88**	**0**	**0**	**0**	**36**	**8.3**		**0**	**0.0**	**89**	**54**	**12:57**	**6**	**0**	**0**	**0**	**0**	**0**	**0**	**0**

WHL West Second All-Star Team (1994)

Signed as a free agent by **Edmonton**, June 2, 1994. Traded to **Ottawa** by **Edmonton** for Frantisek Musil, March 9, 1998. Signed as a free agent by **Anaheim**, July 27, 1998. Signed as a free agent by **Edmonton**, July 5, 2000.

FERRARO, Chris
(fuh-RAHR-oh, KRIHS)

Center. Shoots right. 5'9", 175 lbs. Born, Port Jefferson, NY, January 24, 1973. NY Rangers' 4th choice, 85th overall, in 1992 Entry Draft.

			Regular Season																Playoffs							
Season	Club	League	GP	G	A	Pts	PIM	PP	SH	GW	S	%	+/-	TF	F%	H	SB	Min	GP	G	A	Pts	PIM	PP	SH	GW
1990-91	Dubuque	USHL	45	53	44	97	84												8	3	9	12	12			
1991-92	Dubuque	USHL	20	30	19	49	52																			
	Waterloo	USHL	18	19	31	50	54												4	5	6	11	14			
1992-93	U. of Maine	H-East	39	25	26	51	46																			
1993-94	U. of Maine	H-East	4	0	1	1	8																			
	Team USA	Nat-Tm	48	8	34	42	58																			
1994-95	Atlanta Knights	IHL	54	13	14	27	72																			
	Binghamton	AHL	13	6	4	10	38												10	2	3	5	16			
1995-96	**NY Rangers**	**NHL**	**2**	**1**	**0**	**1**	**0**	**1**	**0**	**0**	**4**	**25.0**	**–3**													
	Binghamton	AHL	77	32	67	99	208												4	4	2	6	13			
1996-97	**NY Rangers**	**NHL**	**12**	**1**	**1**	**2**	**6**	**0**	**0**	**0**	**23**	**4.3**	**1**													
	Binghamton	AHL	53	29	34	63	94																			
1997-98	**Pittsburgh**	**NHL**	**46**	**3**	**4**	**7**	**43**	**0**	**0**	**0**	**42**	**7.1**	**–2**													
1998-99	**Edmonton**	**NHL**	**2**	**1**	**0**	**1**	**0**	**0**	**0**	**0**	**1**	**100.0**	**1**	**19**	**52.6**	**0**	**0**	**8:33**								
	Hamilton	AHL	72	35	41	76	104												11	8	5	13	20			
99-2000	**NY Islanders**	**NHL**	**11**	**1**	**3**	**4**	**8**	**0**	**0**	**0**	**15**	**6.7**	**1**	**92**	**50.0**	**8**	**7**	**9:30**								
	Providence	AHL	21	9	9	18	32																			
	Chicago Wolves	IHL	25	7	18	25	40												16	5	8	13	14			
2000-01	Albany	AHL	74	24	42	66	111																			
2001-02	**Washington**	**NHL**	**1**	**0**	**1**	**1**	**0**	**0**	**0**	**0**	**4**	**0.0**	**0**	**2**	**0.0**	**4**	**0**	**15:17**								
	Portland Pirates	AHL	2	1	1	2	6																			
	NHL Totals		**74**	**7**	**9**	**16**	**57**	**1**	**0**	**0**	**89**	**7.9**		**113**	**49.6**	**12**	**7**	**9:47**								

Claimed on waivers by **Pittsburgh** from **NY Rangers**, October 1, 1997. Signed as a free agent by **Edmonton**, August 13, 1998. Signed as a free agent by **NY Islanders**, July 22, 1999. Signed as a free agent by **New Jersey**, July 20, 2000. Traded to **Washington** by **New Jersey** for future considerations, August 22, 2001. • Missed majority of 2001-02 season after being granted personal leave of absence by Washington, October 15, 2001.

FERRARO, Peter

(fuh-RAHR-oh, PEE-tuhr) **WSH.**

Right wing. Shoots right. 5'10", 180 lbs. Born, Port Jefferson, NY, January 24, 1973. NY Rangers' 1st choice, 24th overall, in 1992 Entry Draft.

			Regular Season																Playoffs							
Season	Club	League	GP	G	A	Pts	PIM	PP	SH	GW	S	%	+/-	TF	F%	H	SB	Min	GP	G	A	Pts	PIM	PP	SH	GW
1990-91	Dubuque	USHL	29	21	31	52	83	....	....	....	....	....	....	....	....	....	....	....	8	7	5	12	10	....	....	....
1991-92	Dubuque	USHL	21	25	25	50	92	....	....	....	....	....	....	....	....	....	....	....	....	....	....	....	....	....	....	....
	Waterloo	USHL	21	23	28	51	76	....	....	....	....	....	....	....	....	....	....	....	4	8	5	13	16	....	....	....
1992-93	U. of Maine	H-East	36	18	32	50	106	....	....	....	....	....	....	....	....	....	....	....	....	....	....	....	....	....	....	....
1993-94	U. of Maine	H-East	4	3	6	9	16	....	....	....	....	....	....	....	....	....	....	....	....	....	....	....	....	....	....	....
	Team USA	Nat-Tm	60	30	34	64	87	....	....	....	....	....	....	....	....	....	....	....	....	....	....	....	....	....	....	....
	United States	Olympics	8	6	0	6	6	....	....	....	....	....	....	....	....	....	....	....	....	....	....	....	....	....	....	....
1994-95	Atlanta Knights	IHL	61	15	24	39	118	....	....	....	....	....	....	....	....	....	....	....	....	....	....	....	....	....	....	....
	Binghamton	AHL	12	2	6	8	67	....	....	....	....	....	....	....	....	....	....	....	11	4	3	7	51	....	....	....
1995-96	**NY Rangers**	**NHL**	**5**	**0**	**1**	**1**	**0**	**0**	**0**	**0**	**6**	**0.0**	**-5**	....	....	....	....	....	....	....	....	....	....	....	....	....
	Binghamton	AHL	68	48	53	101	157	....	....	....	....	....	....	....	....	....	....	....	4	1	6	7	22	....	....	....
1996-97	**NY Rangers**	**NHL**	**2**	**0**	**0**	**0**	**0**	**0**	**0**	**0**	**3**	**0.0**	**0**	....	....	....	....	....	**2**	**0**	**0**	**0**	**0**	**0**	**0**	**0**
	Binghamton	AHL	75	38	39	77	171	....	....	....	....	....	....	....	....	....	....	....	4	3	1	4	18	....	....	....
1997-98	**Pittsburgh**	**NHL**	**29**	**3**	**4**	**7**	**12**	**0**	**0**	**0**	**34**	**8.8**	**-2**	....	....	....	....	....	....	....	....	....	....	....	....	....
	NY Rangers	**NHL**	**1**	**0**	**0**	**0**	**2**	**0**	**0**	**0**	**3**	**0.0**	**-2**	....	....	....	....	....	....	....	....	....	....	....	....	....
	Hartford	AHL	36	17	23	40	54	....	....	....	....	....	....	....	....	....	....	....	15	8	6	14	59	....	....	....
1998-99	**Boston**	**NHL**	**46**	**6**	**8**	**14**	**44**	**1**	**0**	**1**	**61**	**9.8**	**10**	**70**	**37.1**	**52**	**21**	**10:12**	....	....	....	....	....	....	....	....
	Providence	AHL	16	15	10	25	14	....	....	....	....	....	....	....	....	....	....	....	19	9	12	21	38	....	....	....
99-2000	**Boston**	**NHL**	**5**	**0**	**1**	**1**	**0**	**0**	**0**	**0**	**3**	**0.0**	**-1**	**19**	**47.4**	**8**	**1**	**8:11**	....	....	....	....	....	....	....	....
	Providence	AHL	48	21	25	46	98	....	....	....	....	....	....	....	....	....	....	....	13	5	7	12	14	....	....	....
2000-01	Providence	AHL	78	26	45	71	109	....	....	....	....	....	....	....	....	....	....	....	17	4	5	9	34	....	....	....
2001-02	**Washington**	**NHL**	**4**	**0**	**1**	**1**	**0**	**0**	**0**	**0**	**3**	**0.0**	**-1**	**0**	**0.0**	**6**	**2**	**12:54**	....	....	....	....	....	....	....	....
	Portland Pirates	AHL	67	21	37	58	119	....	....	....	....	....	....	....	....	....	....	....	....	....	....	....	....	....	....	....
	NHL Totals		**92**	**9**	**15**	**24**	**58**	**1**	**0**	**1**	**113**	**8.0**		**89**	**39.3**	**66**	**24**	**10:13**	**2**	**0**	**0**	**0**	**0**	**0**	**0**	**0**

AHL First All-Star Team (1996) • Won Jack A. Butterfield Trophy (Playoff MVP - AHL) (1999)

Claimed on waivers by **Pittsburgh** from **NY Rangers**, October 1, 1997. Claimed on waivers by **NY Rangers** from **Pittsburgh**, January 9, 1998. Signed as a free agent by **Boston**, August 5, 1998. Claimed by **Atlanta** from **Boston** in Expansion Draft, June 25, 1999. Traded to **Boston** by **Atlanta** for Randy Robitaille, June 25, 1999. Signed as a free agent by **Washington**, August 1, 2001.

FERRARO, Ray

(fuh-RAHR-oh, RAY)

Center. Shoots left. 5'9", 200 lbs. Born, Trail, B.C., August 23, 1964. Hartford's 5th choice, 88th overall, in 1982 Entry Draft.

			Regular Season																Playoffs							
Season	Club	League	GP	G	A	Pts	PIM	PP	SH	GW	S	%	+/-	TF	F%	H	SB	Min	GP	G	A	Pts	PIM	PP	SH	GW
1981-82	Penticton	BCJHL	48	65	70	135	90	....	....	....	....	....	....	....	....	....	....	....	....	....	....	....	....	....	....	....
1982-83	Portland	WHL	50	41	49	90	39	....	....	....	....	....	....	....	....	....	....	....	14	14	10	24	13	....	....	....
1983-84	Brandon	WHL	72	*108	84	*192	84	....	....	....	....	....	....	....	....	....	....	....	11	13	15	28	20	....	....	....
1984-85	**Hartford**	**NHL**	**44**	**11**	**17**	**28**	**40**	**6**	**0**	**2**	**59**	**18.6**	**-1**	....	....	....	....	....	....	....	....	....	....	....	....	....
	Binghamton	AHL	37	20	13	33	29	....	....	....	....	....	....	....	....	....	....	....	....	....	....	....	....	....	....	....
1985-86	**Hartford**	**NHL**	**76**	**30**	**47**	**77**	**57**	**14**	**0**	**0**	**132**	**22.7**	**10**	....	....	....	....	....	**10**	**3**	**6**	**9**	**4**	**3**	**0**	**0**
1986-87	**Hartford**	**NHL**	**80**	**27**	**32**	**59**	**42**	**14**	**0**	**2**	**96**	**28.1**	**-9**	....	....	....	....	....	**6**	**1**	**1**	**2**	**8**	**0**	**0**	**0**
1987-88	**Hartford**	**NHL**	**68**	**21**	**29**	**50**	**81**	**6**	**0**	**2**	**105**	**20.0**	**1**	....	....	....	....	....	**6**	**1**	**1**	**2**	**6**	**1**	**0**	**0**
1988-89	**Hartford**	**NHL**	**80**	**41**	**35**	**76**	**86**	**11**	**0**	**7**	**169**	**24.3**	**1**	....	....	....	....	....	**4**	**2**	**0**	**2**	**4**	**0**	**0**	**0**
1989-90	**Hartford**	**NHL**	**79**	**25**	**29**	**54**	**109**	**7**	**0**	**4**	**138**	**18.1**	**-15**	....	....	....	....	....	**7**	**0**	**3**	**3**	**2**	**0**	**0**	**0**
1990-91	**Hartford**	**NHL**	**15**	**2**	**5**	**7**	**18**	**1**	**0**	**0**	**18**	**11.1**	**-1**	....	....	....	....	....	....	....	....	....	....	....	....	....
	NY Islanders	**NHL**	**61**	**19**	**16**	**35**	**52**	**5**	**0**	**1**	**91**	**20.9**	**-11**	....	....	....	....	....	....	....	....	....	....	....	....	....
1991-92	**NY Islanders**	**NHL**	**80**	**40**	**40**	**80**	**92**	**7**	**0**	**4**	**154**	**26.0**	**25**	....	....	....	....	....	....	....	....	....	....	....	....	....
1992-93	**NY Islanders**	**NHL**	**46**	**14**	**13**	**27**	**40**	**3**	**0**	**1**	**72**	**19.4**	**0**	....	....	....	....	....	**18**	**13**	**7**	**20**	**18**	**0**	**0**	**0**
	Capital District	AHL	1	0	2	2	2	....	....	....	....	....	....	....	....	....	....	....	....	....	....	....	....	....	....	....
1993-94	**NY Islanders**	**NHL**	**82**	**21**	**32**	**53**	**83**	**5**	**0**	**3**	**136**	**15.4**	**1**	....	....	....	....	....	**4**	**1**	**0**	**1**	**6**	**0**	**0**	**0**
1994-95	**NY Islanders**	**NHL**	**47**	**22**	**21**	**43**	**30**	**2**	**0**	**1**	**94**	**23.4**	**1**	....	....	....	....	....	....	....	....	....	....	....	....	....
1995-96	**NY Rangers**	**NHL**	**65**	**25**	**29**	**54**	**82**	**8**	**0**	**4**	**160**	**15.6**	**13**	....	....	....	....	....	....	....	....	....	....	....	....	....
	Los Angeles	**NHL**	**11**	**4**	**2**	**6**	**10**	**1**	**0**	**0**	**18**	**22.2**	**-13**	....	....	....	....	....	....	....	....	....	....	....	....	....
1996-97	**Los Angeles**	**NHL**	**81**	**25**	**21**	**46**	**112**	**11**	**0**	**2**	**152**	**16.4**	**-22**	....	....	....	....	....	....	....	....	....	....	....	....	....
1997-98	**Los Angeles**	**NHL**	**40**	**6**	**9**	**15**	**42**	**0**	**0**	**2**	**45**	**13.3**	**-10**	....	....	....	....	....	**3**	**0**	**1**	**1**	**2**	**0**	**0**	**0**
1998-99	**Los Angeles**	**NHL**	**65**	**13**	**18**	**31**	**59**	**4**	**0**	**4**	**84**	**15.5**	**0**	**979**	**47.8**	**58**	**24**	**14:34**	....	....	....	....	....	....	....	....
99-2000	**Atlanta**	**NHL**	**81**	**19**	**25**	**44**	**88**	**10**	**0**	**3**	**170**	**11.2**	**-33**	**1390**	**51.2**	**111**	**26**	**16:10**	....	....	....	....	....	....	....	....
2000-01	**Atlanta**	**NHL**	**81**	**29**	**47**	**76**	**91**	**11**	**0**	**2**	**172**	**16.9**	**-11**	**1705**	**48.0**	**78**	**43**	**18:19**	....	....	....	....	....	....	....	....
2001-02	**Atlanta**	**NHL**	**61**	**8**	**19**	**27**	**66**	**2**	**0**	**0**	**72**	**11.1**	**-32**	**933**	**46.7**	**56**	**25**	**15:59**	....	....	....	....	....	....	....	....
	St. Louis	**NHL**	**15**	**6**	**4**	**10**	**8**	**2**	**0**	**1**	**27**	**22.2**	**2**	**167**	**47.3**	**20**	**2**	**14:05**	**10**	**0**	**3**	**3**	**4**	**0**	**0**	**0**
	NHL Totals		**1258**	**408**	**490**	**898**	**1288**	**130**	**0**	**45**	**2164**	**18.9**		**5174**	**48.6**	**323**	**120**	**16:15**	**68**	**21**	**22**	**43**	**54**	**4**	**0**	**0**

WHL East First All-Star Team (1984) • WHL MVP (1984) • Played in NHL All-Star Game (1992)

Traded to **NY Islanders** by **Hartford** for Doug Crossman, November 13, 1990. Signed as a free agent by **NY Rangers**, August 9, 1995. Traded to **LA Kings** by **NY Rangers** with Ian Laperriere, Mattias Norstrom, Nathan LaFayette and NY Rangers' 4th round choice (Sean Blanchard) in 1997 Entry Draft for Marty McSorley, Jari Kurri and Shane Churla, March 14, 1996. Signed as a free agent by **Atlanta**, August 9, 1999. Traded to **St. Louis** by **Atlanta** for Carolina's 4th round choice (previously acquired, Atlanta selected Lane Manson) in 2002 Entry Draft, March 18, 2002. • Officially announced retirement, August 2, 2002.

FINLEY, Jeff

(FIHN-lee, JEHF) **ST.L.**

Defense. Shoots left. 6'2", 205 lbs. Born, Edmonton, Alta., April 14, 1967. NY Islanders' 4th choice, 55th overall, in 1985 Entry Draft.

			Regular Season																Playoffs							
Season	Club	League	GP	G	A	Pts	PIM	PP	SH	GW	S	%	+/-	TF	F%	H	SB	Min	GP	G	A	Pts	PIM	PP	SH	GW
1983-84	Summerland	BCJHL	49	0	21	21	14	....	....	....	....	....	....	....	....	....	....	....	....	....	....	....	....	....	....	....
	Portland	WHL	5	0	0	0	5	....	....	....	....	....	....	....	....	....	....	....	5	0	1	1	4	....	....	....
1984-85	Portland	WHL	69	6	44	50	57	....	....	....	....	....	....	....	....	....	....	....	6	1	2	3	2	....	....	....
1985-86	Portland	WHL	70	11	59	70	83	....	....	....	....	....	....	....	....	....	....	....	15	1	7	8	16	....	....	....
1986-87	Portland	WHL	72	13	53	66	113	....	....	....	....	....	....	....	....	....	....	....	20	1	*21	22	27	....	....	....
1987-88	**NY Islanders**	**NHL**	**10**	**0**	**5**	**5**	**15**	**0**	**0**	**0**	**9**	**0.0**	**5**	....	....	....	....	....	**1**	**0**	**0**	**0**	**2**	**0**	**0**	**0**
	Springfield	AHL	52	5	18	23	50	....	....	....	....	....	....	....	....	....	....	....	....	....	....	....	....	....	....	....
1988-89	**NY Islanders**	**NHL**	**4**	**0**	**0**	**0**	**6**	**0**	**0**	**0**	**1**	**0.0**	**1**	....	....	....	....	....	....	....	....	....	....	....	....	....
	Springfield	AHL	65	3	16	19	55	....	....	....	....	....	....	....	....	....	....	....	....	....	....	....	....	....	....	....
1989-90	**NY Islanders**	**NHL**	**11**	**0**	**1**	**1**	**0**	**0**	**0**	**0**	**7**	**0.0**	**0**	....	....	....	....	....	**5**	**0**	**2**	**2**	**2**	**0**	**0**	**0**
	Springfield	AHL	57	1	15	16	41	....	....	....	....	....	....	....	....	....	....	....	13	1	4	5	23	....	....	....
1990-91	**NY Islanders**	**NHL**	**11**	**0**	**0**	**0**	**4**	**0**	**0**	**0**	**0**	**0.0**	**-1**	....	....	....	....	....	....	....	....	....	....	....	....	....
	Capital District	AHL	67	10	34	44	34	....	....	....	....	....	....	....	....	....	....	....	....	....	....	....	....	....	....	....
1991-92	**NY Islanders**	**NHL**	**51**	**1**	**10**	**11**	**26**	**0**	**0**	**0**	**25**	**4.0**	**-6**	....	....	....	....	....	....	....	....	....	....	....	....	....
	Capital District	AHL	20	1	9	10	6	....	....	....	....	....	....	....	....	....	....	....	....	....	....	....	....	....	....	....
1992-93	Capital District	AHL	61	6	29	35	34	....	....	....	....	....	....	....	....	....	....	....	4	0	1	1	0	....	....	....
1993-94	**Philadelphia**	**NHL**	**55**	**1**	**8**	**9**	**24**	**0**	**0**	**0**	**43**	**2.3**	**16**	....	....	....	....	....	....	....	....	....	....	....	....	....
1994-95	Hershey Bears	AHL	36	2	9	11	33	....	....	....	....	....	....	....	....	....	....	....	6	0	1	1	8	....	....	....
1995-96	**Winnipeg**	**NHL**	**65**	**1**	**5**	**6**	**81**	**0**	**0**	**0**	**27**	**3.7**	**-2**	....	....	....	....	....	**6**	**0**	**0**	**0**	**4**	**0**	**0**	**0**
	Springfield	AHL	14	3	12	15	22	....	....	....	....	....	....	....	....	....	....	....	....	....	....	....	....	....	....	....
1996-97	**Phoenix**	**NHL**	**65**	**3**	**7**	**10**	**40**	**1**	**0**	**1**	**38**	**7.9**	**-8**	....	....	....	....	....	**1**	**0**	**0**	**0**	**2**	**0**	**0**	**0**
1997-98	**NY Rangers**	**NHL**	**63**	**1**	**6**	**7**	**55**	**0**	**0**	**0**	**32**	**3.1**	**-3**	....	....	....	....	....	....	....	....	....	....	....	....	....
1998-99	**NY Rangers**	**NHL**	**2**	**0**	**0**	**0**	**0**	**0**	**0**	**0**	**0**	**0.0**	**-1**	**0**	**0.0**	**2**	**2**	**11:40**	....	....	....	....	....	....	....	....
	Hartford	AHL	42	2	10	12	28	....	....	....	....	....	....	....	....	....	....	....	....	....	....	....	....	....	....	....
	St. Louis	**NHL**	**30**	**1**	**2**	**3**	**20**	**0**	**0**	**0**	**16**	**6.3**	**12**	**0**	**0.0**	**35**	**28**	**17:36**	**13**	**1**	**2**	**3**	**8**	**0**	**0**	**1**
99-2000	**St. Louis**	**NHL**	**74**	**2**	**8**	**10**	**38**	**0**	**0**	**2**	**31**	**6.5**	**26**	**1**	**100.0**	**121**	**67**	**17:49**	**7**	**0**	**2**	**2**	**6**	**0**	**0**	**0**
2000-01	**St. Louis**	**NHL**	**72**	**2**	**8**	**10**	**38**	**0**	**0**	**0**	**35**	**5.7**	**7**	**1**	**0.0**	**85**	**83**	**18:53**	**2**	**0**	**0**	**0**	**0**	**0**	**0**	**0**
2001-02	**St. Louis**	**NHL**	**78**	**0**	**6**	**6**	**30**	**0**	**0**	**0**	**39**	**0.0**	**12**	**0**	**0.0**	**96**	**125**	**18:32**	**10**	**0**	**0**	**0**	**8**	**0**	**0**	**0**
	NHL Totals		**591**	**12**	**66**	**78**	**377**	**1**	**0**	**3**	**303**	**4.0**		**2**	**50.0**	**339**	**305**	**18:15**	**45**	**1**	**6**	**7**	**30**	**0**	**0**	**1**

Rights traded to **Ottawa** by **NY Islanders** for Chris Luongo, June 30, 1993. Signed as a free agent by **Philadelphia**, July 30, 1993. Traded to **Winnipeg** by **Philadelphia** for Russ Romaniuk, June 27, 1995. Transferred to **Phoenix** after **Winnipeg** franchise relocated, July 1, 1996. Signed as a free agent by **NY Rangers**, August 18, 1997. Traded to **St. Louis** by **NY Rangers** with Geoff Smith for future considerations (Chris Kenady, February 22, 1999), February 13, 1999.

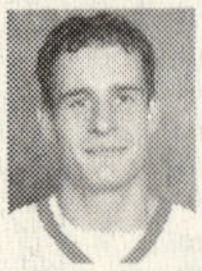

FISCHER, Jiri

(FIH-shuhr, YIH-ree) **DET.**

Defense. Shoots left. 6'5", 225 lbs. Born, Horovice, Czech., July 31, 1980. Detroit's 1st choice, 25th overall, in 1998 Entry Draft.

			Regular Season																Playoffs							
Season	Club	League	GP	G	A	Pts	PIM	PP	SH	GW	S	%	+/-	TF	F%	H	SB	Min	GP	G	A	Pts	PIM	PP	SH	GW
1995-96	Kladno Jr.	Czech-Jr.	39	6	10	16	....	....	....	....	....	....	....	....	....	....	....	....	....	....	....	....	....	....	....	....
1996-97	Kladno Jr.	Czech-Jr.	38	7	21	28	....	....	....	....	....	....	....	....	....	....	....	....	....	....	....	....	....	....	....	....
1997-98	Hull Olympiques	QMJHL	70	3	19	22	112	....	....	....	....	....	....	....	....	....	....	....	11	1	4	5	16	....	....	....
1998-99	Hull Olympiques	QMJHL	65	22	56	78	141	....	....	....	....	....	....	....	....	....	....	....	23	6	17	23	44	....	....	....
99-2000	**Detroit**	**NHL**	**52**	**0**	**8**	**8**	**45**	**0**	**0**	**0**	**41**	**0.0**	**1**	**0**	**0.0**	**68**	**25**	**10:51**	....	....	....	....	....	....	....	....
	Cincinnati	AHL	7	0	2	2	10	....	....	....	....	....	....	....	....	....	....	....	....	....	....	....	....	....	....	....
2000-01	**Detroit**	**NHL**	**55**	**1**	**8**	**9**	**59**	**0**	**0**	**0**	**64**	**1.6**	**3**	**0**	**0.0**	**131**	**37**	**16:46**	**5**	**0**	**0**	**0**	**9**	**0**	**0**	**0**
	Cincinnati	AHL	18	2	6	8	22	....	....	....	....	....	....	....	....	....	....	....	....	....	....	....	....	....	....	....
2001-02♦	**Detroit**	**NHL**	**80**	**2**	**8**	**10**	**67**	**0**	**0**	**1**	**103**	**1.9**	**17**	**0**	**0.0**	**171**	**117**	**17:10**	**22**	**3**	**3**	**6**	**30**	**0**	**0**	**1**
	NHL Totals		**187**	**3**	**24**	**27**	**171**	**0**	**0**	**1**	**208**	**1.4**		**0**	**0.0**	**370**	**179**	**15:18**	**27**	**3**	**3**	**6**	**39**	**0**	**0**	**1**

QMJHL First All-Star Team (1999)

FISHER, Mike

(FIH-shuhr, MIGHK) **OTT.**

Center. Shoots right. 6'1", 193 lbs. Born, Peterborough, Ont., June 5, 1980. Ottawa's 2nd choice, 44th overall, in 1998 Entry Draft.

			Regular Season																Playoffs							
Season	Club	League	GP	G	A	Pts	PIM	PP	SH	GW	S	%	+/-	TF	F%	H	SB	Min	GP	G	A	Pts	PIM	PP	SH	GW
1996-97	Peterborough	OPJHL	51	26	30	56	35	....	....	....	....	....	....	....	....	....	....	....	....	....	....	....	....	....	....	....
1997-98	Sudbury Wolves	OHL	66	24	25	49	65	....	....	....	....	....	....	....	....	....	....	....	9	2	2	4	13	....	....	....
1998-99	Sudbury Wolves	OHL	68	41	65	106	55	....	....	....	....	....	....	....	....	....	....	....	4	2	1	3	4	....	....	....
99-2000	**Ottawa**	**NHL**	**32**	**4**	**5**	**9**	**15**	**0**	**0**	**1**	**49**	**8.2**	**-6**	**356**	**47.8**	**76**	**13**	**12:57**	....	....	....	....	....	....	....	....
2000-01	**Ottawa**	**NHL**	**60**	**7**	**12**	**19**	**46**	**0**	**0**	**3**	**83**	**8.4**	**-1**	**709**	**50.2**	**129**	**32**	**11:38**	**4**	**0**	**1**	**1**	**4**	**0**	**0**	**0**
2001-02	**Ottawa**	**NHL**	**58**	**15**	**9**	**24**	**55**	**0**	**3**	**4**	**123**	**12.2**	**8**	**848**	**48.7**	**137**	**39**	**14:05**	**10**	**2**	**1**	**3**	**0**	**0**	**0**	**0**
	NHL Totals		**150**	**26**	**26**	**52**	**116**	**0**	**3**	**8**	**255**	**10.2**		**1913**	**49.1**	**342**	**84**	**12:52**	**14**	**2**	**2**	**4**	**4**	**0**	**0**	**0**

• Missed majority of 1999-2000 season recovering from knee injury suffered in game vs. Boston, December 30, 1999.

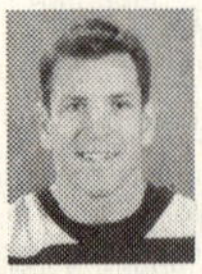

FITZGERALD, Tom

(FIHTZ-jair-uhld, TAWM) **TOR.**

Right wing/Center. Shoots right. 6', 195 lbs. Born, Billerica, MA, August 28, 1968. NY Islanders' 1st choice, 17th overall, in 1986 Entry Draft.

			Regular Season																Playoffs							
Season	Club	League	GP	G	A	Pts	PIM	PP	SH	GW	S	%	+/-	TF	F%	H	SB	Min	GP	G	A	Pts	PIM	PP	SH	GW
1984-85	Austin Mustangs	Hi-School	18	20	21	41	....	....	....	....	....	....	....	....	....	....	....	....	....	....	....	....	....	....	....	....
1985-86	Austin Mustangs	Hi-School	24	35	38	73	....	....	....	....	....	....	....	....	....	....	....	....	....	....	....	....	....	....	....	....
1986-87	Providence	H-East	27	8	14	22	22	....	....	....	....	....	....	....	....	....	....	....	....	....	....	....	....	....	....	....
1987-88	Providence	H-East	36	19	15	34	50	....	....	....	....	....	....	....	....	....	....	....	....	....	....	....	....	....	....	....
1988-89	**NY Islanders**	**NHL**	**23**	**3**	**5**	**8**	**10**	**0**	**0**	**1**	**24**	**12.5**	**1**	....	....	....	....	....	....	....	....	....	....	....	....	....
	Springfield	AHL	61	24	18	42	43	....	....	....	....	....	....	....	....	....	....	....	....	....	....	....	....	....	....	....
1989-90	**NY Islanders**	**NHL**	**19**	**2**	**5**	**7**	**4**	**0**	**0**	**1**	**24**	**8.3**	**-3**	....	....	....	....	....	**4**	**1**	**0**	**1**	**4**	**0**	**0**	**0**
	Springfield	AHL	53	30	23	53	32	....	....	....	....	....	....	....	....	....	....	....	14	2	9	11	13	....	....	....
1990-91	**NY Islanders**	**NHL**	**41**	**5**	**5**	**10**	**24**	**0**	**0**	**2**	**60**	**8.3**	**-9**	....	....	....	....	....	....	....	....	....	....	....	....	....
	Capital District	AHL	27	7	7	14	50	....	....	....	....	....	....	....	....	....	....	....	....	....	....	....	....	....	....	....
1991-92	**NY Islanders**	**NHL**	**45**	**6**	**11**	**17**	**28**	**0**	**2**	**2**	**71**	**8.5**	**-3**	....	....	....	....	....	....	....	....	....	....	....	....	....
	Capital District	AHL	4	1	1	2	4	....	....	....	....	....	....	....	....	....	....	....	....	....	....	....	....	....	....	....
1992-93	**NY Islanders**	**NHL**	**77**	**9**	**18**	**27**	**34**	**0**	**3**	**1**	**83**	**10.8**	**-2**	....	....	....	....	....	**18**	**2**	**5**	**7**	**18**	**0**	**0**	**0**
1993-94	**Florida**	**NHL**	**83**	**18**	**14**	**32**	**54**	**0**	**3**	**1**	**144**	**12.5**	**-3**	....	....	....	....	....	....	....	....	....	....	....	....	....
1994-95	**Florida**	**NHL**	**48**	**3**	**13**	**16**	**31**	**0**	**0**	**0**	**78**	**3.8**	**-3**	....	....	....	....	....	....	....	....	....	....	....	....	....
1995-96	**Florida**	**NHL**	**82**	**13**	**21**	**34**	**75**	**1**	**6**	**2**	**141**	**9.2**	**-3**	....	....	....	....	....	**22**	**4**	**4**	**8**	**34**	**0**	**0**	**2**
1996-97	**Florida**	**NHL**	**71**	**10**	**14**	**24**	**64**	**0**	**2**	**1**	**135**	**7.4**	**7**	....	....	....	....	....	**5**	**0**	**1**	**1**	**0**	**0**	**0**	**0**
1997-98	**Florida**	**NHL**	**69**	**10**	**5**	**15**	**57**	**0**	**1**	**1**	**105**	**9.5**	**-4**	....	....	....	....	....	....	....	....	....	....	....	....	....
	Colorado	**NHL**	**11**	**2**	**1**	**3**	**22**	**0**	**1**	**0**	**14**	**14.3**	**0**	....	....	....	....	....	**7**	**0**	**1**	**1**	**20**	**0**	**0**	**0**
1998-99	**Nashville**	**NHL**	**80**	**13**	**19**	**32**	**48**	**0**	**0**	**1**	**180**	**7.2**	**-18**	**155**	**52.3**	**70**	**49**	**17:17**	....	....	....	....	....	....	....	....
99-2000	**Nashville**	**NHL**	**82**	**13**	**9**	**22**	**66**	**0**	**3**	**1**	**119**	**10.9**	**-18**	**264**	**51.9**	**75**	**43**	**13:57**	....	....	....	....	....	....	....	....
2000-01	**Nashville**	**NHL**	**82**	**9**	**9**	**18**	**71**	**0**	**2**	**2**	**135**	**6.7**	**-5**	**458**	**54.6**	**96**	**42**	**14:58**	....	....	....	....	....	....	....	....
2001-02	**Nashville**	**NHL**	**63**	**7**	**9**	**16**	**33**	**0**	**1**	**0**	**101**	**6.9**	**-4**	**525**	**48.4**	**64**	**30**	**14:15**	....	....	....	....	....	....	....	....
	Chicago	**NHL**	**15**	**1**	**3**	**4**	**6**	**0**	**1**	**0**	**24**	**4.2**	**-3**	**110**	**50.0**	**38**	**9**	**16:16**	**5**	**0**	**0**	**0**	**4**	**0**	**0**	**0**
	NHL Totals		**891**	**124**	**161**	**285**	**627**	**1**	**25**	**16**	**1438**	**8.6**		**1512**	**51.4**	**343**	**173**	**15:12**	**61**	**7**	**11**	**18**	**80**	**0**	**0**	**2**

Claimed by **Florida** from **NY Islanders** in Expansion Draft, June 24, 1993. Traded to **Colorado** by **Florida** for the rights to Mark Parrish and Anaheim's 3rd round choice (previously acquired, Florida selected Lance Ward) in 1998 Entry Draft, March 24, 1998. Signed as a free agent by **Nashville**, July 6, 1998. Traded to **Chicago** by **Nashville** for Chicago's 4th round choice in 2003 Entry Draft and future considerations, March 13, 2002. Signed as a free agent by **Toronto**, July 17, 2002.

FITZPATRICK, Rory

(fitz-PA-trihk, ROHR-ee) **BUF.**

Defense. Shoots right. 6'2", 215 lbs. Born, Rochester, NY, January 11, 1975. Montreal's 2nd choice, 47th overall, in 1993 Entry Draft.

			Regular Season																Playoffs							
Season	Club	League	GP	G	A	Pts	PIM	PP	SH	GW	S	%	+/-	TF	F%	H	SB	Min	GP	G	A	Pts	PIM	PP	SH	GW
1990-91	Rochester	NAJHL	40	0	5	5	....	....	....	....	....	....	....	....	....	....	....	....	....	....	....	....	....	....	....	....
1991-92	Rochester	NAJHL	28	8	28	36	141	....	....	....	....	....	....	....	....	....	....	....	....	....	....	....	....	....	....	....
1992-93	Sudbury Wolves	OHL	58	4	20	24	68	....	....	....	....	....	....	....	....	....	....	....	14	0	0	0	17	....	....	....
1993-94	Sudbury Wolves	OHL	65	12	34	46	112	....	....	....	....	....	....	....	....	....	....	....	10	2	5	7	10	....	....	....
1994-95	Sudbury Wolves	OHL	56	12	36	48	72	....	....	....	....	....	....	....	....	....	....	....	18	3	15	18	21	....	....	....
	Fredericton	AHL	....	....	....	....	....	....	....	....	....	....	....	....	....	....	....	....	10	1	2	3	5	....	....	....
1995-96	**Montreal**	**NHL**	**42**	**0**	**2**	**2**	**18**	**0**	**0**	**0**	**31**	**0.0**	**-7**	....	....	....	....	....	**6**	**1**	**1**	**2**	**0**	**0**	**0**	**0**
	Fredericton	AHL	18	4	6	10	36	....	....	....	....	....	....	....	....	....	....	....	....	....	....	....	....	....	....	....
1996-97	**Montreal**	**NHL**	**6**	**0**	**1**	**1**	**6**	**0**	**0**	**0**	**5**	**0.0**	**-2**	....	....	....	....	....	....	....	....	....	....	....	....	....
	St. Louis	**NHL**	**2**	**0**	**0**	**0**	**2**	**0**	**0**	**0**	**1**	**0.0**	**-2**	....	....	....	....	....	....	....	....	....	....	....	....	....
	Worcester	AHL	49	4	13	17	78	....	....	....	....	....	....	....	....	....	....	....	5	1	2	3	0	....	....	....
1997-98	Worcester	AHL	62	8	22	30	111	....	....	....	....	....	....	....	....	....	....	....	11	0	3	3	26	....	....	....
1998-99	**St. Louis**	**NHL**	**1**	**0**	**0**	**0**	**2**	**0**	**0**	**0**	**0**	**0.0**	**-3**	**0**	**0.0**	**0**	**0**	**4:49**	....	....	....	....	....	....	....	....
	Worcester	AHL	53	5	16	21	82	....	....	....	....	....	....	....	....	....	....	....	4	0	1	1	17	....	....	....
99-2000	Worcester	AHL	28	0	5	5	48	....	....	....	....	....	....	....	....	....	....	....	....	....	....	....	....	....	....	....
	Milwaukee	IHL	27	2	1	3	27	....	....	....	....	....	....	....	....	....	....	....	3	0	2	2	2	....	....	....
2000-01	**Nashville**	**NHL**	**2**	**0**	**0**	**0**	**2**	**0**	**0**	**0**	**0**	**0.0**	**-2**	**0**	**0.0**	**1**	**0**	**9:47**	....	....	....	....	....	....	....	....
	Milwaukee	IHL	22	0	2	2	32	....	....	....	....	....	....	....	....	....	....	....	....	....	....	....	....	....	....	....
	Hamilton	AHL	34	3	17	20	29	....	....	....	....	....	....	....	....	....	....	....	....	....	....	....	....	....	....	....
2001-02	**Buffalo**	**NHL**	**5**	**0**	**0**	**0**	**4**	**0**	**0**	**0**	**2**	**0.0**	**-2**	**0**	**0.0**	**6**	**4**	**11:54**	....	....	....	....	....	....	....	....
	Rochester	AHL	60	4	8	12	83	....	....	....	....	....	....	....	....	....	....	....	2	0	1	1	0	....	....	....
	NHL Totals		**58**	**0**	**3**	**3**	**34**	**0**	**0**	**0**	**39**	**0.0**		**0**	**0.0**	**7**	**4**	**10:29**	**6**	**1**	**1**	**2**	**0**	**0**	**0**	**0**

OHL All-Rookie Team (1993)

Traded to **St. Louis** by **Montreal** with Pierre Turgeon and Craig Conroy for Murray Baron, Shayne Corson and St. Louis' 5th round choice (Gennady Razin) in 1997 Entry Draft, October 29, 1996. Claimed by **Boston** from **St. Louis** in NHL Waiver Draft, October 5, 1998. Claimed on waivers by **St. Louis** from **Boston**, October 7, 1998. Traded to **Nashville** by **St. Louis** for Dan Keczmer, February 9, 2000. Traded to **Edmonton** by **Nashville** for future considerations, January 12, 2001. Signed as a free agent by **Buffalo**, August 14, 2001.

FLEURY, Theoren

(FLUH-ree, THAIR-ihn) **CHI.**

Right wing. Shoots right. 5'6", 180 lbs. Born, Oxbow, Sask., June 29, 1968. Calgary's 9th choice, 166th overall, in 1987 Entry Draft.

			Regular Season																Playoffs							
Season	Club	League	GP	G	A	Pts	PIM	PP	SH	GW	S	%	+/-	TF	F%	H	SB	Min	GP	G	A	Pts	PIM	PP	SH	GW
1983-84	St. James	MJHL	22	33	31	64	88	....	....	....	....	....	....	....	....	....	....	....	....	....	....	....	....	....	....	....
1984-85	Moose Jaw	WHL	71	29	46	75	82	....	....	....	....	....	....	....	....	....	....	....	....	....	....	....	....	....	....	....
1985-86	Moose Jaw	WHL	72	43	65	108	124	....	....	....	....	....	....	....	....	....	....	....	13	7	13	20	16	....	....	....
1986-87	Moose Jaw	WHL	66	61	68	129	110	....	....	....	....	....	....	....	....	....	....	....	9	7	9	16	34	....	....	....
1987-88	Moose Jaw	WHL	65	68	92	*160	235	....	....	....	....	....	....	....	....	....	....	....	....	....	....	....	....	....	....	....
	Salt Lake	IHL	2	3	4	7	7	....	....	....	....	....	....	....	....	....	....	....	8	11	5	16	16	....	....	....
1988-89♦	**Calgary**	**NHL**	**36**	**14**	**20**	**34**	**46**	**5**	**0**	**3**	**89**	**15.7**	**5**	....	....	....	....	....	**22**	**5**	**6**	**11**	**24**	**3**	**0**	**3**
	Salt Lake	IHL	40	37	37	74	81	....	....	....	....	....	....	....	....	....	....	....	....	....	....	....	....	....	....	....
1989-90	**Calgary**	**NHL**	**80**	**31**	**35**	**66**	**157**	**9**	**3**	**6**	**200**	**15.5**	**22**	....	....	....	....	....	**6**	**2**	**3**	**5**	**10**	**0**	**0**	**0**
1990-91	**Calgary**	**NHL**	**79**	**51**	**53**	**104**	**136**	**9**	**7**	**9**	**249**	**20.5**	**48**	....	....	....	....	....	**7**	**2**	**5**	**7**	**14**	**0**	**0**	**1**
1991-92	**Calgary**	**NHL**	**80**	**33**	**40**	**73**	**133**	**11**	**1**	**6**	**225**	**14.7**	**0**	....	....	....	....	....	....	....	....	....	....	....	....	....
1992-93	**Calgary**	**NHL**	**83**	**34**	**66**	**100**	**88**	**12**	**2**	**4**	**250**	**13.6**	**14**	....	....	....	....	....	**6**	**5**	**7**	**12**	**27**	**3**	**1**	**0**
1993-94	**Calgary**	**NHL**	**83**	**40**	**45**	**85**	**186**	**16**	**1**	**6**	**278**	**14.4**	**30**	....	....	....	....	....	**7**	**6**	**4**	**10**	**5**	**1**	**0**	**2**
1994-95	Tappara Tampere	Finland	10	8	9	17	22	....	....	....	....	....	....	....	....	....	....	....	....	....	....	....	....	....	....	....
	Calgary	**NHL**	**47**	**29**	**29**	**58**	**112**	**9**	**2**	**5**	**173**	**16.8**	**6**	....	....	....	....	....	**7**	**7**	**7**	**14**	**2**	**2**	**1**	**0**
1995-96	**Calgary**	**NHL**	**80**	**46**	**50**	**96**	**112**	**17**	**5**	**4**	**353**	**13.0**	**17**	....	....	....	....	....	**4**	**2**	**1**	**3**	**14**	**0**	**0**	**0**
1996-97	**Calgary**	**NHL**	**81**	**29**	**38**	**67**	**104**	**9**	**2**	**3**	**336**	**8.6**	**-12**	....	....	....	....	....	....	....	....	....	....	....	....	....

			Regular Season																Playoffs							
Season	Club	League	GP	G	A	Pts	PIM	PP	SH	GW	S	%	+/-	TF	F%	H	SB	Min	GP	G	A	Pts	PIM	PP	SH	GW
1997-98	**Calgary**	**NHL**	**82**	**27**	**51**	**78**	**197**	**3**	**2**	**4**	**282**	**9.6**	**0**													
	Canada	Olympics	6	1	3	4	2																			
1998-99	**Calgary**	**NHL**	**60**	**30**	**39**	**69**	**68**	**7**	**3**	**3**	**250**	**12.0**	**18**	**517**	**59.2**	**51**	**25**	**23:33**								
	Colorado	**NHL**	**15**	**10**	**14**	**24**	**18**	**1**	**0**	**2**	**51**	**19.6**	**8**	**150**	**58.7**	**12**	**3**	**22:33**	**18**	**5**	**12**	**17**	**20**	**2**	**0**	**0**
99-2000	**NY Rangers**	**NHL**	**80**	**15**	**49**	**64**	**68**	**1**	**0**	**1**	**246**	**6.1**	**–4**	**490**	**56.5**	**69**	**31**	**19:41**								
2000-01	**NY Rangers**	**NHL**	**62**	**30**	**44**	**74**	**122**	**8**	**7**	**3**	**238**	**12.6**	**0**	**139**	**47.5**	**63**	**30**	**21:47**								
2001-02	**NY Rangers**	**NHL**	**82**	**24**	**39**	**63**	**216**	**7**	**0**	**5**	**267**	**9.0**	**0**	**170**	**47.7**	**113**	**51**	**20:48**								
	Canada	Olympics	6	0	2	2	6																			
	NHL Totals		**1030**	**443**	**612**	**1055**	**1763**	**124**	**35**	**64**	**3487**	**12.7**		**1466**	**55.8**	**308**	**140**	**21:21**	**77**	**34**	**45**	**79**	**116**	**11**	**2**	**6**

WHL East First All-Star Team (1987) • WHL East Second All-Star Team (1988) • Shared Alka-Seltzer Plus Award with Marty McSorley (1991) • NHL Second All-Star Team (1995) • Played in NHL All-Star Game (1991, 1992, 1996, 1997, 1998, 1999, 2001)

Traded to **Colorado** by **Calgary** with Chris Dingman for Rene Corbet, Wade Belak, Robyn Regehr and Colorado's 2nd round compensatory choice (Jarret Stoll) in 2000 Entry Draft, February 28, 1999. Signed as a free agent by **NY Rangers**, July 8, 1999. Traded to **San Jose** by **NY Rangers** to complete transaction that sent San Jose's 6th round choice (Kim Hirschovits) in 2002 Entry Draft for NY Rangers' 6th round choice in 2003 Entry Draft (June 23, 2002), June 26, 2002. Signed as a free agent by **Chicago**, August 15, 2002.

FLINN, Ryan

(FLIHN, RIGH-yan) **L.A.**

Left wing. Shoots left. 6'5", 223 lbs. Born, Halifax, N.S., April 20, 1980. New Jersey's 8th choice, 143rd overall, in 1998 Entry Draft.

Season	Club	League	GP	G	A	Pts	PIM	PP	SH	GW	S	%	+/-	TF	F%	H	SB	Min	GP	G	A	Pts	PIM	PP	SH	GW
1996-97	Laval Titan	QMJHL	23	3	2	5	56												2	0	0	0	0			
1997-98	Laval Titan	QMJHL	59	4	12	16	217												15	1	0	1	63			
1998-99	Acadie-Bathurst	QMJHL	44	3	4	7	195												23	2	0	2	37			
99-2000	Halifax	QMJHL	67	14	19	33	365																			
2000-01	Cape Breton	QMJHL	57	16	17	33	280												9	1	1	2	43			
2001-02	Reading Royals	ECHL	20	1	3	4	130																			
	Los Angeles	**NHL**	**10**	**0**	**0**	**0**	**51**	**0**	**0**	**0**	**2**	**0.0**	**0**	**0**	**0.0**	**11**	**0**	**3:29**								
	Manchester	AHL	37	0	1	1	113												1	0	0	0	0			
	NHL Totals		**10**	**0**	**0**	**0**	**51**	**0**	**0**	**0**	**2**	**0.0**		**0**	**0.0**	**11**	**0**	**3:29**								

Traded to **Cape Breton** (QMJHL) by **Halifax** (QMJHL) for Cape Breton's 4th round (Vincent Lambert) and 10th round choice (later traded to Acadie-Bathurst who selected Charles Bergeron) in 2001 QMJHL Priority Draft, October 1, 1999. Signed as a free agent by **LA Kings**, January 8, 2002.

FOCHT, Dan

(FOHKT, DAN) **PHX.**

Defense. Shoots left. 6'6", 242 lbs. Born, Regina, Sask., December 31, 1977. Phoenix's 1st choice, 11th overall, in 1996 Entry Draft.

Season	Club	League	GP	G	A	Pts	PIM	PP	SH	GW	S	%	+/-	TF	F%	H	SB	Min	GP	G	A	Pts	PIM	PP	SH	GW
1994-95	Sask. Blazers	SMHL	33	6	12	18	98																			
1995-96	Tri-City	WHL	63	6	12	18	161												11	1	1	2	23			
1996-97	Tri-City	WHL	28	0	5	5	92																			
	Regina Pats	WHL	22	2	2	4	59												5	0	2	2	8			
	Springfield	AHL	1	0	0	0	2																			
1997-98	Springfield	AHL	61	2	5	7	125												3	0	0	0	4			
1998-99	Mississippi	ECHL	2	0	0	0	6																			
	Springfield	AHL	30	0	2	2	58												3	1	0	1	10			
99-2000	Jokerit Helsinki	Finland	2	0	0	0	0																			
	Mississippi	ECHL	4	0	1	1	0																			
	Springfield	AHL	44	2	9	11	86												5	0	1	1	2			
2000-01	Springfield	AHL	69	0	6	6	156																			
2001-02	**Phoenix**	**NHL**	**8**	**0**	**0**	**0**	**11**	**0**	**0**	**0**	**5**	**0.0**	**0**	**0**	**0.0**	**24**	**7**	**12:25**	**1**	**0**	**1**	**1**	**0**	**0**	**0**	**0**
	Springfield	AHL	56	2	8	10	134																			
	NHL Totals		**8**	**0**	**0**	**0**	**11**	**0**	**0**	**0**	**5**	**0.0**		**0**	**0.0**	**24**	**7**	**12:25**	**1**	**0**	**1**	**1**	**0**	**0**	**0**	**0**

FOOTE, Adam

(FUT, A-duhm) **COL.**

Defense. Shoots right. 6'2", 215 lbs. Born, Toronto, Ont., July 10, 1971. Quebec's 2nd choice, 22nd overall, in 1989 Entry Draft.

Season	Club	League	GP	G	A	Pts	PIM	PP	SH	GW	S	%	+/-	TF	F%	H	SB	Min	GP	G	A	Pts	PIM	PP	SH	GW
1987-88	Brooklin	OMHA	65	25	43	68	108																			
1988-89	Sault Ste. Marie	OHL	66	7	32	39	120																			
1989-90	Sault Ste. Marie	OHL	61	12	43	55	199																			
1990-91	Sault Ste. Marie	OHL	59	18	51	69	93												14	5	12	17	28			
1991-92	**Quebec**	**NHL**	**46**	**2**	**5**	**7**	**44**	**0**	**0**	**0**	**55**	**3.6**	**–4**													
	Halifax Citadels	AHL	6	0	1	1	2																			
1992-93	**Quebec**	**NHL**	**81**	**4**	**12**	**16**	**168**	**0**	**1**	**0**	**54**	**7.4**	**6**						**6**	**0**	**1**	**1**	**2**	**0**	**0**	**0**
1993-94	**Quebec**	**NHL**	**45**	**2**	**6**	**8**	**67**	**0**	**0**	**0**	**42**	**4.8**	**3**													
1994-95	**Quebec**	**NHL**	**35**	**0**	**7**	**7**	**52**	**0**	**0**	**0**	**24**	**0.0**	**17**						**6**	**0**	**1**	**1**	**14**	**0**	**0**	**0**
1995-96♦	**Colorado**	**NHL**	**73**	**5**	**11**	**16**	**88**	**1**	**0**	**1**	**49**	**10.2**	**27**						**22**	**1**	**3**	**4**	**36**	**0**	**0**	**0**
1996-97	**Colorado**	**NHL**	**78**	**2**	**19**	**21**	**135**	**0**	**0**	**0**	**60**	**3.3**	**16**						**17**	**0**	**4**	**4**	**62**	**0**	**0**	**0**
1997-98	**Colorado**	**NHL**	**77**	**3**	**14**	**17**	**124**	**0**	**0**	**1**	**64**	**4.7**	**–3**						**7**	**0**	**0**	**0**	**23**	**0**	**0**	**0**
	Canada	Olympics	6	0	1	1	4																			
1998-99	**Colorado**	**NHL**	**64**	**5**	**16**	**21**	**92**	**3**	**0**	**0**	**83**	**6.0**	**20**	**0**	**0.0**	**125**	**93**	**24:50**	**19**	**2**	**3**	**5**	**24**	**1**	**0**	**0**
99-2000	**Colorado**	**NHL**	**59**	**5**	**13**	**18**	**98**	**1**	**0**	**2**	**63**	**7.9**	**5**	**0**	**0.0**	**156**	**87**	**25:51**	**16**	**0**	**7**	**7**	**28**	**0**	**0**	**0**
2000-01♦	**Colorado**	**NHL**	**35**	**3**	**12**	**15**	**42**	**1**	**1**	**1**	**59**	**5.1**	**6**	**0**	**0.0**	**96**	**43**	**25:22**	**23**	**3**	**4**	**7**	***47**	**1**	**0**	**1**
2001-02	**Colorado**	**NHL**	**55**	**5**	**22**	**27**	**55**	**1**	**1**	**0**	**85**	**5.9**	**7**	**0**	**0.0**	**159**	**95**	**25:59**	**21**	**1**	**6**	**7**	**28**	**0**	**0**	**0**
	Canada	Olympics	6	1	0	1	2																			
	NHL Totals		**648**	**36**	**137**	**173**	**965**	**7**	**3**	**5**	**638**	**5.6**		**0**	**0.0**	**536**	**318**	**25:30**	**137**	**7**	**29**	**36**	**264**	**2**	**0**	**1**

OHL First All-Star Team (1991)

Transferred to **Colorado** after **Quebec** franchise relocated, June 21, 1995. • Missed majority of 2000-01 season recovering from shoulder injury suffered in game vs. Carolina, January 6, 2001.

FORBES, Colin

(FOHRBS, COHL-ihn) **WSH.**

Left wing. Shoots left. 6'3", 205 lbs. Born, New Westminster, B.C., February 16, 1976. Philadelphia's 5th choice, 166th overall, in 1994 Entry Draft.

Season	Club	League	GP	G	A	Pts	PIM	PP	SH	GW	S	%	+/-	TF	F%	H	SB	Min	GP	G	A	Pts	PIM	PP	SH	GW
1993-94	Sherwood Park	AJHL	47	18	22	40	76																			
1994-95	Portland	WHL	72	24	31	55	108												9	1	3	4	10			
1995-96	Portland	WHL	72	33	44	77	137												7	2	5	7	14			
	Hershey Bears	AHL	2	1	0	1	2												4	0	2	2	2			
1996-97	**Philadelphia**	**NHL**	**3**	**1**	**0**	**1**	**0**	**0**	**0**	**0**	**3**	**33.3**	**0**						**3**	**0**	**0**	**0**	**0**	**0**	**0**	**0**
	Philadelphia	AHL	74	21	28	49	108												10	5	5	10	33			
1997-98	**Philadelphia**	**NHL**	**63**	**12**	**7**	**19**	**59**	**2**	**0**	**2**	**93**	**12.9**	**2**						**5**	**0**	**0**	**0**	**2**	**0**	**0**	**0**
	Philadelphia	AHL	13	7	4	11	22																			
1998-99	**Philadelphia**	**NHL**	**66**	**9**	**7**	**16**	**51**	**0**	**0**	**4**	**92**	**9.8**	**0**	**2**	**50.0**	**46**	**10**	**12:35**								
	Tampa Bay	**NHL**	**14**	**3**	**1**	**4**	**10**	**0**	**1**	**0**	**25**	**12.0**	**–5**	**0**	**0.0**	**21**	**4**	**17:30**								
99-2000	**Tampa Bay**	**NHL**	**8**	**0**	**0**	**0**	**18**	**0**	**0**	**0**	**3**	**0.0**	**–4**	**1**	**0.0**	**9**	**0**	**8:53**								
	Ottawa	**NHL**	**45**	**2**	**5**	**7**	**12**	**0**	**0**	**0**	**54**	**3.7**	**–1**	**82**	**47.6**	**70**	**6**	**8:34**	**5**	**1**	**0**	**1**	**14**	**0**	**0**	**0**
2000-01	**Ottawa**	**NHL**	**39**	**0**	**1**	**1**	**31**	**0**	**0**	**0**	**26**	**0.0**	**–3**	**10**	**40.0**	**60**	**3**	**6:10**								
	NY Rangers	**NHL**	**19**	**1**	**4**	**5**	**15**	**0**	**0**	**0**	**20**	**5.0**	**–3**	**1**	**0.0**	**28**	**5**	**7:52**								
2001-02	Utah Grizzlies	AHL	4	0	0	0	21																			
	Washington	**NHL**	**38**	**5**	**3**	**8**	**15**	**0**	**1**	**1**	**49**	**10.2**	**–2**	**253**	**44.7**	**41**	**12**	**11:01**								
	Portland Pirates	AHL	14	4	5	9	18																			
	NHL Totals		**295**	**33**	**28**	**61**	**211**	**2**	**2**	**7**	**365**	**9.0**		**349**	**45.0**	**275**	**40**	**10:13**	**13**	**1**	**0**	**1**	**16**	**0**	**0**	**0**

Traded to **Tampa Bay** by **Philadelphia** with Philadelphia's 4th round choice (Michal Lanicek) in 1999 Entry Draft for Mikael Andersson and Sandy McCarthy, March 20, 1999. Traded to **Ottawa** by **Tampa Bay** for Bruce Gardiner, November 11, 1999. Traded to **NY Rangers** by **Ottawa** for Eric Lacroix, March 1, 2001. Signed as a free agent by **Washington**, January 8, 2002.

FORSBERG, Peter

(FOHRS-buhrg, PEE-tuhr) **COL.**

Center. Shoots left. 6', 205 lbs. Born, Ornskoldsvik, Sweden, July 20, 1973. Philadelphia's 1st choice, 6th overall, in 1991 Entry Draft.

Season	Club	League	GP	G	A	Pts	PIM	PP	SH	GW	S	%	+/-	TF	F%	H	SB	Min	GP	G	A	Pts	PIM	PP	SH	GW
1989-90	MoDo Jr.	Swede-Jr.	30	15	12	27	42																			
	MoDo	Sweden	1	0	1	1	4																			
1990-91	MoDo Jr.	Swede-Jr.	39	38	64	102	56																			
	MoDo	Sweden	23	7	10	17	22																			
1991-92	MoDo	Sweden	39	9	18	27	78																			
1992-93	MoDo Jr.	Swede-Jr.	2	0	3	3	4																			
	MoDo	Sweden	39	23	24	47	92												3	4	1	5	0			

			Regular Season																Playoffs							
Season	Club	League	GP	G	A	Pts	PIM	PP	SH	GW	S	%	+/-	TF	F%	H	SB	Min	GP	G	A	Pts	PIM	PP	SH	GW
1993-94	MoDo	Sweden	39	18	26	44	82	....	....	....	....	....	....	....	....	....	....	....	11	9	7	16	14	....	....	....
	Sweden	Olympics	8	2	6	8	6	....	....	....	....	....	....	....	....	....	....	....	....	....	....	....	....	....	....	....
1994-95	MoDo	Sweden	11	5	9	14	20	....	....	....	....	....	....	....	....	....	....	....	....	....	....	....	....	....	....	....
	Quebec	**NHL**	**47**	**15**	**35**	**50**	**16**	**3**	**0**	**3**	**86**	**17.4**	**17**	....	....	....	....	....	**6**	**2**	**4**	**6**	**4**	**1**	**0**	**0**
1995-96♦	**Colorado**	**NHL**	**82**	**30**	**86**	**116**	**47**	**7**	**3**	**3**	**217**	**13.8**	**26**	....	....	....	....	....	**22**	**10**	**11**	**21**	**18**	**3**	**0**	**1**
1996-97	**Colorado**	**NHL**	**65**	**28**	**58**	**86**	**73**	**5**	**4**	**4**	**188**	**14.9**	**31**	....	....	....	....	....	**14**	**5**	**12**	**17**	**10**	**3**	**0**	**0**
1997-98	**Colorado**	**NHL**	**72**	**25**	**66**	**91**	**94**	**7**	**3**	**7**	**202**	**12.4**	**6**	....	....	....	....	....	**7**	**6**	**5**	**11**	**12**	**2**	**0**	**0**
	Sweden	Olympics	4	1	4	5	6	....	....	....	....	....	....	....	....	....	....	....	....	....	....	....	....	....	....	....
1998-99	**Colorado**	**NHL**	**78**	**30**	**67**	**97**	**108**	**9**	**2**	**7**	**217**	**13.8**	**27**	**895**	**54.4**	**108**	**31**	**23:29**	**19**	**8**	**16**	***24**	**31**	**1**	**1**	**0**
99-2000	**Colorado**	**NHL**	**49**	**14**	**37**	**51**	**52**	**3**	**0**	**2**	**105**	**13.3**	**9**	**519**	**46.6**	**70**	**19**	**20:55**	**16**	**7**	**8**	**15**	**12**	**2**	**1**	**4**
2000-01♦	**Colorado**	**NHL**	**73**	**27**	**62**	**89**	**54**	**12**	**2**	**5**	**178**	**15.2**	**23**	**755**	**46.6**	**85**	**19**	**20:48**	**11**	**4**	**10**	**14**	**6**	**1**	**0**	**2**
2001-02	**Colorado**	**NHL**	....	....	....	....	....	....	....	....	....	....	....	....	....	....	....	....	**20**	**9**	***18**	***27**	**20**	**0**	**0**	**4**
	NHL Totals		**466**	**169**	**411**	**580**	**444**	**46**	**14**	**31**	**1193**	**14.2**		**2169**	**49.8**	**263**	**69**	**21:52**	**115**	**51**	**84**	**135**	**113**	**13**	**2**	**11**

NHL All-Rookie Team (1995) • Won Calder Memorial Trophy (1995) • NHL First All-Star Team (1998, 1999) • Played in NHL All-Star Game (1996, 1998, 1999, 2001)

Traded to **Quebec** by **Philadelphia** with Steve Duchesne, Kerry Huffman, Mike Ricci, Ron Hextall, Philadelphia's 1st round choice (Jocelyn Thibault) in 1993 Entry Draft, $15,000,000 and future considerations (Chris Simon and Philadelphia's 1st round choice (later traded to Toronto - later traded to Washington - Washington selected Nolan Baumgartner) in 1994 Entry Draft, July 21, 1992) for Eric Lindros, June 30, 1992. Transferred to **Colorado** after **Quebec** franchise relocated, June 21, 1995. • Missed entire 2001-02 regular season recovering from spleen injury suffered in game vs. LA Kings, May 10, 2001 and ankle injury suffered in practice, January 10, 2002.

FORTIN, Jean-Francois (fohr-TEHN, ZHAWN-fran-SWUH) WSH.

Defense. Shoots right. 6'2", 205 lbs. Born, Laval, Que., March 15, 1979. Washington's 2nd choice, 35th overall, in 1997 Entry Draft.

Season	Club	League	GP	G	A	Pts	PIM	PP	SH	GW	S	%	+/-	TF	F%	H	SB	Min	GP	G	A	Pts	PIM	PP	SH	GW
1993-94	Laval Laurentide	QAHA	31	8	20	28	32	....	....	....	....	....	....	....	....	....	....	....	....	....	....	....	....	....	....	....
1994-95	Abitibi	QAAA	44	2	14	16	34	....	....	....	....	....	....	....	....	....	....	....	10	2	2	4	....	....	....	....
1995-96	Sherbrooke	QMJHL	69	7	15	22	40	....	....	....	....	....	....	....	....	....	....	....	7	2	6	8	2	....	....	....
1996-97	Sherbrooke	QMJHL	59	7	30	37	89	....	....	....	....	....	....	....	....	....	....	....	2	0	1	1	14	....	....	....
1997-98	Sherbrooke	QMJHL	55	12	25	37	37	....	....	....	....	....	....	....	....	....	....	....	....	....	....	....	....	....	....	....
1998-99	Sherbrooke	QMJHL	64	17	33	50	78	....	....	....	....	....	....	....	....	....	....	....	12	5	13	18	20	....	....	....
99-2000	Portland Pirates	AHL	43	3	5	8	44	....	....	....	....	....	....	....	....	....	....	....	2	0	0	0	0	....	....	....
	Hampton Roads	ECHL	7	0	2	2	0	....	....	....	....	....	....	....	....	....	....	....	....	....	....	....	....	....	....	....
2000-01	Richmond	ECHL	15	0	4	4	2	....	....	....	....	....	....	....	....	....	....	....	....	....	....	....	....	....	....	....
	Portland Pirates	AHL	32	1	7	8	22	....	....	....	....	....	....	....	....	....	....	....	1	0	0	0	0	....	....	....
2001-02	**Washington**	**NHL**	**36**	**1**	**3**	**4**	**20**	**0**	**0**	**0**	**24**	**4.2**	**-1**	**1**	**100.0**	**47**	**35**	**19:25**	....	....	....	....	....	....	....	....
	Portland Pirates	AHL	44	4	9	13	20	....	....	....	....	....	....	....	....	....	....	....	....	....	....	....	....	....	....	....
	NHL Totals		**36**	**1**	**3**	**4**	**20**	**0**	**0**	**0**	**24**	**4.2**		**1**	**100.0**	**47**	**35**	**19:25**	....	....	....	....	....	....	....	....

FRANCIS, Ron (FRAN-sihs, RAWN) CAR.

Center. Shoots left. 6'3", 200 lbs. Born, Sault Ste. Marie, Ont., March 1, 1963. Hartford's 1st choice, 4th overall, in 1981 Entry Draft.

Season	Club	League	GP	G	A	Pts	PIM	PP	SH	GW	S	%	+/-	TF	F%	H	SB	Min	GP	G	A	Pts	PIM	PP	SH	GW
1979-80	Soo Legion	NOHA	45	57	92	149	....	....	....	....	....	....	....	....	....	....	....	....	....	....	....	....	....	....	....	....
1980-81	Sault Ste. Marie	OMJHL	64	26	43	69	33	....	....	....	....	....	....	....	....	....	....	....	19	7	8	15	34	....	....	....
1981-82	Sault Ste. Marie	OHL	25	18	30	48	46	....	....	....	....	....	....	....	....	....	....	....	....	....	....	....	....	....	....	....
	Hartford	**NHL**	**59**	**25**	**43**	**68**	**51**	**12**	**0**	**1**	**163**	**15.3**	**-13**	....	....	....	....	....	....	....	....	....	....	....	....	....
1982-83	**Hartford**	**NHL**	**79**	**31**	**59**	**90**	**60**	**4**	**2**	**4**	**212**	**14.6**	**-25**	....	....	....	....	....	....	....	....	....	....	....	....	....
1983-84	**Hartford**	**NHL**	**72**	**23**	**60**	**83**	**45**	**5**	**0**	**5**	**202**	**11.4**	**-10**	....	....	....	....	....	....	....	....	....	....	....	....	....
1984-85	**Hartford**	**NHL**	**80**	**24**	**57**	**81**	**66**	**4**	**0**	**1**	**195**	**12.3**	**-23**	....	....	....	....	....	....	....	....	....	....	....	....	....
1985-86	**Hartford**	**NHL**	**53**	**24**	**53**	**77**	**24**	**7**	**1**	**4**	**120**	**20.0**	**8**	....	....	....	....	....	**10**	**1**	**2**	**3**	**4**	**0**	**0**	**0**
1986-87	**Hartford**	**NHL**	**75**	**30**	**63**	**93**	**45**	**7**	**0**	**7**	**189**	**15.9**	**10**	....	....	....	....	....	**6**	**2**	**2**	**4**	**6**	**1**	**0**	**0**
1987-88	**Hartford**	**NHL**	**80**	**25**	**50**	**75**	**87**	**11**	**1**	**3**	**172**	**14.5**	**-8**	....	....	....	....	....	**6**	**2**	**5**	**7**	**2**	**1**	**0**	**0**
1988-89	**Hartford**	**NHL**	**69**	**29**	**48**	**77**	**36**	**8**	**0**	**4**	**156**	**18.6**	**4**	....	....	....	....	....	**4**	**0**	**2**	**2**	**0**	**0**	**0**	**0**
1989-90	**Hartford**	**NHL**	**80**	**32**	**69**	**101**	**73**	**15**	**1**	**5**	**170**	**18.8**	**13**	....	....	....	....	....	**7**	**3**	**3**	**6**	**8**	**1**	**0**	**0**
1990-91	**Hartford**	**NHL**	**67**	**21**	**55**	**76**	**51**	**10**	**1**	**6**	**149**	**14.1**	**-2**	....	....	....	....	....	....	....	....	....	....	....	....	....
	♦ Pittsburgh	**NHL**	**14**	**2**	**9**	**11**	**21**	**0**	**0**	**1**	**25**	**8.0**	**0**	....	....	....	....	....	**24**	**7**	**10**	**17**	**24**	**0**	**0**	**4**
1991-92♦	**Pittsburgh**	**NHL**	**70**	**21**	**33**	**54**	**30**	**5**	**1**	**2**	**121**	**17.4**	**-7**	....	....	....	....	....	**21**	**8**	***19**	**27**	**6**	**2**	**0**	**2**
1992-93	**Pittsburgh**	**NHL**	**84**	**24**	**76**	**100**	**68**	**9**	**2**	**4**	**215**	**11.2**	**6**	....	....	....	....	....	**12**	**6**	**11**	**17**	**19**	**1**	**0**	**1**
1993-94	**Pittsburgh**	**NHL**	**82**	**27**	**66**	**93**	**62**	**8**	**0**	**2**	**216**	**12.5**	**-3**	....	....	....	....	....	**6**	**0**	**2**	**2**	**6**	**0**	**0**	**0**
1994-95	**Pittsburgh**	**NHL**	**44**	**11**	***48**	**59**	**18**	**3**	**0**	**1**	**94**	**11.7**	**30**	....	....	....	....	....	**12**	**6**	**13**	**19**	**4**	**2**	**0**	**0**
1995-96	**Pittsburgh**	**NHL**	**77**	**27**	***92**	**119**	**56**	**12**	**1**	**4**	**158**	**17.1**	**25**	....	....	....	....	....	**11**	**3**	**6**	**9**	**4**	**2**	**0**	**1**
1996-97	**Pittsburgh**	**NHL**	**81**	**27**	**63**	**90**	**20**	**10**	**1**	**2**	**183**	**14.8**	**7**	....	....	....	....	....	**5**	**1**	**2**	**3**	**2**	**1**	**0**	**0**
1997-98	**Pittsburgh**	**NHL**	**81**	**25**	**62**	**87**	**20**	**7**	**0**	**5**	**189**	**13.2**	**12**	....	....	....	....	....	**6**	**1**	**5**	**6**	**2**	**0**	**0**	**0**
1998-99	**Carolina**	**NHL**	**82**	**21**	**31**	**52**	**34**	**8**	**0**	**2**	**133**	**15.8**	**-2**	**1589**	**51.5**	**36**	**57**	**21:55**	**3**	**0**	**1**	**1**	**0**	**0**	**0**	**0**
99-2000	**Carolina**	**NHL**	**78**	**23**	**50**	**73**	**18**	**7**	**0**	**4**	**150**	**15.3**	**10**	**1566**	**53.3**	**48**	**53**	**21:58**	....	....	....	....	....	....	....	....
2000-01	**Carolina**	**NHL**	**82**	**15**	**50**	**65**	**32**	**7**	**0**	**4**	**130**	**11.5**	**-15**	**1271**	**57.5**	**37**	**54**	**20:15**	**3**	**0**	**0**	**0**	**0**	**0**	**0**	**0**
2001-02	**Carolina**	**NHL**	**80**	**27**	**50**	**77**	**18**	**14**	**0**	**5**	**165**	**16.4**	**4**	**1136**	**58.9**	**19**	**47**	**20:41**	**23**	**6**	**10**	**16**	**6**	**4**	**0**	**3**
	NHL Totals		**1569**	**514**	**1187**	**1701**	**935**	**173**	**11**	**76**	**3507**	**14.7**		**5562**	**54.9**	**140**	**211**	**21:12**	**159**	**46**	**93**	**139**	**93**	**15**	**0**	**11**

Won Alka-Seltzer Plus Award (1995) • Won Frank J. Selke Trophy (1995) • Won Lady Byng Trophy (1995, 1998, 2002) • Won King Clancy Memorial Trophy (2002) • Played in NHL All-Star Game (1983, 1985, 1990, 1996)

Traded to **Pittsburgh** by **Hartford** with Grant Jennings and Ulf Samuelsson for John Cullen, Jeff Parker and Zarley Zalapski, March 4, 1991. Signed as a free agent by **Carolina**, July 13, 1998.

FREADRICH, Kyle (FREE-drihk, KIGHL)

Left wing. Shoots left. 6'7", 260 lbs. Born, Edmonton, Alta., December 28, 1978. Vancouver's 4th choice, 64th overall, in 1997 Entry Draft.

Season	Club	League	GP	G	A	Pts	PIM	PP	SH	GW	S	%	+/-	TF	F%	H	SB	Min	GP	G	A	Pts	PIM	PP	SH	GW
1995-96	Killam Selects	AAHA	37	11	22	33	176	....	....	....	....	....	....	....	....	....	....	....	....	....	....	....	....	....	....	....
1996-97	Prince George	WHL	12	0	0	0	12	....	....	....	....	....	....	....	....	....	....	....	....	....	....	....	....	....	....	....
	Regina Pats	WHL	50	1	3	4	152	....	....	....	....	....	....	....	....	....	....	....	4	0	0	0	8	....	....	....
1997-98	Regina Pats	WHL	62	6	5	11	259	....	....	....	....	....	....	....	....	....	....	....	9	0	1	1	25	....	....	....
1998-99	Regina Pats	WHL	52	2	2	4	215	....	....	....	....	....	....	....	....	....	....	....	....	....	....	....	....	....	....	....
	Syracuse Crunch	AHL	5	0	0	0	20	....	....	....	....	....	....	....	....	....	....	....	....	....	....	....	....	....	....	....
	Louisiana	ECHL	5	0	0	0	17	....	....	....	....	....	....	....	....	....	....	....	4	0	0	0	2	....	....	....
99-2000	**Tampa Bay**	**NHL**	**10**	**0**	**0**	**0**	**39**	**0**	**0**	**0**	**0**	**0.0**	**-1**	**0**	**0.0**	**4**	**1**	**2:19**	....	....	....	....	....	....	....	....
	Louisiana	ECHL	3	0	0	0	17	....	....	....	....	....	....	....	....	....	....	....	....	....	....	....	....	....	....	....
	Detroit Vipers	IHL	45	0	1	1	203	....	....	....	....	....	....	....	....	....	....	....	....	....	....	....	....	....	....	....
2000-01	**Tampa Bay**	**NHL**	**13**	**0**	**1**	**1**	**36**	**0**	**0**	**0**	**3**	**0.0**	**-1**	**0**	**0.0**	**6**	**2**	**3:32**	....	....	....	....	....	....	....	....
	Detroit Vipers	IHL	29	3	3	6	120	....	....	....	....	....	....	....	....	....	....	....	....	....	....	....	....	....	....	....
2001-02	**NY Rangers**	**NHL**	DID NOT PLAY – INJURED																							
	NHL Totals		**23**	**0**	**1**	**1**	**75**	**0**	**0**	**0**	**3**	**0.0**		**0**	**0.0**	**10**	**3**	**3:00**	....	....	....	....	....	....	....	....

Signed as a free agent by **Tampa Bay**, July 16, 1999. Traded to **NY Rangers** by **Tampa Bay** with Nils Ekman for Tim Taylor, June 30, 2001. • Missed entire 2001-02 season recovering from head injury suffered in training camp, October 1, 2001.

FRIESEN, Jeff (FREE-zuhn, JEHF) N.J.

Left wing. Shoots left. 6', 215 lbs. Born, Meadow Lake, Sask., August 5, 1976. San Jose's 1st choice, 11th overall, in 1994 Entry Draft.

Season	Club	League	GP	G	A	Pts	PIM	PP	SH	GW	S	%	+/-	TF	F%	H	SB	Min	GP	G	A	Pts	PIM	PP	SH	GW
1991-92	Sask. Contacts	SMHL	35	37	51	88	75	....	....	....	....	....	....	....	....	....	....	....	....	....	....	....	....	....	....	....
	Regina Pats	WHL	4	3	1	4	2	....	....	....	....	....	....	....	....	....	....	....	....	....	....	....	....	....	....	....
1992-93	Regina Pats	WHL	70	45	38	83	23	....	....	....	....	....	....	....	....	....	....	....	13	7	10	17	8	....	....	....
1993-94	Regina Pats	WHL	66	51	67	118	48	....	....	....	....	....	....	....	....	....	....	....	4	3	2	5	2	....	....	....
1994-95	Regina Pats	WHL	25	21	23	44	22	....	....	....	....	....	....	....	....	....	....	....	....	....	....	....	....	....	....	....
	San Jose	**NHL**	**48**	**15**	**10**	**25**	**14**	**5**	**1**	**2**	**86**	**17.4**	**-8**	....	....	....	....	....	**11**	**1**	**5**	**6**	**4**	**0**	**0**	**0**
1995-96	**San Jose**	**NHL**	**79**	**15**	**31**	**46**	**42**	**2**	**0**	**0**	**123**	**12.2**	**-19**	....	....	....	....	....	....	....	....	....	....	....	....	....
1996-97	**San Jose**	**NHL**	**82**	**28**	**34**	**62**	**75**	**6**	**2**	**5**	**200**	**14.0**	**-8**	....	....	....	....	....	....	....	....	....	....	....	....	....
1997-98	**San Jose**	**NHL**	**79**	**31**	**32**	**63**	**40**	**7**	**6**	**7**	**186**	**16.7**	**8**	....	....	....	....	....	**6**	**0**	**1**	**1**	**2**	**0**	**0**	**0**
1998-99	**San Jose**	**NHL**	**78**	**22**	**35**	**57**	**42**	**10**	**1**	**3**	**215**	**10.2**	**3**	**24**	**33.3**	**99**	**22**	**19:25**	**6**	**2**	**2**	**4**	**14**	**1**	**0**	**0**
99-2000	**San Jose**	**NHL**	**82**	**26**	**35**	**61**	**47**	**11**	**3**	**7**	**191**	**13.6**	**-2**	**3**	**66.7**	**110**	**37**	**19:48**	**11**	**2**	**2**	**4**	**10**	**0**	**0**	**0**

Season	Club	League	Regular Season																Playoffs							
			GP	G	A	Pts	PIM	PP	SH	GW	S	%	+/-	TF	F%	H	SB	Min	GP	G	A	Pts	PIM	PP	SH	GW
2000-01	San Jose	NHL	64	12	24	36	56	2	0	1	120	10.0	7	7	28.6	107	36	18:51								
	Anaheim	NHL	15	2	10	12	10	2	0	0	29	6.9	−2	43	55.8	19	11	21:28								
2001-02	Anaheim	NHL	81	17	26	43	44	1	1	0	161	10.6	−1	45	48.9	78	22	17:59								
	NHL Totals		608	168	237	405	370	46	14	25	1311	12.8		122	47.5	413	128	19:08	34	5	10	15	30	1	0	0

WHL Rookie of the Year (1993) • Canadian Major Junior Rookie of the Year (1993) • NHL All-Rookie Team (1995)

Traded to **Anaheim** by **San Jose** with Steve Shields and future considerations for Teemu Selanne, March 5, 2001. Traded to **New Jersey** by **Anaheim** with Oleg Tverdovsky and Maxim Balmochnykh for Petr Sykora, Mike Commodore, Jean-Francois Damphousse and Igor Pohanka, July 6, 2002.

GABORIK, Marian

(gah-BOHR-ihk, MAIR-ee-uhn) **MIN.**

Left wing. Shoots left. 6'1", 183 lbs. Born, Trencin, Czech., February 14, 1982. Minnesota's 1st choice, 3rd overall, in 2000 Entry Draft.

Season	Club	League	GP	G	A	Pts	PIM	PP	SH	GW	S	%	+/-	TF	F%	H	SB	Min	GP	G	A	Pts	PIM	PP	SH	GW
1997-98	Dukla Trencin Jr.	Slovak-Jr.	36	37	22	59	28																			
	Dukla Trencin	Slovakia	1	1	0	1	0																			
1998-99	Dukla Trencin	Slovakia	33	11	9	20	6												3	1	0	1	2			
99-2000	Dukla Trencin	Slovakia	50	25	21	46	34												5	1	2	3	2			
2000-01	Minnesota	NHL	71	18	18	36	32	6	0	3	179	10.1	−6	3	33.3	25	14	15:26								
2001-02	Minnesota	NHL	78	30	37	67	34	10	0	4	221	13.6	0	4	25.0	22	26	16:47								
	NHL Totals		149	48	55	103	66	16	0	7	400	12.0		7	28.6	47	40	16:08								

GAGNE, Simon

(GAH-nyay, see-MOHN) **PHI.**

Left wing. Shoots left. 6', 190 lbs. Born, Ste-Foy, Que., February 29, 1980. Philadelphia's 1st choice, 22nd overall, in 1998 Entry Draft.

Season	Club	League	GP	G	A	Pts	PIM	PP	SH	GW	S	%	+/-	TF	F%	H	SB	Min	GP	G	A	Pts	PIM	PP	SH	GW
1995-96	Ste-Foy	QAAA	27	13	9	22	18												15	7	8	15	8			
1996-97	Beauport	QMJHL	51	9	22	31	49																			
1997-98	Quebec Remparts	QMJHL	53	30	39	69	26												12	11	5	16	23			
1998-99	Quebec Remparts	QMJHL	61	50	70	120	42												13	9	8	17	4			
99-2000	Philadelphia	NHL	80	20	28	48	22	8	1	4	159	12.6	11	443	42.2	46	23	14:58	17	5	5	10	2	2	0	1
2000-01	Philadelphia	NHL	69	27	32	59	18	6	0	7	191	14.1	24	21	28.6	41	17	18:05	6	3	0	3	0	2	0	0
2001-02	Philadelphia	NHL	79	33	33	66	32	4	1	7	199	16.6	31	6	83.3	44	26	18:09	5	0	0	0	2	0	0	0
	Canada	Olympics	6	1	3	4	0																			
	NHL Totals		228	80	93	173	72	18	2	18	549	14.6		470	42.1	131	66	17:01	28	8	5	13	4	4	0	1

QMJHL Second All-Star Team (1999) • NHL All-Rookie Team (2000) • Played in NHL ALL-Star Game (2001)

GAGNON, Sean

(gah-NYAWN, SHAWN)

Defense. Shoots left. 6'2", 219 lbs. Born, Sault Ste. Marie, Ont., September 11, 1973.

Season	Club	League	GP	G	A	Pts	PIM	PP	SH	GW	S	%	+/-	TF	F%	H	SB	Min	GP	G	A	Pts	PIM	PP	SH	GW
1990-91	Soo Elks	NOHA	46	21	26	47	218																			
1991-92	Sud. N. Wolves	NOJHA	13	10	13	23	34																			
	Sudbury Wolves	OHL	44	3	4	7	60												5	0	1	1	0			
1992-93	Sudbury Wolves	OHL	6	1	1	2	16																			
	Ottawa 67's	OHL	33	2	10	12	68																			
	Sault Ste. Marie	OHL	24	1	5	6	65												15	2	2	4	25			
1993-94	Sault Ste. Marie	OHL	42	4	12	16	147												14	1	1	2	52			
1994-95	Dayton Bombers	ECHL	68	9	23	32	339												8	0	3	3	69			
1995-96	Dayton Bombers	ECHL	68	7	22	29	326												3	0	1	1	33			
1996-97	Fort Wayne	IHL	72	7	7	14	*457																			
1997-98	Phoenix	NHL	5	0	1	1	14	0	0	0	3	0.0	1													
	Springfield	AHL	54	4	13	17	330												2	0	1	1	17			
1998-99	Phoenix	NHL	2	0	0	0	7	0	0	0	1	0.0	−2	0	0.0	1	2	7:56								
	Springfield	AHL	68	8	14	22	331												3	0	0	0	14			
99-2000	Jokerit Helsinki	Finland	42	3	5	8	183												11	4	1	5	22			
2000-01	Ottawa	NHL	5	0	0	0	13	0	0	0	0	0.0	0	0	0.0	12	4	10:45								
	Grand Rapids	IHL	70	4	16	20	226												10	2	3	5	30			
2001-02	Hartford	AHL	42	3	5	8	200																			
	NHL Totals		12	0	1	1	34	0	0	0	4	0.0		0	0.0	13	6	9:57								

Signed as a free agent by **Phoenix**, May 14, 1997. Signed as a free agent by **Ottawa**, July 7, 2000. Traded to **NY Rangers** by **Ottawa** for Jason Doig and Jeff Ulmer, June 29, 2001.

GAINEY, Steve

(GAY-nee, STEEV) **DAL.**

Left wing. Shoots left. 6'1", 192 lbs. Born, Montreal, Que., January 26, 1979. Dallas' 3rd choice, 77th overall, in 1997 Entry Draft.

Season	Club	League	GP	G	A	Pts	PIM	PP	SH	GW	S	%	+/-	TF	F%	H	SB	Min	GP	G	A	Pts	PIM	PP	SH	GW
1995-96	Kamloops Blazers	WHL	49	1	4	5	40												3	0	0	0	0			
1996-97	Kamloops Blazers	WHL	60	9	18	27	60												2	0	0	0	9			
1997-98	Kamloops Blazers	WHL	68	21	34	55	93												7	1	7	8	15			
1998-99	Kamloops Blazers	WHL	68	30	34	64	155												15	5	4	9	38			
99-2000	Fort Wayne	UHL	1	0	0	0	0																			
	Michigan K-Wings	IHL	58	8	10	18	41																			
2000-01	Dallas	NHL	1	0	0	0	0	0	0	0	0	0.0	0	0	0.0	0	0	2:21								
	Utah Grizzlies	IHL	61	7	7	14	167																			
2001-02	Dallas	NHL	5	0	1	1	7	0	0	0	1	0.0	−1	0	0.0	9	0	7:24								
	Utah Grizzlies	AHL	58	16	18	34	87																			
	NHL Totals		6	0	1	1	7	0	0	0	1	0.0		0	0.0	9	0	6:34								

GALANOV, Maxim

(gah-LAH-nahf, mahx-EEM) **TOR.**

Defense. Shoots left. 6'1", 205 lbs. Born, Krasnoyarsk, USSR, March 13, 1974. NY Rangers' 3rd choice, 61st overall, in 1993 Entry Draft.

Season	Club	League	GP	G	A	Pts	PIM	PP	SH	GW	S	%	+/-	TF	F%	H	SB	Min	GP	G	A	Pts	PIM	PP	SH	GW
1992-93	Lada Togliatti	CIS	41	4	2	6	12												10	1	1	2	12			
1993-94	Lada Togliatti	CIS	7	1	0	1	4												12	1	0	1	8			
1994-95	Lada Togliatti	CIS	45	5	6	11	54												9	0	1	1	12			
1995-96	Binghamton	AHL	72	17	36	53	24												4	1	1	2	0			
1996-97	Binghamton	AHL	73	13	30	43	30												3	0	0	0	2			
1997-98	NY Rangers	NHL	6	0	1	1	2	0	0	0	5	0.0	1													
	Hartford	AHL	61	6	24	30	22												13	3	6	9	2			
1998-99	Pittsburgh	NHL	51	4	3	7	14	2	0	0	44	9.1	−8	1	0.0	32	49	15:13	1	0	0	0	0	0	0	0
99-2000	Atlanta	NHL	40	4	3	7	20	0	0	0	47	8.5	−12	0	0.0	40	56	21:32								
2000-01	Louisville	AHL	9	4	5	9	11																			
	Tampa Bay	NHL	25	0	5	5	8	0	0	0	10	0.0	−5	0	0.0	10	19	16:12								
	Detroit Vipers	IHL	16	0	3	3	6																			
2001-02	Lada Togliatti	Russia	3	0	0	0	0																			
	Cherepovets	Russia	8	0	0	0	4																			
	NHL Totals		122	8	12	20	44	2	0	0	106	7.5		1	0.0	82	124	17:36	1	0	0	0	0	0	0	0

Claimed by **Pittsburgh** from **NY Rangers** in NHL Waiver Draft, October 5, 1998. Claimed by **Atlanta** from **Pittsburgh** in Expansion Draft, June 25, 1999. • Missed majority of 1999-2000 season recovering from hand injury suffered in game vs. NY Rangers, October 17, 1999. Signed as a free agent by **Florida**, September, 2000. Claimed on waivers by **Tampa Bay** from **Florida**, November 1, 2000. Traded to **Toronto** by **Tampa Bay** for Konstantin Kalmikov, February 20, 2001.

GARDINER, Bruce

(gahr-DIHN-uhr, BREWS)

Right wing. Shoots right. 6'1", 193 lbs. Born, Barrie, Ont., February 11, 1972. St. Louis' 6th choice, 131st overall, in 1991 Entry Draft.

Season	Club	League	GP	G	A	Pts	PIM	PP	SH	GW	S	%	+/-	TF	F%	H	SB	Min	GP	G	A	Pts	PIM	PP	SH	GW
1988-89	Barrie Colts	OJHL-B	41	17	28	45	29																			
1989-90	Barrie Colts	OJHL-B	40	19	26	45	89												13	10	11	21	32			
1990-91	Colgate	ECAC	27	4	9	13	72																			
1991-92	Colgate	ECAC	23	7	8	15	77																			
1992-93	Colgate	ECAC	33	17	12	29	64																			
1993-94	Colgate	ECAC	33	23	23	46	68																			
	Peoria Rivermen	IHL	3	0	0	0	0																			
1994-95	P.E.I. Senators	AHL	72	17	20	37	132												7	4	1	5	4			
1995-96	P.E.I. Senators	AHL	38	11	13	24	87												5	2	4	6	4			
1996-97	Ottawa	NHL	67	11	10	21	49	0	1	2	94	11.7	4						7	0	1	1	2	0	0	0
1997-98	Ottawa	NHL	55	7	11	18	50	0	0	0	64	10.9	2						11	1	3	4	2	0	0	1
1998-99	Ottawa	NHL	59	4	8	12	43	0	0	1	70	5.7	6	278	45.7	88	17	12:52	3	0	0	0	4	0	0	0

			Regular Season																Playoffs							
Season	Club	League	GP	G	A	Pts	PIM	PP	SH	GW	S	%	+/-	TF	F%	H	SB	Min	GP	G	A	Pts	PIM	PP	SH	GW
99-2000	Ottawa	NHL	10	0	3	3	4	0	0	0	18	0.0	1	62	59.7	16	2	13:24	....	....	....	....	....	....	....	....
	Tampa Bay	NHL	41	3	6	9	37	0	0	0	30	10.0	−21	330	56.4	52	28	13:36	....	....	....	....	....	....	....	....
2000-01	Columbus	NHL	73	7	15	22	78	0	0	1	60	11.7	−1	505	51.5	143	47	14:36	....	....	....	....	....	....	....	....
2001-02	New Jersey	NHL	7	2	1	3	2	1	0	0	10	20.0	−1	9	44.4	12	5	13:16	....	....	....	....	....	....	....	....
	Albany	AHL	45	5	18	23	71	....	....	....	....	....	....	....	....	....	....	....	....	....	....	....	....	....	....	....
	NHL Totals		312	34	54	88	263	1	1	4	346	9.8		1184	51.9	311	99	13:44	21	1	4	5	8	0	0	1

ECAC Second All-Star Team (1994)

Signed as a free agent by **Ottawa**, June 14, 1994. Traded to **Tampa Bay** by **Ottawa** for Colin Forbes, November 11, 1999. Selected by **Columbus** from **Tampa Bay** in Expansion Draft, June 23, 2000. Signed as a free agent by **New Jersey**, October 21, 2001.

GAUL, Mike

(GAWL, MIGH-kuhl)

Defense. Shoots right. 6'1", 200 lbs. Born, Lachine, Que., April 22, 1973. Los Angeles' 10th choice, 262nd overall, in 1991 Entry Draft.

Season	Club	League	GP	G	A	Pts	PIM	PP	SH	GW	S	%	+/-	TF	F%	H	SB	Min	GP	G	A	Pts	PIM	PP	SH	GW
1989-90	Lac St-Louis	QAAA	39	5	9	14	46	....	....	....	....	....	....	....	....	....	....	....	2	0	1	1	14	....	....	....
1990-91	St. Lawrence	ECAC	31	1	3	4	46	....	....	....	....	....	....	....	....	....	....	....	....	....	....	....	....	....	....	....
1991-92	Laval Titan	QMJHL	50	6	38	44	44	....	....	....	....	....	....	....	....	....	....	....	10	0	2	2	20	....	....	....
1992-93	Laval Titan	QMJHL	57	16	57	73	66	....	....	....	....	....	....	....	....	....	....	....	13	3	10	13	10	....	....	....
1993-94	Laval Titan	QMJHL	22	10	17	27	24	....	....	....	....	....	....	....	....	....	....	....	21	5	15	20	14	....	....	....
1994-95	Phoenix	IHL	4	0	1	1	2	....	....	....	....	....	....	....	....	....	....	....	....	....	....	....	....	....	....	....
	Knoxville	ECHL	68	13	41	54	51	....	....	....	....	....	....	....	....	....	....	....	4	2	1	3	2	....	....	....
1995-96	Knoxville	ECHL	54	13	48	61	44	....	....	....	....	....	....	....	....	....	....	....	....	....	....	....	....	....	....	....
1996-97	ETC Timmendorf	German-2	51	40	52	92	100	....	....	....	....	....	....	....	....	....	....	....	....	....	....	....	....	....	....	....
1997-98	Hershey Bears	AHL	60	12	47	59	69	....	....	....	....	....	....	....	....	....	....	....	7	0	7	7	6	....	....	....
	Mobile Mysticks	ECHL	5	0	7	7	0	....	....	....	....	....	....	....	....	....	....	....	....	....	....	....	....	....	....	....
1998-99	Lowell	AHL	18	3	5	8	14	....	....	....	....	....	....	....	....	....	....	....	....	....	....	....	....	....	....	....
	Colorado	**NHL**	**1**	**0**	**0**	**0**	**0**	**0**	**0**	**0**	**1**	**0.0**	**0**	**0**	**0.0**	**2**	**0**	**10:46**	....	....	....	....	....	....	....	....
	Hershey Bears	AHL	43	9	31	40	22	....	....	....	....	....	....	....	....	....	....	....	5	1	1	2	6	....	....	....
99-2000	Hershey Bears	AHL	65	12	57	69	52	....	....	....	....	....	....	....	....	....	....	....	12	0	8	8	....	....	....	....
2000-01	**Columbus**	**NHL**	**2**	**0**	**0**	**0**	**4**	**0**	**0**	**0**	**3**	**0.0**	**0**	**0**	**0.0**	**2**	**3**	**13:17**	....	....	....	....	....	....	....	....
	Syracuse Crunch	AHL	70	16	45	61	80	....	....	....	....	....	....	....	....	....	....	....	5	1	2	3	8	....	....	....
2001-02	Fribourg	Swiss	44	11	29	40	64	....	....	....	....	....	....	....	....	....	....	....	4	0	1	1	10	....	....	....
	NHL Totals		**3**	**0**	**0**	**0**	**4**	**0**	**0**	**0**	**4**	**0.0**		**0**	**0.0**	**4**	**3**	**12:27**	....	....	....	....	....	....	....	....

QMJHL All-Rookie Team (1992) • AHL Second All-Star Team (2000, 2001)

Signed as a free agent by **NY Islanders**, July 16, 1998. Traded to **Colorado** by **NY Islanders** for Ted Crowley, December 15, 1998. Signed as a free agent by **Columbus**, July 18, 2000.

GAUTHIER, Denis

(GOH-tyay, DEH-nihs) **CGY.**

Defense. Shoots left. 6'2", 210 lbs. Born, Montreal, Que., October 1, 1976. Calgary's 1st choice, 20th overall, in 1995 Entry Draft.

Season	Club	League	GP	G	A	Pts	PIM	PP	SH	GW	S	%	+/-	TF	F%	H	SB	Min	GP	G	A	Pts	PIM	PP	SH	GW
1991-92	St-Jean-de-Rich.	QAHA	STATISTICS NOT AVAILABLE																							
1992-93	Drummondville	QMJHL	61	1	7	8	136	....	....	....	....	....	....	....	....	....	....	....	10	0	5	5	40	....	....	....
1993-94	Drummondville	QMJHL	60	0	7	7	176	....	....	....	....	....	....	....	....	....	....	....	9	2	0	2	41	....	....	....
1994-95	Drummondville	QMJHL	64	9	31	40	190	....	....	....	....	....	....	....	....	....	....	....	4	0	5	5	12	....	....	....
1995-96	Drummondville	QMJHL	53	25	49	74	140	....	....	....	....	....	....	....	....	....	....	....	6	4	4	8	32	....	....	....
	Saint John	AHL	5	2	0	2	8	....	....	....	....	....	....	....	....	....	....	....	16	1	6	7	20	....	....	....
1996-97	Saint John	AHL	73	3	28	31	74	....	....	....	....	....	....	....	....	....	....	....	5	0	0	0	6	....	....	....
1997-98	**Calgary**	**NHL**	**10**	**0**	**0**	**0**	**16**	**0**	**0**	**0**	**3**	**0.0**	**−5**	....	....	....	....	....	....	....	....	....	....	....	....	....
	Saint John	AHL	68	4	20	24	154	....	....	....	....	....	....	....	....	....	....	....	21	0	4	4	83	....	....	....
1998-99	**Calgary**	**NHL**	**55**	**3**	**4**	**7**	**68**	**0**	**0**	**0**	**40**	**7.5**	**3**	**0**	**0.0**	**162**	**51**	**12:41**	....	....	....	....	....	....	....	....
	Saint John	AHL	16	0	3	3	31	....	....	....	....	....	....	....	....	....	....	....	....	....	....	....	....	....	....	....
99-2000	**Calgary**	**NHL**	**39**	**1**	**1**	**2**	**50**	**0**	**0**	**0**	**29**	**3.4**	**−4**	**0**	**0.0**	**168**	**52**	**19:21**	....	....	....	....	....	....	....	....
2000-01	**Calgary**	**NHL**	**62**	**2**	**6**	**8**	**78**	**0**	**0**	**0**	**33**	**6.1**	**3**	**0**	**0.0**	**252**	**62**	**16:37**	....	....	....	....	....	....	....	....
2001-02	**Calgary**	**NHL**	**66**	**5**	**8**	**13**	**91**	**0**	**1**	**2**	**76**	**6.6**	**9**	**0**	**0.0**	**194**	**112**	**19:19**	....	....	....	....	....	....	....	....
	NHL Totals		**232**	**11**	**19**	**30**	**303**	**0**	**1**	**2**	**181**	**6.1**		**0**	**0.0**	**776**	**277**	**16:55**	....	....	....	....	....	....	....	....

QMJHL First All-Star Team (1996) • Canadian Major Junior First All-Star Team (1996) • Missed majority of 1999-2000 season recovering from hip injury suffered in game vs. St. Louis, February 1, 2000.

GAVEY, Aaron

(GAY-vee, AIR-ruhn) **TOR.**

Center. Shoots left. 6'2", 200 lbs. Born, Sudbury, Ont., February 22, 1974. Tampa Bay's 4th choice, 74th overall, in 1992 Entry Draft.

Season	Club	League	GP	G	A	Pts	PIM	PP	SH	GW	S	%	+/-	TF	F%	H	SB	Min	GP	G	A	Pts	PIM	PP	SH	GW
1990-91	Peterborough	OPJHL	42	26	30	56	68	....	....	....	....	....	....	....	....	....	....	....	....	....	....	....	....	....	....	....
1991-92	Sault Ste. Marie	OHL	48	7	11	18	27	....	....	....	....	....	....	....	....	....	....	....	19	5	1	6	10	....	....	....
1992-93	Sault Ste. Marie	OHL	62	45	39	84	116	....	....	....	....	....	....	....	....	....	....	....	18	5	9	14	36	....	....	....
1993-94	Sault Ste. Marie	OHL	60	42	60	102	116	....	....	....	....	....	....	....	....	....	....	....	14	11	10	21	22	....	....	....
1994-95	Atlanta Knights	IHL	66	18	17	35	85	....	....	....	....	....	....	....	....	....	....	....	5	0	1	1	9	....	....	....
1995-96	**Tampa Bay**	**NHL**	**73**	**8**	**4**	**12**	**56**	**1**	**1**	**2**	**65**	**12.3**	**−6**	....	....	....	....	....	**6**	**0**	**0**	**0**	**4**	**0**	**0**	**0**
1996-97	**Tampa Bay**	**NHL**	**16**	**1**	**2**	**3**	**12**	**0**	**0**	**0**	**8**	**12.5**	**−1**	....	....	....	....	....	....	....	....	....	....	....	....	....
	Calgary	**NHL**	**41**	**7**	**9**	**16**	**34**	**3**	**0**	**1**	**54**	**13.0**	**−11**	....	....	....	....	....	....	....	....	....	....	....	....	....
1997-98	**Calgary**	**NHL**	**26**	**2**	**3**	**5**	**24**	**0**	**0**	**1**	**27**	**7.4**	**−5**	....	....	....	....	....	....	....	....	....	....	....	....	....
	Saint John	AHL	8	4	3	7	28	....	....	....	....	....	....	....	....	....	....	....	....	....	....	....	....	....	....	....
1998-99	**Dallas**	**NHL**	**7**	**0**	**0**	**0**	**10**	**0**	**0**	**0**	**4**	**0.0**	**−1**	**43**	**48.8**	**13**	**0**	**8:09**	....	....	....	....	....	....	....	....
	Michigan K-Wings	IHL	67	24	33	57	128	....	....	....	....	....	....	....	....	....	....	....	5	2	3	5	4	....	....	....
99-2000	**Dallas**	**NHL**	**41**	**7**	**6**	**13**	**44**	**1**	**0**	**2**	**39**	**17.9**	**0**	**263**	**51.7**	**88**	**12**	**9:55**	**13**	**1**	**2**	**3**	**10**	**0**	**0**	**1**
	Michigan K-Wings	IHL	28	14	15	29	73	....	....	....	....	....	....	....	....	....	....	....	....	....	....	....	....	....	....	....
2000-01	**Minnesota**	**NHL**	**75**	**10**	**14**	**24**	**52**	**1**	**0**	**2**	**100**	**10.0**	**−8**	**584**	**43.5**	**84**	**53**	**14:00**	....	....	....	....	....	....	....	....
2001-02	**Minnesota**	**NHL**	**71**	**6**	**11**	**17**	**38**	**1**	**0**	**0**	**75**	**8.0**	**−21**	**254**	**42.5**	**50**	**19**	**11:55**	....	....	....	....	....	....	....	....
	NHL Totals		**350**	**41**	**49**	**90**	**270**	**7**	**1**	**8**	**372**	**11.0**		**1144**	**45.4**	**235**	**84**	**12:10**	**19**	**1**	**2**	**3**	**14**	**0**	**0**	**1**

Traded to **Calgary** by **Tampa Bay** for Rick Tabaracci, November 19, 1996. Traded to **Dallas** by **Calgary** for Bob Bassen, July 14, 1998. Traded to **Minnesota** by **Dallas** with Pavel Patera, Dallas' 8th round choice (Eric Johansson) in 2000 Entry Draft and Minnesota's 4th round choice (previously acquired) in 2002 Entry Draft for Brad Lukowich and Minnesota's 3rd (Yared Hagos) and 9th (Dale Sullivan) round choices in 2001 Entry Draft, June 25, 2000. Signed as a free agent by **Toronto**, July 24, 2002.

GELINAS, Martin

(ZHEHL-in-nuh, MAHR-tihn) **CGY.**

Left wing. Shoots left. 5'11", 195 lbs. Born, Shawinigan, Que., June 5, 1970. Los Angeles' 1st choice, 7th overall, in 1988 Entry Draft.

Season	Club	League	GP	G	A	Pts	PIM	PP	SH	GW	S	%	+/-	TF	F%	H	SB	Min	GP	G	A	Pts	PIM	PP	SH	GW
1985-86	Noranda Aces	NOHA	5	1	1	2	0	....	....	....	....	....	....	....	....	....	....	....	....	....	....	....	....	....	....	....
1986-87	L'est Cantonniers	QAAA	41	36	42	78	36	....	....	....	....	....	....	....	....	....	....	....	7	7	5	12	2	....	....	....
1987-88	Hull Olympiques	QMJHL	65	63	68	131	74	....	....	....	....	....	....	....	....	....	....	....	17	15	18	33	32	....	....	....
1988-89	Hull Olympiques	QMJHL	41	38	39	77	31	....	....	....	....	....	....	....	....	....	....	....	9	5	4	9	14	....	....	....
	Edmonton	**NHL**	**6**	**1**	**2**	**3**	**0**	**0**	**0**	**0**	**14**	**7.1**	**−1**	....	....	....	....	....	....	....	....	....	....	....	....	....
1989-90♦	**Edmonton**	**NHL**	**46**	**17**	**8**	**25**	**30**	**5**	**0**	**2**	**71**	**23.9**	**0**	....	....	....	....	....	**20**	**2**	**3**	**5**	**6**	**0**	**0**	**0**
1990-91	**Edmonton**	**NHL**	**73**	**20**	**20**	**40**	**34**	**4**	**0**	**2**	**124**	**16.1**	**−7**	....	....	....	....	....	**18**	**3**	**6**	**9**	**25**	**0**	**0**	**1**
1991-92	**Edmonton**	**NHL**	**68**	**11**	**18**	**29**	**62**	**1**	**0**	**0**	**94**	**11.7**	**14**	....	....	....	....	....	**15**	**1**	**3**	**4**	**10**	**0**	**0**	**0**
1992-93	**Edmonton**	**NHL**	**65**	**11**	**12**	**23**	**30**	**0**	**0**	**1**	**93**	**11.8**	**3**	....	....	....	....	....	....	....	....	....	....	....	....	....
1993-94	**Quebec**	**NHL**	**31**	**6**	**6**	**12**	**8**	**0**	**0**	**0**	**53**	**11.3**	**−2**	....	....	....	....	....	....	....	....	....	....	....	....	....
	Vancouver	**NHL**	**33**	**8**	**8**	**16**	**26**	**3**	**0**	**1**	**54**	**14.8**	**−6**	....	....	....	....	....	**24**	**5**	**4**	**9**	**14**	**2**	**0**	**1**
1994-95	**Vancouver**	**NHL**	**46**	**13**	**10**	**23**	**36**	**1**	**0**	**4**	**75**	**17.3**	**8**	....	....	....	....	....	**3**	**0**	**1**	**1**	**0**	**0**	**0**	**0**
1995-96	**Vancouver**	**NHL**	**81**	**30**	**26**	**56**	**59**	**3**	**4**	**5**	**181**	**16.6**	**8**	....	....	....	....	....	**6**	**1**	**1**	**2**	**12**	**1**	**0**	**0**
1996-97	**Vancouver**	**NHL**	**74**	**35**	**33**	**68**	**42**	**6**	**1**	**3**	**177**	**19.8**	**6**	....	....	....	....	....	....	....	....	....	....	....	....	....
1997-98	**Vancouver**	**NHL**	**24**	**4**	**4**	**8**	**10**	**1**	**1**	**1**	**49**	**8.2**	**−6**	....	....	....	....	....	....	....	....	....	....	....	....	....
	Carolina	**NHL**	**40**	**12**	**14**	**26**	**30**	**2**	**1**	**4**	**98**	**12.2**	**1**	....	....	....	....	....	....	....	....	....	....	....	....	....
1998-99	**Carolina**	**NHL**	**76**	**13**	**15**	**28**	**67**	**0**	**0**	**2**	**111**	**11.7**	**3**	**6**	**50.0**	**70**	**13**	**13:13**	**6**	**0**	**3**	**3**	**2**	**0**	**0**	**0**
99-2000	**Carolina**	**NHL**	**81**	**14**	**16**	**30**	**40**	**3**	**0**	**0**	**139**	**10.1**	**−10**	**5**	**40.0**	**85**	**31**	**13:39**	....	....	....	....	....	....	....	....
2000-01	**Carolina**	**NHL**	**79**	**23**	**29**	**52**	**59**	**6**	**1**	**4**	**170**	**13.5**	**−4**	**6**	**0.0**	**131**	**54**	**17:54**	**6**	**0**	**1**	**1**	**6**	**0**	**0**	**0**
2001-02	**Carolina**	**NHL**	**72**	**13**	**16**	**29**	**30**	**3**	**0**	**1**	**121**	**10.7**	**−1**	**13**	**23.1**	**85**	**28**	**16:05**	**23**	**3**	**4**	**7**	**10**	**0**	**0**	**1**
	NHL Totals		**895**	**231**	**237**	**468**	**563**	**38**	**8**	**30**	**1624**	**14.2**		**30**	**26.7**	**371**	**126**	**15:12**	**121**	**15**	**26**	**41**	**85**	**3**	**0**	**3**

QMJHL First All-Star Team (1988) • QMJHL Offensive Rookie of the Year (1988) • Canadian Major Junior Rookie of the Year (1988) • Won George Parsons Trophy (Memorial Cup Tournament Most Sportsmanlike Player) (1988)

Traded to **Edmonton** by **LA Kings** with Jimmy Carson and LA Kings' 1st round choices in 1989 (later traded to New Jersey - New Jersey selected Jason Miller), 1991 (Martin Rucinsky) and 1993 (Nick Stajduhar) Entry Drafts and cash for Wayne Gretzky, Mike Krushelnyski and Marty McSorley, August 9, 1988. Traded to **Quebec** by **Edmonton** with Edmonton's 6th round choice (Nicholas Checco) in 1993 Entry Draft for Scott Pearson, June 20, 1993. Claimed on waivers by **Vancouver** from **Quebec**, January 15, 1994. Traded to **Carolina** by **Vancouver** with Kirk McLean for Sean Burke, Geoff Sanderson and Enrico Ciccone, January 3, 1998. Signed as a free agent by **Calgary**, July 2, 2002.

			Regular Season																Playoffs							
Season	Club	League	GP	G	A	Pts	PIM	PP	SH	GW	S	%	+/-	TF	F%	H	SB	Min	GP	G	A	Pts	PIM	PP	SH	GW

GERNANDER, Ken

(guhr-NAN-duhr, KEHN) **NYR**

Center. Shoots left. 5'10", 175 lbs. Born, Coleraine, MN, June 30, 1969. Winnipeg's 4th choice, 96th overall, in 1987 Entry Draft.

Season	Club	League	GP	G	A	Pts	PIM	PP	SH	GW	S	%	+/-	TF	F%	H	SB	Min	GP	G	A	Pts	PIM	PP	SH	GW
1985-86	Greenway Raiders	Hi-School	23	14	23	37																				
1986-87	Greenway Raiders	Hi-School	26	35	34	69																				
1987-88	U. of Minnesota	WCHA	44	14	14	28	14																			
1988-89	U. of Minnesota	WCHA	44	9	11	20	2																			
1989-90	U. of Minnesota	WCHA	44	32	17	49	24																			
1990-91	U. of Minnesota	WCHA	44	23	20	43	24																			
1991-92	Fort Wayne	IHL	13	7	6	13	2																			
	Moncton Hawks	AHL	43	8	18	26	9												8	1	1	2	2			
1992-93	Moncton Hawks	AHL	71	18	29	47	20												5	1	4	5	0			
1993-94	Moncton Hawks	AHL	71	22	25	47	12												19	6	1	7	0			
1994-95	Binghamton	AHL	80	28	25	53	24												11	2	2	4	6			
1995-96	**NY Rangers**	**NHL**	**10**	**2**	**3**	**5**	**4**	**2**	**0**	**0**	**10**	**20.0**	**-3**						**6**	**0**	**0**	**0**	**0**	**0**	**0**	**0**
	Binghamton	AHL	63	44	29	73	38																			
1996-97	Binghamton	AHL	46	13	18	31	30												2	0	1	1	0			
	NY Rangers	**NHL**																	**9**	**0**	**0**	**0**	**0**	**0**	**0**	**0**
1997-98	Hartford	AHL	80	35	28	63	26												12	5	6	11	4			
1998-99	Hartford	AHL	70	23	26	49	32												7	1	2	3	2			
99-2000	Hartford	AHL	79	28	29	57	24												23	5	5	10	0			
2000-01	Hartford	AHL	80	22	27	49	39												2	0	0	0	0			
2001-02	Hartford	AHL	75	18	31	49	19												10	1	3	4	4			
	NHL Totals		**10**	**2**	**3**	**5**	**4**	**2**	**0**	**0**	**10**	**20.0**							**15**	**0**	**0**	**0**	**0**	**0**	**0**	**0**

Won Fred Hunt Memorial Trophy (Sportsmanship - AHL) (1996)

Signed as a free agent by **NY Rangers**, July 4, 1994.

GILCHRIST, Brent

(GIHL-chrihst, BREHNT) **NSH.**

Left wing. Shoots left. 5'11", 180 lbs. Born, Moose Jaw, Sask., April 3, 1967. Montreal's 6th choice, 79th overall, in 1985 Entry Draft.

Season	Club	League	GP	G	A	Pts	PIM	PP	SH	GW	S	%	+/-	TF	F%	H	SB	Min	GP	G	A	Pts	PIM	PP	SH	GW
1983-84	Kelowna Wings	WHL	69	16	11	27	16																			
1984-85	Kelowna Wings	WHL	51	35	38	73	58												6	5	2	7	8			
1985-86	Spokane Chiefs	WHL	52	45	45	90	57												9	6	7	13	19			
1986-87	Spokane Chiefs	WHL	46	45	55	100	71												5	2	7	9	6			
	Sherbrooke	AHL																	10	2	7	9	2			
1987-88	Sherbrooke	AHL	77	26	48	74	83												6	1	3	4	6			
1988-89	**Montreal**	**NHL**	**49**	**8**	**16**	**24**	**16**	**0**	**0**	**2**	**68**	**11.8**	**9**						**9**	**1**	**1**	**2**	**10**	**0**	**0**	**0**
	Sherbrooke	AHL	7	6	5	11	7																			
1989-90	**Montreal**	**NHL**	**57**	**9**	**15**	**24**	**28**	**1**	**0**	**0**	**80**	**11.3**	**3**						**8**	**2**	**0**	**2**	**2**	**0**	**0**	**0**
1990-91	**Montreal**	**NHL**	**51**	**6**	**9**	**15**	**10**	**1**	**0**	**1**	**81**	**7.4**	**-3**						**13**	**5**	**3**	**8**	**6**	**0**	**0**	**1**
1991-92	**Montreal**	**NHL**	**79**	**23**	**27**	**50**	**57**	**2**	**0**	**3**	**146**	**15.8**	**29**						**11**	**2**	**4**	**6**	**6**	**1**	**0**	**0**
1992-93	**Edmonton**	**NHL**	**60**	**10**	**10**	**20**	**47**	**2**	**0**	**0**	**94**	**10.6**	**-10**													
	Minnesota	**NHL**	**8**	**0**	**1**	**1**	**2**	**0**	**0**	**0**	**12**	**0.0**	**-2**													
1993-94	**Dallas**	**NHL**	**76**	**17**	**14**	**31**	**31**	**3**	**1**	**5**	**103**	**16.5**	**0**						**9**	**3**	**1**	**4**	**2**	**1**	**0**	**0**
1994-95	**Dallas**	**NHL**	**32**	**9**	**4**	**13**	**16**	**1**	**3**	**1**	**70**	**12.9**	**-3**						**5**	**0**	**1**	**1**	**2**	**0**	**0**	**0**
1995-96	**Dallas**	**NHL**	**77**	**20**	**22**	**42**	**36**	**6**	**1**	**2**	**164**	**12.2**	**-11**													
1996-97	**Dallas**	**NHL**	**67**	**10**	**20**	**30**	**24**	**2**	**0**	**2**	**116**	**8.6**	**6**						**6**	**2**	**2**	**4**	**2**	**0**	**0**	**0**
1997-98♦	**Detroit**	**NHL**	**61**	**13**	**14**	**27**	**40**	**5**	**0**	**3**	**124**	**10.5**	**4**						**15**	**2**	**1**	**3**	**12**	**0**	**0**	**0**
1998-99	**Detroit**	**NHL**	**5**	**1**	**0**	**1**	**0**	**0**	**0**	**1**	**4**	**25.0**	**-1**	28	42.9	1	2	11:58	**3**	**0**	**0**	**0**	**0**	**0**	**0**	**0**
99-2000	**Detroit**	**NHL**	**24**	**4**	**2**	**6**	**24**	**0**	**0**	**0**	**33**	**12.1**	**1**	180	50.0	18	8	11:19	**6**	**0**	**0**	**0**	**6**	**0**	**0**	**0**
2000-01	**Detroit**	**NHL**	**60**	**1**	**8**	**9**	**41**	**0**	**0**	**0**	**75**	**1.3**	**-8**	401	50.9	59	15	11:41	**5**	**0**	**1**	**1**	**0**	**0**	**0**	**0**
2001-02	**Detroit**	**NHL**	**19**	**1**	**1**	**2**	**8**	**0**	**0**	**1**	**24**	**4.2**	**-3**	118	53.4	18	4	9:22								
	Dallas	**NHL**	**26**	**2**	**5**	**7**	**6**	**0**	**0**	**0**	**29**	**6.9**	**-6**	141	48.9	16	9	12:35								
	NHL Totals		**751**	**134**	**168**	**302**	**386**	**23**	**5**	**21**	**1223**	**11.0**		868	50.5	112	38	11:29	90	17	14	31	48	2	0	1

Traded to **Edmonton** by **Montreal** with Shayne Corson and Vladimir Vujtek for Vincent Damphousse and Edmonton's 4th round choice (Adam Wiesel) in 1993 Entry Draft, August 27, 1992. Traded to **Minnesota** by **Edmonton** for Todd Elik, March 5, 1993. Transferred to **Dallas** after **Minnesota** franchise relocated, June 9, 1993. Signed as a free agent by **Detroit**, August 1, 1997. Claimed by **Tampa Bay** from **Detroit** in NHL Waiver Draft, October 5, 1998. Traded to **Detroit** by **Tampa Bay** for future considerations, October 5, 1998. • Missed majority of 1998-99 and 1999-2000 seasons recovering from hernia surgery, September 22, 1998. Claimed on waivers by **Dallas** from **Detroit**, February 13, 2002. Signed as a free agent by **Nashville**, July 11, 2002.

GILL, Hal

(GIHL, HAL) **BOS.**

Defense. Shoots left. 6'7", 230 lbs. Born, Concord, MA, April 6, 1975. Boston's 8th choice, 207th overall, in 1993 Entry Draft.

Season	Club	League	GP	G	A	Pts	PIM	PP	SH	GW	S	%	+/-	TF	F%	H	SB	Min	GP	G	A	Pts	PIM	PP	SH	GW
1992-93	Nashoba High	Hi-School	20	25	25	50																				
1993-94	Providence	H-East	31	1	2	3	26																			
1994-95	Providence	H-East	26	1	3	4	22																			
1995-96	Providence	H-East	39	5	12	17	54																			
1996-97	Providence	H-East	35	5	16	21	52																			
1997-98	**Boston**	**NHL**	**68**	**2**	**4**	**6**	**47**	**0**	**0**	**0**	**56**	**3.6**	**4**						**6**	**0**	**0**	**0**	**4**	**0**	**0**	**0**
	Providence	AHL	4	1	0	1	23																			
1998-99	**Boston**	**NHL**	**80**	**3**	**7**	**10**	**63**	**0**	**0**	**2**	**102**	**2.9**	**-10**	1	100.0	144	102	20:54	**12**	**0**	**0**	**0**	**14**	**0**	**0**	**0**
99-2000	**Boston**	**NHL**	**81**	**3**	**9**	**12**	**51**	**0**	**0**	**0**	**120**	**2.5**	**0**	0	0.0	245	63	17:15								
2000-01	**Boston**	**NHL**	**80**	**1**	**10**	**11**	**71**	**0**	**0**	**0**	**79**	**1.3**	**-2**	0	0.0	206	62	18:21								
2001-02	**Boston**	**NHL**	**79**	**4**	**18**	**22**	**77**	**0**	**0**	**0**	**137**	**2.9**	**16**	0	0.0	152	76	24:13	**6**	**0**	**1**	**1**	**2**	**0**	**0**	**0**
	NHL Totals		**388**	**13**	**48**	**61**	**309**	**0**	**0**	**2**	**494**	**2.6**		1	100.0	747	303	20:09	24	0	1	1	20	0	0	0

GILL, Todd

(GIHL, TAWD)

Defense. Shoots left. 6', 180 lbs. Born, Cardinal, Ont., November 9, 1965. Toronto's 2nd choice, 25th overall, in 1984 Entry Draft.

Season	Club	League	GP	G	A	Pts	PIM	PP	SH	GW	S	%	+/-	TF	F%	H	SB	Min	GP	G	A	Pts	PIM	PP	SH	GW
1980-81	Cardinal Broncos	OHA-B	35	10	14	24	65																			
1981-82	Brockville	OCJHL	48	5	16	21	169																			
1982-83	Windsor	OHL	70	12	24	36	108												3	0	0	0	11			
1983-84	Windsor	OHL	68	9	48	57	184												3	1	1	2	10			
1984-85	Windsor	OHL	53	17	40	57	148												4	0	1	1	14			
	Toronto	**NHL**	**10**	**1**	**0**	**1**	**13**	**0**	**0**	**0**	**9**	**11.1**	**-1**													
1985-86	**Toronto**	**NHL**	**15**	**1**	**2**	**3**	**28**	**0**	**0**	**0**	**9**	**11.1**	**0**						**1**	**0**	**0**	**0**	**0**	**0**	**0**	**0**
	St. Catharines	AHL	58	8	25	33	90												10	1	6	7	17			
1986-87	**Toronto**	**NHL**	**61**	**4**	**27**	**31**	**92**	**1**	**0**	**0**	**51**	**7.8**	**-3**						**13**	**2**	**2**	**4**	**42**	**0**	**0**	**0**
	Newmarket Saints	AHL	11	1	8	9	33																			
1987-88	**Toronto**	**NHL**	**65**	**8**	**17**	**25**	**131**	**1**	**0**	**3**	**109**	**7.3**	**-20**						**6**	**1**	**3**	**4**	**20**	**1**	**0**	**0**
	Newmarket Saints	AHL	2	0	1	1	2																			
1988-89	**Toronto**	**NHL**	**59**	**11**	**14**	**25**	**72**	**0**	**0**	**1**	**92**	**12.0**	**-3**													
1989-90	**Toronto**	**NHL**	**48**	**1**	**14**	**15**	**92**	**0**	**0**	**0**	**44**	**2.3**	**-8**						**5**	**0**	**3**	**3**	**16**	**0**	**0**	**0**
1990-91	**Toronto**	**NHL**	**72**	**2**	**22**	**24**	**113**	**0**	**0**	**0**	**90**	**2.2**	**-4**													
1991-92	**Toronto**	**NHL**	**74**	**2**	**15**	**17**	**91**	**1**	**0**	**0**	**82**	**2.4**	**-22**													
1992-93	**Toronto**	**NHL**	**69**	**11**	**32**	**43**	**66**	**5**	**0**	**2**	**113**	**9.7**	**4**						**21**	**1**	**10**	**11**	**26**	**0**	**0**	**0**
1993-94	**Toronto**	**NHL**	**45**	**4**	**24**	**28**	**44**	**2**	**0**	**1**	**74**	**5.4**	**8**						**18**	**1**	**5**	**6**	**37**	**0**	**0**	**1**
1994-95	**Toronto**	**NHL**	**47**	**7**	**25**	**32**	**64**	**3**	**1**	**2**	**82**	**8.5**	**-8**						**7**	**0**	**3**	**3**	**6**	**0**	**0**	**0**
1995-96	**Toronto**	**NHL**	**74**	**7**	**18**	**25**	**116**	**1**	**0**	**2**	**109**	**6.4**	**-15**						**6**	**0**	**0**	**0**	**24**	**0**	**0**	**0**
1996-97	**San Jose**	**NHL**	**79**	**0**	**21**	**21**	**101**	**0**	**0**	**0**	**101**	**0.0**	**-20**													
1997-98	**San Jose**	**NHL**	**64**	**8**	**13**	**21**	**31**	**4**	**0**	**1**	**100**	**8.0**	**-13**													
	St. Louis	**NHL**	**11**	**5**	**4**	**9**	**10**	**3**	**0**	**1**	**22**	**22.7**	**2**						**10**	**2**	**2**	**4**	**10**	**1**	**1**	**0**
1998-99	**St. Louis**	**NHL**	**28**	**2**	**3**	**5**	**16**	**1**	**0**	**0**	**36**	**5.6**	**-6**	0	0.0	34	16	17:36								
	Detroit	**NHL**	**23**	**2**	**2**	**4**	**11**	**0**	**0**	**1**	**25**	**8.0**	**-4**	0	0.0	32	13	18:45	**2**	**0**	**1**	**1**	**0**	**0**	**0**	**0**
99-2000	**Phoenix**	**NHL**	**41**	**1**	**6**	**7**	**30**	**0**	**0**	**1**	**41**	**2.4**	**-10**	0	0.0	74	37	16:07								
	Detroit	**NHL**	**13**	**2**	**0**	**2**	**15**	**0**	**0**	**1**	**20**	**10.0**	**2**	1	100.0	12	10	15:31	**9**	**0**	**1**	**1**	**4**	**0**	**0**	**0**

			Regular Season																Playoffs							
Season	Club	League	GP	G	A	Pts	PIM	PP	SH	GW	S	%	+/-	TF	F%	H	SB	Min	GP	G	A	Pts	PIM	PP	SH	GW
2000-01	**Detroit**	**NHL**	**68**	**3**	**8**	**11**	**53**	**0**	**1**	**0**	**66**	**4.5**	**17**	**2**	**0.0**	**103**	**70**	**18:36**	**5**	**0**	**0**	**0**	**8**	**0**	**0**	**0**
	Cincinnati	AHL	2	0	1	1	2	....	....	....	....	....	....	....	....	....	....	....	....	....	....	....	....	....	....	....
2001-02	**Colorado**	**NHL**	**36**	**0**	**4**	**4**	**25**	**0**	**0**	**0**	**26**	**0.0**	**3**	**0**	**0.0**	**29**	**24**	**12:44**	....	....	....	....	....	....	....	....
	NHL Totals		**1002**	**82**	**271**	**353**	**1214**	**22**	**2**	**16**	**1301**	**6.3**		**3**	**33.3**	**284**	**170**	**16:48**	**103**	**7**	**30**	**37**	**193**	**2**	**1**	**1**

Traded to **San Jose** by **Toronto** for Jamie Baker and San Jose's 5th round choice (Peter Cava) in 1996 Entry Draft, June 14, 1996. Traded to **St. Louis** by **San Jose** for Joe Murphy, March 24, 1998. Claimed on waivers by **Detroit** from **St. Louis**, December 30, 1998. Signed as a free agent by **Phoenix**, July 21, 1999. Traded to **Detroit** by **Phoenix** for Philippe Audet, March 13, 2000. Signed as a free agent by **Colorado**, July 24, 2001. • Released by **Colorado**, February 12, 2002.

GILMOUR, Doug

(GIHL-mohr, DUHG) **MTL.**

Center/Left wing. Shoots left. 5'11", 177 lbs. Born, Kingston, Ont., June 25, 1963. St. Louis' 4th choice, 134th overall, in 1982 Entry Draft.

Season	Club	League	GP	G	A	Pts	PIM	PP	SH	GW	S	%	+/-	TF	F%	H	SB	Min	GP	G	A	Pts	PIM	PP	SH	GW
1979-80	Kingston	OHA-B	15	2	5	7	26	....	....	....	....	....	....	....	....	....	....	....	....	....	....	....	....	....	....	....
	Belleville	OHA-B	25	9	14	23	18	....	....	....	....	....	....	....	....	....	....	....	....	....	....	....	....	....	....	....
1980-81	Cornwall Royals	QMJHL	51	12	23	35	35	....	....	....	....	....	....	....	....	....	....	....	....	....	....	....	....	....	....	....
1981-82	Cornwall Royals	OHL	67	46	73	119	42	....	....	....	....	....	....	....	....	....	....	....	5	6	9	15	2	....	....	....
1982-83	Cornwall Royals	OHL	68	70	*107	*177	62	....	....	....	....	....	....	....	....	....	....	....	8	8	10	18	16	....	....	....
1983-84	**St. Louis**	**NHL**	**80**	**25**	**28**	**53**	**57**	**3**	**1**	**1**	**157**	**15.9**	**6**	....	....	....	....	....	**11**	**2**	**9**	**11**	**10**	**1**	**0**	**1**
1984-85	**St. Louis**	**NHL**	**78**	**21**	**36**	**57**	**49**	**3**	**1**	**3**	**162**	**13.0**	**3**	....	....	....	....	....	**3**	**1**	**1**	**2**	**2**	**0**	**0**	**0**
1985-86	**St. Louis**	**NHL**	**74**	**25**	**28**	**53**	**41**	**2**	**1**	**5**	**183**	**13.7**	**–3**	....	....	....	....	....	**19**	**9**	**12**	***21**	**25**	**1**	**2**	**2**
1986-87	**St. Louis**	**NHL**	**80**	**42**	**63**	**105**	**58**	**17**	**1**	**2**	**207**	**20.3**	**–2**	....	....	....	....	....	**6**	**2**	**2**	**4**	**16**	**1**	**0**	**1**
1987-88	**St. Louis**	**NHL**	**72**	**36**	**50**	**86**	**59**	**19**	**2**	**4**	**163**	**22.1**	**–13**	....	....	....	....	....	**10**	**3**	**14**	**17**	**18**	**1**	**0**	**0**
1988-89 ♦	**Calgary**	**NHL**	**72**	**26**	**59**	**85**	**44**	**11**	**0**	**5**	**161**	**16.1**	**45**	....	....	....	....	....	**22**	**11**	**11**	**22**	**20**	**3**	**0**	**3**
1989-90	**Calgary**	**NHL**	**78**	**24**	**67**	**91**	**54**	**12**	**1**	**3**	**152**	**15.8**	**20**	....	....	....	....	....	**6**	**3**	**1**	**4**	**8**	**0**	**0**	**1**
1990-91	**Calgary**	**NHL**	**78**	**20**	**61**	**81**	**144**	**2**	**2**	**5**	**135**	**14.8**	**27**	....	....	....	....	....	**7**	**1**	**1**	**2**	**0**	**0**	**0**	**1**
1991-92	**Calgary**	**NHL**	**38**	**11**	**27**	**38**	**46**	**4**	**1**	**1**	**64**	**17.2**	**12**	....	....	....	....	....	....	....	....	....	....	....	....	....
	Toronto	**NHL**	**40**	**15**	**34**	**49**	**32**	**6**	**0**	**3**	**104**	**14.4**	**13**	....	....	....	....	....	....	....	....	....	....	....	....	....
1992-93	**Toronto**	**NHL**	**83**	**32**	**95**	**127**	**100**	**15**	**3**	**2**	**211**	**15.2**	**32**	....	....	....	....	....	**21**	**10**	***25**	**35**	**30**	**4**	**0**	**1**
1993-94	**Toronto**	**NHL**	**83**	**27**	**84**	**111**	**105**	**10**	**1**	**3**	**167**	**16.2**	**25**	....	....	....	....	....	**18**	**6**	**22**	**28**	**42**	**5**	**0**	**1**
1994-95	Rapperswil	Swiss	9	2	13	15	16	....	....	....	....	....	....	....	....	....	....	....	....	....	....	....	....	....	....	....
	Toronto	**NHL**	**44**	**10**	**23**	**33**	**26**	**3**	**0**	**1**	**73**	**13.7**	**–5**	....	....	....	....	....	**7**	**0**	**6**	**6**	**6**	**0**	**0**	**0**
1995-96	**Toronto**	**NHL**	**81**	**32**	**40**	**72**	**77**	**10**	**2**	**3**	**180**	**17.8**	**–5**	....	....	....	....	....	**6**	**1**	**7**	**8**	**12**	**1**	**0**	**0**
1996-97	**Toronto**	**NHL**	**61**	**15**	**45**	**60**	**46**	**2**	**1**	**1**	**103**	**14.6**	**–5**	....	....	....	....	....	....	....	....	....	....	....	....	....
	New Jersey	**NHL**	**20**	**7**	**15**	**22**	**22**	**2**	**0**	**0**	**40**	**17.5**	**7**	....	....	....	....	....	**10**	**0**	**4**	**4**	**14**	**0**	**0**	**0**
1997-98	**New Jersey**	**NHL**	**63**	**13**	**40**	**53**	**68**	**3**	**0**	**5**	**94**	**13.8**	**10**	....	....	....	....	....	**6**	**5**	**2**	**7**	**4**	**1**	**0**	**1**
1998-99	**Chicago**	**NHL**	**72**	**16**	**40**	**56**	**56**	**7**	**1**	**4**	**110**	**14.5**	**–16**	**1619**	**53.6**	**37**	**45**	**22:29**	....	....	....	....	....	....	....	....
99-2000	**Chicago**	**NHL**	**63**	**22**	**34**	**56**	**51**	**8**	**0**	**3**	**100**	**22.0**	**–12**	**941**	**53.7**	**34**	**31**	**19:59**	....	....	....	....	....	....	....	....
	Buffalo	**NHL**	**11**	**3**	**14**	**17**	**12**	**2**	**0**	**0**	**13**	**23.1**	**3**	**35**	**51.4**	**8**	**6**	**18:58**	**5**	**0**	**1**	**1**	**0**	**0**	**0**	**0**
2000-01	**Buffalo**	**NHL**	**71**	**7**	**31**	**38**	**70**	**4**	**0**	**0**	**91**	**7.7**	**3**	**429**	**51.1**	**51**	**36**	**18:02**	**13**	**2**	**4**	**6**	**12**	**1**	**0**	**1**
2001-02	**Montreal**	**NHL**	**70**	**10**	**31**	**41**	**48**	**5**	**0**	**2**	**78**	**12.8**	**–7**	**1241**	**51.1**	**43**	**38**	**18:39**	**12**	**4**	**6**	**10**	**16**	**1**	**0**	**0**
	NHL Totals		**1412**	**439**	**945**	**1384**	**1265**	**150**	**18**	**56**	**2748**	**16.0**		**4265**	**52.6**	**173**	**156**	**19:46**	**182**	**60**	**128**	**188**	**235**	**20**	**2**	**13**

OHL First All-Star Team (1983) • OHL MVP (1983) • Won Frank J. Selke Trophy (1993) • Played in NHL All-Star Game (1993, 1994)

Traded to **Calgary** by **St. Louis** with Mark Hunter, Steve Bozek and Michael Dark for Mike Bullard, Craig Coxe and Tim Corkery, September 6, 1988. Traded to **Toronto** by **Calgary** with Jamie Macoun, Ric Nattress, Kent Manderville and Rick Wamsley for Gary Leeman, Alexander Godynyuk, Jeff Reese, Michel Petit and Craig Berube, January 2, 1992. Traded to **New Jersey** by **Toronto** with Dave Ellett and New Jersey's 3rd round choice (previously acquired, New Jersey selected Andre Lakos) in 1999 Entry Draft for Jason Smith, Steve Sullivan and the rights to Alyn McCauley, February 25, 1997. Signed as a free agent by **Chicago**, July 28, 1998. Traded to **Buffalo** by **Chicago** with J-P Dumont for Michal Grosek, March 10, 2000. Signed as a free agent by **Montreal**, October 5, 2001.

GIONTA, Brian

(jee-OHN-tuh, BRIGH-uhn) **N.J.**

Right wing. Shoots right. 5'7", 175 lbs. Born, Rochester, NY, January 18, 1979. New Jersey's 4th choice, 82nd overall, in 1998 Entry Draft.

Season	Club	League	GP	G	A	Pts	PIM	PP	SH	GW	S	%	+/-	TF	F%	H	SB	Min	GP	G	A	Pts	PIM	PP	SH	GW
1994-95	Rochester	NEJHL	28	*52	37	*89	....	....	....	....	....	....	....	....	....	....	....	....	....	....	....	....	....	....	....	....
1995-96	Niagara Scenics	MTJHL	51	47	44	91	59	....	....	....	....	....	....	....	....	....	....	....	....	....	....	....	....	....	....	....
1996-97	Niagara Scenics	MTJHL	50	57	70	127	101	....	....	....	....	....	....	....	....	....	....	....	6	6	11	17	21	....	....	....
1997-98	Boston College	H-East	40	30	32	62	44	....	....	....	....	....	....	....	....	....	....	....	....	....	....	....	....	....	....	....
1998-99	Boston College	H-East	39	27	33	60	46	....	....	....	....	....	....	....	....	....	....	....	....	....	....	....	....	....	....	....
99-2000	Boston College	H-East	42	*33	23	56	66	....	....	....	....	....	....	....	....	....	....	....	....	....	....	....	....	....	....	....
2000-01	Boston College	H-East	43	*33	21	*54	47	....	....	....	....	....	....	....	....	....	....	....	....	....	....	....	....	....	....	....
2001-02	**New Jersey**	**NHL**	**33**	**4**	**7**	**11**	**8**	**0**	**0**	**0**	**58**	**6.9**	**10**	**36**	**44.4**	**54**	**8**	**13:25**	**6**	**2**	**2**	**4**	**0**	**0**	**1**	**2**
	Albany	AHL	37	9	16	25	18	....	....	....	....	....	....	....	....	....	....	....	....	....	....	....	....	....	....	....
	NHL Totals		**33**	**4**	**7**	**11**	**8**	**0**	**0**	**0**	**58**	**6.9**		**36**	**44.4**	**54**	**8**	**13:25**	**6**	**2**	**2**	**4**	**0**	**0**	**1**	**2**

MTJHL Player of the Year (1997) • Hockey East Rookie of the Year (1998) • Hockey East Second All-Star Team (1998) • NCAA East Second All-American Team (1998) • Hockey East First All-Star Team (1999, 2000, 2001) • NCAA East First All-American Team (1999, 2000, 2001) • Hockey East Player of the Year (2001)

GIRARD, Jonathan

(zhih-RAHR, JAWN-ah-thuhn) **BOS.**

Defense. Shoots right. 5'11", 192 lbs. Born, Joliette, Que., May 27, 1980. Boston's 1st choice, 48th overall, in 1998 Entry Draft.

Season	Club	League	GP	G	A	Pts	PIM	PP	SH	GW	S	%	+/-	TF	F%	H	SB	Min	GP	G	A	Pts	PIM	PP	SH	GW
1995-96	Laval Laurentide	QAAA	39	11	22	33	44	....	....	....	....	....	....	....	....	....	....	....	16	4	11	15	16	....	....	....
1996-97	Laval Titan	QMJHL	39	11	23	34	13	....	....	....	....	....	....	....	....	....	....	....	3	0	3	3	0	....	....	....
1997-98	Laval Titan	QMJHL	64	20	47	67	44	....	....	....	....	....	....	....	....	....	....	....	16	2	16	18	13	....	....	....
1998-99	Acadie-Bathurst	QMJHL	50	9	58	67	60	....	....	....	....	....	....	....	....	....	....	....	23	13	18	31	22	....	....	....
	Boston	**NHL**	**3**	**0**	**0**	**0**	**0**	**0**	**0**	**0**	**3**	**0.0**	**1**	**0**	**0.0**	**0**	**0**	**9:28**	....	....	....	....	....	....	....	....
99-2000	**Boston**	**NHL**	**23**	**1**	**2**	**3**	**2**	**0**	**0**	**0**	**17**	**5.9**	**–1**	**0**	**0.0**	**24**	**4**	**9:32**	....	....	....	....	....	....	....	....
	Moncton Wildcats	QMJHL	26	10	25	35	36	....	....	....	....	....	....	....	....	....	....	....	16	3	15	18	36	....	....	....
	Providence	AHL	5	0	1	1	0	....	....	....	....	....	....	....	....	....	....	....	....	....	....	....	....	....	....	....
2000-01	**Boston**	**NHL**	**31**	**3**	**13**	**16**	**14**	**2**	**0**	**1**	**42**	**7.1**	**2**	**0**	**0.0**	**31**	**17**	**16:32**	....	....	....	....	....	....	....	....
	Providence	AHL	39	3	21	24	6	....	....	....	....	....	....	....	....	....	....	....	17	0	5	5	4	....	....	....
2001-02	**Boston**	**NHL**	**20**	**0**	**3**	**3**	**9**	**0**	**0**	**0**	**28**	**0.0**	**0**	**0**	**0.0**	**19**	**4**	**14:43**	**1**	**0**	**0**	**0**	**2**	**0**	**0**	**0**
	Providence	AHL	59	6	31	37	36	....	....	....	....	....	....	....	....	....	....	....	2	0	0	0	2	....	....	....
	NHL Totals		**77**	**4**	**18**	**22**	**25**	**2**	**0**	**1**	**90**	**4.4**		**0**	**0.0**	**74**	**25**	**13:42**	**1**	**0**	**0**	**0**	**2**	**0**	**0**	**0**

QMJHL All-Rookie Team (1997) • QMJHL Second All-Star Team (1998) • QMJHL First All-Star Team (1999, 2000)

Traded to **Moncton** (QMJHL) by **Acadie-Bathurst** (QMJHL) with Acadie-Bathurst's 4th round choice (Hugh Verpaelst) in 2000 QMJHL Priority Draft and future considerations for David Comeau and Moncton's 1st round choice (Michel Tessier) in 2000 and future considerations (Sergei Kaltygen), January 12, 2000.

GIROUX, Raymond

(zhih-ROO, ray-MAWN) **N.J.**

Defense. Shoots left. 6'1", 190 lbs. Born, North Bay, Ont., July 20, 1976. Philadelphia's 7th choice, 202nd overall, in 1994 Entry Draft.

Season	Club	League	GP	G	A	Pts	PIM	PP	SH	GW	S	%	+/-	TF	F%	H	SB	Min	GP	G	A	Pts	PIM	PP	SH	GW
1992-93	Powassan Hawks	NOJHA	45	8	18	26	117	....	....	....	....	....	....	....	....	....	....	....	....	....	....	....	....	....	....	....
1993-94	Powassan Hawks	NOJHA	36	10	40	50	42	....	....	....	....	....	....	....	....	....	....	....	....	....	....	....	....	....	....	....
1994-95	Yale Bulldogs	ECAC	27	1	3	4	8	....	....	....	....	....	....	....	....	....	....	....	....	....	....	....	....	....	....	....
1995-96	Yale Bulldogs	ECAC	30	3	16	19	36	....	....	....	....	....	....	....	....	....	....	....	....	....	....	....	....	....	....	....
1996-97	Yale Bulldogs	ECAC	32	9	12	21	38	....	....	....	....	....	....	....	....	....	....	....	....	....	....	....	....	....	....	....
1997-98	Yale Bulldogs	ECAC	35	9	*30	39	62	....	....	....	....	....	....	....	....	....	....	....	....	....	....	....	....	....	....	....
1998-99	Lowell	AHL	59	13	19	32	92	....	....	....	....	....	....	....	....	....	....	....	3	1	1	2	0	....	....	....
99-2000	**NY Islanders**	**NHL**	**14**	**0**	**9**	**9**	**10**	**0**	**0**	**0**	**24**	**0.0**	**0**	**9**	**22.2**	**34**	**5**	**14:40**	....	....	....	....	....	....	....	....
	Lowell	AHL	49	12	21	33	34	....	....	....	....	....	....	....	....	....	....	....	7	0	0	0	2	....	....	....
2000-01	HIFK Helsinki	Finland	22	3	9	12	34	....	....	....	....	....	....	....	....	....	....	....	....	....	....	....	....	....	....	....
	AIK Solna	Sweden	9	0	1	1	16	....	....	....	....	....	....	....	....	....	....	....	....	....	....	....	....	....	....	....
	Jokerit Helsinki	Finland	24	4	9	13	16	....	....	....	....	....	....	....	....	....	....	....	5	0	0	0	0	....	....	....
2001-02	**NY Islanders**	**NHL**	**2**	**0**	**0**	**0**	**2**	**0**	**0**	**0**	**2**	**0.0**	**–1**	**0**	**0.0**	**1**	**1**	**12:18**	....	....	....	....	....	....	....	....
	Bridgeport	AHL	79	13	40	53	73	....	....	....	....	....	....	....	....	....	....	....	19	1	7	8	20	....	....	....
	NHL Totals		**16**	**0**	**9**	**9**	**12**	**0**	**0**	**0**	**26**	**0.0**		**9**	**22.2**	**35**	**6**	**14:22**	....	....	....	....	....	....	....	....

ECAC First All-Star Team (1998) • NCAA East First All-American Team (1998)

Rights traded to **NY Islanders** by **Philadelphia** for NY Islanders' 6th round choice (later traded to Montreal - Montreal selected Scott Selig) in 2000 Entry Draft, August 25, 1998. Signed as a free agent by **New Jersey**, July 12, 2002.

GOC, Sascha

(GAWCH, SA-shah) **T.B.**

Defense. Shoots right. 6'2", 225 lbs. Born, Calw, West Germany, April 17, 1979. New Jersey's 5th choice, 159th overall, in 1997 Entry Draft.

Season	Club	League	GP	G	A	Pts	PIM	PP	SH	GW	S	%	+/-	TF	F%	H	SB	Min	Playoffs GP	G	A	Pts	PIM	PP	SH	GW
1995-96	Schwenningen Jr.	Ger.-Jr.	11	3	6	9	77																			
	Schwenningen	Germany	1	0	0	0	0																			
1996-97	Schwenningen	Germany	41	3	1	4	28												5	0	0	0	0			
1997-98	Schwenningen	Germany	49	5	5	10	45																			
1998-99	Albany	AHL	55	1	12	13	24												2	0	0	0	0			
99-2000	Albany	AHL	64	9	22	31	35												5	2	0	2	6			
2000-01	**New Jersey**	**NHL**	**11**	**0**	**0**	**0**	**4**	**0**	**0**	**0**	**7**	**0.0**	**7**	**0**	**0.0**	**14**	**13**	**13:37**								
	Albany	AHL	55	10	29	39	49																			
2001-02	**New Jersey**	**NHL**	**2**	**0**	**0**	**0**	**0**	**0**	**0**	**0**	**2**	**0.0**	**–2**	**0**	**0.0**	**4**	**3**	**19:21**								
	Albany	AHL	10	0	5	5	12																			
	Tampa Bay	**NHL**	**9**	**0**	**0**	**0**	**0**	**0**	**0**	**0**	**2**	**0.0**	**0**	**0**	**0.0**	**7**	**5**	**5:32**								
	Springfield	AHL	36	3	9	12	30																			
	NHL Totals		**22**	**0**	**0**	**0**	**4**	**0**	**0**	**0**	**11**	**0.0**		**0**	**0.0**	**25**	**21**	**10:50**								

Traded to **Tampa Bay** by **New Jersey** with Josef Boumedienne and the rights to Anton But for Andrei Zyuzin, November 9, 2001.

GOLDMANN, Erich

(GOHLD-mahn, AIR-ihk)

Defense. Shoots left. 6'3", 212 lbs. Born, Dingolfing, West Germany, April 7, 1976. Ottawa's 5th choice, 212th overall, in 1996 Entry Draft.

Season	Club	League	GP	G	A	Pts	PIM	PP	SH	GW	S	%	+/-	TF	F%	H	SB	Min	Playoffs GP	G	A	Pts	PIM	PP	SH	GW
1993-94	EV Landshut	German-2	33	0	0	0	4												7	0	0	0	0			
1994-95	Adler Mannheim	Germany	31	0	0	0	22												10	1	0	1	2			
1995-96	Adler Mannheim	Germany	47	0	3	3	40												8	0	0	0	4			
1996-97	Kaufbeurer Adler	Germany	44	2	4	6	58												6	1	0	1	2			
1997-98	Worcester	AHL	31	0	2	2	40																			
	Germany	Olympics	4	0	1	1	27																			
	Detroit Vipers	IHL	3	0	0	0	2																			
	Dayton Bombers	ECHL	3	0	2	2	5												5	0	0	0	8			
1998-99	Hershey Bears	AHL	21	1	1	2	23																			
	Cincinnati	IHL	5	0	1	1	7																			
	Cincinnati	AHL	32	0	2	2	18												3	0	0	0	2			
99-2000	**Ottawa**	**NHL**	**1**	**0**	**0**	**0**	**0**	**0**	**0**	**0**	**0**	**0.0**	**0**	**0**	**0.0**	**0**	**1**	**9:44**								
	Grand Rapids	IHL	26	1	1	2	15																			
	Detroit Vipers	IHL	11	1	0	1	13																			
2000-01	Essen	Germany	58	7	3	10	44																			
2001-02	Moskitos Essen	Germany	59	1	17	18	32																			
	Germany	Olympics	7	0	0	0	27																			
	NHL Totals		**1**	**0**	**0**	**0**	**0**	**0**	**0**	**0**	**0**	**0.0**		**0**	**0.0**	**0**	**1**	**9:44**								

Loaned to **Cincinnati** (AHL) by **Ottawa** with Ratislav Pavlikovsky and Ivan Ciernik, January 12, 1999. Signed as a free agent by **Inglostadt** (Germany), August 17, 2002.

GOLUBOVSKY, Yan

(goh-luh-BOHV-skee, YAN)

Defense. Shoots right. 6'3", 183 lbs. Born, Novosibirsk, USSR, March 9, 1976. Detroit's 1st choice, 23rd overall, in 1994 Entry Draft.

Season	Club	League	GP	G	A	Pts	PIM	PP	SH	GW	S	%	+/-	TF	F%	H	SB	Min	Playoffs GP	G	A	Pts	PIM	PP	SH	GW
1993-94	DynamoMoscow2	CIS-3	10	0	1	1																				
	Russian Penguins	IHL	8	0	0	0	23																			
1994-95	Adirondack	AHL	57	4	2	6	39																			
1995-96	Adirondack	AHL	71	5	16	21	97												3	0	0	0	2			
1996-97	Adirondack	AHL	62	2	11	13	67												4	0	0	0	0			
1997-98	**Detroit**	**NHL**	**12**	**0**	**2**	**2**	**6**	**0**	**0**	**0**	**9**	**0.0**	**1**													
	Adirondack	AHL	52	1	15	16	57												3	0	0	0	2			
1998-99	**Detroit**	**NHL**	**17**	**0**	**1**	**1**	**16**	**0**	**0**	**0**	**10**	**0.0**	**4**	**0**	**0.0**	**9**	**4**	**9:39**								
	Adirondack	AHL	43	2	2	4	32												2	0	0	0	4			
99-2000	**Detroit**	**NHL**	**21**	**1**	**2**	**3**	**8**	**0**	**0**	**0**	**7**	**14.3**	**3**	**0**	**0.0**	**13**	**23**	**8:54**								
2000-01	Cincinnati	AHL	28	4	4	8	16																			
	Florida	**NHL**	**6**	**0**	**2**	**2**	**2**	**0**	**0**	**0**	**4**	**0.0**	**3**	**0**	**0.0**	**6**	**6**	**16:38**								
	Louisville	AHL	30	1	12	13	36																			
2001-02	Magnitogorsk	Russia	9	0	1	1	10																			
	Novosibirsk	Russia-2	31	6	8	14	50												10	1	1	2	12			
	NHL Totals		**56**	**1**	**7**	**8**	**32**	**0**	**0**	**0**	**30**	**3.3**		**0**	**0.0**	**28**	**33**	**10:15**								

Traded to **Florida** by **Detroit** for Igor Larionov, December 28, 2000.

GOMEZ, Scott

(GOH-mehz, SKAWT) **N.J.**

Center. Shoots left. 5'11", 200 lbs. Born, Anchorage, AK, December 23, 1979. New Jersey's 2nd choice, 27th overall, in 1998 Entry Draft.

Season	Club	League	GP	G	A	Pts	PIM	PP	SH	GW	S	%	+/-	TF	F%	H	SB	Min	Playoffs GP	G	A	Pts	PIM	PP	SH	GW
1994-95	East High	Hi-School	28	30	48	78																				
1995-96	East High	Hi-School	27	*56	49	*101																				
	Anchorage	AAHL	40	*70	*67	*137	44																			
1996-97	South Surrey	BCHL	56	48	76	124	94												21	18	23	41	57			
1997-98	Tri-City	WHL	45	12	37	49	57																			
1998-99	Tri-City	WHL	58	30	*78	108	55												10	6	13	19	31			
99-2000♦	**New Jersey**	**NHL**	**82**	**19**	**51**	**70**	**78**	**7**	**0**	**1**	**204**	**9.3**	**14**	**341**	**44.6**	**49**	**12**	**16:21**	**23**	**4**	**6**	**10**	**4**	**1**	**0**	**2**
2000-01	**New Jersey**	**NHL**	**76**	**14**	**49**	**63**	**46**	**2**	**0**	**4**	**155**	**9.0**	**–1**	**1010**	**44.6**	**36**	**15**	**15:46**	**25**	**5**	**9**	**14**	**24**	**0**	**0**	**0**
2001-02	**New Jersey**	**NHL**	**76**	**10**	**38**	**48**	**36**	**1**	**0**	**1**	**156**	**6.4**	**–4**	**628**	**48.7**	**49**	**16**	**16:46**								
	NHL Totals		**234**	**43**	**138**	**181**	**160**	**10**	**0**	**6**	**515**	**8.3**		**1979**	**45.9**	**134**	**43**	**16:18**	**48**	**9**	**15**	**24**	**28**	**1**	**0**	**2**

BCHL All-Rookie Team (1997) • WHL West First All-Star Team (1999) • NHL All-Rookie Team (2000) • Won Calder Memorial Trophy (2000) • Played in NHL All-Star Game (2000)

GONCHAR, Sergei

(gohn-CHAR, SAIR-gay) **WSH.**

Defense. Shoots left. 6'2", 208 lbs. Born, Chelyabinsk, USSR, April 13, 1974. Washington's 1st choice, 14th overall, in 1992 Entry Draft.

Season	Club	League	GP	G	A	Pts	PIM	PP	SH	GW	S	%	+/-	TF	F%	H	SB	Min	Playoffs GP	G	A	Pts	PIM	PP	SH	GW
1991-92	Chelyabinsk	CIS	31	1	0	1	6																			
1992-93	Dynamo Moscow	CIS	31	1	3	4	70												10	0	0	0	12			
1993-94	Dynamo Moscow	CIS	44	4	5	9	36												10	0	3	3	14			
	Portland Pirates	AHL																	2	0	0	0	0			
1994-95	Portland Pirates	AHL	61	10	32	42	67																			
	Washington	**NHL**	**31**	**2**	**5**	**7**	**22**	**0**	**0**	**0**	**38**	**5.3**	**4**						**7**	**2**	**2**	**4**	**2**	**0**	**0**	**1**
1995-96	**Washington**	**NHL**	**78**	**15**	**26**	**41**	**60**	**4**	**0**	**4**	**139**	**10.8**	**25**						**6**	**2**	**4**	**6**	**4**	**1**	**0**	**0**
1996-97	**Washington**	**NHL**	**57**	**13**	**17**	**30**	**36**	**3**	**0**	**3**	**129**	**10.1**	**–11**													
1997-98	Lada Togliatti	Russia	7	3	2	5	4																			
	Lada Togliatti	EuroHL	1	1	0	1	2																			
	Washington	**NHL**	**72**	**5**	**16**	**21**	**66**	**2**	**0**	**0**	**134**	**3.7**	**2**						**21**	**7**	**4**	**11**	**30**	**3**	**1**	**2**
	Russia	Olympics	6	0	2	2	0																			
1998-99	**Washington**	**NHL**	**53**	**21**	**10**	**31**	**57**	**13**	**1**	**3**	**180**	**11.7**	**1**	**0**	**0.0**	**72**	**31**	**23:55**								
99-2000	**Washington**	**NHL**	**73**	**18**	**36**	**54**	**52**	**5**	**0**	**3**	**181**	**9.9**	**26**	**0**	**0.0**	**97**	**44**	**21:46**	**5**	**1**	**0**	**1**	**6**	**0**	**0**	**0**
2000-01	**Washington**	**NHL**	**76**	**19**	**38**	**57**	**70**	**8**	**0**	**2**	**241**	**7.9**	**12**	**1**	**100.0**	**88**	**53**	**22:26**	**6**	**1**	**3**	**4**	**2**	**1**	**0**	**0**
2001-02	**Washington**	**NHL**	**76**	**26**	**33**	**59**	**58**	**7**	**0**	**2**	**216**	**12.0**	**–1**	**1**	**100.0**	**63**	**61**	**23:51**								
	Russia	Olympics	6	0	0	0	2																			
	NHL Totals		**516**	**119**	**181**	**300**	**421**	**42**	**1**	**17**	**1258**	**9.5**		**2**	**100.0**	**320**	**189**	**22:56**	**45**	**13**	**13**	**26**	**44**	**5**	**1**	**3**

NHL Second All-Star Team (2002) • Played in NHL All-Star Game (2001, 2002)

GOREN, Lee

(GOH-rehn, LEE) **BOS.**

Right wing. Shoots right. 6'3", 205 lbs. Born, Winnipeg, Man., December 26, 1977. Boston's 5th choice, 63rd overall, in 1997 Entry Draft.

Season	Club	League	GP	G	A	Pts	PIM	PP	SH	GW	S	%	+/-	TF	F%	H	SB	Min	Playoffs GP	G	A	Pts	PIM	PP	SH	GW
1994-95	Wpg. Warriors	MMHL	31	19	31	50	50																			
1995-96	Minot Top Guns	SJHL	56	25	35	61													12	5	20	25				
	Saskatoon Blades	WHL	2	0	0	0	2																			
1996-97	North Dakota	WCHA	DID NOT PLAY – FRESHMAN																							
1997-98	North Dakota	WCHA	29	3	13	16	26																			
1998-99	North Dakota	WCHA	38	26	19	45	20																			

Season	Club	League	Regular Season GP	G	A	Pts	PIM	PP	SH	GW	S	%	+/-	TF	F%	H	SB	Min	Playoffs GP	G	A	Pts	PIM	PP	SH	GW
99-2000	North Dakota	WCHA	44	*34	29	63	42																			
2000-01	**Boston**	**NHL**	**21**	**2**	**0**	**2**	**7**	**1**	**0**	**0**	**9**	**22.2**	**–3**	**21**	**38.1**	**15**	**0**	**4:24**								
	Providence	AHL	54	15	18	33	72												17	5	2	7	11			
2001-02	Providence	AHL	71	11	26	37	121												2	0	0	0	0			
	NHL Totals		**21**	**2**	**0**	**2**	**7**	**1**	**0**	**0**	**9**	**22.2**		**21**	**38.1**	**15**	**0**	**4:24**								

WCHA Second All-Star Team (2000) • NCAA West Second All-American Team (2000) • NCAA Championship All-Tournament Team (2000) • NCAA Championship Tournament MVP (2000)

• Ruled ineligible to play during 1996-97 season by NCAA due to appearance with **Saskatoon** (WHL) in 1995-96 season.

GOSSELIN, David

(GAH-sih-lihn, DAY-vihd) **DAL.**

Right wing. Shoots right. 6'1", 205 lbs. Born, Levis, Que., June 22, 1977. New Jersey's 4th choice, 78th overall, in 1995 Entry Draft.

Season	Club	League	GP	G	A	Pts	PIM	PP	SH	GW	S	%	+/-	TF	F%	H	SB	Min	Playoffs GP	G	A	Pts	PIM	PP	SH	GW
1992-93	Richelieu	QAAA	40	5	12	17	24												4	0	0	0	2			
1993-94	Richelieu	QAAA	44	26	19	45	62												4	2	1	3	0			
1994-95	Sherbrooke	QMJHL	58	8	8	16	36												7	0	0	0	2			
1995-96	Sherbrooke	QMJHL	55	24	24	48	147												7	2	2	4	4			
1996-97	Sherbrooke	QMJHL	23	11	15	26	52																			
	Chicoutimi	QMJHL	28	16	33	49	65												12	9	7	16	16			
1997-98	Chicoutimi	QMJHL	69	46	64	110	139												6	1	4	5	8			
1998-99	Milwaukee	IHL	74	17	11	28	78												2	0	2	2	2			
99-2000	**Nashville**	**NHL**	**10**	**2**	**1**	**3**	**6**	**0**	**0**	**0**	**14**	**14.3**	**–4**	**0**	**0.0**	**7**	**1**	**9:16**								
	Milwaukee	IHL	70	21	20	41	118												3	0	0	0	10			
2000-01	Milwaukee	IHL	32	5	9	14	56																			
2001-02	**Nashville**	**NHL**	**3**	**0**	**0**	**0**	**5**	**0**	**0**	**0**	**0**	**0.0**	**–1**	**0**	**0.0**	**0**	**0**	**7:05**								
	Milwaukee	AHL	66	11	21	32	112																			
	NHL Totals		**13**	**2**	**1**	**3**	**11**	**0**	**0**	**0**	**14**	**14.3**		**0**	**0.0**	**7**	**1**	**8:46**								

Signed as a free agent by **Nashville**, July 1, 1998. • Missed majority of 2000-01 season recovering from knee injury suffered in game vs. Grand Rapids (IHL), December 28, 2000. Traded to **Dallas** by **Nashville** with Nashville's 5th round choice in 2003 Entry Draft for Ed Belfour and Cameron Mann, June 29, 2002.

GRAND-PIERRE, Jean-Luc

(GRAHN pee-AIR, ZHAHN-LOOK) **CBJ**

Defense. Shoots right. 6'3", 223 lbs. Born, Montreal, Que., February 2, 1977. St. Louis' 6th choice, 179th overall, in 1995 Entry Draft.

Season	Club	League	GP	G	A	Pts	PIM	PP	SH	GW	S	%	+/-	TF	F%	H	SB	Min	Playoffs GP	G	A	Pts	PIM	PP	SH	GW
1992-93	Lac St-Louis	QAAA	1	0	0	0	2																			
1993-94	Beauport	QMJHL	46	1	4	5	27												1	0	0	0	0			
1994-95	Val-d'Or Foreurs	QMJHL	59	10	13	23	126																			
1995-96	Val-d'Or Foreurs	QMJHL	67	13	21	34	209												13	1	4	5	47			
1996-97	Val-d'Or Foreurs	QMJHL	58	9	24	33	186												13	5	8	13	46			
1997-98	Rochester	AHL	75	4	6	10	211												4	0	0	0	2			
1998-99	**Buffalo**	**NHL**	**16**	**0**	**1**	**1**	**17**	**0**	**0**	**0**	**11**	**0.0**	**0**	**0**	**0.0**	**46**	**13**	**13:36**								
	Rochester	AHL	55	5	4	9	90																			
99-2000	**Buffalo**	**NHL**	**11**	**0**	**0**	**0**	**15**	**0**	**0**	**0**	**11**	**0.0**	**–1**	**0**	**0.0**	**29**	**8**	**15:11**	**4**	**0**	**0**	**0**	**4**	**0**	**0**	**0**
	Rochester	AHL	62	5	8	13	124												17	0	1	1	40			
2000-01	**Columbus**	**NHL**	**64**	**1**	**4**	**5**	**73**	**0**	**0**	**0**	**33**	**3.0**	**–6**	**0**	**0.0**	**112**	**53**	**12:51**								
2001-02	**Columbus**	**NHL**	**81**	**2**	**6**	**8**	**90**	**0**	**0**	**0**	**62**	**3.2**	**–28**	**3**	**0.0**	**176**	**105**	**15:20**								
	NHL Totals		**172**	**3**	**11**	**14**	**195**	**0**	**0**	**0**	**117**	**2.6**		**3**	**0.0**	**363**	**179**	**14:14**	**4**	**0**	**0**	**0**	**4**	**0**	**0**	**0**

Traded to **Buffalo** by **St. Louis** with Ottawa's 2nd round choice (previously acquired, Buffalo selected Cory Sarich) in 1996 Entry Draft and St. Louis' 3rd round choice (Maxim Afinogenov) in 1997 Entry Draft for Yuri Khmylev and Buffalo's 8th round choice (Andrei Podkonicky) in 1996 Entry Draft, March 20, 1996. Traded to **Columbus** by **Buffalo** with Matt Davidson, San Jose's 5th round choice (previously acquired, Columbus selected Tyler Kolarik) in 2000 Entry Draft and Buffalo's 5th round choice (later traded to Calgary - later traded to Detroit - Detroit selected Andreas Jamtin) in 2001 Entry Draft to complete Expansion Draft agreement which had Columbus select Geoff Sanderson and Dwayne Roloson from Buffalo, June 23, 2000.

GRATTON, Benoit

(grah-TOHN, BEHN-wah) **MTL.**

Left wing. Shoots left. 5'11", 194 lbs. Born, Montreal, Que., December 28, 1976. Washington's 6th choice, 105th overall, in 1995 Entry Draft.

Season	Club	League	GP	G	A	Pts	PIM	PP	SH	GW	S	%	+/-	TF	F%	H	SB	Min	Playoffs GP	G	A	Pts	PIM	PP	SH	GW
1992-93	Laval Laurentide	QAAA	40	19	38	57	74												13	1	9	10	27			
1993-94	Laval Titan	QMJHL	51	9	14	23	70												20	2	1	3	19			
1994-95	Laval Titan	QMJHL	71	30	58	88	199												20	8	*21	29	42			
1995-96	Laval Titan	QMJHL	38	21	39	60	130																			
	Granby	QMJHL	27	12	46	58	97												21	13	26	39	68			
1996-97	Portland Pirates	AHL	76	6	40	46	140												5	2	1	3	14			
1997-98	**Washington**	**NHL**	**6**	**0**	**1**	**1**	**6**	**0**	**0**	**0**	**5**	**0.0**	**1**													
	Portland Pirates	AHL	58	19	31	50	137												8	4	2	6	24			
1998-99	**Washington**	**NHL**	**16**	**4**	**3**	**7**	**16**	**0**	**0**	**0**	**24**	**16.7**	**–1**	**136**	**54.4**	**26**	**7**	**13:28**								
	Portland Pirates	AHL	64	18	42	60	135																			
99-2000	**Calgary**	**NHL**	**10**	**0**	**2**	**2**	**10**	**0**	**0**	**0**	**4**	**0.0**	**1**	**68**	**63.2**	**8**	**2**	**8:15**								
	Saint John	AHL	65	17	49	66	137												3	0	1	1	4			
2000-01	**Calgary**	**NHL**	**14**	**1**	**3**	**4**	**14**	**0**	**0**	**0**	**13**	**7.7**	**0**	**105**	**63.8**	**13**	**3**	**9:11**								
	Saint John	AHL	53	10	36	46	153																			
2001-02	**Montreal**	**NHL**	**8**	**1**	**0**	**1**	**8**	**0**	**0**	**0**	**8**	**12.5**	**–1**	**98**	**63.3**	**7**	**6**	**9:51**								
	Quebec	AHL	35	10	19	29	70												3	2	3	5	10			
	NHL Totals		**54**	**6**	**9**	**15**	**54**	**0**	**0**	**0**	**54**	**11.1**		**407**	**60.4**	**54**	**18**	**10:32**								

Traded to **Calgary** by **Washington** for Steve Shirreffs, August 18, 1999. Claimed on waivers by **Montreal** from **Calgary**, April 11, 2001.

GRATTON, Chris

(GRA-tuhn, KRIHS) **BUF.**

Center. Shoots left. 6'4", 225 lbs. Born, Brantford, Ont., July 5, 1975. Tampa Bay's 1st choice, 3rd overall, in 1993 Entry Draft.

Season	Club	League	GP	G	A	Pts	PIM	PP	SH	GW	S	%	+/-	TF	F%	H	SB	Min	Playoffs GP	G	A	Pts	PIM	PP	SH	GW
1989-90	Brantford	OJHL-B	1	0	2	2	2																			
1990-91	Brantford	OJHL-B	31	30	30	60	28																			
1991-92	Kingston	OHL	62	27	39	66	37																			
1992-93	Kingston	OHL	58	55	54	109	125												16	11	18	29	42			
1993-94	**Tampa Bay**	**NHL**	**84**	**13**	**29**	**42**	**123**	**5**	**1**	**2**	**161**	**8.1**	**–25**													
1994-95	**Tampa Bay**	**NHL**	**46**	**7**	**20**	**27**	**89**	**2**	**0**	**0**	**91**	**7.7**	**–2**													
1995-96	**Tampa Bay**	**NHL**	**82**	**17**	**21**	**38**	**105**	**7**	**0**	**3**	**183**	**9.3**	**–13**						**6**	**0**	**2**	**2**	**27**	**0**	**0**	**0**
1996-97	**Tampa Bay**	**NHL**	**82**	**30**	**32**	**62**	**201**	**9**	**0**	**4**	**230**	**13.0**	**–28**													
1997-98	**Philadelphia**	**NHL**	**82**	**22**	**40**	**62**	**159**	**5**	**0**	**2**	**182**	**12.1**	**11**						**5**	**2**	**0**	**2**	**10**	**0**	**0**	**0**
1998-99	**Philadelphia**	**NHL**	**26**	**1**	**7**	**8**	**41**	**0**	**0**	**0**	**54**	**1.9**	**–8**	**38**	**42.1**	**35**	**0**	**14:25**								
	Tampa Bay	**NHL**	**52**	**7**	**19**	**26**	**102**	**1**	**0**	**1**	**127**	**5.5**	**–20**	**1032**	**53.9**	**74**	**12**	**18:20**								
99-2000	**Tampa Bay**	**NHL**	**58**	**14**	**27**	**41**	**121**	**4**	**0**	**1**	**168**	**8.3**	**–24**	**1341**	**55.9**	**120**	**13**	**20:03**								
	Buffalo	**NHL**	**14**	**1**	**7**	**8**	**15**	**0**	**0**	**0**	**34**	**2.9**	**1**	**256**	**54.3**	**27**	**8**	**16:40**	**5**	**0**	**1**	**1**	**4**	**0**	**0**	**0**
2000-01	**Buffalo**	**NHL**	**82**	**19**	**21**	**40**	**102**	**5**	**0**	**5**	**156**	**12.2**	**0**	**1161**	**57.3**	**118**	**34**	**14:37**	**13**	**6**	**4**	**10**	**14**	**2**	**0**	**1**
2001-02	**Buffalo**	**NHL**	**82**	**15**	**24**	**39**	**75**	**1**	**0**	**5**	**139**	**10.8**	**0**	**1297**	**53.8**	**117**	**30**	**14:57**								
	NHL Totals		**690**	**146**	**247**	**393**	**1133**	**39**	**1**	**23**	**1525**	**9.6**		**5125**	**55.1**	**491**	**97**	**16:24**	**29**	**8**	**7**	**15**	**55**	**2**	**0**	**1**

OHL All-Rookie Team (1992) • OHL Rookie of the Year (1992)

Signed as a free agent by **Philadelphia**, August 14, 1997. Traded to **Tampa Bay** by **Philadelphia** with Mike Sillinger for Mikael Renberg and Daymond Langkow, December 12, 1998. Traded to **Buffalo** by **Tampa Bay** with Tampa Bay's 2nd round choice (Derek Roy) in 2001 Entry Draft for Cory Sarich, Wayne Primeau, Brian Holzinger and Buffalo's 3rd round choice (Alexander Kharitonov) in 2000 Entry Draft, March 9, 2000.

GRAVES, Adam

(GRAYVS, A-duhm) **S.J.**

Center. Shoots left. 6', 205 lbs. Born, Toronto, Ont., April 12, 1968. Detroit's 2nd choice, 22nd overall, in 1986 Entry Draft.

Season	Club	League	GP	G	A	Pts	PIM	PP	SH	GW	S	%	+/-	TF	F%	H	SB	Min	Playoffs GP	G	A	Pts	PIM	PP	SH	GW
1984-85	King City Dukes	OJHL-B	25	23	33	56	29																			
1985-86	Windsor	OHL	62	27	37	64	35												16	5	11	16	10			
1986-87	Windsor	OHL	66	45	55	100	70												14	9	8	17	32			
	Adirondack	AHL																	5	0	1	1	0			
1987-88	Windsor	OHL	37	28	32	60	107												12	14	18	*32	16			
	Detroit	**NHL**	**9**	**0**	**1**	**1**	**8**	**0**	**0**	**0**	**9**	**0.0**	**–2**													
1988-89	**Detroit**	**NHL**	**56**	**7**	**5**	**12**	**60**	**0**	**0**	**1**	**60**	**11.7**	**–5**						**5**	**0**	**0**	**0**	**4**	**0**	**0**	**0**
	Adirondack	AHL	14	10	11	21	28												14	11	7	18	17			
1989-90	**Detroit**	**NHL**	**13**	**0**	**1**	**1**	**13**	**0**	**0**	**0**	**10**	**0.0**	**–5**													
	♦ **Edmonton**	**NHL**	**63**	**9**	**12**	**21**	**123**	**1**	**0**	**1**	**84**	**10.7**	**5**						**22**	**5**	**6**	**11**	**17**	**0**	**0**	**1**
1990-91	**Edmonton**	**NHL**	**76**	**7**	**18**	**25**	**127**	**2**	**0**	**1**	**126**	**5.6**	**–21**						**18**	**2**	**4**	**6**	**22**	**0**	**0**	**0**

Season	Club	League	GP	G	A	Pts	PIM	PP	SH	GW	S	%	+/-	TF	F%	H	SB	Min	GP	G	A	Pts	PIM	PP	SH	GW
			Regular Season																Playoffs							
1991-92	**NY Rangers**	**NHL**	**80**	**26**	**33**	**59**	**139**	**4**	**4**	**4**	**228**	**11.4**	**19**						**10**	**5**	**3**	**8**	**22**	**1**	**0**	**1**
1992-93	**NY Rangers**	**NHL**	**84**	**36**	**29**	**65**	**148**	**12**	**1**	**6**	**275**	**13.1**	**–4**													
1993-94♦	**NY Rangers**	**NHL**	**84**	**52**	**27**	**79**	**127**	**20**	**4**	**4**	**291**	**17.9**	**27**						**23**	**10**	**7**	**17**	**24**	**3**	**0**	**0**
1994-95	**NY Rangers**	**NHL**	**47**	**17**	**14**	**31**	**51**	**9**	**0**	**3**	**185**	**9.2**	**9**						**10**	**4**	**4**	**8**	**8**	**2**	**0**	**0**
1995-96	**NY Rangers**	**NHL**	**82**	**22**	**36**	**58**	**100**	**9**	**1**	**2**	**266**	**8.3**	**18**						**10**	**7**	**1**	**8**	**4**	**6**	**0**	**2**
1996-97	**NY Rangers**	**NHL**	**82**	**33**	**28**	**61**	**66**	**10**	**4**	**3**	**269**	**12.3**	**10**						**15**	**2**	**1**	**3**	**12**	**1**	**0**	**2**
1997-98	**NY Rangers**	**NHL**	**72**	**23**	**12**	**35**	**41**	**10**	**0**	**2**	**226**	**10.2**	**–30**													
1998-99	**NY Rangers**	**NHL**	**82**	**38**	**15**	**53**	**47**	**14**	**2**	**7**	**239**	**15.9**	**–12**	**347**	**51.3**	**162**	**28**	**20:33**								
99-2000	**NY Rangers**	**NHL**	**77**	**23**	**17**	**40**	**14**	**11**	**0**	**4**	**194**	**11.9**	**–15**	**51**	**49.0**	**166**	**27**	**18:46**								
2000-01	**NY Rangers**	**NHL**	**82**	**10**	**16**	**26**	**77**	**1**	**0**	**1**	**136**	**7.4**	**–16**	**98**	**55.1**	**186**	**23**	**15:44**								
2001-02	**San Jose**	**NHL**	**81**	**17**	**14**	**31**	**51**	**1**	**3**	**1**	**139**	**12.2**	**11**	**24**	**41.7**	**133**	**19**	**15:47**	**12**	**3**	**1**	**4**	**6**	**0**	**0**	**2**
	NHL Totals		**1070**	**320**	**278**	**598**	**1192**	**104**	**19**	**40**	**2737**	**11.7**		**520**	**51.3**	**647**	**97**	**17:42**	**125**	**38**	**27**	**65**	**119**	**13**	**0**	**8**

NHL Second All-Star Team (1994) • Won King Clancy Memorial Trophy (1994) • Won Bill Masterton Memorial Trophy (2001) • Played in NHL All-Star Game (1994)

Traded to **Edmonton** by **Detroit** with Petr Klima, Joe Murphy and Jeff Sharples for Jimmy Carson, Kevin McClelland and Edmonton's 5th round choice (later traded to Montreal - Montreal selected Brad Layzell) in 1991 Entry Draft, November 2, 1989. Signed as a free agent by **NY Rangers**, September 3, 1991. Traded to **San Jose** by **NY Rangers** with future considerations for Mikael Samuelsson and Christian Gosselin, June 24, 2001.

GREEN, Josh

(GREEN, JAWSH) **EDM.**

Left wing. Shoots left. 6'4", 212 lbs. Born, Camrose, Alta., November 16, 1977. Los Angeles' 1st choice, 30th overall, in 1996 Entry Draft.

Season	Club	League	GP	G	A	Pts	PIM	PP	SH	GW	S	%	+/-	TF	F%	H	SB	Min	GP	G	A	Pts	PIM	PP	SH	GW
1992-93	Camrose Kodiacs	ABHL	60	55	45	100	80																			
1993-94	Medicine Hat	WHL	63	22	22	44	43												3	0	0	0	4			
1994-95	Medicine Hat	WHL	68	32	23	55	64												5	5	1	6	2			
1995-96	Medicine Hat	WHL	46	18	25	43	55												5	2	2	4	4			
1996-97	Medicine Hat	WHL	51	25	32	57	61																			
	Swift Current	WHL	23	10	15	25	33												10	9	7	16	19			
1997-98	Swift Current	WHL	5	9	1	10	9																			
	Portland	WHL	26	26	18	44	27																			
	Fredericton	AHL	43	16	15	31	14												4	1	3	4	6			
1998-99	**Los Angeles**	**NHL**	**27**	**1**	**3**	**4**	**8**	**1**	**0**	**0**	**35**	**2.9**	**–5**	**2**	**50.0**	**30**	**1**	**11:44**								
	Springfield	AHL	41	15	15	30	29																			
99-2000	**NY Islanders**	**NHL**	**49**	**12**	**14**	**26**	**41**	**2**	**0**	**3**	**109**	**11.0**	**–7**	**12**	**50.0**	**105**	**8**	**13:36**								
	Lowell	AHL	17	6	2	8	19																			
2000-01	Hamilton	AHL	2	2	0	2	2																			
	Edmonton	**NHL**																	**3**	**0**	**0**	**0**	**0**	**0**	**0**	**0**
2001-02	**Edmonton**	**NHL**	**61**	**10**	**5**	**15**	**52**	**1**	**0**	**1**	**78**	**12.8**	**9**	**18**	**38.9**	**77**	**18**	**10:05**								
	NHL Totals		**137**	**23**	**22**	**45**	**101**	**4**	**0**	**4**	**222**	**10.4**		**32**	**43.8**	**212**	**27**	**11:40**	**3**	**0**	**0**	**0**	**0**	**0**	**0**	**0**

Traded to **Portland** (WHL) by **Swift Current** (WHL) for Tyler Murray, October 1, 1997. Traded to **NY Islanders** by **LA Kings** with Olli Jokinen, Mathieu Biron and LA Kings' 1st round choice (Taylor Pyatt) in 1999 Entry Draft for Ziggy Palffy, Brian Smolinski, Marcel Cousineau and New Jersey's 4th round choice (previously acquired, LA Kings selected Daniel Johansson) in 1999 Entry Draft, June 20, 1999. Traded to **Edmonton** by **NY Islanders** with Eric Brewer and NY Islanders' 2nd round choice (Brad Winchester) in 2000 Entry Draft for Roman Hamrlik, June 24, 2000. • Missed majority of 2000-01 season recovering from shoulder injury suffered in game vs. Detroit, October 10, 2000.

GREEN, Travis

(GREEN, TRA-vihs) **TOR.**

Center. Shoots right. 6'2", 200 lbs. Born, Castlegar, B.C., December 20, 1970. NY Islanders' 2nd choice, 23rd overall, in 1989 Entry Draft.

Season	Club	League	GP	G	A	Pts	PIM	PP	SH	GW	S	%	+/-	TF	F%	H	SB	Min	GP	G	A	Pts	PIM	PP	SH	GW
1985-86	Castlegar Rebels	KIJHL	35	30	40	70	41																			
1986-87	Spokane Chiefs	WHL	64	8	17	25	27												3	0	0	0	0			
1987-88	Spokane Chiefs	WHL	72	33	54	87	42												15	10	10	20	13			
1988-89	Spokane Chiefs	WHL	75	51	51	102	79																			
1989-90	Spokane Chiefs	WHL	50	45	44	89	80																			
	Medicine Hat	WHL	25	15	24	39	19												3	0	0	0	2			
1990-91	Capital District	AHL	73	21	34	55	26																			
1991-92	Capital District	AHL	71	23	27	50	10												7	0	4	4	21			
1992-93	**NY Islanders**	**NHL**	**61**	**7**	**18**	**25**	**43**	**1**	**0**	**0**	**115**	**6.1**	**4**						**12**	**3**	**1**	**4**	**6**	**0**	**0**	**0**
	Capital District	AHL	20	12	11	23	39																			
1993-94	**NY Islanders**	**NHL**	**83**	**18**	**22**	**40**	**44**	**1**	**0**	**2**	**164**	**11.0**	**16**						**4**	**0**	**0**	**0**	**2**	**0**	**0**	**0**
1994-95	**NY Islanders**	**NHL**	**42**	**5**	**7**	**12**	**25**	**0**	**0**	**0**	**59**	**8.5**	**–10**													
1995-96	**NY Islanders**	**NHL**	**69**	**25**	**45**	**70**	**42**	**14**	**1**	**2**	**186**	**13.4**	**–20**													
1996-97	**NY Islanders**	**NHL**	**79**	**23**	**41**	**64**	**38**	**10**	**0**	**3**	**177**	**13.0**	**–5**													
1997-98	**NY Islanders**	**NHL**	**54**	**14**	**12**	**26**	**66**	**8**	**0**	**2**	**99**	**14.1**	**–19**													
	Anaheim	**NHL**	**22**	**5**	**11**	**16**	**16**	**1**	**0**	**0**	**42**	**11.9**	**–10**													
1998-99	**Anaheim**	**NHL**	**79**	**13**	**17**	**30**	**81**	**3**	**1**	**2**	**165**	**7.9**	**–7**	**1325**	**52.8**	**97**	**24**	**17:17**	**4**	**0**	**1**	**1**	**4**	**0**	**0**	**0**
99-2000	**Phoenix**	**NHL**	**78**	**25**	**21**	**46**	**45**	**6**	**0**	**2**	**157**	**15.9**	**–4**	**1322**	**55.6**	**120**	**11**	**16:36**	**5**	**2**	**1**	**3**	**2**	**0**	**0**	**0**
2000-01	**Phoenix**	**NHL**	**69**	**13**	**15**	**28**	**63**	**3**	**0**	**0**	**113**	**11.5**	**–11**	**1135**	**54.9**	**83**	**16**	**16:05**								
2001-02	**Toronto**	**NHL**	**82**	**11**	**23**	**34**	**61**	**3**	**0**	**2**	**119**	**9.2**	**13**	**647**	**54.1**	**99**	**17**	**14:32**	**20**	**3**	**6**	**9**	**34**	**0**	**0**	**1**
	NHL Totals		**718**	**159**	**232**	**391**	**524**	**50**	**2**	**15**	**1396**	**11.4**		**4429**	**54.4**	**399**	**68**	**16:06**	**45**	**8**	**9**	**17**	**48**	**0**	**0**	**1**

Traded to **Anaheim** by **NY Islanders** with Doug Houda and Tony Tuzzolino for Joe Sacco, J-J Daigneault and Mark Janssens, February 6, 1998. Traded to **Phoenix** by **Anaheim** with Anaheim's 1st round choice (Scott Kelman) in 1999 Entry Draft for Oleg Tverdovsky, June 26, 1999. Traded to **Toronto** by **Phoenix** with Robert Reichel and Craig Mills for Danny Markov, June 12, 2001.

GREIG, Mark

(GREG, MAHRK) **PHI.**

Right wing. Shoots right. 5'11", 190 lbs. Born, High River, Alta., January 25, 1970. Hartford's 1st choice, 15th overall, in 1990 Entry Draft.

Season	Club	League	GP	G	A	Pts	PIM	PP	SH	GW	S	%	+/-	TF	F%	H	SB	Min	GP	G	A	Pts	PIM	PP	SH	GW
1985-86	Blackie Bisons	AAHA	31	12	43	55	44																			
1986-87	Cgy. North Stars	AMHL	18	9	28	37	30																			
	Calgary	WHL	5	0	0	0	0																			
1987-88	Lethbridge	WHL	65	9	18	27	38																			
1988-89	Lethbridge	WHL	71	36	72	108	113												8	5	5	10	16			
1989-90	Lethbridge	WHL	65	55	80	135	149												18	11	21	32	35			
1990-91	**Hartford**	**NHL**	**4**	**0**	**0**	**0**	**0**	**0**	**0**	**0**	**1**	**0.0**	**–1**													
	Springfield	AHL	73	32	55	87	73												17	2	6	8	22			
1991-92	**Hartford**	**NHL**	**17**	**0**	**5**	**5**	**6**	**0**	**0**	**0**	**18**	**0.0**	**7**													
	Springfield	AHL	50	20	27	47	38												9	1	1	2	20			
1992-93	**Hartford**	**NHL**	**22**	**1**	**7**	**8**	**27**	**0**	**0**	**0**	**16**	**6.3**	**–11**													
	Springfield	AHL	55	20	38	58	86																			
1993-94	**Hartford**	**NHL**	**31**	**4**	**5**	**9**	**31**	**0**	**0**	**0**	**41**	**9.8**	**–6**													
	Springfield	AHL	4	0	4	4	21																			
	Toronto	**NHL**	**13**	**2**	**2**	**4**	**10**	**0**	**0**	**0**	**14**	**14.3**	**1**													
	St. John's	AHL	9	4	6	10	0												11	4	2	6	26			
1994-95	Saint John	AHL	67	31	50	81	82												2	0	1	1	0			
	Calgary	**NHL**	**8**	**1**	**1**	**2**	**2**	**0**	**0**	**0**	**5**	**20.0**	**1**													
1995-96	Atlanta Knights	IHL	71	25	48	73	104												3	2	1	3	4			
1996-97	Quebec Rafales	IHL	5	1	2	3	0																			
	Houston Aeros	IHL	59	12	30	42	59												13	5	8	13	2			
1997-98	Grand Rapids	IHL	69	26	36	62	103												3	0	4	4	4			
1998-99	**Philadelphia**	**NHL**	**7**	**1**	**3**	**4**	**2**	**0**	**0**	**0**	**9**	**11.1**	**1**	**0**	**0.0**	**5**	**3**	**9:55**	**2**	**0**	**1**	**1**	**0**	**0**	**0**	**0**
	Philadelphia	AHL	67	23	46	69	102												7	1	5	6	14			
99-2000	**Philadelphia**	**NHL**	**11**	**3**	**2**	**5**	**6**	**0**	**0**	**1**	**14**	**21.4**	**0**	**1**	**0.0**	**7**	**1**	**11:19**	**3**	**0**	**0**	**0**	**0**	**0**	**0**	**0**
	Philadelphia	AHL	68	34	48	82	116												5	3	2	5	6			
2000-01	**Philadelphia**	**NHL**	**7**	**1**	**1**	**2**	**4**	**0**	**0**	**0**	**7**	**14.3**	**–2**	**0**	**0.0**	**2**	**0**	**14:23**								
	Philadelphia	AHL	74	31	57	88	98												10	6	5	11	4			
2001-02	Philadelphia	AHL	66	22	37	59	105												5	0	4	4	6			
	NHL Totals		**120**	**13**	**26**	**39**	**88**	**0**	**0**	**1**	**125**	**10.4**		**1**	**0.0**	**14**	**4**	**11:47**	**5**	**0**	**1**	**1**	**0**	**0**	**0**	**0**

WHL East First All-Star Team (1990) • AHL First All-Star Team (2001)

Traded to **Toronto** by **Hartford** with Hartford's 6th round choice (Doug Bonner) in 1995 Entry Draft for Ted Crowley, January 25, 1994. Signed as a free agent by **Calgary**, August 9, 1994. Signed as a free agent by **Philadelphia**, July 28, 1998.

GRENIER, Martin
(GREH-nyay, MAHR-tihn) **PHX.**

Defense. Shoots left. 6'5", 245 lbs. Born, Laval, Que., November 2, 1980. Colorado's 2nd choice, 45th overall, in 1999 Entry Draft.

Season	Club	League	GP	G	A	Pts	PIM	PP	SH	GW	S	%	+/-	TF	F%	H	SB	Min	Playoffs GP	G	A	Pts	PIM	PP	SH	GW
1996-97	Laval Laurentide	QAAA	34	3	16	19	117												13	0	4	4				
1997-98	Quebec Remparts	QMJHL	61	4	11	15	202												14	0	2	2	36			
1998-99	Quebec Remparts	QMJHL	60	7	18	25	*479												13	0	4	4	29			
99-2000	Quebec Remparts	QMJHL	67	11	35	46	302												7	1	4	5	27			
2000-01	Quebec Remparts	QMJHL	26	5	16	21	82																			
	Victoriaville	QMJHL	28	9	19	28	108												13	2	8	10	51			
2001-02	**Phoenix**	**NHL**	**5**	**0**	**0**	**0**	**5**	**0**	**0**	**0**	**1**	**0.0**	**0**	**1**	**100.0**	**2**	**2**	**5:56**								
	Springfield	AHL	69	2	6	8	241																			
	NHL Totals		**5**	**0**	**0**	**0**	**5**	**0**	**0**	**0**	**1**	**0.0**		**1**	**100.0**	**2**	**2**	**5:56**								

Traded to **Victoriaville** (QMJHL) by **Quebec** (QMJHL) for Daniel Masse, Daniel Houle and Victoriaville's 1st round choice (Jeff MacAuley) in 2001 QMJHL Midget Draft, January 12, 2001. Traded to **Boston** by **Colorado** with Brian Rolston, Sami Pahlsson and New Jersey's 1st round choice (previously acquired, Boston selected Martin Samuelsson) in 2000 Entry Draft for Raymond Bourque and Dave Andreychuk, March 6, 2000. Signed as a free agent by **Phoenix**, June 27, 2001.

GRIER, Mike
(GREER, MIGHK) **EDM.**

Right wing. Shoots right. 6'1", 227 lbs. Born, Detroit, MI, January 5, 1975. St. Louis' 7th choice, 219th overall, in 1993 Entry Draft.

Season	Club	League	GP	G	A	Pts	PIM	PP	SH	GW	S	%	+/-	TF	F%	H	SB	Min	Playoffs GP	G	A	Pts	PIM	PP	SH	GW
1992-93	St. Sebastian's	Hi-School	22	16	27	43	32																			
1993-94	Boston University	H-East	39	9	9	18	56																			
1994-95	Boston University	H-East	37	*29	26	55	85																			
1995-96	Boston University	H-East	38	21	25	46	82																			
1996-97	**Edmonton**	**NHL**	**79**	**15**	**17**	**32**	**45**	**4**	**0**	**2**	**89**	**16.9**	**7**						**12**	**3**	**1**	**4**	**4**	**1**	**0**	**1**
1997-98	**Edmonton**	**NHL**	**66**	**9**	**6**	**15**	**73**	**1**	**0**	**1**	**90**	**10.0**	**–3**						**12**	**2**	**2**	**4**	**13**	**0**	**0**	**1**
1998-99	**Edmonton**	**NHL**	**82**	**20**	**24**	**44**	**54**	**3**	**2**	**1**	**143**	**14.0**	**5**	**34**	**20.6**	**188**	**49**	**15:57**	**4**	**1**	**1**	**2**	**6**	**0**	**0**	**0**
99-2000	**Edmonton**	**NHL**	**65**	**9**	**22**	**31**	**68**	**0**	**3**	**2**	**115**	**7.8**	**9**	**32**	**46.8**	**174**	**38**	**15:45**								
2000-01	**Edmonton**	**NHL**	**74**	**20**	**16**	**36**	**20**	**2**	**3**	**2**	**124**	**16.1**	**11**	**36**	**38.9**	**144**	**51**	**16:44**	**6**	**0**	**0**	**0**	**8**	**0**	**0**	**0**
2001-02	**Edmonton**	**NHL**	**82**	**8**	**17**	**25**	**32**	**0**	**2**	**3**	**112**	**7.1**	**1**	**38**	**47.4**	**136**	**50**	**15:01**								
	NHL Totals		**448**	**81**	**102**	**183**	**292**	**10**	**10**	**11**	**673**	**12.0**		**140**	**27.9**	**642**	**188**	**15:52**	**34**	**6**	**4**	**10**	**31**	**1**	**0**	**2**

Hockey East First All-Star Team (1995) • NCAA East First All-American Team (1995)

Rights traded to **Edmonton** by **St. Louis** with Curtis Joseph for St. Louis' 1st round choices in 1996 (previously acquired, St. Louis selected Marty Reasoner) and 1997 (later traded to LA Kings - LA Kings selected Matt Zultek) Entry Drafts, August 4, 1995.

GRIMSON, Stu
(GRIHM-suhn, STOO) **NSH.**

Left wing. Shoots left. 6'4", 240 lbs. Born, Kamloops, B.C., May 20, 1965. Calgary's 8th choice, 143rd overall, in 1985 Entry Draft.

Season	Club	League	GP	G	A	Pts	PIM	PP	SH	GW	S	%	+/-	TF	F%	H	SB	Min	Playoffs GP	G	A	Pts	PIM	PP	SH	GW
1982-83	Regina Pats	WHL	48	0	1	1	105												5	0	0	0	14			
1983-84	Regina Pats	WHL	63	8	8	16	131												21	0	1	1	29			
1984-85	Regina Pats	WHL	71	24	32	56	248												8	1	2	3	14			
1985-86	U. of Manitoba	CWUAA	12	7	4	11	113												8	1	1	2	24			
1986-87	U. of Manitoba	CWUAA	29	8	8	16	67												14	4	2	6	28			
1987-88	Salt Lake	IHL	38	9	5	14	268																			
1988-89	**Calgary**	**NHL**	**1**	**0**	**0**	**0**	**5**	**0**	**0**	**0**	**0**	**0.0**	**0**													
	Salt Lake	IHL	72	9	18	27	397												14	2	3	5	86			
1989-90	**Calgary**	**NHL**	**3**	**0**	**0**	**0**	**17**	**0**	**0**	**0**	**0**	**0.0**	**–1**													
	Salt Lake	IHL	62	8	8	16	319												4	0	0	0	8			
1990-91	**Chicago**	**NHL**	**35**	**0**	**1**	**1**	**183**	**0**	**0**	**0**	**14**	**0.0**	**–3**						**5**	**0**	**0**	**0**	**46**	**0**	**0**	**0**
1991-92	**Chicago**	**NHL**	**54**	**2**	**2**	**4**	**234**	**0**	**0**	**0**	**23**	**8.7**	**–2**						**14**	**0**	**1**	**1**	**10**	**0**	**0**	**0**
	Indianapolis Ice	IHL	5	1	1	2	17																			
1992-93	**Chicago**	**NHL**	**78**	**1**	**1**	**2**	**193**	**1**	**0**	**0**	**14**	**7.1**	**2**						**2**	**0**	**0**	**0**	**4**	**0**	**0**	**0**
1993-94	**Anaheim**	**NHL**	**77**	**1**	**5**	**6**	**199**	**0**	**0**	**0**	**34**	**2.9**	**–6**													
1994-95	**Anaheim**	**NHL**	**31**	**0**	**1**	**1**	**110**	**0**	**0**	**0**	**14**	**0.0**	**–7**													
	Detroit	**NHL**	**11**	**0**	**0**	**0**	**37**	**0**	**0**	**0**	**4**	**0.0**	**–4**						**11**	**1**	**0**	**1**	**26**	**0**	**0**	**0**
1995-96	**Detroit**	**NHL**	**56**	**0**	**1**	**1**	**128**	**0**	**0**	**0**	**19**	**0.0**	**–10**						**2**	**0**	**0**	**0**	**0**	**0**	**0**	**0**
1996-97	**Detroit**	**NHL**	**1**	**0**	**0**	**0**	**0**	**0**	**0**	**0**	**0**	**0.0**	**–1**													
	Hartford	**NHL**	**75**	**2**	**2**	**4**	**218**	**0**	**0**	**0**	**17**	**11.8**	**–7**													
1997-98	**Carolina**	**NHL**	**82**	**3**	**4**	**7**	**204**	**0**	**0**	**1**	**17**	**17.6**	**0**													
1998-99	**Anaheim**	**NHL**	**73**	**3**	**0**	**3**	**158**	**0**	**0**	**1**	**10**	**30.0**	**0**	**0**	**0.0**	**25**	**4**	**3:25**	**3**	**0**	**0**	**0**	**30**	**0**	**0**	**0**
99-2000	**Anaheim**	**NHL**	**50**	**1**	**2**	**3**	**116**	**0**	**0**	**0**	**14**	**7.1**	**0**	**0**	**0.0**	**55**	**2**	**5:13**								
2000-01	**Los Angeles**	**NHL**	**72**	**3**	**2**	**5**	**235**	**0**	**0**	**1**	**26**	**11.5**	**–2**	**0**	**0.0**	**79**	**1**	**5:60**	**5**	**0**	**0**	**0**	**4**	**0**	**0**	**0**
2001-02	**Nashville**	**NHL**	**30**	**1**	**1**	**2**	**76**	**0**	**0**	**0**	**5**	**20.0**	**0**	**0**	**0.0**	**23**	**7**	**6:08**								
	NHL Totals		**729**	**17**	**22**	**39**	**2113**	**1**	**0**	**3**	**211**	**8.1**		**0**	**0.0**	**182**	**14**	**5:00**	**42**	**1**	**1**	**2**	**120**	**0**	**0**	**0**

• Re-entered NHL Entry Draft. Originally Detroit's 11th choice, 193rd overall, in 1983 Entry Draft.

Claimed on waivers by **Chicago** from **Calgary**, October 1, 1990. Claimed by **Anaheim** from **Chicago** in Expansion Draft, June 24, 1993. Traded to **Detroit** by **Anaheim** with Mark Ferner and Anaheim's 6th round choice (Magnus Nilsson) in 1996 Entry Draft for Mike Sillinger and Jason York, April 4, 1995. Claimed on waivers by **Hartford** from **Detroit**, October 13, 1996. Transferred to **Carolina** after **Hartford** franchise relocated, June 25, 1997. Traded to **Anaheim** by **Carolina** with Kevin Haller for Dave Karpa and Anaheim's 4th round choice (later traded to Atlanta - Atlanta selected Blake Robson) in 2000 Entry Draft, August 11, 1998. Signed as a free agent by **LA Kings**, July 6, 2000. Signed as a free agent by **Nashville**, July 2, 2001. • Missed majority of 2001-02 season recovering from head injury suffered in game vs. Anaheim, December 2, 2001.

GRON, Stanislav
(GRAHN, Stan-ih-slav) **N.J.**

Right wing. Shoots left. 6'2", 205 lbs. Born, Bratislava, Czech., October 28, 1978. New Jersey's 2nd choice, 38th overall, in 1997 Entry Draft.

Season	Club	League	GP	G	A	Pts	PIM	PP	SH	GW	S	%	+/-	TF	F%	H	SB	Min	Playoffs GP	G	A	Pts	PIM	PP	SH	GW
1994-95	S. Bratislava Jr.	Slovak-Jr.	40	49	26	75	20																			
1995-96	S. Bratislava Jr.	Slovak-Jr.	43	33	25	58	14																			
	Slov. Bratislava	Slovakia																	1	0	0	0	0			
1996-97	S. Bratislava Jr.	Slovak-Jr.	22	20	16	36																				
	Slov. Bratislava	Slovakia	7	0	0	0																				
1997-98	Seattle	WHL	61	9	29	38	21												5	1	5	6	0			
1998-99	Kootenay Ice	WHL	49	28	18	46	18												7	3	8	11	12			
	Utah Grizzlies	IHL	4	0	3	3	0																			
99-2000	Albany	AHL	65	19	10	29	17												5	1	1	2	2			
2000-01	**New Jersey**	**NHL**	**1**	**0**	**0**	**0**	**0**	**0**	**0**	**0**	**2**	**0.0**	**1**	**2**	**0.0**	**2**	**0**	**9:55**								
	Albany	AHL	61	16	9	25	19																			
2001-02	Albany	AHL	76	13	15	28	34																			
	NHL Totals		**1**	**0**	**0**	**0**	**0**	**0**	**0**	**0**	**2**	**0.0**		**2**	**0.0**	**2**	**0**	**9:55**								

GROSEK, Michal
(GROH-shehk, MIHK-al) **BOS.**

Left wing. Shoots right. 6'2", 207 lbs. Born, Vyskov, Czech., June 1, 1975. Winnipeg's 7th choice, 145th overall, in 1993 Entry Draft.

Season	Club	League	GP	G	A	Pts	PIM	PP	SH	GW	S	%	+/-	TF	F%	H	SB	Min	Playoffs GP	G	A	Pts	PIM	PP	SH	GW
1992-93	AC ZPS Zlin	Czech	17	1	3	4																				
1993-94	Tacoma Rockets	WHL	30	25	20	45	106												7	2	2	4	30			
	Winnipeg	**NHL**	**3**	**1**	**0**	**1**	**0**	**0**	**0**	**0**	**4**	**25.0**	**–1**													
	Moncton Hawks	AHL	20	1	2	3	47												2	0	0	0	0			
1994-95	Springfield	AHL	45	10	22	32	98																			
	Winnipeg	**NHL**	**24**	**2**	**2**	**4**	**21**	**0**	**0**	**1**	**27**	**7.4**	**–3**													
1995-96	**Winnipeg**	**NHL**	**1**	**0**	**0**	**0**	**0**	**0**	**0**	**0**	**1**	**0.0**	**–1**													
	Springfield	AHL	39	16	19	35	68																			
	Buffalo	**NHL**	**22**	**6**	**4**	**10**	**31**	**2**	**0**	**1**	**33**	**18.2**	**0**													
1996-97	**Buffalo**	**NHL**	**82**	**15**	**21**	**36**	**71**	**1**	**0**	**2**	**117**	**12.8**	**25**						**12**	**3**	**3**	**6**	**8**	**0**	**0**	**0**
1997-98	**Buffalo**	**NHL**	**67**	**10**	**20**	**30**	**60**	**2**	**0**	**1**	**114**	**8.8**	**9**						**15**	**6**	**4**	**10**	**28**	**2**	**0**	**3**
1998-99	**Buffalo**	**NHL**	**76**	**20**	**30**	**50**	**102**	**4**	**0**	**3**	**140**	**14.3**	**21**	**5**	**60.0**	**98**	**20**	**17:14**	**13**	**0**	**4**	**4**	**28**	**0**	**0**	**0**
99-2000	**Buffalo**	**NHL**	**61**	**11**	**23**	**34**	**35**	**2**	**0**	**2**	**96**	**11.5**	**12**	**8**	**25.0**	**58**	**13**	**16:17**								
	Chicago	**NHL**	**14**	**2**	**4**	**6**	**12**	**1**	**0**	**0**	**18**	**11.1**	**–1**	**1**	**0.0**	**28**	**1**	**13:06**								
2000-01	**NY Rangers**	**NHL**	**65**	**9**	**11**	**20**	**61**	**2**	**0**	**0**	**84**	**10.7**	**–10**	**14**	**28.6**	**99**	**13**	**11:05**								
	Hartford	AHL	12	8	7	15	12																			

Season	Club	League	Regular Season																Playoffs							
			GP	G	A	Pts	PIM	PP	SH	GW	S	%	+/-	TF	F%	H	SB	Min	GP	G	A	Pts	PIM	PP	SH	GW
2001-02	**NY Rangers**	**NHL**	**15**	**3**	**2**	**5**	**12**	**0**	**0**	**0**	**23**	**13.0**	**–3**	**2**	**0.0**	**25**	**4**	**12:24**	....	....	....	....	....	....	....	....
	Hartford	AHL	48	14	30	44	167	....	....	....	....	....	....	....	....	....	....	....	....	....	....	....	....	....	....	....
	NHL Totals		**430**	**79**	**117**	**196**	**405**	**14**	**0**	**10**	**657**	**12.0**		**30**	**30.0**	**308**	**51**	**14:41**	**40**	**9**	**11**	**20**	**64**	**2**	**0**	**3**

Traded to **Buffalo** by **Winnipeg** with Darryl Shannon for Craig Muni, February 15, 1996. Traded to **Chicago** by **Buffalo** for Doug Gilmour, J-P Dumont and future considerations, March 10, 2000. Traded to **NY Rangers** by **Chicago** with Brad Brown for future considerations, October 5, 2000. Signed as a free agent by **Boston**, July 16, 2002.

GRUDEN, John

(GROO-duhn, JAWN) **OTT.**

Defense. Shoots left. 6', 203 lbs. Born, Virginia, MN, June 4, 1970. Boston's 7th choice, 168th overall, in 1990 Entry Draft.

Season	Club	League	GP	G	A	Pts	PIM	PP	SH	GW	S	%	+/-	TF	F%	H	SB	Min	GP	G	A	Pts	PIM	PP	SH	GW
1989-90	Waterloo	USHL	47	7	39	46	35	....	....	....	....	....	....	....	....	....	....	....	....	....	....	....	....	....	....	....
1990-91	Ferris State	CCHA	37	4	11	15	27	....	....	....	....	....	....	....	....	....	....	....	....	....	....	....	....	....	....	....
1991-92	Ferris State	CCHA	37	9	14	23	24	....	....	....	....	....	....	....	....	....	....	....	....	....	....	....	....	....	....	....
1992-93	Ferris State	CCHA	41	16	14	30	58	....	....	....	....	....	....	....	....	....	....	....	....	....	....	....	....	....	....	....
1993-94	Ferris State	CCHA	38	11	25	36	52	....	....	....	....	....	....	....	....	....	....	....	....	....	....	....	....	....	....	....
	Boston	**NHL**	**7**	**0**	**1**	**1**	**2**	**0**	**0**	**0**	**8**	**0.0**	**–3**	....	....	....	....	....	....	....	....	....	....	....	....	....
1994-95	**Boston**	**NHL**	**38**	**0**	**6**	**6**	**22**	**0**	**0**	**0**	**30**	**0.0**	**3**	....	....	....	....	....	....	....	....	....	....	....	....	....
	Providence	AHL	1	0	1	1	0	....	....	....	....	....	....	....	....	....	....	....	....	....	....	....	....	....	....	....
1995-96	**Boston**	**NHL**	**14**	**0**	**0**	**0**	**4**	**0**	**0**	**0**	**12**	**0.0**	**–3**	....	....	....	....	....	**3**	**0**	**1**	**1**	**0**	**0**	**0**	**0**
	Providence	AHL	39	5	19	24	29	....	....	....	....	....	....	....	....	....	....	....	....	....	....	....	....	....	....	....
1996-97	Providence	AHL	78	18	27	45	52	....	....	....	....	....	....	....	....	....	....	....	10	3	6	9	4	....	....	....
1997-98	Detroit Vipers	IHL	76	13	42	55	74	....	....	....	....	....	....	....	....	....	....	....	21	1	8	9	14	....	....	....
1998-99	**Ottawa**	**NHL**	**13**	**0**	**1**	**1**	**8**	**0**	**0**	**0**	**10**	**0.0**	**0**	**0**	**0.0**	**16**	**4**	**13:07**	....	....	....	....	....	....	....	....
	Detroit Vipers	IHL	59	10	28	38	52	....	....	....	....	....	....	....	....	....	....	....	10	0	1	1	6	....	....	....
99-2000	**Ottawa**	**NHL**	**9**	**0**	**0**	**0**	**4**	**0**	**0**	**0**	**3**	**0.0**	**0**	**0**	**0.0**	**6**	**6**	**16:29**	....	....	....	....	....	....	....	....
	Grand Rapids	IHL	50	5	17	22	24	....	....	....	....	....	....	....	....	....	....	....	12	1	4	5	8	....	....	....
2000-01	Grand Rapids	IHL	34	2	6	8	18	....	....	....	....	....	....	....	....	....	....	....	10	1	4	5	8	....	....	....
2001-02	Grand Rapids	AHL	57	3	14	17	48	....	....	....	....	....	....	....	....	....	....	....	5	1	0	1	2	....	....	....
	NHL Totals		**81**	**0**	**8**	**8**	**40**	**0**	**0**	**0**	**63**	**0.0**		**0**	**0.0**	**22**	**10**	**14:30**	**3**	**0**	**1**	**1**	**0**	**0**	**0**	**0**

CCHA First All-Star Team (1994) • NCAA West First All-American Team (1994) • IHL Second All-Star Team (1998) • AHL First All-Star Team (2002)

Signed as a free agent by **Ottawa**, August 7, 1998. • Missed majority of 2000-01 season recovering from shoulder injury suffered in training camp, October 1, 2000. Signed as a free agent by **Eisbaren Berlin** (Germany), May 3, 2002.

GUERIN, Bill

(GAIR-ihn, BIHL) **DAL.**

Right wing. Shoots right. 6'2", 210 lbs. Born, Worcester, MA, November 9, 1970. New Jersey's 1st choice, 5th overall, in 1989 Entry Draft.

Season	Club	League	GP	G	A	Pts	PIM	PP	SH	GW	S	%	+/-	TF	F%	H	SB	Min	GP	G	A	Pts	PIM	PP	SH	GW
1985-86	Springfield	NEJHL	48	26	19	45	71	....	....	....	....	....	....	....	....	....	....	....	....	....	....	....	....	....	....	....
1986-87	Springfield	NEJHL	32	34	20	54	40	....	....	....	....	....	....	....	....	....	....	....	....	....	....	....	....	....	....	....
1987-88	Springfield	NEJHL	38	31	44	75	146	....	....	....	....	....	....	....	....	....	....	....	....	....	....	....	....	....	....	....
1988-89	Springfield	NEJHL	31	32	35	67	90	....	....	....	....	....	....	....	....	....	....	....	....	....	....	....	....	....	....	....
1989-90	Boston College	H-East	39	14	11	25	54	....	....	....	....	....	....	....	....	....	....	....	....	....	....	....	....	....	....	....
1990-91	Boston College	H-East	38	26	19	45	102	....	....	....	....	....	....	....	....	....	....	....	....	....	....	....	....	....	....	....
1991-92	Team USA	Nat-Tm	46	12	15	27	67	....	....	....	....	....	....	....	....	....	....	....	....	....	....	....	....	....	....	....
	New Jersey	**NHL**	**5**	**0**	**1**	**1**	**9**	**0**	**0**	**0**	**8**	**0.0**	**1**	....	....	....	....	....	**6**	**3**	**0**	**3**	**4**	**0**	**0**	**0**
	Utica Devils	AHL	22	13	10	23	6	....	....	....	....	....	....	....	....	....	....	....	4	1	3	4	14	....	....	....
1992-93	**New Jersey**	**NHL**	**65**	**14**	**20**	**34**	**63**	**0**	**0**	**2**	**123**	**11.4**	**14**	....	....	....	....	....	**5**	**1**	**1**	**2**	**4**	**0**	**0**	**0**
	Utica Devils	AHL	18	10	7	17	47	....	....	....	....	....	....	....	....	....	....	....	....	....	....	....	....	....	....	....
1993-94	**New Jersey**	**NHL**	**81**	**25**	**19**	**44**	**101**	**2**	**0**	**3**	**195**	**12.8**	**14**	....	....	....	....	....	**17**	**2**	**1**	**3**	**35**	**0**	**0**	**1**
1994-95♦	**New Jersey**	**NHL**	**48**	**12**	**13**	**25**	**72**	**4**	**0**	**3**	**96**	**12.5**	**6**	....	....	....	....	....	**20**	**3**	**8**	**11**	**30**	**1**	**0**	**0**
1995-96	**New Jersey**	**NHL**	**80**	**23**	**30**	**53**	**116**	**8**	**0**	**6**	**216**	**10.6**	**7**	....	....	....	....	....	....	....	....	....	....	....	....	....
1996-97	**New Jersey**	**NHL**	**82**	**29**	**18**	**47**	**95**	**7**	**0**	**9**	**177**	**16.4**	**–2**	....	....	....	....	....	**8**	**2**	**1**	**3**	**18**	**1**	**0**	**1**
1997-98	**New Jersey**	**NHL**	**19**	**5**	**5**	**10**	**13**	**1**	**0**	**2**	**48**	**10.4**	**0**	....	....	....	....	....	....	....	....	....	....	....	....	....
	Edmonton	**NHL**	**40**	**13**	**16**	**29**	**80**	**8**	**0**	**2**	**130**	**10.0**	**1**	....	....	....	....	....	**12**	**7**	**1**	**8**	**17**	**4**	**0**	**0**
	United States	Olympics	4	0	3	3	2	....	....	....	....	....	....	....	....	....	....	....	....	....	....	....	....	....	....	....
1998-99	**Edmonton**	**NHL**	**80**	**30**	**34**	**64**	**133**	**13**	**0**	**2**	**261**	**11.5**	**7**	**74**	**40.5**	**131**	**20**	**19:42**	**3**	**0**	**2**	**2**	**2**	**0**	**0**	**0**
99-2000	**Edmonton**	**NHL**	**70**	**24**	**22**	**46**	**123**	**11**	**0**	**2**	**188**	**12.8**	**4**	**13**	**46.2**	**101**	**33**	**18:01**	**5**	**3**	**2**	**5**	**9**	**1**	**0**	**0**
2000-01	**Edmonton**	**NHL**	**21**	**12**	**10**	**22**	**18**	**4**	**0**	**1**	**64**	**18.8**	**11**	**0**	**0.0**	**35**	**9**	**19:49**	....	....	....	....	....	....	....	....
	Boston	**NHL**	**64**	**28**	**35**	**63**	**122**	**7**	**1**	**4**	**225**	**12.4**	**–4**	**36**	**41.7**	**118**	**51**	**22:43**	....	....	....	....	....	....	....	....
2001-02	**Boston**	**NHL**	**78**	**41**	**25**	**66**	**91**	**10**	**1**	**7**	**355**	**11.5**	**–1**	**17**	**52.9**	**100**	**42**	**20:45**	**6**	**4**	**2**	**6**	**6**	**3**	**0**	**0**
	United States	Olympics	6	4	0	4	4	....	....	....	....	....	....	....	....	....	....	....	....	....	....	....	....	....	....	....
	NHL Totals		**733**	**256**	**248**	**504**	**1036**	**75**	**2**	**43**	**2086**	**12.3**		**140**	**42.9**	**485**	**155**	**20:13**	**82**	**25**	**18**	**43**	**125**	**10**	**0**	**2**

NHL Second All-Star Team (2002) • Played in NHL All-Star Game (2001)

Traded to **Edmonton** by **New Jersey** with Valeri Zelepukin for Jason Arnott and Bryan Muir, January 4, 1998. Traded to **Boston** by **Edmonton** for Anson Carter, Boston's 1st (Ales Hemsky) and 2nd (Doug Lynch) round choices in 2001 Entry Draft and future considerations, November 15, 2000. Signed as a free agent by **Dallas**, July 3, 2002.

GUOLLA, Stephen

(GUH-wah-lah, STEEV-vuhn) **N.J.**

Center. Shoots left. 6', 190 lbs. Born, Scarborough, Ont., March 15, 1973. Ottawa's 1st choice, 3rd overall, in 1994 Supplemental Draft.

Season	Club	League	GP	G	A	Pts	PIM	PP	SH	GW	S	%	+/-	TF	F%	H	SB	Min	GP	G	A	Pts	PIM	PP	SH	GW
1988-89	Tor. Red Wings	MTHL	25	14	20	34	....	....	....	....	....	....	....	....	....	....	....	....	....	....	....	....	....	....	....	....
1989-90	Tor. Red Wings	MTHL	40	42	47	89	....	....	....	....	....	....	....	....	....	....	....	....	....	....	....	....	....	....	....	....
1990-91	Wexford Raiders	MTJHL	44	34	44	78	34	....	....	....	....	....	....	....	....	....	....	....	12	12	16	28	....	....	....	....
1991-92	Michigan State	CCHA	33	4	9	13	8	....	....	....	....	....	....	....	....	....	....	....	....	....	....	....	....	....	....	....
1992-93	Michigan State	CCHA	39	19	35	54	6	....	....	....	....	....	....	....	....	....	....	....	....	....	....	....	....	....	....	....
1993-94	Michigan State	CCHA	41	23	46	69	16	....	....	....	....	....	....	....	....	....	....	....	....	....	....	....	....	....	....	....
1994-95	Michigan State	CCHA	40	16	35	51	16	....	....	....	....	....	....	....	....	....	....	....	....	....	....	....	....	....	....	....
1995-96	P.E.I. Senators	AHL	72	32	48	80	28	....	....	....	....	....	....	....	....	....	....	....	3	0	0	0	0	....	....	....
1996-97	**San Jose**	**NHL**	**43**	**13**	**8**	**21**	**14**	**2**	**0**	**1**	**81**	**16.0**	**–10**	....	....	....	....	....	....	....	....	....	....	....	....	....
	Kentucky	AHL	34	22	22	44	10	....	....	....	....	....	....	....	....	....	....	....	4	2	1	3	0	....	....	....
1997-98	**San Jose**	**NHL**	**7**	**1**	**1**	**2**	**0**	**0**	**0**	**0**	**9**	**11.1**	**–2**	....	....	....	....	....	....	....	....	....	....	....	....	....
	Kentucky	AHL	69	37	63	100	45	....	....	....	....	....	....	....	....	....	....	....	3	0	0	0	0	....	....	....
1998-99	**San Jose**	**NHL**	**14**	**2**	**2**	**4**	**6**	**0**	**0**	**1**	**22**	**9.1**	**3**	**172**	**36.6**	**19**	**6**	**13:54**	....	....	....	....	....	....	....	....
	Kentucky	AHL	53	29	47	76	33	....	....	....	....	....	....	....	....	....	....	....	....	....	....	....	....	....	....	....
99-2000	**Tampa Bay**	**NHL**	**46**	**6**	**10**	**16**	**11**	**2**	**0**	**0**	**52**	**11.5**	**2**	**155**	**45.8**	**33**	**8**	**11:26**	....	....	....	....	....	....	....	....
	Atlanta	**NHL**	**20**	**4**	**9**	**13**	**4**	**2**	**0**	**0**	**34**	**11.8**	**–13**	**345**	**42.6**	**19**	**9**	**17:47**	....	....	....	....	....	....	....	....
2000-01	**Atlanta**	**NHL**	**63**	**12**	**16**	**28**	**23**	**2**	**0**	**3**	**96**	**12.5**	**–6**	**859**	**47.7**	**63**	**22**	**14:41**	....	....	....	....	....	....	....	....
2001-02	Albany	AHL	68	25	35	60	27	....	....	....	....	....	....	....	....	....	....	....	....	....	....	....	....	....	....	....
	NHL Totals		**193**	**38**	**46**	**84**	**58**	**8**	**0**	**5**	**294**	**12.9**		**1531**	**45.1**	**134**	**45**	**13:60**	....	....	....	....	....	....	....	....

CCHA Second All-Star Team (1994) • NCAA West Second All-American Team (1994) • AHL Second All-Star Team (1998, 1999) • Won Les Cunningham Award (MVP - AHL) (1998)

Signed as a free agent by **San Jose**, August 22, 1996. Traded to **Tampa Bay** by **San Jose** with Bill Houlder, Shawn Burr and Andrei Zyuzin for Niklas Sundstrom and NY Rangers' 3rd round choice (previously acquired, later traded to Chicago - Chicago selected Igor Radulov) in 2000 Entry Draft, August 4, 1999. Claimed on waivers by **Atlanta** from **Tampa Bay**, March 1, 2000. Signed as a free agent by **New Jersey**, October 21, 2001.

GUREN, Miloslav

(GOO-rihn, MEER-oh-slahf) **MTL.**

Defense. Shoots left. 6'2", 215 lbs. Born, Uherske Hradiste, Czech., September 24, 1976. Montreal's 2nd choice, 60th overall, in 1995 Entry Draft.

Season	Club	League	GP	G	A	Pts	PIM	PP	SH	GW	S	%	+/-	TF	F%	H	SB	Min	GP	G	A	Pts	PIM	PP	SH	GW
1993-94	AC ZPS Zlin	Czech	22	1	5	6	....	....	....	....	....	....	....	....	....	....	....	....	3	0	0	0	....	....	....	....
1994-95	AC ZPS Zlin	Czech	32	3	7	10	10	....	....	....	....	....	....	....	....	....	....	....	12	1	0	1	6	....	....	....
1995-96	AC ZPS Zlin	Czech	28	1	2	3	....	....	....	....	....	....	....	....	....	....	....	....	7	1	0	1	....	....	....	....
1996-97	Fredericton	AHL	79	6	26	32	26	....	....	....	....	....	....	....	....	....	....	....	....	....	....	....	....	....	....	....
1997-98	Fredericton	AHL	78	15	36	51	36	....	....	....	....	....	....	....	....	....	....	....	4	1	2	3	0	....	....	....
1998-99	**Montreal**	**NHL**	**12**	**0**	**1**	**1**	**4**	**0**	**0**	**0**	**11**	**0.0**	**–1**	**0**	**0.0**	**4**	**10**	**12:02**	....	....	....	....	....	....	....	....
	Fredericton	AHL	63	5	16	21	24	....	....	....	....	....	....	....	....	....	....	....	15	4	7	11	10	....	....	....
99-2000	**Montreal**	**NHL**	**24**	**1**	**2**	**3**	**12**	**1**	**0**	**0**	**20**	**5.0**	**–5**	**0**	**0.0**	**20**	**26**	**15:14**	....	....	....	....	....	....	....	....
	Quebec	AHL	29	5	12	17	16	....	....	....	....	....	....	....	....	....	....	....	3	0	0	0	2	....	....	....
2000-01	Quebec	AHL	75	11	40	51	24	....	....	....	....	....	....	....	....	....	....	....	8	4	2	6	6	....	....	....
2001-02	Trinec	Czech	52	2	9	11	44	....	....	....	....	....	....	....	....	....	....	....	6	1	2	3	9	....	....	....
	NHL Totals		**36**	**1**	**3**	**4**	**16**	**1**	**0**	**0**	**31**	**3.2**		**0**	**0.0**	**24**	**36**	**14:10**	....	....	....	....	....	....	....	....

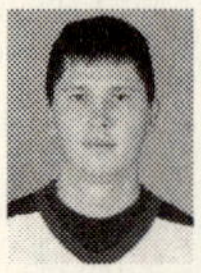

GUSEV, Sergey

(GOO-sehv, SAIR-gay)

Defense. Shoots left. 6'1", 205 lbs. Born, Nizhny Tagil, USSR, July 31, 1975. Dallas' 4th choice, 69th overall, in 1995 Entry Draft.

			Regular Season																Playoffs							
Season	Club	League	GP	G	A	Pts	PIM	PP	SH	GW	S	%	+/-	TF	F%	H	SB	Min	GP	G	A	Pts	PIM	PP	SH	GW
1994-95	CSK VVS Samara	CIS	50	3	5	8	58																			
1995-96	Michigan K-Wings	IHL	73	11	17	28	76																			
1996-97	Michigan K-Wings	IHL	51	7	8	15	44												4	0	4	4	6			
1997-98	**Dallas**	**NHL**	**9**	**0**	**0**	**0**	**2**	**0**	**0**	**0**	**5**	**0.0**	**–5**													
	Michigan K-Wings	IHL	36	3	6	9	36												4	0	2	2	6			
1998-99	**Dallas**	**NHL**	**22**	**1**	**4**	**5**	**6**	**0**	**0**	**1**	**30**	**3.3**	**5**	**0**	**0.0**	**12**	**16**	**12:04**								
	Michigan K-Wings	IHL	12	0	6	6	14																			
	Tampa Bay	**NHL**	**14**	**0**	**3**	**3**	**10**	**0**	**0**	**0**	**16**	**0.0**	**–8**	**0**	**0.0**	**13**	**28**	**21:30**								
99-2000	**Tampa Bay**	**NHL**	**28**	**2**	**3**	**5**	**6**	**1**	**0**	**0**	**23**	**8.7**	**–9**	**0**	**0.0**	**35**	**46**	**17:34**								
2000-01	**Tampa Bay**	**NHL**	**16**	**1**	**0**	**1**	**10**	**0**	**0**	**0**	**13**	**7.7**	**–3**	**0**	**0.0**	**8**	**22**	**13:23**								
	Detroit Vipers	IHL	13	1	4	5	10																			
2001-02	Cherepovets	Russia	41	4	8	12	26												4	0	0	0	6			
	NHL Totals		**89**	**4**	**10**	**14**	**34**	**1**	**0**	**1**	**87**	**4.6**		**0**	**0.0**	**68**	**112**	**15:55**								

Traded to **Tampa Bay** by **Dallas** for Benoit Hogue and Tampa Bay's 6th round choice (Michal Blazek) in 2001 Entry Draft, March 21, 1999. • Missed majority of 1999-2000 season recovering from knee injury suffered in game vs. NY Rangers, December 19, 1999. • Missed majority of 2000-01 and 2001-02 seasons recovering from knee injury originally suffered in game vs. NY Islanders, October 10, 2000.

GUSMANOV, Ravil

(goos-MAN-ohv, ra-VIHL)

Left wing. Shoots left. 6'3", 185 lbs. Born, Naberezhnye Chelny, USSR, July 25, 1972. Winnipeg's 5th choice, 93rd overall, in 1993 Entry Draft.

Season	Club	League	GP	G	A	Pts	PIM	PP	SH	GW	S	%	+/-	TF	F%	H	SB	Min	GP	G	A	Pts	PIM	PP	SH	GW
1990-91	Chelyabinsk	USSR	15	0	0	0	10																			
1991-92	Chelyabinsk	CIS	38	4	4	8	20																			
1992-93	Chelyabinsk	CIS	39	15	8	23	30												8	4	0	4	2			
1993-94	Chelyabinsk	CIS	43	18	9	27	51												6	4	3	7	10			
	Russia	Olympics	8	3	1	4	0																			
1994-95	Springfield	AHL	72	18	15	33	14																			
1995-96	**Winnipeg**	**NHL**	**4**	**0**	**0**	**0**	**0**	**0**	**0**	**0**	**6**	**0.0**	**–3**													
	Springfield	AHL	60	36	32	68	20																			
	Indianapolis Ice	IHL	11	6	10	16	4												5	2	3	5	4			
1996-97	Indianapolis Ice	IHL	60	21	27	48	14																			
	Saint John	AHL	12	4	4	8	2												3	0	1	1	2			
1997-98	Chicago Wolves	IHL	56	27	28	55	26												11	1	3	4	19			
1998-99	Magnitogorsk	Russia	42	14	24	38	28												16	2	8	10	16			
	Magnitogorsk	EuroHL	6	3	3	6	4												6	0	2	2	0			
99-2000	Magnitogorsk	Russia	37	12	13	25	30												11	3	5	8	6			
	Magnitogorsk	EuroHL	6	2	0	2	0												5	0	1	1	6			
2000-01	Magnitogorsk	Russia	43	6	19	25	20												12	1	9	10	10			
2001-02	Houston Aeros	AHL	1	0	0	0	0																			
	Magnitogorsk	Russia	35	8	18	26	22												9	1	1	2	12			
	NHL Totals		**4**	**0**	**0**	**0**	**0**	**0**	**0**	**0**	**6**	**0.0**														

Traded to **Chicago** by **Winnipeg** for Chicago's 4th round choice (later traded to Toronto who selected Vladimir Antipov) in 1996 Entry Draft, March 20, 1996. Traded to **Calgary** by **Chicago** for Marc Hussey, March 18, 1997. Signed as a free agent by **Minnesota**, June 21, 2001.

HAGGERTY, Sean

(HA-guhr-tee, SHAWN)

Left wing. Shoots left. 6'1", 186 lbs. Born, Rye, NY, February 11, 1976. Toronto's 2nd choice, 48th overall, in 1994 Entry Draft.

Season	Club	League	GP	G	A	Pts	PIM	PP	SH	GW	S	%	+/-	TF	F%	H	SB	Min	GP	G	A	Pts	PIM	PP	SH	GW
1990-91	Westminster High	Hi-School	25	20	22	42	....																			
1991-92	Westminster High	Hi-School	25	24	36	60	....																			
1992-93	Boston Jr. Bruins	MBAHL	72	70	111	181	80																			
1993-94	Detroit	OHL	60	31	32	63	21												17	9	10	19	11			
1994-95	Detroit	OHL	61	40	49	89	37												21	13	24	37	18			
1995-96	Detroit	OHL	66	*60	51	111	78												17	15	9	24	30			
	Toronto	**NHL**	**1**	**0**	**0**	**0**	**0**	**0**	**0**	**0**	**0**	**0.0**	**0**													
	Worcester	AHL																	1	0	0	0	2			
1996-97	Kentucky	AHL	77	13	22	35	60												4	1	0	1	4			
1997-98	**NY Islanders**	**NHL**	**5**	**0**	**0**	**0**	**0**	**0**	**0**	**0**	**2**	**0.0**	**–3**													
	Kentucky	AHL	63	33	20	53	64												3	0	2	2	4			
1998-99	Lowell	AHL	77	19	27	46	40												3	0	1	1	0			
99-2000	**NY Islanders**	**NHL**	**5**	**1**	**1**	**2**	**4**	**0**	**0**	**0**	**2**	**50.0**	**3**	**0**	**0.0**	**6**	**3**	**9:38**								
	Kansas City	IHL	76	27	33	60	94																			
2000-01	**Nashville**	**NHL**	**3**	**0**	**1**	**1**	**0**	**0**	**0**	**0**	**2**	**0.0**	**1**	**0**	**0.0**	**4**	**1**	**8:35**								
	Milwaukee	IHL	76	27	23	50	59												5	0	1	1	8			
2001-02	Providence	AHL	69	21	26	47	46												2	0	0	0	4			
	NHL Totals		**14**	**1**	**2**	**3**	**4**	**0**	**0**	**0**	**6**	**16.7**		**0**	**0.0**	**10**	**4**	**9:15**								

OHL All-Rookie Team (1994) • Memorial Cup All-Star Team (1995) • OHL Second All-Star Team (1996) • AHL Second All-Star Team (1998)

Traded to **NY Islanders** by **Toronto** with Darby Hendrickson, Kenny Jonsson and Toronto's 1st round choice (Roberto Luongo) in 1997 Entry Draft for Wendel Clark, Mathieu Schneider and D.J. Smith, March 13, 1996. Claimed on waivers by **Nashville** from **NY Islanders**, May 23, 2000. Signed as a free agent by **Providence** (AHL), October 11, 2001.

HAGMAN, Niklas

(HAG-muhn, NIHK-las) **FLA.**

Left wing. Shoots left. 6', 200 lbs. Born, Espoo, Finland, December 5, 1979. Florida's 3rd choice, 70th overall, in 1999 Entry Draft.

Season	Club	League	GP	G	A	Pts	PIM	PP	SH	GW	S	%	+/-	TF	F%	H	SB	Min	GP	G	A	Pts	PIM	PP	SH	GW
1994-95	HIFK Helsinki-C	Finn-Jr.	28	30	15	45	40												4	2	0	2	6			
1995-96	HIFK Helsinki-B	Finn-Jr.	26	12	21	33	32												4	3	0	3	2			
	HIFK Jr.	Finn-Jr.	12	3	1	4	0																			
1996-97	HIFK Jr.	Finn-Jr.	30	13	12	25	30												4	1	1	2	0			
1997-98	HIFK Jr.	Finn-Jr.	26	9	5	14	16																			
	HIFK Helsinki	Finland	8	1	0	1	0																			
	HIFK Helsinki-B	Finn-Jr.	1	0	1	1	0																			
1998-99	HIFK Jr.	Finn-Jr.	14	4	9	13	43																			
	HIFK Helsinki	Finland	17	1	1	2	14																			
	HIFK Helsinki	EuroHL	1	0	1	1	0																			
	Blues Espoo	Finland	14	1	1	2	2												4	1	0	1	0			
99-2000	Karpat Oulu	Finland-2	41	17	18	35	12												7	4	2	6	....			
2000-01	Karpat Oulu	Finland	56	28	18	46	32												8	3	1	4	0			
2001-02	**Florida**	**NHL**	**78**	**10**	**18**	**28**	**8**	**0**	**1**	**2**	**134**	**7.5**	**–6**	**32**	**28.1**	**52**	**13**	**13:50**								
	Finland	Olympics	4	1	2	3	0																			
	NHL Totals		**78**	**10**	**18**	**28**	**8**	**0**	**1**	**2**	**134**	**7.5**		**32**	**28.1**	**52**	**13**	**13:50**								

HAHL, Riku

(HAHL, REE-koo) **COL.**

Center. Shoots left. 6', 190 lbs. Born, Hameenlinna, Finland, November 1, 1980. Colorado's 9th choice, 183rd overall, in 1999 Entry Draft.

Season	Club	League	GP	G	A	Pts	PIM	PP	SH	GW	S	%	+/-	TF	F%	H	SB	Min	GP	G	A	Pts	PIM	PP	SH	GW
1995-96	HPK-C	Finn-Jr.	32	18	30	48	28																			
1996-97	HPK-B	Finn-Jr.	32	19	24	43	22																			
	HPK Jr.	Finn-Jr.	2	0	1	1	2												6	2	0	2	2			
1997-98	HPK-B	Finn-Jr.	10	5	14	19	6																			
	HPK Jr.	Finn-Jr.	35	13	6	19	12																			
1998-99	HPK Jr.	Finn-Jr.	6	0	2	2	6																			
	HPK Hameenlinna	Finland	28	0	1	1	0												8	0	0	0	2			
99-2000	HPK Jr.	Finn-Jr.	12	1	6	7	8												9	5	4	9	16			
	HPK Hameenlinna	Finland	50	4	3	7	18												8	0	0	0	2			
2000-01	HPK Jr.	Finn-Jr.	2	1	3	4	0																			
	HPK Hameenlinna	Finland	55	3	9	12	32																			
2001-02	**Colorado**	**NHL**	**22**	**2**	**3**	**5**	**14**	**0**	**0**	**1**	**17**	**11.8**	**1**	**94**	**35.1**	**35**	**6**	**9:26**	**21**	**1**	**2**	**3**	**0**	**0**	**0**	**0**
	Hershey Bears	AHL	52	6	17	23	16																			
	NHL Totals		**22**	**2**	**3**	**5**	**14**	**0**	**0**	**1**	**17**	**11.8**		**94**	**35.1**	**35**	**6**	**9:26**	**21**	**1**	**2**	**3**	**0**	**0**	**0**	**0**

HAJT, Chris
(HIGHT, KRIHS) **WSH.**

Defense. Shoots left. 6'3", 206 lbs. Born, Saskatoon, Sask., July 5, 1978. Edmonton's 3rd choice, 32nd overall, in 1996 Entry Draft.

Season	Club	League	GP	G	A	Pts	PIM	PP	SH	GW	S	%	+/-	TF	F%	H	SB	Min	Playoffs GP	G	A	Pts	PIM	PP	SH	GW
1993-94	Amherst Knights	WNYHA	38	8	20	28	16																			
1994-95	Guelph Storm	OHL	57	1	7	8	35												14	0	2	2	9			
1995-96	Guelph Storm	OHL	63	8	27	35	69												16	0	6	6	13			
1996-97	Guelph Storm	OHL	58	11	15	26	62												18	0	8	8	25			
1997-98	Guelph Storm	OHL	44	2	21	23	46												12	1	5	6	11			
1998-99	Hamilton	AHL	64	0	4	4	36																			
99-2000	Hamilton	AHL	54	0	8	8	30												10	0	2	2	0			
2000-01	**Edmonton**	**NHL**	**1**	**0**	**0**	**0**	**0**	**0**	**0**	**0**	**0**	**0.0**	**−1**	**0**	**0.0**	**1**	**1**	**7:38**								
	Hamilton	AHL	70	0	10	10	48																			
2001-02	Hamilton	AHL	39	2	3	5	34												15	1	2	3	8			
	NHL Totals		**1**	**0**	**0**	**0**	**0**	**0**	**0**	**0**	**0**	**0.0**		**0**	**0.0**	**1**	**1**	**7:38**								

OHL Second All-Star Team (1998)

Signed as a free agent by **Washington**, July 23, 2002.

HALKO, Steven
(HAL-koh, STEE-vehn) **CAR.**

Defense. Shoots right. 6'1", 200 lbs. Born, Etobicoke, Ont., March 8, 1974. Hartford's 10th choice, 225th overall, in 1992 Entry Draft.

Season	Club	League	GP	G	A	Pts	PIM	PP	SH	GW	S	%	+/-	TF	F%	H	SB	Min	Playoffs GP	G	A	Pts	PIM	PP	SH	GW
1989-90	Newmarket	OJHL-B	30	3	5	8	16																			
1990-91	Newmarket	OJHL-B	35	2	13	15	37																			
	Markham	OJHL-B	8	4	3	7	2																			
1991-92	Thornhill	MTJHL	44	15	46	61	43																			
1992-93	U. of Michigan	CCHA	39	1	12	13	12																			
1993-94	U. of Michigan	CCHA	41	2	13	15	32																			
1994-95	U. of Michigan	CCHA	39	2	14	16	20																			
1995-96	U. of Michigan	CCHA	43	4	16	20	32																			
1996-97	Springfield	AHL	70	1	5	6	37												11	0	2	2	8			
1997-98	**Carolina**	**NHL**	**18**	**0**	**2**	**2**	**10**	**0**	**0**	**0**	**7**	**0.0**	**−1**													
	New Haven	AHL	65	1	19	20	44												1	0	0	0	0			
1998-99	**Carolina**	**NHL**	**20**	**0**	**3**	**3**	**24**	**0**	**0**	**0**	**6**	**0.0**	**5**	**0**	**0.0**	**32**	**11**	**15:57**	**4**	**0**	**0**	**0**	**2**	**0**	**0**	**0**
	New Haven	AHL	42	2	7	9	58																			
99-2000	**Carolina**	**NHL**	**58**	**0**	**8**	**8**	**25**	**0**	**0**	**0**	**54**	**0.0**	**0**	**1**	**100.0**	**85**	**60**	**16:43**								
2000-01	**Carolina**	**NHL**	**48**	**0**	**1**	**1**	**6**	**0**	**0**	**0**	**24**	**0.0**	**−10**	**0**	**0.0**	**67**	**36**	**13:27**								
2001-02	**Carolina**	**NHL**	**5**	**0**	**1**	**1**	**6**	**0**	**0**	**0**	**0**	**0.0**	**3**	**0**	**0.0**	**4**	**2**	**9:43**								
	Worcester	AHL	43	3	5	8	15												3	0	1	1	0			
	NHL Totals		**149**	**0**	**15**	**15**	**71**	**0**	**0**	**0**	**91**	**0.0**		**1**	**100.0**	**188**	**109**	**15:08**	**4**	**0**	**0**	**0**	**2**	**0**	**0**	**0**

CCHA Second All-Star Team (1995, 1996) • NCAA Championship All-Tournament Team (1996)

Transferred to **Carolina** after **Hartford** franchise relocated, June 25, 1997. Traded to **St. Louis** by **Carolina** with Carolina's 4th round choice (later traded to Atlanta - Atlanta selected Lane Manson) in 2002 Entry Draft for Sean Hill, December 5, 2001. Signed as a free agent by **Carolina**, August 5, 2002.

HALL, Adam
(HAWL, A-dam) **NSH.**

Right wing. Shoots right. 6'3", 205 lbs. Born, Kalamazoo, MI, August 14, 1980. Nashville's 3rd choice, 52nd overall, in 1999 Entry Draft.

Season	Club	League	GP	G	A	Pts	PIM	PP	SH	GW	S	%	+/-	TF	F%	H	SB	Min	Playoffs GP	G	A	Pts	PIM	PP	SH	GW
1996-97	Bramalea Blues	OPJHL	43	9	14	23	92																			
1997-98	Team USA	USDP-18	71	42	23	65	63																			
1998-99	Michigan State	CCHA	36	16	7	23	74																			
99-2000	Michigan State	CCHA	40	*26	13	39	38																			
2000-01	Michigan State	CCHA	42	18	12	30	42																			
2001-02	Michigan State	CCHA	41	19	15	34	36																			
	Nashville	**NHL**	**1**	**0**	**1**	**1**	**0**	**0**	**0**	**0**	**2**	**0.0**	**0**	**0**	**0.0**	**1**	**0**	**14:04**								
	Milwaukee	AHL	6	2	2	4	4																			
	NHL Totals		**1**	**0**	**1**	**1**	**0**	**0**	**0**	**0**	**2**	**0.0**		**0**	**0.0**	**1**	**0**	**14:04**								

CCHA Second All-Star Team (2000)

HALLER, Kevin
(HAHL-her, KEH-vihn) **NYI**

Defense. Shoots left. 6'2", 199 lbs. Born, Trochu, Alta., December 5, 1970. Buffalo's 1st choice, 14th overall, in 1989 Entry Draft.

Season	Club	League	GP	G	A	Pts	PIM	PP	SH	GW	S	%	+/-	TF	F%	H	SB	Min	Playoffs GP	G	A	Pts	PIM	PP	SH	GW
1986-87	Three Hills	AAHA	12	10	11	21	8																			
1987-88	Olds Grizzlies	AJHL	51	13	31	44	58																			
	Regina Pats	WHL	5	0	1	1	2												4	1	1	2	2			
1988-89	Regina Pats	WHL	72	10	31	41	99																			
1989-90	Regina Pats	WHL	58	16	37	53	93												11	2	9	11	16			
	Buffalo	**NHL**	**2**	**0**	**0**	**0**	**0**	**0**	**0**	**0**	**1**	**0.0**	**0**													
1990-91	**Buffalo**	**NHL**	**21**	**1**	**8**	**9**	**20**	**1**	**0**	**0**	**42**	**2.4**	**9**						**6**	**1**	**4**	**5**	**10**	**0**	**0**	**0**
	Rochester	AHL	52	2	8	10	53												10	2	1	3	6			
1991-92	**Buffalo**	**NHL**	**58**	**6**	**15**	**21**	**75**	**2**	**0**	**1**	**76**	**7.9**	**−13**													
	Rochester	AHL	4	0	0	0	18																			
	Montreal	**NHL**	**8**	**2**	**2**	**4**	**17**	**1**	**0**	**0**	**9**	**22.2**	**4**						**9**	**0**	**0**	**0**	**6**	**0**	**0**	**0**
1992-93♦	**Montreal**	**NHL**	**73**	**11**	**14**	**25**	**117**	**6**	**0**	**1**	**126**	**8.7**	**7**						**17**	**1**	**6**	**7**	**16**	**1**	**0**	**0**
1993-94	**Montreal**	**NHL**	**68**	**4**	**9**	**13**	**118**	**0**	**0**	**1**	**72**	**5.6**	**3**						**7**	**1**	**1**	**2**	**19**	**0**	**0**	**0**
1994-95	**Philadelphia**	**NHL**	**36**	**2**	**8**	**10**	**48**	**0**	**0**	**0**	**26**	**7.7**	**16**						**15**	**4**	**4**	**8**	**10**	**0**	**1**	**1**
1995-96	**Philadelphia**	**NHL**	**69**	**5**	**9**	**14**	**92**	**0**	**2**	**2**	**89**	**5.6**	**18**						**6**	**0**	**1**	**1**	**8**	**0**	**0**	**0**
1996-97	**Philadelphia**	**NHL**	**27**	**0**	**5**	**5**	**37**	**0**	**0**	**0**	**34**	**0.0**	**−1**													
	Hartford	**NHL**	**35**	**2**	**6**	**8**	**48**	**0**	**0**	**0**	**43**	**4.7**	**−11**													
1997-98	**Carolina**	**NHL**	**65**	**3**	**5**	**8**	**94**	**0**	**0**	**0**	**67**	**4.5**	**−5**													
1998-99	**Anaheim**	**NHL**	**82**	**1**	**6**	**7**	**122**	**0**	**0**	**0**	**64**	**1.6**	**−1**	**0**	**0.0**	**95**	**110**	**20:39**	**4**	**0**	**0**	**0**	**2**	**0**	**0**	**0**
99-2000	**Anaheim**	**NHL**	**67**	**3**	**5**	**8**	**61**	**0**	**0**	**2**	**50**	**6.0**	**−8**	**1**	**100.0**	**107**	**83**	**18:10**								
2000-01	**NY Islanders**	**NHL**	**30**	**1**	**5**	**6**	**56**	**0**	**0**	**0**	**19**	**5.3**	**5**	**0**	**0.0**	**75**	**30**	**19:07**								
2001-02	**NY Islanders**	**NHL**	**1**	**0**	**0**	**0**	**2**	**0**	**0**	**0**	**2**	**0.0**	**−1**	**0**	**0.0**	**1**	**0**	**17:12**								
	NHL Totals		**642**	**41**	**97**	**138**	**907**	**10**	**2**	**7**	**720**	**5.7**		**1**	**100.0**	**278**	**223**	**19:27**	**64**	**7**	**16**	**23**	**71**	**1**	**1**	**1**

WHL East First All-Star Team (1990)

Traded to **Montreal** by **Buffalo** for Petr Svoboda, March 10, 1992. Traded to **Philadelphia** by **Montreal** for Yves Racine, June 29, 1994. Traded to **Hartford** by **Philadelphia** with Philadelphia's 1st round choice (later traded to San Jose - San Jose selected Scott Hannan) in 1997 Entry Draft and Hartford/Carolina's 7th round choice (previously acquired, Carolina selected Andrew Merrick) in 1997 Entry Draft for Paul Coffey and Hartford/Carolina's 3rd round choice (Kris Mallette) in 1997 Entry Draft, December 15, 1996. Transferred to **Carolina** after **Hartford** franchise relocated, June 25, 1997. Traded to **Anaheim** by **Carolina** with Stu Grimson for Dave Karpa and Anaheim's 4th round choice (later traded to Atlanta - Atlanta selected Blake Robson) in 2000 Entry Draft, August 11, 1998. Signed as a free agent by **NY Islanders**, July 3, 2000. • Missed majority of 2000-01 season recovering from hernia injury suffered in game vs. Ottawa, December 16, 2000. • Missed majority of 2001-02 season recovering from groin injury suffered in game vs. LA Kings, October 19, 2001.

HALPERN, Jeff
(HAL-pehrn, JEHF) **WSH.**

Center. Shoots right. 6', 201 lbs. Born, Potomac, MD, May 3, 1976.

Season	Club	League	GP	G	A	Pts	PIM	PP	SH	GW	S	%	+/-	TF	F%	H	SB	Min	Playoffs GP	G	A	Pts	PIM	PP	SH	GW
1994-95	Stratford	OJHL-B	44	29	54	83	43																			
1995-96	Princeton	ECAC	29	3	11	14	30																			
1996-97	Princeton	ECAC	33	7	24	31	35																			
1997-98	Princeton	ECAC	36	*28	25	*53	46																			
1998-99	Princeton	ECAC	33	*22	22	44	32																			
	Portland Pirates	AHL	6	2	1	3	4																			
99-2000	**Washington**	**NHL**	**79**	**18**	**11**	**29**	**39**	**4**	**4**	**1**	**108**	**16.7**	**21**	**812**	**51.1**	**84**	**44**	**13:14**	**5**	**2**	**1**	**3**	**0**	**1**	**0**	**1**
2000-01	**Washington**	**NHL**	**80**	**21**	**21**	**42**	**60**	**2**	**1**	**5**	**110**	**19.1**	**13**	**1293**	**52.4**	**92**	**49**	**16:08**	**6**	**2**	**3**	**5**	**17**	**1**	**0**	**1**
2001-02	**Washington**	**NHL**	**48**	**5**	**14**	**19**	**29**	**0**	**0**	**4**	**74**	**6.8**	**−9**	**661**	**56.0**	**41**	**17**	**15:19**								
	NHL Totals		**207**	**44**	**46**	**90**	**128**	**6**	**5**	**10**	**292**	**15.1**		**2766**	**52.9**	**217**	**110**	**14:50**	**11**	**4**	**4**	**8**	**17**	**2**	**0**	**2**

ECAC Second All-Star Team (1998, 1999)

Signed as a free agent by **Washington**, March 29, 1999.

HAMEL, Denis
(ha-MEHL, deh-NEE) **BUF.**

Left wing. Shoots left. 6'1", 201 lbs. Born, Lachute, Que., May 10, 1977. St. Louis' 5th choice, 153rd overall, in 1995 Entry Draft.

Season	Club	League	GP	G	A	Pts	PIM	PP	SH	GW	S	%	+/-	TF	F%	H	SB	Min	GP	G	A	Pts	PIM	PP	SH	GW
			Regular Season																Playoffs							
1992-93	Lachute Regents	QAAA	32	18	24	42	....	....	....	....	....	....	....	....	....	....	....	....	....	....	....	....	....	....	....	....
1993-94	Lac St-Louis	QAAA	28	10	11	21	50	....	....	....	....	....	....	....	....	....	....	....	....	....	....	....	....	....	....	....
	Abitibi	QAAA	15	5	7	12	29	....	....	....	....	....	....	....	....	....	....	....	5	0	3	3	16	....	....	....
1994-95	Chicoutimi	QMJHL	66	15	12	27	155	....	....	....	....	....	....	....	....	....	....	....	12	2	0	2	27	....	....	....
1995-96	Chicoutimi	QMJHL	65	40	49	89	199	....	....	....	....	....	....	....	....	....	....	....	17	10	14	24	64	....	....	....
1996-97	Chicoutimi	QMJHL	70	50	50	100	357	....	....	....	....	....	....	....	....	....	....	....	20	15	10	25	58	....	....	....
1997-98	Rochester	AHL	74	10	15	25	98	....	....	....	....	....	....	....	....	....	....	....	4	1	2	3	0	....	....	....
1998-99	Rochester	AHL	74	16	17	33	121	....	....	....	....	....	....	....	....	....	....	....	20	3	4	7	10	....	....	....
99-2000	**Buffalo**	**NHL**	**3**	**1**	**0**	**1**	**0**	**0**	**0**	**0**	**3**	**33.3**	**–1**	**0**	**0.0**	**10**	**0**	**9:45**	....	....	....	....	....	....	....	....
	Rochester	AHL	76	34	24	58	122	....	....	....	....	....	....	....	....	....	....	....	21	6	7	13	49	....	....	....
2000-01	**Buffalo**	**NHL**	**41**	**8**	**3**	**11**	**22**	**1**	**1**	**3**	**55**	**14.5**	**–2**	**171**	**33.9**	**78**	**18**	**10:58**	....	....	....	....	....	....	....	....
2001-02	**Buffalo**	**NHL**	**61**	**2**	**6**	**8**	**28**	**0**	**0**	**0**	**80**	**2.5**	**–1**	**94**	**39.4**	**85**	**26**	**11:00**	....	....	....	....	....	....	....	....
	NHL Totals		**105**	**11**	**9**	**20**	**50**	**1**	**1**	**3**	**138**	**8.0**		**265**	**35.8**	**173**	**44**	**10:57**	....	....	....	....	....	....	....	....

QMJHL All-Rookie Team (1995)

Traded to **Buffalo** by **St. Louis** for Charlie Huddy and Buffalo's 7th round choice (Daniel Corso) in 1996 Entry Draft, March 19, 1996. • Missed majority of 2000-01 season recovering from knee injury suffered in game vs. NY Islanders, January 27, 2001.

HAMRLIK, Roman
(HAHM-reh-lik, ROH-muhn) **NYI**

Defense. Shoots left. 6'2", 200 lbs. Born, Zlin, Czech., April 12, 1974. Tampa Bay's 1st choice, 1st overall, in 1992 Entry Draft.

Season	Club	League	GP	G	A	Pts	PIM	PP	SH	GW	S	%	+/-	TF	F%	H	SB	Min	GP	G	A	Pts	PIM	PP	SH	GW
1990-91	AC ZPS Zlin	Czech	14	2	2	4	18	....	....	....	....	....	....	....	....	....	....	....	....	....	....	....	....	....	....	....
1991-92	AC ZPS Zlin	Czech	34	5	5	10	50	....	....	....	....	....	....	....	....	....	....	....	....	....	....	....	....	....	....	....
1992-93	**Tampa Bay**	**NHL**	**67**	**6**	**15**	**21**	**71**	**1**	**0**	**1**	**113**	**5.3**	**–21**	....	....	....	....	....	....	....	....	....	....	....	....	....
	Atlanta Knights	IHL	2	1	1	2	2	....	....	....	....	....	....	....	....	....	....	....	....	....	....	....	....	....	....	....
1993-94	**Tampa Bay**	**NHL**	**64**	**3**	**18**	**21**	**135**	**0**	**0**	**0**	**158**	**1.9**	**–14**	....	....	....	....	....	....	....	....	....	....	....	....	....
1994-95	AC ZPS Zlin	Czech	2	1	0	1	10	....	....	....	....	....	....	....	....	....	....	....	....	....	....	....	....	....	....	....
	Tampa Bay	**NHL**	**48**	**12**	**11**	**23**	**86**	**7**	**1**	**2**	**134**	**9.0**	**–18**	....	....	....	....	....	....	....	....	....	....	....	....	....
1995-96	**Tampa Bay**	**NHL**	**82**	**16**	**49**	**65**	**103**	**12**	**0**	**2**	**281**	**5.7**	**–24**	....	....	....	....	....	**5**	**0**	**1**	**1**	**4**	**0**	**0**	**0**
1996-97	**Tampa Bay**	**NHL**	**79**	**12**	**28**	**40**	**57**	**6**	**0**	**0**	**238**	**5.0**	**–29**	....	....	....	....	....	....	....	....	....	....	....	....	....
1997-98	**Tampa Bay**	**NHL**	**37**	**3**	**12**	**15**	**22**	**1**	**0**	**0**	**86**	**3.5**	**–18**	....	....	....	....	....	....	....	....	....	....	....	....	....
	Edmonton	**NHL**	**41**	**6**	**20**	**26**	**48**	**4**	**1**	**3**	**112**	**5.4**	**3**	....	....	....	....	....	**12**	**0**	**6**	**6**	**12**	**0**	**0**	**0**
	Czech Republic	Olympics	6	1	0	1	2	....	....	....	....	....	....	....	....	....	....	....	....	....	....	....	....	....	....	....
1998-99	**Edmonton**	**NHL**	**75**	**8**	**24**	**32**	**70**	**3**	**0**	**0**	**172**	**4.7**	**9**	**0**	**0.0**	**144**	**121**	**23:49**	**3**	**0**	**0**	**0**	**2**	**0**	**0**	**0**
99-2000	Zlin	Czech	6	0	3	3	4	....	....	....	....	....	....	....	....	....	....	....	....	....	....	....	....	....	....	....
	Edmonton	**NHL**	**80**	**8**	**37**	**45**	**68**	**5**	**0**	**0**	**180**	**4.4**	**1**	**0**	**0.0**	**122**	**99**	**25:18**	**5**	**0**	**1**	**1**	**4**	**0**	**0**	**0**
2000-01	**NY Islanders**	**NHL**	**76**	**16**	**30**	**46**	**92**	**5**	**1**	**4**	**232**	**6.9**	**–20**	**1**	**100.0**	**156**	**119**	**25:12**	....	....	....	....	....	....	....	....
2001-02	**NY Islanders**	**NHL**	**70**	**11**	**26**	**37**	**78**	**4**	**1**	**1**	**169**	**6.5**	**7**	**1**	**0.0**	**112**	**149**	**25:32**	**7**	**1**	**6**	**7**	**6**	**0**	**0**	**0**
	Czech Republic	Olympics	4	0	1	1	2	....	....	....	....	....	....	....	....	....	....	....	....	....	....	....	....	....	....	....
	NHL Totals		**719**	**101**	**270**	**371**	**830**	**48**	**4**	**13**	**1875**	**5.4**		**2**	**50.0**	**534**	**488**	**24:57**	**32**	**1**	**14**	**15**	**28**	**0**	**0**	**0**

Played in NHL All-Star Game (1996, 1999)

Traded to **Edmonton** by **Tampa Bay** with Paul Comrie for Bryan Marchment, Steve Kelly and Jason Bonsignore, December 30, 1997. Traded to **NY Islanders** by **Edmonton** for Eric Brewer, Josh Green and NY Islanders' 2nd round choice (Brad Winchester) in 2000 Entry Draft, June 24, 2000.

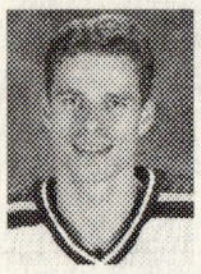

HANDZUS, Michal
(HAHND-zuhs, MEE-chal) **PHI.**

Center. Shoots left. 6'5", 210 lbs. Born, Banska Bystrica, Czech., March 11, 1977. St. Louis' 3rd choice, 101st overall, in 1995 Entry Draft.

Season	Club	League	GP	G	A	Pts	PIM	PP	SH	GW	S	%	+/-	TF	F%	H	SB	Min	GP	G	A	Pts	PIM	PP	SH	GW
1993-94	B. Bystrica Jr.	Slovak-Jr.	40	23	36	59	....	....	....	....	....	....	....	....	....	....	....	....	....	....	....	....	....	....	....	....
1994-95	Banska Bystrica	Slovak-2	22	15	14	29	10	....	....	....	....	....	....	....	....	....	....	....	....	....	....	....	....	....	....	....
1995-96	Banska Bystrica	Slovakia	19	3	1	4	8	....	....	....	....	....	....	....	....	....	....	....	....	....	....	....	....	....	....	....
1996-97	HC SKP PS Poprad	Slovakia	44	15	18	33	....	....	....	....	....	....	....	....	....	....	....	....	....	....	....	....	....	....	....	....
1997-98	Worcester	AHL	69	27	36	63	54	....	....	....	....	....	....	....	....	....	....	....	11	2	6	8	10	....	....	....
1998-99	**St. Louis**	**NHL**	**66**	**4**	**12**	**16**	**30**	**0**	**0**	**0**	**78**	**5.1**	**–9**	**794**	**49.9**	**56**	**38**	**14:48**	**11**	**0**	**2**	**2**	**8**	**0**	**0**	**0**
99-2000	**St. Louis**	**NHL**	**81**	**25**	**28**	**53**	**44**	**3**	**4**	**5**	**166**	**15.1**	**19**	**1243**	**51.5**	**50**	**31**	**17:43**	**7**	**0**	**3**	**3**	**6**	**0**	**0**	**0**
2000-01	**St. Louis**	**NHL**	**36**	**10**	**14**	**24**	**12**	**3**	**2**	**2**	**58**	**17.2**	**11**	**581**	**50.6**	**34**	**17**	**17:60**	....	....	....	....	....	....	....	....
	Phoenix	**NHL**	**10**	**4**	**4**	**8**	**21**	**0**	**1**	**0**	**14**	**28.6**	**5**	**111**	**60.4**	**6**	**1**	**15:26**	....	....	....	....	....	....	....	....
2001-02	**Phoenix**	**NHL**	**79**	**15**	**30**	**45**	**34**	**3**	**1**	**1**	**94**	**16.0**	**–8**	**1227**	**48.7**	**67**	**40**	**16:09**	**5**	**0**	**0**	**0**	**2**	**0**	**0**	**0**
	Slovakia	Olympics	2	1	0	1	6	....	....	....	....	....	....	....	....	....	....	....	....	....	....	....	....	....	....	....
	NHL Totals		**272**	**58**	**88**	**146**	**141**	**9**	**8**	**8**	**410**	**14.1**		**3956**	**50.4**	**213**	**127**	**16:31**	**23**	**0**	**5**	**5**	**16**	**0**	**0**	**0**

Traded to **Phoenix** by **St. Louis** with Ladislav Nagy, the rights to Jeff Taffe and St. Louis' 1st round choice (Ben Eager) in 2002 Entry Draft for Keith Tkachuk, March 13, 2001. Traded to **Philadelphia** by **Phoenix** with Robert Esche for Brian Boucher and Nashville's 3rd round choice (previously acquired, Phoenix selected Joe Callahan) in 2002 Entry Draft, June 12, 2002.

HANKINSON, Casey
(HAN-kihn-suhn, KAY-see) **CHI.**

Left wing. Shoots left. 6'1", 187 lbs. Born, Edina, MN, May 8, 1976. Chicago's 9th choice, 201st overall, in 1995 Entry Draft.

Season	Club	League	GP	G	A	Pts	PIM	PP	SH	GW	S	%	+/-	TF	F%	H	SB	Min	GP	G	A	Pts	PIM	PP	SH	GW
1992-93	Edina Hornets	Hi-School	25	20	26	46	....	....	....	....	....	....	....	....	....	....	....	....	....	....	....	....	....	....	....	....
1993-94	Edina Hornets	Hi-School	24	21	20	41	50	....	....	....	....	....	....	....	....	....	....	....	....	....	....	....	....	....	....	....
1994-95	U. of Minnesota	WCHA	33	7	1	8	86	....	....	....	....	....	....	....	....	....	....	....	....	....	....	....	....	....	....	....
1995-96	U. of Minnesota	WCHA	39	16	19	35	101	....	....	....	....	....	....	....	....	....	....	....	....	....	....	....	....	....	....	....
1996-97	U. of Minnesota	WCHA	42	17	24	41	79	....	....	....	....	....	....	....	....	....	....	....	....	....	....	....	....	....	....	....
1997-98	U. of Minnesota	WCHA	35	10	12	22	81	....	....	....	....	....	....	....	....	....	....	....	....	....	....	....	....	....	....	....
1998-99	Portland Pirates	AHL	72	10	13	23	106	....	....	....	....	....	....	....	....	....	....	....	....	....	....	....	....	....	....	....
99-2000	Cleveland	IHL	82	7	22	29	140	....	....	....	....	....	....	....	....	....	....	....	2	0	0	0	2	....	....	....
2000-01	**Chicago**	**NHL**	**11**	**0**	**1**	**1**	**9**	**0**	**0**	**0**	**15**	**0.0**	**–3**	**0**	**0.0**	**30**	**3**	**9:46**	....	....	....	....	....	....	....	....
	Norfolk Admirals	AHL	69	30	21	51	74	....	....	....	....	....	....	....	....	....	....	....	9	5	4	9	2	....	....	....
2001-02	**Chicago**	**NHL**	**3**	**0**	**0**	**0**	**0**	**0**	**0**	**0**	**1**	**0.0**	**–2**	**6**	**33.3**	**5**	**0**	**8:38**	....	....	....	....	....	....	....	....
	Norfolk Admirals	AHL	72	19	30	49	85	....	....	....	....	....	....	....	....	....	....	....	4	1	2	3	0	....	....	....
	NHL Totals		**14**	**0**	**1**	**1**	**9**	**0**	**0**	**0**	**16**	**0.0**		**6**	**33.3**	**35**	**3**	**9:31**	....	....	....	....	....	....	....	....

HANNAN, Scott
(HAN-nan, SKAWT) **S.J.**

Defense. Shoots left. 6'2", 220 lbs. Born, Richmond, B.C., January 23, 1979. San Jose's 2nd choice, 23rd overall, in 1997 Entry Draft.

Season	Club	League	GP	G	A	Pts	PIM	PP	SH	GW	S	%	+/-	TF	F%	H	SB	Min	GP	G	A	Pts	PIM	PP	SH	GW
1994-95	Surrey Wolves	BCAHA	70	54	54	108	200	....	....	....	....	....	....	....	....	....	....	....	....	....	....	....	....	....	....	....
	Tacoma Rockets	WHL	2	0	0	0	0	....	....	....	....	....	....	....	....	....	....	....	....	....	....	....	....	....	....	....
1995-96	Kelowna Rockets	WHL	69	4	5	9	76	....	....	....	....	....	....	....	....	....	....	....	6	0	1	1	4	....	....	....
1996-97	Kelowna Rockets	WHL	70	17	26	43	101	....	....	....	....	....	....	....	....	....	....	....	6	0	0	0	8	....	....	....
1997-98	Kelowna Rockets	WHL	47	10	30	40	70	....	....	....	....	....	....	....	....	....	....	....	7	2	7	9	14	....	....	....
1998-99	**San Jose**	**NHL**	**5**	**0**	**2**	**2**	**6**	**0**	**0**	**0**	**4**	**0.0**	**0**	**0**	**0.0**	**2**	**0**	**7:15**	....	....	....	....	....	....	....	....
	Kelowna Rockets	WHL	47	15	30	45	92	....	....	....	....	....	....	....	....	....	....	....	6	1	2	3	14	....	....	....
	Kentucky	AHL	2	0	0	0	2	....	....	....	....	....	....	....	....	....	....	....	12	0	2	2	10	....	....	....
99-2000	**San Jose**	**NHL**	**30**	**1**	**2**	**3**	**10**	**0**	**0**	**0**	**28**	**3.6**	**7**	**1**	**0.0**	**27**	**9**	**17:09**	**1**	**0**	**1**	**1**	**0**	**0**	**0**	**0**
	Kentucky	AHL	41	5	12	17	40	....	....	....	....	....	....	....	....	....	....	....	....	....	....	....	....	....	....	....
2000-01	**San Jose**	**NHL**	**75**	**3**	**14**	**17**	**51**	**0**	**0**	**1**	**96**	**3.1**	**10**	**0**	**0.0**	**109**	**67**	**19:02**	**6**	**0**	**1**	**1**	**6**	**0**	**0**	**0**
2001-02	**San Jose**	**NHL**	**75**	**2**	**12**	**14**	**57**	**0**	**0**	**1**	**68**	**2.9**	**10**	**1**	**100.0**	**103**	**76**	**20:19**	**12**	**0**	**2**	**2**	**12**	**0**	**0**	**0**
	NHL Totals		**185**	**6**	**30**	**36**	**124**	**0**	**0**	**2**	**196**	**3.1**		**2**	**50.0**	**241**	**152**	**18:56**	**19**	**0**	**4**	**4**	**18**	**0**	**0**	**0**

WHL West First All-Star Team (1999)

HANSEN, Tavis
(HAN-sehn, TA-vihs)

Center. Shoots right. 6'1", 205 lbs. Born, Prince Albert, Sask., June 17, 1975. Winnipeg's 3rd choice, 58th overall, in 1994 Entry Draft.

Season	Club	League	GP	G	A	Pts	PIM	PP	SH	GW	S	%	+/-	TF	F%	H	SB	Min	GP	G	A	Pts	PIM	PP	SH	GW
1992-93	Shellbrook	SMHL	42	42	63	105	107	....	....	....	....	....	....	....	....	....	....	....	....	....	....	....	....	....	....	....
1993-94	Tacoma Rockets	WHL	71	23	31	54	122	....	....	....	....	....	....	....	....	....	....	....	8	1	3	4	17	....	....	....
1994-95	Tacoma Rockets	WHL	71	32	41	73	142	....	....	....	....	....	....	....	....	....	....	....	4	1	1	2	8	....	....	....
	Winnipeg	**NHL**	**1**	**0**	**0**	**0**	**0**	**0**	**0**	**0**	**0**	**0.0**	**0**	....	....	....	....	....	....	....	....	....	....	....	....	....
1995-96	Springfield	AHL	67	6	16	22	85	....	....	....	....	....	....	....	....	....	....	....	5	1	2	3	2	....	....	....
1996-97	**Phoenix**	**NHL**	**1**	**0**	**0**	**0**	**0**	**0**	**0**	**0**	**0**	**0.0**	**0**	....	....	....	....	....	....	....	....	....	....	....	....	....
	Springfield	AHL	12	3	1	4	23	....	....	....	....	....	....	....	....	....	....	....	....	....	....	....	....	....	....	....

			Regular Season																Playoffs							
Season	Club	League	GP	G	A	Pts	PIM	PP	SH	GW	S	%	+/-	TF	F%	H	SB	Min	GP	G	A	Pts	PIM	PP	SH	GW
1997-98	Springfield	AHL	73	20	14	34	70												4	1	2	3	18			
1998-99	**Phoenix**	**NHL**	**20**	**2**	**1**	**3**	**12**	**0**	**0**	**0**	**14**	**14.3**	**–4**	**5**	**80.0**	**26**	**3**	**8:07**	**2**	**0**	**0**	**0**	**0**	**0**	**0**	**0**
	Springfield	AHL	63	23	11	34	85												3	0	1	1	5			
99-2000	**Phoenix**	**NHL**	**5**	**0**	**0**	**0**	**0**	**0**	**0**	**0**	**2**	**0.0**	**0**	**0**	**0.0**	**2**	**1**	**4:17**								
	Springfield	AHL	59	21	27	48	164												5	2	1	3	4			
2000-01	Springfield	AHL	24	6	10	16	81																			
	Phoenix	**NHL**	**7**	**0**	**0**	**0**	**4**	**0**	**0**	**0**	**2**	**0.0**	**–1**	**1**	**100.0**	**9**	**1**	**5:05**								
2001-02	Hershey Bears	AHL	35	9	6	15	50												8	0	3	3	8			
	NHL Totals		**34**	**2**	**1**	**3**	**16**	**0**	**0**	**0**	**18**	**11.1**		**6**	**83.3**	**37**	**5**	**6:51**	**2**	**0**	**0**	**0**	**0**	**0**	**0**	**0**

Transferred to **Phoenix** after **Winnipeg** franchise relocated, July 1, 1996. • Missed majority of 2000-01 and 2001-02 seasons recovering from arm injury originally suffered in game vs. Hershey (AHL), February 3, 2001. Signed as a free agent by **Hershey** (AHL), January 16, 2002.

HARKINS, Brett

(HAHR-kihns, BREHT)

Left wing. Shoots left. 6'1", 185 lbs. Born, North Ridgeville, OH, July 2, 1970. NY Islanders' 9th choice, 133rd overall, in 1989 Entry Draft.

Season	Club	League	GP	G	A	Pts	PIM	PP	SH	GW	S	%	+/-	TF	F%	H	SB	Min	GP	G	A	Pts	PIM	PP	SH	GW
1986-87	St. Andrew's	Hi-School	30	47	60	107																				
1987-88	Brockville	OCJHL	55	21	55	76	36																			
1988-89	Det. Compuware	NAJHL	38	23	46	69	94																			
1989-90	Bowling Green	CCHA	41	11	43	54	45																			
1990-91	Bowling Green	CCHA	40	22	38	60	30																			
1991-92	Bowling Green	CCHA	34	8	39	47	32																			
1992-93	Bowling Green	CCHA	35	19	28	47	28																			
1993-94	Adirondack	AHL	80	22	47	69	23												10	1	5	6	4			
1994-95	Providence	AHL	80	23	*69	92	32												13	8	14	22	4			
	Boston	**NHL**	**1**	**0**	**1**	**1**	**0**	**0**	**0**	**0**	**1**	**0.0**	**0**													
1995-96	**Florida**	**NHL**	**8**	**0**	**3**	**3**	**6**	**0**	**0**	**0**	**4**	**0.0**	**–2**													
	Carolina	AHL	55	23	*71	94	44																			
1996-97	**Boston**	**NHL**	**44**	**4**	**14**	**18**	**8**	**3**	**0**	**2**	**52**	**7.7**	**–3**													
	Providence	AHL	28	9	31	40	32												10	2	10	12	0			
1997-98	Cleveland	IHL	80	32	62	94	82												10	4	13	17	14			
1998-99	Cleveland	IHL	74	20	67	87	84																			
99-2000	Cleveland	IHL	76	20	50	70	79												9	2	8	10	6			
2000-01	Houston Aeros	IHL	81	16	*64	80	51												7	0	3	3	8			
2001-02	**Columbus**	**NHL**	**25**	**2**	**12**	**14**	**8**	**2**	**0**	**0**	**9**	**22.2**	**–5**	**160**	**43.1**	**9**	**6**	**12:38**								
	Syracuse Crunch	AHL	22	4	20	24	13																			
	NHL Totals		**78**	**6**	**30**	**36**	**22**	**5**	**0**	**2**	**66**	**9.1**		**160**	**43.1**	**9**	**6**	**12:38**								

Signed as a free agent by **Adirondack** (AHL), September, 1993. Signed as a free agent by **Boston**, July 1, 1994. Signed as a free agent by **Florida**, July 24, 1995. Signed as a free agent by **Boston**, September 4, 1996. Signed as a free agent by **Columbus**, May 29, 2001. Signed as a free agent by **Skelleftea** (Sweden), July 4, 2002.

HARLOCK, David

(HAHR-lahk, DAY-vihd) **PHI.**

Defense. Shoots left. 6'2", 215 lbs. Born, Toronto, Ont., March 16, 1971. New Jersey's 2nd choice, 24th overall, in 1990 Entry Draft.

Season	Club	League	GP	G	A	Pts	PIM	PP	SH	GW	S	%	+/-	TF	F%	H	SB	Min	GP	G	A	Pts	PIM	PP	SH	GW
1986-87	Tor. Red Wings	MTHL	86	17	55	72	60																			
1987-88	Tor. Red Wings	MTHL	70	16	56	72	100																			
	Henry Carr	MTJHL	3	0	0	0	4																			
1988-89	St. Michael's B	OJHL-B	25	4	16	20	34												27	3	12	15	14			
1989-90	U. of Michigan	CCHA	42	2	13	15	44																			
1990-91	U. of Michigan	CCHA	39	2	8	10	70																			
1991-92	U. of Michigan	CCHA	44	1	6	7	80																			
1992-93	U. of Michigan	CCHA	38	3	9	12	58																			
1993-94	Team Canada	Nat-Tm	41	0	3	3	28																			
	Canada	Olympics	8	0	0	0	8																			
	Toronto	**NHL**	**6**	**0**	**0**	**0**	**0**	**0**	**0**	**0**	**2**	**0.0**	**–2**													
	St. John's	AHL	10	0	3	3	2												9	0	0	0	6			
1994-95	St. John's	AHL	58	0	6	6	44												5	0	0	0	0			
	Toronto	**NHL**	**1**	**0**	**0**	**0**	**0**	**0**	**0**	**0**	**0**	**0.0**	**–1**													
1995-96	**Toronto**	**NHL**	**1**	**0**	**0**	**0**	**0**	**0**	**0**	**0**	**0**	**0.0**	**0**													
	St. John's	AHL	77	0	12	12	92												4	0	1	1	2			
1996-97	San Antonio	IHL	69	3	10	13	82												9	0	0	0	10			
1997-98	**Washington**	**NHL**	**6**	**0**	**0**	**0**	**4**	**0**	**0**	**0**	**2**	**0.0**	**2**													
	Portland Pirates	AHL	71	3	15	18	66												10	2	2	4	6			
1998-99	**NY Islanders**	**NHL**	**70**	**2**	**6**	**8**	**68**	**0**	**0**	**0**	**35**	**5.7**	**–16**	**0**	**0.0**	**172**	**65**	**18:15**								
99-2000	**Atlanta**	**NHL**	**44**	**0**	**6**	**6**	**36**	**0**	**0**	**0**	**29**	**0.0**	**–8**	**0**	**0.0**	**166**	**55**	**19:41**								
2000-01	**Atlanta**	**NHL**	**65**	**0**	**1**	**1**	**62**	**0**	**0**	**0**	**26**	**0.0**	**–28**	**0**	**0.0**	**186**	**87**	**17:04**								
2001-02	**Atlanta**	**NHL**	**19**	**0**	**1**	**1**	**18**	**0**	**0**	**0**	**12**	**0.0**	**–2**	**0**	**0.0**	**31**	**16**	**14:55**								
	Chicago Wolves	AHL	24	2	9	11	28																			
	Philadelphia	AHL	11	0	4	4	14												5	0	1	1	4			
	NHL Totals		**212**	**2**	**14**	**16**	**188**	**0**	**0**	**0**	**106**	**1.9**		**0**	**0.0**	**555**	**223**	**17:52**								

Signed as a free agent by **Toronto**, August 20, 1993. Signed as a free agent by **Washington**, August 20, 1997. Signed as a free agent by **NY Islanders**, August 24, 1998. Claimed by **Atlanta** from **NY Islanders** in Expansion Draft, June 25, 1999. Traded to **Philadelphia** by **Atlanta** with Atlanta's 3rd and 7th round choices in 2003 Entry Draft for Francis Lessard, March 15, 2002.

HARTIGAN, Mark

(HAHR-tih-guhn, MAHRK) **ATL.**

Right wing. Shoots left. 6', 200 lbs. Born, Fort St. John, B.C., October 15, 1977.

Season	Club	League	GP	G	A	Pts	PIM	PP	SH	GW	S	%	+/-	TF	F%	H	SB	Min	GP	G	A	Pts	PIM	PP	SH	GW
1996-97	Weyburn	SJHL	52	44	32	76																				
1997-98	Weyburn	SJHL	62	*59	46	*105	81												23	17	21	38	10			
1998-99	St. Cloud State	WCHA	DID NOT PLAY – FRESHMAN																							
99-2000	St. Cloud State	WCHA	37	22	20	42	24																			
2000-01	St. Cloud State	WCHA	40	27	21	48	20																			
2001-02	St. Cloud State	WCHA	42	*37	38	75	42																			
	Atlanta	**NHL**	**2**	**0**	**0**	**0**	**2**	**0**	**0**	**0**	**3**	**0.0**	**–2**	**18**	**38.9**	**1**	**1**	**13:16**								
	NHL Totals		**2**	**0**	**0**	**0**	**2**	**0**	**0**	**0**	**3**	**0.0**		**18**	**38.9**	**1**	**1**	**13:16**								

WCHA First All-Star Team (2002) • WCHA Player of the Year (2002)

Signed as a free agent by **Atlanta**, March 27, 2002.

HARTNELL, Scott

(HAHRT-nuhl, SKAWT) **NSH.**

Left wing. Shoots left. 6'2", 208 lbs. Born, Regina, Sask., April 18, 1982. Nashville's 1st choice, 6th overall, in 2000 Entry Draft.

Season	Club	League	GP	G	A	Pts	PIM	PP	SH	GW	S	%	+/-	TF	F%	H	SB	Min	GP	G	A	Pts	PIM	PP	SH	GW
1997-98	Lloydminster	AJHL	56	9	25	34	82												4	2	1	3	8			
	Prince Albert	WHL	1	0	1	1	2																			
1998-99	Prince Albert	WHL	65	10	34	44	104												14	0	5	5	22			
99-2000	Prince Albert	WHL	62	27	55	82	124												6	3	2	5	6			
2000-01	**Nashville**	**NHL**	**75**	**2**	**14**	**16**	**48**	**0**	**0**	**0**	**92**	**2.2**	**–8**	**3**	**33.3**	**100**	**22**	**10:54**								
2001-02	**Nashville**	**NHL**	**75**	**14**	**27**	**41**	**111**	**3**	**0**	**4**	**162**	**8.6**	**5**	**12**	**25.0**	**144**	**23**	**16:58**								
	NHL Totals		**150**	**16**	**41**	**57**	**159**	**3**	**0**	**4**	**254**	**6.3**		**15**	**26.7**	**244**	**45**	**13:56**								

HARVEY, Todd

(HAHR-vee, TAWD) **S.J.**

Center. Shoots right. 6', 200 lbs. Born, Hamilton, Ont., February 17, 1975. Dallas' 1st choice, 9th overall, in 1993 Entry Draft.

Season	Club	League	GP	G	A	Pts	PIM	PP	SH	GW	S	%	+/-	TF	F%	H	SB	Min	GP	G	A	Pts	PIM	PP	SH	GW
1989-90	Cambridge	OJHL-B	41	35	27	62	213																			
1990-91	Cambridge	OJHL-B	35	32	39	71	174																			
1991-92	Detroit	OHL	58	21	43	64	141												7	3	5	8	30			
1992-93	Detroit	OHL	55	50	50	100	83												15	9	12	21	39			
1993-94	Detroit	OHL	49	34	51	85	75												17	10	12	22	26			
1994-95	Detroit	OHL	11	8	14	22	12																			
	Dallas	**NHL**	**40**	**11**	**9**	**20**	**67**	**2**	**0**	**1**	**64**	**17.2**	**–3**						**5**	**0**	**0**	**0**	**8**	**0**	**0**	**0**
1995-96	**Dallas**	**NHL**	**69**	**9**	**20**	**29**	**136**	**3**	**0**	**1**	**101**	**8.9**	**–13**													
	Michigan K-Wings	IHL	5	1	3	4	8																			
1996-97	**Dallas**	**NHL**	**71**	**9**	**22**	**31**	**142**	**1**	**0**	**2**	**99**	**9.1**	**19**						**7**	**0**	**1**	**1**	**10**	**0**	**0**	**0**

			Regular Season																Playoffs							
Season	Club	League	GP	G	A	Pts	PIM	PP	SH	GW	S	%	+/-	TF	F%	H	SB	Min	GP	G	A	Pts	PIM	PP	SH	GW
1997-98	Dallas	NHL	59	9	10	19	104	0	0	1	88	10.2	5													
1998-99	NY Rangers	NHL	37	11	17	28	72	6	0	2	58	19.0	–1	175	50.3	126	22	17:19								
99-2000	NY Rangers	NHL	31	3	3	6	62	0	0	0	31	9.7	–9	173	49.1	94	20	12:21								
	San Jose	NHL	40	8	4	12	78	2	0	0	59	13.6	–2	44	43.2	94	9	12:55	12	1	0	1	8	1	0	0
2000-01	San Jose	NHL	69	10	11	21	72	1	0	2	66	15.2	6	98	40.8	178	7	11:05	6	0	0	0	8	0	0	0
2001-02	San Jose	NHL	69	9	13	22	73	0	0	1	66	13.6	16	223	48.4	172	13	9:47	12	0	2	2	12	0	0	0
	NHL Totals		485	79	109	188	806	15	0	10	632	12.5		713	47.7	664	71	12:07	42	1	3	4	46	1	0	0

OHL All-Rookie Team (1992)

Traded to **NY Rangers** by **Dallas** with Bob Errey and Dallas' 4th round choice (Boyd Kane) in 1998 Entry Draft for Brian Skrudland, Mike Keane and NY Rangers' 6th round choice (Pavel Patera) in 1998 Entry Draft, March 24, 1998. Traded to **San Jose** by **NY Rangers** with NY Rangers' 4th round choice (Dimitri Patzold) in 2001 Entry Draft for Radek Dvorak, December 30, 1999.

HATCHER, Derian

(HAT-chuhr, DAIR-ee-an) **DAL.**

Defense. Shoots left. 6'5", 235 lbs. Born, Sterling Hts., MI, June 4, 1972. Minnesota's 1st choice, 8th overall, in 1990 Entry Draft.

Season	Club	League	GP	G	A	Pts	PIM	PP	SH	GW	S	%	+/-	TF	F%	H	SB	Min	GP	G	A	Pts	PIM	PP	SH	GW
1987-88	Detroit GPD	MNHL	25	5	13	18	52																			
1988-89	Detroit GPD	MNHL	51	19	35	54	100																			
1989-90	North Bay	OHL	64	14	38	52	81												5	2	3	5	8			
1990-91	North Bay	OHL	64	13	49	62	163												10	2	10	12	28			
1991-92	Minnesota	NHL	43	8	4	12	88	0	0	2	51	15.7	7						5	0	2	2	8	0	0	0
1992-93	Minnesota	NHL	67	4	15	19	178	0	0	1	73	5.5	–27													
	Kalamazoo Wings	IHL	2	1	2	3	21																			
1993-94	Dallas	NHL	83	12	19	31	211	2	1	2	132	9.1	19						9	0	2	2	14	0	0	0
1994-95	Dallas	NHL	43	5	11	16	105	2	0	2	74	6.8	3													
1995-96	Dallas	NHL	79	8	23	31	129	2	0	1	125	6.4	–12													
1996-97	Dallas	NHL	63	3	19	22	97	0	0	0	96	3.1	8						7	0	2	2	20	0	0	0
1997-98	Dallas	NHL	70	6	25	31	132	3	0	2	74	8.1	9						17	3	3	6	39	2	0	0
	United States	Olympics	4	0	0	0	0																			
1998-99♦	Dallas	NHL	80	9	21	30	102	3	0	2	125	7.2	21	0	0.0	204	97	24:44	18	1	6	7	24	0	0	0
99-2000	Dallas	NHL	57	2	22	24	68	0	0	0	90	2.2	6	0	0.0	179	82	27:33	23	1	3	4	29	0	0	0
2000-01	Dallas	NHL	80	2	21	23	77	1	0	2	97	2.1	5	0	0.0	250	114	25:53	10	0	1	1	16	0	0	0
2001-02	Dallas	NHL	80	4	21	25	87	1	0	0	111	3.6	12	0	0.0	330	111	26:40								
	NHL Totals		745	63	201	264	1274	14	1	14	1048	6.0		0	0.0	963	404	26:06	89	5	19	24	150	2	0	0

Played in NHL All-Star Game (1997)

Transferred to **Dallas** after **Minnesota** franchise relocated, June 9, 1993.

HAUER, Brett

(HOW-uhr, BREHT)

Defense. Shoots right. 6'2", 210 lbs. Born, Richfield, MN, July 11, 1971. Vancouver's 3rd choice, 71st overall, in 1989 Entry Draft.

Season	Club	League	GP	G	A	Pts	PIM	PP	SH	GW	S	%	+/-	TF	F%	H	SB	Min	GP	G	A	Pts	PIM	PP	SH	GW
1987-88	Richfield High	Hi-School	24	3	3	6																				
1988-89	Richfield High	Hi-School	24	8	15	23	70																			
1989-90	U. Minn-Duluth	WCHA	37	2	6	8	44																			
1990-91	U. Minn-Duluth	WCHA	30	1	7	8	54																			
1991-92	U. Minn-Duluth	WCHA	33	8	14	22	40																			
1992-93	U. Minn-Duluth	WCHA	40	10	46	56	52																			
1993-94	Team USA	Nat-Tm	57	6	14	20	88																			
	United States	Olympics	8	0	0	0	10																			
	Las Vegas	IHL	21	0	7	7	8												1	0	0	0	0			
1994-95	AIK Solna	Swede-2	37	1	3	4	38																			
1995-96	Edmonton	NHL	29	4	2	6	30	2	0	1	53	7.5	–11													
	Cape Breton	AHL	17	3	5	8	29																			
1996-97	Chicago Wolves	IHL	81	10	30	40	50												4	2	0	2	4			
1997-98	Manitoba Moose	IHL	82	13	48	61	58												3	0	0	0	2			
1998-99	Manitoba Moose	IHL	81	15	56	71	66												5	0	5	5	4			
99-2000	Edmonton	NHL	5	0	2	2	2	0	0	0	8	0.0	–2	0	0.0	10	2	14:48								
	Manitoba Moose	IHL	77	13	47	60	92												2	0	1	1	2			
2000-01	Manitoba Moose	IHL	82	17	42	59	52												13	1	9	10	12			
2001-02	Manchester	AHL	29	2	11	13	38																			
	Nashville	NHL	3	0	0	0	6	0	0	0	2	0.0	–3	0	0.0	3	1	10:29								
	Milwaukee	AHL	48	6	21	27	14																			
	NHL Totals		37	4	4	8	38	2	0	1	63	6.3		0	0.0	13	3	13:11								

WCHA First All-Star Team (1993) • NCAA West First All-American Team (1993) • IHL First All-Star Team (1999, 2000, 2001) • Won Govenors' Trophy (Top Defenseman - IHL) (2000, 2001)

Signed as a free agent by **Las Vegas** (IHL), February 15, 1994. Traded to **Edmonton** by **Vancouver** for Edmonton's 7th round choice (Larry Shapley) in 1997 Entry Draft, August 24, 1995. Signed as a free agent by **Manitoba** (IHL), September 15, 1997. Signed as a free agent by **LA Kings**, July 8, 2001. Traded to **Nashville** by **LA Kings** for Rich Brennan, December 19, 2001. Signed as a free agent by **Lausanne** (Swiss), July 11, 2002.

HAVELID, Niclas

(HAHV-lihd, NIHK-lahs) **ANA.**

Defense. Shoots left. 5'11", 196 lbs. Born, Stockholm, Sweden, April 12, 1973. Anaheim's 2nd choice, 83rd overall, in 1999 Entry Draft.

Season	Club	League	GP	G	A	Pts	PIM	PP	SH	GW	S	%	+/-	TF	F%	H	SB	Min	GP	G	A	Pts	PIM	PP	SH	GW
1988-89	Enkopings SK	Swede-3	7	0	1	1	0																			
1989-90	Enkopings SK	Swede-3	24	1	2	3	28																			
1990-91	RA-73	Swede-2	30	2	3	5	22																			
1991-92	AIK Solna	Sweden	10	0	0	0	2																			
1992-93	AIK Solna	Sweden	30	1	2	3	22												3	0	0	0	2			
1993-94	AIK Solna	Swede-2	22	3	9	12	14																			
1994-95	AIK Solna	Swede-2	40	3	7	10	38																			
1995-96	AIK Solna	Sweden	40	5	6	11	30																			
1996-97	AIK Solna	Sweden	49	3	6	9	42												7	1	2	3	8			
1997-98	AIK Solna	Sweden	43	8	4	12	42												10	1	3	4	39			
1998-99	Malmo IF	Sweden	50	10	12	22	42												8	0	4	4	10			
99-2000	Anaheim	NHL	50	2	7	9	20	0	0	2	70	2.9	0	1	0.0	103	66	19:10								
	Cincinnati	AHL	2	0	0	0	0																			
2000-01	Anaheim	NHL	47	4	10	14	34	2	0	1	69	5.8	–6	4	0.0	101	54	21:51								
2001-02	Anaheim	NHL	52	1	2	3	40	0	0	0	45	2.2	–13	1	0.0	58	51	17:01								
	NHL Totals		149	7	19	26	94	2	0	3	184	3.8		6	0.0	262	171	19:16								

HAVLAT, Martin

(HAHV-lat, MAHR-tihn) **OTT.**

Center. Shoots left. 6'1", 190 lbs. Born, Mlada Boleslav, Czech., April 19, 1981. Ottawa's 1st choice, 26th overall, in 1999 Entry Draft.

Season	Club	League	GP	G	A	Pts	PIM	PP	SH	GW	S	%	+/-	TF	F%	H	SB	Min	GP	G	A	Pts	PIM	PP	SH	GW
1997-98	Ytong Brno Jr.	Czech-Jr.	32	38	29	67																				
1998-99	Trinec Jr.	Czech-Jr.	31	28	23	51																				
	Trinec	Czech	24	2	3	5	4												8	0	0	0				
99-2000	Trinec	Czech	46	13	29	42	42												4	0	2	2	8			
2000-01	Ottawa	NHL	73	19	23	42	20	7	0	5	133	14.3	8	40	30.0	68	9	13:47	4	0	0	0	2	0	0	0
2001-02	Ottawa	NHL	72	22	28	50	66	9	0	6	145	15.2	–7	15	40.0	60	4	14:46	12	2	5	7	14	2	0	2
	Czech Republic	Olympics	4	3	1	4	27																			
	NHL Totals		145	41	51	92	86	16	0	11	278	14.7		55	32.7	128	13	14:16	16	2	5	7	16	2	0	2

NHL All-Rookie Team (2001)

HAWGOOD, Greg

(HAW-guhd, GREHG) **DAL.**

Defense. Shoots left. 5'10", 190 lbs. Born, Edmonton, Alta., August 10, 1968. Boston's 9th choice, 202nd overall, in 1986 Entry Draft.

Season	Club	League	GP	G	A	Pts	PIM	PP	SH	GW	S	%	+/-	TF	F%	H	SB	Min	GP	G	A	Pts	PIM	PP	SH	GW
1983-84	Kamloops	WHL	49	10	23	33	39												6	0	2	2	2			
1984-85	Kamloops Blazers	WHL	66	25	40	65	72												15	3	15	18	15			
1985-86	Kamloops Blazers	WHL	71	34	85	119	86												16	9	22	31	16			
1986-87	Kamloops Blazers	WHL	61	30	93	123	139												13	7	16	23	18			
1987-88	Kamloops Blazers	WHL	63	48	85	133	142												16	10	16	26	33			
	Boston	NHL	1	0	0	0	0	0	0	0	1	0.0	–1						3	1	0	1	0	0	0	0
1988-89	Boston	NHL	56	16	24	40	84	5	0	0	132	12.1	4						10	0	2	2	2	0	0	0
	Maine Mariners	AHL	21	2	9	11	41																			

Season	Club	League	GP	G	A	Pts	PIM	PP	SH	GW	S	%	+/-	TF	F%	H	SB	Min	GP	G	A	Pts	PIM	PP	SH	GW
			Regular Season																Playoffs							
1989-90	Boston	NHL	77	11	27	38	76	2	0	1	127	8.7	12						15	1	3	4	12	1	0	0
1990-91	HC Asiago	Italy	2	3	0	3	9																			
	Maine Mariners	AHL	5	0	1	1	13																			
	Edmonton	NHL	6	0	1	1	6	0	0	0	9	0.0	–2													
	Cape Breton	AHL	55	10	32	42	73												4	0	3	3	23			
1991-92	Edmonton	NHL	20	2	11	13	22	0	0	0	24	8.3	19						13	0	3	3	23	0	0	0
	Cape Breton	AHL	56	20	55	75	26												3	2	2	4	0			
1992-93	Edmonton	NHL	29	5	13	18	35	2	0	0	47	10.6	–1													
	Philadelphia	NHL	40	6	22	28	39	5	0	1	91	6.6	–7													
1993-94	Philadelphia	NHL	19	3	12	15	19	3	0	0	37	8.1	2													
	Florida	NHL	33	2	14	16	9	0	0	1	55	3.6	8													
	Pittsburgh	NHL	12	1	2	3	8	1	0	1	20	5.0	–1						1	0	0	0	0	0	0	0
1994-95	Pittsburgh	NHL	21	1	4	5	25	1	0	0	17	5.9	2													
	Cleveland	IHL																	3	1	0	1	4			
1995-96	Las Vegas	IHL	78	20	65	85	101												15	5	11	16	24			
1996-97	San Jose	NHL	63	6	12	18	69	3	0	0	83	7.2	–22													
1997-98	Kolner Haie	Germany	4	0	1	1	16																			
	Kolner Haie	EuroHL	1	0	0	0	2																			
	Houston Aeros	IHL	81	19	52	71	75												4	0	4	4	0			
1998-99	Houston Aeros	IHL	76	17	57	74	90												19	4	8	12	24			
99-2000	Vancouver	NHL	79	5	17	22	26	2	0	0	70	7.1	5	2	50.0	64	57	17:11								
2000-01	Vancouver	NHL	16	2	5	7	6	1	0	1	16	12.5	8	1	0.0	16	5	14:55								
	Kansas City	IHL	46	6	16	22	21																			
2001-02	Dallas	NHL	2	0	0	0	2	0	0	0	1	0.0	0	0	0.0	2	0	6:21								
	Utah Grizzlies	AHL	67	18	43	61	83												5	0	2	2	2			
	NHL Totals		474	60	164	224	426	25	0	5	730	8.2		3	33.3	82	62	16:35	42	2	8	10	37	1	0	0

WHL West First All-Star Team (1986, 1987, 1988) • Canadian Major Junior Defenseman of the Year (1988) • AHL First All-Star Team (1992) • Won Eddie Shore Award (Top Defenseman - AHL) (1992) • IHL First All-Star Team (1996, 1998, 1999) • Won Governors' Trophy (Top Defenseman - IHL) (1996, 1999) • AHL Second All-Star Team (2002)

Traded to **Edmonton** by **Boston** for Vladimir Ruzicka, October 22, 1990. Traded to **Philadelphia** by **Edmonton** with Josef Beranek for Brian Benning, January 16, 1993. Traded to **Florida** by **Philadelphia** for cash, November 30, 1993. Traded to **Pittsburgh** by **Florida** for Jeff Daniels, March 19, 1994. Signed as a free agent by **San Jose**, September 25, 1996. Signed as a free agent by **Vancouver**, September 30, 1999. Signed as a free agent by **Dallas**, July 17, 2001.

HAY, Dwayne

(HAY, DWAYN) **CGY.**

Left wing. Shoots left. 6'1", 203 lbs. Born, London, Ont., February 11, 1977. Washington's 3rd choice, 43rd overall, in 1995 Entry Draft.

Season	Club	League	GP	G	A	Pts	PIM	PP	SH	GW	S	%	+/-	TF	F%	H	SB	Min	GP	G	A	Pts	PIM	PP	SH	GW
1991-92	London	OMHA	86	70	56	126	104																			
1992-93	Listowel	OJHL-B	50	19	33	52	40																			
1993-94	Listowel	OJHL-B	48	10	24	34	56																			
1994-95	Guelph Storm	OHL	65	26	28	54	37												14	5	7	12	6			
1995-96	Guelph Storm	OHL	60	28	30	58	49												16	4	9	13	18			
1996-97	Guelph Storm	OHL	32	17	17	34	21												11	4	6	10	0			
1997-98	Washington	NHL	2	0	0	0	2	0	0	0	1	0.0	0													
	Portland Pirates	AHL	58	6	7	13	35																			
	New Haven	AHL	10	3	2	5	4												2	0	0	0	0			
1998-99	Florida	NHL	9	0	0	0	0	0	0	0	3	0.0	–1	1	0.0	5	0	6:35								
	New Haven	AHL	46	18	17	35	22																			
99-2000	Florida	NHL	6	0	0	0	2	0	0	0	3	0.0	–2	0	0.0	3	1	6:36								
	Louisville	AHL	41	11	20	31	18																			
	Tampa Bay	NHL	13	1	1	2	2	0	0	0	11	9.1	0	1	0.0	15	1	6:04								
2000-01	Calgary	NHL	49	1	3	4	16	0	0	0	39	2.6	–4	6	16.7	92	9	8:24								
2001-02	Saint John	AHL	70	5	12	17	39																			
	NHL Totals		79	2	4	6	22	0	0	0	57	3.5		8	12.5	115	11	7:39								

Traded to **Florida** by **Washington** with future considerations for Esa Tikkanen, March 9, 1998. Traded to **Tampa Bay** by **Florida** with Ryan Johnson for Mike Sillinger, March 14, 2000. Claimed on waivers by **Calgary** from **Tampa Bay**, October 3, 2000.

HEALEY, Paul

(HEE-lee, PAWL) **TOR.**

Right wing. Shoots right. 6'2", 198 lbs. Born, Edmonton, Alta., March 20, 1975. Philadelphia's 7th choice, 192nd overall, in 1993 Entry Draft.

Season	Club	League	GP	G	A	Pts	PIM	PP	SH	GW	S	%	+/-	TF	F%	H	SB	Min	GP	G	A	Pts	PIM	PP	SH	GW
1991-92	Ft. Saskatchewan	AJHL	52	11	19	30	40																			
1992-93	Prince Albert	WHL	72	12	20	32	66																			
1993-94	Prince Albert	WHL	63	23	26	49	70																			
1994-95	Prince Albert	WHL	71	43	50	93	67												12	3	4	7	2			
1995-96	Hershey Bears	AHL	60	7	15	22	35																			
1996-97	Philadelphia	NHL	2	0	0	0	0	0	0	0	0	0.0	0													
	Philadelphia	AHL	64	21	19	40	56												10	4	1	5	10			
1997-98	Philadelphia	NHL	4	0	0	0	12	0	0	0	0	0.0	0													
	Philadelphia	AHL	71	34	18	52	48												20	6	2	8	4			
1998-99	Philadelphia	AHL	72	26	20	46	39												15	4	6	10	11			
99-2000	Milwaukee	IHL	76	21	18	39	28												3	1	2	3	0			
2000-01	Hamilton	AHL	79	39	32	71	34																			
2001-02	Toronto	NHL	21	3	7	10	2	0	0	0	29	10.3	7	5	60.0	20	5	11:03	18	0	1	1	2	0	0	0
	St. John's	AHL	58	27	29	56	30												2	1	1	2	8			
	NHL Totals		27	3	7	10	14	0	0	0	29	10.3		5	60.0	20	5	11:03	18	0	1	1	2	0	0	0

WHL East Second All-Star Team (1995)

Traded to **Nashville** by **Philadelphia** for Matt Henderson, September 27, 1999. Signed as a free agent by **Edmonton**, August 31, 2000. Signed as a free agent by **Toronto**, July 24, 2001.

HEATLEY, Dany

(HEET-lee, DA-nee) **ATL.**

Right wing. Shoots left. 6'3", 210 lbs. Born, Freiburg, West Germany, January 21, 1981. Atlanta's 1st choice, 2nd overall, in 2000 Entry Draft.

Season	Club	League	GP	G	A	Pts	PIM	PP	SH	GW	S	%	+/-	TF	F%	H	SB	Min	GP	G	A	Pts	PIM	PP	SH	GW
1996-97	Calgary Blazers	AMHL	25	30	42	72	26																			
1997-98	Cgy. Buffaloes	AMHL	36	39	42	*91	34												10	10	12	*22	30			
1998-99	Calgary Canucks	AJHL	60	*70	56	*126	91												13	*22	13	*35	6			
99-2000	U. of Wisconsin	WCHA	38	28	28	56	32																			
2000-01	U. of Wisconsin	WCHA	39	24	33	57	74																			
2001-02	Atlanta	NHL	82	26	41	67	56	7	0	4	202	12.9	–19	116	32.8	62	21	19:53								
	NHL Totals		82	26	41	67	56	7	0	4	202	12.9		116	32.8	62	21	19:53								

Air Canada Cup MVP (1997) • AJHL Player of the Year (1999) • Canadian Junior "A" Player of the Year (1999) • WCHA First All-Star Team (2000) • WCHA Rookie of the Year (2000) • NCAA West Second All-American Team (2000) • WCHA Second All-Star Team (2001) • NCAA West First All-American Team (2001) • NHL All-Rookie Team (2002) • Won Calder Memorial Trophy (2002)

HECHT, Jochen

(HEHKHT, YOH-khehn) **BUF.**

Center. Shoots left. 6'1", 200 lbs. Born, Mannheim, West Germany, June 21, 1977. St. Louis' 1st choice, 49th overall, in 1995 Entry Draft.

Season	Club	League	GP	G	A	Pts	PIM	PP	SH	GW	S	%	+/-	TF	F%	H	SB	Min	GP	G	A	Pts	PIM	PP	SH	GW
1993-94	Mannheim Jr.	Ger.-Jr.	28	27	13	40	103																			
1994-95	Adler Mannheim	Germany	43	11	12	23	68												10	5	4	9	12			
1995-96	Adler Mannheim	Germany	44	12	16	28	68												8	3	2	5	6			
1996-97	Adler Mannheim	Germany	46	21	21	42	36												9	3	3	6	4			
1997-98	Adler Mannheim	Germany	44	7	19	26	42												10	1	1	2	14			
	Adler Mannheim	EuroHL	5	0	4	4	8																			
	Germany	Olympics	4	1	0	1	6																			
1998-99	St. Louis	NHL	3	0	0	0	0	0	0	0	4	0.0	–2	19	21.1	1	0	13:16	5	2	0	2	0	0	0	0
	Worcester	AHL	74	21	35	56	48												4	1	1	2	2			
99-2000	St. Louis	NHL	63	13	21	34	28	5	0	1	140	9.3	20	75	49.3	36	6	15:25	7	4	6	10	2	1	0	1
2000-01	St. Louis	NHL	72	19	25	44	48	8	3	1	208	9.1	11	160	43.8	48	18	17:56	15	2	4	6	4	0	0	0
2001-02	Edmonton	NHL	82	16	24	40	60	5	0	3	211	7.6	4	26	53.9	40	24	15:00								
	Germany	Olympics	4	1	1	2	2																			
	NHL Totals		220	48	70	118	136	18	3	5	563	8.5		280	44.6	125	48	16:03	27	8	10	18	6	1	0	1

Traded to **Edmonton** by **St. Louis** with Marty Reasoner and Jan Horacek for Doug Weight and Michel Riesen, July 1, 2001. Traded to **Buffalo** by **Edmonton** for Atlanta's 2nd round choice (previously acquired, Edmonton selected Jeff Deslauriers) in 2002 Entry Draft and Nashville's 2nd round choice (previously acquired, Edmonton selected Jarret Stoll) in 2002 Entry Draft, June 22, 2002.

HEDICAN, Bret

(HEH-dih-kan, BREHT) **CAR.**

Defense. Shoots left. 6'2", 205 lbs. Born, St. Paul, MN, August 10, 1970. St. Louis' 10th choice, 198th overall, in 1988 Entry Draft.

			Regular Season																Playoffs							
Season	Club	League	GP	G	A	Pts	PIM	PP	SH	GW	S	%	+/-	TF	F%	H	SB	Min	GP	G	A	Pts	PIM	PP	SH	GW
1987-88	North St. Paul	Hi-School	23	15	19	34	16																			
1988-89	St. Cloud State	NCAA-3	28	5	3	8	28																			
1989-90	St. Cloud State	NCAA-3	36	4	17	21	37																			
1990-91	St. Cloud State	WCHA	41	21	26	47	26																			
1991-92	Team USA	Nat-Tm	54	1	8	9	59																			
	United States	Olympics	8	0	0	0	4																			
	St. Louis	**NHL**	4	1	0	1	0	0	0	0	1	100.0	1						5	0	0	0	0	0	0	0
1992-93	**St. Louis**	**NHL**	42	0	8	8	30	0	0	0	40	0.0	–2						10	0	0	0	14	0	0	0
	Peoria Rivermen	IHL	19	0	8	8	10																			
1993-94	**St. Louis**	**NHL**	61	0	11	11	64	0	0	0	78	0.0	–8													
	Vancouver	**NHL**	8	0	1	1	0	0	0	0	10	0.0	1						24	1	6	7	16	0	0	0
1994-95	**Vancouver**	**NHL**	45	2	11	13	34	0	0	0	56	3.6	–3						11	0	2	2	6	0	0	0
1995-96	**Vancouver**	**NHL**	77	6	23	29	83	1	0	0	113	5.3	8						6	0	1	1	10	0	0	0
1996-97	**Vancouver**	**NHL**	67	4	15	19	51	2	0	1	93	4.3	–3													
1997-98	**Vancouver**	**NHL**	71	3	24	27	79	1	0	0	84	3.6	3													
1998-99	**Vancouver**	**NHL**	42	2	11	13	34	0	2	0	52	3.8	7	0	0.0	60	32	18:40								
	Florida	**NHL**	25	3	7	10	17	0	0	1	38	7.9	–2	0	0.0	46	42	22:24								
99-2000	**Florida**	**NHL**	76	6	19	25	68	2	0	1	58	10.3	4	0	0.0	128	89	19:36	4	0	0	0	0	0	0	0
2000-01	**Florida**	**NHL**	70	5	15	20	72	4	0	1	104	4.8	–7	0	0.0	144	80	21:49								
2001-02	**Florida**	**NHL**	31	3	7	10	12	0	0	0	46	6.5	–4	0	0.0	64	41	24:27								
	Carolina	**NHL**	26	2	4	6	10	0	0	1	39	5.1	3	0	0.0	35	31	22:56	23	1	4	5	20	0	0	0
	NHL Totals		645	37	156	193	554	10	2	5	812	4.6		0	0.0	477	315	21:10	83	2	13	15	66	0	0	0

WCHA First All-Star Team (1991)

Traded to **Vancouver** by **St. Louis** with Jeff Brown and Nathan Lafayette for Craig Janney, March 21, 1994. Traded to **Florida** by **Vancouver** with Pavel Bure, Brad Ference and Vancouver's 3rd round choice (Robert Fried) in 2000 Entry Draft for Ed Jovanovski, Dave Gagner, Mike Brown, Kevin Weekes and Florida's 1st round choice (Nathan Smith) in 2000 Entry Draft, January 17, 1999. Traded to **Carolina** by **Florida** with Kevyn Adams, Tomas Malec and a conditional 3rd round choice in 2003 Entry Draft for Sandis Ozolinsh and Byron Ritchie, January 16, 2002.

HEINS, Shawn

(HIGHNS, SHAWN) **S.J.**

Defense. Shoots left. 6'4", 210 lbs. Born, Eganville, Ont., December 24, 1973.

Season	Club	League	GP	G	A	Pts	PIM	PP	SH	GW	S	%	+/-	TF	F%	H	SB	Min	GP	G	A	Pts	PIM	PP	SH	GW
1991-92	Peterborough	OHL	49	1	1	2	73												7	0	0	0	5			
1992-93	Peterborough	OHL	5	0	0	0	10																			
	Windsor	OHL	53	7	10	17	107																			
1993-94	Renfrew	NOJHA	32	16	34	50	250																			
1994-95	Renfrew	NOJHA	35	30	49	79	188																			
1995-96	Mobile Mysticks	ECHL	62	7	20	27	152																			
	Cape Breton	AHL	1	0	0	0	0																			
1996-97	Mobile Mysticks	ECHL	56	6	17	23	253												3	0	2	2	2			
	Kansas City	IHL	6	0	0	0	9																			
1997-98	Kansas City	IHL	82	22	28	50	303												11	1	0	1	49			
1998-99	Team Canada	Nat-Tm	36	5	16	21	66																			
	San Jose	**NHL**	5	0	0	0	13	0	0	0	4	0.0	0	0	0.0	4	1	13:38								
	Kentucky	AHL	18	2	2	4	108												12	2	7	9	10			
99-2000	**San Jose**	**NHL**	1	0	0	0	2	0	0	0	1	0.0	–1	0	0.0	1	2	10:57								
	Kentucky	AHL	69	11	52	63	238												9	3	3	6	44			
2000-01	**San Jose**	**NHL**	38	3	4	7	57	2	0	0	45	6.7	2	0	0.0	54	13	10:15	2	0	0	0	0	0	0	0
2001-02	**San Jose**	**NHL**	17	0	2	2	24	0	0	0	20	0.0	1	0	0.0	27	5	9:46								
	NHL Totals		61	3	6	9	96	2	0	0	70	4.3		0	0.0	86	21	10:24	2	0	0	0	0	0	0	0

AHL First All-Star Team (2000)

Signed as a free agent by **San Jose**, January 5, 1997. • Missed majority of 2000-01 season recovering from head injury suffered in game vs. Chicago, February 14, 2001. • Missed majority of 2001-02 season recovering from knee (December 4, 2001 vs. Calgary) and jaw (January 19, 2002 vs. Colorado) injuries.

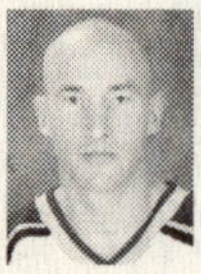

HEINZE, Steve

(HIGHNS, STEEV) **L.A.**

Right wing. Shoots right. 5'11", 202 lbs. Born, Lawrence, MA, January 30, 1970. Boston's 2nd choice, 60th overall, in 1988 Entry Draft.

Season	Club	League	GP	G	A	Pts	PIM	PP	SH	GW	S	%	+/-	TF	F%	H	SB	Min	GP	G	A	Pts	PIM	PP	SH	GW
1986-87	Lawrence School	Hi-School	23	26	24	50																				
1987-88	Lawrence School	Hi-School	23	30	25	55																				
1988-89	Boston College	H-East	36	26	23	49	26																			
1989-90	Boston College	H-East	40	27	36	63	41																			
1990-91	Boston College	H-East	35	21	26	47	35																			
1991-92	Team USA	Nat-Tm	49	18	15	33	38																			
	United States	Olympics	8	1	3	4	8																			
	Boston	**NHL**	14	3	4	7	6	0	0	2	29	10.3	–1						7	0	3	3	17	0	0	0
1992-93	**Boston**	**NHL**	73	18	13	31	24	0	2	4	146	12.3	20						4	1	1	2	2	0	0	0
1993-94	**Boston**	**NHL**	77	10	11	21	32	0	2	1	183	5.5	–2						13	2	3	5	7	0	0	0
1994-95	**Boston**	**NHL**	36	7	9	16	23	0	1	0	70	10.0	0						5	0	0	0	0	0	0	0
1995-96	**Boston**	**NHL**	76	16	12	28	43	0	1	3	129	12.4	–3						5	1	1	2	4	0	1	0
1996-97	**Boston**	**NHL**	30	17	8	25	27	4	2	2	96	17.7	–8													
1997-98	**Boston**	**NHL**	61	26	20	46	54	9	0	6	160	16.3	8						6	0	0	0	6	0	0	0
1998-99	**Boston**	**NHL**	73	22	18	40	30	9	0	3	146	15.1	7	2	0.0	91	13	15:48	12	4	3	7	0	2	0	0
99-2000	**Boston**	**NHL**	75	12	13	25	36	2	0	2	145	8.3	–8	8	12.5	129	10	14:57								
2000-01	**Columbus**	**NHL**	65	22	20	42	38	14	0	3	125	17.6	–19	38	29.0	71	12	18:49								
	Buffalo	**NHL**	14	5	7	12	8	1	0	1	19	26.3	6	0	0.0	20	3	15:24	13	3	4	7	10	3	0	0
2001-02	**Los Angeles**	**NHL**	73	15	16	31	46	8	0	4	123	12.2	–15	4	50.0	61	7	15:47	4	0	0	0	2	0	0	0
	NHL Totals		667	173	151	324	367	47	8	31	1371	12.6		52	26.9	372	45	16:13	69	11	15	26	48	5	1	0

Hockey East First All-Star Team (1990) • NCAA East First All-American Team (1990)

Selected by **Columbus** from **Boston** in Expansion Draft, June 23, 2000. Traded to **Buffalo** by **Columbus** for Buffalo's 3rd round choice (Per Mars) in 2001 Entry Draft, March 13, 2001. Signed as a free agent by **LA Kings**, July 4, 2001.

HEISTEN, Barrett

(HIGH-stehn, BAIR-reht) **DAL.**

Left wing. Shoots left. 6'1", 200 lbs. Born, Anchorage, AK, March 19, 1980. Buffalo's 1st choice, 20th overall, in 1999 Entry Draft.

Season	Club	League	GP	G	A	Pts	PIM	PP	SH	GW	S	%	+/-	TF	F%	H	SB	Min	GP	G	A	Pts	PIM	PP	SH	GW
1996-97	Anchorage	AAHL	39	35	29	64																				
1997-98	Team USA	USDP-18	50	11	26	37	245																			
1998-99	U. of Maine	H-East	34	12	16	28	72																			
99-2000	U. of Maine	H-East	37	13	24	37	86																			
2000-01	Seattle	WHL	58	20	57	77	61												9	2	6	8	20			
2001-02	**NY Rangers**	**NHL**	10	0	0	0	2	0	0	0	7	0.0	–4	7	28.6	7	2	7:36								
	Hartford	AHL	49	9	9	18	60																			
	Utah Grizzlies	AHL	12	5	1	6	14												5	1	0	1	4			
	NHL Totals		10	0	0	0	2	0	0	0	7	0.0		7	28.6	7	2	7:36								

• Left **University of Maine** (H-East) and signed with **Seattle** (WHL) who had selected him 80th overall in 1998 WHL Bantam Draft, August 7, 2000. Signed as a free agent by **NY Rangers**, June 16, 2001. Traded to **Dallas** by **NY Rangers** with Manny Malhotra for Martin Rucinsky and Roman Lyashenko, March 12, 2002.

HEJDUK, Milan

(HAY-dook, MEE-lan) **COL.**

Right wing. Shoots right. 5'11", 185 lbs. Born, Usti-nad-Labem, Czech., February 14, 1976. Quebec's 6th choice, 87th overall, in 1994 Entry Draft.

Season	Club	League	GP	G	A	Pts	PIM	PP	SH	GW	S	%	+/-	TF	F%	H	SB	Min	GP	G	A	Pts	PIM	PP	SH	GW
1993-94	HC Pardubice	Czech	22	6	3	9													10	5	1	6				
1994-95	HC Pardubice	Czech	43	11	13	24	6												6	3	1	4	0			
1995-96	Pardubice	Czech	37	13	7	20																				
1996-97	Pardubice	Czech	51	27	11	38	10												10	6	0	6	27			
1997-98	Pardubice	Czech	48	26	19	45	20												3	0	0	0	2			
	Czech Republic	Olympics	4	0	0	0	2																			
1998-99	**Colorado**	**NHL**	82	14	34	48	26	4	0	5	178	7.9	8	2	50.0	50	30	15:45	16	6	6	12	4	1	0	3
99-2000	**Colorado**	**NHL**	82	36	36	72	16	13	0	9	228	15.8	14	3	100.0	46	40	19:58	17	5	4	9	6	3	0	1
2000-01♦	**Colorado**	**NHL**	80	41	38	79	36	12	1	9	213	19.2	32	3	33.3	88	33	19:52	23	7	*16	23	6	4	0	1

Season	Club	League	Regular Season																Playoffs							
			GP	G	A	Pts	PIM	PP	SH	GW	S	%	+/-	TF	F%	H	SB	Min	GP	G	A	Pts	PIM	PP	SH	GW
2001-02	**Colorado**	**NHL**	**62**	**21**	**23**	**44**	**24**	**7**	**1**	**5**	**139**	**15.1**	**0**	**5**	**40.0**	**54**	**33**	**20:11**	**16**	**3**	**3**	**6**	**4**	**1**	**0**	**0**
	Czech Republic	Olympics	4	1	0	1	0																			
	NHL Totals		**306**	**112**	**131**	**243**	**102**	**36**	**2**	**28**	**758**	**14.8**		**13**	**53.8**	**238**	**136**	**18:51**	**72**	**21**	**29**	**50**	**20**	**9**	**0**	**5**

NHL All-Rookie Team (1999) • Played in NHL All-Star Game (2000, 2001)

Rights transferred to **Colorado** after **Quebec** franchise relocated, June 21, 1995.

HELENIUS, Sami (huh-LEHN-ee-uhs, SA-mee) DAL.

Defense. Shoots left. 6'6", 230 lbs. Born, Helsinki, Finland, January 22, 1974. Calgary's 5th choice, 102nd overall, in 1992 Entry Draft.

Season	Club	League	GP	G	A	Pts	PIM	PP	SH	GW	S	%	+/-	TF	F%	H	SB	Min	GP	G	A	Pts	PIM	PP	SH	GW
1990-91	Jokerit Jr.	Finn-Jr.	2	0	0	0	6																			
1991-92	Jokerit Jr.	Finn-Jr.	14	3	3	6	24																			
	Jokerit Helsinki	Finland-2	13	4	4	8	24																			
1992-93	Jokerit Jr.	Finn-Jr.	13	2	3	5	18																			
	Vantaa HT	Finland-2	21	3	2	5	50																			
	Jokerit Helsinki	Finland	1	0	0	0	0																			
1993-94	Reipas Lahti Jr.	Finn-Jr.	11	3	4	7	48																			
	Reipas Lahti	Finland	37	2	3	5	46																			
1994-95	Saint John	AHL	69	2	5	7	217																			
1995-96	Saint John	AHL	68	0	3	3	231												10	0	0	0	9			
1996-97	**Calgary**	**NHL**	**3**	**0**	**1**	**1**	**0**	**0**	**0**	**0**	**1**	**0.0**	**1**													
	Saint John	AHL	72	5	10	15	218												2	0	0	0	0			
1997-98	Saint John	AHL	63	1	2	3	185																			
	Las Vegas	IHL	10	0	1	1	19												4	0	0	0	25			
1998-99	**Calgary**	**NHL**	**4**	**0**	**0**	**0**	**8**	**0**	**0**	**0**	**1**	**0.0**	**–2**	**0**	**0.0**	**6**	**6**	**10:16**								
	Las Vegas	IHL	42	2	3	5	193																			
	Tampa Bay	**NHL**	**4**	**1**	**0**	**1**	**15**	**0**	**1**	**0**	**3**	**33.3**	**–3**	**0**	**0.0**	**6**	**3**	**16:53**								
	Chicago Wolves	IHL	4	0	0	0	11																			
	Hershey Bears	AHL	8	0	0	0	29												5	0	0	0	16			
99-2000	**Colorado**	**NHL**	**33**	**0**	**0**	**0**	**46**	**0**	**0**	**0**	**6**	**0.0**	**–5**	**0**	**0.0**	**34**	**19**	**7:04**								
	Hershey Bears	AHL	12	0	1	1	31												9	0	0	0	40			
2000-01	**Dallas**	**NHL**	**57**	**1**	**2**	**3**	**99**	**0**	**0**	**0**	**18**	**5.6**	**1**	**0**	**0.0**	**84**	**25**	**10:39**	**1**	**0**	**0**	**0**	**0**	**0**	**0**	**0**
2001-02	**Dallas**	**NHL**	**39**	**0**	**0**	**0**	**58**	**0**	**0**	**0**	**18**	**0.0**	**–4**	**0**	**0.0**	**54**	**22**	**9:41**								
	NHL Totals		**140**	**2**	**3**	**5**	**226**	**0**	**1**	**0**	**47**	**4.3**		**0**	**0.0**	**184**	**75**	**9:41**	**1**	**0**	**0**	**0**	**0**	**0**	**0**	**0**

Traded to **Tampa Bay** by **Calgary** for future considerations, January 29, 1999. Traded to **Colorado** by **Tampa Bay** for future considerations, March 23, 1999. Signed as a free agent by **Dallas**, July 12, 2000. Signed as a free agent by **Jokerit Helsinki** (Finland) with Dallas retaining NHL rights, May 15, 2002.

HELMER, Bryan (HEHL-muhr, BRIGH-uhn) VAN.

Defense. Shoots right. 6'1", 200 lbs. Born, Sault Ste. Marie, Ont., July 15, 1972.

Season	Club	League	GP	G	A	Pts	PIM	PP	SH	GW	S	%	+/-	TF	F%	H	SB	Min	GP	G	A	Pts	PIM	PP	SH	GW
1989-90	Wellington Dukes	MTJHL	44	4	20	24	204																			
	Belleville Bulls	OHL	6	0	1	1	0																			
1990-91	Wellington Dukes	MTJHL	50	11	14	25	109																			
1991-92	Wellington Dukes	MTJHL	42	17	31	48	66												3	2	1	3	0			
1992-93	Wellington Dukes	MTJHL	48	21	54	75	84												9	4	8	12	22			
1993-94	Albany	AHL	65	4	19	23	79												5	0	0	0	9			
1994-95	Albany	AHL	77	7	36	43	101												7	1	0	1	0			
1995-96	Albany	AHL	80	14	30	44	107												4	2	0	2	6			
1996-97	Albany	AHL	77	12	27	39	113												16	1	7	8	10			
1997-98	Albany	AHL	80	14	49	63	101												13	4	9	13	18			
1998-99	**Phoenix**	**NHL**	**11**	**0**	**0**	**0**	**23**	**0**	**0**	**0**	**11**	**0.0**	**2**	**0**	**0.0**	**1**	**2**	**7:43**								
	Las Vegas	IHL	8	1	3	4	28																			
	St. Louis	**NHL**	**29**	**0**	**4**	**4**	**19**	**0**	**0**	**0**	**38**	**0.0**	**3**	**1**	**100.0**	**34**	**28**	**19:08**								
	Worcester	AHL	16	7	8	15	18												4	0	0	0	12			
99-2000	**St. Louis**	**NHL**	**15**	**1**	**1**	**2**	**10**	**1**	**0**	**1**	**19**	**5.3**	**–3**	**0**	**0.0**	**16**	**8**	**16:15**								
	Worcester	AHL	54	10	25	35	124												9	1	4	5	10			
2000-01	**Vancouver**	**NHL**	**20**	**2**	**4**	**6**	**18**	**0**	**0**	**0**	**28**	**7.1**	**0**	**0**	**0.0**	**16**	**21**	**16:51**								
	Kansas City	IHL	42	4	15	19	76																			
2001-02	**Vancouver**	**NHL**	**40**	**5**	**5**	**10**	**53**	**2**	**0**	**1**	**43**	**11.6**	**10**	**0**	**0.0**	**45**	**20**	**12:04**	**6**	**0**	**0**	**0**	**0**	**0**	**0**	**0**
	Manitoba Moose	AHL	34	6	18	24	69																			
	NHL Totals		**115**	**8**	**14**	**22**	**123**	**3**	**0**	**2**	**139**	**5.8**		**1**	**100.0**	**112**	**79**	**14:49**	**6**	**0**	**0**	**0**	**0**	**0**	**0**	**0**

AHL First All-Star Team (1998)

Signed as a free agent by **New Jersey**, July 10, 1994. Signed as a free agent by **Phoenix**, July 17, 1998. Claimed on waivers by **St. Louis** from **Phoenix**, December 19, 1998. Signed as a free agent by **Vancouver**, August 21, 2000.

HENDERSON, Jay (HEHN-duhr-SOHN, JAY) BOS.

Left wing. Shoots left. 5'11", 190 lbs. Born, Edmonton, Alta., September 17, 1978. Boston's 12th choice, 246th overall, in 1997 Entry Draft.

Season	Club	League	GP	G	A	Pts	PIM	PP	SH	GW	S	%	+/-	TF	F%	H	SB	Min	GP	G	A	Pts	PIM	PP	SH	GW
1993-94	Sherwood Park	AMBHL	31	12	21	33	36																			
1994-95	Red Deer Rebels	WHL	54	3	9	12	80																			
1995-96	Red Deer Rebels	WHL	71	15	13	28	139												10	1	1	2	11			
1996-97	Edmonton Ice	WHL	66	28	32	60	127																			
1997-98	Edmonton Ice	WHL	72	49	45	94	130																			
	Providence	AHL	11	3	1	4	11																			
1998-99	**Boston**	**NHL**	**4**	**0**	**0**	**0**	**2**	**0**	**0**	**0**	**4**	**0.0**	**–1**	**0**	**0.0**	**1**	**1**	**5:39**								
	Providence	AHL	55	7	9	16	172												2	0	0	0	2			
99-2000	**Boston**	**NHL**	**16**	**1**	**3**	**4**	**9**	**0**	**0**	**0**	**18**	**5.6**	**1**	**2**	**0.0**	**16**	**2**	**5:22**								
	Providence	AHL	60	18	27	45	200												14	1	2	3	16			
2000-01	**Boston**	**NHL**	**13**	**0**	**0**	**0**	**26**	**0**	**0**	**0**	**12**	**0.0**	**–1**	**3**	**100.0**	**9**	**2**	**6:58**								
	Providence	AHL	41	9	7	16	121												1	0	0	0	2			
2001-02	**Boston**	**NHL**	DID NOT PLAY – INJURED																							
	NHL Totals		**33**	**1**	**3**	**4**	**37**	**0**	**0**	**0**	**34**	**2.9**		**5**	**60.0**	**26**	**5**	**6:02**								

• Missed entire 2001-02 season recovering from knee injury originally suffered in pre-season game vs. Detroit, September 21, 2001.

HENDERSON, Matt (HEHN-duhr-SOHN, MAT)

Right wing. Shoots left. 6'1", 200 lbs. Born, White Bear Lake, MN, June 22, 1974.

Season	Club	League	GP	G	A	Pts	PIM	PP	SH	GW	S	%	+/-	TF	F%	H	SB	Min	GP	G	A	Pts	PIM	PP	SH	GW
1993-94	St. Paul Vulcans	USHL	48	27	24	51																				
1994-95	North Dakota	WCHA	19	1	3	4	16																			
1995-96	North Dakota	WCHA	36	9	10	19	34												2	0	1	1	0			
1996-97	North Dakota	WCHA	42	14	17	31	71												7	5	4	9	10			
1997-98	North Dakota	WCHA	38	24	14	38	74												5	2	2	4	4			
1998-99	**Nashville**	**NHL**	**2**	**0**	**0**	**0**	**2**	**0**	**0**	**0**	**0**	**0.0**	**–1**	**0**	**0.0**	**4**	**0**	**6:23**								
	Milwaukee	IHL	77	19	19	38	117												2	0	0	0	0			
99-2000	Philadelphia	AHL	51	4	8	12	37												5	0	0	0	4			
	Trenton Titans	ECHL	16	2	4	6	47																			
2000-01	Norfolk Admirals	AHL	78	14	24	38	80												9	1	1	2	16			
2001-02	**Chicago**	**NHL**	**4**	**0**	**1**	**1**	**0**	**0**	**0**	**0**	**3**	**0.0**	**–1**	**0**	**0.0**	**8**	**0**	**7:08**								
	Norfolk Admirals	AHL	74	21	28	49	90												4	0	0	0	6			
	NHL Totals		**6**	**0**	**1**	**1**	**2**	**0**	**0**	**0**	**3**	**0.0**		**0**	**0.0**	**12**	**0**	**6:53**								

NCAA Championship All-Tournament Team (1997) • NCAA Championship Tournament MVP (1997)

Signed as a free agent by **Nashville**, July 14, 1998. Traded to **Philadelphia** by **Nashville** for Paul Healey, September 27, 1999. Signed as a free agent by **Chicago**, September 20, 2001.

HENDRICKSON, Darby
(HEHN-drihk-SOHN, DAHR-bee) **MIN.**

Center. Shoots left. 6'1", 195 lbs. Born, Richfield, MN, August 28, 1972. Toronto's 3rd choice, 73rd overall, in 1990 Entry Draft.

			Regular Season																Playoffs							
Season	Club	League	GP	G	A	Pts	PIM	PP	SH	GW	S	%	+/-	TF	F%	H	SB	Min	GP	G	A	Pts	PIM	PP	SH	GW
1987-88	Richfield High	Hi-School	22	12	9	21	10																			
1988-89	Richfield High	Hi-School	22	22	20	42	12																			
1989-90	Richfield High	Hi-School	24	23	27	50	49																			
1990-91	Richfield High	Hi-School	27	32	29	61	....																			
1991-92	U. of Minnesota	WCHA	41	25	28	53	61																			
1992-93	U. of Minnesota	WCHA	31	12	15	27	35																			
1993-94	Team USA	Nat-Tm	59	12	16	28	30																			
	United States	Olympics	8	0	0	0	6																			
	Toronto	**NHL**																	2	0	0	0	0	0	0	0
	St. John's	AHL	6	4	1	5	4												3	1	1	2	0			
1994-95	St. John's	AHL	59	16	20	36	48																			
	Toronto	**NHL**	8	0	1	1	4	0	0	0	4	0.0	0													
1995-96	**Toronto**	**NHL**	46	6	6	12	47	0	0	0	43	14.0	-2													
	NY Islanders	**NHL**	16	1	4	5	33	0	0	1	30	3.3	-6													
1996-97	**Toronto**	**NHL**	64	11	6	17	47	0	1	0	105	10.5	-20													
	St. John's	AHL	12	5	4	9	21																			
1997-98	**Toronto**	**NHL**	80	8	4	12	67	0	0	0	115	7.0	-20													
1998-99	**Toronto**	**NHL**	35	2	3	5	30	0	0	0	34	5.9	-4	278	46.0	35	8	10:16								
	Vancouver	**NHL**	27	2	2	4	22	1	0	0	36	5.6	-15	427	46.8	24	23	17:15								
99-2000	**Vancouver**	**NHL**	40	5	4	9	14	0	1	1	39	12.8	-3	407	46.2	26	27	11:32								
	Syracuse Crunch	AHL	20	5	8	13	16																			
2000-01	**Minnesota**	**NHL**	72	18	11	29	36	3	1	1	114	15.8	1	1119	45.2	44	44	15:50								
2001-02	**Minnesota**	**NHL**	68	9	15	24	50	2	2	1	79	11.4	-22	1206	47.7	25	52	16:43								
	NHL Totals		456	62	56	118	350	6	5	4	599	10.4		3437	46.5	154	154	14:44	2	0	0	0	0	0	0	0

Minnesota High School Player of the Year (1991) • WCHA Rookie of the Year (1992)

Traded to **NY Islanders** by **Toronto** with Sean Haggerty, Kenny Jonsson and Toronto's 1st round choice (Roberto Luongo) in 1997 Entry Draft for Wendel Clark, Mathieu Schneider and D.J. Smith, March 13, 1996. Traded to **Toronto** by **NY Islanders** for Toronto's 5th round choice (Jiri Dopita) in 1998 Entry Draft, October 11, 1996. Traded to **Vancouver** by **Toronto** for Chris McAllister, February 16, 1999. Selected by **Minnesota** from **Vancouver** in Expansion Draft, June 23, 2000.

HENTUNEN, Jukka
(HEHN-too-nehn, YOO-kuh) **NSH.**

Right wing. Shoots right. 5'10", 194 lbs. Born, Joroinen, Finland, May 3, 1974. Calgary's 7th choice, 176th overall, in 2000 Entry Draft.

Season	Club	League	GP	G	A	Pts	PIM	PP	SH	GW	S	%	+/-	TF	F%	H	SB	Min	GP	G	A	Pts	PIM	PP	SH	GW
1993-94	Kiekko Warkaus	Finland-3	16	7	6	13	10												8	4	1	5	2			
1994-95	Kiekko Warkaus	Finland-3	29	23	23	46	28																			
1995-96	Diskos Jyvaskyla	Finland-2	43	23	18	41	14												10	5	6	11	4			
1996-97	Hermes Kokkola	Finland-2	35	10	13	23	43												3	1	0	1	0			
1997-98	Hermes Kokkola	Finland-2	49	19	16	35	36												3	3	3	6	0			
1998-99	Hermes Kokkola	Finland-2	1	0	0	0	0																			
	HPK Hameenlinna	Finland	41	13	21	34	32												8	1	4	5	12			
99-2000	HPK Hameenlinna	Finland	53	17	28	45	76												8	4	2	6	12			
2000-01	Jokerit Helsinki	Finland	56	27	28	55	24												5	1	0	1	4			
2001-02	**Calgary**	**NHL**	28	2	3	5	4	1	0	0	38	5.3	-9	0	0.0	41	6	11:09								
	Saint John	AHL	9	3	3	6	0																			
	Nashville	**NHL**	10	2	2	4	0	0	0	1	12	16.7	0	0	0.0	10	2	12:46								
	NHL Totals		38	4	5	9	4	1	0	1	50	8.0		0	0.0	51	8	11:35								

Traded to **Nashville** by **Calgary** for a conditional choice in 2003 Entry Draft, March 17, 2002. Signed as a free agent by

HERPERGER, Chris
(HUHR-puhr-GEHR, KRIHS) **ATL.**

Left wing. Shoots left. 6', 190 lbs. Born, Esterhazy, Sask., February 24, 1974. Philadelphia's 9th choice, 223rd overall, in 1992 Entry Draft.

Season	Club	League	GP	G	A	Pts	PIM	PP	SH	GW	S	%	+/-	TF	F%	H	SB	Min	GP	G	A	Pts	PIM	PP	SH	GW
1990-91	Swift Current	SMHL	STATISTICS NOT AVAILABLE																							
	Swift Current	WHL	10	0	1	1	5																			
1991-92	Swift Current	WHL	72	14	19	33	44												8	0	1	1	9			
1992-93	Swift Current	WHL	20	9	7	16	31																			
	Seattle	WHL	46	20	11	31	30												5	1	1	2	6			
1993-94	Seattle	WHL	71	44	51	95	110												9	12	10	22	12			
1994-95	Seattle	WHL	59	49	52	101	106												4	4	0	4	6			
	Hershey Bears	AHL	4	0	0	0	0																			
1995-96	Hershey Bears	AHL	46	8	12	20	36																			
	Baltimore	AHL	21	2	3	5	17												9	2	3	5	6			
1996-97	Baltimore	AHL	67	19	22	41	88												3	0	0	0	0			
1997-98	Team Canada	Nat-Tm	63	20	30	50	102																			
1998-99	Indianapolis Ice	IHL	79	19	29	48	81												7	0	4	4	4			
99-2000	**Chicago**	**NHL**	9	0	0	0	5	0	0	0	2	0.0	-2	52	55.8	7	1	7:39								
	Cleveland	IHL	73	22	26	48	122												9	3	3	6	8			
2000-01	**Chicago**	**NHL**	61	10	15	25	20	0	1	3	76	13.2	0	678	56.2	28	21	13:07								
	Norfolk Admirals	AHL	9	1	4	5	9																			
2001-02	**Ottawa**	**NHL**	72	4	9	13	43	0	0	0	81	4.9	4	769	49.3	41	37	11:19								
	NHL Totals		142	14	24	38	68	0	1	3	159	8.8		1499	52.6	76	59	11:52								

WHL West Second All-Star Team (1995)

Traded to **Anaheim** by **Philadelphia** with Winnipeg/Phoenix's 7th round choice (previously acquired, Anaheim selected Tony Mohagen) in 1997 Entry Draft for Bob Corkum, February 6, 1996. Signed as a free agent by **Chicago**, September 2, 1998. Signed as a free agent by **Ottawa**, July 13, 2001. Signed as a free agent by **Atlanta**, August 1, 2002.

HERR, Matt
(HUHR, MAT) **BOS.**

Center. Shoots left. 6'2", 204 lbs. Born, Hackensack, NJ, May 26, 1976. Washington's 4th choice, 93rd overall, in 1994 Entry Draft.

Season	Club	League	GP	G	A	Pts	PIM	PP	SH	GW	S	%	+/-	TF	F%	H	SB	Min	GP	G	A	Pts	PIM	PP	SH	GW
1990-91	Hotchkiss High	Hi-School	26	9	5	14	....																			
1991-92	Hotchkiss High	Hi-School	25	17	16	33	....																			
1992-93	Hotchkiss High	Hi-School	24	48	30	78	....																			
1993-94	Hotchkiss High	Hi-School	24	28	19	47	....																			
1994-95	U. of Michigan	CCHA	37	11	8	19	51												3	1	0	1	4			
1995-96	U. of Michigan	CCHA	40	18	13	31	55												7	0	4	4	0			
1996-97	U. of Michigan	CCHA	43	29	23	52	67												6	2	2	4	8			
1997-98	U. of Michigan	CCHA	31	14	17	31	62																			
1998-99	**Washington**	**NHL**	30	2	2	4	8	1	0	0	40	5.0	-7	176	52.8	42	10	11:05								
	Portland Pirates	AHL	46	15	14	29	29																			
99-2000	Portland Pirates	AHL	77	22	21	43	51												4	1	1	2	4			
2000-01	**Washington**	**NHL**	22	2	3	5	17	0	0	1	20	10.0	3	2	100.0	30	1	8:09								
	Portland Pirates	AHL	40	21	13	34	58																			
	Philadelphia	AHL	11	2	4	6	18												9	2	1	3	8			
2001-02	**Florida**	**NHL**	3	0	0	0	0	0	0	0	1	0.0	-2	15	53.3	1	1	6:27								
	Hershey Bears	AHL	61	18	16	34	68												7	1	2	3	15			
	NHL Totals		55	4	5	9	25	1	0	1	61	6.6		193	53.4	73	12	9:39								

Traded to **Philadelphia** by **Washington** for Dean Melanson, March 13, 2001. Signed as a free agent by **Florida**, August 21, 2001. Signed as a free agent by **Boston**, July 18, 2002.

HEWARD, Jamie
(HEW-uhrd, JAY-mee)

Defense. Shoots right. 6'2", 207 lbs. Born, Regina, Sask., March 30, 1971. Pittsburgh's 1st choice, 16th overall, in 1989 Entry Draft.

Season	Club	League	GP	G	A	Pts	PIM	PP	SH	GW	S	%	+/-	TF	F%	H	SB	Min	GP	G	A	Pts	PIM	PP	SH	GW
1987-88	Regina Pats	WHL	68	10	17	27	17												4	1	1	2	2			
1988-89	Regina Pats	WHL	52	31	28	59	29																			
1989-90	Regina Pats	WHL	72	14	44	58	42												11	2	2	4	10			
1990-91	Regina Pats	WHL	71	23	61	84	41												8	2	9	11	6			
1991-92	Muskegon	IHL	54	6	21	27	37												14	1	4	5	4			
1992-93	Cleveland	IHL	58	9	18	27	64																			
1993-94	Cleveland	IHL	73	8	16	24	72																			
1994-95	Team Canada	Nat-Tm	51	11	35	46	32																			

Season	Club	League	GP	G	A	Pts	PIM	PP	SH	GW	S	%	+/-	TF	F%	H	SB	Min	GP	G	A	Pts	PIM	PP	SH	GW
			Regular Season																Playoffs							
1995-96	**Toronto**	**NHL**	**5**	**0**	**0**	**0**	**0**	**0**	**0**	**0**	**8**	**0.0**	**–1**													
	St. John's	AHL	73	22	34	56	33												3	1	1	2	6			
1996-97	**Toronto**	**NHL**	**20**	**1**	**4**	**5**	**6**	**0**	**0**	**0**	**23**	**4.3**	**–6**													
	St. John's	AHL	27	8	19	27	26												9	1	3	4	6			
1997-98	Philadelphia	AHL	72	17	48	65	54												20	3	16	19	10			
1998-99	**Nashville**	**NHL**	**63**	**6**	**12**	**18**	**44**	**4**	**0**	**1**	**124**	**4.8**	**–24**	**0**	**0.0**	**80**	**35**	**16:12**								
99-2000	**NY Islanders**	**NHL**	**54**	**6**	**11**	**17**	**26**	**2**	**0**	**1**	**92**	**6.5**	**–9**	**0**	**0.0**	**54**	**74**	**19:58**								
2000-01	**Columbus**	**NHL**	**69**	**11**	**16**	**27**	**33**	**9**	**0**	**1**	**108**	**10.2**	**3**	**0**	**0.0**	**53**	**38**	**14:21**								
2001-02	**Columbus**	**NHL**	**28**	**1**	**2**	**3**	**7**	**0**	**0**	**0**	**38**	**2.6**	**–9**	**1**	**100.0**	**23**	**16**	**14:04**								
	Syracuse Crunch	AHL	14	3	10	13	6												10	0	4	4	6			
	NHL Totals		**239**	**25**	**45**	**70**	**116**	**15**	**0**	**3**	**393**	**6.4**		**1**	**100.0**	**210**	**163**	**16:17**								

WHL East First All-Star Team (1991) • AHL First All-Star Team (1996, 1998) • Won Eddie Shore Award (Top Defenseman - AHL) (1998)

Signed as a free agent by **Toronto**, May 4, 1995. Signed as a free agent by **Philadelphia**, July 31, 1997. Signed as a free agent by **Nashville**, August 10, 1998. Signed as a free agent by **NY Islanders**, July 27, 1999. Claimed on waivers by **Columbus** from **NY Islanders**, May 26, 2000. Signed as a free agent by **Geneve-Servette** (Swiss-2), April 17, 2002.

HIGGINS, Matt

(HIH-gihns, MAT)

Center. Shoots left. 6'2", 190 lbs. Born, Calgary, Alta., October 29, 1977. Montreal's 1st choice, 18th overall, in 1996 Entry Draft.

Season	Club	League	GP	G	A	Pts	PIM	PP	SH	GW	S	%	+/-	TF	F%	H	SB	Min	GP	G	A	Pts	PIM	PP	SH	GW
1992-93	Vernon	BCAHA	70	53	76	129	54																			
1993-94	Moose Jaw	WHL	64	6	10	16	10																			
1994-95	Moose Jaw	WHL	72	36	34	70	26												10	1	2	3	2			
1995-96	Moose Jaw	WHL	67	30	33	63	43																			
1996-97	Moose Jaw	WHL	71	33	57	90	51												12	3	5	8	2			
1997-98	**Montreal**	**NHL**	**1**	**0**	**0**	**0**	**0**	**0**	**0**	**0**	**1**	**0.0**	**–1**													
	Fredericton	AHL	50	5	22	27	12												4	1	2	3	2			
1998-99	**Montreal**	**NHL**	**25**	**1**	**0**	**1**	**0**	**0**	**0**	**0**	**12**	**8.3**	**–2**	**108**	**45.4**	**9**	**4**	**5:41**								
	Fredericton	AHL	11	3	4	7	6												5	0	2	2	0			
99-2000	**Montreal**	**NHL**	**25**	**0**	**2**	**2**	**4**	**0**	**0**	**0**	**9**	**0.0**	**–6**	**145**	**48.3**	**12**	**13**	**7:57**								
	Quebec	AHL	29	1	15	16	21																			
2000-01	**Montreal**	**NHL**	**6**	**0**	**0**	**0**	**2**	**0**	**0**	**0**	**3**	**0.0**	**–2**	**40**	**47.5**	**5**	**3**	**9:27**								
	Quebec	AHL	66	10	18	28	18												8	0	1	1	4			
2001-02	Bridgeport	AHL	43	13	19	32	24												15	1	0	1	6			
	NHL Totals		**57**	**1**	**2**	**3**	**6**	**0**	**0**	**0**	**25**	**4.0**		**293**	**47.1**	**26**	**20**	**7:06**								

Signed as a free agent by **Bridgeport** (AHL), December 26, 2001.

HILBERT, Andy

(HIHL-buhrt, AN-dee) **BOS.**

Center. Shoots left. 5'11", 190 lbs. Born, Howell, MI, February 6, 1981. Boston's 3rd choice, 37th overall, in 2000 Entry Draft.

Season	Club	League	GP	G	A	Pts	PIM	PP	SH	GW	S	%	+/-	TF	F%	H	SB	Min	GP	G	A	Pts	PIM	PP	SH	GW
1997-98	Team USA	USDP-18	75	34	30	64	148																			
1998-99	Team USA	USDP-18	46	23	35	58	140																			
99-2000	U. of Michigan	CCHA	35	17	15	32	39																			
2000-01	U. of Michigan	CCHA	42	26	38	64	72																			
2001-02	**Boston**	**NHL**	**6**	**1**	**0**	**1**	**2**	**0**	**0**	**0**	**11**	**9.1**	**–2**	**4**	**50.0**	**4**	**2**	**11:34**								
	Providence	AHL	72	26	27	53	74												2	0	0	0	2			
	NHL Totals		**6**	**1**	**0**	**1**	**2**	**0**	**0**	**0**	**11**	**9.1**		**4**	**50.0**	**4**	**2**	**11:34**								

CCHA First All-Star Team (2001) • NCAA West First All-American Team (2001) • AHL All-Rookie Team (2002)

HILL, Sean

(HIHL, SHAWN) **CAR.**

Defense. Shoots right. 6', 203 lbs. Born, Duluth, MN, February 14, 1970. Montreal's 9th choice, 167th overall, in 1988 Entry Draft.

Season	Club	League	GP	G	A	Pts	PIM	PP	SH	GW	S	%	+/-	TF	F%	H	SB	Min	GP	G	A	Pts	PIM	PP	SH	GW
1986-87	Lakefield Chiefs	OJHL-C	3	1	1	2	14																			
1987-88	East Duluth	Hi-School	24	10	17	27	....																			
1988-89	U. of Wisconsin	WCHA	45	2	23	25	69																			
1989-90	U. of Wisconsin	WCHA	42	14	39	53	78																			
1990-91	U. of Wisconsin	WCHA	37	19	32	51	122																			
	Montreal	**NHL**																	**1**	**0**	**0**	**0**	**0**	**0**	**0**	**0**
	Fredericton	AHL																	3	0	2	2	2			
1991-92	Fredericton	AHL	42	7	20	27	65												7	1	3	4	6			
	Team USA	Nat-Tm	12	4	3	7	16																			
	United States	Olympics	8	2	0	2	6																			
	Montreal	**NHL**																	**4**	**1**	**0**	**1**	**2**	**0**	**0**	**0**
1992-93♦	**Montreal**	**NHL**	**31**	**2**	**6**	**8**	**54**	**1**	**0**	**1**	**37**	**5.4**	**–5**						**3**	**0**	**0**	**0**	**4**	**0**	**0**	**0**
	Fredericton	AHL	6	1	3	4	10																			
1993-94	**Anaheim**	**NHL**	**68**	**7**	**20**	**27**	**78**	**2**	**1**	**1**	**165**	**4.2**	**–12**													
1994-95	**Ottawa**	**NHL**	**45**	**1**	**14**	**15**	**30**	**1**	**0**	**0**	**107**	**0.9**	**–11**													
1995-96	**Ottawa**	**NHL**	**80**	**7**	**14**	**21**	**94**	**2**	**0**	**2**	**157**	**4.5**	**–26**													
1996-97	**Ottawa**	**NHL**	**5**	**0**	**0**	**0**	**4**	**0**	**0**	**0**	**9**	**0.0**	**1**													
1997-98	**Ottawa**	**NHL**	**13**	**1**	**1**	**2**	**6**	**0**	**0**	**0**	**16**	**6.3**	**–3**													
	Carolina	**NHL**	**42**	**0**	**5**	**5**	**48**	**0**	**0**	**0**	**37**	**0.0**	**–2**													
1998-99	**Carolina**	**NHL**	**54**	**0**	**10**	**10**	**48**	**0**	**0**	**0**	**44**	**0.0**	**9**	**0**	**0.0**	**194**	**79**	**19:02**								
99-2000	**Carolina**	**NHL**	**62**	**13**	**31**	**44**	**59**	**8**	**0**	**2**	**150**	**8.7**	**3**	**1**	**0.0**	**246**	**94**	**24:31**								
2000-01	**St. Louis**	**NHL**	**48**	**1**	**10**	**11**	**51**	**0**	**0**	**0**	**47**	**2.1**	**5**	**1**	**0.0**	**111**	**49**	**17:23**	**15**	**0**	**1**	**1**	**12**	**0**	**0**	**0**
2001-02	**St. Louis**	**NHL**	**23**	**0**	**3**	**3**	**28**	**0**	**0**	**0**	**29**	**0.0**	**1**	**0**	**0.0**	**67**	**24**	**15:56**								
	Carolina	**NHL**	**49**	**7**	**23**	**30**	**61**	**4**	**0**	**2**	**116**	**6.0**	**–1**	**1**	**0.0**	**157**	**71**	**23:58**	**23**	**4**	**4**	**8**	**20**	**4**	**0**	**1**
	NHL Totals		**520**	**39**	**137**	**176**	**561**	**17**	**1**	**8**	**914**	**4.3**		**3**	**0.0**	**775**	**317**	**20:52**	**46**	**5**	**5**	**10**	**38**	**4**	**0**	**1**

WCHA Second All-Star Team (1990, 1991) • NCAA West Second All-American Team (1991)

Claimed by **Anaheim** from **Montreal** in Expansion Draft, June 24, 1993. Traded to **Ottawa** by **Anaheim** with Anaheim's 9th round choice (Frederic Cassivi) in 1994 Entry Draft for Ottawa's 3rd round choice (later traded to Tampa Bay - Tampa Bay selected Vadim Epanchintsev) in 1994 Entry Draft, June 29, 1994. • Missed remainder of 1996-97 season recovering from knee injury suffered in game vs. New Jersey, October 18, 1996. Traded to **Carolina** by **Ottawa** for Chris Murray, November 18, 1997. Signed as a free agent by **St. Louis**, July 1, 2000. Traded to **Carolina** by **St. Louis** for Steve Halko and Carolina's 4th round choice (later traded to Atlanta - Atlanta selected Lane Manson) in 2002 Entry Draft, December 5, 2001.

HINOTE, Dan

(HIGH-noht, DAN) **COL.**

Right wing. Shoots right. 6', 190 lbs. Born, Leesburg, FL, January 30, 1977. Colorado's 9th choice, 167th overall, in 1996 Entry Draft.

Season	Club	League	GP	G	A	Pts	PIM	PP	SH	GW	S	%	+/-	TF	F%	H	SB	Min	GP	G	A	Pts	PIM	PP	SH	GW
1993-94	Elk River Elks	Hi-School	STATISTICS NOT AVAILABLE																							
1994-95	Army	NCAA	33	20	24	44	20																			
1995-96	Army	NCAA	34	21	24	45	22																			
1996-97	Oshawa Generals	OHL	60	15	13	28	58												18	4	5	9	8			
1997-98	Oshawa Generals	OHL	35	12	15	27	39												5	2	2	4	7			
	Hershey Bears	AHL	24	1	4	5	25																			
1998-99	Hershey Bears	AHL	65	4	16	20	95												5	3	1	4	6			
99-2000	**Colorado**	**NHL**	**27**	**1**	**3**	**4**	**10**	**0**	**0**	**0**	**14**	**7.1**	**0**	**132**	**51.5**	**47**	**8**	**7:51**								
	Hershey Bears	AHL	55	28	31	59	96												14	4	5	9	19			
2000-01♦	**Colorado**	**NHL**	**76**	**5**	**10**	**15**	**51**	**1**	**0**	**1**	**69**	**7.2**	**1**	**506**	**49.8**	**199**	**38**	**10:21**	**23**	**2**	**4**	**6**	**21**	**0**	**0**	**0**
2001-02	**Colorado**	**NHL**	**58**	**6**	**6**	**12**	**39**	**0**	**1**	**3**	**75**	**8.0**	**8**	**267**	**51.3**	**182**	**19**	**12:27**	**19**	**1**	**2**	**3**	**9**	**0**	**0**	**0**
	NHL Totals		**161**	**12**	**19**	**31**	**100**	**1**	**1**	**4**	**158**	**7.6**		**905**	**50.5**	**428**	**65**	**10:41**	**42**	**3**	**6**	**9**	**30**	**0**	**0**	**0**

HLAVAC, Jan

(huh-LAH-vahch, YAHN) **VAN.**

Left wing. Shoots left. 6', 185 lbs. Born, Prague, Czech., September 20, 1976. NY Islanders' 2nd choice, 28th overall, in 1995 Entry Draft.

Season	Club	League	GP	G	A	Pts	PIM	PP	SH	GW	S	%	+/-	TF	F%	H	SB	Min	GP	G	A	Pts	PIM	PP	SH	GW
1993-94	Sparta Praha Jr.	Czech-Jr.	27	12	15	27	....																			
	HC Sparta Praha	Czech	9	1	1	2																				
1994-95	HC Sparta Praha	Czech	38	7	6	13	18												5	0	2	2	0			
1995-96	HC Sparta Praha	Czech	34	8	5	13	....												12	1	2	3	....			
1996-97	HC Sparta Praha	Czech	38	8	13	21	24												10	5	2	7	2			
	HC Sparta Praha	EuroHL	3	4	0	4	6																			
1997-98	HC Sparta Praha	Czech	48	17	30	47	40												5	1	0	1	2			
	HC Sparta Praha	EuroHL	5	0	3	3	4																			

			Regular Season																Playoffs							
Season	Club	League	GP	G	A	Pts	PIM	PP	SH	GW	S	%	+/-	TF	F%	H	SB	Min	GP	G	A	Pts	PIM	PP	SH	GW
1998-99	HC Sparta Praha	Czech	49	*33	20	53	52												6	1	3	4				
	HC Sparta Praha	EuroHL	5	4	2	6	0												1	1	1	2	0			
99-2000	**NY Rangers**	**NHL**	**67**	**19**	**23**	**42**	**16**	**6**	**0**	**2**	**134**	**14.2**	**3**	**6**	**33.3**	**39**	**17**	**15:09**								
	Hartford	AHL	3	1	0	1	0																			
2000-01	**NY Rangers**	**NHL**	**79**	**28**	**36**	**64**	**20**	**5**	**0**	**6**	**195**	**14.4**	**3**	**0**	**0.0**	**68**	**25**	**16:38**								
2001-02	**Philadelphia**	**NHL**	**31**	**7**	**3**	**10**	**8**	**0**	**0**	**1**	**62**	**11.3**	**5**	**0**	**0.0**	**19**	**4**	**12:36**								
	Vancouver	**NHL**	**46**	**9**	**12**	**21**	**10**	**1**	**0**	**2**	**70**	**12.9**	**4**	**2**	**100.0**	**36**	**12**	**14:46**	**5**	**0**	**1**	**1**	**0**	**0**	**0**	**0**
	NHL Totals		**223**	**63**	**74**	**137**	**54**	**12**	**0**	**11**	**461**	**13.7**		**8**	**50.0**	**162**	**58**	**15:15**	**5**	**0**	**1**	**1**	**0**	**0**	**0**	**0**

Traded to **Calgary** by **NY Islanders** for Jorgen Jonsson, July 14, 1998. Rights traded to **NY Rangers** by **Calgary** with Calgary's 1st (Jamie Lundmark) and 3rd (later traded back to Calgary - Calgary selected Craig Andersson) round choices in 1999 Entry Draft for Marc Savard and NY Rangers' 1st round choice (Oleg Saprykin) in 1999 Entry Draft, June 26, 1999. Traded to **Philadelphia** by **NY Rangers** with Kim Johnsson, Pavel Brendl and NY Rangers' 3rd round choice in 2003 Entry Draft for Eric Lindros, August 20, 2001. Traded to **Vancouver** by **Philadelphia** with Tampa Bay's 3rd round choice (previously acquired, Vancouver selected Brett Skinner) in 2002 Entry Draft for Donald Brashear and Vancouver's 6th round choice (later traded to Columbus - Columbus selected Jaroslav Balastik) in 2002 Entry Draft, December 17, 2001.

HNIDY, Shane

(NIGH-dee, SHAYN) **OTT.**

Defense. Shoots right. 6'2", 210 lbs. Born, Neepawa, Man., November 8, 1975. Buffalo's 7th choice, 173rd overall, in 1994 Entry Draft.

Season	Club	League	GP	G	A	Pts	PIM	PP	SH	GW	S	%	+/-	TF	F%	H	SB	Min	GP	G	A	Pts	PIM	PP	SH	GW
1990-91	Yellowhead Pass	MMHL	36	9	11	20	92																			
1991-92	Swift Current	WHL	56	1	3	4	11												4	0	0	0	0			
1992-93	Swift Current	WHL	45	5	12	17	62																			
	Prince Albert	WHL	27	2	10	12	43																			
1993-94	Prince Albert	WHL	69	7	26	33	113																			
1994-95	Prince Albert	WHL	72	5	29	34	169												15	4	7	11	29			
1995-96	Prince Albert	WHL	58	11	42	53	100												18	4	11	15	34			
1996-97	Baton Rouge	ECHL	21	3	10	13	50																			
	Saint John	AHL	44	2	12	14	112																			
1997-98	Grand Rapids	IHL	77	6	12	18	210												3	0	2	2	23			
1998-99	Adirondack	AHL	68	9	20	29	121												3	0	1	1	0			
99-2000	Cincinnati	AHL	68	9	19	28	153																			
2000-01	**Ottawa**	**NHL**	**52**	**3**	**2**	**5**	**84**	**0**	**0**	**1**	**47**	**6.4**	**8**	**0**	**0.0**	**90**	**44**	**13:05**	**1**	**0**	**0**	**0**	**0**	**0**	**0**	**0**
	Grand Rapids	IHL	2	0	0	0	2																			
2001-02	**Ottawa**	**NHL**	**33**	**1**	**1**	**2**	**57**	**0**	**0**	**0**	**34**	**2.9**	**–10**	**0**	**0.0**	**54**	**36**	**16:56**	**12**	**1**	**1**	**2**	**12**	**0**	**0**	**0**
	NHL Totals		**85**	**4**	**3**	**7**	**141**	**0**	**0**	**1**	**81**	**4.9**		**0**	**0.0**	**144**	**80**	**14:35**	**13**	**1**	**1**	**2**	**12**	**0**	**0**	**0**

Signed as a free agent by **Detroit**, August 6, 1998. Traded to **Ottawa** by **Detroit** for Ottawa's 8th round choice (Todd Jackson) in 2000 Entry Draft, June 25, 2000. • Missed majority of 2001-02 season recovering from ankle injury suffered in game vs. Boston, December 26, 2001.

HOGLUND, Jonas

(HOHG-lund, YOH-nuhs) **TOR.**

Right wing. Shoots right. 6'3", 215 lbs. Born, Hammaro, Sweden, August 29, 1972. Calgary's 11th choice, 222nd overall, in 1992 Entry Draft.

Season	Club	League	GP	G	A	Pts	PIM	PP	SH	GW	S	%	+/-	TF	F%	H	SB	Min	GP	G	A	Pts	PIM	PP	SH	GW
1990-91	Farjestad	Sweden	40	5	5	10	4												8	1	0	1	0			
1991-92	Farjestad	Sweden	40	14	11	25	6												6	2	4	6	2			
1992-93	Farjestad	Sweden	40	13	13	26	14												3	1	0	1	0			
1993-94	Farjestad	Sweden	22	7	2	9	10																			
1994-95	Farjestad	Sweden	40	14	12	26	16												4	3	2	5	0			
1995-96	Farjestad	Sweden	40	*32	11	43	18												8	2	1	3	6			
1996-97	**Calgary**	**NHL**	**68**	**19**	**16**	**35**	**12**	**3**	**0**	**6**	**189**	**10.1**	**–4**													
1997-98	**Calgary**	**NHL**	**50**	**6**	**8**	**14**	**16**	**0**	**0**	**0**	**124**	**4.8**	**–9**													
	Montreal	**NHL**	**28**	**6**	**5**	**11**	**6**	**4**	**0**	**0**	**62**	**9.7**	**2**						**10**	**2**	**0**	**2**	**0**	**0**	**0**	**0**
1998-99	**Montreal**	**NHL**	**74**	**8**	**10**	**18**	**16**	**1**	**0**	**0**	**122**	**6.6**	**–5**	**17**	**29.4**	**41**	**15**	**11:50**								
99-2000	**Toronto**	**NHL**	**82**	**29**	**27**	**56**	**10**	**9**	**1**	**3**	**215**	**13.5**	**–2**	**4**	**50.0**	**55**	**24**	**17:10**	**12**	**2**	**4**	**6**	**2**	**0**	**0**	**0**
2000-01	**Toronto**	**NHL**	**82**	**23**	**26**	**49**	**14**	**5**	**0**	**5**	**196**	**11.7**	**1**	**4**	**50.0**	**53**	**23**	**15:08**	**10**	**0**	**0**	**0**	**4**	**0**	**0**	**0**
2001-02	**Toronto**	**NHL**	**82**	**13**	**34**	**47**	**26**	**1**	**1**	**4**	**199**	**6.5**	**11**	**1**	**0.0**	**36**	**15**	**15:34**	**20**	**4**	**6**	**10**	**2**	**3**	**0**	**1**
	NHL Totals		**466**	**104**	**126**	**230**	**100**	**23**	**2**	**18**	**1107**	**9.4**		**26**	**34.6**	**185**	**77**	**15:00**	**52**	**8**	**10**	**18**	**8**	**3**	**0**	**1**

Traded to **Montreal** by **Calgary** with Zarley Zalapski for Valeri Bure and Montreal's 4th round choice (Shaun Sutter) in 1998 Entry Draft, February 1, 1998. Signed as a free agent by **Toronto**, July 13, 1999.

HOGUE, Benoit

(HOHG, BEHN-wah)

Center. Shoots left. 5'10", 194 lbs. Born, Repentigny, Que., October 28, 1966. Buffalo's 2nd choice, 35th overall, in 1985 Entry Draft.

Season	Club	League	GP	G	A	Pts	PIM	PP	SH	GW	S	%	+/-	TF	F%	H	SB	Min	GP	G	A	Pts	PIM	PP	SH	GW
1982-83	Mtl-Bourassa	QAAA	40	20	20	40	34												10	2	1	3	4			
1983-84	St-Jean Castors	QMJHL	59	14	11	25	42																			
1984-85	St-Jean Castors	QMJHL	63	46	44	90	92																			
1985-86	St-Jean Castors	QMJHL	65	54	54	108	115												9	6	4	10	26			
1986-87	Rochester	AHL	52	14	20	34	52												12	5	4	9	8			
1987-88	**Buffalo**	**NHL**	**3**	**1**	**1**	**2**	**0**	**0**	**0**	**1**	**3**	**33.3**	**3**													
	Rochester	AHL	62	24	31	55	141												7	6	1	7	46			
1988-89	**Buffalo**	**NHL**	**69**	**14**	**30**	**44**	**120**	**1**	**2**	**0**	**114**	**12.3**	**–5**						**5**	**0**	**0**	**0**	**17**	**0**	**0**	**0**
1989-90	**Buffalo**	**NHL**	**45**	**11**	**7**	**18**	**79**	**1**	**0**	**1**	**73**	**15.1**	**0**						**3**	**0**	**0**	**0**	**10**	**0**	**0**	**0**
1990-91	**Buffalo**	**NHL**	**76**	**19**	**28**	**47**	**76**	**1**	**0**	**2**	**134**	**14.2**	**–8**						**5**	**3**	**1**	**4**	**10**	**0**	**0**	**0**
1991-92	**Buffalo**	**NHL**	**3**	**0**	**1**	**1**	**0**	**0**	**0**	**0**	**6**	**0.0**	**0**													
	NY Islanders	**NHL**	**72**	**30**	**45**	**75**	**67**	**8**	**0**	**5**	**143**	**21.0**	**30**													
1992-93	**NY Islanders**	**NHL**	**70**	**33**	**42**	**75**	**108**	**5**	**3**	**5**	**147**	**22.4**	**13**						**18**	**6**	**6**	**12**	**31**	**0**	**0**	**0**
1993-94	**NY Islanders**	**NHL**	**83**	**36**	**33**	**69**	**73**	**9**	**5**	**3**	**218**	**16.5**	**–7**						**4**	**0**	**1**	**1**	**4**	**0**	**0**	**0**
1994-95	**NY Islanders**	**NHL**	**33**	**6**	**4**	**10**	**34**	**1**	**0**	**1**	**50**	**12.0**	**0**													
	Toronto	**NHL**	**12**	**3**	**3**	**6**	**0**	**1**	**0**	**1**	**16**	**18.8**	**0**						**7**	**0**	**0**	**0**	**6**	**0**	**0**	**0**
1995-96	**Toronto**	**NHL**	**44**	**12**	**25**	**37**	**68**	**3**	**0**	**5**	**94**	**12.8**	**6**													
	Dallas	**NHL**	**34**	**7**	**20**	**27**	**36**	**2**	**0**	**0**	**61**	**11.5**	**4**													
1996-97	**Dallas**	**NHL**	**73**	**19**	**24**	**43**	**54**	**5**	**0**	**5**	**131**	**14.5**	**8**						**7**	**2**	**2**	**4**	**6**	**1**	**0**	**0**
1997-98	**Dallas**	**NHL**	**53**	**6**	**16**	**22**	**35**	**3**	**0**	**1**	**55**	**10.9**	**7**						**17**	**4**	**2**	**6**	**16**	**1**	**0**	**2**
1998-99	**Tampa Bay**	**NHL**	**62**	**11**	**14**	**25**	**50**	**2**	**0**	**3**	**101**	**10.9**	**–12**	**63**	**42.9**	**89**	**24**	**16:23**								
	♦ **Dallas**	**NHL**	**12**	**1**	**3**	**4**	**4**	**0**	**0**	**0**	**20**	**5.0**	**2**	**52**	**46.2**	**31**	**4**	**15:04**	**14**	**0**	**2**	**2**	**16**	**0**	**0**	**0**
99-2000	**Phoenix**	**NHL**	**27**	**3**	**10**	**13**	**10**	**0**	**0**	**0**	**39**	**7.7**	**–1**	**24**	**37.5**	**56**	**11**	**15:45**	**5**	**1**	**2**	**3**	**2**	**0**	**0**	**0**
2000-01	**Dallas**	**NHL**	**34**	**3**	**7**	**10**	**26**	**0**	**0**	**0**	**35**	**8.6**	**–1**	**143**	**44.8**	**70**	**5**	**12:28**	**7**	**1**	**0**	**1**	**6**	**0**	**0**	**1**
2001-02	**Dallas**	**NHL**	**32**	**3**	**3**	**6**	**24**	**0**	**0**	**0**	**20**	**15.0**	**–4**	**23**	**34.8**	**46**	**7**	**10:37**								
	Boston	**NHL**	**17**	**4**	**4**	**8**	**9**	**0**	**0**	**1**	**17**	**23.5**	**–3**	**148**	**52.7**	**30**	**9**	**13:38**								
	Washington	**NHL**	**9**	**0**	**1**	**1**	**4**	**0**	**0**	**0**	**5**	**0.0**	**2**	**9**	**33.3**	**6**	**3**	**8:52**								
	NHL Totals		**863**	**222**	**321**	**543**	**877**	**42**	**10**	**34**	**1482**	**15.0**		**462**	**46.1**	**328**	**63**	**13:58**	**92**	**17**	**16**	**33**	**124**	**2**	**0**	**3**

Traded to **NY Islanders** by **Buffalo** with Pierre Turgeon, Uwe Krupp and Dave McLlwain for Pat LaFontaine, Randy Hillier, Randy Wood and NY Islanders' 4th round choice (Dean Melanson) in 1992 Entry Draft, October 25, 1991. Traded to **Toronto** by **NY Islanders** with NY Islanders' 3rd round choice (Ryan Pepperall) in 1995 Entry Draft and 5th round choice (Brandon Sugden) in 1996 Entry Draft for Eric Fichaud, April 6, 1995. Traded to **Dallas** by **Toronto** with Randy Wood for Dave Gagner and Dallas' 6th round choice (Dmitri Yakushin) in 1996 Entry Draft, January 29, 1996. Signed as a free agent by **Tampa Bay**, August 19, 1998. Traded to **Dallas** by **Tampa Bay** with Tampa Bay's 6th round choice (Michal Blazek) in 2001 Entry Draft for Sergey Gusev, March 21, 1999. Signed as a free agent by **Phoenix**, February 3, 2000. Signed as a free agent by **Dallas**, January 5, 2001. Traded to **Boston** by **Dallas** for future considerations, January 12, 2002. Claimed on waivers by **Washington** from **Boston**, March 19, 2002.

HOLDEN, Josh

(HOHL-dehn, JAWSH) **TOR.**

Center. Shoots left. 6', 190 lbs. Born, Calgary, Alta., January 18, 1978. Vancouver's 1st choice, 12th overall, in 1996 Entry Draft.

Season	Club	League	GP	G	A	Pts	PIM	PP	SH	GW	S	%	+/-	TF	F%	H	SB	Min	GP	G	A	Pts	PIM	PP	SH	GW
1993-94	Cgy. Buffaloes	AMHL	34	14	15	29	82																			
1994-95	Regina Pats	WHL	62	20	23	43	45												4	3	1	4	0			
1995-96	Regina Pats	WHL	70	57	55	112	105												11	4	5	9	23			
1996-97	Regina Pats	WHL	58	49	49	98	148												5	3	2	5	10			
1997-98	Regina Pats	WHL	56	41	58	99	134												2	2	2	4	10			
1998-99	**Vancouver**	**NHL**	**30**	**2**	**4**	**6**	**10**	**1**	**0**	**0**	**44**	**4.5**	**–10**	**269**	**39.0**	**38**	**6**	**12:44**								
	Syracuse Crunch	AHL	38	14	15	29	48																			
99-2000	**Vancouver**	**NHL**	**6**	**1**	**5**	**6**	**2**	**0**	**0**	**0**	**5**	**20.0**	**2**	**42**	**42.9**	**14**	**2**	**10:25**								
	Syracuse Crunch	AHL	45	19	32	51	113												4	1	0	1	10			
2000-01	**Vancouver**	**NHL**	**10**	**1**	**0**	**1**	**0**	**0**	**0**	**0**	**12**	**8.3**	**0**	**85**	**35.3**	**21**	**2**	**9:27**								
	Kansas City	IHL	60	27	26	53	136																			

			Regular Season																Playoffs							
Season	Club	League	GP	G	A	Pts	PIM	PP	SH	GW	S	%	+/-	TF	F%	H	SB	Min	GP	G	A	Pts	PIM	PP	SH	GW
2001-02	**Carolina**	**NHL**	**8**	**0**	**0**	**0**	**2**	**0**	**0**	**0**	**3**	**0.0**	**0**	**41**	**41.5**	**9**	**2**	**5:17**	....	....	....	....	....	....	....	....
	Manitoba Moose	AHL	68	16	17	33	187	....	....	....	....	....	....	....	....	....	....	....	7	1	1	2	4	....	....	....
	NHL Totals		**54**	**4**	**9**	**13**	**14**	**1**	**0**	**0**	**64**	**6.3**		**437**	**38.9**	**82**	**12**	**10:46**	....	....	....	....	....	....	....	....

WHL East Second All-Star Team (1998)

Claimed by **Carolina** from **Vancouver** in Waiver Draft, September 28, 2001. Claimed on waivers by **Vancouver** from **Carolina**, October 25, 2001. Traded to **Toronto** by **Vancouver** for Jeff Farkas, June 23, 2002.

HOLIK, Bobby (HOH-leek, BAWB-ee) NYR

Center. Shoots right. 6'4", 230 lbs. Born, Jihlava, Czech., January 1, 1971. Hartford's 1st choice, 10th overall, in 1989 Entry Draft.

Season	Club	League	GP	G	A	Pts	PIM	PP	SH	GW	S	%	+/-	TF	F%	H	SB	Min	GP	G	A	Pts	PIM	PP	SH	GW
1987-88	Dukla Jihlava	Czech	31	5	9	14	16	....	....	....	....	....	....	....	....	....	....	....	....	....	....	....	....	....	....	....
1988-89	Dukla Jihlava	Czech	24	7	10	17	32	....	....	....	....	....	....	....	....	....	....	....	....	....	....	....	....	....	....	....
1989-90	Dukla Jihlava	Czech	42	15	26	41	....	....	....	....	....	....	....	....	....	....	....	....	....	....	....	....	....	....	....	....
1990-91	**Hartford**	**NHL**	**78**	**21**	**22**	**43**	**113**	**8**	**0**	**3**	**173**	**12.1**	**–3**	....	....	....	....	....	**6**	**0**	**0**	**0**	**7**	**0**	**0**	**0**
1991-92	**Hartford**	**NHL**	**76**	**21**	**24**	**45**	**44**	**1**	**0**	**2**	**207**	**10.1**	**4**	....	....	....	....	....	**7**	**0**	**1**	**1**	**6**	**0**	**0**	**0**
1992-93	**New Jersey**	**NHL**	**61**	**20**	**19**	**39**	**76**	**7**	**0**	**4**	**180**	**11.1**	**–6**	....	....	....	....	....	**5**	**1**	**1**	**2**	**6**	**0**	**0**	**0**
	Utica Devils	AHL	1	0	0	0	2	....	....	....	....	....	....	....	....	....	....	....	....	....	....	....	....	....	....	....
1993-94	**New Jersey**	**NHL**	**70**	**13**	**20**	**33**	**72**	**2**	**0**	**3**	**130**	**10.0**	**28**	....	....	....	....	....	**20**	**0**	**3**	**3**	**6**	**0**	**0**	**0**
1994-95♦	**New Jersey**	**NHL**	**48**	**10**	**10**	**20**	**18**	**0**	**0**	**2**	**84**	**11.9**	**9**	....	....	....	....	....	**20**	**4**	**4**	**8**	**22**	**2**	**0**	**1**
1995-96	**New Jersey**	**NHL**	**63**	**13**	**17**	**30**	**58**	**1**	**0**	**1**	**157**	**8.3**	**9**	....	....	....	....	....	....	....	....	....	....	....	....	....
1996-97	**New Jersey**	**NHL**	**82**	**23**	**39**	**62**	**54**	**5**	**0**	**6**	**192**	**12.0**	**24**	....	....	....	....	....	**10**	**2**	**3**	**5**	**4**	**1**	**0**	**0**
1997-98	**New Jersey**	**NHL**	**82**	**29**	**36**	**65**	**100**	**8**	**0**	**8**	**238**	**12.2**	**23**	....	....	....	....	....	**5**	**0**	**0**	**0**	**8**	**0**	**0**	**0**
1998-99	**New Jersey**	**NHL**	**78**	**27**	**37**	**64**	**119**	**5**	**0**	**8**	**253**	**10.7**	**16**	**1350**	**53.6**	**217**	**24**	**17:34**	**7**	**0**	**7**	**7**	**6**	**0**	**0**	**0**
99-2000♦	**New Jersey**	**NHL**	**79**	**23**	**23**	**46**	**106**	**7**	**0**	**4**	**257**	**8.9**	**7**	**1390**	**55.6**	**139**	**17**	**16:53**	**23**	**3**	**7**	**10**	**14**	**0**	**0**	**1**
2000-01	**New Jersey**	**NHL**	**80**	**15**	**35**	**50**	**97**	**3**	**0**	**3**	**206**	**7.3**	**19**	**1365**	**56.0**	**213**	**20**	**15:49**	**25**	**6**	**10**	**16**	**37**	**1**	**0**	**3**
2001-02	**New Jersey**	**NHL**	**81**	**25**	**29**	**54**	**97**	**6**	**0**	**3**	**270**	**9.3**	**7**	**1594**	**54.5**	**217**	**28**	**17:42**	**6**	**4**	**1**	**5**	**2**	**1**	**0**	**0**
	NHL Totals		**878**	**240**	**311**	**551**	**954**	**53**	**0**	**47**	**2347**	**10.2**		**5699**	**54.9**	**786**	**89**	**16:59**	**134**	**20**	**37**	**57**	**118**	**5**	**0**	**5**

Played in NHL All-Star Game (1998, 1999)

Traded to **New Jersey** by **Hartford** with Hartford's 2nd round choice (Jay Pandolfo) in 1993 Entry Draft for Sean Burke and Eric Weinrich, August 28, 1992. Signed as a free agent by **NY Rangers**, July 1, 2002.

HOLLAND, Jason (HAWL-land, JAY-suhn) L.A.

Defense. Shoots right. 6'3", 209 lbs. Born, Morinville, Alta., April 30, 1976. NY Islanders' 2nd choice, 38th overall, in 1994 Entry Draft.

Season	Club	League	GP	G	A	Pts	PIM	PP	SH	GW	S	%	+/-	TF	F%	H	SB	Min	GP	G	A	Pts	PIM	PP	SH	GW
1991-92	St. Albert	AMHL	38	9	29	38	94	....	....	....	....	....	....	....	....	....	....	....	....	....	....	....	....	....	....	....
1992-93	St. Albert	AMHL	31	11	25	36	36	....	....	....	....	....	....	....	....	....	....	....	....	....	....	....	....	....	....	....
	Kamloops Blazers	WHL	4	0	0	0	2	....	....	....	....	....	....	....	....	....	....	....	....	....	....	....	....	....	....	....
1993-94	Kamloops Blazers	WHL	59	14	15	29	80	....	....	....	....	....	....	....	....	....	....	....	18	2	3	5	4	....	....	....
1994-95	Kamloops Blazers	WHL	71	9	32	41	65	....	....	....	....	....	....	....	....	....	....	....	21	2	7	9	9	....	....	....
1995-96	Kamloops Blazers	WHL	63	24	33	57	98	....	....	....	....	....	....	....	....	....	....	....	16	4	9	13	22	....	....	....
1996-97	**NY Islanders**	**NHL**	**4**	**1**	**0**	**1**	**0**	**0**	**0**	**0**	**3**	**33.3**	**1**	....	....	....	....	....	....	....	....	....	....	....	....	....
	Kentucky	AHL	72	14	25	39	46	....	....	....	....	....	....	....	....	....	....	....	4	0	2	2	0	....	....	....
1997-98	**NY Islanders**	**NHL**	**8**	**0**	**0**	**0**	**4**	**0**	**0**	**0**	**6**	**0.0**	**–4**	....	....	....	....	....	....	....	....	....	....	....	....	....
	Kentucky	AHL	50	10	16	26	29	....	....	....	....	....	....	....	....	....	....	....	....	....	....	....	....	....	....	....
	Rochester	AHL	9	0	4	4	10	....	....	....	....	....	....	....	....	....	....	....	4	0	3	3	4	....	....	....
1998-99	**Buffalo**	**NHL**	**3**	**0**	**0**	**0**	**8**	**0**	**0**	**0**	**2**	**0.0**	**–1**	**0**	**0.0**	**0**	**2**	**10:58**	....	....	....	....	....	....	....	....
	Rochester	AHL	74	4	25	29	36	....	....	....	....	....	....	....	....	....	....	....	20	2	5	7	8	....	....	....
99-2000	**Buffalo**	**NHL**	**9**	**0**	**1**	**1**	**0**	**0**	**0**	**0**	**8**	**0.0**	**0**	**0**	**0.0**	**5**	**3**	**15:31**	**1**	**0**	**0**	**0**	**0**	**0**	**0**	**0**
	Rochester	AHL	54	3	15	18	24	....	....	....	....	....	....	....	....	....	....	....	12	1	0	1	2	....	....	....
2000-01	Rochester	AHL	63	4	19	23	45	....	....	....	....	....	....	....	....	....	....	....	4	1	0	1	0	....	....	....
2001-02	**Los Angeles**	**NHL**	**3**	**0**	**0**	**0**	**0**	**0**	**0**	**0**	**1**	**0.0**	**–1**	**0**	**0.0**	**3**	**1**	**15:50**	....	....	....	....	....	....	....	....
	Manchester	AHL	65	9	18	27	39	....	....	....	....	....	....	....	....	....	....	....	5	1	0	1	5	....	....	....
	NHL Totals		**27**	**1**	**1**	**2**	**12**	**0**	**0**	**0**	**20**	**5.0**		**0**	**0.0**	**8**	**6**	**14:40**	**1**	**0**	**0**	**0**	**0**	**0**	**0**	**0**

Won Warwick Trophy (MVP - AMHL) (1993) • WHL West First All-Star Team (1996)

Traded to **Buffalo** by **NY Islanders** with Paul Kruse for Jason Dawe, March 24, 1998. Signed as a free agent by **LA Kings**, August 23, 2001.

HOLMSTROM, Tomas (HOHLM-struhm, TAW-mas) DET.

Left wing. Shoots left. 6', 200 lbs. Born, Pitea, Sweden, January 23, 1973. Detroit's 9th choice, 257th overall, in 1994 Entry Draft.

Season	Club	League	GP	G	A	Pts	PIM	PP	SH	GW	S	%	+/-	TF	F%	H	SB	Min	GP	G	A	Pts	PIM	PP	SH	GW
1989-90	Pitea HC	Swede-2	9	1	0	1	4	....	....	....	....	....	....	....	....	....	....	....	....	....	....	....	....	....	....	....
1990-91	Pitea HC	Swede-2	26	5	4	9	16	....	....	....	....	....	....	....	....	....	....	....	....	....	....	....	....	....	....	....
1991-92	Pitea HC	Swede-2	31	15	12	27	44	....	....	....	....	....	....	....	....	....	....	....	....	....	....	....	....	....	....	....
1992-93	Pitea HC	Swede-2	32	17	15	32	30	....	....	....	....	....	....	....	....	....	....	....	....	....	....	....	....	....	....	....
1993-94	Bodens IK	Swede-2	34	23	16	39	86	....	....	....	....	....	....	....	....	....	....	....	9	3	3	6	24	....	....	....
1994-95	Lulea HF	Sweden	40	14	14	28	56	....	....	....	....	....	....	....	....	....	....	....	8	1	2	3	20	....	....	....
1995-96	Lulea HF	Sweden	34	12	11	23	78	....	....	....	....	....	....	....	....	....	....	....	11	6	2	8	22	....	....	....
1996-97♦	**Detroit**	**NHL**	**47**	**6**	**3**	**9**	**33**	**3**	**0**	**0**	**53**	**11.3**	**–10**	....	....	....	....	....	**1**	**0**	**0**	**0**	**0**	**0**	**0**	**0**
	Adirondack	AHL	6	3	1	4	7	....	....	....	....	....	....	....	....	....	....	....	....	....	....	....	....	....	....	....
1997-98♦	**Detroit**	**NHL**	**57**	**5**	**17**	**22**	**44**	**1**	**0**	**1**	**48**	**10.4**	**6**	....	....	....	....	....	**22**	**7**	**12**	**19**	**16**	**2**	**0**	**0**
1998-99	**Detroit**	**NHL**	**82**	**13**	**21**	**34**	**69**	**5**	**0**	**4**	**100**	**13.0**	**–11**	**0**	**0.0**	**94**	**9**	**12:22**	**10**	**4**	**3**	**7**	**4**	**2**	**0**	**1**
99-2000	**Detroit**	**NHL**	**72**	**13**	**22**	**35**	**43**	**4**	**0**	**1**	**71**	**18.3**	**4**	**0**	**0.0**	**68**	**9**	**12:06**	**9**	**3**	**1**	**4**	**16**	**1**	**0**	**1**
2000-01	**Detroit**	**NHL**	**73**	**16**	**24**	**40**	**40**	**9**	**0**	**2**	**74**	**21.6**	**–12**	**2**	**50.0**	**75**	**8**	**11:41**	**6**	**1**	**3**	**4**	**8**	**1**	**0**	**0**
2001-02♦	**Detroit**	**NHL**	**69**	**8**	**18**	**26**	**58**	**6**	**0**	**1**	**79**	**10.1**	**–12**	**2**	**0.0**	**73**	**12**	**12:23**	**23**	**8**	**3**	**11**	**8**	**3**	**0**	**2**
	Sweden	Olympics	4	1	0	1	0	....	....	....	....	....	....	....	....	....	....	....	....	....	....	....	....	....	....	....
	NHL Totals		**400**	**61**	**105**	**166**	**287**	**28**	**0**	**9**	**425**	**14.4**		**4**	**25.0**	**310**	**38**	**12:08**	**71**	**23**	**22**	**45**	**52**	**9**	**0**	**4**

HOLZINGER, Brian (HOHL-zihn-guhr, BRIGH-uhn) T.B.

Center. Shoots right. 5'11", 190 lbs. Born, Parma, OH, October 10, 1972. Buffalo's 7th choice, 124th overall, in 1991 Entry Draft.

Season	Club	League	GP	G	A	Pts	PIM	PP	SH	GW	S	%	+/-	TF	F%	H	SB	Min	GP	G	A	Pts	PIM	PP	SH	GW
1988-89	Padua High	Hi-School	35	73	65	138	....	....	....	....	....	....	....	....	....	....	....	....	....	....	....	....	....	....	....	....
1989-90	Det. Compuware	NAJHL	44	36	37	73	....	....	....	....	....	....	....	....	....	....	....	....	....	....	....	....	....	....	....	....
1990-91	Det. Compuware	NAJHL	37	45	41	86	16	....	....	....	....	....	....	....	....	....	....	....	....	....	....	....	....	....	....	....
1991-92	Bowling Green	CCHA	30	14	8	22	36	....	....	....	....	....	....	....	....	....	....	....	....	....	....	....	....	....	....	....
1992-93	Bowling Green	CCHA	41	31	26	57	44	....	....	....	....	....	....	....	....	....	....	....	....	....	....	....	....	....	....	....
1993-94	Bowling Green	CCHA	38	22	15	37	24	....	....	....	....	....	....	....	....	....	....	....	....	....	....	....	....	....	....	....
1994-95	Bowling Green	CCHA	38	35	33	68	42	....	....	....	....	....	....	....	....	....	....	....	....	....	....	....	....	....	....	....
	Buffalo	**NHL**	**4**	**0**	**3**	**3**	**0**	**0**	**0**	**0**	**3**	**0.0**	**2**	....	....	....	....	....	**4**	**2**	**1**	**3**	**2**	**1**	**0**	**0**
1995-96	**Buffalo**	**NHL**	**58**	**10**	**10**	**20**	**37**	**5**	**0**	**1**	**71**	**14.1**	**–21**	....	....	....	....	....	....	....	....	....	....	....	....	....
	Rochester	AHL	17	10	11	21	14	....	....	....	....	....	....	....	....	....	....	....	19	10	14	24	10	....	....	....
1996-97	**Buffalo**	**NHL**	**81**	**22**	**29**	**51**	**54**	**2**	**2**	**6**	**142**	**15.5**	**9**	....	....	....	....	....	**12**	**2**	**5**	**7**	**8**	**0**	**1**	**0**
1997-98	**Buffalo**	**NHL**	**69**	**14**	**21**	**35**	**36**	**4**	**2**	**1**	**116**	**12.1**	**–2**	....	....	....	....	....	**15**	**4**	**7**	**11**	**18**	**1**	**1**	**0**
1998-99	**Buffalo**	**NHL**	**81**	**17**	**17**	**34**	**45**	**5**	**0**	**2**	**143**	**11.9**	**2**	**852**	**50.4**	**72**	**23**	**16:31**	**21**	**3**	**5**	**8**	**33**	**1**	**0**	**0**
99-2000	**Buffalo**	**NHL**	**59**	**7**	**17**	**24**	**30**	**0**	**1**	**2**	**81**	**8.6**	**4**	**839**	**45.7**	**60**	**18**	**14:38**	....	....	....	....	....	....	....	....
	Tampa Bay	**NHL**	**14**	**3**	**3**	**6**	**21**	**1**	**1**	**0**	**23**	**13.0**	**–7**	**119**	**46.2**	**16**	**3**	**16:10**	....	....	....	....	....	....	....	....
2000-01	**Tampa Bay**	**NHL**	**70**	**11**	**25**	**36**	**64**	**3**	**0**	**2**	**87**	**12.6**	**–9**	**775**	**47.4**	**40**	**40**	**16:03**	....	....	....	....	....	....	....	....
2001-02	**Tampa Bay**	**NHL**	**23**	**1**	**2**	**3**	**4**	**0**	**0**	**0**	**20**	**5.0**	**–4**	**54**	**55.6**	**22**	**4**	**9:19**	....	....	....	....	....	....	....	....
	NHL Totals		**459**	**85**	**127**	**212**	**291**	**20**	**6**	**14**	**686**	**12.4**		**2639**	**47.9**	**210**	**88**	**15:15**	**52**	**11**	**18**	**29**	**61**	**3**	**2**	**0**

CCHA Second All-Star Team (1993) • CCHA First All-Star Team (1995) • CCHA Player of the Year (1995) • NCAA West First All-American Team (1995) • Won Hobey Baker Memorial Award (Top U.S. Collegiate Player) (1995)

Traded to **Tampa Bay** by **Buffalo** with Cory Sarich, Wayne Primeau and Buffalo's 3rd round choice (Alexander Kharitonov) in 2000 Entry Draft for Chris Gratton and Tampa Bay's 2nd round choice (Derek Roy) in 2001 Entry Draft, March 9, 2000. • Missed majority of 2001-02 season recovering from shoulder injury suffered in game vs. Florida, October 7, 2001.

HORCOFF, Shawn (HOHR-cuhf, SHAWN) EDM.

Center. Shoots left. 6'1", 202 lbs. Born, Trail, B.C., September 17, 1978. Edmonton's 3rd choice, 99th overall, in 1998 Entry Draft.

			Regular Season																Playoffs							
Season	Club	League	GP	G	A	Pts	PIM	PP	SH	GW	S	%	+/-	TF	F%	H	SB	Min	GP	G	A	Pts	PIM	PP	SH	GW
1994-95	Trail Smokies	RMJHL	47	50	46	96	26	....	....	....	....	....	....	....	....	....	....	....	....	....	....	....	....	....	....	....
1995-96	Chilliwack	BCJHL	58	49	96	*146	44	....	....	....	....	....	....	....	....	....	....	....	....	....	....	....	....	....	....	....
1996-97	Michigan State	CCHA	40	10	13	23	20	....	....	....	....	....	....	....	....	....	....	....	....	....	....	....	....	....	....	....
1997-98	Michigan State	CCHA	34	14	13	27	50	....	....	....	....	....	....	....	....	....	....	....	....	....	....	....	....	....	....	....
1998-99	Michigan State	CCHA	39	12	25	37	70	....	....	....	....	....	....	....	....	....	....	....	....	....	....	....	....	....	....	....
99-2000	Michigan State	CCHA	42	14	*51	*65	50	....	....	....	....	....	....	....	....	....	....	....	....	....	....	....	....	....	....	....
2000-01	**Edmonton**	**NHL**	**49**	**9**	**7**	**16**	**10**	**0**	**0**	**2**	**42**	**21.4**	**8**	**122**	**41.8**	**9**	**4**	**9:14**	**5**	**0**	**0**	**0**	**0**	**0**	**0**	**0**
	Hamilton	AHL	24	10	18	28	19	....	....	....	....	....	....	....	....	....	....	....	....	....	....	....	....	....	....	....
2001-02	**Edmonton**	**NHL**	**61**	**8**	**14**	**22**	**18**	**0**	**0**	**0**	**57**	**14.0**	**3**	**454**	**46.3**	**24**	**14**	**11:20**	....	....	....	....	....	....	....	....
	Hamilton	AHL	2	1	2	3	6	....	....	....	....	....	....	....	....	....	....	....	....	....	....	....	....	....	....	....
	NHL Totals		**110**	**17**	**21**	**38**	**28**	**0**	**0**	**2**	**99**	**17.2**		**576**	**45.3**	**33**	**18**	**10:23**	**5**	**0**	**0**	**0**	**0**	**0**	**0**	**0**

BCHL Player of the Year (1996) • Won Brett Hull Trophy (Top Scorer - BCHL) (1996) • BCHL First All-Star Team (1996) • CCHA First All-Star Team (2000) • CCHA Player of the Year (2000) • NCAA West First All-American Team (2000)

HORDICHUK, Darcy (HOHR-dih-chuhk, DAHR-see) PHX.

Left wing. Shoots left. 6'1", 215 lbs. Born, Kamsack, Sask., August 10, 1980. Atlanta's 9th choice, 180th overall, in 2000 Entry Draft.

			Regular Season																Playoffs							
Season	Club	League	GP	G	A	Pts	PIM	PP	SH	GW	S	%	+/-	TF	F%	H	SB	Min	GP	G	A	Pts	PIM	PP	SH	GW
1996-97	Yorkton Mallers	SMHL	57	6	15	21	230	....	....	....	....	....	....	....	....	....	....	....	....	....	....	....	....	....	....	....
	Calgary Hitmen	WHL	3	0	0	0	2	....	....	....	....	....	....	....	....	....	....	....	....	....	....	....	....	....	....	....
1997-98	Dauphin Kings	MJHL	58	12	21	33	279	....	....	....	....	....	....	....	....	....	....	....	....	....	....	....	....	....	....	....
1998-99	Saskatoon Blades	WHL	66	3	2	5	246	....	....	....	....	....	....	....	....	....	....	....	....	....	....	....	....	....	....	....
99-2000	Saskatoon Blades	WHL	63	6	8	14	269	....	....	....	....	....	....	....	....	....	....	....	11	4	2	6	43	....	....	....
2000-01	**Atlanta**	**NHL**	**11**	**0**	**0**	**0**	**38**	**0**	**0**	**0**	**6**	**0.0**	**–3**	**0**	**0.0**	**34**	**2**	**7:18**	....	....	....	....	....	....	....	....
	Orlando	IHL	69	7	3	10	*369	....	....	....	....	....	....	....	....	....	....	....	16	3	3	6	*41	....	....	....
2001-02	**Atlanta**	**NHL**	**33**	**1**	**1**	**2**	**127**	**0**	**0**	**0**	**8**	**12.5**	**–5**	**4**	**25.0**	**70**	**4**	**6:03**	....	....	....	....	....	....	....	....
	Chicago Wolves	AHL	34	5	4	9	127	....	....	....	....	....	....	....	....	....	....	....	....	....	....	....	....	....	....	....
	Phoenix	**NHL**	**1**	**0**	**0**	**0**	**14**	**0**	**0**	**0**	**0**	**0.0**	**0**	**0**	**0.0**	**3**	**0**	**7:18**	....	....	....	....	....	....	....	....
	NHL Totals		**45**	**1**	**1**	**2**	**179**	**0**	**0**	**0**	**14**	**7.1**		**4**	**25.0**	**107**	**6**	**6:23**	....	....	....	....	....	....	....	....

Traded to **Phoenix** by **Atlanta** with Atlanta's 4th (Lance Monych) and 5th (John Zeiler) round choices in 2002 Entry Draft for Kiril Safronov, the rights to Ruslan Zainullin and Phoenix's 4th round choice (Patrick Dwyer) in 2002 Entry Draft, March 19, 2002.

HOSSA, Marcel (HOH-sah, MAHR-sehl) MTL.

Center. Shoots left. 6'2", 211 lbs. Born, Ilava, Czech., October 12, 1981. Montreal's 2nd choice, 16th overall, in 2000 Entry Draft.

			Regular Season																Playoffs							
Season	Club	League	GP	G	A	Pts	PIM	PP	SH	GW	S	%	+/-	TF	F%	H	SB	Min	GP	G	A	Pts	PIM	PP	SH	GW
1996-97	Dukla Trencin Jr.	Slovak-Jr.	45	30	21	51	30	....	....	....	....	....	....	....	....	....	....	....	....	....	....	....	....	....	....	....
1997-98	Dukla Trencin Jr.	Slovak-Jr.	39	11	38	49	44	....	....	....	....	....	....	....	....	....	....	....	....	....	....	....	....	....	....	....
1998-99	Portland	WHL	70	7	14	21	66	....	....	....	....	....	....	....	....	....	....	....	2	0	0	0	2	....	....	....
99-2000	Portland	WHL	60	24	29	53	58	....	....	....	....	....	....	....	....	....	....	....	....	....	....	....	....	....	....	....
2000-01	Portland	WHL	58	34	56	90	58	....	....	....	....	....	....	....	....	....	....	....	16	5	7	12	14	....	....	....
2001-02	**Montreal**	**NHL**	**10**	**3**	**1**	**4**	**2**	**0**	**0**	**0**	**20**	**15.0**	**2**	**0**	**0.0**	**4**	**2**	**11:09**	....	....	....	....	....	....	....	....
	Quebec	AHL	50	17	15	32	24	....	....	....	....	....	....	....	....	....	....	....	3	0	0	0	4	....	....	....
	NHL Totals		**10**	**3**	**1**	**4**	**2**	**0**	**0**	**0**	**20**	**15.0**		**0**	**0.0**	**4**	**2**	**11:09**	....	....	....	....	....	....	....	....

WHL West Second All-Star Team (2001)

HOSSA, Marian (HOH-sah, MAIR-ee-an) OTT.

Left wing. Shoots left. 6'1", 199 lbs. Born, Stara Lubovna, Czech., January 12, 1979. Ottawa's 1st choice, 12th overall, in 1997 Entry Draft.

			Regular Season																Playoffs							
Season	Club	League	GP	G	A	Pts	PIM	PP	SH	GW	S	%	+/-	TF	F%	H	SB	Min	GP	G	A	Pts	PIM	PP	SH	GW
1995-96	Dukla Trencin Jr.	Slovak-Jr.	53	42	49	91	26	....	....	....	....	....	....	....	....	....	....	....	....	....	....	....	....	....	....	....
1996-97	Dukla Trencin	Slovakia	46	25	19	44	33	....	....	....	....	....	....	....	....	....	....	....	7	5	5	10	....	....	....	....
1997-98	Portland	WHL	53	45	40	85	50	....	....	....	....	....	....	....	....	....	....	....	16	13	6	19	6	....	....	....
	Ottawa	**NHL**	**7**	**0**	**1**	**1**	**0**	**0**	**0**	**0**	**10**	**0.0**	**–1**	....	....	....	....	....	....	....	....	....	....	....	....	....
1998-99	**Ottawa**	**NHL**	**60**	**15**	**15**	**30**	**37**	**1**	**0**	**2**	**124**	**12.1**	**18**	**4**	**25.0**	**59**	**6**	**13:59**	**4**	**0**	**2**	**2**	**4**	**0**	**0**	**0**
99-2000	**Ottawa**	**NHL**	**78**	**29**	**27**	**56**	**32**	**5**	**0**	**4**	**240**	**12.1**	**5**	**7**	**57.1**	**100**	**18**	**17:12**	**6**	**0**	**0**	**0**	**2**	**0**	**0**	**0**
2000-01	**Ottawa**	**NHL**	**81**	**32**	**43**	**75**	**44**	**11**	**2**	**7**	**249**	**12.9**	**19**	**14**	**42.9**	**97**	**32**	**18:01**	**4**	**1**	**1**	**2**	**4**	**0**	**0**	**0**
2001-02	Dukla Trencin	Slovakia	8	3	4	7	16	....	....	....	....	....	....	....	....	....	....	....	....	....	....	....	....	....	....	....
	Ottawa	**NHL**	**80**	**31**	**35**	**66**	**50**	**9**	**1**	**4**	**278**	**11.2**	**11**	**12**	**33.3**	**82**	**18**	**18:29**	**12**	**4**	**6**	**10**	**2**	**1**	**0**	**0**
	Slovakia	Olympics	2	4	2	6	0	....	....	....	....	....	....	....	....	....	....	....	....	....	....	....	....	....	....	....
	NHL Totals		**306**	**107**	**121**	**228**	**163**	**26**	**3**	**17**	**901**	**11.9**		**37**	**40.5**	**338**	**74**	**17:07**	**26**	**5**	**9**	**14**	**12**	**1**	**0**	**0**

WHL West First All-Star Team (1998) • Canadian Major Junior First All-Star Team (1998) • Memorial Cup All-Star Team (1998) • NHL All-Rookie Team (1999) • Played in NHL All-Star Game (2001)

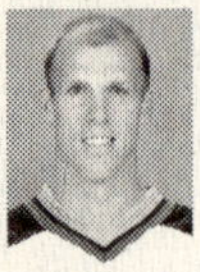

HOUDA, Doug (HOO-duh, DUHG) BUF.

Defense. Shoots right. 6'2", 208 lbs. Born, Blairmore, Alta., June 3, 1966. Detroit's 2nd choice, 28th overall, in 1984 Entry Draft.

			Regular Season																Playoffs							
Season	Club	League	GP	G	A	Pts	PIM	PP	SH	GW	S	%	+/-	TF	F%	H	SB	Min	GP	G	A	Pts	PIM	PP	SH	GW
1982-83	Calgary	WHL	71	5	23	28	99	....	....	....	....	....	....	....	....	....	....	....	16	1	3	4	44	....	....	....
1983-84	Calgary	WHL	69	6	30	36	195	....	....	....	....	....	....	....	....	....	....	....	4	0	0	0	7	....	....	....
1984-85	Calgary	WHL	65	20	54	74	182	....	....	....	....	....	....	....	....	....	....	....	8	3	4	7	29	....	....	....
	Kalamazoo Wings	IHL	....	....	....	....	....	....	....	....	....	....	....	....	....	....	....	....	7	0	2	2	10	....	....	....
1985-86	Calgary	WHL	16	4	10	14	60	....	....	....	....	....	....	....	....	....	....	....	....	....	....	....	....	....	....	....
	Medicine Hat	WHL	35	9	23	32	80	....	....	....	....	....	....	....	....	....	....	....	25	4	19	23	64	....	....	....
	Detroit	**NHL**	**6**	**0**	**0**	**0**	**4**	**0**	**0**	**0**	**5**	**0.0**	**–7**	....	....	....	....	....	....	....	....	....	....	....	....	....
1986-87	Adirondack	AHL	77	6	23	29	142	....	....	....	....	....	....	....	....	....	....	....	11	1	8	9	50	....	....	....
1987-88	**Detroit**	**NHL**	**11**	**1**	**1**	**2**	**10**	**0**	**0**	**0**	**10**	**10.0**	**0**	....	....	....	....	....	....	....	....	....	....	....	....	....
	Adirondack	AHL	71	10	32	42	169	....	....	....	....	....	....	....	....	....	....	....	11	0	3	3	44	....	....	....
1988-89	**Detroit**	**NHL**	**57**	**2**	**11**	**13**	**67**	**0**	**0**	**0**	**38**	**5.3**	**17**	....	....	....	....	....	**6**	**0**	**1**	**1**	**0**	**0**	**0**	**0**
	Adirondack	AHL	7	0	3	3	8	....	....	....	....	....	....	....	....	....	....	....	....	....	....	....	....	....	....	....
1989-90	**Detroit**	**NHL**	**73**	**2**	**9**	**11**	**127**	**0**	**0**	**0**	**59**	**3.4**	**–5**	....	....	....	....	....	....	....	....	....	....	....	....	....
1990-91	**Detroit**	**NHL**	**22**	**0**	**4**	**4**	**43**	**0**	**0**	**0**	**21**	**0.0**	**–2**	....	....	....	....	....	....	....	....	....	....	....	....	....
	Adirondack	AHL	38	9	17	26	67	....	....	....	....	....	....	....	....	....	....	....	....	....	....	....	....	....	....	....
	Hartford	**NHL**	**19**	**1**	**2**	**3**	**41**	**0**	**0**	**0**	**21**	**4.8**	**–3**	....	....	....	....	....	**6**	**0**	**0**	**0**	**8**	**0**	**0**	**0**
1991-92	**Hartford**	**NHL**	**56**	**3**	**6**	**9**	**125**	**1**	**0**	**1**	**40**	**7.5**	**–2**	....	....	....	....	....	**6**	**0**	**2**	**2**	**13**	**0**	**0**	**0**
1992-93	**Hartford**	**NHL**	**60**	**2**	**6**	**8**	**167**	**0**	**0**	**0**	**43**	**4.7**	**–19**	....	....	....	....	....	....	....	....	....	....	....	....	....
1993-94	**Hartford**	**NHL**	**7**	**0**	**0**	**0**	**23**	**0**	**0**	**0**	**1**	**0.0**	**–4**	....	....	....	....	....	....	....	....	....	....	....	....	....
	Los Angeles	**NHL**	**54**	**2**	**6**	**8**	**165**	**0**	**0**	**0**	**31**	**6.5**	**–15**	....	....	....	....	....	....	....	....	....	....	....	....	....
1994-95	**Buffalo**	**NHL**	**28**	**1**	**2**	**3**	**68**	**0**	**0**	**0**	**21**	**4.8**	**1**	....	....	....	....	....	....	....	....	....	....	....	....	....
1995-96	**Buffalo**	**NHL**	**38**	**1**	**3**	**4**	**52**	**0**	**0**	**0**	**21**	**4.8**	**3**	....	....	....	....	....	....	....	....	....	....	....	....	....
	Rochester	AHL	21	1	6	7	41	....	....	....	....	....	....	....	....	....	....	....	19	3	5	8	30	....	....	....
1996-97	**NY Islanders**	**NHL**	**70**	**2**	**8**	**10**	**99**	**0**	**0**	**0**	**29**	**6.9**	**1**	....	....	....	....	....	....	....	....	....	....	....	....	....
	Utah Grizzlies	IHL	3	0	0	0	7	....	....	....	....	....	....	....	....	....	....	....	....	....	....	....	....	....	....	....
1997-98	**NY Islanders**	**NHL**	**31**	**1**	**2**	**3**	**47**	**0**	**0**	**0**	**15**	**6.7**	**–6**	....	....	....	....	....	....	....	....	....	....	....	....	....
	Anaheim	**NHL**	**24**	**1**	**2**	**3**	**52**	**0**	**1**	**0**	**9**	**11.1**	**–5**	....	....	....	....	....	....	....	....	....	....	....	....	....
1998-99	**Detroit**	**NHL**	**3**	**0**	**1**	**1**	**0**	**0**	**0**	**0**	**1**	**0.0**	**–2**	**0**	**0.0**	**4**	**1**	**6:51**	....	....	....	....	....	....	....	....
	Adirondack	AHL	73	7	21	28	122	....	....	....	....	....	....	....	....	....	....	....	3	0	1	1	4	....	....	....
99-2000	**Buffalo**	**NHL**	**1**	**0**	**0**	**0**	**12**	**0**	**0**	**0**	**0**	**0.0**	**0**	**0**	**0.0**	**5**	**0**	**9:10**	....	....	....	....	....	....	....	....
	Rochester	AHL	79	7	17	24	175	....	....	....	....	....	....	....	....	....	....	....	21	1	8	9	39	....	....	....
2000-01	Rochester	AHL	43	6	20	26	106	....	....	....	....	....	....	....	....	....	....	....	4	0	0	0	4	....	....	....
2001-02	Rochester	AHL	64	6	22	28	170	....	....	....	....	....	....	....	....	....	....	....	2	0	0	0	4	....	....	....
	NHL Totals		**560**	**19**	**63**	**82**	**1102**	**1**	**1**	**1**	**365**	**5.2**		**0**	**0.0**	**9**	**1**	**7:26**	**18**	**0**	**3**	**3**	**21**	**0**	**0**	**0**

WHL East Second All-Star Team (1985) • AHL First All-Star Team (1988)

Traded to **Hartford** by **Detroit** for Doug Crossman, February 20, 1991. Traded to **LA Kings** by **Hartford** for Marc Potvin, November 3, 1993. Traded to **Buffalo** by **LA Kings** for Sean O'Donnell, July 26, 1994. Signed as a free agent by **NY Islanders**, October 26, 1996. Traded to **Anaheim** by **NY Islanders** with Travis Green and Tony Tuzzolino for Joe Sacco, J-J Daigneault and Mark Janssens, February 6, 1998. Traded to **Detroit** by **Anaheim** for future considerations, October 9, 1998. Signed as a free agent by **Buffalo**, July 13, 1999.

HOULDER, Bill

(HOHL-duhr, BIHL) **NSH.**

Defense. Shoots left. 6'2", 217 lbs. Born, Thunder Bay, Ont., March 11, 1967. Washington's 4th choice, 82nd overall, in 1985 Entry Draft.

			Regular Season																Playoffs							
Season	Club	League	GP	G	A	Pts	PIM	PP	SH	GW	S	%	+/-	TF	F%	H	SB	Min	GP	G	A	Pts	PIM	PP	SH	GW
1983-84	T. Bay Beavers	TBJHL	23	4	18	22	37																			
1984-85	North Bay	OHL	66	4	20	24	37												8	0	0	0	2			
1985-86	North Bay	OHL	59	5	30	35	97												10	1	6	7	12			
1986-87	North Bay	OHL	62	17	51	68	68												22	4	19	23	20			
1987-88	**Washington**	**NHL**	**30**	**1**	**2**	**3**	**10**	**0**	**0**	**0**	**20**	**5.0**	**–2**													
	Fort Wayne	IHL	43	10	14	24	32																			
1988-89	**Washington**	**NHL**	**8**	**0**	**3**	**3**	**4**	**0**	**0**	**0**	**5**	**0.0**	**7**													
	Baltimore	AHL	65	10	36	46	50																			
1989-90	**Washington**	**NHL**	**41**	**1**	**11**	**12**	**28**	**0**	**0**	**0**	**49**	**2.0**	**8**													
	Baltimore	AHL	26	3	7	10	12												7	0	2	2	2			
1990-91	**Buffalo**	**NHL**	**7**	**0**	**2**	**2**	**4**	**0**	**0**	**0**	**7**	**0.0**	**–2**													
	Rochester	AHL	69	13	53	66	28												15	5	13	18	4			
1991-92	**Buffalo**	**NHL**	**10**	**1**	**0**	**1**	**8**	**0**	**0**	**0**	**18**	**5.6**	**–2**													
	Rochester	AHL	42	8	26	34	16												16	5	6	11	4			
1992-93	**Buffalo**	**NHL**	**15**	**3**	**5**	**8**	**6**	**0**	**0**	**0**	**29**	**10.3**	**5**						**8**	**0**	**2**	**2**	**4**	**0**	**0**	**0**
	San Diego Gulls	IHL	64	24	48	72	39																			
1993-94	**Anaheim**	**NHL**	**80**	**14**	**25**	**39**	**40**	**3**	**0**	**3**	**187**	**7.5**	**–18**													
1994-95	**St. Louis**	**NHL**	**41**	**5**	**13**	**18**	**20**	**1**	**0**	**0**	**59**	**8.5**	**16**						**4**	**1**	**1**	**2**	**0**	**0**	**0**	**0**
1995-96	**Tampa Bay**	**NHL**	**61**	**5**	**23**	**28**	**22**	**3**	**0**	**0**	**90**	**5.6**	**1**						**6**	**0**	**1**	**1**	**4**	**0**	**0**	**0**
1996-97	**Tampa Bay**	**NHL**	**79**	**4**	**21**	**25**	**30**	**0**	**0**	**2**	**116**	**3.4**	**16**													
1997-98	**San Jose**	**NHL**	**82**	**7**	**25**	**32**	**48**	**4**	**0**	**2**	**102**	**6.9**	**13**						**6**	**1**	**2**	**3**	**2**	**0**	**0**	**0**
1998-99	**San Jose**	**NHL**	**76**	**9**	**23**	**32**	**40**	**7**	**0**	**5**	**115**	**7.8**	**8**	**0**	**0.0**	**70**	**83**	**22:08**	**6**	**3**	**0**	**3**	**4**	**3**	**0**	**0**
99-2000	**Tampa Bay**	**NHL**	**14**	**1**	**2**	**3**	**2**	**1**	**0**	**0**	**21**	**4.8**	**–3**	**1**	**100.0**	**12**	**26**	**21:29**								
	Nashville	**NHL**	**57**	**2**	**12**	**14**	**24**	**1**	**0**	**1**	**68**	**2.9**	**–6**	**1**	**100.0**	**56**	**71**	**22:36**								
2000-01	**Nashville**	**NHL**	**81**	**4**	**12**	**16**	**40**	**0**	**1**	**1**	**78**	**5.1**	**–7**	**2**	**0.0**	**66**	**76**	**21:12**								
2001-02	**Nashville**	**NHL**	**82**	**0**	**8**	**8**	**40**	**0**	**0**	**0**	**44**	**0.0**	**–1**	**3**	**0.0**	**53**	**124**	**21:33**								
	NHL Totals		**764**	**57**	**187**	**244**	**366**	**20**	**1**	**14**	**1008**	**5.7**		**7**	**28.6**	**257**	**380**	**21:48**	**30**	**5**	**6**	**11**	**14**	**3**	**0**	**0**

AHL First All-Star Team (1991) • Won Governor's Trophy (Top Defenseman - IHL) (1993) • IHL First All-Star Team (1993)

Traded to **Buffalo** by **Washington** for Shawn Anderson, September 30, 1990. Claimed by **Anaheim** from **Buffalo** in Expansion Draft, June 24, 1993. Traded to **St. Louis** by **Anaheim** for Jason Marshall, August 29, 1994. Signed as a free agent by **Tampa Bay**, July 26, 1995. Signed as a free agent by **San Jose**, July 16, 1997. Traded to **Tampa Bay** by **San Jose** with Andrei Zyuzin, Shawn Burr and Steve Guolla for Niklas Sundstrom and NY Rangers' 3rd round choice (previously acquired, later traded to Chicago - Chicago selected Igor Radulov) in 2000 Entry Draft, August 4, 1999. Claimed on waivers by **Nashville** from **Tampa Bay**, November 10, 1999.

HOUSLEY, Phil

(HOWZ-lee, FIHL) **CHI.**

Defense. Shoots left. 5'10", 185 lbs. Born, St. Paul, MN, March 9, 1964. Buffalo's 1st choice, 6th overall, in 1982 Entry Draft.

Season	Club	League	GP	G	A	Pts	PIM	PP	SH	GW	S	%	+/-	TF	F%	H	SB	Min	GP	G	A	Pts	PIM	PP	SH	GW
1980-81	St. Paul Vulcans	USHL	6	7	7	14	6												10	5	5	10	0			
1981-82	South St. Paul	Hi-School	22	31	34	65	18																			
1982-83	**Buffalo**	**NHL**	**77**	**19**	**47**	**66**	**39**	**11**	**0**	**2**	**183**	**10.4**	**–4**						**10**	**3**	**4**	**7**	**2**	**1**	**0**	**0**
1983-84	**Buffalo**	**NHL**	**75**	**31**	**46**	**77**	**33**	**13**	**2**	**6**	**234**	**13.2**	**3**						**3**	**0**	**0**	**0**	**6**	**0**	**0**	**0**
1984-85	**Buffalo**	**NHL**	**73**	**16**	**53**	**69**	**28**	**3**	**0**	**4**	**188**	**8.5**	**15**						**5**	**3**	**2**	**5**	**2**	**0**	**0**	**0**
1985-86	**Buffalo**	**NHL**	**79**	**15**	**47**	**62**	**54**	**7**	**0**	**2**	**180**	**8.3**	**–9**													
1986-87	**Buffalo**	**NHL**	**78**	**21**	**46**	**67**	**57**	**8**	**1**	**2**	**202**	**10.4**	**–2**													
1987-88	**Buffalo**	**NHL**	**74**	**29**	**37**	**66**	**96**	**6**	**0**	**1**	**231**	**12.6**	**–17**						**6**	**2**	**4**	**6**	**6**	**1**	**0**	**0**
1988-89	**Buffalo**	**NHL**	**72**	**26**	**44**	**70**	**47**	**5**	**0**	**3**	**178**	**14.6**	**6**						**5**	**1**	**3**	**4**	**2**	**0**	**0**	**0**
1989-90	**Buffalo**	**NHL**	**80**	**21**	**60**	**81**	**32**	**8**	**1**	**4**	**201**	**10.4**	**11**						**6**	**1**	**4**	**5**	**4**	**1**	**0**	**0**
1990-91	**Winnipeg**	**NHL**	**78**	**23**	**53**	**76**	**24**	**12**	**1**	**3**	**206**	**11.2**	**–13**													
1991-92	**Winnipeg**	**NHL**	**74**	**23**	**63**	**86**	**92**	**11**	**0**	**4**	**234**	**9.8**	**–5**						**7**	**1**	**4**	**5**	**0**	**1**	**0**	**1**
1992-93	**Winnipeg**	**NHL**	**80**	**18**	**79**	**97**	**52**	**6**	**0**	**2**	**249**	**7.2**	**–14**						**6**	**0**	**7**	**7**	**2**	**0**	**0**	**0**
1993-94	**St. Louis**	**NHL**	**26**	**7**	**15**	**22**	**12**	**4**	**0**	**1**	**60**	**11.7**	**–5**						**4**	**2**	**1**	**3**	**4**	**2**	**0**	**0**
1994-95	Zurcher SC	Swiss	10	6	8	14	34																			
	Calgary	**NHL**	**43**	**8**	**35**	**43**	**18**	**3**	**0**	**0**	**135**	**5.9**	**17**						**7**	**0**	**9**	**9**	**0**	**0**	**0**	**0**
1995-96	**Calgary**	**NHL**	**59**	**16**	**36**	**52**	**22**	**6**	**0**	**1**	**155**	**10.3**	**–2**													
	New Jersey	**NHL**	**22**	**1**	**15**	**16**	**8**	**0**	**0**	**0**	**50**	**2.0**	**–4**													
1996-97	**Washington**	**NHL**	**77**	**11**	**29**	**40**	**24**	**3**	**1**	**2**	**167**	**6.6**	**–10**													
1997-98	**Washington**	**NHL**	**64**	**6**	**25**	**31**	**24**	**4**	**1**	**0**	**116**	**5.2**	**–10**						**18**	**0**	**4**	**4**	**4**	**0**	**0**	**0**
1998-99	**Calgary**	**NHL**	**79**	**11**	**43**	**54**	**52**	**4**	**0**	**1**	**193**	**5.7**	**14**	**0**	**0.0**	**21**	**52**	**20:52**								
99-2000	**Calgary**	**NHL**	**78**	**11**	**44**	**55**	**24**	**5**	**0**	**2**	**176**	**6.3**	**–12**	**1**	**0.0**	**23**	**50**	**23:29**								
2000-01	**Calgary**	**NHL**	**69**	**4**	**30**	**34**	**24**	**0**	**0**	**0**	**115**	**3.5**	**–15**	**0**	**0.0**	**25**	**40**	**18:10**								
2001-02	**Chicago**	**NHL**	**80**	**15**	**24**	**39**	**34**	**8**	**0**	**6**	**218**	**6.9**	**–3**	**0**	**0.0**	**43**	**55**	**21:46**	**5**	**0**	**1**	**1**	**4**	**0**	**0**	**0**
	United States	Olympics	6	1	4	5	0																			
	NHL Totals		**1437**	**332**	**871**	**1203**	**796**	**127**	**7**	**46**	**3671**	**9.0**		**1**	**0.0**	**112**	**197**	**21:10**	**82**	**13**	**43**	**56**	**36**	**6**	**0**	**1**

NHL All-Rookie Team (1983) • NHL Second All-Star Team (1992) • Played in NHL All-Star Game (1984, 1989, 1990, 1991, 1992, 1993, 2000)

Traded to **Winnipeg** by **Buffalo** with Scott Arniel, Jeff Parker and Buffalo's 1st round choice (Keith Tkachuk) in 1990 Entry Draft for Dale Hawerchuk and Winnipeg's 1st round choice (Brad May) in 1990 Entry Draft, June 16, 1990. Traded to **St. Louis** by **Winnipeg** for Nelson Emerson and Stephane Quintal, September 24, 1993. Traded to **Calgary** by **St. Louis** with St. Louis' 2nd round choices in 1996 (Steve Begin) and 1997 (John Tripp) Entry Drafts for Al MacInnis and Calgary's 4th round choice (Didier Tremblay) in 1997 Entry Draft, July 4, 1994. Traded to **New Jersey** by **Calgary** with Dan Keczmer for Tommy Albelin, Cale Hulse and Jocelyn Lemieux, February 26, 1996. Signed as a free agent by **Washington**, July 22, 1996. Claimed on waivers by **Calgary** from **Washington**, July 21, 1998. Claimed by **Chicago** from **Calgary** in Waiver Draft, September 28, 2001.

HRDINA, Jan

(huhr-DEE-nah, YAN) **PIT.**

Center. Shoots right. 6', 206 lbs. Born, Hradec Kralove, Czech., February 5, 1976. Pittsburgh's 4th choice, 128th overall, in 1995 Entry Draft.

Season	Club	League	GP	G	A	Pts	PIM	PP	SH	GW	S	%	+/-	TF	F%	H	SB	Min	GP	G	A	Pts	PIM	PP	SH	GW
1993-94	H. Kralove Jr.	Czech-Jr.	10	1	6	7	0																			
	Hradec Kralove	Czech	23	1	5	6													4	0	1	1				
1994-95	Seattle	WHL	69	41	59	100	79												4	0	1	1	8			
1995-96	Seattle	WHL	30	19	28	47	37																			
	Spokane Chiefs	WHL	18	10	16	26	25												18	5	14	19	49			
1996-97	Cleveland	IHL	68	23	31	54	82												13	1	2	3	8			
1997-98	Syracuse Crunch	AHL	72	20	24	44	82												5	1	3	4	10			
1998-99	**Pittsburgh**	**NHL**	**82**	**13**	**29**	**42**	**40**	**3**	**0**	**2**	**94**	**13.8**	**–2**	**1461**	**56.7**	**104**	**26**	**16:26**	**13**	**4**	**1**	**5**	**12**	**1**	**0**	**1**
99-2000	**Pittsburgh**	**NHL**	**70**	**13**	**33**	**46**	**43**	**3**	**0**	**1**	**84**	**15.5**	**13**	**1392**	**53.7**	**57**	**24**	**18:47**	**9**	**4**	**8**	**12**	**2**	**1**	**0**	**0**
2000-01	**Pittsburgh**	**NHL**	**78**	**15**	**28**	**43**	**48**	**3**	**0**	**1**	**89**	**16.9**	**19**	**1067**	**53.8**	**47**	**24**	**15:56**	**18**	**2**	**5**	**7**	**8**	**0**	**0**	**0**
2001-02	**Pittsburgh**	**NHL**	**79**	**24**	**33**	**57**	**50**	**6**	**0**	**6**	**115**	**20.9**	**–7**	**667**	**50.4**	**41**	**26**	**19:51**								
	Czech Republic	Olympics	4	0	0	0	0																			
	NHL Totals		**309**	**65**	**123**	**188**	**181**	**15**	**0**	**10**	**382**	**17.0**		**4587**	**54.2**	**249**	**100**	**17:43**	**40**	**10**	**14**	**24**	**22**	**2**	**0**	**1**

HRKAC, Tony

(HUHR-kuhz, TOH-nee) **ATL.**

Center. Shoots left. 5'10", 190 lbs. Born, Thunder Bay, Ont., July 7, 1966. St. Louis' 2nd choice, 32nd overall, in 1984 Entry Draft.

Season	Club	League	GP	G	A	Pts	PIM	PP	SH	GW	S	%	+/-	TF	F%	H	SB	Min	GP	G	A	Pts	PIM	PP	SH	GW
1983-84	Orillia	OPJHL	42	*52	54	*106	20																			
1984-85	North Dakota	WCHA	36	18	36	54	16																			
1985-86	Team Canada	Nat-Tm	62	19	30	49	36																			
1986-87	North Dakota	WCHA	48	46	70	116	48																			
	St. Louis	**NHL**																	**3**	**0**	**0**	**0**	**0**	**0**	**0**	**0**
1987-88	**St. Louis**	**NHL**	**67**	**11**	**37**	**48**	**22**	**2**	**1**	**3**	**86**	**12.8**	**5**						**10**	**6**	**1**	**7**	**4**	**3**	**1**	**1**
1988-89	**St. Louis**	**NHL**	**70**	**17**	**28**	**45**	**8**	**5**	**0**	**1**	**133**	**12.8**	**–10**						**4**	**1**	**1**	**2**	**0**	**0**	**0**	**1**
1989-90	**St. Louis**	**NHL**	**28**	**5**	**12**	**17**	**8**	**1**	**0**	**0**	**41**	**12.2**	**1**													
	Quebec	**NHL**	**22**	**4**	**8**	**12**	**2**	**2**	**0**	**0**	**29**	**13.8**	**–5**													
	Halifax Citadels	AHL	20	12	21	33	4												6	5	9	14	4			
1990-91	**Quebec**	**NHL**	**70**	**16**	**32**	**48**	**16**	**6**	**0**	**0**	**122**	**13.1**	**–22**													
	Halifax Citadels	AHL	3	4	1	5	2																			
1991-92	**San Jose**	**NHL**	**22**	**2**	**10**	**12**	**4**	**0**	**0**	**0**	**31**	**6.5**	**–2**													
	Chicago	**NHL**	**18**	**1**	**2**	**3**	**6**	**0**	**0**	**0**	**22**	**4.5**	**4**						**3**	**0**	**0**	**0**	**2**	**0**	**0**	**0**
1992-93	Indianapolis Ice	IHL	80	45	*87	*132	70												5	0	2	2	2			
1993-94	**St. Louis**	**NHL**	**36**	**6**	**5**	**11**	**8**	**1**	**1**	**1**	**43**	**14.0**	**–11**						**4**	**0**	**0**	**0**	**0**	**0**	**0**	**0**
	Peoria Rivermen	IHL	45	30	51	81	25												1	1	2	3	0			
1994-95	Milwaukee	IHL	71	24	67	91	26												15	4	9	13	16			
1995-96	Milwaukee	IHL	43	14	28	42	18												5	1	3	4	4			

Season	Club	League	GP	G	A	Pts	PIM	PP	SH	GW	S	%	+/-	TF	F%	H	SB	Min	GP	G	A	Pts	PIM	PP	SH	GW
								Regular Season											Playoffs							
1996-97	Milwaukee	IHL	81	27	61	88	20	....	....	....	....	....	....	....	....	....	....	....	3	1	1	2	2	....	....	....
1997-98	**Dallas**	**NHL**	**13**	**5**	**3**	**8**	**0**	**3**	**0**	**0**	**14**	**35.7**	**0**	....	....	....	....	....	....	....	....	....	....	....	....	....
	Michigan K-Wings	IHL	20	7	15	22	6	....	....	....	....	....	....	....	....	....	....	....	....	....	....	....	....	....	....	....
	Edmonton	**NHL**	**36**	**8**	**11**	**19**	**10**	**4**	**0**	**1**	**43**	**18.6**	**3**	....	....	....	....	....	**12**	**0**	**3**	**3**	**2**	**0**	**0**	**0**
1998-99♦	**Dallas**	**NHL**	**69**	**13**	**14**	**27**	**26**	**2**	**0**	**2**	**67**	**19.4**	**2**	**666**	**48.0**	**49**	**13**	**12:02**	**5**	**0**	**2**	**2**	**4**	**0**	**0**	**0**
99-2000	**NY Islanders**	**NHL**	**7**	**0**	**2**	**2**	**0**	**0**	**0**	**0**	**2**	**0.0**	**–1**	**34**	**35.3**	**2**	**0**	**11:22**	....	....	....	....	....	....	....	....
	Anaheim	**NHL**	**60**	**4**	**7**	**11**	**8**	**1**	**0**	**0**	**37**	**10.8**	**–2**	**536**	**50.8**	**22**	**11**	**9:04**	....	....	....	....	....	....	....	....
2000-01	**Anaheim**	**NHL**	**80**	**13**	**25**	**38**	**29**	**0**	**0**	**1**	**88**	**14.8**	**0**	**1072**	**50.7**	**22**	**17**	**13:46**	....	....	....	....	....	....	....	....
2001-02	**Atlanta**	**NHL**	**80**	**18**	**26**	**44**	**12**	**5**	**1**	**2**	**101**	**17.8**	**–12**	**935**	**47.5**	**21**	**34**	**17:29**	....	....	....	....	....	....	....	....
	NHL Totals		**678**	**123**	**222**	**345**	**159**	**32**	**3**	**11**	**859**	**14.3**		**3243**	**49.1**	**116**	**75**	**13:22**	**41**	**7**	**7**	**14**	**12**	**3**	**1**	**2**

WCHA First All-Star Team (1987) • WCHA Player of the Year (1987) • NCAA West First All-American Team (1987) • NCAA Championship All-Tournament Team (1987) • NCAA Championship Tournament MVP (1987) • Won Hobey Baker Memorial Award (Top U.S. Collegiate Player) (1987) • Won James Gatschene Memorial Trophy (MVP - IHL) (1993) • Won Leo P. Lamoureux Memorial Trophy (Top Scorer - IHL) (1993) • IHL First All-Star Team (1993)

Traded to **Quebec** by **St. Louis** with Greg Millen for Jeff Brown, December 13, 1989. Traded to **San Jose** by **Quebec** for Greg Paslawski, May 31, 1991. Traded to **Chicago** by **San Jose** for Chicago's 6th round choice (Fredrik Oduya) in 1993 Entry Draft, February 7, 1992. Signed as a free agent by **St. Louis**, July 30, 1993. Signed as a free agent by **Dallas**, August 12, 1997. Claimed on waivers by **Edmonton** from **Dallas**, January 6, 1998. Traded to **Pittsburgh** by **Edmonton** with Bobby Dollas for Josef Beranek, June 16, 1998. Claimed by **Nashville** from **Pittsburgh** in Expansion Draft, June 26, 1998. Traded to **Dallas** by **Nashville** for future considerations, July 9, 1998. Signed as a free agent by **NY Islanders**, July 29, 1999. Traded to **Anaheim** by **NY Islanders** with Dean Malkoc for Ted Drury, October 29, 1999. Signed as a free agent by **Atlanta**, July 25, 2001.

HUBACEK, Petr

(HOO-buh-chehk, PEE-tuhr) **NSH.**

Center. Shoots right. 6'2", 183 lbs. Born, Brno, Czech., September 2, 1979. Philadelphia's 11th choice, 243rd overall, in 1998 Entry Draft.

Season	Club	League	GP	G	A	Pts	PIM	PP	SH	GW	S	%	+/-	TF	F%	H	SB	Min	GP	G	A	Pts	PIM	PP	SH	GW
1997-98	Kometa Brno Jr.	Czech-Jr.	17	9	5	14	....	....	....	....	....	....	....	....	....	....	....	....	....	....	....	....	....	....	....	....
	HC Kometa Brno	Czech-2	48	6	10	16	....	....	....	....	....	....	....	....	....	....	....	....	....	....	....	....	....	....	....	....
1998-99	HC Vitkovice	Czech	25	0	4	4	2	....	....	....	....	....	....	....	....	....	....	....	4	0	0	0		....	....	....
99-2000	HC Vitkovice	Czech	48	11	12	23	81	....	....	....	....	....	....	....	....	....	....	....	....	....	....	....	....	....	....	....
2000-01	**Philadelphia**	**NHL**	**6**	**1**	**0**	**1**	**2**	**0**	**0**	**0**	**5**	**20.0**	**–1**	**39**	**25.6**	**1**	**3**	**11:20**	....	....	....	....	....	....	....	....
	Philadelphia	AHL	62	3	9	12	29	....	....	....	....	....	....	....	....	....	....	....	9	0	1	1	6	....	....	....
2001-02	Philadelphia	AHL	22	1	6	7	8	....	....	....	....	....	....	....	....	....	....	....	....	....	....	....	....	....	....	....
	Milwaukee	AHL	14	2	0	2	0	....	....	....	....	....	....	....	....	....	....	....	....	....	....	....	....	....	....	....
	NHL Totals		**6**	**1**	**0**	**1**	**2**	**0**	**0**	**0**	**5**	**20.0**		**39**	**25.6**	**1**	**3**	**11:20**	....	....	....	....	....	....	....	....

Traded to **Nashville** by **Philadelphia** with Jason Beckett for Yves Sarault and a conditional choice in 2003 Entry Draft, January 11, 2002. Signed as a free agent by **Zlin** (Czech) with Nashville retaining NHL rights, August 4, 2002.

HULBIG, Joe

(HUHL-bihg, JOH)

Left wing. Shoots left. 6'3", 215 lbs. Born, Norwood, MA, September 29, 1973. Edmonton's 1st choice, 13th overall, in 1992 Entry Draft.

Season	Club	League	GP	G	A	Pts	PIM	PP	SH	GW	S	%	+/-	TF	F%	H	SB	Min	GP	G	A	Pts	PIM	PP	SH	GW
1989-90	St. Sebastian's	Hi-School	30	13	12	25	....	....	....	....	....	....	....	....	....	....	....	....	....	....	....	....	....	....	....	....
1990-91	St. Sebastian's	Hi-School	30	23	19	42	....	....	....	....	....	....	....	....	....	....	....	....	....	....	....	....	....	....	....	....
1991-92	St. Sebastian's	Hi-School	17	19	24	43	30	....	....	....	....	....	....	....	....	....	....	....	....	....	....	....	....	....	....	....
1992-93	Providence	H-East	26	3	13	16	22	....	....	....	....	....	....	....	....	....	....	....	....	....	....	....	....	....	....	....
1993-94	Providence	H-East	28	6	4	10	36	....	....	....	....	....	....	....	....	....	....	....	....	....	....	....	....	....	....	....
1994-95	Providence	H-East	37	14	21	35	36	....	....	....	....	....	....	....	....	....	....	....	....	....	....	....	....	....	....	....
1995-96	Providence	H-East	31	14	22	36	56	....	....	....	....	....	....	....	....	....	....	....	....	....	....	....	....	....	....	....
1996-97	**Edmonton**	**NHL**	**6**	**0**	**0**	**0**	**0**	**0**	**0**	**0**	**4**	**0.0**	**–1**	....	....	....	....	....	**6**	**0**	**1**	**1**	**2**	**0**	**0**	**0**
	Hamilton	AHL	73	18	28	46	59	....	....	....	....	....	....	....	....	....	....	....	16	6	10	16	6	....	....	....
1997-98	**Edmonton**	**NHL**	**17**	**2**	**2**	**4**	**2**	**0**	**0**	**1**	**8**	**25.0**	**–1**	....	....	....	....	....	....	....	....	....	....	....	....	....
	Hamilton	AHL	46	15	16	31	52	....	....	....	....	....	....	....	....	....	....	....	3	0	1	1	2	....	....	....
1998-99	**Edmonton**	**NHL**	**1**	**0**	**0**	**0**	**2**	**0**	**0**	**0**	**2**	**0.0**	**1**	**0**	**0.0**	**0**	**1**	**8:20**	....	....	....	....	....	....	....	....
	Hamilton	AHL	76	22	24	46	68	....	....	....	....	....	....	....	....	....	....	....	11	4	2	6	18	....	....	....
99-2000	**Boston**	**NHL**	**24**	**2**	**2**	**4**	**8**	**0**	**0**	**0**	**15**	**13.3**	**–8**	**2**	**0.0**	**52**	**4**	**8:18**	....	....	....	....	....	....	....	....
	Providence	AHL	15	4	5	9	17	....	....	....	....	....	....	....	....	....	....	....	....	....	....	....	....	....	....	....
2000-01	**Boston**	**NHL**	**7**	**0**	**0**	**0**	**4**	**0**	**0**	**0**	**0**	**0.0**	**–3**	**0**	**0.0**	**15**	**0**	**5:28**	....	....	....	....	....	....	....	....
	Providence	AHL	36	4	11	15	19	....	....	....	....	....	....	....	....	....	....	....	15	2	2	4	20	....	....	....
2001-02	Providence	AHL	54	8	10	18	41	....	....	....	....	....	....	....	....	....	....	....	....	....	....	....	....	....	....	....
	Worcester	AHL	7	0	3	3	2	....	....	....	....	....	....	....	....	....	....	....	3	1	0	1	7	....	....	....
	NHL Totals		**55**	**4**	**4**	**8**	**16**	**0**	**0**	**1**	**29**	**13.8**		**2**	**0.0**	**67**	**5**	**7:41**	**6**	**0**	**1**	**1**	**2**	**0**	**0**	**0**

Signed as a free agent by **Boston**, July 23, 1999. • Missed majority of 2000-01 season recovering from head injury suffered in game vs. Ottawa, November 9, 2000.

HULL, Brett

(HUHL, BREHT) **DET.**

Right wing. Shoots right. 5'11", 203 lbs. Born, Belleville, Ont., August 9, 1964. Calgary's 6th choice, 117th overall, in 1984 Entry Draft.

Season	Club	League	GP	G	A	Pts	PIM	PP	SH	GW	S	%	+/-	TF	F%	H	SB	Min	GP	G	A	Pts	PIM	PP	SH	GW
1982-83	Penticton	BCJHL	50	48	56	104	27	....	....	....	....	....	....	....	....	....	....	....	....	....	....	....	....	....	....	....
1983-84	Penticton	BCJHL	56	*105	83	*188	20	....	....	....	....	....	....	....	....	....	....	....	....	....	....	....	....	....	....	....
1984-85	U. Minn-Duluth	WCHA	48	32	28	60	24	....	....	....	....	....	....	....	....	....	....	....	....	....	....	....	....	....	....	....
1985-86	U. Minn-Duluth	WCHA	42	52	32	84	46	....	....	....	....	....	....	....	....	....	....	....	....	....	....	....	....	....	....	....
	Calgary	**NHL**	....	....	....	....	....	....	....	....	....	....	....	....	....	....	....	....	**2**	**0**	**0**	**0**	**0**	**0**	**0**	**0**
1986-87	**Calgary**	**NHL**	**5**	**1**	**0**	**1**	**0**	**0**	**0**	**1**	**5**	**20.0**	**–1**	....	....	....	....	....	**4**	**2**	**1**	**3**	**0**	**0**	**0**	**0**
	Moncton	AHL	67	50	42	92	16	....	....	....	....	....	....	....	....	....	....	....	3	2	2	4	2	....	....	....
1987-88	**Calgary**	**NHL**	**52**	**26**	**24**	**50**	**12**	**4**	**0**	**3**	**153**	**17.0**	**10**	....	....	....	....	....	....	....	....	....	....	....	....	....
	St. Louis	**NHL**	**13**	**6**	**8**	**14**	**4**	**2**	**0**	**0**	**58**	**10.3**	**4**	....	....	....	....	....	**10**	**7**	**2**	**9**	**4**	**4**	**0**	**3**
1988-89	**St. Louis**	**NHL**	**78**	**41**	**43**	**84**	**33**	**16**	**0**	**6**	**305**	**13.4**	**–17**	....	....	....	....	....	**10**	**5**	**5**	**10**	**6**	**1**	**0**	**2**
1989-90	**St. Louis**	**NHL**	**80**	***72**	**41**	**113**	**24**	**27**	**0**	**12**	**385**	**18.7**	**–1**	....	....	....	....	....	**12**	**13**	**8**	**21**	**17**	**7**	**0**	**3**
1990-91	**St. Louis**	**NHL**	**78**	***86**	**45**	**131**	**22**	**29**	**0**	**11**	**389**	**22.1**	**23**	....	....	....	....	....	**13**	**11**	**8**	**19**	**4**	**3**	**0**	**2**
1991-92	**St. Louis**	**NHL**	**73**	***70**	**39**	**109**	**48**	**20**	**5**	**9**	**408**	**17.2**	**–2**	....	....	....	....	....	**6**	**4**	**4**	**8**	**4**	**1**	**1**	**1**
1992-93	**St. Louis**	**NHL**	**80**	**54**	**47**	**101**	**41**	**29**	**0**	**2**	**390**	**13.8**	**–27**	....	....	....	....	....	**11**	**8**	**5**	**13**	**2**	**5**	**0**	**2**
1993-94	**St. Louis**	**NHL**	**81**	**57**	**40**	**97**	**38**	**25**	**3**	**6**	**392**	**14.5**	**–3**	....	....	....	....	....	**4**	**2**	**1**	**3**	**0**	**1**	**0**	**0**
1994-95	**St. Louis**	**NHL**	**48**	**29**	**21**	**50**	**10**	**9**	**3**	**6**	**200**	**14.5**	**13**	....	....	....	....	....	**7**	**6**	**2**	**8**	**0**	**2**	**0**	**0**
1995-96	**St. Louis**	**NHL**	**70**	**43**	**40**	**83**	**30**	**16**	**5**	**6**	**327**	**13.1**	**4**	....	....	....	....	....	**13**	**6**	**5**	**11**	**10**	**2**	**1**	**1**
1996-97	**St. Louis**	**NHL**	**77**	**42**	**40**	**82**	**10**	**12**	**2**	**6**	**302**	**13.9**	**–9**	....	....	....	....	....	**6**	**2**	**7**	**9**	**2**	**0**	**0**	**0**
1997-98	**St. Louis**	**NHL**	**66**	**27**	**45**	**72**	**26**	**10**	**0**	**6**	**211**	**12.8**	**–1**	....	....	....	....	....	**10**	**3**	**3**	**6**	**2**	**1**	**0**	**1**
	United States	Olympics	4	2	1	3	0	....	....	....	....	....	....	....	....	....	....	....	....	....	....	....	....	....	....	....
1998-99♦	**Dallas**	**NHL**	**60**	**32**	**26**	**58**	**30**	**15**	**0**	**11**	**192**	**16.7**	**19**	**12**	**50.0**	**9**	**18**	**17:24**	**22**	**8**	**7**	**15**	**4**	**3**	**0**	**2**
99-2000	**Dallas**	**NHL**	**79**	**24**	**35**	**59**	**43**	**11**	**0**	**3**	**223**	**10.8**	**–21**	**10**	**40.0**	**27**	**18**	**18:37**	**23**	***11**	***13**	***24**	**4**	**3**	**0**	**4**
2000-01	**Dallas**	**NHL**	**79**	**39**	**40**	**79**	**18**	**11**	**0**	**8**	**219**	**17.8**	**10**	**10**	**30.0**	**31**	**20**	**17:53**	**10**	**2**	**5**	**7**	**6**	**1**	**0**	**0**
2001-02♦	**Detroit**	**NHL**	**82**	**30**	**33**	**63**	**35**	**7**	**1**	**4**	**247**	**12.1**	**18**	**5**	**20.0**	**33**	**25**	**18:49**	**23**	***10**	**8**	**18**	**4**	**3**	**2**	**2**
	United States	Olympics	6	3	5	8	6	....	....	....	....	....	....	....	....	....	....	....	....	....	....	....	....	....	....	....
	NHL Totals		**1101**	**679**	**567**	**1246**	**424**	**243**	**19**	**100**	**4406**	**15.4**		**37**	**37.8**	**100**	**81**	**18:14**	**186**	**100**	**84**	**184**	**69**	**37**	**4**	**23**

• BCJHL Interior Division First All-Star Team (1983, 1984) • WCHA Freshman of the Year (1985) • WCHA First All-Star Team (1986) • AHL First All-Star Team (1987) • Won Dudley "Red" Garrett Memorial Trophy (Top Rookie - AHL) (1987) • NHL First All-Star Team (1990, 1991, 1992) • Won Lady Byng Trophy (1990) • Won Dodge Ram Tough Award (1990, 1991) • Won Hart Memorial Trophy (1991) • Won Lester B. Pearson Award (1991) • Won ProSet/NHL Player of the Year Award (1991) • Played in NHL All-Star Game (1989, 1990, 1992, 1993, 1994, 1996, 1997, 2001)

Traded to **St. Louis** by **Calgary** with Steve Bozek for Rob Ramage and Rick Wamsley, March 7, 1988. Signed as a free agent by **Dallas**, July 3, 1998. Signed as a free agent by **Detroit**, August 22, 2001.

HULL, Jody

(HUHL, JOH-dee)

Right wing. Shoots right. 6'2", 200 lbs. Born, Petrolia, Ont., February 2, 1969. Hartford's 1st choice, 18th overall, in 1987 Entry Draft.

Season	Club	League	GP	G	A	Pts	PIM	PP	SH	GW	S	%	+/-	TF	F%	H	SB	Min	GP	G	A	Pts	PIM	PP	SH	GW
1984-85	Cambridge	OJHL-B	38	13	17	30	39	....	....	....	....	....	....	....	....	....	....	....	....	....	....	....	....	....	....	....
1985-86	Peterborough	OHL	61	20	22	42	29	....	....	....	....	....	....	....	....	....	....	....	16	1	5	6	4	....	....	....
1986-87	Peterborough	OHL	49	18	34	52	22	....	....	....	....	....	....	....	....	....	....	....	12	4	9	13	14	....	....	....
1987-88	Peterborough	OHL	60	50	44	94	33	....	....	....	....	....	....	....	....	....	....	....	12	10	8	18	8	....	....	....
1988-89	**Hartford**	**NHL**	**60**	**16**	**18**	**34**	**10**	**6**	**0**	**2**	**82**	**19.5**	**6**	....	....	....	....	....	**1**	**0**	**0**	**0**	**2**	**0**	**0**	**0**
1989-90	**Hartford**	**NHL**	**38**	**7**	**10**	**17**	**21**	**2**	**0**	**0**	**46**	**15.2**	**–6**	....	....	....	....	....	**5**	**0**	**1**	**1**	**2**	**0**	**0**	**0**
	Binghamton	AHL	21	7	10	17	6	....	....	....	....	....	....	....	....	....	....	....	....	....	....	....	....	....	....	....
1990-91	**NY Rangers**	**NHL**	**47**	**5**	**8**	**13**	**10**	**0**	**0**	**0**	**57**	**8.8**	**2**	....	....	....	....	....	....	....	....	....	....	....	....	....
1991-92	**NY Rangers**	**NHL**	**3**	**0**	**0**	**0**	**2**	**0**	**0**	**0**	**4**	**0.0**	**–4**	....	....	....	....	....	....	....	....	....	....	....	....	....
	Binghamton	AHL	69	34	31	65	28	....	....	....	....	....	....	....	....	....	....	....	11	5	2	7	4	....	....	....
1992-93	**Ottawa**	**NHL**	**69**	**13**	**21**	**34**	**14**	**5**	**1**	**0**	**134**	**9.7**	**–24**	....	....	....	....	....	....	....	....	....	....	....	....	....
1993-94	**Florida**	**NHL**	**69**	**13**	**13**	**26**	**8**	**0**	**1**	**5**	**100**	**13.0**	**6**	....	....	....	....	....	....	....	....	....	....	....	....	....
1994-95	**Florida**	**NHL**	**46**	**11**	**8**	**19**	**8**	**0**	**0**	**4**	**63**	**17.5**	**–1**	....	....	....	....	....	....	....	....	....	....	....	....	....
1995-96	**Florida**	**NHL**	**78**	**20**	**17**	**37**	**25**	**2**	**0**	**3**	**120**	**16.7**	**5**	....	....	....	....	....	**14**	**3**	**2**	**5**	**0**	**0**	**0**	**0**

Season	Club	League	GP	G	A	Pts	PIM	PP	SH	GW	S	%	+/-	TF	F%	H	SB	Min	GP	G	A	Pts	PIM	PP	SH	GW
								Regular Season											Playoffs							
1996-97	**Florida**	**NHL**	**67**	**10**	**6**	**16**	**4**	**0**	**1**	**2**	**92**	**10.9**	**1**	....	....	....	....	....	**5**	**0**	**0**	**0**	**0**	**0**	**0**	**0**
1997-98	**Florida**	**NHL**	**21**	**2**	**0**	**2**	**4**	**0**	**1**	**0**	**23**	**8.7**	**1**	....	....	....	....	....	....	....	....	....	....	....	....	....
	Tampa Bay	**NHL**	**28**	**2**	**4**	**6**	**4**	**0**	**0**	**2**	**28**	**7.1**	**2**	....	....	....	....	....	....	....	....	....	....	....	....	....
1998-99	**Philadelphia**	**NHL**	**72**	**3**	**11**	**14**	**12**	**0**	**0**	**1**	**73**	**4.1**	**–2**	**15**	**53.3**	**34**	**28**	**12:59**	**6**	**0**	**0**	**0**	**4**	**0**	**0**	**0**
99-2000	Orlando	IHL	1	0	0	0	0	....	....	....	....	....	....	....	....	....	....	....	....	....	....	....	....	....	....	....
	Philadelphia	**NHL**	**67**	**10**	**3**	**13**	**4**	**0**	**2**	**2**	**63**	**15.9**	**8**	**36**	**41.7**	**28**	**44**	**11:58**	**18**	**0**	**1**	**1**	**0**	**0**	**0**	**0**
2000-01	**Philadelphia**	**NHL**	**71**	**7**	**8**	**15**	**10**	**0**	**2**	**2**	**78**	**9.0**	**–1**	**83**	**36.1**	**41**	**34**	**13:13**	**6**	**0**	**0**	**0**	**4**	**0**	**0**	**0**
2001-02	**Ottawa**	**NHL**	**24**	**2**	**2**	**4**	**6**	**0**	**0**	**1**	**12**	**16.7**	**0**	**10**	**30.0**	**13**	**5**	**10:10**	**12**	**1**	**1**	**2**	**2**	**0**	**0**	**0**
	Grand Rapids	AHL	3	2	1	3	2	....	....	....	....	....	....	....	....	....	....	....	....	....	....	....	....	....	....	....
	NHL Totals		**760**	**121**	**129**	**250**	**142**	**15**	**8**	**24**	**975**	**12.4**		**144**	**38.9**	**116**	**111**	**12:28**	**67**	**4**	**5**	**9**	**14**	**0**	**0**	**0**

OHL Second All-Star Team (1988)

Traded to **NY Rangers** by **Hartford** for Carey Wilson and NY Rangers' 3rd round choice (Michael Nylander) in the 1991 Entry Draft, July 9, 1990. Traded to **Ottawa** by **NY Rangers** for future considerations, July 28, 1992. Signed as a free agent by **Florida**, August 10, 1993. Traded to **Tampa Bay** by **Florida** with Mark Fitzpatrick for Dino Ciccarelli and Jeff Norton, January 15, 1998. Signed as a free agent by **Philadelphia**, October 7, 1998. Claimed by **Atlanta** from **Philadelphia** in Expansion Draft, June 25, 1999. Traded to **Philadelphia** by **Atlanta** for cash, October 15, 1999. Signed as a free agent by **Ottawa**, January 24, 2002.

HULSE, Cale

(HUHLS, KAYL) **NSH.**

Defense. Shoots right. 6'3", 220 lbs. Born, Edmonton, Alta., November 10, 1973. New Jersey's 3rd choice, 66th overall, in 1992 Entry Draft.

Season	Club	League	GP	G	A	Pts	PIM	PP	SH	GW	S	%	+/-	TF	F%	H	SB	Min	GP	G	A	Pts	PIM	PP	SH	GW
1990-91	Calgary Royals	AJHL	49	3	23	26	220	....	....	....	....	....	....	....	....	....	....	....	....	....	....	....	....	....	....	....
1991-92	Portland	WHL	70	4	18	22	230	....	....	....	....	....	....	....	....	....	....	....	6	0	2	2	27	....	....	....
1992-93	Portland	WHL	72	10	26	36	284	....	....	....	....	....	....	....	....	....	....	....	16	4	4	8	65	....	....	....
1993-94	Albany	AHL	79	7	14	21	186	....	....	....	....	....	....	....	....	....	....	....	5	0	3	3	11	....	....	....
1994-95	Albany	AHL	77	5	13	18	215	....	....	....	....	....	....	....	....	....	....	....	12	1	1	2	17	....	....	....
1995-96	**New Jersey**	**NHL**	**8**	**0**	**0**	**0**	**15**	**0**	**0**	**0**	**5**	**0.0**	**–2**	....	....	....	....	....	....	....	....	....	....	....	....	....
	Albany	AHL	42	4	23	27	107	....	....	....	....	....	....	....	....	....	....	....	....	....	....	....	....	....	....	....
	Calgary	**NHL**	**3**	**0**	**0**	**0**	**5**	**0**	**0**	**0**	**4**	**0.0**	**3**	....	....	....	....	....	**1**	**0**	**0**	**0**	**0**	**0**	**0**	**0**
	Saint John	AHL	13	2	7	9	39	....	....	....	....	....	....	....	....	....	....	....	....	....	....	....	....	....	....	....
1996-97	**Calgary**	**NHL**	**63**	**1**	**6**	**7**	**91**	**0**	**1**	**0**	**58**	**1.7**	**–2**	....	....	....	....	....	....	....	....	....	....	....	....	....
1997-98	**Calgary**	**NHL**	**79**	**5**	**22**	**27**	**169**	**1**	**1**	**0**	**117**	**4.3**	**1**	....	....	....	....	....	....	....	....	....	....	....	....	....
1998-99	**Calgary**	**NHL**	**73**	**3**	**9**	**12**	**117**	**0**	**0**	**0**	**83**	**3.6**	**–8**	**1**	**0.0**	**113**	**74**	**16:38**	....	....	....	....	....	....	....	....
99-2000	**Calgary**	**NHL**	**47**	**1**	**6**	**7**	**47**	**0**	**0**	**0**	**41**	**2.4**	**–11**	**1**	**100.0**	**85**	**37**	**12:38**	....	....	....	....	....	....	....	....
2000-01	**Nashville**	**NHL**	**82**	**1**	**7**	**8**	**128**	**0**	**0**	**1**	**93**	**1.1**	**–5**	**0**	**0.0**	**232**	**71**	**20:05**	....	....	....	....	....	....	....	....
2001-02	**Nashville**	**NHL**	**63**	**0**	**2**	**2**	**121**	**0**	**0**	**0**	**70**	**0.0**	**–18**	**0**	**0.0**	**147**	**62**	**18:51**	....	....	....	....	....	....	....	....
	NHL Totals		**418**	**11**	**52**	**63**	**693**	**1**	**2**	**1**	**471**	**2.3**		**2**	**50.0**	**577**	**244**	**17:31**	**1**	**0**	**0**	**0**	**0**	**0**	**0**	**0**

Traded to **Calgary** by **New Jersey** with Tommy Albelin and Jocelyn Lemieux for Phil Housley and Dan Keczmer, February 26, 1996. Traded to **Nashville** by **Calgary** with Calgary's 3rd round choice (Denis Platonov) in 2001 Entry Draft for Sergei Krivokrasov, March 14, 2000.

HUML, Ivan

(HUH-muhl, ee-VAHN) **BOS.**

Left wing. Shoots left. 6'2", 195 lbs. Born, Kladno, Czech., September 6, 1981. Boston's 4th choice, 59th overall, in 2000 Entry Draft.

Season	Club	League	GP	G	A	Pts	PIM	PP	SH	GW	S	%	+/-	TF	F%	H	SB	Min	GP	G	A	Pts	PIM	PP	SH	GW
1996-97	Kladno Jr.	Czech-Jr.	37	16	3	19	....	....	....	....	....	....	....	....	....	....	....	....	....	....	....	....	....	....	....	....
1997-98	Kladno Jr.	Czech-Jr.	46	37	24	61	....	....	....	....	....	....	....	....	....	....	....	....	....	....	....	....	....	....	....	....
	Kladno	Czech	1	0	0	0	0	....	....	....	....	....	....	....	....	....	....	....	....	....	....	....	....	....	....	....
1998-99	Kladno Jr.	Czech-Jr.	18	6	6	12	....	....	....	....	....	....	....	....	....	....	....	....	....	....	....	....	....	....	....	....
	Langley Hornets	BCHL	33	23	17	40	41	....	....	....	....	....	....	....	....	....	....	....	....	....	....	....	....	....	....	....
99-2000	Langley Hornets	BCHL	49	53	51	104	72	....	....	....	....	....	....	....	....	....	....	....	....	....	....	....	....	....	....	....
2000-01	Providence	AHL	79	13	6	19	28	....	....	....	....	....	....	....	....	....	....	....	17	0	0	0	2	....	....	....
2001-02	**Boston**	**NHL**	**1**	**0**	**1**	**1**	**0**	**0**	**0**	**0**	**2**	**0.0**	**2**	**0**	**0.0**	**1**	**1**	**15:43**	....	....	....	....	....	....	....	....
	Providence	AHL	76	28	19	47	75	....	....	....	....	....	....	....	....	....	....	....	2	0	0	0	0	....	....	....
	NHL Totals		**1**	**0**	**1**	**1**	**0**	**0**	**0**	**0**	**2**	**0.0**		**0**	**0.0**	**1**	**1**	**15:43**	....	....	....	....	....	....	....	....

HUNTER, Trent

(HUHN-tuhr, TREHNT) **NYI**

Right wing. Shoots right. 6'3", 191 lbs. Born, Red Deer, Alta., July 5, 1980. Anaheim's 4th choice, 150th overall, in 1998 Entry Draft.

Season	Club	League	GP	G	A	Pts	PIM	PP	SH	GW	S	%	+/-	TF	F%	H	SB	Min	GP	G	A	Pts	PIM	PP	SH	GW
1996-97	Red Deer	AMHL	42	30	25	55	50	....	....	....	....	....	....	....	....	....	....	....	....	....	....	....	....	....	....	....
1997-98	Prince George	WHL	60	13	14	27	34	....	....	....	....	....	....	....	....	....	....	....	8	1	0	1	4	....	....	....
1998-99	Prince George	WHL	50	18	20	38	34	....	....	....	....	....	....	....	....	....	....	....	7	2	5	7	2	....	....	....
99-2000	Prince George	WHL	67	46	49	95	47	....	....	....	....	....	....	....	....	....	....	....	13	7	15	22	6	....	....	....
2000-01	Springfield	AHL	57	18	17	35	14	....	....	....	....	....	....	....	....	....	....	....	....	....	....	....	....	....	....	....
2001-02	Bridgeport	AHL	80	30	35	65	30	....	....	....	....	....	....	....	....	....	....	....	17	8	11	19	6	....	....	....
	NY Islanders	**NHL**	....	....	....	....	....	....	....	....	....	....	....	....	....	....	....	....	**4**	**1**	**1**	**2**	**2**	**0**	**0**	**0**
	NHL Totals		....	....	....	....	....	....	....	....	....	....	....	....	....	....	....	....	**4**	**1**	**1**	**2**	**2**	**0**	**0**	**0**

WHL West First All-Star Team (2000)

Traded to **NY Islanders** by **Anaheim** for Columbus' 4th round choice (previously acquired, Anaheim selected Jonas Ronnqvist) in 2000 Entry Draft, May 23, 2000.

HURLBUT, Mike

(HUHRL-buht, MIGHK)

Defense. Shoots left. 6'2", 206 lbs. Born, Massena, NY, October 7, 1966. NY Rangers' 1st choice, 5th overall, in 1988 Supplemental Draft.

Season	Club	League	GP	G	A	Pts	PIM	PP	SH	GW	S	%	+/-	TF	F%	H	SB	Min	GP	G	A	Pts	PIM	PP	SH	GW
1983-84	Massena High	Hi-School	27	22	31	53	15	....	....	....	....	....	....	....	....	....	....	....	....	....	....	....	....	....	....	....
1984-85	Northfield Prep	Hi-School	34	20	27	47	30	....	....	....	....	....	....	....	....	....	....	....	....	....	....	....	....	....	....	....
1985-86	St. Lawrence	ECAC	25	2	10	12	40	....	....	....	....	....	....	....	....	....	....	....	....	....	....	....	....	....	....	....
1986-87	St. Lawrence	ECAC	35	8	15	23	44	....	....	....	....	....	....	....	....	....	....	....	....	....	....	....	....	....	....	....
1987-88	St. Lawrence	ECAC	38	6	12	18	18	....	....	....	....	....	....	....	....	....	....	....	....	....	....	....	....	....	....	....
1988-89	St. Lawrence	ECAC	36	8	25	33	30	....	....	....	....	....	....	....	....	....	....	....	....	....	....	....	....	....	....	....
	Denver Rangers	IHL	8	0	2	2	13	....	....	....	....	....	....	....	....	....	....	....	4	1	2	3	2	....	....	....
1989-90	Flint Spirits	IHL	74	3	34	37	38	....	....	....	....	....	....	....	....	....	....	....	3	0	1	1	2	....	....	....
1990-91	San Diego Gulls	IHL	2	1	0	1	0	....	....	....	....	....	....	....	....	....	....	....	....	....	....	....	....	....	....	....
	Binghamton	AHL	33	2	11	13	27	....	....	....	....	....	....	....	....	....	....	....	3	0	1	1	0	....	....	....
1991-92	Binghamton	AHL	79	16	39	55	64	....	....	....	....	....	....	....	....	....	....	....	11	2	7	9	8	....	....	....
1992-93	**NY Rangers**	**NHL**	**23**	**1**	**8**	**9**	**16**	**1**	**0**	**0**	**26**	**3.8**	**4**	....	....	....	....	....	....	....	....	....	....	....	....	....
	Binghamton	AHL	45	11	25	36	46	....	....	....	....	....	....	....	....	....	....	....	14	2	5	7	12	....	....	....
1993-94	**Quebec**	**NHL**	**1**	**0**	**0**	**0**	**0**	**0**	**0**	**0**	**1**	**0.0**	**–1**	....	....	....	....	....	....	....	....	....	....	....	....	....
	Cornwall Aces	AHL	77	13	33	46	100	....	....	....	....	....	....	....	....	....	....	....	13	3	7	10	12	....	....	....
1994-95	Cornwall Aces	AHL	74	11	49	60	69	....	....	....	....	....	....	....	....	....	....	....	3	1	0	1	15	....	....	....
1995-96	Minnesota Moose	IHL	22	1	4	5	22	....	....	....	....	....	....	....	....	....	....	....	....	....	....	....	....	....	....	....
	Houston Aeros	IHL	38	3	12	15	33	....	....	....	....	....	....	....	....	....	....	....	....	....	....	....	....	....	....	....
1996-97	Houston Aeros	IHL	70	11	24	35	62	....	....	....	....	....	....	....	....	....	....	....	13	5	8	13	12	....	....	....
1997-98	**Buffalo**	**NHL**	**3**	**0**	**0**	**0**	**2**	**0**	**0**	**0**	**3**	**0.0**	**–1**	....	....	....	....	....	....	....	....	....	....	....	....	....
	Rochester	AHL	45	10	20	30	48	....	....	....	....	....	....	....	....	....	....	....	4	1	1	2	2	....	....	....
1998-99	**Buffalo**	**NHL**	**1**	**0**	**0**	**0**	**0**	**0**	**0**	**0**	**2**	**0.0**	**2**	**0**	**0.0**	**1**	**2**	**17:53**	....	....	....	....	....	....	....	....
	Rochester	AHL	72	15	39	54	46	....	....	....	....	....	....	....	....	....	....	....	20	4	5	9	12	....	....	....
99-2000	**Buffalo**	**NHL**	**1**	**0**	**0**	**0**	**2**	**0**	**0**	**0**	**1**	**0.0**	**1**	**0**	**0.0**	**0**	**1**	**12:25**	....	....	....	....	....	....	....	....
	Rochester	AHL	74	10	29	39	83	....	....	....	....	....	....	....	....	....	....	....	21	5	6	11	14	....	....	....
2000-01	Rochester	AHL	53	6	26	32	36	....	....	....	....	....	....	....	....	....	....	....	4	1	0	1	6	....	....	....
2001-02	Rochester	AHL	44	3	9	12	44	....	....	....	....	....	....	....	....	....	....	....	2	0	0	0	0	....	....	....
	NHL Totals		**29**	**1**	**8**	**9**	**20**	**1**	**0**	**0**	**33**	**3.0**		**0**	**0.0**	**1**	**3**	**15:09**	....	....	....	....	....	....	....	....

ECAC First All-Star Team (1989) • NCAA East First All-American Team (1989) • AHL Second All-Star Team (1995)

Traded to **Quebec** by **NY Rangers** for Alexander Karpovtsev, September 7, 1993. Signed as a free agent by **Buffalo**, September 9, 1997.

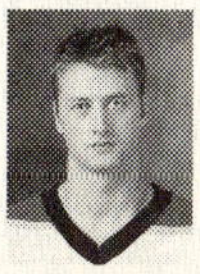

HUSELIUS, Kristian

(hoo-SAY-lee-oos, KRIHST-yan) **FLA.**

Right wing. Shoots left. 6'1", 190 lbs. Born, Osterhaninge, Sweden, November 10, 1978. Florida's 2nd choice, 47th overall, in 1997 Entry Draft.

Season	Club	League	GP	G	A	Pts	PIM	PP	SH	GW	S	%	+/-	TF	F%	H	SB	Min	GP	G	A	Pts	PIM	PP	SH	GW
1994-95	Hammarby Jr.	Swede-Jr.	17	6	2	8	2	....	....	....	....	....	....	....	....	....	....	....	....	....	....	....	....	....	....	....
1995-96	Hammarby Jr.	Swede-Jr.	25	13	8	21	14	....	....	....	....	....	....	....	....	....	....	....	....	....	....	....	....	....	....	....
	Hammarby	Swede-2	6	1	0	1	0	....	....	....	....	....	....	....	....	....	....	....	....	....	....	....	....	....	....	....
1996-97	Farjestad	Sweden	13	2	0	2	4	....	....	....	....	....	....	....	....	....	....	....	5	1	0	1	0	....	....	....
1997-98	Farjestad	Sweden	34	2	1	3	2	....	....	....	....	....	....	....	....	....	....	....	11	0	0	0	0	....	....	....
	Farjestad	EuroHL	5	2	3	5	0	....	....	....	....	....	....	....	....	....	....	....	....	....	....	....	....	....	....	....

			Regular Season																Playoffs							
Season	Club	League	GP	G	A	Pts	PIM	PP	SH	GW	S	%	+/-	TF	F%	H	SB	Min	GP	G	A	Pts	PIM	PP	SH	GW
1998-99	Farjestad	Sweden	28	4	4	8	4																			
	Farjestad	EuroHL	6	2	2	4	8												1	0	0	0	0			
	Vastra Frolunda	Sweden	20	2	2	4	2												4	1	0	1	0			
99-2000	Vastra Frolunda	Sweden	50	21	23	44	20												5	2	2	4	8			
2000-01	Vastra Frolunda	Sweden	49	*32	*35	*67	26												5	4	5	9	14			
2001-02	**Florida**	**NHL**	**79**	**23**	**22**	**45**	**14**	**6**	**1**	**3**	**169**	**13.6**	**–4**	**14**	**21.4**	**13**	**21**	**16:55**								
	NHL Totals		**79**	**23**	**22**	**45**	**14**	**6**	**1**	**3**	**169**	**13.6**		**14**	**21.4**	**13**	**21**	**16:55**								

NHL All-Rookie Team (2002)

HYVONEN, Hannes
(HOO-voh-nuhn, HAH-nuhs) **FLA.**

Right wing. Shoots right. 6'2", 200 lbs. Born, Oulu, Finland, August 29, 1975. San Jose's 7th choice, 257th overall, in 1999 Entry Draft.

Season	Club	League	GP	G	A	Pts	PIM	PP	SH	GW	S	%	+/-	TF	F%	H	SB	Min	GP	G	A	Pts	PIM	PP	SH	GW
1993-94	Karpat Oulu Jr.	Finn-Jr.	35	15	13	28	26												3	0	0	0	0			
	Karpat Oulu	Finland-2	3	3	1	4	2																			
1994-95	TPS Turku Jr.	Finn-Jr.	10	8	2	10	64																			
	Kiekko-67 Jr.	Finn-Jr.	1	1	0	1	0																			
	Kiekko-67 Turku	Finland-2	16	4	2	6	10																			
	TPS Turku	Finland	9	4	3	7	16												5	0	0	0	7			
1995-96	Kiekko-67 Turku	Finland-2	2	1	0	1	8																			
	TPS Turku	Finland	30	11	5	16	49												7	0	1	1	28			
1996-97	TPS Turku	Finland	41	10	5	15	48												10	4	2	6	14			
1997-98	TPS Turku	Finland	29	2	6	8	71												2	0	0	0	0			
1998-99	Blues Espoo	Finland	52	23	18	41	74												4	2	1	3	2			
99-2000	Blues Espoo	Finland	18	5	2	7	*89																			
	HIFK Helsinki	Finland	22	2	2	4	*100												9	4	0	4	8			
2000-01	HIFK Helsinki	Finland	56	14	12	26	34												5	0	0	0	8			
2001-02	**San Jose**	**NHL**	**6**	**0**	**0**	**0**	**0**	**0**	**0**	**0**	**4**	**0.0**	**–2**	**0**	**0.0**	**12**	**1**	**5:40**								
	Cleveland Barons	AHL	67	24	18	42	136																			
	NHL Totals		**6**	**0**	**0**	**0**	**0**	**0**	**0**	**0**	**4**	**0.0**		**0**	**0.0**	**12**	**1**	**5:40**								

Traded to **Florida** by **San Jose** for future considerations, July 16, 2002.

IGINLA, Jarome
(ih-GIHN-lah, jah-ROHM) **CGY.**

Right wing. Shoots right. 6'1", 200 lbs. Born, Edmonton, Alta., July 1, 1977. Dallas' 1st choice, 11th overall, in 1995 Entry Draft.

Season	Club	League	GP	G	A	Pts	PIM	PP	SH	GW	S	%	+/-	TF	F%	H	SB	Min	GP	G	A	Pts	PIM	PP	SH	GW
1991-92	St. Albert	AMHL	36	26	30	56	22																			
1992-93	St. Albert	AMHL	36	34	53	87	20																			
1993-94	Kamloops Blazers	WHL	48	6	23	29	33												19	3	6	9	10			
1994-95	Kamloops Blazers	WHL	72	33	38	71	111												21	7	11	18	34			
1995-96	Kamloops Blazers	WHL	63	63	73	136	120												16	16	13	29	44			
	Calgary	**NHL**																	**2**	**1**	**1**	**2**	**0**	**0**	**0**	**0**
1996-97	**Calgary**	**NHL**	**82**	**21**	**29**	**50**	**37**	**8**	**1**	**3**	**169**	**12.4**	**–4**													
1997-98	**Calgary**	**NHL**	**70**	**13**	**19**	**32**	**29**	**0**	**2**	**1**	**154**	**8.4**	**–10**													
1998-99	**Calgary**	**NHL**	**82**	**28**	**23**	**51**	**58**	**7**	**0**	**4**	**211**	**13.3**	**1**	**111**	**51.4**	**119**	**25**	**16:30**								
99-2000	**Calgary**	**NHL**	**77**	**29**	**34**	**63**	**26**	**12**	**0**	**4**	**256**	**11.3**	**0**	**278**	**52.9**	**133**	**29**	**18:24**								
2000-01	**Calgary**	**NHL**	**77**	**31**	**40**	**71**	**62**	**10**	**0**	**4**	**229**	**13.5**	**–2**	**638**	**51.7**	**99**	**34**	**19:58**								
2001-02	**Calgary**	**NHL**	**82**	***52**	**44**	***96**	**77**	**16**	**1**	**7**	**311**	**16.7**	**27**	**308**	**55.2**	**98**	**25**	**22:22**								
	Canada	Olympics	6	3	1	4	0																			
	NHL Totals		**470**	**174**	**189**	**363**	**289**	**53**	**4**	**23**	**1330**	**13.1**		**1335**	**52.7**	**449**	**113**	**19:19**	**2**	**1**	**1**	**2**	**0**	**0**	**0**	**0**

Won George Parsons Trophy (Memorial Cup Tournament Most Sportsmanlike Player) (1995) • WHL West First All-Star Team (1996) • Canadian Major Junior First All-Star Team (1996) • NHL All-Rookie Team (1997) • NHL First All-Star Team (2002) • Won Maurice "Rocket" Richard Trophy (2002) • Won Art Ross Trophy (2002) • Won Lester B. Pearson Award (2002) • Played in NHL All-Star Game (2002)

Traded to **Calgary** by **Dallas** with Corey Millen for Joe Nieuwendyk, December 19, 1995.

IGNATJEV, Victor
(ihg-NYAT-ee-ehv, VIHK-tohr)

Defense. Shoots left. 6'4", 215 lbs. Born, Riga, USSR, April 26, 1970. San Jose's 11th choice, 243rd overall, in 1992 Entry Draft.

Season	Club	League	GP	G	A	Pts	PIM	PP	SH	GW	S	%	+/-	TF	F%	H	SB	Min	GP	G	A	Pts	PIM	PP	SH	GW
1989-90	Dynamo Riga	USSR	40	0	0	0	26																			
1990-91	Dynamo Riga	USSR	10	0	0	0	2																			
1991-92	Riga Stars	CIS	22	4	5	9	22																			
1992-93	Kansas City	IHL	64	5	16	21	68												4	1	2	3	24			
1993-94	Kansas City	IHL	67	1	24	25	123																			
1994-95	Oklahoma City	CHL	47	11	35	46	66																			
	Denver Grizzlies	IHL	23	2	11	13	4												17	3	8	11	8			
1995-96	Utah Grizzlies	IHL	73	9	29	38	67												21	3	8	11	22			
1996-97	Long Beach	IHL	82	16	53	69	112												16	3	4	7	26			
1997-98	Long Beach	IHL	71	12	33	45	102												17	3	11	14	16			
1998-99	**Pittsburgh**	**NHL**	**11**	**0**	**1**	**1**	**6**	**0**	**0**	**0**	**15**	**0.0**	**–3**	**0**	**0.0**	**10**	**6**	**12:14**	**1**	**0**	**0**	**0**	**2**	**0**	**0**	**0**
99-2000	Nurnberg	Germany	60	3	15	18	56																			
	EHC Nurnberg	EuroHL	4	0	3	3	22												2	0	0	0	6			
2000-01	Leksands IF	Sweden	39	0	1	1	32																			
2001-02	Cherepovets	Russia	26	0	3	3	34												1	0	0	0	0			
	Latvia	Olympics	4	0	0	0	4																			
	NHL Totals		**11**	**0**	**1**	**1**	**6**	**0**	**0**	**0**	**15**	**0.0**		**0**	**0.0**	**10**	**6**	**12:14**	**1**	**0**	**0**	**0**	**2**	**0**	**0**	**0**

IHL Second All-Star Team (1997)

Signed as a free agent by **Pittsburgh**, August 11, 1998. • Missed majority of 1998-99 season recovering from shoulder surgery, November, 1998.

ISBISTER, Brad
(IHZ-bihs-tuhr, BRAD) **NYI**

Left wing. Shoots right. 6'4", 227 lbs. Born, Edmonton, Alta., May 7, 1977. Winnipeg's 4th choice, 67th overall, in 1995 Entry Draft.

Season	Club	League	GP	G	A	Pts	PIM	PP	SH	GW	S	%	+/-	TF	F%	H	SB	Min	GP	G	A	Pts	PIM	PP	SH	GW
1992-93	Calgary Canucks	ABHL	35	24	25	49	74																			
1993-94	Portland	WHL	64	7	10	17	45												10	0	2	2	0			
1994-95	Portland	WHL	67	16	20	36	123																			
1995-96	Portland	WHL	71	45	44	89	184												7	2	4	6	20			
1996-97	Portland	WHL	24	15	18	33	45												6	2	1	3	16			
	Springfield	AHL	7	3	1	4	14												9	1	2	3	10			
1997-98	**Phoenix**	**NHL**	**66**	**9**	**8**	**17**	**102**	**1**	**0**	**1**	**115**	**7.8**	**4**						**5**	**0**	**0**	**0**	**2**	**0**	**0**	**0**
	Springfield	AHL	9	8	2	10	36																			
1998-99	**Phoenix**	**NHL**	**32**	**4**	**4**	**8**	**46**	**0**	**0**	**2**	**48**	**8.3**	**1**	**3**	**0.0**	**39**	**3**	**11:33**								
	Springfield	AHL	4	1	1	2	12																			
	Las Vegas	IHL	2	0	0	0	9																			
99-2000	**NY Islanders**	**NHL**	**64**	**22**	**20**	**42**	**100**	**9**	**0**	**1**	**135**	**16.3**	**–18**	**55**	**54.6**	**134**	**15**	**16:58**								
2000-01	**NY Islanders**	**NHL**	**51**	**18**	**14**	**32**	**59**	**7**	**1**	**4**	**129**	**14.0**	**–19**	**255**	**45.9**	**119**	**16**	**19:26**								
2001-02	**NY Islanders**	**NHL**	**79**	**17**	**21**	**38**	**113**	**4**	**0**	**2**	**142**	**12.0**	**1**	**71**	**45.1**	**163**	**15**	**15:18**	**3**	**1**	**1**	**2**	**17**	**1**	**0**	**1**
	NHL Totals		**292**	**70**	**67**	**137**	**420**	**21**	**1**	**10**	**569**	**12.3**		**384**	**46.6**	**455**	**49**	**16:10**	**8**	**1**	**1**	**2**	**19**	**1**	**0**	**1**

WHL West Second All-Star Team (1997)

Rights transferred to **Phoenix** after **Winnipeg** franchise relocated, July 1, 1996. Traded to **NY Islanders** by **Phoenix** with Phoenix's 3rd round choice (Brian Collins) in 1999 Entry Draft for Robert Reichel, NY Islanders' 3rd round choice (Jason Jaspers) in 1999 Entry Draft and Ottawa's 4th round choice (previously acquired, Phoenix selected Preston Mizzi) in 1999 Entry Draft, March 20, 1999.

JACKMAN, Barret
(JAK-man, BAIR-reht) **ST.L.**

Defense. Shoots left. 6'1", 200 lbs. Born, Trail, B.C., March 5, 1981. St. Louis' 1st choice, 17th overall, in 1999 Entry Draft.

Season	Club	League	GP	G	A	Pts	PIM	PP	SH	GW	S	%	+/-	TF	F%	H	SB	Min	GP	G	A	Pts	PIM	PP	SH	GW
1996-97	Beaver Valley	VIJHL	32	22	25	47	180																			
1997-98	Regina Pats	WHL	68	2	11	13	224												9	0	3	3	32			
1998-99	Regina Pats	WHL	70	8	36	44	259																			
99-2000	Regina Pats	WHL	53	9	37	46	175												6	1	1	2	19			
	Worcester	AHL																	2	0	0	0	13			
2000-01	Regina Pats	WHL	43	9	27	36	138												6	0	3	3	8			

			Regular Season																Playoffs							
Season	Club	League	GP	G	A	Pts	PIM	PP	SH	GW	S	%	+/-	TF	F%	H	SB	Min	GP	G	A	Pts	PIM	PP	SH	GW
2001-02	**St. Louis**	**NHL**	**1**	**0**	**0**	**0**	**0**	**0**	**0**	**0**	**1**	**0.0**	**0**	**0**	**0.0**	**4**	**0**	**18:56**	**1**	**0**	**0**	**0**	**2**	**0**	**0**	**0**
	Worcester	AHL	75	2	12	14	266	....	....	....	....	....	....	....	....	....	....	....	3	0	1	1	4	....	....	....
	NHL Totals		**1**	**0**	**0**	**0**	**0**	**0**	**0**	**0**	**1**	**0.0**		**0**	**0.0**	**4**	**0**	**18:56**	**1**	**0**	**0**	**0**	**2**	**0**	**0**	**0**

WHL East Second All-Star Team (2000) • AHL All-Rookie Team (2002)

JACKMAN, Richard

(JAK-man, RIH-chuhrd) **TOR.**

Defense. Shoots right. 6'2", 192 lbs. Born, Toronto, Ont., June 28, 1978. Dallas' 1st choice, 5th overall, in 1996 Entry Draft.

Season	Club	League	GP	G	A	Pts	PIM	PP	SH	GW	S	%	+/-	TF	F%	H	SB	Min	GP	G	A	Pts	PIM	PP	SH	GW
1993-94	Mississauga Sens	MTHL	81	35	53	88	156	....	....	....	....	....	....	....	....	....	....	....	....	....	....	....	....	....	....	....
1994-95	Mississauga Sens	MTHL	53	20	37	57	120	....	....	....	....	....	....	....	....	....	....	....	....	....	....	....	....	....	....	....
	Richmond Hill	OJHL	10	2	9	11	16	....	....	....	....	....	....	....	....	....	....	....	....	....	....	....	....	....	....	....
1995-96	Sault Ste. Marie	OHL	66	13	29	42	97	....	....	....	....	....	....	....	....	....	....	....	4	1	0	1	15	....	....	....
1996-97	Sault Ste. Marie	OHL	53	13	34	47	116	....	....	....	....	....	....	....	....	....	....	....	10	2	6	8	24	....	....	....
1997-98	Sault Ste. Marie	OHL	60	33	40	73	111	....	....	....	....	....	....	....	....	....	....	....	....	....	....	....	....	....	....	....
	Michigan K-Wings	IHL	14	1	5	6	10	....	....	....	....	....	....	....	....	....	....	....	4	0	0	0	10	....	....	....
1998-99	Michigan K-Wings	IHL	71	13	17	30	106	....	....	....	....	....	....	....	....	....	....	....	5	0	4	4	6	....	....	....
99-2000	**Dallas**	**NHL**	**22**	**1**	**2**	**3**	**6**	**1**	**0**	**0**	**16**	**6.3**	**–1**	**0**	**0.0**	**18**	**10**	**8:06**	....	....	....	....	....	....	....	....
	Michigan K-Wings	IHL	50	3	16	19	51	....	....	....	....	....	....	....	....	....	....	....	....	....	....	....	....	....	....	....
2000-01	**Dallas**	**NHL**	**16**	**0**	**0**	**0**	**18**	**0**	**0**	**0**	**10**	**0.0**	**–6**	**0**	**0.0**	**23**	**8**	**8:51**	....	....	....	....	....	....	....	....
	Utah Grizzlies	IHL	57	9	19	28	24	....	....	....	....	....	....	....	....	....	....	....	....	....	....	....	....	....	....	....
2001-02	**Boston**	**NHL**	**2**	**0**	**0**	**0**	**2**	**0**	**0**	**0**	**4**	**0.0**	**–1**	**0**	**0.0**	**2**	**3**	**13:26**	....	....	....	....	....	....	....	....
	Providence	AHL	9	0	1	1	8	....	....	....	....	....	....	....	....	....	....	....	2	0	0	0	2	....	....	....
	NHL Totals		**40**	**1**	**2**	**3**	**26**	**1**	**0**	**0**	**30**	**3.3**		**0**	**0.0**	**43**	**21**	**8:40**	....	....	....	....	....	....	....	....

OHL All-Rookie Team (1996) • OHL Second All-Star Team (1998)

Traded to **Boston** by **Dallas** for Cameron Mann, June 23, 2001. • Missed majority of 2001-02 season recovering from shoulder injury suffered in game vs. St. Louis, October 21, 2001. Traded to **Toronto** by **Boston** for the rights to Kris Vernarsky, May 13, 2002.

JACKSON, Dane

(JAK-sohn, DAYN)

Right wing. Shoots right. 6'1", 200 lbs. Born, Castlegar, B.C., May 17, 1970. Vancouver's 3rd choice, 44th overall, in 1988 Entry Draft.

Season	Club	League	GP	G	A	Pts	PIM	PP	SH	GW	S	%	+/-	TF	F%	H	SB	Min	GP	G	A	Pts	PIM	PP	SH	GW
1987-88	Vernon Lakers	BCJHL	49	24	30	54	95	....	....	....	....	....	....	....	....	....	....	....	13	7	10	17	49	....	....	....
1988-89	North Dakota	WCHA	30	4	5	9	33	....	....	....	....	....	....	....	....	....	....	....	....	....	....	....	....	....	....	....
1989-90	North Dakota	WCHA	44	15	11	26	56	....	....	....	....	....	....	....	....	....	....	....	....	....	....	....	....	....	....	....
1990-91	North Dakota	WCHA	37	17	9	26	79	....	....	....	....	....	....	....	....	....	....	....	....	....	....	....	....	....	....	....
1991-92	North Dakota	WCHA	39	23	19	42	81	....	....	....	....	....	....	....	....	....	....	....	....	....	....	....	....	....	....	....
1992-93	Hamilton Canucks	AHL	68	23	20	43	59	....	....	....	....	....	....	....	....	....	....	....	....	....	....	....	....	....	....	....
1993-94	**Vancouver**	**NHL**	**12**	**5**	**1**	**6**	**9**	**0**	**0**	**0**	**18**	**27.8**	**3**	....	....	....	....	....	....	....	....	....	....	....	....	....
	Hamilton Canucks	AHL	60	25	35	60	75	....	....	....	....	....	....	....	....	....	....	....	4	2	2	4	16	....	....	....
1994-95	Syracuse Crunch	AHL	78	30	28	58	162	....	....	....	....	....	....	....	....	....	....	....	....	....	....	....	....	....	....	....
	Vancouver	**NHL**	**3**	**1**	**0**	**1**	**4**	**0**	**0**	**0**	**6**	**16.7**	**0**	....	....	....	....	....	**6**	**0**	**0**	**0**	**10**	**0**	**0**	**0**
1995-96	**Buffalo**	**NHL**	**22**	**5**	**4**	**9**	**41**	**0**	**0**	**1**	**20**	**25.0**	**3**	....	....	....	....	....	....	....	....	....	....	....	....	....
	Rochester	AHL	50	27	19	46	132	....	....	....	....	....	....	....	....	....	....	....	19	4	6	10	53	....	....	....
1996-97	Rochester	AHL	78	24	34	58	111	....	....	....	....	....	....	....	....	....	....	....	10	7	4	11	14	....	....	....
1997-98	**NY Islanders**	**NHL**	**8**	**1**	**1**	**2**	**4**	**0**	**0**	**1**	**5**	**20.0**	**1**	....	....	....	....	....	....	....	....	....	....	....	....	....
	Rochester	AHL	28	10	13	23	55	....	....	....	....	....	....	....	....	....	....	....	3	2	2	4	4	....	....	....
1998-99	Lowell	AHL	80	16	27	43	103	....	....	....	....	....	....	....	....	....	....	....	3	0	1	1	16	....	....	....
99-2000	Rochester	AHL	21	6	9	15	8	....	....	....	....	....	....	....	....	....	....	....	....	....	....	....	....	....	....	....
2000-01	Rochester	AHL	69	16	12	28	104	....	....	....	....	....	....	....	....	....	....	....	4	1	1	2	4	....	....	....
2001-02	Manchester	AHL	76	16	21	37	93	....	....	....	....	....	....	....	....	....	....	....	5	1	0	1	9	....	....	....
	NHL Totals		**45**	**12**	**6**	**18**	**58**	**0**	**0**	**2**	**49**	**24.5**		....	....	....	....	....	**6**	**0**	**0**	**0**	**10**	**0**	**0**	**0**

Signed as a free agent by **Buffalo**, September 20, 1995. Signed as a free agent by **NY Islanders**, July 21, 1997. Signed as a free agent by **Rochester** (AHL), August 29, 1999. • Missed majority of 1999-2000 season recovering from knee injury suffered in game vs. Springfield (AHL), January 21, 2000. Signed as a free agent by **Manchester** (AHL), September 7, 2001.

JAGR, Jaromir

(YAH-guhr, YAIR-oh-MEER) **WSH.**

Right wing. Shoots left. 6'2", 234 lbs. Born, Kladno, Czech., February 15, 1972. Pittsburgh's 1st choice, 5th overall, in 1990 Entry Draft.

Season	Club	League	GP	G	A	Pts	PIM	PP	SH	GW	S	%	+/-	TF	F%	H	SB	Min	GP	G	A	Pts	PIM	PP	SH	GW
1984-85	Kladno Jr.	Czech-Jr.	34	24	17	41	....	....	....	....	....	....	....	....	....	....	....	....	....	....	....	....	....	....	....	....
1985-86	Kladno Jr.	Czech-Jr.	36	41	29	70	....	....	....	....	....	....	....	....	....	....	....	....	....	....	....	....	....	....	....	....
1986-87	Kladno Jr.	Czech-Jr.	30	35	35	70	....	....	....	....	....	....	....	....	....	....	....	....	....	....	....	....	....	....	....	....
1987-88	Kladno Jr.	Czech-Jr.	35	57	27	84	....	....	....	....	....	....	....	....	....	....	....	....	....	....	....	....	....	....	....	....
1988-89	Kladno	Czech	29	3	3	6	4	....	....	....	....	....	....	....	....	....	....	....	10	5	7	12	0	....	....	....
1989-90	Poldi Kladno	Czech	42	22	28	50	....	....	....	....	....	....	....	....	....	....	....	....	9	*8	2	10	....	....	....	....
1990-91♦	**Pittsburgh**	**NHL**	**80**	**27**	**30**	**57**	**42**	**7**	**0**	**4**	**136**	**19.9**	**–4**	....	....	....	....	....	**24**	**3**	**10**	**13**	**6**	**1**	**0**	**1**
1991-92♦	**Pittsburgh**	**NHL**	**70**	**32**	**37**	**69**	**34**	**4**	**0**	**4**	**194**	**16.5**	**12**	....	....	....	....	....	**21**	**11**	**13**	**24**	**6**	**2**	**0**	**4**
1992-93	**Pittsburgh**	**NHL**	**81**	**34**	**60**	**94**	**61**	**10**	**1**	**9**	**242**	**14.0**	**30**	....	....	....	....	....	**12**	**5**	**4**	**9**	**23**	**1**	**0**	**1**
1993-94	**Pittsburgh**	**NHL**	**80**	**32**	**67**	**99**	**61**	**9**	**0**	**6**	**298**	**10.7**	**15**	....	....	....	....	....	**6**	**2**	**4**	**6**	**16**	**0**	**0**	**1**
1994-95	HC Kladno	Czech	11	8	14	22	10	....	....	....	....	....	....	....	....	....	....	....	....	....	....	....	....	....	....	....
	HC Bolzano	Euroliga	5	8	8	16	4	....	....	....	....	....	....	....	....	....	....	....	....	....	....	....	....	....	....	....
	HC Bolzano	Italy	1	0	0	0	0	....	....	....	....	....	....	....	....	....	....	....	....	....	....	....	....	....	....	....
	EHC Schalke	German-3	1	1	10	11	0	....	....	....	....	....	....	....	....	....	....	....	....	....	....	....	....	....	....	....
	Pittsburgh	**NHL**	**48**	**32**	**38**	***70**	**37**	**8**	**3**	**7**	**192**	**16.7**	**23**	....	....	....	....	....	**12**	**10**	**5**	**15**	**6**	**2**	**1**	**1**
1995-96	**Pittsburgh**	**NHL**	**82**	**62**	**87**	**149**	**96**	**20**	**1**	**12**	**403**	**15.4**	**31**	....	....	....	....	....	**18**	**11**	**12**	**23**	**18**	**5**	**1**	**1**
1996-97	**Pittsburgh**	**NHL**	**63**	**47**	**48**	**95**	**40**	**11**	**2**	**6**	**234**	**20.1**	**22**	....	....	....	....	....	**5**	**4**	**4**	**8**	**4**	**2**	**0**	**0**
1997-98	**Pittsburgh**	**NHL**	**77**	**35**	***67**	***102**	**64**	**7**	**0**	**8**	**262**	**13.4**	**17**	....	....	....	....	....	**6**	**4**	**5**	**9**	**2**	**1**	**0**	**0**
	Czech Republic	Olympics	6	1	4	5	2	....	....	....	....	....	....	....	....	....	....	....	....	....	....	....	....	....	....	....
1998-99	**Pittsburgh**	**NHL**	**81**	**44**	***83**	***127**	**66**	**10**	**1**	**7**	**343**	**12.8**	**17**	**4**	**50.0**	**27**	**23**	**25:51**	**9**	**5**	**7**	**12**	**16**	**1**	**0**	**1**
99-2000	**Pittsburgh**	**NHL**	**63**	**42**	**54**	***96**	**50**	**10**	**0**	**5**	**290**	**14.5**	**25**	**9**	**22.2**	**19**	**12**	**23:12**	**11**	**8**	**8**	**16**	**6**	**2**	**0**	**4**
2000-01	**Pittsburgh**	**NHL**	**81**	**52**	***69**	***121**	**42**	**14**	**1**	**10**	**317**	**16.4**	**19**	**2**	**0.0**	**25**	**27**	**23:19**	**16**	**2**	**10**	**12**	**18**	**2**	**0**	**0**
2001-02	**Washington**	**NHL**	**69**	**31**	**48**	**79**	**30**	**10**	**0**	**3**	**197**	**15.7**	**0**	**2**	**50.0**	**20**	**13**	**21:43**	....	....	....	....	....	....	....	....
	Czech Republic	Olympics	4	2	3	5	4	....	....	....	....	....	....	....	....	....	....	....	....	....	....	....	....	....	....	....
	NHL Totals		**875**	**470**	**688**	**1158**	**623**	**120**	**9**	**81**	**3108**	**15.1**		**17**	**29.4**	**91**	**75**	**23:37**	**140**	**65**	**82**	**147**	**121**	**19**	**2**	**14**

NHL All-Rookie Team (1991) • NHL First All-Star Team (1995, 1996, 1998, 1999, 2000, 2001) • Won Art Ross Trophy (1995, 1998, 1999, 2000, 2001) • NHL Second All-Star Team (1997) • Won Lester B. Pearson Award (1999, 2000) • Won Hart Trophy (1999) • Played in NHL All-Star Game (1992, 1993, 1996, 1998, 1999, 2000, 2002)

Traded to **Washington** by **Pittsburgh** with Frantisek Kucera for Kris Beech, Michal Sivek, Ross Lupaschuk and future considerations, July 11, 2001.

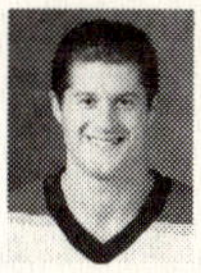

JAKOPIN, John

(JA-koh-pihn, JAWN)

Defense. Shoots right. 6'5", 239 lbs. Born, Toronto, Ont., May 16, 1975. Detroit's 4th choice, 97th overall, in 1993 Entry Draft.

Season	Club	League	GP	G	A	Pts	PIM	PP	SH	GW	S	%	+/-	TF	F%	H	SB	Min	GP	G	A	Pts	PIM	PP	SH	GW
1992-93	St. Michael's B	OJHL-B	45	9	21	30	42	....	....	....	....	....	....	....	....	....	....	....	13	3	2	5	4	....	....	....
1993-94	Merrimack	H-East	36	2	8	10	64	....	....	....	....	....	....	....	....	....	....	....	....	....	....	....	....	....	....	....
1994-95	Merrimack	H-East	37	4	10	14	42	....	....	....	....	....	....	....	....	....	....	....	....	....	....	....	....	....	....	....
1995-96	Merrimack	H-East	32	10	15	25	68	....	....	....	....	....	....	....	....	....	....	....	....	....	....	....	....	....	....	....
1996-97	Merrimack	H-East	31	4	12	16	68	....	....	....	....	....	....	....	....	....	....	....	....	....	....	....	....	....	....	....
	Adirondack	AHL	3	0	0	0	9	....	....	....	....	....	....	....	....	....	....	....	....	....	....	....	....	....	....	....
1997-98	**Florida**	**NHL**	**2**	**0**	**0**	**0**	**4**	**0**	**0**	**0**	**1**	**0.0**	**–3**	....	....	....	....	....	....	....	....	....	....	....	....	....
	New Haven	AHL	60	2	18	20	151	....	....	....	....	....	....	....	....	....	....	....	3	0	0	0	0	....	....	....
1998-99	**Florida**	**NHL**	**3**	**0**	**0**	**0**	**0**	**0**	**0**	**0**	**0**	**0.0**	**–1**	**0**	**0.0**	**8**	**2**	**13:32**	....	....	....	....	....	....	....	....
	New Haven	AHL	60	2	7	9	154	....	....	....	....	....	....	....	....	....	....	....	....	....	....	....	....	....	....	....
99-2000	**Florida**	**NHL**	**17**	**0**	**0**	**0**	**26**	**0**	**0**	**0**	**1**	**0.0**	**–2**	**0**	**0.0**	**37**	**10**	**11:58**	....	....	....	....	....	....	....	....
	Louisville	AHL	23	4	6	10	47	....	....	....	....	....	....	....	....	....	....	....	....	....	....	....	....	....	....	....
2000-01	**Florida**	**NHL**	**60**	**1**	**2**	**3**	**62**	**0**	**0**	**0**	**23**	**4.3**	**–4**	**2**	**50.0**	**181**	**46**	**12:54**	....	....	....	....	....	....	....	....
	Louisville	AHL	8	0	1	1	21	....	....	....	....	....	....	....	....	....	....	....	....	....	....	....	....	....	....	....
2001-02	**Pittsburgh**	**NHL**	**19**	**0**	**4**	**4**	**42**	**0**	**0**	**0**	**3**	**0.0**	**2**	**0**	**0.0**	**28**	**11**	**8:16**	....	....	....	....	....	....	....	....
	Wilkes-Barre	AHL	30	3	5	8	90	....	....	....	....	....	....	....	....	....	....	....	....	....	....	....	....	....	....	....
	NHL Totals		**101**	**1**	**6**	**7**	**134**	**0**	**0**	**0**	**28**	**3.6**		**2**	**50.0**	**254**	**69**	**11:52**	....	....	....	....	....	....	....	....

Signed as a free agent by **Florida**, May 14, 1997. • Missed majority of 1999-2000 season recovering from groin injury suffered in game vs. Carolina, February 1, 2000. Claimed on waivers by **Pittsburgh** from **Florida**, October 3, 2001.

JARDINE, Ryan (JAHR-dighn, RIGH-yan) **FLA.**

Left wing. Shoots left. 6', 210 lbs. Born, Ottawa, Ont., March 15, 1980. Florida's 4th choice, 89th overall, in 1998 Entry Draft.

			Regular Season																Playoffs							
Season	Club	League	GP	G	A	Pts	PIM	PP	SH	GW	S	%	+/-	TF	F%	H	SB	Min	GP	G	A	Pts	PIM	PP	SH	GW
1996-97	Kanata Valley	OCJHL	52	30	27	57	76																			
1997-98	Sault Ste. Marie	OHL	65	28	32	60	16																			
1998-99	Sault Ste. Marie	OHL	68	27	34	61	56												5	0	1	1	6			
99-2000	Sault Ste. Marie	OHL	65	43	34	77	58												17	11	8	19	16			
2000-01	Louisville	AHL	77	12	14	26	38																			
2001-02	**Florida**	**NHL**	**8**	**0**	**2**	**2**	**2**	**0**	**0**	**0**	**6**	**0.0**	**0**	**2**	**100.0**	**12**	**1**	**8:55**								
	Utah Grizzlies	AHL	64	16	16	32	56												4	1	1	2	0			
	NHL Totals		**8**	**0**	**2**	**2**	**2**	**0**	**0**	**0**	**6**	**0.0**		**2**	**100.0**	**12**	**1**	**8:55**								

OHL All-Rookie Team (1998)

JARVENTIE, Martti (yar-VEHN-tee-eh, MAHR-tee) **MTL.**

Defense. Shoots left. 5'11", 196 lbs. Born, Tampere, Finland, April 4, 1976. Montreal's 5th choice, 109th overall, in 2001 Entry Draft.

Season	Club	League	GP	G	A	Pts	PIM	PP	SH	GW	S	%	+/-	TF	F%	H	SB	Min	GP	G	A	Pts	PIM	PP	SH	GW
1992-93	Ilves Tampere-B	Finn-Jr.	27	1	4	5	92																			
	Ilves Jr.	Finn-Jr.	12	0	1	1	6																			
1993-94	Ilves Tampere-B	Finn-Jr.	5	2	2	4	14																			
	Ilves Jr.	Finn-Jr.	36	7	6	13	34												6	1	2	3				
1994-95	Ilves Jr.	Finn-Jr.	9	2	2	4	26																			
	Ilves Tampere	Finland-2	7	0	1	1	0																			
	Ilves Tampere	Finland	37	1	6	7	18																			
1995-96	Ilves Jr.	Finn-Jr.	3	2	2	4	4																			
	Ilves Tampere	Finland	21	2	1	3	30																			
	KooVee Tampere	Finland-2	3	0	0	0	2																			
	Lukko Rauma	Finland	15	0	1	1	10																			
1996-97	Ilves Jr.	Finn-Jr.	2	0	0	0	2																			
	Ilves Tampere	Finland	44	2	11	13	34												6	0	1	1	0			
1997-98	Ilves Tampere	Finland	37	2	5	7	22												9	2	2	4	14			
1998-99	Ilves Tampere	Finland	43	2	4	6	56												4	0	0	0	0			
99-2000	Ilves Tampere	Finland	50	14	14	28	77												3	1	1	2	2			
2000-01	TPS Turku	Finland	56	5	14	19	71												10	1	2	3	4			
2001-02	**Montreal**	**NHL**	**1**	**0**	**0**	**0**	**0**	**0**	**0**	**0**	**0**	**0.0**	**2**	**0**	**0.0**	**3**	**0**	**12:57**								
	Quebec	AHL	59	7	14	21	18																			
	NHL Totals		**1**	**0**	**0**	**0**	**0**	**0**	**0**	**0**	**0**	**0.0**		**0**	**0.0**	**3**	**0**	**12:57**								

JASPERS, Jason (JAS-puhrs, JAY-suhn) **PHX.**

Center/Left wing. Shoots left. 5'11", 200 lbs. Born, Thunder Bay, Ont., April 8, 1981. Phoenix's 4th choice, 71st overall, in 1999 Entry Draft.

Season	Club	League	GP	G	A	Pts	PIM	PP	SH	GW	S	%	+/-	TF	F%	H	SB	Min	GP	G	A	Pts	PIM	PP	SH	GW
1996-97	Thunder Bay	TBAHA	70	51	69	120	67																			
1997-98	Thunder Bay	TBAHA	72	45	75	120	90																			
1998-99	Sudbury Wolves	OHL	68	28	33	61	81												4	2	1	3	13			
99-2000	Sudbury Wolves	OHL	68	46	61	107	107												12	4	6	10	27			
2000-01	Sudbury Wolves	OHL	63	42	42	84	77												12	3	16	19	18			
2001-02	**Phoenix**	**NHL**	**4**	**0**	**1**	**1**	**4**	**0**	**0**	**0**	**1**	**0.0**	**–1**	**14**	**35.7**	**6**	**2**	**8:07**								
	Springfield	AHL	71	25	23	48	55																			
	NHL Totals		**4**	**0**	**1**	**1**	**4**	**0**	**0**	**0**	**1**	**0.0**		**14**	**35.7**	**6**	**2**	**8:07**								

OHL Second All-Star Team (2000)

JILLSON, Jeff (JIHL-sohn, JEHF) **S.J.**

Defense. Shoots right. 6'3", 220 lbs. Born, North Smithfield, RI, July 24, 1980. San Jose's 1st choice, 14th overall, in 1999 Entry Draft.

Season	Club	League	GP	G	A	Pts	PIM	PP	SH	GW	S	%	+/-	TF	F%	H	SB	Min	GP	G	A	Pts	PIM	PP	SH	GW
1995-96	Mount St. Charles	Hi-School	15	8	7	15	15												5	1	1	2	4			
1996-97	Mount St. Charles	Hi-School	15	16	14	30	20												4	0	4	4	6			
1997-98	Mount St. Charles	Hi-School	15	10	13	23	32												5	4	5	9	6			
1998-99	U. of Michigan	CCHA	38	5	19	24	71																			
99-2000	U. of Michigan	CCHA	38	8	26	34	115																			
2000-01	U. of Michigan	CCHA	43	10	20	30	74																			
2001-02	**San Jose**	**NHL**	**48**	**5**	**13**	**18**	**29**	**3**	**0**	**2**	**47**	**10.6**	**2**	**0**	**0.0**	**65**	**39**	**14:36**	**4**	**0**	**0**	**0**	**0**	**0**	**0**	**0**
	Cleveland Barons	AHL	27	2	13	15	45																			
	NHL Totals		**48**	**5**	**13**	**18**	**29**	**3**	**0**	**2**	**47**	**10.6**		**0**	**0.0**	**65**	**39**	**14:36**	**4**	**0**	**0**	**0**	**0**	**0**	**0**	**0**

Rhode Island All-State First All-Star Team (1996, 1997, 1998) • CCHA All-Rookie Team (1999) • CCHA First All-Star Team (2000, 2001) • NCAA West First All-American Team (2000) • NCAA West Second All-American Team (2001)

JOHANSSON, Andreas (yoh-HAHN-suhn, ahn-DRAY-uhs)

Center. Shoots left. 6', 202 lbs. Born, Hofors, Sweden, May 19, 1973. NY Islanders' 7th choice, 136th overall, in 1991 Entry Draft.

Season	Club	League	GP	G	A	Pts	PIM	PP	SH	GW	S	%	+/-	TF	F%	H	SB	Min	GP	G	A	Pts	PIM	PP	SH	GW
1987-88	Bofors IK	Swede-3	1	0	0	0	0																			
1988-89	Bofors IK	Swede-3	28	19	11	30																				
1989-90	Falu IF	Swede-2	21	3	1	4	14																			
1990-91	Falu IF	Swede-2	31	12	10	22	38																			
1991-92	Farjestad	Sweden	30	3	1	4	10												6	0	0	0	4			
1992-93	Farjestad	Sweden	38	4	7	11	38												2	0	0	0	0			
1993-94	Farjestad	Sweden	37	11	16	27	24												3	1	4	5	2			
1994-95	Farjestad	Sweden	36	9	10	19	42												4	0	0	0	10			
1995-96	**NY Islanders**	**NHL**	**3**	**0**	**1**	**1**	**0**	**0**	**0**	**0**	**6**	**0.0**	**1**													
	Worcester	AHL	29	5	5	10	32																			
	Utah Grizzlies	IHL	22	4	13	17	28												12	0	5	5	6			
1996-97	**NY Islanders**	**NHL**	**15**	**2**	**2**	**4**	**0**	**1**	**0**	**0**	**21**	**9.5**	**–6**													
	Pittsburgh	**NHL**	**27**	**2**	**7**	**9**	**20**	**0**	**0**	**0**	**38**	**5.3**	**–6**													
	Cleveland	IHL	10	2	4	6	42												11	1	5	6	8			
1997-98	**Pittsburgh**	**NHL**	**50**	**5**	**10**	**15**	**20**	**0**	**1**	**0**	**49**	**10.2**	**4**						**1**	**0**	**0**	**0**	**0**	**0**	**0**	**0**
	Sweden	Olympics	3	0	0	0	2																			
1998-99	**Ottawa**	**NHL**	**69**	**21**	**16**	**37**	**34**	**7**	**0**	**6**	**144**	**14.6**	**1**	**9**	**22.2**	**48**	**8**	**14:39**	**2**	**0**	**0**	**0**	**0**	**0**	**0**	**0**
99-2000	**Tampa Bay**	**NHL**	**12**	**2**	**3**	**5**	**8**	**0**	**0**	**0**	**11**	**18.2**	**1**	**0**	**0.0**	**8**	**2**	**10:50**								
	Calgary	**NHL**	**28**	**3**	**7**	**10**	**14**	**1**	**0**	**0**	**47**	**6.4**	**–3**	**5**	**20.0**	**22**	**8**	**13:33**								
2000-01	SC Bern	Swiss	40	15	29	44	94												7	5	4	9	0			
2001-02	**NY Rangers**	**NHL**	**70**	**14**	**10**	**24**	**46**	**3**	**0**	**1**	**108**	**13.0**	**6**	**299**	**43.1**	**64**	**19**	**16:19**								
	NHL Totals		**274**	**49**	**56**	**105**	**142**	**12**	**1**	**7**	**424**	**11.6**		**313**	**42.2**	**142**	**37**	**14:52**	**3**	**0**	**0**	**0**	**0**	**0**	**0**	**0**

Traded to **Pittsburgh** by **NY Islanders** with Darius Kasparaitis for Bryan Smolinski, November 17, 1996. Signed as a free agent by **Ottawa**, September 29, 1998. Traded to **Tampa Bay** by **Ottawa** for Rob Zamuner and Tampa Bay's 2nd round choice (later traded to Philadelphia - later traded back to Tampa Bay - later traded to Dallas - Dallas selected Tobias Stephan) in 2002 Entry Draft, June 29, 1999. Traded to **Calgary** by **Tampa Bay** for Nils Ekman and Calgary's 4th round choice (later traded to NY Islanders - NY Islanders selected Vladimir Gorbunov) in 2000 Entry Draft, November 13, 1999. • Missed majority of 1999-2000 season recovering from back injury suffered in game vs. Vancouver, January 2, 2000. Claimed by **NY Rangers** from **Calgary** in Waiver Draft, September 29, 2000.

JOHANSSON, Calle (yoh-HAHN-suhn, KAL-ee) **WSH.**

Defense. Shoots left. 5'11", 203 lbs. Born, Goteborg, Sweden, February 14, 1967. Buffalo's 1st choice, 14th overall, in 1985 Entry Draft.

Season	Club	League	GP	G	A	Pts	PIM	PP	SH	GW	S	%	+/-	TF	F%	H	SB	Min	GP	G	A	Pts	PIM	PP	SH	GW
1981-82	KBA-67	Swede-3	27	3	3	6																				
1982-83	KBA-67	Swede-3	29	12	11	23																				
1983-84	Vastra Frolunda	Sweden	28	4	4	8	10																			
1984-85	Vastra Frolunda	Swede-2	30	8	13	21	16																			
1985-86	Bjorkloven	Sweden	17	1	2	3	4																			
1986-87	Bjorkloven	Sweden	30	2	13	15	20												6	1	3	4	6			
1987-88	**Buffalo**	**NHL**	**71**	**4**	**38**	**42**	**37**	**2**	**0**	**0**	**93**	**4.3**	**12**						**6**	**0**	**1**	**1**	**0**	**0**	**0**	**0**
1988-89	**Buffalo**	**NHL**	**47**	**2**	**11**	**13**	**33**	**0**	**0**	**1**	**53**	**3.8**	**–7**													
	Washington	**NHL**	**12**	**1**	**7**	**8**	**4**	**1**	**0**	**0**	**22**	**4.5**	**1**						**6**	**1**	**2**	**3**	**0**	**1**	**0**	**0**
1989-90	**Washington**	**NHL**	**70**	**8**	**31**	**39**	**25**	**4**	**0**	**2**	**103**	**7.8**	**7**						**15**	**1**	**6**	**7**	**4**	**0**	**0**	**0**
1990-91	**Washington**	**NHL**	**80**	**11**	**41**	**52**	**23**	**2**	**1**	**2**	**128**	**8.6**	**–2**						**10**	**2**	**7**	**9**	**8**	**1**	**0**	**0**
1991-92	**Washington**	**NHL**	**80**	**14**	**42**	**56**	**49**	**5**	**2**	**2**	**119**	**11.8**	**2**						**7**	**0**	**5**	**5**	**4**	**0**	**0**	**0**

Season	Club	League	GP	G	A	Pts	PIM	PP	SH	GW	S	%	+/-	TF	F%	H	SB	Min	GP	G	A	Pts	PIM	PP	SH	GW
			Regular Season																Playoffs							
1992-93	**Washington**	**NHL**	**77**	**7**	**38**	**45**	**56**	**6**	**0**	**0**	**133**	**5.3**	**3**						**6**	**0**	**5**	**5**	**4**	**0**	**0**	**0**
1993-94	**Washington**	**NHL**	**84**	**9**	**33**	**42**	**59**	**4**	**0**	**1**	**141**	**6.4**	**3**						**6**	**1**	**3**	**4**	**4**	**0**	**0**	**1**
1994-95	EHC Kloten	Swiss	5	1	2	3	8																			
	Washington	**NHL**	**46**	**5**	**26**	**31**	**35**	**4**	**0**	**2**	**112**	**4.5**	**–6**						**7**	**3**	**1**	**4**	**0**	**1**	**0**	**0**
1995-96	**Washington**	**NHL**	**78**	**10**	**25**	**35**	**50**	**4**	**0**	**0**	**182**	**5.5**	**13**													
1996-97	**Washington**	**NHL**	**65**	**6**	**11**	**17**	**16**	**2**	**0**	**0**	**133**	**4.5**	**–2**													
1997-98	**Washington**	**NHL**	**73**	**15**	**20**	**35**	**30**	**10**	**1**	**1**	**163**	**9.2**	**–11**						**21**	**2**	**8**	**10**	**16**	**0**	**0**	**0**
	Sweden	Olympics	4	0	0	0	2																			
1998-99	**Washington**	**NHL**	**67**	**8**	**21**	**29**	**22**	**2**	**0**	**2**	**145**	**5.5**	**10**	**0**	**0.0**	**51**	**140**	**23:58**								
99-2000	**Washington**	**NHL**	**82**	**7**	**25**	**32**	**24**	**1**	**0**	**3**	**138**	**5.1**	**13**	**0**	**0.0**	**67**	**161**	**23:55**	**5**	**1**	**2**	**3**	**0**	**1**	**0**	**0**
2000-01	**Washington**	**NHL**	**76**	**7**	**29**	**36**	**26**	**5**	**0**	**0**	**154**	**4.5**	**11**	**0**	**0.0**	**41**	**136**	**23:44**	**6**	**1**	**2**	**3**	**2**	**0**	**0**	**0**
2001-02	**Washington**	**NHL**	**11**	**2**	**0**	**2**	**8**	**0**	**0**	**1**	**18**	**11.1**	**–4**	**0**	**0.0**	**1**	**16**	**21:14**								
	NHL Totals		**1019**	**116**	**398**	**514**	**497**	**52**	**4**	**17**	**1837**	**6.3**		**0**	**0.0**	**160**	**453**	**23:45**	**95**	**12**	**42**	**54**	**42**	**4**	**0**	**1**

NHL All-Rookie Team (1988)

Traded to **Washington** by **Buffalo** with Buffalo's 2nd round choice (Byron Dafoe) in 1989 Entry Draft for Clint Malarchuk, Grant Ledyard and Washington's 6th round choice (Brian Holzinger) in 1991 Entry Draft, March 7, 1989. • Missed majority of 2001-02 season recovering from rotator cuff injury suffered in game vs. Atlanta, November 10, 2001.

JOHNSON, Craig

(JAWN-suhn, KRAYG) **L.A.**

Left wing. Shoots left. 6'2", 200 lbs. Born, St. Paul, MN, March 18, 1972. St. Louis' 1st choice, 33rd overall, in 1990 Entry Draft.

Season	Club	League	GP	G	A	Pts	PIM	PP	SH	GW	S	%	+/-	TF	F%	H	SB	Min	GP	G	A	Pts	PIM	PP	SH	GW
1987-88	Hill-Murray	Hi-School	28	14	20	34	4																			
1988-89	Hill-Murray	Hi-School	24	22	30	52	10																			
1989-90	Hill-Murray	Hi-School	23	15	36	51	0																			
1990-91	U. of Minnesota	WCHA	33	13	18	31	34																			
1991-92	U. of Minnesota	WCHA	41	17	38	55	66																			
1992-93	U. of Minnesota	WCHA	42	22	24	46	70																			
	Jacksonville	SunHL	23	2	9	11	38																			
1993-94	Team USA	Nat-Tm	54	25	26	51	64																			
	United States	Olympics	8	0	4	4	4																			
1994-95	**St. Louis**	**NHL**	**15**	**3**	**3**	**6**	**6**	**0**	**0**	**0**	**19**	**15.8**	**4**						**1**	**0**	**0**	**0**	**2**	**0**	**0**	**0**
	Peoria Rivermen	IHL	16	2	6	8	25												9	0	4	4	10			
1995-96	**St. Louis**	**NHL**	**49**	**8**	**7**	**15**	**30**	**1**	**0**	**0**	**69**	**11.6**	**–4**													
	Worcester	AHL	5	3	0	3	2																			
	Los Angeles	**NHL**	**11**	**5**	**4**	**9**	**6**	**3**	**0**	**0**	**28**	**17.9**	**–4**													
1996-97	**Los Angeles**	**NHL**	**31**	**4**	**3**	**7**	**26**	**1**	**0**	**0**	**30**	**13.3**	**–7**													
1997-98	**Los Angeles**	**NHL**	**74**	**17**	**21**	**38**	**42**	**6**	**0**	**2**	**125**	**13.6**	**9**						**4**	**1**	**0**	**1**	**4**	**0**	**0**	**0**
1998-99	**Los Angeles**	**NHL**	**69**	**7**	**12**	**19**	**32**	**2**	**0**	**2**	**94**	**7.4**	**–12**	**2**	**50.0**	**70**	**11**	**12:02**								
99-2000	**Los Angeles**	**NHL**	**76**	**9**	**14**	**23**	**28**	**1**	**0**	**1**	**106**	**8.5**	**–10**	**9**	**55.6**	**82**	**18**	**13:56**	**4**	**1**	**0**	**1**	**2**	**0**	**0**	**0**
2000-01	**Los Angeles**	**NHL**	**26**	**4**	**5**	**9**	**16**	**0**	**0**	**0**	**36**	**11.1**	**0**	**2**	**100.0**	**26**	**3**	**10:40**								
2001-02	**Los Angeles**	**NHL**	**72**	**13**	**14**	**27**	**24**	**4**	**1**	**3**	**102**	**12.7**	**14**	**11**	**36.4**	**62**	**17**	**14:15**	**7**	**1**	**2**	**3**	**2**	**0**	**0**	**1**
	NHL Totals		**423**	**70**	**83**	**153**	**210**	**18**	**1**	**8**	**609**	**11.5**		**24**	**50.0**	**240**	**49**	**13:08**	**16**	**3**	**2**	**5**	**10**	**0**	**0**	**1**

Traded to **LA Kings** by **St. Louis** with Patrice Tardif, Roman Vopat, St. Louis 5th round choice (Peter Hogan) in 1996 Entry Draft and 1st round choice (Matt Zultek) in 1997 Entry Draft for Wayne Gretzky, February 27, 1996. • Missed majority of 2000-01 season recovering from ankle injury suffered in game vs. San Jose, December 26, 2000.

JOHNSON, Greg

(JAWN-suhn, GREHG) **NSH.**

Center. Shoots left. 5'11", 202 lbs. Born, Thunder Bay, Ont., March 16, 1971. Philadelphia's 1st choice, 33rd overall, in 1989 Entry Draft.

Season	Club	League	GP	G	A	Pts	PIM	PP	SH	GW	S	%	+/-	TF	F%	H	SB	Min	GP	G	A	Pts	PIM	PP	SH	GW
1988-89	Thunder Bay	USHL	47	32	64	96	4												12	5	13	18	0			
1989-90	North Dakota	WCHA	44	17	38	55	11																			
1990-91	North Dakota	WCHA	38	18	*61	79	6																			
1991-92	North Dakota	WCHA	39	20	*54	74	8																			
1992-93	North Dakota	WCHA	34	19	45	64	18																			
	Team Canada	Nat-Tm	23	6	14	20	2																			
1993-94	**Detroit**	**NHL**	**52**	**6**	**11**	**17**	**22**	**1**	**1**	**0**	**48**	**12.5**	**–7**						**7**	**2**	**2**	**4**	**2**	**1**	**0**	**0**
	Adirondack	AHL	3	2	4	6	0												4	0	4	4	2			
	Canada	Olympics	8	0	3	3	0																			
1994-95	**Detroit**	**NHL**	**22**	**3**	**5**	**8**	**14**	**2**	**0**	**0**	**32**	**9.4**	**1**						**1**	**0**	**0**	**0**	**0**	**0**	**0**	**0**
1995-96	**Detroit**	**NHL**	**60**	**18**	**22**	**40**	**30**	**5**	**0**	**2**	**87**	**20.7**	**6**						**13**	**3**	**1**	**4**	**8**	**0**	**0**	**0**
1996-97	**Detroit**	**NHL**	**43**	**6**	**10**	**16**	**12**	**0**	**0**	**0**	**56**	**10.7**	**–5**													
	Pittsburgh	**NHL**	**32**	**7**	**9**	**16**	**14**	**1**	**0**	**0**	**52**	**13.5**	**–13**						**5**	**1**	**0**	**1**	**2**	**0**	**0**	**0**
1997-98	**Pittsburgh**	**NHL**	**5**	**1**	**0**	**1**	**2**	**0**	**0**	**0**	**4**	**25.0**	**0**													
	Chicago	**NHL**	**69**	**11**	**22**	**33**	**38**	**4**	**0**	**3**	**85**	**12.9**	**–2**													
1998-99	**Nashville**	**NHL**	**68**	**16**	**34**	**50**	**24**	**2**	**3**	**0**	**120**	**13.3**	**–8**	**1441**	**53.6**	**28**	**36**	**19:26**								
99-2000	**Nashville**	**NHL**	**82**	**11**	**33**	**44**	**40**	**2**	**0**	**1**	**133**	**8.3**	**–15**	**1684**	**50.8**	**17**	**41**	**19:13**								
2000-01	**Nashville**	**NHL**	**82**	**15**	**17**	**32**	**46**	**1**	**0**	**4**	**97**	**15.5**	**–6**	**1583**	**51.8**	**27**	**32**	**17:49**								
2001-02	**Nashville**	**NHL**	**82**	**18**	**26**	**44**	**38**	**3**	**0**	**2**	**145**	**12.4**	**–14**	**1764**	**51.8**	**23**	**32**	**19:34**								
	NHL Totals		**597**	**112**	**189**	**301**	**280**	**21**	**4**	**12**	**859**	**13.0**		**6472**	**51.9**	**95**	**141**	**18:60**	**26**	**6**	**3**	**9**	**12**	**1**	**0**	**0**

WCHA First All-Star Team (1991, 1992, 1993) • NCAA West First All-American Team (1991, 1993) • NCAA West Second All-American Team (1992)

Traded to **Detroit** by **Philadelphia** with Philadelphia's 5th round choice (Frederic Deschenes) in 1994 Entry Draft for Jim Cummins and Philadelphia's 4th round choice (previously acquired by Detroit - later traded to Boston - Boston selected Charles Paquette) in 1993 Entry Draft, June 20, 1993. Traded to **Pittsburgh** by **Detroit** for Tomas Sandstrom, January 27, 1997. Traded to **Chicago** by **Pittsburgh** for Tuomas Gronman, October 27, 1997. Claimed by **Nashville** from **Chicago** in Expansion Draft, June 26, 1998.

JOHNSON, Matt

(JAWN-suhn, MAT) **MIN.**

Left wing. Shoots left. 6'5", 232 lbs. Born, Welland, Ont., November 23, 1975. Los Angeles' 2nd choice, 33rd overall, in 1994 Entry Draft.

Season	Club	League	GP	G	A	Pts	PIM	PP	SH	GW	S	%	+/-	TF	F%	H	SB	Min	GP	G	A	Pts	PIM	PP	SH	GW
1991-92	Welland	OJHL-B	38	6	19	25	214																			
	Ajax Axemen	MTJHL	1	0	0	0	0																			
1992-93	Peterborough	OHL	66	8	17	25	211												16	1	1	2	56			
1993-94	Peterborough	OHL	50	13	24	37	233																			
1994-95	Peterborough	OHL	14	1	2	3	43																			
	Los Angeles	**NHL**	**14**	**1**	**0**	**1**	**102**	**0**	**0**	**0**	**4**	**25.0**	**0**													
1995-96	**Los Angeles**	**NHL**	**1**	**0**	**0**	**0**	**5**	**0**	**0**	**0**	**1**	**0.0**	**0**													
	Phoenix	IHL	29	4	4	8	87																			
1996-97	**Los Angeles**	**NHL**	**52**	**1**	**3**	**4**	**194**	**0**	**0**	**0**	**20**	**5.0**	**–4**													
1997-98	**Los Angeles**	**NHL**	**66**	**2**	**4**	**6**	**249**	**0**	**0**	**0**	**18**	**11.1**	**–8**						**4**	**0**	**0**	**0**	**6**	**0**	**0**	**0**
1998-99	**Los Angeles**	**NHL**	**49**	**2**	**1**	**3**	**131**	**0**	**0**	**0**	**14**	**14.3**	**–5**	**1**	**0.0**	**62**	**4**	**5:55**								
99-2000	**Atlanta**	**NHL**	**64**	**2**	**5**	**7**	**144**	**0**	**0**	**0**	**54**	**3.7**	**–11**	**1**	**100.0**	**133**	**8**	**8:25**								
2000-01	**Minnesota**	**NHL**	**50**	**1**	**1**	**2**	**137**	**0**	**0**	**0**	**21**	**4.8**	**–6**	**1**	**100.0**	**103**	**8**	**7:43**								
2001-02	**Minnesota**	**NHL**	**60**	**4**	**0**	**4**	**183**	**0**	**0**	**1**	**23**	**17.4**	**–13**	**1**	**100.0**	**131**	**15**	**7:23**								
	NHL Totals		**356**	**13**	**14**	**27**	**1145**	**0**	**0**	**1**	**155**	**8.4**		**4**	**75.0**	**429**	**35**	**7:26**	**4**	**0**	**0**	**0**	**6**	**0**	**0**	**0**

OHL All-Rookie Team (1993)

Claimed by **Atlanta** from **Los Angeles** in Expansion Draft, June 25, 1999. Traded to **Minnesota** by **Atlanta** for San Jose's 3rd round choice (previously acquired, later traded to Pittsburgh, later traded to Columbus - Columbus selected Aaron Johnson) in 2001 Entry Draft, September 29, 2000.

JOHNSON, Mike

(JAWN-suhn, MIGHK) **PHX.**

Right wing. Shoots right. 6'2", 200 lbs. Born, Scarborough, Ont., October 3, 1974.

Season	Club	League	GP	G	A	Pts	PIM	PP	SH	GW	S	%	+/-	TF	F%	H	SB	Min	GP	G	A	Pts	PIM	PP	SH	GW
1991-92	Hillcrest	MTHL	45	43	66	109													20	10	19	29				
1992-93	Aurora Eagles	MTJHL	48	25	40	65	18												7	7	15	22				
1993-94	Bowling Green	CCHA	38	6	14	20	18																			
1994-95	Bowling Green	CCHA	37	16	33	49	35																			
1995-96	Bowling Green	CCHA	30	12	19	31	22																			
1996-97	Bowling Green	CCHA	38	30	32	62	46																			
	Toronto	**NHL**	**13**	**2**	**2**	**4**	**4**	**0**	**1**	**1**	**27**	**7.4**	**–2**													
1997-98	**Toronto**	**NHL**	**82**	**15**	**32**	**47**	**24**	**5**	**0**	**0**	**143**	**10.5**	**–4**													
1998-99	**Toronto**	**NHL**	**79**	**20**	**24**	**44**	**35**	**5**	**3**	**2**	**149**	**13.4**	**13**	**15**	**53.3**	**70**	**17**	**16:16**	**17**	**3**	**2**	**5**	**4**	**0**	**0**	**1**
99-2000	**Toronto**	**NHL**	**52**	**11**	**14**	**25**	**23**	**2**	**1**	**3**	**89**	**12.4**	**8**	**2**	**50.0**	**58**	**6**	**15:22**								
	Tampa Bay	**NHL**	**28**	**10**	**12**	**22**	**4**	**4**	**0**	**0**	**43**	**23.3**	**–2**	**5**	**60.0**	**26**	**11**	**20:33**								

			Regular Season																Playoffs							
Season	Club	League	GP	G	A	Pts	PIM	PP	SH	GW	S	%	+/-	TF	F%	H	SB	Min	GP	G	A	Pts	PIM	PP	SH	GW
2000-01	Tampa Bay	NHL	64	11	27	38	38	3	1	0	107	10.3	-10	2	0.0	67	22	18:13	....	....	....	....	....	....	....	....
	Phoenix	NHL	12	2	3	5	4	1	0	0	17	11.8	0	0	0.0	14	1	12:12	....	....	....	....	....	....	....	....
2001-02	Phoenix	NHL	57	5	22	27	28	1	2	0	73	6.8	14	13	30.8	92	23	15:49	5	1	1	2	6	0	0	0
	NHL Totals		387	76	136	212	160	21	8	6	648	11.7		37	43.2	327	80	16:41	22	4	3	7	10	0	0	1

NHL All-Rookie Team (1998)

Signed as a free agent by **Toronto**, March 16, 1997. Traded to **Tampa Bay** by **Toronto** with Marek Posmyk, Toronto's 5th (Pavel Sedov) and 6th (Aaron Gionet) round choices in 2000 Entry Draft and future considerations for Darcy Tucker, Tampa Bay's 4th round choice (Miguel Delisle) in 2000 Entry Draft and future considerations, February 9, 2000. Traded to **Phoenix** by **Tampa Bay** with Paul Mara, Ruslan Zainullin and NY Islanders' 2nd round choice (previously acquired, Phoenix selected Matthew Spiller) in 2001 Entry Draft for Nikolai Khabibulin and Stan Neckar, March 5, 2001.

JOHNSON, Ryan

(JAWN-suhn, RIGH-yuhn) **FLA.**

Center. Shoots left. 6'1", 200 lbs. Born, Thunder Bay, Ont., June 14, 1976. Florida's 4th choice, 36th overall, in 1994 Entry Draft.

Season	Club	League	GP	G	A	Pts	PIM	PP	SH	GW	S	%	+/-	TF	F%	H	SB	Min	GP	G	A	Pts	PIM	PP	SH	GW
1992-93	Thunder Bay	TBAHA	60	25	33	58	....	....	....	....	....	....	....	....	....	....	....	....	....	....	....	....	....	....	....	....
1993-94	Thunder Bay	USHL	48	14	36	50	28	....	....	....	....	....	....	....	....	....	....	....	....	....	....	....	....	....	....	....
1994-95	North Dakota	WCHA	38	6	22	28	39	....	....	....	....	....	....	....	....	....	....	....	....	....	....	....	....	....	....	....
1995-96	North Dakota	WCHA	21	2	17	19	14	....	....	....	....	....	....	....	....	....	....	....	....	....	....	....	....	....	....	....
	Team Canada	Nat-Tm	28	5	12	17	14	....	....	....	....	....	....	....	....	....	....	....	....	....	....	....	....	....	....	....
1996-97	Carolina	AHL	79	18	24	42	28	....	....	....	....	....	....	....	....	....	....	....	....	....	....	....	....	....	....	....
1997-98	Florida	NHL	10	0	2	2	0	0	0	0	6	0.0	-4	....	....	....	....	....	....	....	....	....	....	....	....	....
	New Haven	AHL	64	19	48	67	12	....	....	....	....	....	....	....	....	....	....	....	3	0	1	1	0	....	....	....
1998-99	Florida	NHL	1	1	0	1	0	0	0	0	1	100.0	0	16	37.5	1	0	15:26	....	....	....	....	....	....	....	....
	New Haven	AHL	37	8	19	27	18	....	....	....	....	....	....	....	....	....	....	....	....	....	....	....	....	....	....	....
99-2000	Florida	NHL	66	4	12	16	14	0	0	0	44	9.1	1	684	51.8	127	29	11:47	....	....	....	....	....	....	....	....
	Tampa Bay	NHL	14	0	2	2	2	0	0	0	5	0.0	-9	117	53.0	28	6	11:02	....	....	....	....	....	....	....	....
2000-01	Tampa Bay	NHL	80	7	14	21	44	1	0	0	71	9.9	-20	951	48.9	177	68	15:47	....	....	....	....	....	....	....	....
2001-02	Florida	NHL	29	1	3	4	10	0	0	0	24	4.2	-5	336	47.9	40	21	13:00	....	....	....	....	....	....	....	....
	NHL Totals		200	13	33	46	70	1	0	0	151	8.6		2104	49.8	373	124	13:37	....	....	....	....	....	....	....	....

Traded to **Tampa Bay** by **Florida** with Dwayne Hay for Mike Sillinger, March 14, 2000. Traded to **Florida** by **Tampa Bay** with Tampa Bay's 6th round choice in 2003 Entry Draft for Vaclav Prospal, July 10, 2001. • Missed majority of 2001-02 season recovering from head injury suffered in game vs. St. Louis, December 22, 2001.

JOHNSSON, Kim

(YAWN-suhn, KIHM) **PHI.**

Defense. Shoots left. 6'1", 205 lbs. Born, Malmo, Sweden, March 16, 1976. NY Rangers' 15th choice, 286th overall, in 1994 Entry Draft.

Season	Club	League	GP	G	A	Pts	PIM	PP	SH	GW	S	%	+/-	TF	F%	H	SB	Min	GP	G	A	Pts	PIM	PP	SH	GW
1993-94	Malmo IF Jr.	Swede-Jr.	14	5	3	8	14	....	....	....	....	....	....	....	....	....	....	....	....	....	....	....	....	....	....	....
	Malmo IF	Sweden	2	0	0	0	0	....	....	....	....	....	....	....	....	....	....	....	....	....	....	....	....	....	....	....
1994-95	Malmo IF Jr.	Swede-Jr.	29	6	15	21	40	....	....	....	....	....	....	....	....	....	....	....	....	....	....	....	....	....	....	....
	Malmo IF	Sweden	13	0	0	0	4	....	....	....	....	....	....	....	....	....	....	....	1	0	0	0	0	....	....	....
1995-96	Malmo IF	Sweden	38	2	0	2	30	....	....	....	....	....	....	....	....	....	....	....	4	0	1	1	8	....	....	....
1996-97	Malmo IF	Sweden	49	4	9	13	42	....	....	....	....	....	....	....	....	....	....	....	4	0	0	0	2	....	....	....
1997-98	Malmo IF	Sweden	45	5	9	14	29	....	....	....	....	....	....	....	....	....	....	....	....	....	....	....	....	....	....	....
1998-99	Malmo IF	Sweden	49	9	8	17	76	....	....	....	....	....	....	....	....	....	....	....	8	2	3	5	12	....	....	....
99-2000	NY Rangers	NHL	76	6	15	21	46	1	0	1	101	5.9	-13	0	0.0	61	116	18:06	....	....	....	....	....	....	....	....
2000-01	NY Rangers	NHL	75	5	21	26	40	4	0	0	104	4.8	-3	0	0.0	68	122	21:16	....	....	....	....	....	....	....	....
2001-02	Philadelphia	NHL	82	11	30	41	42	5	0	1	150	7.3	12	0	0.0	94	118	23:02	5	0	0	0	2	0	0	0
	Sweden	Olympics	4	1	1	2	0	....	....	....	....	....	....	....	....	....	....	....	....	....	....	....	....	....	....	....
	NHL Totals		233	22	66	88	128	10	0	2	355	6.2		0	0.0	223	356	20:51	5	0	0	0	2	0	0	0

Traded to **Philadelphia** by **NY Rangers** with Jan Hlavac, Pavel Brendl and NY Rangers' 3rd round choice in 2003 Entry Draft for Eric Lindros, August 20, 2001.

JOKINEN, Olli

(YOH-kih-nihn, OH-lee) **FLA.**

Center. Shoots left. 6'3", 205 lbs. Born, Kuopio, Finland, December 5, 1978. Los Angeles' 1st choice, 3rd overall, in 1997 Entry Draft.

Season	Club	League	GP	G	A	Pts	PIM	PP	SH	GW	S	%	+/-	TF	F%	H	SB	Min	GP	G	A	Pts	PIM	PP	SH	GW
1992-93	KalPa Kuopio-C	Finn-Jr.	14	8	3	11	12	....	....	....	....	....	....	....	....	....	....	....	....	....	....	....	....	....	....	....
1993-94	KalPa Kuopio-C	Finn-Jr.	31	27	25	52	62	....	....	....	....	....	....	....	....	....	....	....	....	....	....	....	....	....	....	....
1994-95	KalPa Kuopio-B	Finn-Jr.	12	9	14	23	46	....	....	....	....	....	....	....	....	....	....	....	....	....	....	....	....	....	....	....
	KalPa Kuopio Jr.	Finn-Jr.	6	0	1	1	6	....	....	....	....	....	....	....	....	....	....	....	....	....	....	....	....	....	....	....
1995-96	KalPa Kuopio Jr.	Finn-Jr.	25	20	14	34	47	....	....	....	....	....	....	....	....	....	....	....	7	4	4	8	20	....	....	....
	KalPa Kuopio	Finland	15	1	1	2	2	....	....	....	....	....	....	....	....	....	....	....	....	....	....	....	....	....	....	....
1996-97	HIFK Jr.	Finn-Jr.	2	1	0	1	6	....	....	....	....	....	....	....	....	....	....	....	....	....	....	....	....	....	....	....
	HIFK Helsinki	Finland	50	14	27	41	88	....	....	....	....	....	....	....	....	....	....	....	....	....	....	....	....	....	....	....
1997-98	Los Angeles	NHL	8	0	0	0	6	0	0	0	12	0.0	-5	....	....	....	....	....	....	....	....	....	....	....	....	....
	HIFK Helsinki	Finland	30	11	28	39	8	....	....	....	....	....	....	....	....	....	....	....	9	*7	2	9	2	....	....	....
1998-99	Los Angeles	NHL	66	9	12	21	44	3	1	1	87	10.3	-10	779	43.9	109	26	14:42	....	....	....	....	....	....	....	....
	Springfield	AHL	9	3	6	9	6	....	....	....	....	....	....	....	....	....	....	....	....	....	....	....	....	....	....	....
99-2000	NY Islanders	NHL	82	11	10	21	80	1	2	3	138	8.0	0	841	46.1	156	27	16:15	....	....	....	....	....	....	....	....
2000-01	Florida	NHL	78	6	10	16	106	0	0	0	121	5.0	-22	638	42.3	98	21	13:23	....	....	....	....	....	....	....	....
2001-02	Florida	NHL	80	9	20	29	98	3	1	0	153	5.9	-16	1222	45.2	101	27	18:05	....	....	....	....	....	....	....	....
	Finland	Olympics	4	2	1	3	0	....	....	....	....	....	....	....	....	....	....	....	....	....	....	....	....	....	....	....
	NHL Totals		314	35	52	87	334	7	4	4	511	6.8		3480	44.6	464	101	15:40	....	....	....	....	....	....	....	....

Traded to **NY Islanders** by **LA Kings** with Josh Green, Mathieu Biron and LA Kings' 1st round choice (Taylor Pyatt) in 1999 Entry Draft for Ziggy Palffy, Brian Smolinski, Marcel Cousineau and New Jersey's 4th round choice (previously acquired, LA Kings selected Daniel Johansson) in 1999 Entry Draft, June 20, 1999. Traded to **Florida** by **NY Islanders** with Roberto Luongo for Mark Parrish and Oleg Kvasha, June 24, 2000.

JONES, Ty

(JOHNZ, TIGH) **CHI.**

Right wing. Shoots right. 6'3", 218 lbs. Born, Richland, WA, February 22, 1979. Chicago's 2nd choice, 16th overall, in 1997 Entry Draft.

Season	Club	League	GP	G	A	Pts	PIM	PP	SH	GW	S	%	+/-	TF	F%	H	SB	Min	GP	G	A	Pts	PIM	PP	SH	GW
1993-94	Alaska All-Stars	AAHL	64	84	104	188	126	....	....	....	....	....	....	....	....	....	....	....	....	....	....	....	....	....	....	....
1994-95	Alaska All-Stars	AAHL	42	33	35	68	98	....	....	....	....	....	....	....	....	....	....	....	....	....	....	....	....	....	....	....
1995-96	Spokane Chiefs	WHL	34	1	0	1	77	....	....	....	....	....	....	....	....	....	....	....	3	0	0	0	6	....	....	....
1996-97	Spokane Chiefs	WHL	67	20	34	54	202	....	....	....	....	....	....	....	....	....	....	....	9	2	4	6	10	....	....	....
1997-98	Spokane Chiefs	WHL	60	36	48	84	161	....	....	....	....	....	....	....	....	....	....	....	18	2	14	16	35	....	....	....
1998-99	Spokane Chiefs	WHL	26	15	12	27	98	....	....	....	....	....	....	....	....	....	....	....	....	....	....	....	....	....	....	....
	Kamloops Blazers	WHL	20	3	16	19	84	....	....	....	....	....	....	....	....	....	....	....	14	5	3	8	22	....	....	....
	Chicago	NHL	8	0	0	0	12	0	0	0	3	0.0	-1	0	0.0	5	1	7:53	....	....	....	....	....	....	....	....
99-2000	Cleveland	IHL	10	1	1	2	34	....	....	....	....	....	....	....	....	....	....	....	....	....	....	....	....	....	....	....
	Florida	ECHL	48	11	26	37	81	....	....	....	....	....	....	....	....	....	....	....	5	1	1	2	17	....	....	....
2000-01	Norfolk Admirals	AHL	64	11	17	28	114	....	....	....	....	....	....	....	....	....	....	....	....	....	....	....	....	....	....	....
2001-02	Norfolk Admirals	AHL	55	6	14	20	172	....	....	....	....	....	....	....	....	....	....	....	4	0	0	0	2	....	....	....
	NHL Totals		8	0	0	0	12	0	0	0	3	0.0		0	0.0	5	1	7:53	....	....	....	....	....	....	....	....

Traded to **Kamloops** (WHL) by **Spokane** (WHL) for Ryan Thorpe, February 1, 1999.

JONSSON, Hans

(YAWN-suhn, HANS) **PIT.**

Defense. Shoots left. 6'1", 205 lbs. Born, Jarved, Sweden, August 2, 1973. Pittsburgh's 11th choice, 286th overall, in 1993 Entry Draft.

Season	Club	League	GP	G	A	Pts	PIM	PP	SH	GW	S	%	+/-	TF	F%	H	SB	Min	GP	G	A	Pts	PIM	PP	SH	GW
1991-92	Hasums IF	Swede-2	13	4	6	10	10	....	....	....	....	....	....	....	....	....	....	....	....	....	....	....	....	....	....	....
	MoDo	Sweden	6	0	1	1	4	....	....	....	....	....	....	....	....	....	....	....	....	....	....	....	....	....	....	....
1992-93	MoDo	Sweden	40	2	2	4	24	....	....	....	....	....	....	....	....	....	....	....	3	0	1	1	2	....	....	....
1993-94	MoDo	Sweden	23	4	1	5	18	....	....	....	....	....	....	....	....	....	....	....	10	0	1	1	12	....	....	....
1994-95	MoDo	Sweden	39	4	6	10	30	....	....	....	....	....	....	....	....	....	....	....	....	....	....	....	....	....	....	....
1995-96	MoDo	Sweden	36	10	6	16	30	....	....	....	....	....	....	....	....	....	....	....	8	2	1	3	24	....	....	....
1996-97	MoDo	Sweden	27	7	5	12	18	....	....	....	....	....	....	....	....	....	....	....	....	....	....	....	....	....	....	....
1997-98	MoDo	Sweden	40	8	6	14	40	....	....	....	....	....	....	....	....	....	....	....	8	1	1	2	12	....	....	....
1998-99	MoDo	Sweden	41	3	4	7	40	....	....	....	....	....	....	....	....	....	....	....	13	2	4	6	22	....	....	....
99-2000	Pittsburgh	NHL	68	3	11	14	12	0	1	1	49	6.1	-5	0	0.0	79	97	18:34	11	0	1	1	6	0	0	0
2000-01	Pittsburgh	NHL	58	4	18	22	22	2	0	0	44	9.1	11	0	0.0	72	96	18:27	16	0	0	0	8	0	0	0
2001-02	Pittsburgh	NHL	53	2	5	7	22	2	0	0	37	5.4	-12	0	0.0	70	75	18:12	....	....	....	....	....	....	....	....
	NHL Totals		179	9	34	43	56	4	1	1	130	6.9		0	0.0	221	268	18:25	27	0	1	1	14	0	0	0

JONSSON, Jorgen

(YAWN-suhn, YOHR-gahn)

Left wing. Shoots left. 6', 185 lbs. Born, Angelholm, Sweden, September 29, 1972. Calgary's 11th choice, 227th overall, in 1994 Entry Draft.

			Regular Season																Playoffs							
Season	Club	League	GP	G	A	Pts	PIM	PP	SH	GW	S	%	+/-	TF	F%	H	SB	Min	GP	G	A	Pts	PIM	PP	SH	GW
1990-91	Rogle	Swede-2	21	4	2	6	2												12	2	1	3	2			
1991-92	Rogle	Swede-2	27	1	8	9	6												5	0	0	0	0			
1992-93	Rogle	Sweden	40	17	11	28	28																			
1993-94	Rogle	Sweden	40	17	14	31	46																			
	Sweden	Olympics	6	0	0	0	0																			
1994-95	Rogle	Sweden	22	4	6	10	18																			
1995-96	Farjestad	Sweden	39	11	15	26	36												8	0	4	4	6			
1996-97	Farjestad	Sweden	49	12	21	33	58												14	9	5	14	14			
	Farjestad	EuroHL	4	2	1	3	2																			
1997-98	Farjestad	Sweden	45	22	25	47	53												12	2	*9	11	12			
	Farjestad	EuroHL	7	2	4	6	6																			
	Sweden	Olympics	1	0	0	0	0																			
1998-99	Farjestad	Sweden	48	17	24	41	44												4	0	2	2	4			
	Farjestad	EuroHL	5	2	4	6	4												2	1	0	1	4			
99-2000	**NY Islanders**	**NHL**	**68**	**11**	**17**	**28**	**16**	**1**	**2**	**0**	**95**	**11.6**	**-6**	**642**	**44.2**	**72**	**37**	**16:07**								
	Anaheim	**NHL**	**13**	**1**	**2**	**3**	**0**	**0**	**0**	**1**	**21**	**4.8**	**-2**	**118**	**35.6**	**15**	**8**	**12:51**								
2000-01	Farjestad	Sweden	50	20	26	46	32												15	5	12	17	12			
2001-02	Farjestad	Sweden	50	*22	17	39	20												10	5	1	6	16			
	Sweden	Olympics	4	0	0	0	4																			
	NHL Totals		**81**	**12**	**19**	**31**	**16**	**1**	**2**	**1**	**116**	**10.3**		**760**	**42.9**	**87**	**45**	**15:35**								

Traded to **NY Islanders** by **Calgary** for Jan Hlavac, July 14, 1998. Traded to **Anaheim** by **NY Islanders** for Johan Davidsson and future considerations, March 11, 2000.

JONSSON, Kenny

(YAWN-suhn, KEHN-nee) **NYI**

Defense. Shoots left. 6'3", 217 lbs. Born, Angelholm, Sweden, October 6, 1974. Toronto's 1st choice, 12th overall, in 1993 Entry Draft.

			Regular Season																Playoffs							
Season	Club	League	GP	G	A	Pts	PIM	PP	SH	GW	S	%	+/-	TF	F%	H	SB	Min	GP	G	A	Pts	PIM	PP	SH	GW
1991-92	Rogle	Swede-2	30	4	11	15	24												5	0	0	0	0			
1992-93	Rogle Jr.	Swede-Jr.	2	1	2	3	25																			
	Rogle	Sweden	39	3	10	13	42																			
1993-94	Rogle	Sweden	36	4	13	17	40												3	1	1	2	2			
	Sweden	Olympics	3	1	0	1	0																			
1994-95	Rogle	Sweden	8	3	1	4	20																			
	St. John's	AHL	10	2	5	7	2																			
	Toronto	**NHL**	**39**	**2**	**7**	**9**	**16**	**0**	**0**	**1**	**50**	**4.0**	**-8**						**4**	**0**	**0**	**0**	**0**	**0**	**0**	**0**
1995-96	**Toronto**	**NHL**	**50**	**4**	**22**	**26**	**22**	**3**	**0**	**1**	**90**	**4.4**	**12**													
	NY Islanders	**NHL**	**16**	**0**	**4**	**4**	**10**	**0**	**0**	**0**	**40**	**0.0**	**-5**													
1996-97	**NY Islanders**	**NHL**	**81**	**3**	**18**	**21**	**24**	**1**	**0**	**0**	**92**	**3.3**	**10**													
1997-98	**NY Islanders**	**NHL**	**81**	**14**	**26**	**40**	**58**	**6**	**0**	**2**	**108**	**13.0**	**-2**													
1998-99	**NY Islanders**	**NHL**	**63**	**8**	**18**	**26**	**34**	**6**	**0**	**0**	**91**	**8.8**	**-18**	**0**	**0.0**	**57**	**90**	**24:59**								
99-2000	**NY Islanders**	**NHL**	**65**	**1**	**24**	**25**	**32**	**1**	**0**	**0**	**84**	**1.2**	**-15**	**0**	**0.0**	**51**	**113**	**24:29**								
2000-01	**NY Islanders**	**NHL**	**65**	**8**	**21**	**29**	**30**	**5**	**0**	**0**	**91**	**8.8**	**-22**	**0**	**0.0**	**47**	**114**	**24:04**								
2001-02	**NY Islanders**	**NHL**	**76**	**10**	**22**	**32**	**26**	**2**	**1**	**0**	**107**	**9.3**	**15**	**2**	**0.0**	**76**	**141**	**25:34**	**5**	**1**	**2**	**3**	**4**	**1**	**0**	**0**
	Sweden	Olympics	3	1	0	1	2																			
	NHL Totals		**536**	**50**	**162**	**212**	**252**	**24**	**1**	**4**	**753**	**6.6**		**2**	**0.0**	**231**	**458**	**24:48**	**9**	**1**	**2**	**3**	**4**	**1**	**0**	**0**

• NHL All-Rookie Team (1995)

Traded to **NY Islanders** by **Toronto** with Sean Haggerty, Darby Hendrickson and Toronto's 1st round choice (Roberto Luongo) in 1997 Entry Draft for Wendel Clark, Mathieu Schneider and D.J. Smith, March 13, 1996.

JOSEPH, Chris

(JOH-sehf, KRIHS)

Defense. Shoots right. 6'3", 212 lbs. Born, Burnaby, B.C., September 10, 1969. Pittsburgh's 1st choice, 5th overall, in 1987 Entry Draft.

			Regular Season																Playoffs							
Season	Club	League	GP	G	A	Pts	PIM	PP	SH	GW	S	%	+/-	TF	F%	H	SB	Min	GP	G	A	Pts	PIM	PP	SH	GW
1984-85	Burnaby Beavers	BCAHA	52	18	48	66	52																			
1985-86	Seattle	WHL	72	4	8	12	50												5	0	3	3	12			
1986-87	Seattle	WHL	67	13	45	58	155																			
1987-88	**Pittsburgh**	**NHL**	**17**	**0**	**4**	**4**	**12**	**0**	**0**	**0**	**13**	**0.0**	**2**													
	Edmonton	**NHL**	**7**	**0**	**4**	**4**	**6**	**0**	**0**	**0**	**1**	**0.0**	**-3**													
	Seattle	WHL	23	5	14	19	49																			
	Nova Scotia	AHL	8	0	2	2	8												4	0	0	0	9			
1988-89	**Edmonton**	**NHL**	**44**	**4**	**5**	**9**	**54**	**0**	**0**	**0**	**36**	**11.1**	**-9**													
	Cape Breton	AHL	5	1	1	2	18																			
1989-90	**Edmonton**	**NHL**	**4**	**0**	**2**	**2**	**2**	**0**	**0**	**0**	**5**	**0.0**	**-2**													
	Cape Breton	AHL	61	10	20	30	69												6	2	1	3	4			
1990-91	**Edmonton**	**NHL**	**49**	**5**	**17**	**22**	**59**	**2**	**0**	**0**	**74**	**6.8**	**3**													
1991-92	**Edmonton**	**NHL**	**7**	**0**	**0**	**0**	**8**	**0**	**0**	**0**	**5**	**0.0**	**-1**						**5**	**1**	**3**	**4**	**2**	**0**	**0**	**0**
	Cape Breton	AHL	63	14	29	43	72												5	0	2	2	8			
1992-93	**Edmonton**	**NHL**	**33**	**2**	**10**	**12**	**48**	**1**	**0**	**0**	**49**	**4.1**	**-9**													
1993-94	**Edmonton**	**NHL**	**10**	**1**	**1**	**2**	**28**	**1**	**0**	**0**	**25**	**4.0**	**-8**													
	Tampa Bay	**NHL**	**66**	**10**	**19**	**29**	**108**	**7**	**0**	**0**	**154**	**6.5**	**-13**													
1994-95	**Pittsburgh**	**NHL**	**33**	**5**	**10**	**15**	**46**	**3**	**0**	**0**	**73**	**6.8**	**3**						**10**	**1**	**1**	**2**	**12**	**0**	**0**	**0**
1995-96	**Pittsburgh**	**NHL**	**70**	**5**	**14**	**19**	**71**	**0**	**0**	**1**	**94**	**5.3**	**6**						**15**	**1**	**0**	**1**	**8**	**0**	**0**	**0**
1996-97	**Vancouver**	**NHL**	**63**	**3**	**13**	**16**	**62**	**2**	**0**	**1**	**99**	**3.0**	**-21**													
1997-98	**Philadelphia**	**NHL**	**15**	**1**	**0**	**1**	**19**	**0**	**0**	**1**	**20**	**5.0**	**1**						**1**	**0**	**0**	**0**	**2**	**0**	**0**	**0**
	Philadelphia	AHL	6	2	3	5	2																			
1998-99	**Philadelphia**	**NHL**	**2**	**0**	**0**	**0**	**2**	**0**	**0**	**0**	**1**	**0.0**	**0**	**0**	**0.0**	**0**	**0**	**9:36**								
	Cincinnati	IHL	27	11	19	30	38																			
	Philadelphia	AHL	51	9	29	38	26												16	3	10	13	8			
99-2000	**Vancouver**	**NHL**	**38**	**2**	**9**	**11**	**6**	**1**	**0**	**0**	**73**	**2.7**	**-4**	**0**	**0.0**	**17**	**24**	**17:14**								
	Phoenix	**NHL**	**9**	**0**	**0**	**0**	**0**	**0**	**0**	**0**	**13**	**0.0**	**-5**	**0**	**0.0**	**8**	**9**	**13:20**								
2000-01	**Phoenix**	**NHL**	**24**	**1**	**1**	**2**	**16**	**0**	**1**	**0**	**33**	**3.0**	**-4**	**0**	**0.0**	**11**	**20**	**12:59**								
	Atlanta	**NHL**	**19**	**0**	**3**	**3**	**20**	**0**	**0**	**0**	**25**	**0.0**	**-7**	**0**	**0.0**	**33**	**26**	**20:48**								
2001-02	TPS Turku	Finland	32	6	9	15	105												8	1	1	2	*31			
	NHL Totals		**510**	**39**	**112**	**151**	**567**	**17**	**1**	**3**	**793**	**4.9**		**0**	**0.0**	**69**	**79**	**16:19**	**31**	**3**	**4**	**7**	**24**	**0**	**0**	**0**

WHL West Second All-Star Team (1987)

Traded to **Edmonton** by **Pittsburgh** with Craig Simpson, Dave Hannan and Moe Mantha for Paul Coffey, Dave Hunter and Wayne Van Dorp, November 24, 1987. Traded to **Tampa Bay** by **Edmonton** for Bob Beers, November 11, 1993. Claimed by **Pittsburgh** from **Tampa Bay** in Waiver Draft, January 18, 1995. Claimed by **Vancouver** from **Pittsburgh** in Waiver Draft, September 30, 1996. Signed as a free agent by **Philadelphia**, September 11, 1997. Signed as a free agent by **Ottawa**, August 18, 1999. Claimed by **Vancouver** from **Ottawa** in Waiver Draft, September 27, 1999. Claimed on waivers by **Phoenix** from **Vancouver**, March 14, 2000. Claimed on waivers by **Atlanta** from **Phoenix**, February 14, 2001.

JOVANOVSKI, Ed

(joh-van-OHV-skee, EHD) **VAN.**

Defense. Shoots left. 6'2", 210 lbs. Born, Windsor, Ont., June 26, 1976. Florida's 1st choice, 1st overall, in 1994 Entry Draft.

			Regular Season																Playoffs							
Season	Club	League	GP	G	A	Pts	PIM	PP	SH	GW	S	%	+/-	TF	F%	H	SB	Min	GP	G	A	Pts	PIM	PP	SH	GW
1991-92	Windsor	OMHA	50	25	40	65	88																			
1992-93	Windsor Bulldogs	OJHL-B	48	7	46	53	88																			
1993-94	Windsor	OHL	62	15	36	51	221												4	0	0	0	15			
1994-95	Windsor	OHL	50	23	42	65	198												9	2	7	9	39			
1995-96	**Florida**	**NHL**	**70**	**10**	**11**	**21**	**137**	**2**	**0**	**2**	**116**	**8.6**	**-3**						**22**	**1**	**8**	**9**	**52**	**0**	**0**	**0**
1996-97	**Florida**	**NHL**	**61**	**7**	**16**	**23**	**172**	**3**	**0**	**1**	**80**	**8.8**	**-1**						**5**	**0**	**0**	**0**	**4**	**0**	**0**	**0**
1997-98	**Florida**	**NHL**	**81**	**9**	**14**	**23**	**158**	**2**	**1**	**3**	**142**	**6.3**	**-12**													
1998-99	**Florida**	**NHL**	**41**	**3**	**13**	**16**	**82**	**1**	**0**	**1**	**68**	**4.4**	**-4**	**0**	**0.0**	**88**	**36**	**22:35**								
	Vancouver	**NHL**	**31**	**2**	**9**	**11**	**44**	**0**	**0**	**0**	**41**	**4.9**	**-5**	**0**	**0.0**	**68**	**35**	**21:16**								
99-2000	**Vancouver**	**NHL**	**75**	**5**	**21**	**26**	**54**	**1**	**0**	**1**	**109**	**4.6**	**-3**	**0**	**0.0**	**167**	**104**	**24:03**								
2000-01	**Vancouver**	**NHL**	**79**	**12**	**35**	**47**	**102**	**4**	**0**	**2**	**193**	**6.2**	**-1**	**0**	**0.0**	**172**	**122**	**24:57**	**4**	**1**	**1**	**2**	**0**	**0**	**0**	**0**
2001-02	**Vancouver**	**NHL**	**82**	**17**	**31**	**48**	**101**	**7**	**1**	**3**	**202**	**8.4**	**-7**	**0**	**0.0**	**138**	**101**	**25:11**	**6**	**1**	**4**	**5**	**8**	**1**	**0**	**0**
	Canada	Olympics	6	0	3	3	4																			
	NHL Totals		**520**	**65**	**150**	**215**	**850**	**20**	**2**	**13**	**951**	**6.8**		**0**	**0.0**	**633**	**398**	**24:06**	**37**	**3**	**13**	**16**	**64**	**1**	**0**	**0**

OHL All-Rookie Team (1994) • OHL Second All-Star Team (1994) • OHL First All-Star Team (1995) • NHL All-Rookie Team (1996) • Played in NHL All-Star Game (2001, 2002)

Traded to **Vancouver** by **Florida** with Dave Gagner, Mike Brown, Kevin Weekes and Florida's 1st round choice (Nathan Smith) in 2000 Entry Draft for Pavel Bure, Bret Hedican, Brad Ference and Vancouver's 3rd round choice (Robert Fried) in 2000 Entry Draft, January 17, 1999.

JUNEAU, Joe (ZHOO-noh, JOH) MTL.

Center. Shoots left. 6', 195 lbs. Born, Pont-Rouge, Que., January 5, 1968. Boston's 3rd choice, 81st overall, in 1988 Entry Draft.

			Regular Season																Playoffs							
Season	Club	League	GP	G	A	Pts	PIM	PP	SH	GW	S	%	+/-	TF	F%	H	SB	Min	GP	G	A	Pts	PIM	PP	SH	GW
1983-84	Ste-Foy	QAAA	30	3	7	10	24	....	....	....	....	....	....	....	....	....	....	....	12	3	11	14	4	....	....	....
1984-85	Ste-Foy	QAAA	41	25	46	71	60	....	....	....	....	....	....	....	....	....	....	....	13	9	15	24	20	....	....	....
1985-86	Levis-Lauzon	CEGEP	STATISTICS NOT AVAILABLE																							
1986-87	Levis-Lauzon	CEGEP	38	27	57	84	....	....	....	....	....	....	....	....	....	....	....	....	....	....	....	....	....	....	....	....
1987-88	RPI Engineers	ECAC	31	16	29	45	18	....	....	....	....	....	....	....	....	....	....	....	....	....	....	....	....	....	....	....
1988-89	RPI Engineers	ECAC	30	12	23	35	40	....	....	....	....	....	....	....	....	....	....	....	....	....	....	....	....	....	....	....
1989-90	RPI Engineers	ECAC	34	18	*52	*70	31	....	....	....	....	....	....	....	....	....	....	....	....	....	....	....	....	....	....	....
1990-91	RPI Engineers	ECAC	29	23	40	63	68	....	....	....	....	....	....	....	....	....	....	....	....	....	....	....	....	....	....	....
1991-92	Team Canada	Nat-Tm	60	20	49	69	35	....	....	....	....	....	....	....	....	....	....	....	....	....	....	....	....	....	....	....
	Canada	Olympics	8	6	*9	*15	4	....	....	....	....	....	....	....	....	....	....	....	....	....	....	....	....	....	....	....
	Boston	**NHL**	**14**	**5**	**14**	**19**	**4**	**2**	**0**	**0**	**38**	**13.2**	**6**	....	....	....	....	....	**15**	**4**	**8**	**12**	**21**	**2**	**0**	**0**
1992-93	**Boston**	**NHL**	**84**	**32**	**70**	**102**	**33**	**9**	**0**	**3**	**229**	**14.0**	**23**	....	....	....	....	....	**4**	**2**	**4**	**6**	**6**	**2**	**0**	**0**
1993-94	**Boston**	**NHL**	**63**	**14**	**58**	**72**	**35**	**4**	**0**	**2**	**142**	**9.9**	**11**	....	....	....	....	....	....	....	....	....	....	....	....	....
	Washington	**NHL**	**11**	**5**	**8**	**13**	**6**	**2**	**0**	**0**	**22**	**22.7**	**0**	....	....	....	....	....	**11**	**4**	**5**	**9**	**6**	**2**	**0**	**1**
1994-95	**Washington**	**NHL**	**44**	**5**	**38**	**43**	**8**	**3**	**0**	**0**	**70**	**7.1**	**–1**	....	....	....	....	....	**7**	**2**	**6**	**8**	**2**	**0**	**0**	**0**
1995-96	**Washington**	**NHL**	**80**	**14**	**50**	**64**	**30**	**7**	**2**	**2**	**176**	**8.0**	**–3**	....	....	....	....	....	**5**	**0**	**7**	**7**	**6**	**0**	**0**	**0**
1996-97	**Washington**	**NHL**	**58**	**15**	**27**	**42**	**8**	**9**	**1**	**3**	**124**	**12.1**	**–11**	....	....	....	....	....	....	....	....	....	....	....	....	....
1997-98	**Washington**	**NHL**	**56**	**9**	**22**	**31**	**26**	**4**	**1**	**1**	**87**	**10.3**	**–8**	....	....	....	....	....	**21**	**7**	**10**	**17**	**8**	**1**	**1**	**4**
1998-99	**Washington**	**NHL**	**63**	**14**	**27**	**41**	**20**	**2**	**1**	**3**	**142**	**9.9**	**–3**	**437**	**48.1**	**34**	**26**	**19:28**	....	....	....	....	....	....	....	....
	Buffalo	**NHL**	**9**	**1**	**1**	**2**	**2**	**0**	**0**	**0**	**8**	**12.5**	**–1**	**8**	**12.5**	**3**	**2**	**17:11**	**20**	**3**	**8**	**11**	**10**	**0**	**1**	**0**
99-2000	**Ottawa**	**NHL**	**65**	**13**	**24**	**37**	**22**	**2**	**0**	**2**	**126**	**10.3**	**3**	**830**	**51.6**	**27**	**25**	**18:28**	**6**	**2**	**1**	**3**	**0**	**0**	**0**	**0**
2000-01	**Phoenix**	**NHL**	**69**	**10**	**23**	**33**	**28**	**5**	**0**	**3**	**100**	**10.0**	**–2**	**210**	**50.5**	**27**	**14**	**17:25**	....	....	....	....	....	....	....	....
2001-02	**Montreal**	**NHL**	**70**	**8**	**28**	**36**	**10**	**1**	**0**	**1**	**96**	**8.3**	**–3**	**1337**	**48.6**	**30**	**42**	**18:37**	**12**	**1**	**4**	**5**	**6**	**0**	**0**	**0**
	NHL Totals		**686**	**145**	**390**	**535**	**232**	**50**	**5**	**20**	**1360**	**10.7**		**2822**	**49.4**	**121**	**109**	**18:26**	**101**	**25**	**53**	**78**	**65**	**7**	**2**	**5**

NCAA East First All-American Team (1990) • ECAC Second All-Star Team (1991) • NCAA East Second All-American Team (1991) • NHL All-Rookie Team (1993)

Traded to **Washington** by **Boston** for Al Iafrate, March 21, 1994. Traded to **Buffalo** by **Washington** with Washington's 3rd round choice (Tim Preston) in 1999 Entry Draft for Alexei Tezikov and Buffalo's 4th round compensatory choice (later traded to Calgary - Calgary selected Levente Szuper) in 2000 Entry Draft, March 22, 1999. Signed as a free agent by **Ottawa**, October 25, 1999. Selected by **Minnesota** from **Ottawa** in Expansion Draft, June 23, 2000. Traded to **Phoenix** by **Minnesota** for the rights to Rickard Wallin, June 23, 2000. Traded to **Montreal** by **Phoenix** for future considerations, June 15, 2001.

KABERLE, Frantisek (KA-buhr-lay, FRAN-tih-sehk) ATL.

Defense. Shoots left. 6', 190 lbs. Born, Kladno, Czech., November 8, 1973. Los Angeles' 3rd choice, 76th overall, in 1999 Entry Draft.

			Regular Season																Playoffs							
Season	Club	League	GP	G	A	Pts	PIM	PP	SH	GW	S	%	+/-	TF	F%	H	SB	Min	GP	G	A	Pts	PIM	PP	SH	GW
1991-92	Poldi Kladno	Czech	37	1	4	5	8	....	....	....	....	....	....	....	....	....	....	....	8	0	1	1	0	....	....	....
1992-93	Poldi Kladno	Czech	40	4	5	9	....	....	....	....	....	....	....	....	....	....	....	....	9	2	4	6	....	....	....	....
1993-94	HC Kladno	Czech	41	4	16	20	....	....	....	....	....	....	....	....	....	....	....	....	11	1	1	2	....	....	....	....
1994-95	HC Kladno	Czech	40	7	17	24	20	....	....	....	....	....	....	....	....	....	....	....	8	0	3	3	12	....	....	....
1995-96	MoDo	Sweden	40	5	7	12	34	....	....	....	....	....	....	....	....	....	....	....	8	0	1	1	0	....	....	....
1996-97	MoDo	Sweden	50	3	11	14	28	....	....	....	....	....	....	....	....	....	....	....	....	....	....	....	....	....	....	....
1997-98	MoDo	Sweden	46	5	4	9	22	....	....	....	....	....	....	....	....	....	....	....	9	1	1	2	4	....	....	....
1998-99	MoDo	Sweden	45	15	18	33	4	....	....	....	....	....	....	....	....	....	....	....	13	2	5	7	8	....	....	....
99-2000	**Los Angeles**	**NHL**	**37**	**0**	**9**	**9**	**4**	**0**	**0**	**0**	**41**	**0.0**	**3**	**0**	**0.0**	**29**	**31**	**17:04**	....	....	....	....	....	....	....	....
	Long Beach	IHL	18	2	8	10	8	....	....	....	....	....	....	....	....	....	....	....	....	....	....	....	....	....	....	....
	Atlanta	**NHL**	**14**	**1**	**6**	**7**	**6**	**0**	**1**	**0**	**35**	**2.9**	**–13**	**0**	**0.0**	**25**	**23**	**24:39**	....	....	....	....	....	....	....	....
	Lowell	AHL	4	0	2	2	0	....	....	....	....	....	....	....	....	....	....	....	....	....	....	....	....	....	....	....
2000-01	**Atlanta**	**NHL**	**51**	**4**	**11**	**15**	**18**	**1**	**0**	**1**	**99**	**4.0**	**11**	**1**	**0.0**	**35**	**62**	**22:17**	....	....	....	....	....	....	....	....
2001-02	**Atlanta**	**NHL**	**61**	**5**	**20**	**25**	**24**	**1**	**0**	**0**	**82**	**6.1**	**–11**	**0**	**0.0**	**49**	**87**	**21:35**	....	....	....	....	....	....	....	....
	NHL Totals		**163**	**10**	**46**	**56**	**52**	**2**	**1**	**1**	**257**	**3.9**		**1**	**0.0**	**138**	**203**	**21:02**	....	....	....	....	....	....	....	....

Traded to **Atlanta** by **Los Angeles** with Donald Audette for Kelly Buchberger and Nelson Emerson, March 13, 2000.

KABERLE, Tomas (KA-buhr-lay, TAW-mas) TOR.

Defense. Shoots left. 6'2", 200 lbs. Born, Rakovnik, Czech., March 2, 1978. Toronto's 13th choice, 204th overall, in 1996 Entry Draft.

			Regular Season																Playoffs							
Season	Club	League	GP	G	A	Pts	PIM	PP	SH	GW	S	%	+/-	TF	F%	H	SB	Min	GP	G	A	Pts	PIM	PP	SH	GW
1994-95	HC Kladno Jr.	Czech-Jr.	37	7	10	17	....	....	....	....	....	....	....	....	....	....	....	....	....	....	....	....	....	....	....	....
	HC Kladno	Czech	4	0	1	1	0	....	....	....	....	....	....	....	....	....	....	....	....	....	....	....	....	....	....	....
1995-96	Kladno Jr.	Czech-Jr.	23	6	13	19	....	....	....	....	....	....	....	....	....	....	....	....	....	....	....	....	....	....	....	....
	HC Poldi Kladno	Czech	23	0	1	1	2	....	....	....	....	....	....	....	....	....	....	....	2	0	0	0	0	....	....	....
1996-97	HC Poldi Kladno	Czech	49	0	5	5	26	....	....	....	....	....	....	....	....	....	....	....	3	0	0	0	0	....	....	....
1997-98	Kladno	Czech	47	4	19	23	12	....	....	....	....	....	....	....	....	....	....	....	....	....	....	....	....	....	....	....
	St. John's	AHL	2	0	0	0	0	....	....	....	....	....	....	....	....	....	....	....	....	....	....	....	....	....	....	....
1998-99	**Toronto**	**NHL**	**57**	**4**	**18**	**22**	**12**	**0**	**0**	**2**	**71**	**5.6**	**3**	**0**	**0.0**	**27**	**46**	**18:42**	**14**	**0**	**3**	**3**	**2**	**0**	**0**	**0**
99-2000	**Toronto**	**NHL**	**82**	**7**	**33**	**40**	**24**	**2**	**0**	**0**	**82**	**8.5**	**3**	**0**	**0.0**	**86**	**106**	**22:55**	**12**	**1**	**4**	**5**	**0**	**0**	**0**	**1**
2000-01	**Toronto**	**NHL**	**82**	**6**	**39**	**45**	**24**	**0**	**0**	**1**	**96**	**6.3**	**10**	**2**	**0.0**	**57**	**124**	**22:41**	**11**	**1**	**3**	**4**	**0**	**0**	**0**	**1**
2001-02	Kladno	Czech	9	1	7	8	4	....	....	....	....	....	....	....	....	....	....	....	....	....	....	....	....	....	....	....
	Toronto	**NHL**	**69**	**10**	**29**	**39**	**2**	**5**	**0**	**3**	**85**	**11.8**	**5**	**2**	**100.0**	**80**	**95**	**25:00**	**20**	**2**	**8**	**10**	**16**	**0**	**0**	**0**
	Czech Republic	Olympics	4	0	1	1	2	....	....	....	....	....	....	....	....	....	....	....	....	....	....	....	....	....	....	....
	NHL Totals		**290**	**27**	**119**	**146**	**62**	**7**	**0**	**6**	**334**	**8.1**		**4**	**50.0**	**250**	**371**	**22:31**	**57**	**4**	**18**	**22**	**18**	**0**	**0**	**2**

Played in NHL All-Star Game (2002)

Signed as a restricted free agent by **Kladno** (Czech) with **Toronto** retaining NHL rights, September 29, 2001. Re-signed by **Toronto**, October 30, 2001.

KALININ, Dmitri (kah-LIHN-ihn, DIH-mih-TREE) BUF.

Defense. Shoots left. 6'3", 215 lbs. Born, Chelyabinsk, USSR, July 22, 1980. Buffalo's 1st choice, 18th overall, in 1998 Entry Draft.

			Regular Season																Playoffs							
Season	Club	League	GP	G	A	Pts	PIM	PP	SH	GW	S	%	+/-	TF	F%	H	SB	Min	GP	G	A	Pts	PIM	PP	SH	GW
1995-96	Chelyabinsk Jr.	CIS-Jr.	30	10	10	20	60	....	....	....	....	....	....	....	....	....	....	....	....	....	....	....	....	....	....	....
	Chelyabinsk	CIS	20	0	3	3	10	....	....	....	....	....	....	....	....	....	....	....	....	....	....	....	....	....	....	....
1996-97	Chelyabinsk 2	Russia-3	20	0	0	0	10	....	....	....	....	....	....	....	....	....	....	....	....	....	....	....	....	....	....	....
	Chelyabinsk	Russia	2	0	0	0	0	....	....	....	....	....	....	....	....	....	....	....	2	0	0	0	0	....	....	....
1997-98	Chelyabinsk	Russia	26	0	2	2	24	....	....	....	....	....	....	....	....	....	....	....	....	....	....	....	....	....	....	....
1998-99	Moncton Wildcats	QMJHL	39	7	18	25	44	....	....	....	....	....	....	....	....	....	....	....	4	1	1	2	0	....	....	....
	Rochester	AHL	3	0	1	1	14	....	....	....	....	....	....	....	....	....	....	....	7	0	0	0	6	....	....	....
99-2000	**Buffalo**	**NHL**	**4**	**0**	**0**	**0**	**4**	**0**	**0**	**0**	**3**	**0.0**	**0**	**0**	**0.0**	**2**	**0**	**16:53**	....	....	....	....	....	....	....	....
	Rochester	AHL	75	2	19	21	52	....	....	....	....	....	....	....	....	....	....	....	21	2	9	11	8	....	....	....
2000-01	**Buffalo**	**NHL**	**79**	**4**	**18**	**22**	**38**	**2**	**0**	**0**	**88**	**4.5**	**–2**	**1**	**100.0**	**88**	**101**	**19:50**	**13**	**0**	**2**	**2**	**4**	**0**	**0**	**0**
2001-02	**Buffalo**	**NHL**	**58**	**2**	**11**	**13**	**26**	**0**	**0**	**0**	**67**	**3.0**	**–6**	**0**	**0.0**	**83**	**69**	**18:03**	....	....	....	....	....	....	....	....
	NHL Totals		**141**	**6**	**29**	**35**	**68**	**2**	**0**	**0**	**158**	**3.8**		**1**	**100.0**	**173**	**170**	**19:05**	**13**	**0**	**2**	**2**	**4**	**0**	**0**	**0**

KALLIO, Tomi (KAL-ee-oh, TAW-mee) ATL.

Left wing. Shoots left. 6', 190 lbs. Born, Turku, Finland, January 27, 1977. Colorado's 4th choice, 81st overall, in 1995 Entry Draft.

			Regular Season																Playoffs							
Season	Club	League	GP	G	A	Pts	PIM	PP	SH	GW	S	%	+/-	TF	F%	H	SB	Min	GP	G	A	Pts	PIM	PP	SH	GW
1992-93	TPS Turku-C	Finn-Jr.	39	39	34	73	18	....	....	....	....	....	....	....	....	....	....	....	....	....	....	....	....	....	....	....
1993-94	TPS Turku-B	Finn-Jr.	10	5	6	11	14	....	....	....	....	....	....	....	....	....	....	....	1	0	1	1	0	....	....	....
	TPS Turku Jr.	Finn-Jr.	33	9	7	16	16	....	....	....	....	....	....	....	....	....	....	....	6	0	1	1	2	....	....	....
1994-95	TPS Turku-B	Finn-Jr.	1	2	0	2	0	....	....	....	....	....	....	....	....	....	....	....	....	....	....	....	....	....	....	....
	TPS Turku Jr.	Finn-Jr.	14	5	12	17	24	....	....	....	....	....	....	....	....	....	....	....	....	....	....	....	....	....	....	....
	Kiekko-67 Turku	Finland-2	25	8	5	13	16	....	....	....	....	....	....	....	....	....	....	....	7	3	1	4	6	....	....	....
1995-96	TPS Turku Jr.	Finn-Jr.	8	8	3	11	14	....	....	....	....	....	....	....	....	....	....	....	....	....	....	....	....	....	....	....
	Kiekko-67 Turku	Finland-2	29	10	11	21	28	....	....	....	....	....	....	....	....	....	....	....	....	....	....	....	....	....	....	....
	TPS Turku	Finland	8	2	3	5	10	....	....	....	....	....	....	....	....	....	....	....	4	0	0	0	2	....	....	....
1996-97	TPS Turku	Finland	47	9	10	19	18	....	....	....	....	....	....	....	....	....	....	....	8	2	0	2	2	....	....	....
	TPS Turku	EuroHL	6	2	0	2	25	....	....	....	....	....	....	....	....	....	....	....	4	0	0	0	0	....	....	....
1997-98	TPS Turku	Finland	47	10	10	20	8	....	....	....	....	....	....	....	....	....	....	....	4	0	2	2	0	....	....	....
	TPS Turku	EuroHL	6	0	1	1	2	....	....	....	....	....	....	....	....	....	....	....	....	....	....	....	....	....	....	....
1998-99	TPS Turku	Finland	54	15	21	36	20	....	....	....	....	....	....	....	....	....	....	....	10	3	4	7	6	....	....	....
99-2000	TPS Turku	Finland	50	26	27	53	40	....	....	....	....	....	....	....	....	....	....	....	11	4	*9	13	4	....	....	....
	TPS Turku	EuroHL	5	2	1	3	0	....	....	....	....	....	....	....	....	....	....	....	5	5	3	8	2	....	....	....
2000-01	**Atlanta**	**NHL**	**56**	**14**	**13**	**27**	**22**	**2**	**0**	**2**	**115**	**12.2**	**–3**	**8**	**12.5**	**45**	**10**	**16:17**	....	....	....	....	....	....	....	....

			Regular Season																Playoffs							
Season	Club	League	GP	G	A	Pts	PIM	PP	SH	GW	S	%	+/-	TF	F%	H	SB	Min	GP	G	A	Pts	PIM	PP	SH	GW
2001-02	**Atlanta**	**NHL**	**60**	**8**	**14**	**22**	**12**	**1**	**0**	**0**	**102**	**7.8**	**–8**	**3**	**33.3**	**33**	**4**	**14:45**								
	Finland	Olympics	4	1	2	3	2																			
	NHL Totals		**116**	**22**	**27**	**49**	**34**	**3**	**0**	**2**	**217**	**10.1**		**11**	**18.2**	**78**	**14**	**15:29**								

Claimed by **Atlanta** from **Colorado** in Expansion Draft, June 25, 1999.

KAMENSKY, Valeri

(kah-MEHN-skee, VAL-uhr-ee)

Left wing. Shoots right. 6'2", 198 lbs. Born, Voskresensk, USSR, April 18, 1966. Quebec's 8th choice, 129th overall, in 1988 Entry Draft.

Season	Club	League	GP	G	A	Pts	PIM	PP	SH	GW	S	%	+/-	TF	F%	H	SB	Min	GP	G	A	Pts	PIM	PP	SH	GW
1982-83	Voskresensk	USSR	5	0	0	0	0																			
1983-84	Voskresensk	USSR	20	2	2	4	6																			
1984-85	Voskresensk	USSR	45	9	3	12	24																			
1985-86	CSKA Moscow	USSR	40	15	9	24	8																			
1986-87	CSKA Moscow	USSR	37	13	8	21	16																			
1987-88	CSKA Moscow	USSR	51	26	20	46	40																			
	Soviet Union	Olympics	8	4	2	6	4																			
1988-89	CSKA Moscow	USSR	40	18	10	28	30																			
1989-90	CSKA Moscow	USSR	45	19	18	37	40																			
1990-91	CSKA Moscow	USSR	46	20	26	46	66																			
1991-92	**Quebec**	**NHL**	**23**	**7**	**14**	**21**	**14**	**2**	**0**	**1**	**42**	**16.7**	**–1**													
1992-93	**Quebec**	**NHL**	**32**	**15**	**22**	**37**	**14**	**2**	**3**	**0**	**94**	**16.0**	**13**						**6**	**0**	**1**	**1**	**6**	**0**	**0**	**0**
1993-94	**Quebec**	**NHL**	**76**	**28**	**37**	**65**	**42**	**6**	**0**	**1**	**170**	**16.5**	**12**													
1994-95	HC Ambri-Piotta	Swiss	12	13	6	19	2																			
	Quebec	**NHL**	**40**	**10**	**20**	**30**	**22**	**5**	**1**	**5**	**70**	**14.3**	**3**						**2**	**1**	**0**	**1**	**0**	**0**	**0**	**0**
1995-96♦	**Colorado**	**NHL**	**81**	**38**	**47**	**85**	**85**	**18**	**1**	**5**	**220**	**17.3**	**14**						**22**	**10**	**12**	**22**	**28**	**3**	**0**	**2**
1996-97	**Colorado**	**NHL**	**68**	**28**	**38**	**66**	**38**	**8**	**0**	**4**	**165**	**17.0**	**5**						**17**	**8**	**14**	**22**	**16**	**5**	**0**	**2**
1997-98	**Colorado**	**NHL**	**75**	**26**	**40**	**66**	**60**	**8**	**0**	**4**	**173**	**15.0**	**–2**						**7**	**2**	**3**	**5**	**18**	**1**	**0**	**0**
	Russia	Olympics	6	1	2	3	0																			
1998-99	**Colorado**	**NHL**	**65**	**14**	**30**	**44**	**28**	**2**	**0**	**2**	**123**	**11.4**	**1**	**4**	**25.0**	**32**	**9**	**17:35**	**10**	**4**	**5**	**9**	**4**	**1**	**0**	**1**
99-2000	**NY Rangers**	**NHL**	**58**	**13**	**19**	**32**	**24**	**3**	**0**	**1**	**88**	**14.8**	**–13**	**5**	**0.0**	**32**	**14**	**14:57**								
2000-01	**NY Rangers**	**NHL**	**65**	**14**	**20**	**34**	**36**	**6**	**0**	**1**	**129**	**10.9**	**–18**	**2**	**50.0**	**49**	**15**	**15:19**								
2001-02	**Dallas**	**NHL**	**24**	**3**	**6**	**9**	**2**	**0**	**0**	**1**	**32**	**9.4**	**3**	**0**	**0.0**	**14**	**3**	**9:38**								
	New Jersey	**NHL**	**30**	**4**	**8**	**12**	**18**	**0**	**0**	**0**	**40**	**10.0**	**–2**	**3**	**0.0**	**24**	**6**	**12:54**	**2**	**0**	**0**	**0**	**0**	**0**	**0**	**0**
	NHL Totals		**637**	**200**	**301**	**501**	**383**	**60**	**5**	**25**	**1346**	**14.9**		**14**	**14.3**	**151**	**47**	**14:58**	**66**	**25**	**35**	**60**	**72**	**10**	**0**	**5**

• Played in NHL All-Star Game (1998)

• Missed majority of 1991-92 season recovering from ankle injury suffered in game vs. Tampa Bay, October 27, 1991. Transferred to **Colorado** after **Quebec** franchise relocated, June 21, 1995. Signed as a free agent by **NY Rangers**, July 7, 1999. Signed as a free agent by **Dallas**, July 5, 2001. Traded to **New Jersey** by **Dallas** for Andre Lakos and future considerations, January 16, 2002.

KAPANEN, Niko

(KA-pah-nehn, NEE-KOH) **DAL.**

Center. Shoots left. 5'9", 180 lbs. Born, Hattula, Finland, April 29, 1978. Dallas' 5th choice, 173rd overall, in 1998 Entry Draft.

Season	Club	League	GP	G	A	Pts	PIM	PP	SH	GW	S	%	+/-	TF	F%	H	SB	Min	GP	G	A	Pts	PIM	PP	SH	GW
1992-93	HPK-C	Finn-Jr.	14	14	6	20	2																			
1993-94	HPK-C	Finn-Jr.	2	0	1	1	0																			
	HPK Jr.	Finn-Jr.	31	17	33	50	34																			
1994-95	HPK-B	Finn-Jr.	37	19	44	63	40																			
1995-96	HPK-B	Finn-Jr.	10	6	6	12	8																			
	HPK Jr.	Finn-Jr.	26	15	22	37	34																			
	HPK Hameenlinna	Finland	7	1	0	1	0																			
1996-97	HPK Jr.	Finn-Jr.	5	1	7	8	2																			
	HPK Hameenlinna	Finland	41	6	9	15	12												2	0	1	1	2			
	HPK Hameenlinna	EuroHL	6	3	0	3	4												10	4	5	9	2			
1997-98	HPK Jr.	Finn-Jr.	2	1	1	2	0												1	0	0	0	0			
	HPK Hameenlinna	Finland	48	8	18	26	44																			
1998-99	HPK Hameenlinna	Finland	53	14	29	43	49												8	3	4	7	4			
99-2000	HPK Hameenlinna	Finland	53	20	28	48	40												8	1	9	10	4			
2000-01	TPS Turku	Finland	56	11	21	32	20												10	2	1	3	4			
2001-02	**Dallas**	**NHL**	**9**	**0**	**1**	**1**	**2**	**0**	**0**	**0**	**3**	**0.0**	**–1**	**59**	**40.7**	**4**	**1**	**9:44**								
	Utah Grizzlies	AHL	59	13	28	41	40												5	2	1	3	0			
	NHL Totals		**9**	**0**	**1**	**1**	**2**	**0**	**0**	**0**	**3**	**0.0**		**59**	**40.7**	**4**	**1**	**9:44**								

KAPANEN, Sami

(KA-pah-nehn, SA-mee) **CAR.**

Left wing. Shoots left. 5'10", 195 lbs. Born, Vantaa, Finland, June 14, 1973. Hartford's 4th choice, 87th overall, in 1995 Entry Draft.

Season	Club	League	GP	G	A	Pts	PIM	PP	SH	GW	S	%	+/-	TF	F%	H	SB	Min	GP	G	A	Pts	PIM	PP	SH	GW
1989-90	KalPa Kuopio Jr.	Finn-Jr.	30	14	13	27	4																			
1990-91	KalPa Kuopio Jr.	Finn-Jr.	31	9	27	36	10																			
	KalPa Kuopio	Finland	14	1	2	3	2												8	2	1	3	2			
1991-92	KalPa Kuopio Jr.	Finn-Jr.	8	1	3	4	12																			
	KalPa Kuopio	Finland	42	15	10	25	8																			
1992-93	KalPa Kuopio Jr.	Finn-Jr.	7	11	14	25	2																			
	KalPa Kuopio	Finland	37	4	17	21	12																			
1993-94	KalPa Kuopio	Finland	48	23	32	55	16																			
	Finland	Olympics	8	1	0	1	2																			
1994-95	HIFK Helsinki	Finland	49	14	28	42	42												3	0	0	0	0			
1995-96	**Hartford**	**NHL**	**35**	**5**	**4**	**9**	**6**	**0**	**0**	**0**	**46**	**10.9**	**0**													
	Springfield	AHL	28	14	17	31	4												3	1	2	3	0			
1996-97	**Hartford**	**NHL**	**45**	**13**	**12**	**25**	**2**	**3**	**0**	**2**	**82**	**15.9**	**6**													
1997-98	**Carolina**	**NHL**	**81**	**26**	**37**	**63**	**16**	**4**	**0**	**5**	**190**	**13.7**	**9**													
	Finland	Olympics	6	0	1	1	0																			
1998-99	**Carolina**	**NHL**	**81**	**24**	**35**	**59**	**10**	**5**	**0**	**7**	**254**	**9.4**	**–1**	**10**	**50.0**	**123**	**28**	**19:25**	**5**	**1**	**1**	**2**	**0**	**0**	**0**	**0**
99-2000	**Carolina**	**NHL**	**76**	**24**	**24**	**48**	**12**	**7**	**0**	**5**	**229**	**10.5**	**10**	**2**	**50.0**	**91**	**33**	**19:53**								
2000-01	**Carolina**	**NHL**	**82**	**20**	**37**	**57**	**24**	**7**	**0**	**4**	**223**	**9.0**	**–12**	**6**	**16.7**	**155**	**38**	**18:56**	**6**	**2**	**3**	**5**	**0**	**1**	**0**	**0**
2001-02	**Carolina**	**NHL**	**77**	**27**	**42**	**69**	**23**	**11**	**0**	**4**	**248**	**10.9**	**9**	**7**	**14.3**	**122**	**40**	**20:38**	**23**	**1**	**8**	**9**	**6**	**0**	**0**	**0**
	Finland	Olympics	4	1	2	3	4																			
	NHL Totals		**477**	**139**	**191**	**330**	**93**	**37**	**0**	**27**	**1272**	**10.9**		**25**	**32.0**	**491**	**139**	**19:42**	**34**	**4**	**12**	**16**	**6**	**1**	**0**	**0**

Played in NHL All-Star Game (2000, 2002)

Transferred to **Carolina** after **Hartford** franchise relocated, June 25, 1997.

KARALAHTI, Jere

(kar-ah-LAHKH-tee, YEH-reh) **NSH.**

Defense. Shoots right. 6'2", 210 lbs. Born, Helsinki, Finland, March 25, 1975. Los Angeles' 7th choice, 146th overall, in 1993 Entry Draft.

Season	Club	League	GP	G	A	Pts	PIM	PP	SH	GW	S	%	+/-	TF	F%	H	SB	Min	GP	G	A	Pts	PIM	PP	SH	GW
1991-92	HIFK Jr.	Finn-Jr.	30	12	5	17	36												1	0	0	0	2			
1992-93	HIFK Helsinki-B	Finn-Jr.	7	3	1	4	4																			
	HIFK Jr.	Finn-Jr.	30	2	13	15	49												2	0	0	0	0			
1993-94	HIFK Jr.	Finn-Jr.	3	0	0	0	0																			
	HIFK Helsinki	Finland	46	1	10	11	36												3	0	0	0	6			
1994-95	HIFK Jr.	Finn-Jr.	1	0	0	0	8																			
	HIFK Helsinki	Finland	37	1	7	8	42												3	0	0	0	0			
1995-96	HIFK Jr.	Finn-Jr.	3	1	2	3	2																			
	HIFK Helsinki	Finland	36	4	6	10	102												3	0	0	0	4			
1996-97	HIFK Helsinki	Finland	18	3	5	8	20																			
1997-98	HIFK Helsinki	Finland	43	14	16	30	32												9	2	0	2	8			
1998-99	HIFK Helsinki	Finland	49	11	22	33	65												11	1	1	2	10			
	HIFK Helsinki	EuroHL	6	2	1	3	2												4		2					
99-2000	HIFK Helsinki	Finland	13	2	2	4	55																			
	Los Angeles	**NHL**	**48**	**6**	**10**	**16**	**18**	**4**	**0**	**1**	**69**	**8.7**	**3**	**0**	**0.0**	**108**	**28**	**17:05**	**4**	**0**	**1**	**1**	**2**	**0**	**0**	**0**
	Long Beach	IHL	10	0	3	3	4																			
2000-01	**Los Angeles**	**NHL**	**56**	**2**	**7**	**9**	**38**	**0**	**0**	**0**	**26**	**7.7**	**8**	**0**	**0.0**	**159**	**65**	**17:04**	**13**	**0**	**0**	**0**	**18**	**0**	**0**	**0**
2001-02	**Los Angeles**	**NHL**	**30**	**0**	**1**	**1**	**29**	**0**	**0**	**0**	**19**	**0.0**	**–5**	**0**	**0.0**	**60**	**20**	**16:02**								
	Nashville	**NHL**	**15**	**0**	**1**	**1**	**12**	**0**	**0**	**0**	**8**	**0.0**	**–1**	**0**	**0.0**	**54**	**13**	**18:45**								
	NHL Totals		**149**	**8**	**19**	**27**	**97**	**4**	**0**	**1**	**122**	**6.6**		**0**	**0.0**	**381**	**126**	**17:02**	**17**	**0**	**1**	**1**	**20**	**0**	**0**	**0**

Traded to **Nashville** by **LA Kings** with a conditional choice in 2003 Entry Draft for Cliff Ronning, March 16, 2002.

			Regular Season																Playoffs							
Season	Club	League	GP	G	A	Pts	PIM	PP	SH	GW	S	%	+/-	TF	F%	H	SB	Min	GP	G	A	Pts	PIM	PP	SH	GW

KARIYA, Paul
(kah-REE-ah, PAWL) **ANA.**

Left wing. Shoots left. 5'10", 176 lbs. Born, Vancouver, B.C., October 16, 1974. Anaheim's 1st choice, 4th overall, in 1993 Entry Draft.

Season	Club	League	GP	G	A	Pts	PIM	PP	SH	GW	S	%	+/-	TF	F%	H	SB	Min	GP	G	A	Pts	PIM	PP	SH	GW
1990-91	Penticton	BCJHL	54	45	67	112	8	....	....	....	....	....	....	....	....	....	....	....	....	....	....	....	....	....	....	....
1991-92	Penticton	BCJHL	40	46	86	132	18	....	....	....	....	....	....	....	....	....	....	....	....	....	....	....	....	....	....	....
1992-93	U. of Maine	H-East	39	25	*75	*100	12	....	....	....	....	....	....	....	....	....	....	....	....	....	....	....	....	....	....	....
1993-94	U. of Maine	H-East	12	8	16	24	4	....	....	....	....	....	....	....	....	....	....	....	....	....	....	....	....	....	....	....
	Team Canada	Nat-Tm	23	7	34	41	2	....	....	....	....	....	....	....	....	....	....	....	....	....	....	....	....	....	....	....
	Canada	Olympics	8	3	4	7	2	....	....	....	....	....	....	....	....	....	....	....	....	....	....	....	....	....	....	....
1994-95	**Anaheim**	**NHL**	**47**	**18**	**21**	**39**	**4**	**7**	**1**	**3**	**134**	**13.4**	**–17**	....	....	....	....	....	....	....	....	....	....	....	....	....
1995-96	**Anaheim**	**NHL**	**82**	**50**	**58**	**108**	**20**	**20**	**3**	**9**	**349**	**14.3**	**9**	....	....	....	....	....	....	....	....	....	....	....	....	....
1996-97	**Anaheim**	**NHL**	**69**	**44**	**55**	**99**	**6**	**15**	**3**	**10**	**340**	**12.9**	**36**	....	....	....	....	....	**11**	**7**	**6**	**13**	**4**	**4**	**0**	**1**
1997-98	**Anaheim**	**NHL**	**22**	**17**	**14**	**31**	**23**	**3**	**0**	**2**	**103**	**16.5**	**12**	....	....	....	....	....	....	....	....	....	....	....	....	....
1998-99	**Anaheim**	**NHL**	**82**	**39**	**62**	**101**	**40**	**11**	**2**	**4**	**429**	**9.1**	**17**	**91**	**48.4**	**35**	**65**	**25:32**	**3**	**1**	**3**	**4**	**0**	**0**	**0**	**0**
99-2000	**Anaheim**	**NHL**	**74**	**42**	**44**	**86**	**24**	**11**	**3**	**3**	**324**	**13.0**	**22**	**99**	**39.4**	**23**	**39**	**24:22**	....	....	....	....	....	....	....	....
2000-01	**Anaheim**	**NHL**	**66**	**33**	**34**	**67**	**20**	**18**	**3**	**3**	**230**	**14.3**	**–9**	**149**	**44.3**	**19**	**29**	**23:02**	....	....	....	....	....	....	....	....
2001-02	**Anaheim**	**NHL**	**82**	**32**	**25**	**57**	**28**	**11**	**0**	**8**	**289**	**11.1**	**–15**	**94**	**41.5**	**23**	**27**	**22:13**	....	....	....	....	....	....	....	....
	Canada	Olympics	6	3	1	4	0	....	....	....	....	....	....	....	....	....	....	....	....	....	....	....	....	....	....	....
	NHL Totals		**524**	**275**	**313**	**588**	**165**	**96**	**15**	**42**	**2198**	**12.5**		**433**	**43.4**	**100**	**160**	**23:49**	**14**	**8**	**9**	**17**	**4**	**4**	**0**	**1**

Hockey East First All-Star Team (1993) • Hockey East Rookie of the Year (1993) • Hockey East Player of the Year (1993) • NCAA East First All-American Team (1993) • NCAA Championship All-Tournament Team (1993) • Won Hobey Baker Memorial Award (Top U.S. Collegiate Player) (1993) • NHL All-Rookie Team (1995) • Won Lady Byng Trophy (1996, 1997) • NHL First All-Star Team (1996, 1997, 1999) • NHL Second All-Star Team (2000) • Played in NHL All-Star Game (1996, 1997, 1999, 2000, 2001, 2002)

• Missed majority of 1997-98 season after failing to come to contract terms with **Anaheim** and recovering from head injury suffered in game vs. San Jose, February 1, 1998.

KARIYA, Steve
(kah-REE-ah, STEEV) **VAN.**

Left wing. Shoots right. 5'8", 170 lbs. Born, North Vancouver, B.C., December 22, 1977.

Season	Club	League	GP	G	A	Pts	PIM	PP	SH	GW	S	%	+/-	TF	F%	H	SB	Min	GP	G	A	Pts	PIM	PP	SH	GW
1994-95	Nanaimo Clippers	BCJHL	60	36	60	96	4	....	....	....	....	....	....	....	....	....	....	....	....	....	....	....	....	....	....	....
1995-96	U. of Maine	H-East	39	7	16	23	8	....	....	....	....	....	....	....	....	....	....	....	....	....	....	....	....	....	....	....
1996-97	U. of Maine	H-East	35	19	31	50	10	....	....	....	....	....	....	....	....	....	....	....	....	....	....	....	....	....	....	....
1997-98	U. of Maine	H-East	35	25	25	50	22	....	....	....	....	....	....	....	....	....	....	....	....	....	....	....	....	....	....	....
1998-99	U. of Maine	H-East	41	27	38	65	24	....	....	....	....	....	....	....	....	....	....	....	....	....	....	....	....	....	....	....
99-2000	**Vancouver**	**NHL**	**45**	**8**	**11**	**19**	**22**	**0**	**0**	**0**	**41**	**19.5**	**9**	**4**	**50.0**	**29**	**4**	**12:38**	....	....	....	....	....	....	....	....
	Syracuse Crunch	AHL	29	18	23	41	22	....	....	....	....	....	....	....	....	....	....	....	4	2	1	3	0	....	....	....
2000-01	**Vancouver**	**NHL**	**17**	**1**	**6**	**7**	**8**	**1**	**0**	**0**	**22**	**4.5**	**–1**	**0**	**0.0**	**5**	**2**	**11:42**	....	....	....	....	....	....	....	....
	Kansas City	IHL	43	15	29	44	51	....	....	....	....	....	....	....	....	....	....	....	....	....	....	....	....	....	....	....
2001-02	**Vancouver**	**NHL**	**3**	**0**	**1**	**1**	**2**	**0**	**0**	**0**	**0**	**0.0**	**–2**	**0**	**0.0**	**2**	**2**	**12:17**	....	....	....	....	....	....	....	....
	Manitoba Moose	AHL	67	25	37	62	48	....	....	....	....	....	....	....	....	....	....	....	7	1	2	3	2	....	....	....
	NHL Totals		**65**	**9**	**18**	**27**	**32**	**1**	**0**	**0**	**63**	**14.3**		**4**	**50.0**	**36**	**8**	**12:23**	....	....	....	....	....	....	....	....

BCJHL First Team All-Star (1995) • BCJHL Most Sportsmanlike Player (1995) • Hockey East First All-Star Team (1999) • NCAA East First All-American Team (1999)

Signed as a free agent by **Vancouver**, April 21, 1999.

KARLSSON, Andreas
(KARLS-uhn, AN-dray-uhs) **ATL.**

Center. Shoots left. 6'4", 205 lbs. Born, Ludvika, Sweden, August 19, 1975. Calgary's 8th choice, 148th overall, in 1993 Entry Draft.

Season	Club	League	GP	G	A	Pts	PIM	PP	SH	GW	S	%	+/-	TF	F%	H	SB	Min	GP	G	A	Pts	PIM	PP	SH	GW
1992-93	Leksands IF	Sweden	13	0	0	0	6	....	....	....	....	....	....	....	....	....	....	....	....	....	....	....	....	....	....	....
1993-94	Leksands IF	Sweden	21	0	0	0	10	....	....	....	....	....	....	....	....	....	....	....	3	0	0	0	0	....	....	....
1994-95	Leksands IF Jr.	Swede-Jr.	3	3	3	6	0	....	....	....	....	....	....	....	....	....	....	....	....	....	....	....	....	....	....	....
	Leksands IF	Sweden	24	7	8	15	0	....	....	....	....	....	....	....	....	....	....	....	4	0	1	1	0	....	....	....
1995-96	Leksands IF Jr.	Swede-Jr.	2	4	1	5	6	....	....	....	....	....	....	....	....	....	....	....	....	....	....	....	....	....	....	....
	Leksands IF	Sweden	40	10	13	23	10	....	....	....	....	....	....	....	....	....	....	....	....	....	....	....	....	....	....	....
1996-97	Leksands IF	Sweden	49	13	11	24	39	....	....	....	....	....	....	....	....	....	....	....	9	2	0	2	2	....	....	....
1997-98	Leksands IF	Sweden	33	9	14	23	20	....	....	....	....	....	....	....	....	....	....	....	4	1	0	1	0	....	....	....
	Leksands IF	EuroHL	6	2	3	5	2	....	....	....	....	....	....	....	....	....	....	....	....	....	....	....	....	....	....	....
1998-99	Leksands IF	Sweden	49	18	15	33	18	....	....	....	....	....	....	....	....	....	....	....	4	1	0	1	6	....	....	....
	Leksands IF	EuroHL	6	1	3	4	2	....	....	....	....	....	....	....	....	....	....	....	2	1	1	2	2	....	....	....
99-2000	**Atlanta**	**NHL**	**51**	**5**	**9**	**14**	**14**	**1**	**0**	**0**	**74**	**6.8**	**–17**	**552**	**46.7**	**52**	**17**	**13:00**	....	....	....	....	....	....	....	....
	Orlando	IHL	18	5	5	10	6	....	....	....	....	....	....	....	....	....	....	....	....	....	....	....	....	....	....	....
2000-01	**Atlanta**	**NHL**	**60**	**5**	**11**	**16**	**16**	**0**	**1**	**0**	**83**	**6.0**	**–2**	**743**	**48.6**	**47**	**28**	**12:54**	....	....	....	....	....	....	....	....
2001-02	**Atlanta**	**NHL**	**42**	**1**	**7**	**8**	**20**	**0**	**0**	**0**	**41**	**2.4**	**–8**	**386**	**45.1**	**28**	**18**	**12:30**	....	....	....	....	....	....	....	....
	Chicago Wolves	AHL	16	6	14	20	11	....	....	....	....	....	....	....	....	....	....	....	23	7	14	21	6	....	....	....
	NHL Totals		**153**	**11**	**27**	**38**	**50**	**1**	**1**	**0**	**198**	**5.6**		**1681**	**47.2**	**127**	**63**	**12:49**	....	....	....	....	....	....	....	....

Traded to **Atlanta** by **Calgary** for future considerations, June 25, 1999.

KARPA, Dave
(KAHR-puh, DAYV) **NYR**

Defense. Shoots right. 6'1", 210 lbs. Born, Regina, Sask., May 7, 1971. Quebec's 4th choice, 68th overall, in 1991 Entry Draft.

Season	Club	League	GP	G	A	Pts	PIM	PP	SH	GW	S	%	+/-	TF	F%	H	SB	Min	GP	G	A	Pts	PIM	PP	SH	GW
1988-89	Notre Dame	SJHL	41	16	37	53	....	....	....	....	....	....	....	....	....	....	....	....	....	....	....	....	....	....	....	....
1989-90	Notre Dame	SJHL	43	9	19	28	271	....	....	....	....	....	....	....	....	....	....	....	....	....	....	....	....	....	....	....
1990-91	Ferris State	CCHA	41	6	19	25	109	....	....	....	....	....	....	....	....	....	....	....	....	....	....	....	....	....	....	....
1991-92	Ferris State	CCHA	34	7	12	19	124	....	....	....	....	....	....	....	....	....	....	....	....	....	....	....	....	....	....	....
	Quebec	**NHL**	**4**	**0**	**0**	**0**	**14**	**0**	**0**	**0**	**2**	**0.0**	**2**	....	....	....	....	....	....	....	....	....	....	....	....	....
	Halifax Citadels	AHL	2	0	0	0	4	....	....	....	....	....	....	....	....	....	....	....	....	....	....	....	....	....	....	....
1992-93	**Quebec**	**NHL**	**12**	**0**	**1**	**1**	**13**	**0**	**0**	**0**	**2**	**0.0**	**–6**	....	....	....	....	....	**3**	**0**	**0**	**0**	**0**	**0**	**0**	**0**
	Halifax Citadels	AHL	71	4	27	31	167	....	....	....	....	....	....	....	....	....	....	....	....	....	....	....	....	....	....	....
1993-94	**Quebec**	**NHL**	**60**	**5**	**12**	**17**	**148**	**2**	**0**	**0**	**48**	**10.4**	**0**	....	....	....	....	....	....	....	....	....	....	....	....	....
	Cornwall Aces	AHL	1	0	0	0	0	....	....	....	....	....	....	....	....	....	....	....	12	2	2	4	27	....	....	....
1994-95	Cornwall Aces	AHL	6	0	2	2	19	....	....	....	....	....	....	....	....	....	....	....	....	....	....	....	....	....	....	....
	Quebec	**NHL**	**2**	**0**	**0**	**0**	**0**	**0**	**0**	**0**	**1**	**0.0**	**–1**	....	....	....	....	....	....	....	....	....	....	....	....	....
	Anaheim	**NHL**	**26**	**1**	**5**	**6**	**91**	**0**	**0**	**0**	**32**	**3.1**	**0**	....	....	....	....	....	....	....	....	....	....	....	....	....
1995-96	**Anaheim**	**NHL**	**72**	**3**	**16**	**19**	**270**	**0**	**1**	**1**	**62**	**4.8**	**–3**	....	....	....	....	....	....	....	....	....	....	....	....	....
1996-97	**Anaheim**	**NHL**	**69**	**2**	**11**	**13**	**210**	**0**	**0**	**1**	**90**	**2.2**	**11**	....	....	....	....	....	**8**	**1**	**1**	**2**	**20**	**0**	**0**	**1**
1997-98	**Anaheim**	**NHL**	**78**	**1**	**11**	**12**	**217**	**0**	**0**	**0**	**64**	**1.6**	**–3**	....	....	....	....	....	....	....	....	....	....	....	....	....
1998-99	**Carolina**	**NHL**	**33**	**0**	**2**	**2**	**55**	**0**	**0**	**0**	**21**	**0.0**	**1**	**0**	**0.0**	**62**	**45**	**16:55**	**2**	**0**	**0**	**0**	**2**	**0**	**0**	**0**
99-2000	**Carolina**	**NHL**	**27**	**1**	**4**	**5**	**52**	**0**	**0**	**0**	**24**	**4.2**	**9**	**0**	**0.0**	**68**	**33**	**17:21**	....	....	....	....	....	....	....	....
	Cincinnati	IHL	39	1	8	9	147	....	....	....	....	....	....	....	....	....	....	....	....	....	....	....	....	....	....	....
2000-01	**Carolina**	**NHL**	**80**	**4**	**6**	**10**	**159**	**2**	**0**	**0**	**69**	**5.8**	**–19**	**0**	**0.0**	**193**	**188**	**20:01**	**6**	**0**	**0**	**0**	**17**	**0**	**0**	**0**
2001-02	**NY Rangers**	**NHL**	**75**	**1**	**10**	**11**	**131**	**0**	**0**	**1**	**53**	**1.9**	**–9**	**0**	**0.0**	**124**	**153**	**16:54**	....	....	....	....	....	....	....	....
	NHL Totals		**538**	**18**	**78**	**96**	**1360**	**4**	**1**	**3**	**468**	**3.8**		**0**	**0.0**	**447**	**419**	**18:07**	**19**	**1**	**1**	**2**	**39**	**0**	**0**	**1**

Traded to **Anaheim** by **Quebec** for Anaheim's 4th round choice (later traded to St. Louis - St. Louis selected Jan Horacek) in 1997 Entry Draft, March 9, 1995. Traded to **Carolina** by **Anaheim** with Anaheim's 4th round choice (later traded to Atlanta - Atlanta selected Blake Robson) in 2000 Entry Draft for Stu Grimson and Kevin Haller, August 11, 1998. Signed as a free agent by **NY Rangers**, July 1, 2001.

KARPOVTSEV, Alexander
(kar-POHV-tzehv, al-ehx-AN-duhr) **CHI.**

Defense. Shoots right. 6'3", 215 lbs. Born, Moscow, USSR, April 7, 1970. Quebec's 7th choice, 158th overall, in 1990 Entry Draft.

Season	Club	League	GP	G	A	Pts	PIM	PP	SH	GW	S	%	+/-	TF	F%	H	SB	Min	GP	G	A	Pts	PIM	PP	SH	GW
1989-90	Dynamo Moscow	USSR	35	1	1	2	27	....	....	....	....	....	....	....	....	....	....	....	....	....	....	....	....	....	....	....
1990-91	Dynamo Moscow	USSR	40	0	5	5	15	....	....	....	....	....	....	....	....	....	....	....	....	....	....	....	....	....	....	....
1991-92	Dynamo Moscow	CIS	35	4	2	6	26	....	....	....	....	....	....	....	....	....	....	....	....	....	....	....	....	....	....	....
1992-93	Dynamo Moscow	CIS	36	3	11	14	100	....	....	....	....	....	....	....	....	....	....	....	7	2	1	3	0	....	....	....
1993-94♦	**NY Rangers**	**NHL**	**67**	**3**	**15**	**18**	**58**	**1**	**0**	**1**	**78**	**3.8**	**12**	....	....	....	....	....	**17**	**0**	**4**	**4**	**12**	**0**	**0**	**0**
1994-95	Dynamo Moscow	CIS	13	0	2	2	10	....	....	....	....	....	....	....	....	....	....	....	....	....	....	....	....	....	....	....
	NY Rangers	**NHL**	**47**	**4**	**8**	**12**	**30**	**1**	**0**	**1**	**82**	**4.9**	**–4**	....	....	....	....	....	**8**	**1**	**0**	**1**	**0**	**0**	**0**	**0**
1995-96	**NY Rangers**	**NHL**	**40**	**2**	**16**	**18**	**26**	**1**	**0**	**1**	**71**	**2.8**	**12**	....	....	....	....	....	**6**	**0**	**1**	**1**	**4**	**0**	**0**	**0**
1996-97	**NY Rangers**	**NHL**	**77**	**9**	**29**	**38**	**59**	**6**	**1**	**0**	**84**	**10.7**	**1**	....	....	....	....	....	**13**	**1**	**3**	**4**	**20**	**1**	**0**	**0**
1997-98	**NY Rangers**	**NHL**	**47**	**3**	**7**	**10**	**38**	**1**	**0**	**1**	**46**	**6.5**	**–1**	....	....	....	....	....	....	....	....	....	....	....	....	....
1998-99	**NY Rangers**	**NHL**	**2**	**1**	**0**	**1**	**0**	**0**	**0**	**0**	**4**	**25.0**	**1**	**0**	**0.0**	**2**	**3**	**22:38**	....	....	....	....	....	....	....	....
	Toronto	**NHL**	**56**	**2**	**25**	**27**	**52**	**1**	**0**	**1**	**61**	**3.3**	**38**	**0**	**0.0**	**79**	**103**	**20:58**	**14**	**1**	**3**	**4**	**12**	**1**	**0**	**0**
99-2000	**Toronto**	**NHL**	**69**	**3**	**14**	**17**	**54**	**3**	**0**	**0**	**51**	**5.9**	**9**	**2**	**0.0**	**88**	**129**	**20:14**	**11**	**0**	**3**	**3**	**4**	**0**	**0**	**0**

Season	Club	League	GP	G	A	Pts	PIM	PP	SH	GW	S	%	+/-	TF	F%	H	SB	Min	GP	G	A	Pts	PIM	PP	SH	GW
			Regular Season																Playoffs							
2000-01	Dynamo Moscow	Russia	5	0	1	1	0	....	....	....	....	....	....	....	....	....	....	....	....	....	....	....	....	....	....	....
	Chicago	**NHL**	**53**	**2**	**13**	**15**	**39**	**1**	**0**	**0**	**52**	**3.8**	**–4**	**0**	**0.0**	**47**	**110**	**20:28**	....	....	....	....	....	....	....	....
2001-02	**Chicago**	**NHL**	**65**	**1**	**9**	**10**	**40**	**0**	**1**	**0**	**40**	**2.5**	**10**	**3**	**33.3**	**91**	**191**	**20:40**	**5**	**1**	**0**	**1**	**0**	**0**	**0**	**1**
	NHL Totals		**523**	**30**	**136**	**166**	**396**	**15**	**2**	**5**	**569**	**5.3**		**5**	**20.0**	**307**	**536**	**20:35**	**74**	**4**	**14**	**18**	**52**	**2**	**0**	**1**

Traded to **NY Rangers** by **Quebec** for Mike Hurlbut, September 7, 1993. Traded to **Toronto** by **NY Rangers** with NY Rangers' 4th round choice (Mirko Murovic) in 1999 Entry Draft for Mathieu Schneider, October 14, 1998. Traded to **Chicago** by **Toronto** with Toronto's 4th round choice (Vladimir Gusev) in 2001 Entry Draft for Bryan McCabe, October 2, 2000.

KASPARAITIS, Darius

(KAZ-puhr-IGH-tihz, DAIR-ee-uhs) **NYR**

Defense. Shoots left. 5'11", 212 lbs. Born, Elektrenai, USSR, October 16, 1972. NY Islanders' 1st choice, 5th overall, in 1992 Entry Draft.

Season	Club	League	GP	G	A	Pts	PIM	PP	SH	GW	S	%	+/-	TF	F%	H	SB	Min	GP	G	A	Pts	PIM	PP	SH	GW
1988-89	Dynamo Moscow	USSR	3	0	0	0	0	....	....	....	....	....	....	....	....	....	....	....	....	....	....	....	....	....	....	....
1989-90	Dynamo Moscow	USSR	1	0	0	0	0	....	....	....	....	....	....	....	....	....	....	....	....	....	....	....	....	....	....	....
1990-91	Dynamo Moscow	USSR	17	0	1	1	10	....	....	....	....	....	....	....	....	....	....	....	....	....	....	....	....	....	....	....
1991-92	Dynamo Moscow	CIS	31	2	10	12	14	....	....	....	....	....	....	....	....	....	....	....	....	....	....	....	....	....	....	....
1992-93	Dynamo Moscow	CIS	7	1	3	4	8	....	....	....	....	....	....	....	....	....	....	....	....	....	....	....	....	....	....	....
	NY Islanders	**NHL**	**79**	**4**	**17**	**21**	**166**	**0**	**0**	**0**	**92**	**4.3**	**15**	....	....	....	....	....	**18**	**0**	**5**	**5**	**31**	**0**	**0**	**0**
1993-94	**NY Islanders**	**NHL**	**76**	**1**	**10**	**11**	**142**	**0**	**0**	**0**	**81**	**1.2**	**–6**	....	....	....	....	....	**4**	**0**	**0**	**0**	**8**	**0**	**0**	**0**
1994-95	**NY Islanders**	**NHL**	**13**	**0**	**1**	**1**	**22**	**0**	**0**	**0**	**8**	**0.0**	**–11**	....	....	....	....	....	....	....	....	....	....	....	....	....
1995-96	**NY Islanders**	**NHL**	**46**	**1**	**7**	**8**	**93**	**0**	**0**	**0**	**34**	**2.9**	**–12**	....	....	....	....	....	....	....	....	....	....	....	....	....
1996-97	**NY Islanders**	**NHL**	**18**	**0**	**5**	**5**	**16**	**0**	**0**	**0**	**12**	**0.0**	**–7**	....	....	....	....	....	....	....	....	....	....	....	....	....
	Pittsburgh	**NHL**	**57**	**2**	**16**	**18**	**84**	**0**	**0**	**0**	**46**	**4.3**	**24**	....	....	....	....	....	**5**	**0**	**0**	**0**	**6**	**0**	**0**	**0**
1997-98	**Pittsburgh**	**NHL**	**81**	**4**	**8**	**12**	**127**	**0**	**2**	**0**	**71**	**5.6**	**3**	....	....	....	....	....	**5**	**0**	**0**	**0**	**8**	**0**	**0**	**0**
	Russia	Olympics	6	0	2	2	6	....	....	....	....	....	....	....	....	....	....	....	....	....	....	....	....	....	....	....
1998-99	**Pittsburgh**	**NHL**	**48**	**1**	**4**	**5**	**70**	**0**	**0**	**0**	**32**	**3.1**	**12**	**0**	**0.0**	**173**	**48**	**16:01**	....	....	....	....	....	....	....	....
99-2000	**Pittsburgh**	**NHL**	**73**	**3**	**12**	**15**	**146**	**1**	**0**	**1**	**76**	**3.9**	**–12**	**0**	**0.0**	**261**	**119**	**18:07**	**11**	**1**	**1**	**2**	**10**	**0**	**0**	**0**
2000-01	**Pittsburgh**	**NHL**	**77**	**3**	**16**	**19**	**111**	**1**	**0**	**0**	**81**	**3.7**	**11**	**0**	**0.0**	**351**	**124**	**19:14**	**17**	**1**	**1**	**2**	**26**	**0**	**0**	**1**
2001-02	**Pittsburgh**	**NHL**	**69**	**2**	**12**	**14**	**123**	**0**	**0**	**0**	**75**	**2.7**	**–1**	**1**	**0.0**	**316**	**136**	**20:32**	....	....	....	....	....	....	....	....
	Russia	Olympics	6	1	0	1	4	....	....	....	....	....	....	....	....	....	....	....	....	....	....	....	....	....	....	....
	Colorado	**NHL**	**11**	**0**	**0**	**0**	**19**	**0**	**0**	**0**	**6**	**0.0**	**1**	**0**	**0.0**	**57**	**17**	**19:44**	**21**	**0**	**3**	**3**	**18**	**0**	**0**	**0**
	NHL Totals		**648**	**21**	**108**	**129**	**1119**	**2**	**2**	**1**	**614**	**3.4**		**1**	**0.0**	**1158**	**444**	**18:44**	**81**	**2**	**10**	**12**	**107**	**0**	**0**	**1**

Traded to **Pittsburgh** by **NY Islanders** with Andreas Johansson for Bryan Smolinski, November 17, 1996. Traded to **Colorado** by **Pittsburgh** for Ville Niemenen and Rick Berry, March 19, 2002. Signed as a free agent by **NY Rangers**, July 2, 2002.

KAVANAGH, Pat

(KA-vuh-naw, PAT) **VAN.**

Right wing. Shoots right. 6'3", 192 lbs. Born, Ottawa, Ont., March 14, 1979. Philadelphia's 2nd choice, 50th overall, in 1997 Entry Draft.

Season	Club	League	GP	G	A	Pts	PIM	PP	SH	GW	S	%	+/-	TF	F%	H	SB	Min	GP	G	A	Pts	PIM	PP	SH	GW
1995-96	Kanata Valley	OCJHL	54	19	16	35	99	....	....	....	....	....	....	....	....	....	....	....	....	....	....	....	....	....	....	....
1996-97	Peterborough	OHL	43	6	8	14	53	....	....	....	....	....	....	....	....	....	....	....	11	1	1	2	12	....	....	....
1997-98	Peterborough	OHL	66	10	16	26	85	....	....	....	....	....	....	....	....	....	....	....	4	1	0	1	6	....	....	....
1998-99	Peterborough	OHL	68	26	43	69	118	....	....	....	....	....	....	....	....	....	....	....	5	0	5	5	10	....	....	....
99-2000	Syracuse Crunch	AHL	68	12	8	20	56	....	....	....	....	....	....	....	....	....	....	....	4	0	0	0	0	....	....	....
2000-01	Kansas City	IHL	78	26	15	41	86	....	....	....	....	....	....	....	....	....	....	....	....	....	....	....	....	....	....	....
	Vancouver	**NHL**	....	....	....	....	....	....	....	....	....	....	....	....	....	....	....	....	**3**	**0**	**0**	**0**	**2**	**0**	**0**	**0**
2001-02	Manitoba Moose	AHL	70	13	19	32	100	....	....	....	....	....	....	....	....	....	....	....	7	1	0	1	6	....	....	....
	NHL Totals		....	....	....	....	....	....	....	....	....	....	....	....	....	....	....	....	**3**	**0**	**0**	**0**	**2**	**0**	**0**	**0**

Traded to **Vancouver** by **Philadelphia** for Vancouver's 6th round choice (Konstantin Rudenko) in 1999 Entry Draft, June 1, 1999.

KEANE, Mike

(KEEN, MIGHK) **COL.**

Right wing. Shoots right. 5'10", 185 lbs. Born, Winnipeg, Man., May 29, 1967.

Season	Club	League	GP	G	A	Pts	PIM	PP	SH	GW	S	%	+/-	TF	F%	H	SB	Min	GP	G	A	Pts	PIM	PP	SH	GW
1983-84	Wpg. Monarchs	MMHL	21	17	19	36	59	....	....	....	....	....	....	....	....	....	....	....	....	....	....	....	....	....	....	....
	Winnipeg	WHL	1	0	0	0	0	....	....	....	....	....	....	....	....	....	....	....	....	....	....	....	....	....	....	....
1984-85	Moose Jaw	WHL	65	17	26	43	141	....	....	....	....	....	....	....	....	....	....	....	....	....	....	....	....	....	....	....
1985-86	Moose Jaw	WHL	67	34	49	83	162	....	....	....	....	....	....	....	....	....	....	....	13	6	8	14	9	....	....	....
1986-87	Moose Jaw	WHL	53	25	45	70	107	....	....	....	....	....	....	....	....	....	....	....	9	3	9	12	11	....	....	....
	Sherbrooke	AHL	....	....	....	....	....	....	....	....	....	....	....	....	....	....	....	....	9	2	2	4	16	....	....	....
1987-88	Sherbrooke	AHL	78	25	43	68	70	....	....	....	....	....	....	....	....	....	....	....	6	1	1	2	18	....	....	....
1988-89	**Montreal**	**NHL**	**69**	**16**	**19**	**35**	**69**	**5**	**0**	**1**	**90**	**17.8**	**9**	....	....	....	....	....	**21**	**4**	**3**	**7**	**17**	**2**	**0**	**0**
1989-90	**Montreal**	**NHL**	**74**	**9**	**15**	**24**	**78**	**1**	**0**	**1**	**92**	**9.8**	**0**	....	....	....	....	....	**11**	**0**	**1**	**1**	**8**	**0**	**0**	**0**
1990-91	**Montreal**	**NHL**	**73**	**13**	**23**	**36**	**50**	**2**	**1**	**2**	**109**	**11.9**	**6**	....	....	....	....	....	**12**	**3**	**2**	**5**	**6**	**0**	**0**	**0**
1991-92	**Montreal**	**NHL**	**67**	**11**	**30**	**41**	**64**	**2**	**0**	**2**	**116**	**9.5**	**16**	....	....	....	....	....	**8**	**1**	**1**	**2**	**16**	**0**	**0**	**0**
1992-93 ♦	**Montreal**	**NHL**	**77**	**15**	**45**	**60**	**95**	**0**	**0**	**1**	**120**	**12.5**	**29**	....	....	....	....	....	**19**	**2**	**13**	**15**	**6**	**0**	**0**	**0**
1993-94	**Montreal**	**NHL**	**80**	**16**	**30**	**46**	**119**	**6**	**2**	**2**	**129**	**12.4**	**6**	....	....	....	....	....	**6**	**3**	**1**	**4**	**4**	**0**	**0**	**0**
1994-95	**Montreal**	**NHL**	**48**	**10**	**10**	**20**	**15**	**1**	**0**	**0**	**75**	**13.3**	**5**	....	....	....	....	....	....	....	....	....	....	....	....	....
1995-96	**Montreal**	**NHL**	**18**	**0**	**7**	**7**	**6**	**0**	**0**	**0**	**17**	**0.0**	**–6**	....	....	....	....	....	....	....	....	....	....	....	....	....
	♦ **Colorado**	**NHL**	**55**	**10**	**10**	**20**	**40**	**0**	**2**	**2**	**67**	**14.9**	**1**	....	....	....	....	....	**22**	**3**	**2**	**5**	**16**	**0**	**0**	**1**
1996-97	**Colorado**	**NHL**	**81**	**10**	**17**	**27**	**63**	**0**	**1**	**1**	**91**	**11.0**	**2**	....	....	....	....	....	**17**	**3**	**1**	**4**	**24**	**0**	**0**	**1**
1997-98	**NY Rangers**	**NHL**	**70**	**8**	**10**	**18**	**47**	**2**	**0**	**0**	**113**	**7.1**	**–12**	....	....	....	....	....	....	....	....	....	....	....	....	....
	Dallas	**NHL**	**13**	**2**	**3**	**5**	**5**	**0**	**0**	**1**	**15**	**13.3**	**0**	....	....	....	....	....	**17**	**4**	**4**	**8**	**0**	**0**	**1**	**1**
1998-99 ♦	**Dallas**	**NHL**	**81**	**6**	**23**	**29**	**62**	**1**	**1**	**1**	**106**	**5.7**	**–2**	**11**	**27.3**	**126**	**36**	**13:57**	**23**	**5**	**2**	**7**	**6**	**0**	**1**	**1**
99-2000	**Dallas**	**NHL**	**81**	**13**	**21**	**34**	**41**	**0**	**4**	**3**	**85**	**15.3**	**9**	**10**	**50.0**	**163**	**54**	**16:10**	**23**	**2**	**4**	**6**	**14**	**0**	**0**	**0**
2000-01	**Dallas**	**NHL**	**67**	**10**	**14**	**24**	**35**	**1**	**0**	**1**	**64**	**15.6**	**4**	**25**	**60.0**	**90**	**41**	**15:28**	**10**	**3**	**2**	**5**	**4**	**0**	**0**	**0**
2001-02	**St. Louis**	**NHL**	**56**	**4**	**6**	**10**	**22**	**1**	**0**	**0**	**47**	**8.5**	**–2**	**49**	**30.6**	**58**	**23**	**14:19**	....	....	....	....	....	....	....	....
	Colorado	**NHL**	**22**	**2**	**5**	**7**	**16**	**0**	**0**	**0**	**26**	**7.7**	**–2**	**42**	**31.0**	**39**	**23**	**16:55**	**18**	**1**	**4**	**5**	**8**	**0**	**0**	**0**
	NHL Totals		**1032**	**155**	**288**	**443**	**827**	**22**	**11**	**18**	**1362**	**11.4**		**137**	**37.2**	**476**	**177**	**15:09**	**207**	**34**	**40**	**74**	**129**	**2**	**2**	**4**

Signed as a free agent by **Montreal**, September 25, 1985. Traded to **Colorado** by **Montreal** with Patrick Roy for Andrei Kovalenko, Martin Rucinsky and Jocelyn Thibault, December 6, 1995. Signed as a free agent by **NY Rangers**, July 30, 1997. Traded to **Dallas** by **NY Rangers** with Brian Skrudland and NY Rangers' 6th round choice (Pavel Patera) in 1998 Entry Draft for Todd Harvey, Bob Errey and Dallas' 4th round choice (Boyd Kane) in 1998 Entry Draft, March 24, 1998. Signed as a free agent by **St. Louis**, July 10, 2001. Traded to **Colorado** by **St. Louis** for Shjon Podein, February 11, 2002.

KEEFE, Sheldon

(KEEF, SHEHL-duhn) **T.B.**

Right wing. Shoots right. 5'11", 185 lbs. Born, Brampton, Ont., September 17, 1980. Tampa Bay's 1st choice, 47th overall, in 1999 Entry Draft.

Season	Club	League	GP	G	A	Pts	PIM	PP	SH	GW	S	%	+/-	TF	F%	H	SB	Min	GP	G	A	Pts	PIM	PP	SH	GW
1995-96	Tor. Young Nats	MTHL	45	66	71	137	....	....	....	....	....	....	....	....	....	....	....	....	....	....	....	....	....	....	....	....
1996-97	Quinte Hawks	MTJHL	44	21	23	44	41	....	....	....	....	....	....	....	....	....	....	....	....	....	....	....	....	....	....	....
	Bramalea Blues	OPJHL	8	0	3	3	4	....	....	....	....	....	....	....	....	....	....	....	....	....	....	....	....	....	....	....
1997-98	Caledon	MTJHL	43	41	40	81	117	....	....	....	....	....	....	....	....	....	....	....	13	15	8	23	....	....	....	....
1998-99	St. Michael's	OHL	38	37	37	74	80	....	....	....	....	....	....	....	....	....	....	....	....	....	....	....	....	....	....	....
	Barrie Colts	OHL	28	14	28	42	60	....	....	....	....	....	....	....	....	....	....	....	10	5	5	10	31	....	....	....
99-2000	Barrie Colts	OHL	66	48	*73	*121	95	....	....	....	....	....	....	....	....	....	....	....	25	10	13	23	41	....	....	....
2000-01	**Tampa Bay**	**NHL**	**49**	**4**	**0**	**4**	**38**	**0**	**0**	**0**	**32**	**12.5**	**–13**	**1**	**0.0**	**77**	**8**	**8:00**	....	....	....	....	....	....	....	....
	Detroit Vipers	IHL	13	7	5	12	23	....	....	....	....	....	....	....	....	....	....	....	....	....	....	....	....	....	....	....
2001-02	**Tampa Bay**	**NHL**	**39**	**6**	**7**	**13**	**16**	**0**	**0**	**1**	**52**	**11.5**	**–11**	**70**	**45.7**	**76**	**10**	**13:00**	....	....	....	....	....	....	....	....
	Springfield	AHL	24	9	9	18	26	....	....	....	....	....	....	....	....	....	....	....	....	....	....	....	....	....	....	....
	NHL Totals		**88**	**10**	**7**	**17**	**54**	**0**	**0**	**1**	**84**	**11.9**		**71**	**45.1**	**153**	**18**	**10:13**	....	....	....	....	....	....	....	....

OHL All-Rookie Team (1999) • OHL Rookie of the Year (1999) • Won Eddie Powers Memorial Trophy (Top Scorer - OHL) (2000) • OHL Second All-Star Team (2000) • Canadian Major Junior First All-Star Team (2000) • Memorial Cup All-Star Team (2000)

Traded to **Barrie** (OHL) by **St. Michael's** (OHL) with Mike Jefferson, Ryan Barnes and Shawn Cation for Keith Delaney, Darryl Bootland, Adam DeLeew and Brad Pierce, January 11, 1999.

KELLEHER, Chris

(KEH-leh-huhr, KRIHS) **BOS.**

Defense. Shoots left. 6'1", 210 lbs. Born, Cambridge, MA, March 23, 1975. Pittsburgh's 5th choice, 130th overall, in 1993 Entry Draft.

Season	Club	League	GP	G	A	Pts	PIM	PP	SH	GW	S	%	+/-	TF	F%	H	SB	Min	GP	G	A	Pts	PIM	PP	SH	GW
1990-91	Belmont Hill	Hi-School	20	4	23	27	14	....	....	....	....	....	....	....	....	....	....	....	....	....	....	....	....	....	....	....
1991-92	St. Sebastian's	Hi-School	28	7	27	34	12	....	....	....	....	....	....	....	....	....	....	....	....	....	....	....	....	....	....	....
1992-93	St. Sebastian's	Hi-School	25	8	30	38	16	....	....	....	....	....	....	....	....	....	....	....	....	....	....	....	....	....	....	....
1993-94	St. Sebastian's	Hi-School	24	10	21	31	....	....	....	....	....	....	....	....	....	....	....	....	....	....	....	....	....	....	....	....
1994-95	Boston University	H-East	35	3	17	20	62	....	....	....	....	....	....	....	....	....	....	....	....	....	....	....	....	....	....	....
1995-96	Boston University	H-East	37	7	18	25	43	....	....	....	....	....	....	....	....	....	....	....	....	....	....	....	....	....	....	....
1996-97	Boston University	H-East	39	10	24	34	54	....	....	....	....	....	....	....	....	....	....	....	....	....	....	....	....	....	....	....
1997-98	Boston University	H-East	37	4	26	30	40	....	....	....	....	....	....	....	....	....	....	....	....	....	....	....	....	....	....	....

			Regular Season															Playoffs							
Season	Club	League	GP	G	A	Pts	PIM	PP	SH	GW	S	%	+/-	TF	F%	H	SB	Min	GP	G	A	Pts	PIM	PP	SH	GW
1998-99	Syracuse Crunch	AHL	45	1	4	5	43																			
99-2000	Wilkes-Barre	AHL	67	0	12	12	40																			
2000-01	Wilkes-Barre	AHL	66	6	13	19	37												21	7	18	*25	4			
2001-02	**Boston**	**NHL**	**1**	**0**	**0**	**0**	**0**	**0**	**0**	**0**	**0**	**0.0**	**0**	**0**	**0.0**	**1**	**0**	**6:07**								
	Providence	AHL	31	6	13	19	14												2	0	1	1	0			
	NHL Totals		**1**	**0**	**0**	**0**	**0**	**0**	**0**	**0**	**0**	**0.0**		**0**	**0.0**	**1**	**0**	**6:07**								

NCAA East Second All-American Team (1997, 1998) • Hockey East Second All-Star Team (1998)

Signed as a free agent by **Boston**, July 24, 2001.

KELLY, Steve

(KEHL-lee, STEEV) **L.A.**

Center. Shoots left. 6'2", 210 lbs. Born, Vancouver, B.C., October 26, 1976. Edmonton's 1st choice, 6th overall, in 1995 Entry Draft.

Season	Club	League	GP	G	A	Pts	PIM	PP	SH	GW	S	%	+/-	TF	F%	H	SB	Min	GP	G	A	Pts	PIM	PP	SH	GW
1991-92	Westbank	BCAHA	30	25	60	85	75																			
1992-93	Prince Albert	WHL	65	11	9	20	75																			
1993-94	Prince Albert	WHL	65	19	42	61	106																			
1994-95	Prince Albert	WHL	68	31	41	72	153												15	7	9	16	35			
1995-96	Prince Albert	WHL	70	27	74	101	203												18	13	18	31	47			
1996-97	**Edmonton**	**NHL**	**8**	**1**	**0**	**1**	**6**	**0**	**0**	**1**	**6**	**16.7**	**–1**						**6**	**0**	**0**	**0**	**2**	**0**	**0**	**0**
	Hamilton	AHL	48	9	29	38	111												11	3	3	6	24			
1997-98	**Edmonton**	**NHL**	**19**	**0**	**2**	**2**	**8**	**0**	**0**	**0**	**5**	**0.0**	**–4**													
	Hamilton	AHL	11	2	8	10	18																			
	Tampa Bay	**NHL**	**24**	**2**	**1**	**3**	**15**	**1**	**0**	**0**	**17**	**11.8**	**–9**													
	Milwaukee	IHL	5	0	1	1	19																			
	Cleveland	IHL	5	1	1	2	29												1	0	1	1	0			
1998-99	**Tampa Bay**	**NHL**	**34**	**1**	**3**	**4**	**27**	**0**	**0**	**1**	**15**	**6.7**	**–15**	**11**	**54.5**	**12**	**13**	**10:51**								
	Cleveland	IHL	18	6	7	13	36																			
99-2000	Detroit Vipers	IHL	1	0	0	0	4																			
	♦ **New Jersey**	**NHL**	**1**	**0**	**0**	**0**	**0**	**0**	**0**	**0**	**0**	**0.0**	**0**	**0**	**0.0**	**0**	**0**	**4:28**	**10**	**0**	**0**	**0**	**4**	**0**	**0**	**0**
	Albany	AHL	76	21	36	57	131												3	1	1	2	2			
2000-01	**New Jersey**	**NHL**	**24**	**2**	**2**	**4**	**21**	**0**	**0**	**0**	**18**	**11.1**	**0**	**87**	**48.3**	**21**	**10**	**9:58**								
	Los Angeles	**NHL**	**11**	**1**	**0**	**1**	**4**	**0**	**0**	**0**	**4**	**25.0**	**0**	**51**	**39.2**	**8**	**5**	**6:44**	**8**	**0**	**0**	**0**	**2**	**0**	**0**	**0**
2001-02	**Los Angeles**	**NHL**	**8**	**0**	**1**	**1**	**2**	**0**	**0**	**0**	**0**	**0.0**	**–1**	**44**	**36.4**	**5**	**5**	**6:52**	**1**	**0**	**0**	**0**	**0**	**0**	**0**	**0**
	Manchester	AHL	49	10	21	31	88												5	1	8	9	4			
	NHL Totals		**129**	**7**	**9**	**16**	**83**	**1**	**0**	**2**	**65**	**10.8**		**193**	**43.5**	**46**	**33**	**9:31**	**25**	**0**	**0**	**0**	**8**	**0**	**0**	**0**

Traded to **Tampa Bay** by **Edmonton** with Bryan Marchment and Jason Bonsignore for Roman Hamrlik and Paul Comrie, December 30, 1997. Traded to **New Jersey** by **Tampa Bay** for New Jersey's 7th round choice (Brian Eklund) in 2000 Entry Draft, October 7, 1999. Traded to **LA Kings** by **New Jersey** to complete transaction that sent Bob Corkum to New Jersey (February 23, 2001), February 27, 2001.
• Healthy scratch for majority of 2000-01 season.

KENADY, Chris

(KEHN-a-dee, KRIHS)

Right wing. Shoots right. 6'2", 195 lbs. Born, Mound, MN, April 10, 1973. St. Louis' 8th choice, 175th overall, in 1991 Entry Draft.

Season	Club	League	GP	G	A	Pts	PIM	PP	SH	GW	S	%	+/-	TF	F%	H	SB	Min	GP	G	A	Pts	PIM	PP	SH	GW
1990-91	St. Paul Vulcans	USHL	45	16	20	36	57																			
1991-92	U. of Denver	WCHA	36	8	5	13	56																			
1992-93	U. of Denver	WCHA	38	8	16	24	95																			
1993-94	U. of Denver	WCHA	37	14	11	25	125																			
1994-95	U. of Denver	WCHA	39	21	17	38	113																			
1995-96	Worcester	AHL	43	9	10	19	58												2	0	0	0	0			
1996-97	Worcester	AHL	73	23	26	49	131												5	0	1	1	2			
1997-98	**St. Louis**	**NHL**	**5**	**0**	**2**	**2**	**0**	**0**	**0**	**0**	**3**	**0.0**	**1**													
	Worcester	AHL	63	23	22	45	84												11	1	5	6	26			
1998-99	Utah Grizzlies	IHL	35	7	6	13	68																			
	Long Beach	IHL	19	1	6	7	47																			
	Hartford	AHL	22	2	6	8	52												2	0	1	1	6			
99-2000	**NY Rangers**	**NHL**	**2**	**0**	**0**	**0**	**0**	**0**	**0**	**0**	**1**	**0.0**	**–1**	**0**	**0.0**	**6**	**0**	**7:27**								
	Hartford	AHL	71	15	16	31	196												21	8	3	11	40			
2000-01	Louisville	AHL	20	2	1	3	36																			
	Hartford	AHL	42	5	12	17	58												5	2	0	2	7			
2001-02	Long Beach	WCHL	68	35	34	69	161												5	4	0	4	4			
	NHL Totals		**7**	**0**	**2**	**2**	**0**	**0**	**0**	**0**	**4**	**0.0**		**0**	**0.0**	**6**	**0**	**7:27**								

WCHL Second All-Star Team (2002)

Traded to **Long Beach** (IHL) by **Utah** (IHL) with Rene Chapdelaine for Andy Roach and John Byce with St. Louis retaining NHL rights, January 9, 1999. Traded to **NY Rangers** by **St. Louis** to complete transaction that sent Jeff Finley and Geoff Smith to St. Louis (February 13, 1999), February 22, 1999. Signed as a free agent by **Long Beach** (WCHL), September 17, 2001.

KHARITONOV, Alexander

(khar-ih-TOH-nahf, al-ehx-AN-duhr)

Left wing. Shoots right. 5'9", 169 lbs. Born, Moscow, USSR, March 30, 1976. Tampa Bay's 3rd choice, 81st overall, in 2000 Entry Draft.

Season	Club	League	GP	G	A	Pts	PIM	PP	SH	GW	S	%	+/-	TF	F%	H	SB	Min	GP	G	A	Pts	PIM	PP	SH	GW
1993-94	Vyatich Ryazan	CIS-3	44	19	8	27	10																			
1994-95	Dynamo Moscow	CIS	10	0	0	0	4																			
1995-96	DynamoMoscow2	CIS-2	3	1	3	4	4																			
	HC Lipetsk	CIS-2	64	30	22	52	44																			
1996-97	DynamoMoscow2	Russia-3	3	1	1	2	0																			
	Dynamo Moscow	Russia	36	11	9	20	12												4	2	0	2	2			
	Dynamo Moscow	EuroHL	6	0	1	1	4												4	2	1	3	2			
1997-98	Dynamo Moscow	Russia	44	19	16	35	20																			
	Dynamo Moscow	EuroHL	6	5	4	9	2												2	0	0	0	0			
1998-99	Dynamo Moscow	Russia	42	8	6	14	24												16	4	3	7	2			
	Dynamo Moscow	EuroHL	5	0	1	1	2												6	0	0	0	4			
99-2000	Dynamo Moscow	Russia	35	14	20	34	26												17	*8	4	12	10			
2000-01	**Tampa Bay**	**NHL**	**66**	**7**	**15**	**22**	**8**	**0**	**0**	**0**	**103**	**6.8**	**–9**	**1**	**100.0**	**20**	**12**	**11:40**								
2001-02	**NY Islanders**	**NHL**	**5**	**0**	**0**	**0**	**4**	**0**	**0**	**0**	**5**	**0.0**	**–1**	**0**	**0.0**	**2**	**0**	**3:57**								
	Bridgeport	AHL	2	1	0	1	0																			
	Avangard Omsk	Russia	29	2	6	8	16												4	1	0	1	2			
	NHL Totals		**71**	**7**	**15**	**22**	**12**	**0**	**0**	**0**	**108**	**6.5**		**1**	**100.0**	**22**	**12**	**11:07**								

Traded to **NY Islanders** by **Tampa Bay** with Adrian Aucoin for Mathieu Biron and NY Islanders' 2nd round choice (later traded to Washington - later traded to Vancouver - Vancouver selected Denis Grot) in 2002 Entry Draft, June 22, 2001.

KHAVANOV, Alexander

(khuh-VAN-ahf, al-ehx-AN-duhr) **ST.L.**

Defense. Shoots left. 6', 187 lbs. Born, Ryazan, USSR, January 30, 1972. St. Louis' 8th choice, 232nd overall, in 1999 Entry Draft.

Season	Club	League	GP	G	A	Pts	PIM	PP	SH	GW	S	%	+/-	TF	F%	H	SB	Min	GP	G	A	Pts	PIM	PP	SH	GW
1992-93	Birmingham Bulls	ECHL	19	0	3	3	14																			
	Raleigh Icecaps	ECHL	17	0	6	6	8																			
1993-94	St. Petersburg	CIS	41	1	2	3	24																			
1994-95	St. Petersburg	CIS	49	7	0	7	32												3	0	0	0	0			
1995-96	St. Petersburg	CIS	32	1	5	6	41																			
	HPK Hameenlinna	Finland	16	0	2	2	4												9	0	0	0	0			
1996-97	Cherepovets	Russia	39	3	8	11	56												3	1	0	1	4			
1997-98	Cherepovets	Russia	44	3	5	8	46																			
1998-99	Dynamo Moscow	Russia	40	2	7	9	14												16	1	5	6	35			
	Dynamo Moscow	EuroHL	5	0	1	1	2												6	0	0	0	4			
99-2000	Dynamo Moscow	Russia	38	5	12	17	49												17	0	3	3	4			
	Dynamo Moscow	EuroHL	6	2	0	2	0																			
2000-01	**St. Louis**	**NHL**	**74**	**7**	**16**	**23**	**52**	**2**	**0**	**0**	**92**	**7.6**	**16**	**0**	**0.0**	**122**	**89**	**20:54**	**15**	**3**	**2**	**5**	**14**	**1**	**0**	**0**
2001-02	**St. Louis**	**NHL**	**81**	**3**	**21**	**24**	**55**	**0**	**0**	**0**	**87**	**3.4**	**9**	**0**	**0.0**	**88**	**92**	**17:13**	**4**	**0**	**0**	**0**	**2**	**0**	**0**	**0**
	NHL Totals		**155**	**10**	**37**	**47**	**107**	**2**	**0**	**0**	**179**	**5.6**		**0**	**0.0**	**210**	**181**	**18:59**	**19**	**3**	**2**	**5**	**16**	**1**	**0**	**0**

KHRISTICH, Dmitri

(KRIH-stihch, dih-MEE-tree)

Left wing/Center. Shoots right. 6'2", 195 lbs. Born, Kiev, USSR, July 23, 1969. Washington's 6th choice, 120th overall, in 1988 Entry Draft.

			Regular Season																Playoffs							
Season	Club	League	GP	G	A	Pts	PIM	PP	SH	GW	S	%	+/-	TF	F%	H	SB	Min	GP	G	A	Pts	PIM	PP	SH	GW
1985-86	Sokol Kiev	USSR	4	0	0	0	0																			
1986-87	Sokol Kiev	USSR	20	3	0	3	4																			
1987-88	Sokol Kiev	USSR	37	9	1	10	18																			
1988-89	Sokol Kiev	USSR	42	17	10	27	15																			
1989-90	Sokol Kiev	USSR	47	14	22	36	32																			
1990-91	Sokol Kiev	USSR	28	10	12	22	20																			
	Washington	NHL	40	13	14	27	21	1	0	0	77	16.9	-1						11	1	3	4	6	0	0	0
	Baltimore	AHL	3	0	0	0	0																			
1991-92	Washington	NHL	80	36	37	73	35	14	1	7	188	19.1	24						7	3	2	5	15	3	0	1
1992-93	Washington	NHL	64	31	35	66	28	9	1	1	127	24.4	29						6	2	5	7	2	1	0	0
1993-94	Washington	NHL	83	29	29	58	73	10	0	4	195	14.9	-2						11	2	3	5	10	0	0	0
1994-95	Washington	NHL	48	12	14	26	41	8	0	2	92	13.0	0						7	1	4	5	0	0	0	0
1995-96	Los Angeles	NHL	76	27	37	64	44	12	0	3	204	13.2	0													
1996-97	Los Angeles	NHL	75	19	37	56	38	3	0	2	135	14.1	8													
1997-98	Boston	NHL	82	29	37	66	42	13	2	1	144	20.1	25						6	2	2	4	2	2	0	0
1998-99	Boston	NHL	79	29	42	71	48	13	1	6	144	20.1	11	76	44.7	80	39	19:46	12	3	4	7	6	0	0	1
99-2000	Toronto	NHL	53	12	18	30	24	3	0	0	79	15.2	8	84	45.2	52	19	16:10	12	1	2	3	0	1	0	0
2000-01	Toronto	NHL	27	3	6	9	8	2	0	0	23	13.0	8	129	41.9	21	8	15:01								
	Washington	NHL	43	10	19	29	8	4	0	4	54	18.5	-8	3	33.3	25	19	14:53	3	0	0	0	0	0	0	0
2001-02	Washington	NHL	61	9	12	21	12	3	0	2	54	16.7	2	621	46.4	37	39	14:00								
	Ukraine	Olympics	2	2	0	2	0																			
	NHL Totals		811	259	337	596	422	95	5	32	1516	17.1		913	45.5	215	124	16:25	75	15	25	40	41	7	0	2

Played in NHL All-Star Game (1997, 1999)

Traded to **LA Kings** by **Washington** with Byron Dafoe for LA Kings' 1st round choice (Alexandre Volchkov) and Dallas' 4th round choice (previously acquired, Washington selected Justin Davis) in 1996 Entry Draft, July 8, 1995. Traded to **Boston** by **LA Kings** with Byron Dafoe for Jozef Stumpel, Sandy Moger and Boston's 4th round choice (later traded to New Jersey - New Jersey selected Pierre Dagenais) in 1998 Entry Draft, August 29, 1997. Traded to **Toronto** by **Boston** for Toronto's 2nd round choice (Ivan Huml) in 2000 Entry Draft, October 20, 1999. Traded to **Washington** by **Toronto** for Tampa Bay's 3rd round choice (previously acquired, Toronto selected Brendan Bell) in 2001 Entry Draft, December 11, 2000.

KILGER, Chad

(KIHL-guhr, CHAD) **MTL.**

Center. Shoots left. 6'4", 224 lbs. Born, Cornwall, Ont., November 27, 1976. Anaheim's 1st choice, 4th overall, in 1995 Entry Draft.

Season	Club	League	GP	G	A	Pts	PIM	PP	SH	GW	S	%	+/-	TF	F%	H	SB	Min	GP	G	A	Pts	PIM	PP	SH	GW
1992-93	Cornwall Colts	OCJHL	55	30	36	66	26												6	0	0	0	0			
1993-94	Kingston	OHL	66	17	35	52	23												6	7	2	9	8			
1994-95	Kingston	OHL	65	42	53	95	95												6	5	2	7	10			
1995-96	Anaheim	NHL	45	5	7	12	22	0	0	1	38	13.2	-2													
	Winnipeg	NHL	29	2	3	5	12	0	0	0	19	10.5	-2						4	1	0	1	0	0	0	1
1996-97	Phoenix	NHL	24	4	3	7	13	1	0	0	30	13.3	-5													
	Springfield	AHL	52	17	28	45	36												16	5	7	12	56			
1997-98	Phoenix	NHL	10	0	1	1	4	0	0	0	9	0.0	-2													
	Springfield	AHL	35	14	14	28	33																			
	Chicago	NHL	22	3	8	11	6	2	0	1	23	13.0	2													
1998-99	Chicago	NHL	64	14	11	25	30	2	1	1	68	20.6	-1	488	56.6	124	27	14:03								
	Edmonton	NHL	13	1	1	2	4	0	0	0	13	7.7	-3	82	53.7	33	3	11:22	4	0	0	0	4	0	0	0
99-2000	Edmonton	NHL	40	3	2	5	18	0	0	0	32	9.4	-6	269	48.0	60	7	8:33	3	0	0	0	0	0	0	0
	Hamilton	AHL	7	4	2	6	4																			
2000-01	Edmonton	NHL	34	5	2	7	17	1	0	0	28	17.9	-7	391	53.5	53	12	8:18								
	Montreal	NHL	43	9	16	25	34	1	1	1	75	12.0	-1	319	52.4	148	26	17:57								
2001-02	Montreal	NHL	75	8	15	23	27	0	1	2	87	9.2	-7	357	53.8	169	43	13:14	12	0	1	1	9	0	0	0
	NHL Totals		399	54	69	123	187	7	3	6	422	12.8		1906	53.4	587	118	12:46	23	1	1	2	13	0	0	1

Traded to **Winnipeg** by **Anaheim** with Oleg Tverdovsky and Anaheim's 3rd round choice (Per-Anton Lundstrom) in 1996 Entry Draft for Teemu Selanne, Marc Chouinard and Winnipeg's 4th round choice (later traded to Toronto - later traded to Montreal - Montreal selected Kim Staal) in 1996 Entry Draft, February 7, 1996. Transferred to **Phoenix** after **Winnipeg** franchise relocated, July 1, 1996. Traded to **Chicago** by **Phoenix** with Jayson More for Keith Carney and Jim Cummins, March 4, 1998. Traded to **Edmonton** by **Chicago** with Daniel Cleary, Ethan Moreau and Christian Laflamme for Boris Mironov, Dean McAmmond and Jonas Elofsson, March 20, 1999. Traded to **Montreal** by **Edmonton** for Sergei Zholtok, December 18, 2000.

KING, Derek

(KIHNG, DAIR-ehk) **DET.**

Left wing. Shoots left. 6'1", 203 lbs. Born, Hamilton, Ont., February 11, 1967. NY Islanders' 2nd choice, 13th overall, in 1985 Entry Draft.

Season	Club	League	GP	G	A	Pts	PIM	PP	SH	GW	S	%	+/-	TF	F%	H	SB	Min	GP	G	A	Pts	PIM	PP	SH	GW
1982-83	Ham. Mtn. A's	OPJHL	8	1	2	3	0																			
1983-84	Ham. Mtn. A's	OPJHL	37	10	14	24	142																			
1984-85	Sault Ste. Marie	OHL	63	35	38	73	106												16	3	13	16	11			
1985-86	Sault Ste. Marie	OHL	25	12	17	29	33																			
	Oshawa Generals	OHL	19	8	13	21	15												6	3	2	5	13			
1986-87	Oshawa Generals	OHL	57	53	53	106	74												17	14	10	24	40			
	NY Islanders	NHL	2	0	0	0	0	0	0	0	5	0.0	0													
1987-88	NY Islanders	NHL	55	12	24	36	30	1	0	4	94	12.8	7						5	0	2	2	2	0	0	0
	Springfield	AHL	10	7	6	13	6																			
1988-89	NY Islanders	NHL	60	14	29	43	14	4	0	0	103	13.6	10													
	Springfield	AHL	4	4	0	4	0																			
1989-90	NY Islanders	NHL	46	13	27	40	20	5	0	1	91	14.3	2						4	0	0	0	4	0	0	0
	Springfield	AHL	21	11	12	23	33																			
1990-91	NY Islanders	NHL	66	19	26	45	44	2	0	2	130	14.6	1													
1991-92	NY Islanders	NHL	80	40	38	78	46	21	0	6	189	21.2	-10													
1992-93	NY Islanders	NHL	77	38	38	76	47	21	0	7	201	18.9	-4						18	3	11	14	14	0	0	0
1993-94	NY Islanders	NHL	78	30	40	70	59	10	0	7	171	17.5	18						4	0	1	1	0	0	0	0
1994-95	NY Islanders	NHL	43	10	16	26	41	7	0	0	118	8.5	-5													
1995-96	NY Islanders	NHL	61	12	20	32	23	5	1	0	154	7.8	-10													
1996-97	NY Islanders	NHL	70	23	30	53	20	5	0	3	153	15.0	-6													
	Hartford	NHL	12	3	3	6	2	1	0	0	28	10.7	0													
1997-98	Toronto	NHL	77	21	25	46	43	4	0	3	166	12.7	-7													
1998-99	Toronto	NHL	81	24	28	52	20	8	0	4	150	16.0	15	1	0.0	36	20	14:02	16	1	3	4	4	0	0	0
99-2000	Toronto	NHL	3	0	0	0	2	0	0	0	4	0.0	-2	0	0.0	3	0	11:29								
	St. Louis	NHL	19	2	7	9	6	1	0	0	29	6.9	0	0	0.0	5	11	12:44								
	Grand Rapids	IHL	52	19	30	49	25												17	7	8	15	8			
2000-01	Grand Rapids	IHL	76	32	51	*83	19												10	5	5	10	4			
2001-02	Munchen Barons	Germany	60	19	26	45	22												9	2	4	6	4			
	NHL Totals		830	261	351	612	417	95	1	37	1786	14.6		1	0.0	44	31	13:43	47	4	17	21	24	0	0	0

OHL Rookie of the Year (1985) • OHL First All-Star Team (1987) • IHL Second All-Star Team (2001) • Shared Leo P. Lamoureux Memorial Trophy (Top Scorer - IHL) with Steve Larouche (2001)

Traded to **Hartford** by **NY Islanders** for Hartford's 5th round choice (Adam Edinger) in 1997 Entry Draft, March 18, 1997. Signed as a free agent by **Toronto**, July 4, 1997. Traded to **St. Louis** by **Toronto** for Tyler Harlton and future considerations, October 20, 1999. Signed as a free agent by **Ottawa**, August 10, 2000. Signed as a free agent with **Munchen** (Germany), July 13, 2001. Signed as a free agent by **Detroit**, July 24, 2002.

KIPRUSOFF, Marko

(KIHP-ruh-sohf)

Defense. Shoots left. 6'1", 195 lbs. Born, Turku, Finland, June 6, 1972. Montreal's 4th choice, 70th overall, in 1994 Entry Draft.

Season	Club	League	GP	G	A	Pts	PIM	PP	SH	GW	S	%	+/-	TF	F%	H	SB	Min	GP	G	A	Pts	PIM	PP	SH	GW
1988-89	TPS Turku Jr.	Finn-Jr.	19	3	5	8	4																			
1989-90	TPS Turku Jr.	Finn-Jr.	23	4	9	13	2																			
1990-91	TPS Turku Jr.	Finn-Jr.	17	2	9	11	2																			
	TuTo Turku	Finland-2	22	4	8	12	0																			
	TPS Turku	Finland	3	0	0	0	0																			
1991-92	TPS Turku	Finland	23	0	2	2	0																			
	Kiekko-67 Turku	Finland-2	4	0	0	0	2																			
	HPK Hameenlinna	Finland	3	0	0	0	0																			
1992-93	Kiekko-67 Turku	Finland-2	1	0	1	1	2																			
	TPS Turku	Finland	43	3	7	10	14												12	2	3	5	6			
	Team Finland	Nat-Tm	7	0	0	0	0																			
1993-94	TPS Turku	Finland	48	5	19	24	8												11	0	6	6	4			
	Team Finland	Nat-Tm	13	0	2	2	10																			
	Finland	Olympics	8	3	3	6	4																			

Season	Club	League	GP	G	A	Pts	PIM	PP	SH	GW	S	%	+/-	TF	F%	H	SB	Min	Playoffs GP	G	A	Pts	PIM	PP	SH	GW
1994-95	TPS Turku	Finland	50	10	21	31	16												13	0	9	9	2			
	Team Finland	Nat-Tm	15	2	2	4	2																			
1995-96	**Montreal**	**NHL**	**24**	**0**	**4**	**4**	**8**	**0**	**0**	**0**	**36**	**0.0**	**-3**													
	Fredericton	AHL	28	4	10	14	2												10	2	5	7	2			
1996-97	Malmo IF	Sweden	50	10	18	28	24												4	0	0	0	0			
	Team Finland	Nat-Tm	25	5	7	12	6																			
1997-98	Malmo IF	Sweden	46	7	16	23	23																			
	Team Finland	Nat-Tm	16	2	6	8	2																			
1998-99	TPS Turku	Finland	49	15	22	37	12												10	3	6	9	0			
99-2000	TPS Turku	Finland	53	6	27	33	10												11	0	3	3	0			
2000-01	EHC Kloten	Swiss	43	6	20	26	10												9	2	7	9	2			
2001-02	**NY Islanders**	**NHL**	**27**	**0**	**6**	**6**	**4**	**0**	**0**	**0**	**17**	**0.0**	**0**	**0**	**0.0**	**7**	**20**	**14:01**								
	Bridgeport	AHL	9	0	2	2	0																			
	TPS Turku	Finland	11	0	2	2	4												6	1	0	1	2			
	NHL Totals		**51**	**0**	**10**	**10**	**12**	**0**	**0**	**0**	**53**	**0.0**		**0**	**0.0**	**7**	**20**	**14:01**								

Finnish League First All-Star Team (1994, 1995)
Signed as a free agent by **NY Islanders**, June 15, 2001.

KJELLBERG, Patric
(SHEHL-buhrg, PA-trihk) **ANA.**

Right wing. Shoots left. 6'2", 210 lbs. Born, Trelleborg, Sweden, June 17, 1969. Montreal's 4th choice, 83rd overall, in 1988 Entry Draft.

Season	Club	League	GP	G	A	Pts	PIM	PP	SH	GW	S	%	+/-	TF	F%	H	SB	Min	Playoffs GP	G	A	Pts	PIM	PP	SH	GW
1985-86	Falu IF	Swede-2	5	0	2	2	0																			
1986-87	Falu IF	Swede-2	32	11	13	24	16																			
1987-88	Falu IF	Swede-2	29	15	10	25	6																			
1988-89	AIK Solna	Sweden	25	7	9	16	8																			
1989-90	AIK Solna	Sweden	33	8	16	24	6												3	1	0	1	0			
1990-91	AIK Solna	Sweden	38	4	11	15	18																			
1991-92	AIK Solna	Sweden	40	20	13	33	14												3	1	0	1	2			
	Sweden	Olympics	8	1	3	4	0																			
1992-93	**Montreal**	**NHL**	**7**	**0**	**0**	**0**	**2**	**0**	**0**	**0**	**7**	**0.0**	**-3**													
	Fredericton	AHL	41	10	27	37	14												5	2	2	4	0			
1993-94	HV 71 Jonkoping	Sweden	40	11	17	28	18																			
	Sweden	Olympics	8	0	1	1	2																			
1994-95	HV 71 Jonkoping	Sweden	29	5	15	20	12																			
1995-96	Djurgarden	Sweden	40	9	7	16	10												4	0	2	2	2			
1996-97	Djurgarden	Sweden	49	29	11	40	18												4	2	3	5	4			
1997-98	Djurgarden	Sweden	46	*30	18	48	16												15	7	3	10	12			
1998-99	**Nashville**	**NHL**	**71**	**11**	**20**	**31**	**24**	**2**	**0**	**2**	**103**	**10.7**	**-13**	**83**	**37.3**	**44**	**18**	**17:41**								
99-2000	**Nashville**	**NHL**	**82**	**23**	**23**	**46**	**14**	**9**	**0**	**3**	**129**	**17.8**	**-11**	**22**	**31.8**	**44**	**24**	**18:12**								
2000-01	**Nashville**	**NHL**	**81**	**14**	**31**	**45**	**12**	**5**	**0**	**2**	**139**	**10.1**	**-2**	**9**	**33.3**	**21**	**20**	**17:39**								
2001-02	**Nashville**	**NHL**	**12**	**1**	**3**	**4**	**6**	**0**	**0**	**0**	**19**	**5.3**	**-3**	**1**	**0.0**	**5**	**5**	**16:51**								
	Anaheim	**NHL**	**65**	**7**	**8**	**15**	**10**	**4**	**0**	**0**	**69**	**10.1**	**-9**	**24**	**37.5**	**28**	**14**	**17:07**								
	NHL Totals		**318**	**56**	**85**	**141**	**68**	**20**	**0**	**7**	**466**	**12.0**		**139**	**36.0**	**142**	**81**	**17:40**								

Signed as a free agent by **Nashville**, June 27, 1998. Traded to **Anaheim** by **Nashville** for Petr Tenkrat, November 1, 2001.

KLATT, Trent
(KLAT, TREHNT) **VAN.**

Right wing. Shoots right. 6'1", 210 lbs. Born, Robbinsdale, MN, January 30, 1971. Washington's 5th choice, 82nd overall, in 1989 Entry Draft.

Season	Club	League	GP	G	A	Pts	PIM	PP	SH	GW	S	%	+/-	TF	F%	H	SB	Min	Playoffs GP	G	A	Pts	PIM	PP	SH	GW
1986-87	Osseo Orioles	Hi-School	22	9	27	36																				
1987-88	Osseo Orioles	Hi-School	22	19	17	36																				
1988-89	Osseo Orioles	Hi-School	22	24	39	63																				
1989-90	U. of Minnesota	WCHA	38	22	14	36	16																			
1990-91	U. of Minnesota	WCHA	39	16	28	44	58																			
1991-92	U. of Minnesota	WCHA	41	27	36	63	76																			
	Minnesota	**NHL**	**1**	**0**	**0**	**0**	**0**	**0**	**0**	**0**	**1**	**0.0**	**0**						**6**	**0**	**0**	**0**	**2**	**0**	**0**	**0**
1992-93	**Minnesota**	**NHL**	**47**	**4**	**19**	**23**	**38**	**1**	**0**	**0**	**69**	**5.8**	**2**													
	Kalamazoo Wings	IHL	31	8	11	19	18																			
1993-94	**Dallas**	**NHL**	**61**	**14**	**24**	**38**	**30**	**3**	**0**	**2**	**86**	**16.3**	**13**						**9**	**2**	**1**	**3**	**4**	**1**	**0**	**0**
	Kalamazoo Wings	IHL	6	3	2	5	4																			
1994-95	**Dallas**	**NHL**	**47**	**12**	**10**	**22**	**26**	**5**	**0**	**3**	**91**	**13.2**	**-2**						**5**	**1**	**0**	**1**	**0**	**1**	**0**	**0**
1995-96	**Dallas**	**NHL**	**22**	**4**	**4**	**8**	**23**	**0**	**0**	**1**	**37**	**10.8**	**0**													
	Michigan K-Wings	IHL	2	1	2	3	5																			
	Philadelphia	**NHL**	**49**	**3**	**8**	**11**	**21**	**0**	**0**	**1**	**64**	**4.7**	**2**						**12**	**4**	**1**	**5**	**0**	**0**	**0**	**0**
1996-97	**Philadelphia**	**NHL**	**76**	**24**	**21**	**45**	**20**	**5**	**5**	**5**	**131**	**18.3**	**9**						**19**	**4**	**3**	**7**	**12**	**0**	**0**	**2**
1997-98	**Philadelphia**	**NHL**	**82**	**14**	**28**	**42**	**16**	**5**	**0**	**3**	**143**	**9.8**	**2**						**5**	**0**	**0**	**0**	**0**	**0**	**0**	**0**
1998-99	**Philadelphia**	**NHL**	**2**	**0**	**0**	**0**	**0**	**0**	**0**	**0**	**2**	**0.0**	**0**	**0**	**0.0**	**3**	**2**	**11:11**								
	Vancouver	**NHL**	**73**	**4**	**10**	**14**	**12**	**0**	**0**	**0**	**58**	**6.9**	**-3**	**37**	**32.4**	**73**	**29**	**11:21**								
99-2000	**Vancouver**	**NHL**	**47**	**10**	**10**	**20**	**26**	**8**	**0**	**0**	**100**	**10.0**	**-8**	**19**	**63.2**	**129**	**14**	**16:04**								
	Syracuse Crunch	AHL	24	13	10	23	6																			
2000-01	**Vancouver**	**NHL**	**77**	**13**	**20**	**33**	**31**	**3**	**0**	**1**	**140**	**9.3**	**8**	**75**	**50.7**	**148**	**26**	**13:33**	**4**	**3**	**0**	**3**	**0**	**2**	**0**	**0**
2001-02	**Vancouver**	**NHL**	**34**	**8**	**7**	**15**	**10**	**2**	**1**	**3**	**67**	**11.9**	**9**	**101**	**54.5**	**77**	**19**	**15:27**								
	NHL Totals		**618**	**110**	**161**	**271**	**253**	**32**	**6**	**19**	**989**	**11.1**		**232**	**50.4**	**430**	**90**	**13:38**	**60**	**14**	**5**	**19**	**18**	**4**	**0**	**2**

Minnesota High School Player of the Year (1989)

Traded to **Minnesota** by **Washington** with Steve Maltais for Shawn Chambers, June 21, 1991. Transferred to **Dallas** after **Minnesota** franchise relocated, June 9, 1993. Traded to **Philadelphia** by **Dallas** for Brent Fedyk, December 13, 1995. Traded to **Vancouver** by **Philadelphia** for Vancouver's 6th round choice (later traded to Atlanta - Atlanta selected Jeff Dwyer) in 2000 Entry Draft, October 19, 1998. • Missed majority of 2001-02 season recovering from abdominal injury suffered in game vs. Ottawa, November 20, 2001.

KLEE, Ken
(KLEE, KEHN) **WSH.**

Defense. Shoots right. 6', 210 lbs. Born, Indianapolis, IN, April 24, 1971. Washington's 11th choice, 177th overall, in 1990 Entry Draft.

Season	Club	League	GP	G	A	Pts	PIM	PP	SH	GW	S	%	+/-	TF	F%	H	SB	Min	Playoffs GP	G	A	Pts	PIM	PP	SH	GW
1988-89	St. Michael's B	OJHL-B	40	9	23	32	64												27	5	12	17	54			
1989-90	Bowling Green	CCHA	39	0	5	5	52																			
1990-91	Bowling Green	CCHA	37	7	28	35	50																			
1991-92	Bowling Green	CCHA	10	0	1	1	14																			
1992-93	Baltimore	AHL	77	4	14	18	93												7	0	1	1	15			
1993-94	Portland Pirates	AHL	65	2	9	11	87												17	1	2	3	14			
1994-95	Portland Pirates	AHL	49	5	7	12	89																			
	Washington	**NHL**	**23**	**3**	**1**	**4**	**41**	**0**	**0**	**0**	**18**	**16.7**	**2**						**7**	**0**	**0**	**0**	**4**	**0**	**0**	**0**
1995-96	**Washington**	**NHL**	**66**	**8**	**3**	**11**	**60**	**0**	**1**	**2**	**76**	**10.5**	**-1**						**1**	**0**	**0**	**0**	**0**	**0**	**0**	**0**
1996-97	**Washington**	**NHL**	**80**	**3**	**8**	**11**	**115**	**0**	**0**	**2**	**108**	**2.8**	**-5**													
1997-98	**Washington**	**NHL**	**51**	**4**	**2**	**6**	**46**	**0**	**0**	**1**	**44**	**9.1**	**-3**						**9**	**1**	**0**	**1**	**10**	**0**	**0**	**0**
1998-99	**Washington**	**NHL**	**78**	**7**	**13**	**20**	**80**	**0**	**0**	**1**	**132**	**5.3**	**-9**	**0**	**0.0**	**248**	**71**	**19:07**								
99-2000	**Washington**	**NHL**	**80**	**7**	**13**	**20**	**79**	**0**	**0**	**2**	**113**	**6.2**	**8**	**0**	**0.0**	**307**	**134**	**20:29**	**5**	**0**	**1**	**1**	**10**	**0**	**0**	**0**
2000-01	**Washington**	**NHL**	**54**	**2**	**4**	**6**	**60**	**0**	**0**	**0**	**58**	**3.4**	**-5**	**0**	**0.0**	**127**	**40**	**17:15**	**6**	**0**	**1**	**1**	**8**	**0**	**0**	**0**
2001-02	**Washington**	**NHL**	**68**	**8**	**8**	**16**	**38**	**2**	**0**	**3**	**85**	**9.4**	**4**	**2**	**0.0**	**105**	**88**	**19:24**								
	NHL Totals		**500**	**42**	**52**	**94**	**519**	**2**	**1**	**11**	**634**	**6.6**		**2**	**0.0**	**787**	**333**	**19:13**	**28**	**1**	**2**	**3**	**32**	**0**	**0**	**0**

KLEMM, Jon
(KLEHM, JAWN) **CHI.**

Defense. Shoots right. 6'2", 200 lbs. Born, Cranbrook, B.C., January 8, 1970.

Season	Club	League	GP	G	A	Pts	PIM	PP	SH	GW	S	%	+/-	TF	F%	H	SB	Min	Playoffs GP	G	A	Pts	PIM	PP	SH	GW
1986-87	Cranbrook Colts	KIJHL	59	20	51	71	54																			
1987-88	Seattle	WHL	68	6	7	13	24																			
1988-89	Seattle	WHL	2	1	1	2	0																			
	Spokane Chiefs	WHL	66	6	34	40	42																			
1989-90	Spokane Chiefs	WHL	66	3	28	31	100												6	1	1	2	5			
1990-91	Spokane Chiefs	WHL	72	7	58	65	65												15	3	6	9	8			
1991-92	**Quebec**	**NHL**	**4**	**0**	**1**	**1**	**0**	**0**	**0**	**0**	**2**	**0.0**	**2**													
	Halifax Citadels	AHL	70	6	13	19	40																			
1992-93	Halifax Citadels	AHL	80	3	20	23	32																			

			Regular Season																Playoffs							
Season	Club	League	GP	G	A	Pts	PIM	PP	SH	GW	S	%	+/-	TF	F%	H	SB	Min	GP	G	A	Pts	PIM	PP	SH	GW
1993-94	Quebec	NHL	7	0	0	0	4	0	0	0	11	0.0	–1													
	Cornwall Aces	AHL	66	4	26	30	78												13	1	2	3	6			
1994-95	Cornwall Aces	AHL	65	6	13	19	84																			
	Quebec	NHL	4	1	0	1	2	0	0	0	5	20.0	3													
1995-96♦	Colorado	NHL	56	3	12	15	20	0	1	1	61	4.9	12						15	2	1	3	0	1	0	0
1996-97	Colorado	NHL	80	9	15	24	37	1	2	1	103	8.7	12						17	1	1	2	6	0	0	0
1997-98	Colorado	NHL	67	6	8	14	30	0	0	0	60	10.0	–3						4	0	0	0	0	0	0	0
1998-99	Colorado	NHL	39	1	2	3	31	0	0	0	28	3.6	4	14	35.7	38	21	13:43	19	0	1	1	10	0	0	0
99-2000	Colorado	NHL	73	5	7	12	34	0	0	0	64	7.8	26	17	47.1	129	80	17:22	17	2	1	3	9	0	0	0
2000-01♦	Colorado	NHL	78	4	11	15	54	2	0	2	97	4.1	22	1	100.0	175	92	19:56	22	1	2	3	16	0	0	1
2001-02	Chicago	NHL	82	4	16	20	42	2	0	1	111	3.6	–3	1	0.0	243	111	23:50	5	0	1	1	4	0	0	0
	NHL Totals		490	33	72	105	254	5	3	5	542	6.1		33	42.4	585	304	19:32	99	6	7	13	45	1	0	1

WHL West Second All-Star Team (1991)

Signed as a free agent by **Quebec**, May 14, 1991. Transferred to **Colorado** after **Quebec** franchise relocated, June 21, 1995. • Missed majority of 1998-99 season recovering from knee injury suffered in game vs. Phoenix, November 10, 1998. Signed as a free agent by **Chicago**, July 1, 2001.

KLESLA, Rostislav

(KLEHS-luh, RAHS-tih-slav) **CBJ**

Defense. Shoots left. 6'3", 206 lbs. Born, Novy Jicin, Czech., March 21, 1982. Columbus' 1st choice, 4th overall, in 2000 Entry Draft.

Season	Club	League	GP	G	A	Pts	PIM	PP	SH	GW	S	%	+/-	TF	F%	H	SB	Min	GP	G	A	Pts	PIM	PP	SH	GW
1997-98	HC Opava Jr.	Czech-Jr.	38	11	18	29	87												8	2	2	4	0			
1998-99	Sioux City	USHL	54	4	12	16	100												5	2	0	2	2			
99-2000	Brampton	OHL	67	16	29	45	174												6	1	1	2	21			
2000-01	Columbus	NHL	8	2	0	2	6	0	0	0	10	20.0	–1	0	0.0	7	9	18:25								
	Brampton	OHL	45	18	36	54	59												9	2	9	11	26			
2001-02	Columbus	NHL	75	8	8	16	74	1	0	0	102	7.8	–6	0	0.0	83	175	18:52								
	NHL Totals		83	10	8	18	80	1	0	0	112	8.9		0	0.0	90	184	18:49								

OHL All-Rookie Team (2000) • Canadian Major Junior All-Rookie Team (2000) • OHL First All-Star Team (2001) • NHL All-Rookie Team (2002)

Returned to **Brampton** (OHL) by **Columbus**, October 28, 2000.

KLOUCEK, Tomas

(KLOH-chehk, TAW-mahsh) **NYR**

Defense. Shoots left. 6'3", 203 lbs. Born, Prague, Czech., March 7, 1980. NY Rangers' 6th choice, 131st overall, in 1998 Entry Draft.

Season	Club	League	GP	G	A	Pts	PIM	PP	SH	GW	S	%	+/-	TF	F%	H	SB	Min	GP	G	A	Pts	PIM	PP	SH	GW
1995-96	Slavia Praha Jr.	Czech-Jr.	40	2	8	10																				
1996-97	Slavia Praha Jr.	Czech-Jr.	43	4	14	18	44																			
1997-98	Slavia Praha Jr.	Czech-Jr.	43	1	9	10																				
1998-99	Cape Breton	QMJHL	59	4	17	21	162												2	0	0	0	4			
99-2000	Hartford	AHL	73	2	8	10	113												23	0	4	4	18			
2000-01	NY Rangers	NHL	43	1	4	5	74	0	0	0	22	4.5	–3	0	0.0	127	53	16:43								
	Hartford	AHL	21	0	2	2	44																			
2001-02	NY Rangers	NHL	52	1	3	4	137	0	0	0	21	4.8	–2	1	0.0	162	35	11:58								
	Hartford	AHL	9	0	2	2	27												10	1	1	2	8			
	NHL Totals		95	2	7	9	211	0	0	0	43	4.7		1	0.0	289	88	14:07								

KNUBLE, Mike

(kuh-NOO-buhl, MIGHK) **BOS.**

Right wing. Shoots right. 6'3", 208 lbs. Born, Toronto, Ont., July 4, 1972. Detroit's 4th choice, 76th overall, in 1991 Entry Draft.

Season	Club	League	GP	G	A	Pts	PIM	PP	SH	GW	S	%	+/-	TF	F%	H	SB	Min	GP	G	A	Pts	PIM	PP	SH	GW
1988-89	East Kentwood	Hi-School	28	52	37	89	60																			
1989-90	East Kentwood	Hi-School	29	63	40	103	40																			
1990-91	Kalamazoo	NAJHL	36	18	24	42	30																			
1991-92	U. of Michigan	CCHA	43	7	8	15	48																			
1992-93	U. of Michigan	CCHA	39	26	16	42	57																			
1993-94	U. of Michigan	CCHA	41	32	26	58	71																			
1994-95	U. of Michigan	CCHA	34	*38	22	60	62																			
	Adirondack	AHL																	3	0	0	0	0			
1995-96	Adirondack	AHL	80	22	23	45	59												3	1	0	1	0			
1996-97	Detroit	NHL	9	1	0	1	0	0	0	0	10	10.0	–1													
	Adirondack	AHL	68	28	35	63	54																			
1997-98♦	Detroit	NHL	53	7	6	13	16	0	0	0	54	13.0	2						3	0	1	1	0	0	0	0
1998-99	NY Rangers	NHL	82	15	20	35	26	3	0	1	113	13.3	–7	1	100.0	180	32	14:52								
99-2000	NY Rangers	NHL	59	9	5	14	18	1	0	1	50	18.0	–5	9	55.6	101	18	10:39								
	Boston	NHL	14	3	3	6	8	1	0	1	28	10.7	–2	3	0.0	25	8	19:29								
2000-01	Boston	NHL	82	7	13	20	37	0	1	1	92	7.6	0	115	31.3	167	28	10:34								
2001-02	Boston	NHL	54	8	6	14	42	0	0	2	77	10.4	9	27	44.4	67	6	9:45	2	0	0	0	0	0	0	0
	NHL Totals		353	50	53	103	147	5	1	6	424	11.8		155	34.8	540	92	12:04	5	0	1	1	0	0	0	0

CCHA Second All-Star Team (1994, 1995) • NCAA West Second All-American Team (1995)

Traded to **NY Rangers** by **Detroit** for NY Rangers' 2nd round choice (Tomas Kopecky) in 2000 Entry Draft, October 1, 1998. Traded to **Boston** by **NY Rangers** for Rob DiMaio, March 10, 2000.

KNUTSEN, Espen

(kuh-NOOT-suhn, EHS-pehn) **CBJ**

Center. Shoots left. 5'11", 188 lbs. Born, Oslo, Norway, January 12, 1972. Hartford's 9th choice, 204th overall, in 1990 Entry Draft.

Season	Club	League	GP	G	A	Pts	PIM	PP	SH	GW	S	%	+/-	TF	F%	H	SB	Min	GP	G	A	Pts	PIM	PP	SH	GW
1988-89	Valerengen Jr.	Nor-Jr.	36	14	7	21	18																			
1989-90	Valerengen	Norway	40	25	28	53	44																			
1990-91	Valerengen	Norway	31	30	24	54	42												5	3	4	7				
1991-92	Valerengen	Norway	30	28	26	54	37												8	7	8	15				
1992-93	Valerengen	Norway	13	11	13	24	4																			
1993-94	Valerengen	Norway	38	32	26	58	20																			
	Norway	Olympics	7	1	3	4	2																			
1994-95	Djurgarden	Sweden	30	6	14	20	18												3	0	1	1	0			
1995-96	Djurgarden	Sweden	32	10	23	33	50												4	1	0	1	2			
1996-97	Djurgarden	Sweden	39	16	33	49	20												4	2	4	6	6			
1997-98	Anaheim	NHL	19	3	0	3	6	1	0	0	21	14.3	–10													
	Cincinnati	AHL	41	4	13	17	18																			
1998-99	Djurgarden	Sweden	39	18	24	42	32												4	0	1	1	2			
	Djurgarden	EuroHL	4	2	2	4	2																			
99-2000	Djurgarden	Sweden	48	18	35	53	65												13	5	*16	*21	2			
2000-01	Columbus	NHL	66	11	42	53	30	2	0	0	62	17.7	–3	125	52.8	57	27	15:59								
2001-02	Columbus	NHL	77	11	31	42	47	5	2	1	102	10.8	–28	500	46.4	48	43	20:15								
	NHL Totals		162	25	73	98	83	8	2	1	185	13.5		625	47.7	105	70	18:17								

Played in NHL All-Star Game (2002)

Rights traded to **Anaheim** by **Hartford** for Kevin Brown, October 1, 1996. Traded to **Columbus** by **Anaheim** for Columbus' 4th round choice (Anaheim selected Vladmir Korsunov) in 2001 Entry Draft, May 25, 2000.

KOEHLER, Greg

(KEE-luhr, GREHG) **NSH.**

Center. Shoots left. 6'2", 195 lbs. Born, Scarborough, Ont., February 27, 1975.

Season	Club	League	GP	G	A	Pts	PIM	PP	SH	GW	S	%	+/-	TF	F%	H	SB	Min	GP	G	A	Pts	PIM	PP	SH	GW
1992-93	Niagara Falls	OJHL-B	40	24	19	43	125																			
1993-94	North York	MTJHL	49	27	47	74	179																			
1994-95	North York	MTJHL	47	28	43	71	126																			
1995-96	Brampton	MTJHL	49	33	64	97	87																			
1996-97	U. Mass-Lowell	H-East	37	16	20	36	49																			
1997-98	U. Mass-Lowell	H-East	33	20	17	37	62																			
	New Haven	AHL	3	0	0	0	2																			
1998-99	New Haven	AHL	26	4	0	4	29																			
	Florida	ECHL	29	13	14	27	62												6	2	3	5	12			
99-2000	Cincinnati	IHL	74	12	13	25	157												8	0	3	3	14			
2000-01	Carolina	NHL	1	0	0	0	0	0	0	0	0	0.0	0	0	0.0	0	0	0:46								
	Cincinnati	IHL	80	35	36	71	122												5	2	2	4	6			

			Regular Season																Playoffs							
Season	Club	League	GP	G	A	Pts	PIM	PP	SH	GW	S	%	+/-	TF	F%	H	SB	Min	GP	G	A	Pts	PIM	PP	SH	GW
2001-02	Lowell	AHL	56	18	18	36	58																			
	Philadelphia	AHL	22	8	4	12	34												5	1	2	3	6			
	NHL Totals		**1**	**0**	**0**	**0**	**0**	**0**	**0**	**0**	**0**	**0.0**		**0**	**0.0**	**0**	**0**	**0:46**								

Hockey East Rookie of the Year (1997) • Hockey East All-Rookie Team (1997) • IHL Second All-Star Team (2001)

Signed as a free agent by **Carolina**, March 31, 1998. Traded to **Philadelphia** by **Carolina** for Jesse Boulerice, February 13, 2002. Signed as a free agent by **Nashville**, July 15, 2002.

KOHN, Ladislav

(KOHN, LA-dih-slahf)

Right wing. Shoots left. 5'11", 194 lbs. Born, Uherske Hradiste, Czech., March 4, 1975. Calgary's 9th choice, 175th overall, in 1994 Entry Draft.

Season	Club	League	GP	G	A	Pts	PIM	PP	SH	GW	S	%	+/-	TF	F%	H	SB	Min	GP	G	A	Pts	PIM	PP	SH	GW
1993-94	Brandon	WHL	2	0	0	0	0																			
	Swift Current	WHL	69	33	35	68	68												7	5	4	9	8			
1994-95	Swift Current	WHL	65	32	60	92	122												6	2	6	8	14			
	Saint John	AHL	1	0	0	0	0																			
1995-96	**Calgary**	**NHL**	**5**	**1**	**0**	**1**	**2**	**0**	**0**	**0**	**8**	**12.5**	**–1**													
	Saint John	AHL	73	28	45	73	97												16	6	5	11	12			
1996-97	Saint John	AHL	76	28	29	57	81												5	0	0	0	0			
1997-98	**Calgary**	**NHL**	**4**	**0**	**1**	**1**	**0**	**0**	**0**	**0**	**2**	**0.0**	**2**													
	Saint John	AHL	65	25	31	56	90												21	14	6	20	20			
1998-99	**Toronto**	**NHL**	**16**	**1**	**3**	**4**	**4**	**0**	**0**	**0**	**23**	**4.3**	**1**	**16**	**18.8**	**15**	**3**	**12:34**	**2**	**0**	**0**	**0**	**5**	**0**	**0**	**0**
	St. John's	AHL	61	27	42	69	90																			
99-2000	**Anaheim**	**NHL**	**77**	**5**	**16**	**21**	**27**	**1**	**0**	**1**	**123**	**4.1**	**–17**	**15**	**33.3**	**121**	**16**	**12:06**								
2000-01	**Anaheim**	**NHL**	**51**	**4**	**3**	**7**	**42**	**0**	**1**	**0**	**86**	**4.7**	**–15**	**45**	**26.7**	**68**	**18**	**10:59**								
	Atlanta	**NHL**	**26**	**3**	**4**	**7**	**44**	**0**	**1**	**0**	**43**	**7.0**	**–12**	**16**	**37.5**	**33**	**14**	**13:38**								
2001-02	Cincinnati	AHL	4	0	0	0	9																			
	Detroit	**NHL**	**4**	**0**	**0**	**0**	**4**	**0**	**0**	**0**	**2**	**0.0**	**0**	**0**	**0.0**	**1**	**0**	**5:13**								
	Blues Espoo	Finland	40	22	13	35	103												3	0	0	0	14			
	NHL Totals		**183**	**14**	**27**	**41**	**123**	**1**	**2**	**1**	**287**	**4.9**		**92**	**28.3**	**238**	**51**	**11:53**	**2**	**0**	**0**	**0**	**5**	**0**	**0**	**0**

Traded to **Toronto** by **Calgary** for David Cooper, July 2, 1998. Claimed by **Atlanta** from **Toronto** in Waiver Draft, September 27, 1999. Traded to **Anaheim** by **Atlanta** for Anaheim's 8th round choice (Evan Nielsen) in 2000 Entry Draft, September 27, 1999. Traded to **Atlanta** by **Anaheim** for Sergei Vyshedkevich and Scott Langkow, February 9, 2001. Signed as a free agent by **Detroit** with player option to return to Finland, October 22, 2001.

KOIVU, Saku

(KOI-voo, SA-koo) **MTL.**

Center. Shoots left. 5'10", 181 lbs. Born, Turku, Finland, November 23, 1974. Montreal's 1st choice, 21st overall, in 1993 Entry Draft.

Season	Club	League	GP	G	A	Pts	PIM	PP	SH	GW	S	%	+/-	TF	F%	H	SB	Min	GP	G	A	Pts	PIM	PP	SH	GW
1990-91	TPS Turku-B	Finn-Jr.	24	20	28	48	26																			
1991-92	TPS Turku-B	Finn-Jr.	12	3	7	10	6																			
	TPS Turku Jr.	Finn-Jr.	34	25	28	53	57												8	5	*9	*14	6			
1992-93	TPS Turku	Finland	46	3	7	10	28												11	3	2	5	2			
1993-94	TPS Turku	Finland	47	23	30	53	42												11	4	8	12	16			
	Finland	Olympics	8	4	3	7	12																			
1994-95	TPS Turku	Finland	45	27	*47	*74	73												13	*7	10	17	16			
1995-96	**Montreal**	**NHL**	**82**	**20**	**25**	**45**	**40**	**8**	**3**	**2**	**136**	**14.7**	**–7**						**6**	**3**	**1**	**4**	**8**	**0**	**0**	**0**
1996-97	**Montreal**	**NHL**	**50**	**17**	**39**	**56**	**38**	**5**	**0**	**3**	**135**	**12.6**	**7**						**5**	**1**	**3**	**4**	**10**	**0**	**0**	**0**
1997-98	**Montreal**	**NHL**	**69**	**14**	**43**	**57**	**48**	**2**	**2**	**3**	**145**	**9.7**	**8**						**6**	**2**	**3**	**5**	**2**	**1**	**0**	**0**
	Finland	Olympics	6	2	*8	*10	4																			
1998-99	**Montreal**	**NHL**	**65**	**14**	**30**	**44**	**38**	**4**	**2**	**0**	**145**	**9.7**	**–7**	**1427**	**52.6**	**53**	**12**	**20:02**								
99-2000	**Montreal**	**NHL**	**24**	**3**	**18**	**21**	**14**	**1**	**0**	**0**	**53**	**5.7**	**7**	**495**	**52.9**	**21**	**3**	**19:13**								
2000-01	**Montreal**	**NHL**	**54**	**17**	**30**	**47**	**40**	**7**	**0**	**3**	**113**	**15.0**	**2**	**1092**	**47.6**	**44**	**12**	**21:23**								
2001-02	**Montreal**	**NHL**	**3**	**0**	**2**	**2**	**0**	**0**	**0**	**0**	**2**	**0.0**	**0**	**13**	**61.5**	**2**	**1**	**13:57**	**12**	**4**	**6**	**10**	**4**	**1**	**0**	**1**
	NHL Totals		**347**	**85**	**187**	**272**	**218**	**27**	**7**	**11**	**729**	**11.7**		**3027**	**50.9**	**120**	**28**	**20:16**	**29**	**10**	**13**	**23**	**24**	**2**	**0**	**1**

Won Bill Masterton Memorial Trophy (2002) • Played in NHL All-Star Game (1998) • Missed majority of 1999-2000 season recovering from shoulder injury suffered in game vs. NY Rangers, October 30, 1999. • Missed majority of 2001-02 season recovering from non-Hodgkins lymphoma, September 6, 2001.

KOLANOS, Krys

(koh-LA-nohs, KRIHS) **PHX.**

Center. Shoots right. 6'2", 205 lbs. Born, Calgary, Alta., July 27, 1981. Phoenix's 1st choice, 19th overall, in 2000 Entry Draft.

Season	Club	League	GP	G	A	Pts	PIM	PP	SH	GW	S	%	+/-	TF	F%	H	SB	Min	GP	G	A	Pts	PIM	PP	SH	GW
1996-97	Calgary Flames	AAHA	24	24	35	59	....																			
1997-98	Cgy. Buffaloes	AMHL	34	34	43	77	29																			
1998-99	Calgary Royals	AJHL	58	43	67	110	98																			
99-2000	Boston College	H-East	42	16	16	32	48																			
2000-01	Boston College	H-East	41	25	25	50	54																			
2001-02	**Phoenix**	**NHL**	**57**	**11**	**11**	**22**	**48**	**0**	**0**	**5**	**81**	**13.6**	**6**	**703**	**46.4**	**59**	**9**	**13:05**	**2**	**0**	**0**	**0**	**6**	**0**	**0**	**0**
	NHL Totals		**57**	**11**	**11**	**22**	**48**	**0**	**0**	**5**	**81**	**13.6**		**703**	**46.4**	**59**	**9**	**13:05**	**2**	**0**	**0**	**0**	**6**	**0**	**0**	**0**

AJHL First All-Star Team (1999) • AJHL Rookie of the Year (1999) • Hockey East All-Rookie Team (2000) • Hockey East Second All-Star Team (2001) • NCAA East Second All-American Team (2001) • NCAA Championship All-Tournament Team (2001)

KOLARIK, Pavel

(koh-LAHR-ihk, PAH-vehl)

Defense. Shoots left. 6'1", 207 lbs. Born, Vyskov, Czech., October 24, 1972. Boston's 11th choice, 268th overall, in 2000 Entry Draft.

Season	Club	League	GP	G	A	Pts	PIM	PP	SH	GW	S	%	+/-	TF	F%	H	SB	Min	GP	G	A	Pts	PIM	PP	SH	GW
1996-97	HC Slavia Praha	Czech	27	1	2	3	10												3	0	0	0	2			
1997-98	HC Slavia Praha	Czech	51	0	4	4	24												5	0	0	0	2			
1998-99	HC Slavia Praha	Czech	51	1	8	9	44																			
99-2000	HC Slavia Praha	Czech	52	5	3	8	38																			
2000-01	**Boston**	**NHL**	**10**	**0**	**0**	**0**	**4**	**0**	**0**	**0**	**1**	**0.0**	**–2**	**0**	**0.0**	**7**	**3**	**8:49**								
	Providence	AHL	51	5	6	11	14												17	0	4	4	4			
2001-02	Providence	AHL	47	3	4	7	10												2	0	0	0	2			
	Boston	**NHL**	**13**	**0**	**0**	**0**	**6**	**0**	**0**	**0**	**6**	**0.0**	**0**	**0**	**0.0**	**4**	**4**	**6:45**								
	NHL Totals		**23**	**0**	**0**	**0**	**10**	**0**	**0**	**0**	**7**	**0.0**		**0**	**0.0**	**11**	**7**	**7:39**								

KOLNIK, Juraj

(KOHL-nihk, YEW-igh) **NYI**

Right wing. Shoots right. 5'10", 182 lbs. Born, Nitra, Czech., November 13, 1980. NY Islanders' 7th choice, 101st overall, in 1999 Entry Draft.

Season	Club	League	GP	G	A	Pts	PIM	PP	SH	GW	S	%	+/-	TF	F%	H	SB	Min	GP	G	A	Pts	PIM	PP	SH	GW
1997-98	Nitra Jr.	Slovak-Jr.	26	28	16	44	50																			
	Nitra	Slovakia	28	1	3	4	6																			
1998-99	Quebec Remparts	QMJHL	12	6	5	11	6																			
	Rimouski Oceanic	QMJHL	50	36	37	73	34												11	9	6	15	6			
99-2000	Rimouski Oceanic	QMJHL	47	53	53	106	53												14	10	17	27	16			
2000-01	**NY Islanders**	**NHL**	**29**	**4**	**3**	**7**	**12**	**0**	**0**	**0**	**38**	**10.5**	**–8**	**1**	**100.0**	**49**	**6**	**10:28**								
	Lowell	AHL	25	2	6	8	18																			
	Springfield	AHL	29	15	20	35	20																			
2001-02	**NY Islanders**	**NHL**	**7**	**2**	**0**	**2**	**0**	**1**	**0**	**0**	**10**	**20.0**	**–2**	**1**	**0.0**	**17**	**0**	**7:57**								
	Bridgeport	AHL	67	18	30	48	40												20	7	14	21	17			
	NHL Totals		**36**	**6**	**3**	**9**	**12**	**1**	**0**	**0**	**48**	**12.5**		**2**	**50.0**	**66**	**6**	**9:59**								

Memorial Cup All-Star Team (2000)

Traded to **Rimouski** (QMJHL) by **Quebec** (QMJHL) for Quebec's 3rd round choice (Noye Tyler) in 1999 QMJHL Priority Draft, October 30, 1998.

KOMARNISKI, Zenith

(KOH-mahr-NIHS-kee, ZEE-nihth) **VAN.**

Defense. Shoots left. 6', 200 lbs. Born, Edmonton, Alta., August 13, 1978. Vancouver's 2nd choice, 75th overall, in 1996 Entry Draft.

Season	Club	League	GP	G	A	Pts	PIM	PP	SH	GW	S	%	+/-	TF	F%	H	SB	Min	GP	G	A	Pts	PIM	PP	SH	GW
1993-94	Ft. Saskatchewan	AMHL	32	14	32	46	42																			
1994-95	Tri-City	WHL	66	5	19	24	110												17	1	2	3	47			
1995-96	Tri-City	WHL	42	5	21	26	85																			
1996-97	Tri-City	WHL	58	12	44	56	112																			
1997-98	Tri-City	WHL	3	0	4	4	18																			
	Spokane Chiefs	WHL	43	7	20	27	90												18	4	6	10	49			
1998-99	Syracuse Crunch	AHL	58	9	19	28	89																			
99-2000	**Vancouver**	**NHL**	**18**	**1**	**1**	**2**	**8**	**0**	**0**	**0**	**21**	**4.8**	**–1**	**0**	**0.0**	**40**	**29**	**16:11**								
	Syracuse Crunch	AHL	42	4	12	16	130												4	2	0	2	6			

			Regular Season																Playoffs							
Season	Club	League	GP	G	A	Pts	PIM	PP	SH	GW	S	%	+/-	TF	F%	H	SB	Min	GP	G	A	Pts	PIM	PP	SH	GW
2000-01	Kansas City	IHL	70	7	22	29	191	....	....	....	....	....	....	....	....	....	....	....	....	....	....	....	....	....	....	....
2001-02	Manitoba Moose	AHL	77	5	20	25	153	....	....	....	....	....	....	....	....	....	....	....	7	0	2	2	13	....	....	....
	NHL Totals		18	1	1	2	8	0	0	0	21	4.8		0	0.0	40	29	16:11	....	....	....	....	....	....	....	....

WHL West First All-Star Team (1997)
Traded to **Spokane** (WHL) by **Tri-City** (WHL) for Blake Evans, October 25, 1997.

KONOWALCHUK, Steve
(kahn-uh-WAHL-chuhk, STEEV) **WSH.**

Center. Shoots left. 6'1", 207 lbs. Born, Salt Lake City, UT, November 11, 1972. Washington's 5th choice, 58th overall, in 1991 Entry Draft.

Season	Club	League	GP	G	A	Pts	PIM	PP	SH	GW	S	%	+/-	TF	F%	H	SB	Min	GP	G	A	Pts	PIM	PP	SH	GW
1989-90	Prince Albert	SMHL	36	30	28	58	22	....	....	....	....	....	....	....	....	....	....	....	....	....	....	....	....	....	....	....
1990-91	Portland	WHL	72	43	49	92	78	....	....	....	....	....	....	....	....	....	....	....	....	....	....	....	....	....	....	....
1991-92	Portland	WHL	64	51	53	104	95	....	....	....	....	....	....	....	....	....	....	....	6	3	6	9	12	....	....	....
	Washington	**NHL**	**1**	**0**	**0**	**0**	**0**	**0**	**0**	**0**	**1**	**0.0**	**0**	....	....	....	....	....	....	....	....	....	....	....	....	....
	Baltimore	AHL	3	1	1	2	0	....	....	....	....	....	....	....	....	....	....	....	....	....	....	....	....	....	....	....
1992-93	**Washington**	**NHL**	**36**	**4**	**7**	**11**	**16**	**1**	**0**	**1**	**34**	**11.8**	**4**	....	....	....	....	....	**2**	**0**	**1**	**1**	**0**	**0**	**0**	**0**
	Baltimore	AHL	37	18	28	46	74	....	....	....	....	....	....	....	....	....	....	....	....	....	....	....	....	....	....	....
1993-94	**Washington**	**NHL**	**62**	**12**	**14**	**26**	**33**	**0**	**0**	**0**	**63**	**19.0**	**9**	....	....	....	....	....	**11**	**0**	**1**	**1**	**10**	**0**	**0**	**0**
	Portland Pirates	AHL	8	11	4	15	4	....	....	....	....	....	....	....	....	....	....	....	....	....	....	....	....	....	....	....
1994-95	**Washington**	**NHL**	**46**	**11**	**14**	**25**	**44**	**3**	**3**	**3**	**88**	**12.5**	**7**	....	....	....	....	....	**7**	**2**	**5**	**7**	**12**	**0**	**1**	**0**
1995-96	**Washington**	**NHL**	**70**	**23**	**22**	**45**	**92**	**7**	**1**	**3**	**197**	**11.7**	**13**	....	....	....	....	....	**2**	**0**	**2**	**2**	**0**	**0**	**0**	**0**
1996-97	**Washington**	**NHL**	**78**	**17**	**25**	**42**	**67**	**2**	**1**	**3**	**155**	**11.0**	**-3**	....	....	....	....	....	....	....	....	....	....	....	....	....
1997-98	**Washington**	**NHL**	**80**	**10**	**24**	**34**	**80**	**2**	**0**	**2**	**131**	**7.6**	**9**	....	....	....	....	....	....	....	....	....	....	....	....	....
1998-99	**Washington**	**NHL**	**45**	**12**	**12**	**24**	**26**	**4**	**1**	**2**	**98**	**12.2**	**0**	**124**	**51.6**	**125**	**13**	**17:50**	....	....	....	....	....	....	....	....
99-2000	**Washington**	**NHL**	**82**	**16**	**27**	**43**	**80**	**3**	**0**	**1**	**146**	**11.0**	**19**	**147**	**49.7**	**245**	**39**	**17:36**	**5**	**1**	**0**	**1**	**2**	**0**	**1**	**0**
2000-01	**Washington**	**NHL**	**82**	**24**	**23**	**47**	**87**	**6**	**0**	**5**	**163**	**14.7**	**8**	**91**	**55.0**	**203**	**35**	**17:04**	**6**	**2**	**3**	**5**	**14**	**2**	**0**	**0**
2001-02	**Washington**	**NHL**	**28**	**2**	**12**	**14**	**23**	**0**	**0**	**0**	**36**	**5.6**	**-2**	**64**	**56.3**	**51**	**18**	**16:29**	....	....	....	....	....	....	....	....
	NHL Totals		**610**	**131**	**180**	**311**	**548**	**28**	**6**	**20**	**1112**	**11.8**		**426**	**52.3**	**624**	**105**	**17:20**	**33**	**5**	**12**	**17**	**38**	**2**	**2**	**0**

WHL West First All-Star Team (1992) • WHL MVP (1992)
• Missed majority of 2001-02 season recovering from shoulder injury suffered in game vs. LA Kings, October 16, 2001.

KOROLEV, Evgeny
(KOH-roh-lehv, ehv-GEHN-ee) **NYI**

Defense. Shoots left. 6'1", 214 lbs. Born, Moscow, USSR, July 24, 1978. NY Islanders' 6th choice, 182nd overall, in 1998 Entry Draft.

Season	Club	League	GP	G	A	Pts	PIM	PP	SH	GW	S	%	+/-	TF	F%	H	SB	Min	GP	G	A	Pts	PIM	PP	SH	GW
1995-96	Peterborough	OHL	60	2	12	14	60	....	....	....	....	....	....	....	....	....	....	....	6	0	0	0	2	....	....	....
1996-97	Peterborough	OHL	64	5	17	22	60	....	....	....	....	....	....	....	....	....	....	....	11	1	1	2	8	....	....	....
1997-98	Peterborough	OHL	37	5	21	26	39	....	....	....	....	....	....	....	....	....	....	....	....	....	....	....	....	....	....	....
	London Knights	OHL	27	4	10	14	36	....	....	....	....	....	....	....	....	....	....	....	15	2	7	9	29	....	....	....
1998-99	Roanoke Express	ECHL	2	0	1	1	0	....	....	....	....	....	....	....	....	....	....	....	....	....	....	....	....	....	....	....
	Lowell	AHL	54	2	6	8	48	....	....	....	....	....	....	....	....	....	....	....	2	0	1	1	0	....	....	....
99-2000	**NY Islanders**	**NHL**	**17**	**1**	**2**	**3**	**8**	**0**	**0**	**0**	**7**	**14.3**	**-10**	**0**	**0.0**	**34**	**19**	**16:12**	....	....	....	....	....	....	....	....
	Lowell	AHL	57	1	10	11	61	....	....	....	....	....	....	....	....	....	....	....	6	0	0	0	4	....	....	....
2000-01	**NY Islanders**	**NHL**	**8**	**0**	**0**	**0**	**6**	**0**	**0**	**0**	**11**	**0.0**	**0**	**0**	**0.0**	**17**	**11**	**16:40**	....	....	....	....	....	....	....	....
	Chicago Wolves	IHL	4	0	1	1	0	....	....	....	....	....	....	....	....	....	....	....	....	....	....	....	....	....	....	....
	Louisville	AHL	36	2	14	16	68	....	....	....	....	....	....	....	....	....	....	....	....	....	....	....	....	....	....	....
2001-02	**NY Islanders**	**NHL**	**17**	**0**	**2**	**2**	**6**	**0**	**0**	**0**	**9**	**0.0**	**0**	**0**	**0.0**	**8**	**9**	**10:22**	**2**	**0**	**0**	**0**	**0**	**0**	**0**	**0**
	Bridgeport	AHL	53	5	8	13	30	....	....	....	....	....	....	....	....	....	....	....	....	....	....	....	....	....	....	....
	NHL Totals		**42**	**1**	**4**	**5**	**20**	**0**	**0**	**0**	**27**	**3.7**		**0**	**0.0**	**59**	**39**	**13:56**	**2**	**0**	**0**	**0**	**0**	**0**	**0**	**0**

• Re-entered NHL Entry Draft. Originally NY Islanders' 9th choice, 192nd overall, in 1996 Entry Draft.
Traded to **London** (OHL) by **Peterborough** (OHL) for Duncan Dalmo, January 10, 1998.

KOROLEV, Igor
(KOH-roh-lehv, EE-gohr) **CHI.**

Center/Left wing. Shoots left. 6'1", 190 lbs. Born, Moscow, USSR, September 6, 1970. St. Louis' 1st choice, 38th overall, in 1992 Entry Draft.

Season	Club	League	GP	G	A	Pts	PIM	PP	SH	GW	S	%	+/-	TF	F%	H	SB	Min	GP	G	A	Pts	PIM	PP	SH	GW
1988-89	Dynamo Moscow	USSR	1	0	0	0	2	....	....	....	....	....	....	....	....	....	....	....	....	....	....	....	....	....	....	....
1989-90	Dynamo Moscow	USSR	17	3	2	5	2	....	....	....	....	....	....	....	....	....	....	....	....	....	....	....	....	....	....	....
1990-91	Dynamo Moscow	USSR	38	12	4	16	12	....	....	....	....	....	....	....	....	....	....	....	....	....	....	....	....	....	....	....
1991-92	Dynamo Moscow	CIS	39	15	12	27	16	....	....	....	....	....	....	....	....	....	....	....	....	....	....	....	....	....	....	....
1992-93	Dynamo Moscow	CIS	5	1	2	3	4	....	....	....	....	....	....	....	....	....	....	....	....	....	....	....	....	....	....	....
	St. Louis	**NHL**	**74**	**4**	**23**	**27**	**20**	**2**	**0**	**0**	**76**	**5.3**	**-1**	....	....	....	....	....	**3**	**0**	**0**	**0**	**0**	**0**	**0**	**0**
1993-94	**St. Louis**	**NHL**	**73**	**6**	**10**	**16**	**40**	**0**	**0**	**1**	**93**	**6.5**	**-12**	....	....	....	....	....	**2**	**0**	**0**	**0**	**0**	**0**	**0**	**0**
1994-95	Dynamo Moscow	CIS	13	4	6	10	18	....	....	....	....	....	....	....	....	....	....	....	....	....	....	....	....	....	....	....
	Winnipeg	**NHL**	**45**	**8**	**22**	**30**	**10**	**1**	**0**	**1**	**85**	**9.4**	**1**	....	....	....	....	....	....	....	....	....	....	....	....	....
1995-96	**Winnipeg**	**NHL**	**73**	**22**	**29**	**51**	**42**	**8**	**0**	**5**	**165**	**13.3**	**1**	....	....	....	....	....	**6**	**0**	**3**	**3**	**0**	**0**	**0**	**0**
1996-97	**Phoenix**	**NHL**	**41**	**3**	**7**	**10**	**28**	**2**	**0**	**0**	**41**	**7.3**	**-5**	....	....	....	....	....	**1**	**0**	**0**	**0**	**0**	**0**	**0**	**0**
	Michigan K-Wings	IHL	4	2	2	4	0	....	....	....	....	....	....	....	....	....	....	....	....	....	....	....	....	....	....	....
	Phoenix	IHL	4	2	6	8	4	....	....	....	....	....	....	....	....	....	....	....	....	....	....	....	....	....	....	....
1997-98	**Toronto**	**NHL**	**78**	**17**	**22**	**39**	**22**	**6**	**3**	**5**	**97**	**17.5**	**-18**	....	....	....	....	....	....	....	....	....	....	....	....	....
1998-99	**Toronto**	**NHL**	**66**	**13**	**34**	**47**	**46**	**1**	**0**	**2**	**99**	**13.1**	**11**	**973**	**42.0**	**23**	**13**	**18:06**	**1**	**0**	**0**	**0**	**0**	**0**	**0**	**0**
99-2000	**Toronto**	**NHL**	**80**	**20**	**26**	**46**	**22**	**5**	**3**	**4**	**101**	**19.8**	**12**	**964**	**41.3**	**42**	**26**	**17:56**	**12**	**0**	**4**	**4**	**6**	**0**	**0**	**0**
2000-01	**Toronto**	**NHL**	**73**	**10**	**19**	**29**	**28**	**2**	**0**	**0**	**78**	**12.8**	**3**	**569**	**42.5**	**30**	**19**	**15:41**	**11**	**0**	**0**	**0**	**0**	**0**	**0**	**0**
2001-02	**Chicago**	**NHL**	**82**	**9**	**20**	**29**	**20**	**0**	**1**	**1**	**78**	**11.5**	**-5**	**1166**	**41.1**	**30**	**32**	**16:49**	**5**	**0**	**1**	**1**	**0**	**0**	**0**	**0**
	NHL Totals		**685**	**112**	**212**	**324**	**278**	**27**	**7**	**19**	**913**	**12.3**		**3672**	**41.6**	**125**	**90**	**17:07**	**41**	**0**	**8**	**8**	**6**	**0**	**0**	**0**

Claimed by **Winnipeg** from **St. Louis** in NHL Waiver Draft, January 18, 1995. Transferred to **Phoenix** after **Winnipeg** franchise relocated, July 1, 1996. Signed as a free agent by **Toronto**, September 29, 1997. Traded to **Chicago** by **Toronto** for Philadelphia's 3rd round choice (previously acquired, Toronto selected Nicolas Corbeil) in 2001 Entry Draft, June 23, 2001.

KOROLYUK, Alexander
(koh-roh-LYUHK, al-ehx-AN-duhr) **S.J.**

Right wing. Shoots left. 5'9", 195 lbs. Born, Moscow, USSR, January 15, 1976. San Jose's 6th choice, 141st overall, in 1994 Entry Draft.

Season	Club	League	GP	G	A	Pts	PIM	PP	SH	GW	S	%	+/-	TF	F%	H	SB	Min	GP	G	A	Pts	PIM	PP	SH	GW
1993-94	Krylja Sovetov	CIS	22	4	4	8	20	....	....	....	....	....	....	....	....	....	....	....	3	1	0	1	4	....	....	....
1994-95	Krylja Sovetov	CIS	52	16	13	29	62	....	....	....	....	....	....	....	....	....	....	....	4	1	2	3	4	....	....	....
1995-96	Krylja Sovetov	CIS	50	30	19	49	77	....	....	....	....	....	....	....	....	....	....	....	....	....	....	....	....	....	....	....
1996-97	Krylja Sovetov	Russia	17	8	5	13	46	....	....	....	....	....	....	....	....	....	....	....	....	....	....	....	....	....	....	....
	Manitoba Moose	IHL	42	20	16	36	71	....	....	....	....	....	....	....	....	....	....	....	....	....	....	....	....	....	....	....
1997-98	**San Jose**	**NHL**	**19**	**2**	**3**	**5**	**6**	**1**	**0**	**0**	**23**	**8.7**	**-5**	....	....	....	....	....	....	....	....	....	....	....	....	....
	Kentucky	AHL	44	16	23	39	96	....	....	....	....	....	....	....	....	....	....	....	3	0	0	0	0	....	....	....
1998-99	**San Jose**	**NHL**	**55**	**12**	**18**	**30**	**26**	**2**	**0**	**0**	**96**	**12.5**	**3**	**4**	**50.0**	**66**	**7**	**13:53**	**6**	**1**	**3**	**4**	**2**	**0**	**0**	**1**
	Kentucky	AHL	23	9	13	22	16	....	....	....	....	....	....	....	....	....	....	....	....	....	....	....	....	....	....	....
99-2000	**San Jose**	**NHL**	**57**	**14**	**21**	**35**	**35**	**3**	**0**	**1**	**124**	**11.3**	**4**	**1**	**100.0**	**47**	**12**	**13:36**	**9**	**0**	**3**	**3**	**6**	**0**	**0**	**0**
2000-01	Ak Bars Kazan	Russia	6	0	5	5	4	....	....	....	....	....	....	....	....	....	....	....	....	....	....	....	....	....	....	....
	San Jose	**NHL**	**70**	**12**	**13**	**25**	**41**	**3**	**0**	**1**	**140**	**8.6**	**2**	**30**	**33.3**	**49**	**10**	**11:56**	**2**	**0**	**0**	**0**	**0**	**0**	**0**	**0**
2001-02	**San Jose**	**NHL**	**32**	**3**	**7**	**10**	**14**	**0**	**0**	**1**	**49**	**6.1**	**2**	**10**	**30.0**	**16**	**5**	**12:16**	....	....	....	....	....	....	....	....
	NHL Totals		**233**	**43**	**62**	**105**	**122**	**9**	**0**	**3**	**432**	**10.0**		**45**	**35.6**	**178**	**34**	**12:56**	**17**	**1**	**6**	**7**	**8**	**0**	**0**	**1**

• Spent majority of 2001-02 season on practice roster, October 4, 2001. Signed as a free agent by **AK Bars Kazan** (Russia) with San Jose retaining NHL rights, July 9, 2002.

KOSTOPOULOS, Tom
(kaw-STAWP-oh-lihs, TAWM) **PIT.**

Right wing. Shoots right. 6', 200 lbs. Born, Mississauga, Ont., January 24, 1979. Pittsburgh's 9th choice, 204th overall, in 1999 Entry Draft.

Season	Club	League	GP	G	A	Pts	PIM	PP	SH	GW	S	%	+/-	TF	F%	H	SB	Min	GP	G	A	Pts	PIM	PP	SH	GW
1995-96	Brampton	OPJHL	24	9	9	18	28	....	....	....	....	....	....	....	....	....	....	....	....	....	....	....	....	....	....	....
1996-97	London Knights	OHL	64	13	12	25	67	....	....	....	....	....	....	....	....	....	....	....	....	....	....	....	....	....	....	....
1997-98	London Knights	OHL	66	24	26	50	108	....	....	....	....	....	....	....	....	....	....	....	16	6	4	10	26	....	....	....
1998-99	London Knights	OHL	66	27	60	87	114	....	....	....	....	....	....	....	....	....	....	....	25	19	16	35	32	....	....	....
99-2000	Wilkes-Barre	AHL	76	26	32	58	121	....	....	....	....	....	....	....	....	....	....	....	....	....	....	....	....	....	....	....
2000-01	Wilkes-Barre	AHL	80	16	36	52	120	....	....	....	....	....	....	....	....	....	....	....	21	3	9	12	6	....	....	....
2001-02	**Pittsburgh**	**NHL**	**11**	**1**	**2**	**3**	**9**	**0**	**0**	**0**	**8**	**12.5**	**-1**	**0**	**0.0**	**20**	**3**	**12:03**	....	....	....	....	....	....	....	....
	Wilkes-Barre	AHL	70	27	26	53	112	....	....	....	....	....	....	....	....	....	....	....	....	....	....	....	....	....	....	....
	NHL Totals		**11**	**1**	**2**	**3**	**9**	**0**	**0**	**0**	**8**	**12.5**		**0**	**0.0**	**20**	**3**	**12:03**	....	....	....	....	....	....	....	....

			Regular Season																Playoffs							
Season	Club	League	GP	G	A	Pts	PIM	PP	SH	GW	S	%	+/-	TF	F%	H	SB	Min	GP	G	A	Pts	PIM	PP	SH	GW

KOTALIK, Ales

(KOH-tahl-eek, AL-ehsh) **BUF.**

Right wing. Shoots right. 6'1", 217 lbs. Born, Jindrichuv Hradec, Czech., December 23, 1978. Buffalo's 7th choice, 164th overall, in 1998 Entry Draft.

Season	Club	League	GP	G	A	Pts	PIM	PP	SH	GW	S	%	+/-	TF	F%	H	SB	Min	GP	G	A	Pts	PIM	PP	SH	GW
1993-94	C. Budejovice Jr.	Czech-Jr.	28	12	12	24																				
1994-95	C. Budejovice Jr.	Czech-Jr.	36	26	17	43																				
1995-96	C. Budejovice Jr.	Czech-Jr.	28	6	7	13																				
1996-97	C. Budejovice Jr.	Czech-Jr.	36	15	16	31	24																			
1997-98	Ceske Budejovice	Czech	47	9	7	16	14																			
1998-99	Ceske Budejovice	Czech	41	8	13	21	16												3	0	0	0				
99-2000	Ceske Budejovice	Czech	43	7	12	19	34												3	0	1	1	6			
2000-01	Ceske Budejovice	Czech	52	19	29	48	54																			
2001-02	**Buffalo**	**NHL**	**13**	**1**	**3**	**4**	**2**	**0**	**0**	**0**	**21**	**4.8**	**–1**	**11**	**27.3**	**24**	**0**	**12:35**								
	Rochester	AHL	68	18	25	43	55												1	0	0	0	0			
	NHL Totals		**13**	**1**	**3**	**4**	**2**	**0**	**0**	**0**	**21**	**4.8**		**11**	**27.3**	**24**	**0**	**12:35**								

KOVALCHUK, Ilya

(koh-vuhl-CHOOK, ILL-yah) **ATL.**

Left wing. Shoots right. 6'1", 220 lbs. Born, Tver, USSR, April 15, 1983. Atlanta's 1st choice, 1st overall, in 2001 Entry Draft.

Season	Club	League	GP	G	A	Pts	PIM	PP	SH	GW	S	%	+/-	TF	F%	H	SB	Min	GP	G	A	Pts	PIM	PP	SH	GW
99-2000	Spartak Moscow	Russia-2	49	12	5	17	75																			
	Spartak Moscow 2	Russia-3	2	2	1	3	14																			
2000-01	Spartak Moscow	Russia-2	51	42	22	64	112																			
2001-02	**Atlanta**	**NHL**	**65**	**29**	**22**	**51**	**28**	**7**	**0**	**4**	**184**	**15.8**	**–19**	**6**	**16.7**	**34**	**11**	**18:32**								
	Russia	Olympics	6	1	2	3	14																			
	NHL Totals		**65**	**29**	**22**	**51**	**28**	**7**	**0**	**4**	**184**	**15.8**		**6**	**16.7**	**34**	**11**	**18:32**								

NHL All-Rookie Team (2002)

KOVALENKO, Andrei

(koh-vah-LEHN-koh, AWN-dray)

Right wing. Shoots left. 5'10", 200 lbs. Born, Balakovo, USSR, June 7, 1970. Quebec's 6th choice, 148th overall, in 1990 Entry Draft.

Season	Club	League	GP	G	A	Pts	PIM	PP	SH	GW	S	%	+/-	TF	F%	H	SB	Min	GP	G	A	Pts	PIM	PP	SH	GW
1987-88	Torpedo Gorky	USSR	2	1	0	1	0																			
1988-89	SKA MVO Kalinin	USSR-2	30	8	7	15	29																			
	CSKA Moscow	USSR	10	1	0	1	0																			
1989-90	CSKA Moscow	USSR	48	8	5	13	20																			
1990-91	CSKA Moscow	USSR	45	13	8	21	26																			
1991-92	CSKA Moscow	CIS	44	19	13	32	32																			
	Russia	Olympics	8	1	1	2	2																			
1992-93	CSKA Moscow	CIS	3	3	1	4	4																			
	Quebec	**NHL**	**81**	**27**	**41**	**68**	**57**	**8**	**1**	**4**	**153**	**17.6**	**13**						**4**	**1**	**0**	**1**	**2**	**0**	**0**	**0**
1993-94	**Quebec**	**NHL**	**58**	**16**	**17**	**33**	**46**	**5**	**0**	**4**	**92**	**17.4**	**–5**													
1994-95	Lada Togliatti	CIS	11	9	2	11	14																			
	Quebec	**NHL**	**45**	**14**	**10**	**24**	**31**	**1**	**0**	**3**	**63**	**22.2**	**–4**						**6**	**0**	**1**	**1**	**2**	**0**	**0**	**0**
1995-96	**Colorado**	**NHL**	**26**	**11**	**11**	**22**	**16**	**3**	**0**	**3**	**46**	**23.9**	**11**													
	Montreal	**NHL**	**51**	**17**	**17**	**34**	**33**	**3**	**0**	**3**	**85**	**20.0**	**9**						**6**	**0**	**0**	**0**	**6**	**0**	**0**	**0**
1996-97	**Edmonton**	**NHL**	**74**	**32**	**27**	**59**	**81**	**14**	**0**	**2**	**163**	**19.6**	**–5**						**12**	**4**	**3**	**7**	**6**	**3**	**0**	**0**
1997-98	**Edmonton**	**NHL**	**59**	**6**	**17**	**23**	**28**	**1**	**0**	**2**	**89**	**6.7**	**–14**						**1**	**0**	**0**	**0**	**2**	**0**	**0**	**0**
	Russia	Olympics	6	4	1	5	14																			
1998-99	**Edmonton**	**NHL**	**43**	**13**	**14**	**27**	**30**	**2**	**0**	**3**	**75**	**17.3**	**–4**	**0**	**0.0**	**40**	**7**	**16:19**								
	Philadelphia	**NHL**	**13**	**0**	**1**	**1**	**2**	**0**	**0**	**0**	**8**	**0.0**	**–5**	**0**	**0.0**	**12**	**1**	**8:02**								
	Carolina	**NHL**	**18**	**6**	**6**	**12**	**0**	**1**	**0**	**1**	**21**	**28.6**	**3**	**1**	**0.0**	**40**	**2**	**13:53**	**4**	**0**	**2**	**2**	**2**	**0**	**0**	**0**
99-2000	**Carolina**	**NHL**	**76**	**15**	**24**	**39**	**38**	**2**	**0**	**3**	**114**	**13.2**	**–13**	**4**	**75.0**	**143**	**21**	**14:59**								
2000-01	**Boston**	**NHL**	**76**	**16**	**21**	**37**	**27**	**7**	**1**	**3**	**119**	**13.4**	**–14**	**17**	**29.4**	**123**	**20**	**15:27**								
2001-02	Yaroslavl	Russia	51	*27	20	47	62												9	4	3	7	28			
	NHL Totals		**620**	**173**	**206**	**379**	**389**	**47**	**2**	**31**	**1028**	**16.8**		**22**	**36.4**	**358**	**51**	**14:54**	**33**	**5**	**6**	**11**	**20**	**3**	**0**	**0**

Transferred to **Colorado** after **Quebec** franchise relocated, June 21, 1995. Traded to **Montreal** by **Colorado** with Martin Rucinsky and Jocelyn Thibault for Patrick Roy and Mike Keane, December 6, 1995. Traded to **Edmonton** by **Montreal** for Scott Thornton, September 6, 1996. Traded to **Philadelphia** by **Edmonton** for Alexandre Daigle, January 29, 1999. Traded to **Carolina** by **Philadelphia** for Adam Burt, March 6, 1999. Signed as a free agent by **Boston**, July 25, 2000.

KOVALEV, Alexei

(koh-VAH-lehv, al-EHX-ay) **PIT.**

Right wing. Shoots left. 6'1", 220 lbs. Born, Togliatti, USSR, February 24, 1973. NY Rangers' 1st choice, 15th overall, in 1991 Entry Draft.

Season	Club	League	GP	G	A	Pts	PIM	PP	SH	GW	S	%	+/-	TF	F%	H	SB	Min	GP	G	A	Pts	PIM	PP	SH	GW
1989-90	Dynamo Moscow	USSR	1	0	0	0	0																			
1990-91	Dynamo Moscow	USSR	18	1	2	3	4																			
1991-92	Dynamo Moscow	CIS	33	16	9	25	20																			
	Russia	Olympics	8	1	2	3	14																			
1992-93	**NY Rangers**	**NHL**	**65**	**20**	**18**	**38**	**79**	**3**	**0**	**3**	**134**	**14.9**	**–10**													
	Binghamton	AHL	13	13	11	24	35												9	3	5	8	14			
1993-94♦	**NY Rangers**	**NHL**	**76**	**23**	**33**	**56**	**154**	**7**	**0**	**3**	**184**	**12.5**	**18**						**23**	**9**	**12**	**21**	**18**	**5**	**0**	**2**
1994-95	Lada Togliatti	CIS	12	8	8	16	49																			
	NY Rangers	**NHL**	**48**	**13**	**15**	**28**	**30**	**1**	**1**	**1**	**103**	**12.6**	**–6**						**10**	**4**	**7**	**11**	**10**	**0**	**0**	**0**
1995-96	**NY Rangers**	**NHL**	**81**	**24**	**34**	**58**	**98**	**8**	**1**	**7**	**206**	**11.7**	**5**						**11**	**3**	**4**	**7**	**14**	**0**	**0**	**1**
1996-97	**NY Rangers**	**NHL**	**45**	**13**	**22**	**35**	**42**	**1**	**0**	**0**	**110**	**11.8**	**11**													
1997-98	**NY Rangers**	**NHL**	**73**	**23**	**30**	**53**	**44**	**8**	**0**	**3**	**173**	**13.3**	**–22**													
1998-99	**NY Rangers**	**NHL**	**14**	**3**	**4**	**7**	**12**	**1**	**0**	**1**	**35**	**8.6**	**–6**	**18**	**44.4**	**13**	**5**	**19:53**								
	Pittsburgh	**NHL**	**63**	**20**	**26**	**46**	**37**	**5**	**1**	**4**	**156**	**12.8**	**8**	**226**	**43.4**	**82**	**40**	**20:30**	**10**	**5**	**7**	**12**	**14**	**0**	**0**	**1**
99-2000	**Pittsburgh**	**NHL**	**82**	**26**	**40**	**66**	**94**	**9**	**2**	**4**	**254**	**10.2**	**–3**	**306**	**47.4**	**90**	**25**	**22:53**	**11**	**1**	**5**	**6**	**10**	**0**	**0**	**0**
2000-01	**Pittsburgh**	**NHL**	**79**	**44**	**51**	**95**	**96**	**12**	**2**	**9**	**307**	**14.3**	**12**	**255**	**40.0**	**104**	**27**	**23:35**	**18**	**5**	**5**	**10**	**16**	**1**	**0**	**0**
2001-02	**Pittsburgh**	**NHL**	**67**	**32**	**44**	**76**	**80**	**8**	**1**	**3**	**266**	**12.0**	**2**	**179**	**45.3**	**76**	**26**	**24:03**								
	Russia	Olympics	6	3	1	4	4																			
	NHL Totals		**693**	**241**	**317**	**558**	**766**	**63**	**8**	**38**	**1928**	**12.5**		**984**	**44.1**	**365**	**123**	**22:41**	**83**	**27**	**40**	**67**	**82**	**6**	**0**	**4**

Played in NHL All-Star Game (2001)

Traded to **Pittsburgh** by **NY Rangers** with Harry York for Petr Nedved, Chris Tamer and Sean Pronger, November 25, 1998.

KOZLOV, Viktor

(KAHS-lahf, VIHK-tohr) **FLA.**

Center. Shoots right. 6'5", 225 lbs. Born, Togliatti, USSR, February 14, 1975. San Jose's 1st choice, 6th overall, in 1993 Entry Draft.

Season	Club	League	GP	G	A	Pts	PIM	PP	SH	GW	S	%	+/-	TF	F%	H	SB	Min	GP	G	A	Pts	PIM	PP	SH	GW
1990-91	Lada Togliatti	USSR-2	2	2	0	2	0																			
1991-92	Lada Togliatti	CIS	3	0	0	0	0																			
1992-93	Dynamo Moscow	CIS	30	6	5	11	4												10	3	0	3	0			
1993-94	Dynamo Moscow	CIS	42	16	9	25	14												7	3	2	5	0			
1994-95	Dynamo Moscow	CIS	3	1	1	2	2																			
	San Jose	**NHL**	**16**	**2**	**0**	**2**	**2**	**0**	**0**	**0**	**23**	**8.7**	**–5**													
	Kansas City	IHL	4	1	1	2	0												13	4	5	9	12			
1995-96	**San Jose**	**NHL**	**62**	**6**	**13**	**19**	**6**	**1**	**0**	**0**	**107**	**5.6**	**–15**													
	Kansas City	IHL	15	4	7	11	12																			
1996-97	**San Jose**	**NHL**	**78**	**16**	**25**	**41**	**40**	**4**	**0**	**4**	**184**	**8.7**	**–16**													
1997-98	**San Jose**	**NHL**	**18**	**5**	**2**	**7**	**2**	**2**	**0**	**0**	**51**	**9.8**	**–2**													
	Florida	**NHL**	**46**	**12**	**11**	**23**	**14**	**3**	**2**	**0**	**114**	**10.5**	**–1**													
1998-99	**Florida**	**NHL**	**65**	**16**	**35**	**51**	**24**	**5**	**1**	**1**	**209**	**7.7**	**13**	**985**	**41.2**	**32**	**30**	**19:03**								
99-2000	**Florida**	**NHL**	**80**	**17**	**53**	**70**	**16**	**6**	**0**	**2**	**223**	**7.6**	**24**	**1616**	**42.9**	**62**	**33**	**19:27**	**4**	**0**	**1**	**1**	**0**	**0**	**0**	**0**
2000-01	**Florida**	**NHL**	**51**	**14**	**23**	**37**	**10**	**6**	**0**	**2**	**139**	**10.1**	**–4**	**817**	**41.6**	**50**	**22**	**18:23**								
2001-02	**Florida**	**NHL**	**50**	**9**	**18**	**27**	**20**	**6**	**0**	**1**	**143**	**6.3**	**–16**	**840**	**43.1**	**54**	**23**	**19:54**								
	NHL Totals		**466**	**97**	**180**	**277**	**134**	**33**	**3**	**10**	**1193**	**8.1**		**4258**	**42.3**	**198**	**108**	**19:13**	**4**	**0**	**1**	**1**	**0**	**0**	**0**	**0**

Played in NHL All-Star Game (2000)

Traded to **Florida** by **San Jose** with Florida's 5th round choice (previously acquired, Florida selected Jaroslav Spacek) in 1998 Entry Draft for Dave Lowry and Florida's 1st round choice (later traded to Tampa Bay - Tampa Bay selected Vincent Lecavalier) in 1998 Entry Draft, November 13, 1997.

KOZLOV, Vyacheslav

(KAHS-lahf, VYACH-ih-slav) **ATL.**

Center. Shoots left. 5'10", 185 lbs. Born, Voskresensk, USSR, May 3, 1972. Detroit's 2nd choice, 45th overall, in 1990 Entry Draft.

			Regular Season																Playoffs							
Season	Club	League	GP	G	A	Pts	PIM	PP	SH	GW	S	%	+/-	TF	F%	H	SB	Min	GP	G	A	Pts	PIM	PP	SH	GW
1987-88	Voskresensk	USSR	2	0	0	0	0																			
1988-89	Voskresensk	USSR	14	0	1	1	2																			
1989-90	Voskresensk	USSR	45	14	12	26	38																			
1990-91	Voskresensk	USSR	45	11	13	24	46																			
1991-92	CSKA Moscow	CIS	11	6	5	11	12																			
	Detroit	**NHL**	**7**	**0**	**2**	**2**	**2**	**0**	**0**	**0**	**9**	**0.0**	**-2**													
1992-93	**Detroit**	**NHL**	**17**	**4**	**1**	**5**	**14**	**0**	**0**	**0**	**26**	**15.4**	**-1**						**4**	**0**	**2**	**2**	**2**	**0**	**0**	**0**
	Adirondack	AHL	45	23	36	59	54												4	1	1	2	4			
1993-94	**Detroit**	**NHL**	**77**	**34**	**39**	**73**	**50**	**8**	**2**	**6**	**202**	**16.8**	**27**						**7**	**2**	**5**	**7**	**12**	**0**	**0**	**0**
	Adirondack	AHL	3	0	1	1	15																			
1994-95	CSKA Moscow	CIS	10	3	4	7	14																			
	Detroit	**NHL**	**46**	**13**	**20**	**33**	**45**	**5**	**0**	**3**	**97**	**13.4**	**12**						**18**	**9**	**7**	**16**	**10**	**1**	**0**	**4**
1995-96	**Detroit**	**NHL**	**82**	**36**	**37**	**73**	**70**	**9**	**0**	**7**	**237**	**15.2**	**33**						**19**	**5**	**7**	**12**	**10**	**2**	**0**	**1**
1996-97♦	**Detroit**	**NHL**	**75**	**23**	**22**	**45**	**46**	**3**	**0**	**6**	**211**	**10.9**	**21**						**20**	**8**	**5**	**13**	**14**	**4**	**0**	**2**
1997-98♦	**Detroit**	**NHL**	**80**	**25**	**27**	**52**	**46**	**6**	**0**	**1**	**221**	**11.3**	**14**						**22**	**6**	**8**	**14**	**10**	**1**	**0**	**4**
1998-99	**Detroit**	**NHL**	**79**	**29**	**29**	**58**	**45**	**6**	**1**	**4**	**209**	**13.9**	**10**	**38**	**36.8**	**41**	**20**	**16:02**	**10**	**6**	**1**	**7**	**4**	**3**	**0**	**0**
99-2000	**Detroit**	**NHL**	**72**	**18**	**18**	**36**	**28**	**4**	**0**	**3**	**165**	**10.9**	**11**	**28**	**35.7**	**35**	**20**	**15:30**	**8**	**2**	**1**	**3**	**12**	**1**	**0**	**1**
2000-01	**Detroit**	**NHL**	**72**	**20**	**18**	**38**	**30**	**4**	**0**	**5**	**187**	**10.7**	**9**	**51**	**47.1**	**50**	**19**	**14:43**	**6**	**4**	**1**	**5**	**2**	**2**	**0**	**0**
2001-02	**Buffalo**	**NHL**	**38**	**9**	**13**	**22**	**16**	**3**	**0**	**1**	**68**	**13.2**	**0**	**24**	**41.7**	**20**	**9**	**16:31**								
	NHL Totals		**645**	**211**	**226**	**437**	**392**	**48**	**3**	**36**	**1632**	**12.9**		**141**	**41.1**	**146**	**68**	**15:36**	**114**	**42**	**37**	**79**	**76**	**14**	**0**	**12**

Traded to **Buffalo** by **Detroit** with Detroit's 1st round choice (later traded to Columbus - later traded to Atlanta - Atlanta selected Jim Slater) in 2002 Entry Draft and future considerations for Dominik Hasek, July 1, 2001. • Missed majority of 2001-02 season recovering from Achilles tendon injury suffered in game vs. Columbus, December 31, 2001. Traded to **Atlanta** by **Buffalo** with Buffalo's 2nd round choice in 2003 Entry Draft for Atlanta's 2nd and 3rd round choices in 2003 Entry Draft, June 22, 2002.

KRAFT, Milan

(KRAFT, MIH-lan) **PIT.**

Center. Shoots right. 6'3", 211 lbs. Born, Plzen, Czech., January 7, 1980. Pittsburgh's 1st choice, 23rd overall, in 1998 Entry Draft.

Season	Club	League	GP	G	A	Pts	PIM	PP	SH	GW	S	%	+/-	TF	F%	H	SB	Min	GP	G	A	Pts	PIM	PP	SH	GW
1995-96	HC ZKZ Plzen Jr.	Czech-Jr.	49	54	41	95																				
1996-97	HC ZKZ Plzen Jr.	Czech-Jr.	29	24	12	36																				
	HC ZKZ Plzen	Czech	9	0	1	1	2																			
1997-98	Plzen Jr.	Czech-Jr.	24	22	21	43	12																			
	Plzen	Czech	16	0	5	5	0												1	0	0	0	0			
1998-99	Prince Albert	WHL	68	40	46	86	32												14	7	13	20	6			
99-2000	Prince Albert	WHL	56	34	35	69	42												6	4	1	5	4			
2000-01	**Pittsburgh**	**NHL**	**42**	**7**	**7**	**14**	**8**	**1**	**1**	**1**	**63**	**11.1**	**-6**	**427**	**37.9**	**10**	**3**	**11:41**	**8**	**0**	**0**	**0**	**2**	**0**	**0**	**0**
	Wilkes-Barre	AHL	40	21	23	44	27												14	12	7	19	6			
2001-02	**Pittsburgh**	**NHL**	**68**	**8**	**8**	**16**	**16**	**1**	**0**	**2**	**103**	**7.8**	**-9**	**766**	**44.7**	**21**	**18**	**12:29**								
	Wilkes-Barre	AHL	8	4	4	8	10																			
	NHL Totals		**110**	**15**	**15**	**30**	**24**	**2**	**1**	**3**	**166**	**9.0**		**1193**	**42.2**	**31**	**21**	**12:11**	**8**	**0**	**0**	**0**	**2**	**0**	**0**	**0**

KRAJICEK, Lukas

(KRIGH-ee-chehk, LOO-kahsh) **FLA.**

Defense. Shoots left. 6'2", 182 lbs. Born, Prostejov, Czech., March 11, 1983. Florida's 2nd choice, 24th overall, in 2001 Entry Draft.

Season	Club	League	GP	G	A	Pts	PIM	PP	SH	GW	S	%	+/-	TF	F%	H	SB	Min	GP	G	A	Pts	PIM	PP	SH	GW
1998-99	Zlin Jr.	Czech-Jr.	48	8	18	26	40																			
99-2000	Det. Compuware	NAJHL	53	5	22	27	61												5	0	1	1	18			
2000-01	Peterborough	OHL	61	8	27	35	53												7	0	5	5	0			
2001-02	**Florida**	**NHL**	**5**	**0**	**0**	**0**	**0**	**0**	**0**	**0**	**3**	**0.0**	**0**	**0**	**0.0**	**4**	**2**	**13:23**								
	Peterborough	OHL	55	10	32	42	56												6	0	5	5	6			
	NHL Totals		**5**	**0**	**0**	**0**	**0**	**0**	**0**	**0**	**3**	**0.0**		**0**	**0.0**	**4**	**2**	**13:23**								

OHL All-Rookie Team (2001)

• Returned to **Peterborough** (OHL) by **Florida**, October 28, 2001.

KRAVCHUK, Igor

(krahv-CHOOK, EE-gohr)

Defense. Shoots left. 6'1", 218 lbs. Born, Ufa, USSR, September 13, 1966. Chicago's 5th choice, 71st overall, in 1991 Entry Draft.

Season	Club	League	GP	G	A	Pts	PIM	PP	SH	GW	S	%	+/-	TF	F%	H	SB	Min	GP	G	A	Pts	PIM	PP	SH	GW
1984-85	Ufa	USSR-2	50	3	2	5	22																			
1985-86	Ufa	USSR	21	2	2	4	6																			
1986-87	Ufa	USSR	22	0	1	1	8																			
1987-88	CSKA Moscow	USSR	48	1	8	9	12																			
	Soviet Union	Olympics	6	1	0	1	0																			
1988-89	CSKA Moscow	USSR	22	3	3	6	2																			
1989-90	CSKA Moscow	USSR	48	1	3	4	16																			
1990-91	CSKA Moscow	USSR	41	6	5	11	16																			
1991-92	CSKA Moscow	CIS	30	3	8	11	6																			
	Russia	Olympics	8	3	2	5	6																			
	Chicago	**NHL**	**18**	**1**	**8**	**9**	**4**	**0**	**0**	**1**	**40**	**2.5**	**-3**						**18**	**2**	**6**	**8**	**8**	**1**	**0**	**0**
1992-93	**Chicago**	**NHL**	**38**	**6**	**9**	**15**	**30**	**3**	**0**	**0**	**101**	**5.9**	**11**													
	Edmonton	**NHL**	**17**	**4**	**8**	**12**	**2**	**1**	**0**	**0**	**42**	**9.5**	**-8**													
1993-94	**Edmonton**	**NHL**	**81**	**12**	**38**	**50**	**16**	**5**	**0**	**2**	**197**	**6.1**	**-12**													
1994-95	**Edmonton**	**NHL**	**36**	**7**	**11**	**18**	**29**	**3**	**1**	**0**	**93**	**7.5**	**-15**													
1995-96	**Edmonton**	**NHL**	**26**	**4**	**4**	**8**	**10**	**3**	**0**	**0**	**59**	**6.8**	**-13**													
	St. Louis	**NHL**	**40**	**3**	**12**	**15**	**24**	**0**	**0**	**1**	**114**	**2.6**	**-6**						**10**	**1**	**5**	**6**	**4**	**0**	**0**	**1**
1996-97	**St. Louis**	**NHL**	**82**	**4**	**24**	**28**	**35**	**1**	**0**	**0**	**142**	**2.8**	**7**						**2**	**0**	**0**	**0**	**2**	**0**	**0**	**0**
1997-98	**Ottawa**	**NHL**	**81**	**8**	**27**	**35**	**8**	**3**	**1**	**1**	**191**	**4.2**	**-19**						**11**	**2**	**3**	**5**	**4**	**0**	**0**	**0**
	Russia	Olympics	6	0	2	2	2																			
1998-99	**Ottawa**	**NHL**	**79**	**4**	**21**	**25**	**32**	**3**	**0**	**0**	**171**	**2.3**	**14**	**0**	**0.0**	**89**	**115**	**23:51**	**4**	**0**	**0**	**0**	**0**	**0**	**0**	**0**
99-2000	**Ottawa**	**NHL**	**64**	**6**	**12**	**18**	**20**	**5**	**0**	**1**	**126**	**4.8**	**-5**	**0**	**0.0**	**57**	**78**	**20:41**	**6**	**1**	**1**	**2**	**0**	**0**	**0**	**0**
2000-01	**Ottawa**	**NHL**	**15**	**1**	**5**	**6**	**14**	**0**	**0**	**1**	**13**	**7.7**	**4**	**0**	**0.0**	**11**	**14**	**20:41**								
	Calgary	**NHL**	**37**	**0**	**8**	**8**	**4**	**0**	**0**	**0**	**54**	**0.0**	**-12**	**0**	**0.0**	**23**	**49**	**23:43**								
2001-02	**Calgary**	**NHL**	**78**	**4**	**22**	**26**	**19**	**1**	**0**	**1**	**135**	**3.0**	**3**	**0**	**0.0**	**39**	**73**	**18:32**								
	Russia	Olympics	6	0	2	2	0																			
	NHL Totals		**692**	**64**	**209**	**273**	**247**	**28**	**2**	**8**	**1478**	**4.3**		**0**	**0.0**	**219**	**329**	**21:24**	**51**	**6**	**15**	**21**	**18**	**1**	**0**	**1**

Played in NHL All-Star Game (1999)

Traded to **Edmonton** by **Chicago** with Dean McAmmond for Joe Murphy, February 24, 1993. Traded to **St. Louis** by **Edmonton** with Ken Sutton for Jeff Norton and Donald Dufresne, January 4, 1996. Traded to **Ottawa** by **St. Louis** for Steve Duchesne, August 25, 1997. Claimed on waivers by **Calgary** from **Ottawa**, November 10, 2000.

KRESTANOVICH, Jordan

(KREH-sta-noh-vihtch, JOHR-dan) **COL.**

Left wing. Shoots left. 6'1", 170 lbs. Born, Langley, B.C., June 14, 1981. Colorado's 7th choice, 152nd overall, in 1999 Entry Draft.

Season	Club	League	GP	G	A	Pts	PIM	PP	SH	GW	S	%	+/-	TF	F%	H	SB	Min	GP	G	A	Pts	PIM	PP	SH	GW
1996-97	Surrey Chiefs	BCAHA	55	79	81	160																				
1997-98	Calgary Hitmen	WHL	22	1	0	1	0												13	0	0	0	0			
1998-99	Calgary Hitmen	WHL	62	6	13	19	10												20	3	8	11	4			
99-2000	Calgary Hitmen	WHL	72	19	24	43	22												13	7	7	14	4			
	Hershey Bears	AHL																	1	0	0	0	0			
2000-01	Calgary Hitmen	WHL	70	40	60	100	32												12	8	4	12	8			
	Hershey Bears	AHL																	2	0	0	0	0			
2001-02	**Colorado**	**NHL**	**8**	**0**	**2**	**2**	**0**	**0**	**0**	**0**	**6**	**0.0**	**1**	**0**	**0.0**	**5**	**1**	**8:34**								
	Hershey Bears	AHL	68	12	22	34	18												8	1	0	1	0			
	NHL Totals		**8**	**0**	**2**	**2**	**0**	**0**	**0**	**0**	**6**	**0.0**		**0**	**0.0**	**5**	**1**	**8:34**								

KRIVOKRASOV, Sergei
(krih-vuh-KRA-sahf, SAIR-gay)

Right wing. Shoots left. 5'11", 185 lbs. Born, Angarsk, USSR, April 15, 1974. Chicago's 1st choice, 12th overall, in 1992 Entry Draft.

Season	Club	League	Regular Season																Playoffs							
			GP	G	A	Pts	PIM	PP	SH	GW	S	%	+/-	TF	F%	H	SB	Min	GP	G	A	Pts	PIM	PP	SH	GW
1990-91	CSKA Moscow	USSR	41	4	0	4	8																			
1991-92	CSKA Moscow	CIS	42	10	8	18	35																			
1992-93	**Chicago**	**NHL**	**4**	**0**	**0**	**0**	**2**	**0**	**0**	**0**	**0**	**0.0**	**–2**													
	Indianapolis Ice	IHL	78	36	33	69	157												5	3	1	4	2			
1993-94	**Chicago**	**NHL**	**9**	**1**	**0**	**1**	**4**	**0**	**0**	**0**	**7**	**14.3**	**–2**													
	Indianapolis Ice	IHL	53	19	26	45	145																			
1994-95	Indianapolis Ice	IHL	29	12	15	27	41																			
	Chicago	**NHL**	**41**	**12**	**7**	**19**	**33**	**6**	**0**	**2**	**72**	**16.7**	**9**						**10**	**0**	**0**	**0**	**8**	**0**	**0**	**0**
1995-96	**Chicago**	**NHL**	**46**	**6**	**10**	**16**	**32**	**0**	**0**	**1**	**52**	**11.5**	**10**						**5**	**1**	**0**	**1**	**2**	**0**	**0**	**1**
	Indianapolis Ice	IHL	9	4	5	9	28																			
1996-97	**Chicago**	**NHL**	**67**	**13**	**11**	**24**	**42**	**2**	**0**	**3**	**104**	**12.5**	**–1**						**6**	**1**	**0**	**1**	**4**	**0**	**0**	**0**
1997-98	**Chicago**	**NHL**	**58**	**10**	**13**	**23**	**33**	**1**	**0**	**2**	**127**	**7.9**	**–1**													
	Russia	Olympics	6	0	0	0	4																			
1998-99	**Nashville**	**NHL**	**70**	**25**	**23**	**48**	**42**	**10**	**0**	**6**	**208**	**12.0**	**–5**	**0**	**0.0**	**19**	**7**	**16:08**								
99-2000	**Nashville**	**NHL**	**63**	**9**	**17**	**26**	**40**	**3**	**0**	**2**	**132**	**6.8**	**–7**	**1**	**0.0**	**24**	**5**	**13:08**								
	Calgary	**NHL**	**12**	**1**	**10**	**11**	**4**	**0**	**0**	**0**	**27**	**3.7**	**2**	**0**	**0.0**	**3**	**2**	**13:22**								
2000-01	**Minnesota**	**NHL**	**54**	**7**	**15**	**22**	**20**	**2**	**0**	**1**	**107**	**6.5**	**–1**	**3**	**0.0**	**37**	**13**	**13:05**								
2001-02	**Minnesota**	**NHL**	**9**	**1**	**1**	**2**	**17**	**0**	**0**	**1**	**13**	**7.7**	**–1**	**0**	**0.0**	**6**	**2**	**12:13**								
	Anaheim	**NHL**	**17**	**1**	**2**	**3**	**19**	**0**	**0**	**0**	**38**	**2.6**	**–1**	**5**	**60.0**	**12**	**2**	**12:05**								
	Cincinnati	AHL	15	3	5	8	27												1	0	0	0	2			
	NHL Totals		**450**	**86**	**109**	**195**	**288**	**24**	**0**	**18**	**887**	**9.7**		**9**	**33.3**	**101**	**31**	**13:57**	**21**	**2**	**0**	**2**	**14**	**0**	**0**	**1**

Played in NHL All-Star Game (1999)

Traded to **Nashville** by **Chicago** for future considerations, June 26, 1998. Traded to **Calgary** by **Nashville** for Cale Hulse and Calgary's 3rd round choice (Denis Platonov) in 2001 Entry Draft, March 14, 2000. Selected by **Minnesota** from **Calgary** in Expansion Draft, June 23, 2000. Traded to **Anaheim** by **Minnesota** for Anaheim's 7th round choice (Niklas Eckerblom) in 2002 Entry Draft and a conditional choice in 2003 Entry Draft, November 1, 2001.

KROG, Jason
(KRAWG, JAY-suhn) **ANA.**

Center. Shoots right. 5'11", 191 lbs. Born, Fernie, B.C., October 9, 1975.

Season	Club	League	GP	G	A	Pts	PIM	PP	SH	GW	S	%	+/-	TF	F%	H	SB	Min	GP	G	A	Pts	PIM	PP	SH	GW
1992-93	Chilliwack	BCJHL	52	30	27	57	52																			
1993-94	Chilliwack	BCJHL	42	19	36	55	20																			
1994-95	Chilliwack	BCJHL	60	47	81	128	36																			
1995-96	New Hampshire	H-East	34	4	16	20	20																			
1996-97	New Hampshire	H-East	39	23	*44	*67	28																			
1997-98	New Hampshire	H-East	38	*33	33	66	44																			
1998-99	New Hampshire	H-East	41	*34	*51	*85	38																			
99-2000	**NY Islanders**	**NHL**	**17**	**2**	**4**	**6**	**6**	**1**	**0**	**0**	**22**	**9.1**	**–1**	**81**	**53.1**	**15**	**6**	**10:03**								
	Lowell	AHL	45	6	21	27	22																			
	Providence	AHL	11	9	8	17	4												6	2	2	4	0			
2000-01	**NY Islanders**	**NHL**	**9**	**0**	**3**	**3**	**0**	**0**	**0**	**0**	**7**	**0.0**	**4**	**60**	**48.3**	**6**	**4**	**10:32**								
	Lowell	AHL	26	11	16	27	6																			
	Springfield	AHL	24	7	23	30	4																			
2001-02	**NY Islanders**	**NHL**	**2**	**0**	**0**	**0**	**0**	**0**	**0**	**0**	**0**	**0.0**	**0**	**13**	**46.2**	**0**	**0**	**6:40**								
	Bridgeport	AHL	64	26	36	62	13												20	10	13	23	8			
	NHL Totals		**28**	**2**	**7**	**9**	**6**	**1**	**0**	**0**	**29**	**6.9**		**154**	**50.6**	**21**	**10**	**9:58**								

Hockey East All-Star Team (1997) • NCAA East Second All-American Team (1997) • Hockey East First All-Star Team (1998, 1999) • Hockey East Player of the Year (1999) • NCAA East First All-American Team (1999) • NCAA Championship All-Tournament Team (1999) • Won Hobey Baker Memorial Award (Top U.S. Collegiate Player) (1999)

Signed as a free agent by **NY Islanders**, May 14, 1999. Loaned to **Providence** (AHL) by **NY Islanders**, March 1, 2000. Signed as a free agent by **Anaheim**, July 17, 2002.

KRON, Robert
(KROHN, RAW-buhrt)

Left wing. Shoots left. 5'11", 185 lbs. Born, Brno, Czech., February 27, 1967. Vancouver's 5th choice, 88th overall, in 1985 Entry Draft.

Season	Club	League	GP	G	A	Pts	PIM	PP	SH	GW	S	%	+/-	TF	F%	H	SB	Min	GP	G	A	Pts	PIM	PP	SH	GW
1983-84	Ingstav Brno	Czech-2	3	0	1	1	0																			
1984-85	Zetor Brno	Czech	40	6	8	14	6																			
1985-86	Zetor Brno	Czech	44	5	6	11																				
1986-87	Zetor Brno	Czech	34	18	11	29	10																			
1987-88	Zetor Brno	Czech	44	14	7	21	30																			
1988-89	Dukla Trencin	Czech	43	28	19	47	26																			
1989-90	Dukla Trencin	Czech	39	22	22	44																				
1990-91	**Vancouver**	**NHL**	**76**	**12**	**20**	**32**	**21**	**2**	**3**	**0**	**124**	**9.7**	**–11**													
1991-92	**Vancouver**	**NHL**	**36**	**2**	**2**	**4**	**2**	**0**	**0**	**0**	**49**	**4.1**	**–9**						**11**	**1**	**2**	**3**	**2**	**0**	**1**	**0**
1992-93	**Vancouver**	**NHL**	**32**	**10**	**11**	**21**	**14**	**0**	**2**	**2**	**60**	**16.7**	**10**													
	Hartford	**NHL**	**13**	**4**	**2**	**6**	**4**	**2**	**0**	**0**	**37**	**10.8**	**–5**													
1993-94	**Hartford**	**NHL**	**77**	**24**	**26**	**50**	**8**	**2**	**1**	**3**	**194**	**12.4**	**0**													
1994-95	**Hartford**	**NHL**	**37**	**10**	**8**	**18**	**10**	**3**	**1**	**1**	**88**	**11.4**	**–3**													
1995-96	**Hartford**	**NHL**	**77**	**22**	**28**	**50**	**6**	**8**	**1**	**3**	**203**	**10.8**	**–1**													
1996-97	**Hartford**	**NHL**	**68**	**10**	**12**	**22**	**10**	**2**	**0**	**4**	**182**	**5.5**	**–18**													
1997-98	**Carolina**	**NHL**	**81**	**16**	**20**	**36**	**12**	**4**	**0**	**2**	**175**	**9.1**	**–8**													
1998-99	**Carolina**	**NHL**	**75**	**9**	**16**	**25**	**10**	**3**	**1**	**2**	**134**	**6.7**	**–13**	**244**	**38.9**	**88**	**27**	**16:14**	**5**	**2**	**0**	**2**	**0**	**0**	**0**	**1**
99-2000	**Carolina**	**NHL**	**81**	**13**	**27**	**40**	**8**	**2**	**1**	**3**	**134**	**9.7**	**–4**	**717**	**43.0**	**48**	**21**	**15:14**								
2000-01	**Columbus**	**NHL**	**59**	**8**	**11**	**19**	**10**	**0**	**1**	**1**	**134**	**6.0**	**4**	**264**	**43.9**	**34**	**22**	**17:08**								
2001-02	**Columbus**	**NHL**	**59**	**4**	**11**	**15**	**4**	**1**	**0**	**0**	**92**	**4.3**	**–14**	**174**	**40.2**	**21**	**19**	**15:59**								
	Syracuse Crunch	AHL	6	2	4	6	5																			
	NHL Totals		**771**	**144**	**194**	**338**	**119**	**31**	**11**	**21**	**1606**	**9.0**		**1399**	**42.1**	**191**	**89**	**16:04**	**16**	**3**	**2**	**5**	**2**	**0**	**1**	**1**

Traded to **Hartford** by **Vancouver** with Vancouver's 3rd round choice (Marek Malik) in 1993 Entry Draft and future considerations (Jim Sandlak, May 17, 1993) for Murray Craven and Vancouver's 5th round choice (previously acquired, Vancouver selected Scott Walker) in 1993 Entry Draft, March 22, 1993. Transferred to **Carolina** after **Hartford** franchise relocated, June 25, 1997. Selected by **Columbus** from **Carolina** in Expansion Draft, June 23, 2000.

KRUPP, Uwe
(KROOP, OO-VAY) **ATL.**

Defense. Shoots right. 6'6", 235 lbs. Born, Cologne, West Germany, June 24, 1965. Buffalo's 13th choice, 223rd overall, in 1983 Entry Draft.

Season	Club	League	GP	G	A	Pts	PIM	PP	SH	GW	S	%	+/-	TF	F%	H	SB	Min	GP	G	A	Pts	PIM	PP	SH	GW
1982-83	Kolner EC	Germany	11	0	0	0	0																			
1983-84	Kolner EC	Germany	26	0	4	4	22																			
1984-85	Kolner EC	Germany	31	7	7	14													9	4	1	5				
1985-86	Kolner EC	Germany	35	6	18	24	83												10	4	3	7				
1986-87	**Buffalo**	**NHL**	**26**	**1**	**4**	**5**	**23**	**0**	**0**	**0**	**34**	**2.9**	**–9**													
	Rochester	AHL	42	3	19	22	50												17	1	11	12	16			
1987-88	**Buffalo**	**NHL**	**75**	**2**	**9**	**11**	**151**	**0**	**0**	**0**	**84**	**2.4**	**–1**						**6**	**0**	**0**	**0**	**15**	**0**	**0**	**0**
1988-89	**Buffalo**	**NHL**	**70**	**5**	**13**	**18**	**55**	**0**	**1**	**0**	**51**	**9.8**	**0**						**5**	**0**	**1**	**1**	**4**	**0**	**0**	**0**
1989-90	**Buffalo**	**NHL**	**74**	**3**	**20**	**23**	**85**	**0**	**1**	**1**	**69**	**4.3**	**15**						**6**	**0**	**0**	**0**	**4**	**0**	**0**	**0**
1990-91	**Buffalo**	**NHL**	**74**	**12**	**32**	**44**	**66**	**6**	**0**	**0**	**138**	**8.7**	**14**						**6**	**1**	**1**	**2**	**6**	**1**	**0**	**0**
1991-92	**Buffalo**	**NHL**	**8**	**2**	**0**	**2**	**6**	**0**	**0**	**0**	**13**	**15.4**	**0**													
	NY Islanders	**NHL**	**59**	**6**	**29**	**35**	**43**	**2**	**0**	**0**	**115**	**5.2**	**13**													
1992-93	**NY Islanders**	**NHL**	**80**	**9**	**29**	**38**	**67**	**2**	**0**	**2**	**116**	**7.8**	**6**						**18**	**1**	**5**	**6**	**12**	**0**	**0**	**0**
1993-94	**NY Islanders**	**NHL**	**41**	**7**	**14**	**21**	**30**	**3**	**0**	**0**	**82**	**8.5**	**11**						**4**	**0**	**1**	**1**	**4**	**0**	**0**	**0**
1994-95	EV Landshut	Germany	5	1	2	3	6																			
	Quebec	**NHL**	**44**	**6**	**17**	**23**	**20**	**3**	**0**	**1**	**102**	**5.9**	**14**						**5**	**0**	**2**	**2**	**2**	**0**	**0**	**0**
1995-96♦	**Colorado**	**NHL**	**6**	**0**	**3**	**3**	**4**	**0**	**0**	**0**	**9**	**0.0**	**4**						**22**	**4**	**12**	**16**	**33**	**1**	**0**	**2**
1996-97	**Colorado**	**NHL**	**60**	**4**	**17**	**21**	**48**	**2**	**0**	**1**	**107**	**3.7**	**12**													
1997-98	**Colorado**	**NHL**	**78**	**9**	**22**	**31**	**38**	**5**	**0**	**2**	**149**	**6.0**	**21**						**7**	**0**	**1**	**1**	**4**	**0**	**0**	**0**
	Germany	Olympics	2	0	2	2	4																			
1998-99	**Detroit**	**NHL**	**22**	**3**	**2**	**5**	**6**	**0**	**0**	**0**	**32**	**9.4**	**0**	**0**	**0.0**	**41**	**37**	**21:23**								

			Regular Season																Playoffs							
Season	Club	League	GP	G	A	Pts	PIM	PP	SH	GW	S	%	+/-	TF	F%	H	SB	Min	GP	G	A	Pts	PIM	PP	SH	GW
99-2000	**Detroit**	**NHL**	DID NOT PLAY – INJURED																							
2000-01	**Detroit**	**NHL**	DID NOT PLAY – INJURED																							
2001-02	**Detroit**	**NHL**	**8**	**0**	**1**	**1**	**8**	**0**	**0**	**0**	**9**	**0.0**	**–1**	**0**	**0.0**	**16**	**14**	**17:03**	**2**	**0**	**0**	**0**	**2**	**0**	**0**	**0**
	NHL Totals		**725**	**69**	**212**	**281**	**650**	**23**	**2**	**7**	**1110**	**6.2**		**0**	**0.0**	**57**	**51**	**20:14**	**81**	**6**	**23**	**29**	**86**	**2**	**0**	**2**

Played in NHL All-Star Game (1991)

Traded to **NY Islanders** by **Buffalo** with Pierre Turgeon, Benoit Hogue and Dave McLlwain for Pat LaFontaine, Randy Hillier, Randy Wood and NY Islanders' 4th round choice (Dean Melanson) in 1992 Entry Draft, October 25, 1991. Traded to **Quebec** by **NY Islanders** with NY Islanders' 1st round choice (Wade Belak) in 1994 Entry Draft for Ron Sutter and Quebec's 1st round choice (Brett Lindros) in 1994 Entry Draft, June 28, 1994. Transferred to **Colorado** after **Quebec** franchise relocated, June 21, 1995. Claimed by **Nashville** from **Colorado** in Expansion Draft, June 26, 1998. Signed as a free agent by **Detroit**, July 7, 1998. • Missed remainder of 1998-99 season and entire 1999-2000 and 2000-01 seasons recovering from back injury suffered prior to game vs. Phoenix, December 19, 1998. Signed as a free agent by **Atlanta**, July 19, 2002.

KUBA, Filip
(KOO-bah, FIHL-ihp) **MIN.**

Defense. Shoots left. 6'3", 205 lbs. Born, Ostrava, Czech., December 29, 1976. Florida's 8th choice, 192nd overall, in 1995 Entry Draft.

Season	Club	League	GP	G	A	Pts	PIM	PP	SH	GW	S	%	+/-	TF	F%	H	SB	Min	GP	G	A	Pts	PIM	PP	SH	GW
1994-95	HC Vitkovice Jr.	Czech-Jr.	35	10	15	25	....	....	....	....	....	....	....	....	....	....	....	....	....	....	....	....	....	....	....	....
	HC Vitkovice	Czech	....	....	....	....	....	....	....	....	....	....	....	....	....	....	....	....	4	0	0	0	2	....	....	....
1995-96	HC Vitkovice	Czech	19	0	1	1	....	....	....	....	....	....	....	....	....	....	....	....	....	....	....	....	....	....	....	....
1996-97	Carolina	AHL	51	0	12	12	38	....	....	....	....	....	....	....	....	....	....	....	....	....	....	....	....	....	....	....
1997-98	New Haven	AHL	77	4	13	17	58	....	....	....	....	....	....	....	....	....	....	....	3	1	1	2	0	....	....	....
1998-99	**Florida**	**NHL**	**5**	**0**	**1**	**1**	**0**	**0**	**0**	**0**	**5**	**0.0**	**2**	**0**	**0.0**	**7**	**6**	**22:29**	....	....	....	....	....	....	....	....
	Kentucky	AHL	45	2	8	10	33	....	....	....	....	....	....	....	....	....	....	....	10	0	1	1	4	....	....	....
99-2000	**Florida**	**NHL**	**13**	**1**	**5**	**6**	**2**	**1**	**0**	**1**	**16**	**6.3**	**–3**	**0**	**0.0**	**10**	**13**	**13:52**	....	....	....	....	....	....	....	....
	Houston Aeros	IHL	27	3	6	9	13	....	....	....	....	....	....	....	....	....	....	....	11	1	2	3	4	....	....	....
2000-01	**Minnesota**	**NHL**	**75**	**9**	**21**	**30**	**28**	**4**	**0**	**4**	**141**	**6.4**	**–6**	**1**	**0.0**	**81**	**139**	**24:16**	....	....	....	....	....	....	....	....
2001-02	**Minnesota**	**NHL**	**62**	**5**	**19**	**24**	**32**	**3**	**0**	**1**	**101**	**5.0**	**–6**	**0**	**0.0**	**52**	**94**	**25:30**	....	....	....	....	....	....	....	....
	NHL Totals		**155**	**15**	**46**	**61**	**62**	**8**	**0**	**6**	**263**	**5.7**		**1**	**0.0**	**150**	**252**	**23:50**	....	....	....	....	....	....	....	....

Traded to **Calgary** by **Florida** for Rocky Thompson, March 16, 2000. Selected by **Minnesota** from **Calgary** in Expansion Draft, June 23, 2000.

KUBINA, Pavel
(koo-BEE-nuh, PAH-vehl) **T.B.**

Defense. Shoots right. 6'4", 230 lbs. Born, Celadna, Czech., April 15, 1977. Tampa Bay's 6th choice, 179th overall, in 1996 Entry Draft.

Season	Club	League	GP	G	A	Pts	PIM	PP	SH	GW	S	%	+/-	TF	F%	H	SB	Min	GP	G	A	Pts	PIM	PP	SH	GW
1993-94	HC Vitkovice Jr.	Czech-Jr.	35	4	3	7	....	....	....	....	....	....	....	....	....	....	....	....	....	....	....	....	....	....	....	....
	HC Vitkovice	Czech	1	0	0	0	....	....	....	....	....	....	....	....	....	....	....	....	....	....	....	....	....	....	....	....
1994-95	HC Vitkovice Jr.	Czech-Jr.	20	6	10	16	....	....	....	....	....	....	....	....	....	....	....	....	....	....	....	....	....	....	....	....
	HC Vitkovice	Czech	8	2	0	2	10	....	....	....	....	....	....	....	....	....	....	....	4	0	0	0	0	....	....	....
1995-96	HC Vitkovice Jr.	Czech-Jr.	16	5	10	15	....	....	....	....	....	....	....	....	....	....	....	....	....	....	....	....	....	....	....	....
	HC Vitkovice	Czech	33	3	4	7	32	....	....	....	....	....	....	....	....	....	....	....	4	0	0	0	0	....	....	....
1996-97	HC Vitkovice	Czech	1	0	0	0	0	....	....	....	....	....	....	....	....	....	....	....	....	....	....	....	....	....	....	....
	Moose Jaw	WHL	61	12	32	44	116	....	....	....	....	....	....	....	....	....	....	....	11	2	5	7	27	....	....	....
1997-98	**Tampa Bay**	**NHL**	**10**	**1**	**2**	**3**	**22**	**0**	**0**	**0**	**8**	**12.5**	**–1**	....	....	....	....	....	....	....	....	....	....	....	....	....
	Adirondack	AHL	55	4	8	12	86	....	....	....	....	....	....	....	....	....	....	....	1	1	0	1	14	....	....	....
1998-99	**Tampa Bay**	**NHL**	**68**	**9**	**12**	**21**	**80**	**3**	**1**	**1**	**119**	**7.6**	**–33**	**2**	**0.0**	**156**	**82**	**22:47**	....	....	....	....	....	....	....	....
	Cleveland	IHL	6	2	2	4	16	....	....	....	....	....	....	....	....	....	....	....	....	....	....	....	....	....	....	....
99-2000	**Tampa Bay**	**NHL**	**69**	**8**	**18**	**26**	**93**	**6**	**0**	**3**	**128**	**6.3**	**–19**	**0**	**0.0**	**121**	**78**	**22:32**	....	....	....	....	....	....	....	....
2000-01	**Tampa Bay**	**NHL**	**70**	**11**	**19**	**30**	**103**	**6**	**1**	**1**	**128**	**8.6**	**–14**	**2**	**0.0**	**105**	**131**	**24:06**	....	....	....	....	....	....	....	....
2001-02	**Tampa Bay**	**NHL**	**82**	**11**	**23**	**34**	**106**	**5**	**2**	**3**	**189**	**5.8**	**–22**	**1**	**100.0**	**94**	**125**	**23:39**	....	....	....	....	....	....	....	....
	Czech Republic	Olympics	4	0	1	1	0	....	....	....	....	....	....	....	....	....	....	....	....	....	....	....	....	....	....	....
	NHL Totals		**299**	**40**	**74**	**114**	**404**	**20**	**4**	**8**	**572**	**7.0**		**5**	**20.0**	**476**	**416**	**23:17**	....	....	....	....	....	....	....	....

KUCERA, Frantisek
(koo-CHAIR-uh, FRAN-tih-sehk)

Defense. Shoots right. 6'2", 205 lbs. Born, Prague, Czech., February 3, 1968. Chicago's 3rd choice, 77th overall, in 1986 Entry Draft.

Season	Club	League	GP	G	A	Pts	PIM	PP	SH	GW	S	%	+/-	TF	F%	H	SB	Min	GP	G	A	Pts	PIM	PP	SH	GW
1985-86	Sparta CKD Praha	Czech	15	0	0	0	....	....	....	....	....	....	....	....	....	....	....	....	....	....	....	....	....	....	....	....
1986-87	Sparta CKD Praha	Czech	40	5	2	7	14	....	....	....	....	....	....	....	....	....	....	....	....	....	....	....	....	....	....	....
1987-88	Sparta CKD Praha	Czech	46	7	2	9	30	....	....	....	....	....	....	....	....	....	....	....	....	....	....	....	....	....	....	....
1988-89	Dukla Jihlava	Czech	45	10	9	19	28	....	....	....	....	....	....	....	....	....	....	....	....	....	....	....	....	....	....	....
1989-90	Dukla Jihlava	Czech	42	8	10	18	....	....	....	....	....	....	....	....	....	....	....	....	1	1	0	1	....	....	....	....
1990-91	**Chicago**	**NHL**	**40**	**2**	**12**	**14**	**32**	**1**	**0**	**0**	**65**	**3.1**	**3**	....	....	....	....	....	....	....	....	....	....	....	....	....
	Indianapolis Ice	IHL	35	8	19	27	23	....	....	....	....	....	....	....	....	....	....	....	7	0	1	1	15	....	....	....
1991-92	**Chicago**	**NHL**	**61**	**3**	**10**	**13**	**36**	**1**	**0**	**1**	**82**	**3.7**	**3**	....	....	....	....	....	**6**	**0**	**0**	**0**	**0**	**0**	**0**	**0**
	Indianapolis Ice	IHL	7	1	2	3	4	....	....	....	....	....	....	....	....	....	....	....	....	....	....	....	....	....	....	....
1992-93	**Chicago**	**NHL**	**71**	**5**	**14**	**19**	**59**	**1**	**0**	**1**	**117**	**4.3**	**7**	....	....	....	....	....	....	....	....	....	....	....	....	....
1993-94	**Chicago**	**NHL**	**60**	**4**	**13**	**17**	**34**	**2**	**0**	**0**	**90**	**4.4**	**9**	....	....	....	....	....	....	....	....	....	....	....	....	....
	Hartford	**NHL**	**16**	**1**	**3**	**4**	**14**	**1**	**0**	**0**	**32**	**3.1**	**–12**	....	....	....	....	....	....	....	....	....	....	....	....	....
1994-95	HC Sparta Praha	Czech	16	1	2	3	14	....	....	....	....	....	....	....	....	....	....	....	....	....	....	....	....	....	....	....
	Hartford	**NHL**	**48**	**3**	**17**	**20**	**30**	**0**	**0**	**1**	**73**	**4.1**	**3**	....	....	....	....	....	....	....	....	....	....	....	....	....
1995-96	**Hartford**	**NHL**	**30**	**2**	**6**	**8**	**10**	**0**	**0**	**1**	**43**	**4.7**	**–3**	....	....	....	....	....	....	....	....	....	....	....	....	....
	Vancouver	**NHL**	**24**	**1**	**0**	**1**	**10**	**0**	**0**	**0**	**34**	**2.9**	**5**	....	....	....	....	....	**6**	**0**	**1**	**1**	**0**	**0**	**0**	**0**
1996-97	**Vancouver**	**NHL**	**2**	**0**	**0**	**0**	**0**	**0**	**0**	**0**	**3**	**0.0**	**0**	....	....	....	....	....	....	....	....	....	....	....	....	....
	Syracuse Crunch	AHL	42	6	29	35	36	....	....	....	....	....	....	....	....	....	....	....	....	....	....	....	....	....	....	....
	Houston Aeros	IHL	12	0	3	3	20	....	....	....	....	....	....	....	....	....	....	....	....	....	....	....	....	....	....	....
	Philadelphia	**NHL**	**2**	**0**	**0**	**0**	**2**	**0**	**0**	**0**	**2**	**0.0**	**–2**	....	....	....	....	....	....	....	....	....	....	....	....	....
	Philadelphia	AHL	9	1	5	6	2	....	....	....	....	....	....	....	....	....	....	....	10	1	6	7	20	....	....	....
1997-98	HC Sparta Praha	Czech	43	8	12	20	49	....	....	....	....	....	....	....	....	....	....	....	9	3	1	4	*53	....	....	....
	HC Sparta Praha	EuroHL	4	0	1	1	2	....	....	....	....	....	....	....	....	....	....	....	....	....	....	....	....	....	....	....
	Czech Republic	Olympics	6	0	0	0	0	....	....	....	....	....	....	....	....	....	....	....	....	....	....	....	....	....	....	....
1998-99	HC Sparta Praha	Czech	42	3	12	15	92	....	....	....	....	....	....	....	....	....	....	....	8	0	2	2	....	....	....	....
	HC Sparta Praha	EuroHL	6	0	2	2	10	....	....	....	....	....	....	....	....	....	....	....	2	0	0	0	2	....	....	....
99-2000	HC Sparta Praha	Czech	51	7	26	33	40	....	....	....	....	....	....	....	....	....	....	....	9	1	9	10	4	....	....	....
	HC Sparta Praha	EuroHL	6	0	2	2	4	....	....	....	....	....	....	....	....	....	....	....	4	0	1	1	2	....	....	....
2000-01	**Columbus**	**NHL**	**48**	**2**	**5**	**7**	**12**	**0**	**0**	**0**	**51**	**3.9**	**–5**	**0**	**0.0**	**43**	**69**	**17:42**	....	....	....	....	....	....	....	....
	Pittsburgh	**NHL**	**7**	**0**	**2**	**2**	**0**	**0**	**0**	**0**	**9**	**0.0**	**–2**	**0**	**0.0**	**10**	**10**	**16:28**	....	....	....	....	....	....	....	....
2001-02	HC Sparta Praha	Czech	10	2	2	4	2	....	....	....	....	....	....	....	....	....	....	....	....	....	....	....	....	....	....	....
	Washington	**NHL**	**56**	**1**	**13**	**14**	**12**	**0**	**0**	**0**	**67**	**1.5**	**7**	**0**	**0.0**	**54**	**91**	**20:39**	....	....	....	....	....	....	....	....
	NHL Totals		**465**	**24**	**95**	**119**	**251**	**6**	**0**	**4**	**668**	**3.6**		**0**	**0.0**	**107**	**170**	**19:07**	**12**	**0**	**1**	**1**	**0**	**0**	**0**	**0**

Traded to **Hartford** by **Chicago** with Jocelyn Lemieux for Gary Suter, Randy Cunneyworth and Hartford's 3rd round choice (later traded to Vancouver - Vancouver selected Larry Courville) in 1995 Entry Draft, March 11, 1994. Traded to **Vancouver** by **Hartford** with Jim Dowd and Hartford's 2nd round choice (Ryan Bonni) in 1997 Entry Draft for Jeff Brown and Vancouver's 3rd round choice (later traded to Calgary - Calgary selected Paul Manning) in 1998 Entry Draft, December 19, 1995. Traded to **Philadelphia** by **Vancouver** for future considerations, March 18, 1997. Signed as a free agent by **Columbus**, July 7, 2000. Traded to **Pittsburgh** by **Columbus** for Pittsburgh's 6th round choice (Columbus selected Scott Horvath) in 2001 Entry Draft, March 13, 2001. Traded to **Washington** by **Pittsburgh** with Jaromir Jagr for Kris Beech, Michal Sivek, Ross Lupaschuk and future considerations, July 11, 2001.

KUDROC, Kristian
(KOO-drawch, KRIHS-tan) **T.B.**

Defense. Shoots right. 6'6", 255 lbs. Born, Michalovce, Czech., May 21, 1981. NY Islanders' 4th choice, 28th overall, in 1999 Entry Draft.

Season	Club	League	GP	G	A	Pts	PIM	PP	SH	GW	S	%	+/-	TF	F%	H	SB	Min	GP	G	A	Pts	PIM	PP	SH	GW
1997-98	Michalovce Jr.	Slovak-Jr.	47	7	4	11	66	....	....	....	....	....	....	....	....	....	....	....	....	....	....	....	....	....	....	....
	Michalovce	Slovak-2	4	0	0	0	0	....	....	....	....	....	....	....	....	....	....	....	....	....	....	....	....	....	....	....
1998-99	Michalovce	Slovak-2	17	0	3	3	12	....	....	....	....	....	....	....	....	....	....	....	....	....	....	....	....	....	....	....
99-2000	Quebec Remparts	QMJHL	57	9	22	31	172	....	....	....	....	....	....	....	....	....	....	....	11	2	5	7	29	....	....	....
2000-01	**Tampa Bay**	**NHL**	**22**	**2**	**2**	**4**	**36**	**0**	**0**	**1**	**12**	**16.7**	**0**	**0**	**0.0**	**44**	**11**	**9:09**	....	....	....	....	....	....	....	....
	Detroit Vipers	IHL	44	4	3	7	80	....	....	....	....	....	....	....	....	....	....	....	....	....	....	....	....	....	....	....
2001-02	**Tampa Bay**	**NHL**	**2**	**0**	**0**	**0**	**0**	**0**	**0**	**0**	**0**	**0.0**	**0**	**0**	**0.0**	**2**	**1**	**7:49**	....	....	....	....	....	....	....	....
	Springfield	AHL	55	0	8	8	126	....	....	....	....	....	....	....	....	....	....	....	....	....	....	....	....	....	....	....
	Philadelphia	AHL	10	0	3	3	14	....	....	....	....	....	....	....	....	....	....	....	5	1	1	2	21	....	....	....
	NHL Totals		**24**	**2**	**2**	**4**	**36**	**0**	**0**	**1**	**12**	**16.7**		**0**	**0.0**	**46**	**12**	**9:02**	....	....	....	....	....	....	....	....

Traded to **Tampa Bay** by **NY Islanders** with Kevin Weekes and NY Islanders' 2nd round choice (later traded to Phoenix - Phoenix selected Matthew Spiller) in 2001 Entry Draft for Tampa Bay's 1st round choice (Raffi Torres) in 2000 Entry Draft, Calgary's 4th round choice (previously acquired, NY Islanders selected Vladimir Gorbunov) in 2000 Entry Draft and NY Islanders' 7th round choice (previously acquired, NY Islanders selected Ryan Caldwell) in 2000 Entry Draft, June 24, 2000.

KULTANEN, Jarno
(kuhl-TAH-nuhn, YAR-noh) **BOS.**

Defense. Shoots left. 6'2", 198 lbs. Born, Luumaki, Finland, January 8, 1973. Boston's 8th choice, 174th overall, in 2000 Entry Draft.

			Regular Season																Playoffs							
Season	Club	League	GP	G	A	Pts	PIM	PP	SH	GW	S	%	+/-	TF	F%	H	SB	Min	GP	G	A	Pts	PIM	PP	SH	GW
1991-92	KooKoo Jr.	Finn-Jr.	22	7	17	24	28																			
	KooKoo Kouvola	Finland-2	1	0	0	0	0																			
1992-93	KooKoo Jr.	Finn-Jr.	11	6	5	11	8																			
	KooKoo Kouvola	Finland-2	27	1	1	2	33																			
	Centers	Finland-2	1	1	0	1	0																			
1993-94	KooKoo Jr.	Finn-Jr.	3	1	0	1	2																			
	KooKoo Kouvola	Finland-2	45	7	10	17	42																			
1994-95	KalPa Kuopio	Finland	47	5	12	17	26												3	0	0	0	8			
1995-96	KalPa Kuopio	Finland	49	4	10	14	42																			
1996-97	HIFK Helsinki	Finland	24	1	3	4	6																			
1997-98	HIFK Helsinki	Finland	25	0	1	1	20												8	0	1	1	2			
1998-99	HIFK Helsinki	Finland	51	6	6	12	53												10	1	0	1	27			
	HIFK Helsinki	EuroHL	5	1	0	1	4												4	0	1	1	2			
99-2000	HIFK Helsinki	Finland	46	6	8	14	51												9	0	0	0	6			
	HIFK Helsinki	EuroHL	5	2	2	4	31												2	0	0	0	2			
2000-01	**Boston**	**NHL**	**62**	**2**	**8**	**10**	**26**	**0**	**0**	**1**	**76**	**2.6**	**-3**	**0**	**0.0**	**87**	**53**	**19:29**								
2001-02	**Boston**	**NHL**	**38**	**0**	**3**	**3**	**33**	**0**	**0**	**0**	**31**	**0.0**	**-1**	**0**	**0.0**	**31**	**19**	**12:58**								
	NHL Totals		**100**	**2**	**11**	**13**	**59**	**0**	**0**	**1**	**107**	**1.9**		**0**	**0.0**	**118**	**72**	**17:00**								

• Missed majority of 2001-02 season recovering from knee injury originally suffered in game vs. Minnesota, October 8, 2001.

KURTZ, Justin
(KUHRTZ, JUHS-tihn) **VAN.**

Defense. Shoots left. 6', 188 lbs. Born, Winnipeg, Manitoba, January 14, 1977. Winnipeg's 5th choice, 84th overall, in 1995 Entry Draft.

			Regular Season																Playoffs							
Season	Club	League	GP	G	A	Pts	PIM	PP	SH	GW	S	%	+/-	TF	F%	H	SB	Min	GP	G	A	Pts	PIM	PP	SH	GW
1993-94	Brandon	WHL	63	3	13	16	37												14	1	3	4	24			
1994-95	Brandon	WHL	65	8	34	42	75												18	2	2	4	26			
1995-96	Brandon	WHL	53	19	55	74	107												18	4	6	10	37			
1996-97	Brandon	WHL	59	24	44	68	85												6	3	2	5	4			
1997-98	Las Vegas	IHL	63	11	11	22	62																			
	Saint John	AHL	4	0	0	0	6												6	0	0	0	0			
1998-99	Louisiana	ECHL	6	1	4	5	6																			
	Manitoba Moose	IHL	38	4	6	10	40												5	0	0	0	4			
99-2000	Manitoba Moose	IHL	66	3	15	18	94												2	0	0	0	0			
2000-01	Manitoba Moose	IHL	69	8	22	30	62												13	2	3	5	12			
2001-02	**Vancouver**	**NHL**	**27**	**3**	**5**	**8**	**14**	**2**	**0**	**0**	**30**	**10.0**	**-4**	**0**	**0.0**	**30**	**17**	**15:58**								
	Manitoba Moose	AHL	52	10	29	39	64												7	0	4	4	4			
	NHL Totals		**27**	**3**	**5**	**8**	**14**	**2**	**0**	**0**	**30**	**10.0**		**0**	**0.0**	**30**	**17**	**15:58**								

WHL East Second All-Star Team (1995-96)

Signed as a free agent by **Vancouver**, October 15, 2001.

KUTLAK, Zdenek
(KUHT-lak, zuh-DEHN-ehk) **BOS.**

Defense. Shoots left. 6'3", 207 lbs. Born, Budejovice, Czech., February 13, 1980. Boston's 10th choice, 237th overall, in 2000 Entry Draft.

			Regular Season																Playoffs							
Season	Club	League	GP	G	A	Pts	PIM	PP	SH	GW	S	%	+/-	TF	F%	H	SB	Min	GP	G	A	Pts	PIM	PP	SH	GW
1996-97	C. Budejovice Jr.	Czech-Jr.	45	8	11	19	20																			
1997-98	C. Budejovice Jr.	Czech-Jr.	43	1	6	7	30																			
1998-99	C. Budejovice Jr.	Czech-Jr.	31	6	14	20	20																			
	Ceske Budejovice	Czech	22	1	3	4	4												3	0	0	0	0			
99-2000	C. Budejovice Jr.	Czech-Jr.	8	4	2	6	26																			
	JindrichuvHradec	Czech-3	4	1	1	2	0																			
	IHC Pisek	Czech-2	3	1	0	1	0												2	0	0	0	2			
	Ceske Budejovice	Czech	28	1	0	1	2												1	0	0	0	0			
2000-01	**Boston**	**NHL**	**10**	**0**	**2**	**2**	**4**	**0**	**0**	**0**	**7**	**0.0**	**-3**	**0**	**0.0**	**9**	**4**	**16:05**								
	Providence	AHL	62	4	5	9	16																			
2001-02	Providence	AHL	80	5	15	20	73												2	0	0	0	0			
	NHL Totals		**10**	**0**	**2**	**2**	**4**	**0**	**0**	**0**	**7**	**0.0**		**0**	**0.0**	**9**	**4**	**16:05**								

KUZNETSOV, Maxim
(kooz-NEHT-zahv, MAX-ihm) **DET.**

Defense. Shoots left. 6'5", 198 lbs. Born, Pavlodar, USSR, March 24, 1977. Detroit's 1st choice, 26th overall, in 1995 Entry Draft.

			Regular Season																Playoffs							
Season	Club	League	GP	G	A	Pts	PIM	PP	SH	GW	S	%	+/-	TF	F%	H	SB	Min	GP	G	A	Pts	PIM	PP	SH	GW
1994-95	Dynamo Moscow	CIS	11	0	0	0	8																			
1995-96	Dynamo Moscow	CIS	9	1	1	2	22												4	0	0	0	0			
1996-97	Dynamo Moscow	Russia	23	0	2	2	16																			
	Adirondack	AHL	2	0	1	1	6												2	0	0	0	0			
1997-98	Adirondack	AHL	51	5	5	10	43												3	0	1	1	4			
1998-99	Adirondack	AHL	60	0	4	4	30												3	0	0	0	0			
99-2000	Cincinnati	AHL	47	2	9	11	36																			
2000-01	**Detroit**	**NHL**	**25**	**1**	**2**	**3**	**23**	**0**	**0**	**0**	**17**	**5.9**	**-1**	**1**	**0.0**	**54**	**7**	**9:28**								
2001-02	**Detroit**	**NHL**	**39**	**1**	**2**	**3**	**40**	**0**	**0**	**0**	**27**	**3.7**	**0**	**0**	**0.0**	**71**	**19**	**11:51**								
	Cincinnati	AHL	7	1	0	1	4																			
	NHL Totals		**64**	**2**	**4**	**6**	**63**	**0**	**0**	**0**	**44**	**4.5**		**1**	**0.0**	**125**	**26**	**10:55**								

• Missed majority of 2000-01 season recovering from knee injury originally suffered in game vs. Vancouver, November 24, 2000.

KUZNIK, Greg
(kooz-NIHK, GREHG)

Defense. Shoots left. 6', 185 lbs. Born, Prince George, B.C., June 12, 1978. Hartford's 7th choice, 171st overall, in 1996 Entry Draft.

			Regular Season																Playoffs							
Season	Club	League	GP	G	A	Pts	PIM	PP	SH	GW	S	%	+/-	TF	F%	H	SB	Min	GP	G	A	Pts	PIM	PP	SH	GW
1994-95	Royal City	BCJHL	45	2	13	15	83																			
1995-96	Seattle	WHL	70	2	13	15	149												5	0	0	0	6			
1996-97	Seattle	WHL	70	4	9	13	161												14	0	2	2	26			
1997-98	Seattle	WHL	72	5	12	17	197												5	0	0	0	4			
1998-99	New Haven	AHL	27	1	0	1	33																			
	Florida	ECHL	50	6	8	14	110												5	1	0	1	0			
99-2000	Cincinnati	IHL	46	0	3	3	53												4	0	0	0	4			
	Dayton Bombers	ECHL	7	1	0	1	16																			
	Florida	ECHL	9	1	4	5	6												3	0	0	0	2			
2000-01	**Carolina**	**NHL**	**1**	**0**	**0**	**0**	**0**	**0**	**0**	**0**	**0**	**0.0**	**0**	**0**	**0.0**	**0**	**1**	**7:32**								
	Cincinnati	IHL	73	0	7	7	72												5	0	1	1	4			
2001-02	Lowell	AHL	58	3	8	11	40												5	0	0	0	8			
	NHL Totals		**1**	**0**	**0**	**0**	**0**	**0**	**0**	**0**	**0**	**0.0**		**0**	**0.0**	**0**	**1**	**7:32**								

Transferred to **Carolina** after **Hartford** franchise relocated, June 25, 1997.

KVASHA, Oleg
(kuh-VAH-shah, OH-lehg) **NYI**

Center/Left wing. Shoots right. 6'5", 230 lbs. Born, Moscow, USSR, July 26, 1978. Florida's 3rd choice, 65th overall, in 1996 Entry Draft.

			Regular Season																Playoffs							
Season	Club	League	GP	G	A	Pts	PIM	PP	SH	GW	S	%	+/-	TF	F%	H	SB	Min	GP	G	A	Pts	PIM	PP	SH	GW
1995-96	CSKA Moscow	CIS	38	2	3	5	14												2	0	0	0	0			
1996-97	H.C. CSKA	Russia	44	20	22	42	115																			
1997-98	New Haven	AHL	57	13	16	29	46												3	2	1	3	0			
1998-99	**Florida**	**NHL**	**68**	**12**	**13**	**25**	**45**	**4**	**0**	**2**	**138**	**8.7**	**5**	**373**	**28.4**	**21**	**13**	**12:48**								
99-2000	**Florida**	**NHL**	**78**	**5**	**20**	**25**	**34**	**2**	**0**	**0**	**110**	**4.5**	**3**	**553**	**34.9**	**22**	**22**	**11:24**	**4**	**0**	**0**	**0**	**0**	**0**	**0**	**0**
2000-01	**NY Islanders**	**NHL**	**62**	**11**	**9**	**20**	**46**	**0**	**0**	**0**	**118**	**9.3**	**-15**	**627**	**43.7**	**33**	**22**	**14:56**								
2001-02	**NY Islanders**	**NHL**	**71**	**13**	**25**	**38**	**80**	**2**	**0**	**3**	**119**	**10.9**	**-4**	**456**	**47.4**	**61**	**23**	**14:12**	**7**	**0**	**1**	**1**	**6**	**0**	**0**	**0**
	Russia	Olympics	5	0	0	0	0																			
	NHL Totals		**279**	**41**	**67**	**108**	**205**	**8**	**0**	**5**	**485**	**8.5**		**2009**	**39.3**	**137**	**80**	**13:15**	**11**	**0**	**1**	**1**	**6**	**0**	**0**	**0**

Traded to **NY Islanders** by **Florida** with Mark Parrish for Roberto Luongo and Olli Jokinen, June 24, 2000.

KWIATKOWSKI, Joel

(KWEE-at-KOW-skee, JOHL) **OTT.**

Defense. Shoots left. 6'2", 210 lbs. Born, Kindersley, Sask., March 22, 1977. Dallas' 7th choice, 194th overall, in 1996 Entry Draft.

			Regular Season																Playoffs							
Season	Club	League	GP	G	A	Pts	PIM	PP	SH	GW	S	%	+/-	TF	F%	H	SB	Min	GP	G	A	Pts	PIM	PP	SH	GW
1994-95	North Battleford	SJHL	51	3	14	17	89	....	....	....	....	....	....	....	....	....	....	....	....	....	....	....	....	....	....	....
	Tacoma Rockets	WHL	70	4	13	17	66	....	....	....	....	....	....	....	....	....	....	....	4	0	0	0	2	....	....	....
1995-96	Kelowna Rockets	WHL	40	6	17	23	85	....	....	....	....	....	....	....	....	....	....	....	....	....	....	....	....	....	....	....
	Prince George	WHL	32	6	11	17	48	....	....	....	....	....	....	....	....	....	....	....	....	....	....	....	....	....	....	....
1996-97	Prince George	WHL	72	15	37	52	94	....	....	....	....	....	....	....	....	....	....	....	15	4	2	6	24	....	....	....
1997-98	Prince George	WHL	62	21	43	64	65	....	....	....	....	....	....	....	....	....	....	....	11	3	6	9	6	....	....	....
1998-99	Cincinnati	AHL	80	12	21	33	48	....	....	....	....	....	....	....	....	....	....	....	3	2	0	2	0	....	....	....
99-2000	Cincinnati	AHL	70	4	22	26	28	....	....	....	....	....	....	....	....	....	....	....	....	....	....	....	....	....	....	....
2000-01	**Ottawa**	**NHL**	**4**	**1**	**0**	**1**	**0**	**0**	**0**	**0**	**2**	**50.0**	**1**	**0**	**0.0**	**3**	**4**	**12:04**	....	....	....	....	....	....	....	....
	Grand Rapids	IHL	77	4	17	21	58	....	....	....	....	....	....	....	....	....	....	....	10	1	0	1	4	....	....	....
2001-02	**Ottawa**	**NHL**	**11**	**0**	**0**	**0**	**12**	**0**	**0**	**0**	**9**	**0.0**	**5**	**0**	**0.0**	**13**	**4**	**13:41**	....	....	....	....	....	....	....	....
	Grand Rapids	AHL	65	8	21	29	94	....	....	....	....	....	....	....	....	....	....	....	5	1	2	3	12	....	....	....
	NHL Totals		**15**	**1**	**0**	**1**	**12**	**0**	**0**	**0**	**11**	**9.1**		**0**	**0.0**	**16**	**8**	**13:15**	....	....	....	....	....	....	....	....

WHL West Second All-Star Team (1997) • WHL West First All-Star Team (1998)

Signed as a free agent by **Anaheim**, June 18, 1998. Traded to **Ottawa** by **Anaheim** for Patrick Traverse, June 12, 2000.

LAAKSONEN, Antti

(lah-AHK-soh-nehn, AHN-tee) **MIN.**

Left wing. Shoots left. 6', 180 lbs. Born, Tammela, Finland, October 3, 1973. Boston's 10th choice, 191st overall, in 1997 Entry Draft.

Season	Club	League	GP	G	A	Pts	PIM	PP	SH	GW	S	%	+/-	TF	F%	H	SB	Min	GP	G	A	Pts	PIM	PP	SH	GW
1991-92	FoPS Forssa Jr.	Finn-Jr.	24	19	23	42	22	....	....	....	....	....	....	....	....	....	....	....	....	....	....	....	....	....	....	....
	FoPS Forssa	Finland-2	41	16	15	31	8	....	....	....	....	....	....	....	....	....	....	....	....	....	....	....	....	....	....	....
1992-93	FoPS Forssa Jr.	Finn-Jr.	9	5	3	8	10	....	....	....	....	....	....	....	....	....	....	....	....	....	....	....	....	....	....	....
	FoPS Forssa	Finland-2	34	11	19	30	36	....	....	....	....	....	....	....	....	....	....	....	....	....	....	....	....	....	....	....
	HPK Jr.	Finn-Jr.	1	1	1	2	0	....	....	....	....	....	....	....	....	....	....	....	....	....	....	....	....	....	....	....
	HPK Hameenlinna	Finland	2	0	0	0	0	....	....	....	....	....	....	....	....	....	....	....	....	....	....	....	....	....	....	....
1993-94	U. of Denver	WCHA	36	12	9	21	38	....	....	....	....	....	....	....	....	....	....	....	....	....	....	....	....	....	....	....
1994-95	U. of Denver	WCHA	40	17	18	35	42	....	....	....	....	....	....	....	....	....	....	....	....	....	....	....	....	....	....	....
1995-96	U. of Denver	WCHA	39	25	28	53	71	....	....	....	....	....	....	....	....	....	....	....	....	....	....	....	....	....	....	....
1996-97	U. of Denver	WCHA	39	21	17	38	63	....	....	....	....	....	....	....	....	....	....	....	....	....	....	....	....	....	....	....
1997-98	Providence	AHL	38	3	2	5	14	....	....	....	....	....	....	....	....	....	....	....	....	....	....	....	....	....	....	....
	Charlotte	ECHL	15	4	3	7	12	....	....	....	....	....	....	....	....	....	....	....	6	0	3	3	0	....	....	....
1998-99	**Boston**	**NHL**	**11**	**1**	**2**	**3**	**2**	**0**	**0**	**0**	**8**	**12.5**	**–1**	**0**	**0.0**	**5**	**3**	**9:20**	....	....	....	....	....	....	....	....
	Providence	AHL	66	25	33	58	52	....	....	....	....	....	....	....	....	....	....	....	19	7	2	9	28	....	....	....
99-2000	**Boston**	**NHL**	**27**	**6**	**3**	**9**	**2**	**0**	**0**	**1**	**23**	**26.1**	**3**	**3**	**66.7**	**22**	**2**	**7:50**	....	....	....	....	....	....	....	....
	Providence	AHL	40	10	12	22	57	....	....	....	....	....	....	....	....	....	....	....	14	5	4	9	4	....	....	....
2000-01	**Minnesota**	**NHL**	**82**	**12**	**16**	**28**	**24**	**0**	**2**	**1**	**129**	**9.3**	**–7**	**15**	**26.7**	**106**	**32**	**16:27**	....	....	....	....	....	....	....	....
2001-02	**Minnesota**	**NHL**	**82**	**16**	**17**	**33**	**22**	**0**	**0**	**1**	**104**	**15.4**	**–5**	**19**	**42.1**	**67**	**37**	**16:20**	....	....	....	....	....	....	....	....
	NHL Totals		**202**	**35**	**38**	**73**	**50**	**0**	**2**	**3**	**264**	**13.3**		**37**	**37.8**	**200**	**74**	**14:52**	....	....	....	....	....	....	....	....

WCHA Second All-Star Team (1996)

Signed as a free agent by **Minnesota**, July 14, 2000.

LACHANCE, Scott

(lah-CHANTS, SKAWT) **CBJ**

Defense. Shoots left. 6'1", 215 lbs. Born, Charlottesville, VA, October 22, 1972. NY Islanders' 1st choice, 4th overall, in 1991 Entry Draft.

Season	Club	League	GP	G	A	Pts	PIM	PP	SH	GW	S	%	+/-	TF	F%	H	SB	Min	GP	G	A	Pts	PIM	PP	SH	GW
1988-89	Springfield	NEJHL	36	8	28	36	20	....	....	....	....	....	....	....	....	....	....	....	....	....	....	....	....	....	....	....
1989-90	Springfield	NEJHL	34	25	41	66	62	....	....	....	....	....	....	....	....	....	....	....	....	....	....	....	....	....	....	....
1990-91	Boston University	H-East	31	5	19	24	48	....	....	....	....	....	....	....	....	....	....	....	....	....	....	....	....	....	....	....
1991-92	Team USA	Nat-Tm	36	1	10	11	34	....	....	....	....	....	....	....	....	....	....	....	....	....	....	....	....	....	....	....
	United States	Olympics	8	0	1	1	6	....	....	....	....	....	....	....	....	....	....	....	....	....	....	....	....	....	....	....
	NY Islanders	**NHL**	**17**	**1**	**4**	**5**	**9**	**0**	**0**	**0**	**20**	**5.0**	**13**	....	....	....	....	....	....	....	....	....	....	....	....	....
1992-93	**NY Islanders**	**NHL**	**75**	**7**	**17**	**24**	**67**	**0**	**1**	**2**	**62**	**11.3**	**–1**	....	....	....	....	....	....	....	....	....	....	....	....	....
1993-94	**NY Islanders**	**NHL**	**74**	**3**	**11**	**14**	**70**	**0**	**0**	**1**	**59**	**5.1**	**–5**	....	....	....	....	....	**3**	**0**	**0**	**0**	**0**	**0**	**0**	**0**
1994-95	**NY Islanders**	**NHL**	**26**	**6**	**7**	**13**	**26**	**3**	**0**	**0**	**56**	**10.7**	**2**	....	....	....	....	....	....	....	....	....	....	....	....	....
1995-96	**NY Islanders**	**NHL**	**55**	**3**	**10**	**13**	**54**	**1**	**0**	**0**	**81**	**3.7**	**–19**	....	....	....	....	....	....	....	....	....	....	....	....	....
1996-97	**NY Islanders**	**NHL**	**81**	**3**	**11**	**14**	**47**	**1**	**0**	**0**	**97**	**3.1**	**–7**	....	....	....	....	....	....	....	....	....	....	....	....	....
1997-98	**NY Islanders**	**NHL**	**63**	**2**	**11**	**13**	**45**	**1**	**0**	**0**	**62**	**3.2**	**–11**	....	....	....	....	....	....	....	....	....	....	....	....	....
1998-99	**NY Islanders**	**NHL**	**59**	**1**	**8**	**9**	**30**	**1**	**0**	**0**	**37**	**2.7**	**–19**	**0**	**0.0**	**67**	**92**	**21:34**	....	....	....	....	....	....	....	....
	Montreal	**NHL**	**17**	**1**	**1**	**2**	**11**	**0**	**0**	**0**	**22**	**4.5**	**–2**	**0**	**0.0**	**19**	**47**	**22:29**	....	....	....	....	....	....	....	....
99-2000	**Montreal**	**NHL**	**57**	**0**	**6**	**6**	**22**	**0**	**0**	**0**	**41**	**0.0**	**–4**	**0**	**0.0**	**95**	**86**	**17:47**	....	....	....	....	....	....	....	....
2000-01	**Vancouver**	**NHL**	**76**	**3**	**11**	**14**	**46**	**0**	**0**	**0**	**55**	**5.5**	**5**	**0**	**0.0**	**139**	**136**	**19:26**	**2**	**0**	**1**	**1**	**2**	**0**	**0**	**0**
2001-02	**Vancouver**	**NHL**	**81**	**1**	**10**	**11**	**50**	**0**	**0**	**0**	**48**	**2.1**	**15**	**2**	**100.0**	**120**	**115**	**20:02**	**6**	**1**	**1**	**2**	**4**	**0**	**0**	**1**
	NHL Totals		**681**	**31**	**107**	**138**	**477**	**7**	**1**	**3**	**640**	**4.8**		**2**	**100.0**	**440**	**476**	**19:53**	**11**	**1**	**2**	**3**	**6**	**0**	**0**	**1**

Played in NHL All-Star Game (1997)

Traded to **Montreal** by **NY Islanders** for Montreal's 3rd round choice (Mattias Weinhandl) in 1999 Entry Draft, March 9, 1999. Signed as a free agent by **Vancouver**, August 13, 2000. Signed as a free agent by **Columbus**, July 4, 2002.

LaCOUTURE, Dan

(LA-koo-TUHR, DAN) **PIT.**

Left wing. Shoots left. 6'2", 208 lbs. Born, Hyannis, MA, April 18, 1977. NY Islanders' 2nd choice, 29th overall, in 1996 Entry Draft.

Season	Club	League	GP	G	A	Pts	PIM	PP	SH	GW	S	%	+/-	TF	F%	H	SB	Min	GP	G	A	Pts	PIM	PP	SH	GW
1992-93	Natick Redmen	Hi-School	20	38	34	72	46	....	....	....	....	....	....	....	....	....	....	....	....	....	....	....	....	....	....	....
1993-94	Natick Redmen	Hi-School	21	52	49	101	58	....	....	....	....	....	....	....	....	....	....	....	....	....	....	....	....	....	....	....
1994-95	Springfield	NEJHL	52	44	56	100	98	....	....	....	....	....	....	....	....	....	....	....	....	....	....	....	....	....	....	....
1995-96	Springfield	NAJHL	41	36	41	77	87	....	....	....	....	....	....	....	....	....	....	....	13	12	13	25	23	....	....	....
1996-97	Boston University	H-East	31	13	12	25	18	....	....	....	....	....	....	....	....	....	....	....	....	....	....	....	....	....	....	....
1997-98	Hamilton	AHL	77	15	10	25	31	....	....	....	....	....	....	....	....	....	....	....	5	1	0	1	0	....	....	....
1998-99	**Edmonton**	**NHL**	**3**	**0**	**0**	**0**	**0**	**0**	**0**	**0**	**0**	**0.0**	**1**	**0**	**0.0**	**3**	**0**	**6:30**	....	....	....	....	....	....	....	....
	Hamilton	AHL	72	17	14	31	73	....	....	....	....	....	....	....	....	....	....	....	9	2	1	3	2	....	....	....
99-2000	**Edmonton**	**NHL**	**5**	**0**	**0**	**0**	**10**	**0**	**0**	**0**	**2**	**0.0**	**0**	**0**	**0.0**	**7**	**1**	**7:02**	**1**	**0**	**0**	**0**	**0**	**0**	**0**	**0**
	Hamilton	AHL	70	23	17	40	85	....	....	....	....	....	....	....	....	....	....	....	6	2	1	3	0	....	....	....
2000-01	**Edmonton**	**NHL**	**37**	**2**	**4**	**6**	**29**	**0**	**0**	**1**	**22**	**9.1**	**–2**	**5**	**20.0**	**44**	**5**	**7:06**	....	....	....	....	....	....	....	....
	Pittsburgh	**NHL**	**11**	**0**	**0**	**0**	**14**	**0**	**0**	**0**	**1**	**0.0**	**0**	**1**	**100.0**	**14**	**2**	**5:57**	**5**	**0**	**0**	**0**	**2**	**0**	**0**	**0**
2001-02	**Pittsburgh**	**NHL**	**82**	**6**	**11**	**17**	**71**	**0**	**1**	**0**	**77**	**7.8**	**–19**	**21**	**38.1**	**76**	**36**	**13:16**	....	....	....	....	....	....	....	....
	NHL Totals		**138**	**8**	**15**	**23**	**124**	**0**	**1**	**1**	**102**	**7.8**		**27**	**37.0**	**144**	**44**	**10:40**	**6**	**0**	**0**	**0**	**2**	**0**	**0**	**0**

Traded to **Edmonton** by **NY Islanders** for Mariusz Czerkawski, August 25, 1997. Traded to **Pittsburgh** by **Edmonton** for Sven Butenschon, March 13, 2001.

LAFLAMME, Christian

(lah-FLAM, KRIHS-tan) **ST.L.**

Defense. Shoots right. 6'1", 210 lbs. Born, St-Charles, Que., November 24, 1976. Chicago's 2nd choice, 45th overall, in 1995 Entry Draft.

Season	Club	League	GP	G	A	Pts	PIM	PP	SH	GW	S	%	+/-	TF	F%	H	SB	Min	GP	G	A	Pts	PIM	PP	SH	GW
1991-92	Ste-Foy	QAAA	42	5	27	32	100	....	....	....	....	....	....	....	....	....	....	....	8	1	2	3	14	....	....	....
1992-93	Verdun	QMJHL	69	2	17	19	85	....	....	....	....	....	....	....	....	....	....	....	3	0	2	2	6	....	....	....
1993-94	Verdun	QMJHL	72	4	34	38	85	....	....	....	....	....	....	....	....	....	....	....	4	0	3	3	4	....	....	....
1994-95	Beauport	QMJHL	67	6	41	47	82	....	....	....	....	....	....	....	....	....	....	....	8	1	4	5	6	....	....	....
1995-96	Beauport	QMJHL	41	13	23	36	63	....	....	....	....	....	....	....	....	....	....	....	20	7	17	24	32	....	....	....
1996-97	**Chicago**	**NHL**	**4**	**0**	**1**	**1**	**2**	**0**	**0**	**0**	**3**	**0.0**	**3**	....	....	....	....	....	....	....	....	....	....	....	....	....
	Indianapolis Ice	IHL	62	5	15	20	60	....	....	....	....	....	....	....	....	....	....	....	4	1	1	2	16	....	....	....
1997-98	**Chicago**	**NHL**	**72**	**0**	**11**	**11**	**59**	**0**	**0**	**0**	**75**	**0.0**	**14**	....	....	....	....	....	....	....	....	....	....	....	....	....
1998-99	**Chicago**	**NHL**	**62**	**2**	**11**	**13**	**70**	**0**	**0**	**0**	**53**	**3.8**	**0**	**0**	**0.0**	**154**	**58**	**18:51**	....	....	....	....	....	....	....	....
	Portland Pirates	AHL	2	0	1	1	2	....	....	....	....	....	....	....	....	....	....	....	....	....	....	....	....	....	....	....
	Edmonton	**NHL**	**11**	**0**	**1**	**1**	**0**	**0**	**0**	**0**	**15**	**0.0**	**–3**	**0**	**0.0**	**22**	**14**	**16:33**	**4**	**0**	**1**	**1**	**2**	**0**	**0**	**0**
99-2000	**Edmonton**	**NHL**	**50**	**0**	**5**	**5**	**32**	**0**	**0**	**0**	**18**	**0.0**	**–4**	**5**	**40.0**	**113**	**33**	**13:40**	....	....	....	....	....	....	....	....
	Montreal	**NHL**	**15**	**0**	**2**	**2**	**8**	**0**	**0**	**0**	**6**	**0.0**	**–5**	**0**	**0.0**	**28**	**11**	**14:32**	....	....	....	....	....	....	....	....
2000-01	**Montreal**	**NHL**	**39**	**0**	**3**	**3**	**42**	**0**	**0**	**0**	**16**	**0.0**	**–11**	**1**	**0.0**	**74**	**27**	**12:04**	....	....	....	....	....	....	....	....

			Regular Season																Playoffs							
Season	Club	League	GP	G	A	Pts	PIM	PP	SH	GW	S	%	+/-	TF	F%	H	SB	Min	GP	G	A	Pts	PIM	PP	SH	GW
2001-02	**St. Louis**	**NHL**	**8**	**0**	**1**	**1**	**4**	**0**	**0**	**0**	**6**	**0.0**	**3**	**0**	**0.0**	**12**	**5**	**13:54**	....	....	....	....	....	....	....	....
	Worcester	AHL	62	2	17	19	52	....	....	....	....	....	....	....	....	....	....	....	....	....	....	....	....	....	....	....
	NHL Totals		**261**	**2**	**35**	**37**	**217**	**0**	**0**	**0**	**192**	**1.0**		**6**	**33.3**	**403**	**148**	**15:19**	**4**	**0**	**1**	**1**	**2**	**0**	**0**	**0**

QMJHL All-Rookie Team (1993) • QMJHL Second All-Star Team (1995)

Traded to **Edmonton** by **Chicago** with Daniel Cleary, Ethan Moreau and Chad Kilger for Boris Mironov, Dean McAmmond and Jonas Elofsson, March 20, 1999. Traded to **Montreal** by **Edmonton** with Matthieu Descoteaux for Igor Ulanov and Alain Nasreddine, March 9, 2000. • Missed majority of 2000-01 season recovering from groin injury suffered in game vs. Calgary, December 13, 2000. Signed as a free agent by **St. Louis**, August 21, 2001.

LAKOVIC, Sasha

(LA-koh-vik, SA-shuh)

Right wing. Shoots left. 6', 220 lbs. Born, Vancouver, B.C., September 7, 1971.

Season	Club	League	GP	G	A	Pts	PIM	PP	SH	GW	S	%	+/-	TF	F%	H	SB	Min	GP	G	A	Pts	PIM	PP	SH	GW
1991-92	Kelowna Spartans	BCJHL	4	1	0	1	14	....	....	....	....	....	....	....	....	....	....	....	....	....	....	....	....	....	....	....
	Bellingham	BCJHL	24	8	3	11	67	....	....	....	....	....	....	....	....	....	....	....	....	....	....	....	....	....	....	....
1992-93	Chatham Wheels	ColHL	28	7	5	12	235	....	....	....	....	....	....	....	....	....	....	....	....	....	....	....	....	....	....	....
	Columbus Chill	ECHL	27	7	9	16	162	....	....	....	....	....	....	....	....	....	....	....	....	....	....	....	....	....	....	....
	Binghamton	AHL	3	0	0	0	0	....	....	....	....	....	....	....	....	....	....	....	....	....	....	....	....	....	....	....
	Brantford Smoke	ColHL	....	....	....	....	....	....	....	....	....	....	....	....	....	....	....	....	5	2	1	3	66	....	....	....
1993-94	Toledo Storm	ECHL	24	5	10	15	198	....	....	....	....	....	....	....	....	....	....	....	....	....	....	....	....	....	....	....
	Chatham Wheels	ColHL	13	11	7	18	61	....	....	....	....	....	....	....	....	....	....	....	....	....	....	....	....	....	....	....
1994-95	Tulsa Oilers	CHL	40	20	24	44	214	....	....	....	....	....	....	....	....	....	....	....	5	1	3	4	88	....	....	....
1995-96	Las Vegas	IHL	49	1	2	3	416	....	....	....	....	....	....	....	....	....	....	....	13	1	1	2	*57	....	....	....
1996-97	**Calgary**	**NHL**	**19**	**0**	**1**	**1**	**54**	**0**	**0**	**0**	**10**	**0.0**	**–1**	....	....	....	....	....	....	....	....	....	....	....	....	....
	Saint John	AHL	18	1	8	9	182	....	....	....	....	....	....	....	....	....	....	....	....	....	....	....	....	....	....	....
	Las Vegas	IHL	10	0	0	0	81	....	....	....	....	....	....	....	....	....	....	....	2	0	0	0	14	....	....	....
1997-98	**New Jersey**	**NHL**	**2**	**0**	**0**	**0**	**5**	**0**	**0**	**0**	**2**	**0.0**	**0**	....	....	....	....	....	....	....	....	....	....	....	....	....
	Albany	AHL	30	7	6	13	158	....	....	....	....	....	....	....	....	....	....	....	13	3	4	7	*84	....	....	....
1998-99	**New Jersey**	**NHL**	**16**	**0**	**3**	**3**	**59**	**0**	**0**	**0**	**10**	**0.0**	**0**	**0**	**0.0**	**41**	**1**	**6:20**	....	....	....	....	....	....	....	....
	Albany	AHL	10	1	1	2	93	....	....	....	....	....	....	....	....	....	....	....	....	....	....	....	....	....	....	....
99-2000	Albany	AHL	51	10	16	26	144	....	....	....	....	....	....	....	....	....	....	....	5	0	0	0	14	....	....	....
2000-01	Rochester	AHL	51	3	9	12	161	....	....	....	....	....	....	....	....	....	....	....	4	1	1	2	32	....	....	....
	Long Beach	WCHL	8	3	6	9	29	....	....	....	....	....	....	....	....	....	....	....	....	....	....	....	....	....	....	....
2001-02	Bakersfield	WCHL	30	5	13	18	147	....	....	....	....	....	....	....	....	....	....	....	....	....	....	....	....	....	....	....
	Anchorage Aces	WCHL	....	....	....	....	....	....	....	....	....	....	....	....	....	....	....	....	2	0	1	1	6	....	....	....
	NHL Totals		**37**	**0**	**4**	**4**	**118**	**0**	**0**	**0**	**22**	**0.0**		**0**	**0.0**	**41**	**1**	**6:20**	....	....	....	....	....	....	....	....

Signed as a free agent by **Calgary**, October 10, 1996. Signed as a free agent by **New Jersey**, September 24, 1997.

LAMBERT, Denny

(lahm-BAIR, DEH-nee) **ANA.**

Left wing. Shoots left. 5'11", 211 lbs. Born, Wawa, Ont., January 7, 1970.

Season	Club	League	GP	G	A	Pts	PIM	PP	SH	GW	S	%	+/-	TF	F%	H	SB	Min	GP	G	A	Pts	PIM	PP	SH	GW
1986-87	Soo Legion	NOHA	22	8	13	21	129	....	....	....	....	....	....	....	....	....	....	....	....	....	....	....	....	....	....	....
1987-88	Soo Thunderbirds	NOJHA	32	25	27	52	184	....	....	....	....	....	....	....	....	....	....	....	....	....	....	....	....	....	....	....
1988-89	Sault Ste. Marie	OHL	61	14	15	29	203	....	....	....	....	....	....	....	....	....	....	....	....	....	....	....	....	....	....	....
1989-90	Sault Ste. Marie	OHL	61	23	29	52	276	....	....	....	....	....	....	....	....	....	....	....	....	....	....	....	....	....	....	....
1990-91	Sault Ste. Marie	OHL	59	28	39	67	169	....	....	....	....	....	....	....	....	....	....	....	14	7	9	16	48	....	....	....
1991-92	San Diego Gulls	IHL	71	17	14	31	229	....	....	....	....	....	....	....	....	....	....	....	3	0	0	0	10	....	....	....
	St. Thomas	ColHL	5	2	6	8	9	....	....	....	....	....	....	....	....	....	....	....	....	....	....	....	....	....	....	....
1992-93	San Diego Gulls	IHL	56	18	12	30	277	....	....	....	....	....	....	....	....	....	....	....	14	1	1	2	44	....	....	....
1993-94	San Diego Gulls	IHL	79	13	14	27	314	....	....	....	....	....	....	....	....	....	....	....	6	1	0	1	55	....	....	....
1994-95	San Diego Gulls	IHL	75	25	35	60	222	....	....	....	....	....	....	....	....	....	....	....	....	....	....	....	....	....	....	....
	Anaheim	**NHL**	**13**	**1**	**3**	**4**	**4**	**0**	**0**	**0**	**14**	**7.1**	**3**	....	....	....	....	....	....	....	....	....	....	....	....	....
1995-96	**Anaheim**	**NHL**	**33**	**0**	**8**	**8**	**55**	**0**	**0**	**0**	**28**	**0.0**	**–2**	....	....	....	....	....	....	....	....	....	....	....	....	....
	Baltimore	AHL	44	14	28	42	126	....	....	....	....	....	....	....	....	....	....	....	12	3	9	12	39	....	....	....
1996-97	**Ottawa**	**NHL**	**80**	**4**	**16**	**20**	**217**	**0**	**0**	**1**	**58**	**6.9**	**–4**	....	....	....	....	....	**6**	**0**	**1**	**1**	**9**	**0**	**0**	**0**
1997-98	**Ottawa**	**NHL**	**72**	**9**	**10**	**19**	**250**	**0**	**0**	**1**	**76**	**11.8**	**4**	....	....	....	....	....	**11**	**0**	**0**	**0**	**19**	**0**	**0**	**0**
1998-99	**Nashville**	**NHL**	**76**	**5**	**11**	**16**	**218**	**1**	**0**	**0**	**66**	**7.6**	**–3**	**1**	**100.0**	**57**	**17**	**10:20**	....	....	....	....	....	....	....	....
99-2000	**Atlanta**	**NHL**	**73**	**5**	**6**	**11**	***219**	**2**	**0**	**0**	**83**	**6.0**	**–17**	**5**	**0.0**	**127**	**22**	**11:39**	....	....	....	....	....	....	....	....
2000-01	**Atlanta**	**NHL**	**67**	**1**	**7**	**8**	**215**	**0**	**0**	**0**	**44**	**2.3**	**–5**	**18**	**38.9**	**79**	**11**	**9:08**	....	....	....	....	....	....	....	....
2001-02	**Anaheim**	**NHL**	**73**	**2**	**5**	**7**	**213**	**0**	**0**	**0**	**51**	**3.9**	**1**	**8**	**37.5**	**60**	**6**	**7:00**	....	....	....	....	....	....	....	....
	NHL Totals		**487**	**27**	**66**	**93**	**1391**	**3**	**0**	**2**	**420**	**6.4**		**32**	**34.4**	**323**	**56**	**9:33**	**17**	**0**	**1**	**1**	**28**	**0**	**0**	**0**

Signed as a free agent by **Anaheim**, August 16, 1993. Signed as a free agent by **Ottawa**, July 29, 1996. Claimed by **Nashville** from **Ottawa** in Expansion Draft, June 26, 1998. Traded to **Atlanta** by **Nashville** for the rights to Randy Robitaille, August 16, 1999. Traded to **Anaheim** by **Atlanta** with Atlanta's 9th round choice (Francois Caron) in 2002 Entry Draft for Anaheim's 8th round choice (Tyler Boldt) in 2002 Entry Draft, July 2, 2001.

LANDRY, Eric

(LAN-dree, AIR-ihk) **MTL.**

Center. Shoots left. 5'10", 184 lbs. Born, Gatineau, Que., January 20, 1975.

Season	Club	League	GP	G	A	Pts	PIM	PP	SH	GW	S	%	+/-	TF	F%	H	SB	Min	GP	G	A	Pts	PIM	PP	SH	GW
1992-93	Abitibi	QAAA	40	15	11	26	98	....	....	....	....	....	....	....	....	....	....	....	1	0	0	0	19	....	....	....
1993-94	St-Hyacinthe	QMJHL	69	42	34	76	128	....	....	....	....	....	....	....	....	....	....	....	7	4	2	6	13	....	....	....
1994-95	St-Hyacinthe	QMJHL	68	38	36	74	249	....	....	....	....	....	....	....	....	....	....	....	5	2	1	3	10	....	....	....
1995-96	Cape Breton	AHL	74	19	33	52	187	....	....	....	....	....	....	....	....	....	....	....	....	....	....	....	....	....	....	....
1996-97	Hamilton	AHL	74	15	17	32	139	....	....	....	....	....	....	....	....	....	....	....	22	6	7	13	43	....	....	....
1997-98	**Calgary**	**NHL**	**12**	**1**	**0**	**1**	**4**	**0**	**0**	**0**	**7**	**14.3**	**–2**	....	....	....	....	....	....	....	....	....	....	....	....	....
	Saint John	AHL	61	17	21	38	194	....	....	....	....	....	....	....	....	....	....	....	20	4	6	10	58	....	....	....
1998-99	**Calgary**	**NHL**	**3**	**0**	**1**	**1**	**0**	**0**	**0**	**0**	**1**	**0.0**	**1**	**15**	**46.7**	**8**	**0**	**9:54**	....	....	....	....	....	....	....	....
	Saint John	AHL	56	19	22	41	158	....	....	....	....	....	....	....	....	....	....	....	7	2	5	7	12	....	....	....
99-2000	Kentucky	AHL	79	35	31	66	170	....	....	....	....	....	....	....	....	....	....	....	9	3	6	9	2	....	....	....
2000-01	**Montreal**	**NHL**	**51**	**4**	**7**	**11**	**43**	**2**	**0**	**0**	**54**	**7.4**	**–9**	**510**	**52.8**	**94**	**10**	**9:19**	....	....	....	....	....	....	....	....
	Quebec	AHL	27	14	18	32	90	....	....	....	....	....	....	....	....	....	....	....	9	4	4	8	35	....	....	....
2001-02	**Montreal**	**NHL**	**2**	**0**	**1**	**1**	**0**	**0**	**0**	**0**	**0**	**0.0**	**2**	**1**	**0.0**	**3**	**0**	**7:34**	....	....	....	....	....	....	....	....
	Quebec	AHL	63	32	43	75	125	....	....	....	....	....	....	....	....	....	....	....	3	1	1	2	16	....	....	....
	NHL Totals		**68**	**5**	**9**	**14**	**47**	**2**	**0**	**0**	**62**	**8.1**		**526**	**52.5**	**105**	**10**	**9:17**	....	....	....	....	....	....	....	....

QMJHL All-Rookie Team (1994)

Signed as a free agent by **Calgary**, August 20, 1997. Traded to **San Jose** by **Calgary** for Fredrik Oduya, July 12, 1999. Signed as a free agent by **Montreal**, July 7, 2000.

LANG, Robert

(LANG, RAW-buhrt) **WSH.**

Center. Shoots right. 6'2", 216 lbs. Born, Teplice, Czech., December 19, 1970. Los Angeles' 6th choice, 133rd overall, in 1990 Entry Draft.

Season	Club	League	GP	G	A	Pts	PIM	PP	SH	GW	S	%	+/-	TF	F%	H	SB	Min	GP	G	A	Pts	PIM	PP	SH	GW
1988-89	CHZ Litvinov	Czech	7	3	2	5	0	....	....	....	....	....	....	....	....	....	....	....	....	....	....	....	....	....	....	....
1989-90	CHZ Litvinov	Czech	32	8	7	15	....	....	....	....	....	....	....	....	....	....	....	....	8	3	3	6	....	....	....	....
1990-91	CHZ Litvinov	Czech	56	26	26	52	38	....	....	....	....	....	....	....	....	....	....	....	....	....	....	....	....	....	....	....
1991-92	CHZ Litvinov	Czech	43	12	31	43	34	....	....	....	....	....	....	....	....	....	....	....	....	....	....	....	....	....	....	....
	Czechoslovakia	Olympics	8	5	8	13	8	....	....	....	....	....	....	....	....	....	....	....	....	....	....	....	....	....	....	....
1992-93	**Los Angeles**	**NHL**	**11**	**0**	**5**	**5**	**2**	**0**	**0**	**0**	**3**	**0.0**	**–3**	....	....	....	....	....	....	....	....	....	....	....	....	....
	Phoenix	IHL	38	9	21	30	20	....	....	....	....	....	....	....	....	....	....	....	....	....	....	....	....	....	....	....
1993-94	**Los Angeles**	**NHL**	**32**	**9**	**10**	**19**	**10**	**0**	**0**	**0**	**41**	**22.0**	**7**	....	....	....	....	....	....	....	....	....	....	....	....	....
	Phoenix	IHL	44	11	24	35	34	....	....	....	....	....	....	....	....	....	....	....	....	....	....	....	....	....	....	....
1994-95	Litvinov	Czech	16	4	19	23	28	....	....	....	....	....	....	....	....	....	....	....	....	....	....	....	....	....	....	....
	Los Angeles	**NHL**	**36**	**4**	**8**	**12**	**4**	**0**	**0**	**0**	**38**	**10.5**	**–7**	....	....	....	....	....	....	....	....	....	....	....	....	....
1995-96	**Los Angeles**	**NHL**	**68**	**6**	**16**	**22**	**10**	**0**	**2**	**0**	**71**	**8.5**	**–15**	....	....	....	....	....	....	....	....	....	....	....	....	....
1996-97	HC Sparta Praha	Czech	38	14	27	41	30	....	....	....	....	....	....	....	....	....	....	....	5	1	2	3	4	....	....	....
	HC Sparta Praha	EuroHL	4	2	2	4	0	....	....	....	....	....	....	....	....	....	....	....	4	2	1	3	2	....	....	....
1997-98	**Boston**	**NHL**	**3**	**0**	**0**	**0**	**2**	**0**	**0**	**0**	**2**	**0.0**	**1**	....	....	....	....	....	....	....	....	....	....	....	....	....
	Pittsburgh	**NHL**	**51**	**9**	**13**	**22**	**14**	**1**	**1**	**2**	**64**	**14.1**	**6**	....	....	....	....	....	**6**	**0**	**3**	**3**	**2**	**0**	**0**	**0**
	Czech Republic	Olympics	6	0	3	3	0	....	....	....	....	....	....	....	....	....	....	....	....	....	....	....	....	....	....	....
	Houston Aeros	IHL	9	1	7	8	4	....	....	....	....	....	....	....	....	....	....	....	....	....	....	....	....	....	....	....
1998-99	**Pittsburgh**	**NHL**	**72**	**21**	**23**	**44**	**24**	**7**	**0**	**3**	**137**	**15.3**	**–10**	**964**	**44.8**	**84**	**22**	**16:24**	**12**	**0**	**2**	**2**	**0**	**0**	**0**	**0**
99-2000	**Pittsburgh**	**NHL**	**78**	**23**	**42**	**65**	**14**	**13**	**0**	**5**	**142**	**16.2**	**–9**	**1433**	**50.7**	**60**	**51**	**19:22**	**11**	**3**	**3**	**6**	**0**	**2**	**0**	**0**
2000-01	**Pittsburgh**	**NHL**	**82**	**32**	**48**	**80**	**28**	**10**	**0**	**2**	**177**	**18.1**	**20**	**1348**	**43.9**	**51**	**38**	**20:24**	**16**	**4**	**4**	**8**	**4**	**0**	**0**	**0**

			Regular Season																Playoffs							
Season	Club	League	GP	G	A	Pts	PIM	PP	SH	GW	S	%	+/-	TF	F%	H	SB	Min	GP	G	A	Pts	PIM	PP	SH	GW
2001-02	Pittsburgh	NHL	62	18	32	50	16	5	1	3	175	10.3	9	1172	46.3	28	39	22:56								
	Czech Republic	Olympics	4	1	2	3	2																			
	NHL Totals		495	122	197	319	124	36	4	15	850	14.4		4917	46.6	223	150	19:41	45	7	12	19	6	2	0	0

Signed as a free agent by **Pittsburgh**, September 2, 1997. Claimed by **Boston** from **Pittsburgh** in NHL Waiver Draft, September 28, 1997. Claimed on waivers by **Pittsburgh** from **Boston**, October 25, 1997. Signed as a free agent by **Washington**, July 1, 2002.

LANGDON, Darren

(LAING-duhn, DAIR-uhn) **CAR.**

Left wing. Shoots left. 6'1", 205 lbs. Born, Deer Lake, Nfld., January 8, 1971.

Season	Club	League	GP	G	A	Pts	PIM	PP	SH	GW	S	%	+/-	TF	F%	H	SB	Min	GP	G	A	Pts	PIM	PP	SH	GW
1991-92	Summerside	MJrHL	44	34	49	83	441																			
1992-93	Binghamton	AHL	18	3	4	7	115												8	0	1	1	14			
	Dayton Bombers	ECHL	54	23	22	45	429												3	0	1	1	40			
1993-94	Binghamton	AHL	54	2	7	9	327																			
1994-95	Binghamton	AHL	55	6	14	20	296												11	1	3	4	*84			
	NY Rangers	NHL	18	1	1	2	62	0	0	0	6	16.7	0													
1995-96	NY Rangers	NHL	64	7	4	11	175	0	0	1	29	24.1	2						2	0	0	0	0	0	0	0
	Binghamton	AHL	1	0	0	0	12																			
1996-97	NY Rangers	NHL	60	3	6	9	195	0	0	1	24	12.5	−1						10	0	0	0	2	0	0	0
1997-98	NY Rangers	NHL	70	3	3	6	197	0	0	0	15	20.0	0													
1998-99	NY Rangers	NHL	44	0	0	0	80	0	0	0	8	0.0	−3	0	0.0	37	5	3:33								
99-2000	NY Rangers	NHL	21	0	1	1	26	0	0	0	13	0.0	−2	0	0.0	22	2	5:36								
2000-01	Carolina	NHL	54	0	2	2	94	0	0	0	6	0.0	−4	2	100.0	36	6	3:21	4	0	0	0	12	0	0	0
2001-02	Carolina	NHL	58	2	1	3	106	0	0	1	12	16.7	2	1	100.0	47	4	4:11								
	NHL Totals		389	16	18	34	935	0	0	3	113	14.2		3	100.0	142	17	3:56	16	0	0	0	14	0	0	0

Signed as a free agent by **NY Rangers**, August 16, 1993. • Missed majority of 1999-2000 season recovering from hernia injury suffered in game vs. New Jersey, December 1, 1999. Traded to **Carolina** by **NY Rangers** with Rob DiMaio for Sandy McCarthy and Carolina's 4th round choice (Bryce Lampman) in 2001 Entry Draft, August 4, 2000.

LANGENBRUNNER, Jamie

(lan-gehn-BRUH-nuhr, JAY-mee) **N.J.**

Right wing. Shoots right. 6'1", 200 lbs. Born, Duluth, MN, July 24, 1975. Dallas' 2nd choice, 35th overall, in 1993 Entry Draft.

Season	Club	League	GP	G	A	Pts	PIM	PP	SH	GW	S	%	+/-	TF	F%	H	SB	Min	GP	G	A	Pts	PIM	PP	SH	GW
1990-91	Cloquet High	Hi-School	20	6	16	22	8																			
1991-92	Cloquet High	Hi-School	23	16	23	39	24																			
1992-93	Cloquet High	Hi-School	27	27	62	89	18																			
1993-94	Peterborough	OHL	62	33	58	91	53												7	4	6	10	2			
1994-95	Peterborough	OHL	62	42	57	99	84												11	8	14	22	12			
	Dallas	NHL	2	0	0	0	2	0	0	0	1	0.0	0													
	Kalamazoo Wings	IHL																	11	1	3	4	2			
1995-96	Dallas	NHL	12	2	2	4	6	1	0	0	15	13.3	−2													
	Michigan K-Wings	IHL	59	25	40	65	129												10	3	10	13	8			
1996-97	Dallas	NHL	76	13	26	39	51	3	0	3	112	11.6	−2						5	1	1	2	14	0	0	1
1997-98	Dallas	NHL	81	23	29	52	61	8	0	6	159	14.5	9						16	1	4	5	14	0	0	1
	United States	Olympics	3	0	0	0	4																			
1998-99♦	Dallas	NHL	75	12	33	45	62	4	0	1	145	8.3	10	217	46.1	129	21	15:51	23	10	7	17	16	4	0	3
99-2000	Dallas	NHL	65	18	21	39	68	4	2	6	153	11.8	16	40	50.0	117	10	17:33	15	1	7	8	18	1	0	0
2000-01	Dallas	NHL	53	12	18	30	57	3	2	4	104	11.5	4	316	45.3	78	26	16:30	10	2	2	4	6	0	0	1
2001-02	Dallas	NHL	68	10	16	26	54	0	1	2	132	7.6	−11	120	45.0	98	26	15:45								
	New Jersey	NHL	14	3	3	6	23	0	0	2	31	9.7	2	2	50.0	13	6	15:27	5	0	1	1	8	0	0	0
	NHL Totals		446	93	148	241	384	23	5	24	852	10.9		695	45.8	435	89	16:20	74	15	22	37	76	5	0	6

Traded to **New Jersey** by **Dallas** with Joe Nieuwendyk for Jason Arnott, Randy McKay and New Jersey's 1st round choice (later traded to Columbus - later traded to Buffalo - Buffalo selected Dan Paille) in 2002 Entry Draft, March 19, 2002.

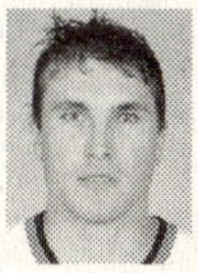

LANGFELD, Josh

(LANG-fehld, JAWSH) **OTT.**

Right wing. Shoots right. 6'3", 216 lbs. Born, Fridley, MN, July 17, 1977. Ottawa's 3rd choice, 66th overall, in 1997 Entry Draft.

Season	Club	League	GP	G	A	Pts	PIM	PP	SH	GW	S	%	+/-	TF	F%	H	SB	Min	GP	G	A	Pts	PIM	PP	SH	GW
1995-96	Great Falls	AFJHL	45	45	40	85	105																			
1996-97	Lincoln Stars	USHL	38	35	23	58	100												14	8	*13	*21	42			
1997-98	U. of Michigan	CCHA	46	19	17	36	66																			
1998-99	U. of Michigan	CCHA	41	21	14	35	84																			
99-2000	U. of Michigan	CCHA	39	9	21	30	56																			
2000-01	U. of Michigan	CCHA	42	16	12	28	44																			
2001-02	Ottawa	NHL	1	0	0	0	2	0	0	0	5	0.0	0	0	0.0	1	0	8:15								
	Grand Rapids	AHL	68	21	16	37	29												5	2	0	2	0			
	NHL Totals		1	0	0	0	2	0	0	0	5	0.0		0	0.0	1	0	8:15								

NCAA Championship All-Tournament Team (1998)

LANGKOW, Daymond

(LAING-kow, DAY-muhn) **PHX.**

Center. Shoots left. 5'11", 180 lbs. Born, Edmonton, Alta, September 27, 1976. Tampa Bay's 1st choice, 5th overall, in 1995 Entry Draft.

Season	Club	League	GP	G	A	Pts	PIM	PP	SH	GW	S	%	+/-	TF	F%	H	SB	Min	GP	G	A	Pts	PIM	PP	SH	GW
1991-92	Edmonton Pats	AMHL	35	36	45	81	100																			
	Tri-City	WHL	1	0	0	0	0																			
1992-93	Tri-City	WHL	64	22	42	64	100												4	1	0	1	4			
1993-94	Tri-City	WHL	61	40	43	83	174												4	2	2	4	15			
1994-95	Tri-City	WHL	72	*67	73	*140	142												17	12	15	27	52			
1995-96	Tri-City	WHL	48	30	61	91	103												11	14	13	27	20			
	Tampa Bay	NHL	4	0	1	1	0	0	0	0	4	0.0	−1													
1996-97	Tampa Bay	NHL	79	15	13	28	35	3	1	1	170	8.8	1													
	Adirondack	AHL	2	1	1	2	0																			
1997-98	Tampa Bay	NHL	68	8	14	22	62	2	0	1	156	5.1	−9													
1998-99	Tampa Bay	NHL	22	4	6	10	15	1	0	1	40	10.0	0	399	48.4	20	8	17:10								
	Cleveland	IHL	4	1	1	2	18																			
	Philadelphia	NHL	56	10	13	23	24	3	1	1	109	9.2	−8	738	48.0	35	12	15:12	6	0	2	2	2	0	0	0
99-2000	Philadelphia	NHL	82	18	32	50	56	5	0	7	222	8.1	1	1263	45.1	78	41	16:57	16	5	5	10	23	1	1	2
2000-01	Philadelphia	NHL	71	13	41	54	50	3	0	2	190	6.8	12	1181	47.2	61	32	18:38	6	2	4	6	2	1	0	0
2001-02	Phoenix	NHL	80	27	35	62	36	6	3	2	171	15.8	18	1379	46.2	88	54	19:11	5	1	0	1	0	0	0	0
	NHL Totals		462	95	155	250	278	23	5	15	1062	8.9		4960	46.6	282	147	17:37	33	8	11	19	27	2	1	2

WHL West First All-Star Team (1995) • Canadian Major Junior First All-Star Team (1995) • WHL West Second All-Star Team (1996)

Traded to **Philadelphia** by **Tampa Bay** with Mikael Renberg for Chris Gratton and Mike Sillinger, December 12, 1998. Traded to **Phoenix** by **Philadelphia** for Phoenix's 2nd round choice (later traded to Tampa Bay - later traded to San Jose - San Jose selected Dan Spang) in 2002 Entry Draft and 1st round choice in 2003 Entry Draft, July 2, 2001.

LAPERRIERE, Ian

(luh-PAIR-ee-YAIR, EE-ihn) **L.A.**

Center. Shoots right. 6'1", 201 lbs. Born, Montreal, Que., January 19, 1974. St. Louis' 6th choice, 158th overall, in 1992 Entry Draft.

Season	Club	League	GP	G	A	Pts	PIM	PP	SH	GW	S	%	+/-	TF	F%	H	SB	Min	GP	G	A	Pts	PIM	PP	SH	GW
1989-90	Mtl-Bourassa	QAAA	22	4	10	14	10												3	0	1	1	6			
1990-91	Drummondville	QMJHL	65	19	29	48	117												14	2	9	11	48			
1991-92	Drummondville	QMJHL	70	28	49	77	160												4	2	2	4	9			
1992-93	Drummondville	QMJHL	60	44	*96	140	188												10	6	13	19	20			
1993-94	Drummondville	QMJHL	62	41	72	113	150												9	4	6	10	35			
	St. Louis	NHL	1	0	0	0	0	0	0	0	1	0.0	0													
	Peoria Rivermen	IHL																	5	1	3	4	2			
1994-95	Peoria Rivermen	IHL	51	16	32	48	111																			
	St. Louis	NHL	37	13	14	27	85	1	0	1	53	24.5	12						7	0	4	4	21	0	0	0
1995-96	St. Louis	NHL	33	3	6	9	87	1	0	1	31	9.7	−4													
	Worcester	AHL	3	2	1	3	22																			
	NY Rangers	NHL	28	1	2	3	53	0	0	0	21	4.8	−5													
	Los Angeles	NHL	10	2	3	5	15	0	0	0	18	11.1	−2													
1996-97	Los Angeles	NHL	62	8	15	23	102	0	1	2	84	9.5	−25													
1997-98	Los Angeles	NHL	77	6	15	21	131	0	1	1	74	8.1	0						4	1	0	1	6	0	0	0
1998-99	Los Angeles	NHL	72	3	10	13	138	0	0	1	62	4.8	−5	643	47.3	149	60	11:47								
99-2000	Los Angeles	NHL	79	9	13	22	185	0	0	1	87	10.3	−14	1111	53.7	181	60	13:15	4	0	0	0	2	0	0	0

			Regular Season																Playoffs							
Season	Club	League	GP	G	A	Pts	PIM	PP	SH	GW	S	%	+/-	TF	F%	H	SB	Min	GP	G	A	Pts	PIM	PP	SH	GW
2000-01	Los Angeles	NHL	79	8	10	18	141	0	0	0	60	13.3	5	297	51.9	209	29	12:02	13	1	2	3	12	0	0	0
2001-02	Los Angeles	NHL	81	8	14	22	125	0	0	3	89	9.0	5	134	49.3	213	45	13:45	7	0	1	1	9	0	0	0
	NHL Totals		559	61	102	163	1062	2	2	10	580	10.5		2185	51.3	752	194	12:44	35	2	7	9	50	0	0	0

QMJHL Second All-Star Team (1993)

Traded to **NY Rangers** by **St. Louis** for Stephane Matteau, December 28, 1995. Traded to **LA Kings** by **NY Rangers** with Ray Ferraro, Mattias Norstrom, Nathan Lafayette and NY Rangers' 4th round choice (Sean Blanchard) in 1997 Entry Draft for Marty McSorley, Jari Kurri and Shane Churla, March 14, 1996.

LAPLANTE, Darryl

(LA-plawnt, DAIR-ihl) **BOS.**

Center. Shoots left. 6', 198 lbs. Born, Calgary, Alta., March 28, 1977. Detroit's 3rd choice, 58th overall, in 1995 Entry Draft.

Season	Club	League	GP	G	A	Pts	PIM	PP	SH	GW	S	%	+/-	TF	F%	H	SB	Min	GP	G	A	Pts	PIM	PP	SH	GW
1992-93	Cgy. AA Royals	ABHL	32	20	26	46	60																			
1993-94	Calgary Royals	AMHL	35	24	27	51	50																			
1994-95	Moose Jaw	WHL	71	22	24	46	66												10	2	2	4	7			
1995-96	Moose Jaw	WHL	72	42	40	82	76																			
1996-97	Moose Jaw	WHL	69	38	42	80	79												12	2	4	6	15			
1997-98	**Detroit**	**NHL**	**2**	**0**	**0**	**0**	**0**	**0**	**0**	**0**	**2**	**0.0**	**0**													
	Adirondack	AHL	77	15	10	25	51												3	0	1	1	4			
1998-99	**Detroit**	**NHL**	**3**	**0**	**0**	**0**	**0**	**0**	**0**	**0**	**0**	**0.0**	**0**	**0**	**0.0**	**0**	**1**	**1:59**								
	Adirondack	AHL	71	17	15	32	96												3	0	1	1	0			
99-2000	**Detroit**	**NHL**	**30**	**0**	**6**	**6**	**10**	**0**	**0**	**0**	**19**	**0.0**	**–2**	**58**	**53.5**	**44**	**9**	**9:35**								
	Cincinnati	AHL	35	13	9	22	47																			
2000-01	Cleveland	IHL	67	6	19	25	43												4	0	1	1	6			
2001-02	Houston Aeros	AHL	31	11	6	17	57																			
	Providence	AHL	10	0	0	0	4												2	0	1	1	4			
	NHL Totals		**35**	**0**	**6**	**6**	**10**	**0**	**0**	**0**	**21**	**0.0**		**58**	**53.4**	**44**	**10**	**8:53**								

Selected by **Minnesota** from **Detroit** in Expansion Draft, June 23, 2000. Traded to **Boston** by **Minnesota** for Greg Crozier, March 19, 2002.

LAPOINTE, Claude

(luh-POYNT, KLOHD) **NYI**

Center. Shoots left. 5'9", 188 lbs. Born, Lachine, Que., October 11, 1968. Quebec's 12th choice, 234th overall, in 1988 Entry Draft.

Season	Club	League	GP	G	A	Pts	PIM	PP	SH	GW	S	%	+/-	TF	F%	H	SB	Min	GP	G	A	Pts	PIM	PP	SH	GW
1983-84	Lac St-Louis	QAAA	42	28	29	57	42												8	3	7	10	8			
1984-85	Lac St-Louis	QAAA	42	20	32	52	66												11	4	8	12	16			
1985-86	Trois-Rivieres	QMJHL	63	14	32	46	70												9	5	6	11	4			
1986-87	Trois-Rivieres	QMJHL	70	47	57	104	123																			
1987-88	Laval Titan	QMJHL	69	37	83	120	143												13	2	17	19	53			
1988-89	Laval Titan	QMJHL	63	32	72	104	158												17	5	14	19	66			
1989-90	Halifax Citadels	AHL	63	18	19	37	51												6	1	1	2	34			
1990-91	**Quebec**	**NHL**	**13**	**2**	**2**	**4**	**4**	**0**	**0**	**0**	**7**	**28.6**	**3**													
	Halifax Citadels	AHL	43	17	17	34	46																			
1991-92	**Quebec**	**NHL**	**78**	**13**	**20**	**33**	**86**	**0**	**2**	**2**	**95**	**13.7**	**–8**													
1992-93	**Quebec**	**NHL**	**74**	**10**	**26**	**36**	**98**	**0**	**0**	**1**	**91**	**11.0**	**5**						**6**	**2**	**4**	**6**	**8**	**0**	**0**	**0**
1993-94	**Quebec**	**NHL**	**59**	**11**	**17**	**28**	**70**	**1**	**1**	**1**	**73**	**15.1**	**2**													
1994-95	**Quebec**	**NHL**	**29**	**4**	**8**	**12**	**41**	**0**	**0**	**0**	**40**	**10.0**	**5**						**5**	**0**	**0**	**0**	**8**	**0**	**0**	**0**
1995-96	**Colorado**	**NHL**	**3**	**0**	**0**	**0**	**0**	**0**	**0**	**0**	**0**	**0.0**	**–1**													
	Calgary	**NHL**	**32**	**4**	**5**	**9**	**20**	**0**	**2**	**1**	**44**	**9.1**	**2**						**2**	**0**	**0**	**0**	**0**	**0**	**0**	**0**
	Saint John	AHL	12	5	3	8	10																			
1996-97	**NY Islanders**	**NHL**	**73**	**13**	**5**	**18**	**49**	**0**	**3**	**3**	**80**	**16.3**	**–12**													
	Utah Grizzlies	IHL	9	7	6	13	14																			
1997-98	**NY Islanders**	**NHL**	**78**	**10**	**10**	**20**	**47**	**0**	**1**	**3**	**82**	**12.2**	**–9**													
1998-99	**NY Islanders**	**NHL**	**82**	**14**	**23**	**37**	**62**	**2**	**2**	**1**	**134**	**10.4**	**–19**	**1218**	**56.6**	**168**	**60**	**19:21**								
99-2000	**NY Islanders**	**NHL**	**76**	**15**	**16**	**31**	**60**	**2**	**1**	**3**	**129**	**11.6**	**–22**	**1284**	**54.0**	**147**	**71**	**19:39**								
2000-01	**NY Islanders**	**NHL**	**80**	**9**	**23**	**32**	**56**	**1**	**1**	**1**	**94**	**9.6**	**–2**	**1074**	**50.4**	**133**	**79**	**18:40**								
2001-02	**NY Islanders**	**NHL**	**80**	**9**	**12**	**21**	**60**	**0**	**3**	**0**	**74**	**12.2**	**–9**	**907**	**54.6**	**123**	**37**	**13:05**	**7**	**0**	**0**	**0**	**14**	**0**	**0**	**0**
	NHL Totals		**757**	**114**	**167**	**281**	**653**	**6**	**16**	**16**	**943**	**12.1**		**4483**	**53.9**	**571**	**247**	**17:41**	**20**	**2**	**4**	**6**	**30**	**0**	**0**	**0**

Transferred to **Colorado** after **Quebec** franchise relocated, June 21, 1995. Traded to **Calgary** by **Colorado** for Calgary's 7th round choice (Samual Pahlsson) in 1996 Entry Draft, November 1, 1995. Signed as a free agent by **NY Islanders**, August 14, 1996.

LAPOINTE, Martin

(luh-POYNT, MAHR-tihn) **BOS.**

Right wing. Shoots right. 5'11", 200 lbs. Born, Ville St-Pierre, Que., September 12, 1973. Detroit's 1st choice, 10th overall, in 1991 Entry Draft.

Season	Club	League	GP	G	A	Pts	PIM	PP	SH	GW	S	%	+/-	TF	F%	H	SB	Min	GP	G	A	Pts	PIM	PP	SH	GW
1988-89	Lac St-Louis	QAAA	42	39	45	84	46												3	6	2	8	4			
1989-90	Laval Titan	QMJHL	65	42	54	96	77												14	8	17	25	54			
1990-91	Laval Titan	QMJHL	64	44	54	98	66												13	7	14	21	26			
1991-92	Laval Titan	QMJHL	31	25	30	55	84												10	4	10	14	32			
	Detroit	**NHL**	**4**	**0**	**1**	**1**	**5**	**0**	**0**	**0**	**2**	**0.0**	**2**						**3**	**0**	**1**	**1**	**4**	**0**	**0**	**0**
	Adirondack	AHL																	8	2	2	4	4			
1992-93	Laval Titan	QMJHL	35	38	51	89	41												13	*13	*17	*30	22			
	Detroit	**NHL**	**3**	**0**	**0**	**0**	**0**	**0**	**0**	**0**	**2**	**0.0**	**–2**													
	Adirondack	AHL	8	1	2	3	9																			
1993-94	**Detroit**	**NHL**	**50**	**8**	**8**	**16**	**55**	**2**	**0**	**0**	**45**	**17.8**	**7**						**4**	**0**	**0**	**0**	**6**	**0**	**0**	**0**
	Adirondack	AHL	28	25	21	46	47												4	1	1	2	8			
1994-95	Adirondack	AHL	39	29	16	45	80																			
	Detroit	**NHL**	**39**	**4**	**6**	**10**	**73**	**0**	**0**	**1**	**46**	**8.7**	**1**						**2**	**0**	**1**	**1**	**8**	**0**	**0**	**0**
1995-96	**Detroit**	**NHL**	**58**	**6**	**3**	**9**	**93**	**1**	**0**	**0**	**76**	**7.9**	**0**						**11**	**1**	**2**	**3**	**12**	**0**	**0**	**0**
1996-97♦	**Detroit**	**NHL**	**78**	**16**	**17**	**33**	**167**	**5**	**1**	**1**	**149**	**10.7**	**–14**						**20**	**4**	**8**	**12**	**60**	**1**	**0**	**1**
1997-98♦	**Detroit**	**NHL**	**79**	**15**	**19**	**34**	**106**	**4**	**0**	**3**	**154**	**9.7**	**0**						**21**	**9**	**6**	**15**	**20**	**2**	**1**	**1**
1998-99	**Detroit**	**NHL**	**77**	**16**	**13**	**29**	**141**	**7**	**1**	**4**	**153**	**10.5**	**7**	**217**	**47.9**	**167**	**16**	**15:06**	**10**	**0**	**2**	**2**	**20**	**0**	**0**	**0**
99-2000	**Detroit**	**NHL**	**82**	**16**	**25**	**41**	**121**	**1**	**1**	**2**	**127**	**12.6**	**17**	**287**	**54.4**	**207**	**19**	**14:43**	**9**	**3**	**1**	**4**	**20**	**2**	**0**	**1**
2000-01	**Detroit**	**NHL**	**82**	**27**	**30**	**57**	**127**	**13**	**0**	**8**	**181**	**14.9**	**3**	**461**	**53.2**	**259**	**17**	**16:06**	**6**	**0**	**1**	**1**	**8**	**0**	**0**	**0**
2001-02	**Boston**	**NHL**	**68**	**17**	**23**	**40**	**101**	**4**	**0**	**2**	**141**	**12.1**	**12**	**222**	**53.6**	**182**	**29**	**17:22**	**6**	**1**	**2**	**3**	**12**	**1**	**0**	**1**
	NHL Totals		**620**	**125**	**145**	**270**	**989**	**37**	**3**	**21**	**1076**	**11.6**		**1187**	**52.6**	**815**	**81**	**15:46**	**92**	**18**	**24**	**42**	**170**	**6**	**1**	**4**

QMJHL First All-Star Team (1990, 1993) • QMJHL Offensive Rookie of the Year) (1990) • QMJHL Second All-Star Team (1991) • Memorial Cup All-Star Team (1993)

Signed as a free agent by **Boston**, July 2, 2001.

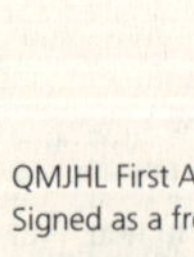

LARAQUE, Georges

(luh-RAK, zhawrzh) **EDM.**

Right wing. Shoots right. 6'3", 240 lbs. Born, Montreal, Que., December 7, 1976. Edmonton's 2nd choice, 31st overall, in 1995 Entry Draft.

Season	Club	League	GP	G	A	Pts	PIM	PP	SH	GW	S	%	+/-	TF	F%	H	SB	Min	GP	G	A	Pts	PIM	PP	SH	GW
1991-92	Mtl-Bourassa	QAHA	28	20	20	40	30																			
1992-93	Mtl-Bourassa	QAAA	37	8	20	28	50												3	1	2	3	2			
1993-94	St-Jean Lynx	QMJHL	70	11	11	22	142												4	0	0	0	7			
1994-95	St-Jean Lynx	QMJHL	62	19	22	41	259												7	1	1	2	42			
1995-96	Laval Titan	QMJHL	11	8	13	21	76																			
	St-Hyacinthe	QMJHL	8	3	4	7	59																			
	Granby	QMJHL	22	9	7	16	125												18	7	6	13	104			
1996-97	Hamilton	AHL	73	14	20	34	179												15	1	3	4	12			
1997-98	**Edmonton**	**NHL**	**11**	**0**	**0**	**0**	**59**	**0**	**0**	**0**	**4**	**0.0**	**–4**													
	Hamilton	AHL	46	10	20	30	154												3	0	0	0	11			
1998-99	**Edmonton**	**NHL**	**39**	**3**	**2**	**5**	**57**	**0**	**0**	**0**	**17**	**17.6**	**–1**	**0**	**0.0**	**32**	**4**	**5:31**	**4**	**0**	**0**	**0**	**2**	**0**	**0**	**0**
	Hamilton	AHL	25	6	8	14	93																			
99-2000	**Edmonton**	**NHL**	**76**	**8**	**8**	**16**	**123**	**0**	**0**	**0**	**56**	**14.3**	**5**	**0**	**0.0**	**84**	**21**	**8:28**	**5**	**0**	**1**	**1**	**6**	**0**	**0**	**0**
2000-01	**Edmonton**	**NHL**	**82**	**13**	**16**	**29**	**148**	**1**	**0**	**1**	**73**	**17.8**	**5**	**0**	**0.0**	**95**	**19**	**9:03**	**6**	**1**	**1**	**2**	**8**	**0**	**0**	**0**
2001-02	**Edmonton**	**NHL**	**80**	**5**	**14**	**19**	**157**	**1**	**0**	**1**	**95**	**5.3**	**6**	**0**	**0.0**	**73**	**13**	**9:48**								
	NHL Totals		**288**	**29**	**40**	**69**	**544**	**2**	**0**	**2**	**245**	**11.8**		**0**	**0.0**	**284**	**57**	**8:36**	**15**	**1**	**2**	**3**	**16**	**0**	**0**	**0**

LARIONOV, Igor
(LAIR-ee-AH-nohv, EE-gohr) **DET.**

Center. Shoots left. 5'9", 170 lbs. Born, Voskresensk, USSR, December 3, 1960. Vancouver's 11th choice, 214th overall, in 1985 Entry Draft.

			Regular Season																Playoffs							
Season	Club	League	GP	G	A	Pts	PIM	PP	SH	GW	S	%	+/-	TF	F%	H	SB	Min	GP	G	A	Pts	PIM	PP	SH	GW
1977-78	Voskresensk	USSR	6	3	0	3	4	....	....	....	....	....	....	....	....	....	....	....	....	....	....	....	....	....	....	....
1978-79	Voskresensk	USSR	32	3	4	7	12	....	....	....	....	....	....	....	....	....	....	....	....	....	....	....	....	....	....	....
1979-80	Voskresensk	USSR	42	11	7	18	24	....	....	....	....	....	....	....	....	....	....	....	....	....	....	....	....	....	....	....
1980-81	Voskresensk	USSR	43	22	23	45	36	....	....	....	....	....	....	....	....	....	....	....	....	....	....	....	....	....	....	....
1981-82	CSKA Moscow	USSR	46	31	22	53	6	....	....	....	....	....	....	....	....	....	....	....	....	....	....	....	....	....	....	....
1982-83	CSKA Moscow	USSR	44	20	19	39	20	....	....	....	....	....	....	....	....	....	....	....	....	....	....	....	....	....	....	....
1983-84	CSKA Moscow	USSR	43	15	26	41	30	....	....	....	....	....	....	....	....	....	....	....	....	....	....	....	....	....	....	....
	Soviet Union	Olympics	6	1	4	5	6	....	....	....	....	....	....	....	....	....	....	....	....	....	....	....	....	....	....	....
1984-85	CSKA Moscow	USSR	40	18	28	46	20	....	....	....	....	....	....	....	....	....	....	....	....	....	....	....	....	....	....	....
1985-86	CSKA Moscow	USSR	40	21	31	52	33	....	....	....	....	....	....	....	....	....	....	....	....	....	....	....	....	....	....	....
1986-87	CSKA Moscow	USSR	39	20	26	46	34	....	....	....	....	....	....	....	....	....	....	....	....	....	....	....	....	....	....	....
1987-88	CSKA Moscow	USSR	51	25	32	57	54	....	....	....	....	....	....	....	....	....	....	....	....	....	....	....	....	....	....	....
	Soviet Union	Olympics	8	4	*9	13	4	....	....	....	....	....	....	....	....	....	....	....	....	....	....	....	....	....	....	....
1988-89	CSKA Moscow	USSR	31	15	12	27	22	....	....	....	....	....	....	....	....	....	....	....	....	....	....	....	....	....	....	....
1989-90	**Vancouver**	**NHL**	**74**	**17**	**27**	**44**	**20**	**8**	**0**	**2**	**118**	**14.4**	**–5**	....	....	....	....	....	....	....	....	....	....	....	....	....
1990-91	**Vancouver**	**NHL**	**64**	**13**	**21**	**34**	**14**	**1**	**1**	**0**	**66**	**19.7**	**–3**	....	....	....	....	....	**6**	**1**	**0**	**1**	**6**	**0**	**0**	**0**
1991-92	**Vancouver**	**NHL**	**72**	**21**	**44**	**65**	**54**	**10**	**3**	**4**	**97**	**21.6**	**7**	....	....	....	....	....	**13**	**3**	**7**	**10**	**4**	**1**	**0**	**0**
1992-93	HC Lugano	Swiss	24	10	19	29	44	....	....	....	....	....	....	....	....	....	....	....	8	3	15	18	0	....	....	....
1993-94	**San Jose**	**NHL**	**60**	**18**	**38**	**56**	**40**	**3**	**2**	**2**	**72**	**25.0**	**20**	....	....	....	....	....	**14**	**5**	**13**	**18**	**10**	**0**	**0**	**0**
1994-95	**San Jose**	**NHL**	**33**	**4**	**20**	**24**	**14**	**0**	**0**	**1**	**69**	**5.8**	**–3**	....	....	....	....	....	**11**	**1**	**8**	**9**	**2**	**0**	**0**	**0**
1995-96	**San Jose**	**NHL**	**4**	**1**	**1**	**2**	**0**	**1**	**0**	**0**	**5**	**20.0**	**–6**	....	....	....	....	....	....	....	....	....	....	....	....	....
	Detroit	**NHL**	**69**	**21**	**50**	**71**	**34**	**9**	**1**	**5**	**108**	**19.4**	**37**	....	....	....	....	....	**19**	**6**	**7**	**13**	**6**	**3**	**0**	**2**
1996-97♦	**Detroit**	**NHL**	**64**	**12**	**42**	**54**	**26**	**2**	**1**	**4**	**95**	**12.6**	**31**	....	....	....	....	....	**20**	**4**	**8**	**12**	**8**	**3**	**0**	**1**
1997-98♦	**Detroit**	**NHL**	**69**	**8**	**39**	**47**	**40**	**3**	**0**	**2**	**93**	**8.6**	**14**	....	....	....	....	....	**22**	**3**	**10**	**13**	**12**	**0**	**0**	**0**
1998-99	**Detroit**	**NHL**	**75**	**14**	**49**	**63**	**48**	**4**	**2**	**2**	**83**	**16.9**	**13**	**867**	**49.8**	**13**	**16**	**17:20**	**7**	**0**	**2**	**2**	**0**	**0**	**0**	**0**
99-2000	**Detroit**	**NHL**	**79**	**9**	**38**	**47**	**28**	**3**	**0**	**4**	**69**	**13.0**	**13**	**729**	**43.6**	**22**	**18**	**16:05**	**9**	**1**	**2**	**3**	**6**	**1**	**0**	**0**
2000-01	**Florida**	**NHL**	**26**	**5**	**6**	**11**	**10**	**2**	**0**	**0**	**15**	**33.3**	**–11**	**299**	**48.8**	**7**	**8**	**16:33**	....	....	....	....	....	....	....	....
	Detroit	**NHL**	**39**	**4**	**25**	**29**	**28**	**2**	**0**	**1**	**31**	**12.9**	**6**	**311**	**44.4**	**8**	**6**	**16:36**	**6**	**1**	**3**	**4**	**2**	**1**	**0**	**0**
2001-02♦	**Detroit**	**NHL**	**70**	**11**	**32**	**43**	**50**	**4**	**0**	**1**	**50**	**22.0**	**–5**	**653**	**43.3**	**12**	**16**	**14:28**	**18**	**5**	**6**	**11**	**4**	**0**	**0**	**1**
	Russia	Olympics	6	0	3	3	4	....	....	....	....	....	....	....	....	....	....	....	....	....	....	....	....	....	....	....
	NHL Totals		**798**	**158**	**432**	**590**	**406**	**52**	**10**	**28**	**971**	**16.3**		**2859**	**46.1**	**62**	**64**	**16:08**	**145**	**30**	**66**	**96**	**60**	**9**	**0**	**4**

Played in NHL All-Star Game (1998)

Claimed by **San Jose** from **Vancouver** in NHL Waiver Draft, October 4, 1992. Traded to **Detroit** by **San Jose** for Ray Sheppard, October 24, 1995. Signed as a free agent by **Florida**, July 1, 2000. Traded to **Detroit** by **Florida** for Yan Golubovsky, December 28, 2000.

LAROCQUE, Mario
(luh-RAWK, MAIR-ee-oh)

Defense. Shoots left. 6'2", 182 lbs. Born, Montreal, Que., April 24, 1978. Tampa Bay's 1st choice, 16th overall, in 1996 Entry Draft.

			Regular Season																Playoffs							
Season	Club	League	GP	G	A	Pts	PIM	PP	SH	GW	S	%	+/-	TF	F%	H	SB	Min	GP	G	A	Pts	PIM	PP	SH	GW
1994-95	Mtl-Bourassa	QAAA	43	0	6	6	153	....	....	....	....	....	....	....	....	....	....	....	....	....	....	....	....	....	....	....
1995-96	Hull Olympiques	QMJHL	68	7	19	26	196	....	....	....	....	....	....	....	....	....	....	....	14	2	5	7	16	....	....	....
1996-97	Hull Olympiques	QMJHL	64	14	36	50	155	....	....	....	....	....	....	....	....	....	....	....	14	2	6	8	36	....	....	....
1997-98	Sherbrooke	QMJHL	28	6	10	16	125	....	....	....	....	....	....	....	....	....	....	....	....	....	....	....	....	....	....	....
1998-99	**Tampa Bay**	**NHL**	**5**	**0**	**0**	**0**	**16**	**0**	**0**	**0**	**3**	**0.0**	**–4**	**0**	**0.0**	**8**	**1**	**12:33**	....	....	....	....	....	....	....	....
	Cleveland	IHL	59	5	7	12	202	....	....	....	....	....	....	....	....	....	....	....	....	....	....	....	....	....	....	....
99-2000	Detroit Vipers	IHL	60	0	5	5	234	....	....	....	....	....	....	....	....	....	....	....	....	....	....	....	....	....	....	....
2000-01	Detroit Vipers	IHL	71	2	1	3	233	....	....	....	....	....	....	....	....	....	....	....	....	....	....	....	....	....	....	....
2001-02	Rochester	AHL	75	8	7	15	219	....	....	....	....	....	....	....	....	....	....	....	2	0	0	0	0	....	....	....
	NHL Totals		**5**	**0**	**0**	**0**	**16**	**0**	**0**	**0**	**3**	**0.0**		**0**	**0.0**	**8**	**1**	**12:33**	....	....	....	....	....	....	....	....

QMJHL All-Rookie Team (1996)

Signed as a free agent by **Buffalo**, August 7, 2001.

LARSEN, Brad
(LARH-sehn, BRAD) **COL.**

Left wing. Shoots left. 6', 200 lbs. Born, Nakusp, B.C., June 28, 1977. Colorado's 5th choice, 87th overall, in 1997 Entry Draft.

			Regular Season																Playoffs							
Season	Club	League	GP	G	A	Pts	PIM	PP	SH	GW	S	%	+/-	TF	F%	H	SB	Min	GP	G	A	Pts	PIM	PP	SH	GW
1992-93	Nelson	RMJHL	42	31	37	68	164	....	....	....	....	....	....	....	....	....	....	....	....	....	....	....	....	....	....	....
1993-94	Swift Current	WHL	64	15	18	33	32	....	....	....	....	....	....	....	....	....	....	....	7	1	2	3	4	....	....	....
1994-95	Swift Current	WHL	62	24	33	57	73	....	....	....	....	....	....	....	....	....	....	....	6	0	1	1	2	....	....	....
1995-96	Swift Current	WHL	51	30	47	77	67	....	....	....	....	....	....	....	....	....	....	....	6	3	2	5	13	....	....	....
1996-97	Swift Current	WHL	61	36	46	82	61	....	....	....	....	....	....	....	....	....	....	....	....	....	....	....	....	....	....	....
1997-98	**Colorado**	**NHL**	**1**	**0**	**0**	**0**	**0**	**0**	**0**	**0**	**0**	**0.0**	**0**	....	....	....	....	....	....	....	....	....	....	....	....	....
	Hershey Bears	AHL	65	12	10	22	80	....	....	....	....	....	....	....	....	....	....	....	7	3	2	5	2	....	....	....
1998-99	Hershey Bears	AHL	18	3	4	7	11	....	....	....	....	....	....	....	....	....	....	....	5	0	1	1	6	....	....	....
99-2000	Hershey Bears	AHL	52	13	26	39	66	....	....	....	....	....	....	....	....	....	....	....	14	5	2	7	29	....	....	....
2000-01	**Colorado**	**NHL**	**9**	**0**	**0**	**0**	**0**	**0**	**0**	**0**	**3**	**0.0**	**1**	**14**	**57.1**	**19**	**4**	**9:17**	....	....	....	....	....	....	....	....
	Hershey Bears	AHL	67	21	25	46	93	....	....	....	....	....	....	....	....	....	....	....	10	1	3	4	6	....	....	....
2001-02	**Colorado**	**NHL**	**50**	**2**	**7**	**9**	**47**	**1**	**0**	**0**	**38**	**5.3**	**4**	**71**	**54.9**	**126**	**13**	**8:07**	**21**	**1**	**1**	**2**	**13**	**0**	**0**	**0**
	NHL Totals		**60**	**2**	**7**	**9**	**47**	**1**	**0**	**0**	**41**	**4.9**		**85**	**55.3**	**145**	**17**	**8:18**	**21**	**1**	**1**	**2**	**13**	**0**	**0**	**0**

• Re-entered NHL Entry Draft. Originally Ottawa's 3rd choice, 53rd overall, in 1995 Entry Draft.

WHL East Second All-Star Team (1997)

Rights traded to **Colorado** by **Ottawa** for Janne Laukkanen, January 26, 1996. • Missed majority of 1998-99 season recovering from abdominal injury suffered in game vs. Albany (AHL), November 20, 1998.

LAUKKANEN, Janne
(LOW-kah-nehn, YAN-nee) **PIT.**

Defense. Shoots left. 6'1", 196 lbs. Born, Lahti, Finland, March 19, 1970. Quebec's 8th choice, 156th overall, in 1991 Entry Draft.

			Regular Season																Playoffs							
Season	Club	League	GP	G	A	Pts	PIM	PP	SH	GW	S	%	+/-	TF	F%	H	SB	Min	GP	G	A	Pts	PIM	PP	SH	GW
1986-87	K. Reipas Jr.	Finn-Jr.	1	0	0	0	0	....	....	....	....	....	....	....	....	....	....	....	....	....	....	....	....	....	....	....
1987-88	K. Reipas-B	Finn-Jr.	20	5	5	10	48	....	....	....	....	....	....	....	....	....	....	....	....	....	....	....	....	....	....	....
1988-89	Army Jr.	Finn-Jr.	6	0	1	1	6	....	....	....	....	....	....	....	....	....	....	....	....	....	....	....	....	....	....	....
	Hockey-Reipas	Finland-2	33	1	7	8	24	....	....	....	....	....	....	....	....	....	....	....	....	....	....	....	....	....	....	....
1989-90	H. Reipas Jr.	Finn-Jr.	2	2	2	4	2	....	....	....	....	....	....	....	....	....	....	....	....	....	....	....	....	....	....	....
	Hockey-Reipas	Finland-2	44	8	22	30	60	....	....	....	....	....	....	....	....	....	....	....	....	....	....	....	....	....	....	....
1990-91	Reipas Lahti	Finland	44	8	14	22	56	....	....	....	....	....	....	....	....	....	....	....	....	....	....	....	....	....	....	....
1991-92	HPK Hameenlinna	Finland	43	5	14	19	62	....	....	....	....	....	....	....	....	....	....	....	....	....	....	....	....	....	....	....
	Finland	Olympics	8	0	1	1	6	....	....	....	....	....	....	....	....	....	....	....	....	....	....	....	....	....	....	....
1992-93	HPK Hameenlinna	Finland	47	8	21	29	76	....	....	....	....	....	....	....	....	....	....	....	12	1	4	5	10	....	....	....
1993-94	HPK Hameenlinna	Finland	48	5	24	29	46	....	....	....	....	....	....	....	....	....	....	....	....	....	....	....	....	....	....	....
	Finland	Olympics	8	0	2	2	12	....	....	....	....	....	....	....	....	....	....	....	....	....	....	....	....	....	....	....
	Ceske Budejovice	Czech	....	....	....	....	....	....	....	....	....	....	....	....	....	....	....	....	3	0	1	1	0	....	....	....
1994-95	Cornwall Aces	AHL	55	8	26	34	41	....	....	....	....	....	....	....	....	....	....	....	....	....	....	....	....	....	....	....
	Quebec	**NHL**	**11**	**0**	**3**	**3**	**4**	**0**	**0**	**0**	**12**	**0.0**	**3**	....	....	....	....	....	**6**	**1**	**0**	**1**	**2**	**0**	**0**	**0**
1995-96	**Colorado**	**NHL**	**3**	**1**	**0**	**1**	**0**	**1**	**0**	**0**	**4**	**25.0**	**–1**	....	....	....	....	....	....	....	....	....	....	....	....	....
	Cornwall Aces	AHL	35	7	20	27	60	....	....	....	....	....	....	....	....	....	....	....	....	....	....	....	....	....	....	....
	Ottawa	**NHL**	**20**	**0**	**2**	**2**	**14**	**0**	**0**	**0**	**31**	**0.0**	**0**	....	....	....	....	....	....	....	....	....	....	....	....	....
1996-97	**Ottawa**	**NHL**	**76**	**3**	**18**	**21**	**76**	**2**	**0**	**0**	**109**	**2.8**	**–14**	....	....	....	....	....	**7**	**0**	**1**	**1**	**6**	**0**	**0**	**0**
1997-98	**Ottawa**	**NHL**	**60**	**4**	**17**	**21**	**64**	**2**	**0**	**2**	**69**	**5.8**	**–15**	....	....	....	....	....	**11**	**2**	**2**	**4**	**8**	**1**	**0**	**1**
	Finland	Olympics	6	0	0	0	4	....	....	....	....	....	....	....	....	....	....	....	....	....	....	....	....	....	....	....
1998-99	**Ottawa**	**NHL**	**50**	**1**	**11**	**12**	**40**	**0**	**0**	**0**	**46**	**2.2**	**18**	**0**	**0.0**	**87**	**80**	**18:37**	**4**	**0**	**0**	**0**	**4**	**0**	**0**	**0**
99-2000	**Ottawa**	**NHL**	**60**	**1**	**11**	**12**	**55**	**0**	**0**	**0**	**62**	**1.6**	**14**	**0**	**0.0**	**108**	**108**	**19:47**	....	....	....	....	....	....	....	....
	Pittsburgh	**NHL**	**11**	**1**	**7**	**8**	**12**	**1**	**0**	**0**	**19**	**5.3**	**3**	**0**	**0.0**	**20**	**9**	**16:55**	**11**	**2**	**4**	**6**	**10**	**1**	**0**	**1**
2000-01	**Pittsburgh**	**NHL**	**50**	**3**	**17**	**20**	**34**	**0**	**0**	**0**	**58**	**5.2**	**9**	**0**	**0.0**	**96**	**64**	**18:28**	**18**	**2**	**2**	**4**	**14**	**1**	**0**	**0**
2001-02	**Pittsburgh**	**NHL**	**47**	**6**	**7**	**13**	**28**	**3**	**0**	**1**	**66**	**9.1**	**–18**	**0**	**0.0**	**69**	**51**	**17:51**	....	....	....	....	....	....	....	....
	NHL Totals		**388**	**20**	**93**	**113**	**327**	**9**	**0**	**3**	**476**	**4.2**		**0**	**0.0**	**380**	**312**	**18:39**	**57**	**7**	**9**	**16**	**44**	**3**	**0**	**2**

Transferred to **Colorado** after **Quebec** franchise relocated, June 21, 1995. Traded to **Ottawa** by **Colorado** for the rights to Brad Larsen, January 26, 1996. Traded to **Pittsburgh** by **Ottawa** with Ron Tugnutt for Tom Barrasso, March 14, 2000.

LAUS, Paul
(LOWZ, PAWL) **FLA.**

Defense. Shoots right. 6'1", 215 lbs. Born, Beamsville, Ont., September 26, 1970. Pittsburgh's 2nd choice, 37th overall, in 1989 Entry Draft.

			Regular Season																Playoffs							
Season	Club	League	GP	G	A	Pts	PIM	PP	SH	GW	S	%	+/-	TF	F%	H	SB	Min	GP	G	A	Pts	PIM	PP	SH	GW
1986-87	St. Catharines	OJHL-B	40	1	8	9	56																			
1987-88	Hamilton	OHL	56	1	9	10	171												14	0	0	0	28			
1988-89	Niagara Falls	OHL	49	1	10	11	225												15	0	5	5	56			
1989-90	Niagara Falls	OHL	60	13	35	48	231												16	6	16	22	71			
1990-91	Albany Choppers	IHL	7	0	0	0	7																			
	Knoxville	ECHL	20	6	12	18	83																			
	Muskegon	IHL	35	3	4	7	103												4	0	0	0	13			
1991-92	Muskegon	IHL	75	0	21	21	248												14	2	5	7	70			
1992-93	Cleveland	IHL	76	8	18	26	427												4	1	0	1	27			
1993-94	**Florida**	**NHL**	**39**	**2**	**0**	**2**	**109**	**0**	**0**	**1**	**15**	**13.3**	**9**													
1994-95	**Florida**	**NHL**	**37**	**0**	**7**	**7**	**138**	**0**	**0**	**0**	**18**	**0.0**	**12**													
1995-96	**Florida**	**NHL**	**78**	**3**	**6**	**9**	**236**	**0**	**0**	**0**	**45**	**6.7**	**-2**						**21**	**2**	**6**	**8**	***62**	**0**	**0**	**0**
1996-97	**Florida**	**NHL**	**77**	**0**	**12**	**12**	**313**	**0**	**0**	**0**	**63**	**0.0**	**13**						**5**	**0**	**1**	**1**	**4**	**0**	**0**	**0**
1997-98	**Florida**	**NHL**	**77**	**0**	**11**	**11**	**293**	**0**	**0**	**0**	**64**	**0.0**	**-5**													
1998-99	**Florida**	**NHL**	**75**	**1**	**9**	**10**	**218**	**0**	**0**	**0**	**54**	**1.9**	**-1**	**0**	**0.0**	**93**	**20**	**11:09**								
99-2000	**Florida**	**NHL**	**77**	**3**	**8**	**11**	**172**	**0**	**0**	**0**	**44**	**6.8**	**-1**	**1**	**0.0**	**106**	**15**	**7:37**	**4**	**0**	**0**	**0**	**8**	**0**	**0**	**0**
2000-01	**Florida**	**NHL**	**25**	**1**	**2**	**3**	**66**	**0**	**0**	**0**	**18**	**5.6**	**5**	**0**	**0.0**	**63**	**13**	**14:22**								
2001-02	**Florida**	**NHL**	**45**	**4**	**3**	**7**	**157**	**0**	**1**	**0**	**39**	**10.3**	**1**	**0**	**0.0**	**106**	**20**	**14:10**								
	NHL Totals		**530**	**14**	**58**	**72**	**1702**	**0**	**1**	**1**	**360**	**3.9**		**1**	**0.0**	**368**	**68**	**10:54**	**30**	**2**	**7**	**9**	**74**	**0**	**0**	**0**

Claimed by **Florida** from **Pittsburgh** in Expansion Draft, June 24, 1993. • Missed majority of 2000-01 season recovering from hernia injury suffered in game vs. Carolina, November 15, 2000.

LAW, Kirby
(LAW, KUHR-bee) **PHI.**

Right wing. Shoots right. 6'1", 185 lbs. Born, McCreary, Man., March 11, 1977.

			Regular Season																Playoffs							
Season	Club	League	GP	G	A	Pts	PIM	PP	SH	GW	S	%	+/-	TF	F%	H	SB	Min	GP	G	A	Pts	PIM	PP	SH	GW
1991-92	McCreary	MAHA	60	89	103	192	60																			
1992-93	Dauphin Kings	MJHL	48	20	15	35	8																			
1993-94	Saskatoon Blades	WHL	66	9	11	20	39												16	0	0	0	6			
1994-95	Saskatoon Blades	WHL	46	10	15	25	44																			
	Lethbridge	WHL	24	4	10	14	38																			
1995-96	Lethbridge	WHL	71	17	45	62	133												4	0	0	0	12			
1996-97	Lethbridge	WHL	72	39	52	91	200												19	4	14	18	60			
1997-98	Brandon	WHL	49	34	44	78	153												9	3	3	6	41			
1998-99	Orlando	IHL	67	18	13	31	136																			
	Adirondack	AHL	11	2	3	5	40												3	1	0	1	2			
99-2000	Louisville	AHL	66	31	21	52	173																			
	Orlando	IHL	1	1	0	1	0																			
	Philadelphia	AHL	12	1	4	5	6												5	2	0	2	2			
2000-01	**Philadelphia**	**NHL**	**1**	**0**	**0**	**0**	**0**	**0**	**0**	**0**	**0**	**0.0**	**-1**	**0**	**0.0**	**0**	**0**	**3:23**								
	Philadelphia	AHL	78	27	34	61	150												10	1	6	7	16			
2001-02	Philadelphia	AHL	71	18	24	42	102												5	0	0	0	0			
	NHL Totals		**1**	**0**	**0**	**0**	**0**	**0**	**0**	**0**	**0**	**0.0**		**0**	**0.0**	**0**	**0**	**3:23**								

Traded to **Brandon** (WHL) by **Lethbridge** (WHL) for Jason Boyd, September 1, 1997. Signed as a free agent by **Atlanta**, July 27, 1999. Traded to **Philadelphia** by **Atlanta** for Vancouver's 6th round choice (previously acquired, Atlanta selected Jeff Dwyer) in 2000 Entry Draft and Philadelphia's 6th round choice (Pasi Nurminen) in 2001 Entry Draft, March 14, 2000.

LECAVALIER, Vincent
(luh-KAV-uhl-YAY, VIHN-sihnt) **T.B.**

Center. Shoots left. 6'4", 205 lbs. Born, Ile Bizard, Que., April 21, 1980. Tampa Bay's 1st choice, 1st overall, in 1998 Entry Draft.

			Regular Season																Playoffs							
Season	Club	League	GP	G	A	Pts	PIM	PP	SH	GW	S	%	+/-	TF	F%	H	SB	Min	GP	G	A	Pts	PIM	PP	SH	GW
1995-96	Notre Dame	SMHL	22	52	52	104																				
1996-97	Rimouski Oceanic	QMJHL	64	42	61	103	38												4	4	3	7	2			
1997-98	Rimouski Oceanic	QMJHL	58	44	71	115	117												18	*15	*26	*41	46			
1998-99	**Tampa Bay**	**NHL**	**82**	**13**	**15**	**28**	**23**	**2**	**0**	**2**	**125**	**10.4**	**-19**	**953**	**40.3**	**52**	**15**	**13:40**								
99-2000	**Tampa Bay**	**NHL**	**80**	**25**	**42**	**67**	**43**	**6**	**0**	**3**	**166**	**15.1**	**-25**	**1288**	**44.4**	**116**	**19**	**19:18**								
2000-01	**Tampa Bay**	**NHL**	**68**	**23**	**28**	**51**	**66**	**7**	**0**	**3**	**165**	**13.9**	**-26**	**1278**	**44.9**	**72**	**22**	**19:57**								
2001-02	**Tampa Bay**	**NHL**	**76**	**20**	**17**	**37**	**61**	**5**	**0**	**3**	**164**	**12.2**	**-18**	**931**	**41.5**	**85**	**18**	**17:09**								
	NHL Totals		**306**	**81**	**102**	**183**	**193**	**20**	**0**	**11**	**620**	**13.1**		**4450**	**43.1**	**325**	**74**	**17:24**								

QMJHL All-Rookie Team (1997) • QMJHL Offensive Rookie of the Year (1997) • QMJHL First All-Star Team (1998) • Canadian Major Junior First All-Star Team (1998) • Canadian Major Junior Rookie of the Year (1997)

LeCLAIR, John
(luh-KLAIR, JAWN) **PHI.**

Left wing. Shoots left. 6'3", 226 lbs. Born, St. Albans, VT, July 5, 1969. Montreal's 2nd choice, 33rd overall, in 1987 Entry Draft.

			Regular Season																Playoffs							
Season	Club	League	GP	G	A	Pts	PIM	PP	SH	GW	S	%	+/-	TF	F%	H	SB	Min	GP	G	A	Pts	PIM	PP	SH	GW
1985-86	Bellows	Hi-School	22	41	28	69	14																			
1986-87	Bellows	Hi-School	23	44	40	84	14																			
1987-88	U. of Vermont	ECAC	31	12	22	34	62																			
1988-89	U. of Vermont	ECAC	18	9	12	21	40																			
1989-90	U. of Vermont	ECAC	10	10	6	16	38																			
1990-91	U. of Vermont	ECAC	33	25	20	45	58																			
	Montreal	**NHL**	**10**	**2**	**5**	**7**	**2**	**0**	**0**	**1**	**12**	**16.7**	**1**						**3**	**0**	**0**	**0**	**0**	**0**	**0**	**0**
1991-92	**Montreal**	**NHL**	**59**	**8**	**11**	**19**	**14**	**3**	**0**	**0**	**73**	**11.0**	**5**						**8**	**1**	**1**	**2**	**4**	**0**	**0**	**0**
	Fredericton	AHL	8	7	7	14	10												2	0	0	0	4			
1992-93♦	**Montreal**	**NHL**	**72**	**19**	**25**	**44**	**33**	**2**	**0**	**2**	**139**	**13.7**	**11**						**20**	**4**	**6**	**10**	**14**	**0**	**0**	**3**
1993-94	**Montreal**	**NHL**	**74**	**19**	**24**	**43**	**32**	**1**	**0**	**1**	**153**	**12.4**	**17**						**7**	**2**	**1**	**3**	**8**	**1**	**0**	**0**
1994-95	**Montreal**	**NHL**	**9**	**1**	**4**	**5**	**10**	**1**	**0**	**0**	**18**	**5.6**	**-1**													
	Philadelphia	**NHL**	**37**	**25**	**24**	**49**	**20**	**5**	**0**	**7**	**113**	**22.1**	**21**						**15**	**5**	**7**	**12**	**4**	**1**	**0**	**1**
1995-96	**Philadelphia**	**NHL**	**82**	**51**	**46**	**97**	**64**	**19**	**0**	**10**	**270**	**18.9**	**21**						**11**	**6**	**5**	**11**	**6**	**4**	**0**	**1**
1996-97	**Philadelphia**	**NHL**	**82**	**50**	**47**	**97**	**58**	**10**	**0**	**5**	**324**	**15.4**	**44**						**19**	**9**	**12**	**21**	**10**	**4**	**0**	**3**
1997-98	**Philadelphia**	**NHL**	**82**	**51**	**36**	**87**	**32**	**16**	**0**	**9**	**303**	**16.8**	**30**						**5**	**1**	**1**	**2**	**8**	**1**	**0**	**1**
	United States	Olympics	4	0	1	1	0																			
1998-99	**Philadelphia**	**NHL**	**76**	**43**	**47**	**90**	**30**	**16**	**0**	**7**	**246**	**17.5**	**36**	**7**	**14.3**	**86**	**10**	**21:03**	**6**	**3**	**0**	**3**	**12**	**2**	**0**	**0**
99-2000	**Philadelphia**	**NHL**	**82**	**40**	**37**	**77**	**36**	**13**	**0**	**7**	**249**	**16.1**	**8**	**7**	**28.6**	**106**	**26**	**20:18**	**18**	**6**	**7**	**13**	**6**	**4**	**0**	**2**
2000-01	**Philadelphia**	**NHL**	**16**	**7**	**5**	**12**	**0**	**3**	**0**	**2**	**48**	**14.6**	**2**	**0**	**0.0**	**15**	**3**	**19:06**	**6**	**1**	**2**	**3**	**2**	**0**	**0**	**0**
2001-02	**Philadelphia**	**NHL**	**82**	**25**	**26**	**51**	**30**	**4**	**0**	**6**	**220**	**11.4**	**5**	**4**	**75.0**	**91**	**10**	**17:30**	**5**	**0**	**0**	**0**	**2**	**0**	**0**	**0**
	United States	Olympics	6	*6	1	7	2																			
	NHL Totals		**763**	**341**	**337**	**678**	**361**	**93**	**0**	**57**	**2168**	**15.7**		**18**	**33.3**	**298**	**49**	**19:33**	**123**	**38**	**42**	**80**	**76**	**17**	**0**	**11**

ECAC Second All-Star Team (1991) • NHL First All-Star Team (1995, 1998) • NHL Second All-Star Team (1996, 1997, 1999) • Won Bud Light Plus/Minus Award (1997) • Won Bud Ice Plus/Minus Award (1999) • Played in NHL All-Star Game (1996, 1997, 1998, 1999, 2000)

• Missed majority of 1989-90 season recovering from knee surgery, January 20, 1990. Traded to **Philadelphia** by **Montreal** with Eric Desjardins and Gilbert Dionne for Mark Recchi and Philadelphia's 3rd round choice (Martin Hohenberger) in 1995 Entry Draft, February 9, 1995. • Missed majority of 2000-01 season recovering from back injury suffered in game vs. Boston, October 7, 2000.

LECLERC, Mike
(luh-KLUHRK, MIGHK) **ANA.**

Left wing. Shoots left. 6'2", 208 lbs. Born, Winnipeg, Man., November 10, 1976. Anaheim's 3rd choice, 55th overall, in 1995 Entry Draft.

			Regular Season																Playoffs							
Season	Club	League	GP	G	A	Pts	PIM	PP	SH	GW	S	%	+/-	TF	F%	H	SB	Min	GP	G	A	Pts	PIM	PP	SH	GW
1991-92	St. Boniface	MJHL	43	16	12	28	25																			
	Victoria Cougars	WHL	2	0	0	0	0																			
1992-93	Victoria Cougars	WHL	70	4	11	15	118																			
1993-94	Victoria Cougars	WHL	68	29	11	40	112																			
1994-95	Prince George	WHL	43	20	36	56	78																			
	Brandon	WHL	23	5	8	13	50												18	10	6	16	33			
1995-96	Brandon	WHL	71	58	53	111	161												19	6	19	25	25			
1996-97	**Anaheim**	**NHL**	**5**	**1**	**1**	**2**	**0**	**0**	**0**	**1**	**3**	**33.3**	**2**						**1**	**0**	**0**	**0**	**0**	**0**	**0**	**0**
	Baltimore	AHL	71	29	27	56	134																			
1997-98	**Anaheim**	**NHL**	**7**	**0**	**0**	**0**	**6**	**0**	**0**	**0**	**11**	**0.0**	**-6**													
	Cincinnati	AHL	48	18	22	40	83																			
1998-99	**Anaheim**	**NHL**	**7**	**0**	**0**	**0**	**4**	**0**	**0**	**0**	**1**	**0.0**	**-2**	**0**	**0.0**	**9**	**2**	**5:52**	**1**	**0**	**0**	**0**	**0**	**0**	**0**	**0**
	Cincinnati	AHL	65	25	28	53	153												3	0	1	1	19			
99-2000	**Anaheim**	**NHL**	**69**	**8**	**11**	**19**	**70**	**0**	**0**	**2**	**105**	**7.6**	**-15**	**1**	**0.0**	**145**	**11**	**12:08**								

Season	Club	League	GP	G	A	Pts	PIM	PP	SH	GW	S	%	+/-	TF	F%	H	SB	Min	GP	G	A	Pts	PIM	PP	SH	GW
			Regular Season																Playoffs							
2000-01	**Anaheim**	**NHL**	**54**	**15**	**20**	**35**	**26**	**3**	**0**	**3**	**130**	**11.5**	**-1**	**5**	**20.0**	**87**	**23**	**17:35**								
2001-02	**Anaheim**	**NHL**	**82**	**20**	**24**	**44**	**107**	**8**	**0**	**4**	**178**	**11.2**	**-12**	**10**	**50.0**	**119**	**23**	**17:23**								
	NHL Totals		**224**	**44**	**56**	**100**	**213**	**11**	**0**	**10**	**428**	**10.3**		**16**	**37.5**	**360**	**59**	**15:21**	**2**	**0**	**0**	**0**	**0**	**0**	**0**	**0**

WHL East Second All-Star Team (1996)

LEDYARD, Grant

(LEHD-yahrd, GRANT)

Defense. Shoots left. 6'2", 195 lbs. Born, Winnipeg, Man., November 19, 1961.

Season	Club	League	GP	G	A	Pts	PIM	PP	SH	GW	S	%	+/-	TF	F%	H	SB	Min	GP	G	A	Pts	PIM	PP	SH	GW
1979-80	Fort Garry Blues	MJHL	49	13	24	37	90																			
1980-81	Saskatoon Blades	WHL	71	9	28	37	148																			
1981-82	Fort Garry Blues	MJHL	63	25	45	70	150																			
1982-83	Tulsa Oilers	CHL	80	13	29	42	115																			
1983-84	Tulsa Oilers	CHL	58	9	17	26	71												9	5	4	9	10			
1984-85	**NY Rangers**	**NHL**	**42**	**8**	**12**	**20**	**53**	**1**	**0**	**1**	**91**	**8.8**	**8**						**3**	**0**	**2**	**2**	**4**	**0**	**0**	**0**
	New Haven	AHL	36	6	20	26	18																			
1985-86	**NY Rangers**	**NHL**	**27**	**2**	**9**	**11**	**20**	**0**	**0**	**0**	**57**	**3.5**	**-7**													
	Los Angeles	**NHL**	**52**	**7**	**18**	**25**	**78**	**4**	**0**	**2**	**113**	**6.2**	**-22**													
1986-87	**Los Angeles**	**NHL**	**67**	**14**	**23**	**37**	**93**	**5**	**0**	**1**	**144**	**9.7**	**-40**						**5**	**0**	**0**	**0**	**10**	**0**	**0**	**0**
1987-88	**Los Angeles**	**NHL**	**23**	**1**	**7**	**8**	**52**	**1**	**0**	**0**	**40**	**2.5**	**-7**													
	New Haven	AHL	3	2	1	3	4																			
	Washington	**NHL**	**21**	**4**	**3**	**7**	**14**	**1**	**0**	**1**	**41**	**9.8**	**-4**						**14**	**1**	**0**	**1**	**30**	**0**	**0**	**0**
1988-89	**Washington**	**NHL**	**61**	**3**	**11**	**14**	**43**	**1**	**0**	**1**	**81**	**3.7**	**1**													
	Buffalo	**NHL**	**13**	**1**	**5**	**6**	**8**	**0**	**0**	**1**	**25**	**4.0**	**1**						**5**	**1**	**2**	**3**	**2**	**0**	**0**	**0**
1989-90	**Buffalo**	**NHL**	**67**	**2**	**13**	**15**	**37**	**0**	**0**	**1**	**91**	**2.2**	**2**													
1990-91	**Buffalo**	**NHL**	**60**	**8**	**23**	**31**	**46**	**2**	**1**	**1**	**118**	**6.8**	**13**						**6**	**3**	**3**	**6**	**10**	**0**	**0**	**0**
1991-92	**Buffalo**	**NHL**	**50**	**5**	**16**	**21**	**45**	**0**	**0**	**0**	**87**	**5.7**	**-4**													
1992-93	**Buffalo**	**NHL**	**50**	**2**	**14**	**16**	**45**	**1**	**0**	**0**	**79**	**2.5**	**-2**						**8**	**0**	**0**	**0**	**8**	**0**	**0**	**0**
	Rochester	AHL	5	0	2	2	8																			
1993-94	**Dallas**	**NHL**	**84**	**9**	**37**	**46**	**42**	**6**	**0**	**1**	**177**	**5.1**	**7**						**9**	**1**	**2**	**3**	**6**	**0**	**0**	**1**
1994-95	**Dallas**	**NHL**	**38**	**5**	**13**	**18**	**20**	**4**	**0**	**0**	**79**	**6.3**	**6**						**3**	**0**	**0**	**0**	**2**	**0**	**0**	**0**
1995-96	**Dallas**	**NHL**	**73**	**5**	**19**	**24**	**20**	**2**	**0**	**1**	**123**	**4.1**	**-15**													
1996-97	**Dallas**	**NHL**	**67**	**1**	**15**	**16**	**61**	**0**	**0**	**0**	**99**	**1.0**	**31**						**7**	**0**	**2**	**2**	**0**	**0**	**0**	**0**
1997-98	**Vancouver**	**NHL**	**49**	**2**	**13**	**15**	**14**	**1**	**0**	**0**	**57**	**3.5**	**-2**													
	Boston	**NHL**	**22**	**2**	**7**	**9**	**6**	**1**	**0**	**0**	**33**	**6.1**	**-2**						**6**	**0**	**0**	**0**	**2**	**0**	**0**	**0**
1998-99	**Boston**	**NHL**	**47**	**4**	**8**	**12**	**33**	**1**	**0**	**2**	**47**	**8.5**	**-8**	**1**	**0.0**	**65**	**49**	**17:31**	**2**	**0**	**0**	**0**	**2**	**0**	**0**	**0**
99-2000	**Ottawa**	**NHL**	**40**	**2**	**4**	**6**	**8**	**0**	**0**	**1**	**42**	**4.8**	**-3**	**0**	**0.0**	**58**	**31**	**14:45**	**6**	**0**	**0**	**0**	**16**	**0**	**0**	**0**
2000-01	**Tampa Bay**	**NHL**	**14**	**2**	**2**	**4**	**12**	**0**	**0**	**0**	**12**	**16.7**	**-5**	**0**	**0.0**	**20**	**14**	**18:15**								
	Dallas	**NHL**	**8**	**0**	**1**	**1**	**4**	**0**	**0**	**0**	**7**	**0.0**	**3**	**0**	**0.0**	**12**	**3**	**15:26**	**9**	**0**	**1**	**1**	**4**	**0**	**0**	**0**
2001-02	**Tampa Bay**	**NHL**	**53**	**1**	**3**	**4**	**12**	**0**	**0**	**0**	**27**	**3.7**	**-5**	**2**	**50.0**	**50**	**44**	**14:36**								
	NHL Totals		**1028**	**90**	**276**	**366**	**766**	**31**	**1**	**14**	**1670**	**5.4**		**3**	**33.3**	**205**	**141**	**15:50**	**83**	**6**	**12**	**18**	**96**	**0**	**0**	**1**

Won Bob Gassoff Trophy (Most Improved Defenseman - CHL) (1984)

Signed as a free agent by **NY Rangers**, July 7, 1982. Traded to **LA Kings** by **NY Rangers** with Rollie Melanson for LA Kings' 4th round choice (Mike Sullivan) in 1987 Entry Draft and Brian MacLellan, December 9, 1985. Traded to **Washington** by **LA Kings** for Craig Laughlin, February 9, 1988. Traded to **Buffalo** by **Washington** with Clint Malarchuk and Washington's 6th round choice (Brian Holzinger) in 1991 Entry Draft for Calle Johansson and Buffalo's 2nd round choice (Byron Dafoe) in 1989 Entry Draft, March 7, 1989. Signed as a free agent by **Dallas**, August 12, 1993. Signed as a free agent by **Vancouver**, July 17, 1997. Traded to **Boston** by **Vancouver** for Boston's 8th round choice (Curtis Valentine) in 1998 Entry Draft, March 3, 1998. Signed as a free agent by **Ottawa**, November 16, 1999. Signed as a free agent by **Tampa Bay**, January 31, 2001. Traded to **Dallas** by **Tampa Bay** for Dallas' 7th round choice (Tampa Bay selected Jeremy Van Hoof) in 2001 Entry Draft, March 13, 2001. Signed as a free agent by **Tampa Bay**, July 13, 2001.

LEEB, Brad

(LEEB, BRAD) **VAN.**

Right wing. Shoots right. 5'11", 180 lbs. Born, Red Deer, Alta., August 27, 1979.

Season	Club	League	GP	G	A	Pts	PIM	PP	SH	GW	S	%	+/-	TF	F%	H	SB	Min	GP	G	A	Pts	PIM	PP	SH	GW
1994-95	Red Deer	AMHL	36	31	14	45	93																			
	Red Deer Rebels	WHL	3	0	0	0	4																			
1995-96	Red Deer Rebels	WHL	38	3	6	9	30												10	2	0	2	11			
1996-97	Red Deer Rebels	WHL	70	15	20	35	76												16	3	3	6	6			
1997-98	Red Deer Rebels	WHL	63	23	23	46	88												3	2	0	2	2			
1998-99	Red Deer Rebels	WHL	64	32	47	79	84												9	5	9	14	10			
99-2000	**Vancouver**	**NHL**	**2**	**0**	**0**	**0**	**2**	**0**	**0**	**0**	**3**	**0.0**	**-2**	**0**	**0.0**	**2**	**1**	**12:07**								
	Syracuse Crunch	AHL	61	19	18	37	50												4	0	0	0	6			
2000-01	Kansas City	IHL	53	18	16	34	53																			
2001-02	**Vancouver**	**NHL**	**2**	**0**	**0**	**0**	**0**	**0**	**0**	**0**	**1**	**0.0**	**1**	**0**	**0.0**	**2**	**0**	**9:35**								
	Manitoba Moose	AHL	60	17	15	32	45																			
	NHL Totals		**4**	**0**	**0**	**0**	**2**	**0**	**0**	**0**	**4**	**0.0**		**0**	**0.0**	**4**	**1**	**9:35**								

WHL East Second All-Star Team (1999)

Signed as a free agent by **Vancouver**, October 8, 1999.

LEEB, Greg

(LEEB, GREHG)

Center. Shoots left. 5'9", 165 lbs. Born, Red Deer, Alta., May 31, 1977.

Season	Club	League	GP	G	A	Pts	PIM	PP	SH	GW	S	%	+/-	TF	F%	H	SB	Min	GP	G	A	Pts	PIM	PP	SH	GW
1993-94	Red Deer Royals	AMHL	36	19	30	49	24																			
1994-95	Spokane Chiefs	WHL	72	21	34	55	48												11	5	10	15	10			
1995-96	Spokane Chiefs	WHL	64	33	21	54	54												18	1	7	8	16			
1996-97	Spokane Chiefs	WHL	72	27	59	86	69												9	3	3	6	4			
1997-98	Spokane Chiefs	WHL	68	46	50	96	54												18	10	10	20	10			
1998-99	Michigan K-Wings	IHL	77	16	27	43	18												5	0	3	3	4			
99-2000	Michigan K-Wings	IHL	73	9	17	26	76																			
2000-01	**Dallas**	**NHL**	**2**	**0**	**0**	**0**	**0**	**0**	**0**	**0**	**4**	**0.0**	**-1**	**17**	**47.1**	**2**	**0**	**6:56**								
	Utah Grizzlies	IHL	78	25	40	65	36																			
2001-02	Hamilton	AHL	79	14	17	31	36												15	0	2	2	6			
	NHL Totals		**2**	**0**	**0**	**0**	**0**	**0**	**0**	**0**	**4**	**0.0**		**17**	**47.1**	**2**	**0**	**6:56**								

WHL West Second All-Star Team (1998)

Signed as a free agent by **Dallas**, July 24, 1998. Signed as a free agent by **Edmonton**, July 17, 2001. Signed as a free agent by **Augsburg** (Germany), July 24, 2002.

LEETCH, Brian

(LEECH, BRIGH-uhn) **NYR**

Defense. Shoots left. 6'1", 190 lbs. Born, Corpus Christi, TX, March 3, 1968. NY Rangers' 1st choice, 9th overall, in 1986 Entry Draft.

Season	Club	League	GP	G	A	Pts	PIM	PP	SH	GW	S	%	+/-	TF	F%	H	SB	Min	GP	G	A	Pts	PIM	PP	SH	GW
1983-84	Avon Old Farms	Hi-School	28	52	49	101	24																			
1984-85	Avon Old Farms	Hi-School	26	30	46	76	15																			
1985-86	Avon Old Farms	Hi-School	28	40	44	84	18																			
1986-87	Boston College	H-East	37	9	38	47	10																			
1987-88	Team USA	Nat-Tm	50	13	61	74	38																			
	United States	Olympics	6	1	5	6	4																			
	NY Rangers	**NHL**	**17**	**2**	**12**	**14**	**0**	**1**	**0**	**1**	**40**	**5.0**	**5**													
1988-89	**NY Rangers**	**NHL**	**68**	**23**	**48**	**71**	**50**	**8**	**3**	**1**	**268**	**8.6**	**8**						**4**	**3**	**2**	**5**	**2**	**2**	**0**	**0**
1989-90	**NY Rangers**	**NHL**	**72**	**11**	**45**	**56**	**26**	**5**	**0**	**2**	**222**	**5.0**	**-18**													
1990-91	**NY Rangers**	**NHL**	**80**	**16**	**72**	**88**	**42**	**6**	**0**	**4**	**206**	**7.8**	**2**						**6**	**1**	**3**	**4**	**0**	**0**	**0**	**0**
1991-92	**NY Rangers**	**NHL**	**80**	**22**	**80**	**102**	**26**	**10**	**1**	**3**	**245**	**9.0**	**25**						**13**	**4**	**11**	**15**	**4**	**1**	**1**	**0**
1992-93	**NY Rangers**	**NHL**	**36**	**6**	**30**	**36**	**26**	**2**	**1**	**1**	**150**	**4.0**	**2**													
1993-94♦	**NY Rangers**	**NHL**	**84**	**23**	**56**	**79**	**67**	**17**	**1**	**4**	**328**	**7.0**	**28**						**23**	**11**	***23**	***34**	**6**	**4**	**0**	**4**
1994-95	**NY Rangers**	**NHL**	**48**	**9**	**32**	**41**	**18**	**3**	**0**	**2**	**182**	**4.9**	**0**						**10**	**6**	**8**	**14**	**8**	**3**	**0**	**1**
1995-96	**NY Rangers**	**NHL**	**82**	**15**	**70**	**85**	**30**	**7**	**0**	**3**	**276**	**5.4**	**12**						**11**	**1**	**6**	**7**	**4**	**1**	**0**	**0**
1996-97	**NY Rangers**	**NHL**	**82**	**20**	**58**	**78**	**40**	**9**	**0**	**2**	**256**	**7.8**	**31**						**15**	**2**	**8**	**10**	**6**	**1**	**0**	**1**
1997-98	**NY Rangers**	**NHL**	**76**	**17**	**33**	**50**	**32**	**11**	**0**	**2**	**230**	**7.4**	**-36**													
	United States	Olympics	4	1	1	2	0																			
1998-99	**NY Rangers**	**NHL**	**82**	**13**	**42**	**55**	**42**	**4**	**0**	**1**	**184**	**7.1**	**-7**	**0**	**0.0**	**173**	**212**	**29:52**								
99-2000	**NY Rangers**	**NHL**	**50**	**7**	**19**	**26**	**20**	**3**	**0**	**2**	**124**	**5.6**	**-16**	**0**	**0.0**	**82**	**94**	**26:57**								
2000-01	**NY Rangers**	**NHL**	**82**	**21**	**58**	**79**	**34**	**10**	**1**	**3**	**241**	**8.7**	**-18**	**0**	**0.0**	**112**	**177**	**29:21**								

Season	Club	League	GP	G	A	Pts	PIM	PP	SH	GW	S	%	+/-	TF	F%	H	SB	Min	GP	G	A	Pts	PIM	PP	SH	GW
			Regular Season																Playoffs							
2001-02	NY Rangers	NHL	82	10	45	55	28	1	0	3	202	5.0	14	0	0.0	104	155	25:52								
	United States	Olympics	6	0	5	5	0																			
	NHL Totals		1021	215	700	915	481	97	7	34	3154	6.8		0	0.0	471	638	28:07	82	28	61	89	30	12	1	6

Hockey East First All-Star Team (1987) • Hockey East Rookie of the Year (1987) • Hockey East Player of the Year (1987) • NCAA East First All-American Team (1987) • NHL All-Rookie Team (1989) • Won Calder Memorial Trophy (1989) • NHL Second All-Star Team (1991, 1994, 1996) • Won James Norris Memorial Trophy (1992, 1997) • NHL First All-Star Team (1992, 1997) • Won Conn Smythe Trophy (1994) • Played in NHL All-Star Game (1990, 1991, 1992, 1994, 1996, 1997, 1998, 2001, 2002)

LEFEBVRE, Guillaume (luh-FAYV, GEE-ohm) PHI.

Left wing. Shoots left. 6'1", 195 lbs. Born, Amos, Que., May 7, 1981. Philadelphia's 6th choice, 227th overall, in 2000 Entry Draft.

Season	Club	League	GP	G	A	Pts	PIM	PP	SH	GW	S	%	+/-	TF	F%	H	SB	Min	GP	G	A	Pts	PIM	PP	SH	GW
1996-97	Amos Forestiers	QAAA	40	7	12	19	14																			
1997-98	Amos Forestiers	QAAA	42	12	16	28	100												6	4	5	9				
1998-99	Shawinigan	QMJHL	40	3	1	4	49																			
	Cape Breton	QMJHL	24	2	7	9	13												5	0	1	1	0			
99-2000	Cape Breton	QMJHL	44	26	28	54	82																			
	Quebec Remparts	QMJHL	2	3	1	4	0																			
	Rouyn-Noranda	QMJHL	25	4	11	15	39												11	4	0	4	25			
2000-01	Rouyn-Noranda	QMJHL	61	24	43	67	160												9	3	1	4	22			
	Philadelphia	AHL																	9	0	1	1	2			
2001-02	Philadelphia	NHL	3	0	0	0	0	0	0	0	3	0.0	-1	0	0.0	2	0	5:55								
	Philadelphia	AHL	78	19	15	34	111												5	0	0	0	4			
	NHL Totals		3	0	0	0	0	0	0	0	3	0.0		0	0.0	2	0	5:55								

Traded to **Cape Breton** (QMJHL) by **Quebec** (QMJHL) with Chris Lyness for Jean-Philippe Cote, Stuart MacRea and future considerations, January 5, 2000. Traded to **Rouyn-Noranda** (QMJHL) by **Quebec** (QMJHL) with future considerations for Mike Ribeiro, January 9, 2000.

LEFEBVRE, Sylvain (luh-FAYV, SIHL-veh) NYR

Defense. Shoots left. 6'2", 205 lbs. Born, Richmond, Que., October 14, 1967.

Season	Club	League	GP	G	A	Pts	PIM	PP	SH	GW	S	%	+/-	TF	F%	H	SB	Min	GP	G	A	Pts	PIM	PP	SH	GW
1983-84	Cantons	QAAA	1	0	0	0	0												2	1	0	1	2			
1984-85	Laval Voisins	QMJHL	66	7	5	12	31																			
1985-86	Laval Titan	QMJHL	71	8	17	25	48												14	1	0	1	25			
1986-87	Laval Titan	QMJHL	70	10	36	46	44												15	1	6	7	12			
1987-88	Sherbrooke	AHL	79	3	24	27	73												6	2	3	5	4			
1988-89	Sherbrooke	AHL	77	15	32	47	119												6	1	3	4	4			
1989-90	Montreal	NHL	68	3	10	13	61	0	0	0	89	3.4	18						6	0	0	0	2	0	0	0
1990-91	Montreal	NHL	63	5	18	23	30	1	0	1	76	6.6	-11						11	1	0	1	6	0	0	0
1991-92	Montreal	NHL	69	3	14	17	91	0	0	0	85	3.5	9						2	0	0	0	2	0	0	0
1992-93	Toronto	NHL	81	2	12	14	90	0	0	0	81	2.5	8						21	3	3	6	20	0	0	0
1993-94	Toronto	NHL	84	2	9	11	79	0	0	0	96	2.1	33						18	0	3	3	16	0	0	0
1994-95	Quebec	NHL	48	2	11	13	17	0	0	0	81	2.5	13						6	0	2	2	2	0	0	0
1995-96♦	Colorado	NHL	75	5	11	16	49	2	0	0	115	4.3	26						22	0	5	5	12	0	0	0
1996-97	Colorado	NHL	71	2	11	13	30	1	0	0	77	2.6	12						17	0	0	0	25	0	0	0
1997-98	Colorado	NHL	81	0	10	10	48	0	0	0	66	0.0	2						7	0	0	0	4	0	0	0
1998-99	Colorado	NHL	76	2	18	20	48	0	0	0	64	3.1	18	0	0.0	120	79	20:56	19	0	1	1	12	0	0	0
99-2000	NY Rangers	NHL	82	2	10	12	43	0	0	0	67	3.0	-13	0	0.0	181	142	18:36								
2000-01	NY Rangers	NHL	71	2	13	15	55	0	0	0	39	5.1	3	0	0.0	200	103	18:10								
2001-02	NY Rangers	NHL	41	0	5	5	23	0	0	0	20	0.0	-3	0	0.0	108	69	16:38								
	Hartford	AHL	15	0	5	5	11																			
	NHL Totals		910	30	152	182	664	4	0	1	956	3.1		0	0.0	609	393	18:50	129	4	14	18	101	0	0	0

AHL Second All-Star Team (1989)

Signed as a free agent by **Montreal**, September 24, 1986. Traded to **Toronto** by **Montreal** for Toronto's 3rd round choice (Martin Belanger) in 1994 Entry Draft, August 20, 1992. Traded to **Quebec** by **Toronto** with Wendel Clark, Landon Wilson and Toronto's 1st round choice (Jeffrey Kealty) in 1994 Entry Draft for Mats Sundin, Garth Butcher, Todd Warriner and Philadelphia's 1st round choice (previously acquired by Quebec - later traded to Washington - Washington selected Nolan Baumgartner) in 1994 Entry Draft, June 28, 1994. Transferred to **Colorado** after **Quebec** franchise relocated, June 21, 1995. Signed as a free agent by **NY Rangers**, July 22, 1999.

LEGWAND, David (LEHG-wuhnd, DAY-vihd) NSH.

Center. Shoots left. 6'2", 190 lbs. Born, Detroit, MI, August 17, 1980. Nashville's 1st choice, 2nd overall, in 1998 Entry Draft.

Season	Club	League	GP	G	A	Pts	PIM	PP	SH	GW	S	%	+/-	TF	F%	H	SB	Min	GP	G	A	Pts	PIM	PP	SH	GW
1996-97	Det. Compuware	MNHL	44	21	41	62	58																			
1997-98	Plymouth Whalers	OHL	59	54	51	105	56												15	8	12	20	24			
1998-99	Plymouth Whalers	OHL	55	31	49	80	65												11	3	8	11	8			
	Nashville	NHL	1	0	0	0	0	0	0	0	2	0.0	0	9	55.6	0	1	12:50								
99-2000	Nashville	NHL	71	13	15	28	30	4	0	2	111	11.7	-6	637	41.6	48	27	14:43								
2000-01	Nashville	NHL	81	13	28	41	38	3	0	3	172	7.6	1	888	40.3	50	23	15:14								
2001-02	Nashville	NHL	63	11	19	30	54	1	1	1	121	9.1	1	843	40.5	43	25	16:25								
	NHL Totals		216	37	62	99	122	8	1	6	406	9.1		2377	40.8	141	76	15:24								

OHL All-Rookie Team (1998) • OHL First All-Star Team (1998) • OHL Rookie of the Year (1998) • OHL MVP (1998) • Canadian Major Junior Rookie of the Year (1998)

LEHTINEN, Jere (LEH-tih-nehn, YUH-ree) DAL.

Right wing. Shoots right. 6', 200 lbs. Born, Espoo, Finland, June 24, 1973. Minnesota's 3rd choice, 88th overall, in 1992 Entry Draft.

Season	Club	League	GP	G	A	Pts	PIM	PP	SH	GW	S	%	+/-	TF	F%	H	SB	Min	GP	G	A	Pts	PIM	PP	SH	GW
1989-90	Kiekko-67 Jr.	Finn-Jr.	32	23	23	46	6												5	0	3	3	0			
1990-91	Kiekko Espoo Jr.	Finn-Jr.	3	3	1	4	0																			
	Kiekko Espoo	Finland-2	32	15	9	24	12																			
1991-92	Kiekko-67 Jr.	Finn-Jr.	8	5	4	9	2																			
	Kiekko Espoo	Finland-2	43	32	17	49	6																			
1992-93	Kiekko-67 Jr.	Finn-Jr.	4	5	3	8	8																			
	Kiekko Espoo	Finland	45	13	14	27	6																			
1993-94	TPS Turku	Finland	42	19	20	39	6												11	*11	2	13	*2			
	Finland	Olympics	8	3	0	3	0																			
1994-95	TPS Turku	Finland	39	19	23	42	33												13	*8	6	14	4			
1995-96	Dallas	NHL	57	6	22	28	16	0	0	1	109	5.5	5													
	Michigan K-Wings	IHL	1	1	0	1	0																			
1996-97	Dallas	NHL	63	16	27	43	2	3	1	2	134	11.9	26						7	2	2	4	0	0	0	0
1997-98	Dallas	NHL	72	23	19	42	20	7	2	6	201	11.4	19						12	3	5	8	2	1	0	0
	Finland	Olympics	6	4	2	6	2																			
1998-99♦	Dallas	NHL	74	20	32	52	18	7	1	2	173	11.6	29	9	33.3	72	47	19:36	23	10	3	13	2	1	1	0
99-2000	Dallas	NHL	17	3	5	8	0	0	0	1	29	10.3	1	0	0.0	17	6	17:31	13	1	5	6	2	0	0	0
2000-01	Dallas	NHL	74	20	25	45	24	7	0	1	148	13.5	14	7	28.6	94	41	19:17	10	1	0	1	2	0	0	0
2001-02	Dallas	NHL	73	25	24	49	14	7	1	4	198	12.6	27	18	22.2	75	45	19:50								
	Finland	Olympics	4	1	2	3	2																			
	NHL Totals		430	113	154	267	94	31	5	17	992	11.4		34	26.5	258	139	19:25	65	17	15	32	8	2	1	0

Won Frank J. Selke Trophy (1998, 1999) • Played in NHL All-Star Game (1998)

Rights transferred to **Dallas** after **Minnesota** franchise relocated, June 9, 1993. • Missed majority of 1999-2000 season recovering from leg injury suffered in game vs. Nashville, October 16, 1999.

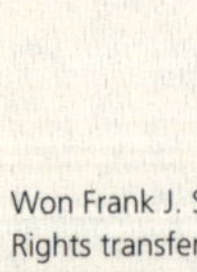

LEMIEUX, Claude (lehm-YOO, KLOHD) PHX.

Right wing. Shoots right. 6'1", 226 lbs. Born, Buckingham, Que., July 16, 1965. Montreal's 2nd choice, 26th overall, in 1983 Entry Draft.

Season	Club	League	GP	G	A	Pts	PIM	PP	SH	GW	S	%	+/-	TF	F%	H	SB	Min	GP	G	A	Pts	PIM	PP	SH	GW
1981-82	Richelieu	QAAA	48	24	48	72	96												8	10	13	23	14			
1982-83	Trois-Rivieres	QMJHL	62	28	38	66	187												4	1	0	1	30			
1983-84	Verdun Juniors	QMJHL	51	41	45	86	225												9	8	12	20	63			
	Montreal	NHL	8	1	1	2	12	0	0	0	7	14.3	-2													
	Nova Scotia	AHL																	2	1	0	1	0			
1984-85	Verdun	QMJHL	52	58	66	124	152												14	23	17	40	38			
	Montreal	NHL	1	0	1	1	7	0	0	0	0	0.0	1													
1985-86♦	Montreal	NHL	10	1	2	3	22	1	0	0	16	6.3	-6						20	10	6	16	68	4	0	4
	Sherbrooke	AHL	58	21	32	53	145																			
1986-87	Montreal	NHL	76	27	26	53	156	5	0	1	184	14.7	0						17	4	9	13	41	2	0	0
1987-88	Montreal	NHL	78	31	30	61	137	6	0	3	241	12.9	16						11	3	2	5	20	0	0	2

			Regular Season																Playoffs							
Season	**Club**	**League**	**GP**	**G**	**A**	**Pts**	**PIM**	**PP**	**SH**	**GW**	**S**	**%**	**+/-**	**TF**	**F%**	**H**	**SB**	**Min**	**GP**	**G**	**A**	**Pts**	**PIM**	**PP**	**SH**	**GW**
1988-89	**Montreal**	**NHL**	**69**	**29**	**22**	**51**	**136**	**7**	**0**	**3**	**220**	**13.2**	**14**	....	....	....	....	....	**18**	**4**	**3**	**7**	**58**	**0**	**0**	**1**
1989-90	**Montreal**	**NHL**	**39**	**8**	**10**	**18**	**106**	**3**	**0**	**1**	**104**	**7.7**	**-8**	....	....	....	....	....	**11**	**1**	**3**	**4**	**38**	**0**	**0**	**1**
1990-91	**New Jersey**	**NHL**	**78**	**30**	**17**	**47**	**105**	**10**	**0**	**2**	**271**	**11.1**	**-8**	....	....	....	....	....	**7**	**4**	**0**	**4**	**34**	**2**	**0**	**1**
1991-92	**New Jersey**	**NHL**	**74**	**41**	**27**	**68**	**109**	**13**	**1**	**8**	**296**	**13.9**	**9**	....	....	....	....	....	**7**	**4**	**3**	**7**	**26**	**1**	**0**	**0**
1992-93	**New Jersey**	**NHL**	**77**	**30**	**51**	**81**	**155**	**13**	**0**	**3**	**311**	**9.6**	**3**	....	....	....	....	....	**5**	**2**	**0**	**2**	**19**	**1**	**0**	**0**
1993-94	**New Jersey**	**NHL**	**79**	**18**	**26**	**44**	**86**	**5**	**0**	**5**	**181**	**9.9**	**13**	....	....	....	....	....	**20**	**7**	**11**	**18**	**44**	**0**	**0**	**2**
1994-95♦	**New Jersey**	**NHL**	**45**	**6**	**13**	**19**	**86**	**1**	**0**	**1**	**117**	**5.1**	**2**	....	....	....	....	....	**20**	***13**	**3**	**16**	**20**	**0**	**0**	**3**
1995-96♦	**Colorado**	**NHL**	**79**	**39**	**32**	**71**	**117**	**9**	**2**	**10**	**315**	**12.4**	**14**	....	....	....	....	....	**19**	**5**	**7**	**12**	**55**	**3**	**0**	**0**
1996-97	**Colorado**	**NHL**	**45**	**11**	**17**	**28**	**43**	**5**	**0**	**4**	**168**	**6.5**	**-4**	....	....	....	....	....	**17**	***13**	**10**	**23**	**32**	**4**	**0**	**4**
1997-98	**Colorado**	**NHL**	**78**	**26**	**27**	**53**	**115**	**11**	**1**	**1**	**261**	**10.0**	**-7**	....	....	....	....	....	**7**	**3**	**3**	**6**	**8**	**1**	**0**	**1**
1998-99	**Colorado**	**NHL**	**82**	**27**	**24**	**51**	**102**	**11**	**0**	**8**	**292**	**9.2**	**0**	**43**	**41.9**	**110**	**26**	**21:14**	**19**	**3**	**11**	**14**	**26**	**1**	**0**	**0**
99-2000	**Colorado**	**NHL**	**13**	**3**	**6**	**9**	**4**	**0**	**0**	**0**	**36**	**8.3**	**0**	**2**	**50.0**	**20**	**2**	**18:40**	....	....	....	....	....	....	....	....
	♦ New Jersey	**NHL**	**70**	**17**	**21**	**38**	**86**	**7**	**0**	**3**	**221**	**7.7**	**-3**	**49**	**28.6**	**116**	**15**	**17:55**	**23**	**4**	**6**	**10**	**28**	**1**	**0**	**0**
2000-01	**Phoenix**	**NHL**	**46**	**10**	**16**	**26**	**58**	**2**	**0**	**1**	**99**	**10.1**	**1**	**28**	**21.4**	**68**	**11**	**17:12**	....	....	....	....	....	....	....	....
2001-02	**Phoenix**	**NHL**	**82**	**16**	**25**	**41**	**70**	**4**	**1**	**3**	**174**	**9.2**	**-5**	**49**	**28.6**	**98**	**11**	**17:12**	**5**	**0**	**0**	**0**	**2**	**0**	**0**	**0**
	NHL Totals		**1129**	**371**	**394**	**765**	**1712**	**113**	**5**	**57**	**3514**	**10.6**		**171**	**31.0**	**412**	**65**	**18:34**	**226**	**80**	**77**	**157**	**519**	**20**	**0**	**19**

QMJHL Second All-Star Team (1984) • QMJHL First All-Star Team (1985) • Won Conn Smythe Trophy (1995)

• Missed majority of 1989-90 season recovering from abdominal injury suffered in game vs. Boston, October 9, 1989. Traded to **New Jersey** by **Montreal** for Sylvain Turgeon, September 4, 1990. Traded to **NY Islanders** by **New Jersey** for Steve Thomas, October 3, 1995. Traded to **Colorado** by **NY Islanders** for Wendel Clark, October 3, 1995. Traded to **New Jersey** by **Colorado** with Colorado's 1st (David Hale) and 2nd (Matt DeMarchi) round choices in 2000 Entry Draft for Brian Rolston and New Jersey's 1st round choice (later traded to Boston - Boston selected Martin Samuelsson) in 2000 Entry Draft, November 3, 1999. Signed as a free agent by **Phoenix**, December 5, 2000.

LEMIEUX, Mario

(lehm-YOO, MAHR-ee-oh) **PIT.**

Center. Shoots right. 6'4", 230 lbs. Born, Montreal, Que., October 5, 1965. Pittsburgh's 1st choice, 1st overall, in 1984 Entry Draft.

Season	Club	League	GP	G	A	Pts	PIM	PP	SH	GW	S	%	+/-	TF	F%	H	SB	Min	GP	G	A	Pts	PIM	PP	SH	GW
1980-81	Mtl-Concordia	QAAA	47	62	62	124	127	....	....	....	....	....	....	....	....	....	....	....	3	2	5	7	8	....	....	....
1981-82	Laval Voisins	QMJHL	64	30	66	96	22	....	....	....	....	....	....	....	....	....	....	....	18	5	9	14	31	....	....	....
1982-83	Laval Voisins	QMJHL	66	84	100	184	76	....	....	....	....	....	....	....	....	....	....	....	12	14	18	32	18	....	....	....
1983-84	Laval Voisins	QMJHL	70	*133	*149	*282	92	....	....	....	....	....	....	....	....	....	....	....	14	*29	*23	*52	29	....	....	....
1984-85	**Pittsburgh**	**NHL**	**73**	**43**	**57**	**100**	**54**	**11**	**0**	**2**	**209**	**20.6**	**-35**	....	....	....	....	....	....	....	....	....	....	....	....	....
1985-86	**Pittsburgh**	**NHL**	**79**	**48**	**93**	**141**	**43**	**17**	**0**	**4**	**276**	**17.4**	**-6**	....	....	....	....	....	....	....	....	....	....	....	....	....
1986-87	**Pittsburgh**	**NHL**	**63**	**54**	**53**	**107**	**57**	**19**	**0**	**4**	**267**	**20.2**	**13**	....	....	....	....	....	....	....	....	....	....	....	....	....
1987-88	**Pittsburgh**	**NHL**	**77**	***70**	**98**	***168**	**92**	**22**	**10**	**7**	**382**	**18.3**	**23**	....	....	....	....	....	....	....	....	....	....	....	....	....
1988-89	**Pittsburgh**	**NHL**	**76**	***85**	***114**	***199**	**100**	**31**	**13**	**8**	**313**	**27.2**	**41**	....	....	....	....	....	**11**	**12**	**7**	**19**	**16**	**7**	**1**	**0**
1989-90	**Pittsburgh**	**NHL**	**59**	**45**	**78**	**123**	**78**	**14**	**3**	**4**	**226**	**19.9**	**-18**	....	....	....	....	....	....	....	....	....	....	....	....	....
1990-91♦	**Pittsburgh**	**NHL**	**26**	**19**	**26**	**45**	**30**	**6**	**1**	**2**	**89**	**21.3**	**8**	....	....	....	....	....	**23**	**16**	***28**	***44**	**16**	**6**	**2**	**0**
1991-92♦	**Pittsburgh**	**NHL**	**64**	**44**	**87**	***131**	**94**	**12**	**4**	**5**	**249**	**17.7**	**27**	....	....	....	....	....	**15**	***16**	**18**	***34**	**2**	**8**	**2**	**5**
1992-93	**Pittsburgh**	**NHL**	**60**	**69**	**91**	***160**	**38**	**16**	**6**	**10**	**286**	**24.1**	**55**	....	....	....	....	....	**11**	**8**	**10**	**18**	**10**	**3**	**1**	**1**
1993-94	**Pittsburgh**	**NHL**	**22**	**17**	**20**	**37**	**32**	**7**	**0**	**4**	**92**	**18.5**	**-2**	....	....	....	....	....	**6**	**4**	**3**	**7**	**2**	**1**	**0**	**0**
1994-95	**Pittsburgh**	**NHL**	DID NOT PLAY																							
1995-96	**Pittsburgh**	**NHL**	**70**	***69**	***92**	***161**	**54**	**31**	**8**	**8**	**338**	**20.4**	**10**	....	....	....	....	....	**18**	**11**	**16**	**27**	**33**	**3**	**1**	**2**
1996-97	**Pittsburgh**	**NHL**	**76**	**50**	***72**	***122**	**65**	**15**	**3**	**7**	**327**	**15.3**	**27**	....	....	....	....	....	**5**	**3**	**3**	**6**	**4**	**0**	**0**	**0**
1997-98			OUT OF HOCKEY – RETIRED																							
1998-99			OUT OF HOCKEY – RETIRED																							
99-2000			OUT OF HOCKEY – RETIRED																							
2000-01	**Pittsburgh**	**NHL**	**43**	**35**	**41**	**76**	**18**	**16**	**1**	**5**	**171**	**20.5**	**15**	**852**	**52.1**	**29**	**23**	**24:20**	**18**	**6**	**11**	**17**	**4**	**1**	**0**	**3**
2001-02	**Pittsburgh**	**NHL**	**24**	**6**	**25**	**31**	**14**	**2**	**0**	**0**	**75**	**8.0**	**0**	**354**	**44.6**	**5**	**7**	**22:16**	....	....	....	....	....	....	....	....
	Canada	Olympics	5	2	4	6	0	....	....	....	....	....	....	....	....	....	....	....	....	....	....	....	....	....	....	....
	NHL Totals		**812**	**654**	**947**	**1601**	**769**	**219**	**49**	**70**	**3300**	**19.8**		**1206**	**49.9**	**34**	**30**	**23:36**	**107**	**76**	**96**	**172**	**87**	**29**	**7**	**11**

• Inducted into Hockey Hall of Fame (1997)

QMJHL Second All-Star Team (1983) • QMJHL First All-Star Team (1984) • QMJHL MVP (1984) • Canadian Major Junior Player of the Year (1984) • NHL All-Rookie Team (1985) • Won Calder Memorial Trophy (1985) • NHL Second All-Star Team (1986, 1987, 1992, 2001) • Won Lester B. Pearson Award (1986, 1988, 1993, 1996) • Canada Cup All-Star Team (1987) • NHL First All-Star Team (1988, 1989, 1993, 1996, 1997) • Won Dodge Performance of the Year Award (1988) • Won Dodge Performer of the Year Award (1988, 1989) • Won Art Ross Trophy (1988, 1989, 1992, 1993, 1996, 1997) • Won Hart Trophy (1988, 1993, 1996) • Won Dodge Ram Tough Award (1989) • Won Conn Smythe Trophy (1991, 1992) • Won ProSet/NHL Player of the Year Award (1992) • Won Alka-Seltzer Plus Award (1993) • Won Bill Masterton Memorial Trophy (1993) • Won Lester Patrick Trophy (2000) • Played in NHL All-Star Game (1985, 1986, 1988, 1989, 1990, 1992, 1996, 1997, 2001, 2002)

• Missed remainder of 1989-90 and majority of 1990-91 seasons recovering from back injury suffered in game vs. NY Rangers, February 14, 1989. • Missed most of 1992-93 season after being diagnosed with Hodgkin's Disease, January 12, 1993. • Missed majority of 1993-94 season recovering from back injury originally suffered in game vs. Chicago, November 11, 1993. • Missed entire 1994-95 season recovering from effects of treatment for Hodgkin's Disease and back injury suffered in game vs. NY Rangers, March 12, 1994. • Became third player (Gordie Howe, Guy Lafleur) to appear in NHL game after being inducted into Hockey Hall-of-Fame, December 27, 2000. • Missed majority of 2001-02 season recovering from hip injury originally suffered in game vs. Anaheim, October 6, 2001.

LEROUX, Francois

(leh-ROO, FRAN-swuh) **PIT.**

Defense. Shoots left. 6'6", 247 lbs. Born, Ste-Adele, Que., April 18, 1970. Edmonton's 1st choice, 19th overall, in 1988 Entry Draft.

Season	Club	League	GP	G	A	Pts	PIM	PP	SH	GW	S	%	+/-	TF	F%	H	SB	Min	GP	G	A	Pts	PIM	PP	SH	GW
1986-87	Laval Laurentide	QAAA	42	5	11	16	76	....	....	....	....	....	....	....	....	....	....	....	8	0	1	1	12	....	....	....
1987-88	St-Jean Castors	QMJHL	58	3	8	11	143	....	....	....	....	....	....	....	....	....	....	....	7	2	0	2	21	....	....	....
1988-89	St-Jean Castors	QMJHL	57	8	34	42	185	....	....	....	....	....	....	....	....	....	....	....	....	....	....	....	....	....	....	....
	Edmonton	**NHL**	**2**	**0**	**0**	**0**	**0**	**0**	**0**	**0**	**0**	**0.0**	**1**	....	....	....	....	....	....	....	....	....	....	....	....	....
1989-90	Victoriaville	QMJHL	54	4	33	37	169	....	....	....	....	....	....	....	....	....	....	....	....	....	....	....	....	....	....	....
	Edmonton	**NHL**	**3**	**0**	**1**	**1**	**0**	**0**	**0**	**0**	**0**	**0.0**	**-2**	....	....	....	....	....	....	....	....	....	....	....	....	....
1990-91	**Edmonton**	**NHL**	**1**	**0**	**2**	**2**	**0**	**0**	**0**	**0**	**1**	**0.0**	**1**	....	....	....	....	....	....	....	....	....	....	....	....	....
	Cape Breton	AHL	71	2	7	9	124	....	....	....	....	....	....	....	....	....	....	....	4	0	1	1	19	....	....	....
1991-92	**Edmonton**	**NHL**	**4**	**0**	**0**	**0**	**7**	**0**	**0**	**0**	**0**	**0.0**	**-1**	....	....	....	....	....	....	....	....	....	....	....	....	....
	Cape Breton	AHL	61	7	22	29	114	....	....	....	....	....	....	....	....	....	....	....	5	0	0	0	8	....	....	....
1992-93	**Edmonton**	**NHL**	**1**	**0**	**0**	**0**	**4**	**0**	**0**	**0**	**0**	**0.0**	**0**	....	....	....	....	....	....	....	....	....	....	....	....	....
	Cape Breton	AHL	55	10	24	34	139	....	....	....	....	....	....	....	....	....	....	....	16	0	5	5	29	....	....	....
1993-94	**Ottawa**	**NHL**	**23**	**0**	**1**	**1**	**70**	**0**	**0**	**0**	**8**	**0.0**	**-4**	....	....	....	....	....	....	....	....	....	....	....	....	....
	P.E.I. Senators	AHL	25	4	6	10	52	....	....	....	....	....	....	....	....	....	....	....	....	....	....	....	....	....	....	....
1994-95	P.E.I. Senators	AHL	45	4	14	18	137	....	....	....	....	....	....	....	....	....	....	....	....	....	....	....	....	....	....	....
	Pittsburgh	**NHL**	**40**	**0**	**2**	**2**	**114**	**0**	**0**	**0**	**19**	**0.0**	**7**	....	....	....	....	....	**12**	**0**	**2**	**2**	**14**	**0**	**0**	**0**
1995-96	**Pittsburgh**	**NHL**	**66**	**2**	**9**	**11**	**161**	**0**	**0**	**0**	**43**	**4.7**	**2**	....	....	....	....	....	**18**	**1**	**1**	**2**	**20**	**0**	**0**	**1**
1996-97	**Pittsburgh**	**NHL**	**59**	**0**	**3**	**3**	**81**	**0**	**0**	**0**	**5**	**0.0**	**-3**	....	....	....	....	....	**3**	**0**	**0**	**0**	**0**	**0**	**0**	**0**
1997-98	**Colorado**	**NHL**	**50**	**1**	**2**	**3**	**140**	**0**	**0**	**0**	**14**	**7.1**	**-3**	....	....	....	....	....	....	....	....	....	....	....	....	....
1998-99	Grand Rapids	IHL	13	1	1	2	22	....	....	....	....	....	....	....	....	....	....	....	....	....	....	....	....	....	....	....
99-2000	Springfield	AHL	64	3	6	9	162	....	....	....	....	....	....	....	....	....	....	....	5	0	0	0	2	....	....	....
2000-01	Springfield	AHL	65	4	6	10	180	....	....	....	....	....	....	....	....	....	....	....	....	....	....	....	....	....	....	....
2001-02	Berlin Capitals	Germany	56	1	10	11	110	....	....	....	....	....	....	....	....	....	....	....	....	....	....	....	....	....	....	....
	NHL Totals		**249**	**3**	**20**	**23**	**577**	**0**	**0**	**0**	**90**	**3.3**							**33**	**1**	**3**	**4**	**34**	**0**	**0**	**1**

Claimed on waivers by **Ottawa** from **Edmonton**, October 6, 1993. Claimed by **Pittsburgh** from **Ottawa** in Waiver Draft, January 18, 1995. Traded to **Colorado** by **Pittsburgh** for Colorado's 3rd round choice (David Cameron) in 1998 Entry Draft, September 28, 1997. Signed as a free agent by **Grand Rapids** (IHL), February 18, 1999. Signed as a free agent by **Phoenix**, July 20, 1999. Signed as a free agent by **Berlin Capitals** (Germany), July 17, 2001. Signed as a free agent by **Pittsburgh**, July 16, 2002.

LEROUX, Jean-Yves

(leh-ROO, ZHAWN-EEV)

Left wing. Shoots left. 6'2", 211 lbs. Born, Montreal, Que., June 24, 1976. Chicago's 2nd choice, 40th overall, in 1994 Entry Draft.

Season	Club	League	GP	G	A	Pts	PIM	PP	SH	GW	S	%	+/-	TF	F%	H	SB	Min	GP	G	A	Pts	PIM	PP	SH	GW
1991-92	Mtl-Bourassa	QAAA	35	14	31	45	62	....	....	....	....	....	....	....	....	....	....	....	7	1	3	4	8	....	....	....
1992-93	Beauport	QMJHL	62	20	25	45	33	....	....	....	....	....	....	....	....	....	....	....	....	....	....	....	....	....	....	....
1993-94	Beauport	QMJHL	45	14	25	39	43	....	....	....	....	....	....	....	....	....	....	....	15	7	6	13	33	....	....	....
1994-95	Beauport	QMJHL	59	19	33	52	125	....	....	....	....	....	....	....	....	....	....	....	17	4	6	10	39	....	....	....
1995-96	Beauport	QMJHL	54	41	41	82	176	....	....	....	....	....	....	....	....	....	....	....	20	5	18	23	20	....	....	....
1996-97	**Chicago**	**NHL**	**1**	**0**	**1**	**1**	**5**	**0**	**0**	**0**	**0**	**0.0**	**1**	....	....	....	....	....	....	....	....	....	....	....	....	....
	Indianapolis Ice	IHL	69	14	17	31	112	....	....	....	....	....	....	....	....	....	....	....	4	1	0	1	2	....	....	....
1997-98	**Chicago**	**NHL**	**66**	**6**	**7**	**13**	**55**	**0**	**0**	**0**	**57**	**10.5**	**-2**	....	....	....	....	....	....	....	....	....	....	....	....	....
1998-99	**Chicago**	**NHL**	**40**	**3**	**5**	**8**	**21**	**0**	**0**	**0**	**47**	**6.4**	**-7**	**10**	**50.0**	**98**	**8**	**12:43**	....	....	....	....	....	....	....	....
	Chicago Wolves	IHL	....	....	....	....	....	....	....	....	....	....	....	....	....	....	....	....	10	1	1	2	18	....	....	....
99-2000	**Chicago**	**NHL**	**54**	**3**	**5**	**8**	**43**	**0**	**0**	**1**	**36**	**8.3**	**-10**	**7**	**0.0**	**96**	**9**	**10:44**	....	....	....	....	....	....	....	....
2000-01	**Chicago**	**NHL**	**59**	**4**	**4**	**8**	**22**	**1**	**0**	**0**	**60**	**6.7**	**-9**	**1**	**0.0**	**98**	**11**	**10:11**	....	....	....	....	....	....	....	....
2001-02	Norfolk Admirals	AHL	59	8	16	24	94	....	....	....	....	....	....	....	....	....	....	....	4	1	0	1	0	....	....	....
	NHL Totals		**220**	**16**	**22**	**38**	**146**	**1**	**0**	**1**	**200**	**8.0**		**18**	**27.8**	**292**	**28**	**11:02**	....	....	....	....	....	....	....	....

QMJHL All-Rookie Team (1993) • QMJHL Second All-Star Team (1994)

LESCHYSHYN, Curtis — (luh-SIH-shuhn, KUHR-tihs) — OTT.

Defense. Shoots left. 6'1", 220 lbs. Born, Thompson, Man., September 21, 1969. Quebec's 1st choice, 3rd overall, in 1988 Entry Draft.

			Regular Season																Playoffs							
Season	Club	League	GP	G	A	Pts	PIM	PP	SH	GW	S	%	+/-	TF	F%	H	SB	Min	GP	G	A	Pts	PIM	PP	SH	GW
1985-86	Sask. Blazers	SMHL	34	9	34	43	52																			
	Saskatoon Blades	WHL	1	0	0	0	0																			
1986-87	Saskatoon Blades	WHL	70	14	26	40	107												11	1	5	6	14			
1987-88	Saskatoon Blades	WHL	56	14	41	55	86												10	2	5	7	16			
1988-89	**Quebec**	**NHL**	71	4	9	13	71	1	1	0	58	6.9	–32													
1989-90	**Quebec**	**NHL**	68	2	6	8	44	1	0	0	42	4.8	–41													
1990-91	**Quebec**	**NHL**	55	3	7	10	49	2	0	1	57	5.3	–19													
1991-92	**Quebec**	**NHL**	42	5	12	17	42	3	0	1	61	8.2	–28													
	Halifax Citadels	AHL	6	0	2	2	4																			
1992-93	**Quebec**	**NHL**	82	9	23	32	61	4	0	2	73	12.3	25						6	1	1	2	6	1	0	0
1993-94	**Quebec**	**NHL**	72	5	17	22	65	3	0	2	97	5.2	–2													
1994-95	**Quebec**	**NHL**	44	2	13	15	20	0	0	0	43	4.7	29						3	0	1	1	4	0	0	0
1995-96♦	**Colorado**	**NHL**	77	4	15	19	73	0	0	1	76	5.3	32						17	1	2	3	8	0	0	0
1996-97	**Colorado**	**NHL**	11	0	5	5	6	0	0	0	8	0.0	1													
	Washington	**NHL**	2	0	0	0	2	0	0	0	0	0.0	0													
	Hartford	**NHL**	64	4	13	17	30	1	1	1	94	4.3	–19													
1997-98	**Carolina**	**NHL**	73	2	10	12	45	1	0	1	53	3.8	–2													
1998-99	**Carolina**	**NHL**	65	2	7	9	50	0	0	0	35	5.7	–1	0	0.0	208	103	19:18	6	0	0	0	6	0	0	0
99-2000	**Carolina**	**NHL**	53	0	2	2	14	0	0	0	31	0.0	–19	0	0.0	164	85	17:47								
2000-01	**Minnesota**	**NHL**	54	2	3	5	19	1	0	1	43	4.7	–2	0	0.0	160	93	19:31								
	Ottawa	**NHL**	11	0	4	4	0	0	0	0	8	0.0	7	0	0.0	32	14	19:04	4	0	0	0	0	0	0	0
2001-02	**Ottawa**	**NHL**	79	1	9	10	44	0	0	0	59	1.7	–5	0	0.0	171	133	18:28	12	0	1	1	0	0	0	0
	NHL Totals		923	45	155	200	635	17	2	10	838	5.4		0	0.0	735	428	18:47	48	2	5	7	24	1	0	0

WHL East First All-Star Team (1988)

Transferred to **Colorado** after **Quebec** franchise relcoated, June 21, 1995. Traded to **Washington** by **Colorado** with Chris Simon for Keith Jones, Washington's 1st (Scott Parker) and 4th (later traded back to Washington - Washington selected Krys Barch) round choices in 1998 Entry Draft, November 2, 1996. Traded to **Hartford** by **Washington** for Andrei Nikolishin, November 9, 1996. Transferred to **Carolina** after **Hartford** franchise relocated, June 25, 1997. Selected by **Minnesota** from **Carolina** in Expansion Draft, June 23, 2000. Traded to **Ottawa** by **Minnesota** for Ottawa's 3rd round choice (Stephane Veilleux) in 2001 Entry Draft and future considerations, March 13, 2001.

LESSARD, Francis — (leh-SAHR, FRAN-sihs) — ATL.

Defense. Shoots right. 6'2", 220 lbs. Born, Montreal, Que., May 30, 1979. Carolina's 3rd choice, 80th overall, in 1997 Entry Draft.

			Regular Season																Playoffs							
Season	Club	League	GP	G	A	Pts	PIM	PP	SH	GW	S	%	+/-	TF	F%	H	SB	Min	GP	G	A	Pts	PIM	PP	SH	GW
1995-96	Laval Laurentide	QAAA	41	5	7	12	73																			
1996-97	Val-d'Or Foreurs	QMJHL	66	1	9	10	287																			
1997-98	Val-d'Or Foreurs	QMJHL	63	3	20	23	338												19	1	6	7	*101			
1998-99	Drummondville	QMJHL	53	12	36	48	295																			
99-2000	Philadelphia	AHL	78	4	8	12	416												5	0	1	1	7			
2000-01	Philadelphia	AHL	64	3	7	10	330												10	0	0	0	33			
2001-02	Philadelphia	AHL	60	0	6	6	251																			
	Atlanta	**NHL**	5	0	0	0	26	0	0	0	2	0.0	0	0	0.0	12	2	12:45								
	Chicago Wolves	AHL	7	2	1	3	34												15	0	1	1	40			
	NHL Totals		5	0	0	0	26	0	0	0	2	0.0		0	0.0	12	2	12:45								

Memorial Cup All-Star Team (1998)

Traded to **Philadelphia** by **Carolina** for Philadelphia's 8th round choice (Antti Jokella) in 1999 Entry Draft, May 25, 1999. Traded to **Atlanta** by **Philadelphia** for David Harlock and Atlanta's 3rd and 7th round choices in 2003 Entry Draft, March 15, 2002.

LETANG, Alan — (leh-TANG, A-luhn) — NYI

Defense. Shoots left. 6'1", 205 lbs. Born, Renfrew, Ont., September 4, 1975. Montreal's 10th choice, 203rd overall, in 1993 Entry Draft.

			Regular Season																Playoffs							
Season	Club	League	GP	G	A	Pts	PIM	PP	SH	GW	S	%	+/-	TF	F%	H	SB	Min	GP	G	A	Pts	PIM	PP	SH	GW
1990-91	Ottawa Valley	OMHA	32	3	26	29	16																			
1991-92	Cornwall Royals	OHL	47	1	4	5	16												6	0	0	0	2			
1992-93	Newmarket Royals	OHL	66	1	25	26	14												6	0	3	3	2			
1993-94	Newmarket Royals	OHL	58	3	21	24	30																			
1994-95	Sarnia Sting	OHL	62	5	36	41	35												4	2	2	4	6			
1995-96	Fredericton	AHL	71	0	26	26	40												10	0	3	3	4			
1996-97	Fredericton	AHL	60	2	9	11	8																			
1997-98	Kaufbeurer Adler	Germany	15	1	5	6	8																			
	SC Langnau	Swiss-2	11	4	3	7	6																			
	Augsburg	Germany	17	0	1	1	4																			
1998-99	Team Canada	Nat-Tm	41	3	9	12	20																			
	EV Zug	Swiss																	9	0	4	4	4			
	Michigan K-Wings	IHL	12	3	3	6	0												5	0	2	2	0			
99-2000	**Dallas**	**NHL**	8	0	0	0	2	0	0	0	1	0.0	–5	0	0.0	2	4	9:57								
	Michigan K-Wings	IHL	51	1	12	13	30																			
2000-01	Utah Grizzlies	IHL	79	6	24	30	26																			
2001-02	**Calgary**	**NHL**	2	0	0	0	0	0	0	0	0	0.0	–2	0	0.0	0	1	11:06								
	Saint John	AHL	61	4	24	28	33																			
	NHL Totals		10	0	0	0	2	0	0	0	1	0.0		0	0.0	2	5	10:10								

Signed as a free agent by **Dallas**, March 22, 1999. Signed as a free agent by **Calgary**, August 22, 2001. Signed as a free agent by **NY Islanders**, July 18, 2002.

LETOWSKI, Trevor — (leh-TOW-skee, TREH-vuhr) — VAN.

Center. Shoots right. 5'10", 176 lbs. Born, Thunder Bay, Ont., April 5, 1977. Phoenix's 6th choice, 174th overall, in 1996 Entry Draft.

			Regular Season																Playoffs							
Season	Club	League	GP	G	A	Pts	PIM	PP	SH	GW	S	%	+/-	TF	F%	H	SB	Min	GP	G	A	Pts	PIM	PP	SH	GW
1993-94	T. Bay Kings	TBMHL	64	41	60	101	48																			
1994-95	Sarnia Sting	OHL	66	22	19	41	33												4	0	1	1	9			
1995-96	Sarnia Sting	OHL	66	36	63	99	66												10	9	5	14	10			
1996-97	Sarnia Sting	OHL	55	35	73	108	51												12	9	12	21	20			
1997-98	Springfield	AHL	75	11	20	31	26												4	1	0	1	2			
1998-99	**Phoenix**	**NHL**	14	2	2	4	2	0	0	0	8	25.0	1	49	55.1	4	3	6:01								
	Springfield	AHL	67	32	35	67	46												3	1	0	1	2			
99-2000	**Phoenix**	**NHL**	82	19	20	39	20	3	4	3	125	15.2	2	692	47.7	24	33	16:03	5	1	1	2	4	0	0	0
2000-01	**Phoenix**	**NHL**	77	7	15	22	32	0	1	3	110	6.4	–2	726	46.1	24	35	16:20								
2001-02	**Phoenix**	**NHL**	33	2	6	8	4	0	0	0	43	4.7	2	250	52.4	9	11	14:27								
	Vancouver	**NHL**	42	7	10	17	15	1	0	0	65	10.8	2	111	44.1	15	7	12:47	6	0	1	1	8	0	0	0
	NHL Totals		248	37	53	90	73	4	5	6	351	10.5		1828	47.7	76	89	14:48	11	1	2	3	12	0	0	0

Traded to **Vancouver** by **Phoenix** with Todd Warriner, Tyler Bouck and Phoenix's 3rd round choice in 2003 Entry Draft for Drake Berehowsky and Denis Pederson, December 28, 2001.

LIDSTROM, Nicklas — (LID-struhm, NIHK-las) — DET.

Defense. Shoots left. 6'2", 185 lbs. Born, Vasteras, Sweden, April 28, 1970. Detroit's 3rd choice, 53rd overall, in 1989 Entry Draft.

			Regular Season																Playoffs							
Season	Club	League	GP	G	A	Pts	PIM	PP	SH	GW	S	%	+/-	TF	F%	H	SB	Min	GP	G	A	Pts	PIM	PP	SH	GW
1987-88	Vasteras IK	Swede-2	3	0	0	0	0												5	0	0	0	6			
1988-89	Vasteras IK	Sweden	34	1	6	7	4												5	0	2	2	0			
1989-90	Vasteras IK	Sweden	39	8	8	16	14												2	0	1	1	2			
1990-91	Vasteras IK	Sweden	38	4	19	23	2												4	0	0	0	4			
1991-92	**Detroit**	**NHL**	80	11	49	60	22	5	0	1	168	6.5	36						11	1	2	3	0	1	0	0
1992-93	**Detroit**	**NHL**	84	7	34	41	28	3	0	2	156	4.5	7						7	1	0	1	0	1	0	0
1993-94	**Detroit**	**NHL**	84	10	46	56	26	4	0	3	200	5.0	43						7	3	2	5	0	1	1	0
1994-95	Vasteras IK	Sweden	13	2	10	12	4																			
	Detroit	**NHL**	43	10	16	26	6	7	0	0	90	11.1	15						18	4	12	16	8	3	0	2
1995-96	**Detroit**	**NHL**	81	17	50	67	20	8	1	1	211	8.1	29						19	5	9	14	10	1	0	0
1996-97♦	**Detroit**	**NHL**	79	15	42	57	30	8	0	1	214	7.0	11						20	2	6	8	2	0	0	0
1997-98♦	**Detroit**	**NHL**	80	17	42	59	18	7	1	1	205	8.3	22						22	6	13	19	8	2	0	2
	Sweden	Olympics	4	1	1	2	2																			
1998-99	**Detroit**	**NHL**	81	14	43	57	14	6	2	3	205	6.8	14	0	0.0	53	88	26:31	10	2	9	11	4	2	0	0
99-2000	**Detroit**	**NHL**	81	20	53	73	18	9	4	3	218	9.2	19	0	0.0	59	95	28:45	9	2	4	6	4	1	0	0
2000-01	**Detroit**	**NHL**	82	15	56	71	18	8	0	0	272	5.5	9	0	0.0	65	94	28:27	6	1	7	8	0	0	0	0

Season	Club	League	GP	G	A	Pts	PIM	PP	SH	GW	S	%	+/-	TF	F%	H	SB	Min	GP	G	A	Pts	PIM	PP	SH	GW
			Regular Season																Playoffs							
2001-02♦	Detroit	NHL	78	9	50	59	20	6	0	0	215	4.2	13	0	0.0	46	103	28:49	23	5	11	16	2	2	1	2
	Sweden	Olympics	4	1	5	6	0																			
	NHL Totals		853	145	481	626	220	71	8	15	2154	6.7		0	0.0	223	380	28:07	152	32	75	107	38	14	2	6

NHL All-Rookie Team (1992) • NHL First All-Star Team (1998, 1999, 2000, 2001, 2002) • Won James Norris Memorial Trophy (2001, 2002) • Won Conn Smythe Trophy (2002) • Played in NHL All-Star Game (1996, 1998, 1999, 2000, 2001, 2002)

LILJA, Andreas

(LIHL-yuh, an-DRAY-uhs) **L.A.**

Defense. Shoots left. 6'3", 222 lbs. Born, Landskrona, Sweden, July 13, 1975. Los Angeles' 2nd choice, 54th overall, in 2000 Entry Draft.

Season	Club	League	GP	G	A	Pts	PIM	PP	SH	GW	S	%	+/-	TF	F%	H	SB	Min	GP	G	A	Pts	PIM	PP	SH	GW
1993-94	Malmo IF Jr.	Swede-Jr.	14	3	7	10	38																			
1994-95	Malmo IF Jr.	Swede-Jr.	30	7	13	20	82																			
	Malmo IF	Sweden	3	0	0	0	2																			
1995-96	Malmo IF Jr.	Swede-Jr.	3	0	1	1	6																			
	Malmo IF	Sweden	40	1	5	6	63												5	0	1	1	2			
1996-97	Malmo IF	Sweden	47	1	0	1	22												4	0	0	0	10			
1997-98	Malmo IF	Swede-2	11	6	5	11	24																			
	Malmo IF	Sweden	10	0	0	0	0																			
	Mora IK	Swede-2	13	1	4	5	30												4	1	0	1	14			
1998-99	Malmo IF	Sweden	41	0	3	3	44												1	0	0	0	4			
99-2000	Malmo IF	Sweden	49	8	11	19	88												6	0	0	0	8			
2000-01	Los Angeles	NHL	2	0	0	0	4	0	0	0	1	0.0	−2	0	0.0	6	0	12:22	1	0	0	0	0	0	0	0
	Lowell	AHL	61	7	29	36	149												4	0	6	6	6			
2001-02	Los Angeles	NHL	26	1	4	5	22	1	0	0	12	8.3	3	0	0.0	39	11	11:27	5	0	0	0	6	0	0	0
	Manchester	AHL	4	0	1	1	4																			
	NHL Totals		28	1	4	5	26	1	0	0	13	7.7		0	0.0	45	11	11:31	6	0	0	0	6	0	0	0

• Spent majority of 2001-02 season on practice roster, October 4, 2001.

LIND, Juha

(LIHND, YOO-huh)

Center. Shoots left. 5'11", 185 lbs. Born, Helsinki, Finland, January 2, 1974. Minnesota's 6th choice, 178th overall, in 1992 Entry Draft.

Season	Club	League	GP	G	A	Pts	PIM	PP	SH	GW	S	%	+/-	TF	F%	H	SB	Min	GP	G	A	Pts	PIM	PP	SH	GW
1990-91	Jokerit Jr.	Finn-Jr.	8	1	1	2	0																			
1991-92	Jokerit-B	Finn-Jr.	14	9	16	25	2												14	7	8	15	8			
1992-93	Vantaa HT	Finland-2	25	8	12	20	8																			
	Jokerit Helsinki	Finland-2	3	2	4	6	2																			
	Jokerit Helsinki	Finland	6	0	0	0	2												1	0	0	0	0			
1993-94	Jokerit Helsinki	Finland-2	11	6	7	13	4																			
	Jokerit Helsinki	Finland	47	17	11	28	37												11	2	5	7	4			
1994-95	Jokerit Jr.	Finn-Jr.	3	2	1	3	2																			
	Jokerit Helsinki	Finland	50	10	8	18	12												11	1	2	3	6			
1995-96	Jokerit Helsinki	Finland	50	15	22	37	32												11	4	5	9	4			
1996-97	Jokerit Helsinki	Finland	50	16	22	38	28												9	5	3	8	0			
	Jokerit Helsinki	EuroHL	6	4	1	5	6												2	1	0	1	0			
1997-98	Dallas	NHL	39	2	3	5	6	0	0	0	27	7.4	4						15	2	2	4	8	0	0	1
	Michigan K-Wings	IHL	8	2	2	4	2																			
	Finland	Olympics	6	0	1	1	6																			
1998-99	Jokerit Helsinki	Finland	50	20	19	39	22												3	3	1	4	2			
	Jokerit Helsinki	EuroHL	6	6	2	8	14												2	0	2	2	0			
99-2000	Dallas	NHL	34	3	4	7	6	0	0	0	36	8.3	−1	8	50.0	52	4	10:49								
	Montreal	NHL	13	1	2	3	4	0	0	0	6	16.7	−2	0	0.0	18	3	9:37								
2000-01	Montreal	NHL	47	3	4	7	4	0	0	2	36	8.3	−4	13	46.2	37	19	7:49								
	Quebec	AHL	3	1	1	2	0																			
2001-02	Sodertalje SK	Sweden	41	16	10	26	26																			
	Finland	Olympics	4	0	0	0	0																			
	NHL Totals		133	9	13	22	20	0	0	2	105	8.6		21	47.6	107	26	9:09	15	2	2	4	8	0	0	1

Rights transferred to **Dallas** after **Minnesota** franchise relocated, June 9, 1993. Traded to **Montreal** by **Dallas** for Scott Thornton, January 22, 2000.

LINDEN, Trevor

(LIHND-dehn, TREH-vohr) **VAN.**

Center/Right wing. Shoots right. 6'4", 215 lbs. Born, Medicine Hat, Alta., April 11, 1970. Vancouver's 1st choice, 2nd overall, in 1988 Entry Draft.

Season	Club	League	GP	G	A	Pts	PIM	PP	SH	GW	S	%	+/-	TF	F%	H	SB	Min	GP	G	A	Pts	PIM	PP	SH	GW
1985-86	Medicine Hat	AMHL	40	14	22	36	14																			
	Medicine Hat	WHL	5	2	0	2	0																			
1986-87	Medicine Hat	WHL	72	14	22	36	59												20	5	4	9	17			
1987-88	Medicine Hat	WHL	67	46	64	110	76												16	*13	12	25	19			
1988-89	Vancouver	NHL	80	30	29	59	41	10	1	2	186	16.1	−10						7	3	4	7	8	2	1	0
1989-90	Vancouver	NHL	73	21	30	51	43	6	2	3	171	12.3	−17													
1990-91	Vancouver	NHL	80	33	37	70	65	16	2	4	229	14.4	−25						6	0	7	7	2	0	0	0
1991-92	Vancouver	NHL	80	31	44	75	101	6	1	6	201	15.4	3						13	4	8	12	6	2	0	1
1992-93	Vancouver	NHL	84	33	39	72	64	8	0	3	209	15.8	19						12	5	8	13	16	2	0	1
1993-94	Vancouver	NHL	84	32	29	61	73	10	2	3	234	13.7	6						24	12	13	25	18	5	1	1
1994-95	Vancouver	NHL	48	18	22	40	40	9	0	1	129	14.0	−5						11	2	6	8	12	1	0	0
1995-96	Vancouver	NHL	82	33	47	80	42	12	1	2	202	16.3	6						6	4	4	8	6	2	0	0
1996-97	Vancouver	NHL	49	9	31	40	27	2	2	2	84	10.7	5													
1997-98	Vancouver	NHL	42	7	14	21	49	2	0	1	74	9.5	−13													
	NY Islanders	NHL	25	10	7	17	33	3	2	1	59	16.9	−1													
	Canada	Olympics	6	1	0	1	10																			
1998-99	NY Islanders	NHL	82	18	29	47	32	8	1	1	167	10.8	−14	261	50.2	144	47	21:29								
99-2000	Montreal	NHL	50	13	17	30	34	4	0	3	87	14.9	−3	860	56.3	88	29	17:51								
2000-01	Montreal	NHL	57	12	21	33	52	6	0	3	96	12.5	−2	1142	52.7	93	37	20:47								
	Washington	NHL	12	3	1	4	8	0	0	0	30	10.0	2	75	60.0	14	7	18:03	6	0	4	4	14	0	0	0
2001-02	Washington	NHL	16	1	2	3	6	1	0	0	19	5.3	−2	71	49.3	17	3	16:06								
	Vancouver	NHL	64	12	22	34	65	2	0	2	122	9.8	−3	1190	53.4	155	30	19:23	6	1	4	5	0	0	0	0
	NHL Totals		1008	316	421	737	775	105	14	37	2299	13.7		3599	53.7	511	153	19:46	91	31	58	89	82	14	2	3

WHL East Second All-Star Team (1988) • NHL All-Rookie Team (1989) • Won King Clancy Memorial Trophy (1997) • Played in NHL All-Star Game (1991, 1992)

Traded to **NY Islanders** by **Vancouver** for Todd Bertuzzi, Bryan McCabe and NY Islanders' 3rd round choice (Jarkko Ruutu) in 1998 Entry Draft, February 6, 1998. Traded to **Montreal** by **NY Islanders** for Montreal's 1st round choice (Branislav Mezei) in 1999 Entry Draft, May 29, 1999. Traded to **Washington** by **Montreal** with Dainius Zubrus and New Jersey's 2nd round choice (previously acquired, later traded to Tampa Bay - Tampa Bay selected Andreas Holmqvist) in 2001 Entry Draft for Richard Zednik, Jan Bulis and Washington's 1st round choice (Alexander Perezhogin) in 2001 Entry Draft, March 13, 2001. Traded to **Vancouver** by **Washington** with NY Islanders' 2nd round choice (previously acquired, Vancouver selected Denis Grot) in 2002 Entry Draft for Vancouver's 1st round choice (Boyd Gordon) in 2002 Entry Draft and 3rd round choice in 2003 Entry Draft, November 10, 2001.

LINDGREN, Mats

(LIHND-gruhn, MAHTS)

Center/Left wing. Shoots left. 6'2", 202 lbs. Born, Skelleftea, Sweden, October 1, 1974. Winnipeg's 1st choice, 15th overall, in 1993 Entry Draft.

Season	Club	League	GP	G	A	Pts	PIM	PP	SH	GW	S	%	+/-	TF	F%	H	SB	Min	GP	G	A	Pts	PIM	PP	SH	GW
1990-91	Skelleftea AIK	Swede-2	10	0	1	1	0																			
1991-92	Skelleftea AIK	Swede-2	29	14	18	32	12												3	3	2	5	2			
1992-93	Skelleftea AIK	Swede-2	32	20	18	38	18												3	0	0	0	2			
1993-94	Farjestad	Sweden	22	11	6	17	26																			
1994-95	Farjestad	Sweden	37	17	15	32	20												3	0	0	0	4			
1995-96	Cape Breton	AHL	13	7	5	12	6																			
1996-97	Edmonton	NHL	69	11	14	25	12	2	3	1	71	15.5	−7						12	0	4	4	0	0	0	0
	Hamilton	AHL	9	6	7	13	6																			
1997-98	Edmonton	NHL	82	13	13	26	42	1	3	3	131	9.9	0						12	1	1	2	10	0	0	0
	Sweden	Olympics	4	0	0	0	2																			
1998-99	Edmonton	NHL	48	5	12	17	22	0	1	0	53	9.4	4	363	47.9	44	16	11:31								
	NY Islanders	NHL	12	5	3	8	2	3	0	1	30	16.7	2	180	48.9	18	4	20:08								
99-2000	NY Islanders	NHL	43	9	7	16	24	1	0	1	68	13.2	0	551	49.2	54	25	19:17								

			Regular Season																Playoffs							
Season	Club	League	GP	G	A	Pts	PIM	PP	SH	GW	S	%	+/-	TF	F%	H	SB	Min	GP	G	A	Pts	PIM	PP	SH	GW
2000-01	NY Islanders	NHL	20	3	4	7	10	0	2	0	34	8.8	4	176	48.3	35	3	15:14								
2001-02	NY Islanders	NHL	59	3	12	15	16	0	0	1	35	8.6	0	237	59.1	68	7	8:21								
	NHL Totals		333	49	65	114	128	7	9	7	422	11.6		1507	50.3	219	55	13:18	24	1	5	6	10	0	0	0

Traded to **Edmonton** by **Winnipeg** with Boris Mironov, Winnipeg's 1st round choice (Jason Bonsignore) in 1994 Entry Draft and Florida's 4th round choice (previously acquired, Edmonton selected Adam Copeland) in 1994 Entry Draft for Dave Manson and St. Louis' 6th round choice (previously acquired, Winnipeg selected Chris Kibermanis) in 1994 Entry Draft, March 15, 1994. Traded to **NY Islanders** by **Edmonton** with Edmonton's 8th round choice (Radek Martinek) in 1999 Entry Draft for Tommy Salo, March 20, 1999. • Missed majority of 2000-01 season recovering from shoulder injury suffered in game vs. Anaheim, November 25, 2000.

LINDROS, Eric

(LIHND-rahz, AIR-ihk) **NYR**

Center. Shoots right. 6'4", 240 lbs. Born, London, Ont., February 28, 1973. Quebec's 1st choice, 1st overall, in 1991 Entry Draft.

Season	Club	League	GP	G	A	Pts	PIM	PP	SH	GW	S	%	+/-	TF	F%	H	SB	Min	GP	G	A	Pts	PIM	PP	SH	GW
1988-89	St. Michael's B	OJHL-B	37	24	43	67	193												27	23	25	48	155			
1989-90	Det. Compuware	NAJHL	14	23	29	52	123																			
	Oshawa Generals	OHL	25	17	19	36	61												17	18	18	36	76			
1990-91	Oshawa Generals	OHL	57	*71	78	*149	189												16	*18	20	*38	*93			
1991-92	Oshawa Generals	OHL	13	9	22	31	54																			
	Team Canada	Nat-Tm	24	19	16	35	34																			
	Canada	Olympics	8	5	6	11	5																			
1992-93	Philadelphia	NHL	61	41	34	75	147	8	1	5	180	22.8	28													
1993-94	Philadelphia	NHL	65	44	53	97	103	13	2	9	197	22.3	16													
1994-95	Philadelphia	NHL	46	29	41	*70	60	7	0	4	144	20.1	27						12	4	11	15	18	0	0	1
1995-96	Philadelphia	NHL	73	47	68	115	163	15	0	4	294	16.0	26						12	6	6	12	43	3	0	2
1996-97	Philadelphia	NHL	52	32	47	79	136	9	0	7	198	16.2	31						19	12	14	*26	40	4	0	1
1997-98	Philadelphia	NHL	63	30	41	71	134	10	1	4	202	14.9	14						5	1	2	3	17	0	0	0
	Canada	Olympics	6	2	3	5	2																			
1998-99	Philadelphia	NHL	71	40	53	93	120	10	1	2	242	16.5	35	1529	60.0	117	17	22:56								
99-2000	Philadelphia	NHL	55	27	32	59	83	10	1	2	187	14.4	11	1318	57.8	130	29	22:02	2	1	0	1	0	0	0	0
2000-01	Philadelphia	NHL	DID NOT PLAY																							
2001-02	NY Rangers	NHL	72	37	36	73	138	12	1	4	196	18.9	19	1707	54.4	145	43	21:03								
	Canada	Olympics	6	1	0	1	8																			
	NHL Totals		558	327	405	732	1084	94	7	41	1840	17.8		4554	57.3	392	89	21:60	50	24	33	57	118	7	0	4

• Memorial Cup All-Star Team (1990) OHL First All-Star Team (1991) • OHL MVP (1991) • Canadian Major Junior Player of the Year (1991) • NHL All-Rookie Team (1993) • NHL First All-Star Team (1995) • Won Lester B. Pearson Award (1995) • Won Hart Trophy (1995) • NHL Second All-Star Team (1996) • Played in NHL All-Star Game (1994, 1996, 1997, 1998, 1999, 2000)

Rights traded to **Oshawa** by **Sault Ste. Marie** for Mike DeCoff, Jason Denomme, Mike Lenarduzzi and Oshawa's 2nd round choices in 1990 (Drew Bannister) and 1991 (Dave Roach) OHL Priority Drafts, December 18, 1989. Traded to **Philadelphia** by **Quebec** for Peter Forsberg, Steve Duchesne, Kerry Huffman, Mike Ricci, Ron Hextall, Philadelphia's 1st round choice (Jocelyn Thibault) in 1993 Entry Draft, $15,000,000 and future considerations (Chris Simon and Philadelphia's 1st round choice (later traded to Toronto - later traded to Washington - Washington selected Nolan Baumgartner) in 1994 Entry Draft, July 21, 1992), June 30, 1992. • Missed entire 2000-01 season recovering from head injury suffered in game vs. New Jersey, May 26, 2000 and contract dispute with Philadelphia Flyers management. Traded to **NY Rangers** by **Philadelphia** for Kim Johnsson, Jan Havac, Pavel Brendl and NY Rangers' 3rd round choice in 2003 Entry Draft, August 20, 2001.

LINDSAY, Bill

(LIHND-see, BIHL) **MTL.**

Left wing. Shoots left. 6', 195 lbs. Born, Fernie, B.C., May 17, 1971. Quebec's 6th choice, 103rd overall, in 1991 Entry Draft.

Season	Club	League	GP	G	A	Pts	PIM	PP	SH	GW	S	%	+/-	TF	F%	H	SB	Min	GP	G	A	Pts	PIM	PP	SH	GW
1988-89	Vernon Lakers	BCJHL	56	24	29	53	166																			
1989-90	Tri-City	WHL	72	40	45	85	84												7	3	0	3	17			
1990-91	Tri-City	WHL	63	46	47	93	151												5	3	6	9	10			
1991-92	Tri-City	WHL	42	34	59	93	81												3	2	3	5	16			
	Quebec	NHL	23	2	4	6	14	0	0	1	35	5.7	–6													
1992-93	Quebec	NHL	44	4	9	13	16	0	0	0	58	6.9	0													
	Halifax Citadels	AHL	20	11	13	24	18																			
1993-94	Florida	NHL	84	6	6	12	97	0	0	0	90	6.7	–2													
1994-95	Florida	NHL	48	10	9	19	46	0	1	0	63	15.9	1													
1995-96	Florida	NHL	73	12	22	34	57	0	3	2	118	10.2	13						22	5	5	10	18	0	1	1
1996-97	Florida	NHL	81	11	23	34	120	0	1	3	168	6.5	1						3	0	1	1	8	0	0	0
1997-98	Florida	NHL	82	12	16	28	80	0	2	5	150	8.0	–2													
1998-99	Florida	NHL	75	12	15	27	92	0	1	2	135	8.9	–1	57	40.4	149	20	13:37								
99-2000	Calgary	NHL	80	8	12	20	86	0	0	2	147	5.4	–7	28	39.3	122	50	12:55								
2000-01	Calgary	NHL	52	1	9	10	97	0	0	0	57	1.8	–8	9	55.6	105	6	10:32								
	San Jose	NHL	16	0	4	4	29	0	0	0	14	0.0	2	2	0.0	49	5	9:16	6	0	0	0	16	0	0	0
2001-02	Florida	NHL	63	4	7	11	117	0	0	1	63	6.3	–11	124	44.4	192	9	9:40								
	Montreal	NHL	13	1	3	4	23	0	0	0	14	7.1	0	26	53.9	32	5	10:10	11	2	2	4	2	0	0	0
	NHL Totals		734	83	139	222	874	0	8	16	1112	7.5		246	43.9	649	95	11:41	42	7	8	15	44	0	1	1

WHL West Second All-Star Team (1992)

Claimed by **Florida** from **Quebec** in Expansion Draft, June 24, 1993. Traded to **Calgary** by **Florida** for Todd Simpson, September 30, 1999. Traded to **San Jose** by **Calgary** for Minnesota's 8th round choice (previously acquired, Calgary selected Joe Campbell) in 2001 Entry Draft, March 6, 2001. Signed as a free agent by **Florida**, August 23, 2001. Claimed on waivers by **Montreal** from **Florida**, March 19, 2002.

LING, David

(LIHNG, DAY-vihd) **CBJ**

Right wing. Shoots right. 5'10", 204 lbs. Born, Halifax, N.S., January 9, 1975. Quebec's 9th choice, 179th overall, in 1993 Entry Draft.

Season	Club	League	GP	G	A	Pts	PIM	PP	SH	GW	S	%	+/-	TF	F%	H	SB	Min	GP	G	A	Pts	PIM	PP	SH	GW
1991-92	Charlottetown	MJrHL	30	33	42	75	270																			
	St. Michael's B	OJHL-B	8	5	14	19	25																			
1992-93	Kingston	OHL	64	17	46	63	275												16	3	12	15	*72			
1993-94	Kingston	OHL	61	37	40	77	*254												6	4	2	6	16			
1994-95	Kingston	OHL	62	*61	74	135	136												6	7	8	15	12			
1995-96	Saint John	AHL	75	24	32	56	179												9	0	5	5	12			
1996-97	Saint John	AHL	5	0	2	2	19																			
	Montreal	NHL	2	0	0	0	0	0	0	0	0	0.0	0													
	Fredericton	AHL	48	22	36	58	229																			
1997-98	Montreal	NHL	1	0	0	0	0	0	0	0	1	0.0	–1													
	Fredericton	AHL	67	25	41	66	148																			
	Indianapolis Ice	IHL	12	8	6	14	30												5	4	1	5	31			
1998-99	Kansas City	IHL	82	30	42	72	112												3	1	0	1	20			
99-2000	Kansas City	IHL	82	35	48	83	210																			
2000-01	Utah Grizzlies	IHL	79	15	28	43	202																			
2001-02	Columbus	NHL	5	0	0	0	7	0	0	0	5	0.0	–1	1	0.0	11	2	9:47								
	Syracuse Crunch	AHL	71	19	41	60	240												10	5	5	10	16			
	NHL Totals		8	0	0	0	7	0	0	0	6	0.0		1	0.0	11	2	9:47								

OHL First All-Star Team (1995) • OHL MVP (1995) • Canadian Major Junior First All-Star Team (1995) • Canadian Major Junior Player of the Year (1995) • IHL First All-Star Team (2000)

Rights transferred to **Colorado** after **Quebec** franchise relocated, June 21, 1995. Traded to **Calgary** by **Colorado** with Colorado's 9th round choice (Steve Shirreffs) in 1995 Entry Draft for Calgary's 9th round choice (Chris George) in 1995 Entry Draft, July 7, 1995. Traded to **Montreal** by **Calgary** with Calgary's 6th round choice (Gordie Dwyer) in 1998 Entry Draft for Scott Fraser, October 24, 1996. Traded to **Chicago** by **Montreal** for Martin Gendron, March 14, 1998. Signed as a free agent by **Kansas City** (IHL) with Chicago retaining NHL rights, September 3, 1998. Traded to **Dallas** by **Chicago** for future considerations, August 11, 2000. Signed as a free agent by **Columbus**, July 7, 2001.

LINTNER, Richard

(LIHNT-nuhr, RIH-chahrd) **NYR**

Defense. Shoots right. 6'3", 212 lbs. Born, Trencin, Czech., November 15, 1977. Phoenix's 4th choice, 119th overall, in 1996 Entry Draft.

Season	Club	League	GP	G	A	Pts	PIM	PP	SH	GW	S	%	+/-	TF	F%	H	SB	Min	GP	G	A	Pts	PIM	PP	SH	GW
1994-95	Dukla Trencin Jr.	Slovak-Jr.	42	12	13	25	20																			
1995-96	Dukla Trencin Jr.	Slovak-Jr.	30	15	17	32	210																			
	Dukla Trencin	Slovakia	2	0	0	0	0																			
1996-97	Nova Ves	Slovakia	35	2	1	3																				
	MoDo Jr.	Swede-Jr.	20	6	11	17																				
1997-98	Springfield	AHL	71	6	9	15	61												3	1	1	2	4			
1998-99	MoDo Jr.	Swede-Jr.	2	1	0	1	0																			
	Springfield	AHL	8	0	1	1	16																			
	Milwaukee	IHL	66	9	16	25	75																			
99-2000	Nashville	NHL	33	1	5	6	22	0	0	0	58	1.7	–6	0	0.0	44	15	14:51								
	Milwaukee	IHL	31	13	8	21	37																			
2000-01	Nashville	NHL	50	3	5	8	22	1	0	0	81	3.7	2	0	0.0	48	26	13:05								

Season	Club	League	GP	G	A	Pts	PIM	PP	SH	GW	S	%	+/-	TF	F%	H	SB	Min	GP	G	A	Pts	PIM	PP	SH	GW
							Regular Season															Playoffs				
2001-02	Dukla Trencin	Slovakia	6	2	0	2	0																			
	MoDo	Sweden	28	12	9	21	85												14	3	0	3	8			
	Slovakia	Olympics	4	1	1	2	0																			
	NHL Totals		**83**	**4**	**10**	**14**	**44**	**1**	**0**	**0**	**139**	**2.9**		**0**	**0.0**	**92**	**41**	**13:47**								

Traded to **Nashville** by **Phoenix** with Cliff Ronning for future considerations, October 31, 1998. Signed as a free agent by **Dukla Trencin** (Czech) with Nashville retaining NHL rights, October 17, 2001. Traded to **NY Rangers** by **Nashville** for Peter Smrek, March 19, 2002.

LOW, Reed

(LOH, REED) **ST.L.**

Right wing. Shoots right. 6'3", 222 lbs. Born, Moose Jaw, Sask., June 21, 1976. St. Louis' 7th choice, 177th overall, in 1996 Entry Draft.

Season	Club	League	GP	G	A	Pts	PIM	PP	SH	GW	S	%	+/-	TF	F%	H	SB	Min	GP	G	A	Pts	PIM	PP	SH	GW
1994-95	Minot Top Guns	SJHL	STATISTICS NOT AVAILABLE																							
	Regina Pats	WHL	2	0	0	0	5																			
1995-96	Moose Jaw	WHL	61	12	7	19	221																			
1996-97	Moose Jaw	WHL	62	16	11	27	228												12	2	1	3	50			
1997-98	Worcester	AHL	17	1	1	2	75												3	0	0	0	0			
	Baton Rouge	ECHL	39	4	2	6	145																			
1998-99	Worcester	AHL	77	5	6	11	239												4	0	0	0	2			
99-2000	Worcester	AHL	80	12	16	28	203												9	1	3	4	16			
2000-01	**St. Louis**	**NHL**	**56**	**1**	**5**	**6**	**159**	**0**	**0**	**0**	**31**	**3.2**	**4**	**2**	**50.0**	**71**	**4**	**6:17**								
2001-02	**St. Louis**	**NHL**	**58**	**0**	**5**	**5**	**160**	**0**	**0**	**0**	**25**	**0.0**	**–3**	**0**	**0.0**	**78**	**4**	**5:22**								
	NHL Totals		**114**	**1**	**10**	**11**	**319**	**0**	**0**	**0**	**56**	**1.8**		**2**	**50.0**	**149**	**8**	**5:49**								

LOWRY, Dave

(LOW-ree, DAYV) **CGY.**

Left wing. Shoots left. 6'1", 200 lbs. Born, Sudbury, Ont., February 14, 1965. Vancouver's 6th choice, 114th overall, in 1983 Entry Draft.

Season	Club	League	GP	G	A	Pts	PIM	PP	SH	GW	S	%	+/-	TF	F%	H	SB	Min	GP	G	A	Pts	PIM	PP	SH	GW
1981-82	Nepean	OMHA	60	50	64	114	46																			
1982-83	London Knights	OHL	42	11	16	27	48												3	0	0	0	14			
1983-84	London Knights	OHL	66	29	47	76	125												8	6	6	12	41			
1984-85	London Knights	OHL	61	60	60	120	94												8	6	5	11	10			
1985-86	**Vancouver**	**NHL**	**73**	**10**	**8**	**18**	**143**	**1**	**0**	**1**	**66**	**15.2**	**–21**						**3**	**0**	**0**	**0**	**0**	**0**	**0**	**0**
1986-87	**Vancouver**	**NHL**	**70**	**8**	**10**	**18**	**176**	**0**	**0**	**1**	**74**	**10.8**	**–23**													
1987-88	**Vancouver**	**NHL**	**22**	**1**	**3**	**4**	**38**	**0**	**0**	**0**	**14**	**7.1**	**–2**													
	Fredericton	AHL	46	18	27	45	59												14	7	3	10	72			
1988-89	**St. Louis**	**NHL**	**21**	**3**	**3**	**6**	**11**	**0**	**1**	**0**	**22**	**13.6**	**1**						**10**	**0**	**5**	**5**	**4**	**0**	**0**	**0**
	Peoria Rivermen	IHL	58	31	35	66	45																			
1989-90	**St. Louis**	**NHL**	**78**	**19**	**6**	**25**	**75**	**0**	**2**	**1**	**98**	**19.4**	**1**						**12**	**2**	**1**	**3**	**39**	**0**	**0**	**0**
1990-91	**St. Louis**	**NHL**	**79**	**19**	**21**	**40**	**168**	**0**	**2**	**5**	**123**	**15.4**	**19**						**13**	**1**	**4**	**5**	**35**	**0**	**0**	**0**
1991-92	**St. Louis**	**NHL**	**75**	**7**	**13**	**20**	**77**	**0**	**0**	**1**	**85**	**8.2**	**–11**						**6**	**0**	**1**	**1**	**20**	**0**	**0**	**0**
1992-93	**St. Louis**	**NHL**	**58**	**5**	**8**	**13**	**101**	**0**	**0**	**0**	**59**	**8.5**	**–18**						**11**	**2**	**0**	**2**	**14**	**0**	**1**	**0**
1993-94	**Florida**	**NHL**	**80**	**15**	**22**	**37**	**64**	**3**	**0**	**3**	**122**	**12.3**	**–4**													
1994-95	**Florida**	**NHL**	**45**	**10**	**10**	**20**	**25**	**2**	**0**	**3**	**70**	**14.3**	**–3**													
1995-96	**Florida**	**NHL**	**63**	**10**	**14**	**24**	**36**	**0**	**0**	**1**	**83**	**12.0**	**–2**						**22**	**10**	**7**	**17**	**39**	**4**	**0**	**2**
1996-97	**Florida**	**NHL**	**77**	**15**	**14**	**29**	**51**	**2**	**0**	**2**	**96**	**15.6**	**2**						**5**	**0**	**0**	**0**	**0**	**0**	**0**	**0**
1997-98	**Florida**	**NHL**	**7**	**0**	**0**	**0**	**2**	**0**	**0**	**0**	**4**	**0.0**	**–1**													
	San Jose	**NHL**	**50**	**4**	**4**	**8**	**51**	**0**	**0**	**1**	**47**	**8.5**	**0**						**6**	**0**	**0**	**0**	**18**	**0**	**0**	**0**
1998-99	**San Jose**	**NHL**	**61**	**6**	**9**	**15**	**24**	**2**	**0**	**0**	**58**	**10.3**	**–5**	**6**	**50.0**	**70**	**7**	**9:14**	**1**	**0**	**0**	**0**	**0**	**0**	**0**	**0**
99-2000	**San Jose**	**NHL**	**32**	**1**	**4**	**5**	**18**	**0**	**0**	**0**	**25**	**4.0**	**1**	**1**	**100.0**	**81**	**3**	**9:11**	**12**	**1**	**2**	**3**	**6**	**0**	**0**	**0**
2000-01	**Calgary**	**NHL**	**79**	**18**	**17**	**35**	**47**	**5**	**0**	**5**	**108**	**16.7**	**–2**	**20**	**20.0**	**121**	**35**	**15:57**								
2001-02	**Calgary**	**NHL**	**62**	**7**	**6**	**13**	**51**	**2**	**1**	**1**	**74**	**9.5**	**–20**	**15**	**26.7**	**79**	**28**	**14:58**								
	NHL Totals		**1032**	**158**	**172**	**330**	**1158**	**17**	**6**	**25**	**1228**	**12.9**		**42**	**28.6**	**351**	**73**	**13:01**	**101**	**16**	**20**	**36**	**175**	**4**	**1**	**2**

OHL First All-Star Team (1985)

Traded to **St. Louis** by **Vancouver** for Ernie Vargas, September 29, 1988. Claimed by **Florida** from **St. Louis** in Expansion Draft, June 24, 1993. Traded to **San Jose** by **Florida** with Florida's 1st round choice (later traded to Tampa Bay - Tampa Bay selected Vincent Lecavalier) in 1998 Entry Draft for Viktor Kozlov and Florida's 5th round choice (previously acquired, Florida selected Jaroslav Spacek) in 1998 Entry Draft, November 13, 1997. • Missed majority of 1999-2000 season recovering from shoulder injury suffered in game vs. Montreal, November 23, 1999. Signed as a free agent by **Calgary**, July 24, 2000.

LUKOWICH, Brad

(loo-KUH-which, BRAD) **T.B.**

Defense. Shoots left. 6'1", 200 lbs. Born, Cranbrook, B.C., August 12, 1976. NY Islanders' 4th choice, 90th overall, in 1994 Entry Draft.

Season	Club	League	GP	G	A	Pts	PIM	PP	SH	GW	S	%	+/-	TF	F%	H	SB	Min	GP	G	A	Pts	PIM	PP	SH	GW
1992-93	Cranbrook Colts	RMJHL	54	21	41	62	162																			
	Kamloops Blazers	WHL	1	0	0	0	0																			
1993-94	Kamloops Blazers	WHL	42	5	11	16	166												16	0	1	1	35			
1994-95	Kamloops Blazers	WHL	63	10	35	45	125												18	0	7	7	21			
1995-96	Kamloops Blazers	WHL	65	14	55	69	114												13	2	10	12	29			
1996-97	Michigan K-Wings	IHL	69	2	6	8	77												4	0	1	1	2			
1997-98	**Dallas**	**NHL**	**4**	**0**	**1**	**1**	**2**	**0**	**0**	**0**	**2**	**0.0**	**–2**													
	Michigan K-Wings	IHL	60	6	27	33	104												4	0	4	4	14			
1998-99	**Dallas**	**NHL**	**14**	**1**	**2**	**3**	**19**	**0**	**0**	**0**	**8**	**12.5**	**3**	**0**	**0.0**	**32**	**12**	**16:18**	**8**	**0**	**1**	**1**	**4**	**0**	**0**	**0**
	Michigan K-Wings	IHL	67	8	21	29	95																			
99-2000	**Dallas**	**NHL**	**60**	**3**	**1**	**4**	**50**	**0**	**0**	**1**	**33**	**9.1**	**–14**	**1**	**0.0**	**96**	**42**	**11:44**								
2000-01	**Dallas**	**NHL**	**80**	**4**	**10**	**14**	**76**	**0**	**0**	**2**	**43**	**9.3**	**28**	**1**	**100.0**	**204**	**70**	**14:48**	**10**	**1**	**0**	**1**	**4**	**0**	**0**	**0**
2001-02	**Dallas**	**NHL**	**66**	**1**	**6**	**7**	**40**	**0**	**0**	**0**	**56**	**1.8**	**–1**	**0**	**0.0**	**131**	**49**	**13:14**								
	NHL Totals		**224**	**9**	**20**	**29**	**187**	**0**	**0**	**3**	**142**	**6.3**		**2**	**50.0**	**463**	**173**	**13:35**	**18**	**1**	**1**	**2**	**8**	**0**	**0**	**0**

Traded to **Dallas** by **NY Islanders** for Dallas' 3rd round choice (Robert Schnabel) in 1997 Entry Draft, June 1, 1996. Traded to **Minnesota** by **Dallas** with Manny Fernandez for Minnesota's 3rd round choice (Joel Lundqvist) in 2000 Entry Draft and 4th round choice (later traded back to Minnesota - later traded to LA Kings - LA Kings selected Aaron Rome) in 2002 Entry Draft, June 12, 2000. Traded to **Dallas** by **Minnesota** with Minnesota's 3rd (Yared Hagos) and 9th (Dale Sullivan) round choices in 2001 Entry Draft for Aaron Gavey, Pavel Patera, Dallas' 8th round choice (Eric Johansson) in 2000 Entry Draft and Minnesota's 4th round choice (previously acquired by Dallas - later traded to LA Kings - LA Kings selected Aaron Rome) in 2002 Entry Draft, June 25, 2000. Traded to **Tampa Bay** by **Dallas** with Dallas' 7th round choice in 2003 Entry Draft for Tampa Bay's 2nd round choice (previously acquired, later traded back to Tampa Bay - later traded to Dallas - Dallas selected Tobias Stephan) in 2002 Entry Draft, June 22, 2002.

LUMME, Jyrki

(LOO-may, YUHR-kee) **TOR.**

Defense. Shoots left. 6'1", 209 lbs. Born, Tampere, Finland, July 16, 1966. Montreal's 3rd choice, 57th overall, in 1986 Entry Draft.

Season	Club	League	GP	G	A	Pts	PIM	PP	SH	GW	S	%	+/-	TF	F%	H	SB	Min	GP	G	A	Pts	PIM	PP	SH	GW
1983-84	KooVee Jr.	Finn-Jr.	28	5	4	9	61																			
1984-85	KooVee Tampere	Finland-3	30	6	4	10	44																			
1985-86	Ilves Jr.	Finn-Jr.	6	3	3	6	6												4	0	0	0	8			
	Ilves Tampere	Finland	31	1	4	5	4																			
1986-87	Ilves Jr.	Finn-Jr.	1	0	1	1	6												1	1	0	1	6			
	Ilves Tampere	Finland	43	12	12	24	52												4	0	1	1	2			
1987-88	Ilves Tampere	Finland	43	8	22	30	75																			
	Finland	Olympics	6	0	1	1	2																			
1988-89	**Montreal**	**NHL**	**21**	**1**	**3**	**4**	**10**	**1**	**0**	**0**	**18**	**5.6**	**3**													
	Sherbrooke	AHL	26	4	11	15	10												6	1	3	4	4			
1989-90	**Montreal**	**NHL**	**54**	**1**	**19**	**20**	**41**	**0**	**0**	**0**	**79**	**1.3**	**17**													
	Vancouver	**NHL**	**11**	**3**	**7**	**10**	**8**	**0**	**0**	**1**	**30**	**10.0**	**0**													
1990-91	**Vancouver**	**NHL**	**80**	**5**	**27**	**32**	**59**	**1**	**0**	**0**	**157**	**3.2**	**–15**						**6**	**2**	**3**	**5**	**0**	**1**	**1**	**0**
1991-92	**Vancouver**	**NHL**	**75**	**12**	**32**	**44**	**65**	**3**	**1**	**1**	**106**	**11.3**	**25**						**13**	**2**	**3**	**5**	**4**	**1**	**0**	**1**
1992-93	**Vancouver**	**NHL**	**74**	**8**	**36**	**44**	**55**	**3**	**2**	**1**	**123**	**6.5**	**30**						**12**	**0**	**5**	**5**	**6**	**0**	**0**	**0**
1993-94	**Vancouver**	**NHL**	**83**	**13**	**42**	**55**	**50**	**1**	**3**	**3**	**161**	**8.1**	**3**						**24**	**2**	**11**	**13**	**16**	**2**	**0**	**1**
1994-95	Ilves Tampere	Finland	12	4	4	8	24																			
	Vancouver	**NHL**	**36**	**5**	**12**	**17**	**26**	**3**	**0**	**1**	**78**	**6.4**	**4**						**11**	**2**	**6**	**8**	**8**	**1**	**0**	**0**
1995-96	**Vancouver**	**NHL**	**80**	**17**	**37**	**54**	**50**	**8**	**0**	**2**	**192**	**8.9**	**–9**						**6**	**1**	**3**	**4**	**2**	**1**	**0**	**0**
1996-97	**Vancouver**	**NHL**	**66**	**11**	**24**	**35**	**32**	**5**	**0**	**2**	**107**	**10.3**	**8**													
1997-98	**Vancouver**	**NHL**	**74**	**9**	**21**	**30**	**34**	**4**	**0**	**1**	**117**	**7.7**	**–25**													
	Finland	Olympics	6	1	0	1	16																			
1998-99	**Phoenix**	**NHL**	**60**	**7**	**21**	**28**	**34**	**1**	**0**	**4**	**121**	**5.8**	**5**	**0**	**0.0**	**26**	**71**	**23:20**	**7**	**0**	**1**	**1**	**6**	**0**	**0**	**0**
99-2000	**Phoenix**	**NHL**	**74**	**8**	**32**	**40**	**44**	**4**	**0**	**3**	**142**	**5.6**	**9**	**0**	**0.0**	**51**	**105**	**23:36**	**5**	**0**	**1**	**1**	**2**	**0**	**0**	**0**
2000-01	**Phoenix**	**NHL**	**58**	**4**	**21**	**25**	**44**	**0**	**0**	**0**	**77**	**5.2**	**3**	**0**	**0.0**	**35**	**78**	**21:44**								

			Regular Season																Playoffs							
Season	Club	League	GP	G	A	Pts	PIM	PP	SH	GW	S	%	+/-	TF	F%	H	SB	Min	GP	G	A	Pts	PIM	PP	SH	GW
2001-02	Dallas	NHL	15	0	1	1	4	0	0	0	12	0.0	−5	0	0.0	8	11	13:40	....	....	....	....	....	....	....	....
	Toronto	NHL	51	4	8	12	18	1	2	1	61	6.6	13	0	0.0	47	72	19:57	14	0	0	0	4	0	0	0
	Finland	Olympics	4	0	1	1	0	....	....	....	....	....	....	....	....	....	....	....	....	....	....	....	....	....	....	....
	NHL Totals		912	108	343	451	574	35	8	20	1581	6.8		0	0.0	167	337	21:49	98	9	33	42	48	6	1	2

Traded to **Vancouver** by **Montreal** for St. Louis' 2nd round choice (previously acquired, Montreal selected Craig Darby) in 1991 Entry Draft, March 6, 1990. Signed as a free agent by **Phoenix**, July 3, 1998. Traded to **Dallas** by **Phoenix** for Tyler Bouck, June 23, 2001. Traded to **Toronto** by **Dallas** for Dave Manson, November 21, 2001.

LYASHENKO, Roman

(LIGH-a-SHEHN-koh, ROH-muhn) **NYR**

Center. Shoots right. 6', 189 lbs. Born, Murmansk, Russia, May 2, 1979. Dallas' 2nd choice, 52nd overall, in 1997 Entry Draft.

Season	Club	League	GP	G	A	Pts	PIM	PP	SH	GW	S	%	+/-	TF	F%	H	SB	Min	GP	G	A	Pts	PIM	PP	SH	GW
1995-96	Yaroslavl	CIS	60	7	10	17	12	....	....	....	....	....	....	....	....	....	....	....	....	....	....	....	....	....	....	....
1996-97	Yaroslavl 2	Russia-3	2	1	1	2	8	....	....	....	....	....	....	....	....	....	....	....	....	....	....	....	....	....	....	....
	Yaroslavl	Russia	42	5	7	12	16	....	....	....	....	....	....	....	....	....	....	....	9	3	0	3	6	....	....	....
1997-98	Yaroslavl	Russia	46	7	6	13	28	....	....	....	....	....	....	....	....	....	....	....	....	....	....	....	....	....	....	....
	Yaroslavl	EuroHL	10	1	1	2	2	....	....	....	....	....	....	....	....	....	....	....	....	....	....	....	....	....	....	....
1998-99	Yaroslavl	Russia	42	10	9	19	51	....	....	....	....	....	....	....	....	....	....	....	9	0	4	4	8	....	....	....
99-2000	Dallas	NHL	58	6	6	12	10	0	0	1	51	11.8	−2	339	45.7	46	15	10:56	16	2	1	3	0	0	0	2
	Michigan K-Wings	IHL	9	3	2	5	8	....	....	....	....	....	....	....	....	....	....	....	....	....	....	....	....	....	....	....
2000-01	Dallas	NHL	60	6	3	9	45	0	0	1	48	12.5	−1	418	43.3	54	10	9:35	1	0	0	0	0	0	0	0
	Utah Grizzlies	IHL	6	0	1	1	2	....	....	....	....	....	....	....	....	....	....	....	....	....	....	....	....	....	....	....
2001-02	Dallas	NHL	4	0	0	0	0	0	0	0	3	0.0	−2	25	44.0	0	0	7:42	....	....	....	....	....	....	....	....
	Utah Grizzlies	AHL	58	11	25	36	37	....	....	....	....	....	....	....	....	....	....	....	....	....	....	....	....	....	....	....
	NY Rangers	NHL	15	2	0	2	0	0	0	0	13	15.4	0	99	41.4	6	2	7:50	....	....	....	....	....	....	....	....
	Hartford	AHL	....	....	....	....	....	....	....	....	....	....	....	....	....	....	....	....	4	1	1	2	2	....	....	....
	NHL Totals		137	14	9	23	55	0	0	2	115	12.2		881	44.0	106	27	9:55	17	2	1	3	0	0	0	2

Traded to **NY Rangers** by **Dallas** with Martin Rucinsky for Manny Malhotra and Barrett Heisten, March 12, 2002.

LYDMAN, Toni

(LEED-man, TOH-nee) **CGY.**

Defense. Shoots left. 6'1", 200 lbs. Born, Lahti, Finland, September 25, 1977. Calgary's 5th choice, 89th overall, in 1996 Entry Draft.

Season	Club	League	GP	G	A	Pts	PIM	PP	SH	GW	S	%	+/-	TF	F%	H	SB	Min	GP	G	A	Pts	PIM	PP	SH	GW
1992-93	Reipas Lahti-C	Finn-Jr.	36	10	9	19	22	....	....	....	....	....	....	....	....	....	....	....	....	....	....	....	....	....	....	....
1993-94	Reipas Lahti-B	Finn-Jr.	9	3	1	4	4	....	....	....	....	....	....	....	....	....	....	....	....	....	....	....	....	....	....	....
	Reipas Lahti Jr.	Finn-Jr.	1	0	0	0	0	....	....	....	....	....	....	....	....	....	....	....	....	....	....	....	....	....	....	....
1994-95	Reipas Lahti-B	Finn-Jr.	9	7	4	11	12	....	....	....	....	....	....	....	....	....	....	....	....	....	....	....	....	....	....	....
	Reipas Lahti-B	Finn-Jr.	26	6	4	10	10	....	....	....	....	....	....	....	....	....	....	....	....	....	....	....	....	....	....	....
1995-96	Reipas Lahti Jr.	Finn-Jr.	9	2	2	4	6	....	....	....	....	....	....	....	....	....	....	....	....	....	....	....	....	....	....	....
	Reipas Lahti	Finland	39	5	2	7	30	....	....	....	....	....	....	....	....	....	....	....	3	0	1	1	0	....	....	....
1996-97	Tappara Tampere	Finland	49	1	2	3	65	....	....	....	....	....	....	....	....	....	....	....	3	0	0	0	6	....	....	....
1997-98	Tappara Tampere	Finland	48	4	10	14	48	....	....	....	....	....	....	....	....	....	....	....	4	0	2	2	0	....	....	....
1998-99	HIFK Helsinki	Finland	42	4	7	11	36	....	....	....	....	....	....	....	....	....	....	....	11	0	3	3	2	....	....	....
	HIFK Helsinki	EuroHL	6	0	2	2	29	....	....	....	....	....	....	....	....	....	....	....	4	1	....	....	....	....	....	....
99-2000	HIFK Helsinki	Finland	46	4	18	22	36	....	....	....	....	....	....	....	....	....	....	....	9	0	4	4	6	....	....	....
2000-01	Calgary	NHL	62	3	16	19	30	1	0	0	80	3.8	−7	0	0.0	60	61	20:36	....	....	....	....	....	....	....	....
2001-02	Calgary	NHL	79	6	22	28	52	1	0	0	126	4.8	−8	0	0.0	81	102	21:10	....	....	....	....	....	....	....	....
	NHL Totals		141	9	38	47	82	2	0	0	206	4.4		0	0.0	141	163	20:55	....	....	....	....	....	....	....	....

MacDONALD, Craig

(MAK-DAWN-uhld, KRAYG) **CAR.**

Center. Shoots left. 6'2", 195 lbs. Born, Antigonish, N.S., April 7, 1977. Hartford's 3rd choice, 88th overall, in 1996 Entry Draft.

Season	Club	League	GP	G	A	Pts	PIM	PP	SH	GW	S	%	+/-	TF	F%	H	SB	Min	GP	G	A	Pts	PIM	PP	SH	GW
1994-95	Lawrence School	Hi-School	30	25	52	77	10	....	....	....	....	....	....	....	....	....	....	....	....	....	....	....	....	....	....	....
1995-96	Harvard Crimson	ECAC	34	7	10	17	10	....	....	....	....	....	....	....	....	....	....	....	....	....	....	....	....	....	....	....
1996-97	Harvard Crimson	ECAC	32	6	10	16	20	....	....	....	....	....	....	....	....	....	....	....	....	....	....	....	....	....	....	....
1997-98	Team Canada	Nat-Tm	58	18	29	47	38	....	....	....	....	....	....	....	....	....	....	....	....	....	....	....	....	....	....	....
1998-99	Carolina	NHL	11	0	0	0	0	0	0	0	5	0.0	0	2	100.0	6	1	2:29	1	0	0	0	0	0	0	0
	New Haven	AHL	62	17	31	48	77	....	....	....	....	....	....	....	....	....	....	....	....	....	....	....	....	....	....	....
99-2000	Cincinnati	IHL	78	12	24	36	76	....	....	....	....	....	....	....	....	....	....	....	11	4	1	5	8	....	....	....
2000-01	Cincinnati	IHL	82	20	28	48	104	....	....	....	....	....	....	....	....	....	....	....	5	0	1	1	6	....	....	....
2001-02	Carolina	NHL	12	1	1	2	0	0	0	0	15	6.7	−1	19	47.4	17	1	10:11	4	0	0	0	2	0	0	0
	Lowell	AHL	64	19	22	41	61	....	....	....	....	....	....	....	....	....	....	....	....	....	....	....	....	....	....	....
	NHL Totals		23	1	1	2	0	0	0	0	20	5.0		21	52.4	23	2	6:30	5	0	0	0	2	0	0	0

Rights transferred to **Carolina** after **Hartford** franchise relocated, June 25, 1997

MacINNIS, Al

(MAK-IHN-his, AL) **ST.L.**

Defense. Shoots right. 6'2", 209 lbs. Born, Inverness, N.S., July 11, 1963. Calgary's 1st choice, 15th overall, in 1981 Entry Draft.

Season	Club	League	GP	G	A	Pts	PIM	PP	SH	GW	S	%	+/-	TF	F%	H	SB	Min	GP	G	A	Pts	PIM	PP	SH	GW
1979-80	Regina Blues	SJHL	59	20	28	48	110	....	....	....	....	....	....	....	....	....	....	....	....	....	....	....	....	....	....	....
1980-81	Kitchener	OMJHL	47	11	28	39	59	....	....	....	....	....	....	....	....	....	....	....	18	4	12	16	20	....	....	....
1981-82	Kitchener	OHL	59	25	50	75	145	....	....	....	....	....	....	....	....	....	....	....	15	5	10	15	44	....	....	....
	Calgary	NHL	2	0	0	0	0	0	0	0	2	0.0	0	....	....	....	....	....	....	....	....	....	....	....	....	....
1982-83	Kitchener	OHL	51	38	46	84	67	....	....	....	....	....	....	....	....	....	....	....	8	3	8	11	9	....	....	....
	Calgary	NHL	14	1	3	4	9	0	0	0	7	14.3	0	....	....	....	....	....	....	....	....	....	....	....	....	....
1983-84	Calgary	NHL	51	11	34	45	42	7	0	2	160	6.9	0	....	....	....	....	....	11	2	12	14	13	2	0	1
	Colorado Flames	CHL	19	5	14	19	22	....	....	....	....	....	....	....	....	....	....	....	....	....	....	....	....	....	....	....
1984-85	Calgary	NHL	67	14	52	66	75	8	0	0	259	5.4	7	....	....	....	....	....	4	1	2	3	8	1	0	0
1985-86	Calgary	NHL	77	11	57	68	76	4	0	0	241	4.6	38	....	....	....	....	....	21	4	*15	19	30	2	0	0
1986-87	Calgary	NHL	79	20	56	76	97	7	0	2	262	7.6	20	....	....	....	....	....	4	1	0	1	0	1	0	0
1987-88	Calgary	NHL	80	25	58	83	114	7	2	2	245	10.2	13	....	....	....	....	....	7	3	6	9	18	2	0	0
1988-89♦	Calgary	NHL	79	16	58	74	126	8	0	3	277	5.8	38	....	....	....	....	....	22	7	*24	*31	46	5	0	4
1989-90	Calgary	NHL	79	28	62	90	82	14	1	3	304	9.2	20	....	....	....	....	....	6	2	3	5	8	1	0	0
1990-91	Calgary	NHL	78	28	75	103	90	17	0	1	305	9.2	42	....	....	....	....	....	7	2	3	5	8	2	0	0
1991-92	Calgary	NHL	72	20	57	77	83	11	0	0	304	6.6	13	....	....	....	....	....	....	....	....	....	....	....	....	....
1992-93	Calgary	NHL	50	11	43	54	61	7	0	4	201	5.5	15	....	....	....	....	....	6	1	6	7	10	1	0	0
1993-94	Calgary	NHL	75	28	54	82	95	12	1	5	324	8.6	35	....	....	....	....	....	7	2	6	8	12	1	0	0
1994-95	St. Louis	NHL	32	8	20	28	43	2	0	0	110	7.3	19	....	....	....	....	....	7	1	5	6	10	0	0	0
1995-96	St. Louis	NHL	82	17	44	61	88	9	1	1	317	5.4	5	....	....	....	....	....	13	3	4	7	20	1	0	0
1996-97	St. Louis	NHL	72	13	30	43	65	6	1	1	296	4.4	2	....	....	....	....	....	6	1	2	3	4	1	0	0
1997-98	St. Louis	NHL	71	19	30	49	80	9	1	2	227	8.4	6	....	....	....	....	....	8	2	6	8	12	1	0	0
	Canada	Olympics	6	2	0	2	2	....	....	....	....	....	....	....	....	....	....	....	....	....	....	....	....	....	....	....
1998-99	St. Louis	NHL	82	20	42	62	70	11	1	2	314	6.4	33	0	0.0	56	128	29:07	13	4	8	12	20	2	0	0
99-2000	St. Louis	NHL	61	11	28	39	34	6	0	7	245	4.5	20	0	0.0	44	72	26:07	7	1	3	4	14	1	0	0
2000-01	St. Louis	NHL	59	12	42	54	52	6	1	3	218	5.5	23	0	0.0	40	73	26:32	15	2	8	10	18	2	0	0
2001-02	St. Louis	NHL	71	11	35	46	52	6	0	4	231	4.8	3	0	0.0	59	89	26:56	10	0	7	7	4	0	0	0
	Canada	Olympics	6	0	0	0	8	....	....	....	....	....	....	....	....	....	....	....	....	....	....	....	....	....	....	....
	NHL Totals		1333	324	880	1204	1434	157	9	42	4849	6.7		0	0.0	199	362	27:19	174	39	120	159	255	26	0	5

OHL First All-Star Team (1982, 1983) • NHL Second All-Star Team (1987, 1989, 1994) • Won Conn Smythe Trophy (1989) • NHL First All-Star Team (1990, 1991, 1999) • Won James Norris Memorial Trophy (1999) • Played in NHL All-Star Game (1985, 1988, 1990, 1991, 1992, 1994, 1996, 1997, 1998, 1999, 2000)

Traded to **St. Louis** by **Calgary** with Calgary's 4th round choice (Didier Tremblay) in 1997 Entry Draft for Phil Housley and St. Louis' 2nd round choices in 1996 (Steve Begin) and 1997 (John Tripp) Entry Drafts, July 4, 1994.

MacKENZIE, Derek

(muh-KEHN-zee, DAIR-ehk) **ATL.**

Center. Shoots left. 5'11", 175 lbs. Born, Sudbury, Ont., June 11, 1981. Atlanta's 6th choice, 128th overall, in 1999 Entry Draft.

Season	Club	League	GP	G	A	Pts	PIM	PP	SH	GW	S	%	+/-	TF	F%	H	SB	Min	GP	G	A	Pts	PIM	PP	SH	GW
1996-97	Rayside-Balfour	NOJHA	40	23	32	55	40	....	....	....	....	....	....	....	....	....	....	....	....	....	....	....	....	....	....	....
1997-98	Sudbury Wolves	OHL	59	9	11	20	26	....	....	....	....	....	....	....	....	....	....	....	....	....	....	....	....	....	....	....
1998-99	Sudbury Wolves	OHL	68	22	65	87	74	....	....	....	....	....	....	....	....	....	....	....	4	2	4	6	2	....	....	....
99-2000	Sudbury Wolves	OHL	68	24	33	57	110	....	....	....	....	....	....	....	....	....	....	....	12	5	9	14	16	....	....	....
2000-01	Sudbury Wolves	OHL	62	40	49	89	89	....	....	....	....	....	....	....	....	....	....	....	12	6	8	14	16	....	....	....

Season	Club	League	GP	G	A	Pts	PIM	PP	SH	GW	S	%	+/-	TF	F%	H	SB	Min	GP	G	A	Pts	PIM	PP	SH	GW
			Regular Season																Playoffs							
2001-02	**Atlanta**	**NHL**	**1**	**0**	**0**	**0**	**2**	**0**	**0**	**0**	**1**	**0.0**	**–1**	**16**	**56.3**	**0**	**0**	**13:51**								
	Chicago Wolves	AHL	68	13	12	25	80												25	4	2	6	20			
	NHL Totals		**1**	**0**	**0**	**0**	**2**	**0**	**0**	**0**	**1**	**0.0**		**16**	**56.3**	**0**	**0**	**13:51**								

MacLEAN, Don

(MAK-layn, DAWN) **CBJ**

Center. Shoots left. 6'2", 199 lbs. Born, Sydney, N.S., January 14, 1977. Los Angeles' 2nd choice, 33rd overall, in 1995 Entry Draft.

Season	Club	League	GP	G	A	Pts	PIM	PP	SH	GW	S	%	+/-	TF	F%	H	SB	Min	GP	G	A	Pts	PIM	PP	SH	GW
1992-93	Halifax Hawks	NSMHL	27	15	25	40	34																			
1993-94	Halifax Hawks	NSMHL	25	35	35	70	151																			
1994-95	Beauport	QMJHL	64	15	27	42	37												17	4	4	8	6			
1995-96	Beauport	QMJHL	1	0	1	1	0																			
	Laval Titan	QMJHL	21	17	11	28	29																			
	Hull Olympiques	QMJHL	39	26	34	60	44												17	6	7	13	14			
1996-97	Hull Olympiques	QMJHL	69	34	47	81	67												14	11	10	21	39			
1997-98	**Los Angeles**	**NHL**	**22**	**5**	**2**	**7**	**4**	**2**	**0**	**0**	**25**	**20.0**	**–1**													
	Fredericton	AHL	39	9	5	14	32												4	1	3	4	2			
1998-99	Springfield	AHL	41	5	14	19	31																			
	Grand Rapids	IHL	28	6	13	19	8																			
99-2000	Lowell	AHL	40	11	17	28	18																			
	St. John's	AHL	21	14	12	26	8																			
2000-01	**Toronto**	**NHL**	**3**	**0**	**1**	**1**	**2**	**0**	**0**	**0**	**2**	**0.0**	**–2**	**33**	**54.6**	**3**	**1**	**9:48**								
	St. John's	AHL	61	26	34	60	48												4	2	1	3	2			
2001-02	St. John's	AHL	75	33	*54	*87	49												9	5	5	10	6			
	Toronto	**NHL**																	**3**	**0**	**0**	**0**	**0**	**0**	**0**	**0**
	NHL Totals		**25**	**5**	**3**	**8**	**6**	**2**	**0**	**0**	**27**	**18.5**		**33**	**54.5**	**3**	**1**	**9:48**	**3**	**0**	**0**	**0**	**0**	**0**	**0**	**0**

Won John P. Sollenberger Trophy (Top Scorer - AHL) (2002)

Traded to **Toronto** by **Los Angeles** for Craig Charron, February 23, 2000. Signed as a free agent by **Columbus**, July 17, 2002.

MacLEAN, John

(MAK-layn, JAWN)

Right wing. Shoots right. 6', 200 lbs. Born, Oshawa, Ont., November 20, 1964. New Jersey's 1st choice, 6th overall, in 1983 Entry Draft.

Season	Club	League	GP	G	A	Pts	PIM	PP	SH	GW	S	%	+/-	TF	F%	H	SB	Min	GP	G	A	Pts	PIM	PP	SH	GW
1980-81	Oshawa	OJHL-B	41	35	35	70	151																			
1981-82	Oshawa Generals	OHL	67	17	22	39	197												12	3	6	9	63			
1982-83	Oshawa Generals	OHL	66	47	51	98	138												17	*18	20	*38	35			
1983-84	Oshawa Generals	OHL	30	23	36	59	58												7	2	5	7	18			
	New Jersey	**NHL**	**23**	**1**	**0**	**1**	**10**	**0**	**0**	**0**	**22**	**4.5**	**–7**													
1984-85	**New Jersey**	**NHL**	**61**	**13**	**20**	**33**	**44**	**1**	**0**	**4**	**92**	**14.1**	**–11**													
1985-86	**New Jersey**	**NHL**	**74**	**21**	**36**	**57**	**112**	**1**	**0**	**4**	**139**	**15.1**	**–3**													
1986-87	**New Jersey**	**NHL**	**80**	**31**	**36**	**67**	**120**	**9**	**0**	**4**	**197**	**15.7**	**–23**													
1987-88	**New Jersey**	**NHL**	**76**	**23**	**16**	**39**	**147**	**12**	**0**	**4**	**204**	**11.3**	**–10**						**20**	**7**	**11**	**18**	**60**	**2**	**0**	**2**
1988-89	**New Jersey**	**NHL**	**74**	**42**	**45**	**87**	**122**	**14**	**0**	**4**	**266**	**15.8**	**26**													
1989-90	**New Jersey**	**NHL**	**80**	**41**	**38**	**79**	**80**	**10**	**3**	**11**	**322**	**12.7**	**17**						**6**	**4**	**1**	**5**	**12**	**2**	**1**	**0**
1990-91	**New Jersey**	**NHL**	**78**	**45**	**33**	**78**	**150**	**19**	**2**	**7**	**292**	**15.4**	**8**						**7**	**5**	**3**	**8**	**20**	**1**	**0**	**0**
1991-92	**New Jersey**	**NHL**	DID NOT PLAY – INJURED																							
1992-93	**New Jersey**	**NHL**	**80**	**24**	**24**	**48**	**102**	**7**	**1**	**3**	**195**	**12.3**	**–6**						**5**	**0**	**1**	**1**	**10**	**0**	**0**	**0**
1993-94	**New Jersey**	**NHL**	**80**	**37**	**33**	**70**	**95**	**8**	**0**	**4**	**277**	**13.4**	**30**						**20**	**6**	**10**	**16**	**22**	**2**	**0**	**1**
1994-95♦	**New Jersey**	**NHL**	**46**	**17**	**12**	**29**	**32**	**2**	**1**	**0**	**139**	**12.2**	**13**						**20**	**5**	**13**	**18**	**14**	**2**	**0**	**0**
1995-96	**New Jersey**	**NHL**	**76**	**20**	**28**	**48**	**91**	**3**	**3**	**3**	**237**	**8.4**	**3**													
1996-97	**New Jersey**	**NHL**	**80**	**29**	**25**	**54**	**49**	**5**	**0**	**6**	**254**	**11.4**	**11**						**10**	**4**	**5**	**9**	**4**	**2**	**1**	**1**
1997-98	**New Jersey**	**NHL**	**26**	**3**	**8**	**11**	**14**	**1**	**0**	**1**	**74**	**4.1**	**–6**													
	San Jose	**NHL**	**51**	**13**	**19**	**32**	**28**	**5**	**0**	**2**	**139**	**9.4**	**0**						**6**	**2**	**3**	**5**	**4**	**1**	**0**	**0**
1998-99	**NY Rangers**	**NHL**	**82**	**28**	**27**	**55**	**46**	**11**	**1**	**2**	**231**	**12.1**	**5**	**20**	**35.0**	**120**	**30**	**20:43**								
99-2000	**NY Rangers**	**NHL**	**77**	**18**	**24**	**42**	**52**	**6**	**2**	**3**	**158**	**11.4**	**–2**	**40**	**55.0**	**48**	**43**	**14:44**								
2000-01	**NY Rangers**	**NHL**	**2**	**0**	**0**	**0**	**0**	**0**	**0**	**0**	**0**	**0.0**	**–2**	**1**	**0.0**	**2**	**0**	**10:14**								
	Manitoba Moose	IHL	32	6	12	18	28																			
	Dallas	**NHL**	**28**	**4**	**2**	**6**	**17**	**1**	**0**	**0**	**41**	**9.8**	**0**	**1**	**100.0**	**34**	**11**	**12:45**	**10**	**2**	**1**	**3**	**6**	**0**	**0**	**0**
2001-02	**Dallas**	**NHL**	**20**	**3**	**3**	**6**	**17**	**0**	**0**	**1**	**40**	**7.5**	**–1**	**3**	**100.0**	**32**	**5**	**12:34**								
	Utah Grizzlies	AHL	5	0	1	1	4																			
	NHL Totals		**1194**	**413**	**429**	**842**	**1328**	**115**	**13**	**63**	**3319**	**12.4**		**65**	**50.8**	**236**	**89**	**16:34**	**104**	**35**	**48**	**83**	**152**	**12**	**2**	**4**

Memorial Cup All-Star Team (1983) • Played in NHL All-Star Game (1989, 1991)

• Missed entire 1991-92 season recovering from knee surgery, June, 1991. Traded to **San Jose** by **New Jersey** with Ken Sutton for Doug Bodger and Dody Wood, December 7, 1997. Signed as a free agent by **NY Rangers**, July 22, 1998. Traded to **Dallas** by **NY Rangers** for future considerations, February 5, 2001. Signed as a free agent by **Dallas**, February 26, 2002. • Officially announced retirement, June 7, 2002.

MADDEN, John

(MA-dehn, JAWN) **N.J.**

Left wing. Shoots left. 5'11", 190 lbs. Born, Barrie, Ont., May 4, 1975.

Season	Club	League	GP	G	A	Pts	PIM	PP	SH	GW	S	%	+/-	TF	F%	H	SB	Min	GP	G	A	Pts	PIM	PP	SH	GW
1989-90	Alliston Hornets	OJHL-C	31	24	25	49	26																			
1990-91	Alliston Hornets	OJHL-C	14	15	21	36	10																			
	Barrie Colts	OJHL-B	1	0	0	0	0																			
1991-92	Barrie Colts	OJHL-B	42	50	54	104	46												13	10	9	19	14			
1992-93	Barrie Colts	OJHL-B	43	49	75	124	62																			
1993-94	U. of Michigan	CCHA	36	6	11	17	14																			
1994-95	U. of Michigan	CCHA	39	21	22	43	8																			
1995-96	U. of Michigan	CCHA	43	27	30	57	45																			
1996-97	U. of Michigan	CCHA	42	26	37	63	56																			
1997-98	Albany	AHL	74	20	36	56	40												13	3	13	16	14			
1998-99	**New Jersey**	**NHL**	**4**	**0**	**1**	**1**	**0**	**0**	**0**	**0**	**4**	**0.0**	**–2**	**0**	**0.0**	**3**	**1**	**9:13**								
	Albany	AHL	75	38	60	98	44												5	2	2	4	6			
99-2000♦	**New Jersey**	**NHL**	**74**	**16**	**9**	**25**	**6**	**0**	**6**	**3**	**115**	**13.9**	**7**	**770**	**47.5**	**92**	**25**	**11:40**	**20**	**3**	**4**	**7**	**0**	**0**	**1**	**2**
2000-01	**New Jersey**	**NHL**	**80**	**23**	**15**	**38**	**12**	**0**	**3**	**4**	**163**	**14.1**	**24**	**974**	**46.6**	**69**	**39**	**15:35**	**25**	**4**	**3**	**7**	**6**	**0**	**0**	**0**
2001-02	**New Jersey**	**NHL**	**82**	**15**	**8**	**23**	**25**	**0**	**0**	**2**	**170**	**8.8**	**6**	**1001**	**47.0**	**86**	**46**	**15:36**	**6**	**0**	**0**	**0**	**0**	**0**	**0**	**0**
	NHL Totals		**240**	**54**	**33**	**87**	**43**	**0**	**9**	**9**	**452**	**11.9**		**2745**	**47.0**	**250**	**111**	**14:17**	**51**	**7**	**7**	**14**	**6**	**0**	**1**	**2**

CCHA First All-Star Team (1997) • NCAA West First All-American Team (1997) • Won Frank J. Selke Trophy (2001)

Signed as a free agent by **New Jersey**, June 26, 1997.

MAIR, Adam

(MAIR, A-duhm) **BUF.**

Center. Shoots right. 6'2", 215 lbs. Born, Hamilton, Ont., February 15, 1979. Toronto's 2nd choice, 84th overall, in 1997 Entry Draft.

Season	Club	League	GP	G	A	Pts	PIM	PP	SH	GW	S	%	+/-	TF	F%	H	SB	Min	GP	G	A	Pts	PIM	PP	SH	GW
1994-95	Ohsweken	OJHL-B	39	21	23	44	91																			
1995-96	Owen Sound	OHL	62	12	15	27	63												6	0	0	0	2			
1996-97	Owen Sound	OHL	65	16	35	51	113												4	1	0	1	2			
1997-98	Owen Sound	OHL	56	25	27	52	179												11	6	3	9	31			
1998-99	Owen Sound	OHL	43	23	41	64	109												16	10	10	20	*47			
	St. John's	AHL																	3	1	0	1	6			
	Toronto	**NHL**																	**5**	**1**	**0**	**1**	**14**	**0**	**0**	**0**
99-2000	**Toronto**	**NHL**	**8**	**1**	**0**	**1**	**6**	**0**	**0**	**0**	**7**	**14.3**	**–1**	**9**	**33.3**	**14**	**1**	**11:33**	**5**	**0**	**0**	**0**	**8**	**0**	**0**	**0**
	St. John's	AHL	66	22	27	49	124																			
2000-01	**Toronto**	**NHL**	**16**	**0**	**2**	**2**	**14**	**0**	**0**	**0**	**17**	**0.0**	**3**	**56**	**51.8**	**25**	**0**	**9:00**								
	St. John's	AHL	47	18	27	45	69																			
	Los Angeles	**NHL**	**10**	**0**	**0**	**0**	**6**	**0**	**0**	**0**	**5**	**0.0**	**–3**	**21**	**61.9**	**21**	**4**	**6:14**								
2001-02	**Los Angeles**	**NHL**	**18**	**1**	**1**	**2**	**57**	**0**	**0**	**0**	**10**	**10.0**	**1**	**31**	**58.1**	**31**	**1**	**7:11**								
	Manchester	AHL	27	10	9	19	48												5	5	1	6	10			
	NHL Totals		**52**	**2**	**3**	**5**	**83**	**0**	**0**	**0**	**39**	**5.1**		**117**	**53.8**	**91**	**6**	**8:14**	**10**	**1**	**0**	**1**	**22**	**0**	**0**	**0**

Traded to **LA Kings** by **Toronto** with Toronto's 2nd round choice (Mike Cammalleri) in 2001 Entry Draft for Aki Berg, March 13, 2001. Traded to **Buffalo** by **LA Kings** with LA Kings' 5th round choice in 2003 Entry Draft for Erik Rasmussen, July 24, 2002.

MALAKHOV, Vladimir

(mah-LAH-kahf, vla-DIH-meer) **NYR**

Defense. Shoots left. 6'4", 230 lbs. Born, Sverdlovsk, USSR, August 30, 1968. NY Islanders' 12th choice, 191st overall, in 1989 Entry Draft.

Season	Club	League	GP	G	A	Pts	PIM	PP	SH	GW	S	%	+/-	TF	F%	H	SB	Min	GP	G	A	Pts	PIM	PP	SH	GW
			Regular Season																Playoffs							
1986-87	Spartak Moscow	USSR	22	0	1	1	12	....	....	....	....	....	....	....	....	....	....	....	....	....	....	....	....	....	....	....
1987-88	Spartak Moscow	USSR	28	2	2	4	26	....	....	....	....	....	....	....	....	....	....	....	....	....	....	....	....	....	....	....
1988-89	CSKA Moscow	USSR	34	6	2	8	16	....	....	....	....	....	....	....	....	....	....	....	....	....	....	....	....	....	....	....
1989-90	CSKA Moscow	USSR	48	2	10	12	34	....	....	....	....	....	....	....	....	....	....	....	....	....	....	....	....	....	....	....
1990-91	CSKA Moscow	USSR	46	5	13	18	22	....	....	....	....	....	....	....	....	....	....	....	....	....	....	....	....	....	....	....
1991-92	CSKA Moscow	CIS	40	1	9	10	12	....	....	....	....	....	....	....	....	....	....	....	....	....	....	....	....	....	....	....
	Russia	Olympics	8	3	0	3	4	....	....	....	....	....	....	....	....	....	....	....	....	....	....	....	....	....	....	....
1992-93	**NY Islanders**	**NHL**	**64**	**14**	**38**	**52**	**59**	**7**	**0**	**0**	**178**	**7.9**	**14**	....	....	....	....	....	**17**	**3**	**6**	**9**	**12**	**0**	**0**	**0**
	Capital District	AHL	3	2	1	3	11	....	....	....	....	....	....	....	....	....	....	....	....	....	....	....	....	....	....	....
1993-94	**NY Islanders**	**NHL**	**76**	**10**	**47**	**57**	**80**	**4**	**0**	**2**	**235**	**4.3**	**29**	....	....	....	....	....	**4**	**0**	**0**	**0**	**6**	**0**	**0**	**0**
1994-95	**NY Islanders**	**NHL**	**26**	**3**	**13**	**16**	**32**	**1**	**0**	**0**	**61**	**4.9**	**–1**	....	....	....	....	....	....	....	....	....	....	....	....	....
	Montreal	**NHL**	**14**	**1**	**4**	**5**	**14**	**0**	**0**	**0**	**30**	**3.3**	**–2**	....	....	....	....	....	....	....	....	....	....	....	....	....
1995-96	**Montreal**	**NHL**	**61**	**5**	**23**	**28**	**79**	**2**	**0**	**0**	**122**	**4.1**	**7**	....	....	....	....	....	....	....	....	....	....	....	....	....
1996-97	**Montreal**	**NHL**	**65**	**10**	**20**	**30**	**43**	**5**	**0**	**1**	**177**	**5.6**	**3**	....	....	....	....	....	**5**	**0**	**0**	**0**	**6**	**0**	**0**	**0**
1997-98	**Montreal**	**NHL**	**74**	**13**	**31**	**44**	**70**	**8**	**0**	**2**	**166**	**7.8**	**16**	....	....	....	....	....	**9**	**3**	**4**	**7**	**10**	**2**	**0**	**0**
1998-99	**Montreal**	**NHL**	**62**	**13**	**21**	**34**	**77**	**8**	**0**	**3**	**143**	**9.1**	**–7**	**0**	**0.0**	**74**	**88**	**23:29**	....	....	....	....	....	....	....	....
99-2000	**Montreal**	**NHL**	**7**	**0**	**0**	**0**	**4**	**0**	**0**	**0**	**7**	**0.0**	**0**	**0**	**0.0**	**6**	**9**	**21:20**	....	....	....	....	....	....	....	....
	♦ **New Jersey**	**NHL**	**17**	**1**	**4**	**5**	**19**	**1**	**0**	**1**	**11**	**9.1**	**1**	**0**	**0.0**	**20**	**13**	**20:18**	**23**	**1**	**4**	**5**	**18**	**1**	**0**	**0**
2000-01	**NY Rangers**	**NHL**	**3**	**0**	**2**	**2**	**4**	**0**	**0**	**0**	**6**	**0.0**	**0**	**0**	**0.0**	**4**	**7**	**19:07**	....	....	....	....	....	....	....	....
2001-02	**NY Rangers**	**NHL**	**81**	**6**	**22**	**28**	**83**	**1**	**0**	**0**	**145**	**4.1**	**10**	**1**	**100.0**	**199**	**132**	**22:48**	....	....	....	....	....	....	....	....
	Russia	Olympics	6	1	3	4	4	....	....	....	....	....	....	....	....	....	....	....	....	....	....	....	....	....	....	....
	NHL Totals		**550**	**76**	**225**	**301**	**564**	**37**	**0**	**9**	**1281**	**5.9**		**1**	**100.0**	**303**	**249**	**22:41**	**58**	**7**	**14**	**21**	**52**	**3**	**0**	**0**

NHL All-Rookie Team (1993)

Traded to **Montreal** by **NY Islanders** with Pierre Turgeon for Kirk Muller, Mathieu Schneider and Craig Darby, April 5, 1995. • Missed majority of 1999-2000 season recovering from knee injury suffered in exhibition game vs. Boston, September 27, 1999. Traded to **New Jersey** by **Montreal** for Sheldon Souray, Josh DeWolf and New Jersey's 2nd round choice (later traded to Washington - later traded to Tampa Bay - Tampa Bay selected Andreas Holmqvist) in 2001 Entry Draft, March 1, 2000. Signed as a free agent by **NY Rangers**, July 10, 2000. • Missed majority of 2000-01 season recovering from knee injury suffered in game vs. Montreal, November 11, 2000.

MALGUNAS, Stewart

(mal-GOO-nuhs, STEW-ahrt)

Defense. Shoots left. 6', 200 lbs. Born, Prince George, B.C., April 21, 1970. Detroit's 3rd choice, 66th overall, in 1990 Entry Draft.

Season	Club	League	GP	G	A	Pts	PIM	PP	SH	GW	S	%	+/-	TF	F%	H	SB	Min	GP	G	A	Pts	PIM	PP	SH	GW
1985-86	Prince George	BCAHA	49	10	25	35	85	....	....	....	....	....	....	....	....	....	....	....	....	....	....	....	....	....	....	....
1986-87	Prince George	BCAHA	50	11	31	42	102	....	....	....	....	....	....	....	....	....	....	....	....	....	....	....	....	....	....	....
1987-88	Prince George	PCJHL	48	12	34	46	99	....	....	....	....	....	....	....	....	....	....	....	....	....	....	....	....	....	....	....
	New Westminster	WHL	6	0	0	0	0	....	....	....	....	....	....	....	....	....	....	....	....	....	....	....	....	....	....	....
1988-89	Seattle	WHL	72	11	41	52	51	....	....	....	....	....	....	....	....	....	....	....	....	....	....	....	....	....	....	....
1989-90	Seattle	WHL	63	15	48	63	116	....	....	....	....	....	....	....	....	....	....	....	13	2	9	11	32	....	....	....
1990-91	Adirondack	AHL	78	5	19	24	70	....	....	....	....	....	....	....	....	....	....	....	2	0	0	0	4	....	....	....
1991-92	Adirondack	AHL	69	4	28	32	82	....	....	....	....	....	....	....	....	....	....	....	18	2	6	8	28	....	....	....
1992-93	Adirondack	AHL	45	3	12	15	39	....	....	....	....	....	....	....	....	....	....	....	11	3	3	6	8	....	....	....
1993-94	**Philadelphia**	**NHL**	**67**	**1**	**3**	**4**	**86**	**0**	**0**	**0**	**54**	**1.9**	**2**	....	....	....	....	....	....	....	....	....	....	....	....	....
1994-95	**Philadelphia**	**NHL**	**4**	**0**	**0**	**0**	**4**	**0**	**0**	**0**	**1**	**0.0**	**–1**	....	....	....	....	....	....	....	....	....	....	....	....	....
	Hershey Bears	AHL	32	3	5	8	28	....	....	....	....	....	....	....	....	....	....	....	6	2	1	3	31	....	....	....
1995-96	**Winnipeg**	**NHL**	**29**	**0**	**1**	**1**	**32**	**0**	**0**	**0**	**13**	**0.0**	**–10**	....	....	....	....	....	....	....	....	....	....	....	....	....
	Washington	**NHL**	**1**	**0**	**0**	**0**	**0**	**0**	**0**	**0**	**0**	**0.0**	**0**	....	....	....	....	....	....	....	....	....	....	....	....	....
	Portland Pirates	AHL	16	2	5	7	18	....	....	....	....	....	....	....	....	....	....	....	13	1	3	4	19	....	....	....
1996-97	**Washington**	**NHL**	**6**	**0**	**0**	**0**	**2**	**0**	**0**	**0**	**3**	**0.0**	**2**	....	....	....	....	....	....	....	....	....	....	....	....	....
	Portland Pirates	AHL	68	6	12	18	59	....	....	....	....	....	....	....	....	....	....	....	5	0	0	0	8	....	....	....
1997-98	**Washington**	**NHL**	**8**	**0**	**0**	**0**	**12**	**0**	**0**	**0**	**5**	**0.0**	**1**	....	....	....	....	....	....	....	....	....	....	....	....	....
	Portland Pirates	AHL	69	14	25	39	73	....	....	....	....	....	....	....	....	....	....	....	9	1	1	2	19	....	....	....
1998-99	**Washington**	**NHL**	**10**	**0**	**0**	**0**	**6**	**0**	**0**	**0**	**2**	**0.0**	**–5**	**0**	**0.0**	**12**	**9**	**9:02**	....	....	....	....	....	....	....	....
	Portland Pirates	AHL	33	2	10	12	49	....	....	....	....	....	....	....	....	....	....	....	....	....	....	....	....	....	....	....
	Detroit Vipers	IHL	9	0	2	2	10	....	....	....	....	....	....	....	....	....	....	....	11	0	1	1	21	....	....	....
99-2000	Utah Grizzlies	IHL	34	4	9	13	55	....	....	....	....	....	....	....	....	....	....	....	....	....	....	....	....	....	....	....
	Calgary	**NHL**	**4**	**0**	**1**	**1**	**2**	**0**	**0**	**0**	**0**	**0.0**	**1**	**0**	**0.0**	**7**	**4**	**12:57**	....	....	....	....	....	....	....	....
2000-01	Hershey Bears	AHL	25	0	2	2	39	....	....	....	....	....	....	....	....	....	....	....	11	0	1	1	14	....	....	....
2001-02	Frankfurt Lions	Germany	50	4	11	15	108	....	....	....	....	....	....	....	....	....	....	....	....	....	....	....	....	....	....	....
	NHL Totals		**129**	**1**	**5**	**6**	**144**	**0**	**0**	**0**	**78**	**1.3**		**0**	**0.0**	**19**	**13**	**10:09**	....	....	....	....	....	....	....	....

WHL West First All-Star Team (1990)

Traded to **Philadelphia** by **Detroit** for Philadelphia's 5th round choice (David Arsenault) in 1995 Entry Draft, September 9, 1993. Signed as a free agent by **Winnipeg**, August 9, 1995. Traded to **Washington** by **Winnipeg** for Denis Chasse, February 15, 1996. Traded to **Nashville** by **Washington** for future considerations, February 2, 2000. Claimed on waivers by **Calgary** from **Nashville**, February 3, 2000. • Missed majority of 1999-2000 season recovering from head injury suffered in game vs. Los Angeles, February 14, 2000. Signed as a free agent by **Colorado**, August 29, 2000. Signed as a free agent by **Frankfurt Lions** (Germany), March 24, 2001.

MALHOTRA, Manny

(mal-HOH-truh, MAHN-ee) **DAL.**

Center. Shoots left. 6'2", 215 lbs. Born, Mississauga, Ont., May 18, 1980. NY Rangers' 1st choice, 7th overall, in 1998 Entry Draft.

Season	Club	League	GP	G	A	Pts	PIM	PP	SH	GW	S	%	+/-	TF	F%	H	SB	Min	GP	G	A	Pts	PIM	PP	SH	GW
1995-96	Mississauga Reps	MTHL	54	27	44	71	62	....	....	....	....	....	....	....	....	....	....	....	....	....	....	....	....	....	....	....
1996-97	Guelph Storm	OHL	61	16	28	44	26	....	....	....	....	....	....	....	....	....	....	....	18	7	7	14	11	....	....	....
1997-98	Guelph Storm	OHL	57	16	35	51	29	....	....	....	....	....	....	....	....	....	....	....	12	7	6	13	8	....	....	....
1998-99	**NY Rangers**	**NHL**	**73**	**8**	**8**	**16**	**13**	**1**	**0**	**2**	**61**	**13.1**	**–2**	**588**	**43.9**	**115**	**17**	**8:36**	....	....	....	....	....	....	....	....
99-2000	**NY Rangers**	**NHL**	**27**	**0**	**0**	**0**	**4**	**0**	**0**	**0**	**18**	**0.0**	**–6**	**132**	**44.7**	**37**	**7**	**6:42**	....	....	....	....	....	....	....	....
	Guelph Storm	OHL	5	2	2	4	4	....	....	....	....	....	....	....	....	....	....	....	6	0	2	2	4	....	....	....
	Hartford	AHL	12	1	5	6	2	....	....	....	....	....	....	....	....	....	....	....	23	1	2	3	10	....	....	....
2000-01	**NY Rangers**	**NHL**	**50**	**4**	**8**	**12**	**31**	**0**	**0**	**2**	**46**	**8.7**	**–10**	**248**	**44.4**	**73**	**17**	**9:03**	....	....	....	....	....	....	....	....
	Hartford	AHL	28	5	6	11	69	....	....	....	....	....	....	....	....	....	....	....	5	0	0	0	0	....	....	....
2001-02	**NY Rangers**	**NHL**	**56**	**7**	**6**	**13**	**42**	**0**	**1**	**1**	**41**	**17.1**	**–1**	**310**	**42.9**	**106**	**29**	**10:14**	....	....	....	....	....	....	....	....
	Dallas	**NHL**	**16**	**1**	**0**	**1**	**5**	**0**	**0**	**0**	**19**	**5.3**	**–3**	**121**	**48.8**	**40**	**8**	**10:37**	....	....	....	....	....	....	....	....
	NHL Totals		**222**	**20**	**22**	**42**	**95**	**1**	**1**	**5**	**185**	**10.8**		**1399**	**44.2**	**371**	**78**	**9:02**	....	....	....	....	....	....	....	....

Memorial Cup All-Star Team (1998) • Won George Parsons Trophy (Memorial Cup Tournament Most Sportsmanlike Player) (1998)

Traded to **Dallas** by **NY Rangers** with Barrett Heisten for Martin Rucinsky and Roman Lyashenko, March 12, 2002.

MALIK, Marek

(MAW-leck, MAIR-ehk) **CAR.**

Defense. Shoots left. 6'5", 215 lbs. Born, Ostrava, Czech., June 24, 1975. Hartford's 2nd choice, 72nd overall, in 1993 Entry Draft.

Season	Club	League	GP	G	A	Pts	PIM	PP	SH	GW	S	%	+/-	TF	F%	H	SB	Min	GP	G	A	Pts	PIM	PP	SH	GW
1992-93	TJ Vitkovice Jr.	Czech-Jr.	20	5	10	15	16	....	....	....	....	....	....	....	....	....	....	....	....	....	....	....	....	....	....	....
1993-94	HC Vitkovice	Czech	38	3	3	6	0	....	....	....	....	....	....	....	....	....	....	....	3	0	1	1	0	....	....	....
1994-95	Springfield	AHL	58	11	30	41	91	....	....	....	....	....	....	....	....	....	....	....	....	....	....	....	....	....	....	....
	Hartford	**NHL**	**1**	**0**	**1**	**1**	**0**	**0**	**0**	**0**	**0**	**0.0**	**1**	....	....	....	....	....	....	....	....	....	....	....	....	....
1995-96	**Hartford**	**NHL**	**7**	**0**	**0**	**0**	**4**	**0**	**0**	**0**	**2**	**0.0**	**–3**	....	....	....	....	....	....	....	....	....	....	....	....	....
	Springfield	AHL	68	8	14	22	135	....	....	....	....	....	....	....	....	....	....	....	8	1	3	4	20	....	....	....
1996-97	**Hartford**	**NHL**	**47**	**1**	**5**	**6**	**50**	**0**	**0**	**1**	**33**	**3.0**	**5**	....	....	....	....	....	....	....	....	....	....	....	....	....
	Springfield	AHL	3	0	3	3	4	....	....	....	....	....	....	....	....	....	....	....	....	....	....	....	....	....	....	....
1997-98	Malmo IF	Sweden	37	1	5	6	21	....	....	....	....	....	....	....	....	....	....	....	....	....	....	....	....	....	....	....
1998-99	HC Vitkovice	Czech	1	1	0	1	6	....	....	....	....	....	....	....	....	....	....	....	....	....	....	....	....	....	....	....
	Carolina	**NHL**	**52**	**2**	**9**	**11**	**36**	**1**	**0**	**0**	**36**	**5.6**	**–6**	**0**	**0.0**	**101**	**76**	**21:14**	**4**	**0**	**0**	**0**	**4**	**0**	**0**	**0**
	New Haven	AHL	21	2	8	10	28	....	....	....	....	....	....	....	....	....	....	....	....	....	....	....	....	....	....	....
99-2000	**Carolina**	**NHL**	**57**	**4**	**10**	**14**	**63**	**0**	**0**	**1**	**57**	**7.0**	**13**	**0**	**0.0**	**44**	**64**	**18:00**	....	....	....	....	....	....	....	....
2000-01	**Carolina**	**NHL**	**61**	**6**	**14**	**20**	**34**	**1**	**0**	**1**	**72**	**8.3**	**–4**	**0**	**0.0**	**75**	**87**	**19:36**	**3**	**0**	**0**	**0**	**6**	**0**	**0**	**0**
2001-02	**Carolina**	**NHL**	**82**	**4**	**19**	**23**	**88**	**0**	**0**	**0**	**91**	**4.4**	**8**	**0**	**0.0**	**84**	**105**	**20:29**	**23**	**0**	**3**	**3**	**18**	**0**	**0**	**0**
	NHL Totals		**307**	**17**	**58**	**75**	**275**	**2**	**0**	**3**	**291**	**5.8**		**0**	**0.0**	**304**	**332**	**19:52**	**30**	**0**	**3**	**3**	**28**	**0**	**0**	**0**

Transferred to **Carolina** after **Hartford** franchise relocated, June 25, 1997.

MALTAIS, Steve

(MAHL-tay, STEEV)

Left wing. Shoots left. 6'2", 205 lbs. Born, Arvida, Que., January 25, 1969. Washington's 2nd choice, 57th overall, in 1987 Entry Draft.

			Regular Season																Playoffs							
Season	Club	League	GP	G	A	Pts	PIM	PP	SH	GW	S	%	+/-	TF	F%	H	SB	Min	GP	G	A	Pts	PIM	PP	SH	GW
1985-86	Wexford Hawks	MTHL	33	35	19	54	38	....	....	....	....	....	....	....	....	....	....	....	....	....	....	....	....	....	....	....
	Wexford Raiders	MTJHL	1	1	0	1	0	....	....	....	....	....	....	....	....	....	....	....	....	....	....	....	....	....	....	....
1986-87	Cornwall Royals	OHL	65	32	12	44	29	....	....	....	....	....	....	....	....	....	....	....	5	0	0	0	2	....	....	....
1987-88	Cornwall Royals	OHL	59	39	46	85	30	....	....	....	....	....	....	....	....	....	....	....	11	9	6	15	33	....	....	....
1988-89	Cornwall Royals	OHL	58	53	70	123	67	....	....	....	....	....	....	....	....	....	....	....	18	14	16	30	16	....	....	....
	Fort Wayne	IHL	....	....	....	....	....	....	....	....	....	....	....	....	....	....	....	....	4	2	1	3	0	....	....	....
1989-90	**Washington**	**NHL**	**8**	**0**	**0**	**0**	**2**	**0**	**0**	**0**	**11**	**0.0**	**-2**	....	....	....	....	....	**1**	**0**	**0**	**0**	**0**	**0**	**0**	**0**
	Baltimore	AHL	67	29	37	66	54	....	....	....	....	....	....	....	....	....	....	....	12	6	10	16	6	....	....	....
1990-91	**Washington**	**NHL**	**7**	**0**	**0**	**0**	**2**	**0**	**0**	**0**	**3**	**0.0**	**-1**	....	....	....	....	....	....	....	....	....	....	....	....	....
	Baltimore	AHL	73	36	43	79	97	....	....	....	....	....	....	....	....	....	....	....	6	1	4	5	10	....	....	....
1991-92	**Minnesota**	**NHL**	**12**	**2**	**1**	**3**	**2**	**0**	**0**	**0**	**6**	**33.3**	**-1**	....	....	....	....	....	....	....	....	....	....	....	....	....
	Kalamazoo Wings	IHL	48	25	31	56	51	....	....	....	....	....	....	....	....	....	....	....	....	....	....	....	....	....	....	....
	Halifax Citadels	AHL	10	3	3	6	0	....	....	....	....	....	....	....	....	....	....	....	....	....	....	....	....	....	....	....
1992-93	**Tampa Bay**	**NHL**	**63**	**7**	**13**	**20**	**35**	**4**	**0**	**1**	**96**	**7.3**	**-20**	....	....	....	....	....	....	....	....	....	....	....	....	....
	Atlanta Knights	IHL	16	14	10	24	22	....	....	....	....	....	....	....	....	....	....	....	....	....	....	....	....	....	....	....
1993-94	**Detroit**	**NHL**	**4**	**0**	**1**	**1**	**0**	**0**	**0**	**0**	**2**	**0.0**	**-1**	....	....	....	....	....	....	....	....	....	....	....	....	....
	Adirondack	AHL	73	35	49	84	79	....	....	....	....	....	....	....	....	....	....	....	12	5	11	16	14	....	....	....
1994-95	Chicago Wolves	IHL	79	*57	40	97	145	....	....	....	....	....	....	....	....	....	....	....	3	1	1	2	0	....	....	....
1995-96	Chicago Wolves	IHL	81	56	66	122	161	....	....	....	....	....	....	....	....	....	....	....	9	7	7	14	20	....	....	....
1996-97	Chicago Wolves	IHL	81	*60	54	114	62	....	....	....	....	....	....	....	....	....	....	....	4	2	0	2	4	....	....	....
1997-98	Chicago Wolves	IHL	82	*46	57	103	120	....	....	....	....	....	....	....	....	....	....	....	22	8	11	19	28	....	....	....
1998-99	Chicago Wolves	IHL	82	*56	44	100	164	....	....	....	....	....	....	....	....	....	....	....	10	4	6	10	2	....	....	....
99-2000	Chicago Wolves	IHL	82	*44	46	*90	78	....	....	....	....	....	....	....	....	....	....	....	16	9	4	13	14	....	....	....
2000-01	**Columbus**	**NHL**	**26**	**0**	**3**	**3**	**12**	**0**	**0**	**0**	**30**	**0.0**	**-9**	**2**	**0.0**	**24**	**6**	**11:57**	....	....	....	....	....	....	....	....
	Chicago Wolves	IHL	50	25	26	51	57	....	....	....	....	....	....	....	....	....	....	....	16	7	10	17	2	....	....	....
2001-02	Chicago Wolves	AHL	67	31	32	63	62	....	....	....	....	....	....	....	....	....	....	....	25	*12	10	22	22	....	....	....
	NHL Totals		**120**	**9**	**18**	**27**	**53**	**4**	**0**	**1**	**148**	**6.1**		**2**	**0.0**	**24**	**6**	**11:57**	**1**	**0**	**0**	**0**	**0**	**0**	**0**	**0**

OHL Second All-Star Team (1989) • IHL First All-Star Team (1995, 1999, 2000) • IHL Second All-Star Team (1996, 1997) • Won Leo P. Lamoureux Memorial Trophy (Top Scorer - IHL) (2000)

Traded to **Minnesota** by **Washington** with Trent Klatt for Shawn Chambers, June 21, 1991. Traded to **Quebec** by **Minnesota** for Kip Miller, March 8, 1992. Claimed by **Tampa Bay** from **Quebec** in Expansion Draft, June 18, 1992. Traded to **Detroit** by **Tampa Bay** for Dennis Vial, June 8, 1993. Signed as a free agent by **Chicago** (IHL), September 8, 1994. Signed as a free agent by **Columbus**, October 6, 2000.

MALTBY, Kirk

(MAHLT-bee, KUHRK) **DET.**

Right wing. Shoots right. 6', 180 lbs. Born, Guelph, Ont., December 22, 1972. Edmonton's 4th choice, 65th overall, in 1992 Entry Draft.

Season	Club	League	GP	G	A	Pts	PIM	PP	SH	GW	S	%	+/-	TF	F%	H	SB	Min	GP	G	A	Pts	PIM	PP	SH	GW
1988-89	Cambridge	OJHL-B	48	28	18	46	138	....	....	....	....	....	....	....	....	....	....	....	....	....	....	....	....	....	....	....
1989-90	Owen Sound	OHL	61	12	15	27	90	....	....	....	....	....	....	....	....	....	....	....	12	1	6	7	15	....	....	....
1990-91	Owen Sound	OHL	66	34	32	66	100	....	....	....	....	....	....	....	....	....	....	....	....	....	....	....	....	....	....	....
1991-92	Owen Sound	OHL	66	50	41	91	99	....	....	....	....	....	....	....	....	....	....	....	5	3	3	6	18	....	....	....
1992-93	Cape Breton	AHL	73	22	23	45	130	....	....	....	....	....	....	....	....	....	....	....	16	3	3	6	45	....	....	....
1993-94	**Edmonton**	**NHL**	**68**	**11**	**8**	**19**	**74**	**0**	**1**	**1**	**89**	**12.4**	**-2**	....	....	....	....	....	....	....	....	....	....	....	....	....
1994-95	**Edmonton**	**NHL**	**47**	**8**	**3**	**11**	**49**	**0**	**2**	**1**	**73**	**11.0**	**-11**	....	....	....	....	....	....	....	....	....	....	....	....	....
1995-96	**Edmonton**	**NHL**	**49**	**2**	**6**	**8**	**61**	**0**	**0**	**1**	**51**	**3.9**	**-16**	....	....	....	....	....	....	....	....	....	....	....	....	....
	Cape Breton	AHL	4	1	2	3	6	....	....	....	....	....	....	....	....	....	....	....	....	....	....	....	....	....	....	....
	Detroit	**NHL**	**6**	**1**	**0**	**1**	**6**	**0**	**0**	**0**	**4**	**25.0**	**0**	....	....	....	....	....	**8**	**0**	**1**	**1**	**4**	**0**	**0**	**0**
1996-97♦	**Detroit**	**NHL**	**66**	**3**	**5**	**8**	**75**	**0**	**0**	**0**	**62**	**4.8**	**3**	....	....	....	....	....	**20**	**5**	**2**	**7**	**24**	**0**	**1**	**1**
1997-98♦	**Detroit**	**NHL**	**65**	**14**	**9**	**23**	**89**	**2**	**1**	**3**	**106**	**13.2**	**11**	....	....	....	....	....	**22**	**3**	**1**	**4**	**30**	**0**	**1**	**0**
1998-99	**Detroit**	**NHL**	**53**	**8**	**6**	**14**	**34**	**0**	**1**	**2**	**76**	**10.5**	**-6**	**10**	**40.0**	**129**	**33**	**13:13**	**10**	**1**	**0**	**1**	**8**	**0**	**0**	**1**
99-2000	**Detroit**	**NHL**	**41**	**6**	**8**	**14**	**24**	**0**	**2**	**1**	**71**	**8.5**	**1**	**2**	**50.0**	**90**	**34**	**13:30**	**8**	**0**	**1**	**1**	**4**	**0**	**0**	**0**
2000-01	**Detroit**	**NHL**	**79**	**12**	**7**	**19**	**22**	**1**	**3**	**3**	**119**	**10.1**	**16**	**14**	**35.7**	**217**	**54**	**14:17**	**6**	**0**	**0**	**0**	**6**	**0**	**0**	**0**
2001-02♦	**Detroit**	**NHL**	**82**	**9**	**15**	**24**	**40**	**0**	**1**	**5**	**108**	**8.3**	**15**	**38**	**47.4**	**182**	**53**	**13:23**	**23**	**3**	**3**	**6**	**32**	**0**	**2**	**0**
	NHL Totals		**556**	**74**	**67**	**141**	**474**	**3**	**11**	**17**	**759**	**9.7**		**64**	**43.8**	**618**	**174**	**13:39**	**97**	**12**	**8**	**20**	**108**	**0**	**4**	**2**

Traded to **Detroit** by **Edmonton** for Dan McGillis, March 20, 1996. • Missed majority of 1999-2000 season recovering from hernia injury suffered in game vs. Dallas, October 5, 1999.

MANDERVILLE, Kent

(MAN-duhr-VIHL, KEHNT) **PIT.**

Center. Shoots left. 6'3", 200 lbs. Born, Edmonton, Alta., April 12, 1971. Calgary's 1st choice, 24th overall, in 1989 Entry Draft.

Season	Club	League	GP	G	A	Pts	PIM	PP	SH	GW	S	%	+/-	TF	F%	H	SB	Min	GP	G	A	Pts	PIM	PP	SH	GW
1987-88	Notre Dame	SMHL	32	22	18	40	42	....	....	....	....	....	....	....	....	....	....	....	....	....	....	....	....	....	....	....
1988-89	Notre Dame	SJHL	58	39	36	75	165	....	....	....	....	....	....	....	....	....	....	....	....	....	....	....	....	....	....	....
1989-90	Cornell Big Red	ECAC	26	11	15	26	28	....	....	....	....	....	....	....	....	....	....	....	....	....	....	....	....	....	....	....
1990-91	Cornell Big Red	ECAC	28	17	14	31	60	....	....	....	....	....	....	....	....	....	....	....	....	....	....	....	....	....	....	....
1991-92	Team Canada	Nat-Tm	63	16	24	40	78	....	....	....	....	....	....	....	....	....	....	....	....	....	....	....	....	....	....	....
	Canada	Olympics	8	1	2	3	0	....	....	....	....	....	....	....	....	....	....	....	....	....	....	....	....	....	....	....
	Toronto	**NHL**	**15**	**0**	**4**	**4**	**0**	**0**	**0**	**0**	**14**	**0.0**	**1**	....	....	....	....	....	....	....	....	....	....	....	....	....
	St. John's	AHL	....	....	....	....	....	....	....	....	....	....	....	....	....	....	....	....	12	5	9	14	14	....	....	....
1992-93	**Toronto**	**NHL**	**18**	**1**	**1**	**2**	**17**	**0**	**0**	**1**	**15**	**6.7**	**-9**	....	....	....	....	....	**18**	**1**	**0**	**1**	**8**	**0**	**0**	**0**
	St. John's	AHL	56	19	28	47	86	....	....	....	....	....	....	....	....	....	....	....	2	0	2	2	0	....	....	....
1993-94	**Toronto**	**NHL**	**67**	**7**	**9**	**16**	**63**	**0**	**0**	**1**	**81**	**8.6**	**5**	....	....	....	....	....	**12**	**1**	**0**	**1**	**4**	**0**	**1**	**0**
1994-95	**Toronto**	**NHL**	**36**	**0**	**1**	**1**	**22**	**0**	**0**	**0**	**43**	**0.0**	**-2**	....	....	....	....	....	**7**	**0**	**0**	**0**	**6**	**0**	**0**	**0**
1995-96	**Edmonton**	**NHL**	**37**	**3**	**5**	**8**	**38**	**0**	**2**	**0**	**63**	**4.8**	**-5**	....	....	....	....	....	....	....	....	....	....	....	....	....
	St. John's	AHL	27	16	12	28	26	....	....	....	....	....	....	....	....	....	....	....	....	....	....	....	....	....	....	....
1996-97	**Hartford**	**NHL**	**44**	**6**	**5**	**11**	**18**	**0**	**0**	**1**	**51**	**11.8**	**3**	....	....	....	....	....	....	....	....	....	....	....	....	....
	Springfield	AHL	23	5	20	25	18	....	....	....	....	....	....	....	....	....	....	....	....	....	....	....	....	....	....	....
1997-98	**Carolina**	**NHL**	**77**	**4**	**4**	**8**	**31**	**0**	**0**	**0**	**80**	**5.0**	**-6**	....	....	....	....	....	....	....	....	....	....	....	....	....
1998-99	**Carolina**	**NHL**	**81**	**5**	**11**	**16**	**38**	**0**	**0**	**0**	**71**	**7.0**	**9**	**609**	**46.3**	**140**	**28**	**8:07**	**6**	**0**	**0**	**0**	**2**	**0**	**0**	**0**
99-2000	**Carolina**	**NHL**	**56**	**1**	**4**	**5**	**12**	**0**	**0**	**1**	**45**	**2.2**	**-8**	**399**	**47.1**	**96**	**26**	**8:15**	....	....	....	....	....	....	....	....
	Philadelphia	**NHL**	**13**	**0**	**3**	**3**	**4**	**0**	**0**	**0**	**17**	**0.0**	**2**	**148**	**54.1**	**21**	**2**	**11:56**	**18**	**0**	**1**	**1**	**22**	**0**	**0**	**0**
2000-01	**Philadelphia**	**NHL**	**82**	**5**	**10**	**15**	**47**	**0**	**3**	**2**	**136**	**3.7**	**-2**	**813**	**47.9**	**108**	**42**	**12:38**	**6**	**1**	**2**	**3**	**2**	**0**	**0**	**0**
2001-02	**Philadelphia**	**NHL**	**34**	**2**	**5**	**7**	**8**	**0**	**0**	**0**	**39**	**5.1**	**2**	**178**	**43.8**	**34**	**8**	**8:51**	....	....	....	....	....	....	....	....
	Pittsburgh	**NHL**	**4**	**1**	**0**	**1**	**4**	**0**	**0**	**0**	**4**	**25.0**	**1**	**57**	**45.6**	**3**	**4**	**12:43**	....	....	....	....	....	....	....	....
	NHL Totals		**564**	**35**	**62**	**97**	**302**	**0**	**5**	**6**	**659**	**5.3**		**2204**	**47.3**	**402**	**110**	**9:51**	**67**	**3**	**3**	**6**	**44**	**0**	**1**	**0**

ECAC Rookie of the Year (1990)

Traded to **Toronto** by **Calgary** with Doug Gilmour, Jamie Macoun, Rick Wamsley and Ric Nattress for Gary Leeman, Alexander Godynyuk, Jeff Reese, Michel Petit and Craig Berube, January 2, 1992. Traded to **Edmonton** by **Toronto** for Peter White and Edmonton's 4th round choice (Jason Sessa) in 1996 Entry Draft, December 4, 1995. Signed as a free agent by **Hartford**, October 2, 1996. Transferred to **Carolina** after **Hartford** franchise relocated, June 25, 1997. Traded to **Philadelphia** by **Carolina** for Sandy McCarthy, March 14, 2000. • Missed majority of 2001-02 season recovering from ankle injury originally suffered in game vs. Montreal, October 27, 2001. Traded to **Pittsburgh** by **Philadelphia** for Billy Tibbetts, March 17, 2002.

MANLOW, Eric

(MAN-low, AIR-ihk) **NYI**

Center. Shoots left. 6', 180 lbs. Born, Belleville, Ont., April 7, 1975. Chicago's 2nd choice, 50th overall, in 1993 Entry Draft.

Season	Club	League	GP	G	A	Pts	PIM	PP	SH	GW	S	%	+/-	TF	F%	H	SB	Min	GP	G	A	Pts	PIM	PP	SH	GW
1990-91	Peterborough	OMHA	59	67	51	118	90	....	....	....	....	....	....	....	....	....	....	....	....	....	....	....	....	....	....	....
	Peterborough	OJHL-B	1	0	0	0	0	....	....	....	....	....	....	....	....	....	....	....	....	....	....	....	....	....	....	....
1991-92	Kitchener	OHL	59	12	20	32	17	....	....	....	....	....	....	....	....	....	....	....	14	2	5	7	10	....	....	....
1992-93	Kitchener	OHL	53	26	21	47	31	....	....	....	....	....	....	....	....	....	....	....	4	0	1	1	2	....	....	....
1993-94	Kitchener	OHL	49	28	32	60	25	....	....	....	....	....	....	....	....	....	....	....	3	0	1	1	4	....	....	....
1994-95	Kitchener	OHL	44	25	29	54	26	....	....	....	....	....	....	....	....	....	....	....	....	....	....	....	....	....	....	....
	Detroit	OHL	16	4	16	20	11	....	....	....	....	....	....	....	....	....	....	....	21	11	10	21	18	....	....	....
1995-96	Indianapolis Ice	IHL	75	6	11	17	32	....	....	....	....	....	....	....	....	....	....	....	4	0	1	1	4	....	....	....
1996-97	Baltimore	AHL	36	6	6	12	13	....	....	....	....	....	....	....	....	....	....	....	3	0	0	0	0	....	....	....
	Columbus Chill	ECHL	32	18	18	36	20	....	....	....	....	....	....	....	....	....	....	....	....	....	....	....	....	....	....	....
1997-98	Indianapolis Ice	IHL	60	8	11	19	25	....	....	....	....	....	....	....	....	....	....	....	3	1	0	1	0	....	....	....
1998-99	Long Beach	IHL	51	9	19	28	30	....	....	....	....	....	....	....	....	....	....	....	8	0	0	0	8	....	....	....
	Florida	ECHL	18	8	15	23	11	....	....	....	....	....	....	....	....	....	....	....	....	....	....	....	....	....	....	....
99-2000	Florida	ECHL	26	14	24	38	24	....	....	....	....	....	....	....	....	....	....	....	....	....	....	....	....	....	....	....
	Providence	AHL	46	17	16	33	14	....	....	....	....	....	....	....	....	....	....	....	14	6	8	14	8	....	....	....
2000-01	**Boston**	**NHL**	**8**	**0**	**1**	**1**	**2**	**0**	**0**	**0**	**3**	**0.0**	**0**	**61**	**50.8**	**3**	**1**	**7:26**	....	....	....	....	....	....	....	....
	Providence	AHL	60	16	51	67	18	....	....	....	....	....	....	....	....	....	....	....	17	6	7	13	6	....	....	....

Season	Club	League	Regular Season GP	G	A	Pts	PIM	PP	SH	GW	S	%	+/-	TF	F%	H	SB	Min	Playoffs GP	G	A	Pts	PIM	PP	SH	GW
2001-02	Boston	NHL	3	0	0	0	0	0	0	0	2	0.0	0	16	31.3	1	0	6:05	....	....	....	....	....	....	....	....
	Providence	AHL	70	13	35	48	30	....	....	....	....	....	....	....	....	....	....	....	2	0	0	0	2	....	....	....
	NHL Totals		11	0	1	1	2	0	0	0	5	0.0		77	46.8	4	1	7:04	....	....	....	....	....	....	....	....

Signed as a free agent by **Providence** (AHL), January 24, 2000. Signed as a free agent by **Boston**, July 11, 2000. Signed as a free agent by **NY Islanders**, July 21, 2002.

MANN, Cameron

(MAN, CAM-uhr-ROHN) **NSH.**

Right wing. Shoots right. 6', 195 lbs. Born, Thompson, Man., April 20, 1977. Boston's 5th choice, 99th overall, in 1995 Entry Draft.

Season	Club	League	GP	G	A	Pts	PIM	PP	SH	GW	S	%	+/-	TF	F%	H	SB	Min	GP	G	A	Pts	PIM	PP	SH	GW
1992-93	Kenora Thistles	NOJHA	35	23	24	47	49	....	....	....	....	....	....	....	....	....	....	....	....	....	....	....	....	....	....	....
1993-94	Peterborough	OPJHL	16	3	14	17	23	....	....	....	....	....	....	....	....	....	....	....	....	....	....	....	....	....	....	....
	Peterborough	OHL	49	8	17	25	18	....	....	....	....	....	....	....	....	....	....	....	7	1	1	2	2	....	....	....
1994-95	Peterborough	OHL	64	19	24	43	40	....	....	....	....	....	....	....	....	....	....	....	11	3	8	11	4	....	....	....
1995-96	Peterborough	OHL	66	42	60	102	108	....	....	....	....	....	....	....	....	....	....	....	24	*27	16	*43	33	....	....	....
1996-97	Peterborough	OHL	51	33	50	83	91	....	....	....	....	....	....	....	....	....	....	....	11	10	18	28	16	....	....	....
1997-98	Boston	NHL	9	0	1	1	4	0	0	0	6	0.0	1	....	....	....	....	....	....	....	....	....	....	....	....	....
	Providence	AHL	71	21	26	47	99	....	....	....	....	....	....	....	....	....	....	....	....	....	....	....	....	....	....	....
1998-99	Boston	NHL	33	5	2	7	17	1	0	1	42	11.9	0	22	36.4	28	4	10:40	1	0	0	0	0	0	0	0
	Providence	AHL	43	21	25	46	65	....	....	....	....	....	....	....	....	....	....	....	11	7	7	14	4	....	....	....
99-2000	Boston	NHL	32	8	4	12	13	1	0	0	48	16.7	-6	16	25.0	28	2	12:50	....	....	....	....	....	....	....	....
	Providence	AHL	29	7	12	19	45	....	....	....	....	....	....	....	....	....	....	....	11	6	7	13	0	....	....	....
2000-01	Boston	NHL	15	1	3	4	6	0	0	0	17	5.9	0	1	0.0	11	0	8:29	....	....	....	....	....	....	....	....
	Providence	AHL	39	24	23	47	59	....	....	....	....	....	....	....	....	....	....	....	....	....	....	....	....	....	....	....
2001-02	Utah Grizzlies	AHL	61	19	32	51	83	....	....	....	....	....	....	....	....	....	....	....	5	0	0	0	6	....	....	....
	NHL Totals		89	14	10	24	40	2	0	1	113	12.4		39	30.8	67	6	11:08	1	0	0	0	0	0	0	0

OHL First All-Star Team (1996, 1997) • Memorial Cup All-Star Team (1996) • Won Stafford Smythe Memorial Trophy (Memorial Cup Tournament MVP) (1996)

Traded to **Dallas** by **Boston** for Richard Jackman, June 23, 2001. Traded to **Nashville** by **Dallas** with Ed Belfour for David Gosselin and Nashville's 5th round choice in 2003 Entry Draft, June 29, 2002.

MANSON, Dave

(MAN-suhn, DAIV)

Defense. Shoots left. 6'2", 200 lbs. Born, Prince Albert, Sask., January 27, 1967. Chicago's 1st choice, 11th overall, in 1985 Entry Draft.

Season	Club	League	GP	G	A	Pts	PIM	PP	SH	GW	S	%	+/-	TF	F%	H	SB	Min	GP	G	A	Pts	PIM	PP	SH	GW
1982-83	Prince Albert	SMHL	28	11	11	22	170	....	....	....	....	....	....	....	....	....	....	....	....	....	....	....	....	....	....	....
	Prince Albert	WHL	6	0	1	1	9	....	....	....	....	....	....	....	....	....	....	....	....	....	....	....	....	....	....	....
1983-84	Prince Albert	WHL	70	2	7	9	233	....	....	....	....	....	....	....	....	....	....	....	5	0	0	0	4	....	....	....
1984-85	Prince Albert	WHL	72	8	30	38	247	....	....	....	....	....	....	....	....	....	....	....	13	1	0	1	34	....	....	....
1985-86	Prince Albert	WHL	70	14	34	48	177	....	....	....	....	....	....	....	....	....	....	....	20	1	8	9	63	....	....	....
1986-87	Chicago	NHL	63	1	8	9	146	0	0	0	42	2.4	-2	....	....	....	....	....	3	0	0	0	10	0	0	0
1987-88	Chicago	NHL	54	1	6	7	185	0	0	0	47	2.1	-12	....	....	....	....	....	5	0	0	0	27	0	0	0
	Saginaw Hawks	IHL	6	0	3	3	37	....	....	....	....	....	....	....	....	....	....	....	....	....	....	....	....	....	....	....
1988-89	Chicago	NHL	79	18	36	54	352	8	1	0	224	8.0	5	....	....	....	....	....	16	0	8	8	84	0	0	0
1989-90	Chicago	NHL	59	5	23	28	301	1	0	1	126	4.0	4	....	....	....	....	....	20	2	4	6	46	1	0	0
1990-91	Chicago	NHL	75	14	15	29	191	6	1	2	154	9.1	20	....	....	....	....	....	6	0	1	1	36	0	0	0
1991-92	Edmonton	NHL	79	15	32	47	220	7	0	2	206	7.3	9	....	....	....	....	....	16	3	9	12	44	1	0	0
1992-93	Edmonton	NHL	83	15	30	45	210	9	1	1	244	6.1	-28	....	....	....	....	....	....	....	....	....	....	....	....	....
1993-94	Edmonton	NHL	57	3	13	16	140	0	0	0	144	2.1	-4	....	....	....	....	....	....	....	....	....	....	....	....	....
	Winnipeg	NHL	13	1	4	5	51	1	0	0	36	2.8	-10	....	....	....	....	....	....	....	....	....	....	....	....	....
1994-95	Winnipeg	NHL	44	3	15	18	139	2	0	1	104	2.9	-20	....	....	....	....	....	....	....	....	....	....	....	....	....
1995-96	Winnipeg	NHL	82	7	23	30	205	3	0	0	189	3.7	8	....	....	....	....	....	6	2	1	3	30	0	0	1
1996-97	Phoenix	NHL	66	3	17	20	164	2	0	0	153	2.0	-25	....	....	....	....	....	....	....	....	....	....	....	....	....
	Montreal	NHL	9	1	1	2	23	0	0	0	22	4.5	-1	....	....	....	....	....	5	0	0	0	17	0	0	0
1997-98	Montreal	NHL	81	4	30	34	122	2	0	0	148	2.7	22	....	....	....	....	....	10	0	1	1	14	0	0	0
1998-99	Montreal	NHL	11	0	2	2	48	0	0	0	11	0.0	-3	0	0.0	14	8	17:19	....	....	....	....	....	....	....	....
	Chicago	NHL	64	6	15	21	107	2	0	0	134	4.5	4	0	0.0	127	51	22:37	....	....	....	....	....	....	....	....
99-2000	Chicago	NHL	37	0	7	7	40	0	0	0	45	0.0	2	0	0.0	46	27	17:29	....	....	....	....	....	....	....	....
	Dallas	NHL	26	1	2	3	22	0	0	0	21	4.8	10	0	0.0	34	6	13:16	23	0	0	0	33	0	0	0
2000-01	Toronto	NHL	74	4	7	11	93	0	0	0	70	5.7	13	0	0.0	124	61	15:48	2	0	0	0	2	0	0	0
2001-02	Toronto	NHL	13	0	1	1	10	0	0	0	12	0.0	3	0	0.0	21	12	17:34	....	....	....	....	....	....	....	....
	Dallas	NHL	34	0	1	1	23	0	0	0	24	0.0	-1	0	0.0	33	18	11:03	....	....	....	....	....	....	....	....
	Utah Grizzlies	AHL	2	0	0	0	2	....	....	....	....	....	....	....	....	....	....	....	....	....	....	....	....	....	....	....
	NHL Totals		1103	102	288	390	2792	43	3	7	2156	4.7		0	0.0	399	183	17:00	112	7	24	31	343	2	0	1

WHL East Second All-Star Team (1986) • Played in NHL All-Star Game (1989, 1993)

Traded to **Edmonton** by **Chicago** with Chicago's 3rd round choice (Kirk Maltby) in 1992 Entry Draft for Steve Smith, October 2, 1991. Traded to **Winnipeg** by **Edmonton** with St. Louis' 6th round choice (previously acquired, Winnipeg selected Chris Kibermanis) in 1994 Entry Draft for Boris Mironov, Mats Lindgren, Winnipeg's 1st round choice (Jason Bonsignore) in 1994 Entry Draft and Florida's 4th round choice (previously acquired, Edmonton selected Adam Copeland) in 1994 Entry Draft, March 15, 1994. Transferred to **Phoenix** after **Winnipeg** franchise relocated, July 1, 1996. Traded to **Montreal** by **Phoenix** for Murray Baron and Chris Murray, March 18, 1997. Traded to **Chicago** by **Montreal** with Jocelyn Thibault and Brad Brown for Jeff Hackett, Eric Weinrich, Alain Nasreddine and Tampa Bay's 4th round choice (previously acquired, Montreal selected Chris Dyment) in 1999 Entry Draft, November 16, 1998. Traded to **Dallas** by **Chicago** with Sylvain Cote for Kevin Dean, Derek Plante and Dallas' 2nd round choice (Matt Keith) in 2001 Entry Draft, February 8, 2000. Signed as a free agent by **Toronto**, August 16, 2000. Traded to **Dallas** by **Toronto** for Jyrki Lumme, November 21, 2001.

MARA, Paul

(MAIR-uh, PAWL) **PHX.**

Defense. Shoots left. 6'4", 210 lbs. Born, Ridgewood, NJ, September 7, 1979. Tampa Bay's 1st choice, 7th overall, in 1997 Entry Draft.

Season	Club	League	GP	G	A	Pts	PIM	PP	SH	GW	S	%	+/-	TF	F%	H	SB	Min	GP	G	A	Pts	PIM	PP	SH	GW
1994-95	Belmont Hill	Hi-School	28	5	17	22	28	....	....	....	....	....	....	....	....	....	....	....	....	....	....	....	....	....	....	....
1995-96	Belmont Hill	Hi-School	28	18	20	38	40	....	....	....	....	....	....	....	....	....	....	....	....	....	....	....	....	....	....	....
1996-97	Sudbury Wolves	OHL	44	9	34	43	61	....	....	....	....	....	....	....	....	....	....	....	....	....	....	....	....	....	....	....
1997-98	Sudbury Wolves	OHL	25	8	18	26	79	....	....	....	....	....	....	....	....	....	....	....	....	....	....	....	....	....	....	....
	Plymouth Whalers	OHL	25	8	15	23	30	....	....	....	....	....	....	....	....	....	....	....	15	3	14	17	30	....	....	....
1998-99	Plymouth Whalers	OHL	52	13	41	54	95	....	....	....	....	....	....	....	....	....	....	....	11	5	7	12	28	....	....	....
	Tampa Bay	NHL	1	1	1	2	0	1	0	0	1	100.0	-3	0	0.0	1	3	19:34	....	....	....	....	....	....	....	....
99-2000	Tampa Bay	NHL	54	7	11	18	73	4	0	1	78	9.0	-27	0	0.0	62	49	22:13	....	....	....	....	....	....	....	....
	Detroit Vipers	IHL	15	3	5	8	22	....	....	....	....	....	....	....	....	....	....	....	....	....	....	....	....	....	....	....
2000-01	Tampa Bay	NHL	46	6	10	16	40	2	0	1	58	10.3	-17	0	0.0	45	56	23:06	....	....	....	....	....	....	....	....
	Detroit Vipers	IHL	10	3	3	6	22	....	....	....	....	....	....	....	....	....	....	....	....	....	....	....	....	....	....	....
	Phoenix	NHL	16	0	4	4	14	0	0	0	20	0.0	1	0	0.0	17	10	19:22	....	....	....	....	....	....	....	....
2001-02	Phoenix	NHL	75	7	17	24	58	2	0	0	112	6.3	-6	2	100.0	129	65	21:34	5	0	0	0	4	0	0	0
	NHL Totals		192	21	43	64	185	9	0	2	269	7.8		2	100.0	254	183	21:56	5	0	0	0	4	0	0	0

Traded to **Windsor** (OHL) by **Sudbury** (OHL) with Steve Valiquette for Glenn Crawford, Kip Brennan and future considerations, December 16, 1997. Traded to **Plymouth** (OHL) by **Windsor** (OHL) with Rick Smith and future considerations for Luc Rioux, Andy Burnham and Plymouth's 1st pick (later forwarded to Sudbury - Sudbury selected Alexei Shalashenko) in CHL Import Draft, December 17, 1997. Traded to **Phoenix** by **Tampa Bay** with Mike Johnson, Ruslan Zainullin and NY Islanders' 2nd round choice (previously acquired, Phoenix selected Matthew Spiller) in 2001 Entry Draft for Nikolai Khabibulin and Stan Neckar, March 5, 2001.

MARCHANT, Todd

(mahr-SHAHNT, TAWD) **EDM.**

Center. Shoots left. 5'10", 178 lbs. Born, Buffalo, NY, August 12, 1973. NY Rangers' 8th choice, 164th overall, in 1993 Entry Draft.

Season	Club	League	GP	G	A	Pts	PIM	PP	SH	GW	S	%	+/-	TF	F%	H	SB	Min	GP	G	A	Pts	PIM	PP	SH	GW
1990-91	Niagara Scenics	NAJHL	37	31	47	78	....	....	....	....	....	....	....	....	....	....	....	....	....	....	....	....	....	....	....	....
1991-92	Clarkson Knights	ECAC	32	20	12	32	32	....	....	....	....	....	....	....	....	....	....	....	....	....	....	....	....	....	....	....
1992-93	Clarkson Knights	ECAC	33	18	28	46	38	....	....	....	....	....	....	....	....	....	....	....	....	....	....	....	....	....	....	....
1993-94	Team USA	Nat-Tm	59	28	39	67	48	....	....	....	....	....	....	....	....	....	....	....	....	....	....	....	....	....	....	....
	United States	Olympics	8	1	1	2	6	....	....	....	....	....	....	....	....	....	....	....	....	....	....	....	....	....	....	....
	NY Rangers	NHL	1	0	0	0	0	0	0	0	1	0.0	-1	....	....	....	....	....	....	....	....	....	....	....	....	....
	Binghamton	AHL	8	2	7	9	6	....	....	....	....	....	....	....	....	....	....	....	....	....	....	....	....	....	....	....
	Edmonton	NHL	3	0	1	1	2	0	0	0	5	0.0	-1	....	....	....	....	....	....	....	....	....	....	....	....	....
	Cape Breton	AHL	3	1	4	5	2	....	....	....	....	....	....	....	....	....	....	....	5	1	1	2	0	....	....	....
1994-95	Cape Breton	AHL	38	22	25	47	25	....	....	....	....	....	....	....	....	....	....	....	....	....	....	....	....	....	....	....
	Edmonton	NHL	45	13	14	27	32	3	2	2	95	13.7	-3	....	....	....	....	....	....	....	....	....	....	....	....	....
1995-96	Edmonton	NHL	81	19	19	38	66	2	3	2	221	8.6	-19	....	....	....	....	....	....	....	....	....	....	....	....	....
1996-97	Edmonton	NHL	79	14	19	33	44	0	4	3	202	6.9	11	....	....	....	....	....	12	4	2	6	12	0	3	1
1997-98	Edmonton	NHL	76	14	21	35	71	2	1	3	194	7.2	9	....	....	....	....	....	12	1	1	2	10	0	0	0
1998-99	Edmonton	NHL	82	14	22	36	65	3	1	2	183	7.7	3	1449	50.0	133	57	16:47	4	1	1	2	12	0	0	0
99-2000	Edmonton	NHL	82	17	23	40	70	0	1	0	170	10.0	7	1593	52.9	92	58	17:08	3	1	0	1	2	0	0	0

			Regular Season																Playoffs							
Season	Club	League	GP	G	A	Pts	PIM	PP	SH	GW	S	%	+/-	TF	F%	H	SB	Min	GP	G	A	Pts	PIM	PP	SH	GW
2000-01	Edmonton	NHL	71	13	26	39	51	0	4	2	113	11.5	1	1549	53.8	85	73	17:54	6	0	0	0	4	0	0	0
2001-02	Edmonton	NHL	82	12	22	34	41	0	3	1	124	9.7	7	1523	52.4	91	74	16:58								
	NHL Totals		602	116	167	283	442	10	19	15	1308	8.9		6114	52.3	401	262	17:10	37	7	4	11	40	0	3	1

ECAC Second All-Star Team (1993)

Traded to **Edmonton** by **NY Rangers** for Craig MacTavish, March 21, 1994.

MARCHMENT, Bryan

(MAHRCH-mehnt, BRIGH-uhn) **S.J.**

Defense. Shoots left. 6'1", 200 lbs. Born, Scarborough, Ont., May 1, 1969. Winnipeg's 1st choice, 16th overall, in 1987 Entry Draft.

Season	Club	League	GP	G	A	Pts	PIM	PP	SH	GW	S	%	+/-	TF	F%	H	SB	Min	GP	G	A	Pts	PIM	PP	SH	GW
1984-85	Tor. Young Nats	MTHL	69	14	35	49	229																			
1985-86	Belleville Bulls	OHL	57	5	15	20	225												21	0	7	7	83			
1986-87	Belleville Bulls	OHL	52	6	38	44	238												6	0	4	4	17			
1987-88	Belleville Bulls	OHL	56	7	51	58	200												6	1	3	4	19			
1988-89	Belleville Bulls	OHL	43	14	36	50	118												5	0	1	1	12			
	Winnipeg	**NHL**	**2**	**0**	**0**	**0**	**2**	**0**	**0**	**0**	**1**	**0.0**	**0**													
1989-90	**Winnipeg**	**NHL**	**7**	**0**	**2**	**2**	**28**	**0**	**0**	**0**	**5**	**0.0**	**0**													
	Moncton Hawks	AHL	56	4	19	23	217																			
1990-91	**Winnipeg**	**NHL**	**28**	**2**	**2**	**4**	**91**	**0**	**0**	**0**	**24**	**8.3**	**–5**													
	Moncton Hawks	AHL	33	2	11	13	101																			
1991-92	**Chicago**	**NHL**	**58**	**5**	**10**	**15**	**168**	**2**	**0**	**0**	**55**	**9.1**	**–4**						**16**	**1**	**0**	**1**	**36**	**0**	**0**	**0**
1992-93	**Chicago**	**NHL**	**78**	**5**	**15**	**20**	**313**	**1**	**0**	**1**	**75**	**6.7**	**15**						**4**	**0**	**0**	**0**	**12**	**0**	**0**	**0**
1993-94	**Chicago**	**NHL**	**13**	**1**	**4**	**5**	**42**	**0**	**0**	**0**	**18**	**5.6**	**–2**													
	Hartford	**NHL**	**42**	**3**	**7**	**10**	**124**	**0**	**1**	**1**	**74**	**4.1**	**–12**													
1994-95	**Edmonton**	**NHL**	**40**	**1**	**5**	**6**	**184**	**0**	**0**	**0**	**57**	**1.8**	**–11**													
1995-96	**Edmonton**	**NHL**	**78**	**3**	**15**	**18**	**202**	**0**	**0**	**0**	**96**	**3.1**	**–7**													
1996-97	**Edmonton**	**NHL**	**71**	**3**	**13**	**16**	**132**	**1**	**0**	**0**	**89**	**3.4**	**13**						**3**	**0**	**0**	**0**	**4**	**0**	**0**	**0**
1997-98	**Edmonton**	**NHL**	**27**	**0**	**4**	**4**	**58**	**0**	**0**	**0**	**23**	**0.0**	**–2**													
	Tampa Bay	**NHL**	**22**	**2**	**4**	**6**	**43**	**0**	**0**	**0**	**20**	**10.0**	**–3**													
	San Jose	**NHL**	**12**	**0**	**3**	**3**	**43**	**0**	**0**	**0**	**13**	**0.0**	**2**						**6**	**0**	**0**	**0**	**10**	**0**	**0**	**0**
1998-99	**San Jose**	**NHL**	**59**	**2**	**6**	**8**	**101**	**0**	**0**	**0**	**49**	**4.1**	**–7**	**0**	**0.0**	**108**	**64**	**17:43**	**6**	**0**	**0**	**0**	**4**	**0**	**0**	**0**
99-2000	**San Jose**	**NHL**	**49**	**0**	**4**	**4**	**72**	**0**	**0**	**0**	**51**	**0.0**	**3**	**0**	**0.0**	**127**	**51**	**18:55**	**11**	**2**	**1**	**3**	**12**	**0**	**0**	**0**
2000-01	**San Jose**	**NHL**	**75**	**7**	**11**	**18**	**204**	**0**	**1**	**3**	**73**	**9.6**	**15**	**1**	**100.0**	**229**	**90**	**18:12**	**5**	**0**	**1**	**1**	**2**	**0**	**0**	**0**
2001-02	**San Jose**	**NHL**	**72**	**2**	**20**	**22**	**178**	**0**	**0**	**0**	**68**	**2.9**	**22**	**0**	**0.0**	**157**	**69**	**18:47**	**12**	**1**	**1**	**2**	**10**	**0**	**0**	**0**
	NHL Totals		**733**	**36**	**125**	**161**	**1985**	**4**	**2**	**5**	**791**	**4.6**		**1**	**100.0**	**621**	**274**	**18:23**	**63**	**4**	**3**	**7**	**90**	**0**	**0**	**0**

OHL Second All-Star Team (1989)

Traded to **Chicago** by **Winnipeg** with Chris Norton for Troy Murray and Warren Rychel, July 22, 1991. Traded to **Hartford** by **Chicago** with Steve Larmer for Eric Weinrich and Patrick Poulin, November 2, 1993. Transferred to **Edmonton** from **Hartford** as compensation for Hartford's signing of free agent Steven Rice, August 30, 1994. Traded to **Tampa Bay** by **Edmonton** with Steve Kelly and Jason Bonsignore for Roman Hamrlik and Paul Comrie, December 30, 1997. Traded to **San Jose** by **Tampa Bay** with David Shaw and Tampa Bay's 1st round choice (later traded to Nashville - Nashville selected David Legwand) in 1998 Entry Draft for Andrei Nazarov and Florida's 1st round choice (previously acquired, Tampa Bay selected Vincent Lecavallier) in 1998 Entry Draft, March 24, 1998.

MARHA, Josef

(MAHR-hah, JOH-sehf) **CHI.**

Center. Shoots left. 6', 176 lbs. Born, Havlickuv Brod, Czech., June 2, 1976. Quebec's 3rd choice, 35th overall, in 1994 Entry Draft.

Season	Club	League	GP	G	A	Pts	PIM	PP	SH	GW	S	%	+/-	TF	F%	H	SB	Min	GP	G	A	Pts	PIM	PP	SH	GW
1991-92	Dukla Jihlava Jr.	Czech-Jr.	25	12	13	25	0																			
1992-93	Dukla Jihlava	Czech	7	2	2	4																				
1993-94	HC Dukla Jihlava	Czech	41	7	2	9													3	0	1	1				
1994-95	HC Dukla Jihlava	Czech	35	3	7	10	6																			
1995-96	**Colorado**	**NHL**	**2**	**0**	**1**	**1**	**0**	**0**	**0**	**0**	**2**	**0.0**	**1**													
	Cornwall Aces	AHL	74	18	30	48	30												8	1	2	3	10			
1996-97	**Colorado**	**NHL**	**6**	**0**	**1**	**1**	**0**	**0**	**0**	**0**	**6**	**0.0**	**0**													
	Hershey Bears	AHL	67	23	49	72	44												19	6	*16	*22	10			
1997-98	**Colorado**	**NHL**	**11**	**2**	**5**	**7**	**4**	**0**	**0**	**0**	**10**	**20.0**	**0**													
	Hershey Bears	AHL	55	6	46	52	30																			
	Anaheim	**NHL**	**12**	**7**	**4**	**11**	**0**	**3**	**0**	**0**	**21**	**33.3**	**4**													
1998-99	**Anaheim**	**NHL**	**10**	**0**	**1**	**1**	**0**	**0**	**0**	**0**	**13**	**0.0**	**–4**	**107**	**40.2**	**3**	**1**	**12:02**								
	Cincinnati	AHL	3	1	0	1	4																			
	Chicago	**NHL**	**22**	**2**	**5**	**7**	**4**	**1**	**0**	**1**	**32**	**6.3**	**5**	**275**	**50.9**	**6**	**10**	**14:37**								
	Portland Pirates	AHL	8	0	8	8	2																			
99-2000	**Chicago**	**NHL**	**81**	**10**	**12**	**22**	**18**	**2**	**1**	**3**	**91**	**11.0**	**–10**	**1110**	**46.3**	**26**	**31**	**13:20**								
2000-01	**Chicago**	**NHL**	**15**	**0**	**3**	**3**	**6**	**0**	**0**	**0**	**17**	**0.0**	**–4**	**196**	**50.0**	**2**	**3**	**12:46**								
	Norfolk Admirals	AHL	60	18	28	46	44												9	1	8	9	6			
2001-02	HC Davos	Swiss	44	19	15	34	38												16	6	9	15	8			
	NHL Totals		**159**	**21**	**32**	**53**	**32**	**6**	**1**	**4**	**192**	**10.9**		**1688**	**47.1**	**37**	**45**	**13:23**								

Rights transferred to **Colorado** after **Quebec** franchise relocated, June 21, 1995. Traded to **Anaheim** by **Colorado** for Warren Rychel and Anaheim's 4th round choice (Sanny Lindstrom) in 1999 Entry Draft, March 24, 1998. Traded to **Chicago** by **Anaheim** for Chicago's 4th round choice (Alexandr Chagodayev) in 1999 Entry Draft, January 28, 1999. Signed as a free agent by **HC Davos** (Swiss) with **Chicago** retaining NHL rights, June 13, 2001.

MARKOV, Andrei

(MAHR-kahf, AHN-dray) **MTL.**

Defense. Shoots left. 6', 208 lbs. Born, Voskresensk, USSR, December 20, 1978. Montreal's 6th choice, 162nd overall, in 1998 Entry Draft.

Season	Club	League	GP	G	A	Pts	PIM	PP	SH	GW	S	%	+/-	TF	F%	H	SB	Min	GP	G	A	Pts	PIM	PP	SH	GW
1995-96	Voskresensk	CIS	38	0	0	0	14																			
1996-97	Voskresensk	Russia	43	8	4	12	32												2	1	1	2	0			
1997-98	Voskresensk	Russia	43	10	5	15	83																			
1998-99	Dynamo Moscow	Russia	38	10	11	21	32												16	3	6	9	6			
	Dynamo Moscow	EuroHL	12	7	5	12	12												6	2	2	4	4			
99-2000	Dynamo Moscow	Russia	29	11	12	23	28												17	4	3	7	8			
2000-01	**Montreal**	**NHL**	**63**	**6**	**17**	**23**	**18**	**2**	**0**	**0**	**82**	**7.3**	**–6**	**2**	**50.0**	**38**	**55**	**16:53**								
	Quebec	AHL	14	0	5	5	4												7	1	1	2	2			
2001-02	**Montreal**	**NHL**	**56**	**5**	**19**	**24**	**24**	**2**	**0**	**1**	**73**	**6.8**	**–1**	**0**	**0.0**	**31**	**57**	**17:15**	**12**	**1**	**3**	**4**	**8**	**0**	**0**	**1**
	Quebec	AHL	12	4	6	10	7																			
	NHL Totals		**119**	**11**	**36**	**47**	**42**	**4**	**0**	**1**	**155**	**7.1**		**2**	**50.0**	**69**	**112**	**17:04**	**12**	**1**	**3**	**4**	**8**	**0**	**0**	**1**

MARKOV, Danny

(MAHR-kahf, DA-nee) **PHX.**

Defense. Shoots left. 6'1", 190 lbs. Born, Moscow, USSR, July 30, 1976. Toronto's 7th choice, 223rd overall, in 1995 Entry Draft.

Season	Club	League	GP	G	A	Pts	PIM	PP	SH	GW	S	%	+/-	TF	F%	H	SB	Min	GP	G	A	Pts	PIM	PP	SH	GW
1993-94	Spartak Moscow	CIS	13	1	0	1	6												1	0	0	0	0			
1994-95	Spartak Moscow	CIS	39	0	1	1	36																			
1995-96	Spartak Moscow	CIS	38	2	0	2	12												2	0	0	0	2			
1996-97	Spartak Moscow	Russia	39	3	6	9	41																			
	St. John's	AHL	10	2	4	6	18												11	2	6	8	14			
1997-98	**Toronto**	**NHL**	**25**	**2**	**5**	**7**	**28**	**1**	**0**	**0**	**15**	**13.3**	**0**													
	St. John's	AHL	52	3	23	26	124												2	0	1	1	0			
1998-99	**Toronto**	**NHL**	**57**	**4**	**8**	**12**	**47**	**0**	**0**	**0**	**34**	**11.8**	**5**	**0**	**0.0**	**92**	**66**	**18:41**	**17**	**0**	**6**	**6**	**18**	**0**	**0**	**0**
99-2000	**Toronto**	**NHL**	**59**	**0**	**10**	**10**	**28**	**0**	**0**	**0**	**38**	**0.0**	**13**	**1**	**0.0**	**97**	**100**	**20:08**	**12**	**0**	**3**	**3**	**10**	**0**	**0**	**0**
2000-01	**Toronto**	**NHL**	**59**	**3**	**13**	**16**	**34**	**1**	**0**	**2**	**49**	**6.1**	**6**	**0**	**0.0**	**123**	**73**	**19:02**	**11**	**1**	**1**	**2**	**12**	**0**	**0**	**0**
2001-02	**Phoenix**	**NHL**	**72**	**6**	**30**	**36**	**67**	**4**	**0**	**1**	**103**	**5.8**	**–7**	**0**	**0.0**	**208**	**80**	**22:55**								
	Russia	Olympics	5	0	1	1	0																			
	NHL Totals		**272**	**15**	**66**	**81**	**204**	**6**	**0**	**3**	**239**	**6.3**		**1**	**0.0**	**520**	**319**	**20:21**	**40**	**1**	**10**	**11**	**40**	**0**	**0**	**0**

Traded to **Phoenix** by **Toronto** for Robert Reichel, Travis Green and Craig Mills, June 12, 2001.

MARLEAU, Patrick

(mahr-LOH, PAT-rihk) **S.J.**

Center. Shoots left. 6'2", 210 lbs. Born, Aneroid, Sask., September 15, 1979. San Jose's 1st choice, 2nd overall, in 1997 Entry Draft.

Season	Club	League	GP	G	A	Pts	PIM	PP	SH	GW	S	%	+/-	TF	F%	H	SB	Min	GP	G	A	Pts	PIM	PP	SH	GW
1993-94	Swift Current	SMHL	53	72	95	167																				
1994-95	Swift Current	SMHL	31	30	22	52	18																			
1995-96	Seattle	WHL	72	32	42	74	22												5	3	4	7	4			
1996-97	Seattle	WHL	71	51	74	125	37												15	7	16	23	12			
1997-98	**San Jose**	**NHL**	**74**	**13**	**19**	**32**	**14**	**1**	**0**	**2**	**90**	**14.4**	**5**						**5**	**0**	**1**	**1**	**0**	**0**	**0**	**0**
1998-99	**San Jose**	**NHL**	**81**	**21**	**24**	**45**	**24**	**4**	**0**	**4**	**134**	**15.7**	**10**	**1121**	**43.4**	**59**	**15**	**15:11**	**6**	**2**	**1**	**3**	**4**	**2**	**0**	**0**

Season	Club	League	Regular Season GP	G	A	Pts	PIM	PP	SH	GW	S	%	+/-	TF	F%	H	SB	Min	Playoffs GP	G	A	Pts	PIM	PP	SH	GW
99-2000	**San Jose**	**NHL**	**81**	**17**	**23**	**40**	**36**	**3**	**0**	**3**	**161**	**10.6**	**-9**	**851**	**42.0**	**74**	**13**	**14:11**	**5**	**1**	**1**	**2**	**2**	**1**	**0**	**0**
2000-01	**San Jose**	**NHL**	**81**	**25**	**27**	**52**	**22**	**5**	**0**	**6**	**146**	**17.1**	**7**	**1088**	**44.8**	**85**	**15**	**16:17**	**6**	**2**	**0**	**2**	**4**	**0**	**0**	**0**
2001-02	**San Jose**	**NHL**	**79**	**21**	**23**	**44**	**40**	**3**	**0**	**5**	**121**	**17.4**	**9**	**897**	**47.3**	**103**	**9**	**14:04**	**12**	**6**	**5**	**11**	**6**	**1**	**0**	**3**
	NHL Totals		**396**	**97**	**116**	**213**	**136**	**16**	**0**	**20**	**652**	**14.9**		**3957**	**44.4**	**321**	**52**	**14:56**	**34**	**11**	**8**	**19**	**16**	**4**	**0**	**3**

WHL West First All-Star Team (1997)

MARSHALL, Grant

(MAHR-shahl, GRANT) **CBJ**

Right wing. Shoots right. 6'1", 200 lbs. Born, Mississauga, Ont., June 9, 1973. Toronto's 2nd choice, 23rd overall, in 1992 Entry Draft.

Season	Club	League	GP	G	A	Pts	PIM	PP	SH	GW	S	%	+/-	TF	F%	H	SB	Min	GP	G	A	Pts	PIM	PP	SH	GW
1989-90	Tor. Young Nats	MTHL	39	15	28	43	56	....	....	....	....	....	....	....	....	....	....	....	....	....	....	....	....	....	....	....
1990-91	Ottawa 67's	OHL	26	6	11	17	25	....	....	....	....	....	....	....	....	....	....	....	1	0	0	0	0	....	....	....
1991-92	Ottawa 67's	OHL	61	32	51	83	132	....	....	....	....	....	....	....	....	....	....	....	11	6	11	17	11	....	....	....
1992-93	Ottawa 67's	OHL	30	14	29	43	83	....	....	....	....	....	....	....	....	....	....	....	....	....	....	....	....	....	....	....
	Newmarket Royals	OHL	31	11	25	36	89	....	....	....	....	....	....	....	....	....	....	....	7	4	7	11	20	....	....	....
	St. John's	AHL	2	0	0	0	0	....	....	....	....	....	....	....	....	....	....	....	2	0	0	0	2	....	....	....
1993-94	St. John's	AHL	67	11	29	40	155	....	....	....	....	....	....	....	....	....	....	....	11	1	5	6	17	....	....	....
1994-95	Kalamazoo Wings	IHL	61	17	29	46	96	....	....	....	....	....	....	....	....	....	....	....	16	9	3	12	27	....	....	....
	Dallas	**NHL**	**2**	**0**	**1**	**1**	**0**	**0**	**0**	**0**	**0**	**0.0**	**1**	....	....	....	....	....	....	....	....	....	....	....	....	....
1995-96	**Dallas**	**NHL**	**70**	**9**	**19**	**28**	**111**	**0**	**0**	**0**	**62**	**14.5**	**0**	....	....	....	....	....	....	....	....	....	....	....	....	....
1996-97	**Dallas**	**NHL**	**56**	**6**	**4**	**10**	**98**	**0**	**0**	**0**	**0**	**0.0**	**5**	....	....	....	....	....	**5**	**0**	**2**	**2**	**8**	**0**	**0**	**0**
1997-98	**Dallas**	**NHL**	**72**	**9**	**10**	**19**	**96**	**3**	**0**	**1**	**91**	**9.9**	**-2**	....	....	....	....	....	**17**	**0**	**2**	**2**	***47**	**0**	**0**	**0**
1998-99♦	**Dallas**	**NHL**	**82**	**13**	**18**	**31**	**85**	**2**	**0**	**4**	**112**	**11.6**	**1**	**2**	**50.0**	**172**	**12**	**12:39**	**14**	**0**	**3**	**3**	**20**	**0**	**0**	**0**
99-2000	**Dallas**	**NHL**	**45**	**2**	**6**	**8**	**38**	**1**	**0**	**0**	**43**	**4.7**	**-5**	**3**	**0.0**	**111**	**6**	**11:19**	**14**	**0**	**1**	**1**	**4**	**0**	**0**	**0**
2000-01	**Dallas**	**NHL**	**75**	**13**	**24**	**37**	**64**	**4**	**0**	**1**	**93**	**14.0**	**1**	**16**	**56.3**	**186**	**13**	**11:05**	**9**	**0**	**0**	**0**	**0**	**0**	**0**	**0**
2001-02	**Columbus**	**NHL**	**81**	**15**	**18**	**33**	**86**	**6**	**0**	**4**	**152**	**9.9**	**-20**	**28**	**39.3**	**220**	**38**	**15:27**	....	....	....	....	....	....	....	....
	NHL Totals		**483**	**67**	**100**	**167**	**578**	**16**	**0**	**10**	**553**	**12.1**		**49**	**42.9**	**689**	**69**	**12:50**	**59**	**0**	**8**	**8**	**79**	**0**	**0**	**0**

• Missed majority of 1990-91 season recovering from neck injury suffered in game vs. Sudbury (OHL), December 4, 1990. Transferred to **Dallas** from **Toronto** with Peter Zezel as compensation for Toronto's signing of free agent Mike Craig, August 10, 1994. Traded to **Columbus** by **Dallas** for Columbus' 3rd round choice in 2002 Entry Draft, August 29, 2001.

MARSHALL, Jason

(MAHR-shahl, JAY-suhn) **MIN.**

Defense. Shoots right. 6'2", 200 lbs. Born, Cranbrook, B.C., February 22, 1971. St. Louis' 1st choice, 9th overall, in 1989 Entry Draft.

Season	Club	League	GP	G	A	Pts	PIM	PP	SH	GW	S	%	+/-	TF	F%	H	SB	Min	GP	G	A	Pts	PIM	PP	SH	GW
1987-88	Columbia Valley	RMJHL	40	4	28	32	150	....	....	....	....	....	....	....	....	....	....	....	....	....	....	....	....	....	....	....
1988-89	Vernon Lakers	BCJHL	48	10	30	40	197	....	....	....	....	....	....	....	....	....	....	....	31	6	6	12	14	....	....	....
1989-90	Team Canada	Nat-Tm	73	1	11	12	57	....	....	....	....	....	....	....	....	....	....	....	....	....	....	....	....	....	....	....
1990-91	Tri-City	WHL	59	10	34	44	236	....	....	....	....	....	....	....	....	....	....	....	7	1	2	3	20	....	....	....
	Peoria Rivermen	IHL	....	....	....	....	....	....	....	....	....	....	....	....	....	....	....	....	18	0	1	1	48	....	....	....
1991-92	**St. Louis**	**NHL**	**2**	**1**	**0**	**1**	**4**	**0**	**0**	**0**	**2**	**50.0**	**0**	....	....	....	....	....	....	....	....	....	....	....	....	....
	Peoria Rivermen	IHL	78	4	18	22	178	....	....	....	....	....	....	....	....	....	....	....	10	0	1	1	16	....	....	....
1992-93	Peoria Rivermen	IHL	77	4	16	20	229	....	....	....	....	....	....	....	....	....	....	....	4	0	0	0	20	....	....	....
1993-94	Team Canada	Nat-Tm	41	3	10	13	60	....	....	....	....	....	....	....	....	....	....	....	....	....	....	....	....	....	....	....
	Peoria Rivermen	IHL	20	1	1	2	72	....	....	....	....	....	....	....	....	....	....	....	3	2	0	2	2	....	....	....
1994-95	San Diego Gulls	IHL	80	7	18	25	218	....	....	....	....	....	....	....	....	....	....	....	5	0	1	1	8	....	....	....
	Anaheim	**NHL**	**1**	**0**	**0**	**0**	**0**	**0**	**0**	**0**	**1**	**0.0**	**-2**	....	....	....	....	....	....	....	....	....	....	....	....	....
1995-96	**Anaheim**	**NHL**	**24**	**0**	**1**	**1**	**42**	**0**	**0**	**0**	**9**	**0.0**	**3**	....	....	....	....	....	....	....	....	....	....	....	....	....
	Baltimore	AHL	57	1	13	14	150	....	....	....	....	....	....	....	....	....	....	....	....	....	....	....	....	....	....	....
1996-97	**Anaheim**	**NHL**	**73**	**1**	**9**	**10**	**140**	**0**	**0**	**0**	**34**	**2.9**	**6**	....	....	....	....	....	**7**	**0**	**1**	**1**	**4**	**0**	**0**	**0**
1997-98	**Anaheim**	**NHL**	**72**	**3**	**6**	**9**	**189**	**1**	**0**	**0**	**68**	**4.4**	**-8**	....	....	....	....	....	....	....	....	....	....	....	....	....
1998-99	**Anaheim**	**NHL**	**72**	**1**	**7**	**8**	**142**	**0**	**0**	**0**	**63**	**1.6**	**-5**	**0**	**0.0**	**150**	**95**	**19:06**	**4**	**1**	**0**	**1**	**10**	**1**	**0**	**0**
99-2000	**Anaheim**	**NHL**	**55**	**0**	**3**	**3**	**88**	**0**	**0**	**0**	**41**	**0.0**	**-10**	**2**	**50.0**	**143**	**57**	**16:33**	....	....	....	....	....	....	....	....
2000-01	**Anaheim**	**NHL**	**50**	**3**	**4**	**7**	**105**	**2**	**1**	**1**	**38**	**7.9**	**-12**	**1**	**0.0**	**122**	**39**	**14:36**	....	....	....	....	....	....	....	....
	Washington	**NHL**	**5**	**0**	**0**	**0**	**17**	**0**	**0**	**0**	**5**	**0.0**	**-1**	**0**	**0.0**	**16**	**4**	**11:48**	....	....	....	....	....	....	....	....
2001-02	**Minnesota**	**NHL**	**80**	**5**	**6**	**11**	**148**	**1**	**0**	**0**	**73**	**6.8**	**-8**	**0**	**0.0**	**108**	**85**	**16:18**	....	....	....	....	....	....	....	....
	NHL Totals		**434**	**14**	**36**	**50**	**875**	**4**	**1**	**1**	**334**	**4.2**		**3**	**33.3**	**539**	**280**	**16:43**	**11**	**1**	**1**	**2**	**14**	**1**	**0**	**0**

Traded to **Anaheim** by **St. Louis** for Bill Houlder, August 29, 1994. Traded to **Washington** by **Anaheim** for Alexei Tezikov and Edmonton's 4th round choice (previously acquired, Anaheim selected Brandon Rogers) in 2001 Entry Draft, March 13, 2001. Signed as a free agent by **Minnesota**, July 2, 2001.

MARTINEK, Radek

(MAHR-tih-nehk, RA-dehk) **NYI**

Defense. Shoots right. 6'1", 200 lbs. Born, Havlickuv Brod, Czech., August 31, 1976. NY Islanders' 12th choice, 228th overall, in 1999 Entry Draft.

Season	Club	League	GP	G	A	Pts	PIM	PP	SH	GW	S	%	+/-	TF	F%	H	SB	Min	GP	G	A	Pts	PIM	PP	SH	GW
1996-97	Ceske Budejovice	Czech	52	3	5	8	40	....	....	....	....	....	....	....	....	....	....	....	5	0	1	1	2	....	....	....
1997-98	Ceske Budejovice	Czech	42	2	7	9	36	....	....	....	....	....	....	....	....	....	....	....	....	....	....	....	....	....	....	....
1998-99	Ceske Budejovice	Czech	52	12	13	25	50	....	....	....	....	....	....	....	....	....	....	....	3	0	2	2	....	....	....	....
99-2000	Ceske Budejovice	Czech	45	5	18	23	24	....	....	....	....	....	....	....	....	....	....	....	3	0	0	0	6	....	....	....
2000-01	Ceske Budejovice	Czech	44	8	10	18	45	....	....	....	....	....	....	....	....	....	....	....	....	....	....	....	....	....	....	....
2001-02	**NY Islanders**	**NHL**	**23**	**1**	**4**	**5**	**16**	**0**	**0**	**1**	**25**	**4.0**	**5**	**0**	**0.0**	**37**	**24**	**21:07**	....	....	....	....	....	....	....	....
	NHL Totals		**23**	**1**	**4**	**5**	**16**	**0**	**0**	**1**	**25**	**4.0**		**0**	**0.0**	**37**	**24**	**21:07**	....	....	....	....	....	....	....	....

• Missed majority of 2001-02 season recovering from knee injury originally suffered in game vs. NY Rangers, November 11, 2001.

MARTINS, Steve

(MAHR-tihns, STEEV) **OTT.**

Center. Shoots left. 5'9", 175 lbs. Born, Gatineau, Que., April 13, 1972. Hartford's 1st choice, 5th overall, in 1994 Supplemental Draft.

Season	Club	League	GP	G	A	Pts	PIM	PP	SH	GW	S	%	+/-	TF	F%	H	SB	Min	GP	G	A	Pts	PIM	PP	SH	GW
1988-89	L'Outaouais	QAAA	38	18	33	51	70	....	....	....	....	....	....	....	....	....	....	....	....	....	....	....	....	....	....	....
1989-90	Choate-Rosemary	Hi-School	STATISTICS NOT AVAILABLE																							
1990-91	Choate-Rosemary	Hi-School	STATISTICS NOT AVAILABLE																							
1991-92	Harvard Crimson	ECAC	20	13	14	27	26	....	....	....	....	....	....	....	....	....	....	....	....	....	....	....	....	....	....	....
1992-93	Harvard Crimson	ECAC	18	6	8	14	40	....	....	....	....	....	....	....	....	....	....	....	....	....	....	....	....	....	....	....
1993-94	Harvard Crimson	ECAC	32	25	35	60	*93	....	....	....	....	....	....	....	....	....	....	....	....	....	....	....	....	....	....	....
1994-95	Harvard Crimson	ECAC	28	15	23	38	93	....	....	....	....	....	....	....	....	....	....	....	....	....	....	....	....	....	....	....
1995-96	**Hartford**	**NHL**	**23**	**1**	**3**	**4**	**8**	**0**	**0**	**0**	**27**	**3.7**	**-3**	....	....	....	....	....	....	....	....	....	....	....	....	....
	Springfield	AHL	30	9	20	29	10	....	....	....	....	....	....	....	....	....	....	....	....	....	....	....	....	....	....	....
1996-97	**Hartford**	**NHL**	**2**	**0**	**1**	**1**	**0**	**0**	**0**	**0**	**2**	**0.0**	**0**	....	....	....	....	....	....	....	....	....	....	....	....	....
	Springfield	AHL	63	12	31	43	78	....	....	....	....	....	....	....	....	....	....	....	17	1	3	4	26	....	....	....
1997-98	**Carolina**	**NHL**	**3**	**0**	**0**	**0**	**0**	**0**	**0**	**0**	**0**	**0.0**	**0**	....	....	....	....	....	....	....	....	....	....	....	....	....
	Chicago Wolves	IHL	78	20	41	61	122	....	....	....	....	....	....	....	....	....	....	....	21	6	14	20	28	....	....	....
1998-99	**Ottawa**	**NHL**	**36**	**4**	**3**	**7**	**10**	**1**	**0**	**1**	**27**	**14.8**	**4**	**191**	**56.0**	**23**	**2**	**8:28**	....	....	....	....	....	....	....	....
	Detroit Vipers	IHL	4	1	6	7	16	....	....	....	....	....	....	....	....	....	....	....	....	....	....	....	....	....	....	....
99-2000	**Ottawa**	**NHL**	**2**	**1**	**0**	**1**	**0**	**0**	**0**	**0**	**3**	**33.3**	**-1**	**3**	**0.0**	**2**	**2**	**11:10**	....	....	....	....	....	....	....	....
	Tampa Bay	**NHL**	**57**	**5**	**7**	**12**	**37**	**0**	**1**	**1**	**62**	**8.1**	**-11**	**806**	**50.9**	**56**	**36**	**13:27**	....	....	....	....	....	....	....	....
2000-01	**Tampa Bay**	**NHL**	**20**	**1**	**1**	**2**	**13**	**0**	**0**	**0**	**18**	**5.6**	**-9**	**184**	**52.2**	**19**	**3**	**9:35**	....	....	....	....	....	....	....	....
	Detroit Vipers	IHL	8	5	4	9	4	....	....	....	....	....	....	....	....	....	....	....	....	....	....	....	....	....	....	....
	NY Islanders	**NHL**	**39**	**1**	**3**	**4**	**20**	**0**	**1**	**0**	**28**	**3.6**	**-7**	**302**	**58.0**	**22**	**17**	**9:41**	....	....	....	....	....	....	....	....
	Chicago Wolves	IHL	5	1	2	3	0	....	....	....	....	....	....	....	....	....	....	....	16	1	6	7	22	....	....	....
2001-02	**Ottawa**	**NHL**	**14**	**1**	**0**	**1**	**4**	**0**	**0**	**0**	**11**	**9.1**	**1**	**121**	**55.4**	**7**	**6**	**9:33**	**2**	**0**	**0**	**0**	**0**	**0**	**0**	**0**
	Grand Rapids	AHL	51	10	21	31	66	....	....	....	....	....	....	....	....	....	....	....	3	0	0	0	0	....	....	....
	NHL Totals		**196**	**14**	**18**	**32**	**92**	**1**	**2**	**2**	**178**	**7.9**		**1607**	**53.2**	**129**	**66**	**10:42**	**2**	**0**	**0**	**0**	**0**	**0**	**0**	**0**

ECAC First All-Star Team (1994) • ECAC Player of the Year (1994) • NCAA East First All-American Team (1994) • NCAA Final Four All-Tournament Team (1994)

Transferred to **Carolina** after **Hartford** franchise relocated, June 25, 1997. Signed as a free agent by **Ottawa**, July 20, 1998. Claimed on waivers by **Tampa Bay** from **Ottawa**, October 29, 1999. Traded to **NY Islanders** by **Tampa Bay** for future considerations, January 3, 2001. Signed as a free agent by **Ottawa**, August 30, 2001.

MATTE, Christian

(MA-tay, KRIH-stan)

Right wing. Shoots right. 6', 190 lbs. Born, Hull, Que., January 20, 1975. Quebec's 8th choice, 153rd overall, in 1993 Entry Draft.

Season	Club	League	GP	G	A	Pts	PIM	PP	SH	GW	S	%	+/-	TF	F%	H	SB	Min	GP	G	A	Pts	PIM	PP	SH	GW
1991-92	Abitibi	QAAA	42	18	27	45	30	....	....	....	....	....	....	....	....	....	....	....	4	1	0	1	0	....	....	....
1992-93	Granby Bisons	QMJHL	68	17	36	53	59	....	....	....	....	....	....	....	....	....	....	....	....	....	....	....	....	....	....	....
1993-94	Granby Bisons	QMJHL	59	50	47	97	103	....	....	....	....	....	....	....	....	....	....	....	7	5	5	10	12	....	....	....
	Cornwall Aces	AHL	1	0	0	0	0	....	....	....	....	....	....	....	....	....	....	....	....	....	....	....	....	....	....	....
1994-95	Granby Bisons	QMJHL	66	50	66	116	86	....	....	....	....	....	....	....	....	....	....	....	13	11	7	18	12	....	....	....
	Cornwall Aces	AHL	....	....	....	....	....	....	....	....	....	....	....	....	....	....	....	....	3	0	1	1	2	....	....	....

Season	Club	League	GP	G	A	Pts	PIM	PP	SH	GW	S	%	+/-	TF	F%	H	SB	Min	GP	G	A	Pts	PIM	PP	SH	GW
			Regular Season																Playoffs							
1995-96	Cornwall Aces	AHL	64	20	32	52	51												7	1	1	2	6			
1996-97	**Colorado**	**NHL**	**5**	**1**	**1**	**2**	**0**	**0**	**0**	**0**	**6**	**16.7**	**1**													
	Hershey Bears	AHL	49	18	18	36	78												22	8	3	11	25			
1997-98	**Colorado**	**NHL**	**5**	**0**	**0**	**0**	**6**	**0**	**0**	**0**	**5**	**0.0**	**0**													
	Hershey Bears	AHL	71	33	40	73	109												7	3	2	5	4			
1998-99	**Colorado**	**NHL**	**7**	**1**	**1**	**2**	**0**	**0**	**0**	**0**	**9**	**11.1**	**–2**	**13**	**30.8**	**4**	**1**	**7:45**								
	Hershey Bears	AHL	60	31	47	78	48												5	2	1	3	8			
99-2000	**Colorado**	**NHL**	**5**	**0**	**1**	**1**	**4**	**0**	**0**	**0**	**1**	**0.0**	**–2**	**2**	**0.0**	**6**	**1**	**8:37**								
	Hershey Bears	AHL	73	43	*61	*104	85												14	8	6	14	10			
2000-01	**Minnesota**	**NHL**	**3**	**0**	**0**	**0**	**2**	**0**	**0**	**0**	**8**	**0.0**	**0**	**0**	**0.0**	**2**	**0**	**13:21**								
	Cleveland	IHL	58	*38	29	67	59												4	1	1	2	0			
2001-02	Rochester	AHL	72	22	29	51	48												2	0	0	0	0			
	NHL Totals		**25**	**2**	**3**	**5**	**12**	**0**	**0**	**0**	**29**	**6.9**		**15**	**26.7**	**12**	**2**	**9:09**								

QMJHL All-Rookie Team (1993) • QMJHL Second All-Star Team (1994) • AHL First All-Star Team (2000) • Won John P. Sollenberger Trophy (Top Scorer - AHL) (2000)

Rights transferred to **Colorado** after **Quebec** franchise relocated, June 21, 1995. Signed as a free agent by **Minnesota**, July 11, 2000. Signed as a free agent by **Buffalo**, August 2, 2001. Signed as a free agent by **Zurich** (Swiss), June 14, 2002.

MATTEAU, Stephane

(mah-TOH, STEH-fan) **FLA.**

Left wing. Shoots left. 6'4", 215 lbs. Born, Rouyn-Noranda, Que., September 2, 1969. Calgary's 2nd choice, 25th overall, in 1987 Entry Draft.

Season	Club	League	GP	G	A	Pts	PIM	PP	SH	GW	S	%	+/-	TF	F%	H	SB	Min	GP	G	A	Pts	PIM	PP	SH	GW
1985-86	Hull Olympiques	QMJHL	60	6	8	14	19												4	0	0	0	0			
1986-87	Hull Olympiques	QMJHL	69	27	48	75	113												8	3	7	10	8			
1987-88	Hull Olympiques	QMJHL	57	17	40	57	179												18	5	14	19	94			
1988-89	Hull Olympiques	QMJHL	59	44	45	89	202												9	8	6	14	30			
	Salt Lake	IHL																	9	0	4	4	13			
1989-90	Salt Lake	IHL	81	23	35	58	130												10	6	3	9	38			
1990-91	**Calgary**	**NHL**	**78**	**15**	**19**	**34**	**93**	**0**	**1**	**1**	**114**	**13.2**	**17**						**5**	**0**	**1**	**1**	**0**	**0**	**0**	**0**
1991-92	**Calgary**	**NHL**	**4**	**1**	**0**	**1**	**19**	**0**	**0**	**0**	**7**	**14.3**	**2**													
	Chicago	**NHL**	**20**	**5**	**8**	**13**	**45**	**1**	**0**	**0**	**31**	**16.1**	**3**						**18**	**4**	**6**	**10**	**24**	**1**	**1**	**0**
1992-93	**Chicago**	**NHL**	**79**	**15**	**18**	**33**	**98**	**2**	**0**	**4**	**95**	**15.8**	**6**						**3**	**0**	**1**	**1**	**2**	**0**	**0**	**0**
1993-94	**Chicago**	**NHL**	**65**	**15**	**16**	**31**	**55**	**2**	**0**	**2**	**113**	**13.3**	**10**													
	◆ **NY Rangers**	**NHL**	**12**	**4**	**3**	**7**	**2**	**1**	**0**	**0**	**22**	**18.2**	**5**						**23**	**6**	**3**	**9**	**20**	**1**	**0**	**2**
1994-95	**NY Rangers**	**NHL**	**41**	**3**	**5**	**8**	**25**	**0**	**0**	**0**	**37**	**8.1**	**–8**						**9**	**0**	**1**	**1**	**10**	**0**	**0**	**0**
1995-96	**NY Rangers**	**NHL**	**32**	**4**	**2**	**6**	**22**	**1**	**0**	**0**	**39**	**10.3**	**–4**													
	St. Louis	**NHL**	**46**	**7**	**13**	**20**	**65**	**3**	**0**	**2**	**70**	**10.0**	**–4**						**11**	**0**	**2**	**2**	**8**	**0**	**0**	**0**
1996-97	**St. Louis**	**NHL**	**74**	**16**	**20**	**36**	**50**	**1**	**2**	**2**	**98**	**16.3**	**11**						**5**	**0**	**0**	**0**	**0**	**0**	**0**	**0**
1997-98	**San Jose**	**NHL**	**73**	**15**	**14**	**29**	**60**	**1**	**0**	**2**	**79**	**19.0**	**4**						**4**	**0**	**1**	**1**	**0**	**0**	**0**	**0**
1998-99	**San Jose**	**NHL**	**68**	**8**	**15**	**23**	**73**	**0**	**0**	**0**	**72**	**11.1**	**2**	**13**	**38.5**	**58**	**20**	**13:34**	**5**	**0**	**0**	**0**	**6**	**0**	**0**	**0**
99-2000	**San Jose**	**NHL**	**69**	**12**	**12**	**24**	**61**	**0**	**0**	**3**	**73**	**16.4**	**–3**	**8**	**50.0**	**83**	**14**	**11:52**	**10**	**0**	**2**	**2**	**8**	**0**	**0**	**0**
2000-01	**San Jose**	**NHL**	**80**	**13**	**19**	**32**	**32**	**1**	**0**	**3**	**81**	**16.0**	**5**	**61**	**37.7**	**94**	**24**	**10:55**	**6**	**1**	**3**	**4**	**0**	**0**	**0**	**0**
2001-02	**San Jose**	**NHL**	**55**	**7**	**4**	**11**	**15**	**1**	**1**	**0**	**43**	**16.3**	**4**	**12**	**50.0**	**58**	**11**	**9:09**	**10**	**1**	**2**	**3**	**2**	**0**	**0**	**0**
	NHL Totals		**796**	**140**	**168**	**308**	**715**	**14**	**4**	**19**	**974**	**14.4**		**94**	**40.4**	**293**	**69**	**11:28**	**109**	**12**	**22**	**34**	**80**	**2**	**1**	**2**

• Missed majority of 1991-92 season recovering from thigh injury suffered in game vs. LA Kings, October 10, 1991. Traded to **Chicago** by **Calgary** for Trent Yawney, December 16, 1991. Traded to **NY Rangers** by **Chicago** with Brian Noonan for Tony Amonte and the rights to Matt Oates, March 21, 1994. Traded to **St. Louis** by **NY Rangers** for Ian Laperriere, December 28, 1995. Traded to **San Jose** by **St. Louis** for Darren Turcotte, July 24, 1997. Signed as a free agent by **Florida**, August 2, 2002.

MATTEUCCI, Mike

(ma-TEW-chee, MIGHK) **N.J.**

Defense. Shoots left. 6'2", 210 lbs. Born, Trail, B.C., December 27, 1971.

Season	Club	League	GP	G	A	Pts	PIM	PP	SH	GW	S	%	+/-	TF	F%	H	SB	Min	GP	G	A	Pts	PIM	PP	SH	GW
1991-92	Estevan Bruins	SJHL	STATISTICS NOT AVAILABLE																							
1992-93	Lake Superior	CCHA	19	1	3	4	16																			
1993-94	Lake Superior	CCHA	45	6	11	17	64																			
1994-95	Lake Superior	CCHA	38	3	11	14	52																			
1995-96	Lake Superior	CCHA	40	3	13	16	82																			
	Los Angeles	IHL	4	0	0	0	7																			
1996-97	Long Beach	IHL	81	4	4	8	254												18	0	1	1	42			
1997-98	Long Beach	IHL	79	1	7	8	258												17	0	2	2	57			
1998-99	Long Beach	IHL	79	3	9	12	253												8	0	1	1	12			
99-2000	Long Beach	IHL	64	0	4	4	170												6	0	0	0	16			
2000-01	**Minnesota**	**NHL**	**3**	**0**	**0**	**0**	**2**	**0**	**0**	**0**	**3**	**0.0**	**–2**	**1**	**100.0**	**6**	**0**	**11:29**								
	Cleveland	IHL	69	0	7	7	189												4	0	0	0	15			
2001-02	**Minnesota**	**NHL**	**3**	**0**	**0**	**0**	**2**	**0**	**0**	**0**	**0**	**0.0**	**1**	**0**	**0.0**	**5**	**4**	**10:06**								
	Houston Aeros	AHL	69	3	10	13	128												14	1	1	2	33			
	NHL Totals		**6**	**0**	**0**	**0**	**4**	**0**	**0**	**0**	**3**	**0.0**		**1**	**100.0**	**11**	**4**	**10:48**								

Signed as a free agent by **Edmonton**, September 10, 1998. Traded to **Boston** by **Edmonton** for Kay Whitmore, December 29, 1999. Signed as a free agent by **Minnesota**, July 20, 2000. Signed as a free agent by **New Jersey**, July 12, 2002.

MATVICHUK, Richard

(MAT-vih-chuhk, RIH-chahrd) **DAL.**

Defense. Shoots left. 6'2", 215 lbs. Born, Edmonton, Alta., February 5, 1973. Minnesota's 1st choice, 8th overall, in 1991 Entry Draft.

Season	Club	League	GP	G	A	Pts	PIM	PP	SH	GW	S	%	+/-	TF	F%	H	SB	Min	GP	G	A	Pts	PIM	PP	SH	GW
1988-89	Ft. Saskatchewan	AJHL	58	7	36	43	147																			
1989-90	Saskatoon Blades	WHL	56	8	24	32	126												10	2	8	10	16			
1990-91	Saskatoon Blades	WHL	68	13	36	49	117																			
1991-92	Saskatoon Blades	WHL	58	14	40	54	126												22	1	9	10	61			
1992-93	**Minnesota**	**NHL**	**53**	**2**	**3**	**5**	**26**	**1**	**0**	**0**	**51**	**3.9**	**–8**													
	Kalamazoo Wings	IHL	3	0	1	1	6																			
1993-94	**Dallas**	**NHL**	**25**	**0**	**3**	**3**	**22**	**0**	**0**	**0**	**18**	**0.0**	**1**						**7**	**1**	**1**	**2**	**12**	**1**	**0**	**0**
	Kalamazoo Wings	IHL	43	8	17	25	84																			
1994-95	**Dallas**	**NHL**	**14**	**0**	**2**	**2**	**14**	**0**	**0**	**0**	**21**	**0.0**	**–7**						**5**	**0**	**2**	**2**	**4**	**0**	**0**	**0**
	Kalamazoo Wings	IHL	17	0	6	6	16																			
1995-96	**Dallas**	**NHL**	**73**	**6**	**16**	**22**	**71**	**0**	**0**	**1**	**81**	**7.4**	**4**													
1996-97	**Dallas**	**NHL**	**57**	**5**	**7**	**12**	**87**	**0**	**2**	**0**	**83**	**6.0**	**1**						**7**	**0**	**1**	**1**	**20**	**0**	**0**	**0**
1997-98	**Dallas**	**NHL**	**74**	**3**	**15**	**18**	**63**	**0**	**0**	**0**	**71**	**4.2**	**7**						**16**	**1**	**1**	**2**	**14**	**0**	**0**	**0**
1998-99◆	**Dallas**	**NHL**	**64**	**3**	**9**	**12**	**51**	**1**	**0**	**0**	**54**	**5.6**	**23**	**0**	**0.0**	**186**	**153**	**21:19**	**22**	**1**	**5**	**6**	**20**	**0**	**0**	**0**
99-2000	**Dallas**	**NHL**	**70**	**4**	**21**	**25**	**42**	**0**	**0**	**1**	**73**	**5.5**	**7**	**0**	**0.0**	**208**	**150**	**24:27**	**23**	**2**	**5**	**7**	**14**	**0**	**0**	**0**
2000-01	**Dallas**	**NHL**	**78**	**4**	**16**	**20**	**62**	**2**	**0**	**1**	**85**	**4.7**	**5**	**1**	**100.0**	**233**	**145**	**22:53**	**10**	**0**	**0**	**0**	**14**	**0**	**0**	**0**
2001-02	**Dallas**	**NHL**	**82**	**9**	**12**	**21**	**52**	**4**	**0**	**2**	**109**	**8.3**	**11**	**1**	**0.0**	**237**	**169**	**23:47**								
	NHL Totals		**590**	**36**	**104**	**140**	**490**	**8**	**2**	**5**	**646**	**5.6**		**2**	**50.0**	**864**	**617**	**23:10**	**90**	**5**	**15**	**20**	**98**	**1**	**0**	**0**

WHL East First All-Star Team (1992)

Transferred to **Dallas** after **Minnesota** franchise relocated, June 9, 1993.

MAY, Brad

(MAY, BRAD) **PHX.**

Left wing. Shoots left. 6'1", 207 lbs. Born, Toronto, Ont., November 29, 1971. Buffalo's 1st choice, 14th overall, in 1990 Entry Draft.

Season	Club	League	GP	G	A	Pts	PIM	PP	SH	GW	S	%	+/-	TF	F%	H	SB	Min	GP	G	A	Pts	PIM	PP	SH	GW
1987-88	Markham	OMHA	31	22	37	59	58																			
	Markham	MTJHL	6	1	1	2	21																			
1988-89	Niagara Falls	OHL	65	8	14	22	304												17	0	1	1	55			
1989-90	Niagara Falls	OHL	61	32	58	90	223												16	9	13	22	64			
1990-91	Niagara Falls	OHL	34	37	32	69	93												14	11	14	25	53			
1991-92	**Buffalo**	**NHL**	**69**	**11**	**6**	**17**	**309**	**1**	**0**	**3**	**82**	**13.4**	**–12**						**7**	**1**	**4**	**5**	**2**	**0**	**0**	**1**
1992-93	**Buffalo**	**NHL**	**82**	**13**	**13**	**26**	**242**	**0**	**0**	**1**	**114**	**11.4**	**3**						**8**	**1**	**1**	**2**	**14**	**0**	**0**	**1**
1993-94	**Buffalo**	**NHL**	**84**	**18**	**27**	**45**	**171**	**3**	**0**	**3**	**166**	**10.8**	**–6**						**7**	**0**	**2**	**2**	**9**	**0**	**0**	**0**
1994-95	**Buffalo**	**NHL**	**33**	**3**	**3**	**6**	**87**	**1**	**0**	**0**	**42**	**7.1**	**5**						**4**	**0**	**0**	**0**	**2**	**0**	**0**	**0**
1995-96	**Buffalo**	**NHL**	**79**	**15**	**29**	**44**	**295**	**3**	**0**	**4**	**168**	**8.9**	**6**													
1996-97	**Buffalo**	**NHL**	**42**	**3**	**4**	**7**	**106**	**1**	**0**	**1**	**75**	**4.0**	**–8**						**10**	**1**	**1**	**2**	**32**	**0**	**0**	**0**
1997-98	**Buffalo**	**NHL**	**36**	**4**	**7**	**11**	**113**	**0**	**0**	**0**	**41**	**9.8**	**2**													
	Vancouver	**NHL**	**27**	**9**	**3**	**12**	**41**	**4**	**0**	**2**	**56**	**16.1**	**0**													
1998-99	**Vancouver**	**NHL**	**66**	**6**	**11**	**17**	**102**	**1**	**0**	**1**	**91**	**6.6**	**–14**	**8**	**12.5**	**109**	**14**	**13:04**								
99-2000	**Vancouver**	**NHL**	**59**	**9**	**7**	**16**	**90**	**0**	**0**	**3**	**66**	**13.6**	**–2**	**3**	**0.0**	**117**	**10**	**10:24**								

Season	Club	League	GP	G	A	Pts	PIM	PP	SH	GW	S	%	+/-	TF	F%	H	SB	Min	GP	G	A	Pts	PIM	PP	SH	GW
			Regular Season																Playoffs							
2000-01	**Phoenix**	**NHL**	**62**	**11**	**14**	**25**	**107**	**0**	**0**	**0**	**83**	**13.3**	**10**	**3**	**33.3**	**146**	**11**	**11:11**	....	....	....	....	....	....	....	....
2001-02	**Phoenix**	**NHL**	**72**	**10**	**12**	**22**	**95**	**1**	**0**	**3**	**105**	**9.5**	**11**	**3**	**0.0**	**125**	**6**	**12:04**	**5**	**0**	**0**	**0**	**0**	**0**	**0**	**0**
	NHL Totals		**711**	**112**	**136**	**248**	**1758**	**15**	**0**	**21**	**1089**	**10.3**		**17**	**11.8**	**497**	**41**	**11:44**	**41**	**3**	**8**	**11**	**59**	**0**	**0**	**2**

OHL Second All-Star Team (1990, 1991)

• Missed majority of 1990-91 season recovering from knee injury suffered at Team Canada Juniors evaluation camp, August 21, 1990. Traded to **Vancouver** by **Buffalo** with Buffalo's 3rd round choice (later traded to Tampa Bay - Tampa Bay selected Jimmie Olvestad) in 1999 Entry Draft for Geoff Sanderson, February 4, 1998. Traded to **Phoenix** by **Vancouver** for future considerations, June 24, 2000.

MAYERS, Jamal

(MAI-uhrz, JUH-MAHL) **ST.L.**

Center. Shoots right. 6'1", 212 lbs. Born, Toronto, Ont., October 24, 1974. St. Louis' 3rd choice, 89th overall, in 1993 Entry Draft.

Season	Club	League	GP	G	A	Pts	PIM	PP	SH	GW	S	%	+/-	TF	F%	H	SB	Min	GP	G	A	Pts	PIM	PP	SH	GW
1990-91	Thornhill	MTJHL	44	12	24	36	78	....	....	....	....	....	....	....	....	....	....	....	....	....	....	....	....	....	....	....
1991-92	Thornhill	MTJHL	56	38	69	107	36	....	....	....	....	....	....	....	....	....	....	....	....	....	....	....	....	....	....	....
1992-93	West-Michigan	CCHA	38	8	17	25	26	....	....	....	....	....	....	....	....	....	....	....	....	....	....	....	....	....	....	....
1993-94	West-Michigan	CCHA	40	17	32	49	40	....	....	....	....	....	....	....	....	....	....	....	....	....	....	....	....	....	....	....
1994-95	West-Michigan	CCHA	39	13	32	45	40	....	....	....	....	....	....	....	....	....	....	....	....	....	....	....	....	....	....	....
1995-96	West-Michigan	CCHA	38	17	22	39	75	....	....	....	....	....	....	....	....	....	....	....	....	....	....	....	....	....	....	....
1996-97	**St. Louis**	**NHL**	**6**	**0**	**1**	**1**	**2**	**0**	**0**	**0**	**7**	**0.0**	**–3**	....	....	....	....	....	....	....	....	....	....	....	....	....
	Worcester	AHL	62	12	14	26	104	....	....	....	....	....	....	....	....	....	....	....	5	4	5	9	4	....	....	....
1997-98	Worcester	AHL	61	19	24	43	117	....	....	....	....	....	....	....	....	....	....	....	11	3	4	7	10	....	....	....
1998-99	**St. Louis**	**NHL**	**34**	**4**	**5**	**9**	**40**	**0**	**0**	**0**	**48**	**8.3**	**–3**	**2**	**50.0**	**56**	**3**	**8:08**	**11**	**0**	**1**	**1**	**8**	**0**	**0**	**0**
	Worcester	AHL	20	9	7	16	34	....	....	....	....	....	....	....	....	....	....	....	....	....	....	....	....	....	....	....
99-2000	**St. Louis**	**NHL**	**79**	**7**	**10**	**17**	**90**	**0**	**0**	**0**	**99**	**7.1**	**0**	**77**	**52.0**	**150**	**11**	**9:46**	**7**	**0**	**4**	**4**	**2**	**0**	**0**	**0**
2000-01	**St. Louis**	**NHL**	**77**	**8**	**13**	**21**	**117**	**0**	**0**	**0**	**132**	**6.1**	**–3**	**273**	**51.3**	**126**	**7**	**11:04**	**15**	**2**	**3**	**5**	**8**	**0**	**0**	**0**
2001-02	**St. Louis**	**NHL**	**77**	**9**	**8**	**17**	**99**	**0**	**1**	**0**	**105**	**8.6**	**9**	**761**	**52.6**	**132**	**25**	**11:36**	**10**	**3**	**0**	**3**	**2**	**0**	**0**	**2**
	NHL Totals		**273**	**28**	**37**	**65**	**348**	**0**	**1**	**0**	**391**	**7.2**		**1113**	**52.2**	**464**	**46**	**10:28**	**43**	**5**	**8**	**13**	**20**	**0**	**0**	**2**

McALLISTER, Chris

(mih-KAL-ihs-tuhr, KRIHS) **PHI.**

Defense. Shoots left. 6'8", 240 lbs. Born, Saskatoon, Sask., June 16, 1975. Vancouver's 1st choice, 40th overall, in 1995 Entry Draft.

Season	Club	League	GP	G	A	Pts	PIM	PP	SH	GW	S	%	+/-	TF	F%	H	SB	Min	GP	G	A	Pts	PIM	PP	SH	GW
1992-93	Saskatoon Royals	NSJHL	40	14	14	28	224	....	....	....	....	....	....	....	....	....	....	....	....	....	....	....	....	....	....	....
	Saskatoon Blades	WHL	4	0	0	0	2	....	....	....	....	....	....	....	....	....	....	....	....	....	....	....	....	....	....	....
1993-94	Humboldt Broncos	SJHL	50	3	5	8	150	....	....	....	....	....	....	....	....	....	....	....	....	....	....	....	....	....	....	....
	Saskatoon Blades	WHL	2	0	0	0	5	....	....	....	....	....	....	....	....	....	....	....	....	....	....	....	....	....	....	....
1994-95	Saskatoon Blades	WHL	65	2	8	10	134	....	....	....	....	....	....	....	....	....	....	....	10	0	0	0	28	....	....	....
1995-96	Syracuse Crunch	AHL	68	0	2	2	142	....	....	....	....	....	....	....	....	....	....	....	16	0	0	0	34	....	....	....
1996-97	Syracuse Crunch	AHL	43	3	1	4	108	....	....	....	....	....	....	....	....	....	....	....	3	0	0	0	6	....	....	....
1997-98	**Vancouver**	**NHL**	**36**	**1**	**2**	**3**	**106**	**0**	**0**	**0**	**15**	**6.7**	**–12**	....	....	....	....	....	....	....	....	....	....	....	....	....
	Syracuse Crunch	AHL	23	0	1	1	71	....	....	....	....	....	....	....	....	....	....	....	5	0	0	0	21	....	....	....
1998-99	**Vancouver**	**NHL**	**28**	**1**	**1**	**2**	**63**	**0**	**0**	**0**	**6**	**16.7**	**–7**	**0**	**0.0**	**19**	**5**	**5:53**	....	....	....	....	....	....	....	....
	Syracuse Crunch	AHL	5	0	0	0	24	....	....	....	....	....	....	....	....	....	....	....	....	....	....	....	....	....	....	....
	Toronto	**NHL**	**20**	**0**	**2**	**2**	**39**	**0**	**0**	**0**	**12**	**0.0**	**4**	**0**	**0.0**	**37**	**17**	**13:59**	**6**	**0**	**1**	**1**	**4**	**0**	**0**	**0**
99-2000	**Toronto**	**NHL**	**36**	**0**	**3**	**3**	**68**	**0**	**0**	**0**	**12**	**0.0**	**–4**	**0**	**0.0**	**65**	**26**	**12:02**	....	....	....	....	....	....	....	....
2000-01	**Philadelphia**	**NHL**	**60**	**2**	**2**	**4**	**124**	**0**	**0**	**0**	**33**	**6.1**	**1**	**0**	**0.0**	**80**	**74**	**11:36**	**2**	**0**	**0**	**0**	**0**	**0**	**0**	**0**
2001-02	**Philadelphia**	**NHL**	**42**	**0**	**5**	**5**	**113**	**0**	**0**	**0**	**26**	**0.0**	**–7**	**0**	**0.0**	**55**	**40**	**9:12**	....	....	....	....	....	....	....	....
	NHL Totals		**222**	**4**	**15**	**19**	**513**	**0**	**0**	**0**	**104**	**3.8**		**0**	**0.0**	**256**	**162**	**10:32**	**8**	**0**	**1**	**1**	**4**	**0**	**0**	**0**

Traded to **Toronto** by **Vancouver** for Darby Hendrickson, February 16, 1999. Traded to **Philadelphia** by **Toronto** for the rights to Regan Kelly, September 26, 2000.

McALPINE, Chris

(mih-KAL-pighn, KRIHS)

Defense. Shoots right. 6', 210 lbs. Born, Roseville, MN, December 1, 1971. New Jersey's 10th choice, 137th overall, in 1990 Entry Draft.

Season	Club	League	GP	G	A	Pts	PIM	PP	SH	GW	S	%	+/-	TF	F%	H	SB	Min	GP	G	A	Pts	PIM	PP	SH	GW
1989-90	Roseville High	Hi-School	25	15	13	28	....	....	....	....	....	....	....	....	....	....	....	....	....	....	....	....	....	....	....	....
1990-91	U. of Minnesota	WCHA	38	7	9	16	112	....	....	....	....	....	....	....	....	....	....	....	....	....	....	....	....	....	....	....
1991-92	U. of Minnesota	WCHA	39	3	9	12	126	....	....	....	....	....	....	....	....	....	....	....	....	....	....	....	....	....	....	....
1992-93	U. of Minnesota	WCHA	41	14	9	23	82	....	....	....	....	....	....	....	....	....	....	....	....	....	....	....	....	....	....	....
1993-94	U. of Minnesota	WCHA	36	12	18	30	121	....	....	....	....	....	....	....	....	....	....	....	....	....	....	....	....	....	....	....
1994-95	Albany	AHL	48	4	18	22	49	....	....	....	....	....	....	....	....	....	....	....	....	....	....	....	....	....	....	....
	♦ New Jersey	**NHL**	**24**	**0**	**3**	**3**	**17**	**0**	**0**	**0**	**19**	**0.0**	**4**	....	....	....	....	....	....	....	....	....	....	....	....	....
1995-96	Albany	AHL	57	5	14	19	72	....	....	....	....	....	....	....	....	....	....	....	4	0	0	0	13	....	....	....
1996-97	Albany	AHL	44	1	9	10	48	....	....	....	....	....	....	....	....	....	....	....	....	....	....	....	....	....	....	....
	St. Louis	**NHL**	**15**	**0**	**0**	**0**	**24**	**0**	**0**	**0**	**3**	**0.0**	**–2**	....	....	....	....	....	**4**	**0**	**1**	**1**	**0**	**0**	**0**	**0**
1997-98	**St. Louis**	**NHL**	**54**	**3**	**7**	**10**	**36**	**0**	**0**	**0**	**35**	**8.6**	**14**	....	....	....	....	....	**10**	**0**	**0**	**0**	**16**	**0**	**0**	**0**
1998-99	**St. Louis**	**NHL**	**51**	**1**	**1**	**2**	**50**	**0**	**0**	**0**	**56**	**1.8**	**–10**	**0**	**0.0**	**75**	**42**	**12:51**	**13**	**0**	**0**	**0**	**2**	**0**	**0**	**0**
99-2000	**St. Louis**	**NHL**	**21**	**1**	**1**	**2**	**14**	**0**	**0**	**0**	**25**	**4.0**	**1**	**0**	**0.0**	**28**	**14**	**11:41**	....	....	....	....	....	....	....	....
	Worcester	AHL	10	1	4	5	4	....	....	....	....	....	....	....	....	....	....	....	....	....	....	....	....	....	....	....
	Tampa Bay	**NHL**	**10**	**1**	**1**	**2**	**10**	**0**	**0**	**0**	**5**	**20.0**	**–5**	**0**	**0.0**	**15**	**12**	**18:23**	....	....	....	....	....	....	....	....
	Detroit Vipers	IHL	8	0	0	0	6	....	....	....	....	....	....	....	....	....	....	....	....	....	....	....	....	....	....	....
	Atlanta	**NHL**	**3**	**0**	**0**	**0**	**2**	**0**	**0**	**0**	**4**	**0.0**	**–4**	**0**	**0.0**	**7**	**4**	**19:40**	....	....	....	....	....	....	....	....
2000-01	**Chicago**	**NHL**	**50**	**0**	**6**	**6**	**32**	**0**	**0**	**0**	**61**	**0.0**	**5**	**2**	**50.0**	**74**	**60**	**17:41**	....	....	....	....	....	....	....	....
	Norfolk Admirals	AHL	13	4	7	11	6	....	....	....	....	....	....	....	....	....	....	....	....	....	....	....	....	....	....	....
2001-02	**Chicago**	**NHL**	**40**	**0**	**3**	**3**	**36**	**0**	**0**	**0**	**40**	**0.0**	**8**	**0**	**0.0**	**86**	**43**	**15:22**	**1**	**0**	**0**	**0**	**0**	**0**	**0**	**0**
	Norfolk Admirals	AHL	8	0	4	4	4	....	....	....	....	....	....	....	....	....	....	....	....	....	....	....	....	....	....	....
	NHL Totals		**268**	**6**	**22**	**28**	**221**	**0**	**0**	**0**	**248**	**2.4**		**2**	**50.0**	**285**	**175**	**15:06**	**28**	**0**	**1**	**1**	**18**	**0**	**0**	**0**

WCHA First All-Star Team (1994) • NCAA West Second All-American Team (1994)

Traded to **St. Louis** by **New Jersey** with New Jersey's 9th round choice (James Desmarais) in 1999 Entry Draft for Peter Zezel, February 11, 1997. Traded to **Tampa Bay** by **St. Louis** with Rich Parent for Stephane Richer, January 13, 2000. Traded to **Atlanta** by **Tampa Bay** for Mikko Kuparinen, March 11, 2000. Signed as a free agent by **Chicago**, July 27, 2000.

McAMMOND, Dean

(MIHK-AM-uhnd, DEEN) **CGY.**

Center. Shoots left. 5'11", 200 lbs. Born, Grand Cache, Alta., June 15, 1973. Chicago's 1st choice, 22nd overall, in 1991 Entry Draft.

Season	Club	League	GP	G	A	Pts	PIM	PP	SH	GW	S	%	+/-	TF	F%	H	SB	Min	GP	G	A	Pts	PIM	PP	SH	GW
1988-89	St. Albert	AMHL	36	33	44	77	132	....	....	....	....	....	....	....	....	....	....	....	....	....	....	....	....	....	....	....
1989-90	Prince Albert	WHL	53	11	11	22	49	....	....	....	....	....	....	....	....	....	....	....	14	2	3	5	18	....	....	....
1990-91	Prince Albert	WHL	71	33	35	68	108	....	....	....	....	....	....	....	....	....	....	....	2	0	1	1	6	....	....	....
1991-92	Prince Albert	WHL	63	37	54	91	189	....	....	....	....	....	....	....	....	....	....	....	10	12	11	23	26	....	....	....
	Chicago	**NHL**	**5**	**0**	**2**	**2**	**0**	**0**	**0**	**0**	**4**	**0.0**	**–2**	....	....	....	....	....	**3**	**0**	**0**	**0**	**2**	**0**	**0**	**0**
1992-93	Prince Albert	WHL	30	19	29	48	44	....	....	....	....	....	....	....	....	....	....	....	....	....	....	....	....	....	....	....
	Swift Current	WHL	18	10	13	23	24	....	....	....	....	....	....	....	....	....	....	....	17	*16	19	35	20	....	....	....
1993-94	**Edmonton**	**NHL**	**45**	**6**	**21**	**27**	**16**	**2**	**0**	**0**	**52**	**11.5**	**12**	....	....	....	....	....	....	....	....	....	....	....	....	....
	Cape Breton	AHL	28	9	12	21	38	....	....	....	....	....	....	....	....	....	....	....	....	....	....	....	....	....	....	....
1994-95	**Edmonton**	**NHL**	**6**	**0**	**0**	**0**	**0**	**0**	**0**	**0**	**3**	**0.0**	**–1**	....	....	....	....	....	....	....	....	....	....	....	....	....
1995-96	**Edmonton**	**NHL**	**53**	**15**	**15**	**30**	**23**	**4**	**0**	**0**	**79**	**19.0**	**6**	....	....	....	....	....	....	....	....	....	....	....	....	....
	Cape Breton	AHL	22	9	15	24	55	....	....	....	....	....	....	....	....	....	....	....	....	....	....	....	....	....	....	....
1996-97	**Edmonton**	**NHL**	**57**	**12**	**17**	**29**	**28**	**4**	**0**	**6**	**106**	**11.3**	**–15**	....	....	....	....	....	....	....	....	....	....	....	....	....
1997-98	**Edmonton**	**NHL**	**77**	**19**	**31**	**50**	**46**	**8**	**0**	**3**	**128**	**14.8**	**9**	....	....	....	....	....	**12**	**1**	**4**	**5**	**12**	**0**	**0**	**0**
1998-99	**Edmonton**	**NHL**	**65**	**9**	**16**	**25**	**36**	**1**	**0**	**0**	**122**	**7.4**	**5**	**26**	**38.5**	**116**	**21**	**14:15**	....	....	....	....	....	....	....	....
	Chicago	**NHL**	**12**	**1**	**4**	**5**	**2**	**0**	**0**	**1**	**16**	**6.3**	**3**	**37**	**48.6**	**22**	**4**	**15:43**	....	....	....	....	....	....	....	....
99-2000	**Chicago**	**NHL**	**76**	**14**	**18**	**32**	**72**	**1**	**0**	**1**	**118**	**11.9**	**11**	**257**	**39.7**	**121**	**35**	**16:25**	....	....	....	....	....	....	....	....
2000-01	**Chicago**	**NHL**	**61**	**10**	**16**	**26**	**43**	**1**	**0**	**1**	**95**	**10.5**	**4**	**23**	**43.5**	**105**	**20**	**15:30**	....	....	....	....	....	....	....	....
	Philadelphia	**NHL**	**10**	**1**	**1**	**2**	**0**	**1**	**0**	**0**	**17**	**5.9**	**–1**	**65**	**46.2**	**12**	**2**	**11:60**	**4**	**0**	**0**	**0**	**2**	**0**	**0**	**0**
2001-02	**Calgary**	**NHL**	**73**	**21**	**30**	**51**	**60**	**7**	**0**	**4**	**152**	**13.8**	**2**	**143**	**55.2**	**65**	**29**	**18:56**	....	....	....	....	....	....	....	....
	NHL Totals		**540**	**108**	**171**	**279**	**326**	**29**	**0**	**16**	**892**	**12.1**		**551**	**45.2**	**441**	**111**	**16:12**	**19**	**1**	**4**	**5**	**16**	**0**	**0**	**0**

Traded to **Edmonton** by **Chicago** with Igor Kravchuk for Joe Murphy, February 24, 1993. Traded to **Chicago** by **Edmonton** with Boris Mironov and Jonas Elofsson for Chad Kilger, Daniel Cleary, Ethan Moreau and Christian Laflamme, March 20, 1999. Traded to **Philadelphia** by **Chicago** for Philadelphia's 3rd round choice (later traded to Toronto - Toronto selected Nicolas Corbeil) in 2001 Entry Draft, March 13, 2001. Traded to **Calgary** by **Philadelphia** for Calgary's 4th round choice (Rosario Ruggeri) in 2002 Entry Draft, June 24, 2001.

McCABE, Bryan

(mih-KAYB, BRIGH-uhn) **TOR.**

Defense. Shoots left. 6'2", 213 lbs. Born, St. Catharines, Ont., June 8, 1975. NY Islanders' 2nd choice, 40th overall, in 1993 Entry Draft.

Season	Club	League	GP	G	A	Pts	PIM	PP	SH	GW	S	%	+/-	TF	F%	H	SB	Min	GP	G	A	Pts	PIM	PP	SH	GW
			Regular Season																Playoffs							
1990-91	Calgary Canucks	AMHL	33	14	34	48	55																			
1991-92	Medicine Hat	WHL	68	6	24	30	157												4	0	0	0	6			
1992-93	Medicine Hat	WHL	14	0	13	13	83																			
	Spokane Chiefs	WHL	46	3	44	47	134												6	1	5	6	28			
1993-94	Spokane Chiefs	WHL	64	22	62	84	218												3	0	4	4	4			
1994-95	Spokane Chiefs	WHL	42	14	39	53	115																			
	Brandon	WHL	20	6	10	16	38												18	4	13	17	59			
1995-96	**NY Islanders**	**NHL**	**82**	**7**	**16**	**23**	**156**	**3**	**0**	**1**	**130**	**5.4**	**-24**													
1996-97	**NY Islanders**	**NHL**	**82**	**8**	**20**	**28**	**165**	**2**	**1**	**2**	**117**	**6.8**	**-2**													
1997-98	**NY Islanders**	**NHL**	**56**	**3**	**9**	**12**	**145**	**1**	**0**	**0**	**81**	**3.7**	**9**													
	Vancouver	**NHL**	**26**	**1**	**11**	**12**	**64**	**0**	**1**	**0**	**42**	**2.4**	**10**													
1998-99	**Vancouver**	**NHL**	**69**	**7**	**14**	**21**	**120**	**1**	**2**	**0**	**98**	**7.1**	**-11**	**1**	**0.0**	**107**	**117**	**24:13**								
99-2000	**Chicago**	**NHL**	**79**	**6**	**19**	**25**	**139**	**2**	**0**	**2**	**119**	**5.0**	**-8**	**1**	**0.0**	**196**	**109**	**23:23**								
2000-01	**Toronto**	**NHL**	**82**	**5**	**24**	**29**	**123**	**3**	**0**	**2**	**159**	**3.1**	**16**	**0**	**0.0**	**202**	**110**	**23:49**	**11**	**2**	**3**	**5**	**16**	**1**	**0**	**0**
2001-02	**Toronto**	**NHL**	**82**	**17**	**26**	**43**	**129**	**8**	**0**	**1**	**157**	**10.8**	**16**	**1**	**0.0**	**250**	**122**	**24:34**	**20**	**5**	**5**	**10**	**30**	**3**	**0**	**1**
	NHL Totals		**558**	**54**	**139**	**193**	**1041**	**20**	**4**	**8**	**903**	**6.0**		**3**	**0.0**	**755**	**458**	**23:60**	**31**	**7**	**8**	**15**	**46**	**4**	**0**	**1**

WHL West Second All-Star Team (1993) • WHL West First All-Star Team (1994) • WHL East First All-Star Team (1995) • Memorial Cup All-Star Team (1995)

Traded to **Vancouver** by **NY Islanders** with Todd Bertuzzi and NY Islanders' 3rd round choice (Jarkko Ruutu) in 1998 Entry Draft for Trevor Linden, February 6, 1998. Traded to **Chicago** by **Vancouver** with Vancouver's 1st round choice (Pavel Vorobiev) in 2000 Entry Draft for Chicago's 1st round choice (later traded to Tampa Bay - later traded to NY Rangers - NY Rangers selected Pavel Brendl) in 1999 Entry Draft, June 25, 1999. Traded to **Toronto** by **Chicago** for Alexander Karpovtsev and Toronto's 4th round choice (Vladimir Gusev) in 2001 Entry Draft, October 2, 2000.

McCARTHY, Sandy

(mih-KAHR-thee, SAN-dee) **NYR**

Right wing. Shoots right. 6'3", 225 lbs. Born, Toronto, Ont., June 15, 1972. Calgary's 3rd choice, 52nd overall, in 1991 Entry Draft.

Season	Club	League	GP	G	A	Pts	PIM	PP	SH	GW	S	%	+/-	TF	F%	H	SB	Min	GP	G	A	Pts	PIM	PP	SH	GW
1987-88	Midland	OJHL-C	18	2	1	3	70																			
1988-89	Hawkesbury	OCJHL	42	4	11	15	139																			
1989-90	Laval Titan	QMJHL	65	10	11	21	269												14	3	3	6	60			
1990-91	Laval Titan	QMJHL	68	21	19	40	297												13	6	5	11	67			
1991-92	Laval Titan	QMJHL	62	39	51	90	326												8	4	5	9	81			
1992-93	Salt Lake	IHL	77	18	20	38	220																			
1993-94	**Calgary**	**NHL**	**79**	**5**	**5**	**10**	**173**	**0**	**0**	**0**	**39**	**12.8**	**-3**						**7**	**0**	**0**	**0**	**34**	**0**	**0**	**0**
1994-95	**Calgary**	**NHL**	**37**	**5**	**3**	**8**	**101**	**0**	**0**	**2**	**29**	**17.2**	**1**						**6**	**0**	**1**	**1**	**17**	**0**	**0**	**0**
1995-96	**Calgary**	**NHL**	**75**	**9**	**7**	**16**	**173**	**3**	**0**	**1**	**98**	**9.2**	**-8**						**4**	**0**	**0**	**0**	**10**	**0**	**0**	**0**
1996-97	**Calgary**	**NHL**	**33**	**3**	**5**	**8**	**113**	**1**	**0**	**1**	**38**	**7.9**	**-8**													
1997-98	**Calgary**	**NHL**	**52**	**8**	**5**	**13**	**170**	**1**	**0**	**1**	**68**	**11.8**	**-18**													
	Tampa Bay	**NHL**	**14**	**0**	**5**	**5**	**71**	**0**	**0**	**0**	**26**	**0.0**	**-1**													
1998-99	**Tampa Bay**	**NHL**	**67**	**5**	**7**	**12**	**135**	**1**	**0**	**0**	**89**	**5.6**	**-22**	**0**	**0.0**	**118**	**13**	**11:02**								
	Philadelphia	**NHL**	**13**	**0**	**1**	**1**	**25**	**0**	**0**	**0**	**18**	**0.0**	**-2**	**2**	**50.0**	**26**	**2**	**11:09**	**6**	**0**	**1**	**1**	**0**	**0**	**0**	**0**
99-2000	**Philadelphia**	**NHL**	**58**	**6**	**5**	**11**	**111**	**1**	**0**	**0**	**68**	**8.8**	**-5**	**4**	**0.0**	**121**	**5**	**10:27**								
	Carolina	**NHL**	**13**	**0**	**0**	**0**	**9**	**0**	**0**	**0**	**12**	**0.0**	**2**	**0**	**0.0**	**20**	**1**	**7:24**								
2000-01	**NY Rangers**	**NHL**	**81**	**11**	**10**	**21**	**171**	**0**	**0**	**2**	**95**	**11.6**	**3**	**4**	**0.0**	**206**	**29**	**10:25**								
2001-02	**NY Rangers**	**NHL**	**82**	**10**	**13**	**23**	**171**	**1**	**0**	**3**	**90**	**11.1**	**-8**	**6**	**50.0**	**137**	**29**	**8:29**								
	NHL Totals		**604**	**62**	**66**	**128**	**1423**	**8**	**0**	**10**	**670**	**9.3**		**16**	**25.0**	**628**	**79**	**9:57**	**23**	**0**	**2**	**2**	**61**	**0**	**0**	**0**

Traded to **Tampa Bay** by **Calgary** with Calgary's 3rd (Brad Richards) and 5th (Curtis Rich) round choices in 1998 Entry Draft for Jason Wiemer, March 24, 1998. Traded to **Philadelphia** by **Tampa Bay** with Mikael Andersson for Colin Forbes and Philadelphia's 4th round choice (Michal Lanicek) in 1999 Entry Draft, March 20, 1999. Traded to **Carolina** by **Philadelphia** for Kent Manderville, March 14, 2000. Traded to **NY Rangers** by **Carolina** with Carolina's' 4th round choice (Bryce Lampman) in 2001 Entry Draft for Darren Langdon and Rob DiMaio, August 4, 2000.

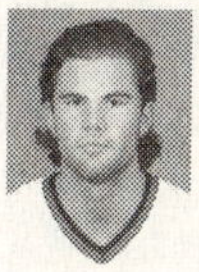

McCARTHY, Steve

(mih-KAHR-thee, STEEV) **CHI.**

Defense. Shoots left. 6', 197 lbs. Born, Trail, B.C., February 3, 1981. Chicago's 1st choice, 23rd overall, in 1999 Entry Draft.

Season	Club	League	GP	G	A	Pts	PIM	PP	SH	GW	S	%	+/-	TF	F%	H	SB	Min	GP	G	A	Pts	PIM	PP	SH	GW
1996-97	Trail Smokies	BCHL	57	25	52	77	81																			
	Edmonton Ice	WHL	2	0	0	0	0																			
1997-98	Edmonton Ice	WHL	58	11	29	40	59																			
1998-99	Kootenay Ice	WHL	57	19	33	52	79												6	0	5	5	8			
99-2000	**Chicago**	**NHL**	**5**	**1**	**1**	**2**	**4**	**1**	**0**	**0**	**4**	**25.0**	**0**	**0**	**0.0**	**4**	**2**	**15:09**								
	Kootenay Ice	WHL	37	13	23	36	36																			
2000-01	**Chicago**	**NHL**	**44**	**0**	**5**	**5**	**8**	**0**	**0**	**0**	**32**	**0.0**	**-7**	**0**	**0.0**	**33**	**42**	**14:47**								
	Norfolk Admirals	AHL	7	0	4	4	2																			
2001-02	**Chicago**	**NHL**	**3**	**0**	**0**	**0**	**2**	**0**	**0**	**0**	**2**	**0.0**	**-1**	**0**	**0.0**	**2**	**2**	**11:48**								
	Norfolk Admirals	AHL	77	7	21	28	37												2	0	3	3	2			
	NHL Totals		**52**	**1**	**6**	**7**	**14**	**1**	**0**	**0**	**38**	**2.6**		**0**	**0.0**	**39**	**46**	**14:39**								

Returned to **Kootenay** (WHL) by **Chicago**, October 18, 1999.

McCARTY, Darren

(mih-KAHR-tee, DAIR-ehn) **DET.**

Right wing. Shoots right. 6'1", 210 lbs. Born, Burnaby, B.C., April 1, 1972. Detroit's 2nd choice, 46th overall, in 1992 Entry Draft.

Season	Club	League	GP	G	A	Pts	PIM	PP	SH	GW	S	%	+/-	TF	F%	H	SB	Min	GP	G	A	Pts	PIM	PP	SH	GW
1988-89	Peterboro B's	OJHL-B	34	18	17	35	135																			
1989-90	Belleville Bulls	OHL	63	12	15	27	142												11	1	1	2	21			
1990-91	Belleville Bulls	OHL	60	30	37	67	151												6	2	2	4	13			
1991-92	Belleville Bulls	OHL	65	*55	72	127	177												5	1	4	5	13			
1992-93	Adirondack	AHL	73	17	19	36	278												11	0	1	1	33			
1993-94	**Detroit**	**NHL**	**67**	**9**	**17**	**26**	**181**	**0**	**0**	**2**	**81**	**11.1**	**12**						**7**	**2**	**2**	**4**	**8**	**0**	**0**	**0**
1994-95	**Detroit**	**NHL**	**31**	**5**	**8**	**13**	**88**	**1**	**0**	**2**	**27**	**18.5**	**5**						**18**	**3**	**2**	**5**	**14**	**0**	**0**	**0**
1995-96	**Detroit**	**NHL**	**63**	**15**	**14**	**29**	**158**	**8**	**0**	**1**	**102**	**14.7**	**14**						**19**	**3**	**2**	**5**	**20**	**0**	**0**	**1**
1996-97♦	**Detroit**	**NHL**	**68**	**19**	**30**	**49**	**126**	**5**	**0**	**6**	**171**	**11.1**	**14**						**20**	**3**	**4**	**7**	**34**	**0**	**0**	**2**
1997-98♦	**Detroit**	**NHL**	**71**	**15**	**22**	**37**	**157**	**5**	**1**	**2**	**166**	**9.0**	**0**						**22**	**3**	**8**	**11**	**34**	**0**	**0**	**1**
1998-99	**Detroit**	**NHL**	**69**	**14**	**26**	**40**	**108**	**6**	**0**	**1**	**140**	**10.0**	**10**	**15**	**33.3**	**217**	**34**	**17:04**	**10**	**1**	**1**	**2**	**23**	**0**	**0**	**0**
99-2000	**Detroit**	**NHL**	**24**	**6**	**6**	**12**	**48**	**0**	**0**	**1**	**40**	**15.0**	**1**	**1**	**0.0**	**129**	**7**	**13:40**	**9**	**0**	**1**	**1**	**12**	**0**	**0**	**0**
2000-01	**Detroit**	**NHL**	**72**	**12**	**10**	**22**	**123**	**1**	**1**	**3**	**118**	**10.2**	**-5**	**26**	**53.9**	**235**	**24**	**13:26**	**6**	**1**	**0**	**1**	**2**	**0**	**0**	**0**
2001-02♦	**Detroit**	**NHL**	**62**	**5**	**7**	**12**	**98**	**0**	**0**	**1**	**74**	**6.8**	**2**	**26**	**38.5**	**170**	**12**	**11:47**	**23**	**4**	**4**	**8**	**34**	**0**	**0**	**1**
	NHL Totals		**527**	**100**	**140**	**240**	**1087**	**26**	**2**	**19**	**919**	**10.9**		**68**	**42.6**	**751**	**77**	**14:07**	**134**	**20**	**24**	**44**	**181**	**0**	**0**	**5**

OHL First All-Star Team (1992)

• Missed majority of 1999-2000 season recovering from hernia injury suffered in game vs. Dallas, November 10, 1999.

McCAULEY, Alyn

(mih-KAW-lee, AL-ihn) **TOR.**

Center. Shoots left. 5'11", 190 lbs. Born, Brockville, Ont., May 29, 1977. New Jersey's 5th choice, 79th overall, in 1995 Entry Draft.

Season	Club	League	GP	G	A	Pts	PIM	PP	SH	GW	S	%	+/-	TF	F%	H	SB	Min	GP	G	A	Pts	PIM	PP	SH	GW
1991-92	Kingston	OCJHL	37	5	17	22	6																			
1992-93	Kingston	OCJHL	38	31	29	60	18																			
1993-94	Ottawa 67's	OHL	38	13	23	36	10												13	5	14	19	4			
1994-95	Ottawa 67's	OHL	65	16	38	54	20																			
1995-96	Ottawa 67's	OHL	55	34	48	82	24												2	0	0	0	0			
1996-97	Ottawa 67's	OHL	50	*56	56	112	16												22	14	22	36	14			
	St. John's	AHL																	3	0	1	1	0			
1997-98	**Toronto**	**NHL**	**60**	**6**	**10**	**16**	**6**	**0**	**0**	**1**	**77**	**7.8**	**-7**													
1998-99	**Toronto**	**NHL**	**39**	**9**	**15**	**24**	**2**	**1**	**0**	**1**	**76**	**11.8**	**7**	**591**	**46.4**	**10**	**4**	**15:10**								
99-2000	**Toronto**	**NHL**	**45**	**5**	**5**	**10**	**10**	**1**	**0**	**0**	**41**	**12.2**	**-6**	**450**	**47.8**	**22**	**11**	**10:46**	**5**	**0**	**0**	**0**	**6**	**0**	**0**	**0**
	St. John's	AHL	5	1	1	2	0																			
2000-01	**Toronto**	**NHL**	**14**	**1**	**0**	**1**	**0**	**0**	**0**	**0**	**13**	**7.7**	**0**	**139**	**46.8**	**5**	**3**	**10:28**	**10**	**0**	**0**	**0**	**2**	**0**	**0**	**0**
	St. John's	AHL	47	16	28	44	12																			
2001-02	**Toronto**	**NHL**	**82**	**6**	**10**	**16**	**18**	**0**	**1**	**1**	**95**	**6.3**	**10**	**951**	**48.1**	**38**	**26**	**11:24**	**20**	**5**	**10**	**15**	**4**	**1**	**0**	**2**
	NHL Totals		**240**	**27**	**40**	**67**	**36**	**2**	**1**	**3**	**302**	**8.9**		**2131**	**47.4**	**75**	**44**	**11:59**	**35**	**5**	**10**	**15**	**12**	**1**	**0**	**2**

OHL First All-Star Team (1996, 1997) • OHL MVP (1996, 1997) • Canadian Major Junior First All-Star Team (1997) • Canadian Major Junior Player of the Year (1997)

Rights traded to **Toronto** by **New Jersey** with Jason Smith and Steve Sullivan for Doug Gilmour, Dave Ellett and New Jersey's 3rd round choice (previously acquired, New Jersey selected Andre Lakos) in 1999 Entry Draft, February 25, 1997.

McDONALD, Andy

(mihk-DAW-nuhld, AN-dee) **ANA.**

Center. Shoots left. 5'10", 186 lbs. Born, Strathroy, Ont., August 25, 1977.

Season	Club	League	GP	G	A	Pts	PIM	PP	SH	GW	S	%	+/-	TF	F%	H	SB	Min	Playoffs GP	G	A	Pts	PIM	PP	SH	GW
1993-94	Strathroy	OJHL-B	7	2	2	4	0																			
1994-95	Strathroy	OJHL-B	50	32	41	73	24																			
1995-96	Strathroy	OJHL-B	52	31	56	87	103																			
1996-97	Colgate	ECAC	33	9	10	19	16																			
1997-98	Colgate	ECAC	35	13	19	32	26																			
1998-99	Colgate	ECAC	35	20	26	46	42																			
99-2000	Colgate	ECAC	34	25	*33	*58	49																			
2000-01	**Anaheim**	**NHL**	**16**	**1**	**0**	**1**	**6**	**0**	**0**	**0**	**21**	**4.8**	**0**	**139**	**48.9**	**13**	**4**	**11:11**								
	Cincinnati	AHL	46	15	25	40	21												3	0	1	1	2			
2001-02	**Anaheim**	**NHL**	**53**	**7**	**21**	**28**	**10**	**2**	**0**	**3**	**79**	**8.9**	**2**	**818**	**53.7**	**52**	**14**	**15:59**								
	Cincinnati	AHL	21	7	25	32	6																			
	NHL Totals		**69**	**8**	**21**	**29**	**16**	**2**	**0**	**3**	**100**	**8.0**		**957**	**53.0**	**65**	**18**	**14:52**								

OJHL-B Player of the Year (1996) • ECAC Second All-Star Team (1999) • ECAC First All-Star Team (2000) • NCAA East First All-American Team (2000)

Signed as a free agent by **Anaheim**, April 3, 2000.

McEACHERN, Shawn

(muh-GEH-kruhn, SHAWN) **ATL.**

Left wing. Shoots left. 5'11", 200 lbs. Born, Waltham, MA, February 28, 1969. Pittsburgh's 6th choice, 110th overall, in 1987 Entry Draft.

Season	Club	League	GP	G	A	Pts	PIM	PP	SH	GW	S	%	+/-	TF	F%	H	SB	Min	Playoffs GP	G	A	Pts	PIM	PP	SH	GW
1985-86	Matignon	Hi-School	20	32	20	52																				
1986-87	Matignon	Hi-School	16	29	28	57																				
1987-88	Matignon	Hi-School	22	52	40	92																				
1988-89	Boston University	H-East	36	20	28	48	32																			
1989-90	Boston University	H-East	43	25	31	56	78																			
1990-91	Boston University	H-East	41	34	48	82	43																			
1991-92	Team USA	Nat-Tm	57	26	23	49	38																			
	United States	Olympics	8	1	0	1	10																			
	♦ **Pittsburgh**	**NHL**	**15**	**0**	**4**	**4**	**0**	**0**	**0**	**0**	**14**	**0.0**	**1**						**19**	**2**	**7**	**9**	**4**	**0**	**0**	**0**
1992-93	**Pittsburgh**	**NHL**	**84**	**28**	**33**	**61**	**46**	**7**	**0**	**6**	**196**	**14.3**	**21**						**12**	**3**	**2**	**5**	**10**	**0**	**0**	**1**
1993-94	**Los Angeles**	**NHL**	**49**	**8**	**13**	**21**	**24**	**0**	**3**	**0**	**81**	**9.9**	**1**													
	Pittsburgh	**NHL**	**27**	**12**	**9**	**21**	**10**	**0**	**2**	**1**	**78**	**15.4**	**13**						**6**	**1**	**0**	**1**	**2**	**0**	**0**	**0**
1994-95	Kiekko Espoo	Finland	8	1	3	4	6																			
	Pittsburgh	**NHL**	**44**	**13**	**13**	**26**	**22**	**1**	**2**	**1**	**97**	**13.4**	**4**						**11**	**0**	**2**	**2**	**8**	**0**	**0**	**0**
1995-96	**Boston**	**NHL**	**82**	**24**	**29**	**53**	**34**	**3**	**2**	**3**	**238**	**10.1**	**−5**						**5**	**2**	**1**	**3**	**8**	**0**	**0**	**0**
1996-97	**Ottawa**	**NHL**	**65**	**11**	**20**	**31**	**18**	**0**	**1**	**2**	**150**	**7.3**	**−5**						**7**	**2**	**0**	**2**	**8**	**1**	**0**	**0**
1997-98	**Ottawa**	**NHL**	**81**	**24**	**24**	**48**	**42**	**8**	**2**	**4**	**229**	**10.5**	**1**						**11**	**0**	**4**	**4**	**8**	**0**	**0**	**0**
1998-99	**Ottawa**	**NHL**	**77**	**31**	**25**	**56**	**46**	**7**	**0**	**4**	**223**	**13.9**	**8**	**441**	**48.5**	**37**	**23**	**18:45**	**4**	**2**	**0**	**2**	**6**	**1**	**0**	**0**
99-2000	**Ottawa**	**NHL**	**69**	**29**	**22**	**51**	**24**	**10**	**0**	**4**	**219**	**13.2**	**2**	**54**	**50.0**	**45**	**13**	**17:50**	**6**	**0**	**3**	**3**	**4**	**0**	**0**	**0**
2000-01	**Ottawa**	**NHL**	**82**	**32**	**40**	**72**	**62**	**9**	**0**	**1**	**231**	**13.9**	**10**	**420**	**48.8**	**82**	**23**	**18:25**	**4**	**0**	**2**	**2**	**2**	**0**	**0**	**0**
2001-02	**Ottawa**	**NHL**	**80**	**15**	**31**	**46**	**52**	**5**	**0**	**3**	**196**	**7.7**	**9**	**210**	**52.4**	**62**	**24**	**17:40**	**12**	**0**	**4**	**4**	**2**	**0**	**0**	**0**
	NHL Totals		**755**	**227**	**263**	**490**	**380**	**50**	**12**	**29**	**1952**	**11.6**		**1125**	**49.4**	**226**	**83**	**18:11**	**97**	**12**	**25**	**37**	**62**	**2**	**0**	**1**

Hockey East Second All-Star Team (1990) • Hockey East First All-Star Team (1991) • NCAA East First All-American Team (1991)

Traded to **LA Kings** by **Pittsburgh** for Marty McSorley, August 27, 1993. Traded to **Pittsburgh** by **LA Kings** with Tomas Sandstrom for Marty McSorley and Jim Paek, February 16, 1994. Traded to **Boston** by **Pittsburgh** with Kevin Stevens for Glen Murray, Bryan Smolinski and Boston's 3rd round choice (Boyd Kane) in 1996 Entry Draft, August 2, 1995. Traded to **Ottawa** by **Boston** for Trent McCleary and Ottawa's 3rd round choice (Eric Naud) in 1996 Entry Draft, June 22, 1996. Traded to **Atlanta** by **Ottawa** with Ottawa's 6th round choice in 2004 Entry Draft for Brian Pothier, June 29, 2002. Traded to **Ottawa** by **Atlanta** for Shawn McEachern and Ottawa's 6th round choice in 2004 Entry Draft, June 29, 2002.

McGILLIS, Dan

(MIHK-gihl-his, DAN) **PHI.**

Defense. Shoots left. 6'2", 230 lbs. Born, Hawkesbury, Ont., July 1, 1972. Detroit's 10th choice, 238th overall, in 1992 Entry Draft.

Season	Club	League	GP	G	A	Pts	PIM	PP	SH	GW	S	%	+/-	TF	F%	H	SB	Min	Playoffs GP	G	A	Pts	PIM	PP	SH	GW
1989-90	Hawkesbury	OCJHL	55	2	1	3	52																			
1990-91	Hawkesbury	OCJHL	56	8	22	30	92																			
1991-92	Hawkesbury	OCJHL	36	5	19	24	106																			
1992-93	Northeastern	H-East	35	5	12	17	42																			
1993-94	Northeastern	H-East	38	4	25	29	82																			
1994-95	Northeastern	H-East	34	9	22	31	70																			
1995-96	Northeastern	H-East	34	12	24	36	50																			
1996-97	**Edmonton**	**NHL**	**73**	**6**	**16**	**22**	**52**	**2**	**1**	**2**	**139**	**4.3**	**2**						**12**	**0**	**5**	**5**	**24**	**0**	**0**	**0**
1997-98	**Edmonton**	**NHL**	**67**	**10**	**15**	**25**	**74**	**5**	**0**	**3**	**119**	**8.4**	**−17**													
	Philadelphia	**NHL**	**13**	**1**	**5**	**6**	**35**	**1**	**0**	**0**	**18**	**5.6**	**−4**						**5**	**1**	**2**	**3**	**10**	**1**	**0**	**0**
1998-99	**Philadelphia**	**NHL**	**78**	**8**	**37**	**45**	**61**	**6**	**0**	**4**	**164**	**4.9**	**16**	**0**	**0.0**	**220**	**71**	**21:41**	**6**	**0**	**1**	**1**	**12**	**0**	**0**	**0**
99-2000	**Philadelphia**	**NHL**	**68**	**4**	**14**	**18**	**55**	**3**	**0**	**1**	**128**	**3.1**	**16**	**0**	**0.0**	**264**	**63**	**20:04**	**18**	**2**	**6**	**8**	**12**	**0**	**0**	**0**
2000-01	**Philadelphia**	**NHL**	**82**	**14**	**35**	**49**	**86**	**4**	**0**	**4**	**207**	**6.8**	**13**	**1**	**0.0**	**292**	**127**	**23:23**	**6**	**1**	**0**	**1**	**6**	**0**	**1**	**0**
2001-02	**Philadelphia**	**NHL**	**75**	**5**	**14**	**19**	**46**	**2**	**0**	**1**	**147**	**3.4**	**17**	**0**	**0.0**	**246**	**75**	**21:04**	**5**	**1**	**0**	**1**	**8**	**1**	**0**	**0**
	NHL Totals		**456**	**48**	**136**	**184**	**409**	**23**	**1**	**15**	**922**	**5.2**		**1**	**0.0**	**1022**	**336**	**21:38**	**52**	**5**	**14**	**19**	**72**	**2**	**1**	**0**

Hockey East First All-Star Team (1995, 1996) • NCAA East First All-American Team (1996)

Traded to **Edmonton** by **Detroit** for Kirk Maltby, March 20, 1996. Traded to **Philadelphia** by **Edmonton** with Edmonton's 2nd round choice (Jason Beckett) in 1998 Entry Draft for Janne Niinimaa, March 24, 1998.

McINNIS, Marty

(MAK-ih-nihs, MAHR-tee) **BOS.**

Left wing. Shoots right. 5'11", 187 lbs. Born, Hingham, MA, June 2, 1970. NY Islanders' 10th choice, 163rd overall, in 1988 Entry Draft.

Season	Club	League	GP	G	A	Pts	PIM	PP	SH	GW	S	%	+/-	TF	F%	H	SB	Min	Playoffs GP	G	A	Pts	PIM	PP	SH	GW
1986-87	Milton Academy	Hi-School	25	21	19	40																				
1987-88	Milton Academy	Hi-School	25	26	25	51																				
1988-89	Boston College	H-East	39	13	19	32	8																			
1989-90	Boston College	H-East	41	24	29	53	43																			
1990-91	Boston College	H-East	38	21	36	57	40																			
1991-92	Team USA	Nat-Tm	54	15	19	34	20																			
	United States	Olympics	8	5	2	7	4																			
	NY Islanders	**NHL**	**15**	**3**	**5**	**8**	**0**	**0**	**0**	**0**	**24**	**12.5**	**6**													
1992-93	**NY Islanders**	**NHL**	**56**	**10**	**20**	**30**	**24**	**0**	**1**	**0**	**60**	**16.7**	**7**						**3**	**0**	**1**	**1**	**0**	**0**	**0**	**0**
	Capital District	AHL	10	4	12	16	2																			
1993-94	**NY Islanders**	**NHL**	**81**	**25**	**31**	**56**	**24**	**3**	**5**	**3**	**136**	**18.4**	**31**						**4**	**0**	**0**	**0**	**0**	**0**	**0**	**0**
1994-95	**NY Islanders**	**NHL**	**41**	**9**	**7**	**16**	**8**	**0**	**0**	**1**	**68**	**13.2**	**−1**													
1995-96	**NY Islanders**	**NHL**	**74**	**12**	**34**	**46**	**39**	**2**	**0**	**1**	**167**	**7.2**	**−11**													
1996-97	**NY Islanders**	**NHL**	**70**	**20**	**22**	**42**	**20**	**4**	**1**	**4**	**163**	**12.3**	**−7**													
	Calgary	**NHL**	**10**	**3**	**4**	**7**	**2**	**1**	**0**	**0**	**19**	**15.8**	**−1**													
1997-98	**Calgary**	**NHL**	**75**	**19**	**25**	**44**	**34**	**5**	**4**	**0**	**128**	**14.8**	**1**													
1998-99	**Calgary**	**NHL**	**6**	**1**	**1**	**2**	**6**	**0**	**0**	**0**	**7**	**14.3**	**−1**	**28**	**32.1**	**5**	**0**	**13:51**								
	Anaheim	**NHL**	**75**	**18**	**34**	**52**	**36**	**11**	**1**	**5**	**139**	**12.9**	**−14**	**390**	**46.9**	**52**	**15**	**18:59**	**4**	**2**	**0**	**2**	**2**	**2**	**0**	**0**
99-2000	**Anaheim**	**NHL**	**62**	**10**	**18**	**28**	**26**	**2**	**1**	**2**	**129**	**7.8**	**−4**	**355**	**50.1**	**60**	**9**	**19:04**								
2000-01	**Anaheim**	**NHL**	**75**	**20**	**22**	**42**	**40**	**10**	**0**	**1**	**136**	**14.7**	**−21**	**605**	**47.1**	**57**	**17**	**18:30**								
2001-02	**Anaheim**	**NHL**	**60**	**9**	**14**	**23**	**25**	**2**	**0**	**0**	**131**	**6.9**	**−14**	**150**	**48.7**	**34**	**8**	**15:18**								
	Boston	**NHL**	**19**	**2**	**3**	**5**	**8**	**0**	**0**	**1**	**26**	**7.7**	**−1**	**204**	**41.2**	**11**	**2**	**17:42**	**6**	**0**	**1**	**1**	**0**	**0**	**0**	**0**
	NHL Totals		**719**	**161**	**240**	**401**	**292**	**40**	**13**	**18**	**1333**	**12.1**		**1732**	**46.9**	**219**	**51**	**17:57**	**17**	**2**	**2**	**4**	**2**	**2**	**0**	**0**

Traded to **Calgary** by **NY Islanders** with Tyrone Garner and Calgary's 6th round choice (previously acquired, Calgary selected Ilja Demidov) in 1997 Entry Draft for Robert Reichel, March 18, 1997. Traded to **Chicago** by **Calgary** with Eric Andersson and Jamie Allison for Jeff Shantz and Steve Dubinsky, October 27, 1998. Traded to **Anaheim** by **Chicago** for Toronto's 4th round choice (previously acquired, later traded to Washington - Washington selected Ryan Vanbuskirk) in 2000 Entry Draft, October 27, 1998. Traded to **Boston by Anaheim** for Boston's 3rd round choice (later traded to Nashville - later traded to Detroit - Detroit selected Valtteri Filppula) in 2002 Entry Draft, March 6, 2002.

McKAY, Randy (mih-KAY, RAN-dee) MTL.

Right wing. Shoots right. 6'2", 210 lbs. Born, Montreal, Que., January 25, 1967. Detroit's 6th choice, 113th overall, in 1985 Entry Draft.

Season	Club	League	GP	G	A	Pts	PIM	PP	SH	GW	S	%	+/-	TF	F%	H	SB	Min	PO GP	PO G	PO A	PO Pts	PO PIM	PO PP	PO SH	PO GW
			Regular Season																Playoffs							
1983-84	Lac St-Louis	QAAA	38	18	28	46	62												11	6	10	16	8			
1984-85	Michigan Tech	WCHA	25	4	5	9	32																			
1985-86	Michigan Tech	WCHA	40	12	22	34	46																			
1986-87	Michigan Tech	WCHA	39	5	11	16	46																			
1987-88	Michigan Tech	WCHA	41	17	24	41	70																			
	Adirondack	AHL	10	0	3	3	12												6	0	4	4	0			
1988-89	**Detroit**	**NHL**	**3**	**0**	**0**	**0**	**0**	**0**	**0**	**0**	**2**	**0.0**	**–1**						**2**	**0**	**0**	**0**	**2**	**0**	**0**	**0**
	Adirondack	AHL	58	29	34	63	170												14	4	7	11	60			
1989-90	**Detroit**	**NHL**	**33**	**3**	**6**	**9**	**51**	**0**	**0**	**0**	**33**	**9.1**	**1**													
	Adirondack	AHL	36	16	23	39	99												6	3	0	3	35			
1990-91	**Detroit**	**NHL**	**47**	**1**	**7**	**8**	**183**	**0**	**0**	**0**	**22**	**4.5**	**–15**						**5**	**0**	**1**	**1**	**41**	**0**	**0**	**0**
1991-92	**New Jersey**	**NHL**	**80**	**17**	**16**	**33**	**246**	**2**	**0**	**1**	**111**	**15.3**	**6**						**7**	**1**	**3**	**4**	**10**	**1**	**0**	**0**
1992-93	**New Jersey**	**NHL**	**73**	**11**	**11**	**22**	**206**	**1**	**0**	**2**	**94**	**11.7**	**0**						**5**	**0**	**0**	**0**	**16**	**0**	**0**	**0**
1993-94	**New Jersey**	**NHL**	**78**	**12**	**15**	**27**	**244**	**0**	**0**	**1**	**77**	**15.6**	**24**						**20**	**1**	**2**	**3**	**24**	**0**	**0**	**0**
1994-95♦	**New Jersey**	**NHL**	**33**	**5**	**7**	**12**	**44**	**0**	**0**	**0**	**44**	**11.4**	**10**						**19**	**8**	**4**	**12**	**11**	**2**	**0**	**2**
1995-96	**New Jersey**	**NHL**	**76**	**11**	**10**	**21**	**145**	**3**	**0**	**3**	**97**	**11.3**	**7**													
1996-97	**New Jersey**	**NHL**	**77**	**9**	**18**	**27**	**109**	**0**	**0**	**2**	**92**	**9.8**	**15**						**10**	**1**	**1**	**2**	**0**	**0**	**0**	**0**
1997-98	**New Jersey**	**NHL**	**74**	**24**	**24**	**48**	**86**	**8**	**0**	**5**	**141**	**17.0**	**30**						**6**	**0**	**1**	**1**	**0**	**0**	**0**	**0**
1998-99	**New Jersey**	**NHL**	**70**	**17**	**20**	**37**	**143**	**3**	**0**	**5**	**136**	**12.5**	**10**	**1**	**0.0**	**124**	**11**	**16:06**	**7**	**3**	**2**	**5**	**2**	**0**	**0**	**1**
99-2000♦	**New Jersey**	**NHL**	**67**	**16**	**23**	**39**	**80**	**3**	**0**	**4**	**116**	**13.8**	**8**	**2**	**0.0**	**144**	**16**	**15:39**	**23**	**0**	**6**	**6**	**9**	**0**	**0**	**0**
2000-01	**New Jersey**	**NHL**	**77**	**23**	**20**	**43**	**50**	**12**	**0**	**5**	**120**	**19.2**	**3**	**2**	**50.0**	**114**	**15**	**13:58**	**19**	**6**	**3**	**9**	**8**	**2**	**0**	**1**
2001-02	**New Jersey**	**NHL**	**55**	**6**	**7**	**13**	**65**	**3**	**0**	**1**	**63**	**9.5**	**2**	**2**	**0.0**	**128**	**12**	**13:15**								
	Dallas	**NHL**	**14**	**1**	**4**	**5**	**7**	**0**	**0**	**0**	**10**	**10.0**	**2**	**1**	**0.0**	**44**	**5**	**12:02**								
	NHL Totals		**857**	**156**	**188**	**344**	**1659**	**35**	**0**	**29**	**1158**	**13.5**		**8**	**12.5**	**554**	**59**	**14:40**	**123**	**20**	**23**	**43**	**123**	**5**	**0**	**4**

Transferred to **New Jersey** by **Detroit** with Dave Barr as compensation for Detroit's signing of free agent Troy Crowder, September 9, 1991. Traded to **Dallas** by **New Jersey** Jason Arnott and New Jersey's 1st round choice (later traded to Columbus - later traded to Buffalo - Buffalo selected Dan Paille) in 2002 Entry Draft for Joe Nieuwendyk and Jamie Langenbrunner, March 19, 2002. Signed as a free agent by **Montreal**, July 4, 2002.

McKEE, Jay (mih-KEE, JAY) BUF.

Defense. Shoots left. 6'4", 212 lbs. Born, Kingston, Ont., September 8, 1977. Buffalo's 1st choice, 14th overall, in 1995 Entry Draft.

Season	Club	League	GP	G	A	Pts	PIM	PP	SH	GW	S	%	+/-	TF	F%	H	SB	Min	PO GP	PO G	PO A	PO Pts	PO PIM	PO PP	PO SH	PO GW
1991-92	Mimico Monarchs	OJHL-B	39	7	9	16	67																			
	Kingston	MTJHL	1	0	0	0	0																			
1992-93	Ernestown Jets	OJHL-C	36	0	17	17	37																			
	Kingston	MTJHL	2	0	0	0	0																			
1993-94	Sudbury Wolves	OHL	51	0	1	1	51												3	0	0	0	0			
1994-95	Sudbury Wolves	OHL	39	6	6	12	91																			
	Niagara Falls	OHL	26	3	13	16	60												6	2	3	5	10			
1995-96	Niagara Falls	OHL	64	5	41	46	129												10	1	5	6	16			
	Buffalo	**NHL**	**1**	**0**	**1**	**1**	**2**	**0**	**0**	**0**	**2**	**0.0**	**1**													
	Rochester	AHL	4	0	1	1	15																			
1996-97	**Buffalo**	**NHL**	**43**	**1**	**9**	**10**	**35**	**0**	**0**	**0**	**29**	**3.4**	**3**						**3**	**0**	**0**	**0**	**0**	**0**	**0**	**0**
	Rochester	AHL	7	2	5	7	4																			
1997-98	**Buffalo**	**NHL**	**56**	**1**	**13**	**14**	**42**	**0**	**0**	**0**	**55**	**1.8**	**–1**						**1**	**0**	**0**	**0**	**0**	**0**	**0**	**0**
	Rochester	AHL	13	1	7	8	11																			
1998-99	**Buffalo**	**NHL**	**72**	**0**	**6**	**6**	**75**	**0**	**0**	**0**	**57**	**0.0**	**20**	**0**	**0.0**	**204**	**129**	**20:28**	**21**	**0**	**3**	**3**	**24**	**0**	**0**	**0**
99-2000	**Buffalo**	**NHL**	**78**	**5**	**12**	**17**	**50**	**1**	**0**	**1**	**84**	**6.0**	**5**	**0**	**0.0**	**153**	**170**	**20:58**	**1**	**0**	**0**	**0**	**0**	**0**	**0**	**0**
2000-01	**Buffalo**	**NHL**	**74**	**1**	**10**	**11**	**76**	**0**	**0**	**0**	**62**	**1.6**	**9**	**2**	**0.0**	**190**	**133**	**19:24**	**8**	**1**	**0**	**1**	**6**	**0**	**0**	**1**
2001-02	**Buffalo**	**NHL**	**81**	**2**	**11**	**13**	**43**	**0**	**0**	**1**	**50**	**4.0**	**18**	**0**	**0.0**	**143**	**175**	**19:26**								
	NHL Totals		**405**	**10**	**62**	**72**	**323**	**1**	**0**	**2**	**339**	**2.9**		**2**	**0.0**	**690**	**607**	**20:03**	**34**	**1**	**3**	**4**	**30**	**0**	**0**	**1**

OHL Second All-Star Team (1996)

McKENNA, Steve (mih-KEHN-ah, STEEV) PIT.

Left wing. Shoots left. 6'8", 255 lbs. Born, Toronto, Ont., August 21, 1973.

Season	Club	League	GP	G	A	Pts	PIM	PP	SH	GW	S	%	+/-	TF	F%	H	SB	Min	PO GP	PO G	PO A	PO Pts	PO PIM	PO PP	PO SH	PO GW
1991-92	Cambridge	OJHL-B	48	21	23	44	173																			
1992-93	Notre Dame	SJHL	STATISTICS NOT AVAILABLE																							
1993-94	Merrimack	H-East	37	1	2	3	74																			
1994-95	Merrimack	H-East	37	1	9	10	74																			
1995-96	Merrimack	H-East	33	3	11	14	67																			
1996-97	**Los Angeles**	**NHL**	**9**	**0**	**0**	**0**	**37**	**0**	**0**	**0**	**6**	**0.0**	**1**													
	Phoenix	IHL	66	6	5	11	187																			
1997-98	**Los Angeles**	**NHL**	**62**	**4**	**4**	**8**	**150**	**1**	**0**	**0**	**42**	**9.5**	**–9**						**3**	**0**	**1**	**1**	**8**	**0**	**0**	**0**
	Fredericton	AHL	6	2	1	3	48																			
1998-99	**Los Angeles**	**NHL**	**20**	**1**	**0**	**1**	**36**	**0**	**0**	**0**	**12**	**8.3**	**–3**	**0**	**0.0**	**35**	**4**	**8:24**								
99-2000	**Los Angeles**	**NHL**	**46**	**0**	**5**	**5**	**125**	**0**	**0**	**0**	**14**	**0.0**	**3**	**1**	**0.0**	**39**	**5**	**4:53**								
2000-01	**Minnesota**	**NHL**	**20**	**1**	**1**	**2**	**19**	**0**	**0**	**0**	**12**	**8.3**	**0**	**0**	**0.0**	**26**	**5**	**7:59**								
	Pittsburgh	**NHL**	**34**	**0**	**0**	**0**	**100**	**0**	**0**	**0**	**7**	**0.0**	**–4**	**0**	**0.0**	**20**	**2**	**3:16**								
2001-02	**NY Rangers**	**NHL**	**54**	**2**	**1**	**3**	**144**	**1**	**0**	**1**	**17**	**11.8**	**0**	**2**	**0.0**	**33**	**6**	**3:59**								
	Hartford	AHL	3	0	0	0	11																			
	NHL Totals		**245**	**8**	**11**	**19**	**611**	**2**	**0**	**1**	**110**	**7.3**		**3**	**0.0**	**153**	**22**	**5:03**	**3**	**0**	**1**	**1**	**8**	**0**	**0**	**0**

Signed as a free agent by **LA Kings**, May 23, 1996. Selected by **Minnesota** from **LA Kings** in Expansion Draft, June 23, 2000. Traded to **Pittsburgh** by **Minnesota** for Roman Simicek, January 13, 2001. Signed as a free agent by **NY Rangers**, August 28, 2001. Signed as a free agent by **Pittsburgh**, July 12, 2002.

McKENZIE, Jim (MIHK-ehn-zee, JIHM) N.J.

Left wing. Shoots left. 6'4", 230 lbs. Born, Gull Lake, Sask., November 3, 1969. Hartford's 3rd choice, 73rd overall, in 1989 Entry Draft.

Season	Club	League	GP	G	A	Pts	PIM	PP	SH	GW	S	%	+/-	TF	F%	H	SB	Min	PO GP	PO G	PO A	PO Pts	PO PIM	PO PP	PO SH	PO GW
1985-86	Moose Jaw	SMHL	36	18	26	44	89																			
	Moose Jaw	WHL	3	0	2	2	0																			
1986-87	Moose Jaw	WHL	65	5	3	8	125												9	0	0	0	7			
1987-88	Moose Jaw	WHL	62	1	17	18	134																			
1988-89	Victoria Cougars	WHL	67	15	27	42	176												8	1	4	5	30			
1989-90	**Hartford**	**NHL**	**5**	**0**	**0**	**0**	**4**	**0**	**0**	**0**	**0**	**0.0**	**0**													
	Binghamton	AHL	56	4	12	16	149																			
1990-91	**Hartford**	**NHL**	**41**	**4**	**3**	**7**	**108**	**0**	**0**	**0**	**16**	**25.0**	**–7**						**6**	**0**	**0**	**0**	**8**	**0**	**0**	**0**
	Springfield	AHL	24	3	4	7	102																			
1991-92	**Hartford**	**NHL**	**67**	**5**	**1**	**6**	**87**	**0**	**0**	**0**	**34**	**14.7**	**–6**													
1992-93	**Hartford**	**NHL**	**64**	**3**	**6**	**9**	**202**	**0**	**0**	**0**	**36**	**8.3**	**–10**													
1993-94	**Hartford**	**NHL**	**26**	**1**	**2**	**3**	**67**	**0**	**0**	**0**	**9**	**11.1**	**–6**													
	Dallas	**NHL**	**34**	**2**	**3**	**5**	**63**	**0**	**0**	**1**	**18**	**11.1**	**4**													
	Pittsburgh	**NHL**	**11**	**0**	**0**	**0**	**16**	**0**	**0**	**0**	**6**	**0.0**	**–5**						**3**	**0**	**0**	**0**	**0**	**0**	**0**	**0**
1994-95	**Pittsburgh**	**NHL**	**39**	**2**	**1**	**3**	**63**	**0**	**0**	**1**	**16**	**12.5**	**–7**						**5**	**0**	**0**	**0**	**4**	**0**	**0**	**0**
1995-96	**Winnipeg**	**NHL**	**73**	**4**	**2**	**6**	**202**	**0**	**0**	**0**	**28**	**14.3**	**–4**						**1**	**0**	**0**	**0**	**2**	**0**	**0**	**0**
1996-97	**Phoenix**	**NHL**	**65**	**5**	**3**	**8**	**200**	**0**	**0**	**1**	**38**	**13.2**	**–5**						**7**	**0**	**0**	**0**	**2**	**0**	**0**	**0**
1997-98	**Phoenix**	**NHL**	**64**	**3**	**4**	**7**	**146**	**0**	**0**	**0**	**35**	**8.6**	**–7**						**1**	**0**	**0**	**0**	**0**	**0**	**0**	**0**
1998-99	**Anaheim**	**NHL**	**73**	**5**	**4**	**9**	**99**	**1**	**0**	**1**	**59**	**8.5**	**–18**	**8**	**50.0**	**85**	**8**	**10:22**	**4**	**0**	**0**	**0**	**4**	**0**	**0**	**0**
99-2000	**Anaheim**	**NHL**	**31**	**3**	**3**	**6**	**48**	**0**	**0**	**0**	**22**	**13.6**	**–5**	**2**	**0.0**	**50**	**2**	**10:26**								
	Washington	**NHL**	**30**	**1**	**2**	**3**	**16**	**0**	**0**	**0**	**10**	**10.0**	**0**	**0**	**0.0**	**27**	**5**	**6:22**	**1**	**0**	**0**	**0**	**0**	**0**	**0**	**0**
2000-01	**New Jersey**	**NHL**	**53**	**2**	**2**	**4**	**119**	**0**	**0**	**0**	**32**	**6.3**	**0**	**0**	**0.0**	**64**	**4**	**7:56**	**3**	**0**	**0**	**0**	**2**	**0**	**0**	**0**
2001-02	**New Jersey**	**NHL**	**67**	**3**	**5**	**8**	**123**	**1**	**0**	**0**	**33**	**9.1**	**0**	**3**	**0.0**	**75**	**10**	**7:19**	**6**	**0**	**0**	**0**	**2**	**0**	**0**	**0**
	NHL Totals		**743**	**43**	**41**	**84**	**1563**	**2**	**0**	**4**	**392**	**11.0**		**13**	**30.8**	**301**	**29**	**8:35**	**37**	**0**	**0**	**0**	**24**	**0**	**0**	**0**

Traded to **Florida** by **Hartford** for Alexander Godynyuk, December 16, 1993. Traded to **Dallas** by **Florida** for Dallas' 4th round choice (later traded to Ottawa - Ottawa selected Kevin Bolibruck) in 1995 Entry Draft, December 16, 1993. Traded to **Pittsburgh** by **Dallas** for Mike Needham, March 21, 1994. Signed as a free agent by **NY Islanders**, August 2, 1995. Claimed by **Winnipeg** from **NY Islanders** in NHL Waiver Draft, October 2, 1995. Transferred to **Phoenix** after **Winnipeg** franchise relocated, July 1, 1996. Traded to **Anaheim** by **Phoenix** for Jean-Francois Jomphe, June 18, 1998. Claimed on waivers by **Washington** from **Anaheim**, January 20, 2000. Signed as a free agent by **New Jersey**, July 3, 2000.

McLAREN, Kyle

(mih-KLAIR-uhn, KIGHL) **BOS.**

Defense. Shoots left. 6'4", 230 lbs. Born, Humboldt, Sask., June 18, 1977. Boston's 1st choice, 9th overall, in 1995 Entry Draft.

			Regular Season																Playoffs							
Season	Club	League	GP	G	A	Pts	PIM	PP	SH	GW	S	%	+/-	TF	F%	H	SB	Min	GP	G	A	Pts	PIM	PP	SH	GW
1992-93	Lethbridge	AMHL	60	28	28	56	84																			
1993-94	Tacoma Rockets	WHL	62	1	9	10	53												6	1	4	5	6			
1994-95	Tacoma Rockets	WHL	47	13	19	32	68												4	1	1	2	4			
1995-96	**Boston**	**NHL**	**74**	**5**	**12**	**17**	**73**	**0**	**0**	**0**	**74**	**6.8**	**16**						**5**	**0**	**0**	**0**	**14**	**0**	**0**	**0**
1996-97	**Boston**	**NHL**	**58**	**5**	**9**	**14**	**54**	**0**	**0**	**1**	**68**	**7.4**	**–9**													
1997-98	**Boston**	**NHL**	**66**	**5**	**20**	**25**	**56**	**2**	**0**	**0**	**101**	**5.0**	**13**						**6**	**1**	**0**	**1**	**4**	**1**	**0**	**0**
1998-99	**Boston**	**NHL**	**52**	**6**	**18**	**24**	**48**	**3**	**0**	**0**	**97**	**6.2**	**1**	**0**	**0.0**	**205**	**69**	**23:25**	**12**	**0**	**3**	**3**	**10**	**0**	**0**	**0**
99-2000	**Boston**	**NHL**	**71**	**8**	**11**	**19**	**67**	**2**	**0**	**3**	**142**	**5.6**	**–4**	**5**	**40.0**	**282**	**155**	**23:18**								
2000-01	**Boston**	**NHL**	**58**	**5**	**12**	**17**	**53**	**2**	**0**	**0**	**91**	**5.5**	**–5**	**4**	**50.0**	**156**	**131**	**24:14**								
2001-02	**Boston**	**NHL**	**38**	**0**	**8**	**8**	**19**	**0**	**0**	**0**	**57**	**0.0**	**–4**	**1**	**0.0**	**73**	**59**	**19:21**	**4**	**0**	**0**	**0**	**20**	**0**	**0**	**0**
	NHL Totals		**417**	**34**	**90**	**124**	**370**	**9**	**0**	**4**	**630**	**5.4**		**10**	**40.0**	**716**	**414**	**22:53**	**27**	**1**	**3**	**4**	**48**	**1**	**0**	**0**

NHL All-Rookie Team (1996)

• Missed majority of 2001-02 season recovering from chest (October 10, 2001 vs. Minnesota) and wrist (December 26, 2001 vs. Ottawa) injuries.

MELANSON, Dean

(meh-LAHN-suhn, DEEN) **WSH.**

Defense. Shoots right. 5'11", 190 lbs. Born, Antigonish, N.S., November 19, 1973. Buffalo's 4th choice, 80th overall, in 1992 Entry Draft.

Season	Club	League	GP	G	A	Pts	PIM	PP	SH	GW	S	%	+/-	TF	F%	H	SB	Min	GP	G	A	Pts	PIM	PP	SH	GW
1989-90	Antigonish	MJrHL	STATISTICS NOT AVAILABLE																							
1990-91	St-Hyacinthe	QMJHL	69	10	17	27	110												4	0	1	1	2			
1991-92	St-Hyacinthe	QMJHL	42	8	19	27	158												6	1	2	3	25			
1992-93	St-Hyacinthe	QMJHL	57	13	29	42	253																			
	Rochester	AHL	8	0	1	1	6												14	1	6	7	18			
1993-94	Rochester	AHL	80	1	21	22	138												4	0	1	1	2			
1994-95	Rochester	AHL	43	4	7	11	84																			
	Buffalo	**NHL**	**5**	**0**	**0**	**0**	**4**	**0**	**0**	**0**	**1**	**0.0**	**–1**													
1995-96	Rochester	AHL	70	3	13	16	204												14	3	3	6	22			
1996-97	Quebec Rafales	IHL	72	3	21	24	95												7	0	2	2	12			
1997-98	Rochester	AHL	73	7	9	16	228												4	0	2	2	0			
1998-99	Rochester	AHL	79	7	27	34	192												17	3	2	5	32			
99-2000	Philadelphia	AHL	58	11	25	36	178												4	2	3	5	10			
2000-01	Philadelphia	AHL	15	1	4	5	48																			
	Chicago Wolves	IHL	42	1	7	8	80																			
	Portland Pirates	AHL	13	1	4	5	14												2	0	0	0	10			
2001-02	**Washington**	**NHL**	**4**	**0**	**0**	**0**	**4**	**0**	**0**	**0**	**6**	**0.0**	**1**	**0**	**0.0**	**7**	**1**	**12:36**								
	Portland Pirates	AHL	70	2	14	16	140																			
	NHL Totals		**9**	**0**	**0**	**0**	**8**	**0**	**0**	**0**	**7**	**0.0**		**0**	**0.0**	**7**	**1**	**12:36**								

QMJHL All-Rookie Team (1991)

Signed as a free agent by **Philadelphia**, July 22, 1999. Traded to **Washington** by **Philadelphia** for Matt Herr, March 13, 2001.

MELICHAR, Josef

(mehl-ee-KHAHR, YOH-sehf) **PIT.**

Defense. Shoots left. 6'2", 221 lbs. Born, Ceske Budejovice, Czech., January 20, 1979. Pittsburgh's 3rd choice, 71st overall, in 1997 Entry Draft.

Season	Club	League	GP	G	A	Pts	PIM	PP	SH	GW	S	%	+/-	TF	F%	H	SB	Min	GP	G	A	Pts	PIM	PP	SH	GW
1995-96	C. Budejovice Jr.	Czech-Jr.	38	3	4	7																				
1996-97	C. Budejovice Jr.	Czech-Jr.	41	2	3	5	10																			
1997-98	Tri-City	WHL	67	9	24	33	154																			
1998-99	Tri-City	WHL	65	8	28	36	125												11	1	0	1	15			
99-2000	Wilkes-Barre	AHL	80	3	9	12	126																			
2000-01	**Pittsburgh**	**NHL**	**18**	**0**	**2**	**2**	**21**	**0**	**0**	**0**	**9**	**0.0**	**–5**	**0**	**0.0**	**33**	**12**	**14:54**								
	Wilkes-Barre	AHL	46	2	5	7	69												21	0	5	5	6			
2001-02	**Pittsburgh**	**NHL**	**60**	**0**	**3**	**3**	**68**	**0**	**0**	**0**	**46**	**0.0**	**–1**	**0**	**0.0**	**93**	**60**	**16:46**								
	NHL Totals		**78**	**0**	**5**	**5**	**89**	**0**	**0**	**0**	**55**	**0.0**		**0**	**0.0**	**126**	**72**	**16:20**								

MELLANBY, Scott

(MEH-lihn-bee, SKAWT) **ST.L.**

Right wing. Shoots right. 6'1", 205 lbs. Born, Montreal, Que., June 11, 1966. Philadelphia's 2nd choice, 27th overall, in 1984 Entry Draft.

Season	Club	League	GP	G	A	Pts	PIM	PP	SH	GW	S	%	+/-	TF	F%	H	SB	Min	GP	G	A	Pts	PIM	PP	SH	GW
1982-83	Don Mills	MTHL	72	66	52	118	38																			
1983-84	Henry Carr	MTJHL	39	37	37	74	97																			
1984-85	U. of Wisconsin	WCHA	40	14	24	38	60																			
1985-86	U. of Wisconsin	WCHA	32	21	23	44	89																			
	Philadelphia	**NHL**	**2**	**0**	**0**	**0**	**0**	**0**	**0**	**0**	**0**	**0.0**	**–1**													
1986-87	**Philadelphia**	**NHL**	**71**	**11**	**21**	**32**	**94**	**1**	**0**	**0**	**118**	**9.3**	**8**						**24**	**5**	**5**	**10**	**46**	**0**	**0**	**1**
1987-88	**Philadelphia**	**NHL**	**75**	**25**	**26**	**51**	**185**	**7**	**0**	**2**	**190**	**13.2**	**–7**						**7**	**0**	**1**	**1**	**16**	**0**	**0**	**0**
1988-89	**Philadelphia**	**NHL**	**76**	**21**	**29**	**50**	**183**	**11**	**0**	**3**	**202**	**10.4**	**–13**						**19**	**4**	**5**	**9**	**28**	**0**	**0**	**0**
1989-90	**Philadelphia**	**NHL**	**57**	**6**	**17**	**23**	**77**	**0**	**0**	**1**	**104**	**5.8**	**–4**													
1990-91	**Philadelphia**	**NHL**	**74**	**20**	**21**	**41**	**155**	**5**	**0**	**6**	**165**	**12.1**	**8**													
1991-92	**Edmonton**	**NHL**	**80**	**23**	**27**	**50**	**197**	**7**	**0**	**5**	**159**	**14.5**	**5**						**16**	**2**	**1**	**3**	**29**	**1**	**0**	**1**
1992-93	**Edmonton**	**NHL**	**69**	**15**	**17**	**32**	**147**	**6**	**0**	**3**	**114**	**13.2**	**–4**													
1993-94	**Florida**	**NHL**	**80**	**30**	**30**	**60**	**149**	**17**	**0**	**4**	**204**	**14.7**	**0**													
1994-95	**Florida**	**NHL**	**48**	**13**	**12**	**25**	**90**	**4**	**0**	**5**	**130**	**10.0**	**–16**													
1995-96	**Florida**	**NHL**	**79**	**32**	**38**	**70**	**160**	**19**	**0**	**3**	**225**	**14.2**	**4**						**22**	**3**	**6**	**9**	**44**	**2**	**0**	**0**
1996-97	**Florida**	**NHL**	**82**	**27**	**29**	**56**	**170**	**9**	**1**	**4**	**221**	**12.2**	**7**						**5**	**0**	**2**	**2**	**4**	**0**	**0**	**0**
1997-98	**Florida**	**NHL**	**79**	**15**	**24**	**39**	**127**	**6**	**0**	**1**	**188**	**8.0**	**–14**													
1998-99	**Florida**	**NHL**	**67**	**18**	**27**	**45**	**85**	**4**	**0**	**3**	**136**	**13.2**	**5**	**11**	**27.3**	**64**	**15**	**16:14**								
99-2000	**Florida**	**NHL**	**77**	**18**	**28**	**46**	**126**	**6**	**0**	**2**	**134**	**13.4**	**14**	**20**	**60.0**	**80**	**25**	**14:51**	**4**	**0**	**1**	**1**	**2**	**0**	**0**	**0**
2000-01	**Florida**	**NHL**	**40**	**4**	**9**	**13**	**46**	**1**	**0**	**0**	**58**	**6.9**	**–13**	**4**	**50.0**	**36**	**11**	**14:59**								
	St. Louis	**NHL**	**23**	**7**	**1**	**8**	**25**	**2**	**0**	**0**	**37**	**18.9**	**0**	**1**	**0.0**	**26**	**4**	**15:00**	**15**	**3**	**3**	**6**	**17**	**2**	**0**	**0**
2001-02	**St. Louis**	**NHL**	**64**	**15**	**26**	**41**	**93**	**8**	**0**	**2**	**137**	**10.9**	**–5**	**3**	**0.0**	**119**	**17**	**15:40**	**10**	**7**	**3**	**10**	**18**	**4**	**0**	**1**
	NHL Totals		**1143**	**300**	**382**	**682**	**2109**	**113**	**1**	**44**	**2522**	**11.9**		**39**	**43.6**	**325**	**72**	**15:25**	**122**	**24**	**27**	**51**	**204**	**9**	**0**	**3**

Played in NHL All-Star Game (1996)

Traded to **Edmonton** by **Philadelphia** with Craig Fisher and Craig Berube for Dave Brown, Corey Foster and Jari Kurri, May 30, 1991. Claimed by **Florida** from **Edmonton** in Expansion Draft, June 24, 1993. Traded to **St. Louis** by **Florida** for rights to Dave Morisset and St. Louis' 5th round choice (Vince Bellissimo) in 2002 Entry Draft, February 9, 2001.

MELOCHE, Eric

(muh-LAWSH, AIR-ihk) **PIT.**

Right wing. Shoots right. 5'10", 195 lbs. Born, Montreal, Que., May 1, 1976. Pittsburgh's 7th choice, 186th overall, in 1996 Entry Draft.

Season	Club	League	GP	G	A	Pts	PIM	PP	SH	GW	S	%	+/-	TF	F%	H	SB	Min	GP	G	A	Pts	PIM	PP	SH	GW
1994-95	Cornwall Colts	OCJHL	40	7	15	22	51																			
1995-96	Cornwall Colts	OCJHL	64	68	53	121	162																			
1996-97	Ohio State	CCHA	39	12	11	23	78																			
1997-98	Ohio State	CCHA	42	26	22	48	86																			
1998-99	Ohio State	CCHA	35	11	16	27	87																			
99-2000	Ohio State	CCHA	36	20	11	31	*138																			
2000-01	Wilkes-Barre	AHL	79	20	20	40	72												21	6	10	16	17			
2001-02	**Pittsburgh**	**NHL**	**23**	**0**	**1**	**1**	**8**	**0**	**0**	**0**	**29**	**0.0**	**–7**	**4**	**0.0**	**38**	**9**	**10:10**								
	Wilkes-Barre	AHL	55	13	14	27	91																			
	NHL Totals		**23**	**0**	**1**	**1**	**8**	**0**	**0**	**0**	**29**	**0.0**		**4**	**0.0**	**38**	**9**	**10:10**								

MESSIER, Eric

(MEHS-see-ay, AIR-ihk) **COL.**

Left wing. Shoots left. 6'2", 200 lbs. Born, Drummondville, Que., October 29, 1973.

Season	Club	League	GP	G	A	Pts	PIM	PP	SH	GW	S	%	+/-	TF	F%	H	SB	Min	GP	G	A	Pts	PIM	PP	SH	GW
1990-91	Swift Textile	QAHA	STATISTICS NOT AVAILABLE																							
	Mtl-Bourassa	QAAA	3	0	1	1	0												2	0	0	0	0			
1991-92	Trois-Rivieres	QMJHL	58	2	10	12	28												15	2	2	4	13			
1992-93	Sherbrooke	QMJHL	51	4	17	21	82												15	0	4	4	18			
1993-94	Sherbrooke	QMJHL	67	4	24	28	69												12	1	7	8	14			
1994-95	U. Quebec T-R	OUAA	13	8	5	13	20												4	0	3	3	8			
1995-96	Cornwall Aces	AHL	72	5	9	14	111												8	1	1	2	20			

			Regular Season																Playoffs							
Season	**Club**	**League**	**GP**	**G**	**A**	**Pts**	**PIM**	**PP**	**SH**	**GW**	**S**	**%**	**+/-**	**TF**	**F%**	**H**	**SB**	**Min**	**GP**	**G**	**A**	**Pts**	**PIM**	**PP**	**SH**	**GW**
1996-97	**Colorado**	**NHL**	21	0	0	0	4	0	0	0	11	0.0	7	....	....	....	....	....	6	0	0	0	4	0	0	0
	Hershey Bears	AHL	55	16	26	42	69	....	....	....	....	....	....	....	....	....	....	....	9	3	8	11	14	....	....	....
1997-98	**Colorado**	**NHL**	62	4	12	16	20	0	0	0	66	6.1	4	....	....	....	....	....	....	....	....	....	....	....	....	....
1998-99	**Colorado**	**NHL**	31	4	2	6	14	1	0	1	30	13.3	0	0	0.0	29	22	13:43	3	0	0	0	0	0	0	0
	Hershey Bears	AHL	6	1	3	4	4	....	....	....	....	....	....	....	....	....	....	....	....	....	....	....	....	....	....	....
99-2000	**Colorado**	**NHL**	61	3	6	9	24	1	0	0	28	10.7	0	4	25.0	52	26	10:27	14	0	1	1	4	0	0	0
2000-01♦	**Colorado**	**NHL**	64	5	7	12	26	0	0	1	60	8.3	−3	9	44.4	142	27	12:16	23	2	2	4	14	0	0	0
2001-02	**Colorado**	**NHL**	74	5	10	15	26	0	0	3	84	6.0	−5	8	25.0	170	85	15:15	21	1	2	3	0	0	0	0
	NHL Totals		**313**	**21**	**37**	**58**	**114**	**2**	**0**	**5**	**279**	**7.5**		**21**	**33.3**	**393**	**160**	**12:56**	**67**	**3**	**5**	**8**	**22**	**0**	**0**	**0**

QMJHL Second All-Star Team (1994)

Signed as a free agent by **Colorado**, June 14, 1995. • Missed majority of 1998-99 season recovering from elbow injury suffered in game vs. Ottawa, October 10, 1998.

MESSIER, Mark

(MEHS-see-ay, MAHRK) **NYR**

Center. Shoots left. 6'1", 210 lbs. Born, Edmonton, Alta., January 18, 1961. Edmonton's 2nd choice, 48th overall, in 1979 Entry Draft.

Season	Club	League	GP	G	A	Pts	PIM	PP	SH	GW	S	%	+/-	TF	F%	H	SB	Min	GP	G	A	Pts	PIM	PP	SH	GW
1976-77	Spruce Grove	AJHL	57	27	39	66	91	....	....	....	....	....	....	....	....	....	....	....	....	....	....	....	....	....	....	....
1977-78	St. Albert	AJHL	54	25	49	74	194	....	....	....	....	....	....	....	....	....	....	....	....	....	....	....	....	....	....	....
	Portland	WHL	....	....	....	....	....	....	....	....	....	....	....	....	....	....	....	....	7	4	1	5	2	....	....	....
1978-79	St. Albert	AJHL	17	15	18	33	64	....	....	....	....	....	....	....	....	....	....	....	....	....	....	....	....	....	....	....
	Indianapolis	WHA	5	0	0	0	0	....	....	....	....	....	....	....	....	....	....	....	....	....	....	....	....	....	....	....
	Cincinnati	WHA	47	1	10	11	58	....	....	....	....	....	....	....	....	....	....	....	....	....	....	....	....	....	....	....
1979-80	**Edmonton**	**NHL**	75	12	21	33	120	1	1	1	113	10.6	−10	....	....	....	....	....	3	1	2	3	2	0	1	0
	Houston Apollos	CHL	4	0	3	3	4	....	....	....	....	....	....	....	....	....	....	....	....	....	....	....	....	....	....	....
1980-81	**Edmonton**	**NHL**	72	23	40	63	102	4	0	1	179	12.8	−12	....	....	....	....	....	9	2	5	7	13	0	0	0
1981-82	**Edmonton**	**NHL**	78	50	38	88	119	10	0	3	235	21.3	21	....	....	....	....	....	5	1	2	3	8	0	0	0
1982-83	**Edmonton**	**NHL**	77	48	58	106	72	12	1	2	237	20.3	19	....	....	....	....	....	15	15	6	21	14	4	2	0
1983-84♦	**Edmonton**	**NHL**	73	37	64	101	165	7	4	7	219	16.9	40	....	....	....	....	....	19	8	18	26	19	1	1	2
1984-85♦	**Edmonton**	**NHL**	55	23	31	54	57	4	5	1	136	16.9	8	....	....	....	....	....	18	12	13	25	12	1	1	1
1985-86	**Edmonton**	**NHL**	63	35	49	84	68	10	5	7	201	17.4	36	....	....	....	....	....	10	4	6	10	18	0	2	0
1986-87♦	**Edmonton**	**NHL**	77	37	70	107	73	7	4	5	208	17.8	21	....	....	....	....	....	21	12	16	28	16	1	2	1
1987-88♦	**Edmonton**	**NHL**	77	37	74	111	103	12	3	7	182	20.3	21	....	....	....	....	....	19	11	23	34	29	7	1	0
1988-89	**Edmonton**	**NHL**	72	33	61	94	130	6	6	4	164	20.1	−5	....	....	....	....	....	7	1	11	12	8	0	0	0
1989-90♦	**Edmonton**	**NHL**	79	45	84	129	79	13	6	3	211	21.3	19	....	....	....	....	....	22	9	*22	*31	20	1	1	1
1990-91	**Edmonton**	**NHL**	53	12	52	64	34	3	1	2	109	11.0	15	....	....	....	....	....	18	4	11	15	16	1	0	0
1991-92	**NY Rangers**	**NHL**	79	35	72	107	76	12	4	6	212	16.5	31	....	....	....	....	....	11	7	7	14	6	2	2	0
1992-93	**NY Rangers**	**NHL**	75	25	66	91	72	7	2	2	215	11.6	−6	....	....	....	....	....	....	....	....	....	....	....	....	....
1993-94♦	**NY Rangers**	**NHL**	76	26	58	84	76	6	2	5	216	12.0	25	....	....	....	....	....	23	12	18	30	33	2	1	4
1994-95	**NY Rangers**	**NHL**	46	14	39	53	40	3	3	2	126	11.1	8	....	....	....	....	....	10	3	10	13	8	2	0	1
1995-96	**NY Rangers**	**NHL**	74	47	52	99	122	14	1	5	241	19.5	29	....	....	....	....	....	11	4	7	11	16	2	0	1
1996-97	**NY Rangers**	**NHL**	71	36	48	84	88	7	5	9	227	15.9	12	....	....	....	....	....	15	3	9	12	6	0	0	1
1997-98	**Vancouver**	**NHL**	82	22	38	60	58	8	2	2	139	15.8	−10	....	....	....	....	....	....	....	....	....	....	....	....	....
1998-99	**Vancouver**	**NHL**	59	13	35	48	33	4	2	2	97	13.4	−12	1536	53.9	29	35	22:36	....	....	....	....	....	....	....	....
99-2000	**Vancouver**	**NHL**	66	17	37	54	30	6	0	4	131	13.0	−15	1684	56.8	50	26	21:12	....	....	....	....	....	....	....	....
2000-01	**NY Rangers**	**NHL**	82	24	43	67	89	12	3	2	131	18.3	−25	1879	55.5	78	47	19:14	....	....	....	....	....	....	....	....
2001-02	**NY Rangers**	**NHL**	41	7	16	23	32	2	0	2	69	10.1	−1	787	53.6	35	27	18:31	....	....	....	....	....	....	....	....
	NHL Totals		**1602**	**658**	**1146**	**1804**	**1838**	**170**	**60**	**84**	**3998**	**16.5**		**5886**	**55.2**	**192**	**135**	**20:26**	**236**	**109**	**186**	**295**	**244**	**24**	**14**	**12**

NHL First All-Star Team (1982, 1983, 1990, 1992) • NHL Second All-Star Team (1984) • Won Conn Smythe Trophy (1984) • Won Lester B. Pearson Award (1990, 1992) • Won Hart Trophy (1990, 1992) • Played in NHL All-Star Game (1982, 1983, 1984, 1986, 1988, 1989, 1990, 1991, 1992, 1994, 1996, 1997, 1998, 2000)

Signed as an underage free agent by **Indianapolis** (WHA) to a 10-game tryout contract, November 5, 1978. Signed as a free agent by **Cincinnati** (WHA) after **Indianapolis** (WHA) franchise folded, December, 1978. Traded to **NY Rangers** by **Edmonton** with future considerations (Jeff Beukeboom for David Shaw, November 12, 1991) for Bernie Nicholls, Steven Rice and Louie DeBrusk, October 4, 1991. Signed as a free agent by **Vancouver**, July 30, 1997. Signed as a free agent by **NY Rangers**, July 13, 2000. • Missed majority of 2001-02 season recovering from back injury suffered in game vs. Toronto, December 8, 2001.

METROPOLIT, Glen

(MEH-troh-poh-LIHT, GLEHN) **WSH.**

Right wing. Shoots right. 5'10", 200 lbs. Born, Toronto, Ont., June 25, 1974.

Season	Club	League	GP	G	A	Pts	PIM	PP	SH	GW	S	%	+/-	TF	F%	H	SB	Min	GP	G	A	Pts	PIM	PP	SH	GW
1992-93	Richmond Hill	MTJHL	43	27	36	63	36	....	....	....	....	....	....	....	....	....	....	....	....	....	....	....	....	....	....	....
1993-94	Richmond Hill	MTJHL	49	38	62	100	83	....	....	....	....	....	....	....	....	....	....	....	....	....	....	....	....	....	....	....
1994-95	Vernon Vipers	BCJHL	60	43	74	117	92	....	....	....	....	....	....	....	....	....	....	....	....	....	....	....	....	....	....	....
1995-96	Nashville	ECHL	58	30	31	61	62	....	....	....	....	....	....	....	....	....	....	....	5	3	8	11	2	....	....	....
	Atlanta Knights	IHL	1	0	0	0	0	....	....	....	....	....	....	....	....	....	....	....	....	....	....	....	....	....	....	....
1996-97	Pensacola	ECHL	54	35	47	82	45	....	....	....	....	....	....	....	....	....	....	....	12	9	16	25	28	....	....	....
	Quebec Rafales	IHL	22	5	4	9	14	....	....	....	....	....	....	....	....	....	....	....	5	0	0	0	2	....	....	....
1997-98	Grand Rapids	IHL	79	20	35	55	90	....	....	....	....	....	....	....	....	....	....	....	3	1	1	2	0	....	....	....
1998-99	Grand Rapids	IHL	77	28	53	81	92	....	....	....	....	....	....	....	....	....	....	....	....	....	....	....	....	....	....	....
99-2000	**Washington**	**NHL**	30	6	13	19	4	1	0	1	57	10.5	5	37	46.0	48	7	13:17	2	0	0	0	2	0	0	0
	Portland Pirates	AHL	48	18	42	60	73	....	....	....	....	....	....	....	....	....	....	....	1	1	0	1	0	....	....	....
2000-01	**Washington**	**NHL**	15	1	5	6	10	0	0	0	20	5.0	−2	3	33.3	13	0	11:50	1	0	0	0	0	0	0	0
	Portland Pirates	AHL	51	25	42	67	59	....	....	....	....	....	....	....	....	....	....	....	....	....	....	....	....	....	....	....
2001-02	**Tampa Bay**	**NHL**	2	0	0	0	0	0	0	0	1	0.0	−2	2	50.0	3	0	10:26	....	....	....	....	....	....	....	....
	Washington	**NHL**	33	1	16	17	6	0	0	0	51	2.0	3	145	49.7	20	8	14:24	....	....	....	....	....	....	....	....
	Portland Pirates	AHL	32	17	22	39	20	....	....	....	....	....	....	....	....	....	....	....	....	....	....	....	....	....	....	....
	NHL Totals		**80**	**8**	**34**	**42**	**20**	**1**	**0**	**1**	**129**	**6.2**		**187**	**48.7**	**84**	**15**	**13:24**	**3**	**0**	**0**	**0**	**2**	**0**	**0**	**0**

Signed as a free agent by **Washington**, July 19, 1999. Claimed by **Tampa Bay** from **Washington** in Waiver Draft, September 28, 2001. Claimed on waivers by **Washington** from **Tampa Bay**, October 20, 2001.

MEZEI, Branislav

(MEH-tzay, BRAN-ih-slav) **FLA.**

Defense. Shoots left. 6'5", 236 lbs. Born, Nitra, Czech., October 8, 1980. NY Islanders' 3rd choice, 10th overall, in 1999 Entry Draft.

Season	Club	League	GP	G	A	Pts	PIM	PP	SH	GW	S	%	+/-	TF	F%	H	SB	Min	GP	G	A	Pts	PIM	PP	SH	GW
1996-97	HC Nitra Jr.	Slovak-Jr.	40	8	17	25	42	....	....	....	....	....	....	....	....	....	....	....	....	....	....	....	....	....	....	....
1997-98	Belleville Bulls	OHL	53	3	5	8	58	....	....	....	....	....	....	....	....	....	....	....	8	0	2	2	8	....	....	....
1998-99	Belleville Bulls	OHL	60	5	18	23	90	....	....	....	....	....	....	....	....	....	....	....	18	0	4	4	29	....	....	....
99-2000	Belleville Bulls	OHL	58	7	21	28	99	....	....	....	....	....	....	....	....	....	....	....	6	0	3	3	10	....	....	....
2000-01	**NY Islanders**	**NHL**	42	1	4	5	53	0	0	0	29	3.4	−5	0	0.0	127	31	14:48	....	....	....	....	....	....	....	....
	Lowell	AHL	20	0	3	3	28	....	....	....	....	....	....	....	....	....	....	....	....	....	....	....	....	....	....	....
2001-02	**NY Islanders**	**NHL**	24	0	2	2	12	0	0	0	4	0.0	2	0	0.0	39	18	8:28	....	....	....	....	....	....	....	....
	Bridgeport	AHL	59	1	9	10	137	....	....	....	....	....	....	....	....	....	....	....	20	0	1	1	48	....	....	....
	NHL Totals		**66**	**1**	**6**	**7**	**65**	**0**	**0**	**0**	**33**	**3.0**		**0**	**0.0**	**166**	**49**	**12:30**	....	....	....	....	....	....	....	....

OHL First All-Star Team (2000)

Traded to **Florida** by **NY Islanders** for Jason Wiemer, July 3, 2002.

MIKA, Petr

(MEE-kah, PEE-tuhr)

Left wing. Shoots right. 6'4", 194 lbs. Born, Prague, Czech., February 12, 1979. NY Islanders' 6th choice, 85th overall, in 1997 Entry Draft.

Season	Club	League	GP	G	A	Pts	PIM	PP	SH	GW	S	%	+/-	TF	F%	H	SB	Min	GP	G	A	Pts	PIM	PP	SH	GW
1995-96	Slavia Praha Jr.	Czech-Jr.	26	5	12	17	....	....	....	....	....	....	....	....	....	....	....	....	....	....	....	....	....	....	....	....
	HC Slavia Praha	Czech	1	0	0	0	0	....	....	....	....	....	....	....	....	....	....	....	....	....	....	....	....	....	....	....
1996-97	H+S Beroun	Czech-2	9	1	0	1	....	....	....	....	....	....	....	....	....	....	....	....	....	....	....	....	....	....	....	....
	HC Slavia Praha	Czech	20	1	2	3	6	....	....	....	....	....	....	....	....	....	....	....	....	....	....	....	....	....	....	....
	Slavia Praha Jr.	Czech-Jr.	15	8	0	8	....	....	....	....	....	....	....	....	....	....	....	....	....	....	....	....	....	....	....	....
1997-98	Ottawa 67's	OHL	41	10	8	18	28	....	....	....	....	....	....	....	....	....	....	....	....	....	....	....	....	....	....	....
1998-99	HC Slavia Praha	Czech	49	6	5	11	57	....	....	....	....	....	....	....	....	....	....	....	....	....	....	....	....	....	....	....
99-2000	**NY Islanders**	**NHL**	3	0	0	0	0	0	0	0	1	0.0	−1	0	0.0	3	2	4:30	....	....	....	....	....	....	....	....
	Lowell	AHL	50	8	9	17	20	....	....	....	....	....	....	....	....	....	....	....	6	0	0	0	0	....	....	....
2000-01	Lowell	AHL	13	0	1	1	7	....	....	....	....	....	....	....	....	....	....	....	....	....	....	....	....	....	....	....
	Springfield	AHL	27	0	2	2	8	....	....	....	....	....	....	....	....	....	....	....	....	....	....	....	....	....	....	....
2001-02	Bridgeport	AHL	1	0	0	0	0	....	....	....	....	....	....	....	....	....	....	....	....	....	....	....	....	....	....	....
	HC Slavia Praha	Czech	29	10	7	17	16	....	....	....	....	....	....	....	....	....	....	....	9	0	1	1	4	....	....	....
	NHL Totals		**3**	**0**	**0**	**0**	**0**	**0**	**0**	**0**	**1**	**0.0**		**0**	**0.0**	**3**	**2**	**4:30**	....	....	....	....	....	....	....	....

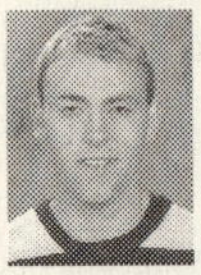

MILLAR, Craig

(MIHL-uhr, KRAYG)

Defense. Shoots left. 6'2", 212 lbs. Born, Winnipeg, Man., July 12, 1976. Buffalo's 10th choice, 225th overall, in 1994 Entry Draft.

Season	Club	League	GP	G	A	Pts	PIM	PP	SH	GW	S	%	+/-	TF	F%	H	SB	Min	PO GP	PO G	PO A	PO Pts	PO PIM	PO PP	PO SH	PO GW
			Regular Season																Playoffs							
1991-92	Wpg. Mavericks	MMHL	STATISTICS NOT AVAILABLE					....	....	....	....	....	....	....	....	....	....	....	....	....	....	....	....	....	....	....
1992-93	Swift Current	WHL	43	2	1	3	8	....	....	....	....	....	....	....	....	....	....	....	....	....	....	....	....	....	....	....
1993-94	Swift Current	WHL	66	2	9	11	53	....	....	....	....	....	....	....	....	....	....	....	7	0	3	3	4	....	....	....
1994-95	Swift Current	WHL	72	8	42	50	80	....	....	....	....	....	....	....	....	....	....	....	6	1	1	2	10	....	....	....
1995-96	Swift Current	WHL	72	31	46	77	151	....	....	....	....	....	....	....	....	....	....	....	6	1	0	1	22	....	....	....
1996-97	Rochester	AHL	64	7	18	25	65	....	....	....	....	....	....	....	....	....	....	....	....	....	....	....	....	....	....	....
	Edmonton	**NHL**	**1**	**0**	**0**	**0**	**2**	**0**	**0**	**0**	**1**	**0.0**	**0**	....	....	....	....	....	....	....	....	....	....	....	....	....
	Hamilton	AHL	10	1	3	4	10	....	....	....	....	....	....	....	....	....	....	....	22	4	4	8	21	....	....	....
1997-98	**Edmonton**	**NHL**	**11**	**4**	**0**	**4**	**8**	**1**	**0**	**0**	**10**	**40.0**	**–3**	....	....	....	....	....	....	....	....	....	....	....	....	....
	Hamilton	AHL	60	10	22	32	113	....	....	....	....	....	....	....	....	....	....	....	9	3	1	4	22	....	....	....
1998-99	**Edmonton**	**NHL**	**24**	**0**	**2**	**2**	**19**	**0**	**0**	**0**	**18**	**0.0**	**–6**	**0**	**0.0**	**23**	**24**	**15:09**	....	....	....	....	....	....	....	....
	Hamilton	AHL	43	3	17	20	38	....	....	....	....	....	....	....	....	....	....	....	11	1	5	6	18	....	....	....
99-2000	**Nashville**	**NHL**	**57**	**3**	**11**	**14**	**28**	**0**	**0**	**1**	**50**	**6.0**	**–6**	**1**	**0.0**	**56**	**38**	**16:53**	....	....	....	....	....	....	....	....
	Milwaukee	IHL	8	1	5	6	6	....	....	....	....	....	....	....	....	....	....	....	....	....	....	....	....	....	....	....
2000-01	**Nashville**	**NHL**	**5**	**0**	**0**	**0**	**6**	**0**	**0**	**0**	**2**	**0.0**	**1**	**0**	**0.0**	**6**	**1**	**12:04**	....	....	....	....	....	....	....	....
	Grand Rapids	IHL	12	1	2	3	2	....	....	....	....	....	....	....	....	....	....	....	....	....	....	....	....	....	....	....
	Tampa Bay	**NHL**	**16**	**1**	**1**	**2**	**10**	**0**	**0**	**1**	**10**	**10.0**	**–8**	**0**	**0.0**	**4**	**13**	**15:01**	....	....	....	....	....	....	....	....
	Detroit Vipers	IHL	11	0	2	2	32	....	....	....	....	....	....	....	....	....	....	....	....	....	....	....	....	....	....	....
2001-02	Slov. Bratislava	Slovakia	10	0	0	0	8	....	....	....	....	....	....	....	....	....	....	....	....	....	....	....	....	....	....	....
	EV Regensburg	German-2	18	5	5	10	122	....	....	....	....	....	....	....	....	....	....	....	10	3	5	8	38	....	....	....
	NHL Totals		**114**	**8**	**14**	**22**	**73**	**1**	**0**	**2**	**91**	**8.8**		**1**	**0.0**	**89**	**76**	**15:57**	....	....	....	....	....	....	....	....

WHL East First All-Star Team (1996)

Traded to **Edmonton** by **Buffalo** with Barrie Moore for Miroslav Satan, March 18, 1997. Traded to **Nashville** by **Edmonton** for Detroit's 3rd round choice (previously acquired, Edmonton selected Mike Comrie) in 1999 Entry Draft, June 26, 1999. Claimed on waivers by **Tampa Bay** from **Nashville**, October 25, 2000. Traded to **Ottawa** by **Tampa Bay** for John Emmons, March 13, 2001.

MILLER, Aaron

(MIHL-luhr, AIR-ruhn) **L.A.**

Defense. Shoots right. 6'3", 200 lbs. Born, Buffalo, NY, August 11, 1971. NY Rangers' 6th choice, 88th overall, in 1989 Entry Draft.

Season	Club	League	GP	G	A	Pts	PIM	PP	SH	GW	S	%	+/-	TF	F%	H	SB	Min	PO GP	PO G	PO A	PO Pts	PO PIM	PO PP	PO SH	PO GW
1987-88	Niagara Scenics	NAJHL	30	4	9	13	2	....	....	....	....	....	....	....	....	....	....	....	....	....	....	....	....	....	....	....
1988-89	Niagara Scenics	NAJHL	59	24	38	62	60	....	....	....	....	....	....	....	....	....	....	....	....	....	....	....	....	....	....	....
1989-90	U. of Vermont	ECAC	31	1	15	16	24	....	....	....	....	....	....	....	....	....	....	....	....	....	....	....	....	....	....	....
1990-91	U. of Vermont	ECAC	30	3	7	10	22	....	....	....	....	....	....	....	....	....	....	....	....	....	....	....	....	....	....	....
1991-92	U. of Vermont	ECAC	31	3	16	19	28	....	....	....	....	....	....	....	....	....	....	....	....	....	....	....	....	....	....	....
1992-93	U. of Vermont	ECAC	30	4	13	17	16	....	....	....	....	....	....	....	....	....	....	....	....	....	....	....	....	....	....	....
1993-94	**Quebec**	**NHL**	**1**	**0**	**0**	**0**	**0**	**0**	**0**	**0**	**0**	**0.0**	**–1**	....	....	....	....	....	....	....	....	....	....	....	....	....
	Cornwall Aces	AHL	64	4	10	14	49	....	....	....	....	....	....	....	....	....	....	....	13	0	2	2	10	....	....	....
1994-95	Cornwall Aces	AHL	76	4	18	22	69	....	....	....	....	....	....	....	....	....	....	....	....	....	....	....	....	....	....	....
	Quebec	**NHL**	**9**	**0**	**3**	**3**	**6**	**0**	**0**	**0**	**12**	**0.0**	**2**	....	....	....	....	....	....	....	....	....	....	....	....	....
1995-96	**Colorado**	**NHL**	**5**	**0**	**0**	**0**	**0**	**0**	**0**	**0**	**2**	**0.0**	**0**	....	....	....	....	....	....	....	....	....	....	....	....	....
	Cornwall Aces	AHL	62	4	23	27	77	....	....	....	....	....	....	....	....	....	....	....	8	0	1	1	6	....	....	....
1996-97	**Colorado**	**NHL**	**56**	**5**	**12**	**17**	**15**	**0**	**0**	**3**	**47**	**10.6**	**15**	....	....	....	....	....	**17**	**1**	**2**	**3**	**10**	**0**	**0**	**0**
1997-98	**Colorado**	**NHL**	**55**	**2**	**2**	**4**	**51**	**0**	**0**	**0**	**29**	**6.9**	**0**	....	....	....	....	....	**7**	**0**	**0**	**0**	**8**	**0**	**0**	**0**
1998-99	**Colorado**	**NHL**	**76**	**5**	**13**	**18**	**42**	**1**	**0**	**2**	**87**	**5.7**	**3**	**0**	**0.0**	**115**	**128**	**21:49**	**19**	**1**	**5**	**6**	**10**	**0**	**0**	**0**
99-2000	**Colorado**	**NHL**	**53**	**1**	**7**	**8**	**36**	**0**	**0**	**0**	**44**	**2.3**	**3**	**0**	**0.0**	**64**	**76**	**19:05**	**17**	**1**	**1**	**2**	**6**	**0**	**0**	**0**
2000-01	**Colorado**	**NHL**	**56**	**4**	**9**	**13**	**29**	**0**	**0**	**0**	**49**	**8.2**	**19**	**0**	**0.0**	**69**	**62**	**18:25**	....	....	....	....	....	....	....	....
	Los Angeles	**NHL**	**13**	**0**	**5**	**5**	**14**	**0**	**0**	**0**	**10**	**0.0**	**3**	**1**	**0.0**	**41**	**26**	**22:44**	**13**	**0**	**1**	**1**	**6**	**0**	**0**	**0**
2001-02	**Los Angeles**	**NHL**	**74**	**5**	**12**	**17**	**54**	**0**	**1**	**3**	**75**	**6.7**	**14**	**0**	**0.0**	**153**	**147**	**22:21**	**7**	**0**	**0**	**0**	**0**	**0**	**0**	**0**
	United States	Olympics	6	0	0	0	4	....	....	....	....	....	....	....	....	....	....	....	....	....	....	....	....	....	....	....
	NHL Totals		**398**	**22**	**63**	**85**	**247**	**1**	**1**	**8**	**355**	**6.2**		**1**	**0.0**	**442**	**439**	**20:46**	**80**	**3**	**9**	**12**	**40**	**0**	**0**	**0**

ECAC First All-Star Team (1993) • NCAA East Second All-American Team (1993)

Traded to **Quebec** by **NY Rangers** with NY Rangers' 5th round choice (Bill Lindsay) in 1991 Entry Draft for Joe Cirella, January 17, 1991. Transferred to **Colorado** after **Quebec** franchise relocated, June 21, 1995. Traded to **LA Kings** by **Colorado** with Adam Deadmarsh, a player to be named later (Jared Aulin, March 22, 2001), Colorado's 1st round choice (Dave Steckel) in 2001 Entry Draft and future considerations for Rob Blake and Steve Reinprecht, February 21, 2001.

MILLER, Kevin

(MIHL-luhr, KEH-vihn)

Center. Shoots right. 5'11", 184 lbs. Born, Lansing, MI, September 2, 1965. NY Rangers' 10th choice, 202nd overall, in 1984 Entry Draft.

Season	Club	League	GP	G	A	Pts	PIM	PP	SH	GW	S	%	+/-	TF	F%	H	SB	Min	PO GP	PO G	PO A	PO Pts	PO PIM	PO PP	PO SH	PO GW
1983-84	Redford Royals	GLJHL	44	28	57	85	....	....	....	....	....	....	....	....	....	....	....	....	....	....	....	....	....	....	....	....
1984-85	Michigan State	CCHA	44	11	29	40	84	....	....	....	....	....	....	....	....	....	....	....	....	....	....	....	....	....	....	....
1985-86	Michigan State	CCHA	45	19	52	71	112	....	....	....	....	....	....	....	....	....	....	....	....	....	....	....	....	....	....	....
1986-87	Michigan State	CCHA	42	25	56	81	63	....	....	....	....	....	....	....	....	....	....	....	....	....	....	....	....	....	....	....
1987-88	Michigan State	CCHA	9	6	3	9	18	....	....	....	....	....	....	....	....	....	....	....	....	....	....	....	....	....	....	....
	Team USA	Nat-Tm	48	31	32	63	33	....	....	....	....	....	....	....	....	....	....	....	....	....	....	....	....	....	....	....
	United States	Olympics	5	1	3	4	4	....	....	....	....	....	....	....	....	....	....	....	....	....	....	....	....	....	....	....
1988-89	**NY Rangers**	**NHL**	**24**	**3**	**5**	**8**	**2**	**0**	**0**	**1**	**40**	**7.5**	**–1**	....	....	....	....	....	....	....	....	....	....	....	....	....
	Denver Rangers	IHL	55	29	47	76	19	....	....	....	....	....	....	....	....	....	....	....	4	2	1	3	2	....	....	....
1989-90	**NY Rangers**	**NHL**	**16**	**0**	**5**	**5**	**2**	**0**	**0**	**0**	**9**	**0.0**	**–1**	....	....	....	....	....	**1**	**0**	**0**	**0**	**0**	**0**	**0**	**0**
	Flint Spirits	IHL	48	19	23	42	41	....	....	....	....	....	....	....	....	....	....	....	....	....	....	....	....	....	....	....
1990-91	**NY Rangers**	**NHL**	**63**	**17**	**27**	**44**	**63**	**1**	**2**	**3**	**113**	**15.0**	**1**	....	....	....	....	....	....	....	....	....	....	....	....	....
	Detroit	**NHL**	**11**	**5**	**2**	**7**	**4**	**0**	**1**	**0**	**23**	**21.7**	**–4**	....	....	....	....	....	**7**	**3**	**2**	**5**	**20**	**0**	**1**	**0**
1991-92	**Detroit**	**NHL**	**80**	**20**	**26**	**46**	**53**	**3**	**1**	**4**	**130**	**15.4**	**6**	....	....	....	....	....	**9**	**0**	**2**	**2**	**4**	**0**	**0**	**0**
1992-93	**Washington**	**NHL**	**10**	**0**	**3**	**3**	**35**	**0**	**0**	**0**	**10**	**0.0**	**–4**	....	....	....	....	....	....	....	....	....	....	....	....	....
	St. Louis	**NHL**	**72**	**24**	**22**	**46**	**65**	**8**	**3**	**4**	**153**	**15.7**	**6**	....	....	....	....	....	**10**	**0**	**3**	**3**	**11**	**0**	**0**	**0**
1993-94	**St. Louis**	**NHL**	**75**	**23**	**25**	**48**	**83**	**6**	**3**	**5**	**154**	**14.9**	**6**	....	....	....	....	....	**3**	**1**	**0**	**1**	**4**	**0**	**1**	**0**
1994-95	**St. Louis**	**NHL**	**15**	**2**	**5**	**7**	**0**	**0**	**0**	**0**	**19**	**10.5**	**4**	....	....	....	....	....	....	....	....	....	....	....	....	....
	San Jose	**NHL**	**21**	**6**	**7**	**13**	**13**	**1**	**1**	**2**	**41**	**14.6**	**0**	....	....	....	....	....	**6**	**0**	**0**	**0**	**2**	**0**	**0**	**0**
1995-96	**San Jose**	**NHL**	**68**	**22**	**20**	**42**	**41**	**2**	**2**	**2**	**146**	**15.1**	**–8**	....	....	....	....	....	....	....	....	....	....	....	....	....
	Pittsburgh	**NHL**	**13**	**6**	**5**	**11**	**4**	**1**	**0**	**0**	**33**	**18.2**	**4**	....	....	....	....	....	**18**	**3**	**2**	**5**	**8**	**0**	**0**	**0**
1996-97	**Chicago**	**NHL**	**69**	**14**	**17**	**31**	**41**	**5**	**1**	**2**	**139**	**10.1**	**–10**	....	....	....	....	....	**6**	**0**	**1**	**1**	**0**	**0**	**0**	**0**
1997-98	**Chicago**	**NHL**	**37**	**4**	**7**	**11**	**8**	**0**	**0**	**1**	**37**	**10.8**	**–4**	....	....	....	....	....	....	....	....	....	....	....	....	....
	Indianapolis Ice	IHL	26	11	11	22	41	....	....	....	....	....	....	....	....	....	....	....	2	1	1	2	0	....	....	....
1998-99	**NY Islanders**	**NHL**	**33**	**1**	**5**	**6**	**13**	**0**	**0**	**0**	**37**	**2.7**	**–5**	**114**	**49.1**	**42**	**11**	**10:19**	....	....	....	....	....	....	....	....
	Chicago Wolves	IHL	30	11	20	31	8	....	....	....	....	....	....	....	....	....	....	....	10	2	7	9	22	....	....	....
99-2000	**Ottawa**	**NHL**	**9**	**3**	**2**	**5**	**2**	**1**	**0**	**2**	**11**	**27.3**	**1**	**34**	**41.2**	**9**	**2**	**8:10**	**1**	**0**	**0**	**0**	**0**	**0**	**0**	**0**
	Grand Rapids	IHL	63	20	34	54	51	....	....	....	....	....	....	....	....	....	....	....	17	*11	7	*18	30	....	....	....
2000-01	HC Davos	Swiss	36	*29	27	56	61	....	....	....	....	....	....	....	....	....	....	....	4	3	0	3	2	....	....	....
2001-02	HC Davos	Swiss	43	23	18	41	78	....	....	....	....	....	....	....	....	....	....	....	16	4	10	14	12	....	....	....
	NHL Totals		**616**	**150**	**183**	**333**	**429**	**28**	**14**	**26**	**1095**	**13.7**		**148**	**47.3**	**51**	**13**	**9:51**	**61**	**7**	**10**	**17**	**49**	**0**	**2**	**0**

Traded to **Detroit** by **NY Rangers** with Jim Cummins and Dennis Vial for Joe Kocur and Per Djoos, March 5, 1991. Traded to **Washington** by **Detroit** for Dino Ciccarelli, June 20, 1992. Traded to **St. Louis** by **Washington** for Paul Cavallini, November 2, 1992. Traded to **San Jose** by **St. Louis** for Todd Elik, March 23, 1995. Traded to **Pittsburgh** by **San Jose** for Pittsburgh's 5th round choice (later traded to Boston - Boston selected Elias Abrahamsson) in 1996 Entry Draft , March 20, 1996. Signed as a free agent by **Chicago**, July 18, 1996. Signed as a free agent by **NY Islanders**, October 9, 1998. Signed as a free agent by **Ottawa**, August 24, 1999. Signed as a free agent by **HC Davos** (Switz.), July 26, 2000.

MILLER, Kip

(MIHL-luhr, KIHP) **WSH.**

Center. Shoots left. 5'10", 190 lbs. Born, Lansing, MI, June 11, 1969. Quebec's 4th choice, 72nd overall, in 1987 Entry Draft.

Season	Club	League	GP	G	A	Pts	PIM	PP	SH	GW	S	%	+/-	TF	F%	H	SB	Min	PO GP	PO G	PO A	PO Pts	PO PIM	PO PP	PO SH	PO GW
1984-85	Det. Compuware	MNHL	65	69	63	132	....	....	....	....	....	....	....	....	....	....	....	....	....	....	....	....	....	....	....	....
1985-86	Det. Compuware	GLJHL	30	25	28	53	....	....	....	....	....	....	....	....	....	....	....	....	....	....	....	....	....	....	....	....
1986-87	Michigan State	CCHA	41	20	19	39	92	....	....	....	....	....	....	....	....	....	....	....	....	....	....	....	....	....	....	....
1987-88	Michigan State	CCHA	39	16	25	41	51	....	....	....	....	....	....	....	....	....	....	....	....	....	....	....	....	....	....	....
1988-89	Michigan State	CCHA	47	32	45	77	94	....	....	....	....	....	....	....	....	....	....	....	....	....	....	....	....	....	....	....
1989-90	Michigan State	CCHA	45	*48	53	*101	60	....	....	....	....	....	....	....	....	....	....	....	....	....	....	....	....	....	....	....
1990-91	**Quebec**	**NHL**	**13**	**4**	**3**	**7**	**7**	**0**	**0**	**0**	**16**	**25.0**	**–1**	....	....	....	....	....	....	....	....	....	....	....	....	....
	Halifax Citadels	AHL	66	36	33	69	40	....	....	....	....	....	....	....	....	....	....	....	....	....	....	....	....	....	....	....
1991-92	**Quebec**	**NHL**	**36**	**5**	**10**	**15**	**12**	**1**	**0**	**2**	**46**	**10.9**	**–21**	....	....	....	....	....	....	....	....	....	....	....	....	....
	Halifax Citadels	AHL	24	9	17	26	8	....	....	....	....	....	....	....	....	....	....	....	....	....	....	....	....	....	....	....
	Minnesota	**NHL**	**3**	**1**	**2**	**3**	**2**	**1**	**0**	**0**	**3**	**33.3**	**–1**	....	....	....	....	....	....	....	....	....	....	....	....	....
	Kalamazoo Wings	IHL	6	1	8	9	4	....	....	....	....	....	....	....	....	....	....	....	12	3	9	12	12	....	....	....

			Regular Season																Playoffs							
Season	Club	League	GP	G	A	Pts	PIM	PP	SH	GW	S	%	+/-	TF	F%	H	SB	Min	GP	G	A	Pts	PIM	PP	SH	GW
1992-93	Kalamazoo Wings	IHL	61	17	39	56	59																			
1993-94	San Jose	NHL	11	2	2	4	6	0	0	0	21	9.5	–1													
	Kansas City	IHL	71	38	54	92	51																			
1994-95	Denver Grizzlies	IHL	71	46	60	106	54												17	*15	14	29	8			
	NY Islanders	NHL	8	0	1	1	0	0	0	0	11	0.0	1													
1995-96	Chicago	NHL	10	1	4	5	2	0	0	0	12	8.3	1													
	Indianapolis Ice	IHL	73	32	59	91	46												5	2	6	8	2			
1996-97	Chicago Wolves	IHL	43	11	41	52	32																			
	Indianapolis Ice	IHL	37	17	24	41	18												4	2	2	4	2			
1997-98	Utah Grizzlies	IHL	72	38	59	97	30												4	3	2	5	10			
	NY Islanders	NHL	9	1	3	4	2	0	0	0	11	9.1	–2													
1998-99	Pittsburgh	NHL	77	19	23	42	22	1	0	4	125	15.2	1	150	44.7	67	25	16:55	13	2	7	9	19	1	0	0
99-2000	Pittsburgh	NHL	44	4	15	19	10	0	0	1	50	8.0	–1	132	40.2	23	13	14:18								
	Anaheim	NHL	30	6	17	23	4	2	0	1	32	18.8	1	7	42.9	28	1	13:44								
2000-01	Pittsburgh	NHL	33	3	8	11	6	1	0	0	38	7.9	0	61	50.8	18	6	9:46								
	Grand Rapids	IHL	34	16	19	35	12												10	5	8	13	2			
2001-02	Grand Rapids	AHL	41	21	35	56	27																			
	NY Islanders	NHL	37	7	17	24	6	2	0	1	52	13.5	2	119	58.8	14	4	14:07	7	4	2	6	2	2	0	1
	NHL Totals		311	53	105	158	79	8	0	9	417	12.7		469	47.8	150	49	14:26	20	6	9	15	21	3	0	1

CCHA First All-Star Team (1989, 1990) • CCHA Player of the Year (1990) • NCAA West First All-American Team (1989, 1990) • Won Hobey Baker Memorial Award (Top U.S. Collegiate Player) (1990)

Traded to **Minnesota** by **Quebec** for Steve Maltais, March 8, 1992. Signed as a free agent by **San Jose**, August 10, 1993. Signed as a free agent by **NY Islanders**, July 7, 1994. Signed as a free agent by **Chicago**, July 21, 1995. Signed as a free agent by **NY Islanders**, November 26, 1997. Claimed by **Pittsburgh** from **NY Islanders** in NHL Waiver Draft, October 5, 1998. Traded to **Anaheim** by **Pittsburgh** for Anaheim's 9th round choice (Roman Simicek) in 2000 Entry Draft, January 29, 2000. Signed as a free agent by **Pittsburgh**, September 24, 2000. Signed as a free agent by **Grand Rapids** (AHL), May 31, 2001. Signed as a free agent by **NY Islanders**, January 16, 2002. Signed as a free agent by **Washington**, July 9, 2002.

MILLEY, Norm

(MIHL-lee, NOHR-man) **BUF.**

Right wing. Shoots right. 6', 200 lbs. Born, Toronto, Ont., February 14, 1980. Buffalo's 3rd choice, 47th overall, in 1998 Entry Draft.

Season	Club	League	GP	G	A	Pts	PIM	PP	SH	GW	S	%	+/-	TF	F%	H	SB	Min	GP	G	A	Pts	PIM	PP	SH	GW
1995-96	Tor. Red Wings	MTHL	42	42	36	78																				
	St. Michael's B	OJHL-B	5	2	1	3	0																			
1996-97	Sudbury Wolves	OHL	61	30	32	62	15																			
1997-98	Sudbury Wolves	OHL	62	33	41	74	48												10	0	1	1	4			
1998-99	Sudbury Wolves	OHL	68	52	68	120	47												4	2	3	5	4			
99-2000	Sudbury Wolves	OHL	68	*52	60	112	47												12	8	11	19	6			
2000-01	Rochester	AHL	77	20	27	47	56												4	0	0	0	2			
2001-02	Buffalo	NHL	5	0	1	1	0	0	0	0	10	0.0	0	1	0.0	6	3	13:12								
	Rochester	AHL	74	20	18	38	20												2	0	3	3	6			
	NHL Totals		5	0	1	1	0	0	0	0	10	0.0		1	0.0	6	3	13:12								

OHL All-Rookie Team (1997) • OHL Second All-Star Team (1999) • OHL First All-Star Team (2000) • Canadian Major Junior First All-Star Team (2000)

MILLS, Craig

(MIHLS, KRAYG) **TOR.**

Right wing. Shoots right. 6', 190 lbs. Born, Toronto, Ont., August 27, 1976. Winnipeg's 5th choice, 108th overall, in 1994 Entry Draft.

Season	Club	League	GP	G	A	Pts	PIM	PP	SH	GW	S	%	+/-	TF	F%	H	SB	Min	GP	G	A	Pts	PIM	PP	SH	GW
1992-93	St. Michael's B	OJHL-B	44	9	21	30	42												15	1	6	7	8			
1993-94	Belleville Bulls	OHL	63	15	18	33	88												12	2	1	3	11			
1994-95	Belleville Bulls	OHL	62	39	41	80	104												13	7	9	16	8			
1995-96	Belleville Bulls	OHL	48	10	19	29	113												14	4	5	9	32			
	Winnipeg	NHL	4	0	2	2	0	0	0	0	0	0.0	0						1	0	0	0	0	0	0	0
	Springfield	AHL																	2	0	0	0	0			
1996-97	Indianapolis Ice	IHL	80	12	7	19	199												4	0	0	0	4			
1997-98	Chicago	NHL	20	0	3	3	34	0	0	0	5	0.0	1													
	Indianapolis Ice	IHL	42	8	11	19	119												5	0	0	0	27			
1998-99	Chicago	NHL	7	0	0	0	2	0	0	0	1	0.0	–2	0	0.0	4	0	5:48								
	Chicago Wolves	IHL	5	0	0	0	14																			
	Portland Pirates	AHL	48	7	11	18	59																			
	Indianapolis Ice	IHL	12	2	3	5	14												6	1	0	1	5			
99-2000	Springfield	AHL	78	10	13	23	151												5	2	1	3	6			
2000-01	Springfield	AHL	64	8	5	13	131																			
2001-02	St. John's	AHL	74	10	21	31	137												11	3	3	6	22			
	NHL Totals		31	0	5	5	36	0	0	0	6	0.0		0	0.0	4	0	5:48	1	0	0	0	0	0	0	0

Canadian Major Junior Humanitarian Player of the Year (1996)

Rights transferred to **Phoenix** after **Winnipeg** franchise relocated, July 1, 1996. Traded to **Chicago** by **Phoenix** with Alexei Zhamnov and Phoenix's 1st round choice (Ty Jones) in 1997 Entry Draft for Jeremy Roenick, August 16, 1996. Traded to **Phoenix** by **Chicago** for cash, September 11, 1999. Traded to **Toronto** by **Phoenix** with Robert Reichel and Travis Green for Danny Markov, June 12, 2001.

MIRONOV, Boris

(mih-RAWN-ohv, BOHR-ihs) **CHI.**

Defense. Shoots right. 6'3", 223 lbs. Born, Moscow, USSR, March 21, 1972. Winnipeg's 2nd choice, 27th overall, in 1992 Entry Draft.

Season	Club	League	GP	G	A	Pts	PIM	PP	SH	GW	S	%	+/-	TF	F%	H	SB	Min	GP	G	A	Pts	PIM	PP	SH	GW
1988-89	CSKA Moscow	USSR	1	0	0	0	0																			
1989-90	CSKA Moscow	USSR	7	0	0	0	0																			
1990-91	CSKA Moscow	USSR	36	1	5	6	16																			
1991-92	CSKA Moscow	CIS	36	2	1	3	22																			
1992-93	CSKA Moscow	CIS	19	0	5	5	20																			
1993-94	Winnipeg	NHL	65	7	22	29	96	5	0	0	122	5.7	–29													
	Edmonton	NHL	14	0	2	2	14	0	0	0	23	0.0	–4													
1994-95	Edmonton	NHL	29	1	7	8	40	0	0	0	48	2.1	–9													
	Cape Breton	AHL	4	2	5	7	23																			
1995-96	Edmonton	NHL	78	8	24	32	101	7	0	1	158	5.1	–23													
1996-97	Edmonton	NHL	55	6	26	32	85	2	0	1	147	4.1	2						12	2	8	10	16	2	0	0
1997-98	Edmonton	NHL	81	16	30	46	100	10	1	1	203	7.9	–8						12	3	3	6	27	1	0	1
	Russia	Olympics	6	0	2	2	2																			
1998-99	Edmonton	NHL	63	11	29	40	104	5	0	4	138	8.0	6	0	0.0	142	103	25:55								
	Chicago	NHL	12	0	9	9	27	0	0	0	35	0.0	7	0	0.0	32	17	24:17								
99-2000	Chicago	NHL	58	9	28	37	72	4	2	1	144	6.3	–3	1	100.0	121	60	24:53								
2000-01	Assat Pori	Finland	66	5	17	22	42																			
2001-02	Chicago	NHL	64	4	14	18	68	0	0	1	129	3.1	15	0	0.0	191	62	22:43	1	0	0	0	2	0	0	0
	Russia	Olympics	6	1	0	1	2																			
	NHL Totals		519	62	191	253	707	33	3	9	1147	5.4		1	100.0	486	242	24:28	25	5	11	16	45	3	0	1

NHL All-Rookie Team (1994)

Traded to **Edmonton** by **Winnipeg** with Mats Lindgren, Winnipeg's 1st round choice (Jason Bonsignore) in 1994 Entry Draft and Florida's 4th round choice (previously acquired, Edmonton selected Adam Copeland) in 1994 Entry Draft for Dave Manson and St. Louis' 6th round choice (previously acquired, Winnipeg selected Chris Kibermanis) in 1994 Entry Draft, March 15, 1994. Traded to **Chicago** by **Edmonton** with Dean McAmmond and Jonas Elofsson for Chad Kilger, Daniel Cleary, Ethan Moreau and Christian Laflamme, March 20, 1999.

MIRONOV, Dmitri

(mih-RAWN-ohv, dih-MEE-tree)

Defense. Shoots right. 6'4", 224 lbs. Born, Moscow, USSR, December 25, 1965. Toronto's 7th choice, 160th overall, in 1991 Entry Draft.

Season	Club	League	GP	G	A	Pts	PIM	PP	SH	GW	S	%	+/-	TF	F%	H	SB	Min	GP	G	A	Pts	PIM	PP	SH	GW
1985-86	CSKA Moscow	USSR	9	0	1	1	8																			
1986-87	CSKA Moscow	USSR	20	1	3	4	10																			
1987-88	Krylja Sovetov	USSR	44	12	6	18	30																			
1988-89	Krylja Sovetov	USSR	44	5	6	11	44																			
1989-90	Krylja Sovetov	USSR	45	4	11	15	34																			
1990-91	Krylja Sovetov	USSR	45	16	12	28	22																			
1991-92	Krylja Sovetov	CIS	35	15	16	31	62																			
	Russia	Olympics	8	3	1	4	6																			
	Toronto	NHL	7	1	0	1	0	0	0	1	7	14.3	–4													
1992-93	Toronto	NHL	59	7	24	31	40	4	0	1	105	6.7	–1						14	1	2	3	2	1	0	0
1993-94	Toronto	NHL	76	9	27	36	78	3	0	1	147	6.1	5						18	6	9	15	6	6	0	0
1994-95	Toronto	NHL	33	5	12	17	28	2	0	0	68	7.4	6						6	2	1	3	2	1	0	0
1995-96	Pittsburgh	NHL	72	3	31	34	88	1	0	1	86	3.5	19						15	0	1	1	10	0	0	0
1996-97	Pittsburgh	NHL	15	1	5	6	24	0	0	1	19	5.3	–4													
	Anaheim	NHL	62	12	34	46	77	3	1	1	158	7.6	20						11	1	10	11	10	1	0	0

Season	Club	League	GP	G	A	Pts	PIM	PP	SH	GW	S	%	+/-	TF	F%	H	SB	Min	Playoffs GP	G	A	Pts	PIM	PP	SH	GW
1997-98	Anaheim	NHL	66	6	30	36	115	2	0	1	142	4.2	−7													
	♦ Detroit	NHL	11	2	5	7	4	1	0	0	28	7.1	0						7	0	3	3	14	0	0	0
	Russia	Olympics	6	0	3	3	0																			
1998-99	Washington	NHL	46	2	14	16	80	2	0	0	86	2.3	−5	0	0.0	47	37	19:51								
99-2000	Washington	NHL	73	3	19	22	28	1	0	0	99	3.0	7	1	0.0	93	52	20:22	4	0	0	0	4	0	0	0
2000-01	Washington	NHL	36	3	5	8	6	1	0	1	33	9.1	−7	1	0.0	26	32	16:32								
	Houston Aeros	IHL	3	2	0	2	2																			
2001-02	Washington	NHL	DID NOT PLAY – INJURED																							
	NHL Totals		556	54	206	260	568	20	1	7	978	5.5		2	0.0	166	121	19:19	75	10	26	36	48	9	0	0

Played in NHL All-Star Game (1998)

Traded to **Pittsburgh** by **Toronto** with Toronto's 2nd round choice (later traded to New Jersey - New Jersey selected Josh DeWolf) in 1996 Entry Draft for Larry Murphy, July 8, 1995. Traded to **Anaheim** by **Pittsburgh** with Shawn Antoski for Alex Hicks and Fredrik Olausson, November 19, 1996. Traded to **Detroit** by **Anaheim** for Jamie Pushor and Detroit's 4th round choice (Viktor Wallin) in 1998 Entry Draft, March 24, 1998. Signed as a free agent by **Washington**, July 29, 1998. • Missed majority of 2000-01 season and entire 2001-02 season recovering from back injury suffered in game vs. Tampa Bay, January 23, 2001.

MITCHELL, Willie (MIHT-chehl, WIHL-lee) MIN.

Defense. Shoots left. 6'3", 205 lbs. Born, Port McNeill, B.C., April 23, 1977. New Jersey's 12th choice, 199th overall, in 1996 Entry Draft.

Season	Club	League	GP	G	A	Pts	PIM	PP	SH	GW	S	%	+/-	TF	F%	H	SB	Min	Playoffs GP	G	A	Pts	PIM	PP	SH	GW
1993-94	Notre Dame	SMHL	31	4	11	15	81																			
1994-95	Kelowna Spartans	BCHL	42	3	8	11	71																			
1995-96	Melfort Mustangs	SJHL	19	2	6	8													14	0	2	2	12			
1996-97	Melfort Mustangs	SJHL	64	14	42	56	227												4	0	1	1	23			
1997-98	Clarkson Knights	ECAC	34	9	17	26	105																			
1998-99	Clarkson Knights	ECAC	34	10	19	29	40																			
	Albany	AHL	6	1	3	4	29																			
99-2000	New Jersey	NHL	2	0	0	0	0	0	0	0	2	0.0	1	0	0.0	1	3	16:04								
	Albany	AHL	63	5	14	19	71												5	1	2	3	4			
2000-01	New Jersey	NHL	16	0	2	2	29	0	0	0	14	0.0	0	0	0.0	12	16	14:52								
	Albany	AHL	41	3	13	16	94																			
	Minnesota	NHL	17	1	7	8	11	0	0	0	16	6.3	4	0	0.0	32	24	20:49								
2001-02	Minnesota	NHL	68	3	10	13	68	0	0	1	67	4.5	−16	0	0.0	94	110	21:25								
	NHL Totals		103	4	19	23	108	0	0	1	99	4.0		0	0.0	139	153	20:12								

SJHL First All-Star Team (1997) • Won SJHL Top Defenseman Award (1997) • ECAC Rookie of the Year (Shared with Erik Cole) (1998) • ECAC Second All-Star Team (1998) • ECAC First All-Star Team (1999) • NCAA East Second All-American Team (1999)

Traded to **Minnesota** by **New Jersey** for Sean O'Donnell, March 4, 2001.

MODANO, Mike (moh-DA-noh, MIGHK) DAL.

Center. Shoots left. 6'3", 205 lbs. Born, Livonia, MI, June 7, 1970. Minnesota's 1st choice, 1st overall, in 1988 Entry Draft.

Season	Club	League	GP	G	A	Pts	PIM	PP	SH	GW	S	%	+/-	TF	F%	H	SB	Min	Playoffs GP	G	A	Pts	PIM	PP	SH	GW
1985-86	Det. Compuware	MNHL	69	66	65	131	32																			
1986-87	Prince Albert	WHL	70	32	30	62	96												8	1	4	5	4			
1987-88	Prince Albert	WHL	65	47	80	127	80												9	7	11	18	18			
1988-89	Prince Albert	WHL	41	39	66	105	74																			
	Minnesota	NHL																	2	0	0	0	0	0	0	0
1989-90	Minnesota	NHL	80	29	46	75	63	12	0	2	172	16.9	−7						7	1	1	2	12	0	0	0
1990-91	Minnesota	NHL	79	28	36	64	65	9	0	2	232	12.1	2						23	8	12	20	16	3	0	1
1991-92	Minnesota	NHL	76	33	44	77	46	5	0	8	256	12.9	−9						7	3	2	5	4	1	0	0
1992-93	Minnesota	NHL	82	33	60	93	83	9	0	7	307	10.7	−7													
1993-94	Dallas	NHL	76	50	43	93	54	18	0	4	281	17.8	−8						9	7	3	10	16	2	0	2
1994-95	Dallas	NHL	30	12	17	29	8	4	1	0	100	12.0	7													
1995-96	Dallas	NHL	78	36	45	81	63	8	4	4	320	11.3	−12													
1996-97	Dallas	NHL	80	35	48	83	42	9	5	9	291	12.0	43						7	4	1	5	0	1	1	2
1997-98	Dallas	NHL	52	21	38	59	32	7	5	2	191	11.0	25						17	4	10	14	12	1	0	1
	United States	Olympics	4	2	0	2	0																			
1998-99♦	Dallas	NHL	77	34	47	81	44	6	4	7	224	15.2	29	1572	51.1	15	33	20:50	23	5	*18	23	16	1	1	1
99-2000	Dallas	NHL	77	38	43	81	48	11	1	8	188	20.2	0	1763	51.4	16	43	22:55	23	10	*13	23	10	4	0	2
2000-01	Dallas	NHL	81	33	51	84	52	8	3	7	208	15.9	26	1791	52.0	27	36	22:24	9	3	4	7	0	2	0	0
2001-02	Dallas	NHL	78	34	43	77	38	6	2	5	219	15.5	14	1710	53.7	23	39	22:27								
	United States	Olympics	6	0	*6	6	4																			
	NHL Totals		946	416	561	977	638	112	25	65	2989	13.9		6836	52.1	81	151	22:09	127	45	64	109	86	15	2	9

WHL East First All-Star Team (1989) • NHL All-Rookie Team (1990) • NHL Second All-Star Team (2000) • Played in NHL All-Star Game (1993, 1998, 1999, 2000)

Transferred to **Dallas** after **Minnesota** franchise relocated, June 9, 1993.

MODIN, Fredrik (moh-DEEN, FREHD-rihk) T.B.

Left wing. Shoots left. 6'4", 225 lbs. Born, Sundsvall, Sweden, October 8, 1974. Toronto's 3rd choice, 64th overall, in 1994 Entry Draft.

Season	Club	League	GP	G	A	Pts	PIM	PP	SH	GW	S	%	+/-	TF	F%	H	SB	Min	Playoffs GP	G	A	Pts	PIM	PP	SH	GW
1991-92	Timra IK	Swede-2	11	1	0	1	0																			
1992-93	Timra IK	Swede-2	30	5	7	12	12												5	1	0	1	0			
1993-94	Timra IK	Swede-2	30	16	15	31	36												2	0	1	1	6			
1994-95	Brynas IF Gavle	Sweden	38	9	10	19	33												14	4	4	8	6			
1995-96	Brynas IF Gavle	Sweden	22	4	8	12	22																			
1996-97	Toronto	NHL	76	6	7	13	24	0	0	0	85	7.1	−14													
1997-98	Toronto	NHL	74	16	16	32	32	1	0	4	137	11.7	−5													
1998-99	Toronto	NHL	67	16	15	31	35	1	0	3	108	14.8	14	2	50.0	77	12	13:34	8	0	0	0	6	0	0	0
99-2000	Tampa Bay	NHL	80	22	26	48	18	3	0	5	167	13.2	−26	6	50.0	123	25	15:32								
2000-01	Tampa Bay	NHL	76	32	24	56	48	8	0	4	217	14.7	−1	21	42.9	65	32	17:15								
2001-02	Tampa Bay	NHL	54	14	17	31	27	2	0	4	141	9.9	0	25	40.0	36	27	19:05								
	NHL Totals		427	106	105	211	184	15	0	20	855	12.4		54	42.6	301	96	16:13	8	0	0	0	6	0	0	0

Played in NHL All-Star Game (2001)

Traded to **Tampa Bay** by **Toronto** for Cory Cross and Tampa Bay's 7th round choice (Ivan Kolozvary) in 2001 Entry Draft, October 1, 1999.

MODRY, Jaroslav (MOH-dree, YAHRO-slahv) L.A.

Defense. Shoots left. 6'2", 220 lbs. Born, Ceske Budejovice, Czech., February 27, 1971. New Jersey's 11th choice, 179th overall, in 1990 Entry Draft.

Season	Club	League	GP	G	A	Pts	PIM	PP	SH	GW	S	%	+/-	TF	F%	H	SB	Min	Playoffs GP	G	A	Pts	PIM	PP	SH	GW
1987-88	Ceske Budejovice	Czech	3	0	0	0	0																			
1988-89	Ceske Budejovice	Czech	28	0	1	1	8																			
1989-90	Ceske Budejovice	Czech	41	2	2	4																				
1990-91	Dukla Trencin	Czech	33	1	9	10	6																			
1991-92	Ceske Budejovice	Czech-2	14	4	10	14																				
	Dukla Trencin	Czech	18	0	4	4	6																			
1992-93	Utica Devils	AHL	80	7	35	42	62												5	0	2	2	2			
1993-94	New Jersey	NHL	41	2	15	17	18	2	0	0	35	5.7	10													
	Albany	AHL	19	1	5	6	25																			
1994-95	Ceske Budejovice	Czech	19	1	3	4	30																			
	New Jersey	NHL	11	0	0	0	0	0	0	0	10	0.0	−1													
	Albany	AHL	18	5	6	11	14												14	3	3	6	4			
1995-96	Ottawa	NHL	64	4	14	18	38	1	0	1	89	4.5	−17													
	Los Angeles	NHL	9	0	3	3	6	0	0	0	17	0.0	−4													
1996-97	Los Angeles	NHL	30	3	3	6	25	1	1	0	32	9.4	−13													
	Phoenix	IHL	23	3	12	15	17																			
	Utah Grizzlies	IHL	11	1	4	5	20												7	0	1	1	6			
1997-98	Utah Grizzlies	IHL	74	12	21	33	72												4	0	2	2	6			
1998-99	Los Angeles	NHL	5	0	1	1	0	0	0	0	11	0.0	1	0	0.0	6	7	26:00								
	Long Beach	IHL	64	6	29	35	44												8	4	2	6	4			
99-2000	Los Angeles	NHL	26	5	4	9	18	5	0	1	32	15.6	−2	0	0.0	20	24	19:13	2	0	0	0	2	0	0	0
	Long Beach	IHL	11	2	4	6	8																			

								Regular Season											Playoffs							
Season	Club	League	GP	G	A	Pts	PIM	PP	SH	GW	S	%	+/-	TF	F%	H	SB	Min	GP	G	A	Pts	PIM	PP	SH	GW
2000-01	Los Angeles	NHL	63	4	15	19	48	0	0	0	72	5.6	16	0	0.0	63	62	18:22	10	1	0	1	4	1	0	1
2001-02	Los Angeles	NHL	80	4	38	42	65	4	0	0	119	3.4	–4	0	0.0	70	82	19:31	7	0	2	2	0	0	0	0
	NHL Totals		**329**	**22**	**93**	**115**	**218**	**13**	**1**	**2**	**417**	**5.3**		**0**	**0.0**	**159**	**175**	**19:15**	**19**	**1**	**2**	**3**	**6**	**1**	**0**	**1**

Played in NHL All-Star Game (2002)

Traded to **Ottawa** by **New Jersey** for Ottawa's 4th round choice (Alyn McCauley) in 1995 Entry Draft, July 8, 1995. Traded to **LA Kings** by **Ottawa** with Ottawa's 8th round choice (Stephen Valiquette) in 1996 Entry Draft for Kevin Brown, March 20, 1996.

MOGILNY, Alexander

(moh-GIHL-nee, al-ehx-AN-duhr) **TOR.**

Right wing. Shoots left. 6', 200 lbs. Born, Khabarovsk, USSR, February 18, 1969. Buffalo's 4th choice, 89th overall, in 1988 Entry Draft.

Season	Club	League	GP	G	A	Pts	PIM	PP	SH	GW	S	%	+/-	TF	F%	H	SB	Min	GP	G	A	Pts	PIM	PP	SH	GW
1986-87	CSKA Moscow	USSR	28	15	1	16	4																			
1987-88	CSKA Moscow	USSR	39	12	8	20	14																			
	Soviet Union	Olympics	6	3	2	5	2																			
1988-89	CSKA Moscow	USSR	31	11	11	22	24																			
1989-90	**Buffalo**	**NHL**	**65**	**15**	**28**	**43**	**16**	**4**	**0**	**2**	**130**	**11.5**	**8**						**4**	**0**	**1**	**1**	**2**	**0**	**0**	**0**
1990-91	**Buffalo**	**NHL**	**62**	**30**	**34**	**64**	**16**	**3**	**3**	**5**	**201**	**14.9**	**14**						**6**	**0**	**6**	**6**	**2**	**0**	**0**	**0**
1991-92	**Buffalo**	**NHL**	**67**	**39**	**45**	**84**	**73**	**15**	**0**	**2**	**236**	**16.5**	**7**						**2**	**0**	**2**	**2**	**0**	**0**	**0**	**0**
1992-93	**Buffalo**	**NHL**	**77**	***76**	**51**	**127**	**40**	**27**	**0**	**11**	**360**	**21.1**	**7**						**7**	**7**	**3**	**10**	**6**	**2**	**0**	**0**
1993-94	**Buffalo**	**NHL**	**66**	**32**	**47**	**79**	**22**	**17**	**0**	**7**	**258**	**12.4**	**8**						**7**	**4**	**2**	**6**	**6**	**1**	**0**	**0**
1994-95	Spartak Moscow	CIS	1	0	1	1	0																			
	Buffalo	**NHL**	**44**	**19**	**28**	**47**	**36**	**12**	**0**	**2**	**148**	**12.8**	**0**						**5**	**3**	**2**	**5**	**2**	**0**	**0**	**0**
1995-96	**Vancouver**	**NHL**	**79**	**55**	**52**	**107**	**16**	**10**	**5**	**6**	**292**	**18.8**	**14**						**6**	**1**	**8**	**9**	**8**	**0**	**0**	**0**
1996-97	**Vancouver**	**NHL**	**76**	**31**	**42**	**73**	**18**	**7**	**1**	**4**	**174**	**17.8**	**9**													
1997-98	**Vancouver**	**NHL**	**51**	**18**	**27**	**45**	**36**	**5**	**4**	**1**	**118**	**15.3**	**–6**													
1998-99	**Vancouver**	**NHL**	**59**	**14**	**31**	**45**	**58**	**3**	**2**	**1**	**110**	**12.7**	**0**	**47**	**23.4**	**34**	**10**	**20:35**								
99-2000	**Vancouver**	**NHL**	**47**	**21**	**17**	**38**	**16**	**3**	**1**	**1**	**126**	**16.7**	**7**	**9**	**11.1**	**37**	**12**	**19:34**								
	♦ New Jersey	**NHL**	**12**	**3**	**3**	**6**	**4**	**2**	**0**	**0**	**35**	**8.6**	**–4**	**0**	**0.0**	**4**	**3**	**17:04**	**23**	**4**	**3**	**7**	**4**	**2**	**0**	**1**
2000-01	**New Jersey**	**NHL**	**75**	**43**	**40**	**83**	**43**	**12**	**0**	**7**	**240**	**17.9**	**10**	**11**	**36.4**	**86**	**8**	**16:53**	**25**	**5**	**11**	**16**	**8**	**1**	**0**	**2**
2001-02	**Toronto**	**NHL**	**66**	**24**	**33**	**57**	**8**	**5**	**0**	**4**	**188**	**12.8**	**1**	**5**	**40.0**	**51**	**10**	**17:38**	**20**	**8**	**3**	**11**	**8**	**2**	**0**	**2**
	NHL Totals		**846**	**420**	**478**	**898**	**402**	**125**	**16**	**53**	**2616**	**16.1**		**72**	**25.0**	**212**	**43**	**18:25**	**105**	**32**	**41**	**73**	**46**	**8**	**0**	**5**

NHL Second All-Star Team (1993, 1996) • Played in NHL All-Star Game (1992, 1993, 1994, 1996)

Traded to **Vancouver** by **Buffalo** with Buffalo's 5th round choice (Todd Norman) in 1995 Entry Draft for Michael Peca, Mike Wilson and Vancouver's 1st round choice (Jay McKee) in 1995 Entry Draft, July 8, 1995. Traded to **New Jersey** by **Vancouver** for Brendan Morrison and Denis Pederson, March 14, 2000. Signed as a free agent by **Toronto**, July 3, 2001.

MONTADOR, Steve

(MAWN-tuh-dohr, STEEV) **CGY.**

Defense. Shoots right. 6', 200 lbs. Born, Vancouver, B.C., December 21, 1979.

Season	Club	League	GP	G	A	Pts	PIM	PP	SH	GW	S	%	+/-	TF	F%	H	SB	Min	GP	G	A	Pts	PIM	PP	SH	GW
1995-96	St. Michael's B	OPJHL	46	3	16	19	145																			
1996-97	North Bay	OHL	63	7	28	35	129																			
1997-98	North Bay	OHL	37	5	16	21	54																			
	Erie Otters	OHL	26	3	17	20	35												7	1	1	2	9			
1998-99	Erie Otters	OHL	61	9	33	42	114												5	0	2	2	4			
99-2000	Peterborough	OHL	64	14	42	56	97												5	0	2	2	4			
	Saint John Flames	AHL																	2	0	0	0	0			
2000-01	Saint John	AHL	58	1	6	7	95												19	0	8	8	13			
2001-02	**Calgary**	**NHL**	**11**	**1**	**2**	**3**	**26**	**0**	**0**	**0**	**10**	**10.0**	**–2**	**0**	**0.0**	**5**	**4**	**12:12**								
	Saint John	AHL	67	9	16	25	107																			
	NHL Totals		**11**	**1**	**2**	**3**	**26**	**0**	**0**	**0**	**10**	**10.0**		**0**	**0.0**	**5**	**4**	**12:12**								

Signed as a free agent by **Calgary**, April 10, 2000.

MONTGOMERY, Jim

(mawnt-GUHM-uhr-ee, JIHM) **DAL.**

Center. Shoots right. 5'10", 180 lbs. Born, Montreal, Que., June 30, 1969.

Season	Club	League	GP	G	A	Pts	PIM	PP	SH	GW	S	%	+/-	TF	F%	H	SB	Min	GP	G	A	Pts	PIM	PP	SH	GW
1988-89	Pembroke	OCJHL	50	53	*101	154	112																			
1989-90	U. of Maine	H-East	45	26	34	60	35																			
1990-91	U. of Maine	H-East	43	24	*57	81	44																			
1991-92	U. of Maine	H-East	37	21	44	65	46																			
1992-93	U. of Maine	H-East	45	32	63	95	40																			
1993-94	**St. Louis**	**NHL**	**67**	**6**	**14**	**20**	**44**	**0**	**0**	**1**	**67**	**9.0**	**–1**													
	Peoria Rivermen	IHL	12	7	8	15	10																			
1994-95	**Montreal**	**NHL**	**5**	**0**	**0**	**0**	**2**	**0**	**0**	**0**	**3**	**0.0**	**–2**													
	Philadelphia	**NHL**	**8**	**1**	**1**	**2**	**6**	**0**	**0**	**0**	**10**	**10.0**	**–2**						**7**	**1**	**0**	**1**	**2**	**0**	**0**	**0**
	Hershey Bears	AHL	16	8	6	14	14												6	3	2	5	25			
1995-96	**Philadelphia**	**NHL**	**5**	**1**	**2**	**3**	**9**	**0**	**0**	**0**	**4**	**25.0**	**1**						**1**	**0**	**0**	**0**	**0**	**0**	**0**	**0**
	Hershey Bears	AHL	78	34	*71	105	95												4	3	2	5	6			
1996-97	Kolner Haie	Germany	50	12	35	47	111												4	0	1	1	6			
	Kolner Haie	EuroHL	6	0	1	1	16																			
1997-98	Philadelphia	AHL	68	19	43	62	75												20	*13	16	29	55			
1998-99	Philadelphia	AHL	78	29	58	87	89												16	4	11	15	20			
99-2000	Philadelphia	AHL	13	3	9	12	22																			
	Manitoba Moose	IHL	67	18	28	46	111																			
2000-01	**San Jose**	**NHL**	**28**	**1**	**6**	**7**	**19**	**1**	**0**	**0**	**17**	**5.9**		**0**	**0.0**	**0**	**0**	**0:00**								
	Kentucky	AHL	55	22	52	74	44												3	1	2	3	5			
2001-02	**Dallas**	**NHL**	**8**	**0**	**2**	**2**	**0**	**0**	**0**	**0**	**8**	**0.0**	**–1**	**38**	**63.2**	**6**	**2**	**8:35**								
	Utah Grizzlies	AHL	71	28	43	71	90												5	0	1	1	23			
	NHL Totals		**121**	**9**	**25**	**34**	**80**	**1**	**0**	**1**	**109**	**8.3**		**38**	**63.2**	**6**	**2**	**8:35**	**8**	**1**	**0**	**1**	**2**	**0**	**0**	**0**

Hockey East Second All-Star Team (1991, 1992) • Hockey East First All-Star Team (1993) • NCAA East Second All-American Team (1993) • NCAA Championship All-Tournament Team (1993) • NCAA Championship Tournament MVP (1993) • AHL Second All-Star Team (1996)

Signed as a free agent by **St. Louis**, June 2, 1993. Traded to **Montreal** by **St. Louis** for Guy Carbonneau, August 19, 1994. Claimed on waivers by **Philadelphia** from **Montreal**, February 10, 1995. Signed as a free agent by **San Jose**, August 15, 2000. Signed as a free agent by **Dallas**, July 10, 2001.

MOORE, Barrie

(MOOR, BAIR-ee)

Left wing. Shoots left. 5'11", 198 lbs. Born, London, Ont., May 22, 1975. Buffalo's 7th choice, 220th overall, in 1993 Entry Draft.

Season	Club	League	GP	G	A	Pts	PIM	PP	SH	GW	S	%	+/-	TF	F%	H	SB	Min	GP	G	A	Pts	PIM	PP	SH	GW
1990-91	Strathroy	OJHL-B	24	9	10	19	14																			
1991-92	Sudbury Wolves	OHL	62	15	38	53	57												11	0	7	7	12			
1992-93	Sudbury Wolves	OHL	57	13	26	39	71												14	4	3	7	19			
1993-94	Sudbury Wolves	OHL	65	36	49	85	69												10	3	5	8	14			
1994-95	Sudbury Wolves	OHL	60	47	42	89	67												18	*15	14	29	24			
1995-96	**Buffalo**	**NHL**	**3**	**0**	**0**	**0**	**0**	**0**	**0**	**0**	**3**	**0.0**	**0**													
	Rochester	AHL	64	26	30	56	40												18	3	6	9	18			
1996-97	**Buffalo**	**NHL**	**31**	**2**	**6**	**8**	**18**	**1**	**0**	**0**	**42**	**4.8**	**1**													
	Rochester	AHL	32	14	15	29	14																			
	Edmonton	**NHL**	**4**	**0**	**0**	**0**	**0**	**0**	**0**	**0**	**1**	**0.0**	**0**													
	Hamilton	AHL	9	5	2	7	0												22	2	6	8	15			
1997-98	Hamilton	AHL	70	22	29	51	64												8	0	1	1	4			
1998-99	Indianapolis Ice	IHL	43	9	10	19	18																			
	Portland Pirates	AHL	23	3	7	10	4																			
99-2000	**Washington**	**NHL**	**1**	**0**	**0**	**0**	**0**	**0**	**0**	**0**	**2**	**0.0**	**0**	**0**	**0.0**	**0**	**0**	**9:50**								
	Portland Pirates	AHL	80	18	33	51	50												4	0	0	0	6			
2000-01	Manitoba Moose	IHL	2	0	0	0	0																			
	Manchester Storm	Britain	32	11	16	27	26												6	3	3	6	2			
2001-02	Columbia Inferno	ECHL	48	20	24	44	34												5	0	2	2	6			
	Portland Pirates	AHL	14	2	3	5	2																			
	NHL Totals		**39**	**2**	**6**	**8**	**18**	**1**	**0**	**0**	**48**	**4.2**		**0**	**0.0**	**0**	**0**	**9:50**								

Traded to **Edmonton** by **Buffalo** with Craig Millar for Miroslav Satan, March 18, 1997. Rights traded to **Washington** by **Edmonton** for Brad Church, February 3, 1999. Selected by **Columbus** from **Washington** in Expansion Draft, June 23, 2000. Signed as a free agent by **Columbia** (ECHL), April 10, 2001.

MOORE, Steve
(MOOR, STEEV) **COL.**

Center. Shoots right. 6'2", 205 lbs. Born, Windsor, Ont., September 22, 1978. Colorado's 7th choice, 53rd overall, in 1998 Entry Draft.

			Regular Season																Playoffs							
Season	Club	League	GP	G	A	Pts	PIM	PP	SH	GW	S	%	+/-	TF	F%	H	SB	Min	GP	G	A	Pts	PIM	PP	SH	GW
1995-96	Thornhill	MTJHL	50	25	27	52	57												18	4	5	9				
1996-97	Thornhill	MTJHL	50	34	52	86	52												13	10	11	21	2			
1997-98	Harvard Crimson	ECAC	33	10	23	33	46																			
1998-99	Harvard Crimson	ECAC	30	18	13	31	34																			
99-2000	Harvard Crimson	ECAC	27	10	16	26	53																			
2000-01	Harvard Crimson	ECAC	32	7	26	33	43																			
2001-02	**Colorado**	**NHL**	**8**	**0**	**0**	**0**	**4**	**0**	**0**	**0**	**5**	**0.0**	**−4**	**33**	**51.5**	**13**	**1**	**7:05**								
	Hershey Bears	AHL	68	10	17	27	31												8	0	1	1	6			
	NHL Totals		**8**	**0**	**0**	**0**	**4**	**0**	**0**	**0**	**5**	**0.0**		**33**	**51.5**	**13**	**1**	**7:05**								

MORAN, Brad
(moh-RAN, BRAD) **CBJ**

Center. Shoots left. 5'11", 187 lbs. Born, Abbotsford, B.C., March 20, 1979. Buffalo's 8th choice, 191st overall, in 1998 Entry Draft.

Season	Club	League	GP	G	A	Pts	PIM	PP	SH	GW	S	%	+/-	TF	F%	H	SB	Min	GP	G	A	Pts	PIM	PP	SH	GW
1994-95	Abbotsford Hawks	BCAHA	56	66	93	159	40																			
1995-96	Calgary Hitmen	WHL	70	13	31	44	28																			
1996-97	Calgary Hitmen	WHL	72	30	36	66	61																			
1997-98	Calgary Hitmen	WHL	72	53	49	102	64												18	10	8	18	20			
1998-99	Calgary Hitmen	WHL	71	60	58	118	96												21	17	*25	42	26			
99-2000	Calgary Hitmen	WHL	72	48	*72	*120	84												13	7	15	22	18			
2000-01	Syracuse Crunch	AHL	71	11	19	30	30												5	3	4	7	2			
2001-02	**Columbus**	**NHL**	**3**	**0**	**0**	**0**	**0**	**0**	**0**	**0**	**2**	**0.0**	**0**	**22**	**40.9**	**0**	**1**	**7:39**								
	Syracuse Crunch	AHL	64	25	24	49	51												10	5	8	13	2			
	NHL Totals		**3**	**0**	**0**	**0**	**0**	**0**	**0**	**0**	**2**	**0.0**		**22**	**40.9**	**0**	**1**	**7:39**								

WHL East First All-Star Team (1999, 2000)

Signed as a free agent by **Columbus**, June 5, 2000.

MORAN, Ian
(moh-RAN, EE-an) **PIT.**

Defense. Shoots right. 6', 200 lbs. Born, Cleveland, OH, August 24, 1972. Pittsburgh's 5th choice, 107th overall, in 1990 Entry Draft.

Season	Club	League	GP	G	A	Pts	PIM	PP	SH	GW	S	%	+/-	TF	F%	H	SB	Min	GP	G	A	Pts	PIM	PP	SH	GW
1987-88	Belmont Hill	Hi-School	25	3	13	16	15																			
1988-89	Belmont Hill	Hi-School	23	7	25	32	8																			
1989-90	Belmont Hill	Hi-School	23	10	36	46																				
1990-91	Belmont Hill	Hi-School	23	7	44	51	12																			
1991-92	Boston College	H-East	30	2	16	18	44																			
1992-93	Boston College	H-East	31	8	12	20	32																			
1993-94	Team USA	Nat-Tm	50	8	15	23	69																			
	Cleveland	IHL	33	5	13	18	39																			
1994-95	Cleveland	IHL	64	7	31	38	94												4	0	1	1	2			
	Pittsburgh	**NHL**																	**8**	**0**	**0**	**0**	**0**	**0**	**0**	**0**
1995-96	**Pittsburgh**	**NHL**	**51**	**1**	**1**	**2**	**47**	**0**	**0**	**0**	**44**	**2.3**	**−1**													
1996-97	**Pittsburgh**	**NHL**	**36**	**4**	**5**	**9**	**22**	**0**	**0**	**0**	**50**	**8.0**	**−11**						**5**	**1**	**2**	**3**	**4**	**0**	**0**	**0**
	Cleveland	IHL	36	6	23	29	26																			
1997-98	**Pittsburgh**	**NHL**	**37**	**1**	**6**	**7**	**19**	**0**	**0**	**1**	**33**	**3.0**	**0**						**6**	**0**	**0**	**0**	**2**	**0**	**0**	**0**
1998-99	**Pittsburgh**	**NHL**	**62**	**4**	**5**	**9**	**37**	**0**	**1**	**0**	**65**	**6.2**	**1**	**32**	**34.4**	**48**	**98**	**16:34**	**13**	**0**	**2**	**2**	**8**	**0**	**0**	**0**
99-2000	**Pittsburgh**	**NHL**	**73**	**4**	**8**	**12**	**28**	**0**	**0**	**0**	**58**	**6.9**	**−10**	**210**	**33.8**	**48**	**62**	**11:21**	**11**	**0**	**1**	**1**	**2**	**0**	**0**	**0**
2000-01	**Pittsburgh**	**NHL**	**40**	**3**	**4**	**7**	**28**	**0**	**0**	**1**	**73**	**4.1**	**5**	**4**	**25.0**	**74**	**34**	**17:42**	**18**	**0**	**1**	**1**	**4**	**0**	**0**	**0**
2001-02	**Pittsburgh**	**NHL**	**64**	**2**	**8**	**10**	**54**	**0**	**0**	**1**	**94**	**2.1**	**−11**	**2**	**100.0**	**46**	**105**	**20:00**								
	NHL Totals		**363**	**19**	**37**	**56**	**235**	**0**	**1**	**3**	**417**	**4.6**		**248**	**34.3**	**216**	**299**	**16:05**	**61**	**1**	**6**	**7**	**20**	**0**	**0**	**0**

Hockey East Rookie of the Year (Shared with Craig Darby) (1992) • Hockey East All-Rookie Team (1992)

• Missed majority of 1997-98 season recovering from knee injury suffered in training camp, September 30, 1997. • Missed majority of 2000-01 season recovering from hand injury originally suffered in game vs. Edmonton, November 11, 2000.

MORAVEC, David
(muh-RAHV-ehts, DAY-vihd) **BUF.**

Right wing. Shoots left. 6', 180 lbs. Born, Vitkovice, Czech., March 24, 1973. Buffalo's 9th choice, 218th overall, in 1998 Entry Draft.

Season	Club	League	GP	G	A	Pts	PIM	PP	SH	GW	S	%	+/-	TF	F%	H	SB	Min	GP	G	A	Pts	PIM	PP	SH	GW
1994-95	HC Vitkovice	Czech	38	4	13	17	12												6	1	7	8	0			
1995-96	HC Vitkovice	Czech	37	6	5	11	14												4	0	0	0	4			
1996-97	HC Vitkovice	Czech	52	18	22	40	30												9	6	3	9	0			
1997-98	HC Vitkovice	Czech	51	*38	26	64	28												11	6	9	15	8			
1998-99	HC Vitkovice	Czech	50	21	22	43	44												4	1	1	2				
99-2000	**Buffalo**	**NHL**	**1**	**0**	**0**	**0**	**0**	**0**	**0**	**0**	**2**	**0.0**	**−1**	**2**	**50.0**	**0**	**0**	**15:15**								
	HC Vitkovice	Czech	38	11	18	29	34																			
2000-01	HC Vitkovice	Czech	51	15	20	35	34												10	4	6	10	4			
2001-02	HC Vitkovice	Czech	46	18	26	44	32												14	7	7	14	2			
	NHL Totals		**1**	**0**	**0**	**0**	**0**	**0**	**0**	**0**	**2**	**0.0**		**2**	**50.0**	**0**	**0**	**15:15**								

MOREAU, Ethan
(moh-ROH, EE-than) **EDM.**

Left wing. Shoots left. 6'2", 211 lbs. Born, Huntsville, Ont., September 22, 1975. Chicago's 1st choice, 14th overall, in 1994 Entry Draft.

Season	Club	League	GP	G	A	Pts	PIM	PP	SH	GW	S	%	+/-	TF	F%	H	SB	Min	GP	G	A	Pts	PIM	PP	SH	GW
1990-91	Orillia Terriers	OPJHL	42	17	22	39	26												12	6	6	12	18			
1991-92	Niagara Falls	OHL	62	20	35	55	39												17	4	6	10	4			
1992-93	Niagara Falls	OHL	65	32	41	73	69												4	0	3	3	4			
1993-94	Niagara Falls	OHL	59	44	54	98	100																			
1994-95	Niagara Falls	OHL	39	25	41	66	69																			
	Sudbury Wolves	OHL	23	13	17	30	22												18	6	12	18	26			
1995-96	**Chicago**	**NHL**	**8**	**0**	**1**	**1**	**4**	**0**	**0**	**0**	**1**	**0.0**	**1**													
	Indianapolis Ice	IHL	71	21	20	41	126												5	4	0	4	8			
1996-97	**Chicago**	**NHL**	**82**	**15**	**16**	**31**	**123**	**0**	**0**	**1**	**114**	**13.2**	**13**						**6**	**1**	**0**	**1**	**9**	**0**	**0**	**0**
1997-98	**Chicago**	**NHL**	**54**	**9**	**9**	**18**	**73**	**2**	**0**	**0**	**87**	**10.3**	**0**													
1998-99	**Chicago**	**NHL**	**66**	**9**	**6**	**15**	**84**	**0**	**0**	**1**	**80**	**11.3**	**−5**	**3**	**33.3**	**113**	**15**	**12:30**								
	Edmonton	**NHL**	**14**	**1**	**5**	**6**	**8**	**0**	**0**	**1**	**16**	**6.3**	**2**	**1**	**0.0**	**26**	**9**	**11:47**	**4**	**0**	**3**	**3**	**6**	**0**	**0**	**0**
99-2000	**Edmonton**	**NHL**	**73**	**17**	**10**	**27**	**62**	**1**	**0**	**3**	**106**	**16.0**	**8**	**8**	**62.5**	**158**	**35**	**15:07**	**5**	**0**	**1**	**1**	**0**	**0**	**0**	**0**
2000-01	**Edmonton**	**NHL**	**68**	**9**	**10**	**19**	**90**	**0**	**1**	**3**	**97**	**9.3**	**−6**	**2**	**0.0**	**148**	**46**	**14:11**	**4**	**0**	**0**	**0**	**2**	**0**	**0**	**0**
2001-02	**Edmonton**	**NHL**	**80**	**11**	**5**	**16**	**81**	**0**	**2**	**1**	**129**	**8.5**	**4**	**11**	**54.6**	**134**	**51**	**12:43**								
	NHL Totals		**445**	**71**	**62**	**133**	**525**	**3**	**3**	**10**	**630**	**11.3**		**25**	**48.0**	**579**	**156**	**13:32**	**19**	**1**	**4**	**5**	**17**	**0**	**0**	**0**

OHL All-Rookie Team (1992)

Traded to **Edmonton** by **Chicago** with Daniel Cleary, Chad Kilger and Christian Laflamme for Boris Mironov, Dean McAmmond and Jonas Elofsson, March 20, 1999.

MORGAN, Jason
(MOHR-gan, JAY-son) **CGY.**

Center. Shoots left. 6'1", 200 lbs. Born, St. John's, Nfld., October 9, 1976. Los Angeles' 5th choice, 118th overall, in 1995 Entry Draft.

Season	Club	League	GP	G	A	Pts	PIM	PP	SH	GW	S	%	+/-	TF	F%	H	SB	Min	GP	G	A	Pts	PIM	PP	SH	GW
1992-93	Kit. Rangers	OMHA	69	44	40	84	85																			
1993-94	Kitchener	OHL	65	6	15	21	16												5	1	0	1	0			
1994-95	Kitchener	OHL	35	3	15	18	25																			
	Kingston	OHL	20	0	3	3	14												6	0	2	2	0			
1995-96	Kingston	OHL	66	16	38	54	50												6	1	2	3	0			
1996-97	**Los Angeles**	**NHL**	**3**	**0**	**0**	**0**	**0**	**0**	**0**	**0**	**4**	**0.0**	**−3**													
	Phoenix	IHL	57	3	6	9	29																			
	Mississippi	ECHL	6	3	0	3	0												3	1	1	2	6			
1997-98	**Los Angeles**	**NHL**	**11**	**1**	**0**	**1**	**4**	**0**	**0**	**0**	**5**	**20.0**	**−7**													
	Springfield	AHL	58	13	22	35	66												3	1	0	1	18			
1998-99	Long Beach	IHL	13	4	6	10	18																			
	Springfield	AHL	46	6	16	22	51												3	0	0	0	6			
99-2000	Cincinnati	IHL	15	1	3	4	14																			
	Florida	ECHL	48	14	25	39	79												5	2	2	4	16			

Season	Club	League	GP	G	A	Pts	PIM	PP	SH	GW	S	%	+/-	TF	F%	H	SB	Min	GP	G	A	Pts	PIM	PP	SH	GW
			Regular Season																Playoffs							
2000-01	Florida	ECHL	37	15	22	37	41	....	....	....	....	....	....	....	....	....	....	....	5	2	3	5	17	....	....	....
	Hamilton	AHL	11	2	0	2	10	....	....	....	....	....	....	....	....	....	....	....	....	....	....	....	....	....	....	....
	Springfield	AHL	16	1	4	5	19	....	....	....	....	....	....	....	....	....	....	....	....	....	....	....	....	....	....	....
	Saint John	AHL	....	....	....	....	....	....	....	....	....	....	....	....	....	....	....	....	6	0	1	1	2	....	....	....
2001-02	Saint John	AHL	76	17	20	37	69	....	....	....	....	....	....	....	....	....	....	....	....	....	....	....	....	....	....	....
	NHL Totals		**14**	**1**	**0**	**1**	**4**	**0**	**0**	**0**	**9**	**11.1**		....	....	....	....	....	....	....	....	....	....	....	....	....

Signed to tryout contract by **Saint John** (AHL), April 22, 2001. Signed as a free agent by **Saint John** (AHL), August 28, 2001. Signed as a free agent by **Calgary**, July 11, 2002.

MORISSET, Dave

(moh-rih-SEHT, DAYV) **FLA.**

Right wing. Shoots right. 6'2", 195 lbs. Born, Langley, B.C., April 6, 1981. St. Louis' 2nd choice, 65th overall, in 2000 Entry Draft.

Season	Club	League	GP	G	A	Pts	PIM	PP	SH	GW	S	%	+/-	TF	F%	H	SB	Min	GP	G	A	Pts	PIM	PP	SH	GW
1997-98	Seattle	WHL	58	6	2	8	104	....	....	....	....	....	....	....	....	....	....	....	5	1	0	1	6	....	....	....
1998-99	Seattle	WHL	17	4	0	4	31	....	....	....	....	....	....	....	....	....	....	....	11	1	1	2	22	....	....	....
99-2000	Seattle	WHL	60	23	34	57	69	....	....	....	....	....	....	....	....	....	....	....	7	3	4	7	12	....	....	....
2000-01	Seattle	WHL	61	32	36	68	95	....	....	....	....	....	....	....	....	....	....	....	9	4	2	6	12	....	....	....
2001-02	**Florida**	**NHL**	**4**	**0**	**0**	**0**	**5**	**0**	**0**	**0**	**2**	**0.0**	**–7**	**1**	**0.0**	**12**	**0**	**11:21**	....	....	....	....	....	....	....	....
	Bridgeport	AHL	62	9	10	19	46	....	....	....	....	....	....	....	....	....	....	....	19	0	1	1	13	....	....	....
	NHL Totals		**4**	**0**	**0**	**0**	**5**	**0**	**0**	**0**	**2**	**0.0**		**1**	**0.0**	**12**	**0**	**11:21**	....	....	....	....	....	....	....	....

• Missed majority of 1998-99 season recovering from shoulder injury suffered in practice, November, 1998. • Rights traded to **Florida** by **St. Louis** with St. Louis' 5th round choice (Vince Bellissimo) in 2002 Entry Draft for Scott Mellanby, February 9, 2001.

MORO, Marc

(MOH-roh, MAHRK) **TOR.**

Defense. Shoots left. 6'1", 220 lbs. Born, Toronto, Ont., July 17, 1977. Ottawa's 2nd choice, 27th overall, in 1995 Entry Draft.

Season	Club	League	GP	G	A	Pts	PIM	PP	SH	GW	S	%	+/-	TF	F%	H	SB	Min	GP	G	A	Pts	PIM	PP	SH	GW
1992-93	Mississauga Reps	MTHL	42	9	18	27	56	....	....	....	....	....	....	....	....	....	....	....	....	....	....	....	....	....	....	....
	Mississauga Sens	MTJHL	2	0	0	0	0	....	....	....	....	....	....	....	....	....	....	....	....	....	....	....	....	....	....	....
1993-94	Kingston	MTJHL	12	0	2	2	10	....	....	....	....	....	....	....	....	....	....	....	....	....	....	....	....	....	....	....
	Kingston	OHL	43	0	3	3	81	....	....	....	....	....	....	....	....	....	....	....	....	....	....	....	....	....	....	....
1994-95	Kingston	OHL	64	4	12	16	255	....	....	....	....	....	....	....	....	....	....	....	6	0	0	0	23	....	....	....
1995-96	Kingston	OHL	66	4	17	21	261	....	....	....	....	....	....	....	....	....	....	....	6	0	0	0	12	....	....	....
	P.E.I. Senators	AHL	2	0	0	0	7	....	....	....	....	....	....	....	....	....	....	....	2	0	0	0	4	....	....	....
1996-97	Kingston	OHL	37	4	8	12	97	....	....	....	....	....	....	....	....	....	....	....	....	....	....	....	....	....	....	....
	Sault Ste. Marie	OHL	26	0	5	5	74	....	....	....	....	....	....	....	....	....	....	....	11	1	6	7	38	....	....	....
1997-98	**Anaheim**	**NHL**	**1**	**0**	**0**	**0**	**0**	**0**	**0**	**0**	**0**	**0.0**	**0**	....	....	....	....	....	....	....	....	....	....	....	....	....
	Cincinnati	AHL	74	1	6	7	181	....	....	....	....	....	....	....	....	....	....	....	....	....	....	....	....	....	....	....
1998-99	Milwaukee	IHL	80	0	5	5	264	....	....	....	....	....	....	....	....	....	....	....	2	0	0	0	4	....	....	....
99-2000	**Nashville**	**NHL**	**8**	**0**	**0**	**0**	**40**	**0**	**0**	**0**	**3**	**0.0**	**–3**	**0**	**0.0**	**19**	**2**	**10:55**	....	....	....	....	....	....	....	....
	Milwaukee	IHL	64	5	5	10	203	....	....	....	....	....	....	....	....	....	....	....	....	....	....	....	....	....	....	....
2000-01	**Nashville**	**NHL**	**6**	**0**	**0**	**0**	**12**	**0**	**0**	**0**	**1**	**0.0**	**1**	**0**	**0.0**	**7**	**0**	**3:34**	....	....	....	....	....	....	....	....
	Milwaukee	IHL	68	2	9	11	190	....	....	....	....	....	....	....	....	....	....	....	5	1	0	1	10	....	....	....
2001-02	**Nashville**	**NHL**	**13**	**0**	**0**	**0**	**23**	**0**	**0**	**0**	**7**	**0.0**	**–3**	**0**	**0.0**	**33**	**5**	**12:02**	....	....	....	....	....	....	....	....
	Milwaukee	AHL	41	1	8	9	81	....	....	....	....	....	....	....	....	....	....	....	....	....	....	....	....	....	....	....
	Toronto	**NHL**	**2**	**0**	**0**	**0**	**2**	**0**	**0**	**0**	**0**	**0.0**	**0**	**0**	**0.0**	**7**	**0**	**11:23**	....	....	....	....	....	....	....	....
	St. John's	AHL	7	1	0	1	21	....	....	....	....	....	....	....	....	....	....	....	....	....	....	....	....	....	....	....
	NHL Totals		**30**	**0**	**0**	**0**	**77**	**0**	**0**	**0**	**11**	**0.0**		**0**	**0.0**	**66**	**7**	**9:56**	....	....	....	....	....	....	....	....

Rights traded to **Anaheim** by **Ottawa** with Ted Drury for Jason York and Shaun Van Allen, October 1, 1996. Traded to **Nashville** by **Anaheim** with Chris Mason for Dominic Roussel, October 5, 1998. Traded to **Toronto** by **Nashville** for D.J. Smith and Marty Wilford, March 1, 2002.

MOROZOV, Aleksey

(moh-ROH-zohv, ah-LEHK-see) **PIT.**

Right wing. Shoots left. 6'1", 202 lbs. Born, Moscow, USSR, February 16, 1977. Pittsburgh's 1st choice, 24th overall, in 1995 Entry Draft.

Season	Club	League	GP	G	A	Pts	PIM	PP	SH	GW	S	%	+/-	TF	F%	H	SB	Min	GP	G	A	Pts	PIM	PP	SH	GW
1993-94	Krylja Sovetov	CIS	7	0	0	0	0	....	....	....	....	....	....	....	....	....	....	....	3	0	0	0	2	....	....	....
1994-95	Krylja Sovetov	CIS	48	15	12	27	53	....	....	....	....	....	....	....	....	....	....	....	4	0	3	3	0	....	....	....
1995-96	Krylja Sovetov	CIS	47	13	9	22	26	....	....	....	....	....	....	....	....	....	....	....	....	....	....	....	....	....	....	....
1996-97	Krylja Sovetov	Russia	44	21	11	32	32	....	....	....	....	....	....	....	....	....	....	....	2	0	1	1	2	....	....	....
1997-98	Krylja Sovetov	Russia	6	2	1	3	4	....	....	....	....	....	....	....	....	....	....	....	....	....	....	....	....	....	....	....
	Pittsburgh	**NHL**	**76**	**13**	**13**	**26**	**8**	**2**	**0**	**3**	**80**	**16.3**	**–4**	....	....	....	....	....	**6**	**0**	**1**	**1**	**2**	**0**	**0**	**0**
	Russia	Olympics	6	2	2	4	0	....	....	....	....	....	....	....	....	....	....	....	....	....	....	....	....	....	....	....
1998-99	**Pittsburgh**	**NHL**	**67**	**9**	**10**	**19**	**14**	**0**	**0**	**0**	**75**	**12.0**	**5**	**7**	**42.9**	**44**	**33**	**11:50**	**10**	**1**	**1**	**2**	**0**	**0**	**0**	**0**
99-2000	**Pittsburgh**	**NHL**	**68**	**12**	**19**	**31**	**14**	**0**	**1**	**0**	**101**	**11.9**	**12**	**27**	**33.3**	**33**	**20**	**13:51**	**5**	**0**	**0**	**0**	**0**	**0**	**0**	**0**
2000-01	**Pittsburgh**	**NHL**	**66**	**5**	**14**	**19**	**6**	**0**	**0**	**1**	**72**	**6.9**	**–8**	**19**	**42.1**	**23**	**25**	**10:41**	**18**	**3**	**3**	**6**	**6**	**0**	**1**	**0**
2001-02	**Pittsburgh**	**NHL**	**72**	**20**	**29**	**49**	**16**	**7**	**0**	**3**	**162**	**12.3**	**–7**	**1**	**100.0**	**38**	**30**	**16:42**	....	....	....	....	....	....	....	....
	NHL Totals		**349**	**59**	**85**	**144**	**58**	**9**	**1**	**7**	**490**	**12.0**		**54**	**38.9**	**138**	**108**	**13:21**	**39**	**4**	**5**	**9**	**8**	**0**	**1**	**0**

MORRIS, Derek

(MOH-rihs, DAIR-ihk) **CGY.**

Defense. Shoots right. 5'11", 200 lbs. Born, Edmonton, Alta., August 24, 1978. Calgary's 1st choice, 13th overall, in 1996 Entry Draft.

Season	Club	League	GP	G	A	Pts	PIM	PP	SH	GW	S	%	+/-	TF	F%	H	SB	Min	GP	G	A	Pts	PIM	PP	SH	GW
1994-95	Red Deer	AMHL	31	6	35	41	74	....	....	....	....	....	....	....	....	....	....	....	....	....	....	....	....	....	....	....
1995-96	Regina Pats	WHL	67	8	44	52	70	....	....	....	....	....	....	....	....	....	....	....	11	1	7	8	26	....	....	....
1996-97	Regina Pats	WHL	67	18	57	75	180	....	....	....	....	....	....	....	....	....	....	....	5	0	3	3	9	....	....	....
	Saint John	AHL	7	0	3	3	7	....	....	....	....	....	....	....	....	....	....	....	5	0	3	3	7	....	....	....
1997-98	**Calgary**	**NHL**	**82**	**9**	**20**	**29**	**88**	**5**	**1**	**1**	**120**	**7.5**	**1**	....	....	....	....	....	....	....	....	....	....	....	....	....
1998-99	**Calgary**	**NHL**	**71**	**7**	**27**	**34**	**73**	**3**	**0**	**2**	**150**	**4.7**	**4**	**0**	**0.0**	**93**	**78**	**20:44**	....	....	....	....	....	....	....	....
99-2000	**Calgary**	**NHL**	**78**	**9**	**29**	**38**	**80**	**3**	**0**	**2**	**193**	**4.7**	**2**	**0**	**0.0**	**127**	**119**	**24:51**	....	....	....	....	....	....	....	....
2000-01	**Calgary**	**NHL**	**51**	**5**	**23**	**28**	**56**	**3**	**1**	**4**	**142**	**3.5**	**–15**	**0**	**0.0**	**52**	**79**	**25:51**	....	....	....	....	....	....	....	....
	Saint John	AHL	3	1	2	3	2	....	....	....	....	....	....	....	....	....	....	....	....	....	....	....	....	....	....	....
2001-02	**Calgary**	**NHL**	**61**	**4**	**30**	**34**	**88**	**2**	**0**	**1**	**166**	**2.4**	**–4**	**1**	**100.0**	**81**	**85**	**24:40**	....	....	....	....	....	....	....	....
	NHL Totals		**343**	**34**	**129**	**163**	**385**	**16**	**2**	**10**	**771**	**4.4**		**1**	**100.0**	**353**	**361**	**23:53**	....	....	....	....	....	....	....	....

WHL East First All-Star Team (1997) • NHL All-Rookie Team (1998)

MORRISON, Brendan

(MOHR-ih-suhn, BREHN-duhn) **VAN.**

Center. Shoots left. 5'11", 190 lbs. Born, Pitt Meadows, B.C., August 15, 1975. New Jersey's 3rd choice, 39th overall, in 1993 Entry Draft.

Season	Club	League	GP	G	A	Pts	PIM	PP	SH	GW	S	%	+/-	TF	F%	H	SB	Min	GP	G	A	Pts	PIM	PP	SH	GW
1990-91	Ridge Meadows	BCAHA	77	126	127	253	88	....	....	....	....	....	....	....	....	....	....	....	....	....	....	....	....	....	....	....
1991-92	Ridge Meadows	BCAHA	55	56	111	167	56	....	....	....	....	....	....	....	....	....	....	....	....	....	....	....	....	....	....	....
1992-93	Penticton	BCJHL	56	35	59	94	45	....	....	....	....	....	....	....	....	....	....	....	....	....	....	....	....	....	....	....
1993-94	U. of Michigan	CCHA	38	20	28	48	24	....	....	....	....	....	....	....	....	....	....	....	5	2	7	9	2	....	....	....
1994-95	U. of Michigan	CCHA	39	23	*53	*76	42	....	....	....	....	....	....	....	....	....	....	....	5	1	11	12	6	....	....	....
1995-96	U. of Michigan	CCHA	35	28	44	*72	41	....	....	....	....	....	....	....	....	....	....	....	7	6	9	15	4	....	....	....
1996-97	U. of Michigan	CCHA	43	31	*57	*88	52	....	....	....	....	....	....	....	....	....	....	....	6	6	8	14	8	....	....	....
1997-98	**New Jersey**	**NHL**	**11**	**5**	**4**	**9**	**0**	**0**	**0**	**1**	**19**	**26.3**	**3**	....	....	....	....	....	**3**	**0**	**1**	**1**	**0**	**0**	**0**	**0**
	Albany	AHL	72	35	49	84	44	....	....	....	....	....	....	....	....	....	....	....	8	3	4	7	19	....	....	....
1998-99	**New Jersey**	**NHL**	**76**	**13**	**33**	**46**	**18**	**5**	**0**	**2**	**111**	**11.7**	**–4**	**920**	**51.1**	**63**	**19**	**13:55**	**7**	**0**	**2**	**2**	**0**	**0**	**0**	**0**
99-2000	Trebic	Czech-2	2	0	0	0	0	....	....	....	....	....	....	....	....	....	....	....	....	....	....	....	....	....	....	....
	Pardubice	Czech	6	5	2	7	2	....	....	....	....	....	....	....	....	....	....	....	....	....	....	....	....	....	....	....
	New Jersey	**NHL**	**44**	**5**	**21**	**26**	**8**	**2**	**0**	**1**	**79**	**6.3**	**8**	**572**	**51.1**	**53**	**18**	**16:09**	....	....	....	....	....	....	....	....
	Vancouver	**NHL**	**12**	**2**	**7**	**9**	**10**	**0**	**0**	**0**	**17**	**11.8**	**4**	**48**	**54.2**	**11**	**3**	**14:41**	....	....	....	....	....	....	....	....
2000-01	**Vancouver**	**NHL**	**82**	**16**	**38**	**54**	**42**	**3**	**2**	**3**	**179**	**8.9**	**2**	**1685**	**50.1**	**40**	**41**	**18:22**	**4**	**1**	**2**	**3**	**0**	**1**	**0**	**0**
2001-02	**Vancouver**	**NHL**	**82**	**23**	**44**	**67**	**26**	**6**	**0**	**4**	**183**	**12.6**	**18**	**1307**	**49.9**	**54**	**35**	**19:21**	**6**	**0**	**2**	**2**	**6**	**0**	**0**	**0**
	NHL Totals		**307**	**64**	**147**	**211**	**104**	**16**	**2**	**11**	**588**	**10.9**		**4532**	**50.4**	**221**	**116**	**17:01**	**20**	**1**	**7**	**8**	**6**	**1**	**0**	**0**

CCHA Rookie of the Year (1994) • CCHA First All-Star Team (1995, 1996, 1997) • NCAA West First All-American Team (1995, 1996, 1997) • CCHA Player of the Year (1996, 1997) • NCAA Championship All-Tournament Team (1996) • NCAA Championship Tournament MVP (1996) • Won Hobey Baker Memorial Award (Top U.S. Collegiate Player) (1997)

Traded to **Vancouver** by **New Jersey** with Denis Pederson for Alexander Mogilny, March 14, 2000.

MORROW, Brenden

(MOHR-rohw, BREHN-dehn) **DAL.**

Left wing. Shoots left. 5'11", 200 lbs. Born, Carlyle, Sask., January 16, 1979. Dallas' 1st choice, 25th overall, in 1997 Entry Draft.

			Regular Season																Playoffs							
Season	Club	League	GP	G	A	Pts	PIM	PP	SH	GW	S	%	+/-	TF	F%	H	SB	Min	GP	G	A	Pts	PIM	PP	SH	GW
1994-95	Estevan	SMBHL	60	117	72	189	45																			
1995-96	Portland	WHL	65	13	12	25	61												7	0	0	0	8			
1996-97	Portland	WHL	71	39	49	88	178												6	2	1	3	4			
1997-98	Portland	WHL	68	34	52	86	184												16	10	8	18	65			
1998-99	Portland	WHL	61	41	44	85	248												4	0	4	4	18			
99-2000	Dallas	NHL	64	14	19	33	81	3	0	3	113	12.4	8	25	48.0	170	24	15:51	21	2	4	6	22	1	0	0
	Michigan K-Wings	IHL	9	2	0	2	18																			
2000-01	Dallas	NHL	82	20	24	44	128	7	0	6	121	16.5	18	22	45.5	230	20	15:29	10	0	3	3	12	0	0	0
2001-02	Dallas	NHL	72	17	18	35	109	4	0	3	102	16.7	12	39	41.0	214	46	16:52								
	NHL Totals		218	51	61	112	318	14	0	12	336	15.2		86	44.2	614	90	16:03	31	2	7	9	34	1	0	0

WHL West First All-Star Team (1999)

MOTTAU, Mike

(MAW-tuh, MIGHK) **NYR**

Defense. Shoots left. 6', 192 lbs. Born, Quincy, MA, March 19, 1978. NY Rangers' 10th choice, 182nd overall, in 1997 Entry Draft.

			Regular Season																Playoffs							
Season	Club	League	GP	G	A	Pts	PIM	PP	SH	GW	S	%	+/-	TF	F%	H	SB	Min	GP	G	A	Pts	PIM	PP	SH	GW
1994-95	Thayer Academy	Hi-School	29	7	19	26																				
1995-96	Thayer Academy	Hi-School	31	6	20	26	14																			
1996-97	Boston College	H-East	38	5	18	23	77																			
1997-98	Boston College	H-East	40	13	36	49	50																			
1998-99	Boston College	H-East	43	3	39	42	44																			
	United States	WC-A	3	2	1	3	0																			
99-2000	Boston College	H-East	42	6	37	43	61																			
2000-01	NY Rangers	NHL	18	0	3	3	13	0	0	0	17	0.0	−6	0	0.0	19	17	15:18								
	Hartford	AHL	61	10	33	43	45												5	0	1	1	19			
2001-02	NY Rangers	NHL	1	0	0	0	0	0	0	0	0	0.0	0	0	0.0	0	0	6:20								
	Hartford	AHL	80	9	42	51	56												10	0	5	5	4			
	NHL Totals		19	0	3	3	13	0	0	0	17	0.0		0	0.0	19	17	14:49								

Hockey East First All-Star Team (1998, 2000) • NCAA East Second All-American Team (1998) • NCAA Championship All-Tournament Team (1998, 2000) • Hockey East Second All-Star Team (1999) • NCAA East First All-American Team (1999, 2000) • Hockey East Player of the Year (Shared award with Ty Conklin) (2000) • Won Hobey Baker Memorial Award (Top U.S. Collegiate Player) (2000)

MOWERS, Mark

(MAHW-uhrs, MAHRK) **DET.**

Center. Shoots right. 5'11", 187 lbs. Born, Whitesboro, NY, February 16, 1974.

			Regular Season																Playoffs							
Season	Club	League	GP	G	A	Pts	PIM	PP	SH	GW	S	%	+/-	TF	F%	H	SB	Min	GP	G	A	Pts	PIM	PP	SH	GW
1992-93	Saginaw Gears	NAJHL	39	31	39	70																				
1993-94	Dubuque	USHL	47	51	31	82	80																			
1994-95	New Hampshire	H-East	36	13	23	36	16																			
1995-96	New Hampshire	H-East	34	21	26	47	18																			
1996-97	New Hampshire	H-East	39	26	32	58	52																			
1997-98	New Hampshire	H-East	35	25	31	56	32																			
1998-99	Nashville	NHL	30	0	6	6	4	0	0	0	24	0.0	−4	241	49.0	19	2	9:22								
	Milwaukee	IHL	51	14	22	36	24												1	0	0	0	0			
99-2000	Nashville	NHL	41	4	5	9	10	0	0	0	50	8.0	0	312	45.2	30	11	10:58								
	Milwaukee	IHL	23	11	15	26	34																			
2000-01	Milwaukee	IHL	63	25	25	50	54												5	1	2	3	2			
2001-02	Nashville	NHL	14	1	2	3	2	0	0	0	5	20.0	−2	24	33.3	13	1	8:31								
	Milwaukee	AHL	45	19	20	39	34																			
	NHL Totals		85	5	13	18	16	0	0	0	79	6.3		577	46.3	62	14	9:60								

Hockey East Rookie of the Year (1995) • Hockey East Second All-Star Team (1998) • NCAA East First All-American Team (1998) • Won Ken McKenzie Trophy (U.S.- Born Rookie of the Year - IHL) (1999)

Signed as a free agent by **Nashville**, June 11, 1998. Signed as a free agent by **Detroit**, August 5, 2002.

MUCKALT, Bill

(MUH-kawlt, BIHL) **MIN.**

Right wing. Shoots right. 6'1", 200 lbs. Born, Surrey, B.C., July 15, 1974. Vancouver's 9th choice, 221st overall, in 1994 Entry Draft.

			Regular Season																Playoffs							
Season	Club	League	GP	G	A	Pts	PIM	PP	SH	GW	S	%	+/-	TF	F%	H	SB	Min	GP	G	A	Pts	PIM	PP	SH	GW
1991-92	Merritt	BCJHL	55	14	11	25	75																			
1992-93	Merritt	BCJHL	59	31	43	74	80																			
1993-94	Merritt	BCJHL	43	58	51	109	99																			
	Kelowna Spartans	BCJHL	15	12	10	22	20																			
1994-95	U. of Michigan	CCHA	39	19	18	37	42												5	1	1	2	6			
1995-96	U. of Michigan	CCHA	41	28	30	58	34												7	5	6	11	6			
1996-97	U. of Michigan	CCHA	36	26	38	64	69												6	5	9	14	2			
1997-98	U. of Michigan	CCHA	46	32	*35	*67	94																			
1998-99	Vancouver	NHL	73	16	20	36	98	4	2	1	119	13.4	−9	68	55.9	66	23	15:24								
99-2000	Vancouver	NHL	33	4	8	12	17	1	0	1	53	7.5	6	6	50.0	38	6	14:34								
	NY Islanders	NHL	12	4	3	7	4	0	0	0	26	15.4	5	8	50.0	20	1	12:23								
2000-01	NY Islanders	NHL	60	11	15	26	33	1	0	2	90	12.2	−4	7	14.3	94	9	13:43								
2001-02	Ottawa	NHL	70	0	8	8	46	0	0	0	73	0.0	−3	13	69.2	92	25	9:46								
	NHL Totals		248	35	54	89	198	6	2	4	361	9.7		102	53.9	310	64	13:09								

CCHA First All-Star Team (1998) • NCAA West First All-American Team (1998)

Traded to **NY Islanders** by **Vancouver** with Kevin Weekes and Dave Scatchard for Felix Potvin and NY Islanders' compensatory 2nd (later traded to New Jersey - New Jersey selected Teemu Laine) and 3rd (Thatcher Bell) round choices in 2000 Entry Draft, December 19, 1999. • Missed majority of 1999-2000 season recovering from shoulder injury suffered in game vs. Tampa Bay, January 13, 2000. Traded to **Ottawa** by **NY Islanders** with Zdeno Chara and NY Islanders' 1st round choice (Jason Spezza) in 2001 Entry Draft for Alexei Yashin, June 23, 2001. Signed as a free agent by **Minnesota**, July 3, 2002.

MUIR, Bryan

(MEWR, BRIGH-uhn) **COL.**

Defense. Shoots left. 6'4", 220 lbs. Born, Winnipeg, Man., June 8, 1973.

			Regular Season																Playoffs							
Season	Club	League	GP	G	A	Pts	PIM	PP	SH	GW	S	%	+/-	TF	F%	H	SB	Min	GP	G	A	Pts	PIM	PP	SH	GW
1991-92	Wexford Raiders	MTJHL	44	3	19	22	35																			
1992-93	New Hampshire	H-East	26	1	2	3	24																			
1993-94	New Hampshire	H-East	40	0	4	4	48																			
1994-95	New Hampshire	H-East	28	9	9	18	46																			
1995-96	Team Canada	Nat-Tm	42	6	12	18	38																			
	Edmonton	NHL	5	0	0	0	6	0	0	0	4	0.0	−4													
1996-97	Hamilton	AHL	75	8	16	24	80												14	0	5	5	12			
	Edmonton	NHL																	5	0	0	0	4	0	0	0
1997-98	Edmonton	NHL	7	0	0	0	17	0	0	0	6	0.0	0													
	Hamilton	AHL	28	3	10	13	62																			
	Albany	AHL	41	3	10	13	67												13	3	0	3	12			
1998-99	New Jersey	NHL	1	0	0	0	0	0	0	0	4	0.0	0	0	0.0	0	0	9:54								
	Albany	AHL	10	0	0	0	29																			
	Chicago	NHL	53	1	4	5	50	0	0	0	78	1.3	1	0	0.0	82	59	18:49								
	Portland Pirates	AHL	2	1	1	2	2																			
99-2000	Chicago	NHL	11	2	3	5	13	0	1	0	19	10.5	−1	0	0.0	17	18	17:54								
	Tampa Bay	NHL	30	1	1	2	32	0	0	0	32	3.1	−8	1	100.0	41	28	19:29								
2000-01	Tampa Bay	NHL	10	0	3	3	15	0	0	0	14	0.0	−7	1	0.0	16	24	18:34								
	Detroit Vipers	IHL	21	5	7	12	36																			
	Colorado	NHL	8	0	0	0	4	0	0	0	3	0.0	0	0	0.0	5	4	8:14	3	0	0	0	0	0	0	0
	Hershey Bears	AHL	26	5	8	13	50																			
2001-02	Colorado	NHL	22	1	1	2	9	0	0	0	26	3.8	1	0	0.0	29	9	10:20	21	0	0	0	2	0	0	0
	Hershey Bears	AHL	59	10	16	26	133																			
	NHL Totals		147	5	12	17	146	0	1	0	186	2.7		2	50.0	190	142	16:48	29	0	0	0	6	0	0	0

Signed to five-game amateur try-out contract by **Edmonton**, February 29, 1996. Signed as a free agent by **Edmonton**, April 30, 1996. Traded to **New Jersey** by **Edmonton** with Jason Arnott for Valeri Zelepukin and Bill Guerin, January 4, 1998. Traded to **Chicago** by **New Jersey** for Chicago's 3rd round choice (Michael Rupp) in 2000 Entry Draft. November 13, 1998. Traded to **Tampa Bay** by **Chicago** with Reid Simpson for Michael Nylander, November 12, 1999. • Missed majority of 1999-2000 season recovering from leg injury suffered in game vs. Atlanta, November 17, 1999. Traded to **Colorado** by **Tampa Bay** for Colorado's 8th round choice (Dmitri Bezrukov) in 2001 Entry Draft, January 23, 2001.

MULLER, Kirk

(MUHL-luhr, KUHRK) **DAL.**

Left wing. Shoots left. 6', 205 lbs. Born, Kingston, Ont., February 8, 1966. New Jersey's 1st choice, 2nd overall, in 1984 Entry Draft.

			Regular Season																Playoffs							
Season	Club	League	GP	G	A	Pts	PIM	PP	SH	GW	S	%	+/-	TF	F%	H	SB	Min	GP	G	A	Pts	PIM	PP	SH	GW
1980-81	Kingston	OHA-B	42	17	37	54	5																			
	Kingston	OMJHL	2	0	0	0	0																			
1981-82	Kingston	OHL	67	12	39	51	27												4	5	1	6	4			
1982-83	Guelph Platers	OHL	66	52	60	112	41																			
1983-84	Guelph Platers	OHL	49	31	63	94	27																			
	Canada	Olympics	6	2	1	3	0																			
1984-85	**New Jersey**	**NHL**	**80**	**17**	**37**	**54**	**69**	**9**	**1**	**0**	**157**	**10.8**	**–31**													
1985-86	**New Jersey**	**NHL**	**77**	**25**	**41**	**66**	**45**	**5**	**1**	**1**	**168**	**14.9**	**–20**													
1986-87	**New Jersey**	**NHL**	**79**	**26**	**50**	**76**	**75**	**10**	**1**	**4**	**193**	**13.5**	**–7**													
1987-88	**New Jersey**	**NHL**	**80**	**37**	**57**	**94**	**114**	**17**	**2**	**1**	**215**	**17.2**	**19**						**20**	**4**	**8**	**12**	**37**	**0**	**0**	**0**
1988-89	**New Jersey**	**NHL**	**80**	**31**	**43**	**74**	**119**	**12**	**1**	**4**	**182**	**17.0**	**–23**													
1989-90	**New Jersey**	**NHL**	**80**	**30**	**56**	**86**	**74**	**9**	**0**	**6**	**200**	**15.0**	**–1**						**6**	**1**	**3**	**4**	**11**	**0**	**0**	**0**
1990-91	**New Jersey**	**NHL**	**80**	**19**	**51**	**70**	**76**	**7**	**0**	**3**	**221**	**8.6**	**1**						**7**	**0**	**2**	**2**	**10**	**0**	**0**	**0**
1991-92	**Montreal**	**NHL**	**78**	**36**	**41**	**77**	**86**	**15**	**1**	**7**	**191**	**18.8**	**15**						**11**	**4**	**3**	**7**	**31**	**2**	**1**	**1**
1992-93♦	**Montreal**	**NHL**	**80**	**37**	**57**	**94**	**77**	**12**	**0**	**4**	**231**	**16.0**	**8**						**20**	**10**	**7**	**17**	**18**	**3**	**0**	**3**
1993-94	**Montreal**	**NHL**	**76**	**23**	**34**	**57**	**96**	**9**	**2**	**3**	**168**	**13.7**	**–1**						**7**	**6**	**2**	**8**	**4**	**3**	**0**	**2**
1994-95	**Montreal**	**NHL**	**33**	**8**	**11**	**19**	**33**	**3**	**0**	**1**	**81**	**9.9**	**–21**													
	NY Islanders	**NHL**	**12**	**3**	**5**	**8**	**14**	**1**	**1**	**1**	**16**	**18.8**	**3**													
1995-96	**NY Islanders**	**NHL**	**15**	**4**	**3**	**7**	**15**	**0**	**0**	**0**	**23**	**17.4**	**–10**													
	Toronto	**NHL**	**36**	**9**	**16**	**25**	**42**	**7**	**0**	**1**	**79**	**11.4**	**–3**						**6**	**3**	**2**	**5**	**0**	**2**	**0**	**0**
1996-97	**Toronto**	**NHL**	**66**	**20**	**17**	**37**	**85**	**9**	**1**	**3**	**153**	**13.1**	**–23**													
	Florida	**NHL**	**10**	**1**	**2**	**3**	**4**	**1**	**0**	**1**	**21**	**4.8**	**–2**						**5**	**1**	**2**	**3**	**4**	**1**	**0**	**0**
1997-98	**Florida**	**NHL**	**70**	**8**	**21**	**29**	**54**	**1**	**0**	**3**	**115**	**7.0**	**–14**													
1998-99	**Florida**	**NHL**	**82**	**4**	**11**	**15**	**49**	**0**	**0**	**1**	**107**	**3.7**	**–11**	**1157**	**49.1**	**68**	**39**	**14:28**								
99-2000	**Dallas**	**NHL**	**47**	**7**	**15**	**22**	**24**	**3**	**0**	**2**	**57**	**12.3**	**–3**	**443**	**48.5**	**77**	**19**	**16:24**	**23**	**2**	**3**	**5**	**18**	**0**	**0**	**1**
2000-01	**Dallas**	**NHL**	**55**	**1**	**9**	**10**	**26**	**0**	**0**	**0**	**54**	**1.9**	**–4**	**539**	**50.8**	**102**	**14**	**12:22**	**10**	**1**	**3**	**4**	**12**	**0**	**0**	**1**
2001-02	**Dallas**	**NHL**	**78**	**10**	**20**	**30**	**28**	**4**	**0**	**1**	**111**	**9.0**	**–12**	**511**	**49.7**	**115**	**18**	**13:48**								
	NHL Totals		**1294**	**356**	**597**	**953**	**1205**	**134**	**11**	**47**	**2743**	**13.0**		**2650**	**49.5**	**362**	**90**	**14:10**	**115**	**32**	**35**	**67**	**145**	**11**	**1**	**8**

Played in NHL All-Star Game (1985, 1986, 1988, 1990, 1992, 1993)

Traded to **Montreal** by **New Jersey** with Rollie Melanson for Stephane Richer and Tom Chorske, September 20, 1991. Traded to **NY Islanders** by **Montreal** with Mathieu Schneider and Craig Darby for Pierre Turgeon and Vladimir Malakhov, April 5, 1995. Traded to **Toronto** by **NY Islanders** with Don Beaupre to complete transaction that sent Damian Rhodes and Ken Belanger to NY Islanders (January 23, 1996), January 23, 1996. Traded to **Florida** by **Toronto** for Jason Podollan, March 18, 1997. Signed as a free agent by **Dallas**, December 15, 1999. Claimed by **Columbus** from **Dallas** in Waiver Draft, September 28, 2001. Traded to **Dallas** by **Columbus** for the rights to Evgeny Petrochinin, September 28, 2001.

MURPHY, Gord

(MUHR-fee, GOHRD)

Defense. Shoots right. 6'2", 195 lbs. Born, Willowdale, Ont., March 23, 1967. Philadelphia's 10th choice, 189th overall, in 1985 Entry Draft.

Season	Club	League	GP	G	A	Pts	PIM	PP	SH	GW	S	%	+/-	TF	F%	H	SB	Min	GP	G	A	Pts	PIM	PP	SH	GW
1983-84	Don Mills	MTHL	65	24	42	66	130																			
1984-85	Oshawa Generals	OHL	59	3	12	15	25																			
1985-86	Oshawa Generals	OHL	64	7	15	22	56												6	1	1	2	6			
1986-87	Oshawa Generals	OHL	56	7	30	37	95												24	6	16	22	22			
1987-88	Hershey Bears	AHL	62	8	20	28	44												12	0	8	8	12			
1988-89	**Philadelphia**	**NHL**	**75**	**4**	**31**	**35**	**68**	**3**	**0**	**1**	**116**	**3.4**	**–3**						**19**	**2**	**7**	**9**	**13**	**1**	**0**	**1**
1989-90	**Philadelphia**	**NHL**	**75**	**14**	**27**	**41**	**95**	**4**	**0**	**1**	**160**	**8.8**	**–7**													
1990-91	**Philadelphia**	**NHL**	**80**	**11**	**31**	**42**	**58**	**6**	**0**	**2**	**203**	**5.4**	**–7**													
1991-92	**Philadelphia**	**NHL**	**31**	**2**	**8**	**10**	**33**	**0**	**0**	**0**	**50**	**4.0**	**–4**													
	Boston	**NHL**	**42**	**3**	**6**	**9**	**51**	**0**	**0**	**0**	**82**	**3.7**	**2**						**15**	**1**	**0**	**1**	**12**	**0**	**0**	**0**
1992-93	**Boston**	**NHL**	**49**	**5**	**12**	**17**	**62**	**3**	**0**	**2**	**68**	**7.4**	**–13**													
	Providence	AHL	2	1	3	4	2																			
1993-94	**Florida**	**NHL**	**84**	**14**	**29**	**43**	**71**	**9**	**0**	**2**	**172**	**8.1**	**–11**													
1994-95	**Florida**	**NHL**	**46**	**6**	**16**	**22**	**24**	**5**	**0**	**0**	**94**	**6.4**	**–14**													
1995-96	**Florida**	**NHL**	**70**	**8**	**22**	**30**	**30**	**4**	**0**	**0**	**125**	**6.4**	**5**						**14**	**0**	**4**	**4**	**6**	**0**	**0**	**0**
1996-97	**Florida**	**NHL**	**80**	**8**	**15**	**23**	**51**	**2**	**0**	**0**	**137**	**5.8**	**3**						**5**	**0**	**5**	**5**	**4**	**0**	**0**	**0**
1997-98	**Florida**	**NHL**	**79**	**6**	**11**	**17**	**46**	**3**	**0**	**0**	**123**	**4.9**	**–3**													
1998-99	**Florida**	**NHL**	**51**	**0**	**7**	**7**	**16**	**0**	**0**	**0**	**56**	**0.0**	**4**	**1**	**0.0**	**44**	**58**	**19:57**								
99-2000	**Atlanta**	**NHL**	**58**	**1**	**10**	**11**	**38**	**0**	**0**	**0**	**74**	**1.4**	**–26**	**0**	**0.0**	**96**	**132**	**22:41**								
2000-01	**Atlanta**	**NHL**	**27**	**3**	**11**	**14**	**12**	**2**	**0**	**0**	**44**	**6.8**	**–11**	**0**	**0.0**	**22**	**47**	**20:44**								
2001-02	**Boston**	**NHL**	**15**	**0**	**2**	**2**	**13**	**0**	**0**	**0**	**10**	**0.0**	**2**	**0**	**0.0**	**12**	**12**	**14:32**								
	Providence	AHL	8	0	3	3	6																			
	NHL Totals		**862**	**85**	**238**	**323**	**668**	**41**	**0**	**8**	**1514**	**5.6**		**1**	**0.0**	**174**	**249**	**20:36**	**53**	**3**	**16**	**19**	**35**	**1**	**0**	**1**

Traded to **Boston** by **Philadelphia** with Brian Dobbin, Philadelphia's 3rd round choice (Sergei Zholtok) in 1992 Entry Draft and 4th round choice (Charles Paquette) in 1993 Entry Draft, for Garry Galley, Wes Walz and Boston's 3rd round choice (Milos Holan) in 1993 Entry Draft, January 2, 1992. Traded to **Dallas** by **Boston** for future considerations (Jon Casey to Boston for Andy Moog, June 25, 1993), June 20, 1993. Claimed by **Florida** from **Dallas** in Expansion Draft, June 24, 1993. Traded to **Atlanta** by **Florida** with Herbert Vasiljevs, Daniel Tjarnqvist and Ottawa's 6th round choice (previously acquired, later traded to Dallas - Dallas selected Justin Cox) in 1999 Entry Draft for Trevor Kidd, June 25, 1999. • Missed majority of 2000-01 season recovering from shoulder injury suffered in game vs. NY Rangers, October 7, 2000. Signed as a free agent by **Boston**, January 29, 2002. • Officially announced retirement, March 19, 2002. Named Assistant Coach by **Columbus**, July 17, 2002.

MURRAY, Glen

(MUHR-ray, GLEHN) **BOS.**

Right wing. Shoots right. 6'3", 225 lbs. Born, Halifax, N.S., November 1, 1972. Boston's 1st choice, 18th overall, in 1991 Entry Draft.

Season	Club	League	GP	G	A	Pts	PIM	PP	SH	GW	S	%	+/-	TF	F%	H	SB	Min	GP	G	A	Pts	PIM	PP	SH	GW
1988-89	Bridgewater	NSMHL	45	50	56	106	62																			
1989-90	Sudbury Wolves	OHL	62	8	28	36	17												7	0	0	0	4			
1990-91	Sudbury Wolves	OHL	66	27	38	65	82												5	8	4	12	10			
1991-92	Sudbury Wolves	OHL	54	37	47	84	93												11	7	4	11	18			
	Boston	**NHL**	**5**	**3**	**1**	**4**	**0**	**1**	**0**	**0**	**20**	**15.0**	**2**						**15**	**4**	**2**	**6**	**10**	**1**	**0**	**0**
1992-93	**Boston**	**NHL**	**27**	**3**	**4**	**7**	**8**	**2**	**0**	**1**	**28**	**10.7**	**–6**													
	Providence	AHL	48	30	26	56	42												6	1	4	5	4			
1993-94	**Boston**	**NHL**	**81**	**18**	**13**	**31**	**48**	**0**	**0**	**4**	**114**	**15.8**	**–1**						**13**	**4**	**5**	**9**	**14**	**0**	**0**	**0**
1994-95	**Boston**	**NHL**	**35**	**5**	**2**	**7**	**46**	**0**	**0**	**2**	**64**	**7.8**	**–11**						**2**	**0**	**0**	**0**	**2**	**0**	**0**	**0**
1995-96	**Pittsburgh**	**NHL**	**69**	**14**	**15**	**29**	**57**	**0**	**0**	**2**	**100**	**14.0**	**4**						**18**	**2**	**6**	**8**	**10**	**0**	**0**	**1**
1996-97	**Pittsburgh**	**NHL**	**66**	**11**	**11**	**22**	**24**	**3**	**0**	**1**	**127**	**8.7**	**–19**													
	Los Angeles	**NHL**	**11**	**5**	**3**	**8**	**8**	**0**	**0**	**0**	**26**	**19.2**	**–2**													
1997-98	**Los Angeles**	**NHL**	**81**	**29**	**31**	**60**	**54**	**7**	**3**	**7**	**193**	**15.0**	**6**						**4**	**2**	**0**	**2**	**6**	**0**	**0**	**0**
1998-99	**Los Angeles**	**NHL**	**61**	**16**	**15**	**31**	**36**	**3**	**3**	**3**	**173**	**9.2**	**–14**	**12**	**25.0**	**63**	**15**	**20:33**								
99-2000	**Los Angeles**	**NHL**	**78**	**29**	**33**	**62**	**60**	**10**	**1**	**2**	**202**	**14.4**	**13**	**15**	**80.0**	**85**	**24**	**18:30**	**4**	**0**	**0**	**0**	**2**	**0**	**0**	**0**
2000-01	**Los Angeles**	**NHL**	**64**	**18**	**21**	**39**	**32**	**3**	**1**	**1**	**138**	**13.0**	**9**	**7**	**42.9**	**88**	**16**	**18:12**	**13**	**4**	**3**	**7**	**4**	**1**	**0**	**1**
2001-02	**Los Angeles**	**NHL**	**9**	**6**	**5**	**11**	**0**	**4**	**0**	**2**	**34**	**17.6**	**5**	**1**	**100.0**	**13**	**1**	**19:08**								
	Boston	**NHL**	**73**	**35**	**25**	**60**	**40**	**5**	**0**	**7**	**212**	**16.5**	**26**	**39**	**23.1**	**57**	**20**	**19:50**	**6**	**1**	**4**	**5**	**4**	**0**	**0**	**0**
	NHL Totals		**660**	**192**	**179**	**371**	**413**	**38**	**8**	**32**	**1431**	**13.4**		**74**	**37.8**	**306**	**76**	**19:14**	**75**	**17**	**20**	**37**	**52**	**2**	**0**	**2**

Traded to **Pittsburgh** by **Boston** with Bryan Smolinski and Boston's 3rd round choice (Boyd Kane) in 1996 Entry Draft for Kevin Stevens and Shawn McEachern, August 2, 1995. Traded to **LA Kings** by **Pittsburgh** for Ed Olczyk, March 18, 1997. Traded to **Boston** by **LA Kings** with Jozef Stumpel for Jason Allison and Mikko Eloranta, October 24, 2001.

MURRAY, Marty

(MUHR-ray, MAHR-tee) **PHI.**

Center. Shoots left. 5'9", 180 lbs. Born, Deloraine, Man., February 16, 1975. Calgary's 5th choice, 96th overall, in 1993 Entry Draft.

Season	Club	League	GP	G	A	Pts	PIM	PP	SH	GW	S	%	+/-	TF	F%	H	SB	Min	GP	G	A	Pts	PIM	PP	SH	GW
1990-91	S-W Cougars	MMHL	36	46	47	93	50																			
1991-92	Brandon	WHL	68	20	36	56	22																			
1992-93	Brandon	WHL	67	29	65	94	50												4	1	3	4	0			
1993-94	Brandon	WHL	64	43	71	114	33												14	6	14	20	14			
1994-95	Brandon	WHL	65	40	*88	128	53												18	9	*20	29	16			
1995-96	**Calgary**	**NHL**	**15**	**3**	**3**	**6**	**0**	**2**	**0**	**0**	**22**	**13.6**	**–4**													
	Saint John	AHL	58	25	31	56	20												14	2	4	6	4			
1996-97	**Calgary**	**NHL**	**2**	**0**	**0**	**0**	**4**	**0**	**0**	**0**	**2**	**0.0**	**0**													
	Saint John	AHL	67	19	39	58	40												5	2	3	5	4			
1997-98	**Calgary**	**NHL**	**2**	**0**	**0**	**0**	**2**	**0**	**0**	**0**	**2**	**0.0**	**1**													
	Saint John	AHL	41	10	30	40	16												21	10	10	20	12			
1998-99	EC Villacher SV	Alpenliga	33	26	41	67	12																			
	EC Villacher SV	Austria	17	13	17	30	6												6	1	4	5	0			
99-2000	Kolner Haie	Germany	56	12	47	59	28												10	4	3	7	2			

			Regular Season																Playoffs							
Season	Club	League	GP	G	A	Pts	PIM	PP	SH	GW	S	%	+/-	TF	F%	H	SB	Min	GP	G	A	Pts	PIM	PP	SH	GW
2000-01	**Calgary**	**NHL**	**7**	**0**	**0**	**0**	**0**	**0**	**0**	**0**	**6**	**0.0**	**–2**	**88**	**55.7**	**5**	**4**	**14:28**	....	....	....	....	....	....	....	....
	Saint John	AHL	56	24	52	76	36	....	....	....	....	....	....	....	....	....	....	....	19	4	16	20	18	....	....	....
2001-02	**Philadelphia**	**NHL**	**74**	**12**	**15**	**27**	**10**	**1**	**1**	**2**	**109**	**11.0**	**10**	**913**	**50.7**	**35**	**40**	**13:56**	**5**	**0**	**1**	**1**	**0**	**0**	**0**	**0**
	Philadelphia	AHL	3	0	3	3	2	....	....	....	....	....	....	....	....	....	....	....	....	....	....	....	....	....	....	....
	NHL Totals		**100**	**15**	**18**	**33**	**16**	**3**	**1**	**2**	**141**	**10.6**		**1001**	**51.1**	**40**	**44**	**13:59**	**5**	**0**	**1**	**1**	**0**	**0**	**0**	**0**

WHL East First All-Star Team (1994, 1995) • Canadian Major Junior Second All-Star Team (1994) • WHL MVP (1995)

Signed as a free agent by **Philadelphia**, July 9, 2001.

MURRAY, Rem

(MUHR-ray, REHM) **NYR**

Center/Left wing. Shoots left. 6'2", 195 lbs. Born, Stratford, Ont., October 9, 1972. Los Angeles' 5th choice, 135th overall, in 1992 Entry Draft.

Season	Club	League	GP	G	A	Pts	PIM	PP	SH	GW	S	%	+/-	TF	F%	H	SB	Min	GP	G	A	Pts	PIM	PP	SH	GW
1989-90	Stratford	OJHL-B	46	19	32	51	48	....	....	....	....	....	....	....	....	....	....	....	....	....	....	....	....	....	....	....
1990-91	Stratford	OJHL-B	48	39	59	98	39	....	....	....	....	....	....	....	....	....	....	....	....	....	....	....	....	....	....	....
1991-92	Michigan State	CCHA	41	12	36	48	16	....	....	....	....	....	....	....	....	....	....	....	....	....	....	....	....	....	....	....
1992-93	Michigan State	CCHA	40	22	35	57	24	....	....	....	....	....	....	....	....	....	....	....	....	....	....	....	....	....	....	....
1993-94	Michigan State	CCHA	41	16	38	54	18	....	....	....	....	....	....	....	....	....	....	....	....	....	....	....	....	....	....	....
1994-95	Michigan State	CCHA	40	20	36	56	21	....	....	....	....	....	....	....	....	....	....	....	....	....	....	....	....	....	....	....
1995-96	Cape Breton	AHL	79	31	59	90	40	....	....	....	....	....	....	....	....	....	....	....	....	....	....	....	....	....	....	....
1996-97	**Edmonton**	**NHL**	**82**	**11**	**20**	**31**	**16**	**1**	**0**	**2**	**85**	**12.9**	**9**	....	....	....	....	....	**12**	**1**	**2**	**3**	**4**	**0**	**0**	**0**
1997-98	**Edmonton**	**NHL**	**61**	**9**	**9**	**18**	**39**	**2**	**2**	**0**	**59**	**15.3**	**–9**	....	....	....	....	....	**11**	**1**	**4**	**5**	**2**	**0**	**0**	**0**
1998-99	**Edmonton**	**NHL**	**78**	**21**	**18**	**39**	**20**	**4**	**1**	**4**	**116**	**18.1**	**4**	**1013**	**48.1**	**68**	**30**	**15:50**	**4**	**1**	**1**	**2**	**2**	**0**	**0**	**0**
99-2000	**Edmonton**	**NHL**	**44**	**9**	**5**	**14**	**8**	**2**	**0**	**3**	**65**	**13.8**	**–2**	**303**	**50.5**	**22**	**14**	**14:16**	**5**	**0**	**1**	**1**	**2**	**0**	**0**	**0**
2000-01	**Edmonton**	**NHL**	**82**	**15**	**21**	**36**	**24**	**1**	**3**	**3**	**122**	**12.3**	**5**	**694**	**49.3**	**42**	**29**	**15:21**	**6**	**2**	**0**	**2**	**6**	**1**	**0**	**0**
2001-02	**Edmonton**	**NHL**	**69**	**7**	**17**	**24**	**14**	**0**	**2**	**1**	**84**	**8.3**	**5**	**825**	**50.9**	**33**	**45**	**14:27**	....	....	....	....	....	....	....	....
	NY Rangers	**NHL**	**11**	**1**	**2**	**3**	**4**	**0**	**0**	**0**	**14**	**7.1**	**–9**	**151**	**51.7**	**7**	**4**	**15:51**	....	....	....	....	....	....	....	....
	NHL Totals		**427**	**73**	**92**	**165**	**125**	**10**	**8**	**13**	**545**	**13.4**		**2986**	**49.6**	**172**	**122**	**15:07**	**38**	**5**	**8**	**13**	**16**	**1**	**0**	**0**

CCHA Second All-Star Team (1995)

Signed as a free agent by **Edmonton**, September 19, 1995. Traded to **NY Rangers** by **Edmonton** with Tom Poti for Mike York and NY Rangers' 4th round choice (Ivan Koltsov) in 2002 Entry Draft, March 19, 2002.

MURRAY, Rob

(MUHR-ray, RAWB) **CGY.**

Center. Shoots right. 6'1", 180 lbs. Born, Toronto, Ont., April 4, 1967. Washington's 3rd choice, 61st overall, in 1985 Entry Draft.

Season	Club	League	GP	G	A	Pts	PIM	PP	SH	GW	S	%	+/-	TF	F%	H	SB	Min	GP	G	A	Pts	PIM	PP	SH	GW
1983-84	Mississauga Reps	MTHL	35	18	36	54	32	....	....	....	....	....	....	....	....	....	....	....	....	....	....	....	....	....	....	....
1984-85	Peterborough	OHL	63	12	9	21	155	....	....	....	....	....	....	....	....	....	....	....	17	2	7	9	45	....	....	....
1985-86	Peterborough	OHL	52	14	18	32	125	....	....	....	....	....	....	....	....	....	....	....	16	1	2	3	50	....	....	....
1986-87	Peterborough	OHL	62	17	37	54	204	....	....	....	....	....	....	....	....	....	....	....	3	1	4	5	8	....	....	....
1987-88	Fort Wayne	IHL	80	12	21	33	139	....	....	....	....	....	....	....	....	....	....	....	6	0	2	2	16	....	....	....
1988-89	Baltimore	AHL	80	11	23	34	235	....	....	....	....	....	....	....	....	....	....	....	....	....	....	....	....	....	....	....
1989-90	**Washington**	**NHL**	**41**	**2**	**7**	**9**	**58**	**0**	**0**	**0**	**29**	**6.9**	**–10**	....	....	....	....	....	**9**	**0**	**0**	**0**	**18**	**0**	**0**	**0**
	Baltimore	AHL	23	5	4	9	63	....	....	....	....	....	....	....	....	....	....	....	....	....	....	....	....	....	....	....
1990-91	**Washington**	**NHL**	**17**	**0**	**3**	**3**	**19**	**0**	**0**	**0**	**8**	**0.0**	**0**	....	....	....	....	....	....	....	....	....	....	....	....	....
	Baltimore	AHL	48	6	20	26	177	....	....	....	....	....	....	....	....	....	....	....	4	0	0	0	12	....	....	....
1991-92	**Winnipeg**	**NHL**	**9**	**0**	**1**	**1**	**18**	**0**	**0**	**0**	**2**	**0.0**	**–2**	....	....	....	....	....	....	....	....	....	....	....	....	....
	Moncton Hawks	AHL	60	16	15	31	247	....	....	....	....	....	....	....	....	....	....	....	8	0	1	1	56	....	....	....
1992-93	**Winnipeg**	**NHL**	**10**	**1**	**0**	**1**	**6**	**0**	**0**	**1**	**4**	**25.0**	**0**	....	....	....	....	....	....	....	....	....	....	....	....	....
	Moncton Hawks	AHL	56	16	21	37	147	....	....	....	....	....	....	....	....	....	....	....	3	0	0	0	6	....	....	....
1993-94	**Winnipeg**	**NHL**	**6**	**0**	**0**	**0**	**2**	**0**	**0**	**0**	**1**	**0.0**	**0**	....	....	....	....	....	....	....	....	....	....	....	....	....
	Moncton Hawks	AHL	69	25	32	57	280	....	....	....	....	....	....	....	....	....	....	....	21	2	3	5	60	....	....	....
1994-95	Springfield	AHL	78	16	38	54	373	....	....	....	....	....	....	....	....	....	....	....	....	....	....	....	....	....	....	....
	Winnipeg	**NHL**	**10**	**0**	**2**	**2**	**2**	**0**	**0**	**0**	**5**	**0.0**	**1**	....	....	....	....	....	....	....	....	....	....	....	....	....
1995-96	**Winnipeg**	**NHL**	**1**	**0**	**0**	**0**	**2**	**0**	**0**	**0**	**1**	**0.0**	**–1**	....	....	....	....	....	....	....	....	....	....	....	....	....
	Springfield	AHL	74	10	28	38	263	....	....	....	....	....	....	....	....	....	....	....	10	1	6	7	32	....	....	....
1996-97	Springfield	AHL	78	16	27	43	234	....	....	....	....	....	....	....	....	....	....	....	17	2	3	5	66	....	....	....
1997-98	Springfield	AHL	80	7	30	37	255	....	....	....	....	....	....	....	....	....	....	....	4	0	2	2	2	....	....	....
1998-99	**Phoenix**	**NHL**	**13**	**1**	**2**	**3**	**4**	**0**	**0**	**0**	**11**	**9.1**	**2**	**28**	**46.4**	**11**	**9**	**8:18**	....	....	....	....	....	....	....	....
	Springfield	AHL	68	6	19	25	197	....	....	....	....	....	....	....	....	....	....	....	3	0	0	0	4	....	....	....
99-2000	Springfield	AHL	22	1	3	4	70	....	....	....	....	....	....	....	....	....	....	....	....	....	....	....	....	....	....	....
	Hamilton	AHL	55	11	20	31	100	....	....	....	....	....	....	....	....	....	....	....	10	2	3	5	4	....	....	....
2000-01	Philadelphia	AHL	46	3	6	9	65	....	....	....	....	....	....	....	....	....	....	....	....	....	....	....	....	....	....	....
	Springfield	AHL	30	3	2	5	43	....	....	....	....	....	....	....	....	....	....	....	....	....	....	....	....	....	....	....
2001-02	Saint John	AHL	80	7	14	21	97	....	....	....	....	....	....	....	....	....	....	....	....	....	....	....	....	....	....	....
	NHL Totals		**107**	**4**	**15**	**19**	**111**	**0**	**0**	**1**	**61**	**6.6**		**28**	**46.4**	**11**	**9**	**8:18**	**9**	**0**	**0**	**0**	**18**	**0**	**0**	**0**

Claimed by **Minnesota** from **Washington** in Expansion Draft, May 30, 1991. Traded to **Winnipeg** by **Minnesota** with future considerations for Winnipeg's 7th round choice (Geoff Finch) in 1991 Entry Draft and future considerations, May 31, 1991. Transferred to **Phoenix** after **Winnipeg** franchise relocated, July 1, 1996. Traded to **Edmonton** by **Phoenix** for Eric Houde, November 30, 1999. Signed as a free agent by **Philadelphia**, July 24, 2000. Signed as a free agent by **Calgary**, August 2, 2001. Signed as a free agent by **Springfield** (AHL), August 1, 2002.

MUSIL, Frantisek

(moo-SIHL, FRAN-tih-sehk)

Defense. Shoots left. 6'3", 215 lbs. Born, Pardubice, Czech., December 17, 1964. Minnesota's 3rd choice, 38th overall, in 1983 Entry Draft.

Season	Club	League	GP	G	A	Pts	PIM	PP	SH	GW	S	%	+/-	TF	F%	H	SB	Min	GP	G	A	Pts	PIM	PP	SH	GW
1980-81	Tesla Pardubice	Czech	2	0	0	0	0	....	....	....	....	....	....	....	....	....	....	....	....	....	....	....	....	....	....	....
1981-82	Tesla Pardubice	Czech	35	1	3	4	34	....	....	....	....	....	....	....	....	....	....	....	....	....	....	....	....	....	....	....
1982-83	Tesla Pardubice	Czech	33	1	2	3	44	....	....	....	....	....	....	....	....	....	....	....	....	....	....	....	....	....	....	....
1983-84	Tesla Pardubice	Czech	37	4	8	12	72	....	....	....	....	....	....	....	....	....	....	....	....	....	....	....	....	....	....	....
1984-85	Dukla Jihlava	Czech	44	4	6	10	76	....	....	....	....	....	....	....	....	....	....	....	....	....	....	....	....	....	....	....
1985-86	Dukla Jihlava	Czech	34	4	7	11	42	....	....	....	....	....	....	....	....	....	....	....	....	....	....	....	....	....	....	....
1986-87	**Minnesota**	**NHL**	**72**	**2**	**9**	**11**	**148**	**0**	**0**	**0**	**83**	**2.4**	**0**	....	....	....	....	....	....	....	....	....	....	....	....	....
1987-88	**Minnesota**	**NHL**	**80**	**9**	**8**	**17**	**213**	**1**	**1**	**0**	**78**	**11.5**	**–2**	....	....	....	....	....	....	....	....	....	....	....	....	....
1988-89	**Minnesota**	**NHL**	**55**	**1**	**19**	**20**	**54**	**0**	**0**	**1**	**78**	**1.3**	**4**	....	....	....	....	....	**5**	**1**	**1**	**2**	**4**	**0**	**0**	**0**
1989-90	**Minnesota**	**NHL**	**56**	**2**	**8**	**10**	**109**	**0**	**0**	**1**	**78**	**2.6**	**0**	....	....	....	....	....	**4**	**0**	**0**	**0**	**14**	**0**	**0**	**0**
1990-91	**Minnesota**	**NHL**	**8**	**0**	**2**	**2**	**23**	**0**	**0**	**0**	**5**	**0.0**	**0**	....	....	....	....	....	....	....	....	....	....	....	....	....
	Calgary	**NHL**	**67**	**7**	**14**	**21**	**160**	**2**	**0**	**1**	**68**	**10.3**	**12**	....	....	....	....	....	**7**	**0**	**0**	**0**	**10**	**0**	**0**	**0**
1991-92	**Calgary**	**NHL**	**78**	**4**	**8**	**12**	**103**	**1**	**1**	**0**	**71**	**5.6**	**12**	....	....	....	....	....	....	....	....	....	....	....	....	....
1992-93	**Calgary**	**NHL**	**80**	**6**	**10**	**16**	**131**	**0**	**0**	**1**	**87**	**6.9**	**28**	....	....	....	....	....	**6**	**1**	**1**	**2**	**7**	**0**	**0**	**0**
1993-94	**Calgary**	**NHL**	**75**	**1**	**8**	**9**	**50**	**0**	**0**	**0**	**65**	**1.5**	**38**	....	....	....	....	....	**7**	**0**	**1**	**1**	**4**	**0**	**0**	**0**
1994-95	HC Sparta Praha	Czech	19	1	4	5	50	....	....	....	....	....	....	....	....	....	....	....	....	....	....	....	....	....	....	....
	HC Saxonia	German-2	1	0	0	0	2	....	....	....	....	....	....	....	....	....	....	....	....	....	....	....	....	....	....	....
	Calgary	**NHL**	**35**	**0**	**5**	**5**	**61**	**0**	**0**	**0**	**18**	**0.0**	**6**	....	....	....	....	....	**5**	**0**	**1**	**1**	**0**	**0**	**0**	**0**
1995-96	Karlovy Vary	Czech-2	16	7	4	11	16	....	....	....	....	....	....	....	....	....	....	....	....	....	....	....	....	....	....	....
	Ottawa	**NHL**	**65**	**1**	**3**	**4**	**85**	**0**	**0**	**0**	**37**	**2.7**	**–10**	....	....	....	....	....	....	....	....	....	....	....	....	....
1996-97	**Ottawa**	**NHL**	**57**	**0**	**5**	**5**	**58**	**0**	**0**	**0**	**24**	**0.0**	**6**	....	....	....	....	....	....	....	....	....	....	....	....	....
1997-98	Indianapolis Ice	IHL	52	5	8	13	122	....	....	....	....	....	....	....	....	....	....	....	....	....	....	....	....	....	....	....
	Detroit Vipers	IHL	9	0	0	0	6	....	....	....	....	....	....	....	....	....	....	....	....	....	....	....	....	....	....	....
	Edmonton	**NHL**	**17**	**1**	**2**	**3**	**8**	**0**	**1**	**1**	**8**	**12.5**	**1**	....	....	....	....	....	**7**	**0**	**0**	**0**	**6**	**0**	**0**	**0**
1998-99	**Edmonton**	**NHL**	**39**	**0**	**3**	**3**	**34**	**0**	**0**	**0**	**9**	**0.0**	**0**	**0**	**0.0**	**60**	**56**	**14:21**	**1**	**0**	**0**	**0**	**2**	**0**	**0**	**0**
99-2000	**Edmonton**	**NHL**	DID NOT PLAY – INJURED																							
2000-01	**Edmonton**	**NHL**	**13**	**0**	**2**	**2**	**4**	**0**	**0**	**0**	**0**	**0.0**	**–2**	**0**	**0.0**	**9**	**17**	**11:09**	....	....	....	....	....	....	....	....
2001-02	HC Dukla Jihlava	Czech-2	3	0	1	1	54	....	....	....	....	....	....	....	....	....	....	....	13	0	2	2	47	....	....	....
	NHL Totals		**797**	**34**	**106**	**140**	**1241**	**4**	**3**	**5**	**709**	**4.8**		**0**	**0.0**	**69**	**73**	**13:33**	**42**	**2**	**4**	**6**	**47**	**0**	**0**	**0**

Traded to **Calgary** by **Minnesota** for Brian Glynn, October 26, 1990. Traded to **Ottawa** by **Calgary** for Ottawa's 4th round choice (Chris St. Croix) in 1997 Entry Draft, October 7, 1995. Traded to **Edmonton** by **Ottawa** for Scott Ferguson, March 9, 1998. • Missed entire 1999-2000 season and start of 2000-01 season recovering from spinal cord injury suffered in practice, October 2, 1999. • Missed majority of 2000-01 season recovering from neck injury suffered in game vs. Columbus, January 7, 2001.

MYRVOLD, Anders (MYOOR-vohld, AN-duhrs) FLA.

Defense. Shoots left. 6'2", 200 lbs. Born, Lorenskog, Norway, August 12, 1975. Quebec's 6th choice, 127th overall, in 1993 Entry Draft.

Season	Club	League	GP	G	A	Pts	PIM	PP	SH	GW	S	%	+/-	TF	F%	H	SB	Min	GP	G	A	Pts	PIM	PP	SH	GW
			Regular Season																Playoffs							
1991-92	Storhamr IL	Norway	1	0	0	0	4	....	....	....	....	....	....	....	....	....	....	....	....	....	....	....	....	....	....	....
1992-93	Farjestad	Sweden	2	0	0	0	0	....	....	....	....	....	....	....	....	....	....	....	....	....	....	....	....	....	....	....
1993-94	Grums IK	Swede-2	24	1	0	1	59	....	....	....	....	....	....	....	....	....	....	....	2	1	0	1	5	....	....	....
1994-95	Laval Titan	QMJHL	64	14	50	64	173	....	....	....	....	....	....	....	....	....	....	....	20	4	10	14	68	....	....	....
	Cornwall Aces	AHL	....	....	....	....	....	....	....	....	....	....	....	....	....	....	....	....	3	0	1	1	2	....	....	....
1995-96	**Colorado**	**NHL**	**4**	**0**	**1**	**1**	**6**	**0**	**0**	**0**	**4**	**0.0**	**–2**	....	....	....	....	....	....	....	....	....	....	....	....	....
	Cornwall Aces	AHL	70	5	24	29	125	....	....	....	....	....	....	....	....	....	....	....	5	1	0	1	19	....	....	....
1996-97	Hershey Bears	AHL	20	0	3	3	16	....	....	....	....	....	....	....	....	....	....	....	....	....	....	....	....	....	....	....
	Boston	**NHL**	**9**	**0**	**2**	**2**	**4**	**0**	**0**	**0**	**8**	**0.0**	**–1**	....	....	....	....	....	....	....	....	....	....	....	....	....
	Providence	AHL	53	6	15	21	107	....	....	....	....	....	....	....	....	....	....	....	10	0	1	1	6	....	....	....
1997-98	Providence	AHL	75	4	21	25	91	....	....	....	....	....	....	....	....	....	....	....	....	....	....	....	....	....	....	....
1998-99	Djurgarden	Sweden	29	3	4	7	52	....	....	....	....	....	....	....	....	....	....	....	....	....	....	....	....	....	....	....
	Djurgarden	EuroHL	3	0	1	1	4	....	....	....	....	....	....	....	....	....	....	....	....	....	....	....	....	....	....	....
	AIK Solna	Sweden	19	1	3	4	24	....	....	....	....	....	....	....	....	....	....	....	....	....	....	....	....	....	....	....
99-2000	AIK Solna	Sweden	49	1	3	4	87	....	....	....	....	....	....	....	....	....	....	....	....	....	....	....	....	....	....	....
2000-01	**NY Islanders**	**NHL**	**12**	**0**	**1**	**1**	**0**	**0**	**0**	**0**	**8**	**0.0**	**–2**	**0**	**0.0**	**10**	**4**	**9:26**	....	....	....	....	....	....	....	....
	Springfield	AHL	69	5	25	30	129	....	....	....	....	....	....	....	....	....	....	....	....	....	....	....	....	....	....	....
2001-02	Hartford	AHL	19	3	3	6	28	....	....	....	....	....	....	....	....	....	....	....	....	....	....	....	....	....	....	....
	Fribourg	Swiss	6	0	0	0	16	....	....	....	....	....	....	....	....	....	....	....	4	0	1	1	6	....	....	....
	NHL Totals		**25**	**0**	**4**	**4**	**10**	**0**	**0**	**0**	**20**	**0.0**		**0**	**0.0**	**10**	**4**	**9:26**	....	....	....	....	....	....	....	....

QMJHL All-Rookie Team (1995)

Rights transferred to **Colorado** after **Quebec** franchise relocated, June 21, 1995. Traded to **Boston** by **Colorado** with Landon Wilson for Boston's 1st round choice (Robyn Regehr) in 1998 Entry Draft, November 22, 1996. Signed as a free agent by **NY Islanders**, August 28, 2000. Signed as a free agent by **Fribourg** (Swiss), January 10, 2002. Signed as a free agent by **Florida**, July 28, 2002.

NABOKOV, Dmitri (na-BAW-kahv, dih-MEE-tree)

Center/Left wing. Shoots right. 6'2", 209 lbs. Born, Novosibirsk, USSR, January 4, 1977. Chicago's 1st choice, 19th overall, in 1995 Entry Draft.

Season	Club	League	GP	G	A	Pts	PIM	PP	SH	GW	S	%	+/-	TF	F%	H	SB	Min	GP	G	A	Pts	PIM	PP	SH	GW
1993-94	Krylja Sovetov	CIS	17	0	2	2	6	....	....	....	....	....	....	....	....	....	....	....	3	0	0	0	0	....	....	....
1994-95	Krylja Sovetov	CIS	49	15	12	27	32	....	....	....	....	....	....	....	....	....	....	....	4	5	0	5	6	....	....	....
1995-96	Krylja Sovetov	CIS	50	12	14	26	51	....	....	....	....	....	....	....	....	....	....	....	....	....	....	....	....	....	....	....
1996-97	Krylja Sovetov	Russia	1	0	0	0	0	....	....	....	....	....	....	....	....	....	....	....	....	....	....	....	....	....	....	....
	Regina Pats	WHL	50	39	56	95	61	....	....	....	....	....	....	....	....	....	....	....	5	2	3	5	2	....	....	....
	Indianapolis Ice	IHL	2	0	0	0	0	....	....	....	....	....	....	....	....	....	....	....	....	....	....	....	....	....	....	....
1997-98	**Chicago**	**NHL**	**25**	**7**	**4**	**11**	**10**	**3**	**0**	**2**	**34**	**20.6**	**–1**	....	....	....	....	....	....	....	....	....	....	....	....	....
	Indianapolis Ice	IHL	46	6	15	21	16	....	....	....	....	....	....	....	....	....	....	....	5	2	1	3	0	....	....	....
1998-99	**NY Islanders**	**NHL**	**4**	**0**	**2**	**2**	**2**	**0**	**0**	**0**	**4**	**0.0**	**4**	**0**	**0.0**	**3**	**0**	**11:38**	....	....	....	....	....	....	....	....
	Lowell	AHL	73	17	25	42	46	....	....	....	....	....	....	....	....	....	....	....	3	0	1	1	0	....	....	....
99-2000	**NY Islanders**	**NHL**	**26**	**4**	**7**	**11**	**16**	**0**	**0**	**0**	**40**	**10.0**	**–8**	**12**	**25.0**	**36**	**10**	**13:33**	....	....	....	....	....	....	....	....
	Lowell	AHL	51	8	26	34	42	....	....	....	....	....	....	....	....	....	....	....	6	1	2	3	2	....	....	....
2000-01	Lada Togliatti	Russia	24	8	5	13	40	....	....	....	....	....	....	....	....	....	....	....	5	0	0	0	10	....	....	....
2001-02	Novokuznetsk	Russia	30	9	11	20	14	....	....	....	....	....	....	....	....	....	....	....	....	....	....	....	....	....	....	....
	NHL Totals		**55**	**11**	**13**	**24**	**28**	**3**	**0**	**2**	**78**	**14.1**		**12**	**25.0**	**39**	**10**	**13:18**	....	....	....	....	....	....	....	....

WHL East Second All-Star Team (1997)

Traded to **NY Islanders** by **Chicago** for J-P Dumont and Chicago's 5th round choice (later traded to Philadelphia - Philadelphia selected Francis Belanger) in 1998 Entry Draft, June 1, 1998.

NAGY, Ladislav (NA-gee, LA-dih-slahv) PHX.

Center. Shoots left. 5'11", 194 lbs. Born, Saca, Czech., June 1, 1979. St. Louis' 6th choice, 177th overall, in 1997 Entry Draft.

Season	Club	League	GP	G	A	Pts	PIM	PP	SH	GW	S	%	+/-	TF	F%	H	SB	Min	GP	G	A	Pts	PIM	PP	SH	GW
1996-97	HC Kosice Jr.	Slovak-Jr.	45	29	30	59	105	....	....	....	....	....	....	....	....	....	....	....	....	....	....	....	....	....	....	....
	HK Dragon Presov	Slovak-2	11	6	5	11	....	....	....	....	....	....	....	....	....	....	....	....	....	....	....	....	....	....	....	....
1997-98	HC Kosice	Slovakia	29	19	15	34	41	....	....	....	....	....	....	....	....	....	....	....	11	2	4	6	6	....	....	....
1998-99	Halifax	QMJHL	63	71	55	126	148	....	....	....	....	....	....	....	....	....	....	....	5	3	3	6	18	....	....	....
	Worcester	AHL	....	....	....	....	....	....	....	....	....	....	....	....	....	....	....	....	3	2	2	4	0	....	....	....
99-2000	**St. Louis**	**NHL**	**11**	**2**	**4**	**6**	**2**	**1**	**0**	**0**	**15**	**13.3**	**2**	**6**	**33.3**	**6**	**3**	**12:19**	**6**	**1**	**1**	**2**	**0**	**0**	**0**	**0**
	Worcester	AHL	69	23	28	51	67	....	....	....	....	....	....	....	....	....	....	....	2	1	0	1	0	....	....	....
2000-01	**St. Louis**	**NHL**	**40**	**8**	**8**	**16**	**20**	**2**	**0**	**2**	**59**	**13.6**	**–2**	**28**	**50.0**	**20**	**5**	**13:03**	....	....	....	....	....	....	....	....
	Worcester	AHL	20	6	14	20	36	....	....	....	....	....	....	....	....	....	....	....	....	....	....	....	....	....	....	....
	Phoenix	**NHL**	**6**	**0**	**1**	**1**	**2**	**0**	**0**	**0**	**5**	**0.0**	**0**	**0**	**0.0**	**4**	**0**	**12:38**	....	....	....	....	....	....	....	....
2001-02	**Phoenix**	**NHL**	**74**	**23**	**19**	**42**	**50**	**5**	**0**	**5**	**187**	**12.3**	**6**	**17**	**47.1**	**60**	**11**	**15:04**	**5**	**0**	**0**	**0**	**21**	**0**	**0**	**0**
	NHL Totals		**131**	**33**	**32**	**65**	**74**	**8**	**0**	**7**	**266**	**12.4**		**51**	**47.1**	**90**	**19**	**14:06**	**11**	**1**	**1**	**2**	**21**	**0**	**0**	**0**

Traded to **Phoenix** by **St. Louis** with Michal Handzus, the rights to Jeff Taffe and St. Louis' 1st round choice (Ben Eager) in 2002 Entry Draft for Keith Tkachuk, March 13, 2001.

NAMESTNIKOV, John (nah-MEST-nih-kov, JAWN)

Defense. Shoots right. 5'11", 190 lbs. Born, Arzamis-Ig, USSR, October 9, 1971. Vancouver's 5th choice, 117th overall, in 1991 Entry Draft.

Season	Club	League	GP	G	A	Pts	PIM	PP	SH	GW	S	%	+/-	TF	F%	H	SB	Min	GP	G	A	Pts	PIM	PP	SH	GW
1988-89	Torpedo Gorky	USSR	2	0	0	0	2	....	....	....	....	....	....	....	....	....	....	....	....	....	....	....	....	....	....	....
1989-90	Torpedo Gorky	USSR	23	0	0	0	25	....	....	....	....	....	....	....	....	....	....	....	....	....	....	....	....	....	....	....
1990-91	Nizhny Novgorod	USSR	42	1	2	3	49	....	....	....	....	....	....	....	....	....	....	....	....	....	....	....	....	....	....	....
1991-92	CSKA Moscow	CIS	42	1	1	2	47	....	....	....	....	....	....	....	....	....	....	....	....	....	....	....	....	....	....	....
1992-93	CSKA Moscow	CIS	42	5	5	10	68	....	....	....	....	....	....	....	....	....	....	....	....	....	....	....	....	....	....	....
1993-94	**Vancouver**	**NHL**	**17**	**0**	**5**	**5**	**10**	**0**	**0**	**0**	**11**	**0.0**	**–2**	....	....	....	....	....	....	....	....	....	....	....	....	....
	Hamilton Canucks	AHL	59	7	27	34	97	....	....	....	....	....	....	....	....	....	....	....	4	0	2	2	19	....	....	....
1994-95	Syracuse Crunch	AHL	59	11	22	33	59	....	....	....	....	....	....	....	....	....	....	....	....	....	....	....	....	....	....	....
	Vancouver	**NHL**	**16**	**0**	**3**	**3**	**4**	**0**	**0**	**0**	**18**	**0.0**	**2**	....	....	....	....	....	**1**	**0**	**0**	**0**	**2**	**0**	**0**	**0**
1995-96	Syracuse Crunch	AHL	59	13	34	47	85	....	....	....	....	....	....	....	....	....	....	....	15	1	8	9	16	....	....	....
	Vancouver	**NHL**	....	....	....	....	....	....	....	....	....	....	....	....	....	....	....	....	**1**	**0**	**0**	**0**	**0**	**0**	**0**	**0**
1996-97	**Vancouver**	**NHL**	**2**	**0**	**0**	**0**	**4**	**0**	**0**	**0**	**1**	**0.0**	**–1**	....	....	....	....	....	....	....	....	....	....	....	....	....
	Syracuse Crunch	AHL	55	9	37	46	73	....	....	....	....	....	....	....	....	....	....	....	3	2	0	2	0	....	....	....
1997-98	**NY Islanders**	**NHL**	**6**	**0**	**1**	**1**	**4**	**0**	**0**	**0**	**2**	**0.0**	**–1**	....	....	....	....	....	....	....	....	....	....	....	....	....
	Utah Grizzlies	IHL	62	6	19	25	48	....	....	....	....	....	....	....	....	....	....	....	4	1	0	1	2	....	....	....
1998-99	Lowell	AHL	42	12	14	26	42	....	....	....	....	....	....	....	....	....	....	....	....	....	....	....	....	....	....	....
99-2000	Hartford	AHL	33	1	9	10	14	....	....	....	....	....	....	....	....	....	....	....	....	....	....	....	....	....	....	....
	Nashville	**NHL**	**2**	**0**	**0**	**0**	**2**	**0**	**0**	**0**	**3**	**0.0**	**0**	**0**	**0.0**	**5**	**2**	**15:48**	....	....	....	....	....	....	....	....
	Milwaukee	IHL	12	2	3	5	17	....	....	....	....	....	....	....	....	....	....	....	3	0	0	0	0	....	....	....
2000-01	Milwaukee	IHL	56	7	22	29	36	....	....	....	....	....	....	....	....	....	....	....	3	0	1	1	0	....	....	....
2001-02	Lada Togliatti	Russia	50	6	7	13	56	....	....	....	....	....	....	....	....	....	....	....	4	0	1	1	2	....	....	....
	NHL Totals		**43**	**0**	**9**	**9**	**24**	**0**	**0**	**0**	**35**	**0.0**		**0**	**0.0**	**5**	**2**	....	**2**	**0**	**0**	**0**	**2**	**0**	**0**	**0**

Signed as a free agent by **NY Islanders**, July 21, 1997. Signed as a free agent by **NY Rangers**, August 9, 1999. Claimed by **Vancouver** from **NY Rangers** in Waiver Draft, September 27, 1999. Claimed on waivers by **NY Rangers** from **Vancouver**, October 5, 1999. Traded to **Nashville** by **NY Rangers** for Jason Dawe, February 3, 2000.

NASH, Tyson (NASH, TIGH-sohn) ST.L.

Left wing. Shoots left. 6', 185 lbs. Born, Edmonton, Alta., March 11, 1975. Vancouver's 10th choice, 247th overall, in 1994 Entry Draft.

Season	Club	League	GP	G	A	Pts	PIM	PP	SH	GW	S	%	+/-	TF	F%	H	SB	Min	GP	G	A	Pts	PIM	PP	SH	GW
1990-91	Sherwood Park	AMHL	40	17	28	43	63	....	....	....	....	....	....	....	....	....	....	....	....	....	....	....	....	....	....	....
1991-92	Kamloops Blazers	WHL	33	1	6	7	62	....	....	....	....	....	....	....	....	....	....	....	4	0	0	0	0	....	....	....
1992-93	Kamloops Blazers	WHL	61	10	16	26	78	....	....	....	....	....	....	....	....	....	....	....	13	3	2	5	32	....	....	....
1993-94	Kamloops Blazers	WHL	65	20	36	56	135	....	....	....	....	....	....	....	....	....	....	....	16	3	4	7	12	....	....	....
1994-95	Kamloops Blazers	WHL	63	34	41	75	70	....	....	....	....	....	....	....	....	....	....	....	21	10	7	17	30	....	....	....
1995-96	Syracuse Crunch	AHL	50	4	7	11	58	....	....	....	....	....	....	....	....	....	....	....	4	0	0	0	11	....	....	....
	Raleigh IceCaps	ECHL	6	1	1	2	8	....	....	....	....	....	....	....	....	....	....	....	....	....	....	....	....	....	....	....
1996-97	Syracuse Crunch	AHL	77	17	17	34	105	....	....	....	....	....	....	....	....	....	....	....	3	0	2	2	0	....	....	....
1997-98	Syracuse Crunch	AHL	74	20	20	40	184	....	....	....	....	....	....	....	....	....	....	....	5	0	2	2	28	....	....	....
1998-99	**St. Louis**	**NHL**	**2**	**0**	**0**	**0**	**5**	**0**	**0**	**0**	**1**	**0.0**	**–1**	**0**	**0.0**	**8**	**0**	**7:44**	**1**	**0**	**0**	**0**	**2**	**0**	**0**	**0**
	Worcester	AHL	55	14	22	36	143	....	....	....	....	....	....	....	....	....	....	....	4	4	1	5	27	....	....	....
99-2000	**St. Louis**	**NHL**	**66**	**4**	**9**	**13**	**150**	**0**	**1**	**1**	**68**	**5.9**	**6**	**0**	**0.0**	**193**	**7**	**8:35**	**6**	**1**	**0**	**1**	**24**	**0**	**0**	**0**

Season	Club	League	GP	G	A	Pts	PIM	PP	SH	GW	S	%	+/-	TF	F%	H	SB	Min	GP	G	A	Pts	PIM	PP	SH	GW
			Regular Season																Playoffs							
2000-01	**St. Louis**	**NHL**	**57**	**8**	**7**	**15**	**110**	**0**	**1**	**0**	**113**	**7.1**	**8**	**2**	**50.0**	**149**	**10**	**12:29**								
2001-02	**St. Louis**	**NHL**	**64**	**6**	**7**	**13**	**100**	**0**	**0**	**1**	**66**	**9.1**	**2**	**14**	**35.7**	**152**	**15**	**10:03**	**9**	**0**	**1**	**1**	**20**	**0**	**0**	**0**
	NHL Totals		**189**	**18**	**23**	**41**	**365**	**0**	**2**	**2**	**248**	**7.3**		**16**	**37.5**	**502**	**32**	**10:15**	**16**	**1**	**1**	**2**	**46**	**0**	**0**	**0**

Signed as a free agent by **St. Louis**, July 14, 1998.

NASLUND, Markus

(NAZ-luhnd, MAHR-kuhs) **VAN.**

Right wing. Shoots left. 5'11", 195 lbs. Born, Ornskoldsvik, Sweden, July 30, 1973. Pittsburgh's 1st choice, 16th overall, in 1991 Entry Draft.

Season	Club	League	GP	G	A	Pts	PIM	PP	SH	GW	S	%	+/-	TF	F%	H	SB	Min	GP	G	A	Pts	PIM	PP	SH	GW
1988-89	Ornskoldsviks IF	Swede-3	14	7	6	13																				
1989-90	MoDo Jr.	Swede-Jr.	33	43	35	78	20																			
1990-91	MoDo	Sweden	32	10	9	19	14																			
1991-92	MoDo	Sweden	39	22	18	40	54																			
1992-93	MoDo Jr.	Swede-Jr.	2	4	1	5	2																			
	MoDo	Sweden	39	22	17	39	67												3	3	2	5	0			
1993-94	**Pittsburgh**	**NHL**	**71**	**4**	**7**	**11**	**27**	**1**	**0**	**0**	**80**	**5.0**	**−3**													
	Cleveland	IHL	5	1	6	7	4																			
1994-95	**Pittsburgh**	**NHL**	**14**	**2**	**2**	**4**	**2**	**0**	**0**	**0**	**13**	**15.4**	**0**													
	Cleveland	IHL	7	3	4	7	6												4	1	3	4	8			
1995-96	**Pittsburgh**	**NHL**	**66**	**19**	**33**	**52**	**36**	**3**	**0**	**4**	**125**	**15.2**	**17**													
	Vancouver	**NHL**	**10**	**3**	**0**	**3**	**6**	**1**	**0**	**1**	**19**	**15.8**	**3**						**6**	**1**	**2**	**3**	**8**	**1**	**0**	**0**
1996-97	**Vancouver**	**NHL**	**78**	**21**	**20**	**41**	**30**	**4**	**0**	**4**	**120**	**17.5**	**−15**													
1997-98	**Vancouver**	**NHL**	**76**	**14**	**20**	**34**	**56**	**2**	**1**	**0**	**106**	**13.2**	**5**													
1998-99	**Vancouver**	**NHL**	**80**	**36**	**30**	**66**	**74**	**15**	**2**	**3**	**205**	**17.6**	**−13**	**14**	**57.1**	**40**	**20**	**19:57**								
99-2000	**Vancouver**	**NHL**	**82**	**27**	**38**	**65**	**64**	**6**	**2**	**3**	**271**	**10.0**	**−5**	**13**	**46.2**	**50**	**28**	**20:13**								
2000-01	**Vancouver**	**NHL**	**72**	**41**	**34**	**75**	**58**	**18**	**1**	**5**	**277**	**14.8**	**−2**	**6**	**50.0**	**37**	**11**	**19:03**								
2001-02	**Vancouver**	**NHL**	**81**	**40**	**50**	**90**	**50**	**8**	**0**	**6**	**302**	**13.2**	**22**	**5**	**20.0**	**36**	**11**	**19:31**	**6**	**1**	**1**	**2**	**2**	**0**	**0**	**0**
	Sweden	Olympics	4	2	1	3	0																			
	NHL Totals		**630**	**207**	**234**	**441**	**403**	**58**	**6**	**26**	**1518**	**13.6**		**38**	**47.4**	**163**	**70**	**19:42**	**12**	**2**	**3**	**5**	**10**	**1**	**0**	**0**

NHL First All-Star Team (2002) • Played in NHL All-Star Game (1999, 2001, 2002)

Traded to **Vancouver** by **Pittsburgh** for Alek Stojanov, March 20, 1996.

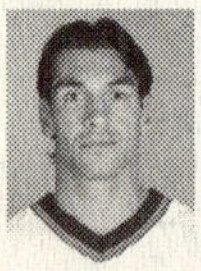

NASREDDINE, Alain

(NAS-ruh-deen, AL-eh)

Defense. Shoots left. 6'1", 201 lbs. Born, Montreal, Que., July 10, 1975. Florida's 8th choice, 135th overall, in 1993 Entry Draft.

Season	Club	League	GP	G	A	Pts	PIM	PP	SH	GW	S	%	+/-	TF	F%	H	SB	Min	GP	G	A	Pts	PIM	PP	SH	GW
1990-91	Mtl-Bourassa	QAAA	35	10	25	35	50																			
1991-92	Drummondville	QMJHL	61	1	9	10	78												4	0	0	0	17			
1992-93	Drummondville	QMJHL	64	0	14	14	137												10	0	1	1	36			
1993-94	Chicoutimi	QMJHL	60	3	24	27	218												26	2	10	12	118			
1994-95	Chicoutimi	QMJHL	67	8	31	39	342												13	3	5	8	40			
1995-96	Carolina	AHL	63	0	5	5	245																			
1996-97	Carolina	AHL	26	0	4	4	109																			
	Indianapolis Ice	IHL	49	0	2	2	248												4	1	1	2	27			
1997-98	Indianapolis Ice	IHL	75	1	12	13	258												5	0	2	2	12			
1998-99	**Chicago**	**NHL**	**7**	**0**	**0**	**0**	**19**	**0**	**0**	**0**	**2**	**0.0**	**−2**	**0**	**0.0**	**5**	**1**	**12:11**								
	Portland Pirates	AHL	7	0	1	1	36																			
	Montreal	**NHL**	**8**	**0**	**0**	**0**	**33**	**0**	**0**	**0**	**1**	**0.0**	**1**	**0**	**0.0**	**7**	**1**	**8:12**								
	Fredericton	AHL	38	0	10	10	108												15	0	3	3	39			
99-2000	Quebec	AHL	59	1	6	7	178																			
	Hamilton	AHL	11	0	0	0	12												10	1	1	2	14			
2000-01	Hamilton	AHL	74	4	14	18	164																			
2001-02	Hamilton	AHL	79	7	10	17	154												12	1	3	4	22			
	NHL Totals		**15**	**0**	**0**	**0**	**52**	**0**	**0**	**0**	**3**	**0.0**		**0**	**0.0**	**12**	**2**	**10:04**								

QMJHL Second All-Star Team (1995)

Traded to **Chicago** by **Florida** for Ivan Droppa, December 18, 1996. Traded to **Montreal** by **Chicago** with Jeff Hackett, Eric Weinrich and Tampa Bay's 4th round choice (previously acquired, Montreal selected Chris Dyment) in 1999 Entry Draft for Jocelyn Thibault, Dave Manson and Brad Brown, November 16, 1998. Traded to **Edmonton** by **Montreal** with Igor Ulanov for Christian Laflamme and Matthieu Descoteaux, March 9, 2000.

NAZAROV, Andrei

(nah-ZAH-rohv, AWN-dray) **PHX.**

Left wing. Shoots right. 6'5", 241 lbs. Born, Chelyabinsk, USSR, May 22, 1974. San Jose's 2nd choice, 10th overall, in 1992 Entry Draft.

Season	Club	League	GP	G	A	Pts	PIM	PP	SH	GW	S	%	+/-	TF	F%	H	SB	Min	GP	G	A	Pts	PIM	PP	SH	GW
1991-92	Dynamo Moscow	CIS	2	1	0	1	2																			
1992-93	Dynamo Moscow	CIS	42	8	2	10	79												10	1	1	2	8			
1993-94	Dynamo Moscow	CIS	6	2	2	4	0																			
	San Jose	**NHL**	**1**	**0**	**0**	**0**	**0**	**0**	**0**	**0**	**0**	**0.0**	**0**													
	Kansas City	IHL	71	15	18	33	64																			
1994-95	Kansas City	IHL	43	15	10	25	55																			
	San Jose	**NHL**	**26**	**3**	**5**	**8**	**94**	**0**	**0**	**0**	**19**	**15.8**	**−1**						**6**	**0**	**0**	**0**	**9**	**0**	**0**	**0**
1995-96	**San Jose**	**NHL**	**42**	**7**	**7**	**14**	**62**	**2**	**0**	**1**	**55**	**12.7**	**−15**													
	Kansas City	IHL	27	4	6	10	118												2	0	0	0	2			
1996-97	**San Jose**	**NHL**	**60**	**12**	**15**	**27**	**222**	**1**	**0**	**1**	**116**	**10.3**	**−4**													
	Kentucky	AHL	3	1	2	3	4																			
1997-98	**San Jose**	**NHL**	**40**	**1**	**1**	**2**	**112**	**0**	**0**	**0**	**31**	**3.2**	**−4**													
	Tampa Bay	**NHL**	**14**	**1**	**1**	**2**	**58**	**0**	**0**	**0**	**19**	**5.3**	**−9**													
1998-99	**Tampa Bay**	**NHL**	**26**	**2**	**0**	**2**	**43**	**0**	**0**	**0**	**18**	**11.1**	**−5**	**4**	**50.0**	**26**	**2**	**8:13**								
	Calgary	**NHL**	**36**	**5**	**9**	**14**	**30**	**0**	**0**	**2**	**53**	**9.4**	**1**	**0**	**0.0**	**38**	**10**	**14:31**								
99-2000	**Calgary**	**NHL**	**76**	**10**	**22**	**32**	**78**	**1**	**0**	**1**	**110**	**9.1**	**3**	**2**	**100.0**	**79**	**16**	**11:44**								
2000-01	**Anaheim**	**NHL**	**16**	**1**	**0**	**1**	**29**	**0**	**0**	**0**	**13**	**7.7**	**−9**	**2**	**50.0**	**19**	**3**	**8:42**								
	Boston	**NHL**	**63**	**1**	**4**	**5**	**200**	**0**	**0**	**0**	**50**	**2.0**	**−14**	**14**	**21.4**	**94**	**13**	**8:12**								
2001-02	**Boston**	**NHL**	**47**	**0**	**2**	**2**	**164**	**0**	**0**	**0**	**18**	**0.0**	**−2**	**1**	**100.0**	**9**	**2**	**3:10**								
	Phoenix	**NHL**	**30**	**6**	**3**	**9**	**51**	**0**	**0**	**0**	**38**	**15.8**	**7**	**1**	**100.0**	**24**	**3**	**7:35**	**3**	**0**	**0**	**0**	**2**	**0**	**0**	**0**
	NHL Totals		**477**	**49**	**69**	**118**	**1143**	**4**	**0**	**5**	**540**	**9.1**		**24**	**41.7**	**289**	**49**	**9:03**	**9**	**0**	**0**	**0**	**11**	**0**	**0**	**0**

Traded to **Tampa Bay** by **San Jose** with Florida's 1st round choice (previously acquired, Tampa Bay selected Vincent Lecavalier) in 1998 Entry Draft for Bryan Marchment, David Shaw and Tampa Bay's 1st round choice (later traded to Nashville - Nashville selected David Legwand) in 1998 Entry Draft, March 24, 1998. Traded to **Calgary** by **Tampa Bay** for Michael Nylander, January 19, 1999. Traded to **Anaheim** by **Calgary** with Calgary's 2nd round choice (later traded to Phoenix - later traded back to Calgary - Calgary selected Andrei Taratukhin) in 2001 Entry Draft for Jordan Leopold, September 26, 2000. Traded to **Boston** by **Anaheim** with Patrick Traverse for Sami Pahlsson, November 18, 2000. Traded to **Phoenix** by **Boston** for Phoenix's 5th round choice (Peter Hamerlik) in 2002 Entry Draft, January 25, 2002.

NDUR, Rumun

(nih-DOOR, ROO-muhn)

Defense. Shoots left. 6'2", 222 lbs. Born, Zaria, Nigeria, July 7, 1975. Buffalo's 3rd choice, 69th overall, in 1994 Entry Draft.

Season	Club	League	GP	G	A	Pts	PIM	PP	SH	GW	S	%	+/-	TF	F%	H	SB	Min	GP	G	A	Pts	PIM	PP	SH	GW
1990-91	Belmont Bombers	OJHL-D	36	3	11	14	70																			
1991-92	Sarnia	OJHL-B	30	2	5	7	46																			
	Clearwater	OJHL-C	4	0	4	4	4																			
1992-93	Guelph Platers	OJHL-B	24	7	8	15	202																			
	Guelph Storm	OHL	22	1	3	4	30												4	0	1	1	4			
1993-94	Guelph Storm	OHL	61	6	33	39	176												9	4	1	5	24			
1994-95	Guelph Storm	OHL	63	10	21	31	187												14	0	4	4	28			
1995-96	Rochester	AHL	73	2	12	14	306												17	1	2	3	33			
1996-97	**Buffalo**	**NHL**	**2**	**0**	**0**	**0**	**2**	**0**	**0**	**0**	**0**	**0.0**	**1**													
	Rochester	AHL	68	5	11	16	282												10	3	1	4	21			
1997-98	**Buffalo**	**NHL**	**1**	**0**	**0**	**0**	**2**	**0**	**0**	**0**	**0**	**0.0**	**−1**													
	Rochester	AHL	50	1	12	13	207												4	0	2	2	16			
1998-99	**Buffalo**	**NHL**	**8**	**0**	**0**	**0**	**16**	**0**	**0**	**0**	**1**	**0.0**	**1**	**0**	**0.0**	**11**	**2**	**10:58**								
	NY Rangers	**NHL**	**31**	**1**	**3**	**4**	**46**	**0**	**0**	**0**	**21**	**4.8**	**−2**	**0**	**0.0**	**58**	**11**	**11:58**								
	Hartford	AHL	6	0	1	1	4																			
99-2000	Hartford	AHL	2	0	0	0	0																			
	Atlanta	**NHL**	**27**	**1**	**0**	**1**	**71**	**0**	**0**	**0**	**6**	**16.7**	**−17**	**0**	**0.0**	**64**	**18**	**13:12**								

Season	Club	League	GP	G	A	Pts	PIM	PP	SH	GW	S	%	+/-	TF	F%	H	SB	Min	GP	G	A	Pts	PIM	PP	SH	GW
			Regular Season																Playoffs							
2000-01	Orlando	IHL	16	0	0	0	50																			
	Norfolk Admirals	AHL	21	1	2	3	93												9	1	0	1	26			
2001-02	Norfolk Admirals	AHL	52	4	9	13	133												2	0	0	0	2			
	NHL Totals		**69**	**2**	**3**	**5**	**137**	**0**	**0**	**0**	**28**	**7.1**		**0**	**0.0**	**133**	**31**	**12:21**								

Claimed on waivers by **NY Rangers** from **Buffalo**, December 18, 1998. Claimed on waivers by **Atlanta** from **NY Rangers**, December 11, 1999. Signed as a free agent by **Chicago**, August 7, 2001.

NECKAR, Stan

(NEHTS-kahzh, STAN) **T.B.**

Defense. Shoots left. 6'1", 214 lbs. Born, Ceske Budejovice, Czech., December 22, 1975. Ottawa's 2nd choice, 29th overall, in 1994 Entry Draft.

Season	Club	League	GP	G	A	Pts	PIM	PP	SH	GW	S	%	+/-	TF	F%	H	SB	Min	GP	G	A	Pts	PIM	PP	SH	GW
1991-92	C. Budejovice Jr.	Czech-Jr.	18	1	3	4																				
1992-93	Ceske Budejovice	Czech	42	2	9	11	12																			
1993-94	Ceske Budejovice	Czech	12	3	2	5	2												3	0	0	0				
1994-95	Detroit Vipers	IHL	15	2	2	4	15																			
	Ottawa	**NHL**	**48**	**1**	**3**	**4**	**37**	**0**	**0**	**0**	**34**	**2.9**	**-20**													
1995-96	**Ottawa**	**NHL**	**82**	**3**	**9**	**12**	**54**	**1**	**0**	**0**	**57**	**5.3**	**-16**													
1996-97	**Ottawa**	**NHL**	**5**	**0**	**0**	**0**	**2**	**0**	**0**	**0**	**3**	**0.0**	**2**													
1997-98	**Ottawa**	**NHL**	**60**	**2**	**2**	**4**	**31**	**0**	**0**	**0**	**43**	**4.7**	**-14**						**9**	**0**	**0**	**0**	**2**	**0**	**0**	**0**
1998-99	**Ottawa**	**NHL**	**3**	**0**	**2**	**2**	**0**	**0**	**0**	**0**	**2**	**0.0**	**-1**	**0**	**0.0**	**6**	**3**	**15:53**								
	NY Rangers	**NHL**	**18**	**0**	**0**	**0**	**8**	**0**	**0**	**0**	**8**	**0.0**	**-1**	**0**	**0.0**	**30**	**27**	**13:46**								
	Phoenix	**NHL**	**11**	**0**	**1**	**1**	**10**	**0**	**0**	**0**	**6**	**0.0**	**3**	**0**	**0.0**	**26**	**10**	**15:08**	**6**	**0**	**1**	**1**	**4**	**0**	**0**	**0**
99-2000	**Phoenix**	**NHL**	**66**	**2**	**8**	**10**	**36**	**0**	**0**	**0**	**34**	**5.9**	**1**	**0**	**0.0**	**119**	**54**	**14:27**	**5**	**0**	**0**	**0**	**0**	**0**	**0**	**0**
2000-01	**Phoenix**	**NHL**	**53**	**2**	**2**	**4**	**63**	**0**	**0**	**1**	**16**	**12.5**	**-2**	**0**	**0.0**	**75**	**56**	**15:26**								
	Tampa Bay	**NHL**	**16**	**0**	**2**	**2**	**8**	**0**	**0**	**0**	**10**	**0.0**	**-1**	**0**	**0.0**	**27**	**15**	**18:33**								
2001-02	**Tampa Bay**	**NHL**	**77**	**1**	**7**	**8**	**24**	**0**	**1**	**0**	**38**	**2.6**	**-18**	**0**	**0.0**	**33**	**102**	**20:31**								
	NHL Totals		**439**	**11**	**36**	**47**	**273**	**1**	**1**	**1**	**251**	**4.4**		**0**	**0.0**	**316**	**267**	**16:51**	**20**	**0**	**1**	**1**	**6**	**0**	**0**	**0**

Traded to **NY Rangers** by **Ottawa** for Bill Berg and NY Rangers' 2nd round choice (later traded to Anaheim - Anaheim selected Jordan Leopold) in 1999 Entry Draft, November 27, 1998. Traded to **Phoenix** by **NY Rangers** for Jason Doig and Phoenix's 6th round choice (Jay Dardis) in 1999 Entry Draft, March 23, 1999. Traded to **Tampa Bay** by **Phoenix** with Nikolai Khabibulin for Mike Johnson, Paul Mara, Ruslan Zainullin and NY Islanders' 2nd round choice (previously acquired, Phoenix selected Matthew Spiller) in 2001 Entry Draft, March 5, 2001.

NEDOROST, Andrej

(NEHD-ohr-ohst, awn-DRAY) **CBJ**

Center. Shoots left. 6', 192 lbs. Born, Trencin, Czech., April 30, 1980. Columbus' 10th choice, 286th overall, in 2000 Entry Draft.

Season	Club	League	GP	G	A	Pts	PIM	PP	SH	GW	S	%	+/-	TF	F%	H	SB	Min	GP	G	A	Pts	PIM	PP	SH	GW
1996-96	Dukla Trencin Jr.	Slovak-Jr.	40	50	35	85																				
1996-97	Dukla Trencin Jr.	Slovak-Jr.	45	15	16	31																				
1997-98	Dukla Trencin Jr.	Slovak-Jr.	45	27	22	49	61																			
	Dukla Trencin	Slovakia	1	0	0	0	0																			
1998-99	Essen Jr.	Ger.-Jr.	17	37	18	55	43																			
	Essen	German-2	30	3	5	8	22																			
99-2000	Essen	Germany	66	7	5	12	44																			
2000-01	Plzen	Czech	33	10	8	18	22																			
2001-02	**Columbus**	**NHL**	**7**	**0**	**2**	**2**	**2**	**0**	**0**	**0**	**12**	**0.0**	**-3**	**1**	**100.0**	**6**	**0**	**12:56**								
	Syracuse Crunch	AHL	37	5	13	18	28												10	1	3	4	4			
	NHL Totals		**7**	**0**	**2**	**2**	**2**	**0**	**0**	**0**	**12**	**0.0**		**1**	**100.0**	**6**	**0**	**12:56**								

NEDOROST, Vaclav

(neh-DOHR-uhst, VA-tslav) **COL.**

Center. Shoots left. 6'1", 190 lbs. Born, Budejovice, Czech., March 16, 1982. Colorado's 1st choice, 14th overall, in 2000 Entry Draft.

Season	Club	League	GP	G	A	Pts	PIM	PP	SH	GW	S	%	+/-	TF	F%	H	SB	Min	GP	G	A	Pts	PIM	PP	SH	GW
1997-98	C. Budejovice Jr.	Czech-Jr.	43	30	23	53	20																			
1998-99	C. Budejovice Jr.	Czech-Jr.	39	6	15	21	20																			
	Ceske Budejovice	Czech	7	0	2	2	0																			
99-2000	C. Budejovice Jr.	Czech-Jr.	14	4	7	11	4																			
	Ceske Budejovice	Czech	38	8	6	14	6												3	0	0	0	0			
2000-01	Ceske Budejovice	Czech	36	3	12	15	14																			
2001-02	**Colorado**	**NHL**	**25**	**2**	**2**	**4**	**2**	**1**	**0**	**0**	**22**	**9.1**	**-4**	**62**	**45.2**	**6**	**7**	**10:15**								
	Hershey Bears	AHL	49	12	22	34	16												7	2	3	5	2			
	NHL Totals		**25**	**2**	**2**	**4**	**2**	**1**	**0**	**0**	**22**	**9.1**		**62**	**45.2**	**6**	**7**	**10:15**								

NEDVED, Petr

(NEHD-VEHD, PEE-tuhr) **NYR**

Center. Shoots left. 6'3", 195 lbs. Born, Liberec, Czech., December 9, 1971. Vancouver's 1st choice, 2nd overall, in 1990 Entry Draft.

Season	Club	League	GP	G	A	Pts	PIM	PP	SH	GW	S	%	+/-	TF	F%	H	SB	Min	GP	G	A	Pts	PIM	PP	SH	GW
1988-89	CHZ Litvinov Jr.	Czech-Jr.	20	32	19	51	12																			
1989-90	Seattle	WHL	71	65	80	145	80												11	4	9	13	2			
1990-91	**Vancouver**	**NHL**	**61**	**10**	**6**	**16**	**20**	**1**	**0**	**0**	**97**	**10.3**	**-21**						**6**	**0**	**1**	**1**	**0**	**0**	**0**	**0**
1991-92	**Vancouver**	**NHL**	**77**	**15**	**22**	**37**	**36**	**5**	**0**	**1**	**99**	**15.2**	**-3**						**10**	**1**	**4**	**5**	**16**	**0**	**0**	**0**
1992-93	**Vancouver**	**NHL**	**84**	**38**	**33**	**71**	**96**	**2**	**1**	**3**	**149**	**25.5**	**20**						**12**	**2**	**3**	**5**	**2**	**0**	**0**	**0**
1993-94	Team Canada	Nat-Tm	17	19	12	31	16																			
	Canada	Olympics	8	5	1	6	6																			
	St. Louis	**NHL**	**19**	**6**	**14**	**20**	**8**	**2**	**0**	**0**	**63**	**9.5**	**2**						**4**	**0**	**1**	**1**	**4**	**0**	**0**	**0**
1994-95	**NY Rangers**	**NHL**	**46**	**11**	**12**	**23**	**26**	**1**	**0**	**3**	**123**	**8.9**	**-1**						**10**	**3**	**2**	**5**	**6**	**2**	**0**	**0**
1995-96	**Pittsburgh**	**NHL**	**80**	**45**	**54**	**99**	**68**	**8**	**1**	**5**	**204**	**22.1**	**37**						**18**	**10**	**10**	**20**	**16**	**4**	**0**	**2**
1996-97	**Pittsburgh**	**NHL**	**74**	**33**	**38**	**71**	**66**	**12**	**3**	**4**	**189**	**17.5**	**-2**						**5**	**1**	**2**	**3**	**12**	**0**	**1**	**0**
1997-98	Stadion Liberec	Czech-2	2	0	3	3																				
	TJ Novy Jicin	Czech-3	7	9	16	25																				
	HC Sparta Praha	Czech	5	2	3	5	8												6	0	2	2	52			
	Las Vegas	IHL	3	3	3	6	4																			
1998-99	Las Vegas	IHL	13	8	10	18	32																			
	NY Rangers	**NHL**	**56**	**20**	**27**	**47**	**50**	**9**	**1**	**3**	**153**	**13.1**	**-6**	**1069**	**52.5**	**58**	**34**	**20:31**								
99-2000	**NY Rangers**	**NHL**	**76**	**24**	**44**	**68**	**40**	**6**	**2**	**4**	**201**	**11.9**	**2**	**1354**	**54.0**	**55**	**33**	**19:54**								
2000-01	**NY Rangers**	**NHL**	**79**	**32**	**46**	**78**	**54**	**9**	**1**	**5**	**230**	**13.9**	**10**	**1349**	**49.7**	**48**	**40**	**20:16**								
2001-02	Liberec	Czech-2	1	3	0	3	2																			
	NY Rangers	**NHL**	**78**	**21**	**25**	**46**	**36**	**6**	**1**	**3**	**175**	**12.0**	**-8**	**1402**	**52.1**	**60**	**50**	**19:22**								
	NHL Totals		**730**	**255**	**321**	**576**	**500**	**61**	**10**	**31**	**1683**	**15.2**		**5174**	**52.0**	**221**	**157**	**19:59**	**65**	**17**	**23**	**40**	**56**	**6**	**1**	**2**

WHL Rookie of the Year (1990) • Canadian Major Junior Rookie of the Year (1990)

Signed as a free agent by **St. Louis**, March 5, 1994. Traded to **NY Rangers** by **St. Louis** for Esa Tikkanen and Doug Lidster, July 24, 1994. Traded to **Pittsburgh** by **NY Rangers** with Sergei Zubov for Luc Robitaille and Ulf Samuelsson, August 31, 1995. Traded to **NY Rangers** by **Pittsburgh** with Chris Tamer and Sean Pronger for Alexei Kovalev and Harry York, November 25, 1998.

NEIL, Christopher

(NEEL, KRIHS-toh-fuhr) **OTT.**

Right wing. Shoots right. 6', 213 lbs. Born, Markdale, Ont., June 18, 1979. Ottawa's 7th choice, 161st overall, in 1998 Entry Draft.

Season	Club	League	GP	G	A	Pts	PIM	PP	SH	GW	S	%	+/-	TF	F%	H	SB	Min	GP	G	A	Pts	PIM	PP	SH	GW
1995-96	Orangeville	OJHL-B	43	15	15	30	50																			
1996-97	North Bay	OHL	65	13	16	29	150																			
1997-98	North Bay	OHL	59	26	29	55	231																			
1998-99	North Bay	OHL	66	26	46	72	215												4	1	0	1	15			
99-2000	Mobile Mysticks	ECHL	4	0	2	2	39																			
	Grand Rapids	IHL	51	9	10	19	301												8	0	2	2	24			
2000-01	Grand Rapids	IHL	78	15	21	36	354												10	2	2	4	22			
2001-02	**Ottawa**	**NHL**	**72**	**10**	**7**	**17**	**231**	**1**	**0**	**0**	**56**	**17.9**	**5**	**0**	**0.0**	**163**	**13**	**8:22**	**12**	**0**	**0**	**0**	**12**	**0**	**0**	**0**
	NHL Totals		**72**	**10**	**7**	**17**	**231**	**1**	**0**	**0**	**56**	**17.9**		**0**	**0.0**	**163**	**13**	**8:22**	**12**	**0**	**0**	**0**	**12**	**0**	**0**	**0**

NELSON, Jeff

(NEHL-sohn, JEHF)

Center. Shoots left. 5'11", 190 lbs. Born, Prince Albert, Sask., December 18, 1972. Washington's 4th choice, 36th overall, in 1991 Entry Draft.

Season	Club	League	GP	G	A	Pts	PIM	PP	SH	GW	S	%	+/-	TF	F%	H	SB	Min	GP	G	A	Pts	PIM	PP	SH	GW
1987-88	Prince Albert	SMHL	31	24	32	56	32																			
1988-89	Prince Albert	WHL	71	30	57	87	74												4	0	3	3	4			
1989-90	Prince Albert	WHL	72	28	69	97	79												14	2	11	13	10			
1990-91	Prince Albert	WHL	72	46	74	120	58												3	1	1	2	4			
1991-92	Prince Albert	WHL	64	48	65	113	84												9	7	14	21	18			
1992-93	Baltimore	AHL	72	14	38	52	12												7	1	3	4	2			
1993-94	Portland Pirates	AHL	80	34	73	107	92												17	10	5	15	20			

			Regular Season																Playoffs							
Season	Club	League	GP	G	A	Pts	PIM	PP	SH	GW	S	%	+/-	TF	F%	H	SB	Min	GP	G	A	Pts	PIM	PP	SH	GW
1994-95	Portland Pirates	AHL	64	33	50	83	57												7	1	4	5	8			
	Washington	NHL	10	1	0	1	2	0	0	0	4	25.0	–2													
1995-96	Washington	NHL	33	0	7	7	16	0	0	0	21	0.0	3						3	0	0	0	4	0	0	0
	Portland Pirates	AHL	39	15	32	47	62																			
1996-97	Grand Rapids	IHL	82	34	55	89	85												5	0	4	4	4			
1997-98	Milwaukee	IHL	52	20	34	54	30												10	2	7	9	15			
1998-99	Nashville	NHL	9	2	1	3	2	0	0	0	8	25.0	–1	138	55.1	4	8	16:09								
	Milwaukee	IHL	70	20	31	51	66												2	0	0	0	0			
99-2000	Portland Pirates	AHL	73	24	30	54	38												1	0	0	0	0			
2000-01	Portland Pirates	AHL	80	18	37	55	63												3	0	2	2	6			
2001-02	Schwenningen	Germany	60	13	14	27	60																			
	NHL Totals		52	3	8	11	20	0	0	0	33	9.1		138	55.1	4	8	16:09	3	0	0	0	4	0	0	0

Canadian Major Junior Scholastic Player of the Year (1989, 1990) • WHL East Second All-Star Team (1991, 1992)

Signed as a free agent by **Grand Rapids** (IHL), September 9, 1996. Signed as a free agent by **Nashville**, August 19, 1998. Traded to **Washington** by **Nashville** for cash, June 21, 1999. Signed as a free agent by **Schwenningen** (Germany), July 17, 2001.

NEMCHINOV, Sergei (nehm-CHEE-nahf, SAIR-gay)

Left wing. Shoots left. 6'1", 205 lbs. Born, Moscow, USSR, January 14, 1964. NY Rangers' 14th choice, 244th overall, in 1990 Entry Draft.

Season	Club	League	GP	G	A	Pts	PIM	PP	SH	GW	S	%	+/-	TF	F%	H	SB	Min	GP	G	A	Pts	PIM	PP	SH	GW
1981-82	Krylja Sovetov	USSR	15	1	0	1	0																			
1982-83	CSKA Moscow	USSR	11	0	0	0	2																			
1983-84	CSKA Moscow	USSR	20	6	5	11	4																			
1984-85	CSKA Moscow	USSR	31	2	4	6	4																			
1985-86	Krylja Sovetov	USSR	39	7	12	19	28																			
1986-87	Krylja Sovetov	USSR	40	13	9	22	24																			
1987-88	Krylja Sovetov	USSR	48	17	11	28	26																			
1988-89	Krylja Sovetov	USSR	43	15	14	29	28																			
1989-90	Krylja Sovetov	USSR	48	17	16	33	34																			
1990-91	Krylja Sovetov	USSR	46	21	24	45	30																			
1991-92	NY Rangers	NHL	73	30	28	58	15	2	0	5	124	24.2	19						13	1	4	5	8	0	0	0
1992-93	NY Rangers	NHL	81	23	31	54	34	0	1	3	144	16.0	15													
1993-94♦	NY Rangers	NHL	76	22	27	49	36	4	0	6	144	15.3	13						23	2	5	7	6	0	0	0
1994-95	NY Rangers	NHL	47	7	6	13	16	0	0	3	67	10.4	–6						10	4	5	9	2	0	0	1
1995-96	NY Rangers	NHL	78	17	15	32	38	0	0	2	118	14.4	9						6	0	1	1	2	0	0	0
1996-97	NY Rangers	NHL	63	6	13	19	12	1	0	1	90	6.7	5													
	Vancouver	NHL	6	2	3	5	4	0	0	1	7	28.6	4													
1997-98	NY Islanders	NHL	74	10	19	29	24	2	1	1	94	10.6	3													
	Russia	Olympics	6	1	0	1	0																			
1998-99	NY Islanders	NHL	67	8	8	16	22	1	0	0	61	13.1	–17	606	42.4	52	44	14:21								
	New Jersey	NHL	10	4	0	4	6	1	0	1	13	30.8	4	38	52.6	13	0	15:36	4	0	0	0	0	0	0	0
99-2000♦	New Jersey	NHL	53	10	16	26	18	0	1	1	55	18.2	1	453	45.7	47	19	13:42	21	3	2	5	2	1	0	0
2000-01	New Jersey	NHL	65	8	22	30	16	1	0	2	70	11.4	11	649	48.1	57	16	13:13	25	1	3	4	4	0	0	0
2001-02	New Jersey	NHL	68	5	5	10	10	0	0	1	49	10.2	–9	497	42.7	50	9	11:09	3	0	0	0	0	0	0	0
	NHL Totals		761	152	193	345	251	12	3	27	1036	14.7		2243	44.9	219	88	13:10	105	11	20	31	24	1	0	1

Traded to **Vancouver** by **NY Rangers** with Brian Noonan for Esa Tikkanen and Russ Courtnall, March 8, 1997. Signed as a free agent by **NY Islanders**, July 10, 1997. Traded to **New Jersey** by **NY Islanders** for New Jersey's 4th round choice (later traded to Los Angeles - Los Angeles selected Daniel Johansson) in 1999 Entry Draft, March 22, 1999.

NEMECEK, Jan (NEHM-eh-chehk, YAHN) L.A.

Defense. Shoots Left. 6'1", 220 lbs. Born, Pisek, Czech., February 14, 1976. Los Angeles' 7th choice, 215th overall, in 1994 Entry Draft.

Season	Club	League	GP	G	A	Pts	PIM	PP	SH	GW	S	%	+/-	TF	F%	H	SB	Min	GP	G	A	Pts	PIM	PP	SH	GW
1992-93	Ceske Budejovice	Czech	15	0	0	0																				
1993-94	Ceske Budejovice	Czech	16	0	1	1	16																			
1994-95	Hull Olympiques	QMJHL	49	10	16	26	48												21	5	9	14	10			
1995-96	Hull Olympiques	QMJHL	57	17	49	66	58												17	2	13	15	10			
1996-97	Mississippi	ECHL	20	3	9	12	16												3	0	0	0	4			
	Phoenix	IHL	24	1	1	2	2																			
1997-98	Fredericton	AHL	65	7	24	31	43												2	0	0	0	0			
1998-99	Los Angeles	NHL	6	1	0	1	4	0	0	1	8	12.5	–1	0	0.0	2	4	16:42								
	Long Beach	IHL	66	5	16	21	42																			
99-2000	Los Angeles	NHL	1	0	0	0	0	0	0	0	0	0.0	0	0	0.0	0	0	9:36								
	Long Beach	IHL	71	9	15	24	22												6	1	0	1	4			
2000-01	Nurnberg	Germany	60	5	16	21	18												4	1	0	1	0			
2001-02	Nurnberg	Germany	60	6	18	24	18												4	0	3	3	2			
	NHL Totals		7	1	0	1	4	0	0	1	8	12.5		0	0.0	2	4	15:41								

QMJHL Second All-Star Team (1996)

NEMIROVSKY, David (neh-mih-ROHV-skee, DAY-vihd) TOR.

Right wing. Shoots right. 6'2", 205 lbs. Born, Toronto, Ont., August 1, 1976. Florida's 5th choice, 84th overall, in 1994 Entry Draft.

Season	Club	League	GP	G	A	Pts	PIM	PP	SH	GW	S	%	+/-	TF	F%	H	SB	Min	GP	G	A	Pts	PIM	PP	SH	GW
1991-92	Pickering	MTJHL	14	3	10	13	5																			
	Weston Dukes	MTJHL	23	6	13	19	2																			
1992-93	Weston Dukes	MTJHL	2	0	3	3	0																			
	North York	MTJHL	40	19	23	42	27																			
1993-94	Ottawa 67's	OHL	64	21	31	52	18												17	10	10	20	2			
1994-95	Ottawa 67's	OHL	59	27	29	56	25																			
1995-96	Sarnia Sting	OHL	26	18	27	45	14												10	8	8	16	6			
	Florida	NHL	9	0	2	2	2	0	0	0	6	0.0	–1													
	Carolina	AHL	5	1	2	3	0																			
1996-97	Florida	NHL	39	7	7	14	32	1	0	0	53	13.2	1						3	1	0	1	0	0	0	0
	Carolina	AHL	34	21	21	42	18																			
1997-98	Florida	NHL	41	9	12	21	8	2	0	1	62	14.5	–3													
	New Haven	AHL	29	10	15	25	10												1	1	0	1	0			
1998-99	Florida	NHL	2	0	1	1	0	0	0	0	2	0.0	1	0	0.0	0	0	8:58								
	Fort Wayne	IHL	44	22	13	35	24																			
	St. John's	AHL	22	3	9	12	18												5	4	1	5	0			
99-2000	St. John's	AHL	57	18	25	43	69																			
2000-01	St. John's	AHL	9	1	2	3	10																			
	HV 71 Jonkoping	Sweden	26	7	11	18	45																			
2001-02	Yaroslavl	Russia	10	2	2	4	4																			
	Ilves Tampere	Finland	20	12	18	30	34												3	1	0	1	8			
	NHL Totals		91	16	22	38	42	3	0	1	123	13.0		0	0.0	0	0	8:58	3	1	0	1	0	0	0	0

Traded to **Toronto** by **Florida** for Jeff Ware, February 17, 1999.

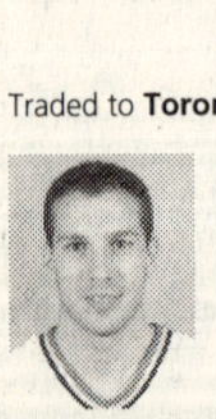

NICHOL, Scott (NIH-KOHL, SKAWT) CGY.

Center. Shoots right. 5'8", 160 lbs. Born, Edmonton, Alta., December 31, 1974. Buffalo's 9th choice, 272nd overall, in 1993 Entry Draft.

Season	Club	League	GP	G	A	Pts	PIM	PP	SH	GW	S	%	+/-	TF	F%	H	SB	Min	GP	G	A	Pts	PIM	PP	SH	GW
1991-92	Calgary Flames	AMHL	23	26	16	42	132																			
1992-93	Portland	WHL	67	31	33	64	146												16	8	8	16	41			
1993-94	Portland	WHL	65	40	53	93	144												10	3	8	11	16			
1994-95	Rochester	AHL	71	11	16	27	136												5	0	3	3	14			
1995-96	Buffalo	NHL	2	0	0	0	10	0	0	0	4	0.0	0													
	Rochester	AHL	62	14	18	32	170												19	7	6	13	36			
1996-97	Rochester	AHL	68	22	21	43	133												10	2	1	3	26			
1997-98	Buffalo	NHL	3	0	0	0	4	0	0	0	5	0.0	0													
	Rochester	AHL	35	13	7	20	113																			
1998-99	Rochester	AHL	52	13	20	33	120																			
99-2000	Rochester	AHL	37	7	11	18	141																			

Season	Club	League	GP	G	A	Pts	PIM	PP	SH	GW	S	%	+/-	TF	F%	H	SB	Min	GP	G	A	Pts	PIM	PP	SH	GW
			Regular Season																Playoffs							
2000-01	Detroit Vipers	IHL	67	7	24	31	198	....	....	....	....	....	....	....	....	....	....	....	....	....	....	....	....	....	....	....
2001-02	**Calgary**	**NHL**	**60**	**8**	**9**	**17**	**107**	**2**	**1**	**0**	**49**	**16.3**	**−9**	**458**	**53.1**	**109**	**31**	**12:41**	....	....	....	....	....	....	....	....
	NHL Totals		**65**	**8**	**9**	**17**	**121**	**2**	**1**	**0**	**58**	**13.8**		**458**	**53.1**	**109**	**31**	**12:41**	....	....	....	....	....	....	....	....

• Missed majority of 1999-2000 season recovering from knee injury suffered in game vs. Saint John (AHL), February 16, 2000. Signed as a free agent by **Calgary**, July 1, 2001.

NICKULAS, Eric

(NICK-luhs, AIR-ihk)

Center. Shoots right. 5'11", 200 lbs. Born, Hyannis, MA, March 25, 1975. Boston's 3rd choice, 99th overall, in 1994 Entry Draft.

Season	Club	League	GP	G	A	Pts	PIM	PP	SH	GW	S	%	+/-	TF	F%	H	SB	Min	GP	G	A	Pts	PIM	PP	SH	GW
1991-92	Barnstable	Hi-School	24	30	25	55	....	....	....	....	....	....	....	....	....	....	....	....	....	....	....	....	....	....	....	....
1992-93	Tabor Academy	Hi-School	28	25	25	50	....	....	....	....	....	....	....	....	....	....	....	....	....	....	....	....	....	....	....	....
1993-94	Cushing Academy	Hi-School	25	46	36	82	....	....	....	....	....	....	....	....	....	....	....	....	....	....	....	....	....	....	....	....
1994-95	New Hampshire	H-East	33	15	9	24	32	....	....	....	....	....	....	....	....	....	....	....	....	....	....	....	....	....	....	....
1995-96	New Hampshire	H-East	34	26	12	38	66	....	....	....	....	....	....	....	....	....	....	....	....	....	....	....	....	....	....	....
1996-97	New Hampshire	H-East	39	29	22	51	80	....	....	....	....	....	....	....	....	....	....	....	....	....	....	....	....	....	....	....
1997-98	Orlando	IHL	76	22	9	31	77	....	....	....	....	....	....	....	....	....	....	....	6	0	0	0	10	....	....	....
1998-99	**Boston**	**NHL**	**2**	**0**	**0**	**0**	**0**	**0**	**0**	**0**	**0**	**0.0**	**0**	**0**	**0.0**	**0**	**0**	**3:27**	**1**	**0**	**0**	**0**	**2**	**0**	**0**	**0**
	Providence	AHL	75	31	27	58	83	....	....	....	....	....	....	....	....	....	....	....	18	8	12	20	33	....	....	....
99-2000	**Boston**	**NHL**	**20**	**5**	**6**	**11**	**12**	**1**	**0**	**0**	**28**	**17.9**	**−1**	**4**	**50.0**	**31**	**3**	**11:13**	....	....	....	....	....	....	....	....
	Providence	AHL	40	6	6	12	37	....	....	....	....	....	....	....	....	....	....	....	12	2	3	5	20	....	....	....
2000-01	**Boston**	**NHL**	**7**	**0**	**0**	**0**	**4**	**0**	**0**	**0**	**6**	**0.0**	**−2**	**1**	**100.0**	**17**	**0**	**7:07**	....	....	....	....	....	....	....	....
	Providence	AHL	62	20	23	43	100	....	....	....	....	....	....	....	....	....	....	....	12	4	4	8	24	....	....	....
2001-02	Worcester	AHL	54	11	25	36	48	....	....	....	....	....	....	....	....	....	....	....	3	0	1	1	2	....	....	....
	NHL Totals		**29**	**5**	**6**	**11**	**16**	**1**	**0**	**0**	**34**	**14.7**		**5**	**60.0**	**48**	**3**	**9:41**	**1**	**0**	**0**	**0**	**2**	**0**	**0**	**0**

Won Ken McKenzie Trophy (U.S.- Born Rookie of the Year - IHL) (1998)

Signed as a free agent by **Worcester** (AHL), November 10, 2001. Signed as a free agent by **St. Louis**, July 16, 2002.

NIEDERMAYER, Rob

(NEE-duhr-MIGH-uhr, RAWB) **CGY.**

Center. Shoots left. 6'2", 204 lbs. Born, Cassiar, B.C., December 28, 1974. Florida's 1st choice, 5th overall, in 1993 Entry Draft.

Season	Club	League	GP	G	A	Pts	PIM	PP	SH	GW	S	%	+/-	TF	F%	H	SB	Min	GP	G	A	Pts	PIM	PP	SH	GW
1989-90	Cranbrook	BCAHA	35	42	40	82	30	....	....	....	....	....	....	....	....	....	....	....	....	....	....	....	....	....	....	....
1990-91	Medicine Hat	WHL	71	24	26	50	8	....	....	....	....	....	....	....	....	....	....	....	12	3	7	10	2	....	....	....
1991-92	Medicine Hat	WHL	71	32	46	78	77	....	....	....	....	....	....	....	....	....	....	....	4	2	3	5	2	....	....	....
1992-93	Medicine Hat	WHL	52	43	34	77	67	....	....	....	....	....	....	....	....	....	....	....	....	....	....	....	....	....	....	....
1993-94	**Florida**	**NHL**	**65**	**9**	**17**	**26**	**51**	**3**	**0**	**2**	**67**	**13.4**	**−11**	....	....	....	....	....	....	....	....	....	....	....	....	....
1994-95	Medicine Hat	WHL	13	9	15	24	14	....	....	....	....	....	....	....	....	....	....	....	....	....	....	....	....	....	....	....
	Florida	**NHL**	**48**	**4**	**6**	**10**	**36**	**1**	**0**	**0**	**58**	**6.9**	**−13**	....	....	....	....	....	....	....	....	....	....	....	....	....
1995-96	**Florida**	**NHL**	**82**	**26**	**35**	**61**	**107**	**11**	**0**	**6**	**155**	**16.8**	**1**	....	....	....	....	....	**22**	**5**	**3**	**8**	**12**	**2**	**0**	**2**
1996-97	**Florida**	**NHL**	**60**	**14**	**24**	**38**	**54**	**3**	**0**	**2**	**136**	**10.3**	**4**	....	....	....	....	....	**5**	**2**	**1**	**3**	**6**	**1**	**0**	**0**
1997-98	**Florida**	**NHL**	**33**	**8**	**7**	**15**	**41**	**5**	**0**	**2**	**64**	**12.5**	**−9**	....	....	....	....	....	....	....	....	....	....	....	....	....
1998-99	**Florida**	**NHL**	**82**	**18**	**33**	**51**	**50**	**6**	**1**	**3**	**142**	**12.7**	**−13**	**1895**	**47.1**	**152**	**38**	**21:17**	....	....	....	....	....	....	....	....
99-2000	**Florida**	**NHL**	**81**	**10**	**23**	**33**	**46**	**1**	**0**	**4**	**135**	**7.4**	**−5**	**1632**	**47.9**	**211**	**40**	**19:04**	**4**	**1**	**0**	**1**	**6**	**0**	**0**	**0**
2000-01	**Florida**	**NHL**	**67**	**12**	**20**	**32**	**50**	**3**	**1**	**0**	**115**	**10.4**	**−12**	**997**	**45.0**	**149**	**49**	**20:30**	....	....	....	....	....	....	....	....
2001-02	**Calgary**	**NHL**	**57**	**6**	**14**	**20**	**49**	**1**	**2**	**1**	**87**	**6.9**	**−15**	**777**	**48.4**	**71**	**28**	**18:01**	....	....	....	....	....	....	....	....
	NHL Totals		**575**	**107**	**179**	**286**	**484**	**34**	**4**	**20**	**959**	**11.2**		**5301**	**47.2**	**583**	**155**	**19:50**	**31**	**8**	**4**	**12**	**24**	**3**	**0**	**2**

WHL East First All-Star Team (1993)

• Missed majority of 1997-98 season recovering from thumb (vs. Boston, November 26, 1997) and head (vs. Buffalo, March 19, 1998) injuries. Traded to **Calgary** by **Florida** with Philadelphia's 2nd round choice (previously acquired, Calgary selected Andrei Medvedev) in 2001 Entry Draft for Valeri Bure and Jason Wiemer, June 23, 2001.

NIEDERMAYER, Scott

(NEE-duhr-MIGH-uhr, SKAWT) **N.J.**

Defense. Shoots left. 6'1", 200 lbs. Born, Edmonton, Alta., August 31, 1973. New Jersey's 1st choice, 3rd overall, in 1991 Entry Draft.

Season	Club	League	GP	G	A	Pts	PIM	PP	SH	GW	S	%	+/-	TF	F%	H	SB	Min	GP	G	A	Pts	PIM	PP	SH	GW
1988-89	Cranbrook	BCAHA	62	55	37	92	100	....	....	....	....	....	....	....	....	....	....	....	....	....	....	....	....	....	....	....
1989-90	Kamloops Blazers	WHL	64	14	55	69	64	....	....	....	....	....	....	....	....	....	....	....	17	2	14	16	35	....	....	....
1990-91	Kamloops Blazers	WHL	57	26	56	82	52	....	....	....	....	....	....	....	....	....	....	....	....	....	....	....	....	....	....	....
1991-92	Kamloops Blazers	WHL	35	7	32	39	61	....	....	....	....	....	....	....	....	....	....	....	17	9	14	23	28	....	....	....
	New Jersey	**NHL**	**4**	**0**	**1**	**1**	**2**	**0**	**0**	**0**	**4**	**0.0**	**1**	....	....	....	....	....	....	....	....	....	....	....	....	....
1992-93	**New Jersey**	**NHL**	**80**	**11**	**29**	**40**	**47**	**5**	**0**	**0**	**131**	**8.4**	**8**	....	....	....	....	....	**5**	**0**	**3**	**3**	**2**	**0**	**0**	**0**
1993-94	**New Jersey**	**NHL**	**81**	**10**	**36**	**46**	**42**	**5**	**0**	**2**	**135**	**7.4**	**34**	....	....	....	....	....	**20**	**2**	**2**	**4**	**8**	**1**	**0**	**0**
1994-95♦	**New Jersey**	**NHL**	**48**	**4**	**15**	**19**	**18**	**4**	**0**	**0**	**52**	**7.7**	**19**	....	....	....	....	....	**20**	**4**	**7**	**11**	**10**	**2**	**0**	**1**
1995-96	**New Jersey**	**NHL**	**79**	**8**	**25**	**33**	**46**	**6**	**0**	**0**	**179**	**4.5**	**5**	....	....	....	....	....	....	....	....	....	....	....	....	....
1996-97	**New Jersey**	**NHL**	**81**	**5**	**30**	**35**	**64**	**3**	**0**	**3**	**159**	**3.1**	**−4**	....	....	....	....	....	**10**	**2**	**4**	**6**	**6**	**2**	**0**	**1**
1997-98	**New Jersey**	**NHL**	**81**	**14**	**43**	**57**	**27**	**11**	**0**	**1**	**175**	**8.0**	**5**	....	....	....	....	....	**6**	**0**	**2**	**2**	**4**	**0**	**0**	**0**
1998-99	Utah Grizzlies	IHL	5	0	2	2	0	....	....	....	....	....	....	....	....	....	....	....	....	....	....	....	....	....	....	....
	New Jersey	**NHL**	**72**	**11**	**35**	**46**	**26**	**1**	**1**	**3**	**161**	**6.8**	**16**	**13**	**15.4**	**99**	**49**	**24:40**	**7**	**1**	**3**	**4**	**18**	**1**	**0**	**0**
99-2000♦	**New Jersey**	**NHL**	**71**	**7**	**31**	**38**	**48**	**1**	**0**	**0**	**109**	**6.4**	**19**	**8**	**37.5**	**105**	**82**	**24:21**	**22**	**5**	**2**	**7**	**10**	**0**	**2**	**1**
2000-01	**New Jersey**	**NHL**	**57**	**6**	**29**	**35**	**22**	**1**	**0**	**5**	**87**	**6.9**	**14**	**5**	**0.0**	**61**	**56**	**23:19**	**21**	**0**	**6**	**6**	**14**	**0**	**0**	**0**
2001-02	**New Jersey**	**NHL**	**76**	**11**	**22**	**33**	**30**	**2**	**0**	**6**	**129**	**8.5**	**12**	**1**	**100.0**	**64**	**71**	**24:17**	**6**	**0**	**2**	**2**	**6**	**0**	**0**	**0**
	Canada	Olympics	6	1	1	2	4	....	....	....	....	....	....	....	....	....	....	....	....	....	....	....	....	....	....	....
	NHL Totals		**730**	**87**	**296**	**383**	**372**	**39**	**1**	**20**	**1321**	**6.6**		**27**	**22.2**	**329**	**258**	**24:12**	**117**	**14**	**31**	**45**	**78**	**6**	**2**	**3**

WHL West First All-Star Team (1991, 1992) • Canadian Major Junior Scholastic Player of the Year (1991) • Memorial Cup All-Star Team (1992) • Won Stafford Smythe Memorial Trophy (Memorial Cup Tournament MVP) (1992) • NHL All-Rookie Team (1993) • NHL Second All-Star Team (1998) • Played in NHL All-Star Game (1998, 2001)

Signed to 25-game try-out contract by **Utah** (IHL) with **New Jersey** retaining NHL rights, October 19, 1998.

NIELSEN, Chris

(NEEL-sehn, KRIHS) **CBJ**

Center. Shoots right. 6'2", 204 lbs. Born, Moshi, Tanzania, February 16, 1980. NY Islanders' 2nd choice, 36th overall, in 1998 Entry Draft.

Season	Club	League	GP	G	A	Pts	PIM	PP	SH	GW	S	%	+/-	TF	F%	H	SB	Min	GP	G	A	Pts	PIM	PP	SH	GW
1995-96	S-W Cougars	MMHL	39	37	35	72	59	....	....	....	....	....	....	....	....	....	....	....	....	....	....	....	....	....	....	....
	Calgary Hitmen	WHL	6	0	0	0	0	....	....	....	....	....	....	....	....	....	....	....	....	....	....	....	....	....	....	....
1996-97	Calgary Hitmen	WHL	62	11	19	30	39	....	....	....	....	....	....	....	....	....	....	....	....	....	....	....	....	....	....	....
1997-98	Calgary Hitmen	WHL	68	22	29	51	31	....	....	....	....	....	....	....	....	....	....	....	18	2	4	6	10	....	....	....
1998-99	Calgary Hitmen	WHL	70	22	24	46	45	....	....	....	....	....	....	....	....	....	....	....	21	11	5	16	28	....	....	....
99-2000	Calgary Hitmen	WHL	62	38	31	69	86	....	....	....	....	....	....	....	....	....	....	....	13	14	9	23	20	....	....	....
2000-01	**Columbus**	**NHL**	**29**	**4**	**5**	**9**	**4**	**0**	**0**	**1**	**36**	**11.1**	**4**	**18**	**55.6**	**33**	**3**	**10:30**	....	....	....	....	....	....	....	....
	Syracuse Crunch	AHL	47	10	11	21	24	....	....	....	....	....	....	....	....	....	....	....	5	2	2	4	4	....	....	....
2001-02	**Columbus**	**NHL**	**23**	**2**	**3**	**5**	**4**	**0**	**0**	**1**	**28**	**7.1**	**−3**	**6**	**33.3**	**24**	**4**	**10:47**	....	....	....	....	....	....	....	....
	Syracuse Crunch	AHL	47	12	12	24	18	....	....	....	....	....	....	....	....	....	....	....	10	2	2	4	2	....	....	....
	NHL Totals		**52**	**6**	**8**	**14**	**8**	**0**	**0**	**2**	**64**	**9.4**		**24**	**50.0**	**57**	**7**	**10:38**	....	....	....	....	....	....	....	....

Traded to **Columbus** by **NY Islanders** for Columbus' 4th (later traded to Anaheim - Anaheim selected Jonas Ronnqvist) and 9th (Dmitri Altarev) round choices in 2000 Entry Draft, May 11, 2000.

NIEMI, Antti-Jussi

(nee-mee, AN-tee-YOO-see)

Defense. Shoots left. 6'1", 195 lbs. Born, Vantaa, Finland, September 22, 1977. Ottawa's 2nd choice, 81st overall, in 1996 Entry Draft.

Season	Club	League	GP	G	A	Pts	PIM	PP	SH	GW	S	%	+/-	TF	F%	H	SB	Min	GP	G	A	Pts	PIM	PP	SH	GW
1992-93	Jokerit-C	Finn-Jr.	38	9	23	32	54	....	....	....	....	....	....	....	....	....	....	....	....	....	....	....	....	....	....	....
	Jokerit-B	Finn-Jr.	2	0	0	0	2	....	....	....	....	....	....	....	....	....	....	....	....	....	....	....	....	....	....	....
1993-94	Jokerit-B	Finn-Jr.	13	1	1	2	8	....	....	....	....	....	....	....	....	....	....	....	....	....	....	....	....	....	....	....
	Jokerit Jr.	Finn-Jr.	33	0	3	3	26	....	....	....	....	....	....	....	....	....	....	....	....	....	....	....	....	....	....	....
1994-95	Jokerit-B	Finn-Jr.	10	3	5	8	42	....	....	....	....	....	....	....	....	....	....	....	....	....	....	....	....	....	....	....
	Jokerit Jr.	Finn-Jr.	24	4	8	12	74	....	....	....	....	....	....	....	....	....	....	....	....	....	....	....	....	....	....	....
1995-96	Jokerit Jr.	Finn-Jr.	34	11	18	29	56	....	....	....	....	....	....	....	....	....	....	....	8	0	4	4	39	....	....	....
	Haukat Jarvenpaa	Finland-2	4	0	2	2	8	....	....	....	....	....	....	....	....	....	....	....	....	....	....	....	....	....	....	....
	Jokerit Helsinki	Finland	6	0	2	2	6	....	....	....	....	....	....	....	....	....	....	....	1	0	0	0	0	....	....	....
1996-97	Jokerit Helsinki	Finland	44	2	9	11	38	....	....	....	....	....	....	....	....	....	....	....	9	0	2	2	2	....	....	....
1997-98	Jokerit Helsinki	Finland	46	2	6	8	24	....	....	....	....	....	....	....	....	....	....	....	8	0	1	1	0	....	....	....
	Jokerit Helsinki	EuroHL	6	0	1	1	6	....	....	....	....	....	....	....	....	....	....	....	....	....	....	....	....	....	....	....
1998-99	Jokerit Helsinki	Finland	53	3	7	10	107	....	....	....	....	....	....	....	....	....	....	....	3	0	0	0	2	....	....	....
	Jokerit Helsinki	EuroHL	6	2	2	4	4	....	....	....	....	....	....	....	....	....	....	....	2	0	1	1	0	....	....	....
99-2000	Jokerit Helsinki	Finland	53	8	8	16	79	....	....	....	....	....	....	....	....	....	....	....	11	0	3	3	6	....	....	....

Season	Club	League	Regular Season																Playoffs							
			GP	G	A	Pts	PIM	PP	SH	GW	S	%	+/-	TF	F%	H	SB	Min	GP	G	A	Pts	PIM	PP	SH	GW
2000-01	Anaheim	NHL	28	1	1	2	22	0	0	0	18	5.6	–6	0	0.0	50	32	15:56								
	Cincinnati	AHL	36	3	8	11	26																			
2001-02	Anaheim	NHL	1	0	0	0	0	0	0	0	0	0.0	–1	0	0.0	1	0	9:53								
	Cincinnati	AHL	39	10	9	19	25												3	1	1	2	2			
	NHL Totals		29	1	1	2	22	0	0	0	18	5.6		0	0.0	51	32	15:44								

Rights traded to **Anaheim** by **Ottawa** with Ted Donato for Patrick Lalime, June 18, 1999. Signed as a free agent by **Jokerit Helsinki** (Sweden), August 13, 2002.

NIEMINEN, Ville

(nee-EHM-ih-nehn, VIHL-ee) **PIT.**

Left wing. Shoots left. 6', 200 lbs. Born, Tampere, Finland, April 6, 1977. Colorado's 4th choice, 78th overall, in 1997 Entry Draft.

Season	Club	League	GP	G	A	Pts	PIM	PP	SH	GW	S	%	+/-	TF	F%	H	SB	Min	GP	G	A	Pts	PIM	PP	SH	GW
1994-95	Tappara Jr.	Finn-Jr.	16	11	21	32	47																			
	Tappara Tampere	Finland	16	0	0	0	0																			
1995-96	Tappara Jr.	Finn-Jr.	20	20	23	43	63																			
	KooKoo Kouvola	Finland-2	7	2	1	3	4																			
	Tappara Tampere	Finland	4	0	1	1	8																			
1996-97	Tappara Tampere	Finland	49	10	13	23	120												3	1	0	1	8			
1997-98	Hershey Bears	AHL	74	14	22	36	85																			
1998-99	Hershey Bears	AHL	67	24	19	43	127												3	0	1	1	0			
99-2000	Colorado	NHL	1	0	0	0	0	0	0	0	2	0.0	0	0	0.0	0	0	10:12								
	Hershey Bears	AHL	74	21	30	51	54												9	2	4	6	6			
2000-01♦	Colorado	NHL	50	14	8	22	38	2	0	3	68	20.6	8	3	33.3	62	14	12:26	23	4	6	10	20	3	0	1
	Hershey Bears	AHL	28	10	11	21	48																			
2001-02	Colorado	NHL	53	10	14	24	30	1	0	5	72	13.9	1	8	62.5	92	18	12:41								
	Finland	Olympics	4	0	1	1	2																			
	Pittsburgh	NHL	13	1	2	3	8	0	0	0	11	9.1	–2	0	0.0	20	2	16:10								
	NHL Totals		117	25	24	49	76	3	0	8	153	16.3		11	54.5	174	34	12:57	23	4	6	10	20	3	0	1

Traded to **Pittsburgh** by **Colorado** with Rick Berry for Darius Kasparaitis, March 19, 2002.

NIEUWENDYK, Joe

(NOO-ihn-DIGHK, JOH) **N.J.**

Center. Shoots left. 6'1", 205 lbs. Born, Oshawa, Ont., September 10, 1966. Calgary's 2nd choice, 27th overall, in 1985 Entry Draft.

Season	Club	League	GP	G	A	Pts	PIM	PP	SH	GW	S	%	+/-	TF	F%	H	SB	Min	GP	G	A	Pts	PIM	PP	SH	GW
1983-84	Pickering	MTJHL	38	30	28	58	35																			
1984-85	Cornell Big Red	ECAC	29	21	24	45	30																			
1985-86	Cornell Big Red	ECAC	29	26	28	54	67																			
1986-87	Cornell Big Red	ECAC	23	26	26	52	26																			
	Calgary	NHL	9	5	1	6	0	2	0	1	16	31.3	0						6	2	2	4	0	0	0	0
1987-88	Calgary	NHL	75	51	41	92	23	31	3	8	212	24.1	20						8	3	4	7	2	1	0	0
1988-89♦	Calgary	NHL	77	51	31	82	40	19	3	11	215	23.7	26						22	10	4	14	10	6	0	1
1989-90	Calgary	NHL	79	45	50	95	40	18	0	3	226	19.9	32						6	4	6	10	4	1	0	0
1990-91	Calgary	NHL	79	45	40	85	36	22	4	1	222	20.3	19						7	4	1	5	10	2	0	0
1991-92	Calgary	NHL	69	22	34	56	55	7	0	2	137	16.1	–1													
1992-93	Calgary	NHL	79	38	37	75	52	14	0	6	208	18.3	9						6	3	6	9	10	1	0	0
1993-94	Calgary	NHL	64	36	39	75	51	14	1	7	191	18.8	19						6	2	2	4	0	1	0	0
1994-95	Calgary	NHL	46	21	29	50	33	3	0	4	122	17.2	11						5	4	3	7	0	2	0	1
1995-96	Dallas	NHL	52	14	18	32	41	8	0	3	138	10.1	–17													
1996-97	Dallas	NHL	66	30	21	51	32	8	0	2	173	17.3	–5						7	2	2	4	6	0	0	0
1997-98	Dallas	NHL	73	39	30	69	30	14	0	11	203	19.2	16						1	1	0	1	0	0	0	0
	Canada	Olympics	6	2	3	5	2																			
1998-99♦	Dallas	NHL	67	28	27	55	34	8	0	8	157	17.8	11	1170	63.2	42	9	15:33	23	*11	10	21	19	3	0	6
99-2000	Dallas	NHL	48	15	19	34	26	7	0	2	110	13.6	–1	924	59.1	17	8	16:15	23	7	3	10	18	3	0	2
2000-01	Dallas	NHL	69	29	23	52	30	12	0	4	166	17.5	5	1262	57.2	34	21	16:11	7	4	0	4	4	1	0	1
2001-02	Dallas	NHL	67	23	24	47	18	6	0	5	157	14.6	–2	1345	59.6	28	13	16:59								
	Canada	Olympics	6	1	1	2	0																			
	New Jersey	NHL	14	2	9	11	4	0	0	1	32	6.3	2	275	55.3	13	4	16:22	5	0	1	1	0	0	0	0
	NHL Totals		1033	494	473	967	545	193	11	79	2685	18.4		4976	59.5	134	55	16:15	132	57	44	101	83	21	0	11

ECAC Rookie of the Year (1985) • ECAC First All-Star Team (1986, 1987) • NCAA East First All-American Team (1986, 1987) • ECAC Player of the Year (1987) • Won Calder Memorial Trophy (1988) • NHL All-Rookie Team (1988) • Won Dodge Ram Tough Award (1988) • Won King Clancy Memorial Trophy (1995) • Won Conn Smythe Trophy (1999) • Played in NHL All-Star Game (1988, 1989, 1990, 1994)

Traded to **Dallas** by **Calgary** for Corey Millen and Jarome Iginla, December 19, 1995. Traded to **New Jersey** by **Dallas** with Jamie Langenbrunner for Jason Arnott, Randy McKay and New Jersey's 1st round choice (later traded to Columbus - later traded to Buffalo - Buffalo selected Dan Paille) in 2002 Entry Draft, March 19, 2002.

NIINIMAA, Janne

(nihn-EE-mah, YAH-nee) **EDM.**

Defense. Shoots left. 6'1", 220 lbs. Born, Raahe, Finland, May 22, 1975. Philadelphia's 1st choice, 36th overall, in 1993 Entry Draft.

Season	Club	League	GP	G	A	Pts	PIM	PP	SH	GW	S	%	+/-	TF	F%	H	SB	Min	GP	G	A	Pts	PIM	PP	SH	GW
1990-91	Karpat Oulu Jr.	Finn-Jr.	3	1	0	1	2																			
1991-92	Karpat Oulu Jr.	Finn-Jr.	3	0	0	0	4																			
	Karpat Oulu	Finland-2	41	2	11	13	49																			
1992-93	Karpat Oulu Jr.	Finn-Jr.	10	3	9	12	16																			
	KKP Kiimimki	Finland-3	1	0	2	2	4																			
	Karpat Oulu	Finland-2	29	2	3	5	14																			
1993-94	Jokerit Jr.	Finn-Jr.	10	2	6	8	41																			
	Jokerit Helsinki	Finland	45	3	8	11	24												12	1	1	2	4			
1994-95	Jokerit Jr.	Finn-Jr.	3	1	2	3	4																			
	Jokerit Helsinki	Finland	42	7	10	17	36												10	1	4	5	35			
1995-96	Jokerit Helsinki	Finland	49	5	15	20	79												11	0	2	2	12			
	Jokerit Jr.	Finn-Jr.																	2	3	4	7	6			
1996-97	Philadelphia	NHL	77	4	40	44	58	1	0	2	141	2.8	12						19	1	12	13	16	1	0	1
1997-98	Philadelphia	NHL	66	3	31	34	56	2	0	1	115	2.6	6													
	Edmonton	NHL	11	1	8	9	6	1	0	0	19	5.3	7						11	1	1	2	12	0	0	1
	Finland	Olympics	6	0	3	3	8																			
1998-99	Edmonton	NHL	81	4	24	28	88	2	0	1	142	2.8	7	1	0.0	144	122	23:54	4	0	0	0	2	0	0	0
99-2000	Edmonton	NHL	81	8	25	33	89	2	2	0	133	6.0	14	0	0.0	124	107	24:28	5	0	2	2	2	0	0	0
2000-01	Edmonton	NHL	82	12	34	46	90	8	0	1	122	9.8	6	0	0.0	137	129	25:20	6	0	2	2	6	0	0	0
2001-02	Edmonton	NHL	81	5	39	44	80	1	0	2	119	4.2	13	0	0.0	115	123	26:02								
	Finland	Olympics	4	0	3	3	2																			
	NHL Totals		479	37	201	238	467	17	2	7	791	4.7		1	0.0	520	481	24:56	45	2	17	19	38	1	0	2

NHL All-Rookie Team (1997) • Played in NHL All-Star Game (2001)

Traded to **Edmonton** by **Philadelphia** for Dan McGillis and Edmonton's 2nd round choice (Jason Beckett) in 1998 Entry Draft, March 24, 1998.

NIKOLISHIN, Andrei

(nee-koh-LEE-shin, AWN-dray) **WSH.**

Center. Shoots left. 6', 213 lbs. Born, Vorkuta, USSR, March 25, 1973. Hartford's 2nd choice, 47th overall, in 1992 Entry Draft.

Season	Club	League	GP	G	A	Pts	PIM	PP	SH	GW	S	%	+/-	TF	F%	H	SB	Min	GP	G	A	Pts	PIM	PP	SH	GW
1990-91	Dynamo Moscow	USSR	2	0	0	0	0																			
1991-92	Dynamo Moscow	CIS	18	1	0	1	4																			
1992-93	Dynamo Moscow	CIS	42	5	7	12	30												10	2	1	3	8			
1993-94	Dynamo Moscow	CIS	41	8	12	20	30												9	1	3	4	4			
	Russia	Olympics	8	2	5	7	6																			
1994-95	Dynamo Moscow	CIS	12	7	2	9	6																			
	Hartford	NHL	39	8	10	18	10	1	1	0	57	14.0	7													
1995-96	Hartford	NHL	61	14	37	51	34	4	1	3	83	16.9	–2													
1996-97	Hartford	NHL	12	2	5	7	2	0	0	0	25	8.0	–2													
	Washington	NHL	59	7	14	21	30	1	0	0	73	9.6	5													
1997-98	Washington	NHL	38	6	10	16	14	1	0	1	40	15.0	1						21	1	13	14	12	1	0	0
	Portland Pirates	AHL	2	0	0	0	2																			
1998-99	Dynamo Moscow	Russia	4	0	0	0	0																			
	Washington	NHL	73	8	27	35	28	0	1	1	121	6.6	0	1354	52.5	70	31	17:34								
99-2000	Washington	NHL	76	11	14	25	28	0	2	2	98	11.2	6	1190	54.7	61	39	15:44	5	0	2	2	4	0	0	0
2000-01	Washington	NHL	81	13	25	38	34	4	0	2	145	9.0	9	1214	55.4	45	32	15:32	6	0	0	0	2	0	0	0

Season	Club	League	GP	G	A	Pts	PIM	PP	SH	GW	S	%	+/-	TF	F%	H	SB	Min	GP	G	A	Pts	PIM	PP	SH	GW
			Regular Season																Playoffs							
2001-02	Washington	NHL	80	13	23	36	40	1	0	0	143	9.1	–1	1445	55.7	30	33	17:23								
	Russia	Olympics	6	0	1	1	6																			
	NHL Totals		519	82	165	247	220	12	5	9	785	10.4		5203	54.6	206	135	16:32	32	1	15	16	18	1	0	0

Traded to **Washington** by **Hartford** for Curtis Leschyshyn, November 9, 1996.

NILSON, Marcus (NIHL-suhn, MAHR-kuhs) FLA.

Right wing. Shoots right. 6'2", 195 lbs. Born, Balsta, Sweden, March 1, 1978. Florida's 1st choice, 20th overall, in 1996 Entry Draft.

Season	Club	League	GP	G	A	Pts	PIM	PP	SH	GW	S	%	+/-	TF	F%	H	SB	Min	GP	G	A	Pts	PIM	PP	SH	GW
1994-95	Djurgarden Jr.	Swede-Jr.	24	7	8	15	22																			
1995-96	Djurgarden Jr.	Swede-Jr.	25	19	17	36	46												2	1	1	2	12			
	Djurgarden	Sweden	12	0	0	0	0												1	0	0	0	0			
1996-97	Djurgarden	Sweden	37	0	3	3	33												4	0	0	0	0			
1997-98	Djurgarden	Sweden	41	4	7	11	18												15	2	1	3	16			
1998-99	Florida	NHL	8	1	1	2	5	0	0	1	7	14.3	2	6	50.0	6	3	12:24								
	New Haven	AHL	69	8	25	33	10																			
99-2000	Florida	NHL	9	0	2	2	2	0	0	0	6	0.0	2	14	64.3	5	3	7:56								
	Louisville	AHL	64	9	23	32	52												4	0	0	0	2			
2000-01	Florida	NHL	78	12	24	36	74	0	0	2	141	8.5	–3	169	40.8	67	29	15:46								
2001-02	Florida	NHL	81	14	19	33	55	6	1	2	147	9.5	–14	539	43.8	49	40	16:31								
	NHL Totals		176	27	46	73	136	6	1	5	301	9.0		728	43.5	127	75	15:34								

NOLAN, Owen (NOH-lan, OH-wehn) S.J.

Right wing. Shoots right. 6'1", 210 lbs. Born, Belfast, Ireland, February 12, 1972. Quebec's 1st choice, 1st overall, in 1990 Entry Draft.

Season	Club	League	GP	G	A	Pts	PIM	PP	SH	GW	S	%	+/-	TF	F%	H	SB	Min	GP	G	A	Pts	PIM	PP	SH	GW
1987-88	Thorold	OMHA	28	53	32	85	24																			
	Thorold	OJHL-B	3	1	0	1	2																			
1988-89	Cornwall Royals	OHL	62	34	25	59	213												18	5	11	16	41			
1989-90	Cornwall Royals	OHL	58	51	59	110	240												6	7	5	12	26			
1990-91	Quebec	NHL	59	3	10	13	109	0	0	0	54	5.6	–19													
	Halifax Citadels	AHL	6	4	4	8	11																			
1991-92	Quebec	NHL	75	42	31	73	183	17	0	0	190	22.1	–9													
1992-93	Quebec	NHL	73	36	41	77	185	15	0	4	241	14.9	–1						5	1	0	1	2	0	0	0
1993-94	Quebec	NHL	6	2	2	4	8	0	0	0	15	13.3	2													
1994-95	Quebec	NHL	46	30	19	49	46	13	2	8	137	21.9	21						6	2	3	5	6	0	0	0
1995-96	Colorado	NHL	9	4	4	8	9	4	0	0	23	17.4	–3													
	San Jose	NHL	72	29	32	61	137	12	1	2	184	15.8	–30													
1996-97	San Jose	NHL	72	31	32	63	155	10	0	3	225	13.8	–19													
1997-98	San Jose	NHL	75	14	27	41	144	3	1	1	192	7.3	–2						6	2	2	4	26	2	0	1
1998-99	San Jose	NHL	78	19	26	45	129	6	2	3	207	9.2	16	657	49.3	174	15	19:09	6	1	1	2	6	0	0	0
99-2000	San Jose	NHL	78	44	40	84	110	18	4	6	261	16.9	–1	357	50.7	209	29	21:07	10	8	2	10	6	2	2	3
2000-01	San Jose	NHL	57	24	25	49	75	10	1	4	191	12.6	0	407	46.9	116	30	21:49	6	1	1	2	8	0	0	1
2001-02	San Jose	NHL	75	23	43	66	93	8	2	2	217	10.6	7	545	47.0	133	31	19:23	12	3	6	9	8	0	0	0
	Canada	Olympics	6	0	3	3	2																			
	NHL Totals		775	301	332	633	1383	116	13	33	2137	14.1		1966	48.4	632	105	20:16	51	18	15	33	62	4	2	5

OHL Rookie of the Year (1989) • OHL First All-Star Team (1990) • Played in NHL All-Star Game (1992, 1996, 1997, 2000, 2002)

• Missed majority of 1993-94 season recovering from shoulder injury suffered in game vs. Tampa Bay, November 13, 1993. Transferred to **Colorado** after **Quebec** franchise relocated, June 21, 1995. Traded to **San Jose** by **Colorado** for Sandis Ozolinsh, October 26, 1995.

NORDSTROM, Peter (NOHRD-struhm, PEE-tuhr) BOS.

Center. Shoots left. 6'1", 200 lbs. Born, Munkfors, Sweden, July 26, 1974. Boston's 3rd choice, 78th overall, in 1998 Entry Draft.

Season	Club	League	GP	G	A	Pts	PIM	PP	SH	GW	S	%	+/-	TF	F%	H	SB	Min	GP	G	A	Pts	PIM	PP	SH	GW
1989-90	IFK Munkfors	Swede-3	21	2	1	3	8																			
1990-91	IFK Munkfors	Swede-3	32	10	18	28	20																			
1991-92	IFK Munkfors	Swede-3	31	12	20	32	42																			
1992-93	IFK Munkfors	Swede-3	35	19	11	30	44																			
1993-94	IFK Munkfors	Swede-3	31	17	26	43	87																			
1994-95	IFK Munkfors	Swede-2	21	8	17	25	30																			
	Leksands IF	Sweden	13	1	0	1	0																			
1995-96	Farjestad	Sweden	40	6	5	11	36												8	0	3	3	12			
1996-97	Farjestad	Sweden	44	9	5	14	32												14	1	2	3	6			
1997-98	Farjestad	Sweden	45	6	19	25	46												12	5	7	*12	8			
1998-99	Farjestad	Sweden	21	4	4	8	14												4	1	1	2	2			
	Farjestad	EuroHL	2	1	0	1	2																			
	Boston	NHL	2	0	0	0	0	0	0	0	0	0.0	–1	0	0.0	1	0	8:04								
	Providence	AHL	13	2	1	3	2																			
99-2000	Farjestad	Sweden	45	8	14	22	48												7	0	3	3	2			
2000-01	Farjestad	Sweden	49	7	15	22	59												16	2	6	8	30			
2001-02	Farjestad	Sweden	45	5	18	23	56												9	3	3	6	4			
	NHL Totals		2	0	0	0	0	0	0	0	0	0.0		0	0.0	1	0	8:04								

NORSTROM, Mattias (NOHR-struhm, MAT-tee-ahs) L.A.

Defense. Shoots left. 6'2", 211 lbs. Born, Stockholm, Sweden, January 2, 1972. NY Rangers' 2nd choice, 48th overall, in 1992 Entry Draft.

Season	Club	League	GP	G	A	Pts	PIM	PP	SH	GW	S	%	+/-	TF	F%	H	SB	Min	GP	G	A	Pts	PIM	PP	SH	GW
1990-91	Mora IK	Swede-2	9	1	1	2	6												1	0	0	0	2			
1991-92	AIK Solna	Sweden	39	4	3	7	28												3	0	2	2	2			
1992-93	AIK Solna	Sweden	22	0	1	1	16																			
1993-94	NY Rangers	NHL	9	0	2	2	6	0	0	0	3	0.0	0													
	Binghamton	AHL	55	1	9	10	70																			
1994-95	Binghamton	AHL	63	9	10	19	91																			
	NY Rangers	NHL	9	0	3	3	2	0	0	0	4	0.0	2						3	0	0	0	0	0	0	0
1995-96	NY Rangers	NHL	25	2	1	3	22	0	0	0	17	11.8	5													
	Los Angeles	NHL	11	0	1	1	18	0	0	0	17	0.0	–8													
1996-97	Los Angeles	NHL	80	1	21	22	84	0	0	0	106	0.9	–4													
1997-98	Los Angeles	NHL	73	1	12	13	90	0	0	0	61	1.6	14						4	0	0	0	2	0	0	0
	Sweden	Olympics	4	0	1	1	2																			
1998-99	Los Angeles	NHL	78	2	5	7	36	0	1	0	61	3.3	–10	1	0.0	236	157	20:20								
99-2000	Los Angeles	NHL	82	1	13	14	66	0	0	0	62	1.6	22	0	0.0	261	127	21:49	4	0	0	0	6	0	0	0
2000-01	Los Angeles	NHL	82	0	18	18	60	0	0	0	59	0.0	10	2	0.0	249	137	21:50	13	0	2	2	18	0	0	0
2001-02	Los Angeles	NHL	79	2	9	11	38	0	0	0	42	4.8	–2	0	0.0	157	140	23:01	7	0	0	0	4	0	0	0
	Sweden	Olympics	4	0	0	0	0																			
	NHL Totals		528	9	85	94	422	0	1	0	432	2.1		3	0.0	903	561	21:45	31	0	2	2	30	0	0	0

Played in NHL All-Star Game (1999)

Traded to **LA Kings** by **NY Rangers** with Ray Ferraro, Ian Laperriere, Nathan Lafayette and NY Rangers' 4th round choice (Sean Blanchard) in 1997 Entry Draft for Marty McSorley, Jari Kurri and Shane Churla, March 14, 1996.

NORTON, Brad (NOHR-tohn, BRAD)

Defense. Shoots left. 6'4", 225 lbs. Born, Cambridge, MA, February 13, 1975. Edmonton's 9th choice, 215th overall, in 1993 Entry Draft.

Season	Club	League	GP	G	A	Pts	PIM	PP	SH	GW	S	%	+/-	TF	F%	H	SB	Min	GP	G	A	Pts	PIM	PP	SH	GW
1992-93	Cushing Academy	Hi-School	31	10	26	36	....																			
1993-94	Cushing Academy	Hi-School	STATISTICS NOT AVAILABLE																							
1994-95	U. Mass-Amherst	H-East	30	0	6	6	89																			
1995-96	U. Mass-Amherst	H-East	34	4	12	16	99																			
1996-97	U. Mass-Amherst	H-East	35	2	16	18	88																			
1997-98	U. Mass-Amherst	H-East	20	2	13	15	28																			
	Detroit Vipers	IHL	33	1	4	5	56												22	0	2	2	87			
1998-99	Hamilton	AHL	58	1	8	9	134												11	0	1	1	6			
99-2000	Hamilton	AHL	40	5	12	17	104												10	1	4	5	26			
2000-01	Hamilton	AHL	46	3	15	18	114																			

			Regular Season																Playoffs							
Season	Club	League	GP	G	A	Pts	PIM	PP	SH	GW	S	%	+/-	TF	F%	H	SB	Min	GP	G	A	Pts	PIM	PP	SH	GW
2001-02	**Florida**	**NHL**	**22**	**0**	**2**	**2**	**45**	**0**	**0**	**0**	**6**	**0.0**	**−2**	**1**	**0.0**	**34**	**10**	**9:15**								
	Hershey Bears	AHL	40	0	10	10	62												2	0	0	0	6			
	NHL Totals		**22**	**0**	**2**	**2**	**45**	**0**	**0**	**0**	**6**	**0.0**		**1**	**0.0**	**34**	**10**	**9:15**								

Signed as a free agent by **Florida**, July 27, 2001.

NORTON, Jeff

(NOHR-tohn, JEHF)

Defense. Shoots left. 6'2", 195 lbs. Born, Acton, MA, November 25, 1965. NY Islanders' 3rd choice, 62nd overall, in 1984 Entry Draft.

Season	Club	League	GP	G	A	Pts	PIM	PP	SH	GW	S	%	+/-	TF	F%	H	SB	Min	GP	G	A	Pts	PIM	PP	SH	GW
1983-84	Cushing Academy	Hi-School	21	22	33	55																				
1984-85	U. of Michigan	CCHA	37	8	16	24	103																			
1985-86	U. of Michigan	CCHA	37	15	30	45	99																			
1986-87	U. of Michigan	CCHA	39	12	36	48	92																			
1987-88	Team USA	Nat-Tm	54	7	22	29	52																			
	United States	Olympics	6	0	4	4	4																			
	NY Islanders	**NHL**	**15**	**1**	**6**	**7**	**14**	**1**	**0**	**1**	**18**	**5.6**	**3**						**3**	**0**	**2**	**2**	**13**	**0**	**0**	**0**
1988-89	**NY Islanders**	**NHL**	**69**	**1**	**30**	**31**	**74**	**1**	**0**	**0**	**126**	**0.8**	**−24**													
1989-90	**NY Islanders**	**NHL**	**60**	**4**	**49**	**53**	**65**	**4**	**0**	**0**	**104**	**3.8**	**−9**						**4**	**1**	**3**	**4**	**17**	**0**	**0**	**0**
1990-91	**NY Islanders**	**NHL**	**44**	**3**	**25**	**28**	**16**	**2**	**1**	**0**	**87**	**3.4**	**−13**													
1991-92	**NY Islanders**	**NHL**	**28**	**1**	**18**	**19**	**18**	**0**	**1**	**0**	**34**	**2.9**	**2**													
1992-93	**NY Islanders**	**NHL**	**66**	**12**	**38**	**50**	**45**	**5**	**0**	**0**	**127**	**9.4**	**−3**						**10**	**1**	**1**	**2**	**4**	**0**	**0**	**0**
1993-94	**San Jose**	**NHL**	**64**	**7**	**33**	**40**	**36**	**1**	**0**	**0**	**92**	**7.6**	**16**						**14**	**1**	**5**	**6**	**20**	**0**	**0**	**0**
1994-95	**San Jose**	**NHL**	**20**	**1**	**9**	**10**	**39**	**0**	**0**	**0**	**21**	**4.8**	**1**													
	St. Louis	**NHL**	**28**	**2**	**18**	**20**	**33**	**0**	**0**	**1**	**27**	**7.4**	**21**						**7**	**1**	**1**	**2**	**11**	**0**	**0**	**0**
1995-96	**St. Louis**	**NHL**	**36**	**4**	**7**	**11**	**26**	**0**	**0**	**1**	**33**	**12.1**	**4**													
	Edmonton	**NHL**	**30**	**4**	**16**	**20**	**16**	**1**	**0**	**1**	**52**	**7.7**	**5**													
1996-97	**Edmonton**	**NHL**	**62**	**2**	**11**	**13**	**42**	**0**	**0**	**0**	**68**	**2.9**	**−7**													
	Tampa Bay	**NHL**	**13**	**0**	**5**	**5**	**16**	**0**	**0**	**0**	**13**	**0.0**	**0**													
1997-98	**Tampa Bay**	**NHL**	**37**	**4**	**6**	**10**	**26**	**4**	**0**	**0**	**41**	**9.8**	**−25**													
	Florida	**NHL**	**19**	**0**	**7**	**7**	**18**	**0**	**0**	**0**	**20**	**0.0**	**−7**													
1998-99	**Florida**	**NHL**	**3**	**0**	**0**	**0**	**2**	**0**	**0**	**0**	**2**	**0.0**	**0**	**0**	**0.0**	**1**	**3**	**17:26**								
	San Jose	**NHL**	**69**	**4**	**18**	**22**	**42**	**2**	**0**	**1**	**68**	**5.9**	**2**	**0**	**0.0**	**37**	**86**	**20:55**	**6**	**0**	**7**	**7**	**10**	**0**	**0**	**0**
99-2000	**San Jose**	**NHL**	**62**	**0**	**20**	**20**	**49**	**0**	**0**	**0**	**45**	**0.0**	**−2**	**0**	**0.0**	**29**	**74**	**19:32**	**12**	**0**	**1**	**1**	**7**	**0**	**0**	**0**
2000-01	**Pittsburgh**	**NHL**	**32**	**2**	**10**	**12**	**20**	**1**	**0**	**1**	**17**	**11.8**	**8**	**0**	**0.0**	**12**	**51**	**18:30**								
	San Jose	**NHL**	**10**	**0**	**1**	**1**	**8**	**0**	**0**	**0**	**5**	**0.0**	**4**	**0**	**0.0**	**5**	**14**	**19:18**	**6**	**0**	**1**	**1**	**2**	**0**	**0**	**0**
2001-02	**Florida**	**NHL**	**29**	**0**	**4**	**4**	**8**	**0**	**0**	**0**	**13**	**0.0**	**−5**	**0**	**0.0**	**25**	**48**	**20:35**								
	Boston	**NHL**	**3**	**0**	**1**	**1**	**2**	**0**	**0**	**0**	**2**	**0.0**	**0**	**0**	**0.0**	**1**	**3**	**15:46**	**3**	**0**	**0**	**0**	**5**	**0**	**0**	**0**
	NHL Totals		**799**	**52**	**332**	**384**	**615**	**22**	**2**	**6**	**1015**	**5.1**		**0**	**0.0**	**110**	**279**	**19:53**	**65**	**4**	**21**	**25**	**89**	**0**	**0**	**0**

CCHA Second All-Star Team (1987)

• Missed majority of 1991-92 season recovering from wrist injury suffered in game vs. Buffalo, January 3, 1992. Traded to **San Jose** by **NY Islanders** for San Jose's 3rd round choice (Jason Strudwick) in 1994 Entry Draft, June 20, 1993. Traded to **St. Louis** by **San Jose** with San Jose's 3rd round choice (later traded to Colorado - Colorado selected Rick Berry) in 1997 Entry Draft for Craig Janney and cash, March 6, 1995. Traded to **Edmonton** by **St. Louis** with Donald Dufresne for Igor Kravchuk and Ken Sutton, January 4, 1996. Traded to **Tampa Bay** by **Edmonton** for Drew Bannister and Tampa Bay's 6th round choice (Peter Sarno) in 1997 Entry Draft, March 18, 1997. Traded to **Florida** by **Tampa Bay** with Dino Ciccarelli for Mark Fitzpatrick and Jody Hull, January 15, 1998. Traded to **San Jose** by **Florida** for Alex Hicks and San Jose's 5th round choice (later traded to NY Islanders - NY Islanders selected Adam Johnson) in 1999 Entry Draft, November 11, 1998. Signed as a free agent by **Pittsburgh**, November 14, 2000. Traded to **San Jose** by **Pittsburgh** for Bobby Dollas and Johan Hedberg, March 12, 2001. Signed as a free agent by **Florida**, July 18, 2001. • Missed majority of 2001-02 season recovering from knee injury suffered in game vs. Toronto, December 28, 2001. Traded to **Boston** by **Florida** for Boston's 6th round choice (Mikael Vuorio) in 2002 Entry Draft, March 19, 2002.

NOVOSELTSEV, Ivan

(noh-voh-SEHLT-sehv, ee-VAHN) **FLA.**

Left wing. Shoots left. 6'1", 210 lbs. Born, Golitsino, USSR, January 23, 1979. Florida's 5th choice, 95th overall, in 1997 Entry Draft.

Season	Club	League	GP	G	A	Pts	PIM	PP	SH	GW	S	%	+/-	TF	F%	H	SB	Min	GP	G	A	Pts	PIM	PP	SH	GW
1995-96	Krylja Sovetov	CIS	1	0	0	0	2																			
1996-97	Krylja Sovetov 2	Russia-3	19	5	3	8	39																			
	Krylja Sovetov	Russia	30	0	3	3	18												2	0	0	0	4			
1997-98	Sarnia Sting	OHL	53	26	22	48	41												5	1	1	2	8			
1998-99	Sarnia Sting	OHL	68	57	39	96	45												5	2	4	6	6			
99-2000	**Florida**	**NHL**	**14**	**2**	**1**	**3**	**8**	**2**	**0**	**0**	**8**	**25.0**	**−3**	**1**	**100.0**	**3**	**3**	**10:29**								
	Louisville	AHL	47	14	21	35	22												4	1	0	1	6			
2000-01	**Florida**	**NHL**	**38**	**3**	**6**	**9**	**16**	**0**	**0**	**0**	**34**	**8.8**	**−5**	**1**	**0.0**	**27**	**4**	**10:44**								
	Louisville	AHL	34	2	10	12	8																			
2001-02	**Florida**	**NHL**	**70**	**13**	**16**	**29**	**44**	**1**	**1**	**5**	**109**	**11.9**	**−10**	**27**	**44.4**	**52**	**18**	**14:53**								
	NHL Totals		**122**	**18**	**23**	**41**	**68**	**3**	**1**	**5**	**151**	**11.9**		**29**	**44.8**	**82**	**25**	**13:05**								

OHL First All-Star Team (1999)

NUMMELIN, Petteri

(NOO-muh-lihn, PEH-tuh-ree) **CBJ**

Defense. Shoots left. 5'10", 196 lbs. Born, Turku, Finland, November 25, 1972. Columbus' 3rd choice, 133rd overall, in 2000 Entry Draft.

Season	Club	League	GP	G	A	Pts	PIM	PP	SH	GW	S	%	+/-	TF	F%	H	SB	Min	GP	G	A	Pts	PIM	PP	SH	GW
1988-89	TPS Turku Jr.	Finn-Jr.	11	2	3	5	2																			
1989-90	TPS Turku Jr.	Finn-Jr.	33	6	14	20	45																			
1990-91	TPS Turku Jr.	Finn-Jr.	35	20	16	36	28																			
	Kiekko-67 Turku	Finland-2	2	0	2	2	4																			
1991-92	Kiekko-67 Jr.	Finn-Jr.	13	16	15	31	28																			
	Kiekko-67 Turku	Finland-2	41	12	24	36	36																			
1992-93	TPS Turku Jr.	Finn-Jr.	1	1	0	1	0																			
	TPS Turku	Finland	3	0	0	0	8																			
	Reipas Lahti	Finland	20	5	8	13	20																			
	Kiekko-67 Turku	Finland-2	28	14	15	29	18																			
1993-94	TPS Turku	Finland	44	14	24	38	20												11	0	3	3	4			
1994-95	TPS Turku	Finland	48	10	17	27	32												11	4	3	7	0			
1995-96	Vastra Frolunda	Sweden	32	7	11	18	26												12	2	7	9	4			
1996-97	Vastra Frolunda	Sweden	44	20	14	34	39												2	0	1	1	0			
1997-98	HC Davos	Swiss	33	13	17	30	24												17	8	14	22	2			
1998-99	HC Davos	Swiss	44	11	42	53	22												4	0	2	2	2			
99-2000	HC Davos	Swiss	40	15	23	38	20												5	0	3	3	0			
2000-01	**Columbus**	**NHL**	**61**	**4**	**12**	**16**	**10**	**2**	**0**	**0**	**99**	**4.0**	**−11**	**1**	**0.0**	**14**	**39**	**17:14**								
2001-02	HC Lugano	Swiss	35	4	18	22	6												13	6	9	15	2			
	NHL Totals		**61**	**4**	**12**	**16**	**10**	**2**	**0**	**0**	**99**	**4.0**		**1**	**0.0**	**14**	**39**	**17:14**								

NUMMINEN, Teppo

(NOO-mih-nehn, TEH-poh) **PHX.**

Defense. Shoots right. 6'2", 199 lbs. Born, Tampere, Finland, July 3, 1968. Winnipeg's 2nd choice, 29th overall, in 1986 Entry Draft.

Season	Club	League	GP	G	A	Pts	PIM	PP	SH	GW	S	%	+/-	TF	F%	H	SB	Min	GP	G	A	Pts	PIM	PP	SH	GW
1984-85	Tappara Jr.	Finn-Jr.	30	14	17	31	10																			
	Whitby Lawmen	OPJHL	16	3	9	12	0																			
1985-86	Tappara Jr.	Finn-Jr.	2	0	0	0	0												3	0	1	1	2			
	Tappara Tampere	Finland	31	2	4	6	6												8	0	0	0	0			
1986-87	Tappara Tampere	Finland	44	9	9	18	16												9	4	1	5	4			
1987-88	Tappara Tampere	Finland	40	10	10	20	29												10	6	6	12	6			
	Finland	Olympics	6	1	4	5	0																			
1988-89	**Winnipeg**	**NHL**	**69**	**1**	**14**	**15**	**36**	**0**	**1**	**0**	**85**	**1.2**	**−11**													
1989-90	**Winnipeg**	**NHL**	**79**	**11**	**32**	**43**	**20**	**1**	**0**	**1**	**105**	**10.5**	**−4**						**7**	**1**	**2**	**3**	**10**	**0**	**0**	**0**
1990-91	**Winnipeg**	**NHL**	**80**	**8**	**25**	**33**	**28**	**3**	**0**	**0**	**151**	**5.3**	**−15**													
1991-92	**Winnipeg**	**NHL**	**80**	**5**	**34**	**39**	**32**	**4**	**0**	**1**	**143**	**3.5**	**15**						**7**	**0**	**0**	**0**	**0**	**0**	**0**	**0**
1992-93	**Winnipeg**	**NHL**	**66**	**7**	**30**	**37**	**33**	**3**	**1**	**0**	**103**	**6.8**	**4**						**6**	**1**	**1**	**2**	**2**	**1**	**0**	**0**
1993-94	**Winnipeg**	**NHL**	**57**	**5**	**18**	**23**	**28**	**4**	**0**	**1**	**89**	**5.6**	**−23**													
1994-95	TuTo Turku	Finland	12	3	8	11	4																			
	Winnipeg	**NHL**	**42**	**5**	**16**	**21**	**16**	**2**	**0**	**0**	**86**	**5.8**	**12**													
1995-96	**Winnipeg**	**NHL**	**74**	**11**	**43**	**54**	**22**	**6**	**0**	**3**	**165**	**6.7**	**−4**						**6**	**0**	**0**	**0**	**2**	**0**	**0**	**0**
1996-97	**Phoenix**	**NHL**	**82**	**2**	**25**	**27**	**28**	**0**	**0**	**0**	**135**	**1.5**	**−3**						**7**	**3**	**3**	**6**	**0**	**1**	**0**	**1**
1997-98	**Phoenix**	**NHL**	**82**	**11**	**40**	**51**	**30**	**6**	**0**	**2**	**126**	**8.7**	**25**						**1**	**0**	**0**	**0**	**0**	**0**	**0**	**0**
	Finland	Olympics	6	1	1	2	2																			
1998-99	**Phoenix**	**NHL**	**82**	**10**	**30**	**40**	**30**	**1**	**0**	**0**	**156**	**6.4**	**3**	**2**	**0.0**	**73**	**99**	**24:26**	**7**	**2**	**1**	**3**	**4**	**2**	**0**	**0**
99-2000	**Phoenix**	**NHL**	**79**	**8**	**34**	**42**	**16**	**2**	**0**	**2**	**126**	**6.3**	**21**	**1**	**0.0**	**75**	**96**	**23:37**	**5**	**1**	**1**	**2**	**0**	**0**	**0**	**0**
2000-01	**Phoenix**	**NHL**	**72**	**5**	**26**	**31**	**36**	**1**	**0**	**2**	**109**	**4.6**	**9**	**0**	**0.0**	**42**	**102**	**24:28**								

Season	Club	League	GP	G	A	Pts	PIM	PP	SH	GW	S	%	+/-	TF	F%	H	SB	Min	GP	G	A	Pts	PIM	PP	SH	GW
			Regular Season																Playoffs							
2001-02	**Phoenix**	**NHL**	**76**	**13**	**35**	**48**	**20**	**4**	**0**	**6**	**117**	**11.1**	**13**	**0**	**0.0**	**66**	**106**	**23:51**	**4**	**0**	**0**	**0**	**2**	**0**	**0**	**0**
	Finland	Olympics	4	0	1	1	0																			
	NHL Totals		**1020**	**102**	**402**	**504**	**375**	**37**	**2**	**18**	**1696**	**6.0**		**3**	**0.0**	**256**	**403**	**24:05**	**50**	**8**	**8**	**16**	**20**	**4**	**0**	**1**

Played in NHL All-Star Game (1999, 2000, 2001)

Transferred to **Phoenix** after **Winnipeg** franchise relocated, July 1, 1996.

NURMINEN, Kai

(NUHR-mih-nehn, KIGH)

Left wing. Shoots left. 6'1", 190 lbs. Born, Turku, Finland, March 29, 1969. Los Angeles' 9th choice, 193rd overall, in 1996 Entry Draft.

Season	Club	League	GP	G	A	Pts	PIM	PP	SH	GW	S	%	+/-	TF	F%	H	SB	Min	GP	G	A	Pts	PIM	PP	SH	GW
1986-87	TPS Turku Jr.	Finn-Jr.	2	0	1	1	0																			
1987-88	TPS Turku Jr.	Finn-Jr.	32	9	7	16	16																			
1988-89	TPS Turku Jr.	Finn-Jr.	22	13	10	23	14																			
1989-90	TPS Turku Jr.	Finn-Jr.	22	13	10	23	14																			
1990-91	TuTo Turku	Finland-2	33	26	20	46	14																			
1991-92	Kiekko-67 Turku	Finland-2	44	44	19	63	34																			
1992-93	Kiekko-67 Turku	Finland-2	8	6	4	10	2																			
	TPS Turku	Finland	31	4	6	10	13												7	1	2	3	0			
1993-94	TPS Turku	Finland	45	23	12	35	20												11	0	3	3	4			
1994-95	HPK Hameenlinna	Finland	49	*30	25	55	40																			
1995-96	HV 71 Jonkoping	Sweden	40	31	24	55	30												4	3	1	4	8			
1996-97	**Los Angeles**	**NHL**	**67**	**16**	**11**	**27**	**22**	**4**	**0**	**1**	**112**	**14.3**	**–3**													
1997-98	Vastra Frolunda	Sweden	23	9	7	16	24																			
	Jokerit Helsinki	Finland	20	7	9	16	30												8	5	3	8	4			
1998-99	HC Davos	Swiss	42	26	14	40	26																			
99-2000	TPS Turku	Finland	54	*41	37	*78	40												10	5	*9	*14	0			
2000-01	**Minnesota**	**NHL**	**2**	**1**	**0**	**1**	**2**	**0**	**0**	**0**	**1**	**100.0**	**–1**	**0**	**0.0**	**1**	**0**	**15:34**								
	Cleveland	IHL	74	28	46	74	34												1	0	0	0	0			
2001-02	TPS Turku	Finland	52	15	18	33	42												8	0	2	2	2			
	NHL Totals		**69**	**17**	**11**	**28**	**24**	**4**	**0**	**1**	**113**	**15.0**		**0**	**0.0**	**1**	**0**	**15:34**								

Signed as a free agent by **Minnesota**, May 24, 2000.

NYLANDER, Michael

(NEE-lan-duhr, MIGH-kuhl) **CHI.**

Center. Shoots left. 6'1", 195 lbs. Born, Stockholm, Sweden, October 3, 1972. Hartford's 4th choice, 59th overall, in 1991 Entry Draft.

Season	Club	League	GP	G	A	Pts	PIM	PP	SH	GW	S	%	+/-	TF	F%	H	SB	Min	GP	G	A	Pts	PIM	PP	SH	GW
1989-90	Huddinge IK	Swede-2	31	7	15	22	4												5	3	0	3	0			
1990-91	Huddinge IK	Swede-2	33	14	20	34	10												2	0	0	0	0			
1991-92	AIK Solna	Sweden	40	11	17	28	30												3	1	4	5	4			
1992-93	**Hartford**	**NHL**	**59**	**11**	**22**	**33**	**36**	**3**	**0**	**1**	**85**	**12.9**	**–7**													
	Springfield	AHL																	3	3	3	6	2			
1993-94	**Hartford**	**NHL**	**58**	**11**	**33**	**44**	**24**	**4**	**0**	**1**	**74**	**14.9**	**–2**													
	Springfield	AHL	4	0	9	9	0																			
	Calgary	**NHL**	**15**	**2**	**9**	**11**	**6**	**0**	**0**	**0**	**21**	**9.5**	**10**						**3**	**0**	**0**	**0**	**0**	**0**	**0**	**0**
1994-95	JyP HT Jyvaskyla	Finland	16	11	19	30	63																			
	Calgary	**NHL**	**6**	**0**	**1**	**1**	**2**	**0**	**0**	**0**	**2**	**0.0**	**1**						**6**	**0**	**6**	**6**	**2**	**0**	**0**	**0**
1995-96	**Calgary**	**NHL**	**73**	**17**	**38**	**55**	**20**	**4**	**0**	**6**	**163**	**10.4**	**0**						**4**	**0**	**0**	**0**	**0**	**0**	**0**	**0**
1996-97	HC Lugano	Swiss	36	12	43	55	28												8	3	8	11	8			
1997-98	**Calgary**	**NHL**	**65**	**13**	**23**	**36**	**24**	**0**	**0**	**2**	**117**	**11.1**	**10**													
	Sweden	Olympics	4	0	0	0	6																			
1998-99	**Calgary**	**NHL**	**9**	**2**	**3**	**5**	**2**	**1**	**0**	**0**	**7**	**28.6**	**1**	**25**	**60.0**	**1**	**0**	**11:10**								
	Tampa Bay	**NHL**	**24**	**2**	**7**	**9**	**6**	**0**	**0**	**0**	**26**	**7.7**	**–10**	**75**	**44.0**	**4**	**6**	**13:29**								
99-2000	**Tampa Bay**	**NHL**	**11**	**1**	**2**	**3**	**4**	**1**	**0**	**0**	**10**	**10.0**	**–3**	**35**	**57.1**	**0**	**0**	**10:32**								
	Chicago	**NHL**	**66**	**23**	**28**	**51**	**26**	**4**	**0**	**2**	**112**	**20.5**	**9**	**561**	**46.9**	**9**	**11**	**16:39**								
2000-01	**Chicago**	**NHL**	**82**	**25**	**39**	**64**	**32**	**4**	**0**	**5**	**176**	**14.2**	**7**	**1036**	**48.3**	**12**	**22**	**18:52**								
2001-02	**Chicago**	**NHL**	**82**	**15**	**46**	**61**	**50**	**6**	**0**	**2**	**158**	**9.5**	**28**	**974**	**50.2**	**17**	**13**	**15:33**	**5**	**0**	**3**	**3**	**2**	**0**	**0**	**0**
	Sweden	Olympics	4	1	2	3	0																			
	NHL Totals		**550**	**122**	**251**	**373**	**232**	**27**	**0**	**19**	**951**	**12.8**		**2706**	**48.8**	**43**	**52**	**16:17**	**18**	**0**	**9**	**9**	**4**	**0**	**0**	**0**

Traded to **Calgary** by **Hartford** with James Patrick and Zarley Zalapski for Gary Suter, Paul Ranheim and Ted Drury, March 10, 1994. • Missed majority of 1994-95 season recovering from wrist injury suffered in game vs. St. Louis, January 24, 1995. Traded to **Tampa Bay** by **Calgary** for Andrei Nazarov, January 19, 1999. Traded to **Chicago** by **Tampa Bay** for Bryan Muir and Reid Simpson, November 12, 1999.

OATES, Adam

(OHTS, A-duhm) **ANA.**

Center. Shoots right. 5'11", 190 lbs. Born, Weston, Ont., August 27, 1962.

Season	Club	League	GP	G	A	Pts	PIM	PP	SH	GW	S	%	+/-	TF	F%	H	SB	Min	GP	G	A	Pts	PIM	PP	SH	GW
1979-80	Port Credit	OHA-B	34	30	36	66	41																			
	Markham Waxers	OHA-B	9	1	6	7	2																			
1980-81	Markham Waxers	OHA-B	43	36	53	89	89																			
1981-82	Markham Waxers	OJHL-B	40	59	110	169																				
1982-83	RPI Engineers	ECAC	22	9	33	42	8																			
1983-84	RPI Engineers	ECAC	38	26	57	83	15																			
1984-85	RPI Engineers	ECAC	38	31	60	91	29																			
1985-86	**Detroit**	**NHL**	**38**	**9**	**11**	**20**	**10**	**1**	**0**	**1**	**49**	**18.4**	**–24**													
	Adirondack	AHL	34	18	28	46	4												17	7	14	21	4			
1986-87	**Detroit**	**NHL**	**76**	**15**	**32**	**47**	**21**	**4**	**0**	**1**	**138**	**10.9**	**0**						**16**	**4**	**7**	**11**	**6**	**0**	**0**	**1**
1987-88	**Detroit**	**NHL**	**63**	**14**	**40**	**54**	**20**	**3**	**0**	**3**	**111**	**12.6**	**16**						**16**	**8**	**12**	**20**	**6**	**4**	**0**	**1**
1988-89	**Detroit**	**NHL**	**69**	**16**	**62**	**78**	**14**	**2**	**0**	**1**	**127**	**12.6**	**–1**						**6**	**0**	**8**	**8**	**2**	**0**	**0**	**0**
1989-90	**St. Louis**	**NHL**	**80**	**23**	**79**	**102**	**30**	**6**	**2**	**3**	**168**	**13.7**	**9**						**12**	**2**	**12**	**14**	**4**	**1**	**0**	**0**
1990-91	**St. Louis**	**NHL**	**61**	**25**	**90**	**115**	**29**	**3**	**1**	**3**	**139**	**18.0**	**15**						**13**	**7**	**13**	**20**	**10**	**2**	**0**	**1**
1991-92	**St. Louis**	**NHL**	**54**	**10**	**59**	**69**	**12**	**3**	**0**	**3**	**118**	**8.5**	**–4**													
	Boston	**NHL**	**26**	**10**	**20**	**30**	**10**	**3**	**0**	**1**	**73**	**13.7**	**–5**						**15**	**5**	**14**	**19**	**4**	**3**	**0**	**2**
1992-93	**Boston**	**NHL**	**84**	**45**	***97**	**142**	**32**	**24**	**1**	**11**	**254**	**17.7**	**15**						**4**	**0**	**9**	**9**	**4**	**0**	**0**	**0**
1993-94	**Boston**	**NHL**	**77**	**32**	**80**	**112**	**45**	**16**	**2**	**3**	**197**	**16.2**	**10**						**13**	**3**	**9**	**12**	**8**	**2**	**0**	**0**
1994-95	**Boston**	**NHL**	**48**	**12**	**41**	**53**	**8**	**4**	**1**	**2**	**109**	**11.0**	**–11**						**5**	**1**	**0**	**1**	**2**	**1**	**0**	**0**
1995-96	**Boston**	**NHL**	**70**	**25**	**67**	**92**	**18**	**7**	**1**	**2**	**183**	**13.7**	**16**						**5**	**2**	**5**	**7**	**2**	**0**	**1**	**0**
1996-97	**Boston**	**NHL**	**63**	**18**	**52**	**70**	**10**	**2**	**2**	**4**	**138**	**13.0**	**–3**													
	Washington	**NHL**	**17**	**4**	**8**	**12**	**4**	**1**	**0**	**1**	**22**	**18.2**	**–2**													
1997-98	**Washington**	**NHL**	**82**	**18**	**58**	**76**	**36**	**3**	**2**	**3**	**121**	**14.9**	**6**						**21**	**6**	**11**	**17**	**8**	**1**	**1**	**1**
1998-99	**Washington**	**NHL**	**59**	**12**	**42**	**54**	**22**	**3**	**0**	**0**	**79**	**15.2**	**–1**	**1330**	**59.2**	**17**	**22**	**20:34**								
99-2000	**Washington**	**NHL**	**82**	**15**	**56**	**71**	**14**	**5**	**0**	**6**	**93**	**16.1**	**13**	**2176**	**56.8**	**19**	**45**	**22:20**	**5**	**0**	**3**	**3**	**4**	**0**	**0**	**0**
2000-01	**Washington**	**NHL**	**81**	**13**	***69**	**82**	**28**	**5**	**0**	**4**	**72**	**18.1**	**–9**	**1836**	**58.9**	**16**	**49**	**20:60**	**6**	**0**	**0**	**0**	**0**	**0**	**0**	**0**
2001-02	**Washington**	**NHL**	**66**	**11**	***57**	**68**	**22**	**3**	**0**	**1**	**85**	**12.9**	**–2**	**1642**	**56.5**	**3**	**29**	**22:06**								
	Philadelphia	**NHL**	**14**	**3**	***7**	**10**	**6**	**0**	**0**	**0**	**17**	**17.6**	**–2**	**323**	**55.7**	**0**	**6**	**20:51**	**5**	**0**	**2**	**2**	**0**	**0**	**0**	**0**
	NHL Totals		**1210**	**330**	**1027**	**1357**	**391**	**98**	**12**	**53**	**2293**	**14.4**		**7307**	**57.6**	**55**	**151**	**21:31**	**142**	**38**	**105**	**143**	**60**	**14**	**2**	**6**

ECAC Second All-Star Team (1984) • NCAA East First All-American Team (1984, 1985) • ECAC First All-Star Team (1985) • NCAA Championship All-Tournament Team (1985) • NHL Second All-Star Team (1991) • Played in NHL All-Star Game (1991, 1992, 1993, 1994, 1997)

Signed as a free agent by **Detroit**, June 28, 1985. Traded to **St. Louis** by **Detroit** with Paul MacLean for Bernie Federko and Tony McKegney, June 15, 1989. Traded to **Boston** by **St. Louis** for Craig Janney and Stephane Quintal, February 7, 1992. Traded to **Washington** by **Boston** with Bill Ranford and Rick Tocchet for Jim Carey, Anson Carter, Jason Allison and Washington's 3rd round choice (Lee Goren) in 1997 Entry Draft, March 1, 1997. Traded to **Philadelphia** by **Washington** for Maxime Ouellet and Philadelphia's 1st (later traded to Dallas - Dallas selected Martin Vagner), 2nd (Maxime Daigneault) and 3rd (Derek Krestanovich) round choices in 2002 Entry Draft, March 19, 2002. Signed as a free agent by **Anaheim**, July 1, 2002.

OBSUT, Jaroslav

(OHB-suht, YAHR-oh-slahv) **VAN.**

Defense. Shoots left. 6'1", 200 lbs. Born, Presov, Czech., September 3, 1976. Winnipeg's 9th choice, 188th overall, in 1995 Entry Draft.

Season	Club	League	GP	G	A	Pts	PIM	PP	SH	GW	S	%	+/-	TF	F%	H	SB	Min	GP	G	A	Pts	PIM	PP	SH	GW
1994-95	North Battleford	SJHL	55	21	30	51	126																			
1995-96	Swift Current	WHL	72	10	11	21	57												6	0	0	0	2			
1996-97	Edmonton Ice	WHL	13	2	9	11	4																			
	Medicine Hat	WHL	50	8	26	34	42												4	0	2	2	2			
	Toledo Storm	ECHL	3	1	0	1	0												5	0	1	1	6			
1997-98	Raleigh Icecaps	ECHL	60	6	26	32	46																			
	Syracuse Crunch	AHL	4	0	1	1	4																			

			Regular Season																Playoffs							
Season	Club	League	GP	G	A	Pts	PIM	PP	SH	GW	S	%	+/-	TF	F%	H	SB	Min	GP	G	A	Pts	PIM	PP	SH	GW
1998-99	Augusta Lynx	ECHL	41	11	25	36	42	....	....	....	....	....	....	....	....	....	....	....	....	....	....	....	....	....	....	....
	Manitoba Moose	IHL	2	0	0	0	0	....	....	....	....	....	....	....	....	....	....	....	....	....	....	....	....	....	....	....
	Worcester	AHL	31	2	8	10	14	....	....	....	....	....	....	....	....	....	....	....	4	0	1	1	2	....	....	....
99-2000	Worcester	AHL	7	0	2	2	4	....	....	....	....	....	....	....	....	....	....	....	....	....	....	....	....	....	....	....
2000-01	**St. Louis**	**NHL**	4	0	0	0	2	0	0	0	3	0.0	1	0	0.0	6	4	18:31	....	....	....	....	....	....	....	....
	Peoria Rivermen	ECHL	3	0	4	4	2	....	....	....	....	....	....	....	....	....	....	....	....	....	....	....	....	....	....	....
	Worcester	AHL	47	9	12	21	20	....	....	....	....	....	....	....	....	....	....	....	7	0	1	1	4	....	....	....
2001-02	**Colorado**	**NHL**	3	0	0	0	0	0	0	0	3	0.0	0	0	0.0	9	0	9:47	....	....	....	....	....	....	....	....
	Hershey Bears	AHL	58	3	22	25	48	....	....	....	....	....	....	....	....	....	....	....	8	0	2	2	2	....	....	....
	Slovakia	Olympics	4	0	0	0	2	....	....	....	....	....	....	....	....	....	....	....	....	....	....	....	....	....	....	....
	NHL Totals		7	0	0	0	2	0	0	0	6	0.0		0	0.0	15	4	14:46	....	....	....	....	....	....	....	....

Signed as a free agent by **St. Louis**, April 26, 1999. • Missed majority of 1999-2000 season recovering from knee injury suffered in practice, October, 1999. Signed as a free agent by **Colorado**, August 11, 2001. Signed as a free agent by **Vancouver**, July 10, 2002.

ODELEIN, Lyle (OH-duh-LIGHN, LIGHL) CHI.

Defense. Shoots right. 5'11", 210 lbs. Born, Quill Lake, Sask., July 21, 1968. Montreal's 8th choice, 141st overall, in 1986 Entry Draft.

Season	Club	League	GP	G	A	Pts	PIM	PP	SH	GW	S	%	+/-	TF	F%	H	SB	Min	GP	G	A	Pts	PIM	PP	SH	GW
1984-85	Regina Pat Cdns.	SMHL	26	12	13	25	30	....	....	....	....	....	....	....	....	....	....	....	....	....	....	....	....	....	....	....
1985-86	Moose Jaw	WHL	67	9	37	46	117	....	....	....	....	....	....	....	....	....	....	....	13	1	6	7	34	....	....	....
1986-87	Moose Jaw	WHL	59	9	50	59	70	....	....	....	....	....	....	....	....	....	....	....	9	2	5	7	26	....	....	....
1987-88	Moose Jaw	WHL	63	15	43	58	166	....	....	....	....	....	....	....	....	....	....	....	....	....	....	....	....	....	....	....
1988-89	Sherbrooke	AHL	33	3	4	7	120	....	....	....	....	....	....	....	....	....	....	....	3	0	2	2	5	....	....	....
	Peoria Rivermen	IHL	36	2	8	10	116	....	....	....	....	....	....	....	....	....	....	....	....	....	....	....	....	....	....	....
1989-90	**Montreal**	**NHL**	8	0	2	2	33	0	0	0	1	0.0	−1	....	....	....	....	....	....	....	....	....	....	....	....	....
	Sherbrooke	AHL	68	7	24	31	265	....	....	....	....	....	....	....	....	....	....	....	12	6	5	11	79	....	....	....
1990-91	**Montreal**	**NHL**	52	0	2	2	259	0	0	0	25	0.0	7	....	....	....	....	....	12	0	0	0	54	0	0	0
1991-92	**Montreal**	**NHL**	71	1	7	8	212	0	0	0	43	2.3	15	....	....	....	....	....	7	0	0	0	11	0	0	0
1992-93♦	**Montreal**	**NHL**	83	2	14	16	205	0	0	0	79	2.5	35	....	....	....	....	....	20	1	5	6	30	0	0	0
1993-94	**Montreal**	**NHL**	79	11	29	40	276	6	0	2	116	9.5	8	....	....	....	....	....	7	0	0	0	17	0	0	0
1994-95	**Montreal**	**NHL**	48	3	7	10	152	0	0	0	74	4.1	−13	....	....	....	....	....	....	....	....	....	....	....	....	....
1995-96	**Montreal**	**NHL**	79	3	14	17	230	0	1	0	74	4.1	8	....	....	....	....	....	6	1	1	2	6	0	1	0
1996-97	**New Jersey**	**NHL**	79	3	13	16	110	1	0	2	93	3.2	16	....	....	....	....	....	10	2	2	4	19	1	0	0
1997-98	**New Jersey**	**NHL**	79	4	19	23	171	1	0	0	76	5.3	11	....	....	....	....	....	6	1	1	2	21	1	0	1
1998-99	**New Jersey**	**NHL**	70	5	26	31	114	1	0	0	101	5.0	6	0	0.0	51	71	19:53	7	0	3	3	10	0	0	0
99-2000	**New Jersey**	**NHL**	57	1	15	16	104	0	0	1	59	1.7	−10	0	0.0	44	63	16:42	....	....	....	....	....	....	....	....
	Phoenix	**NHL**	16	1	7	8	19	1	0	0	30	3.3	1	0	0.0	13	28	21:45	5	0	0	0	16	0	0	0
2000-01	**Columbus**	**NHL**	81	3	14	17	118	1	0	0	104	2.9	−16	0	0.0	80	142	21:31	....	....	....	....	....	....	....	....
2001-02	**Columbus**	**NHL**	65	2	14	16	89	0	0	0	76	2.6	−28	0	0.0	44	113	22:12	....	....	....	....	....	....	....	....
	Chicago	**NHL**	12	0	2	2	4	0	0	0	10	0.0	0	0	0.0	14	18	24:23	4	0	1	1	25	0	0	0
	NHL Totals		879	39	185	224	2096	11	1	5	961	4.1		0	0.0	246	435	20:30	84	5	13	18	209	2	1	1

Traded to **New Jersey** by **Montreal** for Stephane Richer, August 22, 1996. Traded to **Phoenix** by **New Jersey** for Deron Quint and Phoenix's 3rd round choice (later traded back to Phoenix - Phoenix selected Beat Forster) in 2001 Entry Draft, March 7, 2000. Selected by **Columbus** from **Phoenix** in Expansion Draft, June 23, 2000. Traded to **Chicago** by **Columbus** for Jaroslav Spacek and Chicago's 2nd round choice in 2003 Entry Draft, March 19, 2002.

ODGERS, Jeff (AWD-juhrs, JEHF) ATL.

Right wing. Shoots right. 5'11", 200 lbs. Born, Spy Hill, Sask., May 31, 1969.

Season	Club	League	GP	G	A	Pts	PIM	PP	SH	GW	S	%	+/-	TF	F%	H	SB	Min	GP	G	A	Pts	PIM	PP	SH	GW
1985-86	Sask. Blazers	SMHL	36	27	29	56	74	....	....	....	....	....	....	....	....	....	....	....	....	....	....	....	....	....	....	....
1986-87	Brandon	WHL	70	7	14	21	150	....	....	....	....	....	....	....	....	....	....	....	....	....	....	....	....	....	....	....
1987-88	Brandon	WHL	70	17	18	35	202	....	....	....	....	....	....	....	....	....	....	....	4	1	1	2	14	....	....	....
1988-89	Brandon	WHL	71	31	29	60	277	....	....	....	....	....	....	....	....	....	....	....	....	....	....	....	....	....	....	....
1989-90	Brandon	WHL	64	37	28	65	209	....	....	....	....	....	....	....	....	....	....	....	....	....	....	....	....	....	....	....
1990-91	Kansas City	IHL	77	12	19	31	318	....	....	....	....	....	....	....	....	....	....	....	....	....	....	....	....	....	....	....
1991-92	**San Jose**	**NHL**	61	7	4	11	217	0	0	0	64	10.9	−21	....	....	....	....	....	....	....	....	....	....	....	....	....
	Kansas City	IHL	12	2	2	4	56	....	....	....	....	....	....	....	....	....	....	....	4	2	1	3	0	....	....	....
1992-93	**San Jose**	**NHL**	66	12	15	27	253	6	0	0	100	12.0	−26	....	....	....	....	....	....	....	....	....	....	....	....	....
1993-94	**San Jose**	**NHL**	81	13	8	21	222	7	0	0	73	17.8	−13	....	....	....	....	....	11	0	0	0	11	0	0	0
1994-95	**San Jose**	**NHL**	48	4	3	7	117	0	0	1	47	8.5	−8	....	....	....	....	....	11	1	1	2	23	0	0	0
1995-96	**San Jose**	**NHL**	78	12	4	16	192	0	0	1	84	14.3	−4	....	....	....	....	....	....	....	....	....	....	....	....	....
1996-97	**Boston**	**NHL**	80	7	8	15	197	1	0	1	84	8.3	−15	....	....	....	....	....	....	....	....	....	....	....	....	....
1997-98	Providence	AHL	4	0	0	0	31	....	....	....	....	....	....	....	....	....	....	....	....	....	....	....	....	....	....	....
	Colorado	**NHL**	68	5	8	13	213	0	0	0	47	10.6	5	....	....	....	....	....	6	0	0	0	25	0	0	0
1998-99	**Colorado**	**NHL**	75	2	3	5	259	1	0	0	39	5.1	−3	8	37.5	56	6	5:07	15	1	0	1	14	0	0	1
99-2000	**Colorado**	**NHL**	62	1	2	3	162	0	0	1	29	3.4	−7	2	50.0	64	8	5:23	4	0	0	0	0	0	0	0
2000-01	**Atlanta**	**NHL**	82	6	7	13	226	0	0	1	67	9.0	−8	2	50.0	109	31	8:39	....	....	....	....	....	....	....	....
2001-02	**Atlanta**	**NHL**	46	4	4	8	135	0	0	1	34	11.8	−3	5	20.0	55	2	8:37	....	....	....	....	....	....	....	....
	NHL Totals		747	73	66	139	2193	15	0	6	668	10.9		17	35.3	284	47	6:53	47	2	1	3	73	0	0	1

Signed as a free agent by **San Jose**, September 3, 1991. Traded to **Boston** by **San Jose** with Pittsburgh's 5th round choice (previously acquired, Boston selected Elias Abrahamsson) in 1996 Entry Draft for Al Iafrate, June 21, 1996. Signed as a free agent by **Colorado**, October 24, 1997. Selected by **Minnesota** from **Colorado** in Expansion Draft, June 23, 2000. Claimed by **Atlanta** from **Minnesota** in Waiver Draft, September 29, 2000.

ODJICK, Gino (OH-jihk, GEE-noh) MTL.

Left wing. Shoots left. 6'3", 224 lbs. Born, Maniwaki, Que., September 7, 1970. Vancouver's 5th choice, 86th overall, in 1990 Entry Draft.

Season	Club	League	GP	G	A	Pts	PIM	PP	SH	GW	S	%	+/-	TF	F%	H	SB	Min	GP	G	A	Pts	PIM	PP	SH	GW
1987-88	Hawkesbury	OCJHL	40	2	4	6	167	....	....	....	....	....	....	....	....	....	....	....	....	....	....	....	....	....	....	....
1988-89	Laval Titan	QMJHL	50	9	15	24	278	....	....	....	....	....	....	....	....	....	....	....	16	0	9	9	129	....	....	....
1989-90	Laval Titan	QMJHL	51	12	26	38	280	....	....	....	....	....	....	....	....	....	....	....	13	6	5	11	110	....	....	....
1990-91	**Vancouver**	**NHL**	45	7	1	8	296	0	0	0	39	17.9	−6	....	....	....	....	....	6	0	0	0	18	0	0	0
	Milwaukee	IHL	17	7	3	10	102	....	....	....	....	....	....	....	....	....	....	....	....	....	....	....	....	....	....	....
1991-92	**Vancouver**	**NHL**	65	4	6	10	348	0	0	0	68	5.9	−1	....	....	....	....	....	4	0	0	0	6	0	0	0
1992-93	**Vancouver**	**NHL**	75	4	13	17	370	0	0	1	79	5.1	3	....	....	....	....	....	1	0	0	0	0	0	0	0
1993-94	**Vancouver**	**NHL**	76	16	13	29	271	4	0	5	121	13.2	13	....	....	....	....	....	10	0	0	0	18	0	0	0
1994-95	**Vancouver**	**NHL**	23	4	5	9	109	0	0	0	35	11.4	−3	....	....	....	....	....	5	0	0	0	47	0	0	0
1995-96	**Vancouver**	**NHL**	55	3	4	7	181	0	0	0	59	5.1	−16	....	....	....	....	....	6	3	1	4	6	0	0	2
1996-97	**Vancouver**	**NHL**	70	5	8	13	*371	1	0	0	85	5.9	−5	....	....	....	....	....	....	....	....	....	....	....	....	....
1997-98	**Vancouver**	**NHL**	35	3	2	5	181	0	0	1	36	8.3	−3	....	....	....	....	....	....	....	....	....	....	....	....	....
	NY Islanders	**NHL**	13	0	0	0	31	0	0	0	16	0.0	1	....	....	....	....	....	....	....	....	....	....	....	....	....
1998-99	**NY Islanders**	**NHL**	23	4	3	7	133	1	0	2	28	14.3	−2	2	0.0	20	2	9:51	....	....	....	....	....	....	....	....
99-2000	**NY Islanders**	**NHL**	46	5	10	15	90	0	0	3	91	5.5	−7	3	33.3	56	8	12:10	....	....	....	....	....	....	....	....
	Philadelphia	**NHL**	13	3	1	4	10	0	0	1	24	12.5	2	0	0.0	11	2	9:03	....	....	....	....	....	....	....	....
2000-01	**Philadelphia**	**NHL**	17	1	3	4	28	0	0	0	21	4.8	0	3	0.0	12	3	8:33	....	....	....	....	....	....	....	....
	Montreal	**NHL**	13	1	0	1	44	0	0	0	11	9.1	0	3	66.7	13	3	7:04	....	....	....	....	....	....	....	....
2001-02	**Montreal**	**NHL**	36	4	4	8	104	0	0	0	40	10.0	3	6	33.3	23	8	7:37	12	1	0	1	47	0	0	0
	Quebec	AHL	13	2	1	3	40	....	....	....	....	....	....	....	....	....	....	....	....	....	....	....	....	....	....	....
	NHL Totals		605	64	73	137	2567	6	0	13	753	8.5		17	29.4	135	26	9:34	44	4	1	5	142	0	0	2

Traded to **NY Islanders** by **Vancouver** for Jason Strudwick, March 23, 1998. Traded to **Philadelphia** by **NY Islanders** for Mikael Andersson and Carolina's 5th round choice (previously acquired, NY Islanders selected Kristofer Ottosson) in 2000 Entry Draft, February 15, 2000. Traded to **Montreal** by **Philadelphia** for P.J. Stock and Montreal's 6th round choice (Dennis Seidenberg) in 2001 Entry Draft, December 7, 2000. • Missed majority of 2000-01 season recovering from wrist injury suffered in game vs. Carolina, January 16, 2001.

O'DONNELL, Sean (oh-DOHN-ehl, SHAWN) BOS.

Defense. Shoots left. 6'3", 230 lbs. Born, Ottawa, Ont., October 13, 1971. Buffalo's 6th choice, 123rd overall, in 1991 Entry Draft.

Season	Club	League	GP	G	A	Pts	PIM	PP	SH	GW	S	%	+/-	TF	F%	H	SB	Min	GP	G	A	Pts	PIM	PP	SH	GW
1987-88	Kanata Valley	OCJHL	54	4	25	29	96	....	....	....	....	....	....	....	....	....	....	....	....	....	....	....	....	....	....	....
1988-89	Sudbury Wolves	OHL	56	1	9	10	49	....	....	....	....	....	....	....	....	....	....	....	....	....	....	....	....	....	....	....
1989-90	Sudbury Wolves	OHL	64	7	19	26	84	....	....	....	....	....	....	....	....	....	....	....	7	1	2	3	8	....	....	....
1990-91	Sudbury Wolves	OHL	66	8	23	31	114	....	....	....	....	....	....	....	....	....	....	....	5	1	4	5	10	....	....	....
1991-92	Rochester	AHL	73	4	9	13	193	....	....	....	....	....	....	....	....	....	....	....	16	1	2	3	21	....	....	....
1992-93	Rochester	AHL	74	3	18	21	203	....	....	....	....	....	....	....	....	....	....	....	17	1	6	7	38	....	....	....
1993-94	Rochester	AHL	64	2	10	12	242	....	....	....	....	....	....	....	....	....	....	....	4	0	1	1	21	....	....	....
1994-95	Phoenix	IHL	61	2	18	20	132	....	....	....	....	....	....	....	....	....	....	....	9	0	1	1	21	....	....	....
	Los Angeles	**NHL**	15	0	2	2	49	0	0	0	12	0.0	−2	....	....	....	....	....	....	....	....	....	....	....	....	....

			Regular Season																Playoffs							
Season	Club	League	GP	G	A	Pts	PIM	PP	SH	GW	S	%	+/-	TF	F%	H	SB	Min	GP	G	A	Pts	PIM	PP	SH	GW
1995-96	**Los Angeles**	**NHL**	**71**	**2**	**5**	**7**	**127**	**0**	**0**	**0**	**65**	**3.1**	**3**	....	....	....	....	....	....	....	....	....	....	....	....	....
1996-97	**Los Angeles**	**NHL**	**55**	**5**	**12**	**17**	**144**	**2**	**0**	**0**	**68**	**7.4**	**–13**	....	....	....	....	....	....	....	....	....	....	....	....	....
1997-98	**Los Angeles**	**NHL**	**80**	**2**	**15**	**17**	**179**	**0**	**0**	**1**	**71**	**2.8**	**7**	....	....	....	....	....	**4**	**1**	**0**	**1**	**36**	**0**	**0**	**0**
1998-99	**Los Angeles**	**NHL**	**80**	**1**	**13**	**14**	**186**	**0**	**0**	**0**	**64**	**1.6**	**1**	**0**	**0.0**	**133**	**71**	**19:10**	....	....	....	....	....	....	....	....
99-2000	**Los Angeles**	**NHL**	**80**	**2**	**12**	**14**	**114**	**0**	**0**	**1**	**51**	**3.9**	**4**	**0**	**0.0**	**154**	**80**	**17:41**	**4**	**1**	**0**	**1**	**4**	**0**	**0**	**0**
2000-01	**Minnesota**	**NHL**	**63**	**4**	**12**	**16**	**128**	**1**	**0**	**2**	**58**	**6.9**	**–2**	**12**	**50.0**	**81**	**72**	**23:00**	....	....	....	....	....	....	....	....
	New Jersey	**NHL**	**17**	**0**	**1**	**1**	**33**	**0**	**0**	**0**	**9**	**0.0**	**2**	**0**	**0.0**	**14**	**13**	**16:27**	**23**	**1**	**2**	**3**	**41**	**0**	**0**	**0**
2001-02	**Boston**	**NHL**	**80**	**3**	**22**	**25**	**89**	**1**	**0**	**2**	**112**	**2.7**	**27**	**0**	**0.0**	**102**	**81**	**24:50**	**6**	**0**	**2**	**2**	**4**	**0**	**0**	**0**
	NHL Totals		**541**	**19**	**94**	**113**	**1049**	**4**	**0**	**6**	**510**	**3.7**		**12**	**50.0**	**484**	**317**	**20:49**	**37**	**3**	**4**	**7**	**85**	**0**	**0**	**0**

Traded to **LA Kings** by **Buffalo** for Doug Houda, July 26, 1994. Selected by **Minnesota** from **LA Kings** in Expansion Draft, June 23, 2000. Traded to **New Jersey** by **Minnesota** for Willie Mitchell, March 4, 2001. Signed as a free agent by **Boston**, July 2, 2001.

OHLUND, Mattias

(OH-luhnd, MAT-tee-ahs) **VAN.**

Defense. Shoots left. 6'2", 220 lbs. Born, Pitea, Sweden, September 9, 1976. Vancouver's 1st choice, 13th overall, in 1994 Entry Draft.

Season	Club	League	GP	G	A	Pts	PIM	PP	SH	GW	S	%	+/-	TF	F%	H	SB	Min	GP	G	A	Pts	PIM	PP	SH	GW
1992-93	Pitea HC	Swede-2	22	0	6	6	16	....	....	....	....	....	....	....	....	....	....	....	....	....	....	....	....	....	....	....
1993-94	Pitea HC	Swede-2	28	7	10	17	62	....	....	....	....	....	....	....	....	....	....	....	....	....	....	....	....	....	....	....
1994-95	Lulea HF	Sweden	34	6	10	16	34	....	....	....	....	....	....	....	....	....	....	....	9	4	0	4	16	....	....	....
1995-96	Lulea HF	Sweden	38	4	10	14	26	....	....	....	....	....	....	....	....	....	....	....	13	1	0	1	47	....	....	....
1996-97	Lulea HF	Sweden	47	7	9	16	38	....	....	....	....	....	....	....	....	....	....	....	10	1	2	3	8	....	....	....
	Lulea HF	EuroHL	6	0	3	3	0	....	....	....	....	....	....	....	....	....	....	....	....	....	....	....	....	....	....	....
1997-98	**Vancouver**	**NHL**	**77**	**7**	**23**	**30**	**76**	**1**	**0**	**0**	**172**	**4.1**	**3**	....	....	....	....	....	....	....	....	....	....	....	....	....
	Sweden	Olympics	4	0	1	1	4	....	....	....	....	....	....	....	....	....	....	....	....	....	....	....	....	....	....	....
1998-99	**Vancouver**	**NHL**	**74**	**9**	**26**	**35**	**83**	**2**	**1**	**1**	**129**	**7.0**	**–19**	**0**	**0.0**	**133**	**154**	**26:04**	....	....	....	....	....	....	....	....
99-2000	**Vancouver**	**NHL**	**42**	**4**	**16**	**20**	**24**	**2**	**1**	**1**	**63**	**6.3**	**6**	**0**	**0.0**	**91**	**85**	**27:41**	....	....	....	....	....	....	....	....
2000-01	**Vancouver**	**NHL**	**65**	**8**	**20**	**28**	**46**	**1**	**1**	**4**	**136**	**5.9**	**–16**	**0**	**0.0**	**125**	**146**	**25:00**	**4**	**1**	**3**	**4**	**6**	**1**	**0**	**0**
2001-02	**Vancouver**	**NHL**	**81**	**10**	**26**	**36**	**56**	**4**	**1**	**3**	**193**	**5.2**	**16**	**0**	**0.0**	**164**	**118**	**25:17**	**6**	**1**	**1**	**2**	**6**	**0**	**0**	**0**
	Sweden	Olympics	4	0	2	2	2	....	....	....	....	....	....	....	....	....	....	....	....	....	....	....	....	....	....	....
	NHL Totals		**339**	**38**	**111**	**149**	**285**	**10**	**4**	**9**	**693**	**5.5**		**0**	**0.0**	**513**	**503**	**25:49**	**10**	**2**	**4**	**6**	**12**	**1**	**0**	**0**

NHL All-Rookie Team (1998) • Played in NHL All-Star Game (1999)

OLAUSSON, Fredrik

(OHL-ah-suhn, FREHD-rihk) **ANA.**

Defense. Shoots right. 6'2", 198 lbs. Born, Dadesjo, Sweden, October 5, 1966. Winnipeg's 4th choice, 81st overall, in 1985 Entry Draft.

Season	Club	League	GP	G	A	Pts	PIM	PP	SH	GW	S	%	+/-	TF	F%	H	SB	Min	GP	G	A	Pts	PIM	PP	SH	GW
1982-83	Nybro SK	Swede-2	31	4	4	8	12	....	....	....	....	....	....	....	....	....	....	....	....	....	....	....	....	....	....	....
1983-84	Nybro SK	Swede-2	28	8	14	22	32	....	....	....	....	....	....	....	....	....	....	....	....	....	....	....	....	....	....	....
1984-85	Farjestad	Sweden	29	5	12	17	22	....	....	....	....	....	....	....	....	....	....	....	3	1	0	1	0	....	....	....
1985-86	Farjestad	Sweden	33	4	12	16	22	....	....	....	....	....	....	....	....	....	....	....	8	3	2	5	6	....	....	....
1986-87	**Winnipeg**	**NHL**	**72**	**7**	**29**	**36**	**24**	**1**	**0**	**2**	**119**	**5.9**	**–3**	....	....	....	....	....	**10**	**2**	**3**	**5**	**4**	**1**	**0**	**0**
1987-88	**Winnipeg**	**NHL**	**38**	**5**	**10**	**15**	**18**	**2**	**0**	**2**	**65**	**7.7**	**3**	....	....	....	....	....	**5**	**1**	**1**	**2**	**0**	**0**	**0**	**0**
1988-89	**Winnipeg**	**NHL**	**75**	**15**	**47**	**62**	**32**	**4**	**0**	**1**	**178**	**8.4**	**6**	....	....	....	....	....	....	....	....	....	....	....	....	....
1989-90	**Winnipeg**	**NHL**	**77**	**9**	**46**	**55**	**32**	**3**	**0**	**0**	**147**	**6.1**	**–1**	....	....	....	....	....	**7**	**0**	**2**	**2**	**2**	**0**	**0**	**0**
1990-91	**Winnipeg**	**NHL**	**71**	**12**	**29**	**41**	**24**	**5**	**0**	**0**	**168**	**7.1**	**–22**	....	....	....	....	....	....	....	....	....	....	....	....	....
1991-92	**Winnipeg**	**NHL**	**77**	**20**	**42**	**62**	**34**	**13**	**1**	**2**	**227**	**8.8**	**–31**	....	....	....	....	....	**7**	**1**	**5**	**6**	**4**	**1**	**0**	**0**
1992-93	**Winnipeg**	**NHL**	**68**	**16**	**41**	**57**	**22**	**11**	**0**	**3**	**165**	**9.7**	**–4**	....	....	....	....	....	**6**	**0**	**2**	**2**	**2**	**0**	**0**	**0**
1993-94	**Winnipeg**	**NHL**	**18**	**2**	**5**	**7**	**10**	**1**	**0**	**0**	**41**	**4.9**	**–3**	....	....	....	....	....	....	....	....	....	....	....	....	....
	Edmonton	**NHL**	**55**	**9**	**19**	**28**	**20**	**6**	**0**	**1**	**85**	**10.6**	**–4**	....	....	....	....	....	....	....	....	....	....	....	....	....
1994-95	EC Ehrwald	Austria	10	4	3	7	8	....	....	....	....	....	....	....	....	....	....	....	....	....	....	....	....	....	....	....
	Edmonton	**NHL**	**33**	**0**	**10**	**10**	**20**	**0**	**0**	**0**	**52**	**0.0**	**–4**	....	....	....	....	....	....	....	....	....	....	....	....	....
1995-96	**Edmonton**	**NHL**	**20**	**0**	**6**	**6**	**14**	**0**	**0**	**0**	**20**	**0.0**	**–14**	....	....	....	....	....	....	....	....	....	....	....	....	....
	Anaheim	**NHL**	**36**	**2**	**16**	**18**	**24**	**1**	**0**	**0**	**63**	**3.2**	**7**	....	....	....	....	....	....	....	....	....	....	....	....	....
1996-97	**Anaheim**	**NHL**	**20**	**2**	**9**	**11**	**8**	**1**	**0**	**0**	**35**	**5.7**	**–5**	....	....	....	....	....	....	....	....	....	....	....	....	....
	Pittsburgh	**NHL**	**51**	**7**	**20**	**27**	**24**	**2**	**0**	**3**	**75**	**9.3**	**21**	....	....	....	....	....	**4**	**0**	**1**	**1**	**0**	**0**	**0**	**0**
1997-98	**Pittsburgh**	**NHL**	**76**	**6**	**27**	**33**	**42**	**2**	**0**	**1**	**89**	**6.7**	**13**	....	....	....	....	....	**6**	**0**	**3**	**3**	**2**	**0**	**0**	**0**
1998-99	**Anaheim**	**NHL**	**74**	**16**	**40**	**56**	**30**	**10**	**0**	**2**	**121**	**13.2**	**17**	**0**	**0.0**	**41**	**56**	**19:47**	**4**	**0**	**2**	**2**	**4**	**0**	**0**	**0**
99-2000	**Anaheim**	**NHL**	**70**	**15**	**19**	**34**	**28**	**8**	**0**	**1**	**120**	**12.5**	**–13**	**0**	**0.0**	**62**	**45**	**20:01**	....	....	....	....	....	....	....	....
2000-01	SC Bern	Swiss	43	12	15	27	28	....	....	....	....	....	....	....	....	....	....	....	4	1	4	5	0	....	....	....
2001-02♦	**Detroit**	**NHL**	**47**	**2**	**13**	**15**	**22**	**0**	**1**	**1**	**61**	**3.3**	**9**	**0**	**0.0**	**24**	**35**	**16:37**	**21**	**2**	**4**	**6**	**10**	**1**	**0**	**1**
	Sweden	Olympics	4	0	0	0	2	....	....	....	....	....	....	....	....	....	....	....	....	....	....	....	....	....	....	....
	NHL Totals		**978**	**145**	**428**	**573**	**428**	**70**	**2**	**19**	**1831**	**7.9**		**0**	**0.0**	**127**	**136**	**19:05**	**70**	**6**	**23**	**29**	**28**	**3**	**0**	**1**

Traded to **Edmonton** by **Winnipeg** with Winnipeg's 7th round choice (Curtis Sheptak) in 1994 Entry Draft for Edmonton's 3rd round choice (Tavis Hansen) in 1994 Entry Draft, December 6, 1993. Claimed on waivers by **Anaheim** from **Edmonton**, January 16, 1996. Traded to **Pittsburgh** by **Anaheim** with Alex Hicks for Shawn Antoski and Dmitri Mironov, November 19, 1996. Signed as a free agent by **Anaheim**, August 28, 1998. Signed as a free agent by **Detroit**, May 24, 2001. Signed as a free agent by **Anaheim**, July 12, 2002.

OLIVER, David

(AWL-ih-vuhr, DAY-vihd) **DAL.**

Right wing. Shoots right. 6', 190 lbs. Born, Sechelt, B.C., April 17, 1971. Edmonton's 7th choice, 144th overall, in 1991 Entry Draft.

Season	Club	League	GP	G	A	Pts	PIM	PP	SH	GW	S	%	+/-	TF	F%	H	SB	Min	GP	G	A	Pts	PIM	PP	SH	GW
1988-89	Vernon Lakers	BCJHL	58	41	38	79	38	....	....	....	....	....	....	....	....	....	....	....	....	....	....	....	....	....	....	....
1989-90	Vernon Lakers	BCJHL	58	51	48	99	22	....	....	....	....	....	....	....	....	....	....	....	....	....	....	....	....	....	....	....
1990-91	U. of Michigan	CCHA	27	13	11	24	34	....	....	....	....	....	....	....	....	....	....	....	....	....	....	....	....	....	....	....
1991-92	U. of Michigan	CCHA	44	31	27	58	32	....	....	....	....	....	....	....	....	....	....	....	....	....	....	....	....	....	....	....
1992-93	U. of Michigan	CCHA	40	35	20	55	18	....	....	....	....	....	....	....	....	....	....	....	....	....	....	....	....	....	....	....
1993-94	U. of Michigan	CCHA	41	28	40	68	16	....	....	....	....	....	....	....	....	....	....	....	....	....	....	....	....	....	....	....
1994-95	Cape Breton	AHL	32	11	18	29	8	....	....	....	....	....	....	....	....	....	....	....	....	....	....	....	....	....	....	....
	Edmonton	**NHL**	**44**	**16**	**14**	**30**	**20**	**10**	**0**	**0**	**79**	**20.3**	**–11**	....	....	....	....	....	....	....	....	....	....	....	....	....
1995-96	**Edmonton**	**NHL**	**80**	**20**	**19**	**39**	**34**	**14**	**0**	**0**	**131**	**15.3**	**–22**	....	....	....	....	....	....	....	....	....	....	....	....	....
1996-97	**Edmonton**	**NHL**	**17**	**1**	**2**	**3**	**4**	**0**	**0**	**0**	**22**	**4.5**	**–8**	....	....	....	....	....	....	....	....	....	....	....	....	....
	NY Rangers	**NHL**	**14**	**2**	**1**	**3**	**4**	**0**	**0**	**0**	**13**	**15.4**	**3**	....	....	....	....	....	**3**	**0**	**0**	**0**	**0**	**0**	**0**	**0**
1997-98	Houston Aeros	IHL	78	38	27	65	60	....	....	....	....	....	....	....	....	....	....	....	4	3	0	3	4	....	....	....
1998-99	**Ottawa**	**NHL**	**17**	**2**	**5**	**7**	**4**	**0**	**0**	**0**	**18**	**11.1**	**1**	**3**	**33.3**	**8**	**2**	**10:34**	....	....	....	....	....	....	....	....
	Houston Aeros	IHL	37	18	17	35	30	....	....	....	....	....	....	....	....	....	....	....	19	10	6	16	22	....	....	....
99-2000	**Phoenix**	**NHL**	**9**	**1**	**0**	**1**	**2**	**1**	**0**	**0**	**6**	**16.7**	**0**	**0**	**0.0**	**11**	**1**	**7:38**	....	....	....	....	....	....	....	....
	Houston Aeros	IHL	45	16	11	27	40	....	....	....	....	....	....	....	....	....	....	....	11	3	4	7	8	....	....	....
2000-01	**Ottawa**	**NHL**	**7**	**0**	**0**	**0**	**2**	**0**	**0**	**0**	**2**	**0.0**	**0**	**0**	**0.0**	**8**	**0**	**5:37**	....	....	....	....	....	....	....	....
	Grand Rapids	IHL	51	14	17	31	35	....	....	....	....	....	....	....	....	....	....	....	10	6	2	8	8	....	....	....
2001-02	Munchen Barons	Germany	59	20	14	34	30	....	....	....	....	....	....	....	....	....	....	....	9	2	2	4	6	....	....	....
	NHL Totals		**188**	**42**	**41**	**83**	**70**	**25**	**0**	**0**	**271**	**15.5**		**3**	**33.3**	**27**	**3**	**8:43**	**3**	**0**	**0**	**0**	**0**	**0**	**0**	**0**

CCHA Second All-Star Team (1993) • CCHA First All-Star Team (1994) • CCHA Player of the Year (1994) • NCAA West First All-American Team (1994)

Claimed on waivers by **NY Rangers** from **Edmonton**, February 21, 1997. Signed as a free agent by **Ottawa**, July 2, 1998. Signed as a free agent by **Phoenix**, July 20, 1999. Signed as a free agent by **Ottawa**, August 2, 2000. Signed as a free agent by **Dallas**, July 30, 2002.

OLIWA, Krzysztof

(oh-LEE-vuh, KHRIH-stahf) **NYR**

Left wing. Shoots left. 6'5", 235 lbs. Born, Tychy, Poland, April 12, 1973. New Jersey's 4th choice, 65th overall, in 1993 Entry Draft.

Season	Club	League	GP	G	A	Pts	PIM	PP	SH	GW	S	%	+/-	TF	F%	H	SB	Min	GP	G	A	Pts	PIM	PP	SH	GW
1990-91	GKS Katowski Jr.	Pol.-Jr.	5	4	4	8	10	....	....	....	....	....	....	....	....	....	....	....	....	....	....	....	....	....	....	....
1991-92	GKS Tychy	Poland	10	3	7	10	6	....	....	....	....	....	....	....	....	....	....	....	....	....	....	....	....	....	....	....
1992-93	Welland Cougars	OJHL-B	30	13	21	34	127	....	....	....	....	....	....	....	....	....	....	....	....	....	....	....	....	....	....	....
1993-94	Albany	AHL	33	2	4	6	151	....	....	....	....	....	....	....	....	....	....	....	....	....	....	....	....	....	....	....
	Raleigh IceCaps	ECHL	15	0	2	2	65	....	....	....	....	....	....	....	....	....	....	....	9	0	0	0	35	....	....	....
1994-95	Albany	AHL	20	1	1	2	77	....	....	....	....	....	....	....	....	....	....	....	....	....	....	....	....	....	....	....
	Saint John	AHL	14	1	4	5	79	....	....	....	....	....	....	....	....	....	....	....	....	....	....	....	....	....	....	....
	Raleigh IceCaps	ECHL	5	0	2	2	32	....	....	....	....	....	....	....	....	....	....	....	....	....	....	....	....	....	....	....
	Detroit Vipers	IHL	4	0	1	1	24	....	....	....	....	....	....	....	....	....	....	....	....	....	....	....	....	....	....	....
1995-96	Albany	AHL	51	5	11	16	217	....	....	....	....	....	....	....	....	....	....	....	....	....	....	....	....	....	....	....
	Raleigh IceCaps	ECHL	9	1	0	1	53	....	....	....	....	....	....	....	....	....	....	....	....	....	....	....	....	....	....	....
1996-97	**New Jersey**	**NHL**	**1**	**0**	**0**	**0**	**5**	**0**	**0**	**0**	**0**	**0.0**	**–1**	....	....	....	....	....	....	....	....	....	....	....	....	....
	Albany	AHL	60	13	14	27	322	....	....	....	....	....	....	....	....	....	....	....	15	7	1	8	49	....	....	....
1997-98	**New Jersey**	**NHL**	**73**	**2**	**3**	**5**	**295**	**0**	**0**	**2**	**53**	**3.8**	**3**	....	....	....	....	....	**6**	**0**	**0**	**0**	**23**	**0**	**0**	**0**
1998-99	**New Jersey**	**NHL**	**64**	**5**	**7**	**12**	**240**	**0**	**0**	**1**	**59**	**8.5**	**4**	**1**	**0.0**	**117**	**9**	**7:02**	**1**	**0**	**0**	**0**	**2**	**0**	**0**	**0**
99-2000♦	**New Jersey**	**NHL**	**69**	**6**	**10**	**16**	**184**	**1**	**0**	**2**	**61**	**9.8**	**–2**	**3**	**66.7**	**84**	**4**	**6:45**	....	....	....	....	....	....	....	....

Season	Club	League	GP	G	A	Pts	PIM	PP	SH	GW	S	%	+/-	TF	F%	H	SB	Min	GP	G	A	Pts	PIM	PP	SH	GW
			Regular Season																Playoffs							
2000-01	**Columbus**	**NHL**	**10**	**0**	**2**	**2**	**34**	**0**	**0**	**0**	**5**	**0.0**	**1**	**0**	**0.0**	**10**	**0**	**5:17**	....	....	....	....	....	....	....	....
	Pittsburgh	**NHL**	**26**	**1**	**2**	**3**	**131**	**0**	**0**	**0**	**17**	**5.9**	**–4**	**1**	**100.0**	**13**	**4**	**4:58**	**5**	**0**	**0**	**0**	**16**	**0**	**0**	**0**
2001-02	**Pittsburgh**	**NHL**	**57**	**0**	**2**	**2**	**150**	**0**	**0**	**0**	**31**	**0.0**	**–5**	**0**	**0.0**	**37**	**2**	**5:35**	....	....	....	....	....	....	....	....
	NHL Totals		**300**	**14**	**26**	**40**	**1039**	**1**	**0**	**5**	**226**	**6.2**		**5**	**60.0**	**261**	**19**	**6:16**	**12**	**0**	**0**	**0**	**41**	**0**	**0**	**0**

• Born Krzystof Graboski

Traded to **Columbus** by **New Jersey** with future considerations (Deron Quint, June 23, 2000) for Columbus' 3rd round choice (Brandon Nolan) in 2001 Entry Draft and future considerations (Turner Stevenson, June 23, 2000), June 12, 2000. • Missed majority of 2000-2001 season recovering from arm injury originally suffered in game vs. Detroit, October 28, 2000. Traded to **Pittsburgh** by **Columbus** for San Jose's 3rd round choice (previously acquired, Columbus selected Aaron Johnson) in 2001 Entry Draft, January 14, 2001. Traded to **NY Rangers** by **Pittsburgh** for future considerations, June 23, 2002.

OLVESTAD, Jimmie
(OHL-vuh-stahd, JIHM-mee) **T.B.**

Left wing. Shoots left. 6'1", 189 lbs. Born, Stockholm, Sweden, February 16, 1980. Tampa Bay's 4th choice, 88th overall, in 1999 Entry Draft.

Season	Club	League	GP	G	A	Pts	PIM	PP	SH	GW	S	%	+/-	TF	F%	H	SB	Min	GP	G	A	Pts	PIM	PP	SH	GW
1996-97	Huddinge IK Jr.	Swede-Jr.	40	15	16	31	....	....	....	....	....	....	....	....	....	....	....	....	....	....	....	....	....	....	....	....
1997-98	Djurgarden Jr.	Swede-Jr.	10	3	3	6	10	....	....	....	....	....	....	....	....	....	....	....	....	....	....	....	....	....	....	....
	Huddinge IK	Swede-2	11	0	0	0	6	....	....	....	....	....	....	....	....	....	....	....	....	....	....	....	....	....	....	....
1998-99	Djurgarden	Sweden	44	2	4	6	18	....	....	....	....	....	....	....	....	....	....	....	4	0	0	0	8	....	....	....
99-2000	Djurgarden	Sweden	50	6	3	9	34	....	....	....	....	....	....	....	....	....	....	....	13	1	2	3	12	....	....	....
2000-01	Djurgarden	Sweden	50	7	8	15	79	....	....	....	....	....	....	....	....	....	....	....	16	7	2	9	14	....	....	....
2001-02	**Tampa Bay**	**NHL**	**74**	**3**	**11**	**14**	**24**	**0**	**0**	**0**	**99**	**3.0**	**3**	**14**	**28.6**	**64**	**16**	**14:00**	....	....	....	....	....	....	....	....
	NHL Totals		**74**	**3**	**11**	**14**	**24**	**0**	**0**	**0**	**99**	**3.0**		**14**	**28.6**	**64**	**16**	**13:60**	....	....	....	....	....	....	....	....

O'NEILL, Jeff
(OH-NEEL, JEHF) **CAR.**

Center. Shoots right. 6'1", 190 lbs. Born, Richmond Hill, Ont., February 23, 1976. Hartford's 1st choice, 5th overall, in 1994 Entry Draft.

Season	Club	League	GP	G	A	Pts	PIM	PP	SH	GW	S	%	+/-	TF	F%	H	SB	Min	GP	G	A	Pts	PIM	PP	SH	GW
1990-91	Richmond Hill	OMHA	78	56	134	190	....	....	....	....	....	....	....	....	....	....	....	....	....	....	....	....	....	....	....	....
1991-92	Thornhill	MTJHL	43	27	*53	80	48	....	....	....	....	....	....	....	....	....	....	....	....	....	....	....	....	....	....	....
1992-93	Guelph Storm	OHL	65	32	47	79	88	....	....	....	....	....	....	....	....	....	....	....	5	2	2	4	6	....	....	....
1993-94	Guelph Storm	OHL	66	45	81	126	95	....	....	....	....	....	....	....	....	....	....	....	9	2	11	13	31	....	....	....
1994-95	Guelph Storm	OHL	57	43	81	124	56	....	....	....	....	....	....	....	....	....	....	....	14	8	18	26	34	....	....	....
1995-96	**Hartford**	**NHL**	**65**	**8**	**19**	**27**	**40**	**1**	**0**	**1**	**65**	**12.3**	**–3**	....	....	....	....	....	....	....	....	....	....	....	....	....
1996-97	**Hartford**	**NHL**	**72**	**14**	**16**	**30**	**40**	**2**	**1**	**2**	**101**	**13.9**	**–24**	....	....	....	....	....	....	....	....	....	....	....	....	....
	Springfield	AHL	1	0	0	0	0	....	....	....	....	....	....	....	....	....	....	....	....	....	....	....	....	....	....	....
1997-98	**Carolina**	**NHL**	**74**	**19**	**20**	**39**	**67**	**7**	**1**	**4**	**114**	**16.7**	**–8**	....	....	....	....	....	....	....	....	....	....	....	....	....
1998-99	**Carolina**	**NHL**	**75**	**16**	**15**	**31**	**66**	**4**	**0**	**2**	**121**	**13.2**	**3**	**941**	**45.6**	**207**	**19**	**16:44**	**6**	**0**	**1**	**1**	**0**	**0**	**0**	**0**
99-2000	**Carolina**	**NHL**	**80**	**25**	**38**	**63**	**72**	**4**	**0**	**7**	**189**	**13.2**	**–9**	**1337**	**49.5**	**165**	**33**	**19:20**	....	....	....	....	....	....	....	....
2000-01	**Carolina**	**NHL**	**82**	**41**	**26**	**67**	**106**	**17**	**0**	**5**	**242**	**16.9**	**–18**	**726**	**50.0**	**263**	**16**	**18:20**	**6**	**1**	**2**	**3**	**10**	**0**	**0**	**1**
2001-02	**Carolina**	**NHL**	**76**	**31**	**33**	**64**	**63**	**11**	**0**	**6**	**272**	**11.4**	**–5**	**831**	**56.7**	**177**	**17**	**19:44**	**22**	**8**	**5**	**13**	**27**	**3**	**0**	**1**
	NHL Totals		**524**	**154**	**167**	**321**	**454**	**46**	**2**	**27**	**1104**	**13.9**		**3835**	**50.2**	**812**	**85**	**18:33**	**34**	**9**	**8**	**17**	**37**	**3**	**0**	**2**

OHL All-Rookie Team (1993) • OHL Rookie of the Year (1993) • OHL First All-Star Team (1995)

Transferred to **Carolina** after **Hartford** franchise relocated, June 25, 1997.

ORSZAGH, Vladimir
(OHR-sahg, VLAD-ih-meer) **NSH.**

Right wing. Shoots left. 5'11", 193 lbs. Born, Banska Bystrica, Czech., May 24, 1977. NY Islanders' 4th choice, 106th overall, in 1995 Entry Draft.

Season	Club	League	GP	G	A	Pts	PIM	PP	SH	GW	S	%	+/-	TF	F%	H	SB	Min	GP	G	A	Pts	PIM	PP	SH	GW
1993-94	B. Bystrica Jr.	Slovak-Jr.	38	38	27	65	....	....	....	....	....	....	....	....	....	....	....	....	....	....	....	....	....	....	....	....
1994-95	Banska Bystrica	Slovak-2	38	18	12	30	....	....	....	....	....	....	....	....	....	....	....	....	....	....	....	....	....	....	....	....
1995-96	Banska Bystrica	Slovakia	31	9	5	14	22	....	....	....	....	....	....	....	....	....	....	....	....	....	....	....	....	....	....	....
1996-97	Utah Grizzlies	IHL	68	12	15	27	30	....	....	....	....	....	....	....	....	....	....	....	3	0	1	1	4	....	....	....
1997-98	**NY Islanders**	**NHL**	**11**	**0**	**1**	**1**	**2**	**0**	**0**	**0**	**9**	**0.0**	**–3**	....	....	....	....	....	....	....	....	....	....	....	....	....
	Utah Grizzlies	IHL	62	13	10	23	60	....	....	....	....	....	....	....	....	....	....	....	4	2	0	2	0	....	....	....
1998-99	**NY Islanders**	**NHL**	**12**	**1**	**0**	**1**	**6**	**0**	**0**	**0**	**4**	**20.0**	**2**	**0**	**0.0**	**8**	**2**	**6:36**	....	....	....	....	....	....	....	....
	Lowell	AHL	68	18	23	41	57	....	....	....	....	....	....	....	....	....	....	....	3	2	2	4	2	....	....	....
99-2000	**NY Islanders**	**NHL**	**11**	**2**	**1**	**3**	**4**	**0**	**0**	**0**	**16**	**12.5**	**1**	**0**	**0.0**	**16**	**6**	**11:54**	....	....	....	....	....	....	....	....
	Lowell	AHL	55	8	12	20	22	....	....	....	....	....	....	....	....	....	....	....	7	3	3	6	2	....	....	....
2000-01	Djurgarden	Sweden	50	23	13	36	62	....	....	....	....	....	....	....	....	....	....	....	16	*7	3	10	20	....	....	....
2001-02	**Nashville**	**NHL**	**79**	**15**	**21**	**36**	**56**	**5**	**0**	**3**	**113**	**13.3**	**–15**	**10**	**20.0**	**124**	**33**	**16:04**	....	....	....	....	....	....	....	....
	NHL Totals		**113**	**18**	**23**	**41**	**68**	**5**	**0**	**3**	**142**	**12.7**		**10**	**20.0**	**148**	**41**	**15:33**	....	....	....	....	....	....	....	....

Signed as a free agent by **Nashville**, May 30, 2001.

O'SULLIVAN, Chris
(oh-SUHL-lih-van, KRIHS) **ANA.**

Defense. Shoots left. 6'2", 205 lbs. Born, Dorchester, MA, May 15, 1974. Calgary's 2nd choice, 30th overall, in 1992 Entry Draft.

Season	Club	League	GP	G	A	Pts	PIM	PP	SH	GW	S	%	+/-	TF	F%	H	SB	Min	GP	G	A	Pts	PIM	PP	SH	GW
1991-92	Catholic Memorial	Hi-School	26	26	23	49	65	....	....	....	....	....	....	....	....	....	....	....	....	....	....	....	....	....	....	....
1992-93	Boston University	H-East	5	0	2	2	4	....	....	....	....	....	....	....	....	....	....	....	....	....	....	....	....	....	....	....
1993-94	Boston University	H-East	32	5	18	23	25	....	....	....	....	....	....	....	....	....	....	....	....	....	....	....	....	....	....	....
1994-95	Boston University	H-East	40	23	33	56	48	....	....	....	....	....	....	....	....	....	....	....	....	....	....	....	....	....	....	....
1995-96	Boston University	H-East	37	12	35	47	50	....	....	....	....	....	....	....	....	....	....	....	....	....	....	....	....	....	....	....
1996-97	**Calgary**	**NHL**	**27**	**2**	**8**	**10**	**2**	**1**	**0**	**1**	**41**	**4.9**	**0**	....	....	....	....	....	....	....	....	....	....	....	....	....
	Saint John	AHL	29	3	8	11	17	....	....	....	....	....	....	....	....	....	....	....	5	0	4	4	0	....	....	....
1997-98	**Calgary**	**NHL**	**12**	**0**	**2**	**2**	**10**	**0**	**0**	**0**	**12**	**0.0**	**4**	....	....	....	....	....	....	....	....	....	....	....	....	....
	Saint John	AHL	32	4	10	14	2	....	....	....	....	....	....	....	....	....	....	....	21	2	17	19	18	....	....	....
1998-99	**Calgary**	**NHL**	**10**	**0**	**1**	**1**	**2**	**0**	**0**	**0**	**10**	**0.0**	**–1**	**1**	**0.0**	**4**	**1**	**9:07**	....	....	....	....	....	....	....	....
	Saint John	AHL	41	7	29	36	24	....	....	....	....	....	....	....	....	....	....	....	....	....	....	....	....	....	....	....
	Hartford	AHL	10	1	4	5	0	....	....	....	....	....	....	....	....	....	....	....	7	1	3	4	11	....	....	....
99-2000	**Vancouver**	**NHL**	**11**	**0**	**5**	**5**	**2**	**0**	**0**	**0**	**16**	**0.0**	**2**	**0**	**0.0**	**4**	**2**	**17:47**	....	....	....	....	....	....	....	....
	Syracuse Crunch	AHL	59	18	47	65	24	....	....	....	....	....	....	....	....	....	....	....	4	0	1	1	0	....	....	....
2000-01	Cincinnati	AHL	60	9	40	49	31	....	....	....	....	....	....	....	....	....	....	....	4	0	3	3	0	....	....	....
2001-02	Kloten Flyers	Swiss	39	7	16	23	34	....	....	....	....	....	....	....	....	....	....	....	6	1	2	3	12	....	....	....
	NHL Totals		**60**	**2**	**16**	**18**	**16**	**1**	**0**	**1**	**79**	**2.5**		**1**	**0.0**	**8**	**3**	**13:39**	....	....	....	....	....	....	....	....

Hockey East First All-Star Team (1995) • NCAA East Second All-American Team (1995) • NCAA Championship All-Tournament Team (1995) • NCAA Championship Tournament MVP (1995)

• Missed majority of 1992-93 season recovering from neck injury suffered in game vs. Boston College (H-East), November 11, 1992. Traded to **NY Rangers** by **Calgary** for Lee Sorochan, March 23, 1999. Signed as a free agent by **Vancouver**, August 20, 1999. Signed as a free agent by **Anaheim**, July 20, 2000. Signed as a free agent by **Kloten Flyers** (Swiss), June 22, 2001. Signed as a free agent by **Anaheim**, July 22, 2002.

OZOLINSH, Sandis
(OH-zoh-LIHNCH, SAN-dihz) **FLA.**

Defense. Shoots left. 6'3", 215 lbs. Born, Riga, Latvia, August 3, 1972. San Jose's 3rd choice, 30th overall, in 1991 Entry Draft.

Season	Club	League	GP	G	A	Pts	PIM	PP	SH	GW	S	%	+/-	TF	F%	H	SB	Min	GP	G	A	Pts	PIM	PP	SH	GW
1990-91	Dynamo Riga	USSR	44	0	3	3	51	....	....	....	....	....	....	....	....	....	....	....	....	....	....	....	....	....	....	....
1991-92	Riga Stars	CIS	30	6	0	6	42	....	....	....	....	....	....	....	....	....	....	....	....	....	....	....	....	....	....	....
	Kansas City	IHL	34	6	9	15	20	....	....	....	....	....	....	....	....	....	....	....	15	2	5	7	22	....	....	....
1992-93	**San Jose**	**NHL**	**37**	**7**	**16**	**23**	**40**	**2**	**0**	**0**	**83**	**8.4**	**–9**	....	....	....	....	....	....	....	....	....	....	....	....	....
1993-94	**San Jose**	**NHL**	**81**	**26**	**38**	**64**	**24**	**4**	**0**	**3**	**157**	**16.6**	**16**	....	....	....	....	....	**14**	**0**	**10**	**10**	**8**	**0**	**0**	**0**
1994-95	**San Jose**	**NHL**	**48**	**9**	**16**	**25**	**30**	**3**	**1**	**2**	**83**	**10.8**	**–6**	....	....	....	....	....	**11**	**3**	**2**	**5**	**6**	**1**	**0**	**0**
1995-96	San Francisco	IHL	2	1	0	1	0	....	....	....	....	....	....	....	....	....	....	....	....	....	....	....	....	....	....	....
	San Jose	**NHL**	**7**	**1**	**3**	**4**	**4**	**1**	**0**	**0**	**21**	**4.8**	**2**	....	....	....	....	....	....	....	....	....	....	....	....	....
	♦ Colorado	**NHL**	**66**	**13**	**37**	**50**	**50**	**7**	**1**	**1**	**145**	**9.0**	**0**	....	....	....	....	....	**22**	**5**	**14**	**19**	**16**	**2**	**0**	**1**
1996-97	**Colorado**	**NHL**	**80**	**23**	**45**	**68**	**88**	**13**	**0**	**4**	**232**	**9.9**	**4**	....	....	....	....	....	**17**	**4**	**13**	**17**	**24**	**2**	**0**	**1**
1997-98	**Colorado**	**NHL**	**66**	**13**	**38**	**51**	**65**	**9**	**0**	**2**	**135**	**9.6**	**–12**	....	....	....	....	....	**7**	**0**	**7**	**7**	**14**	**0**	**0**	**0**
1998-99	**Colorado**	**NHL**	**39**	**7**	**25**	**32**	**22**	**4**	**0**	**3**	**81**	**8.6**	**10**	**0**	**0.0**	**34**	**23**	**22:06**	**19**	**4**	**8**	**12**	**22**	**3**	**0**	**1**
99-2000	**Colorado**	**NHL**	**82**	**16**	**36**	**52**	**46**	**6**	**0**	**1**	**210**	**7.6**	**17**	**0**	**0.0**	**71**	**47**	**22:41**	**17**	**5**	**5**	**10**	**20**	**3**	**0**	**1**
2000-01	**Carolina**	**NHL**	**72**	**12**	**32**	**44**	**71**	**4**	**2**	**2**	**145**	**8.3**	**–25**	**0**	**0.0**	**84**	**71**	**22:12**	**6**	**0**	**2**	**2**	**5**	**0**	**0**	**0**

Season	Club	League	GP	G	A	Pts	PIM	PP	SH	GW	S	%	+/-	TF	F%	H	SB	Min	GP	G	A	Pts	PIM	PP	SH	GW
			Regular Season																Playoffs							
2001-02	**Carolina**	**NHL**	**46**	**4**	**19**	**23**	**34**	**1**	**0**	**0**	**71**	**5.6**	**–4**	**0**	**0.0**	**41**	**42**	**19:30**	....	....	....	....	....	....	....	....
	Florida	**NHL**	**37**	**10**	**19**	**29**	**24**	**2**	**0**	**1**	**101**	**9.9**	**–3**	**0**	**0.0**	**77**	**58**	**30:30**	....	....	....	....	....	....	....	....
	Latvia	Olympics	1	0	4	4	0	....	....	....	....	....	....	....	....	....	....	....	....	....	....	....	....	....	....	....
	NHL Totals		**661**	**141**	**324**	**465**	**498**	**56**	**4**	**19**	**1464**	**9.6**		**0**	**0.0**	**307**	**241**	**22:60**	**113**	**21**	**61**	**82**	**115**	**11**	**0**	**4**

NHL First All-Star Team (1997) • Played in NHL All-Star Game (1994, 1997, 1998, 2000, 2001, 2002)

• Missed majority of 1992-93 season recovering from knee injury suffered in game vs. Philadelphia, December 30, 1992. Traded to **Colorado** by **San Jose** for Owen Nolan, October 26, 1995. Traded to **Carolina** by **Colorado** with Columbus' 2nd round choice (previously acquired, Carolina selected Tomas Kurka) in 2000 Entry Draft for Nolan Pratt, Carolina's 1st (Vaclav Nedorost) and 2nd (Jared Aulin) round choices in 2000 Entry Draft and Philadelphia's 2nd round choice (previously acquired, Colorado selected Agris Saviels) in 2000 Entry Draft, June 24, 2000. Traded to **Florida** by **Carolina** with Byron Ritchie for Bret Hedican, Kevyn Adams, Tomas Malec and a conditional 3rd round choice in 2003 Entry Draft, January 16, 2002.

PAHLSSON, Sami

(PAWL-suhn, SAM-ee) **ANA.**

Center. Shoots left. 5'11", 212 lbs. Born, Ornskoldsvik, Sweden, December 17, 1977. Colorado's 10th choice, 176th overall, in 1996 Entry Draft.

Season	Club	League	GP	G	A	Pts	PIM	PP	SH	GW	S	%	+/-	TF	F%	H	SB	Min	GP	G	A	Pts	PIM	PP	SH	GW
1992-93	Ange IK	Swede-4	9	0	0	0	0	....	....	....	....	....	....	....	....	....	....	....	....	....	....	....	....	....	....	....
1993-94	Ange IK	Swede-4	STATISTICS NOT AVAILABLE																							
1994-95	MoDo Jr.	Swede-Jr.	30	10	11	21	26	....	....	....	....	....	....	....	....	....	....	....	....	....	....	....	....	....	....	....
	MoDo	Sweden	1	0	0	0	0	....	....	....	....	....	....	....	....	....	....	....	....	....	....	....	....	....	....	....
1995-96	MoDo Jr.	Swede-Jr.	5	2	6	8	2	....	....	....	....	....	....	....	....	....	....	....	....	....	....	....	....	....	....	....
	MoDo	Sweden	36	1	3	4	8	....	....	....	....	....	....	....	....	....	....	....	4	0	0	0	0	....	....	....
1996-97	MoDo	Sweden	49	8	9	17	83	....	....	....	....	....	....	....	....	....	....	....	....	....	....	....	....	....	....	....
1997-98	MoDo	Sweden	23	6	11	17	24	....	....	....	....	....	....	....	....	....	....	....	9	3	0	3	6	....	....	....
1998-99	MoDo	Sweden	50	17	17	34	44	....	....	....	....	....	....	....	....	....	....	....	13	3	3	6	10	....	....	....
99-2000	MoDo	Sweden	47	16	11	27	67	....	....	....	....	....	....	....	....	....	....	....	13	3	3	6	8	....	....	....
	MoDo	EuroHL	4	1	0	1	0	....	....	....	....	....	....	....	....	....	....	....	3	1	1	2	2	....	....	....
2000-01	**Boston**	**NHL**	**17**	**1**	**1**	**2**	**6**	**0**	**0**	**0**	**13**	**7.7**	**–5**	**239**	**40.2**	**33**	**4**	**14:19**	....	....	....	....	....	....	....	....
	Anaheim	**NHL**	**59**	**3**	**4**	**7**	**14**	**1**	**1**	**1**	**46**	**6.5**	**–9**	**867**	**45.1**	**65**	**25**	**14:14**	....	....	....	....	....	....	....	....
2001-02	**Anaheim**	**NHL**	**80**	**6**	**14**	**20**	**26**	**1**	**1**	**0**	**99**	**6.1**	**–16**	**1201**	**49.8**	**127**	**31**	**16:24**	....	....	....	....	....	....	....	....
	NHL Totals		**156**	**10**	**19**	**29**	**46**	**2**	**2**	**1**	**158**	**6.3**		**2307**	**47.0**	**225**	**60**	**15:21**	....	....	....	....	....	....	....	....

Traded to **Boston** by **Colorado** with Brian Rolston, Martin Grenier and New Jersey's 1st round choice (previously acquired, Boston selected Martin Samuelsson) in 2000 Entry Draft for Raymond Bourque and Dave Andreychuk, March 6, 2000. Traded to **Anaheim** by **Boston** for Patrick Traverse and Andrei Nazarov, November 18, 2000.

PALFFY, Ziggy

(PAHL-fee, ZIHG-gee) **L.A.**

Right wing. Shoots left. 5'10", 183 lbs. Born, Skalica, Czech., May 5, 1972. NY Islanders' 2nd choice, 26th overall, in 1991 Entry Draft.

Season	Club	League	GP	G	A	Pts	PIM	PP	SH	GW	S	%	+/-	TF	F%	H	SB	Min	GP	G	A	Pts	PIM	PP	SH	GW
1990-91	AC Nitra	Czech	50	34	16	50	18	....	....	....	....	....	....	....	....	....	....	....	....	....	....	....	....	....	....	....
1991-92	Dukla Trencin	Czech	45	41	33	74	36	....	....	....	....	....	....	....	....	....	....	....	....	....	....	....	....	....	....	....
1992-93	Dukla Trencin	Czech	43	38	41	79	....	....	....	....	....	....	....	....	....	....	....	....	....	....	....	....	....	....	....	....
1993-94	**NY Islanders**	**NHL**	**5**	**0**	**0**	**0**	**0**	**0**	**0**	**0**	**5**	**0.0**	**–6**	....	....	....	....	....	....	....	....	....	....	....	....	....
	Salt Lake	IHL	57	25	32	57	83	....	....	....	....	....	....	....	....	....	....	....	....	....	....	....	....	....	....	....
	Slovakia	Olympics	8	3	*7	*10	8	....	....	....	....	....	....	....	....	....	....	....	....	....	....	....	....	....	....	....
1994-95	Denver Grizzlies	IHL	33	20	23	43	40	....	....	....	....	....	....	....	....	....	....	....	....	....	....	....	....	....	....	....
	NY Islanders	**NHL**	**33**	**10**	**7**	**17**	**6**	**1**	**0**	**1**	**75**	**13.3**	**3**	....	....	....	....	....	....	....	....	....	....	....	....	....
1995-96	**NY Islanders**	**NHL**	**81**	**43**	**44**	**87**	**56**	**17**	**1**	**6**	**257**	**16.7**	**–17**	....	....	....	....	....	....	....	....	....	....	....	....	....
1996-97	Dukla Trencin	Slovakia	1	0	0	0	....	....	....	....	....	....	....	....	....	....	....	....	....	....	....	....	....	....	....	....
	NY Islanders	**NHL**	**80**	**48**	**42**	**90**	**43**	**6**	**4**	**6**	**292**	**16.4**	**21**	....	....	....	....	....	....	....	....	....	....	....	....	....
1997-98	**NY Islanders**	**NHL**	**82**	**45**	**42**	**87**	**34**	**17**	**2**	**5**	**277**	**16.2**	**–2**	....	....	....	....	....	....	....	....	....	....	....	....	....
1998-99	HK 36 Skalica	Slovakia	9	11	8	19	6	....	....	....	....	....	....	....	....	....	....	....	....	....	....	....	....	....	....	....
	NY Islanders	**NHL**	**50**	**22**	**28**	**50**	**34**	**5**	**2**	**1**	**168**	**13.1**	**–6**	**1**	**100.0**	**28**	**34**	**22:04**	....	....	....	....	....	....	....	....
99-2000	**Los Angeles**	**NHL**	**64**	**27**	**39**	**66**	**32**	**4**	**0**	**3**	**186**	**14.5**	**18**	**7**	**42.9**	**63**	**19**	**19:39**	**4**	**2**	**0**	**2**	**0**	**0**	**0**	**0**
2000-01	**Los Angeles**	**NHL**	**73**	**38**	**51**	**89**	**20**	**12**	**4**	**8**	**217**	**17.5**	**22**	**5**	**80.0**	**54**	**34**	**19:46**	**13**	**3**	**5**	**8**	**8**	**0**	**0**	**0**
2001-02	**Los Angeles**	**NHL**	**63**	**32**	**27**	**59**	**26**	**15**	**1**	**6**	**161**	**19.9**	**5**	**3**	**66.7**	**32**	**36**	**20:18**	**7**	**4**	**5**	**9**	**0**	**0**	**0**	**0**
	Slovakia	Olympics	1	0	0	0	0	....	....	....	....	....	....	....	....	....	....	....	....	....	....	....	....	....	....	....
	NHL Totals		**531**	**265**	**280**	**545**	**251**	**77**	**14**	**36**	**1638**	**16.2**		**16**	**62.5**	**177**	**123**	**20:20**	**24**	**9**	**10**	**19**	**8**	**0**	**0**	**0**

Played in NHL All-Star Game (1998, 2001, 2002)

Traded to **LA Kings** by **NY Islanders** with Brian Smolinski, Marcel Cousineau and New Jersey's 4th round choice (previously acquired, LA Kings selected Daniel Johansson) in 1999 Entry Draft for Olli Jokinen, Josh Green, Mathieu Biron and LA Kings' 1st round choice (Taylor Pyatt) in 1999 Entry Draft, June 20, 1999.

PANDOLFO, Jay

(pan-DAHL-foh, JAY) **N.J.**

Left wing. Shoots left. 6'1", 190 lbs. Born, Winchester, MA, December 27, 1974. New Jersey's 2nd choice, 32nd overall, in 1993 Entry Draft.

Season	Club	League	GP	G	A	Pts	PIM	PP	SH	GW	S	%	+/-	TF	F%	H	SB	Min	GP	G	A	Pts	PIM	PP	SH	GW
1989-90	Burlington Prep	Hi-School	23	33	30	63	18	....	....	....	....	....	....	....	....	....	....	....	....	....	....	....	....	....	....	....
1990-91	Burlington Prep	Hi-School	20	19	27	46	10	....	....	....	....	....	....	....	....	....	....	....	....	....	....	....	....	....	....	....
1991-92	Burlington Prep	Hi-School	20	35	34	69	14	....	....	....	....	....	....	....	....	....	....	....	....	....	....	....	....	....	....	....
1992-93	Boston University	H-East	37	16	22	38	16	....	....	....	....	....	....	....	....	....	....	....	....	....	....	....	....	....	....	....
1993-94	Boston University	H-East	37	17	25	42	27	....	....	....	....	....	....	....	....	....	....	....	....	....	....	....	....	....	....	....
1994-95	Boston University	H-East	20	7	13	20	6	....	....	....	....	....	....	....	....	....	....	....	....	....	....	....	....	....	....	....
1995-96	Boston University	H-East	39	*38	29	67	6	....	....	....	....	....	....	....	....	....	....	....	....	....	....	....	....	....	....	....
	Albany	AHL	5	3	1	4	0	....	....	....	....	....	....	....	....	....	....	....	3	0	0	0	0	....	....	....
1996-97	**New Jersey**	**NHL**	**46**	**6**	**8**	**14**	**6**	**0**	**0**	**1**	**61**	**9.8**	**–1**	....	....	....	....	....	**6**	**0**	**1**	**1**	**0**	**0**	**0**	**0**
	Albany	AHL	12	3	9	12	0	....	....	....	....	....	....	....	....	....	....	....	....	....	....	....	....	....	....	....
1997-98	**New Jersey**	**NHL**	**23**	**1**	**3**	**4**	**4**	**0**	**0**	**0**	**23**	**4.3**	**–4**	....	....	....	....	....	**3**	**0**	**2**	**2**	**0**	**0**	**0**	**0**
	Albany	AHL	51	18	19	37	24	....	....	....	....	....	....	....	....	....	....	....	....	....	....	....	....	....	....	....
1998-99	**New Jersey**	**NHL**	**70**	**14**	**13**	**27**	**10**	**1**	**1**	**4**	**100**	**14.0**	**3**	**10**	**40.0**	**103**	**39**	**15:13**	**7**	**1**	**0**	**1**	**0**	**0**	**0**	**0**
99-2000♦	**New Jersey**	**NHL**	**71**	**7**	**8**	**15**	**4**	**0**	**0**	**0**	**86**	**8.1**	**0**	**19**	**47.4**	**91**	**22**	**13:25**	**23**	**0**	**5**	**5**	**0**	**0**	**0**	**0**
2000-01	**New Jersey**	**NHL**	**63**	**4**	**12**	**16**	**16**	**0**	**0**	**0**	**57**	**7.0**	**3**	**15**	**53.3**	**77**	**16**	**14:05**	**25**	**1**	**4**	**5**	**4**	**0**	**0**	**0**
2001-02	**New Jersey**	**NHL**	**65**	**4**	**10**	**14**	**15**	**0**	**1**	**0**	**72**	**5.6**	**12**	**12**	**41.7**	**87**	**40**	**13:59**	**6**	**0**	**0**	**0**	**0**	**0**	**0**	**0**
	NHL Totals		**338**	**36**	**54**	**90**	**55**	**1**	**2**	**5**	**399**	**9.0**		**56**	**46.4**	**358**	**117**	**14:11**	**70**	**2**	**12**	**14**	**4**	**0**	**0**	**0**

Hockey East First All-Star Team (1996) • Hockey East Player of the Year (1996) • NCAA East First All-American Team (1996)

PANKEWICZ, Greg

(PAN-kuh-wihts, GREHG)

Right wing. Shoots right. 6', 185 lbs. Born, Drayton Valley, Alta., November 6, 1970.

Season	Club	League	GP	G	A	Pts	PIM	PP	SH	GW	S	%	+/-	TF	F%	H	SB	Min	GP	G	A	Pts	PIM	PP	SH	GW
1988-89	Sherwood Park	AJHL	53	26	18	44	307	....	....	....	....	....	....	....	....	....	....	....	....	....	....	....	....	....	....	....
1989-90	Regina Pats	WHL	63	14	24	38	136	....	....	....	....	....	....	....	....	....	....	....	10	1	3	4	19	....	....	....
1990-91	Regina Pats	WHL	72	39	41	80	134	....	....	....	....	....	....	....	....	....	....	....	8	4	7	11	12	....	....	....
1991-92	Knoxville	ECHL	59	41	39	80	214	....	....	....	....	....	....	....	....	....	....	....	....	....	....	....	....	....	....	....
1992-93	New Haven	AHL	62	23	20	43	163	....	....	....	....	....	....	....	....	....	....	....	....	....	....	....	....	....	....	....
1993-94	**Ottawa**	**NHL**	**3**	**0**	**0**	**0**	**2**	**0**	**0**	**0**	**3**	**0.0**	**–1**	....	....	....	....	....	....	....	....	....	....	....	....	....
	P.E.I. Senators	AHL	69	33	29	62	241	....	....	....	....	....	....	....	....	....	....	....	....	....	....	....	....	....	....	....
1994-95	P.E.I. Senators	AHL	75	37	30	67	161	....	....	....	....	....	....	....	....	....	....	....	6	1	1	2	24	....	....	....
1995-96	Portland Pirates	AHL	28	9	12	21	99	....	....	....	....	....	....	....	....	....	....	....	....	....	....	....	....	....	....	....
	Chicago Wolves	IHL	45	9	16	25	164	....	....	....	....	....	....	....	....	....	....	....	5	4	0	4	8	....	....	....
1996-97	Manitoba Moose	IHL	79	32	34	66	222	....	....	....	....	....	....	....	....	....	....	....	....	....	....	....	....	....	....	....
1997-98	Manitoba Moose	IHL	76	42	34	76	246	....	....	....	....	....	....	....	....	....	....	....	3	0	0	0	6	....	....	....
1998-99	**Calgary**	**NHL**	**18**	**0**	**3**	**3**	**20**	**0**	**0**	**0**	**10**	**0.0**	**0**	**3**	**66.7**	**17**	**2**	**7:26**	....	....	....	....	....	....	....	....
	Saint John	AHL	30	10	14	24	84	....	....	....	....	....	....	....	....	....	....	....	....	....	....	....	....	....	....	....
	Kentucky	AHL	10	2	3	5	7	....	....	....	....	....	....	....	....	....	....	....	11	4	1	5	10	....	....	....
99-2000	Houston Aeros	IHL	62	22	19	41	134	....	....	....	....	....	....	....	....	....	....	....	5	2	1	3	18	....	....	....
2000-01	Houston Aeros	IHL	74	22	24	46	231	....	....	....	....	....	....	....	....	....	....	....	7	1	1	2	10	....	....	....
2001-02	Pensacola	ECHL	63	39	46	85	306	....	....	....	....	....	....	....	....	....	....	....	3	2	0	2	8	....	....	....
	NHL Totals		**21**	**0**	**3**	**3**	**22**	**0**	**0**	**0**	**13**	**0.0**		**3**	**66.7**	**17**	**2**	**7:26**	....	....	....	....	....	....	....	....

ECHL Second All-Star Team (2002)

Signed as a free agent by **Ottawa**, May 27, 1993. Signed as a free agent by **Calgary**, September 1, 1998. Traded to **San Jose** by **Calgary** for cash, March 23, 1999. Signed as a free agent by **Houston** (IHL), August 31, 2000. Signed as a free agent by **Pensacola** (ECHL), August 21, 2001.

PAPINEAU, Justin

(PA-pee-noh, JUHS-tihn) **ST.L.**

Center. Shoots left. 5'10", 178 lbs. Born, Ottawa, Ont., January 15, 1980. St. Louis' 3rd choice, 75th overall, in 2000 Entry Draft.

			Regular Season																Playoffs							
Season	Club	League	GP	G	A	Pts	PIM	PP	SH	GW	S	%	+/-	TF	F%	H	SB	Min	GP	G	A	Pts	PIM	PP	SH	GW
1995-96	Ottawa Jr. Sens	OCJHL	52	31	19	50	51	....	....	....	....	....	....	....	....	....	....	....	....	....	....	....	....	....	....	....
1996-97	Belleville Bulls	OHL	50	10	32	42	32	....	....	....	....	....	....	....	....	....	....	....	....	....	....	....	....	....	....	....
1997-98	Belleville Bulls	OHL	66	41	53	94	34	....	....	....	....	....	....	....	....	....	....	....	10	5	9	14	6	....	....	....
1998-99	Belleville Bulls	OHL	68	52	47	99	28	....	....	....	....	....	....	....	....	....	....	....	21	*21	*30	*51	20	....	....	....
99-2000	Belleville Bulls	OHL	60	40	36	76	52	....	....	....	....	....	....	....	....	....	....	....	16	4	12	16	16	....	....	....
2000-01	Worcester	AHL	43	7	22	29	33	....	....	....	....	....	....	....	....	....	....	....	11	7	3	10	8	....	....	....
2001-02	**St. Louis**	**NHL**	**1**	**0**	**0**	**0**	**0**	**0**	**0**	**0**	**0**	**0.0**	**–2**	**7**	**42.9**	**0**	**1**	**8:40**	....	....	....	....	....	....	....	....
	Worcester	AHL	75	*38	38	76	86	....	....	....	....	....	....	....	....	....	....	....	3	1	2	3	4	....	....	....
	NHL Totals		**1**	**0**	**0**	**0**	**0**	**0**	**0**	**0**	**0**	**0.0**		**7**	**42.9**	**0**	**1**	**8:40**	....	....	....	....	....	....	....	....

• Re-entered NHL Entry Draft. Originally Los Angeles' 2nd choice, 46th overall, in 1998 Entry Draft.

PARK, Richard

(PAHRK, RIH-chahrd) **MIN.**

Center. Shoots right. 5'11", 190 lbs. Born, Seoul, South Korea, May 27, 1976. Pittsburgh's 2nd choice, 50th overall, in 1994 Entry Draft.

Season	Club	League	GP	G	A	Pts	PIM	PP	SH	GW	S	%	+/-	TF	F%	H	SB	Min	GP	G	A	Pts	PIM	PP	SH	GW
1991-92	Tor. Young Nats	MTHL	76	49	58	107	91	....	....	....	....	....	....	....	....	....	....	....	....	....	....	....	....	....	....	....
1992-93	Belleville Bulls	OHL	66	23	38	61	38	....	....	....	....	....	....	....	....	....	....	....	5	0	0	0	14	....	....	....
1993-94	Belleville Bulls	OHL	59	27	49	76	70	....	....	....	....	....	....	....	....	....	....	....	12	3	5	8	18	....	....	....
1994-95	Belleville Bulls	OHL	45	28	51	79	35	....	....	....	....	....	....	....	....	....	....	....	16	9	18	27	12	....	....	....
	Pittsburgh	**NHL**	**1**	**0**	**1**	**1**	**2**	**0**	**0**	**0**	**4**	**0.0**	**1**	....	....	....	....	....	**3**	**0**	**0**	**0**	**2**	**0**	**0**	**0**
1995-96	Belleville Bulls	OHL	6	7	6	13	2	....	....	....	....	....	....	....	....	....	....	....	14	18	12	30	10	....	....	....
	Pittsburgh	**NHL**	**56**	**4**	**6**	**10**	**36**	**0**	**1**	**1**	**62**	**6.5**	**3**	....	....	....	....	....	**1**	**0**	**0**	**0**	**0**	**0**	**0**	**0**
1996-97	**Pittsburgh**	**NHL**	**1**	**0**	**0**	**0**	**0**	**0**	**0**	**0**	**1**	**0.0**	**–1**	....	....	....	....	....	....	....	....	....	....	....	....	....
	Cleveland	IHL	50	12	15	27	30	....	....	....	....	....	....	....	....	....	....	....	....	....	....	....	....	....	....	....
	Anaheim	**NHL**	**11**	**1**	**1**	**2**	**10**	**0**	**0**	**0**	**9**	**11.1**	**0**	....	....	....	....	....	**11**	**0**	**1**	**1**	**2**	**0**	**0**	**0**
1997-98	**Anaheim**	**NHL**	**15**	**0**	**2**	**2**	**8**	**0**	**0**	**0**	**14**	**0.0**	**–3**	....	....	....	....	....	....	....	....	....	....	....	....	....
	Cincinnati	AHL	56	17	26	43	36	....	....	....	....	....	....	....	....	....	....	....	....	....	....	....	....	....	....	....
1998-99	**Philadelphia**	**NHL**	**7**	**0**	**0**	**0**	**0**	**0**	**0**	**0**	**5**	**0.0**	**–1**	**15**	**53.3**	**2**	**0**	**9:21**	....	....	....	....	....	....	....	....
	Philadelphia	AHL	75	41	42	83	33	....	....	....	....	....	....	....	....	....	....	....	16	9	6	15	4	....	....	....
99-2000	Utah Grizzlies	IHL	82	28	32	60	36	....	....	....	....	....	....	....	....	....	....	....	5	1	0	1	0	....	....	....
2000-01	Cleveland	IHL	75	27	21	48	29	....	....	....	....	....	....	....	....	....	....	....	4	0	2	2	4	....	....	....
2001-02	**Minnesota**	**NHL**	**63**	**10**	**15**	**25**	**10**	**2**	**1**	**2**	**115**	**8.7**	**–1**	**79**	**41.8**	**21**	**23**	**16:28**	....	....	....	....	....	....	....	....
	Houston Aeros	AHL	13	4	10	14	6	....	....	....	....	....	....	....	....	....	....	....	....	....	....	....	....	....	....	....
	NHL Totals		**154**	**15**	**25**	**40**	**66**	**2**	**2**	**3**	**210**	**7.1**		**94**	**43.6**	**23**	**23**	**15:45**	**15**	**0**	**1**	**1**	**4**	**0**	**0**	**0**

OHL All-Rookie Team (1993) • AHL Second All-Star Team (1999)

Traded to **Anaheim** by **Pittsburgh** for Roman Oksiuta, March 18, 1997. Signed as a free agent by **Philadelphia**, August 24, 1998. Signed as a free agent by **Utah** (IHL), September 22, 1999. Signed as a free agent by **Minnesota**, June 6, 2000.

PARKER, Scott

(PAR-kuhr, SKAWT) **COL.**

Right wing. Shoots right. 6'5", 230 lbs. Born, Hanford, CA, January 29, 1978. Colorado's 4th choice, 20th overall, in 1998 Entry Draft.

Season	Club	League	GP	G	A	Pts	PIM	PP	SH	GW	S	%	+/-	TF	F%	H	SB	Min	GP	G	A	Pts	PIM	PP	SH	GW
1993-94	Alaska	AAHL	34	8	12	20	86	....	....	....	....	....	....	....	....	....	....	....	....	....	....	....	....	....	....	....
1994-95	Spokane Braves	KIJHL	43	7	21	28	128	....	....	....	....	....	....	....	....	....	....	....	....	....	....	....	....	....	....	....
1995-96	Kelowna Rockets	WHL	64	3	4	7	159	....	....	....	....	....	....	....	....	....	....	....	6	0	0	0	12	....	....	....
1996-97	Kelowna Rockets	WHL	68	18	8	26	*330	....	....	....	....	....	....	....	....	....	....	....	6	0	2	2	4	....	....	....
1997-98	Kelowna Rockets	WHL	71	30	22	52	243	....	....	....	....	....	....	....	....	....	....	....	7	6	0	6	23	....	....	....
1998-99	**Colorado**	**NHL**	**27**	**0**	**0**	**0**	**71**	**0**	**0**	**0**	**3**	**0.0**	**–3**	**1**	**0.0**	**10**	**0**	**1:37**	....	....	....	....	....	....	....	....
	Hershey Bears	AHL	32	4	3	7	143	....	....	....	....	....	....	....	....	....	....	....	4	0	0	0	6	....	....	....
99-2000	Hershey Bears	AHL	68	12	7	19	206	....	....	....	....	....	....	....	....	....	....	....	11	1	1	2	56	....	....	....
2000-01 ♦	**Colorado**	**NHL**	**69**	**2**	**3**	**5**	**155**	**0**	**0**	**1**	**35**	**5.7**	**–2**	**2**	**0.0**	**104**	**5**	**5:42**	**4**	**0**	**0**	**0**	**2**	**0**	**0**	**0**
2001-02	**Colorado**	**NHL**	**63**	**1**	**4**	**5**	**154**	**0**	**0**	**0**	**32**	**3.1**	**0**	**0**	**0.0**	**84**	**6**	**5:50**	....	....	....	....	....	....	....	....
	NHL Totals		**159**	**3**	**7**	**10**	**380**	**0**	**0**	**1**	**70**	**4.3**		**3**	**0.0**	**198**	**11**	**5:04**	**4**	**0**	**0**	**0**	**2**	**0**	**0**	**0**

• Re-entered NHL Entry Draft. Originally New Jersey's 6th choice, 63rd overall, in 1996 Entry Draft.

PARRISH, Mark

(PAIR-ihsh, MAHRK) **NYI**

Right wing. Shoots right. 5'11", 200 lbs. Born, Edina, MN, February 2, 1977. Colorado's 3rd choice, 79th overall, in 1996 Entry Draft.

Season	Club	League	GP	G	A	Pts	PIM	PP	SH	GW	S	%	+/-	TF	F%	H	SB	Min	GP	G	A	Pts	PIM	PP	SH	GW
1994-95	Jefferson High	Hi-School	27	40	20	60	42	....	....	....	....	....	....	....	....	....	....	....	....	....	....	....	....	....	....	....
1995-96	St. Cloud State	WCHA	39	15	13	28	30	....	....	....	....	....	....	....	....	....	....	....	....	....	....	....	....	....	....	....
1996-97	St. Cloud State	WCHA	35	*27	15	42	60	....	....	....	....	....	....	....	....	....	....	....	....	....	....	....	....	....	....	....
1997-98	Seattle	WHL	54	54	38	92	29	....	....	....	....	....	....	....	....	....	....	....	5	2	3	5	2	....	....	....
	New Haven	AHL	1	1	0	1	2	....	....	....	....	....	....	....	....	....	....	....	....	....	....	....	....	....	....	....
1998-99	**Florida**	**NHL**	**73**	**24**	**13**	**37**	**25**	**5**	**0**	**5**	**129**	**18.6**	**–6**	**1**	**0.0**	**56**	**5**	**13:59**	....	....	....	....	....	....	....	....
	New Haven	AHL	2	1	0	1	0	....	....	....	....	....	....	....	....	....	....	....	....	....	....	....	....	....	....	....
99-2000	**Florida**	**NHL**	**81**	**26**	**18**	**44**	**39**	**6**	**0**	**3**	**152**	**17.1**	**1**	**8**	**75.0**	**113**	**8**	**14:04**	**4**	**0**	**1**	**1**	**0**	**0**	**0**	**0**
2000-01	**NY Islanders**	**NHL**	**70**	**17**	**13**	**30**	**28**	**6**	**0**	**3**	**123**	**13.8**	**–27**	**3**	**33.3**	**176**	**13**	**15:27**	....	....	....	....	....	....	....	....
2001-02	**NY Islanders**	**NHL**	**78**	**30**	**30**	**60**	**32**	**9**	**1**	**6**	**162**	**18.5**	**10**	**10**	**40.0**	**132**	**14**	**16:48**	**7**	**2**	**1**	**3**	**6**	**2**	**0**	**0**
	NHL Totals		**302**	**97**	**74**	**171**	**124**	**26**	**1**	**17**	**566**	**17.1**		**22**	**50.0**	**477**	**40**	**15:05**	**11**	**2**	**2**	**4**	**6**	**2**	**0**	**0**

NCAA West Second All-American Team (1997) • WHL West First All-Star Team (1998) • Played in NHL All-Star Game (2002)

Rights traded to **Florida** by **Colorado** with Anaheim's 3rd round choice (previously acquired, Florida selected Lance Ward) in 1998 Entry Draft for Tom Fitzgerald, March 24, 1998. Traded to **NY Islanders** by **Florida** with Oleg Kvasha for Roberto Luongo and Olli Jokinen, June 24, 2000.

PARSSINEN, Timo

(pahr-SIH-nehn, TEE-moh)

Left wing. Shoots left. 5'10", 176 lbs. Born, Lohjan mlk., Finland, January 19, 1977. Anaheim's 4th choice, 102nd overall, in 2001 Entry Draft.

Season	Club	League	GP	G	A	Pts	PIM	PP	SH	GW	S	%	+/-	TF	F%	H	SB	Min	GP	G	A	Pts	PIM	PP	SH	GW
1994-95	TuTo Turku Jr.	Finn-Jr.	14	9	6	15	33	....	....	....	....	....	....	....	....	....	....	....	....	....	....	....	....	....	....	....
	TuTo Turku	Finland	1	1	0	1	0	....	....	....	....	....	....	....	....	....	....	....	....	....	....	....	....	....	....	....
1995-96	TuTo Turku Jr.	Finn-Jr.	14	18	15	33	12	....	....	....	....	....	....	....	....	....	....	....	11	12	6	18	36	....	....	....
	TuTo Turku	Finland	7	0	0	0	4	....	....	....	....	....	....	....	....	....	....	....	....	....	....	....	....	....	....	....
1996-97	TuTo Turku	Finland-2	39	21	29	50	30	....	....	....	....	....	....	....	....	....	....	....	....	....	....	....	....	....	....	....
	TuTo Turku Jr.	Finn-Jr.	....	....	....	....	....	....	....	....	....	....	....	....	....	....	....	....	9	2	14	16	36	....	....	....
1997-98	Hermes Kokkola	Finland-2	46	29	39	68	66	....	....	....	....	....	....	....	....	....	....	....	3	1	0	1	4	....	....	....
1998-99	HPK Hameenlinna	Finland	46	15	24	39	46	....	....	....	....	....	....	....	....	....	....	....	8	2	3	5	8	....	....	....
99-2000	HPK Hameenlinna	Finland	53	25	27	52	60	....	....	....	....	....	....	....	....	....	....	....	8	5	5	10	8	....	....	....
2000-01	HPK Hameenlinna	Finland	54	18	31	49	48	....	....	....	....	....	....	....	....	....	....	....	....	....	....	....	....	....	....	....
2001-02	**Anaheim**	**NHL**	**17**	**0**	**3**	**3**	**2**	**0**	**0**	**0**	**16**	**0.0**	**0**	**1**	**100.0**	**3**	**2**	**11:44**	....	....	....	....	....	....	....	....
	Cincinnati	AHL	49	14	24	38	22	....	....	....	....	....	....	....	....	....	....	....	3	0	2	2	0	....	....	....
	NHL Totals		**17**	**0**	**3**	**3**	**2**	**0**	**0**	**0**	**16**	**0.0**		**1**	**100.0**	**3**	**2**	**11:44**	....	....	....	....	....	....	....	....

PATERA, Pavel

(puh-TEHR-uh, PAH-vehl)

Center. Shoots left. 6'1", 172 lbs. Born, Kladno, Czech., September 6, 1971. Dallas' 4th choice, 153rd overall, in 1998 Entry Draft.

Season	Club	League	GP	G	A	Pts	PIM	PP	SH	GW	S	%	+/-	TF	F%	H	SB	Min	GP	G	A	Pts	PIM	PP	SH	GW
1990-91	Poldi Kladno	Czech	3	0	0	0	....	....	....	....	....	....	....	....	....	....	....	....	....	....	....	....	....	....	....	....
1991-92	Poldi Kladno	Czech	38	12	13	25	26	....	....	....	....	....	....	....	....	....	....	....	8	8	4	12	0	....	....	....
1992-93	Poldi Kladno	Czech	42	9	23	32	....	....	....	....	....	....	....	....	....	....	....	....	....	....	....	....	....	....	....	....
1993-94	HC Kladno	Czech	43	21	39	60	....	....	....	....	....	....	....	....	....	....	....	....	11	5	10	15	....	....	....	....
1994-95	HC Kladno	Czech	43	26	49	75	24	....	....	....	....	....	....	....	....	....	....	....	11	5	7	12	6	....	....	....
1995-96	HC Poldi Kladno	Czech	40	24	31	55	38	....	....	....	....	....	....	....	....	....	....	....	8	3	1	4	34	....	....	....
1996-97	AIK Solna	Sweden	50	19	24	43	44	....	....	....	....	....	....	....	....	....	....	....	7	2	3	5	6	....	....	....
1997-98	AIK Solna	Sweden	46	8	17	25	50	....	....	....	....	....	....	....	....	....	....	....	....	....	....	....	....	....	....	....
1998-99	Vsetin	Czech	52	16	37	53	58	....	....	....	....	....	....	....	....	....	....	....	12	5	*10	15	....	....	....	....
99-2000	**Dallas**	**NHL**	**12**	**1**	**4**	**5**	**4**	**0**	**0**	**0**	**18**	**5.6**	**–1**	**61**	**49.2**	**2**	**1**	**14:04**	....	....	....	....	....	....	....	....
	Vsetin	Czech	29	8	14	22	36	....	....	....	....	....	....	....	....	....	....	....	9	3	4	7	8	....	....	....
2000-01	**Minnesota**	**NHL**	**20**	**1**	**3**	**4**	**4**	**0**	**0**	**0**	**14**	**7.1**	**–8**	**184**	**44.0**	**7**	**15**	**14:27**	....	....	....	....	....	....	....	....
	Cleveland	IHL	54	8	44	52	22	....	....	....	....	....	....	....	....	....	....	....	....	....	....	....	....	....	....	....

Season	Club	League	GP	G	A	Pts	PIM	PP	SH	GW	S	%	+/-	TF	F%	H	SB	Min	GP	G	A	Pts	PIM	PP	SH	GW
			Regular Season																Playoffs							
2001-02	Kladno	Czech	3	0	2	2	0																			
	Avangard Omsk	Russia	28	5	7	12	32												11	1	3	4	6			
	Czech Republic	Olympics	4	0	0	0	0																			
	NHL Totals		**32**	**2**	**7**	**9**	**8**	**0**	**0**	**0**	**32**	**6.3**		**245**	**45.3**	**9**	**16**	**14:19**								

Traded to **Minnesota** by **Dallas** with Aaron Gavey, Dallas' 8th round choice (Eric Johansson) in 2000 Entry Draft and Minnesota's 4th round choice (previously acquired) in 2002 Entry Draft for Brad Lukowich and Minnesota's 3rd (Yared Hagos) and 9th (Dale Sullivan) round choices in 2001 Entry Draft, June 25, 2000.

PATRICK, James

(PAT-rihk, JAYMS) **BUF.**

Defense. Shoots right. 6'2", 202 lbs. Born, Winnipeg, Man., June 14, 1963. NY Rangers' 1st choice, 9th overall, in 1981 Entry Draft.

Season	Club	League	GP	G	A	Pts	PIM	PP	SH	GW	S	%	+/-	TF	F%	H	SB	Min	GP	G	A	Pts	PIM	PP	SH	GW
1980-81	Prince Albert	SJHL	59	21	61	82	162																			
1981-82	North Dakota	WCHA	42	5	24	29	26																			
1982-83	North Dakota	WCHA	36	12	36	48	29																			
1983-84	Team Canada	Nat-Tm	63	7	24	31	52																			
	Canada	Olympics	7	0	3	3	4																			
	NY Rangers	NHL	12	1	7	8	2	0	0	0	15	6.7	6						5	0	3	3	2	0	0	0
1984-85	NY Rangers	NHL	75	8	28	36	71	4	1	1	101	7.9	−17						3	0	0	0	4	0	0	0
1985-86	NY Rangers	NHL	75	14	29	43	88	2	1	1	131	10.7	14						16	1	5	6	34	0	0	0
1986-87	NY Rangers	NHL	78	10	45	55	62	5	0	0	143	7.0	13						6	1	2	3	2	1	0	1
1987-88	NY Rangers	NHL	70	17	45	62	52	9	0	1	187	9.1	16													
1988-89	NY Rangers	NHL	68	11	36	47	41	6	0	2	147	7.5	3						4	0	1	1	2	0	0	0
1989-90	NY Rangers	NHL	73	14	43	57	50	9	0	0	136	10.3	4						10	3	8	11	0	2	0	1
1990-91	NY Rangers	NHL	74	10	49	59	58	6	0	2	138	7.2	−5						6	0	0	0	6	0	0	0
1991-92	NY Rangers	NHL	80	14	57	71	54	6	0	1	148	9.5	34						13	0	7	7	12	0	0	0
1992-93	NY Rangers	NHL	60	5	21	26	61	3	0	0	99	5.1	1													
1993-94	NY Rangers	NHL	6	0	3	3	2	0	0	0	6	0.0	1													
	Hartford	NHL	47	8	20	28	32	4	1	2	65	12.3	−12													
	Calgary	NHL	15	2	2	4	6	1	0	0	20	10.0	6						7	0	1	1	6	0	0	0
1994-95	Calgary	NHL	43	0	10	10	14	0	0	0	43	0.0	−3						5	0	1	1	0	0	0	0
1995-96	Calgary	NHL	80	3	32	35	30	1	0	0	116	2.6	3						4	0	0	0	2	0	0	0
1996-97	Calgary	NHL	19	3	1	4	6	1	0	0	22	13.6	2													
1997-98	Calgary	NHL	60	6	11	17	26	1	0	1	57	10.5	−2													
1998-99	Buffalo	NHL	45	1	7	8	16	0	0	0	31	3.2	12	0	0.0	23	38	14:45	20	0	1	1	12	0	0	0
99-2000	Buffalo	NHL	66	5	8	13	22	0	0	3	40	12.5	8	0	0.0	44	61	15:59	5	0	1	1	2	0	0	0
2000-01	Buffalo	NHL	54	4	9	13	12	1	0	0	48	8.3	9	0	0.0	28	64	17:16	13	1	2	3	2	0	0	0
2001-02	Buffalo	NHL	56	5	8	13	16	1	0	0	45	11.1	3	0	0.0	35	70	16:27								
	NHL Totals		**1156**	**141**	**471**	**612**	**721**	**60**	**3**	**14**	**1738**	**8.1**		**0**	**0.0**	**130**	**233**	**16:10**	**117**	**6**	**32**	**38**	**86**	**3**	**0**	**2**

WCHA Second All-Star Team (1982) • WCHA Freshman of the Year (1982) • NCAA Chamionship All-Tournament Team (1982) • WCHA First All-Star Team (1983) • NCAA West All American Team (1983)

Traded to **Hartford** by **NY Rangers** with Darren Turcotte for Steve Larmer, Nick Kypreos, Barry Richter and Hartford's 6th round choice (Yuri Litvinov) in 1994 Entry Draft, November 2, 1993. Traded to **Calgary** by **Hartford** with Zarley Zalapski and Michael Nylander for Gary Suter, Paul Ranheim and Ted Drury, March 10, 1994. • Missed majority of 1996-97 season recovering from knee injury originally suffered in game vs. Pittsburgh, October 24, 1996. Signed as a free agent by **Buffalo**, October 7, 1998.

PAYER, Serge

(pie-YAY, SAIRZH) **FLA.**

Center. Shoots left. 6', 203 lbs. Born, Rockland, Ont., May 7, 1979.

Season	Club	League	GP	G	A	Pts	PIM	PP	SH	GW	S	%	+/-	TF	F%	H	SB	Min	GP	G	A	Pts	PIM	PP	SH	GW
1994-95	Cumberland Colts	OMHA	42	37	46	83	55																			
1995-96	Kitchener	OHL	66	8	16	24	18												12	0	2	2	2			
1996-97	Kitchener	OHL	63	7	16	23	27												13	1	3	4	2			
1997-98	Kitchener	OHL	44	20	21	41	51												6	3	0	3	7			
1998-99	Kitchener	OHL	40	18	19	37	22																			
99-2000	Kitchener	OHL	44	10	26	36	53												5	0	3	3	6			
2000-01	Florida	NHL	43	5	1	6	21	0	1	0	34	14.7	0	97	42.3	26	9	7:27								
	Louisville	AHL	32	6	6	12	15																			
2001-02	Utah Grizzlies	AHL	20	6	2	8	9																			
	NHL Totals		**43**	**5**	**1**	**6**	**21**	**0**	**1**	**0**	**34**	**14.7**		**97**	**42.3**	**26**	**9**	**7:27**								

• Missed remainder of 1998-99 and majority of 1999-2000 seasons recovering from Guillian-Barre Syndrome, January 25, 1999. • Signed as a free agent by **Florida**, September 30, 1997. • Missed majority of 2001-02 season recovering from back injury originally suffered in training camp, September, 2001.

PEAT, Stephen

(PEET, STEEV-vuhn) **WSH.**

Defense. Shoots right. 6'3", 210 lbs. Born, Princeton, B.C., March 10, 1980. Anaheim's 2nd choice, 32nd overall, in 1998 Entry Draft.

Season	Club	League	GP	G	A	Pts	PIM	PP	SH	GW	S	%	+/-	TF	F%	H	SB	Min	GP	G	A	Pts	PIM	PP	SH	GW
1995-96	Langley Thunder	BCJHL	59	5	15	20	112																			
	Red Deer Rebels	WHL	1	0	0	0	0																			
1996-97	Red Deer Rebels	WHL	68	3	14	17	161												16	0	2	2	22			
1997-98	Red Deer Rebels	WHL	63	6	12	18	189												5	0	0	0	8			
1998-99	Red Deer Rebels	WHL	31	2	6	8	98																			
	Tri-City	WHL	,5	0	0	0	19																			
99-2000	Tri-City	WHL	12	0	2	2	48																			
	Calgary Hitmen	WHL	23	0	8	8	100												13	0	1	1	33			
2000-01	Portland Pirates	AHL	6	0	0	0	16																			
2001-02	Washington	NHL	38	2	2	4	85	0	0	0	11	18.2	−1	0	0.0	37	2	5:04								
	Portland Pirates	AHL	17	2	2	4	57																			
	NHL Totals		**38**	**2**	**2**	**4**	**85**	**0**	**0**	**0**	**11**	**18.2**		**0**	**0.0**	**37**	**2**	**5:04**								

Traded to **Tri-City** (WHL) by **Red Deer** (WHL) for Regan Darby and Jarrett Thompson, January 4, 1999. Traded to **Calgary** (WHL) by **Tri-City** (WHL) with Toni Bader for Jeff Feniak, Eric Clark and Matthew Ireland, December 6, 1999. • Missed majority of 1999-00 season recovering from injuries sustained off-ice, February 8, 2000. Rights traded to **Washington** by **Anaheim** for Washington's 4th round choice (later traded to Montreal - later traded to Pittsburgh - Pittsburgh selected Michel Ouellet) in 2000 Entry Draft, June 1, 2000. • Missed majority of 2000-01 recovering from groin injury originally suffered in training camp, September 29, 2000.

PECA, Michael

(PEH-kuh, MIGHK-uhl) **NYI**

Center. Shoots right. 5'11", 190 lbs. Born, Toronto, Ont., March 26, 1974. Vancouver's 2nd choice, 40th overall, in 1992 Entry Draft.

Season	Club	League	GP	G	A	Pts	PIM	PP	SH	GW	S	%	+/-	TF	F%	H	SB	Min	GP	G	A	Pts	PIM	PP	SH	GW
1989-90	Tor. Young Nats	MTHL	39	42	53	95	40																			
1990-91	Sudbury Wolves	OHL	62	14	27	41	24												5	1	0	1	7			
1991-92	Sudbury Wolves	OHL	39	16	34	50	61																			
	Ottawa 67's	OHL	27	8	17	25	32												11	6	10	16	6			
1992-93	Ottawa 67's	OHL	55	38	64	102	80																			
	Hamilton Canucks	AHL	9	6	3	9	11																			
1993-94	Ottawa 67's	OHL	55	50	63	113	101												17	7	22	29	30			
	Vancouver	NHL	4	0	0	0	2	0	0	0	5	0.0	−1													
1994-95	Syracuse Crunch	AHL	35	10	24	34	75																			
	Vancouver	NHL	33	6	6	12	30	2	0	1	46	13.0	−6						5	0	1	1	8	0	0	0
1995-96	Buffalo	NHL	68	11	20	31	67	4	3	1	109	10.1	−1													
1996-97	Buffalo	NHL	79	20	29	49	80	5	6	4	137	14.6	26						10	0	2	2	8	0	0	0
1997-98	Buffalo	NHL	61	18	22	40	57	6	5	1	132	13.6	12						13	3	2	5	8	0	0	1
1998-99	Buffalo	NHL	82	27	29	56	81	10	0	8	199	13.6	7	1855	49.4	181	64	20:44	21	5	8	13	18	2	1	0
99-2000	Buffalo	NHL	73	20	21	41	67	2	0	3	144	13.9	6	1604	48.6	132	76	19:57	5	0	1	1	4	0	0	0
2000-01	Buffalo	NHL	DID NOT PLAY																							
2001-02	NY Islanders	NHL	80	25	35	60	62	3	6	5	168	14.9	19	1804	52.4	132	107	20:14	5	1	0	1	2	0	0	0
	Canada	Olympics	6	0	2	2	2																			
	NHL Totals		**480**	**127**	**162**	**289**	**446**	**32**	**20**	**23**	**940**	**13.5**		**5263**	**50.2**	**445**	**247**	**20:19**	**59**	**9**	**14**	**23**	**48**	**2**	**1**	**1**

Won Frank J. Selke Trophy (1997, 2002)

Traded to **Buffalo** by **Vancouver** with Mike Wilson and Vancouver's 1st round choice (Jay McKee) in 1995 Entry Draft for Alexander Mogilny and Buffalo's 5th round choice (Todd Norman) in 1995 Entry Draft, July 8, 1995. • Missed entire 2000-01 season after failing to come to contract terms with **Buffalo**. Rights traded to **NY Islanders** by **Buffalo** for Tim Connolly and Taylor Pyatt, June 24, 2001.

PEDERSON, Denis

(PEE-duhr-suhn, DEH-nihs) **NSH.**

Center. Shoots right. 6'2", 205 lbs. Born, Prince Albert, Sask., September 10, 1975. New Jersey's 1st choice, 13th overall, in 1993 Entry Draft.

			Regular Season																Playoffs							
Season	Club	League	GP	G	A	Pts	PIM	PP	SH	GW	S	%	+/-	TF	F%	H	SB	Min	GP	G	A	Pts	PIM	PP	SH	GW
1990-91	Prince Albert	SMHL	30	25	17	42	84	....	....	....	....	....	....	....	....	....	....	....	....	....	....	....	....	....	....	....
1991-92	Prince Albert	SMHL	21	33	25	58	40	....	....	....	....	....	....	....	....	....	....	....	....	....	....	....	....	....	....	....
	Prince Albert	WHL	10	0	0	0	6	....	....	....	....	....	....	....	....	....	....	....	7	0	1	1	13	....	....	....
1992-93	Prince Albert	WHL	72	33	40	73	134	....	....	....	....	....	....	....	....	....	....	....	....	....	....	....	....	....	....	....
1993-94	Prince Albert	WHL	71	53	45	98	157	....	....	....	....	....	....	....	....	....	....	....	....	....	....	....	....	....	....	....
1994-95	Prince Albert	WHL	63	30	38	68	122	....	....	....	....	....	....	....	....	....	....	....	15	11	14	25	14	....	....	....
	Albany	AHL	....	....	....	....	....	....	....	....	....	....	....	....	....	....	....	....	3	0	0	0	2	....	....	....
1995-96	**New Jersey**	**NHL**	**10**	**3**	**1**	**4**	**0**	**1**	**0**	**2**	**6**	**50.0**	**–1**	....	....	....	....	....	....	....	....	....	....	....	....	....
	Albany	AHL	68	28	43	71	104	....	....	....	....	....	....	....	....	....	....	....	4	1	2	3	0	....	....	....
1996-97	**New Jersey**	**NHL**	**70**	**12**	**20**	**32**	**62**	**3**	**0**	**3**	**106**	**11.3**	**7**	....	....	....	....	....	**9**	**0**	**0**	**0**	**2**	**0**	**0**	**0**
	Albany	AHL	3	1	3	4	7	....	....	....	....	....	....	....	....	....	....	....	....	....	....	....	....	....	....	....
1997-98	**New Jersey**	**NHL**	**80**	**15**	**13**	**28**	**97**	**7**	**0**	**1**	**135**	**11.1**	**–6**	....	....	....	....	....	**6**	**1**	**1**	**2**	**2**	**0**	**1**	**0**
1998-99	**New Jersey**	**NHL**	**76**	**11**	**12**	**23**	**66**	**3**	**0**	**1**	**145**	**7.6**	**–10**	**540**	**42.8**	**105**	**39**	**15:19**	**3**	**0**	**1**	**1**	**0**	**0**	**0**	**0**
99-2000	**New Jersey**	**NHL**	**35**	**3**	**3**	**6**	**16**	**0**	**0**	**0**	**41**	**7.3**	**–7**	**125**	**48.8**	**41**	**9**	**10:42**	....	....	....	....	....	....	....	....
	Vancouver	**NHL**	**12**	**3**	**2**	**5**	**2**	**0**	**0**	**1**	**15**	**20.0**	**1**	**70**	**45.7**	**15**	**3**	**12:41**	....	....	....	....	....	....	....	....
2000-01	**Vancouver**	**NHL**	**61**	**4**	**8**	**12**	**65**	**0**	**1**	**3**	**70**	**5.7**	**0**	**351**	**42.7**	**71**	**24**	**11:42**	**4**	**0**	**1**	**1**	**4**	**0**	**0**	**0**
2001-02	**Vancouver**	**NHL**	**29**	**1**	**5**	**6**	**31**	**0**	**0**	**0**	**24**	**4.2**	**–2**	**142**	**47.2**	**25**	**9**	**8:18**	....	....	....	....	....	....	....	....
	Phoenix	**NHL**	**19**	**1**	**1**	**2**	**20**	**0**	**0**	**0**	**18**	**5.6**	**–2**	**202**	**49.0**	**30**	**7**	**10:32**	**5**	**0**	**2**	**2**	**0**	**0**	**0**	**0**
	NHL Totals		**392**	**53**	**65**	**118**	**359**	**14**	**1**	**11**	**560**	**9.5**		**1430**	**44.8**	**287**	**91**	**12:16**	**27**	**1**	**5**	**6**	**8**	**0**	**1**	**0**

WHL East Second All-Star Team (1994)

Traded to **Vancouver** by **New Jersey** with Brendan Morrison for Alexander Mogilny, March 14, 2000. Traded to **Phoenix** by **Vancouver** with Drake Berehowsky for Todd Warriner, Trevor Letowski, Tyler Bouck and Phoenix's 3rd round choice in 2003 Entry Draft, December 28, 2001. Signed as a free agent by **Nashville**, July 24, 2002.

PELLERIN, Scott

(PEHL-ih-rihn, SKAWT) **DAL.**

Left wing. Shoots left. 5'11", 190 lbs. Born, Shediac, N.B., January 9, 1970. New Jersey's 4th choice, 47th overall, in 1989 Entry Draft.

			Regular Season																Playoffs							
Season	Club	League	GP	G	A	Pts	PIM	PP	SH	GW	S	%	+/-	TF	F%	H	SB	Min	GP	G	A	Pts	PIM	PP	SH	GW
1985-86	Moncton Flyers	NBAHA	45	65	34	99	34	....	....	....	....	....	....	....	....	....	....	....	....	....	....	....	....	....	....	....
1986-87	Notre Dame	SMHL	72	62	68	130	98	....	....	....	....	....	....	....	....	....	....	....	....	....	....	....	....	....	....	....
1987-88	Notre Dame	SJHL	57	37	49	86	139	....	....	....	....	....	....	....	....	....	....	....	....	....	....	....	....	....	....	....
1988-89	U. of Maine	H-East	45	29	33	62	92	....	....	....	....	....	....	....	....	....	....	....	....	....	....	....	....	....	....	....
1989-90	U. of Maine	H-East	42	22	34	56	68	....	....	....	....	....	....	....	....	....	....	....	....	....	....	....	....	....	....	....
1990-91	U. of Maine	H-East	43	23	25	48	60	....	....	....	....	....	....	....	....	....	....	....	....	....	....	....	....	....	....	....
1991-92	U. of Maine	H-East	37	*32	25	57	54	....	....	....	....	....	....	....	....	....	....	....	....	....	....	....	....	....	....	....
	Utica Devils	AHL	....	....	....	....	....	....	....	....	....	....	....	....	....	....	....	....	3	1	0	1	0	....	....	....
1992-93	**New Jersey**	**NHL**	**45**	**10**	**11**	**21**	**41**	**1**	**2**	**0**	**60**	**16.7**	**–1**	....	....	....	....	....	....	....	....	....	....	....	....	....
	Utica Devils	AHL	27	15	18	33	33	....	....	....	....	....	....	....	....	....	....	....	2	0	1	1	0	....	....	....
1993-94	**New Jersey**	**NHL**	**1**	**0**	**0**	**0**	**2**	**0**	**0**	**0**	**0**	**0.0**	**0**	....	....	....	....	....	....	....	....	....	....	....	....	....
	Albany	AHL	73	28	46	74	84	....	....	....	....	....	....	....	....	....	....	....	5	2	1	3	11	....	....	....
1994-95	Albany	AHL	74	23	33	56	95	....	....	....	....	....	....	....	....	....	....	....	14	6	4	10	8	....	....	....
1995-96	**New Jersey**	**NHL**	**6**	**2**	**1**	**3**	**0**	**0**	**0**	**0**	**9**	**22.2**	**1**	....	....	....	....	....	....	....	....	....	....	....	....	....
	Albany	AHL	75	35	47	82	142	....	....	....	....	....	....	....	....	....	....	....	4	0	3	3	10	....	....	....
1996-97	**St. Louis**	**NHL**	**54**	**8**	**10**	**18**	**35**	**0**	**2**	**2**	**76**	**10.5**	**12**	....	....	....	....	....	**6**	**0**	**0**	**0**	**6**	**0**	**0**	**0**
	Worcester	AHL	24	10	16	26	37	....	....	....	....	....	....	....	....	....	....	....	....	....	....	....	....	....	....	....
1997-98	**St. Louis**	**NHL**	**80**	**8**	**21**	**29**	**62**	**1**	**1**	**0**	**96**	**8.3**	**14**	....	....	....	....	....	**10**	**0**	**2**	**2**	**10**	**0**	**0**	**0**
1998-99	**St. Louis**	**NHL**	**80**	**20**	**21**	**41**	**42**	**0**	**5**	**4**	**138**	**14.5**	**1**	**6**	**66.7**	**90**	**50**	**17:18**	**8**	**1**	**0**	**1**	**4**	**0**	**0**	**0**
99-2000	**St. Louis**	**NHL**	**80**	**8**	**15**	**23**	**48**	**0**	**2**	**2**	**120**	**6.7**	**9**	**6**	**16.7**	**132**	**33**	**14:47**	**7**	**0**	**0**	**0**	**2**	**0**	**0**	**0**
2000-01	**Minnesota**	**NHL**	**58**	**11**	**28**	**39**	**45**	**2**	**2**	**2**	**117**	**9.4**	**6**	**47**	**31.9**	**111**	**36**	**18:40**	....	....	....	....	....	....	....	....
	Carolina	**NHL**	**19**	**0**	**5**	**5**	**6**	**0**	**0**	**0**	**21**	**0.0**	**–4**	**52**	**44.2**	**66**	**6**	**14:14**	**6**	**0**	**0**	**0**	**4**	**0**	**0**	**0**
2001-02	**Boston**	**NHL**	**35**	**1**	**5**	**6**	**6**	**0**	**0**	**0**	**41**	**2.4**	**–6**	**24**	**37.5**	**38**	**8**	**11:38**	....	....	....	....	....	....	....	....
	Dallas	**NHL**	**33**	**3**	**5**	**8**	**15**	**0**	**0**	**0**	**22**	**13.6**	**–5**	**8**	**25.0**	**35**	**9**	**9:13**	....	....	....	....	....	....	....	....
	NHL Totals		**491**	**71**	**122**	**193**	**302**	**4**	**14**	**10**	**700**	**10.1**		**143**	**37.8**	**472**	**142**	**15:11**	**37**	**1**	**2**	**3**	**26**	**0**	**0**	**0**

Hockey East Rookie of the Year (Shared with Rob Gaudreau (1989) • Hockey East First All-Star Team (1992) • Hockey East Player of the Year (1992) • NCAA East First All-American Team (1992) • Won Hobey Baker Memorial Award (Top U.S. Collegiate Player) (1992)

Signed as a free agent by **St. Louis**, July 10, 1996. Selected by **Minnesota** from **St. Louis** in Expansion Draft, June 23, 2000. Traded to **Carolina** by **Minnesota** for Askhat Rakhmatullin, Carolina's 3rd round choice (later traded to NY Rangers - NY Rangers selected Garth Murray) in 2001 Entry Draft and Carolina's compensatory 5th round choice (Armands Berzins) in 2002 Entry Draft, March 1, 2001. Signed as a free agent by **Boston**, July 26, 2001. Claimed on waivers by **Dallas** from **Boston**, January 12, 2002.

PELTONEN, Ville

(PEHL-TOH-nen, VIHL-ee)

Left wing. Shoots left. 5'11", 188 lbs. Born, Vantaa, Finland, May 24, 1973. San Jose's 4th choice, 58th overall, in 1993 Entry Draft.

			Regular Season																Playoffs							
Season	Club	League	GP	G	A	Pts	PIM	PP	SH	GW	S	%	+/-	TF	F%	H	SB	Min	GP	G	A	Pts	PIM	PP	SH	GW
1990-91	HIFK Jr.	Finn-Jr.	36	21	16	37	16	....	....	....	....	....	....	....	....	....	....	....	7	2	3	5	10	....	....	....
1991-92	HIFK Jr.	Finn-Jr.	37	28	23	51	28	....	....	....	....	....	....	....	....	....	....	....	4	0	2	2	0	....	....	....
	HIFK Helsinki	Finland	6	0	0	0	0	....	....	....	....	....	....	....	....	....	....	....	....	....	....	....	....	....	....	....
1992-93	HIFK Jr.	Finn-Jr.	2	4	2	6	4	....	....	....	....	....	....	....	....	....	....	....	....	....	....	....	....	....	....	....
	HIFK Helsinki	Finland	46	13	24	37	16	....	....	....	....	....	....	....	....	....	....	....	4	0	2	2	2	....	....	....
1993-94	HIFK Helsinki	Finland	43	16	22	38	14	....	....	....	....	....	....	....	....	....	....	....	3	0	0	0	2	....	....	....
	Finland	Olympics	8	4	3	7	0	....	....	....	....	....	....	....	....	....	....	....	....	....	....	....	....	....	....	....
1994-95	HIFK Helsinki	Finland	45	20	16	36	16	....	....	....	....	....	....	....	....	....	....	....	3	0	0	0	0	....	....	....
1995-96	**San Jose**	**NHL**	**31**	**2**	**11**	**13**	**14**	**0**	**0**	**0**	**58**	**3.4**	**–7**	....	....	....	....	....	....	....	....	....	....	....	....	....
	Kansas City	IHL	29	5	13	18	8	....	....	....	....	....	....	....	....	....	....	....	....	....	....	....	....	....	....	....
1996-97	**San Jose**	**NHL**	**28**	**2**	**3**	**5**	**0**	**1**	**0**	**0**	**35**	**5.7**	**–8**	....	....	....	....	....	....	....	....	....	....	....	....	....
	Kentucky	AHL	40	22	30	52	21	....	....	....	....	....	....	....	....	....	....	....	....	....	....	....	....	....	....	....
1997-98	Vastra Frolunda	Sweden	45	22	29	*51	44	....	....	....	....	....	....	....	....	....	....	....	7	4	2	6	0	....	....	....
	Finland	Olympics	6	2	1	3	6	....	....	....	....	....	....	....	....	....	....	....	....	....	....	....	....	....	....	....
1998-99	**Nashville**	**NHL**	**14**	**5**	**5**	**10**	**2**	**1**	**0**	**0**	**31**	**16.1**	**1**	**0**	**0.0**	**4**	**4**	**15:54**	....	....	....	....	....	....	....	....
99-2000	**Nashville**	**NHL**	**79**	**6**	**22**	**28**	**22**	**2**	**0**	**2**	**125**	**4.8**	**–1**	**1**	**100.0**	**41**	**14**	**14:41**	....	....	....	....	....	....	....	....
2000-01	**Nashville**	**NHL**	**23**	**3**	**1**	**4**	**2**	**0**	**0**	**0**	**38**	**7.9**	**–7**	**2**	**0.0**	**13**	**6**	**11:40**	....	....	....	....	....	....	....	....
	Milwaukee	IHL	53	27	33	60	26	....	....	....	....	....	....	....	....	....	....	....	5	2	1	3	6	....	....	....
2001-02	Jokerit Helsinki	Finland	30	11	18	29	8	....	....	....	....	....	....	....	....	....	....	....	....	....	....	....	....	....	....	....
	NHL Totals		**175**	**18**	**42**	**60**	**40**	**4**	**0**	**2**	**287**	**6.3**		**3**	**33.3**	**58**	**24**	**14:14**	....	....	....	....	....	....	....	....

• IHL Second All-Star Team (2001)

Traded to **Nashville** by **San Jose** for Nashville's 5th round choice (later traded to Phoenix - Phoenix selected Josh Blackburn) in 1998 Entry Draft, June 26, 1998. • Missed majority of 1998-99 season recovering from shoulder surgery, December 10, 1998.

PELUSO, Mike

(puh-LOO-soh, MIGHK) **CHI.**

Right wing. Shoots right. 6'1", 208 lbs. Born, Bismark, ND, September 2, 1974. Calgary's 12th choice, 253rd overall, in 1994 Entry Draft.

			Regular Season																Playoffs							
Season	Club	League	GP	G	A	Pts	PIM	PP	SH	GW	S	%	+/-	TF	F%	H	SB	Min	GP	G	A	Pts	PIM	PP	SH	GW
1993-94	Omaha Lancers	USHL	48	36	29	65	77	....	....	....	....	....	....	....	....	....	....	....	....	....	....	....	....	....	....	....
1994-95	U. Minn-Duluth	WCHA	38	11	23	34	38	....	....	....	....	....	....	....	....	....	....	....	....	....	....	....	....	....	....	....
1995-96	U. Minn-Duluth	WCHA	38	25	19	44	64	....	....	....	....	....	....	....	....	....	....	....	....	....	....	....	....	....	....	....
1996-97	U. Minn-Duluth	WCHA	37	20	20	40	53	....	....	....	....	....	....	....	....	....	....	....	....	....	....	....	....	....	....	....
1997-98	U. Minn-Duluth	WCHA	40	24	21	45	100	....	....	....	....	....	....	....	....	....	....	....	....	....	....	....	....	....	....	....
1998-99	Portland Pirates	AHL	26	7	6	13	6	....	....	....	....	....	....	....	....	....	....	....	....	....	....	....	....	....	....	....
99-2000	Portland Pirates	AHL	71	25	29	54	86	....	....	....	....	....	....	....	....	....	....	....	4	2	0	2	0	....	....	....
2000-01	Portland Pirates	AHL	19	12	10	22	17	....	....	....	....	....	....	....	....	....	....	....	....	....	....	....	....	....	....	....
	Worcester	AHL	44	17	23	40	22	....	....	....	....	....	....	....	....	....	....	....	11	3	3	6	4	....	....	....
2001-02	**Chicago**	**NHL**	**37**	**4**	**2**	**6**	**19**	**0**	**0**	**1**	**45**	**8.9**	**–3**	**1**	**0.0**	**60**	**8**	**9:17**	....	....	....	....	....	....	....	....
	Norfolk Admirals	AHL	29	18	9	27	4	....	....	....	....	....	....	....	....	....	....	....	4	1	0	1	0	....	....	....
	NHL Totals		**37**	**4**	**2**	**6**	**19**	**0**	**0**	**1**	**45**	**8.9**		**1**	**0.0**	**60**	**8**	**9:17**	....	....	....	....	....	....	....	....

WCHA Second All-Star Team (1997)

Signed as a free agent by **Washington**, October 9, 1998. Traded to **St. Louis** by **Washington** for Derek Bekar, November 29, 2000. Signed as a free agent by **Chicago**, August 1, 2001.

Season	Club	League	GP	G	A	Pts	PIM	PP	SH	GW	S	%	+/-	TF	F%	H	SB	Min	Playoffs GP	G	A	Pts	PIM	PP	SH	GW

PERREAULT, Yanic

(puh-ROH, YAH-nihk) **MTL.**

Center. Shoots left. 5'11", 185 lbs. Born, Sherbrooke, Que., April 4, 1971. Toronto's 1st choice, 47th overall, in 1991 Entry Draft.

Season	Club	League	GP	G	A	Pts	PIM	PP	SH	GW	S	%	+/-	TF	F%	H	SB	Min	GP	G	A	Pts	PIM	PP	SH	GW
1987-88	L'est Cantonniers	QAAA	42	*70	57	*127	14												8	12	10	22	6			
1988-89	Trois-Rivieres	QMJHL	70	53	55	108	48																			
1989-90	Trois-Rivieres	QMJHL	63	51	63	114	75												7	6	5	11	19			
1990-91	Trois-Rivieres	QMJHL	67	*87	98	*185	103												6	4	7	11	6			
1991-92	St. John's	AHL	62	38	38	76	19												16	7	8	15	4			
1992-93	St. John's	AHL	79	49	46	95	56												9	4	5	9	2			
1993-94	**Toronto**	**NHL**	**13**	**3**	**3**	**6**	**0**	**2**	**0**	**0**	**24**	**12.5**	**1**													
	St. John's	AHL	62	45	60	105	38												11	*12	6	18	14			
1994-95	Phoenix	IHL	68	51	48	99	52																			
	Los Angeles	**NHL**	**26**	**2**	**5**	**7**	**20**	**0**	**0**	**1**	**43**	**4.7**	**3**													
1995-96	**Los Angeles**	**NHL**	**78**	**25**	**24**	**49**	**16**	**8**	**3**	**7**	**175**	**14.3**	**–11**													
1996-97	**Los Angeles**	**NHL**	**41**	**11**	**14**	**25**	**20**	**1**	**1**	**0**	**98**	**11.2**	**0**													
1997-98	**Los Angeles**	**NHL**	**79**	**28**	**20**	**48**	**32**	**3**	**2**	**3**	**206**	**13.6**	**6**						**4**	**1**	**2**	**3**	**6**	**1**	**0**	**0**
1998-99	**Los Angeles**	**NHL**	**64**	**10**	**17**	**27**	**30**	**2**	**2**	**1**	**113**	**8.8**	**–3**	**1024**	**56.5**	**35**	**33**	**15:24**								
	Toronto	**NHL**	**12**	**7**	**8**	**15**	**12**	**2**	**1**	**2**	**28**	**25.0**	**10**	**164**	**62.8**	**9**	**1**	**13:20**	**17**	**3**	**6**	**9**	**6**	**0**	**0**	**2**
99-2000	**Toronto**	**NHL**	**58**	**18**	**27**	**45**	**22**	**5**	**0**	**4**	**114**	**15.8**	**3**	**987**	**61.8**	**23**	**10**	**15:18**	**1**	**0**	**1**	**1**	**0**	**0**	**0**	**0**
2000-01	**Toronto**	**NHL**	**76**	**24**	**28**	**52**	**52**	**5**	**0**	**2**	**134**	**17.9**	**0**	**1055**	**62.7**	**22**	**21**	**14:01**	**11**	**2**	**3**	**5**	**4**	**1**	**0**	**1**
2001-02	**Montreal**	**NHL**	**82**	**27**	**29**	**56**	**40**	**6**	**0**	**7**	**156**	**17.3**	**–3**	**1485**	**61.3**	**18**	**30**	**16:47**	**11**	**3**	**5**	**8**	**0**	**2**	**0**	**1**
	NHL Totals		**529**	**155**	**175**	**330**	**244**	**34**	**9**	**27**	**1091**	**14.2**		**4715**	**60.7**	**107**	**95**	**15:20**	**44**	**9**	**17**	**26**	**16**	**4**	**0**	**4**

QMJHL All-Rookie Team (1989) • QMJHL Offensive Rookie of the Year (1989) • Canadian Major Junior Rookie of the Year (1989) • QMJHL First All-Star Team (1991) • QMJHL MVP (1991)

Traded to **LA Kings** by **Toronto** for LA Kings' 4th round choice (later traded to Philadelphia - later traded to LA Kings - LA Kings selected Mikael Simons) in 1996 Entry Draft, July 11, 1994. Traded to **Toronto** by **Los Angeles** for Jason Podollan and Toronto's 3rd round choice (Cory Campbell) in 1999 Entry Draft, March 23, 1999. Signed as a free agent by **Montreal**, July 4, 2001.

PERROTT, Nathan

(PEHR-roht, NAY-than) **NSH.**

Right wing. Shoots right. 6', 225 lbs. Born, Owen Sound, Ont., December 8, 1976. New Jersey's 2nd choice, 44th overall, in 1995 Entry Draft.

Season	Club	League	GP	G	A	Pts	PIM	PP	SH	GW	S	%	+/-	TF	F%	H	SB	Min	GP	G	A	Pts	PIM	PP	SH	GW
1992-93	Walkerton	OJHL-C	25	6	13	19	45																			
1993-94	St. Mary's	OJHL-B	41	11	26	37	249																			
1994-95	Oshawa Generals	OHL	63	18	28	46	233												2	1	1	2	9			
1995-96	Oshawa Generals	OHL	59	30	32	62	158												5	2	3	5	8			
	Albany	AHL	4	0	0	0	12																			
1996-97	Oshawa Generals	OHL	5	1	0	1	17																			
	Sault Ste. Marie	OHL	37	18	23	41	120												11	5	5	10	60			
1997-98	Indianapolis Ice	IHL	31	4	3	7	76																			
	Jacksonville	ECHL	30	6	8	14	135																			
1998-99	Indianapolis Ice	IHL	72	14	11	25	307												7	3	1	4	45			
99-2000	Cleveland	IHL	65	12	9	21	248												9	2	1	3	19			
2000-01	Norfolk Admirals	AHL	73	11	17	28	268												9	2	0	2	18			
2001-02	Norfolk Admirals	AHL	2	0	0	0	5																			
	Nashville	**NHL**	**22**	**1**	**2**	**3**	**74**	**0**	**0**	**1**	**7**	**14.3**	**–1**	**1**	**0.0**	**26**	**2**	**4:55**								
	Milwaukee	AHL	56	6	10	16	190																			
	NHL Totals		**22**	**1**	**2**	**3**	**74**	**0**	**0**	**1**	**7**	**14.3**		**1**	**0.0**	**26**	**2**	**4:55**								

Signed as a free agent by **Chicago**, August 27, 1997. Traded to **Nashville** by **Chicago** for future considerations, October 9, 2001.

PERSSON, Ricard

(PAIR-suhn, RIH-kahrd)

Defense. Shoots left. 6'1", 201 lbs. Born, Ostersund, Sweden, August 24, 1969. New Jersey's 2nd choice, 23rd overall, in 1987 Entry Draft.

Season	Club	League	GP	G	A	Pts	PIM	PP	SH	GW	S	%	+/-	TF	F%	H	SB	Min	GP	G	A	Pts	PIM	PP	SH	GW
1984-85	Ostersunds IK	Swede-2	13	0	3	3	6																			
1985-86	Ostersunds IK	Swede-2	24	2	2	4	16																			
1986-87	Ostersunds IK	Swede-2	31	10	11	21	28																			
1987-88	Leksands IF	Sweden	31	2	0	2	8												2	0	1	1	2			
1988-89	Leksands IF	Sweden	33	2	4	6	28												9	0	1	1	6			
1989-90	Leksands IF	Sweden	43	9	10	19	62												3	0	0	0	6			
1990-91	Leksands IF	Sweden	37	6	9	15	42																			
1991-92	Leksands IF	Sweden	21	0	7	7	28																			
1992-93	Leksands IF	Sweden	36	7	15	22	63												2	0	2	2	0			
1993-94	Malmo IF	Sweden	40	11	9	20	38												11	2	0	2	12			
1994-95	Malmo IF	Sweden	31	3	13	16	38												9	0	2	2	8			
	Albany	AHL	3	0	0	0	0												9	3	5	8	7			
1995-96	**New Jersey**	**NHL**	**12**	**2**	**1**	**3**	**8**	**1**	**0**	**0**	**41**	**4.9**	**5**													
	Albany	AHL	67	15	31	46	59												4	0	0	0	7			
1996-97	**New Jersey**	**NHL**	**1**	**0**	**0**	**0**	**0**	**0**	**0**	**0**	**2**	**0.0**	**0**													
	Albany	AHL	13	1	4	5	8																			
	St. Louis	**NHL**	**53**	**4**	**8**	**12**	**45**	**1**	**0**	**0**	**68**	**5.9**	**–2**						**6**	**0**	**0**	**0**	**27**	**0**	**0**	**0**
1997-98	**St. Louis**	**NHL**	**1**	**0**	**0**	**0**	**0**	**0**	**0**	**0**	**0**	**0.0**	**0**													
	Worcester	AHL	32	2	16	18	58												10	3	7	10	24			
1998-99	**St. Louis**	**NHL**	**54**	**1**	**12**	**13**	**94**	**0**	**0**	**0**	**52**	**1.9**	**4**	**1**	**100.0**	**33**	**81**	**19:58**	**13**	**0**	**3**	**3**	**17**	**0**	**0**	**0**
	Worcester	AHL	19	6	4	10	42																			
99-2000	**St. Louis**	**NHL**	**41**	**0**	**8**	**8**	**38**	**0**	**0**	**0**	**30**	**0.0**	**–2**	**0**	**0.0**	**17**	**32**	**12:24**	**3**	**1**	**0**	**1**	**0**	**0**	**0**	**0**
	Worcester	AHL	2	0	1	1	0																			
2000-01	**Ottawa**	**NHL**	**33**	**1**	**8**	**9**	**35**	**0**	**0**	**1**	**43**	**2.3**	**8**	**0**	**0.0**	**26**	**60**	**17:06**	**2**	**0**	**0**	**0**	**0**	**0**	**0**	**0**
2001-02	**Ottawa**	**NHL**	**34**	**2**	**7**	**9**	**42**	**0**	**0**	**0**	**35**	**5.7**	**3**	**0**	**0.0**	**44**	**29**	**14:34**	**2**	**0**	**0**	**0**	**15**	**0**	**0**	**0**
	NHL Totals		**229**	**10**	**44**	**54**	**262**	**2**	**0**	**1**	**271**	**3.7**		**1**	**100.0**	**120**	**202**	**16:20**	**26**	**1**	**3**	**4**	**59**	**0**	**0**	**0**

Traded to **St. Louis** by **New Jersey** with Mike Peluso for Ken Sutton and St. Louis' 2nd round choice (Brett Clouthier) in 1999 Entry Draft, November 26, 1996. Signed as a free agent by **Ottawa**, July 12, 2000. • Missed majority of 2000-01 season recovering from ankle injury originally suffered in game vs. Pittsburgh, October 25, 2001. • Spent majority of 2001-02 season on **Ottawa** practice roster. Signed as a free agent by **Eisbaren Berlin** (Germany), August 8, 2002.

PETERSEN, Toby

(PEE-tuhr-sohn, TOH-bee) **PIT.**

Center. Shoots left. 5'9", 197 lbs. Born, Minneapolis, MN, October 27, 1978. Pittsburgh's 9th choice, 244th overall, in 1998 Entry Draft.

Season	Club	League	GP	G	A	Pts	PIM	PP	SH	GW	S	%	+/-	TF	F%	H	SB	Min	GP	G	A	Pts	PIM	PP	SH	GW
1995-96	Jefferson High	Hi-School	25	29	30	59	....																			
1996-97	Colorado College	WCHA	40	17	21	38	18																			
1997-98	Colorado College	WCHA	40	16	17	33	34																			
1998-99	Colorado College	WCHA	21	12	12	24	2																			
99-2000	Colorado College	WCHA	37	14	19	33	8																			
2000-01	**Pittsburgh**	**NHL**	**12**	**2**	**6**	**8**	**4**	**0**	**0**	**1**	**25**	**8.0**	**3**	**39**	**35.9**	**5**	**2**	**13:22**								
	Wilkes-Barre	AHL	73	26	41	67	22												21	7	6	13	4			
2001-02	**Pittsburgh**	**NHL**	**79**	**8**	**10**	**18**	**4**	**1**	**1**	**0**	**116**	**6.9**	**–15**	**338**	**45.6**	**34**	**35**	**12:16**								
	NHL Totals		**91**	**10**	**16**	**26**	**8**	**1**	**1**	**1**	**141**	**7.1**		**377**	**44.6**	**39**	**37**	**12:25**								

Minnesota High School All-State, All-Metro and All-Conference Player of the Year (1996) • WCHA All-Rookie Team (1997)

PETERSON, Brent

(PEE-tuhr-sohn, BREHNT)

Left wing. Shoots left. 6'3", 200 lbs. Born, Calgary, Alta., July 20, 1972. Tampa Bay's 1st choice, 3rd overall, in 1993 Supplemental Draft.

Season	Club	League	GP	G	A	Pts	PIM	PP	SH	GW	S	%	+/-	TF	F%	H	SB	Min	GP	G	A	Pts	PIM	PP	SH	GW
1990-91	Thunder Bay	USHL	48	27	40	67	10												10	8	9	17	4			
1991-92	Michigan Tech	WCHA	39	11	9	20	18																			
1992-93	Michigan Tech	WCHA	37	24	18	42	32																			
1993-94	Michigan Tech	WCHA	43	25	21	46	30																			
1994-95	Michigan Tech	WCHA	39	20	16	36	27																			
1995-96	Atlanta Knights	IHL	69	9	19	28	33												3	0	0	0	0			
1996-97	**Tampa Bay**	**NHL**	**17**	**2**	**0**	**2**	**4**	**0**	**0**	**0**	**11**	**18.2**	**–4**													
	Adirondack	AHL	52	22	23	45	56												4	3	1	4	2			
1997-98	**Tampa Bay**	**NHL**	**19**	**5**	**0**	**5**	**2**	**0**	**0**	**0**	**15**	**33.3**	**–2**													
	Milwaukee	IHL	63	20	39	59	48												8	5	3	8	22			
1998-99	**Tampa Bay**	**NHL**	**20**	**2**	**1**	**3**	**0**	**0**	**0**	**0**	**16**	**12.5**	**–2**	**1**	**100.0**	**14**	**3**	**8:58**								
	Cleveland	IHL	18	6	7	13	31																			
	Grand Rapids	IHL	17	7	5	12	14																			

			Regular Season																Playoffs							
Season	Club	League	GP	G	A	Pts	PIM	PP	SH	GW	S	%	+/-	TF	F%	H	SB	Min	GP	G	A	Pts	PIM	PP	SH	GW
99-2000	Milwaukee	IHL	66	8	24	32	62												3	3	2	5	4			
2000-01	Langnau	Swiss	10	3	2	5	8																			
	Kassel Huskies	Germany	28	7	14	21	40												8	2	2	4	8			
2001-02	Kassel Huskies	Germany	51	8	15	23	46												7	1	2	3	10			
	NHL Totals		56	9	1	10	6	0	0	0	42	21.4		1	100.0	14	3	8:58								

Traded to **Pittsburgh** by **Tampa Bay** for cash, March 18, 1999. Signed as a free agent by **Nashville**, July 24, 1999.

PETROV, Oleg

(PEH-trahf, OH-lehg) **MTL.**

Right wing. Shoots left. 5'9", 172 lbs. Born, Moscow, USSR, April 18, 1971. Montreal's 9th choice, 127th overall, in 1991 Entry Draft.

Season	Club	League	GP	G	A	Pts	PIM	PP	SH	GW	S	%	+/-	TF	F%	H	SB	Min	GP	G	A	Pts	PIM	PP	SH	GW
1989-90	CSKA Moscow	USSR	30	4	7	11	4																			
1990-91	CSKA Moscow	USSR	43	7	4	11	8																			
1991-92	CSKA Moscow	CIS	42	10	16	26	8																			
1992-93	**Montreal**	**NHL**	**9**	**2**	**1**	**3**	**10**	**0**	**0**	**1**	**20**	**10.0**	**2**						**1**	**0**	**0**	**0**	**0**	**0**	**0**	**0**
	Fredericton	AHL	55	26	29	55	36												5	4	1	5	0			
1993-94	**Montreal**	**NHL**	**55**	**12**	**15**	**27**	**2**	**1**	**0**	**1**	**107**	**11.2**	**7**						**2**	**0**	**0**	**0**	**0**	**0**	**0**	**0**
	Fredericton	AHL	23	8	20	28	18																			
1994-95	**Montreal**	**NHL**	**12**	**2**	**3**	**5**	**4**	**0**	**0**	**0**	**26**	**7.7**	**-7**													
	Fredericton	AHL	17	7	11	18	12												17	5	6	11	10			
1995-96	**Montreal**	**NHL**	**36**	**4**	**7**	**11**	**23**	**0**	**0**	**2**	**44**	**9.1**	**-9**						**5**	**0**	**1**	**1**	**0**	**0**	**0**	**0**
	Fredericton	AHL	22	12	18	30	71												6	2	6	8	0			
1996-97	HC Ambri-Piotta	Swiss	45	24	28	52	44																			
	HC Merano	Italy	12	5	12	17	4																			
1997-98	HC Ambri-Piotta	Swiss	40	30	*63	*93	60												14	11	11	22	40			
1998-99	HC Ambri-Piotta	Swiss	45	35	*52	*87	52												15	9	11	*20	32			
99-2000	**Montreal**	**NHL**	**44**	**2**	**24**	**26**	**8**	**1**	**0**	**0**	**96**	**2.1**	**10**	**9**	**33.3**	**36**	**10**	**15:51**								
	Quebec	AHL	16	7	7	14	4																			
2000-01	**Montreal**	**NHL**	**81**	**17**	**30**	**47**	**24**	**4**	**2**	**1**	**158**	**10.8**	**-11**	**8**	**25.0**	**81**	**27**	**18:33**								
2001-02	**Montreal**	**NHL**	**75**	**24**	**17**	**41**	**12**	**3**	**1**	**6**	**152**	**15.8**	**-4**	**4**	**25.0**	**60**	**20**	**18:56**	**12**	**1**	**5**	**6**	**2**	**0**	**0**	**1**
	NHL Totals		**312**	**63**	**97**	**160**	**83**	**9**	**3**	**11**	**603**	**10.4**		**21**	**28.6**	**177**	**57**	**18:06**	**20**	**1**	**6**	**7**	**2**	**0**	**0**	**1**

NHL All-Rookie Team (1994)

Signed as a free agent by **Montreal**, July 15, 1999.

PETROVICKY, Robert

(PEHT-roh-vih-kee, RAW-buhrt)

Center. Shoots left. 5'11", 172 lbs. Born, Kosice, Czech., October 26, 1973. Hartford's 1st choice, 9th overall, in 1992 Entry Draft.

Season	Club	League	GP	G	A	Pts	PIM	PP	SH	GW	S	%	+/-	TF	F%	H	SB	Min	GP	G	A	Pts	PIM	PP	SH	GW
1990-91	Dukla Trencin	Czech	33	9	14	23	12																			
1991-92	Dukla Trencin	Czech	46	25	36	61	28																			
1992-93	**Hartford**	**NHL**	**42**	**3**	**6**	**9**	**45**	**0**	**0**	**0**	**41**	**7.3**	**-10**													
	Springfield	AHL	16	5	3	8	39												15	5	6	11	14			
1993-94	Dukla Trencin	Slovakia	1	0	0	0	0																			
	Hartford	**NHL**	**33**	**6**	**5**	**11**	**39**	**1**	**0**	**0**	**33**	**18.2**	**-1**													
	Springfield	AHL	30	16	8	24	39												4	0	2	2	4			
	Slovakia	Olympics	8	1	6	7	18																			
1994-95	Springfield	AHL	74	30	52	82	121																			
	Hartford	**NHL**	**2**	**0**	**0**	**0**	**0**	**0**	**0**	**0**	**1**	**0.0**	**0**													
1995-96	Springfield	AHL	9	4	8	12	18																			
	Detroit Vipers	IHL	12	5	3	8	16																			
	Dallas	**NHL**	**5**	**1**	**1**	**2**	**0**	**1**	**0**	**1**	**3**	**33.3**	**1**													
	Michigan K-Wings	IHL	50	23	23	46	63												7	3	1	4	16			
1996-97	**St. Louis**	**NHL**	**44**	**7**	**12**	**19**	**10**	**0**	**0**	**1**	**54**	**13.0**	**2**						**2**	**0**	**0**	**0**	**0**	**0**	**0**	**0**
	Worcester	AHL	12	5	4	9	19																			
1997-98	Worcester	AHL	65	27	34	61	97												10	3	4	7	12			
	Slovakia	Olympics	4	2	1	3	0																			
1998-99	**Tampa Bay**	**NHL**	**28**	**3**	**4**	**7**	**6**	**0**	**0**	**0**	**32**	**9.4**	**-8**	**35**	**34.3**	**21**	**8**	**10:33**								
	Grand Rapids	IHL	49	26	32	58	87																			
99-2000	**Tampa Bay**	**NHL**	**43**	**7**	**10**	**17**	**14**	**1**	**0**	**0**	**50**	**14.0**	**2**	**38**	**47.4**	**32**	**5**	**10:05**								
	Grand Rapids	IHL	7	5	3	8	4																			
2000-01	**NY Islanders**	**NHL**	**11**	**0**	**0**	**0**	**4**	**0**	**0**	**0**	**3**	**0.0**	**-1**	**15**	**46.7**	**1**	**0**	**3:26**								
	Chicago Wolves	IHL	23	13	10	23	22																			
	MoDo	Sweden	7	3	2	5	10												7	1	1	2	6			
2001-02	HC Ambri-Piotta	Swiss	38	23	23	46	32												5	2	2	4	18			
	Slovakia	Olympics	4	1	1	2	2																			
	NHL Totals		**208**	**27**	**38**	**65**	**118**	**3**	**0**	**2**	**217**	**12.4**		**88**	**42.0**	**54**	**13**	**9:21**	**2**	**0**	**0**	**0**	**0**	**0**	**0**	**0**

Traded to **Dallas** by **Hartford** for Dan Kesa, November 29, 1995. Signed as a free agent by **St. Louis**, September 6, 1996. Signed as a free agent by **Tampa Bay**, February 15, 1999. Signed as a free agent by **NY Islanders**, July 28, 2000.

PETROVICKY, Ronald

(PEHT-roh-vih-kee, RAW-nohld) **CGY.**

Right wing. Shoots right. 5'11", 190 lbs. Born, Zilina, Czech., February 15, 1977. Calgary's 9th choice, 228th overall, in 1996 Entry Draft.

Season	Club	League	GP	G	A	Pts	PIM	PP	SH	GW	S	%	+/-	TF	F%	H	SB	Min	GP	G	A	Pts	PIM	PP	SH	GW
1993-94	Dukla Trencin Jr.	Slovak-Jr.	36	28	27	55	42																			
	Dukla Trencin	Slovakia	1	0	0	0	0																			
1994-95	Tri-City	WHL	39	4	11	15	86																			
	Prince George	WHL	21	4	6	10	37																			
1995-96	Prince George	WHL	39	19	21	40	61																			
1996-97	Prince George	WHL	72	32	37	69	119												15	4	9	13	31			
1997-98	Regina Pats	WHL	71	64	49	113	168												9	2	4	6	11			
1998-99	Saint John	AHL	78	12	21	33	114												7	1	2	3	19			
99-2000	Saint John	AHL	67	23	33	56	131												3	1	1	2	6			
2000-01	**Calgary**	**NHL**	**30**	**4**	**5**	**9**	**54**	**1**	**0**	**1**	**30**	**13.3**	**0**	**7**	**42.9**	**68**	**11**	**11:33**								
2001-02	**Calgary**	**NHL**	**77**	**5**	**7**	**12**	**85**	**1**	**0**	**1**	**78**	**6.4**	**0**	**28**	**46.4**	**150**	**22**	**11:42**								
	NHL Totals		**107**	**9**	**12**	**21**	**139**	**2**	**0**	**2**	**108**	**8.3**		**35**	**45.7**	**218**	**33**	**11:39**								

WHL East Second All-Star Team (1998)

• Missed majority of 2000-01 season recovering from wrist injury suffered in game vs. Detroit, October 5, 2000.

PETTINGER, Matt

(PEH-tihn-juhr, MAT) **WSH.**

Left wing. Shoots left. 6'1", 205 lbs. Born, Edmonton, Alta, October 22, 1980. Washington's 2nd choice, 43rd overall, in 2000 Entry Draft.

Season	Club	League	GP	G	A	Pts	PIM	PP	SH	GW	S	%	+/-	TF	F%	H	SB	Min	GP	G	A	Pts	PIM	PP	SH	GW
1994-95	Victoria Racquet	BCAHA	55	52	48	100	41																			
1995-96	Victoria Racquet	BCAHA	60	80	65	145	45																			
1996-97	Victoria Salsa	BCHL	49	22	14	36	31																			
1997-98	Victoria Salsa	BCHL	55	22	20	42	56												7	5	1	6	8			
1998-99	U. of Denver	WCHA	33	6	14	20	44																			
99-2000	U. of Denver	WCHA	19	2	6	8	49																			
	Calgary Hitmen	WHL	27	14	6	20	41												11	2	6	8	30			
2000-01	**Washington**	**NHL**	**10**	**0**	**0**	**0**	**2**	**0**	**0**	**0**	**6**	**0.0**	**-1**	**2**	**50.0**	**13**	**1**	**7:47**								
	Portland Pirates	AHL	64	19	17	36	92												2	0	0	0	4			
2001-02	**Washington**	**NHL**	**61**	**7**	**3**	**10**	**44**	**1**	**0**	**1**	**73**	**9.6**	**-8**	**5**	**20.0**	**99**	**8**	**9:39**								
	Portland Pirates	AHL	9	3	3	6	24																			
	NHL Totals		**71**	**7**	**3**	**10**	**46**	**1**	**0**	**1**	**79**	**8.9**		**7**	**28.6**	**112**	**9**	**9:23**								

Left **U. of Denver** (WCHA) and signed as a free agent with **Calgary** (WHL), January 10, 2000.

PHILLIPS, Chris
(FIHL-ihps, KRIHS) **OTT.**

Defense. Shoots left. 6'3", 215 lbs. Born, Calgary, Alta., March 9, 1978. Ottawa's 1st choice, 1st overall, in 1996 Entry Draft.

			Regular Season																Playoffs							
Season	Club	League	GP	G	A	Pts	PIM	PP	SH	GW	S	%	+/-	TF	F%	H	SB	Min	GP	G	A	Pts	PIM	PP	SH	GW
1993-94	Fort McMurray	AJHL	56	6	16	22	72												10	0	3	3	16			
1994-95	Fort McMurray	AJHL	48	16	32	48	127												11	4	2	6	10			
1995-96	Prince Albert	WHL	61	10	30	40	97												18	2	12	14	30			
1996-97	Prince Albert	WHL	32	3	23	26	58																			
	Lethbridge	WHL	26	4	18	22	28												19	4	*21	25	20			
1997-98	**Ottawa**	**NHL**	**72**	**5**	**11**	**16**	**38**	**2**	**0**	**2**	**107**	**4.7**	**2**						**11**	**0**	**2**	**2**	**2**	**0**	**0**	**0**
1998-99	**Ottawa**	**NHL**	**34**	**3**	**3**	**6**	**32**	**2**	**0**	**0**	**51**	**5.9**	**-5**	**0**	**0.0**	**53**	**28**	**18:06**	**3**	**0**	**0**	**0**	**0**	**0**	**0**	**0**
99-2000	**Ottawa**	**NHL**	**65**	**5**	**14**	**19**	**39**	**0**	**0**	**1**	**96**	**5.2**	**12**	**0**	**0.0**	**143**	**50**	**16:50**	**6**	**0**	**1**	**1**	**4**	**0**	**0**	**0**
2000-01	**Ottawa**	**NHL**	**73**	**2**	**12**	**14**	**31**	**2**	**0**	**0**	**77**	**2.6**	**8**	**1**	**0.0**	**136**	**102**	**21:28**	**1**	**1**	**0**	**1**	**0**	**0**	**0**	**0**
2001-02	**Ottawa**	**NHL**	**63**	**6**	**16**	**22**	**29**	**1**	**0**	**1**	**103**	**5.8**	**5**	**0**	**0.0**	**137**	**43**	**19:31**	**12**	**0**	**0**	**0**	**12**	**0**	**0**	**0**
	NHL Totals		**307**	**21**	**56**	**77**	**169**	**7**	**0**	**4**	**434**	**4.8**		**1**	**0.0**	**469**	**223**	**19:10**	**33**	**1**	**3**	**4**	**18**	**0**	**0**	**0**

WHL East First All-Star Team (1997) • Canadian Major Junior First All-Star Team (1997)

• Missed majority of 1998-99 season recovering from ankle injury suffered in game vs. Buffalo, December 30, 1998.

PICARD, Michel
(PEE-cahr, mih-SHEHL) **DET.**

Left wing. Shoots left. 5'11", 190 lbs. Born, Beauport, Que., November 7, 1969. Hartford's 8th choice, 178th overall, in 1989 Entry Draft.

			Regular Season																Playoffs							
Season	Club	League	GP	G	A	Pts	PIM	PP	SH	GW	S	%	+/-	TF	F%	H	SB	Min	GP	G	A	Pts	PIM	PP	SH	GW
1985-86	Ste-Foy	QAAA	42	53	34	87																				
1986-87	Trois-Rivieres	QMJHL	66	33	35	68	53																			
1987-88	Trois-Rivieres	QMJHL	69	40	55	95	71																			
1988-89	Trois-Rivieres	QMJHL	66	59	81	140	170												4	1	3	4	2			
1989-90	Binghamton	AHL	67	16	24	40	98																			
1990-91	**Hartford**	**NHL**	**5**	**1**	**0**	**1**	**2**	**0**	**0**	**0**	**7**	**14.3**	**-2**													
	Springfield	AHL	77	*56	40	96	61												18	8	13	21	18			
1991-92	**Hartford**	**NHL**	**25**	**3**	**5**	**8**	**6**	**1**	**0**	**0**	**41**	**7.3**	**-2**													
	Springfield	AHL	40	21	17	38	44												11	2	0	2	34			
1992-93	**San Jose**	**NHL**	**25**	**4**	**0**	**4**	**24**	**2**	**0**	**0**	**32**	**12.5**	**-17**													
	Kansas City	IHL	33	7	10	17	51												12	3	2	5	20			
1993-94	Portland Pirates	AHL	61	41	44	85	99												17	11	10	21	22			
1994-95	P.E.I. Senators	AHL	57	32	57	89	58												8	4	4	8	6			
	Ottawa	**NHL**	**24**	**5**	**8**	**13**	**14**	**1**	**0**	**0**	**33**	**15.2**	**-1**													
1995-96	**Ottawa**	**NHL**	**17**	**2**	**6**	**8**	**10**	**0**	**0**	**1**	**21**	**9.5**	**-1**													
	P.E.I. Senators	AHL	55	37	45	82	79												5	5	1	6	2			
1996-97	Vastra Frolunda	Sweden	3	0	1	1	0																			
	Grand Rapids	IHL	82	46	55	101	58												5	2	0	2	10			
1997-98	Grand Rapids	IHL	58	28	41	69	42																			
	St. Louis	**NHL**	**16**	**1**	**8**	**9**	**29**	**0**	**0**	**0**	**19**	**5.3**	**3**													
1998-99	**St. Louis**	**NHL**	**45**	**11**	**11**	**22**	**16**	**0**	**0**	**2**	**69**	**15.9**	**5**	**1**	**100.0**	**9**	**6**	**14:20**	**5**	**0**	**0**	**0**	**2**	**0**	**0**	**0**
	Grand Rapids	IHL	6	2	2	4	2																			
99-2000	**Edmonton**	**NHL**	**2**	**0**	**0**	**0**	**2**	**0**	**0**	**0**	**2**	**0.0**	**0**	**0**	**0.0**	**4**	**0**	**9:56**								
	Grand Rapids	IHL	65	33	35	68	50												17	8	10	*18	4			
2000-01	**Philadelphia**	**NHL**	**7**	**1**	**4**	**5**	**0**	**1**	**0**	**0**	**12**	**8.3**	**6**	**0**	**0.0**	**4**	**2**	**14:14**								
	Philadelphia	AHL	72	31	39	70	22												10	4	5	9	4			
2001-02	Adler Mannheim	Germany	60	24	28	52	30												12	*7	6	13	4			
	NHL Totals		**166**	**28**	**42**	**70**	**103**	**5**	**0**	**3**	**236**	**11.9**		**1**	**100.0**	**17**	**8**	**14:09**	**5**	**0**	**0**	**0**	**2**	**0**	**0**	**0**

QMJHL Second All-Star Team (1989) • AHL First All-Star Team (1991, 1995) • AHL Second All-Star Team (1994) • IHL First All-Star Team (1997)

Traded to **San Jose** by **Hartford** for future considerations (Yvon Corriveau, January 21, 1993), October 9, 1992. Signed as a free agent by **Portland** (AHL), September, 1993. Signed as a free agent by **Ottawa**, June 16, 1994. Traded to **Washington** by **Ottawa** for cash, May 21, 1996. Signed as a free agent by **St. Louis**, January 5, 1998. Signed as a free agent by **Edmonton**, December 2, 1999. Signed as a free agent by **Philadelphia**, August 14, 2000. Signed as a free agent by **Detroit**, July 15, 2002.

PILAR, Karel
(PEE-lahr, KAH-rehl) **TOR.**

Defense. Shoots right. 6'3", 210 lbs. Born, Prague, Czech., December 23, 1977. Toronto's 2nd choice, 39th overall, in 2001 Entry Draft.

			Regular Season																Playoffs							
Season	Club	League	GP	G	A	Pts	PIM	PP	SH	GW	S	%	+/-	TF	F%	H	SB	Min	GP	G	A	Pts	PIM	PP	SH	GW
99-2000	Litvinov	Czech	49	2	12	14	61												7	0	0	0	4			
2000-01	Litvinov	Czech	52	12	26	38	52												4	1	1	2	25			
2001-02	**Toronto**	**NHL**	**23**	**1**	**3**	**4**	**8**	**0**	**0**	**0**	**32**	**3.1**	**3**	**0**	**0.0**	**39**	**7**	**16:03**	**11**	**0**	**4**	**4**	**12**	**0**	**0**	**0**
	St. John's	AHL	52	10	14	24	26																			
	NHL Totals		**23**	**1**	**3**	**4**	**8**	**0**	**0**	**0**	**32**	**3.1**		**0**	**0.0**	**39**	**7**	**16:03**	**11**	**0**	**4**	**4**	**12**	**0**	**0**	**0**

PILON, Rich
(PEE-lahn, RITCH) **ST.L.**

Defense. Shoots left. 6'2", 220 lbs. Born, Saskatoon, Sask., April 30, 1968. NY Islanders' 9th choice, 143rd overall, in 1986 Entry Draft.

			Regular Season																Playoffs							
Season	Club	League	GP	G	A	Pts	PIM	PP	SH	GW	S	%	+/-	TF	F%	H	SB	Min	GP	G	A	Pts	PIM	PP	SH	GW
1984-85	Prince Albert	SMHL	26	3	11	14	41																			
1985-86	Prince Albert	SMHL	35	3	28	31	142																			
	Prince Albert	WHL	6	0	0	0	0																			
1986-87	Prince Albert	WHL	68	4	21	25	192												7	1	6	7	17			
1987-88	Prince Albert	WHL	65	13	34	47	177												9	0	6	6	38			
1988-89	**NY Islanders**	**NHL**	**62**	**0**	**14**	**14**	**242**	**0**	**0**	**0**	**47**	**0.0**	**-9**													
1989-90	**NY Islanders**	**NHL**	**14**	**0**	**2**	**2**	**31**	**0**	**0**	**0**	**5**	**0.0**	**2**													
1990-91	**NY Islanders**	**NHL**	**60**	**1**	**4**	**5**	**126**	**0**	**0**	**0**	**33**	**3.0**	**-12**													
1991-92	**NY Islanders**	**NHL**	**65**	**1**	**6**	**7**	**183**	**0**	**0**	**0**	**27**	**3.7**	**-1**													
1992-93	**NY Islanders**	**NHL**	**44**	**1**	**3**	**4**	**164**	**0**	**0**	**0**	**20**	**5.0**	**-4**						**15**	**0**	**0**	**0**	**50**	**0**	**0**	**0**
	Capital District	AHL	6	0	1	1	8																			
1993-94	**NY Islanders**	**NHL**	**28**	**1**	**4**	**5**	**75**	**0**	**0**	**0**	**20**	**5.0**	**-4**													
	Salt Lake	IHL	2	0	0	0	8																			
1994-95	**NY Islanders**	**NHL**	**20**	**1**	**1**	**2**	**40**	**0**	**0**	**0**	**11**	**9.1**	**-3**													
1995-96	**NY Islanders**	**NHL**	**27**	**0**	**3**	**3**	**72**	**0**	**0**	**0**	**7**	**0.0**	**-9**													
1996-97	**NY Islanders**	**NHL**	**52**	**1**	**4**	**5**	**179**	**0**	**0**	**0**	**17**	**5.9**	**4**													
1997-98	**NY Islanders**	**NHL**	**76**	**0**	**7**	**7**	**291**	**0**	**0**	**0**	**37**	**0.0**	**1**													
1998-99	**NY Islanders**	**NHL**	**52**	**0**	**4**	**4**	**88**	**0**	**0**	**0**	**27**	**0.0**	**-8**	**0**	**0.0**	**155**	**49**	**16:35**								
99-2000	**NY Islanders**	**NHL**	**9**	**0**	**2**	**2**	**34**	**0**	**0**	**0**	**0**	**0.0**	**-2**	**0**	**0.0**	**22**	**17**	**16:21**								
	NY Rangers	**NHL**	**45**	**0**	**4**	**4**	**36**	**0**	**0**	**0**	**16**	**0.0**	**0**	**0**	**0.0**	**135**	**67**	**16:29**								
2000-01	**NY Rangers**	**NHL**	**69**	**2**	**9**	**11**	**175**	**0**	**0**	**0**	**24**	**8.3**	**-2**	**0**	**0.0**	**247**	**135**	**16:52**								
2001-02	**St. Louis**	**NHL**	**8**	**0**	**2**	**2**	**9**	**0**	**0**	**0**	**3**	**0.0**	**-1**	**0**	**0.0**	**18**	**0**	**12:45**								
	NHL Totals		**631**	**8**	**69**	**77**	**1745**	**0**	**0**	**0**	**294**	**2.7**		**0**	**0.0**	**577**	**268**	**16:29**	**15**	**0**	**0**	**0**	**50**	**0**	**0**	**0**

WHL East Second All-Star Team (1988)

• Missed majority of 1989-90 season recovering from eye injury suffered in game vs. Detroit, November 4, 1989. • Missed majority of 1993-94 season recovering from shoulder injury originally suffered in game vs. Boston, November 13, 1993. • Missed majority of 1994-95 and 1995-96 seasons recovering from wrist injury suffered in game vs. Quebec, April 18, 1995. Claimed on waivers by **NY Rangers** from **NY Islanders**, December 1, 1999. Traded to **San Jose** by **NY Rangers** for San Jose's 7th round choice (Joseph Crabb) in 2002 Entry Draft, June 29, 2001. Signed as a free agent by **St. Louis**, July 10, 2001. • Missed majority of 2001-02 season recovering from wrist injury suffered in game vs. NY Rangers, October 25, 2001.

PIROS, Kamil
(PIH-ruhsh, KA-mihl) **ATL.**

Center. Shoots left. 6', 200 lbs. Born, Most, Czech., November 20, 1978. Buffalo's 9th choice, 212th overall, in 1997 Entry Draft.

			Regular Season																Playoffs							
Season	Club	League	GP	G	A	Pts	PIM	PP	SH	GW	S	%	+/-	TF	F%	H	SB	Min	GP	G	A	Pts	PIM	PP	SH	GW
1993-94	Most Jr.	Czech-Jr.	16	12	10	22																				
	Litvinov Jr.	Czech-Jr.	22	6	13	19																				
1994-95	Litvinov Jr.	Czech-Jr.	40	27	16	43																				
1995-96	Litvinov Jr.	Czech-Jr.	42	16	13	29																				
1996-97	Litvinov Jr.	Czech-Jr.	3	2	1	3																				
	Litvinov	Czech	38	4	9	13	10																			
1997-98	Litvinov	Czech	14	0	1	1	2																			
	HC Vitkovice	Czech	26	2	9	11	14																			
1998-99	Litvinov	Czech	41	7	9	16	10																			
99-2000	Litvinov	Czech	40	8	8	16	18												7	0	3	3	2			
2000-01	Litvinov	Czech	48	11	13	24	28												6	1	1	2	2			

			Regular Season																Playoffs							
Season	Club	League	GP	G	A	Pts	PIM	PP	SH	GW	S	%	+/-	TF	F%	H	SB	Min	GP	G	A	Pts	PIM	PP	SH	GW
2001-02	**Atlanta**	**NHL**	**8**	**0**	**1**	**1**	**4**	**0**	**0**	**0**	**4**	**0.0**	**–2**	**81**	**32.1**	**5**	**1**	**12:12**								
	Chicago Wolves	AHL	64	19	30	49	16												25	6	11	17	6			
	NHL Totals		**8**	**0**	**1**	**1**	**4**	**0**	**0**	**0**	**4**	**0.0**		**81**	**32.1**	**5**	**1**	**12:12**								

Rights traded to **Atlanta** by **Buffalo** with Buffalo's 4th round choice (later traded to St. Louis - St. Louis selected Igor Valeyev) in 2001 Entry Draft for Donald Audette, March 13, 2001.

PISA, Ales

(PEE-sha, al-EHSH) **EDM.**

Defense. Shoots left. 6', 195 lbs. Born, Pardibuce, Czech., January 2, 1977. Edmonton's 10th choice, 272nd overall, in 2001 Entry Draft.

Season	Club	League	GP	G	A	Pts	PIM	PP	SH	GW	S	%	+/-	TF	F%	H	SB	Min	GP	G	A	Pts	PIM	PP	SH	GW
1993-94	HC Pardubice	Czech	2	0	0	0	0																			
1994-95	HC Pardubice	Czech	22	0	0	0	18												2	0	0	0	0			
1995-96	Pardubice	Czech	33	0	4	4	82																			
1996-97	Pardubice	Czech	41	4	2	6	70												6	0	2	2	6			
1997-98	Pardubice	Czech	50	4	9	13	107												3	0	0	0	4			
1998-99	Pardubice	Czech	48	7	12	19	74												3	0	1	1				
99-2000	Pardubice	Czech	51	5	11	16	58												1	0	0	0	4			
2000-01	Pardubice	Czech	47	10	13	23	75												7	2	2	4	4			
2001-02	**Edmonton**	**NHL**	**2**	**0**	**0**	**0**	**2**	**0**	**0**	**0**	**3**	**0.0**	**0**	**0**	**0.0**	**2**	**1**	**15:02**								
	Hamilton	AHL	52	6	12	18	62												14	1	4	5	8			
	NHL Totals		**2**	**0**	**0**	**0**	**2**	**0**	**0**	**0**	**3**	**0.0**		**0**	**0.0**	**2**	**1**	**15:02**								

PITLICK, Lance

(PIHT-lihk, LANS)

Defense. Shoots right. 6', 205 lbs. Born, Minneapolis, MN, November 5, 1967. Minnesota's 10th choice, 180th overall, in 1986 Entry Draft.

Season	Club	League	GP	G	A	Pts	PIM	PP	SH	GW	S	%	+/-	TF	F%	H	SB	Min	GP	G	A	Pts	PIM	PP	SH	GW
1984-85	Cooper Hawks	Hi-School	23	8	4	12																				
1985-86	Cooper Hawks	Hi-School	21	17	8	25	247																			
1986-87	U. of Minnesota	WCHA	45	0	9	9	88																			
1987-88	U. of Minnesota	WCHA	38	3	9	12	76																			
1988-89	U. of Minnesota	WCHA	47	4	9	13	95																			
1989-90	U. of Minnesota	WCHA	14	3	2	5	26																			
1990-91	Hershey Bears	AHL	64	6	15	21	75												3	0	0	0	9			
1991-92	Hershey Bears	AHL	4	0	0	0	6												3	0	0	0	4			
1992-93	Hershey Bears	AHL	53	5	10	15	77																			
1993-94	Hershey Bears	AHL	58	4	13	17	93												11	1	0	1	11			
1994-95	P.E.I. Senators	AHL	61	8	19	27	55												11	1	4	5	10			
	Ottawa	**NHL**	**15**	**0**	**1**	**1**	**6**	**0**	**0**	**0**	**11**	**0.0**	**–5**													
1995-96	**Ottawa**	**NHL**	**28**	**1**	**6**	**7**	**20**	**0**	**0**	**0**	**13**	**7.7**	**–8**													
	P.E.I. Senators	AHL	29	4	10	14	39												5	0	0	0	0			
1996-97	**Ottawa**	**NHL**	**66**	**5**	**5**	**10**	**91**	**0**	**0**	**1**	**54**	**9.3**	**2**						**7**	**0**	**0**	**0**	**4**	**0**	**0**	**0**
1997-98	**Ottawa**	**NHL**	**69**	**2**	**7**	**9**	**50**	**0**	**0**	**0**	**66**	**3.0**	**8**						**11**	**0**	**1**	**1**	**17**	**0**	**0**	**0**
1998-99	**Ottawa**	**NHL**	**50**	**3**	**6**	**9**	**33**	**0**	**0**	**0**	**34**	**8.8**	**7**	**0**	**0.0**	**120**	**64**	**16:39**	**2**	**0**	**0**	**0**	**0**	**0**	**0**	**0**
99-2000	**Florida**	**NHL**	**62**	**3**	**5**	**8**	**44**	**0**	**0**	**1**	**26**	**11.5**	**7**	**0**	**0.0**	**142**	**91**	**16:33**	**4**	**0**	**1**	**1**	**0**	**0**	**0**	**0**
2000-01	**Florida**	**NHL**	**68**	**1**	**2**	**3**	**42**	**0**	**0**	**0**	**24**	**4.2**	**–5**	**0**	**0.0**	**161**	**64**	**13:35**								
2001-02	**Florida**	**NHL**	**35**	**1**	**1**	**2**	**12**	**0**	**0**	**0**	**15**	**6.7**	**–14**	**0**	**0.0**	**50**	**15**	**9:22**								
	NHL Totals		**393**	**16**	**33**	**49**	**298**	**0**	**0**	**2**	**243**	**6.6**		**0**	**0.0**	**473**	**234**	**14:28**	**24**	**0**	**2**	**2**	**21**	**0**	**0**	**0**

Signed as a free agent by **Philadelphia**, September 5, 1990. Signed as a free agent by **Ottawa**, June 22, 1994. Signed as a free agent by **Florida**, July 21, 1999. • Spent majority of 2001-02 season on practice roster, October 4, 2001.

PITTIS, Domenic

(PIH-THIS, DOHM-ihn-ihk) **NSH.**

Center. Shoots left. 5'11", 190 lbs. Born, Calgary, Alta., October 1, 1974. Pittsburgh's 2nd choice, 52nd overall, in 1993 Entry Draft.

Season	Club	League	GP	G	A	Pts	PIM	PP	SH	GW	S	%	+/-	TF	F%	H	SB	Min	GP	G	A	Pts	PIM	PP	SH	GW
1990-91	Calgary Flames	AMHL	35	23	54	77	43																			
1991-92	Lethbridge	WHL	65	6	17	23	18												5	0	2	2	4			
1992-93	Lethbridge	WHL	66	46	73	119	69												4	3	3	6	8			
1993-94	Lethbridge	WHL	72	58	69	127	93												8	4	11	15	16			
1994-95	Cleveland	IHL	62	18	32	50	66												3	0	2	2	2			
1995-96	Cleveland	IHL	74	10	28	38	100												3	0	0	0	2			
1996-97	**Pittsburgh**	**NHL**	**1**	**0**	**0**	**0**	**0**	**0**	**0**	**0**	**0**	**0.0**	**–1**													
	Long Beach	IHL	65	23	43	66	91												18	5	9	14	26			
1997-98	Syracuse Crunch	AHL	75	23	41	64	90												5	1	3	4	4			
1998-99	**Buffalo**	**NHL**	**3**	**0**	**0**	**0**	**2**	**0**	**0**	**0**	**1**	**0.0**	**0**	**19**	**42.1**	**4**	**1**	**8:35**								
	Rochester	AHL	76	38	66	*104	108												20	7	*14	*21	40			
99-2000	**Buffalo**	**NHL**	**7**	**1**	**0**	**1**	**6**	**0**	**0**	**0**	**6**	**16.7**	**1**	**65**	**44.6**	**7**	**0**	**11:27**								
	Rochester	AHL	53	17	48	65	85												21	4	*26	*30	28			
2000-01	**Edmonton**	**NHL**	**47**	**4**	**5**	**9**	**49**	**0**	**0**	**2**	**42**	**9.5**	**–5**	**367**	**54.8**	**57**	**23**	**10:46**	**3**	**0**	**0**	**0**	**2**	**0**	**0**	**0**
2001-02	**Edmonton**	**NHL**	**22**	**0**	**6**	**6**	**8**	**0**	**0**	**0**	**18**	**0.0**	**–2**	**52**	**55.8**	**14**	**9**	**11:23**								
	NHL Totals		**80**	**5**	**11**	**16**	**65**	**0**	**0**	**2**	**67**	**7.5**		**503**	**53.1**	**82**	**33**	**10:55**	**3**	**0**	**0**	**0**	**2**	**0**	**0**	**0**

WHL East Second All-Star Team (1994) • Won John P. Sollenberger Trophy (Top Scorer - AHL) (1999)

Signed as a free agent by **Buffalo**, August 10, 1998. Signed as a free agent by **Edmonton**, July 25, 2000. • Missed majority of 2001-02 season recovering from head injury suffered in game vs. Nashvile, February 28, 2002. Signed as a free agent by **Nashville**, July 24, 2002.

PLANTE, Dan

(PLAHNT, DAN)

Right wing. Shoots right. 5'11", 202 lbs. Born, Hayward, WI, October 5, 1971. NY Islanders' 3rd choice, 48th overall, in 1990 Entry Draft.

Season	Club	League	GP	G	A	Pts	PIM	PP	SH	GW	S	%	+/-	TF	F%	H	SB	Min	GP	G	A	Pts	PIM	PP	SH	GW
1988-89	Edina Hornets	Hi-School	27	10	26	36	23																			
1989-90	Edina Hornets	Hi-School	24	8	18	26	12																			
1990-91	U. of Wisconsin	WCHA	33	1	2	3	54																			
1991-92	U. of Wisconsin	WCHA	36	13	13	26	107																			
1992-93	U. of Wisconsin	WCHA	42	26	31	57	142																			
1993-94	**NY Islanders**	**NHL**	**12**	**0**	**1**	**1**	**4**	**0**	**0**	**0**	**9**	**0.0**	**–2**						**1**	**1**	**0**	**1**	**2**	**0**	**0**	**0**
	Salt Lake	IHL	66	7	17	24	148																			
1994-95	Denver Grizzlies	IHL	2	0	0	0	4																			
1995-96	**NY Islanders**	**NHL**	**73**	**5**	**3**	**8**	**50**	**0**	**2**	**0**	**103**	**4.9**	**–22**													
1996-97	**NY Islanders**	**NHL**	**67**	**4**	**9**	**13**	**75**	**0**	**2**	**0**	**61**	**6.6**	**–6**													
1997-98	**NY Islanders**	**NHL**	**7**	**0**	**1**	**1**	**6**	**0**	**0**	**0**	**7**	**0.0**	**–1**													
	Utah Grizzlies	IHL	73	22	27	49	125												4	0	2	2	14			
1998-99	Chicago Wolves	IHL	81	21	12	33	119												10	1	5	6	10			
99-2000	Chicago Wolves	IHL	79	11	11	22	71												16	3	5	8	14			
2000-01	Chicago Wolves	IHL	76	15	11	26	58												16	1	0	1	27			
2001-02	Chicago Wolves	AHL	70	11	15	26	44												6	0	0	0	15			
	NHL Totals		**159**	**9**	**14**	**23**	**135**	**0**	**4**	**0**	**180**	**5.0**							**1**	**1**	**0**	**1**	**2**	**0**	**0**	**0**

• Missed majority of 1994-95 season recovering from knee injury suffered in game vs. Houston (IHL), October 2, 1994. Signed as a free agent by **Chicago** (IHL), July 21, 1999.

PLANTE, Derek

(PLAHNT, DAIR-ehk)

Center. Shoots left. 5'11", 181 lbs. Born, Cloquet, MN, January 17, 1971. Buffalo's 7th choice, 161st overall, in 1989 Entry Draft.

Season	Club	League	GP	G	A	Pts	PIM	PP	SH	GW	S	%	+/-	TF	F%	H	SB	Min	GP	G	A	Pts	PIM	PP	SH	GW
1987-88	Cloquet High	Hi-School	23	16	25	41																				
1988-89	Cloquet High	Hi-School	24	30	33	63																				
1989-90	U. Minn-Duluth	WCHA	28	10	11	21	12																			
1990-91	U. Minn-Duluth	WCHA	36	23	20	43	6																			
1991-92	U. Minn-Duluth	WCHA	37	27	36	63	28																			
1992-93	U. Minn-Duluth	WCHA	37	*36	*56	*92	30																			
1993-94	**Buffalo**	**NHL**	**77**	**21**	**35**	**56**	**24**	**8**	**1**	**2**	**147**	**14.3**	**4**						**7**	**1**	**0**	**1**	**0**	**0**	**0**	**0**
1994-95	**Buffalo**	**NHL**	**47**	**3**	**19**	**22**	**12**	**2**	**0**	**0**	**94**	**3.2**	**–4**													
1995-96	**Buffalo**	**NHL**	**76**	**23**	**33**	**56**	**28**	**4**	**0**	**5**	**203**	**11.3**	**–4**													
1996-97	**Buffalo**	**NHL**	**82**	**27**	**26**	**53**	**24**	**5**	**0**	**6**	**191**	**14.1**	**14**						**12**	**4**	**6**	**10**	**4**	**1**	**0**	**2**
1997-98	**Buffalo**	**NHL**	**72**	**13**	**21**	**34**	**26**	**5**	**0**	**1**	**150**	**8.7**	**8**						**11**	**0**	**3**	**3**	**10**	**0**	**0**	**0**
1998-99	**Buffalo**	**NHL**	**41**	**4**	**11**	**15**	**12**	**0**	**0**	**0**	**66**	**6.1**	**3**	**598**	**46.2**	**10**	**24**	**15:31**								
	♦ Dallas	**NHL**	**10**	**2**	**3**	**5**	**4**	**1**	**0**	**0**	**24**	**8.3**	**1**	**113**	**58.4**	**9**	**4**	**13:42**	**6**	**1**	**0**	**1**	**4**	**0**	**0**	**0**

Season	Club	League	Regular Season																Playoffs							
			GP	G	A	Pts	PIM	PP	SH	GW	S	%	+/-	TF	F%	H	SB	Min	GP	G	A	Pts	PIM	PP	SH	GW
99-2000	**Dallas**	**NHL**	16	1	1	2	2	1	0	0	17	5.9	–4	119	51.3	5	7	10:05								
	Michigan K-Wings	IHL	13	0	4	4	2																			
	Chicago	**NHL**	17	1	1	2	2	0	0	0	14	7.1	–1	98	38.8	6	3	7:48								
	Chicago Wolves	IHL	4	2	1	3	2												8	3	1	4	6			
2000-01	**Philadelphia**	**NHL**	12	1	2	3	4	0	0	0	20	5.0	0	177	48.6	2	10	16:30	5	0	1	1	0	0	0	0
	Philadelphia	AHL	57	18	35	53	19																			
2001-02	Munchen Barons	Germany	60	20	38	58	22												9	3	7	10	10			
	NHL Totals		450	96	152	248	138	26	1	14	926	10.4		1105	47.7	32	48	13:11	41	6	10	16	18	1	0	2

WCHA Second All-Star Team (1992) • WCHA First All-Star Team (1993) • WCHA Player of the Year (1993) • NCAA West First All-American Team (1993)

Traded to **Dallas** by **Buffalo** for Dallas' 2nd round choice (Michael Zigomanis) in 1999 Entry Draft, March 23, 1999. Traded to **Chicago** by **Dallas** with Kevin Dean and Dallas' 2nd round choice (Matt Keith) in 2001 Entry Draft for Sylvain Cote and Dave Manson, February 8, 2000. Signed as a free agent by **Philadelphia**, July 26, 2000.

PLETKA, Vaclav

(PLEHT-kuh, VA-tslav) **PHI.**

Right wing. Shoots left. 5'11", 182 lbs. Born, Mlada Boleslav, Czech., June 8, 1979. Philadelphia's 5th choice, 208th overall, in 1999 Entry Draft.

Season	Club	League	GP	G	A	Pts	PIM	PP	SH	GW	S	%	+/-	TF	F%	H	SB	Min	GP	G	A	Pts	PIM	PP	SH	GW
1995-96	Mlada Boleslav Jr.	Czech-Jr.	35	22	17	39																				
1996-97	Mlada Boleslav Jr.	Czech-Jr.	37	28	13	41																				
1997-98	Mlada Boleslav Jr.	Czech-Jr.	36	23	25	48																				
1998-99	Trinec Jr.	Czech-Jr.	20	11	5	16																				
	Trinec	Czech	50	15	11	26	20												10	2	1	3				
99-2000	Trinec	Czech	51	28	21	49	62												4	2	2	4	2			
2000-01	Philadelphia	AHL	71	20	21	41	51												10	1	3	4	4			
2001-02	**Philadelphia**	**NHL**	1	0	0	0	0	0	0	0	2	0.0	0	0	0.0	0	1	11:34								
	Philadelphia	AHL	61	20	19	39	43																			
	NHL Totals		1	0	0	0	0	0	0	0	2	0.0		0	0.0	0	1	11:34								

Signed as a free agent by **HC Trinec** (Czech) with Philadelphia retaining NHL rights, July 28, 2002.

POAPST, Steve

(POHPST, STEEV) **CHI.**

Defense. Shoots left. 6', 200 lbs. Born, Cornwall, Ont., January 3, 1969.

Season	Club	League	GP	G	A	Pts	PIM	PP	SH	GW	S	%	+/-	TF	F%	H	SB	Min	GP	G	A	Pts	PIM	PP	SH	GW
1986-87	Smiths Falls	OCJHL	54	10	27	37	94																			
1987-88	Colgate	ECAC	32	3	13	16	22																			
1988-89	Colgate	ECAC	30	0	5	5	38																			
1989-90	Colgate	ECAC	38	4	15	19	54																			
1990-91	Colgate	ECAC	32	6	15	21	43																			
1991-92	Hampton Roads	ECHL	55	8	20	28	29												14	1	4	5	12			
1992-93	Hampton Roads	ECHL	63	10	35	45	57												4	0	1	1	4			
	Baltimore	AHL	7	0	1	1	4												7	0	3	3	6			
1993-94	Portland Pirates	AHL	78	14	21	35	47												12	0	3	3	8			
1994-95	Portland Pirates	AHL	71	8	22	30	60												7	0	1	1	16			
1995-96	**Washington**	**NHL**	3	1	0	1	0	0	0	1	2	50.0	–1						6	0	0	0	0	0	0	0
	Portland Pirates	AHL	70	10	24	34	79												20	2	6	8	16			
1996-97	Portland Pirates	AHL	47	1	20	21	34												5	0	1	1	6			
1997-98	Portland Pirates	AHL	76	8	29	37	46												10	2	3	5	8			
1998-99	**Washington**	**NHL**	22	0	0	0	8	0	0	0	11	0.0	–8	0	0.0	48	6	12:26								
	Portland Pirates	AHL	54	3	21	24	36																			
99-2000	Portland Pirates	AHL	58	0	14	14	20												3	1	0	1	2			
2000-01	**Chicago**	**NHL**	36	2	3	5	12	0	0	0	27	7.4	3	0	0.0	57	48	17:03								
	Norfolk Admirals	AHL	37	1	8	9	14																			
2001-02	**Chicago**	**NHL**	56	1	7	8	30	0	0	0	48	2.1	6	0	0.0	147	43	14:52	5	0	0	0	0	0	0	0
	NHL Totals		117	4	10	14	50	0	0	1	88	4.5		0	0.0	252	97	15:05	11	0	0	0	0	0	0	0

ECHL First All-Star Team (1993)

Signed as a free agent by **Washington**, February 4, 1995. Signed as a free agent by **Chicago**, July 27, 2000.

PODEIN, Shjon

(poh-DEEN, SHAWN) **ST.L.**

Left wing. Shoots left. 6'2", 200 lbs. Born, Rochester, MN, March 5, 1968. Edmonton's 9th choice, 166th overall, in 1988 Entry Draft.

Season	Club	League	GP	G	A	Pts	PIM	PP	SH	GW	S	%	+/-	TF	F%	H	SB	Min	GP	G	A	Pts	PIM	PP	SH	GW
1985-86	John Marshall	Hi-School	25	34	30	64																				
1986-87	American Int'l.	NCAA-2	6	0	1	1	0																			
1987-88	U. Minn-Duluth	WCHA	30	4	4	8	48																			
1988-89	U. Minn-Duluth	WCHA	36	7	5	12	46																			
1989-90	U. Minn-Duluth	WCHA	35	21	18	39	36																			
1990-91	Cape Breton	AHL	63	14	15	29	65												4	0	0	0	5			
1991-92	Cape Breton	AHL	80	30	24	54	46												5	3	1	4	2			
1992-93	**Edmonton**	**NHL**	40	13	6	19	25	2	1	1	64	20.3	–2													
	Cape Breton	AHL	38	18	21	39	32												9	2	2	4	29			
1993-94	**Edmonton**	**NHL**	28	3	5	8	8	0	0	0	26	11.5	3													
	Cape Breton	AHL	5	4	4	8	4																			
1994-95	**Philadelphia**	**NHL**	44	3	7	10	33	0	0	1	48	6.3	–2						15	1	3	4	10	0	0	0
1995-96	**Philadelphia**	**NHL**	79	15	10	25	89	0	4	4	115	13.0	25						12	1	2	3	50	0	0	1
1996-97	**Philadelphia**	**NHL**	82	14	18	32	41	0	0	4	153	9.2	7						19	4	3	7	16	0	0	1
1997-98	**Philadelphia**	**NHL**	82	11	13	24	53	1	1	2	126	8.7	8						5	0	0	0	10	0	0	0
1998-99	**Philadelphia**	**NHL**	14	1	0	1	0	0	0	0	26	3.8	–2	2	50.0	13	4	11:52								
	Colorado	**NHL**	41	2	6	8	24	0	0	0	49	4.1	–3	21	42.9	41	24	11:49	19	1	1	2	12	0	0	0
99-2000	**Colorado**	**NHL**	75	11	8	19	29	0	1	3	104	10.6	12	17	64.7	84	60	13:31	17	5	0	5	8	0	0	1
2000-01♦	**Colorado**	**NHL**	82	15	17	32	68	0	0	3	137	10.9	7	69	44.9	116	47	14:23	23	2	3	5	14	0	0	1
2001-02	**Colorado**	**NHL**	41	6	6	12	39	0	1	2	43	14.0	0	15	13.3	66	25	13:00								
	St. Louis	**NHL**	23	2	4	6	2	0	0	0	24	8.3	2	14	21.4	44	13	15:09	10	0	0	0	6	0	0	0
	NHL Totals		631	96	100	196	411	3	8	20	915	10.5		138	41.3	364	173	13:30	120	14	12	26	126	0	0	4

Won King Clancy Memorial Trophy (2001)

Signed as a free agent by **Philadelphia**, July 27, 1994. Traded to **Colorado** by **Philadelphia** for Keith Jones, November 12, 1998. Traded to **St. Louis** by **Colorado** for Mike Keane, February 11, 2002.

PODKONICKY, Andrej

(pohd-koh-NIHTZ-kee, AWN-dray)

Center. Shoots left. 6'2", 202 lbs. Born, Zvolen, Czech., May 9, 1978. St. Louis' 8th choice, 196th overall, in 1996 Entry Draft.

Season	Club	League	GP	G	A	Pts	PIM	PP	SH	GW	S	%	+/-	TF	F%	H	SB	Min	GP	G	A	Pts	PIM	PP	SH	GW
1994-95	HKm Zvolen	Slovak-2	17	0	4	4	6																			
1995-96	HKm Zvolen	Slovak-2	38	18	12	30	18																			
1996-97	Portland	WHL	71	25	46	71	127												6	1	1	2	8			
1997-98	Portland	WHL	64	30	44	74	81												16	4	12	16	20			
1998-99	Worcester	AHL	61	19	24	43	52												4	0	0	0	4			
99-2000	Worcester	AHL	77	16	25	41	68												9	2	5	7	6			
2000-01	Worcester	AHL	16	2	3	5	15																			
	Florida	**NHL**	6	1	0	1	2	0	0	0	5	20.0	0	34	55.9	7	1	6:04								
	Louisville	AHL	41	6	10	16	31																			
2001-02	HIFK Helsinki	Finland	23	3	6	9	41																			
	Slov. Bratislava	Slovakia	16	11	2	13	2												19	4	7	11	49			
	NHL Totals		6	1	0	1	2	0	0	0	5	20.0		34	55.9	7	1	6:04								

Memorial Cup All-Star Team (1998) • Won Ed Chynoweth Award (Memorial Cup Tournament Top Scorer) (1998)

Traded to **Florida** by **St. Louis** for Eric Boguniecki, December 17, 2000. Signed as a free agent by **HIFK Helsinki** (Finland), August 14, 2001.

PODOLLAN, Jason

(poh-DOH-luhn, JAY-suhn)

Right wing. Shoots right. 6'1", 198 lbs. Born, Vernon, B.C., February 18, 1976. Florida's 3rd choice, 31st overall, in 1994 Entry Draft.

			Regular Season																Playoffs							
Season	Club	League	GP	G	A	Pts	PIM	PP	SH	GW	S	%	+/-	TF	F%	H	SB	Min	GP	G	A	Pts	PIM	PP	SH	GW
1990-91	Sherwood Park	AMHL	61	105	111	216	133	....	....	....	....	....	....	....	....	....	....	....	....	....	....	....	....	....	....	....
1991-92	Penticton	BCJHL	59	20	26	46	66	....	....	....	....	....	....	....	....	....	....	....	....	....	....	....	....	....	....	....
	Spokane Chiefs	WHL	2	0	0	0	2	....	....	....	....	....	....	....	....	....	....	....	10	3	1	4	16	....	....	....
1992-93	Spokane Chiefs	WHL	72	36	33	69	108	....	....	....	....	....	....	....	....	....	....	....	10	4	4	8	14	....	....	....
1993-94	Spokane Chiefs	WHL	69	29	37	66	108	....	....	....	....	....	....	....	....	....	....	....	3	3	0	3	2	....	....	....
1994-95	Spokane Chiefs	WHL	72	43	41	84	102	....	....	....	....	....	....	....	....	....	....	....	11	5	7	12	18	....	....	....
	Cincinnati	IHL	....	....	....	....	....	....	....	....	....	....	....	....	....	....	....	....	3	0	0	0	2	....	....	....
1995-96	Spokane Chiefs	WHL	56	37	25	62	103	....	....	....	....	....	....	....	....	....	....	....	18	*21	12	33	28	....	....	....
1996-97	**Florida**	**NHL**	**19**	**1**	**1**	**2**	**4**	**1**	**0**	**0**	**20**	**5.0**	**–3**	....	....	....	....	....	....	....	....	....	....	....	....	....
	Carolina	AHL	39	21	25	46	36	....	....	....	....	....	....	....	....	....	....	....	....	....	....	....	....	....	....	....
	Toronto	**NHL**	**10**	**0**	**3**	**3**	**6**	**0**	**0**	**0**	**10**	**0.0**	**–2**	....	....	....	....	....	....	....	....	....	....	....	....	....
	St. John's	AHL	....	....	....	....	....	....	....	....	....	....	....	....	....	....	....	....	11	2	3	5	6	....	....	....
1997-98	St. John's	AHL	70	30	31	61	116	....	....	....	....	....	....	....	....	....	....	....	4	1	0	1	10	....	....	....
1998-99	**Toronto**	**NHL**	**4**	**0**	**0**	**0**	**0**	**0**	**0**	**0**	**2**	**0.0**	**0**	**0**	**0.0**	**5**	**0**	**6:29**	....	....	....	....	....	....	....	....
	St. John's	AHL	68	42	26	68	65	....	....	....	....	....	....	....	....	....	....	....	....	....	....	....	....	....	....	....
	Los Angeles	**NHL**	**6**	**0**	**0**	**0**	**5**	**0**	**0**	**0**	**7**	**0.0**	**–3**	**0**	**0.0**	**13**	**1**	**10:15**	....	....	....	....	....	....	....	....
	Long Beach	IHL	8	5	3	8	2	....	....	....	....	....	....	....	....	....	....	....	6	1	2	3	4	....	....	....
99-2000	**Los Angeles**	**NHL**	**1**	**0**	**1**	**1**	**2**	**0**	**0**	**0**	**2**	**0.0**	**0**	**0**	**0.0**	**3**	**0**	**16:13**	....	....	....	....	....	....	....	....
	Lowell	AHL	71	29	26	55	91	....	....	....	....	....	....	....	....	....	....	....	4	0	0	0	4	....	....	....
2000-01	Detroit Vipers	IHL	63	15	16	31	98	....	....	....	....	....	....	....	....	....	....	....	....	....	....	....	....	....	....	....
	Manitoba Moose	IHL	16	5	2	7	10	....	....	....	....	....	....	....	....	....	....	....	4	0	0	0	2	....	....	....
2001-02	**NY Islanders**	**NHL**	**1**	**0**	**0**	**0**	**2**	**0**	**0**	**0**	**1**	**0.0**	**0**	**0**	**0.0**	**3**	**1**	**4:10**	....	....	....	....	....	....	....	....
	Bridgeport	AHL	65	21	24	45	63	....	....	....	....	....	....	....	....	....	....	....	20	5	4	9	32	....	....	....
	NHL Totals		**41**	**1**	**5**	**6**	**19**	**1**	**0**	**0**	**42**	**2.4**		**0**	**0.0**	**24**	**2**	**8:59**	....	....	....	....	....	....	....	....

WHL West Second All-Star Team (1996)

Traded to **Toronto** by **Florida** for Kirk Muller, March 18, 1997. Traded to **Los Angeles** by **Toronto** with Toronto's 3rd round choice (Cory Campbell) in 1999 Entry Draft for Yanic Perreault, March 23, 1999. Claimed by **Tampa Bay** from **LA Kings** in Waiver Draft, September 29, 2000. Signed as a free agent by **NY Islanders**, August 24, 2001. Signed as a free agent by **Mannheim** (Germany), July 29, 2002.

PONIKAROVSKY, Alexei

(poh-NIH-kahr-ohv-skee, al-EHX-ay) **TOR.**

Right wing. Shoots left. 6'4", 196 lbs. Born, Kiev, USSR, April 9, 1980. Toronto's 4th choice, 87th overall, in 1998 Entry Draft.

			Regular Season																Playoffs							
Season	Club	League	GP	G	A	Pts	PIM	PP	SH	GW	S	%	+/-	TF	F%	H	SB	Min	GP	G	A	Pts	PIM	PP	SH	GW
1995-96	Dyno. Moscow Jr.	CIS-Jr.	70	14	10	24	20	....	....	....	....	....	....	....	....	....	....	....	....	....	....	....	....	....	....	....
1996-97	Dyno. Moscow Jr.	Russia-Jr.	60	12	15	27	30	....	....	....	....	....	....	....	....	....	....	....	....	....	....	....	....	....	....	....
	DynamoMoscow2	Russia-3	2	0	0	0	2	....	....	....	....	....	....	....	....	....	....	....	....	....	....	....	....	....	....	....
1997-98	Dynamo Moscow	Russia	24	1	2	3	30	....	....	....	....	....	....	....	....	....	....	....	....	....	....	....	....	....	....	....
1998-99	Krylja Sovetov	Russia	13	2	1	3	2	....	....	....	....	....	....	....	....	....	....	....	....	....	....	....	....	....	....	....
	Dynamo Moscow	Russia	....	....	....	....	....	....	....	....	....	....	....	....	....	....	....	....	3	0	0	0	2	....	....	....
99-2000	THC Tver	Russia-2	29	8	14	22	26	....	....	....	....	....	....	....	....	....	....	....	....	....	....	....	....	....	....	....
	Dynamo Moscow	Russia	19	1	0	1	8	....	....	....	....	....	....	....	....	....	....	....	1	0	0	0	0	....	....	....
	Dynamo Moscow	EuroHL	2	0	2	2	0	....	....	....	....	....	....	....	....	....	....	....	....	....	....	....	....	....	....	....
2000-01	**Toronto**	**NHL**	**22**	**1**	**3**	**4**	**14**	**0**	**0**	**0**	**21**	**4.8**	**–1**	**7**	**28.6**	**18**	**7**	**8:32**	....	....	....	....	....	....	....	....
	St. John's	AHL	49	12	24	36	44	....	....	....	....	....	....	....	....	....	....	....	4	0	0	0	4	....	....	....
2001-02	**Toronto**	**NHL**	**8**	**2**	**0**	**2**	**0**	**0**	**0**	**1**	**8**	**25.0**	**2**	**2**	**50.0**	**4**	**0**	**8:03**	**10**	**0**	**0**	**0**	**4**	**0**	**0**	**0**
	St. John's	AHL	72	21	27	48	74	....	....	....	....	....	....	....	....	....	....	....	5	2	1	3	8	....	....	....
	Ukraine	Olympics	4	1	1	2	6	....	....	....	....	....	....	....	....	....	....	....	....	....	....	....	....	....	....	....
	NHL Totals		**30**	**3**	**3**	**6**	**14**	**0**	**0**	**1**	**29**	**10.3**		**9**	**33.3**	**22**	**7**	**8:24**	**10**	**0**	**0**	**0**	**4**	**0**	**0**	**0**

POPOVIC, Peter

(puh-PUH-vihch, PEE-tuhr)

Defense. Shoots left. 6'6", 243 lbs. Born, Koping, Sweden, February 10, 1968. Montreal's 5th choice, 93rd overall, in 1988 Entry Draft.

			Regular Season																Playoffs							
Season	Club	League	GP	G	A	Pts	PIM	PP	SH	GW	S	%	+/-	TF	F%	H	SB	Min	GP	G	A	Pts	PIM	PP	SH	GW
1986-87	Vasteras IK Jr.	Swede-Jr.	10	0	1	1	2	....	....	....	....	....	....	....	....	....	....	....	....	....	....	....	....	....	....	....
	Vasteras IK	Swede-2	24	1	2	3	10	....	....	....	....	....	....	....	....	....	....	....	12	2	8	10	6	....	....	....
1987-88	Vasteras IK	Swede-2	28	3	17	20	16	....	....	....	....	....	....	....	....	....	....	....	15	1	4	5	20	....	....	....
1988-89	Vasteras IK	Sweden	39	3	7	10	52	....	....	....	....	....	....	....	....	....	....	....	5	0	1	1	16	....	....	....
1989-90	Vasteras IK	Sweden	30	2	10	12	24	....	....	....	....	....	....	....	....	....	....	....	2	0	1	1	2	....	....	....
1990-91	Vasteras IK	Sweden	40	3	2	5	62	....	....	....	....	....	....	....	....	....	....	....	4	0	0	0	4	....	....	....
1991-92	Vasteras IK	Sweden	34	7	10	17	30	....	....	....	....	....	....	....	....	....	....	....	....	....	....	....	....	....	....	....
1992-93	Vasteras IK	Sweden	39	6	12	18	46	....	....	....	....	....	....	....	....	....	....	....	3	0	1	1	2	....	....	....
1993-94	**Montreal**	**NHL**	**47**	**2**	**12**	**14**	**26**	**1**	**0**	**0**	**58**	**3.4**	**10**	....	....	....	....	....	**6**	**0**	**1**	**1**	**0**	**0**	**0**	**0**
1994-95	Vasteras IK	Sweden	11	0	3	3	10	....	....	....	....	....	....	....	....	....	....	....	....	....	....	....	....	....	....	....
	Montreal	**NHL**	**33**	**0**	**5**	**5**	**8**	**0**	**0**	**0**	**23**	**0.0**	**–10**	....	....	....	....	....	....	....	....	....	....	....	....	....
1995-96	**Montreal**	**NHL**	**76**	**2**	**12**	**14**	**69**	**0**	**0**	**0**	**59**	**3.4**	**21**	....	....	....	....	....	**6**	**0**	**2**	**2**	**4**	**0**	**0**	**0**
1996-97	**Montreal**	**NHL**	**78**	**1**	**13**	**14**	**32**	**0**	**0**	**0**	**82**	**1.2**	**9**	....	....	....	....	....	**3**	**0**	**0**	**0**	**2**	**0**	**0**	**0**
1997-98	**Montreal**	**NHL**	**69**	**2**	**6**	**8**	**38**	**0**	**0**	**0**	**40**	**5.0**	**–6**	....	....	....	....	....	**10**	**1**	**1**	**2**	**2**	**0**	**0**	**0**
1998-99	**NY Rangers**	**NHL**	**68**	**1**	**4**	**5**	**40**	**0**	**0**	**0**	**64**	**1.6**	**–12**	**2**	**0.0**	**112**	**178**	**20:41**	....	....	....	....	....	....	....	....
99-2000	**Pittsburgh**	**NHL**	**54**	**1**	**5**	**6**	**30**	**0**	**0**	**0**	**23**	**4.3**	**–8**	**2**	**0.0**	**59**	**85**	**16:12**	**10**	**0**	**0**	**0**	**10**	**0**	**0**	**0**
2000-01	**Boston**	**NHL**	**60**	**1**	**6**	**7**	**48**	**0**	**0**	**0**	**34**	**2.9**	**–5**	**1**	**0.0**	**104**	**102**	**18:41**	....	....	....	....	....	....	....	....
2001-02	Sodertalje SK	Sweden	50	3	18	21	52	....	....	....	....	....	....	....	....	....	....	....	....	....	....	....	....	....	....	....
	NHL Totals		**485**	**10**	**63**	**73**	**291**	**1**	**0**	**0**	**383**	**2.6**		**5**	**0.0**	**275**	**365**	**18:42**	**35**	**1**	**4**	**5**	**18**	**0**	**0**	**0**

Traded to **NY Rangers** by **Montreal** for Sylvain Blouin and NY Rangers' 6th round choice (later traded to Phoenix - Phoenix selected Erik Lewerstrom) in 1999 Entry Draft, June 30, 1998. Traded to **Pittsburgh** by **NY Rangers** for Kevin Hatcher, September 30, 1999. Signed as a free agent by **Boston**, July 2, 2000.

POSMYK, Marek

(PAWZ-mihk, MAHR-ehk) **T.B.**

Defense. Shoots right. 6'5", 228 lbs. Born, Jihlava, Czech., September 15, 1978. Toronto's 1st choice, 36th overall, in 1996 Entry Draft.

			Regular Season																Playoffs							
Season	Club	League	GP	G	A	Pts	PIM	PP	SH	GW	S	%	+/-	TF	F%	H	SB	Min	GP	G	A	Pts	PIM	PP	SH	GW
1994-95	Dukla Jihlava Jr.	Czech-Jr.	16	1	3	4	....	....	....	....	....	....	....	....	....	....	....	....	....	....	....	....	....	....	....	....
1995-96	Dukla Jihlava Jr.	Czech-Jr.	16	6	5	11	....	....	....	....	....	....	....	....	....	....	....	....	....	....	....	....	....	....	....	....
	HC Dukla Jihlava	Czech	18	1	2	3	....	....	....	....	....	....	....	....	....	....	....	....	1	0	0	0	....	....	....	....
1996-97	HC Dukla Jihlava	Czech	24	1	7	8	44	....	....	....	....	....	....	....	....	....	....	....	....	....	....	....	....	....	....	....
	St. John's	AHL	2	0	0	0	2	....	....	....	....	....	....	....	....	....	....	....	....	....	....	....	....	....	....	....
1997-98	Sarnia Sting	OHL	48	8	16	24	94	....	....	....	....	....	....	....	....	....	....	....	5	0	2	2	6	....	....	....
	St. John's	AHL	3	0	0	0	4	....	....	....	....	....	....	....	....	....	....	....	....	....	....	....	....	....	....	....
1998-99	St. John's	AHL	41	1	0	1	36	....	....	....	....	....	....	....	....	....	....	....	....	....	....	....	....	....	....	....
99-2000	St. John's	AHL	38	1	6	7	57	....	....	....	....	....	....	....	....	....	....	....	....	....	....	....	....	....	....	....
	Tampa Bay	**NHL**	**18**	**1**	**2**	**3**	**20**	**0**	**0**	**0**	**22**	**4.5**	**1**	**1**	**100.0**	**21**	**13**	**13:12**	....	....	....	....	....	....	....	....
	Detroit Vipers	IHL	1	0	1	1	0	....	....	....	....	....	....	....	....	....	....	....	....	....	....	....	....	....	....	....
2000-01	**Tampa Bay**	**NHL**	**1**	**0**	**0**	**0**	**0**	**0**	**0**	**0**	**0**	**0.0**	**–1**	**0**	**0.0**	**0**	**0**	**13:27**	....	....	....	....	....	....	....	....
	Detroit Vipers	IHL	49	7	14	21	58	....	....	....	....	....	....	....	....	....	....	....	....	....	....	....	....	....	....	....
2001-02	Zlin	Czech	45	8	6	14	147	....	....	....	....	....	....	....	....	....	....	....	6	0	0	0	10	....	....	....
	NHL Totals		**19**	**1**	**2**	**3**	**20**	**0**	**0**	**0**	**22**	**4.5**		**1**	**100.0**	**21**	**13**	**13:12**	....	....	....	....	....	....	....	....

Traded to **Tampa Bay** by **Toronto** with Mike Johnson, Toronto's 5th (Pavel Sedov) and 6th (Aaron Gionet) round choices in 2000 Entry Draft and future considerations for Darcy Tucker, Tampa Bay's 4th round choice (Miguel Delisle) in 2000 Entry Draft and future considerations, February 9, 2000. Signed as a free agent by **Zlin** (Czech) with Tampa Bay retaining NHL rights, August 12, 2001.

POTHIER, Brian

(PAW-tee-ay, BRIGH-uhn) **OTT.**

Defense. Shoots right. 6', 195 lbs. Born, New Bedford, MA, April 15, 1977.

			Regular Season																Playoffs							
Season	Club	League	GP	G	A	Pts	PIM	PP	SH	GW	S	%	+/-	TF	F%	H	SB	Min	GP	G	A	Pts	PIM	PP	SH	GW
1995-96	Northfield High	Hi-School	27	11	22	33	36	....	....	....	....	....	....	....	....	....	....	....	....	....	....	....	....	....	....	....
1996-97	RPI Engineers	ECAC	34	1	11	12	42	....	....	....	....	....	....	....	....	....	....	....	....	....	....	....	....	....	....	....
1997-98	RPI Engineers	ECAC	35	2	9	11	28	....	....	....	....	....	....	....	....	....	....	....	....	....	....	....	....	....	....	....
1998-99	RPI Engineers	ECAC	37	5	13	18	36	....	....	....	....	....	....	....	....	....	....	....	....	....	....	....	....	....	....	....
99-2000	RPI Engineers	ECAC	36	9	24	33	44	....	....	....	....	....	....	....	....	....	....	....	....	....	....	....	....	....	....	....
2000-01	**Atlanta**	**NHL**	**3**	**0**	**0**	**0**	**2**	**0**	**0**	**0**	**0**	**0.0**	**4**	**0**	**0.0**	**2**	**6**	**20:38**	....	....	....	....	....	....	....	....
	Orlando	IHL	76	12	29	41	69	....	....	....	....	....	....	....	....	....	....	....	16	3	5	8	11	....	....	....

Season	Club	League	GP	G	A	Pts	PIM	PP	SH	GW	S	%	+/-	TF	F%	H	SB	Min	GP	G	A	Pts	PIM	PP	SH	GW
			Regular Season																Playoffs							
2001-02	Atlanta	NHL	33	3	6	9	22	1	0	1	65	4.6	–19	0	0.0	51	36	21:41								
	Chicago Wolves	AHL	39	6	13	19	30																			
	NHL Totals		36	3	6	9	24	1	0	1	65	4.6		0	0.0	53	42	21:36								

ECAC Second All-Star Team (2000) • ECAC All-Tournament Team (2000) • NCAA East Second All-American Team (2000) • Won Garry F. Longman Memorial Trophy (Top Rookie - IHL) (2001)

Signed as a free agent by **Atlanta**, March 27, 2000. Traded to **Ottawa** by **Atlanta** for Shawn McEachern and Ottawa's 6th round choice in 2004 Entry Draft, June 29, 2002.

POTI, Tom (POH-tee, TAWM) NYR

Defense. Shoots left. 6'3", 215 lbs. Born, Worcester, MA, March 22, 1977. Edmonton's 4th choice, 59th overall, in 1996 Entry Draft.

Season	Club	League	GP	G	A	Pts	PIM	PP	SH	GW	S	%	+/-	TF	F%	H	SB	Min	GP	G	A	Pts	PIM	PP	SH	GW
1992-93	St. Peter's High	Hi-School	55	25	46	71																				
1993-94	Cushing Academy	Hi-School	30	10	35	45																				
1994-95	Cushing Academy	Hi-School	36	17	54	71	35																			
	Central-Mass	MBAHL	8	8	10	18																				
1995-96	Cushing Academy	Hi-School	29	14	59	73	18																			
1996-97	Boston University	H-East	38	4	17	21	54																			
1997-98	Boston University	H-East	38	13	29	42	60																			
1998-99	Edmonton	NHL	73	5	16	21	42	2	0	3	94	5.3	10	0	0.0	35	82	19:33	4	0	1	1	2	0	0	0
99-2000	Edmonton	NHL	76	9	26	35	65	2	1	1	125	7.2	8	0	0.0	38	119	24:10	5	0	1	1	0	0	0	0
2000-01	Edmonton	NHL	81	12	20	32	60	6	0	3	161	7.5	–4	0	0.0	37	114	22:44	6	0	2	2	2	0	0	0
2001-02	Edmonton	NHL	55	1	16	17	42	1	0	0	100	1.0	–6	0	0.0	29	62	24:32								
	United States	Olympics	6	0	1	1	4																			
	NY Rangers	NHL	11	1	7	8	2	1	0	1	9	11.1	–4	0	0.0	8	10	21:45								
	NHL Totals		296	28	85	113	211	12	1	8	489	5.7		0	0.0	147	387	22:37	15	0	4	4	4	0	0	0

NCAA Championship All-Tournament Team (1997) • Hockey East First All-Star Team (1998) • NCAA East First All-American Team (1998) • NHL All-Rookie Team (1999)

Traded to **NY Rangers** by **Edmonton** with Rem Murray for Mike York and NY Rangers' 4th round choice (Ivan Koltsov) in 2002 Entry Draft, March 19, 2002.

POULIN, Patrick (poo-LIHN, PAT-rihk)

Center. Shoots left. 6'1", 216 lbs. Born, Vanier, Que., April 23, 1973. Hartford's 1st choice, 9th overall, in 1991 Entry Draft.

Season	Club	League	GP	G	A	Pts	PIM	PP	SH	GW	S	%	+/-	TF	F%	H	SB	Min	GP	G	A	Pts	PIM	PP	SH	GW
1988-89	Ste-Foy	QAAA	42	28	42	70	44												13	*13	23	*36	24			
1989-90	St-Hyacinthe	QMJHL	60	25	26	51	55												12	1	9	10	5			
1990-91	St-Hyacinthe	QMJHL	56	32	38	70	82												4	0	2	2	23			
1991-92	St-Hyacinthe	QMJHL	56	52	86	*138	58												5	2	2	4	4			
	Hartford	NHL	1	0	0	0	2	0	0	0	0	0.0	–1						7	2	1	3	0	1	0	0
	Springfield	AHL																	1	0	0	0	0			
1992-93	Hartford	NHL	81	20	31	51	37	4	0	2	160	12.5	–19													
1993-94	Hartford	NHL	9	2	1	3	11	1	0	0	13	15.4	–8													
	Chicago	NHL	58	12	13	25	40	1	0	3	83	14.5	0						4	0	0	0	0	0	0	0
1994-95	Chicago	NHL	45	15	15	30	53	4	0	2	77	19.5	13						16	4	1	5	8	1	0	0
1995-96	Chicago	NHL	38	7	8	15	16	1	0	0	40	17.5	7													
	Indianapolis Ice	IHL	1	0	1	1	0																			
	Tampa Bay	NHL	8	0	1	1	0	0	0	0	11	0.0	0						2	0	0	0	0	0	0	0
1996-97	Tampa Bay	NHL	73	12	14	26	56	2	3	1	124	9.7	–16													
1997-98	Tampa Bay	NHL	44	2	7	9	19	0	0	0	49	4.1	–3													
	Montreal	NHL	34	4	6	10	8	0	1	1	39	10.3	–1						3	0	0	0	0	0	0	0
1998-99	Montreal	NHL	81	8	17	25	21	0	1	1	87	9.2	6	112	30.4	92	29	13:30								
99-2000	Montreal	NHL	82	10	5	15	17	0	1	2	82	12.2	–15	35	34.3	79	38	12:34								
2000-01	Montreal	NHL	52	9	11	20	13	0	0	4	65	13.8	1	11	9.1	77	34	13:16								
2001-02	Montreal	NHL	28	0	5	5	6	0	0	0	15	0.0	6	4	0.0	32	12	9:13								
	Quebec	AHL	31	12	7	19	6												3	0	2	2	0			
	NHL Totals		634	101	134	235	299	13	6	16	845	12.0		162	29.0	280	113	12:38	32	6	2	8	8	2	0	0

QMJHL All-Rookie Team (1990) • QMJHL First All-Star Team (1992) • Canadian Major Junior Player of the Year (1992)

Traded to **Chicago** by **Hartford** with Eric Weinrich for Steve Larmer and Bryan Marchment, November 2, 1993. Traded to **Tampa Bay** by **Chicago** with Igor Ulanov and Chicago's 2nd round choice (later traded to New Jersey - New Jersey selected Pierre Dagenais) in 1996 Entry Draft for Enrico Ciccone and Tampa Bay's 2nd round choice (Jeff Paul) in 1996 Entry Draft, March 20, 1996. Traded to **Montreal** by **Tampa Bay** with Mick Vukota and Igor Ulanov for Stephane Richer, Darcy Tucker and David Wilkie, January 15, 1998.

PRATT, Nolan (PRAT, NOH-lan) T.B.

Defense. Shoots left. 6'3", 200 lbs. Born, Fort McMurray, Alta., August 14, 1975. Hartford's 4th choice, 115th overall, in 1993 Entry Draft.

Season	Club	League	GP	G	A	Pts	PIM	PP	SH	GW	S	%	+/-	TF	F%	H	SB	Min	GP	G	A	Pts	PIM	PP	SH	GW
1991-92	Bonnyville	AJHL	33	3	7	10	57																			
	Portland	WHL	22	2	9	11	13												6	1	3	4	12			
1992-93	Portland	WHL	70	4	19	23	97												16	2	7	9	31			
1993-94	Portland	WHL	72	4	32	36	105												10	1	2	3	14			
1994-95	Portland	WHL	72	6	37	43	196												9	1	6	7	10			
1995-96	Springfield	AHL	62	2	6	8	72												2	0	0	0	0			
	Richmond	ECHL	4	1	0	1	2																			
1996-97	Hartford	NHL	9	0	2	2	6	0	0	0	4	0.0	0													
	Springfield	AHL	66	1	18	19	127												17	0	3	3	18			
1997-98	Carolina	NHL	23	0	2	2	44	0	0	0	11	0.0	–2													
	New Haven	AHL	54	3	15	18	135																			
1998-99	Carolina	NHL	61	1	14	15	95	0	0	1	46	2.2	15	0	0.0	121	59	16:45	3	0	0	0	2	0	0	0
99-2000	Carolina	NHL	64	3	1	4	90	0	0	1	47	6.4	–22	0	0.0	155	74	19:10								
2000-01	Colorado	NHL	46	1	2	3	40	0	0	1	26	3.8	2	1	0.0	41	33	9:50								
2001-02	Tampa Bay	NHL	46	0	3	3	51	0	0	0	38	0.0	–4	1	0.0	67	65	18:25								
	NHL Totals		249	5	24	29	326	0	0	3	172	2.9		2	0.0	384	231	16:21	3	0	0	0	2	0	0	0

Transferred to **Carolina** after **Hartford** franchise relocated, June 25, 1997. Traded to **Colorado** by **Carolina** with Carolina's 1st (Vaclav Nedorost) and 2nd (Jared Aulin) round choices in 2000 Entry Draft and Philadelphia's 2nd round choice (previously acquired, Colorado selected Agris Saviels) in 2000 Entry Draft for Sandis Ozolinsh and Columbus' 2nd round choice (previously acquired, Carolina selected Tomas Kurka) in 2000 Entry Draft, June 24, 2000. Traded to **Tampa Bay** by **Colorado** for LA Kings' 6th round choice (previously acquired, Colorado selected Scott Horvath) in 2001 Entry Draft, June 24, 2001.

PRIMEAU, Keith (PREE-moh, KEETH) PHI.

Center. Shoots left. 6'5", 220 lbs. Born, Toronto, Ont., November 24, 1971. Detroit's 1st choice, 3rd overall, in 1990 Entry Draft.

Season	Club	League	GP	G	A	Pts	PIM	PP	SH	GW	S	%	+/-	TF	F%	H	SB	Min	GP	G	A	Pts	PIM	PP	SH	GW
1986-87	Whitby Flyers	OMHA	65	69	80	149	116																			
1987-88	Hamilton	OJHL-B	19	19	17	36	16																			
	Hamilton	OHL	47	6	6	12	69												11	0	2	2	2			
1988-89	Niagara Falls	OHL	48	20	35	55	56												17	9	16	25	12			
1989-90	Niagara Falls	OHL	65	*57	70	*127	97												16	*16	17	*33	49			
1990-91	Detroit	NHL	58	3	12	15	106	0	0	1	33	9.1	–12						5	1	1	2	25	0	0	0
	Adirondack	AHL	6	3	5	8	8																			
1991-92	Detroit	NHL	35	6	10	16	83	0	0	0	27	22.2	9						11	0	0	0	14	0	0	0
	Adirondack	AHL	42	21	24	45	89												9	1	7	8	27			
1992-93	Detroit	NHL	73	15	17	32	152	4	1	2	75	20.0	–6						7	0	2	2	26	0	0	0
1993-94	Detroit	NHL	78	31	42	73	173	7	3	4	155	20.0	34						7	0	2	2	6	0	0	0
1994-95	Detroit	NHL	45	15	27	42	99	1	0	3	96	15.6	17						17	4	5	9	45	2	0	0
1995-96	Detroit	NHL	74	27	25	52	168	6	2	7	150	18.0	19						17	1	4	5	28	0	0	0
1996-97	Hartford	NHL	75	26	25	51	161	6	3	2	169	15.4	–3													
1997-98	Carolina	NHL	81	26	37	63	110	7	3	2	180	14.4	19													
	Canada	Olympics	6	2	1	3	4																			
1998-99	Carolina	NHL	78	30	32	62	75	9	1	5	178	16.9	8	1823	53.5	231	62	21:21	6	0	3	3	6	0	0	0
99-2000	Philadelphia	NHL	23	7	10	17	31	1	0	1	51	13.7	10	478	54.2	37	18	17:38	18	2	11	13	13	0	0	1
2000-01	Philadelphia	NHL	71	34	39	73	76	11	0	4	165	20.6	17	1811	53.5	151	27	19:58	4	0	3	3	8	0	0	0
2001-02	Philadelphia	NHL	75	19	29	48	128	5	0	3	151	12.6	–3	1595	51.3	131	31	18:00	5	0	0	0	6	0	0	0
	NHL Totals		766	239	305	544	1362	57	13	34	1430	16.7		5707	52.9	550	138	19:35	97	8	31	39	177	2	0	1

OHL Second All-Star Team (1990) • Played in NHL All-Star Game (1999)

Traded to **Hartford** by **Detroit** with Paul Coffey and Detroit's 1st round choice (Nikos Tselios) in 1997 Entry Draft for Brendan Shanahan and Brian Glynn, October 9, 1996. Transferred to **Carolina** after **Hartford** franchise relocated, June 25, 1997. • Missed majority of 1999-2000 season after failing to come to contract terms with **Carolina**. Traded to **Philadelphia** by **Carolina** with Carolina's 5th round choice (later traded to NY Islanders - NY Islanders selected Kristofer Ottosson) in 2000 Entry Draft for Rod Brind'Amour, Jean-Marc Pelletier and Philadelphia's 2nd round choice (later traded to Colorado - Colorado selected Agris Saviels) in 2000 Entry Draft, January 23, 2000.

PRIMEAU, Wayne
(PREE-moh, WAYN) **PIT.**

Center. Shoots left. 6'3", 220 lbs. Born, Scarborough, Ont., June 4, 1976. Buffalo's 1st choice, 17th overall, in 1994 Entry Draft.

			Regular Season																Playoffs							
Season	Club	League	GP	G	A	Pts	PIM	PP	SH	GW	S	%	+/-	TF	F%	H	SB	Min	GP	G	A	Pts	PIM	PP	SH	GW
1991-92	Whitby Flyers	OMHA	63	36	50	86	96	....	....	....	....	....	....	....	....	....	....	....	....	....	....	....	....	....	....	....
1992-93	Owen Sound	OHL	66	10	27	37	108	....	....	....	....	....	....	....	....	....	....	....	8	1	4	5	0	....	....	....
1993-94	Owen Sound	OHL	65	25	50	75	75	....	....	....	....	....	....	....	....	....	....	....	9	1	6	7	8	....	....	....
1994-95	Owen Sound	OHL	66	34	62	96	84	....	....	....	....	....	....	....	....	....	....	....	10	4	9	13	15	....	....	....
	Buffalo	**NHL**	**1**	**1**	**0**	**1**	**0**	**0**	**0**	**1**	**2**	**50.0**	**–2**	....	....	....	....	....	....	....	....	....	....	....	....	....
1995-96	Owen Sound	OHL	28	15	29	44	52	....	....	....	....	....	....	....	....	....	....	....	....	....	....	....	....	....	....	....
	Oshawa Generals	OHL	24	12	13	25	33	....	....	....	....	....	....	....	....	....	....	....	3	2	3	5	2	....	....	....
	Buffalo	**NHL**	**2**	**0**	**0**	**0**	**0**	**0**	**0**	**0**	**0**	**0.0**	**0**	....	....	....	....	....	....	....	....	....	....	....	....	....
	Rochester	AHL	8	2	3	5	6	....	....	....	....	....	....	....	....	....	....	....	17	3	1	4	11	....	....	....
1996-97	**Buffalo**	**NHL**	**45**	**2**	**4**	**6**	**64**	**1**	**0**	**0**	**25**	**8.0**	**–2**	....	....	....	....	....	**9**	**0**	**0**	**0**	**6**	**0**	**0**	**0**
	Rochester	AHL	24	9	5	14	27	....	....	....	....	....	....	....	....	....	....	....	1	0	0	0	0	....	....	....
1997-98	**Buffalo**	**NHL**	**69**	**6**	**6**	**12**	**87**	**2**	**0**	**1**	**51**	**11.8**	**9**	....	....	....	....	....	**14**	**1**	**3**	**4**	**6**	**0**	**0**	**0**
1998-99	**Buffalo**	**NHL**	**67**	**5**	**8**	**13**	**38**	**0**	**0**	**0**	**55**	**9.1**	**–6**	**529**	**48.6**	**74**	**15**	**10:19**	**19**	**3**	**4**	**7**	**6**	**1**	**0**	**0**
99-2000	**Buffalo**	**NHL**	**41**	**5**	**7**	**12**	**38**	**2**	**0**	**1**	**40**	**12.5**	**–8**	**430**	**45.6**	**47**	**8**	**11:03**	....	....	....	....	....	....	....	....
	Tampa Bay	**NHL**	**17**	**2**	**3**	**5**	**25**	**0**	**0**	**0**	**35**	**5.7**	**–4**	**290**	**45.5**	**26**	**9**	**14:21**	....	....	....	....	....	....	....	....
2000-01	**Tampa Bay**	**NHL**	**47**	**2**	**13**	**15**	**77**	**0**	**0**	**0**	**47**	**4.3**	**–17**	**630**	**52.2**	**75**	**21**	**14:11**	....	....	....	....	....	....	....	....
	Pittsburgh	**NHL**	**28**	**1**	**6**	**7**	**54**	**0**	**0**	**0**	**30**	**3.3**	**0**	**318**	**51.3**	**39**	**17**	**12:45**	**18**	**1**	**3**	**4**	**2**	**0**	**0**	**0**
2001-02	**Pittsburgh**	**NHL**	**33**	**3**	**7**	**10**	**18**	**0**	**1**	**0**	**28**	**10.7**	**–1**	**519**	**53.2**	**54**	**19**	**12:38**	....	....	....	....	....	....	....	....
	NHL Totals		**350**	**27**	**54**	**81**	**401**	**5**	**1**	**3**	**313**	**8.6**		**2716**	**49.8**	**315**	**89**	**12:09**	**60**	**5**	**10**	**15**	**20**	**1**	**0**	**0**

Traded to **Tampa Bay** by **Buffalo** with Cory Sarich, Brian Holzinger and Buffalo's 3rd round choice (Alexander Kharitonov) in 2000 Entry Draft for Chris Gratton and Tampa Bay's 2nd round choice (Derek Roy) in 2001 Entry Draft, March 9, 2000. Traded to **Pittsburgh** by **Tampa Bay** for Matthew Barnaby, February 1, 2001. • Missed majority of 2001-02 season recovering from knee injury suffered in game vs. Buffalo, January 8, 2002.

PROBERT, Bob
(PROH-buhrt, BAWB) **CHI.**

Left wing. Shoots left. 6'3", 225 lbs. Born, Windsor, Ont., June 5, 1965. Detroit's 3rd choice, 46th overall, in 1983 Entry Draft.

Season	Club	League	GP	G	A	Pts	PIM	PP	SH	GW	S	%	+/-	TF	F%	H	SB	Min	GP	G	A	Pts	PIM	PP	SH	GW
1981-82	Windsor Club 240	OMHA	55	60	40	100	40	....	....	....	....	....	....	....	....	....	....	....	....	....	....	....	....	....	....	....
1982-83	Brantford	OHL	51	12	16	28	133	....	....	....	....	....	....	....	....	....	....	....	8	2	2	4	23	....	....	....
1983-84	Brantford	OHL	65	35	28	63	189	....	....	....	....	....	....	....	....	....	....	....	6	0	3	3	16	....	....	....
1984-85	Hamilton	OHL	4	0	1	1	21	....	....	....	....	....	....	....	....	....	....	....	....	....	....	....	....	....	....	....
	Sault Ste. Marie	OHL	44	20	52	72	172	....	....	....	....	....	....	....	....	....	....	....	15	6	11	17	60	....	....	....
1985-86	**Detroit**	**NHL**	**44**	**8**	**13**	**21**	**186**	**3**	**0**	**0**	**46**	**17.4**	**–14**	....	....	....	....	....	....	....	....	....	....	....	....	....
	Adirondack	AHL	32	12	15	27	152	....	....	....	....	....	....	....	....	....	....	....	10	2	3	5	68	....	....	....
1986-87	**Detroit**	**NHL**	**63**	**13**	**11**	**24**	**221**	**2**	**0**	**0**	**56**	**23.2**	**–6**	....	....	....	....	....	**16**	**3**	**4**	**7**	**63**	**1**	**0**	**1**
	Adirondack	AHL	7	1	4	5	15	....	....	....	....	....	....	....	....	....	....	....	....	....	....	....	....	....	....	....
1987-88	**Detroit**	**NHL**	**74**	**29**	**33**	**62**	***398**	**15**	**0**	**5**	**126**	**23.0**	**16**	....	....	....	....	....	**16**	**8**	**13**	**21**	**51**	**5**	**0**	**1**
1988-89	**Detroit**	**NHL**	**25**	**4**	**2**	**6**	**106**	**1**	**0**	**0**	**23**	**17.4**	**–11**	....	....	....	....	....	....	....	....	....	....	....	....	....
1989-90	**Detroit**	**NHL**	**4**	**3**	**0**	**3**	**21**	**0**	**0**	**1**	**12**	**25.0**	**0**	....	....	....	....	....	....	....	....	....	....	....	....	....
1990-91	**Detroit**	**NHL**	**55**	**16**	**23**	**39**	**315**	**4**	**0**	**3**	**88**	**18.2**	**–3**	....	....	....	....	....	**6**	**1**	**2**	**3**	**50**	**0**	**0**	**0**
1991-92	**Detroit**	**NHL**	**63**	**20**	**24**	**44**	**276**	**8**	**0**	**1**	**96**	**20.8**	**16**	....	....	....	....	....	**11**	**1**	**6**	**7**	**28**	**0**	**0**	**0**
1992-93	**Detroit**	**NHL**	**80**	**14**	**29**	**43**	**292**	**6**	**0**	**3**	**128**	**10.9**	**–9**	....	....	....	....	....	**7**	**0**	**3**	**3**	**10**	**0**	**0**	**0**
1993-94	**Detroit**	**NHL**	**66**	**7**	**10**	**17**	**275**	**1**	**0**	**0**	**105**	**6.7**	**–1**	....	....	....	....	....	**7**	**1**	**1**	**2**	**8**	**0**	**0**	**0**
1994-95	**Chicago**	**NHL**	DID NOT PLAY – SUSPENDED																							
1995-96	**Chicago**	**NHL**	**78**	**19**	**21**	**40**	**237**	**1**	**0**	**3**	**97**	**19.6**	**15**	....	....	....	....	....	**10**	**0**	**2**	**2**	**23**	**0**	**0**	**0**
1996-97	**Chicago**	**NHL**	**82**	**9**	**14**	**23**	**326**	**1**	**0**	**3**	**111**	**8.1**	**–3**	....	....	....	....	....	**6**	**2**	**1**	**3**	**41**	**0**	**0**	**0**
1997-98	**Chicago**	**NHL**	**14**	**2**	**1**	**3**	**48**	**2**	**0**	**0**	**18**	**11.1**	**–7**	....	....	....	....	....	....	....	....	....	....	....	....	....
1998-99	**Chicago**	**NHL**	**78**	**7**	**14**	**21**	**206**	**0**	**0**	**3**	**87**	**8.0**	**–11**	**45**	**51.1**	**105**	**12**	**10:38**	....	....	....	....	....	....	....	....
99-2000	**Chicago**	**NHL**	**69**	**4**	**11**	**15**	**114**	**0**	**0**	**0**	**38**	**10.5**	**10**	**32**	**40.6**	**87**	**15**	**10:28**	....	....	....	....	....	....	....	....
2000-01	**Chicago**	**NHL**	**79**	**7**	**12**	**19**	**103**	**1**	**0**	**0**	**45**	**15.6**	**–13**	**2**	**0.0**	**117**	**13**	**10:50**	....	....	....	....	....	....	....	....
2001-02	**Chicago**	**NHL**	**61**	**1**	**3**	**4**	**176**	**0**	**0**	**1**	**33**	**3.0**	**–9**	**1**	**0.0**	**64**	**8**	**6:44**	**2**	**0**	**0**	**0**	**0**	**0**	**0**	**0**
	NHL Totals		**935**	**163**	**221**	**384**	**3300**	**45**	**0**	**23**	**1109**	**14.7**		**80**	**45.0**	**373**	**48**	**9:49**	**81**	**16**	**32**	**48**	**274**	**6**	**0**	**2**

Played in NHL All-Star Game (1988)

Signed as a free agent by **Chicago**, July 23, 1994. • Suspended for entire 1994-95 season for violating NHL substance abuse policy, September 2, 1994. • Missed majority of 1997-98 season recovering from rotator cuff injury suffered in game vs. Detroit, November 16, 1997.

PROCHAZKA, Martin
(pro-HAHS-kah, MAHR-tihn)

Right wing. Shoots right. 5'11", 180 lbs. Born, Slany, Czech., March 3, 1972. Toronto's 6th choice, 135th overall, in 1991 Entry Draft.

Season	Club	League	GP	G	A	Pts	PIM	PP	SH	GW	S	%	+/-	TF	F%	H	SB	Min	GP	G	A	Pts	PIM	PP	SH	GW
1989-90	Poldi Kladno	Czech	49	18	12	30	....	....	....	....	....	....	....	....	....	....	....	....	....	....	....	....	....	....	....	....
1990-91	Poldi Kladno	Czech	50	19	10	29	21	....	....	....	....	....	....	....	....	....	....	....	....	....	....	....	....	....	....	....
1991-92	Dukla Jihlava	Czech	44	18	11	29	2	....	....	....	....	....	....	....	....	....	....	....	....	....	....	....	....	....	....	....
1992-93	Poldi Kladno	Czech	46	26	12	38	....	....	....	....	....	....	....	....	....	....	....	....	....	....	....	....	....	....	....	....
1993-94	HC Kladno	Czech	43	24	16	40	0	....	....	....	....	....	....	....	....	....	....	....	2	2	0	2		....	....	....
1994-95	HC Kladno	Czech	41	25	33	58	18	....	....	....	....	....	....	....	....	....	....	....	11	8	4	12	4	....	....	....
1995-96	HC Poldi Kladno	Czech	37	15	27	42	....	....	....	....	....	....	....	....	....	....	....	....	8	2	4	6	....	....	....	....
1996-97	AIK Solna	Sweden	49	16	23	39	38	....	....	....	....	....	....	....	....	....	....	....	7	2	3	5	8	....	....	....
1997-98	**Toronto**	**NHL**	**29**	**2**	**4**	**6**	**8**	**0**	**0**	**0**	**40**	**5.0**	**–1**	....	....	....	....	....	....	....	....	....	....	....	....	....
	Czech Republic	Olympics	6	1	1	2	0	....	....	....	....	....	....	....	....	....	....	....	....	....	....	....	....	....	....	....
1998-99	Vsetin	Czech	47	20	29	49	12	....	....	....	....	....	....	....	....	....	....	....	12	*10	9	*19	....	....	....	....
99-2000	**Atlanta**	**NHL**	**3**	**0**	**1**	**1**	**0**	**0**	**0**	**0**	**5**	**0.0**	**–1**	**0**	**0.0**	**2**	**2**	**16:37**	....	....	....	....	....	....	....	....
	Vsetin	Czech	31	10	10	20	16	....	....	....	....	....	....	....	....	....	....	....	9	2	0	2	0	....	....	....
2000-01	HC Vitkovice	Czech	32	15	16	31	12	....	....	....	....	....	....	....	....	....	....	....	10	4	2	6	0	....	....	....
2001-02	HC Vitkovice	Czech	20	6	11	17	4	....	....	....	....	....	....	....	....	....	....	....	....	....	....	....	....	....	....	....
	Avangard Omsk	Russia	31	8	6	14	6	....	....	....	....	....	....	....	....	....	....	....	11	2	1	3	2	....	....	....
	NHL Totals		**32**	**2**	**5**	**7**	**8**	**0**	**0**	**0**	**45**	**4.4**		**0**	**0.0**	**2**	**2**	**16:37**	....	....	....	....	....	....	....	....

Traded to **Atlanta** by **Toronto** for Atlanta's 6th round choice (Maxim Kondratjev) in 2001 Entry Draft, July 15, 1999.

PRONGER, Chris
(PRAHN-guhr, KRIHS) **ST.L.**

Defense. Shoots left. 6'6", 220 lbs. Born, Dryden, Ont., October 10, 1974. Hartford's 1st choice, 2nd overall, in 1993 Entry Draft.

Season	Club	League	GP	G	A	Pts	PIM	PP	SH	GW	S	%	+/-	TF	F%	H	SB	Min	GP	G	A	Pts	PIM	PP	SH	GW
1990-91	Stratford	OJHL-B	48	15	37	52	132	....	....	....	....	....	....	....	....	....	....	....	....	....	....	....	....	....	....	....
1991-92	Peterborough	OHL	63	17	45	62	90	....	....	....	....	....	....	....	....	....	....	....	10	1	8	9	28	....	....	....
1992-93	Peterborough	OHL	61	15	62	77	108	....	....	....	....	....	....	....	....	....	....	....	21	15	25	40	51	....	....	....
1993-94	**Hartford**	**NHL**	**81**	**5**	**25**	**30**	**113**	**2**	**0**	**0**	**174**	**2.9**	**–3**	....	....	....	....	....	....	....	....	....	....	....	....	....
1994-95	**Hartford**	**NHL**	**43**	**5**	**9**	**14**	**54**	**3**	**0**	**1**	**94**	**5.3**	**–12**	....	....	....	....	....	....	....	....	....	....	....	....	....
1995-96	**St. Louis**	**NHL**	**78**	**7**	**18**	**25**	**110**	**3**	**1**	**1**	**138**	**5.1**	**–18**	....	....	....	....	....	**13**	**1**	**5**	**6**	**16**	**0**	**0**	**0**
1996-97	**St. Louis**	**NHL**	**79**	**11**	**24**	**35**	**143**	**4**	**0**	**0**	**147**	**7.5**	**15**	....	....	....	....	....	**6**	**1**	**1**	**2**	**22**	**0**	**0**	**0**
1997-98	**St. Louis**	**NHL**	**81**	**9**	**27**	**36**	**180**	**1**	**0**	**2**	**145**	**6.2**	**47**	....	....	....	....	....	**10**	**1**	**9**	**10**	**26**	**0**	**0**	**0**
	Canada	Olympics	6	0	0	0	4	....	....	....	....	....	....	....	....	....	....	....	....	....	....	....	....	....	....	....
1998-99	**St. Louis**	**NHL**	**67**	**13**	**33**	**46**	**113**	**8**	**0**	**0**	**172**	**7.6**	**3**	**0**	**0.0**	**132**	**119**	**30:36**	**13**	**1**	**4**	**5**	**28**	**1**	**0**	**0**
99-2000	**St. Louis**	**NHL**	**79**	**14**	**48**	**62**	**92**	**8**	**0**	**3**	**192**	**7.3**	**52**	**1**	**0.0**	**106**	**185**	**30:14**	**7**	**3**	**4**	**7**	**32**	**2**	**0**	**2**
2000-01	**St. Louis**	**NHL**	**51**	**8**	**39**	**47**	**75**	**4**	**0**	**0**	**121**	**6.6**	**21**	**0**	**0.0**	**56**	**80**	**27:45**	**15**	**1**	**7**	**8**	**32**	**0**	**0**	**0**
2001-02	**St. Louis**	**NHL**	**78**	**7**	**40**	**47**	**120**	**4**	**1**	**3**	**204**	**3.4**	**23**	**0**	**0.0**	**115**	**161**	**29:28**	**9**	**1**	**7**	**8**	**24**	**0**	**0**	**0**
	Canada	Olympics	6	0	1	1	2	....	....	....	....	....	....	....	....	....	....	....	....	....	....	....	....	....	....	....
	NHL Totals		**637**	**79**	**263**	**342**	**1000**	**37**	**2**	**10**	**1387**	**5.7**		**1**	**0.0**	**409**	**545**	**29:39**	**73**	**9**	**37**	**46**	**180**	**3**	**0**	**2**

OHL All-Rookie Team (1992) • OHL First All-Star Team (1993) • Canadian Major Junior First All-Star Team (1993) • Canadian Major Junior Defenseman of the Year (1993) • NHL All-Rookie Team (1994) • NHL Second All-Star Team (1998) • Won Bud Ice Plus/Minus Award (1998) • NHL First All-Star Team (2000) • Won Bud Light Plus/Minus Award (2000) • Won James Norris Memorial Trophy (2000) • Won Hart Trophy (2000) • Played in NHL All-Star Game (1999, 2000, 2002)

Traded to **St. Louis** by **Hartford** for Brendan Shanahan, July 27, 1995.

			Regular Season																Playoffs							
Season	Club	League	GP	G	A	Pts	PIM	PP	SH	GW	S	%	+/-	TF	F%	H	SB	Min	GP	G	A	Pts	PIM	PP	SH	GW

PRONGER, Sean

(PRAHN-guhr, SHAWN) **CBJ**

Center. Shoots left. 6'3", 209 lbs. Born, Thunder Bay, Ont., November 30, 1972. Vancouver's 3rd choice, 51st overall, in 1991 Entry Draft.

Season	Club	League	GP	G	A	Pts	PIM	PP	SH	GW	S	%	+/-	TF	F%	H	SB	Min	GP	G	A	Pts	PIM	PP	SH	GW
1988-89	Kenora Boise	NOJHA	33	38	30	68	....	....	....	....	....	....	....	....	....	....	....	....	....	....	....	....	....	....	....	....
1989-90	Thunder Bay	USHL	48	18	34	52	61	....	....	....	....	....	....	....	....	....	....	....	....	....	....	....	....	....	....	....
1990-91	Bowling Green	CCHA	40	3	7	10	30	....	....	....	....	....	....	....	....	....	....	....	....	....	....	....	....	....	....	....
1991-92	Bowling Green	CCHA	34	9	7	16	28	....	....	....	....	....	....	....	....	....	....	....	....	....	....	....	....	....	....	....
1992-93	Bowling Green	CCHA	39	23	23	46	35	....	....	....	....	....	....	....	....	....	....	....	....	....	....	....	....	....	....	....
1993-94	Bowling Green	CCHA	38	17	17	34	38	....	....	....	....	....	....	....	....	....	....	....	....	....	....	....	....	....	....	....
1994-95	Knoxville	ECHL	34	18	23	41	55	....	....	....	....	....	....	....	....	....	....	....	....	....	....	....	....	....	....	....
	Greensboro	ECHL	2	0	2	2	0	....	....	....	....	....	....	....	....	....	....	....	....	....	....	....	....	....	....	....
	San Diego Gulls	IHL	8	0	0	0	2	....	....	....	....	....	....	....	....	....	....	....	....	....	....	....	....	....	....	....
1995-96	**Anaheim**	**NHL**	**7**	**0**	**1**	**1**	**6**	**0**	**0**	**0**	**3**	**0.0**	**0**	....	....	....	....	....	....	....	....	....	....	....	....	....
	Baltimore	AHL	72	16	17	33	61	....	....	....	....	....	....	....	....	....	....	....	12	3	7	10	16	....	....	....
1996-97	**Anaheim**	**NHL**	**39**	**7**	**7**	**14**	**20**	**1**	**0**	**1**	**43**	**16.3**	**6**	....	....	....	....	....	**9**	**0**	**2**	**2**	**4**	**0**	**0**	**0**
	Baltimore	AHL	41	26	17	43	17	....	....	....	....	....	....	....	....	....	....	....	....	....	....	....	....	....	....	....
1997-98	**Anaheim**	**NHL**	**62**	**5**	**15**	**20**	**30**	**1**	**0**	**2**	**68**	**7.4**	**-9**	....	....	....	....	....	....	....	....	....	....	....	....	....
	Pittsburgh	**NHL**	**5**	**1**	**0**	**1**	**2**	**0**	**0**	**1**	**5**	**20.0**	**-1**	....	....	....	....	....	**5**	**0**	**0**	**0**	**4**	**0**	**0**	**0**
1998-99	**Pittsburgh**	**NHL**	**2**	**0**	**0**	**0**	**0**	**0**	**0**	**0**	**3**	**0.0**	**0**	**4**	**75.0**	**2**	**0**	**8:48**	....	....	....	....	....	....	....	....
	Houston Aeros	IHL	16	11	7	18	32	....	....	....	....	....	....	....	....	....	....	....	....	....	....	....	....	....	....	....
	NY Rangers	**NHL**	**14**	**0**	**3**	**3**	**4**	**0**	**0**	**0**	**3**	**0.0**	**-3**	**15**	**20.0**	**12**	**1**	**6:28**	....	....	....	....	....	....	....	....
	Los Angeles	**NHL**	**13**	**0**	**1**	**1**	**4**	**0**	**0**	**0**	**8**	**0.0**	**2**	**20**	**35.0**	**13**	**5**	**11:00**	....	....	....	....	....	....	....	....
99-2000	**Boston**	**NHL**	**11**	**0**	**1**	**1**	**13**	**0**	**0**	**0**	**7**	**0.0**	**-4**	**116**	**46.6**	**28**	**3**	**10:47**	....	....	....	....	....	....	....	....
	Providence	AHL	51	11	18	29	26	....	....	....	....	....	....	....	....	....	....	....	....	....	....	....	....	....	....	....
	Manitoba Moose	IHL	14	3	5	8	21	....	....	....	....	....	....	....	....	....	....	....	2	0	1	1	2	....	....	....
2000-01	Manitoba Moose	IHL	82	18	21	39	85	....	....	....	....	....	....	....	....	....	....	....	13	3	6	9	2	....	....	....
2001-02	**Columbus**	**NHL**	**26**	**3**	**1**	**4**	**4**	**0**	**0**	**0**	**25**	**12.0**	**-4**	**167**	**52.1**	**25**	**13**	**11:26**	....	....	....	....	....	....	....	....
	Syracuse Crunch	AHL	54	23	26	49	53	....	....	....	....	....	....	....	....	....	....	....	8	4	1	5	10	....	....	....
	NHL Totals		**179**	**16**	**29**	**45**	**83**	**2**	**0**	**4**	**165**	**9.7**		**322**	**47.8**	**80**	**22**	**10:06**	**14**	**0**	**2**	**2**	**8**	**0**	**0**	**0**

Signed as a free agent by **Anaheim**, February 14, 1995. Traded to **Pittsburgh** by **Anaheim** for the rights to Patrick Lalime, March 24, 1998. Traded to **NY Rangers** by **Pittsburgh** with Chris Tamer and Petr Nedved for Alexei Kovalev and Harry York, November 25, 1998. Traded to **Los Angeles** by **NY Rangers** for Eric Lacroix, February 12, 1999. Signed as a free agent by **Boston**, August 25, 1999. Traded to **Manitoba** (IHL) by **Providence** (AHL) with Keith McCambridge for Terry Hollinger, March 16, 2000 with Boston retaining Pronger's NHL rights. Traded to **NY Islanders** by **Boston** for future considerations, December 5, 2000. Claimed on waivers by **Columbus** from **NY Islanders**, May 18, 2001.

PROSPAL, Vaclav

(PRAWS-pahl, VAHT-slahv) **T.B.**

Center. Shoots left. 6'2", 195 lbs. Born, Ceske Budejovice, Czech., February 17, 1975. Philadelphia's 2nd choice, 71st overall, in 1993 Entry Draft.

Season	Club	League	GP	G	A	Pts	PIM	PP	SH	GW	S	%	+/-	TF	F%	H	SB	Min	GP	G	A	Pts	PIM	PP	SH	GW
1991-92	C. Budejovice Jr.	Czech-Jr.	36	16	16	32	12	....	....	....	....	....	....	....	....	....	....	....	....	....	....	....	....	....	....	....
1992-93	C. Budejovice Jr.	Czech-Jr.	32	26	31	57	24	....	....	....	....	....	....	....	....	....	....	....	....	....	....	....	....	....	....	....
1993-94	Hershey Bears	AHL	55	14	21	35	38	....	....	....	....	....	....	....	....	....	....	....	2	0	0	0	2	....	....	....
1994-95	Hershey Bears	AHL	69	13	32	45	36	....	....	....	....	....	....	....	....	....	....	....	2	1	0	1	4	....	....	....
1995-96	Hershey Bears	AHL	68	15	36	51	59	....	....	....	....	....	....	....	....	....	....	....	5	2	4	6	2	....	....	....
1996-97	**Philadelphia**	**NHL**	**18**	**5**	**10**	**15**	**4**	**0**	**0**	**0**	**35**	**14.3**	**3**	....	....	....	....	....	**5**	**1**	**3**	**4**	**4**	**0**	**0**	**0**
	Philadelphia	AHL	63	32	63	95	70	....	....	....	....	....	....	....	....	....	....	....	....	....	....	....	....	....	....	....
1997-98	**Philadelphia**	**NHL**	**41**	**5**	**13**	**18**	**17**	**4**	**0**	**0**	**60**	**8.3**	**-10**	....	....	....	....	....	....	....	....	....	....	....	....	....
	Ottawa	**NHL**	**15**	**1**	**6**	**7**	**4**	**0**	**0**	**0**	**28**	**3.6**	**-1**	....	....	....	....	....	**6**	**0**	**0**	**0**	**0**	**0**	**0**	**0**
1998-99	**Ottawa**	**NHL**	**79**	**10**	**26**	**36**	**58**	**2**	**0**	**3**	**114**	**8.8**	**8**	**997**	**56.2**	**202**	**20**	**13:03**	**4**	**0**	**0**	**0**	**0**	**0**	**0**	**0**
99-2000	**Ottawa**	**NHL**	**79**	**22**	**33**	**55**	**40**	**5**	**0**	**4**	**204**	**10.8**	**-2**	**1331**	**49.6**	**92**	**26**	**16:26**	**6**	**0**	**4**	**4**	**4**	**0**	**0**	**0**
2000-01	**Ottawa**	**NHL**	**40**	**1**	**12**	**13**	**12**	**0**	**0**	**0**	**68**	**1.5**	**1**	**501**	**50.1**	**37**	**9**	**12:57**	....	....	....	....	....	....	....	....
	Florida	**NHL**	**34**	**4**	**12**	**16**	**10**	**1**	**0**	**0**	**68**	**5.9**	**-2**	**487**	**54.6**	**58**	**14**	**16:36**	....	....	....	....	....	....	....	....
2001-02	**Tampa Bay**	**NHL**	**81**	**18**	**37**	**55**	**38**	**7**	**0**	**2**	**166**	**10.8**	**-11**	**555**	**52.8**	**58**	**37**	**17:31**	....	....	....	....	....	....	....	....
	NHL Totals		**387**	**66**	**149**	**215**	**183**	**19**	**0**	**9**	**743**	**8.9**		**3871**	**52.4**	**447**	**106**	**15:26**	**21**	**1**	**7**	**8**	**8**	**0**	**0**	**0**

AHL First All-Star Team (1997)

Traded to **Ottawa** by **Philadelphia** with Pat Falloon and Dallas' 2nd round choice (previously acquired, Ottawa selected Chris Bala) in 1998 Entry Draft for Alexandre Daigle, January 17, 1998. Traded to **Florida** by **Ottawa** for future considerations, January 20, 2001. Traded to **Tampa Bay** by **Florida** for Ryan Johnson and Tampa Bay's 6th round choice in 2003 Entry Draft, July 10, 2001.

PRPIC, Joel

(puhr-PIHCH, JOHL) **S.J.**

Center. Shoots left. 6'6", 225 lbs. Born, Sudbury, Ont., September 25, 1974. Boston's 9th choice, 233rd overall, in 1993 Entry Draft.

Season	Club	League	GP	G	A	Pts	PIM	PP	SH	GW	S	%	+/-	TF	F%	H	SB	Min	GP	G	A	Pts	PIM	PP	SH	GW
1992-93	Waterloo Siskins	OJHL-B	45	17	43	60	160	....	....	....	....	....	....	....	....	....	....	....	....	....	....	....	....	....	....	....
1993-94	St. Lawrence	ECAC	31	2	4	6	90	....	....	....	....	....	....	....	....	....	....	....	....	....	....	....	....	....	....	....
1994-95	St. Lawrence	ECAC	32	7	10	17	62	....	....	....	....	....	....	....	....	....	....	....	....	....	....	....	....	....	....	....
1995-96	St. Lawrence	ECAC	32	3	10	13	77	....	....	....	....	....	....	....	....	....	....	....	....	....	....	....	....	....	....	....
1996-97	St. Lawrence	ECAC	34	10	8	18	57	....	....	....	....	....	....	....	....	....	....	....	....	....	....	....	....	....	....	....
1997-98	**Boston**	**NHL**	**1**	**0**	**0**	**0**	**2**	**0**	**0**	**0**	**0**	**0.0**	**0**	....	....	....	....	....	....	....	....	....	....	....	....	....
	Providence	AHL	73	17	18	35	53	....	....	....	....	....	....	....	....	....	....	....	....	....	....	....	....	....	....	....
1998-99	Providence	AHL	75	14	16	30	163	....	....	....	....	....	....	....	....	....	....	....	18	4	6	10	48	....	....	....
99-2000	**Boston**	**NHL**	**14**	**0**	**3**	**3**	**0**	**0**	**0**	**0**	**13**	**0.0**	**-6**	**117**	**52.1**	**10**	**2**	**6:47**	....	....	....	....	....	....	....	....
	Providence	AHL	70	9	20	29	143	....	....	....	....	....	....	....	....	....	....	....	14	3	4	7	58	....	....	....
2000-01	**Colorado**	**NHL**	**3**	**0**	**0**	**0**	**2**	**0**	**0**	**0**	**0**	**0.0**	**0**	**20**	**60.0**	**7**	**1**	**9:47**	....	....	....	....	....	....	....	....
	Hershey Bears	AHL	74	16	23	39	128	....	....	....	....	....	....	....	....	....	....	....	12	1	1	2	26	....	....	....
2001-02	Cleveland Barons	AHL	80	10	38	48	174	....	....	....	....	....	....	....	....	....	....	....	....	....	....	....	....	....	....	....
	NHL Totals		**18**	**0**	**3**	**3**	**4**	**0**	**0**	**0**	**13**	**0.0**		**137**	**53.3**	**17**	**3**	**7:19**	....	....	....	....	....	....	....	....

Signed as a free agent by **Colorado**, August, 2000. Signed as a free agent by **San Jose**, August 15, 2001.

PURINTON, Dale

(PUHR-ihn-TOHN, DAYL) **NYR**

Defense. Shoots left. 6'3", 214 lbs. Born, Fort Wayne, IN, October 11, 1976. NY Rangers' 5th choice, 117th overall, in 1995 Entry Draft.

Season	Club	League	GP	G	A	Pts	PIM	PP	SH	GW	S	%	+/-	TF	F%	H	SB	Min	GP	G	A	Pts	PIM	PP	SH	GW
1992-93	Moose Jaw	SMHL	34	1	16	17	107	....	....	....	....	....	....	....	....	....	....	....	....	....	....	....	....	....	....	....
	Moose Jaw	WHL	2	0	0	0	2	....	....	....	....	....	....	....	....	....	....	....	....	....	....	....	....	....	....	....
1993-94	Vernon Vipers	BCJHL	42	1	6	7	194	....	....	....	....	....	....	....	....	....	....	....	....	....	....	....	....	....	....	....
1994-95	Tacoma Rockets	WHL	65	0	8	8	291	....	....	....	....	....	....	....	....	....	....	....	3	0	0	0	13	....	....	....
1995-96	Kelowna Rockets	WHL	22	1	4	5	88	....	....	....	....	....	....	....	....	....	....	....	....	....	....	....	....	....	....	....
	Lethbridge	WHL	37	3	6	9	144	....	....	....	....	....	....	....	....	....	....	....	4	1	1	2	25	....	....	....
1996-97	Lethbridge	WHL	51	6	26	32	254	....	....	....	....	....	....	....	....	....	....	....	18	3	5	8	*88	....	....	....
1997-98	Hartford	AHL	17	0	0	0	95	....	....	....	....	....	....	....	....	....	....	....	....	....	....	....	....	....	....	....
	Charlotte	ECHL	34	3	5	8	186	....	....	....	....	....	....	....	....	....	....	....	....	....	....	....	....	....	....	....
1998-99	Hartford	AHL	45	1	3	4	306	....	....	....	....	....	....	....	....	....	....	....	7	0	2	2	24	....	....	....
99-2000	**NY Rangers**	**NHL**	**1**	**0**	**0**	**0**	**7**	**0**	**0**	**0**	**1**	**0.0**	**-1**	**0**	**0.0**	**1**	**1**	**12:45**	....	....	....	....	....	....	....	....
	Hartford	AHL	62	4	4	8	415	....	....	....	....	....	....	....	....	....	....	....	23	0	3	3	*87	....	....	....
2000-01	**NY Rangers**	**NHL**	**42**	**0**	**2**	**2**	**180**	**0**	**0**	**0**	**13**	**0.0**	**5**	**0**	**0.0**	**46**	**30**	**9:33**	....	....	....	....	....	....	....	....
	Hartford	AHL	11	0	1	1	75	....	....	....	....	....	....	....	....	....	....	....	....	....	....	....	....	....	....	....
2001-02	**NY Rangers**	**NHL**	**40**	**0**	**4**	**4**	**113**	**0**	**0**	**0**	**11**	**0.0**	**4**	**0**	**0.0**	**36**	**20**	**7:37**	....	....	....	....	....	....	....	....
	NHL Totals		**83**	**0**	**6**	**6**	**300**	**0**	**0**	**0**	**25**	**0.0**		**0**	**0.0**	**83**	**51**	**8:40**	....	....	....	....	....	....	....	....

PUSHOR, Jamie

(PUH-shohr, JAY-mee) **PIT.**

Defense. Shoots right. 6'3", 218 lbs. Born, Lethbridge, Alta., February 11, 1973. Detroit's 2nd choice, 32nd overall, in 1991 Entry Draft.

Season	Club	League	GP	G	A	Pts	PIM	PP	SH	GW	S	%	+/-	TF	F%	H	SB	Min	GP	G	A	Pts	PIM	PP	SH	GW
1988-89	Lethbridge	AMHL	37	1	8	9	20	....	....	....	....	....	....	....	....	....	....	....	....	....	....	....	....	....	....	....
	Lethbridge	WHL	2	0	0	0	0	....	....	....	....	....	....	....	....	....	....	....	....	....	....	....	....	....	....	....
1989-90	Lethbridge	AMHL	35	6	27	33	92	....	....	....	....	....	....	....	....	....	....	....	....	....	....	....	....	....	....	....
	Lethbridge	WHL	10	0	2	2	2	....	....	....	....	....	....	....	....	....	....	....	16	0	0	0	63	....	....	....
1990-91	Lethbridge	WHL	71	1	13	14	202	....	....	....	....	....	....	....	....	....	....	....	....	....	....	....	....	....	....	....
1991-92	Lethbridge	WHL	49	2	15	17	232	....	....	....	....	....	....	....	....	....	....	....	5	0	0	0	33	....	....	....
1992-93	Lethbridge	WHL	72	6	22	28	200	....	....	....	....	....	....	....	....	....	....	....	4	0	1	1	9	....	....	....
1993-94	Adirondack	AHL	73	1	17	18	124	....	....	....	....	....	....	....	....	....	....	....	12	0	0	0	22	....	....	....
1994-95	Adirondack	AHL	58	2	11	13	129	....	....	....	....	....	....	....	....	....	....	....	4	0	1	1	0	....	....	....
1995-96	**Detroit**	**NHL**	**5**	**0**	**1**	**1**	**17**	**0**	**0**	**0**	**6**	**0.0**	**2**	....	....	....	....	....	....	....	....	....	....	....	....	....
	Adirondack	AHL	65	2	16	18	126	....	....	....	....	....	....	....	....	....	....	....	3	0	0	0	5	....	....	....
1996-97♦	**Detroit**	**NHL**	**75**	**4**	**7**	**11**	**129**	**0**	**0**	**0**	**63**	**6.3**	**1**	....	....	....	....	....	**5**	**0**	**1**	**1**	**5**	**0**	**0**	**0**

Season	Club	League	GP	G	A	Pts	PIM	PP	SH	GW	S	%	+/-	TF	F%	H	SB	Min	GP	G	A	Pts	PIM	PP	SH	GW
			Regular Season																Playoffs							
1997-98	**Detroit**	**NHL**	**54**	**2**	**5**	**7**	**71**	**0**	**0**	**0**	**43**	**4.7**	**2**	....	....	....	....	....	....	....	....	....	....	....	....	....
	Anaheim	**NHL**	**10**	**0**	**2**	**2**	**10**	**0**	**0**	**0**	**8**	**0.0**	**1**	....	....	....	....	....	....	....	....	....	....	....	....	....
1998-99	**Anaheim**	**NHL**	**70**	**1**	**2**	**3**	**112**	**0**	**0**	**0**	**75**	**1.3**	**–20**	**0**	**0.0**	**110**	**153**	**19:16**	**4**	**0**	**0**	**0**	**6**	**0**	**0**	**0**
99-2000	**Dallas**	**NHL**	**62**	**0**	**8**	**8**	**53**	**0**	**0**	**0**	**27**	**0.0**	**0**	**0**	**0.0**	**99**	**43**	**11:36**	**5**	**0**	**0**	**0**	**5**	**0**	**0**	**0**
2000-01	**Columbus**	**NHL**	**75**	**3**	**10**	**13**	**94**	**0**	**1**	**0**	**64**	**4.7**	**7**	**0**	**0.0**	**133**	**156**	**20:48**	....	....	....	....	....	....	....	....
2001-02	**Columbus**	**NHL**	**61**	**0**	**6**	**6**	**54**	**0**	**0**	**0**	**46**	**0.0**	**–10**	**0**	**0.0**	**94**	**103**	**17:06**	....	....	....	....	....	....	....	....
	Pittsburgh	**NHL**	**15**	**0**	**2**	**2**	**30**	**0**	**0**	**0**	**14**	**0.0**	**–3**	**0**	**0.0**	**38**	**14**	**19:03**	....	....	....	....	....	....	....	....
	NHL Totals		**427**	**10**	**43**	**53**	**570**	**0**	**1**	**0**	**346**	**2.9**		**0**	**0.0**	**474**	**469**	**17:31**	**14**	**0**	**1**	**1**	**16**	**0**	**0**	**0**

Traded to **Anaheim** by **Detroit** with Detroit's 4th round choice (Viktor Wallin) in 1998 Entry Draft for Dmitri Mironov, March 24, 1998. Claimed by **Atlanta** from **Anaheim** in Expansion Draft, June 25, 1999. Traded to **Dallas** by **Atlanta** for Jason Botterill, July 15, 1999. Selected by **Columbus** from **Dallas** in Expansion Draft, June 23, 2000. Traded to **Pittsburgh** by **Columbus** for Pittsburgh's 4th round choice in 2003 Entry Draft, March 15, 2002.

PYATT, Taylor
(PIGH-at, TAY-lohr) **BUF.**

Left wing. Shoots left. 6'4", 222 lbs. Born, Thunder Bay, Ont., August 19, 1981. NY Islanders' 2nd choice, 8th overall, in 1999 Entry Draft.

Season	Club	League	GP	G	A	Pts	PIM	PP	SH	GW	S	%	+/-	TF	F%	H	SB	Min	GP	G	A	Pts	PIM	PP	SH	GW
1996-97	Thunder Bay	TBAHA	60	52	61	113	72	....	....	....	....	....	....	....	....	....	....	....	....	....	....	....	....	....	....	....
1997-98	Sudbury Wolves	OHL	58	14	17	31	104	....	....	....	....	....	....	....	....	....	....	....	10	3	1	4	8	....	....	....
1998-99	Sudbury Wolves	OHL	68	37	38	75	95	....	....	....	....	....	....	....	....	....	....	....	4	0	4	4	6	....	....	....
99-2000	Sudbury Wolves	OHL	68	40	49	89	98	....	....	....	....	....	....	....	....	....	....	....	12	8	7	15	25	....	....	....
2000-01	**NY Islanders**	**NHL**	**78**	**4**	**14**	**18**	**39**	**1**	**0**	**2**	**86**	**4.7**	**–17**	**1**	**0.0**	**116**	**28**	**12:14**	....	....	....	....	....	....	....	....
2001-02	**Buffalo**	**NHL**	**48**	**10**	**10**	**20**	**35**	**0**	**0**	**0**	**61**	**16.4**	**4**	**0**	**0.0**	**61**	**20**	**13:30**	....	....	....	....	....	....	....	....
	Rochester	AHL	27	6	4	10	36	....	....	....	....	....	....	....	....	....	....	....	....	....	....	....	....	....	....	....
	NHL Totals		**126**	**14**	**24**	**38**	**74**	**1**	**0**	**2**	**147**	**9.5**		**1**	**0.0**	**177**	**48**	**12:43**	....	....	....	....	....	....	....	....

OHL First All-Star Team (2000)

Traded to **Buffalo** by **NY Islanders** with Tim Connolly for Michael Peca, June 24, 2001.

QUINT, Deron
(KWIHNT, DAIR-ohn)

Defense. Shoots left. 6'2", 219 lbs. Born, Durham, NH, March 12, 1976. Winnipeg's 1st choice, 30th overall, in 1994 Entry Draft.

Season	Club	League	GP	G	A	Pts	PIM	PP	SH	GW	S	%	+/-	TF	F%	H	SB	Min	GP	G	A	Pts	PIM	PP	SH	GW
1990-91	Cardigan High	Hi-School	31	67	54	121	....	....	....	....	....	....	....	....	....	....	....	....	....	....	....	....	....	....	....	....
1991-92	Cardigan High	Hi-School	21	111	58	169	....	....	....	....	....	....	....	....	....	....	....	....	....	....	....	....	....	....	....	....
1992-93	Tabor Academy	Hi-School	28	15	26	41	30	....	....	....	....	....	....	....	....	....	....	....	1	0	2	2	0	....	....	....
1993-94	Seattle	WHL	63	15	29	44	47	....	....	....	....	....	....	....	....	....	....	....	9	4	12	16	8	....	....	....
1994-95	Seattle	WHL	65	29	60	89	82	....	....	....	....	....	....	....	....	....	....	....	3	1	2	3	6	....	....	....
1995-96	**Winnipeg**	**NHL**	**51**	**5**	**13**	**18**	**22**	**2**	**0**	**0**	**97**	**5.2**	**–2**	....	....	....	....	....	....	....	....	....	....	....	....	....
	Springfield	AHL	11	2	3	5	4	....	....	....	....	....	....	....	....	....	....	....	10	2	3	5	6	....	....	....
	Seattle	WHL	....	....	....	....	....	....	....	....	....	....	....	....	....	....	....	....	5	4	1	5	6	....	....	....
1996-97	**Phoenix**	**NHL**	**27**	**3**	**11**	**14**	**4**	**1**	**0**	**0**	**63**	**4.8**	**–4**	....	....	....	....	....	**7**	**0**	**2**	**2**	**0**	**0**	**0**	**0**
	Springfield	AHL	43	6	18	24	20	....	....	....	....	....	....	....	....	....	....	....	12	2	7	9	4	....	....	....
1997-98	**Phoenix**	**NHL**	**32**	**4**	**7**	**11**	**16**	**1**	**0**	**1**	**61**	**6.6**	**–6**	....	....	....	....	....	....	....	....	....	....	....	....	....
	Springfield	AHL	8	1	7	8	10	....	....	....	....	....	....	....	....	....	....	....	1	0	0	0	0	....	....	....
1998-99	**Phoenix**	**NHL**	**60**	**5**	**8**	**13**	**20**	**2**	**0**	**0**	**94**	**5.3**	**–10**	**0**	**0.0**	**53**	**32**	**16:12**	....	....	....	....	....	....	....	....
99-2000	**Phoenix**	**NHL**	**50**	**3**	**7**	**10**	**22**	**0**	**0**	**1**	**88**	**3.4**	**0**	**0**	**0.0**	**61**	**31**	**16:39**	....	....	....	....	....	....	....	....
	New Jersey	**NHL**	**4**	**1**	**0**	**1**	**2**	**0**	**0**	**0**	**6**	**16.7**	**–2**	**0**	**0.0**	**6**	**3**	**16:26**	....	....	....	....	....	....	....	....
2000-01	**Columbus**	**NHL**	**57**	**7**	**16**	**23**	**16**	**3**	**0**	**0**	**148**	**4.7**	**–19**	**1**	**100.0**	**37**	**68**	**24:06**	....	....	....	....	....	....	....	....
	Syracuse Crunch	AHL	21	5	15	20	30	....	....	....	....	....	....	....	....	....	....	....	....	....	....	....	....	....	....	....
2001-02	**Columbus**	**NHL**	**75**	**7**	**18**	**25**	**26**	**3**	**0**	**1**	**169**	**4.1**	**–34**	**0**	**0.0**	**44**	**91**	**22:01**	....	....	....	....	....	....	....	....
	NHL Totals		**356**	**35**	**80**	**115**	**128**	**12**	**0**	**3**	**726**	**4.8**		**1**	**100.0**	**201**	**225**	**19:54**	**7**	**0**	**2**	**2**	**0**	**0**	**0**	**0**

WHL West First All-Star Team (1995)

Transferred to **Phoenix** after **Winnipeg** franchise relocated, July 1, 1996. Traded to **New Jersey** by **Phoenix** with Phoenix's 3rd round choice (later traded back to Phoenix - Phoenix selected Beat Forster) in 2001 Entry Draft for Lyle Odelein, March 7, 2000. Traded to **Columbus** by **New Jersey** to complete transaction that sent Krzysztof Oliwa to Columbus (June 12, 2000) and Turner Stevenson to New Jersey (June 23, 2000), June 23, 2000.

QUINTAL, Stephane
(KAYN-tahl, STEH-fan) **MTL.**

Defense. Shoots right. 6'3", 231 lbs. Born, Boucherville, Que., October 22, 1968. Boston's 2nd choice, 14th overall, in 1987 Entry Draft.

Season	Club	League	GP	G	A	Pts	PIM	PP	SH	GW	S	%	+/-	TF	F%	H	SB	Min	GP	G	A	Pts	PIM	PP	SH	GW
1984-85	Richelieu	QAAA	41	1	10	11	68	....	....	....	....	....	....	....	....	....	....	....	9	0	5	5	27	....	....	....
1985-86	Granby Bisons	QMJHL	67	2	17	19	144	....	....	....	....	....	....	....	....	....	....	....	....	....	....	....	....	....	....	....
1986-87	Granby Bisons	QMJHL	67	13	41	54	178	....	....	....	....	....	....	....	....	....	....	....	8	0	9	9	10	....	....	....
1987-88	Hull Olympiques	QMJHL	38	13	23	36	138	....	....	....	....	....	....	....	....	....	....	....	19	7	12	19	30	....	....	....
1988-89	**Boston**	**NHL**	**26**	**0**	**1**	**1**	**29**	**0**	**0**	**0**	**23**	**0.0**	**–5**	....	....	....	....	....	....	....	....	....	....	....	....	....
	Maine Mariners	AHL	16	4	10	14	28	....	....	....	....	....	....	....	....	....	....	....	....	....	....	....	....	....	....	....
1989-90	**Boston**	**NHL**	**38**	**2**	**2**	**4**	**22**	**0**	**0**	**0**	**43**	**4.7**	**–11**	....	....	....	....	....	....	....	....	....	....	....	....	....
	Maine Mariners	AHL	37	4	16	20	27	....	....	....	....	....	....	....	....	....	....	....	....	....	....	....	....	....	....	....
1990-91	**Boston**	**NHL**	**45**	**2**	**6**	**8**	**89**	**1**	**0**	**0**	**54**	**3.7**	**2**	....	....	....	....	....	**3**	**0**	**1**	**1**	**7**	**0**	**0**	**0**
	Maine Mariners	AHL	23	1	5	6	30	....	....	....	....	....	....	....	....	....	....	....	....	....	....	....	....	....	....	....
1991-92	**Boston**	**NHL**	**49**	**4**	**10**	**14**	**77**	**0**	**0**	**0**	**52**	**7.7**	**–8**	....	....	....	....	....	....	....	....	....	....	....	....	....
	St. Louis	**NHL**	**26**	**0**	**6**	**6**	**32**	**0**	**0**	**0**	**19**	**0.0**	**–3**	....	....	....	....	....	**4**	**1**	**2**	**3**	**6**	**1**	**0**	**0**
1992-93	**St. Louis**	**NHL**	**75**	**1**	**10**	**11**	**100**	**0**	**1**	**0**	**81**	**1.2**	**–6**	....	....	....	....	....	**9**	**0**	**0**	**0**	**8**	**0**	**0**	**0**
1993-94	**Winnipeg**	**NHL**	**81**	**8**	**18**	**26**	**119**	**1**	**1**	**1**	**154**	**5.2**	**–25**	....	....	....	....	....	....	....	....	....	....	....	....	....
1994-95	**Winnipeg**	**NHL**	**43**	**6**	**17**	**23**	**78**	**3**	**0**	**2**	**107**	**5.6**	**0**	....	....	....	....	....	....	....	....	....	....	....	....	....
1995-96	**Montreal**	**NHL**	**68**	**2**	**14**	**16**	**117**	**0**	**1**	**1**	**104**	**1.9**	**–4**	....	....	....	....	....	**6**	**0**	**1**	**1**	**6**	**0**	**0**	**0**
1996-97	**Montreal**	**NHL**	**71**	**7**	**15**	**22**	**100**	**1**	**0**	**0**	**139**	**5.0**	**1**	....	....	....	....	....	**5**	**0**	**1**	**1**	**6**	**0**	**0**	**0**
1997-98	**Montreal**	**NHL**	**71**	**6**	**10**	**16**	**97**	**0**	**0**	**0**	**88**	**6.8**	**13**	....	....	....	....	....	**9**	**0**	**2**	**2**	**4**	**0**	**0**	**0**
1998-99	**Montreal**	**NHL**	**82**	**8**	**19**	**27**	**84**	**1**	**1**	**4**	**159**	**5.0**	**–23**	**0**	**0.0**	**98**	**125**	**22:06**	....	....	....	....	....	....	....	....
99-2000	**NY Rangers**	**NHL**	**75**	**2**	**14**	**16**	**77**	**0**	**0**	**1**	**102**	**2.0**	**–10**	**0**	**0.0**	**133**	**125**	**19:04**	....	....	....	....	....	....	....	....
2000-01	**Chicago**	**NHL**	**72**	**1**	**18**	**19**	**60**	**0**	**0**	**0**	**109**	**0.9**	**–9**	**0**	**0.0**	**85**	**90**	**22:30**	....	....	....	....	....	....	....	....
2001-02	**Montreal**	**NHL**	**75**	**6**	**10**	**16**	**87**	**1**	**0**	**1**	**85**	**7.1**	**–7**	**0**	**0.0**	**115**	**117**	**18:52**	**12**	**1**	**3**	**4**	**12**	**0**	**0**	**0**
	NHL Totals		**897**	**55**	**170**	**225**	**1168**	**8**	**4**	**10**	**1319**	**4.2**		**0**	**0.0**	**431**	**457**	**20:39**	**48**	**2**	**10**	**12**	**49**	**1**	**0**	**0**

QMJHL First All-Star Team (1987)

Traded to **St. Louis** by **Boston** with Craig Janney for Adam Oates, February 7, 1992. Traded to **Winnipeg** by **St. Louis** with Nelson Emerson for Phil Housley, September 24, 1993. Traded to **Montreal** by **Winnipeg** for Montreal's 2nd round choice (Jason Doig) in 1995 Entry Draft, July 8, 1995. Signed as a free agent by **NY Rangers**, July 13, 1999. Claimed on waivers by **Chicago** from **NY Rangers**, October 5, 2000. Traded to **Montreal** by **Chicago** for Montreal's 4th round choice (Brent MacLellan) in 2001 Entry Draft, June 23, 2001.

RACHUNEK, Karel
(ra-KHOO-nehk, KAH-rehl) **OTT.**

Defense. Shoots right. 6'2", 202 lbs. Born, Gottwaldov, Czech., August 27, 1979. Ottawa's 8th choice, 229th overall, in 1997 Entry Draft.

Season	Club	League	GP	G	A	Pts	PIM	PP	SH	GW	S	%	+/-	TF	F%	H	SB	Min	GP	G	A	Pts	PIM	PP	SH	GW
1995-96	AC ZPS Zlin Jr.	Czech-Jr.	38	8	11	19	....	....	....	....	....	....	....	....	....	....	....	....	....	....	....	....	....	....	....	....
1996-97	AC ZPS Zlin Jr.	Czech-Jr.	27	2	11	13	....	....	....	....	....	....	....	....	....	....	....	....	....	....	....	....	....	....	....	....
1997-98	Zlin	Czech	27	1	2	3	16	....	....	....	....	....	....	....	....	....	....	....	....	....	....	....	....	....	....	....
1998-99	Zlin	Czech	39	3	9	12	88	....	....	....	....	....	....	....	....	....	....	....	6	0	0	0	....	....	....	....
99-2000	**Ottawa**	**NHL**	**6**	**0**	**0**	**0**	**2**	**0**	**0**	**0**	**3**	**0.0**	**0**	**0**	**0.0**	**10**	**5**	**8:03**	....	....	....	....	....	....	....	....
	Grand Rapids	IHL	62	6	20	26	64	....	....	....	....	....	....	....	....	....	....	....	9	0	5	5	6	....	....	....
2000-01	**Ottawa**	**NHL**	**71**	**3**	**30**	**33**	**60**	**3**	**0**	**0**	**77**	**3.9**	**17**	**0**	**0.0**	**173**	**98**	**20:54**	**3**	**0**	**0**	**0**	**0**	**0**	**0**	**0**
2001-02	**Ottawa**	**NHL**	**51**	**3**	**15**	**18**	**24**	**1**	**0**	**2**	**55**	**5.5**	**7**	**2**	**0.0**	**108**	**62**	**19:19**	....	....	....	....	....	....	....	....
	NHL Totals		**128**	**6**	**45**	**51**	**86**	**4**	**0**	**2**	**135**	**4.4**		**2**	**0.0**	**291**	**165**	**19:40**	**3**	**0**	**0**	**0**	**0**	**0**	**0**	**0**

RADIVOJEVIC, Branko
(ra-dih-VOI-uh-vihch, BRAN-koh) **PHX.**

Right wing. Shoots right. 6'1", 209 lbs. Born, Piestany, Czech., November 24, 1980. Colorado's 3rd choice, 93rd overall, in 1999 Entry Draft.

Season	Club	League	GP	G	A	Pts	PIM	PP	SH	GW	S	%	+/-	TF	F%	H	SB	Min	GP	G	A	Pts	PIM	PP	SH	GW
1997-98	Dukla Trencin Jr.	Slovak-Jr.	52	30	31	61	50	....	....	....	....	....	....	....	....	....	....	....	....	....	....	....	....	....	....	....
	Dukla Trencin	Slovakia	1	0	0	0	2	....	....	....	....	....	....	....	....	....	....	....	....	....	....	....	....	....	....	....
1998-99	Belleville Bulls	OHL	68	20	38	58	61	....	....	....	....	....	....	....	....	....	....	....	21	7	17	24	18	....	....	....
99-2000	Belleville Bulls	OHL	59	23	49	72	86	....	....	....	....	....	....	....	....	....	....	....	16	5	8	13	32	....	....	....
2000-01	Belleville Bulls	OHL	61	34	70	104	77	....	....	....	....	....	....	....	....	....	....	....	10	6	10	16	18	....	....	....

			Regular Season																Playoffs							
Season	Club	League	GP	G	A	Pts	PIM	PP	SH	GW	S	%	+/-	TF	F%	H	SB	Min	GP	G	A	Pts	PIM	PP	SH	GW
2001-02	**Phoenix**	**NHL**	**18**	**4**	**2**	**6**	**4**	**0**	**0**	**1**	**19**	**21.1**	**1**	**0**	**0.0**	**26**	**3**	**9:22**	**1**	**0**	**0**	**0**	**2**	**0**	**0**	**0**
	Springfield	AHL	62	18	21	39	64	....	....	....	....	....	....	....	....	....	....	....	....	....	....	....	....	....	....	....
	NHL Totals		**18**	**4**	**2**	**6**	**4**	**0**	**0**	**1**	**19**	**21.1**		**0**	**0.0**	**26**	**3**	**9:22**	**1**	**0**	**0**	**0**	**2**	**0**	**0**	**0**

OHL First All-Star Team (2001)

Signed as a free agent by **Phoenix**, June 19, 2001.

RAFALSKI, Brian

(ra-FAWL-skee, BRIGH-uhn) **N.J.**

Defense. Shoots right. 5'9", 190 lbs. Born, Dearborn, MI, September 28, 1973.

Season	Club	League	GP	G	A	Pts	PIM	PP	SH	GW	S	%	+/-	TF	F%	H	SB	Min	GP	G	A	Pts	PIM	PP	SH	GW
1990-91	Madison Capitols	USHL	47	12	11	23	28	....	....	....	....	....	....	....	....	....	....	....	....	....	....	....	....	....	....	....
1991-92	U. of Wisconsin	WCHA	34	3	14	17	34	....	....	....	....	....	....	....	....	....	....	....	....	....	....	....	....	....	....	....
1992-93	U. of Wisconsin	WCHA	32	0	13	13	10	....	....	....	....	....	....	....	....	....	....	....	....	....	....	....	....	....	....	....
1993-94	U. of Wisconsin	WCHA	37	6	17	23	26	....	....	....	....	....	....	....	....	....	....	....	....	....	....	....	....	....	....	....
1994-95	U. of Wisconsin	WCHA	43	11	34	45	48	....	....	....	....	....	....	....	....	....	....	....	....	....	....	....	....	....	....	....
1995-96	Brynas IF Gavle	Sweden	40	4	14	18	26	....	....	....	....	....	....	....	....	....	....	....	9	0	1	1	2	....	....	....
1996-97	HPK Hameenlinna	Finland	49	11	24	35	26	....	....	....	....	....	....	....	....	....	....	....	10	6	5	11	4	....	....	....
1997-98	HIFK Helsinki	Finland	40	13	10	23	20	....	....	....	....	....	....	....	....	....	....	....	9	5	6	11	0	....	....	....
1998-99	HIFK Helsinki	Finland	53	19	34	53	18	....	....	....	....	....	....	....	....	....	....	....	11	5	*9	*14	4	....	....	....
	HIFK Helsinki	EuroHL	6	4	6	10	10	....	....	....	....	....	....	....	....	....	....	....	4	1	0	1	2	....	....	....
99-2000♦	**New Jersey**	**NHL**	**75**	**5**	**27**	**32**	**28**	**1**	**0**	**1**	**128**	**3.9**	**21**	**1**	**0.0**	**102**	**68**	**18:51**	**23**	**2**	**6**	**8**	**8**	**0**	**0**	**1**
2000-01	**New Jersey**	**NHL**	**78**	**9**	**43**	**52**	**26**	**6**	**0**	**1**	**142**	**6.3**	**36**	**2**	**100.0**	**82**	**55**	**21:41**	**25**	**7**	**11**	**18**	**7**	**1**	**0**	**3**
2001-02	**New Jersey**	**NHL**	**76**	**7**	**40**	**47**	**18**	**2**	**0**	**4**	**125**	**5.6**	**15**	**0**	**0.0**	**61**	**73**	**22:08**	**6**	**3**	**2**	**5**	**4**	**3**	**0**	**0**
	United States	Olympics	6	1	2	3	2	....	....	....	....	....	....	....	....	....	....	....	....	....	....	....	....	....	....	....
	NHL Totals		**229**	**21**	**110**	**131**	**72**	**9**	**0**	**6**	**395**	**5.3**		**3**	**66.7**	**245**	**196**	**20:54**	**54**	**12**	**19**	**31**	**19**	**4**	**0**	**4**

WCHA First All-Star Team (1995) • NCAA West First All-American Team (1995) • NHL All-Rookie Team (2000)

Signed as a free agent by **New Jersey**, May 7, 1999.

RAGNARSSON, Marcus

(RAG-nahr-suhn, MAHR-kuhs) **S.J.**

Defense. Shoots left. 6'1", 215 lbs. Born, Ostervala, Sweden, August 13, 1971. San Jose's 5th choice, 99th overall, in 1992 Entry Draft.

Season	Club	League	GP	G	A	Pts	PIM	PP	SH	GW	S	%	+/-	TF	F%	H	SB	Min	GP	G	A	Pts	PIM	PP	SH	GW
1986-87	Ostervala IF	Swede-3	28	1	6	7	....	....	....	....	....	....	....	....	....	....	....	....	....	....	....	....	....	....	....	....
1987-88	Ostervala IF	Swede-3	25	3	12	15	....	....	....	....	....	....	....	....	....	....	....	....	....	....	....	....	....	....	....	....
1988-89	Ostervala IF	Swede-3	30	15	14	29	....	....	....	....	....	....	....	....	....	....	....	....	....	....	....	....	....	....	....	....
1989-90	Nacka HK	Swede-2	9	2	3	5	4	....	....	....	....	....	....	....	....	....	....	....	....	....	....	....	....	....	....	....
	Djurgarden	Sweden	13	0	2	2	0	....	....	....	....	....	....	....	....	....	....	....	1	0	0	0	0	....	....	....
1990-91	Djurgarden	Sweden	35	4	1	5	12	....	....	....	....	....	....	....	....	....	....	....	7	0	0	0	6	....	....	....
1991-92	Djurgarden	Sweden	40	8	5	13	14	....	....	....	....	....	....	....	....	....	....	....	10	0	1	1	4	....	....	....
1992-93	Djurgarden	Sweden	35	3	3	6	53	....	....	....	....	....	....	....	....	....	....	....	6	0	3	3	8	....	....	....
1993-94	Djurgarden	Sweden	19	0	4	4	24	....	....	....	....	....	....	....	....	....	....	....	....	....	....	....	....	....	....	....
1994-95	Djurgarden	Sweden	38	7	9	16	20	....	....	....	....	....	....	....	....	....	....	....	3	0	0	0	4	....	....	....
1995-96	**San Jose**	**NHL**	**71**	**8**	**31**	**39**	**42**	**4**	**0**	**0**	**94**	**8.5**	**–24**	....	....	....	....	....	....	....	....	....	....	....	....	....
1996-97	**San Jose**	**NHL**	**69**	**3**	**14**	**17**	**63**	**2**	**0**	**0**	**57**	**5.3**	**–18**	....	....	....	....	....	....	....	....	....	....	....	....	....
1997-98	**San Jose**	**NHL**	**79**	**5**	**20**	**25**	**65**	**3**	**0**	**2**	**91**	**5.5**	**–11**	....	....	....	....	....	**6**	**0**	**0**	**0**	**4**	**0**	**0**	**0**
	Sweden	Olympics	3	0	1	1	0	....	....	....	....	....	....	....	....	....	....	....	....	....	....	....	....	....	....	....
1998-99	**San Jose**	**NHL**	**74**	**0**	**13**	**13**	**66**	**0**	**0**	**0**	**87**	**0.0**	**7**	**3**	**66.7**	**99**	**66**	**21:55**	**6**	**0**	**1**	**1**	**6**	**0**	**0**	**0**
99-2000	**San Jose**	**NHL**	**63**	**3**	**13**	**16**	**38**	**0**	**0**	**0**	**60**	**5.0**	**13**	**0**	**0.0**	**92**	**76**	**22:52**	**12**	**0**	**3**	**3**	**10**	**0**	**0**	**0**
2000-01	**San Jose**	**NHL**	**68**	**3**	**12**	**15**	**44**	**1**	**0**	**0**	**74**	**4.1**	**2**	**0**	**0.0**	**168**	**60**	**23:36**	**5**	**0**	**1**	**1**	**8**	**0**	**0**	**0**
2001-02	**San Jose**	**NHL**	**70**	**5**	**15**	**20**	**44**	**2**	**0**	**3**	**68**	**7.4**	**4**	**4**	**25.0**	**102**	**70**	**22:37**	**12**	**1**	**3**	**4**	**12**	**0**	**0**	**0**
	Sweden	Olympics	4	0	2	2	2	....	....	....	....	....	....	....	....	....	....	....	....	....	....	....	....	....	....	....
	NHL Totals		**494**	**27**	**118**	**145**	**362**	**12**	**0**	**5**	**531**	**5.1**		**7**	**42.9**	**461**	**272**	**22:44**	**41**	**1**	**8**	**9**	**40**	**0**	**0**	**0**

Played in NHL All-Star Game (2001)

RALPH, Brad

(RALF, BRAD) **PHX.**

Left wing. Shoots left. 6'2", 206 lbs. Born, Ottawa, Ont., October 17, 1980. Phoenix's 3rd choice, 53rd overall, in 1999 Entry Draft.

Season	Club	League	GP	G	A	Pts	PIM	PP	SH	GW	S	%	+/-	TF	F%	H	SB	Min	GP	G	A	Pts	PIM	PP	SH	GW
1995-96	Kanata Valley	OCJHL	19	6	1	7	19	....	....	....	....	....	....	....	....	....	....	....	....	....	....	....	....	....	....	....
1996-97	Kanata Valley	OCJHL	44	13	13	26	63	....	....	....	....	....	....	....	....	....	....	....	....	....	....	....	....	....	....	....
1997-98	Oshawa Generals	OHL	59	20	17	37	45	....	....	....	....	....	....	....	....	....	....	....	7	2	1	3	8	....	....	....
1998-99	Oshawa Generals	OHL	67	31	44	75	93	....	....	....	....	....	....	....	....	....	....	....	14	7	7	14	10	....	....	....
99-2000	Oshawa Generals	OHL	56	28	35	63	68	....	....	....	....	....	....	....	....	....	....	....	5	1	1	2	4	....	....	....
2000-01	**Phoenix**	**NHL**	**1**	**0**	**0**	**0**	**0**	**0**	**0**	**0**	**0**	**0.0**	**0**	**0**	**0.0**	**1**	**0**	**5:16**	....	....	....	....	....	....	....	....
	Springfield	AHL	50	5	13	18	23	....	....	....	....	....	....	....	....	....	....	....	....	....	....	....	....	....	....	....
2001-02	Mississippi	ECHL	16	2	3	5	12	....	....	....	....	....	....	....	....	....	....	....	....	....	....	....	....	....	....	....
	NHL Totals		**1**	**0**	**0**	**0**	**0**	**0**	**0**	**0**	**0**	**0.0**		**0**	**0.0**	**1**	**0**	**5:16**	....	....	....	....	....	....	....	....

• Missed majority of 2001-02 season recovering from shoulder injury suffered in game vs. Greenville (ECHL), December 8, 2001.

RANHEIM, Paul

(RAN-highm, PAWL) **PHI.**

Left wing. Shoots right. 6'1", 210 lbs. Born, St. Louis, MO, January 25, 1966. Calgary's 3rd choice, 38th overall, in 1984 Entry Draft.

Season	Club	League	GP	G	A	Pts	PIM	PP	SH	GW	S	%	+/-	TF	F%	H	SB	Min	GP	G	A	Pts	PIM	PP	SH	GW
1982-83	Edina Hornets	Hi-School	26	12	25	37	4	....	....	....	....	....	....	....	....	....	....	....	....	....	....	....	....	....	....	....
1983-84	Edina Hornets	Hi-School	26	16	24	40	6	....	....	....	....	....	....	....	....	....	....	....	....	....	....	....	....	....	....	....
1984-85	U. of Wisconsin	WCHA	42	11	11	22	40	....	....	....	....	....	....	....	....	....	....	....	....	....	....	....	....	....	....	....
1985-86	U. of Wisconsin	WCHA	33	17	17	34	34	....	....	....	....	....	....	....	....	....	....	....	....	....	....	....	....	....	....	....
1986-87	U. of Wisconsin	WCHA	42	24	35	59	54	....	....	....	....	....	....	....	....	....	....	....	....	....	....	....	....	....	....	....
1987-88	U. of Wisconsin	WCHA	44	36	26	62	63	....	....	....	....	....	....	....	....	....	....	....	....	....	....	....	....	....	....	....
1988-89	**Calgary**	**NHL**	**5**	**0**	**0**	**0**	**0**	**0**	**0**	**0**	**4**	**0.0**	**–3**	....	....	....	....	....	....	....	....	....	....	....	....	....
	Salt Lake	IHL	75	*68	29	97	16	....	....	....	....	....	....	....	....	....	....	....	14	5	5	10	8	....	....	....
1989-90	**Calgary**	**NHL**	**80**	**26**	**28**	**54**	**23**	**1**	**3**	**4**	**197**	**13.2**	**27**	....	....	....	....	....	**6**	**1**	**3**	**4**	**2**	**0**	**0**	**0**
1990-91	**Calgary**	**NHL**	**39**	**14**	**16**	**30**	**4**	**2**	**0**	**2**	**108**	**13.0**	**20**	....	....	....	....	....	**7**	**2**	**2**	**4**	**0**	**0**	**0**	**0**
1991-92	**Calgary**	**NHL**	**80**	**23**	**20**	**43**	**32**	**1**	**3**	**3**	**159**	**14.5**	**16**	....	....	....	....	....	....	....	....	....	....	....	....	....
1992-93	**Calgary**	**NHL**	**83**	**21**	**22**	**43**	**26**	**3**	**4**	**1**	**179**	**11.7**	**–4**	....	....	....	....	....	**6**	**0**	**1**	**1**	**0**	**0**	**0**	**0**
1993-94	**Calgary**	**NHL**	**67**	**10**	**14**	**24**	**20**	**0**	**2**	**2**	**110**	**9.1**	**–7**	....	....	....	....	....	....	....	....	....	....	....	....	....
	Hartford	**NHL**	**15**	**0**	**3**	**3**	**2**	**0**	**0**	**0**	**21**	**0.0**	**–11**	....	....	....	....	....	....	....	....	....	....	....	....	....
1994-95	**Hartford**	**NHL**	**47**	**6**	**14**	**20**	**10**	**0**	**0**	**1**	**73**	**8.2**	**–3**	....	....	....	....	....	....	....	....	....	....	....	....	....
1995-96	**Hartford**	**NHL**	**73**	**10**	**20**	**30**	**14**	**0**	**1**	**1**	**126**	**7.9**	**–2**	....	....	....	....	....	....	....	....	....	....	....	....	....
1996-97	**Hartford**	**NHL**	**67**	**10**	**11**	**21**	**18**	**0**	**3**	**1**	**96**	**10.4**	**–13**	....	....	....	....	....	....	....	....	....	....	....	....	....
1997-98	**Carolina**	**NHL**	**73**	**5**	**9**	**14**	**28**	**0**	**1**	**2**	**77**	**6.5**	**–11**	....	....	....	....	....	....	....	....	....	....	....	....	....
1998-99	**Carolina**	**NHL**	**78**	**9**	**10**	**19**	**39**	**0**	**2**	**1**	**67**	**13.4**	**4**	**10**	**50.0**	**72**	**32**	**9:02**	**6**	**0**	**0**	**0**	**2**	**0**	**0**	**0**
99-2000	**Carolina**	**NHL**	**79**	**9**	**13**	**22**	**6**	**0**	**0**	**2**	**98**	**9.2**	**–14**	**78**	**46.2**	**107**	**31**	**11:19**	....	....	....	....	....	....	....	....
2000-01	**Philadelphia**	**NHL**	**80**	**10**	**7**	**17**	**14**	**0**	**2**	**0**	**123**	**8.1**	**2**	**18**	**55.6**	**51**	**38**	**12:57**	**6**	**0**	**2**	**2**	**2**	**0**	**0**	**0**
2001-02	**Philadelphia**	**NHL**	**79**	**5**	**4**	**9**	**36**	**1**	**0**	**0**	**75**	**6.7**	**5**	**40**	**40.0**	**32**	**30**	**9:15**	**5**	**0**	**0**	**0**	**0**	**0**	**0**	**0**
	NHL Totals		**945**	**158**	**191**	**349**	**272**	**8**	**21**	**20**	**1513**	**10.4**		**146**	**45.9**	**262**	**131**	**10:39**	**36**	**3**	**8**	**11**	**6**	**0**	**0**	**0**

WCHA Second All-Star Team (1987) • WCHA First All-Star Team (1988) • NCAA West First All-American Team (1988) • IHL Second All-Star Team (1989) • Won Ken McKenzie Trophy (U.S.- Born Rookie of the Year - IHL) (1989)• Won Garry F. Longman Memorial Trophy (Top Rookie - IHL) (1989)

• Missed majority of 1990-91 season recovering from ankle injury suffered in game vs. Minnesota, December 11, 1990. Traded to **Hartford** by **Calgary** with Gary Suter and Ted Drury for James Patrick, Zarley Zalapski and Michael Nylander, March 10, 1994. Transferred to **Carolina** after **Hartford** franchise relocated, June 25, 1997. Traded to **Philadelphia** by **Carolina** for Philadelphia's 8th round choice (later traded to Tampa Bay - Tampa Bay selected Darren Reid) in 2002 Entry Draft, May 31, 2000.

RASMUSSEN, Erik

(RAS-moo-suhn, AIR-ihk) **L.A.**

Center/Left wing. Shoots left. 6'3", 208 lbs. Born, Minneapolis, MN, March 28, 1977. Buffalo's 1st choice, 7th overall, in 1996 Entry Draft.

Season	Club	League	GP	G	A	Pts	PIM	PP	SH	GW	S	%	+/-	TF	F%	H	SB	Min	GP	G	A	Pts	PIM	PP	SH	GW
1992-93	St. Louis Park	Hi-School	23	16	24	40	50	....	....	....	....	....	....	....	....	....	....	....	....	....	....	....	....	....	....	....
1993-94	St. Louis Park	Hi-School	18	25	18	43	80	....	....	....	....	....	....	....	....	....	....	....	....	....	....	....	....	....	....	....
1994-95	St. Louis Park	Hi-School	23	19	33	52	80	....	....	....	....	....	....	....	....	....	....	....	....	....	....	....	....	....	....	....
1995-96	U. of Minnesota	WCHA	40	16	32	48	55	....	....	....	....	....	....	....	....	....	....	....	....	....	....	....	....	....	....	....
1996-97	U. of Minnesota	WCHA	34	15	12	27	*123	....	....	....	....	....	....	....	....	....	....	....	....	....	....	....	....	....	....	....
1997-98	**Buffalo**	**NHL**	**21**	**2**	**3**	**5**	**14**	**0**	**0**	**0**	**28**	**7.1**	**2**	....	....	....	....	....	....	....	....	....	....	....	....	....
	Rochester	AHL	53	9	14	23	83	....	....	....	....	....	....	....	....	....	....	....	1	0	0	0	5	....	....	....

			Regular Season																Playoffs							
Season	Club	League	GP	G	A	Pts	PIM	PP	SH	GW	S	%	+/-	TF	F%	H	SB	Min	GP	G	A	Pts	PIM	PP	SH	GW
1998-99	**Buffalo**	**NHL**	**42**	**3**	**7**	**10**	**37**	**0**	**0**	**0**	**40**	**7.5**	**6**	**67**	**40.3**	**104**	**18**	**12:22**	**21**	**2**	**4**	**6**	**18**	**0**	**0**	**1**
	Rochester	AHL	37	12	14	26	47	....	....	....	....	....	....	....	....	....	....	....	....	....	....	....	....	....	....	....
99-2000	**Buffalo**	**NHL**	**67**	**8**	**6**	**14**	**43**	**0**	**0**	**2**	**76**	**10.5**	**1**	**130**	**44.6**	**175**	**12**	**11:27**	**3**	**0**	**0**	**0**	**4**	**0**	**0**	**0**
2000-01	**Buffalo**	**NHL**	**82**	**12**	**19**	**31**	**51**	**1**	**0**	**3**	**95**	**12.6**	**0**	**565**	**43.7**	**207**	**36**	**13:47**	**3**	**0**	**1**	**1**	**0**	**0**	**0**	**0**
2001-02	**Buffalo**	**NHL**	**69**	**8**	**11**	**19**	**34**	**0**	**0**	**2**	**89**	**9.0**	**–1**	**236**	**39.8**	**164**	**19**	**13:03**	....	....	....	....	....	....	....	....
	NHL Totals		**281**	**33**	**46**	**79**	**179**	**1**	**0**	**7**	**328**	**10.1**		**998**	**42.7**	**650**	**85**	**12:45**	**27**	**2**	**5**	**7**	**22**	**0**	**0**	**1**

Minnesota High School Player of the Year (1995)

Traded to **LA Kings** by **Buffalo** for Adam Mair and LA Kings' 5th round choice in 2003 Entry Draft, July 24, 2002.

RATCHUK, Peter

(RAT-chuhk, PEE-tuhr) **BUF.**

Defense. Shoots left. 6'1", 185 lbs. Born, Buffalo, NY, September 10, 1977. Colorado's 1st choice, 25th overall, in 1996 Entry Draft.

Season	Club	League	GP	G	A	Pts	PIM	PP	SH	GW	S	%	+/-	TF	F%	H	SB	Min	GP	G	A	Pts	PIM	PP	SH	GW
1994-95	Lawrence School	Hi-School	31	8	15	23	18	....	....	....	....	....	....	....	....	....	....	....	....	....	....	....	....	....	....	....
1995-96	Shattuck High	Hi-School	35	22	28	50	24	....	....	....	....	....	....	....	....	....	....	....	....	....	....	....	....	....	....	....
1996-97	Bowling Green	CCHA	35	9	12	21	14	....	....	....	....	....	....	....	....	....	....	....	....	....	....	....	....	....	....	....
1997-98	Hull Olympiques	QMJHL	60	23	31	54	34	....	....	....	....	....	....	....	....	....	....	....	11	3	6	9	8	....	....	....
1998-99	**Florida**	**NHL**	**24**	**1**	**1**	**2**	**10**	**0**	**0**	**0**	**34**	**2.9**	**–1**	**0**	**0.0**	**16**	**14**	**13:55**	....	....	....	....	....	....	....	....
	New Haven	AHL	53	7	20	27	44	....	....	....	....	....	....	....	....	....	....	....	....	....	....	....	....	....	....	....
99-2000	Louisville	AHL	76	9	17	26	64	....	....	....	....	....	....	....	....	....	....	....	4	1	2	3	0	....	....	....
2000-01	**Florida**	**NHL**	**8**	**0**	**0**	**0**	**0**	**0**	**0**	**0**	**11**	**0.0**	**–1**	**1**	**100.0**	**9**	**1**	**13:09**	....	....	....	....	....	....	....	....
	Louisville	AHL	64	5	13	18	85	....	....	....	....	....	....	....	....	....	....	....	....	....	....	....	....	....	....	....
2001-02	Wilkes-Barre	AHL	75	16	23	39	55	....	....	....	....	....	....	....	....	....	....	....	....	....	....	....	....	....	....	....
	NHL Totals		**32**	**1**	**1**	**2**	**10**	**0**	**0**	**0**	**45**	**2.2**		**1**	**100.0**	**25**	**15**	**13:43**	....	....	....	....	....	....	....	....

Signed as a free agent by **Florida**, June 15, 1998. Signed as a free agent by **Pittsburgh**, August 14, 2001. Signed as a free agent by **Buffalo**, August 7, 2002.

RATHJE, Mike

(RATH-jee, MIGHK) **S.J.**

Defense. Shoots left. 6'5", 245 lbs. Born, Mannville, Alta., May 11, 1974. San Jose's 1st choice, 3rd overall, in 1992 Entry Draft.

Season	Club	League	GP	G	A	Pts	PIM	PP	SH	GW	S	%	+/-	TF	F%	H	SB	Min	GP	G	A	Pts	PIM	PP	SH	GW
1989-90	Sherwood Park	AMHL	33	6	11	17	30	....	....	....	....	....	....	....	....	....	....	....	6	1	1	2	2	....	....	....
1990-91	Medicine Hat	WHL	64	1	16	17	28	....	....	....	....	....	....	....	....	....	....	....	12	0	4	4	2	....	....	....
1991-92	Medicine Hat	WHL	67	11	23	34	99	....	....	....	....	....	....	....	....	....	....	....	4	0	1	1	2	....	....	....
1992-93	Medicine Hat	WHL	57	12	37	49	103	....	....	....	....	....	....	....	....	....	....	....	10	3	3	6	12	....	....	....
	Kansas City	IHL	....	....	....	....	....	....	....	....	....	....	....	....	....	....	....	....	5	0	0	0	12	....	....	....
1993-94	**San Jose**	**NHL**	**47**	**1**	**9**	**10**	**59**	**1**	**0**	**0**	**30**	**3.3**	**–9**	....	....	....	....	....	**1**	**0**	**0**	**0**	**0**	**0**	**0**	**0**
	Kansas City	IHL	6	0	2	2	0	....	....	....	....	....	....	....	....	....	....	....	....	....	....	....	....	....	....	....
1994-95	Kansas City	IHL	6	0	1	1	7	....	....	....	....	....	....	....	....	....	....	....	....	....	....	....	....	....	....	....
	San Jose	**NHL**	**42**	**2**	**7**	**9**	**29**	**0**	**0**	**0**	**38**	**5.3**	**–1**	....	....	....	....	....	**11**	**5**	**2**	**7**	**4**	**5**	**0**	**0**
1995-96	**San Jose**	**NHL**	**27**	**0**	**7**	**7**	**14**	**0**	**0**	**0**	**26**	**0.0**	**–16**	....	....	....	....	....	....	....	....	....	....	....	....	....
	Kansas City	IHL	36	6	11	17	34	....	....	....	....	....	....	....	....	....	....	....	....	....	....	....	....	....	....	....
1996-97	**San Jose**	**NHL**	**31**	**0**	**8**	**8**	**21**	**0**	**0**	**0**	**22**	**0.0**	**–1**	....	....	....	....	....	....	....	....	....	....	....	....	....
1997-98	**San Jose**	**NHL**	**81**	**3**	**12**	**15**	**59**	**1**	**0**	**0**	**61**	**4.9**	**–4**	....	....	....	....	....	**6**	**1**	**0**	**1**	**6**	**1**	**0**	**0**
1998-99	**San Jose**	**NHL**	**82**	**5**	**9**	**14**	**36**	**2**	**0**	**1**	**67**	**7.5**	**15**	**0**	**0.0**	**100**	**64**	**20:07**	**6**	**0**	**0**	**0**	**4**	**0**	**0**	**0**
99-2000	**San Jose**	**NHL**	**66**	**2**	**14**	**16**	**31**	**0**	**0**	**0**	**46**	**4.3**	**–2**	**0**	**0.0**	**82**	**54**	**22:11**	**12**	**1**	**3**	**4**	**8**	**0**	**0**	**0**
2000-01	**San Jose**	**NHL**	**81**	**0**	**11**	**11**	**48**	**0**	**0**	**0**	**89**	**0.0**	**7**	**0**	**0.0**	**144**	**98**	**22:20**	**6**	**0**	**1**	**1**	**4**	**0**	**0**	**0**
2001-02	**San Jose**	**NHL**	**52**	**5**	**12**	**17**	**48**	**4**	**0**	**0**	**56**	**8.9**	**23**	**0**	**0.0**	**102**	**52**	**21:31**	**12**	**1**	**3**	**4**	**6**	**1**	**0**	**0**
	NHL Totals		**509**	**18**	**89**	**107**	**345**	**8**	**0**	**1**	**435**	**4.1**		**0**	**0.0**	**428**	**268**	**21:30**	**54**	**8**	**9**	**17**	**32**	**7**	**0**	**0**

WHL East Second All-Star Team (1992, 1993)

• Missed majority of 1996-97 season recovering from groin injury suffered in game vs. Dallas, November 8, 1996.

RAY, Rob

(RAY, RAWB) **BUF.**

Right wing. Shoots left. 6', 217 lbs. Born, Stirling, Ont., June 8, 1968. Buffalo's 5th choice, 97th overall, in 1988 Entry Draft.

Season	Club	League	GP	G	A	Pts	PIM	PP	SH	GW	S	%	+/-	TF	F%	H	SB	Min	GP	G	A	Pts	PIM	PP	SH	GW
1983-84	Trenton Bobcats	OJHL-B	40	11	10	21	57	....	....	....	....	....	....	....	....	....	....	....	....	....	....	....	....	....	....	....
1984-85	Whitby Lawmen	OPJHL	35	5	10	15	318	....	....	....	....	....	....	....	....	....	....	....	....	....	....	....	....	....	....	....
1985-86	Cornwall Royals	OHL	53	6	13	19	253	....	....	....	....	....	....	....	....	....	....	....	6	0	0	0	26	....	....	....
1986-87	Cornwall Royals	OHL	46	17	20	37	158	....	....	....	....	....	....	....	....	....	....	....	5	1	1	2	16	....	....	....
1987-88	Cornwall Royals	OHL	61	11	41	52	179	....	....	....	....	....	....	....	....	....	....	....	11	2	3	5	33	....	....	....
1988-89	Rochester	AHL	74	11	18	29	*446	....	....	....	....	....	....	....	....	....	....	....	....	....	....	....	....	....	....	....
1989-90	**Buffalo**	**NHL**	**27**	**2**	**1**	**3**	**99**	**0**	**0**	**0**	**20**	**10.0**	**–2**	....	....	....	....	....	....	....	....	....	....	....	....	....
	Rochester	AHL	43	2	13	15	335	....	....	....	....	....	....	....	....	....	....	....	17	1	3	4	115	....	....	....
1990-91	**Buffalo**	**NHL**	**66**	**8**	**8**	**16**	***350**	**0**	**0**	**1**	**54**	**14.8**	**–11**	....	....	....	....	....	**6**	**1**	**1**	**2**	**56**	**0**	**0**	**1**
	Rochester	AHL	8	1	1	2	15	....	....	....	....	....	....	....	....	....	....	....	....	....	....	....	....	....	....	....
1991-92	**Buffalo**	**NHL**	**63**	**5**	**3**	**8**	**354**	**0**	**0**	**0**	**29**	**17.2**	**–9**	....	....	....	....	....	**7**	**0**	**0**	**0**	**2**	**0**	**0**	**0**
1992-93	**Buffalo**	**NHL**	**68**	**3**	**2**	**5**	**211**	**1**	**0**	**0**	**28**	**10.7**	**–3**	....	....	....	....	....	....	....	....	....	....	....	....	....
1993-94	**Buffalo**	**NHL**	**82**	**3**	**4**	**7**	**274**	**0**	**0**	**0**	**34**	**8.8**	**2**	....	....	....	....	....	**7**	**1**	**0**	**1**	**43**	**0**	**0**	**0**
1994-95	**Buffalo**	**NHL**	**47**	**0**	**3**	**3**	**173**	**0**	**0**	**0**	**7**	**0.0**	**–4**	....	....	....	....	....	**5**	**0**	**0**	**0**	**14**	**0**	**0**	**0**
1995-96	**Buffalo**	**NHL**	**71**	**3**	**6**	**9**	**287**	**0**	**0**	**0**	**21**	**14.3**	**–8**	....	....	....	....	....	....	....	....	....	....	....	....	....
1996-97	**Buffalo**	**NHL**	**82**	**7**	**3**	**10**	**286**	**0**	**0**	**1**	**45**	**15.6**	**3**	....	....	....	....	....	**12**	**0**	**1**	**1**	**28**	**0**	**0**	**0**
1997-98	**Buffalo**	**NHL**	**63**	**2**	**4**	**6**	**234**	**1**	**0**	**1**	**19**	**10.5**	**2**	....	....	....	....	....	**10**	**0**	**0**	**0**	**24**	**0**	**0**	**0**
1998-99	**Buffalo**	**NHL**	**76**	**0**	**4**	**4**	***261**	**0**	**0**	**0**	**23**	**0.0**	**–2**	**0**	**0.0**	**60**	**4**	**5:11**	**5**	**1**	**0**	**1**	**0**	**0**	**0**	**1**
99-2000	**Buffalo**	**NHL**	**69**	**1**	**3**	**4**	**158**	**0**	**0**	**0**	**17**	**5.9**	**0**	**0**	**0.0**	**55**	**4**	**4:13**	....	....	....	....	....	....	....	....
2000-01	**Buffalo**	**NHL**	**63**	**4**	**6**	**10**	**210**	**0**	**0**	**1**	**33**	**12.1**	**2**	**1**	**100.0**	**83**	**4**	**5:38**	**3**	**0**	**0**	**0**	**2**	**0**	**0**	**0**
2001-02	**Buffalo**	**NHL**	**71**	**2**	**3**	**5**	**200**	**0**	**0**	**0**	**23**	**8.7**	**–3**	**0**	**0.0**	**78**	**6**	**5:08**	....	....	....	....	....	....	....	....
	NHL Totals		**848**	**40**	**50**	**90**	**3097**	**2**	**0**	**4**	**353**	**11.3**		**1**	**100.0**	**276**	**18**	**5:02**	**55**	**3**	**2**	**5**	**169**	**0**	**0**	**2**

Won King Clancy Memorial Trophy (1999)

REASONER, Marty

(REE-sohn-uhr, MAHR-tee) **EDM.**

Center. Shoots left. 6'1", 190 lbs. Born, Rochester, NY, February 26, 1977. St. Louis' 1st choice, 14th overall, in 1996 Entry Draft.

Season	Club	League	GP	G	A	Pts	PIM	PP	SH	GW	S	%	+/-	TF	F%	H	SB	Min	GP	G	A	Pts	PIM	PP	SH	GW
1993-94	Deerfield	Hi-School	22	27	25	52	....	....	....	....	....	....	....	....	....	....	....	....	....	....	....	....	....	....	....	....
1994-95	Deerfield	Hi-School	26	25	32	57	14	....	....	....	....	....	....	....	....	....	....	....	....	....	....	....	....	....	....	....
1995-96	Boston College	H-East	34	16	29	45	32	....	....	....	....	....	....	....	....	....	....	....	....	....	....	....	....	....	....	....
1996-97	Boston College	H-East	35	20	24	44	31	....	....	....	....	....	....	....	....	....	....	....	....	....	....	....	....	....	....	....
1997-98	Boston College	H-East	42	*33	40	*73	56	....	....	....	....	....	....	....	....	....	....	....	....	....	....	....	....	....	....	....
1998-99	**St. Louis**	**NHL**	**22**	**3**	**7**	**10**	**8**	**1**	**0**	**0**	**33**	**9.1**	**2**	**224**	**53.6**	**19**	**1**	**13:55**	....	....	....	....	....	....	....	....
	Worcester	AHL	44	17	22	39	24	....	....	....	....	....	....	....	....	....	....	....	4	2	1	3	6	....	....	....
99-2000	**St. Louis**	**NHL**	**32**	**10**	**14**	**24**	**20**	**3**	**0**	**0**	**51**	**19.6**	**9**	**379**	**49.6**	**26**	**5**	**15:20**	**7**	**2**	**1**	**3**	**4**	**1**	**0**	**0**
	Worcester	AHL	44	23	28	51	39	....	....	....	....	....	....	....	....	....	....	....	....	....	....	....	....	....	....	....
2000-01	**St. Louis**	**NHL**	**41**	**4**	**9**	**13**	**14**	**0**	**0**	**0**	**65**	**6.2**	**–5**	**454**	**53.1**	**31**	**14**	**14:00**	**10**	**3**	**1**	**4**	**0**	**0**	**0**	**1**
	Worcester	AHL	34	17	18	35	25	....	....	....	....	....	....	....	....	....	....	....	....	....	....	....	....	....	....	....
2001-02	**Edmonton**	**NHL**	**52**	**6**	**5**	**11**	**41**	**3**	**0**	**2**	**66**	**9.1**	**0**	**470**	**55.5**	**34**	**8**	**11:44**	....	....	....	....	....	....	....	....
	NHL Totals		**147**	**23**	**35**	**58**	**83**	**7**	**0**	**2**	**215**	**10.7**		**1527**	**53.0**	**110**	**28**	**13:29**	**17**	**5**	**2**	**7**	**4**	**1**	**0**	**1**

Hockey East Rookie of the Year (1996) • Hockey East First All-Star Team (1997, 1998) • NCAA East First All-American Team (1998) • NCAA Championship All-Tournament Team (1998)

Traded to **Edmonton** by **St. Louis** with Jochen Hecht and Jan Horacek for Doug Weight and Michel Riesen, July 1, 2001.

RECCHI, Mark

(REH-kee, MAHRK) **PHI.**

Right wing. Shoots left. 5'10", 185 lbs. Born, Kamloops, B.C., February 1, 1968. Pittsburgh's 4th choice, 67th overall, in 1988 Entry Draft.

Season	Club	League	GP	G	A	Pts	PIM	PP	SH	GW	S	%	+/-	TF	F%	H	SB	Min	GP	G	A	Pts	PIM	PP	SH	GW
1984-85	Langley Eagles	BCJHL	51	26	39	65	39	....	....	....	....	....	....	....	....	....	....	....	....	....	....	....	....	....	....	....
	New Westminster	WHL	4	1	0	1	0	....	....	....	....	....	....	....	....	....	....	....	....	....	....	....	....	....	....	....
1985-86	New Westminster	WHL	72	21	40	61	55	....	....	....	....	....	....	....	....	....	....	....	....	....	....	....	....	....	....	....
1986-87	Kamloops Blazers	WHL	40	26	50	76	63	....	....	....	....	....	....	....	....	....	....	....	13	3	16	19	17	....	....	....
1987-88	Kamloops Blazers	WHL	62	61	*93	154	75	....	....	....	....	....	....	....	....	....	....	....	17	10	*21	*31	18	....	....	....
1988-89	**Pittsburgh**	**NHL**	**15**	**1**	**1**	**2**	**0**	**0**	**0**	**0**	**11**	**9.1**	**–2**	....	....	....	....	....	....	....	....	....	....	....	....	....
	Muskegon	IHL	63	50	49	99	86	....	....	....	....	....	....	....	....	....	....	....	14	7	*14	*21	28	....	....	....
1989-90	**Pittsburgh**	**NHL**	**74**	**30**	**37**	**67**	**44**	**6**	**2**	**4**	**143**	**21.0**	**6**	....	....	....	....	....	....	....	....	....	....	....	....	....
	Muskegon	IHL	4	7	4	11	2	....	....	....	....	....	....	....	....	....	....	....	....	....	....	....	....	....	....	....
1990-91 ♦	**Pittsburgh**	**NHL**	**78**	**40**	**73**	**113**	**48**	**12**	**0**	**9**	**184**	**21.7**	**0**	....	....	....	....	....	**24**	**10**	**24**	**34**	**33**	**5**	**0**	**2**

			Regular Season																Playoffs							
Season	Club	League	GP	G	A	Pts	PIM	PP	SH	GW	S	%	+/-	TF	F%	H	SB	Min	GP	G	A	Pts	PIM	PP	SH	GW
1991-92	**Pittsburgh**	**NHL**	**58**	**33**	**37**	**70**	**78**	**16**	**1**	**4**	**156**	**21.2**	**–16**													
	Philadelphia	**NHL**	**22**	**10**	**17**	**27**	**18**	**4**	**0**	**1**	**54**	**18.5**	**–5**													
1992-93	**Philadelphia**	**NHL**	**84**	**53**	**70**	**123**	**95**	**15**	**4**	**6**	**274**	**19.3**	**1**													
1993-94	**Philadelphia**	**NHL**	**84**	**40**	**67**	**107**	**46**	**11**	**0**	**5**	**217**	**18.4**	**–2**													
1994-95	**Philadelphia**	**NHL**	**10**	**2**	**3**	**5**	**12**	**1**	**0**	**2**	**17**	**11.8**	**–6**													
	Montreal	**NHL**	**39**	**14**	**29**	**43**	**16**	**8**	**0**	**1**	**104**	**13.5**	**–3**													
1995-96	**Montreal**	**NHL**	**82**	**28**	**50**	**78**	**69**	**11**	**2**	**6**	**191**	**14.7**	**20**						**6**	**3**	**3**	**6**	**0**	**3**	**0**	**0**
1996-97	**Montreal**	**NHL**	**82**	**34**	**46**	**80**	**58**	**7**	**2**	**3**	**202**	**16.8**	**–1**						**5**	**4**	**2**	**6**	**2**	**0**	**0**	**0**
1997-98	**Montreal**	**NHL**	**82**	**32**	**42**	**74**	**51**	**9**	**1**	**6**	**216**	**14.8**	**11**						**10**	**4**	**8**	**12**	**6**	**0**	**0**	**2**
	Canada	Olympics	5	0	2	2	0																			
1998-99	**Montreal**	**NHL**	**61**	**12**	**35**	**47**	**28**	**3**	**0**	**2**	**152**	**7.9**	**–4**	**239**	**44.8**	**76**	**23**	**20:37**								
	Philadelphia	**NHL**	**10**	**4**	**2**	**6**	**6**	**0**	**0**	**0**	**19**	**21.1**	**–3**	**4**	**25.0**	**21**	**1**	**19:30**	**6**	**0**	**1**	**1**	**2**	**0**	**0**	**0**
99-2000	**Philadelphia**	**NHL**	**82**	**28**	***63**	**91**	**50**	**7**	**1**	**5**	**223**	**12.6**	**20**	**353**	**49.6**	**95**	**39**	**21:43**	**18**	**6**	**12**	**18**	**6**	**2**	**0**	**1**
2000-01	**Philadelphia**	**NHL**	**69**	**27**	**50**	**77**	**33**	**7**	**1**	**8**	**191**	**14.1**	**15**	**138**	**42.8**	**76**	**18**	**21:40**	**6**	**2**	**2**	**4**	**2**	**1**	**0**	**1**
2001-02	**Philadelphia**	**NHL**	**80**	**22**	**42**	**64**	**46**	**7**	**2**	**4**	**205**	**10.7**	**5**	**82**	**53.7**	**94**	**26**	**20:40**	**4**	**0**	**0**	**0**	**2**	**0**	**0**	**0**
	NHL Totals		**1012**	**410**	**664**	**1074**	**698**	**124**	**16**	**66**	**2559**	**16.0**		**816**	**47.3**	**362**	**107**	**21:08**	**79**	**29**	**52**	**81**	**53**	**11**	**0**	**6**

WHL West First All-Star Team (1988) • IHL Second All-Star Team (1989) • NHL Second All-Star Team (1992) • Played in NHL All-Star Game (1991, 1993, 1994, 1997, 1998, 1999, 2000)

Traded to **Philadelphia** by **Pittsburgh** with Brian Benning and LA Kings' 1st round choice (previously acquired, Philadelphia selected Jason Bowen) in 1992 Entry Draft for Rick Tocchet, Kjell Samuelsson, Ken Wregget and Philadelphia's 3rd round choice (Dave Roche) in 1993 Entry Draft, February 19, 1992. Traded to **Montreal** by **Philadelphia** with Philadelphia's 3rd round choice (Martin Hohenberger) in 1995 Entry Draft for Eric Desjardins, Gilbert Dionne and John LeClair, February 9, 1995. Traded to **Philadelphia** by **Montreal** for Danius Zubrus, Philadelphia's 2nd round choice (Matt Carkner) in 1999 Entry Draft and NY Islanders' 6th round choice (previously acquired, Montreal selected Scott Selig) in 2000 Entry Draft, March 10, 1999.

REDDEN, Wade

(REH-duhn, WAYD) **OTT.**

Defense. Shoots left. 6'2", 205 lbs. Born, Lloydminster, Sask., June 12, 1977. NY Islanders' 1st choice, 2nd overall, in 1995 Entry Draft.

Season	Club	League	GP	G	A	Pts	PIM	PP	SH	GW	S	%	+/-	TF	F%	H	SB	Min	GP	G	A	Pts	PIM	PP	SH	GW
1992-93	Lloydminster	AJHL	34	4	11	15	64																			
1993-94	Brandon	WHL	63	4	35	39	98												14	2	4	6	10			
1994-95	Brandon	WHL	64	14	46	60	83												18	5	10	15	8			
1995-96	Brandon	WHL	51	9	45	54	55												19	5	10	15	19			
1996-97	**Ottawa**	**NHL**	**82**	**6**	**24**	**30**	**41**	**2**	**0**	**1**	**102**	**5.9**	**1**						**7**	**1**	**3**	**4**	**2**	**0**	**0**	**0**
1997-98	**Ottawa**	**NHL**	**80**	**8**	**14**	**22**	**27**	**3**	**0**	**2**	**103**	**7.8**	**17**						**9**	**0**	**2**	**2**	**2**	**0**	**0**	**0**
1998-99	**Ottawa**	**NHL**	**72**	**8**	**21**	**29**	**54**	**3**	**0**	**1**	**127**	**6.3**	**7**	**0**	**0.0**	**83**	**73**	**23:27**	**4**	**1**	**2**	**3**	**2**	**1**	**0**	**0**
99-2000	**Ottawa**	**NHL**	**81**	**10**	**26**	**36**	**49**	**3**	**0**	**2**	**163**	**6.1**	**–1**	**0**	**0.0**	**119**	**103**	**23:43**								
2000-01	**Ottawa**	**NHL**	**78**	**10**	**37**	**47**	**49**	**4**	**0**	**0**	**159**	**6.3**	**22**	**0**	**0.0**	**149**	**120**	**25:17**	**4**	**0**	**0**	**0**	**0**	**0**	**0**	**0**
2001-02	**Ottawa**	**NHL**	**79**	**9**	**25**	**34**	**48**	**4**	**1**	**1**	**156**	**5.8**	**22**	**1**	**0.0**	**131**	**108**	**25:06**	**12**	**3**	**2**	**5**	**6**	**1**	**0**	**1**
	NHL Totals		**472**	**51**	**147**	**198**	**268**	**19**	**1**	**7**	**810**	**6.3**		**1**	**0.0**	**482**	**404**	**24:24**	**36**	**5**	**9**	**14**	**12**	**2**	**0**	**1**

WHL Rookie of the Year (1994) • WHL East Second All-Star Team (1995) • WHL East First All-Star Team (1996) • Memorial Cup All-Star Team (1996) • Played in NHL All-Star Game (2002)

Traded to **Ottawa** by **NY Islanders** with Damian Rhodes for Don Beaupre, Martin Straka and Bryan Berard, January 23, 1996.

REEKIE, Joe

(REE-kee, JOH)

Defense. Shoots left. 6'3", 220 lbs. Born, Victoria, B.C., February 22, 1965. Buffalo's 6th choice, 119th overall, in 1985 Entry Draft.

Season	Club	League	GP	G	A	Pts	PIM	PP	SH	GW	S	%	+/-	TF	F%	H	SB	Min	GP	G	A	Pts	PIM	PP	SH	GW
1981-82	Nepean Raiders	OCJHL	16	2	5	7	4																			
1982-83	Pembroke	OCJHL	7	2	1	3	16																			
	North Bay	OHL	59	2	9	11	49												8	0	1	1	11			
1983-84	North Bay	OHL	9	1	0	1	18																			
	Cornwall Royals	OHL	53	6	27	33	166												3	0	0	0	4			
1984-85	Cornwall Royals	OHL	65	19	63	82	134												9	4	13	17	18			
1985-86	**Buffalo**	**NHL**	**3**	**0**	**0**	**0**	**14**	**0**	**0**	**0**	**1**	**0.0**	**–2**													
	Rochester	AHL	77	3	25	28	178																			
1986-87	**Buffalo**	**NHL**	**56**	**1**	**8**	**9**	**82**	**0**	**0**	**0**	**56**	**1.8**	**6**													
	Rochester	AHL	22	0	6	6	52																			
1987-88	**Buffalo**	**NHL**	**30**	**1**	**4**	**5**	**68**	**0**	**0**	**0**	**23**	**4.3**	**–3**						**2**	**0**	**0**	**0**	**4**	**0**	**0**	**0**
1988-89	**Buffalo**	**NHL**	**15**	**1**	**3**	**4**	**26**	**1**	**0**	**0**	**14**	**7.1**	**6**													
	Rochester	AHL	21	1	2	3	56																			
1989-90	**NY Islanders**	**NHL**	**31**	**1**	**8**	**9**	**43**	**0**	**0**	**1**	**22**	**4.5**	**13**													
	Springfield	AHL	15	1	4	5	24																			
1990-91	**NY Islanders**	**NHL**	**66**	**3**	**16**	**19**	**96**	**0**	**0**	**2**	**70**	**4.3**	**17**													
	Capital District	AHL	2	1	0	1	0																			
1991-92	**NY Islanders**	**NHL**	**54**	**4**	**12**	**16**	**85**	**0**	**0**	**0**	**59**	**6.8**	**15**													
	Capital District	AHL	3	2	2	4	2																			
1992-93	**Tampa Bay**	**NHL**	**42**	**2**	**11**	**13**	**69**	**0**	**0**	**0**	**53**	**3.8**	**2**													
1993-94	**Tampa Bay**	**NHL**	**73**	**1**	**11**	**12**	**127**	**0**	**0**	**0**	**88**	**1.1**	**8**													
	Washington	**NHL**	**12**	**0**	**5**	**5**	**29**	**0**	**0**	**0**	**10**	**0.0**	**7**						**11**	**2**	**1**	**3**	**29**	**0**	**1**	**1**
1994-95	**Washington**	**NHL**	**48**	**1**	**6**	**7**	**97**	**0**	**0**	**0**	**52**	**1.9**	**10**						**7**	**0**	**0**	**0**	**2**	**0**	**0**	**0**
1995-96	**Washington**	**NHL**	**78**	**3**	**7**	**10**	**149**	**0**	**0**	**0**	**52**	**5.8**	**7**													
1996-97	**Washington**	**NHL**	**65**	**1**	**8**	**9**	**107**	**0**	**0**	**0**	**65**	**1.5**	**8**													
1997-98	**Washington**	**NHL**	**68**	**2**	**8**	**10**	**70**	**0**	**0**	**1**	**59**	**3.4**	**15**						**21**	**1**	**2**	**3**	**20**	**0**	**0**	**0**
1998-99	**Washington**	**NHL**	**73**	**0**	**10**	**10**	**68**	**0**	**0**	**0**	**81**	**0.0**	**11**	**0**	**0.0**	**182**	**101**	**21:53**								
99-2000	**Washington**	**NHL**	**59**	**0**	**7**	**7**	**50**	**0**	**0**	**0**	**32**	**0.0**	**21**	**0**	**0.0**	**99**	**77**	**17:32**	**5**	**0**	**1**	**1**	**2**	**0**	**0**	**0**
2000-01	**Washington**	**NHL**	**74**	**2**	**9**	**11**	**77**	**0**	**0**	**1**	**59**	**3.4**	**14**	**1**	**0.0**	**122**	**88**	**18:49**	**4**	**0**	**0**	**0**	**4**	**0**	**0**	**0**
2001-02	**Washington**	**NHL**	**38**	**2**	**4**	**6**	**41**	**0**	**0**	**0**	**26**	**7.7**	**–7**	**0**	**0.0**	**50**	**35**	**15:06**								
	Chicago	**NHL**	**17**	**0**	**2**	**2**	**28**	**0**	**0**	**0**	**7**	**0.0**	**2**	**0**	**0.0**	**30**	**18**	**14:00**	**1**	**0**	**0**	**0**	**2**	**0**	**0**	**0**
	NHL Totals		**902**	**25**	**139**	**164**	**1326**	**1**	**0**	**5**	**829**	**3.0**		**1**	**0.0**	**483**	**319**	**18:32**	**51**	**3**	**4**	**7**	**63**	**0**	**1**	**1**

• Re-entered NHL Entry Draft. Originally Hartford's 8th choice, 128th overall, in 1983 Entry Draft.

• Missed majority of 1987-88 and 1988-89 seasons recovering from knee injury originally suffered in game vs. Toronto, November 11, 1987. Traded to **NY Islanders** by **Buffalo** for NY Islanders' 6th round choice (Bill Pye) in 1989 Entry Draft, June 17, 1989. Claimed by **Tampa Bay** from **NY Islanders** in Expansion Draft, June 18, 1992. Traded to **Washington** by **Tampa Bay** for Enrico Ciccone, Washington's 3rd round choice (later traded to Anaheim - Anaheim selected Craig Reichert) in 1994 Entry Draft and the return of conditional draft choice transferred in the Pat Elynuik trade, March 21, 1994. Traded to **Chicago** by **Washington** for Chicago's 4th round choice (Petr Dvorak) in 2002 Entry Draft, January 17, 2002.

REGEHR, Robyn

(reh-GUHR, RAW-bihn) **CGY.**

Defense. Shoots left. 6'2", 210 lbs. Born, Recife, Brazil, April 19, 1980. Colorado's 3rd choice, 19th overall, in 1998 Entry Draft.

Season	Club	League	GP	G	A	Pts	PIM	PP	SH	GW	S	%	+/-	TF	F%	H	SB	Min	GP	G	A	Pts	PIM	PP	SH	GW
1995-96	Prince Albert	SMHL	59	8	24	32	157																			
1996-97	Kamloops Blazers	WHL	64	4	19	23	96												5	0	1	1	18			
1997-98	Kamloops Blazers	WHL	65	4	10	14	120												5	0	3	3	8			
1998-99	Kamloops Blazers	WHL	54	12	20	32	130												12	1	4	5	21			
99-2000	**Calgary**	**NHL**	**57**	**5**	**7**	**12**	**46**	**2**	**0**	**0**	**64**	**7.8**	**–2**	**0**	**0.0**	**135**	**62**	**18:24**								
	Saint John	AHL	5	0	0	0	0																			
2000-01	**Calgary**	**NHL**	**71**	**1**	**3**	**4**	**70**	**0**	**0**	**0**	**62**	**1.6**	**–7**	**1**	**0.0**	**171**	**104**	**19:43**								
2001-02	**Calgary**	**NHL**	**77**	**2**	**6**	**8**	**93**	**0**	**0**	**0**	**82**	**2.4**	**–24**	**0**	**0.0**	**161**	**145**	**20:54**								
	NHL Totals		**205**	**8**	**16**	**24**	**209**	**2**	**0**	**0**	**208**	**3.8**		**1**	**0.0**	**467**	**311**	**19:48**								

WHL West First All-Star Team (1999)

Traded to **Calgary** by **Colorado** with Rene Corbet, Wade Belak and Colorado's 2nd round compensatory choice (Jarret Stoll) in 2000 Entry Draft for Theoren Fleury and Chris Dingman, February 28, 1999.

REICHEL, Robert

(RIGH-khul, RAW-buhrt) **TOR.**

Center. Shoots left. 5'10", 185 lbs. Born, Litvinov, Czech., June 25, 1971. Calgary's 5th choice, 70th overall, in 1989 Entry Draft.

Season	Club	League	GP	G	A	Pts	PIM	PP	SH	GW	S	%	+/-	TF	F%	H	SB	Min	GP	G	A	Pts	PIM	PP	SH	GW
1987-88	CHZ Litvinov	Czech	36	17	10	27	8																			
1988-89	CHZ Litvinov	Czech	44	23	25	48	32																			
1989-90	CHZ Litvinov	Czech	44	*43	28	*71													8	6	6	12				
1990-91	**Calgary**	**NHL**	**66**	**19**	**22**	**41**	**22**	**3**	**0**	**3**	**131**	**14.5**	**17**						**6**	**1**	**1**	**2**	**0**	**1**	**0**	**0**
1991-92	**Calgary**	**NHL**	**77**	**20**	**34**	**54**	**32**	**8**	**0**	**3**	**181**	**11.0**	**1**													
1992-93	**Calgary**	**NHL**	**80**	**40**	**48**	**88**	**54**	**12**	**0**	**5**	**238**	**16.8**	**25**						**6**	**2**	**4**	**6**	**2**	**2**	**0**	**0**
1993-94	**Calgary**	**NHL**	**84**	**40**	**53**	**93**	**58**	**14**	**0**	**6**	**249**	**16.1**	**20**						**7**	**0**	**5**	**5**	**0**	**0**	**0**	**0**
1994-95	Frankfurt Lions	Germany	21	19	24	43	41																			
	Calgary	**NHL**	**48**	**18**	**17**	**35**	**28**	**5**	**0**	**2**	**160**	**11.3**	**–2**						**7**	**2**	**4**	**6**	**4**	**0**	**0**	**1**
1995-96	Frankfurt Lions	Germany	46	47	54	101	84												3	1	3	4	0			

Season	Club	League	GP	G	A	Pts	PIM	PP	SH	GW	S	%	+/-	TF	F%	H	SB	Min	GP	G	A	Pts	PIM	PP	SH	GW
			Regular Season																Playoffs							
1996-97	Calgary	NHL	70	16	27	43	22	6	0	3	181	8.8	–2													
	NY Islanders	NHL	12	5	14	19	4	0	1	0	33	15.2	7													
1997-98	NY Islanders	NHL	82	25	40	65	32	8	0	2	201	12.4	–11													
	Czech Republic	Olympics	6	3	0	3	0																			
1998-99	NY Islanders	NHL	70	19	37	56	50	5	1	1	186	10.2	–15	1241	51.7	52	19	19:36								
	Phoenix	NHL	13	7	6	13	4	3	0	3	50	14.0	2	241	48.5	10	4	20:03	7	1	3	4	2	0	0	0
99-2000	Litvinov	Czech	45	25	32	57	24												7	3	4	7	2			
2000-01	Litvinov	Czech	49	23	33	56	72												5	1	2	3	4			
2001-02	Toronto	NHL	78	20	31	51	26	1	0	3	152	13.2	7	1045	49.6	44	24	15:11	18	0	3	3	4	0	0	0
	Czech Republic	Olympics	4	1	0	1	2																			
	NHL Totals		680	229	329	558	332	65	2	31	1762	13.0		2527	50.5	106	47	17:30	51	6	20	26	12	3	0	1

Traded to **NY Islanders** by **Calgary** for Marty McInnis, Tyrone Garner and Calgary's 6th round choice (previously acquired, Calgary selected Ilja Demidov) in 1997 Entry Draft, March 18, 1997. Traded to **Phoenix** by **NY Islanders** with NY Islanders' 3rd round choice (Jason Jaspers) in 1999 Entry Draft and Ottawa's 4th round choice (previously acquired, Phoenix selected Preston Mizzi) in 1999 Entry Draft for Brad Isbister and Phoenix's 3rd round choice (Brian Collins) in 1999 Entry Draft, March 20, 1999. Traded to **Toronto** by **Phoenix** with Travis Green and Craig Mills for Danny Markov, June 12, 2001.

REICHERT, Craig

(RIGH-kuhrt, KRAYG) **EDM.**

Right wing. Shoots right. 6'1", 200 lbs. Born, Winnipeg, Man., May 11, 1974. Anaheim's 3rd choice, 67th overall, in 1994 Entry Draft.

Season	Club	League	GP	G	A	Pts	PIM	PP	SH	GW	S	%	+/-	TF	F%	H	SB	Min	GP	G	A	Pts	PIM	PP	SH	GW
1990-91	Cgy. Buffaloes	AMHL	47	32	36	68	54												13	13	28	41	27			
1991-92	Spokane Chiefs	WHL	68	13	20	33	56												4	1	0	1	4			
1992-93	Red Deer Rebels	WHL	66	32	33	65	62												4	3	1	4	2			
1993-94	Red Deer Rebels	WHL	72	52	67	119	153												4	2	2	4	8			
1994-95	San Diego Gulls	IHL	49	4	12	16	28																			
1995-96	Baltimore	AHL	68	10	17	27	50												1	0	0	0	0			
1996-97	Anaheim	NHL	3	0	0	0	0	0	0	0	3	0.0	–2													
	Baltimore	AHL	77	22	53	75	54												3	0	2	2	0			
1997-98	Cincinnati	AHL	78	28	59	87	28																			
1998-99	Cincinnati	AHL	72	28	41	69	56												3	2	0	2	0			
99-2000	Louisville	AHL	72	16	42	58	41												4	1	1	2	2			
2000-01	Dusseldorfer EG	Germany	60	12	23	35	76																			
2001-02	Hamilton	AHL	49	10	12	22	26																			
	NHL Totals		3	0	0	0	0	0	0	0	3	0.0														

Signed as a free agent by **Florida**, July 21, 1999. Signed as a free agent by **Edmonton**, June 11, 2001.

REINPRECHT, Steve

(REIGHN-prehkt, STEEV) **COL.**

Center. Shoots left. 6', 190 lbs. Born, Edmonton, Alta., May 7, 1976.

Season	Club	League	GP	G	A	Pts	PIM	PP	SH	GW	S	%	+/-	TF	F%	H	SB	Min	GP	G	A	Pts	PIM	PP	SH	GW
1993-94	Edmonton SSAC	AMHL	71	48	77	125	....																			
1994-95	St. Albert	AJHL	56	35	44	79	14																			
1995-96	St. Albert	AJHL	32	24	36	60																				
1996-97	U. of Wisconsin	WCHA	38	11	9	20	12																			
1997-98	U. of Wisconsin	WCHA	41	19	24	43	18																			
1998-99	U. of Wisconsin	WCHA	38	16	17	33	14																			
99-2000	U. of Wisconsin	WCHA	37	26	40	*66	14																			
	Los Angeles	NHL	1	0	0	0	2	0	0	0	0	0.0	0	6	50.0	0	0	6:01								
2000-01	Los Angeles	NHL	59	12	17	29	12	3	2	3	72	16.7	11	676	41.4	65	34	12:39								
	♦ Colorado	NHL	21	3	4	7	2	0	0	0	28	10.7	–1	209	51.2	23	14	15:38	22	2	3	5	2	0	0	0
2001-02	Colorado	NHL	67	19	27	46	18	4	0	3	111	17.1	14	413	52.1	92	32	16:32	21	7	5	12	8	0	0	2
	NHL Totals		148	34	48	82	34	7	2	6	211	16.1		1304	46.4	180	80	14:47	43	9	8	17	10	0	0	2

WCHA Second All-Star Team (1998) • WCHA First All-Star Team (2000) • WCHA Player of the Year (2000) • NCAA West First All-American Team (2000)

Signed as a free agent by **LA Kings**, March 31, 2000. Traded to **Colorado** by **LA Kings** with Rob Blake for Adam Deadmarsh, Aaron Miller, a player to be named later (Jared Aulin, March 22, 2001), Colorado's 1st round choice (Dave Steckel) in 2001 Entry Draft and future considerations, February 21, 2001.

REIRDEN, Todd

(REER-dehn, TAWD) **ANA.**

Defense. Shoots left. 6'5", 225 lbs. Born, Deerfield, IL, June 25, 1971. New Jersey's 14th choice, 242nd overall, in 1990 Entry Draft.

Season	Club	League	GP	G	A	Pts	PIM	PP	SH	GW	S	%	+/-	TF	F%	H	SB	Min	GP	G	A	Pts	PIM	PP	SH	GW
1987-88	Deerfield	Hi-School	22	19	32	51	....																			
1988-89	Tabor Academy	Hi-School	22	6	16	22	....																			
1989-90	Tabor Academy	Hi-School	22	10	28	38	....																			
1990-91	Bowling Green	CCHA	28	1	5	6	22																			
1991-92	Bowling Green	CCHA	33	8	7	15	34																			
1992-93	Bowling Green	CCHA	41	8	17	25	48																			
1993-94	Bowling Green	CCHA	38	7	23	30	56																			
1994-95	Albany	AHL	2	0	1	1	2																			
	Raleigh Icecaps	ECHL	26	2	13	15	33																			
	Tallahassee	ECHL	43	5	25	30	61												13	2	5	7	40			
1995-96	Tallahassee	ECHL	7	1	3	4	10																			
	Jacksonville	ECHL	15	1	10	11	41												1	0	2	2	4			
	Chicago Wolves	IHL	31	0	2	2	39												9	0	2	2	16			
1996-97	Chicago Wolves	IHL	57	3	10	13	108																			
	San Antonio	IHL	23	2	5	7	51												9	0	1	1	17			
1997-98	San Antonio	IHL	70	5	14	19	132																			
	Fort Wayne	IHL	11	2	2	4	16												4	0	2	2	4			
1998-99	Edmonton	NHL	17	2	3	5	20	0	0	0	26	7.7	–1	0	0.0	18	18	17:17								
	Hamilton	AHL	58	9	25	34	84												11	0	5	5	6			
99-2000	St. Louis	NHL	56	4	21	25	32	0	0	1	77	5.2	18	1	0.0	59	49	18:18	4	0	1	1	0	0	0	0
2000-01	St. Louis	NHL	38	2	4	6	43	1	0	0	58	3.4	–2	2	50.0	48	46	16:50	1	0	0	0	0	0	0	0
	Worcester	AHL	7	2	6	8	20																			
2001-02	Atlanta	NHL	65	3	5	8	82	1	0	0	85	3.5	–25	1	0.0	77	123	18:02								
	NHL Totals		176	11	33	44	177	2	0	1	246	4.5		4	25.0	202	236	17:47	5	0	1	1	0	0	0	0

Signed as a free agent by **Edmonton**, September 17, 1998. Claimed on waivers by **St. Louis** from **Edmonton**, September 30, 1999. Signed as a free agent by **Atlanta**, July 16, 2001. Signed as a free agent by **Anaheim**, July 17, 2002.

RENBERG, Mikael

(REHN-buhrg, MIHK-al) **TOR.**

Right wing. Shoots left. 6'2", 218 lbs. Born, Pitea, Sweden, May 5, 1972. Philadelphia's 3rd choice, 40th overall, in 1990 Entry Draft.

Season	Club	League	GP	G	A	Pts	PIM	PP	SH	GW	S	%	+/-	TF	F%	H	SB	Min	GP	G	A	Pts	PIM	PP	SH	GW
1988-89	Pitea HC	Swede-2	12	6	3	9	....																			
1989-90	Pitea HC	Swede-2	29	15	19	34	....																			
1990-91	Lulea HF	Sweden	29	11	6	17	12												5	1	1	2	4			
1991-92	Lulea HF	Sweden	38	8	15	23	20												2	0	0	0	0			
1992-93	Lulea HF	Sweden	39	19	13	32	61												11	4	4	8	4			
1993-94	Philadelphia	NHL	83	38	44	82	36	9	0	1	195	19.5	8													
1994-95	Lulea HF	Sweden	10	9	4	13	16																			
	Philadelphia	NHL	47	26	31	57	20	8	0	4	143	18.2	20						15	6	7	13	6	2	0	0
1995-96	Philadelphia	NHL	51	23	20	43	45	9	0	4	198	11.6	8						11	3	6	9	14	1	0	0
1996-97	Philadelphia	NHL	77	22	37	59	65	1	0	4	249	8.8	36						18	5	6	11	4	2	0	0
1997-98	Tampa Bay	NHL	68	16	22	38	34	6	3	0	175	9.1	–37													
	Sweden	Olympics	4	1	2	3	4																			
1998-99	Tampa Bay	NHL	20	4	8	12	4	2	0	0	42	9.5	–2	2	100.0	1	4	15:32								
	Philadelphia	NHL	46	11	15	26	14	4	0	2	112	9.8	7	1	0.0	8	5	16:00	6	0	1	1	0	0	0	0
99-2000	Philadelphia	NHL	62	8	21	29	30	3	0	1	106	7.5	–1	3	33.3	23	10	13:27								
	Phoenix	NHL	10	2	4	6	2	0	0	0	16	12.5	0	0	0.0	2	1	15:31	5	1	2	3	4	0	0	1
2000-01	Lulea HF	Sweden	48	22	32	54	36												11	6	5	11	35			

			Regular Season																Playoffs							
Season	Club	League	GP	G	A	Pts	PIM	PP	SH	GW	S	%	+/-	TF	F%	H	SB	Min	GP	G	A	Pts	PIM	PP	SH	GW
2001-02	**Toronto**	**NHL**	**71**	**14**	**38**	**52**	**36**	**4**	**0**	**3**	**130**	**10.8**	**11**	**6**	**33.3**	**29**	**17**	**14:35**	**3**	**0**	**0**	**0**	**2**	**0**	**0**	**0**
	Sweden	Olympics	4	1	0	1	4	....	....	....	....	....	....	....	....	....	....	....	....	....	....	....	....	....	....	....
	NHL Totals		**535**	**164**	**240**	**404**	**286**	**46**	**3**	**19**	**1366**	**12.0**		**12**	**41.7**	**63**	**37**	**14:42**	**58**	**15**	**22**	**37**	**30**	**5**	**0**	**1**

NHL All-Rookie Team (1994)

Traded to **Tampa Bay** by **Philadelphia** with Karl Dykhuis for Philadelphia's 1st round choices in 1998 (previously acquired, Tampa Bay selected Simon Gagne), 1999 (Maxime Ouellet), 2000 (Justin Williams) and 2001 (later traded to Ottawa - Ottawa selected Tim Gleason) Entry Drafts, August 20, 1997. Traded to **Philadelphia** by **Tampa Bay** with Daymond Langkow for Chris Gratton and Mike Sillinger, December 12, 1998. Traded to **Phoenix** by **Philadelphia** for Rick Tocchet, March 8, 2000. Traded to **Toronto** by **Phoenix** for Sergei Berezin, June 23, 2001.

RHEAUME, Pascal (RAY-awm, PAS-kal) ATL.

Left wing. Shoots left. 6'1", 210 lbs. Born, Quebec City, Que., June 21, 1973.

Season	Club	League	GP	G	A	Pts	PIM	PP	SH	GW	S	%	+/-	TF	F%	H	SB	Min	GP	G	A	Pts	PIM	PP	SH	GW
1990-91	Ste-Foy	QAAA	37	20	38	58	25	....	....	....	....	....	....	....	....	....	....	....	7	7	1	8	6	....	....	....
1991-92	Trois-Rivieres	QMJHL	65	17	20	37	84	....	....	....	....	....	....	....	....	....	....	....	14	5	4	9	23	....	....	....
1992-93	Sherbrooke	QMJHL	65	28	34	62	88	....	....	....	....	....	....	....	....	....	....	....	14	6	5	11	31	....	....	....
1993-94	Albany	AHL	55	17	18	35	43	....	....	....	....	....	....	....	....	....	....	....	5	0	1	1	0	....	....	....
1994-95	Albany	AHL	78	19	25	44	46	....	....	....	....	....	....	....	....	....	....	....	14	3	6	9	19	....	....	....
1995-96	Albany	AHL	68	26	42	68	50	....	....	....	....	....	....	....	....	....	....	....	4	1	2	3	2	....	....	....
1996-97	**New Jersey**	**NHL**	**2**	**1**	**0**	**1**	**0**	**0**	**0**	**0**	**5**	**20.0**	**1**	....	....	....	....	....	....	....	....	....	....	....	....	....
	Albany	AHL	51	22	23	45	40	....	....	....	....	....	....	....	....	....	....	....	16	2	8	10	16	....	....	....
1997-98	**St. Louis**	**NHL**	**48**	**6**	**9**	**15**	**35**	**1**	**0**	**0**	**45**	**13.3**	**4**	....	....	....	....	....	**10**	**1**	**3**	**4**	**8**	**1**	**0**	**0**
1998-99	**St. Louis**	**NHL**	**60**	**9**	**18**	**27**	**24**	**2**	**0**	**0**	**85**	**10.6**	**10**	**21**	**71.4**	**105**	**14**	**13:19**	**5**	**1**	**0**	**1**	**4**	**0**	**0**	**0**
99-2000	**St. Louis**	**NHL**	**7**	**1**	**1**	**2**	**6**	**0**	**0**	**0**	**5**	**20.0**	**–2**	**2**	**0.0**	**9**	**1**	**10:08**	....	....	....	....	....	....	....	....
	Worcester	AHL	7	1	1	2	4	....	....	....	....	....	....	....	....	....	....	....	....	....	....	....	....	....	....	....
2000-01	**St. Louis**	**NHL**	**8**	**2**	**0**	**2**	**5**	**2**	**0**	**0**	**16**	**12.5**	**–1**	**7**	**42.9**	**24**	**0**	**11:56**	**3**	**0**	**1**	**1**	**0**	**0**	**0**	**0**
	Worcester	AHL	56	23	35	58	63	....	....	....	....	....	....	....	....	....	....	....	11	2	4	6	2	....	....	....
2001-02	**Chicago**	**NHL**	**19**	**0**	**2**	**2**	**4**	**0**	**0**	**0**	**19**	**0.0**	**–1**	**165**	**51.5**	**26**	**5**	**9:22**	....	....	....	....	....	....	....	....
	Atlanta	**NHL**	**42**	**11**	**9**	**20**	**25**	**6**	**0**	**2**	**61**	**18.0**	**–3**	**510**	**45.9**	**81**	**28**	**14:27**	....	....	....	....	....	....	....	....
	NHL Totals		**186**	**30**	**39**	**69**	**99**	**11**	**0**	**2**	**236**	**12.7**		**705**	**47.8**	**245**	**48**	**12:52**	**18**	**2**	**4**	**6**	**12**	**1**	**0**	**0**

Signed as a free agent by **New Jersey**, October 1, 1993. Claimed by **St. Louis** from **New Jersey** in NHL Waiver Draft, September 28, 1997. • Missed majority of 1999-2000 season recovering from shoulder surgery, August, 1999. Signed as a free agent by **Chicago**, July 31, 2001. Claimed on waivers by **Atlanta** from **Chicago**, November 14, 2001.

RIBEIRO, Mike (rih-bee-AIR-roh, MIGHK) MTL.

Center. Shoots left. 6', 177 lbs. Born, Montreal, Que., February 10, 1980. Montreal's 2nd choice, 45th overall, in 1998 Entry Draft.

Season	Club	League	GP	G	A	Pts	PIM	PP	SH	GW	S	%	+/-	TF	F%	H	SB	Min	GP	G	A	Pts	PIM	PP	SH	GW
1996-97	Mtl-Bourassa	QAAA	43	32	57	89	48	....	....	....	....	....	....	....	....	....	....	....	16	15	23	38	14	....	....	....
1997-98	Rouyn-Noranda	QMJHL	67	40	*85	125	55	....	....	....	....	....	....	....	....	....	....	....	6	3	1	4	0	....	....	....
1998-99	Rouyn-Noranda	QMJHL	69	*67	*100	*167	137	....	....	....	....	....	....	....	....	....	....	....	11	5	11	16	12	....	....	....
	Fredericton	AHL	....	....	....	....	....	....	....	....	....	....	....	....	....	....	....	....	5	0	1	1	2	....	....	....
99-2000	**Montreal**	**NHL**	**19**	**1**	**1**	**2**	**2**	**1**	**0**	**0**	**18**	**5.6**	**–6**	**95**	**34.7**	**15**	**5**	**10:40**	....	....	....	....	....	....	....	....
	Quebec	AHL	3	0	0	0	2	....	....	....	....	....	....	....	....	....	....	....	....	....	....	....	....	....	....	....
	Rouyn-Noranda	QMJHL	2	1	3	4	0	....	....	....	....	....	....	....	....	....	....	....	....	....	....	....	....	....	....	....
	Quebec Remparts	QMJHL	21	17	28	45	30	....	....	....	....	....	....	....	....	....	....	....	11	3	20	23	38	....	....	....
2000-01	**Montreal**	**NHL**	**2**	**0**	**0**	**0**	**2**	**0**	**0**	**0**	**3**	**0.0**	**0**	**11**	**18.2**	**2**	**0**	**10:38**	....	....	....	....	....	....	....	....
	Quebec	AHL	74	26	40	66	44	....	....	....	....	....	....	....	....	....	....	....	9	1	5	6	23	....	....	....
2001-02	**Montreal**	**NHL**	**43**	**8**	**10**	**18**	**12**	**3**	**0**	**0**	**48**	**16.7**	**–11**	**141**	**44.0**	**44**	**7**	**13:55**	....	....	....	....	....	....	....	....
	Quebec	AHL	23	9	14	23	36	....	....	....	....	....	....	....	....	....	....	....	3	0	3	3	0	....	....	....
	NHL Totals		**64**	**9**	**11**	**20**	**16**	**4**	**0**	**0**	**69**	**13.0**		**247**	**39.3**	**61**	**12**	**12:51**	....	....	....	....	....	....	....	....

QMJHL Second All-Star Team (1998) • QMJHL First All-Star Team (1999) • Canadian Major Junior First All-Star Team (1999)

Assigned to **Rouyn-Noranda** (QMJHL) by **Montreal**, January 5, 2000. Traded to **Quebec** (QMJHL) by **Rouyn-Noranda** (QMJHL) for Guillaume Lefebvre and future considerations, January 9, 2000.

RICCI, Mike (REE-CHEE, MIGHK) S.J.

Center. Shoots left. 6', 185 lbs. Born, Scarborough, Ont., October 27, 1971. Philadelphia's 1st choice, 4th overall, in 1990 Entry Draft.

Season	Club	League	GP	G	A	Pts	PIM	PP	SH	GW	S	%	+/-	TF	F%	H	SB	Min	GP	G	A	Pts	PIM	PP	SH	GW
1986-87	Toronto Marlies	MTHL	38	39	42	81	27	....	....	....	....	....	....	....	....	....	....	....	....	....	....	....	....	....	....	....
1987-88	Peterborough	OHL	41	24	37	61	20	....	....	....	....	....	....	....	....	....	....	....	8	5	5	10	4	....	....	....
1988-89	Peterborough	OHL	60	54	52	106	43	....	....	....	....	....	....	....	....	....	....	....	17	19	16	35	18	....	....	....
1989-90	Peterborough	OHL	60	52	64	116	39	....	....	....	....	....	....	....	....	....	....	....	12	5	7	12	26	....	....	....
1990-91	**Philadelphia**	**NHL**	**68**	**21**	**20**	**41**	**64**	**9**	**0**	**4**	**121**	**17.4**	**–8**	....	....	....	....	....	....	....	....	....	....	....	....	....
1991-92	**Philadelphia**	**NHL**	**78**	**20**	**36**	**56**	**93**	**11**	**2**	**0**	**149**	**13.4**	**–10**	....	....	....	....	....	....	....	....	....	....	....	....	....
1992-93	**Quebec**	**NHL**	**77**	**27**	**51**	**78**	**123**	**12**	**1**	**10**	**142**	**19.0**	**8**	....	....	....	....	....	**6**	**0**	**6**	**6**	**8**	**0**	**0**	**0**
1993-94	**Quebec**	**NHL**	**83**	**30**	**21**	**51**	**113**	**13**	**3**	**6**	**138**	**21.7**	**–9**	....	....	....	....	....	....	....	....	....	....	....	....	....
1994-95	**Quebec**	**NHL**	**48**	**15**	**21**	**36**	**40**	**9**	**0**	**1**	**73**	**20.5**	**5**	....	....	....	....	....	**6**	**1**	**3**	**4**	**8**	**0**	**0**	**0**
1995-96♦	**Colorado**	**NHL**	**62**	**6**	**21**	**27**	**52**	**3**	**0**	**1**	**73**	**8.2**	**1**	....	....	....	....	....	**22**	**6**	**11**	**17**	**18**	**3**	**0**	**1**
1996-97	**Colorado**	**NHL**	**63**	**13**	**19**	**32**	**59**	**5**	**0**	**3**	**74**	**17.6**	**–3**	....	....	....	....	....	**17**	**2**	**4**	**6**	**17**	**0**	**0**	**1**
1997-98	**Colorado**	**NHL**	**6**	**0**	**4**	**4**	**2**	**0**	**0**	**0**	**5**	**0.0**	**0**	....	....	....	....	....	....	....	....	....	....	....	....	....
	San Jose	**NHL**	**59**	**9**	**14**	**23**	**30**	**5**	**0**	**2**	**86**	**10.5**	**–4**	....	....	....	....	....	**6**	**1**	**3**	**4**	**6**	**0**	**0**	**0**
1998-99	**San Jose**	**NHL**	**82**	**13**	**26**	**39**	**68**	**2**	**1**	**2**	**98**	**13.3**	**1**	**1465**	**49.6**	**73**	**41**	**15:23**	**6**	**2**	**3**	**5**	**10**	**1**	**0**	**0**
99-2000	**San Jose**	**NHL**	**82**	**20**	**24**	**44**	**60**	**10**	**0**	**5**	**134**	**14.9**	**14**	**1522**	**50.7**	**94**	**52**	**16:52**	**12**	**5**	**1**	**6**	**2**	**3**	**0**	**1**
2000-01	**San Jose**	**NHL**	**81**	**22**	**22**	**44**	**60**	**9**	**2**	**4**	**141**	**15.6**	**3**	**1631**	**51.4**	**107**	**46**	**17:60**	**6**	**0**	**3**	**3**	**0**	**0**	**0**	**0**
2001-02	**San Jose**	**NHL**	**79**	**19**	**34**	**53**	**44**	**5**	**2**	**0**	**115**	**16.5**	**9**	**1501**	**49.0**	**100**	**35**	**17:04**	**12**	**4**	**6**	**10**	**4**	**0**	**0**	**1**
	NHL Totals		**868**	**215**	**313**	**528**	**808**	**93**	**11**	**38**	**1349**	**15.9**		**6119**	**50.2**	**374**	**174**	**16:49**	**93**	**21**	**40**	**61**	**73**	**7**	**0**	**4**

OHL Second All-Star Team (1989) • OHL First All-Star Team (1990) • OHL MVP (1990) • Canadian Major Junior Player of the Year (1990) • OHL First All-Star Team (1990)

Traded to **Quebec** by **Philadelphia** with Steve Duchesne, Peter Forsberg, Kerry Huffman, Ron Hextall, Philadelphia's 1st round choice (Jocelyn Thibault) in 1993 Entry Draft, $15,000,000 and future considerations (Chris Simon and Philadelphia's 1st round choice (later traded to Toronto - later traded to Washington - Washington selected Nolan Baumgartner) in 1994 Entry Draft, July 21, 1992) for Eric Lindros, June 30, 1992. Transferred to **Colorado** after **Quebec** franchise relocated, June 21, 1995. Traded to **San Jose** by **Colorado** with Colorado's 2nd round choice (later traded to Buffalo - Buffalo selected Jaroslav Kristek) in 1998 Entry Draft for Shean Donovan and San Jose's 1st round choice (Alex Tanguay) in 1998 Entry Draft, November 21, 1997.

RICHARDS, Brad (RIH-chahrds, BRAD) T.B.

Left wing. Shoots left. 6'1", 198 lbs. Born, Montague, P.E.I., May 2, 1980. Tampa Bay's 2nd choice, 64th overall, in 1998 Entry Draft.

Season	Club	League	GP	G	A	Pts	PIM	PP	SH	GW	S	%	+/-	TF	F%	H	SB	Min	GP	G	A	Pts	PIM	PP	SH	GW
1996-97	Notre Dame	SJHL	63	39	48	87	73	....	....	....	....	....	....	....	....	....	....	....	....	....	....	....	....	....	....	....
1997-98	Rimouski Oceanic	QMJHL	68	33	82	115	44	....	....	....	....	....	....	....	....	....	....	....	19	8	24	32	2	....	....	....
1998-99	Rimouski Oceanic	QMJHL	59	39	92	131	55	....	....	....	....	....	....	....	....	....	....	....	11	9	12	21	6	....	....	....
99-2000	Rimouski Oceanic	QMJHL	63	*71	*115	*186	69	....	....	....	....	....	....	....	....	....	....	....	12	13	*24	*37	16	....	....	....
2000-01	**Tampa Bay**	**NHL**	**82**	**21**	**41**	**62**	**14**	**7**	**0**	**3**	**179**	**11.7**	**–10**	**955**	**41.4**	**14**	**28**	**16:54**	....	....	....	....	....	....	....	....
2001-02	**Tampa Bay**	**NHL**	**82**	**20**	**42**	**62**	**13**	**5**	**0**	**0**	**251**	**8.0**	**–18**	**911**	**41.2**	**24**	**30**	**19:48**	....	....	....	....	....	....	....	....
	NHL Totals		**164**	**41**	**83**	**124**	**27**	**12**	**0**	**3**	**430**	**9.5**		**1866**	**41.3**	**38**	**58**	**18:21**	....	....	....	....	....	....	....	....

SJHL Rookie of the Year (1997) • QMJHL First All-Star Team (2000) • Canadian Major Junior First All-Star Team (2000) • Canadian Major Junior Player of the Year (2000) • Memorial Cup All-Star Team (2000) • Won Stafford Smythe Memorial Trophy (Memorial Cup Tournament MVP) (2000) • NHL All-Rookie Team (2001)

RICHARDS, Travis (RIH-chuhrds, TRA-vihs)

Defense. Shoots left. 6'1", 195 lbs. Born, Crystal, MN, March 22, 1970. Minnesota's 6th choice, 169th overall, in 1988 Entry Draft.

Season	Club	League	GP	G	A	Pts	PIM	PP	SH	GW	S	%	+/-	TF	F%	H	SB	Min	GP	G	A	Pts	PIM	PP	SH	GW
1986-87	Armstrong	Hi-School	22	6	16	22	20	....	....	....	....	....	....	....	....	....	....	....	....	....	....	....	....	....	....	....
1987-88	Armstrong	Hi-School	24	14	14	28	....	....	....	....	....	....	....	....	....	....	....	....	....	....	....	....	....	....	....	....
1988-89	Armstrong	Hi-School	STATISTICS NOT AVAILABLE																							
1989-90	U. of Minnesota	WCHA	45	4	24	28	38	....	....	....	....	....	....	....	....	....	....	....	....	....	....	....	....	....	....	....
1990-91	U. of Minnesota	WCHA	45	9	25	34	28	....	....	....	....	....	....	....	....	....	....	....	....	....	....	....	....	....	....	....
1991-92	U. of Minnesota	WCHA	41	10	22	32	65	....	....	....	....	....	....	....	....	....	....	....	....	....	....	....	....	....	....	....
1992-93	U. of Minnesota	WCHA	42	12	26	38	52	....	....	....	....	....	....	....	....	....	....	....	....	....	....	....	....	....	....	....
1993-94	Team USA	Nat-Tm	51	1	11	12	38	....	....	....	....	....	....	....	....	....	....	....	....	....	....	....	....	....	....	....
	United States	Olympics	8	0	0	0	2	....	....	....	....	....	....	....	....	....	....	....	....	....	....	....	....	....	....	....
	Kalamazoo Wings	IHL	19	2	10	12	20	....	....	....	....	....	....	....	....	....	....	....	4	1	1	2	0	....	....	....
1994-95	Kalamazoo Wings	IHL	63	4	16	20	53	....	....	....	....	....	....	....	....	....	....	....	15	1	5	6	12	....	....	....
	Dallas	**NHL**	**2**	**0**	**0**	**0**	**0**	**0**	**0**	**0**	**1**	**0.0**	**0**	....	....	....	....	....	....	....	....	....	....	....	....	....
1995-96	**Dallas**	**NHL**	**1**	**0**	**0**	**0**	**2**	**0**	**0**	**0**	**0**	**0.0**	**–1**	....	....	....	....	....	....	....	....	....	....	....	....	....
	Michigan K-Wings	IHL	65	8	15	23	55	....	....	....	....	....	....	....	....	....	....	....	9	2	2	4	4	....	....	....
1996-97	Grand Rapids	IHL	77	10	13	23	83	....	....	....	....	....	....	....	....	....	....	....	5	1	3	4	2	....	....	....
1997-98	Grand Rapids	IHL	81	12	20	32	70	....	....	....	....	....	....	....	....	....	....	....	3	1	1	2	4	....	....	....

			Regular Season																Playoffs							
Season	Club	League	GP	G	A	Pts	PIM	PP	SH	GW	S	%	+/-	TF	F%	H	SB	Min	GP	G	A	Pts	PIM	PP	SH	GW
1998-99	Grand Rapids	IHL	82	9	23	32	84																			
99-2000	Grand Rapids	IHL	71	5	23	28	47												17	1	7	8	18			
2000-01	Grand Rapids	IHL	75	5	30	35	42												10	0	5	5	8			
2001-02	Grand Rapids	AHL	72	6	23	29	36												5	0	1	1	2			
	NHL Totals		3	0	0	0	2	0	0	0	1	0.0														

WCHA Second All-Star Team (1992, 1993) • IHL First All-Star Team (1995, 1996) • Won Governors' Trophy (Outstanding Defenseman - IHL) (1995) • IHL Second All-Star Team (2001)

Rights transferred to **Dallas** after **Minnesota** franchise relocated, June 9, 1993. Signed as a free agent by **Ottawa**, July 13, 2001.

RICHARDSON, Luke
(RIH-chahrd-sohn, LEWK) **CBJ**

Defense. Shoots left. 6'4", 210 lbs. Born, Ottawa, Ont., March 26, 1969. Toronto's 1st choice, 7th overall, in 1987 Entry Draft.

Season	Club	League	GP	G	A	Pts	PIM	PP	SH	GW	S	%	+/-	TF	F%	H	SB	Min	GP	G	A	Pts	PIM	PP	SH	GW
1984-85	Ottawa Knights	OMHA	35	5	26	31	72																			
1985-86	Peterborough	OHL	63	6	18	24	57												16	2	1	3	50			
1986-87	Peterborough	OHL	59	13	32	45	70												12	0	5	5	24			
1987-88	Toronto	NHL	78	4	6	10	90	0	0	0	49	8.2	–25						2	0	0	0	0	0	0	0
1988-89	Toronto	NHL	55	2	7	9	106	0	0	0	59	3.4	–15													
1989-90	Toronto	NHL	67	4	14	18	122	0	0	0	80	5.0	–1						5	0	0	0	22	0	0	0
1990-91	Toronto	NHL	78	1	9	10	238	0	0	0	68	1.5	–28													
1991-92	Edmonton	NHL	75	2	19	21	118	0	0	0	85	2.4	–9						16	0	5	5	45	0	0	0
1992-93	Edmonton	NHL	82	3	10	13	142	0	2	0	78	3.8	–18													
1993-94	Edmonton	NHL	69	2	6	8	131	0	0	0	92	2.2	–13													
1994-95	Edmonton	NHL	46	3	10	13	40	1	1	1	51	5.9	–6													
1995-96	Edmonton	NHL	82	2	9	11	108	0	0	0	61	3.3	–27													
1996-97	Edmonton	NHL	82	1	11	12	91	0	0	0	67	1.5	9						12	0	2	2	14	0	0	0
1997-98	Philadelphia	NHL	81	2	3	5	139	2	0	0	57	3.5	7						5	0	0	0	0	0	0	0
1998-99	Philadelphia	NHL	78	0	6	6	106	0	0	0	49	0.0	–3	0	0.0	116	94	16:33								
99-2000	Philadelphia	NHL	74	2	5	7	140	0	0	1	50	4.0	14	0	0.0	125	109	16:11	18	0	1	1	41	0	0	0
2000-01	Philadelphia	NHL	82	2	6	8	131	0	1	0	75	2.7	23	1	0.0	148	125	20:42	6	0	0	0	4	0	0	0
2001-02	Philadelphia	NHL	72	1	8	9	102	0	0	0	65	1.5	18	0	0.0	113	83	18:17	5	0	0	0	4	0	0	0
	NHL Totals		1101	31	129	160	1804	3	4	2	986	3.1		1	0.0	502	411	17:59	69	0	8	8	130	0	0	0

Traded to **Edmonton** by **Toronto** with Vincent Damphousse, Peter Ing and Scott Thornton for Grant Fuhr, Glenn Anderson and Craig Berube, September 19, 1991. Signed as a free agent by **Philadelphia**, July 23, 1997. Signed as a free agent by **Columbus**, July 4, 2002.

RICHER, Stephane
(REE-shay, STEH-fan)

Right wing. Shoots right. 6'2", 215 lbs. Born, Ripon, Que., June 7, 1966. Montreal's 3rd choice, 29th overall, in 1984 Entry Draft.

Season	Club	League	GP	G	A	Pts	PIM	PP	SH	GW	S	%	+/-	TF	F%	H	SB	Min	GP	G	A	Pts	PIM	PP	SH	GW
1982-83	Laval Insulaires	QAAA	48	47	54	101	86																			
1983-84	Granby Bisons	QMJHL	67	39	37	76	58												3	1	1	2	4			
1984-85	Granby Bisons	QMJHL	30	30	27	57	31																			
	Chicoutimi	QMJHL	27	31	32	63	40												12	13	13	26	25			
	Montreal	NHL	1	0	0	0	0	0	0	0	0	0.0	0													
	Sherbrooke	AHL																	9	6	3	9	10			
1985-86 ♦	Montreal	NHL	65	21	16	37	50	5	0	2	112	18.8	1						16	4	1	5	23	3	0	1
1986-87	Montreal	NHL	57	20	19	39	80	4	0	3	109	18.3	11						5	3	2	5	0	0	0	1
	Sherbrooke	AHL	12	10	4	14	11																			
1987-88	Montreal	NHL	72	50	28	78	72	16	0	11	263	19.0	12						8	7	5	12	6	1	0	2
1988-89	Montreal	NHL	68	25	35	60	61	11	0	6	214	11.7	4						21	6	5	11	14	2	0	3
1989-90	Montreal	NHL	75	51	40	91	46	9	0	8	269	19.0	35						9	7	3	10	2	1	0	1
1990-91	Montreal	NHL	75	31	30	61	53	9	0	4	221	14.0	0						13	9	5	14	6	1	0	1
1991-92	New Jersey	NHL	74	29	35	64	25	5	1	6	240	12.1	–1						7	1	2	3	0	0	0	0
1992-93	New Jersey	NHL	78	38	35	73	44	7	1	7	286	13.3	–1						5	2	2	4	2	1	0	0
1993-94	New Jersey	NHL	80	36	36	72	16	7	3	9	217	16.6	31						20	7	5	12	6	3	0	2
1994-95 ♦	New Jersey	NHL	45	23	16	39	10	1	2	5	133	17.3	8						19	6	15	21	2	3	1	2
1995-96	New Jersey	NHL	73	20	12	32	30	3	4	3	192	10.4	–8													
1996-97	Montreal	NHL	63	22	24	46	32	2	0	2	126	17.5	0						5	0	0	0	0	0	0	0
1997-98	Montreal	NHL	14	5	4	9	5	2	0	0	24	20.8	1													
	Tampa Bay	NHL	26	9	11	20	36	3	0	2	71	12.7	–7													
1998-99	Tampa Bay	NHL	64	12	21	33	22	3	2	1	139	8.6	–10	67	31.3	32	11	16:47								
99-2000	Tampa Bay	NHL	20	7	5	12	4	1	0	0	47	14.9	2	7	57.1	14	6	14:31								
	Detroit Vipers	IHL	2	0	0	0	0																			
	St. Louis	NHL	36	8	17	25	14	4	0	1	63	12.7	7	0	0.0	21	3	13:27	3	1	0	1	0	0	0	0
2000-01			OUT OF HOCKEY – RETIRED																							
2001-02	Pittsburgh	NHL	58	13	12	25	14	1	0	2	107	12.1	–8	8	50.0	17	14	14:04								
	New Jersey	NHL	10	1	2	3	0	0	0	0	16	6.3	–1	0	0.0	6	0	12:09	3	0	0	0	0	0	0	0
	NHL Totals		1054	421	398	819	614	93	13	72	2849	14.8		82	35.4	90	34	14:49	134	53	45	98	61	15	1	13

QMJHL Offensive Rookie of the Year (1984) • QMJHL Second All-Star Team (1985) • Played in NHL All-Star Game (1990)

Traded to **New Jersey** by **Montreal** with Tom Chorske for Kirk Muller and Rollie Melanson, September 20, 1991. Traded to **Montreal** by **New Jersey** for Lyle Odelein, August 22, 1996. Traded to **Tampa Bay** by **Montreal** with Darcy Tucker and David Wilkie for Patrick Poulin, Mick Vukota and Igor Ulanov, January 15, 1998. Traded to **St. Louis** by **Tampa Bay** for Rich Parent and Chris McAlpine, January 13, 2000. Signed as a free agent by **Washington**, August 25, 2000. Signed as a free agent by **Pittsburgh**, October 2, 2001. Traded to **New Jersey** by **Pittsburgh** for a conditional choice in 2003 Entry Draft, March 19, 2002.

RICHTER, Barry
(RIHK-tuhr, BAIR-ree)

Defense. Shoots left. 6'2", 200 lbs. Born, Madison, WI, September 11, 1970. Hartford's 2nd choice, 32nd overall, in 1988 Entry Draft.

Season	Club	League	GP	G	A	Pts	PIM	PP	SH	GW	S	%	+/-	TF	F%	H	SB	Min	GP	G	A	Pts	PIM	PP	SH	GW
1986-87	Culver Eagles	Hi-School	39	15	30	45																				
1987-88	Culver Eagles	Hi-School	35	24	29	53	18																			
1988-89	Culver Eagles	Hi-School	19	21	29	50	16																			
1989-90	U. of Wisconsin	WCHA	42	13	23	36	36																			
1990-91	U. of Wisconsin	WCHA	43	15	20	35	42																			
1991-92	U. of Wisconsin	WCHA	39	10	25	35	62																			
1992-93	U. of Wisconsin	WCHA	42	14	32	46	74																			
1993-94	Team USA	Nat-Tm	56	7	16	23	50																			
	United States	Olympics	8	0	3	3	4																			
	Binghamton	AHL	21	0	9	9	12																			
1994-95	Binghamton	AHL	73	15	41	56	54												11	4	5	9	12			
1995-96	NY Rangers	NHL	4	0	1	1	0	0	0	0	3	0.0	2													
	Binghamton	AHL	69	20	61	81	64												3	0	3	3	0			
1996-97	Boston	NHL	50	5	13	18	32	1	0	0	79	6.3	–7													
	Providence	AHL	19	2	6	8	4												10	4	4	8	4			
1997-98	Providence	AHL	75	16	29	45	47																			
1998-99	NY Islanders	NHL	72	6	18	24	34	0	0	2	111	5.4	–4	0	0.0	52	49	21:08								
99-2000	Montreal	NHL	23	0	2	2	8	0	0	0	13	0.0	–5	1	100.0	20	13	12:15								
	Quebec	AHL	2	0	0	0	0																			
	Manitoba Moose	IHL	19	5	4	9	6												2	1	1	2	0			
2000-01	Montreal	NHL	2	0	0	0	2	0	0	0	0	0.0	–1	0	0.0	0	0	10:44								
	Quebec	AHL	68	4	47	51	45												6	0	3	3	2			
2001-02	Linkopings HC	Sweden	44	5	12	17	82																			
	NHL Totals		151	11	34	45	76	1	0	2	206	5.3		1	100.0	72	62	18:49								

NCAA Championship All-Tournament Team (1992) • WCHA First All-Star Team (1993) • NCAA West First All-American Team (1993) • AHL First All-Star Team (1996) • Won Eddie Shore Award (Top Defenseman - AHL) (1996)

Traded to **NY Rangers** by **Hartford** with Steve Larmer, Nick Kypreos and Hartford's 6th round choice (Yuri Litvinov) in 1994 Entry Draft for Darren Turcotte and James Patrick, November 2, 1993. Signed as a free agent by **Boston**, July 19, 1996. Signed as a free agent by **NY Islanders**, August 17, 1998. Signed as a free agent by **Montreal**, August 20, 1999. Loaned to **Manitoba** (IHL) by **Montreal** for loan of Patrice Tardif to **Quebec** (AHL), March 3, 2000.

RIESEN, Michel
(REE-sehn, MEE-shehl)

Right wing. Shoots right. 6'2", 190 lbs. Born, Oberbalm, Switz., April 11, 1979. Edmonton's 1st choice, 14th overall, in 1997 Entry Draft.

			Regular Season																Playoffs							
Season	Club	League	GP	G	A	Pts	PIM	PP	SH	GW	S	%	+/-	TF	F%	H	SB	Min	GP	G	A	Pts	PIM	PP	SH	GW
1994-95	EHC Biel-Bienne	Swiss	12	0	2	2	0	....	....	....	....	....	....	....	....	....	....	....	6	2	0	2	0	....	....	....
1995-96	EHC Biel-Bienne	Swiss-2	34	9	6	15	2	....	....	....	....	....	....	....	....	....	....	....	3	1	0	1	0	....	....	....
1996-97	EHC Biel-Bienne	Swiss-2	38	16	16	32	49	....	....	....	....	....	....	....	....	....	....	....	....	....	....	....	....	....	....	....
1997-98	HC Davos	Swiss	32	16	9	25	8	....	....	....	....	....	....	....	....	....	....	....	18	5	5	10	4	....	....	....
1998-99	Hamilton	AHL	60	6	17	23	6	....	....	....	....	....	....	....	....	....	....	....	3	0	0	0	0	....	....	....
99-2000	Hamilton	AHL	73	29	31	60	20	....	....	....	....	....	....	....	....	....	....	....	10	3	5	8	4	....	....	....
2000-01	**Edmonton**	**NHL**	**12**	**0**	**1**	**1**	**4**	**0**	**0**	**0**	**16**	**0.0**	**2**	**0**	**0.0**	**1**	**1**	**9:54**	....	....	....	....	....	....	....	....
	Hamilton	AHL	69	26	28	54	14	....	....	....	....	....	....	....	....	....	....	....	....	....	....	....	....	....	....	....
2001-02	HC Davos	Swiss	35	12	12	24	28	....	....	....	....	....	....	....	....	....	....	....	16	7	8	15	*52	....	....	....
	NHL Totals		**12**	**0**	**1**	**1**	**4**	**0**	**0**	**0**	**16**	**0.0**		**0**	**0.0**	**1**	**1**	**9:54**	....	....	....	....	....	....	....	....

Traded to **St. Louis** by **Edmonton** with Doug Weight for Marty Reasoner, Jochen Hecht and Jan Horacek, July 1, 2001. Signed as a free agent by **HC Davos** (Switz.), October 2, 2001.

RITA, Jani
(REETA, YA-nee) **EDM.**

Left wing. Shoots left. 6'1", 206 lbs. Born, Helsinki, Finland, July 25, 1981. Edmonton's 1st choice, 13th overall, in 1999 Entry Draft.

			Regular Season																Playoffs							
Season	Club	League	GP	G	A	Pts	PIM	PP	SH	GW	S	%	+/-	TF	F%	H	SB	Min	GP	G	A	Pts	PIM	PP	SH	GW
1994-95	Jokerit-C	Finn-Jr.	7	3	0	3	0	....	....	....	....	....	....	....	....	....	....	....	6	1	0	1	0	....	....	....
1995-96	Jokerit-C	Finn-Jr.	12	10	3	13	2	....	....	....	....	....	....	....	....	....	....	....	....	....	....	....	....	....	....	....
	Jokerit-B	Finn-Jr.	5	0	0	0	0	....	....	....	....	....	....	....	....	....	....	....	....	....	....	....	....	....	....	....
1996-97	Jokerit-B	Finn-Jr.	27	22	7	29	4	....	....	....	....	....	....	....	....	....	....	....	....	....	....	....	....	....	....	....
1997-98	Jokerit-B	Finn-Jr.	7	7	4	11	2	....	....	....	....	....	....	....	....	....	....	....	....	....	....	....	....	....	....	....
	Jokerit Jr.	Finn-Jr.	36	15	9	24	2	....	....	....	....	....	....	....	....	....	....	....	8	4	1	5	0	....	....	....
	Jokerit Helsinki	Finland	....	....	....	....	....	....	....	....	....	....	....	....	....	....	....	....	1	0	0	0	0	....	....	....
1998-99	Jokerit Jr.	Finn-Jr.	20	9	13	22	8	....	....	....	....	....	....	....	....	....	....	....	....	....	....	....	....	....	....	....
	Jokerit Helsinki	Finland	41	3	2	5	39	....	....	....	....	....	....	....	....	....	....	....	....	....	....	....	....	....	....	....
	Jokerit Helsinki	EuroHL	3	0	0	0	0	....	....	....	....	....	....	....	....	....	....	....	....	....	....	....	....	....	....	....
99-2000	Jokerit Jr.	Finn-Jr.	1	1	0	1	0	....	....	....	....	....	....	....	....	....	....	....	....	....	....	....	....	....	....	....
	Jokerit Helsinki	Finland	49	6	3	9	10	....	....	....	....	....	....	....	....	....	....	....	11	1	0	1	0	....	....	....
2000-01	Jokerit Jr.	Finn-Jr.	3	3	2	5	0	....	....	....	....	....	....	....	....	....	....	....	....	....	....	....	....	....	....	....
	Jokerit Helsinki	Finland	50	5	10	15	18	....	....	....	....	....	....	....	....	....	....	....	5	0	0	0	2	....	....	....
2001-02	**Edmonton**	**NHL**	**1**	**0**	**0**	**0**	**0**	**0**	**0**	**0**	**0**	**0.0**	**0**	**0**	**0.0**	**0**	**0**	**6:09**	....	....	....	....	....	....	....	....
	Hamilton	AHL	76	25	17	42	32	....	....	....	....	....	....	....	....	....	....	....	15	8	4	12	0	....	....	....
	NHL Totals		**1**	**0**	**0**	**0**	**0**	**0**	**0**	**0**	**0**	**0.0**		**0**	**0.0**	**0**	**0**	**6:09**	....	....	....	....	....	....	....	....

RITCHIE, Byron
(RIHT-chee, BIGH-rohn) **FLA.**

Center. Shoots left. 5'10", 195 lbs. Born, Burnaby, B.C., April 24, 1977. Hartford's 6th choice, 165th overall, in 1995 Entry Draft.

			Regular Season																Playoffs							
Season	Club	League	GP	G	A	Pts	PIM	PP	SH	GW	S	%	+/-	TF	F%	H	SB	Min	GP	G	A	Pts	PIM	PP	SH	GW
1992-93	North Delta	BCAHA	60	102	151	253	147	....	....	....	....	....	....	....	....	....	....	....	....	....	....	....	....	....	....	....
1993-94	Lethbridge	WHL	44	4	11	15	44	....	....	....	....	....	....	....	....	....	....	....	6	0	0	0	14	....	....	....
1994-95	Lethbridge	WHL	58	22	28	50	132	....	....	....	....	....	....	....	....	....	....	....	....	....	....	....	....	....	....	....
1995-96	Lethbridge	WHL	66	55	51	106	163	....	....	....	....	....	....	....	....	....	....	....	4	0	2	2	4	....	....	....
	Springfield	AHL	6	2	1	3	4	....	....	....	....	....	....	....	....	....	....	....	8	0	3	3	0	....	....	....
1996-97	Lethbridge	WHL	63	50	76	126	115	....	....	....	....	....	....	....	....	....	....	....	18	*16	12	*28	28	....	....	....
1997-98	New Haven	AHL	65	13	18	31	97	....	....	....	....	....	....	....	....	....	....	....	....	....	....	....	....	....	....	....
1998-99	**Carolina**	**NHL**	**3**	**0**	**0**	**0**	**0**	**0**	**0**	**0**	**0**	**0.0**	**0**	**5**	**20.0**	**1**	**0**	**3:25**	....	....	....	....	....	....	....	....
	New Haven	AHL	66	24	33	57	139	....	....	....	....	....	....	....	....	....	....	....	....	....	....	....	....	....	....	....
99-2000	**Carolina**	**NHL**	**26**	**0**	**2**	**2**	**17**	**0**	**0**	**0**	**13**	**0.0**	**−10**	**155**	**49.7**	**33**	**7**	**7:24**	....	....	....	....	....	....	....	....
	Cincinnati	IHL	34	8	13	21	81	....	....	....	....	....	....	....	....	....	....	....	10	1	6	7	32	....	....	....
2000-01	Cincinnati	IHL	77	31	35	66	166	....	....	....	....	....	....	....	....	....	....	....	5	3	2	5	10	....	....	....
2001-02	**Carolina**	**NHL**	**4**	**0**	**0**	**0**	**2**	**0**	**0**	**0**	**5**	**0.0**	**0**	**9**	**44.4**	**5**	**1**	**11:30**	....	....	....	....	....	....	....	....
	Lowell	AHL	43	25	30	55	38	....	....	....	....	....	....	....	....	....	....	....	....	....	....	....	....	....	....	....
	Florida	**NHL**	**31**	**5**	**6**	**11**	**34**	**2**	**0**	**0**	**55**	**9.1**	**−2**	**324**	**50.9**	**41**	**10**	**12:26**	....	....	....	....	....	....	....	....
	NHL Totals		**64**	**5**	**8**	**13**	**53**	**2**	**0**	**0**	**73**	**6.8**		**493**	**50.1**	**80**	**18**	**9:54**	....	....	....	....	....	....	....	....

WHL East Second All-Star Team (1996, 1997)

Rights transferred to **Carolina** after **Hartford** franchise relocated, June 25, 1997. Traded to **Florida** by **Carolina** with Sandis Ozolinsh for Bret Hedican, Kevyn Adams, Tomas Malec and a conditional 3rd round choice in 2003 Entry Draft, January 16, 2002.

RIVERS, Jamie
(RIH-vuhrs, JAY-mee)

Defense. Shoots left. 6'1", 200 lbs. Born, Ottawa, Ont., March 16, 1975. St. Louis' 2nd choice, 63rd overall, in 1993 Entry Draft.

			Regular Season																Playoffs							
Season	Club	League	GP	G	A	Pts	PIM	PP	SH	GW	S	%	+/-	TF	F%	H	SB	Min	GP	G	A	Pts	PIM	PP	SH	GW
1989-90	Ottawa South	OMHA	50	26	46	72	46	....	....	....	....	....	....	....	....	....	....	....	....	....	....	....	....	....	....	....
1990-91	Ottawa Jr. Sens	OCJHL	55	4	30	34	74	....	....	....	....	....	....	....	....	....	....	....	....	....	....	....	....	....	....	....
1991-92	Sudbury Wolves	OHL	55	3	13	16	20	....	....	....	....	....	....	....	....	....	....	....	8	0	0	0	0	....	....	....
1992-93	Sudbury Wolves	OHL	62	12	43	55	20	....	....	....	....	....	....	....	....	....	....	....	14	7	19	26	4	....	....	....
1993-94	Sudbury Wolves	OHL	65	32	*89	121	58	....	....	....	....	....	....	....	....	....	....	....	10	1	9	10	14	....	....	....
1994-95	Sudbury Wolves	OHL	46	9	56	65	30	....	....	....	....	....	....	....	....	....	....	....	18	7	26	33	22	....	....	....
1995-96	**St. Louis**	**NHL**	**3**	**0**	**0**	**0**	**2**	**0**	**0**	**0**	**5**	**0.0**	**−1**	....	....	....	....	....	....	....	....	....	....	....	....	....
	Worcester	AHL	75	7	45	52	130	....	....	....	....	....	....	....	....	....	....	....	4	0	1	1	4	....	....	....
1996-97	**St. Louis**	**NHL**	**15**	**2**	**5**	**7**	**6**	**1**	**0**	**0**	**9**	**22.2**	**−4**	....	....	....	....	....	....	....	....	....	....	....	....	....
	Worcester	AHL	63	8	35	43	83	....	....	....	....	....	....	....	....	....	....	....	5	1	2	3	14	....	....	....
1997-98	**St. Louis**	**NHL**	**59**	**2**	**4**	**6**	**36**	**1**	**0**	**1**	**53**	**3.8**	**5**	....	....	....	....	....	....	....	....	....	....	....	....	....
1998-99	**St. Louis**	**NHL**	**76**	**2**	**5**	**7**	**47**	**1**	**0**	**0**	**78**	**2.6**	**−3**	**0**	**0.0**	**131**	**63**	**14:10**	**9**	**1**	**1**	**2**	**2**	**1**	**0**	**1**
99-2000	**NY Islanders**	**NHL**	**75**	**1**	**16**	**17**	**84**	**1**	**0**	**0**	**95**	**1.1**	**−4**	**0**	**0.0**	**219**	**112**	**19:39**	....	....	....	....	....	....	....	....
2000-01	**Ottawa**	**NHL**	**45**	**2**	**4**	**6**	**44**	**0**	**0**	**0**	**41**	**4.9**	**6**	**0**	**0.0**	**115**	**41**	**14:01**	**1**	**0**	**0**	**0**	**4**	**0**	**0**	**0**
	Grand Rapids	IHL	2	0	0	0	2	....	....	....	....	....	....	....	....	....	....	....	....	....	....	....	....	....	....	....
2001-02	**Ottawa**	**NHL**	**2**	**0**	**0**	**0**	**4**	**0**	**0**	**0**	**3**	**0.0**	**−3**	**0**	**0.0**	**3**	**0**	**11:37**	....	....	....	....	....	....	....	....
	Boston	**NHL**	**64**	**4**	**2**	**6**	**45**	**1**	**0**	**1**	**48**	**8.3**	**6**	**39**	**33.3**	**77**	**27**	**8:27**	**3**	**0**	**0**	**0**	**0**	**0**	**0**	**0**
	NHL Totals		**339**	**13**	**36**	**49**	**268**	**5**	**0**	**2**	**332**	**3.9**		**39**	**33.3**	**545**	**243**	**14:18**	**13**	**1**	**1**	**2**	**6**	**1**	**0**	**1**

OHL First All-Star Team (1994) • Canadian Major Junior Second All-Star Team (1994) • OHL Second All-Star Team (1995) • AHL Second All-Star Team (1997)

Claimed by **NY Islanders** from **St. Louis** in Waiver Draft, September 27, 1999. Signed as a free agent by **Ottawa**, November 30, 2000. Claimed on waivers by **Boston** from **Ottawa**, October 13, 2001.

RIVET, Craig
(rih-VAY, KRAYG) **MTL.**

Defense. Shoots right. 6'2", 207 lbs. Born, North Bay, Ont., September 13, 1974. Montreal's 4th choice, 68th overall, in 1992 Entry Draft.

			Regular Season																Playoffs							
Season	Club	League	GP	G	A	Pts	PIM	PP	SH	GW	S	%	+/-	TF	F%	H	SB	Min	GP	G	A	Pts	PIM	PP	SH	GW
1990-91	Barrie Colts	OJHL-B	42	9	17	26	55	....	....	....	....	....	....	....	....	....	....	....	....	....	....	....	....	....	....	....
1991-92	Kingston	OHL	66	5	21	26	97	....	....	....	....	....	....	....	....	....	....	....	....	....	....	....	....	....	....	....
1992-93	Kingston	OHL	64	19	55	74	117	....	....	....	....	....	....	....	....	....	....	....	16	5	7	12	39	....	....	....
1993-94	Kingston	OHL	61	12	52	64	100	....	....	....	....	....	....	....	....	....	....	....	6	0	3	3	6	....	....	....
	Fredericton	AHL	4	0	2	2	2	....	....	....	....	....	....	....	....	....	....	....	....	....	....	....	....	....	....	....
1994-95	Fredericton	AHL	78	5	27	32	126	....	....	....	....	....	....	....	....	....	....	....	12	0	4	4	17	....	....	....
	Montreal	**NHL**	**5**	**0**	**1**	**1**	**5**	**0**	**0**	**0**	**2**	**0.0**	**2**	....	....	....	....	....	....	....	....	....	....	....	....	....
1995-96	**Montreal**	**NHL**	**19**	**1**	**4**	**5**	**54**	**0**	**0**	**0**	**9**	**11.1**	**4**	....	....	....	....	....	....	....	....	....	....	....	....	....
	Fredericton	AHL	49	5	18	23	189	....	....	....	....	....	....	....	....	....	....	....	6	0	0	0	12	....	....	....
1996-97	**Montreal**	**NHL**	**35**	**0**	**4**	**4**	**54**	**0**	**0**	**0**	**24**	**0.0**	**7**	....	....	....	....	....	**5**	**0**	**1**	**1**	**14**	**0**	**0**	**0**
	Fredericton	AHL	23	3	12	15	99	....	....	....	....	....	....	....	....	....	....	....	....	....	....	....	....	....	....	....
1997-98	**Montreal**	**NHL**	**61**	**0**	**2**	**2**	**93**	**0**	**0**	**0**	**26**	**0.0**	**−3**	....	....	....	....	....	**5**	**0**	**0**	**0**	**2**	**0**	**0**	**0**
1998-99	**Montreal**	**NHL**	**66**	**2**	**8**	**10**	**66**	**0**	**0**	**0**	**39**	**5.1**	**−3**	**0**	**0.0**	**78**	**40**	**14:20**	....	....	....	....	....	....	....	....
99-2000	**Montreal**	**NHL**	**61**	**3**	**14**	**17**	**76**	**0**	**0**	**1**	**71**	**4.2**	**11**	**0**	**0.0**	**80**	**59**	**19:03**	....	....	....	....	....	....	....	....
2000-01	**Montreal**	**NHL**	**26**	**1**	**2**	**3**	**36**	**0**	**0**	**0**	**22**	**4.5**	**−8**	**0**	**0.0**	**38**	**31**	**19:04**	....	....	....	....	....	....	....	....
2001-02	**Montreal**	**NHL**	**82**	**8**	**17**	**25**	**76**	**0**	**0**	**0**	**90**	**8.9**	**1**	**1**	**0.0**	**97**	**137**	**19:00**	**12**	**0**	**3**	**3**	**4**	**0**	**0**	**0**
	NHL Totals		**355**	**15**	**52**	**67**	**460**	**0**	**0**	**1**	**283**	**5.3**		**1**	**0.0**	**293**	**267**	**17:43**	**22**	**0**	**4**	**4**	**20**	**0**	**0**	**0**

• Missed majority of 2000-01 season recovering from shoulder injury suffered in game vs. Vancouver, October 30, 2000.

ROBERTS, Gary

(RAW-buhrts, GAIR-ree) **TOR.**

Left wing. Shoots left. 6'1", 190 lbs. Born, North York, Ont., May 23, 1966. Calgary's 1st choice, 12th overall, in 1984 Entry Draft.

			Regular Season																Playoffs							
Season	Club	League	GP	G	A	Pts	PIM	PP	SH	GW	S	%	+/-	TF	F%	H	SB	Min	GP	G	A	Pts	PIM	PP	SH	GW
1980-81	Hamilton	OHA-B	3	0	1	1	0																			
1981-82	Whitby	OMHA	44	55	31	86	133																			
1982-83	Ottawa 67's	OHL	53	12	8	20	83												5	1	0	1	19			
1983-84	Ottawa 67's	OHL	48	27	30	57	144												13	10	7	17	62			
1984-85	Ottawa 67's	OHL	59	44	62	106	186												5	2	8	10	10			
	Moncton	AHL	7	4	2	6	7																			
1985-86	Ottawa 67's	OHL	24	26	25	51	83																			
	Guelph Platers	OHL	23	18	15	33	65												20	18	13	31	43			
1986-87	**Calgary**	**NHL**	**32**	**5**	**10**	**15**	**85**	**0**	**0**	**0**	**38**	**13.2**	**6**						**2**	**0**	**0**	**0**	**4**	**0**	**0**	**0**
	Moncton	AHL	38	20	18	38	72																			
1987-88	**Calgary**	**NHL**	**74**	**13**	**15**	**28**	**282**	**0**	**0**	**1**	**118**	**11.0**	**24**						**9**	**2**	**3**	**5**	**29**	**0**	**0**	**0**
1988-89♦	**Calgary**	**NHL**	**71**	**22**	**16**	**38**	**250**	**0**	**1**	**2**	**123**	**17.9**	**32**						**22**	**5**	**7**	**12**	**57**	**0**	**0**	**0**
1989-90	**Calgary**	**NHL**	**78**	**39**	**33**	**72**	**222**	**5**	**0**	**5**	**175**	**22.3**	**31**						**6**	**2**	**5**	**7**	**41**	**0**	**0**	**0**
1990-91	**Calgary**	**NHL**	**80**	**22**	**31**	**53**	**252**	**0**	**0**	**3**	**132**	**16.7**	**15**						**7**	**1**	**3**	**4**	**18**	**0**	**0**	**0**
1991-92	**Calgary**	**NHL**	**76**	**53**	**37**	**90**	**207**	**15**	**0**	**2**	**196**	**27.0**	**32**													
1992-93	**Calgary**	**NHL**	**58**	**38**	**41**	**79**	**172**	**8**	**3**	**4**	**166**	**22.9**	**32**						**5**	**1**	**6**	**7**	**43**	**1**	**0**	**0**
1993-94	**Calgary**	**NHL**	**73**	**41**	**43**	**84**	**145**	**12**	**3**	**5**	**202**	**20.3**	**37**						**7**	**2**	**6**	**8**	**24**	**1**	**0**	**1**
1994-95	**Calgary**	**NHL**	**8**	**2**	**2**	**4**	**43**	**2**	**0**	**0**	**20**	**10.0**	**1**													
1995-96	**Calgary**	**NHL**	**35**	**22**	**20**	**42**	**78**	**9**	**0**	**5**	**84**	**26.2**	**15**													
1996-97	**Calgary**	**NHL**	DID NOT PLAY – INJURED																							
1997-98	**Carolina**	**NHL**	**61**	**20**	**29**	**49**	**103**	**4**	**0**	**2**	**106**	**18.9**	**3**													
1998-99	**Carolina**	**NHL**	**77**	**14**	**28**	**42**	**178**	**1**	**1**	**4**	**138**	**10.1**	**2**	**15**	**46.7**	**260**	**16**	**19:36**	**6**	**1**	**1**	**2**	**8**	**0**	**0**	**0**
99-2000	**Carolina**	**NHL**	**69**	**23**	**30**	**53**	**62**	**12**	**0**	**1**	**150**	**15.3**	**–10**	**7**	**28.6**	**212**	**27**	**18:31**								
2000-01	**Toronto**	**NHL**	**82**	**29**	**24**	**53**	**109**	**8**	**2**	**3**	**138**	**21.0**	**16**	**13**	**46.2**	**206**	**17**	**17:08**	**11**	**2**	**9**	**11**	**0**	**0**	**0**	**0**
2001-02	**Toronto**	**NHL**	**69**	**21**	**27**	**48**	**63**	**6**	**2**	**2**	**122**	**17.2**	**–4**	**6**	**33.3**	**151**	**9**	**17:23**	**19**	**7**	**12**	**19**	**56**	**3**	**0**	**1**
	NHL Totals		**943**	**364**	**386**	**750**	**2251**	**82**	**12**	**39**	**1908**	**19.1**		**41**	**41.5**	**829**	**69**	**18:09**	**94**	**23**	**52**	**75**	**280**	**5**	**0**	**2**

OHL Second All-Star Team (1985, 1986) • Won Bill Masterton Memorial Trophy (1996) • Played in NHL All-Star Game (1992, 1993)

• Missed remainder of 1994-95 and majority of 1995-96 seasons recovering from neck injury suffered in game vs. Toronto, February 4, 1995. • Missed remainder of 1995-96 and entire 1996-97 seasons recovering from neck injury suffered in game vs. Vancouver, April 3, 1996. Traded to **Carolina** by **Calgary** with Trevor Kidd for Andrew Cassels and Jean-Sebastien Giguere, August 25, 1997. Signed as a free agent by **Toronto**, July 4, 2000.

ROBERTSSON, Bert

(ROH-behrt-suhn, BUHRT)

Defense. Shoots left. 6'3", 205 lbs. Born, Sodertalje, Sweden, June 30, 1974. Vancouver's 8th choice, 254th overall, in 1993 Entry Draft.

Season	Club	League	GP	G	A	Pts	PIM	PP	SH	GW	S	%	+/-	TF	F%	H	SB	Min	GP	G	A	Pts	PIM	PP	SH	GW
1992-93	Sodertalje Jr.	Swede-Jr.	12	1	5	6	20																			
	Sodertalje SK	Swede-2	23	2	1	3	24												2	0	0	0	0			
1993-94	Sodertalje SK	Swede-2	28	0	1	1	12												1	0	0	0	0			
1994-95	Sodertalje Jr.	Swede-Jr.	11	2	3	5	4																			
	Sodertalje SK	Swede-2	23	1	2	3	24												3	0	1	1	2			
1995-96	Syracuse Crunch	AHL	65	1	7	8	109												16	0	1	1	26			
1996-97	Syracuse Crunch	AHL	80	4	9	13	132												3	1	0	1	4			
1997-98	**Vancouver**	**NHL**	**30**	**2**	**4**	**6**	**24**	**0**	**0**	**0**	**19**	**10.5**	**2**													
	Syracuse Crunch	AHL	42	5	9	14	87												3	0	0	0	6			
1998-99	**Vancouver**	**NHL**	**39**	**2**	**2**	**4**	**13**	**0**	**0**	**0**	**13**	**15.4**	**–7**	**0**	**0.0**	**28**	**6**	**7:26**								
	Syracuse Crunch	AHL	8	1	0	1	21																			
99-2000	**Edmonton**	**NHL**	**52**	**0**	**4**	**4**	**34**	**0**	**0**	**0**	**31**	**0.0**	**–3**	**3**	**0.0**	**62**	**28**	**10:49**	**5**	**0**	**0**	**0**	**0**	**0**	**0**	**0**
	Hamilton	AHL	6	0	3	3	12																			
2000-01	**NY Rangers**	**NHL**	**2**	**0**	**0**	**0**	**4**	**0**	**0**	**0**	**0**	**0.0**	**–1**	**0**	**0.0**	**4**	**0**	**5:37**								
	Hartford	AHL	27	7	5	12	27																			
	Houston Aeros	IHL	14	0	0	0	26																			
	Milwaukee	IHL	10	0	0	0	0												5	0	1	1	2			
2001-02	Milwaukee	AHL	14	0	1	1	18																			
	Cincinnati	AHL	16	0	2	2	39																			
	Wilkes-Barre	AHL	13	0	0	0	15																			
	NHL Totals		**123**	**4**	**10**	**14**	**75**	**0**	**0**	**0**	**63**	**6.3**		**3**	**0.0**	**94**	**34**	**9:17**	**5**	**0**	**0**	**0**	**0**	**0**	**0**	**0**

Signed as a free agent by **Edmonton**, August 19, 1999. Selected by **Columbus** from **Edmonton** in Expansion Draft, June 23, 2000. Traded to **NY Rangers** by **Columbus** for Jean-Francois Labbe, November 9, 2000. Traded to **Nashville** by **NY Rangers** for Ryan Tobler, March 7, 2001. Traded to **Anaheim** by **Nashville** for Jay Legault, December 4, 2001. Traded to **Pittsburgh** by **Anaheim** for Mark Moore, March 8, 2002. Signed as a free agent by **Ilves** (Finland), August 8, 2002.

ROBIDAS, Stephane

(ROH-bih-dah, STEH-fan) **MTL.**

Defense. Shoots right. 5'11", 189 lbs. Born, Sherbrooke, Que., March 3, 1977. Montreal's 7th choice, 164th overall, in 1995 Entry Draft.

Season	Club	League	GP	G	A	Pts	PIM	PP	SH	GW	S	%	+/-	TF	F%	H	SB	Min	GP	G	A	Pts	PIM	PP	SH	GW
1992-93	Magog	QAAA	41	3	12	15	16												5	1	1	2	2			
1993-94	Shawinigan	QMJHL	67	3	18	21	33												1	0	0	0	0			
1994-95	Shawinigan	QMJHL	71	13	56	69	44												15	7	12	19	4			
1995-96	Shawinigan	QMJHL	67	23	56	79	53												6	1	5	6	10			
1996-97	Shawinigan	QMJHL	67	24	51	75	59												7	4	6	10	14			
1997-98	Fredericton	AHL	79	10	21	31	50												4	0	2	2	0			
1998-99	Fredericton	AHL	79	8	33	41	59												15	1	5	6	10			
99-2000	**Montreal**	**NHL**	**1**	**0**	**0**	**0**	**0**	**0**	**0**	**0**	**0**	**0.0**	**0**	**0**	**0.0**	**1**	**1**	**15:54**								
	Quebec	AHL	76	14	31	45	36												3	0	1	1	0			
2000-01	**Montreal**	**NHL**	**65**	**6**	**6**	**12**	**14**	**1**	**0**	**0**	**77**	**7.8**	**0**	**1**	**100.0**	**115**	**83**	**20:44**								
2001-02	**Montreal**	**NHL**	**56**	**1**	**10**	**11**	**14**	**1**	**0**	**0**	**68**	**1.5**	**–25**	**3**	**33.3**	**107**	**87**	**18:58**	**2**	**0**	**0**	**0**	**4**	**0**	**0**	**0**
	NHL Totals		**122**	**7**	**16**	**23**	**28**	**2**	**0**	**0**	**145**	**4.8**		**4**	**50.0**	**223**	**171**	**19:53**	**2**	**0**	**0**	**0**	**4**	**0**	**0**	**0**

QMJHL First All-Star Team (1996, 1997)

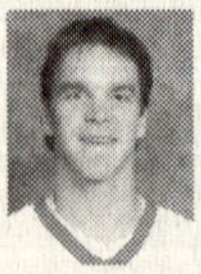

ROBITAILLE, Luc

(ROH-buh-tigh, LEWK) **DET.**

Left wing. Shoots left. 6'1", 215 lbs. Born, Montreal, Que., February 17, 1966. Los Angeles' 9th choice, 171st overall, in 1984 Entry Draft.

Season	Club	League	GP	G	A	Pts	PIM	PP	SH	GW	S	%	+/-	TF	F%	H	SB	Min	GP	G	A	Pts	PIM	PP	SH	GW
1982-83	Mtl-Bourassa	QAAA	48	36	57	93	28												7	9	6	15	14			
1983-84	Hull Olympiques	QMJHL	70	32	53	85	48																			
1984-85	Hull Olympiques	QMJHL	64	55	94	149	115												5	4	2	6	27			
1985-86	Hull Olympiques	QMJHL	63	68	123	191	91												15	17	27	44	28			
1986-87	**Los Angeles**	**NHL**	**79**	**45**	**39**	**84**	**28**	**18**	**0**	**3**	**199**	**22.6**	**–18**						**5**	**1**	**4**	**5**	**2**	**0**	**0**	**0**
1987-88	**Los Angeles**	**NHL**	**80**	**53**	**58**	**111**	**82**	**17**	**0**	**6**	**220**	**24.1**	**–9**						**5**	**2**	**5**	**7**	**18**	**2**	**0**	**1**
1988-89	**Los Angeles**	**NHL**	**78**	**46**	**52**	**98**	**65**	**10**	**0**	**4**	**237**	**19.4**	**5**						**11**	**2**	**6**	**8**	**10**	**0**	**0**	**1**
1989-90	**Los Angeles**	**NHL**	**80**	**52**	**49**	**101**	**38**	**20**	**0**	**7**	**210**	**24.8**	**8**						**10**	**5**	**5**	**10**	**12**	**1**	**0**	**1**
1990-91	**Los Angeles**	**NHL**	**76**	**45**	**46**	**91**	**68**	**11**	**0**	**5**	**229**	**19.7**	**28**						**12**	**12**	**4**	**16**	**22**	**5**	**0**	**2**
1991-92	**Los Angeles**	**NHL**	**80**	**44**	**63**	**107**	**95**	**26**	**0**	**6**	**240**	**18.3**	**–4**						**6**	**3**	**4**	**7**	**12**	**1**	**0**	**1**
1992-93	**Los Angeles**	**NHL**	**84**	**63**	**62**	**125**	**100**	**24**	**2**	**8**	**265**	**23.8**	**18**						**24**	**9**	**13**	**22**	**28**	**4**	**0**	**2**
1993-94	**Los Angeles**	**NHL**	**83**	**44**	**42**	**86**	**86**	**24**	**0**	**3**	**267**	**16.5**	**–20**													
1994-95	**Pittsburgh**	**NHL**	**46**	**23**	**19**	**42**	**37**	**5**	**0**	**3**	**109**	**21.1**	**10**						**12**	**7**	**4**	**11**	**26**	**0**	**0**	**2**
1995-96	**NY Rangers**	**NHL**	**77**	**23**	**46**	**69**	**80**	**11**	**0**	**4**	**223**	**10.3**	**13**						**11**	**1**	**5**	**6**	**8**	**0**	**0**	**0**
1996-97	**NY Rangers**	**NHL**	**69**	**24**	**24**	**48**	**48**	**5**	**0**	**4**	**200**	**12.0**	**16**						**15**	**4**	**7**	**11**	**4**	**0**	**0**	**0**
1997-98	**Los Angeles**	**NHL**	**57**	**16**	**24**	**40**	**66**	**5**	**0**	**7**	**130**	**12.3**	**5**						**4**	**1**	**2**	**3**	**6**	**0**	**0**	**0**
1998-99	**Los Angeles**	**NHL**	**82**	**39**	**35**	**74**	**54**	**11**	**0**	**7**	**292**	**13.4**	**–1**	**7**	**57.1**	**40**	**22**	**19:11**								
99-2000	**Los Angeles**	**NHL**	**71**	**36**	**38**	**74**	**68**	**13**	**0**	**7**	**221**	**16.3**	**11**	**10**	**30.0**	**29**	**12**	**18:34**	**4**	**2**	**2**	**4**	**6**	**0**	**0**	**0**
2000-01	**Los Angeles**	**NHL**	**82**	**37**	**51**	**88**	**66**	**16**	**1**	**4**	**235**	**15.7**	**10**	**12**	**33.3**	**42**	**11**	**18:42**	**13**	**4**	**3**	**7**	**10**	**1**	**0**	**1**
2001-02♦	**Detroit**	**NHL**	**81**	**30**	**20**	**50**	**38**	**13**	**0**	**5**	**190**	**15.8**	**–2**	**12**	**41.7**	**35**	**6**	**14:51**	**23**	**4**	**5**	**9**	**10**	**1**	**0**	**1**
	NHL Totals		**1205**	**620**	**668**	**1288**	**1019**	**229**	**3**	**83**	**3467**	**17.9**		**41**	**39.0**	**146**	**51**	**17:49**	**155**	**57**	**69**	**126**	**174**	**15**	**0**	**12**

QMJHL Second All-Star Team (1985) • QMJHL First All-Star Team (1986) • Canadian Major Junior Player of the Year (1986) • NHL All-Rookie Team (1987) • NHL Second All-Star Team (1987, 1992, 2001) • Won Calder Memorial Trophy (1987) • NHL First All-Star Team (1988, 1989, 1990, 1991, 1993) • Played in NHL All-Star Game (1988, 1989, 1990, 1991, 1992, 1993, 1999, 2001)

Traded to **Pittsburgh** by **LA Kings** for Rick Tocchet and Pittsburgh's 2nd round choice (Pavel Rosa) in 1995 Entry Draft, July 29, 1994. Traded to **NY Rangers** by **Pittsburgh** with Ulf Samuelsson for Petr Nedved and Sergei Zubov, August 31, 1995. Traded to **LA Kings** by **NY Rangers** for Kevin Stevens, August 28, 1997. Signed as a free agent by **Detroit**, July 5, 2001.

ROBITAILLE, Randy
(ROH-buh-tigh, RAN-dee) **PIT.**

Center. Shoots left. 5'11", 196 lbs. Born, Ottawa, Ont., October 12, 1975.

			Regular Season																Playoffs							
Season	Club	League	GP	G	A	Pts	PIM	PP	SH	GW	S	%	+/-	TF	F%	H	SB	Min	GP	G	A	Pts	PIM	PP	SH	GW
1993-94	Ottawa Jr. Sens	OCJHL	57	33	55	88	31																			
1994-95	Ottawa Jr. Sens	OCJHL	54	48	77	*125	111																			
1995-96	Miami-Ohio	CCHA	36	14	31	45	26																			
1996-97	Miami-Ohio	CCHA	39	27	34	61	44																			
	Boston	**NHL**	**1**	**0**	**0**	**0**	**0**	**0**	**0**	**0**	**0**	**0.0**	**0**													
1997-98	**Boston**	**NHL**	**4**	**0**	**0**	**0**	**0**	**0**	**0**	**0**	**5**	**0.0**	**–2**													
	Providence	AHL	48	15	29	44	16																			
1998-99	**Boston**	**NHL**	**4**	**0**	**2**	**2**	**0**	**0**	**0**	**0**	**5**	**0.0**	**–1**	**24**	**25.0**	**0**	**0**	**10:11**	**1**	**0**	**0**	**0**	**0**	**0**	**0**	**0**
	Providence	AHL	74	28	*74	102	34												19	6	*14	20	20			
99-2000	**Nashville**	**NHL**	**69**	**11**	**14**	**25**	**10**	**2**	**0**	**1**	**113**	**9.7**	**–13**	**528**	**51.5**	**30**	**14**	**12:52**								
2000-01	Milwaukee	IHL	19	10	23	33	4																			
	Nashville	**NHL**	**62**	**9**	**17**	**26**	**12**	**5**	**0**	**0**	**121**	**7.4**	**–11**	**481**	**48.4**	**30**	**13**	**14:10**								
2001-02	**Los Angeles**	**NHL**	**18**	**4**	**3**	**7**	**17**	**2**	**0**	**0**	**30**	**13.3**	**–9**	**60**	**65.0**	**5**	**3**	**12:54**								
	Manchester	AHL	6	7	3	10	0																			
	Pittsburgh	**NHL**	**40**	**10**	**20**	**30**	**16**	**3**	**0**	**1**	**91**	**11.0**	**–14**	**599**	**51.4**	**13**	**9**	**18:08**								
	NHL Totals		**198**	**34**	**56**	**90**	**55**	**12**	**0**	**2**	**365**	**9.3**		**1692**	**50.7**	**78**	**39**	**14:20**	**1**	**0**	**0**	**0**	**0**	**0**	**0**	**0**

OCJHL First All-Star Team (1995) • CCHA First All-Star Team (1997) • NCAA West First All-American Team (1997) • AHL First All-Star Team (1999) • Won Les Cunningham Award (MVP - AHL) (1999)

Signed as a free agent by **Boston**, March 27, 1997. Traded to **Atlanta** by **Boston** for Peter Ferraro, June 25, 1999. Traded to **Nashville** by **Atlanta** for Denny Lambert, August 16, 1999. Signed as a free agent by **LA Kings**, July 6, 2001. Claimed on waivers by **Pittsburgh** from **LA Kings**, January 4, 2002.

ROCHE, Dave
(ROHSH, DAYV) **N.J.**

Left wing. Shoots left. 6'4", 230 lbs. Born, Lindsay, Ont., June 13, 1975. Pittsburgh's 3rd choice, 62nd overall, in 1993 Entry Draft.

Season	Club	League	GP	G	A	Pts	PIM	PP	SH	GW	S	%	+/-	TF	F%	H	SB	Min	GP	G	A	Pts	PIM	PP	SH	GW
1990-91	Peterborough	OJHL-B	40	22	17	39	86																			
1991-92	Peterborough	OHL	62	10	17	27	134												10	0	0	0	34			
1992-93	Peterborough	OHL	56	40	60	100	105												21	14	15	29	42			
1993-94	Peterborough	OHL	34	15	22	37	127																			
	Windsor	OHL	29	14	20	34	73												4	1	1	2	15			
1994-95	Windsor	OHL	66	55	59	114	180												10	9	6	15	16			
1995-96	**Pittsburgh**	**NHL**	**71**	**7**	**7**	**14**	**130**	**0**	**0**	**1**	**65**	**10.8**	**–5**						**16**	**2**	**7**	**9**	**26**	**0**	**0**	**0**
1996-97	**Pittsburgh**	**NHL**	**61**	**5**	**5**	**10**	**155**	**2**	**0**	**0**	**53**	**9.4**	**–13**													
	Cleveland	IHL	18	5	5	10	25												13	6	3	9	*87			
1997-98	Syracuse Crunch	AHL	73	12	20	32	307												5	2	0	2	10			
1998-99	**Calgary**	**NHL**	**36**	**3**	**3**	**6**	**44**	**1**	**0**	**2**	**30**	**10.0**	**–1**	**0**	**0.0**	**35**	**9**	**7:15**								
	Saint John	AHL	7	0	3	3	6																			
99-2000	**Calgary**	**NHL**	**2**	**0**	**0**	**0**	**5**	**0**	**0**	**0**	**3**	**0.0**	**–1**	**0**	**0.0**	**5**	**0**	**8:33**								
	Saint John	AHL	67	22	21	43	130												3	0	1	1	8			
2000-01	Saint John	AHL	79	32	26	58	179												19	3	6	9	43			
2001-02	Cincinnati	AHL	29	6	7	13	41																			
	NY Islanders	**NHL**	**1**	**0**	**0**	**0**	**0**	**0**	**0**	**0**	**0**	**0.0**	**0**	**1**	**0.0**	**0**	**0**	**2:51**								
	Bridgeport	AHL	48	25	14	39	64												20	3	0	3	20			
	NHL Totals		**171**	**15**	**15**	**30**	**334**	**3**	**0**	**3**	**151**	**9.9**		**1**	**0.0**	**40**	**9**	**7:12**	**16**	**2**	**7**	**9**	**26**	**0**	**0**	**0**

OHL First All-Star Team (1995)

Traded to **Calgary** by **Pittsburgh** with Ken Wregget for German Titov and Todd Hlushko, June 17, 1998. Signed as a free agent by **NY Islanders**, August 17, 2001. Traded to **Anaheim** by **NY Islanders** for Jim Cummins, January 14, 2002. Traded to **NY Islanders** by **Anaheim** for Ben Guite and the rights to Bjorn Mellin, March 19, 2002. Signed as a free agent by **New Jersey**, August, 2002.

ROCHE, Travis
(ROHSH, TRA-vihs) **MIN.**

Defense. Shoots right. 6'1", 190 lbs. Born, Grand Cache, Alta, June 17, 1978.

Season	Club	League	GP	G	A	Pts	PIM	PP	SH	GW	S	%	+/-	TF	F%	H	SB	Min	GP	G	A	Pts	PIM	PP	SH	GW
1996-97	Trail	BCHL	49	17	40	57	159																			
1997-98	Trail	BCHL	38	11	31	42	104												11	0	8	8	21			
1998-99	North Dakota	WCHA	DID NOT PLAY – FRESHMAN																							
99-2000	North Dakota	WCHA	42	6	22	28	60																			
2000-01	North Dakota	WCHA	42	11	38	49	42																			
	Minnesota	**NHL**	**1**	**0**	**0**	**0**	**0**	**0**	**0**	**0**	**0**	**0.0**	**0**	**0**	**0.0**	**0**	**1**	**15:22**								
2001-02	**Minnesota**	**NHL**	**4**	**0**	**0**	**0**	**2**	**0**	**0**	**0**	**1**	**0.0**	**–1**	**0**	**0.0**	**3**	**3**	**12:30**								
	Houston Aeros	AHL	60	13	21	34	107												12	2	3	5	6			
	NHL Totals		**5**	**0**	**0**	**0**	**2**	**0**	**0**	**0**	**1**	**0.0**		**0**	**0.0**	**3**	**4**	**13:04**								

BCHL Second All-Star Team (1997) • Won BCHL Rookie of the Year Award (1997) • Won BCHL Playoff MVP Award (1997) • BCHL First All-Star Team (1998) • Won BCHL Best Defenseman Award (1998) • WCHA All-Rookie Team (2000) • WCHA First All-Star Team (2001) • NCAA West First All-American Team (2001) • NCAA Championship All-Tournament Team (2001)

Signed as a free agent by **Minnesota**, April 8, 2001.

ROENICK, Jeremy
(ROH-nihk, JAIR-eh-mee) **PHI.**

Center. Shoots right. 6'1", 207 lbs. Born, Boston, MA, January 17, 1970. Chicago's 1st choice, 8th overall, in 1988 Entry Draft.

Season	Club	League	GP	G	A	Pts	PIM	PP	SH	GW	S	%	+/-	TF	F%	H	SB	Min	GP	G	A	Pts	PIM	PP	SH	GW
1986-87	Thayer Academy	Hi-School	24	31	34	65	....																			
1987-88	Thayer Academy	Hi-School	24	34	50	84	....																			
1988-89	Hull Olympiques	QMJHL	28	34	36	70	14																			
	Chicago	**NHL**	**20**	**9**	**9**	**18**	**4**	**2**	**0**	**0**	**52**	**17.3**	**4**						**10**	**1**	**3**	**4**	**7**	**1**	**0**	**1**
1989-90	**Chicago**	**NHL**	**78**	**26**	**40**	**66**	**54**	**6**	**0**	**4**	**173**	**15.0**	**2**						**20**	**11**	**7**	**18**	**8**	**4**	**0**	**1**
1990-91	**Chicago**	**NHL**	**79**	**41**	**53**	**94**	**80**	**15**	**4**	**10**	**194**	**21.1**	**38**						**6**	**3**	**5**	**8**	**4**	**1**	**0**	**1**
1991-92	**Chicago**	**NHL**	**80**	**53**	**50**	**103**	**98**	**22**	**3**	**13**	**234**	**22.6**	**23**						**18**	**12**	**10**	**22**	**12**	**4**	**0**	**3**
1992-93	**Chicago**	**NHL**	**84**	**50**	**57**	**107**	**86**	**22**	**3**	**3**	**255**	**19.6**	**15**						**4**	**1**	**2**	**3**	**2**	**0**	**0**	**0**
1993-94	**Chicago**	**NHL**	**84**	**46**	**61**	**107**	**125**	**24**	**5**	**5**	**281**	**16.4**	**21**						**6**	**1**	**6**	**7**	**2**	**0**	**0**	**1**
1994-95	Kolner Haie	Germany	3	3	1	4	2																			
	Chicago	**NHL**	**33**	**10**	**24**	**34**	**14**	**5**	**0**	**1**	**93**	**10.8**	**5**						**8**	**1**	**2**	**3**	**16**	**0**	**0**	**0**
1995-96	**Chicago**	**NHL**	**66**	**32**	**35**	**67**	**109**	**12**	**4**	**2**	**171**	**18.7**	**9**						**10**	**5**	**7**	**12**	**2**	**1**	**0**	**1**
1996-97	**Phoenix**	**NHL**	**72**	**29**	**40**	**69**	**115**	**10**	**3**	**7**	**228**	**12.7**	**–7**						**6**	**2**	**4**	**6**	**4**	**0**	**0**	**0**
1997-98	**Phoenix**	**NHL**	**79**	**24**	**32**	**56**	**103**	**6**	**1**	**3**	**182**	**13.2**	**5**						**6**	**5**	**3**	**8**	**4**	**2**	**2**	**2**
	United States	Olympics	4	0	1	1	6																			
1998-99	**Phoenix**	**NHL**	**78**	**24**	**48**	**72**	**130**	**4**	**0**	**3**	**203**	**11.8**	**7**	**956**	**47.6**	**154**	**30**	**20:10**	**1**	**0**	**0**	**0**	**0**	**0**	**0**	**0**
99-2000	**Phoenix**	**NHL**	**75**	**34**	**44**	**78**	**102**	**6**	**3**	**12**	**192**	**17.7**	**11**	**925**	**50.1**	**125**	**28**	**20:51**	**5**	**2**	**2**	**4**	**10**	**1**	**0**	**0**
2000-01	**Phoenix**	**NHL**	**80**	**30**	**46**	**76**	**114**	**13**	**0**	**7**	**192**	**15.6**	**–1**	**888**	**49.1**	**135**	**25**	**20:60**								
2001-02	**Philadelphia**	**NHL**	**75**	**21**	**46**	**67**	**74**	**5**	**0**	**3**	**167**	**12.6**	**32**	**1329**	**49.1**	**151**	**29**	**18:14**	**5**	**0**	**0**	**0**	**14**	**0**	**0**	**0**
	United States	Olympics	6	1	4	5	2																			
	NHL Totals		**983**	**429**	**585**	**1014**	**1208**	**152**	**26**	**73**	**2617**	**16.4**		**4098**	**49.0**	**565**	**112**	**20:05**	**105**	**44**	**51**	**95**	**85**	**14**	**2**	**10**

• QMJHL Second All-Star Team (1989) • Played in NHL All-Star Game (1991, 1992, 1993, 1994, 1999, 2000, 2002)

Traded to **Phoenix** by **Chicago** for Alexei Zhamnov, Craig Mills and Phoenix's 1st round choice (Ty Jones) in 1997 Entry Draft, August 16, 1996. Signed as a free agent by **Philadelphia**, July 2, 2001.

ROEST, Stacy
(ROOST, STAY-see)

Center. Shoots right. 5'9", 185 lbs. Born, Lethbridge, Alta., March 15, 1974.

Season	Club	League	GP	G	A	Pts	PIM	PP	SH	GW	S	%	+/-	TF	F%	H	SB	Min	GP	G	A	Pts	PIM	PP	SH	GW
1990-91	Lethbridge	AMHL	34	22	50	72	38																			
	Medicine Hat	WHL	5	1	2	3	0												12	5	5	10	4			
1991-92	Medicine Hat	WHL	72	22	43	65	20												4	2	1	3	0			
1992-93	Medicine Hat	WHL	72	33	73	106	30												10	3	10	13	6			
1993-94	Medicine Hat	WHL	72	48	72	120	48												3	1	0	1	4			
1994-95	Medicine Hat	WHL	69	37	78	115	32												5	2	7	9	2			
	Adirondack	AHL	3	0	0	0	0																			
1995-96	Adirondack	AHL	76	16	39	55	40												3	0	0	0	0			
1996-97	Adirondack	AHL	78	25	41	66	30												4	1	1	2	0			
1997-98	Adirondack	AHL	80	34	58	92	30												3	2	1	3	6			
1998-99	**Detroit**	**NHL**	**59**	**4**	**8**	**12**	**14**	**0**	**0**	**1**	**50**	**8.0**	**–7**	**234**	**57.3**	**72**	**7**	**8:04**								
	Adirondack	AHL	2	0	1	1	0																			
99-2000	**Detroit**	**NHL**	**49**	**7**	**9**	**16**	**12**	**1**	**0**	**1**	**56**	**12.5**	**–1**	**397**	**55.4**	**54**	**12**	**9:56**	**3**	**0**	**0**	**0**	**0**	**0**	**0**	**0**

			Regular Season																Playoffs							
Season	Club	League	GP	G	A	Pts	PIM	PP	SH	GW	S	%	+/-	TF	F%	H	SB	Min	GP	G	A	Pts	PIM	PP	SH	GW
2000-01	Minnesota	NHL	76	7	20	27	20	1	0	1	125	5.6	3	731	57.1	51	16	14:01								
2001-02	Minnesota	NHL	58	10	11	21	8	1	4	2	98	10.2	–3	504	55.0	33	25	15:50								
	NHL Totals		242	28	48	76	54	3	4	5	329	8.5		1866	56.2	210	60	12:10	3	0	0	0	0	0	0	0

WHL East First All-Star Team (1994) • WHL East Second All-Star Team (1995)

Signed as a free agent by **Detroit**, June 9, 1997. Selected by **Minnesota** from **Detroit** in Expansion Draft, June 23, 2000.

ROHLOFF, Todd

(ROH-lawf, TAWD) **WSH.**

Defense. Shoots left. 6'3", 213 lbs. Born, Grand Rapids, IL, January 16, 1974.

Season	Club	League	GP	G	A	Pts	PIM	PP	SH	GW	S	%	+/-	TF	F%	H	SB	Min	GP	G	A	Pts	PIM	PP	SH	GW
1993-94	St. Paul Vulcans	USHL	47	4	22	26																				
1994-95	Miami-Ohio	CCHA	38	1	6	7	22																			
1995-96	Miami-Ohio	CCHA	23	2	4	6	24																			
1996-97	Miami-Ohio	CCHA	38	2	12	14	48																			
1997-98	Miami-Ohio	CCHA	17	2	5	7	38																			
	Indianapolis Ice	IHL	5	0	1	1	6												1	0	0	0	0			
1998-99	Portland Pirates	AHL	58	1	6	7	58																			
	Indianapolis Ice	IHL	12	2	0	2	8												5	1	1	2	6			
99-2000	Cleveland	IHL	77	1	13	14	88												9	0	0	0	6			
2000-01	Portland Pirates	AHL	58	3	8	11	59												3	0	0	0	2			
2001-02	Washington	NHL	16	0	1	1	14	0	0	0	6	0.0	–2	0	0.0	24	14	14:11								
	Portland Pirates	AHL	17	1	3	4	22																			
	NHL Totals		16	0	1	1	14	0	0	0	6	0.0		0	0.0	24	14	14:11								

Signed as a free agent by **Chicago**, March 24, 1998. Signed as a free agent by **Washington**, July 21, 2000. • Missed majority of 2001-02 season recovering from ankle injury suffered in off-season, September, 2001.

ROLSTON, Brian

(ROHL-stuhn, BRIGH-uhn) **BOS.**

Center. Shoots left. 6'2", 205 lbs. Born, Flint, MI, February 21, 1973. New Jersey's 2nd choice, 11th overall, in 1991 Entry Draft.

Season	Club	League	GP	G	A	Pts	PIM	PP	SH	GW	S	%	+/-	TF	F%	H	SB	Min	GP	G	A	Pts	PIM	PP	SH	GW
1989-90	Det. Compuware	NAJHL	40	36	37	73	57																			
1990-91	Det. Compuware	NAJHL	36	49	46	95	14																			
1991-92	Lake Superior	CCHA	37	14	23	37	14																			
1992-93	Lake Superior	CCHA	39	33	31	64	20																			
1993-94	Team USA	Nat-Tm	41	20	28	48	36																			
	United States	Olympics	8	7	0	7	8																			
	Albany	AHL	17	5	5	10	8												5	1	2	3	0			
1994-95	Albany	AHL	18	9	11	20	10																			
	♦ New Jersey	NHL	40	7	11	18	17	2	0	3	92	7.6	5						6	2	1	3	4	1	0	0
1995-96	New Jersey	NHL	58	13	11	24	8	3	1	4	139	9.4	9													
1996-97	New Jersey	NHL	81	18	27	45	20	2	2	3	237	7.6	6						10	4	1	5	6	1	2	0
1997-98	New Jersey	NHL	76	16	14	30	16	0	2	1	185	8.6	7						6	1	0	1	2	0	1	0
1998-99	New Jersey	NHL	82	24	33	57	14	5	5	3	210	11.4	11	51	45.1	70	20	18:49	7	1	0	1	2	0	1	0
99-2000	New Jersey	NHL	11	3	1	4	0	1	0	2	33	9.1	–2	37	37.8	13	2	19:09								
	Colorado	NHL	50	8	10	18	12	1	0	3	107	7.5	–6	65	41.5	28	19	16:18								
	Boston	NHL	16	5	4	9	6	3	0	1	66	7.6	–4	265	41.1	22	11	22:13								
2000-01	Boston	NHL	77	19	39	58	28	5	0	4	286	6.6	6	666	45.7	71	31	19:19								
2001-02	Boston	NHL	82	31	31	62	30	6	9	7	331	9.4	11	1289	46.6	63	27	20:24	6	4	1	5	0	1	1	0
	United States	Olympics	6	0	3	3	0																			
	NHL Totals		573	144	181	325	151	28	19	31	1686	8.5		2373	45.4	267	110	19:08	35	12	3	15	14	3	5	0

NCAA Championship All-Tournament Team (1992, 1993) • CCHA First All-Star Team (1993) • NCAA West Second All-American Team (1993)

Traded to **Colorado** by **New Jersey** with New Jersey's 1st round choice (later traded to Boston - Boston selected Martin Samuelsson) in 2000 Entry Draft for Claude Lemieux and Colorado's 1st (David Hale) and 2nd (Matt DeMarchi) round choices in 2000 Entry Draft, November 3, 1999. Traded to **Boston** by **Colorado** with Martin Grenier, Sami Pahlsson and New Jersey's 1st round choice (previously acquired, Boston selected Martin Samuelsson) in 2000 Entry Draft for Raymond Bourque and Dave Andreychuk, March 6, 2000.

RONNING, Cliff

(RAWN-ihng, KLIHF) **MIN.**

Center. Shoots left. 5'8", 165 lbs. Born, Burnaby, B.C., October 1, 1965. St. Louis' 9th choice, 134th overall, in 1984 Entry Draft.

Season	Club	League	GP	G	A	Pts	PIM	PP	SH	GW	S	%	+/-	TF	F%	H	SB	Min	GP	G	A	Pts	PIM	PP	SH	GW
1982-83	New Westminster	BCJHL	52	83	68	151	22																			
1983-84	New Westminster	WHL	71	69	67	136	10												9	8	13	21	10			
1984-85	New Westminster	WHL	70	*89	108	*197	20												11	10	14	24	4			
1985-86	Team Canada	Nat-Tm	71	*55	*63	*118	53																			
	St. Louis	NHL																	5	1	1	2	2	1	0	0
1986-87	Team Canada	Nat-Tm	26	17	16	33	12																			
	St. Louis	NHL	42	11	14	25	6	2	0	2	68	16.2	–1						4	0	1	1	0	0	0	0
1987-88	St. Louis	NHL	26	5	8	13	12	1	0	1	38	13.2	6													
1988-89	St. Louis	NHL	64	24	31	55	18	16	0	1	150	16.0	3						7	1	3	4	0	1	0	0
	Peoria Rivermen	IHL	12	11	20	31	8																			
1989-90	HC Asiago	Italy	36	67	49	116	25												6	7	12	19	4			
1990-91	St. Louis	NHL	48	14	18	32	10	5	0	2	81	17.3	2													
	Vancouver	NHL	11	6	6	12	0	2	0	0	32	18.8	–2						6	6	3	9	12	2	0	2
1991-92	Vancouver	NHL	80	24	47	71	42	6	0	2	216	11.1	18						13	8	5	13	6	1	0	1
1992-93	Vancouver	NHL	79	29	56	85	30	10	0	2	209	13.9	19						12	2	9	11	6	0	0	0
1993-94	Vancouver	NHL	76	25	43	68	42	10	0	4	197	12.7	7						24	5	10	15	16	2	0	2
1994-95	Vancouver	NHL	41	6	19	25	27	3	0	2	93	6.5	–4						11	3	5	8	2	1	0	2
1995-96	Vancouver	NHL	79	22	45	67	42	5	0	1	187	11.8	16						6	0	2	2	6	0	0	0
1996-97	Phoenix	NHL	69	19	32	51	26	8	0	2	171	11.1	–9						7	0	7	7	12	0	0	0
1997-98	Phoenix	NHL	80	11	44	55	36	3	0	0	197	5.6	5						6	1	3	4	4	0	0	0
1998-99	Phoenix	NHL	7	2	5	7	2	2	0	1	18	11.1	3	81	51.9	0	1	15:22								
	Nashville	NHL	72	18	35	53	40	8	0	3	239	7.5	–6	1129	47.7	11	35	19:42								
99-2000	Nashville	NHL	82	26	36	62	34	7	0	2	248	10.5	–13	611	45.7	8	20	18:11								
2000-01	Nashville	NHL	80	19	43	62	28	6	0	4	237	8.0	4	331	45.0	13	13	17:23								
2001-02	Nashville	NHL	67	18	31	49	24	4	0	0	164	11.0	0	98	51.0	6	20	16:44								
	Los Angeles	NHL	14	1	4	5	8	1	0	0	35	2.9	0	14	50.0	6	4	16:53	4	0	1	1	2	0	0	0
	NHL Totals		1017	280	517	797	427	99	0	29	2580	10.9		2264	47.1	44	93	17:54	105	27	50	77	68	8	0	7

BCJHL Coastal Division First All-Star Team (1983) • WHL Rookie of the Year (1984) • WHL West First All-Star Team (1985) • WHL MVP (1985)

Traded to **Vancouver** by **St. Louis** with Geoff Courtnall, Robert Dirk, Sergio Momesso and St. Louis' 5th round choice (Brian Loney) in 1992 Entry Draft for Dan Quinn and Garth Butcher, March 5, 1991. Signed as a free agent by **Phoenix**, July 1, 1996. Traded to **Nashville** by **Phoenix** with Richard Lintner for future considerations, October 31, 1998. Traded to **LA Kings** by **Nashville** for Jere Karalahti and a conditional choice in 2003 Entry Draft, March 16, 2002. Traded to **Minnesota** by **LA Kings** for Minnesota's 4th round choice (Aaron Rome) in 2002 Entry Draft, June 22, 2002.

RONNQVIST, Jonas

(RAWN-kvihst, YOH-nuhs)

Right wing. Shoots right. 6'2", 200 lbs. Born, Kalix, Sweden, August 22, 1973. Anaheim's 3rd choice, 98th overall, in 2000 Entry Draft.

Season	Club	League	GP	G	A	Pts	PIM	PP	SH	GW	S	%	+/-	TF	F%	H	SB	Min	GP	G	A	Pts	PIM	PP	SH	GW
1991-92	Bodens IK	Swede-2	6	1	1	2	2												1	0	0	0	0			
1992-93	Bodens IK	Swede-2	35	10	4	14	16																			
1993-94	Bodens IK	Swede-2	34	15	10	25	24												9	1	1	2	6			
1994-95	Bodens IK	Swede-2	35	10	15	5	48												8	1	0	1	0			
1995-96	Bodens IK	Swede-2	26	12	10	22	26																			
1996-97	Bodens IK	Swede-2	32	15	14	29	48																			
1997-98	Lulea HF	Sweden	40	6	8	14	24												3	0	0	0	2			
	Lulea HF	EuroHL	5	0	1	1	0																			
1998-99	Lulea HF	Sweden	41	5	8	13	30												3	0	2	2	29			
99-2000	Lulea HF	Sweden	49	15	24	39	42												8	3	3	6	4			
2000-01	Anaheim	NHL	38	0	4	4	14	0	0	0	30	0.0	–7	101	44.6	19	6	10:37								
	Cincinnati	AHL	13	3	2	5	6																			
2001-02	Cincinnati	AHL	74	10	18	28	30												3	0	0	0	0			
	NHL Totals		38	0	4	4	14	0	0	0	30	0.0		101	44.6	19	6	10:37								

			Regular Season																Playoffs							
Season	Club	League	GP	G	A	Pts	PIM	PP	SH	GW	S	%	+/-	TF	F%	H	SB	Min	GP	G	A	Pts	PIM	PP	SH	GW

ROSA, Pavel
(ROHZA, PAH-vehl) **L.A.**

Right wing. Shoots right. 6', 195 lbs. Born, Most, Czech., June 7, 1977. Los Angeles' 3rd choice, 50th overall, in 1995 Entry Draft.

Season	Club	League	GP	G	A	Pts	PIM	PP	SH	GW	S	%	+/-	TF	F%	H	SB	Min	GP	G	A	Pts	PIM	PP	SH	GW
1994-95	Litvinov Jr.	Czech-Jr.	40	56	42	98																				
	Litvinov	Czech	2	0	0	0	0												1	0	0	0	0			
1995-96	Hull Olympiques	QMJHL	61	46	70	116	39												18	14	22	36	25			
1996-97	Hull Olympiques	QMJHL	68	*63	*90	*153	66												14	18	13	31	16			
1997-98	Fredericton	AHL	1	0	0	0	0																			
	Long Beach	IHL	2	0	1	1	0												1	1	1	2	0			
1998-99	**Los Angeles**	**NHL**	**29**	**4**	**12**	**16**	**6**	**0**	**0**	**0**	**61**	**6.6**	**0**	**0**	**0.0**	**24**	**6**	**13:34**								
	Long Beach	IHL	31	17	13	30	28												6	1	2	3	0			
99-2000	**Los Angeles**	**NHL**	**3**	**0**	**0**	**0**	**0**	**0**	**0**	**0**	**1**	**0.0**	**–1**	**0**	**0.0**	**6**	**1**	**11:21**								
	Long Beach	IHL	74	22	31	53	76												6	2	2	4	4			
2000-01	HPK Hameenlinna	Finland	54	25	25	50	53																			
2001-02	Jokerit Helsinki	Finland	46	21	22	43	37												12	3	5	8	18			
	NHL Totals		**32**	**4**	**12**	**16**	**6**	**0**	**0**	**0**	**62**	**6.5**		**0**	**0.0**	**30**	**7**	**13:22**								

QMJHL All-Rookie Team (1996) • QMJHL Offensive Rookie of the Year (1996) • QMJHL First All-Star Team (1997) • Canadian Major Junior First All-Star Team (1997)
• Missed majority of 1997-98 season recovering from head injury originally suffered in training camp, September, 1997.

ROSSITER, Kyle
(RAWS-ih-tuhr, KIGHL) **FLA.**

Defense. Shoots left. 6'3", 217 lbs. Born, Edmonton, Alta., June 9, 1980. Florida's 1st choice, 30th overall, in 1998 Entry Draft.

Season	Club	League	GP	G	A	Pts	PIM	PP	SH	GW	S	%	+/-	TF	F%	H	SB	Min	GP	G	A	Pts	PIM	PP	SH	GW
1995-96	Edmonton SSAC	AMHL	34	5	19	24	116																			
1996-97	Spokane Chiefs	WHL	50	0	2	2	65												9	0	0	0	6			
1997-98	Spokane Chiefs	WHL	61	6	16	22	190												15	0	3	3	28			
1998-99	Spokane Chiefs	WHL	71	4	17	21	206																			
99-2000	Spokane Chiefs	WHL	63	11	22	33	155												15	1	4	5	25			
2000-01	Louisville	AHL	78	2	5	7	110																			
2001-02	**Florida**	**NHL**	**2**	**0**	**0**	**0**	**2**	**0**	**0**	**0**	**0**	**0.0**	**–1**	**0**	**0.0**	**4**	**2**	**15:27**								
	Utah Grizzlies	AHL	74	3	7	10	88												5	0	1	1	0			
	NHL Totals		**2**	**0**	**0**	**0**	**2**	**0**	**0**	**0**	**0**	**0.0**		**0**	**0.0**	**4**	**2**	**15:27**								

Canadian Major Junior Scholastic Player of the Year (1998)

ROY, Andre
(WAH, AHN-dray) **T.B.**

Left wing. Shoots left. 6'4", 213 lbs. Born, Port Chester, NY, February 8, 1975. Boston's 5th choice, 151st overall, in 1994 Entry Draft.

Season	Club	League	GP	G	A	Pts	PIM	PP	SH	GW	S	%	+/-	TF	F%	H	SB	Min	GP	G	A	Pts	PIM	PP	SH	GW
1992-93	Nord Selects	QAHA	STATISTICS NOT AVAILABLE																							
1993-94	Goulbourn Royals	OJHL-C	9	9	13	22	98																			
	Beauport	QMJHL	33	6	7	13	125																			
	Chicoutimi	QMJHL	32	4	14	18	152												25	3	6	9	94			
1994-95	Chicoutimi	QMJHL	20	15	8	23	90																			
	Drummondville	QMJHL	34	18	13	31	233												4	2	0	2	34			
1995-96	**Boston**	**NHL**	**3**	**0**	**0**	**0**	**0**	**0**	**0**	**0**	**0**	**0.0**	**0**													
	Providence	AHL	58	7	8	15	167												1	0	0	0	10			
1996-97	**Boston**	**NHL**	**10**	**0**	**2**	**2**	**12**	**0**	**0**	**0**	**12**	**0.0**	**–5**													
	Providence	AHL	50	17	11	28	234																			
1997-98	Providence	AHL	36	3	11	14	154																			
	Charlotte	ECHL	27	10	8	18	132												7	2	3	5	34			
1998-99	Fort Wayne	IHL	65	15	6	21	*395												2	0	0	0	11			
99-2000	**Ottawa**	**NHL**	**73**	**4**	**3**	**7**	**145**	**0**	**0**	**1**	**39**	**10.3**	**3**	**3**	**33.3**	**109**	**5**	**6:29**	**5**	**0**	**0**	**0**	**2**	**0**	**0**	**0**
2000-01	**Ottawa**	**NHL**	**64**	**3**	**5**	**8**	**169**	**0**	**0**	**0**	**33**	**9.1**	**1**	**2**	**50.0**	**96**	**3**	**4:36**	**2**	**0**	**0**	**0**	**16**	**0**	**0**	**0**
2001-02	**Ottawa**	**NHL**	**56**	**6**	**8**	**14**	**148**	**0**	**0**	**0**	**60**	**10.0**	**3**	**1**	**100.0**	**95**	**6**	**8:23**								
	Tampa Bay	**NHL**	**9**	**1**	**1**	**2**	**63**	**0**	**0**	**0**	**6**	**16.7**	**–5**	**0**	**0.0**	**20**	**0**	**8:58**								
	NHL Totals		**215**	**14**	**19**	**33**	**537**	**0**	**0**	**1**	**150**	**9.3**		**6**	**50.0**	**320**	**14**	**6:31**	**7**	**0**	**0**	**0**	**18**	**0**	**0**	**0**

Signed as a free agent by **Ottawa**, April 28, 1999. Traded to **Tampa Bay** by **Ottawa** with Ottawa's 6th round choice (Paul Ranger) in 2002 Entry Draft for Juha Ylonen, March 15, 2002.

ROYER, Gaetan
(ROI-ay, GAY-tan)

Right wing. Shoots right. 6'3", 210 lbs. Born, Donnacona, Que., March 13, 1976.

Season	Club	League	GP	G	A	Pts	PIM	PP	SH	GW	S	%	+/-	TF	F%	H	SB	Min	GP	G	A	Pts	PIM	PP	SH	GW
1994-95	Sherbrooke	QMJHL	65	11	25	36	194												7	0	2	2	6			
1995-96	Sherbrooke	QMJHL	36	25	26	51	174																			
	Beauport	QMJHL	25	11	10	21	59												19	5	9	14	47			
1996-97	Jacksonville	ECHL	28	7	8	15	149																			
	Indianapolis Ice	IHL	29	2	4	6	60																			
1997-98	Grand Rapids	IHL	52	12	6	18	177																			
1998-99	Saint John	AHL	15	1	0	1	36												7	0	1	1	8			
99-2000	Michigan K-Wings	IHL	20	6	2	8	64																			
2000-01	Saint John	AHL	58	8	4	12	134												14	0	0	0	16			
2001-02	Muskegon Fury	UHL	1	0	0	0	7																			
	Tampa Bay	**NHL**	**3**	**0**	**0**	**0**	**2**	**0**	**0**	**0**	**0**	**0.0**	**–1**	**0**	**0.0**	**4**	**0**	**4:57**								
	Pensacola	ECHL	33	13	10	23	129																			
	Springfield	AHL	40	6	4	10	112																			
	NHL Totals		**3**	**0**	**0**	**0**	**2**	**0**	**0**	**0**	**0**	**0.0**		**0**	**0.0**	**4**	**0**	**4:57**								

Signed as a free agent by **Calgary**, September 12, 2000. Signed as a free agent by **Tampa Bay**, October 23, 2001.

ROZSIVAL, Michal
(roh-ZIH-vahl, mee-KHUHL) **PIT.**

Defense. Shoots right. 6'1", 208 lbs. Born, Vlasim, Czech., September 3, 1978. Pittsburgh's 5th choice, 105th overall, in 1996 Entry Draft.

Season	Club	League	GP	G	A	Pts	PIM	PP	SH	GW	S	%	+/-	TF	F%	H	SB	Min	GP	G	A	Pts	PIM	PP	SH	GW
1994-95	Dukla Jihlava Jr.	Czech-Jr.	31	8	13	21																				
1995-96	HC Dukla Jihlava	Czech	36	3	4	7																				
1996-97	Swift Current	WHL	63	8	31	39	80												10	0	6	6	15			
1997-98	Swift Current	WHL	71	14	55	69	122												12	0	5	5	33			
1998-99	Syracuse Crunch	AHL	49	3	22	25	72																			
99-2000	**Pittsburgh**	**NHL**	**75**	**4**	**17**	**21**	**48**	**1**	**0**	**1**	**73**	**5.5**	**11**	**1**	**0.0**	**124**	**72**	**19:01**	**2**	**0**	**0**	**0**	**4**	**0**	**0**	**0**
2000-01	**Pittsburgh**	**NHL**	**30**	**1**	**4**	**5**	**26**	**0**	**0**	**0**	**17**	**5.9**	**3**	**1**	**100.0**	**37**	**38**	**17:06**								
	Wilkes-Barre	AHL	29	8	8	16	32												21	3	*19	22	23			
2001-02	**Pittsburgh**	**NHL**	**79**	**9**	**20**	**29**	**47**	**4**	**0**	**4**	**89**	**10.1**	**–6**	**0**	**0.0**	**89**	**116**	**20:01**								
	NHL Totals		**184**	**14**	**41**	**55**	**121**	**5**	**0**	**5**	**179**	**7.8**		**2**	**50.0**	**250**	**226**	**19:08**	**2**	**0**	**0**	**0**	**4**	**0**	**0**	**0**

WHL East First All-Star Team (1998)

RUCCHIN, Steve
(ROO-chihn, STEEV) **ANA.**

Center. Shoots left. 6'2", 211 lbs. Born, Thunder Bay, Ont., July 4, 1971. Anaheim's 1st choice, 2nd overall, in 1994 Supplemental Draft.

Season	Club	League	GP	G	A	Pts	PIM	PP	SH	GW	S	%	+/-	TF	F%	H	SB	Min	GP	G	A	Pts	PIM	PP	SH	GW
1989-90	Banting High	Hi-School	STATISTICS NOT AVAILABLE																							
	Thamesford	OJHL-D	2	1	2	3	0																			
1990-91	Western Ontario	OUAA	34	13	16	29	14																			
1991-92	Western Ontario	OUAA	37	28	34	62	36																			
1992-93	Western Ontario	OUAA	34	22	26	48	16																			
1993-94	Western Ontario	OUAA	35	30	23	53	30																			
1994-95	San Diego Gulls	IHL	41	11	15	26	14																			
	Anaheim	**NHL**	**43**	**6**	**11**	**17**	**23**	**0**	**0**	**1**	**59**	**10.2**	**7**													
1995-96	**Anaheim**	**NHL**	**64**	**19**	**25**	**44**	**12**	**8**	**1**	**4**	**113**	**16.8**	**3**													
1996-97	**Anaheim**	**NHL**	**79**	**19**	**48**	**67**	**24**	**6**	**1**	**2**	**153**	**12.4**	**26**						**8**	**1**	**2**	**3**	**10**	**0**	**0**	**0**
1997-98	**Anaheim**	**NHL**	**72**	**17**	**36**	**53**	**13**	**8**	**1**	**3**	**131**	**13.0**	**8**													
1998-99	**Anaheim**	**NHL**	**69**	**23**	**39**	**62**	**22**	**5**	**1**	**5**	**145**	**15.9**	**11**	**1845**	**52.3**	**43**	**72**	**22:33**	**4**	**0**	**3**	**3**	**0**	**0**	**0**	**0**
99-2000	**Anaheim**	**NHL**	**71**	**19**	**38**	**57**	**16**	**10**	**0**	**2**	**131**	**14.5**	**9**	**1996**	**53.4**	**69**	**69**	**22:12**								

			Regular Season																Playoffs							
Season	Club	League	GP	G	A	Pts	PIM	PP	SH	GW	S	%	+/-	TF	F%	H	SB	Min	GP	G	A	Pts	PIM	PP	SH	GW
2000-01	**Anaheim**	**NHL**	**16**	**3**	**5**	**8**	**0**	**2**	**0**	**0**	**19**	**15.8**	**–5**	**289**	**51.6**	**13**	**5**	**18:47**								
2001-02	**Anaheim**	**NHL**	**38**	**7**	**16**	**23**	**6**	**4**	**0**	**1**	**57**	**12.3**	**–3**	**808**	**52.2**	**28**	**18**	**19:17**								
	NHL Totals		**452**	**113**	**218**	**331**	**116**	**43**	**4**	**18**	**808**	**14.0**		**4938**	**52.7**	**153**	**164**	**21:28**	**12**	**1**	**5**	**6**	**10**	**0**	**0**	**0**

• Missed majority of 2000-01 season recovering from jaw injury originally suffered in game vs. Colorado, November 15, 2000. • Missed majority of 2001-02 season recovering from leg injury suffered in game vs. San Jose, November 16, 2001.

RUCINSKI, Mike

(roo-SIHN-skee, MIGHK)

Defense. Shoots left. 5'11", 179 lbs. Born, Trenton, MI, March 30, 1975. Hartford's 8th choice, 217th overall, in 1995 Entry Draft.

Season	Club	League	GP	G	A	Pts	PIM	PP	SH	GW	S	%	+/-	TF	F%	H	SB	Min	GP	G	A	Pts	PIM	PP	SH	GW
1991-92	Det. Caesars	MNHL	29	4	15	19	38																			
1992-93	Detroit	OHL	66	6	13	19	59												15	0	4	4	12			
1993-94	Detroit	OHL	66	2	26	28	58												17	0	7	7	15			
1994-95	Detroit	OHL	64	9	18	27	61												21	3	3	6	8			
1995-96	Detroit	OHL	51	10	26	36	65												11	2	4	6	14			
1996-97	Richmond	ECHL	61	20	23	43	85												8	2	6	8	18			
	Springfield	AHL	6	0	1	1	0																			
1997-98	**Carolina**	**NHL**	**9**	**0**	**1**	**1**	**2**	**0**	**0**	**0**	**3**	**0.0**	**0**													
	New Haven	AHL	65	5	17	22	50												1	0	0	0	0			
	Cleveland	IHL	2	0	0	0	4																			
1998-99	**Carolina**	**NHL**	**15**	**0**	**1**	**1**	**8**	**0**	**0**	**0**	**8**	**0.0**	**1**	**0**	**0.0**	**8**	**4**	**10:29**								
	New Haven	AHL	23	2	6	8	27																			
	Florida	ECHL	16	2	5	7	13																			
	Charlotte	ECHL	16	6	10	16	4																			
99-2000	Cincinnati	IHL	66	3	10	13	34												11	0	0	0	28			
2000-01	**Carolina**	**NHL**	**2**	**0**	**0**	**0**	**0**	**0**	**0**	**0**	**2**	**0.0**	**0**	**0**	**0.0**	**1**	**1**	**13:18**								
	Cincinnati	IHL	79	1	22	23	46												5	0	0	0	2			
2001-02	Lowell	AHL	63	5	11	16	33																			
	Albany	AHL	19	0	2	2	4																			
	NHL Totals		**26**	**0**	**2**	**2**	**10**	**0**	**0**	**0**	**13**	**0.0**		**0**	**0.0**	**9**	**5**	**10:49**								

Rights transferred to **Carolina** after **Hartford** franchise relocated, June 25, 1997. Traded to **New Jersey** by **Carolina** for Ted Drury, March 4, 2002.

RUCINSKY, Martin

(roo-SHIHN-skee, MAHR-tihn)

Left wing. Shoots left. 6'1", 205 lbs. Born, Most, Czech., March 11, 1971. Edmonton's 2nd choice, 20th overall, in 1991 Entry Draft.

Season	Club	League	GP	G	A	Pts	PIM	PP	SH	GW	S	%	+/-	TF	F%	H	SB	Min	GP	G	A	Pts	PIM	PP	SH	GW
1988-89	CHZ Litvinov	Czech	3	1	0	1	2																			
1989-90	CHZ Litvinov	Czech	39	12	6	18	....												8	5	3	8				
1990-91	CHZ Litvinov	Czech	56	24	20	44	69																			
1991-92	**Edmonton**	**NHL**	**2**	**0**	**0**	**0**	**0**	**0**	**0**	**0**	**1**	**0.0**	**–3**													
	Cape Breton	AHL	35	11	12	23	34																			
	Quebec	**NHL**	**4**	**1**	**1**	**2**	**2**	**0**	**0**	**0**	**4**	**25.0**	**1**													
	Halifax Citadels	AHL	7	1	1	2	6																			
1992-93	**Quebec**	**NHL**	**77**	**18**	**30**	**48**	**51**	**4**	**0**	**1**	**133**	**13.5**	**16**						**6**	**1**	**1**	**2**	**4**	**1**	**0**	**0**
1993-94	**Quebec**	**NHL**	**60**	**9**	**23**	**32**	**58**	**4**	**0**	**1**	**96**	**9.4**	**4**													
1994-95	Litvinov	Czech	13	12	10	22	54																			
	Quebec	**NHL**	**20**	**3**	**6**	**9**	**14**	**0**	**0**	**0**	**32**	**9.4**	**5**													
1995-96	HC Petra Vsetin	Czech	1	1	1	2	0																			
	Colorado	**NHL**	**22**	**4**	**11**	**15**	**14**	**0**	**0**	**1**	**39**	**10.3**	**10**													
	Montreal	**NHL**	**56**	**25**	**35**	**60**	**54**	**9**	**2**	**3**	**142**	**17.6**	**8**													
1996-97	**Montreal**	**NHL**	**70**	**28**	**27**	**55**	**62**	**6**	**3**	**3**	**172**	**16.3**	**1**						**5**	**0**	**0**	**0**	**4**	**0**	**0**	**0**
1997-98	**Montreal**	**NHL**	**78**	**21**	**32**	**53**	**84**	**5**	**3**	**3**	**192**	**10.9**	**13**						**10**	**3**	**0**	**3**	**4**	**1**	**0**	**0**
	Czech Republic	Olympics	6	3	1	4	4																			
1998-99	Litvinov	Czech	3	2	2	4	0																			
	Montreal	**NHL**	**73**	**17**	**17**	**34**	**50**	**5**	**0**	**1**	**180**	**9.4**	**–25**	**12**	**50.0**	**75**	**18**	**18:12**								
99-2000	**Montreal**	**NHL**	**80**	**25**	**24**	**49**	**70**	**7**	**1**	**4**	**242**	**10.3**	**1**	**31**	**54.8**	**116**	**21**	**18:54**								
2000-01	**Montreal**	**NHL**	**57**	**16**	**22**	**38**	**66**	**5**	**1**	**4**	**141**	**11.3**	**–5**	**5**	**40.0**	**82**	**18**	**19:11**								
2001-02	**Montreal**	**NHL**	**18**	**2**	**6**	**8**	**12**	**1**	**0**	**0**	**41**	**4.9**	**–1**	**5**	**80.0**	**24**	**3**	**16:15**								
	Dallas	**NHL**	**42**	**6**	**11**	**17**	**24**	**2**	**0**	**1**	**63**	**9.5**	**3**	**7**	**57.1**	**43**	**7**	**14:33**								
	Czech Republic	Olympics	4	0	3	3	2																			
	NY Rangers	**NHL**	**15**	**3**	**10**	**13**	**6**	**0**	**0**	**1**	**24**	**12.5**	**6**	**3**	**33.3**	**19**	**3**	**16:39**								
	NHL Totals		**674**	**178**	**255**	**433**	**567**	**48**	**10**	**23**	**1502**	**11.9**		**63**	**54.0**	**359**	**70**	**17:51**	**21**	**4**	**1**	**5**	**12**	**2**	**0**	**0**

Played in NHL All-Star Game (2000)

Traded to **Quebec** by **Edmonton** for Ron Tugnutt and Brad Zavisha, March 10, 1992. Transferred to **Colorado** after **Quebec** franchise relocated, June 21, 1995. Traded to **Montreal** by **Colorado** with Andrei Kovalenko and Jocelyn Thibault for Patrick Roy and Mike Keane, December 6, 1995. Traded to **Dallas** by **Montreal** with Benoit Brunet for Donald Audette and Shaun Van Allen, November 21, 2001. Traded to **NY Rangers** by **Dallas** with Roman Lyashenko for Manny Malhotra and Barrett Heisten, March 12, 2002.

RUMBLE, Darren

(RUHM-buhl, DAIR-rehn)

Defense. Shoots left. 6'1", 200 lbs. Born, Barrie, Ont., January 23, 1969. Philadelphia's 1st choice, 20th overall, in 1987 Entry Draft.

Season	Club	League	GP	G	A	Pts	PIM	PP	SH	GW	S	%	+/-	TF	F%	H	SB	Min	GP	G	A	Pts	PIM	PP	SH	GW
1985-86	Barrie Colts	OJHL-B	46	14	32	46	91																			
1986-87	Kitchener	OHL	64	11	32	43	44												4	0	1	1	9			
1987-88	Kitchener	OHL	55	15	50	65	64																			
1988-89	Kitchener	OHL	46	11	28	39	25												5	1	0	1	2			
1989-90	Hershey Bears	AHL	57	2	13	15	31																			
1990-91	**Philadelphia**	**NHL**	**3**	**1**	**0**	**1**	**0**	**0**	**0**	**0**	**2**	**50.0**	**1**													
	Hershey Bears	AHL	73	6	35	41	48												3	0	5	5	2			
1991-92	Hershey Bears	AHL	79	12	54	66	118												6	0	3	3	2			
1992-93	**Ottawa**	**NHL**	**69**	**3**	**13**	**16**	**61**	**0**	**0**	**0**	**92**	**3.3**	**–24**													
	New Haven	AHL	2	1	0	1	0																			
1993-94	**Ottawa**	**NHL**	**70**	**6**	**9**	**15**	**116**	**0**	**0**	**0**	**95**	**6.3**	**–50**													
	P.E.I. Senators	AHL	3	2	0	2	0																			
1994-95	P.E.I. Senators	AHL	70	7	46	53	77												11	0	6	6	4			
1995-96	**Philadelphia**	**NHL**	**5**	**0**	**0**	**0**	**4**	**0**	**0**	**0**	**7**	**0.0**	**0**													
	Hershey Bears	AHL	58	13	37	50	83												5	0	0	0	6			
1996-97	**Philadelphia**	**NHL**	**10**	**0**	**0**	**0**	**0**	**0**	**0**	**0**	**9**	**0.0**	**–2**													
	Philadelphia	AHL	72	18	44	62	83												7	0	3	3	19			
1997-98	Adler Mannheim	Germany	21	2	7	9	18																			
	Adler Mannheim	EuroHL	4	0	1	1	4																			
	San Antonio	IHL	46	7	22	29	47																			
1998-99	Utah Grizzlies	IHL	10	1	4	5	10																			
	Grand Rapids	IHL	53	6	22	28	44																			
99-2000	Grand Rapids	IHL	29	3	10	13	20																			
	Worcester	AHL	39	0	17	17	31												9	0	2	2	6			
2000-01	**St. Louis**	**NHL**	**12**	**0**	**4**	**4**	**27**	**0**	**0**	**0**	**11**	**0.0**	**7**	**0**	**0.0**	**25**	**11**	**18:12**								
	Worcester	AHL	53	6	24	30	65												8	0	1	1	10			
2001-02	Worcester	AHL	60	3	29	32	48												3	0	4	4	2			
	NHL Totals		**169**	**10**	**26**	**36**	**208**	**0**	**0**	**0**	**216**	**4.6**		**0**	**0.0**	**25**	**11**	**18:12**								

AHL Second All-Star Team (1995) • AHL First All-Star Team (1997) • Won Eddie Shore Award (Top Defenseman - AHL) (1997)

Claimed by **Ottawa** from **Philadelphia** in Expansion Draft, June 18, 1992. Signed as a free agent by **Philadelphia**, July 31, 1995. Signed as a free agent by **St. Louis**, February 1, 2000.

RUUTU, Jarkko

(ROO-too, YAHR-koh) **VAN.**

Left wing. Shoots left. 6'2", 194 lbs. Born, Vantaa, Finland, August 23, 1975. Vancouver's 3rd choice, 68th overall, in 1998 Entry Draft.

Season	Club	League	GP	G	A	Pts	PIM	PP	SH	GW	S	%	+/-	TF	F%	H	SB	Min	GP	G	A	Pts	PIM	PP	SH	GW
1991-92	HIFK Jr.	Finn-Jr.	1	0	0	0	0																			
1992-93	HIFK Helsinki-B	Finn-Jr.	33	26	21	47	53																			
	HIFK Jr.	Finn-Jr.	1	0	0	0	0																			
1993-94	HIFK Jr.	Finn-Jr.	19	9	12	21	44																			
1994-95	HIFK Jr.	Finn-Jr.	35	26	22	48	117																			
1995-96	Michigan Tech	WCHA	39	12	10	22	96																			
1996-97	HIFK Helsinki	Finland	48	11	10	21	*155																			
1997-98	HIFK Helsinki	Finland	37	10	10	20	87												8	*7	4	11	10			

								Regular Season											Playoffs							
Season	Club	League	GP	G	A	Pts	PIM	PP	SH	GW	S	%	+/-	TF	F%	H	SB	Min	GP	G	A	Pts	PIM	PP	SH	GW
1998-99	HIFK Helsinki	Finland	25	10	4	14	136	....	....	....	....	....	....	....	....	....	....	....	9	0	2	2	43	....	....	....
	HIFK Helsinki	EuroHL	5	1	2	3	8	....	....	....	....	....	....	....	....	....	....	....	....	....	....	....	....	....	....	....
99-2000	**Vancouver**	**NHL**	**8**	**0**	**1**	**1**	**6**	**0**	**0**	**0**	**4**	**0.0**	**–1**	**0**	**0.0**	**14**	**0**	**8:47**	....	....	....	....	....	....	....	....
	Syracuse Crunch	AHL	65	26	32	58	164	....	....	....	....	....	....	....	....	....	....	....	4	3	1	4	8	....	....	....
2000-01	**Vancouver**	**NHL**	**21**	**3**	**3**	**6**	**32**	**0**	**1**	**0**	**23**	**13.0**	**1**	**0**	**0.0**	**65**	**7**	**10:39**	**4**	**0**	**1**	**1**	**8**	**0**	**0**	**0**
	Kansas City	IHL	46	11	18	29	111	....	....	....	....	....	....	....	....	....	....	....	....	....	....	....	....	....	....	....
2001-02	**Vancouver**	**NHL**	**49**	**2**	**7**	**9**	**74**	**0**	**0**	**0**	**37**	**5.4**	**–1**	**5**	**0.0**	**130**	**9**	**10:11**	**1**	**0**	**0**	**0**	**0**	**0**	**0**	**0**
	Finland	Olympics	4	0	0	0	4	....	....	....	....	....	....	....	....	....	....	....	....	....	....	....	....	....	....	....
	NHL Totals		**78**	**5**	**11**	**16**	**112**	**0**	**1**	**0**	**64**	**7.8**		**5**	**0.0**	**209**	**16**	**10:10**	**5**	**0**	**1**	**1**	**8**	**0**	**0**	**0**

RYCROFT, Mark

(RIGH-krawft, MAHRK) **ST.L.**

Right wing. Shoots right. 5'11", 197 lbs. Born, Nanaimo, B.C., July 12, 1978.

Season	Club	League	GP	G	A	Pts	PIM	PP	SH	GW	S	%	+/-	TF	F%	H	SB	Min	GP	G	A	Pts	PIM	PP	SH	GW
1993-94	Penticton Ice	BCAHA	60	47	65	112	100	....	....	....	....	....	....	....	....	....	....	....	....	....	....	....	....	....	....	....
1994-95	Penticton Ice	BCAHA	43	33	43	76	90	....	....	....	....	....	....	....	....	....	....	....	....	....	....	....	....	....	....	....
1995-96	Nanaimo Clippers	BCHL	60	17	28	45	28	....	....	....	....	....	....	....	....	....	....	....	....	....	....	....	....	....	....	....
1996-97	Nanaimo Clippers	BCHL	58	32	35	67	79	....	....	....	....	....	....	....	....	....	....	....	....	....	....	....	....	....	....	....
1997-98	U. of Denver	WCHA	35	15	17	32	28	....	....	....	....	....	....	....	....	....	....	....	....	....	....	....	....	....	....	....
1998-99	U. of Denver	WCHA	41	19	18	37	36	....	....	....	....	....	....	....	....	....	....	....	....	....	....	....	....	....	....	....
99-2000	U. of Denver	WCHA	41	17	17	34	87	....	....	....	....	....	....	....	....	....	....	....	....	....	....	....	....	....	....	....
2000-01	Worcester	AHL	71	24	26	50	68	....	....	....	....	....	....	....	....	....	....	....	11	2	5	7	4	....	....	....
2001-02	**St. Louis**	**NHL**	**9**	**0**	**3**	**3**	**4**	**0**	**0**	**0**	**14**	**0.0**	**0**	**1**	**0.0**	**16**	**2**	**9:50**	....	....	....	....	....	....	....	....
	Worcester	AHL	66	12	19	31	68	....	....	....	....	....	....	....	....	....	....	....	3	0	1	1	0	....	....	....
	NHL Totals		**9**	**0**	**3**	**3**	**4**	**0**	**0**	**0**	**14**	**0.0**		**1**	**0.0**	**16**	**2**	**9:50**	....	....	....	....	....	....	....	....

WCHA All-Rookie Team (1998)

Signed as a free agent by **St. Louis**, May 15, 2000.

SACCO, Joe

(SAK-oh, JOH)

Right wing. Shoots left. 6'1", 190 lbs. Born, Medford, MA, February 4, 1969. Toronto's 4th choice, 71st overall, in 1987 Entry Draft.

Season	Club	League	GP	G	A	Pts	PIM	PP	SH	GW	S	%	+/-	TF	F%	H	SB	Min	GP	G	A	Pts	PIM	PP	SH	GW
1985-86	Medford	Hi-School	20	30	30	60	....	....	....	....	....	....	....	....	....	....	....	....	....	....	....	....	....	....	....	....
1986-87	Medford	Hi-School	21	22	32	54	....	....	....	....	....	....	....	....	....	....	....	....	....	....	....	....	....	....	....	....
1987-88	Boston University	H-East	34	16	20	36	40	....	....	....	....	....	....	....	....	....	....	....	....	....	....	....	....	....	....	....
1988-89	Boston University	H-East	33	21	19	40	66	....	....	....	....	....	....	....	....	....	....	....	....	....	....	....	....	....	....	....
1989-90	Boston University	H-East	44	28	24	52	70	....	....	....	....	....	....	....	....	....	....	....	....	....	....	....	....	....	....	....
1990-91	**Toronto**	**NHL**	**20**	**0**	**5**	**5**	**2**	**0**	**0**	**0**	**20**	**0.0**	**–5**	....	....	....	....	....	....	....	....	....	....	....	....	....
	Newmarket Saints	AHL	49	18	17	35	24	....	....	....	....	....	....	....	....	....	....	....	....	....	....	....	....	....	....	....
1991-92	Team USA	Nat-Tm	50	11	26	37	61	....	....	....	....	....	....	....	....	....	....	....	....	....	....	....	....	....	....	....
	United States	Olympics	8	0	2	2	0	....	....	....	....	....	....	....	....	....	....	....	....	....	....	....	....	....	....	....
	Toronto	**NHL**	**17**	**7**	**4**	**11**	**4**	**0**	**0**	**1**	**40**	**17.5**	**8**	....	....	....	....	....	....	....	....	....	....	....	....	....
	St. John's	AHL	....	....	....	....	....	....	....	....	....	....	....	....	....	....	....	....	1	1	1	2	0	....	....	....
1992-93	**Toronto**	**NHL**	**23**	**4**	**4**	**8**	**8**	**0**	**0**	**0**	**38**	**10.5**	**–4**	....	....	....	....	....	....	....	....	....	....	....	....	....
	St. John's	AHL	37	14	16	30	45	....	....	....	....	....	....	....	....	....	....	....	7	6	4	10	2	....	....	....
1993-94	**Anaheim**	**NHL**	**84**	**19**	**18**	**37**	**61**	**3**	**1**	**2**	**206**	**9.2**	**–11**	....	....	....	....	....	....	....	....	....	....	....	....	....
1994-95	**Anaheim**	**NHL**	**41**	**10**	**8**	**18**	**23**	**2**	**0**	**0**	**77**	**13.0**	**–8**	....	....	....	....	....	....	....	....	....	....	....	....	....
1995-96	**Anaheim**	**NHL**	**76**	**13**	**14**	**27**	**40**	**1**	**2**	**2**	**132**	**9.8**		....	....	....	....	....	....	....	....	....	....	....	....	....
1996-97	**Anaheim**	**NHL**	**77**	**12**	**17**	**29**	**35**	**1**	**1**	**2**	**131**	**9.2**	**1**	....	....	....	....	....	**11**	**2**	**0**	**2**	**2**	**0**	**0**	**0**
1997-98	**Anaheim**	**NHL**	**55**	**8**	**11**	**19**	**24**	**0**	**2**	**2**	**90**	**8.9**	**–1**	....	....	....	....	....	....	....	....	....	....	....	....	....
	NY Islanders	**NHL**	**25**	**3**	**3**	**6**	**10**	**0**	**0**	**0**	**32**	**9.4**	**1**	....	....	....	....	....	....	....	....	....	....	....	....	....
1998-99	**NY Islanders**	**NHL**	**73**	**3**	**0**	**3**	**45**	**0**	**1**	**2**	**84**	**3.6**	**–24**	**32**	**56.3**	**58**	**27**	**9:59**	....	....	....	....	....	....	....	....
99-2000	**Washington**	**NHL**	**79**	**7**	**16**	**23**	**50**	**0**	**0**	**1**	**117**	**6.0**	**7**	**9**	**33.3**	**146**	**40**	**11:51**	**5**	**0**	**0**	**0**	**4**	**0**	**0**	**0**
2000-01	**Washington**	**NHL**	**69**	**7**	**7**	**14**	**48**	**0**	**0**	**0**	**81**	**8.6**	**4**	**1**	**100.0**	**97**	**41**	**11:10**	**6**	**0**	**0**	**0**	**2**	**0**	**0**	**0**
2001-02	**Washington**	**NHL**	**65**	**0**	**7**	**7**	**51**	**0**	**0**	**0**	**57**	**0.0**	**–13**	**12**	**16.7**	**38**	**22**	**7:38**	....	....	....	....	....	....	....	....
	NHL Totals		**704**	**93**	**114**	**207**	**401**	**7**	**7**	**12**	**1105**	**8.4**		**54**	**44.4**	**339**	**130**	**10:15**	**22**	**2**	**0**	**2**	**8**	**0**	**0**	**0**

Claimed by **Anaheim** from **Toronto** in Expansion Draft, June 24, 1993. Traded to **NY Islanders** by **Anaheim** with J-J Daigneault and Mark Janssens for Travis Green, Doug Houda and Tony Tuzzolino, February 6, 1998. Signed as a free agent by **Washington**, August 9, 1999.

SAFRONOV, Kirill

(sah-FRAW-nawf, kih-RIHL) **ATL.**

Defense. Shoots left. 6'1", 210 lbs. Born, Leningrad, USSR, February 26, 1981. Phoenix's 2nd choice, 19th overall, in 1999 Entry Draft.

Season	Club	League	GP	G	A	Pts	PIM	PP	SH	GW	S	%	+/-	TF	F%	H	SB	Min	GP	G	A	Pts	PIM	PP	SH	GW
1996-97	St. Petersburg 2	Russia-3	9	0	0	0	6	....	....	....	....	....	....	....	....	....	....	....	....	....	....	....	....	....	....	....
	St. Petersburg	Russia	1	0	0	0	0	....	....	....	....	....	....	....	....	....	....	....	....	....	....	....	....	....	....	....
1997-98	St. Petersburg 2	Russia-3	34	4	3	7	36	....	....	....	....	....	....	....	....	....	....	....	....	....	....	....	....	....	....	....
	St. Petersburg	Russia	9	0	1	1	4	....	....	....	....	....	....	....	....	....	....	....	1	0	0	0	0	....	....	....
1998-99	St. Petersburg 2	Russia-4	4	2	1	3	2	....	....	....	....	....	....	....	....	....	....	....	....	....	....	....	....	....	....	....
	St. Petersburg	Russia	45	1	3	4	32	....	....	....	....	....	....	....	....	....	....	....	....	....	....	....	....	....	....	....
99-2000	Quebec Remparts	QMJHL	55	11	32	43	95	....	....	....	....	....	....	....	....	....	....	....	11	2	4	6	14	....	....	....
2000-01	Springfield	AHL	65	5	13	18	77	....	....	....	....	....	....	....	....	....	....	....	....	....	....	....	....	....	....	....
2001-02	**Phoenix**	**NHL**	**1**	**0**	**0**	**0**	**0**	**0**	**0**	**0**	**0**	**0.0**	**–2**	**0**	**0.0**	**1**	**0**	**6:08**	....	....	....	....	....	....	....	....
	Springfield	AHL	68	3	19	22	26	....	....	....	....	....	....	....	....	....	....	....	....	....	....	....	....	....	....	....
	Atlanta	**NHL**	**2**	**0**	**0**	**0**	**2**	**0**	**0**	**0**	**2**	**0.0**	**–3**	**0**	**0.0**	**10**	**2**	**21:11**	....	....	....	....	....	....	....	....
	Chicago Wolves	AHL	8	0	2	2	2	....	....	....	....	....	....	....	....	....	....	....	25	2	6	8	8	....	....	....
	NHL Totals		**3**	**0**	**0**	**0**	**2**	**0**	**0**	**0**	**2**	**0.0**		**0**	**0.0**	**11**	**2**	**16:10**	....	....	....	....	....	....	....	....

Traded to **Atlanta** by **Phoenix** with the rights to Ruslan Zainullin and Phoenix's 5th round choice (Patrick Dwyer) in 2002 Entry Draft for Darcy Hordichuk and Atlanta's 4th (Lance Monych) and 5th (John Zeiler) round choices in 2002 Entry Draft, March 19, 2002.

ST-JACQUES, Bruno

(SAINT ZHAWK, BREW-noh) **PHI.**

Defense. Shoots left. 6'2", 210 lbs. Born, Montreal, Que., August 22, 1980. Philadelphia's 12th choice, 253rd overall, in 1998 Entry Draft.

Season	Club	League	GP	G	A	Pts	PIM	PP	SH	GW	S	%	+/-	TF	F%	H	SB	Min	GP	G	A	Pts	PIM	PP	SH	GW
1996-97	Mtl-Bourassa	QAAA	40	5	8	13	....	....	....	....	....	....	....	....	....	....	....	....	16	0	7	7	....	....	....	....
1997-98	Baie-Comeau	QMJHL	63	1	11	12	140	....	....	....	....	....	....	....	....	....	....	....	....	....	....	....	....	....	....	....
1998-99	Baie-Comeau	QMJHL	49	8	13	21	85	....	....	....	....	....	....	....	....	....	....	....	....	....	....	....	....	....	....	....
99-2000	Baie-Comeau	QMJHL	60	8	28	36	120	....	....	....	....	....	....	....	....	....	....	....	6	0	2	2	10	....	....	....
	Philadelphia	AHL	3	0	1	1	0	....	....	....	....	....	....	....	....	....	....	....	1	0	0	0	0	....	....	....
2000-01	Philadelphia	AHL	45	1	16	17	83	....	....	....	....	....	....	....	....	....	....	....	10	1	0	1	16	....	....	....
2001-02	**Philadelphia**	**NHL**	**7**	**0**	**0**	**0**	**2**	**0**	**0**	**0**	**4**	**0.0**	**4**	**0**	**0.0**	**13**	**9**	**13:51**	....	....	....	....	....	....	....	....
	Philadelphia	AHL	55	3	11	14	59	....	....	....	....	....	....	....	....	....	....	....	4	0	0	0	0	....	....	....
	NHL Totals		**7**	**0**	**0**	**0**	**2**	**0**	**0**	**0**	**4**	**0.0**		**0**	**0.0**	**13**	**9**	**13:51**	....	....	....	....	....	....	....	....

ST-LOUIS, Martin

(sehn-loo-EE, mahr-TEHN) **T.B.**

Right wing. Shoots left. 5'9", 185 lbs. Born, Laval, Que., June 18, 1975.

Season	Club	League	GP	G	A	Pts	PIM	PP	SH	GW	S	%	+/-	TF	F%	H	SB	Min	GP	G	A	Pts	PIM	PP	SH	GW
1991-92	Laval Laurentide	QAAA	42	29	*74	*103	38	....	....	....	....	....	....	....	....	....	....	....	12	7	15	22	16	....	....	....
1992-93	Hawkesbury	OCJHL	31	37	50	87	70	....	....	....	....	....	....	....	....	....	....	....	....	....	....	....	....	....	....	....
1993-94	U. of Vermont	ECAC	33	15	36	51	24	....	....	....	....	....	....	....	....	....	....	....	....	....	....	....	....	....	....	....
1994-95	U. of Vermont	ECAC	35	23	48	71	36	....	....	....	....	....	....	....	....	....	....	....	....	....	....	....	....	....	....	....
1995-96	U. of Vermont	ECAC	35	29	56	85	38	....	....	....	....	....	....	....	....	....	....	....	....	....	....	....	....	....	....	....
1996-97	U. of Vermont	ECAC	36	24	*36	60	65	....	....	....	....	....	....	....	....	....	....	....	....	....	....	....	....	....	....	....
1997-98	Cleveland	IHL	56	16	34	50	24	....	....	....	....	....	....	....	....	....	....	....	....	....	....	....	....	....	....	....
	Saint John	AHL	25	15	11	26	20	....	....	....	....	....	....	....	....	....	....	....	20	5	15	20	16	....	....	....
1998-99	**Calgary**	**NHL**	**13**	**1**	**1**	**2**	**10**	**0**	**0**	**0**	**14**	**7.1**	**–2**	**0**	**0.0**	**8**	**4**	**8:15**	....	....	....	....	....	....	....	....
	Saint John	AHL	53	28	34	62	30	....	....	....	....	....	....	....	....	....	....	....	7	4	4	8	2	....	....	....
99-2000	**Calgary**	**NHL**	**56**	**3**	**15**	**18**	**22**	**0**	**0**	**1**	**73**	**4.1**	**–5**	**3**	**0.0**	**47**	**45**	**14:41**	....	....	....	....	....	....	....	....
	Saint John	AHL	17	15	11	26	14	....	....	....	....	....	....	....	....	....	....	....	....	....	....	....	....	....	....	....
2000-01	**Tampa Bay**	**NHL**	**78**	**18**	**22**	**40**	**12**	**3**	**3**	**4**	**141**	**12.8**	**–4**	**48**	**41.7**	**45**	**53**	**15:14**	....	....	....	....	....	....	....	....
2001-02	**Tampa Bay**	**NHL**	**53**	**16**	**19**	**35**	**20**	**6**	**1**	**2**	**105**	**15.2**	**4**	**33**	**39.4**	**32**	**32**	**18:41**	....	....	....	....	....	....	....	....
	NHL Totals		**200**	**38**	**57**	**95**	**64**	**9**	**4**	**7**	**333**	**11.4**		**84**	**39.3**	**132**	**134**	**15:33**	....	....	....	....	....	....	....	....

ECAC First All-Star Team (1995, 1996, 1997) • ECAC Player of the Year (1995) • NCAA East First All-American Team (1995, 1996, 1997) • NCAA Championship All-Tournament Team (1996)

Signed as a free agent by **Calgary**, February 19, 1998. Signed as a free agent by **Tampa Bay**, July 31, 2000.

SAKIC, Joe
(SAK-ihk, JOH) **COL.**

Center. Shoots left. 5'11", 195 lbs. Born, Burnaby, B.C., July 7, 1969. Quebec's 2nd choice, 15th overall, in 1987 Entry Draft.

Season	Club	League	GP	G	A	Pts	PIM	PP	SH	GW	S	%	+/-	TF	F%	H	SB	Min	GP	G	A	Pts	PIM	PP	SH	GW
			Regular Season																Playoffs							
1985-86	Burnaby	BCAHA	80	83	73	156	96																			
	Lethbridge	WHL	3	0	0	0	0																			
1986-87	Swift Current	WHL	72	60	73	133	31												4	0	1	1	0			
1987-88	Swift Current	WHL	64	*78	82	*160	64												10	11	13	24	12			
1988-89	**Quebec**	**NHL**	**70**	**23**	**39**	**62**	**24**	**10**	**0**	**2**	**148**	**15.5**	**–36**													
1989-90	**Quebec**	**NHL**	**80**	**39**	**63**	**102**	**27**	**8**	**1**	**2**	**234**	**16.7**	**–40**													
1990-91	**Quebec**	**NHL**	**80**	**48**	**61**	**109**	**24**	**12**	**3**	**7**	**245**	**19.6**	**–26**													
1991-92	**Quebec**	**NHL**	**69**	**29**	**65**	**94**	**20**	**6**	**3**	**1**	**217**	**13.4**	**5**													
1992-93	**Quebec**	**NHL**	**78**	**48**	**57**	**105**	**40**	**20**	**2**	**4**	**264**	**18.2**	**–3**						**6**	**3**	**3**	**6**	**2**	**1**	**0**	**0**
1993-94	**Quebec**	**NHL**	**84**	**28**	**64**	**92**	**18**	**10**	**1**	**9**	**279**	**10.0**	**–8**													
1994-95	**Quebec**	**NHL**	**47**	**19**	**43**	**62**	**30**	**3**	**2**	**5**	**157**	**12.1**	**7**						**6**	**4**	**1**	**5**	**0**	**1**	**1**	**1**
1995-96♦	**Colorado**	**NHL**	**82**	**51**	**69**	**120**	**44**	**17**	**6**	**7**	**339**	**15.0**	**14**						**22**	***18**	**16**	***34**	**14**	**6**	**0**	**6**
1996-97	**Colorado**	**NHL**	**65**	**22**	**52**	**74**	**34**	**10**	**2**	**5**	**261**	**8.4**	**–10**						**17**	**8**	***17**	**25**	**14**	**3**	**0**	**0**
1997-98	**Colorado**	**NHL**	**64**	**27**	**36**	**63**	**50**	**12**	**1**	**2**	**254**	**10.6**	**0**						**6**	**2**	**3**	**5**	**6**	**0**	**1**	**2**
	Canada	Olympics	4	1	2	3	4																			
1998-99	**Colorado**	**NHL**	**73**	**41**	**55**	**96**	**29**	**12**	**5**	**6**	**255**	**16.1**	**23**	**1723**	**51.4**	**31**	**47**	**25:35**	**19**	**6**	**13**	**19**	**8**	**1**	**1**	**1**
99-2000	**Colorado**	**NHL**	**60**	**28**	**53**	**81**	**28**	**5**	**1**	**5**	**242**	**11.6**	**30**	**1392**	**53.8**	**19**	**26**	**23:16**	**17**	**2**	**7**	**9**	**8**	**2**	**0**	**0**
2000-01♦	**Colorado**	**NHL**	**82**	**54**	**64**	**118**	**30**	**19**	**3**	**12**	**332**	**16.3**	**45**	**2292**	**53.0**	**44**	**54**	**23:01**	**21**	***13**	**13**	***26**	**6**	**5**	**0**	**3**
2001-02	**Colorado**	**NHL**	**82**	**26**	**53**	**79**	**18**	**9**	**1**	**4**	**260**	**10.0**	**12**	**2148**	**52.2**	**64**	**52**	**22:01**	**21**	**9**	**10**	**19**	**4**	**4**	**0**	**1**
	Canada	Olympics	6	4	3	7	0																			
	NHL Totals		**1016**	**483**	**774**	**1257**	**416**	**153**	**31**	**71**	**3487**	**13.9**		**7555**	**52.6**	**158**	**179**	**23:25**	**135**	**65**	**83**	**148**	**62**	**23**	**3**	**14**

WHL East Second All-Star Team (1987) • WHL East Rookie of the Year (1987) • WHL East MVP (1987) • WHL East First All-Star Team (1988) • WHL MVP (1988) • Canadian Major Junior Player of the Year (1988) • Shared Bob Clarke Trophy (Top Scorer - WHL) with Theo Fleury (1988) • Won Conn Smythe Trophy (1996) • NHL First All-Star Team (2001, 2002) • Won Lady Byng Trophy (2001) • Won Hart Trophy (2001) • Won Lester B. Pearson Award (2001) • Played in NHL All-Star Game (1990, 1991, 1992, 1993, 1994, 1996, 1998, 2000, 2001, 2002)

Transferred to **Colorado** after **Quebec** franchise relocated, June 21, 1995.

SALEI, Ruslan
(sah-LAY, roos-LAHN) **ANA.**

Defense. Shoots left. 6'1", 205 lbs. Born, Minsk, USSR, November 2, 1974. Anaheim's 1st choice, 9th overall, in 1996 Entry Draft.

Season	Club	League	GP	G	A	Pts	PIM	PP	SH	GW	S	%	+/-	TF	F%	H	SB	Min	GP	G	A	Pts	PIM	PP	SH	GW
1992-93	Dynamo Minsk	CIS	9	1	0	1	10																			
1993-94	Tivali Minsk	CIS	39	2	3	5	50																			
1994-95	Tivali Minsk	CIS	51	4	2	6	44																			
1995-96	Las Vegas	IHL	76	7	23	30	123												15	3	7	10	18			
1996-97	**Anaheim**	**NHL**	**30**	**0**	**1**	**1**	**37**	**0**	**0**	**0**	**14**	**0.0**	**–8**													
	Baltimore	AHL	12	1	4	5	12																			
	Las Vegas	IHL	8	0	2	2	24												3	2	1	3	6			
1997-98	**Anaheim**	**NHL**	**66**	**5**	**10**	**15**	**70**	**1**	**0**	**0**	**104**	**4.8**	**7**													
	Cincinnati	AHL	6	3	6	9	14																			
	Belarus	Olympics	7	1	0	1	4																			
1998-99	**Anaheim**	**NHL**	**74**	**2**	**14**	**16**	**65**	**1**	**0**	**0**	**123**	**1.6**	**1**	**0**	**0.0**	**154**	**105**	**22:03**	**3**	**0**	**0**	**0**	**4**	**0**	**0**	**0**
99-2000	**Anaheim**	**NHL**	**71**	**5**	**5**	**10**	**94**	**1**	**0**	**0**	**116**	**4.3**	**3**	**0**	**0.0**	**205**	**107**	**20:21**								
2000-01	**Anaheim**	**NHL**	**50**	**1**	**5**	**6**	**70**	**0**	**0**	**0**	**73**	**1.4**	**–14**	**0**	**0.0**	**146**	**78**	**20:40**								
2001-02	**Anaheim**	**NHL**	**82**	**4**	**7**	**11**	**97**	**0**	**0**	**1**	**96**	**4.2**	**–10**	**0**	**0.0**	**191**	**109**	**21:25**								
	Belarus	Olympics	6	2	1	3	4																			
	NHL Totals		**373**	**17**	**42**	**59**	**433**	**3**	**0**	**1**	**526**	**3.2**		**0**	**0.0**	**696**	**399**	**21:11**	**3**	**0**	**0**	**0**	**4**	**0**	**0**	**0**

SALO, Sami
(SA-loh, SA-mee) **OTT.**

Defense. Shoots right. 6'3", 215 lbs. Born, Turku, Finland, September 2, 1974. Ottawa's 7th choice, 239th overall, in 1996 Entry Draft.

Season	Club	League	GP	G	A	Pts	PIM	PP	SH	GW	S	%	+/-	TF	F%	H	SB	Min	GP	G	A	Pts	PIM	PP	SH	GW
1991-92	Kiekko-67 Jr.	Finn-Jr.	23	4	5	9	26																			
1992-93	Kiekko-67-B	Finn-Jr.	21	9	4	13	4																			
	Kiekko-67 Jr.	Finn-Jr.	13	6	2	8	2																			
1993-94	TPS Turku Jr.	Finn-Jr.	36	7	13	20	16												7	0	1	1	10			
1994-95	TPS Turku Jr.	Finn-Jr.	14	1	3	4	6																			
	Kiekko-67 Turku	Finland-2	19	4	2	6	4																			
	TPS Turku	Finland	7	1	2	3	8												1	0	0	0	0			
1995-96	TPS Turku	Finland	47	7	14	21	32												11	1	3	4	8			
1996-97	TPS Turku	Finland	48	9	6	15	10												10	2	3	5	4			
	TPS Turku	EuroHL	6	0	2	2	6												2	0	0	0	2			
1997-98	Jokerit Helsinki	Finland	35	3	5	8	10												8	0	1	1	2			
	Jokerit Helsinki	EuroHL	6	1	1	2	2																			
1998-99	**Ottawa**	**NHL**	**61**	**7**	**12**	**19**	**24**	**2**	**0**	**1**	**106**	**6.6**	**20**	**0**	**0.0**	**90**	**58**	**19:42**	**4**	**0**	**0**	**0**	**0**	**0**	**0**	**0**
	Detroit Vipers	IHL	5	0	2	2	0																			
99-2000	**Ottawa**	**NHL**	**37**	**6**	**8**	**14**	**2**	**3**	**0**	**1**	**85**	**7.1**	**6**	**0**	**0.0**	**52**	**29**	**20:18**	**6**	**1**	**1**	**2**	**0**	**1**	**0**	**0**
2000-01	**Ottawa**	**NHL**	**31**	**2**	**16**	**18**	**10**	**1**	**0**	**0**	**61**	**3.3**	**9**	**0**	**0.0**	**57**	**42**	**19:44**	**4**	**0**	**0**	**0**	**0**	**0**	**0**	**0**
2001-02	**Ottawa**	**NHL**	**66**	**4**	**14**	**18**	**14**	**1**	**1**	**2**	**122**	**3.3**	**1**	**0**	**0.0**	**84**	**68**	**19:52**	**12**	**2**	**1**	**3**	**4**	**0**	**0**	**0**
	Finland	Olympics	4	0	0	0	0																			
	NHL Totals		**195**	**19**	**50**	**69**	**50**	**7**	**1**	**4**	**374**	**5.1**		**0**	**0.0**	**283**	**197**	**19:53**	**26**	**3**	**2**	**5**	**4**	**1**	**0**	**0**

NHL All-Rookie Team (1999) • Missed majority of 1999-2000 season recovering from wrist injury originally suffered in game vs. Philadelphia, November 28, 1999. • Missed majority of 2000-01 season recovering from shoulder injury suffered in game vs. Atlanta, December 14, 2000.

SALOMONSSON, Andreas
(sal-oh-MAWN-suhn, an-DRAY-uhs) **N.J.**

Right wing. Shoots left. 6'1", 200 lbs. Born, Ornskoldsvik, Sweden, December 19, 1973. New Jersey's 8th choice, 163rd overall, in 2001 Entry Draft.

Season	Club	League	GP	G	A	Pts	PIM	PP	SH	GW	S	%	+/-	TF	F%	H	SB	Min	GP	G	A	Pts	PIM	PP	SH	GW
1990-91	MoDo	Sweden	2	0	0	0	0																			
1991-92	MoDo	Sweden	20	1	1	2	26																			
1992-93	MoDo Jr.	Swede-Jr.	10	4	16	20	8																			
	MoDo	Sweden	33	1	1	2	0																			
1993-94	MoDo	Sweden	38	15	8	23	33												11	1	2	3	4			
1994-95	MoDo	Sweden	40	5	9	14	34																			
1995-96	MoDo	Sweden	38	13	6	19	22												7	0	4	4	8			
1996-97	MoDo	Sweden	23	6	6	12	22																			
	Ratingen	Germany	21	5	1	6	49																			
1997-98	MoDo	Sweden	45	6	15	21	69												8	4	3	7	4			
1998-99	MoDo	Sweden	46	13	13	26	60												13	4	5	9	12			
99-2000	MoDo	EuroHL	5	2	2	4	4												3	1	1	2	4			
	MoDo	Sweden	48	9	15	24	38												12	0	4	4	12			
2000-01	Djurgarden	Sweden	48	10	12	22	46												13	3	4	7	31			
2001-02	**New Jersey**	**NHL**	**39**	**4**	**5**	**9**	**22**	**1**	**0**	**1**	**58**	**6.9**	**–12**	**6**	**33.3**	**50**	**8**	**13:30**	**4**	**0**	**1**	**1**	**0**	**0**	**0**	**0**
	Albany	AHL	19	3	10	13	6																			
	NHL Totals		**39**	**4**	**5**	**9**	**22**	**1**	**0**	**1**	**58**	**6.9**		**6**	**33.3**	**50**	**8**	**13:30**	**4**	**0**	**1**	**1**	**0**	**0**	**0**	**0**

SALVADOR, Bryce
(SAL-vuh-dohr, BRIGHS) **ST.L.**

Defense. Shoots left. 6'2", 215 lbs. Born, Brandon, Man., February 11, 1976. Tampa Bay's 6th choice, 138th overall, in 1994 Entry Draft.

Season	Club	League	GP	G	A	Pts	PIM	PP	SH	GW	S	%	+/-	TF	F%	H	SB	Min	GP	G	A	Pts	PIM	PP	SH	GW
1991-92	Brandon	MAHA	52	6	23	29	38																			
1992-93	Lethbridge	WHL	64	1	4	5	29												4	0	0	0	0			
1993-94	Lethbridge	WHL	61	4	14	18	36												9	0	1	1	2			
1994-95	Lethbridge	WHL	67	1	9	10	88																			
1995-96	Lethbridge	WHL	56	4	12	16	75												3	0	1	1	2			
1996-97	Lethbridge	WHL	63	8	32	40	81												19	0	7	7	14			
1997-98	Worcester	AHL	46	2	8	10	74												11	0	1	1	45			
1998-99	Worcester	AHL	69	5	13	18	129												4	0	1	1	2			
99-2000	Worcester	AHL	55	0	13	13	53												9	0	1	1	2			

Season	Club	League	GP	G	A	Pts	PIM	PP	SH	GW	S	%	+/-	TF	F%	H	SB	Min	GP	G	A	Pts	PIM	PP	SH	GW
			Regular Season																Playoffs							
2000-01	St. Louis	NHL	75	2	8	10	69	0	0	1	60	3.3	–4	1	0.0	142	87	16:38	14	2	0	2	18	0	0	1
2001-02	St. Louis	NHL	66	5	7	12	78	1	0	2	37	13.5	3	0	0.0	138	69	16:55	10	0	1	1	4	0	0	0
	NHL Totals		**141**	**7**	**15**	**22**	**147**	**1**	**0**	**3**	**97**	**7.2**		**1**	**0.0**	**280**	**156**	**16:46**	**24**	**2**	**1**	**3**	**22**	**0**	**0**	**1**

Signed as a free agent by **St. Louis**, December 16, 1996.

SAMSONOV, Sergei

(sam-SAWN-nahf, SAIR-gay) **BOS.**

Left wing. Shoots right. 5'8", 180 lbs. Born, Moscow, USSR, October 27, 1978. Boston's 2nd choice, 8th overall, in 1997 Entry Draft.

Season	Club	League	GP	G	A	Pts	PIM	PP	SH	GW	S	%	+/-	TF	F%	H	SB	Min	GP	G	A	Pts	PIM	PP	SH	GW
1994-95	CSKA Moscow Jr.	CIS-Jr.	50	110	72	182	....																			
	CSKA Moscow	CIS	13	2	2	4	14												2	0	0	0	0			
1995-96	CSKA Moscow	CIS	51	21	17	38	12												3	1	1	2	4			
1996-97	Detroit Vipers	IHL	73	29	35	64	18												19	8	4	12	12			
1997-98	**Boston**	**NHL**	**81**	**22**	**25**	**47**	**8**	**7**	**0**	**3**	**159**	**13.8**	**9**						**6**	**2**	**5**	**7**	**0**	**0**	**0**	**1**
1998-99	**Boston**	**NHL**	**79**	**25**	**26**	**51**	**18**	**6**	**0**	**8**	**160**	**15.6**	**–6**	**0**	**0.0**	**39**	**10**	**16:23**	**11**	**3**	**1**	**4**	**0**	**0**	**0**	**0**
99-2000	**Boston**	**NHL**	**77**	**19**	**26**	**45**	**4**	**6**	**0**	**3**	**145**	**13.1**	**–6**	**3**	**0.0**	**29**	**9**	**16:32**								
2000-01	**Boston**	**NHL**	**82**	**29**	**46**	**75**	**18**	**3**	**0**	**3**	**215**	**13.5**	**6**	**14**	**42.9**	**41**	**19**	**19:23**								
2001-02	**Boston**	**NHL**	**74**	**29**	**41**	**70**	**27**	**3**	**0**	**4**	**192**	**15.1**	**21**	**1**	**0.0**	**29**	**8**	**18:47**	**6**	**2**	**2**	**4**	**0**	**0**	**0**	**0**
	Russia	Olympics	6	1	2	3	4																			
	NHL Totals		**393**	**124**	**164**	**288**	**75**	**25**	**0**	**21**	**871**	**14.2**		**18**	**33.3**	**138**	**46**	**17:47**	**23**	**7**	**8**	**15**	**0**	**0**	**0**	**1**

Won Garry F. Longman Memorial Trophy (Top Rookie - IHL) (1997) • NHL All-Rookie Team (1998) • Won Calder Memorial Trophy (1998) • Played in NHL All-Star Game (2001)

SAMUELSSON, Mikael

(SAM-yuhl-suhn, MIH-kigh-ehl) **NYR**

Right wing. Shoots left. 6'1", 195 lbs. Born, Mariefred, Sweden, December 23, 1976. San Jose's 7th choice, 145th overall, in 1998 Entry Draft.

Season	Club	League	GP	G	A	Pts	PIM	PP	SH	GW	S	%	+/-	TF	F%	H	SB	Min	GP	G	A	Pts	PIM	PP	SH	GW
1994-95	Sodertalje Jr.	Swede-Jr.	30	8	6	14	12																			
1995-96	Sodertalje Jr.	Swede-Jr.	22	13	12	25	20																			
	Sodertalje SK	Swede-2	18	5	1	6	0												4	0	0	0	0			
1996-97	Sodertalje Jr.	Swede-Jr.	2	2	1	3	....																			
	Sodertalje SK	Sweden	29	3	2	5	10												10	0	0	0	4			
1997-98	IK Nykopings	Swede-2	10	5	1	6	14																			
	Sodertalje SK	Sweden	41	11	9	20	66																			
1998-99	Sodertalje SK	Swede-2	18	13	10	23	26												10	2	2	4	12			
	Vastra Frolunda	Sweden	27	0	5	5	10																			
99-2000	Brynas IF Gavle	Sweden	40	4	3	7	76												11	7	2	9	6			
	Brynas IF Gavle	EuroHL	4	0	2	2	4																			
2000-01	**San Jose**	**NHL**	**4**	**0**	**0**	**0**	**0**	**0**	**0**	**0**	**3**	**0.0**	**0**	**0**	**0.0**	**2**	**0**	**4:41**								
	Kentucky	AHL	66	32	46	78	58												3	1	0	1	0			
2001-02	**NY Rangers**	**NHL**	**67**	**6**	**10**	**16**	**23**	**1**	**2**	**1**	**94**	**6.4**	**10**	**5**	**40.0**	**59**	**31**	**11:52**								
	Hartford	AHL	8	3	6	9	12																			
	NHL Totals		**71**	**6**	**10**	**16**	**23**	**1**	**2**	**1**	**97**	**6.2**		**5**	**40.0**	**61**	**31**	**11:27**								

Traded to **NY Rangers** by **San Jose** with Christian Gosselin for Adam Graves and future considerations, June 24, 2001.

SANDERSON, Geoff

(SAN-duhr-sohn, JEHF) **CBJ**

Left wing. Shoots left. 6', 190 lbs. Born, Hay River, N.W.T., February 1, 1972. Hartford's 2nd choice, 36th overall, in 1990 Entry Draft.

Season	Club	League	GP	G	A	Pts	PIM	PP	SH	GW	S	%	+/-	TF	F%	H	SB	Min	GP	G	A	Pts	PIM	PP	SH	GW
1987-88	St. Albert	AMHL	45	65	55	120	175																			
1988-89	Swift Current	WHL	58	17	11	28	16												12	3	5	8	6			
1989-90	Swift Current	WHL	70	32	62	94	56												4	1	4	5	8			
1990-91	Swift Current	WHL	70	62	50	112	57												3	1	2	3	4			
	Hartford	**NHL**	**2**	**1**	**0**	**1**	**0**	**0**	**0**	**0**	**2**	**50.0**	**–2**						**3**	**0**	**0**	**0**	**0**	**0**	**0**	**0**
	Springfield	AHL																	1	0	0	0	2			
1991-92	**Hartford**	**NHL**	**64**	**13**	**18**	**31**	**18**	**2**	**0**	**1**	**98**	**13.3**	**5**						**7**	**1**	**0**	**1**	**2**	**0**	**0**	**0**
1992-93	**Hartford**	**NHL**	**82**	**46**	**43**	**89**	**28**	**21**	**2**	**4**	**271**	**17.0**	**–21**													
1993-94	**Hartford**	**NHL**	**82**	**41**	**26**	**67**	**42**	**15**	**1**	**6**	**266**	**15.4**	**–13**													
1994-95	HPK Hameenlinna	Finland	12	6	4	10	24																			
	Hartford	**NHL**	**46**	**18**	**14**	**32**	**24**	**4**	**0**	**4**	**170**	**10.6**	**–10**													
1995-96	**Hartford**	**NHL**	**81**	**34**	**31**	**65**	**40**	**6**	**0**	**7**	**314**	**10.8**	**0**													
1996-97	**Hartford**	**NHL**	**82**	**36**	**31**	**67**	**29**	**12**	**1**	**4**	**297**	**12.1**	**–9**													
1997-98	**Carolina**	**NHL**	**40**	**7**	**10**	**17**	**14**	**2**	**0**	**0**	**96**	**7.3**	**–4**													
	Vancouver	**NHL**	**9**	**0**	**3**	**3**	**4**	**0**	**0**	**0**	**29**	**0.0**	**–1**													
	Buffalo	**NHL**	**26**	**4**	**5**	**9**	**20**	**0**	**0**	**2**	**72**	**5.6**	**6**						**14**	**3**	**1**	**4**	**4**	**1**	**0**	**1**
1998-99	**Buffalo**	**NHL**	**75**	**12**	**18**	**30**	**22**	**1**	**0**	**1**	**155**	**7.7**	**8**	**4**	**50.0**	**43**	**9**	**12:55**	**19**	**4**	**6**	**10**	**14**	**0**	**0**	**1**
99-2000	**Buffalo**	**NHL**	**67**	**13**	**13**	**26**	**22**	**4**	**0**	**3**	**136**	**9.6**	**4**	**3**	**100.0**	**38**	**17**	**12:54**	**5**	**0**	**2**	**2**	**8**	**0**	**0**	**0**
2000-01	**Columbus**	**NHL**	**68**	**30**	**26**	**56**	**46**	**9**	**0**	**7**	**199**	**15.1**	**4**	**726**	**49.0**	**56**	**17**	**16:37**								
2001-02	**Columbus**	**NHL**	**42**	**11**	**5**	**16**	**12**	**5**	**0**	**2**	**112**	**9.8**	**–15**	**322**	**42.9**	**30**	**6**	**16:50**								
	NHL Totals		**766**	**266**	**243**	**509**	**321**	**81**	**4**	**41**	**2217**	**12.0**		**1055**	**47.3**	**167**	**49**	**14:34**	**48**	**8**	**9**	**17**	**28**	**1**	**0**	**2**

Played in NHL All-Star Game (1994, 1997)

Transferred to **Carolina** after **Hartford** franchise relocated, June 25, 1997. Traded to **Vancouver** by **Carolina** with Sean Burke and Enrico Ciccone for Kirk McLean and Martin Gelinas, January 3, 1998. Traded to **Buffalo** by **Vancouver** for Brad May and Buffalo's 3rd round choice (later traded to Tampa Bay - Tampa Bay selected Jimmie Olvestad) in 1999 Entry Draft, February 4, 1998. Selected by **Columbus** from **Buffalo** in Expansion Draft, June 23, 2000.

SAPRYKIN, Oleg

(sah-PRIH-kihn, OH-lehg) **CGY.**

Center. Shoots left. 6', 195 lbs. Born, Moscow, USSR, February 12, 1981. Calgary's 1st choice, 11th overall, in 1999 Entry Draft.

Season	Club	League	GP	G	A	Pts	PIM	PP	SH	GW	S	%	+/-	TF	F%	H	SB	Min	GP	G	A	Pts	PIM	PP	SH	GW
1997-98	H.C. CSKA 2	Russia-3	15	0	3	3	6																			
	H.C. CSKA	Russia	20	0	2	2	8																			
1998-99	Seattle	WHL	66	47	46	93	107												11	5	11	16	36			
99-2000	**Calgary**	**NHL**	**4**	**0**	**1**	**1**	**2**	**0**	**0**	**0**	**2**	**0.0**	**–4**	**0**	**0.0**	**7**	**2**	**12:35**								
	Seattle	WHL	48	30	36	66	91												6	3	3	6	37			
2000-01	**Calgary**	**NHL**	**59**	**9**	**14**	**23**	**43**	**2**	**0**	**0**	**95**	**9.5**	**4**	**2**	**50.0**	**34**	**12**	**12:10**								
2001-02	**Calgary**	**NHL**	**3**	**0**	**0**	**0**	**0**	**0**	**0**	**0**	**9**	**0.0**	**–2**	**0**	**0.0**	**1**	**0**	**13:25**								
	Saint John	AHL	52	5	19	24	53																			
	NHL Totals		**66**	**9**	**15**	**24**	**45**	**2**	**0**	**0**	**106**	**8.5**		**2**	**50.0**	**42**	**14**	**12:15**								

WHL West Second All-Star Team (1999, 2000)

• Returned to **Seattle** (WHL) by **Calgary**, October 18, 1999.

SARAULT, Yves

(sah-ROH, EEV)

Left wing. Shoots left. 6'1", 190 lbs. Born, Valleyfield, Que., December 23, 1972. Montreal's 4th choice, 61st overall, in 1991 Entry Draft.

Season	Club	League	GP	G	A	Pts	PIM	PP	SH	GW	S	%	+/-	TF	F%	H	SB	Min	GP	G	A	Pts	PIM	PP	SH	GW
1987-88	Lac St-Louis	QAAA	4	0	0	0	4																			
1988-89	Lac St-Louis	QAAA	42	23	30	53	64												3	2	3	5	4			
1989-90	Victoriaville	QMJHL	70	12	28	40	140												16	0	3	3	26			
1990-91	St-Jean Lynx	QMJHL	56	22	24	46	113																			
1991-92	St-Jean Lynx	QMJHL	50	28	38	66	96																			
	Trois-Rivieres	QMJHL	18	15	14	29	12												15	10	10	20	18			
1992-93	Fredericton	AHL	59	14	17	31	41												3	0	1	1	2			
	Wheeling	ECHL	2	1	3	4	0																			
1993-94	Fredericton	AHL	60	13	14	27	72																			
1994-95	Fredericton	AHL	69	24	21	45	96												13	2	1	3	33			
	Montreal	**NHL**	**8**	**0**	**1**	**1**	**0**	**0**	**0**	**0**	**9**	**0.0**	**–1**													
1995-96	**Montreal**	**NHL**	**14**	**0**	**0**	**0**	**4**	**0**	**0**	**0**	**14**	**0.0**	**–7**													
	Calgary	**NHL**	**11**	**2**	**1**	**3**	**4**	**0**	**0**	**1**	**12**	**16.7**	**–2**													
	Saint John	AHL	26	10	12	22	34												16	6	2	8	33			
1996-97	**Colorado**	**NHL**	**28**	**2**	**1**	**3**	**6**	**0**	**0**	**0**	**41**	**4.9**	**0**						**5**	**0**	**0**	**0**	**2**	**0**	**0**	**0**
	Hershey Bears	AHL	6	2	3	5	8																			
1997-98	**Colorado**	**NHL**	**2**	**1**	**0**	**1**	**0**	**0**	**0**	**0**	**1**	**100.0**	**1**													
	Hershey Bears	AHL	63	23	36	59	43												7	1	2	3	14			
1998-99	**Ottawa**	**NHL**	**11**	**0**	**1**	**1**	**4**	**0**	**0**	**0**	**7**	**0.0**	**1**	**0**	**0.0**	**15**	**1**	**7:15**								
	Detroit Vipers	IHL	36	11	12	23	52												11	7	2	9	40			

			Regular Season																Playoffs							
Season	Club	League	GP	G	A	Pts	PIM	PP	SH	GW	S	%	+/-	TF	F%	H	SB	Min	GP	G	A	Pts	PIM	PP	SH	GW
99-2000	**Ottawa**	**NHL**	**11**	**0**	**2**	**2**	**7**	**0**	**0**	**0**	**13**	**0.0**	**–3**	**1**	**0.0**	**5**	**1**	**9:03**	....	....	....	....	....	....	....	....
	Grand Rapids	IHL	62	17	26	43	77	....	....	....	....	....	....	....	....	....	....	....	17	7	4	11	32	....	....	....
2000-01	**Atlanta**	**NHL**	**20**	**5**	**4**	**9**	**26**	**2**	**0**	**0**	**44**	**11.4**	**–9**	**2**	**50.0**	**30**	**4**	**13:08**	....	....	....	....	....	....	....	....
	Orlando	IHL	35	17	17	34	42	....	....	....	....	....	....	....	....	....	....	....	....	....	....	....	....	....	....	....
2001-02	**Nashville**	**NHL**	**1**	**0**	**0**	**0**	**0**	**0**	**0**	**0**	**0**	**0.0**	**0**	**0**	**0.0**	**0**	**0**	**5:15**	....	....	....	....	....	....	....	....
	Milwaukee	AHL	27	5	5	10	24	....	....	....	....	....	....	....	....	....	....	....	....	....	....	....	....	....	....	....
	Philadelphia	AHL	7	0	2	2	9	....	....	....	....	....	....	....	....	....	....	....	5	0	0	0	6	....	....	....
	NHL Totals		**106**	**10**	**10**	**20**	**51**	**2**	**0**	**1**	**141**	**7.1**		**3**	**33.3**	**50**	**6**	**10:24**	**5**	**0**	**0**	**0**	**2**	**0**	**0**	**0**

QMJHL Second All-Star Team (1992)

Traded to **Calgary** by **Montreal** with Craig Ferguson for Calgary's 8th round choice (Petr Kubos) in 1997 Entry Draft, November 26, 1995. Signed as a free agent by **Colorado**, September 13, 1996. Signed as a free agent by **Ottawa**, August 7, 1998. Signed as a free agent by **Atlanta**, July 20, 2000. Claimed on waivers by **Nashville** from **Atlanta**, June 19, 2001. Traded to **Philadelphia** by **Nashville** with a conditional choice in 2003 Entry Draft for Petr Hubacek and Jason Beckett, January 11, 2002.

SARICH, Cory

(SAHR-ihch, KOH-ree) **T.B.**

Defense. Shoots right. 6'3", 204 lbs. Born, Saskatoon, Sask., August 16, 1978. Buffalo's 2nd choice, 27th overall, in 1996 Entry Draft.

Season	Club	League	GP	G	A	Pts	PIM	PP	SH	GW	S	%	+/-	TF	F%	H	SB	Min	GP	G	A	Pts	PIM	PP	SH	GW
1994-95	Sask. Contacts	SMHL	31	5	22	27	99	....	....	....	....	....	....	....	....	....	....	....	....	....	....	....	....	....	....	....
	Saskatoon Blades	WHL	6	0	0	0	4	....	....	....	....	....	....	....	....	....	....	....	3	0	1	1	0	....	....	....
1995-96	Saskatoon Blades	WHL	59	5	18	23	54	....	....	....	....	....	....	....	....	....	....	....	3	0	0	0	4	....	....	....
1996-97	Saskatoon Blades	WHL	58	6	27	33	158	....	....	....	....	....	....	....	....	....	....	....	....	....	....	....	....	....	....	....
1997-98	Saskatoon Blades	WHL	33	5	24	29	90	....	....	....	....	....	....	....	....	....	....	....	....	....	....	....	....	....	....	....
	Seattle	WHL	13	3	16	19	47	....	....	....	....	....	....	....	....	....	....	....	0	0	0	0	0	....	....	....
1998-99	**Buffalo**	**NHL**	**4**	**0**	**0**	**0**	**0**	**0**	**0**	**0**	**2**	**0.0**	**3**	**0**	**0.0**	**7**	**0**	**13:11**	....	....	....	....	....	....	....	....
	Rochester	AHL	77	3	26	29	82	....	....	....	....	....	....	....	....	....	....	....	20	2	4	6	14	....	....	....
99-2000	**Buffalo**	**NHL**	**42**	**0**	**4**	**4**	**35**	**0**	**0**	**0**	**49**	**0.0**	**2**	**0**	**0.0**	**108**	**31**	**17:42**	....	....	....	....	....	....	....	....
	Rochester	AHL	15	0	6	6	44	....	....	....	....	....	....	....	....	....	....	....	....	....	....	....	....	....	....	....
	Tampa Bay	**NHL**	**17**	**0**	**2**	**2**	**42**	**0**	**0**	**0**	**20**	**0.0**	**–8**	**0**	**0.0**	**44**	**12**	**20:42**	....	....	....	....	....	....	....	....
2000-01	**Tampa Bay**	**NHL**	**73**	**1**	**8**	**9**	**106**	**0**	**0**	**1**	**66**	**1.5**	**–25**	**3**	**0.0**	**139**	**84**	**18:44**	....	....	....	....	....	....	....	....
	Detroit Vipers	IHL	3	0	2	2	2	....	....	....	....	....	....	....	....	....	....	....	....	....	....	....	....	....	....	....
2001-02	**Tampa Bay**	**NHL**	**72**	**0**	**11**	**11**	**105**	**0**	**0**	**0**	**55**	**0.0**	**–4**	**2**	**50.0**	**112**	**59**	**16:06**	....	....	....	....	....	....	....	....
	Springfield	AHL	2	0	0	0	0	....	....	....	....	....	....	....	....	....	....	....	....	....	....	....	....	....	....	....
	NHL Totals		**208**	**1**	**25**	**26**	**288**	**0**	**0**	**1**	**192**	**0.5**		**5**	**20.0**	**410**	**186**	**17:40**	....	....	....	....	....	....	....	....

WHL West Second All-Star Team (1998)

Traded to **Tampa Bay** by **Buffalo** with Wayne Primeau, Brian Holzinger and Buffalo's 3rd round choice (Alexander Kharitonov) in 2000 Entry Draft for Chris Gratton and Tampa Bay's 2nd round choice (Derek Roy) in 2001 Entry Draft, March 9, 2000.

SATAN, Miroslav

(SHA-tuhn, MEER-oh-slahv) **BUF.**

Left wing. Shoots left. 6'3", 190 lbs. Born, Topolcany, Czech., October 22, 1974. Edmonton's 6th choice, 111th overall, in 1993 Entry Draft.

Season	Club	League	GP	G	A	Pts	PIM	PP	SH	GW	S	%	+/-	TF	F%	H	SB	Min	GP	G	A	Pts	PIM	PP	SH	GW
1991-92	Topolcany Jr.	Czech-Jr.	31	30	22	52	....	....	....	....	....	....	....	....	....	....	....	....	....	....	....	....	....	....	....	....
	VTJ Topolcany	Czech-2	9	2	1	3	6	....	....	....	....	....	....	....	....	....	....	....	....	....	....	....	....	....	....	....
1992-93	Dukla Trencin	Czech	38	11	6	17	....	....	....	....	....	....	....	....	....	....	....	....	....	....	....	....	....	....	....	....
1993-94	Dukla Trencin	Slovakia	30	32	16	48	16	....	....	....	....	....	....	....	....	....	....	....	....	....	....	....	....	....	....	....
	Slovakia	Olympics	8	*9	0	9	0	....	....	....	....	....	....	....	....	....	....	....	....	....	....	....	....	....	....	....
1994-95	Cape Breton	AHL	25	24	16	40	15	....	....	....	....	....	....	....	....	....	....	....	....	....	....	....	....	....	....	....
	Detroit Vipers	IHL	8	1	3	4	4	....	....	....	....	....	....	....	....	....	....	....	....	....	....	....	....	....	....	....
	San Diego Gulls	IHL	6	0	2	2	6	....	....	....	....	....	....	....	....	....	....	....	....	....	....	....	....	....	....	....
1995-96	**Edmonton**	**NHL**	**62**	**18**	**17**	**35**	**22**	**6**	**0**	**4**	**113**	**15.9**	**0**	....	....	....	....	....	....	....	....	....	....	....	....	....
1996-97	**Edmonton**	**NHL**	**64**	**17**	**11**	**28**	**22**	**5**	**0**	**2**	**90**	**18.9**	**–4**	....	....	....	....	....	....	....	....	....	....	....	....	....
	Buffalo	**NHL**	**12**	**8**	**2**	**10**	**4**	**2**	**0**	**1**	**29**	**27.6**	**1**	....	....	....	....	....	**7**	**0**	**0**	**0**	**0**	**0**	**0**	**0**
1997-98	**Buffalo**	**NHL**	**79**	**22**	**24**	**46**	**34**	**9**	**0**	**4**	**139**	**15.8**	**2**	....	....	....	....	....	**14**	**5**	**4**	**9**	**4**	**4**	**0**	**1**
1998-99	**Buffalo**	**NHL**	**81**	**40**	**26**	**66**	**44**	**13**	**3**	**6**	**208**	**19.2**	**24**	**9**	**55.6**	**42**	**24**	**20:49**	**12**	**3**	**5**	**8**	**2**	**1**	**0**	**1**
99-2000	Dukla Trencin	Slovakia	3	2	8	10	2	....	....	....	....	....	....	....	....	....	....	....	....	....	....	....	....	....	....	....
	Buffalo	**NHL**	**81**	**33**	**34**	**67**	**32**	**5**	**3**	**5**	**265**	**12.5**	**16**	**7**	**14.3**	**45**	**28**	**20:35**	**5**	**3**	**2**	**5**	**0**	**0**	**0**	**0**
2000-01	**Buffalo**	**NHL**	**82**	**29**	**33**	**62**	**36**	**8**	**2**	**4**	**206**	**14.1**	**5**	**11**	**36.4**	**36**	**30**	**19:56**	**13**	**3**	**10**	**13**	**8**	**1**	**0**	**0**
2001-02	**Buffalo**	**NHL**	**82**	**37**	**36**	**73**	**33**	**15**	**5**	**5**	**267**	**13.9**	**14**	**4**	**50.0**	**33**	**31**	**21:10**	....	....	....	....	....	....	....	....
	Slovakia	Olympics	2	0	1	1	0	....	....	....	....	....	....	....	....	....	....	....	....	....	....	....	....	....	....	....
	NHL Totals		**543**	**204**	**183**	**387**	**227**	**63**	**13**	**31**	**1317**	**15.5**		**31**	**38.7**	**156**	**113**	**20:37**	**51**	**14**	**21**	**35**	**14**	**6**	**0**	**2**

Played in NHL All-Star Game (2000)

Traded to **Buffalo** by **Edmonton** for Barrie Moore and Craig Millar, March 18, 1997.

SAVAGE, Andre

(SA-vahj, AWN-dray) **PHI.**

Center. Shoots right. 6', 195 lbs. Born, Ottawa, Ont., May 27, 1975.

Season	Club	League	GP	G	A	Pts	PIM	PP	SH	GW	S	%	+/-	TF	F%	H	SB	Min	GP	G	A	Pts	PIM	PP	SH	GW
1992-93	Gloucester	OCJHL	54	34	34	68	38	....	....	....	....	....	....	....	....	....	....	....	....	....	....	....	....	....	....	....
1993-94	Gloucester	OCJHL	57	43	74	117	44	....	....	....	....	....	....	....	....	....	....	....	....	....	....	....	....	....	....	....
1994-95	Michigan Tech	WCHA	39	7	17	24	56	....	....	....	....	....	....	....	....	....	....	....	....	....	....	....	....	....	....	....
1995-96	Michigan Tech	WCHA	38	13	27	40	42	....	....	....	....	....	....	....	....	....	....	....	....	....	....	....	....	....	....	....
1996-97	Michigan Tech	WCHA	37	18	20	38	34	....	....	....	....	....	....	....	....	....	....	....	....	....	....	....	....	....	....	....
1997-98	Michigan Tech	WCHA	33	14	27	41	34	....	....	....	....	....	....	....	....	....	....	....	....	....	....	....	....	....	....	....
1998-99	**Boston**	**NHL**	**6**	**1**	**0**	**1**	**0**	**0**	**0**	**0**	**8**	**12.5**	**2**	**32**	**65.6**	**3**	**1**	**9:31**	....	....	....	....	....	....	....	....
	Providence	AHL	63	27	42	69	54	....	....	....	....	....	....	....	....	....	....	....	5	0	1	1	0	....	....	....
99-2000	**Boston**	**NHL**	**43**	**7**	**13**	**20**	**10**	**2**	**0**	**1**	**70**	**10.0**	**–8**	**619**	**55.1**	**50**	**10**	**14:40**	....	....	....	....	....	....	....	....
	Providence	AHL	30	15	17	32	22	....	....	....	....	....	....	....	....	....	....	....	14	6	7	13	22	....	....	....
2000-01	**Boston**	**NHL**	**1**	**0**	**0**	**0**	**0**	**0**	**0**	**0**	**1**	**0.0**	**0**	**3**	**100.0**	**0**	**0**	**4:38**	....	....	....	....	....	....	....	....
	Providence	AHL	35	13	15	28	47	....	....	....	....	....	....	....	....	....	....	....	17	3	4	7	18	....	....	....
2001-02	Manitoba Moose	AHL	76	35	26	61	115	....	....	....	....	....	....	....	....	....	....	....	6	2	3	5	16	....	....	....
	NHL Totals		**50**	**8**	**13**	**21**	**10**	**2**	**0**	**1**	**79**	**10.1**		**654**	**55.8**	**53**	**11**	**13:51**	....	....	....	....	....	....	....	....

WCHA First All-Star Team (1998)

Signed as a free agent by **Boston**, June 18, 1998. Signed as a free agent by **Vancouver**, August 2, 2001. Signed as a free agent by **Philadelphia**, August 20, 2002.

SAVAGE, Brian

(SA-vuhj, BRIGH-uhn) **PHX.**

Right wing. Shoots left. 6'1", 205 lbs. Born, Sudbury, Ont., February 24, 1971. Montreal's 11th choice, 171st overall, in 1991 Entry Draft.

Season	Club	League	GP	G	A	Pts	PIM	PP	SH	GW	S	%	+/-	TF	F%	H	SB	Min	GP	G	A	Pts	PIM	PP	SH	GW
1989-90	Sud. Cub Wolves	NOJHA	32	45	40	85	61	....	....	....	....	....	....	....	....	....	....	....	....	....	....	....	....	....	....	....
1990-91	Miami-Ohio	CCHA	28	5	6	11	26	....	....	....	....	....	....	....	....	....	....	....	....	....	....	....	....	....	....	....
1991-92	Miami-Ohio	CCHA	40	24	16	40	43	....	....	....	....	....	....	....	....	....	....	....	....	....	....	....	....	....	....	....
1992-93	Miami-Ohio	CCHA	38	*37	21	58	44	....	....	....	....	....	....	....	....	....	....	....	....	....	....	....	....	....	....	....
1993-94	Team Canada	Nat-Tm	51	20	26	46	38	....	....	....	....	....	....	....	....	....	....	....	....	....	....	....	....	....	....	....
	Canada	Olympics	8	2	2	4	6	....	....	....	....	....	....	....	....	....	....	....	....	....	....	....	....	....	....	....
	Montreal	**NHL**	**3**	**1**	**0**	**1**	**0**	**0**	**0**	**0**	**3**	**33.3**	**0**	....	....	....	....	....	**3**	**0**	**2**	**2**	**0**	**0**	**0**	**0**
	Fredericton	AHL	17	12	15	27	4	....	....	....	....	....	....	....	....	....	....	....	....	....	....	....	....	....	....	....
1994-95	**Montreal**	**NHL**	**37**	**12**	**7**	**19**	**27**	**0**	**0**	**0**	**64**	**18.8**	**5**	....	....	....	....	....	....	....	....	....	....	....	....	....
1995-96	**Montreal**	**NHL**	**75**	**25**	**8**	**33**	**28**	**4**	**0**	**4**	**150**	**16.7**	**–8**	....	....	....	....	....	**6**	**0**	**2**	**2**	**2**	**0**	**0**	**0**
1996-97	**Montreal**	**NHL**	**81**	**23**	**37**	**60**	**39**	**5**	**0**	**2**	**219**	**10.5**	**–14**	....	....	....	....	....	**5**	**1**	**1**	**2**	**0**	**0**	**0**	**0**
1997-98	**Montreal**	**NHL**	**64**	**26**	**17**	**43**	**36**	**8**	**0**	**7**	**152**	**17.1**	**11**	....	....	....	....	....	**9**	**0**	**2**	**2**	**6**	**0**	**0**	**0**
1998-99	**Montreal**	**NHL**	**54**	**16**	**10**	**26**	**20**	**5**	**0**	**4**	**124**	**12.9**	**–14**	**70**	**44.3**	**57**	**14**	**16:30**	....	....	....	....	....	....	....	....
99-2000	**Montreal**	**NHL**	**38**	**17**	**12**	**29**	**19**	**6**	**1**	**5**	**107**	**15.9**	**–4**	**67**	**47.8**	**27**	**12**	**17:56**	....	....	....	....	....	....	....	....
2000-01	**Montreal**	**NHL**	**62**	**21**	**24**	**45**	**26**	**12**	**0**	**1**	**172**	**12.2**	**–13**	**30**	**56.7**	**70**	**9**	**18:50**	....	....	....	....	....	....	....	....
2001-02	**Montreal**	**NHL**	**47**	**14**	**15**	**29**	**30**	**7**	**0**	**2**	**117**	**12.0**	**–14**	**3**	**0.0**	**41**	**13**	**18:12**	....	....	....	....	....	....	....	....
	Phoenix	**NHL**	**30**	**6**	**6**	**12**	**8**	**2**	**0**	**2**	**47**	**12.8**	**1**	**4**	**50.0**	**35**	**9**	**15:02**	**5**	**0**	**0**	**0**	**0**	**0**	**0**	**0**
	NHL Totals		**491**	**161**	**136**	**297**	**233**	**49**	**1**	**27**	**1155**	**13.9**		**174**	**47.1**	**230**	**57**	**17:31**	**28**	**1**	**7**	**8**	**8**	**0**	**0**	**0**

CCHA First All-Star Team (1993) • CCHA Player of the Year (1993) • NCAA West Second All-American Team (1993)

• Missed majority of 1999-2000 season recovering from neck injury suffered in game vs. LA Kings, November 20, 1999. Traded to **Phoenix** by **Montreal** with Montreal's 3rd round choice (Matt Jones) in 2002 Entry Draft and future considerations for Sergei Berezin, January 25, 2002.

SAVAGE, Reggie
(SA-vuhj, REH-jee)

Center. Shoots left. 5'10", 197 lbs. Born, Montreal, Que., May 1, 1970. Washington's 1st choice, 15th overall, in 1988 Entry Draft.

			Regular Season																Playoffs							
Season	Club	League	GP	G	A	Pts	PIM	PP	SH	GW	S	%	+/-	TF	F%	H	SB	Min	GP	G	A	Pts	PIM	PP	SH	GW
1985-86	Richelieu	QAAA	40	38	26	64	....	....	....	....	....	....	....	....	....	....	....	....	7	10	1	11	....	....	....	....
1986-87	Richelieu	QAAA	42	82	57	139	44	....	....	....	....	....	....	....	....	....	....	....	9	10	9	19	10	....	....	....
1987-88	Victoriaville	QMJHL	68	68	54	122	77	....	....	....	....	....	....	....	....	....	....	....	5	2	3	5	8	....	....	....
1988-89	Victoriaville	QMJHL	54	58	55	113	178	....	....	....	....	....	....	....	....	....	....	....	16	15	13	28	52	....	....	....
1989-90	Victoriaville	QMJHL	63	51	43	94	79	....	....	....	....	....	....	....	....	....	....	....	16	13	10	23	40	....	....	....
1990-91	**Washington**	**NHL**	**1**	**0**	**0**	**0**	**0**	**0**	**0**	**0**	**2**	**0.0**	**−1**	....	....	....	....	....	....	....	....	....	....	....	....	....
	Baltimore	AHL	62	32	29	61	10	....	....	....	....	....	....	....	....	....	....	....	6	1	1	2	6	....	....	....
1991-92	Baltimore	AHL	77	42	28	70	51	....	....	....	....	....	....	....	....	....	....	....	....	....	....	....	....	....	....	....
1992-93	**Washington**	**NHL**	**16**	**2**	**3**	**5**	**12**	**2**	**0**	**0**	**20**	**10.0**	**−4**	....	....	....	....	....	....	....	....	....	....	....	....	....
	Baltimore	AHL	40	37	18	55	28	....	....	....	....	....	....	....	....	....	....	....	....	....	....	....	....	....	....	....
1993-94	**Quebec**	**NHL**	**17**	**3**	**4**	**7**	**16**	**1**	**0**	**0**	**25**	**12.0**	**3**	....	....	....	....	....	....	....	....	....	....	....	....	....
	Cornwall Aces	AHL	33	21	13	34	56	....	....	....	....	....	....	....	....	....	....	....	....	....	....	....	....	....	....	....
1994-95	Cornwall Aces	AHL	34	13	7	20	56	....	....	....	....	....	....	....	....	....	....	....	14	5	6	11	40	....	....	....
1995-96	Atlanta Knights	IHL	66	22	14	36	118	....	....	....	....	....	....	....	....	....	....	....	....	....	....	....	....	....	....	....
	Syracuse Crunch	AHL	10	9	5	14	28	....	....	....	....	....	....	....	....	....	....	....	16	9	6	15	54	....	....	....
1996-97	Springfield	AHL	68	32	25	57	103	....	....	....	....	....	....	....	....	....	....	....	17	6	7	13	24	....	....	....
1997-98	Kansas City	IHL	51	6	10	16	60	....	....	....	....	....	....	....	....	....	....	....	....	....	....	....	....	....	....	....
	San Antonio	IHL	22	6	12	18	24	....	....	....	....	....	....	....	....	....	....	....	....	....	....	....	....	....	....	....
	Orlando	IHL	10	5	5	10	18	....	....	....	....	....	....	....	....	....	....	....	17	2	9	11	60	....	....	....
1998-99	Asiago Hockey	Alpenliga	27	25	27	52	69	....	....	....	....	....	....	....	....	....	....	....	....	....	....	....	....	....	....	....
	Asiago Hockey	Italy	16	18	15	33	8	....	....	....	....	....	....	....	....	....	....	....	2	1	0	1	22	....	....	....
99-2000	Syracuse Crunch	AHL	78	36	34	70	135	....	....	....	....	....	....	....	....	....	....	....	4	0	0	0	8	....	....	....
2000-01	Syracuse Crunch	AHL	78	37	24	61	90	....	....	....	....	....	....	....	....	....	....	....	5	0	2	2	16	....	....	....
2001-02	EHC Biel-Bienne	Swiss-2	31	17	10	27	24	....	....	....	....	....	....	....	....	....	....	....	....	....	....	....	....	....	....	....
	NHL Totals		**34**	**5**	**7**	**12**	**28**	**3**	**0**	**0**	**47**	**10.6**		....	....	....	....	....	....	....	....	....	....	....	....	....

Traded to **Quebec** by **Washington** with Paul MacDermid for Mike Hough, June 20, 1993. Signed as a free agent by **Phoenix**, August 28, 1996. Signed as a free agent by **Vancouver**, June 17, 1999. Signed as a free agent by **Columbus**, June 2, 2000. Signed as a free agent by **Biel-Bienne** (Swiss-2), June 21, 2001.

SAVARD, Marc
(sa-VAHR, MAHRK) **CGY.**

Center. Shoots left. 5'10", 185 lbs. Born, Ottawa, Ont., July 17, 1977. NY Rangers' 3rd choice, 91st overall, in 1995 Entry Draft.

Season	Club	League	GP	G	A	Pts	PIM	PP	SH	GW	S	%	+/-	TF	F%	H	SB	Min	GP	G	A	Pts	PIM	PP	SH	GW
1992-93	Metcalfe Jets	OJHL-B	36	*44	55	*99	38	....	....	....	....	....	....	....	....	....	....	....	....	....	....	....	....	....	....	....
1993-94	Oshawa Generals	OHL	61	18	39	57	20	....	....	....	....	....	....	....	....	....	....	....	5	4	3	7	8	....	....	....
1994-95	Oshawa Generals	OHL	66	43	96	*139	78	....	....	....	....	....	....	....	....	....	....	....	7	5	6	11	8	....	....	....
1995-96	Oshawa Generals	OHL	48	28	59	87	77	....	....	....	....	....	....	....	....	....	....	....	5	4	5	9	6	....	....	....
1996-97	Oshawa Generals	OHL	64	43	*87	*130	94	....	....	....	....	....	....	....	....	....	....	....	18	13	*24	*37	20	....	....	....
1997-98	**NY Rangers**	**NHL**	**28**	**1**	**5**	**6**	**4**	**0**	**0**	**0**	**32**	**3.1**	**−4**	....	....	....	....	....	....	....	....	....	....	....	....	....
	Hartford	AHL	58	21	53	74	66	....	....	....	....	....	....	....	....	....	....	....	15	8	19	27	24	....	....	....
1998-99	**NY Rangers**	**NHL**	**70**	**9**	**36**	**45**	**38**	**4**	**0**	**1**	**116**	**7.8**	**−7**	**956**	**48.4**	**30**	**18**	**14:35**	....	....	....	....	....	....	....	....
	Hartford	AHL	9	3	10	13	16	....	....	....	....	....	....	....	....	....	....	....	7	1	12	13	16	....	....	....
99-2000	**Calgary**	**NHL**	**78**	**22**	**31**	**53**	**56**	**4**	**0**	**3**	**184**	**12.0**	**−2**	**1021**	**49.6**	**44**	**33**	**16:36**	....	....	....	....	....	....	....	....
2000-01	**Calgary**	**NHL**	**77**	**23**	**42**	**65**	**46**	**10**	**1**	**5**	**197**	**11.7**	**−12**	**1050**	**53.1**	**48**	**19**	**19:13**	....	....	....	....	....	....	....	....
2001-02	**Calgary**	**NHL**	**56**	**14**	**19**	**33**	**48**	**7**	**0**	**3**	**140**	**10.0**	**−18**	**577**	**54.8**	**16**	**7**	**17:20**	....	....	....	....	....	....	....	....
	NHL Totals		**309**	**69**	**133**	**202**	**192**	**25**	**1**	**12**	**669**	**10.3**		**3604**	**51.1**	**138**	**77**	**16:58**	....	....	....	....	....	....	....	....

OHL Second All-Star Team (1995)

Traded to **Calgary** by **NY Rangers** with NY Rangers 1st round choice (Oleg Saprykin) in 1999 Entry Draft for the rights to Jan Hlavac and Calgary's 1st (Jamie Lundmark) and 3rd (later traded back to Calgary - Calgary selected Craig Andersson) round choices in 1999 Entry Draft, June 26, 1999.

SAWYER, Kevin
(SOI-yuhr, KEH-vihn) **ANA.**

Left wing. Shoots left. 6'2", 212 lbs. Born, Christina Lake, B.C., February 21, 1974.

Season	Club	League	GP	G	A	Pts	PIM	PP	SH	GW	S	%	+/-	TF	F%	H	SB	Min	GP	G	A	Pts	PIM	PP	SH	GW
1991-92	Grand Forks	KIJHL	24	9	11	20	200	....	....	....	....	....	....	....	....	....	....	....	....	....	....	....	....	....	....	....
	Kelowna Spartans	BCJHL	3	0	0	0	9	....	....	....	....	....	....	....	....	....	....	....	....	....	....	....	....	....	....	....
	Vernon Lakers	BCJHL	12	0	1	1	18	....	....	....	....	....	....	....	....	....	....	....	....	....	....	....	....	....	....	....
	Penticton	BCJHL	3	0	0	0	13	....	....	....	....	....	....	....	....	....	....	....	....	....	....	....	....	....	....	....
1992-93	Spokane Chiefs	WHL	62	4	3	7	274	....	....	....	....	....	....	....	....	....	....	....	8	1	1	2	13	....	....	....
1993-94	Spokane Chiefs	WHL	60	10	15	25	350	....	....	....	....	....	....	....	....	....	....	....	3	0	1	1	6	....	....	....
1994-95	Spokane Chiefs	WHL	54	7	9	16	365	....	....	....	....	....	....	....	....	....	....	....	11	2	0	2	58	....	....	....
	Peoria Rivermen	IHL	....	....	....	....	....	....	....	....	....	....	....	....	....	....	....	....	2	0	0	0	12	....	....	....
1995-96	**St. Louis**	**NHL**	**6**	**0**	**0**	**0**	**23**	**0**	**0**	**0**	**1**	**0.0**	**−2**	....	....	....	....	....	....	....	....	....	....	....	....	....
	Worcester	AHL	41	3	4	7	268	....	....	....	....	....	....	....	....	....	....	....	....	....	....	....	....	....	....	....
	Boston	**NHL**	**2**	**0**	**0**	**0**	**5**	**0**	**0**	**0**	**0**	**0.0**	**1**	....	....	....	....	....	....	....	....	....	....	....	....	....
	Providence	AHL	4	0	0	0	29	....	....	....	....	....	....	....	....	....	....	....	4	0	1	1	9	....	....	....
1996-97	**Boston**	**NHL**	**2**	**0**	**0**	**0**	**0**	**0**	**0**	**0**	**0**	**0.0**	**0**	....	....	....	....	....	....	....	....	....	....	....	....	....
	Providence	AHL	60	8	9	17	367	....	....	....	....	....	....	....	....	....	....	....	6	0	0	0	32	....	....	....
1997-98	Michigan K-Wings	IHL	60	2	5	7	*398	....	....	....	....	....	....	....	....	....	....	....	3	0	0	0	23	....	....	....
1998-99	Worcester	AHL	70	8	14	22	299	....	....	....	....	....	....	....	....	....	....	....	4	0	1	1	4	....	....	....
99-2000	**Phoenix**	**NHL**	**3**	**0**	**0**	**0**	**12**	**0**	**0**	**0**	**0**	**0.0**	**1**	**0**	**0.0**	**4**	**0**	**2:20**	....	....	....	....	....	....	....	....
	Springfield	AHL	56	4	8	12	321	....	....	....	....	....	....	....	....	....	....	....	4	0	0	0	6	....	....	....
2000-01	**Anaheim**	**NHL**	**9**	**0**	**1**	**1**	**27**	**0**	**0**	**0**	**6**	**0.0**	**−1**	**0**	**0.0**	**12**	**0**	**6:31**	....	....	....	....	....	....	....	....
	Cincinnati	AHL	41	2	12	14	211	....	....	....	....	....	....	....	....	....	....	....	....	....	....	....	....	....	....	....
2001-02	**Anaheim**	**NHL**	**57**	**1**	**1**	**2**	**221**	**0**	**0**	**0**	**29**	**3.4**	**−4**	**3**	**0.0**	**53**	**1**	**5:47**	....	....	....	....	....	....	....	....
	NHL Totals		**79**	**1**	**2**	**3**	**288**	**0**	**0**	**0**	**36**	**2.8**		**3**	**0.0**	**69**	**1**	**5:44**	....	....	....	....	....	....	....	....

Signed as a free agent by **St. Louis**, February 28, 1995. Traded to **Boston** by **St. Louis** with Steve Staios for Steve Leach, March 8, 1996. Signed as a free agent by **Dallas**, August 19, 1997. Signed as a free agent by **St. Louis**, September 4, 1998. Signed as a free agent by **Phoenix**, August 15, 1999. Signed as a free agent by **Anaheim**, July 13, 2000.

SCATCHARD, Dave
(SKAT-chuhrd, DAYV) **NYI**

Center. Shoots right. 6'2", 224 lbs. Born, Hinton, Alta., February 20, 1976. Vancouver's 3rd choice, 42nd overall, in 1994 Entry Draft.

Season	Club	League	GP	G	A	Pts	PIM	PP	SH	GW	S	%	+/-	TF	F%	H	SB	Min	GP	G	A	Pts	PIM	PP	SH	GW
1991-92	Salmon Arm	BCAHA	65	98	100	198	167	....	....	....	....	....	....	....	....	....	....	....	....	....	....	....	....	....	....	....
1992-93	Kimberley	RMJHL	51	20	23	43	61	....	....	....	....	....	....	....	....	....	....	....	....	....	....	....	....	....	....	....
1993-94	Portland	WHL	47	9	11	20	46	....	....	....	....	....	....	....	....	....	....	....	10	2	1	3	4	....	....	....
1994-95	Portland	WHL	71	20	30	50	148	....	....	....	....	....	....	....	....	....	....	....	8	0	3	3	21	....	....	....
1995-96	Portland	WHL	59	19	28	47	146	....	....	....	....	....	....	....	....	....	....	....	7	1	8	9	14	....	....	....
	Syracuse Crunch	AHL	1	0	0	0	0	....	....	....	....	....	....	....	....	....	....	....	15	2	5	7	29	....	....	....
1996-97	Syracuse Crunch	AHL	26	8	7	15	65	....	....	....	....	....	....	....	....	....	....	....	....	....	....	....	....	....	....	....
1997-98	**Vancouver**	**NHL**	**76**	**13**	**11**	**24**	**165**	**0**	**0**	**1**	**85**	**15.3**	**−4**	....	....	....	....	....	....	....	....	....	....	....	....	....
1998-99	**Vancouver**	**NHL**	**82**	**13**	**13**	**26**	**140**	**0**	**2**	**2**	**130**	**10.0**	**−12**	**1007**	**56.3**	**147**	**33**	**13:46**	....	....	....	....	....	....	....	....
99-2000	**Vancouver**	**NHL**	**21**	**0**	**4**	**4**	**24**	**0**	**0**	**0**	**25**	**0.0**	**−3**	**190**	**59.5**	**40**	**6**	**10:12**	....	....	....	....	....	....	....	....
	NY Islanders	**NHL**	**44**	**12**	**14**	**26**	**93**	**0**	**1**	**1**	**103**	**11.7**	**0**	**710**	**55.8**	**123**	**21**	**13:42**	....	....	....	....	....	....	....	....
2000-01	**NY Islanders**	**NHL**	**81**	**21**	**24**	**45**	**114**	**4**	**0**	**5**	**176**	**11.9**	**−9**	**1322**	**55.1**	**199**	**22**	**16:50**	....	....	....	....	....	....	....	....
2001-02	**NY Islanders**	**NHL**	**80**	**12**	**15**	**27**	**111**	**3**	**1**	**4**	**117**	**10.3**	**−4**	**788**	**53.8**	**181**	**19**	**12:31**	**7**	**1**	**1**	**2**	**22**	**0**	**0**	**0**
	NHL Totals		**384**	**71**	**81**	**152**	**647**	**7**	**4**	**13**	**636**	**11.2**		**4017**	**55.5**	**690**	**101**	**13:60**	**7**	**1**	**1**	**2**	**22**	**0**	**0**	**0**

Traded to **NY Islanders** by **Vancouver** with Kevin Weekes and Bill Muckalt for Felix Potvin and NY Islanders' compensatory 2nd (later traded to New Jersey - New Jersey selected Teemu Laine) and 3rd (Thatcher Bell) round choices in 2000 Entry Draft, December 19, 1999.

SCHAEFER, Peter
(SHAY-fuhr, PEE-tuhr) **VAN.**

Left wing. Shoots left. 5'11", 195 lbs. Born, Yellow Grass, Sask., July 12, 1977. Vancouver's 3rd choice, 66th overall, in 1995 Entry Draft.

Season	Club	League	GP	G	A	Pts	PIM	PP	SH	GW	S	%	+/-	TF	F%	H	SB	Min	GP	G	A	Pts	PIM	PP	SH	GW
1993-94	Yorkton Mallers	SMHL	32	27	14	41	133	....	....	....	....	....	....	....	....	....	....	....	....	....	....	....	....	....	....	....
	Brandon	WHL	2	1	0	1	0	....	....	....	....	....	....	....	....	....	....	....	....	....	....	....	....	....	....	....
1994-95	Brandon	WHL	68	27	32	59	34	....	....	....	....	....	....	....	....	....	....	....	18	5	3	8	18	....	....	....
1995-96	Brandon	WHL	69	47	61	108	53	....	....	....	....	....	....	....	....	....	....	....	19	10	13	23	5	....	....	....
1996-97	Brandon	WHL	61	49	74	123	85	....	....	....	....	....	....	....	....	....	....	....	6	1	4	5	4	....	....	....
	Syracuse Crunch	AHL	5	0	3	3	0	....	....	....	....	....	....	....	....	....	....	....	3	1	3	4	14	....	....	....
1997-98	Syracuse Crunch	AHL	73	19	44	63	41	....	....	....	....	....	....	....	....	....	....	....	5	2	1	3	2	....	....	....

			Regular Season																Playoffs							
Season	Club	League	GP	G	A	Pts	PIM	PP	SH	GW	S	%	+/-	TF	F%	H	SB	Min	GP	G	A	Pts	PIM	PP	SH	GW
1998-99	**Vancouver**	**NHL**	**25**	**4**	**4**	**8**	**8**	**1**	**0**	**1**	**24**	**16.7**	**-1**	**6**	**0.0**	**27**	**6**	**13:21**								
	Syracuse Crunch	AHL	41	10	19	29	66																			
99-2000	**Vancouver**	**NHL**	**71**	**16**	**15**	**31**	**20**	**2**	**2**	**4**	**101**	**15.8**	**0**	**21**	**19.1**	**56**	**32**	**15:28**								
	Syracuse Crunch	AHL	2	0	0	0	2																			
2000-01	**Vancouver**	**NHL**	**82**	**16**	**20**	**36**	**22**	**3**	**4**	**2**	**163**	**9.8**	**4**	**25**	**32.0**	**64**	**46**	**16:18**	**3**	**0**	**0**	**0**	**0**	**0**	**0**	**0**
2001-02	TPS Turku	Finland	33	16	15	31	93												8	1	2	3	2			
	NHL Totals		**178**	**36**	**39**	**75**	**50**	**6**	**6**	**7**	**288**	**12.5**		**52**	**23.1**	**147**	**84**	**15:33**	**3**	**0**	**0**	**0**	**0**	**0**	**0**	**0**

WHL East First All-Star Team (1996, 1997) • Canadian Major Junior First All-Star Team (1997)

Signed as a free agent by **TPS Turku** (Finland) with Vancouver retaining NHL rights, October 18, 2001.

SCHASTLIVY, Petr
(schust-LEE-vee, PEH-tuhr) **OTT.**

Left wing. Shoots left. 6'1", 204 lbs. Born, Angarsk, USSR, April 18, 1979. Ottawa's 5th choice, 101st overall, in 1998 Entry Draft.

Season	Club	League	GP	G	A	Pts	PIM	PP	SH	GW	S	%	+/-	TF	F%	H	SB	Min	GP	G	A	Pts	PIM	PP	SH	GW
1997-98	Yaroslavl	Russia	47	15	9	24	34																			
	Yaroslavl	Russia	4	0	0	0	0																			
1998-99	Yaroslavl	Russia	40	6	1	7	28												6	0	0	0	2			
99-2000	**Ottawa**	**NHL**	**13**	**2**	**5**	**7**	**2**	**1**	**0**	**1**	**22**	**9.1**	**4**	**0**	**0.0**	**2**	**2**	**12:18**	**1**	**0**	**0**	**0**	**0**	**0**	**0**	**0**
	Grand Rapids	IHL	46	16	12	28	10												17	8	7	15	6			
2000-01	**Ottawa**	**NHL**	**17**	**3**	**2**	**5**	**6**	**0**	**0**	**0**	**32**	**9.4**	**-1**	**0**	**0.0**	**4**	**2**	**11:20**								
	Grand Rapids	IHL	43	10	14	24	10												7	4	4	8	0			
2001-02	**Ottawa**	**NHL**	**1**	**0**	**1**	**1**	**0**	**0**	**0**	**0**	**0**	**0.0**	**1**	**0**	**0.0**	**0**	**0**	**3:21**								
	Grand Rapids	AHL	31	22	13	35	10																			
	NHL Totals		**31**	**5**	**8**	**13**	**8**	**1**	**0**	**1**	**54**	**9.3**		**0**	**0.0**	**6**	**4**	**11:29**	**1**	**0**	**0**	**0**	**0**	**0**	**0**	**0**

• Missed majority of 2001-02 season recovering from knee injury suffered in game vs. Chicago, December 31, 2001.

SCHNABEL, Robert
(SHNAH-buhl, RAW-buhrt) **NSH.**

Defense. Shoots left. 6'5", 230 lbs. Born, Prague, Czech., November 10, 1978. Phoenix's 7th choice, 129th overall, in 1998 Entry Draft.

Season	Club	League	GP	G	A	Pts	PIM	PP	SH	GW	S	%	+/-	TF	F%	H	SB	Min	GP	G	A	Pts	PIM	PP	SH	GW
1994-95	Slavia Praha Jr.	Czech-Jr.	35	11	6	17	14																			
1995-96	Slavia Praha Jr.	Czech-Jr.	38	3	5	8	....																			
1996-97	Slavia Praha Jr.	Czech-Jr.	36	5	2	7	....																			
	HC Slavia Praha	Czech	4	0	0	0	4												1	0	0	0	0			
1997-98	Red Deer Rebels	WHL	61	1	22	23	143												5	0	0	0	16			
1998-99	Red Deer Rebels	WHL	1	0	0	0	2																			
	Springfield	AHL	77	1	7	8	155												3	1	0	1	4			
99-2000	Springfield	AHL	40	2	8	10	133												5	0	0	0	4			
2000-01	Springfield	AHL	22	1	2	3	38																			
	Timra IK	Sweden	16	0	2	2	72																			
2001-02	**Nashville**	**NHL**	**1**	**0**	**0**	**0**	**0**	**0**	**0**	**0**	**0**	**0.0**	**0**	**0**	**0.0**	**1**	**2**	**7:16**								
	Milwaukee	AHL	67	2	7	9	130																			
	NHL Totals		**1**	**0**	**0**	**0**	**0**	**0**	**0**	**0**	**0**	**0.0**		**0**	**0.0**	**1**	**2**	**7:16**								

• Re-entered NHL Entry Draft. Originally NY Islanders' 5th choice, 79th overall, in 1997 Entry Draft.

Claimed on waivers by **Nashville** from **Phoenix**, January 2, 2001.

SCHNEIDER, Mathieu
(SHNIGH-duhr, MA-thew) **L.A.**

Defense. Shoots left. 5'10", 192 lbs. Born, New York, NY, June 12, 1969. Montreal's 4th choice, 44th overall, in 1987 Entry Draft.

Season	Club	League	GP	G	A	Pts	PIM	PP	SH	GW	S	%	+/-	TF	F%	H	SB	Min	GP	G	A	Pts	PIM	PP	SH	GW
1985-86	Mount St. Charles	Hi-School	19	3	27	30	....																			
1986-87	Cornwall Royals	OHL	63	7	29	36	75												5	0	0	0	22			
1987-88	Cornwall Royals	OHL	48	21	40	61	83												11	2	6	8	14			
	Montreal	**NHL**	**4**	**0**	**0**	**0**	**2**	**0**	**0**	**0**	**2**	**0.0**	**-1**													
	Sherbrooke	AHL	....	....	....	....	....												3	0	3	3	12			
1988-89	Cornwall Royals	OHL	59	16	57	73	96												18	7	20	27	30			
1989-90	**Montreal**	**NHL**	**44**	**7**	**14**	**21**	**25**	**5**	**0**	**1**	**84**	**8.3**	**2**						**9**	**1**	**3**	**4**	**31**	**1**	**0**	**0**
	Sherbrooke	AHL	28	6	13	19	20																			
1990-91	**Montreal**	**NHL**	**69**	**10**	**20**	**30**	**63**	**5**	**0**	**3**	**164**	**6.1**	**7**						**13**	**2**	**7**	**9**	**18**	**1**	**0**	**0**
1991-92	**Montreal**	**NHL**	**78**	**8**	**24**	**32**	**72**	**2**	**0**	**1**	**194**	**4.1**	**10**						**10**	**1**	**4**	**5**	**6**	**1**	**0**	**0**
1992-93♦	**Montreal**	**NHL**	**60**	**13**	**31**	**44**	**91**	**3**	**0**	**2**	**169**	**7.7**	**8**						**11**	**1**	**2**	**3**	**16**	**0**	**0**	**0**
1993-94	**Montreal**	**NHL**	**75**	**20**	**32**	**52**	**62**	**11**	**0**	**4**	**193**	**10.4**	**15**						**1**	**0**	**0**	**0**	**0**	**0**	**0**	**0**
1994-95	**Montreal**	**NHL**	**30**	**5**	**15**	**20**	**49**	**2**	**0**	**0**	**82**	**6.1**	**-3**													
	NY Islanders	**NHL**	**13**	**3**	**6**	**9**	**30**	**1**	**0**	**2**	**36**	**8.3**	**-5**													
1995-96	**NY Islanders**	**NHL**	**65**	**11**	**36**	**47**	**93**	**7**	**0**	**1**	**155**	**7.1**	**-18**													
	Toronto	**NHL**	**13**	**2**	**5**	**7**	**10**	**0**	**0**	**0**	**36**	**5.6**	**-2**						**6**	**0**	**4**	**4**	**8**	**0**	**0**	**0**
1996-97	**Toronto**	**NHL**	**26**	**5**	**7**	**12**	**20**	**1**	**0**	**1**	**63**	**7.9**	**3**													
1997-98	**Toronto**	**NHL**	**76**	**11**	**26**	**37**	**44**	**4**	**1**	**1**	**181**	**6.1**	**-12**													
	United States	Olympics	4	0	0	0	6																			
1998-99	**NY Rangers**	**NHL**	**75**	**10**	**24**	**34**	**71**	**5**	**0**	**2**	**159**	**6.3**	**-19**	**0**	**0.0**	**182**	**149**	**24:35**								
99-2000	**NY Rangers**	**NHL**	**80**	**10**	**20**	**30**	**78**	**3**	**0**	**1**	**228**	**4.4**	**-6**	**0**	**0.0**	**178**	**183**	**22:31**								
2000-01	**Los Angeles**	**NHL**	**73**	**16**	**35**	**51**	**56**	**7**	**1**	**2**	**183**	**8.7**	**0**	**0**	**0.0**	**179**	**137**	**23:04**	**13**	**0**	**9**	**9**	**10**	**0**	**0**	**0**
2001-02	**Los Angeles**	**NHL**	**55**	**7**	**23**	**30**	**68**	**4**	**0**	**0**	**123**	**5.7**	**3**	**0**	**0.0**	**107**	**86**	**22:25**	**7**	**0**	**1**	**1**	**18**	**0**	**0**	**0**
	NHL Totals		**836**	**138**	**318**	**456**	**834**	**60**	**2**	**21**	**2052**	**6.7**		**0**	**0.0**	**646**	**555**	**23:11**	**70**	**5**	**30**	**35**	**107**	**3**	**0**	**0**

OHL First All-Star Team (1988, 1989) • Played in NHL All-Star Game (1996)

Traded to **NY Islanders** by **Montreal** with Kirk Muller and Craig Darby for Pierre Turgeon and Vladimir Malakhov, April 5, 1995. Traded to **Toronto** by **NY Islanders** with Wendel Clark and D.J. Smith for Darby Hendrickson, Sean Haggerty, Kenny Jonsson and Toronto's 1st round choice (Roberto Luongo) in 1997 Entry Draft, March 13, 1996. • Missed majority of 1996-97 season recovering from groin injury suffered in game vs. St. Louis, December 27, 1996. Rights traded to **NY Rangers** by **Toronto** for Alexander Karpovtsev and NY Rangers' 4th round choice (Mirko Murovic) in 1999 Entry Draft, October 14, 1998. Selected by **Columbus** from **NY Rangers** in Expansion Draft, June 23, 2000. Signed as a free agent by **LA Kings**, August 14, 2000.

SCHULTZ, Nick
(SHULTZ, NIHK) **MIN.**

Defense. Shoots left. 6', 187 lbs. Born, Regina, Sask., August 25, 1982. Minnesota's 2nd choice, 33rd overall, in 2000 Entry Draft.

Season	Club	League	GP	G	A	Pts	PIM	PP	SH	GW	S	%	+/-	TF	F%	H	SB	Min	GP	G	A	Pts	PIM	PP	SH	GW
1997-98	Yorkton Mallers	SMHL	59	10	30	40	74																			
1998-99	Prince Albert	WHL	58	5	18	23	37												14	0	7	7	0			
99-2000	Prince Albert	WHL	72	11	33	44	38												6	0	3	3	2			
2000-01	Prince Albert	WHL	59	17	30	47	120																			
	Cleveland	IHL	4	1	1	2	2												3	0	1	1	0			
2001-02	**Minnesota**	**NHL**	**52**	**4**	**6**	**10**	**14**	**1**	**0**	**1**	**47**	**8.5**	**0**	**0**	**0.0**	**36**	**48**	**16:08**								
	Houston Aeros	AHL																	14	1	5	6	2			
	NHL Totals		**52**	**4**	**6**	**10**	**14**	**1**	**0**	**1**	**47**	**8.5**		**0**	**0.0**	**36**	**48**	**16:08**								

SCHULTZ, Ray
(SHUHLTZ, RAY) **NYI**

Defense. Shoots left. 6'2", 215 lbs. Born, Red Deer, Alta., November 14, 1976. Ottawa's 8th choice, 184th overall, in 1995 Entry Draft.

Season	Club	League	GP	G	A	Pts	PIM	PP	SH	GW	S	%	+/-	TF	F%	H	SB	Min	GP	G	A	Pts	PIM	PP	SH	GW
1993-94	Edmonton SSAC	AMHL	31	3	24	27	94																			
	Tri-City	WHL	3	0	0	0	11																			
1994-95	Tri-City	WHL	63	1	8	9	209												11	0	0	0	16			
1995-96	Calgary Hitmen	WHL	66	3	17	20	282																			
1996-97	Calgary Hitmen	WHL	32	3	17	20	141																			
	Kelowna Rockets	WHL	23	3	11	14	63												6	0	2	2	12			
1997-98	**NY Islanders**	**NHL**	**13**	**0**	**1**	**1**	**45**	**0**	**0**	**0**	**4**	**0.0**	**3**													
	Kentucky	AHL	51	2	4	6	179												1	0	0	0	25			
1998-99	**NY Islanders**	**NHL**	**4**	**0**	**0**	**0**	**7**	**0**	**0**	**0**	**2**	**0.0**	**-2**	**1**	**0.0**	**5**	**1**	**15:21**								
	Lowell	AHL	54	0	3	3	184												1	0	0	0	4			
99-2000	**NY Islanders**	**NHL**	**9**	**0**	**1**	**1**	**30**	**0**	**0**	**0**	**2**	**0.0**	**1**	**0**	**0.0**	**21**	**5**	**14:18**								
	Kansas City	IHL	65	5	5	10	208																			
2000-01	**NY Islanders**	**NHL**	**13**	**0**	**2**	**2**	**40**	**0**	**0**	**0**	**3**	**0.0**	**-1**	**0**	**0.0**	**27**	**9**	**10:50**								
	Lowell	AHL	13	0	1	1	33																			
	Cleveland	IHL	44	3	5	8	127												3	1	0	1	16			

			Regular Season																Playoffs							
Season	Club	League	GP	G	A	Pts	PIM	PP	SH	GW	S	%	+/-	TF	F%	H	SB	Min	GP	G	A	Pts	PIM	PP	SH	GW
2001-02	**NY Islanders**	**NHL**	**2**	**0**	**0**	**0**	**5**	**0**	**0**	**0**	**0**	**0.0**	**–1**	**0**	**0.0**	**1**	**0**	**3:22**	**2**	**0**	**0**	**0**	**2**	**0**	**0**	**0**
	Bridgeport	AHL	69	0	15	15	205												19	1	3	4	18			
	NHL Totals		**41**	**0**	**4**	**4**	**127**	**0**	**0**	**0**	**11**	**0.0**		**1**	**0.0**	**54**	**15**	**12:03**	**2**	**0**	**0**	**0**	**2**	**0**	**0**	**0**

Signed as a free agent by **NY Islanders**, June 9, 1997.

SCOTT, Richard

(SKAWT, RIH-churd) **NYR**

Left wing. Shoots left. 6'2", 195 lbs. Born, Orillia, Ont., August 1, 1978.

Season	Club	League	GP	G	A	Pts	PIM	PP	SH	GW	S	%	+/-	TF	F%	H	SB	Min	GP	G	A	Pts	PIM	PP	SH	GW
1996-97	Orillia Terriers	OPJHL	10	0	0	0	23																			
1997-98	Couchiching	OPJHL	45	13	19	32	166																			
1998-99	Oshawa Generals	OHL	54	12	12	24	193																			
99-2000	Charlotte	ECHL	55	1	5	6	317																			
2000-01	Charlotte	ECHL	4	1	1	2	22																			
	Hartford	AHL	64	2	5	7	320												1	0	0	0	0			
2001-02	**NY Rangers**	**NHL**	**5**	**0**	**0**	**0**	**5**	**0**	**0**	**0**	**1**	**0.0**	**0**	**0**	**0.0**	**1**	**0**	**2:25**								
	Hartford	AHL	39	2	3	5	211																			
	NHL Totals		**5**	**0**	**0**	**0**	**5**	**0**	**0**	**0**	**1**	**0.0**		**0**	**0.0**	**1**	**0**	**2:25**								

Signed as a free agent by **NY Rangers**, May 8, 2001.

SCOVILLE, Darrel

(SKO-vihl, DAIR-uhl) **CBJ**

Defense. Shoots left. 6'3", 215 lbs. Born, Swift Current, Sask., October 13, 1975.

Season	Club	League	GP	G	A	Pts	PIM	PP	SH	GW	S	%	+/-	TF	F%	H	SB	Min	GP	G	A	Pts	PIM	PP	SH	GW
1994-95	Lebret Eagles	SJHL	STATISTICS NOT AVAILABLE																							
1995-96	Merrimack	H-East	34	6	20	26	54																			
1996-97	Merrimack	H-East	35	7	16	23	71																			
1997-98	Merrimack	H-East	38	4	26	30	84																			
1998-99	Saint John	AHL	61	1	7	8	66												7	1	2	3	13			
99-2000	**Calgary**	**NHL**	**6**	**0**	**0**	**0**	**2**	**0**	**0**	**0**	**1**	**0.0**	**1**	**0**	**0.0**	**11**	**1**	**9:18**								
	Saint John	AHL	64	11	25	36	99												3	1	2	3	0			
2000-01	Saint John	AHL	76	11	32	47	125												11	2	6	8	8			
2001-02	Syracuse Crunch	AHL	51	5	16	21	60												10	0	0	0	6			
	NHL Totals		**6**	**0**	**0**	**0**	**2**	**0**	**0**	**0**	**1**	**0.0**		**0**	**0.0**	**11**	**1**	**9:18**								

Hockey East All-Rookie Team (1996)

Signed as a free agent by **Calgary**, June 12, 1998. Signed as a free agent by **Columbus**, July 10, 2001.

SEDIN, Daniel

(suh-DEEN, DAN-yehl) **VAN.**

Left wing. Shoots left. 6'1", 200 lbs. Born, Ornskoldsvik, Sweden, September 26, 1980. Vancouver's 1st choice, 2nd overall, in 1999 Entry Draft.

Season	Club	League	GP	G	A	Pts	PIM	PP	SH	GW	S	%	+/-	TF	F%	H	SB	Min	GP	G	A	Pts	PIM	PP	SH	GW
1996-97	MoDo Jr.	Swede-Jr.	26	26	14	40																				
1997-98	MoDo Jr.	Swede-Jr.	4	3	3	6	4																			
	MoDo	Sweden	45	4	8	12	26												9	0	0	0	2			
1998-99	MoDo	Sweden	50	21	21	42	20												13	4	8	12	14			
99-2000	MoDo	Sweden	50	19	26	45	28												13	*8	6	14	18			
	MoDo	EuroHL	4	3	3	6	0												2	0	0	0	0			
2000-01	**Vancouver**	**NHL**	**75**	**20**	**14**	**34**	**24**	**10**	**0**	**3**	**127**	**15.7**	**–3**	**10**	**60.0**	**22**	**7**	**12:60**	**4**	**1**	**2**	**3**	**0**	**0**	**0**	**0**
2001-02	**Vancouver**	**NHL**	**79**	**9**	**23**	**32**	**32**	**4**	**0**	**2**	**117**	**7.7**	**1**	**18**	**33.3**	**43**	**12**	**12:22**	**6**	**0**	**1**	**1**	**0**	**0**	**0**	**0**
	NHL Totals		**154**	**29**	**37**	**66**	**56**	**14**	**0**	**5**	**244**	**11.9**		**28**	**42.9**	**65**	**19**	**12:40**	**10**	**1**	**3**	**4**	**0**	**0**	**0**	**0**

SEDIN, Henrik

(suh-DEEN, HEHN-rihk) **VAN.**

Center. Shoots left. 6'2", 200 lbs. Born, Ornskoldsvik, Sweden, September 26, 1980. Vancouver's 2nd choice, 3rd overall, in 1999 Entry Draft.

Season	Club	League	GP	G	A	Pts	PIM	PP	SH	GW	S	%	+/-	TF	F%	H	SB	Min	GP	G	A	Pts	PIM	PP	SH	GW
1996-97	MoDo Jr.	Swede-Jr.	26	14	22	36																				
1997-98	MoDo Jr.	Swede-Jr.	8	4	7	11	6																			
	MoDo	Sweden	39	1	4	5	8												7	0	0	0	0			
1998-99	MoDo	Sweden	49	12	22	34	32												13	2	8	10	6			
99-2000	MoDo	Sweden	50	9	38	47	22												13	5	9	14	2			
2000-01	**Vancouver**	**NHL**	**82**	**9**	**20**	**29**	**38**	**2**	**0**	**1**	**98**	**9.2**	**–2**	**1020**	**44.1**	**21**	**11**	**13:31**	**4**	**0**	**4**	**4**	**0**	**0**	**0**	**0**
2001-02	**Vancouver**	**NHL**	**82**	**16**	**20**	**36**	**36**	**3**	**0**	**1**	**78**	**20.5**	**9**	**785**	**47.4**	**52**	**22**	**12:48**	**6**	**3**	**0**	**3**	**0**	**0**	**0**	**1**
	NHL Totals		**164**	**25**	**40**	**65**	**74**	**5**	**0**	**2**	**176**	**14.2**		**1805**	**45.5**	**73**	**33**	**13:10**	**10**	**3**	**4**	**7**	**0**	**0**	**0**	**1**

SEKERAS, Lubomir

(SHE-kuhr-ahsh, LOO-boh-mihr) **MIN.**

Defense. Shoots left. 6', 183 lbs. Born, Trencin, Czech., November 18, 1968. Minnesota's 8th choice, 232nd overall, in 2000 Entry Draft.

Season	Club	League	GP	G	A	Pts	PIM	PP	SH	GW	S	%	+/-	TF	F%	H	SB	Min	GP	G	A	Pts	PIM	PP	SH	GW
1987-88	Dukla Trencin Jr.	Czech-Jr.	STATISTICS NOT AVAILABLE																							
	Dukla Trencin	Czech																	9	0	0	0	0			
1988-89	Dukla Trencin	Czech	16	2	5	7	22												11	0	4	4	0			
1989-90	Dukla Trencin	Czech	44	6	8	14													9	0	2	2				
1990-91	Dukla Trencin	Czech	52	6	16	22													6	0	1	1				
1991-92	Dukla Trencin	Czech	30	2	6	8	32												13	1	1	2	0			
1992-93	Dukla Trencin	Czech	40	5	19	24	48												11	4	9	13	0			
1993-94	Dukla Trencin	Slovakia	36	9	12	21	46												9	2	4	6	10			
1994-95	Dukla Trencin	Slovakia	36	11	11	22	24												9	2	7	9	8			
1995-96	Trinec	Czech	40	11	13	24	44												3	0	0	0	0			
1996-97	Trinec	Czech	52	14	21	35	56												4	1	0	1	2			
1997-98	Trinec	Czech	50	11	33	44	42												13	2	10	12	4			
1998-99	Trinec	Czech	50	8	15	23	38												10	2	6	8				
99-2000	Trinec	Czech	52	7	24	31	36												4	0	2	2	2			
2000-01	**Minnesota**	**NHL**	**80**	**11**	**23**	**34**	**52**	**4**	**0**	**2**	**102**	**10.8**	**–8**	**0**	**0.0**	**59**	**70**	**21:13**								
2001-02	**Minnesota**	**NHL**	**69**	**4**	**20**	**24**	**38**	**4**	**0**	**1**	**82**	**4.9**	**–7**	**0**	**0.0**	**67**	**81**	**22:37**								
	NHL Totals		**149**	**15**	**43**	**58**	**90**	**8**	**0**	**3**	**184**	**8.2**		**0**	**0.0**	**126**	**151**	**21:52**								

SELANNE, Teemu

(SEH-lahn-nay, TEE-moo) **S.J.**

Right wing. Shoots right. 6', 204 lbs. Born, Helsinki, Finland, July 3, 1970. Winnipeg's 1st choice, 10th overall, in 1988 Entry Draft.

Season	Club	League	GP	G	A	Pts	PIM	PP	SH	GW	S	%	+/-	TF	F%	H	SB	Min	GP	G	A	Pts	PIM	PP	SH	GW
1986-87	Jokerit Jr.	Finn-Jr.	33	10	12	22	8																			
1987-88	Jokerit Jr.	Finn-Jr.	33	*43	23	*66	18												5	4	3	7	2			
	Jokerit Helsinki	Finland-2	5	1	1	2	0																			
1988-89	Army Jr.	Finn-Jr.	3	3	1	4	2																			
	Jokerit Jr.	Finn-Jr.	3	8	8	16	4																			
	Jokerit Helsinki	Finland-2	34	35	33	68	12												5	7	3	10	4			
1989-90	Jokerit Helsinki	Finland	11	4	8	12	0																			
1990-91	Jokerit Jr.	Finn-Jr.	1	0	0	0	0																			
	Jokerit Helsinki	Finland	42	33	25	58	12																			
1991-92	Jokerit Helsinki	Finland	44	*39	23	62	20												10	*10	7	*17	18			
	Finland	Olympics	8	7	4	11	6																			
1992-93	**Winnipeg**	**NHL**	**84**	***76**	**56**	**132**	**45**	**24**	**0**	**7**	**387**	**19.6**	**8**						**6**	**4**	**2**	**6**	**2**	**2**	**0**	**2**
1993-94	**Winnipeg**	**NHL**	**51**	**25**	**29**	**54**	**22**	**11**	**0**	**2**	**191**	**13.1**	**–23**													
1994-95	Jokerit Helsinki	Finland	20	7	12	19	6																			
	Winnipeg	**NHL**	**45**	**22**	**26**	**48**	**2**	**8**	**2**	**1**	**167**	**13.2**	**1**													
1995-96	**Winnipeg**	**NHL**	**51**	**24**	**48**	**72**	**18**	**6**	**1**	**4**	**163**	**14.7**	**3**													
	Anaheim	**NHL**	**28**	**16**	**20**	**36**	**4**	**3**	**0**	**1**	**104**	**15.4**	**2**													
1996-97	**Anaheim**	**NHL**	**78**	**51**	**58**	**109**	**34**	**11**	**1**	**8**	**273**	**18.7**	**28**						**11**	**7**	**3**	**10**	**4**	**3**	**0**	**1**
1997-98	**Anaheim**	**NHL**	**73**	***52**	**34**	**86**	**30**	**10**	**1**	**10**	**268**	**19.4**	**12**													
	Finland	Olympics	5	4	6	*10	8																			
1998-99	**Anaheim**	**NHL**	**75**	***47**	**60**	**107**	**30**	**25**	**0**	**7**	**281**	**16.7**	**18**	**5**	**20.0**	**27**	**16**	**22:47**	**4**	**2**	**2**	**4**	**2**	**1**	**0**	**0**
99-2000	**Anaheim**	**NHL**	**79**	**33**	**52**	**85**	**12**	**8**	**0**	**6**	**236**	**14.0**	**6**	**13**	**23.1**	**43**	**19**	**22:44**								
2000-01	**Anaheim**	**NHL**	**61**	**26**	**33**	**59**	**36**	**10**	**0**	**5**	**202**	**12.9**	**–8**	**4**	**50.0**	**36**	**10**	**21:51**								
	San Jose	**NHL**	**12**	**7**	**6**	**13**	**0**	**2**	**0**	**2**	**31**	**22.6**	**1**	**4**	**75.0**	**2**	**0**	**18:14**	**6**	**0**	**2**	**2**	**2**	**0**	**0**	**0**

Season	Club	League	GP	G	A	Pts	PIM	PP	SH	GW	S	%	+/-	TF	F%	H	SB	Min	GP	G	A	Pts	PIM	PP	SH	GW
			Regular Season																Playoffs							
2001-02	**San Jose**	**NHL**	**82**	**29**	**25**	**54**	**40**	**9**	**1**	**8**	**202**	**14.4**	**-11**	**12**	**25.0**	**18**	**12**	**16:58**	**12**	**5**	**3**	**8**	**2**	**2**	**0**	**1**
	Finland	Olympics	4	3	0	3	2	....	....	....	....	....	....	....	....	....	....	....	....	....	....	....	....	....	....	....
	NHL Totals		**719**	**408**	**447**	**855**	**273**	**127**	**6**	**61**	**2505**	**16.3**		**38**	**31.6**	**126**	**57**	**20:52**	**39**	**18**	**12**	**30**	**12**	**8**	**0**	**4**

Won Calder Memorial Trophy (1993) • NHL First All-Star Team (1993, 1997) • NHL All-Rookie Team (1993) • NHL Second All-Star Team (1998, 1999) • Won Maurice "Rocket" Richard Trophy (1999) • Played in NHL All-Star Game (1993, 1994, 1996, 1997, 1998, 1999, 2000, 2002)

• Missed majority of 1989-90 season recovering from leg injury suffered in game vs. HIFK Helsinki, October 19, 1989. Traded to **Anaheim** by **Winnipeg** with Marc Chouinard and Winnipeg's 4th round choice (later traded to Toronto - later traded to Montreal - Montreal selected Kim Staal) in 1996 Entry Draft for Chad Kilger, Oleg Tverdovsky and Anaheim's 3rd round choice (Per-Anton Lundstrom) in 1996 Entry Draft, February 7, 1996. Traded to **San Jose** by **Anaheim** for Jeff Friesen, Steve Shields and future considerations, March 5, 2001.

SELIVANOV, Alex

(seh-lih-VAH-nohv, AL-ehx)

Right wing. Shoots left. 6', 208 lbs. Born, Moscow, USSR, March 23, 1971. Philadelphia's 4th choice, 140th overall, in 1994 Entry Draft.

Season	Club	League	GP	G	A	Pts	PIM	PP	SH	GW	S	%	+/-	TF	F%	H	SB	Min	GP	G	A	Pts	PIM	PP	SH	GW
1988-89	Spartak Moscow	USSR	1	0	0	0	0	....	....	....	....	....	....	....	....	....	....	....	....	....	....	....	....	....	....	....
1989-90	Spartak Moscow	USSR	4	0	0	0	0	....	....	....	....	....	....	....	....	....	....	....	....	....	....	....	....	....	....	....
1990-91	Spartak Moscow	USSR	21	3	1	4	6	....	....	....	....	....	....	....	....	....	....	....	....	....	....	....	....	....	....	....
1991-92	Spartak Moscow	CIS	31	6	7	13	16	....	....	....	....	....	....	....	....	....	....	....	....	....	....	....	....	....	....	....
1992-93	Spartak Moscow	CIS	42	12	19	31	66	....	....	....	....	....	....	....	....	....	....	....	3	2	0	2	2	....	....	....
1993-94	Spartak Moscow	CIS	45	30	11	41	50	....	....	....	....	....	....	....	....	....	....	....	6	5	1	6	2	....	....	....
1994-95	Atlanta Knights	IHL	4	0	3	3	2	....	....	....	....	....	....	....	....	....	....	....	....	....	....	....	....	....	....	....
	Chicago Wolves	IHL	14	4	1	5	8	....	....	....	....	....	....	....	....	....	....	....	....	....	....	....	....	....	....	....
	Tampa Bay	**NHL**	**43**	**10**	**6**	**16**	**14**	**4**	**0**	**3**	**94**	**10.6**	**-2**	....	....	....	....	....	....	....	....	....	....	....	....	....
1995-96	**Tampa Bay**	**NHL**	**79**	**31**	**21**	**52**	**93**	**13**	**0**	**5**	**215**	**14.4**	**3**	....	....	....	....	....	**6**	**2**	**2**	**4**	**6**	**0**	**0**	**1**
1996-97	**Tampa Bay**	**NHL**	**69**	**15**	**18**	**33**	**61**	**3**	**0**	**4**	**187**	**8.0**	**-3**	....	....	....	....	....	....	....	....	....	....	....	....	....
1997-98	**Tampa Bay**	**NHL**	**70**	**16**	**19**	**35**	**85**	**4**	**0**	**3**	**206**	**7.8**	**-38**	....	....	....	....	....	....	....	....	....	....	....	....	....
1998-99	**Tampa Bay**	**NHL**	**43**	**6**	**13**	**19**	**18**	**1**	**0**	**0**	**120**	**5.0**	**-8**	**0**	**0.0**	**40**	**5**	**15:44**	....	....	....	....	....	....	....	....
	Cleveland	IHL	2	0	1	1	4	....	....	....	....	....	....	....	....	....	....	....	....	....	....	....	....	....	....	....
	Edmonton	**NHL**	**29**	**8**	**6**	**14**	**24**	**1**	**0**	**1**	**57**	**14.0**	**0**	**5**	**60.0**	**20**	**5**	**13:28**	**2**	**0**	**1**	**1**	**2**	**0**	**0**	**0**
99-2000	**Edmonton**	**NHL**	**67**	**27**	**20**	**47**	**46**	**10**	**0**	**5**	**122**	**22.1**	**2**	**5**	**0.0**	**29**	**6**	**14:28**	**5**	**0**	**0**	**0**	**8**	**0**	**0**	**0**
2000-01	**Columbus**	**NHL**	**59**	**8**	**11**	**19**	**38**	**5**	**0**	**2**	**104**	**7.7**	**-11**	**13**	**30.8**	**34**	**9**	**14:06**	....	....	....	....	....	....	....	....
2001-02	Frankfurt Lions	Germany	58	26	35	61	87	....	....	....	....	....	....	....	....	....	....	....	....	....	....	....	....	....	....	....
	NHL Totals		**459**	**121**	**114**	**235**	**379**	**41**	**0**	**23**	**1105**	**11.0**		**23**	**30.4**	**123**	**25**	**14:29**	**13**	**2**	**3**	**5**	**16**	**0**	**0**	**1**

Traded to **Tampa Bay** by **Philadelphia** for Philadelphia's 4th round choice (previously acquired, Philadelphia selected Radovan Somik) in 1995 Entry Draft, September 6, 1994. Traded to **Edmonton** by **Tampa Bay** for Alexandre Daigle, January 29, 1999. Signed as a free agent by **Columbus**, November 27, 2000. Signed as a free agent by **Frankfurt** (Germany), August 16, 2001.

SELLARS, Luke

(SEHL-lahrs, LEWK) **ATL.**

Defense. Shoots left. 6'1", 205 lbs. Born, Toronto, Ont., May 21, 1981. Atlanta's 2nd choice, 30th overall, in 1999 Entry Draft.

Season	Club	League	GP	G	A	Pts	PIM	PP	SH	GW	S	%	+/-	TF	F%	H	SB	Min	GP	G	A	Pts	PIM	PP	SH	GW
1997-98	Wexford Raiders	MTJHL	46	2	18	20	155	....	....	....	....	....	....	....	....	....	....	....	....	....	....	....	....	....	....	....
1998-99	Ottawa 67's	OHL	56	4	19	23	87	....	....	....	....	....	....	....	....	....	....	....	9	1	2	3	7	....	....	....
99-2000	Ottawa 67's	OHL	56	8	34	42	147	....	....	....	....	....	....	....	....	....	....	....	11	4	6	10	28	....	....	....
2000-01	Ottawa 67's	OHL	59	9	21	30	136	....	....	....	....	....	....	....	....	....	....	....	18	4	10	14	47	....	....	....
2001-02	**Atlanta**	**NHL**	**1**	**0**	**0**	**0**	**2**	**0**	**0**	**0**	**0**	**0.0**	**0**	**0**	**0.0**	**1**	**1**	**3:27**	....	....	....	....	....	....	....	....
	Chicago Wolves	AHL	31	2	4	6	87	....	....	....	....	....	....	....	....	....	....	....	....	....	....	....	....	....	....	....
	Greenville	ECHL	21	2	6	8	61	....	....	....	....	....	....	....	....	....	....	....	17	7	6	13	44	....	....	....
	NHL Totals		**1**	**0**	**0**	**0**	**2**	**0**	**0**	**0**	**0**	**0.0**		**0**	**0.0**	**1**	**1**	**3:27**	....	....	....	....	....	....	....	....

OHL All-Rookie Team (1999)

SELMSER, Sean

(SEHLM-suhr, SHAWN) **EDM.**

Left wing. Shoots left. 6'1", 195 lbs. Born, Calgary, Alta., November 10, 1974. Pittsburgh's 7th choice, 182nd overall, in 1993 Entry Draft.

Season	Club	League	GP	G	A	Pts	PIM	PP	SH	GW	S	%	+/-	TF	F%	H	SB	Min	GP	G	A	Pts	PIM	PP	SH	GW
1991-92	Cgy. Buffaloes	AMHL	32	21	24	45	131	....	....	....	....	....	....	....	....	....	....	....	....	....	....	....	....	....	....	....
1992-93	Red Deer Rebels	WHL	70	13	27	40	216	....	....	....	....	....	....	....	....	....	....	....	4	0	0	0	10	....	....	....
1993-94	Red Deer Rebels	WHL	71	25	25	50	201	....	....	....	....	....	....	....	....	....	....	....	4	1	0	1	14	....	....	....
1994-95	Red Deer Rebels	WHL	33	11	17	28	65	....	....	....	....	....	....	....	....	....	....	....	....	....	....	....	....	....	....	....
1995-96	Hampton Roads	ECHL	70	23	31	54	211	....	....	....	....	....	....	....	....	....	....	....	3	2	0	2	8	....	....	....
	Portland Pirates	AHL	6	4	1	5	28	....	....	....	....	....	....	....	....	....	....	....	5	2	3	5	11	....	....	....
1996-97	Team Canada	Nat-Tm	59	20	18	38	150	....	....	....	....	....	....	....	....	....	....	....	....	....	....	....	....	....	....	....
	Manitoba Moose	IHL	4	0	2	2	12	....	....	....	....	....	....	....	....	....	....	....	....	....	....	....	....	....	....	....
1997-98	Team Canada	Nat-Tm	52	8	22	30	122	....	....	....	....	....	....	....	....	....	....	....	....	....	....	....	....	....	....	....
	Portland Pirates	AHL	....	....	....	....	....	....	....	....	....	....	....	....	....	....	....	....	1	0	1	1	0	....	....	....
1998-99	Fort Wayne	IHL	80	9	15	24	200	....	....	....	....	....	....	....	....	....	....	....	2	0	0	0	2	....	....	....
99-2000	Hamilton	AHL	72	14	12	26	151	....	....	....	....	....	....	....	....	....	....	....	10	3	4	7	10	....	....	....
2000-01	**Columbus**	**NHL**	**1**	**0**	**0**	**0**	**5**	**0**	**0**	**0**	**2**	**0.0**	**0**	**0**	**0.0**	**1**	**0**	**11:03**	....	....	....	....	....	....	....	....
	Syracuse Crunch	AHL	75	11	15	26	151	....	....	....	....	....	....	....	....	....	....	....	5	0	0	0	14	....	....	....
2001-02	Hamilton	AHL	80	10	16	26	103	....	....	....	....	....	....	....	....	....	....	....	15	1	5	6	8	....	....	....
	NHL Totals		**1**	**0**	**0**	**0**	**5**	**0**	**0**	**0**	**2**	**0.0**		**0**	**0.0**	**1**	**0**	**11:03**	....	....	....	....	....	....	....	....

Signed as a free agent by **Columbus**, August 3, 2000. Signed as a free agent by **Edmonton**, August 15, 2001.

SHANAHAN, Brendan

(SHAN-na-HAN, BREHN-duhn) **DET.**

Left wing. Shoots right. 6'3", 218 lbs. Born, Mimico, Ont., January 23, 1969. New Jersey's 1st choice, 2nd overall, in 1987 Entry Draft.

Season	Club	League	GP	G	A	Pts	PIM	PP	SH	GW	S	%	+/-	TF	F%	H	SB	Min	GP	G	A	Pts	PIM	PP	SH	GW
1984-85	Mississauga Reps	MTHL	36	20	21	41	26	....	....	....	....	....	....	....	....	....	....	....	....	....	....	....	....	....	....	....
	Dixie Beehives	MTJHL	1	0	0	0	0	....	....	....	....	....	....	....	....	....	....	....	....	....	....	....	....	....	....	....
1985-86	London Knights	OHL	59	28	34	62	70	....	....	....	....	....	....	....	....	....	....	....	5	5	5	10	5	....	....	....
1986-87	London Knights	OHL	56	39	53	92	92	....	....	....	....	....	....	....	....	....	....	....	....	....	....	....	....	....	....	....
1987-88	**New Jersey**	**NHL**	**65**	**7**	**19**	**26**	**131**	**2**	**0**	**2**	**72**	**9.7**	**-20**	....	....	....	....	....	**12**	**2**	**1**	**3**	**44**	**1**	**0**	**0**
1988-89	**New Jersey**	**NHL**	**68**	**22**	**28**	**50**	**115**	**9**	**0**	**0**	**152**	**14.5**	**2**	....	....	....	....	....	....	....	....	....	....	....	....	....
1989-90	**New Jersey**	**NHL**	**73**	**30**	**42**	**72**	**137**	**8**	**0**	**5**	**196**	**15.3**	**15**	....	....	....	....	....	**6**	**3**	**3**	**6**	**20**	**1**	**0**	**1**
1990-91	**New Jersey**	**NHL**	**75**	**29**	**37**	**66**	**141**	**7**	**0**	**2**	**195**	**14.9**	**4**	....	....	....	....	....	**7**	**3**	**5**	**8**	**12**	**2**	**0**	**0**
1991-92	**St. Louis**	**NHL**	**80**	**33**	**36**	**69**	**171**	**13**	**0**	**2**	**215**	**15.3**	**-3**	....	....	....	....	....	**6**	**2**	**3**	**5**	**14**	**1**	**0**	**0**
1992-93	**St. Louis**	**NHL**	**71**	**51**	**43**	**94**	**174**	**18**	**0**	**8**	**232**	**22.0**	**10**	....	....	....	....	....	**11**	**4**	**3**	**7**	**18**	**2**	**0**	**0**
1993-94	**St. Louis**	**NHL**	**81**	**52**	**50**	**102**	**211**	**15**	**7**	**8**	**397**	**13.1**	**-9**	....	....	....	....	....	**4**	**2**	**5**	**7**	**4**	**0**	**0**	**0**
1994-95	Dusseldorfer EG	Germany	3	5	3	8	4	....	....	....	....	....	....	....	....	....	....	....	....	....	....	....	....	....	....	....
	St. Louis	**NHL**	**45**	**20**	**21**	**41**	**136**	**6**	**2**	**6**	**153**	**13.1**	**7**	....	....	....	....	....	**5**	**4**	**5**	**9**	**14**	**1**	**0**	**1**
1995-96	**Hartford**	**NHL**	**74**	**44**	**34**	**78**	**125**	**17**	**2**	**6**	**280**	**15.7**	**2**	....	....	....	....	....	....	....	....	....	....	....	....	....
1996-97	**Hartford**	**NHL**	**2**	**1**	**0**	**1**	**0**	**0**	**1**	**0**	**13**	**7.7**	**1**	....	....	....	....	....	....	....	....	....	....	....	....	....
	♦ Detroit	**NHL**	**79**	**46**	**41**	**87**	**131**	**20**	**2**	**7**	**323**	**14.2**	**31**	....	....	....	....	....	**20**	**9**	**8**	**17**	**43**	**2**	**0**	**2**
1997-98♦	**Detroit**	**NHL**	**75**	**28**	**29**	**57**	**154**	**15**	**1**	**9**	**266**	**10.5**	**6**	....	....	....	....	....	**20**	**5**	**4**	**9**	**22**	**3**	**0**	**2**
	Canada	Olympics	6	2	0	2	0	....	....	....	....	....	....	....	....	....	....	....	....	....	....	....	....	....	....	....
1998-99	**Detroit**	**NHL**	**81**	**31**	**27**	**58**	**123**	**5**	**0**	**5**	**288**	**10.8**	**2**	**18**	**44.4**	**119**	**35**	**17:31**	**10**	**3**	**7**	**10**	**6**	**1**	**0**	**1**
99-2000	**Detroit**	**NHL**	**78**	**41**	**37**	**78**	**105**	**13**	**1**	**9**	**283**	**14.5**	**24**	**24**	**50.0**	**104**	**34**	**18:35**	**9**	**3**	**2**	**5**	**10**	**0**	**0**	**0**
2000-01	**Detroit**	**NHL**	**81**	**31**	**45**	**76**	**81**	**15**	**1**	**7**	**278**	**11.2**	**9**	**115**	**43.5**	**105**	**24**	**18:22**	**2**	**2**	**2**	**4**	**0**	**0**	**0**	**1**
2001-02♦	**Detroit**	**NHL**	**80**	**37**	**38**	**75**	**118**	**12**	**3**	**7**	**277**	**13.4**	**23**	**70**	**47.1**	**61**	**29**	**18:55**	**23**	**8**	**11**	**19**	**20**	**1**	**0**	**2**
	Canada	Olympics	6	0	1	1	0	....	....	....	....	....	....	....	....	....	....	....	....	....	....	....	....	....	....	....
	NHL Totals		**1108**	**503**	**527**	**1030**	**2053**	**175**	**20**	**83**	**3620**	**13.9**		**227**	**45.4**	**389**	**122**	**18:20**	**135**	**50**	**59**	**109**	**227**	**15**	**0**	**10**

NHL First All-Star Team (1994, 2000) • NHL Second All-Star Team (2002) • Played in NHL All-Star Game (1994, 1996, 1997, 1998, 1999, 2000, 2002)

Signed as a free agent by **St. Louis**, July 25, 1991. Traded to **Hartford** by **St. Louis** for Chris Pronger, July 27, 1995. Traded to **Detroit** by **Hartford** with Brian Glynn for Paul Coffey, Keith Primeau and Detroit's 1st round choice (Nikos Tselios) in 1997 Entry Draft, October 9, 1996.

SHANNON, Darryl

(SHA-nohn, DAIR-ihl)

Defense. Shoots left. 6'2", 208 lbs. Born, Barrie, Ont., June 21, 1968. Toronto's 2nd choice, 36th overall, in 1986 Entry Draft.

Season	Club	League	GP	G	A	Pts	PIM	PP	SH	GW	S	%	+/-	TF	F%	H	SB	Min	GP	G	A	Pts	PIM	PP	SH	GW
1983-84	Alliston Hornets	OJHL-C	30	18	22	40	70	....	....	....	....	....	....	....	....	....	....	....	....	....	....	....	....	....	....	....
1984-85	Richmond Hill	OJHL	1	0	0	0	0	....	....	....	....	....	....	....	....	....	....	....	....	....	....	....	....	....	....	....
	Barrie Colts	OJHL-B	39	5	23	28	50	....	....	....	....	....	....	....	....	....	....	....	....	....	....	....	....	....	....	....
1985-86	Windsor	OHL	57	6	21	27	52	....	....	....	....	....	....	....	....	....	....	....	16	5	6	11	22	....	....	....
1986-87	Windsor	OHL	64	23	27	50	83	....	....	....	....	....	....	....	....	....	....	....	14	4	8	12	18	....	....	....
1987-88	Windsor	OHL	60	16	67	83	116	....	....	....	....	....	....	....	....	....	....	....	12	3	8	11	17	....	....	....

			Regular Season																Playoffs							
Season	Club	League	GP	G	A	Pts	PIM	PP	SH	GW	S	%	+/-	TF	F%	H	SB	Min	GP	G	A	Pts	PIM	PP	SH	GW
1988-89	**Toronto**	**NHL**	**14**	**1**	**3**	**4**	**6**	**0**	**0**	**0**	**16**	**6.3**	**5**													
	Newmarket Saints	AHL	61	5	24	29	37												5	0	3	3	10			
1989-90	**Toronto**	**NHL**	**10**	**0**	**1**	**1**	**12**	**0**	**0**	**0**	**16**	**0.0**	**–10**													
	Newmarket Saints	AHL	47	4	15	19	58																			
1990-91	**Toronto**	**NHL**	**10**	**0**	**1**	**1**	**0**	**0**	**0**	**0**	**3**	**0.0**	**1**													
	Newmarket Saints	AHL	47	2	14	16	51																			
1991-92	**Toronto**	**NHL**	**48**	**2**	**8**	**10**	**23**	**1**	**0**	**0**	**50**	**4.0**	**–17**													
1992-93	**Toronto**	**NHL**	**16**	**0**	**0**	**0**	**11**	**0**	**0**	**0**	**10**	**0.0**	**–5**													
	St. John's	AHL	7	1	1	2	4																			
1993-94	**Winnipeg**	**NHL**	**20**	**0**	**4**	**4**	**18**	**0**	**0**	**0**	**14**	**0.0**	**–6**													
	Moncton Hawks	AHL	37	1	10	11	62												20	1	7	8	32			
1994-95	**Winnipeg**	**NHL**	**40**	**5**	**9**	**14**	**48**	**0**	**1**	**0**	**42**	**11.9**	**1**													
1995-96	**Winnipeg**	**NHL**	**48**	**2**	**7**	**9**	**72**	**0**	**0**	**0**	**34**	**5.9**	**5**													
	Buffalo	**NHL**	**26**	**2**	**6**	**8**	**20**	**0**	**0**	**0**	**25**	**8.0**	**10**													
1996-97	**Buffalo**	**NHL**	**82**	**4**	**19**	**23**	**112**	**1**	**0**	**1**	**94**	**4.3**	**23**						**12**	**2**	**3**	**5**	**8**	**1**	**0**	**0**
1997-98	**Buffalo**	**NHL**	**76**	**3**	**19**	**22**	**56**	**1**	**0**	**1**	**85**	**3.5**	**26**						**15**	**2**	**4**	**6**	**8**	**0**	**1**	**0**
1998-99	**Buffalo**	**NHL**	**71**	**3**	**12**	**15**	**52**	**1**	**0**	**0**	**80**	**3.8**	**28**	**1**	**0.0**	**103**	**111**	**20:10**	**2**	**0**	**0**	**0**	**0**	**0**	**0**	**0**
99-2000	**Atlanta**	**NHL**	**49**	**5**	**13**	**18**	**65**	**1**	**0**	**1**	**66**	**7.6**	**–14**	**1**	**0.0**	**86**	**84**	**21:21**								
	Calgary	**NHL**	**27**	**1**	**8**	**9**	**22**	**0**	**0**	**0**	**46**	**2.2**	**–13**	**0**	**0.0**	**32**	**44**	**23:33**								
2000-01	**Montreal**	**NHL**	**7**	**0**	**1**	**1**	**6**	**0**	**0**	**0**	**6**	**0.0**	**–4**	**0**	**0.0**	**11**	**13**	**18:03**								
	Quebec	AHL	4	0	1	1	4																			
2001-02	Krefeld Pinguine	Germany	38	1	18	19	49												3	1	1	2	0			
	NHL Totals		**544**	**28**	**111**	**139**	**523**	**5**	**1**	**3**	**587**	**4.8**		**2**	**0.0**	**232**	**252**	**21:03**	**29**	**4**	**7**	**11**	**16**	**1**	**1**	**0**

OHL Second All-Star Team (1987) • OHL First All-Star Team (1988)

Signed as a free agent by **Winnipeg**, June 30, 1993. Traded to **Buffalo** by **Winnipeg** with Michal Grosek for Craig Muni, February 15, 1996. Claimed by **Atlanta** from **Buffalo** in Expansion Draft, June 25, 1999. Traded to **Calgary** by **Atlanta** with Jason Botterill for Hnat Domenichelli and Dmitri Vlasenkov, February 11, 2000. Signed as a free agent by **Montreal**, September 25, 2000. Signed as a free agent by **Krefeld** (Germany), November 17, 2001.

SHANTZ, Jeff

(SHAWNTS, JEHF) **CGY.**

Center. Shoots right. 6', 195 lbs. Born, Duchess, Alta., October 10, 1973. Chicago's 2nd choice, 36th overall, in 1992 Entry Draft.

Season	Club	League	GP	G	A	Pts	PIM	PP	SH	GW	S	%	+/-	TF	F%	H	SB	Min	GP	G	A	Pts	PIM	PP	SH	GW
1989-90	Medicine Hat	AMHL	36	18	31	49	30																			
	Regina Pats	WHL	1	0	0	0	0																			
1990-91	Regina Pats	WHL	69	16	21	37	22												8	2	2	4	2			
1991-92	Regina Pats	WHL	72	39	50	89	35																			
1992-93	Regina Pats	WHL	64	29	54	83	75												13	2	12	14	14			
1993-94	**Chicago**	**NHL**	**52**	**3**	**13**	**16**	**30**	**0**	**0**	**0**	**56**	**5.4**	**–14**						**6**	**0**	**0**	**0**	**6**	**0**	**0**	**0**
	Indianapolis Ice	IHL	19	5	9	14	20																			
1994-95	Indianapolis Ice	IHL	32	9	15	24	20																			
	Chicago	**NHL**	**45**	**6**	**12**	**18**	**33**	**0**	**2**	**0**	**58**	**10.3**	**11**						**16**	**3**	**1**	**4**	**2**	**0**	**0**	**0**
1995-96	**Chicago**	**NHL**	**78**	**6**	**14**	**20**	**24**	**1**	**2**	**0**	**72**	**8.3**	**12**						**10**	**2**	**3**	**5**	**6**	**0**	**0**	**0**
1996-97	**Chicago**	**NHL**	**69**	**9**	**21**	**30**	**28**	**0**	**1**	**1**	**86**	**10.5**	**11**						**6**	**0**	**4**	**4**	**6**	**0**	**0**	**0**
1997-98	**Chicago**	**NHL**	**61**	**11**	**20**	**31**	**36**	**1**	**2**	**2**	**69**	**15.9**	**0**													
1998-99	**Chicago**	**NHL**	**7**	**1**	**0**	**1**	**4**	**0**	**0**	**0**	**5**	**20.0**	**–1**	**72**	**38.9**	**16**	**1**	**15:14**								
	Calgary	**NHL**	**69**	**12**	**17**	**29**	**40**	**1**	**1**	**3**	**77**	**15.6**	**15**	**1112**	**48.4**	**83**	**35**	**16:47**								
99-2000	**Calgary**	**NHL**	**74**	**13**	**18**	**31**	**30**	**6**	**0**	**1**	**112**	**11.6**	**–13**	**1576**	**51.1**	**43**	**71**	**18:15**								
2000-01	**Calgary**	**NHL**	**73**	**5**	**15**	**20**	**58**	**0**	**0**	**0**	**88**	**5.7**	**–7**	**876**	**53.5**	**59**	**38**	**14:51**								
2001-02	Saint John	AHL	2	0	1	1	0																			
	Calgary	**NHL**	**40**	**3**	**3**	**6**	**23**	**2**	**0**	**0**	**37**	**8.1**	**–3**	**298**	**53.0**	**47**	**9**	**11:26**								
	NHL Totals		**568**	**69**	**133**	**202**	**306**	**11**	**8**	**7**	**660**	**10.5**		**3934**	**50.8**	**248**	**154**	**15:48**	**38**	**5**	**8**	**13**	**20**	**0**	**0**	**0**

WHL East First All-Star Team (1993)

Traded to **Calgary** by **Chicago** with Steve Dubinsky for Marty McInnis, Jamie Allison and Eric Andersson, October 27, 1998.

SHARIFIJANOV, Vadim

(shah-rih-FYAH-nohv, VA-dihm) **VAN.**

Right wing. Shoots left. 6', 205 lbs. Born, Ufa, USSR, December 23, 1975. New Jersey's 1st choice, 25th overall, in 1994 Entry Draft.

Season	Club	League	GP	G	A	Pts	PIM	PP	SH	GW	S	%	+/-	TF	F%	H	SB	Min	GP	G	A	Pts	PIM	PP	SH	GW
1992-93	Ufa	CIS	37	6	4	10	16												2	1	0	1	0			
1993-94	Ufa	CIS	46	10	6	16	36												5	3	0	3	4			
1994-95	CSKA Moscow	CIS	34	7	3	10	26												2	0	0	0	0			
	Albany	AHL	1	1	1	2	0												9	3	3	6	10			
1995-96	Albany	AHL	69	14	28	42	28																			
1996-97	**New Jersey**	**NHL**	**2**	**0**	**0**	**0**	**0**	**0**	**0**	**0**	**4**	**0.0**	**0**													
	Albany	AHL	70	14	27	41	89												10	3	3	6	6			
1997-98	Albany	AHL	72	23	27	50	69												12	4	9	13	6			
1998-99	**New Jersey**	**NHL**	**53**	**11**	**16**	**27**	**28**	**1**	**0**	**2**	**71**	**15.5**	**11**	**2**	**50.0**	**43**	**10**	**13:39**	**4**	**0**	**0**	**0**	**0**	**0**	**0**	**0**
	Albany	AHL	2	1	1	2	0																			
99-2000	**New Jersey**	**NHL**	**20**	**3**	**4**	**7**	**8**	**0**	**0**	**0**	**20**	**15.0**	**–6**	**0**	**0.0**	**19**	**0**	**10:58**								
	Vancouver	**NHL**	**17**	**2**	**1**	**3**	**14**	**1**	**0**	**0**	**26**	**7.7**	**–7**	**3**	**33.3**	**13**	**0**	**13:11**								
2000-01	Kansas City	IHL	70	20	43	63	48																			
2001-02	Lada Togliatti	Russia	4	0	0	0	12																			
	Cherepovets	Russia	26	9	3	12	2												2	0	0	0	6			
	NHL Totals		**92**	**16**	**21**	**37**	**50**	**2**	**0**	**2**	**121**	**13.2**		**5**	**40.0**	**75**	**10**	**12:58**	**4**	**0**	**0**	**0**	**0**	**0**	**0**	**0**

Traded to **Vancouver** by **New Jersey** with New Jersey's 3rd round choice (Tim Branham) in 2000 Entry Draft for NY Islanders' compensatory 2nd round choice (previously acquired, New Jersey selected Teemu Laine) in 2000 Entry Draft and Atlanta's 3rd round choice (previously acquired, New Jersey selected Max Birbraer) in 2000 Entry Draft, January 14, 2000.

SHEARER, Rob

(SHEER-uhr, RAWB)

Center. Shoots right. 5'10", 190 lbs. Born, Kitchener, Ont., October 19, 1976.

Season	Club	League	GP	G	A	Pts	PIM	PP	SH	GW	S	%	+/-	TF	F%	H	SB	Min	GP	G	A	Pts	PIM	PP	SH	GW
1992-93	Kitchener Lions	OMHA	27	27	24	51																				
1993-94	Windsor	OHL	66	17	25	42	46												4	0	2	2	6			
1994-95	Windsor	OHL	59	28	28	56	48												10	4	4	8	10			
1995-96	Windsor	OHL	63	40	53	93	74												7	6	3	9	8			
1996-97	Hershey Bears	AHL	78	12	16	28	88												23	0	4	4	9			
1997-98	Hershey Bears	AHL	79	30	30	60	44												7	0	5	5	6			
1998-99	Hershey Bears	AHL	77	24	42	66	43												3	0	0	0	6			
99-2000	Hershey Bears	AHL	70	21	46	67	55												14	1	7	8	10			
2000-01	**Colorado**	**NHL**	**2**	**0**	**0**	**0**	**0**	**0**	**0**	**0**	**0**	**0.0**	**–2**	**12**	**41.7**	**3**	**2**	**6:45**								
	Hershey Bears	AHL	73	18	34	52	64												12	1	3	4	0			
2001-02	TPS Turku	Finland	56	24	16	40	56												8	1	2	3	10			
	NHL Totals		**2**	**0**	**0**	**0**	**0**	**0**	**0**	**0**	**0**	**0.0**		**12**	**41.7**	**3**	**2**	**6:45**								

Signed as a free agent by **Colorado**, October 5, 1995.

SHELLEY, Jody

(SHEH-lee, JOH-dee) **CBJ**

Left wing. Shoots left. 6'4", 225 lbs. Born, Thompson, Man., February 7, 1976.

Season	Club	League	GP	G	A	Pts	PIM	PP	SH	GW	S	%	+/-	TF	F%	H	SB	Min	GP	G	A	Pts	PIM	PP	SH	GW
1994-95	Halifax	QMJHL	72	10	12	22	194												7	0	1	1	12			
1995-96	Halifax	QMJHL	50	13	19	32	319												6	0	2	2	36			
1996-97	Halifax	QMJHL	58	25	19	44	*448												17	6	6	12	*123			
1997-98	Dalhousie	AUAA	19	6	11	17	145																			
	Saint John	AHL	18	1	1	2	50																			
1998-99	Saint John	AHL	8	0	0	0	46																			
	Johnstown Chiefs	ECHL	52	12	17	29	325																			
99-2000	Johnstown Chiefs	ECHL	36	9	17	26	256																			
	Saint John	AHL	22	1	4	5	93												3	0	0	0	2			
2000-01	Syracuse Crunch	AHL	69	1	7	8	*357												5	0	0	0	21			
	Columbus	**NHL**	**1**	**0**	**0**	**0**	**10**	**0**	**0**	**0**	**0**	**0.0**	**0**	**0**	**0.0**	**0**	**0**	**1:33**								

			Regular Season																Playoffs							
Season	Club	League	GP	G	A	Pts	PIM	PP	SH	GW	S	%	+/-	TF	F%	H	SB	Min	GP	G	A	Pts	PIM	PP	SH	GW
2001-02	Columbus	NHL	52	3	3	6	206	0	0	0	35	8.6	1	0	0.0	84	10	6:32	....	....	....	....	....	....	....	....
	Syracuse Crunch	AHL	22	3	5	8	165	....	....	....	....	....	....	....	....	....	....	....	....	....	....	....	....	....	....	....
	NHL Totals		53	3	3	6	216	0	0	0	35	8.6		0	0.0	84	10	6:26	....	....	....	....	....	....	....	....

Signed as a free agent by **Calgary**, September 1, 1998. Signed as a free agent by **Syracuse** (AHL), September 15, 2000. Signed as a free agent by **Columbus**, January 31, 2001.

SHVIDKI, Denis
(SHVIHD-kee, DEH-nihs) **FLA.**

Right wing. Shoots left. 6'2", 215 lbs. Born, Kharkov, USSR, November 21, 1980. Florida's 1st choice, 12th overall, in 1999 Entry Draft.

Season	Club	League	GP	G	A	Pts	PIM	PP	SH	GW	S	%	+/-	TF	F%	H	SB	Min	GP	G	A	Pts	PIM	PP	SH	GW
1996-97	Yaroslavl 2	Russia-3	35	21	12	33	32	....	....	....	....	....	....	....	....	....	....	....	....	....	....	....	....	....	....	....
	Yaroslavl	Russia	17	3	2	5	6	....	....	....	....	....	....	....	....	....	....	....	....	....	....	....	....	....	....	....
1997-98	Yaroslavl 2	Russia-2	32	20	13	33	20	....	....	....	....	....	....	....	....	....	....	....	....	....	....	....	....	....	....	....
	Yaroslavl	Russia	15	1	1	2	2	....	....	....	....	....	....	....	....	....	....	....	....	....	....	....	....	....	....	....
1998-99	Barrie Colts	OHL	61	35	59	94	8	....	....	....	....	....	....	....	....	....	....	....	12	7	9	16	2	....	....	....
99-2000	Barrie Colts	OHL	61	41	65	106	55	....	....	....	....	....	....	....	....	....	....	....	9	3	1	4	2	....	....	....
2000-01	Florida	NHL	43	6	10	16	16	0	0	1	28	21.4	6	4	50.0	22	8	10:21	....	....	....	....	....	....	....	....
	Louisville	AHL	34	15	11	26	20	....	....	....	....	....	....	....	....	....	....	....	....	....	....	....	....	....	....	....
2001-02	Florida	NHL	8	1	2	3	2	0	0	0	11	9.1	–4	1	0.0	5	3	11:57	....	....	....	....	....	....	....	....
	Utah Grizzlies	AHL	8	2	4	6	2	....	....	....	....	....	....	....	....	....	....	....	....	....	....	....	....	....	....	....
	NHL Totals		51	7	12	19	18	0	0	1	39	17.9		5	40.0	27	11	10:36	....	....	....	....	....	....	....	....

OHL All-Rookie Team (1999) • OHL Second All-Star Team (1999) • Missed majority of 2001-02 season recovering from head injury suffered in game vs. Philadelphia, October 4, 2001.

SILLINGER, Mike
(sih-LIHN-juhr, MIGHK) **CBJ**

Center. Shoots right. 5'11", 196 lbs. Born, Regina, Sask., June 29, 1971. Detroit's 1st choice, 11th overall, in 1989 Entry Draft.

Season	Club	League	GP	G	A	Pts	PIM	PP	SH	GW	S	%	+/-	TF	F%	H	SB	Min	GP	G	A	Pts	PIM	PP	SH	GW
1986-87	Regina Kings	SMHL	31	83	51	134	....	....	....	....	....	....	....	....	....	....	....	....	....	....	....	....	....	....	....	....
1987-88	Regina Pats	WHL	67	18	25	43	17	....	....	....	....	....	....	....	....	....	....	....	4	2	2	4	0	....	....	....
1988-89	Regina Pats	WHL	72	53	78	131	52	....	....	....	....	....	....	....	....	....	....	....	....	....	....	....	....	....	....	....
1989-90	Regina Pats	WHL	70	57	72	129	41	....	....	....	....	....	....	....	....	....	....	....	11	12	10	22	2	....	....	....
	Adirondack	AHL	....	....	....	....	....	....	....	....	....	....	....	....	....	....	....	....	1	0	0	0	0	....	....	....
1990-91	Regina Pats	WHL	57	50	66	116	42	....	....	....	....	....	....	....	....	....	....	....	8	6	9	15	4	....	....	....
	Detroit	NHL	3	0	1	1	0	0	0	0	6	0.0	–2	....	....	....	....	....	3	0	1	1	0	0	0	0
1991-92	Adirondack	AHL	64	25	41	66	26	....	....	....	....	....	....	....	....	....	....	....	15	9	*19	*28	12	....	....	....
	Detroit	NHL	....	....	....	....	....	....	....	....	....	....	....	....	....	....	....	....	8	2	2	4	2	0	0	0
1992-93	Detroit	NHL	51	4	17	21	16	0	0	0	47	8.5	0	....	....	....	....	....	....	....	....	....	....	....	....	....
	Adirondack	AHL	15	10	20	30	31	....	....	....	....	....	....	....	....	....	....	....	11	5	13	18	10	....	....	....
1993-94	Detroit	NHL	62	8	21	29	10	0	1	1	91	8.8	2	....	....	....	....	....	....	....	....	....	....	....	....	....
1994-95	CE Wien	Austria	13	13	14	27	10	....	....	....	....	....	....	....	....	....	....	....	....	....	....	....	....	....	....	....
	Detroit	NHL	13	2	6	8	2	0	0	0	11	18.2	3	....	....	....	....	....	....	....	....	....	....	....	....	....
	Anaheim	NHL	15	2	5	7	6	2	0	0	28	7.1	1	....	....	....	....	....	....	....	....	....	....	....	....	....
1995-96	Anaheim	NHL	62	13	21	34	32	7	0	2	143	9.1	–20	....	....	....	....	....	....	....	....	....	....	....	....	....
	Vancouver	NHL	12	1	3	4	6	0	1	0	16	6.3	2	....	....	....	....	....	6	0	0	0	2	0	0	0
1996-97	Vancouver	NHL	78	17	20	37	25	3	3	2	112	15.2	–3	....	....	....	....	....	....	....	....	....	....	....	....	....
1997-98	Vancouver	NHL	48	10	9	19	34	1	2	1	56	17.9	–14	....	....	....	....	....	....	....	....	....	....	....	....	....
	Philadelphia	NHL	27	11	11	22	16	1	2	0	40	27.5	3	....	....	....	....	....	3	1	0	1	0	0	0	0
1998-99	Philadelphia	NHL	25	0	3	3	8	0	0	0	23	0.0	–9	229	62.9	12	5	10:42	....	....	....	....	....	....	....	....
	Tampa Bay	NHL	54	8	2	10	28	0	2	0	69	11.6	–20	320	57.8	70	35	13:57	....	....	....	....	....	....	....	....
99-2000	Tampa Bay	NHL	67	19	25	44	86	6	3	1	126	15.1	–29	493	56.0	69	39	19:42	....	....	....	....	....	....	....	....
	Florida	NHL	13	4	4	8	16	2	0	1	20	20.0	–1	248	61.3	15	14	19:33	4	2	1	3	2	0	0	0
2000-01	Florida	NHL	55	13	21	34	44	1	0	2	100	13.0	–12	1028	59.7	74	29	18:52	....	....	....	....	....	....	....	....
	Ottawa	NHL	13	3	4	7	4	0	0	0	19	15.8	1	215	63.3	14	3	14:31	4	0	0	0	2	0	0	0
2001-02	Columbus	NHL	80	20	23	43	54	8	0	5	150	13.3	–35	2024	57.0	64	43	20:51	....	....	....	....	....	....	....	....
	NHL Totals		678	135	196	331	387	31	14	15	1057	12.8		4557	58.4	318	168	17:53	28	5	4	9	8	0	0	0

WHL East Second All-Star Team (1990) • WHL East First All-Star Team (1991)

Traded to **Anaheim** by **Detroit** with Jason York for Stu Grimson, Mark Ferner and Anaheim's 6th round choice (Magnus Nilsson) in 1996 Entry Draft, April 4, 1995. Traded to **Vancouver** by **Anaheim** for Roman Oksiuta, March 15, 1996. Traded to **Philadelphia** by **Vancouver** for Philadelphia's 5th round choice (later traded back to Philadelphia - Philadelphia selected Garrett Prosofsky) in 1998 Entry Draft, February 5, 1998. Traded to **Tampa Bay** by **Philadelphia** with Chris Gratton for Mikael Renberg and Daymond Langkow, December 12, 1998. Traded to **Florida** by **Tampa Bay** for Ryan Johnson and Dwayne Hay, March 14, 2000. Traded to **Ottawa** by **Florida** for future considerations, March 13, 2001. Signed as a free agent by **Columbus**, July 7, 2001.

SIM, Jonathan
(SIHM, JAWN-ah-thuhn) **DAL.**

Center. Shoots left. 5'10", 190 lbs. Born, New Glasgow, N.S., September 29, 1977. Dallas' 2nd choice, 70th overall, in 1996 Entry Draft.

Season	Club	League	GP	G	A	Pts	PIM	PP	SH	GW	S	%	+/-	TF	F%	H	SB	Min	GP	G	A	Pts	PIM	PP	SH	GW
1994-95	Laval Titan	QMJHL	9	0	1	1	6	....	....	....	....	....	....	....	....	....	....	....	....	....	....	....	....	....	....	....
	Sarnia Sting	OHL	25	9	12	21	19	....	....	....	....	....	....	....	....	....	....	....	4	3	2	5	2	....	....	....
1995-96	Sarnia Sting	OHL	63	56	46	102	130	....	....	....	....	....	....	....	....	....	....	....	10	8	7	15	26	....	....	....
1996-97	Sarnia Sting	OHL	64	*56	39	95	109	....	....	....	....	....	....	....	....	....	....	....	12	9	5	14	32	....	....	....
1997-98	Sarnia Sting	OHL	59	44	50	94	95	....	....	....	....	....	....	....	....	....	....	....	5	1	4	5	14	....	....	....
1998-99♦	Dallas	NHL	7	1	0	1	12	0	0	0	8	12.5	1	6	50.0	15	0	11:26	4	0	0	0	0	0	0	0
	Michigan K-Wings	IHL	68	24	27	51	91	....	....	....	....	....	....	....	....	....	....	....	5	3	1	4	18	....	....	....
99-2000	Dallas	NHL	25	5	3	8	10	2	0	1	44	11.4	4	4	75.0	29	2	10:51	7	1	0	1	6	0	0	0
	Michigan K-Wings	IHL	35	14	16	30	65	....	....	....	....	....	....	....	....	....	....	....	....	....	....	....	....	....	....	....
2000-01	Dallas	NHL	15	0	3	3	6	0	0	0	18	0.0	–2	1	100.0	28	0	8:47	....	....	....	....	....	....	....	....
	Utah Grizzlies	IHL	39	16	13	29	44	....	....	....	....	....	....	....	....	....	....	....	....	....	....	....	....	....	....	....
2001-02	Dallas	NHL	26	3	0	3	10	1	0	0	43	7.0	–3	3	0.0	53	8	9:30	....	....	....	....	....	....	....	....
	Utah Grizzlies	AHL	31	21	6	27	63	....	....	....	....	....	....	....	....	....	....	....	....	....	....	....	....	....	....	....
	NHL Totals		73	9	6	15	38	3	0	1	113	8.0		14	50.0	125	10	9:60	11	1	0	1	6	0	0	0

OHL Second All-Star Team (1998)

SIMICEK, Roman
(SIH-mih-chehk, ROH-muhn)

Center. Shoots left. 6'1", 190 lbs. Born, Ostrava, Czech., November 4, 1971. Pittsburgh's 9th choice, 273rd overall, in 2000 Entry Draft.

Season	Club	League	GP	G	A	Pts	PIM	PP	SH	GW	S	%	+/-	TF	F%	H	SB	Min	GP	G	A	Pts	PIM	PP	SH	GW
1990-91	TJ Vitkovice	Czech	35	2	4	6	....	....	....	....	....	....	....	....	....	....	....	....	....	....	....	....	....	....	....	....
1991-92	TJ Vitkovice	Czech	33	6	12	18	34	....	....	....	....	....	....	....	....	....	....	....	12	2	7	9	....	....	....	....
1992-93	TJ Vitkovice	Czech	38	8	11	19	52	....	....	....	....	....	....	....	....	....	....	....	14	5	8	13	....	....	....	....
1993-94	HC Vitkovice	Czech	40	18	16	34	78	....	....	....	....	....	....	....	....	....	....	....	5	0	2	2	....	....	....	....
1994-95	HC Vitkovice	Czech	41	11	14	25	100	....	....	....	....	....	....	....	....	....	....	....	6	1	3	4	8	....	....	....
1995-96	HC Vitkovice	Czech	39	9	11	20	38	....	....	....	....	....	....	....	....	....	....	....	4	2	0	2	8	....	....	....
1996-97	HC Vitkovice	Czech	49	18	19	37	48	....	....	....	....	....	....	....	....	....	....	....	9	4	4	8	22	....	....	....
1997-98	HC Vitkovice	Czech	40	16	27	43	71	....	....	....	....	....	....	....	....	....	....	....	9	2	4	6	6	....	....	....
	HC Vitkovice	EuroHL	4	1	2	3	4	....	....	....	....	....	....	....	....	....	....	....	....	....	....	....	....	....	....	....
1998-99	HPK Hameenlinna	Finland	49	24	27	51	75	....	....	....	....	....	....	....	....	....	....	....	8	2	5	7	18	....	....	....
99-2000	HPK Hameenlinna	Finland	23	10	17	27	50	....	....	....	....	....	....	....	....	....	....	....	8	2	4	6	10	....	....	....
2000-01	Pittsburgh	NHL	29	3	6	9	30	1	0	1	19	15.8	–5	203	45.8	12	2	9:28	....	....	....	....	....	....	....	....
	Minnesota	NHL	28	2	4	6	21	2	0	0	14	14.3	–4	4	0.0	26	2	12:40	....	....	....	....	....	....	....	....
2001-02	Minnesota	NHL	6	2	0	2	8	0	0	0	4	50.0	1	0	0.0	7	1	13:16	....	....	....	....	....	....	....	....
	Houston Aeros	AHL	49	12	14	26	61	....	....	....	....	....	....	....	....	....	....	....	4	1	0	1	2	....	....	....
	NHL Totals		63	7	10	17	59	3	0	1	37	18.9		207	44.9	45	5	11:15	....	....	....	....	....	....	....	....

Traded to **Minnesota** by **Pittsburgh** for Steve McKenna, January 13, 2001.

SIMON, Ben
(SIGH-mohn, BEN) **ATL.**

Center. Shoots left. 6', 195 lbs. Born, Shaker Heights, OH, June 14, 1978. Chicago's 5th choice, 110th overall, in 1997 Entry Draft.

Season	Club	League	GP	G	A	Pts	PIM	PP	SH	GW	S	%	+/-	TF	F%	H	SB	Min	GP	G	A	Pts	PIM	PP	SH	GW
1992-93	Shaker Heights	Hi-School	25	15	21	36	....	....	....	....	....	....	....	....	....	....	....	....	....	....	....	....	....	....	....	....
1993-94	Shaker Heights	Hi-School	24	45	41	86	....	....	....	....	....	....	....	....	....	....	....	....	....	....	....	....	....	....	....	....
1994-95	Shaker Heights	Hi-School	25	61	68	129	....	....	....	....	....	....	....	....	....	....	....	....	....	....	....	....	....	....	....	....
1995-96	Cleveland Barons	NAJHL	45	38	33	71	....	....	....	....	....	....	....	....	....	....	....	....	5	7	13	20	....	....	....	....
1996-97	U. of Notre Dame	CCHA	30	4	15	19	79	....	....	....	....	....	....	....	....	....	....	....	....	....	....	....	....	....	....	....
1997-98	U. of Notre Dame	CCHA	37	9	28	37	91	....	....	....	....	....	....	....	....	....	....	....	....	....	....	....	....	....	....	....
1998-99	U. of Notre Dame	CCHA	37	18	24	42	65	....	....	....	....	....	....	....	....	....	....	....	....	....	....	....	....	....	....	....
99-2000	U. of Notre Dame	CCHA	40	13	19	32	53	....	....	....	....	....	....	....	....	....	....	....	....	....	....	....	....	....	....	....

			Regular Season																Playoffs							
Season	Club	League	GP	G	A	Pts	PIM	PP	SH	GW	S	%	+/-	TF	F%	H	SB	Min	GP	G	A	Pts	PIM	PP	SH	GW
2000-01	Orlando	IHL	77	8	12	20	47												16	6	5	11	20			
2001-02	**Atlanta**	**NHL**	6	0	0	0	6	0	0	0	7	0.0	1	32	40.6	16	2	9:20								
	Chicago Wolves	AHL	74	11	23	34	56												25	2	3	5	24			
	NHL Totals		6	0	0	0	6	0	0	0	7	0.0		32	40.6	16	2	9:20								

NAJHL First All-Star Team (1996) • NAJHL Rookie of the Year (1996) • CCHA Second All-Star Team (1999)

Rights traded to **Atlanta** by **Chicago** for Atlanta's 9th round choice (Peter Flache) in 2000 Entry Draft, June 25, 2000.

SIMON, Chris

(SIGH-mohn, KRIHS) **WSH.**

Left wing. Shoots left. 6'4", 235 lbs. Born, Wawa, Ont., January 30, 1972. Philadelphia's 2nd choice, 25th overall, in 1990 Entry Draft.

Season	Club	League	GP	G	A	Pts	PIM	PP	SH	GW	S	%	+/-	TF	F%	H	SB	Min	GP	G	A	Pts	PIM	PP	SH	GW
1986-87	Wawa Flyers	NOHA	36	12	20	32	108																			
1987-88	Soo Thunderbirds	NOJHA	55	42	36	78	172																			
1988-89	Ottawa 67's	OHL	36	4	2	6	31																			
1989-90	Ottawa 67's	OHL	57	36	38	74	146												3	2	1	3	4			
1990-91	Ottawa 67's	OHL	20	16	6	22	69												17	5	9	14	59			
1991-92	Ottawa 67's	OHL	2	1	1	2	24																			
	Sault Ste. Marie	OHL	31	19	25	44	143												11	5	8	13	49			
1992-93	**Quebec**	**NHL**	16	1	1	2	67	0	0	1	15	6.7	−2						5	0	0	0	26	0	0	0
	Halifax Citadels	AHL	36	12	6	18	131																			
1993-94	**Quebec**	**NHL**	37	4	4	8	132	0	0	1	39	10.3	−2													
1994-95	**Quebec**	**NHL**	29	3	9	12	106	0	0	0	33	9.1	14						6	1	1	2	19	0	0	1
1995-96♦	**Colorado**	**NHL**	64	16	18	34	250	4	0	1	105	15.2	10						12	1	2	3	11	0	0	0
1996-97	**Washington**	**NHL**	42	9	13	22	165	3	0	1	89	10.1	−1													
1997-98	**Washington**	**NHL**	28	7	10	17	38	4	0	1	71	9.9	−1						18	1	0	1	26	0	0	0
1998-99	**Washington**	**NHL**	23	3	7	10	48	0	0	0	29	10.3	−4	2	50.0	53	5	12:08								
99-2000	**Washington**	**NHL**	75	29	20	49	146	7	0	5	201	14.4	11	7	28.6	141	19	15:32	4	2	0	2	24	0	0	0
2000-01	**Washington**	**NHL**	60	10	10	20	109	4	0	2	123	8.1	−12	3	33.3	69	9	14:34	6	0	1	1	4	0	0	0
2001-02	**Washington**	**NHL**	82	14	17	31	137	1	0	1	121	11.6	−8	7	28.6	94	18	12:11								
	NHL Totals		456	96	109	205	1198	23	0	13	826	11.6		19	31.6	357	51	13:49	51	5	4	9	110	0	0	1

• Missed majority of 1990-91 season recovering from shoulder surgery, October, 1990. Traded to **Quebec** by **Philadelphia** with Philadelphia's 1st round choice (later traded to Toronto - later traded to Washington - Washington selected Nolan Baumgartner) in 1994 Entry Draft to complete transaction that sent Eric Lindros to Philadelphia (June 30, 1992), July 21, 1992. Transferred to **Colorado** after **Quebec** franchise relocated, June 21, 1995. Traded to **Washington** by **Colorado** with Curtis Leschyshyn for Keith Jones and Washington's 1st (Scott Parker) and 4th (later traded back to Washington - Washington selected Krys Barch) round choices in 1998 Entry Drarft, November 2, 1996.

SIMPSON, Reid

(SIHMP-sohn, REED)

Left wing. Shoots left. 6'2", 216 lbs. Born, Flin Flon, Man., May 21, 1969. Philadelphia's 3rd choice, 72nd overall, in 1989 Entry Draft.

Season	Club	League	GP	G	A	Pts	PIM	PP	SH	GW	S	%	+/-	TF	F%	H	SB	Min	GP	G	A	Pts	PIM	PP	SH	GW
1984-85	Flin Flon	MMHL	50	60	70	130	100																			
1985-86	Flin Flon	MJHL	40	20	21	41	200																			
	New Westminster	WHL	2	0	0	0	0																			
1986-87	Prince Albert	WHL	47	3	8	11	105												8	2	3	5	13			
1987-88	Prince Albert	WHL	72	13	14	27	164												10	1	0	1	43			
1988-89	Prince Albert	WHL	59	26	29	55	264												4	2	1	3	30			
1989-90	Prince Albert	WHL	29	15	17	32	121												14	4	7	11	34			
	Hershey Bears	AHL	28	2	2	4	175																			
1990-91	Hershey Bears	AHL	54	9	15	24	183												1	0	0	0	0			
1991-92	**Philadelphia**	**NHL**	1	0	0	0	0	0	0	0	0	0.0	0													
	Hershey Bears	AHL	60	11	7	18	145																			
1992-93	**Minnesota**	**NHL**	1	0	0	0	5	0	0	0	0	0.0	0													
	Kalamazoo Wings	IHL	45	5	5	10	193																			
1993-94	Kalamazoo Wings	IHL	5	0	0	0	16																			
	Albany	AHL	37	9	5	14	135												5	1	1	2	18			
1994-95	Albany	AHL	70	18	25	43	268												14	1	8	9	13			
	New Jersey	**NHL**	9	0	0	0	27	0	0	0	5	0.0	−1													
1995-96	**New Jersey**	**NHL**	23	1	5	6	79	0	0	0	8	12.5	2													
	Albany	AHL	6	1	3	4	17																			
1996-97	**New Jersey**	**NHL**	27	0	4	4	60	0	0	0	17	0.0	0						5	0	0	0	29	0	0	0
	Albany	AHL	3	0	0	0	10																			
1997-98	**New Jersey**	**NHL**	6	0	0	0	16	0	0	0	5	0.0	−2													
	Chicago	**NHL**	38	3	2	5	102	1	0	0	19	15.8	−1													
1998-99	**Chicago**	**NHL**	53	5	4	9	145	1	0	0	23	21.7	2	5	40.0	37	1	5:56								
99-2000	Cleveland	IHL	12	2	2	4	56																			
	Tampa Bay	**NHL**	26	1	0	1	103	0	0	0	13	7.7	−3	1	100.0	30	5	4:33								
2000-01	**St. Louis**	**NHL**	38	2	1	3	96	0	0	1	23	8.7	−3	1	0.0	48	5	6:57	5	0	0	0	2	0	0	0
2001-02	**Montreal**	**NHL**	25	1	1	2	63	0	0	1	9	11.1	0	0	0.0	26	1	4:22								
	Nashville	**NHL**	26	5	0	5	69	0	0	0	13	38.5	−1	5	60.0	26	3	5:45								
	Milwaukee	AHL	2	1	0	1	37																			
	NHL Totals		273	18	17	35	765	2	0	2	135	13.3		12	50.0	167	15	5:41	10	0	0	0	31	0	0	0

Signed as a free agent by **Minnesota**, December 14, 1992. Transferred to **Dallas** after **Minnesota** franchise relocated, June 9, 1993. Traded to **New Jersey** by **Dallas** with Roy Mitchell for future considerations, March 21, 1994. Traded to **Chicago** by **New Jersey** for Chicago's 4th round choice (Mikko Jokela) in 1998 Entry Draft and future considerations, January 8, 1998. Traded to **Tampa Bay** by **Chicago** with Bryan Muir for Michael Nylander, November 12, 1999. • Missed majority of 1999-2000 season recovering from jaw injury suffered in game vs. NY Islanders, January 13, 2000. Signed as a free agent by **St. Louis**, August 24, 2000. • Missed majority of 2000-01 season recovering from groin injury originally suffered in game vs. Nashville, November 24, 2000. Signed as a free agent by **Montreal**, September 10, 2001. Claimed on waivers by **Nashville** from **Montreal**, January 28, 2002.

SIMPSON, Todd

(SIHMP-sohn, TAWD) **PHX.**

Defense. Shoots left. 6'3", 215 lbs. Born, North Vancouver, B.C., May 28, 1973.

Season	Club	League	GP	G	A	Pts	PIM	PP	SH	GW	S	%	+/-	TF	F%	H	SB	Min	GP	G	A	Pts	PIM	PP	SH	GW
1989-90	Don Mills	MTHL	42	36	48	84	36																			
1990-91	Port Colborne	OJHL-B	45	40	43	83	120																			
1991-92	Brown U.	ECAC	18	1	4	5	38																			
1992-93	Tri-City	WHL	69	5	18	23	196												4	0	0	0	13			
1993-94	Tri-City	WHL	12	2	3	5	32																			
	Saskatoon Blades	WHL	51	7	19	26	175												16	1	5	6	42			
1994-95	Saint John	AHL	80	3	10	13	321												5	0	0	0	4			
1995-96	**Calgary**	**NHL**	6	0	0	0	32	0	0	0	3	0.0	0													
	Saint John	AHL	66	4	13	17	277												16	2	3	5	32			
1996-97	**Calgary**	**NHL**	82	1	13	14	208	0	0	0	85	1.2	−14													
1997-98	**Calgary**	**NHL**	53	1	5	6	109	0	0	1	51	2.0	−10													
1998-99	**Calgary**	**NHL**	73	2	8	10	151	0	0	0	52	3.8	18	1	100.0	91	62	17:19								
99-2000	**Florida**	**NHL**	82	1	6	7	202	0	0	0	50	2.0	5	0	0.0	112	77	16:35	4	0	0	0	4	0	0	0
2000-01	**Florida**	**NHL**	25	1	3	4	74	0	0	1	26	3.8	0	0	0.0	40	17	16:29								
	Phoenix	**NHL**	13	0	1	1	12	0	0	0	9	0.0	−4	0	0.0	9	5	13:56								
2001-02	**Phoenix**	**NHL**	67	2	13	15	152	0	0	0	51	3.9	20	0	0.0	130	39	17:20	5	0	2	2	6	0	0	0
	NHL Totals		401	8	49	57	940	0	0	2	327	2.4		1	100.0	382	200	16:50	9	0	2	2	10	0	0	0

Signed as free agent by **Calgary**, July 6, 1994. Traded to **Florida** by **Calgary** for Bill Lindsay, September 30, 1999. • Missed majority of 2000-01 season recovering from head injury suffered in game vs. NY Islanders, December 6, 2000. Traded to **Phoenix** by **Florida** for Phoenix's 2nd round choice (later traded to New Jersey - New Jersey selected Tuomas Pihlman) in 2001 Entry Draft, March 13, 2001.

SKALDE, Jarrod

(SKAHL-dee, JAIR-ruhd) **PHI.**

Center. Shoots left. 6', 185 lbs. Born, Niagara Falls, Ont., February 26, 1971. New Jersey's 3rd choice, 26th overall, in 1989 Entry Draft.

Season	Club	League	GP	G	A	Pts	PIM	PP	SH	GW	S	%	+/-	TF	F%	H	SB	Min	GP	G	A	Pts	PIM	PP	SH	GW
1986-87	Fort Erie	OJHL-B	41	27	34	61	36																			
1987-88	Oshawa Generals	OHL	60	12	16	28	24												7	2	1	3	2			
1988-89	Oshawa Generals	OHL	65	38	38	76	36												6	1	5	6	2			
1989-90	Oshawa Generals	OHL	62	40	52	92	66												17	10	7	17	6			
1990-91	Oshawa Generals	OHL	15	8	14	22	14																			
	Belleville Bulls	OHL	40	30	52	82	21												6	9	6	15	10			
	New Jersey	**NHL**	1	0	1	1	0	0	0	0	2	0.0	0													
	Utica Devils	AHL	3	3	2	5	0																			
1991-92	**New Jersey**	**NHL**	15	2	4	6	4	0	0	2	25	8.0	−1													
	Utica Devils	AHL	62	20	20	40	56												4	3	1	4	8			

Season	Club	League	GP	G	A	Pts	PIM	PP	SH	GW	S	%	+/-	TF	F%	H	SB	Min	Playoffs GP	G	A	Pts	PIM	PP	SH	GW
1992-93	**New Jersey**	**NHL**	**11**	**0**	**2**	**2**	**4**	**0**	**0**	**0**	**11**	**0.0**	**–3**													
	Utica Devils	AHL	59	21	39	60	76												5	0	2	2	19			
	Cincinnati	IHL	4	1	2	3	4																			
1993-94	**Anaheim**	**NHL**	**20**	**5**	**4**	**9**	**10**	**2**	**0**	**2**	**25**	**20.0**	**–3**													
	San Diego Gulls	IHL	57	25	38	63	79												9	3	12	15	10			
1994-95	Las Vegas	IHL	74	34	41	75	103												9	2	4	6	8			
1995-96	Baltimore	AHL	11	2	6	8	55																			
	Calgary	**NHL**	**1**	**0**	**0**	**0**	**0**	**0**	**0**	**0**	**0**	**0.0**	**0**													
	Saint John	AHL	68	27	40	67	98												16	4	9	13	6			
1996-97	Saint John	AHL	65	32	36	68	94												3	0	0	0	14			
1997-98	**San Jose**	**NHL**	**22**	**4**	**6**	**10**	**14**	**0**	**0**	**0**	**30**	**13.3**	**–2**													
	Kentucky	AHL	6	2	6	8	10																			
	Chicago	**NHL**	**4**	**0**	**1**	**1**	**2**	**0**	**0**	**0**	**4**	**0.0**	**0**													
	Indianapolis Ice	IHL	2	0	2	2	0																			
	Dallas	**NHL**	**1**	**0**	**0**	**0**	**0**	**0**	**0**	**0**	**0**	**0.0**	**0**													
	Chicago	**NHL**	**3**	**0**	**0**	**0**	**2**	**0**	**0**	**0**	**0**	**0.0**	**0**													
	Kentucky	AHL	17	3	9	12	38												3	3	0	3	6			
1998-99	**San Jose**	**NHL**	**17**	**1**	**1**	**2**	**4**	**0**	**0**	**0**	**17**	**5.9**	**–6**	**191**	**52.4**	**17**	**1**	**10:07**								
	Kentucky	AHL	54	17	40	57	75												12	4	5	9	16			
99-2000	Utah Grizzlies	IHL	77	25	54	79	98												5	0	1	1	10			
2000-01	**Atlanta**	**NHL**	**19**	**1**	**2**	**3**	**20**	**0**	**0**	**0**	**24**	**4.2**	**–8**	**291**	**47.1**	**22**	**5**	**13:57**								
	Orlando	IHL	60	14	40	54	56												15	3	6	9	20			
2001-02	Chicago Wolves	AHL	64	15	37	52	71																			
	Philadelphia	**NHL**	**1**	**0**	**0**	**0**	**2**	**0**	**0**	**0**	**5**	**0.0**	**0**	**15**	**60.0**	**1**	**0**	**12:25**								
	Philadelphia	AHL	16	4	4	8	23												4	0	2	2	4			
	NHL Totals		**115**	**13**	**21**	**34**	**62**	**2**	**0**	**4**	**143**	**9.1**		**497**	**49.5**	**40**	**6**	**12:09**								

OHL Second All-Star Team (1991) • IHL First All-Star Team (2000)

Claimed by **Anaheim** from **New Jersey** in Expansion Draft, June 24, 1993. Traded to **Calgary** by **Anaheim** for Bobby Marshall, October 30, 1995. Signed as a free agent by **San Jose**, August 14, 1997. Claimed on waivers by **Chicago** from **San Jose**, January 8, 1998. Claimed on waivers by **San Jose** from **Chicago**, January 23, 1998. Claimed on waivers by **Dallas** from **San Jose**, January 27, 1998. Claimed on waivers by **Chicago** from **Dallas**, February 10, 1998. Claimed on waivers by **San Jose** from **Chicago**, March 6, 1998. Signed as a free agent by **Atlanta**, July 21, 2000. Traded to **Philadelphia** by **Atlanta** for Joe DiPenta, March 5, 2002.

SKOPINTSEV, Andrei

(skuh-PIHN-sehf, AWN-dray)

Defense. Shoots right. 6', 185 lbs. Born, Elektrostal, USSR, September 28, 1971. Tampa Bay's 7th choice, 153rd overall, in 1997 Entry Draft.

Season	Club	League	GP	G	A	Pts	PIM	PP	SH	GW	S	%	+/-	TF	F%	H	SB	Min	Playoffs GP	G	A	Pts	PIM	PP	SH	GW
1989-90	Krylja Sovetov	USSR	20	0	0	0	10																			
1990-91	Krylja Sovetov	USSR	16	0	1	1	2																			
1991-92	Krylja Sovetov	CIS	36	1	1	2	14																			
1992-93	Krylja Sovetov	CIS	12	1	0	1	4												7	1	0	1	2			
1993-94	Krylja Sovetov	CIS	43	4	8	12	14												3	1	0	1	0			
1994-95	Krylja Sovetov	CIS	52	8	12	20	55												4	1	1	2	0			
1995-96	Augsburg	Germany	46	10	20	30	32												7	3	2	5	22			
1996-97	TPS Turku	Finland	46	3	6	9	80												10	1	1	2	4			
	TPS Turku	EuroHL	5	0	1	1	4																			
1997-98	TPS Turku	Finland	48	2	9	11	8												4	0	1	1	4			
	TPS Turku	EuroHL	5	0	1	1	4																			
1998-99	**Tampa Bay**	**NHL**	**19**	**1**	**1**	**2**	**10**	**0**	**0**	**0**	**17**	**5.9**	**1**	**0**	**0.0**	**13**	**15**	**16:05**								
	Cleveland	IHL	19	3	2	5	8																			
99-2000	**Tampa Bay**	**NHL**	**4**	**0**	**0**	**0**	**6**	**0**	**0**	**0**	**0**	**0.0**	**–4**	**0**	**0.0**	**8**	**7**	**14:07**								
	Detroit Vipers	IHL	51	4	15	19	44																			
2000-01	**Atlanta**	**NHL**	**17**	**1**	**3**	**4**	**16**	**0**	**0**	**1**	**10**	**10.0**	**–7**	**0**	**0.0**	**21**	**33**	**18:13**								
	Orlando	IHL	25	0	6	6	22																			
2001-02	Dynamo Moscow	Russia	43	3	6	9	50												3	0	0	0	0			
	NHL Totals		**40**	**2**	**4**	**6**	**32**	**0**	**0**	**1**	**27**	**7.4**		**0**	**0.0**	**42**	**55**	**16:48**								

Signed as a free agent by **Atlanta**, September 7, 2000. • Missed majority of 2000-01 season recovering from abdominal injury suffered in game vs. NY Islanders, December 29, 2000. Signed as a free agent by **Dynamo Moscow** (Russia), July 2, 2001.

SKOULA, Martin

(SHKOH-la, MAHR-tihn) **COL.**

Defense. Shoots left. 6'2", 195 lbs. Born, Litomerice, Czech., October 28, 1979. Colorado's 2nd choice, 17th overall, in 1998 Entry Draft.

Season	Club	League	GP	G	A	Pts	PIM	PP	SH	GW	S	%	+/-	TF	F%	H	SB	Min	Playoffs GP	G	A	Pts	PIM	PP	SH	GW
1995-96	Litvinov Jr.	Czech-Jr.	38	0	4	4																				
	Litvinov	Czech																	1	0	0	0	0			
1996-97	Litvinov Jr.	Czech-Jr.	38	2	9	11																				
	Litvinov	Czech	1	0	0	0	0																			
1997-98	Barrie Colts	OHL	66	8	36	44	36												6	1	3	4	4			
1998-99	Barrie Colts	OHL	67	13	46	59	46												12	3	10	13	13			
	Hershey Bears	AHL																	1	0	0	0	0			
99-2000	**Colorado**	**NHL**	**80**	**3**	**13**	**16**	**20**	**2**	**0**	**0**	**66**	**4.5**	**5**	**0**	**0.0**	**100**	**45**	**18:15**	**17**	**0**	**2**	**2**	**4**	**0**	**0**	**0**
2000-01♦	**Colorado**	**NHL**	**82**	**8**	**17**	**25**	**38**	**3**	**0**	**2**	**108**	**7.4**	**8**	**1**	**100.0**	**113**	**52**	**20:41**	**23**	**1**	**4**	**5**	**8**	**0**	**0**	**0**
2001-02	**Colorado**	**NHL**	**82**	**10**	**21**	**31**	**42**	**5**	**0**	**1**	**100**	**10.0**	**–3**	**0**	**0.0**	**112**	**85**	**22:18**	**21**	**0**	**6**	**6**	**2**	**0**	**0**	**0**
	Czech Republic	Olympics	4	0	0	0	0																			
	NHL Totals		**244**	**21**	**51**	**72**	**100**	**10**	**0**	**3**	**274**	**7.7**		**1**	**100.0**	**325**	**182**	**20:26**	**61**	**1**	**12**	**13**	**14**	**0**	**0**	**0**

OHL All-Rookie Team (1998) • OHL Second All-Star Team (1999)

SKRASTINS, Karlis

(SKRAS-tinsh, KAR-lihs) **NSH.**

Defense. Shoots left. 6'1", 212 lbs. Born, Riga, USSR, July 9, 1974. Nashville's 8th choice, 230th overall, in 1998 Entry Draft.

Season	Club	League	GP	G	A	Pts	PIM	PP	SH	GW	S	%	+/-	TF	F%	H	SB	Min	Playoffs GP	G	A	Pts	PIM	PP	SH	GW
1992-93	Pardaugava Riga	CIS	40	3	5	8	16												2	0	0	0	0			
1993-94	Pardaugava Riga	CIS	42	7	5	12	18												2	1	0	1	4			
1994-95	Pardaugava Riga	CIS	52	4	14	18	69																			
1995-96	TPS Turku	Finland	50	4	11	15	32												11	2	2	4	10			
1996-97	TPS Turku	Finland	50	2	8	10	20												12	0	4	4	2			
	TPS Turku	EuroHL	6	0	1	1	4												4	0	0	0	14			
1997-98	TPS Turku	Finland	48	4	15	19	67												4	0	0	0	0			
	TPS Turku	EuroHL	6	0	1	1	6																			
1998-99	**Nashville**	**NHL**	**2**	**0**	**1**	**1**	**0**	**0**	**0**	**0**	**0**	**0.0**	**0**	**0**	**0.0**	**1**	**1**	**11:47**								
	Milwaukee	IHL	75	8	36	44	47												2	0	1	1	2			
99-2000	**Nashville**	**NHL**	**59**	**5**	**6**	**11**	**20**	**1**	**0**	**2**	**51**	**9.8**	**–7**	**0**	**0.0**	**104**	**110**	**20:51**								
	Milwaukee	IHL	19	3	8	11	10																			
2000-01	**Nashville**	**NHL**	**82**	**1**	**11**	**12**	**30**	**0**	**0**	**1**	**66**	**1.5**	**–12**	**0**	**0.0**	**146**	**160**	**19:12**								
2001-02	**Nashville**	**NHL**	**82**	**4**	**13**	**17**	**36**	**0**	**0**	**1**	**84**	**4.8**	**–12**	**0**	**0.0**	**161**	**137**	**20:29**								
	Latvia	Olympics	1	0	0	0	0																			
	NHL Totals		**225**	**10**	**31**	**41**	**86**	**1**	**0**	**4**	**201**	**5.0**		**0**	**0.0**	**412**	**408**	**20:02**								

SKRBEK, Pavel

(SKUHR-behk, PAH-vehl) **NSH.**

Defense. Shoots left. 6'3", 217 lbs. Born, Kladno, Czech., August 9, 1978. Pittsburgh's 2nd choice, 28th overall, in 1996 Entry Draft.

Season	Club	League	GP	G	A	Pts	PIM	PP	SH	GW	S	%	+/-	TF	F%	H	SB	Min	Playoffs GP	G	A	Pts	PIM	PP	SH	GW
1994-95	HC Kladno Jr.	Czech-Jr.	29	7	6	13																				
1995-96	Kladno Jr.	Czech-Jr.	29	10	12	22																				
	HC Poldi Kladno	Czech	13	0	1	1													5	0	0	0				
1996-97	HC Poldi Kladno	Czech	35	1	5	6	26												3	0	0	0	4			
1997-98	Kladno	Czech	47	4	10	14	126																			
1998-99	**Pittsburgh**	**NHL**	**4**	**0**	**0**	**0**	**2**	**0**	**0**	**0**	**1**	**0.0**	**2**	**0**	**0.0**	**6**	**3**	**14:21**								
	Syracuse Crunch	AHL	64	6	16	22	38																			
99-2000	Wilkes-Barre	AHL	51	7	16	23	50																			
	Milwaukee	IHL	6	0	0	0	0																			
2000-01	**Nashville**	**NHL**	**5**	**0**	**0**	**0**	**4**	**0**	**0**	**0**	**2**	**0.0**	**1**	**0**	**0.0**	**2**	**5**	**11:44**								
	Milwaukee	IHL	54	2	22	24	55												5	0	0	0	2			

Season	Club	League	Regular Season																Playoffs							
			GP	G	A	Pts	PIM	PP	SH	GW	S	%	+/-	TF	F%	H	SB	Min	GP	G	A	Pts	PIM	PP	SH	GW
2001-02	**Nashville**	**NHL**	**3**	**0**	**0**	**0**	**2**	**0**	**0**	**0**	**0**	**0.0**	**−2**	**0**	**0.0**	**2**	**2**	**10:18**	....	....	....	....	....	....	....	....
	Kladno	Czech	24	2	3	5	30	....	....	....	....	....	....	....	....	....	....	....	....	....	....	....	....	....	....	....
	NHL Totals		**12**	**0**	**0**	**0**	**8**	**0**	**0**	**0**	**3**	**0.0**		**0**	**0.0**	**10**	**10**	**12:15**	....	....	....	....	....	....	....	....

Traded to **Nashville** by **Pittsburgh** for Bob Boughner, March 13, 2000. Assigned to **Kladno** (Czech) by **Nashville**, October 31, 2001.

SLANEY, John

(SLAY-nee, JAWN) **PHI.**

Defense. Shoots left. 6', 189 lbs. Born, St. John's, Nfld., February 2, 1972. Washington's 1st choice, 9th overall, in 1990 Entry Draft.

Season	Club	League	GP	G	A	Pts	PIM	PP	SH	GW	S	%	+/-	TF	F%	H	SB	Min	GP	G	A	Pts	PIM	PP	SH	GW
1987-88	St. John's	NFAHA	65	41	69	110	70	....	....	....	....	....	....	....	....	....	....	....	....	....	....	....	....	....	....	....
1988-89	Cornwall Royals	OHL	66	16	43	59	23	....	....	....	....	....	....	....	....	....	....	....	18	8	16	24	10	....	....	....
1989-90	Cornwall Royals	OHL	64	38	59	97	68	....	....	....	....	....	....	....	....	....	....	....	6	0	8	8	11	....	....	....
1990-91	Cornwall Royals	OHL	34	21	25	46	28	....	....	....	....	....	....	....	....	....	....	....	....	....	....	....	....	....	....	....
1991-92	Cornwall Royals	OHL	34	19	41	60	43	....	....	....	....	....	....	....	....	....	....	....	6	3	8	11	0	....	....	....
	Baltimore	AHL	6	2	4	6	0	....	....	....	....	....	....	....	....	....	....	....	....	....	....	....	....	....	....	....
1992-93	Baltimore	AHL	79	20	46	66	60	....	....	....	....	....	....	....	....	....	....	....	7	0	7	7	8	....	....	....
1993-94	**Washington**	**NHL**	**47**	**7**	**9**	**16**	**27**	**3**	**0**	**1**	**70**	**10.0**	**3**	....	....	....	....	....	**11**	**1**	**1**	**2**	**2**	**1**	**0**	**0**
	Portland Pirates	AHL	29	14	13	27	17	....	....	....	....	....	....	....	....	....	....	....	....	....	....	....	....	....	....	....
1994-95	**Washington**	**NHL**	**16**	**0**	**3**	**3**	**6**	**0**	**0**	**0**	**21**	**0.0**	**−3**	....	....	....	....	....	....	....	....	....	....	....	....	....
	Portland Pirates	AHL	8	3	10	13	4	....	....	....	....	....	....	....	....	....	....	....	7	1	3	4	4	....	....	....
1995-96	**Colorado**	**NHL**	**7**	**0**	**3**	**3**	**4**	**0**	**0**	**0**	**12**	**0.0**	**2**	....	....	....	....	....	....	....	....	....	....	....	....	....
	Cornwall Aces	AHL	5	0	4	4	2	....	....	....	....	....	....	....	....	....	....	....	....	....	....	....	....	....	....	....
	Los Angeles	**NHL**	**31**	**6**	**11**	**17**	**10**	**3**	**1**	**0**	**63**	**9.5**	**5**	....	....	....	....	....	....	....	....	....	....	....	....	....
1996-97	**Los Angeles**	**NHL**	**32**	**3**	**11**	**14**	**4**	**1**	**0**	**1**	**60**	**5.0**	**−10**	....	....	....	....	....	....	....	....	....	....	....	....	....
	Phoenix	IHL	35	9	25	34	8	....	....	....	....	....	....	....	....	....	....	....	....	....	....	....	....	....	....	....
1997-98	**Phoenix**	**NHL**	**55**	**3**	**14**	**17**	**24**	**1**	**0**	**1**	**74**	**4.1**	**−3**	....	....	....	....	....	....	....	....	....	....	....	....	....
	Las Vegas	IHL	5	2	2	4	10	....	....	....	....	....	....	....	....	....	....	....	....	....	....	....	....	....	....	....
1998-99	**Nashville**	**NHL**	**46**	**2**	**12**	**14**	**14**	**0**	**0**	**1**	**84**	**2.4**	**−12**	**0**	**0.0**	**54**	**49**	**20:39**	....	....	....	....	....	....	....	....
	Milwaukee	IHL	7	0	1	1	0	....	....	....	....	....	....	....	....	....	....	....	....	....	....	....	....	....	....	....
99-2000	**Pittsburgh**	**NHL**	**29**	**1**	**4**	**5**	**10**	**1**	**0**	**0**	**27**	**3.7**	**−10**	**35**	**40.0**	**22**	**8**	**12:13**	**2**	**1**	**0**	**1**	**2**	**1**	**0**	**0**
	Wilkes-Barre	AHL	49	30	30	60	25	....	....	....	....	....	....	....	....	....	....	....	....	....	....	....	....	....	....	....
2000-01	Wilkes-Barre	AHL	40	12	38	50	4	....	....	....	....	....	....	....	....	....	....	....	....	....	....	....	....	....	....	....
	Philadelphia	AHL	25	6	11	17	10	....	....	....	....	....	....	....	....	....	....	....	10	2	6	8	6	....	....	....
2001-02	**Philadelphia**	**NHL**	**1**	**0**	**0**	**0**	**0**	**0**	**0**	**0**	**1**	**0.0**	**2**	**0**	**0.0**	**2**	**1**	**23:33**	**1**	**0**	**0**	**0**	**0**	**0**	**0**	**0**
	Philadelphia	AHL	64	20	39	59	26	....	....	....	....	....	....	....	....	....	....	....	5	2	1	3	0	....	....	....
	NHL Totals		**264**	**22**	**67**	**89**	**99**	**9**	**1**	**4**	**412**	**5.3**		**35**	**40.0**	**78**	**58**	**17:28**	**14**	**2**	**1**	**3**	**4**	**2**	**0**	**0**

OHL First All-Star Team (1990) • Canadian Major Junior Defenseman of the Year (1990) • OHL Second All-Star Team (1991) • AHL First All-Star Team (2001, 2002) • Won Eddie Shore Award (Top Defenseman - AHL) (2001, 2002)

Traded to **Colorado** by **Washington** for Philadelphia's 3rd round choice (previously acquired, Washington selected Shawn McNeil) in 1996 Entry Draft, July 12, 1995. Traded to **LA Kings** by **Colorado** for Winnipeg's 6th round choice (previously acquired, Colorado selected Brian Willsie) in 1996 Entry Draft, December 28, 1995. Signed as a free agent by **Phoenix**, August 19, 1997. Claimed by **Nashville** from **Phoenix** in Expansion Draft, June 26, 1998. Signed as a free agent by **Pittsburgh**, September 30, 1999. Traded to **Philadelphia** by **Pittsburgh** for Kevin Stevens, January 14, 2001.

SLEGR, Jiri

(SLAY-guhr, YEE-ree)

Defense. Shoots left. 6', 216 lbs. Born, Jihlava, Czech., May 30, 1971. Vancouver's 3rd choice, 23rd overall, in 1990 Entry Draft.

Season	Club	League	GP	G	A	Pts	PIM	PP	SH	GW	S	%	+/-	TF	F%	H	SB	Min	GP	G	A	Pts	PIM	PP	SH	GW
1987-88	CHZ Litvinov	Czech	4	1	1	2	0	....	....	....	....	....	....	....	....	....	....	....	....	....	....	....	....	....	....	....
1988-89	CHZ Litvinov	Czech	8	0	0	0	4	....	....	....	....	....	....	....	....	....	....	....	....	....	....	....	....	....	....	....
1989-90	CHZ Litvinov	Czech	51	4	15	19	....	....	....	....	....	....	....	....	....	....	....	....	....	....	....	....	....	....	....	....
1990-91	HC CHZ Litvinov	Czech	47	11	36	47	26	....	....	....	....	....	....	....	....	....	....	....	....	....	....	....	....	....	....	....
1991-92	Litvinov	Czech	42	9	23	32	38	....	....	....	....	....	....	....	....	....	....	....	....	....	....	....	....	....	....	....
	Czechoslovakia	Olympics	8	1	1	2	14	....	....	....	....	....	....	....	....	....	....	....	....	....	....	....	....	....	....	....
1992-93	**Vancouver**	**NHL**	**41**	**4**	**22**	**26**	**109**	**2**	**0**	**0**	**89**	**4.5**	**16**	....	....	....	....	....	**5**	**0**	**3**	**3**	**4**	**0**	**0**	**0**
	Hamilton Canucks	AHL	21	4	14	18	42	....	....	....	....	....	....	....	....	....	....	....	....	....	....	....	....	....	....	....
1993-94	**Vancouver**	**NHL**	**78**	**5**	**33**	**38**	**86**	**1**	**0**	**0**	**160**	**3.1**	**0**	....	....	....	....	....	....	....	....	....	....	....	....	....
1994-95	Litvinov	Czech	11	3	10	13	80	....	....	....	....	....	....	....	....	....	....	....	....	....	....	....	....	....	....	....
	Vancouver	**NHL**	**19**	**1**	**5**	**6**	**32**	**0**	**0**	**1**	**42**	**2.4**	**0**	....	....	....	....	....	....	....	....	....	....	....	....	....
	Edmonton	**NHL**	**12**	**1**	**5**	**6**	**14**	**1**	**0**	**0**	**27**	**3.7**	**−5**	....	....	....	....	....	....	....	....	....	....	....	....	....
1995-96	**Edmonton**	**NHL**	**57**	**4**	**13**	**17**	**74**	**0**	**1**	**1**	**91**	**4.4**	**−1**	....	....	....	....	....	....	....	....	....	....	....	....	....
	Cape Breton	AHL	4	1	2	3	4	....	....	....	....	....	....	....	....	....	....	....	....	....	....	....	....	....	....	....
1996-97	Litvinov	Czech	1	0	0	0	0	....	....	....	....	....	....	....	....	....	....	....	....	....	....	....	....	....	....	....
	Sodertalje SK	Sweden	30	4	14	18	62	....	....	....	....	....	....	....	....	....	....	....	10	4	2	6	32	....	....	....
1997-98	**Pittsburgh**	**NHL**	**73**	**5**	**12**	**17**	**109**	**1**	**1**	**0**	**131**	**3.8**	**10**	....	....	....	....	....	**6**	**0**	**4**	**4**	**2**	**0**	**0**	**0**
	Czech Republic	Olympics	6	1	0	1	8	....	....	....	....	....	....	....	....	....	....	....	....	....	....	....	....	....	....	....
1998-99	**Pittsburgh**	**NHL**	**63**	**3**	**20**	**23**	**86**	**1**	**0**	**0**	**91**	**3.3**	**13**	**2**	**0.0**	**78**	**59**	**18:42**	**13**	**1**	**3**	**4**	**12**	**0**	**0**	**1**
99-2000	**Pittsburgh**	**NHL**	**74**	**11**	**20**	**31**	**82**	**0**	**0**	**2**	**144**	**7.6**	**20**	**3**	**66.7**	**106**	**66**	**21:22**	**10**	**2**	**3**	**5**	**19**	**0**	**0**	**1**
2000-01	**Pittsburgh**	**NHL**	**42**	**5**	**10**	**15**	**60**	**0**	**1**	**1**	**67**	**7.5**	**−9**	**0**	**0.0**	**63**	**29**	**17:37**	....	....	....	....	....	....	....	....
	Atlanta	**NHL**	**33**	**3**	**16**	**19**	**36**	**2**	**0**	**0**	**78**	**3.8**	**−1**	**1**	**0.0**	**49**	**32**	**21:34**	....	....	....	....	....	....	....	....
2001-02	**Atlanta**	**NHL**	**38**	**3**	**5**	**8**	**51**	**1**	**0**	**0**	**56**	**5.4**	**−21**	**0**	**0.0**	**81**	**48**	**21:21**	....	....	....	....	....	....	....	....
	♦ Detroit	**NHL**	**8**	**0**	**1**	**1**	**8**	**0**	**0**	**0**	**11**	**0.0**	**1**	**0**	**0.0**	**14**	**3**	**19:16**	**1**	**0**	**0**	**0**	**2**	**0**	**0**	**0**
	NHL Totals		**538**	**45**	**162**	**207**	**747**	**9**	**3**	**5**	**987**	**4.6**		**6**	**33.3**	**391**	**237**	**20:04**	**35**	**3**	**13**	**16**	**39**	**0**	**0**	**2**

Czechoslovakian First All-Star Team (1991)

Traded to **Edmonton** by **Vancouver** for Roman Oksiuta, April 7, 1995. Traded to **Pittsburgh** by **Edmonton** for Pittsburgh's 3rd round choice (later traded to New Jersey - New Jersey selected Brian Gionta) in 1998 Entry Draft, August 12, 1997. Traded to **Atlanta** by **Pittsburgh** for San Jose's 3rd round choice (previously acquired, later traded to Columbus - Columbus selected Aaron Johnson) in 2001 Entry Draft, January 14, 2001. Traded to **Detroit** by **Atlanta** for Yuri Butsayev and Detroit's 3rd round choice (later traded to Columbus - Columbus selected Jeff Genovy) in 2002 Entry Draft, March 19, 2002.

SLOAN, Blake

(SLOHN, BLAYK) **CGY.**

Right wing. Shoots right. 5'10", 196 lbs. Born, Park Ridge, IL, July 27, 1975.

Season	Club	League	GP	G	A	Pts	PIM	PP	SH	GW	S	%	+/-	TF	F%	H	SB	Min	GP	G	A	Pts	PIM	PP	SH	GW
1992-93	Tabor Academy	Hi-School	33	7	15	22	....	....	....	....	....	....	....	....	....	....	....	....	....	....	....	....	....	....	....	....
1993-94	U. of Michigan	CCHA	38	2	4	6	48	....	....	....	....	....	....	....	....	....	....	....	....	....	....	....	....	....	....	....
1994-95	U. of Michigan	CCHA	39	2	15	17	60	....	....	....	....	....	....	....	....	....	....	....	....	....	....	....	....	....	....	....
1995-96	U. of Michigan	CCHA	41	6	24	30	55	....	....	....	....	....	....	....	....	....	....	....	....	....	....	....	....	....	....	....
1996-97	U. of Michigan	CCHA	41	2	15	17	52	....	....	....	....	....	....	....	....	....	....	....	....	....	....	....	....	....	....	....
1997-98	Houston Aeros	IHL	70	2	13	15	86	....	....	....	....	....	....	....	....	....	....	....	2	0	0	0	0	....	....	....
1998-99♦	**Dallas**	**NHL**	**14**	**0**	**0**	**0**	**10**	**0**	**0**	**0**	**7**	**0.0**	**−1**	**0**	**0.0**	**27**	**3**	**9:01**	**19**	**0**	**2**	**2**	**8**	**0**	**0**	**0**
	Houston Aeros	IHL	62	8	10	18	76	....	....	....	....	....	....	....	....	....	....	....	....	....	....	....	....	....	....	....
99-2000	**Dallas**	**NHL**	**67**	**4**	**13**	**17**	**50**	**0**	**0**	**1**	**78**	**5.1**	**11**	**2**	**50.0**	**165**	**25**	**13:30**	**16**	**0**	**0**	**0**	**12**	**0**	**0**	**0**
2000-01	**Dallas**	**NHL**	**33**	**2**	**2**	**4**	**4**	**0**	**0**	**1**	**29**	**6.9**	**−2**	**4**	**25.0**	**79**	**13**	**9:56**	....	....	....	....	....	....	....	....
	Houston Aeros	IHL	20	7	4	11	18	....	....	....	....	....	....	....	....	....	....	....	....	....	....	....	....	....	....	....
	Columbus	**NHL**	**14**	**1**	**0**	**1**	**13**	**0**	**0**	**0**	**16**	**6.3**	**−2**	**7**	**28.6**	**40**	**12**	**13:19**	....	....	....	....	....	....	....	....
2001-02	**Columbus**	**NHL**	**60**	**2**	**7**	**9**	**46**	**0**	**0**	**0**	**49**	**4.1**	**−18**	**3**	**66.7**	**142**	**26**	**10:56**	....	....	....	....	....	....	....	....
	Calgary	**NHL**	**7**	**0**	**2**	**2**	**4**	**0**	**0**	**0**	**7**	**0.0**	**1**	**2**	**0.0**	**13**	**3**	**12:16**	....	....	....	....	....	....	....	....
	NHL Totals		**195**	**9**	**24**	**33**	**127**	**0**	**0**	**2**	**186**	**4.8**		**18**	**33.3**	**466**	**82**	**11:43**	**35**	**0**	**2**	**2**	**20**	**0**	**0**	**0**

Signed as a free agent by **Dallas**, March 10, 1998. Claimed on waivers by **Columbus** from **Dallas**, March 13, 2001. Traded to **Calgary** by **Columbus** for Jamie Allison, March 19, 2002.

SMEHLIK, Richard

(SHMEH-lihk, RIH-chahrd) **ATL.**

Defense. Shoots left. 6'4", 222 lbs. Born, Ostrava, Czech., January 23, 1970. Buffalo's 3rd choice, 97th overall, in 1990 Entry Draft.

Season	Club	League	GP	G	A	Pts	PIM	PP	SH	GW	S	%	+/-	TF	F%	H	SB	Min	GP	G	A	Pts	PIM	PP	SH	GW
1988-89	TJ Vitkovice	Czech	38	2	5	7	12	....	....	....	....	....	....	....	....	....	....	....	....	....	....	....	....	....	....	....
1989-90	TJ Vitkovice	Czech	44	4	3	7	....	....	....	....	....	....	....	....	....	....	....	....	7	1	1	2	....	....	....	....
1990-91	Dukla Jihlava	Czech	58	4	3	7	22	....	....	....	....	....	....	....	....	....	....	....	....	....	....	....	....	....	....	....
1991-92	TJ Vitkovice	Czech	47	9	10	19	42	....	....	....	....	....	....	....	....	....	....	....	....	....	....	....	....	....	....	....
	Czechoslovakia	Olympics	8	0	1	1	2	....	....	....	....	....	....	....	....	....	....	....	....	....	....	....	....	....	....	....
1992-93	**Buffalo**	**NHL**	**80**	**4**	**27**	**31**	**59**	**0**	**0**	**0**	**82**	**4.9**	**9**	....	....	....	....	....	**8**	**0**	**4**	**4**	**2**	**0**	**0**	**0**
1993-94	**Buffalo**	**NHL**	**84**	**14**	**27**	**41**	**69**	**3**	**3**	**1**	**106**	**13.2**	**22**	....	....	....	....	....	**7**	**0**	**2**	**2**	**10**	**0**	**0**	**0**
1994-95	HC Vitkovice	Czech	13	5	2	7	12	....	....	....	....	....	....	....	....	....	....	....	....	....	....	....	....	....	....	....
	Buffalo	**NHL**	**39**	**4**	**7**	**11**	**46**	**0**	**1**	**1**	**49**	**8.2**	**5**	....	....	....	....	....	**5**	**0**	**0**	**0**	**2**	**0**	**0**	**0**
1995-96	**Buffalo Sabres**	**NHL**	DID NOT PLAY – INJURED																							
1996-97	**Buffalo**	**NHL**	**62**	**11**	**19**	**30**	**43**	**2**	**0**	**1**	**100**	**11.0**	**19**	....	....	....	....	....	**12**	**0**	**2**	**2**	**4**	**0**	**0**	**0**
1997-98	**Buffalo**	**NHL**	**72**	**3**	**17**	**20**	**62**	**0**	**1**	**0**	**90**	**3.3**	**11**	....	....	....	....	....	**15**	**0**	**2**	**2**	**6**	**0**	**0**	**0**
	Czech Republic	Olympics	6	0	1	1	4	....	....	....	....	....	....	....	....	....	....	....	....	....	....	....	....	....	....	....

			Regular Season																Playoffs							
Season	Club	League	GP	G	A	Pts	PIM	PP	SH	GW	S	%	+/-	TF	F%	H	SB	Min	GP	G	A	Pts	PIM	PP	SH	GW
1998-99	**Buffalo**	**NHL**	72	3	11	14	44	0	0	0	61	4.9	–9	0	0.0	107	87	21:50	21	0	3	3	10	0	0	0
99-2000	**Buffalo**	**NHL**	64	2	9	11	50	0	0	0	67	3.0	13	0	0.0	65	52	21:04	5	1	0	1	0	0	0	0
2000-01	**Buffalo**	**NHL**	56	3	12	15	4	0	0	1	40	7.5	6	0	0.0	86	60	19:25	10	0	1	1	4	0	0	0
2001-02	**Buffalo**	**NHL**	60	3	6	9	22	0	0	1	52	5.8	–9	0	0.0	73	74	20:22	....	....	....	....	....	....	....	....
	Czech Republic	Olympics	4	0	0	0	0	....	....	....	....	....	....	....	....	....	....	....	....	....	....	....	....	....	....	....
	NHL Totals		589	47	135	182	399	5	5	5	647	7.3		0	0.0	331	273	20:45	83	1	14	15	38	0	0	0

• Missed entire 1995-96 season recovering from knee surgery, August 11, 1995. Signed as a free agent by **Atlanta**, July 11, 2002.

SMITH, Brandon

(SMIHTH, BRAN-duhn) **NYI**

Defense. Shoots left. 6'1", 209 lbs. Born, Hazelton, B.C., February 25, 1973.

Season	Club	League	GP	G	A	Pts	PIM	PP	SH	GW	S	%	+/-	TF	F%	H	SB	Min	GP	G	A	Pts	PIM	PP	SH	GW
1989-90	Portland	WHL	59	2	17	19	16	....	....	....	....	....	....	....	....	....	....	....	....	....	....	....	....	....	....	....
1990-91	Portland	WHL	17	8	5	13	8	....	....	....	....	....	....	....	....	....	....	....	....	....	....	....	....	....	....	....
1991-92	Portland	WHL	70	12	32	44	63	....	....	....	....	....	....	....	....	....	....	....	....	....	....	....	....	....	....	....
1992-93	Portland	WHL	72	20	54	74	38	....	....	....	....	....	....	....	....	....	....	....	16	4	9	13	6	....	....	....
1993-94	Portland	WHL	72	19	63	82	47	....	....	....	....	....	....	....	....	....	....	....	10	2	10	12	8	....	....	....
1994-95	Dayton Bombers	ECHL	60	16	49	65	57	....	....	....	....	....	....	....	....	....	....	....	4	2	3	5	0	....	....	....
	Minnesota Moose	IHL	1	0	0	0	0	....	....	....	....	....	....	....	....	....	....	....	....	....	....	....	....	....	....	....
	Adirondack	AHL	14	1	2	3	7	....	....	....	....	....	....	....	....	....	....	....	3	0	0	0	2	....	....	....
1995-96	Adirondack	AHL	48	4	13	17	22	....	....	....	....	....	....	....	....	....	....	....	3	0	1	1	2	....	....	....
1996-97	Adirondack	AHL	80	8	26	34	30	....	....	....	....	....	....	....	....	....	....	....	4	0	0	0	0	....	....	....
1997-98	Adirondack	AHL	64	9	27	36	26	....	....	....	....	....	....	....	....	....	....	....	1	0	1	1	0	....	....	....
1998-99	**Boston**	**NHL**	5	0	0	0	0	0	0	0	2	0.0	2	0	0.0	9	2	9:38	....	....	....	....	....	....	....	....
	Providence	AHL	72	16	46	62	32	....	....	....	....	....	....	....	....	....	....	....	19	1	9	10	12	....	....	....
99-2000	**Boston**	**NHL**	22	2	4	6	10	0	0	0	24	8.3	–4	0	0.0	28	25	19:29	....	....	....	....	....	....	....	....
	Providence	AHL	55	8	30	38	20	....	....	....	....	....	....	....	....	....	....	....	14	1	11	12	2	....	....	....
2000-01	**Boston**	**NHL**	3	1	0	1	0	1	0	0	2	50.0	–1	0	0.0	1	1	6:46	....	....	....	....	....	....	....	....
	Providence	AHL	63	11	28	39	30	....	....	....	....	....	....	....	....	....	....	....	17	0	5	5	6	....	....	....
2001-02	Cleveland Barons	AHL	59	6	29	35	26	....	....	....	....	....	....	....	....	....	....	....	....	....	....	....	....	....	....	....
	NHL Totals		30	3	4	7	10	1	0	0	28	10.7		0	0.0	38	28	16:34	....	....	....	....	....	....	....	....

WHL West Second All-Star Team (1993, 1994) • ECHL First All-Star Team (1995) • Won ECHL Top Defenseman Award (1995) • AHL First All-Star Team (1999)

Signed as a free agent by **Detroit**, July 22, 1997. Signed as a free agent by **Boston**, August 5, 1998. Signed as a free agent by **San Jose**, July 23, 2001. Signed as a free agent by **NY Islanders**, August 3, 2002.

SMITH, D.J.

(SMIHTH, DEE-JAY) **COL.**

Defense. Shoots left. 6'2", 205 lbs. Born, Windsor, Ont., May 13, 1977. NY Islanders' 3rd choice, 41st overall, in 1995 Entry Draft.

Season	Club	League	GP	G	A	Pts	PIM	PP	SH	GW	S	%	+/-	TF	F%	H	SB	Min	GP	G	A	Pts	PIM	PP	SH	GW
1992-93	Belle River	OJHL-C	40	5	18	23	39	....	....	....	....	....	....	....	....	....	....	....	....	....	....	....	....	....	....	....
	Windsor Bulldogs	OJHL-B	1	0	0	0	0	....	....	....	....	....	....	....	....	....	....	....	....	....	....	....	....	....	....	....
1993-94	Windsor Bulldogs	OJHL-B	51	8	34	42	267	....	....	....	....	....	....	....	....	....	....	....	....	....	....	....	....	....	....	....
1994-95	Windsor	OHL	61	4	13	17	201	....	....	....	....	....	....	....	....	....	....	....	10	1	3	4	41	....	....	....
1995-96	Windsor	OHL	64	14	45	59	260	....	....	....	....	....	....	....	....	....	....	....	7	1	7	8	23	....	....	....
	St. John's	AHL	1	0	0	0	0	....	....	....	....	....	....	....	....	....	....	....	....	....	....	....	....	....	....	....
1996-97	Windsor	OHL	63	15	52	67	190	....	....	....	....	....	....	....	....	....	....	....	5	1	7	8	11	....	....	....
	Toronto	**NHL**	8	0	1	1	7	0	0	0	4	0.0	–5	....	....	....	....	....	....	....	....	....	....	....	....	....
	St. John's	AHL	....	....	....	....	....	....	....	....	....	....	....	....	....	....	....	....	1	0	0	0	0	....	....	....
1997-98	St. John's	AHL	65	4	11	15	237	....	....	....	....	....	....	....	....	....	....	....	4	0	0	0	4	....	....	....
1998-99	St. John's	AHL	79	7	28	35	216	....	....	....	....	....	....	....	....	....	....	....	5	0	1	1	0	....	....	....
99-2000	**Toronto**	**NHL**	3	0	0	0	5	0	0	0	2	0.0	–1	0	0.0	7	1	12:37	....	....	....	....	....	....	....	....
	St. John's	AHL	74	6	22	28	197	....	....	....	....	....	....	....	....	....	....	....	....	....	....	....	....	....	....	....
2000-01	St. John's	AHL	59	7	12	19	106	....	....	....	....	....	....	....	....	....	....	....	4	0	0	0	11	....	....	....
2001-02	St. John's	AHL	59	6	10	16	152	....	....	....	....	....	....	....	....	....	....	....	....	....	....	....	....	....	....	....
	Hershey Bears	AHL	14	0	3	3	33	....	....	....	....	....	....	....	....	....	....	....	8	1	0	1	33	....	....	....
	NHL Totals		11	0	1	1	12	0	0	0	6	0.0		0	0.0	7	1	12:37	....	....	....	....	....	....	....	....

OHL Second All-Star Team (1997)

Traded to **Toronto** by **NY Islanders** with Wendel Clark and Mathieu Schneider for Darby Hendrickson, Sean Haggerty, Kenny Jonsson and Toronto's 1st round choice (Roberto Luongo) in 1997 Entry Draft, March 13, 1996. Traded to **Nashville** by **Toronto** with Marty Wilford for Marc Moro, March 1, 2002. Traded to **Colorado** by **Nashville** for Tampa Bay's 9th round choice (previously acquired, Nashville selected Matt Davis) in 2002 Entry Draft, March 1, 2002.

SMITH, Jason

(SMIHTH, JAY-suhn) **EDM.**

Defense. Shoots right. 6'3", 210 lbs. Born, Calgary, Alta., November 2, 1973. New Jersey's 1st choice, 18th overall, in 1992 Entry Draft.

Season	Club	League	GP	G	A	Pts	PIM	PP	SH	GW	S	%	+/-	TF	F%	H	SB	Min	GP	G	A	Pts	PIM	PP	SH	GW
1990-91	Calgary Canucks	AJHL	45	3	15	18	69	....	....	....	....	....	....	....	....	....	....	....	....	....	....	....	....	....	....	....
	Regina Pats	WHL	2	0	0	0	7	....	....	....	....	....	....	....	....	....	....	....	4	0	0	0	2	....	....	....
1991-92	Regina Pats	WHL	62	9	29	38	138	....	....	....	....	....	....	....	....	....	....	....	....	....	....	....	....	....	....	....
1992-93	Regina Pats	WHL	64	14	52	66	175	....	....	....	....	....	....	....	....	....	....	....	13	4	8	12	39	....	....	....
	Utica Devils	AHL	....	....	....	....	....	....	....	....	....	....	....	....	....	....	....	....	1	0	0	0	2	....	....	....
1993-94	**New Jersey**	**NHL**	41	0	5	5	43	0	0	0	47	0.0	7	....	....	....	....	....	6	0	0	0	7	0	0	0
	Albany	AHL	20	6	3	9	31	....	....	....	....	....	....	....	....	....	....	....	....	....	....	....	....	....	....	....
1994-95	Albany	AHL	7	0	2	2	15	....	....	....	....	....	....	....	....	....	....	....	11	2	2	4	19	....	....	....
	New Jersey	**NHL**	2	0	0	0	0	0	0	0	5	0.0	–3	....	....	....	....	....	....	....	....	....	....	....	....	....
1995-96	**New Jersey**	**NHL**	64	2	1	3	86	0	0	0	52	3.8	5	....	....	....	....	....	....	....	....	....	....	....	....	....
1996-97	**New Jersey**	**NHL**	57	1	2	3	38	0	0	0	48	2.1	–8	....	....	....	....	....	....	....	....	....	....	....	....	....
	Toronto	**NHL**	21	0	5	5	16	0	0	0	26	0.0	–4	....	....	....	....	....	....	....	....	....	....	....	....	....
1997-98	**Toronto**	**NHL**	81	3	13	16	100	0	0	0	97	3.1	–5	....	....	....	....	....	....	....	....	....	....	....	....	....
1998-99	**Toronto**	**NHL**	60	2	11	13	40	0	0	0	53	3.8	–9	0	0.0	113	67	17:31	....	....	....	....	....	....	....	....
	Edmonton	**NHL**	12	1	1	2	11	0	0	0	15	6.7	0	0	0.0	35	26	20:26	4	0	1	1	4	0	0	0
99-2000	**Edmonton**	**NHL**	80	3	11	14	60	0	0	1	96	3.1	16	1	100.0	260	167	21:15	5	0	1	1	4	0	0	0
2000-01	**Edmonton**	**NHL**	82	5	15	20	120	1	1	0	140	3.6	14	1	0.0	254	210	21:40	6	0	2	2	6	0	0	0
2001-02	**Edmonton**	**NHL**	74	5	13	18	103	0	1	1	85	5.9	14	0	0.0	220	161	21:00	....	....	....	....	....	....	....	....
	NHL Totals		574	22	77	99	617	1	2	2	664	3.3		2	50.0	882	631	20:33	21	0	4	4	21	0	0	0

WHL East First All-Star Team (1993) • Canadian Major Junior First All-Star Team (1993)

• Missed majority of 1994-95 season recovering from knee injury suffered in practice, November 5, 1994. Traded to **Toronto** by **New Jersey** with Steve Sullivan and the rights to Alyn McCauley for Doug Gilmour, Dave Ellett and New Jersey's 4th round choice (previously acquired, New Jersey selected Andre Lakos) in 1999 Entry Draft, February 25, 1997. Traded to **Edmonton** by **Toronto** for Edmonton's 4th round choice (Jonathon Zion) in 1999 Entry Draft and 2nd round choice (Kris Vernarsky) in 2000 Entry Draft, March 23, 1999.

SMITH, Mark

(SMIHTH, MAHRK) **S.J.**

Center. Shoots left. 5'10", 205 lbs. Born, Edmonton, Alta., October 24, 1977. San Jose's 7th choice, 219th overall, in 1997 Entry Draft.

Season	Club	League	GP	G	A	Pts	PIM	PP	SH	GW	S	%	+/-	TF	F%	H	SB	Min	GP	G	A	Pts	PIM	PP	SH	GW
1993-94	Nipawin Hawks	SJHL	62	14	12	26	44	....	....	....	....	....	....	....	....	....	....	....	....	....	....	....	....	....	....	....
1994-95	Lethbridge	WHL	49	3	4	7	25	....	....	....	....	....	....	....	....	....	....	....	....	....	....	....	....	....	....	....
1995-96	Lethbridge	WHL	71	11	24	35	59	....	....	....	....	....	....	....	....	....	....	....	4	2	0	2	2	....	....	....
1996-97	Lethbridge	WHL	62	19	38	57	125	....	....	....	....	....	....	....	....	....	....	....	19	7	13	20	51	....	....	....
1997-98	Lethbridge	WHL	70	42	67	109	206	....	....	....	....	....	....	....	....	....	....	....	3	0	2	2	18	....	....	....
	Kentucky	AHL	2	0	0	0	0	....	....	....	....	....	....	....	....	....	....	....	....	....	....	....	....	....	....	....
1998-99	Kentucky	AHL	78	18	21	39	101	....	....	....	....	....	....	....	....	....	....	....	12	2	7	9	16	....	....	....
99-2000	Kentucky	AHL	79	21	45	66	153	....	....	....	....	....	....	....	....	....	....	....	9	0	5	5	22	....	....	....
2000-01	**San Jose**	**NHL**	42	2	2	4	51	0	0	0	39	5.1	2	308	52.9	52	12	8:48	....	....	....	....	....	....	....	....
	Kentucky	AHL	6	2	6	8	23	....	....	....	....	....	....	....	....	....	....	....	....	....	....	....	....	....	....	....
2001-02	**San Jose**	**NHL**	49	3	3	6	72	0	0	1	40	7.5	–1	368	54.1	68	5	8:03	....	....	....	....	....	....	....	....
	NHL Totals		91	5	5	10	123	0	0	1	79	6.3		676	53.6	120	17	8:24	....	....	....	....	....	....	....	....

WHL East Second All-Star Team (1998)

SMITH, Nick

(SMIHTH, NIHK)

Center. Shoots left. 6'2", 196 lbs. Born, Hamilton, Ont., March 23, 1979. Florida's 4th choice, 74th overall, in 1997 Entry Draft.

			Regular Season																Playoffs							
Season	Club	League	GP	G	A	Pts	PIM	PP	SH	GW	S	%	+/-	TF	F%	H	SB	Min	GP	G	A	Pts	PIM	PP	SH	GW
1995-96	Shelburne Wolves	MTJHL	42	13	18	31	12																			
1996-97	Barrie Colts	OHL	63	10	18	28	15												9	3	8	11	13			
1997-98	Barrie Colts	OHL	63	13	21	34	21												6	1	2	3	4			
1998-99	Barrie Colts	OHL	68	19	34	53	18												12	3	8	11	8			
99-2000	Louisville	AHL	53	8	4	12	8												4	0	0	0	0			
	Port Huron	UHL	2	1	1	2	0																			
2000-01	Louisville	AHL	23	1	2	3	25																			
2001-02	**Florida**	**NHL**	**15**	**0**	**0**	**0**	**0**	**0**	**0**	**0**	**2**	**0.0**	**–1**	**53**	**47.2**	**11**	**1**	**4:09**								
	Bridgeport	AHL	22	3	6	9	4																			
	Saint John	AHL	41	8	9	17	10																			
	NHL Totals		**15**	**0**	**0**	**0**	**0**	**0**	**0**	**0**	**2**	**0.0**		**53**	**47.2**	**11**	**1**	**4:09**								

• Missed majority of 2000-01 season recovering from knee injury suffered in training camp, September 25, 2000.

SMITH, Wyatt

(SMIHTH, WIGH-uht) **NSH.**

Center. Shoots left. 5'11", 208 lbs. Born, Thief River Falls, MN, February 13, 1977. Phoenix's 6th choice, 233rd overall, in 1997 Entry Draft.

Season	Club	League	GP	G	A	Pts	PIM	PP	SH	GW	S	%	+/-	TF	F%	H	SB	Min	GP	G	A	Pts	PIM	PP	SH	GW
1994-95	Warroad Warriors	Hi-School	28	29	31	60	28																			
1995-96	U. of Minnesota	WCHA	32	4	5	9	32																			
1996-97	U. of Minnesota	WCHA	38	16	14	30	44																			
1997-98	U. of Minnesota	WCHA	39	24	23	47	62																			
1998-99	U. of Minnesota	WCHA	43	23	20	43	37																			
99-2000	**Phoenix**	**NHL**	**2**	**0**	**0**	**0**	**0**	**0**	**0**	**0**	**0**	**0.0**	**–2**	**20**	**30.0**	**1**	**1**	**11:39**								
	Springfield	AHL	60	14	26	40	26												5	2	3	5	13			
2000-01	**Phoenix**	**NHL**	**42**	**3**	**7**	**10**	**13**	**0**	**1**	**0**	**40**	**7.5**	**7**	**335**	**40.9**	**15**	**29**	**12:20**								
	Springfield	AHL	18	5	7	12	11																			
2001-02	**Phoenix**	**NHL**	**10**	**0**	**0**	**0**	**0**	**0**	**0**	**0**	**4**	**0.0**	**–5**	**81**	**48.2**	**12**	**4**	**10:36**								
	Springfield	AHL	69	23	32	55	69																			
	NHL Totals		**54**	**3**	**7**	**10**	**13**	**0**	**1**	**0**	**44**	**6.8**		**436**	**41.7**	**28**	**34**	**11:59**								

Signed as a free agent by **Nashville**, July 15, 2002.

SMOLINSKI, Bryan

(smoh-LIHN-skee, BRIGH-uhn) **L.A.**

Center/Right wing. Shoots right. 6'1", 208 lbs. Born, Toledo, OH, December 27, 1971. Boston's 1st choice, 21st overall, in 1990 Entry Draft.

Season	Club	League	GP	G	A	Pts	PIM	PP	SH	GW	S	%	+/-	TF	F%	H	SB	Min	GP	G	A	Pts	PIM	PP	SH	GW
1987-88	Det. Caesars	MNHL	80	43	77	120																				
1988-89	Stratford	OJHL-B	46	32	62	94	132																			
1989-90	Michigan State	CCHA	35	9	13	22	34																			
1990-91	Michigan State	CCHA	35	9	12	21	24																			
1991-92	Michigan State	CCHA	41	28	33	61	55																			
1992-93	Michigan State	CCHA	40	31	37	*68	93																			
	Boston	**NHL**	**9**	**1**	**3**	**4**	**0**	**0**	**0**	**0**	**10**	**10.0**	**3**						**4**	**1**	**0**	**1**	**2**	**0**	**0**	**0**
1993-94	**Boston**	**NHL**	**83**	**31**	**20**	**51**	**82**	**4**	**3**	**5**	**179**	**17.3**	**4**						**13**	**5**	**4**	**9**	**4**	**2**	**0**	**0**
1994-95	**Boston**	**NHL**	**44**	**18**	**13**	**31**	**31**	**6**	**0**	**5**	**121**	**14.9**	**–3**						**5**	**0**	**1**	**1**	**4**	**0**	**0**	**0**
1995-96	**Pittsburgh**	**NHL**	**81**	**24**	**40**	**64**	**69**	**8**	**2**	**1**	**229**	**10.5**	**6**						**18**	**5**	**4**	**9**	**10**	**0**	**0**	**1**
1996-97	Detroit Vipers	IHL	6	5	7	12	10																			
	NY Islanders	**NHL**	**64**	**28**	**28**	**56**	**25**	**9**	**0**	**1**	**183**	**15.3**	**9**													
1997-98	**NY Islanders**	**NHL**	**81**	**13**	**30**	**43**	**34**	**3**	**0**	**4**	**203**	**6.4**	**–16**													
1998-99	**NY Islanders**	**NHL**	**82**	**16**	**24**	**40**	**49**	**7**	**0**	**3**	**223**	**7.2**	**–7**	**1011**	**48.3**	**105**	**50**	**19:19**								
99-2000	**Los Angeles**	**NHL**	**79**	**20**	**36**	**56**	**48**	**2**	**0**	**0**	**160**	**12.5**	**2**	**1545**	**50.9**	**129**	**58**	**18:35**	**4**	**0**	**0**	**0**	**2**	**0**	**0**	**0**
2000-01	**Los Angeles**	**NHL**	**78**	**27**	**32**	**59**	**40**	**5**	**3**	**5**	**183**	**14.8**	**10**	**952**	**48.7**	**155**	**54**	**18:32**	**13**	**1**	**5**	**6**	**14**	**0**	**0**	**0**
2001-02	**Los Angeles**	**NHL**	**80**	**13**	**25**	**38**	**56**	**4**	**1**	**0**	**187**	**7.0**	**7**	**1316**	**45.7**	**135**	**57**	**19:23**	**7**	**2**	**0**	**2**	**2**	**1**	**0**	**0**
	NHL Totals		**681**	**191**	**251**	**442**	**434**	**48**	**9**	**24**	**1678**	**11.4**		**4824**	**48.5**	**524**	**219**	**18:58**	**64**	**14**	**14**	**28**	**38**	**3**	**0**	**1**

CCHA First All-Star Team (1993) • NCAA West First All-American Team (1993)

Traded to **Pittsburgh** by **Boston** with Glen Murray and Boston's 3rd round choice (Boyd Kane) in 1996 Entry Draft for Kevin Stevens and Shawn McEachern, August 2, 1995. Traded to **NY Islanders** by **Pittsburgh** for Darius Kasparaitis and Andreas Johansson, November 17, 1996. Traded to **LA Kings** by **NY Islanders** with Ziggy Palffy, Marcel Cousineau and New Jersey's 4th round choice (previously acquired, LA Kings selected Daniel Johansson) in 1999 Entry Draft for Olli Jokinen, Josh Green, Mathieu Biron and LA Kings' 1st round choice (Taylor Pyatt) in 1999 Entry Draft, June 20, 1999.

SMREK, Peter

(SMUHR-ehk, PEE-tuhr) **NSH.**

Defense. Shoots left. 6'1", 215 lbs. Born, Martin, Czech., February 16, 1979. St. Louis' 2nd choice, 85th overall, in 1999 Entry Draft.

Season	Club	League	GP	G	A	Pts	PIM	PP	SH	GW	S	%	+/-	TF	F%	H	SB	Min	GP	G	A	Pts	PIM	PP	SH	GW
1996-97	Martin	Slovakia	12	1	0	1													3	0	0	0				
1997-98	Martin Jr.	Slovak-Jr.	19	7	6	13	32																			
	Martin	Slovakia	23	0	5	5	24												1	0	0	0	0			
1998-99	Des Moines	USHL	52	6	26	32	59												14	2	7	9	8			
99-2000	Peoria Rivermen	ECHL	4	1	1	2	2																			
	Worcester	AHL	64	5	19	24	26												2	0	0	0	4			
2000-01	**St. Louis**	**NHL**	**6**	**2**	**0**	**2**	**2**	**0**	**0**	**1**	**5**	**40.0**	**1**	**0**	**0.0**	**9**	**4**	**13:01**								
	Worcester	AHL	50	2	7	9	71																			
	NY Rangers	**NHL**	**14**	**0**	**3**	**3**	**12**	**0**	**0**	**0**	**9**	**0.0**	**1**	**0**	**0.0**	**18**	**24**	**16:47**								
	Hartford	AHL																	5	0	2	2	2			
2001-02	**NY Rangers**	**NHL**	**8**	**0**	**1**	**1**	**4**	**0**	**0**	**0**	**2**	**0.0**	**–7**	**0**	**0.0**	**7**	**8**	**14:28**								
	Hartford	AHL	50	2	5	7	36																			
	Slovakia	Olympics	4	0	0	0	0																			
	Milwaukee	AHL	8	0	2	2	4																			
	NHL Totals		**28**	**2**	**4**	**6**	**18**	**0**	**0**	**1**	**16**	**12.5**		**0**	**0.0**	**34**	**36**	**15:19**								

Traded to **NY Rangers** by **St. Louis** for Alexei Gusarov, March 5, 2001. Traded to **Nashville** by **NY Rangers** for Richard Lintner, March 19, 2002.

SMYTH, Brad

(SMIHTH, BRAD) **OTT.**

Right wing. Shoots right. 6', 195 lbs. Born, Ottawa, Ont., March 13, 1973.

Season	Club	League	GP	G	A	Pts	PIM	PP	SH	GW	S	%	+/-	TF	F%	H	SB	Min	GP	G	A	Pts	PIM	PP	SH	GW
1989-90	Nepean	OMHA	55	53	36	89	105																			
1990-91	London Knights	OHL	29	2	6	8	22																			
1991-92	London Knights	OHL	58	17	18	35	93												10	2	0	2	8			
1992-93	London Knights	OHL	66	54	55	109	118												12	7	8	15	25			
1993-94	Cincinnati	IHL	30	7	3	10	54																			
	Birmingham Bulls	ECHL	29	26	30	56	38												10	8	8	16	19			
1994-95	Springfield	AHL	3	0	0	0	7																			
	Birmingham Bulls	ECHL	36	33	35	68	52												3	5	2	7	0			
	Cincinnati	IHL	26	2	11	13	34												1	0	0	0	2			
1995-96	**Florida**	**NHL**	**7**	**1**	**1**	**2**	**4**	**1**	**0**	**0**	**12**	**8.3**	**–3**													
	Carolina	AHL	68	*68	58	*126	80																			
1996-97	**Florida**	**NHL**	**8**	**1**	**0**	**1**	**2**	**0**	**0**	**0**	**10**	**10.0**	**–3**													
	Los Angeles	**NHL**	**44**	**8**	**8**	**16**	**74**	**0**	**0**	**1**	**74**	**10.8**	**–7**													
	Phoenix	IHL	3	5	2	7	0																			
1997-98	**Los Angeles**	**NHL**	**9**	**1**	**3**	**4**	**4**	**0**	**0**	**0**	**12**	**8.3**	**–1**													
	NY Rangers	**NHL**	**1**	**0**	**0**	**0**	**0**	**0**	**0**	**0**	**1**	**0.0**	**0**													
	Hartford	AHL	57	29	33	62	79												15	12	8	20	11			
1998-99	**Nashville**	**NHL**	**3**	**0**	**0**	**0**	**6**	**0**	**0**	**0**	**5**	**0.0**	**–1**	**0**	**0.0**	**1**	**0**	**9:54**								
	Milwaukee	IHL	34	11	16	27	21																			
	Hartford	AHL	36	25	19	44	48												7	6	0	6	14			
99-2000	Hartford	AHL	80	39	37	76	62												23	*13	10	23	8			

			Regular Season																Playoffs							
Season	Club	League	GP	G	A	Pts	PIM	PP	SH	GW	S	%	+/-	TF	F%	H	SB	Min	GP	G	A	Pts	PIM	PP	SH	GW
2000-01	NY Rangers	NHL	4	1	0	1	4	0	0	0	10	10.0	0	0	0.0	2	0	13:57	....	....	....	....	....	....	....	....
	Hartford	AHL	77	*50	29	79	110	....	....	....	....	....	....	....	....	....	....	....	5	2	3	5	8	....	....	....
2001-02	Hartford	AHL	79	34	48	82	90	....	....	....	....	....	....	....	....	....	....	....	10	3	8	11	14	....	....	....
	NHL Totals		76	12	12	24	94	1	0	1	124	9.7		0	0.0	3	0	12:13	....	....	....	....	....	....	....	....

AHL First All-Star Team (1996, 2001, 2002) • Won John B. Sollenberger Trophy (Top Scorer - AHL) (1996) • Won Les Cunningham Award (MVP - AHL) (1996)

Signed as a free agent by **Florida**, October 4, 1993. Traded to **LA Kings** by **Florida** for LA Kings' 3rd round choice (Vratislav Cech) in 1997 Entry Draft, November 28, 1996. Traded to **NY Rangers** by **LA Kings** for future considerations, November 14, 1997. Signed as a free agent by **Nashville**, July 16, 1998. Traded to **NY Rangers** by **Nashville** for future considerations, May 3, 1999. Signed as a free agent by **Ottawa**, August 1, 2002.

SMYTH, Ryan
(SMIHTH, RIGH-uhn) **EDM.**

Left wing. Shoots left. 6'1", 195 lbs. Born, Banff, Alta., February 21, 1976. Edmonton's 2nd choice, 6th overall, in 1994 Entry Draft.

Season	Club	League	GP	G	A	Pts	PIM	PP	SH	GW	S	%	+/-	TF	F%	H	SB	Min	GP	G	A	Pts	PIM	PP	SH	GW
1990-91	Banff Blazers	ABHL	25	100	50	150	....	....	....	....	....	....	....	....	....	....	....	....	....	....	....	....	....	....	....	....
	Lethbridge	AMHL	34	8	21	29	....	....	....	....	....	....	....	....	....	....	....	....	....	....	....	....	....	....	....	....
1991-92	Caronport	SMHL	35	55	61	116	98	....	....	....	....	....	....	....	....	....	....	....	....	....	....	....	....	....	....	....
	Moose Jaw	WHL	2	0	0	0	0	....	....	....	....	....	....	....	....	....	....	....	....	....	....	....	....	....	....	....
1992-93	Moose Jaw	WHL	64	19	14	33	59	....	....	....	....	....	....	....	....	....	....	....	....	....	....	....	....	....	....	....
1993-94	Moose Jaw	WHL	72	50	55	105	88	....	....	....	....	....	....	....	....	....	....	....	....	....	....	....	....	....	....	....
1994-95	Moose Jaw	WHL	50	41	45	86	66	....	....	....	....	....	....	....	....	....	....	....	10	6	9	15	22	....	....	....
	Edmonton	**NHL**	**3**	**0**	**0**	**0**	**0**	**0**	**0**	**0**	**2**	**0.0**	**–1**	....	....	....	....	....	....	....	....	....	....	....	....	....
1995-96	**Edmonton**	**NHL**	**48**	**2**	**9**	**11**	**28**	**1**	**0**	**0**	**65**	**3.1**	**–10**	....	....	....	....	....	....	....	....	....	....	....	....	....
	Cape Breton	AHL	9	6	5	11	4	....	....	....	....	....	....	....	....	....	....	....	....	....	....	....	....	....	....	....
1996-97	**Edmonton**	**NHL**	**82**	**39**	**22**	**61**	**76**	**20**	**0**	**4**	**265**	**14.7**	**–7**	....	....	....	....	....	**12**	**5**	**5**	**10**	**12**	**1**	**0**	**2**
1997-98	**Edmonton**	**NHL**	**65**	**20**	**13**	**33**	**44**	**10**	**0**	**2**	**205**	**9.8**	**–24**	....	....	....	....	....	**12**	**1**	**3**	**4**	**16**	**1**	**0**	**0**
1998-99	**Edmonton**	**NHL**	**71**	**13**	**18**	**31**	**62**	**6**	**0**	**2**	**161**	**8.1**	**0**	**5**	**20.0**	**84**	**20**	**14:26**	**3**	**3**	**0**	**3**	**0**	**2**	**0**	**0**
99-2000	**Edmonton**	**NHL**	**82**	**28**	**26**	**54**	**58**	**11**	**0**	**4**	**238**	**11.8**	**–2**	**24**	**54.2**	**84**	**34**	**19:12**	**5**	**1**	**0**	**1**	**6**	**0**	**1**	**0**
2000-01	**Edmonton**	**NHL**	**82**	**31**	**39**	**70**	**58**	**11**	**0**	**6**	**245**	**12.7**	**10**	**17**	**35.3**	**89**	**35**	**19:58**	**6**	**3**	**4**	**7**	**4**	**0**	**0**	**0**
2001-02	**Edmonton**	**NHL**	**61**	**15**	**35**	**50**	**48**	**7**	**1**	**5**	**150**	**10.0**	**7**	**12**	**41.7**	**45**	**32**	**19:27**	....	....	....	....	....	....	....	....
	Canada	Olympics	6	0	1	1	0	....	....	....	....	....	....	....	....	....	....	....	....	....	....	....	....	....	....	....
	NHL Totals		**494**	**148**	**162**	**310**	**374**	**66**	**1**	**23**	**1331**	**11.1**		**58**	**43.1**	**302**	**121**	**18:19**	**38**	**13**	**12**	**25**	**38**	**4**	**1**	**2**

WHL East Second All-Star Team (1995)

SNYDER, Dan
(SHNIGH-duhr, DAN) **ATL.**

Center. Shoots left. 6', 185 lbs. Born, Elmira, Ont., February 23, 1978.

Season	Club	League	GP	G	A	Pts	PIM	PP	SH	GW	S	%	+/-	TF	F%	H	SB	Min	GP	G	A	Pts	PIM	PP	SH	GW
1994-95	Elmira	OJHL-B	43	8	17	25	46	....	....	....	....	....	....	....	....	....	....	....	....	....	....	....	....	....	....	....
1995-96	Owen Sound	OHL	63	8	17	25	78	....	....	....	....	....	....	....	....	....	....	....	6	1	2	3	4	....	....	....
1996-97	Owen Sound	OHL	57	17	29	46	96	....	....	....	....	....	....	....	....	....	....	....	4	2	3	5	8	....	....	....
1997-98	Owen Sound	OHL	46	23	33	56	74	....	....	....	....	....	....	....	....	....	....	....	10	2	3	5	16	....	....	....
1998-99	Owen Sound	OHL	64	27	67	94	110	....	....	....	....	....	....	....	....	....	....	....	16	8	5	13	30	....	....	....
99-2000	Orlando	IHL	71	12	13	25	123	....	....	....	....	....	....	....	....	....	....	....	6	1	2	3	4	....	....	....
2000-01	**Atlanta**	**NHL**	**2**	**0**	**0**	**0**	**0**	**0**	**0**	**0**	**2**	**0.0**	**0**	**12**	**50.0**	**6**	**1**	**6:55**	....	....	....	....	....	....	....	....
	Orlando	IHL	78	13	30	43	127	....	....	....	....	....	....	....	....	....	....	....	16	7	3	10	20	....	....	....
2001-02	**Atlanta**	**NHL**	**11**	**1**	**1**	**2**	**30**	**0**	**0**	**0**	**7**	**14.3**	**–3**	**122**	**43.4**	**20**	**7**	**10:57**	....	....	....	....	....	....	....	....
	Chicago Wolves	AHL	56	11	24	35	115	....	....	....	....	....	....	....	....	....	....	....	22	7	10	17	25	....	....	....
	NHL Totals		**13**	**1**	**1**	**2**	**30**	**0**	**0**	**0**	**9**	**11.1**		**134**	**44.0**	**26**	**8**	**10:20**	....	....	....	....	....	....	....	....

Signed as a free agent by **Atlanta**, June 28, 1999.

SONNENBERG, Martin
(SOHN-nehn-BUHRG, MAHR-tihn) **CGY.**

Left wing. Shoots left. 6', 184 lbs. Born, Wetaskiwin, Alta., January 23, 1978.

Season	Club	League	GP	G	A	Pts	PIM	PP	SH	GW	S	%	+/-	TF	F%	H	SB	Min	GP	G	A	Pts	PIM	PP	SH	GW
1994-95	Leduc Oil Barons	AMHL	35	28	40	68	34	....	....	....	....	....	....	....	....	....	....	....	....	....	....	....	....	....	....	....
1995-96	Saskatoon Blades	WHL	58	8	7	15	24	....	....	....	....	....	....	....	....	....	....	....	3	0	0	0	2	....	....	....
1996-97	Saskatoon Blades	WHL	72	38	26	64	79	....	....	....	....	....	....	....	....	....	....	....	....	....	....	....	....	....	....	....
1997-98	Saskatoon Blades	WHL	72	40	52	92	87	....	....	....	....	....	....	....	....	....	....	....	6	1	3	4	9	....	....	....
1998-99	**Pittsburgh**	**NHL**	**44**	**1**	**1**	**2**	**19**	**0**	**0**	**0**	**12**	**8.3**	**–2**	**2**	**0.0**	**33**	**5**	**4:00**	**7**	**0**	**0**	**0**	**0**	**0**	**0**	**0**
	Syracuse Crunch	AHL	36	15	9	24	31	....	....	....	....	....	....	....	....	....	....	....	....	....	....	....	....	....	....	....
99-2000	**Pittsburgh**	**NHL**	**14**	**1**	**2**	**3**	**0**	**1**	**0**	**0**	**19**	**5.3**	**0**	**7**	**28.6**	**10**	**7**	**7:26**	....	....	....	....	....	....	....	....
	Wilkes-Barre	AHL	62	20	33	53	109	....	....	....	....	....	....	....	....	....	....	....	....	....	....	....	....	....	....	....
2000-01	Wilkes-Barre	AHL	73	14	18	32	89	....	....	....	....	....	....	....	....	....	....	....	21	4	3	7	6	....	....	....
2001-02	Wilkes-Barre	AHL	78	20	30	50	127	....	....	....	....	....	....	....	....	....	....	....	....	....	....	....	....	....	....	....
	NHL Totals		**58**	**2**	**3**	**5**	**19**	**1**	**0**	**0**	**31**	**6.5**		**9**	**22.2**	**43**	**12**	**4:50**	**7**	**0**	**0**	**0**	**0**	**0**	**0**	**0**

Signed as a free agent by **Pittsburgh**, October 9, 1998. Signed as a free agent by **Calgary**, July 9, 2002.

SOPEL, Brent
(SOH-puhl, BREHNT) **VAN.**

Defense. Shoots right. 6'1", 205 lbs. Born, Calgary, Alta., January 7, 1977. Vancouver's 6th choice, 144th overall, in 1995 Entry Draft.

Season	Club	League	GP	G	A	Pts	PIM	PP	SH	GW	S	%	+/-	TF	F%	H	SB	Min	GP	G	A	Pts	PIM	PP	SH	GW
1992-93	Sask. Legion	SMHL	36	7	17	24	95	....	....	....	....	....	....	....	....	....	....	....	....	....	....	....	....	....	....	....
1993-94	Sask. Blazers	SMHL	34	9	30	39	180	....	....	....	....	....	....	....	....	....	....	....	....	....	....	....	....	....	....	....
	Saskatoon Blades	WHL	11	2	2	4	2	....	....	....	....	....	....	....	....	....	....	....	....	....	....	....	....	....	....	....
1994-95	Saskatoon Blades	WHL	22	1	10	11	31	....	....	....	....	....	....	....	....	....	....	....	....	....	....	....	....	....	....	....
	Swift Current	WHL	41	4	19	23	50	....	....	....	....	....	....	....	....	....	....	....	3	0	3	3	0	....	....	....
1995-96	Swift Current	WHL	71	13	48	61	87	....	....	....	....	....	....	....	....	....	....	....	6	1	2	3	4	....	....	....
	Syracuse Crunch	AHL	1	0	0	0	0	....	....	....	....	....	....	....	....	....	....	....	....	....	....	....	....	....	....	....
1996-97	Swift Current	WHL	62	15	41	56	109	....	....	....	....	....	....	....	....	....	....	....	10	5	11	16	32	....	....	....
	Syracuse Crunch	AHL	2	0	0	0	0	....	....	....	....	....	....	....	....	....	....	....	3	0	0	0	0	....	....	....
1997-98	Syracuse Crunch	AHL	76	10	33	43	70	....	....	....	....	....	....	....	....	....	....	....	5	0	7	7	12	....	....	....
1998-99	**Vancouver**	**NHL**	**5**	**1**	**0**	**1**	**4**	**1**	**0**	**0**	**5**	**20.0**	**–1**	**0**	**0.0**	**4**	**1**	**11:58**	....	....	....	....	....	....	....	....
	Syracuse Crunch	AHL	53	10	21	31	59	....	....	....	....	....	....	....	....	....	....	....	....	....	....	....	....	....	....	....
99-2000	**Vancouver**	**NHL**	**18**	**2**	**4**	**6**	**12**	**0**	**0**	**1**	**11**	**18.2**	**9**	**0**	**0.0**	**29**	**13**	**10:31**	....	....	....	....	....	....	....	....
	Syracuse Crunch	AHL	50	6	25	31	67	....	....	....	....	....	....	....	....	....	....	....	4	0	2	2	8	....	....	....
2000-01	**Vancouver**	**NHL**	**52**	**4**	**10**	**14**	**10**	**0**	**0**	**1**	**57**	**7.0**	**4**	**0**	**0.0**	**93**	**44**	**16:01**	**4**	**0**	**0**	**0**	**2**	**0**	**0**	**0**
	Kansas City	IHL	4	0	1	1	0	....	....	....	....	....	....	....	....	....	....	....	....	....	....	....	....	....	....	....
2001-02	**Vancouver**	**NHL**	**66**	**8**	**17**	**25**	**44**	**1**	**0**	**3**	**116**	**6.9**	**21**	**0**	**0.0**	**89**	**74**	**19:01**	**6**	**0**	**2**	**2**	**2**	**0**	**0**	**0**
	NHL Totals		**141**	**15**	**31**	**46**	**70**	**2**	**0**	**5**	**189**	**7.9**		**0**	**0.0**	**215**	**132**	**16:35**	**10**	**0**	**2**	**2**	**4**	**0**	**0**	**0**

SOURAY, Sheldon
(SUHR-ee, SHEHL-dohn) **MTL.**

Defense. Shoots left. 6'4", 223 lbs. Born, Elk Point, Alta., July 13, 1976. New Jersey's 3rd choice, 71st overall, in 1994 Entry Draft.

Season	Club	League	GP	G	A	Pts	PIM	PP	SH	GW	S	%	+/-	TF	F%	H	SB	Min	GP	G	A	Pts	PIM	PP	SH	GW
1990-91	Bonnyville	AAHA	30	15	20	35	100	....	....	....	....	....	....	....	....	....	....	....	....	....	....	....	....	....	....	....
1991-92	Quesnel	BCAHA	20	5	15	20	200	....	....	....	....	....	....	....	....	....	....	....	....	....	....	....	....	....	....	....
	Alberta Cycle	AMHL	11	0	5	5	67	....	....	....	....	....	....	....	....	....	....	....	....	....	....	....	....	....	....	....
1992-93	Ft. Saskatchewan	AJHL	35	0	12	12	125	....	....	....	....	....	....	....	....	....	....	....	....	....	....	....	....	....	....	....
	Tri-City	WHL	2	0	0	0	0	....	....	....	....	....	....	....	....	....	....	....	....	....	....	....	....	....	....	....
1993-94	Tri-City	WHL	42	3	6	9	122	....	....	....	....	....	....	....	....	....	....	....	....	....	....	....	....	....	....	....
1994-95	Tri-City	WHL	40	2	24	26	140	....	....	....	....	....	....	....	....	....	....	....	....	....	....	....	....	....	....	....
	Prince George	WHL	11	2	3	5	23	....	....	....	....	....	....	....	....	....	....	....	....	....	....	....	....	....	....	....
	Albany	AHL	7	0	2	2	8	....	....	....	....	....	....	....	....	....	....	....	....	....	....	....	....	....	....	....
1995-96	Prince George	WHL	32	9	18	27	91	....	....	....	....	....	....	....	....	....	....	....	....	....	....	....	....	....	....	....
	Kelowna Rockets	WHL	27	7	20	27	94	....	....	....	....	....	....	....	....	....	....	....	6	0	5	5	2	....	....	....
	Albany	AHL	6	0	2	2	12	....	....	....	....	....	....	....	....	....	....	....	4	0	1	1	4	....	....	....
1996-97	Albany	AHL	70	2	11	13	160	....	....	....	....	....	....	....	....	....	....	....	16	2	3	5	47	....	....	....
1997-98	**New Jersey**	**NHL**	**60**	**3**	**7**	**10**	**85**	**0**	**0**	**1**	**74**	**4.1**	**18**	....	....	....	....	....	**3**	**0**	**1**	**1**	**2**	**0**	**0**	**0**
	Albany	AHL	6	0	0	0	8	....	....	....	....	....	....	....	....	....	....	....	....	....	....	....	....	....	....	....
1998-99	**New Jersey**	**NHL**	**70**	**1**	**7**	**8**	**110**	**0**	**0**	**0**	**101**	**1.0**	**5**	**0**	**0.0**	**109**	**57**	**14:56**	**2**	**0**	**1**	**1**	**0**	**0**	**0**	**0**
99-2000	**New Jersey**	**NHL**	**52**	**0**	**8**	**8**	**70**	**0**	**0**	**0**	**74**	**0.0**	**–6**	**0**	**0.0**	**97**	**51**	**17:12**	....	....	....	....	....	....	....	....
	Montreal	**NHL**	**19**	**3**	**0**	**3**	**44**	**0**	**0**	**0**	**39**	**7.7**	**7**	**0**	**0.0**	**23**	**34**	**19:18**	....	....	....	....	....	....	....	....

			Regular Season																Playoffs							
Season	Club	League	GP	G	A	Pts	PIM	PP	SH	GW	S	%	+/-	TF	F%	H	SB	Min	GP	G	A	Pts	PIM	PP	SH	GW
2000-01	Montreal	NHL	52	3	8	11	95	0	0	2	103	2.9	−11	0	0.0	72	78	20:36								
2001-02	Montreal	NHL	34	3	5	8	62	1	0	0	56	5.4	−5	1	100.0	67	44	18:11	12	0	1	1	16	0	0	0
	NHL Totals		287	13	35	48	466	1	0	3	447	2.9		1	100.0	368	264	17:36	17	0	3	3	18	0	0	0

WHL West Second All-Star Team (1996)

Traded to **Montreal** by **New Jersey** with Josh DeWolf and New Jersey's 2nd round choice (later traded to Washington - later traded to Tampa Bay - Tampa Bay selected Andreas Holmqvist) in 2001 Entry Draft for Vladimir Malakhov, March 1, 2000. • Missed majority of 2001-02 season recovering from wrist injury originally suffered in game vs. Tampa Bay, November 17, 2001.

SPACEK, Jaroslav

(SPAH-chehk, YA-roh-slahv) **CBJ**

Defense. Shoots left. 5'11", 206 lbs. Born, Rokycany, Czech., February 11, 1974. Florida's 5th choice, 117th overall, in 1998 Entry Draft.

Season	Club	League	GP	G	A	Pts	PIM	PP	SH	GW	S	%	+/-	TF	F%	H	SB	Min	GP	G	A	Pts	PIM	PP	SH	GW
1992-93	HC Skoda Plzen	Czech	16	1	3	4																				
1993-94	HC Skoda Plzen	Czech	34	2	6	8																				
1994-95	Plzen	Czech	38	4	8	12	14												3	1	0	1	2			
1995-96	HC ZKZ Plzen	Czech	40	3	10	13	42												3	0	1	1	4			
1996-97	HC ZKZ Plzen	Czech	52	9	29	38	44																			
1997-98	Farjestad	Sweden	45	10	16	26	63												12	2	5	7	14			
	Farjestad	EuroHL	6	2	3	5	2																			
1998-99	Florida	NHL	63	3	12	15	28	2	1	0	92	3.3	15	1	100.0	78	41	19:27								
	New Haven	AHL	14	4	8	12	15																			
99-2000	Florida	NHL	82	10	26	36	53	4	0	1	111	9.0	7	1	0.0	122	98	22:40	4	0	0	0	0	0	0	0
2000-01	Florida	NHL	12	2	1	3	8	1	0	0	21	9.5	−4	0	0.0	23	11	19:12								
	Chicago	NHL	50	5	18	23	20	2	0	1	85	5.9	7	0	0.0	63	65	21:31								
2001-02	Chicago	NHL	60	3	10	13	29	0	0	1	64	4.7	5	0	0.0	91	45	16:25								
	Czech Republic	Olympics	4	0	0	0	0																			
	Columbus	NHL	14	2	3	5	24	1	1	1	29	6.9	−9	0	0.0	28	25	23:35								
	NHL Totals		281	25	70	95	162	10	2	4	402	6.2		2	50.0	405	285	20:18	4	0	0	0	0	0	0	0

Traded to **Chicago** by **Florida** for Anders Eriksson, November 6, 2000. Traded to **Columbus** by **Chicago** with Chicago's 2nd round choice in 2003 Entry Draft for Lyle Odelein, March 19, 2002.

SPANHEL, Martin

(SPAN-hehl, MAHR-tihn) **CBJ**

Right wing. Shoots left. 6'2", 206 lbs. Born, Zlin, Czech., July 1, 1977. Philadelphia's 6th choice, 152nd overall, in 1995 Entry Draft.

Season	Club	League	GP	G	A	Pts	PIM	PP	SH	GW	S	%	+/-	TF	F%	H	SB	Min	GP	G	A	Pts	PIM	PP	SH	GW
1994-95	AC ZPS Zlin Jr.	Czech-Jr.	33	25	16	41	0																			
	AC ZPS Zlin	Czech	1	0	0	0	0																			
1995-96	Lethbridge	WHL	6	1	0	1	0																			
	Moose Jaw	WHL	61	4	12	16	33																			
1996-97	AC ZPS Zlin	Czech	22	3	6	9	20																			
1997-98	Zlin	Czech	40	7	9	16	70																			
1998-99	Plzen	Czech	49	12	12	24	60												5	2	1	3	27			
99-2000	Plzen	Czech	52	21	27	48	86												7	1	4	5	12			
2000-01	Columbus	NHL	6	1	0	1	2	0	0	0	8	12.5	−1	1	0.0	8	0	12:29								
	Syracuse Crunch	AHL	67	11	13	24	75												2	0	0	0	8			
2001-02	Columbus	NHL	4	1	0	1	2	0	0	0	6	16.7	−2	0	0.0	1	0	11:12								
	Syracuse Crunch	AHL	50	7	12	19	43																			
	NHL Totals		10	2	0	2	4	0	0	0	14	14.3		1	0.0	9	0	11:58								

Traded to **San Jose** by **Philadelphia** with Philadelphia's 1st round choice (later traded to Buffalo - later traded to Phoenix - Phoenix selected Daniel Briere) in 1996 Entry Draft and Philadelphia's 4th round choice (later traded to Buffalo - Buffalo selected Mike Martone) in 1996 Entry Draft for Pat Falloon, November 16, 1995. Traded to **Buffalo** by **San Jose** with Vaclav Varada and Philadelphia's 1st (previously acquired by San Jose - later traded to Phoenix - Phoenix selected Daniel Briere) and 4th (previously acquired, Buffalo selected Mike Martone) round choices in 1996 Entry Draft for Doug Bodger, November 16, 1995. Signed as a free agent by **Columbus**, May 30, 2000. Signed as a free agent by **Sparta Praha** (Czech) with Columbus retaining NHL rights, July 26, 2002.

STAIOS, Steve

(STAY-uhs, STEEV) **EDM.**

Defense. Shoots right. 6'1", 200 lbs. Born, Hamilton, Ont., July 28, 1973. St. Louis' 1st choice, 27th overall, in 1991 Entry Draft.

Season	Club	League	GP	G	A	Pts	PIM	PP	SH	GW	S	%	+/-	TF	F%	H	SB	Min	GP	G	A	Pts	PIM	PP	SH	GW
1988-89	Hamilton Huskies	OMHA	58	13	39	52	78																			
1989-90	Hamilton	OJHL-B	40	9	27	36	66																			
1990-91	Niagara Falls	OHL	66	17	29	46	115												12	2	3	5	10			
1991-92	Niagara Falls	OHL	65	11	42	53	122												17	7	8	15	27			
1992-93	Niagara Falls	OHL	12	4	14	18	30																			
	Sudbury Wolves	OHL	53	13	44	57	67												11	5	6	11	22			
1993-94	Peoria Rivermen	IHL	38	3	9	12	42																			
1994-95	Peoria Rivermen	IHL	60	3	13	16	64												6	0	0	0	10			
1995-96	Peoria Rivermen	IHL	6	0	1	1	14																			
	Worcester	AHL	57	1	11	12	114																			
	Boston	NHL	12	0	0	0	4	0	0	0	4	0.0	−5						3	0	0	0	0	0	0	0
	Providence	AHL	7	1	4	5	8																			
1996-97	Boston	NHL	54	3	8	11	71	0	0	0	56	5.4	−26													
	Vancouver	NHL	9	0	6	6	20	0	0	0	10	0.0	2													
1997-98	Vancouver	NHL	77	3	4	7	134	0	0	1	45	6.7	−3													
1998-99	Vancouver	NHL	57	0	2	2	54	0	0	0	33	0.0	−12	4	25.0	58	8	6:53								
99-2000	Atlanta	NHL	27	2	3	5	66	0	0	0	38	5.3	−5	2	50.0	60	14	13:01								
2000-01	Atlanta	NHL	70	9	13	22	137	4	0	0	156	5.8	−23	1	0.0	102	108	21:45								
2001-02	Edmonton	NHL	73	5	5	10	108	0	0	1	101	5.0	10	0	0.0	85	86	18:05								
	NHL Totals		379	22	41	63	594	4	0	2	443	5.0		7	28.6	305	216	15:48	3	0	0	0	0	0	0	0

Traded to **Boston** by **St. Louis** with Kevin Sawyer for Steve Leach, March 8, 1996. Claimed on waivers by **Vancouver** from **Boston**, March 18, 1997. Claimed by **Atlanta** from **Vancouver** in Expansion Draft, June 25, 1999. • Missed majority of 1999-2000 season recovering from knee injury originally suffered in game vs. Colorado, October 23, 1999. Traded to **New Jersey** by **Atlanta** for New Jersey's 9th round choice (Simon Gamache) in 2000 Entry Draft, June 12, 2000. Traded to **Atlanta** by **New Jersey** for future considerations, July 10, 2000. Signed as a free agent by **Edmonton**, July 12, 2001.

STAPLETON, Mike

(STAY-puhl-TOHN, MIGHK)

Center. Shoots right. 5'10", 183 lbs. Born, Sarnia, Ont., May 5, 1966. Chicago's 7th choice, 132nd overall, in 1984 Entry Draft.

Season	Club	League	GP	G	A	Pts	PIM	PP	SH	GW	S	%	+/-	TF	F%	H	SB	Min	GP	G	A	Pts	PIM	PP	SH	GW
1982-83	Strathroy Blades	OJHL-B	40	39	38	77	99												3	1	2	3	4			
1983-84	Cornwall Royals	OHL	70	24	45	69	94												3	1	2	3	4			
1984-85	Cornwall Royals	OHL	56	41	44	85	68												9	2	4	6	23			
1985-86	Cornwall Royals	OHL	56	39	64	103	74												6	2	3	5	2			
1986-87	Team Canada	Nat-Tm	21	2	4	6	4																			
	Chicago	NHL	39	3	6	9	6	0	0	0	54	5.6	−9						4	0	0	0	2	0	0	0
1987-88	Chicago	NHL	53	2	9	11	59	0	0	1	50	4.0	−10													
	Saginaw Hawks	IHL	31	11	19	30	52												10	5	6	11	10			
1988-89	Chicago	NHL	7	0	1	1	7	0	0	0	6	0.0	−1													
	Saginaw Hawks	IHL	69	21	47	68	162												6	1	3	4	4			
1989-90	Arvika HC	Swede-3	30	15	18	33																				
	Indianapolis Ice	IHL	16	5	10	15	6												13	9	10	19	38			
1990-91	Chicago	NHL	7	0	1	1	2	0	0	0	6	0.0	0													
	Indianapolis Ice	IHL	75	29	52	81	76												7	1	4	5	0			
1991-92	Chicago	NHL	19	4	4	8	8	1	0	0	32	12.5	0													
	Indianapolis Ice	IHL	59	18	40	58	65																			
1992-93	Pittsburgh	NHL	78	4	9	13	10	0	1	1	78	5.1	−8						4	0	0	0	0	0	0	0
1993-94	Pittsburgh	NHL	58	7	4	11	18	3	0	0	59	11.9	−4													
	Edmonton	NHL	23	5	9	14	28	1	0	0	43	11.6	−1													
1994-95	Edmonton	NHL	46	6	11	17	21	3	0	2	59	10.2	−12													
1995-96	Winnipeg	NHL	58	10	14	24	37	3	1	0	91	11.0	−4						6	0	0	0	21	0	0	0
1996-97	Phoenix	NHL	55	4	11	15	36	2	0	1	74	5.4	−4						7	0	0	0	14	0	0	0
1997-98	Phoenix	NHL	64	5	5	10	36	1	1	1	69	7.2	−4						6	0	0	0	2	0	0	0
1998-99	Phoenix	NHL	76	9	9	18	34	0	2	2	106	8.5	−6	345	46.1	67	40	13:40	7	1	0	1	0	0	0	0
99-2000	Atlanta	NHL	62	10	12	22	30	4	0	1	146	6.8	−29	717	48.0	72	44	17:18								

			Regular Season																Playoffs							
Season	Club	League	GP	G	A	Pts	PIM	PP	SH	GW	S	%	+/-	TF	F%	H	SB	Min	GP	G	A	Pts	PIM	PP	SH	GW
2000-01	NY Islanders	NHL	34	1	4	5	2	0	0	0	22	4.5	–5	192	45.3	32	13	9:09	....	....	....	....	....	....	....	....
	Vancouver	NHL	18	1	2	3	8	1	0	0	9	11.1	–6	104	41.4	13	4	9:58	....	....	....	....	....	....	....	....
2001-02	Blues Espoo	Finland	41	18	19	37	42	....	....	....	....	....	....	....	....	....	....	....	3	0	0	0	4	....	....	....
	NHL Totals		697	71	111	182	342	19	5	9	904	7.9		1358	46.6	184	101	13:42	34	1	0	1	39	0	0	0

Signed as a free agent by **Pittsburgh**, September 30, 1992. Claimed on waivers by **Edmonton** from **Pittsburgh**, February 19, 1994. Signed as a free agent by **Winnipeg**, August 18, 1995. Transferred to **Phoenix** after **Winnipeg** franchise relocated, July 1, 1996. Claimed by **Atlanta** from **Phoenix** in Expansion Draft, June 25, 1999. Signed as a free agent by **NY Islanders**, July 3, 2000. Traded to **Vancouver** by **NY Islanders** for Vancouver's 9th round choice (later traded to Washington - Washington selected Robert Muller) in 2001 Entry Draft, December 28, 2000. Signed as a free agent by **Blues Espoo** (Finland), October 23, 2001.

STEFAN, Patrik

(SHTEH-fan, PAT-rihk) **ATL.**

Center. Shoots left. 6'3", 205 lbs. Born, Pribram, Czech., September 16, 1980. Atlanta's 1st choice, 1st overall, in 1999 Entry Draft.

Season	Club	League	GP	G	A	Pts	PIM	PP	SH	GW	S	%	+/-	TF	F%	H	SB	Min	GP	G	A	Pts	PIM	PP	SH	GW
1996-97	HC Sparta Praha	Czech	5	0	1	1	2	....	....	....	....	....	....	....	....	....	....	....	7	1	0	1	0	....	....	....
1997-98	HC Sparta Praha	Czech	27	2	6	8	16	....	....	....	....	....	....	....	....	....	....	....	....	....	....	....	....	....	....	....
	Long Beach	IHL	25	5	10	15	10	....	....	....	....	....	....	....	....	....	....	....	10	1	1	2	2	....	....	....
1998-99	Long Beach	IHL	33	11	24	35	26	....	....	....	....	....	....	....	....	....	....	....	....	....	....	....	....	....	....	....
99-2000	Atlanta	NHL	72	5	20	25	30	1	0	0	117	4.3	–20	988	41.4	58	22	14:49	....	....	....	....	....	....	....	....
2000-01	Atlanta	NHL	66	10	21	31	22	0	0	1	93	10.8	–3	834	42.9	32	35	14:07	....	....	....	....	....	....	....	....
2001-02	Atlanta	NHL	59	7	16	23	22	0	1	0	67	10.4	–4	628	41.9	27	26	15:58	....	....	....	....	....	....	....	....
	Chicago Wolves	AHL	5	3	0	3	0	....	....	....	....	....	....	....	....	....	....	....	....	....	....	....	....	....	....	....
	NHL Totals		197	22	57	79	74	1	1	1	277	7.9		2450	42.0	117	83	14:56	....	....	....	....	....	....	....	....

STEVENS, Kevin

(STEE-vehns, KEH-vihn)

Left wing. Shoots left. 6'3", 230 lbs. Born, Brockton, MA, April 15, 1965. Los Angeles' 6th choice, 112th overall, in 1983 Entry Draft.

Season	Club	League	GP	G	A	Pts	PIM	PP	SH	GW	S	%	+/-	TF	F%	H	SB	Min	GP	G	A	Pts	PIM	PP	SH	GW
1982-83	Silver Lake	Hi-School	18	24	27	51	....	....	....	....	....	....	....	....	....	....	....	....	....	....	....	....	....	....	....	....
1983-84	Boston College	ECAC	37	6	14	20	36	....	....	....	....	....	....	....	....	....	....	....	....	....	....	....	....	....	....	....
1984-85	Boston College	H-East	40	13	23	36	36	....	....	....	....	....	....	....	....	....	....	....	....	....	....	....	....	....	....	....
1985-86	Boston College	H-East	42	17	27	44	56	....	....	....	....	....	....	....	....	....	....	....	....	....	....	....	....	....	....	....
1986-87	Boston College	H-East	39	35	35	70	54	....	....	....	....	....	....	....	....	....	....	....	....	....	....	....	....	....	....	....
1987-88	Team USA	Nat-Tm	44	22	23	45	52	....	....	....	....	....	....	....	....	....	....	....	....	....	....	....	....	....	....	....
	United States	Olympics	5	1	3	4	2	....	....	....	....	....	....	....	....	....	....	....	....	....	....	....	....	....	....	....
	Pittsburgh	NHL	16	5	2	7	8	2	0	0	22	22.7	–6	....	....	....	....	....	....	....	....	....	....	....	....	....
1988-89	Pittsburgh	NHL	24	12	3	15	19	4	0	3	52	23.1	–8	....	....	....	....	....	11	3	7	10	16	0	0	0
	Muskegon	IHL	45	24	41	65	113	....	....	....	....	....	....	....	....	....	....	....	....	....	....	....	....	....	....	....
1989-90	Pittsburgh	NHL	76	29	41	70	171	12	0	1	179	16.2	–13	....	....	....	....	....	....	....	....	....	....	....	....	....
1990-91♦	Pittsburgh	NHL	80	40	46	86	133	18	0	6	253	15.8	–1	....	....	....	....	....	24	*17	16	33	53	7	0	4
1991-92♦	Pittsburgh	NHL	80	54	69	123	254	19	0	4	325	16.6	8	....	....	....	....	....	21	13	15	28	28	4	0	3
1992-93	Pittsburgh	NHL	72	55	56	111	177	26	0	5	326	16.9	17	....	....	....	....	....	12	5	11	16	22	4	0	0
1993-94	Pittsburgh	NHL	83	41	47	88	155	21	0	4	284	14.4	–24	....	....	....	....	....	6	1	1	2	10	0	0	0
1994-95	Pittsburgh	NHL	27	15	12	27	51	6	0	4	80	18.8	0	....	....	....	....	....	12	4	7	11	21	3	0	1
1995-96	Boston	NHL	41	10	13	23	49	3	0	1	101	9.9	1	....	....	....	....	....	....	....	....	....	....	....	....	....
	Los Angeles	NHL	20	3	10	13	22	3	0	0	69	4.3	–11	....	....	....	....	....	....	....	....	....	....	....	....	....
1996-97	Los Angeles	NHL	69	14	20	34	96	4	0	1	175	8.0	–27	....	....	....	....	....	....	....	....	....	....	....	....	....
1997-98	NY Rangers	NHL	80	14	27	41	130	5	0	3	144	9.7	–7	....	....	....	....	....	....	....	....	....	....	....	....	....
1998-99	NY Rangers	NHL	81	23	20	43	64	8	0	3	136	16.9	–10	11	45.5	176	27	15:12	....	....	....	....	....	....	....	....
99-2000	NY Rangers	NHL	38	3	5	8	43	1	0	0	44	6.8	–7	11	54.6	78	20	12:29	....	....	....	....	....	....	....	....
2000-01	Philadelphia	NHL	23	2	7	9	18	0	0	0	31	6.5	–2	0	0.0	45	5	12:43	....	....	....	....	....	....	....	....
	Pittsburgh	NHL	32	8	15	23	55	2	0	0	76	10.5	–4	5	20.0	73	5	18:15	17	3	3	6	20	2	0	1
2001-02	Pittsburgh	NHL	32	1	4	5	25	0	0	0	34	2.9	–9	5	20.0	47	4	10:50	....	....	....	....	....	....	....	....
	NHL Totals		874	329	397	726	1470	134	0	35	2331	14.1		32	40.6	419	61	14:13	103	46	60	106	170	20	0	9

Hockey East First All-Star Team (1987) • NCAA East Second All-American Team (1987) • NHL Second All-Star Team (1991, 1993) • NHL First All-Star Team (1992) • Played in NHL All-Star Game (1991, 1992, 1993)

Rights traded to **Pittsburgh** by **LA Kings** for Anders Hakansson, September 9, 1983. Traded to **Boston** by **Pittsburgh** with Shawn McEachern for Glen Murray, Bryan Smolinski and Boston's 3rd round choice (Boyd Kane) in 1996 Entry Draft, August 2, 1995. Traded to **LA Kings** by **Boston** for Rick Tocchet, January 25, 1996. Traded to **NY Rangers** by **LA Kings** for Luc Robitaille, August 28, 1997. • Missed majority of 1999-2000 season after entering NHL/NHLPA substance abuse program, January 23, 2000. Signed as a free agent by **Philadelphia**, July 7, 2000. Traded to **Pittsburgh** by **Philadelphia** for John Slaney, January 14, 2001. • Missed majority of 2001-02 season for personal reasons, January 30, 2002.

STEVENS, Scott

(STEE-vehns, SKAWT) **N.J.**

Defense. Shoots left. 6'2", 215 lbs. Born, Kitchener, Ont., April 1, 1964. Washington's 1st choice, 5th overall, in 1982 Entry Draft.

Season	Club	League	GP	G	A	Pts	PIM	PP	SH	GW	S	%	+/-	TF	F%	H	SB	Min	GP	G	A	Pts	PIM	PP	SH	GW
1980-81	Kitchener	OHA-B	39	7	33	40	82	....	....	....	....	....	....	....	....	....	....	....	....	....	....	....	....	....	....	....
	Kitchener	OMJHL	1	0	0	0	0	....	....	....	....	....	....	....	....	....	....	....	....	....	....	....	....	....	....	....
1981-82	Kitchener	OHL	68	6	36	42	158	....	....	....	....	....	....	....	....	....	....	....	15	1	10	11	71	....	....	....
1982-83	Washington	NHL	77	9	16	25	195	0	0	0	121	7.4	14	....	....	....	....	....	4	1	0	1	26	0	0	0
1983-84	Washington	NHL	78	13	32	45	201	7	0	2	155	8.4	26	....	....	....	....	....	8	1	8	9	21	1	0	0
1984-85	Washington	NHL	80	21	44	65	221	16	0	5	170	12.4	19	....	....	....	....	....	5	0	1	1	20	0	0	0
1985-86	Washington	NHL	73	15	38	53	165	3	0	2	121	12.4	0	....	....	....	....	....	9	3	8	11	12	2	0	2
1986-87	Washington	NHL	77	10	51	61	283	2	0	0	165	6.1	13	....	....	....	....	....	7	0	5	5	19	0	0	0
1987-88	Washington	NHL	80	12	60	72	184	5	1	2	231	5.2	14	....	....	....	....	....	13	1	11	12	46	0	0	0
1988-89	Washington	NHL	80	7	61	68	225	6	0	3	195	3.6	1	....	....	....	....	....	6	1	4	5	11	0	0	0
1989-90	Washington	NHL	56	11	29	40	154	7	0	0	143	7.7	1	....	....	....	....	....	15	2	7	9	25	1	0	0
1990-91	St. Louis	NHL	78	5	44	49	150	1	0	1	160	3.1	23	....	....	....	....	....	13	0	3	3	36	0	0	0
1991-92	New Jersey	NHL	68	17	42	59	124	7	1	2	156	10.9	24	....	....	....	....	....	7	2	1	3	29	2	0	1
1992-93	New Jersey	NHL	81	12	45	57	120	8	0	1	146	8.2	14	....	....	....	....	....	5	2	2	4	10	1	0	0
1993-94	New Jersey	NHL	83	18	60	78	112	5	1	4	215	8.4	53	....	....	....	....	....	20	2	9	11	42	2	0	1
1994-95♦	New Jersey	NHL	48	2	20	22	56	1	0	1	111	1.8	4	....	....	....	....	....	20	1	7	8	24	0	0	1
1995-96	New Jersey	NHL	82	5	23	28	100	2	1	1	174	2.9	7	....	....	....	....	....	....	....	....	....	....	....	....	....
1996-97	New Jersey	NHL	79	5	19	24	70	0	0	1	166	3.0	26	....	....	....	....	....	10	0	4	4	2	0	0	0
1997-98	New Jersey	NHL	80	4	22	26	80	1	0	1	94	4.3	19	....	....	....	....	....	6	1	0	1	8	0	0	0
	Canada	Olympics	6	0	0	0	2	....	....	....	....	....	....	....	....	....	....	....	....	....	....	....	....	....	....	....
1998-99	New Jersey	NHL	75	5	22	27	64	0	0	1	111	4.5	29	1	0.0	187	149	24:11	7	2	1	3	10	2	0	0
99-2000♦	New Jersey	NHL	78	8	21	29	103	0	1	1	133	6.0	30	0	0.0	169	130	23:23	23	3	8	11	6	0	0	2
2000-01	New Jersey	NHL	81	9	22	31	71	3	0	2	171	5.3	40	0	0.0	175	142	24:37	25	1	7	8	37	0	0	0
2001-02	New Jersey	NHL	82	1	16	17	44	0	0	1	121	0.8	15	0	0.0	184	171	23:19	6	0	0	0	4	0	0	0
	NHL Totals		1516	189	687	876	2722	74	5	31	3059	6.2		1	0.0	715	592	23:52	209	23	86	109	388	11	0	7

NHL All-Rookie Team (1983) • NHL First All-Star Team (1988, 1994) • NHL Second All-Star Team (1992, 1997, 2001) • Won Alka-Seltzer Plus Award (1994) • Won Conn Smythe Trophy (2000) • Played in NHL All-Star Game (1985, 1989, 1991, 1992, 1993, 1994, 1996, 1997, 1998, 1999, 2000, 2001)

Signed as a free agent by **St. Louis**, July 16, 1990. Transferred to **New Jersey** from **St. Louis** as compensation for St. Louis' signing of free agent Brendan Shanahan, September 3, 1991.

STEVENSON, Jeremy

(STEE-vehn-sohn, JAIR-eh-mee)

Left wing. Shoots left. 6'2", 218 lbs. Born, San Bernardino, CA, July 28, 1974. Anaheim's 10th choice, 262nd overall, in 1994 Entry Draft.

Season	Club	League	GP	G	A	Pts	PIM	PP	SH	GW	S	%	+/-	TF	F%	H	SB	Min	GP	G	A	Pts	PIM	PP	SH	GW
1989-90	Elliot Lake	NOHA	61	39	26	65	203	....	....	....	....	....	....	....	....	....	....	....	....	....	....	....	....	....	....	....
1990-91	Cornwall Royals	OHL	58	13	20	33	124	....	....	....	....	....	....	....	....	....	....	....	....	....	....	....	....	....	....	....
1991-92	Cornwall Royals	OHL	63	15	23	38	176	....	....	....	....	....	....	....	....	....	....	....	6	3	1	4	4	....	....	....
1992-93	Newmarket Royals	OHL	54	28	28	56	144	....	....	....	....	....	....	....	....	....	....	....	5	5	1	6	28	....	....	....
1993-94	Newmarket Royals	OHL	9	2	4	6	27	....	....	....	....	....	....	....	....	....	....	....	....	....	....	....	....	....	....	....
	Sault Ste. Marie	OHL	48	18	19	37	183	....	....	....	....	....	....	....	....	....	....	....	14	1	1	2	23	....	....	....
1994-95	Greensboro	ECHL	43	14	13	27	231	....	....	....	....	....	....	....	....	....	....	....	17	6	11	17	64	....	....	....
1995-96	Anaheim	NHL	3	0	1	1	12	0	0	0	1	0.0	1	....	....	....	....	....	....	....	....	....	....	....	....	....
	Baltimore	AHL	60	11	10	21	295	....	....	....	....	....	....	....	....	....	....	....	12	4	2	6	23	....	....	....
1996-97	Anaheim	NHL	5	0	0	0	14	0	0	0	1	0.0	–1	....	....	....	....	....	....	....	....	....	....	....	....	....
	Baltimore	AHL	25	8	8	16	125	....	....	....	....	....	....	....	....	....	....	....	3	0	0	0	8	....	....	....
1997-98	Anaheim	NHL	45	3	5	8	101	0	0	1	43	7.0	–4	....	....	....	....	....	....	....	....	....	....	....	....	....
	Cincinnati	AHL	10	5	0	5	34	....	....	....	....	....	....	....	....	....	....	....	....	....	....	....	....	....	....	....
1998-99	Cincinnati	AHL	22	4	4	8	83	....	....	....	....	....	....	....	....	....	....	....	3	1	0	1	2	....	....	....
99-2000	Anaheim	NHL	3	0	0	0	7	0	0	0	2	0.0	–1	0	0.0	4	1	6:50	....	....	....	....	....	....	....	....
	Cincinnati	AHL	41	11	14	25	100	....	....	....	....	....	....	....	....	....	....	....	....	....	....	....	....	....	....	....

Season	Club	League	GP	G	A	Pts	PIM	PP	SH	GW	S	%	+/-	TF	F%	H	SB	Min	GP	G	A	Pts	PIM	PP	SH	GW
			Regular Season																Playoffs							
2000-01	Nashville	NHL	8	1	0	1	39	0	0	0	6	16.7	–1	0	0.0	8	1	6:24	....	....	....	....	....	....	....	....
	Milwaukee	IHL	60	16	13	29	262	....	....	....	....	....	....	....	....	....	....	....	5	2	0	2	12	....	....	....
2001-02	Nashville	NHL	4	0	0	0	9	0	0	0	0	0.0	0	0	0.0	5	1	5:53	....	....	....	....	....	....	....	....
	Milwaukee	AHL	53	12	7	19	192	....	....	....	....	....	....	....	....	....	....	....	....	....	....	....	....	....	....	....
	NHL Totals		68	4	6	10	182	0	0	1	53	7.5		0	0.0	17	3	6:21	....	....	....	....	....	....	....	....

• Re-entered NHL Entry Draft. Originally Winnipeg's 3rd choice, 60th overall, in 1992 Entry Draft.
Signed as a free agent by **Nashville**, September 25, 2000.

STEVENSON, Turner

(STEE-vehn-sohn, TUHR-nuhr) **N.J.**

Right wing. Shoots right. 6'3", 230 lbs. Born, Prince George, B.C., May 18, 1972. Montreal's 1st choice, 12th overall, in 1990 Entry Draft.

Season	Club	League	GP	G	A	Pts	PIM	PP	SH	GW	S	%	+/-	TF	F%	H	SB	Min	GP	G	A	Pts	PIM	PP	SH	GW
1987-88	Prince George	BCAHA	53	45	46	91	127	....	....	....	....	....	....	....	....	....	....	....	....	....	....	....	....	....	....	....
1988-89	Seattle	WHL	69	15	12	27	84	....	....	....	....	....	....	....	....	....	....	....	....	....	....	....	....	....	....	....
1989-90	Seattle	WHL	62	29	32	61	276	....	....	....	....	....	....	....	....	....	....	....	13	3	2	5	35	....	....	....
1990-91	Seattle	WHL	57	36	27	63	222	....	....	....	....	....	....	....	....	....	....	....	6	1	5	6	15	....	....	....
	Fredericton	AHL	....	....	....	....	....	....	....	....	....	....	....	....	....	....	....	....	4	0	0	0	5	....	....	....
1991-92	Seattle	WHL	58	20	32	52	264	....	....	....	....	....	....	....	....	....	....	....	15	9	3	12	55	....	....	....
1992-93	**Montreal**	**NHL**	**1**	**0**	**0**	**0**	**0**	**0**	**0**	**0**	**1**	**0.0**	**–1**	....	....	....	....	....	....	....	....	....	....	....	....	....
	Fredericton	AHL	79	25	34	59	102	....	....	....	....	....	....	....	....	....	....	....	5	2	3	5	11	....	....	....
1993-94	**Montreal**	**NHL**	**2**	**0**	**0**	**0**	**2**	**0**	**0**	**0**	**0**	**0.0**	**–2**	....	....	....	....	....	**3**	**0**	**2**	**2**	**0**	**0**	**0**	**0**
	Fredericton	AHL	66	19	28	47	155	....	....	....	....	....	....	....	....	....	....	....	....	....	....	....	....	....	....	....
1994-95	Fredericton	AHL	37	12	12	24	109	....	....	....	....	....	....	....	....	....	....	....	....	....	....	....	....	....	....	....
	Montreal	**NHL**	**41**	**6**	**1**	**7**	**86**	**0**	**0**	**1**	**35**	**17.1**	**0**	....	....	....	....	....	....	....	....	....	....	....	....	....
1995-96	**Montreal**	**NHL**	**80**	**9**	**16**	**25**	**167**	**0**	**0**	**2**	**101**	**8.9**	**–2**	....	....	....	....	....	**6**	**0**	**1**	**1**	**2**	**0**	**0**	**0**
1996-97	**Montreal**	**NHL**	**65**	**8**	**13**	**21**	**97**	**1**	**0**	**0**	**76**	**10.5**	**–14**	....	....	....	....	....	**5**	**1**	**1**	**2**	**2**	**0**	**0**	**0**
1997-98	**Montreal**	**NHL**	**63**	**4**	**6**	**10**	**110**	**1**	**0**	**0**	**43**	**9.3**	**–8**	....	....	....	....	....	**10**	**3**	**4**	**7**	**12**	**0**	**0**	**0**
1998-99	**Montreal**	**NHL**	**69**	**10**	**17**	**27**	**88**	**0**	**0**	**2**	**102**	**9.8**	**6**	**29**	**37.9**	**142**	**20**	**12:57**	....	....	....	....	....	....	....	....
99-2000	**Montreal**	**NHL**	**64**	**8**	**13**	**21**	**61**	**0**	**0**	**2**	**94**	**8.5**	**–1**	**5**	**20.0**	**160**	**18**	**13:10**	....	....	....	....	....	....	....	....
2000-01	**New Jersey**	**NHL**	**69**	**8**	**18**	**26**	**97**	**2**	**0**	**1**	**92**	**8.7**	**11**	**0**	**0.0**	**141**	**13**	**11:00**	**23**	**1**	**3**	**4**	**20**	**0**	**0**	**1**
2001-02	**New Jersey**	**NHL**	**21**	**0**	**2**	**2**	**25**	**0**	**0**	**0**	**33**	**0.0**	**–3**	**0**	**0.0**	**60**	**4**	**11:38**	**1**	**0**	**0**	**0**	**4**	**0**	**0**	**0**
	NHL Totals		**475**	**53**	**86**	**139**	**733**	**4**	**0**	**8**	**577**	**9.2**		**34**	**35.3**	**503**	**55**	**12:17**	**48**	**5**	**11**	**16**	**40**	**0**	**0**	**1**

WHL West First All-Star Team (1992) • Memorial Cup All-Star Team (1992).

Selected by **Columbus** from **Montreal** in Expansion Draft, June 23, 2000. Traded to **New Jersey** by **Columbus** to complete transaction that sent Krzysztof Oliwa (June 12, 2000) and Deron Quint (June 23, 2000) to **Columbus**, June 23, 2000. • Missed majority of 2001-02 season recovering from knee injury suffered in game vs. Vancouver, December 29, 2001.

STILLMAN, Cory

(STIHL-mahn, KOHR-ee) **ST.L.**

Center. Shoots left. 6', 194 lbs. Born, Peterborough, Ont., December 20, 1973. Calgary's 1st choice, 6th overall, in 1992 Entry Draft.

Season	Club	League	GP	G	A	Pts	PIM	PP	SH	GW	S	%	+/-	TF	F%	H	SB	Min	GP	G	A	Pts	PIM	PP	SH	GW
1989-90	Peterboro B's	OJHL-B	41	30	*54	84	76	....	....	....	....	....	....	....	....	....	....	....	....	....	....	....	....	....	....	....
1990-91	Windsor	OHL	64	31	70	101	31	....	....	....	....	....	....	....	....	....	....	....	11	3	6	9	8	....	....	....
1991-92	Windsor	OHL	53	29	61	90	59	....	....	....	....	....	....	....	....	....	....	....	7	2	4	6	8	....	....	....
1992-93	Peterborough	OHL	61	25	55	80	55	....	....	....	....	....	....	....	....	....	....	....	18	3	8	11	18	....	....	....
1993-94	Saint John	AHL	79	35	48	83	52	....	....	....	....	....	....	....	....	....	....	....	7	2	4	6	16	....	....	....
1994-95	Saint John	AHL	63	28	53	81	70	....	....	....	....	....	....	....	....	....	....	....	5	0	2	2	2	....	....	....
	Calgary	**NHL**	**10**	**0**	**2**	**2**	**2**	**0**	**0**	**0**	**7**	**0.0**	**1**	....	....	....	....	....	....	....	....	....	....	....	....	....
1995-96	**Calgary**	**NHL**	**74**	**16**	**19**	**35**	**41**	**4**	**1**	**3**	**132**	**12.1**	**–5**	....	....	....	....	....	**2**	**1**	**1**	**2**	**0**	**0**	**0**	**0**
1996-97	**Calgary**	**NHL**	**58**	**6**	**20**	**26**	**14**	**2**	**0**	**0**	**112**	**5.4**	**–6**	....	....	....	....	....	....	....	....	....	....	....	....	....
1997-98	**Calgary**	**NHL**	**72**	**27**	**22**	**49**	**40**	**9**	**4**	**1**	**178**	**15.2**	**–9**	....	....	....	....	....	....	....	....	....	....	....	....	....
1998-99	**Calgary**	**NHL**	**76**	**27**	**30**	**57**	**38**	**9**	**3**	**5**	**175**	**15.4**	**7**	**535**	**46.5**	**128**	**33**	**16:19**	....	....	....	....	....	....	....	....
99-2000	**Calgary**	**NHL**	**37**	**12**	**9**	**21**	**12**	**6**	**0**	**3**	**59**	**20.3**	**–9**	**283**	**54.4**	**39**	**27**	**17:45**	....	....	....	....	....	....	....	....
2000-01	**Calgary**	**NHL**	**66**	**21**	**24**	**45**	**45**	**7**	**0**	**4**	**148**	**14.2**	**–6**	**346**	**43.9**	**62**	**35**	**18:50**	....	....	....	....	....	....	....	....
	St. Louis	**NHL**	**12**	**3**	**4**	**7**	**6**	**3**	**0**	**0**	**26**	**11.5**	**–2**	**36**	**61.1**	**16**	**2**	**18:37**	**15**	**3**	**5**	**8**	**8**	**1**	**0**	**1**
2001-02	**St. Louis**	**NHL**	**80**	**23**	**22**	**45**	**36**	**6**	**0**	**4**	**140**	**16.4**	**8**	**196**	**46.4**	**112**	**26**	**15:03**	**9**	**0**	**2**	**2**	**2**	**0**	**0**	**0**
	NHL Totals		**485**	**135**	**152**	**287**	**234**	**46**	**8**	**20**	**977**	**13.8**		**1396**	**47.9**	**357**	**123**	**16:51**	**26**	**4**	**8**	**12**	**10**	**1**	**0**	**1**

OHL Rookie of the Year (1991)

• Missed majority of 1999-2000 season recovering from shoulder injury suffered in game vs. Philadelphia, December 27, 1999. Traded to **St. Louis by Calgary** for Craig Conroy and St. Louis' 7th round choice (David Moss) in 2001 Entry Draft, March 13, 2001.

STOCK, P.J.

(STAWK, PEE-JAY) **BOS.**

Left wing. Shoots left. 5'10", 190 lbs. Born, Victoriaville, Que., May 26, 1975.

Season	Club	League	GP	G	A	Pts	PIM	PP	SH	GW	S	%	+/-	TF	F%	H	SB	Min	GP	G	A	Pts	PIM	PP	SH	GW
1992-93	Pembroke	OCJHL	55	10	38	48	189	....	....	....	....	....	....	....	....	....	....	....	....	....	....	....	....	....	....	....
1993-94	Pembroke	OCJHL	52	25	48	73	262	....	....	....	....	....	....	....	....	....	....	....	....	....	....	....	....	....	....	....
1994-95	Victoriaville	QMJHL	70	9	46	55	386	....	....	....	....	....	....	....	....	....	....	....	4	0	0	0	60	....	....	....
1995-96	Victoriaville	QMJHL	67	19	43	62	432	....	....	....	....	....	....	....	....	....	....	....	12	5	4	9	79	....	....	....
1996-97	St. FX University	AUAA	27	11	20	31	110	....	....	....	....	....	....	....	....	....	....	....	3	0	4	4	14	....	....	....
1997-98	Hartford	AHL	41	8	8	16	202	....	....	....	....	....	....	....	....	....	....	....	11	1	3	4	79	....	....	....
	NY Rangers	**NHL**	**38**	**2**	**3**	**5**	**114**	**0**	**0**	**1**	**9**	**22.2**	**4**	....	....	....	....	....	....	....	....	....	....	....	....	....
1998-99	**NY Rangers**	**NHL**	**5**	**0**	**0**	**0**	**6**	**0**	**0**	**0**	**0**	**0.0**	**–1**	**8**	**50.0**	**8**	**0**	**2:42**	....	....	....	....	....	....	....	....
	Hartford	AHL	55	4	14	18	250	....	....	....	....	....	....	....	....	....	....	....	6	0	1	1	35	....	....	....
99-2000	**NY Rangers**	**NHL**	**11**	**0**	**1**	**1**	**11**	**0**	**0**	**0**	**2**	**0.0**	**1**	**63**	**31.8**	**14**	**1**	**6:13**	....	....	....	....	....	....	....	....
	Hartford	AHL	64	13	23	36	290	....	....	....	....	....	....	....	....	....	....	....	23	1	11	12	69	....	....	....
2000-01	**Montreal**	**NHL**	**20**	**1**	**2**	**3**	**32**	**0**	**0**	**0**	**9**	**11.1**	**–1**	**83**	**53.0**	**19**	**4**	**5:31**	....	....	....	....	....	....	....	....
	Philadelphia	**NHL**	**31**	**1**	**3**	**4**	**78**	**0**	**0**	**0**	**18**	**5.6**	**–2**	**12**	**50.0**	**37**	**0**	**7:48**	**2**	**0**	**0**	**0**	**0**	**0**	**0**	**0**
	Philadelphia	AHL	9	1	2	3	37	....	....	....	....	....	....	....	....	....	....	....	....	....	....	....	....	....	....	....
2001-02	**Boston**	**NHL**	**58**	**0**	**3**	**3**	**122**	**0**	**0**	**0**	**12**	**0.0**	**–2**	**120**	**45.8**	**38**	**1**	**5:09**	**6**	**1**	**0**	**1**	**19**	**0**	**0**	**0**
	NHL Totals		**163**	**4**	**12**	**16**	**363**	**0**	**0**	**1**	**50**	**8.0**		**286**	**45.1**	**116**	**6**	**5:52**	**8**	**1**	**0**	**1**	**19**	**0**	**0**	**0**

Signed as a free agent by **NY Rangers**, November 18, 1997. Signed as a free agent by **Montreal**, July 7, 2000. Traded to **Philadelphia** by **Montreal** with Montreal's 6th round choice (Dennis Seidenberg) in 2001 Entry Draft for Gino Odjick, December 7, 2000. Signed as a free agent by **NY Rangers**, August 23, 2001. Claimed by **Boston** from **NY Rangers** in Waiver Draft, September 28, 2001.

STRAKA, Martin

(STRAH-kuh, MAHR-tihn) **PIT.**

Center. Shoots left. 5'9", 178 lbs. Born, Plzen, Czech., September 3, 1972. Pittsburgh's 1st choice, 19th overall, in 1992 Entry Draft.

Season	Club	League	GP	G	A	Pts	PIM	PP	SH	GW	S	%	+/-	TF	F%	H	SB	Min	GP	G	A	Pts	PIM	PP	SH	GW
1989-90	TJ Skoda Plzen	Czech	1	0	3	3	....	....	....	....	....	....	....	....	....	....	....	....	....	....	....	....	....	....	....	....
1990-91	HC Skoda Plzen	Czech	47	7	24	31	6	....	....	....	....	....	....	....	....	....	....	....	....	....	....	....	....	....	....	....
1991-92	HC Skoda Plzen	Czech	50	27	28	55	20	....	....	....	....	....	....	....	....	....	....	....	....	....	....	....	....	....	....	....
1992-93	**Pittsburgh**	**NHL**	**42**	**3**	**13**	**16**	**29**	**0**	**0**	**1**	**28**	**10.7**	**2**	....	....	....	....	....	**11**	**2**	**1**	**3**	**2**	**0**	**0**	**0**
	Cleveland	IHL	4	4	3	7	0	....	....	....	....	....	....	....	....	....	....	....	....	....	....	....	....	....	....	....
1993-94	**Pittsburgh**	**NHL**	**84**	**30**	**34**	**64**	**24**	**2**	**0**	**6**	**130**	**23.1**	**24**	....	....	....	....	....	**6**	**1**	**0**	**1**	**2**	**0**	**0**	**0**
1994-95	Plzen	Czech	19	10	11	21	18	....	....	....	....	....	....	....	....	....	....	....	....	....	....	....	....	....	....	....
	Pittsburgh	**NHL**	**31**	**4**	**12**	**16**	**16**	**0**	**0**	**0**	**36**	**11.1**	**0**	....	....	....	....	....	....	....	....	....	....	....	....	....
	Ottawa	**NHL**	**6**	**1**	**1**	**2**	**0**	**0**	**0**	**0**	**13**	**7.7**	**–1**	....	....	....	....	....	....	....	....	....	....	....	....	....
1995-96	**Ottawa**	**NHL**	**43**	**9**	**16**	**25**	**29**	**5**	**0**	**1**	**63**	**14.3**	**–14**	....	....	....	....	....	....	....	....	....	....	....	....	....
	NY Islanders	**NHL**	**22**	**2**	**10**	**12**	**6**	**0**	**0**	**0**	**18**	**11.1**	**–6**	....	....	....	....	....	....	....	....	....	....	....	....	....
	Florida	**NHL**	**12**	**2**	**4**	**6**	**6**	**1**	**0**	**0**	**17**	**11.8**	**1**	....	....	....	....	....	**13**	**2**	**2**	**4**	**2**	**0**	**0**	**0**
1996-97	**Florida**	**NHL**	**55**	**7**	**22**	**29**	**12**	**2**	**0**	**1**	**94**	**7.4**	**9**	....	....	....	....	....	**4**	**0**	**0**	**0**	**0**	**0**	**0**	**0**
1997-98	**Pittsburgh**	**NHL**	**75**	**19**	**23**	**42**	**28**	**4**	**3**	**4**	**117**	**16.2**	**–1**	....	....	....	....	....	**6**	**2**	**0**	**2**	**2**	**0**	**1**	**0**
	Czech Republic	Olympics	6	1	2	3	0	....	....	....	....	....	....	....	....	....	....	....	....	....	....	....	....	....	....	....
1998-99	**Pittsburgh**	**NHL**	**80**	**35**	**48**	**83**	**26**	**5**	**4**	**4**	**177**	**19.8**	**12**	**845**	**43.6**	**75**	**59**	**23:35**	**13**	**6**	**9**	**15**	**6**	**1**	**0**	**0**
99-2000	**Pittsburgh**	**NHL**	**71**	**20**	**39**	**59**	**26**	**3**	**1**	**2**	**146**	**13.7**	**24**	**651**	**42.9**	**62**	**35**	**23:58**	**11**	**3**	**9**	**12**	**10**	**1**	**0**	**0**
2000-01	**Pittsburgh**	**NHL**	**82**	**27**	**68**	**95**	**38**	**7**	**1**	**4**	**185**	**14.6**	**19**	**331**	**43.2**	**58**	**39**	**23:01**	**18**	**5**	**8**	**13**	**8**	**3**	**0**	**2**
2001-02	**Pittsburgh**	**NHL**	**13**	**5**	**4**	**9**	**0**	**1**	**0**	**1**	**33**	**15.2**	**3**	**5**	**60.0**	**5**	**6**	**18:00**	....	....	....	....	....	....	....	....
	NHL Totals		**616**	**164**	**294**	**458**	**240**	**30**	**9**	**24**	**1057**	**15.5**		**1832**	**43.3**	**200**	**139**	**23:13**	**82**	**21**	**29**	**50**	**32**	**5**	**1**	**2**

Played in NHL All-Star Game (1999)

Traded to **Ottawa** by **Pittsburgh** for Troy Murray and Norm Maciver, April 7, 1995. Traded to **NY Islanders** by **Ottawa** with Don Beaupre and Bryan Berard for Damian Rhodes and Wade Redden, January 23, 1996. Claimed on waivers by **Florida** from **NY Islanders**, March 15, 1996. Signed as a free agent by **Pittsburgh**, August 6, 1997. •. Missed majority of 2001-02 season recovering from leg injury originally suffered in game vs. Florida, October 28, 2001.

STRUDWICK, Jason (STRUHD-wihk, JAY-suhn) CHI.

Defense. Shoots left. 6'3", 215 lbs. Born, Edmonton, Alta., July 17, 1975. NY Islanders' 3rd choice, 63rd overall, in 1994 Entry Draft.

			Regular Season																Playoffs							
Season	Club	League	GP	G	A	Pts	PIM	PP	SH	GW	S	%	+/-	TF	F%	H	SB	Min	GP	G	A	Pts	PIM	PP	SH	GW
1991-92	Edmonton Legion	AMHL	35	3	8	11	67	....	....	....	....	....	....	....	....	....	....	....	....	....	....	....	....	....	....	....
1992-93	Edmonton Pats	AMHL	33	8	20	28	135	....	....	....	....	....	....	....	....	....	....	....	....	....	....	....	....	....	....	....
1993-94	Kamloops Blazers	WHL	61	6	8	14	118	....	....	....	....	....	....	....	....	....	....	....	19	0	4	4	24	....	....	....
1994-95	Kamloops Blazers	WHL	72	3	11	14	183	....	....	....	....	....	....	....	....	....	....	....	21	1	1	2	39	....	....	....
1995-96	**NY Islanders**	**NHL**	**1**	**0**	**0**	**0**	**7**	**0**	**0**	**0**	**0**	**0.0**	**0**	....	....	....	....	....	....	....	....	....	....	....	....	....
	Worcester	AHL	60	2	7	9	119	....	....	....	....	....	....	....	....	....	....	....	4	0	1	1	0	....	....	....
1996-97	Kentucky	AHL	80	1	9	10	198	....	....	....	....	....	....	....	....	....	....	....	4	0	0	0	0	....	....	....
1997-98	**NY Islanders**	**NHL**	**17**	**0**	**1**	**1**	**36**	**0**	**0**	**0**	**3**	**0.0**	**1**	....	....	....	....	....	....	....	....	....	....	....	....	....
	Kentucky	AHL	39	3	1	4	87	....	....	....	....	....	....	....	....	....	....	....	....	....	....	....	....	....	....	....
	Vancouver	**NHL**	**11**	**0**	**1**	**1**	**29**	**0**	**0**	**0**	**5**	**0.0**	**–3**	....	....	....	....	....	....	....	....	....	....	....	....	....
	Syracuse Crunch	AHL	....	....	....	....	....	....	....	....	....	....	....	....	....	....	....	....	3	0	0	0	6	....	....	....
1998-99	**Vancouver**	**NHL**	**65**	**0**	**3**	**3**	**114**	**0**	**0**	**0**	**25**	**0.0**	**–19**	**0**	**0.0**	**62**	**52**	**12:49**	....	....	....	....	....	....	....	....
99-2000	**Vancouver**	**NHL**	**63**	**1**	**3**	**4**	**64**	**0**	**0**	**0**	**18**	**5.6**	**–13**	**0**	**0.0**	**105**	**87**	**15:12**	....	....	....	....	....	....	....	....
2000-01	**Vancouver**	**NHL**	**60**	**1**	**4**	**5**	**64**	**0**	**0**	**1**	**21**	**4.8**	**16**	**0**	**0.0**	**90**	**43**	**9:59**	**2**	**0**	**0**	**0**	**0**	**0**	**0**	**0**
2001-02	**Vancouver**	**NHL**	**44**	**2**	**4**	**6**	**96**	**0**	**0**	**0**	**13**	**15.4**	**4**	**0**	**0.0**	**68**	**18**	**9:45**	....	....	....	....	....	....	....	....
	NHL Totals		**261**	**4**	**16**	**20**	**410**	**0**	**0**	**1**	**85**	**4.7**		**0**	**0.0**	**325**	**200**	**12:09**	**2**	**0**	**0**	**0**	**0**	**0**	**0**	**0**

Traded to **Vancouver** by **NY Islanders** for Gino Odjick, March 23, 1998. Signed as a free agent by **Chicago**, July 15, 2002.

STUART, Brad (STEW-ahrt, BRAD) S.J.

Defense. Shoots left. 6'2", 215 lbs. Born, Rocky Mountain House, Alta., November 6, 1979. San Jose's 1st choice, 3rd overall, in 1998 Entry Draft.

			Regular Season																Playoffs							
Season	Club	League	GP	G	A	Pts	PIM	PP	SH	GW	S	%	+/-	TF	F%	H	SB	Min	GP	G	A	Pts	PIM	PP	SH	GW
1995-96	Red Deer	AMHL	35	12	25	37	83	....	....	....	....	....	....	....	....	....	....	....	....	....	....	....	....	....	....	....
	Regina Pats	WHL	3	0	0	0	0	....	....	....	....	....	....	....	....	....	....	....	....	....	....	....	....	....	....	....
1996-97	Regina Pats	WHL	57	7	36	43	58	....	....	....	....	....	....	....	....	....	....	....	5	0	4	4	14	....	....	....
1997-98	Regina Pats	WHL	72	20	45	65	82	....	....	....	....	....	....	....	....	....	....	....	9	3	4	7	10	....	....	....
1998-99	Regina Pats	WHL	29	10	19	29	43	....	....	....	....	....	....	....	....	....	....	....	....	....	....	....	....	....	....	....
	Calgary Hitmen	WHL	30	11	22	33	26	....	....	....	....	....	....	....	....	....	....	....	21	8	15	23	59	....	....	....
99-2000	**San Jose**	**NHL**	**82**	**10**	**26**	**36**	**32**	**5**	**1**	**3**	**133**	**7.5**	**3**	**0**	**0.0**	**123**	**57**	**20:24**	**12**	**1**	**0**	**1**	**6**	**1**	**0**	**0**
2000-01	**San Jose**	**NHL**	**77**	**5**	**18**	**23**	**56**	**1**	**0**	**2**	**119**	**4.2**	**10**	**0**	**0.0**	**160**	**70**	**20:06**	**5**	**1**	**0**	**1**	**0**	**0**	**0**	**0**
2001-02	**San Jose**	**NHL**	**82**	**6**	**23**	**29**	**39**	**2**	**0**	**2**	**96**	**6.3**	**13**	**0**	**0.0**	**197**	**113**	**21:41**	**12**	**0**	**3**	**3**	**8**	**0**	**0**	**0**
	NHL Totals		**241**	**21**	**67**	**88**	**127**	**8**	**1**	**7**	**348**	**6.0**		**0**	**0.0**	**480**	**240**	**20:44**	**29**	**2**	**3**	**5**	**14**	**1**	**0**	**0**

WHL East Second All-Star Team (1998) • WHL East First All-Star Team (1999) • Canadian Major Junior First All-Star Team (1999) • Canadian Major Junior Defenseman of the Year (1999) • NHL All-Rookie Team (2000)

Traded to **Calgary** (WHL) by **Regina** (WHL) for Donald Choukalos and Ryan Geremia, January 15, 1999.

STUMPEL, Jozef (STUM-puhl, JOH-zehf) BOS.

Center. Shoots right. 6'3", 225 lbs. Born, Nitra, Czech., July 20, 1972. Boston's 2nd choice, 40th overall, in 1991 Entry Draft.

			Regular Season																Playoffs							
Season	Club	League	GP	G	A	Pts	PIM	PP	SH	GW	S	%	+/-	TF	F%	H	SB	Min	GP	G	A	Pts	PIM	PP	SH	GW
1989-90	Plastika Nitra	Czech-2	38	12	11	23	....	....	....	....	....	....	....	....	....	....	....	....	....	....	....	....	....	....	....	....
1990-91	AC Nitra	Czech	49	23	22	45	14	....	....	....	....	....	....	....	....	....	....	....	....	....	....	....	....	....	....	....
1991-92	Kolner EC	Germany	33	19	18	37	35	....	....	....	....	....	....	....	....	....	....	....	4	1	1	2	0	....	....	....
	Boston	**NHL**	**4**	**1**	**0**	**1**	**0**	**0**	**0**	**0**	**3**	**33.3**	**1**	....	....	....	....	....	....	....	....	....	....	....	....	....
1992-93	**Boston**	**NHL**	**13**	**1**	**3**	**4**	**4**	**0**	**0**	**0**	**8**	**12.5**	**–3**	....	....	....	....	....	....	....	....	....	....	....	....	....
	Providence	AHL	56	31	61	92	26	....	....	....	....	....	....	....	....	....	....	....	6	4	4	8	0	....	....	....
1993-94	**Boston**	**NHL**	**59**	**8**	**15**	**23**	**14**	**0**	**0**	**1**	**62**	**12.9**	**4**	....	....	....	....	....	**13**	**1**	**7**	**8**	**4**	**0**	**0**	**0**
	Providence	AHL	17	5	12	17	4	....	....	....	....	....	....	....	....	....	....	....	....	....	....	....	....	....	....	....
1994-95	Kolner Haie	Germany	25	16	23	39	18	....	....	....	....	....	....	....	....	....	....	....	....	....	....	....	....	....	....	....
	Boston	**NHL**	**44**	**5**	**13**	**18**	**8**	**1**	**0**	**2**	**46**	**10.9**	**4**	....	....	....	....	....	**5**	**0**	**0**	**0**	**0**	**0**	**0**	**0**
1995-96	**Boston**	**NHL**	**76**	**18**	**36**	**54**	**14**	**5**	**0**	**2**	**158**	**11.4**	**–8**	....	....	....	....	....	**5**	**1**	**2**	**3**	**0**	**0**	**0**	**0**
1996-97	**Boston**	**NHL**	**78**	**21**	**55**	**76**	**14**	**6**	**0**	**1**	**168**	**12.5**	**–22**	....	....	....	....	....	....	....	....	....	....	....	....	....
1997-98	**Los Angeles**	**NHL**	**77**	**21**	**58**	**79**	**53**	**4**	**0**	**2**	**162**	**13.0**	**17**	....	....	....	....	....	**4**	**1**	**2**	**3**	**2**	**0**	**0**	**0**
1998-99	**Los Angeles**	**NHL**	**64**	**13**	**21**	**34**	**10**	**1**	**0**	**1**	**131**	**9.9**	**–18**	**1484**	**54.0**	**67**	**30**	**19:44**	....	....	....	....	....	....	....	....
99-2000	**Los Angeles**	**NHL**	**57**	**17**	**41**	**58**	**10**	**3**	**0**	**7**	**126**	**13.5**	**23**	**1088**	**50.6**	**77**	**10**	**19:16**	**4**	**0**	**4**	**4**	**8**	**0**	**0**	**0**
2000-01	Slov. Bratislava	Slovakia	9	2	4	6	16	....	....	....	....	....	....	....	....	....	....	....	....	....	....	....	....	....	....	....
	Los Angeles	**NHL**	**63**	**16**	**39**	**55**	**14**	**9**	**0**	**6**	**95**	**16.8**	**20**	**1278**	**52.7**	**70**	**21**	**19:35**	**13**	**3**	**5**	**8**	**10**	**2**	**0**	**1**
2001-02	**Los Angeles**	**NHL**	**9**	**1**	**3**	**4**	**4**	**0**	**0**	**0**	**7**	**14.3**	**1**	**164**	**48.2**	**7**	**6**	**20:06**	....	....	....	....	....	....	....	....
	Boston	**NHL**	**72**	**7**	**47**	**54**	**14**	**1**	**0**	**3**	**93**	**7.5**	**21**	**1346**	**49.7**	**39**	**23**	**18:36**	**6**	**0**	**2**	**2**	**0**	**0**	**0**	**0**
	Slovakia	Olympics	2	2	1	3	0	....	....	....	....	....	....	....	....	....	....	....	....	....	....	....	....	....	....	....
	NHL Totals		**616**	**129**	**331**	**460**	**159**	**30**	**0**	**25**	**1059**	**12.2**		**5360**	**51.8**	**260**	**90**	**19:18**	**50**	**6**	**22**	**28**	**24**	**2**	**0**	**1**

Traded to **LA Kings** by **Boston** with Sandy Moger and Boston's 4th round choice (later traded to New Jersey - New Jersey selected Pierre Dagenais) in 1998 Entry Draft for Dmitri Kristich and Byron Dafoe, August 29, 1997. Traded to **Boston** by **LA Kings** with Glen Murray for Jason Allison and Mikko Eloranta, October 24, 2001.

STURM, Marco (STURHM, MAHR-koh) S.J.

Center. Shoots left. 6', 195 lbs. Born, Dingolfing, West Germany, September 8, 1978. San Jose's 2nd choice, 21st overall, in 1996 Entry Draft.

			Regular Season																Playoffs							
Season	Club	League	GP	G	A	Pts	PIM	PP	SH	GW	S	%	+/-	TF	F%	H	SB	Min	GP	G	A	Pts	PIM	PP	SH	GW
1995-96	EV Landshut	Germany	47	12	20	32	50	....	....	....	....	....	....	....	....	....	....	....	11	1	3	4	18	....	....	....
1996-97	EV Landshut	Germany	46	16	27	43	40	....	....	....	....	....	....	....	....	....	....	....	7	1	4	5	6	....	....	....
1997-98	**San Jose**	**NHL**	**74**	**10**	**20**	**30**	**40**	**2**	**0**	**3**	**118**	**8.5**	**–2**	....	....	....	....	....	**2**	**0**	**0**	**0**	**0**	**0**	**0**	**0**
	Germany	Olympics	2	0	0	0	0	....	....	....	....	....	....	....	....	....	....	....	....	....	....	....	....	....	....	....
1998-99	**San Jose**	**NHL**	**78**	**16**	**22**	**38**	**52**	**3**	**2**	**3**	**140**	**11.4**	**7**	**576**	**45.0**	**98**	**37**	**15:23**	**6**	**2**	**2**	**4**	**4**	**0**	**0**	**1**
99-2000	**San Jose**	**NHL**	**74**	**12**	**15**	**27**	**22**	**2**	**4**	**3**	**120**	**10.0**	**4**	**183**	**45.4**	**102**	**28**	**14:07**	**12**	**1**	**3**	**4**	**6**	**0**	**0**	**0**
2000-01	**San Jose**	**NHL**	**81**	**14**	**18**	**32**	**28**	**2**	**3**	**5**	**153**	**9.2**	**9**	**517**	**40.2**	**101**	**46**	**16:06**	**6**	**0**	**2**	**2**	**0**	**0**	**0**	**0**
2001-02	**San Jose**	**NHL**	**77**	**21**	**20**	**41**	**32**	**4**	**3**	**5**	**174**	**12.1**	**23**	**105**	**47.6**	**88**	**30**	**15:39**	**12**	**3**	**2**	**5**	**2**	**0**	**0**	**0**
	Germany	Olympics	5	0	1	1	0	....	....	....	....	....	....	....	....	....	....	....	....	....	....	....	....	....	....	....
	NHL Totals		**384**	**73**	**95**	**168**	**174**	**13**	**12**	**19**	**705**	**10.4**		**1381**	**43.4**	**389**	**141**	**15:20**	**38**	**6**	**9**	**15**	**12**	**0**	**0**	**1**

Played in NHL All-Star Game (1999)

SUCHY, Radoslav (soo-KHEE, RAD-oh-slav) PHX.

Defense. Shoots left. 6'2", 198 lbs. Born, Kezmarok, Czech., April 7, 1976.

			Regular Season																Playoffs							
Season	Club	League	GP	G	A	Pts	PIM	PP	SH	GW	S	%	+/-	TF	F%	H	SB	Min	GP	G	A	Pts	PIM	PP	SH	GW
1993-94	Poprad Jr.	Slovak-Jr.	30	11	12	23	16	....	....	....	....	....	....	....	....	....	....	....	....	....	....	....	....	....	....	....
	SKP PS Poprad	Slovakia	3	0	0	0	0	....	....	....	....	....	....	....	....	....	....	....	....	....	....	....	....	....	....	....
1994-95	Sherbrooke	QMJHL	69	12	32	44	30	....	....	....	....	....	....	....	....	....	....	....	7	0	3	3	2	....	....	....
1995-96	Sherbrooke	QMJHL	68	15	53	68	68	....	....	....	....	....	....	....	....	....	....	....	7	0	3	3	2	....	....	....
1996-97	Sherbrooke	QMJHL	32	6	34	40	14	....	....	....	....	....	....	....	....	....	....	....	....	....	....	....	....	....	....	....
	Chicoutimi	QMJHL	28	5	24	29	26	....	....	....	....	....	....	....	....	....	....	....	19	6	15	21	12	....	....	....
1997-98	Las Vegas	IHL	26	1	4	5	10	....	....	....	....	....	....	....	....	....	....	....	....	....	....	....	....	....	....	....
	Springfield	AHL	41	6	15	21	16	....	....	....	....	....	....	....	....	....	....	....	4	0	1	1	2	....	....	....
1998-99	Springfield	AHL	69	4	32	36	10	....	....	....	....	....	....	....	....	....	....	....	3	0	1	1	0	....	....	....
99-2000	**Phoenix**	**NHL**	**60**	**0**	**6**	**6**	**16**	**0**	**0**	**0**	**36**	**0.0**	**2**	**0**	**0.0**	**57**	**55**	**15:09**	**5**	**0**	**1**	**1**	**0**	**0**	**0**	**0**
	Springfield	AHL	2	0	1	1	0	....	....	....	....	....	....	....	....	....	....	....	....	....	....	....	....	....	....	....
2000-01	**Phoenix**	**NHL**	**72**	**0**	**10**	**10**	**22**	**0**	**0**	**0**	**33**	**0.0**	**1**	**0**	**0.0**	**80**	**112**	**17:09**	....	....	....	....	....	....	....	....
2001-02	**Phoenix**	**NHL**	**81**	**4**	**13**	**17**	**10**	**1**	**0**	**0**	**49**	**8.2**	**25**	**1**	**100.0**	**99**	**96**	**18:05**	**5**	**1**	**0**	**1**	**0**	**0**	**0**	**0**
	NHL Totals		**213**	**4**	**29**	**33**	**48**	**1**	**0**	**0**	**118**	**3.4**		**1**	**100.0**	**236**	**263**	**16:57**	**10**	**1**	**1**	**2**	**0**	**0**	**0**	**0**

QMJHL All-Rookie Team (1995) • QMJHL Second All-Star Team (1997) • Won George Parsons Trophy (Memorial Cup Tournament Most Sportsmanlike Player) (1997)

Signed as a free agent by **Phoenix**, September 26, 1997.

SULLIVAN, Mike

(SUH-lih-van, MIGHK)

Center. Shoots left. 6'2", 204 lbs. Born, Marshfield, MA, February 27, 1968. NY Rangers' 4th choice, 69th overall, in 1987 Entry Draft.

			Regular Season																Playoffs							
Season	Club	League	GP	G	A	Pts	PIM	PP	SH	GW	S	%	+/-	TF	F%	H	SB	Min	GP	G	A	Pts	PIM	PP	SH	GW
1985-86	B.C. High Irish	Hi-School	22	26	33	59	....	....	....	....	....	....	....	....	....	....	....	....	....	....	....	....	....	....	....	....
1986-87	Boston University	H-East	37	13	18	31	18	....	....	....	....	....	....	....	....	....	....	....	....	....	....	....	....	....	....	....
1987-88	Boston University	H-East	30	18	22	40	30	....	....	....	....	....	....	....	....	....	....	....	....	....	....	....	....	....	....	....
1988-89	Boston University	H-East	36	19	17	36	30	....	....	....	....	....	....	....	....	....	....	....	....	....	....	....	....	....	....	....
1989-90	Boston University	H-East	38	11	20	31	26	....	....	....	....	....	....	....	....	....	....	....	....	....	....	....	....	....	....	....
1990-91	San Diego Gulls	IHL	74	12	23	35	27	....	....	....	....	....	....	....	....	....	....	....	....	....	....	....	....	....	....	....
1991-92	**San Jose**	**NHL**	**64**	**8**	**11**	**19**	**15**	**1**	**0**	**1**	**72**	**11.1**	**–18**	....	....	....	....	....	....	....	....	....	....	....	....	....
	Kansas City	IHL	10	2	8	10	8	....	....	....	....	....	....	....	....	....	....	....	....	....	....	....	....	....	....	....
1992-93	**San Jose**	**NHL**	**81**	**6**	**8**	**14**	**30**	**0**	**2**	**0**	**95**	**6.3**	**–42**	....	....	....	....	....	....	....	....	....	....	....	....	....
1993-94	**San Jose**	**NHL**	**26**	**2**	**2**	**4**	**4**	**0**	**2**	**1**	**21**	**9.5**	**–3**	....	....	....	....	....	....	....	....	....	....	....	....	....
	Kansas City	IHL	6	3	3	6	0	....	....	....	....	....	....	....	....	....	....	....	....	....	....	....	....	....	....	....
	Calgary	**NHL**	**19**	**2**	**3**	**5**	**6**	**0**	**2**	**0**	**27**	**7.4**	**2**	....	....	....	....	....	**7**	**1**	**1**	**2**	**8**	**0**	**1**	**0**
	Saint John	AHL	5	2	0	2	4	....	....	....	....	....	....	....	....	....	....	....	....	....	....	....	....	....	....	....
1994-95	**Calgary**	**NHL**	**38**	**4**	**7**	**11**	**14**	**0**	**0**	**2**	**31**	**12.9**	**–2**	....	....	....	....	....	**7**	**3**	**5**	**8**	**2**	**0**	**1**	**1**
1995-96	**Calgary**	**NHL**	**81**	**9**	**12**	**21**	**24**	**0**	**1**	**1**	**106**	**8.5**	**–6**	....	....	....	....	....	**4**	**0**	**0**	**0**	**0**	**0**	**0**	**0**
1996-97	**Calgary**	**NHL**	**67**	**5**	**6**	**11**	**10**	**0**	**3**	**2**	**64**	**7.8**	**–11**	....	....	....	....	....	....	....	....	....	....	....	....	....
1997-98	**Boston**	**NHL**	**77**	**5**	**13**	**18**	**34**	**0**	**0**	**2**	**83**	**6.0**	**–1**	....	....	....	....	....	**6**	**0**	**1**	**1**	**2**	**0**	**0**	**0**
1998-99	**Phoenix**	**NHL**	**63**	**2**	**4**	**6**	**24**	**0**	**1**	**1**	**66**	**3.0**	**–11**	**74**	**50.0**	**52**	**40**	**12:09**	**5**	**0**	**0**	**0**	**2**	**0**	**0**	**0**
99-2000	**Phoenix**	**NHL**	**79**	**5**	**10**	**15**	**10**	**0**	**2**	**1**	**59**	**8.5**	**–4**	**820**	**51.5**	**77**	**37**	**11:36**	**5**	**0**	**1**	**1**	**0**	**0**	**0**	**0**
2000-01	**Phoenix**	**NHL**	**72**	**5**	**4**	**9**	**16**	**0**	**3**	**0**	**59**	**8.5**	**–6**	**797**	**50.3**	**52**	**65**	**12:48**	....	....	....	....	....	....	....	....
2001-02	**Phoenix**	**NHL**	**42**	**1**	**2**	**3**	**16**	**0**	**0**	**0**	**28**	**3.6**	**–3**	**254**	**48.4**	**30**	**34**	**10:56**	....	....	....	....	....	....	....	....
	NHL Totals		**709**	**54**	**82**	**136**	**203**	**1**	**16**	**11**	**711**	**7.6**		**1945**	**50.5**	**211**	**176**	**11:58**	**34**	**4**	**8**	**12**	**14**	**0**	**2**	**1**

Rights traded to **Minnesota** by **NY Rangers** with Mark Tinordi, Paul Jerrard, the rights to Bret Barnett and LA Kings' 3rd round choice (previously acquired, Minnesota selected Murray Garbutt) in 1989 Entry Draft for Brian Lawton, Igor Liba and the rights to Eric Bennett, October 11, 1988. Signed as a free agent by **San Jose**, August 9, 1991. Claimed on waivers by **Calgary** from **San Jose**, January 6, 1994. Traded to **Boston** by **Calgary** for Boston's 7th round choice (Radek Duda) in 1998 Entry Draft, June 21, 1997. Claimed by **Nashville** from **Boston** in Expansion Draft, June 26, 1998. Traded to **Phoenix** by **Nashville** for Phoenix's 7th round choice (Kyle Kettles) in 1999 Entry Draft, June 30, 1998.

• Officially announced retirement and named Head Coach of **Providence** (AHL), July 29, 2002.

SULLIVAN, Steve

(SUH-lih-van, STEEV) **CHI.**

Right wing. Shoots right. 5'9", 160 lbs. Born, Timmins, Ont., July 6, 1974. New Jersey's 10th choice, 233rd overall, in 1994 Entry Draft.

			Regular Season																Playoffs							
Season	Club	League	GP	G	A	Pts	PIM	PP	SH	GW	S	%	+/-	TF	F%	H	SB	Min	GP	G	A	Pts	PIM	PP	SH	GW
1991-92	Timmins	NOJHA	47	66	55	121	141	....	....	....	....	....	....	....	....	....	....	....	....	....	....	....	....	....	....	....
1992-93	Sault Ste. Marie	OHL	62	36	27	63	44	....	....	....	....	....	....	....	....	....	....	....	16	3	8	11	18	....	....	....
1993-94	Sault Ste. Marie	OHL	63	51	62	113	82	....	....	....	....	....	....	....	....	....	....	....	14	9	16	25	22	....	....	....
1994-95	Albany	AHL	75	31	50	81	124	....	....	....	....	....	....	....	....	....	....	....	14	4	7	11	10	....	....	....
1995-96	**New Jersey**	**NHL**	**16**	**5**	**4**	**9**	**8**	**2**	**0**	**1**	**23**	**21.7**	**3**	....	....	....	....	....	....	....	....	....	....	....	....	....
	Albany	AHL	53	33	42	75	127	....	....	....	....	....	....	....	....	....	....	....	4	3	0	3	6	....	....	....
1996-97	**New Jersey**	**NHL**	**33**	**8**	**14**	**22**	**14**	**2**	**0**	**2**	**63**	**12.7**	**9**	....	....	....	....	....	....	....	....	....	....	....	....	....
	Albany	AHL	15	8	7	15	16	....	....	....	....	....	....	....	....	....	....	....	....	....	....	....	....	....	....	....
	Toronto	**NHL**	**21**	**5**	**11**	**16**	**23**	**1**	**0**	**1**	**45**	**11.1**	**5**	....	....	....	....	....	....	....	....	....	....	....	....	....
1997-98	**Toronto**	**NHL**	**63**	**10**	**18**	**28**	**40**	**1**	**0**	**1**	**112**	**8.9**	**–8**	....	....	....	....	....	....	....	....	....	....	....	....	....
1998-99	**Toronto**	**NHL**	**63**	**20**	**20**	**40**	**28**	**4**	**0**	**5**	**110**	**18.2**	**12**	**685**	**44.4**	**26**	**11**	**14:12**	**13**	**3**	**3**	**6**	**14**	**2**	**0**	**0**
99-2000	**Toronto**	**NHL**	**7**	**0**	**1**	**1**	**4**	**0**	**0**	**0**	**11**	**0.0**	**–1**	**47**	**48.9**	**1**	**0**	**11:52**	....	....	....	....	....	....	....	....
	Chicago	**NHL**	**73**	**22**	**42**	**64**	**52**	**2**	**1**	**6**	**169**	**13.0**	**20**	**692**	**48.0**	**39**	**15**	**18:05**	....	....	....	....	....	....	....	....
2000-01	**Chicago**	**NHL**	**81**	**34**	**41**	**75**	**54**	**6**	**8**	**3**	**204**	**16.7**	**3**	**649**	**42.4**	**34**	**22**	**20:32**	....	....	....	....	....	....	....	....
2001-02	**Chicago**	**NHL**	**78**	**21**	**39**	**60**	**67**	**3**	**0**	**8**	**155**	**13.5**	**23**	**758**	**48.9**	**54**	**20**	**19:10**	**5**	**1**	**0**	**1**	**2**	**0**	**0**	**0**
	NHL Totals		**435**	**125**	**190**	**315**	**290**	**21**	**9**	**27**	**892**	**14.0**		**2831**	**46.1**	**154**	**68**	**18:04**	**18**	**4**	**3**	**7**	**16**	**2**	**0**	**0**

AHL First All-Star Team (1996)

Traded to **Toronto** by **New Jersey** with Jason Smith and the rights to Alyn McCauley for Doug Gilmour, Dave Ellett and New Jersey's 3rd round choice (previously acquired, New Jersey selected Andre Lakos) in 1999 Entry Draft, February 25, 1997. Claimed on waivers by **Chicago** from **Toronto**, October 23, 1999.

SUNDIN, Mats

(sahn-DEEN, MATS) **TOR.**

Center/Right wing. Shoots right. 6'4", 220 lbs. Born, Bromma, Sweden, February 13, 1971. Quebec's 1st choice, 1st overall, in 1989 Entry Draft.

			Regular Season																Playoffs							
Season	Club	League	GP	G	A	Pts	PIM	PP	SH	GW	S	%	+/-	TF	F%	H	SB	Min	GP	G	A	Pts	PIM	PP	SH	GW
1988-89	Nacka HK	Swede-2	25	10	8	18	18	....	....	....	....	....	....	....	....	....	....	....	....	....	....	....	....	....	....	....
1989-90	Djurgarden	Sweden	34	10	8	18	16	....	....	....	....	....	....	....	....	....	....	....	8	7	0	7	4	....	....	....
1990-91	**Quebec**	**NHL**	**80**	**23**	**36**	**59**	**58**	**4**	**0**	**0**	**155**	**14.8**	**–24**	....	....	....	....	....	....	....	....	....	....	....	....	....
1991-92	**Quebec**	**NHL**	**80**	**33**	**43**	**76**	**103**	**8**	**2**	**2**	**231**	**14.3**	**–19**	....	....	....	....	....	....	....	....	....	....	....	....	....
1992-93	**Quebec**	**NHL**	**80**	**47**	**67**	**114**	**96**	**13**	**4**	**9**	**215**	**21.9**	**21**	....	....	....	....	....	**6**	**3**	**1**	**4**	**6**	**1**	**0**	**0**
1993-94	**Quebec**	**NHL**	**84**	**32**	**53**	**85**	**60**	**6**	**2**	**4**	**226**	**14.2**	**1**	....	....	....	....	....	....	....	....	....	....	....	....	....
1994-95	Djurgarden	Sweden	12	7	2	9	14	....	....	....	....	....	....	....	....	....	....	....	....	....	....	....	....	....	....	....
	Toronto	**NHL**	**47**	**23**	**24**	**47**	**14**	**9**	**0**	**4**	**173**	**13.3**	**–5**	....	....	....	....	....	**7**	**5**	**4**	**9**	**4**	**2**	**0**	**1**
1995-96	**Toronto**	**NHL**	**76**	**33**	**50**	**83**	**46**	**7**	**6**	**7**	**301**	**11.0**	**8**	....	....	....	....	....	**6**	**3**	**1**	**4**	**4**	**2**	**0**	**1**
1996-97	**Toronto**	**NHL**	**82**	**41**	**53**	**94**	**59**	**7**	**4**	**8**	**281**	**14.6**	**6**	....	....	....	....	....	....	....	....	....	....	....	....	....
1997-98	**Toronto**	**NHL**	**82**	**33**	**41**	**74**	**49**	**9**	**1**	**5**	**219**	**15.1**	**–3**	....	....	....	....	....	....	....	....	....	....	....	....	....
	Sweden	Olympics	4	3	0	3	4	....	....	....	....	....	....	....	....	....	....	....	....	....	....	....	....	....	....	....
1998-99	**Toronto**	**NHL**	**82**	**31**	**52**	**83**	**58**	**4**	**0**	**6**	**209**	**14.8**	**22**	**1993**	**57.3**	**54**	**17**	**20:41**	**17**	**8**	**8**	**16**	**16**	**3**	**0**	**2**
99-2000	**Toronto**	**NHL**	**73**	**32**	**41**	**73**	**46**	**10**	**2**	**7**	**184**	**17.4**	**16**	**1619**	**50.8**	**48**	**33**	**20:11**	**12**	**3**	**5**	**8**	**10**	**0**	**0**	**1**
2000-01	**Toronto**	**NHL**	**82**	**28**	**46**	**74**	**76**	**9**	**0**	**6**	**226**	**12.4**	**15**	**1870**	**56.6**	**83**	**33**	**19:21**	**11**	**6**	**7**	**13**	**14**	**2**	**1**	**1**
2001-02	**Toronto**	**NHL**	**82**	**41**	**39**	**80**	**94**	**10**	**2**	**9**	**262**	**15.6**	**6**	**1812**	**57.5**	**59**	**32**	**19:20**	**8**	**2**	**5**	**7**	**4**	**0**	**0**	**0**
	Sweden	Olympics	4	5	4	*9	10	....	....	....	....	....	....	....	....	....	....	....	....	....	....	....	....	....	....	....
	NHL Totals		**930**	**397**	**545**	**942**	**759**	**96**	**23**	**67**	**2682**	**14.8**		**7294**	**55.7**	**244**	**115**	**19:53**	**67**	**30**	**31**	**61**	**58**	**10**	**1**	**6**

NHL Second All-Star Team (2002) • Played in NHL All-Star Game (1996, 1997, 1998, 1999, 2000, 2001, 2002)

Traded to **Toronto** by **Quebec** with Garth Butcher, Todd Warriner and Philadelphia's 1st round choice (previously acquired by Quebec - later traded to Washington - Washington selected Nolan Baumgartner) in 1994 Entry Draft for Wendel Clark, Sylvain Lefebvre, Landon Wilson and Toronto's 1st round choice (Jeffrey Kealty) in 1994 Entry Draft, June 28, 1994.

SUNDSTROM, Niklas

(SUHN-struhm, NIHK-las) **S.J.**

Left wing. Shoots left. 6', 190 lbs. Born, Ornskoldsvik, Sweden, June 6, 1975. NY Rangers' 1st choice, 8th overall, in 1993 Entry Draft.

			Regular Season																Playoffs							
Season	Club	League	GP	G	A	Pts	PIM	PP	SH	GW	S	%	+/-	TF	F%	H	SB	Min	GP	G	A	Pts	PIM	PP	SH	GW
1991-92	MoDo	Sweden	9	1	3	4	0	....	....	....	....	....	....	....	....	....	....	....	....	....	....	....	....	....	....	....
1992-93	MoDo Jr.	Swede-Jr.	2	3	1	4	0	....	....	....	....	....	....	....	....	....	....	....	....	....	....	....	....	....	....	....
	MoDo	Sweden	40	7	11	18	18	....	....	....	....	....	....	....	....	....	....	....	3	0	0	0	0	....	....	....
1993-94	MoDo Jr.	Swede-Jr.	3	3	4	7	2	....	....	....	....	....	....	....	....	....	....	....	....	....	....	....	....	....	....	....
	MoDo	Sweden	37	7	12	19	28	....	....	....	....	....	....	....	....	....	....	....	11	4	3	7	2	....	....	....
1994-95	MoDo	Sweden	33	8	13	21	30	....	....	....	....	....	....	....	....	....	....	....	....	....	....	....	....	....	....	....
1995-96	**NY Rangers**	**NHL**	**82**	**9**	**12**	**21**	**14**	**1**	**1**	**2**	**90**	**10.0**	**2**	....	....	....	....	....	**11**	**4**	**3**	**7**	**4**	**1**	**0**	**0**
1996-97	**NY Rangers**	**NHL**	**82**	**24**	**28**	**52**	**20**	**5**	**1**	**4**	**132**	**18.2**	**23**	....	....	....	....	....	**9**	**0**	**5**	**5**	**2**	**0**	**0**	**0**
1997-98	**NY Rangers**	**NHL**	**70**	**19**	**28**	**47**	**24**	**4**	**0**	**1**	**115**	**16.5**	**0**	....	....	....	....	....	....	....	....	....	....	....	....	....
	Sweden	Olympics	4	1	1	2	2	....	....	....	....	....	....	....	....	....	....	....	....	....	....	....	....	....	....	....
1998-99	**NY Rangers**	**NHL**	**81**	**13**	**30**	**43**	**20**	**1**	**2**	**3**	**89**	**14.6**	**–2**	**376**	**40.4**	**118**	**55**	**19:11**	....	....	....	....	....	....	....	....
99-2000	**San Jose**	**NHL**	**79**	**12**	**25**	**37**	**22**	**2**	**1**	**2**	**90**	**13.3**	**9**	**10**	**50.0**	**44**	**12**	**15:10**	**12**	**0**	**2**	**2**	**2**	**0**	**0**	**0**
2000-01	**San Jose**	**NHL**	**82**	**10**	**39**	**49**	**28**	**4**	**1**	**0**	**100**	**10.0**	**10**	**26**	**30.8**	**40**	**16**	**16:44**	**6**	**0**	**3**	**3**	**2**	**0**	**0**	**0**
2001-02	**San Jose**	**NHL**	**73**	**9**	**30**	**39**	**50**	**0**	**1**	**0**	**74**	**12.2**	**7**	**9**	**33.3**	**37**	**16**	**15:57**	**12**	**1**	**6**	**7**	**6**	**0**	**0**	**0**
	Sweden	Olympics	4	1	3	4	0	....	....	....	....	....	....	....	....	....	....	....	....	....	....	....	....	....	....	....
	NHL Totals		**549**	**96**	**192**	**288**	**178**	**17**	**7**	**12**	**690**	**13.9**		**421**	**39.9**	**239**	**99**	**16:47**	**50**	**5**	**19**	**24**	**16**	**1**	**0**	**0**

Traded to **Tampa Bay** by **NY Rangers** with Dan Cloutier and NY Rangers' 1st (Nikita Alexeev) and 3rd (later traded to San Jose - later traded to Chicago - Chicago selected Igor Radulov) round choices in 2000 Entry Draft for Chicago's 1st round choice (previously acquired, NY Rangers selected Pavel Brendl) in 1999 Entry Draft, June 26, 1999. Traded to **San Jose** by **Tampa Bay** with NY Rangers' 3rd round choice (previously acquired, later traded to Chicago - Chicago selected Igor Radulov) in 2000 Entry Draft for Bill Houlder, Andrei Zyuzin, Shawn Burr and Steve Guolla, August 4, 1999.

SUSHINSKY, Maxim
(soo-SHIHN-skee, max-EEM) **MIN.**

Right wing. Shoots left. 5'8", 165 lbs. Born, Leningrad, USSR, July 1, 1974. Minnesota's 4th choice, 132nd overall, in 2000 Entry Draft.

			Regular Season																Playoffs							
Season	Club	League	GP	G	A	Pts	PIM	PP	SH	GW	S	%	+/-	TF	F%	H	SB	Min	GP	G	A	Pts	PIM	PP	SH	GW
1990-91	SKA Leningrad 2	USSR-3	8	1	0	1	0	....	....	....	....	....	....	....	....	....	....	....	....	....	....	....	....	....	....	....
	SKA Leningrad	USSR	4	0	1	1	0	....	....	....	....	....	....	....	....	....	....	....	....	....	....	....	....	....	....	....
1991-92	St. Petersburg 2	CIS-3	20	14	1	15	38	....	....	....	....	....	....	....	....	....	....	....	....	....	....	....	....	....	....	....
	St. Petersburg	CIS-2	45	5	3	8	16	....	....	....	....	....	....	....	....	....	....	....	....	....	....	....	....	....	....	....
1992-93	St. Petersburg	CIS	23	2	3	5	22	....	....	....	....	....	....	....	....	....	....	....	6	2	1	3	2	....	....	....
1993-94	St. Petersburg	CIS	45	7	4	11	26	....	....	....	....	....	....	....	....	....	....	....	....	....	....	....	....	....	....	....
1994-95	St. Petersburg	CIS	52	11	11	22	57	....	....	....	....	....	....	....	....	....	....	....	3	1	0	1	6	....	....	....
1995-96	St. Petersburg	CIS	49	21	15	36	43	....	....	....	....	....	....	....	....	....	....	....	2	0	0	0	0	....	....	....
1996-97	Avangard Omsk	Russia	39	20	16	36	24	....	....	....	....	....	....	....	....	....	....	....	5	3	1	4	0	....	....	....
1997-98	Avangard Omsk	Russia	20	6	11	17	6	....	....	....	....	....	....	....	....	....	....	....	....	....	....	....	....	....	....	....
1998-99	Avangard Omsk	Russia	41	15	16	31	46	....	....	....	....	....	....	....	....	....	....	....	5	3	1	4	10	....	....	....
99-2000	Avangard Omsk	Russia	37	19	24	43	58	....	....	....	....	....	....	....	....	....	....	....	8	2	5	7	6	....	....	....
2000-01	**Minnesota**	**NHL**	**30**	**7**	**4**	**11**	**29**	**3**	**0**	**0**	**62**	**11.3**	**–7**	**2**	**50.0**	**17**	**2**	**14:01**	....	....	....	....	....	....	....	....
	Avangard Omsk	Russia	12	5	3	8	14	....	....	....	....	....	....	....	....	....	....	....	13	*9	4	13	12	....	....	....
2001-02	Avangard Omsk	Russia	46	19	*32	*51	60	....	....	....	....	....	....	....	....	....	....	....	11	*6	*11	*17	18	....	....	....
	NHL Totals		**30**	**7**	**4**	**11**	**29**	**3**	**0**	**0**	**62**	**11.3**		**2**	**50.0**	**17**	**2**	**14:01**	....	....	....	....	....	....	....	....

Signed as a free agent by **Avangard Omsk** (Russia) with Minnesota retaining NHL rights, January 9, 2001.

SUTER, Gary
(SOO-tuhr, GAIR-ee) **S.J.**

Defense. Shoots left. 6', 215 lbs. Born, Madison, WI, June 24, 1964. Calgary's 9th choice, 180th overall, in 1984 Entry Draft.

			Regular Season																Playoffs							
Season	Club	League	GP	G	A	Pts	PIM	PP	SH	GW	S	%	+/-	TF	F%	H	SB	Min	GP	G	A	Pts	PIM	PP	SH	GW
1981-82	Dubuque	USHL	18	3	4	7	32	....	....	....	....	....	....	....	....	....	....	....	....	....	....	....	....	....	....	....
1982-83	Dubuque	USHL	41	9	30	39	112	....	....	....	....	....	....	....	....	....	....	....	....	....	....	....	....	....	....	....
1983-84	U. of Wisconsin	WCHA	35	4	18	22	32	....	....	....	....	....	....	....	....	....	....	....	....	....	....	....	....	....	....	....
1984-85	U. of Wisconsin	WCHA	39	12	39	51	110	....	....	....	....	....	....	....	....	....	....	....	....	....	....	....	....	....	....	....
1985-86	**Calgary**	**NHL**	**80**	**18**	**50**	**68**	**141**	**9**	**0**	**4**	**195**	**9.2**	**11**	....	....	....	....	....	**10**	**2**	**8**	**10**	**8**	**0**	**0**	**1**
1986-87	**Calgary**	**NHL**	**68**	**9**	**39**	**48**	**70**	**4**	**0**	**0**	**152**	**5.9**	**–10**	....	....	....	....	....	**6**	**0**	**3**	**3**	**10**	**0**	**0**	**0**
1987-88	**Calgary**	**NHL**	**75**	**21**	**70**	**91**	**124**	**6**	**1**	**3**	**204**	**10.3**	**39**	....	....	....	....	....	**9**	**1**	**9**	**10**	**6**	**0**	**1**	**0**
1988-89♦	**Calgary**	**NHL**	**63**	**13**	**49**	**62**	**78**	**8**	**0**	**1**	**216**	**6.0**	**26**	....	....	....	....	....	**5**	**0**	**3**	**3**	**10**	**0**	**0**	**0**
1989-90	**Calgary**	**NHL**	**76**	**16**	**60**	**76**	**97**	**5**	**0**	**1**	**211**	**7.6**	**4**	....	....	....	....	....	**6**	**0**	**1**	**1**	**14**	**0**	**0**	**0**
1990-91	**Calgary**	**NHL**	**79**	**12**	**58**	**70**	**102**	**6**	**0**	**1**	**258**	**4.7**	**26**	....	....	....	....	....	**7**	**1**	**6**	**7**	**12**	**1**	**0**	**0**
1991-92	**Calgary**	**NHL**	**70**	**12**	**43**	**55**	**128**	**4**	**0**	**0**	**189**	**6.3**	**1**	....	....	....	....	....	....	....	....	....	....	....	....	....
1992-93	**Calgary**	**NHL**	**81**	**23**	**58**	**81**	**112**	**10**	**1**	**2**	**263**	**8.7**	**–1**	....	....	....	....	....	**6**	**2**	**3**	**5**	**8**	**0**	**1**	**0**
1993-94	**Calgary**	**NHL**	**25**	**4**	**9**	**13**	**20**	**2**	**1**	**0**	**51**	**7.8**	**–3**	....	....	....	....	....	....	....	....	....	....	....	....	....
	Chicago	**NHL**	**16**	**2**	**3**	**5**	**18**	**2**	**0**	**0**	**35**	**5.7**	**–9**	....	....	....	....	....	**6**	**3**	**2**	**5**	**6**	**2**	**0**	**0**
1994-95	**Chicago**	**NHL**	**48**	**10**	**27**	**37**	**42**	**5**	**0**	**0**	**144**	**6.9**	**14**	....	....	....	....	....	**12**	**2**	**5**	**7**	**10**	**1**	**0**	**0**
1995-96	**Chicago**	**NHL**	**82**	**20**	**47**	**67**	**80**	**12**	**2**	**4**	**242**	**8.3**	**3**	....	....	....	....	....	**10**	**3**	**3**	**6**	**8**	**2**	**0**	**1**
1996-97	**Chicago**	**NHL**	**82**	**7**	**21**	**28**	**70**	**3**	**0**	**0**	**225**	**3.1**	**–4**	....	....	....	....	....	**6**	**1**	**4**	**5**	**8**	**0**	**0**	**0**
1997-98	**Chicago**	**NHL**	**73**	**14**	**28**	**42**	**74**	**5**	**2**	**0**	**199**	**7.0**	**1**	....	....	....	....	....	....	....	....	....	....	....	....	....
	United States	Olympics	4	0	0	0	2	....	....	....	....	....	....	....	....	....	....	....	....	....	....	....	....	....	....	....
1998-99	**San Jose**	**NHL**	**1**	**0**	**0**	**0**	**0**	**0**	**0**	**0**	**1**	**0.0**	**0**	**0**	**0.0**	**0**	**0**	**12:56**	....	....	....	....	....	....	....	....
99-2000	**San Jose**	**NHL**	**76**	**6**	**28**	**34**	**52**	**2**	**1**	**0**	**175**	**3.4**	**7**	**0**	**0.0**	**118**	**102**	**23:30**	**12**	**2**	**5**	**7**	**12**	**1**	**0**	**1**
2000-01	**San Jose**	**NHL**	**68**	**10**	**24**	**34**	**84**	**4**	**0**	**1**	**157**	**6.4**	**8**	**2**	**50.0**	**126**	**68**	**21:27**	**1**	**0**	**0**	**0**	**0**	**0**	**0**	**0**
2001-02	**San Jose**	**NHL**	**82**	**6**	**27**	**33**	**57**	**3**	**0**	**1**	**174**	**3.4**	**13**	**0**	**0.0**	**99**	**69**	**20:07**	**12**	**0**	**4**	**4**	**8**	**0**	**0**	**0**
	United States	Olympics	6	0	1	1	4	....	....	....	....	....	....	....	....	....	....	....	....	....	....	....	....	....	....	....
	NHL Totals		**1145**	**203**	**641**	**844**	**1349**	**90**	**8**	**18**	**3091**	**6.6**		**2**	**50.0**	**343**	**239**	**21:37**	**108**	**17**	**56**	**73**	**120**	**7**	**2**	**3**

NHL All-Rookie Team (1986) • Won Calder Memorial Trophy (1986) • NHL Second All-Star Team (1988) • Played in NHL All-Star Game (1986, 1988, 1989, 1991)

Traded to **Hartford** by **Calgary** with Paul Ranheim and Ted Drury for James Patrick, Zarley Zalapski and Michael Nylander, March 10, 1994. Traded to **Chicago** by **Hartford** with Randy Cunneyworth and Hartford's 3rd round choice (later traded to Vancouver - Vancouver selected Larry Courville) in 1995 Entry Draft for Frantisek Kucera and Jocelyn Lemieux, March 11, 1994. Signed as a free agent by **San Jose**, July 1, 1998. • Missed majority of 1998-99 season recovering from tricep muscle injury suffered in game vs. Dallas, October 24, 1998.

SUTHERBY, Brian
(SUH-thur-bee, BRIGH-uhn) **WSH.**

Center. Shoots left. 6'2", 180 lbs. Born, Edmonton, Alta., March 1, 1982. Washington's 1st choice, 26th overall, in 2000 Entry Draft.

			Regular Season																Playoffs							
Season	Club	League	GP	G	A	Pts	PIM	PP	SH	GW	S	%	+/-	TF	F%	H	SB	Min	GP	G	A	Pts	PIM	PP	SH	GW
1997-98	CAC Cement	AMHL	36	36	23	59	60	....	....	....	....	....	....	....	....	....	....	....	....	....	....	....	....	....	....	....
1998-99	Moose Jaw	WHL	66	9	12	21	47	....	....	....	....	....	....	....	....	....	....	....	11	0	1	1	0	....	....	....
99-2000	Moose Jaw	WHL	47	18	17	35	102	....	....	....	....	....	....	....	....	....	....	....	4	1	1	2	12	....	....	....
2000-01	Moose Jaw	WHL	59	34	43	77	138	....	....	....	....	....	....	....	....	....	....	....	4	2	1	3	10	....	....	....
2001-02	**Washington**	**NHL**	**7**	**0**	**0**	**0**	**2**	**0**	**0**	**0**	**3**	**0.0**	**–3**	**39**	**35.9**	**3**	**1**	**7:17**	....	....	....	....	....	....	....	....
	Moose Jaw	WHL	36	18	27	45	75	....	....	....	....	....	....	....	....	....	....	....	12	7	5	12	33	....	....	....
	NHL Totals		**7**	**0**	**0**	**0**	**2**	**0**	**0**	**0**	**3**	**0.0**		**39**	**35.9**	**3**	**1**	**7:17**	....	....	....	....	....	....	....	....

• Returned to **Moose Jaw** (WHL) by **Washington**, October 22, 2001.

SUTTON, Andy
(SUH-tohn, AN-dee) **ATL.**

Defense. Shoots left. 6'6", 245 lbs. Born, Edmonton, Alta., March 10, 1975.

			Regular Season																Playoffs							
Season	Club	League	GP	G	A	Pts	PIM	PP	SH	GW	S	%	+/-	TF	F%	H	SB	Min	GP	G	A	Pts	PIM	PP	SH	GW
1991-92	Gananoque	OJHL-B	36	11	9	20	....	....	....	....	....	....	....	....	....	....	....	....	14	9	21	30	....	....	....	....
1992-93	Gananoque	OJHL-B	38	14	9	23	....	....	....	....	....	....	....	....	....	....	....	....	12	16	13	29	....	....	....	....
1993-94	St. Michael's B	OJHL-B	48	17	23	40	161	....	....	....	....	....	....	....	....	....	....	....	3	0	0	0	20	....	....	....
1994-95	Michigan Tech	WCHA	19	2	1	3	42	....	....	....	....	....	....	....	....	....	....	....	....	....	....	....	....	....	....	....
1995-96	Michigan Tech	WCHA	33	2	2	4	58	....	....	....	....	....	....	....	....	....	....	....	....	....	....	....	....	....	....	....
1996-97	Michigan Tech	WCHA	32	2	7	9	73	....	....	....	....	....	....	....	....	....	....	....	....	....	....	....	....	....	....	....
1997-98	Michigan Tech	WCHA	38	16	24	40	97	....	....	....	....	....	....	....	....	....	....	....	....	....	....	....	....	....	....	....
	Kentucky	AHL	7	0	0	0	33	....	....	....	....	....	....	....	....	....	....	....	....	....	....	....	....	....	....	....
1998-99	**San Jose**	**NHL**	**31**	**0**	**3**	**3**	**65**	**0**	**0**	**0**	**24**	**0.0**	**–4**	**0**	**0.0**	**50**	**14**	**12:58**	....	....	....	....	....	....	....	....
	Kentucky	AHL	21	5	10	15	53	....	....	....	....	....	....	....	....	....	....	....	5	0	0	0	23	....	....	....
99-2000	**San Jose**	**NHL**	**40**	**1**	**1**	**2**	**80**	**0**	**0**	**0**	**29**	**3.4**	**–5**	**0**	**0.0**	**93**	**22**	**12:57**	....	....	....	....	....	....	....	....
	Kentucky	AHL	3	0	1	1	0	....	....	....	....	....	....	....	....	....	....	....	....	....	....	....	....	....	....	....
2000-01	**Minnesota**	**NHL**	**69**	**3**	**4**	**7**	**131**	**2**	**0**	**0**	**64**	**4.7**	**–11**	**3**	**33.3**	**123**	**49**	**12:55**	....	....	....	....	....	....	....	....
2001-02	**Minnesota**	**NHL**	**19**	**2**	**4**	**6**	**35**	**1**	**0**	**0**	**21**	**9.5**	**–4**	**2**	**0.0**	**30**	**5**	**10:57**	....	....	....	....	....	....	....	....
	Atlanta	**NHL**	**24**	**0**	**4**	**4**	**46**	**0**	**0**	**0**	**20**	**0.0**	**0**	**0**	**0.0**	**58**	**36**	**15:25**	....	....	....	....	....	....	....	....
	NHL Totals		**183**	**6**	**16**	**22**	**357**	**3**	**0**	**0**	**158**	**3.8**		**5**	**20.0**	**354**	**126**	**13:03**	....	....	....	....	....	....	....	....

WCHA Second All-Star Team (1998)

Signed as a free agent by **San Jose**, March 20, 1998. Traded to **Minnesota** by **San Jose** with San Jose's 7th round choice (Peter Bartos) in 2000 Entry Draft and 3rd round choice (later traded to Atlanta - later traded to Pittsburgh - later traded to Columbus - Columbus selected Aaron Johnson) in 2001 Entry Draft for Minnesota's 8th round choice (later traded to Calgary - Calgary selected Joe Campbell) in 2001 Entry Draft and future considerations, June 12, 2000. Traded to **Atlanta** by **Minnesota** for Hnat Domenichelli, January 22, 2002.

SUTTON, Ken
(SUH-tohn, KEHN) **NYI**

Defense. Shoots left. 6'1", 205 lbs. Born, Edmonton, Alta., November 5, 1969. Buffalo's 4th choice, 98th overall, in 1989 Entry Draft.

			Regular Season																Playoffs							
Season	Club	League	GP	G	A	Pts	PIM	PP	SH	GW	S	%	+/-	TF	F%	H	SB	Min	GP	G	A	Pts	PIM	PP	SH	GW
1987-88	Calgary Canucks	AJHL	53	13	43	56	228	....	....	....	....	....	....	....	....	....	....	....	....	....	....	....	....	....	....	....
1988-89	Saskatoon Blades	WHL	71	22	31	53	104	....	....	....	....	....	....	....	....	....	....	....	8	2	5	7	12	....	....	....
1989-90	Rochester	AHL	57	5	14	19	83	....	....	....	....	....	....	....	....	....	....	....	11	1	6	7	15	....	....	....
1990-91	**Buffalo**	**NHL**	**15**	**3**	**6**	**9**	**13**	**2**	**0**	**0**	**26**	**11.5**	**2**	....	....	....	....	....	**6**	**0**	**1**	**1**	**2**	**0**	**0**	**0**
	Rochester	AHL	62	7	24	31	65	....	....	....	....	....	....	....	....	....	....	....	3	1	1	2	14	....	....	....
1991-92	**Buffalo**	**NHL**	**64**	**2**	**18**	**20**	**71**	**0**	**0**	**0**	**81**	**2.5**	**5**	....	....	....	....	....	**7**	**0**	**2**	**2**	**4**	**0**	**0**	**0**
1992-93	**Buffalo**	**NHL**	**63**	**8**	**14**	**22**	**30**	**1**	**0**	**2**	**77**	**10.4**	**–3**	....	....	....	....	....	**8**	**3**	**1**	**4**	**8**	**0**	**0**	**0**
1993-94	**Buffalo**	**NHL**	**78**	**4**	**20**	**24**	**71**	**1**	**0**	**0**	**95**	**4.2**	**–6**	....	....	....	....	....	**4**	**0**	**0**	**0**	**2**	**0**	**0**	**0**
1994-95	**Buffalo**	**NHL**	**12**	**1**	**2**	**3**	**30**	**0**	**0**	**1**	**12**	**8.3**	**–2**	....	....	....	....	....	....	....	....	....	....	....	....	....
	Edmonton	**NHL**	**12**	**3**	**1**	**4**	**12**	**0**	**0**	**0**	**28**	**10.7**	**–1**	....	....	....	....	....	....	....	....	....	....	....	....	....
1995-96	**Edmonton**	**NHL**	**32**	**0**	**8**	**8**	**39**	**0**	**0**	**0**	**38**	**0.0**	**–12**	....	....	....	....	....	....	....	....	....	....	....	....	....
	St. Louis	**NHL**	**6**	**0**	**0**	**0**	**4**	**0**	**0**	**0**	**3**	**0.0**	**–1**	....	....	....	....	....	**1**	**0**	**0**	**0**	**0**	**0**	**0**	**0**
	Worcester	AHL	32	4	16	20	60	....	....	....	....	....	....	....	....	....	....	....	4	0	2	2	21	....	....	....
1996-97	Manitoba Moose	IHL	20	3	10	13	48	....	....	....	....	....	....	....	....	....	....	....	....	....	....	....	....	....	....	....
	Albany	AHL	61	6	13	19	79	....	....	....	....	....	....	....	....	....	....	....	16	4	8	12	55	....	....	....

Season	Club	League	GP	G	A	Pts	PIM	PP	SH	GW	S	%	+/-	TF	F%	H	SB	Min	GP	G	A	Pts	PIM	PP	SH	GW
			Regular Season																Playoffs							
1997-98	**New Jersey**	**NHL**	**13**	**0**	**0**	**0**	**6**	**0**	**0**	**0**	**5**	**0.0**	**1**													
	Albany	AHL	10	0	7	7	15																			
	San Jose	**NHL**	**8**	**0**	**0**	**0**	**15**	**0**	**0**	**0**	**7**	**0.0**	**-4**													
1998-99	**New Jersey**	**NHL**	**5**	**1**	**0**	**1**	**0**	**0**	**0**	**0**	**5**	**20.0**	**1**	**0**	**0.0**	**3**	**1**	**13:02**								
	Albany	AHL	75	13	42	55	118												5	0	2	2	12			
99-2000♦	**New Jersey**	**NHL**	**6**	**0**	**2**	**2**	**2**	**0**	**0**	**0**	**10**	**0.0**	**2**	**0**	**0.0**	**4**	**5**	**17:20**								
	Albany	AHL	57	5	16	21	129																			
2000-01	**New Jersey**	**NHL**	**53**	**1**	**7**	**8**	**37**	**0**	**0**	**0**	**35**	**2.9**	**9**	**0**	**0.0**	**40**	**71**	**15:44**	**6**	**0**	**0**	**0**	**13**	**0**	**0**	**0**
2001-02	**NY Islanders**	**NHL**	**21**	**0**	**2**	**2**	**8**	**0**	**0**	**0**	**22**	**0.0**	**-5**	**0**	**0.0**	**21**	**40**	**12:44**								
	Bridgeport	AHL	28	1	10	11	51												20	2	8	10	24			
	NHL Totals		**388**	**23**	**80**	**103**	**338**	**4**	**0**	**3**	**444**	**5.2**		**0**	**0.0**	**68**	**117**	**14:57**	**32**	**3**	**4**	**7**	**29**	**0**	**0**	**0**

Memorial Cup All-Star Team (1989) • AHL First All-Star Team (1999) • Won Eddie Shore Award (Top Defenseman - AHL) (1999)

Traded to **Edmonton** by **Buffalo** for Scott Pearson, April 7, 1995. Traded to **St. Louis** by **Edmonton** with Igor Kravchuk for Jeff Norton and Donald Dufresne, January 4, 1996. Traded to **New Jersey** by **St. Louis** with St. Louis' 2nd round choice (Brett Clouthier) in 1999 Entry Draft for Mike Peluso and Ricard Persson, November 26, 1996. Traded to **San Jose** by **New Jersey** with John MacLean for Doug Bodger and Dody Wood, December 7, 1997. Traded to **New Jersey** by **San Jose** for future considerations, August 26, 1998. Claimed by **Washington** from **New Jersey** in Waiver Draft, September 27, 1999. Traded to **New Jersey** by **Washington** for future considerations, October 5, 1999. Signed as a free agent by **NY Islanders**, July 5, 2001.

SVARTVADET, Per

(svahrt-VAH-deht, PAIR) **ATL.**

Center. Shoots left. 6'1", 195 lbs. Born, Solleftea, Sweden, May 17, 1975. Dallas' 5th choice, 139th overall, in 1993 Entry Draft.

Season	Club	League	GP	G	A	Pts	PIM	PP	SH	GW	S	%	+/-	TF	F%	H	SB	Min	GP	G	A	Pts	PIM	PP	SH	GW
1990-91	Solleftea HK	Swede-2	5	1	0	1	0																			
1991-92	MoDo Jr.	Swede-Jr.	30	17	19	36	36																			
1992-93	MoDo Jr.	Swede-Jr.	14	5	10	15	18																			
	MoDo	Sweden	2	0	0	0	0																			
1993-94	MoDo Jr.	Swede-Jr.	12	7	12	19	6																			
	MoDo	Sweden	36	2	1	3	4												11	0	0	0	6			
1994-95	MoDo	Sweden	40	6	9	15	31																			
1995-96	MoDo	Sweden	40	9	14	23	26												8	2	3	5	0			
1996-97	MoDo	Sweden	50	7	18	25	38																			
1997-98	MoDo	Sweden	46	6	12	18	28												7	3	2	5	2			
1998-99	MoDo	Sweden	50	9	23	32	30												13	3	6	9	6			
99-2000	**Atlanta**	**NHL**	**38**	**3**	**4**	**7**	**6**	**0**	**0**	**0**	**36**	**8.3**	**-8**	**452**	**45.6**	**29**	**16**	**13:24**								
	Orlando	IHL	27	4	6	10	10												5	0	1	1	0			
2000-01	**Atlanta**	**NHL**	**69**	**10**	**11**	**21**	**20**	**0**	**2**	**1**	**98**	**10.2**	**-6**	**902**	**42.5**	**41**	**34**	**14:42**								
2001-02	**Atlanta**	**NHL**	**78**	**3**	**12**	**15**	**24**	**0**	**0**	**0**	**80**	**3.8**	**-12**	**347**	**45.0**	**62**	**38**	**13:22**								
	NHL Totals		**185**	**16**	**27**	**43**	**50**	**0**	**2**	**1**	**214**	**7.5**		**1701**	**43.8**	**132**	**88**	**13:52**								

Traded to **Atlanta** by **Dallas** for Ottawa's 6th round choice (previously acquired, Dallas selected Justin Cox) in 1999 Entry Draft, June 26, 1999.

SVEHLA, Robert

(SHVEH-lah, RAW-buhrt) **TOR.**

Defense. Shoots right. 6'1", 210 lbs. Born, Martin, Czech., January 2, 1969. Calgary's 4th choice, 78th overall, in 1992 Entry Draft.

Season	Club	League	GP	G	A	Pts	PIM	PP	SH	GW	S	%	+/-	TF	F%	H	SB	Min	GP	G	A	Pts	PIM	PP	SH	GW
1989-90	Dukla Trencin	Czech	29	4	3	7	....																			
1990-91	Dukla Trencin	Czech	52	16	9	25	62																			
1991-92	Dukla Trencin	Czech	51	23	28	51	74																			
	Czechoslovakia	Olympics	8	2	1	3	8																			
1992-93	Malmo IF	Sweden	40	19	10	29	86												6	0	1	1	14			
1993-94	Malmo IF	Sweden	37	14	25	39	*127												10	5	1	6	23			
	Slovakia	Olympics	8	2	4	6	26																			
1994-95	Malmo IF	Sweden	32	11	13	24	83												9	2	3	5	6			
	Florida	**NHL**	**5**	**1**	**1**	**2**	**0**	**1**	**0**	**0**	**6**	**16.7**	**3**													
1995-96	**Florida**	**NHL**	**81**	**8**	**49**	**57**	**94**	**7**	**0**	**0**	**146**	**5.5**	**-3**						**22**	**0**	**6**	**6**	**32**	**0**	**0**	**0**
1996-97	**Florida**	**NHL**	**82**	**13**	**32**	**45**	**86**	**5**	**0**	**3**	**159**	**8.2**	**2**						**5**	**1**	**4**	**5**	**4**	**1**	**0**	**0**
1997-98	**Florida**	**NHL**	**79**	**9**	**34**	**43**	**113**	**3**	**0**	**0**	**144**	**6.3**	**-3**													
	Slovakia	Olympics	2	0	1	1	0																			
1998-99	**Florida**	**NHL**	**80**	**8**	**29**	**37**	**83**	**4**	**0**	**0**	**157**	**5.1**	**-13**	**2**	**0.0**	**101**	**92**	**24:45**								
99-2000	**Florida**	**NHL**	**82**	**9**	**40**	**49**	**64**	**3**	**0**	**1**	**143**	**6.3**	**23**	**1**	**0.0**	**97**	**169**	**24:33**	**4**	**0**	**1**	**1**	**4**	**0**	**0**	**0**
2000-01	**Florida**	**NHL**	**82**	**6**	**22**	**28**	**76**	**0**	**0**	**0**	**121**	**5.0**	**-8**	**1**	**0.0**	**354**	**130**	**25:34**								
2001-02	**Florida**	**NHL**	**82**	**7**	**22**	**29**	**87**	**3**	**0**	**0**	**119**	**5.9**	**-19**	**1**	**0.0**	**386**	**132**	**25:41**								
	NHL Totals		**573**	**61**	**229**	**290**	**603**	**26**	**0**	**4**	**995**	**6.1**		**5**	**0.0**	**938**	**523**	**25:08**	**31**	**1**	**11**	**12**	**40**	**1**	**0**	**0**

Played in NHL All-Star Game (1997)

Traded to **Florida** by **Calgary** with Magnus Svensson for Florida's 3rd round choice (Dmitri Vlasenkov) in 1996 Entry Draft and 4th round choice (Ryan Ready) in 1997 Entry Draft, September 29, 1994. Traded to **Toronto** by **Florida** for Dmitry Yushkevich, July 18, 2002.

SVOBODA, Jaroslav

(svah-BOH-duh, YAR-oh-slawf) **CAR.**

Left wing. Shoots left. 6'2", 190 lbs. Born, Cervenka, Czech., June 1, 1980. Carolina's 8th choice, 208th overall, in 1998 Entry Draft.

Season	Club	League	GP	G	A	Pts	PIM	PP	SH	GW	S	%	+/-	TF	F%	H	SB	Min	GP	G	A	Pts	PIM	PP	SH	GW
1995-96	HC Olomouc Jr.	Czech-Jr.	40	13	15	28	....																			
1996-97	HC Olomouc Jr.	Czech-Jr.	39	19	14	33	....																			
1997-98	HC Olomouc Jr.	Czech-Jr.	36	14	21	35	....																			
	HC Olomouc	Czech-2	13	0	1	1	....																			
1998-99	Kootenay Ice	WHL	54	26	33	59	46												7	2	2	4	11			
99-2000	Kootenay Ice	WHL	56	23	43	66	97												21	*15	13	*28	51			
2000-01	Cincinnati	IHL	52	4	10	14	25																			
2001-02	Lowell	AHL	66	12	16	28	58																			
	Carolina	**NHL**	**10**	**2**	**2**	**4**	**2**	**0**	**0**	**0**	**12**	**16.7**	**0**	**1**	**0.0**	**12**	**1**	**9:23**	**23**	**1**	**4**	**5**	**28**	**1**	**0**	**1**
	NHL Totals		**10**	**2**	**2**	**4**	**2**	**0**	**0**	**0**	**12**	**16.7**		**1**	**0.0**	**12**	**1**	**9:23**	**23**	**1**	**4**	**5**	**28**	**1**	**0**	**1**

SVOBODA, Petr

(svah-BOH-duh, PEE-tuhr) **TOR.**

Defense. Shoots right. 6'3", 200 lbs. Born, Jihlava, Czech., June 20, 1980. Toronto's 2nd choice, 35th overall, in 1998 Entry Draft.

Season	Club	League	GP	G	A	Pts	PIM	PP	SH	GW	S	%	+/-	TF	F%	H	SB	Min	GP	G	A	Pts	PIM	PP	SH	GW
1995-96	Dukla Jihlava Jr.	Czech-Jr.	38	4	12	16	50																			
1996-97	Dukla Jihlava Jr.	Czech-Jr.	29	1	3	4	....																			
1997-98	Dukla Jihlava Jr.	Czech-Jr.	12	0	2	2	....																			
	Havlickuv Brod	Czech-2	18	1	2	3	16																			
	HC Dukla Jihlava	Czech	1	0	0	0	0																			
1998-99	HC Dukla Jihlava	Czech	40	1	5	6	28																			
	Dukla Jihlava	EuroHL	5	1	0	1	4																			
99-2000	Trinec	Czech	46	1	2	3	44												3	0	0	0	0			
2000-01	**Toronto**	**NHL**	**18**	**1**	**2**	**3**	**10**	**1**	**0**	**0**	**17**	**5.9**	**-5**	**0**	**0.0**	**46**	**23**	**0:00**								
	St. John's	AHL	38	7	7	14	48												4	0	0	0	4			
2001-02	St. John's	AHL	74	3	10	13	58												11	0	1	1	6			
	NHL Totals		**18**	**1**	**2**	**3**	**10**	**1**	**0**	**0**	**17**	**5.9**		**0**	**0.0**	**46**	**23**									

SWANSON, Brian

(SWAHN-suhn, BRIGH-uhn) **EDM.**

Center. Shoots left. 5'10", 185 lbs. Born, Eagle River, AK, March 24, 1976. San Jose's 5th choice, 115th overall, in 1994 Entry Draft.

Season	Club	League	GP	G	A	Pts	PIM	PP	SH	GW	S	%	+/-	TF	F%	H	SB	Min	GP	G	A	Pts	PIM	PP	SH	GW
1991-92	Anchorage	AAHL	50	35	40	75	10																			
1992-93	Anchorage	AAHL	45	40	50	90	12																			
1993-94	Omaha Lancers	USHL	47	38	42	80	40																			
1994-95	Omaha Lancers	USHL	33	14	35	49	12																			
1995-96	Colorado College	WCHA	40	26	33	59	24																			
1996-97	Colorado College	WCHA	43	19	32	51	47																			
1997-98	Colorado College	WCHA	42	18	*38	*56	26																			
1998-99	Colorado College	WCHA	42	25	*41	66	28																			
	Hartford	AHL	4	0	0	0	4																			
99-2000	Hamilton	AHL	69	19	40	59	18												10	2	5	7	6			
2000-01	**Edmonton**	**NHL**	**16**	**1**	**1**	**2**	**6**	**0**	**0**	**0**	**8**	**12.5**	**-1**	**150**	**44.0**	**3**	**4**	**10:55**								
	Hamilton	AHL	49	18	29	47	20																			

Season	Club	League	GP	G	A	Pts	PIM	PP	SH	GW	S	%	+/-	TF	F%	H	SB	Min	Playoffs GP	G	A	Pts	PIM	PP	SH	GW
2001-02	**Edmonton**	**NHL**	**8**	**1**	**1**	**2**	**0**	**0**	**0**	**0**	**7**	**14.3**	**−1**	**59**	**54.2**	**0**	**1**	**10:10**	....	....	....	....	....	....	....	....
	Hamilton	AHL	65	34	39	73	26	....	....	....	....	....	....	....	....	....	....	....	15	7	6	13	6	....	....	....
	NHL Totals		**24**	**2**	**2**	**4**	**6**	**0**	**0**	**0**	**15**	**13.3**		**209**	**46.9**	**3**	**5**	**10:40**	....	....	....	....	....	....	....	....

USHL First All-Star Team (1994) • USHL Second Team All-Star (1995) • WCHA Second All-Star Team (1996) • WCHA Rookie of the Year (1996) • WCHA First All-Star Team (1997, 1998, 1999) • NCAA West Second All-American Team (1998) • NCAA West First All-American Team (1999) • AHL Second All-Star Team (2002)

Traded to **NY Rangers** by **San Jose** with Jayson More and San Jose's 4th round choice (later traded back to San Jose - San Jose selected Adam Colagiacomo) in 1997 Entry Draft for Marty McSorley, August 20, 1996. Signed as a free agent by **Edmonton**, August 19, 1999.

SWEENEY, Don
(SWEE-nee, DAWN) **BOS.**

Defense. Shoots left. 5'10", 185 lbs. Born, St. Stephen, N.B., August 17, 1966. Boston's 8th choice, 166th overall, in 1984 Entry Draft.

Season	Club	League	GP	G	A	Pts	PIM	PP	SH	GW	S	%	+/-	TF	F%	H	SB	Min	Playoffs GP	G	A	Pts	PIM	PP	SH	GW
1983-84	South St. Paul	Hi-School	22	33	26	59	....	....	....	....	....	....	....	....	....	....	....	....	....	....	....	....	....	....	....	....
1984-85	Harvard Crimson	ECAC	29	3	7	10	30	....	....	....	....	....	....	....	....	....	....	....	....	....	....	....	....	....	....	....
1985-86	Harvard Crimson	ECAC	31	4	5	9	12	....	....	....	....	....	....	....	....	....	....	....	....	....	....	....	....	....	....	....
1986-87	Harvard Crimson	ECAC	34	7	4	11	22	....	....	....	....	....	....	....	....	....	....	....	....	....	....	....	....	....	....	....
1987-88	Harvard Crimson	ECAC	30	6	23	29	37	....	....	....	....	....	....	....	....	....	....	....	....	....	....	....	....	....	....	....
	Maine Mariners	AHL	....	....	....	....	....	....	....	....	....	....	....	....	....	....	....	....	6	1	3	4	0	....	....	....
1988-89	**Boston**	**NHL**	**36**	**3**	**5**	**8**	**20**	**0**	**0**	**0**	**35**	**8.6**	**−6**	....	....	....	....	....	....	....	....	....	....	....	....	....
	Maine Mariners	AHL	42	8	17	25	24	....	....	....	....	....	....	....	....	....	....	....	....	....	....	....	....	....	....	....
1989-90	**Boston**	**NHL**	**58**	**3**	**5**	**8**	**58**	**0**	**0**	**0**	**49**	**6.1**	**11**	....	....	....	....	....	**21**	**1**	**5**	**6**	**18**	**1**	**0**	**0**
	Maine Mariners	AHL	11	0	8	8	8	....	....	....	....	....	....	....	....	....	....	....	....	....	....	....	....	....	....	....
1990-91	**Boston**	**NHL**	**77**	**8**	**13**	**21**	**67**	**0**	**1**	**3**	**102**	**7.8**	**2**	....	....	....	....	....	**19**	**3**	**0**	**3**	**25**	**0**	**0**	**0**
1991-92	**Boston**	**NHL**	**75**	**3**	**11**	**14**	**74**	**0**	**0**	**1**	**92**	**3.3**	**−9**	....	....	....	....	....	**15**	**0**	**0**	**0**	**10**	**0**	**0**	**0**
1992-93	**Boston**	**NHL**	**84**	**7**	**27**	**34**	**68**	**0**	**1**	**0**	**107**	**6.5**	**34**	....	....	....	....	....	**4**	**0**	**0**	**0**	**4**	**0**	**0**	**0**
1993-94	**Boston**	**NHL**	**75**	**6**	**15**	**21**	**50**	**1**	**2**	**2**	**136**	**4.4**	**29**	....	....	....	....	....	**12**	**2**	**1**	**3**	**4**	**0**	**0**	**1**
1994-95	**Boston**	**NHL**	**47**	**3**	**19**	**22**	**24**	**1**	**0**	**2**	**102**	**2.9**	**6**	....	....	....	....	....	**5**	**0**	**0**	**0**	**4**	**0**	**0**	**0**
1995-96	**Boston**	**NHL**	**77**	**4**	**24**	**28**	**42**	**2**	**0**	**3**	**142**	**2.8**	**−4**	....	....	....	....	....	**5**	**0**	**2**	**2**	**6**	**0**	**0**	**0**
1996-97	**Boston**	**NHL**	**82**	**3**	**23**	**26**	**39**	**0**	**0**	**0**	**113**	**2.7**	**−5**	....	....	....	....	....	....	....	....	....	....	....	....	....
1997-98	**Boston**	**NHL**	**59**	**1**	**15**	**16**	**24**	**0**	**0**	**0**	**55**	**1.8**	**12**	....	....	....	....	....	....	....	....	....	....	....	....	....
1998-99	**Boston**	**NHL**	**81**	**2**	**10**	**12**	**64**	**0**	**0**	**0**	**79**	**2.5**	**14**	**0**	**0.0**	**205**	**85**	**19:31**	**11**	**3**	**0**	**3**	**6**	**1**	**0**	**0**
99-2000	**Boston**	**NHL**	**81**	**1**	**13**	**14**	**48**	**0**	**0**	**0**	**82**	**1.2**	**−14**	**1**	**0.0**	**301**	**84**	**21:08**	....	....	....	....	....	....	....	....
2000-01	**Boston**	**NHL**	**72**	**2**	**10**	**12**	**26**	**1**	**0**	**1**	**60**	**3.3**	**−1**	**0**	**0.0**	**182**	**82**	**19:16**	....	....	....	....	....	....	....	....
2001-02	**Boston**	**NHL**	**81**	**3**	**15**	**18**	**35**	**1**	**0**	**0**	**70**	**4.3**	**22**	**0**	**0.0**	**144**	**102**	**20:09**	**6**	**0**	**1**	**1**	**2**	**0**	**0**	**0**
	NHL Totals		**985**	**49**	**205**	**254**	**639**	**6**	**4**	**12**	**1224**	**4.0**		**1**	**0.0**	**832**	**353**	**20:02**	**98**	**9**	**9**	**18**	**79**	**2**	**0**	**1**

ECAC First All-Star Team (1988) • NCAA East All-American Team (1988)

SYDOR, Darryl
(sih-DOHR, DAIR-ihl) **DAL.**

Defense. Shoots left. 6'1", 205 lbs. Born, Edmonton, Alta., May 13, 1972. Los Angeles' 1st choice, 7th overall, in 1990 Entry Draft.

Season	Club	League	GP	G	A	Pts	PIM	PP	SH	GW	S	%	+/-	TF	F%	H	SB	Min	Playoffs GP	G	A	Pts	PIM	PP	SH	GW
1985-86	Genstar Cement	AAHA	34	20	17	37	60	....	....	....	....	....	....	....	....	....	....	....	....	....	....	....	....	....	....	....
1986-87	Genstar Cement	AAHA	36	15	20	35	60	....	....	....	....	....	....	....	....	....	....	....	....	....	....	....	....	....	....	....
1987-88	Edmonton Mets	AJHL	38	10	11	21	54	....	....	....	....	....	....	....	....	....	....	....	....	....	....	....	....	....	....	....
1988-89	Kamloops Blazers	WHL	65	12	14	26	86	....	....	....	....	....	....	....	....	....	....	....	15	1	4	5	19	....	....	....
1989-90	Kamloops Blazers	WHL	67	29	66	95	129	....	....	....	....	....	....	....	....	....	....	....	17	2	9	11	28	....	....	....
1990-91	Kamloops Blazers	WHL	66	27	78	105	88	....	....	....	....	....	....	....	....	....	....	....	12	3	*22	25	10	....	....	....
1991-92	Kamloops Blazers	WHL	29	9	39	48	33	....	....	....	....	....	....	....	....	....	....	....	17	3	15	18	18	....	....	....
	Los Angeles	**NHL**	**18**	**1**	**5**	**6**	**22**	**0**	**0**	**0**	**18**	**5.6**	**−3**	....	....	....	....	....	....	....	....	....	....	....	....	....
1992-93	**Los Angeles**	**NHL**	**80**	**6**	**23**	**29**	**63**	**0**	**0**	**1**	**112**	**5.4**	**−2**	....	....	....	....	....	**24**	**3**	**8**	**11**	**16**	**2**	**0**	**0**
1993-94	**Los Angeles**	**NHL**	**84**	**8**	**27**	**35**	**94**	**1**	**0**	**0**	**146**	**5.5**	**−9**	....	....	....	....	....	....	....	....	....	....	....	....	....
1994-95	**Los Angeles**	**NHL**	**48**	**4**	**19**	**23**	**36**	**3**	**0**	**0**	**96**	**4.2**	**−2**	....	....	....	....	....	....	....	....	....	....	....	....	....
1995-96	**Los Angeles**	**NHL**	**58**	**1**	**11**	**12**	**34**	**1**	**0**	**0**	**84**	**1.2**	**−11**	....	....	....	....	....	....	....	....	....	....	....	....	....
	Dallas	**NHL**	**26**	**2**	**6**	**8**	**41**	**1**	**0**	**0**	**33**	**6.1**	**−1**	....	....	....	....	....	....	....	....	....	....	....	....	....
1996-97	**Dallas**	**NHL**	**82**	**8**	**40**	**48**	**51**	**2**	**0**	**2**	**142**	**5.6**	**37**	....	....	....	....	....	**7**	**0**	**2**	**2**	**0**	**0**	**0**	**0**
1997-98	**Dallas**	**NHL**	**79**	**11**	**35**	**46**	**51**	**4**	**1**	**1**	**166**	**6.6**	**17**	....	....	....	....	....	**17**	**0**	**5**	**5**	**14**	**0**	**0**	**0**
1998-99♦	**Dallas**	**NHL**	**74**	**14**	**34**	**48**	**50**	**9**	**0**	**2**	**163**	**8.6**	**−1**	**1**	**100.0**	**83**	**83**	**21:16**	**23**	**3**	**9**	**12**	**16**	**1**	**0**	**1**
99-2000	**Dallas**	**NHL**	**74**	**8**	**26**	**34**	**32**	**5**	**0**	**1**	**132**	**6.1**	**6**	**1**	**0.0**	**85**	**76**	**23:09**	**23**	**1**	**6**	**7**	**6**	**0**	**0**	**0**
2000-01	**Dallas**	**NHL**	**81**	**10**	**37**	**47**	**34**	**8**	**0**	**1**	**140**	**7.1**	**5**	**1**	**0.0**	**72**	**86**	**21:25**	**10**	**1**	**3**	**4**	**0**	**1**	**0**	**0**
2001-02	**Dallas**	**NHL**	**78**	**4**	**29**	**33**	**50**	**2**	**0**	**0**	**183**	**2.2**	**3**	**0**	**0.0**	**85**	**83**	**21:07**	....	....	....	....	....	....	....	....
	NHL Totals		**782**	**77**	**292**	**369**	**558**	**36**	**1**	**8**	**1415**	**5.4**		**3**	**33.3**	**325**	**328**	**21:43**	**104**	**8**	**33**	**41**	**52**	**4**	**0**	**1**

WHL West First All-Star Team (1990, 1991, 1992) • Played in NHL All-Star Game (1998, 1999)

Traded to **Dallas** by **LA Kings** with LA Kings' 5th round choice (Ryan Christie) in 1996 Entry Draft for Shane Churla and Doug Zmolek, February 17, 1996.

SYKORA, Michal
(SEE-koh-ra, MIHK-al)

Defense. Shoots left. 6'5", 225 lbs. Born, Pardubice, Czech., July 5, 1973. San Jose's 6th choice, 123rd overall, in 1992 Entry Draft.

Season	Club	League	GP	G	A	Pts	PIM	PP	SH	GW	S	%	+/-	TF	F%	H	SB	Min	Playoffs GP	G	A	Pts	PIM	PP	SH	GW
1990-91	Pardubice Jr.	Czech-Jr.	40	17	26	43	45	....	....	....	....	....	....	....	....	....	....	....	....	....	....	....	....	....	....	....
	Tesla Pardubice	Czech	2	0	0	0	....	....	....	....	....	....	....	....	....	....	....	....	....	....	....	....	....	....	....	....
1991-92	Tacoma Rockets	WHL	61	13	23	36	46	....	....	....	....	....	....	....	....	....	....	....	4	0	2	2	2	....	....	....
1992-93	Tacoma Rockets	WHL	70	23	50	73	73	....	....	....	....	....	....	....	....	....	....	....	7	4	8	12	2	....	....	....
1993-94	**San Jose**	**NHL**	**22**	**1**	**4**	**5**	**14**	**0**	**0**	**0**	**22**	**4.5**	**−4**	....	....	....	....	....	....	....	....	....	....	....	....	....
	Kansas City	IHL	47	5	11	16	30	....	....	....	....	....	....	....	....	....	....	....	....	....	....	....	....	....	....	....
1994-95	Kansas City	IHL	36	1	10	11	30	....	....	....	....	....	....	....	....	....	....	....	....	....	....	....	....	....	....	....
	San Jose	**NHL**	**16**	**0**	**4**	**4**	**10**	**0**	**0**	**0**	**6**	**0.0**	**6**	....	....	....	....	....	....	....	....	....	....	....	....	....
1995-96	**San Jose**	**NHL**	**79**	**4**	**16**	**20**	**54**	**1**	**0**	**0**	**80**	**5.0**	**−14**	....	....	....	....	....	....	....	....	....	....	....	....	....
1996-97	**San Jose**	**NHL**	**35**	**2**	**5**	**7**	**59**	**1**	**0**	**0**	**39**	**5.1**	**0**	....	....	....	....	....	....	....	....	....	....	....	....	....
	Chicago	**NHL**	**28**	**1**	**9**	**10**	**10**	**0**	**0**	**0**	**38**	**2.6**	**4**	....	....	....	....	....	**1**	**0**	**0**	**0**	**0**	**0**	**0**	**0**
1997-98	**Chicago**	**NHL**	**28**	**1**	**3**	**4**	**12**	**0**	**0**	**0**	**35**	**2.9**	**−10**	....	....	....	....	....	....	....	....	....	....	....	....	....
	Indianapolis Ice	IHL	6	0	0	0	4	....	....	....	....	....	....	....	....	....	....	....	....	....	....	....	....	....	....	....
	Pardubice	Czech	1	1	0	1	2	....	....	....	....	....	....	....	....	....	....	....	....	....	....	....	....	....	....	....
1998-99	HC Sparta Praha	Czech	26	4	9	13	38	....	....	....	....	....	....	....	....	....	....	....	8	2	0	2	....	....	....	....
	HC Sparta Praha	EuroHL	2	2	2	4	4	....	....	....	....	....	....	....	....	....	....	....	2	2	2	4	0	....	....	....
	Tampa Bay	**NHL**	**10**	**1**	**2**	**3**	**0**	**0**	**0**	**1**	**24**	**4.2**	**−7**	**0**	**0.0**	**10**	**8**	**16:20**	....	....	....	....	....	....	....	....
99-2000	HC Sparta Praha	Czech	48	11	14	25	89	....	....	....	....	....	....	....	....	....	....	....	9	5	3	8	8	....	....	....
2000-01	**Philadelphia**	**NHL**	**49**	**5**	**11**	**16**	**26**	**1**	**1**	**1**	**71**	**7.0**	**9**	**1**	**0.0**	**45**	**58**	**16:27**	**6**	**0**	**1**	**1**	**0**	**0**	**0**	**0**
2001-02	Pardubice	Czech	49	7	11	18	111	....	....	....	....	....	....	....	....	....	....	....	6	2	2	4	26	....	....	....
	Czech Republic	Olympics	4	0	0	0	0	....	....	....	....	....	....	....	....	....	....	....	....	....	....	....	....	....	....	....
	NHL Totals		**267**	**15**	**54**	**69**	**185**	**3**	**1**	**2**	**315**	**4.8**		**1**	**0.0**	**55**	**66**	**16:26**	**7**	**0**	**1**	**1**	**0**	**0**	**0**	**0**

WHL West First All-Star Team (1993)

Traded to **Chicago** by **San Jose** with Chris Terreri and Ulf Dahlen for Ed Belfour, January 25, 1997. Traded to **Tampa Bay** by **Chicago** for Mark Fitzpatrick and Tampa Bay's 4th round choice (later traded to Montreal - Montreal selected Chris Dyment) in 1999 Entry Draft, July 17, 1998. Signed as a free agent by **Philadelphia**, July 6, 2000.

SYKORA, Petr
(SEE-koh-ra, PEE-tuhr) **WSH.**

Center. Shoots right. 6'3", 206 lbs. Born, Pardubice, Czech., December 21, 1978. Detroit's 2nd choice, 76th overall, in 1997 Entry Draft.

Season	Club	League	GP	G	A	Pts	PIM	PP	SH	GW	S	%	+/-	TF	F%	H	SB	Min	Playoffs GP	G	A	Pts	PIM	PP	SH	GW
1994-95	HC Pardubice Jr.	Czech-Jr.	38	35	33	68	....	....	....	....	....	....	....	....	....	....	....	....	....	....	....	....	....	....	....	....
1995-96	HC Pardubice Jr.	Czech-Jr.	16	26	17	43	....	....	....	....	....	....	....	....	....	....	....	....	....	....	....	....	....	....	....	....
1996-97	HC Pardubice Jr.	Czech-Jr.	12	14	4	18	....	....	....	....	....	....	....	....	....	....	....	....	....	....	....	....	....	....	....	....
	Pardubice	Czech	29	1	3	4	4	....	....	....	....	....	....	....	....	....	....	....	....	....	....	....	....	....	....	....
1997-98	Pardubice	Czech	39	4	5	9	8	....	....	....	....	....	....	....	....	....	....	....	3	0	0	0	....	....	....	....
1998-99	**Nashville**	**NHL**	**2**	**0**	**0**	**0**	**0**	**0**	**0**	**0**	**2**	**0.0**	**−1**	**11**	**45.5**	**0**	**0**	**8:19**	....	....	....	....	....	....	....	....
	Milwaukee	IHL	73	14	15	29	50	....	....	....	....	....	....	....	....	....	....	....	2	1	1	2	0	....	....	....
99-2000	Milwaukee	IHL	3	0	1	1	2	....	....	....	....	....	....	....	....	....	....	....	....	....	....	....	....	....	....	....
	Pardubice	Czech	36	7	13	20	49	....	....	....	....	....	....	....	....	....	....	....	3	0	0	0	2	....	....	....
2000-01	Pardubice	Czech	47	26	18	44	42	....	....	....	....	....	....	....	....	....	....	....	7	5	3	8	6	....	....	....
2001-02	Pardubice	Czech	32	18	8	26	72	....	....	....	....	....	....	....	....	....	....	....	6	1	2	3	26	....	....	....
	NHL Totals		**2**	**0**	**0**	**0**	**0**	**0**	**0**	**0**	**2**	**0.0**		**11**	**45.5**	**0**	**0**	**8:19**	....	....	....	....	....	....	....	....

Traded to **Nashville** by **Detroit** with Detroit's 3rd round choice (later traded to Edmonton - Edmonton selected Mike Comrie) and 4th round compensatory choice (Alexander Krevsun) in 1999 Entry Draft for Doug Brown, July 14, 1998. Traded to **Washington** by **Nashville** for a conditional 3rd round choice in 2003 Entry Draft, June 22, 2002.

SYKORA, Petr
(SEE-koh-ra, PEE-tuhr) **ANA.**

Right wing. Shoots left. 6', 190 lbs. Born, Plzen, Czech., November 19, 1976. New Jersey's 1st choice, 18th overall, in 1995 Entry Draft.

			Regular Season																Playoffs							
Season	Club	League	GP	G	A	Pts	PIM	PP	SH	GW	S	%	+/-	TF	F%	H	SB	Min	GP	G	A	Pts	PIM	PP	SH	GW
1991-92	Plzen Jr.	Czech-Jr.	30	50	50	100																				
1992-93	HC Skoda Plzen	Czech	19	12	5	17																				
1993-94	HC Skoda Plzen	Czech	37	10	16	26													4	0	1	1				
	Cleveland	IHL	13	4	5	9	8																			
1994-95	Detroit Vipers	IHL	29	12	17	29	16																			
1995-96	**New Jersey**	**NHL**	**63**	**18**	**24**	**42**	**32**	**8**	**0**	**3**	**128**	**14.1**	**7**													
	Albany	AHL	5	4	1	5	0																			
1996-97	**New Jersey**	**NHL**	**19**	**1**	**2**	**3**	**4**	**0**	**0**	**0**	**26**	**3.8**	**–8**						**2**	**0**	**0**	**0**	**2**	**0**	**0**	**0**
	Albany	AHL	43	20	25	45	48												4	1	4	5	2			
1997-98	**New Jersey**	**NHL**	**58**	**16**	**20**	**36**	**22**	**3**	**1**	**4**	**130**	**12.3**	**0**						**2**	**0**	**0**	**0**	**0**	**0**	**0**	**0**
	Albany	AHL	2	4	1	5	0																			
1998-99	**New Jersey**	**NHL**	**80**	**29**	**43**	**72**	**22**	**15**	**0**	**7**	**222**	**13.1**	**16**	**33**	**33.3**	**65**	**16**	**16:14**	**7**	**3**	**3**	**6**	**4**	**0**	**0**	**1**
99-2000♦	**New Jersey**	**NHL**	**79**	**25**	**43**	**68**	**26**	**5**	**1**	**4**	**222**	**11.3**	**24**	**47**	**61.7**	**33**	**22**	**17:06**	**23**	**9**	**8**	**17**	**10**	**1**	**0**	**3**
2000-01	**New Jersey**	**NHL**	**73**	**35**	**46**	**81**	**32**	**9**	**2**	**3**	**249**	**14.1**	**36**	**15**	**33.3**	**19**	**21**	**17:44**	**25**	**10**	**12**	**22**	**12**	**2**	**2**	**2**
2001-02	**New Jersey**	**NHL**	**73**	**21**	**27**	**48**	**44**	**4**	**0**	**4**	**194**	**10.8**	**12**	**1**	**0.0**	**35**	**17**	**17:51**	**4**	**0**	**1**	**1**	**0**	**0**	**0**	**0**
	Czech Republic	Olympics	4	1	0	1	0																			
	NHL Totals		**445**	**145**	**205**	**350**	**182**	**44**	**4**	**25**	**1171**	**12.4**		**96**	**16.7**	**152**	**76**	**16:57**	**63**	**22**	**24**	**46**	**28**	**3**	**2**	**6**

NHL All-Rookie Team (1996)
Traded to **Anaheim** by **New Jersey** with Mike Commodore, Jean-Francois Damphousse and Igor Pohanka for Jeff Friesen, Oleg Tverdovsky and Maxim Balmochnykh, July 6, 2002.

TALLINDER, Henrik
(tah-LIHN-duhr, HEHN-rihk) **BUF.**

Defense. Shoots left. 6'3", 210 lbs. Born, Stockholm, Sweden, January 10, 1979. Buffalo's 2nd choice, 48th overall, in 1997 Entry Draft.

Season	Club	League	GP	G	A	Pts	PIM	PP	SH	GW	S	%	+/-	TF	F%	H	SB	Min	GP	G	A	Pts	PIM	PP	SH	GW
1996-97	AIK Solna Jr.	Swede-Jr.	40	4	13	17	55																			
	AIK Solna	Sweden	1	0	0	0	0																			
1997-98	AIK Solna	Sweden	34	0	0	0	26																			
1998-99	AIK Solna	Sweden	36	0	0	0	30																			
99-2000	AIK Solna	Sweden	50	0	2	2	59																			
2000-01	TPS Turku	Finland	56	5	9	14	62												10	2	1	3	8			
2001-02	**Buffalo**	**NHL**	**2**	**0**	**0**	**0**	**0**	**0**	**0**	**0**	**4**	**0.0**	**–1**	**0**	**0.0**	**1**	**3**	**18:10**								
	Rochester	AHL	73	6	14	20	26												2	0	0	0	0			
	NHL Totals		**2**	**0**	**0**	**0**	**0**	**0**	**0**	**0**	**4**	**0.0**		**0**	**0.0**	**1**	**3**	**18:10**								

TAMER, Chris
(TAY-muhr, KRIHS) **ATL.**

Defense. Shoots left. 6'2", 205 lbs. Born, Dearborn, MI, November 17, 1970. Pittsburgh's 3rd choice, 68th overall, in 1990 Entry Draft.

Season	Club	League	GP	G	A	Pts	PIM	PP	SH	GW	S	%	+/-	TF	F%	H	SB	Min	GP	G	A	Pts	PIM	PP	SH	GW
1987-88	Redford Royals	NAJHL	40	10	20	30	217																			
1988-89	Redford Royals	NAJHL	31	6	13	19	79																			
1989-90	U. of Michigan	CCHA	42	2	7	9	147																			
1990-91	U. of Michigan	CCHA	45	8	19	27	130																			
1991-92	U. of Michigan	CCHA	43	4	15	19	125																			
1992-93	U. of Michigan	CCHA	39	5	18	23	113																			
1993-94	**Pittsburgh**	**NHL**	**12**	**0**	**0**	**0**	**9**	**0**	**0**	**0**	**10**	**0.0**	**3**						**5**	**0**	**0**	**0**	**2**	**0**	**0**	**0**
	Cleveland	IHL	53	1	2	3	160																			
1994-95	Cleveland	IHL	48	4	10	14	204																			
	Pittsburgh	**NHL**	**36**	**2**	**0**	**2**	**82**	**0**	**0**	**0**	**26**	**7.7**	**0**						**4**	**0**	**0**	**0**	**18**	**0**	**0**	**0**
1995-96	**Pittsburgh**	**NHL**	**70**	**4**	**10**	**14**	**153**	**0**	**0**	**1**	**75**	**5.3**	**20**						**18**	**0**	**7**	**7**	**24**	**0**	**0**	**0**
1996-97	**Pittsburgh**	**NHL**	**45**	**2**	**4**	**6**	**131**	**0**	**1**	**0**	**56**	**3.6**	**–25**						**4**	**0**	**0**	**0**	**4**	**0**	**0**	**0**
1997-98	**Pittsburgh**	**NHL**	**79**	**0**	**7**	**7**	**181**	**0**	**0**	**0**	**55**	**0.0**	**4**						**6**	**0**	**1**	**1**	**4**	**0**	**0**	**0**
1998-99	**Pittsburgh**	**NHL**	**11**	**0**	**0**	**0**	**32**	**0**	**0**	**0**	**2**	**0.0**	**–2**	**0**	**0.0**	**6**	**8**	**5:59**								
	NY Rangers	**NHL**	**52**	**1**	**5**	**6**	**92**	**0**	**0**	**1**	**46**	**2.2**	**–12**	**0**	**0.0**	**88**	**77**	**15:25**								
99-2000	**Atlanta**	**NHL**	**69**	**2**	**8**	**10**	**91**	**0**	**0**	**0**	**61**	**3.3**	**–32**	**5**	**40.0**	**129**	**102**	**18:29**								
2000-01	**Atlanta**	**NHL**	**82**	**4**	**13**	**17**	**128**	**0**	**1**	**1**	**90**	**4.4**	**–1**	**1**	**0.0**	**170**	**134**	**19:42**								
2001-02	**Atlanta**	**NHL**	**78**	**3**	**3**	**6**	**111**	**0**	**1**	**0**	**66**	**4.5**	**–11**	**0**	**0.0**	**130**	**120**	**18:30**								
	NHL Totals		**534**	**18**	**50**	**68**	**1010**	**0**	**3**	**3**	**487**	**3.7**		**6**	**33.3**	**523**	**441**	**17:49**	**37**	**0**	**8**	**8**	**52**	**0**	**0**	**0**

Traded to **NY Rangers** by **Pittsburgh** with Petr Nedved and Sean Pronger for Alexei Kovalev and Harry York, November 25, 1998. Claimed by **Atlanta** from **NY Rangers** in Expansion Draft, June 25, 1999.

TANABE, David
(tuh-NA-bee, DAY-vihd) **CAR.**

Defense. Shoots right. 6'1", 190 lbs. Born, White Bear Lake, MN, July 19, 1980. Carolina's 1st choice, 16th overall, in 1999 Entry Draft.

Season	Club	League	GP	G	A	Pts	PIM	PP	SH	GW	S	%	+/-	TF	F%	H	SB	Min	GP	G	A	Pts	PIM	PP	SH	GW
1996-97	Hill-Murray	Hi-School	28	12	14	26																				
1997-98	Team USA	USDP-18	73	8	21	29	96																			
1998-99	U. of Wisconsin	WCHA	35	10	12	22	44																			
99-2000	**Carolina**	**NHL**	**31**	**4**	**0**	**4**	**14**	**3**	**0**	**0**	**28**	**14.3**	**–4**	**0**	**0.0**	**11**	**10**	**12:53**								
	Cincinnati	IHL	32	0	13	13	14												11	1	4	5	6			
2000-01	**Carolina**	**NHL**	**74**	**7**	**22**	**29**	**42**	**5**	**0**	**1**	**130**	**5.4**	**–9**	**0**	**0.0**	**52**	**50**	**17:55**	**6**	**2**	**0**	**2**	**12**	**2**	**0**	**0**
2001-02	**Carolina**	**NHL**	**78**	**1**	**15**	**16**	**35**	**0**	**0**	**0**	**113**	**0.9**	**–13**	**0**	**0.0**	**56**	**56**	**18:27**	**1**	**0**	**1**	**1**	**0**	**0**	**0**	**0**
	NHL Totals		**183**	**12**	**37**	**49**	**91**	**8**	**0**	**1**	**271**	**4.4**		**0**	**0.0**	**119**	**116**	**17:18**	**7**	**2**	**1**	**3**	**12**	**2**	**0**	**0**

WCHA All-Rookie Team (1999)

TANGUAY, Alex
(TAN-guay, AL-ehx) **COL.**

Center. Shoots left. 6', 190 lbs. Born, Ste-Justine, Que., November 21, 1979. Colorado's 1st choice, 12th overall, in 1998 Entry Draft.

Season	Club	League	GP	G	A	Pts	PIM	PP	SH	GW	S	%	+/-	TF	F%	H	SB	Min	GP	G	A	Pts	PIM	PP	SH	GW
1994-95	Cap-d-Madeleine	QAAA	1	0	1	1	0																			
1995-96	Cap-d-Madeleine	QAAA	44	29	34	63	64												5	2	4	6	14			
1996-97	Halifax	QMJHL	70	27	41	68	60												12	5	8	13	8			
1997-98	Halifax	QMJHL	51	47	38	85	32												5	7	6	13	4			
1998-99	Halifax	QMJHL	31	27	34	61	30												5	1	2	3	2			
	Hershey Bears	AHL	5	1	2	3	2												5	0	2	2	0			
99-2000	**Colorado**	**NHL**	**76**	**17**	**34**	**51**	**22**	**5**	**0**	**3**	**74**	**23.0**	**6**	**11**	**45.5**	**72**	**23**	**15:38**	**17**	**2**	**1**	**3**	**2**	**1**	**0**	**1**
2000-01♦	**Colorado**	**NHL**	**82**	**27**	**50**	**77**	**37**	**7**	**1**	**3**	**135**	**20.0**	**35**	**30**	**43.3**	**102**	**22**	**17:51**	**23**	**6**	**15**	**21**	**8**	**1**	**0**	**2**
2001-02	**Colorado**	**NHL**	**70**	**13**	**35**	**48**	**36**	**7**	**0**	**2**	**90**	**14.4**	**8**	**37**	**40.5**	**57**	**24**	**18:20**	**19**	**5**	**8**	**13**	**0**	**3**	**0**	**0**
	NHL Totals		**228**	**57**	**119**	**176**	**95**	**19**	**1**	**8**	**299**	**19.1**		**78**	**42.3**	**231**	**69**	**17:16**	**59**	**13**	**24**	**37**	**10**	**5**	**0**	**3**

QMJHL All-Rookie Team (1997)

TAPPER, Brad
(TA-puhr, BRAD) **ATL.**

Center. Shoots right. 6', 185 lbs. Born, Scarborough, Ont., April 28, 1978.

Season	Club	League	GP	G	A	Pts	PIM	PP	SH	GW	S	%	+/-	TF	F%	H	SB	Min	GP	G	A	Pts	PIM	PP	SH	GW
1996-97	Wexford Raiders	MTJHL	50	42	70	112	169																			
1997-98	RPI Engineers	ECAC	34	14	11	25	62																			
1998-99	RPI Engineers	ECAC	35	20	20	40	60																			
99-2000	RPI Engineers	ECAC	37	*31	20	51	81																			
2000-01	**Atlanta**	**NHL**	**16**	**2**	**3**	**5**	**6**	**0**	**0**	**0**	**21**	**9.5**	**1**	**0**	**0.0**	**11**	**2**	**12:44**								
	Orlando	IHL	45	7	9	16	39												2	0	0	0	2			
2001-02	**Atlanta**	**NHL**	**20**	**2**	**4**	**6**	**43**	**0**	**0**	**0**	**34**	**5.9**	**–3**	**3**	**66.7**	**21**	**7**	**13:21**								
	Chicago Wolves	AHL	50	14	12	26	62												19	3	4	7	42			
	NHL Totals		**36**	**4**	**7**	**11**	**49**	**0**	**0**	**0**	**55**	**7.3**		**3**	**66.7**	**32**	**9**	**13:05**								

ECAC First All-Star Team (2000) • NCAA East Second All-American Team (2000)
Signed as a free agent by **Atlanta**, April 11, 2000.

TARNSTROM, Dick
(TAHRN-struhm, DIHK) **PIT.**

Defense. Shoots left. 6'2", 200 lbs. Born, Sundbyberg, Sweden, January 20, 1975. NY Islanders' 12th choice, 272nd overall, in 1994 Entry Draft.

Season	Club	League	GP	G	A	Pts	PIM	PP	SH	GW	S	%	+/-	TF	F%	H	SB	Min	PO GP	PO G	PO A	PO Pts	PO PIM	PO PP	PO SH	PO GW
1992-93	AIK Solna	Sweden	3	0	0	0	0	....	....	....	....	....	....	....	....	....	....	....	....	....	....	....	....	....	....	....
1993-94	AIK Solna	Sweden	33	1	4	5	....	....	....	....	....	....	....	....	....	....	....	....	....	....	....	....	....	....	....	....
1994-95	AIK Solna	Swede-2	37	8	4	12	26	....	....	....	....	....	....	....	....	....	....	....	....	....	....	....	....	....	....	....
1995-96	AIK Solna	Sweden	40	0	5	5	32	....	....	....	....	....	....	....	....	....	....	....	....	....	....	....	....	....	....	....
1996-97	AIK Solna	Sweden	49	5	3	8	38	....	....	....	....	....	....	....	....	....	....	....	7	0	1	1	6	....	....	....
1997-98	AIK Solna	Sweden	45	2	12	14	30	....	....	....	....	....	....	....	....	....	....	....	....	....	....	....	....	....	....	....
1998-99	AIK Solna	Sweden	47	9	14	23	36	....	....	....	....	....	....	....	....	....	....	....	....	....	....	....	....	....	....	....
99-2000	AIK Solna	Sweden	42	7	15	22	20	....	....	....	....	....	....	....	....	....	....	....	....	....	....	....	....	....	....	....
2000-01	AIK Solna	Sweden	50	10	18	28	28	....	....	....	....	....	....	....	....	....	....	....	5	0	0	0	8	....	....	....
2001-02	**NY Islanders**	**NHL**	**62**	**3**	**16**	**19**	**38**	**0**	**0**	**0**	**59**	**5.1**	**-12**	**0**	**0.0**	**54**	**77**	**17:39**	**5**	**0**	**0**	**0**	**2**	**0**	**0**	**0**
	Bridgeport	AHL	9	0	2	2	2	....	....	....	....	....	....	....	....	....	....	....	....	....	....	....	....	....	....	....
	NHL Totals		**62**	**3**	**16**	**19**	**38**	**0**	**0**	**0**	**59**	**5.1**		**0**	**0.0**	**54**	**77**	**17:39**	**5**	**0**	**0**	**0**	**2**	**0**	**0**	**0**

Claimed on waivers by **Pittsburgh** from **NY Islanders**, August 6, 2002.

TAYLOR, Chris
(TAY-lohr, KRIHS) **BUF.**

Center. Shoots left. 6'2", 192 lbs. Born, Stratford, Ont., March 6, 1972. NY Islanders' 2nd choice, 27th overall, in 1990 Entry Draft.

Season	Club	League	GP	G	A	Pts	PIM	PP	SH	GW	S	%	+/-	TF	F%	H	SB	Min	PO GP	PO G	PO A	PO Pts	PO PIM	PO PP	PO SH	PO GW
1987-88	Stratford	OJHL-B	52	28	37	65	112	....	....	....	....	....	....	....	....	....	....	....	....	....	....	....	....	....	....	....
1988-89	London Knights	OHL	62	7	16	23	52	....	....	....	....	....	....	....	....	....	....	....	15	0	2	2	15	....	....	....
1989-90	London Knights	OHL	66	45	60	105	60	....	....	....	....	....	....	....	....	....	....	....	6	3	2	5	6	....	....	....
1990-91	London Knights	OHL	65	50	78	128	50	....	....	....	....	....	....	....	....	....	....	....	7	4	8	12	6	....	....	....
1991-92	London Knights	OHL	66	48	74	122	57	....	....	....	....	....	....	....	....	....	....	....	10	8	16	24	9	....	....	....
1992-93	Capital District	AHL	77	19	43	62	32	....	....	....	....	....	....	....	....	....	....	....	4	0	1	1	2	....	....	....
1993-94	Salt Lake	IHL	79	21	20	41	38	....	....	....	....	....	....	....	....	....	....	....	....	....	....	....	....	....	....	....
1994-95	Denver Grizzlies	IHL	78	38	48	86	47	....	....	....	....	....	....	....	....	....	....	....	14	7	6	13	10	....	....	....
	NY Islanders	**NHL**	**10**	**0**	**3**	**3**	**2**	**0**	**0**	**0**	**13**	**0.0**	**1**	....	....	....	....	....	....	....	....	....	....	....	....	....
1995-96	**NY Islanders**	**NHL**	**11**	**0**	**1**	**1**	**2**	**0**	**0**	**0**	**4**	**0.0**	**1**	....	....	....	....	....	....	....	....	....	....	....	....	....
	Utah Grizzlies	IHL	50	18	23	41	60	....	....	....	....	....	....	....	....	....	....	....	22	5	11	16	26	....	....	....
1996-97	**NY Islanders**	**NHL**	**1**	**0**	**0**	**0**	**0**	**0**	**0**	**0**	**1**	**0.0**	**0**	....	....	....	....	....	....	....	....	....	....	....	....	....
	Utah Grizzlies	IHL	71	27	40	67	24	....	....	....	....	....	....	....	....	....	....	....	7	1	2	3	0	....	....	....
1997-98	Utah Grizzlies	IHL	79	28	56	84	66	....	....	....	....	....	....	....	....	....	....	....	4	0	2	2	6	....	....	....
1998-99	**Boston**	**NHL**	**37**	**3**	**5**	**8**	**12**	**0**	**1**	**0**	**60**	**5.0**	**-3**	**512**	**53.7**	**52**	**16**	**14:24**	....	....	....	....	....	....	....	....
	Providence	AHL	21	6	11	17	6	....	....	....	....	....	....	....	....	....	....	....	....	....	....	....	....	....	....	....
	Las Vegas	IHL	14	3	12	15	2	....	....	....	....	....	....	....	....	....	....	....	....	....	....	....	....	....	....	....
99-2000	**Buffalo**	**NHL**	**11**	**1**	**1**	**2**	**2**	**0**	**0**	**0**	**15**	**6.7**	**-2**	**125**	**45.6**	**15**	**3**	**10:54**	**2**	**0**	**0**	**0**	**2**	**0**	**0**	**0**
	Rochester	AHL	49	21	28	49	21	....	....	....	....	....	....	....	....	....	....	....	....	....	....	....	....	....	....	....
2000-01	**Buffalo**	**NHL**	**14**	**0**	**2**	**2**	**6**	**0**	**0**	**0**	**21**	**0.0**	**1**	**138**	**50.7**	**28**	**3**	**11:08**	....	....	....	....	....	....	....	....
	Rochester	AHL	45	20	24	44	25	....	....	....	....	....	....	....	....	....	....	....	....	....	....	....	....	....	....	....
2001-02	Rochester	AHL	77	21	45	66	66	....	....	....	....	....	....	....	....	....	....	....	2	0	1	1	0	....	....	....
	NHL Totals		**84**	**4**	**12**	**16**	**24**	**0**	**1**	**0**	**114**	**3.5**		**775**	**51.9**	**95**	**22**	**13:02**	**2**	**0**	**0**	**0**	**2**	**0**	**0**	**0**

Signed as a free agent by **LA Kings**, July 25, 1997. Signed as a free agent by **Boston**, August 5, 1998. Signed as a free agent by **Buffalo**, August 13, 1999.

TAYLOR, Tim
(TAY-lohr, TIHM) **T.B.**

Center. Shoots left. 6'1", 189 lbs. Born, Stratford, Ont., February 6, 1969. Washington's 2nd choice, 36th overall, in 1988 Entry Draft.

Season	Club	League	GP	G	A	Pts	PIM	PP	SH	GW	S	%	+/-	TF	F%	H	SB	Min	PO GP	PO G	PO A	PO Pts	PO PIM	PO PP	PO SH	PO GW
1985-86	Stratford	OJHL-B	1	0	0	0	0	....	....	....	....	....	....	....	....	....	....	....	....	....	....	....	....	....	....	....
1986-87	Stratford	OJHL-B	31	25	26	51	51	....	....	....	....	....	....	....	....	....	....	....	....	....	....	....	....	....	....	....
	London Knights	OHL	34	7	9	16	11	....	....	....	....	....	....	....	....	....	....	....	....	....	....	....	....	....	....	....
1987-88	London Knights	OHL	64	46	50	96	66	....	....	....	....	....	....	....	....	....	....	....	12	9	9	18	26	....	....	....
1988-89	London Knights	OHL	61	34	80	114	93	....	....	....	....	....	....	....	....	....	....	....	21	*21	25	*46	58	....	....	....
1989-90	Baltimore	AHL	79	31	36	67	124	....	....	....	....	....	....	....	....	....	....	....	9	2	2	4	13	....	....	....
1990-91	Baltimore	AHL	79	25	42	67	75	....	....	....	....	....	....	....	....	....	....	....	5	0	1	1	4	....	....	....
1991-92	Baltimore	AHL	65	9	18	27	131	....	....	....	....	....	....	....	....	....	....	....	....	....	....	....	....	....	....	....
1992-93	Baltimore	AHL	41	15	16	31	49	....	....	....	....	....	....	....	....	....	....	....	....	....	....	....	....	....	....	....
	Hamilton Canucks	AHL	36	15	22	37	37	....	....	....	....	....	....	....	....	....	....	....	....	....	....	....	....	....	....	....
1993-94	**Detroit**	**NHL**	**1**	**1**	**0**	**1**	**0**	**0**	**0**	**0**	**4**	**25.0**	**-1**	....	....	....	....	....	....	....	....	....	....	....	....	....
	Adirondack	AHL	79	36	*81	*117	86	....	....	....	....	....	....	....	....	....	....	....	12	2	10	12	12	....	....	....
1994-95	**Detroit**	**NHL**	**22**	**0**	**4**	**4**	**16**	**0**	**0**	**0**	**21**	**0.0**	**3**	....	....	....	....	....	**6**	**0**	**1**	**1**	**12**	**0**	**0**	**0**
1995-96	**Detroit**	**NHL**	**72**	**11**	**14**	**25**	**39**	**1**	**1**	**4**	**81**	**13.6**	**11**	....	....	....	....	....	**18**	**0**	**4**	**4**	**4**	**0**	**0**	**0**
1996-97♦	**Detroit**	**NHL**	**44**	**3**	**4**	**7**	**52**	**0**	**1**	**0**	**44**	**6.8**	**-6**	....	....	....	....	....	**2**	**0**	**0**	**0**	**0**	**0**	**0**	**0**
1997-98	**Boston**	**NHL**	**79**	**20**	**11**	**31**	**57**	**1**	**3**	**0**	**127**	**15.7**	**-16**	....	....	....	....	....	**6**	**0**	**0**	**0**	**10**	**0**	**0**	**0**
1998-99	**Boston**	**NHL**	**49**	**4**	**7**	**11**	**55**	**0**	**0**	**1**	**76**	**5.3**	**-10**	**834**	**58.3**	**93**	**16**	**15:56**	**12**	**0**	**3**	**3**	**8**	**0**	**0**	**0**
99-2000	**NY Rangers**	**NHL**	**76**	**9**	**11**	**20**	**72**	**0**	**0**	**2**	**79**	**11.4**	**-4**	**1276**	**58.9**	**94**	**41**	**14:09**	....	....	....	....	....	....	....	....
2000-01	**NY Rangers**	**NHL**	**38**	**2**	**5**	**7**	**16**	**0**	**0**	**1**	**34**	**5.9**	**-6**	**292**	**59.3**	**34**	**18**	**8:57**	....	....	....	....	....	....	....	....
2001-02	**Tampa Bay**	**NHL**	**48**	**4**	**4**	**8**	**25**	**0**	**1**	**0**	**50**	**8.0**	**-2**	**559**	**54.6**	**21**	**29**	**13:29**	....	....	....	....	....	....	....	....
	NHL Totals		**429**	**54**	**60**	**114**	**332**	**2**	**6**	**8**	**516**	**10.5**		**2961**	**57.9**	**242**	**104**	**13:28**	**44**	**0**	**8**	**8**	**34**	**0**	**0**	**0**

AHL First All-Star Team (1994) • Won John B. Sollenberger Trophy (Top Scorer - AHL) (1994)

Traded to **Vancouver** by **Washington** for Eric Murano, January 29, 1993. Signed as a free agent by **Detroit**, July 28, 1993. Claimed by **Boston** from **Detroit** in NHL Waiver Draft, September 28, 1997. Signed as a free agent by **NY Rangers**, July 30, 1999. • Missed majority of 2000-01 season recovering from abdominal injury suffered in game vs. Phoenix, January 4, 2001. Traded to **Tampa Bay** by **NY Rangers** for Kyle Freadrich and Nils Ekman, June 30, 2001.

TENKRAT, Petr
(TEHN-krat, PEE-tuhr) **NSH.**

Right wing. Shoots right. 5'11", 200 lbs. Born, Kladno, Czech., May 31, 1977. Anaheim's 6th choice, 230th overall, in 1999 Entry Draft.

Season	Club	League	GP	G	A	Pts	PIM	PP	SH	GW	S	%	+/-	TF	F%	H	SB	Min	PO GP	PO G	PO A	PO Pts	PO PIM	PO PP	PO SH	PO GW
1994-95	HC Kladno	Czech	1	0	0	0	0	....	....	....	....	....	....	....	....	....	....	....	....	....	....	....	....	....	....	....
1995-96	HC Poldi Kladno	Czech	20	0	4	4	4	....	....	....	....	....	....	....	....	....	....	....	3	0	1	1	0	....	....	....
1996-97	HC Poldi Kladno	Czech	43	5	9	14	6	....	....	....	....	....	....	....	....	....	....	....	3	0	1	1	0	....	....	....
1997-98	Kladno	Czech	52	9	10	19	24	....	....	....	....	....	....	....	....	....	....	....	....	....	....	....	....	....	....	....
1998-99	Kladno	Czech	50	21	14	35	32	....	....	....	....	....	....	....	....	....	....	....	....	....	....	....	....	....	....	....
99-2000	HPK Hameenlinna	Finland	32	20	9	29	31	....	....	....	....	....	....	....	....	....	....	....	....	....	....	....	....	....	....	....
	Ilves Tampere	Finland	22	15	5	20	44	....	....	....	....	....	....	....	....	....	....	....	3	1	1	2	14	....	....	....
2000-01	**Anaheim**	**NHL**	**46**	**5**	**9**	**14**	**16**	**0**	**0**	**2**	**79**	**6.3**	**-11**	**0**	**0.0**	**33**	**3**	**12:48**	....	....	....	....	....	....	....	....
	Cincinnati	AHL	25	9	9	18	24	....	....	....	....	....	....	....	....	....	....	....	4	3	2	5	0	....	....	....
2001-02	**Anaheim**	**NHL**	**9**	**0**	**0**	**0**	**6**	**0**	**0**	**0**	**13**	**0.0**	**-6**	**1**	**0.0**	**4**	**2**	**11:47**	....	....	....	....	....	....	....	....
	Cincinnati	AHL	3	2	3	5	2	....	....	....	....	....	....	....	....	....	....	....	....	....	....	....	....	....	....	....
	Nashville	**NHL**	**58**	**8**	**16**	**24**	**28**	**0**	**1**	**2**	**82**	**9.8**	**-4**	**7**	**28.6**	**34**	**4**	**12:00**	....	....	....	....	....	....	....	....
	Milwaukee	AHL	4	0	0	0	2	....	....	....	....	....	....	....	....	....	....	....	....	....	....	....	....	....	....	....
	NHL Totals		**113**	**13**	**25**	**38**	**50**	**0**	**1**	**4**	**174**	**7.5**		**8**	**25.0**	**71**	**9**	**12:18**	....	....	....	....	....	....	....	....

Traded to **Nashville** by **Anaheim** for Patrick Kjellberg, November 1, 2001.

TETARENKO, Joey
(teh-tar-EHN-koh, JOH-ee) **FLA.**

Right wing. Shoots right. 6'2", 215 lbs. Born, Prince Albert, Sask., March 3, 1978. Florida's 4th choice, 82nd overall, in 1996 Entry Draft.

Season	Club	League	GP	G	A	Pts	PIM	PP	SH	GW	S	%	+/-	TF	F%	H	SB	Min	PO GP	PO G	PO A	PO Pts	PO PIM	PO PP	PO SH	PO GW
1993-94	North Battleford	SMHL	36	6	13	19	75	....	....	....	....	....	....	....	....	....	....	....	....	....	....	....	....	....	....	....
1994-95	Portland	WHL	59	0	1	1	134	....	....	....	....	....	....	....	....	....	....	....	9	0	0	0	8	....	....	....
1995-96	Portland	WHL	71	4	11	15	190	....	....	....	....	....	....	....	....	....	....	....	7	0	1	1	17	....	....	....
1996-97	Portland	WHL	68	8	18	26	182	....	....	....	....	....	....	....	....	....	....	....	2	0	0	0	2	....	....	....
1997-98	Portland	WHL	49	2	12	14	148	....	....	....	....	....	....	....	....	....	....	....	16	0	2	2	30	....	....	....
1998-99	New Haven	AHL	65	4	10	14	154	....	....	....	....	....	....	....	....	....	....	....	....	....	....	....	....	....	....	....
99-2000	Louisville	AHL	57	3	11	14	136	....	....	....	....	....	....	....	....	....	....	....	4	0	0	0	2	....	....	....
2000-01	**Florida**	**NHL**	**29**	**3**	**1**	**4**	**44**	**0**	**0**	**0**	**21**	**14.3**	**-1**	**0**	**0.0**	**30**	**5**	**6:12**	....	....	....	....	....	....	....	....
	Louisville	AHL	29	1	4	5	74	....	....	....	....	....	....	....	....	....	....	....	....	....	....	....	....	....	....	....
2001-02	**Florida**	**NHL**	**38**	**1**	**0**	**1**	**123**	**0**	**0**	**0**	**10**	**10.0**	**-5**	**0**	**0.0**	**25**	**2**	**5:08**	....	....	....	....	....	....	....	....
	NHL Totals		**67**	**4**	**1**	**5**	**167**	**0**	**0**	**0**	**31**	**12.9**		**0**	**0.0**	**55**	**7**	**5:35**	....	....	....	....	....	....	....	....

• Missed majority of 2001-02 season recovering from jaw injury suffered in game vs. NY Rangers, November 3, 2001.

TEZIKOV, Alexei

(TEH-zih-kahf, al-EHX-ay) **VAN.**

Defense. Shoots left. 6'1", 208 lbs. Born, Togliatti, USSR, June 22, 1978. Buffalo's 7th choice, 115th overall, in 1996 Entry Draft.

Season	Club	League	Regular Season GP	G	A	Pts	PIM	PP	SH	GW	S	%	+/-	TF	F%	H	SB	Min	Playoffs GP	G	A	Pts	PIM	PP	SH	GW
1995-96	Lada Togliatti	CIS	14	0	0	0	8																			
1996-97	Lada Togliatti	Russia	7	0	0	0	4																			
	Nizhny Novgorod	Russia	5	0	2	2	2																			
1997-98	Moncton Wildcats	QMJHL	60	15	33	48	144												10	3	8	11	20			
1998-99	Moncton Wildcats	QMJHL	25	9	21	30	52																			
	Rochester	AHL	31	3	7	10	41																			
	Washington	**NHL**	**5**	**0**	**0**	**0**	**0**	**0**	**0**	**0**	**4**	**0.0**	**–1**	**0**	**0.0**	**9**	**3**	**18:11**								
	Cincinnati	IHL	5	0	0	0	2												3	0	0	0	10			
99-2000	**Washington**	**NHL**	**23**	**1**	**1**	**2**	**2**	**1**	**0**	**1**	**18**	**5.6**	**–2**	**0**	**0.0**	**17**	**9**	**10:30**								
	Portland Pirates	AHL	53	6	9	15	70																			
2000-01	Portland Pirates	AHL	58	7	24	31	58																			
	Cincinnati	AHL	13	2	6	8	8												4	0	1	1	0			
2001-02	**Vancouver**	**NHL**	**2**	**0**	**0**	**0**	**0**	**0**	**0**	**0**	**0**	**0.0**	**2**	**0**	**0.0**	**0**	**1**	**11:46**								
	Manitoba Moose	AHL	28	5	6	11	16												3	0	0	0	2			
	NHL Totals		**30**	**1**	**1**	**2**	**2**	**1**	**0**	**1**	**22**	**4.5**		**0**	**0.0**	**26**	**13**	**11:52**								

QMJHL Second All-Star Team (1998)

Traded to **Washington** by **Buffalo** with Buffalo's 4th round compensatory choice (later traded to Calgary - Calgary selected Levente Szuper) in 2000 Entry Draft for Joe Juneau and Washington's 3rd round choice (Tim Preston) in 1999 Entry Draft, March 22, 1999. Traded to **Anaheim** by **Washington** with Edmonton's 4th round choice (previously acquired, Anaheim selected Brandon Rogers) in 2001 Entry Draft for Jason Marshall, March 13, 2001. Claimed on waivers by **Vancouver** from **Anaheim**, October 30, 2001. • Spent majority of 2001-02 season on practice roster, November 20, 2001.

THERIEN, Chris

(TEH-ree-ehn, KRIHS) **PHI.**

Defense. Shoots left. 6'5", 235 lbs. Born, Ottawa, Ont., December 14, 1971. Philadelphia's 7th choice, 47th overall, in 1990 Entry Draft.

Season	Club	League	Regular Season GP	G	A	Pts	PIM	PP	SH	GW	S	%	+/-	TF	F%	H	SB	Min	Playoffs GP	G	A	Pts	PIM	PP	SH	GW
1988-89	Ottawa Jr. Sens	OCJHL	8	3	1	4	22																			
1989-90	Ottawa Jr. Sens	OCJHL	3	0	2	2	2																			
	Northfield Prep	Hi-School	31	35	37	72	54																			
1990-91	Providence	H-East	36	4	18	22	36																			
1991-92	Providence	H-East	36	16	25	41	38																			
1992-93	Providence	H-East	33	8	11	19	52																			
	Team Canada	Nat-Tm	8	1	4	5	8																			
1993-94	Team Canada	Nat-Tm	59	7	15	22	46																			
	Canada	Olympics	4	0	0	0	4																			
	Hershey Bears	AHL	6	0	0	0	2																			
1994-95	Hershey Bears	AHL	34	3	13	16	27																			
	Philadelphia	**NHL**	**48**	**3**	**10**	**13**	**38**	**1**	**0**	**0**	**53**	**5.7**	**8**						**15**	**0**	**0**	**0**	**10**	**0**	**0**	**0**
1995-96	**Philadelphia**	**NHL**	**82**	**6**	**17**	**23**	**89**	**3**	**0**	**1**	**123**	**4.9**	**16**						**12**	**0**	**0**	**0**	**18**	**0**	**0**	**0**
1996-97	**Philadelphia**	**NHL**	**71**	**2**	**22**	**24**	**64**	**0**	**0**	**0**	**107**	**1.9**	**27**						**19**	**1**	**6**	**7**	**6**	**0**	**0**	**1**
1997-98	**Philadelphia**	**NHL**	**78**	**3**	**16**	**19**	**80**	**1**	**0**	**1**	**102**	**2.9**	**5**						**5**	**0**	**1**	**1**	**4**	**0**	**0**	**0**
1998-99	**Philadelphia**	**NHL**	**74**	**3**	**15**	**18**	**48**	**1**	**0**	**0**	**115**	**2.6**	**16**	**0**	**0.0**	**167**	**103**	**20:45**	**6**	**0**	**0**	**0**	**6**	**0**	**0**	**0**
99-2000	**Philadelphia**	**NHL**	**80**	**4**	**9**	**13**	**66**	**1**	**0**	**1**	**126**	**3.2**	**11**	**0**	**0.0**	**171**	**137**	**20:12**	**18**	**0**	**1**	**1**	**12**	**0**	**0**	**0**
2000-01	**Philadelphia**	**NHL**	**73**	**2**	**12**	**14**	**48**	**1**	**0**	**0**	**103**	**1.9**	**22**	**0**	**0.0**	**196**	**84**	**20:38**	**6**	**1**	**0**	**1**	**8**	**0**	**0**	**0**
2001-02	**Philadelphia**	**NHL**	**77**	**4**	**10**	**14**	**30**	**0**	**2**	**3**	**105**	**3.8**	**16**	**0**	**0.0**	**207**	**103**	**18:42**	**5**	**0**	**0**	**0**	**2**	**0**	**0**	**0**
	NHL Totals		**583**	**27**	**111**	**138**	**463**	**8**	**2**	**6**	**834**	**3.2**		**0**	**0.0**	**741**	**427**	**20:04**	**86**	**2**	**8**	**10**	**66**	**0**	**0**	**1**

Hockey East Second All-Star Team (1993) • NHL All-Rookie Team (1995)

THOMAS, Scott

(TAW-mas, SKAWT)

Right wing. Shoots right. 6'2", 200 lbs. Born, Buffalo, NY, January 18, 1970. Buffalo's 2nd choice, 56th overall, in 1989 Entry Draft.

Season	Club	League	Regular Season GP	G	A	Pts	PIM	PP	SH	GW	S	%	+/-	TF	F%	H	SB	Min	Playoffs GP	G	A	Pts	PIM	PP	SH	GW
1987-88	Nichols High	Hi-School	16	23	39	62	62																			
1988-89	Nichols High	Hi-School	17	38	52	90																				
1989-90	Clarkson Knights	ECAC	34	19	13	32	95																			
1990-91	Clarkson Knights	ECAC	40	28	14	42	89																			
1991-92	Clarkson Knights	ECAC	29	22	20	42	57																			
	Rochester	AHL																	9	0	1	1	17			
1992-93	**Buffalo**	**NHL**	**7**	**1**	**1**	**2**	**15**	**0**	**0**	**0**	**4**	**25.0**	**2**													
	Rochester	AHL	65	32	27	59	38												17	8	5	13	6			
1993-94	**Buffalo**	**NHL**	**32**	**2**	**2**	**4**	**8**	**1**	**0**	**0**	**26**	**7.7**	**–6**													
	Rochester	AHL	11	4	5	9	0																			
1994-95	Rochester	AHL	55	21	25	46	115												5	4	0	4	4			
1995-96	Cincinnati	IHL	78	32	28	60	54												17	*13	2	15	4			
1996-97	Cincinnati	IHL	71	32	29	61	46												3	0	0	0	0			
1997-98	Detroit Vipers	IHL	44	11	16	27	18																			
	Manitoba Moose	IHL	26	12	4	16	8												3	0	1	1	2			
1998-99	Manitoba Moose	IHL	78	45	25	70	32												5	3	4	7	4			
99-2000	Long Beach	IHL	52	15	16	31	18												6	2	1	3	6			
2000-01	**Los Angeles**	**NHL**	**24**	**3**	**1**	**4**	**9**	**0**	**0**	**0**	**16**	**18.8**	**0**	**0**	**0.0**	**32**	**5**	**8:07**	**12**	**1**	**0**	**1**	**4**	**1**	**0**	**0**
	Manitoba Moose	IHL	22	9	14	23	21												3	1	2	3	0			
2001-02	Manchester	AHL	56	14	29	43	6												5	1	0	1	4			
	NHL Totals		**63**	**6**	**4**	**10**	**32**	**1**	**0**	**0**	**46**	**13.0**		**0**	**0.0**	**32**	**5**	**8:07**	**12**	**1**	**0**	**1**	**4**	**1**	**0**	**0**

Signed as a free agent by **LA Kings**, July 30, 1999.

THOMAS, Steve

(TAW-mas, STEEV) **CHI.**

Right wing. Shoots left. 5'10", 185 lbs. Born, Stockport, England, July 15, 1963.

Season	Club	League	Regular Season GP	G	A	Pts	PIM	PP	SH	GW	S	%	+/-	TF	F%	H	SB	Min	Playoffs GP	G	A	Pts	PIM	PP	SH	GW
1980-81	Markham Waxers	OHA-B	42	22	25	47	76																			
	Toronto	OMJHL	1	0	0	0	0																			
1981-82	Markham Waxers	OJHL	48	68	57	125	113																			
	Toronto	OHL	1	0	0	0	0																			
1982-83	Toronto	OHL	61	18	20	38	42																			
1983-84	Toronto	OHL	70	51	54	105	77																			
1984-85	**Toronto**	**NHL**	**18**	**1**	**1**	**2**	**2**	**0**	**0**	**0**	**26**	**3.8**	**–13**													
	St. Catharines	AHL	64	42	48	90	56																			
1985-86	**Toronto**	**NHL**	**65**	**20**	**37**	**57**	**36**	**5**	**0**	**5**	**197**	**10.2**	**–15**						**10**	**6**	**8**	**14**	**9**	**3**	**0**	**0**
	St. Catharines	AHL	19	18	14	32	35																			
1986-87	**Toronto**	**NHL**	**78**	**35**	**27**	**62**	**114**	**3**	**0**	**7**	**245**	**14.3**	**–3**						**13**	**2**	**3**	**5**	**13**	**0**	**0**	**0**
1987-88	**Chicago**	**NHL**	**30**	**13**	**13**	**26**	**40**	**5**	**0**	**3**	**69**	**18.8**	**1**						**3**	**1**	**2**	**3**	**6**	**0**	**0**	**0**
1988-89	**Chicago**	**NHL**	**45**	**21**	**19**	**40**	**69**	**8**	**0**	**0**	**124**	**16.9**	**–2**						**12**	**3**	**5**	**8**	**10**	**1**	**0**	**2**
1989-90	**Chicago**	**NHL**	**76**	**40**	**30**	**70**	**91**	**13**	**0**	**7**	**235**	**17.0**	**–3**						**20**	**7**	**6**	**13**	**33**	**1**	**0**	**3**
1990-91	**Chicago**	**NHL**	**69**	**19**	**35**	**54**	**129**	**2**	**0**	**3**	**192**	**9.9**	**8**						**6**	**1**	**2**	**3**	**15**	**0**	**0**	**0**
1991-92	**Chicago**	**NHL**	**11**	**2**	**6**	**8**	**26**	**0**	**0**	**1**	**35**	**5.7**	**–3**													
	NY Islanders	**NHL**	**71**	**28**	**42**	**70**	**71**	**3**	**0**	**2**	**210**	**13.3**	**11**													
1992-93	**NY Islanders**	**NHL**	**79**	**37**	**50**	**87**	**111**	**12**	**0**	**7**	**264**	**14.0**	**3**						**18**	**9**	**8**	**17**	**37**	**0**	**0**	**0**
1993-94	**NY Islanders**	**NHL**	**78**	**42**	**33**	**75**	**139**	**17**	**0**	**5**	**249**	**16.9**	**–9**						**4**	**1**	**0**	**1**	**8**	**1**	**0**	**0**
1994-95	**NY Islanders**	**NHL**	**47**	**11**	**15**	**26**	**60**	**3**	**0**	**2**	**133**	**8.3**	**–14**													
1995-96	**New Jersey**	**NHL**	**81**	**26**	**35**	**61**	**98**	**6**	**0**	**6**	**192**	**13.5**	**–2**													
1996-97	**New Jersey**	**NHL**	**57**	**15**	**19**	**34**	**46**	**1**	**0**	**2**	**124**	**12.1**	**9**						**10**	**1**	**1**	**2**	**18**	**0**	**0**	**0**
1997-98	**New Jersey**	**NHL**	**55**	**14**	**10**	**24**	**32**	**3**	**0**	**4**	**111**	**12.6**	**4**						**6**	**0**	**3**	**3**	**2**	**0**	**0**	**0**
1998-99	**Toronto**	**NHL**	**78**	**28**	**45**	**73**	**33**	**11**	**0**	**7**	**209**	**13.4**	**26**	**4**	**25.0**	**65**	**25**	**18:23**	**17**	**6**	**3**	**9**	**12**	**2**	**0**	**1**
99-2000	**Toronto**	**NHL**	**81**	**26**	**37**	**63**	**68**	**9**	**0**	**9**	**151**	**17.2**	**1**	**8**	**50.0**	**62**	**20**	**16:19**	**12**	**6**	**3**	**9**	**10**	**0**	**0**	**1**
2000-01	**Toronto**	**NHL**	**57**	**8**	**26**	**34**	**46**	**1**	**0**	**1**	**140**	**5.7**	**0**	**3**	**33.3**	**54**	**17**	**16:02**	**11**	**6**	**3**	**9**	**4**	**4**	**0**	**0**
2001-02	**Chicago**	**NHL**	**34**	**11**	**4**	**15**	**17**	**3**	**0**	**0**	**66**	**16.7**	**0**	**0**	**0.0**	**46**	**12**	**17:21**	**5**	**1**	**1**	**2**	**0**	**0**	**0**	**0**
	NHL Totals		**1110**	**397**	**484**	**881**	**1228**	**105**	**0**	**71**	**2972**	**13.4**		**15**	**40.0**	**227**	**74**	**17:02**	**147**	**50**	**48**	**98**	**177**	**12**	**0**	**7**

AHL First All-Star Team (1985) • Won Dudley "Red" Garrett Memorial Trophy (Top Rookie - AHL) (1985)

Signed as a free agent by **Toronto,** May 12, 1984. Traded to **Chicago** by **Toronto** with Rick Vaive and Bob McGill for Al Secord and Ed Olczyk, September 3, 1987. Traded to **NY Islanders** by **Chicago** with Adam Creighton for Brent Sutter and Brad Lauer, October 25, 1991. Traded to **New Jersey** by **NY Islanders** for Claude Lemieux, October 3, 1995. Signed as a free agent by **Toronto**, July 30, 1998. Signed as a free agent by **Chicago**, July 17, 2001. • Missed majority of 2001-02 season recovering from ankle injury suffered in game vs. Calgary, November 15, 2001.

THOMPSON, Brent

(TAWM-suhn, BREHNT) **COL.**

Defense. Shoots left. 6'2", 205 lbs. Born, Calgary, Alta., January 9, 1971. Los Angeles' 1st choice, 39th overall, in 1989 Entry Draft.

Season	Club	League	GP	G	A	Pts	PIM	PP	SH	GW	S	%	+/-	TF	F%	H	SB	Min	GP	G	A	Pts	PIM	PP	SH	GW
			Regular Season																Playoffs							
1987-88	Cgy. North Stars	AMHL	25	0	13	13	33																			
1988-89	Medicine Hat	WHL	72	3	10	13	160												3	0	0	0	2			
1989-90	Medicine Hat	WHL	68	10	35	45	167												3	0	1	1	14			
1990-91	Medicine Hat	WHL	51	5	40	45	87												12	1	7	8	16			
	Phoenix	IHL																	4	0	1	1	6			
1991-92	**Los Angeles**	**NHL**	**27**	**0**	**5**	**5**	**89**	**0**	**0**	**0**	**18**	**0.0**	**–7**						**4**	**0**	**0**	**0**	**4**	**0**	**0**	**0**
	Phoenix	IHL	42	4	13	17	139																			
1992-93	**Los Angeles**	**NHL**	**30**	**0**	**4**	**4**	**76**	**0**	**0**	**0**	**18**	**0.0**	**–4**													
	Phoenix	IHL	22	0	5	5	112																			
1993-94	**Los Angeles**	**NHL**	**24**	**1**	**0**	**1**	**81**	**0**	**0**	**0**	**9**	**11.1**	**–1**													
	Phoenix	IHL	26	1	11	12	118																			
1994-95	**Winnipeg**	**NHL**	**29**	**0**	**0**	**0**	**78**	**0**	**0**	**0**	**16**	**0.0**	**–17**													
1995-96	**Winnipeg**	**NHL**	**10**	**0**	**1**	**1**	**21**	**0**	**0**	**0**	**7**	**0.0**	**–2**													
	Springfield	AHL	58	2	10	12	203												10	1	4	5	*55			
1996-97	**Phoenix**	**NHL**	**1**	**0**	**0**	**0**	**7**	**0**	**0**	**0**	**0**	**0.0**	**–1**													
	Springfield	AHL	64	2	15	17	215												17	0	2	2	31			
	Phoenix	IHL	12	0	1	1	67																			
1997-98	Hartford	AHL	77	4	15	19	308												15	0	4	4	25			
1998-99	Hartford	AHL	76	3	15	18	265												7	0	0	0	23			
99-2000	Louisville	AHL	67	4	22	26	311												3	0	0	0	11			
2000-01	Louisville	AHL	59	1	9	10	170																			
	Hershey Bears	AHL	15	0	1	1	44												12	0	0	0	10			
2001-02	Hershey Bears	AHL	79	8	16	24	178												8	0	0	0	21			
	NHL Totals		**121**	**1**	**10**	**11**	**352**	**0**	**0**	**0**	**68**	**1.5**							**4**	**0**	**0**	**0**	**4**	**0**	**0**	**0**

WHL East Second All-Star Team (1991)

Traded to **Winnipeg** by **LA Kings** with cash for the rights to Ruslan Batyrshin and Winnipeg's 2nd round choice (Marian Cisar) in 1996 Entry Draft, August 8, 1994. Transferred to **Phoenix** after **Winnipeg** franchise relocated, July 1, 1996. Signed as a free agent by **NY Rangers**, August 26, 1997. Signed as a free agent by **Florida**, July 27, 1999. Traded to **Colorado** by **Florida** for future considerations, March 3, 2001.

THOMPSON, Rocky

(TAWM-suhn, RAW-kee) **FLA.**

Right wing. Shoots right. 6'2", 205 lbs. Born, Calgary, Alta., August 8, 1977. Calgary's 3rd choice, 72nd overall, in 1995 Entry Draft.

Season	Club	League	GP	G	A	Pts	PIM	PP	SH	GW	S	%	+/-	TF	F%	H	SB	Min	GP	G	A	Pts	PIM	PP	SH	GW
1992-93	Spruce Grove	AMHL	65	13	50	63	295																			
1993-94	Medicine Hat	WHL	68	1	4	5	166												3	0	0	0	2			
1994-95	Medicine Hat	WHL	63	1	6	7	220												5	0	0	0	17			
1995-96	Medicine Hat	WHL	71	9	20	29	260												5	2	3	5	26			
	Saint John	AHL	4	0	0	0	33																			
1996-97	Medicine Hat	WHL	47	6	9	15	170																			
	Swift Current	WHL	22	3	5	8	90												10	1	2	3	22			
1997-98	**Calgary**	**NHL**	**12**	**0**	**0**	**0**	**61**	**0**	**0**	**0**	**3**	**0.0**	**0**													
	Saint John	AHL	51	3	0	3	187												18	1	1	2	47			
1998-99	**Calgary**	**NHL**	**3**	**0**	**0**	**0**	**25**	**0**	**0**	**0**	**0**	**0.0**	**0**	**0**	**0.0**	**0**	**0**	**2:01**								
	Saint John	AHL	27	2	2	4	108																			
99-2000	Saint John	AHL	53	2	8	10	125																			
	Louisville	AHL	3	0	1	1	54												4	0	0	0	4			
2000-01	**Florida**	**NHL**	**4**	**0**	**0**	**0**	**19**	**0**	**0**	**0**	**0**	**0.0**	**0**	**0**	**0.0**	**4**	**0**	**1:28**								
	Louisville	AHL	55	3	5	8	193																			
2001-02	**Florida**	**NHL**	**6**	**0**	**0**	**0**	**12**	**0**	**0**	**0**	**1**	**0.0**	**0**	**1**	**0.0**	**7**	**0**	**4:03**								
	Hershey Bears	AHL	42	0	3	3	143												8	1	0	1	19			
	NHL Totals		**25**	**0**	**0**	**0**	**117**	**0**	**0**	**0**	**4**	**0.0**		**1**	**0.0**	**11**	**0**	**2:47**								

Traded to **Florida** by **Calgary** for Filip Kuba, March 16, 2000.

THORNTON, Joe

(THOHRN-tuhn, JOH) **BOS.**

Center. Shoots left. 6'4", 220 lbs. Born, London, Ont., July 2, 1979. Boston's 1st choice, 1st overall, in 1997 Entry Draft.

Season	Club	League	GP	G	A	Pts	PIM	PP	SH	GW	S	%	+/-	TF	F%	H	SB	Min	GP	G	A	Pts	PIM	PP	SH	GW
1993-94	Elgin-Middlesex	OMHA	67	*83	*85	*168	45																			
	St. Thomas Stars	OJHL-B	6	2	6	8	2																			
1994-95	St. Thomas Stars	OJHL-B	50	40	64	104	53																			
1995-96	Sault Ste. Marie	OHL	66	30	46	76	53												4	1	1	2	11			
1996-97	Sault Ste. Marie	OHL	59	41	81	122	123												11	11	8	19	24			
1997-98	**Boston**	**NHL**	**55**	**3**	**4**	**7**	**19**	**0**	**0**	**1**	**33**	**9.1**	**–6**						**6**	**0**	**0**	**0**	**9**	**0**	**0**	**0**
1998-99	**Boston**	**NHL**	**81**	**16**	**25**	**41**	**69**	**7**	**0**	**1**	**128**	**12.5**	**3**	**1073**	**48.7**	**124**	**11**	**15:21**	**11**	**3**	**6**	**9**	**4**	**2**	**0**	**2**
99-2000	**Boston**	**NHL**	**81**	**23**	**37**	**60**	**82**	**5**	**0**	**3**	**171**	**13.5**	**–5**	**1861**	**49.5**	**134**	**14**	**21:18**								
2000-01	**Boston**	**NHL**	**72**	**37**	**34**	**71**	**107**	**19**	**1**	**5**	**181**	**20.4**	**–4**	**1651**	**52.1**	**108**	**20**	**21:45**								
2001-02	**Boston**	**NHL**	**66**	**22**	**46**	**68**	**127**	**6**	**0**	**5**	**152**	**14.5**	**7**	**1341**	**49.1**	**60**	**15**	**19:59**	**6**	**2**	**4**	**6**	**10**	**0**	**0**	**0**
	NHL Totals		**355**	**101**	**146**	**247**	**404**	**37**	**1**	**15**	**665**	**15.2**		**5926**	**50.0**	**426**	**60**	**19:31**	**23**	**5**	**10**	**15**	**23**	**2**	**0**	**2**

OHL All-Rookie Team (1996) • OHL Rookie of the Year (1996) • Canadian Major Junior Rookie of the Year (1996) • OHL Second All-Star Team (1997) • Played in NHL All-Star Game (2002)

THORNTON, Scott

(THOHRN-tuhn, SKAWT) **S.J.**

Center. Shoots left. 6'3", 220 lbs. Born, London, Ont., January 9, 1971. Toronto's 1st choice, 3rd overall, in 1989 Entry Draft.

Season	Club	League	GP	G	A	Pts	PIM	PP	SH	GW	S	%	+/-	TF	F%	H	SB	Min	GP	G	A	Pts	PIM	PP	SH	GW
1986-87	London Diamonds	OJHL-B	31	10	7	17	10																			
1987-88	Belleville Bulls	OHL	62	11	19	30	54												6	0	1	1	2			
1988-89	Belleville Bulls	OHL	59	28	34	62	103												5	1	1	2	6			
1989-90	Belleville Bulls	OHL	47	21	28	49	91												11	2	10	12	15			
1990-91	Belleville Bulls	OHL	3	2	1	3	2												6	0	7	7	14			
	Toronto	**NHL**	**33**	**1**	**3**	**4**	**30**	**0**	**0**	**0**	**31**	**3.2**	**–15**													
	Newmarket Saints	AHL	5	1	0	1	4																			
1991-92	**Edmonton**	**NHL**	**15**	**0**	**1**	**1**	**43**	**0**	**0**	**0**	**11**	**0.0**	**–6**						**1**	**0**	**0**	**0**	**0**	**0**	**0**	**0**
	Cape Breton	AHL	49	9	14	23	40												5	1	0	1	8			
1992-93	**Edmonton**	**NHL**	**9**	**0**	**1**	**1**	**0**	**0**	**0**	**0**	**7**	**0.0**	**–4**													
	Cape Breton	AHL	58	23	27	50	102												16	1	2	3	35			
1993-94	**Edmonton**	**NHL**	**61**	**4**	**7**	**11**	**104**	**0**	**0**	**0**	**65**	**6.2**	**–15**													
	Cape Breton	AHL	2	1	1	2	31																			
1994-95	**Edmonton**	**NHL**	**47**	**10**	**12**	**22**	**89**	**0**	**1**	**1**	**69**	**14.5**	**–4**													
1995-96	**Edmonton**	**NHL**	**77**	**9**	**9**	**18**	**149**	**0**	**2**	**3**	**95**	**9.5**	**–25**													
1996-97	**Montreal**	**NHL**	**73**	**10**	**10**	**20**	**128**	**1**	**1**	**1**	**110**	**9.1**	**–19**						**5**	**1**	**0**	**1**	**2**	**0**	**0**	**0**
1997-98	**Montreal**	**NHL**	**67**	**6**	**9**	**15**	**158**	**1**	**0**	**1**	**51**	**11.8**	**0**						**9**	**0**	**2**	**2**	**10**	**0**	**0**	**0**
1998-99	**Montreal**	**NHL**	**47**	**7**	**4**	**11**	**87**	**1**	**0**	**1**	**56**	**12.5**	**–2**	**466**	**52.8**	**74**	**13**	**12:24**								
99-2000	**Montreal**	**NHL**	**35**	**2**	**3**	**5**	**70**	**0**	**0**	**1**	**36**	**5.6**	**–7**	**253**	**51.8**	**43**	**13**	**12:40**								
	Dallas	**NHL**	**30**	**6**	**3**	**9**	**38**	**1**	**0**	**0**	**47**	**12.8**	**–5**	**14**	**14.3**	**66**	**7**	**13:03**	**23**	**2**	**7**	**9**	**28**	**0**	**0**	**1**
2000-01	**San Jose**	**NHL**	**73**	**19**	**17**	**36**	**114**	**4**	**0**	**1**	**159**	**11.9**	**4**	**29**	**41.4**	**167**	**19**	**13:54**	**6**	**3**	**0**	**3**	**8**	**0**	**0**	**1**
2001-02	**San Jose**	**NHL**	**77**	**26**	**16**	**42**	**116**	**6**	**0**	**5**	**144**	**18.1**	**11**	**18**	**61.1**	**156**	**15**	**13:31**	**12**	**3**	**3**	**6**	**6**	**0**	**0**	**0**
	NHL Totals		**644**	**100**	**95**	**195**	**1126**	**14**	**4**	**14**	**881**	**11.4**		**780**	**51.5**	**506**	**67**	**13:15**	**56**	**9**	**12**	**21**	**54**	**0**	**0**	**2**

Traded to **Edmonton** by **Toronto** with Vincent Damphousse, Peter Ing and Luke Richardson for Grant Fuhr, Glenn Anderson and Craig Berube, September 19, 1991. Traded to **Montreal** by **Edmonton** for Andrei Kovalenko, September 6, 1996. Traded to **Dallas** by **Montreal** for Juha Lind, January 22, 2000. Signed as a free agent by **San Jose**, July 1, 2000.

TIBBETTS, Billy

(TIH-buhts, BIHL-ee) **PHI.**

Right wing. Shoots right. 6'2", 215 lbs. Born, Boston, MA, October 14, 1974.

Season	Club	League	GP	G	A	Pts	PIM	PP	SH	GW	S	%	+/-	TF	F%	H	SB	Min	GP	G	A	Pts	PIM	PP	SH	GW
1992-93	Boston Jr. Bruins	NEJHL	73	60	80	140	150																			
1993-94	Sioux City	USHL	7	1	4	5	27																			
	London Knights	OHL	14	6	6	12	49																			
	Tri-City	WHL	9	0	2	2	39																			
1994-95	Birmingham Bulls	ECHL	2	0	1	1	18																			
1995-96	Johnstown Chiefs	ECHL	58	37	31	68	300																			

			Regular Season																Playoffs							
Season	Club	League	GP	G	A	Pts	PIM	PP	SH	GW	S	%	+/-	TF	F%	H	SB	Min	GP	G	A	Pts	PIM	PP	SH	GW
1996/00			DID NOT PLAY																						
2000-01	**Pittsburgh**	**NHL**	**29**	**1**	**2**	**3**	**79**	**0**	**0**	**0**	**16**	**6.3**	**–2**	**115**	**28.7**	**46**	**9**	**7:14**								
	Wilkes-Barre	AHL	38	14	24	38	185													12	4	6	10	55			
2001-02	**Pittsburgh**	**NHL**	**33**	**1**	**5**	**6**	**109**	**0**	**0**	**1**	**42**	**2.4**	**–13**	**87**	**36.8**	**64**	**7**	**12:05**								
	Wilkes-Barre	AHL	24	13	17	30	193																			
	Philadelphia	**NHL**	**9**	**0**	**1**	**1**	**69**	**0**	**0**	**0**	**6**	**0.0**	**–3**	**25**	**32.0**	**10**	**0**	**6:41**								
	NHL Totals		**71**	**2**	**8**	**10**	**257**	**0**	**0**	**1**	**64**	**3.1**		**227**	**32.2**	**120**	**16**	**9:25**								

• Missed 1996-97 through 1999-2000 seasons serving prison sentence that commenced July 12, 1996. Signed as a free agent by **Pittsburgh**, April 10, 2000. Traded to **Philadelphia** by **Pittsburgh** for Kent Manderville, March 17, 2002.

TILEY, Brad

(TIHL-ee, BRAD) **PHI.**

Defense. Shoots left. 6'1", 185 lbs. Born, Markdale, Ont., July 5, 1971. Boston's 4th choice, 84th overall, in 1991 Entry Draft.

Season	Club	League	GP	G	A	Pts	PIM	PP	SH	GW	S	%	+/-	TF	F%	H	SB	Min	GP	G	A	Pts	PIM	PP	SH	GW
1987-88	Owen Sound	OJHL-B	45	18	25	43	69																		
1988-89	Sault Ste. Marie	OHL	50	4	11	15	31																		
1989-90	Sault Ste. Marie	OHL	66	9	32	41	47																		
1990-91	Sault Ste. Marie	OHL	66	11	55	66	29													14	4	15	19	12			
1991-92	Maine Mariners	AHL	62	7	22	29	36																		
1992-93	Phoenix	IHL	46	11	27	38	35																		
	Binghamton	AHL	26	6	10	16	19													8	0	1	1	2			
1993-94	Binghamton	AHL	29	6	10	16	6																		
	Phoenix	IHL	35	8	15	23	21																		
1994-95	Detroit Vipers	IHL	56	7	19	26	32																		
	Fort Wayne	IHL	14	1	6	7	2													3	1	2	3	0			
1995-96	Orlando	IHL	69	11	23	34	82													23	2	4	6	16			
1996-97	Phoenix	IHL	66	8	28	36	34																		
	Long Beach	IHL	3	1	0	1	2																		
1997-98	**Phoenix**	**NHL**	**1**	**0**	**0**	**0**	**0**																		
	Springfield	AHL	60	10	31	41	36													4	0	4	4	2			
1998-99	**Phoenix**	**NHL**	**8**	**0**	**0**	**0**	**0**	**0**	**0**	**0**	**1**	**0.0**	**–1**	**0**	**0.0**	**5**	**3**	**11:29**	**1**	**0**	**0**	**0**	**0**	**0**	**0**	**0**
	Springfield	AHL	69	9	35	44	14													1	0	0	0	0			
99-2000	Springfield	AHL	80	14	54	68	51													5	0	4	4	2			
2000-01	**Philadelphia**	**NHL**	**2**	**0**	**0**	**0**	**0**	**0**	**0**	**0**	**1**	**0.0**	**–1**	**0**	**0.0**	**1**	**1**	**15:46**								
	Philadelphia	AHL	56	11	19	30	10													10	1	2	3	2			
2001-02	Philadelphia	AHL	56	6	15	21	14																		
	NHL Totals		**11**	**0**	**0**	**0**	**0**	**0**	**0**	**0**	**2**	**0.0**		**0**	**0.0**	**6**	**4**	**12:20**	**1**	**0**	**0**	**0**	**0**	**0**	**0**	**0**

Memorial Cup All-Star Team (1991) • AHL First All-Star Team (2000) • Won Eddie Shore Award (Top Defenseman - AHL) (2000)

Signed as a free agent by **NY Rangers**, September 4, 1992. Traded to **LA Kings** by **NY Rangers** for LA Kings' 11th round choice (Jamie Butt) in 1994 Entry Draft, January 28, 1994. Signed as a free agent by **Phoenix**, September 4, 1997. Signed as a free agent by **Philadelphia**, July 14, 2000.

TIMANDER, Mattias

(tih-MAHN-duhr, MA-tee-uhs) **NYI**

Defense. Shoots left. 6'2", 230 lbs. Born, Solleftea, Sweden, April 16, 1974. Boston's 7th choice, 208th overall, in 1992 Entry Draft.

Season	Club	League	GP	G	A	Pts	PIM	PP	SH	GW	S	%	+/-	TF	F%	H	SB	Min	GP	G	A	Pts	PIM	PP	SH	GW
1992-93	MoDo Jr.	Swede-Jr.	4	0	0	0	0																		
	Husums IF	Swede-2	27	4	9	13	22																		
	MoDo	Sweden	1	0	0	0	0																		
1993-94	MoDo Jr.	Swede-Jr.	3	2	2	4	10																		
	MoDo	Sweden	23	2	2	4	6													11	2	0	2	10			
1994-95	MoDo	Sweden	39	8	9	17	24																		
1995-96	MoDo	Sweden	37	4	10	14	34													7	1	1	2	8			
1996-97	**Boston**	**NHL**	**41**	**1**	**8**	**9**	**14**	**0**	**0**	**0**	**62**	**1.6**	**–9**													
	Providence	AHL	32	3	11	14	20													10	1	1	2	12			
1997-98	**Boston**	**NHL**	**23**	**1**	**1**	**2**	**6**	**0**	**0**	**0**	**17**	**5.9**	**–9**													
	Providence	AHL	31	3	7	10	25																		
1998-99	**Boston**	**NHL**	**22**	**0**	**6**	**6**	**10**	**0**	**0**	**0**	**22**	**0.0**	**4**	**0**	**0.0**	**14**	**13**	**12:54**	**4**	**1**	**1**	**2**	**2**	**0**	**0**	**0**
	Providence	AHL	43	2	22	24	24																		
99-2000	**Boston**	**NHL**	**60**	**0**	**8**	**8**	**22**	**0**	**0**	**0**	**39**	**0.0**	**–11**	**0**	**0.0**	**50**	**63**	**12:29**								
	Hershey Bears	AHL	1	0	0	0	2																		
2000-01	**Columbus**	**NHL**	**76**	**2**	**9**	**11**	**24**	**0**	**0**	**1**	**68**	**2.9**	**–8**	**2**	**100.0**	**51**	**129**	**21:02**								
2001-02	**Columbus**	**NHL**	**78**	**4**	**7**	**11**	**44**	**1**	**0**	**0**	**68**	**5.9**	**–34**	**1**	**0.0**	**58**	**134**	**19:52**								
	NHL Totals		**300**	**8**	**39**	**47**	**120**	**1**	**0**	**1**	**276**	**2.9**		**3**	**66.7**	**173**	**339**	**17:43**	**4**	**1**	**1**	**2**	**2**	**0**	**0**	**0**

Selected by **Columbus** from **Boston** in Expansion Draft, June 23, 2000. Traded to **NY Islanders** by **Columbus** for NY Islanders' 4th round choice (Jekabs Redlihs) in 2002 Entry Draft, June 22, 2002.

TIMONEN, Kimmo

(TEEM-oh-nehn, KEE-moh) **NSH.**

Defense. Shoots left. 5'10", 196 lbs. Born, Kuopio, Finland, March 18, 1975. Los Angeles' 11th choice, 250th overall, in 1993 Entry Draft.

Season	Club	League	GP	G	A	Pts	PIM	PP	SH	GW	S	%	+/-	TF	F%	H	SB	Min	GP	G	A	Pts	PIM	PP	SH	GW
1990-91	KalPa Kuopio Jr.	Finn-Jr.	4	0	1	1	2																		
1991-92	KalPa Kuopio Jr.	Finn-Jr.	32	7	10	17	4																		
	KalPa Kuopio	Finland	5	0	0	0	0																		
1992-93	KalPa Kuopio Jr.	Finn-Jr.	16	9	15	24	10																		
	KalPa Kuopio	Finland	33	0	2	2	4																		
1993-94	KalPa Kuopio Jr.	Finn-Jr.	5	4	7	11	0																		
	KalPa Kuopio	Finland	46	6	7	13	55																		
1994-95	TPS Turku Jr.	Finn-Jr.	1	0	0	0	0																		
	TPS Turku	Finland	45	3	4	7	10													13	0	1	1	6			
1995-96	TPS Turku	Finland	48	3	21	24	22													9	1	2	3	12			
1996-97	TPS Turku	Finland	50	10	14	24	18													12	2	7	9	8			
	TPS Turku	EuroHL	6	1	0	1	27													4	0	1	1	0			
1997-98	HIFK Helsinki	Finland	45	10	15	25	59													9	3	4	7	8			
	Finland	Olympics	6	0	1	1	2																		
1998-99	**Nashville**	**NHL**	**50**	**4**	**8**	**12**	**30**	**1**	**0**	**0**	**75**	**5.3**	**–4**	**0**	**0.0**	**68**	**34**	**19:04**								
	Milwaukee	IHL	29	2	13	15	22																		
99-2000	**Nashville**	**NHL**	**51**	**8**	**25**	**33**	**26**	**2**	**1**	**2**	**97**	**8.2**	**–5**	**0**	**0.0**	**48**	**34**	**21:06**								
2000-01	**Nashville**	**NHL**	**82**	**12**	**13**	**25**	**50**	**6**	**0**	**3**	**151**	**7.9**	**–6**	**2**	**50.0**	**86**	**71**	**23:11**								
2001-02	**Nashville**	**NHL**	**82**	**13**	**29**	**42**	**28**	**9**	**0**	**1**	**154**	**8.4**	**2**	**0**	**0.0**	**97**	**106**	**24:12**								
	Finland	Olympics	4	0	1	1	2																		
	NHL Totals		**265**	**37**	**75**	**112**	**134**	**18**	**1**	**6**	**477**	**7.8**		**2**	**50.0**	**299**	**245**	**22:19**								

Traded to **Nashville** by **LA Kings** with Jan Vopat for future considerations, June 26, 1998.

TITOV, German

(TEE-tahf, GUHR-mihn) **ANA.**

Center. Shoots left. 6'1", 203 lbs. Born, Moscow, USSR, October 16, 1965. Calgary's 10th choice, 252nd overall, in 1993 Entry Draft.

Season	Club	League	GP	G	A	Pts	PIM	PP	SH	GW	S	%	+/-	TF	F%	H	SB	Min	GP	G	A	Pts	PIM	PP	SH	GW
1986-87	Voskresensk	USSR	23	1	0	1	10																		
1987-88	Voskresensk	USSR	39	6	5	11	10																		
1988-89	Voskresensk	USSR	44	10	3	13	24																		
1989-90	Voskresensk	USSR	44	6	14	20	19																		
1990-91	Voskresensk	USSR	45	13	11	24	28																		
1991-92	Voskresensk	CIS	42	18	13	31	35																		
1992-93	TPS Turku	Finland	47	25	19	44	49													12	5	12	17	10			
1993-94	**Calgary**	**NHL**	**76**	**27**	**18**	**45**	**28**	**8**	**3**	**2**	**153**	**17.6**	**20**						**7**	**2**	**1**	**3**	**4**	**1**	**0**	**0**
1994-95	TPS Turku	Finland	14	6	6	12	20																		
	Calgary	**NHL**	**40**	**12**	**12**	**24**	**16**	**3**	**2**	**3**	**88**	**13.6**	**6**						**7**	**5**	**3**	**8**	**10**	**0**	**1**	**0**
1995-96	**Calgary**	**NHL**	**82**	**28**	**39**	**67**	**24**	**13**	**2**	**2**	**214**	**13.1**	**9**						**4**	**0**	**2**	**2**	**0**	**0**	**0**	**0**
1996-97	**Calgary**	**NHL**	**79**	**22**	**30**	**52**	**36**	**12**	**0**	**4**	**192**	**11.5**	**–12**													
1997-98	**Calgary**	**NHL**	**68**	**18**	**22**	**40**	**38**	**6**	**1**	**2**	**133**	**13.5**	**–1**													
	Russia	Olympics	6	1	0	1	6																		
1998-99	**Pittsburgh**	**NHL**	**72**	**11**	**45**	**56**	**34**	**3**	**1**	**3**	**113**	**9.7**	**18**	**41**	**39.0**	**67**	**50**	**19:31**	**11**	**3**	**5**	**8**	**4**	**0**	**0**	**0**
99-2000	**Pittsburgh**	**NHL**	**63**	**17**	**25**	**42**	**34**	**4**	**2**	**3**	**111**	**15.3**	**–3**	**102**	**37.3**	**58**	**26**	**20:20**								
	Edmonton	**NHL**	**7**	**0**	**4**	**4**	**4**	**0**	**0**	**0**	**11**	**0.0**	**2**	**0**	**0.0**	**5**	**4**	**13:33**	**5**	**1**	**1**	**2**	**0**	**0**	**0**	**0**

Season	Club	League	GP	G	A	Pts	PIM	PP	SH	GW	S	%	+/-	TF	F%	H	SB	Min	GP	G	A	Pts	PIM	PP	SH	GW
			Regular Season																Playoffs							
2000-01	**Anaheim**	**NHL**	**71**	**9**	**11**	**20**	**61**	**1**	**0**	**0**	**78**	**11.5**	**−21**	**226**	**38.9**	**60**	**25**	**15:53**								
2001-02	**Anaheim**	**NHL**	**66**	**13**	**14**	**27**	**36**	**1**	**0**	**2**	**63**	**20.6**	**4**	**120**	**46.7**	**40**	**12**	**12:29**								
	NHL Totals		**624**	**157**	**220**	**377**	**311**	**51**	**11**	**21**	**1156**	**13.6**		**489**	**40.5**	**230**	**117**	**16:58**	**34**	**11**	**12**	**23**	**18**	**1**	**1**	**0**

Traded to **Pittsburgh** by **Calgary** with Todd Hlushko for Ken Wregget and Dave Roche, June 17, 1998. Traded to **Edmonton** by **Pittsburgh** for Josef Beranek, March 14, 2000. Signed as a free agent by **Anaheim**, July 1, 2000.

TJARNQVIST, Daniel (TUH-yahrn-kvihst, DAN-yehl) ATL.

Defense. Shoots left. 6'2", 190 lbs. Born, Umea, Sweden, October 14, 1976. Florida's 5th choice, 88th overall, in 1995 Entry Draft.

Season	Club	League	GP	G	A	Pts	PIM	PP	SH	GW	S	%	+/-	TF	F%	H	SB	Min	GP	G	A	Pts	PIM	PP	SH	GW
1994-95	Rogle	Sweden	18	0	1	1	2																			
	Rogle	Swede-Q	15	2	3	5	0																			
1995-96	Rogle	Sweden	22	1	7	8	6																			
1996-97	Jokerit Helsinki	Finland	44	3	8	11	4												9	0	3	3	4			
	Jokerit Helsinki	EuroHL	6	1	1	2	2																			
1997-98	Djurgarden	Sweden	40	5	9	14	12												15	1	1	2	2			
1998-99	Djurgarden	Sweden	40	4	3	7	16												4	0	0	0	2			
99-2000	Djurgarden	Sweden	42	3	16	19	8												5	0	0	0	2			
2000-01	Djurgarden	Sweden	45	9	17	26	26												16	6	5	11	2			
2001-02	**Atlanta**	**NHL**	**75**	**2**	**16**	**18**	**14**	**1**	**0**	**0**	**68**	**2.9**	**−22**	**4**	**25.0**	**78**	**90**	**21:32**								
	NHL Totals		**75**	**2**	**16**	**18**	**14**	**1**	**0**	**0**	**68**	**2.9**		**4**	**25.0**	**78**	**90**	**21:32**								

Traded to **Atlanta** by **Florida** with Gord Murphy, Herbert Vasiljevs and Ottawa's 6th round choice (previously acquired, later traded to Dallas - Dallas selected Justin Cox) in 1999 Entry Draft for Trevor Kidd, June 25, 1999.

TKACHUK, Keith (kuh-CHUK, KEETH) ST.L.

Left wing. Shoots left. 6'2", 225 lbs. Born, Melrose, MA, March 28, 1972. Winnipeg's 1st choice, 19th overall, in 1990 Entry Draft.

Season	Club	League	GP	G	A	Pts	PIM	PP	SH	GW	S	%	+/-	TF	F%	H	SB	Min	GP	G	A	Pts	PIM	PP	SH	GW
1988-89	Malden	Hi-School	21	30	16	46																				
1989-90	Malden	Hi-School	6	12	14	26																				
1990-91	Boston University	H-East	36	17	23	40	70																			
1991-92	Team USA	Nat-Tm	45	10	10	20	141																			
	United States	Olympics	8	1	1	2	12																			
	Winnipeg	**NHL**	**17**	**3**	**5**	**8**	**28**	**2**	**0**	**0**	**22**	**13.6**	**0**						**7**	**3**	**0**	**3**	**30**	**0**	**0**	**0**
1992-93	**Winnipeg**	**NHL**	**83**	**28**	**23**	**51**	**201**	**12**	**0**	**2**	**199**	**14.1**	**−13**						**6**	**4**	**0**	**4**	**14**	**1**	**0**	**0**
1993-94	**Winnipeg**	**NHL**	**84**	**41**	**40**	**81**	**255**	**22**	**3**	**3**	**218**	**18.8**	**−12**													
1994-95	**Winnipeg**	**NHL**	**48**	**22**	**29**	**51**	**152**	**7**	**2**	**2**	**129**	**17.1**	**−4**													
1995-96	**Winnipeg**	**NHL**	**76**	**50**	**48**	**98**	**156**	**20**	**2**	**6**	**249**	**20.1**	**11**						**6**	**1**	**2**	**3**	**22**	**0**	**0**	**0**
1996-97	**Phoenix**	**NHL**	**81**	***52**	**34**	**86**	**228**	**9**	**2**	**7**	**296**	**17.6**	**−1**						**7**	**6**	**0**	**6**	**7**	**2**	**0**	**0**
1997-98	**Phoenix**	**NHL**	**69**	**40**	**26**	**66**	**147**	**11**	**0**	**8**	**232**	**17.2**	**9**						**6**	**3**	**3**	**6**	**10**	**0**	**0**	**0**
	United States	Olympics	4	0	2	2	6																			
1998-99	**Phoenix**	**NHL**	**68**	**36**	**32**	**68**	**151**	**11**	**2**	**7**	**258**	**14.0**	**22**	**770**	**47.7**	**102**	**20**	**20:59**	**7**	**1**	**3**	**4**	**13**	**1**	**0**	**0**
99-2000	**Phoenix**	**NHL**	**50**	**22**	**21**	**43**	**82**	**5**	**1**	**1**	**183**	**12.0**	**7**	**500**	**50.4**	**100**	**9**	**19:21**	**5**	**1**	**1**	**2**	**4**	**1**	**0**	**0**
2000-01	**Phoenix**	**NHL**	**64**	**29**	**42**	**71**	**108**	**15**	**0**	**4**	**230**	**12.6**	**6**	**646**	**51.9**	**86**	**18**	**20:11**								
	St. Louis	**NHL**	**12**	**6**	**2**	**8**	**14**	**2**	**0**	**1**	**41**	**14.6**	**−3**	**87**	**54.0**	**25**	**3**	**19:39**	**15**	**2**	**7**	**9**	**20**	**2**	**0**	**1**
2001-02	**St. Louis**	**NHL**	**73**	**38**	**37**	**75**	**117**	**13**	**0**	**7**	**244**	**15.6**	**21**	**88**	**43.2**	**155**	**20**	**19:38**	**10**	**5**	**5**	**10**	**18**	**1**	**0**	**0**
	United States	Olympics	5	2	0	2	2																			
	NHL Totals		**725**	**367**	**339**	**706**	**1639**	**129**	**12**	**48**	**2301**	**15.9**		**2091**	**49.7**	**468**	**70**	**20:03**	**69**	**26**	**21**	**47**	**138**	**8**	**0**	**1**

NHL Second All-Star Team (1995, 1998) • Played in NHL All-Star Game (1997, 1998, 1999)

Transferred to **Phoenix** after **Winnipeg** franchise relocated, July 1, 1996. Traded to **St. Louis** by **Phoenix** for Michal Handzus, Ladislav Nagy, the rights to Jeff Taffe and St. Louis' 1st round choice (Ben Eager) in 2002 Entry Draft, March 13, 2001.

TKACZUK, Daniel (kuh-CHUK, DAN-yehl)

Center. Shoots left. 6'1", 197 lbs. Born, Toronto, Ont., June 10, 1979. Calgary's 1st choice, 6th overall, in 1997 Entry Draft.

Season	Club	League	GP	G	A	Pts	PIM	PP	SH	GW	S	%	+/-	TF	F%	H	SB	Min	GP	G	A	Pts	PIM	PP	SH	GW
1994-95	Mississauga Reps	MTHL	53	65	66	131	20																			
1995-96	Barrie Colts	OHL	61	22	39	61	38												7	1	2	3	8			
1996-97	Barrie Colts	OHL	62	45	48	93	49												9	7	2	9	2			
1997-98	Barrie Colts	OHL	57	35	40	75	38												6	2	3	5	8			
1998-99	Barrie Colts	OHL	58	43	62	105	58												12	7	8	15	10			
99-2000	Saint John	AHL	80	25	41	66	56												3	0	0	0	0			
2000-01	**Calgary**	**NHL**	**19**	**4**	**7**	**11**	**14**	**1**	**0**	**0**	**34**	**11.8**	**1**	**197**	**43.2**	**16**	**5**	**12:14**								
	Saint John	AHL	50	15	21	36	48												14	10	9	19	4			
2001-02	Worcester	AHL	75	10	27	37	37												3	1	1	2	2			
	NHL Totals		**19**	**4**	**7**	**11**	**14**	**1**	**0**	**0**	**34**	**11.8**		**197**	**43.1**	**16**	**5**	**12:14**								

OHL First All-Star Team (1999)

Traded to **St. Louis** by **Calgary** with Fred Brathwaite, Sergei Varlamov and Calgary's 9th round choice (Grant Jacobsen) in 2001 Entry Draft for Roman Turek and St. Louis' 4th round choice (Yegor Shastin) in 2001 Entry Draft, June 23, 2001.

TOBLER, Ryan (TOH-bluhr, RIGH-uhn) T.B.

Left wing. Shoots left. 6'3", 227 lbs. Born, Calgary, Alta., May 13, 1976.

Season	Club	League	GP	G	A	Pts	PIM	PP	SH	GW	S	%	+/-	TF	F%	H	SB	Min	GP	G	A	Pts	PIM	PP	SH	GW
1993-94	Calgary Royals	AJHL	56	32	17	49	195																			
1994-95	Saskatoon Blades	WHL	61	11	19	30	81												10	1	2	3	8			
1995-96	Calgary Hitmen	WHL	16	10	3	13	8																			
	Swift Current	WHL	25	17	11	28	31												6	1	1	2	2			
1996-97	Swift Current	WHL	39	10	17	27	40																			
	Moose Jaw	WHL	24	6	15	21	16												12	1	6	7	16			
1997-98	Lake Charles	WPHL	66	22	34	56	204												4	2	3	5	18			
	Utah Grizzlies	IHL	3	1	0	1	2																			
1998-99	Adirondack	AHL	64	9	18	27	157												3	0	0	0	2			
99-2000	Milwaukee	IHL	78	19	28	47	293												2	0	0	0	0			
2000-01	Milwaukee	IHL	49	7	9	16	196																			
	Hartford	AHL	13	1	5	6	71												5	0	0	0	2			
2001-02	Springfield	AHL	73	17	24	41	215																			
	Tampa Bay	**NHL**	**4**	**0**	**0**	**0**	**5**	**0**	**0**	**0**	**1**	**0.0**	**−2**	**0**	**0.0**	**1**	**1**	**4:08**								
	NHL Totals		**4**	**0**	**0**	**0**	**5**	**0**	**0**	**0**	**1**	**0.0**		**0**	**0.0**	**1**	**1**	**4:08**								

Signed as a free agent by **Nashville**, May 1, 2000. Traded to **NY Rangers** by **Nashville** for Bert Robertsson, March 7, 2001. Signed as a free agent by **Tampa Bay**, August 21, 2001.

TOCCHET, Rick (TAH-keht, RIHK)

Right wing. Shoots right. 6', 210 lbs. Born, Scarborough, Ont., April 9, 1964. Philadelphia's 5th choice, 125th overall, in 1983 Entry Draft.

Season	Club	League	GP	G	A	Pts	PIM	PP	SH	GW	S	%	+/-	TF	F%	H	SB	Min	GP	G	A	Pts	PIM	PP	SH	GW
1980-81	St. Michael's	MTHL	41	28	46	74																				
	St. Michael's B	OJHL-B	5	1	1	2	2																			
1981-82	Sault Ste. Marie	OHL	59	7	15	22	184												11	1	1	2	28			
1982-83	Sault Ste. Marie	OHL	66	32	34	66	146												16	4	13	17	67			
1983-84	Sault Ste. Marie	OHL	64	44	64	108	209												16	*22	14	*36	41			
1984-85	**Philadelphia**	**NHL**	**75**	**14**	**25**	**39**	**181**	**0**	**0**	**0**	**112**	**12.5**	**6**						**19**	**3**	**4**	**7**	**72**	**0**	**0**	**2**
1985-86	**Philadelphia**	**NHL**	**69**	**14**	**21**	**35**	**284**	**3**	**0**	**1**	**107**	**13.1**	**12**						**5**	**1**	**2**	**3**	**26**	**0**	**0**	**0**
1986-87	**Philadelphia**	**NHL**	**69**	**21**	**28**	**49**	**288**	**1**	**1**	**5**	**147**	**14.3**	**16**						**26**	**11**	**10**	**21**	**72**	**0**	**1**	**2**
1987-88	**Philadelphia**	**NHL**	**65**	**31**	**33**	**64**	**299**	**10**	**2**	**3**	**182**	**17.0**	**3**						**5**	**1**	**4**	**5**	**55**	**0**	**1**	**0**
1988-89	**Philadelphia**	**NHL**	**66**	**45**	**36**	**81**	**183**	**16**	**1**	**5**	**220**	**20.5**	**−1**						**16**	**6**	**6**	**12**	**69**	**2**	**0**	**1**
1989-90	**Philadelphia**	**NHL**	**75**	**37**	**59**	**96**	**196**	**15**	**1**	**0**	**269**	**13.8**	**4**													
1990-91	**Philadelphia**	**NHL**	**70**	**40**	**31**	**71**	**150**	**8**	**0**	**5**	**217**	**18.4**	**2**													
1991-92	**Philadelphia**	**NHL**	**42**	**13**	**16**	**29**	**102**	**4**	**0**	**1**	**107**	**12.1**	**3**													
	♦ **Pittsburgh**	**NHL**	**19**	**14**	**16**	**30**	**49**	**4**	**1**	**1**	**59**	**23.7**	**12**						**14**	**6**	**13**	**19**	**24**	**3**	**0**	**1**
1992-93	**Pittsburgh**	**NHL**	**80**	**48**	**61**	**109**	**252**	**20**	**4**	**5**	**240**	**20.0**	**28**						**12**	**7**	**6**	**13**	**24**	**1**	**0**	**0**
1993-94	**Pittsburgh**	**NHL**	**51**	**14**	**26**	**40**	**134**	**5**	**1**	**2**	**150**	**9.3**	**−15**						**6**	**2**	**3**	**5**	**20**	**1**	**0**	**1**
1994-95	**Los Angeles**	**NHL**	**36**	**18**	**17**	**35**	**70**	**7**	**1**	**3**	**95**	**18.9**	**−8**													
1995-96	**Los Angeles**	**NHL**	**44**	**13**	**23**	**36**	**117**	**4**	**0**	**0**	**100**	**13.0**	**3**													
	Boston	**NHL**	**27**	**16**	**8**	**24**	**64**	**6**	**0**	**3**	**85**	**18.8**	**7**						**5**	**4**	**0**	**4**	**21**	**3**	**0**	**1**

			Regular Season																Playoffs							
Season	Club	League	GP	G	A	Pts	PIM	PP	SH	GW	S	%	+/-	TF	F%	H	SB	Min	GP	G	A	Pts	PIM	PP	SH	GW
1996-97	Boston	NHL	40	16	14	30	67	3	0	1	120	13.3	–3													
	Washington	NHL	13	5	5	10	31	1	0	1	37	13.5	0													
1997-98	Phoenix	NHL	68	26	19	45	157	8	0	6	161	16.1	1						6	6	2	8	25	3	0	0
1998-99	Phoenix	NHL	81	26	30	56	147	6	1	5	178	14.6	5	4	25.0	115	13	18:34	7	0	3	3	8	0	0	0
99-2000	Phoenix	NHL	64	12	17	29	67	2	0	1	107	11.2	–5	1	0.0	103	17	15:54								
	Philadelphia	NHL	16	3	3	6	23	2	0	0	23	13.0	4	0	0.0	26	3	14:47	18	5	6	11	*49	2	0	1
2000-01	Philadelphia	NHL	60	14	22	36	83	5	0	2	76	18.4	10	1	0.0	68	8	15:20	6	0	1	1	6	0	0	0
2001-02	Philadelphia	NHL	14	0	2	2	28	0	0	0	10	0.0	–2	1	100.0	13	0	9:13								
	NHL Totals		1144	440	512	952	2972	130	13	50	2802	15.7		7	28.6	325	41	16:12	145	52	60	112	471	15	2	9

Played in NHL All-Star Game (1989, 1990, 1991, 1993)

Traded to **Pittsburgh** by **Philadelphia** with Kjell Samuelsson, Ken Wregget and Philadelphia's 3rd round choice (Dave Roche) in 1993 Entry Draft for Mark Recchi, Brian Benning and LA Kings' 1st round choice (previously acquired, Philadelphia selected Jason Bowen) in 1992 Entry Draft, February 19, 1992 Traded to **LA Kings** by **Pittsburgh** with Pittsburgh's 2nd round choice (Pavel Rosa) in 1995 Entry Draft for Luc Robitaille, July 29, 1994. Traded to **Boston** by **LA Kings** for Kevin Stevens, January 25, 1996. Traded to **Washington** by **Boston** with Bill Ranford and Adam Oates for Jim Carey, Anson Carter, Jason Allison and Washington's 3rd round choice (Lee Goren) in 1997 Entry Draft, March 1, 1997. Signed as a free agent by **Phoenix**, July 23, 1997. Traded to **Philadelphia** by **Phoenix** for Mikael Renberg, March 8, 2000. • Missed majority of 2001-02 season recovering from knee injury suffered in training camp, October 1, 2001.

TOMS, Jeff (TAWMS, JEHF) FLA.

Left wing. Shoots left. 6'5", 200 lbs. Born, Swift Current, Sask., June 4, 1974. New Jersey's 10th choice, 210th overall, in 1992 Entry Draft.

Season	Club	League	GP	G	A	Pts	PIM	PP	SH	GW	S	%	+/-	TF	F%	H	SB	Min	GP	G	A	Pts	PIM	PP	SH	GW
1990-91	Oakville	OMHA	58	34	47	81	72																			
1991-92	Sault Ste. Marie	OHL	36	9	5	14	0												16	0	1	1	2			
1992-93	Sault Ste. Marie	OHL	59	16	23	39	20												16	4	4	8	7			
1993-94	Sault Ste. Marie	OHL	64	52	45	97	19												14	11	4	15	2			
1994-95	Atlanta Knights	IHL	40	7	8	15	10												4	0	0	0	4			
1995-96	Tampa Bay	NHL	1	0	0	0	0	0	0	0	1	0.0	0													
	Atlanta Knights	IHL	68	16	18	34	18												1	0	0	0	0			
1996-97	Tampa Bay	NHL	34	2	8	10	10	0	0	1	53	3.8	2													
	Adirondack	AHL	37	11	16	27	8												4	1	2	3	0			
1997-98	Tampa Bay	NHL	13	1	2	3	7	0	0	0	14	7.1	–6													
	Washington	NHL	33	3	4	7	8	0	0	1	55	5.5	–11						1	0	0	0	0	0	0	0
1998-99	Washington	NHL	21	1	5	6	2	0	0	0	30	3.3	0	92	54.3	9	1	13:35								
	Portland Pirates	AHL	20	3	7	10	8																			
99-2000	Washington	NHL	20	1	2	3	4	0	0	1	18	5.6	–1	17	52.9	8	2	8:26								
	Portland Pirates	AHL	33	16	21	37	16												4	1	1	2	2			
2000-01	NY Islanders	NHL	39	2	4	6	10	0	0	0	37	5.4	–7	172	42.4	30	4	9:44								
	Springfield	AHL	5	6	5	11	0																			
	NY Rangers	NHL	15	1	1	2	0	0	0	0	12	8.3	–3	21	33.3	1	2	6:34								
	Hartford	AHL	12	4	9	13	2												5	6	0	6	2			
2001-02	NY Rangers	NHL	38	7	4	11	10	2	0	0	62	11.3	–4	243	45.3	14	11	10:31								
	Hartford	AHL	9	6	6	12	4																			
	Pittsburgh	NHL	14	2	1	3	4	0	0	0	22	9.1	–5	71	40.9	6	1	12:19								
	NHL Totals		228	20	31	51	55	2	0	3	304	6.6		616	45.1	68	21	10:14	1	0	0	0	0	0	0	0

Traded to **Tampa Bay** by **New Jersey** for Vancouver's 4th round choice (previously acquired by Tampa Bay - later traded to New Jersey - later traded to Calgary - Calgary selected Ryan Duthie) in 1994 Entry Draft, May 31, 1994. Claimed by on waivers by **Washington** from **Tampa Bay**, November 19, 1997. Signed as a free agent by **NY Islanders**, July 27, 2000. Claimed on waivers by **NY Rangers** from **NY Islanders**, January 13, 2001. Claimed on waivers by **Pittsburgh** from **NY Rangers**, March 16, 2002. Signed as a free agent by **Florida**, July 11, 2002.

TORRES, Raffi (TAW-rehs, RA-fee) NYI

Left wing. Shoots left. 6', 218 lbs. Born, Toronto, Ont., October 8, 1981. NY Islanders' 2nd choice, 5th overall, in 2000 Entry Draft.

Season	Club	League	GP	G	A	Pts	PIM	PP	SH	GW	S	%	+/-	TF	F%	H	SB	Min	GP	G	A	Pts	PIM	PP	SH	GW
1997-98	Thornhill	MTJHL	46	17	16	33	90																			
1998-99	Brampton	OHL	62	35	27	62	32																			
99-2000	Brampton	OHL	68	43	48	91	40												6	5	2	7	23			
2000-01	Brampton	OHL	55	33	37	70	76												8	7	4	11	19			
2001-02	NY Islanders	NHL	14	0	1	1	6	0	0	0	9	0.0	2	0	0.0	26	1	7:35								
	Bridgeport	AHL	59	20	10	30	45												20	8	9	17	26			
	NHL Totals		14	0	1	1	6	0	0	0	9	0.0		0	0.0	26	1	7:35								

OHL All-Rookie Team (1999) • OHL Second All-Star Team (2000, 2001)

TRAVERSE, Patrick (tra-VAIRZ, PAT-rihk) MTL.

Defense. Shoots left. 6'4", 207 lbs. Born, Montreal, Que., March 14, 1974. Ottawa's 3rd choice, 50th overall, in 1992 Entry Draft.

Season	Club	League	GP	G	A	Pts	PIM	PP	SH	GW	S	%	+/-	TF	F%	H	SB	Min	GP	G	A	Pts	PIM	PP	SH	GW
1990-91	Mtl-Bourassa	QAAA	42	4	19	23	10												5	0	3	3	2			
1991-92	Shawinigan	QMJHL	59	3	11	14	12												10	0	0	0	4			
1992-93	Shawinigan	QMJHL	53	5	24	29	24																			
	St-Jean Lynx	QMJHL	15	1	6	7	0												4	0	1	1	2			
	New Haven	AHL	2	0	0	0	2																			
1993-94	St-Jean Lynx	QMJHL	66	15	37	52	30												5	0	4	4	4			
	P.E.I. Senators	AHL	3	0	1	1	2																			
1994-95	P.E.I. Senators	AHL	70	5	13	18	19												7	0	2	2	0			
1995-96	Ottawa	NHL	5	0	0	0	2	0	0	0	2	0.0	–1													
	P.E.I. Senators	AHL	55	4	21	25	32												5	1	2	3	2			
1996-97	Worcester	AHL	24	0	4	4	23																			
	Grand Rapids	IHL	10	2	1	3	10												2	0	1	1	2			
1997-98	Hershey Bears	AHL	71	14	15	29	67												7	1	3	4	4			
1998-99	Ottawa	NHL	46	1	9	10	22	0	0	0	35	2.9	12	0	0.0	41	42	14:56								
99-2000	Ottawa	NHL	66	6	17	23	21	1	0	0	73	8.2	17	0	0.0	51	78	18:43	6	0	0	0	2	0	0	0
2000-01	Anaheim	NHL	15	1	0	1	6	0	0	0	7	14.3	–6	0	0.0	15	12	17:19								
	Boston	NHL	37	2	6	8	14	1	0	1	39	5.1	4	0	0.0	39	25	16:38								
	Montreal	NHL	19	2	3	5	10	0	0	0	16	12.5	–8	0	0.0	25	27	21:36								
2001-02	Quebec	AHL	4	0	2	2	4																			
	Montreal	NHL	25	2	3	5	14	2	0	0	24	8.3	–7	0	0.0	18	30	18:14								
	NHL Totals		213	14	38	52	89	4	0	1	196	7.1		0	0.0	189	214	17:37	6	0	0	0	2	0	0	0

Traded to **Anaheim** by **Ottawa** for Joel Kwiatkowski, June 12, 2000. Traded to **Boston** by **Anaheim** with Andrei Nazarov for Sami Pahlsson, November 18, 2000. • Missed majority of 2001-02 season recovering from knee (November 3, 2001 vs. Calgary) and head (January 10, 2002 vs. NY Islanders) injuries. Traded to **Montreal** by **Boston** for Eric Weinrich, February 21, 2001.

TREBIL, Dan (TREH-bihl, DAN-yehl)

Defense. Shoots right. 6'3", 210 lbs. Born, Bloomington, MN, April 10, 1974. New Jersey's 7th choice, 138th overall, in 1992 Entry Draft.

Season	Club	League	GP	G	A	Pts	PIM	PP	SH	GW	S	%	+/-	TF	F%	H	SB	Min	GP	G	A	Pts	PIM	PP	SH	GW
1989-90	Jefferson High	Hi-School	22	3	6	9	10																			
1990-91	Jefferson High	Hi-School	23	4	12	16	8																			
1991-92	Jefferson High	Hi-School	28	7	26	33	6																			
1992-93	U. of Minnesota	WCHA	36	2	11	13	16																			
1993-94	U. of Minnesota	WCHA	42	1	21	22	24																			
1994-95	U. of Minnesota	WCHA	44	10	33	43	10																			
1995-96	U. of Minnesota	WCHA	42	11	35	46	36																			
1996-97	Anaheim	NHL	29	3	3	6	23	0	0	0	30	10.0	5						9	0	1	1	6	0	0	0
	Baltimore	AHL	49	4	20	24	38																			
1997-98	Anaheim	NHL	21	0	1	1	2	0	0	0	11	0.0	–8													
	Cincinnati	AHL	32	5	15	20	21																			
1998-99	Anaheim	NHL	6	0	0	0	0	0	0	0	1	0.0	–2	1	0.0	7	7	15:02	1	0	0	0	2	0	0	0
	Cincinnati	AHL	52	6	15	21	31																			
99-2000	Cincinnati	AHL	52	7	21	28	48																			
	Pittsburgh	NHL	3	1	0	1	0	0	0	0	2	50.0	2	0	0.0	3	1	12:33								

			Regular Season																Playoffs							
Season	Club	League	GP	G	A	Pts	PIM	PP	SH	GW	S	%	+/-	TF	F%	H	SB	Min	GP	G	A	Pts	PIM	PP	SH	GW
2000-01	**Pittsburgh**	**NHL**	**16**	**0**	**0**	**0**	**7**	**0**	**0**	**0**	**17**	**0.0**	**–1**	**0**	**0.0**	**23**	**6**	**12:59**								
	Chicago Wolves	IHL	6	0	2	2	4																			
	St. Louis	**NHL**	**10**	**0**	**0**	**0**	**0**	**0**	**0**	**0**	**9**	**0.0**	**1**	**0**	**0.0**	**7**	**6**	**14:14**								
	Worcester	AHL	14	2	7	9	0												9	0	5	5	2			
2001-02	Hammarby	Swede-2	18	3	8	11	18																			
	NHL Totals		**85**	**4**	**4**	**8**	**32**	**0**	**0**	**0**	**70**	**5.7**		**1**	**0.0**	**40**	**20**	**13:39**	**10**	**0**	**1**	**1**	**8**	**0**	**0**	**0**

WCHA Second All-Star Team (1996) • NCAA West Second All-American Team (1996)

Signed as a free agent by **Anaheim**, May 30, 1996. Traded to **Pittsburgh** by **Anaheim** for Pittsburgh's 5th round choice (Bill Cass) in 2000 Entry Draft, March 14, 2000. Signed as a free agent by **NY Islanders**, July 31, 2000. Traded to **Pittsburgh** by **NY Islanders** for Pittsburgh's 9th round choice (Roman Kuhtinov) in 2001 Entry Draft, November 14, 2000. Traded to **St. Louis** by **Pittsburgh** for Marc Bergevin, December 28, 2000. Signed as a free agent by **Hammarby** (Swede-2), October 14, 2001.

TREMBLAY, Yannick

(TRAHM-blay, YA-nihk) **ATL.**

Defense. Shoots right. 6'2", 200 lbs. Born, Pointe-aux-Trembles, Que., November 15, 1975. Toronto's 4th choice, 145th overall, in 1995 Entry Draft.

Season	Club	League	GP	G	A	Pts	PIM	PP	SH	GW	S	%	+/-	TF	F%	H	SB	Min	GP	G	A	Pts	PIM	PP	SH	GW
1991-92	Mtl-Bourassa	QAAA	35	2	5	7	55												8	0	4	4	2			
1992-93	Mtl-Bourassa	CEGEP	21	2	5	7	10												3	0	0	0	2			
1993-94	St. Thomas	AUAA	25	2	3	5	10																			
1994-95	Beauport	QMJHL	70	10	32	42	22												17	6	8	14	6			
1995-96	Beauport	QMJHL	61	12	33	45	42												20	3	16	19	18			
	St. John's	AHL	3	0	1	1	0																			
1996-97	**Toronto**	**NHL**	**5**	**0**	**0**	**0**	**0**	**0**	**0**	**0**	**2**	**0.0**	**–4**													
	St. John's	AHL	67	7	25	32	34												11	2	9	11	0			
1997-98	**Toronto**	**NHL**	**38**	**2**	**4**	**6**	**6**	**1**	**0**	**0**	**45**	**4.4**	**–6**													
	St. John's	AHL	17	3	7	10	4												4	0	1	1	5			
1998-99	**Toronto**	**NHL**	**35**	**2**	**7**	**9**	**16**	**0**	**0**	**0**	**37**	**5.4**	**0**	**0**	**0.0**	**20**	**28**	**17:39**								
99-2000	**Atlanta**	**NHL**	**75**	**10**	**21**	**31**	**22**	**4**	**1**	**2**	**139**	**7.2**	**–42**	**3**	**0.0**	**117**	**91**	**19:27**								
2000-01	**Atlanta**	**NHL**	**46**	**4**	**8**	**12**	**30**	**1**	**0**	**1**	**102**	**3.9**	**–6**	**1**	**100.0**	**84**	**45**	**20:27**								
2001-02	**Atlanta**	**NHL**	**66**	**9**	**15**	**24**	**47**	**1**	**0**	**1**	**115**	**7.8**	**–15**	**0**	**0.0**	**96**	**92**	**21:50**								
	NHL Totals		**265**	**27**	**55**	**82**	**121**	**7**	**1**	**4**	**440**	**6.1**		**4**	**25.0**	**317**	**256**	**20:05**								

Claimed by **Atlanta** from **Toronto** in Expansion Draft, June 25, 1999.

TREPANIER, Pascal

(TREHP-uhn-yay, PAS-kal) **NSH.**

Defense. Shoots right. 6', 210 lbs. Born, Gaspe, Que., September 4, 1973.

Season	Club	League	GP	G	A	Pts	PIM	PP	SH	GW	S	%	+/-	TF	F%	H	SB	Min	GP	G	A	Pts	PIM	PP	SH	GW
1989-90	Jonquiere Elites	QAAA	40	2	8	10	46																			
1990-91	Hull Olympiques	QMJHL	46	3	3	6	56												4	0	2	2	7			
1991-92	Trois-Rivieres	QMJHL	53	4	18	22	125												15	3	5	8	21			
1992-93	Sherbrooke	QMJHL	59	15	33	48	130												15	5	7	12	36			
1993-94	Sherbrooke	QMJHL	48	16	41	57	67												12	1	8	9	14			
1994-95	Dayton Bombers	ECHL	36	16	28	44	113												9	2	4	6	20			
	Kalamazoo Wings	IHL	14	1	2	3	47																			
	Cornwall Aces	AHL	4	0	0	0	9												14	2	7	9	32			
1995-96	Cornwall Aces	AHL	70	13	20	33	142												8	1	2	3	24			
1996-97	Hershey Bears	AHL	73	14	39	53	151												23	6	13	19	59			
1997-98	**Colorado**	**NHL**	**15**	**0**	**1**	**1**	**18**	**0**	**0**	**0**	**9**	**0.0**	**–2**													
	Hershey Bears	AHL	43	13	18	31	105												7	4	2	6	8			
1998-99	**Anaheim**	**NHL**	**45**	**2**	**4**	**6**	**48**	**0**	**0**	**1**	**49**	**4.1**	**0**	**1**	**0.0**	**65**	**52**	**12:42**								
99-2000	**Anaheim**	**NHL**	**37**	**0**	**4**	**4**	**54**	**0**	**0**	**0**	**33**	**0.0**	**2**	**1**	**0.0**	**61**	**30**	**11:18**								
2000-01	**Anaheim**	**NHL**	**57**	**6**	**4**	**10**	**73**	**3**	**0**	**0**	**86**	**7.0**	**–12**	**1**	**100.0**	**82**	**63**	**16:47**								
2001-02	**Colorado**	**NHL**	**74**	**4**	**9**	**13**	**59**	**2**	**0**	**0**	**87**	**4.6**	**4**	**0**	**0.0**	**107**	**88**	**14:37**	**2**	**0**	**0**	**0**	**0**	**0**	**0**	**0**
	NHL Totals		**228**	**12**	**22**	**34**	**252**	**5**	**0**	**1**	**264**	**4.5**		**3**	**33.3**	**315**	**233**	**14:13**	**2**	**0**	**0**	**0**	**0**	**0**	**0**	**0**

AHL Second All-Star Team (1997)

Signed as a free agent by **Colorado**, August 30, 1995. Claimed by **Anaheim** from **Colorado** in NHL Waiver Draft, October 5, 1998. Signed as a free agent by **Colorado**, September, 2001. Signed as a free agent by **Nashville**, July 16, 2002.

TRNKA, Pavel

(truhn-KAH, PAH-vehl) **ANA.**

Defense. Shoots left. 6'2", 200 lbs. Born, Plzen, Czech., July 27, 1976. Anaheim's 5th choice, 106th overall, in 1994 Entry Draft.

Season	Club	League	GP	G	A	Pts	PIM	PP	SH	GW	S	%	+/-	TF	F%	H	SB	Min	GP	G	A	Pts	PIM	PP	SH	GW
1993-94	HC Skoda Plzen	Czech	12	0	1	1	....																			
1994-95	HC Kladno	Czech	28	0	5	5	24																			
	Plzen	Czech	6	0	0	0	0																			
1995-96	Baltimore	AHL	69	2	6	8	44												6	0	0	0	2			
1996-97	Baltimore	AHL	69	6	14	20	86												3	0	0	0	2			
1997-98	**Anaheim**	**NHL**	**48**	**3**	**4**	**7**	**40**	**1**	**0**	**0**	**46**	**6.5**	**–4**													
	Cincinnati	AHL	23	3	5	8	28																			
1998-99	**Anaheim**	**NHL**	**63**	**0**	**4**	**4**	**60**	**0**	**0**	**0**	**50**	**0.0**	**–6**	**0**	**0.0**	**115**	**41**	**16:06**	**4**	**0**	**1**	**1**	**2**	**0**	**0**	**0**
99-2000	**Anaheim**	**NHL**	**57**	**2**	**15**	**17**	**34**	**0**	**0**	**0**	**54**	**3.7**	**12**	**0**	**0.0**	**152**	**48**	**19:21**								
2000-01	**Anaheim**	**NHL**	**59**	**1**	**7**	**8**	**42**	**0**	**0**	**0**	**59**	**1.7**	**–12**	**0**	**0.0**	**146**	**54**	**20:07**								
2001-02	**Anaheim**	**NHL**	**71**	**2**	**11**	**13**	**66**	**1**	**0**	**0**	**78**	**2.6**	**–5**	**0**	**0.0**	**144**	**44**	**17:03**								
	NHL Totals		**298**	**8**	**41**	**49**	**242**	**2**	**0**	**0**	**287**	**2.8**		**0**	**0.0**	**557**	**187**	**18:04**	**4**	**0**	**1**	**1**	**2**	**0**	**0**	**0**

TRUDEL, Jean-Guy

(TROO-dehl, zhawn-gee) **MIN.**

Left wing. Shoots left. 5'11", 202 lbs. Born, Sudbury, Ont., October 18, 1975.

Season	Club	League	GP	G	A	Pts	PIM	PP	SH	GW	S	%	+/-	TF	F%	H	SB	Min	GP	G	A	Pts	PIM	PP	SH	GW
1991-92	Beauport	QMJHL	35	5	7	12	20																			
1992-93	Beauport	QMJHL	56	1	4	5	20																			
	Verdun	QMJHL	10	1	0	1	0												2	0	0	0	5			
1993-94			DID NOT PLAY																							
1994-95	Hull Olympiques	QMJHL	54	29	42	71	76												19	4	13	17	25			
1995-96	Hull Olympiques	QMJHL	70	50	71	121	96												17	11	18	29	8			
1996-97	Quad City	ColHL	5	8	7	15	4																			
	Chicago Wolves	IHL	6	1	2	3	2																			
	San Antonio	IHL	12	1	5	6	4																			
	Peoria Rivermen	ECHL	37	25	29	54	47												9	9	10	19	22			
1997-98	Peoria Rivermen	ECHL	62	39	74	113	147												3	0	0	0	2			
1998-99	Kansas City	IHL	76	24	25	49	66												3	1	0	1	0			
99-2000	**Phoenix**	**NHL**	**1**	**0**	**0**	**0**	**0**	**0**	**0**	**0**	**0**	**0.0**	**–1**	**0**	**0.0**	**4**	**0**	**4:33**								
	Springfield	AHL	72	34	39	73	80												3	0	1	1	4			
2000-01	Springfield	AHL	80	34	65	99	89																			
2001-02	**Phoenix**	**NHL**	**3**	**0**	**0**	**0**	**2**	**0**	**0**	**0**	**1**	**0.0**	**0**	**0**	**0.0**	**1**	**2**	**8:42**								
	Springfield	AHL	76	22	48	70	83																			
	NHL Totals		**4**	**0**	**0**	**0**	**2**	**0**	**0**	**0**	**1**	**0.0**		**0**	**0.0**	**5**	**2**	**7:40**								

ECHL First All-Star Team (1998) • AHL Second All-Star Team (2000, 2002) • AHL First All-Star Team (2001)

• Sat out 1993-94 season to regain eligibility for U.S. College scholarship. Signed as a free agent by **Phoenix**, July 17, 1999. Signed as a free agent by **Minnesota**, July 16, 2002.

TSELIOS, Nikos

(TSEHL-ee-ohs, NEE-kohs) **CAR.**

Defense. Shoots left. 6'5", 210 lbs. Born, Oak Park, IL, January 20, 1979. Carolina's 1st choice, 22nd overall, in 1997 Entry Draft.

Season	Club	League	GP	G	A	Pts	PIM	PP	SH	GW	S	%	+/-	TF	F%	H	SB	Min	GP	G	A	Pts	PIM	PP	SH	GW
1995-96	Chicago	MEHL	27	5	8	13	40																			
1996-97	Belleville Bulls	OHL	64	9	37	46	61												6	1	1	2	2			
1997-98	Belleville Bulls	OHL	20	2	10	12	16																			
	Plymouth Whalers	OHL	41	8	20	28	27												15	1	8	9	27			
1998-99	Plymouth Whalers	OHL	60	21	39	60	60												11	4	10	14	8			
99-2000	Cincinnati	IHL	80	3	19	22	75												10	0	2	2	4			
2000-01	Cincinnati	IHL	79	7	18	25	98												5	0	3	3	0			

Season	Club	League	Regular Season GP	G	A	Pts	PIM	PP	SH	GW	S	%	+/-	TF	F%	H	SB	Min	Playoffs GP	G	A	Pts	PIM	PP	SH	GW
2001-02	**Carolina**	**NHL**	**2**	**0**	**0**	**0**	**6**	**0**	**0**	**0**	**3**	**0.0**	**−2**	**0**	**0.0**	**3**	**1**	**13:13**								
	Lowell	AHL	70	3	16	19	64																			
	NHL Totals		**2**	**0**	**0**	**0**	**6**	**0**	**0**	**0**	**3**	**0.0**		**0**	**0.0**	**3**	**1**	**13:13**								

OHL All-Rookie Team (1997)

TSYPLAKOV, Vladimir

(tsih-plah-KAHF, vla-DIH-meer)

Left wing. Shoots left. 6'1", 197 lbs. Born, Inta, USSR, April 18, 1969. Los Angeles' 4th choice, 59th overall, in 1995 Entry Draft.

Season	Club	League	GP	G	A	Pts	PIM	PP	SH	GW	S	%	+/-	TF	F%	H	SB	Min	GP	G	A	Pts	PIM	PP	SH	GW
1988-89	Dynamo Minsk	USSR	19	6	1	7	4																			
1989-90	Dynamo Minsk	USSR	47	11	6	17	20																			
1990-91	Dynamo Minsk	USSR	28	6	5	11	14																			
1991-92	Dynamo Minsk	CIS	29	10	9	19	16																			
1992-93	Detroit Falcons	ColHL	44	33	43	76	20												6	5	4	9	6			
	Indianapolis Ice	IHL	11	6	7	13	4												5	1	1	2	2			
1993-94	Fort Wayne	IHL	63	31	32	63	51												14	6	8	14	16			
1994-95	Fort Wayne	IHL	79	38	40	78	39												4	2	4	6	2			
1995-96	**Los Angeles**	**NHL**	**23**	**5**	**5**	**10**	**4**	**0**	**0**	**0**	**40**	**12.5**	**1**													
	Las Vegas	IHL	9	5	6	11	4																			
1996-97	**Los Angeles**	**NHL**	**67**	**16**	**23**	**39**	**12**	**1**	**0**	**2**	**118**	**13.6**	**8**													
1997-98	**Los Angeles**	**NHL**	**73**	**18**	**34**	**52**	**18**	**2**	**0**	**1**	**113**	**15.9**	**15**						**4**	**0**	**1**	**1**	**8**	**0**	**0**	**0**
	Belarus	Olympics	5	1	1	2	2																			
1998-99	**Los Angeles**	**NHL**	**69**	**11**	**12**	**23**	**32**	**0**	**2**	**2**	**111**	**9.9**	**−7**	**1**	**100.0**	**72**	**14**	**16:23**								
99-2000	**Los Angeles**	**NHL**	**29**	**6**	**7**	**13**	**4**	**1**	**0**	**1**	**30**	**20.0**	**6**	**3**	**0.0**	**32**	**5**	**10:39**								
	Buffalo	**NHL**	**34**	**6**	**13**	**19**	**10**	**0**	**0**	**1**	**46**	**13.0**	**17**	**0**	**0.0**	**44**	**4**	**13:41**	**5**	**0**	**1**	**1**	**4**	**0**	**0**	**0**
2000-01	**Buffalo**	**NHL**	**36**	**7**	**7**	**14**	**10**	**0**	**0**	**0**	**39**	**17.9**	**2**	**1**	**0.0**	**29**	**9**	**11:18**	**9**	**1**	**0**	**1**	**4**	**0**	**0**	**0**
2001-02	Ak Bars Kazan	Russia	25	5	14	19	22												11	4	2	6	4			
	Belarus	Olympics	8	1	3	4	4																			
	NHL Totals		**331**	**69**	**101**	**170**	**90**	**4**	**2**	**7**	**497**	**13.9**		**5**	**20.0**	**177**	**32**	**13:45**	**18**	**1**	**2**	**3**	**16**	**0**	**0**	**0**

ColHL First All-Star Team (1993)

Traded to **Buffalo** by **LA Kings** for Buffalo's 8th round choice (Dan Welch) in 2000 Entry Draft, January 24, 2000. • Missed majority of 2000-01 season recovering from knee injury suffered in game vs. Edmonton, October 13, 2000.

TUCKER, Darcy

(TUH-kuhr, DAHR-see) **TOR.**

Center. Shoots left. 5'11", 185 lbs. Born, Castor, Alta., March 15, 1975. Montreal's 8th choice, 151st overall, in 1993 Entry Draft.

Season	Club	League	GP	G	A	Pts	PIM	PP	SH	GW	S	%	+/-	TF	F%	H	SB	Min	GP	G	A	Pts	PIM	PP	SH	GW
1990-91	Red Deer	AMHL	47	70	90	160	48																			
1991-92	Kamloops Blazers	WHL	26	3	10	13	32												9	0	1	1	16			
1992-93	Kamloops Blazers	WHL	67	31	58	89	155												13	7	6	13	34			
1993-94	Kamloops Blazers	WHL	66	52	88	140	143												19	9	*18	*27	43			
1994-95	Kamloops Blazers	WHL	64	64	73	137	94												21	*16	15	*31	19			
1995-96	**Montreal**	**NHL**	**3**	**0**	**0**	**0**	**0**	**0**	**0**	**0**	**1**	**0.0**	**−1**													
	Fredericton	AHL	74	29	64	93	174												7	7	3	10	14			
1996-97	**Montreal**	**NHL**	**73**	**7**	**13**	**20**	**110**	**1**	**0**	**3**	**62**	**11.3**	**−5**						**4**	**0**	**0**	**0**	**0**	**0**	**0**	**0**
1997-98	**Montreal**	**NHL**	**39**	**1**	**5**	**6**	**57**	**0**	**0**	**0**	**19**	**5.3**	**−6**													
	Tampa Bay	**NHL**	**35**	**6**	**8**	**14**	**89**	**1**	**1**	**0**	**44**	**13.6**	**−8**													
1998-99	**Tampa Bay**	**NHL**	**82**	**21**	**22**	**43**	**176**	**8**	**2**	**3**	**178**	**11.8**	**−34**	**1470**	**45.6**	**120**	**49**	**19:24**								
99-2000	**Tampa Bay**	**NHL**	**50**	**14**	**20**	**34**	**108**	**1**	**0**	**2**	**98**	**14.3**	**−15**	**152**	**48.7**	**106**	**19**	**19:58**								
	Toronto	**NHL**	**27**	**7**	**10**	**17**	**55**	**0**	**2**	**3**	**40**	**17.5**	**3**	**11**	**54.6**	**62**	**19**	**16:41**	**12**	**4**	**2**	**6**	**15**	**1**	**0**	**2**
2000-01	**Toronto**	**NHL**	**82**	**16**	**21**	**37**	**141**	**2**	**0**	**4**	**122**	**13.1**	**6**	**413**	**47.0**	**151**	**50**	**16:09**	**11**	**0**	**2**	**2**	**6**	**0**	**0**	**0**
2001-02	**Toronto**	**NHL**	**77**	**24**	**35**	**59**	**92**	**7**	**0**	**5**	**124**	**19.4**	**24**	**138**	**43.5**	**101**	**40**	**16:59**	**17**	**4**	**4**	**8**	**38**	**1**	**0**	**1**
	NHL Totals		**468**	**96**	**134**	**230**	**828**	**20**	**5**	**20**	**688**	**14.0**		**2184**	**46.0**	**540**	**177**	**17:50**	**44**	**8**	**8**	**16**	**59**	**2**	**0**	**3**

WHL West First All-Star Team (1994, 1995) • Canadian Major Junior First All-Star Team (1994) • Memorial Cup All-Star Team (1994, 1995) • Won Stafford Smythe Memorial Trophy (Memorial Cup Tournament MVP) (1994) • Won Dudley "Red" Garrett Memorial Trophy (Top Rookie - AHL) (1996)

Traded to **Tampa Bay** by **Montreal** with Stephane Richer and David Wilkie for Patrick Poulin, Mick Vukota and Igor Ulanov, January 15, 1998. Traded to **Toronto** by **Tampa Bay** with Tampa Bay's 4th round choice (Miguel Delisle) in 2000 Entry Draft and future considerations for Mike Johnson, Marek Posmyk, Toronto's 5th (Pavel Sedov) and 6th (Aaron Gionet) round choices in 2000 Entry Draft and future considerations, February 9, 2000.

TUOMAINEN, Marko

(TOO-oh-migh-nehn, MAHR-koh)

Right wing. Shoots right. 6'3", 230 lbs. Born, Kuopio, Finland, April 25, 1972. Edmonton's 10th choice, 205th overall, in 1992 Entry Draft.

Season	Club	League	GP	G	A	Pts	PIM	PP	SH	GW	S	%	+/-	TF	F%	H	SB	Min	GP	G	A	Pts	PIM	PP	SH	GW
1988-89	KalPa Kuopio Jr.	Finn-Jr.	7	6	6	12	4																			
1989-90	KalPa Kuopio Jr.	Finn-Jr.	36	13	24	37	30																			
	KalPa Kuopio	Finland	5	0	0	0	0																			
1990-91	KalPa Kuopio Jr.	Finn-Jr.	35	36	17	53	61																			
	KalPa Kuopio	Finland	30	2	1	3	2												8	0	0	0	6			
1991-92	Clarkson Knights	ECAC	28	11	12	23	32																			
1992-93	Clarkson Knights	ECAC	35	25	30	55	26																			
1993-94	Clarkson Knights	ECAC	34	23	29	52	60																			
1994-95	Clarkson Knights	ECAC	37	23	38	61	34																			
	Edmonton	**NHL**	**4**	**0**	**0**	**0**	**0**	**0**	**0**	**0**	**5**	**0.0**	**0**													
1995-96	Cape Breton	AHL	58	25	35	60	71																			
1996-97	Hamilton	AHL	79	31	21	52	130												22	7	5	12	4			
1997-98	HIFK Helsinki	Finland	46	13	9	22	20												9	0	3	3	0			
1998-99	HIFK Helsinki	Finland	48	11	17	28	*173												11	1	3	4	12			
	HIFK Helsinki	EuroHL	6	0	1	1	8												4	3	0	3	4			
99-2000	**Los Angeles**	**NHL**	**63**	**9**	**8**	**17**	**80**	**2**	**1**	**1**	**74**	**12.2**	**−12**	**8**	**25.0**	**69**	**9**	**11:30**	**1**	**0**	**0**	**0**	**0**	**0**	**0**	**0**
2000-01	Lowell	AHL	59	28	39	67	73												4	3	3	6	10			
	Los Angeles	**NHL**	**11**	**0**	**1**	**1**	**4**	**0**	**0**	**0**	**12**	**0.0**	**1**	**1**	**100.0**	**10**	**1**	**9:22**								
2001-02	**NY Islanders**	**NHL**	**1**	**0**	**0**	**0**	**0**	**0**	**0**	**0**	**0**	**0.0**	**−1**	**0**	**0.0**	**0**	**0**	**7:11**								
	Bridgeport	AHL	76	11	34	45	82												20	1	2	3	6			
	NHL Totals		**79**	**9**	**9**	**18**	**84**	**2**	**1**	**1**	**91**	**9.9**		**9**	**33.3**	**79**	**10**	**11:08**	**1**	**0**	**0**	**0**	**0**	**0**	**0**	**0**

ECAC First All-Star Team (1993, 1995) • NCAA East Second All-American Team (1995)

Signed as a free agent by **LA Kings**, June 20, 1999. Signed as a free agent by **LA Kings**, January 4, 2001. Signed as a free agent by **NY Islanders**, July 18, 2001.

TURGEON, Pierre

(TUHR-zhaw, PEE-air) **DAL.**

Center. Shoots left. 6'1", 199 lbs. Born, Rouyn, Que., August 28, 1969. Buffalo's 1st choice, 1st overall, in 1987 Entry Draft.

Season	Club	League	GP	G	A	Pts	PIM	PP	SH	GW	S	%	+/-	TF	F%	H	SB	Min	GP	G	A	Pts	PIM	PP	SH	GW
1984-85	Mtl-Bourassa	QAAA	41	49	52	101	26												5	3	8	11	2			
1985-86	Granby Bisons	QMJHL	69	47	67	114	31																			
1986-87	Granby Bisons	QMJHL	58	69	85	154	8												7	9	6	15	15			
1987-88	**Buffalo**	**NHL**	**76**	**14**	**28**	**42**	**34**	**8**	**0**	**3**	**101**	**13.9**	**−8**						**6**	**4**	**3**	**7**	**4**	**3**	**0**	**0**
1988-89	**Buffalo**	**NHL**	**80**	**34**	**54**	**88**	**26**	**19**	**0**	**5**	**182**	**18.7**	**−2**						**5**	**3**	**5**	**8**	**2**	**1**	**0**	**0**
1989-90	**Buffalo**	**NHL**	**80**	**40**	**66**	**106**	**29**	**17**	**1**	**10**	**193**	**20.7**	**10**						**6**	**2**	**4**	**6**	**2**	**0**	**0**	**1**
1990-91	**Buffalo**	**NHL**	**78**	**32**	**47**	**79**	**26**	**13**	**2**	**3**	**174**	**18.4**	**14**						**6**	**3**	**1**	**4**	**6**	**1**	**0**	**0**
1991-92	**Buffalo**	**NHL**	**8**	**2**	**6**	**8**	**4**	**0**	**0**	**0**	**14**	**14.3**	**−1**													
	NY Islanders	**NHL**	**69**	**38**	**49**	**87**	**16**	**13**	**0**	**6**	**193**	**19.7**	**8**													
1992-93	**NY Islanders**	**NHL**	**83**	**58**	**74**	**132**	**26**	**24**	**0**	**10**	**301**	**19.3**	**−1**						**11**	**6**	**7**	**13**	**0**	**0**	**0**	**0**
1993-94	**NY Islanders**	**NHL**	**69**	**38**	**56**	**94**	**18**	**10**	**4**	**6**	**254**	**15.0**	**14**						**4**	**0**	**1**	**1**	**0**	**0**	**0**	**0**
1994-95	**NY Islanders**	**NHL**	**34**	**13**	**14**	**27**	**10**	**3**	**2**	**2**	**93**	**14.0**	**−12**													
	Montreal	**NHL**	**15**	**11**	**9**	**20**	**4**	**2**	**0**	**2**	**67**	**16.4**	**12**													
1995-96	**Montreal**	**NHL**	**80**	**38**	**58**	**96**	**44**	**17**	**1**	**6**	**297**	**12.8**	**19**						**6**	**2**	**4**	**6**	**2**	**0**	**0**	**0**
1996-97	**Montreal**	**NHL**	**9**	**1**	**10**	**11**	**2**	**0**	**0**	**0**	**22**	**4.5**	**4**													
	St. Louis	**NHL**	**69**	**25**	**49**	**74**	**12**	**5**	**0**	**7**	**194**	**12.9**	**4**						**5**	**1**	**1**	**2**	**2**	**1**	**0**	**0**
1997-98	**St. Louis**	**NHL**	**60**	**22**	**46**	**68**	**24**	**6**	**0**	**4**	**140**	**15.7**	**13**						**10**	**4**	**4**	**8**	**2**	**2**	**0**	**0**
1998-99	**St. Louis**	**NHL**	**67**	**31**	**34**	**65**	**36**	**10**	**0**	**5**	**193**	**16.1**	**4**	**1285**	**50.0**	**16**	**24**	**19:07**	**13**	**4**	**9**	**13**	**6**	**0**	**0**	**2**
99-2000	**St. Louis**	**NHL**	**52**	**26**	**40**	**66**	**8**	**8**	**0**	**3**	**139**	**18.7**	**30**	**1016**	**53.2**	**5**	**19**	**19:13**	**7**	**0**	**7**	**7**	**0**	**0**	**0**	**0**

			Regular Season																Playoffs							
Season	Club	League	GP	G	A	Pts	PIM	PP	SH	GW	S	%	+/-	TF	F%	H	SB	Min	GP	G	A	Pts	PIM	PP	SH	GW
2000-01	St. Louis	NHL	79	30	52	82	37	11	0	6	171	17.5	14	1569	49.7	9	22	18:50	15	5	10	15	2	1	0	0
2001-02	Dallas	NHL	66	15	32	47	16	7	0	1	121	12.4	–4	822	48.4	12	19	16:31	....	....	....	....	....	....	....	....
	NHL Totals		**1074**	**468**	**724**	**1192**	**372**	**173**	**10**	**79**	**2849**	**16.4**		**4692**	**50.3**	**42**	**84**	**18:24**	**94**	**34**	**56**	**90**	**28**	**9**	**0**	**3**

QMJHL Offensive Rookie of the Year) (1986) • Won Lady Byng Memorial Trophy (1993) • Played in NHL All-Star Game (1990, 1993, 1994, 1996)

Traded to **NY Islanders** by **Buffalo** with Uwe Krupp, Benoit Hogue and Dave McLlwain for Pat LaFontaine, Randy Hillier, Randy Wood and NY Islanders' 4th round choice (Dean Melanson) in 1992 Entry Draft, October 25, 1991. Traded to **Montreal** by **NY Islanders** with Vladimir Malakhov for Kirk Muller, Mathieu Schneider and Craig Darby, April 5, 1995. Traded to **St. Louis** by **Montreal** with Rory Fitzpatrick and Craig Conroy for Murray Baron, Shayne Corson and St. Louis' 5th round choice (Gennady Razin) in 1997 Entry Draft, October 29, 1996. Signed as a free agent by **Dallas**, July 1, 2001.

TUZZOLINO, Tony (too-zuh-LEE-noh, TOH-nee) **MIN.**

Right wing. Shoots right. 6'2", 208 lbs. Born, Buffalo, NY, October 9, 1975. Quebec's 7th choice, 113th overall, in 1994 Entry Draft.

Season	Club	League	GP	G	A	Pts	PIM	PP	SH	GW	S	%	+/-	TF	F%	H	SB	Min	GP	G	A	Pts	PIM	PP	SH	GW
1989-90	Amherst Knights	NYAHA	29	50	95	145	....	....	....	....	....	....	....	....	....	....	....	....	....	....	....	....	....	....	....	....
1990-91	Buffalo Regals	NAJHL	55	39	47	86	....	....	....	....	....	....	....	....	....	....	....	....	....	....	....	....	....	....	....	....
1991-92	Niagara Scenics	NAJHL	45	19	27	46	82	....	....	....	....	....	....	....	....	....	....	....	....	....	....	....	....	....	....	....
1992-93	Niagara Scenics	NAJHL	50	36	41	77	134	....	....	....	....	....	....	....	....	....	....	....	....	....	....	....	....	....	....	....
1993-94	Michigan State	CCHA	35	4	3	7	46	....	....	....	....	....	....	....	....	....	....	....	....	....	....	....	....	....	....	....
1994-95	Michigan State	CCHA	39	9	18	27	81	....	....	....	....	....	....	....	....	....	....	....	....	....	....	....	....	....	....	....
1995-96	Michigan State	CCHA	41	12	17	29	120	....	....	....	....	....	....	....	....	....	....	....	....	....	....	....	....	....	....	....
1996-97	Michigan State	CCHA	39	14	18	32	120	....	....	....	....	....	....	....	....	....	....	....	....	....	....	....	....	....	....	....
1997-98	Kentucky	AHL	35	9	14	23	83	....	....	....	....	....	....	....	....	....	....	....	....	....	....	....	....	....	....	....
	Anaheim	**NHL**	**1**	**0**	**0**	**0**	**2**	**0**	**0**	**0**	**0**	**0.0**	**–2**	....	....	....	....	....	....	....	....	....	....	....	....	....
	Cincinnati	AHL	13	3	3	6	6	....	....	....	....	....	....	....	....	....	....	....	....	....	....	....	....	....	....	....
1998-99	Cincinnati	AHL	50	4	10	14	55	....	....	....	....	....	....	....	....	....	....	....	....	....	....	....	....	....	....	....
	Cleveland	IHL	15	2	4	6	22	....	....	....	....	....	....	....	....	....	....	....	....	....	....	....	....	....	....	....
99-2000	Cincinnati	AHL	15	0	3	3	8	....	....	....	....	....	....	....	....	....	....	....	....	....	....	....	....	....	....	....
	Huntington	ECHL	20	6	13	19	43	....	....	....	....	....	....	....	....	....	....	....	....	....	....	....	....	....	....	....
	Hartford	AHL	32	3	8	11	41	....	....	....	....	....	....	....	....	....	....	....	19	2	2	4	16	....	....	....
2000-01	Hartford	AHL	47	12	23	35	136	....	....	....	....	....	....	....	....	....	....	....	5	0	2	2	6	....	....	....
	NY Rangers	**NHL**	**6**	**0**	**0**	**0**	**5**	**0**	**0**	**0**	**3**	**0.0**	**–1**	**1**	**0.0**	**11**	**3**	**3:41**	....	....	....	....	....	....	....	....
2001-02	**Boston**	**NHL**	**2**	**0**	**0**	**0**	**0**	**0**	**0**	**0**	**1**	**0.0**	**–1**	**3**	**33.3**	**4**	**0**	**5:32**	....	....	....	....	....	....	....	....
	Providence	AHL	59	10	19	29	123	....	....	....	....	....	....	....	....	....	....	....	1	0	0	0	0	....	....	....
	NHL Totals		**9**	**0**	**0**	**0**	**7**	**0**	**0**	**0**	**4**	**0.0**		**4**	**25.0**	**15**	**3**	**4:09**	....	....	....	....	....	....	....	....

Rights transferred to **Colorado** after **Quebec** franchise relocated, June 21, 1995. Signed as a free agent by **NY Islanders**, April 26, 1997. Traded to **Anaheim** by **NY Islanders** with Travis Green and Doug Houda for Joe Sacco, J-J Daigneault and Mark Janssens, February 6, 1998. Loaned to **Hartford** (AHL) by **Anaheim**, January 25, 2000. Signed as a free agent by **NY Rangers**, February 9, 2001. Signed as a free agent by **Boston**, July 23, 2001. Signed as a free agent by **Minnesota**, July 9, 2002.

TVERDOVSKY, Oleg (tvehr-DOHV-skee, OH-lehg) **N.J.**

Defense. Shoots left. 6'1", 205 lbs. Born, Donetsk, USSR, May 18, 1976. Anaheim's 1st choice, 2nd overall, in 1994 Entry Draft.

Season	Club	League	GP	G	A	Pts	PIM	PP	SH	GW	S	%	+/-	TF	F%	H	SB	Min	GP	G	A	Pts	PIM	PP	SH	GW
1992-93	Krylja Sovetov	CIS	21	0	1	1	6	....	....	....	....	....	....	....	....	....	....	....	6	0	0	0	0	....	....	....
1993-94	Krylja Sovetov	CIS	46	4	10	14	22	....	....	....	....	....	....	....	....	....	....	....	3	1	0	1	2	....	....	....
1994-95	Brandon	WHL	7	1	4	5	4	....	....	....	....	....	....	....	....	....	....	....	....	....	....	....	....	....	....	....
	Anaheim	**NHL**	**36**	**3**	**9**	**12**	**14**	**1**	**1**	**0**	**26**	**11.5**	**–6**	....	....	....	....	....	....	....	....	....	....	....	....	....
1995-96	**Anaheim**	**NHL**	**51**	**7**	**15**	**22**	**35**	**2**	**0**	**0**	**84**	**8.3**	**0**	....	....	....	....	....	....	....	....	....	....	....	....	....
	Winnipeg	**NHL**	**31**	**0**	**8**	**8**	**6**	**0**	**0**	**0**	**35**	**0.0**	**–7**	....	....	....	....	....	**6**	**0**	**1**	**1**	**0**	**0**	**0**	**0**
1996-97	**Phoenix**	**NHL**	**82**	**10**	**45**	**55**	**30**	**3**	**1**	**2**	**144**	**6.9**	**–5**	....	....	....	....	....	**7**	**0**	**1**	**1**	**0**	**0**	**0**	**0**
1997-98	Hamilton	AHL	9	8	6	14	2	....	....	....	....	....	....	....	....	....	....	....	....	....	....	....	....	....	....	....
	Phoenix	**NHL**	**46**	**7**	**12**	**19**	**12**	**4**	**0**	**1**	**83**	**8.4**	**1**	....	....	....	....	....	**6**	**0**	**7**	**7**	**0**	**0**	**0**	**0**
1998-99	**Phoenix**	**NHL**	**82**	**7**	**18**	**25**	**32**	**2**	**0**	**2**	**117**	**6.0**	**11**	**1**	**0.0**	**50**	**53**	**20:48**	**6**	**0**	**2**	**2**	**6**	**0**	**0**	**0**
99-2000	**Anaheim**	**NHL**	**82**	**15**	**36**	**51**	**30**	**5**	**0**	**5**	**153**	**9.8**	**5**	**1**	**0.0**	**79**	**89**	**22:46**	....	....	....	....	....	....	....	....
2000-01	**Anaheim**	**NHL**	**82**	**14**	**39**	**53**	**32**	**8**	**0**	**3**	**188**	**7.4**	**–11**	**0**	**0.0**	**77**	**70**	**24:25**	....	....	....	....	....	....	....	....
2001-02	**Anaheim**	**NHL**	**73**	**6**	**26**	**32**	**31**	**2**	**0**	**1**	**147**	**4.1**	**0**	**0**	**0.0**	**45**	**68**	**22:50**	....	....	....	....	....	....	....	....
	Russia	Olympics	6	1	1	2	0	....	....	....	....	....	....	....	....	....	....	....	....	....	....	....	....	....	....	....
	NHL Totals		**565**	**69**	**208**	**277**	**222**	**27**	**2**	**14**	**977**	**7.1**		**2**	**0.0**	**251**	**280**	**22:42**	**25**	**0**	**11**	**11**	**6**	**0**	**0**	**0**

Played in NHL All-Star Game (1997)

Traded to **Winnipeg** by **Anaheim** with Chad Kilger and Anaheim's 3rd round choice (Per-Anton Lundstrom) in 1996 Entry Draft for Teemu Selanne, Marc Chouinard and Winnipeg's 4th round choice (later traded to Toronto - later traded to Montreal - Montreal selected Kim Staal) in 1996 Entry Draft, February 7, 1996. Transferred to **Phoenix** after **Winnipeg** franchise relocated, July 1, 1996. Traded to **Anaheim** by **Phoenix** for Travis Green and Anaheim's 1st round choice (Scott Kelman) in 1999 Entry Draft, June 26, 1999. Traded to **New Jersey** by **Anaheim** with Jeff Friesen and Maxim Balmochnykh for Petr Sykora, Mike Commodore, Jean-Francois Damphousse and Igor Pohanka, July 6, 2002.

ULANOV, Igor (yoo-LAH-nahf, EE-gohr) **FLA.**

Defense. Shoots left. 6'3", 220 lbs. Born, Krasnokamsk, USSR, October 1, 1969. Winnipeg's 8th choice, 203rd overall, in 1991 Entry Draft.

Season	Club	League	GP	G	A	Pts	PIM	PP	SH	GW	S	%	+/-	TF	F%	H	SB	Min	GP	G	A	Pts	PIM	PP	SH	GW
1990-91	Voskresensk	USSR	41	2	2	4	52	....	....	....	....	....	....	....	....	....	....	....	....	....	....	....	....	....	....	....
1991-92	Voskresensk	CIS	27	1	4	5	24	....	....	....	....	....	....	....	....	....	....	....	....	....	....	....	....	....	....	....
	Winnipeg	**NHL**	**27**	**2**	**9**	**11**	**67**	**0**	**0**	**0**	**23**	**8.7**	**5**	....	....	....	....	....	**7**	**0**	**0**	**0**	**39**	**0**	**0**	**0**
	Moncton Hawks	AHL	3	0	1	1	16	....	....	....	....	....	....	....	....	....	....	....	....	....	....	....	....	....	....	....
1992-93	**Winnipeg**	**NHL**	**56**	**2**	**14**	**16**	**124**	**0**	**0**	**0**	**26**	**7.7**	**6**	....	....	....	....	....	**4**	**0**	**0**	**0**	**4**	**0**	**0**	**0**
	Moncton Hawks	AHL	9	1	3	4	26	....	....	....	....	....	....	....	....	....	....	....	....	....	....	....	....	....	....	....
	Fort Wayne	IHL	3	0	1	1	29	....	....	....	....	....	....	....	....	....	....	....	....	....	....	....	....	....	....	....
1993-94	**Winnipeg**	**NHL**	**74**	**0**	**17**	**17**	**165**	**0**	**0**	**0**	**46**	**0.0**	**–11**	....	....	....	....	....	....	....	....	....	....	....	....	....
1994-95	**Winnipeg**	**NHL**	**19**	**1**	**3**	**4**	**27**	**0**	**0**	**0**	**13**	**7.7**	**–2**	....	....	....	....	....	....	....	....	....	....	....	....	....
	Washington	**NHL**	**3**	**0**	**1**	**1**	**2**	**0**	**0**	**0**	**0**	**0.0**	**3**	....	....	....	....	....	**2**	**0**	**0**	**0**	**4**	**0**	**0**	**0**
1995-96	**Chicago**	**NHL**	**53**	**1**	**8**	**9**	**92**	**0**	**0**	**0**	**24**	**4.2**	**12**	....	....	....	....	....	....	....	....	....	....	....	....	....
	Indianapolis Ice	IHL	1	0	0	0	0	....	....	....	....	....	....	....	....	....	....	....	....	....	....	....	....	....	....	....
	Tampa Bay	**NHL**	**11**	**2**	**1**	**3**	**24**	**0**	**0**	**1**	**13**	**15.4**	**–1**	....	....	....	....	....	**5**	**0**	**0**	**0**	**15**	**0**	**0**	**0**
1996-97	**Tampa Bay**	**NHL**	**59**	**1**	**7**	**8**	**108**	**0**	**0**	**0**	**56**	**1.8**	**2**	....	....	....	....	....	....	....	....	....	....	....	....	....
1997-98	**Tampa Bay**	**NHL**	**45**	**2**	**7**	**9**	**85**	**1**	**0**	**0**	**32**	**6.3**	**–5**	....	....	....	....	....	....	....	....	....	....	....	....	....
	Montreal	**NHL**	**4**	**0**	**1**	**1**	**12**	**0**	**0**	**0**	**4**	**0.0**	**–2**	....	....	....	....	....	**10**	**1**	**4**	**5**	**12**	**0**	**0**	**0**
1998-99	**Montreal**	**NHL**	**76**	**3**	**9**	**12**	**109**	**0**	**0**	**0**	**55**	**5.5**	**–3**	**0**	**0.0**	**136**	**163**	**17:35**	....	....	....	....	....	....	....	....
99-2000	**Montreal**	**NHL**	**43**	**1**	**5**	**6**	**76**	**0**	**0**	**0**	**33**	**3.0**	**–11**	**0**	**0.0**	**79**	**86**	**16:33**	....	....	....	....	....	....	....	....
	Edmonton	**NHL**	**14**	**0**	**3**	**3**	**10**	**0**	**0**	**0**	**6**	**0.0**	**–3**	**0**	**0.0**	**20**	**21**	**16:23**	**5**	**0**	**0**	**0**	**6**	**0**	**0**	**0**
2000-01	**Edmonton**	**NHL**	**67**	**3**	**20**	**23**	**90**	**1**	**0**	**0**	**74**	**4.1**	**15**	**0**	**0.0**	**106**	**172**	**23:01**	**6**	**0**	**0**	**0**	**4**	**0**	**0**	**0**
2001-02	**NY Rangers**	**NHL**	**39**	**0**	**6**	**6**	**53**	**0**	**0**	**0**	**17**	**0.0**	**–4**	**0**	**0.0**	**34**	**100**	**16:19**	....	....	....	....	....	....	....	....
	Hartford	AHL	6	1	1	2	2	....	....	....	....	....	....	....	....	....	....	....	....	....	....	....	....	....	....	....
	Florida	**NHL**	**14**	**0**	**4**	**4**	**11**	**0**	**0**	**0**	**9**	**0.0**	**–3**	**0**	**0.0**	**22**	**20**	**20:50**	....	....	....	....	....	....	....	....
	NHL Totals		**604**	**18**	**115**	**133**	**1055**	**2**	**0**	**1**	**431**	**4.2**		**0**	**0.0**	**397**	**562**	**18:46**	**39**	**1**	**4**	**5**	**84**	**0**	**0**	**0**

Traded to **Washington** by **Winnipeg** with Mike Eagles for Washington's 3rd (later traded to Dallas - Dallas selected Sergey Gusev) and 5th (Brian Elder) round choices in 1995 Entry Draft, April 7, 1995. Traded to **Chicago** by **Washington** for Chicago's 3rd round choice (Dave Weninger) in 1996 Entry Draft, October 17, 1995. Traded to **Tampa Bay** by **Chicago** with Patrick Poulin and Chicago's 2nd round choice (later traded to New Jersey - New Jersey selected Pierre Dagenais) in 1996 Entry Draft for Enrico Ciccone and Tampa Bay's 2nd round choice (Jeff Paul) in 1996 Entry Draft, March 20, 1996. Traded to **Montreal** by **Tampa Bay** with Patrick Poulin and Mick Vukota for Stephane Richer, Darcy Tucker and David Wilkie, January 15, 1998. Traded to **Edmonton** by **Montreal** with Alain Nasreddine for Christian Laflamme and Matthieu Descoteaux, March 9, 2000. Signed as a free agent by **NY Rangers**, July 20, 2001. Traded to **Florida** by **NY Rangers** with Filip Novak, NY Rangers' 1st (later traded to Calgary - Calgary selected Eric Nystrom) and 2nd (Rob Globke) round choices in 2002 Entry Draft and NY Rangers' 4th round choice in 2003 Entry Draft for Pavel Bure and Florida's 2nd round choice (Lee Falardeau) in 2002 Entry Draft, March 18, 2002.

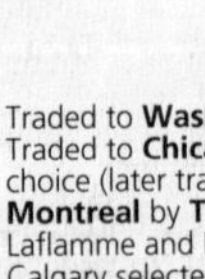

ULMER, Jeff (UHL-muhr, JEHF) **OTT.**

Right wing. Shoots right. 5'11", 195 lbs. Born, Wilcox, Sask., April 27, 1977.

Season	Club	League	GP	G	A	Pts	PIM	PP	SH	GW	S	%	+/-	TF	F%	H	SB	Min	GP	G	A	Pts	PIM	PP	SH	GW
1994-95	Notre Dame	AJHL	63	25	35	60	....	....	....	....	....	....	....	....	....	....	....	....	....	....	....	....	....	....	....	....
1995-96	North Dakota	WCHA	29	5	3	8	26	....	....	....	....	....	....	....	....	....	....	....	....	....	....	....	....	....	....	....
1996-97	North Dakota	WCHA	26	6	11	17	16	....	....	....	....	....	....	....	....	....	....	....	....	....	....	....	....	....	....	....
1997-98	North Dakota	WCHA	32	12	12	24	44	....	....	....	....	....	....	....	....	....	....	....	....	....	....	....	....	....	....	....
1998-99	North Dakota	WCHA	38	16	20	36	46	....	....	....	....	....	....	....	....	....	....	....	....	....	....	....	....	....	....	....
99-2000	Team Canada	Nat-Tm	48	14	25	39	20	....	....	....	....	....	....	....	....	....	....	....	....	....	....	....	....	....	....	....
	Houston Aeros	IHL	5	1	0	1	0	....	....	....	....	....	....	....	....	....	....	....	11	2	4	6	6	....	....	....

Season	Club	League	Regular Season GP	G	A	Pts	PIM	PP	SH	GW	S	%	+/-	TF	F%	H	SB	Min	Playoffs GP	G	A	Pts	PIM	PP	SH	GW
2000-01	**NY Rangers**	**NHL**	**21**	**3**	**0**	**3**	**8**	**0**	**0**	**0**	**22**	**13.6**	**-6**	**16**	**31.3**	**14**	**2**	**10:23**								
	Hartford	AHL	48	11	14	25	34																			
2001-02	Grand Rapids	AHL	73	9	17	26	65												5	0	1	1	11			
	NHL Totals		**21**	**3**	**0**	**3**	**8**	**0**	**0**	**0**	**22**	**13.6**		**16**	**31.3**	**14**	**2**	**10:23**								

Signed as a free agent by **Houston** (IHL), March 30, 2000. Signed as a free agent by **NY Rangers**, July 27, 2000. Traded to **Ottawa** by **NY Rangers** with Jason Doig for Sean Gagnon, June 29, 2001.

VAANANEN, Ossi

(VAN-ih-nehn, AW-see) **PHX.**

Defense. Shoots left. 6'4", 215 lbs. Born, Vantaa, Finland, August 18, 1980. Phoenix's 2nd choice, 43rd overall, in 1998 Entry Draft.

Season	Club	League	GP	G	A	Pts	PIM	PP	SH	GW	S	%	+/-	TF	F%	H	SB	Min	GP	G	A	Pts	PIM	PP	SH	GW
1994-95	Jokerit-C	Finn-Jr.	23	0	1	1	10												6	0	0	0	8			
1995-96	Jokerit-C	Finn-Jr.	12	0	0	0	10																			
	Jokerit-B	Finn-Jr.	1	0	0	0	0												1	0	0	0	0			
1996-97	Jokerit Jr.	Finn-Jr.	17	1	2	3	43																			
1997-98	Jokerit Jr.	Finn-Jr.	31	0	6	6	24																			
1998-99	Jokerit Jr.	Finn-Jr.	12	1	6	7	16																			
	Jokerit Helsinki	EuroHL	5	0	0	0	2												1	0	1	1	2			
	Jokerit Helsinki	Finland	48	0	1	1	42												3	0	1	1	2			
99-2000	Jokerit Helsinki	Finland	49	1	6	7	46												11	1	1	2	2			
2000-01	**Phoenix**	**NHL**	**81**	**4**	**12**	**16**	**90**	**0**	**0**	**2**	**69**	**5.8**	**9**	**0**	**0.0**	**190**	**79**	**19:09**								
2001-02	**Phoenix**	**NHL**	**76**	**2**	**12**	**14**	**74**	**0**	**1**	**0**	**41**	**4.9**	**6**	**0**	**0.0**	**173**	**88**	**20:13**	**5**	**0**	**0**	**0**	**6**	**0**	**0**	**0**
	Finland	Olympics	2	0	1	1	0																			
	NHL Totals		**157**	**6**	**24**	**30**	**164**	**0**	**1**	**2**	**110**	**5.5**		**0**	**0.0**	**363**	**167**	**19:40**	**5**	**0**	**0**	**0**	**6**	**0**	**0**	**0**

VALICEVIC, Rob

(val-IH-seh-VIK, RAWB) **ANA.**

Right wing. Shoots right. 6'1", 198 lbs. Born, Detroit, MI, January 6, 1971. NY Islanders' 6th choice, 114th overall, in 1991 Entry Draft.

Season	Club	League	GP	G	A	Pts	PIM	PP	SH	GW	S	%	+/-	TF	F%	H	SB	Min	GP	G	A	Pts	PIM	PP	SH	GW
1990-91	Det. Compuware	NAJHL	39	31	44	75	54																			
1991-92	Lake Superior	CCHA	32	8	4	12	12																			
1992-93	Lake Superior	CCHA	43	21	20	41	28																			
1993-94	Lake Superior	CCHA	45	18	20	38	46																			
1994-95	Lake Superior	CCHA	37	10	21	31	40																			
1995-96	Louisiana	ECHL	60	42	20	62	85												5	2	3	5	8			
	Springfield	AHL	2	0	0	0	2																			
1996-97	Louisiana	ECHL	8	7	2	9	21																			
	Houston Aeros	IHL	58	11	12	23	42												12	1	3	4	11			
1997-98	Houston Aeros	IHL	72	29	28	57	47												4	2	0	2	2			
1998-99	**Nashville**	**NHL**	**19**	**4**	**2**	**6**	**2**	**0**	**0**	**2**	**23**	**17.4**	**4**	**17**	**29.4**	**8**	**8**	**11:01**								
	Houston Aeros	IHL	57	16	33	49	62												19	7	10	17	8			
99-2000	**Nashville**	**NHL**	**80**	**14**	**11**	**25**	**21**	**2**	**1**	**3**	**113**	**12.4**	**-11**	**50**	**44.0**	**100**	**38**	**14:31**								
2000-01	**Nashville**	**NHL**	**60**	**8**	**6**	**14**	**26**	**1**	**0**	**4**	**62**	**12.9**	**-2**	**35**	**40.0**	**89**	**28**	**14:06**								
2001-02	**Los Angeles**	**NHL**	**17**	**1**	**1**	**2**	**8**	**0**	**0**	**0**	**9**	**11.1**	**-4**	**166**	**41.6**	**16**	**6**	**9:34**								
	Manchester	AHL	59	11	23	34	25												5	1	0	1	4			
	NHL Totals		**176**	**27**	**20**	**47**	**57**	**3**	**1**	**9**	**207**	**13.0**		**268**	**41.0**	**213**	**80**	**13:31**								

Signed as a free agent by **Nashville**, May 28, 1998. Signed as a free agent by **LA Kings**, August 16, 2001. Signed as a free agent by **Anaheim**, July 24, 2002.

VALK, Garry

(VAHLK, GAIR-ee)

Right wing. Shoots left. 6'1", 200 lbs. Born, Edmonton, Alta., November 27, 1967. Vancouver's 5th choice, 108th overall, in 1987 Entry Draft.

Season	Club	League	GP	G	A	Pts	PIM	PP	SH	GW	S	%	+/-	TF	F%	H	SB	Min	GP	G	A	Pts	PIM	PP	SH	GW
1984-85	Sherwood Park	AJHL	53	20	22	42	66																			
1985-86	Sherwood Park	AJHL	40	20	26	46	116																			
1986-87	Sherwood Park	AJHL	59	42	44	86	204																			
1987-88	North Dakota	WCHA	38	23	12	35	64																			
1988-89	North Dakota	WCHA	40	14	17	31	71																			
1989-90	North Dakota	WCHA	43	22	17	39	92																			
1990-91	**Vancouver**	**NHL**	**59**	**10**	**11**	**21**	**67**	**1**	**0**	**1**	**90**	**11.1**	**-23**						**5**	**0**	**0**	**0**	**20**	**0**	**0**	**0**
	Milwaukee	IHL	10	12	4	16	13												3	0	0	0	2			
1991-92	**Vancouver**	**NHL**	**65**	**8**	**17**	**25**	**56**	**2**	**1**	**2**	**93**	**8.6**	**3**						**4**	**0**	**0**	**0**	**5**	**0**	**0**	**0**
1992-93	**Vancouver**	**NHL**	**48**	**6**	**7**	**13**	**77**	**0**	**0**	**2**	**46**	**13.0**	**6**						**7**	**0**	**1**	**1**	**12**	**0**	**0**	**0**
	Hamilton Canucks	AHL	7	3	6	9	6																			
1993-94	**Anaheim**	**NHL**	**78**	**18**	**27**	**45**	**100**	**4**	**1**	**5**	**165**	**10.9**	**8**													
1994-95	**Anaheim**	**NHL**	**36**	**3**	**6**	**9**	**34**	**0**	**0**	**0**	**53**	**5.7**	**-4**													
1995-96	**Anaheim**	**NHL**	**79**	**12**	**12**	**24**	**125**	**1**	**1**	**2**	**108**	**11.1**	**8**													
1996-97	**Anaheim**	**NHL**	**53**	**7**	**7**	**14**	**53**	**0**	**0**	**1**	**68**	**10.3**	**-2**													
	Pittsburgh	**NHL**	**17**	**3**	**4**	**7**	**25**	**0**	**0**	**0**	**32**	**9.4**	**-6**													
1997-98	**Pittsburgh**	**NHL**	**39**	**2**	**1**	**3**	**33**	**0**	**0**	**0**	**32**	**6.3**	**-3**													
1998-99	**Toronto**	**NHL**	**77**	**8**	**21**	**29**	**53**	**1**	**0**	**0**	**93**	**8.6**	**8**	**19**	**36.8**	**99**	**20**	**13:53**	**17**	**3**	**4**	**7**	**22**	**0**	**0**	**1**
99-2000	**Toronto**	**NHL**	**73**	**10**	**14**	**24**	**44**	**0**	**1**	**1**	**91**	**11.0**	**-2**	**10**	**30.0**	**109**	**14**	**12:51**	**12**	**1**	**2**	**3**	**14**	**0**	**0**	**0**
2000-01	**Toronto**	**NHL**	**74**	**8**	**18**	**26**	**46**	**1**	**0**	**2**	**87**	**9.2**	**4**	**30**	**33.3**	**112**	**18**	**11:33**	**5**	**1**	**0**	**1**	**2**	**0**	**0**	**0**
2001-02	**Toronto**	**NHL**	**63**	**5**	**10**	**15**	**28**	**0**	**0**	**0**	**80**	**6.3**	**2**	**6**	**33.3**	**61**	**19**	**10:48**	**11**	**1**	**0**	**1**	**4**	**0**	**0**	**0**
	NHL Totals		**761'**	**100**	**155**	**255**	**741**	**10**	**4**	**16**	**1038**	**9.6**		**65**	**33.8**	**381**	**71**	**12:20**	**61**	**6**	**7**	**13**	**79**	**0**	**0**	**1**

Claimed by **Anaheim** from **Vancouver** in NHL Waiver Draft, October 3, 1993. Traded to **Pittsburgh** by **Anaheim** for J-J Daigneault, February 21, 1997. Signed as a free agent by **Toronto**, October 8, 1998.

VAN ALLEN, Shaun

(VAN-AL-ehn, SHAWN) **OTT.**

Center. Shoots left. 6'1", 205 lbs. Born, Calgary, Alta., August 29, 1967. Edmonton's 5th choice, 105th overall, in 1987 Entry Draft.

Season	Club	League	GP	G	A	Pts	PIM	PP	SH	GW	S	%	+/-	TF	F%	H	SB	Min	GP	G	A	Pts	PIM	PP	SH	GW
1984-85	Swift Current	SJHL	61	12	20	32	136																			
1985-86	Saskatoon Blades	WHL	55	12	11	23	43												13	4	8	12	28			
1986-87	Saskatoon Blades	WHL	72	38	59	97	116												11	4	6	10	24			
1987-88	Milwaukee	IHL	40	14	28	42	34																			
	Nova Scotia	AHL	19	4	10	14	17												4	1	1	2	4			
1988-89	Cape Breton	AHL	76	32	42	74	81																			
1989-90	Cape Breton	AHL	61	25	44	69	83												4	0	2	2	8			
1990-91	**Edmonton**	**NHL**	**2**	**0**	**0**	**0**	**0**	**0**	**0**	**0**	**0**	**0.0**	**0**													
	Cape Breton	AHL	76	25	75	100	182												4	0	1	1	8			
1991-92	Cape Breton	AHL	77	29	*84	*113	80												5	3	7	10	14			
1992-93	**Edmonton**	**NHL**	**21**	**1**	**4**	**5**	**6**	**0**	**0**	**0**	**19**	**5.3**	**-2**													
	Cape Breton	AHL	43	14	62	76	68												15	8	9	17	18			
1993-94	**Anaheim**	**NHL**	**80**	**8**	**25**	**33**	**64**	**2**	**2**	**1**	**104**	**7.7**	**0**													
1994-95	**Anaheim**	**NHL**	**45**	**8**	**21**	**29**	**32**	**1**	**1**	**1**	**68**	**11.8**	**-4**													
1995-96	**Anaheim**	**NHL**	**49**	**8**	**17**	**25**	**41**	**0**	**0**	**2**	**78**	**10.3**	**13**													
1996-97	**Ottawa**	**NHL**	**80**	**11**	**14**	**25**	**35**	**1**	**1**	**2**	**123**	**8.9**	**-8**						**7**	**0**	**1**	**1**	**4**	**0**	**0**	**0**
1997-98	**Ottawa**	**NHL**	**80**	**4**	**15**	**19**	**48**	**0**	**0**	**0**	**104**	**3.8**	**4**						**11**	**0**	**1**	**1**	**10**	**0**	**0**	**0**
1998-99	**Ottawa**	**NHL**	**79**	**6**	**11**	**17**	**30**	**0**	**1**	**0**	**47**	**12.8**	**3**	**656**	**47.4**	**84**	**23**	**11:07**	**4**	**0**	**0**	**0**	**0**	**0**	**0**	**0**
99-2000	**Ottawa**	**NHL**	**75**	**9**	**19**	**28**	**37**	**0**	**2**	**4**	**75**	**12.0**	**20**	**911**	**48.6**	**119**	**35**	**11:49**	**6**	**0**	**1**	**1**	**9**	**0**	**0**	**0**
2000-01	**Dallas**	**NHL**	**59**	**7**	**16**	**23**	**16**	**0**	**2**	**3**	**51**	**13.7**	**5**	**559**	**47.1**	**114**	**29**	**12:05**	**8**	**0**	**2**	**2**	**8**	**0**	**0**	**0**
2001-02	**Dallas**	**NHL**	**19**	**2**	**4**	**6**	**6**	**0**	**0**	**0**	**20**	**10.0**	**-5**	**177**	**53.1**	**31**	**7**	**12:57**								
	Montreal	**NHL**	**54**	**6**	**9**	**15**	**20**	**0**	**1**	**1**	**28**	**21.4**	**5**	**381**	**47.8**	**43**	**25**	**10:08**	**7**	**0**	**1**	**1**	**2**	**0**	**0**	**0**
	NHL Totals		**643**	**70**	**155**	**225**	**335**	**4**	**10**	**14**	**717**	**9.8**		**2684**	**48.2**	**391**	**119**	**11:26**	**43**	**0**	**6**	**6**	**33**	**0**	**0**	**0**

AHL Second All-Star Team (1991) • AHL First All-Star Team (1992) • Won John B. Sollenberger Trophy (Top Scorer - AHL) (1992)

Signed as a free agent by **Anaheim**, July 22, 1993. Traded to **Ottawa** by **Anaheim** with Jason York for Ted Drury and the rights to Marc Moro, October 1, 1996. Signed as a free agent by **Dallas**, July 12, 2000. Traded to **Montreal** by **Dallas** with Donald Audette for Martin Rucinsky and Benoit Brunet, November 21, 2001. Signed as a free agent by **Ottawa**, July 24, 2002.

VANDENBUSSCHE, Ryan
(van-dehn-BUHSH, RIGH-yuhn) **CHI.**

Right wing. Shoots right. 6', 200 lbs. Born, Simcoe, Ont., February 28, 1973. Toronto's 9th choice, 173rd overall, in 1992 Entry Draft.

			Regular Season																Playoffs							
Season	Club	League	GP	G	A	Pts	PIM	PP	SH	GW	S	%	+/-	TF	F%	H	SB	Min	GP	G	A	Pts	PIM	PP	SH	GW
1988-89	Delhi Flames	OJHL-D	3	1	1	2	2																			
1989-90	Norwich	OJHL-C	21	12	10	22	146																			
	Tillsonburg	OJHL-B	24	0	5	5	113																			
1990-91	Massena	OCJHL	10	2	3	5	46																			
	Cornwall Royals	OHL	49	3	8	11	139																			
1991-92	Cornwall Royals	OHL	61	13	15	28	232												6	0	2	2	9			
1992-93	Newmarket Royals	OHL	30	15	12	27	161																			
	Guelph Storm	OHL	29	3	14	17	99												5	1	3	4	13			
	St. John's	AHL	1	0	0	0	0																			
1993-94	St. John's	AHL	44	4	10	14	124																			
	Springfield	AHL	9	1	2	3	29												5	0	0	0	16			
1994-95	St. John's	AHL	53	2	13	15	239												3	0	0	0	17			
1995-96	Binghamton	AHL	68	3	17	20	240												4	0	0	0	9			
1996-97	**NY Rangers**	**NHL**	**11**	**1**	**0**	**1**	**30**	**0**	**0**	**0**	**4**	**25.0**	**–2**													
	Binghamton	AHL	38	8	11	19	133																			
1997-98	**NY Rangers**	**NHL**	**16**	**1**	**0**	**1**	**38**	**0**	**0**	**0**	**2**	**50.0**	**–2**													
	Hartford	AHL	15	2	0	2	45																			
	Chicago	**NHL**	**4**	**0**	**1**	**1**	**5**	**0**	**0**	**0**	**0**	**0.0**	**0**													
	Indianapolis Ice	IHL	3	1	1	2	4																			
1998-99	**Chicago**	**NHL**	**6**	**0**	**0**	**0**	**17**	**0**	**0**	**0**	**3**	**0.0**	**0**	**0**	**0.0**	**13**	**2**	**9:29**								
	Indianapolis Ice	IHL	34	3	10	13	130																			
	Portland Pirates	AHL	37	4	1	5	119																			
99-2000	**Chicago**	**NHL**	**52**	**0**	**1**	**1**	**143**	**0**	**0**	**0**	**19**	**0.0**	**–3**	**3**	**0.0**	**68**	**3**	**5:37**								
2000-01	**Chicago**	**NHL**	**64**	**2**	**5**	**7**	**146**	**0**	**0**	**0**	**24**	**8.3**	**–8**	**2**	**0.0**	**118**	**3**	**7:46**								
2001-02	**Chicago**	**NHL**	**50**	**1**	**2**	**3**	**103**	**0**	**0**	**0**	**22**	**4.5**	**–10**	**2**	**50.0**	**100**	**6**	**6:19**	**1**	**0**	**0**	**0**	**0**	**0**	**0**	**0**
	NHL Totals		**203**	**5**	**9**	**14**	**482**	**0**	**0**	**0**	**74**	**6.8**		**7**	**14.3**	**299**	**14**	**6:46**	**1**	**0**	**0**	**0**	**0**	**0**	**0**	**0**

Signed as a free agent by **NY Rangers**, August 22, 1995. Traded to **Chicago** by **NY Rangers** for Ryan Risidore, March 24, 1998.

VAN IMPE, Darren
(van-IHMP, DAIR-ehn)

Defense. Shoots left. 6'1", 205 lbs. Born, Saskatoon, Sask., May 18, 1973. NY Islanders' 7th choice, 170th overall, in 1993 Entry Draft.

Season	Club	League	GP	G	A	Pts	PIM	PP	SH	GW	S	%	+/-	TF	F%	H	SB	Min	GP	G	A	Pts	PIM	PP	SH	GW
1989-90	Prince Albert	SMHL	32	16	31	47	100																			
	Prince Albert	WHL	1	0	1	1	0																			
1990-91	Prince Albert	WHL	70	15	45	60	57												3	1	1	2	2			
1991-92	Prince Albert	WHL	69	9	37	46	89												8	1	5	6	10			
1992-93	Red Deer Rebels	WHL	54	23	47	70	118												4	2	5	7	16			
1993-94	Red Deer Rebels	WHL	58	20	64	84	125												4	2	4	6	6			
1994-95	San Diego Gulls	IHL	76	6	17	23	74												5	0	0	0	0			
	Anaheim	**NHL**	**1**	**0**	**1**	**1**	**4**	**0**	**0**	**0**	**0**	**0.0**	**0**													
1995-96	**Anaheim**	**NHL**	**16**	**1**	**2**	**3**	**14**	**0**	**0**	**1**	**13**	**7.7**	**8**													
	Baltimore	AHL	63	11	47	58	79																			
1996-97	**Anaheim**	**NHL**	**74**	**4**	**19**	**23**	**90**	**2**	**0**	**0**	**107**	**3.7**	**3**						**9**	**0**	**2**	**2**	**16**	**0**	**0**	**0**
1997-98	**Anaheim**	**NHL**	**19**	**1**	**3**	**4**	**4**	**0**	**0**	**0**	**21**	**4.8**	**–10**													
	Boston	**NHL**	**50**	**2**	**8**	**10**	**36**	**2**	**0**	**0**	**50**	**4.0**	**4**						**6**	**2**	**1**	**3**	**0**	**1**	**0**	**1**
1998-99	**Boston**	**NHL**	**60**	**5**	**15**	**20**	**66**	**4**	**0**	**0**	**92**	**5.4**	**–5**	**0**	**0.0**	**45**	**60**	**17:54**	**11**	**1**	**2**	**3**	**4**	**1**	**0**	**0**
99-2000	**Boston**	**NHL**	**79**	**5**	**23**	**28**	**73**	**4**	**0**	**0**	**97**	**5.2**	**–19**	**0**	**0.0**	**93**	**87**	**19:31**								
2000-01	**Boston**	**NHL**	**31**	**3**	**10**	**13**	**41**	**2**	**0**	**0**	**40**	**7.5**	**–9**	**0**	**0.0**	**37**	**33**	**20:11**								
2001-02	**NY Rangers**	**NHL**	**17**	**1**	**0**	**1**	**12**	**1**	**0**	**0**	**19**	**5.3**	**3**	**1**	**0.0**	**19**	**24**	**14:09**								
	Florida	**NHL**	**36**	**1**	**6**	**7**	**31**	**1**	**0**	**1**	**27**	**3.7**	**3**	**1**	**0.0**	**51**	**33**	**19:17**								
	NY Islanders	**NHL**	**14**	**1**	**2**	**3**	**16**	**0**	**0**	**0**	**14**	**7.1**	**6**	**0**	**0.0**	**12**	**28**	**21:36**	**7**	**0**	**4**	**4**	**8**	**0**	**0**	**0**
	NHL Totals		**397**	**24**	**89**	**113**	**387**	**16**	**0**	**2**	**480**	**5.0**		**2**	**0.0**	**257**	**265**	**18:54**	**33**	**3**	**9**	**12**	**28**	**2**	**0**	**1**

WHL East First All-Star Team (1993, 1994)

Traded to **Anaheim** by **NY Islanders** for Anaheim's 8th round choice (Mike Broda) in 1995 Entry Draft, August 31, 1994. Claimed on waivers by **Boston** from **Anaheim**, November 26, 1997. • Missed majority of 2000-01 season recovering from shoulder injury suffered in game vs. Detroit, December 23, 2000. Claimed on waivers by **NY Rangers** from **Boston**, August 7, 2001. Claimed on waivers by **Florida** from **NY Rangers**, December 18, 2001. Traded to **NY Islanders** by **Florida** for NY Islanders' 5th round choice in 2003 Entry Draft, March 19, 2002.

VAN RYN, Mike
(VAN RIHN, MIGHK) **ST.L.**

Defense. Shoots right. 6'1", 190 lbs. Born, London, Ont., May 14, 1979. New Jersey's 1st choice, 26th overall, in 1998 Entry Draft.

Season	Club	League	GP	G	A	Pts	PIM	PP	SH	GW	S	%	+/-	TF	F%	H	SB	Min	GP	G	A	Pts	PIM	PP	SH	GW
1995-96	London Nationals	OJHL-B	44	9	14	23	24																			
1996-97	London Nationals	OJHL-B	46	14	31	45	32																			
1997-98	U. of Michigan	CCHA	38	4	14	18	44																			
1998-99	U. of Michigan	CCHA	37	10	13	23	52																			
99-2000	Sarnia Sting	OHL	61	6	35	41	34												7	0	5	5	4			
2000-01	**St. Louis**	**NHL**	**1**	**0**	**0**	**0**	**0**	**0**	**0**	**0**	**1**	**0.0**	**–2**	**0**	**0.0**	**1**	**0**	**13:43**								
	Worcester	AHL	37	3	10	13	12												7	1	1	2	2			
2001-02	**St. Louis**	**NHL**	**48**	**2**	**8**	**10**	**18**	**0**	**0**	**1**	**52**	**3.8**	**10**	**0**	**0.0**	**33**	**37**	**16:23**	**9**	**0**	**0**	**0**	**0**	**0**	**0**	**0**
	Worcester	AHL	24	2	7	9	17																			
	NHL Totals		**49**	**2**	**8**	**10**	**18**	**0**	**0**	**1**	**53**	**3.8**		**0**	**0.0**	**34**	**37**	**16:20**	**9**	**0**	**0**	**0**	**0**	**0**	**0**	**0**

OJHL-B First All-Star Team (1997)

Signed as a free agent by **St. Louis**, June 30, 2000. • Missed majority of 2000-01 season recovering from shoulder injury suffered in game vs. Phoenix, October 5, 2000.

VARADA, Vaclav
(vuh-RA-da, VATS-LAV) **BUF.**

Right wing. Shoots left. 6', 208 lbs. Born, Vsetin, Czech., April 26, 1976. San Jose's 4th choice, 89th overall, in 1994 Entry Draft.

Season	Club	League	GP	G	A	Pts	PIM	PP	SH	GW	S	%	+/-	TF	F%	H	SB	Min	GP	G	A	Pts	PIM	PP	SH	GW
1993-94	HC Vitkovice	Czech	24	6	7	13													5	1	1	2				
1994-95	Tacoma Rockets	WHL	68	50	38	88	108												4	4	3	7	11			
1995-96	Kelowna Rockets	WHL	59	39	46	85	100												6	3	3	6	16			
	Buffalo	**NHL**	**1**	**0**	**0**	**0**	**0**	**0**	**0**	**0**	**2**	**0.0**	**0**													
	Rochester	AHL	5	3	0	3	4																			
1996-97	**Buffalo**	**NHL**	**5**	**0**	**0**	**0**	**2**	**0**	**0**	**0**	**2**	**0.0**	**0**													
	Rochester	AHL	53	23	25	48	81												10	1	6	7	27			
1997-98	**Buffalo**	**NHL**	**27**	**5**	**6**	**11**	**15**	**0**	**0**	**1**	**27**	**18.5**	**0**						**15**	**3**	**4**	**7**	**18**	**0**	**0**	**0**
	Rochester	AHL	45	30	26	56	74																			
1998-99	**Buffalo**	**NHL**	**72**	**7**	**24**	**31**	**61**	**1**	**0**	**1**	**123**	**5.7**	**11**	**1**	**0.0**	**185**	**12**	**14:30**	**21**	**5**	**4**	**9**	**14**	**1**	**0**	**0**
99-2000	HC Vitkovice	Czech	5	2	3	5	12																			
	Buffalo	**NHL**	**76**	**10**	**27**	**37**	**62**	**0**	**0**	**0**	**140**	**7.1**	**12**	**1**	**0.0**	**150**	**16**	**14:58**	**5**	**0**	**0**	**0**	**8**	**0**	**0**	**0**
2000-01	**Buffalo**	**NHL**	**75**	**10**	**21**	**31**	**81**	**2**	**0**	**2**	**112**	**8.9**	**–2**	**2**	**0.0**	**134**	**25**	**15:57**	**13**	**0**	**4**	**4**	**8**	**0**	**0**	**0**
2001-02	**Buffalo**	**NHL**	**76**	**7**	**16**	**23**	**82**	**1**	**0**	**1**	**138**	**5.1**	**–7**	**3**	**0.0**	**116**	**30**	**16:33**								
	NHL Totals		**332**	**39**	**94**	**133**	**303**	**4**	**0**	**5**	**544**	**7.2**		**7**	**0.0**	**585**	**83**	**15:30**	**54**	**8**	**12**	**20**	**48**	**1**	**0**	**0**

Traded to **Buffalo** by **San Jose** with Martin Spahnel and Philadelphia's 1st (previously acquired by San Jose - later traded to Phoenix - Phoenix selected Daniel Briere) and 4th (previously acquired, Buffalo selected Mike Martone) round choices in 1996 Entry Draft for Doug Bodger, November 16, 1995.

VARLAMOV, Sergei
(vahr-LAHM-uhf, SAIR-gay) **ST.L.**

Left wing. Shoots left. 5'11", 195 lbs. Born, Kiev, USSR, July 21, 1978.

Season	Club	League	GP	G	A	Pts	PIM	PP	SH	GW	S	%	+/-	TF	F%	H	SB	Min	GP	G	A	Pts	PIM	PP	SH	GW
1994-95	Nelson	RMJHL	26	11	15	26	56																			
1995-96	Swift Current	WHL	55	23	21	44	65																			
1996-97	Swift Current	WHL	72	46	39	85	94												10	3	8	11	10			
	Saint John	AHL	1	0	0	0	0																			
1997-98	Swift Current	WHL	72	*66	53	*119	132												12	10	5	15	28			
	Calgary	**NHL**	**1**	**0**	**0**	**0**	**0**	**0**	**0**	**0**	**0**	**0.0**	**0**													
	Saint John	AHL																	3	0	0	0	0			
1998-99	Saint John	AHL	76	24	33	57	66												7	0	4	4	8			
99-2000	**Calgary**	**NHL**	**7**	**3**	**0**	**3**	**0**	**0**	**0**	**1**	**11**	**27.3**	**0**	**1**	**0.0**	**4**	**1**	**10:22**								
	Saint John	AHL	68	20	21	41	88												3	0	0	0	24			
2000-01	Saint John	AHL	55	21	30	51	56												19	*15	8	23	10			

			Regular Season																Playoffs							
Season	Club	League	GP	G	A	Pts	PIM	PP	SH	GW	S	%	+/-	TF	F%	H	SB	Min	GP	G	A	Pts	PIM	PP	SH	GW
2001-02	St. Louis	NHL	52	5	7	12	26	0	0	0	83	6.0	4	1	0.0	91	10	11:10	1	0	0	0	2	0	0	0
	Ukraine	Olympics	2	0	1	1	14	....	....	....	....	....	....	....	....	....	....	....	....	....	....	....	....	....	....	....
	NHL Totals		60	8	7	15	26	0	0	1	94	8.5		2	0.0	95	11	11:05	1	0	0	0	2	0	0	0

WHL East First All-Star Team (1998) • Canadian Major Junior First All-Star Team (1998) • Canadian Major Junior Player of the Year (1998)

Signed as a free agent by **Calgary**, September 18, 1996. Traded to **St. Louis** by **Calgary** with Fred Brathwaite, Daniel Tkaczuk and Calgary's 9th round choice (Grant Jacobsen) in 2001 Entry Draft for Roman Turek and St. Louis' 4th round choice (Yegor Shastin) in 2001 Entry Draft, June 23, 2001.

VASICEK, Josef

(VAHSH-ih-chehk, YOH-zehf) **CAR.**

Center. Shoots left. 6'4", 200 lbs. Born, Havlickuv Brod, Czech., September 12, 1980. Carolina's 4th choice, 91st overall, in 1998 Entry Draft.

Season	Club	League	GP	G	A	Pts	PIM	PP	SH	GW	S	%	+/-	TF	F%	H	SB	Min	GP	G	A	Pts	PIM	PP	SH	GW
1995-96	HavlickuvBrodJr.	Czech-Jr.	36	25	25	50	....	....	....	....	....	....	....	....	....	....	....	....	....	....	....	....	....	....	....	....
1996-97	Slavia Praha Jr.	Czech-Jr.	37	20	40	60	....	....	....	....	....	....	....	....	....	....	....	....	....	....	....	....	....	....	....	....
1997-98	Slavia Praha Jr.	Czech-Jr.	34	13	20	33	....	....	....	....	....	....	....	....	....	....	....	....	....	....	....	....	....	....	....	....
1998-99	Sault Ste. Marie	OHL	66	21	35	56	30	....	....	....	....	....	....	....	....	....	....	....	5	3	0	3	10	....	....	....
99-2000	Sault Ste. Marie	OHL	54	26	46	72	49	....	....	....	....	....	....	....	....	....	....	....	17	5	15	20	8	....	....	....
2000-01	Carolina	NHL	76	8	13	21	53	1	0	0	103	7.8	–8	786	46.6	69	8	11:49	6	2	0	2	0	0	0	0
	Cincinnati	IHL	....	....	....	....	....	....	....	....	....	....	....	....	....	....	....	....	3	0	0	0	0	....	....	....
2001-02	Carolina	NHL	78	14	17	31	53	3	0	3	117	12.0	–7	878	48.3	80	9	14:11	23	3	2	5	12	0	0	1
	NHL Totals		154	22	30	52	106	4	0	3	220	10.0		1664	47.5	149	17	13:01	29	5	2	7	12	0	0	1

VASILIEV, Alexei

(vah-SEE-lee-ehf, al-EHX-ay)

Defense. Shoots left. 6'1", 192 lbs. Born, Yaroslavl, USSR, September 1, 1977. NY Rangers' 4th choice, 110th overall, in 1995 Entry Draft.

Season	Club	League	GP	G	A	Pts	PIM	PP	SH	GW	S	%	+/-	TF	F%	H	SB	Min	GP	G	A	Pts	PIM	PP	SH	GW
1995-96	Yaroslavl	CIS	40	4	7	11	4	....	....	....	....	....	....	....	....	....	....	....	....	....	....	....	....	....	....	....
1996-97	Yaroslavl	Russia	44	2	8	10	10	....	....	....	....	....	....	....	....	....	....	....	9	1	1	2	8	....	....	....
1997-98	Hartford	AHL	DID NOT PLAY – INJURED																							
1998-99	Hartford	AHL	75	8	19	27	24	....	....	....	....	....	....	....	....	....	....	....	6	0	1	1	2	....	....	....
99-2000	NY Rangers	NHL	1	0	0	0	2	0	0	0	0	0.0	–1	0	0.0	3	0	15:57	....	....	....	....	....	....	....	....
	Hartford	AHL	75	10	28	38	20	....	....	....	....	....	....	....	....	....	....	....	15	3	1	4	2	....	....	....
2000-01	Milwaukee	IHL	69	6	12	18	20	....	....	....	....	....	....	....	....	....	....	....	4	0	0	0	2	....	....	....
2001-02	Yaroslavl	Russia	51	5	11	16	40	....	....	....	....	....	....	....	....	....	....	....	9	1	1	2	2	....	....	....
	NHL Totals		1	0	0	0	2	0	0	0	0	0.0		0	0.0	3	0	....	....	....	....	....	....	....	....	....

• Missed entire 1997-98 season recovering from knee injury suffered in training camp, October, 1997. Traded to **Nashville** by **NY Rangers** for future considerations, September 25, 2000. Signed as a free agent by **Yaroslavl** (Russia), July 18, 2001.

VASILJEVS, Herbert

(vah-SEE-lee-ehf, HUHR-buhrt) **VAN.**

Center. Shoots right. 5'11", 180 lbs. Born, Riga, Latvia, May 27, 1976.

Season	Club	League	GP	G	A	Pts	PIM	PP	SH	GW	S	%	+/-	TF	F%	H	SB	Min	GP	G	A	Pts	PIM	PP	SH	GW
1994-95	Krefelder EV	Germany	42	4	5	9	24	....	....	....	....	....	....	....	....	....	....	....	15	1	4	5	10	....	....	....
1995-96	Guelph Storm	OHL	65	34	33	67	63	....	....	....	....	....	....	....	....	....	....	....	16	6	13	19	6	....	....	....
1996-97	Carolina	AHL	54	13	18	31	30	....	....	....	....	....	....	....	....	....	....	....	....	....	....	....	....	....	....	....
	Port Huron	ColHL	3	3	2	5	4	....	....	....	....	....	....	....	....	....	....	....	....	....	....	....	....	....	....	....
1997-98	New Haven	AHL	76	36	30	66	60	....	....	....	....	....	....	....	....	....	....	....	3	1	0	1	2	....	....	....
1998-99	Florida	NHL	5	0	0	0	2	0	0	0	6	0.0	–1	3	66.7	0	0	11:06	....	....	....	....	....	....	....	....
	Kentucky	AHL	76	28	48	76	66	....	....	....	....	....	....	....	....	....	....	....	12	2	1	3	4	....	....	....
99-2000	Atlanta	NHL	7	1	0	1	4	0	0	0	2	50.0	–3	27	48.2	10	3	9:18	....	....	....	....	....	....	....	....
	Orlando	IHL	73	25	35	60	60	....	....	....	....	....	....	....	....	....	....	....	6	2	2	4	6	....	....	....
2000-01	Atlanta	NHL	21	4	5	9	14	2	0	1	41	9.8	–11	37	40.5	10	6	16:01	....	....	....	....	....	....	....	....
	Orlando	IHL	58	22	26	48	32	....	....	....	....	....	....	....	....	....	....	....	12	8	3	11	14	....	....	....
2001-02	Vancouver	NHL	18	3	2	5	2	2	0	0	17	17.6	0	8	75.0	8	4	9:09	....	....	....	....	....	....	....	....
	Manitoba Moose	AHL	31	12	14	26	10	....	....	....	....	....	....	....	....	....	....	....	....	....	....	....	....	....	....	....
	NHL Totals		51	8	7	15	22	4	0	1	66	12.1		75	48.0	28	13	12:11	....	....	....	....	....	....	....	....

Signed as a free agent by **Florida**, October 3, 1996. Traded to **Atlanta** by **Florida** with Gord Murphy, Daniel Tjarnqvist and Ottawa's 6th round choice (previously acquired, later traded to Dallas - Dallas selected Justin Cox) in 1999 Entry Draft for Trevor Kidd, June 25, 1999. Signed as a free agent by **Vancouver**, August 11, 2001.

VERBEEK, Pat

(vuhr-BEEK, PAT)

Right wing. Shoots right. 5'9", 192 lbs. Born, Sarnia, Ont., May 24, 1964. New Jersey's 3rd choice, 43rd overall, in 1982 Entry Draft.

Season	Club	League	GP	G	A	Pts	PIM	PP	SH	GW	S	%	+/-	TF	F%	H	SB	Min	GP	G	A	Pts	PIM	PP	SH	GW
1979-80	Petrolia Jets	OHA-B	41	17	24	41	85	....	....	....	....	....	....	....	....	....	....	....	....	....	....	....	....	....	....	....
1980-81	Petrolia Jets	OHA-B	42	44	44	88	155	....	....	....	....	....	....	....	....	....	....	....	....	....	....	....	....	....	....	....
1981-82	Sudbury Wolves	OHL	66	37	51	88	180	....	....	....	....	....	....	....	....	....	....	....	....	....	....	....	....	....	....	....
1982-83	Sudbury Wolves	OHL	61	40	67	107	184	....	....	....	....	....	....	....	....	....	....	....	....	....	....	....	....	....	....	....
	New Jersey	NHL	6	3	2	5	8	0	0	0	12	25.0	–2	....	....	....	....	....	....	....	....	....	....	....	....	....
1983-84	New Jersey	NHL	79	20	27	47	158	5	1	2	167	12.0	–19	....	....	....	....	....	....	....	....	....	....	....	....	....
1984-85	New Jersey	NHL	78	15	18	33	162	5	1	1	147	10.2	–24	....	....	....	....	....	....	....	....	....	....	....	....	....
1985-86	New Jersey	NHL	76	25	28	53	79	4	1	0	159	15.7	–24	....	....	....	....	....	....	....	....	....	....	....	....	....
1986-87	New Jersey	NHL	74	35	24	59	120	17	0	5	143	24.5	–23	....	....	....	....	....	....	....	....	....	....	....	....	....
1987-88	New Jersey	NHL	73	46	31	77	227	13	0	8	179	25.7	29	....	....	....	....	....	20	4	8	12	51	2	0	1
1988-89	New Jersey	NHL	77	26	21	47	189	9	0	1	175	14.9	–18	....	....	....	....	....	....	....	....	....	....	....	....	....
1989-90	Hartford	NHL	80	44	45	89	228	14	0	5	219	20.1	1	....	....	....	....	....	7	2	2	4	26	1	0	1
1990-91	Hartford	NHL	80	43	39	82	246	15	0	5	247	17.4	0	....	....	....	....	....	6	3	2	5	40	2	0	0
1991-92	Hartford	NHL	76	22	35	57	243	10	0	3	163	13.5	–16	....	....	....	....	....	7	0	2	2	12	0	0	0
1992-93	Hartford	NHL	84	39	43	82	197	16	0	6	235	16.6	–7	....	....	....	....	....	....	....	....	....	....	....	....	....
1993-94	Hartford	NHL	84	37	38	75	177	15	1	3	226	16.4	–15	....	....	....	....	....	....	....	....	....	....	....	....	....
1994-95	Hartford	NHL	29	7	11	18	53	3	0	0	75	9.3	0	....	....	....	....	....	....	....	....	....	....	....	....	....
	NY Rangers	NHL	19	10	5	15	18	4	0	2	56	17.9	–2	....	....	....	....	....	10	4	6	10	20	3	0	0
1995-96	NY Rangers	NHL	69	41	41	82	129	17	0	6	252	16.3	29	....	....	....	....	....	11	3	6	9	12	1	0	0
1996-97	Dallas	NHL	81	17	36	53	128	5	0	4	172	9.9	3	....	....	....	....	....	7	1	3	4	16	1	0	0
1997-98	Dallas	NHL	82	31	26	57	170	9	0	8	190	16.3	15	....	....	....	....	....	17	3	2	5	26	2	0	1
1998-99♦	Dallas	NHL	78	17	17	34	133	8	0	2	134	12.7	11	1	100.0	130	15	14:34	18	3	4	7	14	0	0	1
99-2000	Detroit	NHL	68	22	26	48	95	7	0	5	138	15.9	22	1	0.0	148	8	16:04	9	1	1	2	2	1	0	1
2000-01	Detroit	NHL	67	15	15	30	73	7	0	0	113	13.3	0	7	57.1	141	13	13:31	5	2	0	2	6	2	0	0
2001-02	Dallas	NHL	64	7	13	20	72	3	0	1	81	8.6	–4	4	0.0	106	6	11:39	....	....	....	....	....	....	....	....
	NHL Totals		1424	522	541	1063	2905	186	4	67	3283	15.9		13	38.5	525	42	14:00	117	26	36	62	225	15	0	5

OHL Rookie of the Year (1982)

Played in NHL All-Star Game (1991, 1996)

Traded to **Hartford** by **New Jersey** for Sylvain Turgeon, June 17, 1989. Traded to **NY Rangers** by **Hartford** for Glen Featherstone, Michael Stewart, NY Rangers' 1st round choice (Jean-Sebastien Giguere) in 1995 Entry Draft and 4th round choice (Steve Wasylko) in 1996 Entry Draft, March 23, 1995. Signed as a free agent by **Dallas**, August 21, 1996. Signed as a free agent by **Detroit**, November 11, 1999. Signed as a free agent by **Dallas**, August 31, 2001.

VIGIER, Jean-Pierre

(vih-ZHAY, zhaw-pee-AIR) **ATL.**

Right wing. Shoots right. 6'1", 190 lbs. Born, Notre Dame de Lourdes, Man., September 11, 1976.

Season	Club	League	GP	G	A	Pts	PIM	PP	SH	GW	S	%	+/-	TF	F%	H	SB	Min	GP	G	A	Pts	PIM	PP	SH	GW
1995-96	Portage Terriers	MJHL	56	32	49	81	....	....	....	....	....	....	....	....	....	....	....	....	....	....	....	....	....	....	....	....
1996-97	North-Michigan	WCHA	36	10	14	24	54	....	....	....	....	....	....	....	....	....	....	....	....	....	....	....	....	....	....	....
1997-98	North-Michigan	CCHA	36	12	15	27	60	....	....	....	....	....	....	....	....	....	....	....	....	....	....	....	....	....	....	....
1998-99	North-Michigan	CCHA	42	21	18	39	80	....	....	....	....	....	....	....	....	....	....	....	....	....	....	....	....	....	....	....
99-2000	North-Michigan	CCHA	39	18	17	35	72	....	....	....	....	....	....	....	....	....	....	....	....	....	....	....	....	....	....	....
	Orlando	IHL	3	1	0	1	0	....	....	....	....	....	....	....	....	....	....	....	....	....	....	....	....	....	....	....
2000-01	Atlanta	NHL	2	0	0	0	0	0	0	0	1	0.0	–2	1	100.0	0	1	9:56	....	....	....	....	....	....	....	....
	Orlando	IHL	78	23	17	40	66	....	....	....	....	....	....	....	....	....	....	....	16	6	6	12	14	....	....	....
2001-02	Atlanta	NHL	15	4	1	5	4	0	0	0	18	22.2	–5	3	66.7	17	7	13:13	....	....	....	....	....	....	....	....
	Chicago Wolves	AHL	62	25	16	41	26	....	....	....	....	....	....	....	....	....	....	....	21	7	7	14	20	....	....	....
	NHL Totals		17	4	1	5	4	0	0	0	19	21.1		4	75.0	17	8	12:50	....	....	....	....	....	....	....	....

CCHA Second All-Star Team (1999) • CCHA All-Tournament Team (1999)

Signed as a free agent by **Atlanta**, April 20, 2000.

VIRTA, Tony
(VIHR-ta, TON-nee) **MIN.**

Right wing. Shoots left. 5'10", 187 lbs. Born, Hameenlinna, Finland, June 28, 1972. Minnesota's 5th choice, 103rd overall, in 2001 Entry Draft.

			Regular Season																Playoffs							
Season	Club	League	GP	G	A	Pts	PIM	PP	SH	GW	S	%	+/-	TF	F%	H	SB	Min	GP	G	A	Pts	PIM	PP	SH	GW
1990-91	HPK-Jr.	Finn-Jr.	28	17	34	51	42	....	....	....	....	....	....	....	....	....	....	....	....	....	....	....	....	....	....	....
1991-92	HPK Hameenlinna	Finland	35	24	31	55	114	....	....	....	....	....	....	....	....	....	....	....	....	....	....	....	....	....	....	....
1992-93	HPK Jr.	Finn-Jr.	2	1	2	3	2	....	....	....	....	....	....	....	....	....	....	....	....	....	....	....	....	....	....	....
	HPK Hameenlinna	Finland	48	9	8	17	35	....	....	....	....	....	....	....	....	....	....	....	12	1	0	1	0	....	....	....
1993-94	HPK Hameenlinna	Finland	47	17	16	33	50	....	....	....	....	....	....	....	....	....	....	....	....	....	....	....	....	....	....	....
1994-95	HPK Hameenlinna	Finland	43	9	24	33	75	....	....	....	....	....	....	....	....	....	....	....	....	....	....	....	....	....	....	....
1995-96	HPK Hameenlinna	Finland	48	15	19	34	12	....	....	....	....	....	....	....	....	....	....	....	9	1	3	4	31	....	....	....
1996-97	Frankfurt Lions	Germany	48	13	13	26	40	....	....	....	....	....	....	....	....	....	....	....	8	1	2	3	56	....	....	....
1997-98	TPS Turku	EuroHL	6	1	0	1	0	....	....	....	....	....	....	....	....	....	....	....	....	....	....	....	....	....	....	....
	TPS Turku	Finland	48	17	15	32	26	....	....	....	....	....	....	....	....	....	....	....	3	0	0	0	0	....	....	....
1998-99	TPS Turku	Finland	54	16	27	43	32	....	....	....	....	....	....	....	....	....	....	....	10	6	6	12	8	....	....	....
99-2000	TPS Turku	EuroHL	5	1	5	6	2	....	....	....	....	....	....	....	....	....	....	....	4	1	2	3	0	....	....	....
	TPS Turku	Finland	53	14	37	51	57	....	....	....	....	....	....	....	....	....	....	....	11	2	7	9	6	....	....	....
2000-01	TPS Turku	Finland	56	27	33	60	24	....	....	....	....	....	....	....	....	....	....	....	10	2	5	7	6	....	....	....
2001-02	**Minnesota**	**NHL**	**8**	**2**	**3**	**5**	**0**	**0**	**0**	**0**	**15**	**13.3**	**0**	**0**	**0.0**	**5**	**1**	**13:23**	....	....	....	....	....	....	....	....
	Houston Aeros	AHL	67	25	33	58	29	....	....	....	....	....	....	....	....	....	....	....	14	1	9	10	10	....	....	....
	NHL Totals		**8**	**2**	**3**	**5**	**0**	**0**	**0**	**0**	**15**	**13.3**		**0**	**0.0**	**5**	**1**	**13:23**	....	....	....	....	....	....	....	....

Finnish League Player of the Year (2001)

VIRTUE, Terry
(VIR-too, TAIR-ee)

Defense. Shoots right. 6', 207 lbs. Born, Scarborough, Ont., August 12, 1970.

			Regular Season																Playoffs							
Season	Club	League	GP	G	A	Pts	PIM	PP	SH	GW	S	%	+/-	TF	F%	H	SB	Min	GP	G	A	Pts	PIM	PP	SH	GW
1988-89	Hobbema Hawks	AJHL	56	6	31	37	339	....	....	....	....	....	....	....	....	....	....	....	....	....	....	....	....	....	....	....
	Victoria Cougars	WHL	8	1	1	2	13	....	....	....	....	....	....	....	....	....	....	....	....	....	....	....	....	....	....	....
1989-90	Victoria Cougars	WHL	24	1	9	10	85	....	....	....	....	....	....	....	....	....	....	....	....	....	....	....	....	....	....	....
	Tri-City	WHL	34	1	10	11	82	....	....	....	....	....	....	....	....	....	....	....	6	0	0	0	30	....	....	....
1990-91	Tri-City	WHL	11	1	8	9	24	....	....	....	....	....	....	....	....	....	....	....	....	....	....	....	....	....	....	....
	Portland	WHL	59	9	44	53	127	....	....	....	....	....	....	....	....	....	....	....	....	....	....	....	....	....	....	....
1991-92	Roanoke Valley	ECHL	38	4	22	26	165	....	....	....	....	....	....	....	....	....	....	....	....	....	....	....	....	....	....	....
	Louisville	ECHL	23	1	15	16	58	....	....	....	....	....	....	....	....	....	....	....	13	0	8	8	49	....	....	....
1992-93	Louisville	ECHL	28	0	17	17	84	....	....	....	....	....	....	....	....	....	....	....	....	....	....	....	....	....	....	....
	Wheeling	ECHL	31	3	15	18	86	....	....	....	....	....	....	....	....	....	....	....	16	3	5	8	18	....	....	....
1993-94	Wheeling	ECHL	34	5	28	33	61	....	....	....	....	....	....	....	....	....	....	....	6	2	2	4	4	....	....	....
	Cape Breton	AHL	26	4	6	10	10	....	....	....	....	....	....	....	....	....	....	....	5	0	0	0	17	....	....	....
1994-95	Worcester	AHL	73	14	25	39	183	....	....	....	....	....	....	....	....	....	....	....	....	....	....	....	....	....	....	....
	Atlanta Knights	IHL	1	0	0	0	2	....	....	....	....	....	....	....	....	....	....	....	....	....	....	....	....	....	....	....
1995-96	Worcester	AHL	76	7	31	38	234	....	....	....	....	....	....	....	....	....	....	....	4	0	0	0	4	....	....	....
1996-97	Worcester	AHL	80	16	26	42	220	....	....	....	....	....	....	....	....	....	....	....	5	0	4	4	8	....	....	....
1997-98	Worcester	AHL	74	8	26	34	233	....	....	....	....	....	....	....	....	....	....	....	11	1	4	5	41	....	....	....
1998-99	**Boston**	**NHL**	**4**	**0**	**0**	**0**	**0**	**0**	**0**	**0**	**2**	**0.0**	**2**	**0**	**0.0**	**4**	**2**	**9:41**	....	....	....	....	....	....	....	....
	Providence	AHL	76	8	48	56	117	....	....	....	....	....	....	....	....	....	....	....	17	2	12	14	29	....	....	....
99-2000	**NY Rangers**	**NHL**	**1**	**0**	**0**	**0**	**0**	**0**	**0**	**0**	**2**	**0.0**	**–2**	**0**	**0.0**	**3**	**1**	**12:32**	....	....	....	....	....	....	....	....
	Hartford	AHL	67	5	22	27	166	....	....	....	....	....	....	....	....	....	....	....	23	3	7	10	51	....	....	....
2000-01	Hartford	AHL	71	5	24	29	166	....	....	....	....	....	....	....	....	....	....	....	5	1	0	1	2	....	....	....
2001-02	Hartford	AHL	76	4	20	24	117	....	....	....	....	....	....	....	....	....	....	....	10	0	1	1	19	....	....	....
	NHL Totals		**5**	**0**	**0**	**0**	**0**	**0**	**0**	**0**	**4**	**0.0**		**0**	**0.0**	**7**	**3**	**10:15**	....	....	....	....	....	....	....	....

AHL Second All-Star Team (1999)

Signed as a free agent by **St. Louis**, January 29, 1996. Signed as a free agent by **Boston**, August 28, 1998. Signed as a free agent by **NY Rangers**, July 29, 1999.

VISHNEVSKI, Vitaly
(vihsh-NEHV-skee, vih-TAL-ee) **ANA.**

Defense. Shoots left. 6'2", 206 lbs. Born, Kharkov, USSR, March 18, 1980. Anaheim's 1st choice, 5th overall, in 1998 Entry Draft.

			Regular Season																Playoffs							
Season	Club	League	GP	G	A	Pts	PIM	PP	SH	GW	S	%	+/-	TF	F%	H	SB	Min	GP	G	A	Pts	PIM	PP	SH	GW
1995-96	Yaroslavl	CIS	40	4	4	8	20	....	....	....	....	....	....	....	....	....	....	....	....	....	....	....	....	....	....	....
1996-97	Yaroslavl 2	Russia-3	45	0	2	2	30	....	....	....	....	....	....	....	....	....	....	....	....	....	....	....	....	....	....	....
1997-98	Yaroslavl	Russia	47	8	9	17	164	....	....	....	....	....	....	....	....	....	....	....	....	....	....	....	....	....	....	....
1998-99	Yaroslavl	Russia	34	3	4	7	38	....	....	....	....	....	....	....	....	....	....	....	10	0	0	0	4	....	....	....
99-2000	**Anaheim**	**NHL**	**31**	**1**	**1**	**2**	**26**	**1**	**0**	**0**	**17**	**5.9**	**0**	**0**	**0.0**	**113**	**28**	**16:38**	....	....	....	....	....	....	....	....
	Cincinnati	AHL	35	1	3	4	45	....	....	....	....	....	....	....	....	....	....	....	....	....	....	....	....	....	....	....
2000-01	**Anaheim**	**NHL**	**76**	**1**	**10**	**11**	**99**	**0**	**0**	**0**	**49**	**2.0**	**–1**	**0**	**0.0**	**286**	**69**	**19:14**	....	....	....	....	....	....	....	....
2001-02	**Anaheim**	**NHL**	**74**	**0**	**3**	**3**	**60**	**0**	**0**	**0**	**54**	**0.0**	**–10**	**0**	**0.0**	**230**	**55**	**17:36**	....	....	....	....	....	....	....	....
	NHL Totals		**181**	**2**	**14**	**16**	**185**	**1**	**0**	**0**	**120**	**1.7**		**0**	**0.0**	**629**	**152**	**18:26**	....	....	....	....	....	....	....	....

VISNOVSKY, Lubomir
(vihsh-NAWV-skee, LOO-boh-mihr) **L.A.**

Defense. Shoots left. 5'10", 183 lbs. Born, Topolcany, Czech., August 11, 1976. Los Angeles' 4th choice, 118th overall, in 2000 Entry Draft.

			Regular Season																Playoffs							
Season	Club	League	GP	G	A	Pts	PIM	PP	SH	GW	S	%	+/-	TF	F%	H	SB	Min	GP	G	A	Pts	PIM	PP	SH	GW
1994-95	Slov. Bratislava	Slovakia	36	11	12	23	10	....	....	....	....	....	....	....	....	....	....	....	9	1	3	4	2	....	....	....
1995-96	Slov. Bratislava	Slovakia	35	8	6	14	22	....	....	....	....	....	....	....	....	....	....	....	13	1	5	6	2	....	....	....
1996-97	Slov. Bratislava	Slovakia	44	11	12	23	....	....	....	....	....	....	....	....	....	....	....	....	2	0	1	1		....	....	....
	Slov. Bratislava	EuroHL	6	3	1	4	2	....	....	....	....	....	....	....	....	....	....	....	2	0	0	0	6	....	....	....
1997-98	Slov. Bratislava	Slovakia	36	7	9	16	16	....	....	....	....	....	....	....	....	....	....	....	11	2	4	6	8	....	....	....
	Slov. Bratislava	EuroHL	6	1	0	1	4	....	....	....	....	....	....	....	....	....	....	....	....	....	....	....	....	....	....	....
1998-99	Slov. Bratislava	Slovakia	40	9	10	19	31	....	....	....	....	....	....	....	....	....	....	....	10	5	5	10	0	....	....	....
	Slov. Bratislava	EuroHL	6	0	3	3	4	....	....	....	....	....	....	....	....	....	....	....	....	....	....	....	....	....	....	....
99-2000	Slov. Bratislava	Slovakia	52	21	24	45	38	....	....	....	....	....	....	....	....	....	....	....	8	5	3	8	16	....	....	....
2000-01	**Los Angeles**	**NHL**	**81**	**7**	**32**	**39**	**36**	**3**	**0**	**3**	**105**	**6.7**	**16**	**0**	**0.0**	**99**	**87**	**16:58**	**8**	**0**	**0**	**0**	**0**	**0**	**0**	**0**
2001-02	**Los Angeles**	**NHL**	**72**	**4**	**17**	**21**	**14**	**1**	**0**	**2**	**95**	**4.2**	**–5**	**0**	**0.0**	**63**	**51**	**16:15**	**4**	**0**	**1**	**1**	**0**	**0**	**0**	**0**
	Slovakia	Olympics	3	1	2	3	0	....	....	....	....	....	....	....	....	....	....	....	....	....	....	....	....	....	....	....
	NHL Totals		**153**	**11**	**49**	**60**	**50**	**4**	**0**	**5**	**200**	**5.5**		**0**	**0.0**	**162**	**138**	**16:38**	**12**	**0**	**1**	**1**	**0**	**0**	**0**	**0**

NHL All-Rookie Team (2001)

VLASAK, Tomas
(VLAH-sahk, TAW-mawsh)

Center. Shoots right. 5'10", 175 lbs. Born, Prague, Czech., February 1, 1975. Los Angeles' 6th choice, 120th overall, in 1993 Entry Draft.

			Regular Season																Playoffs							
Season	Club	League	GP	G	A	Pts	PIM	PP	SH	GW	S	%	+/-	TF	F%	H	SB	Min	GP	G	A	Pts	PIM	PP	SH	GW
1992-93	Slavia IPS Praha	Czech-2	31	17	6	23	....	....	....	....	....	....	....	....	....	....	....	....	....	....	....	....	....	....	....	....
1993-94	Litvinov	Czech	41	16	11	27	0	....	....	....	....	....	....	....	....	....	....	....	4	0	1	1	0	....	....	....
1994-95	Litvinov	Czech	35	6	14	20	4	....	....	....	....	....	....	....	....	....	....	....	4	0	0	0	4	....	....	....
1995-96	Litvinov	Czech	35	10	22	32	....	....	....	....	....	....	....	....	....	....	....	....	15	5	5	10	....	....	....	....
1996-97	Litvinov	EuroHL	6	0	2	2	29	....	....	....	....	....	....	....	....	....	....	....	2	0	2	2	0	....	....	....
	Litvinov	Czech	52	26	34	60	16	....	....	....	....	....	....	....	....	....	....	....	....	....	....	....	....	....	....	....
1997-98	Litvinov	Czech	51	22	22	44	40	....	....	....	....	....	....	....	....	....	....	....	4	1	2	3	2	....	....	....
1998-99	HPK Hameenlinna	Finland	54	28	29	57	36	....	....	....	....	....	....	....	....	....	....	....	8	2	*9	11	0	....	....	....
99-2000	HPK Hameenlinna	Finland	48	24	39	63	63	....	....	....	....	....	....	....	....	....	....	....	8	3	4	7	6	....	....	....
2000-01	**Los Angeles**	**NHL**	**10**	**1**	**3**	**4**	**2**	**0**	**0**	**0**	**15**	**6.7**	**4**	**1**	**0.0**	**5**	**0**	**11:28**	....	....	....	....	....	....	....	....
	Lowell	AHL	5	0	1	1	5	....	....	....	....	....	....	....	....	....	....	....	....	....	....	....	....	....	....	....
	HPK Hameenlinna	Finland	27	6	12	18	10	....	....	....	....	....	....	....	....	....	....	....	....	....	....	....	....	....	....	....
2001-02	HC Ambri-Piotta	Swiss	42	20	23	43	39	....	....	....	....	....	....	....	....	....	....	....	4	1	2	3	2	....	....	....
	NHL Totals		**10**	**1**	**3**	**4**	**2**	**0**	**0**	**0**	**15**	**6.7**		**1**	**0.0**	**5**	**0**	**11:28**	....	....	....	....	....	....	....	....

• Released by **LA Kings**, December 5, 2000.

VON ARX, Reto

(VAWN-ARX, RAY-toh)

Center. Shoots left. 5'10", 190 lbs. Born, Egerkingen, Switz., September 13, 1976. Chicago's 14th choice, 271st overall, in 2000 Entry Draft.

			Regular Season																Playoffs							
Season	Club	League	GP	G	A	Pts	PIM	PP	SH	GW	S	%	+/-	TF	F%	H	SB	Min	GP	G	A	Pts	PIM	PP	SH	GW
1992-93	SC Langnau	Swiss-2	35	11	4	15	28	....	....	....	....	....	....	....	....	....	....	....	5	0	0	0	6	....	....	....
1993-94	SC Langnau	Swiss-3	36	35	31	66	42	....	....	....	....	....	....	....	....	....	....	....	....	....	....	....	....	....	....	....
1994-95	SC Langnau	Swiss-2	36	14	9	23	72	....	....	....	....	....	....	....	....	....	....	....	5	1	2	3	27	....	....	....
1995-96	HC Davos	Swiss	34	4	6	10	59	....	....	....	....	....	....	....	....	....	....	....	5	0	2	2	4	....	....	....
1996-97	HC Davos	Swiss	42	10	17	27	78	....	....	....	....	....	....	....	....	....	....	....	6	1	3	4	4	....	....	....
1997-98	HC Davos	Swiss	39	8	15	23	113	....	....	....	....	....	....	....	....	....	....	....	18	10	6	16	18	....	....	....
1998-99	HC Davos	Swiss	45	21	20	41	76	....	....	....	....	....	....	....	....	....	....	....	6	4	6	10	16	....	....	....
99-2000	HC Davos	Swiss	45	19	26	45	70	....	....	....	....	....	....	....	....	....	....	....	5	2	0	2	10	....	....	....
2000-01	**Chicago**	**NHL**	**19**	**3**	**1**	**4**	**4**	**0**	**0**	**1**	**12**	**25.0**	**–4**	**94**	**42.6**	**1**	**5**	**11:26**	....	....	....	....	....	....	....	....
	Norfolk Admirals	AHL	49	16	26	42	28	....	....	....	....	....	....	....	....	....	....	....	9	1	2	3	8	....	....	....
2001-02	HC Davos	Swiss	31	6	15	21	95	....	....	....	....	....	....	....	....	....	....	....	12	3	8	11	35	....	....	....
	Switzerland	Olympics	2	0	1	1	0	....	....	....	....	....	....	....	....	....	....	....	....	....	....	....	....	....	....	....
	NHL Totals		**19**	**3**	**1**	**4**	**4**	**0**	**0**	**1**	**12**	**25.0**		**94**	**42.6**	**1**	**5**	**11:26**	....	....	....	....	....	....	....	....

Signed as a free agent by **HC Davos** (Swiss) with Chicago retaining NHL rights, October 6, 2001.

VRBATA, Radim

(vuhr-BA-tuh, ra-DEEM) **COL.**

Right wing. Shoots right. 6'1", 185 lbs. Born, Boleslav, Czech., June 13, 1981. Colorado's 10th choice, 212th overall, in 1999 Entry Draft.

			Regular Season																Playoffs							
Season	Club	League	GP	G	A	Pts	PIM	PP	SH	GW	S	%	+/-	TF	F%	H	SB	Min	GP	G	A	Pts	PIM	PP	SH	GW
1997-98	Mlada Boleslav Jr.	Czech-Jr.	35	42	31	73	4	....	....	....	....	....	....	....	....	....	....	....	....	....	....	....	....	....	....	....
1998-99	Hull Olympiques	QMJHL	54	22	38	60	16	....	....	....	....	....	....	....	....	....	....	....	23	6	13	19	6	....	....	....
99-2000	Hull Olympiques	QMJHL	58	29	45	74	26	....	....	....	....	....	....	....	....	....	....	....	15	3	9	12	8	....	....	....
2000-01	Shawinigan	QMJHL	55	56	64	120	67	....	....	....	....	....	....	....	....	....	....	....	10	4	7	11	4	....	....	....
	Hershey Bears	AHL	....	....	....	....	....	....	....	....	....	....	....	....	....	....	....	....	1	0	1	1	2	....	....	....
2001-02	**Colorado**	**NHL**	**52**	**18**	**12**	**30**	**14**	**6**	**0**	**3**	**112**	**16.1**	**7**	**8**	**37.5**	**34**	**23**	**14:32**	**9**	**0**	**0**	**0**	**0**	**0**	**0**	**0**
	Hershey Bears	AHL	20	8	14	22	8	....	....	....	....	....	....	....	....	....	....	....	....	....	....	....	....	....	....	....
	NHL Totals		**52**	**18**	**12**	**30**	**14**	**6**	**0**	**3**	**112**	**16.1**		**8**	**37.5**	**34**	**23**	**14:32**	**9**	**0**	**0**	**0**	**0**	**0**	**0**	**0**

QMJHL First All-Star Team (2001)

VUJTEK, Vladimir

(VYOO-tehk, VLAD-dih-MEER) **PIT.**

Left wing. Shoots left. 6'2", 200 lbs. Born, Ostrava, Czech., February 17, 1972. Montreal's 5th choice, 73rd overall, in 1991 Entry Draft.

			Regular Season																Playoffs							
Season	Club	League	GP	G	A	Pts	PIM	PP	SH	GW	S	%	+/-	TF	F%	H	SB	Min	GP	G	A	Pts	PIM	PP	SH	GW
1988-89	TJ Vitkovice	Czech	3	0	1	1	0	....	....	....	....	....	....	....	....	....	....	....	....	....	....	....	....	....	....	....
1989-90	TJ Vitkovice	Czech	22	3	4	7	....	....	....	....	....	....	....	....	....	....	....	....	7	4	3	7	....	....	....	....
1990-91	TJ Vitkovice	Czech	27	7	4	11	....	....	....	....	....	....	....	....	....	....	....	....	....	....	....	....	....	....	....	....
	Tri-City	WHL	37	26	18	44	74	....	....	....	....	....	....	....	....	....	....	....	7	2	3	5	4	....	....	....
1991-92	Tri-City	WHL	53	41	61	102	114	....	....	....	....	....	....	....	....	....	....	....	....	....	....	....	....	....	....	....
	Montreal	**NHL**	**2**	**0**	**0**	**0**	**0**	**0**	**0**	**0**	**1**	**0.0**	**–1**	....	....	....	....	....	....	....	....	....	....	....	....	....
1992-93	**Edmonton**	**NHL**	**30**	**1**	**10**	**11**	**8**	**0**	**0**	**0**	**49**	**2.0**	**–1**	....	....	....	....	....	....	....	....	....	....	....	....	....
	Cape Breton	AHL	20	10	9	19	14	....	....	....	....	....	....	....	....	....	....	....	1	0	0	0	0	....	....	....
1993-94	**Edmonton**	**NHL**	**40**	**4**	**15**	**19**	**14**	**1**	**0**	**0**	**66**	**6.1**	**–7**	....	....	....	....	....	....	....	....	....	....	....	....	....
1994-95	HC Vitkovice	Czech	18	5	7	12	51	....	....	....	....	....	....	....	....	....	....	....	....	....	....	....	....	....	....	....
	Cape Breton	AHL	30	10	11	21	30	....	....	....	....	....	....	....	....	....	....	....	....	....	....	....	....	....	....	....
	Las Vegas	IHL	1	0	0	0	0	....	....	....	....	....	....	....	....	....	....	....	....	....	....	....	....	....	....	....
1995-96	HC Vitkovice	Czech	26	6	7	13	....	....	....	....	....	....	....	....	....	....	....	....	4	1	1	2	....	....	....	....
1996-97	Assat Pori	Finland	50	27	31	58	48	....	....	....	....	....	....	....	....	....	....	....	4	1	2	3	2	....	....	....
1997-98	**Tampa Bay**	**NHL**	**30**	**2**	**4**	**6**	**16**	**0**	**0**	**1**	**44**	**4.5**	**–2**	....	....	....	....	....	....	....	....	....	....	....	....	....
	Adirondack	AHL	2	1	2	3	0	....	....	....	....	....	....	....	....	....	....	....	....	....	....	....	....	....	....	....
1998-99	HC Vitkovice	Czech	47	20	34	54	77	....	....	....	....	....	....	....	....	....	....	....	....	....	....	....	....	....	....	....
99-2000	**Atlanta**	**NHL**	**3**	**0**	**0**	**0**	**0**	**0**	**0**	**0**	**2**	**0.0**	**0**	**0**	**0.0**	**1**	**0**	**10:52**	....	....	....	....	....	....	....	....
	HC Sparta Praha	Czech	21	12	19	31	14	....	....	....	....	....	....	....	....	....	....	....	8	2	3	5	10	....	....	....
2000-01	HC Sparta Praha	Czech	38	11	18	29	28	....	....	....	....	....	....	....	....	....	....	....	13	3	6	9	4	....	....	....
2001-02	HPK Hameenlinna	Finland	45	19	39	58	38	....	....	....	....	....	....	....	....	....	....	....	8	4	7	11	6	....	....	....
	NHL Totals		**105**	**7**	**29**	**36**	**38**	**1**	**0**	**1**	**162**	**4.3**		**0**	**0.0**	**1**	**0**	**10:52**	....	....	....	....	....	....	....	....

WHL West First All-Star Team (1992)

Traded to **Edmonton** by **Montreal** with Shayne Corson and Brent Gilchrist for Vincent Damphousse and Edmonton's 4th round choice (Adam Wiesel) in 1993 Entry Draft, August 27, 1992. Traded to **Tampa Bay** by **Edmonton** with Edmonton's 3rd round choice (Dmitry Afanasenkov) in 1998 Entry Draft for Brantt Myhres and Toronto's 3rd round choice (previously acquired, Edmonton selected Alex Henry) in 1998 Entry Draft, July 16, 1997. • Missed majority of 1997-98 season recovering from Epstein-Barr Virus, December, 1997. Signed as a free agent by **Atlanta**, July 29, 1999. • Missed majority of 1999-2000 season recovering from facial injuries suffered in exhibition game vs. NY Rangers, September 18, 1999. Signed as a free agent by **Pittsburgh**, July 15, 2002.

VYBORNY, David

(vih-BOHR-nee, DAY-vihd) **CBJ**

Right wing. Shoots left. 5'10", 189 lbs. Born, Jihlava, Czech., June 2, 1975. Edmonton's 3rd choice, 33rd overall, in 1993 Entry Draft.

			Regular Season																Playoffs							
Season	Club	League	GP	G	A	Pts	PIM	PP	SH	GW	S	%	+/-	TF	F%	H	SB	Min	GP	G	A	Pts	PIM	PP	SH	GW
1991-92	HC Sparta Praha	Czech	32	6	9	15	2	....	....	....	....	....	....	....	....	....	....	....	....	....	....	....	....	....	....	....
1992-93	HC Sparta Praha	Czech	52	20	24	44	....	....	....	....	....	....	....	....	....	....	....	....	....	....	....	....	....	....	....	....
1993-94	HC Sparta Praha	Czech	44	15	20	35	0	....	....	....	....	....	....	....	....	....	....	....	6	4	7	11	0	....	....	....
1994-95	Cape Breton	AHL	76	23	38	61	30	....	....	....	....	....	....	....	....	....	....	....	....	....	....	....	....	....	....	....
1995-96	HC Sparta Praha	Czech	40	12	18	30	....	....	....	....	....	....	....	....	....	....	....	....	12	6	5	11	....	....	....	....
1996-97	HC Sparta Praha	Czech	47	20	29	49	14	....	....	....	....	....	....	....	....	....	....	....	10	7	7	14	6	....	....	....
1997-98	MoDo	Sweden	45	16	21	37	34	....	....	....	....	....	....	....	....	....	....	....	9	0	2	2	2	....	....	....
1998-99	HC Sparta Praha	Czech	52	24	*46	*70	22	....	....	....	....	....	....	....	....	....	....	....	8	1	3	4	....	....	....	....
99-2000	HC Sparta Praha	Czech	50	25	38	63	30	....	....	....	....	....	....	....	....	....	....	....	9	3	*8	*11	4	....	....	....
2000-01	**Columbus**	**NHL**	**79**	**13**	**19**	**32**	**22**	**5**	**0**	**1**	**125**	**10.4**	**–9**	**36**	**44.4**	**23**	**17**	**15:25**	....	....	....	....	....	....	....	....
2001-02	**Columbus**	**NHL**	**75**	**13**	**18**	**31**	**6**	**6**	**0**	**2**	**103**	**12.6**	**–14**	**25**	**44.0**	**40**	**18**	**15:27**	....	....	....	....	....	....	....	....
	NHL Totals		**154**	**26**	**37**	**63**	**28**	**11**	**0**	**3**	**228**	**11.4**		**61**	**44.3**	**63**	**35**	**15:26**	....	....	....	....	....	....	....	....

Signed as a free agent by **Columbus**, June 8, 2000.

VYSHEDKEVICH, Sergei

(vee-shehd-KAY-vihch, SAIR-gay)

Defense. Shoots left. 6', 195 lbs. Born, Dedovsk, USSR, January 3, 1975. New Jersey's 3rd choice, 70th overall, in 1995 Entry Draft.

			Regular Season																Playoffs							
Season	Club	League	GP	G	A	Pts	PIM	PP	SH	GW	S	%	+/-	TF	F%	H	SB	Min	GP	G	A	Pts	PIM	PP	SH	GW
1994-95	Dynamo Moscow	CIS	49	6	7	13	67	....	....	....	....	....	....	....	....	....	....	....	14	2	0	2	12	....	....	....
1995-96	Dynamo Moscow	CIS	49	5	4	9	12	....	....	....	....	....	....	....	....	....	....	....	13	1	1	2	6	....	....	....
1996-97	Albany	AHL	65	8	27	35	16	....	....	....	....	....	....	....	....	....	....	....	12	0	6	6	0	....	....	....
1997-98	Albany	AHL	54	12	16	28	12	....	....	....	....	....	....	....	....	....	....	....	13	0	10	10	4	....	....	....
1998-99	Albany	AHL	79	11	38	49	28	....	....	....	....	....	....	....	....	....	....	....	5	0	3	3	0	....	....	....
99-2000	**Atlanta**	**NHL**	**7**	**1**	**3**	**4**	**2**	**1**	**0**	**0**	**5**	**20.0**	**–3**	**0**	**0.0**	**14**	**8**	**23:18**	....	....	....	....	....	....	....	....
	Orlando	IHL	69	11	24	35	32	....	....	....	....	....	....	....	....	....	....	....	6	3	3	6	8	....	....	....
2000-01	**Atlanta**	**NHL**	**23**	**1**	**2**	**3**	**14**	**0**	**0**	**0**	**28**	**3.6**	**–7**	**0**	**0.0**	**19**	**24**	**20:34**	....	....	....	....	....	....	....	....
	Orlando	IHL	10	2	3	5	2	....	....	....	....	....	....	....	....	....	....	....	....	....	....	....	....	....	....	....
	Cincinnati	AHL	17	3	2	5	2	....	....	....	....	....	....	....	....	....	....	....	....	....	....	....	....	....	....	....
2001-02	Dynamo Moscow	Russia	50	5	7	12	56	....	....	....	....	....	....	....	....	....	....	....	3	0	0	0	0	....	....	....
	NHL Totals		**30**	**2**	**5**	**7**	**16**	**1**	**0**	**0**	**33**	**6.1**		**0**	**0.0**	**33**	**32**	**21:13**	....	....	....	....	....	....	....	....

Traded to **Atlanta** by **New Jersey** for future considerations, June 25, 1999. Traded to **Anaheim** by **Atlanta** with Scott Langkow for Ladislav Kohn, February 9, 2001.

WALKER, Scott

(WAH-kuhr, SKAWT) **NSH.**

Center. Shoots right. 5'10", 196 lbs. Born, Cambridge, Ont., July 19, 1973. Vancouver's 4th choice, 124th overall, in 1993 Entry Draft.

			Regular Season																Playoffs							
Season	Club	League	GP	G	A	Pts	PIM	PP	SH	GW	S	%	+/-	TF	F%	H	SB	Min	GP	G	A	Pts	PIM	PP	SH	GW
1989-90	Kitchener	OJHL-B	6	0	5	5	4	....	....	....	....	....	....	....	....	....	....	....	....	....	....	....	....	....	....	....
	Cambridge	OJHL-B	27	7	22	29	87	....	....	....	....	....	....	....	....	....	....	....	....	....	....	....	....	....	....	....
1990-91	Cambridge	OJHL-B	45	10	27	37	241	....	....	....	....	....	....	....	....	....	....	....	....	....	....	....	....	....	....	....
1991-92	Owen Sound	OHL	53	7	31	38	128	....	....	....	....	....	....	....	....	....	....	....	5	0	7	7	8	....	....	....
1992-93	Owen Sound	OHL	57	23	68	91	110	....	....	....	....	....	....	....	....	....	....	....	8	1	5	6	16	....	....	....
1993-94	Hamilton Canucks	AHL	77	10	29	39	272	....	....	....	....	....	....	....	....	....	....	....	4	0	1	1	25	....	....	....
1994-95	Syracuse Crunch	AHL	74	14	38	52	334	....	....	....	....	....	....	....	....	....	....	....	....	....	....	....	....	....	....	....
	Vancouver	**NHL**	**11**	**0**	**1**	**1**	**33**	**0**	**0**	**0**	**8**	**0.0**	**0**	....	....	....	....	....	....	....	....	....	....	....	....	....
1995-96	**Vancouver**	**NHL**	**63**	**4**	**8**	**12**	**137**	**0**	**1**	**1**	**45**	**8.9**	**–7**	....	....	....	....	....	....	....	....	....	....	....	....	....
	Syracuse Crunch	AHL	15	3	12	15	52	....	....	....	....	....	....	....	....	....	....	....	16	9	8	17	39	....	....	....

Season	Club	League	GP	G	A	Pts	PIM	PP	SH	GW	S	%	+/-	TF	F%	H	SB	Min	GP	G	A	Pts	PIM	PP	SH	GW
			Regular Season																Playoffs							
1996-97	Vancouver	NHL	64	3	15	18	132	0	0	0	55	5.5	2	....	....	....	....	....	....	....	....	....	....	....	....	....
1997-98	Vancouver	NHL	59	3	10	13	164	0	1	1	40	7.5	–8	....	....	....	....	....	....	....	....	....	....	....	....	....
1998-99	Nashville	NHL	71	15	25	40	103	0	1	2	96	15.6	0	265	48.3	82	41	16:21	....	....	....	....	....	....	....	....
99-2000	Nashville	NHL	69	7	21	28	90	0	1	0	98	7.1	–16	30	36.7	121	36	15:49	....	....	....	....	....	....	....	....
2000-01	Nashville	NHL	74	25	29	54	66	9	3	1	159	15.7	–2	541	51.4	92	10	19:17	....	....	....	....	....	....	....	....
2001-02	Nashville	NHL	28	4	5	9	18	1	0	0	46	8.7	–13	149	38.9	47	9	18:38	....	....	....	....	....	....	....	....
	NHL Totals		439	61	114	175	743	10	7	5	547	11.2		985	48.2	342	96	17:22	....	....	....	....	....	....	....	....

OHL Second All-Star Team (1993)

Claimed by **Nashville** from **Vancouver** in Expansion Draft, June 26, 1998.

WALLIN, Jesse

(WAHL-ihn, JEH-see) **DET.**

Defense. Shoots left. 6'2", 190 lbs. Born, Saskatoon, Sask., March 10, 1978. Detroit's 1st choice, 26th overall, in 1996 Entry Draft.

Season	Club	League	GP	G	A	Pts	PIM	PP	SH	GW	S	%	+/-	TF	F%	H	SB	Min	GP	G	A	Pts	PIM	PP	SH	GW
1993-94	North Battleford	SMHL	32	1	7	8	41																			
1994-95	Red Deer Rebels	WHL	72	4	20	24	72																			
1995-96	Red Deer Rebels	WHL	70	5	19	24	61												9	0	3	3	4			
1996-97	Red Deer Rebels	WHL	59	6	33	39	70												16	1	4	5	10			
1997-98	Red Deer Rebels	WHL	14	1	6	7	17												5	0	1	1	2			
1998-99	Adirondack	AHL	76	4	12	16	34												3	0	2	2	2			
99-2000	Detroit	NHL	1	0	0	0	0	0	0	0	0	0.0	–2	0	0.0	2	1	19:22								
	Cincinnati	AHL	75	3	14	17	61																			
2000-01	Detroit	NHL	1	0	0	0	2	0	0	0	1	0.0	0	0	0.0	1	1	3:50								
	Cincinnati	AHL	76	2	15	17	50												4	0	1	1	4			
2001-02	Detroit	NHL	15	0	1	1	13	0	0	0	8	0.0	–1	0	0.0	30	7	10:42								
	Cincinnati	AHL	5	1	1	2	2																			
	NHL Totals		17	0	1	1	15	0	0	0	9	0.0		0	0.0	33	9	10:48								

Canadian Major Junior Humanitarian Player of the Year (1997) • Missed majority of 1997-98 season recovering from arm injury suffered in automobile accident (September 10, 1997) and foot injury suffered in game vs. Germany, December 30, 1997 • Missed majority of 2001-02 season recovering from groin injury suffered in training camp, October 1, 2001.

WALLIN, Niclas

(VAH-lihn, NIH-kluhs) **CAR.**

Defense. Shoots left. 6'3", 220 lbs. Born, Boden, Sweden, February 20, 1975. Carolina's 3rd choice, 97th overall, in 2000 Entry Draft.

Season	Club	League	GP	G	A	Pts	PIM	PP	SH	GW	S	%	+/-	TF	F%	H	SB	Min	GP	G	A	Pts	PIM	PP	SH	GW
1994-95	Bodens IK Jr.	Swede-Jr.	30	2	13	15	125																			
	Bodens IK	Swede-2	13	0	0	0	0												2	0	0	0	0			
1995-96	Bodens IK Jr.	Swede-Jr.	2	2	2	4	0																			
	Bodens IK	Swede-2	30	2	7	9	26												2	0	1	1	2			
1996-97	Brynas IF Gavle	Sweden	47	1	1	2	14																			
1997-98	Brynas IF Gavle	Sweden	44	2	3	5	57												3	0	1	1	4			
1998-99	Brynas IF Gavle	Sweden	46	2	4	6	52												14	0	1	1	8			
99-2000	Brynas IF Gavle	Sweden	48	7	9	16	73												11	2	1	3	14			
	Brynas IF Gavle	EuroHL	5	1	1	2	10																			
2000-01	Carolina	NHL	37	2	3	5	21	0	0	0	19	10.5	–11	0	0.0	51	45	14:57	3	0	0	0	2	0	0	0
	Cincinnati	IHL	8	1	2	3	4												3	0	0	0	2			
2001-02	Carolina	NHL	52	1	2	3	36	0	0	0	33	3.0	1	0	0.0	67	32	12:12	23	2	1	3	12	0	0	2
	NHL Totals		89	3	5	8	57	0	0	0	52	5.8		0	0.0	118	77	13:20	26	2	1	3	14	0	0	2

• Missed most of 2000-01 season recovering from shoulder injury suffered in game vs. Florida, January 12, 2001.

WALSER, Derrick

(WAHL-zuhr, DEHR-rihk) **CBJ**

Defense. Shoots left. 5'10", 196 lbs. Born, New Glasgow, N.S., May 12, 1978.

Season	Club	League	GP	G	A	Pts	PIM	PP	SH	GW	S	%	+/-	TF	F%	H	SB	Min	GP	G	A	Pts	PIM	PP	SH	GW
1994-95	Beauport	QMJHL	48	4	18	22	34												12	2	5	7	2			
1995-96	Beauport	QMJHL	69	9	31	40	56												20	2	11	13	16			
1996-97	Beauport	QMJHL	37	13	25	38	26																			
	Rimouski Oceanic	QMJHL	31	15	30	45	44												4	2	2	4	6			
1997-98	Rimouski Oceanic	QMJHL	70	41	69	110	135												18	10	*26	36	49			
1998-99	Saint John	AHL	40	3	7	10	24																			
	Johnstown Chiefs	ECHL	24	8	12	20	29																			
99-2000	Saint John	AHL	14	2	3	5	10																			
	Johnstown Chiefs	ECHL	54	17	26	43	104												7	3	3	6	8			
2000-01	Saint John	AHL	76	19	36	55	36												19	7	9	16	14			
2001-02	Columbus	NHL	2	1	0	1	0	0	0	0	2	50.0	–2	0	0.0	1	0	16:18								
	Syracuse Crunch	AHL	73	23	38	61	70												10	1	5	6	12			
	NHL Totals		2	1	0	1	0	0	0	0	2	50.0		0	0.0	1	0	16:18								

QMJHL First All-Star Team (1997) • Won Emile Bouchard Trophy (Top Defenseman - QMJHL) (1998) • QMJHL First All-Star Team (1998) • Canadian Major Junior First All-Star Team (1998) • Canadian Major Junior Defenseman of the Year (1998)

Signed as a free agent by **Calgary**, October 16, 1998. Signed as a free agent by **Columbus**, September 17, 2001.

WALZ, Wes

(WAHLZ, WEHS) **MIN.**

Center. Shoots right. 5'10", 180 lbs. Born, Calgary, Alta., May 15, 1970. Boston's 3rd choice, 57th overall, in 1989 Entry Draft.

Season	Club	League	GP	G	A	Pts	PIM	PP	SH	GW	S	%	+/-	TF	F%	H	SB	Min	GP	G	A	Pts	PIM	PP	SH	GW
1987-88	Cgy. North Stars	AMHL	35	47	52	99	72																			
	Prince Albert	WHL	1	1	1	2	0																			
1988-89	Lethbridge	WHL	63	29	75	104	32												8	1	5	6	6			
1989-90	Lethbridge	WHL	56	54	86	140	69												19	13	*24	*37	33			
	Boston	NHL	2	1	1	2	0	1	0	0	1	100.0	–1													
1990-91	Boston	NHL	56	8	8	16	32	1	0	1	57	14.0	–14						2	0	0	0	0	0	0	0
	Maine Mariners	AHL	20	8	12	20	19												2	0	0	0	21			
1991-92	Boston	NHL	15	0	3	3	12	0	0	0	17	0.0	–3													
	Maine Mariners	AHL	21	13	11	24	38																			
	Philadelphia	NHL	2	1	0	1	0	0	0	1	2	50.0	1													
	Hershey Bears	AHL	41	13	28	41	37												6	1	2	3	0			
1992-93	Hershey Bears	AHL	78	35	45	80	106																			
1993-94	Calgary	NHL	53	11	27	38	16	1	0	0	79	13.9	20						6	3	0	3	2	0	0	0
	Saint John	AHL	15	6	6	12	14																			
1994-95	Calgary	NHL	39	6	12	18	11	4	0	1	73	8.2	7						1	0	0	0	0	0	0	0
1995-96	Detroit	NHL	2	0	0	0	0	0	0	0	2	0.0	0													
	Adirondack	AHL	38	20	35	55	58																			
1996-97	EV Zug	Swiss	41	24	22	46	67												9	5	1	6	39			
1997-98	EV Zug	Swiss	38	18	34	52	32												20	*16	*12	*28	18			
	EV Zug	EuroHL	5	1	3	4	10																			
1998-99	EV Zug	Swiss	42	22	27	49	75												10	3	9	12	2			
	EV Zug	EuroHL	6	7	5	12	4												2	0	0	0	12			
99-2000	Long Beach	IHL	6	4	3	7	8																			
	HC Lugano	Swiss	13	7	11	18	14												5	3	4	7	4			
2000-01	Minnesota	NHL	82	18	12	30	37	0	7	3	152	11.8	–8	1533	47.2	51	105	16:45								
2001-02	Minnesota	NHL	64	10	20	30	43	0	2	5	97	10.3	0	1231	44.8	43	65	16:42								
	NHL Totals		315	55	83	138	151	7	9	11	480	11.5		2764	46.1	94	170	16:44	9	3	0	3	2	0	0	0

WHL Rookie of the Year (1989) • WHL East First All-Star Team (1990)

Traded to **Philadelphia** by **Boston** with Garry Galley and Boston's 3rd round choice (Milos Holan) in 1993 Entry Draft for Gord Murphy, Brian Dobbin, Philadelphia's 3rd round choice (Sergei Zholtok) in 1992 Entry Draft and 4th round choice (Charles Paquette) in 1993 Entry Draft, January 2, 1992. Signed as a free agent by **Calgary**, August 26, 1993. Signed as a free agent by **Detroit**, September 6, 1995. Signed as a free agent by **Long Beach** (IHL), October 12, 1999. Signed as a free agent by **Minnesota**, June 28, 2000.

WARD, Aaron

(WOHRD, AIR-ruhn) **CAR.**

Defense. Shoots right. 6'2", 200 lbs. Born, Windsor, Ont., January 17, 1973. Winnipeg's 1st choice, 5th overall, in 1991 Entry Draft.

			Regular Season																Playoffs							
Season	Club	League	GP	G	A	Pts	PIM	PP	SH	GW	S	%	+/-	TF	F%	H	SB	Min	GP	G	A	Pts	PIM	PP	SH	GW
1988-89	Nepean Raiders	OCJHL	54	1	14	15	40	....	....	....	....	....	....	....	....	....	....	....	....	....	....	....	....	....	....	....
1989-90	Nepean Raiders	OCJHL	52	6	33	39	85	....	....	....	....	....	....	....	....	....	....	....	....	....	....	....	....	....	....	....
1990-91	U. of Michigan	CCHA	46	8	11	19	126	....	....	....	....	....	....	....	....	....	....	....	....	....	....	....	....	....	....	....
1991-92	U. of Michigan	CCHA	42	7	12	19	64	....	....	....	....	....	....	....	....	....	....	....	....	....	....	....	....	....	....	....
1992-93	U. of Michigan	CCHA	30	5	8	13	73	....	....	....	....	....	....	....	....	....	....	....	....	....	....	....	....	....	....	....
1993-94	**Detroit**	**NHL**	**5**	**1**	**0**	**1**	**4**	**0**	**0**	**0**	**3**	**33.3**	**2**	....	....	....	....	....	....	....	....	....	....	....	....	....
	Adirondack	AHL	58	4	12	16	87	....	....	....	....	....	....	....	....	....	....	....	9	2	6	8	6	....	....	....
1994-95	Adirondack	AHL	76	11	24	35	87	....	....	....	....	....	....	....	....	....	....	....	4	0	1	1	0	....	....	....
	Detroit	**NHL**	**1**	**0**	**1**	**1**	**2**	**0**	**0**	**0**	**0**	**0.0**	**1**	....	....	....	....	....	....	....	....	....	....	....	....	....
1995-96	Adirondack	AHL	74	5	10	15	133	....	....	....	....	....	....	....	....	....	....	....	3	0	0	0	6	....	....	....
1996-97♦	**Detroit**	**NHL**	**49**	**2**	**5**	**7**	**52**	**0**	**0**	**0**	**40**	**5.0**	**–9**	....	....	....	....	....	**19**	**0**	**0**	**0**	**17**	**0**	**0**	**0**
1997-98♦	**Detroit**	**NHL**	**52**	**5**	**5**	**10**	**47**	**0**	**0**	**1**	**47**	**10.6**	**–1**	....	....	....	....	....	....	....	....	....	....	....	....	....
1998-99	**Detroit**	**NHL**	**60**	**3**	**8**	**11**	**52**	**0**	**0**	**0**	**46**	**6.5**	**–5**	**0**	**0.0**	**133**	**37**	**13:55**	**8**	**0**	**1**	**1**	**8**	**0**	**0**	**0**
99-2000	**Detroit**	**NHL**	**36**	**1**	**3**	**4**	**24**	**0**	**0**	**0**	**25**	**4.0**	**–4**	**0**	**0.0**	**87**	**19**	**12:36**	**3**	**0**	**0**	**0**	**0**	**0**	**0**	**0**
2000-01	**Detroit**	**NHL**	**73**	**4**	**5**	**9**	**57**	**0**	**0**	**1**	**48**	**8.3**	**–4**	**0**	**0.0**	**191**	**70**	**16:60**	....	....	....	....	....	....	....	....
2001-02	**Carolina**	**NHL**	**79**	**3**	**11**	**14**	**74**	**0**	**0**	**2**	**69**	**4.3**	**0**	**1**	**100.0**	**162**	**117**	**19:40**	**23**	**1**	**1**	**2**	**22**	**0**	**0**	**0**
	NHL Totals		**355**	**19**	**38**	**57**	**312**	**0**	**0**	**4**	**278**	**6.8**		**1**	**100.0**	**573**	**243**	**16:28**	**53**	**1**	**2**	**3**	**47**	**0**	**0**	**0**

Traded to **Detroit** by **Winnipeg** with Toronto's 4th round choice (previously acquired by Winnipeg - later traded to Detroit - Detroit selected John Jakopin) in 1993 Entry Draft for Paul Ysebaert and future considerations (Alan Kerr, June 18, 1993), June 11, 1993. • Missed majority of 1999-2000 season recovering from shoulder injury suffered in game vs. Vancouver, January 19, 2000. Traded to **Carolina** by **Detroit** for Carolina's 2nd round choice (Jiri Hudler) in 2002 Entry Draft, July 9, 2001.

WARD, Dixon

(WOHRD, DIHX-ohn)

Right wing. Shoots right. 6', 200 lbs. Born, Leduc, Alta., September 23, 1968. Vancouver's 6th choice, 128th overall, in 1988 Entry Draft.

Season	Club	League	GP	G	A	Pts	PIM	PP	SH	GW	S	%	+/-	TF	F%	H	SB	Min	GP	G	A	Pts	PIM	PP	SH	GW
1986-87	Red Deer Rebels	AJHL	59	46	40	86	153	....	....	....	....	....	....	....	....	....	....	....	20	11	11	22	16	....	....	....
1987-88	Red Deer Rebels	AJHL	51	60	71	131	167	....	....	....	....	....	....	....	....	....	....	....	....	....	....	....	....	....	....	....
1988-89	North Dakota	WCHA	37	8	9	17	26	....	....	....	....	....	....	....	....	....	....	....	....	....	....	....	....	....	....	....
1989-90	North Dakota	WCHA	45	35	34	69	44	....	....	....	....	....	....	....	....	....	....	....	....	....	....	....	....	....	....	....
1990-91	North Dakota	WCHA	43	34	35	69	84	....	....	....	....	....	....	....	....	....	....	....	....	....	....	....	....	....	....	....
1991-92	North Dakota	WCHA	38	33	31	64	90	....	....	....	....	....	....	....	....	....	....	....	....	....	....	....	....	....	....	....
1992-93	**Vancouver**	**NHL**	**70**	**22**	**30**	**52**	**82**	**4**	**1**	**0**	**111**	**19.8**	**34**	....	....	....	....	....	**9**	**2**	**3**	**5**	**0**	**2**	**0**	**0**
1993-94	**Vancouver**	**NHL**	**33**	**6**	**1**	**7**	**37**	**2**	**0**	**1**	**46**	**13.0**	**–14**	....	....	....	....	....	....	....	....	....	....	....	....	....
	Los Angeles	**NHL**	**34**	**6**	**2**	**8**	**45**	**2**	**0**	**0**	**44**	**13.6**	**–8**	....	....	....	....	....	....	....	....	....	....	....	....	....
1994-95	**Toronto**	**NHL**	**22**	**0**	**3**	**3**	**31**	**0**	**0**	**0**	**15**	**0.0**	**–4**	....	....	....	....	....	....	....	....	....	....	....	....	....
	St. John's	AHL	6	3	3	6	19	....	....	....	....	....	....	....	....	....	....	....	....	....	....	....	....	....	....	....
	Detroit Vipers	IHL	7	3	6	9	7	....	....	....	....	....	....	....	....	....	....	....	5	3	0	3	7	....	....	....
1995-96	**Buffalo**	**NHL**	**8**	**2**	**2**	**4**	**6**	**0**	**0**	**1**	**12**	**16.7**	**1**	....	....	....	....	....	....	....	....	....	....	....	....	....
	Rochester	AHL	71	38	56	94	74	....	....	....	....	....	....	....	....	....	....	....	19	11	*24	*35	8	....	....	....
1996-97	**Buffalo**	**NHL**	**79**	**13**	**32**	**45**	**36**	**1**	**2**	**4**	**93**	**14.0**	**17**	....	....	....	....	....	**12**	**2**	**3**	**5**	**6**	**0**	**0**	**1**
1997-98	**Buffalo**	**NHL**	**71**	**10**	**13**	**23**	**42**	**0**	**2**	**3**	**99**	**10.1**	**9**	....	....	....	....	....	**15**	**3**	**8**	**11**	**6**	**0**	**0**	**0**
1998-99	**Buffalo**	**NHL**	**78**	**20**	**24**	**44**	**44**	**2**	**1**	**4**	**101**	**19.8**	**10**	**11**	**54.5**	**45**	**49**	**15:47**	**21**	**7**	**5**	**12**	**32**	**0**	**2**	**3**
99-2000	**Buffalo**	**NHL**	**71**	**11**	**9**	**20**	**41**	**1**	**2**	**2**	**101**	**10.9**	**1**	**8**	**50.0**	**48**	**44**	**13:58**	**5**	**0**	**1**	**1**	**2**	**0**	**0**	**0**
2000-01	**Boston**	**NHL**	**63**	**5**	**13**	**18**	**65**	**0**	**0**	**0**	**88**	**5.7**	**–1**	**108**	**48.2**	**62**	**19**	**12:38**	....	....	....	....	....	....	....	....
2001-02	Langnau	Swiss	23	8	19	27	38	....	....	....	....	....	....	....	....	....	....	....	9	3	6	9	2	....	....	....
	NHL Totals		**529**	**95**	**129**	**224**	**429**	**12**	**8**	**15**	**710**	**13.4**		**127**	**48.8**	**155**	**112**	**14:14**	**62**	**14**	**20**	**34**	**46**	**2**	**2**	**4**

WCHA Second All-Star Team (1991, 1992) • Won Jack A. Butterfield Trophy (Playoff MVP - AHL) (1996)

Traded to **LA Kings** by **Vancouver** for Jimmy Carson, January 8, 1994. Traded to **Toronto** by **LA Kings** with Guy Leveque, Kelly Fairchild and Shayne Toporowski for Eric Lacroix, Chris Snell and Toronto's 4th round choice (Eric Belanger) in 1996 Entry Draft, October 3, 1994. Signed as a free agent by **Buffalo**, September 20, 1995. Signed as a free agent by **Boston**, November 8, 2000.

WARD, Ed

(WOHRD, EHD)

Right wing. Shoots right. 6'3", 220 lbs. Born, Edmonton, Alta., November 10, 1969. Quebec's 7th choice, 108th overall, in 1988 Entry Draft.

Season	Club	League	GP	G	A	Pts	PIM	PP	SH	GW	S	%	+/-	TF	F%	H	SB	Min	GP	G	A	Pts	PIM	PP	SH	GW
1986-87	Sherwood Park	AJHL	60	18	28	46	272	....	....	....	....	....	....	....	....	....	....	....	....	....	....	....	....	....	....	....
1987-88	North-Michigan	WCHA	25	0	2	2	40	....	....	....	....	....	....	....	....	....	....	....	....	....	....	....	....	....	....	....
1988-89	North-Michigan	WCHA	42	5	15	20	36	....	....	....	....	....	....	....	....	....	....	....	....	....	....	....	....	....	....	....
1989-90	North-Michigan	WCHA	39	5	11	16	77	....	....	....	....	....	....	....	....	....	....	....	....	....	....	....	....	....	....	....
1990-91	North-Michigan	WCHA	46	13	18	31	109	....	....	....	....	....	....	....	....	....	....	....	....	....	....	....	....	....	....	....
1991-92	Greensboro	ECHL	12	4	8	12	21	....	....	....	....	....	....	....	....	....	....	....	....	....	....	....	....	....	....	....
	Halifax Citadels	AHL	51	7	11	18	65	....	....	....	....	....	....	....	....	....	....	....	....	....	....	....	....	....	....	....
1992-93	Halifax Citadels	AHL	70	13	19	32	56	....	....	....	....	....	....	....	....	....	....	....	....	....	....	....	....	....	....	....
1993-94	**Quebec**	**NHL**	**7**	**1**	**0**	**1**	**5**	**0**	**0**	**0**	**3**	**33.3**	**0**	....	....	....	....	....	....	....	....	....	....	....	....	....
	Cornwall Aces	AHL	60	12	30	42	65	....	....	....	....	....	....	....	....	....	....	....	12	1	3	4	14	....	....	....
1994-95	Cornwall Aces	AHL	56	10	14	24	118	....	....	....	....	....	....	....	....	....	....	....	....	....	....	....	....	....	....	....
	Calgary	**NHL**	**2**	**1**	**1**	**2**	**2**	**0**	**0**	**0**	**1**	**100.0**	**–2**	....	....	....	....	....	....	....	....	....	....	....	....	....
	Saint John	AHL	11	4	5	9	20	....	....	....	....	....	....	....	....	....	....	....	5	1	0	1	10	....	....	....
1995-96	**Calgary**	**NHL**	**41**	**3**	**5**	**8**	**44**	**0**	**0**	**0**	**33**	**9.1**	**–2**	....	....	....	....	....	....	....	....	....	....	....	....	....
	Saint John	AHL	12	1	2	3	45	....	....	....	....	....	....	....	....	....	....	....	16	4	4	8	27	....	....	....
1996-97	**Calgary**	**NHL**	**40**	**5**	**8**	**13**	**49**	**0**	**0**	**1**	**33**	**15.2**	**–3**	....	....	....	....	....	....	....	....	....	....	....	....	....
	Saint John	AHL	1	0	0	0	0	....	....	....	....	....	....	....	....	....	....	....	....	....	....	....	....	....	....	....
	Detroit Vipers	IHL	31	7	6	13	45	....	....	....	....	....	....	....	....	....	....	....	....	....	....	....	....	....	....	....
1997-98	**Calgary**	**NHL**	**64**	**4**	**5**	**9**	**122**	**0**	**0**	**0**	**52**	**7.7**	**–1**	....	....	....	....	....	....	....	....	....	....	....	....	....
1998-99	**Calgary**	**NHL**	**68**	**3**	**5**	**8**	**67**	**0**	**0**	**0**	**56**	**5.4**	**–4**	**6**	**33.3**	**122**	**14**	**8:02**	....	....	....	....	....	....	....	....
99-2000	**Atlanta**	**NHL**	**44**	**5**	**1**	**6**	**44**	**0**	**2**	**0**	**51**	**9.8**	**–5**	**7**	**28.6**	**130**	**15**	**11:26**	....	....	....	....	....	....	....	....
	Anaheim	**NHL**	**8**	**1**	**0**	**1**	**15**	**0**	**0**	**0**	**5**	**20.0**	**–2**	**0**	**0.0**	**13**	**0**	**6:56**	....	....	....	....	....	....	....	....
2000-01	**New Jersey**	**NHL**	**4**	**0**	**1**	**1**	**6**	**0**	**0**	**0**	**4**	**0.0**	**2**	**0**	**0.0**	**2**	**0**	**4:48**	....	....	....	....	....	....	....	....
	Albany	AHL	65	14	19	33	71	....	....	....	....	....	....	....	....	....	....	....	....	....	....	....	....	....	....	....
2001-02	Timra IK	Sweden	34	3	2	5	64	....	....	....	....	....	....	....	....	....	....	....	....	....	....	....	....	....	....	....
	NHL Totals		**278**	**23**	**26**	**49**	**354**	**0**	**2**	**1**	**238**	**9.7**		**13**	**30.8**	**267**	**29**	**9:04**	....	....	....	....	....	....	....	....

Traded to **Calgary** by **Quebec** for Francois Groleau, March 23, 1995. Claimed by **Atlanta** from **Calgary** in Expansion Draft, June 25, 1999. Traded to **Anaheim** by **Atlanta** for Anaheim's 7th round choice (Colin Fitzrandolph) in 2001 Entry Draft, March 14, 2000. Traded to **New Jersey** by **Anaheim** for New Jersey's 7th round choice (Tony Martensson) in 2001 Entry Draft, June 12, 2000.

WARD, Jason

(WOHRD, JAY-suhn) **MTL.**

Right wing. Shoots right. 6'3", 200 lbs. Born, Chapleau, Ont., January 16, 1979. Montreal's 1st choice, 11th overall, in 1997 Entry Draft.

Season	Club	League	GP	G	A	Pts	PIM	PP	SH	GW	S	%	+/-	TF	F%	H	SB	Min	GP	G	A	Pts	PIM	PP	SH	GW
1994-95	Oshawa	OJHL-B	47	30	31	61	75	....	....	....	....	....	....	....	....	....	....	....	....	....	....	....	....	....	....	....
1995-96	Niagara Falls	OHL	64	15	35	50	139	....	....	....	....	....	....	....	....	....	....	....	10	6	4	10	23	....	....	....
1996-97	Erie Otters	OHL	58	25	39	64	137	....	....	....	....	....	....	....	....	....	....	....	5	1	2	3	2	....	....	....
1997-98	Erie Otters	OHL	21	7	9	16	42	....	....	....	....	....	....	....	....	....	....	....	....	....	....	....	....	....	....	....
	Windsor	OHL	26	19	27	46	34	....	....	....	....	....	....	....	....	....	....	....	....	....	....	....	....	....	....	....
	Fredericton	AHL	7	1	0	1	2	....	....	....	....	....	....	....	....	....	....	....	1	0	0	0	2	....	....	....
1998-99	Windsor	OHL	12	8	11	19	25	....	....	....	....	....	....	....	....	....	....	....	....	....	....	....	....	....	....	....
	Plymouth Whalers	OHL	23	14	13	27	28	....	....	....	....	....	....	....	....	....	....	....	11	6	8	14	12	....	....	....
	Fredericton	AHL	....	....	....	....	....	....	....	....	....	....	....	....	....	....	....	....	10	4	2	6	22	....	....	....
99-2000	**Montreal**	**NHL**	**32**	**2**	**1**	**3**	**10**	**1**	**0**	**0**	**24**	**8.3**	**–1**	**86**	**44.2**	**37**	**5**	**9:10**	....	....	....	....	....	....	....	....
	Quebec	AHL	40	14	12	26	30	....	....	....	....	....	....	....	....	....	....	....	3	2	1	3	4	....	....	....
2000-01	**Montreal**	**NHL**	**12**	**0**	**0**	**0**	**12**	**0**	**0**	**0**	**4**	**0.0**	**3**	**2**	**50.0**	**10**	**6**	**8:16**	....	....	....	....	....	....	....	....
	Quebec	AHL	23	7	12	19	69	....	....	....	....	....	....	....	....	....	....	....	....	....	....	....	....	....	....	....
2001-02	Quebec	AHL	78	24	33	57	128	....	....	....	....	....	....	....	....	....	....	....	3	0	0	0	2	....	....	....
	NHL Totals		**44**	**2**	**1**	**3**	**22**	**1**	**0**	**0**	**28**	**7.1**		**88**	**44.3**	**47**	**11**	**8:55**	....	....	....	....	....	....	....	....

Traded to **Windsor** (OHL) by **Erie** (OHL) for Michael Rupp, January 6, 1998. Traded to **Plymouth** (OHL) by **Windsor** (OHL) for Kyle Chapman and Maxim Linnik, January 10, 1999. • Missed majority of 2000-01 season recovering from knee injury suffered in game vs. Carolina, January 16, 2001.

WARD, Lance
(WAWRD, LANTS) **FLA.**

Defense. Shoots left. 6'3", 220 lbs. Born, Lloydminster, Alta., June 2, 1978. Florida's 3rd choice, 63rd overall, in 1998 Entry Draft.

			Regular Season																Playoffs							
Season	Club	League	GP	G	A	Pts	PIM	PP	SH	GW	S	%	+/-	TF	F%	H	SB	Min	GP	G	A	Pts	PIM	PP	SH	GW
1993-94	Lloydminister	AAHA	20	8	12	20	68																			
1994-95	Red Deer Rebels	WHL	28	0	0	0	57																			
1995-96	Red Deer Rebels	WHL	72	4	13	17	127												10	0	4	4	10			
1996-97	Red Deer Rebels	WHL	70	5	34	39	229												16	0	3	3	36			
1997-98	Red Deer Rebels	WHL	71	8	25	33	233												5	0	0	0	16			
1998-99	Miami Matadors	ECHL	6	1	0	1	12																			
	Fort Wayne	IHL	13	0	2	2	28																			
	New Haven	AHL	43	2	5	7	51																			
99-2000	Louisville	AHL	80	4	16	20	190												4	0	0	0	6			
2000-01	**Florida**	**NHL**	**30**	**0**	**2**	**2**	**45**	**0**	**0**	**0**	**17**	**0.0**	**–3**	**0**	**0.0**	**50**	**21**	**15:50**								
	Louisville	AHL	35	3	2	5	78																			
2001-02	**Florida**	**NHL**	**68**	**1**	**4**	**5**	**131**	**0**	**0**	**0**	**39**	**2.6**	**–20**	**1**	**0.0**	**173**	**48**	**14:31**								
	NHL Totals		**98**	**1**	**6**	**7**	**176**	**0**	**0**	**0**	**56**	**1.8**		**1**	**0.0**	**223**	**69**	**14:55**								

• Re-entered NHL Entry Draft. Originally New Jersey's 1st choice, 10th overall, in 1996 Entry Draft.

WARRENER, Rhett
(WAHR-ihn-uhr, REHT) **BUF.**

Defense. Shoots right. 6'2", 217 lbs. Born, Shaunavon, Sask., January 27, 1976. Florida's 2nd choice, 27th overall, in 1994 Entry Draft.

			Regular Season																Playoffs							
Season	Club	League	GP	G	A	Pts	PIM	PP	SH	GW	S	%	+/-	TF	F%	H	SB	Min	GP	G	A	Pts	PIM	PP	SH	GW
1991-92	Sask. Blazers	SMHL	33	6	5	11	71																			
	Saskatoon Blades	WHL	2	0	0	0	0																			
1992-93	Saskatoon Blades	WHL	68	2	17	19	100												9	0	0	0	14			
1993-94	Saskatoon Blades	WHL	61	7	19	26	131												16	0	5	5	33			
1994-95	Saskatoon Blades	WHL	66	13	26	39	137												10	0	3	3	6			
1995-96	**Florida**	**NHL**	**28**	**0**	**3**	**3**	**46**	**0**	**0**	**0**	**19**	**0.0**	**4**						**21**	**0**	**1**	**1**	**0**	**0**	**0**	**0**
	Carolina	AHL	9	0	0	0	4																			
1996-97	**Florida**	**NHL**	**62**	**4**	**9**	**13**	**88**	**1**	**0**	**1**	**58**	**6.9**	**20**						**5**	**0**	**0**	**0**	**0**	**0**	**0**	**0**
1997-98	**Florida**	**NHL**	**79**	**0**	**4**	**4**	**99**	**0**	**0**	**0**	**66**	**0.0**	**–16**													
1998-99	**Florida**	**NHL**	**48**	**0**	**7**	**7**	**64**	**0**	**0**	**0**	**33**	**0.0**	**–1**	**0**	**0.0**	**75**	**38**	**19:01**								
	Buffalo	**NHL**	**13**	**1**	**0**	**1**	**20**	**0**	**0**	**0**	**11**	**9.1**	**3**	**0**	**0.0**	**28**	**12**	**18:13**	**20**	**1**	**3**	**4**	**32**	**0**	**0**	**0**
99-2000	**Buffalo**	**NHL**	**61**	**0**	**3**	**3**	**89**	**0**	**0**	**0**	**68**	**0.0**	**18**	**0**	**0.0**	**131**	**79**	**19:51**	**5**	**0**	**0**	**0**	**2**	**0**	**0**	**0**
2000-01	**Buffalo**	**NHL**	**77**	**3**	**16**	**19**	**78**	**0**	**0**	**2**	**103**	**2.9**	**10**	**0**	**0.0**	**187**	**107**	**20:24**	**13**	**0**	**2**	**2**	**4**	**0**	**0**	**0**
2001-02	**Buffalo**	**NHL**	**65**	**5**	**5**	**10**	**113**	**0**	**0**	**1**	**66**	**7.6**	**15**	**0**	**0.0**	**116**	**91**	**19:39**								
	NHL Totals		**433**	**13**	**47**	**60**	**597**	**1**	**0**	**4**	**424**	**3.1**		**0**	**0.0**	**537**	**327**	**19:44**	**64**	**1**	**6**	**7**	**38**	**0**	**0**	**0**

Traded to **Buffalo** by **Florida** with Florida's 5th round choice (Ryan Miller) in 1999 Entry Draft for Mike Wilson, March 23, 1999.

WARRINER, Todd
(WAHR-ihn-uhr, TAWD) **VAN.**

Left wing. Shoots left. 6'1", 200 lbs. Born, Blenheim, Ont., January 3, 1974. Quebec's 1st choice, 4th overall, in 1992 Entry Draft.

			Regular Season																Playoffs							
Season	Club	League	GP	G	A	Pts	PIM	PP	SH	GW	S	%	+/-	TF	F%	H	SB	Min	GP	G	A	Pts	PIM	PP	SH	GW
1988-89	Blenheim Blades	OJHL-C	10	1	4	5	0																			
1989-90	Chatham	OJHL-B	40	24	21	45	12																			
1990-91	Windsor	OHL	57	36	28	64	26												11	5	6	11	12			
1991-92	Windsor	OHL	50	41	41	82	64												7	5	4	9	6			
1992-93	Windsor	OHL	23	13	21	34	29																			
	Kitchener	OHL	32	19	24	43	35												7	5	14	19	14			
1993-94	Team Canada	Nat-Tm	50	11	20	31	33																			
	Canada	Olympics	4	1	1	2	0																			
	Kitchener	OHL																	1	0	1	1	0			
	Cornwall Aces	AHL																	10	1	4	5	4			
1994-95	St. John's	AHL	46	8	10	18	22												4	1	0	1	2			
	Toronto	**NHL**	**5**	**0**	**0**	**0**	**0**	**0**	**0**	**0**	**1**	**0.0**	**–3**													
1995-96	**Toronto**	**NHL**	**57**	**7**	**8**	**15**	**26**	**1**	**0**	**0**	**79**	**8.9**	**–11**						**6**	**1**	**1**	**2**	**2**	**0**	**0**	**0**
	St. John's	AHL	11	5	6	11	16																			
1996-97	**Toronto**	**NHL**	**75**	**12**	**21**	**33**	**41**	**2**	**2**	**0**	**146**	**8.2**	**–3**													
1997-98	**Toronto**	**NHL**	**45**	**5**	**8**	**13**	**20**	**0**	**0**	**1**	**73**	**6.8**	**5**													
1998-99	**Toronto**	**NHL**	**53**	**9**	**10**	**19**	**28**	**1**	**0**	**1**	**96**	**9.4**	**–6**	**579**	**47.8**	**52**	**10**	**14:06**	**9**	**0**	**0**	**0**	**2**	**0**	**0**	**0**
99-2000	**Toronto**	**NHL**	**18**	**3**	**1**	**4**	**2**	**0**	**0**	**0**	**33**	**9.1**	**6**	**33**	**45.5**	**18**	**5**	**12:33**								
	Tampa Bay	**NHL**	**55**	**11**	**13**	**24**	**34**	**3**	**1**	**0**	**100**	**11.0**	**–14**	**175**	**52.0**	**72**	**20**	**16:28**								
2000-01	**Tampa Bay**	**NHL**	**64**	**10**	**11**	**21**	**46**	**3**	**2**	**1**	**99**	**10.1**	**–13**	**400**	**50.5**	**80**	**22**	**14:43**								
2001-02	**Phoenix**	**NHL**	**18**	**0**	**3**	**3**	**8**	**0**	**0**	**0**	**10**	**0.0**	**–3**	**16**	**62.5**	**27**	**6**	**10:53**								
	Springfield	AHL	2	0	0	0	0																			
	Vancouver	**NHL**	**14**	**2**	**4**	**6**	**12**	**0**	**0**	**0**	**18**	**11.1**	**4**	**4**	**100.0**	**19**	**0**	**9:01**	**6**	**1**	**0**	**1**	**2**	**0**	**0**	**0**
	Manitoba Moose	AHL	30	7	13	20	32																			
	NHL Totals		**404**	**59**	**79**	**138**	**217**	**10**	**5**	**3**	**655**	**9.0**		**1207**	**49.6**	**268**	**63**	**14:09**	**21**	**2**	**1**	**3**	**6**	**0**	**0**	**0**

OHL First All-Star Team (1992)

Traded to **Toronto** by **Quebec** with Mats Sundin, Garth Butcher and Philadelphia's 1st round choice (previously acquired by Quebec - later traded to Washington - Washington selected Nolan Baumgartner) in 1994 Entry Draft for Wendel Clark, Sylvain Lefebvre, Landon Wilson and Toronto's 1st round choice (Jeffrey Kealty) in 1994 Entry Draft, June 28, 1994. Traded to **Tampa Bay** by **Toronto** for Tampa Bay's 3rd round choice (Mikael Tellqvist) in 2000 Entry Draft, November 29, 1999. Traded to **Phoenix** by **Tampa Bay** for Juha Ylonen, June 18, 2001. Traded to **Vancouver** by **Phoenix** with Trevor Letowski, Tyler Bouck and Phoenix's 3rd round choice in 2003 Entry Draft for Drake Berehowsky and Denis Pederson, December 28, 2001.

WASHBURN, Steve
(WAWSH-buhrn, STEEV)

Center. Shoots left. 6'2", 198 lbs. Born, Ottawa, Ont., April 10, 1975. Florida's 5th choice, 78th overall, in 1993 Entry Draft.

			Regular Season																Playoffs							
Season	Club	League	GP	G	A	Pts	PIM	PP	SH	GW	S	%	+/-	TF	F%	H	SB	Min	GP	G	A	Pts	PIM	PP	SH	GW
1990-91	Gloucester	OCJHL	56	21	30	51	47																			
1991-92	Ottawa 67's	OHL	59	5	17	22	10												11	2	3	5	4			
1992-93	Ottawa 67's	OHL	66	20	38	58	54																			
1993-94	Ottawa 67's	OHL	65	30	50	80	88												17	7	16	23	10			
1994-95	Ottawa 67's	OHL	63	43	63	106	72																			
	Cincinnati	IHL	6	3	1	4	0												9	1	3	4	4			
1995-96	**Florida**	**NHL**	**1**	**0**	**1**	**1**	**0**	**0**	**0**	**0**	**1**	**0.0**	**1**						**1**	**0**	**1**	**1**	**0**	**0**	**0**	**0**
	Carolina	AHL	78	29	54	83	45																			
1996-97	**Florida**	**NHL**	**18**	**3**	**6**	**9**	**4**	**1**	**0**	**0**	**21**	**14.3**	**2**													
	Carolina	AHL	60	23	40	63	66																			
1997-98	**Florida**	**NHL**	**58**	**11**	**8**	**19**	**32**	**4**	**0**	**2**	**61**	**18.0**	**–6**													
	New Haven	AHL	6	3	5	8	4												3	2	0	2	15			
1998-99	**Florida**	**NHL**	**4**	**0**	**0**	**0**	**4**	**0**	**0**	**0**	**0**	**0.0**	**–1**	**21**	**38.1**	**1**	**0**	**6:17**								
	New Haven	AHL	10	4	3	7	6																			
	Vancouver	**NHL**	**8**	**0**	**0**	**0**	**2**	**0**	**0**	**0**	**6**	**0.0**	**0**	**30**	**40.0**	**9**	**1**	**8:05**								
	Syracuse Crunch	AHL	13	1	6	7	6																			
99-2000	Milwaukee	IHL	12	0	4	4	16																			
	Philadelphia	**NHL**	**1**	**0**	**0**	**0**	**0**	**0**	**0**	**0**	**1**	**0.0**	**0**	**8**	**12.5**	**7**	**0**	**14:26**								
	Philadelphia	AHL	61	19	52	71	93												5	0	2	2	8			
2000-01	EHC Kloten	Swiss	8	0	6	6	16																			
	Philadelphia	**NHL**	**3**	**0**	**0**	**0**	**0**	**0**	**0**	**0**	**0**	**0.0**	**0**	**16**	**50.0**	**3**	**0**	**7:04**								
	Philadelphia	AHL	46	12	16	28	52												2	0	0	0	2			
2001-02	Iserlohn	Germany	38	7	14	21	59																			
	NHL Totals		**93**	**14**	**15**	**29**	**42**	**5**	**0**	**2**	**90**	**15.6**		**75**	**38.7**	**20**	**1**	**7:50**	**1**	**0**	**1**	**1**	**0**	**0**	**0**	**0**

Claimed on waivers by **Vancouver** from **Florida**, February 18, 1999. Signed as a free agent by **Nashville**, August 11, 1999. Traded to **Philadelphia** by **Nashville** for future considerations, November 16, 1999. Signed as a free agent by **EHC Kloten** (Switz.), July 26, 2000. Signed as a free agent by **Philadelphia**, November 21, 2000.

Regular Season: GP–Min; Playoffs: GP–GW (second group)

WATT, Mike

(WAHT, MIGHK) **CAR.**

Left wing. Shoots left. 6'2", 212 lbs. Born, Seaforth, Ont., March 31, 1976. Edmonton's 3rd choice, 32nd overall, in 1994 Entry Draft.

Season	Club	League	GP	G	A	Pts	PIM	PP	SH	GW	S	%	+/-	TF	F%	H	SB	Min	GP	G	A	Pts	PIM	PP	SH	GW
1990-91	Seaforth	OJHL-D	39	15	23	38	43																			
1991-92	Stratford	OJHL-B	40	5	21	26	103																			
1992-93	Stratford	OJHL-B	45	20	35	55	100																			
1993-94	Stratford	OJHL-B	48	34	34	68	165																			
1994-95	Michigan State	CCHA	39	12	6	18	64																			
1995-96	Michigan State	CCHA	37	17	22	39	60																			
1996-97	Michigan State	CCHA	39	24	17	41	109																			
1997-98	**Edmonton**	**NHL**	**14**	**1**	**2**	**3**	**4**	**0**	**0**	**1**	**14**	**7.1**	**-4**													
	Hamilton	AHL	63	24	25	49	65												9	2	2	4	8			
1998-99	**NY Islanders**	**NHL**	**75**	**8**	**17**	**25**	**12**	**0**	**0**	**4**	**75**	**10.7**	**-2**	**180**	**50.6**	**73**	**29**	**11:15**								
99-2000	**NY Islanders**	**NHL**	**45**	**5**	**6**	**11**	**17**	**0**	**1**	**0**	**49**	**10.2**	**-8**	**81**	**48.2**	**45**	**17**	**12:02**								
	Lowell	AHL	16	6	11	17	6												7	1	1	2	4			
2000-01	**Nashville**	**NHL**	**18**	**1**	**1**	**2**	**8**	**0**	**0**	**1**	**18**	**5.6**	**-2**	**2**	**50.0**	**22**	**7**	**11:02**								
	Milwaukee	IHL	60	20	20	40	48												5	1	2	3	6			
2001-02	Philadelphia	AHL	53	11	13	24	38												5	2	1	3	6			
	NHL Totals		**152**	**15**	**26**	**41**	**41**	**0**	**1**	**6**	**156**	**9.6**		**263**	**49.8**	**140**	**53**	**11:29**								

Traded to **NY Islanders** by **Edmonton** for Eric Fichaud, June 18, 1998. Claimed on waivers by **Nashville** from **NY Islanders**, May 23, 2000. Traded to **Philadelphia** by **Nashville** for Mikhail Chernov, May 24, 2001. Signed as a free agent by **Carolina**, August 7, 2002.

WEAVER, Mike

(WEE-vuhr, MIGHK) **ATL.**

Defense. Shoots right. 5'9", 180 lbs. Born, Bramalea, Ont., May 2, 1978.

Season	Club	League	GP	G	A	Pts	PIM	PP	SH	GW	S	%	+/-	TF	F%	H	SB	Min	GP	G	A	Pts	PIM	PP	SH	GW
1995-96	Bramalea Blues	OPJHL	48	10	39	49	103																			
1996-97	Michigan State	CCHA	39	0	7	7	46																			
1997-98	Michigan State	CCHA	44	4	22	26	68																			
1998-99	Michigan State	CCHA	42	1	6	7	54																			
99-2000	Michigan State	CCHA	26	0	7	7	20																			
2000-01	Orlando	IHL	68	0	8	8	34												16	0	2	2	8			
2001-02	**Atlanta**	**NHL**	**16**	**0**	**1**	**1**	**10**	**0**	**0**	**0**	**9**	**0.0**	**0**	**0**	**0.0**	**23**	**9**	**13:54**								
	Chicago Wolves	AHL	58	2	8	10	67												25	1	3	4	21			
	NHL Totals		**16**	**0**	**1**	**1**	**10**	**0**	**0**	**0**	**9**	**0.0**		**0**	**0.0**	**23**	**9**	**13:54**								

OPJHL Defenseman of the Year (1996) • CCHA All-Tournament Team (1997) • CCHA First All-Star Team (1999, 2000) • NCAA West Second All-American Team (1999, 2000) • Won CCHA Best Defensive Defenseman Award (1999, 2000)

Signed as a free agent by **Atlanta**, June 15, 2000.

WEBB, Steve

(WEHB, STEEV) **NYI**

Right wing. Shoots right. 6', 211 lbs. Born, Peterborough, Ont., April 30, 1975. Buffalo's 8th choice, 176th overall, in 1994 Entry Draft.

Season	Club	League	GP	G	A	Pts	PIM	PP	SH	GW	S	%	+/-	TF	F%	H	SB	Min	GP	G	A	Pts	PIM	PP	SH	GW
1991-92	Peterborough	OJHL-B	37	9	9	18	195																			
1992-93	Windsor	OHL	63	14	25	39	184																			
1993-94	Windsor	OHL	2	0	1	1	9																			
	Peterborough	OHL	33	6	15	21	117												6	1	1	2	20			
1994-95	Peterborough	OHL	42	8	16	24	109												11	3	3	6	22			
1995-96	Muskegon Fury	ColHL	58	18	24	42	263												5	1	2	3	22			
	Detroit Vipers	IHL	4	0	0	0	24																			
1996-97	**NY Islanders**	**NHL**	**41**	**1**	**4**	**5**	**144**	**1**	**0**	**0**	**21**	**4.8**	**-10**													
	Kentucky	AHL	25	6	6	12	103												2	0	0	0	19			
1997-98	**NY Islanders**	**NHL**	**20**	**0**	**0**	**0**	**35**	**0**	**0**	**0**	**6**	**0.0**	**-2**													
	Kentucky	AHL	37	5	13	18	139												3	0	1	1	10			
1998-99	**NY Islanders**	**NHL**	**45**	**0**	**0**	**0**	**32**	**0**	**0**	**0**	**18**	**0.0**	**-10**	**0**	**0.0**	**64**	**1**	**4:13**								
	Lowell	AHL	23	2	4	6	80																			
99-2000	**NY Islanders**	**NHL**	**65**	**1**	**3**	**4**	**103**	**0**	**0**	**0**	**27**	**3.7**	**-4**	**1**	**0.0**	**194**	**12**	**7:00**								
2000-01	**NY Islanders**	**NHL**	**31**	**0**	**2**	**2**	**35**	**0**	**0**	**0**	**8**	**0.0**	**1**	**0**	**0.0**	**88**	**10**	**6:26**								
2001-02	**NY Islanders**	**NHL**	**60**	**2**	**4**	**6**	**104**	**0**	**0**	**0**	**31**	**6.5**	**0**	**1**	**100.0**	**168**	**8**	**6:29**	**7**	**0**	**0**	**0**	**12**	**0**	**0**	**0**
	NHL Totals		**262**	**4**	**13**	**17**	**453**	**1**	**0**	**0**	**111**	**3.6**		**2**	**50.0**	**514**	**31**	**6:08**	**7**	**0**	**0**	**0**	**12**	**0**	**0**	**0**

Signed as a free agent by **NY Islanders**, October 10, 1996. • Missed majority of 2000-01 season recovering from knee injury originally suffered in game vs. Anaheim, November 19, 2000.

WEIGHT, Doug

(WAYT, DUHG) **ST.L.**

Center. Shoots left. 5'11", 200 lbs. Born, Warren, MI, January 21, 1971. NY Rangers' 2nd choice, 34th overall, in 1990 Entry Draft.

Season	Club	League	GP	G	A	Pts	PIM	PP	SH	GW	S	%	+/-	TF	F%	H	SB	Min	GP	G	A	Pts	PIM	PP	SH	GW
1988-89	Bloomfield Jets	NAJHL	34	26	53	79	105																			
1989-90	Lake Superior	CCHA	46	21	48	69	44																			
1990-91	Lake Superior	CCHA	42	29	46	75	86																			
	NY Rangers	**NHL**																	**1**	**0**	**0**	**0**	**0**	**0**	**0**	**0**
1991-92	**NY Rangers**	**NHL**	**53**	**8**	**22**	**30**	**23**	**0**	**0**	**2**	**72**	**11.1**	**-3**						**7**	**2**	**2**	**4**	**0**	**1**	**0**	**0**
	Binghamton	AHL	9	3	14	17	2												4	1	4	5	6			
1992-93	**NY Rangers**	**NHL**	**65**	**15**	**25**	**40**	**55**	**3**	**0**	**1**	**90**	**16.7**	**4**													
	Edmonton	**NHL**	**13**	**2**	**6**	**8**	**10**	**0**	**0**	**0**	**35**	**5.7**	**-2**													
1993-94	**Edmonton**	**NHL**	**84**	**24**	**50**	**74**	**47**	**4**	**1**	**1**	**188**	**12.8**	**-22**													
1994-95	Rosenheim	Germany	8	2	3	5	18																			
	Edmonton	**NHL**	**48**	**7**	**33**	**40**	**69**	**1**	**0**	**1**	**104**	**6.7**	**-17**													
1995-96	**Edmonton**	**NHL**	**82**	**25**	**79**	**104**	**95**	**9**	**0**	**2**	**204**	**12.3**	**-19**													
1996-97	**Edmonton**	**NHL**	**80**	**21**	**61**	**82**	**80**	**4**	**0**	**2**	**235**	**8.9**	**1**						**12**	**3**	**8**	**11**	**8**	**0**	**0**	**0**
1997-98	**Edmonton**	**NHL**	**79**	**26**	**44**	**70**	**69**	**9**	**0**	**4**	**205**	**12.7**	**1**						**12**	**2**	**7**	**9**	**14**	**2**	**0**	**1**
	United States	Olympics	4	0	2	2	2																			
1998-99	**Edmonton**	**NHL**	**43**	**6**	**31**	**37**	**12**	**1**	**0**	**0**	**79**	**7.6**	**-8**	**853**	**49.5**	**43**	**16**	**19:51**	**4**	**1**	**1**	**2**	**15**	**0**	**0**	**0**
99-2000	**Edmonton**	**NHL**	**77**	**21**	**51**	**72**	**54**	**3**	**1**	**4**	**167**	**12.6**	**6**	**1588**	**50.4**	**76**	**44**	**20:35**	**5**	**3**	**2**	**5**	**4**	**2**	**0**	**1**
2000-01	**Edmonton**	**NHL**	**82**	**25**	**65**	**90**	**91**	**8**	**0**	**3**	**188**	**13.3**	**12**	**1514**	**51.3**	**57**	**45**	**22:08**	**6**	**1**	**5**	**6**	**17**	**0**	**0**	**0**
2001-02	**St. Louis**	**NHL**	**61**	**15**	**34**	**49**	**40**	**3**	**0**	**1**	**131**	**11.5**	**20**	**1123**	**49.2**	**55**	**29**	**19:48**	**10**	**1**	**1**	**2**	**4**	**1**	**0**	**1**
	United States	Olympics	6	0	3	3	4																			
	NHL Totals		**767**	**195**	**501**	**696**	**645**	**45**	**2**	**21**	**1698**	**11.5**		**5078**	**50.2**	**231**	**134**	**20:46**	**57**	**13**	**26**	**39**	**62**	**6**	**0**	**3**

CCHA First All-Star Team (1991) • NCAA West Second All-American Team (1991) • Played in NHL All-Star Game (1996, 1998, 2001)

Traded to **Edmonton** by **NY Rangers** for Esa Tikkanen, March 17, 1993. Traded to **St. Louis** by **Edmonton** with Michel Riesen for Marty Reasoner, Jochen Hecht and Jan Horacek, July 1, 2001.

WEINRICH, Eric

(WIGHN-rihk, AIR-ihk) **PHI.**

Defense. Shoots left. 6'1", 213 lbs. Born, Roanoke, VA, December 19, 1966. New Jersey's 3rd choice, 32nd overall, in 1985 Entry Draft.

Season	Club	League	GP	G	A	Pts	PIM	PP	SH	GW	S	%	+/-	TF	F%	H	SB	Min	GP	G	A	Pts	PIM	PP	SH	GW
1983-84	N. Yarmouth	Hi-School	17	23	33	56	....																			
1984-85	N. Yarmouth	Hi-School	20	6	21	27	....																			
1985-86	U. of Maine	H-East	34	0	14	14	26																			
1986-87	U. of Maine	H-East	41	12	32	44	59																			
1987-88	U. of Maine	H-East	8	4	7	11	22																			
	Team USA	Nat-Tm	38	3	9	12	24																			
	United States	Olympics	3	0	0	0	0																			
1988-89	**New Jersey**	**NHL**	**2**	**0**	**0**	**0**	**0**	**0**	**0**	**0**	**3**	**0.0**	**-1**													
	Utica Devils	AHL	80	17	27	44	70												5	0	1	1	4			
1989-90	**New Jersey**	**NHL**	**19**	**2**	**7**	**9**	**11**	**1**	**0**	**1**	**16**	**12.5**	**1**						**6**	**1**	**3**	**4**	**17**	**0**	**0**	**0**
	Utica Devils	AHL	57	12	48	60	38																			
1990-91	**New Jersey**	**NHL**	**76**	**4**	**34**	**38**	**48**	**1**	**0**	**0**	**96**	**4.2**	**10**						**7**	**1**	**2**	**3**	**6**	**1**	**0**	**0**
1991-92	**New Jersey**	**NHL**	**76**	**7**	**25**	**32**	**55**	**5**	**0**	**0**	**97**	**7.2**	**10**						**7**	**0**	**2**	**2**	**4**	**0**	**0**	**0**
1992-93	**Hartford**	**NHL**	**79**	**7**	**29**	**36**	**76**	**0**	**2**	**2**	**104**	**6.7**	**-11**													
1993-94	**Hartford**	**NHL**	**8**	**1**	**1**	**2**	**2**	**1**	**0**	**0**	**10**	**10.0**	**-5**													
	Chicago	**NHL**	**54**	**3**	**23**	**26**	**31**	**1**	**0**	**2**	**105**	**2.9**	**6**						**6**	**0**	**2**	**2**	**6**	**0**	**0**	**0**
1994-95	**Chicago**	**NHL**	**48**	**3**	**10**	**13**	**33**	**1**	**0**	**2**	**50**	**6.0**	**1**						**16**	**1**	**5**	**6**	**4**	**0**	**0**	**0**
1995-96	**Chicago**	**NHL**	**77**	**5**	**10**	**15**	**65**	**0**	**0**	**0**	**76**	**6.6**	**14**						**10**	**1**	**4**	**5**	**10**	**1**	**0**	**0**
1996-97	**Chicago**	**NHL**	**81**	**7**	**25**	**32**	**62**	**1**	**0**	**0**	**115**	**6.1**	**19**						**6**	**0**	**1**	**1**	**4**	**0**	**0**	**0**
1997-98	**Chicago**	**NHL**	**82**	**2**	**21**	**23**	**106**	**0**	**0**	**0**	**85**	**2.4**	**10**													

								Regular Season											Playoffs							
Season	Club	League	GP	G	A	Pts	PIM	PP	SH	GW	S	%	+/-	TF	F%	H	SB	Min	GP	G	A	Pts	PIM	PP	SH	GW
1998-99	**Chicago**	**NHL**	**14**	**1**	**3**	**4**	**12**	**0**	**0**	**0**	**24**	**4.2**	**−13**	**0**	**0.0**	**27**	**13**	**20:12**	....	....	....	....	....	....	....	....
	Montreal	**NHL**	**66**	**6**	**12**	**18**	**77**	**4**	**0**	**1**	**95**	**6.3**	**−12**	**0**	**0.0**	**100**	**117**	**24:44**	....	....	....	....	....	....	....	....
99-2000	**Montreal**	**NHL**	**77**	**4**	**25**	**29**	**39**	**2**	**0**	**0**	**120**	**3.3**	**4**	**0**	**0.0**	**124**	**152**	**25:21**	....	....	....	....	....	....	....	....
2000-01	**Montreal**	**NHL**	**60**	**6**	**19**	**25**	**34**	**2**	**0**	**1**	**81**	**7.4**	**−1**	**1**	**100.0**	**77**	**125**	**24:27**	....	....	....	....	....	....	....	....
	Boston	**NHL**	**22**	**1**	**5**	**6**	**10**	**1**	**0**	**1**	**28**	**3.6**	**−8**	**0**	**0.0**	**33**	**39**	**25:52**	....	....	....	....	....	....	....	....
2001-02	**Philadelphia**	**NHL**	**80**	**4**	**20**	**24**	**26**	**0**	**0**	**2**	**102**	**3.9**	**27**	**0**	**0.0**	**67**	**104**	**21:53**	**5**	**0**	**0**	**0**	**4**	**0**	**0**	**0**
	NHL Totals		**921**	**63**	**269**	**332**	**687**	**20**	**2**	**12**	**1207**	**5.2**		**1**	**100.0**	**428**	**550**	**23:60**	**63**	**4**	**19**	**23**	**55**	**2**	**0**	**0**

Hockey East First All-Star Team (1987) • NCAA East Second All-American Team (1987) • AHL First All-Star Team (1990) • Won Eddie Shore Award (Top Defenseman - AHL) (1990) • NHL All-Rookie Team (1991)

Traded to **Hartford** by **New Jersey** with Sean Burke for Bobby Holik and Hartford's 2nd round choice (Jay Pandolfo) in 1993 Entry Draft, August 28, 1992. Traded to **Chicago** by **Hartford** with Patrick Poulin for Steve Larmer and Bryan Marchment, November 2, 1993. Traded to **Montreal** by **Chicago** with Jeff Hackett, Alain Nasreddine and Tampa Bay's 4th round choice (previously acquired, Montreal selected Chris Dyment) in 1999 Entry Draft for Jocelyn Thibault, Dave Manson and Brad Brown, November 16, 1998. Traded to **Boston** by **Montreal** for Patrick Traverse, February 21, 2001. Signed as a free agent by **Philadelphia**, July 5, 2001.

WEISS, Stephen

(WIGHS, STEEV-ehn) **FLA.**

Center. Shoots left. 5'11", 183 lbs. Born, Toronto, Ont., April 3, 1983. Florida's 1st choice, 4th overall, in 2001 Entry Draft.

Season	Club	League	GP	G	A	Pts	PIM	PP	SH	GW	S	%	+/-	TF	F%	H	SB	Min	GP	G	A	Pts	PIM	PP	SH	GW
1997-98	Tor. Young Nats	MTHL	48	51	58	109	....	....	....	....	....	....	....	....	....	....	....	....	....	....	....	....	....	....	....	....
1998-99	North York	OPJHL	35	15	22	37	10	....	....	....	....	....	....	....	....	....	....	....	....	....	....	....	....	....	....	....
99-2000	Plymouth Whalers	OHL	64	24	42	66	35	....	....	....	....	....	....	....	....	....	....	....	23	8	18	26	18	....	....	....
2000-01	Plymouth Whalers	OHL	62	40	47	87	45	....	....	....	....	....	....	....	....	....	....	....	18	7	16	23	10	....	....	....
2001-02	**Florida**	**NHL**	**7**	**1**	**1**	**2**	**0**	**1**	**0**	**0**	**15**	**6.7**	**0**	**107**	**52.3**	**2**	**4**	**16:14**	....	....	....	....	....	....	....	....
	Plymouth Whalers	OHL	46	25	45	70	69	....	....	....	....	....	....	....	....	....	....	....	6	2	7	9	13	....	....	....
	NHL Totals		**7**	**1**	**1**	**2**	**0**	**1**	**0**	**0**	**15**	**6.7**		**107**	**52.3**	**2**	**4**	**16:14**	....	....	....	....	....	....	....	....

OHL All-Rookie Team (2000) • Returned to **Plymouth** (OHL) by **Florida**, October 21, 2001.

WESLEY, Glen

(WEH-slee, GLEHN) **CAR.**

Defense. Shoots left. 6'1", 205 lbs. Born, Red Deer, Alta., October 2, 1968. Boston's 1st choice, 3rd overall, in 1987 Entry Draft.

Season	Club	League	GP	G	A	Pts	PIM	PP	SH	GW	S	%	+/-	TF	F%	H	SB	Min	GP	G	A	Pts	PIM	PP	SH	GW
1983-84	Red Deer	AJHL	57	9	20	29	40	....	....	....	....	....	....	....	....	....	....	....	....	....	....	....	....	....	....	....
	Portland	WHL	3	1	2	3	0	....	....	....	....	....	....	....	....	....	....	....	....	....	....	....	....	....	....	....
1984-85	Portland	WHL	67	16	52	68	76	....	....	....	....	....	....	....	....	....	....	....	6	1	6	7	8	....	....	....
1985-86	Portland	WHL	69	16	75	91	96	....	....	....	....	....	....	....	....	....	....	....	15	3	11	14	29	....	....	....
1986-87	Portland	WHL	63	16	46	62	72	....	....	....	....	....	....	....	....	....	....	....	20	8	18	26	27	....	....	....
1987-88	**Boston**	**NHL**	**79**	**7**	**30**	**37**	**69**	**1**	**2**	**0**	**158**	**4.4**	**21**	....	....	....	....	....	**23**	**6**	**8**	**14**	**22**	**4**	**1**	**0**
1988-89	**Boston**	**NHL**	**77**	**19**	**35**	**54**	**61**	**8**	**1**	**1**	**181**	**10.5**	**23**	....	....	....	....	....	**10**	**0**	**2**	**2**	**4**	**0**	**0**	**0**
1989-90	**Boston**	**NHL**	**78**	**9**	**27**	**36**	**48**	**5**	**0**	**4**	**166**	**5.4**	**6**	....	....	....	....	....	**21**	**2**	**6**	**8**	**36**	**0**	**0**	**1**
1990-91	**Boston**	**NHL**	**80**	**11**	**32**	**43**	**78**	**5**	**1**	**1**	**199**	**5.5**	**0**	....	....	....	....	....	**19**	**2**	**9**	**11**	**19**	**2**	**0**	**0**
1991-92	**Boston**	**NHL**	**78**	**9**	**37**	**46**	**54**	**4**	**0**	**1**	**211**	**4.3**	**−9**	....	....	....	....	....	**15**	**2**	**4**	**6**	**16**	**0**	**0**	**0**
1992-93	**Boston**	**NHL**	**64**	**8**	**25**	**33**	**47**	**4**	**1**	**0**	**183**	**4.4**	**−2**	....	....	....	....	....	**4**	**0**	**0**	**0**	**0**	**0**	**0**	**0**
1993-94	**Boston**	**NHL**	**81**	**14**	**44**	**58**	**64**	**6**	**1**	**1**	**265**	**5.3**	**1**	....	....	....	....	....	**13**	**3**	**3**	**6**	**12**	**1**	**0**	**0**
1994-95	**Hartford**	**NHL**	**48**	**2**	**14**	**16**	**50**	**1**	**0**	**1**	**125**	**1.6**	**−6**	....	....	....	....	....	....	....	....	....	....	....	....	....
1995-96	**Hartford**	**NHL**	**68**	**8**	**16**	**24**	**88**	**6**	**0**	**1**	**129**	**6.2**	**−9**	....	....	....	....	....	....	....	....	....	....	....	....	....
1996-97	**Hartford**	**NHL**	**68**	**6**	**26**	**32**	**40**	**3**	**1**	**0**	**126**	**4.8**	**0**	....	....	....	....	....	....	....	....	....	....	....	....	....
1997-98	**Carolina**	**NHL**	**82**	**6**	**19**	**25**	**36**	**1**	**0**	**1**	**121**	**5.0**	**7**	....	....	....	....	....	....	....	....	....	....	....	....	....
1998-99	**Carolina**	**NHL**	**74**	**7**	**17**	**24**	**44**	**0**	**0**	**2**	**112**	**6.3**	**14**	**1**	**0.0**	**121**	**94**	**22:31**	**6**	**0**	**0**	**0**	**2**	**0**	**0**	**0**
99-2000	**Carolina**	**NHL**	**78**	**7**	**15**	**22**	**38**	**1**	**0**	**0**	**99**	**7.1**	**−4**	**0**	**0.0**	**116**	**111**	**21:32**	....	....	....	....	....	....	....	....
2000-01	**Carolina**	**NHL**	**71**	**5**	**16**	**21**	**42**	**3**	**0**	**0**	**92**	**5.4**	**−2**	**0**	**0.0**	**132**	**124**	**22:21**	**6**	**0**	**0**	**0**	**0**	**0**	**0**	**0**
2001-02	**Carolina**	**NHL**	**77**	**5**	**13**	**18**	**56**	**1**	**0**	**0**	**88**	**5.7**	**−8**	**0**	**0.0**	**104**	**141**	**20:13**	**22**	**0**	**2**	**2**	**12**	**0**	**0**	**0**
	NHL Totals		**1103**	**123**	**366**	**489**	**815**	**49**	**7**	**13**	**2255**	**5.5**		**1**	**0.0**	**473**	**470**	**21:38**	**139**	**15**	**34**	**49**	**123**	**7**	**1**	**1**

WHL West First All-Star Team (1986, 1987) • NHL All-Rookie Team (1988) • Played in NHL All-Star Game (1989)

Traded to **Hartford** by **Boston** for Hartford's 1st round choices in 1995 (Kyle McLaren), 1996 (Johnathan Aitken) and 1997 (Sergei Samsonov) Entry Drafts, August 26, 1994. Transferred to **Carolina** after **Hartford** franchise relocated, June 25, 1997.

WESTCOTT, Duvie

(WEST-coht, DOO-vee) **CBJ**

Defense. Shoots right. 5'11", 192 lbs. Born, Winnipeg, Man., October 30, 1977.

Season	Club	League	GP	G	A	Pts	PIM	PP	SH	GW	S	%	+/-	TF	F%	H	SB	Min	GP	G	A	Pts	PIM	PP	SH	GW
1996-97	Winnipeg South	MJHL	52	12	47	59	....	....	....	....	....	....	....	....	....	....	....	....	....	....	....	....	....	....	....	....
1997-98	Alaska-Anchorage	WCHA	25	3	5	8	43	....	....	....	....	....	....	....	....	....	....	....	....	....	....	....	....	....	....	....
	Omaha Lancers	USHL	12	3	3	6	31	....	....	....	....	....	....	....	....	....	....	....	14	0	8	8	84	....	....	....
1998-99	St. Cloud State	WCHA	DID NOT PLAY – TRANSFERRED COLLEGES																							
99-2000	St. Cloud State	WCHA	36	1	18	19	67	....	....	....	....	....	....	....	....	....	....	....	....	....	....	....	....	....	....	....
2000-01	St. Cloud State	WCHA	38	10	24	34	116	....	....	....	....	....	....	....	....	....	....	....	....	....	....	....	....	....	....	....
2001-02	**Columbus**	**NHL**	**4**	**0**	**0**	**0**	**2**	**0**	**0**	**0**	**3**	**0.0**	**−2**	**0**	**0.0**	**5**	**1**	**15:08**	....	....	....	....	....	....	....	....
	Syracuse Crunch	AHL	68	4	29	33	99	....	....	....	....	....	....	....	....	....	....	....	10	0	1	1	12	....	....	....
	NHL Totals		**4**	**0**	**0**	**0**	**2**	**0**	**0**	**0**	**3**	**0.0**		**0**	**0.0**	**5**	**1**	**15:08**	....	....	....	....	....	....	....	....

WCHA Second All-Star Team (2001)

Signed as a free agent by **Columbus**, May 10, 2001.

WESTLUND, Tommy

(WEHST-luhnd, TOHM-mee) **CAR.**

Right wing. Shoots right. 6', 210 lbs. Born, Fors, Sweden, December 29, 1974. Carolina's 5th choice, 93rd overall, in 1998 Entry Draft.

Season	Club	League	GP	G	A	Pts	PIM	PP	SH	GW	S	%	+/-	TF	F%	H	SB	Min	GP	G	A	Pts	PIM	PP	SH	GW
1991-92	Avesta BK	Swede-3	27	11	9	20	8	....	....	....	....	....	....	....	....	....	....	....	....	....	....	....	....	....	....	....
1992-93	Avesta BK	Swede-2	32	9	5	14	32	....	....	....	....	....	....	....	....	....	....	....	....	....	....	....	....	....	....	....
1993-94	Avesta BK	Swede-2	31	20	11	31	34	....	....	....	....	....	....	....	....	....	....	....	....	....	....	....	....	....	....	....
1994-95	Avesta BK	Swede-2	32	17	13	30	22	....	....	....	....	....	....	....	....	....	....	....	....	....	....	....	....	....	....	....
1995-96	Brynas IF Gavle	Sweden	18	2	1	3	2	....	....	....	....	....	....	....	....	....	....	....	....	....	....	....	....	....	....	....
	Brynas IF Gavle	Swede-Q	18	10	10	20	4	....	....	....	....	....	....	....	....	....	....	....	8	1	0	1	4	....	....	....
1996-97	Brynas IF Gavle	Sweden	50	21	13	34	16	....	....	....	....	....	....	....	....	....	....	....	....	....	....	....	....	....	....	....
1997-98	Brynas IF Gavle	Sweden	46	29	9	38	45	....	....	....	....	....	....	....	....	....	....	....	3	0	1	1	0	....	....	....
1998-99	New Haven	AHL	50	8	18	26	31	....	....	....	....	....	....	....	....	....	....	....	....	....	....	....	....	....	....	....
99-2000	**Carolina**	**NHL**	**81**	**4**	**8**	**12**	**19**	**0**	**1**	**0**	**67**	**6.0**	**−10**	**336**	**52.7**	**86**	**25**	**10:23**	....	....	....	....	....	....	....	....
2000-01	**Carolina**	**NHL**	**79**	**5**	**3**	**8**	**23**	**0**	**0**	**1**	**47**	**10.6**	**−9**	**340**	**50.9**	**71**	**37**	**9:22**	**6**	**0**	**0**	**0**	**17**	**0**	**0**	**0**
2001-02	**Carolina**	**NHL**	**40**	**0**	**2**	**2**	**6**	**0**	**0**	**0**	**29**	**0.0**	**−8**	**118**	**56.8**	**41**	**21**	**10:38**	**19**	**1**	**0**	**1**	**0**	**0**	**0**	**0**
	NHL Totals		**200**	**9**	**13**	**22**	**48**	**0**	**1**	**1**	**143**	**6.3**		**794**	**52.5**	**198**	**83**	**10:02**	**25**	**1**	**0**	**1**	**17**	**0**	**0**	**0**

• Missed majority of 2001-02 season recovering from back injury originally suffered in game vs. Columbus, November 19, 2001.

WHITE, Brian

(WHIGHT, BRIGH-uhn)

Defense. Shoots right. 6'1", 195 lbs. Born, Winchester, MA, February 7, 1976. Tampa Bay's 11th choice, 268th overall, in 1994 Entry Draft.

Season	Club	League	GP	G	A	Pts	PIM	PP	SH	GW	S	%	+/-	TF	F%	H	SB	Min	GP	G	A	Pts	PIM	PP	SH	GW
1993-94	Arlington	Hi-School	40	14	18	32	81	....	....	....	....	....	....	....	....	....	....	....	....	....	....	....	....	....	....	....
1994-95	U. of Maine	H-East	28	1	1	2	16	....	....	....	....	....	....	....	....	....	....	....	....	....	....	....	....	....	....	....
1995-96	U. of Maine	H-East	39	0	4	4	18	....	....	....	....	....	....	....	....	....	....	....	....	....	....	....	....	....	....	....
1996-97	U. of Maine	H-East	35	4	12	16	36	....	....	....	....	....	....	....	....	....	....	....	....	....	....	....	....	....	....	....
1997-98	U. of Maine	H-East	33	0	12	12	45	....	....	....	....	....	....	....	....	....	....	....	....	....	....	....	....	....	....	....
	Long Beach	IHL	1	0	0	0	0	....	....	....	....	....	....	....	....	....	....	....	....	....	....	....	....	....	....	....
1998-99	**Colorado**	**NHL**	**2**	**0**	**0**	**0**	**0**	**0**	**0**	**0**	**0**	**0.0**	**0**	**0**	**0.0**	**1**	**0**	**0:40**	....	....	....	....	....	....	....	....
	Hershey Bears	AHL	71	4	8	12	41	....	....	....	....	....	....	....	....	....	....	....	4	0	1	1	2	....	....	....
99-2000	Hershey Bears	AHL	79	3	19	22	78	....	....	....	....	....	....	....	....	....	....	....	14	0	3	3	21	....	....	....
2000-01	Hershey Bears	AHL	75	2	9	11	44	....	....	....	....	....	....	....	....	....	....	....	9	0	1	1	12	....	....	....
2001-02	Cincinnati	AHL	73	0	8	8	32	....	....	....	....	....	....	....	....	....	....	....	3	0	0	0	2	....	....	....
	NHL Totals		**2**	**0**	**0**	**0**	**0**	**0**	**0**	**0**	**0**	**0.0**		**0**	**0.0**	**1**	**0**	**0:40**	....	....	....	....	....	....	....	....

Signed as a free agent by **Colorado**, July 7, 1998. Signed as a free agent by **Anaheim**, August 14, 2001.

WHITE, Colin

(WHIGHT, CAWI-ihn) **N.J.**

Defense. Shoots left. 6'4", 215 lbs. Born, New Glasgow, N.S., December 12, 1977. New Jersey's 5th choice, 49th overall, in 1996 Entry Draft.

			Regular Season																Playoffs							
Season	Club	League	GP	G	A	Pts	PIM	PP	SH	GW	S	%	+/-	TF	F%	H	SB	Min	GP	G	A	Pts	PIM	PP	SH	GW
1994-95	Laval Titan	QMJHL	7	0	1	1	32	....	....	....	....	....	....	....	....	....	....	....	....	....	....	....	....	....	....	....
	Hull Olympiques	QMJHL	5	0	1	1	4	....	....	....	....	....	....	....	....	....	....	....	12	0	0	0	23	....	....	....
1995-96	Hull Olympiques	QMJHL	62	2	8	10	303	....	....	....	....	....	....	....	....	....	....	....	18	0	4	4	42	....	....	....
1996-97	Hull Olympiques	QMJHL	63	3	12	15	297	....	....	....	....	....	....	....	....	....	....	....	14	3	12	15	65	....	....	....
1997-98	Albany	AHL	76	3	13	16	235	....	....	....	....	....	....	....	....	....	....	....	13	0	0	0	55	....	....	....
1998-99	Albany	AHL	77	2	12	14	265	....	....	....	....	....	....	....	....	....	....	....	5	0	1	1	8	....	....	....
99-2000♦	**New Jersey**	**NHL**	**21**	**2**	**1**	**3**	**40**	**0**	**0**	**1**	**29**	**6.9**	**3**	**0**	**0.0**	**35**	**26**	**14:45**	**23**	**1**	**5**	**6**	**18**	**0**	**0**	**1**
	Albany	AHL	52	5	21	26	176	....	....	....	....	....	....	....	....	....	....	....	....	....	....	....	....	....	....	....
2000-01	**New Jersey**	**NHL**	**82**	**1**	**19**	**20**	**155**	**0**	**0**	**1**	**114**	**0.9**	**32**	**0**	**0.0**	**163**	**116**	**19:06**	**25**	**0**	**3**	**3**	**42**	**0**	**0**	**0**
2001-02	**New Jersey**	**NHL**	**73**	**2**	**3**	**5**	**133**	**0**	**0**	**0**	**81**	**2.5**	**6**	**0**	**0.0**	**177**	**104**	**20:06**	**6**	**0**	**0**	**0**	**2**	**0**	**0**	**0**
	NHL Totals		**176**	**5**	**23**	**28**	**328**	**0**	**0**	**2**	**224**	**2.2**		**0**	**0.0**	**375**	**246**	**19:00**	**54**	**1**	**8**	**9**	**62**	**0**	**0**	**1**

QMJHL All-Rookie Team (1996) • NHL All-Rookie Team (2001)

WHITE, Peter

(WHIGHT, PEE-tuhr) **CHI.**

Center. Shoots left. 5'11", 200 lbs. Born, Montreal, Que., March 15, 1969. Edmonton's 4th choice, 92nd overall, in 1989 Entry Draft.

Season	Club	League	GP	G	A	Pts	PIM	PP	SH	GW	S	%	+/-	TF	F%	H	SB	Min	GP	G	A	Pts	PIM	PP	SH	GW
1984-85	Lac St-Louis	QAAA	42	16	32	48	18	....	....	....	....	....	....	....	....	....	....	....	11	4	3	7	4	....	....	....
1985-86	Lac St-Louis	QAAA	42	38	62	100	28	....	....	....	....	....	....	....	....	....	....	....	2	3	1	4	2	....	....	....
1986-87	Pembroke	OCJHL	55	20	34	54	20	....	....	....	....	....	....	....	....	....	....	....	....	....	....	....	....	....	....	....
1987-88	Pembroke	OCJHL	56	*90	*136	*226	32	....	....	....	....	....	....	....	....	....	....	....	....	....	....	....	....	....	....	....
1988-89	Michigan State	CCHA	46	20	33	53	17	....	....	....	....	....	....	....	....	....	....	....	....	....	....	....	....	....	....	....
1989-90	Michigan State	CCHA	45	22	40	62	6	....	....	....	....	....	....	....	....	....	....	....	....	....	....	....	....	....	....	....
1990-91	Michigan State	CCHA	37	7	31	38	28	....	....	....	....	....	....	....	....	....	....	....	....	....	....	....	....	....	....	....
1991-92	Michigan State	CCHA	41	26	49	75	32	....	....	....	....	....	....	....	....	....	....	....	....	....	....	....	....	....	....	....
1992-93	Cape Breton	AHL	64	12	28	40	10	....	....	....	....	....	....	....	....	....	....	....	16	3	3	6	12	....	....	....
1993-94	**Edmonton**	**NHL**	**26**	**3**	**5**	**8**	**2**	**0**	**0**	**0**	**17**	**17.6**	**1**	....	....	....	....	....	....	....	....	....	....	....	....	....
	Cape Breton	AHL	45	21	49	70	12	....	....	....	....	....	....	....	....	....	....	....	5	2	3	5	2	....	....	....
1994-95	Cape Breton	AHL	65	36	*69	*105	30	....	....	....	....	....	....	....	....	....	....	....	....	....	....	....	....	....	....	....
	Edmonton	**NHL**	**9**	**2**	**4**	**6**	**0**	**2**	**0**	**0**	**13**	**15.4**	**1**	....	....	....	....	....	....	....	....	....	....	....	....	....
1995-96	**Edmonton**	**NHL**	**26**	**5**	**3**	**8**	**0**	**1**	**0**	**0**	**34**	**14.7**	**–14**	....	....	....	....	....	....	....	....	....	....	....	....	....
	Toronto	**NHL**	**1**	**0**	**0**	**0**	**0**	**0**	**0**	**0**	**0**	**0.0**	**0**	....	....	....	....	....	....	....	....	....	....	....	....	....
	St. John's	AHL	17	6	7	13	6	....	....	....	....	....	....	....	....	....	....	....	....	....	....	....	....	....	....	....
	Atlanta Knights	IHL	36	21	20	41	4	....	....	....	....	....	....	....	....	....	....	....	3	0	3	3	2	....	....	....
1996-97	Philadelphia	AHL	80	*44	61	*105	28	....	....	....	....	....	....	....	....	....	....	....	10	6	8	14	6	....	....	....
1997-98	Philadelphia	AHL	80	27	*78	*105	28	....	....	....	....	....	....	....	....	....	....	....	20	9	9	18	6	....	....	....
1998-99	**Philadelphia**	**NHL**	**3**	**0**	**0**	**0**	**0**	**0**	**0**	**0**	**0**	**0.0**	**0**	**8**	**37.5**	**0**	**0**	**2:02**	....	....	....	....	....	....	....	....
	Philadelphia	AHL	77	31	59	90	20	....	....	....	....	....	....	....	....	....	....	....	16	4	13	17	12	....	....	....
99-2000	**Philadelphia**	**NHL**	**21**	**1**	**5**	**6**	**6**	**0**	**0**	**0**	**24**	**4.2**	**1**	**277**	**54.5**	**9**	**15**	**13:04**	**16**	**0**	**2**	**2**	**0**	**0**	**0**	**0**
	Philadelphia	AHL	62	20	41	61	38	....	....	....	....	....	....	....	....	....	....	....	....	....	....	....	....	....	....	....
2000-01	**Philadelphia**	**NHL**	**77**	**9**	**16**	**25**	**16**	**1**	**0**	**1**	**68**	**13.2**	**1**	**1038**	**54.3**	**23**	**27**	**12:56**	**3**	**0**	**0**	**0**	**0**	**0**	**0**	**0**
2001-02	**Chicago**	**NHL**	**48**	**3**	**3**	**6**	**10**	**1**	**0**	**1**	**21**	**14.3**	**–8**	**513**	**53.0**	**16**	**30**	**10:31**	....	....	....	....	....	....	....	....
	Norfolk Admirals	AHL	24	4	19	23	18	....	....	....	....	....	....	....	....	....	....	....	4	0	1	1	0	....	....	....
	NHL Totals		**211**	**23**	**36**	**59**	**34**	**5**	**0**	**2**	**177**	**13.0**		**1836**	**53.9**	**48**	**72**	**11:57**	**19**	**0**	**2**	**2**	**0**	**0**	**0**	**0**

AHL Second All-Star Team (1995, 1997) • Won John B. Sollenberger Trophy (Top Scorer - AHL) (1995, 1997, 1998)

Traded to **Toronto** by **Edmonton** with Edmonton's 4th round choice (Jason Sessa) in 1996 Entry Draft for Kent Manderville, December 4, 1995. Signed as a free agent by **Philadelphia**, August 19, 1996. Signed as a free agent by **Chicago**, September 10, 2001.

WHITE, Todd

(WHIGHT, TAWD) **OTT.**

Center. Shoots left. 5'10", 194 lbs. Born, Kanata, Ont., May 21, 1975.

Season	Club	League	GP	G	A	Pts	PIM	PP	SH	GW	S	%	+/-	TF	F%	H	SB	Min	GP	G	A	Pts	PIM	PP	SH	GW
1990-91	Powassan Hawks	NOJHA	38	34	38	72	118	....	....	....	....	....	....	....	....	....	....	....	....	....	....	....	....	....	....	....
1991-92	Kanata Valley	OCJHL	55	39	49	88	30	....	....	....	....	....	....	....	....	....	....	....	....	....	....	....	....	....	....	....
1992-93	Kanata Valley	OCJHL	49	51	87	138	46	....	....	....	....	....	....	....	....	....	....	....	....	....	....	....	....	....	....	....
1993-94	Clarkson Knights	ECAC	33	10	12	22	28	....	....	....	....	....	....	....	....	....	....	....	....	....	....	....	....	....	....	....
1994-95	Clarkson Knights	ECAC	34	13	16	29	44	....	....	....	....	....	....	....	....	....	....	....	....	....	....	....	....	....	....	....
1995-96	Clarkson Knights	ECAC	38	29	43	72	36	....	....	....	....	....	....	....	....	....	....	....	....	....	....	....	....	....	....	....
1996-97	Clarkson Knights	ECAC	37	*38	*36	*74	22	....	....	....	....	....	....	....	....	....	....	....	....	....	....	....	....	....	....	....
1997-98	**Chicago**	**NHL**	**7**	**1**	**0**	**1**	**2**	**0**	**0**	**0**	**3**	**33.3**	**0**	....	....	....	....	....	....	....	....	....	....	....	....	....
	Indianapolis Ice	IHL	65	46	36	82	28	....	....	....	....	....	....	....	....	....	....	....	5	2	3	5	4	....	....	....
1998-99	**Chicago**	**NHL**	**35**	**5**	**8**	**13**	**20**	**2**	**0**	**0**	**43**	**11.6**	**–1**	**452**	**46.0**	**27**	**11**	**13:39**	....	....	....	....	....	....	....	....
	Chicago Wolves	IHL	25	11	13	24	8	....	....	....	....	....	....	....	....	....	....	....	10	1	4	5	8	....	....	....
99-2000	**Chicago**	**NHL**	**1**	**0**	**0**	**0**	**0**	**0**	**0**	**0**	**0**	**0.0**	**0**	**9**	**55.6**	**2**	**0**	**13:02**	....	....	....	....	....	....	....	....
	Cleveland	IHL	42	21	30	51	32	....	....	....	....	....	....	....	....	....	....	....	....	....	....	....	....	....	....	....
	Philadelphia	**NHL**	**3**	**1**	**0**	**1**	**0**	**0**	**0**	**0**	**4**	**25.0**	**–1**	**25**	**40.0**	**1**	**1**	**10:29**	....	....	....	....	....	....	....	....
	Philadelphia	AHL	32	19	24	43	12	....	....	....	....	....	....	....	....	....	....	....	5	2	1	3	8	....	....	....
2000-01	**Ottawa**	**NHL**	**16**	**4**	**1**	**5**	**4**	**0**	**0**	**0**	**12**	**33.3**	**5**	**133**	**57.1**	**9**	**6**	**8:33**	**2**	**0**	**0**	**0**	**0**	**0**	**0**	**0**
	Grand Rapids	IHL	64	22	32	54	20	....	....	....	....	....	....	....	....	....	....	....	10	4	4	8	10	....	....	....
2001-02	**Ottawa**	**NHL**	**81**	**20**	**30**	**50**	**24**	**4**	**0**	**1**	**147**	**13.6**	**12**	**1508**	**50.5**	**42**	**43**	**18:22**	**12**	**2**	**2**	**4**	**6**	**0**	**0**	**0**
	NHL Totals		**143**	**31**	**39**	**70**	**50**	**6**	**0**	**1**	**209**	**14.8**		**2127**	**49.8**	**81**	**61**	**15:47**	**14**	**2**	**2**	**4**	**6**	**0**	**0**	**0**

ECAC Second All-Star Team (1996) • NCAA East Second All-American Team (1996) • ECAC First All-Star Team (1997) • NCAA East First All-American Team (1997) • Won Garry F. Longman Memorial Trophy (Top Rookie - IHL) (1998)

Signed as a free agent by **Chicago**, August 27, 1997. Traded to **Philadelphia** by **Chicago** for future considerations, January 26, 2000. Signed as a free agent by **Ottawa**, July 12, 2000.

WHITFIELD, Trent

(WHIHT-feeld, TREHNT) **WSH.**

Center. Shoots left. 5'11", 204 lbs. Born, Estevan, Sask., June 17, 1977. Boston's 5th choice, 100th overall, in 1996 Entry Draft.

Season	Club	League	GP	G	A	Pts	PIM	PP	SH	GW	S	%	+/-	TF	F%	H	SB	Min	GP	G	A	Pts	PIM	PP	SH	GW
1993-94	Sask. Blazers	SMHL	36	26	22	48	42	....	....	....	....	....	....	....	....	....	....	....	....	....	....	....	....	....	....	....
	Spokane Chiefs	WHL	5	1	1	2	0	....	....	....	....	....	....	....	....	....	....	....	....	....	....	....	....	....	....	....
1994-95	Spokane Chiefs	WHL	48	8	17	25	26	....	....	....	....	....	....	....	....	....	....	....	11	7	6	13	5	....	....	....
1995-96	Spokane Chiefs	WHL	72	33	51	84	75	....	....	....	....	....	....	....	....	....	....	....	18	8	10	18	10	....	....	....
1996-97	Spokane Chiefs	WHL	58	34	42	76	74	....	....	....	....	....	....	....	....	....	....	....	9	5	7	12	10	....	....	....
1997-98	Spokane Chiefs	WHL	65	38	44	82	97	....	....	....	....	....	....	....	....	....	....	....	18	9	10	19	15	....	....	....
1998-99	Portland Pirates	AHL	50	10	8	18	20	....	....	....	....	....	....	....	....	....	....	....	....	....	....	....	....	....	....	....
	Hampton Roads	ECHL	19	13	12	25	12	....	....	....	....	....	....	....	....	....	....	....	4	2	0	2	14	....	....	....
99-2000	Portland Pirates	AHL	79	18	35	53	52	....	....	....	....	....	....	....	....	....	....	....	3	1	1	2	2	....	....	....
	Washington	**NHL**	....	....	....	....	....	....	....	....	....	....	....	....	....	....	....	....	**3**	**0**	**0**	**0**	**0**	**0**	**0**	**0**
2000-01	**Washington**	**NHL**	**61**	**2**	**4**	**6**	**35**	**0**	**0**	**0**	**47**	**4.3**	**3**	**520**	**51.9**	**70**	**24**	**9:39**	**5**	**0**	**0**	**0**	**2**	**0**	**0**	**0**
	Portland Pirates	AHL	19	9	11	20	27	....	....	....	....	....	....	....	....	....	....	....	....	....	....	....	....	....	....	....
2001-02	**Washington**	**NHL**	**24**	**0**	**1**	**1**	**28**	**0**	**0**	**0**	**15**	**0.0**	**–3**	**189**	**54.0**	**17**	**12**	**7:06**	....	....	....	....	....	....	....	....
	Portland Pirates	AHL	10	4	4	8	8	....	....	....	....	....	....	....	....	....	....	....	....	....	....	....	....	....	....	....
	NY Rangers	**NHL**	**1**	**0**	**0**	**0**	**0**	**0**	**0**	**0**	**0**	**0.0**	**1**	**18**	**50.0**	**3**	**0**	**12:44**	....	....	....	....	....	....	....	....
	Portland Pirates	AHL	24	10	16	26	16	....	....	....	....	....	....	....	....	....	....	....	....	....	....	....	....	....	....	....
	NHL Totals		**86**	**2**	**5**	**7**	**63**	**0**	**0**	**0**	**62**	**3.2**		**727**	**52.4**	**90**	**36**	**8:58**	**8**	**0**	**0**	**0**	**2**	**0**	**0**	**0**

WHL West First All-Star Team (1997) • WHL West Second All-Star Team (1998)

Signed as a free agent by **Washington**, September 1, 1998. Claimed on waivers by **NY Rangers** from **Washington**, January 16, 2002. Claimed on waivers by **Washington** from **NY Rangers**, February 1, 2002.

WHITNEY, Ray

(WHIHT-nee, RAY) **CBJ**

Left wing. Shoots right. 5'10", 175 lbs. Born, Fort Saskatchewan, Alta., May 8, 1972. San Jose's 2nd choice, 23rd overall, in 1991 Entry Draft.

Season	Club	League	GP	G	A	Pts	PIM	PP	SH	GW	S	%	+/-	TF	F%	H	SB	Min	GP	G	A	Pts	PIM	PP	SH	GW
1987-88	Ft. Saskatchewan	AMHL	71	80	155	235	119	....	....	....	....	....	....	....	....	....	....	....	....	....	....	....	....	....	....	....
1988-89	Spokane Chiefs	WHL	71	17	33	50	16	....	....	....	....	....	....	....	....	....	....	....	....	....	....	....	....	....	....	....
1989-90	Spokane Chiefs	WHL	71	57	56	113	50	....	....	....	....	....	....	....	....	....	....	....	6	3	4	7	6	....	....	....
1990-91	Spokane Chiefs	WHL	72	67	118	*185	36	....	....	....	....	....	....	....	....	....	....	....	15	13	18	*31	12	....	....	....

Season	Club	League	GP	G	A	Pts	PIM	PP	SH	GW	S	%	+/-	TF	F%	H	SB	Min	GP	G	A	Pts	PIM	PP	SH	GW
			Regular Season																Playoffs							
1991-92	Kolner EC	Germany	10	3	6	9	4																			
	Team Canada	Nat-Tm	5	1	0	1	6																			
	San Jose	**NHL**	**2**	**0**	**3**	**3**	**0**	**0**	**0**	**0**	**4**	**0.0**	**–1**													
	San Diego Gulls	IHL	63	36	54	90	12												4	0	0	0	0			
1992-93	**San Jose**	**NHL**	**26**	**4**	**6**	**10**	**4**	**1**	**0**	**0**	**24**	**16.7**	**–14**													
	Kansas City	IHL	46	20	33	53	14												12	5	7	12	2			
1993-94	**San Jose**	**NHL**	**61**	**14**	**26**	**40**	**14**	**1**	**0**	**0**	**82**	**17.1**	**2**						**14**	**0**	**4**	**4**	**8**	**0**	**0**	**0**
1994-95	**San Jose**	**NHL**	**39**	**13**	**12**	**25**	**14**	**4**	**0**	**1**	**67**	**19.4**	**–7**						**11**	**4**	**4**	**8**	**2**	**0**	**0**	**1**
1995-96	**San Jose**	**NHL**	**60**	**17**	**24**	**41**	**16**	**4**	**2**	**2**	**106**	**16.0**	**–23**													
1996-97	**San Jose**	**NHL**	**12**	**0**	**2**	**2**	**4**	**0**	**0**	**0**	**24**	**0.0**	**–6**													
	Kentucky	AHL	9	1	7	8	2																			
	Utah Grizzlies	IHL	43	13	35	48	34												7	3	1	4	6			
1997-98	**Edmonton**	**NHL**	**9**	**1**	**3**	**4**	**0**	**0**	**0**	**0**	**19**	**5.3**	**–1**													
	Florida	**NHL**	**68**	**32**	**29**	**61**	**28**	**12**	**0**	**2**	**156**	**20.5**	**10**													
1998-99	**Florida**	**NHL**	**81**	**26**	**38**	**64**	**18**	**7**	**0**	**6**	**193**	**13.5**	**–3**	**144**	**43.8**	**14**	**7**	**18:20**								
99-2000	**Florida**	**NHL**	**81**	**29**	**42**	**71**	**35**	**5**	**0**	**3**	**198**	**14.6**	**16**	**198**	**49.0**	**17**	**16**	**18:41**	**4**	**1**	**0**	**1**	**4**	**0**	**0**	**0**
2000-01	**Florida**	**NHL**	**43**	**10**	**21**	**31**	**28**	**5**	**0**	**0**	**117**	**8.5**	**–16**	**38**	**39.5**	**6**	**15**	**17:41**								
	Columbus	**NHL**	**3**	**0**	**3**	**3**	**2**	**0**	**0**	**0**	**3**	**0.0**	**–1**	**19**	**36.8**	**2**	**0**	**20:17**								
2001-02	**Columbus**	**NHL**	**67**	**21**	**40**	**61**	**12**	**6**	**0**	**3**	**210**	**10.0**	**–22**	**21**	**47.6**	**14**	**21**	**20:13**								
	NHL Totals		**552**	**167**	**249**	**416**	**175**	**45**	**2**	**17**	**1203**	**13.9**		**420**	**45.7**	**53**	**59**	**18:49**	**29**	**5**	**8**	**13**	**14**	**0**	**0**	**1**

WHL West First All-Star Team (1991) • WHL MVP (1991) • Memorial Cup All-Star Team (1991) • Won George Parsons Trophy (Memorial Cup Tournament Most Sportsmanlike Player) (1991) • Played in NHL All-Star Game (2000)

Signed as a free agent by **Edmonton**, October 1, 1997. Claimed on waivers by **Florida** from **Edmonton**, November 6, 1997. Traded to **Columbus** by **Florida** with future considerations for Kevyn Adams and Coliumbus's 4th round choice (Michael Woodford) in 2001 Entry Draft, March 13, 2001.

WIEMER, Jason

(WEE-muhr, JAY-suhn) **NYI**

Center. Shoots left. 6'1", 225 lbs. Born, Kimberley, B.C., April 14, 1976. Tampa Bay's 1st choice, 8th overall, in 1994 Entry Draft.

Season	Club	League	GP	G	A	Pts	PIM	PP	SH	GW	S	%	+/-	TF	F%	H	SB	Min	GP	G	A	Pts	PIM	PP	SH	GW
1991-92	Kimberley	RMJHL	45	33	33	66	211																			
	Portland	WHL	2	0	1	1	0																			
1992-93	Portland	WHL	68	18	34	52	159												16	7	3	10	27			
1993-94	Portland	WHL	72	45	51	96	236												10	4	4	8	32			
1994-95	Portland	WHL	16	10	14	24	63																			
	Tampa Bay	**NHL**	**36**	**1**	**4**	**5**	**44**	**0**	**0**	**0**	**10**	**10.0**	**–2**													
1995-96	**Tampa Bay**	**NHL**	**66**	**9**	**9**	**18**	**81**	**4**	**0**	**1**	**89**	**10.1**	**–9**						**6**	**1**	**0**	**1**	**28**	**1**	**0**	**0**
1996-97	**Tampa Bay**	**NHL**	**63**	**9**	**5**	**14**	**134**	**2**	**0**	**0**	**103**	**8.7**	**–13**													
	Adirondack	AHL	4	1	0	1	7																			
1997-98	**Tampa Bay**	**NHL**	**67**	**8**	**9**	**17**	**132**	**2**	**0**	**0**	**106**	**7.5**	**–9**													
	Calgary	**NHL**	**12**	**4**	**1**	**5**	**28**	**1**	**0**	**2**	**16**	**25.0**	**–1**													
1998-99	**Calgary**	**NHL**	**78**	**8**	**13**	**21**	**177**	**1**	**0**	**1**	**128**	**6.3**	**–12**	**867**	**40.9**	**171**	**20**	**13:17**								
99-2000	**Calgary**	**NHL**	**64**	**11**	**11**	**22**	**120**	**2**	**0**	**3**	**104**	**10.6**	**–10**	**955**	**47.6**	**149**	**22**	**14:41**								
2000-01	**Calgary**	**NHL**	**65**	**10**	**5**	**15**	**177**	**3**	**0**	**1**	**76**	**13.2**	**–15**	**599**	**51.1**	**143**	**28**	**13:56**								
2001-02	**Florida**	**NHL**	**70**	**11**	**20**	**31**	**178**	**5**	**1**	**1**	**115**	**9.6**	**–4**	**1241**	**44.1**	**168**	**36**	**17:09**								
	NHL Totals		**521**	**71**	**77**	**148**	**1071**	**20**	**1**	**9**	**747**	**9.5**		**3662**	**45.4**	**631**	**106**	**14:44**	**6**	**1**	**0**	**1**	**28**	**1**	**0**	**0**

Traded to **Calgary** by **Tampa Bay** for Sandy McCarthy and Calgary's 3rd (Brad Richards) and 5th (Curtis Rich) round choices in 1998 Entry Draft, March 24, 1998. Traded to **Florida** by **Calgary** with Valeri Bure for Rob Neidermayer and Philadelphia's 2nd round choice (previously acquired, Calgary selected Andrei Medvedev) in 2001 Entry Draft, June 24, 2001. Traded to **NY Islanders** by **Florida** for Branislav Mezei, July 3, 2002.

WILKIE, David

(WIHL-kee, DAY-vihd)

Defense. Shoots right. 6'3", 215 lbs. Born, Ellensburgh, WA, May 30, 1974. Montreal's 1st choice, 20th overall, in 1992 Entry Draft.

Season	Club	League	GP	G	A	Pts	PIM	PP	SH	GW	S	%	+/-	TF	F%	H	SB	Min	GP	G	A	Pts	PIM	PP	SH	GW
1989-90	Seattle	PIJHL	41	21	27	48	59																			
1990-91	Omaha Lancers	USHL	19	2	2	4	18																			
	Seattle	WHL	25	1	1	2	22																			
1991-92	Kamloops Blazers	WHL	71	12	28	40	153												16	6	5	11	19			
1992-93	Kamloops Blazers	WHL	53	11	26	37	109												6	4	2	6	2			
1993-94	Kamloops Blazers	WHL	27	11	18	29	18																			
	Regina Pats	WHL	29	27	21	48	16												4	1	4	5	4			
1994-95	Fredericton	AHL	70	10	43	53	34												1	0	0	0	0			
	Montreal	**NHL**	**1**	**0**	**0**	**0**	**0**	**0**	**0**	**0**	**0**	**0.0**	**0**													
1995-96	**Montreal**	**NHL**	**24**	**1**	**5**	**6**	**10**	**1**	**0**	**0**	**39**	**2.6**	**–10**						**6**	**1**	**2**	**3**	**12**	**0**	**0**	**0**
	Fredericton	AHL	23	5	12	17	20																			
1996-97	**Montreal**	**NHL**	**61**	**6**	**9**	**15**	**63**	**3**	**0**	**0**	**65**	**9.2**	**–9**						**2**	**0**	**0**	**0**	**2**	**0**	**0**	**0**
1997-98	**Montreal**	**NHL**	**5**	**1**	**0**	**1**	**4**	**0**	**0**	**1**	**2**	**50.0**	**–1**													
	Tampa Bay	**NHL**	**29**	**1**	**5**	**6**	**17**	**0**	**0**	**0**	**46**	**2.2**	**–21**													
1998-99	**Tampa Bay**	**NHL**	**46**	**1**	**7**	**8**	**69**	**0**	**0**	**0**	**35**	**2.9**	**–19**	**0**	**0.0**	**49**	**48**	**12:18**								
	Cleveland	IHL	2	0	2	2	0																			
99-2000	Houston Aeros	IHL	57	4	24	28	71												11	1	8	9	10			
	Hartford	AHL	1	0	2	2	0																			
2000-01	**NY Rangers**	**NHL**	**1**	**0**	**0**	**0**	**2**	**0**	**0**	**0**	**1**	**0.0**	**–2**	**0**	**0.0**	**1**	**3**	**11:33**								
	Houston Aeros	IHL	49	8	11	19	29												7	1	1	2	4			
2001-02	Augusta Lynx	ECHL	72	11	34	45	86																			
	NHL Totals		**167**	**10**	**26**	**36**	**165**	**4**	**0**	**1**	**188**	**5.3**		**0**	**0.0**	**50**	**51**	**12:17**	**8**	**1**	**2**	**3**	**14**	**0**	**0**	**0**

Traded to **Tampa Bay** by **Montreal** with Stephane Richer and Darcy Tucker for Patrick Poulin, Mick Vukota and Igor Ulanov, January 15, 1998. Signed as a free agent by **NY Rangers**, September 29, 1999. Signed as a free agent by **Augusta** (ECHL) and named Player-Assistant Coach, August 11, 2001.

WILLIAMS, Jason

(WIHL-yuhms, JAY-suhn) **DET.**

Center. Shoots right. 5'11", 185 lbs. Born, London, Ont., August 11, 1980.

Season	Club	League	GP	G	A	Pts	PIM	PP	SH	GW	S	%	+/-	TF	F%	H	SB	Min	GP	G	A	Pts	PIM	PP	SH	GW
1995-96	Mount Brydges	OJHL-D	36	31	28	59	18																			
1996-97	Peterborough	OHL	60	4	8	12	8												10	1	0	1	2			
1997-98	Peterborough	OHL	55	8	27	35	31												4	0	1	1	2			
1998-99	Peterborough	OHL	68	26	48	74	42												5	1	2	3	2			
99-2000	Peterborough	OHL	66	36	37	75	64												5	2	1	3	2			
2000-01	**Detroit**	**NHL**	**5**	**0**	**3**	**3**	**2**	**0**	**0**	**0**	**7**	**0.0**	**1**	**56**	**39.3**	**15**	**2**	**12:24**	**2**	**0**	**0**	**0**	**0**	**0**	**0**	**0**
	Cincinnati	AHL	76	24	45	69	48												1	0	0	0	2			
2001-02	**Detroit**	**NHL**	**25**	**8**	**2**	**10**	**4**	**4**	**0**	**0**	**32**	**25.0**	**2**	**208**	**47.6**	**35**	**2**	**10:50**	**9**	**0**	**0**	**0**	**2**	**0**	**0**	**0**
	Cincinnati	AHL	52	23	27	50	27												3	0	1	1	6			
	NHL Totals		**30**	**8**	**5**	**13**	**6**	**4**	**0**	**0**	**39**	**20.5**		**264**	**45.8**	**50**	**4**	**11:06**	**11**	**0**	**0**	**0**	**2**	**0**	**0**	**0**

OHL Third Team All-Star (2000)

Signed as a free agent by **Detroit**, September 18, 2000.

WILLIAMS, Justin

(WIHL-yuhms, JUHS-tihn) **PHI.**

Right wing. Shoots right. 6'1", 190 lbs. Born, Cobourg, Ont., October 4, 1981. Philadelphia's 1st choice, 28th overall, in 2000 Entry Draft.

Season	Club	League	GP	G	A	Pts	PIM	PP	SH	GW	S	%	+/-	TF	F%	H	SB	Min	GP	G	A	Pts	PIM	PP	SH	GW
1997-98	Cobourne Colts	OJHL-C	36	32	35	67	26																			
	Cobourg Cougars	OPJHL	17	0	3	3	5																			
1998-99	Plymouth Whalers	OHL	47	4	8	12	28												7	1	2	3	0			
99-2000	Plymouth Whalers	OHL	68	37	46	83	46												23	*14	16	*30	10			
2000-01	**Philadelphia**	**NHL**	**63**	**12**	**13**	**25**	**22**	**0**	**0**	**0**	**99**	**12.1**	**6**	**13**	**53.9**	**20**	**18**	**12:31**								
2001-02	**Philadelphia**	**NHL**	**75**	**17**	**23**	**40**	**32**	**0**	**0**	**1**	**162**	**10.5**	**11**	**16**	**25.0**	**33**	**30**	**14:27**	**5**	**0**	**0**	**0**	**4**	**0**	**0**	**0**
	NHL Totals		**138**	**29**	**36**	**65**	**54**	**0**	**0**	**1**	**261**	**11.1**		**29**	**37.9**	**53**	**48**	**13:34**	**5**	**0**	**0**	**0**	**4**	**0**	**0**	**0**

WILLIS, Shane

(WIH-lihs, SHAYN) **T.B.**

Right wing. Shoots right. 6'1", 190 lbs. Born, Edmonton, Alta., June 13, 1977. Carolina's 4th choice, 88th overall, in 1997 Entry Draft.

Season	Club	League	GP	G	A	Pts	PIM	PP	SH	GW	S	%	+/-	TF	F%	H	SB	Min	GP	G	A	Pts	PIM	PP	SH	GW
			Regular Season																Playoffs							
1992-93	Red Deer	ABHL	36	32	18	50	88																			
1993-94	Red Deer	AMHL	34	40	26	66	103																			
1994-95	Prince Albert	WHL	65	24	19	43	38												13	3	4	7	6			
1995-96	Prince Albert	WHL	69	41	40	81	47												18	11	10	21	18			
1996-97	Prince Albert	WHL	41	34	22	56	63																			
	Lethbridge	WHL	26	22	17	39	24												19	13	11	24	20			
1997-98	Lethbridge	WHL	64	58	54	112	73												4	2	3	5	6			
	New Haven	AHL	1	0	1	1	2																			
1998-99	**Carolina**	**NHL**	**7**	**0**	**0**	**0**	**0**	0	0	0	1	0.0	-2	0	0.0	4	2	2:14								
	New Haven	AHL	73	31	50	81	49																			
99-2000	**Carolina**	**NHL**	**2**	**0**	**0**	**0**	**0**	0	0	0	1	0.0	-1	0	0.0	6	0	5:50								
	Cincinnati	IHL	80	35	25	60	64												11	5	3	8	8			
2000-01	**Carolina**	**NHL**	**73**	**20**	**24**	**44**	**45**	9	0	6	172	11.6	-6	10	20.0	117	16	15:58	**2**	**0**	**0**	**0**	**0**	0	0	0
2001-02	**Carolina**	**NHL**	**59**	**7**	**10**	**17**	**24**	2	0	0	126	5.6	-8	10	50.0	104	10	13:00								
	Tampa Bay	**NHL**	**21**	**4**	**3**	**7**	**6**	0	0	0	29	13.8	0	12	8.3	16	1	11:18								
	NHL Totals		**162**	**31**	**37**	**68**	**75**	11	0	6	329	9.4		32	25.0	247	29	13:34	**2**	**0**	**0**	**0**	**0**	0	0	0

• Re-entered NHL Entry Draft. Originally Tampa Bay's 3rd choice, 56th overall, in 1995 Entry Draft.

WHL East First All-Star Team (1997, 1998) • AHL First All-Star Team (1999) • Won Dudley "Red" Garrett Memorial Trophy (Top Rookie - AHL) (1999) • NHL All-Rookie Team (2001)

Traded to **Tampa Bay** by **Carolina** with Chris Dingman for Kevin Weekes, March 5, 2002.

WILLSIE, Brian

(WIHL-see, BRIGH-uhn) **COL.**

Right wing. Shoots right. 6'1", 195 lbs. Born, London, Ont., March 16, 1978. Colorado's 7th choice, 146th overall, in 1996 Entry Draft.

Season	Club	League	GP	G	A	Pts	PIM	PP	SH	GW	S	%	+/-	TF	F%	H	SB	Min	GP	G	A	Pts	PIM	PP	SH	GW
1993-94	Belmont Bombers	OJHL-D	13	9	5	14	14																			
1994-95	St. Thomas Stars	OJHL-B	45	35	47	82	47																			
1995-96	Guelph Storm	OHL	65	13	21	34	18												16	4	2	6	6			
1996-97	Guelph Storm	OHL	64	37	31	68	37												18	15	4	19	10			
1997-98	Guelph Storm	OHL	57	45	31	76	41												12	9	5	14	18			
1998-99	Hershey Bears	AHL	72	19	10	29	28												3	1	0	1	0			
99-2000	**Colorado**	**NHL**	**1**	**0**	**0**	**0**	**0**	0	0	0	1	0.0	0	0	0.0	0	0	8:16								
	Hershey Bears	AHL	78	20	39	59	44												12	2	6	8	8			
2000-01	Hershey Bears	AHL	48	18	23	41	20												12	7	2	9	14			
2001-02	**Colorado**	**NHL**	**56**	**7**	**7**	**14**	**14**	2	0	1	66	10.6	4	8	12.5	51	10	11:24	**4**	**0**	**1**	**1**	**2**	0	0	0
	NHL Totals		**57**	**7**	**7**	**14**	**14**	2	0	1	67	10.4		8	12.5	51	10	11:21	**4**	**0**	**1**	**1**	**2**	0	0	0

OHL First All-Star Team (1998)

WILM, Clarke

(WIHLM, KLAHRK) **NSH.**

Center. Shoots left. 6', 202 lbs. Born, Central Butte, Sask., October 24, 1976. Calgary's 5th choice, 150th overall, in 1995 Entry Draft.

Season	Club	League	GP	G	A	Pts	PIM	PP	SH	GW	S	%	+/-	TF	F%	H	SB	Min	GP	G	A	Pts	PIM	PP	SH	GW
1991-92	Sask. Blazers	SMHL	36	18	28	46	16																			
	Saskatoon Blades	WHL																	1	0	0	0	0			
1992-93	Saskatoon Blades	WHL	69	14	19	33	71												9	4	2	6	13			
1993-94	Saskatoon Blades	WHL	70	18	32	50	181												16	0	9	9	19			
1994-95	Saskatoon Blades	WHL	71	20	39	59	179												10	6	1	7	21			
1995-96	Saskatoon Blades	WHL	72	49	61	110	83												4	1	1	2	4			
1996-97	Saint John	AHL	62	9	19	28	107												5	2	0	2	15			
1997-98	Saint John	AHL	68	13	26	39	112												21	5	9	14	8			
1998-99	**Calgary**	**NHL**	**78**	**10**	**8**	**18**	**53**	2	2	0	94	10.6	11	609	40.9	94	21	11:32								
99-2000	**Calgary**	**NHL**	**78**	**10**	**12**	**22**	**67**	1	3	0	81	12.3	-6	872	44.4	113	34	12:38								
2000-01	**Calgary**	**NHL**	**81**	**7**	**8**	**15**	**69**	2	0	0	85	8.2	-11	992	51.9	111	54	14:11								
2001-02	**Calgary**	**NHL**	**66**	**4**	**14**	**18**	**61**	0	1	0	83	4.8	-1	995	51.1	73	39	15:00								
	NHL Totals		**303**	**31**	**42**	**73**	**250**	5	6	0	343	9.0		3468	47.8	391	148	13:17								

Signed as a free agent by **Nashville**, July 11, 2002.

WILSON, Landon

(WIHL-sohn, LAN-duhn) **PHX.**

Right wing. Shoots right. 6'3", 232 lbs. Born, St. Louis, MO, March 13, 1975. Toronto's 2nd choice, 19th overall, in 1993 Entry Draft.

Season	Club	League	GP	G	A	Pts	PIM	PP	SH	GW	S	%	+/-	TF	F%	H	SB	Min	GP	G	A	Pts	PIM	PP	SH	GW
1991-92	California	WSJHL	38	50	42	92	135																			
1992-93	Dubuque	USHL	43	29	36	65	284																			
1993-94	North Dakota	WCHA	35	18	15	33	*147																			
1994-95	North Dakota	WCHA	31	7	16	23	141																			
	Cornwall Aces	AHL	8	4	4	8	25												13	3	4	7	68			
1995-96	**Colorado**	**NHL**	**7**	**1**	**0**	**1**	**6**	0	0	0	6	16.7	3													
	Cornwall Aces	AHL	53	21	13	34	154												8	1	3	4	22			
1996-97	**Colorado**	**NHL**	**9**	**1**	**2**	**3**	**23**	0	0	0	7	14.3	1													
	Boston	**NHL**	**40**	**7**	**10**	**17**	**49**	0	0	0	76	9.2	-6													
	Providence	AHL	2	2	1	3	2												10	3	4	7	16			
1997-98	**Boston**	**NHL**	**28**	**1**	**5**	**6**	**7**	0	0	0	26	3.8	3						**1**	**0**	**0**	**0**	**0**	0	0	0
	Providence	AHL	42	18	10	28	146																			
1998-99	**Boston**	**NHL**	**22**	**3**	**3**	**6**	**17**	0	0	0	32	9.4	0	3	0.0	59	2	10:04	**8**	**1**	**1**	**2**	**8**	1	0	1
	Providence	AHL	48	31	22	53	89												11	7	1	8	19			
99-2000	**Boston**	**NHL**	**40**	**1**	**3**	**4**	**18**	0	0	0	67	1.5	-6	14	42.9	84	5	10:09								
	Providence	AHL	17	5	5	10	45												9	2	3	5	38			
2000-01	**Phoenix**	**NHL**	**70**	**18**	**13**	**31**	**92**	2	0	3	123	14.6	3	13	46.2	129	10	11:26								
2001-02	**Phoenix**	**NHL**	**47**	**7**	**12**	**19**	**46**	1	0	0	100	7.0	4	15	53.3	101	9	12:51	**4**	**0**	**0**	**0**	**12**	0	0	0
	Springfield	AHL	2	2	1	3	2																			
	NHL Totals		**263**	**39**	**48**	**87**	**258**	3	0	3	437	8.9		45	44.4	373	26	11:21	**13**	**1**	**1**	**2**	**20**	1	0	1

WCHA Rookie of the Year (1994) • AHL First All-Star Team (1999)

Traded to **Quebec** by **Toronto** with Wendel Clark, Sylvain Lefebvre and Toronto's 1st round choice (Jeffrey Kealty) in 1994 Entry Draft for Mats Sundin, Garth Butcher, Todd Warriner and Philadelphia's 1st round choice (previously acquired by Quebec - later traded to Washington - Washington selected Nolan Baumgartner) in 1994 Entry Draft, June 28, 1994. Transferred to **Colorado** after **Quebec** franchise relocated, June 21, 1995. Traded to **Boston** by **Colorado** with Anders Myrvold for Boston's 1st round choice (Robyn Regehr) in 1998 Entry Draft, November 22, 1996. Signed as a free agent by **Phoenix**, July 7, 2000.

WILSON, Mike

(WIHL-sohn, MIGHK) **PIT.**

Defense. Shoots left. 6'6", 229 lbs. Born, Brampton, Ont., February 26, 1975. Vancouver's 1st choice, 20th overall, in 1993 Entry Draft.

Season	Club	League	GP	G	A	Pts	PIM	PP	SH	GW	S	%	+/-	TF	F%	H	SB	Min	GP	G	A	Pts	PIM	PP	SH	GW
1991-92	Georgetown	OJHL-B	41	9	13	22	65																			
1992-93	Sudbury Wolves	OHL	53	6	7	13	58												14	1	1	2	2			
1993-94	Sudbury Wolves	OHL	60	4	22	26	62												9	1	3	4	8			
1994-95	Sudbury Wolves	OHL	64	13	34	47	46												18	1	8	9	10			
1995-96	**Buffalo**	**NHL**	**58**	**4**	**8**	**12**	**41**	1	0	1	52	7.7	13													
	Rochester	AHL	15	0	5	5	38																			
1996-97	**Buffalo**	**NHL**	**77**	**2**	**9**	**11**	**51**	0	0	1	57	3.5	13						**10**	**0**	**1**	**1**	**2**	0	0	0
1997-98	**Buffalo**	**NHL**	**66**	**4**	**4**	**8**	**48**	0	0	1	52	7.7	13						**15**	**0**	**1**	**1**	**13**	0	0	0
1998-99	Las Vegas	IHL	6	3	1	4	6																			
	Buffalo	**NHL**	**30**	**1**	**2**	**3**	**47**	0	0	1	40	2.5	10	0	0.0	54	19	16:40								
	Florida	**NHL**	**4**	**0**	**0**	**0**	**0**	0	0	0	8	0.0	2	0	0.0	4	6	19:23								
99-2000	**Florida**	**NHL**	**60**	**4**	**16**	**20**	**35**	0	0	2	65	6.2	10	0	0.0	130	62	18:07	**4**	**0**	**0**	**0**	**0**	0	0	0
2000-01	**Florida**	**NHL**	**19**	**0**	**1**	**1**	**25**	0	0	0	26	0.0	-7	0	0.0	54	14	13:20								
	Louisville	AHL	4	0	2	2	5																			

Season	Club	League	Regular Season																Playoffs							
			GP	G	A	Pts	PIM	PP	SH	GW	S	%	+/-	TF	F%	H	SB	Min	GP	G	A	Pts	PIM	PP	SH	GW
2001-02	**Pittsburgh**	**NHL**	21	1	1	2	17	0	0	0	14	7.1	–12	0	0.0	46	18	14:51								
	Wilkes-Barre	AHL	46	3	9	12	59																			
	NHL Totals		335	16	41	57	264	1	0	6	314	5.1		0	0.0	288	119	16:38	29	0	2	2	15	0	0	0

OHL All-Rookie Team (1993)

Traded to **Buffalo** by **Vancouver** with Michael Peca and Vancouver's 1st round choice (Jay McKee) in 1995 Entry Draft for Alexander Mogilny and Buffalo's 5th round choice (Todd Norman) in 1995 Entry Draft, July 8, 1995. Traded to **Florida** by **Buffalo** for Rhett Warrener and Florida's 5th round choice (Ryan Miller) in 1999 Entry Draft, March 23, 1999. • Missed majority of 2000-01 season recovering from shoulder injury suffered in game vs. New Jersey, October 25, 2000. • Missed majority of 2000-01 season recovering from shoulder injury suffered in game vs. New Jersey, October 25, 2000. Signed as a free agent by **Pittsburgh**, July 5, 2001.

WITEHALL, Johan

(WITH-hall, YOH-han)

Left wing. Shoots left. 6'1", 198 lbs. Born, Goteborg, Sweden, January 7, 1972. NY Rangers' 8th choice, 207th overall, in 1998 Entry Draft.

Season	Club	League	GP	G	A	Pts	PIM	PP	SH	GW	S	%	+/-	TF	F%	H	SB	Min	GP	G	A	Pts	PIM	PP	SH	GW
1991-92	Hanhals HF	Swede-2	32	23	14	37	52																			
1992-93	Hanhals HF	Swede-2	29	12	7	19	34																			
1993-94	Hanhals HF	Swede-2	30	13	12	25	66																			
1994-95	Hanhals HF	Swede-3	32	*38	13	*51	44																			
1995-96	Hanhals HF	Swede-3	36	*43	17	60	48																			
1996-97	IK Oskarshamn	Swede-2	32	19	16	35	38																			
1997-98	Leksands IF	Sweden	42	12	4	16	34												2	0	0	0	2			
	Leksands IF	EuroHL	5	3	0	3	2																			
1998-99	**NY Rangers**	**NHL**	4	0	0	0	0	0	0	0	1	0.0	0	1	100.0	5	0	4:06								
	Hartford	AHL	62	14	15	29	56												7	1	2	3	6			
99-2000	**NY Rangers**	**NHL**	9	1	1	2	2	0	0	0	6	16.7	0	1	0.0	13	0	8:26								
	Hartford	AHL	73	17	24	41	65												17	6	7	13	10			
2000-01	**NY Rangers**	**NHL**	15	0	3	3	8	0	0	0	16	0.0	–5	3	66.7	20	2	9:11								
	Hartford	AHL	19	10	8	18	19																			
	Montreal	**NHL**	26	1	1	2	6	0	0	0	18	5.6	0	3	66.7	36	5	10:17								
	Quebec	AHL	1	0	0	0	0												9	3	5	8	6			
2001-02	EHC Chur	Swiss	43	22	16	38	36												9	3	1	4	2			
	NHL Totals		54	2	5	7	16	0	0	0	41	4.9		8	62.5	74	7	9:13								

Claimed on waivers by **Montreal** from **NY Rangers**, January 12, 2001.

WITT, Brendan

(WITH, BREHN-duhn) **WSH.**

Defense. Shoots left. 6'2", 229 lbs. Born, Humboldt, Sask., February 20, 1975. Washington's 1st choice, 11th overall, in 1993 Entry Draft.

Season	Club	League	GP	G	A	Pts	PIM	PP	SH	GW	S	%	+/-	TF	F%	H	SB	Min	GP	G	A	Pts	PIM	PP	SH	GW
1990-91	Sask. Blazers	SMHL	31	5	13	18	42																			
	Seattle	WHL																	1	0	0	0	0			
1991-92	Seattle	WHL	67	3	9	12	212												15	1	1	2	84			
1992-93	Seattle	WHL	70	2	26	28	239												5	1	2	3	30			
1993-94	Seattle	WHL	56	8	31	39	235												9	3	8	11	23			
1994-95			DID NOT PLAY																							
1995-96	**Washington**	**NHL**	48	2	3	5	85	0	0	1	44	4.5	–4													
1996-97	**Washington**	**NHL**	44	3	2	5	88	0	0	0	41	7.3	–20													
	Portland Pirates	AHL	30	2	4	6	56												5	1	0	1	30			
1997-98	**Washington**	**NHL**	64	1	7	8	112	0	0	0	68	1.5	–11						16	1	0	1	14	0	0	0
1998-99	**Washington**	**NHL**	54	2	5	7	87	0	0	0	51	3.9	–6	0	0.0	148	52	15:50								
99-2000	**Washington**	**NHL**	77	1	7	8	114	0	0	0	64	1.6	5	2	50.0	322	105	20:56	3	0	0	0	0	0	0	0
2000-01	**Washington**	**NHL**	72	3	3	6	101	0	0	0	87	3.4	2	1	100.0	207	103	20:41	6	2	0	2	12	1	0	0
2001-02	**Washington**	**NHL**	68	3	7	10	78	0	0	0	81	3.7	–1	1	100.0	168	95	20:03								
	NHL Totals		427	15	34	49	665	0	0	1	436	3.4		4	75.0	845	355	19:38	25	3	0	3	26	1	0	0

WHL West First All-Star Team (1993, 1994) • Canadian Major Junior First All-Star Team (1994)

• Missed entire 1994-95 season after failing to come to contract terms with **Washington**.

WOOLLEY, Jason

(WU-lee, JAY-suhn) **BUF.**

Defense. Shoots left. 6', 203 lbs. Born, Toronto, Ont., July 27, 1969. Washington's 4th choice, 61st overall, in 1989 Entry Draft.

Season	Club	League	GP	G	A	Pts	PIM	PP	SH	GW	S	%	+/-	TF	F%	H	SB	Min	GP	G	A	Pts	PIM	PP	SH	GW
1986-87	St. Michael's B	OJHL-B	35	13	22	35	40																			
1987-88	St. Michael's B	OJHL-B	31	19	37	56	22																			
1988-89	Michigan State	CCHA	47	12	25	37	26																			
1989-90	Michigan State	CCHA	45	10	38	48	26																			
1990-91	Michigan State	CCHA	40	15	44	59	24																			
1991-92	Team Canada	Nat-Tm	60	14	30	44	36																			
	Canada	Olympics	8	0	5	5	4																			
	Washington	**NHL**	1	0	0	0	0	0	0	0	2	0.0	1													
	Baltimore	AHL	15	1	10	11	6																			
1992-93	**Washington**	**NHL**	26	0	2	2	10	0	0	0	11	0.0	3													
	Baltimore	AHL	29	14	27	41	22												1	0	2	2	0			
1993-94	**Washington**	**NHL**	10	1	2	3	4	0	0	0	15	6.7	2						4	1	0	1	4	0	0	1
	Portland Pirates	AHL	41	12	29	41	14												9	2	2	4	4			
1994-95	Detroit Vipers	IHL	48	8	28	36	38																			
	Florida	**NHL**	34	4	9	13	18	1	0	0	76	5.3	–1													
1995-96	**Florida**	**NHL**	52	6	28	34	32	3	0	0	98	6.1	–9						13	2	6	8	14	1	0	1
1996-97	**Florida**	**NHL**	3	0	0	0	2	0	0	0	7	0.0	1													
	Pittsburgh	**NHL**	57	6	30	36	28	2	0	1	79	7.6	3						5	0	3	3	0	0	0	0
1997-98	**Buffalo**	**NHL**	71	9	26	35	35	3	0	2	129	7.0	8						15	2	9	11	12	1	0	1
1998-99	**Buffalo**	**NHL**	80	10	33	43	62	4	0	2	154	6.5	16	0	0.0	68	67	18:43	21	4	11	15	10	2	0	1
99-2000	**Buffalo**	**NHL**	74	8	25	33	52	2	0	2	113	7.1	14	0	0.0	43	57	17:51	5	0	2	2	2	0	0	0
2000-01	**Buffalo**	**NHL**	67	5	18	23	46	4	0	2	92	5.4	0	0	0.0	33	58	17:22	8	1	5	6	2	0	0	1
2001-02	**Buffalo**	**NHL**	59	8	20	28	34	6	0	2	90	8.9	–6	0	0.0	33	55	17:11								
	NHL Totals		534	57	193	250	323	25	0	12	866	6.6		0	0.0	177	237	17:50	71	10	36	46	44	4	0	5

CCHA First All-Star Team (1991) • NCAA West First All-American Team (1991)

Signed as a free agent by **Florida**, February 15, 1995. Traded to **Pittsburgh** by **Florida** with Stu Barnes for Chris Wells, November 19, 1996. Traded to **Buffalo** by **Pittsburgh** for Buffalo's 5th round choice (Robert Scuderi) in 1998 Entry Draft, September 24, 1997.

WORRELL, Peter

(woh-REHL, PEE-tuhr) **FLA.**

Left wing. Shoots left. 6'6", 235 lbs. Born, Pierrefonds, Que., August 18, 1977. Florida's 7th choice, 166th overall, in 1995 Entry Draft.

Season	Club	League	GP	G	A	Pts	PIM	PP	SH	GW	S	%	+/-	TF	F%	H	SB	Min	GP	G	A	Pts	PIM	PP	SH	GW
1993-94	Lac St-Louis	QAAA	1	0	0	0	0												1	0	0	0	0			
1994-95	Hull Olympiques	QMJHL	56	1	8	9	243												21	0	1	1	91			
1995-96	Hull Olympiques	QMJHL	63	23	36	59	464												18	11	8	19	81			
1996-97	Hull Olympiques	QMJHL	62	17	46	63	437												14	3	13	16	83			
1997-98	**Florida**	**NHL**	19	0	0	0	153	0	0	0	15	0.0	–4													
	New Haven	AHL	50	15	12	27	309												1	0	1	1	6			
1998-99	**Florida**	**NHL**	62	4	5	9	258	0	0	2	50	8.0	0	0	0.0	100	5	6:15								
	New Haven	AHL	10	3	1	4	65																			
99-2000	**Florida**	**NHL**	48	3	6	9	169	2	0	1	45	6.7	–7	1	100.0	141	12	8:25	4	1	0	1	8	0	0	0
2000-01	**Florida**	**NHL**	71	3	7	10	248	0	0	0	86	3.5	–10	3	33.3	221	15	9:28								
2001-02	**Florida**	**NHL**	79	4	5	9	*354	0	0	1	65	6.2	–15	9	0.0	226	6	8:45								
	NHL Totals		279	14	23	37	1182	2	0	4	261	5.4		13	15.4	688	38	8:17	4	1	0	1	8	0	0	0

WOTTON, Mark

(WAH-tuhn, MAHRK) **DAL.**

Defense. Shoots left. 6'1", 195 lbs. Born, Foxwarren, Man., November 16, 1973. Vancouver's 11th choice, 237th overall, in 1992 Entry Draft.

Season	Club	League	GP	G	A	Pts	PIM	PP	SH	GW	S	%	+/-	TF	F%	H	SB	Min	GP	G	A	Pts	PIM	PP	SH	GW
1988-89	Foxwarren Blades	MAHA	60	10	30	40	70																			
1989-90	Saskatoon Blades	WHL	51	2	3	5	31												7	1	1	2	15			
1990-91	Saskatoon Blades	WHL	45	4	11	15	37																			
1991-92	Saskatoon Blades	WHL	64	11	25	36	62												21	2	6	8	22			
1992-93	Saskatoon Blades	WHL	71	15	51	66	90												9	6	5	11	18			
1993-94	Saskatoon Blades	WHL	65	12	34	46	108												16	3	12	15	32			

Season	Club	League	GP	G	A	Pts	PIM	PP	SH	GW	S	%	+/-	TF	F%	H	SB	Min	GP	G	A	Pts	PIM	PP	SH	GW
			Regular Season																Playoffs							
1994-95	Syracuse Crunch	AHL	75	12	29	41	50																			
	Vancouver	**NHL**	**1**	**0**	**0**	**0**	**0**	**0**	**0**	**0**	**2**	**0.0**	**1**						**5**	**0**	**0**	**0**	**4**	**0**	**0**	**0**
1995-96	Syracuse Crunch	AHL	80	10	35	45	96												15	1	12	13	20			
1996-97	**Vancouver**	**NHL**	**36**	**3**	**6**	**9**	**19**	**0**	**1**	**0**	**41**	**7.3**	**8**													
	Syracuse Crunch	AHL	27	2	8	10	25												2	0	0	0	4			
1997-98	**Vancouver**	**NHL**	**5**	**0**	**0**	**0**	**6**	**0**	**0**	**0**	**3**	**0.0**	**–2**													
	Syracuse Crunch	AHL	56	12	21	33	80												5	0	0	0	12			
1998-99	Syracuse Crunch	AHL	72	4	31	35	74																			
99-2000	Michigan K-Wings	IHL	70	3	7	10	72																			
2000-01	**Dallas**	**NHL**	**1**	**0**	**0**	**0**	**0**	**0**	**0**	**0**	**0**	**0.0**	**0**	**0**	**0.0**	**0**	**2**	**13:45**								
	Utah Grizzlies	IHL	63	2	2	4	64																			
2001-02	Utah Grizzlies	AHL	57	9	18	27	68												4	0	1	1	6			
	NHL Totals		**43**	**3**	**6**	**9**	**25**	**0**	**1**	**0**	**46**	**6.5**		**0**	**0.0**	**0**	**2**	**13:45**	**5**	**0**	**0**	**0**	**4**	**0**	**0**	**0**

WHL East Second All-Star Team (1994)

Signed as a free agent by **Dallas**, July 9, 1999.

WREN, Bob

(REHN, BAWB) **TOR.**

Center. Shoots left. 5'10", 185 lbs. Born, Preston, Ont., September 16, 1974. Los Angeles' 3rd choice, 94th overall, in 1993 Entry Draft.

Season	Club	League	GP	G	A	Pts	PIM	PP	SH	GW	S	%	+/-	TF	F%	H	SB	Min	GP	G	A	Pts	PIM	PP	SH	GW
1989-90	Guelph Jr. B's	OJHL-B	48	24	36	60	82																			
1990-91	Guelph Jr. B's	OJHL-B	18	17	13	30	51																			
	Kingston	OCJHL	14	10	15	25	34																			
1991-92	Detroit	OHL	62	13	36	49	58												7	3	4	7	19			
1992-93	Detroit	OHL	63	57	88	145	91												15	4	11	15	20			
1993-94	Detroit	OHL	57	45	64	109	81												17	12	18	30	20			
1994-95	Springfield	AHL	61	16	15	31	118																			
	Richmond	ECHL	2	0	1	1	0																			
1995-96	Detroit Vipers	IHL	1	0	0	0	0																			
	Knoxville	ECHL	50	21	35	56	257												8	4	11	15	32			
1996-97	Baltimore	AHL	72	23	36	59	97												3	1	1	2	0			
1997-98	**Anaheim**	**NHL**	**3**	**0**	**0**	**0**	**0**	**0**	**0**	**0**	**4**	**0.0**	**0**													
	Cincinnati	AHL	77	*42	58	100	151																			
1998-99	Cincinnati	AHL	73	27	43	70	102												3	1	2	3	8			
99-2000	Cincinnati	AHL	57	24	38	62	61																			
2000-01	**Anaheim**	**NHL**	**1**	**0**	**0**	**0**	**0**	**0**	**0**	**0**	**0**	**0.0**	**–1**	**0**	**0.0**	**0**	**0**	**11:41**								
	Cincinnati	AHL	70	20	47	67	103												4	4	2	6	2			
2001-02	St. John's	AHL	69	24	49	73	83												11	5	7	12	6			
	Toronto	**NHL**	**1**	**0**	**0**	**0**	**0**	**0**	**0**	**0**	**0**	**0.0**	**1**	**0**	**0.0**	**2**	**0**	**9:33**	**1**	**0**	**0**	**0**	**0**	**0**	**0**	**0**
	NHL Totals		**5**	**0**	**0**	**0**	**0**	**0**	**0**	**0**	**4**	**0.0**		**0**	**0.0**	**2**	**0**	**10:37**	**1**	**0**	**0**	**0**	**0**	**0**	**0**	**0**

OHL Second All-Star Team (1993, 1994)

Signed as a free agent by **Hartford**, September 6, 1994. Signed as a free agent by **Anaheim**, August 1, 1997. Signed as a free agent by **Toronto**, July 24, 2001.

WRIGHT, Jamie

(RIGHT, JAY-mee) **CGY.**

Left wing. Shoots left. 6', 195 lbs. Born, Kitchener, Ont., May 13, 1976. Dallas' 3rd choice, 98th overall, in 1994 Entry Draft.

Season	Club	League	GP	G	A	Pts	PIM	PP	SH	GW	S	%	+/-	TF	F%	H	SB	Min	GP	G	A	Pts	PIM	PP	SH	GW
1991-92	Elmira	OJHL-B	44	17	11	28	46																			
1992-93	Elmira	OJHL-B	47	22	32	54	52																			
1993-94	Guelph Storm	OHL	65	17	15	32	34												8	2	1	3	10			
1994-95	Guelph Storm	OHL	65	43	39	82	36												14	6	8	14	6			
1995-96	Guelph Storm	OHL	55	30	36	66	45												16	10	12	22	35			
1996-97	Michigan K-Wings	IHL	60	6	8	14	34												1	0	0	0	0			
1997-98	**Dallas**	**NHL**	**21**	**4**	**2**	**6**	**2**	**0**	**0**	**2**	**15**	**26.7**	**8**						**5**	**0**	**0**	**0**	**0**	**0**	**0**	**0**
	Michigan K-Wings	IHL	53	15	11	26	31																			
1998-99	**Dallas**	**NHL**	**11**	**0**	**0**	**0**	**0**	**0**	**0**	**0**	**10**	**0.0**	**–3**	**0**	**0.0**	**18**	**4**	**7:37**								
	Michigan K-Wings	IHL	64	16	15	31	92												2	0	0	0	2			
99-2000	**Dallas**	**NHL**	**23**	**1**	**4**	**5**	**16**	**0**	**0**	**0**	**15**	**6.7**	**4**	**2**	**50.0**	**53**	**5**	**9:50**								
	Michigan K-Wings	IHL	49	12	4	16	64																			
2000-01	**Dallas**	**NHL**	**2**	**1**	**0**	**1**	**0**	**0**	**0**	**0**	**4**	**25.0**	**–3**	**1**	**0.0**	**3**	**1**	**10:45**								
	Utah Grizzlies	IHL	74	25	27	52	126																			
2001-02	Saint John	AHL	34	11	13	24	34																			
	Calgary	**NHL**	**44**	**4**	**12**	**16**	**20**	**0**	**0**	**0**	**64**	**6.3**	**6**	**10**	**60.0**	**51**	**14**	**13:44**								
	NHL Totals		**101**	**10**	**18**	**28**	**38**	**0**	**0**	**2**	**108**	**9.3**		**13**	**53.8**	**125**	**24**	**11:42**	**5**	**0**	**0**	**0**	**0**	**0**	**0**	**0**

Signed as a free agent by **Calgary**, August 2, 2001.

WRIGHT, Tyler

(RIGHT, TIGH-luhr) **CBJ**

Center. Shoots right. 6', 190 lbs. Born, Kamsack, Sask., April 6, 1973. Edmonton's 1st choice, 12th overall, in 1991 Entry Draft.

Season	Club	League	GP	G	A	Pts	PIM	PP	SH	GW	S	%	+/-	TF	F%	H	SB	Min	GP	G	A	Pts	PIM	PP	SH	GW
1988-89	Swift Current	SMHL	36	20	13	33	102																			
1989-90	Swift Current	WHL	67	14	18	32	119												4	0	0	0	12			
1990-91	Swift Current	WHL	66	41	51	92	157												3	0	0	0	6			
1991-92	Swift Current	WHL	63	36	46	82	185												8	2	5	7	16			
1992-93	Swift Current	WHL	37	24	41	65	76												17	9	17	26	*49			
	Edmonton	**NHL**	**7**	**1**	**1**	**2**	**19**	**0**	**0**	**0**	**7**	**14.3**	**–4**													
1993-94	**Edmonton**	**NHL**	**5**	**0**	**0**	**0**	**4**	**0**	**0**	**0**	**2**	**0.0**	**–3**													
	Cape Breton	AHL	65	14	27	41	160												5	2	0	2	11			
1994-95	Cape Breton	AHL	70	16	15	31	184																			
	Edmonton	**NHL**	**6**	**1**	**0**	**1**	**14**	**0**	**0**	**0**	**6**	**16.7**	**1**													
1995-96	**Edmonton**	**NHL**	**23**	**1**	**0**	**1**	**33**	**0**	**0**	**0**	**18**	**5.6**	**–7**													
	Cape Breton	AHL	31	6	12	18	158																			
1996-97	**Pittsburgh**	**NHL**	**45**	**2**	**2**	**4**	**70**	**0**	**0**	**2**	**30**	**6.7**	**–7**													
	Cleveland	IHL	10	4	3	7	34												14	4	2	6	44			
1997-98	**Pittsburgh**	**NHL**	**82**	**3**	**4**	**7**	**112**	**1**	**0**	**0**	**46**	**6.5**	**–3**						**6**	**0**	**1**	**1**	**4**	**0**	**0**	**0**
1998-99	**Pittsburgh**	**NHL**	**61**	**0**	**0**	**0**	**90**	**0**	**0**	**0**	**16**	**0.0**	**–2**	**122**	**46.7**	**57**	**3**	**3:46**	**13**	**0**	**0**	**0**	**19**	**0**	**0**	**0**
99-2000	**Pittsburgh**	**NHL**	**50**	**12**	**10**	**22**	**45**	**0**	**0**	**1**	**68**	**17.6**	**4**	**698**	**47.1**	**89**	**13**	**13:24**	**11**	**3**	**1**	**4**	**17**	**0**	**0**	**0**
	Wilkes-Barre	AHL	25	5	15	20	86																			
2000-01	**Columbus**	**NHL**	**76**	**16**	**16**	**32**	**140**	**4**	**1**	**2**	**141**	**11.3**	**–9**	**999**	**45.1**	**197**	**42**	**17:41**								
2001-02	**Columbus**	**NHL**	**77**	**13**	**11**	**24**	**100**	**4**	**0**	**1**	**120**	**10.8**	**–40**	**1036**	**43.9**	**200**	**37**	**17:12**								
	NHL Totals		**432**	**49**	**44**	**93**	**627**	**9**	**1**	**6**	**454**	**10.8**		**2855**	**45.2**	**543**	**95**	**13:31**	**30**	**3**	**2**	**5**	**40**	**0**	**0**	**0**

Traded to **Pittsburgh** by **Edmonton** for Pittsburgh's 7th round choice (Brandon Lafrance) in 1996 Entry Draft, June 22, 1996. Selected by **Columbus** from **Pittsburgh** in Expansion Draft, June 23, 2000.

YACHMENEV, Vitali

(YATCH-muh-nehv, VIH-tal-ee) **NSH.**

Left wing. Shoots left. 5'11", 200 lbs. Born, Chelyabinsk, USSR, January 8, 1975. Los Angeles' 3rd choice, 59th overall, in 1994 Entry Draft.

Season	Club	League	GP	G	A	Pts	PIM	PP	SH	GW	S	%	+/-	TF	F%	H	SB	Min	GP	G	A	Pts	PIM	PP	SH	GW
1990-91	Chelyabinsk Jr.	CIS-Jr.	80	88	60	148	72																			
1991-92	Chelyabinsk Jr.	CIS-Jr.	80	82	70	152	20																			
1992-93	Chelyabinsk	CIS	51	23	20	43	12																			
1993-94	North Bay	OHL	66	*61	52	113	18												18	13	19	32	12			
1994-95	North Bay	OHL	59	53	52	105	8												6	1	8	9	2			
	Phoenix	IHL																	4	1	0	1	0			
1995-96	**Los Angeles**	**NHL**	**80**	**19**	**34**	**53**	**16**	**6**	**1**	**2**	**133**	**14.3**	**–3**													
1996-97	**Los Angeles**	**NHL**	**65**	**10**	**22**	**32**	**10**	**2**	**0**	**2**	**97**	**10.3**	**–9**													
1997-98	**Los Angeles**	**NHL**	**4**	**0**	**1**	**1**	**4**	**0**	**0**	**0**	**4**	**0.0**	**1**													
	Long Beach	IHL	59	23	28	51	14												17	8	9	17	4			
1998-99	**Nashville**	**NHL**	**55**	**7**	**10**	**17**	**10**	**0**	**1**	**2**	**83**	**8.4**	**–10**	**0**	**0.0**	**28**	**21**	**15:06**								
	Milwaukee	IHL	16	7	6	13	0																			
99-2000	**Nashville**	**NHL**	**68**	**16**	**16**	**32**	**12**	**1**	**1**	**3**	**120**	**13.3**	**5**	**8**	**25.0**	**26**	**19**	**15:26**								

			Regular Season																Playoffs							
Season	Club	League	GP	G	A	Pts	PIM	PP	SH	GW	S	%	+/-	TF	F%	H	SB	Min	GP	G	A	Pts	PIM	PP	SH	GW
2000-01	**Nashville**	**NHL**	**78**	**15**	**19**	**34**	**10**	**4**	**1**	**4**	**123**	**12.2**	**-5**	**29**	**37.9**	**40**	**38**	**17:52**								
2001-02	**Nashville**	**NHL**	**75**	**11**	**16**	**27**	**14**	**1**	**2**	**0**	**103**	**10.7**	**-16**	**42**	**33.3**	**36**	**41**	**17:33**								
	NHL Totals		**425**	**78**	**118**	**196**	**76**	**14**	**6**	**13**	**663**	**11.8**		**79**	**34.2**	**130**	**119**	**16:38**								

OHL All-Rookie Team (1994) • OHL Rookie of the Year (1994) • Canadian Major Junior Rookie of the Year (1994)

Traded to **Nashville** by **LA Kings** for future considerations, July 7, 1998.

YAKE, Terry

(YAYK, TAIR-ee)

Center. Shoots right. 5'11", 190 lbs. Born, New Westminster, B.C., October 22, 1968. Hartford's 3rd choice, 81st overall, in 1987 Entry Draft.

Season	Club	League	GP	G	A	Pts	PIM	PP	SH	GW	S	%	+/-	TF	F%	H	SB	Min	GP	G	A	Pts	PIM	PP	SH	GW
1984-85	Brandon	WHL	11	1	1	2	0																			
1985-86	Brandon	WHL	72	26	26	52	49																			
1986-87	Brandon	WHL	71	44	58	102	64																			
1987-88	Brandon	WHL	72	55	85	140	59												4	5	6	11	12			
1988-89	**Hartford**	**NHL**	**2**	**0**	**0**	**0**	**0**	**0**	**0**	**0**	**0**	**0.0**	**1**													
	Binghamton	AHL	75	39	56	95	57																			
1989-90	**Hartford**	**NHL**	**2**	**0**	**1**	**1**	**0**	**0**	**0**	**0**	**2**	**0.0**	**-1**													
	Binghamton	AHL	77	13	42	55	37																			
1990-91	**Hartford**	**NHL**	**19**	**1**	**4**	**5**	**10**	**0**	**0**	**1**	**19**	**5.3**	**-3**						**6**	**1**	**1**	**2**	**16**	**0**	**1**	**0**
	Springfield	AHL	60	35	42	77	56												15	9	9	18	10			
1991-92	**Hartford**	**NHL**	**15**	**1**	**1**	**2**	**4**	**0**	**0**	**0**	**12**	**8.3**	**-2**													
	Springfield	AHL	53	21	34	55	63												8	3	4	7	2			
1992-93	**Hartford**	**NHL**	**66**	**22**	**31**	**53**	**46**	**4**	**1**	**2**	**98**	**22.4**	**3**													
	Springfield	AHL	16	8	14	22	27																			
1993-94	**Anaheim**	**NHL**	**82**	**21**	**31**	**52**	**44**	**5**	**0**	**2**	**188**	**11.2**	**2**													
1994-95	**Toronto**	**NHL**	**19**	**3**	**2**	**5**	**2**	**1**	**0**	**2**	**26**	**11.5**	**1**													
	Denver Grizzlies	IHL	2	0	3	3	2												17	4	11	15	16			
1995-96	Milwaukee	IHL	70	32	56	88	70												5	3	6	9	4			
1996-97	Rochester	AHL	78	34	*67	101	77												10	8	8	16	2			
1997-98	**St. Louis**	**NHL**	**65**	**10**	**15**	**25**	**38**	**3**	**1**	**4**	**60**	**16.7**	**1**						**10**	**2**	**1**	**3**	**6**	**2**	**0**	**1**
1998-99	**St. Louis**	**NHL**	**60**	**9**	**18**	**27**	**34**	**3**	**0**	**4**	**59**	**15.3**	**-9**	**453**	**48.1**	**39**	**17**	**14:50**	**13**	**1**	**2**	**3**	**14**	**1**	**0**	**0**
	Worcester	AHL	24	8	11	19	26																			
99-2000	**St. Louis**	**NHL**	**26**	**4**	**9**	**13**	**22**	**2**	**0**	**2**	**26**	**15.4**	**2**	**129**	**55.8**	**18**	**8**	**13:51**								
	Washington	**NHL**	**35**	**6**	**5**	**11**	**12**	**1**	**0**	**1**	**29**	**20.7**	**2**	**219**	**51.1**	**9**	**8**	**12:36**	**3**	**0**	**0**	**0**	**0**	**0**	**0**	**0**
2000-01	**Washington**	**NHL**	**12**	**0**	**3**	**3**	**8**	**0**	**0**	**0**	**13**	**0.0**	**0**	**32**	**65.6**	**3**	**4**	**12:11**								
	Portland Pirates	AHL	55	11	38	49	47												3	0	1	1	12			
2001-02	Moskitos Essen	Germany	51	19	30	49	78																			
	NHL Totals		**403**	**77**	**120**	**197**	**220**	**19**	**2**	**18**	**532**	**14.5**		**833**	**50.8**	**69**	**37**	**13:49**	**32**	**4**	**4**	**8**	**36**	**3**	**1**	**1**

Claimed by **Anaheim** from **Hartford** in Expansion Draft, June 24, 1993. Traded to **Toronto** by **Anaheim** for David Sacco, September 28, 1994. Loaned to **Denver** (IHL) by **Toronto**, April 5, 1995. Signed as a free agent by **Buffalo**, September 17, 1996. Signed as a free agent by **St. Louis**, July 24, 1997. Claimed by **Atlanta** from **St. Louis** in Expansion Draft, June 25, 1999. Claimed by **St. Louis** from **Atlanta** in Waiver Draft, September 27, 1999. Claimed on waivers by **Washington** from **St. Louis**, January 18, 2000.

YAKUSHIN, Dmitri

(yah-KOO-shihn. DIH-mee-TREE) **TOR.**

Defense. Shoots left. 6', 200 lbs. Born, Kharkov, USSR, January 21, 1978. Toronto's 9th choice, 140th overall, in 1996 Entry Draft.

Season	Club	League	GP	G	A	Pts	PIM	PP	SH	GW	S	%	+/-	TF	F%	H	SB	Min	GP	G	A	Pts	PIM	PP	SH	GW
1995-96	Pembroke	OCJHL	31	8	5	13	62																			
1996-97	Edmonton Ice	WHL	63	3	14	17	103																			
1997-98	Edmonton Ice	WHL	29	1	10	11	41																			
	Regina Pats	WHL	13	0	14	14	16												9	2	8	10	12			
1998-99	St. John's	AHL	71	2	6	8	65												4	0	0	0	0			
99-2000	**Toronto**	**NHL**	**2**	**0**	**0**	**0**	**2**	**0**	**0**	**0**	**1**	**0.0**	**0**	**0**	**0.0**	**3**	**1**	**10:03**								
	St. John's	AHL	64	1	13	14	106																			
2000-01	St. John's	AHL	45	2	0	2	61												1	0	0	0	0			
2001-02	Sokol Kiev	EEHL	14	1	3	4	54																			
	Sokol Kiev	Ukraine	3	1	0	1	4																			
	NHL Totals		**2**	**0**	**0**	**0**	**2**	**0**	**0**	**0**	**1**	**0.0**		**0**	**0.0**	**3**	**1**	**10:03**								

YASHIN, Alexei

(YAH-shin, al-EHX-ay) **NYI**

Center. Shoots right. 6'3", 225 lbs. Born, Sverdlovsk, USSR, November 5, 1973. Ottawa's 1st choice, 2nd overall, in 1992 Entry Draft.

Season	Club	League	GP	G	A	Pts	PIM	PP	SH	GW	S	%	+/-	TF	F%	H	SB	Min	GP	G	A	Pts	PIM	PP	SH	GW
1990-91	Sverdlovsk	USSR	26	2	1	3	10																			
1991-92	Dynamo Moscow	CIS	35	7	5	12	19																			
1992-93	Dynamo Moscow	CIS	27	10	12	22	18												10	7	3	10	18			
1993-94	**Ottawa**	**NHL**	**83**	**30**	**49**	**79**	**22**	**11**	**2**	**3**	**232**	**12.9**	**-49**													
1994-95	Las Vegas	IHL	24	15	20	35	32																			
	Ottawa	**NHL**	**47**	**21**	**23**	**44**	**20**	**11**	**0**	**1**	**154**	**13.6**	**-20**													
1995-96	CSKA Moscow	CIS	4	2	2	4	4																			
	Ottawa	**NHL**	**46**	**15**	**24**	**39**	**28**	**8**	**0**	**1**	**143**	**10.5**	**-15**													
1996-97	**Ottawa**	**NHL**	**82**	**35**	**40**	**75**	**44**	**10**	**0**	**5**	**291**	**12.0**	**-7**						**7**	**1**	**5**	**6**	**2**	**1**	**0**	**0**
1997-98	**Ottawa**	**NHL**	**82**	**33**	**39**	**72**	**24**	**5**	**0**	**6**	**291**	**11.3**	**6**						**11**	**5**	**3**	**8**	**8**	**3**	**0**	**2**
	Russia	Olympics	6	3	3	6	0																			
1998-99	**Ottawa**	**NHL**	**82**	**44**	**50**	**94**	**54**	**19**	**0**	**5**	**337**	**13.1**	**16**	**1428**	**41.9**	**73**	**22**	**22:05**	**4**	**0**	**0**	**0**	**10**	**0**	**0**	**0**
99-2000	**Ottawa**	**NHL**	DID NOT PLAY – SUSPENDED																							
2000-01	**Ottawa**	**NHL**	**82**	**40**	**48**	**88**	**30**	**13**	**2**	**10**	**263**	**15.2**	**10**	**1414**	**43.1**	**50**	**28**	**20:24**	**4**	**0**	**1**	**1**	**0**	**0**	**0**	**0**
2001-02	**NY Islanders**	**NHL**	**78**	**32**	**43**	**75**	**25**	**15**	**0**	**5**	**239**	**13.4**	**-3**	**828**	**46.7**	**47**	**26**	**20:37**	**7**	**3**	**4**	**7**	**2**	**1**	**0**	**0**
	Russia	Olympics	6	1	1	2	0																			
	NHL Totals		**582**	**250**	**316**	**566**	**247**	**92**	**4**	**36**	**1950**	**12.8**		**3670**	**43.5**	**170**	**76**	**21:02**	**33**	**9**	**13**	**22**	**22**	**5**	**0**	**2**

NHL Second All-Star Team (1999) • Played in NHL All-Star Game (1994, 1999, 2002)

• Suspended for entire 1999-2000 season by **Ottawa** for refusing to report to team, November 9, 1999. Traded to **NY Islanders** by **Ottawa** for Bill Muckalt, Zdeno Chara and NY Islanders' 1st round choice (Jason Spezza) in 2001 Entry Draft, June 23, 2001.

YELLE, Stephane

(YEHL, STEH-fan) **COL.**

Center. Shoots left. 6'1", 190 lbs. Born, Ottawa, Ont., May 9, 1974. New Jersey's 9th choice, 186th overall, in 1992 Entry Draft.

Season	Club	League	GP	G	A	Pts	PIM	PP	SH	GW	S	%	+/-	TF	F%	H	SB	Min	GP	G	A	Pts	PIM	PP	SH	GW
1990-91	Cumberland	OJHL-B	33	20	30	50	16																			
1991-92	Oshawa Generals	OHL	55	12	14	26	20												7	2	0	2	1			
1992-93	Oshawa Generals	OHL	66	24	50	74	20												10	2	4	6	4			
1993-94	Oshawa Generals	OHL	66	35	69	104	22												5	1	7	8	2			
1994-95	Cornwall Aces	AHL	40	18	15	33	22												13	7	7	14	8			
1995-96♦	**Colorado**	**NHL**	**71**	**13**	**14**	**27**	**30**	**0**	**2**	**1**	**93**	**14.0**	**15**						**22**	**1**	**4**	**5**	**8**	**0**	**1**	**0**
1996-97	**Colorado**	**NHL**	**79**	**9**	**17**	**26**	**38**	**0**	**1**	**1**	**89**	**10.1**	**1**						**12**	**1**	**6**	**7**	**2**	**0**	**0**	**0**
1997-98	**Colorado**	**NHL**	**81**	**7**	**15**	**22**	**48**	**0**	**1**	**0**	**93**	**7.5**	**-10**						**7**	**1**	**0**	**1**	**12**	**0**	**0**	**0**
1998-99	**Colorado**	**NHL**	**72**	**8**	**7**	**15**	**40**	**1**	**0**	**0**	**99**	**8.1**	**-8**	**1201**	**51.2**	**136**	**72**	**15:15**	**10**	**0**	**1**	**1**	**6**	**0**	**0**	**0**
99-2000	**Colorado**	**NHL**	**79**	**8**	**14**	**22**	**28**	**0**	**1**	**1**	**90**	**8.9**	**9**	**1294**	**52.2**	**140**	**81**	**15:51**	**17**	**1**	**2**	**3**	**4**	**0**	**0**	**0**
2000-01♦	**Colorado**	**NHL**	**50**	**4**	**10**	**14**	**20**	**0**	**1**	**0**	**54**	**7.4**	**-3**	**736**	**56.4**	**92**	**29**	**14:28**	**23**	**1**	**2**	**3**	**8**	**0**	**0**	**1**
2001-02	**Colorado**	**NHL**	**73**	**5**	**12**	**17**	**48**	**0**	**1**	**1**	**71**	**7.0**	**1**	**1036**	**51.8**	**131**	**73**	**14:02**	**20**	**0**	**2**	**2**	**14**	**0**	**0**	**0**
	NHL Totals		**505**	**54**	**89**	**143**	**252**	**1**	**7**	**4**	**589**	**9.2**		**4267**	**52.6**	**499**	**255**	**14:57**	**111**	**5**	**17**	**22**	**54**	**0**	**1**	**1**

Traded to **Quebec** by **New Jersey** with New Jersey's 11th round choice (Steven Low) in 1994 Entry Draft for Quebec's 11th round choice (Mike Hanson) in 1994 Entry Draft, June 1, 1994. Transferred to **Colorado** after **Quebec** franchise relocated, June 21, 1995.

YLONEN, Juha

(YOO-lih-nehn, YOO-hah)

Center. Shoots left. 6'1", 189 lbs. Born, Helsinki, Finland, February 13, 1972. Winnipeg's 3rd choice, 91st overall, in 1991 Entry Draft.

Season	Club	League	GP	G	A	Pts	PIM	PP	SH	GW	S	%	+/-	TF	F%	H	SB	Min	GP	G	A	Pts	PIM	PP	SH	GW
1988-89	Kiekko Espoo Jr.	Finn-Jr.	31	9	14	23	8																			
1989-90	Kiekko Espoo Jr.	Finn-Jr.	4	1	5	6	0												5	1	5	6	0			
	Kiekko Espoo	Finland-2	38	10	17	27	12																			
1990-91	Kiekko Espoo Jr.	Finn-Jr.	5	3	1	4	2																			
	Kiekko Espoo	Finland-2	40	12	21	33	4																			

			Regular Season																Playoffs							
Season	Club	League	GP	G	A	Pts	PIM	PP	SH	GW	S	%	+/-	TF	F%	H	SB	Min	GP	G	A	Pts	PIM	PP	SH	GW
1991-92	HPK Jr.	Finn-Jr.	2	1	2	3	0																			
	HPK Hameenlinna	Finland-2	9	8	14	22	0																			
	HPK Hameenlinna	Finland	43	7	11	18	8																			
1992-93	HPK Hameenlinna	Finland	48	8	18	26	22												12	3	5	8	2			
	HPK Jr.	Finn-Jr.	2	2	1	3	0												1	0	0	0	0			
1993-94	Jokerit Helsinki	Finland	37	5	11	16	2												12	1	3	4	8			
1994-95	Jokerit Helsinki	Finland	50	13	15	28	10												11	3	2	5	0			
1995-96	Jokerit Helsinki	Finland	24	3	13	16	20												11	4	5	9	4			
1996-97	Phoenix	NHL	2	0	0	0	0	0	0	0	2	0.0	0													
	Springfield	AHL	70	20	41	61	6												17	5	*16	21	4			
1997-98	Phoenix	NHL	55	1	11	12	10	0	1	0	60	1.7	-3													
	Finland	Olympics	6	0	0	0	8																			
1998-99	Phoenix	NHL	59	6	17	23	20	2	0	1	66	9.1	18	297	46.5	48	37	15:55	2	0	2	2	2	0	0	0
99-2000	Phoenix	NHL	76	6	23	29	12	0	1	1	82	7.3	-6	753	45.6	52	51	16:43	1	0	0	0	0	0	0	0
2000-01	Phoenix	NHL	69	9	14	23	38	0	1	1	72	12.5	10	163	41.1	42	36	14:18								
2001-02	Tampa Bay	NHL	65	3	10	13	8	0	0	0	75	4.0	-10	427	39.8	32	31	15:45								
	Finland	Olympics	4	0	1	1	2																			
	Ottawa	NHL	15	1	1	2	2	0	1	0	22	4.5	-1	221	49.8	18	10	15:17	12	0	5	5	2	0	0	0
	NHL Totals		341	26	76	102	90	2	4	3	379	6.9		1861	44.5	192	165	15:40	15	0	7	7	4	0	0	0

Rights transferred to **Phoenix** after **Winnipeg** franchise relocated, July 1, 1996. Traded to **Tampa Bay** by **Phoenix** for Todd Warriner, June 18, 2001. Traded to **Ottawa** by **Tampa Bay** for Andre Roy and Ottawa's 6th round choice (Paul Ranger) in 2002 Entry Draft, March 15, 2002.

YONKMAN, Nolan
(YAWK-man, NOH-lan) **WSH.**

Defense. Shoots right. 6'6", 236 lbs. Born, Punnicht, Sask., April 1, 1981. Washington's 5th choice, 37th overall, in 1999 Entry Draft.

Season	Club	League	GP	G	A	Pts	PIM	PP	SH	GW	S	%	+/-	TF	F%	H	SB	Min	GP	G	A	Pts	PIM	PP	SH	GW
1996-97	Naicam Vikings	SAHA	64	15	23	38	36																			
	Kelowna Rockets	WHL	4	0	0	0	0																			
1997-98	Kelowna Rockets	WHL	65	0	2	2	36												7	0	0	0	2			
1998-99	Kelowna Rockets	WHL	61	1	6	7	129												6	0	0	0	6			
99-2000	Kelowna Rockets	WHL	71	5	7	12	153												5	0	0	0	8			
2000-01	Kelowna Rockets	WHL	7	0	1	1	19																			
	Brandon	WHL	51	6	10	16	94												6	0	1	1	12			
2001-02	Washington	NHL	11	1	0	1	4	0	0	0	7	14.3	3	0	0.0	30	11	12:44								
	Portland Pirates	AHL	59	4	3	7	116																			
	NHL Totals		11	1	0	1	4	0	0	0	7	14.3		0	0.0	30	11	12:44								

Traded to **Brandon** (WHL) by **Kelowna** (WHL) for Bart Rushner and Jan Fadrny, October 12, 2000.

YORK, Jason
(YOHRK, JAY-suhn) **ANA.**

Defense. Shoots right. 6'1", 208 lbs. Born, Nepean, Ont., May 20, 1970. Detroit's 6th choice, 129th overall, in 1990 Entry Draft.

Season	Club	League	GP	G	A	Pts	PIM	PP	SH	GW	S	%	+/-	TF	F%	H	SB	Min	GP	G	A	Pts	PIM	PP	SH	GW
1986-87	Smiths Falls	OCJHL	46	6	13	19	86																			
1987-88	Hamilton	OHL	58	4	9	13	110																			
1988-89	Windsor	OHL	65	19	44	63	105																			
1989-90	Windsor	OHL	39	9	30	39	38																			
	Kitchener	OHL	25	11	25	36	17												17	3	19	22	10			
1990-91	Windsor	OHL	66	13	80	93	40												11	3	10	13	12			
1991-92	Adirondack	AHL	49	4	20	24	32												5	0	1	1	0			
1992-93	Detroit	NHL	2	0	0	0	0	0	0	0	1	0.0	0													
	Adirondack	AHL	77	15	40	55	86												11	0	3	3	18			
1993-94	Detroit	NHL	7	1	2	3	2	0	0	0	9	11.1	0													
	Adirondack	AHL	74	10	56	66	98												12	3	11	14	22			
1994-95	Detroit	NHL	10	1	2	3	2	0	0	0	6	16.7	0													
	Adirondack	AHL	5	1	3	4	4																			
	Anaheim	NHL	15	0	8	8	12	0	0	0	22	0.0	4													
1995-96	Anaheim	NHL	79	3	21	24	88	0	0	0	106	2.8	-7													
1996-97	Ottawa	NHL	75	4	17	21	67	1	0	0	121	3.3	-8						7	0	0	0	4	0	0	0
1997-98	Ottawa	NHL	73	3	13	16	62	0	0	0	109	2.8	8						7	1	1	2	7	1	0	0
1998-99	Ottawa	NHL	79	4	31	35	48	2	0	0	177	2.3	17	2	0.0	166	127	23:49	4	1	1	2	4	0	0	0
99-2000	Ottawa	NHL	79	8	22	30	60	1	0	1	159	5.0	-3	0	0.0	158	111	23:20	6	0	2	2	2	0	0	0
2000-01	Ottawa	NHL	74	6	16	22	72	3	0	2	133	4.5	7	0	0.0	167	101	23:49	4	0	0	0	4	0	0	0
2001-02	Anaheim	NHL	74	5	20	25	60	3	0	2	104	4.8	-11	0	0.0	114	62	19:19								
	NHL Totals		567	35	152	187	473	10	0	5	947	3.7		2	0.0	605	401	22:36	28	2	4	6	21	1	0	0

AHL First All-Star Team (1994)

Traded to **Anaheim** by **Detroit** with Mike Sillinger for Stu Grimson, Mark Ferner and Anaheim's 6th round choice (Magnus Nilsson) in 1996 Entry Draft, April 4, 1995. Traded to **Ottawa** by **Anaheim** with Shaun Van Allen for Ted Drury and the rights to Marc Moro, October 1, 1996. Signed as a free agent by **Anaheim**, July 3, 2001.

YORK, Mike
(YOHRK, MIGHK) **EDM.**

Center. Shoots right. 5'10", 185 lbs. Born, Waterford, MI, January 3, 1978. NY Rangers' 7th choice, 136th overall, in 1997 Entry Draft.

Season	Club	League	GP	G	A	Pts	PIM	PP	SH	GW	S	%	+/-	TF	F%	H	SB	Min	GP	G	A	Pts	PIM	PP	SH	GW
1992-93	Michigan	MNHL	50	45	50	95																				
1993-94	Det. Compuware	MNHL	85	136	140	276																				
1994-95	Thornhill	MTJHL	49	39	54	*93	44												11	7	6	13	0			
1995-96	Michigan State	CCHA	39	12	27	39	20																			
1996-97	Michigan State	CCHA	37	18	29	47	42																			
1997-98	Michigan State	CCHA	40	27	34	61	38																			
1998-99	Michigan State	CCHA	42	22	32	*54	41																			
	Hartford	AHL	3	2	2	4	0												6	3	1	4	0			
99-2000	NY Rangers	NHL	82	26	24	50	18	8	0	4	177	14.7	-17	1131	48.1	75	23	15:50								
2000-01	NY Rangers	NHL	79	14	17	31	20	3	2	4	171	8.2	1	1098	46.9	78	41	17:45								
2001-02	NY Rangers	NHL	69	18	39	57	16	2	0	5	188	9.6	8	267	43.1	54	37	20:24								
	United States	Olympics	6	0	1	1	0																			
	Edmonton	NHL	12	2	2	4	0	1	0	1	30	6.7	-1	100	56.0	7	2	16:35								
	NHL Totals		242	60	82	142	54	14	2	14	566	10.6		2596	26.4	214	103	18:49								

MTJHL Bauer Divison Rookie of the Year (1995) • MTJHL Bauer Divison All-Star Team (1995) • CCHA Second All-Star Team (1998) • NCAA West First All-American Team (1998, 1999) • CCHA First All-Star Team (1999) • CCHA Player of the Year (1999) • NHL All-Rookie Team (2000) • Played in NHL All-Star Game (2002)

Traded to **Edmonton** by **NY Rangers** with NY Rangers' 4th round choice (Ivan Koltsov) in 2002 Entry Draft for Tom Poti and Rem Murray, March 19, 2002.

YOUNG, Scott
(YUHNG, SKAWT) **DAL.**

Right wing. Shoots right. 6'1", 200 lbs. Born, Clinton, MA, October 1, 1967. Hartford's 1st choice, 11th overall, in 1986 Entry Draft.

Season	Club	League	GP	G	A	Pts	PIM	PP	SH	GW	S	%	+/-	TF	F%	H	SB	Min	GP	G	A	Pts	PIM	PP	SH	GW
1984-85	St. Mark's	Hi-School	23	28	41	69																				
1985-86	Boston University	H-East	38	16	13	29	31																			
1986-87	Boston University	H-East	33	15	21	36	24																			
1987-88	Team USA	Nat-Tm	56	11	47	58	31																			
	United States	Olympics	6	2	6	8	4																			
	Hartford	NHL	7	0	0	0	2	0	0	0	6	0.0	-6						4	1	0	1	0	0	0	0
1988-89	Hartford	NHL	76	19	40	59	27	6	0	2	203	9.4	-21						4	2	0	2	4	0	0	0
1989-90	Hartford	NHL	80	24	40	64	47	10	2	5	239	10.0	-24						7	2	0	2	2	0	0	0
1990-91	Hartford	NHL	34	6	9	15	8	3	1	2	94	6.4	-9													
	♦ Pittsburgh	NHL	43	11	16	27	33	3	1	3	116	9.5	3						17	1	6	7	2	1	0	0
1991-92	HC Bolzano	Alpenliga	15	19	11	30	14																			
	HC Bolzano	Italy	18	22	17	39	6												5	4	3	7	7			
	Team USA	Nat-Tm	10	2	4	6	21																			
	United States	Olympics	8	2	1	3	2																			
1992-93	Quebec	NHL	82	30	30	60	20	9	6	5	225	13.3	5						6	4	1	5	0	0	0	2
1993-94	Quebec	NHL	76	26	25	51	14	6	1	1	236	11.0	-4													
1994-95	EV Landshut	Germany	4	6	1	7	6																			
	Frankfurt Lions	Germany	1	1	0	1	0																			
	Quebec	NHL	48	18	21	39	14	3	3	0	167	10.8	9						6	3	3	6	2	0	1	0
1995-96 ♦	Colorado	NHL	81	21	39	60	50	7	0	5	229	9.2	2						22	3	12	15	10	0	0	0

			Regular Season																Playoffs							
Season	Club	League	GP	G	A	Pts	PIM	PP	SH	GW	S	%	+/-	TF	F%	H	SB	Min	GP	G	A	Pts	PIM	PP	SH	GW
1996-97	Colorado	NHL	72	18	19	37	14	7	0	0	164	11.0	–5	....	....	....	....	....	17	4	2	6	14	2	0	0
1997-98	Anaheim	NHL	73	13	20	33	22	4	2	1	187	7.0	–13	....	....	....	....	....	....	....	....	....	....	....	....	....
1998-99	St. Louis	NHL	75	24	28	52	27	8	0	4	205	11.7	8	4	25.0	51	18	15:14	13	4	7	11	10	1	0	1
99-2000	St. Louis	NHL	75	24	15	39	18	6	1	7	244	9.8	12	2	50.0	63	15	16:06	6	6	2	8	8	3	0	0
2000-01	St. Louis	NHL	81	40	33	73	30	14	3	7	321	12.5	15	4	25.0	67	29	19:18	15	6	7	13	2	0	2	3
2001-02	St. Louis	NHL	67	19	22	41	26	5	0	1	210	9.0	11	4	0.0	44	25	18:31	10	3	0	3	2	1	1	0
	United States	Olympics	6	4	0	4	2	....	....	....	....	....	....	....	....	....	....	....	....	....	....	....	....	....	....	....
	NHL Totals		970	293	357	650	352	91	20	43	2846	10.3		14	21.4	225	87	17:18	127	39	40	79	56	8	4	6

Hockey East Rookie of the Year (Shared with Al Loring) (1986)

Traded to **Pittsburgh** by **Hartford** for Rob Brown, December 21, 1990. Traded to **Quebec** by **Pittsburgh** for Bryan Fogarty, March 10, 1992. Transferred to **Colorado** after **Quebec** franchise relocated, June 21, 1995. Traded to **Anaheim** by **Colorado** for Anaheim's 3rd round choice (later traded to Florida - Florida selected Lance Ward) in 1998 Entry Draft, September 17, 1997. Signed as a free agent by **St. Louis**, July 28, 1998. Signed as a free agent by **Dallas**, July 5, 2002.

YUSHKEVICH, Dmitry (yoosh-KAY-vihch, dih-MEE-tree) FLA.

Defense. Shoots right. 5'11", 208 lbs. Born, Cherepovets, USSR, November 19, 1971. Philadelphia's 6th choice, 122nd overall, in 1991 Entry Draft.

Season	Club	League	GP	G	A	Pts	PIM	PP	SH	GW	S	%	+/-	TF	F%	H	SB	Min	GP	G	A	Pts	PIM	PP	SH	GW
1988-89	Yaroslavl	USSR	23	2	1	3	8	....	....	....	....	....	....	....	....	....	....	....	....	....	....	....	....	....	....	....
1989-90	Yaroslavl	USSR	41	2	3	5	39	....	....	....	....	....	....	....	....	....	....	....	....	....	....	....	....	....	....	....
1990-91	Yaroslavl	USSR	41	10	4	14	22	....	....	....	....	....	....	....	....	....	....	....	....	....	....	....	....	....	....	....
1991-92	Dynamo Moscow	CIS	35	5	7	12	14	....	....	....	....	....	....	....	....	....	....	....	....	....	....	....	....	....	....	....
	Russia	Olympics	8	1	2	3	4	....	....	....	....	....	....	....	....	....	....	....	....	....	....	....	....	....	....	....
1992-93	Philadelphia	NHL	82	5	27	32	71	1	0	1	155	3.2	12	....	....	....	....	....	....	....	....	....	....	....	....	....
1993-94	Philadelphia	NHL	75	5	25	30	86	1	0	2	136	3.7	–8	....	....	....	....	....	....	....	....	....	....	....	....	....
1994-95	Yaroslavl	CIS	10	3	4	7	8	....	....	....	....	....	....	....	....	....	....	....	....	....	....	....	....	....	....	....
	Philadelphia	NHL	40	5	9	14	47	3	1	1	80	6.3	–4	....	....	....	....	....	15	1	5	6	12	0	0	0
1995-96	Toronto	NHL	69	1	10	11	54	1	0	0	96	1.0	–14	....	....	....	....	....	4	0	0	0	0	0	0	0
1996-97	Toronto	NHL	74	4	10	14	56	1	1	1	99	4.0	–24	....	....	....	....	....	....	....	....	....	....	....	....	....
1997-98	Toronto	NHL	72	0	12	12	78	0	0	0	92	0.0	–13	....	....	....	....	....	....	....	....	....	....	....	....	....
	Russia	Olympics	6	0	0	0	2	....	....	....	....	....	....	....	....	....	....	....	....	....	....	....	....	....	....	....
1998-99	Toronto	NHL	78	6	22	28	88	2	1	0	95	6.3	25	0	0.0	169	107	22:20	17	1	5	6	22	1	0	0
99-2000	Yaroslavl	Russia	7	2	3	5	2	....	....	....	....	....	....	....	....	....	....	....	....	....	....	....	....	....	....	....
	Toronto	NHL	77	3	24	27	55	2	1	1	103	2.9	2	0	0.0	266	160	23:18	12	1	1	2	4	0	0	0
2000-01	Toronto	NHL	81	5	19	24	52	1	0	0	110	4.5	–2	0	0.0	171	174	24:14	11	0	4	4	12	0	0	0
2001-02	Toronto	NHL	55	6	13	19	26	3	0	0	79	7.6	14	1	100.0	150	107	23:25	....	....	....	....	....	....	....	....
	NHL Totals		703	40	171	211	613	15	4	6	1045	3.8		1	100.0	756	548	23:20	59	3	15	18	50	1	0	0

Played in NHL All-Star Game (2000)

Traded to **Toronto** by **Philadelphia** with Philadelphia's 2nd round choice (Francis Larivee) in 1996 Entry Draft for Toronto's 1st round choice (Dainius Zubrus) in 1996 Entry Draft, 2nd round choice (Jean-Marc Pelletier) in 1997 Entry Draft and LA Kings' 4th round choice (previously acquired by Toronto - later traded to LA Kings - LA Kings selected Mikael Simons) in 1996 Entry Draft, August 30, 1995. Traded to **Florida** by **Toronto** for Robert Svehla, July 18, 2002.

YZERMAN, Steve (IGH-zuhr-muhn, STEEV) DET.

Center. Shoots right. 5'11", 185 lbs. Born, Cranbrook, B.C., May 9, 1965. Detroit's 1st choice, 4th overall, in 1983 Entry Draft.

Season	Club	League	GP	G	A	Pts	PIM	PP	SH	GW	S	%	+/-	TF	F%	H	SB	Min	GP	G	A	Pts	PIM	PP	SH	GW
1980-81	Nepean Raiders	OCJHL	50	38	*54	92	44	....	....	....	....	....	....	....	....	....	....	....	....	....	....	....	....	....	....	....
1981-82	Peterborough	OHL	58	21	43	64	65	....	....	....	....	....	....	....	....	....	....	....	6	0	1	1	16	....	....	....
1982-83	Peterborough	OHL	56	42	49	91	33	....	....	....	....	....	....	....	....	....	....	....	4	1	4	5	0	....	....	....
1983-84	Detroit	NHL	80	39	48	87	33	13	0	2	177	22.0	–17	....	....	....	....	....	4	3	3	6	0	1	0	1
1984-85	Detroit	NHL	80	30	59	89	58	9	0	3	231	13.0	–17	....	....	....	....	....	3	2	1	3	2	0	0	0
1985-86	Detroit	NHL	51	14	28	42	16	3	0	3	132	10.6	–24	....	....	....	....	....	....	....	....	....	....	....	....	....
1986-87	Detroit	NHL	80	31	59	90	43	9	1	2	217	14.3	–1	....	....	....	....	....	16	5	13	18	8	1	0	0
1987-88	Detroit	NHL	64	50	52	102	44	10	6	6	242	20.7	30	....	....	....	....	....	3	1	3	4	6	0	0	0
1988-89	Detroit	NHL	80	65	90	155	61	17	3	7	388	16.8	17	....	....	....	....	....	6	5	5	10	2	2	0	0
1989-90	Detroit	NHL	79	62	65	127	79	16	7	8	332	18.7	–6	....	....	....	....	....	....	....	....	....	....	....	....	....
1990-91	Detroit	NHL	80	51	57	108	34	12	6	4	326	15.6	–2	....	....	....	....	....	7	3	3	6	4	1	0	0
1991-92	Detroit	NHL	79	45	58	103	64	9	8	9	295	15.3	26	....	....	....	....	....	11	3	5	8	12	0	1	1
1992-93	Detroit	NHL	84	58	79	137	44	13	7	6	307	18.9	33	....	....	....	....	....	7	4	3	7	4	1	1	1
1993-94	Detroit	NHL	58	24	58	82	36	7	3	3	217	11.1	11	....	....	....	....	....	3	1	3	4	0	0	0	0
1994-95	Detroit	NHL	47	12	26	38	40	4	0	1	134	9.0	6	....	....	....	....	....	15	4	8	12	0	2	0	1
1995-96	Detroit	NHL	80	36	59	95	64	16	2	8	220	16.4	29	....	....	....	....	....	18	8	12	20	4	4	0	1
1996-97♦	Detroit	NHL	81	22	63	85	78	8	0	3	232	9.5	22	....	....	....	....	....	20	7	6	13	4	3	0	2
1997-98♦	Detroit	NHL	75	24	45	69	46	6	2	0	188	12.8	3	....	....	....	....	....	22	6	*18	*24	22	3	1	0
	Canada	Olympics	6	1	1	2	10	....	....	....	....	....	....	....	....	....	....	....	....	....	....	....	....	....	....	....
1998-99	Detroit	NHL	80	29	45	74	42	13	2	4	231	12.6	8	1600	56.9	46	58	21:35	10	9	4	13	0	4	0	2
99-2000	Detroit	NHL	78	35	44	79	34	15	2	6	234	15.0	28	1868	56.8	67	56	21:07	8	0	4	4	0	0	0	0
2000-01	Detroit	NHL	54	18	34	52	18	5	0	7	155	11.6	4	1197	59.6	38	51	22:14	1	0	0	0	0	0	0	0
2001-02♦	Detroit	NHL	52	13	35	48	18	5	1	5	104	12.5	11	1182	58.4	30	43	20:35	23	6	17	23	10	4	0	2
	Canada	Olympics	6	2	4	6	2	....	....	....	....	....	....	....	....	....	....	....	....	....	....	....	....	....	....	....
	NHL Totals		1362	658	1004	1662	852	190	50	87	4362	15.1		5847	57.7	181	208	21:23	177	67	108	175	78	26	3	11

NHL All-Rookie Team (1984) • Won Lester B. Pearson Award (1989) • Won Conn Smythe Trophy (1998) • NHL First All-Star Team (2000) • Won Frank J. Selke Trophy (2000) • Played in NHL All-Star Game (1984, 1988, 1989, 1990, 1991, 1992, 1993, 1997, 2000)

ZAMUNER, Rob (ZAM-nuhr, RAWB) BOS.

Left wing. Shoots left. 6'3", 203 lbs. Born, Oakville, Ont., September 17, 1969. NY Rangers' 3rd choice, 45th overall, in 1989 Entry Draft.

Season	Club	League	GP	G	A	Pts	PIM	PP	SH	GW	S	%	+/-	TF	F%	H	SB	Min	GP	G	A	Pts	PIM	PP	SH	GW
1985-86	Oakville Oaks	OMHA	48	43	50	93	66	....	....	....	....	....	....	....	....	....	....	....	....	....	....	....	....	....	....	....
1986-87	Guelph Jr. B's	OJHL-B	3	6	7	13	15	....	....	....	....	....	....	....	....	....	....	....	....	....	....	....	....	....	....	....
	Guelph Platers	OHL	62	6	15	21	8	....	....	....	....	....	....	....	....	....	....	....	....	....	....	....	....	....	....	....
1987-88	Guelph Platers	OHL	58	20	41	61	18	....	....	....	....	....	....	....	....	....	....	....	....	....	....	....	....	....	....	....
1988-89	Guelph Platers	OHL	66	46	65	111	38	....	....	....	....	....	....	....	....	....	....	....	7	5	5	10	9	....	....	....
1989-90	Flint Spirits	IHL	77	44	35	79	32	....	....	....	....	....	....	....	....	....	....	....	4	1	0	1	6	....	....	....
1990-91	Binghamton	AHL	80	25	58	83	50	....	....	....	....	....	....	....	....	....	....	....	9	7	6	13	35	....	....	....
1991-92	NY Rangers	NHL	9	1	2	3	2	0	0	0	11	9.1	0	....	....	....	....	....	....	....	....	....	....	....	....	....
	Binghamton	AHL	61	19	53	72	42	....	....	....	....	....	....	....	....	....	....	....	11	8	9	17	8	....	....	....
1992-93	Tampa Bay	NHL	84	15	28	43	74	1	0	0	183	8.2	–25	....	....	....	....	....	....	....	....	....	....	....	....	....
1993-94	Tampa Bay	NHL	59	6	6	12	42	0	0	1	109	5.5	–9	....	....	....	....	....	....	....	....	....	....	....	....	....
1994-95	Tampa Bay	NHL	43	9	6	15	24	0	3	1	74	12.2	–3	....	....	....	....	....	....	....	....	....	....	....	....	....
1995-96	Tampa Bay	NHL	72	15	20	35	62	0	3	4	152	9.9	11	....	....	....	....	....	6	2	3	5	10	0	1	0
1996-97	Tampa Bay	NHL	82	17	33	50	56	0	4	3	216	7.9	3	....	....	....	....	....	....	....	....	....	....	....	....	....
1997-98	Tampa Bay	NHL	77	14	12	26	41	0	3	4	126	11.1	–31	....	....	....	....	....	....	....	....	....	....	....	....	....
	Canada	Olympics	6	1	0	1	8	....	....	....	....	....	....	....	....	....	....	....	....	....	....	....	....	....	....	....
1998-99	Tampa Bay	NHL	58	8	11	19	24	1	1	2	89	9.0	–15	34	47.1	52	11	16:26	....	....	....	....	....	....	....	....
99-2000	Ottawa	NHL	57	9	12	21	32	0	1	0	103	8.7	–6	29	37.9	65	19	14:56	6	2	0	2	2	0	0	1
2000-01	Ottawa	NHL	79	19	18	37	52	1	2	4	123	15.4	7	162	35.8	94	24	14:46	4	0	0	0	6	0	0	0
2001-02	Boston	NHL	66	12	13	25	24	1	2	0	98	12.2	6	192	41.7	45	22	13:13	6	0	2	2	4	0	0	0
	NHL Totals		686	125	161	286	433	4	19	19	1284	9.7		417	39.6	256	76	14:47	22	4	5	9	22	0	1	1

Signed as a free agent by **Tampa Bay**, July 13, 1992. Traded to **Ottawa** by **Tampa Bay** with Tampa Bay's 2nd round choice (later traded to Philadelphia - later traded back to Tampa Bay - later traded to Dallas - Dallas selected Tobias Stephan) in 2002 Entry Draft for Andreas Johansson, June 29, 1999. Signed as a free agent by **Boston**, July 6, 2001.

ZEDNIK, Richard (ZEHD-nihk, REE-khahrd) MTL.

Left wing. Shoots left. 6', 200 lbs. Born, Bystrica, Czech., January 6, 1976. Washington's 10th choice, 249th overall, in 1994 Entry Draft.

Season	Club	League	GP	G	A	Pts	PIM	PP	SH	GW	S	%	+/-	TF	F%	H	SB	Min	GP	G	A	Pts	PIM	PP	SH	GW
1993-94	Banska Bystrica	Slovak-2	25	3	6	9	....	....	....	....	....	....	....	....	....	....	....	....	....	....	....	....	....	....	....	....
1994-95	Portland	WHL	65	35	51	86	89	....	....	....	....	....	....	....	....	....	....	....	9	5	5	10	20	....	....	....
1995-96	Portland	WHL	61	44	37	81	154	....	....	....	....	....	....	....	....	....	....	....	7	8	4	12	23	....	....	....
	Washington	NHL	1	0	0	0	0	0	0	0	0	0.0	0	....	....	....	....	....	....	....	....	....	....	....	....	....
	Portland Pirates	AHL	1	1	1	2	0	....	....	....	....	....	....	....	....	....	....	....	21	4	5	9	26	....	....	....
1996-97	Washington	NHL	11	2	1	3	4	1	0	0	21	9.5	–5	....	....	....	....	....	....	....	....	....	....	....	....	....
	Portland Pirates	AHL	56	15	20	35	70	....	....	....	....	....	....	....	....	....	....	....	5	1	0	1	6	....	....	....
1997-98	Washington	NHL	65	17	9	26	28	2	0	2	148	11.5	–2	....	....	....	....	....	17	7	3	10	16	2	0	0
1998-99	Washington	NHL	49	9	8	17	50	1	0	2	115	7.8	–6	2	0.0	81	9	15:08	....	....	....	....	....	....	....	....

			Regular Season																Playoffs							
Season	Club	League	GP	G	A	Pts	PIM	PP	SH	GW	S	%	+/-	TF	F%	H	SB	Min	GP	G	A	Pts	PIM	PP	SH	GW
99-2000	**Washington**	**NHL**	**69**	**19**	**16**	**35**	**54**	**1**	**0**	**2**	**179**	**10.6**	**6**	**11**	**100.0**	**155**	**27**	**15:34**	**5**	**0**	**0**	**0**	**5**	**0**	**0**	**0**
2000-01	**Washington**	**NHL**	**62**	**16**	**19**	**35**	**61**	**4**	**0**	**3**	**155**	**10.3**	**–2**	**1**	**0.0**	**112**	**23**	**15:32**	....	....	....	....	....	....	....	....
	Montreal	**NHL**	**12**	**3**	**6**	**9**	**10**	**1**	**0**	**0**	**23**	**13.0**	**–2**	**0**	**0.0**	**13**	**7**	**18:29**	....	....	....	....	....	....	....	....
2001-02	**Montreal**	**NHL**	**82**	**22**	**22**	**44**	**59**	**4**	**0**	**3**	**249**	**8.8**	**–3**	**10**	**30.0**	**81**	**33**	**17:39**	**4**	**4**	**4**	**8**	**6**	**2**	**0**	**0**
	NHL Totals		**351**	**88**	**81**	**169**	**266**	**14**	**0**	**12**	**890**	**9.9**		**14**	**28.6**	**442**	**99**	**16:14**	**26**	**11**	**7**	**18**	**27**	**4**	**0**	**0**

WHL West Second All-Star Team (1996)

Traded to **Montreal** by **Washington** with Jan Bulis and Washington's 1st round choice (Alexander Perezhogin) in 2001 Entry Draft for Trevor Linden, Dainius Zubrus and New Jersey's 2nd round choice (previously acquired, later traded to Tampa Bay - Tampa Bay selected Andreas Holmqvist) in 2001 Entry Draft, March 13, 2001.

ZELEPUKIN, Valeri

(zeh-leh-POO-kin, VAL-uhr-ee)

Left wing. Shoots left. 6'1", 200 lbs. Born, Voskresensk, USSR, September 17, 1968. New Jersey's 13th choice, 221st overall, in 1990 Entry Draft.

Season	Club	League	GP	G	A	Pts	PIM	PP	SH	GW	S	%	+/-	TF	F%	H	SB	Min	GP	G	A	Pts	PIM	PP	SH	GW
1984-85	Voskresensk	USSR	5	0	0	0	2	....	....	....	....	....	....	....	....	....	....	....	....	....	....	....	....	....	....	....
1985-86	Voskresensk	USSR	33	2	2	4	10	....	....	....	....	....	....	....	....	....	....	....	....	....	....	....	....	....	....	....
1986-87	Voskresensk	USSR	19	1	0	1	4	....	....	....	....	....	....	....	....	....	....	....	....	....	....	....	....	....	....	....
1987-88	CSKA Moscow	USSR	19	3	1	4	8	....	....	....	....	....	....	....	....	....	....	....	....	....	....	....	....	....	....	....
1988-89	CSKA Moscow	USSR	17	2	3	5	2	....	....	....	....	....	....	....	....	....	....	....	....	....	....	....	....	....	....	....
1989-90	Voskresensk	USSR	46	17	14	31	26	....	....	....	....	....	....	....	....	....	....	....	....	....	....	....	....	....	....	....
1990-91	Voskresensk	USSR	34	11	6	17	38	....	....	....	....	....	....	....	....	....	....	....	....	....	....	....	....	....	....	....
1991-92	**New Jersey**	**NHL**	**44**	**13**	**18**	**31**	**28**	**3**	**0**	**3**	**94**	**13.8**	**11**	....	....	....	....	....	**4**	**1**	**1**	**2**	**2**	**0**	**0**	**0**
	Utica Devils	AHL	22	20	9	29	8	....	....	....	....	....	....	....	....	....	....	....	....	....	....	....	....	....	....	....
1992-93	**New Jersey**	**NHL**	**78**	**23**	**41**	**64**	**70**	**5**	**1**	**2**	**174**	**13.2**	**19**	....	....	....	....	....	**5**	**0**	**2**	**2**	**0**	**0**	**0**	**0**
1993-94	**New Jersey**	**NHL**	**82**	**26**	**31**	**57**	**70**	**8**	**0**	**0**	**155**	**16.8**	**36**	....	....	....	....	....	**20**	**5**	**2**	**7**	**14**	**1**	**0**	**0**
1994-95♦	**New Jersey**	**NHL**	**4**	**1**	**2**	**3**	**6**	**0**	**0**	**0**	**6**	**16.7**	**3**	....	....	....	....	....	**18**	**1**	**2**	**3**	**12**	**0**	**0**	**1**
1995-96	**New Jersey**	**NHL**	**61**	**6**	**9**	**15**	**107**	**3**	**0**	**1**	**86**	**7.0**	**–10**	....	....	....	....	....	....	....	....	....	....	....	....	....
1996-97	**New Jersey**	**NHL**	**71**	**14**	**24**	**38**	**36**	**3**	**0**	**2**	**111**	**12.6**	**–10**	....	....	....	....	....	**8**	**3**	**2**	**5**	**2**	**1**	**0**	**1**
1997-98	**New Jersey**	**NHL**	**35**	**2**	**8**	**10**	**32**	**0**	**0**	**0**	**54**	**3.7**	**0**	....	....	....	....	....	....	....	....	....	....	....	....	....
	Edmonton	**NHL**	**33**	**2**	**10**	**12**	**57**	**0**	**0**	**0**	**47**	**4.3**	**–2**	....	....	....	....	....	**8**	**1**	**2**	**3**	**2**	**0**	**0**	**0**
	Russia	Olympics	6	1	2	3	0	....	....	....	....	....	....	....	....	....	....	....	....	....	....	....	....	....	....	....
1998-99	**Philadelphia**	**NHL**	**74**	**16**	**9**	**25**	**48**	**0**	**0**	**5**	**129**	**12.4**	**0**	**0**	**0.0**	**85**	**22**	**14:33**	**4**	**1**	**0**	**1**	**4**	**0**	**0**	**1**
99-2000	**Philadelphia**	**NHL**	**77**	**11**	**21**	**32**	**55**	**2**	**0**	**3**	**125**	**8.8**	**–3**	**0**	**0.0**	**104**	**33**	**14:37**	**18**	**1**	**2**	**3**	**12**	**1**	**0**	**0**
2000-01	**Chicago**	**NHL**	**36**	**3**	**4**	**7**	**18**	**0**	**2**	**0**	**38**	**7.9**	**–14**	**1**	**0.0**	**60**	**11**	**12:07**	....	....	....	....	....	....	....	....
	Norfolk Admirals	AHL	29	10	9	19	28	....	....	....	....	....	....	....	....	....	....	....	9	5	3	8	6	....	....	....
2001-02	Norfolk Admirals	AHL	27	8	10	18	29	....	....	....	....	....	....	....	....	....	....	....	4	0	1	1	2	....	....	....
	NHL Totals		**595**	**117**	**177**	**294**	**527**	**24**	**3**	**16**	**1019**	**11.5**		**1**	**0.0**	**249**	**66**	**14:07**	**85**	**13**	**13**	**26**	**48**	**3**	**0**	**3**

• Missed majority of 1994-95 season recovering from eye injury suffered in practice, January 24, 1995. Traded to **Edmonton** by **New Jersey** with Bill Guerin for Jason Arnott and Bryan Muir, January 4, 1998. Traded to **Philadelphia** by **Edmonton** for Daniel Lacroix, October 5, 1998. Signed as a free agent by **Chicago**, July 18, 2000. • Missed majority of 2001-02 season recovering from knee injury suffered in training camp, October 4, 2001.

ZETTLER, Rob

(ZEHT-luhr, RAWB)

Defense. Shoots left. 6'3", 200 lbs. Born, Sept-Iles, Que., March 8, 1968. Minnesota's 5th choice, 55th overall, in 1986 Entry Draft.

Season	Club	League	GP	G	A	Pts	PIM	PP	SH	GW	S	%	+/-	TF	F%	H	SB	Min	GP	G	A	Pts	PIM	PP	SH	GW
1983-84	Soo Legion	NOHA	40	9	24	33	28	....	....	....	....	....	....	....	....	....	....	....	....	....	....	....	....	....	....	....
1984-85	Sault Ste. Marie	OHL	60	2	14	16	37	....	....	....	....	....	....	....	....	....	....	....	....	....	....	....	....	....	....	....
1985-86	Sault Ste. Marie	OHL	57	5	23	28	92	....	....	....	....	....	....	....	....	....	....	....	....	....	....	....	....	....	....	....
1986-87	Sault Ste. Marie	OHL	64	13	22	35	89	....	....	....	....	....	....	....	....	....	....	....	4	0	0	0	0	....	....	....
1987-88	Sault Ste. Marie	OHL	64	7	41	48	77	....	....	....	....	....	....	....	....	....	....	....	6	2	2	4	9	....	....	....
	Kalamazoo Wings	IHL	2	0	1	1	0	....	....	....	....	....	....	....	....	....	....	....	7	0	2	2	2	....	....	....
1988-89	**Minnesota**	**NHL**	**2**	**0**	**0**	**0**	**0**	**0**	**0**	**0**	**0**	**0.0**	**–1**	....	....	....	....	....	....	....	....	....	....	....	....	....
	Kalamazoo Wings	IHL	80	5	21	26	79	....	....	....	....	....	....	....	....	....	....	....	6	0	1	1	26	....	....	....
1989-90	**Minnesota**	**NHL**	**31**	**0**	**8**	**8**	**45**	**0**	**0**	**0**	**21**	**0.0**	**–7**	....	....	....	....	....	....	....	....	....	....	....	....	....
	Kalamazoo Wings	IHL	41	6	10	16	64	....	....	....	....	....	....	....	....	....	....	....	7	0	0	0	6	....	....	....
1990-91	**Minnesota**	**NHL**	**47**	**1**	**4**	**5**	**119**	**0**	**0**	**0**	**30**	**3.3**	**–10**	....	....	....	....	....	....	....	....	....	....	....	....	....
	Kalamazoo Wings	IHL	1	0	0	0	2	....	....	....	....	....	....	....	....	....	....	....	....	....	....	....	....	....	....	....
1991-92	**San Jose**	**NHL**	**74**	**1**	**8**	**9**	**99**	**0**	**0**	**0**	**72**	**1.4**	**–23**	....	....	....	....	....	....	....	....	....	....	....	....	....
1992-93	**San Jose**	**NHL**	**80**	**0**	**7**	**7**	**150**	**0**	**0**	**0**	**60**	**0.0**	**–50**	....	....	....	....	....	....	....	....	....	....	....	....	....
1993-94	**San Jose**	**NHL**	**42**	**0**	**3**	**3**	**65**	**0**	**0**	**0**	**28**	**0.0**	**–7**	....	....	....	....	....	....	....	....	....	....	....	....	....
	Philadelphia	**NHL**	**33**	**0**	**4**	**4**	**69**	**0**	**0**	**0**	**27**	**0.0**	**–19**	....	....	....	....	....	....	....	....	....	....	....	....	....
1994-95	**Philadelphia**	**NHL**	**32**	**0**	**1**	**1**	**34**	**0**	**0**	**0**	**17**	**0.0**	**–3**	....	....	....	....	....	**1**	**0**	**0**	**0**	**2**	**0**	**0**	**0**
1995-96	**Toronto**	**NHL**	**29**	**0**	**1**	**1**	**48**	**0**	**0**	**0**	**11**	**0.0**	**–1**	....	....	....	....	....	**2**	**0**	**0**	**0**	**0**	**0**	**0**	**0**
1996-97	**Toronto**	**NHL**	**48**	**2**	**12**	**14**	**51**	**0**	**0**	**0**	**31**	**6.5**	**8**	....	....	....	....	....	....	....	....	....	....	....	....	....
	Utah Grizzlies	IHL	30	0	10	10	60	....	....	....	....	....	....	....	....	....	....	....	....	....	....	....	....	....	....	....
1997-98	**Toronto**	**NHL**	**59**	**0**	**7**	**7**	**108**	**0**	**0**	**0**	**28**	**0.0**	**–8**	....	....	....	....	....	....	....	....	....	....	....	....	....
1998-99	**Nashville**	**NHL**	**2**	**0**	**0**	**0**	**2**	**0**	**0**	**0**	**0**	**0.0**	**–2**	**0**	**0.0**	**5**	**1**	**16:16**	....	....	....	....	....	....	....	....
	Utah Grizzlies	IHL	77	2	16	18	136	....	....	....	....	....	....	....	....	....	....	....	....	....	....	....	....	....	....	....
99-2000	**Washington**	**NHL**	**12**	**0**	**2**	**2**	**19**	**0**	**0**	**0**	**15**	**0.0**	**–1**	**0**	**0.0**	**15**	**18**	**15:22**	**5**	**0**	**0**	**0**	**2**	**0**	**0**	**0**
	Portland Pirates	AHL	23	2	2	4	27	....	....	....	....	....	....	....	....	....	....	....	....	....	....	....	....	....	....	....
2000-01	**Washington**	**NHL**	**29**	**0**	**4**	**4**	**55**	**0**	**0**	**0**	**17**	**0.0**	**0**	**0**	**0.0**	**44**	**19**	**12:08**	**6**	**0**	**0**	**0**	**0**	**0**	**0**	**0**
	Portland Pirates	AHL	36	1	9	10	84	....	....	....	....	....	....	....	....	....	....	....	....	....	....	....	....	....	....	....
2001-02	**Washington**	**NHL**	**49**	**1**	**4**	**5**	**56**	**0**	**0**	**0**	**28**	**3.6**	**3**	**0**	**0.0**	**46**	**38**	**14:25**	....	....	....	....	....	....	....	....
	NHL Totals		**569**	**5**	**65**	**70**	**920**	**0**	**0**	**0**	**385**	**1.3**		**0**	**0.0**	**110**	**76**	**13:51**	**14**	**0**	**0**	**0**	**4**	**0**	**0**	**0**

Claimed by **San Jose** from **Minnesota** in Dispersal Draft, May 30, 1991. Traded to **Philadelphia** by **San Jose** for Viacheslav Butsayev, February 1, 1994. Traded to **Toronto** by **Philadelphia** for Toronto's 5th round choice (Per-Ragna Bergqvist) in 1996 Entry Draft, July 8, 1995. Claimed by **Nashville** from **Toronto** in Expansion Draft, June 26, 1998. Signed as a free agent by **Washington**, September 7, 1999. • Missed majority of 1999-2000 season recovering from head injury suffered in game vs. Albany (AHL), October 21, 1999.

ZHAMNOV, Alexei

(ZHAHM-nahf, al-EHX-ay) **CHI.**

Center. Shoots left. 6'1", 200 lbs. Born, Moscow, USSR, October 1, 1970. Winnipeg's 5th choice, 77th overall, in 1990 Entry Draft.

Season	Club	League	GP	G	A	Pts	PIM	PP	SH	GW	S	%	+/-	TF	F%	H	SB	Min	GP	G	A	Pts	PIM	PP	SH	GW
1988-89	Dynamo Moscow	USSR	4	0	0	0	0	....	....	....	....	....	....	....	....	....	....	....	....	....	....	....	....	....	....	....
1989-90	Dynamo Moscow	USSR	43	11	6	17	21	....	....	....	....	....	....	....	....	....	....	....	....	....	....	....	....	....	....	....
1990-91	Dynamo Moscow	USSR	46	16	12	28	24	....	....	....	....	....	....	....	....	....	....	....	....	....	....	....	....	....	....	....
1991-92	Dynamo Moscow	CIS	39	15	21	36	28	....	....	....	....	....	....	....	....	....	....	....	....	....	....	....	....	....	....	....
	Russia	Olympics	8	0	3	3	8	....	....	....	....	....	....	....	....	....	....	....	....	....	....	....	....	....	....	....
1992-93	**Winnipeg**	**NHL**	**68**	**25**	**47**	**72**	**58**	**6**	**1**	**4**	**163**	**15.3**	**7**	....	....	....	....	....	**6**	**0**	**2**	**2**	**2**	**0**	**0**	**0**
1993-94	**Winnipeg**	**NHL**	**61**	**26**	**45**	**71**	**62**	**7**	**0**	**1**	**196**	**13.3**	**–20**	....	....	....	....	....	....	....	....	....	....	....	....	....
1994-95	**Winnipeg**	**NHL**	**48**	**30**	**35**	**65**	**20**	**9**	**0**	**4**	**155**	**19.4**	**5**	....	....	....	....	....	....	....	....	....	....	....	....	....
1995-96	**Winnipeg**	**NHL**	**58**	**22**	**37**	**59**	**65**	**5**	**0**	**2**	**199**	**11.1**	**–4**	....	....	....	....	....	**6**	**2**	**1**	**3**	**8**	**0**	**0**	**0**
1996-97	**Chicago**	**NHL**	**74**	**20**	**42**	**62**	**56**	**6**	**1**	**2**	**208**	**9.6**	**18**	....	....	....	....	....	....	....	....	....	....	....	....	....
1997-98	**Chicago**	**NHL**	**70**	**21**	**28**	**49**	**61**	**6**	**2**	**3**	**193**	**10.9**	**16**	....	....	....	....	....	....	....	....	....	....	....	....	....
	Russia	Olympics	6	2	1	3	2	....	....	....	....	....	....	....	....	....	....	....	....	....	....	....	....	....	....	....
1998-99	**Chicago**	**NHL**	**76**	**20**	**41**	**61**	**50**	**8**	**1**	**2**	**200**	**10.0**	**–10**	**1299**	**48.9**	**40**	**38**	**21:30**	....	....	....	....	....	....	....	....
99-2000	**Chicago**	**NHL**	**71**	**23**	**37**	**60**	**61**	**5**	**0**	**7**	**175**	**13.1**	**7**	**1171**	**45.8**	**28**	**43**	**22:08**	....	....	....	....	....	....	....	....
2000-01	**Chicago**	**NHL**	**63**	**13**	**36**	**49**	**40**	**3**	**1**	**3**	**117**	**11.1**	**–12**	**1486**	**48.1**	**32**	**35**	**21:15**	....	....	....	....	....	....	....	....
2001-02	**Chicago**	**NHL**	**77**	**22**	**45**	**67**	**67**	**6**	**0**	**3**	**173**	**12.7**	**8**	**1634**	**50.0**	**35**	**41**	**22:19**	**5**	**0**	**0**	**0**	**0**	**0**	**0**	**0**
	Russia	Olympics	6	1	0	1	4	....	....	....	....	....	....	....	....	....	....	....	....	....	....	....	....	....	....	....
	NHL Totals		**666**	**222**	**393**	**615**	**540**	**61**	**6**	**31**	**1779**	**12.5**		**5590**	**48.3**	**135**	**157**	**21:49**	**17**	**2**	**3**	**5**	**10**	**0**	**0**	**0**

NHL Second All-Star Team (1995) • Played in NHL All-Star Game (2002)

Traded to **Chicago** by **Phoenix** with Craig Mills and Phoenix's 1st round choice (Ty Jones) in 1997 Entry Draft for Jeremy Roenick, August 16, 1996.

ZHITNIK, Alexei

(ZHIHT-nihk, al-EHX-ay) **BUF.**

Defense. Shoots left. 5'11", 215 lbs. Born, Kiev, USSR, October 10, 1972. Los Angeles' 3rd choice, 81st overall, in 1991 Entry Draft.

Season	Club	League	GP	G	A	Pts	PIM	PP	SH	GW	S	%	+/-	TF	F%	H	SB	Min	GP	G	A	Pts	PIM	PP	SH	GW
1989-90	Sokol Kiev	USSR	31	3	4	7	16	....	....	....	....	....	....	....	....	....	....	....	....	....	....	....	....	....	....	....
1990-91	Sokol Kiev	USSR	46	1	4	5	46	....	....	....	....	....	....	....	....	....	....	....	....	....	....	....	....	....	....	....
1991-92	CSKA Moscow	CIS	44	2	7	9	52	....	....	....	....	....	....	....	....	....	....	....	....	....	....	....	....	....	....	....
	Russia	Olympics	8	1	0	1	0	....	....	....	....	....	....	....	....	....	....	....	....	....	....	....	....	....	....	....
1992-93	**Los Angeles**	**NHL**	**78**	**12**	**36**	**48**	**80**	**5**	**0**	**2**	**136**	**8.8**	**–3**	....	....	....	....	....	**24**	**3**	**9**	**12**	**26**	**2**	**0**	**1**
1993-94	**Los Angeles**	**NHL**	**81**	**12**	**40**	**52**	**101**	**11**	**0**	**1**	**227**	**5.3**	**–11**	....	....	....	....	....	....	....	....	....	....	....	....	....
1994-95	**Los Angeles**	**NHL**	**11**	**2**	**5**	**7**	**27**	**2**	**0**	**0**	**33**	**6.1**	**–3**	....	....	....	....	....	....	....	....	....	....	....	....	....
	Buffalo	**NHL**	**21**	**2**	**5**	**7**	**34**	**1**	**0**	**0**	**33**	**6.1**	**–3**	....	....	....	....	....	**5**	**0**	**1**	**1**	**14**	**0**	**0**	**0**

			Regular Season																Playoffs							
Season	Club	League	GP	G	A	Pts	PIM	PP	SH	GW	S	%	+/-	TF	F%	H	SB	Min	GP	G	A	Pts	PIM	PP	SH	GW
1995-96	Buffalo	NHL	80	6	30	36	58	5	0	0	193	3.1	-25	....	....	....	....	....	....	....	....	....	....	....	....	....
1996-97	Buffalo	NHL	80	7	28	35	95	3	1	0	170	4.1	10	....	....	....	....	....	12	1	0	1	16	0	0	0
1997-98	Buffalo	NHL	78	15	30	45	102	2	3	3	191	7.9	19	....	....	....	....	....	15	0	3	3	36	0	0	0
	Russia	Olympics	6	0	2	2	2	....	....	....	....	....	....	....	....	....	....	....	....	....	....	....	....	....	....	....
1998-99	Buffalo	NHL	81	7	26	33	96	3	1	2	185	3.8	-6	0	0.0	122	109	25:39	21	4	11	15	*52	4	0	2
99-2000	Buffalo	NHL	74	2	11	13	95	1	0	0	139	1.4	-6	0	0.0	117	88	24:48	4	0	0	0	8	0	0	0
2000-01	Buffalo	NHL	78	8	29	37	75	5	0	1	149	5.4	-3	0	0.0	175	70	24:15	13	1	6	7	12	0	0	0
2001-02	Buffalo	NHL	82	1	33	34	80	1	0	0	150	0.7	-1	0	0.0	185	96	25:36	....	....	....	....	....	....	....	....
	NHL Totals		744	74	273	347	843	39	5	9	1606	4.6		0	0.0	599	363	25:06	94	9	30	39	164	6	0	3

Played in NHL All-Star Game (1999, 2002)

Traded to **Buffalo** by **LA Kings** with Robb Stauber, Charlie Huddy and LA Kings' 5th round choice (Marian Menhart) in 1995 Entry Draft for Philippe Boucher, Denis Tsygurov and Grant Fuhr, February 14, 1995.

ZHOLTOK, Sergei

(ZHOL-tok, SAIR-gay) **MIN.**

Center. Shoots right. 6'2", 191 lbs. Born, Riga, Latvia, December 2, 1972. Boston's 2nd choice, 55th overall, in 1992 Entry Draft.

Season	Club	League	GP	G	A	Pts	PIM	PP	SH	GW	S	%	+/-	TF	F%	H	SB	Min	GP	G	A	Pts	PIM	PP	SH	GW
1990-91	Dynamo Riga	USSR	39	4	0	4	16	....	....	....	....	....	....	....	....	....	....	....	....	....	....	....	....	....	....	....
1991-92	Riga Stars	CIS	27	6	3	9	6	....	....	....	....	....	....	....	....	....	....	....	....	....	....	....	....	....	....	....
1992-93	Boston	NHL	1	0	1	1	0	0	0	0	2	0.0	1	....	....	....	....	....	....	....	....	....	....	....	....	....
	Providence	AHL	64	31	35	66	57	....	....	....	....	....	....	....	....	....	....	....	6	3	5	8	4	....	....	....
1993-94	Boston	NHL	24	2	1	3	2	1	0	0	25	8.0	-7	....	....	....	....	....	....	....	....	....	....	....	....	....
	Providence	AHL	54	29	33	62	16	....	....	....	....	....	....	....	....	....	....	....	....	....	....	....	....	....	....	....
1994-95	Providence	AHL	78	23	35	58	42	....	....	....	....	....	....	....	....	....	....	....	13	8	5	13	6	....	....	....
1995-96	Las Vegas	IHL	82	51	50	101	30	....	....	....	....	....	....	....	....	....	....	....	15	7	13	20	6	....	....	....
1996-97	Ottawa	NHL	57	12	16	28	19	5	0	0	96	12.5	2	....	....	....	....	....	7	1	1	2	0	1	0	0
	Las Vegas	IHL	19	13	14	27	20	....	....	....	....	....	....	....	....	....	....	....	....	....	....	....	....	....	....	....
1997-98	Ottawa	NHL	78	10	13	23	16	7	0	1	127	7.9	-7	....	....	....	....	....	11	0	2	2	0	0	0	0
1998-99	Montreal	NHL	70	7	15	22	6	2	0	3	102	6.9	-12	522	49.0	23	12	11:11	....	....	....	....	....	....	....	....
	Fredericton	AHL	7	3	4	7	0	....	....	....	....	....	....	....	....	....	....	....	....	....	....	....	....	....	....	....
99-2000	Montreal	NHL	68	26	12	38	28	9	0	7	163	16.0	2	914	48.9	12	18	17:17	....	....	....	....	....	....	....	....
	Quebec	AHL	1	0	1	1	2	....	....	....	....	....	....	....	....	....	....	....	....	....	....	....	....	....	....	....
2000-01	Montreal	NHL	32	1	10	11	8	0	0	0	78	1.3	-15	318	52.2	6	8	15:38	....	....	....	....	....	....	....	....
	Edmonton	NHL	37	4	16	20	22	1	0	0	61	6.6	8	109	60.6	20	7	12:55	3	0	0	0	0	0	0	0
2001-02	Minnesota	NHL	73	19	20	39	28	10	0	2	146	13.0	-10	823	45.6	14	18	16:04	....	....	....	....	....	....	....	....
	NHL Totals		440	81	104	185	129	35	0	13	800	10.1		2686	48.8	75	63	14:41	21	1	3	4	0	1	0	0

Signed as a free agent by **Las Vegas** (IHL), August 5, 1995. Signed as a free agent by **Ottawa**, July 10, 1996. Signed as a free agent by **Montreal**, September 9, 1998. Traded to **Edmonton** by **Montreal** for Chad Kilger, December 18, 2000. Traded to **Minnesota** by **Edmonton** for Minnesota's 7th round choice (J.F. Dufort) in 2002 Entry Draft, June 29, 2001.

ZUBOV, Sergei

(ZOO-bahf, SAIR-gay) **DAL.**

Defense. Shoots right. 6'1", 200 lbs. Born, Moscow, USSR, July 22, 1970. NY Rangers' 6th choice, 85th overall, in 1990 Entry Draft.

Season	Club	League	GP	G	A	Pts	PIM	PP	SH	GW	S	%	+/-	TF	F%	H	SB	Min	GP	G	A	Pts	PIM	PP	SH	GW
1988-89	CSKA Moscow	USSR	29	1	4	5	10	....	....	....	....	....	....	....	....	....	....	....	....	....	....	....	....	....	....	....
1989-90	CSKA Moscow	USSR	48	6	2	8	16	....	....	....	....	....	....	....	....	....	....	....	....	....	....	....	....	....	....	....
1990-91	CSKA Moscow	USSR	41	6	5	11	12	....	....	....	....	....	....	....	....	....	....	....	....	....	....	....	....	....	....	....
1991-92	CSKA Moscow	CIS	44	4	7	11	8	....	....	....	....	....	....	....	....	....	....	....	....	....	....	....	....	....	....	....
	Russia	Olympics	8	0	1	1	0	....	....	....	....	....	....	....	....	....	....	....	....	....	....	....	....	....	....	....
1992-93	CSKA Moscow	CIS	1	0	1	1	0	....	....	....	....	....	....	....	....	....	....	....	....	....	....	....	....	....	....	....
	NY Rangers	NHL	49	8	23	31	4	3	0	0	93	8.6	-1	....	....	....	....	....	....	....	....	....	....	....	....	....
	Binghamton	AHL	30	7	29	36	14	....	....	....	....	....	....	....	....	....	....	....	11	5	5	10	2	....	....	....
1993-94♦	NY Rangers	NHL	78	12	77	89	39	9	0	1	222	5.4	20	....	....	....	....	....	22	5	14	19	0	2	0	0
	Binghamton	AHL	2	1	2	3	0	....	....	....	....	....	....	....	....	....	....	....	....	....	....	....	....	....	....	....
1994-95	NY Rangers	NHL	38	10	26	36	18	6	0	0	116	8.6	-2	....	....	....	....	....	10	3	8	11	2	1	0	0
1995-96	Pittsburgh	NHL	64	11	55	66	22	3	2	1	141	7.8	28	....	....	....	....	....	18	1	14	15	26	1	0	0
1996-97	Dallas	NHL	78	13	30	43	24	1	0	3	133	9.8	19	....	....	....	....	....	7	0	3	3	2	0	0	0
1997-98	Dallas	NHL	73	10	47	57	16	5	1	2	148	6.8	16	....	....	....	....	....	17	4	5	9	2	3	0	1
1998-99♦	Dallas	NHL	81	10	41	51	20	5	0	3	155	6.5	9	0	0.0	35	41	24:14	23	1	12	13	4	0	0	0
99-2000	Dallas	NHL	77	9	33	42	18	3	1	3	179	5.0	-2	0	0.0	52	58	28:50	18	2	7	9	6	1	1	0
2000-01	Dallas	NHL	79	10	41	51	24	6	0	1	173	5.8	22	0	0.0	49	78	26:37	10	1	5	6	4	0	0	0
2001-02	Dallas	NHL	80	12	32	44	22	8	0	2	198	6.1	-4	0	0.0	45	75	26:46	....	....	....	....	....	....	....	....
	NHL Totals		697	105	405	510	207	49	4	16	1558	6.7		0	0.0	181	252	26:35	125	17	68	85	46	8	1	1

Played in NHL All-Star Game (1998, 1999, 2000)

Traded to **Pittsburgh** by **NY Rangers** with Petr Nedved for Luc Robitaille and Ulf Samuelsson, August 31, 1995. Traded to **Dallas** by **Pittsburgh** for Kevin Hatcher, June 22, 1996.

ZUBRUS, Dainius

(ZOO-bruhs, DAYN-ihs) **WSH.**

Right wing. Shoots left. 6'4", 231 lbs. Born, Elektrenai, USSR, June 16, 1978. Philadelphia's 1st choice, 15th overall, in 1996 Entry Draft.

Season	Club	League	GP	G	A	Pts	PIM	PP	SH	GW	S	%	+/-	TF	F%	H	SB	Min	GP	G	A	Pts	PIM	PP	SH	GW
1995-96	Pembroke	OCJHL	28	19	13	32	73	....	....	....	....	....	....	....	....	....	....	....	....	....	....	....	....	....	....	....
	Caledon	MTJHL	7	3	7	10	2	....	....	....	....	....	....	....	....	....	....	....	17	11	12	23	4	....	....	....
1996-97	Philadelphia	NHL	68	8	13	21	22	1	0	2	71	11.3	3	....	....	....	....	....	19	5	4	9	12	1	0	1
1997-98	Philadelphia	NHL	69	8	25	33	42	1	0	5	101	7.9	29	....	....	....	....	....	5	0	1	1	2	0	0	0
1998-99	Philadelphia	NHL	63	3	5	8	25	0	1	0	49	6.1	-5	29	51.7	54	30	11:00	....	....	....	....	....	....	....	....
	Montreal	NHL	17	3	5	8	4	0	0	1	31	9.7	-3	2	50.0	15	6	16:53	....	....	....	....	....	....	....	....
99-2000	Montreal	NHL	73	14	28	42	54	3	0	1	139	10.1	-1	212	39.2	99	27	17:37	....	....	....	....	....	....	....	....
2000-01	Montreal	NHL	49	12	12	24	30	3	0	0	70	17.1	-7	190	41.1	58	34	18:30	....	....	....	....	....	....	....	....
	Washington	NHL	12	1	1	2	7	1	0	0	13	7.7	-4	0	0.0	15	3	13:05	6	0	0	0	2	0	0	0
2001-02	Washington	NHL	71	17	26	43	38	4	0	3	138	12.3	5	131	37.4	98	43	18:52	....	....	....	....	....	....	....	....
	NHL Totals		422	66	115	181	222	13	1	12	612	10.8		564	40.1	339	143	16:23	30	5	5	10	16	1	0	1

Traded to **Montreal** by **Philadelphia** with Philadelphia's 2nd round choice (Matt Carkner) in 1999 Entry Draft and NY Islanders' 6th round choice (previously acquired, Montreal selected Scott Selig) in 2000 Entry Draft for Mark Recchi, March 10, 1999. Traded to **Washington** by **Montreal** with Trevor Linden and New Jersey's 2nd round choice (previously acquired, later traded to Tampa Bay - Tampa Bay selected Andreas Holmqvist) in 2001 Entry Draft for Richard Zednik, Jan Bulis and Washington's 1st round choice (Alexander Perezhogin) in 2001 Entry Draft, March 13, 2001.

ZYUZIN, Andrei

(ZYOO-zin, AWN-dray) **N.J.**

Defense. Shoots right. 6'1", 215 lbs. Born, Ufa, USSR, January 21, 1978. San Jose's 1st choice, 2nd overall, in 1996 Entry Draft.

Season	Club	League	GP	G	A	Pts	PIM	PP	SH	GW	S	%	+/-	TF	F%	H	SB	Min	GP	G	A	Pts	PIM	PP	SH	GW
1994-95	Ufa	CIS	30	3	0	3	16	....	....	....	....	....	....	....	....	....	....	....	....	....	....	....	....	....	....	....
1995-96	Ufa	CIS	41	6	3	9	24	....	....	....	....	....	....	....	....	....	....	....	....	....	....	....	....	....	....	....
1996-97	Ufa	Russia	32	7	10	17	28	....	....	....	....	....	....	....	....	....	....	....	7	1	1	2	4	....	....	....
1997-98	San Jose	NHL	56	6	7	13	66	2	0	2	72	8.3	8	....	....	....	....	....	6	1	0	1	14	0	0	1
	Kentucky	AHL	17	4	5	9	28	....	....	....	....	....	....	....	....	....	....	....	....	....	....	....	....	....	....	....
1998-99	San Jose	NHL	25	3	1	4	38	2	0	0	44	6.8	5	0	0.0	25	11	15:56	....	....	....	....	....	....	....	....
	Kentucky	AHL	23	2	12	14	42	....	....	....	....	....	....	....	....	....	....	....	....	....	....	....	....	....	....	....
99-2000	Tampa Bay	NHL	34	2	9	11	33	0	0	0	47	4.3	-11	0	0.0	33	27	20:28	....	....	....	....	....	....	....	....
2000-01	Tampa Bay	NHL	64	4	16	20	76	2	1	1	92	4.3	-8	0	0.0	42	66	18:39	....	....	....	....	....	....	....	....
	Detroit Vipers	IHL	2	0	1	1	0	....	....	....	....	....	....	....	....	....	....	....	....	....	....	....	....	....	....	....
2001-02	Tampa Bay	NHL	9	0	2	2	6	0	0	0	14	0.0	-6	0	0.0	7	8	19:41	....	....	....	....	....	....	....	....
	New Jersey	NHL	38	1	2	3	25	1	0	0	47	2.1	1	0	0.0	35	37	15:04	....	....	....	....	....	....	....	....
	Albany	AHL	3	0	1	1	2	....	....	....	....	....	....	....	....	....	....	....	....	....	....	....	....	....	....	....
	NHL Totals		226	16	37	53	244	7	1	3	316	5.1		0	0.0	142	149	17:52	6	1	0	1	14	0	0	1

• Suspended for remainder of 1998-99 season by **San Jose** for leaving team without permission, April 1, 1999. Traded to **Tampa Bay** by **San Jose** with Bill Houlder, Shawn Burr and Steve Guolla for Niklas Sundstrom and NY Rangers' 3rd round choice (previously acquired, later traded to Chicago - Chicago selected Igor Radulov) in 2000 Entry Draft, August 4, 1999. • Missed majority of 1999-2000 season recovering from shoulder injury suffered in game vs. NY Islanders, January 13, 2000. Traded to **New Jersey** by **Tampa Bay** for Josef Boumedienne, Sascha Goc and the rights to Anton But, November 9, 2001.

NHL Goaltenders

David Aebischer · Jean-Sebastien Aubin · Alexander Auld · Tom Barrasso · Ed Belfour · Zac Bierk · Craig Billington · Martin Biron · Dan Blackburn · Brian Boucher

Fred Brathwaite · Martin Brochu · Martin Brodeur · Ilja Bryzgalov · Sean Burke · Roman Cechmanek · Sebastien Centomo · Sebastien Charpentier · Scott Clemmensen · Dan Cloutier

Ty Conklin · Byron Dafoe · Jean-Francois Damphousse · Marc Denis · Patrick DesRochers · Rick DiPietro · Reinhard Divis · Mike Dunham · Robert Esche · Manny Fernandez

Stephane Fiset · Wade Flaherty · Mike Fountain · Mathieu Garon · Jean-Sebastien Giguere · John Grahame · Jeff Hackett · Johan Hedberg · Corey Hirsch · Milan Hnilicka

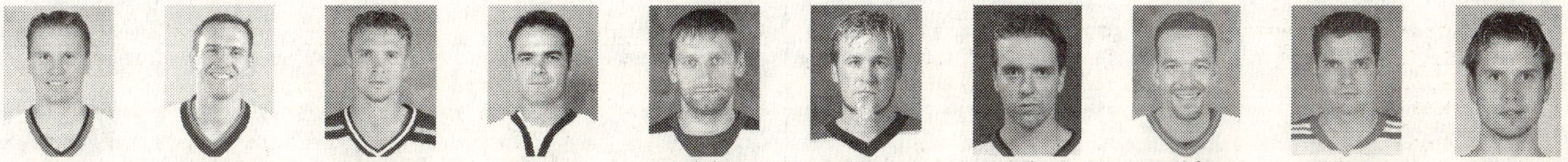

Jani Hurme · Arturs Irbe · Brent Johnson · Curtis Joseph · Nikolai Khabibulin · Trevor Kidd · Miika Kiprusoff · Olaf Kolzig · Jean-Francois Labbe · Simon Lajeunesse

Patrick Lalime · Marc Lamothe · Scott Langkow · Manny Legace · Michael Leighton · Neil Little · Roberto Luongo · Norm Maracle · Jussi Markkanen · Jamie McLennan

Alfie Michaud · Olivier Michaud · Tyler Moss · Evgeni Nabokov · Mika Noronen · Pasi Nurminen · Chris Osgood · Maxime Ouellet · Steve Passmore · Felix Potvin

Andrew Raycroft · Damian Rhodes · Mike Richter · Dwayne Roloson · Patrick Roy · Tommy Salo · Philippe Sauve · Corey Schwab · Steve Shields · Peter Skudra

Garth Snow · Jamie Storr · Jose Theodore · Jocelyn Thibault · Ron Tugnutt · Marty Turco · Roman Turek · Tomas Vokoun · Kevin Weekes · Kay Whitmore

2002-03 Goaltender Register

Note: The 2002-03 Goaltender Register lists every goaltender who appeared in an NHL game in the 2001-02 season, every goaltender drafted in the first five rounds of the 2002 Entry Draft, goaltenders on NHL Reserve Lists and other goaltenders.

Trades and roster changes are current as of August 23, 2002.

To calculate a goaltender's goals-against per game average (**Avg**), divide goals against (**GA**) by minutes played (**Mins**) and multiply this result by **60**.

Abbreviations: Lea – league; **GP** – games played; **W** – wins; **L** – losses; **T** – ties; **GA** – goals against; **SO** – shutouts; **Avg** – goals-against per game average.
♦ – member of Stanley Cup-winning team.

NHL Player Register begins on page 338.
Prospect Register begins on page 269.
League Abbreviations are listed on page 268.

AEBISCHER, David (A-bih-shuhr, DAY-vihd) COL.

Goaltender. Catches left. 6'1", 190 lbs. Born, Fribourg, Switz., February 7, 1978.
(Colorado's 7th choice, 161st overall, in 1997 Entry Draft).

			Regular Season								Playoffs						
Season	**Club**	**Lea**	**GP**	**W**	**L**	**T**	**Mins**	**GA**	**SO**	**Avg**	**GP**	**W**	**L**	**Mins**	**GA**	**SO**	**Avg**
1996-97	Fribourg	Swiss	10	....	....	....	577	34	0	3.54	3	1	2	184	13	0	4.24
1997-98	Chesapeake	ECHL	17	5	7	2	930	52	0	3.35	....	....	....	....	....	....	....
	Wheeling Nailers	ECHL	10	5	3	1	564	30	1	3.19	....	....	....	....	....	....	....
	Hershey Bears	AHL	2	0	0	1	79	5	0	3.76	....	....	....	....	....	....	....
	Fribourg	Swiss	1	1	0	0	60	1	0	1.00	4	....	....	240	17	....	4.25
1998-99	Hershey Bears	AHL	38	17	10	5	1932	79	2	2.45	3	1	2	152	6	0	2.37
99-2000	Hershey Bears	AHL	58	29	23	2	3259	180	1	3.31	14	7	6	788	40	2	3.05
2000-01♦	**Colorado**	**NHL**	**26**	**12**	**7**	**3**	**1393**	**52**	**3**	**2.24**	**1**	**0**	**0**	**1**	**0**	**0**	**0.00**
2001-02	**Colorado**	**NHL**	**21**	**13**	**6**	**0**	**1184**	**37**	**2**	**1.88**	**1**	**0**	**0**	**34**	**1**	**0**	**1.76**
	Switzerland	Olympics	2	1	0	0	81	6	0	4.43	....	....	....	....	....	....	....
	NHL Totals		**47**	**25**	**13**	**3**	**2577**	**89**	**5**	**2.07**	**2**	**0**	**0**	**35**	**1**	**0**	**1.71**

AHONEN, Ari (ah-HOH-nuhn, AH-ree) N.J.

Goaltender. Catches left. 6'1", 190 lbs. Born, Jyvaskyla, Finland, February 6, 1981.
(New Jersey's 1st choice, 27th overall, in 1999 Entry Draft).

			Regular Season								Playoffs						
Season	**Club**	**Lea**	**GP**	**W**	**L**	**T**	**Mins**	**GA**	**SO**	**Avg**	**GP**	**W**	**L**	**Mins**	**GA**	**SO**	**Avg**
1997-98	JYP Jr.	Finn-Jr.	31	....	....	....	1853	64	....	2.09	....	....	....	....	....	....	....
1998-99	JYP Jr.	Finn-Jr.	24	....	....	....	1447	70	....	2.90	....	....	....	....	....	....	....
99-2000	HIFK Helsinki	Finland	24	11	7	1	1347	70	1	3.12	2	0	2	119	7	0	3.53
	HIFK Helsinki	EuroHL	5	4	1	0	285	15	1	3.16	....	....	....	....	....	....	....
2000-01	HIFK Helsinki	Finland	37	18	13	4	2101	97	2	2.77	5	2	3	395	9	1	1.37
2001-02	Albany	AHL	36	6	22	6	2106	106	0	3.02	....	....	....	....	....	....	....

ALBAN, Chad (AL-ban, CHAD) DAL.

Goaltender. Catches left. 5'9", 165 lbs. Born, Kalamazoo, MI, April 27, 1976.

			Regular Season								Playoffs						
Season	**Club**	**Lea**	**GP**	**W**	**L**	**T**	**Mins**	**GA**	**SO**	**Avg**	**GP**	**W**	**L**	**Mins**	**GA**	**SO**	**Avg**
1994-95	Michigan State	CCHA	13	8	2	0	636	29	0	2.73	....	....	....	....	....	....	....
1995-96	Michigan State	CCHA	40	26	13	1	2286	117	0	3.07	....	....	....	....	....	....	....
1996-97	Michigan State	CCHA	39	23	11	4	2272	103	3	2.72	....	....	....	....	....	....	....
1997-98	Michigan State	CCHA	40	31	4	5	2438	64	*6	*1.57	....	....	....	....	....	....	....
1998-99	Mobile Mysticks	ECHL	34	16	14	3	1960	111	1	3.40	2	0	2	119	9	0	4.54
	Houston Aeros	IHL	5	1	3	1	284	14	0	2.96	....	....	....	....	....	....	....
99-2000	Mobile Mysticks	ECHL	39	25	13	1	2334	114	0	2.93	5	2	3	299	20	0	4.01
	Utah Grizzlies	IHL	1	0	1	0	35	3	0	5.16	....	....	....	....	....	....	....
2000-01	Idaho Steelheads	WCHL	20	14	5	1	1121	56	1	3.00	10	6	4	596	33	0	3.32
	Grand Rapids	IHL	3	2	1	0	180	4	1	1.33	....	....	....	....	....	....	....
	Utah Grizzlies	IHL	11	2	4	4	597	23	0	2.31	....	....	....	....	....	....	....
2001-02	Utah Grizzlies	AHL	42	17	15	1	2103	103	2	2.94	....	....	....	....	....	....	....

CCHA Rookie of the Year (1995) • CCHA First All-Star Team (1998) • CCHA Player of the Year (1998) • NCAA West First All-American Team (1998) • Scored goal in game vs. Ferris State (CCHA), February 28, 1998.

Signed as a free agent by **Dallas**, August 30, 2000.

ANDERSSON, Andreas (AN-duhr-suhn, AN-dree-as) ANA.

Goaltender. Catches left. 6', 180 lbs. Born, Jonkoping, Sweden, April 4, 1979.
(Anaheim's 8th choice, 245th overall, in 1998 Entry Draft).

			Regular Season								Playoffs						
Season	**Club**	**Lea**	**GP**	**W**	**L**	**T**	**Mins**	**GA**	**SO**	**Avg**	**GP**	**W**	**L**	**Mins**	**GA**	**SO**	**Avg**
1997-98	HV 71 Jr.	Swede-Jr.	10	....	....	....	600	31	....	3.10	....	....	....	....	....	....	....
	HV 71 Jonkoping	Sweden	7	....	....	....	420	20	....	2.86	....	....	....	....	....	....	....
1998-99	Mora IK Jr.	Swede-Jr.	12	....	....	....	720	28	0	1.92	....	....	....	....	....	....	....
	HV 71 Jonkoping	Sweden	12	....	....	....	633	35	0	3.32	....	....	....	....	....	....	....
99-2000	Tranas AIF	Swede-2	5	....	....	....	297	17	0	3.44	....	....	....	....	....	....	....
	HV 71 Jonkoping	Sweden	1	....	....	....	51	5	0	5.88	....	....	....	....	....	....	....
2000-01	IF Troja-Ljungby	Swede-2	19	....	....	....	1076	66	0	3.68	2	1	1	120	5	0	2.50
2001-02	IF Troja-Ljungby	Swede-2	6	....	....	....	360	27	0	4.50	....	....	....	....	....	....	....

ANDERSSON, Craig (AN-duhr-suhn, KRAYG) CHI.

Goaltender. Catches left. 6'2", 174 lbs. Born, Park Ridge, IL, May 21, 1981.
(Chicago's 4th choice, 73rd overall, in 2001 Entry Draft).

			Regular Season								Playoffs						
Season	**Club**	**Lea**	**GP**	**W**	**L**	**T**	**Mins**	**GA**	**SO**	**Avg**	**GP**	**W**	**L**	**Mins**	**GA**	**SO**	**Avg**
1997-98	Chicago Jets	MEHL	50	....	....	....	2991	143	2	2.86	....	....	....	....	....	....	....
1998-99	Chicago Freeze	NAJHL	14	11	3	0	840	40	0	2.56	....	....	....	....	....	....	....
	Guelph Storm	OHL	21	12	5	1	1006	52	1	3.10	3	0	2	114	9	0	4.74
99-2000	Guelph Storm	OHL	38	12	17	2	1955	117	0	3.59	3	0	1	110	5	0	2.73
2000-01	Guelph Storm	OHL	59	30	19	9	3555	156	3	2.63	4	0	4	240	17	0	4.25
2001-02	Norfolk Admirals	AHL	28	9	13	4	1568	77	2	2.95	1	0	1	21	1	0	2.83

• Re-entered NHL Entry Draft. Originally Calgary's 3rd choice, 77th overall, in 1999 Entry Draft.

OHL First All-Star Team (2001)

ANTILA, Kristian (AN-tih-luh, KRIHS-tan) EDM.

Goaltender. Catches left. 6'3", 207 lbs. Born, Vammala, Finland, January 10, 1980.
(Edmonton's 4th choice, 113th overall, in 1998 Entry Draft).

			Regular Season								Playoffs						
Season	**Club**	**Lea**	**GP**	**W**	**L**	**T**	**Mins**	**GA**	**SO**	**Avg**	**GP**	**W**	**L**	**Mins**	**GA**	**SO**	**Avg**
1997-98	Ilves Jr.	Finn-Jr.	11	....	....	....	564	28	0	2.97	....	....	....	....	....	....	....
1998-99	Ilves Tampere	Finn-Jr.	18	....	....	....	1080	48	0	2.63	5	....	....	300	11	....	2.22
	Ilves Tampere	Finland	5	1	2	0	207	12	0	3.48	....	....	....	....	....	....	....
99-2000	Ilves Jr.	Finn-Jr.	6	....	....	....	360	21	0	3.56	....	....	....	....	....	....	....
	Diskos Jyvaskyla	Finland-2	5	....	....	....	300	23	0	4.67	....	....	....	....	....	....	....
	Ilves Tampere	Finland	25	4	11	4	1239	74	1	3.58	1	0	1	20	4	0	12.00
2000-01	Assat Pori	Finland	42	9	27	7	2417	146	1	3.62	....	....	....	....	....	....	....
2001-02	Assat Pori	Finland	35	7	22	3	1951	112	3	3.44	....	....	....	....	....	....	....

ASKEY, Tom (AS-kee, TAWM) BUF.

Goaltender. Catches left. 6'1", 195 lbs. Born, Kenmore, NY, October 4, 1974.
(Anaheim's 8th choice, 186th overall, in 1993 Entry Draft).

			Regular Season								Playoffs						
Season	**Club**	**Lea**	**GP**	**W**	**L**	**T**	**Mins**	**GA**	**SO**	**Avg**	**GP**	**W**	**L**	**Mins**	**GA**	**SO**	**Avg**
1992-93	Ohio State	CCHA	25	2	19	0	1235	125	0	6.07	....	....	....	....	....	....	....
1993-94	Ohio State	CCHA	27	3	19	4	1488	103	0	4.15	....	....	....	....	....	....	....
1994-95	Ohio State	CCHA	26	4	19	2	1387	121	0	5.23	....	....	....	....	....	....	....
1995-96	Ohio State	CCHA	26	8	11	4	1340	68	0	3.05	....	....	....	....	....	....	....
1996-97	Baltimore	AHL	40	17	18	2	2238	140	1	3.75	3	0	3	137	11	0	4.79
1997-98	**Anaheim**	**NHL**	**7**	**0**	**1**	**2**	**273**	**12**	**0**	**2.64**	....	....	....	....	....	....	....
	Cincinnati	AHL	32	10	16	4	1753	104	3	3.56	....	....	....	....	....	....	....
1998-99	Cincinnati	AHL	53	21	22	3	2893	131	3	2.72	3	0	3	178	13	0	4.38
	Anaheim	**NHL**	....	....	....	....	....	....	....	....	**1**	**0**	**1**	**30**	**2**	**0**	**4.00**
99-2000	Kansas City	IHL	13	3	5	3	658	43	0	3.92	....	....	....	....	....	....	....
	Houston Aeros	IHL	13	4	7	1	727	33	0	2.72	....	....	....	....	....	....	....
2000-01	Rochester	AHL	29	15	8	4	1671	71	1	2.55	....	....	....	....	....	....	....
2001-02	Rochester	AHL	34	16	15	3	2048	86	3	2.52	1	0	1	58	4	0	4.11
	NHL Totals		**7**	**0**	**1**	**2**	**273**	**12**	**0**	**2.64**	**1**	**0**	**1**	**30**	**2**	**0**	**4.00**

CCHA Second All-Star Team (1996) • Shared Harry "Hap" Holmes Memorial Trophy (fewest goals against - AHL) with Mika Noronen (2001)

Signed as a free agent by **Rochester** (AHL), September 29, 2000. Signed as a free agent by **Buffalo**, August 10, 2001.

ASPLUND, Johan (AS-pluhnd, YOH-hahn) NYR

Goaltender. Catches left. 6'1", 180 lbs. Born, Slutskar, Sweden, December 15, 1980.
(NY Rangers' 4th choice, 79th overall, in 1999 Entry Draft).

			Regular Season								Playoffs						
Season	**Club**	**Lea**	**GP**	**W**	**L**	**T**	**Mins**	**GA**	**SO**	**Avg**	**GP**	**W**	**L**	**Mins**	**GA**	**SO**	**Avg**
1998-99	Brynas IF Gavle	Sweden	12	....	....	....	646	32	0	2.97	....	....	....	....	....	....	....
99-2000	Mora IK	Swede-2	3	3	0	0	180	6	0	2.00	....	....	....	....	....	....	....
	Brynas IF Gavle	Sweden	10	....	....	....	622	30	0	2.89	....	....	....	....	....	....	....
2000-01	Brynas IF Gavle	Sweden	29	....	....	....	1761	79	2	2.69	3	0	3	177	11	0	3.73
2001-02	Brynas IF Gavle	Sweden	34	....	....	....	2069	102	1	2.96	3	0	3	188	16	0	5.11

AUBIN, Jean-Sebastien (OH-behn, ZHAWN-suh-BAS-tee-yeh) PIT.

Goaltender. Catches right. 5'11", 180 lbs. Born, Montreal, Que., July 19, 1977.
(Pittsburgh's 2nd choice, 76th overall, in 1995 Entry Draft).

			Regular Season								Playoffs						
Season	Club	Lea	GP	W	L	T	Mins	GA	SO	Avg	GP	W	L	Mins	GA	SO	Avg
1993-94	Mtl-Bourassa	QAAA	27	14	13	0	1524	96	1	3.74	4	1	3	222	19	0	5.14
1994-95	Sherbrooke	QMJHL	27	13	10	1	1287	73	1	3.40	3	1	2	185	11	0	3.57
1995-96	Sherbrooke	QMJHL	40	18	14	2	2140	127	0	3.57	4	1	3	238	23	0	5.55
1996-97	Sherbrooke	QMJHL	4	3	1	0	249	8	0	1.93	1	0	1	60	4	0	4.00
	Moncton Wildcats	QMJHL	22	9	12	0	1252	67	1	3.21	....	....	....	....	....	....	....
	Laval Titan	QMJHL	11	2	6	1	532	41	0	4.62	....	....	....	....	....	....	....
1997-98	Syracuse Crunch	AHL	8	2	4	1	380	26	0	4.10	....	....	....	....	....	....	....
	Dayton Bombers	ECHL	21	15	2	2	1177	59	1	3.01	3	1	1	142	4	0	1.69
1998-99	**Pittsburgh**	**NHL**	**17**	**4**	**3**	**6**	**756**	**28**	**2**	**2.22**	....	....	....	....	....	....	....
	Kansas City	IHL	13	5	7	1	751	41	0	3.28	....	....	....	....	....	....	....
99-2000	**Pittsburgh**	**NHL**	**51**	**23**	**21**	**3**	**2789**	**120**	**2**	**2.58**	....	....	....	....	....	....	....
	Wilkes-Barre	AHL	11	2	8	0	538	39	0	4.35	....	....	....	....	....	....	....
2000-01	**Pittsburgh**	**NHL**	**36**	**20**	**14**	**1**	**2050**	**107**	**0**	**3.13**	**1**	**0**	**0**	**1**	**0**	**0**	**0.00**
2001-02	**Pittsburgh**	**NHL**	**21**	**3**	**12**	**1**	**1094**	**65**	**0**	**3.56**	....	....	....	....	....	....	....
	NHL Totals		**125**	**50**	**50**	**11**	**6689**	**320**	**4**	**2.87**	**1**	**0**	**0**	**1**	**0**	**0**	**0.00**

AULD, Alexander (AWLD, al-ehx-AN-duhr) VAN.

Goaltender. Catches left. 6'4", 197 lbs. Born, Cold Lake, Alta., January 7, 1981.
(Florida's 2nd choice, 40th overall, in 1999 Entry Draft).

			Regular Season								Playoffs						
Season	Club	Lea	GP	W	L	T	Mins	GA	SO	Avg	GP	W	L	Mins	GA	SO	Avg
1996-97	T. Bay Kings	TBMHL	35	....	....	....	2100	46	10	1.35	....	....	....	....	....	....	....
1997-98	Sturgeon Falls	NOJHA	11	4	6	0	611	46	0	4.52	....	....	....	....	....	....	....
	North Bay	OHL	6	0	4	0	206	17	0	4.95	....	....	....	....	....	....	....
1998-99	North Bay	OHL	37	9	20	1	1894	106	1	3.36	3	0	3	170	10	0	3.53
99-2000	North Bay	OHL	55	21	26	6	3047	167	2	3.29	6	2	4	374	12	0	*1.93
2000-01	North Bay	OHL	40	22	11	5	2319	98	1	2.54	4	0	4	240	15	0	3.75
2001-02	**Vancouver**	**NHL**	**1**	**1**	**0**	**0**	**60**	**2**	**0**	**2.00**	....	....	....	....	....	....	....
	Columbia Inferno	ECHL	6	3	1	2	375	12	0	1.92	....	....	....	....	....	....	....
	Manitoba Moose	AHL	21	11	9	0	1104	65	1	3.53	1	0	0	20	0	0	0.00
	NHL Totals		**1**	**1**	**0**	**0**	**60**	**2**	**0**	**2.00**	....	....	....	....	....	....	....

Rights traded to **Vancouver** by **Florida** for Vancouver's compensatory 2nd round choice (later traded to New Jersey - New Jersey selected Tuomas Pihlman) in 2001 Entry Draft and 3rd round choice (later traded to Atlanta - later traded to Buffalo - Buffalo selected John Adams) in 2002 Entry Draft, May 31, 2001.

AYERS, Michael (AY-uhrs, MIGH-kuhl) CHI.

Goaltender. Catches left. 5'11", 183 lbs. Born, Weymouth, MA, January 16, 1980.
(Chicago's 8th choice, 177th overall, in 2000 Entry Draft).

			Regular Season								Playoffs						
Season	Club	Lea	GP	W	L	T	Mins	GA	SO	Avg	GP	W	L	Mins	GA	SO	Avg
1998-99	Trinity Pawling	H.S.					STATISTICS NOT AVAILABLE										
99-2000	Dubuque	USHL	55	16	35	3	3188	196	0	3.69	....	....	....	....	....	....	....
2000-01	New Hampshire	H-East	4	0	0	0	26	2	0	4.68	....	....	....	....	....	....	....
2001-02	New Hampshire	H-East	20	14	3	1	1129	46	1	2.44	....	....	....	....	....	....	....

Hockey East Second All-Star Team (2002)

BACASHIHUA, Jason (bak-ah-SHIH-hu-ah, JAY-suhn) DAL.

Goaltender. Catches left. 5'11", 175 lbs. Born, Garden City, MI, September 20, 1982.
(Dallas' 1st choice, 26th overall, in 2001 Entry Draft).

			Regular Season								Playoffs						
Season	Club	Lea	GP	W	L	T	Mins	GA	SO	Avg	GP	W	L	Mins	GA	SO	Avg
99-2000	Chicago Freeze	NAJHL	41	20	19	2	2432	118	2	2.91	2	0	2	103	12	0	6.97
2000-01	Chicago Freeze	NAJHL	39	24	14	0	2246	121	1	3.23	3	1	2	190	12	0	3.79
2001-02	Plymouth Whalers	OHL	46	26	12	7	2688	105	*5	2.34	6	2	4	360	15	0	2.50
	Utah Grizzlies	AHL	1	0	1	0	61	3	0	2.97	....	....	....	....	....	....	....

BARRASSO, Tom (buh-RAH-soh, TAWM) TOR.

Goaltender. Catches right. 6'3", 210 lbs. Born, Boston, MA, March 31, 1965.
(Buffalo's 1st choice, 5th overall, in 1983 Entry Draft).

			Regular Season								Playoffs						
Season	Club	Lea	GP	W	L	T	Mins	GA	SO	Avg	GP	W	L	Mins	GA	SO	Avg
1981-82	Acton-Boxborough	H.S.	23	....	....	....	1035	32	7	1.86	....	....	....	....	....	....	....
1982-83	Acton-Boxborough	H.S.	23	22	0	1	1035	17	10	0.99	....	....	....	....	....	....	....
1983-84	**Buffalo**	**NHL**	**42**	**26**	**12**	**3**	**2475**	**117**	**2**	**2.84**	**3**	**0**	**2**	**139**	**8**	**0**	**3.45**
1984-85	**Buffalo**	**NHL**	**54**	**25**	**18**	**10**	**3248**	**144**	***5**	***2.66**	**5**	**2**	**3**	**300**	**22**	**0**	**4.40**
	Rochester	AHL	5	3	1	1	267	6	1	1.35	....	....	....	....	....	....	....
1985-86	**Buffalo**	**NHL**	**60**	**29**	**24**	**5**	**3561**	**214**	**2**	**3.61**	....	....	....	....	....	....	....
1986-87	**Buffalo**	**NHL**	**46**	**17**	**23**	**2**	**2501**	**152**	**2**	**3.65**	....	....	....	....	....	....	....
1987-88	**Buffalo**	**NHL**	**54**	**25**	**18**	**8**	**3133**	**173**	**2**	**3.31**	**4**	**1**	**3**	**224**	**16**	**0**	**4.29**
1988-89	**Buffalo**	**NHL**	**10**	**2**	**7**	**0**	**545**	**45**	**0**	**4.95**	....	....	....	....	....	....	....
	Pittsburgh	**NHL**	**44**	**18**	**15**	**7**	**2406**	**162**	**0**	**4.04**	**11**	**7**	**4**	**631**	**40**	**0**	**3.80**
1989-90	**Pittsburgh**	**NHL**	**24**	**7**	**12**	**3**	**1294**	**101**	**0**	**4.68**	....	....	....	....	....	....	....
1990-91 ♦	**Pittsburgh**	**NHL**	**48**	**27**	**16**	**3**	**2754**	**165**	**1**	**3.59**	**20**	**12**	**7**	**1175**	**51**	***1**	***2.60**
1991-92 ♦	**Pittsburgh**	**NHL**	**57**	**25**	**22**	**9**	**3329**	**196**	**1**	**3.53**	***21**	***16**	**5**	***1233**	**58**	**1**	**2.82**
1992-93	**Pittsburgh**	**NHL**	**63**	***43**	**14**	**5**	**3702**	**186**	**4**	**3.01**	**12**	**7**	**5**	**722**	**35**	***2**	**2.91**
1993-94	**Pittsburgh**	**NHL**	**44**	**22**	**15**	**5**	**2482**	**139**	**2**	**3.36**	**6**	**2**	**4**	**356**	**17**	**0**	**2.87**
1994-95	**Pittsburgh**	**NHL**	**2**	**0**	**1**	**1**	**125**	**8**	**0**	**3.84**	**2**	**0**	**1**	**80**	**8**	**0**	**6.00**
1995-96	**Pittsburgh**	**NHL**	**49**	**29**	**16**	**2**	**2799**	**160**	**2**	**3.43**	**10**	**4**	**5**	**558**	**26**	**1**	**2.80**
1996-97	**Pittsburgh**	**NHL**	**5**	**0**	**5**	**0**	**270**	**26**	**0**	**5.78**	....	....	....	....	....	....	....
1997-98	**Pittsburgh**	**NHL**	**63**	**31**	**14**	**13**	**3542**	**122**	**7**	**2.07**	**6**	**2**	**4**	**376**	**17**	**0**	**2.71**
1998-99	**Pittsburgh**	**NHL**	**43**	**19**	**16**	**3**	**2306**	**98**	**4**	**2.55**	**13**	**6**	**7**	**787**	**35**	**1**	**2.67**
99-2000	**Pittsburgh**	**NHL**	**18**	**5**	**7**	**2**	**870**	**46**	**1**	**3.17**	....	....	....	....	....	....	....
	Ottawa	**NHL**	**7**	**3**	**4**	**0**	**418**	**22**	**0**	**3.16**	**6**	**2**	**4**	**372**	**16**	**0**	**2.58**
2000-01							DID NOT PLAY										
2001-02	**Carolina**	**NHL**	**34**	**13**	**12**	**5**	**1908**	**83**	**2**	**2.61**	....	....	....	....	....	....	....
	United States	Olympics	1	1	0	0	60	1	0	1.00	....	....	....	....	....	....	....
	Toronto	**NHL**	**4**	**2**	**2**	**0**	**219**	**10**	**0**	**2.74**	....	....	....	....	....	....	....
	NHL Totals		**771**	**368**	**273**	**86**	**43887**	**2369**	**37**	**3.24**	**119**	**61**	**54**	**6953**	**349**	**6**	**3.01**

NHL All-Rookie Team (1984) • NHL First All-Star Team (1984) • Won Calder Memorial Trophy (1984) • Won Vezina Trophy (1984) • NHL Second All-Star Team (1985, 1993) • Shared William M. Jennings Trophy with Bob Sauve (1985) • Played in NHL All-Star Game (1985)

Traded to **Pittsburgh** by **Buffalo** with Buffalo's 3rd round choice (Joe Dziedzic) in 1990 Entry Draft for Doug Bodger and Darrin Shannon, November 12, 1988. • Missed majority of 1994-95 season recovering from wrist surgery, January 20, 1995. • Missed majority of 1996-97 season recovering from shoulder injury originally suffered in game vs. Montreal, February 5, 1996. Traded to **Ottawa** by **Pittsburgh** for Ron Tugnutt and Janne Laukkanen, March 14, 2000. • Missed entire 2000-01 season for personal reasons. Signed as a free agent by **Carolina**, July 17, 2001. Traded to **Toronto** by **Carolina** for Toronto's 4th round choice in 2003 Entry Draft, March 15, 2002.

BELANGER, Steve (buh-LAWN-zhay, STEEV) PHX.

Goaltender. Catches left. 6' 1", 198 lbs. Born, Anchorage, AK, May 20, 1983.
(Phoenix's 7th choice, 210th overall, in 2001 Entry Draft).

			Regular Season								Playoffs						
Season	Club	Lea	GP	W	L	T	Mins	GA	SO	Avg	GP	W	L	Mins	GA	SO	Avg
99-2000	Team USA	USDP-17	29	11	13	4	1592	85	0	3.20	....	....	....	....	....	....	....
2000-01	Kamloops Blazers	WHL	27	8	14	4	1407	97	1	4.14	2	0	1	84	7	0	5.00
2001-02	Kamloops Blazers	WHL	41	17	17	4	2217	119	0	3.22	2	0	1	68	2	0	1.76

BELFOUR, Ed (BEHL-fohr, EHD) TOR.

Goaltender. Catches left. 5'11", 192 lbs. Born, Carman, Man., April 21, 1965.

			Regular Season								Playoffs						
Season	Club	Lea	GP	W	L	T	Mins	GA	SO	Avg	GP	W	L	Mins	GA	SO	Avg
1983-84	Winkler Flyers	MJHL	14	....	....	....	818	68	0	4.99	....	....	....	....	....	....	....
1984-85	Winkler Flyers	MJHL	34	....	....	....	1973	145	1	4.41	7	3	4	528	41	0	4.66
1985-86	Winkler Flyers	MJHL	33	....	....	....	1943	124	1	3.83	....	....	....	....	....	....	....
1986-87	North Dakota	WCHA	34	29	4	0	2049	81	3	2.37	....	....	....	....	....	....	....
1987-88	Saginaw Hawks	IHL	61	32	25	0	*3446	183	3	3.19	9	4	5	561	33	0	3.53
1988-89	**Chicago**	**NHL**	**23**	**4**	**12**	**3**	**1148**	**74**	**0**	**3.87**	....	....	....	....	....	....	....
	Saginaw Hawks	IHL	29	12	10	0	1760	92	0	3.14	5	2	3	298	14	0	2.82
1989-90	Canada	Nat-Tm	33	13	12	6	1808	93	0	3.09	....	....	....	....	....	....	....
	Chicago	**NHL**	....	....	....	....	....	....	....	....	**9**	**4**	**2**	**409**	**17**	**0**	**2.49**
1990-91	**Chicago**	**NHL**	***74**	***43**	**19**	**7**	***4127**	**170**	**4**	***2.47**	**6**	**2**	**4**	**295**	**20**	**0**	**4.07**
1991-92	**Chicago**	**NHL**	**52**	**21**	**18**	**10**	**2928**	**132**	***5**	**2.70**	**18**	**12**	**4**	**949**	**39**	**1**	***2.47**
1992-93	**Chicago**	**NHL**	***71**	**41**	**18**	**11**	***4106**	**177**	***7**	**2.59**	**4**	**0**	**4**	**249**	**13**	**0**	**3.13**
1993-94	**Chicago**	**NHL**	**70**	**37**	**24**	**6**	**3998**	**178**	***7**	**2.67**	**6**	**2**	**4**	**360**	**15**	**0**	**2.50**
1994-95	**Chicago**	**NHL**	**42**	**22**	**15**	**3**	**2450**	**93**	***5**	**2.28**	**16**	**9**	**7**	**1014**	**37**	**1**	**2.19**
1995-96	**Chicago**	**NHL**	**50**	**22**	**17**	**10**	**2956**	**135**	**1**	**2.74**	**9**	**6**	**3**	**666**	**23**	**1**	**2.07**
1996-97	**Chicago**	**NHL**	**33**	**11**	**15**	**6**	**1966**	**88**	**1**	**2.69**	....	....	....	....	....	....	....
	San Jose	**NHL**	**13**	**3**	**9**	**0**	**757**	**43**	**1**	**3.41**	....	....	....	....	....	....	....
1997-98	**Dallas**	**NHL**	**61**	**37**	**12**	**10**	**3581**	**112**	**9**	***1.88**	**17**	**10**	**7**	**1039**	**31**	**1**	***1.79**
1998-99 ♦	**Dallas**	**NHL**	**61**	**35**	**15**	**9**	**3536**	**117**	**5**	**1.99**	***23**	***16**	**7**	***1544**	**43**	***3**	***1.67**
99-2000	**Dallas**	**NHL**	**62**	**32**	**21**	**7**	**3620**	**127**	**4**	**2.10**	***23**	**14**	**9**	**1443**	**45**	***4**	**1.87**
2000-01	**Dallas**	**NHL**	**63**	**35**	**20**	**7**	**3687**	**144**	**8**	**2.34**	**10**	**4**	**6**	**671**	**25**	**0**	**2.24**
2001-02	**Dallas**	**NHL**	**60**	**21**	**27**	**11**	**3467**	**153**	**1**	**2.65**	....	....	....	....	....	....	....
	Canada	Olympics					DID NOT PLAY – SPARE GOALTENDER										
	NHL Totals		**735**	**364**	**242**	**100**	**42327**	**1743**	**58**	**2.47**	**141**	**79**	**57**	**8639**	**308**	**11**	**2.14**

WCHA First All-Star Team (1987) • NCAA Championship All-Tournament Team (1987) • IHL First All-Star Team (1988) • Shared Garry F. Longman Memorial Trophy (Top Rookie - IHL) with John Cullen (1988) • NHL All-Rookie Team (1991) • NHL First All-Star Team (1991, 1993) • Won Trico Goaltender Award (1991) • Won Calder Memorial Trophy (1991) • Won William M. Jennings Trophy (1991, 1993, 1995) • Won Vezina Trophy (1991, 1993) • NHL Second All-Star Team (1995) • Shared William M. Jennings Trophy with Roman Turek (1999) • Won MBNA Roger Crozier Saving Grace Award (2000) • Played in NHL All-Star Game (1992, 1993, 1996, 1998, 1999)

Signed as a free agent by **Chicago**, September 25, 1987. Traded to **San Jose** by **Chicago** for Chris Terreri, Ulf Dahlen and Michal Sykora, January 25, 1997. Signed as a free agent by **Dallas**, July 2, 1997. Traded to **Nashville** by **Dallas** with Cameron Mann for David Gosselin and Nashville's 5th round choice in 2003 Entry Draft, June 29, 2002. Signed as a free agent by **Toronto**, July 2, 2002.

BENDERA, Shane (behn-DEHR-ah, SHAYN) CBJ

Goaltender. Catches left. 5'11", 170 lbs. Born, St. Albert, Alta., July 13, 1982.
(Columbus' 6th choice, 169th overall, in 2000 Entry Draft).

			Regular Season								Playoffs						
Season	Club	Lea	GP	W	L	T	Mins	GA	SO	Avg	GP	W	L	Mins	GA	SO	Avg
1997-98	Edmonton KC Pats	AMHL	22	8	10	2	1284	82	0	3.84	....	....	....	....	....	....	....
	Red Deer Rebels	WHL	1	0	0	0	8	0	0	0.00	....	....	....	....	....	....	....
1998-99	Bonnyville	AJHL	20	....	....	....	956	70	0	4.38	....	....	....	....	....	....	....
	Red Deer Rebels	WHL	2	0	1	0	72	7	0	5.83	....	....	....	....	....	....	....
99-2000	Red Deer Rebels	WHL	*69	31	27	9	*4003	202	0	3.03	3	0	2	76	15	0	11.84
2000-01	Red Deer Rebels	WHL	45	32	8	2	2603	108	*5	*2.49	*22	*16	6	*1404	43	*4	1.84
2001-02	Red Deer Rebels	WHL	20	11	6	3	1211	46	1	2.28	....	....	....	....	....	....	....
	Kelowna Rockets	WHL	30	13	9	8	1816	79	0	2.61	15	9	6	918	29	*2	*1.90

WHL East Second All-Star Team (2001) • WHL West Second All-Star Team (2002)

Traded to **Kelowna** (WHL) by **Red Deer** (WHL) for Jason Stone and Carsen Germyn, December 18, 2001.

BERKHOEL, Adam (BUHRK-uhl, A-duhm) CHI.

Goaltender. Catches left. 5'11", 173 lbs. Born, St. Paul, MN, May 16, 1981.
(Chicago's 12th choice, 240th overall, in 2000 Entry Draft).

			Regular Season								Playoffs						
Season	Club	Lea	GP	W	L	T	Mins	GA	SO	Avg	GP	W	L	Mins	GA	SO	Avg
1998-99	Stillwater Ponies	H.S.					STATISTICS NOT AVAILABLE										
99-2000	Twin Cities	USHL	49	25	15	7	2848	129	0	2.72	13	7	6	797	43	0	3.24
2000-01	U. of Denver	WCHA	15	7	6	1	745	38	1	3.06	....	....	....	....	....	....	....
2001-02	U. of Denver	WCHA	18	12	4	1	1026	40	1	2.34	....	....	....	....	....	....	....

USHL All-Rookie Team (2000) • USHL Second All-Star Team (2000)

BIERK, Zac (BUHRK, ZAK) PHX.

Goaltender. Catches left. 6'4", 205 lbs. Born, Peterborough, Ont., September 17, 1976.
(Tampa Bay's 8th choice, 212th overall, in 1995 Entry Draft).

			Regular Season								Playoffs						
Season	Club	Lea	GP	W	L	T	Mins	GA	SO	Avg	GP	W	L	Mins	GA	SO	Avg
1993-94	Peterborough	OPJHL	4	....	....	....	205	17	0	4.98	....	....	....	....	....	....	....
	Peterborough	OHL	9	0	4	2	423	37	0	5.22	1	0	0	33	7	0	12.70
1994-95	Peterborough	OHL	35	11	15	5	1779	117	0	3.95	6	2	3	301	24	0	4.78
1995-96	Peterborough	OHL	58	31	16	6	3292	174	2	3.17	*22	*14	7	*1383	83	0	3.60
1996-97	Peterborough	OHL	49	*28	16	0	2744	151	2	3.30	11	6	5	666	35	0	3.15
1997-98	**Tampa Bay**	**NHL**	**13**	**1**	**4**	**1**	**433**	**30**	**0**	**4.16**	....	....	....	....	....	....	....
	Adirondack	AHL	12	1	6	1	557	36	0	3.87	....	....	....	....	....	....	....
1998-99	**Tampa Bay**	**NHL**	**1**	**0**	**1**	**0**	**59**	**2**	**0**	**2.03**	....	....	....	....	....	....	....
	Cleveland	IHL	27	11	12	4	1556	79	0	3.05	....	....	....	....	....	....	....
99-2000	**Tampa Bay**	**NHL**	**12**	**4**	**4**	**1**	**509**	**31**	**0**	**3.65**	....	....	....	....	....	....	....
	Detroit Vipers	IHL	15	4	8	2	846	46	1	3.26	....	....	....	....	....	....	....
2000-01	**Minnesota**	**NHL**	**1**	**0**	**1**	**0**	**60**	**6**	**0**	**6.00**	....	....	....	....	....	....	....
	Cleveland	IHL	49	24	18	5	2785	134	6	2.89	4	0	3	182	10	0	3.29
2001-02	Augusta Lynx	ECHL	30	16	9	3	1748	68	1	2.33	....	....	....	....	....	....	....
	Springfield	AHL	1	0	1	0	20	4	0	12.00	....	....	....	....	....	....	....
	NHL Totals		**27**	**5**	**10**	**2**	**1061**	**69**	**0**	**3.90**	....	....	....	....	....	....	....

CHL Second All-Star Team (1997) • OHL First All-Star Team (1997)

• Missed remainder of 1998-99 season recovering from Meniere's Disease which was diagnosed on March 25, 1999. Selected by **Minnesota** from **Tampa Bay** in Expansion Draft, June 23, 2000. Signed as a free agent by **Phoenix**, August 30, 2001.

BILLINGTON, Craig (BIHL-lihng-tohn, KRAYG) WSH.

Goaltender. Catches left. 5'10", 170 lbs. Born, London, Ont., September 11, 1966.
(New Jersey's 2nd choice, 23rd overall, in 1984 Entry Draft).

			Regular Season								Playoffs						
Season	Club	Lea	GP	W	L	T	Mins	GA	SO	Avg	GP	W	L	Mins	GA	SO	Avg
1982-83	London Diamonds	OJHL-B	23	....	....	....	1338	76	0	3.41	....	....	....	....	....	....	....
1983-84	Belleville Bulls	OHL	44	20	19	0	2335	162	1	4.16	1	0	0	30	3	0	6.00
1984-85	Belleville Bulls	OHL	47	26	19	0	2544	180	1	4.25	14	7	5	761	47	1	3.71
1985-86	Belleville Bulls	OHL	3	2	1	0	180	11	0	3.67	20	9	6	1133	68	0	3.60
	New Jersey	**NHL**	**18**	**4**	**9**	**1**	**901**	**77**	**0**	**5.13**	....	....	....	....	....	....	....
1986-87	**New Jersey**	**NHL**	**22**	**4**	**13**	**2**	**1114**	**89**	**0**	**4.79**	....	....	....	....	....	....	....
	Maine Mariners	AHL	20	9	8	2	1151	70	0	3.65	....	....	....	....	....	....	....
1987-88	Utica Devils	AHL	*59	22	27	8	*3404	208	1	3.67	....	....	....	....	....	....	....
1988-89	**New Jersey**	**NHL**	**3**	**1**	**1**	**0**	**140**	**11**	**0**	**4.71**	....	....	....	....	....	....	....
	Utica Devils	AHL	41	17	18	6	2432	150	2	3.70	4	1	3	220	18	0	4.91
1989-90	Utica Devils	AHL	38	20	13	1	2087	138	0	3.97	....	....	....	....	....	....	....
1990-91	Canada	Nat-Tm	34	17	14	2	1879	110	2	3.51	....	....	....	....	....	....	....
1991-92	**New Jersey**	**NHL**	**26**	**13**	**7**	**1**	**1363**	**69**	**2**	**3.04**	....	....	....	....	....	....	....
1992-93	**New Jersey**	**NHL**	**42**	**21**	**16**	**4**	**2389**	**146**	**2**	**3.67**	**2**	**0**	**1**	**78**	**5**	**0**	**3.85**
1993-94	**Ottawa**	**NHL**	**63**	**11**	**41**	**4**	**3319**	**254**	**0**	**4.59**	....	....	....	....	....	....	....
1994-95	**Ottawa**	**NHL**	**9**	**0**	**6**	**2**	**472**	**32**	**0**	**4.07**	....	....	....	....	....	....	....
	Boston	**NHL**	**8**	**5**	**1**	**0**	**373**	**19**	**0**	**3.06**	**1**	**0**	**0**	**25**	**1**	**0**	**2.40**
1995-96	**Boston**	**NHL**	**27**	**10**	**13**	**3**	**1380**	**79**	**1**	**3.43**	**1**	**0**	**1**	**60**	**6**	**0**	**6.00**
1996-97	**Colorado**	**NHL**	**23**	**11**	**8**	**2**	**1200**	**53**	**1**	**2.65**	**1**	**0**	**0**	**20**	**1**	**0**	**3.00**
1997-98	**Colorado**	**NHL**	**23**	**8**	**7**	**4**	**1162**	**45**	**1**	**2.32**	**1**	**0**	**0**	**1**	**0**	**0**	**0.00**
1998-99	**Colorado**	**NHL**	**21**	**11**	**8**	**1**	**1086**	**52**	**0**	**2.87**	**1**	**0**	**0**	**9**	**1**	**0**	**6.67**
99-2000	**Washington**	**NHL**	**13**	**3**	**6**	**1**	**611**	**28**	**2**	**2.75**	**1**	**0**	**0**	**20**	**1**	**0**	**3.00**
2000-01	**Washington**	**NHL**	**12**	**3**	**5**	**2**	**660**	**27**	**0**	**2.45**	....	....	....	....	....	....	....
2001-02	**Washington**	**NHL**	**17**	**4**	**5**	**3**	**710**	**36**	**0**	**3.04**	....	....	....	....	....	....	....
	NHL Totals		**327**	**109**	**146**	**30**	**16880**	**1017**	**9**	**3.61**	**8**	**0**	**2**	**213**	**15**	**0**	**4.23**

OHL First All-Star Team (1985) • Played in NHL All-Star Game (1993)

Traded to **Ottawa** by **New Jersey** with Troy Mallette and New Jersey's 4th round choice (Cosmo Dupaul) in 1993 Entry Draft for Peter Sidorkiewicz and future considerations (Mike Peluso, June 26, 1993), June 20, 1993. Traded to **Boston** by **Ottawa** for NY Islanders' 8th round choice (previously acquired, Ottawa selected Ray Schultz) in 1995 Entry Draft, April 7, 1995. Signed as a free agent by **Florida**, September 5, 1996. Claimed by **Colorado** from **Florida** in Waiver Draft, September 30, 1996. Traded to **Washington** by **Colorado** for future considerations, July 16, 1999.

BIRON, Martin (BIH-rohn, MAHR-tihn) BUF.

Goaltender. Catches left. 6'2", 168 lbs. Born, Lac-St-Charles, Que., August 15, 1977.
(Buffalo's 2nd choice, 16th overall, in 1995 Entry Draft).

			Regular Season								Playoffs						
Season	Club	Lea	GP	W	L	T	Mins	GA	SO	Avg	GP	W	L	Mins	GA	SO	Avg
1993-94	Trois-Rivieres	QAAA	23	14	8	1	1412	80	1	3.40	2	1	1	112	7	0	3.73
1994-95	Beauport	QMJHL	56	29	16	9	3193	132	3	*2.48	16	8	7	900	37	*4	2.47
1995-96	Beauport	QMJHL	55	29	17	7	3201	152	1	2.85	*19	*12	7	1134	64	0	3.39
	Buffalo	**NHL**	**3**	**0**	**2**	**0**	**119**	**10**	**0**	**5.04**	....	....	....	....	....	....	....
1996-97	Beauport	QMJHL	18	6	9	1	928	61	1	3.94	....	....	....	....	....	....	....
	Hull Olympiques	QMJHL	16	11	4	1	974	43	2	2.65	6	3	1	325	19	0	3.51
1997-98	South Carolina	ECHL	2	0	1	1	86	3	0	2.09	....	....	....	....	....	....	....
	Rochester	AHL	41	14	18	6	2312	113	*5	2.93	4	1	3	239	16	0	4.01
1998-99	**Buffalo**	**NHL**	**6**	**1**	**2**	**1**	**281**	**10**	**0**	**2.14**	....	....	....	....	....	....	....
	Rochester	AHL	52	36	13	3	3129	108	*6	*2.07	*20	12	8	1167	42	1	*2.16
99-2000	**Buffalo**	**NHL**	**41**	**19**	**18**	**2**	**2229**	**90**	**5**	**2.42**	....	....	....	....	....	....	....
	Rochester	AHL	6	6	0	0	344	12	1	2.09	....	....	....	....	....	....	....
2000-01	Rochester	AHL	4	3	1	0	239	4	1	1.00	....	....	....	....	....	....	....
	Buffalo	**NHL**	**18**	**7**	**7**	**1**	**918**	**39**	**2**	**2.55**	....	....	....	....	....	....	....
2001-02	**Buffalo**	**NHL**	**72**	**31**	**28**	**10**	**4085**	**151**	**4**	**2.22**	....	....	....	....	....	....	....
	NHL Totals		**140**	**58**	**57**	**14**	**7632**	**300**	**11**	**2.36**	....	....	....	....	....	....	....

QMJHL All-Rookie Team (1995) • QMJHL Defensive Rookie of the Year (1995) • Canadian Major Junior First All-Star Team (1995) • Canadian Major Junior Goaltender of the Year (1995) • AHL First All-Star Team (1999) • Shared Harry "Hap" Holmes Memorial Trophy (fewest goals against - AHL) with Tom Draper (1999) • Won Baz Bastien Memorial Trophy (Top Goaltender - AHL) (1999)

BLACKBURN, Dan (BLAK-buhrn, DAN) NYR

Goaltender. Catches left. 6', 180 lbs. Born, Montreal, Que., May 20, 1983.
(NY Rangers' 1st choice, 10th overall, in 2001 Entry Draft).

			Regular Season								Playoffs						
Season	Club	Lea	GP	W	L	T	Mins	GA	SO	Avg	GP	W	L	Mins	GA	SO	Avg
1997-98	Bow Valley	AJHL	20	9	6	1	1039	58	1	3.35	3	0	0	97	8	0	4.95
1998-99	Bow Valley	AJHL	38	7	19	6	1941	146	0	4.51	2	0	2	118	8	0	4.07
99-2000	Kootenay Ice	WHL	51	34	8	7	3004	126	3	2.52	*21	*16	5	*1272	43	2	*2.03
2000-01	Kootenay Ice	WHL	50	*33	14	2	2922	135	1	2.77	11	7	4	706	23	1	1.95
2001-02	**NY Rangers**	**NHL**	**31**	**12**	**16**	**0**	**1737**	**95**	**0**	**3.28**	....	....	....	....	....	....	....
	Hartford	AHL	4	2	1	1	244	11	0	2.71	....	....	....	....	....	....	....
	NHL Totals		**31**	**12**	**16**	**0**	**1737**	**95**	**0**	**3.28**	....	....	....	....	....	....	....

AJHL Scholastic Player of the Year (1998, 1999) • WHL East First All-Star Team (2001) • Canadian Major Junior First All-Star Team (2001) • Canadian Major Junior Goaltender of the Year (2001) • NHL All-Rookie Team (2002)

BLACKBURN, Josh (BLAK-buhrn, JAWSH) PHX.

Goaltender. Catches left. 6', 203 lbs. Born, Delrio, TX, November 13, 1978.
(Phoenix's 6th choice, 116th overall, in 1998 Entry Draft).

			Regular Season								Playoffs						
Season	Club	Lea	GP	W	L	T	Mins	GA	SO	Avg	GP	W	L	Mins	GA	SO	Avg
1996-97	Dubuque	USHL	52	15	32	3	2979	185	1	3.72	....	....	....	....	....	....	....
1997-98	Dubuque	USHL	28	11	15	1	1605	94	0	3.51	....	....	....	....	....	....	....
	Lincoln Stars	USHL	17	13	4	0	1004	41	1	2.45	9	4	5	522	29	0	3.33
1998-99	U. of Michigan	CCHA	*42	*25	10	6	*2398	91	3	2.28	....	....	....	....	....	....	....
99-2000	U. of Michigan	CCHA	22	14	4	4	1337	53	1	2.38	....	....	....	....	....	....	....
2000-01	U. of Michigan	CCHA	*45	26	13	5	*2647	101	5	2.29	....	....	....	....	....	....	....
2001-02	U. of Michigan	CCHA	*43	*27	11	5	*2568	96	5	2.24	....	....	....	....	....	....	....

CCHA Second All-Star Team (1999, 2001) • NCAA West First All-American Team (1999)

BOISCLAIR, Daniel (BWUH-klair, DAN-yehl) CAR.

Goaltender. Catches left. 6'2", 185 lbs. Born, Sept-Iles, Que., November 2, 1982.
(Carolina's 5th choice, 181st overall, in 2001 Entry Draft).

			Regular Season								Playoffs						
Season	Club	Lea	GP	W	L	T	Mins	GA	SO	Avg	GP	W	L	Mins	GA	SO	Avg
1998-99	Charles-Lemoyne	QAAA	25	14	7	2	1360	69	0	3.04	....	....	....	....	....	....	....
99-2000	Cape Breton	QMJHL	29	7	14	1	1421	95	0	4.01	3	0	1	86	6	0	4.21
2000-01	Cape Breton	QMJHL	47	16	23	2	2426	161	0	3.98	12	5	6	685	36	0	3.16
2001-02	Cape Breton	QMJHL	25	9	10	2	1151	73	1	3.81	....	....	....	....	....	....	....
	Victoriaville	QMJHL	15	5	5	3	780	42	0	3.23	*19	*13	6	*1107	53	0	2.87

Traded to **Victoriaville** (QMJHL) by **Cape Breton** (QMJHL) with Cape Breton's 5th choice (J-F Gouin) in 2002 QMJHL Priority Draft for Sylvain Swells and Carl McLean, January 7, 2002.

BOUCHER, Brian (BOO-shay, BRIGH-uhn) PHX.

Goaltender. Catches left. 6'2", 190 lbs. Born, Woonsocket, RI, January 2, 1977.
(Philadelphia's 1st choice, 22nd overall, in 1995 Entry Draft).

			Regular Season								Playoffs						
Season	Club	Lea	GP	W	L	T	Mins	GA	SO	Avg	GP	W	L	Mins	GA	SO	Avg
1993-94	Mount St. Charles	H.S.	15	*14	0	1	*504	*8	*9	*0.57	4	*4	0	*180	*6	*1	*1.20
1994-95	Wexford Raiders	MTJHL	8	....	....	....	425	23	0	3.25	....	....	....	....	....	....	....
	Tri-City	WHL	35	17	11	2	1969	108	1	3.29	13	6	5	795	50	0	3.77
1995-96	Tri-City	WHL	55	33	19	2	3183	181	1	3.41	11	6	5	653	37	*2	3.40
1996-97	Tri-City	WHL	41	10	24	6	2458	149	1	3.64	....	....	....	....	....	....	....
1997-98	Philadelphia	AHL	34	16	12	3	1901	101	0	3.19	2	0	0	30	1	0	1.95
1998-99	Philadelphia	AHL	36	20	8	5	2061	89	2	2.59	16	9	7	947	45	0	2.85
99-2000	**Philadelphia**	**NHL**	**35**	**20**	**10**	**3**	**2038**	**65**	**4**	***1.91**	**18**	**11**	**7**	**1183**	**40**	**1**	**2.03**
	Philadelphia	AHL	1	0	0	1	65	3	0	2.77	....	....	....	....	....	....	....
2000-01	**Philadelphia**	**NHL**	**27**	**8**	**12**	**5**	**1470**	**80**	**1**	**3.27**	**1**	**0**	**0**	**37**	**3**	**0**	**4.86**
2001-02	**Philadelphia**	**NHL**	**41**	**18**	**16**	**4**	**2295**	**92**	**2**	**2.41**	**2**	**0**	**1**	**88**	**2**	**0**	**1.36**
	NHL Totals		**103**	**46**	**38**	**12**	**5803**	**237**	**7**	**2.45**	**21**	**11**	**8**	**1308**	**45**	**1**	**2.06**

WHL West Second All-Star Team (1996) • WHL West First All-Star Team (1997) • NHL All-Rookie Team (2000)

Traded to **Phoenix** by **Philadelphia** with Nashville's 3rd round choice (previously acquired, Phoenix selected Joe Callahan) in 2002 Entry Draft for Michal Handzus and Robert Esche, June 12, 2002.

BOUCHER, Nick (BOO-shay, NIHK) PIT.

Goaltender. Catches left. 5'11", 175 lbs. Born, Leduc, Alta., December 29, 1980.
(Pittsburgh's 10th choice, 280th overall, in 2000 Entry Draft).

			Regular Season								Playoffs						
Season	Club	Lea	GP	W	L	T	Mins	GA	SO	Avg	GP	W	L	Mins	GA	SO	Avg
1997-98	Cowichan	BCHL	20	6	7	0	918	65	0	4.25	....	....	....	....	....	....	....
1998-99	Cowichan	BCHL	10	3	5	0	464	40	0	5.17	....	....	....	....	....	....	....
	Ft. Saskatchewan	AJHL	22	11	9	2	1243	70	1	3.38	7	3	4	420	24	3	3.43
99-2000	Dartmouth	ECAC	24	8	12	3	1378	66	1	2.87	....	....	....	....	....	....	....
2000-01	Dartmouth	ECAC	*33	*16	12	4	*1966	84	1	2.56	....	....	....	....	....	....	....
2001-02	Dartmouth	ECAC	21	9	7	4	1187	57	1	2.88	....	....	....	....	....	....	....

BRATHWAITE, Fred (BRAYTH-wayt, FREHD) ST.L.

Goaltender. Catches left. 5'7", 175 lbs. Born, Ottawa, Ont., November 24, 1972.

			Regular Season								Playoffs						
Season	Club	Lea	GP	W	L	T	Mins	GA	SO	Avg	GP	W	L	Mins	GA	SO	Avg
1988-89	Smiths Falls Bears	COJHL	38	16	18	1	2130	187	0	5.27	....	....	....	....	....	....	....
1989-90	Orillia Terriers	OJHL-B	15	....	....	....	782	47	0	3.61	....	....	....	....	....	....	....
	Oshawa Generals	OHL	20	11	2	1	886	43	1	2.91	10	4	2	451	22	0	*2.93
1990-91	Oshawa Generals	OHL	39	25	6	3	1986	112	1	3.38	13	*9	2	677	43	0	3.81
1991-92	Oshawa Generals	OHL	24	12	7	2	1248	81	0	3.89	....	....	....	....	....	....	....
	London Knights	OHL	23	15	6	2	1325	61	*4	2.76	10	5	5	615	36	0	3.51
1992-93	Detroit	OHL	37	23	10	4	2192	134	0	3.67	15	9	6	858	48	1	3.36
1993-94	**Edmonton**	**NHL**	**19**	**3**	**10**	**3**	**982**	**58**	**0**	**3.54**	....	....	....	....	....	....	....
	Cape Breton	AHL	2	1	1	0	119	6	0	3.04	....	....	....	....	....	....	....
1994-95	**Edmonton**	**NHL**	**14**	**2**	**5**	**1**	**601**	**40**	**0**	**3.99**	....	....	....	....	....	....	....
1995-96	**Edmonton**	**NHL**	**7**	**0**	**2**	**0**	**293**	**12**	**0**	**2.46**	....	....	....	....	....	....	....
	Cape Breton	AHL	31	12	16	0	1699	110	1	3.88	....	....	....	....	....	....	....
1996-97	Manitoba Moose	IHL	58	22	22	5	2945	167	1	3.40	....	....	....	....	....	....	....
1997-98	Manitoba Moose	IHL	51	23	18	4	2736	138	1	3.03	2	0	1	72	4	0	3.30
1998-99	Canada	Nat-Tm	24	6	8	3	989	47	2	2.85	....	....	....	....	....	....	....
	Calgary	**NHL**	**28**	**11**	**9**	**7**	**1663**	**68**	**1**	**2.45**	....	....	....	....	....	....	....
99-2000	**Calgary**	**NHL**	**61**	**25**	**25**	**7**	**3448**	**158**	**5**	**2.75**	....	....	....	....	....	....	....
	Saint John	AHL	2	2	0	0	120	4	0	2.00	....	....	....	....	....	....	....
2000-01	**Calgary**	**NHL**	**49**	**15**	**17**	**10**	**2742**	**106**	**5**	**2.32**	....	....	....	....	....	....	....
2001-02	**St. Louis**	**NHL**	**25**	**9**	**11**	**4**	**1446**	**54**	**2**	**2.24**	**1**	**0**	**0**	**1**	**0**	**0**	**0.00**
	NHL Totals		**203**	**65**	**79**	**32**	**11175**	**496**	**13**	**2.66**	**1**	**0**	**0**	**1**	**0**	**0**	**0.00**

• Scored a goal while with Detroit (OHL), April 20, 1993. • Scored a goal while with Manitoba (IHL), November 9, 1996. • Played 6 seconds of playoff game vs. Detroit, May 4, 2002.

Signed as a free agent by **Edmonton**, October 6, 1993. Signed as a free agent by **Calgary**, January 6, 1999. Traded to **St. Louis** by **Calgary** with Daniel Tkaczuk, Sergei Varlamov and Calgary's 9th round choice (Grant Jacobsen) in 2001 Entry Draft for Roman Turek and St. Louis' 4th round choice (Yegor Shastin) in 2001 Entry Draft, June 23, 2001.

BROCHU, Martin (broh-SHOO, MAHR-tihn) VAN.

Goaltender. Catches left. 6', 199 lbs. Born, Anjou, Que., March 10, 1973.

			Regular Season								Playoffs						
Season	Club	Lea	GP	W	L	T	Mins	GA	SO	Avg	GP	W	L	Mins	GA	SO	Avg
1989-90	Mtl-Bourassa	QAAA	27	11	14	1	1471	103	3	4.20	3	1	2	193	10	1	3.10
1990-91	Granby Bisons	QMJHL	16	6	5	0	622	39	....	3.76	....	....	....	....	....	....	....
1991-92	Granby Bisons	QMJHL	52	15	29	2	2772	278	0	4.72	....	....	....	....	....	....	....
1992-93	Hull Olympiques	QMJHL	29	9	15	1	1453	137	0	5.66	2	0	1	69	7	0	6.07
1993-94	Fredericton	AHL	32	10	11	3	1505	76	2	3.03	....	....	....	....	....	....	....
1994-95	Fredericton	AHL	44	18	18	4	2475	145	0	3.51	....	....	....	....	....	....	....
1995-96	Fredericton	AHL	17	6	8	2	986	70	0	4.26	....	....	....	....	....	....	....
	Wheeling	ECHL	19	10	6	2	1060	51	1	2.89	....	....	....	....	....	....	....
	Portland Pirates	AHL	5	2	2	1	287	15	0	3.14	12	7	4	700	28	*2	*2.40
1996-97	Portland Pirates	AHL	55	23	17	7	2962	150	2	3.04	5	2	3	324	13	0	2.41
1997-98	Portland Pirates	AHL	37	16	14	1	1926	96	2	2.99	6	3	2	296	16	0	3.24
1998-99	**Washington**	**NHL**	**2**	**0**	**2**	**0**	**120**	**6**	**0**	**3.00**	....	....	....	....	....	....	....
	Portland Pirates	AHL	20	6	10	3	1164	57	2	2.94	....	....	....	....	....	....	....
	Utah Grizzlies	IHL	5	1	3	1	298	13	0	2.62	....	....	....	....	....	....	....
99-2000	Portland Pirates	AHL	54	32	15	6	3192	116	4	2.18	2	0	2	80	7	0	5.27
2000-01	Saint John	AHL	55	27	19	5	3049	132	2	2.60	19	*14	4	1148	39	*4	2.04
2001-02	**Vancouver**	**NHL**	**6**	**0**	**3**	**0**	**216**	**15**	**0**	**4.17**	....	....	....	....	....	....	....
	Manitoba Moose	AHL	29	10	14	3	1625	91	1	3.36	....	....	....	....	....	....	....
	NHL Totals		**8**	**0**	**5**	**0**	**336**	**21**	**0**	**3.75**	....	....	....	....	....	....	....

AHL First All-Star Team (2000) • Won Baz Bastien Memorial Trophy (Top Goaltender - AHL) (2000) • Won Les Cunningham Award (MVP - AHL) (2000)

Signed as a free agent by **Montreal**, September 22, 1992. Traded to **Washington** by **Montreal** for future considerations, March 15, 1996. Signed as a free agent by **Calgary**, August 25, 2000. Signed as a free agent by **Minnesota**, July 17, 2001. Claimed by **Vancouver** from **Minnesota** in Waiver Draft, September 28, 2001.

BRODEUR, Martin (broh-DOOR, MAHR-tihn) N.J.

Goaltender. Catches left. 6'2", 210 lbs. Born, Montreal, Que., May 6, 1972.
(New Jersey's 1st choice, 20th overall, in 1990 Entry Draft).

			Regular Season								Playoffs						
Season	Club	Lea	GP	W	L	T	Mins	GA	SO	Avg	GP	W	L	Mins	GA	SO	Avg
1988-89	Mtl-Bourassa	QAAA	27	13	12	1	1580	98	0	3.72	3	0	3	210	14	0	3.99
1989-90	St-Hyacinthe	QMJHL	42	23	13	2	2333	156	0	4.01	12	5	7	678	46	0	4.07
1990-91	St-Hyacinthe	QMJHL	52	22	24	4	2946	162	2	3.30	4	0	4	232	16	0	4.14
1991-92	St-Hyacinthe	QMJHL	48	27	16	4	2846	161	2	3.39	5	2	3	317	14	0	2.65
	New Jersey	NHL	4	2	1	0	179	10	0	3.35	1	0	1	32	3	0	5.63
1992-93	Utica Devils	AHL	32	14	13	5	1952	131	0	4.03	4	1	3	258	18	0	4.19
1993-94	New Jersey	NHL	47	27	11	8	2625	105	3	2.40	17	8	9	1171	38	1	1.95
1994-95♦	New Jersey	NHL	40	19	11	6	2184	89	3	2.45	*20	*16	4	*1222	34	*3	*1.67
1995-96	New Jersey	NHL	77	34	30	12	*4433	173	6	2.34	....	....	....	....	....	....	....
1996-97	New Jersey	NHL	67	37	14	13	3838	120	*10	*1.88	10	5	5	659	19	2	*1.73
1997-98	New Jersey	NHL	70	*43	17	8	4128	130	10	1.89	6	2	4	366	12	0	1.97
1998-99	New Jersey	NHL	*70	*39	21	10	*4239	162	4	2.29	7	3	4	425	20	0	2.82
99-2000♦	New Jersey	NHL	72	*43	20	8	4312	161	6	2.24	*23	*16	7	*1450	39	2	*1.61
2000-01	New Jersey	NHL	72	*42	17	11	4297	166	9	2.32	*25	15	10	*1505	52	*4	2.07
2001-02	New Jersey	NHL	*73	38	26	9	*4347	156	4	2.15	6	2	4	381	9	1	1.42
	Canada	Olympics	5	*4	0	1	300	9	0	*1.80	....	....	....	....	....	....	....
	NHL Totals		592	324	168	85	34582	1272	55	2.21	115	67	48	7211	226	13	1.88

QMJHL All-Rookie Team (1990) • QMJHL Second All-Star Team (1992) • NHL All-Rookie Team (1994) • Won Calder Memorial Trophy (1994) • NHL Second All-Star Team (1997, 1998) • Shared William M. Jennings Trophy with Mike Dunham (1997) • Won William M. Jennings Trophy (1998) • Played in NHL All-Star Game (1996, 1997, 1998, 1999, 2000, 2001) • Scored a goal while with New Jersey in playoffs vs. Montreal, April 17, 1997.

BRUCKLER, Bernd (BRUK-luhr, BUHRND) PHI.

Goaltender. Catches left. 6'1", 180 lbs. Born, Graz, Austria, September 26, 1981.
(Philadelphia's 4th choice, 150th overall, in 2001 Entry Draft).

			Regular Season								Playoffs						
Season	Club	Lea	GP	W	L	T	Mins	GA	SO	Avg	GP	W	L	Mins	GA	SO	Avg
1997-98	EC Graz	Austria	2	....	....	....	61	11	0	10.82	....	....	....	....	....	....	....
1998-99	Ponoka	HJHL					Statistics not available										
99-2000	EC Graz	Austria-Jr.					Statistics not available										
2000-01	Tri-City Storm	USHL	28	15	8	3	1624	67	2	2.48	7	3	4	420	....	....	....
2001-02	U. of Wisconsin	WCHA	18	6	8	2	973	50	1	3.08	....	....	....	....	....	....	....

USHL Second All-Star Team (2001) • WCHA All-Rookie Team (2002)

BRUST, Barry (BRUHST, BAIR-ree) MIN.

Goaltender. Catches left. 6'3", 225 lbs. Born, Swan River, Man., August 8, 1983.
(Minnesota's 4th choice, 73rd overall, in 2002 Entry Draft).

			Regular Season								Playoffs						
Season	Club	Lea	GP	W	L	T	Mins	GA	SO	Avg	GP	W	L	Mins	GA	SO	Avg
99-2000	Swan Valley	MJHL					Statistics not available										
2000-01	Spokane Chiefs	WHL	16	4	6	1	777	42	0	3.24	....	....	....	....	....	....	....
2001-02	Spokane Chiefs	WHL	60	28	21	10	3540	152	1	2.58	11	6	5	677	23	0	2.04

WHL West First All-Star Team (2002)

BRYZGALOV, Ilja (breez-GAH-lahf, ihl-YUH) ANA.

Goaltender. Catches left. 6'3", 198 lbs. Born, Togliatti, USSR, June 22, 1980.
(Anaheim's 2nd choice, 44th overall, in 2000 Entry Draft).

			Regular Season								Playoffs						
Season	Club	Lea	GP	W	L	T	Mins	GA	SO	Avg	GP	W	L	Mins	GA	SO	Avg
1997-98	Lada Togliatti 2	Russia-3	8	....	....	....	480	28	....	3.50	....	....	....	....	....	....	....
1998-99	Lada Togliatti 2	Russia-4	20	....	....	....	1200	43	....	2.15	....	....	....	....	....	....	....
99-2000	Spartak Moscow	Russia-2	9	....	....	....	500	21	....	2.52	....	....	....	....	....	....	....
	Lada Togliatti	Russia	14	....	....	....	796	18	3	1.36	7	....	....	407	10	1	1.47
2000-01	Lada Togliatti	Russia	34	....	....	....	1992	61	*8	1.84	5	....	....	249	8	0	1.93
2001-02	Cincinnati	AHL	45	20	16	4	2399	99	4	2.48	....	....	....	....	....	....	....
	Anaheim	NHL	1	0	0	0	32	1	0	1.88	....	....	....	....	....	....	....
	Russia	Olympics					Did not play – spare goaltender										
	NHL Totals		1	0	0	0	32	1	0	1.88	....	....	....	....	....	....	....

BUDAJ, Peter (BOO-digh, PEE-tuhr) COL.

Goaltender. Catches left. 6', 200 lbs. Born, Bystrica, Czech., September 18, 1982.
(Colorado's 1st choice, 63rd overall, in 2001 Entry Draft).

			Regular Season								Playoffs						
Season	Club	Lea	GP	W	L	T	Mins	GA	SO	Avg	GP	W	L	Mins	GA	SO	Avg
99-2000	St. Michael's	OHL	34	6	18	1	1676	112	1	4.01	....	....	....	....	....	....	....
2000-01	St. Michael's	OHL	37	17	12	3	1996	95	3	2.86	11	6	4	621	26	1	2.51
2001-02	St. Michael's	OHL	42	26	9	5	2329	89	2	*2.29	12	5	6	620	34	*1	3.29

OHL Second All-Star Team (2002)

BURKE, Sean (BUHRK, SHAWN) PHX.

Goaltender. Catches left. 6'4", 210 lbs. Born, Windsor, Ont., January 29, 1967.
(New Jersey's 2nd choice, 24th overall, in 1985 Entry Draft).

			Regular Season								Playoffs						
Season	Club	Lea	GP	W	L	T	Mins	GA	SO	Avg	GP	W	L	Mins	GA	SO	Avg
1983-84	St. Michael's B	MTJHL	25	....	....	....	1482	120	0	4.86	....	....	....	....	....	....	....
1984-85	Toronto	OHL	49	25	21	3	2987	211	0	4.24	5	1	3	266	25	0	5.64
1985-86	Toronto	OHL	47	16	27	3	2840	233	0	4.92	4	0	4	238	24	0	6.05
1986-87	Canada	Nat-Tm	42	27	13	2	2550	130	0	3.05	....	....	....	....	....	....	....
1987-88	Canada	Nat-Tm	37	19	9	2	1962	92	1	2.81	....	....	....	....	....	....	....
	Canada	Olympics	4	1	2	1	238	12	0	3.02	....	....	....	....	....	....	....
	New Jersey	NHL	13	10	1	0	689	35	1	3.05	17	9	8	1001	57	*1	3.42
1988-89	New Jersey	NHL	62	22	31	9	3590	230	3	3.84	....	....	....	....	....	....	....
1989-90	New Jersey	NHL	52	22	22	6	2914	175	0	3.60	2	0	2	125	8	0	3.84
1990-91	New Jersey	NHL	35	8	12	8	1870	112	0	3.59	....	....	....	....	....	....	....
1991-92	Canada	Nat-Tm	31	18	6	4	1721	75	1	2.61	....	....	....	....	....	....	....
	Canada	Olympics	7	5	2	0	429	17	0	2.37	....	....	....	....	....	....	....
	San Diego Gulls	IHL	7	4	2	1	424	17	0	2.41	3	0	3	160	13	0	4.88
1992-93	Hartford	NHL	50	16	27	3	2656	184	0	4.16	....	....	....	....	....	....	....
1993-94	Hartford	NHL	47	17	24	5	2750	137	2	2.99	....	....	....	....	....	....	....
1994-95	Hartford	NHL	42	17	19	4	2418	108	0	2.68	....	....	....	....	....	....	....
1995-96	Hartford	NHL	66	28	28	6	3669	190	4	3.11	....	....	....	....	....	....	....
1996-97	Hartford	NHL	51	22	22	6	2985	134	4	2.69	....	....	....	....	....	....	....
1997-98	Carolina	NHL	25	7	11	5	1415	66	1	2.80	....	....	....	....	....	....	....
	Vancouver	NHL	16	2	9	4	838	49	0	3.51	....	....	....	....	....	....	....
	Philadelphia	NHL	11	7	3	0	632	27	1	2.56	5	1	4	283	17	0	3.60
1998-99	Florida	NHL	59	21	24	14	3402	151	3	2.66	....	....	....	....	....	....	....
99-2000	Florida	NHL	7	2	5	0	418	18	0	2.58	....	....	....	....	....	....	....
	Phoenix	NHL	35	17	14	3	2074	88	3	2.55	5	1	4	296	16	0	3.24
2000-01	Phoenix	NHL	62	25	22	13	3644	138	4	2.27	....	....	....	....	....	....	....
2001-02	Phoenix	NHL	60	33	21	6	3587	137	5	2.29	5	1	4	297	13	0	2.63
	NHL Totals		693	276	295	92	39551	1979	31	3.00	34	12	22	2002	111	1	3.33

Played in NHL All-Star Game (1989, 2001, 2002)

Traded to **Hartford** by **New Jersey** with Eric Weinrich for Bobby Holik and Hartford's 2nd round choice (Jay Pandolfo) in 1993 Entry Draft, August 28, 1992. Transferred to **Carolina** after **Hartford** franchise relocated, June 25, 1997. Traded to **Vancouver** by **Carolina** with Geoff Sanderson and Enrico Ciccone for Kirk McLean and Martin Gelinas, January 3, 1998. Traded to **Philadelphia** by **Vancouver** for Garth Snow, March 4, 1998. Signed as a free agent by **Florida**, September 12, 1998. Traded to **Phoenix** by **Florida** with Florida's 5th round choice (Nate Kiser) in 2000 Entry Draft for Mikhail Shtalenkov and Phoenix's 4th round choice (Chris Eade) in 2000 Entry Draft, November 18, 1999.

CARON, Sebastian (KAIR-aw, suh-BAS-tee-yeh) PIT.

Goaltender. Catches left. 6'1", 167 lbs. Born, Amqui, Que., June 25, 1980.
(Pittsburgh's 4th choice, 86th overall, in 1999 Entry Draft).

			Regular Season								Playoffs						
Season	Club	Lea	GP	W	L	T	Mins	GA	SO	Avg	GP	W	L	Mins	GA	SO	Avg
1997-98	TGV Pentagone	QAHA	17	....	....	....	762	48	1	2.84	....	....	....	....	....	....	....
1998-99	Rimouski Oceanic	QMJHL	30	13	10	3	1570	85	0	3.25	2	1	0	68	0	0	0.00
99-2000	Rimouski Oceanic	QMJHL	54	*38	11	3	3040	179	1	3.53	14	*12	2	828	50	0	3.62
2000-01	Wilkes-Barre	AHL	30	12	14	3	1746	103	4	3.54	....	....	....	....	....	....	....
2001-02	Wilkes-Barre	AHL	46	14	22	8	2671	139	1	3.12	....	....	....	....	....	....	....

Memorial Cup All-Star Team (2000) • Won Hap Emms Memorial Trophy (Memorial Cup Tournament Top Goaltender) (2000)

CASSIVI, Frederic (KASS-ih-vee, FREHD-uhr-ihk) ATL.

Goaltender. Catches left. 6'4", 220 lbs. Born, Sorel, Que., June 12, 1975.
(Ottawa's 7th choice, 210th overall, in 1994 Entry Draft).

			Regular Season								Playoffs						
Season	Club	Lea	GP	W	L	T	Mins	GA	SO	Avg	GP	W	L	Mins	GA	SO	Avg
1991-92	Abitibi	QAAA	22	5	17	0	1320	106	0	4.84	3	1	2	180	15	0	5.06
1992-93							Statistics not available										
1993-94	St-Hyacinthe	QMJHL	35	15	13	3	1751	127	1	4.35	....	....	....	....	....	....	....
1994-95	Halifax	QMJHL	24	9	12	1	1362	105	0	4.63	....	....	....	....	....	....	....
	St-Jean Lynx	QMJHL	19	12	6	0	1021	55	1	3.23	5	2	3	258	18	0	4.19
1995-96	Thunder Bay	ColHL	12	6	4	2	715	51	0	4.28	....	....	....	....	....	....	....
	P.E.I. Senators	AHL	41	20	14	3	2347	128	1	3.27	5	2	3	317	24	0	4.54
1996-97	Syracuse Crunch	AHL	55	23	22	8	3069	164	2	3.21	1	0	1	60	3	0	3.01
1997-98	Worcester	AHL	45	20	22	2	2593	140	1	3.24	6	3	3	326	18	0	3.31
1998-99	Cincinnati	IHL	44	21	17	2	2418	123	1	3.05	3	1	2	139	6	0	2.59
99-2000	Hershey Bears	AHL	31	14	9	3	1554	78	1	3.01	2	0	1	63	5	0	4.75
2000-01	Hershey Bears	AHL	49	17	24	3	2620	124	2	2.84	9	7	2	564	14	1	*1.49
2001-02	Hershey Bears	AHL	21	6	10	4	1201	50	0	2.50	....	....	....	....	....	....	....
	Atlanta	NHL	6	2	3	0	307	17	0	3.32	....	....	....	....	....	....	....
	Chicago Wolves	AHL	12	6	2	1	625	26	0	2.50	5	2	2	264	11	0	2.50
	NHL Totals		6	2	3	0	307	17	0	3.32	....	....	....	....	....	....	....

Signed as a free agent by **Colorado**, August 17, 1999. Traded to **Atlanta** by **Colorado** for Brett Clark, January 24, 2002.

CECHMANEK, Roman (chehkh-MAN-ehk, ROH-muhn) PHI.

Goaltender. Catches left. 6'3", 187 lbs. Born, Gottwaldov, Czech., March 2, 1971.
(Philadelphia's 3rd choice, 171st overall, in 2000 Entry Draft).

			Regular Season								Playoffs						
Season	Club	Lea	GP	W	L	T	Mins	GA	SO	Avg	GP	W	L	Mins	GA	SO	Avg
1988-89	TJ Gottwaldov	Czech	1	0	0	0	13	0	0	0.00	....	....	....	....	....	....	....
1989-90	TJ Zlin	Czech	2	0	0	0	89	5	0	3.37	....	....	....	....	....	....	....
1990-91	Dukla Jihlava	Czech	9	....	....	....	447	18	2	2.42	....	....	....	....	....	....	....
1991-92	DS Olomouc	Czech	13	....	....	....	731	54	0	4.43	....	....	....	....	....	....	....
	AC ZPS Zlin	Czech	2	0	1	0	67	8	0	7.73	....	....	....	....	....	....	....
1992-93	Banik Hodonin	Czech-2					Statistics not available										
1993-94	Zbojkova Vsetin	Czech-2					Statistics not available										
1994-95	HC Dadak Vsetin	Czech	41	....	....	....	2413	98	5	2.44	11	....	....	619	23	1	2.23
1995-96	HC Petra Vsetin	Czech	36	....	....	....	2142	77	4	2.16	13	....	....	783	17	2	1.30
1996-97	HC Petra Vsetin	Czech	48	....	....	....	2762	98	3	2.13	10	....	....	602	11	2	1.10
1997-98	HC Petra Vsetin	Czech	41	....	....	....	2306	76	....	*1.98	10	....	....	600	16	1	*1.60
1998-99	Vsetin	Czech	45	....	....	....	2696	77	5	*1.71	12	8	4	*747	23	1	1.85
99-2000	Vsetin	Czech	37	....	....	....	2141	88	0	2.47	9	5	4	545	15	3	1.65
2000-01	Philadelphia	NHL	59	35	15	6	3431	115	10	2.01	6	2	4	347	18	0	3.11
	Philadelphia	AHL	3	1	1	0	160	3	0	1.12	....	....	....	....	....	....	....
2001-02	Philadelphia	NHL	46	24	13	6	2603	89	4	2.05	4	1	3	227	7	1	1.85
	Czech Republic	Olympics					Did not play – spare goaltender										
	NHL Totals		105	59	28	12	6034	204	14	2.03	10	3	7	574	25	1	2.61

NHL Second All-Star Team (2001) • Played in NHL All-Star Game (2001)

CENTOMO, Sebastien (sehn-TOH-moh, suh-BAS-tee-yeh) TOR.

Goaltender. Catches right. 6'1", 193 lbs. Born, Montreal, Que., March 26, 1981.

			Regular Season								Playoffs						
Season	Club	Lea	GP	W	L	T	Mins	GA	SO	Avg	GP	W	L	Mins	GA	SO	Avg
1997-98	Laval Laurentide	QAAA	30	16	11	2	1696	87	0	2.86	....	....	....	....	....	....	....
1998-99	Rouyn-Noranda	QMJHL	32	14	9	4	1658	104	1	3.76	2	0	1	28	5	0	10.71
99-2000	Rouyn-Noranda	QMJHL	50	24	17	3	2758	160	1	3.48	11	6	5	695	41	0	3.54
2000-01	Rouyn-Noranda	QMJHL	46	25	14	4	2599	158	3	3.65	1	0	1	60	4	0	4.00
2001-02	**Toronto**	**NHL**	**1**	**0**	**0**	**0**	**40**	**3**	**0**	**4.50**	....	....	....	....	....	....	....
	Memphis	CHL	19	16	1	0	1035	36	1	2.09	....	....	....	....	....	....	....
	St. John's	AHL	25	12	7	4	1429	60	2	2.52	11	4	6	691	29	2	2.52
	NHL Totals		**1**	**0**	**0**	**0**	**40**	**3**	**0**	**4.50**	....	....	....	....	....	....	....

CHL Rookie of the Year (2002)

Signed as a free agent by **Toronto**, September 10, 1999.

CHARPENTIER, Sebastien (shahr-PUHNT-yay, suh-BAS-tee-yeh) WSH.

Goaltender. Catches left. 5'9", 177 lbs. Born, Drummondville, Que., April 18, 1977.
(Washington's 4th choice, 93rd overall, in 1995 Entry Draft).

			Regular Season								Playoffs						
Season	Club	Lea	GP	W	L	T	Mins	GA	SO	Avg	GP	W	L	Mins	GA	SO	Avg
1991-92	Drummondville	QAHA	14	....	....	....	840	34	2	2.42	....	....	....	....	....	....	....
1992-93	Drummondville	QAHA	20	....	....	....	1215	37	*7	*1.80	....	....	....	....	....	....	....
1993-94	Magog	QAAA	24	14	5	1	1443	75	1	3.16	....	....	....	....	....	....	....
1994-95	Laval Titan	QMJHL	41	25	12	1	2152	99	2	2.76	16	9	4	886	45	0	3.05
1995-96	Laval Titan	QMJHL	18	4	10	0	938	97	0	6.20	....	....	....	....	....	....	....
	Val-d'Or Foreurs	QMJHL	33	21	9	1	1906	87	1	2.74	13	7	5	740	45	0	3.64
1996-97	Shawinigan	QMJHL	*62	*37	17	4	*3480	177	1	3.05	4	2	1	196	13	0	3.98
1997-98	Portland Pirates	AHL	4	1	3	0	229	10	0	2.61	....	....	....	....	....	....	....
	Hampton Roads	ECHL	43	20	16	6	2388	114	0	2.86	18	*14	4	*1183	38	1	*1.93
1998-99	Quad City	UHL	6	0	0	0	4	0	0	0.00	....	....	....	....	....	....	....
	Portland Pirates	AHL	3	0	3	0	180	10	0	3.34	....	....	....	....	....	....	....
99-2000	Portland Pirates	AHL	18	10	4	3	1041	48	0	2.77	3	1	1	183	9	0	2.96
2000-01	Portland Pirates	AHL	34	16	16	1	1978	113	1	3.43	1	0	1	102	3	0	1.76
2001-02	**Washington**	**NHL**	**2**	**1**	**1**	**0**	**122**	**5**	**0**	**2.46**	....	....	....	....	....	....	....
	Portland Pirates	AHL	49	20	18	10	2941	131	3	2.67	....	....	....	....	....	....	....
	NHL Totals		**2**	**1**	**1**	**0**	**122**	**5**	**0**	**2.46**	....	....	....	....	....	....	....

QMJHL Second All-Rookie Team (1995) • ECHL Playoff MVP (1998)

CHIODO, Andy (CHEE-aw-doh, AN-dee) NYI

Goaltender. Catches left. 5'11", 201 lbs. Born, Toronto, Ont., April 25, 1983.
(NY Islanders' 3rd choice, 166th overall, in 2001 Entry Draft).

			Regular Season								Playoffs						
Season	Club	Lea	GP	W	L	T	Mins	GA	SO	Avg	GP	W	L	Mins	GA	SO	Avg
1998-99	Wexford Raiders	OPJHL	26	....	....	....	1519	105	0	4.05	....	....	....	....	....	....	....
99-2000	Wexford Raiders	GTHL	24	....	....	....	1389	89	0	3.84	....	....	....	....	....	....	....
2000-01	St. Michael's	OHL	38	18	12	5	2069	86	*4	2.49	9	2	6	479	30	0	3.76
2001-02	St. Michael's	OHL	33	14	10	3	1743	79	2	2.72	7	3	1	288	17	*1	3.54

CHOUINARD, Mathieu (SHWEE-nuhr, ma-TEW) OTT.

Goaltender. Catches left. 6'1", 211 lbs. Born, Laval, Que., April 11, 1980.
(Ottawa's 2nd choice, 45th overall, in 2000 Entry Draft).

			Regular Season								Playoffs						
Season	Club	Lea	GP	W	L	T	Mins	GA	SO	Avg	GP	W	L	Mins	GA	SO	Avg
1995-96	Amos Forestiers	QAAA	31	14	14	1	1613	114	1	4.24	1	1	2	190	11	0	3.48
1996-97	Shawinigan	QMJHL	17	4	7	1	795	51	0	3.85	4	1	3	264	15	0	3.41
1997-98	Shawinigan	QMJHL	55	*32	18	3	3055	142	2	2.79	6	2	4	348	24	0	4.14
1998-99	Shawinigan	QMJHL	56	36	16	4	3288	150	*5	2.74	6	2	4	392	27	0	4.13
99-2000	Shawinigan	QMJHL	*59	32	20	5	*3339	186	4	3.34	13	7	6	769	41	0	3.20
2000-01	Grand Rapids	IHL	28	17	7	1	1567	69	1	2.64	3	1	1	135	4	0	1.78
2001-02	Grand Rapids	AHL	25	11	12	1	1404	58	2	2.48	....	....	....	....	....	....	....

• Re-entered NHL Entry Draft. Originally Ottawa's 1st choice, 15th overall, in 1998 Entry Draft.

QMJHL First All-Star Team (1999) • Shared Harry "Hap" Holmes Memorial Trophy (fewest goals against - AHL) with Martin Prusek and Simon Lajeunesse (2002)

CHOUKALOS, Donald (koo-KAH-luhs, DAWN-ohld)

Goaltender. Catches left. 6'2", 186 lbs. Born, Calgary, Alta., April 11, 1981.
(Boston's 6th choice, 179th overall, in 1999 Entry Draft).

			Regular Season								Playoffs						
Season	Club	Lea	GP	W	L	T	Mins	GA	SO	Avg	GP	W	L	Mins	GA	SO	Avg
1997-98	Cgy. North Stars	AMHL	11	2	8	1	644	59	0	5.50	....	....	....	....	....	....	....
	Calgary Hitmen	WHL	9	4	1	1	404	21	0	3.12	....	....	....	....	....	....	....
1998-99	Calgary Hitmen	WHL	10	8	1	0	571	20	1	2.10	....	....	....	....	....	....	....
	Regina Pats	WHL	24	6	12	4	1337	98	0	4.40	....	....	....	....	....	....	....
99-2000	Regina Pats	WHL	58	26	24	4	3259	194	2	3.57	7	3	4	385	24	0	3.74
2000-01	Regina Pats	WHL	60	*33	23	2	3376	187	3	3.32	6	1	4	315	18	0	3.43
2001-02	Vancouver Giants	WHL	26	6	13	1	1189	91	0	4.59	....	....	....	....	....	....	....
	New Mexico	CHL	8	4	1	2	420	22	0	3.14	....	....	....	....	....	....	....

Claimed by **Vancouver** (WHL) from **Regina** (WHL) in Expansion Draft, June 1, 2001.

CHOVAN, Jan (HOH-van, YAN) TOR.

Goaltender. Catches left. 5'11", 178 lbs. Born, Bratislava, Czech., September 7, 1983.
(Toronto's 10th choice, 213th overall, in 2001 Entry Draft).

			Regular Season								Playoffs						
Season	Club	Lea	GP	W	L	T	Mins	GA	SO	Avg	GP	W	L	Mins	GA	SO	Avg
1998/00	S. Bratislava Jr.	Slovak-Jr.	STATISTICS NOT AVAILABLE														
2000-01	Belleville Bulls	OHL	39	19	15	2	2074	98	2	2.84	10	6	4	618	38	0	3.69
2001-02	Belleville Bulls	OHL	20	10	7	1	1061	61	0	3.45	....	....	....	....	....	....	....
	London Knights	OHL	10	3	5	1	506	36	0	4.27	2	0	0	33	3	0	5.45

Traded to **London** (OHL) by **Belleville** (OHL) with Alex White for Lubos Velebny and Glen Ridler, December 5, 2001.

CLEMMENSEN, Scott (KLEH-mehn-sehn, SKAWT) N.J.

Goaltender. Catches left. 6'2", 205 lbs. Born, Des Moines, IA, July 23, 1977.
(New Jersey's 7th choice, 215th overall, in 1997 Entry Draft).

			Regular Season								Playoffs						
Season	Club	Lea	GP	W	L	T	Mins	GA	SO	Avg	GP	W	L	Mins	GA	SO	Avg
1995-96	Des Moines	USHL	20	10	7	1	1082	62	0	3.44	....	....	....	....	....	....	....
1996-97	Des Moines	USHL	36	22	9	2	2042	111	1	3.26	4	1	2	200	9	1	2.70
1997-98	Boston College	H-East	37	24	9	4	2205	102	*4	2.78	....	....	....	....	....	....	....
1998-99	Boston College	H-East	*42	26	12	4	*2507	120	1	2.87	....	....	....	....	....	....	....
99-2000	Boston College	H-East	29	19	7	0	1610	59	*5	2.20	....	....	....	....	....	....	....
2000-01	Boston College	H-East	*39	*30	7	2	*2312	82	3	2.13	....	....	....	....	....	....	....
2001-02	**New Jersey**	**NHL**	**2**	**0**	**0**	**0**	**20**	**1**	**0**	**3.00**	....	....	....	....	....	....	....
	Albany	AHL	29	5	19	4	1677	92	0	3.29	....	....	....	....	....	....	....
	NHL Totals		**2**	**0**	**0**	**0**	**20**	**1**	**0**	**3.00**	....	....	....	....	....	....	....

NCAA Championship All-Tournament Team (2001)

CLOUTIER, Dan (KLOO-tyay, DAN) VAN.

Goaltender. Catches left. 6'1", 182 lbs. Born, Mont-Laurier, Que., April 22, 1976.
(NY Rangers' 1st choice, 26th overall, in 1994 Entry Draft).

			Regular Season								Playoffs						
Season	Club	Lea	GP	W	L	T	Mins	GA	SO	Avg	GP	W	L	Mins	GA	SO	Avg
1991-92	St. Thomas Stars	OJHL-B	14	....	....	....	823	80	0	5.83	....	....	....	....	....	....	....
1992-93	Timmins	NOJHA	5	4	0	0	255	10	0	2.35	....	....	....	....	....	....	....
	Sault Ste. Marie	OHL	12	4	6	0	572	44	0	4.62	4	1	2	231	12	0	3.12
1993-94	Sault Ste. Marie	OHL	55	28	14	6	2934	174	*2	3.56	14	*10	4	833	52	0	3.75
1994-95	Sault Ste. Marie	OHL	45	15	26	2	2518	185	1	4.41	....	....	....	....	....	....	....
1995-96	Sault Ste. Marie	OHL	13	9	3	0	641	43	0	4.02	....	....	....	....	....	....	....
	Guelph Storm	OHL	17	12	2	2	1004	35	2	2.09	16	11	5	993	52	*2	3.14
1996-97	Binghamton	AHL	60	23	28	8	3367	199	3	3.55	4	1	3	236	13	0	3.31
1997-98	**NY Rangers**	**NHL**	**12**	**4**	**5**	**1**	**551**	**23**	**0**	**2.50**	....	....	....	....	....	....	....
	Hartford	AHL	24	12	8	3	1417	62	0	2.63	8	5	3	478	24	0	3.01
1998-99	**NY Rangers**	**NHL**	**22**	**6**	**8**	**3**	**1097**	**49**	**0**	**2.68**	....	....	....	....	....	....	....
99-2000	**Tampa Bay**	**NHL**	**52**	**9**	**30**	**3**	**2492**	**145**	**0**	**3.49**	....	....	....	....	....	....	....
2000-01	**Tampa Bay**	**NHL**	**24**	**3**	**13**	**3**	**1005**	**59**	**1**	**3.52**	....	....	....	....	....	....	....
	Detroit Vipers	IHL	1	0	1	0	59	3	0	3.05	....	....	....	....	....	....	....
	Vancouver	**NHL**	**16**	**4**	**6**	**5**	**914**	**37**	**0**	**2.43**	**2**	**0**	**2**	**117**	**9**	**0**	**4.62**
2001-02	**Vancouver**	**NHL**	**62**	**31**	**22**	**5**	**3502**	**142**	**7**	**2.43**	**6**	**2**	**3**	**273**	**16**	**0**	**3.52**
	NHL Totals		**188**	**57**	**84**	**20**	**9561**	**455**	**8**	**2.86**	**8**	**2**	**5**	**390**	**25**	**0**	**3.85**

OHL Second All-Star Team (1996)

Traded to **Tampa Bay** by **NY Rangers** with Niklas Sundstrom and NY Rangers' 1st (Nikita Alexeev) and 3rd (later traded to San Jose - later traded to Chicago - Chicago selected Igor Radulov) round choices in 2000 Entry Draft for Chicago's 1st round choice (previously acquired, NY Rangers selected Pavel Brendl) in 1999 Entry Draft, June 26, 1999. Traded to **Vancouver** by **Tampa Bay** for Adrian Aucoin and Vancouver's 2nd round choice (Alexander Polushin) in 2001 Entry Draft, February 7, 2001.

CLOUTIER, Frederic (KLOO-tyay, FREHD-uhr-ihk) MIN.

Goaltender. Catches right. 6', 165 lbs. Born, St-Georges, Que., May 14, 1981.

			Regular Season								Playoffs						
Season	Club	Lea	GP	W	L	T	Mins	GA	SO	Avg	GP	W	L	Mins	GA	SO	Avg
1996-97	Cap-d-Madeleine	QAAA	24	5	16	2	1400	105	0	4.37	....	....	....	....	....	....	....
1997-98	Levis	QAAA	28	18	9	1	1680	106	0	3.77	3	0	3	180	10	0	3.72
1998-99	Acadie-Bathurst	QMJHL	8	2	3	0	383	30	0	4.70	....	....	....	....	....	....	....
99-2000	Acadie-Bathurst	QMJHL	58	16	34	6	3262	208	3	3.83	....	....	....	....	....	....	....
2000-01	Acadie-Bathurst	QMJHL	58	*42	8	2	3270	136	*6	*2.50	9	5	2	467	24	1	3.08
2001-02	Louisiana	ECHL	36	28	5	3	2155	66	*7	*1.84	1	0	1	115	5	0	2.60

QMJHL First All-Star Team (2001) •QMJHL Top Goaltender (2001) • ECHL All-Rookie Team (2002) • ECHL First All-Star Team (2002) • ECHL Rookie of the Year (2002) • ECHL Top Goaltender (2002) • ECHL MVP (2002)

Signed as a free agent by **Minnesota**, November 23, 2001.

CONKLIN, Ty (KAWN-klihn, TIGH) EDM.

Goaltender. Catches left. 6', 180 lbs. Born, Anchorage, AK, March 30, 1976.

			Regular Season								Playoffs						
Season	Club	Lea	GP	W	L	T	Mins	GA	SO	Avg	GP	W	L	Mins	GA	SO	Avg
1995-96	Green Bay	USHL	30	....	....	....	1727	82	1	2.85	....	....	....	....	....	....	....
1996-97	Alaska-Anchorage	WCHA	DID NOT PLAY – FRESHMAN														
	Green Bay	USHL	30	19	7	1	1609	86	1	3.21	17	8	9	980	56	1	3.43
1997-98	New Hampshire	H-East	DID NOT PLAY – TRANSFERRED COLLEGES														
1998-99	New Hampshire	H-East	22	18	3	1	1338	41	0	*1.84	....	....	....	....	....	....	....
99-2000	New Hampshire	H-East	*37	*22	8	6	*2194	91	2	2.49	....	....	....	....	....	....	....
2000-01	New Hampshire	H-East	34	17	12	5	2048	70	*5	*2.05	....	....	....	....	....	....	....
2001-02	**Edmonton**	**NHL**	**4**	**2**	**0**	**0**	**148**	**4**	**0**	**1.62**	....	....	....	....	....	....	....
	Hamilton	AHL	37	13	12	8	2043	89	1	2.61	7	4	2	416	18	0	2.60
	NHL Totals		**4**	**2**	**0**	**0**	**148**	**4**	**0**	**1.62**	....	....	....	....	....	....	....

USHL Second All-Star Team (1996) • Hockey East All-Rookie Team (1999) • Hockey East Second All-Star Team (1999) • Hockey East First All-Star Team (2000, 2001) • Shared Hockey East Player of the Year Award with Mike Mottau (2000) • NCAA East Second All-American Team (2000) • NCAA East First All-American Team (2001) • Shared Walter Brown Award (New England's Outstanding American-born College player) with Brian Gionta, March 14, 2001.

• Left **Alaska-Anchorage** (WCHA) and returned to **Green Bay** (USHL), November 14, 1996. • First goaltender to be named captain of **New Hampshire** (H-East) since 1961, October 5, 2000. Signed as a free agent by **Edmonton**, April 18, 2001.

CRAWFORD-WEST, Brandon (KRAW-fohrd-WEHST, BRAN-duhn) PIT.

Goaltender. Catches right. 5'11", 185 lbs. Born, San Diego, CA, July 1, 1982.
(Pittsburgh's 9th choice, 250th overall, in 2001 Entry Draft).

			Regular Season								Playoffs						
Season	Club	Lea	GP	W	L	T	Mins	GA	SO	Avg	GP	W	L	Mins	GA	SO	Avg
2000-01	Texas Tornado	NAJHL	43	30	9	3	2534	110	4	2.60	7	6	1	452	6	3	0.80
2001-02	Texas Tornado	NAJHL	50	35	11	2	2825	109	4	2.32	6	3	3	389	13	0	2.01

NAJHL All-Rookie Team (2001) • NAJHL Second All-Star Team (2001)

DAFOE, Byron (duh-FOH, BIGH-ruhn)

Goaltender. Catches left. 5'11", 200 lbs. Born, Sussex, England, February 25, 1971.
(Washington's 2nd choice, 35th overall, in 1989 Entry Draft).

			Regular Season								Playoffs						
Season	Club	Lea	GP	W	L	T	Mins	GA	SO	Avg	GP	W	L	Mins	GA	SO	Avg
1987-88	Juan de Fuca	BCJHL	32				1716	129	0	4.51							
1988-89	Portland	WHL	59	29	24	3	3279	291	1	5.32	*18	10	8	*1091	81	*1	4.45
1989-90	Portland	WHL	40	14	21	3	2265	193	0	5.11							
1990-91	Portland	WHL	8	1	5	1	414	41	0	5.94							
	Prince Albert	WHL	32	13	12	4	1839	124	0	4.05							
1991-92	Baltimore	AHL	33	12	16	4	1847	119	0	3.87							
	New Haven	AHL	7	3	2	1	364	22	0	3.63							
	Hampton Roads	ECHL	10	6	4	0	562	26	0	2.78							
1992-93	Washington	NHL	1	0	0	0	1	0	0	0.00							
	Baltimore	AHL	48	16	20	7	2617	191	1	4.38	5	2	3	241	22	0	5.48
1993-94	Washington	NHL	5	2	2	0	230	13	0	3.39	2	0	2	118	5	0	2.54
	Portland Pirates	AHL	47	24	16	4	2661	148	1	3.34	1	0	0	9	1	0	6.79
1994-95	Washington	NHL	4	1	1	1	187	11	0	3.53	1	0	0	20	1	0	3.00
	Phoenix	IHL	49	25	16	6	2743	169	2	3.70							
	Portland Pirates	AHL	6	5	0	0	330	16	0	2.91	7	3	4	416	29	0	4.18
1995-96	Los Angeles	NHL	47	14	24	8	2666	172	1	3.87							
1996-97	Los Angeles	NHL	40	13	17	5	2162	112	0	3.11							
1997-98	Boston	NHL	65	30	25	9	3693	138	6	2.24	6	2	4	422	14	1	1.99
1998-99	Boston	NHL	68	32	23	11	4001	133	*10	1.99	12	6	6	768	26	2	2.03
99-2000	Boston	NHL	41	13	16	10	2307	114	3	2.96							
2000-01	Boston	NHL	45	22	14	7	2536	101	2	2.39							
2001-02	Boston	NHL	64	35	26	3	3827	141	4	2.21	6	2	4	358	19	0	3.18
	NHL Totals		380	162	148	54	21610	935	26	2.60	27	10	16	1686	65	3	2.31

AHL First All-Star Team (1994) • Shared Harry "Hap" Holmes Memorial Trophy (fewest goals against - AHL) with Olaf Kolzig (1994) • NHL Second All-Star Team (1999)

Traded to **LA Kings** by **Washington** with Dmitri Khristich for LA Kings' 1st round choice (Alexandre Volchkov) and Dallas' 4th round choice (previously acquired, Washington selected Justin Davis) in 1996 Entry Draft, July 8, 1995. Traded to **Boston** by **LA Kings** with Dimitri Khristich for Jozef Stumpel, Sandy Moger and Boston's 4th round choice (later traded to New Jersey - New Jersey selected Pierre Dagenais) in 1998 Entry Draft, August 29, 1997.

DAIGNEAULT, Maxime (DAYN-yoh, mahx-EEM) WSH.

Goaltender. Catches left. 6'1", 185 lbs. Born, St-Jacques-le-Mineur, Que., January 23, 1984.
(Washington's 4th choice, 59th overall, in 2002 Entry Draft).

			Regular Season								Playoffs						
Season	Club	Lea	GP	W	L	T	Mins	GA	SO	Avg	GP	W	L	Mins	GA	SO	Avg
99-2000	Cap-d-Madeleine	QAAA	19	12	3	3	1108	53	3	2.87	18	12	5	945	42	1	2.67
2000-01	Val-d'Or Foreurs	QMJHL	28	14	8	1	1386	82	0	3.55	10	8	1	504	21	0	2.50
2001-02	Val-d'Or Foreurs	QMJHL	61	25	27	5	3270	184	3	3.38	7	3	4	431	23	0	3.20

Memorial Cup All-Star Team (2002) • Won Hap Emms Memorial Trophy (Memorial Cup Tournament Top Goaltender) (2002)

DAMPHOUSSE, Jean-Francois (DAHM-fooz, ZHAWN-fran-SWUH) ANA.

Goaltender. Catches left. 6', 180 lbs. Born, St-Alexis-des-Monts, Que., July 21, 1979.
(New Jersey's 1st choice, 24th overall, in 1997 Entry Draft).

			Regular Season								Playoffs						
Season	Club	Lea	GP	W	L	T	Mins	GA	SO	Avg	GP	W	L	Mins	GA	SO	Avg
1993-94	Ste-Foy	QAHA	18	10	1	0	1078	53	0	2.95	14	10	4	842	50	0	3.52
1994-95	Ste-Foy	QAHA	16				958	48	0	3.01							
	Ste-Foy	QAAA	2	1	0	1	120	8	0	3.84							
1995-96	Ste-Foy	QAAA	32	18	10	1	1629	83	2	3.06							
1996-97	Moncton Wildcats	QMJHL	39	6	25	2	2063	190	0	5.53							
1997-98	Moncton Wildcats	QMJHL	59	24	26	6	3400	174	1	3.07	10	5	5	595	28	0	2.82
1998-99	Moncton Wildcats	QMJHL	40	19	17	2	2163	121	1	3.36	4	0	4	200	12	0	3.60
	Albany	AHL	1	0	1	0	59	3	0	3.06							
99-2000	Augusta Lynx	ECHL	14	4	7	0	676	49	0	4.35							
	Albany	AHL	26	9	11	2	1326	62	0	2.81	2	0	1	62	4	0	3.86
2000-01	Albany	AHL	55	24	23	3	2963	141	1	2.86							
2001-02	New Jersey	NHL	6	1	3	0	294	12	0	2.45							
	Albany	AHL	18	3	11	2	1001	57	0	3.42							
	NHL Totals		6	1	3	0	294	12	0	2.45							

Traded to **Anaheim** by **New Jersey** with Petr Sykora, Mike Commodore and Igor Pohanka for Jeff Friesen, Oleg Tverdovsky and Maxim Balmochnykh, July 6, 2002.

DENIKE, Terry (deh-NIGHK, TEHR-ee) L.A.

Goaltender. Catches left. 6'2", 190 lbs. Born, Burlington, Ont., April 16, 1981.
(Los Angeles' 7th choice, 152nd overall, in 2001 Entry Draft).

			Regular Season								Playoffs						
Season	Club	Lea	GP	W	L	T	Mins	GA	SO	Avg	GP	W	L	Mins	GA	SO	Avg
99-2000	Weyburn	SJHL	38	25	7	5	2171	108	2	2.99							
2000-01	Weyburn	SJHL	43	28	14	0	2528	107	*6	*2.54	17	13	4	1005	47	0	2.81
2001-02	Lake Superior	CCHA	16	3	10	0	700	42	0	3.60							

SJHL Second All-Star Team (2001) • SJHL Dedication and Sportsmanship Award (2001) • SJHL Top Goaltender (2001)

DENIS, Marc (deh-NEE, MAHRK) CBJ

Goaltender. Catches left. 6'1", 190 lbs. Born, Montreal, Que., August 1, 1977.
(Colorado's 1st choice, 25th overall, in 1995 Entry Draft).

			Regular Season								Playoffs						
Season	Club	Lea	GP	W	L	T	Mins	GA	SO	Avg	GP	W	L	Mins	GA	SO	Avg
1992-93	Mtl-Bourassa	QAAA	26				1559	74	5	2.87							
1993-94	Trois-Rivieres	QAAA	36	10	22	3	2093	158	0	4.53	4	1	3	249	20	0	4.83
1994-95	Chicoutimi	QMJHL	32	17	9	1	1688	98	0	3.48	6	4	2	372	19	1	3.06
1995-96	Chicoutimi	QMJHL	51	23	21	4	2951	157	2	3.19	16	8	8	957	69	0	4.33
1996-97	Chicoutimi	QMJHL	41	22	15	2	2323	104	4	*2.69	*21	*11	10	*1229	70	*1	3.42
	Colorado	NHL	1	0	1	0	60	3	0	3.00							
	Hershey Bears	AHL									4	1	0	56	1	0	1.08
1997-98	Hershey Bears	AHL	47	17	23	4	2588	125	1	2.90	6	3	3	346	15	0	2.59
1998-99	Colorado	NHL	4	1	1	1	217	9	0	2.49							
	Hershey Bears	AHL	52	20	23	5	2908	137	4	2.83	3	1	1	143	7	0	2.93
99-2000	Colorado	NHL	23	9	8	3	1203	51	3	2.54							
2000-01	Columbus	NHL	32	6	20	4	1830	99	0	3.25							
2001-02	Columbus	NHL	42	9	24	5	2335	121	1	3.11							
	NHL Totals		102	25	54	13	5645	283	4	3.01							

QMJHL First All-Star Team (1997) • Canadian Major Junior First All-Star Team (1997) • Canadian Major Junior Goaltender of the Year (1997)

Traded to **Columbus** by **Colorado** for Columbus' 2nd round choice (later traded to Carolina - Carolina selected Tomas Kurka) in 2000 Entry Draft, June 7, 2000.

DESLAURIERS, Jeff (duh-LAW-ree-yay, JEHF) EDM.

Goaltender. Catches right. 6'3", 175 lbs. Born, St-Jean-Richelieu, Que., May 15, 1984.
(Edmonton's 2nd choice, 31st overall, in 2002 Entry Draft).

			Regular Season								Playoffs						
Season	Club	Lea	GP	W	L	T	Mins	GA	SO	Avg	GP	W	L	Mins	GA	SO	Avg
2000-01	Gatineau	QAAA	22	10	9	2	1194	61	2	3.07	2	1	0	125	6	0	2.89
2001-02	Chicoutimi	QMJHL	51	28	20	1	2909	170	1	3.51	4	0	3	197	20	0	6.11

DesROCHERS, Patrick (duh-RAWSH-ay, PAT-rihk) PHX.

Goaltender. Catches left. 6'4", 208 lbs. Born, Penetanguishene, Ont., October 27, 1979.
(Phoenix's 1st choice, 14th overall, in 1998 Entry Draft).

			Regular Season								Playoffs						
Season	Club	Lea	GP	W	L	T	Mins	GA	SO	Avg	GP	W	L	Mins	GA	SO	Avg
1994-95	Barrie Colts	OPJHL	26				3205	179	3	3.08							
1995-96	Sarnia Sting	OHL	29	12	6	2	1265	96	0	4.55	3	0	1	71	5	0	4.23
1996-97	Sarnia Sting	OHL	50	22	17	4	2667	154	*4	3.46	11	6	5	576	42	0	4.38
1997-98	Sarnia Sting	OHL	56	26	17	11	3205	179	1	3.35	4	1	2	160	12	0	4.50
1998-99	Sarnia Sting	OHL	8	3	5	0	425	26	0	3.67							
	Kingston	OHL	44	14	22	3	2389	177	1	4.45	5	1	4	323	21	0	3.90
99-2000	Springfield	AHL	52	21	17	7	2710	137	1	3.03	2	1	1	120	7	1	3.50
2000-01	Springfield	AHL	50	17	24	5	2807	156	0	3.33							
2001-02	Phoenix	NHL	5	1	2	1	243	15	0	3.70							
	Springfield	AHL	34	12	18	1	1864	94	2	3.03							
	NHL Totals		5	1	2	1	243	15	0	3.70							

Traded to **Kingston** (OHL) by **Sarnia** (OHL) for Matt Price and Curtis Cruickshank, November 26, 1998.

DiPIETRO, Rick (dee-pee-EHT-roh, RIHK) NYI

Goaltender. Catches right. 5'11", 185 lbs. Born, Winthrop, MA, September 19, 1981.
(NY Islanders' 1st choice, 1st overall, in 2000 Entry Draft).

			Regular Season								Playoffs						
Season	Club	Lea	GP	W	L	T	Mins	GA	SO	Avg	GP	W	L	Mins	GA	SO	Avg
1997-98	Team USA	USDP-18	46	21	19	4	2526	131	2	3.11							
1998-99	Team USA	USDP-18	46	31	11	2	2760	113	2	2.46							
99-2000	Boston University	H-East	29	18	5	5	1790	73	2	2.45							
2000-01	NY Islanders	NHL	20	3	15	1	1083	63	0	3.49							
	Chicago Wolves	IHL	14	4	5	2	778	44	0	3.39							
2001-02	Bridgeport	AHL	59	*30	22	7	3472	134	4	2.32	20	12	8	*1270	45	*3	2.13
	NHL Totals		20	3	15	1	1083	63	0	3.49							

Hockey East Second All-Star Team (2000) • Hockey East Rookie of the Year (2000)

DIVIS, Reinhard (DIH-vihs, RIGHN-hard) ST.L.

Goaltender. Catches left. 5'11", 200 lbs. Born, Vienna, Austria, July 4, 1975.
(St. Louis' 8th choice, 261st overall, in 2000 Entry Draft).

			Regular Season								Playoffs						
Season	Club	Lea	GP	W	L	T	Mins	GA	SO	Avg	GP	W	L	Mins	GA	SO	Avg
1995-96	VEU Feldkirch	Austria	37				2200	85	0	2.32							
1996-97	VEU Feldkirch	Alpenliga	45				2738	105	0	2.30							
	VEU Feldkirch	Austria									11			620	27	0	2.61
1997-98	VEU Feldkirch	Alpenliga	13				779	22	0	1.69							
	VEU Feldkirch	Austria	27				1620	55	0	2.07							
1998-99	VEU Feldkirch	Austria	15				900	58	0	3.86							
99-2000	Leksands IF	Sweden	48				2839	160	3	3.38							
2000-01	Leksands IF	Sweden	41				2451	141	3	3.45							
2001-02	St. Louis	NHL	1	0	0	0	25	0	0	0.00							
	Worcester	AHL	55	28	20	5	3173	137	3	2.59	3	1	2	205	8	0	2.34
	Austria	Olympics	4	1	1	2	238	12	0	3.02							
	NHL Totals		1	0	0	0	25	0	0	0.00							

DUBA, Tomas (DOO-bah, TAW-mash) PIT.

Goaltender. Catches left. 6', 176 lbs. Born, Prague, Czech., July 2, 1981.
(Pittsburgh's 8th choice, 217th overall, in 2001 Entry Draft).

			Regular Season								Playoffs						
Season	Club	Lea	GP	W	L	T	Mins	GA	SO	Avg	GP	W	L	Mins	GA	SO	Avg
1998-99	Sparta Praha Jr.	Czech-Jr.	34				1850	95		3.08							
99-2000	Sparta Praha Jr.	Czech-Jr.	30				1670	73		2.62							
	HC CKD Slany	Czech-3	1	0	1	0	60	5	0	5.00							
2000-01	Sparta Praha Jr.	Czech-Jr.	14				774	41	0	3.18	2			60	6	0	6.00
	Beroun	Czech-2	8				426	18		2.54							
2001-02	SaiPa	Finland	47	10	31	4	2755	152	3	3.31							

DUNHAM, Mike (DUHN-uhm, MIGHK) NSH.

Goaltender. Catches left. 6'3", 200 lbs. Born, Johnson City, NY, June 1, 1972.
(New Jersey's 4th choice, 53rd overall, in 1990 Entry Draft).

			Regular Season								Playoffs						
Season	Club	Lea	GP	W	L	T	Mins	GA	SO	Avg	GP	W	L	Mins	GA	SO	Avg
1987-88	Canterbury School	H.S.	29				1740	69	4	2.38							
1988-89	Canterbury School	H.S.	25				1500	63	2	2.52							
1989-90	Canterbury School	H.S.	32				1558	68	3	1.96							
1990-91	U. of Maine	H-East	23	14	5	2	1275	63	0	*2.96							
1991-92	U. of Maine	H-East	7	6	0	0	382	14	1	2.20							
	United States	Nat-Tm	3	0	1	1	157	10	0	3.82							
1992-93	U. of Maine	H-East	25	*21	1	1	1429	63	0	2.65							
1993-94	United States	Nat-Tm	33	22	9	2	1983	125	2	3.78							
	United States	Olympics	3	0	1	2	180	15	0	5.00							
	Albany	AHL	5	2	2	1	304	26	0	5.12							
1994-95	Albany	AHL	35	20	7	8	2120	99	1	2.80	7	6	1	419	20	1	2.86
1995-96	Albany	AHL	44	30	10	2	2592	109	1	2.52	3	1	2	182	5	1	1.65
1996-97	New Jersey	NHL	26	8	7	1	1013	43	2	2.55							
	Albany	AHL	3	1	1	1	184	12	0	3.91							
1997-98	New Jersey	NHL	15	5	5	3	773	29	1	2.25							
1998-99	Nashville	NHL	44	16	23	3	2472	127	1	3.08							
99-2000	Nashville	NHL	52	19	27	6	3077	146	0	2.85							
	Milwaukee	IHL	1	1	0	0	60	1	0	1.00							
2000-01	Nashville	NHL	48	21	21	4	2810	107	4	2.28							
2001-02	Nashville	NHL	58	23	24	9	3316	144	3	2.61							
	United States	Olympics	1	1	0	0	60	0	*1	0.00							
	NHL Totals		243	92	107	26	13461	596	11	2.66							

Hockey East First All-Star Team (1993) • NCAA East First All-American Team (1993) • Shared Harry "Hap" Holmes Memorial Trophy (fewest goals against - AHL) with Corey Schwab (1995) • Shared Jack A. Butterfield Trophy (Playoff MVP - AHL) with Corey Schwab (1995) • AHL Second All-Star Team (1996) • Shared William M. Jennings Trophy with Martin Brodeur (1997)

Claimed by **Nashville** from **New Jersey** in Expansion Draft, June 26, 1998.

EKLUND, Brian (EHK-luhnd, BRIGH-uhn) T.B.

Goaltender. Catches left. 6'5", 200 lbs. Born, Quincy, MA, May 24, 1980.
(Tampa Bay's 8th choice, 226th overall, in 2000 Entry Draft).

			Regular Season								Playoffs						
Season	Club	Lea	GP	W	L	T	Mins	GA	SO	Avg	GP	W	L	Mins	GA	SO	Avg
1997-98	Archbishop Prep	H.S.	22				1320	40	*6	*1.84							
1998-99	Brown U.	ECAC	8	1	3	0	299	17	0	3.41							
99-2000	Brown U.	ECAC	12	1	6	2	569	28	1	2.95							
2000-01	Brown U.	ECAC	19	2	13	3	1084	62	0	3.43							
2001-02	Brown U.	ECAC	9	3	5	0	454	30	0	3.97							

School Sports Hockey Player of the Year (1998) • HNIB Division 1 Goalie of the Year (1998)

ELLIS, Dan (EHL-ihs, DAN) DAL.

Goaltender. Catches left. 6', 185 lbs. Born, Saskatoon, Sask., June 19, 1980.
(Dallas' 2nd choice, 60th overall, in 2000 Entry Draft).

			Regular Season								Playoffs						
Season	**Club**	**Lea**	**GP**	**W**	**L**	**T**	**Mins**	**GA**	**SO**	**Avg**	**GP**	**W**	**L**	**Mins**	**GA**	**SO**	**Avg**
1998-99	Newmarket	OPJHL	28	24	3	1	1670	63	3	2.25	....	....	....	....	....	....	....
99-2000	Omaha Lancers	USHL	55	*34	16	4	*3274	123	*11	*2.25	4	1	3	238	10	0	2.52
2000-01	Nebraska-Omaha	CCHA	40	21	14	3	2285	95	2	2.49	....	....	....	....	....	....	....
2001-02	Nebraska-Omaha	CCHA	40	20	15	4	2405	97	3	2.42	....	....	....	....	....	....	....

USHL First All-Star Team (2000) • USHL Goaltender of the Year (2000) • USHL Player of the Year (2000) • CCHA Second All-Star Team (2002)

EMERY, Ray (EH-muhr-ee, RAY) OTT.

Goaltender. Catches left. 6'3", 192 lbs. Born, Cayuga, Ont., September 28, 1982.
(Ottawa's 4th choice, 99th overall, in 2001 Entry Draft).

			Regular Season								Playoffs						
Season	**Club**	**Lea**	**GP**	**W**	**L**	**T**	**Mins**	**GA**	**SO**	**Avg**	**GP**	**W**	**L**	**Mins**	**GA**	**SO**	**Avg**
1998-99	Dunnville	OJHL-C	22	3	19	0	1320	140	0	6.37	....	....	....	....	....	....	....
99-2000	Welland Cougars	OJHL-B	23	13	10	1	1323	62	1	2.68	....	....	....	....	....	....	....
	Sault Ste. Marie	OHL	16	9	3	0	716	36	1	3.02	15	8	7	883	33	*3	2.24
2000-01	Sault Ste. Marie	OHL	52	18	29	2	2938	174	1	3.55	....	....	....	....	....	....	....
2001-02	Sault Ste. Marie	OHL	*59	*33	17	9	*3477	158	4	2.73	6	2	4	360	19	*1	3.17

OHL First All-Star Team (2002) • Canadian Major Junior First All-Star Team (2002) • Canadian Major Junior Goaltender of the Year (2002)

ESCHE, Robert (EHSH, RAW-buhrt) PHI.

Goaltender. Catches left. 6'1", 210 lbs. Born, Whitesboro, NY, January 22, 1978.
(Phoenix's 5th choice, 139th overall, in 1996 Entry Draft).

			Regular Season								Playoffs						
Season	**Club**	**Lea**	**GP**	**W**	**L**	**T**	**Mins**	**GA**	**SO**	**Avg**	**GP**	**W**	**L**	**Mins**	**GA**	**SO**	**Avg**
1994-95	Gloucester	COJHL	20	10	6	0	1034	70	0	4.06	....	....	....	....	....	....	....
1995-96	Detroit	OHL	23	13	6	0	1219	76	1	3.74	3	0	2	105	4	0	2.29
1996-97	Detroit	OHL	58	24	28	2	3241	206	2	3.81	5	1	4	317	19	0	3.60
1997-98	Plymouth Whalers	OHL	48	29	13	4	2810	135	3	2.88	15	8	7	869	45	0	3.11
1998-99	**Phoenix**	**NHL**	**3**	**0**	**1**	**0**	**130**	**7**	**0**	**3.23**	....	....	....	....	....	....	....
	Springfield	AHL	55	24	20	6	2957	138	1	2.80	1	0	1	60	4	0	4.02
99-2000	**Phoenix**	**NHL**	**8**	**2**	**5**	**0**	**408**	**23**	**0**	**3.38**	....	....	....	....	....	....	....
	Houston Aeros	IHL	7	4	2	1	419	16	2	2.29	....	....	....	....	....	....	....
	Springfield	AHL	21	9	9	2	1207	61	2	3.03	3	1	2	180	12	0	4.01
2000-01	**Phoenix**	**NHL**	**25**	**10**	**8**	**4**	**1350**	**68**	**2**	**3.02**	....	....	....	....	....	....	....
2001-02	**Phoenix**	**NHL**	**22**	**6**	**10**	**2**	**1145**	**52**	**1**	**2.72**	....	....	....	....	....	....	....
	Springfield	AHL	1	1	0	0	60	0	1	0.00	....	....	....	....	....	....	....
	NHL Totals		**58**	**18**	**24**	**6**	**3033**	**150**	**3**	**2.97**	....	....	....	....	....	....	....

OHL Second All-Star Team (1998)

Traded to **Philadelphia** by **Phoenix** with Michal Handzus for Brian Boucher and Nashville's 3rd round choice (previously acquired, Phoenix selected Joe Callahan) in 2002 Entry Draft, June 12, 2002.

ESSENSA, Bob (EH-sehn-suh, BAWB)

Goaltender. Catches left. 6', 190 lbs. Born, Toronto, Ont., January 14, 1965.
(Winnipeg's 5th choice, 71st overall, in 1983 Entry Draft).

			Regular Season								Playoffs						
Season	**Club**	**Lea**	**GP**	**W**	**L**	**T**	**Mins**	**GA**	**SO**	**Avg**	**GP**	**W**	**L**	**Mins**	**GA**	**SO**	**Avg**
1981-82	Henry Carr	MTJHL	17	....	....	....	948	79	0	4.99	....	....	....	....	....	....	....
1982-83	Henry Carr	MTJHL	31	....	....	....	1840	98	2	3.20	....	....	....	....	....	....	....
	Markham Waxers	MTJHL	1	1	0	0	60	1	0	1.00	....	....	....	....	....	....	....
1983-84	Michigan State	CCHA	17	11	4	0	946	44	2	2.79	....	....	....	....	....	....	....
1984-85	Michigan State	CCHA	18	15	2	0	1059	29	2	1.64	....	....	....	....	....	....	....
1985-86	Michigan State	CCHA	23	17	4	1	1333	74	1	3.33	....	....	....	....	....	....	....
1986-87	Michigan State	CCHA	25	19	3	1	1383	64	2	2.78	....	....	....	....	....	....	....
1987-88	Moncton Hawks	AHL	27	7	11	1	1287	100	1	4.66	....	....	....	....	....	....	....
1988-89	**Winnipeg**	**NHL**	**20**	**6**	**8**	**3**	**1102**	**68**	**1**	**3.70**	....	....	....	....	....	....	....
	Fort Wayne	IHL	22	14	7	0	1287	70	0	3.26	....	....	....	....	....	....	....
1989-90	**Winnipeg**	**NHL**	**36**	**18**	**9**	**5**	**2035**	**107**	**1**	**3.15**	**4**	**2**	**1**	**206**	**12**	**0**	**3.50**
	Moncton Hawks	AHL	6	3	3	0	358	15	0	2.51	....	....	....	....	....	....	....
1990-91	**Winnipeg**	**NHL**	**55**	**19**	**24**	**6**	**2916**	**153**	**4**	**3.15**	....	....	....	....	....	....	....
	Moncton Hawks	AHL	2	1	0	1	125	6	0	2.88	....	....	....	....	....	....	....
1991-92	**Winnipeg**	**NHL**	**47**	**21**	**17**	**6**	**2627**	**126**	***5**	**2.88**	**1**	**0**	**0**	**33**	**3**	**0**	**5.45**
1992-93	**Winnipeg**	**NHL**	**67**	**33**	**26**	**6**	**3855**	**227**	**2**	**3.53**	**6**	**2**	**4**	**367**	**20**	**0**	**3.27**
1993-94	**Winnipeg**	**NHL**	**56**	**19**	**30**	**6**	**3136**	**201**	**1**	**3.85**	....	....	....	....	....	....	....
	Detroit	**NHL**	**13**	**4**	**7**	**2**	**778**	**34**	**1**	**2.62**	**2**	**0**	**2**	**109**	**9**	**0**	**4.95**
1994-95	San Diego Gulls	IHL	16	6	8	1	919	52	0	3.39	1	0	1	59	3	0	3.05
1995-96	Adirondack	AHL	3	1	2	0	179	11	0	3.69	....	....	....	....	....	....	....
	Fort Wayne	IHL	45	24	14	5	2529	122	1	2.89	5	2	3	299	12	0	2.41
1996-97	**Edmonton**	**NHL**	**19**	**4**	**8**	**0**	**868**	**41**	**1**	**2.83**	....	....	....	....	....	....	....
1997-98	**Edmonton**	**NHL**	**16**	**6**	**6**	**1**	**825**	**35**	**0**	**2.55**	**1**	**0**	**0**	**27**	**1**	**0**	**2.22**
1998-99	**Edmonton**	**NHL**	**39**	**12**	**14**	**6**	**2091**	**96**	**0**	**2.75**	....	....	....	....	....	....	....
99-2000	**Phoenix**	**NHL**	**30**	**13**	**10**	**3**	**1573**	**73**	**1**	**2.78**	....	....	....	....	....	....	....
2000-01	**Vancouver**	**NHL**	**39**	**18**	**12**	**3**	**2059**	**92**	**1**	**2.68**	**2**	**0**	**2**	**122**	**6**	**0**	**2.95**
2001-02	**Buffalo**	**NHL**	**9**	**0**	**5**	**0**	**350**	**17**	**0**	**2.91**	....	....	....	....	....	....	....
	NHL Totals		**446**	**173**	**176**	**47**	**24215**	**1270**	**18**	**3.15**	**16**	**4**	**9**	**864**	**51**	**0**	**3.54**

CCHA First All-Star Team (1985) • CCHA Second All-Star Team (1986) • NHL All-Rookie Team (1990)

Traded to **Detroit** by **Winnipeg** with Sergei Bautin for Tim Cheveldae and Dallas Drake, March 8, 1994. Traded to **Edmonton** by **Detroit** for future considerations, June 14, 1996. Signed as a free agent by **Phoenix**, September 5, 1999. Signed as a free agent by **Vancouver**, July 26, 2000. Signed as a free agent by **Buffalo**, August 3, 2001.

FANKHOUSER, Scott (FANK-how-suhr, SKAWT)

Goaltender. Catches left. 6'2", 205 lbs. Born, Bismark, ND, July 1, 1975.
(St. Louis' 8th choice, 276th overall, in 1994 Entry Draft).

			Regular Season								Playoffs						
Season	**Club**	**Lea**	**GP**	**W**	**L**	**T**	**Mins**	**GA**	**SO**	**Avg**	**GP**	**W**	**L**	**Mins**	**GA**	**SO**	**Avg**
1993-94	Loomis-Chaffe	H.S.					STATISTICS NOT AVAILABLE										
1994-95	U. Mass-Lowell	H-East	11	4	4	1	499	37	0	4.44	....	....	....	....	....	....	....
1995-96	Melfort Mustangs	SJHL	45	31	9	4	2544	109	3	2.57	....	....	....	....	....	....	....
1996-97	U. Mass-Lowell	H-East	11	2	4	1	517	38	0	4.41	....	....	....	....	....	....	....
1997-98	U. Mass-Lowell	H-East	16	4	7	2	798	48	0	3.61	....	....	....	....	....	....	....
1998-99	U. Mass-Lowell	H-East	32	16	14	0	1729	80	1	2.78	....	....	....	....	....	....	....
99-2000	**Atlanta**	**NHL**	**16**	**2**	**11**	**2**	**920**	**49**	**0**	**3.20**	....	....	....	....	....	....	....
	Greenville	ECHL	7	6	1	0	419	18	0	2.58	....	....	....	....	....	....	....
	Orlando	IHL	6	2	2	1	320	14	0	2.63	....	....	....	....	....	....	....
	Louisville	AHL	1	0	1	0	59	3	0	3.05	....	....	....	....	....	....	....
2000-01	**Atlanta**	**NHL**	**7**	**2**	**1**	**0**	**260**	**16**	**0**	**3.69**	....	....	....	....	....	....	....
	Orlando	IHL	28	13	12	3	1603	69	1	2.58	1	0	0	37	3	0	4.83
2001-02	Chicago Wolves	AHL	2	1	0	1	125	4	0	1.92	....	....	....	....	....	....	....
	Greenville	ECHL	3	1	2	0	180	11	0	3.67	....	....	....	....	....	....	....
	Hershey Bears	AHL	8	5	2	1	487	19	1	2.34	3	0	0	41	3	0	4.44
	NHL Totals		**23**	**4**	**12**	**2**	**1180**	**65**	**0**	**3.31**	....	....	....	....	....	....	....

SJHL First All-Star Team (1996) • SJHL Playoff MVP (1996) • Shared James Norris Memorial Trophy (fewest goals against - IHL) with Norm Maracle (2001)

Signed as a free agent by **Atlanta**, August 24, 1999.

FERNANDEZ, Manny (fuhr-NAN-dehz, MAN-ee) MIN.

Goaltender. Catches left. 6', 180 lbs. Born, Etobicoke, Ont., August 27, 1974.
(Quebec's 4th choice, 52nd overall, in 1992 Entry Draft).

			Regular Season								Playoffs						
Season	**Club**	**Lea**	**GP**	**W**	**L**	**T**	**Mins**	**GA**	**SO**	**Avg**	**GP**	**W**	**L**	**Mins**	**GA**	**SO**	**Avg**
1990-91	Lac St-Louis	QAAA	20	13	5	0	1176	69	*3	3.52	3	2	1	181	12	0	3.98
1991-92	Laval Titan	QMJHL	31	14	13	2	1593	99	1	3.73	9	3	5	468	39	0	5.00
1992-93	Laval Titan	QMJHL	43	26	14	2	2347	141	1	3.60	13	*12	1	818	42	0	3.08
1993-94	Laval Titan	QMJHL	51	29	14	1	2776	143	*5	3.09	19	14	5	1116	49	*1	*2.63
1994-95	Kalamazoo Wings	IHL	46	21	10	9	2470	115	2	2.79	14	10	2	753	34	1	2.71
	Dallas	**NHL**	**1**	**0**	**1**	**0**	**59**	**3**	**0**	**3.05**	....	....	....	....	....	....	....
1995-96	**Dallas**	**NHL**	**5**	**0**	**1**	**1**	**249**	**19**	**0**	**4.58**	....	....	....	....	....	....	....
	Michigan K-Wings	IHL	47	22	15	9	2664	133	*4	3.00	6	5	1	372	14	0	*2.26
1996-97	Michigan K-Wings	IHL	48	20	24	2	2720	142	2	3.13	4	1	3	277	15	0	3.25
1997-98	**Dallas**	**NHL**	**2**	**1**	**0**	**0**	**69**	**2**	**0**	**1.74**	**1**	**0**	**0**	**2**	**0**	**0**	**0.00**
	Michigan K-Wings	IHL	55	27	17	5	3022	139	5	2.76	2	0	2	88	7	0	4.73
1998-99	**Dallas**	**NHL**	**1**	**0**	**1**	**0**	**60**	**2**	**0**	**2.00**	....	....	....	....	....	....	....
	Houston Aeros	IHL	50	34	6	9	2949	116	2	2.36	*19	*11	8	*1126	49	1	2.61
99-2000	**Dallas**	**NHL**	**24**	**11**	**8**	**3**	**1353**	**48**	**1**	**2.13**	**1**	**0**	**0**	**17**	**1**	**0**	**3.53**
2000-01	**Minnesota**	**NHL**	**42**	**19**	**17**	**4**	**2461**	**92**	**4**	**2.24**	....	....	....	....	....	....	....
2001-02	**Minnesota**	**NHL**	**44**	**12**	**24**	**5**	**2463**	**125**	**1**	**3.05**	....	....	....	....	....	....	....
	NHL Totals		**119**	**43**	**52**	**13**	**6714**	**291**	**6**	**2.60**	**2**	**0**	**0**	**19**	**1**	**0**	**3.16**

QMJHL First All-Star Team (1994) • QMJHL MVP (1994) • IHL Second All-Star Team (1995)

Rights traded to **Dallas** by **Quebec** for Tommy Sjodin and Dallas' 3rd round choice (Chris Drury) in 1994 Entry Draft, February 13, 1994. Traded to **Minnesota** by **Dallas** with Brad Lukowich for Minnesota's 3rd round choice (Joel Lundqvist) in 2000 Entry Draft and 4th round choice (later traded back to Minnesota - later traded to LA Kings - LA Kings selected Aaron Rome) in 2002 Entry Draft, June 12, 2000.

FINLEY, Brian (FIHN-lee, BRIGH-uhn) NSH.

Goaltender. Catches right. 6'3", 205 lbs. Born, Sault Ste. Marie, Ont., July 13, 1981.
(Nashville's 1st choice, 6th overall, in 1999 Entry Draft).

			Regular Season								Playoffs						
Season	**Club**	**Lea**	**GP**	**W**	**L**	**T**	**Mins**	**GA**	**SO**	**Avg**	**GP**	**W**	**L**	**Mins**	**GA**	**SO**	**Avg**
1996-97	Soo Carlucci's	NOBHL	45	....	....	....	1943	109	3	2.38	....	....	....	....	....	....	....
1997-98	Barrie Colts	OHL	41	23	14	1	2154	105	3	2.92	5	1	3	260	13	0	3.00
1998-99	Barrie Colts	OHL	52	*36	10	4	3063	136	3	2.66	5	4	1	323	15	0	2.79
99-2000	Barrie Colts	OHL	47	24	12	6	2540	130	2	3.07	*23	14	8	1353	58	1	2.57
2000-01	Barrie Colts	OHL	16	5	8	0	818	42	0	3.08	....	....	....	....	....	....	....
	Brampton	OHL	11	7	3	1	631	31	0	2.95	9	5	4	503	26	1	3.10
2001-02							DID NOT PLAY – INJURED										

NOBHL Top Goaltender (1997) • OHL All-Rookie Team (1998) • OHL First All-Star Team (1999) • OHL Playoff MVP (2000)

Traded to **Brampton** (OHL) by **Barrie** (OHL) with Mississauga's 7th round choice (Michael Root) in 2001 OHL Midget Draft and Kitchener's 4th round choice (previously acquired, Brampton selected Tyler McCormick) in 2002 OHL Priority Draft for David Chant, Tyler Hanchuk and Matt Grenier, January 10, 2001. • Missed entire 2001-02 season recovering from groin injury originally suffered during 2000-01 season and re-injured in training camp, October 3, 2001.

FISET, Stephane (fih-SEHT, STEH-fan)

Goaltender. Catches left. 6'1", 215 lbs. Born, Montreal, Que., June 17, 1970.
(Quebec's 3rd choice, 24th overall, in 1988 Entry Draft).

			Regular Season								Playoffs						
Season	**Club**	**Lea**	**GP**	**W**	**L**	**T**	**Mins**	**GA**	**SO**	**Avg**	**GP**	**W**	**L**	**Mins**	**GA**	**SO**	**Avg**
1986-87	Mtl-Bourassa	QAAA	30	8	21	1	1689	155	0	5.51	....	....	....	....	....	....	....
1987-88	Victoriaville	QMJHL	40	15	17	4	2221	146	1	3.94	2	0	2	163	10	0	3.68
1988-89	Victoriaville	QMJHL	43	25	14	0	2401	138	1	*3.45	12	*9	2	711	33	0	*2.78
1989-90	**Quebec**	**NHL**	**6**	**0**	**5**	**1**	**342**	**34**	**0**	**5.96**	....	....	....	....	....	....	....
	Victoriaville	QMJHL	24	14	6	3	1383	63	1	*2.73	*14	7	6	*790	49	0	3.72
1990-91	**Quebec**	**NHL**	**3**	**0**	**2**	**1**	**186**	**12**	**0**	**3.87**	....	....	....	....	....	....	....
	Halifax Citadels	AHL	36	10	15	8	1902	131	0	4.13	....	....	....	....	....	....	....
1991-92	**Quebec**	**NHL**	**23**	**7**	**10**	**2**	**1133**	**71**	**1**	**3.76**	....	....	....	....	....	....	....
	Halifax Citadels	AHL	29	8	14	6	1675	110	*3	3.94	....	....	....	....	....	....	....
1992-93	**Quebec**	**NHL**	**37**	**18**	**9**	**4**	**1939**	**110**	**0**	**3.40**	**1**	**0**	**0**	**21**	**1**	**0**	**2.86**
	Halifax Citadels	AHL	3	2	1	0	180	11	0	3.67	....	....	....	....	....	....	....
1993-94	**Quebec**	**NHL**	**50**	**20**	**25**	**4**	**2798**	**158**	**2**	**3.39**	....	....	....	....	....	....	....
	Cornwall Aces	AHL	1	0	1	0	60	4	0	4.00	....	....	....	....	....	....	....
1994-95	**Quebec**	**NHL**	**32**	**17**	**10**	**3**	**1879**	**87**	**2**	**2.78**	**4**	**1**	**2**	**209**	**16**	**0**	**4.59**
1995-96♦	**Colorado**	**NHL**	**37**	**22**	**6**	**7**	**2107**	**103**	**1**	**2.93**	**1**	**0**	**0**	**1**	**0**	**0**	**0.00**
1996-97	**Los Angeles**	**NHL**	**44**	**13**	**24**	**5**	**2482**	**132**	**4**	**3.19**	....	....	....	....	....	....	....
1997-98	**Los Angeles**	**NHL**	**60**	**26**	**25**	**8**	**3497**	**158**	**2**	**2.71**	**2**	**0**	**2**	**93**	**7**	**0**	**4.52**
1998-99	**Los Angeles**	**NHL**	**42**	**18**	**21**	**1**	**2403**	**104**	**3**	**2.60**	....	....	....	....	....	....	....
99-2000	**Los Angeles**	**NHL**	**47**	**20**	**15**	**7**	**2592**	**119**	**1**	**2.75**	**4**	**0**	**3**	**200**	**10**	**0**	**3.00**
2000-01	**Los Angeles**	**NHL**	**7**	**3**	**0**	**1**	**318**	**19**	**0**	**3.58**	**1**	**0**	**0**	**1**	**0**	**0**	**0.00**
	Lowell	AHL	3	1	0	2	190	9	0	2.84	....	....	....	....	....	....	....
2001-02	**Montreal**	**NHL**	**2**	**0**	**1**	**0**	**109**	**7**	**0**	**3.85**	**1**	**0**	**0**	**38**	**3**	**0**	**4.74**
	Manchester	AHL	23	7	7	6	1228	64	0	3.13	....	....	....	....	....	....	....
	NHL Totals		**390**	**164**	**153**	**44**	**21785**	**1114**	**16**	**3.07**	**14**	**1**	**7**	**563**	**37**	**0**	**3.94**

QMJHL First All-Star Team (1989) • Canadian Major Junior Goaltender of the Year (1989)

Transferred to **Colorado** after **Quebec** franchise relocated, June 21, 1995. Traded to **LA Kings** by **Colorado** with Colorado's 1st round choice (Mathieu Biron) in 1998 Entry Draft for Eric Lacroix and LA Kings' 1st round choice (Martin Skoula) in 1998 Entry Draft, June 20, 1996. • Missed majority of 2000-01 season recovering from knee injury suffered in exhibition game vs. Anaheim, September 22, 2000. • Played 12 seconds of playoff game vs. Colorado, April 28, 2001. Traded to **Montreal** by **LA Kings** for future considerations, March 19, 2002.

FISHER, Glenn (FIH-shuhr, GLEHN) EDM.

Goaltender. Catches left. 6'1", 160 lbs. Born, Edmonton, Alta., April 25, 1983.
(Edmonton's 9th choice, 148th overall, in 2002 Entry Draft).

			Regular Season								Playoffs						
Season	**Club**	**Lea**	**GP**	**W**	**L**	**T**	**Mins**	**GA**	**SO**	**Avg**	**GP**	**W**	**L**	**Mins**	**GA**	**SO**	**Avg**
99-2000	Edm. Leafs	AMBHL	16	9	5	2	944	62	0	3.94	....	....	....	....	....	....	....
2000-01	Edm. Leafs	AMHL	19	6	9	3	1116	77	0	4.14	....	....	....	....	....	....	....
2001-02	Ft. Saskatchewan	AJHL	47	....	....	....	2649	196	2	4.44	....	....	....	....	....	....	....

AJHL Rookie of the Year (2002) • Signed Letter of Intent to attend **North Dakota** (WCHA), February 12, 2002.

FLAHERTY, Wade (FLAY-uhr-tee, WAYD) FLA.

Goaltender. Catches left. 6', 170 lbs. Born, Terrace, B.C., January 11, 1968.
(Buffalo's 10th choice, 181st overall, in 1988 Entry Draft).

			Regular Season								Playoffs						
Season	**Club**	**Lea**	**GP**	**W**	**L**	**T**	**Mins**	**GA**	**SO**	**Avg**	**GP**	**W**	**L**	**Mins**	**GA**	**SO**	**Avg**
1984-85	Kelowna Wings	WHL	1	0	0	0	55	5	0	5.45	....	....	....	....	....	....	....
1985-86	Seattle	WHL	9	1	3	0	271	36	0	7.97	....	....	....	....	....	....	....
	Spokane Chiefs	WHL	5	0	3	0	161	21	0	7.83	....	....	....	....	....	....	....
1986-87	Nanaimo Clippers	BCJHL	15	....	....	....	830	53	0	3.83	....	....	....	....	....	....	....
	Victoria Cougars	WHL	3	0	2	0	127	16	0	7.56	....	....	....	....	....	....	....
1987-88	Victoria Cougars	WHL	36	20	15	0	2052	135	0	3.95	5	2	3	300	18	0	3.60
1988-89	Victoria Cougars	WHL	42	21	19	0	2408	180	4	4.49	....	....	....	....	....	....	....
1989-90	Greensboro	ECHL	27	12	10	0	1308	96	0	4.40	....	....	....	....	....	....	....
1990-91	Kansas City	IHL	*56	16	31	4	2990	224	0	4.49	....	....	....	....	....	....	....
1991-92	**San Jose**	**NHL**	**3**	**0**	**3**	**0**	**178**	**13**	**0**	**4.38**	....	....	....	....	....	....	....
	Kansas City	IHL	43	26	14	3	2603	140	1	3.23	1	0	0	1	0	0	0.00
1992-93	**San Jose**	**NHL**	**1**	**0**	**1**	**0**	**60**	**5**	**0**	**5.00**	....	....	....	....	....	....	....
	Kansas City	IHL	*61	*34	19	7	*3642	195	2	3.21	*12	6	6	733	34	*1	2.78
1993-94	Kansas City	IHL	*60	32	19	9	*3564	202	0	3.40	....	....	....	....	....	....	....
1994-95	**San Jose**	**NHL**	**18**	**5**	**6**	**1**	**852**	**44**	**1**	**3.10**	**7**	**2**	**3**	**377**	**31**	**0**	**4.93**
1995-96	**San Jose**	**NHL**	**24**	**3**	**12**	**1**	**1137**	**92**	**0**	**4.85**	....	....	....	....	....	....	....
1996-97	**San Jose**	**NHL**	**7**	**2**	**4**	**0**	**359**	**31**	**0**	**5.18**	....	....	....	....	....	....	....
	Kentucky	AHL	19	8	6	2	1032	54	1	3.14	3	1	2	200	11	0	3.30
1997-98	**NY Islanders**	**NHL**	**16**	**4**	**4**	**3**	**694**	**23**	**3**	**1.99**	....	....	....	....	....	....	....
	Utah Grizzlies	IHL	24	16	5	3	1341	40	3	1.79	....	....	....	....	....	....	....
1998-99	**NY Islanders**	**NHL**	**20**	**5**	**11**	**2**	**1048**	**53**	**0**	**3.03**	....	....	....	....	....	....	....
	Lowell	AHL	5	1	3	1	305	16	0	3.15	....	....	....	....	....	....	....
99-2000	**NY Islanders**	**NHL**	**4**	**0**	**1**	**1**	**182**	**7**	**0**	**2.31**	....	....	....	....	....	....	....
2000-01	**NY Islanders**	**NHL**	**20**	**6**	**10**	**0**	**1017**	**56**	**1**	**3.30**	....	....	....	....	....	....	....
	Tampa Bay	**NHL**	**2**	**0**	**2**	**0**	**118**	**8**	**0**	**4.07**	....	....	....	....	....	....	....
2001-02	**Florida**	**NHL**	**4**	**2**	**1**	**1**	**245**	**12**	**0**	**2.94**	....	....	....	....	....	....	....
	Utah Grizzlies	AHL	45	22	13	5	2351	92	2	2.35	5	2	3	312	11	0	2.12
	NHL Totals		**119**	**27**	**55**	**9**	**5890**	**344**	**5**	**3.50**	**7**	**2**	**3**	**377**	**31**	**0**	**4.93**

WHL West Second All-Star Team (1988) • ECHL Playoff MVP (1990) • Shared James Norris Memorial Trophy (fewest goals against - IHL) with Arturs Irbe (1992) • IHL Second All-Star Team (1993, 1994)

Signed as a free agent by **San Jose**, September 3, 1991. Signed as a free agent by **NY Islanders**, July 22, 1997. Traded to **Tampa Bay** by **NY Islanders** for future considerations, February 16, 2001. Signed as a free agent by **Florida**, August 2, 2001.

FORD, Todd (FOHRD, TAWD) TOR.

Goaltender. Catches left. 6'3", 175 lbs. Born, Calgary, Alta., May 1, 1984.
(Toronto's 3rd choice, 74th overall, in 2002 Entry Draft).

			Regular Season								Playoffs						
Season	**Club**	**Lea**	**GP**	**W**	**L**	**T**	**Mins**	**GA**	**SO**	**Avg**	**GP**	**W**	**L**	**Mins**	**GA**	**SO**	**Avg**
2000-01	Swift Current	WHL	20	12	4	2	1066	53	0	2.98	1	0	0	26	2	0	4.62
2001-02	Swift Current	WHL	37	18	12	3	2003	99	2	2.97	10	5	5	603	28	0	2.79

FOUNTAIN, Mike (FOWN-tehn, MIGHK)

Goaltender. Catches left. 6'1", 180 lbs. Born, North York, Ont., January 26, 1972.
(Vancouver's 3rd choice, 45th overall, in 1992 Entry Draft).

			Regular Season								Playoffs						
Season	**Club**	**Lea**	**GP**	**W**	**L**	**T**	**Mins**	**GA**	**SO**	**Avg**	**GP**	**W**	**L**	**Mins**	**GA**	**SO**	**Avg**
1988-89	Huntsville	OJHL-C	22	*18	3	2	1306	82	0	3.77	....	....	....	....	....	....	....
1989-90	Chatham Maroons	OJHL-B	21	....	....	....	1249	76	0	3.65	....	....	....	....	....	....	....
1990-91	Sault Ste. Marie	OHL	7	5	2	0	380	19	0	3.00	....	....	....	....	....	....	....
	Oshawa Generals	OHL	30	17	5	1	1483	84	0	3.40	8	1	4	292	26	0	5.34
1991-92	Oshawa Generals	OHL	40	18	13	6	2260	149	1	3.96	7	3	4	429	26	0	3.64
1992-93	Canada	Nat-Tm	13	7	5	1	745	37	1	2.98	....	....	....	....	....	....	....
	Hamilton Canucks	AHL	12	2	8	0	618	46	0	4.47	....	....	....	....	....	....	....
1993-94	Hamilton Canucks	AHL	*70	*34	28	6	*4005	241	*4	3.61	3	0	2	146	12	0	4.92
1994-95	Syracuse Crunch	AHL	61	25	29	7	3618	225	2	3.73	....	....	....	....	....	....	....
1995-96	Syracuse Crunch	AHL	54	21	27	3	3060	184	1	3.61	15	8	7	915	57	*2	3.74
1996-97	**Vancouver**	**NHL**	**6**	**2**	**2**	**0**	**245**	**14**	**1**	**3.43**	....	....	....	....	....	....	....
	Syracuse Crunch	AHL	25	8	14	2	1462	78	1	3.20	2	0	2	120	12	0	6.02
1997-98	**Carolina**	**NHL**	**3**	**0**	**3**	**0**	**163**	**10**	**0**	**3.68**	....	....	....	....	....	....	....
	New Haven	AHL	50	25	19	5	2922	139	3	2.85	....	....	....	....	....	....	....
1998-99	New Haven	AHL	51	23	24	3	2989	150	2	3.01	....	....	....	....	....	....	....
99-2000	**Ottawa**	**NHL**	**1**	**0**	**0**	**0**	**16**	**1**	**0**	**3.75**	....	....	....	....	....	....	....
	Grand Rapids	IHL	36	21	7	4	1851	77	3	2.50	1	0	0	20	4	0	12.00
2000-01	**Ottawa**	**NHL**	**1**	**0**	**1**	**0**	**59**	**3**	**0**	**3.05**	....	....	....	....	....	....	....
	Grand Rapids	IHL	*52	*34	10	6	*3005	104	6	2.08	8	5	3	522	21	1	2.41
2001-02	Lada Togliatti	Russia	43	....	....	....	2591	59	*14	*1.37	4	....	....	249	7	0	1.69
	NHL Totals		**11**	**2**	**6**	**0**	**483**	**28**	**1**	**3.48**	....	....	....	....	....	....	....

OHL First All-Star Team (1992) • AHL Second All-Star Team (1994) • IHL Second All-Star Team (2001)

• Recorded shutout (3-0) in NHL debut vs. **New Jersey**, November 14, 1996. Signed as a free agent by **Carolina**, August 19, 1997. Signed as a free agent by **Ottawa**, July 30, 1999.

FRANEK, Petr (FRAH-nehk, PEE-tuhr) COL.

Goaltender. Catches left. 5'11", 185 lbs. Born, Most, Czech., April 6, 1975.
(Quebec's 10th choice, 205th overall, in 1993 Entry Draft).

			Regular Season								Playoffs						
Season	**Club**	**Lea**	**GP**	**W**	**L**	**T**	**Mins**	**GA**	**SO**	**Avg**	**GP**	**W**	**L**	**Mins**	**GA**	**SO**	**Avg**
1992-93	Litvinov	Czech	5	....	....	....	273	15	0	3.29	....	....	....	....	....	....	....
1993-94	Litvinov	Czech	11	....	....	....	535	34	0	3.81	2	0	1	61	10	0	9.83
1994-95	Litvinov	Czech	12	....	....	....	657	47	0	4.29	1	0	0	16	0	0	0.00
1995-96	Litvinov	Czech	36	....	....	....	2096	85	3	2.43	16	....	....	948	47	1	2.97
1996-97	Hershey Bears	AHL	15	4	1	0	457	23	3	3.02	....	....	....	....	....	....	....
	Brantford Smoke	ColHL	6	4	1	0	321	14	0	2.61	....	....	....	....	....	....	....
	Quebec Rafales	IHL	6	3	3	0	357	18	0	3.02	1	0	1	40	4	0	6.00
1997-98	Hershey Bears	AHL	43	19	14	2	2169	98	2	2.71	1	0	1	60	4	0	4.00
1998-99	Utah Grizzlies	IHL	8	1	6	1	446	26	0	3.50	....	....	....	....	....	....	....
	Las Vegas	IHL	37	17	13	2	1879	107	0	3.42	....	....	....	....	....	....	....
99-2000	Nurnberg	Germany	30	....	....	....	1603	73	2	2.73	....	....	....	....	....	....	....
2000-01	HC Karlovy Vary	Czech	44	....	....	....	2507	121	....	2.90	....	....	....	....	....	....	....
2001-02	HC Karlovy Vary	Czech	40	....	....	....	2189	109	....	2.99	....	....	....	....	....	....	....

Rights transferred to **Colorado** after **Quebec** franchise relocated, June 21, 1995.

GARDNER, Greg (GAHRD-nuhr, GREHG)

Goaltender. Catches left. 6', 190 lbs. Born, Mississauga, Ont., November 21, 1975.

			Regular Season								Playoffs						
Season	**Club**	**Lea**	**GP**	**W**	**L**	**T**	**Mins**	**GA**	**SO**	**Avg**	**GP**	**W**	**L**	**Mins**	**GA**	**SO**	**Avg**
1992-93	Caledon	MTJHL	28	....	....	....	1390	73	1	3.15	....	....	....	....	....	....	....
1993-94	Caledon	MTJHL	1	1	0	0	40	5	0	7.50	....	....	....	....	....	....	....
	Thornhill	MTJHL	31	18	11	1	1813	103	1	3.41	....	....	....	....	....	....	....
1994-95	Thornhill	MTJHL	42	....	....	....	2430	117	4	2.89	11	7	4	677	36	1	3.19
1995-96	Thornhill	MTJHL	34	....	....	....	2012	99	2	2.95	18	11	7	1095	59	0	3.23
1996-97	Niagara U.	ECAC-2	17	8	5	2	939	54	0	3.45	....	....	....	....	....	....	....
1997-98	Niagara U.	ECAC-2	25	12	10	3	1454	74	0	3.05	....	....	....	....	....	....	....
1998-99	Niagara U.	CHA	30	15	10	3	1742	78	4	2.69	....	....	....	....	....	....	....
99-2000	Niagara U.	CHA	*41	*29	8	4	*2503	64	*12	*1.53	....	....	....	....	....	....	....
2000-01	Syracuse Crunch	AHL	9	1	5	0	351	28	0	4.78	....	....	....	....	....	....	....
	Dayton Bombers	ECHL	28	14	9	2	1600	70	2	2.62	3	1	2	181	9	0	2.96
2001-02	Dayton Bombers	ECHL	16	11	4	1	965	43	1	2.67	....	....	....	....	....	....	....
	Syracuse Crunch	AHL	18	7	6	3	992	41	2	2.48	1	0	0	1	0	0	0.00

MTJHL East All-Star Team (1994, 1995) • MTJHL East Goaltender of the Year (1995, 1996) • CHA First All-Star Team (2000) • CHA Goaltender of the Year (2000) • CHA Player of the Year (2000)

Signed as a free agent by **Columbus**, May 16, 2000. • Played 24 seconds of playoff game vs. Chicago (AHL), May 7, 2002.

GARNER, Tyrone (GAHR-nuhr, TIGH-rohn)

Goaltender. Catches left. 6'1", 200 lbs. Born, Stoney Creek, Ont., July 27, 1978.
(NY Islanders' 4th choice, 83rd overall, in 1996 Entry Draft).

			Regular Season								Playoffs						
Season	**Club**	**Lea**	**GP**	**W**	**L**	**T**	**Mins**	**GA**	**SO**	**Avg**	**GP**	**W**	**L**	**Mins**	**GA**	**SO**	**Avg**
1994-95	Stoney Creek	OJHL-B	10	2	7	1	589	62	0	6.32	....	....	....	....	....	....	....
	Hamilton	OPJHL	8	....	....	....	419	28	0	4.01	....	....	....	....	....	....	....
1995-96	Oshawa Generals	OHL	32	11	15	4	1697	112	0	3.96	....	....	....	....	....	....	....
1996-97	Oshawa Generals	OHL	9	6	1	0	434	20	0	2.76	3	1	0	88	6	0	4.09
1997-98	Oshawa Generals	OHL	54	23	17	8	2946	162	1	3.30	7	3	4	450	25	0	3.33
1998-99	Oshawa Generals	OHL	44	24	15	3	2496	124	4	2.98	15	9	6	901	57	0	3.80
	Calgary	**NHL**	**3**	**0**	**2**	**0**	**139**	**12**	**0**	**5.18**	....	....	....	....	....	....	....
99-2000	Saint John	AHL	19	4	8	4	940	70	0	4.47	....	....	....	....	....	....	....
	Dayton Bombers	ECHL	3	0	2	0	113	11	0	5.86	....	....	....	....	....	....	....
	Johnstown Chiefs	ECHL	17	8	6	3	971	48	0	2.97	1	0	1	59	2	0	2.03
2000-01	Johnstown Chiefs	ECHL	5	3	1	1	306	15	0	2.94	....	....	....	....	....	....	....
	Greenville	ECHL	35	17	15	3	2114	99	3	2.81	....	....	....	....	....	....	....
2001-02	Greenville	ECHL	29	12	12	5	1763	74	0	2.52	*14	*12	2	803	33	0	2.47
	NHL Totals		**3**	**0**	**2**	**0**	**139**	**12**	**0**	**5.18**	....	....	....	....	....	....	....

OHL Second All-Star Team (1999) • Shared ECHL Playoff MVP Award (2002) with Simon Gamache

Traded to **Calgary** by **NY Islanders** with Marty McInnis and Calgary's 6th round choice (previously acquired, Calgary selected Ilja Demidov) in 1997 Entry Draft for Robert Reichel, March 18, 1997.

GARNETT, Michael (gahr-NEHT, MIGHK-uhl) ATL.

Goaltender. Catches left. 6'1", 185 lbs. Born, Saskatoon, Sask., November 25, 1982.
(Atlanta's 2nd choice, 80th overall, in 2001 Entry Draft).

			Regular Season								Playoffs						
Season	**Club**	**Lea**	**GP**	**W**	**L**	**T**	**Mins**	**GA**	**SO**	**Avg**	**GP**	**W**	**L**	**Mins**	**GA**	**SO**	**Avg**
1997-98	Sask. Contacts	SMHL	3	1	1	0	82	8	0	5.85	....	....	....	....	....	....	....
1998-99	Sask. Contacts	SMHL					STATISTICS NOT AVAILABLE										
99-2000	Kindersley	SJHL	36	....	....	....	2067	140	1	3.57	....	....	....	....	....	....	....
	Red Deer Rebels	WHL	1	0	0	0	14	0	0	0.00	....	....	....	....	....	....	....
2000-01	Red Deer Rebels	WHL	21	14	5	1	1133	39	3	2.07	....	....	....	....	....	....	....
	Saskatoon Blades	WHL	28	7	17	2	1501	83	1	3.32	....	....	....	....	....	....	....
2001-02	Saskatoon Blades	WHL	*67	27	34	4	*3738	205	2	3.29	7	3	4	450	15	0	2.00

Traded to **Saskatoon** (WHL) by **Red Deer** (WHL) with Justin Wallin, Martin Vymazzal and future considerations for Martin Erat, Darcy Robinson and Cam Ornik, January 11, 2001.

GARON, Mathieu (gah-ROHN, MAT-yoo) MTL.

Goaltender. Catches right. 6'2", 192 lbs. Born, Chandler, Que., January 9, 1978.
(Montreal's 2nd choice, 44th overall, in 1996 Entry Draft).

			Regular Season								Playoffs						
Season	**Club**	**Lea**	**GP**	**W**	**L**	**T**	**Mins**	**GA**	**SO**	**Avg**	**GP**	**W**	**L**	**Mins**	**GA**	**SO**	**Avg**
1993-94	Jonquiere Elites	QAAA	17	0	13	0	834	88	0	6.33	....	....	....	....	....	....	....
1994-95	Jonquiere Elites	QAAA	27	13	13	1	1554	94	0	3.63	9	6	2	467	26	0	3.34
1995-96	Victoriaville	QMJHL	51	18	27	0	2709	189	1	4.19	12	7	4	676	38	1	3.39
1996-97	Victoriaville	QMJHL	53	29	18	3	3032	150	*6	2.97	6	2	4	330	23	0	4.18
1997-98	Victoriaville	QMJHL	47	27	18	2	2802	125	5	2.68	6	2	4	345	22	0	3.82
1998-99	Fredericton	AHL	40	14	22	2	2222	114	3	3.08	6	1	1	208	12	0	3.47
99-2000	Quebec	AHL	53	17	28	3	2884	149	2	3.10	1	0	0	20	3	0	8.82
2000-01	**Montreal**	**NHL**	**11**	**4**	**5**	**1**	**589**	**24**	**2**	**2.44**	....	....	....	....	....	....	....
	Quebec	AHL	31	16	13	1	1768	86	1	2.92	8	4	4	459	22	1	2.88
2001-02	**Montreal**	**NHL**	**5**	**1**	**4**	**0**	**261**	**19**	**0**	**4.37**	....	....	....	....	....	....	....
	Quebec	AHL	50	21	15	12	2988	136	2	2.73	3	0	3	198	12	0	3.63
	NHL Totals		**16**	**5**	**9**	**1**	**850**	**43**	**2**	**3.04**	....	....	....	....	....	....	....

QMJHL All-Rookie Team (1996) • QMJHL Defensive Rookie of the Year (1996) • QMJHL First All-Star Team (1998) • Canadian Major Junior First All-Star Team (1998) • Canadian Major Junior Goaltender of the Year (1998)

GERBER, Martin (GUHR-buhr, MAHR-tihn) ANA.

Goaltender. Catches left. 6', 185 lbs. Born, Burgdorf, Switz., September 3, 1974.
(Anaheim's 10th choice, 232nd overall, in 2001 Entry Draft).

			Regular Season								Playoffs						
Season	**Club**	**Lea**	**GP**	**W**	**L**	**T**	**Mins**	**GA**	**SO**	**Avg**	**GP**	**W**	**L**	**Mins**	**GA**	**SO**	**Avg**
1996-97	SC Langnau	Swiss-2	38	....	....	....	2286	121	0	3.18	8	....	....	488	29	0	3.57
1997-98	SC Langnau	Swiss-2	40	....	....	....	2430	141	2	3.48	16	....	....	961	42	0	2.62
1998-99	SC Langnau	Swiss	42	....	....	....	2521	203	1	4.83	11	....	....	664	50	0	4.52
99-2000	SC Langnau	Swiss	44	....	....	....	2652	161	3	3.64	6	....	....	360	13	*2	*2.17
2000-01	Langnau	Swiss	*44	....	....	....	2671	114	3	2.56	5	....	....	319	7	1	1.32
2001-02	Farjestad	Sweden	44	....	....	....	2664	87	*4	*1.96	*10	....	....	*657	18	*2	*1.64
	Switzerland	Olympics	3	1	1	1	158	4	0	1.52	....	....	....	....	....	....	....

GHERSON, Rob (GAIR-suhn, RAWB) WSH.

Goaltender. Catches left. 6'1", 155 lbs. Born, Toronto, Ont., October 8, 1983.
(Washington's 9th choice, 145th overall, in 2002 Entry Draft).

			Regular Season								Playoffs						
Season	**Club**	**Lea**	**GP**	**W**	**L**	**T**	**Mins**	**GA**	**SO**	**Avg**	**GP**	**W**	**L**	**Mins**	**GA**	**SO**	**Avg**
99-2000	Wellington Dukes	MTJHL	31	....	....	....	1810	80	3	2.65	....	....	....	....	....	....	....
2000-01	Sarnia Sting	OHL	41	11	22	5	2230	136	1	3.66	4	0	2	158	9	0	3.42
2001-02	Sarnia Sting	OHL	55	20	28	4	3098	183	1	3.54	5	1	3	270	17	0	3.78

GIGUERE, Jean-Sebastien (ZHEE-gair, ZHAWN-suh-BAS-tee-yeh) ANA.

Goaltender. Catches left. 6'1", 199 lbs. Born, Montreal, Que., May 16, 1977.
(Hartford's 1st choice, 13th overall, in 1995 Entry Draft).

			Regular Season								Playoffs						
Season	Club	Lea	GP	W	L	T	Mins	GA	SO	Avg	GP	W	L	Mins	GA	SO	Avg
1992-93	Laval Laurentide	QAAA	25	12	11	2	1498	76	0	3.02	11	6	5	654	38	0	3.49
1993-94	Verdun	QMJHL	25	13	5	2	1234	66	1	3.21	....	....	....	....	....	....	....
1994-95	Halifax	QMJHL	47	14	27	5	2755	181	2	3.94	7	3	4	417	17	1	*2.45
1995-96	Halifax	QMJHL	55	26	23	2	3230	185	1	3.44	6	1	5	354	24	0	4.07
1996-97	Hartford	NHL	8	1	4	0	394	24	0	3.65	....	....	....	....	....	....	....
	Halifax	QMJHL	50	28	19	3	3014	170	2	3.38	16	9	7	954	58	0	3.65
1997-98	Saint John	AHL	31	16	10	3	1758	72	2	2.46	10	5	3	536	27	0	3.02
1998-99	Calgary	NHL	15	6	7	1	860	46	0	3.21	....	....	....	....	....	....	....
	Saint John	AHL	39	18	16	3	2145	123	3	3.44	7	3	2	304	21	0	4.14
99-2000	Calgary	NHL	7	1	3	1	330	15	0	2.73	....	....	....	....	....	....	....
	Saint John	AHL	41	17	17	3	2243	114	0	3.05	3	0	3	178	9	0	3.03
2000-01	Anaheim	NHL	34	11	17	5	2031	87	4	2.57	....	....	....	....	....	....	....
	Cincinnati	AHL	23	12	7	2	1306	53	0	2.43	....	....	....	....	....	....	....
2001-02	Anaheim	NHL	53	20	25	6	3127	111	4	2.13	....	....	....	....	....	....	....
	NHL Totals		117	39	56	13	6742	283	8	2.52	....	....	....	....	....	....	....

QMJHL Second All-Star Team (1997) • Shared Harry "Hap" Holmes Memorial Trophy (fewest goals against - AHL) with Tyler Moss (1998)

Transferred to **Carolina** after **Hartford** franchise relocated, June 25, 1997. Traded to **Calgary** by **Carolina** with Andrew Cassels for Gary Roberts and Trevor Kidd, August 25, 1997. Traded to **Anaheim** by **Calgary** for Anaheim's 2nd round choice (later traded to Washington - Washington selected Matt Pettinger) in 2000 Entry Draft, June 10, 2000.

GOEHRING, Karl (GAIR-ihng, KAHRL) CBJ

Goaltender. Catches left. 5'8", 160 lbs. Born, Apple Valley, MN, August 23, 1978.

			Regular Season								Playoffs						
Season	Club	Lea	GP	W	L	T	Mins	GA	SO	Avg	GP	W	L	Mins	GA	SO	Avg
1996-97	Fargo-Moorhead	USHL	32	13	18	1	1909	79	*4	*2.48	5	2	3	251	15	1	3.58
1997-98	North Dakota	WCHA	27	23	3	1	1504	57	1	*2.27	....	....	....	....	....	....	....
1998-99	North Dakota	WCHA	31	22	5	2	1774	71	3	2.40	....	....	....	....	....	....	....
99-2000	North Dakota	WCHA	30	19	6	4	1747	55	*8	*1.89	....	....	....	....	....	....	....
2000-01	North Dakota	WCHA	30	16	6	6	1662	66	*3	2.38	....	....	....	....	....	....	....
2001-02	Syracuse Crunch	AHL	15	5	6	3	891	37	1	2.49	....	....	....	....	....	....	....
	Dayton Bombers	ECHL	23	11	9	3	1393	52	2	2.24	*14	9	5	*866	35	1	2.43

WCHA First All-Star Team (1998, 2000) • WCHA Rookie of the Year (1998) • NCAA West First All-American Team (1998, 2000) • WCHA Second All-Star Team (1999)

Signed as a free agent by **Columbus**, May 7, 2001.

GRAHAME, John (GRAY-ham, JAWN) BOS.

Goaltender. Catches left. 6'2", 214 lbs. Born, Denver, CO, August 31, 1975.
(Boston's 7th choice, 229th overall, in 1994 Entry Draft).

			Regular Season								Playoffs						
Season	Club	Lea	GP	W	L	T	Mins	GA	SO	Avg	GP	W	L	Mins	GA	SO	Avg
1993-94	Sioux City	USHL	20	....	....	....	1200	73	0	3.70	....	....	....	....	....	....	....
1994-95	Lake Superior	CCHA	28	16	7	3	1616	75	2	2.79	....	....	....	....	....	....	....
1995-96	Lake Superior	CCHA	29	21	4	2	1558	66	2	2.54	....	....	....	....	....	....	....
1996-97	Lake Superior	CCHA	37	19	13	4	2197	134	3	3.66	....	....	....	....	....	....	....
1997-98	Providence	AHL	55	15	31	4	3053	164	3	3.22	....	....	....	....	....	....	....
1998-99	Providence	AHL	48	*37	9	1	2771	134	3	2.90	19	*15	4	*1209	48	1	2.38
99-2000	Boston	NHL	24	7	10	5	1344	55	2	2.46	....	....	....	....	....	....	....
	Providence	AHL	27	11	13	2	1528	86	1	3.38	13	10	3	839	35	0	2.50
2000-01	Boston	NHL	10	3	4	0	471	28	0	3.57	....	....	....	....	....	....	....
	Providence	AHL	16	4	7	3	893	47	0	3.16	17	8	9	1043	46	2	2.65
2001-02	Boston	NHL	19	8	7	2	1079	52	1	2.89	....	....	....	....	....	....	....
	NHL Totals		53	18	21	7	2894	135	3	2.80	....	....	....	....	....	....	....

GRAHN, Carl (GRAHN, KARL) L.A.

Goaltender. Catches left. 5'11", 169 lbs. Born, Kouvola, Finland, January 8, 1981.
(Los Angeles' 11th choice, 282nd overall, in 2000 Entry Draft).

			Regular Season								Playoffs						
Season	Club	Lea	GP	W	L	T	Mins	GA	SO	Avg	GP	W	L	Mins	GA	SO	Avg
1998-99	KalPa Kuopio Jr.	Finn-Jr.	20	7	10	1	1153	63	1	3.28	....	....	....	....	....	....	....
	KalPa Kuopio	Finland	3	0	2	0	126	16	0	7.63	....	....	....	....	....	....	....
99-2000	KooKoo Jr.	Finn-Jr.	38	....	....	....	2165	124	1	3.44	....	....	....	....	....	....	....
2000-01	KalPa Kuopio	Finland-2	39	20	15	4	2340	105	2	2.69	....	....	....	....	....	....	....
2001-02	KooKoo Kouvola	Finland-2	36	....	....	....	2160	92	0	2.56	10	....	....	600	31	0	3.09

GRUMET-MORRIS, Dov (groo-MAY-MAW-rihs, DAWV) PHI.

Goaltender. Catches left. 6'2", 190 lbs. Born, Evanston, IL, February 28, 1982.
(Philadelphia's 4th choice, 161st overall, in 2002 Entry Draft).

			Regular Season								Playoffs						
Season	Club	Lea	GP	W	L	T	Mins	GA	SO	Avg	GP	W	L	Mins	GA	SO	Avg
2000-01	Danville Wings	NAJHL	27	19	5	2	1547	57	3	2.21	5	2	2	300	17	0	3.40
2001-02	Harvard Crimson	ECAC	21	10	8	1	1226	58	1	2.84	....	....	....	....	....	....	....

GUSTAFSON, Derek (GUHST-ahf-suhn, DEH-rihk) MIN.

Goaltender. Catches left. 5'11", 210 lbs. Born, Gresham, OR, June 21, 1979.

			Regular Season								Playoffs						
Season	Club	Lea	GP	W	L	T	Mins	GA	SO	Avg	GP	W	L	Mins	GA	SO	Avg
1995-96	Seattle Ironmen	BCAHA	16	....	....	....	913	46	0	3.02	....	....	....	....	....	....	....
1996-97	Vernon Vipers	BCHL	23	....	....	....	1241	70	0	3.38	....	....	....	....	....	....	....
1997-98	Vernon Vipers	BCHL	42	27	13	2	2270	144	1	3.81	6	2	1	257	13	0	3.04
1998-99	Vernon Vipers	BCHL	42	39	3	0	2505	94	3	2.25	....	....	....	....	....	....	....
99-2000	St. Lawrence	ECAC	24	17	4	2	1475	51	2	2.07	....	....	....	....	....	....	....
2000-01	Minnesota	NHL	4	1	3	0	239	10	0	2.51	....	....	....	....	....	....	....
	Jackson Bandits	ECHL	7	4	3	0	404	15	1	2.23	....	....	....	....	....	....	....
	Cleveland	IHL	24	14	7	1	1293	59	2	2.74	2	0	1	53	5	0	5.64
2001-02	Minnesota	NHL	1	0	0	0	26	0	0	0.00	....	....	....	....	....	....	....
	Houston Aeros	AHL	38	14	13	6	2016	92	4	2.74	2	0	0	25	1	0	2.37
	NHL Totals		5	1	3	0	265	10	0	2.26	....	....	....	....	....	....	....

BCHL First All-Star Team (1999) • BCHL Interior Top Goaltender (1999) • ECAC Second All-Star Team (2000) • ECAC Rookie of the Year (2000)

Signed as a free agent by **Minnesota**, June 9, 2000.

HACKETT, Jeff (HA-keht, JEHF) MTL.

Goaltender. Catches left. 6'1", 198 lbs. Born, London, Ont., June 1, 1968.
(NY Islanders' 2nd choice, 34th overall, in 1987 Entry Draft).

			Regular Season								Playoffs						
Season	Club	Lea	GP	W	L	T	Mins	GA	SO	Avg	GP	W	L	Mins	GA	SO	Avg
1984-85	London Diamonds	OJHL-B	18	....	....	....	1078	73	1	4.06	....	....	....	....	....	....	....
1985-86	London Diamonds	OJHL-B	19	....	....	....	1150	66	0	3.43	....	....	....	....	....	....	....
1986-87	Oshawa Generals	OHL	31	18	9	2	1672	85	2	3.05	15	8	7	895	40	0	2.68
1987-88	Oshawa Generals	OHL	53	30	21	2	3165	205	0	3.89	7	3	4	438	31	0	4.25
1988-89	NY Islanders	NHL	13	4	7	0	662	39	0	3.53	....	....	....	....	....	....	....
	Springfield	AHL	29	12	14	2	1677	116	0	4.15	....	....	....	....	....	....	....
1989-90	Springfield	AHL	54	24	25	3	3045	187	1	3.68	*17	*10	5	934	60	0	3.85
1990-91	NY Islanders	NHL	30	5	18	1	1508	91	0	3.62	....	....	....	....	....	....	....
1991-92	San Jose	NHL	42	11	27	1	2314	148	0	3.84	....	....	....	....	....	....	....
1992-93	San Jose	NHL	36	2	30	1	2000	176	0	5.28	....	....	....	....	....	....	....
1993-94	Chicago	NHL	22	2	12	3	1084	62	0	3.43	....	....	....	....	....	....	....
1994-95	Chicago	NHL	7	1	3	2	328	13	0	2.38	2	0	0	26	1	0	2.31
1995-96	Chicago	NHL	35	18	11	4	2000	80	4	2.40	1	0	1	60	5	0	5.00
1996-97	Chicago	NHL	41	19	18	4	2473	89	2	2.16	6	2	4	345	25	0	4.35
1997-98	Chicago	NHL	58	21	25	11	3441	126	8	2.20	....	....	....	....	....	....	....
1998-99	Chicago	NHL	10	2	6	1	524	33	0	3.78	....	....	....	....	....	....	....
	Montreal	NHL	53	24	20	9	3091	117	5	2.27	....	....	....	....	....	....	....
99-2000	Montreal	NHL	56	23	25	7	3301	132	3	2.40	....	....	....	....	....	....	....
2000-01	Montreal	NHL	19	4	10	2	998	54	0	3.25	....	....	....	....	....	....	....
2001-02	Montreal	NHL	15	5	5	2	717	38	0	3.18	....	....	....	....	....	....	....
	NHL Totals		437	141	217	48	24441	1198	22	2.94	9	2	5	431	31	0	4.32

Won Jack A. Butterfield Trophy (Playoff MVP - AHL) (1990)

Claimed by **San Jose** from **NY Islanders** in Expansion Draft, May 30, 1991. Traded to **Chicago** by **San Jose** for Chicago's 3rd round choice (Alexei Yegorov) in 1994 Entry Draft, July 13, 1993. Traded to **Montreal** by **Chicago** with Eric Weinrich, Alain Nasreddine and Tampa Bay's 4th round choice (previously acquired, Montreal selected Chris Dyment) in 1999 Entry Draft for Jocelyn Thibault, Dave Manson and Brad Brown, November 16, 1998. • Missed majority of 2000-01 season recovering from hand injury originally suffered in game vs. Minnesota, October 24, 2000. • Missed majority of 2001-02 season recovering from shoulder injury originally suffered in game vs. Buffalo, October 20, 2001.

HAMERLIK, Peter (HAHM-ehr-lik, PEE-tuhr) BOS.

Goaltender. Catches left. 6'1", 194 lbs. Born, Myjava, Czech., January 2, 1982.
(Boston's 4th choice, 153rd overall, in 2002 Entry Draft).

			Regular Season								Playoffs						
Season	Club	Lea	GP	W	L	T	Mins	GA	SO	Avg	GP	W	L	Mins	GA	SO	Avg
1997-98	Skalica Jr.	Slovak-Jr.	49	....	....	....	2969	168	....	3.40	....	....	....	....	....	....	....
1998-99	HK 36 Skalica	Slovakia	1	0	1	0	24	3	0	7.50	....	....	....	....	....	....	....
99-2000	Skalica Jr.	Slovak-Jr.	37	....	....	....	1850	121	....	3.92	....	....	....	....	....	....	....
	HK-36 Skalica	Slovakia	7	....	....	....	286	16	....	3.36	....	....	....	....	....	....	....
2000-01	Kingston	OHL	56	21	21	8	3026	153	*4	3.03	3	0	2	131	13	0	5.95
2001-02	Kingston	OHL	43	12	21	6	2371	144	1	3.64	1	0	1	60	6	0	6.00

• Re-entered NHL Entry Draft. Originally Pittsburgh's 3rd choice, 84th overall, in 2000 Entry Draft.

HARDING, Josh (HAHR-dihng, JAWSH) MIN.

Goaltender. Catches right. 6'1", 170 lbs. Born, Regina, Sask., June 18, 1984.
(Minnesota's 2nd choice, 38th overall, in 2002 Entry Draft).

			Regular Season								Playoffs						
Season	Club	Lea	GP	W	L	T	Mins	GA	SO	Avg	GP	W	L	Mins	GA	SO	Avg
2000-01	Regina Pat Cdns.	SMHL	36	....	....	....	2000	81	3	2.41	....	....	....	....	....	....	....
2001-02	Regina Pats	WHL	42	27	13	1	2389	95	*4	2.39	6	2	4	325	16	0	2.95

WHL East Second All-Star Team (2002)

HASEK, Dominik (HAH-shihk, DOHM-ihn-ihk)

Goaltender. Catches left. 5'11", 180 lbs. Born, Pardubice, Czech., January 29, 1965.
(Chicago's 11th choice, 207th overall, in 1983 Entry Draft).

			Regular Season								Playoffs						
Season	Club	Lea	GP	W	L	T	Mins	GA	SO	Avg	GP	W	L	Mins	GA	SO	Avg
1981-82	HC Pardubice	Czech	12	....	....	....	661	34	....	3.09	....	....	....	....	....	....	....
1982-83	HC Pardubice	Czech	42	....	....	....	2358	105	....	2.67	....	....	....	....	....	....	....
1983-84	HC Pardubice	Czech	40	....	....	....	2304	108	....	2.81	....	....	....	....	....	....	....
1984-85	HC Pardubice	Czech	42	....	....	....	2419	131	....	3.25	....	....	....	....	....	....	....
1985-86	HC Pardubice	Czech	45	....	....	....	2689	138	....	3.08	....	....	....	....	....	....	....
1986-87	HC Pardubice	Czech	43	....	....	....	2515	103	....	2.46	....	....	....	....	....	....	....
1987-88	HC Pardubice	Czech	31	....	....	....	1862	93	....	3.00	....	....	....	....	....	....	....
	Czechoslovakia	Olympics	5	3	2	0	217	18	1	4.98	....	....	....	....	....	....	....
1988-89	HC Pardubice	Czech	42	....	....	....	2507	114	....	2.73	....	....	....	....	....	....	....
1989-90	Dukla Jihlava	Czech	40	....	....	....	2251	80	....	2.13	....	....	....	....	....	....	....
1990-91	Chicago	NHL	5	3	0	1	195	8	0	2.46	3	0	0	69	3	0	2.61
	Indianapolis Ice	IHL	33	20	11	1	1903	80	*5	*2.52	1	1	0	60	3	0	3.00
1991-92	Chicago	NHL	20	10	4	1	1014	44	1	2.60	3	0	2	158	8	0	3.04
	Indianapolis Ice	IHL	20	7	10	3	1162	69	1	3.56	....	....	....	....	....	....	....
1992-93	Buffalo	NHL	28	11	10	4	1429	75	0	3.15	1	1	0	45	1	0	1.33
1993-94	Buffalo	NHL	58	30	20	6	3358	109	*7	*1.95	7	3	4	484	13	2	*1.61
1994-95	HC Pardubice	Czech	2	1	0	1	124	6	0	2.90	....	....	....	....	....	....	....
	Buffalo	NHL	41	19	14	7	2416	85	*5	*2.11	5	1	4	309	18	0	3.50
1995-96	Buffalo	NHL	59	22	30	6	3417	161	2	2.83	....	....	....	....	....	....	....
1996-97	Buffalo	NHL	67	37	20	10	4037	153	5	2.27	3	1	1	153	5	0	1.96
1997-98	Buffalo	NHL	*72	33	23	13	*4220	147	*13	2.09	15	10	5	948	32	1	2.03
	Czech Republic	Olympics	6	*5	1	0	*369	6	*2	*0.97	....	....	....	....	....	....	....
1998-99	Buffalo	NHL	64	30	18	14	3817	119	9	1.87	19	13	6	1217	36	2	1.77
99-2000	Buffalo	NHL	35	15	11	6	2066	76	3	2.21	5	1	4	301	12	0	2.39
2000-01	Buffalo	NHL	67	37	24	4	3904	137	*11	2.11	13	7	6	833	29	1	2.09
2001-02 ♦	Detroit	NHL	65	*41	15	8	3872	140	5	2.17	*23	*16	7	*1455	45	*6	1.86
	Czech Republic	Olympics	4	2	1	1	239	8	0	2.01	....	....	....	....	....	....	....
	NHL Totals		581	288	189	80	33745	1254	61	2.23	97	53	39	5972	202	12	2.03

Czechoslovakian Goaltender of the Year (1986, 1987, 1988, 1989, 1990) • Czechoslovakian Player of the Year (1987, 1989, 1990) • Czechoslovakian First All-Star Team (1988, 1989, 1990) • IHL First All-Star Team (1991) • NHL All-Rookie Team (1992) • NHL First All-Star Team (1994, 1995, 1997, 1998, 1999, 2001) • Shared William M. Jennings Trophy with Grant Fuhr (1994) • Won Vezina Trophy (1994, 1995, 1997, 1998, 1999, 2001) • Won Lester B. Pearson Award (1997, 1998) • Won Hart Trophy (1997, 1998) • Won William M. Jennings Trophy (2001) • Played in NHL All-Star Game (1996, 1997, 1998, 1999, 2001, 2002)

Traded to **Buffalo** by **Chicago** for Stephane Beauregard and Buffalo's 4th round choice (Eric Daze) in 1993 Entry Draft, August 7, 1992. Traded to **Detroit** by **Buffalo** for Vyacheslav Kozlov, Detroit's 1st round choice (later traded to Columbus - later traded to Atlanta - Atlanta selected Jim Slater) in 2002 Entry Draft and future considerations, July 1, 2001. • Officially announced retirement, June 25, 2002.

HAUSER, Adam (HOW-suhr, A-duhm)

Goaltender. Catches left. 6'2", 195 lbs. Born, Bovey, MN, May 27, 1980.
(Edmonton's 4th choice, 81st overall, in 1999 Entry Draft).

			Regular Season								Playoffs						
Season	Club	Lea	GP	W	L	T	Mins	GA	SO	Avg	GP	W	L	Mins	GA	SO	Avg
1996-97	Greenway High	H.S.	25	....	....	....	1496	63	0	2.54	....	....	....	....	....	....	....
1997-98	Team USA	USDP-18	38	19	10	7	2110	94	4	2.67	....	....	....	....	....	....	....
1998-99	U. of Minnesota	WCHA	*40	14	18	8	*2350	136	3	3.47	....	....	....	....	....	....	....
99-2000	U. of Minnesota	WCHA	36	20	14	2	2114	104	1	2.95	....	....	....	....	....	....	....
2000-01	U. of Minnesota	WCHA	40	*26	12	2	2366	101	*3	2.56	....	....	....	....	....	....	....
2001-02	U. of Minnesota	WCHA	35	*23	6	4	2003	80	1	2.40	....	....	....	....	....	....	....

NCAA Championship All-Tournament Team (2002)

HEDBERG, Johan (HEHD-buhrg, YO-han) PIT.

Goaltender. Catches left. 6', 184 lbs. Born, Leksand, Sweden, May 3, 1973.
(Philadelphia's 8th choice, 218th overall, in 1994 Entry Draft).

			Regular Season								Playoffs						
Season	Club	Lea	GP	W	L	T	Mins	GA	SO	Avg	GP	W	L	Mins	GA	SO	Avg
1992-93	Leksands IF	Sweden	10	....	....	....	600	24	....	2.40	....	....	....	....	....	....	....
1993-94	Leksands IF	Sweden	17	....	....	....	1020	48	....	2.82	....	....	....	....	....	....	....
1994-95	Leksands IF	Sweden	17	....	....	....	986	58	....	3.53	....	....	....	....	....	....	....
1995-96	Leksands IF	Sweden	34	....	....	....	2013	95	....	2.83	4	....	....	240	13	....	3.25
1996-97	Leksands IF	Sweden	38	....	....	....	2260	95	3	2.52	8	....	....	581	18	1	1.86
1997-98	Baton Rouge	ECHL	2	1	1	0	100	7	0	4.20	....	....	....	....	....	....	....
	Detroit Vipers	IHL	16	7	2	2	726	32	1	2.64	....	....	....	....	....	....	....
	Manitoba Moose	IHL	14	8	4	1	745	32	1	2.58	2	0	2	105	6	0	3.40
1998-99	Leksands IF	Sweden	*48	....	....	....	*2940	140	0	2.86	4	....	....	255	15	0	3.53
99-2000	Kentucky	AHL	33	18	9	5	1973	88	3	2.68	5	3	2	311	10	1	1.93
2000-01	Manitoba Moose	IHL	46	23	13	7	2697	115	1	2.56	....	....	....	....	....	....	....
	Pittsburgh	**NHL**	**9**	**7**	**1**	**1**	**545**	**24**	**0**	**2.64**	**18**	**9**	**9**	**1123**	**43**	**2**	**2.30**
2001-02	**Pittsburgh**	**NHL**	**66**	**25**	**34**	**7**	**3877**	**178**	**6**	**2.75**	....	....	....	....	....	....	....
	Sweden	Olympics	1	1	0	0	60	1	0	1.00	....	....	....	....	....	....	....
	NHL Totals		**75**	**32**	**35**	**8**	**4422**	**202**	**6**	**2.74**	**18**	**9**	**9**	**1123**	**43**	**2**	**2.30**

Rights traded to **San Jose** by **Philadelphia** for San Jose's 7th round choice (Pavel Kasparik) in 1999 Entry Draft, August 6, 1998. Traded to **Pittsburgh** by **San Jose** with Bobby Dollas for Jeff Norton, March 12, 2001.

HIRSCH, Corey (HUHRSH, KOHR-ee) DAL.

Goaltender. Catches left. 5'10", 175 lbs. Born, Medicine Hat, Alta., August 10, 1972.
(NY Rangers' 7th choice, 169th overall, in 1991 Entry Draft).

			Regular Season								Playoffs						
Season	Club	Lea	GP	W	L	T	Mins	GA	SO	Avg	GP	W	L	Mins	GA	SO	Avg
1987-88	Calgary Canucks	AJHL	32	22	5	0	1538	91	1	3.55	....	....	....	....	....	....	....
1988-89	Kamloops Blazers	WHL	32	11	12	2	1516	106	2	4.20	5	3	2	245	19	0	4.65
1989-90	Kamloops Blazers	WHL	*63	*48	13	0	3608	230	*3	3.82	*17	*14	3	*1043	60	0	*3.45
1990-91	Kamloops Blazers	WHL	38	26	7	1	1970	100	3	*3.05	11	5	6	623	42	0	4.04
1991-92	Kamloops Blazers	WHL	48	35	10	2	2732	124	*5	*2.72	*16	*11	5	954	35	*2	*2.20
1992-93	**NY Rangers**	**NHL**	**4**	**1**	**2**	**1**	**224**	**14**	**0**	**3.75**	....	....	....	....	....	....	....
	Binghamton	AHL	46	*35	4	5	2692	125	1	*2.79	14	7	7	831	46	0	3.32
1993-94	Canada	Nat-Tm	45	24	17	3	2653	124	0	2.80	....	....	....	....	....	....	....
	Canada	Olympics	8	5	2	1	495	18	0	2.18	....	....	....	....	....	....	....
	Binghamton	AHL	10	5	4	1	610	38	0	3.73	....	....	....	....	....	....	....
1994-95	Binghamton	AHL	57	31	20	5	3371	175	0	3.11	....	....	....	....	....	....	....
1995-96	**Vancouver**	**NHL**	**41**	**17**	**14**	**6**	**2338**	**114**	**1**	**2.93**	**6**	**2**	**3**	**338**	**21**	**0**	**3.73**
1996-97	**Vancouver**	**NHL**	**39**	**12**	**20**	**4**	**2127**	**116**	**2**	**3.27**	....	....	....	....	....	....	....
1997-98	**Vancouver**	**NHL**	**1**	**0**	**0**	**0**	**50**	**5**	**0**	**6.00**	....	....	....	....	....	....	....
	Syracuse Crunch	AHL	60	30	22	6	3512	187	1	3.19	5	2	3	297	10	1	*2.02
1998-99	**Vancouver**	**NHL**	**20**	**3**	**8**	**3**	**919**	**48**	**1**	**3.13**	....	....	....	....	....	....	....
	Syracuse Crunch	AHL	5	2	3	0	300	14	0	2.80	....	....	....	....	....	....	....
99-2000	Milwaukee	IHL	19	9	8	1	1098	49	0	2.68	....	....	....	....	....	....	....
	Utah Grizzlies	IHL	17	9	5	1	937	42	3	2.69	2	0	2	121	4	0	1.99
2000-01	Albany	AHL	4	0	4	0	199	19	0	5.72	....	....	....	....	....	....	....
	Washington	**NHL**	**1**	**1**	**0**	**0**	**20**	**0**	**0**	**0.00**	....	....	....	....	....	....	....
	Cincinnati	IHL	13	11	2	0	783	28	1	2.15	....	....	....	....	....	....	....
	Portland Pirates	AHL	36	17	17	2	2142	104	1	2.91	2	0	2	118	7	0	3.55
2001-02	Portland Pirates	AHL	23	6	12	5	1395	62	1	2.67	....	....	....	....	....	....	....
	Philadelphia	AHL	5	2	3	0	299	14	1	2.81	....	....	....	....	....	....	....
	NHL Totals		**106**	**34**	**44**	**14**	**5678**	**297**	**4**	**3.14**	**6**	**2**	**3**	**338**	**21**	**0**	**3.73**

WHL West Second All-Star Team (1990) • WHL West First All-Star Team (1991, 1992) • Canadian Major Junior Goaltender of the Year (1992) • Memorial Cup All-Star Team (1992) • Won Hap Emms Memorial Trophy (Memorial Cup Tournament Top Goaltender) (1992) • AHL First All-Star Team (1993) • Won Dudley "Red" Garrett Memorial Trophy (Top Rookie - AHL) (1993) • Shared Harry "Hap" Holmes Memorial Trophy (fewest goals against - AHL) with Boris Rousson (1993) • NHL All-Rookie Team (1996)

Traded to **Vancouver** by **NY Rangers** for Nathan Lafayette, April 7, 1995. Signed as a free agent by **Nashville**, August 10, 1999. Traded to **Anaheim** by **Nashville** for future considerations, March 14, 2000. Signed as a free agent by **Washington**, October 27, 2000. Signed as a free agent by **Dallas**, August 14, 2002.

HNILICKA, Milan (huh-LEETCH-kuh, MEE-lan) ATL.

Goaltender. Catches left. 5'11", 190 lbs. Born, Kladno, Czech., June 25, 1973.
(NY Islanders' 4th choice, 70th overall, in 1991 Entry Draft).

			Regular Season								Playoffs						
Season	Club	Lea	GP	W	L	T	Mins	GA	SO	Avg	GP	W	L	Mins	GA	SO	Avg
1989-90	Poldi Kladno	Czech	24	....	....	....	1113	70	....	3.77	....	....	....	....	....	....	....
1990-91	Poldi Kladno	Czech	40	....	....	....	2122	98	0	2.80	....	....	....	....	....	....	....
1991-92	Poldi Kladno	Czech	38	....	....	....	2066	128	0	3.73	....	....	....	....	....	....	....
1992-93	Swift Current	WHL	*65	*46	12	2	3679	206	2	3.36	*17	*12	5	*1017	54	*2	3.19
1993-94	Richmond	ECHL	43	18	16	5	2299	155	4	4.05	....	....	....	....	....	....	....
	Salt Lake	IHL	8	5	1	0	378	25	0	3.97	....	....	....	....	....	....	....
1994-95	Denver Grizzlies	IHL	15	9	4	1	798	47	1	3.53	....	....	....	....	....	....	....
1995-96	HC Poldi Kladno	Czech	33	....	....	....	1959	93	1	2.84	8	....	....	493	24	....	2.92
1996-97	HC Poldi Kladno	Czech	48	....	....	....	2736	120	*4	2.63	3	....	....	151	14	0	5.56
1997-98	HC Sparta Praha	Czech	49	....	....	....	2847	99	....	2.09	11	....	....	632	31	....	3.00
1998-99	HC Sparta Praha	Czech	*50	....	....	....	*2877	109	....	2.27	8	....	....	507	13	....	*1.54
99-2000	**NY Rangers**	**NHL**	**2**	**0**	**1**	**0**	**86**	**5**	**0**	**3.49**	....	....	....	....	....	....	....
	Hartford	AHL	36	22	11	0	1979	71	5	*2.15	3	0	1	99	6	0	3.64
2000-01	**Atlanta**	**NHL**	**36**	**12**	**19**	**2**	**1879**	**105**	**2**	**3.35**	....	....	....	....	....	....	....
2001-02	**Atlanta**	**NHL**	**60**	**13**	**33**	**10**	**3367**	**179**	**3**	**3.19**	....	....	....	....	....	....	....
	NHL Totals		**98**	**25**	**53**	**12**	**5332**	**289**	**5**	**3.25**	....	....	....	....	....	....	....

Shared Harry "Hap" Holmes Memorial Trophy (fewest goals against - AHL) with Jean-Francois Labbe (2000)

Signed as a free agent by **NY Rangers**, July 15, 1999. Signed as a free agent by **Atlanta**, July 28, 2000.

HODSON, Jamie (HAWD-suhn, JAY-mee) TOR.

Goaltender. Catches left. 6'2", 206 lbs. Born, Brandon, Man., April 8, 1980.
(Toronto's 3rd choice, 69th overall, in 1998 Entry Draft).

			Regular Season								Playoffs						
Season	Club	Lea	GP	W	L	T	Mins	GA	SO	Avg	GP	W	L	Mins	GA	SO	Avg
1996-97	Yellowhead Chiefs	MMHL	12	....	....	....	720	58	1	4.83	....	....	....	....	....	....	....
1997-98	Brandon	WHL	20	12	2	2	964	52	2	3.24	6	5	0	337	16	0	2.85
1998-99	Brandon	WHL	43	23	12	3	2295	123	4	3.22	5	1	4	275	26	0	5.67
99-2000	Brandon	WHL	39	13	22	3	2321	130	2	3.36	....	....	....	....	....	....	....
2000-01	St. John's	AHL	4	0	2	0	137	13	0	5.71	....	....	....	....	....	....	....
	Brandon	WHL	29	9	17	1	1587	92	0	3.48	1	0	1	59	3	0	3.05
2001-02	South Carolina	ECHL	36	20	9	3	1996	100	1	3.01	1	0	1	60	3	0	3.01

• Returned to **Brandon** (WHL) by **St. John's** (AHL), October 25, 2000.

HODSON, Kevin (HAWD-suhn, KEH-vihn) T.B.

Goaltender. Catches left. 6', 182 lbs. Born, Winnipeg, Man., March 27, 1972.

			Regular Season								Playoffs						
Season	Club	Lea	GP	W	L	T	Mins	GA	SO	Avg	GP	W	L	Mins	GA	SO	Avg
1989-90	Winnipeg South	MJHL	35	....	....	....	1900	115	2	3.40	....	....	....	....	....	....	....
1990-91	Sault Ste. Marie	OHL	30	18	11	0	1638	88	*2	*3.22	10	*9	1	581	28	0	*2.89
1991-92	Sault Ste. Marie	OHL	50	28	12	4	2722	151	0	3.33	18	12	6	1116	54	1	2.90
1992-93	Sault Ste. Marie	OHL	26	18	5	2	1470	76	1	*3.10	14	11	2	755	34	0	2.70
	Indianapolis Ice	IHL	14	5	9	0	777	53	0	4.09	....	....	....	....	....	....	....
1993-94	Adirondack	AHL	37	20	10	5	2082	102	2	2.94	3	0	2	89	10	0	6.77
1994-95	Adirondack	AHL	51	19	22	8	2731	161	1	3.54	4	0	4	237	14	0	3.53
1995-96	**Detroit**	**NHL**	**4**	**2**	**0**	**0**	**163**	**3**	**1**	**1.10**	....	....	....	....	....	....	....
	Adirondack	AHL	32	13	13	2	1654	87	0	3.16	3	0	2	150	8	0	3.21
1996-97 ♦	**Detroit**	**NHL**	**6**	**2**	**2**	**1**	**294**	**8**	**1**	**1.63**	....	....	....	....	....	....	....
	Quebec Rafales	IHL	2	1	1	0	118	7	0	3.54	....	....	....	....	....	....	....
1997-98 ♦	**Detroit**	**NHL**	**21**	**9**	**3**	**3**	**988**	**44**	**2**	**2.67**	**1**	**0**	**0**	**1**	**0**	**0**	**0.00**
1998-99	**Detroit**	**NHL**	**4**	**0**	**2**	**0**	**175**	**9**	**0**	**3.09**	....	....	....	....	....	....	....
	Adirondack	AHL	6	1	3	2	349	19	0	3.27	....	....	....	....	....	....	....
	Tampa Bay	**NHL**	**5**	**2**	**1**	**1**	**238**	**11**	**0**	**2.77**	....	....	....	....	....	....	....
99-2000	**Tampa Bay**	**NHL**	**24**	**2**	**7**	**4**	**769**	**47**	**0**	**3.67**	....	....	....	....	....	....	....
	Detroit Vipers	IHL	9	2	6	0	505	22	0	2.61	....	....	....	....	....	....	....
2000-01			OUT OF HOCKEY – RETIRED														
2001-02	Sault Ste. Marie	OHL	DID NOT PLAY – ASSISTANT COACH														
	NHL Totals		**64**	**17**	**15**	**9**	**2627**	**122**	**4**	**2.79**	**1**	**0**	**0**	**1**	**0**	**0**	**0.00**

Memorial Cup All-Star Team (1993) • Won Hap Emms Memorial Trophy (Memorial Cup Tournament Top Goaltender) (1993)

Signed as a free agent by **Chicago**, August 17, 1992. Signed as a free agent by **Detroit**, June 16, 1993. • Played 16 seconds in playoff game vs. St. Louis, May 17, 1998. Traded to **Tampa Bay** by **Detroit** with San Jose's 2nd round choice (previously acquired, Tampa Bay selected Sheldon Keefe) in 1999 Entry Draft for Wendel Clark and Detroit's 6th round choice (previously acquired, Detroit selected Kent McDonell) in 1999 Entry Draft, March 23, 1999. Traded to **Montreal** by **Tampa Bay** for Montreal's 7th round choice (later traded to Philadelphia - Philadelphia selected John Eichelberger) in 2000 Entry Draft, June 2, 2000. Signed as a free agent by **Tampa Bay**, May 28, 2002.

HOLMQVIST, Johan (HOHLM-kvihst, YOH-han) NYR

Goaltender. Catches left. 6'3", 190 lbs. Born, Tolfta, Sweden, May 24, 1978.
(NY Rangers' 9th choice, 175th overall, in 1997 Entry Draft).

			Regular Season								Playoffs						
Season	Club	Lea	GP	W	L	T	Mins	GA	SO	Avg	GP	W	L	Mins	GA	SO	Avg
1996-97	Brynas IF Gavle	Sweden	2	0	0	0	80	4	0	3.00	....	....	....	....	....	....	....
1997-98	Brynas IF Gavle	Sweden	33	....	....	....	1897	82	....	2.59	3	0	3	180	14	....	4.67
1998-99	Brynas IF Gavle	Sweden	41	....	....	....	2383	111	4	2.79	*14	9	5	*855	34	0	2.39
99-2000	Brynas IF Gavle	Sweden	41	....	....	....	2402	104	4	2.60	11	....	....	671	30	1	2.68
2000-01	**NY Rangers**	**NHL**	**2**	**0**	**2**	**0**	**119**	**10**	**0**	**5.04**	....	....	....	....	....	....	....
	Hartford	AHL	43	19	14	4	2305	111	2	2.89	5	2	3	314	13	0	2.48
2001-02	**NY Rangers**	**NHL**	**1**	**0**	**0**	**0**	**9**	**0**	**0**	**0.00**	....	....	....	....	....	....	....
	Hartford	AHL	48	26	12	6	2734	140	1	3.07	4	1	2	163	12	0	4.41
	NHL Totals		**3**	**0**	**2**	**0**	**128**	**10**	**0**	**4.69**	....	....	....	....	....	....	....

HUET, Cristobal (oo-AY, KRIHS-toh-bahl) L.A.

Goaltender. Catches left. 6', 194 lbs. Born, St-Martin-d'Heres, France, September 3, 1975.
(Los Angeles' 9th choice, 214th overall, in 2001 Entry Draft).

			Regular Season								Playoffs						
Season	Club	Lea	GP	W	L	T	Mins	GA	SO	Avg	GP	W	L	Mins	GA	SO	Avg
1994/98	CSG Grenoble	France	STATISTICS NOT AVAILABLE														
1997-98	France	Olympics	2	1	1	0	120	5	0	2.50	....	....	....	....	....	....	....
1998-99	HC Lugano	Swiss	21	....	....	....	1275	58	1	2.73	10	....	....	628	18	1	*1.72
99-2000	HC Lugano	Swiss	31	....	....	....	1886	50	*8	*1.59	13	....	....	783	29	0	2.22
2000-01	HC Lugano	Swiss	39	....	....	....	2365	77	*6	*1.95	*18	....	....	*1141	39	2	2.05
2001-02	HC Lugano	Swiss	39	....	....	....	2313	107	*4	2.78	1	0	1	60	3	0	3.00
	France	Olympics	3	0	2	1	179	10	0	3.36	....	....	....	....	....	....	....

HURME, Jani (HOOR-meh, YAN-ee) OTT.

Goaltender. Catches left. 6', 187 lbs. Born, Turku, Finland, January 7, 1975.
(Ottawa's 2nd choice, 58th overall, in 1997 Entry Draft).

			Regular Season								Playoffs						
Season	Club	Lea	GP	W	L	T	Mins	GA	SO	Avg	GP	W	L	Mins	GA	SO	Avg
1992-93	TPS Turku Jr.	Finn-Jr.	12	....	....	....	669	47	0	4.22	1	....	....	60	0	1	0.00
1993-94	Kiekko-67 Jr.	Finn-Jr.	18	....	....	....	1082	57	0	3.16	....	....	....	....	....	....	....
	Kiekko-67 Turku	Finland-2	3	....	....	....	190	7	0	2.21	....	....	....	....	....	....	....
	TPS Turku	Finland	1	....	....	....	2	0	0	0.00	....	....	....	....	....	....	....
1994-95	TPS Turku Jr.	Finn-Jr.	2	....	....	....	125	5	0	2.40	....	....	....	....	....	....	....
	Kiekko-67 Jr.	Finn-Jr.	9	....	....	....	540	47	....	5.22	....	....	....	....	....	....	....
	Kiekko-67 Turku	Finland-2	19	....	....	....	1049	53	....	3.03	3	....	....	180	6	....	2.00
1995-96	TPS Turku Jr.	Finn-Jr.	13	....	....	....	777	34	1	2.63	....	....	....	....	....	....	....
	Kiekko-67 Turku	Finland-2	16	....	....	....	968	39	1	2.42	....	....	....	....	....	....	....
	TPS Turku	Finland	16	....	....	....	946	34	2	2.16	10	....	....	545	22	2	2.42
1996-97	TPS Turku	Finland	48	*31	11	6	*2917	101	*6	*2.08	*12	6	6	*722	39	0	3.24
1997-98	Detroit Vipers	IHL	6	2	2	2	290	20	0	4.13	....	....	....	....	....	....	....
	Indianapolis Ice	IHL	29	11	11	3	1506	83	1	3.30	3	1	0	129	10	0	4.62
1998-99	Detroit Vipers	IHL	12	7	3	1	643	26	1	2.43	....	....	....	....	....	....	....
	Cincinnati	IHL	26	14	9	2	1428	81	0	3.40	....	....	....	....	....	....	....
99-2000	**Ottawa**	**NHL**	**1**	**1**	**0**	**0**	**60**	**2**	**0**	**2.00**	....	....	....	....	....	....	....
	Grand Rapids	IHL	52	29	15	4	2948	107	4	2.18	*17	*10	7	*1028	37	1	2.16
2000-01	**Ottawa**	**NHL**	**22**	**12**	**5**	**4**	**1296**	**54**	**2**	**2.50**	....	....	....	....	....	....	....
2001-02	**Ottawa**	**NHL**	**25**	**12**	**9**	**1**	**1309**	**54**	**3**	**2.48**	....	....	....	....	....	....	....
	Finland	Olympics	3	1	2	0	179	9	0	3.01	....	....	....	....	....	....	....
	NHL Totals		**48**	**25**	**14**	**5**	**2665**	**110**	**5**	**2.48**	....	....	....	....	....	....	....

Finnish Elite League Rookie of the Year (1996) • Finnish Elite League Player of the Year (1997) • IHL Second All-Star Team (2000)

IRBE, Arturs (UHR-bay, AHR-tuhrs) CAR.

Goaltender. Catches left. 5'8", 190 lbs. Born, Riga, Latvia, February 2, 1967.
(Minnesota's 11th choice, 196th overall, in 1989 Entry Draft).

			Regular Season								Playoffs						
Season	Club	Lea	GP	W	L	T	Mins	GA	SO	Avg	GP	W	L	Mins	GA	SO	Avg
1986-87	Dynamo Riga	USSR	2	....	....	....	27	1	0	2.22	....	....	....	....	....	....	....
1987-88	Dynamo Riga	USSR	34	....	....	....	1870	86	4	2.76	....	....	....	....	....	....	....
1988-89	Dynamo Riga	USSR	40	....	....	....	2460	116	4	2.83	....	....	....	....	....	....	....
1989-90	Dynamo Riga	USSR	48	....	....	....	2880	115	2	2.40	....	....	....	....	....	....	....
1990-91	Dynamo Riga	USSR	46	....	....	....	2713	133	5	2.94	....	....	....	....	....	....	....
1991-92	**San Jose**	**NHL**	**13**	**2**	**6**	**3**	**645**	**48**	**0**	**4.47**	....	....	....	....	....	....	....
	Kansas City	IHL	32	24	7	1	1955	80	0	*2.46	*15	*12	3	914	44	0	*2.89
1992-93	**San Jose**	**NHL**	**36**	**7**	**26**	**0**	**2074**	**142**	**1**	**4.11**	....	....	....	....	....	....	....
	Kansas City	IHL	6	3	3	0	364	20	0	3.30	....	....	....	....	....	....	....
1993-94	**San Jose**	**NHL**	***74**	**30**	**28**	**16**	***4412**	**209**	**3**	**2.84**	**14**	**7**	**7**	**806**	**50**	**0**	**3.72**
1994-95	**San Jose**	**NHL**	**38**	**14**	**19**	**3**	**2043**	**111**	**4**	**3.26**	**6**	**2**	**4**	**316**	**27**	**0**	**5.13**
1995-96	**San Jose**	**NHL**	**22**	**4**	**12**	**4**	**1112**	**85**	**0**	**4.59**	....	....	....	....	....	....	....
	Kansas City	IHL	4	1	2	1	226	16	0	4.24	....	....	....	....	....	....	....
1996-97	**Dallas**	**NHL**	**35**	**17**	**12**	**3**	**1965**	**88**	**3**	**2.69**	**1**	**0**	**0**	**13**	**0**	**0**	**0.00**
1997-98	**Vancouver**	**NHL**	**41**	**14**	**11**	**6**	**1999**	**91**	**2**	**2.73**	....	....	....	....	....	....	....
1998-99	**Carolina**	**NHL**	**62**	**27**	**20**	**12**	**3643**	**135**	**6**	**2.22**	**6**	**2**	**4**	**408**	**15**	**0**	**2.21**
99-2000	**Carolina**	**NHL**	***75**	**34**	**28**	**9**	**4345**	**175**	**5**	**2.42**	....	....	....	....	....	....	....
2000-01	**Carolina**	**NHL**	***77**	**37**	**29**	**9**	***4406**	**180**	**6**	**2.45**	**6**	**2**	**4**	**360**	**20**	**0**	**3.33**
2001-02	**Carolina**	**NHL**	**51**	**20**	**19**	**11**	**2974**	**126**	**3**	**2.54**	**18**	**10**	**8**	**1078**	**30**	**1**	**1.67**
	Latvia	Olympics	1	0	1	0	60	4	0	4.00	....	....	....	....	....	....	....
	NHL Totals		**524**	**206**	**210**	**76**	**29618**	**1390**	**33**	**2.82**	**51**	**23**	**27**	**2981**	**142**	**1**	**2.86**

IHL First All-Star Team (1992) • Shared James Norris Memorial Trophy (fewest goals against - IHL) with Wade Flaherty (1992) • Played in NHL All-Star Game (1994, 1999)

Claimed by **San Jose** from **Minnesota** in Dispersal Draft, May 30, 1991. Signed as a free agent by **Dallas**, August 19, 1996. Signed as a free agent by **Vancouver**, August 25, 1997. Signed as a free agent by **Carolina**, September 14, 1998.

JOHNSON, Brent (JAWN-suhn, BREHNT) ST.L.

Goaltender. Catches left. 6'2", 200 lbs. Born, Farmington, MI, March 12, 1977.
(Colorado's 5th choice, 129th overall, in 1995 Entry Draft).

			Regular Season								Playoffs						
Season	Club	Lea	GP	W	L	T	Mins	GA	SO	Avg	GP	W	L	Mins	GA	SO	Avg
1993-94	Det. Compuware	MTJHL	18	....	....	....	1024	49	1	3.52	....	....	....	....	....	....	....
1994-95	Owen Sound	OHL	18	3	9	1	904	75	0	4.98	....	....	....	....	....	....	....
1995-96	Owen Sound	OHL	58	24	28	1	3211	243	1	4.54	6	2	4	371	29	0	4.69
1996-97	Owen Sound	OHL	50	20	28	1	2798	201	1	4.31	4	0	4	253	24	0	5.69
1997-98	Worcester	AHL	42	14	15	7	2240	119	0	3.19	6	3	2	332	19	0	3.43
1998-99	**St. Louis**	**NHL**	**6**	**3**	**2**	**0**	**286**	**10**	**0**	**2.10**	....	....	....	....	....	....	....
	Worcester	AHL	49	22	22	4	2925	146	2	2.99	4	1	3	238	12	0	3.02
99-2000	Worcester	AHL	58	24	27	5	3319	161	3	2.91	9	4	5	561	23	1	2.46
2000-01	**St. Louis**	**NHL**	**31**	**19**	**9**	**2**	**1744**	**63**	**4**	**2.17**	**2**	**0**	**1**	**62**	**2**	**0**	**1.94**
2001-02	**St. Louis**	**NHL**	**58**	**34**	**20**	**4**	**3491**	**127**	**5**	**2.18**	**10**	**5**	**5**	**590**	**18**	**3**	**1.83**
	NHL Totals		**95**	**56**	**31**	**6**	**5521**	**200**	**9**	**2.17**	**12**	**5**	**6**	**652**	**20**	**3**	**1.84**

Traded to **St. Louis** by **Colorado** for San Jose's 3rd round choice (previously acquired, Colorado selected Rick Berry) in 1997 Entry Draft, May 30, 1997.

JOKELA, Antti (YOH-keh-luh, AHN-tee) CAR.

Goaltender. Catches left. 5'11", 165 lbs. Born, Rauma, Finland, May 7, 1981.
(Carolina's 8th choice, 237th overall, in 1999 Entry Draft).

			Regular Season								Playoffs						
Season	Club	Lea	GP	W	L	T	Mins	GA	SO	Avg	GP	W	L	Mins	GA	SO	Avg
1998-99	Lukko Rauma-B	Finn-Jr.	8	....	....	....	480	26	0	3.25	....	....	....	....	....	....	....
	Lukko Rauma Jr.	Finn-Jr.	18	....	....	....	1038	66	1	3.81	....	....	....	....	....	....	....
99-2000	Lukko Rauma Jr.	Finn-Jr.	21	....	....	....	1196	71	0	3.56	....	....	....	....	....	....	....
2000-01	Jaa-Kotkat	Finn-Jr.	42	20	20	2	2520	100	4	2.38	....	....	....	....	....	....	....
	Lukko Rauma	Finland	1	0	0	0	19	2	0	6.32	....	....	....	....	....	....	....
2001-02	Jaa-Kotkat	Finland-2	28	....	....	....	1680	100	0	3.51	....	....	....	....	....	....	....
	Jaa-Kotkat Jr.	Finn-Jr.	....	....	....	....	....	....	....	....	4	1	3	240	14	0	3.46

JOSEPH, Curtis (JOH-sehf, KUR-tihs) DET.

Goaltender. Catches left. 5'11", 190 lbs. Born, Keswick, Ont., April 29, 1967.

			Regular Season								Playoffs						
Season	Club	Lea	GP	W	L	T	Mins	GA	SO	Avg	GP	W	L	Mins	GA	SO	Avg
1984-85	King City Dukes	OJHL-B	18	....	....	....	947	76	0	4.82	....	....	....	....	....	....	....
	Newmarket Flyers	OPJHL	2	1	1	0	120	16	0	8.00	....	....	....	....	....	....	....
1985-86	Richmond Hill	OPJHL	33	12	18	0	1716	156	1	5.45	....	....	....	....	....	....	....
1986-87	Richmond Hill	OPJHL	30	14	7	6	1764	128	1	4.35	....	....	....	....	....	....	....
1987-88	Notre Dame	SJHL	36	25	4	7	2174	94	1	2.59	....	....	....	....	....	....	....
1988-89	U. of Wisconsin	WCHA	38	21	11	5	2267	94	1	2.49	....	....	....	....	....	....	....
1989-90	Peoria Rivermen	IHL	23	10	8	2	1241	80	0	3.87	....	....	....	....	....	....	....
	St. Louis	**NHL**	**15**	**9**	**5**	**1**	**852**	**48**	**0**	**3.38**	**6**	**4**	**1**	**327**	**18**	**0**	**3.30**
1990-91	**St. Louis**	**NHL**	**30**	**16**	**10**	**2**	**1710**	**89**	**0**	**3.12**	....	....	....	....	....	....	....
1991-92	**St. Louis**	**NHL**	**60**	**27**	**20**	**10**	**3494**	**175**	**2**	**3.01**	**6**	**2**	**4**	**379**	**23**	**0**	**3.64**
1992-93	**St. Louis**	**NHL**	**68**	**29**	**28**	**9**	**3890**	**196**	**1**	**3.02**	**11**	**7**	**4**	**715**	**27**	***2**	**2.27**
1993-94	**St. Louis**	**NHL**	**71**	**36**	**23**	**11**	**4127**	**213**	**1**	**3.10**	**4**	**0**	**4**	**246**	**15**	**0**	**3.66**
1994-95	**St. Louis**	**NHL**	**36**	**20**	**10**	**1**	**1914**	**89**	**1**	**2.79**	**7**	**3**	**3**	**392**	**24**	**0**	**3.67**
1995-96	Las Vegas	IHL	15	12	2	1	874	29	1	1.99	....	....	....	....	....	....	....
	Edmonton	**NHL**	**34**	**15**	**16**	**2**	**1936**	**111**	**0**	**3.44**	....	....	....	....	....	....	....
1996-97	**Edmonton**	**NHL**	**72**	**32**	**29**	**9**	**4100**	**200**	**6**	**2.93**	**12**	**5**	**7**	**767**	**36**	**2**	**2.82**
1997-98	**Edmonton**	**NHL**	**71**	**29**	**31**	**9**	**4132**	**181**	**8**	**2.63**	**12**	**5**	**7**	**716**	**23**	**3**	**1.93**
1998-99	**Toronto**	**NHL**	**67**	**35**	**24**	**7**	**4001**	**171**	**3**	**2.56**	**17**	**9**	**8**	**1011**	**41**	**1**	**2.43**
99-2000	**Toronto**	**NHL**	**63**	**36**	**20**	**7**	**3801**	**158**	**4**	**2.49**	**12**	**6**	**6**	**729**	**25**	**1**	**2.06**
2000-01	**Toronto**	**NHL**	**68**	**33**	**27**	**8**	**4100**	**163**	**6**	**2.39**	**11**	**7**	**4**	**685**	**24**	**3**	**2.10**
2001-02	**Toronto**	**NHL**	**51**	**29**	**17**	**5**	**3065**	**114**	**4**	**2.23**	**20**	**10**	**10**	**1253**	**48**	**3**	**2.30**
	Canada	Olympics	1	0	1	0	60	5	0	5.00	....	....	....	....	....	....	....
	NHL Totals		**706**	**346**	**260**	**81**	**41122**	**1908**	**36**	**2.78**	**118**	**58**	**58**	**7220**	**304**	**15**	**2.53**

WCHA First All-Star Team (1989) • WCHA Freshman of the Year (1989) • WCHA Player of the Year (1989) • NCAA West Second All-American Team (1989) • Won King Clancy Memorial Trophy (2000) • Played in NHL All-Star Game (1994, 2000)

Signed as a free agent by **St. Louis**, June 16, 1989. Traded to **Edmonton** by **St. Louis** with the rights to Michael Grier for St. Louis' 1st round choices (previously acquired) in 1996 (Marty Reasoner) and 1997 (later traded to LA Kings - LA Kings selected Matt Zultek) Entry Drafts, August 4, 1995. Signed as a free agent by **Toronto**, July 15, 1998. Traded to **Calgary** by **Toronto** for Calgary's 8th round choice in 2004 Entry Draft and future considerations, June 30, 2002. Signed as a free agent by **Detroit**, July 2, 2002.

KALTIAINEN, Matti (kal-tee-AY-nehn, MAT-tee) BOS.

Goaltender. Catches left. 6'2", 216 lbs. Born, Espoo, Finland, April 30, 1982.
(Boston's 3rd choice, 111th overall, in 2001 Entry Draft).

			Regular Season								Playoffs						
Season	Club	Lea	GP	W	L	T	Mins	GA	SO	Avg	GP	W	L	Mins	GA	SO	Avg
1998-99	Blues Espoo Jr.	Finn-Jr.	4	....	....	....	258	10	0	2.33	....	....	....	....	....	....	....
99-2000	Blues Espoo Jr.	Finn-Jr.	23	....	....	....	1337	56	0	2.51	4	2	2	244	15	0	3.93
2000-01	Blues Espoo Jr.	Finn-Jr.	25	....	....	....	1500	65	0	2.60	....	....	....	....	....	....	....
2001-02	Boston College	H-East	18	8	10	0	1080	48	1	2.67	....	....	....	....	....	....	....

KETTLES, Kyle (KEH-tuhls, KIGHL) MIN.

Goaltender. Catches left. 6'3", 180 lbs. Born, Lac du Bonnet, Man., February 19, 1981.
(Nashville's 13th choice, 205th overall, in 1999 Entry Draft).

			Regular Season								Playoffs						
Season	Club	Lea	GP	W	L	T	Mins	GA	SO	Avg	GP	W	L	Mins	GA	SO	Avg
1997-98	Selkirk Steelers	MJHL	32	9	19	1	1613	119	0	4.43	....	....	....	....	....	....	....
	Brandon	WHL	....	....	....	....	....	....	....	....	1	0	0	10	2	0	12.00
1998-99	Selkirk Steelers	MJHL	17	7	8	0	939	65	0	4.15	....	....	....	....	....	....	....
	Neepawa Natives	MJHL	6	2	4	0	361	27	0	4.49	....	....	....	....	....	....	....
99-2000	Medicine Hat	WHL	57	16	33	5	3260	215	1	3.96	....	....	....	....	....	....	....
2000-01	Medicine Hat	WHL	47	15	24	2	2586	183	0	4.25	....	....	....	....	....	....	....
2001-02	Medicine Hat	WHL	5	1	4	0	274	23	0	5.04	....	....	....	....	....	....	....
	Moose Jaw	WHL	54	24	25	4	3089	160	*4	3.11	12	6	6	735	29	1	2.37

Traded to **Moose Jaw** (WHL) by **Medicine Hat** (WHL) for Sean Connors, October 10, 2001. Signed as a free agent by **Minnesota**, April 23, 2002.

KHABIBULIN, Nikolai (khah-bee-BOO-lihn, NIH-koh-ligh) T.B.

Goaltender. Catches left. 6'1", 203 lbs. Born, Sverdlovsk, USSR, January 13, 1973.
(Winnipeg's 8th choice, 204th overall, in 1992 Entry Draft).

			Regular Season								Playoffs						
Season	Club	Lea	GP	W	L	T	Mins	GA	SO	Avg	GP	W	L	Mins	GA	SO	Avg
1991-92	CSKA Moscow	CIS	2	0	0	0	34	2	0	3.53	....	....	....	....	....	....	....
1992-93	CSKA Moscow	CIS	13	....	....	....	491	27	....	3.29	....	....	....	....	....	....	....
1993-94	CSKA Moscow	CIS	46	....	....	....	2625	116	....	2.65	3	....	....	193	11	....	3.42
	Russian Penguins	IHL	12	2	7	2	639	47	0	4.41	....	....	....	....	....	....	....
1994-95	Springfield	AHL	23	9	9	3	1240	80	0	3.87	....	....	....	....	....	....	....
	Winnipeg	**NHL**	**26**	**8**	**9**	**4**	**1339**	**76**	**0**	**3.41**	....	....	....	....	....	....	....
1995-96	**Winnipeg**	**NHL**	**53**	**26**	**20**	**3**	**2914**	**152**	**2**	**3.13**	**6**	**2**	**4**	**359**	**19**	**0**	**3.18**
1996-97	**Phoenix**	**NHL**	**72**	**30**	**33**	**6**	**4091**	**193**	**7**	**2.83**	**7**	**3**	**4**	**426**	**15**	**1**	**2.11**
1997-98	**Phoenix**	**NHL**	**70**	**30**	**28**	**10**	**4026**	**184**	**4**	**2.74**	**4**	**2**	**1**	**185**	**13**	**0**	**4.22**
1998-99	**Phoenix**	**NHL**	**63**	**32**	**23**	**7**	**3657**	**130**	**8**	**2.13**	**7**	**3**	**4**	**449**	**18**	**0**	**2.41**
99-2000	Long Beach	IHL	33	21	11	1	1936	59	5	*1.83	5	2	3	321	15	0	2.81
2000-01	**Tampa Bay**	**NHL**	**2**	**1**	**1**	**0**	**123**	**6**	**0**	**2.93**	....	....	....	....	....	....	....
2001-02	**Tampa Bay**	**NHL**	**70**	**24**	**32**	**10**	**3896**	**153**	**7**	**2.36**	....	....	....	....	....	....	....
	Russia	Olympics	6	3	2	1	*359	14	*1	2.34	....	....	....	....	....	....	....
	NHL Totals		**356**	**151**	**146**	**40**	**20046**	**894**	**28**	**2.68**	**24**	**10**	**13**	**1419**	**65**	**1**	**2.75**

Shared James Gatschene Memorial Trophy (MVP - IHL) with Frederic Chabot (2000) • Played in NHL All-Star Game (1998, 1999, 2002)

Transferred to **Phoenix** after **Winnipeg** franchise relocated, July 1, 1996. • Missed entire 1999-2000 NHL season and majority of 2000-01 season after failing to come to contract terms with **Phoenix**. Signed as a free agent by **Long Beach** (IHL) with **Phoenix** retaining NHL rights, January 14, 2000. Traded to **Tampa Bay** by **Phoenix** with Stan Neckar for Mike Johnson, Paul Mara, Ruslan Zainullin and NY Islanders' 2nd round choice (previously acquired, Phoenix selected Matthew Spiller) in 2001 Entry Draft, March 5, 2001.

KHLOPTONOV, Denis (khloh-POHT-nahv, DEH-nihs) FLA.

Goaltender. Catches left. 6'4", 198 lbs. Born, Moscow, USSR, January 27, 1978.
(Florida's 8th choice, 209th overall, in 1996 Entry Draft).

			Regular Season								Playoffs						
Season	Club	Lea	GP	W	L	T	Mins	GA	SO	Avg	GP	W	L	Mins	GA	SO	Avg
1996-97	HC CSKA	Russia	21	....	....	....	1260	42	0	2.00	....	....	....	....	....	....	....
1997-98	HC CSKA	Russia	20	....	....	....	987	58	....	3.53	....	....	....	....	....	....	....
1998-99	Muskegon Fury	UHL	37	21	8	2	1950	98	1	3.02	4	1	1	166	9	0	3.25
99-2000	CSKA Moscow	Russia	10	....	....	....	540	22	1	2.44	2	....	....	119	7	0	3.53
2000-01	CSKA Moscow	Russia	14	....	....	....	753	26	1	2.07	....	....	....	....	....	....	....
2001-02	Magnitogorsk	Russia	14	....	....	....	572	36	0	3.78	....	....	....	....	....	....	....

KIDD, Trevor (KIHD, TREH-vohr) FLA.

Goaltender. Catches left. 6'2", 210 lbs. Born, Dugald, Man., March 26, 1972.
(Calgary's 1st choice, 11th overall, in 1990 Entry Draft).

			Regular Season								Playoffs						
Season	Club	Lea	GP	W	L	T	Mins	GA	SO	Avg	GP	W	L	Mins	GA	SO	Avg
1987-88	Eastman Selects	MAHA	14	....	....	....	840	66	0	4.72	....	....	....	....	....	....	....
1988-89	Brandon	WHL	32	11	13	1	1509	102	0	4.06	....	....	....	....	....	....	....
1989-90	Brandon	WHL	*63	24	32	2	*3676	254	2	4.15	....	....	....	....	....	....	....
1990-91	Brandon	WHL	30	10	19	1	1730	117	0	4.06	....	....	....	....	....	....	....
	Spokane Chiefs	WHL	14	8	3	0	749	44	0	3.52	15	*14	1	926	32	2	*2.07
1991-92	Canada	Nat-Tm	28	18	4	4	1349	79	2	3.51	....	....	....	....	....	....	....
	Canada	Olympics	1	1	0	0	60	0	1	0.00	....	....	....	....	....	....	....
	Calgary	**NHL**	**2**	**1**	**1**	**0**	**120**	**8**	**0**	**4.00**	....	....	....	....	....	....	....
1992-93	Salt Lake	IHL	29	10	16	1	1696	111	1	3.93	....	....	....	....	....	....	....
1993-94	**Calgary**	**NHL**	**31**	**13**	**7**	**6**	**1614**	**85**	**0**	**3.16**	....	....	....	....	....	....	....
1994-95	**Calgary**	**NHL**	***43**	**22**	**14**	**6**	***2463**	**107**	**3**	**2.61**	**7**	**3**	**4**	**434**	**26**	**1**	**3.59**
1995-96	**Calgary**	**NHL**	**47**	**15**	**21**	**8**	**2570**	**119**	**3**	**2.78**	**2**	**0**	**1**	**83**	**9**	**0**	**6.51**
1996-97	**Calgary**	**NHL**	**55**	**21**	**23**	**6**	**2979**	**141**	**4**	**2.84**	....	....	....	....	....	....	....
1997-98	**Carolina**	**NHL**	**47**	**21**	**21**	**3**	**2685**	**97**	**3**	**2.17**	....	....	....	....	....	....	....
1998-99	**Carolina**	**NHL**	**25**	**7**	**10**	**6**	**1358**	**61**	**2**	**2.70**	....	....	....	....	....	....	....
99-2000	**Florida**	**NHL**	**28**	**14**	**11**	**2**	**1574**	**69**	**1**	**2.63**	....	....	....	....	....	....	....
	Louisville	AHL	1	0	1	0	60	5	0	5.04	....	....	....	....	....	....	....
2000-01	**Florida**	**NHL**	**42**	**10**	**23**	**6**	**2354**	**130**	**1**	**3.31**	....	....	....	....	....	....	....
2001-02	**Florida**	**NHL**	**33**	**4**	**16**	**5**	**1683**	**90**	**1**	**3.21**	....	....	....	....	....	....	....
	NHL Totals		**353**	**128**	**147**	**48**	**19400**	**907**	**18**	**2.81**	**9**	**3**	**5**	**517**	**35**	**1**	**4.06**

WHL East First All-Star Team (1990) • Canadian Major Junior Goaltender of the Year (1990)

Traded to **Carolina** by **Calgary** with Gary Roberts for Andrew Cassels and Jean-Sebastien Giguere, August 25, 1997. Claimed by **Atlanta** from **Carolina** in Expansion Draft, June 25, 1999. Traded to **Florida** by **Atlanta** for Gord Murphy, Herbert Vasiljevs, Daniel Tjarnqvist and Ottawa's 6th round choice (previously acquired, later traded to Dallas - Dallas selected Justin Cox) in 1999 Entry Draft, June 25, 1999.

KIELKUCKI, Marc (keel-KOO-kee, MAHRK) S.J.

Goaltender. Catches left. 6'4", 195 lbs. Born, Brooklyn Park, MN, June 5, 1979.

			Regular Season								Playoffs						
Season	Club	Lea	GP	W	L	T	Mins	GA	SO	Avg	GP	W	L	Mins	GA	SO	Avg
1996-97	Champlin Park	H.S.	STATISTICS NOT AVAILABLE														
1997-98	Air Force	CHA	5	0	1	0	148	17	0	6.89	....	....	....	....	....	....	....
1998-99	Air Force	CHA	33	14	15	2	1729	96	5	3.33	....	....	....	....	....	....	....
99-2000	Air Force	CHA	37	18	16	2	2007	102	3	3.05	....	....	....	....	....	....	....
2000-01	Air Force	CHA	*35	*14	16	4	*2024	96	2	2.85	....	....	....	....	....	....	....
2001-02	Dayton Bombers	ECHL	11	4	4	3	675	30	1	2.67	....	....	....	....	....	....	....

CHA First All-Star Team (2001) • CHA Player of the Year (2001)

Signed as a free agent by **San Jose**, July 5, 2001. • Missed majority of 2001-02 season completing compulsary U.S. military service.

KIPRUSOFF, Miikka (KIHP-ruh-sohf, MEE-kah) S.J.

Goaltender. Catches left. 6'2", 190 lbs. Born, Turku, Finland, October 26, 1976.
(San Jose's 5th choice, 116th overall, in 1995 Entry Draft).

			Regular Season								Playoffs						
Season	Club	Lea	GP	W	L	T	Mins	GA	SO	Avg	GP	W	L	Mins	GA	SO	Avg
1994-95	TPS Turku Jr.	Finn-Jr.	31	....	....	....	1896	92	....	2.91	....	....	....	....	....	....	....
	TPS Turku	Finland	4	....	....	....	240	12	0	3.00	2	....	....	120	7	....	3.50
1995-96	TPS Turku Jr.	Finn-Jr.	3	....	....	....	180	9	....	3.00	....	....	....	....	....	....	....
	Kiekko-67 Turku	Finland-2	5	....	....	....	300	7	....	1.40	....	....	....	....	....	....	....
	TPS Turku	Finland	12	....	....	....	550	38	0	4.14	3	....	....	114	4	....	2.11
1996-97	AIK Solna	Sweden	42	....	....	....	2466	104	3	2.53	7	....	....	420	23	0	3.28
1997-98	AIK Solna	Sweden	42	....	....	....	2457	110	....	2.69	....	....	....	....	....	....	....
1998-99	TPS Turku	Finland	39	*26	6	6	2259	70	4	1.86	10	*9	1	580	15	*3	*1.55
99-2000	Kentucky	AHL	47	23	19	4	2759	114	3	2.48	5	1	3	239	13	0	3.27
2000-01	**San Jose**	**NHL**	**5**	**2**	**1**	**0**	**154**	**5**	**0**	**1.95**	**3**	**1**	**1**	**149**	**5**	**0**	**2.01**
	Kentucky	AHL	36	19	9	6	2038	76	2	2.24	....	....	....	....	....	....	....
2001-02	**San Jose**	**NHL**	**20**	**7**	**6**	**3**	**1037**	**43**	**2**	**2.49**	**1**	**0**	**0**	**8**	**0**	**0**	**0.00**
	Cleveland Barons	AHL	4	4	0	0	242	7	0	1.73	....	....	....	....	....	....	....
	NHL Totals		**25**	**9**	**7**	**3**	**1191**	**48**	**2**	**2.42**	**4**	**1**	**1**	**157**	**5**	**0**	**1.91**

KOCHAN, Dieter (KAH-kuhn, DEE-tuhr) MIN.

Goaltender. Catches left. 6'1", 180 lbs. Born, Saskatoon, Sask., May 11, 1974.
(Vancouver's 3rd choice, 98th overall, in 1993 Entry Draft).

			Regular Season								Playoffs						
Season	Club	Lea	GP	W	L	T	Mins	GA	SO	Avg	GP	W	L	Mins	GA	SO	Avg
1991-92	Sioux City	USHL	23	7	10	0	1131	100	0	5.31	....	....	....	....	....	....	....
1992-93	Kelowna Spartans	BCJHL	44	34	8	0	2582	137	1	3.18	15	12	3	927	48	1	3.10
1993-94	North-Michigan	WCHA	20	9	7	0	985	57	2	3.47	....	....	....	....	....	....	....
1994-95	North-Michigan	WCHA	29	8	17	3	1512	107	0	4.25	....	....	....	....	....	....	....
1995-96	North-Michigan	WCHA	31	7	21	2	1627	123	0	4.54	....	....	....	....	....	....	....
1996-97	North-Michigan	WCHA	26	8	15	2	1528	99	0	3.89	....	....	....	....	....	....	....
1997-98	Louisville	ECHL	18	7	9	2	980	61	1	3.73	....	....	....	....	....	....	....
1998-99	Binghamton	UHL	40	18	16	5	2322	115	2	2.97	4	1	2	208	9	0	2.60
99-2000	Binghamton	UHL	43	29	11	3	2544	110	4	2.59	....	....	....	....	....	....	....
	Orlando	IHL	4	4	0	0	240	4	1	1.00	....	....	....	....	....	....	....
	Springfield	AHL	2	1	1	0	120	5	1	2.50	....	....	....	....	....	....	....
	Tampa Bay	**NHL**	**5**	**1**	**4**	**0**	**238**	**17**	**0**	**4.29**	....	....	....	....	....	....	....
	Grand Rapids	IHL	2	1	0	1	93	1	0	0.64	....	....	....	....	....	....	....
2000-01	**Tampa Bay**	**NHL**	**10**	**0**	**3**	**0**	**314**	**18**	**0**	**3.44**	....	....	....	....	....	....	....
	Detroit Vipers	IHL	49	13	28	3	2606	154	0	3.55	....	....	....	....	....	....	....
2001-02	**Tampa Bay**	**NHL**	**5**	**0**	**3**	**1**	**237**	**16**	**0**	**4.05**	....	....	....	....	....	....	....
	Springfield	AHL	45	21	20	1	2518	112	2	2.67	....	....	....	....	....	....	....
	NHL Totals		**20**	**1**	**10**	**1**	**789**	**51**	**0**	**3.88**	....	....	....	....	....	....	....

UHL Second All-Star Team (2000)

• Scored a goal vs. Winston-Salem (UHL), January 5, 1999. Signed as a free agent by **Tampa Bay**, March 27, 2000. Signed as a free agent by **Minnesota**, August 5, 2002.

KOLZIG, Olaf (KOHL-zihg, OH-lahf) WSH.

Goaltender. Catches left. 6'3", 225 lbs. Born, Johannesburg, South Africa, April 9, 1970.
(Washington's 1st choice, 19th overall, in 1989 Entry Draft).

			Regular Season								Playoffs						
Season	Club	Lea	GP	W	L	T	Mins	GA	SO	Avg	GP	W	L	Mins	GA	SO	Avg
1986-87	Abbotsford	BCAHA	17	5	9	0	857	81	0	5.67	....	....	....	....	....	....	....
1987-88	New Westminster	WHL	15	6	5	0	650	48	1	4.43	3	0	3	149	11	0	4.43
1988-89	Tri-City	WHL	30	16	10	2	1671	97	1	*3.48	....	....	....	....	....	....	....
1989-90	**Washington**	**NHL**	**2**	**0**	**2**	**0**	**120**	**12**	**0**	**6.00**	....	....	....	....	....	....	....
	Tri-City	WHL	48	27	27	3	2504	187	1	4.48	6	4	0	318	27	0	5.09
1990-91	Baltimore	AHL	26	10	12	1	1367	72	0	3.16	....	....	....	....	....	....	....
	Hampton Roads	ECHL	21	11	9	1	1248	71	2	3.41	3	1	2	180	14	0	4.66
1991-92	Baltimore	AHL	28	5	17	2	1503	105	1	4.19	....	....	....	....	....	....	....
	Hampton Roads	ECHL	14	11	3	0	847	41	0	2.90	....	....	....	....	....	....	....
1992-93	**Washington**	**NHL**	**1**	**0**	**0**	**0**	**20**	**2**	**0**	**6.00**	....	....	....	....	....	....	....
	Rochester	AHL	49	25	16	4	2737	168	0	3.68	*17	9	8	*1040	61	0	3.52
1993-94	**Washington**	**NHL**	**7**	**0**	**3**	**0**	**224**	**20**	**0**	**5.36**	....	....	....	....	....	....	....
	Portland Pirates	AHL	29	16	8	5	1725	88	3	3.06	17	*12	5	1035	44	0	*2.55
1994-95	**Washington**	**NHL**	**14**	**2**	**8**	**2**	**724**	**30**	**0**	**2.49**	**2**	**1**	**0**	**44**	**1**	**0**	**1.36**
	Portland Pirates	AHL	2	1	0	1	125	3	0	1.44	....	....	....	....	....	....	....
1995-96	**Washington**	**NHL**	**18**	**4**	**8**	**2**	**897**	**46**	**0**	**3.08**	**5**	**2**	**3**	**341**	**11**	**0**	***1.94**
	Portland Pirates	AHL	5	5	0	0	300	7	1	1.40	....	....	....	....	....	....	....
1996-97	**Washington**	**NHL**	**29**	**8**	**15**	**4**	**1645**	**71**	**2**	**2.59**	....	....	....	....	....	....	....
1997-98	**Washington**	**NHL**	**64**	**33**	**18**	**10**	**3788**	**139**	**5**	**2.20**	**21**	**12**	**9**	**1351**	**44**	***4**	**1.95**
	Germany	Olympics	2	2	0	0	120	2	1	1.00	....	....	....	....	....	....	....
1998-99	**Washington**	**NHL**	**64**	**26**	**31**	**3**	**3586**	**154**	**4**	**2.58**	....	....	....	....	....	....	....
99-2000	**Washington**	**NHL**	**73**	**41**	**20**	**11**	***4371**	**163**	**5**	**2.24**	**5**	**1**	**4**	**284**	**16**	**0**	**3.38**
2000-01	**Washington**	**NHL**	**72**	**37**	**26**	**8**	**4279**	**177**	**5**	**2.48**	**6**	**2**	**4**	**375**	**14**	**1**	**2.24**
2001-02	**Washington**	**NHL**	**71**	**31**	**29**	**8**	**4131**	**192**	**6**	**2.79**	....	....	....	....	....	....	....
	NHL Totals		**415**	**182**	**160**	**48**	**23785**	**1006**	**27**	**2.54**	**39**	**18**	**20**	**2395**	**86**	**5**	**2.15**

WHL West Second All-Star Team (1989) • Shared Harry "Hap" Holmes Memorial Trophy (fewest goals against - AHL) with Byron Dafoe (1994) • Won Jack Butterfield Trophy (Playoff MVP - AHL) (1994) • NHL First All-Star Team (2000) • Won Vezina Trophy (2000) • Played in NHL All-Star Game (1998, 2000)

• Scored a goal while with Tri-City (WHL), November 29, 1989.

KONSTANTINOV, Evgeny (kohn-stahn-TEE-nahf, EHV-jeh-nee) T.B.

Goaltender. Catches left. 6', 176 lbs. Born, Kazan, USSR, March 29, 1981.
(Tampa Bay's 2nd choice, 67th overall, in 1999 Entry Draft).

			Regular Season								Playoffs						
Season	Club	Lea	GP	W	L	T	Mins	GA	SO	Avg	GP	W	L	Mins	GA	SO	Avg
1997-98	Ak Bars Kazan 2	Russia-3	34	....	....	....	2040	129	....	3.79	....	....	....	....	....	....	....
1998-99	Ak Bars Kazan 2	Russia-4	17	....	....	....	1020	38	....	2.24	....	....	....	....	....	....	....
99-2000	Leninogorsk	Russia-2					STATISTICS NOT AVAILABLE										
	Ak Bars Kazan	Russia	2	....	....	....	59	5	0	5.08	....	....	....	....	....	....	....
2000-01	Detroit Vipers	IHL	27	4	15	2	1197	85	0	4.26	....	....	....	....	....	....	....
	Tampa Bay	**NHL**	**1**	**0**	**0**	**0**	**1**	**0**	**0**	**0.00**	....	....	....	....	....	....	....
	Louisiana	ECHL	8	4	4	0	458	21	0	2.75	12	5	6	637	32	0	3.01
2001-02	Pensacola	ECHL	24	10	10	0	1229	71	0	3.47	....	....	....	....	....	....	....
	Springfield	AHL	3	1	2	0	178	5	0	1.69	....	....	....	....	....	....	....
	NHL Totals		**1**	**0**	**0**	**0**	**1**	**0**	**0**	**0.00**	....	....	....	....	....	....	....

• Played 24 seconds of game vs. Colorado, December 8, 2000.

KOOPMANS, Logan (KOOP-manz, LOH-guhn) DET.

Goaltender. Catches . 6'2", 182 lbs. Born, Cranbrook, B.C., May 18, 1984.
(Detroit's 5th choice, 166th overall, in 2002 Entry Draft).

			Regular Season								Playoffs						
Season	Club	Lea	GP	W	L	T	Mins	GA	SO	Avg	GP	W	L	Mins	GA	SO	Avg
99-2000	Lethbridge	WHL	5	1	3	0	282	19	0	4.04	....	....	....	....	....	....	....
2000-01	Columbia Valley	KIJHL	37	....	....	....	2140	144	1	3.90	....	....	....	....	....	....	....
2001-02	Lethbridge	WHL	37	20	12	2	2057	97	3	2.83	4	0	4	237	14	0	3.54

KOSTUR, Matus (KAW-stuhr, ma-TOOSH) N.J.

Goaltender. Catches left. 6'1", 190 lbs. Born, Banska Bystrica, Czech., March 28, 1980.
(New Jersey's 10th choice, 164th overall, in 2000 Entry Draft).

			Regular Season								Playoffs						
Season	Club	Lea	GP	W	L	T	Mins	GA	SO	Avg	GP	W	L	Mins	GA	SO	Avg
1997-98	B. Bystrica Jr.	Slovak-Jr.	36	....	....	....	2152	120	0	3.35	....	....	....	....	....	....	....
1998-99	Banska Bystrica	Slovak-2	3	....	....	....	133	9	0	4.06	....	....	....	....	....	....	....
99-2000	HKm Zvolen	Slovak-2	10	....	....	....	538	32	0	3.57	....	....	....	....	....	....	....
	HKm Zvolen	Slovakia	20	....	....	....	768	36	0	2.81	2	0	0	41	3	0	4.39
2000-01	HC Nitra	Slovak-2	20	18	1	1	1132	24	4	1.53	....	....	....	....	....	....	....
	HKm Zvolen	Slovakia	3	....	....	....	110	11	0	6.00	....	....	....	....	....	....	....
2001-02	HKm Zvolen	Slovakia	39	....	....	....	2163	78	5	2.16	2	....	....	99	5	0	3.03

KOTYK, Seamus (koh-TIHK, SHAY-muhs) S.J.

Goaltender. Catches left. 5'11", 187 lbs. Born, London, Ont., October 7, 1980.
(Boston's 5th choice, 147th overall, in 1999 Entry Draft).

			Regular Season								Playoffs						
Season	Club	Lea	GP	W	L	T	Mins	GA	SO	Avg	GP	W	L	Mins	GA	SO	Avg
1996-97	Stratford	OJHL-B	28	....	....	....	1615	105	0	3.82	....	....	....	....	....	....	....
1997-98	Ottawa 67's	OHL	31	13	5	5	1422	63	4	2.66	7	3	2	332	11	0	1.99
1998-99	Ottawa 67's	OHL	41	26	7	4	2314	92	5	2.39	5	3	2	338	13	0	*2.31
99-2000	Ottawa 67's	OHL	26	12	6	2	1241	65	1	3.14	....	....	....	....	....	....	....
2000-01	Ottawa 67's	OHL	55	24	20	7	3087	141	2	2.74	*20	*16	4	*1157	46	*3	2.39
2001-02	Cleveland Barons	AHL	24	6	11	0	981	61	1	3.73	....	....	....	....	....	....	....

• Missed majority of 1999-2000 season recovering from surgery for cardiac arrhythmia, October 18, 1999. Signed as a free agent by **San Jose**, July 23, 2001.

KOWALSKI, Craig (koh-WAHL-skee, KRAYG) CAR.

Goaltender. Catches left. 5'9", 190 lbs. Born, Warren, MI, January 15, 1981.
(Carolina's 6th choice, 235th overall, in 2000 Entry Draft).

			Regular Season								Playoffs						
Season	Club	Lea	GP	W	L	T	Mins	GA	SO	Avg	GP	W	L	Mins	GA	SO	Avg
1998-99	Det. Compuware	NAJHL	42	*34	7	6	2733	96	3	*2.10	7	*7	0	420	13	1	*1.86
99-2000	Det. Compuware	NAJHL	49	33	12	3	2850	113	4	2.38	5	2	3	334	13	0	2.34
2000-01	North-Michigan	CCHA	19	7	8	4	1078	49	1	2.73	....	....	....	....	....	....	....
2001-02	North-Michigan	CCHA	38	24	11	2	2271	89	4	2.35	....	....	....	....	....	....	....

NAJHL First All-Star Team (1998, 1999) • NAJHL Goaltender of the Year (1999)

KRAHN, Brent (KRAWN, BREHNT) CGY.

Goaltender. Catches left. 6'4", 200 lbs. Born, Winnipeg, Man., April 2, 1982.
(Calgary's 1st choice, 9th overall, in 2000 Entry Draft).

			Regular Season								Playoffs						
Season	Club	Lea	GP	W	L	T	Mins	GA	SO	Avg	GP	W	L	Mins	GA	SO	Avg
1997-98	Pembina Valley	MMHL	22	20	0	1	1265	40	3	1.90	2	2	0	120	2	1	1.00
1998-99	Pembina Valley	MMHL	13	10	3	0	770	30	2	2.34	....	....	....	....	....	....	....
99-2000	Calgary Hitmen	WHL	39	33	6	0	2315	92	4	2.38	5	2	2	266	13	0	2.93
2000-01	Calgary Hitmen	WHL	37	22	10	3	2087	104	1	2.99	....	....	....	....	....	....	....
2001-02	Calgary Hitmen	WHL	18	8	6	2	1033	61	0	3.54	2	1	1	119	6	0	3.03

• Missed majority of 2001-02 season recovering from off-season knee surgery, June, 2001.

LABARBERA, Jason (lah-BAR-buhr-uh, JAY-suhn) NYR

Goaltender. Catches left. 6'2", 205 lbs. Born, Prince George, B.C., January 18, 1980.
(NY Rangers' 3rd choice, 66th overall, in 1998 Entry Draft).

			Regular Season								Playoffs						
Season	Club	Lea	GP	W	L	T	Mins	GA	SO	Avg	GP	W	L	Mins	GA	SO	Avg
1995-96	Prince George	BCAHA	31	....	....	....	1860	83	0	2.68	....	....	....	....	....	....	....
1996-97	Tri-City	WHL	2	1	0	0	63	4	0	3.81	....	....	....	....	....	....	....
	Portland	WHL	9	5	1	1	443	18	0	2.44	....	....	....	....	....	....	....
1997-98	Portland	WHL	23	18	4	0	1305	72	1	3.31	....	....	....	....	....	....	....
1998-99	Portland	WHL	51	18	23	9	2991	170	4	3.41	4	0	4	252	19	0	4.52
99-2000	Portland	WHL	34	8	24	2	2005	123	1	3.68	....	....	....	....	....	....	....
	Spokane Chiefs	WHL	21	12	6	2	1146	50	0	2.62	9	6	1	435	18	1	2.48
2000-01	**NY Rangers**	**NHL**	**1**	**0**	**0**	**0**	**10**	**0**	**0**	**0.00**	....	....	....	....	....	....	....
	Hartford	AHL	4	1	1	0	156	12	0	4.61	....	....	....	....	....	....	....
	Charlotte	ECHL	35	18	10	7	2100	112	1	3.20	2	1	1	143	5	0	2.09
2001-02	Hartford	AHL	20	7	11	1	1058	55	0	3.12	....	....	....	....	....	....	....
	Charlotte	ECHL	13	9	3	1	744	29	0	2.34	4	2	2	212	12	0	3.39
	NHL Totals		**1**	**0**	**0**	**0**	**10**	**0**	**0**	**0.00**	....	....	....	....	....	....	....

Traded to **Spokane** (WHL) by **Portland** (WHL) with Portland's 1st choice (Miroslav Stoic) in 2002 CHL Import Draft for Kris Callway, January 10, 2000.

LABBE, Jean-Francois (lah-BAY, ZHAWN-fran-SWUH) CBJ

Goaltender. Catches left. 5'10", 175 lbs. Born, Sherbrooke, Que., June 15, 1972.

			Regular Season								Playoffs						
Season	Club	Lea	GP	W	L	T	Mins	GA	SO	Avg	GP	W	L	Mins	GA	SO	Avg
1988-89	L'est Cantonniers	QAAA	29	*22	7	0	1764	94	1	3.20	5	1	4	333	19	0	3.42
1989-90	Trois-Rivieres	QMJHL	28	13	10	0	1499	106	1	4.24	3	1	1	132	8	0	3.64
1990-91	Trois-Rivieres	QMJHL	54	*35	14	0	2870	158	5	3.30	5	1	4	230	19	0	4.96
1991-92	Trois-Rivieres	QMJHL	48	*31	13	3	2749	142	1	3.10	*15	*10	3	791	33	*1	*2.50
1992-93	Hull Olympiques	QMJHL	46	26	18	2	2701	156	2	3.46	10	6	3	518	24	*1	*2.78
1993-94	Thunder Bay	ColHL	52	*35	11	4	*2900	150	*2	*3.10	8	7	1	493	18	*2	*2.19
	P.E.I. Senators	AHL	7	4	3	0	389	22	0	3.39	....	....	....	....	....	....	....
1994-95	P.E.I. Senators	AHL	32	13	14	3	1817	94	2	3.10	....	....	....	....	....	....	....
1995-96	Cornwall Aces	AHL	55	25	21	5	2972	144	3	2.91	8	3	5	471	21	1	2.68
1996-97	Hershey Bears	AHL	66	*34	22	9	3811	160	*6	*2.52	*23	*14	8	*1364	59	1	2.60
1997-98	Hamilton	AHL	52	24	17	11	3138	149	2	2.85	7	3	4	413	20	0	2.90
1998-99	Hartford	AHL	*59	28	26	3	*3392	182	2	3.22	7	3	4	447	22	0	2.95
99-2000	**NY Rangers**	**NHL**	**1**	**0**	**1**	**0**	**60**	**3**	**0**	**3.00**	....	....	....	....	....	....	....
	Hartford	AHL	49	27	13	7	2853	120	1	2.52	*22	*15	7	*1320	48	3	2.18
2000-01	Hartford	AHL	8	4	2	1	394	20	0	3.04	....	....	....	....	....	....	....
	Syracuse Crunch	AHL	37	15	15	5	2201	105	2	2.86	5	2	3	323	18	0	3.34
2001-02	**Columbus**	**NHL**	**3**	**1**	**1**	**0**	**117**	**6**	**0**	**3.08**	....	....	....	....	....	....	....
	Syracuse Crunch	AHL	51	27	16	7	2993	109	*9	2.18	10	6	4	596	19	2	*1.91
	NHL Totals		**4**	**1**	**2**	**0**	**177**	**9**	**0**	**3.05**	....	....	....	....	....	....	....

QMJHL First All-Star Team (1992) • ColHL First All-Star Team (1994) • ColHL Rookie of the Year (1994) • ColHL Outstanding Goaltender (1994) • ColHL Playoff MVP (1994) • AHL First All-Star Team (1997) • Won Harry "Hap" Holmes Memorial Trophy (fewest goals against - AHL) (1997) • Won Baz Bastien Memorial Trophy (Top Goaltender - AHL) (1997) • Won Les Cunningham Award (MVP - AHL) (1997) • Shared Harry "Hap" Holmes Memorial Trophy (fewest goals against - AHL) with Milan Hnilicka (2000) • AHL Second All-Star Team (2002) • Scored a goal while with Hartford (AHL) vs. Quebec (AHL), February 5, 2000.

Signed as a free agent by **Ottawa**, May 12, 1994. Traded to **Colorado** by **Ottawa** for future considerations, September 20, 1995. Signed as a free agent by **Edmonton**, September 2, 1997. Signed as a free agent by **NY Rangers**, July 30, 1998. Traded to **Columbus** by **NY Rangers** for Bert Robertsson, November 9, 2000.

LAJEUNESSE, Simon (lah-ZHUH-nehs, SIGH-mohn) OTT.

Goaltender. Catches left. 6', 175 lbs. Born, Quebec, Que., January 22, 1981.
(Ottawa's 2nd choice, 48th overall, in 1999 Entry Draft).

			Regular Season								Playoffs						
Season	Club	Lea	GP	W	L	T	Mins	GA	SO	Avg	GP	W	L	Mins	GA	SO	Avg
1996-97	Cap-d-Madeleine	QAAA	23	15	5	1	1300	89	0	4.11	4	1	3	240	26	0	4.72
1997-98	Moncton Wildcats	QMJHL	19	5	6	3	925	51	1	3.31	2	0	0	1	0	0	0.00
1998-99	Moncton Wildcats	QMJHL	36	18	9	3	1993	98	1	2.95	1	0	0	43	2	0	2.79
99-2000	Moncton Wildcats	QMJHL	55	31	15	1	2922	127	6	2.61	16	9	6	910	56	1	3.69
2000-01	Acadie-Bathurst	QMJHL	36	10	19	2	1879	121	1	3.86							
	Val-d'Or Foreurs	QMJHL	21	16	3	1	1159	54	1	2.80	14	8	4	760	52	0	4.10
2001-02	**Ottawa**	**NHL**	**1**	**0**	**0**	**0**	**24**	**0**	**0**	**0.00**							
	Mobile Mysticks	ECHL	13	7	2	3	755	36	2	2.86							
	Grand Rapids	AHL	26	13	7	5	1534	54	3	2.11	2	0	0	21	1	0	2.82
	NHL Totals		**1**	**0**	**0**	**0**	**24**	**0**	**0**	**0.00**							

QMJHL First All-Star Team (2000) • Shared Harry "Hap" Holmes Memorial Trophy (fewest goals against - AHL) with Martin Prusek and Mathieu Chouinard (2002)

Traded to **Val-d'Or** (QMJHL) by **Acadie-Bathurst** (QMJHL) with Acadie-Bathurst's 4th round choice (Mathieu Curadeau) in 2001 QMJHL Midget Draft for Antoine Bergeron, Jean-Francois Laniel, Eric Labelle and future considerations, January 7, 2001.

LALIME, Patrick (lah-LEEM, PAT-rihk) OTT.

Goaltender. Catches left. 6'3", 185 lbs. Born, St-Bonaventure, Que., July 7, 1974.
(Pittsburgh's 6th choice, 156th overall, in 1993 Entry Draft).

			Regular Season								Playoffs						
Season	Club	Lea	GP	W	L	T	Mins	GA	SO	Avg	GP	W	L	Mins	GA	SO	Avg
1990-91	Abitibi	QAAA	26	9	17	0	1595	151	0	5.81							
1991-92	Shawinigan	QMJHL	6				272	25	0	5.50							
1992-93	Shawinigan	QMJHL	44	10	24	4	2467	192	0	4.67							
1993-94	Shawinigan	QMJHL	48	22	20	0	2733	192	1	4.22	5	1	3	223	25	0	6.73
1994-95	Hampton Roads	ECHL	26	15	7	3	1470	82	2	3.35							
	Cleveland	IHL	23	7	10	4	1230	91	0	4.44							
1995-96	Cleveland	IHL	41	20	12	7	2314	149	0	3.86							
1996-97	**Pittsburgh**	**NHL**	**39**	**21**	**12**	**2**	**2058**	**101**	**3**	**2.94**							
	Cleveland	IHL	14	6	6	2	834	45	1	3.24							
1997-98	Grand Rapids	IHL	31	10	10	9	1749	76	2	2.61	1	0	1	77	4	0	3.11
1998-99	Kansas City	IHL	*66	*39	20	4	*3789	190	2	3.01	3	1	2	179	6	1	2.01
99-2000	**Ottawa**	**NHL**	**38**	**19**	**14**	**3**	**2038**	**79**	**3**	**2.33**							
2000-01	**Ottawa**	**NHL**	**60**	**36**	**19**	**5**	**3607**	**141**	**7**	**2.35**	**4**	**0**	**4**	**251**	**10**	**0**	**2.39**
2001-02	**Ottawa**	**NHL**	**61**	**27**	**24**	**8**	**3583**	**148**	**7**	**2.48**	**12**	**7**	**5**	**778**	**18**	**4**	***1.39**
	NHL Totals		**198**	**103**	**69**	**18**	**11286**	**469**	**20**	**2.49**	**16**	**7**	**9**	**1029**	**28**	**4**	**1.63**

NHL All-Rookie Team (1997) • IHL First All-Star Team (1999)

Rights traded to **Anaheim** by **Pittsburgh** for Sean Pronger, March 24, 1998. Traded to **Ottawa** by **Anaheim** for Ted Donato and the rights to Antti-Jussi Niemi, June 18, 1999.

LAMOTHE, Marc (luh-MAWTH, MAHRK) DET.

Goaltender. Catches left. 6'2", 210 lbs. Born, New Liskeard, Ont., February 27, 1974.
(Montreal's 6th choice, 92nd overall, in 1992 Entry Draft).

			Regular Season								Playoffs						
Season	Club	Lea	GP	W	L	T	Mins	GA	SO	Avg	GP	W	L	Mins	GA	SO	Avg
1990-91	Ottawa Jr. Sens	OCJHL	25	13	7	0	1220	82	1	4.03							
1991-92	Kingston	OHL	42	10	25	2	2378	189	1	4.77							
1992-93	Kingston	OHL	45	23	12	6	2489	162	1	3.91	15	8	5	753	48	1	3.82
1993-94	Kingston	OHL	48	23	20	5	2828	177	*2	3.76	6	2	2	224	12	0	3.21
1994-95	Fredericton	AHL	9	2	5	0	428	32	0	4.48							
	Wheeling	ECHL	13	9	2	1	737	38	0	3.10							
1995-96	Fredericton	AHL	23	5	9	3	1166	73	1	3.76	3	1	2	161	9	0	3.36
1996-97	Indianapolis Ice	IHL	38	20	14	4	2271	100	1	2.64	1	0	0	20	1	0	3.00
1997-98	Indianapolis Ice	IHL	31	18	10	2	1772	72	3	2.44	4	1	3	177	10	0	3.38
1998-99	Indianapolis Ice	IHL	32	9	16	6	1823	115	1	3.78	6	3	3	338	10	*2	1.78
99-2000	**Chicago**	**NHL**	**2**	**1**	**1**	**0**	**116**	**10**	**0**	**5.17**							
	Cleveland	IHL	44	19	18	4	2455	112	2	2.74	4	2	2	325	12	0	2.21
2000-01	Syracuse Crunch	AHL	42	17	15	7	2323	112	2	2.89							
2001-02	Hamilton	AHL	45	22	19	2	2569	102	3	2.38	9	6	3	551	18	0	1.96
	NHL Totals		**2**	**1**	**1**	**0**	**116**	**10**	**0**	**5.17**							

Signed as a free agent by **Chicago**, September 26, 1996. Signed as a free agent by **Edmonton**, August 16, 2001. Signed as a free agent by **Detroit**, August 5, 2002.

LANGKOW, Scott (LAING-kow, SKAWT)

Goaltender. Catches left. 5'11", 190 lbs. Born, Sherwood Park, Alta., April 21, 1975.
(Winnipeg's 2nd choice, 31st overall, in 1993 Entry Draft).

			Regular Season								Playoffs						
Season	Club	Lea	GP	W	L	T	Mins	GA	SO	Avg	GP	W	L	Mins	GA	SO	Avg
1990-91	Sherwood Park	AMHL	32				1920	128	0	4.00							
1991-92	Abbotsford	PIJHL					STATISTICS NOT AVAILABLE										
	Portland	WHL	1	0	0	0	33	2	0	3.46							
1992-93	Portland	WHL	34	24	8	2	2064	119	2	3.46	9	6	3	535	31	0	3.48
1993-94	Portland	WHL	39	27	9	1	2302	121	2	3.15	10	6	4	600	34	0	3.40
1994-95	Portland	WHL	63	20	36	5	*3638	240	1	3.96	8	3	5	510	30	0	3.53
1995-96	**Winnipeg**	**NHL**	**1**	**0**	**0**	**0**	**6**	**0**	**0**	**0.00**							
	Springfield	AHL	39	18	15	6	2329	116	3	2.99	7	4	2	393	23	0	3.51
1996-97	Springfield	AHL	33	15	9	7	1929	85	0	2.64							
1997-98	**Phoenix**	**NHL**	**3**	**0**	**1**	**1**	**137**	**10**	**0**	**4.38**							
	Springfield	AHL	51	30	13	5	2874	128	3	2.67	4	1	3	216	14	0	3.88
1998-99	**Phoenix**	**NHL**	**1**	**0**	**0**	**0**	**35**	**3**	**0**	**5.14**							
	Las Vegas	IHL	27	7	14	2	1402	97	1	4.15							
	Utah Grizzlies	IHL	21	10	9	2	1227	59	1	2.89							
99-2000	**Atlanta**	**NHL**	**15**	**3**	**11**	**0**	**765**	**55**	**0**	**4.31**							
	Orlando	IHL	27	14	8	2	1487	57	4	2.30	6	2	3	381	16	0	2.52
2000-01	Orlando	IHL	4	1	1	1	187	9	0	2.88							
	Mobile Mysticks	ECHL	6	2	4	0	358	23	0	3.86							
	Cincinnati	AHL	15	6	4	4	838	44	1	3.15	3	1	1	142	7	0	2.95
2001-02	Kalamazoo Wings	UHL	53	20	24	7	3021	165	1	3.28							
	NHL Totals		**20**	**3**	**12**	**1**	**943**	**68**	**0**	**4.33**							

WHL West Second All-Star Team (1994, 1995) • Shared Harry "Hap" Holmes Memorial Trophy (fewest goals against - AHL) with Manny Legace (1996) • AHL First All-Star Team (1998) • Won Baz Bastien Memorial Trophy (Top Goaltender - AHL) (1998)

Transferred to **Phoenix** after **Winnipeg** franchise relocated, July 1, 1996. Traded to **Atlanta** by **Phoenix** for future considerations, June 25, 1999. Traded to **Anaheim** by **Atlanta** with Sergei Vyshedkevich for Ladislav Kohn, February 9, 2001.

LANICEK, Michal (LAN-ih-chehk, MIHK-uhl) T.B.

Goaltender. Catches left. 6'1", 172 lbs. Born, Benesov, Czech., July 6, 1981.
(Tampa Bay's 6th choice, 148th overall, in 1999 Entry Draft).

			Regular Season								Playoffs						
Season	Club	Lea	GP	W	L	T	Mins	GA	SO	Avg	GP	W	L	Mins	GA	SO	Avg
1996-97	Slavia Praha Jr.	Czech-Jr.	22				1260	42		2.00							
1997-98	Slavia Praha Jr.	Czech-Jr.	39				2162	75		2.08							
1998-99	Slavia Praha Jr.	Czech-Jr.	43				2412	87		2.16							
99-2000	Slavia Praha Jr.	Czech-Jr.	18				1020	35	1	2.06	6			328	7	0	1.28
	HC Liberec	Czech-2	4				109	11	0	6.06							
	Beroun	Czech-2	7				365	12	0	1.97	1	0	1	70	4	0	3.43
2000-01	Beroun	Czech-2	17				958	39	0	2.44	4	1	3	247	14	0	3.40
2001-02	Muskegon Fury	UHL	19	10	4	1	1000	33	2	1.98							
	Pensacola	ECHL	6	4	1	0	277	17	0	3.69							

LASAK, Jan (LA-shak, YAN) NSH.

Goaltender. Catches left. 6'1", 204 lbs. Born, Zvolen, Czech., April 10, 1979.
(Nashville's 6th choice, 65th overall, in 1999 Entry Draft).

			Regular Season								Playoffs						
Season	Club	Lea	GP	W	L	T	Mins	GA	SO	Avg	GP	W	L	Mins	GA	SO	Avg
1996-97	HKm Zvolen Jr.	Slovak-Jr.	49				2940	111		2.27							
1997-98	HKm Zvolen Jr.	Slovak-Jr.	48				2881	119		2.48							
	HK SKP Zilina	Slovak-2	4				208	12		3.46							
1998-99	HKm Zvolen Jr.	Slovak-Jr.	43				2580	91		2.12							
	HKm Zvolen	Slovakia	8				387	29		4.50							
99-2000	Hampton Roads	ECHL	*59	*36	17	4	*3409	145	0	2.55	10	5	5	610	28	1	2.75
2000-01	Milwaukee	IHL	43	23	17	2	2439	106	1	2.61	3	0	1	60	5	0	4.97
2001-02	**Nashville**	**NHL**	**3**	**0**	**3**	**0**	**177**	**13**	**0**	**4.41**							
	Milwaukee	AHL	34	12	18	3	1981	79	2	2.39							
	Slovakia	Olympics	2	0	1	0	94	6	0	3.81							
	NHL Totals		**3**	**0**	**3**	**0**	**177**	**13**	**0**	**4.41**							

ECHL First All-Star Team (2000) • ECHL Rookie of the Year (2000) • ECHL Top Goaltender (2000)

LECLAIRE, Pascal (lah-CLAIR, pas-CAL) CBJ

Goaltender. Catches left. 6'2", 185 lbs. Born, Repentigny, Que., November 7, 1982.
(Columbus' 1st choice, 8th overall, in 2001 Entry Draft).

			Regular Season								Playoffs						
Season	Club	Lea	GP	W	L	T	Mins	GA	SO	Avg	GP	W	L	Mins	GA	SO	Avg
1997-98	Cap-d-Madeleine	QAAA	26	6	17	3	1580	127	0	4.90							
1998-99	Halifax	QMJHL	33	19	11	1	1828	96	2	3.15	1	0	0	17	2	0	7.06
99-2000	Halifax	QMJHL	31	16	8	4	1729	103	1	3.57	5	1	2	198	12	0	3.65
2000-01	Halifax	QMJHL	35	14	16	5	2111	126	1	3.58	2	0	2	109	10	0	5.49
2001-02	Montreal Rocket	QMJHL	45	15	23	4	2513	138	1	3.29	7	3	4	441	15	0	*2.04

Traded to **Montreal** (QMJHL) by **Halifax** (QMJHL) with Hugo Lemoux for future considerations, June 16, 2002.

LEGACE, Manny (LEH-gah-see, MAN-nee) DET.

Goaltender. Catches left. 5'9", 162 lbs. Born, Toronto, Ont., February 4, 1973.
(Hartford's 5th choice, 188th overall, in 1993 Entry Draft).

			Regular Season								Playoffs						
Season	Club	Lea	GP	W	L	T	Mins	GA	SO	Avg	GP	W	L	Mins	GA	SO	Avg
1987-88	Alliston Hornets	OJHL-C	16	7	9	0	960	83	0	5.17							
1988-89	Vaughan Raiders	MTJHL	23				1303	92	1	4.24							
1989-90	Vaughan Raiders	MTJHL	21	8	11	1	1180	89	1	4.53							
	Thornhill	MTJHL	8	3	3	2	480	30	0	3.75							
1990-91	Niagara Falls	OHL	30	13	11	2	1515	107	0	4.24	4	1	1	119	10	0	5.04
1991-92	Niagara Falls	OHL	43	21	16	3	2384	143	0	3.60	14	8	5	791	56	0	4.25
1992-93	Niagara Falls	OHL	48	22	19	3	2630	171	0	3.90	4	0	4	240	18	0	4.50
1993-94	Canada	Nat-Tm	16	8	6	0	859	36	2	2.51							
1994-95	Springfield	AHL	39	12	17	6	2169	128	2	3.54							
1995-96	Springfield	AHL	37	20	12	4	2196	83	*5	*2.27	4	1	3	220	18	0	4.91
1996-97	Springfield	AHL	36	17	14	5	2119	107	1	3.03	12	9	3	745	25	*2	2.01
	Richmond	ECHL	3	2	1	0	157	8	0	3.05							
1997-98	Springfield	AHL	6	4	2	0	345	16	0	2.78							
	Las Vegas	IHL	41	18	16	4	2106	111	1	3.16	4	1	3	237	16	0	4.05
1998-99	**Los Angeles**	**NHL**	**17**	**2**	**9**	**2**	**899**	**39**	**0**	**2.60**							
	Long Beach	IHL	33	22	8	1	1796	67	2	2.24	6	4	2	338	9	0	*1.60
99-2000	**Detroit**	**NHL**	**4**	**4**	**0**	**0**	**240**	**11**	**0**	**2.75**							
	Manitoba Moose	IHL	42	17	18	5	2409	104	2	2.59	2	0	2	141	7	0	2.97
2000-01	**Detroit**	**NHL**	**39**	**24**	**5**	**5**	**2136**	**73**	**2**	**2.05**							
2001-02♦	**Detroit**	**NHL**	**20**	**10**	**6**	**2**	**1117**	**45**	**1**	**2.42**	**1**	**0**	**0**	**11**	**1**	**0**	**5.45**
	NHL Totals		**80**	**40**	**20**	**9**	**4392**	**168**	**3**	**2.30**	**1**	**0**	**0**	**11**	**1**	**0**	**5.45**

OHL First All-Star Team (1993) • AHL First All-Star Team (1996) • Shared Harry "Hap" Holmes Memorial Trophy (fewest goals against - AHL) with Scott Langkow (1996) • Won Baz Bastien Memorial Trophy (Top Goaltender - AHL) (1996)

Rights transferred to **Carolina** after **Hartford** franchise relocated, June 25, 1997. Traded to **LA Kings** by **Carolina** for future considerations, July 31, 1998. Signed as a free agent by **Detroit**, August 9, 1999. Claimed on waivers by **Vancouver** from **Detroit**, September 30, 1999. Claimed on waivers by **Detroit** from **Vancouver**, October 13, 1999.

LEHTO, Mika (leh-TOH, MEE-kuh) PIT.

Goaltender. Catches left. 5'11", 172 lbs. Born, Vammala, Finland, April 12, 1979.
(Pittsburgh's 8th choice, 224th overall, in 1998 Entry Draft).

			Regular Season								Playoffs						
Season	Club	Lea	GP	W	L	T	Mins	GA	SO	Avg	GP	W	L	Mins	GA	SO	Avg
1997-98	Assat Pori Jr.	Finn-Jr.	36				2160	103	2	2.86							
	Assat Pori	Finland	1	0	0	0	35	1	0	1.71	1	0	0	17	0	0	0.00
1998-99	Assat Pori Jr.	Finn-Jr.	20				1202	68		3.39							
	Assat Pori	Finland	15	4	6	1	773	38	1	2.95							
99-2000	Assat Pori	Finland	23	4	11	3	1099	87	0	4.75							
2000-01	JYP Jyvaskyla	Finland	41	12	20	8	2384	126	1	3.17							
2001-02	JYP Jyvaskyla	Finland	37	10	17	10	2167	107	6	2.96							

LEHTONEN, Kari (LEH-tuh-nehn, KAH-ree) ATL.

Goaltender. Catches left. 6'3", 190 lbs. Born, Helsinki, Finland, November 16, 1983.
(Atlanta's 1st choice, 2nd overall, in 2002 Entry Draft).

			Regular Season								Playoffs						
Season	Club	Lea	GP	W	L	T	Mins	GA	SO	Avg	GP	W	L	Mins	GA	SO	Avg
1998-99	Jokerit-C	Finn-Jr.	17				1020	61	0	3.61							
	Jokerit-B	Finn-Jr.									4	2	2	240	7	0	1.75
99-2000	Jokerit Jr.	Finn-Jr.	33	21	9	3	1974	86	2	2.61	12	9	3	758	14	4	1.11
2000-01	Jokerit Helsinki	Finn-Jr.	31	20	9	1	1799	71	3	2.37	1	0	1	54	4	0	4.44
	Jokerit Helsinki	Finland	4	3	0	0	189	6	0	1.90							
2001-02	Jokerit Jr.	Finn-Jr.	6	5	1	0	360	11	1	1.83							
	Jokerit Helsinki	Finland	23	13	5	3	1242	37	4	1.79	*11	*8	3	*623	18	*3	1.73

LEIGHTON, Michael (LAY-tohn, MIGH-kuhl) CHI.

Goaltender. Catches left. 6'2", 175 lbs. Born, Petrolia, Ont., May 19, 1981.
(Chicago's 5th choice, 165th overall, in 1999 Entry Draft).

			Regular Season								Playoffs						
Season	**Club**	**Lea**	**GP**	**W**	**L**	**T**	**Mins**	**GA**	**SO**	**Avg**	**GP**	**W**	**L**	**Mins**	**GA**	**SO**	**Avg**
1997-98	Petrolia Jets	OJHL-B	30	....	....	....	1583	87	2	3.30	....	....	....	....	....	....	....
1998-99	Windsor	OHL	28	4	17	2	1389	112	0	4.84	3	0	1	80	10	0	7.50
99-2000	Windsor	OHL	42	17	17	2	2272	118	1	3.12	12	5	6	616	32	0	3.12
2000-01	Windsor	OHL	54	32	13	5	3035	138	2	2.73	9	4	5	519	27	1	3.12
2001-02	Norfolk Admirals	AHL	52	27	16	8	3114	111	6	2.14	4	1	2	238	8	0	2.02

AHL All-Rookie Team (2002)

LENEVEU, David (LEH-neh-voo, DAY-vihd) PHX.

Goaltender. Catches left. 6', 170 lbs. Born, Fernie, B.C., May 23, 1983.
(Phoenix's 3rd choice, 46th overall, in 2002 Entry Draft).

			Regular Season								Playoffs						
Season	**Club**	**Lea**	**GP**	**W**	**L**	**T**	**Mins**	**GA**	**SO**	**Avg**	**GP**	**W**	**L**	**Mins**	**GA**	**SO**	**Avg**
99-2000	Fernie	AWJHL	22	15	2	0	1140	48	0	2.49	....	....	....	....	....	....	....
2000-01	Nanaimo Clippers	BCHL	41	....	....	....	2330	127	6	3.29	....	....	....	....	....	....	....
2001-02	Cornell Big Red	ECAC	14	11	2	1	842	21	2	*1.50	....	....	....	....	....	....	....

ECAC All-Rookie Team (2002)

Rights traded by **Calgary** (WHL) to **Vancouver** (WHL) for future considerations, May 30, 2001.

LITTLE, Neil (LIH-tuhl, NEEL) PHI.

Goaltender. Catches left. 6'1", 193 lbs. Born, Medicine Hat, Alta., December 18, 1971.
(Philadelphia's 10th choice, 226th overall, in 1991 Entry Draft).

			Regular Season								Playoffs						
Season	**Club**	**Lea**	**GP**	**W**	**L**	**T**	**Mins**	**GA**	**SO**	**Avg**	**GP**	**W**	**L**	**Mins**	**GA**	**SO**	**Avg**
1989-90	Estevan Bruins	SJHL	46	21	19	4	2707	150	1	3.32	....	....	....	....	....	....	....
1990-91	RPI Engineers	ECAC	18	9	8	0	1032	71	0	4.13	....	....	....	....	....	....	....
1991-92	RPI Engineers	ECAC	28	11	11	3	1532	96	0	3.76	....	....	....	....	....	....	....
1992-93	RPI Engineers	ECAC	*31	*19	9	3	*1801	88	0	2.93	....	....	....	....	....	....	....
1993-94	RPI Engineers	ECAC	27	16	7	4	1570	88	0	3.36	....	....	....	....	....	....	....
	Hershey Bears	AHL	1	0	0	0	18	1	0	3.33	....	....	....	....	....	....	....
1994-95	Hershey Bears	AHL	19	5	7	3	919	60	0	3.91	....	....	....	....	....	....	....
	Johnstown Chiefs	ECHL	16	7	6	1	897	55	0	3.68	3	0	2	145	11	0	4.55
1995-96	Hershey Bears	AHL	48	21	18	6	2680	149	0	3.34	1	0	1	60	4	0	4.02
1996-97	Philadelphia	AHL	54	31	12	7	3007	145	0	2.89	10	6	4	620	20	1	*1.94
1997-98	Philadelphia	AHL	51	*31	11	7	2960	145	0	2.94	*20	*15	5	*1193	48	*3	2.41
1998-99	Grand Rapids	IHL	50	18	21	5	2740	144	3	3.15	....	....	....	....	....	....	....
99-2000	Philadelphia	AHL	51	26	18	2	2830	143	1	3.03	5	2	3	298	15	0	3.02
2000-01	Philadelphia	AHL	*58	22	27	4	3117	148	2	2.85	10	5	5	631	23	1	2.19
2001-02	**Philadelphia**	**NHL**	**1**	**0**	**1**	**0**	**60**	**4**	**0**	**4.00**	....	....	....	....	....	....	....
	Philadelphia	AHL	35	13	15	7	2079	70	2	2.02	5	2	3	298	13	0	2.62
	NHL Totals		**1**	**0**	**1**	**0**	**60**	**4**	**0**	**4.00**	....	....	....	....	....	....	....

ECAC First All-Star Team (1993) • NCAA East Second All-American Team (1993)

LIV, Stefan (LIHV, STEH-fuhn) DET.

Goaltender. Catches left. 6', 172 lbs. Born, Jonkoping, Sweden, December 21, 1980.
(Detroit's 3rd choice, 102nd overall, in 2000 Entry Draft).

			Regular Season								Playoffs						
Season	**Club**	**Lea**	**GP**	**W**	**L**	**T**	**Mins**	**GA**	**SO**	**Avg**	**GP**	**W**	**L**	**Mins**	**GA**	**SO**	**Avg**
1997-98	HV 71 Jr.	Swede-Jr.	17	....	....	....	1020	47	....	2.76	....	....	....	....	....	....	....
1998-99	HV 71 Jonkoping	Sweden					DID NOT PLAY – SPARE GOALTENDER										
99-2000	HV 71 Jr.	Swede-Jr.	10	....	....	....	600	17	2	1.70	....	....	....	....	....	....	....
	Tranas AIF	Swede-2	9	....	....	....	541	20	0	2.17	....	....	....	....	....	....	....
	HV 71 Jonkoping	Sweden	12	....	....	....	716	24	1	2.01	3	....	....	178	12	0	4.04
2000-01	HV 71 Jonkoping	Sweden	*46	....	....	....	*2752	127	4	2.77	....	....	....	....	....	....	....
2001-02	HV 71 Jonkoping	Sweden	38	....	....	....	2184	95	*4	2.61	8	....	....	517	27	0	3.13

LUNDQVIST, Henrik (LUHND-kvihst, HEHN-rihk) NYR

Goaltender. Catches left. 5'11", 167 lbs. Born, Are, Sweden, March 2, 1982.
(NY Rangers' 7th choice, 205th overall, in 2000 Entry Draft).

			Regular Season								Playoffs						
Season	**Club**	**Lea**	**GP**	**W**	**L**	**T**	**Mins**	**GA**	**SO**	**Avg**	**GP**	**W**	**L**	**Mins**	**GA**	**SO**	**Avg**
1998-99	V. Frolunda Jr.	Swede-Jr.	35	....	....	....	2100	95	0	2.73	....	....	....	....	....	....	....
99-2000	V. Frolunda Jr.	Swede-Jr.	30	....	....	....	1726	73	0	2.54	5	4	1	300	7	2	1.40
2000-01	V. Frolunda-18	Swede-Jr.	2	....	....	....	120	5	0	2.50	3	2	1	182	5	0	1.62
	V. Frolunda Jr.	Swede-Jr.	19	....	....	....	1140	50	2	2.64	....	....	....	....	....	....	....
	Molndals IF	Swede-2	7	....	....	....	420	29	0	4.22	....	....	....	....	....	....	....
	Vastra Frolunda	Sweden	4	....	....	....	190	11	0	3.47	....	....	....	....	....	....	....
2001-02	Vastra Frolunda	Sweden	20	....	....	....	1152	52	2	2.71	8	8	0	489	18	*2	2.21

LUONGO, Roberto (loo-WAHN-goh, roh-BUHR-toh) FLA.

Goaltender. Catches left. 6'3", 205 lbs. Born, Montreal, Que., April 4, 1979.
(NY Islanders' 1st choice, 4th overall, in 1997 Entry Draft).

			Regular Season								Playoffs						
Season	**Club**	**Lea**	**GP**	**W**	**L**	**T**	**Mins**	**GA**	**SO**	**Avg**	**GP**	**W**	**L**	**Mins**	**GA**	**SO**	**Avg**
1994-95	Mtl-Bourassa	QAAA	25	10	14	0	94	1465	0	3.85	....	....	....	....	....	....	....
1995-96	Val-d'Or Foreurs	QMJHL	23	6	11	4	1201	74	0	3.70	3	0	1	68	5	0	4.41
1996-97	Val-d'Or Foreurs	QMJHL	60	32	22	2	3305	171	2	3.10	13	8	5	777	44	0	3.40
1997-98	Val-d'Or Foreurs	QMJHL	54	27	20	5	3046	157	*7	3.09	*17	*14	3	*1019	37	*2	*2.18
1998-99	Val-d'Or Foreurs	QMJHL	21	6	10	2	1176	77	1	3.93	....	....	....	....	....	....	....
	Acadie-Bathurst	QMJHL	22	14	7	1	1340	74	0	3.31	*23	*16	6	*1400	64	0	2.74
99-2000	**NY Islanders**	**NHL**	**24**	**7**	**14**	**1**	**1292**	**70**	**1**	**3.25**	....	....	....	....	....	....	....
	Lowell	AHL	26	10	12	4	1517	74	1	2.93	6	3	3	359	18	0	3.01
2000-01	**Florida**	**NHL**	**47**	**12**	**24**	**7**	**2628**	**107**	**5**	**2.44**	....	....	....	....	....	....	....
	Louisville	AHL	3	1	2	0	178	10	0	3.38	....	....	....	....	....	....	....
2001-02	**Florida**	**NHL**	**58**	**16**	**33**	**4**	**3030**	**140**	**4**	**2.77**	....	....	....	....	....	....	....
	NHL Totals		**129**	**35**	**71**	**12**	**6950**	**317**	**10**	**2.74**	....	....	....	....	....	....	....

Traded to **Florida** by **NY Islanders** with Olli Jokinen for Mark Parrish and Oleg Kvasha, June 24, 2000.

MacINTYRE, Drew (MAK-ihn-tighr, DROO) DET.

Goaltender. Catches left. 6', 173 lbs. Born, Charlottetown, P.E.I., June 24, 1983.
(Detroit's 2nd choice, 121st overall, in 2001 Entry Draft).

			Regular Season								Playoffs						
Season	**Club**	**Lea**	**GP**	**W**	**L**	**T**	**Mins**	**GA**	**SO**	**Avg**	**GP**	**W**	**L**	**Mins**	**GA**	**SO**	**Avg**
1998-99	Trenton Sting	OPJHL	20	....	....	....	1173	71	2	3.63	....	....	....	....	....	....	....
99-2000	Sherbrooke	QMJHL	24	10	7	2	1253	67	0	3.21	....	....	....	....	....	....	....
2000-01	Sherbrooke	QMJHL	48	17	22	3	2552	139	4	3.27	4	0	4	238	19	0	4.78
2001-02	Sherbrooke	QMJHL	55	15	34	3	3028	201	1	3.98	....	....	....	....	....	....	....

MAGERS, Marty (MAY-juhrs, MAHR-tee) BUF.

Goaltender. Catches left. 6'1", 180 lbs. Born, Maywood, IL, June 7, 1983.
(Buffalo's 6th choice, 121st overall, in 2002 Entry Draft).

			Regular Season								Playoffs						
Season	**Club**	**Lea**	**GP**	**W**	**L**	**T**	**Mins**	**GA**	**SO**	**Avg**	**GP**	**W**	**L**	**Mins**	**GA**	**SO**	**Avg**
99-2000	Fort Erie	OJHL-B	26	8	10	2	1546	113	0	4.27	....	....	....	....	....	....	....
2000-01	Omaha Lancers	USHL	27	16	5	3	1502	58	3	2.32	2	0	0	71	6	0	5.04
2001-02	Omaha Lancers	USHL	30	21	7	2	1780	48	*10	*1.62	5	2	3	276	11	0	2.39

• Signed Letter of Intent to attend **Michigan Tech** (WCHA), May 8, 2002.

USHL Top Goaltender (2001) • USHL Second All-Star Team (2002)

MALEK, Roman (MAHL-ehk, ROH-muhn) PHI.

Goaltender. Catches left. 5'11", 161 lbs. Born, Prague, Czech., September 25, 1977.
(Philadelphia's 5th choice, 158th overall, in 2001 Entry Draft).

			Regular Season								Playoffs						
Season	**Club**	**Lea**	**GP**	**W**	**L**	**T**	**Mins**	**GA**	**SO**	**Avg**	**GP**	**W**	**L**	**Mins**	**GA**	**SO**	**Avg**
1998-99	HC Slavia Praha	Czech	18	....	....	....	830	51	....	3.69	....	....	....	....	....	....	....
99-2000	HC Slavia Praha	Czech	25	....	....	....	1342	59	....	2.64	....	....	....	....	....	....	....
2000-01	HC Slavia Praha	Czech	46	....	....	....	2550	100	....	2.35	11	....	....	665	28	....	2.53
2001-02	HC Slavia Praha	Czech	33	....	....	....	1967	80	....	2.44	9	....	....	485	22	....	2.72

MANZATO, Daniel (man-ZA-toh, DAN-yehl) CAR.

Goaltender. Catches left. 6', 178 lbs. Born, Fribourg, Switz., January 17, 1984.
(Carolina's 3rd choice, 160th overall, in 2002 Entry Draft).

			Regular Season								Playoffs						
Season	**Club**	**Lea**	**GP**	**W**	**L**	**T**	**Mins**	**GA**	**SO**	**Avg**	**GP**	**W**	**L**	**Mins**	**GA**	**SO**	**Avg**
99-2000	Fribourg Jr.	Swiss-Jr.					STATISTICS NOT AVAILABLE										
2000-01	Fribourg Jr.	Swiss-Jr.	36	....	....	....	2160	32	6	0.91	....	....	....	....	....	....	....
2001-02	Victoriaville	QMJHL	36	20	8	2	1894	102	0	3.23	6	3	0	249	17	0	4.09

MARACLE, Norm (MAHR-ah-kuhl, NOHRM) ATL.

Goaltender. Catches left. 5'9", 195 lbs. Born, Belleville, Ont., October 2, 1974.
(Detroit's 6th choice, 126th overall, in 1993 Entry Draft).

			Regular Season								Playoffs						
Season	**Club**	**Lea**	**GP**	**W**	**L**	**T**	**Mins**	**GA**	**SO**	**Avg**	**GP**	**W**	**L**	**Mins**	**GA**	**SO**	**Avg**
1990-91	Cgy. North Stars	AMHL	29	....	....	....	1740	99	0	3.43	....	....	....	....	....	....	....
1991-92	Saskatoon Blades	WHL	29	13	6	3	1529	87	1	3.41	15	9	5	860	37	0	3.38
1992-93	Saskatoon Blades	WHL	53	27	18	3	1939	160	1	3.27	9	4	5	569	33	0	3.48
1993-94	Saskatoon Blades	WHL	56	*41	13	1	3219	148	2	2.76	16	*11	5	940	48	*1	3.06
1994-95	Adirondack	AHL	39	12	15	2	1997	119	0	3.57	....	....	....	....	....	....	....
1995-96	Adirondack	AHL	54	24	18	6	2949	135	2	2.75	1	0	1	30	4	0	8.11
1996-97	Adirondack	AHL	*68	*34	22	9	*3843	173	5	2.70	4	1	3	192	10	1	3.13
1997-98	**Detroit**	**NHL**	**4**	**2**	**0**	**1**	**178**	**6**	**0**	**2.02**	....	....	....	....	....	....	....
	Adirondack	AHL	*66	27	29	8	*3709	190	1	3.07	3	0	3	180	10	0	3.33
1998-99	**Detroit**	**NHL**	**16**	**6**	**5**	**2**	**821**	**31**	**0**	**2.27**	**2**	**0**	**0**	**58**	**3**	**0**	**3.10**
	Adirondack	AHL	6	3	3	0	359	18	0	3.01	....	....	....	....	....	....	....
99-2000	**Atlanta**	**NHL**	**32**	**4**	**19**	**2**	**1618**	**94**	**1**	**3.49**	....	....	....	....	....	....	....
2000-01	**Atlanta**	**NHL**	**13**	**2**	**8**	**3**	**753**	**43**	**0**	**3.43**	....	....	....	....	....	....	....
	Orlando	IHL	51	33	13	3	2963	100	*8	*2.02	*16	*12	4	*1003	37	1	2.21
2001-02	**Atlanta**	**NHL**	**1**	**0**	**1**	**0**	**60**	**3**	**0**	**3.00**	....	....	....	....	....	....	....
	Chicago Wolves	AHL	51	21	25	4	2919	141	3	2.90	2	0	1	55	4	0	4.36
	NHL Totals		**66**	**14**	**33**	**8**	**3430**	**177**	**1**	**3.10**	**2**	**0**	**0**	**58**	**3**	**0**	**3.10**

Won Warwick Trophy (MVP - AMHL) (1991) • WHL East Second All-Star Team (1993) • WHL East First All-Star Team (1994) • Canadian Major Junior First All-Star Team (1994) • Canadian Major Junior Goaltender of the Year (1994) • AHL Second All-Star Team (1997, 1998) • IHL First All-Star Team (2001) • Shared James Norris Memorial Trophy (fewest goals against - IHL) with Scott Fankhouser (2001) • Won James Gatschene Memorial Trophy (MVP - IHL) (2001) • Won "Bud" Poile Trophy (Playoff MVP - IHL) (2001)

Claimed by **Atlanta** from **Detroit** in Expansion Draft, June 25, 1999.

MARKKANEN, Jussi (MAHR-kah-nehn, YOO-see) EDM.

Goaltender. Catches left. 5'11", 183 lbs. Born, Imatra, Finland, May 8, 1975.
(Edmonton's 5th choice, 133rd overall, in 2001 Entry Draft).

			Regular Season								Playoffs						
Season	**Club**	**Lea**	**GP**	**W**	**L**	**T**	**Mins**	**GA**	**SO**	**Avg**	**GP**	**W**	**L**	**Mins**	**GA**	**SO**	**Avg**
1991-92	SaiPa Jr.	Finn-Jr.	2	....	....	....	120	11	0	5.50	....	....	....	....	....	....	....
1992-93	SaiPa Jr.	Finn-Jr.	7	....	....	....	367	28	....	4.58	....	....	....	....	....	....	....
	SaiPa	Finland-2	16	....	....	....	798	60	....	4.51	....	....	....	....	....	....	....
1993-94	SaiPa	Finland-2	30	....	....	....	1726	97	....	3.37	....	....	....	....	....	....	....
1994-95	SaiPa	Finland-2	43	....	....	....	2493	122	....	2.94	3	....	....	179	5	....	1.68
1995-96	Tappara Jr.	Finn-Jr.	5	....	....	....	298	21	....	4.23	....	....	....	....	....	....	....
	Tappara Tampere	Finland	23	11	8	2	1238	59	1	2.86	....	....	....	....	....	....	....
1996-97	SaiPa	Finland	41	9	24	7	2340	132	0	3.38	....	....	....	....	....	....	....
1997-98	SaiPa	Finland	*48	21	20	5	*2870	138	4	2.89	3	0	3	164	11	0	4.02
1998-99	SaiPa	Finland	48	21	19	4	2633	105	4	2.39	7	3	3	366	21	0	3.44
99-2000	SaiPa	Finland	48	4	23	9	2794	150	2	3.22	....	....	....	....	....	....	....
2000-01	Tappara Tampere	Finland	52	*30	17	5	3076	107	*9	2.09	*10	7	3	*608	18	1	1.78
2001-02	**Edmonton**	**NHL**	**14**	**6**	**4**	**2**	**784**	**24**	**2**	**1.84**	....	....	....	....	....	....	....
	Hamilton	AHL	4	2	2	0	239	9	0	2.26	....	....	....	....	....	....	....
	Finland	Olympics					DID NOT PLAY – SPARE GOALTENDER										
	NHL Totals		**14**	**6**	**4**	**2**	**784**	**24**	**2**	**1.84**	....	....	....	....	....	....	....

MARSTERS, Nathan (MAHR-stuhrs, NAY-thuhn) L.A.

Goaltender. Catches left. 6'4", 190 lbs. Born, Burlington, Ont., January 20, 1980.
(Los Angeles' 5th choice, 165th overall, in 2000 Entry Draft).

			Regular Season								Playoffs						
Season	**Club**	**Lea**	**GP**	**W**	**L**	**T**	**Mins**	**GA**	**SO**	**Avg**	**GP**	**W**	**L**	**Mins**	**GA**	**SO**	**Avg**
1997-98	Bramalea Blues	OPJHL	12	....	....	....	539	25	2	2.78	....	....	....	....	....	....	....
1998-99	Chilliwack	BCHL	9	6	2	0	478	29	0	3.65	....	....	....	....	....	....	....
99-2000	Bramalea Blues	OPJHL	27	....	....	....	1668	98	2	3.53	....	....	....	....	....	....	....
	Chilliwack	BCHL	15	9	6	0	825	63	0	4.58	20	15	5	1187	62	0	3.13
2000-01	RPI Engineers	ECAC	28	14	13	1	1631	64	*4	2.35	....	....	....	....	....	....	....
2001-02	RPI Engineers	ECAC	28	15	9	3	1627	70	1	2.58	....	....	....	....	....	....	....

MASON, Chris (MAY-sohn, KRIHS)

Goaltender. Catches left. 6', 195 lbs. Born, Red Deer, Alta., April 20, 1976.
(New Jersey's 7th choice, 122nd overall, in 1995 Entry Draft).

			Regular Season								Playoffs						
Season	Club	Lea	GP	W	L	T	Mins	GA	SO	Avg	GP	W	L	Mins	GA	SO	Avg
1992-93	Red Deer Chiefs	AMHL	20	....	....	....	1280	76	0	3.35	....	....	....	....	....	....	....
1993-94	Victoria Cougars	WHL	5	1	4	0	237	27	0	6.84	....	....	....	....	....	....	....
1994-95	Prince George	WHL	44	8	30	1	2288	192	1	5.03	....	....	....	....	....	....	....
1995-96	Prince George	WHL	59	16	37	1	3289	236	1	4.31	....	....	....	....	....	....	....
1996-97	Prince George	WHL	50	19	24	4	2851	172	2	3.62	15	9	6	938	44	*1	2.81
1997-98	Cincinnati	AHL	47	13	19	7	2368	136	0	3.45	....	....	....	....	....	....	....
1998-99	**Nashville**	**NHL**	**3**	**0**	**0**	**0**	**69**	**6**	**0**	**5.22**	....	....	....	....	....	....	....
	Milwaukee	IHL	34	15	12	6	1901	92	1	2.90	....	....	....	....	....	....	....
99-2000	Milwaukee	IHL	53	20	21	8	2952	137	2	2.78	3	1	2	252	11	0	2.62
2000-01	**Nashville**	**NHL**	**1**	**0**	**1**	**0**	**59**	**2**	**0**	**2.03**	....	....	....	....	....	....	....
	Milwaukee	IHL	37	17	14	5	2226	87	5	2.35	4	1	3	239	12	0	3.02
2001-02	Milwaukee	AHL	48	17	21	7	2755	116	2	2.53	....	....	....	....	....	....	....
	NHL Totals		**4**	**0**	**1**	**0**	**128**	**8**	**0**	**3.75**	....	....	....	....	....	....	....

Signed as a free agent by **Anaheim**, June 27, 1997. Traded to **Nashville** by **Anaheim** with Marc Moro for Dominic Roussel, October 5, 1998.

McLENNAN, Jamie (muh-KLEH-nuhn, JAY-mee) CGY.

Goaltender. Catches left. 6', 190 lbs. Born, Edmonton, Alta., June 30, 1971.
(NY Islanders' 3rd choice, 48th overall, in 1991 Entry Draft).

			Regular Season								Playoffs						
Season	Club	Lea	GP	W	L	T	Mins	GA	SO	Avg	GP	W	L	Mins	GA	SO	Avg
1987-88	St. Albert	AMHL	21	....	....	....	1224	80	0	3.92	....	....	....	....	....	....	....
1988-89	Spokane Chiefs	WHL	11	....	....	....	578	63	0	6.54	....	....	....	....	....	....	....
	Lethbridge	WHL	7	....	....	....	368	22	0	3.59	....	....	....	....	....	....	....
1989-90	Lethbridge	WHL	34	20	4	2	1690	110	1	3.91	13	6	5	677	44	0	3.90
1990-91	Lethbridge	WHL	56	32	18	4	3230	205	0	3.81	*16	8	8	*970	56	0	3.46
1991-92	Capital District	AHL	18	4	10	2	952	60	1	3.78	....	....	....	....	....	....	....
	Richmond	ECHL	32	16	12	2	1837	114	0	3.72	....	....	....	....	....	....	....
1992-93	Capital District	AHL	38	17	14	6	2171	117	1	3.23	1	0	1	20	5	0	15.00
1993-94	**NY Islanders**	**NHL**	**22**	**8**	**7**	**6**	**1287**	**61**	**0**	**2.84**	**2**	**0**	**1**	**82**	**6**	**0**	**4.39**
	Salt Lake	IHL	24	8	12	2	1320	80	0	3.64	....	....	....	....	....	....	....
1994-95	**NY Islanders**	**NHL**	**21**	**6**	**11**	**2**	**1185**	**67**	**0**	**3.39**	....	....	....	....	....	....	....
	Denver Grizzlies	IHL	4	3	0	1	239	12	0	3.00	11	8	2	640	23	1	*2.15
1995-96	**NY Islanders**	**NHL**	**13**	**3**	**9**	**1**	**636**	**39**	**0**	**3.68**	....	....	....	....	....	....	....
	Utah Grizzlies	IHL	14	9	2	2	728	29	0	2.39	....	....	....	....	....	....	....
	Worcester	AHL	22	14	7	1	1216	57	0	2.81	2	0	2	119	8	0	4.04
1996-97	Worcester	AHL	39	18	13	4	2152	100	2	2.79	4	2	2	262	16	0	3.67
1997-98	**St. Louis**	**NHL**	**30**	**16**	**8**	**2**	**1658**	**60**	**2**	**2.17**	**1**	**0**	**0**	**14**	**1**	**0**	**4.29**
1998-99	**St. Louis**	**NHL**	**33**	**13**	**14**	**4**	**1763**	**70**	**3**	**2.38**	**1**	**0**	**1**	**37**	**0**	**0**	**0.00**
99-2000	**St. Louis**	**NHL**	**19**	**9**	**5**	**2**	**1009**	**33**	**2**	**1.96**	....	....	....	....	....	....	....
2000-01	**Minnesota**	**NHL**	**38**	**5**	**23**	**9**	**2230**	**98**	**2**	**2.64**	....	....	....	....	....	....	....
2001-02	Houston Aeros	AHL	51	25	18	4	2852	130	3	2.74	14	8	6	880	31	2	2.11
	NHL Totals		**176**	**60**	**77**	**26**	**9768**	**428**	**9**	**2.63**	**4**	**0**	**2**	**133**	**7**	**0**	**3.16**

WHL East First All-Star Team (1991) • Won Bill Masterton Memorial Trophy (1998)

Signed as a free agent by **St. Louis**, July 15, 1996. Selected by **Minnesota** from **St. Louis** in Expansion Draft, June 23, 2000. Traded to **Calgary** by **Minnesota** for Calgary's 9th round choice (Mika Hannula) in 2002 Entry Draft, June 22, 2002.

McVICAR, Rob (mihk-VIH-kuhr, RAWB) VAN.

Goaltender. Catches left. 6'4", 195 lbs. Born, Hay River, NWT, January 15, 1982.
(Vancouver's 6th choice, 151st overall, in 2002 Entry Draft).

			Regular Season								Playoffs						
Season	Club	Lea	GP	W	L	T	Mins	GA	SO	Avg	GP	W	L	Mins	GA	SO	Avg
1998-99	Brandon Kings	MMMHL	21	....	....	....	1217	69	0	3.40	....	....	....	....	....	....	....
99-2000	Brandon	WHL	14	5	6	0	687	43	0	3.76	....	....	....	....	....	....	....
2000-01	Brandon	WHL	27	12	10	2	1537	76	0	2.97	5	2	3	324	13	1	2.41
2001-02	Brandon	WHL	55	*33	18	2	3276	151	1	2.77	19	11	8	1255	44	1	2.10

MEDVEDEV, Andrei (mehd-VEH-dehv, AN-dray) CGY.

Goaltender. Catches left. 6', 211 lbs. Born, Moscow, USSR, April 1, 1983.
(Calgary's 3rd choice, 56th overall, in 2001 Entry Draft).

			Regular Season								Playoffs						
Season	Club	Lea	GP	W	L	T	Mins	GA	SO	Avg	GP	W	L	Mins	GA	SO	Avg
1998-99	Spartak Moscow	Russia	2	....	....	....	80	2	1	1.50	....	....	....	....	....	....	....
99-2000	Spartak Moscow	Russia-2	STATISTICS NOT AVAILABLE														
2000-01	Spartak Moscow	Russia-2	11	....	....	....	208	8	0	2.31	....	....	....	....	....	....	....
2001-02	Spartak Moscow 2	Russia-3	STATISTICS NOT AVAILABLE														
	Spartak Moscow	Russia	2	....	....	....	61	4	0	3.93	....	....	....	....	....	....	....

MENSATOR, Lukas (MEHN-suh-tohr, loo-KAHSH) VAN.

Goaltender. Catches left. 5'8", 167 lbs. Born, Sokolov, Czech., August 18, 1984.
(Vancouver's 4th choice, 83rd overall, in 2002 Entry Draft).

			Regular Season								Playoffs						
Season	Club	Lea	GP	W	L	T	Mins	GA	SO	Avg	GP	W	L	Mins	GA	SO	Avg
99-2000	Karlovy Vary 18	Czech-Jr.	42	....	....	....	2406	160	0	3.99	....	....	....	....	....	....	....
	Karlovy Vary Jr.	Czech-Jr.	1	1	0	0	60	3	0	3.00	....	....	....	....	....	....	....
2000-01	Karlovy Vary 18	Czech-Jr.	6	....	....	....	360	15	0	2.50	....	....	....	....	....	....	....
	Karlovy Vary Jr.	Czech-Jr.	19	....	....	....	1085	60	0	3.32	....	....	....	....	....	....	....
2001-02	Karlovy Vary Jr.	Czech-Jr.	31	....	....	....	1809	93	0	3.08	9	....	....	459	17	0	2.22
	Banik Sokolov	Czech-3	3	....	....	....	180	12	0	4.00	....	....	....	....	....	....	....

MEYER, Scott (MIGH-uhr, SKAWT) NYR

Goaltender. Catches left. 6', 185 lbs. Born, White Bear Lake, MN, April 10, 1976.

			Regular Season								Playoffs						
Season	Club	Lea	GP	W	L	T	Mins	GA	SO	Avg	GP	W	L	Mins	GA	SO	Avg
1995-96	Fargo-Moorhead	USHL	27	....	....	....	1620	82	1	3.04	....	....	....	....	....	....	....
1996-97	St. Cloud State	WCHA	1	0	0	0	29	1	0	2.07	....	....	....	....	....	....	....
1997-98	St. Cloud State	WCHA	2	0	1	0	75	4	0	3.21	....	....	....	....	....	....	....
1998-99	St. Cloud State	WCHA	9	2	5	1	464	23	0	2.97	....	....	....	....	....	....	....
99-2000	St. Cloud State	WCHA	32	20	8	3	1922	76	7	2.37	....	....	....	....	....	....	....
2000-01	St. Cloud State	WCHA	36	25	8	1	2096	78	2	2.23	....	....	....	....	....	....	....
2001-02	Charlotte	ECHL	30	14	10	5	1720	78	1	2.72	....	....	....	....	....	....	....
	Hartford	AHL	13	4	4	2	646	23	0	2.14	8	3	4	505	21	0	2.50

WCHA Second All-Star Team (2000) • WCHA First All-Star Team (2001) • NCAA West Second All-American Team (2001)

Signed as a free agent by **NY Rangers**, July 5, 2001.

MICHAUD, Alfie (mee-SHOH, AL-fee) VAN.

Goaltender. Catches left. 5'10", 177 lbs. Born, Selkirk, Man., November 6, 1976.

			Regular Season								Playoffs						
Season	Club	Lea	GP	W	L	T	Mins	GA	SO	Avg	GP	W	L	Mins	GA	SO	Avg
1995-96	Lebret Eagles	SJHL	44	....	....	....	2547	121	2	2.85	....	....	....	....	....	....	....
1996-97	U. of Maine	H-East	29	*17	8	1	1515	78	1	3.09	....	....	....	....	....	....	....
1997-98	U. of Maine	H-East	32	15	12	4	1794	94	2	3.14	....	....	....	....	....	....	....
1998-99	U. of Maine	H-East	37	*28	6	3	2147	83	3	2.32	....	....	....	....	....	....	....
99-2000	**Vancouver**	**NHL**	**2**	**0**	**1**	**0**	**69**	**5**	**0**	**4.35**	....	....	....	....	....	....	....
	Syracuse Crunch	AHL	38	10	17	5	2052	132	0	3.86	....	....	....	....	....	....	....
2000-01	Kansas City	IHL	32	14	14	2	1778	93	1	3.14	....	....	....	....	....	....	....
2001-02	Reading Royals	ECHL	11	5	3	2	606	26	2	2.58	....	....	....	....	....	....	....
	Manitoba Moose	AHL	32	16	10	1	1749	78	4	2.68	7	3	4	424	19	0	2.69
	NHL Totals		**2**	**0**	**1**	**0**	**69**	**5**	**0**	**4.35**	....	....	....	....	....	....	....

NCAA Championship All-Tournament Team (1999) • NCAA Championship Tournament MVP (1999)

Signed as a free agent by **Vancouver**, July 12, 1999.

MICHAUD, Olivier (MEE-shoh, OH-lihv-ee-ay) MTL.

Goaltender. Catches left. 5'11", 160 lbs. Born, Beloeil, Que., September 14, 1983.

			Regular Season								Playoffs						
Season	Club	Lea	GP	W	L	T	Mins	GA	SO	Avg	GP	W	L	Mins	GA	SO	Avg
1998-99	Eclaireur	QAHA	23	14	4	5	1380	58	....	2.50	....	....	....	....	....	....	....
99-2000	Antoine-Girourd	QAAA	7	6	1	0	420	15	1	2.14	....	....	....	....	....	....	....
	Charles Lemonthe	QAAA	16	8	4	2	886	57	0	3.86	16	8	8	1015	29	2	1.71
	Shawinigan	QMJHL	1	0	1	0	49	2	0	2.44	....	....	....	....	....	....	....
2000-01	Shawinigan	QMJHL	21	12	4	4	1096	54	1	2.96	3	1	2	150	6	0	2.41
2001-02	**Montreal**	**NHL**	**1**	**0**	**0**	**0**	**18**	**0**	**0**	**0.00**	....	....	....	....	....	....	....
	Shawinigan	QMJHL	46	29	11	3	2650	108	3	*2.45	12	7	5	744	36	0	2.91
	NHL Totals		**1**	**0**	**0**	**0**	**18**	**0**	**0**	**0.00**	....	....	....	....	....	....	....

Signed as a free agent by **Montreal**, September 18, 2001. • Recalled by **Montreal** from **Shawinigan** (QMJHL) under emergency conditions, October 26, 2001. • Returned to **Shawinigan** (QMJHL) by **Montreal**, November 5, 2001.

MILLER, Ryan (MIHL-luhr, RIGH-uhn) BUF.

Goaltender. Catches left. 6'2", 150 lbs. Born, East Lansing, MI, July 17, 1980.
(Buffalo's 7th choice, 138th overall, in 1999 Entry Draft).

			Regular Season								Playoffs						
Season	Club	Lea	GP	W	L	T	Mins	GA	SO	Avg	GP	W	L	Mins	GA	SO	Avg
1997-98	Sault Ste. Marie	NAJHL	31	17	13	0	1804	72	1	2.39	6	2	4	311	10	0	1.93
1998-99	Sault Ste. Marie	NAJHL	47	31	14	1	2711	104	8	2.30	4	2	2	218	10	1	2.76
99-2000	Michigan State	CCHA	26	16	5	3	1525	39	*8	*1.53	....	....	....	....	....	....	....
2000-01	Michigan State	CCHA	40	*31	5	4	2447	54	*10	*1.32	....	....	....	....	....	....	....
2001-02	Michigan State	CCHA	40	26	9	5	2411	71	*8	*1.77	....	....	....	....	....	....	....

CCHA Second All-Star Team (2000) • CCHA First All-Star Team (2001, 2002) • NCAA West First All-American Team (2001, 2002) • Won Hobey Baker Memorial Award (Top U.S. Collegiate Player) (2001) • CCHA Player of the Year (2002)

MINARD, Mike (mih-NAHRD, MIGHK) TOR.

Goaltender. Catches left. 6'3", 205 lbs. Born, Owen Sound, Ont., November 1, 1976.
(Edmonton's 4th choice, 83rd overall, in 1995 Entry Draft).

			Regular Season								Playoffs						
Season	Club	Lea	GP	W	L	T	Mins	GA	SO	Avg	GP	W	L	Mins	GA	SO	Avg
1992-93	St. Marys	OJHL-B	23	....	....	....	1374	162	0	3.10	....	....	....	....	....	....	....
1993-94	St. Marys	OJHL-B	31	*25	5	0	1710	78	1	*2.74	....	....	....	....	....	....	....
1994-95	Chilliwack	BCJHL	40	....	....	....	2330	136	0	3.50	....	....	....	....	....	....	....
1995-96	Barrie Colts	OHL	1	0	1	0	52	8	0	9.23	....	....	....	....	....	....	....
	Detroit	OHL	42	25	10	4	2314	128	2	3.32	17	9	6	922	55	1	3.58
1996-97	Hamilton	AHL	3	1	1	0	100	7	0	4.20	....	....	....	....	....	....	....
	Wheeling Nailers	ECHL	23	3	7	1	899	69	0	4.60	3	0	2	148	16	0	6.47
1997-98	Hamilton	AHL	2	1	0	0	80	2	0	1.50	....	....	....	....	....	....	....
	Brantford Smoke	UHL	2	1	1	0	74	7	0	5.63	....	....	....	....	....	....	....
	New Orleans	ECHL	11	6	2	0	429	30	0	4.19	....	....	....	....	....	....	....
	Milwaukee	IHL	8	2	2	0	362	19	0	3.15	....	....	....	....	....	....	....
1998-99	Dayton Bombers	ECHL	15	8	5	2	788	42	1	3.20	....	....	....	....	....	....	....
	Milwaukee	IHL	10	3	5	0	531	27	0	3.05	....	....	....	....	....	....	....
	Hamilton	AHL	11	8	3	0	645	30	1	2.79	1	0	0	20	0	0	0.00
99-2000	**Edmonton**	**NHL**	**1**	**1**	**0**	**0**	**60**	**3**	**0**	**3.00**	....	....	....	....	....	....	....
	Hamilton	AHL	38	16	12	5	1987	102	0	3.08	1	0	0	23	0	0	0.00
2000-01	St. John's	AHL	43	23	10	4	2252	91	1	2.42	4	1	3	252	15	0	3.57
2001-02	St. John's	AHL	35	14	11	7	1936	100	1	3.10	....	....	....	....	....	....	....
	NHL Totals		**1**	**1**	**0**	**0**	**60**	**3**	**0**	**3.00**	....	....	....	....	....	....	....

Signed as a free agent by **Toronto**, March 16, 2001.

MORRISON, Mike (MOHR-rihs-ohn, MIGHK) EDM.

Goaltender. Catches right. 6'3", 194 lbs. Born, Medford, MA, July 11, 1979.
(Edmonton's 8th choice, 186th overall, in 1998 Entry Draft).

			Regular Season								Playoffs						
Season	Club	Lea	GP	W	L	T	Mins	GA	SO	Avg	GP	W	L	Mins	GA	SO	Avg
1997-98	Exeter Academy	H.S.	27	15	11	2	1632	64	1	2.35	....	....	....	....	....	....	....
1998-99	U. of Maine	H-East	11	3	0	1	347	10	1	1.73	....	....	....	....	....	....	....
99-2000	U. of Maine	H-East	12	7	2	1	608	27	1	2.67	....	....	....	....	....	....	....
2000-01	U. of Maine	H-East	10	2	3	3	490	16	1	1.96	....	....	....	....	....	....	....
2001-02	U. of Maine	H-East	30	20	3	4	1645	60	2	2.19	....	....	....	....	....	....	....

Hockey East First All-Star Team (2002)

MOSS, Tyler (MAWS, TIGH-luhr) VAN.

Goaltender. Catches right. 6', 185 lbs. Born, Ottawa, Ont., June 29, 1975.
(Tampa Bay's 2nd choice, 29th overall, in 1993 Entry Draft).

			Regular Season								Playoffs						
Season	**Club**	**Lea**	**GP**	**W**	**L**	**T**	**Mins**	**GA**	**SO**	**Avg**	**GP**	**W**	**L**	**Mins**	**GA**	**SO**	**Avg**
1991-92	Nepean Raiders	OCJHL	26	7	12	1	1335	109	0	4.90	....	....	....	....	....	....	....
1992-93	Kingston	OHL	31	13	7	5	1537	97	0	3.79	6	1	2	228	19	0	5.00
1993-94	Kingston	OHL	13	6	4	3	795	42	1	3.17	3	0	2	136	8	0	3.53
1994-95	Kingston	OHL	*57	33	17	5	*3249	164	1	3.03	6	2	4	333	27	0	4.86
1995-96	Atlanta Knights	IHL	40	11	19	4	2030	138	1	4.08	3	0	3	213	11	0	3.10
1996-97	Adirondack	AHL	11	1	5	2	507	42	1	4.97	....	....	....	....	....	....	....
	Grand Rapids	IHL	15	5	6	1	715	35	0	2.94	....	....	....	....	....	....	....
	Muskegon Fury	ColHL	2	1	1	0	119	5	0	2.51	....	....	....	....	....	....	....
	Saint John	AHL	9	6	1	1	534	17	0	1.91	5	2	3	242	15	0	3.72
1997-98	**Calgary**	**NHL**	**6**	**2**	**3**	**1**	**367**	**20**	**0**	**3.27**	....	....	....	....	....	....	....
	Saint John	AHL	39	19	10	7	2194	91	0	2.49	15	8	5	761	37	0	2.91
1998-99	**Calgary**	**NHL**	**11**	**3**	**7**	**0**	**550**	**23**	**0**	**2.51**	....	....	....	....	....	....	....
	Saint John	AHL	9	2	5	1	475	25	0	3.16	....	....	....	....	....	....	....
	Orlando	IHL	9	6	2	1	515	21	1	2.45	17	10	7	1017	53	0	3.13
99-2000	Wilkes-Barre	AHL	4	1	1	1	188	11	0	3.52	....	....	....	....	....	....	....
	Kansas City	IHL	36	18	12	5	2116	105	3	2.98	....	....	....	....	....	....	....
2000-01	**Carolina**	**NHL**	**12**	**1**	**6**	**0**	**557**	**37**	**0**	**3.99**	....	....	....	....	....	....	....
	Cincinnati	IHL	9	5	3	1	506	24	2	2.85	....	....	....	....	....	....	....
2001-02	Lowell	AHL	43	20	16	7	2572	106	1	2.47	....	....	....	....	....	....	....
	NHL Totals		**29**	**6**	**16**	**1**	**1474**	**80**	**0**	**3.26**	....	....	....	....	....	....	....

OHL All-Rookie Team (1993) • OHL First All-Star Team (1995) • Shared Harry "Hap" Holmes Memorial Trophy (fewest goals against - AHL) with Jean-Sebastien Giguere (1998)

Traded to **Calgary** by **Tampa Bay** for Jamie Huscroft, March 18, 1997. Traded to **Pittsburgh** by **Calgary** with Rene Corbet for Brad Werenka, March 14, 2000. Signed as a free agent by **Carolina**, August 9, 2000. Signed as a free agent by **Vancouver**, July 5, 2002.

MULLER, Robert (MEW-luhr, RAW-buhrt) WSH.

Goaltender. Catches left. 5'8", 163 lbs. Born, Rosenheim, West Germany, June 25, 1980.
(Washington's 9th choice, 275th overall, in 2001 Entry Draft).

			Regular Season								Playoffs						
Season	**Club**	**Lea**	**GP**	**W**	**L**	**T**	**Mins**	**GA**	**SO**	**Avg**	**GP**	**W**	**L**	**Mins**	**GA**	**SO**	**Avg**
1996-97	Rosenheim Jr.	Ger.-Jr.					STATISTICS NOT AVAILABLE										
1997-98	EHC Klostersee	German-3					STATISTICS NOT AVAILABLE										
1998-99	Rosenheim	Germany	33	....	....	....	1863	105	1	3.38	....	....	....	....	....	....	....
99-2000	Rosenheim	Germany	39	....	....	....	2228	131	1	3.53	....	....	....	....	....	....	....
2000-01	Adler Mannheim	Germany	23	....	....	....	1195	49	1	2.46	2	....	....	103	2	0	1.17
2001-02	Adler Mannheim	Germany	15	....	....	....	637	26	1	2.45	....	....	....	....	....	....	....
	Germany	Olympics	2	0	1	0	78	4	0	3.07	....	....	....	....	....	....	....

MUNRO, Adam (MUHN-roh, A-duhm) CHI.

Goaltender. Catches left. 6'1", 194 lbs. Born, St. George, Ont., November 12, 1982.
(Chicago's 1st choice, 29th overall, in 2001 Entry Draft).

			Regular Season								Playoffs						
Season	**Club**	**Lea**	**GP**	**W**	**L**	**T**	**Mins**	**GA**	**SO**	**Avg**	**GP**	**W**	**L**	**Mins**	**GA**	**SO**	**Avg**
1997-98	Brantford	OMHA	15	13	2	0	660	20	*4	*1.36	....	....	....	....	....	....	....
1998-99	Brant County	OJHL-B	10	....	....	....	348	30	0	5.17	....	....	....	....	....	....	....
	Bowmanville	OPJHL	14	....	....	....	816	50	0	3.68	....	....	....	....	....	....	....
	Erie Otters	OHL	1	0	0	0	1	0	0	0.00	....	....	....	....	....	....	....
99-2000	Bowmanville	OPJHL	2	2	0	0	125	5	0	2.40	....	....	....	....	....	....	....
	Erie Otters	OHL	22	8	7	1	948	48	1	3.04	1	0	0	5	1	0	12.00
2000-01	Erie Otters	OHL	41	26	6	6	2283	88	*4	2.31	10	6	2	509	27	1	3.18
2001-02	Erie Otters	OHL	43	24	13	1	2277	128	3	3.37	6	4	2	361	17	0	2.83

MURPHY, Dan (MUHR-fee, DAN) PHI.

Goaltender. Catches left. 6'2", 191 lbs. Born, Nanaimo, B.C., May 6, 1974.

			Regular Season								Playoffs						
Season	**Club**	**Lea**	**GP**	**W**	**L**	**T**	**Mins**	**GA**	**SO**	**Avg**	**GP**	**W**	**L**	**Mins**	**GA**	**SO**	**Avg**
1993-94	Nanaimo Clippers	BCJHL	36	....	....	....	2123	190	0	5.31	....	....	....	....	....	....	....
1994-95	Clarkson Knights	ECAC	*37	*23	9	4	*2157	118	0	3.28	....	....	....	....	....	....	....
1995-96	Clarkson Knights	ECAC	*38	25	10	3	2230	100	0	2.69	....	....	....	....	....	....	....
1996-97	Clarkson Knights	ECAC	*37	*27	9	0	*2162	84	*4	2.33	....	....	....	....	....	....	....
1997-98	Clarkson Knights	ECAC	23	10	9	2	1266	48	2	*2.27	....	....	....	....	....	....	....
1998-99	Worcester	AHL	8	2	4	1	410	26	0	3.81	....	....	....	....	....	....	....
	Peoria Rivermen	ECHL	29	16	10	2	1672	92	0	3.30	3	1	2	180	11	0	3.67
99-2000	Que. Citadelles	AHL	33	16	9	1	1573	62	3	2.37	....	....	....	....	....	....	....
	Philadelphia	AHL	5	1	4	0	258	17	0	3.95	....	....	....	....	....	....	....
2000-01	Philadelphia	AHL	2	0	2	0	80	9	0	6.75	....	....	....	....	....	....	....
	Springfield	AHL	16	4	7	2	824	51	0	3.71	....	....	....	....	....	....	....
	Trenton Titans	ECHL	15	7	7	1	862	33	2	2.30	3	1	0	154	8	0	3.08
2001-02	Philadelphia	AHL	4	2	1	0	208	12	0	3.46	....	....	....	....	....	....	....
	Trenton Titans	ECHL	45	*30	10	4	2662	94	3	2.12	7	3	4	422	18	0	2.56

ECAC All-Rookie Team (1995) • ECAC Second All-Star Team (1996) • NCAA East Second All-American Team (1996, 1997) • ECHL Second All-Star Team (2002)

Signed as a free agent by **Philadelphia**, March 21, 2000.

NABOKOV, Evgeni (na-BAW-kahv, ehv-GEH-nee) S.J.

Goaltender. Catches left. 6', 200 lbs. Born, Ust-Kamenogorsk, USSR, July 25, 1975.
(San Jose's 9th choice, 219th overall, in 1994 Entry Draft).

			Regular Season								Playoffs						
Season	**Club**	**Lea**	**GP**	**W**	**L**	**T**	**Mins**	**GA**	**SO**	**Avg**	**GP**	**W**	**L**	**Mins**	**GA**	**SO**	**Avg**
1992-93	Ust-Kamenogorsk	CIS	4	1	0	0	109	5	0	2.75	....	....	....	....	....	....	....
1993-94	Ust-Kamenogorsk	CIS	11	....	....	....	539	29	0	3.23	....	....	....	....	....	....	....
1994-95	Dynamo Moscow	CIS	24	....	....	....	1265	40	....	1.90	13	....	....	810	30	....	2.22
1995-96	Dynamo Moscow	CIS	39	....	....	....	2008	67	5	2.00	6	....	....	298	7	....	1.41
1996-97	Dynamo Moscow	Russia	27	....	....	....	1588	56	2	2.12	4	....	....	255	12	0	2.82
1997-98	Kentucky	AHL	33	10	21	2	1866	122	0	3.92	1	0	0	23	1	0	2.59
1998-99	Kentucky	AHL	43	26	14	1	2429	106	5	2.62	11	6	5	599	30	*2	3.00
99-2000	**San Jose**	**NHL**	**11**	**2**	**2**	**1**	**414**	**15**	**1**	**2.17**	**1**	**0**	**0**	**20**	**0**	**0**	**0.00**
	Cleveland	IHL	20	12	4	3	1164	52	0	2.68	....	....	....	....	....	....	....
	Kentucky	AHL	2	1	1	0	120	3	1	1.50	....	....	....	....	....	....	....
2000-01	**San Jose**	**NHL**	**66**	**32**	**21**	**7**	**3700**	**135**	**6**	**2.19**	**4**	**1**	**3**	**218**	**10**	**1**	**2.75**
2001-02	**San Jose**	**NHL**	**67**	**37**	**24**	**5**	**3901**	**149**	**7**	**2.29**	**12**	**7**	**5**	**712**	**31**	**0**	**2.61**
	NHL Totals		**144**	**71**	**47**	**13**	**8015**	**299**	**14**	**2.24**	**17**	**8**	**8**	**950**	**41**	**1**	**2.59**

NHL All-Rookie Team (2001) • Won Calder Memorial Trophy (2001) • Played in NHL All-Star Game (2001)

• Scored a goal vs. Vancouver, March 10, 2002.

NAUMENKO, Gregg (naw-MEHN-koh, GREHG)

Goaltender. Catches left. 6'1", 201 lbs. Born, Chicago, IL, March 30, 1977.

			Regular Season								Playoffs						
Season	**Club**	**Lea**	**GP**	**W**	**L**	**T**	**Mins**	**GA**	**SO**	**Avg**	**GP**	**W**	**L**	**Mins**	**GA**	**SO**	**Avg**
1995-96	North Iowa	USHL	27	15	12	0	1649	103	1	3.75	4	1	3	239	15	0	3.77
1996-97	North Iowa	USHL	25	11	11	1	1342	85	1	3.80	6	3	2	284	19	0	4.01
1997-98	North Iowa	USHL	38	23	11	3	2171	80	3	2.21	5	4	1	299	11	0	2.21
1998-99	Alaska-Anchorage	WCHA	29	11	13	5	1691	65	1	*2.31	....	....	....	....	....	....	....
99-2000	Cincinnati	AHL	50	17	25	7	2877	143	2	2.98	....	....	....	....	....	....	....
2000-01	**Anaheim**	**NHL**	**2**	**0**	**1**	**0**	**70**	**7**	**0**	**6.00**	....	....	....	....	....	....	....
	Cincinnati	AHL	39	20	12	3	2079	101	2	2.91	2	0	2	123	10	0	4.90
2001-02	Cincinnati	AHL	7	2	4	0	364	15	0	2.47	....	....	....	....	....	....	....
	Augusta Lynx	ECHL	10	3	5	2	546	36	0	3.96	....	....	....	....	....	....	....
	Dayton Bombers	ECHL	23	14	3	5	1347	58	1	2.58	....	....	....	....	....	....	....
	NHL Totals		**2**	**0**	**1**	**0**	**70**	**7**	**0**	**6.00**	....	....	....	....	....	....	....

WCHA First All-Star Team (1999) • WCHA Rookie of the Year (1999)

Signed as a free agent by **Anaheim**, March 31, 1999.

NIITTYMAKI, Antero (NEE-too-mah-kee, AN-tehr-oh) PHI.

Goaltender. Catches left. 6', 176 lbs. Born, Turku, Finland, June 18, 1980.
(Philadelphia's 7th choice, 168th overall, in 1998 Entry Draft).

			Regular Season								Playoffs						
Season	**Club**	**Lea**	**GP**	**W**	**L**	**T**	**Mins**	**GA**	**SO**	**Avg**	**GP**	**W**	**L**	**Mins**	**GA**	**SO**	**Avg**
1998-99	TPS Turku Jr.	Finn-Jr.	35	....	....	....	2095	60	0	1.72	....	....	....	....	....	....	....
99-2000	TPS Turku Jr.	Finn-Jr.	1	1	0	0	60	1	0	1.00	....	....	....	....	....	....	....
	TPS Turku	Finland	32	23	6	2	1899	68	3	2.15	8	6	1	453	13	0	1.72
2000-01	TPS Turku	Finland	21	10	6	1	1112	46	2	2.48	....	....	....	....	....	....	....
2001-02	TPS Turku	Finland	27	16	8	1	1498	46	3	1.84	4	2	2	295	11	0	2.24

NISSINEN, Tuomas (NIHS-ih-nehn, too-OH-muhs) ST.L.

Goaltender. Catches left. 6'1", 176 lbs. Born, Kuopio, Finland, July 17, 1983.
(St. Louis' 2nd choice, 89th overall, in 2001 Entry Draft).

			Regular Season								Playoffs						
Season	**Club**	**Lea**	**GP**	**W**	**L**	**T**	**Mins**	**GA**	**SO**	**Avg**	**GP**	**W**	**L**	**Mins**	**GA**	**SO**	**Avg**
2000-01	KalPa Kuopio Jr.	Finn-Jr.	40	15	14	5	2327	125	2	3.22	1	0	1	60	4	0	4.00
2001-02	KalPa Kuopio Jr.	Finn-Jr.	33	20	11	2	1988	81	3	2.44	1	0	1	59	4	0	4.04
	Kalpa Kuopio	Finland-2	4	....	....	....	240	12	0	2.89	....	....	....	....	....	....	....

NORONEN, Mika (NOH-rah-nehn, MEE-kah) BUF.

Goaltender. Catches left. 6'2", 196 lbs. Born, Tampere, Finland, June 17, 1979.
(Buffalo's 1st choice, 21st overall, in 1997 Entry Draft).

			Regular Season								Playoffs						
Season	**Club**	**Lea**	**GP**	**W**	**L**	**T**	**Mins**	**GA**	**SO**	**Avg**	**GP**	**W**	**L**	**Mins**	**GA**	**SO**	**Avg**
1995-96	Tappara Jr.	Finn-Jr.	16	....	....	....	962	37	2	2.31	....	....	....	....	....	....	....
1996-97	Tappara Tampere	Finland	5	1	3	0	215	17	0	4.73	....	....	....	....	....	....	....
1997-98	Tappara Tampere	Finland	37	14	12	3	1704	83	1	2.92	4	1	2	196	12	0	3.67
1998-99	Tappara Tampere	Finland	43	18	20	5	2494	135	2	3.25	....	....	....	....	....	....	....
99-2000	Rochester	AHL	54	*33	13	4	3089	112	*6	2.18	21	13	8	1235	37	*6	*1.80
2000-01	Rochester	AHL	47	26	15	5	2753	100	4	2.18	4	1	3	250	11	0	2.64
	Buffalo	**NHL**	**2**	**2**	**0**	**0**	**108**	**5**	**0**	**2.78**	....	....	....	....	....	....	....
2001-02	**Buffalo**	**NHL**	**10**	**4**	**3**	**1**	**518**	**23**	**0**	**2.66**	....	....	....	....	....	....	....
	Rochester	AHL	45	16	17	12	2764	115	3	2.50	1	0	1	59	3	0	3.06
	NHL Totals		**12**	**6**	**3**	**1**	**626**	**28**	**0**	**2.68**	....	....	....	....	....	....	....

AHL Second All-Star Team (2000, 2001) • Won Dudley "Red" Garrett Memorial Trophy (Top Rookie - AHL) (2000) • Shared Harry "Hap" Holmes Memorial Trophy (fewest goals against - AHL) with Tom Askey (2001)

NURMINEN, Pasi (NUR-mih-nehn, PAS-ee) ATL.

Goaltender. Catches left. 5'10", 190 lbs. Born, Lahti, Finland, December 17, 1975.
(Atlanta's 6th choice, 189th overall, in 2001 Entry Draft).

			Regular Season								Playoffs						
Season	**Club**	**Lea**	**GP**	**W**	**L**	**T**	**Mins**	**GA**	**SO**	**Avg**	**GP**	**W**	**L**	**Mins**	**GA**	**SO**	**Avg**
1993-94	Reipas Lahti Jr.	Finn-Jr.	14	....	....	....	847	58	0	4.11	....	....	....	....	....	....	....
	Reipas Lahti	Finland	1	....	....	....	30	2	0	4.00	....	....	....	....	....	....	....
1994-95	Reipas Lahti Jr.	Finn-Jr.	9	....	....	....	542	22	....	2.44	....	....	....	....	....	....	....
	Reipas Lahti	Finland-2	9	....	....	....	423	44	0	6.24	....	....	....	....	....	....	....
1995-96	Kettera Imatra	Finland-2	38	....	....	....	2204	146	0	3.97	....	....	....	....	....	....	....
1996-97	Pelicans Lahti	Finland-2	30	....	....	....	1726	69	0	2.40	3	....	....	204	8	....	2.35
1997-98	Pelicans Lahti	Finland-2	35	....	....	....	3044	59	....	1.73	3	....	....	180	4	....	1.33
1998-99	HPK Hameenlinna	Finland	*48	24	17	6	*2810	127	2	2.71	7	3	4	425	24	1	3.39
99-2000	Jokerit Helsinki	Finland	48	24	15	8	2770	104	*6	2.25	*11	*7	4	*719	22	*2	1.84
2000-01	Jokerit Helsinki	Finland	52	*30	13	7	2971	107	5	2.16	5	2	3	308	11	1	2.14
2001-02	**Atlanta**	**NHL**	**9**	**2**	**5**	**0**	**465**	**28**	**0**	**3.61**	....	....	....	....	....	....	....
	Chicago Wolves	AHL	20	9	9	1	1165	57	2	2.93	*21	*15	5	1267	41	2	1.94
	Finland	Olympics	1	1	0	0	60	1	0	1.00	....	....	....	....	....	....	....
	NHL Totals		**9**	**2**	**5**	**0**	**465**	**28**	**0**	**3.61**	....	....	....	....	....	....	....

Won Jack A. Butterfield Trophy (Playoff MVP - AHL) (2002)

OSAER, Phil (OH-shar, FIHL) ST.L.

Goaltender. Catches left. 6'1", 186 lbs. Born, Dearborn, MI, February 10, 1980.
(St. Louis' 6th choice, 203rd overall, in 1999 Entry Draft).

			Regular Season								Playoffs						
Season	**Club**	**Lea**	**GP**	**W**	**L**	**T**	**Mins**	**GA**	**SO**	**Avg**	**GP**	**W**	**L**	**Mins**	**GA**	**SO**	**Avg**
1997-98	Waterloo	USHL	36	12	20	2	2094	107	2	3.07	5	1	4	295	17	0	3.46
1998-99	Ferris State	CCHA	9	2	2	1	399	10	0	1.51	....	....	....	....	....	....	....
99-2000	Ferris State	CCHA	25	13	8	2	1350	49	3	2.18	....	....	....	....	....	....	....
2000-01	Ferris State	CCHA	25	9	12	3	1449	57	3	2.36	....	....	....	....	....	....	....
2001-02	Peoria Rivermen	ECHL	29	16	11	2	1705	69	2	2.43	4	2	2	222	9	1	2.43

CCHA Second All-Star Team (2001)

OSGOOD, Chris (AWS-gud, KRIHS) NYI

Goaltender. Catches left. 5'10", 175 lbs. Born, Peace River, Alta., November 26, 1972.
(Detroit's 3rd choice, 54th overall, in 1991 Entry Draft).

			Regular Season								Playoffs						
Season	**Club**	**Lea**	**GP**	**W**	**L**	**T**	**Mins**	**GA**	**SO**	**Avg**	**GP**	**W**	**L**	**Mins**	**GA**	**SO**	**Avg**
1988-89	Medicine Hat	AMHL	26	....	....	....	1441	88	0	3.66	....	....	....	....	....	....	....
1989-90	Medicine Hat	WHL	57	24	28	2	3094	228	0	4.42	3	0	3	173	17	0	5.91
1990-91	Medicine Hat	WHL	46	23	18	3	2630	173	2	3.95	12	7	5	712	42	0	3.54
1991-92	Medicine Hat	WHL	15	10	3	0	819	44	0	3.22	....	....	....	....	....	....	....
	Brandon	WHL	16	3	10	1	890	60	1	4.04	....	....	....	....	....	....	....
	Seattle	WHL	21	12	7	1	1217	65	1	3.20	15	9	6	904	51	0	3.38
1992-93	Adirondack	AHL	45	19	19	4	2438	159	0	3.91	1	0	1	59	2	0	2.03
1993-94	**Detroit**	**NHL**	**41**	**23**	**8**	**5**	**2206**	**105**	**2**	**2.86**	**6**	**3**	**2**	**307**	**12**	**1**	**2.35**
	Adirondack	AHL	4	3	1	0	239	13	0	3.26	....	....	....	....	....	....	....
1994-95	**Detroit**	**NHL**	**19**	**14**	**5**	**0**	**1087**	**41**	**1**	**2.26**	**2**	**0**	**0**	**68**	**2**	**0**	**1.76**
	Adirondack	AHL	2	1	1	0	120	6	0	3.00	....	....	....	....	....	....	....
1995-96	**Detroit**	**NHL**	**50**	***39**	**6**	**5**	**2933**	**106**	**5**	**2.17**	**15**	**8**	**7**	**936**	**33**	**2**	**2.12**
1996-97 ♦	**Detroit**	**NHL**	**47**	**23**	**13**	**9**	**2769**	**106**	**6**	**2.30**	**2**	**0**	**0**	**47**	**2**	**0**	**2.55**
1997-98 ♦	**Detroit**	**NHL**	**64**	**33**	**20**	**11**	**3807**	**140**	**6**	**2.21**	***22**	***16**	**6**	***1361**	**48**	**2**	**2.12**
1998-99	**Detroit**	**NHL**	**63**	**34**	**25**	**4**	**3691**	**149**	**3**	**2.42**	**6**	**4**	**2**	**358**	**14**	**1**	**2.35**
99-2000	**Detroit**	**NHL**	**53**	**30**	**14**	**8**	**3148**	**126**	**6**	**2.40**	**9**	**5**	**4**	**547**	**18**	**2**	**1.97**
2000-01	**Detroit**	**NHL**	**52**	**25**	**19**	**4**	**2834**	**127**	**1**	**2.69**	**6**	**2**	**4**	**365**	**15**	**1**	**2.47**
2001-02	**NY Islanders**	**NHL**	**66**	**32**	**25**	**6**	**3743**	**156**	**4**	**2.50**	**7**	**3**	**4**	**392**	**17**	**0**	**2.60**
	NHL Totals		**455**	**253**	**135**	**52**	**26218**	**1056**	**34**	**2.42**	**75**	**41**	**29**	**4381**	**161**	**9**	**2.20**

WHL East Second All-Star Team (1991) • NHL Second All-Star Team (1996) • Shared William M. Jennings Trophy with Mike Vernon (1996) • Played in NHL All-Star Game (1996, 1997, 1998)

• Scored a goal while with Medicine Hat (WHL), January 3, 1991. • Scored a goal vs. Hartford, March 6, 1996. Claimed by **NY Islanders** from **Detroit** in Waiver Draft, September 28, 2001.

OUELLET, Maxime (OO-leht, MAX-eem) WSH.

Goaltender. Catches left. 6'2", 195 lbs. Born, Beauport, Que., June 17, 1981.
(Philadelphia's 1st choice, 22nd overall, in 1999 Entry Draft).

			Regular Season								Playoffs						
Season	**Club**	**Lea**	**GP**	**W**	**L**	**T**	**Mins**	**GA**	**SO**	**Avg**	**GP**	**W**	**L**	**Mins**	**GA**	**SO**	**Avg**
1996-97	Ste-Foy	QAAA	29	16	8	0	1470	81	0	2.75	9	4	5	555	31	0	3.37
1997-98	Quebec Remparts	QMJHL	24	12	7	1	1188	66	0	3.33	7	3	1	305	16	0	3.15
1998-99	Quebec Remparts	QMJHL	*59	*40	12	6	*3447	155	3	*2.70	13	6	7	803	41	*1	3.06
99-2000	Quebec Remparts	QMJHL	53	31	16	4	2984	133	2	2.67	11	7	4	638	28	*2	*2.63
2000-01	**Philadelphia**	**NHL**	**2**	**0**	**1**	**0**	**76**	**3**	**0**	**2.37**	....	....	....	....	....	....	....
	Philadelphia	AHL	2	1	0	0	86	4	0	2.78	....	....	....	....	....	....	....
	Rouyn-Noranda	QMJHL	25	18	6	1	1471	65	3	2.65	8	4	4	490	25	0	3.06
2001-02	Philadelphia	AHL	41	16	13	8	2294	104	1	2.72	....	....	....	....	....	....	....
	Portland Pirates	AHL	6	3	3	0	358	17	0	2.85	....	....	....	....	....	....	....
	NHL Totals		**2**	**0**	**1**	**0**	**76**	**3**	**0**	**2.37**	....	....	....	....	....	....	....

QMJHL Second All-Star Team (1999, 2000, 2001) • Won Jacques Plante Trophy (fewest goals against - QMJHL) (1999)

• Returned to **Rouyn-Noranda** (QMJHL) by **Philadelphia**, October 27, 2000. Traded to **Washington** by **Philadelphia** with Philadelphia's 1st (later traded to Dallas - Dallas selected Martin Vagner), 2nd (Maxime Daigneault) and 3rd (Derek Krestanovich) round choices in 2002 Entry Draft for Adam Oates, March 19, 2002.

PANNONI, Nick (puh-NOH-nee, NIHK) DET.

Goaltender. Catches left. 5'11", 155 lbs. Born, Cardston, Alta., November 6, 1982.
(Detroit's 5th choice, 195th overall, in 2001 Entry Draft).

			Regular Season								Playoffs						
Season	**Club**	**Lea**	**GP**	**W**	**L**	**T**	**Mins**	**GA**	**SO**	**Avg**	**GP**	**W**	**L**	**Mins**	**GA**	**SO**	**Avg**
99-2000	Team USA	USDP-18	35	24	8	2	2066	95	2	2.76	....	....	....	....	....	....	....
2000-01	Seattle	WHL	41	17	17	5	2165	150	0	4.16	7	3	4	426	18	0	2.54
2001-02	Seattle	WHL	40	13	21	2	2208	152	1	4.13	11	4	7	655	35	1	3.21

PASSMORE, Steve (PAS-mohr, STEEV) CHI.

Goaltender. Catches left. 5'9", 165 lbs. Born, Thunder Bay, Ont., January 29, 1973.
(Quebec's 10th choice, 196th overall, in 1992 Entry Draft).

			Regular Season								Playoffs						
Season	**Club**	**Lea**	**GP**	**W**	**L**	**T**	**Mins**	**GA**	**SO**	**Avg**	**GP**	**W**	**L**	**Mins**	**GA**	**SO**	**Avg**
1988-89	Tri-City	WHL	1	0	1	0	60	6	0	6.00	....	....	....	....	....	....	....
1989-90	West Island Deltas	BCAHA					STATISTICS NOT AVAILABLE										
	Tri-City	WHL	4	....	....	....	215	17	0	4.74	....	....	....	....	....	....	....
1990-91	Victoria Cougars	WHL	35	3	25	1	1838	190	0	6.20	....	....	....	....	....	....	....
1991-92	Victoria Cougars	WHL	*71	15	50	5	*4228	347	0	4.92	....	....	....	....	....	....	....
1992-93	Victoria Cougars	WHL	43	14	24	2	2402	150	1	3.75	....	....	....	....	....	....	....
	Kamloops Blazers	WHL	25	19	6	0	1479	69	1	2.80	7	4	2	401	22	1	3.29
1993-94	Kamloops Blazers	WHL	36	22	9	2	1927	88	1	*2.74	*18	*11	7	*1099	60	0	3.28
1994-95	Cape Breton	AHL	25	8	13	3	1455	93	0	3.83	....	....	....	....	....	....	....
1995-96	Cape Breton	AHL	2	1	0	0	90	2	0	1.33	....	....	....	....	....	....	....
1996-97	Hamilton	AHL	27	12	12	3	1568	70	1	2.68	22	12	10	1325	61	*2	2.76
	Raleigh IceCaps	ECHL	2	1	1	0	118	13	0	6.56	....	....	....	....	....	....	....
1997-98	San Antonio	IHL	14	3	8	2	736	56	0	4.56	....	....	....	....	....	....	....
	Hamilton	AHL	27	11	10	6	1655	87	2	3.15	3	0	2	132	14	0	6.33
1998-99	**Edmonton**	**NHL**	**6**	**1**	**4**	**1**	**362**	**17**	**0**	**2.82**	....	....	....	....	....	....	....
	Hamilton	AHL	54	24	21	7	3148	117	4	2.23	11	5	6	680	31	0	2.74
99-2000	**Chicago**	**NHL**	**24**	**7**	**12**	**3**	**1388**	**63**	**1**	**2.72**	....	....	....	....	....	....	....
	Cleveland	IHL	2	1	0	1	120	3	1	1.50	....	....	....	....	....	....	....
2000-01	**Los Angeles**	**NHL**	**14**	**3**	**8**	**1**	**718**	**37**	**1**	**3.09**	....	....	....	....	....	....	....
	Lowell	AHL	6	2	4	0	334	24	0	4.32	....	....	....	....	....	....	....
	Chicago	**NHL**	**6**	**0**	**4**	**1**	**340**	**14**	**0**	**2.47**	....	....	....	....	....	....	....
	Chicago Wolves	IHL	6	2	2	2	340	22	0	3.88	....	....	....	....	....	....	....
2001-02	**Chicago**	**NHL**	**23**	**8**	**5**	**4**	**1142**	**43**	**0**	**2.26**	**3**	**0**	**2**	**138**	**6**	**0**	**2.61**
	Norfolk Admirals	AHL	2	2	0	0	120	6	0	3.00	....	....	....	....	....	....	....
	NHL Totals		**73**	**19**	**33**	**10**	**3950**	**174**	**2**	**2.64**	**3**	**0**	**2**	**138**	**6**	**0**	**2.61**

WHL West First All-Star Team (1993, 1994) • Won Fred Hunt Memorial Trophy (Sportsmanship - AHL) (1997) • AHL Second All-Star Team (1999)

Traded to **Edmonton** by **Quebec** for Brad Werenka, March 21, 1994. • Missed majority of the 1995-96 season recovering from blood disorder, October, 1995. Signed as a free agent by **Chicago**, July 8, 1999. Traded to **LA Kings** by **Chicago** for LA Kings' 4th round choice (Olli Malmivaara) in 2000 Entry Draft, May 1, 2000. Traded to **Chicago** by **LA Kings** for Chicago's 8th round choice (Mike Gabinet) in 2001 Entry Draft, February 28, 2001.

PATZOLD, Dimitri (PATZ-ohld, dih-MEE-tree) S.J.

Goaltender. Catches left. 6', 183 lbs. Born, Ust-Kamenogorsk, USSR, February 3, 1983.
(San Jose's 3rd choice, 107th overall, in 2001 Entry Draft).

			Regular Season								Playoffs						
Season	**Club**	**Lea**	**GP**	**W**	**L**	**T**	**Mins**	**GA**	**SO**	**Avg**	**GP**	**W**	**L**	**Mins**	**GA**	**SO**	**Avg**
99-2000	Kolner Haie Jr.	Ger-Jr.	38	....	....	....	2131	73	0	2.06	....	....	....	....	....	....	....
	Kolner Haie-2	German5	16	....	....	....	896	58	0	3.88	....	....	....	....	....	....	....
2000-01	EV Duisberg	German2	6	....	....	....	360	17	0	2.83	....	....	....	....	....	....	....
	TSV Erding Jets	German2	24	....	....	....	1378	89	0	3.88	....	....	....	....	....	....	....
2001-02	Kolner Haie	Germany	7	....	....	....	260	16	0	3.69	....	....	....	....	....	....	....

PEARCE, Joseph (PEERS, JOH-sehf) T.B.

Goaltender. Catches left. 6'5", 215 lbs. Born, Point Pleasant, NJ, June 29, 1982.
(Tampa Bay's 3rd choice, 135th overall, in 2002 Entry Draft).

			Regular Season								Playoffs						
Season	**Club**	**Lea**	**GP**	**W**	**L**	**T**	**Mins**	**GA**	**SO**	**Avg**	**GP**	**W**	**L**	**Mins**	**GA**	**SO**	**Avg**
2000-01	Bismark Bobcats	AWJHL	17	....	....	....	1020	40	1	2.31	....	....	....	....	....	....	....
2001-02	N.H. Jr. Monarchs	EJHL	32	....	....	....	1885	57	1	1.82	....	....	....	....	....	....	....

EJHL First All-Star Team (2002)

• Signed Letter of Intent to attend **Boston College** (H-East), February 14, 2002.

PELLETIER, Jean-Marc (PEHL-tyay, ZHAWN-MAHRK) CAR.

Goaltender. Catches left. 6'3", 200 lbs. Born, Atlanta, GA, March 4, 1978.
(Philadelphia's 1st choice, 30th overall, in 1997 Entry Draft).

			Regular Season								Playoffs						
Season	**Club**	**Lea**	**GP**	**W**	**L**	**T**	**Mins**	**GA**	**SO**	**Avg**	**GP**	**W**	**L**	**Mins**	**GA**	**SO**	**Avg**
1993-94	Richelieu	QAAA	24	14	8	2	1440	91	0	3.79	2	1	0	104	11	0	6.32
1994-95	Richelieu	QAAA	21	15	6	0	1260	71	0	3.36	2	1	1	153	11	0	4.32
1995-96	Cornell Big Red	ECAC	5	1	2	0	179	15	0	5.03	....	....	....	....	....	....	....
1996-97	Cornell Big Red	ECAC	11	5	2	3	679	28	1	2.47	....	....	....	....	....	....	....
1997-98	Rimouski Oceanic	QMJHL	34	17	11	3	1913	118	0	3.70	16	11	3	895	51	1	3.42
1998-99	**Philadelphia**	**NHL**	**1**	**0**	**1**	**0**	**60**	**5**	**0**	**5.00**	....	....	....	....	....	....	....
	Philadelphia	AHL	47	25	16	4	2636	122	2	2.78	1	0	0	27	0	0	0.00
99-2000	Philadelphia	AHL	24	14	10	0	1405	58	3	2.48	....	....	....	....	....	....	....
	Cincinnati	IHL	22	14	4	2	1278	52	2	2.44	3	1	1	160	8	1	3.00
2000-01	Cincinnati	IHL	39	18	14	5	2261	119	2	3.16	5	1	4	318	15	0	2.83
2001-02	Lowell	AHL	40	21	12	4	2284	98	2	2.57	5	2	3	298	13	0	2.62
	NHL Totals		**1**	**0**	**1**	**0**	**60**	**5**	**0**	**5.00**	....	....	....	....	....	....	....

Traded to **Carolina** by **Philadelphia** with Rod Brind'Amour and Philadelphia's 2nd round choice (later traded to Colorado - Colorado selected Argis Saviels) in 2000 Entry Draft for Keith Primeau and Carolina's 5th round choice (later traded to NY Islanders - NY Islanders selected Kristofer Ottosson) in 2000 Entry Draft, January 23, 2000.

PENNER, Andrew (PEH-nuhr, AN-droo) CBJ

Goaltender. Catches left. 6'2", 205 lbs. Born, Scarborough, Ont., December 21, 1982.

			Regular Season								Playoffs						
Season	**Club**	**Lea**	**GP**	**W**	**L**	**T**	**Mins**	**GA**	**SO**	**Avg**	**GP**	**W**	**L**	**Mins**	**GA**	**SO**	**Avg**
1998-99	North York	OPJHL	26	....	....	....	1497	107	0	4.29	....	....	....	....	....	....	....
99-2000	North Bay	OHL	22	3	12	0	1070	79	0	4.43	....	....	....	....	....	....	....
2000-01	North Bay	OHL	32	10	19	1	1787	117	1	3.93	....	....	....	....	....	....	....
2001-02	North Bay	OHL	18	4	8	4	917	52	1	3.40	....	....	....	....	....	....	....
	Guelph Storm	OHL	36	18	12	5	2066	107	0	3.11	9	5	4	546	29	0	3.19

Traded to **Guelph** (OHL) by **North Bay** (OHL) with North Bay's 4th round choice (Mark Verstegg-Lytwyn) in 2002 OHL Priority Draft for Colt King and Jeremy Day, November 22, 2001. Signed as a free agent by **Columbus**, September 17, 2001.

PETRUK, Randy (PEHT-ruhk, RAN-dee) CAR.

Goaltender. Catches right. 5'9", 175 lbs. Born, Cranbrook, B.C., April 23, 1978.
(Colorado's 5th choice, 107th overall, in 1996 Entry Draft).

			Regular Season								Playoffs						
Season	**Club**	**Lea**	**GP**	**W**	**L**	**T**	**Mins**	**GA**	**SO**	**Avg**	**GP**	**W**	**L**	**Mins**	**GA**	**SO**	**Avg**
1993-94	Cranbrook Colts	RMJHL	21	....	....	....	1158	89	0	4.61	....	....	....	....	....	....	....
1994-95	Kamloops Blazers	WHL	27	16	3	4	1462	71	1	2.91	7	5	2	423	19	0	2.70
1995-96	Kamloops Blazers	WHL	52	34	15	1	3071	181	1	3.54	16	9	6	990	58	0	3.52
1996-97	Kamloops Blazers	WHL	*60	25	28	5	*3475	210	0	3.63	....	....	....	....	....	....	....
1997-98	Kamloops Blazers	WHL	57	31	21	1	3097	157	3	3.04	7	3	4	425	21	0	2.96
1998-99	Florida	ECHL	25	13	10	2	1441	66	1	2.75	1	0	1	60	5	0	5.00
	New Haven	AHL	1	0	0	1	65	3	0	2.77	....	....	....	....	....	....	....
99-2000	Florida	ECHL	6	5	0	1	339	19	0	3.36	....	....	....	....	....	....	....
	Cincinnati	IHL	26	13	9	3	1436	84	2	3.51	9	4	5	551	27	1	2.94
2000-01	Cincinnati	IHL	8	4	3	0	420	23	1	3.29	....	....	....	....	....	....	....
	Florida	ECHL	13	5	7	1	742	41	0	3.31	....	....	....	....	....	....	....
2001-02	Florida	ECHL	51	27	18	5	3087	140	3	2.72	5	1	3	259	14	0	3.24

WHL West Second All-Star Team (1998)

Traded to **Carolina** by **Colorado** for Carolina's 5th round choice (Will Magnuson) in 1999 Entry Draft, June 1, 1998.

POTVIN, Felix (PAHT-vihn, FEEL-ihx) L.A.

Goaltender. Catches left. 6'1", 190 lbs. Born, Anjou, Que., June 23, 1971.
(Toronto's 2nd choice, 31st overall, in 1990 Entry Draft).

			Regular Season								Playoffs						
Season	**Club**	**Lea**	**GP**	**W**	**L**	**T**	**Mins**	**GA**	**SO**	**Avg**	**GP**	**W**	**L**	**Mins**	**GA**	**SO**	**Avg**
1987-88	Mtl-Bourassa	QAAA	27	15	7	3	1585	103	3	3.90	6	2	4	341	20	0	3.51
1988-89	Chicoutimi	QMJHL	*65	25	31	1	*3489	271	*2	4.66	....	....	....	....	....	....	....
1989-90	Chicoutimi	QMJHL	*62	*31	26	2	*3478	231	*2	3.99	....	....	....	....	....	....	....
1990-91	Chicoutimi	QMJHL	54	33	15	4	3216	145	*6	*2.70	*16	*11	5	*992	46	0	*2.78
1991-92	**Toronto**	**NHL**	**4**	**0**	**2**	**1**	**210**	**8**	**0**	**2.29**	....	....	....	....	....	....	....
	St. John's	AHL	35	18	10	6	2070	101	2	2.93	11	7	4	642	41	0	3.83
1992-93	**Toronto**	**NHL**	**48**	**25**	**15**	**7**	**2781**	**116**	**2**	***2.50**	***21**	**11**	**10**	***1308**	**62**	**1**	**2.84**
	St. John's	AHL	5	3	0	2	309	18	0	3.50	....	....	....	....	....	....	....
1993-94	**Toronto**	**NHL**	**66**	**34**	**22**	**9**	**3883**	**187**	**3**	**2.89**	**18**	**9**	**9**	**1124**	**46**	**3**	**2.46**
1994-95	**Toronto**	**NHL**	**36**	**15**	**13**	**7**	**2144**	**104**	**0**	**2.91**	**7**	**3**	**4**	**424**	**20**	**1**	**2.83**
1995-96	**Toronto**	**NHL**	**69**	**30**	**26**	**11**	**4009**	**192**	**2**	**2.87**	**6**	**2**	**4**	**350**	**19**	**0**	**3.26**
1996-97	**Toronto**	**NHL**	***74**	**27**	**36**	**7**	***4271**	**224**	**0**	**3.15**	....	....	....	....	....	....	....
1997-98	**Toronto**	**NHL**	**67**	**26**	**33**	**7**	**3864**	**176**	**5**	**2.73**	....	....	....	....	....	....	....
1998-99	**Toronto**	**NHL**	**5**	**3**	**2**	**0**	**299**	**19**	**0**	**3.81**	....	....	....	....	....	....	....
	NY Islanders	**NHL**	**11**	**2**	**7**	**1**	**606**	**37**	**0**	**3.66**	....	....	....	....	....	....	....
99-2000	**NY Islanders**	**NHL**	**22**	**5**	**14**	**3**	**1273**	**68**	**1**	**3.21**	....	....	....	....	....	....	....
	Vancouver	**NHL**	**34**	**12**	**13**	**7**	**1966**	**85**	**0**	**2.59**	....	....	....	....	....	....	....
2000-01	**Vancouver**	**NHL**	**35**	**14**	**17**	**3**	**2006**	**103**	**1**	**3.08**	....	....	....	....	....	....	....
	Los Angeles	**NHL**	**23**	**13**	**5**	**5**	**1410**	**46**	**5**	**1.96**	**13**	**7**	**6**	**812**	**33**	**2**	**2.44**
2001-02	**Los Angeles**	**NHL**	**71**	**31**	**27**	**8**	**4071**	**157**	**6**	**2.31**	**7**	**3**	**4**	**417**	**15**	**1**	**2.16**
	NHL Totals		**565**	**237**	**232**	**76**	**32793**	**1522**	**25**	**2.78**	**72**	**35**	**37**	**4435**	**195**	**8**	**2.64**

QMJHL All-Rookie Team (1989) • QMJHL Second All-Star Team (1990) • QMJHL First All-Star Team (1991) • Canadian Major Junior Goaltender of the Year (1991) • Memorial Cup All-Star Team (1991) • Won Hap Emms Memorial Trophy (Memorial Cup Tournament Top Goaltender) (1991) • AHL First All-Star Team (1992) • Won Dudley "Red" Garrett Memorial Trophy (Top Rookie - AHL) (1992) • Won Baz Bastien Memorial Trophy (Top Goaltender - AHL) (1992) • NHL All-Rookie Team (1993) • Played in NHL All-Star Game (1994, 1996)

Traded to **NY Islanders** by **Toronto** with Toronto's 6th round choice (later traded to Tampa Bay - Tampa Bay selected Fedor Fedorov) in 1999 Entry Draft for Bryan Berard and NY Islanders' 6th round choice (Jan Sochor) in 1999 Entry Draft, January 9, 1999. Traded to **Vancouver** by **NY Islanders** with NY Islanders' compensatory 2nd (later traded to New Jersey - New Jersey selected Teemu Laine) and 3rd (Thatcher Bell) round choices in 2000 Entry Draft for Kevin Weekes, Dave Scatchard and Bill Muckalt, December 19, 1999. Traded to **LA Kings** by **Vancouver** for future considerations, February 15, 2001.

PRUSEK, Martin (PREW-sehk, MAHR-tihn) OTT.

Goaltender. Catches left. 6'1", 176 lbs. Born, Ostrava, Czech., December 11, 1975.
(Ottawa's 6th choice, 164th overall, in 1999 Entry Draft).

			Regular Season								Playoffs						
Season	**Club**	**Lea**	**GP**	**W**	**L**	**T**	**Mins**	**GA**	**SO**	**Avg**	**GP**	**W**	**L**	**Mins**	**GA**	**SO**	**Avg**
1994-95	HC Vitkovice	Czech	5	....	....	....	232	18	....	4.65	....	....	....	....	....	....	....
1995-96	HC Vitkovice	Czech	40	....	....	....	2336	113	1	2.90	4	....	....	250	10	1	2.40
1996-97	HC Vitkovice	Czech	49	....	....	....	2841	109	8	2.30	9	....	....	546	19	1	2.08
1997-98	HC Vitkovice	Czech	50	....	....	....	2901	129	....	2.67	9	....	....	529	26	....	3.00
1998-99	HC Vitkovice	Czech	37	....	....	....	1905	85	....	2.68	4	....	....	250	12	....	2.88
99-2000	HC Vitkovice	Czech	50	....	....	....	2647	132	....	2.99	....	....	....	....	....	....	....
2000-01	HC Vitkovice	Czech	30	....	....	....	1679	64	....	2.29	9	....	....	460	25	....	3.26
2001-02	**Ottawa**	**NHL**	**1**	**0**	**1**	**0**	**62**	**3**	**0**	**2.90**	....	....	....	....	....	....	....
	Grand Rapids	AHL	33	18	8	5	1903	58	4	*1.83	5	2	3	278	10	0	2.16
	NHL Totals		**1**	**0**	**1**	**0**	**62**	**3**	**0**	**2.90**	....	....	....	....	....	....	....

AHL First All-Star Team (2002) • Shared Harry "Hap" Holmes Memorial Trophy (fewest goals against - AHL) with Simon Lajeunesse and Mathieu Chouinard (2002) • Won Baz Bastien Memorial Trophy (Top Goaltender - AHL) (2002)

PUURULA, Joni (pu-u-ROO-luh, YOHN-ee) MTL.

Goaltender. Catches left. 5'11", 180 lbs. Born, Kokkola, Finland, August 4, 1982.
(Montreal's 10th choice, 243rd overall, in 2000 Entry Draft).

			Regular Season								Playoffs						
Season	**Club**	**Lea**	**GP**	**W**	**L**	**T**	**Mins**	**GA**	**SO**	**Avg**	**GP**	**W**	**L**	**Mins**	**GA**	**SO**	**Avg**
1998-99	Junkkarit	Finland-2	12	....	....	....	782	37	0	2.84	....	....	....	....	....	....	....
99-2000	Hermes Kokkola	Finland-2	23	8	12	2	1251	81	1	3.88	....	....	....	....	....	....	....
2000-01	FoPS Forssa	Finland-2	39	....	....	....	2263	142	1	3.76	....	....	....	....	....	....	....
	FoPS Forssa Jr.	Finn-Jr.	....	....	....	....	....	....	....	....	4	1	3	240	8	0	2.00
2001-02	HPK Jr.	Finn-Jr.	2	1	1	0	120	11	0	5.52	....	....	....	....	....	....	....
	FPS Fossa	Finland-2	5	....	....	....	300	18	0	3.54	....	....	....	....	....	....	....
	HPK Hameenlinna	Finland	9	8	0	1	515	18	0	2.10	8	4	3	453	13	0	1.72

RACINE, Jean-Francois (RAY-seen, ZHAWN-fran-SWUH) TOR.

Goaltender. Catches left. 6'3", 183 lbs. Born, St-Hyacinthe, Que., April 27, 1982.
(Toronto's 4th choice, 90th overall, in 2000 Entry Draft).

			Regular Season								Playoffs						
Season	**Club**	**Lea**	**GP**	**W**	**L**	**T**	**Mins**	**GA**	**SO**	**Avg**	**GP**	**W**	**L**	**Mins**	**GA**	**SO**	**Avg**
1998-99	Magog	QAAA	36	19	12	1	2160	107	3	2.98	11	5	6	656	37	0	3.39
99-2000	Moncton Wildcats	QMJHL	10	3	3	1	410	28	0	4.10	....	....	....	....	....	....	....
	Drummondville	QMJHL	20	14	6	0	1152	63	1	3.28	3	0	0	65	5	0	4.60
2000-01	Drummondville	QMJHL	61	27	26	3	3362	189	4	3.37	5	2	3	303	20	0	3.97
2001-02	Drummondville	QMJHL	65	29	30	3	3640	208	2	3.43	12	5	7	720	42	1	3.50

RAYCROFT, Andrew (RAY-krawft, AN-droo) BOS.

Goaltender. Catches left. 6', 174 lbs. Born, Belleville, Ont., May 4, 1980.
(Boston's 4th choice, 135th overall, in 1998 Entry Draft).

			Regular Season								Playoffs						
Season	**Club**	**Lea**	**GP**	**W**	**L**	**T**	**Mins**	**GA**	**SO**	**Avg**	**GP**	**W**	**L**	**Mins**	**GA**	**SO**	**Avg**
1996-97	Wellington Dukes	MTJHL	27	....	....	....	1402	92	0	3.94	....	....	....	....	....	....	....
1997-98	Sudbury Wolves	OHL	33	8	16	5	1802	125	0	4.16	2	0	1	89	8	0	5.39
1998-99	Sudbury Wolves	OHL	45	17	22	5	2528	173	1	4.11	3	0	2	96	13	0	8.13
99-2000	Kingston	OHL	*61	33	20	5	3340	191	0	3.43	5	1	4	300	21	0	4.20
2000-01	**Boston**	**NHL**	**15**	**4**	**6**	**0**	**649**	**32**	**0**	**2.96**	....	....	....	....	....	....	....
	Providence	AHL	26	8	14	4	1459	82	1	3.37	....	....	....	....	....	....	....
2001-02	**Boston**	**NHL**	**1**	**0**	**0**	**1**	**65**	**3**	**0**	**2.77**	....	....	....	....	....	....	....
	Providence	AHL	56	25	24	6	3317	142	4	2.57	2	0	2	119	5	0	2.52
	NHL Totals		**16**	**4**	**6**	**1**	**714**	**35**	**0**	**2.94**	....	....	....	....	....	....	....

OHL First All-Star Team (2000) • Canadian Major Junior First All-Star Team (2000) • Canadian Major Junior Goaltender of the Year (2000)

RHODES, Damian (ROHDZ, DAY-mee-uhn)

Goaltender. Catches left. 5'11", 195 lbs. Born, St. Paul, MN, May 28, 1969.
(Toronto's 6th choice, 112th overall, in 1987 Entry Draft).

			Regular Season								Playoffs						
Season	**Club**	**Lea**	**GP**	**W**	**L**	**T**	**Mins**	**GA**	**SO**	**Avg**	**GP**	**W**	**L**	**Mins**	**GA**	**SO**	**Avg**
1985-86	Richfield High	H.S.	16	....	....	....	720	56	0	3.50	....	....	....	....	....	....	....
1986-87	Richfield High	H.S.	19	....	....	....	673	51	1	4.55	....	....	....	....	....	....	....
1987-88	Michigan Tech	WCHA	29	16	10	1	1625	114	0	4.20	....	....	....	....	....	....	....
1988-89	Michigan Tech	WCHA	37	15	22	0	2216	163	0	4.41	....	....	....	....	....	....	....
1989-90	Michigan Tech	WCHA	25	6	17	0	1358	119	0	6.26	....	....	....	....	....	....	....
1990-91	**Toronto**	**NHL**	**1**	**1**	**0**	**0**	**60**	**1**	**0**	**1.00**	....	....	....	....	....	....	....
	Newmarket Saints	AHL	38	8	24	3	2154	144	1	4.01	....	....	....	....	....	....	....
1991-92	St. John's	AHL	43	20	16	5	2454	148	0	3.62	6	4	1	331	16	0	2.90
1992-93	St. John's	AHL	*52	27	16	8	*3074	184	1	3.59	9	4	5	538	37	0	4.13
1993-94	**Toronto**	**NHL**	**22**	**9**	**7**	**3**	**1213**	**53**	**0**	**2.62**	**1**	**0**	**0**	**1**	**0**	**0**	**0.00**
1994-95	**Toronto**	**NHL**	**13**	**6**	**6**	**1**	**760**	**34**	**0**	**2.68**	....	....	....	....	....	....	....
1995-96	**Toronto**	**NHL**	**11**	**4**	**5**	**1**	**624**	**29**	**0**	**2.79**	....	....	....	....	....	....	....
	Ottawa	**NHL**	**36**	**10**	**22**	**4**	**2123**	**98**	**2**	**2.77**	....	....	....	....	....	....	....
1996-97	**Ottawa**	**NHL**	**50**	**14**	**20**	**14**	**2934**	**133**	**1**	**2.72**	....	....	....	....	....	....	....
1997-98	**Ottawa**	**NHL**	**50**	**19**	**19**	**7**	**2743**	**107**	**5**	**2.34**	**10**	**5**	**5**	**590**	**21**	**0**	**2.14**
1998-99	**Ottawa**	**NHL**	**45**	**22**	**13**	**7**	**2480**	**101**	**3**	**2.44**	**2**	**0**	**2**	**150**	**6**	**0**	**2.40**
99-2000	**Atlanta**	**NHL**	**28**	**5**	**19**	**3**	**1561**	**101**	**1**	**3.88**	....	....	....	....	....	....	....
2000-01	**Atlanta**	**NHL**	**38**	**7**	**19**	**7**	**2072**	**116**	**0**	**3.36**	....	....	....	....	....	....	....
2001-02	**Atlanta**	**NHL**	**15**	**2**	**10**	**1**	**769**	**47**	**0**	**3.67**	....	....	....	....	....	....	....
	NHL Totals		**309**	**99**	**140**	**48**	**17339**	**820**	**12**	**2.84**	**13**	**5**	**7**	**741**	**27**	**0**	**2.19**

• Credited with scoring a goal while with Michigan Tech (WCHA), January 21, 1989. • Played 10 seconds of playoff game vs. San Jose, May 6, 1994. • Credited with scoring a goal vs. New Jersey, January 2, 1999.

Traded to **NY Islanders** by **Toronto** with Ken Belanger for future considerations (Kirk Muller and Don Beaupre, January 23, 1996), January 23, 1996. Traded to **Ottawa** by **NY Islanders** with Wade Redden for Don Beaupre, Martin Straka and Bryan Berard, January 23, 1996. Traded to **Atlanta** by **Ottawa** for future considerations, June 18, 1999.

RICHTER, Mike (RIHK-tuhr, MIGHK) NYR

Goaltender. Catches left. 5'11", 185 lbs. Born, Abington, PA, September 22, 1966.
(NY Rangers' 2nd choice, 28th overall, in 1985 Entry Draft).

			Regular Season								Playoffs						
Season	**Club**	**Lea**	**GP**	**W**	**L**	**T**	**Mins**	**GA**	**SO**	**Avg**	**GP**	**W**	**L**	**Mins**	**GA**	**SO**	**Avg**
1983-84	Philadelphia	NEJHL	36	23	10	3	2160	94	0	2.61	....	....	....	....	....	....	....
1984-85	Northwood Prep	H.S.	24	....	....	....	1374	52	2	2.27	....	....	....	....	....	....	....
1985-86	U. of Wisconsin	WCHA	24	14	9	0	1394	92	1	3.96	....	....	....	....	....	....	....
1986-87	U. of Wisconsin	WCHA	36	19	16	1	2136	126	0	3.54	....	....	....	....	....	....	....
1987-88	United States	Nat-Tm	29	17	7	2	1559	86	0	3.31	....	....	....	....	....	....	....
	United States	Olympics	4	2	2	0	230	15	0	3.91	....	....	....	....	....	....	....
	Colorado Rangers	IHL	22	16	5	0	1298	68	1	3.14	10	5	3	536	35	0	3.92
1988-89	Denver Rangers	IHL	*57	23	26	0	3031	217	1	4.30	4	0	4	210	21	0	6.00
	NY Rangers	**NHL**	....	....	....	....	....	....	....	....	**1**	**0**	**1**	**58**	**4**	**0**	**4.14**
1989-90	**NY Rangers**	**NHL**	**23**	**12**	**5**	**5**	**1320**	**66**	**0**	**3.00**	**6**	**3**	**2**	**330**	**19**	**0**	**3.45**
	Flint Spirits	IHL	13	7	4	2	782	49	0	3.76	....	....	....	....	....	....	....
1990-91	**NY Rangers**	**NHL**	**45**	**21**	**13**	**7**	**2596**	**135**	**0**	**3.12**	**6**	**2**	**4**	**313**	**14**	***1**	**2.68**
1991-92	**NY Rangers**	**NHL**	**41**	**23**	**12**	**2**	**2298**	**119**	**3**	**3.11**	**7**	**4**	**2**	**412**	**24**	**1**	**3.50**
1992-93	**NY Rangers**	**NHL**	**38**	**13**	**19**	**3**	**2105**	**134**	**1**	**3.82**	....	....	....	....	....	....	....
	Binghamton	AHL	5	4	0	1	305	6	0	1.18	....	....	....	....	....	....	....
1993-94♦	**NY Rangers**	**NHL**	**68**	***42**	**12**	**6**	**3710**	**159**	**5**	**2.57**	**23**	***16**	**7**	**1417**	**49**	***4**	**2.07**
1994-95	**NY Rangers**	**NHL**	**35**	**14**	**17**	**2**	**1993**	**97**	**2**	**2.92**	**7**	**2**	**5**	**384**	**23**	**0**	**3.59**
1995-96	**NY Rangers**	**NHL**	**41**	**24**	**13**	**3**	**2396**	**107**	**3**	**2.68**	**11**	**5**	**6**	**661**	**36**	**0**	**3.27**
1996-97	**NY Rangers**	**NHL**	**61**	**33**	**22**	**6**	**3598**	**161**	**4**	**2.68**	**15**	**9**	**6**	**939**	**33**	***3**	**2.11**
1997-98	**NY Rangers**	**NHL**	***72**	**21**	**31**	**15**	**4143**	**184**	**0**	**2.66**	....	....	....	....	....	....	....
	United States	Olympics	4	1	3	0	237	14	0	3.54	....	....	....	....	....	....	....
1998-99	**NY Rangers**	**NHL**	**68**	**27**	**30**	**8**	**3878**	**170**	**4**	**2.63**	....	....	....	....	....	....	....
99-2000	**NY Rangers**	**NHL**	**61**	**22**	**31**	**8**	**3622**	**173**	**0**	**2.87**	....	....	....	....	....	....	....
2000-01	**NY Rangers**	**NHL**	**45**	**20**	**21**	**3**	**2635**	**144**	**0**	**3.28**	....	....	....	....	....	....	....
2001-02	**NY Rangers**	**NHL**	**55**	**24**	**26**	**4**	**3195**	**157**	**2**	**2.95**	....	....	....	....	....	....	....
	United States	Olympics	4	2	1	1	240	9	*1	2.25	....	....	....	....	....	....	....
	NHL Totals		**653**	**296**	**252**	**72**	**37489**	**1806**	**24**	**2.89**	**76**	**41**	**33**	**4514**	**202**	**9**	**2.68**

WCHA Freshman of the Year (1986) • WCHA Second All-Star Team (1987) • Played in NHL All-Star Game (1992, 1994, 2000)

Claimed by **Nashville** from **NY Rangers** in Expansion Draft, June 26, 1998. Signed as a free agent by **NY Rangers**, July 15, 1998. Traded to **Edmonton** by **NY Rangers** for future considerations, June 30, 2002. Signed as a free agent by **NY Rangers**, July 4, 2002.

ROLOSON, Dwayne (ROH-loh-suhn, DWAYN) MIN.

Goaltender. Catches left. 6'1", 178 lbs. Born, Simcoe, Ont., October 12, 1969.

			Regular Season								Playoffs						
Season	**Club**	**Lea**	**GP**	**W**	**L**	**T**	**Mins**	**GA**	**SO**	**Avg**	**GP**	**W**	**L**	**Mins**	**GA**	**SO**	**Avg**
1984-85	Simcoe Penguins	OJHL-C	3	....	....	....	100	21	0	12.60	....	....	....	....	....	....	....
1985-86	Simcoe Rams	OJHL-C	1	....	....	....	60	6	0	6.00	....	....	....	....	....	....	....
1986-87	Norwich	OJHL-C	19	....	....	....	1091	55	0	*3.03	....	....	....	....	....	....	....
1987-88	Belleville	OJHL-B	21	9	6	1	1070	60	*2	3.36	....	....	....	....	....	....	....
1988-89	Thorold	OJHL-B	27	15	6	4	1490	82	0	3.30	....	....	....	....	....	....	....
1989-90	Thorold	OJHL-B	30	18	8	1	1683	108	0	3.85	....	....	....	....	....	....	....
1990-91	U. Mass-Lowell	H-East	15	5	9	0	823	63	0	4.59	....	....	....	....	....	....	....
1991-92	U. Mass-Lowell	H-East	12	3	8	0	660	52	0	4.73	....	....	....	....	....	....	....
1992-93	U. Mass-Lowell	H-East	*39	20	17	2	*2342	150	0	3.84	....	....	....	....	....	....	....
1993-94	U. Mass-Lowell	H-East	*40	*23	10	7	*2305	106	0	2.76	....	....	....	....	....	....	....
1994-95	Saint John	AHL	46	16	21	8	2734	156	1	3.42	5	1	4	298	13	0	2.61
1995-96	Saint John	AHL	67	*33	22	11	4026	190	1	2.83	16	10	6	1027	49	1	2.86
1996-97	**Calgary**	**NHL**	**31**	**9**	**14**	**3**	**1618**	**78**	**1**	**2.89**	....	....	....	....	....	....	....
	Saint John	AHL	8	6	2	0	481	22	1	2.75	....	....	....	....	....	....	....
1997-98	**Calgary**	**NHL**	**39**	**11**	**16**	**8**	**2205**	**110**	**0**	**2.99**	....	....	....	....	....	....	....
	Saint John	AHL	4	3	0	1	245	8	0	1.96	....	....	....	....	....	....	....
1998-99	**Buffalo**	**NHL**	**18**	**6**	**8**	**2**	**911**	**42**	**1**	**2.77**	**4**	**1**	**1**	**139**	**10**	**0**	**4.32**
	Rochester	AHL	2	2	0	0	120	4	0	2.00	....	....	....	....	....	....	....
99-2000	**Buffalo**	**NHL**	**14**	**1**	**7**	**3**	**677**	**32**	**0**	**2.84**	....	....	....	....	....	....	....
2000-01	Worcester	AHL	52	*32	15	5	*3127	113	*6	*2.17	11	6	5	697	23	1	1.98
2001-02	**Minnesota**	**NHL**	**45**	**14**	**20**	**7**	**2506**	**112**	**5**	**2.68**	....	....	....	....	....	....	....
	NHL Totals		**147**	**41**	**65**	**23**	**7917**	**374**	**7**	**2.83**	**4**	**1**	**1**	**139**	**10**	**0**	**4.32**

Hockey East First All-Star Team (1994) • Hockey East Player of the Year (1994) • NCAA East First All-Ameircan Team (1994) • AHL First All-Star Team (2001) • Won Baz Bastien Memorial Trophy (Top Goaltender - AHL) (2001)

Signed as a free agent by **Calgary**, July 4, 1994. Signed as a free agent by **Buffalo**, July 15, 1998. Selected by **Columbus** from **Buffalo** in Expansion Draft, June 23, 2000. Signed as a free agent by **St. Louis**, July 14, 2000. Signed as a free agent by **Minnesota**, July 2, 2001.

ROY, Patrick (WAH, PAT-rihk) COL.

Goaltender. Catches left. 6'2", 185 lbs. Born, Quebec, Que., October 5, 1965.
(Montreal's 4th choice, 51st overall, in 1984 Entry Draft).

			Regular Season								Playoffs						
Season	**Club**	**Lea**	**GP**	**W**	**L**	**T**	**Mins**	**GA**	**SO**	**Avg**	**GP**	**W**	**L**	**Mins**	**GA**	**SO**	**Avg**
1981-82	Ste-Foy	QAAA	40	*27	3	10	2400	156	*3	*2.63	2	2	0	114	2	*1	1.05
1982-83	Granby Bisons	QMJHL	54	13	35	1	2808	293	0	6.26	....	....	....	....	....	....	....
1983-84	Granby Bisons	QMJHL	61	29	29	1	3585	265	0	4.44	4	0	4	244	22	0	5.41
1984-85	Granby Bisons	QMJHL	44	16	25	1	2463	228	0	5.55	....	....	....	....	....	....	....
	Montreal	**NHL**	**1**	**1**	**0**	**0**	**20**	**0**	**0**	**0.00**	....	....	....	....	....	....	....
	Sherbrooke	AHL	1	1	0	0	60	4	0	4.00	13	10	3	*769	37	0	*2.89
1985-86♦	**Montreal**	**NHL**	**47**	**23**	**18**	**3**	**2651**	**148**	**1**	**3.35**	**20**	***15**	**5**	**1218**	**39**	***1**	**1.92**
1986-87	**Montreal**	**NHL**	**46**	**22**	**16**	**6**	**2686**	**131**	**1**	**2.93**	**6**	**4**	**2**	**330**	**22**	**0**	**4.00**
1987-88	**Montreal**	**NHL**	**45**	**23**	**12**	**9**	**2586**	**125**	**3**	**2.90**	**8**	**3**	**4**	**430**	**24**	**0**	**3.35**
1988-89	**Montreal**	**NHL**	**48**	**33**	**5**	**6**	**2744**	**113**	**4**	***2.47**	**19**	**13**	**6**	**1206**	**42**	**2**	***2.09**
1989-90	**Montreal**	**NHL**	**54**	***31**	**16**	**5**	**3173**	**134**	**3**	**2.53**	**11**	**5**	**6**	**641**	**26**	**1**	**2.43**
1990-91	**Montreal**	**NHL**	**48**	**25**	**15**	**6**	**2835**	**128**	**1**	**2.71**	**13**	**7**	**5**	**785**	**40**	**0**	**3.06**
1991-92	**Montreal**	**NHL**	**67**	**36**	**22**	**8**	**3935**	**155**	***5**	***2.36**	**11**	**4**	**7**	**686**	**30**	**1**	**2.62**
1992-93♦	**Montreal**	**NHL**	**62**	**31**	**25**	**5**	**3595**	**192**	**2**	**3.20**	**20**	***16**	**4**	**1293**	**46**	**0**	***2.13**
1993-94	**Montreal**	**NHL**	**68**	**35**	**17**	**11**	**3867**	**161**	***7**	**2.50**	**6**	**3**	**3**	**375**	**16**	**0**	**2.56**
1994-95	**Montreal**	**NHL**	**43**	**17**	**20**	**6**	**2566**	**127**	**1**	**2.97**	....	....	....	....	....	....	....
1995-96	**Montreal**	**NHL**	**22**	**12**	**9**	**1**	**1260**	**62**	**1**	**2.95**	....	....	....	....	....	....	....
♦	**Colorado**	**NHL**	**39**	**22**	**15**	**1**	**2305**	**103**	**1**	**2.68**	***22**	***16**	**6**	***1454**	**51**	***3**	**2.10**
1996-97	**Colorado**	**NHL**	**62**	***38**	**15**	**7**	**3698**	**143**	**7**	**2.32**	**17**	**10**	**7**	**1034**	**38**	***3**	**2.21**
1997-98	**Colorado**	**NHL**	**65**	**31**	**19**	**13**	**3835**	**153**	**4**	**2.39**	**7**	**3**	**4**	**430**	**18**	**0**	**2.51**
	Canada	Olympics	6	4	2	0	*369	9	1	1.46	....	....	....	....	....	....	....
1998-99	**Colorado**	**NHL**	**61**	**32**	**19**	**8**	**3648**	**139**	**5**	**2.29**	**19**	**11**	**8**	**1173**	**52**	**1**	**2.66**
99-2000	**Colorado**	**NHL**	**63**	**32**	**21**	**8**	**3704**	**141**	**2**	**2.28**	**17**	**11**	**6**	**1039**	**31**	**3**	**1.79**
2000-01♦	**Colorado**	**NHL**	**62**	**40**	**13**	**7**	**3585**	**132**	**4**	**2.21**	**23**	***16**	**7**	**1451**	**41**	***4**	***1.70**
2001-02	**Colorado**	**NHL**	**63**	**32**	**23**	**8**	**3773**	**122**	***9**	***1.94**	**21**	**11**	**10**	**1241**	**52**	**3**	**2.51**
	NHL Totals		**966**	**516**	**300**	**118**	**56466**	**2409**	**61**	**2.56**	**240**	**148**	**90**	**14786**	**568**	**22**	**2.30**

NHL All-Rookie Team (1986) • Won Conn Smythe Trophy (1986, 1993, 2001) • Shared William M. Jennings Trophy with Brian Hayward (1987, 1988, 1989) • NHL Second All-Star Team (1988, 1991) • NHL First All-Star Team (1989, 1990, 1992, 2002) • Won Trico Goaltending Award (1989, 1990) • Won Vezina Trophy (1989, 1990, 1992) • Won William M. Jennings Trophy (1992, 2002) • Played in NHL All-Star Game (1988, 1990, 1991, 1992, 1993, 1994, 1997, 1998, 2001, 2002)

Traded to **Colorado** by **Montreal** with Mike Keane for Andrei Kovalenko, Martin Rucinsky and Jocelyn Thibault, December 6, 1995.

RUDKOWSKY, Cody (ruhd-KOW-skee, KOH-dee) ST.L.

Goaltender. Catches left. 6'1", 206 lbs. Born, Willingdon, Alta., July 21, 1978.

			Regular Season								Playoffs						
Season	Club	Lea	GP	W	L	T	Mins	GA	SO	Avg	GP	W	L	Mins	GA	SO	Avg
1995-96	Langley Thunder	BCJHL	23	....	....	....	1172	73	1	3.73	....	....	....	....	....	....	....
	Seattle	WHL	2	0	0	0	21	3	0	8.57	....	....	....	....	....	....	....
1996-97	Seattle	WHL	40	19	16	1	2162	124	0	3.44	1	1	0	30	0	0	0.00
1997-98	Seattle	WHL	53	20	22	3	2805	176	1	3.74	5	1	4	278	18	0	3.88
1998-99	Seattle	WHL	64	34	17	10	3665	177	*7	2.90	11	5	6	637	31	1	2.92
99-2000	Worcester	AHL	28	9	7	6	1405	75	0	3.20	....	....	....	....	....	....	....
	Peoria Rivermen	ECHL	10	6	4	0	599	32	0	3.20	2	1	1	119	6	0	3.02
2000-01	Worcester	AHL	25	13	8	3	1477	66	3	2.68	....	....	....	....	....	....	....
2001-02	Worcester	AHL	21	6	10	2	1108	50	1	2.71	....	....	....	....	....	....	....
	Peoria Rivermen	ECHL	12	5	2	4	709	24	3	2.03	2	0	1	78	4	0	3.08

WHL West First All-Star Team (1999) • Canadian Major Junior First All-Star Team (1999) • Canadian Major Junior Goaltender of the Year (1999)

Signed as a free agent by **St. Louis**, March 25, 1999.

SABOURIN, Dany (SA-boo-rihn, DAN-ee) CGY.

Goaltender. Catches left. 6'2", 182 lbs. Born, Val-d'Or, Que., September 2, 1980.
(Calgary's 5th choice, 108th overall, in 1998 Entry Draft).

			Regular Season								Playoffs						
Season	Club	Lea	GP	W	L	T	Mins	GA	SO	Avg	GP	W	L	Mins	GA	SO	Avg
1996-97	Amos Forestiers	QAAA	24	6	16	0	1440	107	0	4.48	....	....	....	....	....	....	....
1997-98	Sherbrooke	QMJHL	37	15	15	2	1906	128	1	4.03	....	....	....	....	....	....	....
1998-99	Sherbrooke	QMJHL	30	8	13	2	1477	102	1	4.14	1	0	1	49	2	0	2.45
	Saint John	AHL	....	....	....	....	....	....	....	....	1	0	1	57	4	0	4.19
99-2000	Sherbrooke	QMJHL	55	25	22	5	3067	181	1	3.54	5	1	4	324	18	0	3.33
2000-01	Saint John	AHL	1	1	0	0	40	0	0	0.00	....	....	....	....	....	....	....
	Johnstown Chiefs	ECHL	19	4	9	1	903	56	0	3.72	1	0	0	40	2	0	3.00
2001-02	Johnstown Chiefs	ECHL	27	14	10	1	1539	84	0	3.28	3	0	2	137	5	0	2.18

SALFICKY, Dusan (sal-FITZ-kee, DOO-shahn) NYI

Goaltender. Catches left. 6'1", 185 lbs. Born, Chrudim, Czech., March 28, 1972.
(NY Islanders' 2nd choice, 132nd overall, in 2001 Entry Draft).

			Regular Season								Playoffs						
Season	Club	Lea	GP	W	L	T	Mins	GA	SO	Avg	GP	W	L	Mins	GA	SO	Avg
1990-91	Tri-City	WHL	2	1	1	0	119	11	0	5.55	....	....	....	....	....	....	....
	Tesla Pardubice	Czech	18	....	....	....	1000	60	....	3.60	....	....	....	....	....	....	....
1991-92	Tesla Pardubice	Czech-Jr.	STATISTICS NOT AVAILABLE														
1992-93	VTJ Tabor	Czech-2	STATISTICS NOT AVAILABLE														
1993-94	HC Pardubice	Czech	1	0	0	1	59	0	1	0.00	....	....	....	....	....	....	....
1994-95	HC Pardubice	Czech	10	....	....	....	548	23	1	2.52	3	....	....	185	12	0	3.89
1995-96	Pardubice	Czech	8	....	....	....	315	22	0	4.19	....	....	....	....	....	....	....
1996-97	Pardubice	Czech	21	....	....	....	1180	48	0	2.44	3	....	....	134	12	0	5.37
1997-98	Plzen	Czech	50	....	....	....	2939	134	0	2.75	5	2	3	310	14	0	2.71
1998-99	Plzen	Czech	44	....	....	....	2506	100	0	2.39	5	2	3	233	14	0	3.32
99-2000	Plzen	Czech	*52	....	....	....	*3061	108	3	2.12	7	....	....	415	15	....	2.17
2000-01	Plzen	Czech	*52	....	....	....	*3014	132	0	2.63	....	....	....	....	....	....	....
2001-02	Bridgeport	AHL	4	3	1	0	239	4	1	1.00	....	....	....	....	....	....	....
	Pardubice	Czech	13	....	....	....	779	35	....	2.70	6	....	....	367	14	....	2.29

SALO, Tommy (SA-loh,TAW-mee) EDM.

Goaltender. Catches left. 5'11", 173 lbs. Born, Surahammar, Sweden, February 1, 1971.
(NY Islanders' 5th choice, 118th overall, in 1993 Entry Draft).

			Regular Season								Playoffs						
Season	Club	Lea	GP	W	L	T	Mins	GA	SO	Avg	GP	W	L	Mins	GA	SO	Avg
1990-91	Vasteras IK	Sweden	2	....	....	....	100	11	0	6.60	....	....	....	....	....	....	....
1991-92	Vasteras IK Jr.	Swede-Jr.	STATISTICS NOT AVAILABLE														
1992-93	Vasteras IK	Sweden	24	....	....	....	1431	59	2	2.47	2	....	....	120	6	0	3.00
1993-94	Vasteras IK	Sweden	32	....	....	....	1896	106	0	3.35	....	....	....	....	....	....	....
	Sweden	Olympics	6	5	1	0	370	13	1	2.11	....	....	....	....	....	....	....
1994-95	Denver Grizzlies	IHL	*65	*45	14	4	*3810	165	*3	*2.60	8	7	0	390	20	0	3.07
	NY Islanders	**NHL**	**6**	**1**	**5**	**0**	**358**	**18**	**0**	**3.02**	....	....	....	....	....	....	....
1995-96	**NY Islanders**	**NHL**	**10**	**1**	**7**	**1**	**523**	**35**	**0**	**4.02**	....	....	....	....	....	....	....
	Utah Grizzlies	IHL	45	28	15	2	2695	119	*4	2.65	22	*15	7	1342	51	*3	2.28
1996-97	**NY Islanders**	**NHL**	**58**	**20**	**27**	**8**	**3208**	**151**	**5**	**2.82**	....	....	....	....	....	....	....
1997-98	**NY Islanders**	**NHL**	**62**	**23**	**29**	**5**	**3461**	**152**	**4**	**2.64**	....	....	....	....	....	....	....
	Sweden	Olympics	4	2	2	0	238	9	0	2.27	....	....	....	....	....	....	....
1998-99	**NY Islanders**	**NHL**	**51**	**17**	**26**	**7**	**3018**	**132**	**5**	**2.62**	....	....	....	....	....	....	....
	Edmonton	**NHL**	**13**	**8**	**2**	**2**	**700**	**27**	**0**	**2.31**	**4**	**0**	**4**	**296**	**11**	**0**	**2.23**
99-2000	**Edmonton**	**NHL**	**70**	**27**	**28**	**13**	**4164**	**162**	**2**	**2.33**	**5**	**1**	**4**	**297**	**14**	**0**	**2.83**
2000-01	**Edmonton**	**NHL**	**73**	**36**	**25**	**12**	**4364**	**179**	**8**	**2.46**	**6**	**2**	**4**	**406**	**15**	**0**	**2.22**
2001-02	**Edmonton**	**NHL**	**69**	**30**	**28**	**10**	**4035**	**149**	**6**	**2.22**	....	....	....	....	....	....	....
	Sweden	Olympics	3	2	1	0	179	7	0	2.35	....	....	....	....	....	....	....
	NHL Totals		**412**	**163**	**177**	**58**	**23831**	**1005**	**30**	**2.53**	**15**	**3**	**12**	**999**	**40**	**0**	**2.40**

IHL First All-Star Team (1995) • Won Garry F. Longman Memorial Trophy (Top Rookie - IHL) (1995) • Won James Norris Memorial Trophy (fewest goals against - IHL) (1995) • Won James Gatschene Memorial Trophy (MVP - IHL) (1995) • Shared James Norris Memorial Trophy (fewest goals against - IHL) with Mark McArthur (1996) • Won ''Bud'' Poile Trophy (Playoff MVP - IHL) (1996) • Played in NHL All-Star Game (2000, 2002)

Traded to **Edmonton** by **NY Islanders** for Mats Lindgren and Edmonton's 8th round choice (Radek Martinek) in 1999 Entry Draft, March 20, 1999.

SAUVE, Philippe (SOH-vay, FIHL-ihp) COL.

Goaltender. Catches left. 6', 180 lbs. Born, Buffalo, NY, February 27, 1980.
(Colorado's 6th choice, 38th overall, in 1998 Entry Draft).

			Regular Season								Playoffs						
Season	Club	Lea	GP	W	L	T	Mins	GA	SO	Avg	GP	W	L	Mins	GA	SO	Avg
1995-96	Laval Laurentide	QAAA	25	9	10	0	1184	87	1	4.11	15	7	8	900	54	0	3.58
1996-97	Rimouski Oceanic	QMJHL	26	11	9	2	1334	84	0	3.78	1	0	0	14	3	0	12.90
1997-98	Rimouski Oceanic	QMJHL	40	23	16	0	2326	131	1	3.38	7	0	5	262	33	0	7.55
1998-99	Rimouski Oceanic	QMJHL	44	16	19	4	2401	155	0	3.87	11	6	4	595	30	*1	3.03
99-2000	Drummondville	QMJHL	28	12	12	2	1526	106	0	4.17	....	....	....	....	....	....	....
	Hull Olympiques	QMJHL	17	9	7	1	992	57	0	3.45	12	6	6	735	47	0	3.84
2000-01	Hershey Bears	AHL	42	17	18	1	2182	100	3	2.75	3	0	3	218	10	0	2.75
2001-02	Hershey Bears	AHL	55	25	20	6	3130	111	6	2.13	8	3	5	486	21	0	2.59

Canadian Major Junior Humanitarian Player of the Year (1999)

Traded to **Hull** (QMJHL) by **Drummondville** (QMJHL) for Frederic Malette and future considerations, January 17, 2000.

SCHAEFER, Nolan (SHAY-fuhr, NOH-luhn) S.J.

Goaltender. Catches left. 6'1", 175 lbs. Born, Yellow Grass, Sask., January 15, 1980.
(San Jose's 4th choice, 166th overall, in 2000 Entry Draft).

			Regular Season								Playoffs						
Season	Club	Lea	GP	W	L	T	Mins	GA	SO	Avg	GP	W	L	Mins	GA	SO	Avg
1996-97	Yorkton Mallers	SMHL	36	....	....	....	1854	132	0	4.27	....	....	....	....	....	....	....
1997-98	Yorkton Mallers	SMHL	5	....	....	....	239	17	0	4.25	....	....	....	....	....	....	....
	Nipawin Hawks	SJHL	21	12	4	3	1080	42	*3	*2.33	....	....	....	....	....	....	....
1998-99	Nipawin Hawks	SJHL	DID NOT PLAY – INJURED														
99-2000	Providence	H-East	14	6	5	1	778	42	0	3.24	....	....	....	....	....	....	....
2000-01	Providence	H-East	25	15	8	2	1529	63	3	2.47	....	....	....	....	....	....	....
2001-02	Providence	H-East	*35	11	18	5	*2062	113	0	3.29	....	....	....	....	....	....	....

SJHL All-Rookie Team (1998) • Hockey East Second All-Star Team (2001) • NCAA East Second All-American Team (2001)

SCHWAB, Corey (SHWAHB, KOHR-ree) N.J.

Goaltender. Catches left. 6', 180 lbs. Born, North Battleford, Sask., November 4, 1970.
(New Jersey's 12th choice, 200th overall, in 1990 Entry Draft).

			Regular Season								Playoffs						
Season	Club	Lea	GP	W	L	T	Mins	GA	SO	Avg	GP	W	L	Mins	GA	SO	Avg
1988-89	Seattle	WHL	10	2	2	0	386	31	0	4.82	....	....	....	....	....	....	....
1989-90	Seattle	WHL	27	15	2	1	1150	69	1	3.60	3	0	0	49	2	0	2.45
1990-91	Seattle	WHL	*58	32	18	3	*3289	224	0	4.09	6	1	5	382	25	0	3.93
1991-92	Utica Devils	AHL	24	9	12	1	1322	95	0	4.31	....	....	....	....	....	....	....
	Cincinnati	ECHL	8	6	0	1	450	31	0	4.13	9	6	3	540	29	0	3.22
1992-93	Utica Devils	AHL	40	18	16	5	2387	169	*2	4.25	1	0	1	59	6	0	6.10
	Cincinnati	IHL	3	1	2	0	185	17	0	5.51	....	....	....	....	....	....	....
1993-94	Albany	AHL	51	27	21	3	3058	184	0	3.61	5	1	4	298	20	0	4.02
1994-95	Albany	AHL	45	25	10	9	2711	117	3	*2.59	7	6	1	425	19	0	2.68
1995-96	**New Jersey**	**NHL**	**10**	**0**	**3**	**0**	**331**	**12**	**0**	**2.18**	....	....	....	....	....	....	....
	Albany	AHL	5	3	2	0	299	13	0	2.61	....	....	....	....	....	....	....
1996-97	**Tampa Bay**	**NHL**	**31**	**11**	**12**	**1**	**1462**	**74**	**2**	**3.04**	....	....	....	....	....	....	....
1997-98	**Tampa Bay**	**NHL**	**16**	**2**	**9**	**1**	**821**	**40**	**1**	**2.92**	....	....	....	....	....	....	....
1998-99	**Tampa Bay**	**NHL**	**40**	**8**	**25**	**3**	**2146**	**126**	**0**	**3.52**	....	....	....	....	....	....	....
	Cleveland	IHL	8	1	6	1	477	31	0	3.90	....	....	....	....	....	....	....
99-2000	Orlando	IHL	16	9	4	2	868	31	1	2.14	....	....	....	....	....	....	....
	Vancouver	**NHL**	**6**	**2**	**1**	**1**	**269**	**16**	**0**	**3.57**	....	....	....	....	....	....	....
	Syracuse Crunch	AHL	12	7	5	0	720	42	0	3.50	4	1	3	246	11	1	2.69
2000-01	Kansas City	IHL	50	22	24	1	2866	150	2	3.14	....	....	....	....	....	....	....
2001-02	**Toronto**	**NHL**	**30**	**12**	**10**	**5**	**1646**	**75**	**1**	**2.73**	**1**	**0**	**0**	**12**	**0**	**0**	**0.00**
	NHL Totals		**133**	**35**	**60**	**11**	**6675**	**343**	**4**	**3.08**	**1**	**0**	**0**	**12**	**0**	**0**	**0.00**

AHL Second All-Star Team (1995) • Shared Harry ''Hap'' Holmes Memorial Trophy (fewest goals against - AHL) with Mike Dunham (1995) • Shared Jack A. Butterfield Trophy (Playoff MVP - AHL) with Mike Dunham (1995)

Traded to **Tampa Bay** by **New Jersey** for Jeff Reese, Chicago's 2nd round choice (previously acquired, New Jersey selected Pierre Dagenais) in 1996 Entry Draft and Tampa Bay's 8th round choice (Jay Bertsch) in 1996 Entry Draft, June 22, 1996. Claimed by **Atlanta** from **Tampa Bay** in Expansion Draft, June 25, 1999. Traded to **Vancouver** by **Atlanta** for Vancouver's 4th round choice (Carl Mallette) in 2000 Entry Draft, October 29, 1999. Signed as a free agent by **Toronto**, October 1, 2001. Signed as a free agent by **New Jersey**, July 8, 2002.

SCOTT, Travis (SKAWT, TRA-vihs) L.A.

Goaltender. Catches left. 6'2", 185 lbs. Born, Kanata, Ont., September 14, 1975.

			Regular Season								Playoffs						
Season	Club	Lea	GP	W	L	T	Mins	GA	SO	Avg	GP	W	L	Mins	GA	SO	Avg
1991-92	Nepean Raiders	COJHL	19	14	5	0	1065	71	1	4.00	....	....	....	....	....	....	....
1992-93	Nepean Raiders	COJHL	36	19	10	2	1968	133	0	4.05	....	....	....	....	....	....	....
1993-94	Windsor	OHL	45	20	18	0	2312	158	1	4.10	4	0	4	240	16	0	4.00
1994-95	Windsor	OHL	48	26	14	3	2644	147	3	3.34	3	0	1	94	6	1	3.83
1995-96	Oshawa Generals	OHL	31	15	9	4	1763	78	3	2.65	5	1	4	315	23	0	4.38
1996-97	Baton Rouge	ECHL	10	5	2	1	501	22	0	2.63	....	....	....	....	....	....	....
	Worcester	AHL	29	14	10	1	1482	75	1	3.04	....	....	....	....	....	....	....
1997-98	Baton Rouge	ECHL	36	14	11	6	1949	96	1	2.96	....	....	....	....	....	....	....
1998-99	Mississippi	ECHL	44	22	12	5	2337	112	1	2.88	*18	*14	4	*1252	42	3	2.01
99-2000	Lowell	AHL	46	15	23	3	2595	126	3	2.91	1	0	1	60	2	0	2.01
2000-01	**Los Angeles**	**NHL**	**1**	**0**	**0**	**0**	**25**	**3**	**0**	**7.20**	....	....	....	....	....	....	....
	Lowell	AHL	34	16	15	1	1977	83	2	2.52	4	1	2	209	7	1	2.01
2001-02	Manchester	AHL	39	21	12	3	2170	83	6	2.30	5	2	3	327	15	0	2.75
	NHL Totals		**1**	**0**	**0**	**0**	**25**	**3**	**0**	**7.20**	....	....	....	....	....	....	....

ECHL Playoff MVP (1999)

Signed as a free agent by **St. Louis**, December 30, 1996. Signed as a free agent by **LA Kings**, February 18, 2000.

SHIELDS, Steve (SHEELDS, STEEV) BOS.

Goaltender. Catches left. 6'3", 215 lbs. Born, Toronto, Ont., July 19, 1972.
(Buffalo's 5th choice, 101st overall, in 1991 Entry Draft).

			Regular Season								Playoffs						
Season	Club	Lea	GP	W	L	T	Mins	GA	SO	Avg	GP	W	L	Mins	GA	SO	Avg
1989-90	St. Marys	OJHL-B	26	....	....	....	1512	121	0	4.80	....	....	....	....	....	....	....
1990-91	U. of Michigan	CCHA	37	26	6	3	1963	106	0	3.24	....	....	....	....	....	....	....
1991-92	U. of Michigan	CCHA	*37	*27	7	2	*2090	99	1	2.84	....	....	....	....	....	....	....
1992-93	U. of Michigan	CCHA	*39	*30	6	2	2027	75	2	*2.22	....	....	....	....	....	....	....
1993-94	U. of Michigan	CCHA	36	*28	6	1	1961	87	0	2.66	....	....	....	....	....	....	....
1994-95	Rochester	AHL	13	3	8	0	673	53	0	4.72	1	0	0	20	3	0	9.00
	South Carolina	ECHL	21	11	5	2	1158	52	2	2.69	3	0	2	144	11	0	4.58
1995-96	**Buffalo**	**NHL**	**2**	**1**	**0**	**0**	**75**	**4**	**0**	**3.20**	....	....	....	....	....	....	....
	Rochester	AHL	43	20	17	2	2357	140	1	3.56	*19	*15	3	*1127	47	1	2.50
1996-97	**Buffalo**	**NHL**	**13**	**3**	**8**	**2**	**789**	**39**	**0**	**2.97**	**10**	**4**	**6**	**570**	**26**	**1**	**2.74**
	Rochester	AHL	23	14	6	2	1331	60	1	2.70	....	....	....	....	....	....	....
1997-98	**Buffalo**	**NHL**	**16**	**3**	**6**	**4**	**785**	**37**	**0**	**2.83**	....	....	....	....	....	....	....
	Rochester	AHL	1	0	1	0	59	3	0	3.04	....	....	....	....	....	....	....
1998-99	**San Jose**	**NHL**	**37**	**15**	**11**	**8**	**2162**	**80**	**4**	**2.22**	**1**	**0**	**1**	**60**	**6**	**0**	**6.00**
99-2000	**San Jose**	**NHL**	**67**	**27**	**30**	**8**	**3797**	**162**	**4**	**2.56**	**12**	**5**	**7**	**696**	**36**	**0**	**3.10**
2000-01	**San Jose**	**NHL**	**21**	**6**	**8**	**5**	**1135**	**47**	**2**	**2.48**	....	....	....	....	....	....	....
2001-02	**Anaheim**	**NHL**	**33**	**9**	**20**	**2**	**1777**	**79**	**0**	**2.67**	....	....	....	....	....	....	....
	NHL Totals		**189**	**64**	**83**	**29**	**10520**	**448**	**10**	**2.56**	**23**	**9**	**14**	**1326**	**68**	**1**	**3.08**

CCHA First All-Star Team (1993, 1994) • NCAA West Second All-American Team (1993, 1994)

Traded to **San Jose** by **Buffalo** with Buffalo's 4th round choice (Miroslav Zalesak) in 1998 Entry Draft for Kay Whitmore, Colorado's 2nd round choice (previously acquired, Buffalo selected Jaroslav Kristek) in 1998 Entry Draft and San Jose's 5th round choice (later traded to Columbus - Columbus selected Tyler Kolarik) in 2000 Entry Draft, June 18, 1998. Traded to **Anaheim** by **San Jose** with Jeff Friesen and future considerations for Teemu Selanne, March 5, 2001. Traded to **Boston** by **Anaheim** for Boston's 3rd round choice in 2003 Entry Draft, June 25, 2002.

SIGALET, Jordan (SIH-ga-leht, JOHR-duhn) BOS.

Goaltender. Catches left. 6'1", 180 lbs. Born, New Westminster, B.C., February 19, 1981.
(Boston's 6th choice, 209th overall, in 2001 Entry Draft).

			Regular Season								Playoffs						
Season	Club	Lea	GP	W	L	T	Mins	GA	SO	Avg	GP	W	L	Mins	GA	SO	Avg
99-2000	Victoria Salsa	BCHL	33	....	....	....	1980	108	0	3.28	....	....	....	....	....	....	....
2000-01	Victoria Salsa	BCHL	48	23	22	0	2820	142	0	3.03	18	12	5	1060	143	0	2.62
2001-02	Bowling Green	CCHA	13	2	6	2	657	38	0	3.47	....	....	....	....	....	....	....

BCHL All-Rookie Team (2000) • BCHL Second All-Star Team (2000) • BCHL First All-Star Team (2001)

SKUDRA, Peter (SKOO-druh, PEE-tuhr) VAN.

Goaltender. Catches left. 6'1", 189 lbs. Born, Riga, Latvia, April 24, 1973.

			Regular Season								Playoffs						
Season	Club	Lea	GP	W	L	T	Mins	GA	SO	Avg	GP	W	L	Mins	GA	SO	Avg
1992-93	Pardaugava Riga	CIS	27	....	....	....	1498	74	....	2.96	1	....	....	60	5	0	5.00
1993-94	Pardaugava Riga	CIS	14	....	....	....	783	42	....	3.22	1	....	....	55	4	0	4.36
1994-95	Greensboro	ECHL	33	13	9	5	1612	113	0	4.20	6	2	2	341	28	0	4.92
	Memphis	CHL	2	0	1	0	80	8	0	6.00	....	....	....	....	....	....	....
1995-96	Erie Panthers	ECHL	12	3	8	1	681	47	0	4.14	....	....	....	....	....	....	....
	Johnstown Chiefs	ECHL	30	12	11	4	1657	98	0	3.55	....	....	....	....	....	....	....
1996-97	Hamilton	AHL	32	8	16	2	1615	101	0	3.75	....	....	....	....	....	....	....
	Johnstown Chiefs	ECHL	4	2	1	1	200	11	0	3.30	....	....	....	....	....	....	....
1997-98	**Pittsburgh**	**NHL**	**17**	**6**	**4**	**3**	**851**	**26**	**0**	**1.83**	....	....	....	....	....	....	....
	Houston Aeros	IHL	9	5	3	1	499	23	0	2.77	....	....	....	....	....	....	....
	Kansas City	IHL	13	10	3	0	775	37	0	2.86	8	4	4	512	20	1	*2.34
1998-99	**Pittsburgh**	**NHL**	**37**	**15**	**11**	**5**	**1914**	**89**	**3**	**2.79**	....	....	....	....	....	....	....
99-2000	**Pittsburgh**	**NHL**	**20**	**5**	**7**	**3**	**922**	**48**	**1**	**3.12**	**1**	**0**	**0**	**20**	**1**	**0**	**3.00**
2000-01	**Buffalo**	**NHL**	**1**	**0**	**0**	**0**	**0**	**0**	**0**	**0.00**	....	....	....	....	....	....	....
	Rochester	AHL	2	2	0	0	120	5	0	2.50	....	....	....	....	....	....	....
	Boston	**NHL**	**25**	**6**	**12**	**1**	**1116**	**62**	**0**	**3.33**	....	....	....	....	....	....	....
	Providence	AHL	3	3	0	0	180	5	0	1.67	....	....	....	....	....	....	....
2001-02	Hartford	AHL	3	2	1	0	179	8	0	2.69	....	....	....	....	....	....	....
	Vancouver	**NHL**	**23**	**10**	**8**	**2**	**1166**	**47**	**1**	**2.42**	**2**	**0**	**1**	**96**	**5**	**0**	**3.13**
	NHL Totals		**123**	**42**	**42**	**14**	**5969**	**272**	**5**	**2.73**	**3**	**0**	**1**	**116**	**6**	**0**	**3.10**

Signed as a free agent by **Pittsburgh**, September 25, 1997. Signed as a free agent by **Boston**, October 3, 2000. Claimed on waivers by **Buffalo** from **Boston**, October 6, 2000. Claimed on waivers by **Boston** from **Buffalo**, November 14, 2000. Signed as a free agent by **Vancouver**, November 7, 2001.

SMID, Zdenek (SHMIHD, zuh-DEHN-ehk) ATL.

Goaltender. Catches left. 5'10", 172 lbs. Born, Plzen, Czech., February 3, 1980.
(Atlanta's 7th choice, 168th overall, in 2000 Entry Draft).

			Regular Season								Playoffs						
Season	Club	Lea	GP	W	L	T	Mins	GA	SO	Avg	GP	W	L	Mins	GA	SO	Avg
1996-97	HC ZKZ Plzen Jr.	Czech-Jr.	23	....	....	....	1304	48	....	2.21	....	....	....	....	....	....	....
1997-98	Plzen Jr.	Czech-Jr.	30	....	....	....	1601	91	....	3.41	....	....	....	....	....	....	....
1998-99	Karlovy Vary Jr.	Czech-Jr.					Statistics Not Available										
	HC Karlovy Vary	Czech	3	....	....	....	160	11	....	4.13	....	....	....	....	....	....	....
99-2000	Karlovy Vary Jr.	Czech-Jr.	24	....	....	....	1409	54	....	2.30	2	....	....	86	9	....	6.28
	HC Karlovy Vary	Czech	14	....	....	....	650	41	0	3.78	3	....	....	150	8	....	3.20
2000-01	HC Karlovy Vary	Czech	12	....	....	....	630	34	....	3.24	....	....	....	....	....	....	....
2001-02	HPK Hameenlinna	Finland	18	10	4	4	1055	44	2	2.50	....	....	....	....	....	....	....
	Lulea HF	Sweden	1	0	1	0	58	3	0	3.10	1	0	1	40	5	0	7.50

SMITH, Mike (SMIHTH, MIGHK) DAL.

Goaltender. Catches left. 6'3", 189 lbs. Born, Kingston, Ont., March 22, 1982.
(Dallas' 5th choice, 161st overall, in 2001 Entry Draft).

			Regular Season								Playoffs						
Season	Club	Lea	GP	W	L	T	Mins	GA	SO	Avg	GP	W	L	Mins	GA	SO	Avg
1998-99	Kingston	OPJHL	16	....	....	....	906	53	0	3.51	....	....	....	....	....	....	....
99-2000	Kingston	OHL	15	4	5	0	666	42	0	3.78	....	....	....	....	....	....	....
2000-01	Kingston	OHL	3	0	0	2	136	8	0	3.53	....	....	....	....	....	....	....
	Sudbury Wolves	OHL	43	22	13	7	2571	108	3	2.52	12	7	5	735	26	2	*2.12
2001-02	Sudbury Wolves	OHL	53	19	28	5	3082	157	3	3.06	5	1	4	302	15	0	2.98

Traded to **Sudbury** (OHL) by **Kingston** (OHL) for Sudbury's 10th round choice (Billy Burke) in 2002 OHL Midget Draft, October 25, 2000.

SNEE, Brandon (SNEE, BRAN-duhn) NYR

Goaltender. Catches left. 6'1", 195 lbs. Born, Philadelphia, PA, June 10, 1980.
(NY Rangers' 5th choice, 143rd overall, in 2000 Entry Draft).

			Regular Season								Playoffs						
Season	Club	Lea	GP	W	L	T	Mins	GA	SO	Avg	GP	W	L	Mins	GA	SO	Avg
1997-98	The Hill School	H.S.	22	....	....	....	1320	43	0	2.02	....	....	....	....	....	....	....
1998-99	Union College	ECAC	19	1	12	3	1011	59	1	3.50	....	....	....	....	....	....	....
99-2000	Union College	ECAC	*31	8	22	1	1765	114	0	3.87	....	....	....	....	....	....	....
2000-01	Union College	ECAC	*33	12	17	4	1849	96	1	3.11	....	....	....	....	....	....	....
2001-02	Union College	ECAC	28	13	11	4	1629	74	1	2.72	....	....	....	....	....	....	....

SNOW, Garth (SNOH, GAHRTH) NYI

Goaltender. Catches left. 6'3", 200 lbs. Born, Wrentham, MA, July 28, 1969.
(Quebec's 6th choice, 114th overall, in 1987 Entry Draft).

			Regular Season								Playoffs						
Season	Club	Lea	GP	W	L	T	Mins	GA	SO	Avg	GP	W	L	Mins	GA	SO	Avg
1986-87	Mount St. Charles	H.S.	30	....	....	....	1795	53	10	1.77	....	....	....	....	....	....	....
1987-88	Stratford	OJHL-B	30	20	6	0	1642	93	2	3.40	....	....	....	....	....	....	....
1988-89	U. of Maine	H-East	5	2	2	0	241	14	1	3.49	....	....	....	....	....	....	....
1989-90	U. of Maine	H-East					Did Not Play – Academically Ineligible										
1990-91	U. of Maine	H-East	25	*18	4	0	1290	64	2	2.98	....	....	....	....	....	....	....
1991-92	U. of Maine	H-East	31	*25	4	2	1792	73	*2	*2.44	....	....	....	....	....	....	....
1992-93	U. of Maine	H-East	23	*21	0	1	1210	42	1	*2.08	....	....	....	....	....	....	....
1993-94	United States	Nat-Tm	23	13	5	3	1324	71	1	3.22	....	....	....	....	....	....	....
	United States	Olympics	5	1	3	1	299	17	0	3.41	....	....	....	....	....	....	....
	Quebec	**NHL**	**5**	**3**	**2**	**0**	**279**	**16**	**0**	**3.44**	....	....	....	....	....	....	....
	Cornwall Aces	AHL	16	6	5	3	927	51	0	3.30	13	8	5	790	42	0	3.19
1994-95	Cornwall Aces	AHL	*62	*32	20	7	*3558	162	3	2.73	8	4	3	402	14	*2	*2.09
	Quebec	**NHL**	**2**	**1**	**1**	**0**	**119**	**11**	**0**	**5.55**	**1**	**0**	**0**	**9**	**1**	**0**	**6.67**
1995-96	**Philadelphia**	**NHL**	**26**	**12**	**8**	**4**	**1437**	**69**	**0**	**2.88**	**1**	**0**	**0**	**1**	**0**	**0**	**0.00**
1996-97	**Philadelphia**	**NHL**	**35**	**14**	**8**	**8**	**1884**	**79**	**2**	**2.52**	**12**	**8**	**4**	**699**	**33**	**0**	**2.83**
1997-98	**Philadelphia**	**NHL**	**29**	**14**	**9**	**4**	**1651**	**67**	**1**	**2.43**	....	....	....	....	....	....	....
	Vancouver	**NHL**	**12**	**3**	**6**	**0**	**504**	**26**	**0**	**3.10**	....	....	....	....	....	....	....
1998-99	**Vancouver**	**NHL**	**65**	**20**	**31**	**8**	**3501**	**171**	**6**	**2.93**	....	....	....	....	....	....	....
99-2000	**Vancouver**	**NHL**	**32**	**10**	**15**	**3**	**1712**	**76**	**0**	**2.66**	....	....	....	....	....	....	....
2000-01	Wilkes-Barre	AHL	3	2	1	0	178	7	0	2.36	....	....	....	....	....	....	....
	Pittsburgh	**NHL**	**35**	**14**	**15**	**4**	**2032**	**101**	**3**	**2.98**	....	....	....	....	....	....	....
2001-02	**NY Islanders**	**NHL**	**25**	**10**	**7**	**2**	**1217**	**55**	**2**	**2.71**	**1**	**0**	**0**	**26**	**2**	**0**	**4.62**
	NHL Totals		**266**	**101**	**102**	**33**	**14336**	**671**	**14**	**2.81**	**15**	**8**	**4**	**735**	**36**	**0**	**2.94**

Hockey East Second All-Star Team (1992, 1993) • NCAA Championship All-Tournament Team (1993)

Transferred to **Colorado** after **Quebec** franchise relocated, June 21, 1995. Traded to **Philadelphia** by **Colorado** for Philadelphia's 3rd (later traded to Washington - Washington selected Shawn McNeil) and 6th (Kai Fischer) round choices in 1996 Entry Draft, July 12, 1995. Traded to **Vancouver** by **Philadelphia** for Sean Burke, March 4, 1998. Signed as a free agent by **Pittsburgh**, October 10, 2000. Signed as a free agent by **NY Islanders**, July 14, 2001.

STANA, Ratislav (STAN-ah, RAH-tih-slahv) WSH.

Goaltender. Catches left. 6'2", 161 lbs. Born, Kosice, Czech., January 10, 1980.
(Washington's 8th choice, 193rd overall, in 1998 Entry Draft).

			Regular Season								Playoffs						
Season	Club	Lea	GP	W	L	T	Mins	GA	SO	Avg	GP	W	L	Mins	GA	SO	Avg
1997-98	HC Kosice Jr.	Slovak-Jr.	32	....	....	....	1920	56	2	1.75	....	....	....	....	....	....	....
1998-99	Moose Jaw	WHL	36	21	14	1	2131	123	2	3.46	9	4	5	544	30	0	3.31
99-2000	Moose Jaw	WHL	14	4	9	0	730	48	0	3.95	....	....	....	....	....	....	....
	Calgary Hitmen	WHL	16	13	2	1	971	37	1	2.29	9	7	2	526	21	1	2.40
2000-01	Richmond	ECHL	38	15	16	2	2111	90	1	2.56	3	1	2	178	7	1	2.34
2001-02	Richmond	ECHL	36	20	12	3	2098	95	1	2.72	....	....	....	....	....	....	....
	Portland Pirates	AHL	3	1	2	0	180	11	0	3.66	....	....	....	....	....	....	....
	Slovakia	Olympics	1	1	0	0	60	1	0	1.00	....	....	....	....	....	....	....

Traded to **Calgary** (WHL) by **Moose Jaw** (WHL) with Cory Hintz for Sean Connors and Anders Lovdahl, January 10, 2000.

STEPHAN, Tobias (STEH-fan, toh-BEE-uhs) DAL.

Goaltender. Catches left. 6'3", 178 lbs. Born, Zurich, Switz., January 21, 1984.
(Dallas' 3rd choice, 34th overall, in 2002 Entry Draft).

			Regular Season								Playoffs						
Season	Club	Lea	GP	W	L	T	Mins	GA	SO	Avg	GP	W	L	Mins	GA	SO	Avg
2000-01	Kloten Jr.	Swiss-Jr.					Statistics Not Available										
2001-02	EHC Chur	Swiss	23	....	....	....	1396	80	0	3.44	10	....	....	604	39	0	3.87

ST-GERMAIN, David (SAN-zhur-meh, DAH-vee) NYI

Goaltender. Catches left. 5'11", 172 lbs. Born, Charles-Lemoye, Que., March 18, 1980.

			Regular Season								Playoffs						
Season	Club	Lea	GP	W	L	T	Mins	GA	SO	Avg	GP	W	L	Mins	GA	SO	Avg
1996-97	Richelieu Elites	QAHA	35	....	....	....	2133	113	2	3.18	....	....	....	....	....	....	....
1997-98	Val-d'Or Foreurs	QMJHL	18	6	4	2	893	51	0	3.42	2	0	2	119	9	0	4.53
1998-99	Val-d'Or Foreurs	QMJHL	16	7	7	1	833	68	0	4.90	....	....	....	....	....	....	....
	Cape Breton	QMJHL	35	10	23	1	1959	127	1	3.89	5	1	4	299	19	0	3.81
99-2000	Cape Breton	QMJHL	36	10	23	2	1986	147	0	4.44	....	....	....	....	....	....	....
	Baie-Comeau	QMJHL	23	10	11	2	1287	77	2	3.59	6	2	4	361	21	0	3.49
2000-01	Baie-Comeau	QMJHL	53	32	13	4	2913	169	3	3.48	11	6	4	683	29	1	2.55
2001-02	Trenton Titans	ECHL	29	16	6	6	1704	73	1	2.57	....	....	....	....	....	....	....
	Bridgeport	AHL	1	0	1	0	59	3	0	3.04	....	....	....	....	....	....	....

Traded to **Baie-Comeau** (QMJHL) by **Cape Breton** (QMJHL) for Dany Dallaire, January 10, 2000. Signed as a free agent by **NY Islanders**, December 12, 2000.

STORR, Jamie (STOHR, JAY-mee) L.A.

Goaltender. Catches left. 6'2", 195 lbs. Born, Brampton, Ont., December 28, 1975.
(Los Angeles' 1st choice, 7th overall, in 1994 Entry Draft).

			Regular Season								Playoffs						
Season	Club	Lea	GP	W	L	T	Mins	GA	SO	Avg	GP	W	L	Mins	GA	SO	Avg
1990-91	Brampton	MTJHL	24	....	....	....	1145	91	0	4.77	15	....	....	885	60	0	4.07
1991-92	Owen Sound	OHL	34	11	16	1	1732	128	0	4.43	5	1	4	299	28	0	5.62
1992-93	Owen Sound	OHL	41	20	17	3	2362	180	0	4.57	8	4	4	454	35	0	4.63
1993-94	Owen Sound	OHL	35	21	11	1	2004	120	1	3.59	9	4	5	547	44	0	4.83
1994-95	Owen Sound	OHL	17	5	9	2	977	64	0	3.93	....	....	....	....	....	....	....
	Los Angeles	**NHL**	**5**	**1**	**3**	**1**	**263**	**17**	**0**	**3.88**	....	....	....	....	....	....	....
	Windsor	OHL	4	3	1	0	241	8	1	1.99	10	6	3	520	34	1	3.92
1995-96	**Los Angeles**	**NHL**	**5**	**3**	**1**	**0**	**262**	**12**	**0**	**2.75**	....	....	....	....	....	....	....
	Phoenix	IHL	48	22	20	4	2711	139	2	3.08	2	1	1	118	4	1	2.03
1996-97	**Los Angeles**	**NHL**	**5**	**2**	**1**	**1**	**265**	**11**	**0**	**2.49**	....	....	....	....	....	....	....
	Phoenix	IHL	44	16	22	4	2441	147	0	3.61	....	....	....	....	....	....	....
1997-98	**Los Angeles**	**NHL**	**17**	**9**	**5**	**1**	**920**	**34**	**2**	**2.22**	**3**	**0**	**2**	**145**	**9**	**0**	**3.72**
	Long Beach	IHL	11	7	2	1	629	31	0	2.96	....	....	....	....	....	....	....
1998-99	**Los Angeles**	**NHL**	**28**	**12**	**12**	**2**	**1525**	**61**	**4**	**2.40**	....	....	....	....	....	....	....
99-2000	**Los Angeles**	**NHL**	**42**	**18**	**15**	**5**	**2206**	**93**	**1**	**2.53**	**1**	**0**	**1**	**36**	**2**	**0**	**3.33**
2000-01	**Los Angeles**	**NHL**	**45**	**19**	**18**	**6**	**2498**	**114**	**4**	**2.74**	....	....	....	....	....	....	....
2001-02	**Los Angeles**	**NHL**	**19**	**9**	**4**	**3**	**886**	**28**	**2**	**1.90**	**1**	**0**	**0**	**0**	**0**	**0**	**0.00**
	NHL Totals		**166**	**73**	**59**	**19**	**8825**	**370**	**13**	**2.52**	**5**	**0**	**3**	**181**	**11**	**0**	**3.65**

OHL All-Rookie Team (1992) • OHL First All-Star Team (1994) • NHL All-Rookie Team (1998, 1999)

SWANSON, Kevin (SWAHN-suhn, KEH-vihn) VAN.

Goaltender. Catches left. 5'10", 170 lbs. Born, Calgary, Alta., April 18, 1980.
(Vancouver's 6th choice, 189th overall, in 1999 Entry Draft).

			Regular Season								Playoffs						
Season	**Club**	**Lea**	**GP**	**W**	**L**	**T**	**Mins**	**GA**	**SO**	**Avg**	**GP**	**W**	**L**	**Mins**	**GA**	**SO**	**Avg**
1996-97	Red Deer Chiefs	AMHL	20	....	....	....	1284	94	0	4.39	....	....	....	....	....	....	....
1997-98	Prince George	WHL	28	14	11	1	1532	93	0	3.64	....	....	....	....	....	....	....
1998-99	Prince George	WHL	4	1	2	0	180	10	0	3.33	....	....	....	....	....	....	....
	Kelowna Rockets	WHL	50	18	23	3	2507	144	2	3.45	6	2	4	355	14	0	2.37
99-2000	Kelowna Rockets	WHL	68	25	40	3	3943	194	*7	2.95	5	1	4	297	16	0	3.23
2000-01	Kelowna Rockets	WHL	49	27	16	5	2854	148	0	3.11	6	2	4	361	17	1	2.83
2001-02	Manitoba Moose	AHL	1	0	0	0	20	1	0	3.00	....	....	....	....	....	....	....
	Columbia Inferno	ECHL	19	4	8	4	986	44	0	2.68	1	0	0	20	2	0	5.91

WHL West First All-Star Team (2000) • WHL West Second All-Star Team (2001)

Traded to **Kelowna** (WHL) by **Prince George** (WHL) for Justin Hansen, November 1, 1998.

SZUPER, Levente (SHOO-puhr, leh-VEHN-teh) CGY.

Goaltender. Catches left. 5'11", 180 lbs. Born, Budapest, Hungary, June 11, 1980.
(Calgary's 4th choice, 116th overall, in 2000 Entry Draft).

			Regular Season								Playoffs						
Season	**Club**	**Lea**	**GP**	**W**	**L**	**T**	**Mins**	**GA**	**SO**	**Avg**	**GP**	**W**	**L**	**Mins**	**GA**	**SO**	**Avg**
1996-97	Ferencvaros Jr.	Hungary	10	....	....	....	600	9	....	0.90	....	....	....	....	....	....	....
	Ferencvaros	Hungary	30	....	....	....	1660	74	3	2.67	....	....	....	....	....	....	....
1997-98	Krefeld Jr.	Ger.-Jr.	40	....	....	....	2300	103	3	2.69	....	....	....	....	....	....	....
1998-99	Ottawa 67's	OHL	32	22	6	3	1800	70	4	2.33	4	2	2	241	11	*1	2.74
99-2000	Ottawa 67's	OHL	53	31	15	2	2862	122	*5	2.56	11	6	5	680	35	1	3.09
2000-01	Saint John	AHL	34	16	10	2	1750	73	2	2.50	1	1	0	36	0	0	0.00
2001-02	Saint John	AHL	43	15	18	5	2429	98	5	2.42	....	....	....	....	....	....	....

TALLAS, Robbie (TAL-as, RAW-bee) PIT.

Goaltender. Catches left. 6', 170 lbs. Born, Edmonton, Alta., March 20, 1973.

			Regular Season								Playoffs						
Season	**Club**	**Lea**	**GP**	**W**	**L**	**T**	**Mins**	**GA**	**SO**	**Avg**	**GP**	**W**	**L**	**Mins**	**GA**	**SO**	**Avg**
1990-91	Penticton	BCJHL	37	....	....	....	2055	196	0	5.72	....	....	....	....	....	....	....
1991-92	Seattle	WHL	14	4	7	0	708	52	0	4.41	....	....	....	....	....	....	....
	South Surrey	BCJHL	19	6	12	0	1043	112	1	6.44	....	....	....	....	....	....	....
1992-93	Seattle	WHL	58	24	23	3	3151	194	2	3.69	5	1	4	333	18	0	3.24
1993-94	Seattle	WHL	51	23	21	3	2849	188	0	3.96	9	5	4	567	40	0	4.23
1994-95	Charlotte	ECHL	36	21	9	3	2011	114	0	3.40	....	....	....	....	....	....	....
	Providence	AHL	2	1	0	0	82	4	1	2.90	....	....	....	....	....	....	....
1995-96	**Boston**	**NHL**	**1**	**1**	**0**	**0**	**60**	**3**	**0**	**3.00**	....	....	....	....	....	....	....
	Providence	AHL	37	12	16	7	2136	117	1	3.29	2	0	2	135	9	0	4.01
1996-97	**Boston**	**NHL**	**28**	**8**	**12**	**1**	**1244**	**69**	**1**	**3.33**	....	....	....	....	....	....	....
	Providence	AHL	24	9	14	1	1424	83	0	3.50	....	....	....	....	....	....	....
1997-98	**Boston**	**NHL**	**14**	**6**	**3**	**3**	**788**	**24**	**1**	**1.83**	....	....	....	....	....	....	....
	Providence	AHL	10	1	8	1	575	39	0	4.07	....	....	....	....	....	....	....
1998-99	**Boston**	**NHL**	**17**	**7**	**7**	**2**	**987**	**43**	**1**	**2.61**	....	....	....	....	....	....	....
99-2000	**Boston**	**NHL**	**27**	**4**	**13**	**4**	**1363**	**72**	**0**	**3.17**	....	....	....	....	....	....	....
2000-01	**Chicago**	**NHL**	**12**	**2**	**7**	**0**	**627**	**35**	**0**	**3.35**	....	....	....	....	....	....	....
	Chicago Wolves	IHL	3	0	1	0	87	6	0	4.13	....	....	....	....	....	....	....
	Norfolk Admirals	AHL	6	2	2	2	333	12	0	2.16	....	....	....	....	....	....	....
2001-02	Wilkes-Barre	AHL	38	6	25	5	2183	126	0	3.46	....	....	....	....	....	....	....
	NHL Totals		**99**	**28**	**42**	**10**	**5069**	**246**	**3**	**2.91**	....	....	....	....	....	....	....

Signed as a free agent by **Boston**, September 13, 1995. Signed as a free agent by **Chicago**, July 31, 2000. Signed as a free agent by **Pittsburgh**, August 14, 2001.

TARASOV, Vadim (ta-RA-sahf, va-DEEM) MTL.

Goaltender. Catches left. 5'11", 187 lbs. Born, Ust-Kamenogorsk, USSR, December 31, 1976.
(Montreal's 9th choice, 196th overall, in 1999 Entry Draft).

			Regular Season								Playoffs						
Season	**Club**	**Lea**	**GP**	**W**	**L**	**T**	**Mins**	**GA**	**SO**	**Avg**	**GP**	**W**	**L**	**Mins**	**GA**	**SO**	**Avg**
1995-96	Novokuznetsk	CIS	26	....	....	....	1355	60	1	2.66	....	....	....	....	....	....	....
1996-97	Novokuznetsk	Russia	34	....	....	....	1971	87	0	2.65	....	....	....	....	....	....	....
1997-98	Novokuznetsk	Russia	23	....	....	....	1364	61	2	2.68	....	....	....	....	....	....	....
1998-99	Novokuznetsk	Russia	*41	....	....	....	*2346	56	*8	1.43	6	....	....	349	16	0	2.75
99-2000	Novokuznetsk	Russia	28	....	....	....	1583	66	1	2.50	14	....	....	791	26	1	1.97
2000-01	Novokuznetsk	Russia	33	....	....	....	1960	69	4	2.11	....	....	....	....	....	....	....
2001-02	Quebec	AHL	14	7	4	2	801	42	0	3.15	....	....	....	....	....	....	....

Signed as a free agent by **Novokuznetsk** (Russia) with Montreal retaining NHL rights, July 18, 2002.

TELLQVIST, Mikael (TEHL-kvihst, mih-KIGH-ehl) TOR.

Goaltender. Catches left. 5'11", 185 lbs. Born, Sundbyberg, Sweden, September 19, 1979.
(Toronto's 3rd choice, 70th overall, in 2000 Entry Draft).

			Regular Season								Playoffs						
Season	**Club**	**Lea**	**GP**	**W**	**L**	**T**	**Mins**	**GA**	**SO**	**Avg**	**GP**	**W**	**L**	**Mins**	**GA**	**SO**	**Avg**
1997-98	Djurgarden Jr.	Swede-Jr.	23	....	....	....	1380	55	....	2.39	2	0	2	120	8	0	4.00
1998-99	Djurgarden	Sweden	3	1	2	0	124	8	0	3.87	4	....	....	240	11	0	2.75
	Djurgarden	EuroHL	3	2	1	0	180	8	....	2.33	....	....	....	....	....	....	....
99-2000	Huddinge IK	Swede-2	11	4	7	0	660	33	....	3.30	....	....	....	....	....	....	....
	Djurgarden	Sweden	30	....	....	....	1909	66	2	*2.07	*13	....	....	*814	21	*3	*1.55
2000-01	Djurgarden	Sweden	43	....	....	....	2622	91	*5	*2.08	*16	....	....	*1006	45	*1	2.68
2001-02	St. John's	AHL	28	8	11	6	1521	79	0	3.12	1	1	0	15	0	0	0.00
	Sweden	Olympics					DID NOT PLAY – SPARE GOALTENDER										

THEODORE, Jose (TEE-uh-dohr, joh-SAY) MTL.

Goaltender. Catches right. 5'11", 182 lbs. Born, Laval, Que., September 13, 1976.
(Montreal's 2nd choice, 44th overall, in 1994 Entry Draft).

			Regular Season								Playoffs						
Season	**Club**	**Lea**	**GP**	**W**	**L**	**T**	**Mins**	**GA**	**SO**	**Avg**	**GP**	**W**	**L**	**Mins**	**GA**	**SO**	**Avg**
1990-91	Richelieu	QAHA	42	....	....	....	2520	80	0	1.90	....	....	....	....	....	....	....
1991-92	Richelieu	QAAA	24	9	13	2	1440	96	0	3.99	5	2	3	295	26	0	5.28
1992-93	St-Jean Lynx	QMJHL	34	12	16	2	1776	112	0	3.78	3	0	2	175	11	0	3.77
1993-94	St-Jean Lynx	QMJHL	57	20	29	6	3225	194	0	3.61	5	1	4	296	18	0	3.65
1994-95	Hull Olympiques	QMJHL	*58	*32	22	2	*3348	193	5	3.46	*21	*15	6	*1263	59	*1	2.80
	Fredericton	AHL	....	....	....	....	....	....	....	....	1	0	1	60	3	0	3.00
1995-96	**Montreal**	**NHL**	**1**	**0**	**0**	**0**	**9**	**1**	**0**	**6.67**	....	....	....	....	....	....	....
	Hull Olympiques	QMJHL	48	33	11	2	2807	158	0	3.38	5	2	3	299	20	0	4.01
1996-97	**Montreal**	**NHL**	**16**	**5**	**6**	**2**	**821**	**53**	**0**	**3.87**	**2**	**1**	**1**	**168**	**7**	**0**	**2.50**
	Fredericton	AHL	26	12	12	0	1469	87	0	3.55	....	....	....	....	....	....	....
1997-98	Fredericton	AHL	53	20	23	8	3053	145	2	2.85	4	1	3	237	13	0	3.28
	Montreal	**NHL**	....	....	....	....	....	....	....	....	**3**	**0**	**1**	**120**	**1**	**0**	**0.50**
1998-99	**Montreal**	**NHL**	**18**	**4**	**12**	**0**	**913**	**50**	**1**	**3.29**	....	....	....	....	....	....	....
	Fredericton	AHL	27	12	13	2	1609	77	2	2.87	13	8	5	694	35	1	3.03
99-2000	**Montreal**	**NHL**	**30**	**12**	**13**	**2**	**1655**	**58**	**5**	**2.10**	....	....	....	....	....	....	....
2000-01	**Montreal**	**NHL**	**59**	**20**	**29**	**5**	**3298**	**141**	**2**	**2.57**	....	....	....	....	....	....	....
	Quebec	AHL	3	3	0	0	180	9	0	3.00	....	....	....	....	....	....	....
2001-02	**Montreal**	**NHL**	**67**	**30**	**24**	**10**	**3864**	**136**	**7**	**2.11**	**12**	**6**	**6**	**686**	**35**	**0**	**3.06**
	NHL Totals		**191**	**71**	**84**	**19**	**10560**	**439**	**15**	**2.49**	**17**	**7**	**8**	**974**	**43**	**0**	**2.65**

QMJHL Second All-Star Team (1995, 1996) • NHL Second All-Star Team (2002) • Won MBNA Roger Crozier Saving Grace Award (2002) • Won Vezina Trophy (2002) • Won Hart Trophy (2002) • Played in NHL All-Star Game (2002)

• Scored a goal vs. NY Islanders, January 2, 2001.

THIBAULT, Jocelyn (TEE-boh, JAW-seh-lihn) CHI.

Goaltender. Catches left. 5'11", 170 lbs. Born, Montreal, Que., January 12, 1975.
(Quebec's 1st choice, 10th overall, in 1993 Entry Draft).

			Regular Season								Playoffs						
Season	**Club**	**Lea**	**GP**	**W**	**L**	**T**	**Mins**	**GA**	**SO**	**Avg**	**GP**	**W**	**L**	**Mins**	**GA**	**SO**	**Avg**
1990-91	Laval Laurentide	QAAA	20	14	5	0	1178	78	1	3.94	5	2	3	300	20	0	4.00
1991-92	Trois-Rivieres	QMJHL	30	14	7	1	1496	77	0	3.09	3	1	1	110	4	0	2.19
1992-93	Sherbrooke	QMJHL	56	34	14	5	3190	159	3	2.99	15	9	6	882	57	0	3.87
1993-94	**Quebec**	**NHL**	**29**	**8**	**13**	**3**	**1504**	**83**	**0**	**3.31**	....	....	....	....	....	....	....
	Cornwall Aces	AHL	4	4	0	0	240	9	1	2.25	....	....	....	....	....	....	....
1994-95	Sherbrooke	QMJHL	13	6	6	1	776	38	1	2.94	....	....	....	....	....	....	....
	Quebec	**NHL**	**18**	**12**	**2**	**2**	**898**	**35**	**1**	**2.34**	**3**	**1**	**2**	**148**	**8**	**0**	**3.24**
1995-96	**Colorado**	**NHL**	**10**	**3**	**4**	**2**	**558**	**28**	**0**	**3.01**	....	....	....	....	....	....	....
	Montreal	**NHL**	**40**	**23**	**13**	**3**	**2334**	**110**	**3**	**2.83**	**6**	**2**	**4**	**311**	**18**	**0**	**3.47**
1996-97	**Montreal**	**NHL**	**61**	**22**	**24**	**11**	**3397**	**164**	**1**	**2.90**	**3**	**0**	**3**	**179**	**13**	**0**	**4.36**
1997-98	**Montreal**	**NHL**	**47**	**19**	**15**	**8**	**2652**	**109**	**2**	**2.47**	**2**	**0**	**0**	**43**	**4**	**0**	**5.58**
1998-99	**Montreal**	**NHL**	**10**	**3**	**4**	**2**	**529**	**23**	**1**	**2.61**	....	....	....	....	....	....	....
	Chicago	**NHL**	**52**	**21**	**26**	**5**	**3014**	**136**	**4**	**2.71**	....	....	....	....	....	....	....
99-2000	**Chicago**	**NHL**	**60**	**25**	**26**	**7**	**3438**	**158**	**3**	**2.76**	....	....	....	....	....	....	....
2000-01	**Chicago**	**NHL**	**66**	**27**	**32**	**7**	**3844**	**180**	**6**	**2.81**	....	....	....	....	....	....	....
2001-02	**Chicago**	**NHL**	**67**	**33**	**23**	**9**	**3838**	**159**	**6**	**2.49**	**3**	**1**	**2**	**159**	**7**	**0**	**2.64**
	NHL Totals		**460**	**196**	**182**	**59**	**26006**	**1185**	**27**	**2.73**	**17**	**4**	**11**	**840**	**50**	**0**	**3.57**

QMJHL All-Rookie Team (1992) • QMJHL First All-Star Team (1993) • QMJHL MVP (1993) • Canadian Major Junior First All-Star Team (1993) • Canadian Major Junior Goaltender of the Year (1993)

Transferred to **Colorado** after **Quebec** franchise relocated, June 21, 1995. Traded to **Montreal** by **Colorado** with Andrei Kovalenko and Martin Rucinsky for Patrick Roy and Mike Keane, December 6, 1995. Traded to **Chicago** by **Montreal** with Dave Manson and Brad Brown for Jeff Hackett, Eric Weinrich, Alain Nasreddine and Tampa Bay's 4th round choice (previously acquired, Montreal selected Chris Dyment) in 1999 Entry Draft, November 16, 1998.

THOMAS, Tim (TAW-mas, TIHM) BOS.

Goaltender. Catches left. 5'11", 181 lbs. Born, Flint, MI, April 15, 1974.
(Quebec's 11th choice, 217th overall, in 1994 Entry Draft).

			Regular Season								Playoffs						
Season	**Club**	**Lea**	**GP**	**W**	**L**	**T**	**Mins**	**GA**	**SO**	**Avg**	**GP**	**W**	**L**	**Mins**	**GA**	**SO**	**Avg**
1992-93	Davison Academy	H.S.	27	....	....	....	1580	87	....	3.30	....	....	....	....	....	....	....
1993-94	U. of Vermont	ECAC	*33	15	12	6	1864	94	0	3.03	....	....	....	....	....	....	....
1994-95	U. of Vermont	ECAC	34	18	13	2	2010	90	*4	*2.69	....	....	....	....	....	....	....
1995-96	U. of Vermont	ECAC	37	*26	7	4	*2254	88	*3	*2.34	....	....	....	....	....	....	....
1996-97	U. of Vermont	ECAC	36	22	11	3	2158	101	2	2.81	....	....	....	....	....	....	....
1997-98	HIFK Helsinki	Finland	18	13	4	1	1035	28	2	*1.62	*9	*9	0	*551	14	*3	*1.52
	Birmingham Bulls	ECHL	6	4	1	1	360	13	1	2.17	....	....	....	....	....	....	....
	Houston Aeros	IHL	1	0	1	0	59	4	0	4.01	....	....	....	....	....	....	....
1998-99	HIFK Helsinki	Finland	14	8	3	3	833	31	2	2.23	*11	7	4	*658	25	0	2.28
	United States	WC-A	2	....	....	....	98	7	0	4.25	....	....	....	....	....	....	....
	Hamilton	AHL	15	6	8	0	837	45	0	3.23	....	....	....	....	....	....	....
99-2000	Detroit Vipers	IHL	36	10	21	3	2020	120	1	3.56	....	....	....	....	....	....	....
2000-01	AIK Solna	Sweden	43	....	....	....	2542	105	3	2.48	5	....	....	299	20	0	4.01
2001-02	Karpat Oulu	Finland	32	15	12	5	1937	79	4	2.45	3	1	2	180	12	0	4.00

ECAC First All-Star Team (1995, 1996) • NCAA East Second All-American Team (1995) • NCAA East First All-American Team (1996)

Signed as a free agent by **Edmonton**, June 4, 1998. Signed as a free agent by **Boston**, August 8, 2002.

THOMPSON, Billy (TAWMP-suhn, BIHL-lee) FLA.

Goaltender. Catches left. 6'3", 188 lbs. Born, Saskatoon, Sask., September 24, 1982.
(Florida's 7th choice, 136th overall, in 2001 Entry Draft).

			Regular Season								Playoffs						
Season	**Club**	**Lea**	**GP**	**W**	**L**	**T**	**Mins**	**GA**	**SO**	**Avg**	**GP**	**W**	**L**	**Mins**	**GA**	**SO**	**Avg**
1997-98	Sask. Contacts	SMHL	23	14	5	3	1336	65	2	2.92	....	....	....	....	....	....	....
1998-99	Lebret Eagles	SJHL					STATISTICS NOT AVAILABLE										
99-2000	Estevan Bruins	SJHL	31	....	....	....	1763	132	1	4.49	5	1	3	328	17	0	3.11
	Prince George	WHL	1	0	1	0	60	5	0	5.00	....	....	....	....	....	....	....
2000-01	Prince George	WHL	57	24	24	3	3185	178	1	3.35	6	2	4	324	22	0	4.07
2001-02	Prince George	WHL	42	20	17	2	2375	108	2	2.73	7	3	4	402	21	0	3.13

TOIVONEN, Hannu (TOI-voh-nuhn, HA-noo) BOS.

Goaltender. Catches left. 6'2", 198 lbs. Born, Kalvola, Finland, May 18, 1984.
(Boston's 1st choice, 29th overall, in 2002 Entry Draft).

			Regular Season								Playoffs						
Season	**Club**	**Lea**	**GP**	**W**	**L**	**T**	**Mins**	**GA**	**SO**	**Avg**	**GP**	**W**	**L**	**Mins**	**GA**	**SO**	**Avg**
2000-01	HPK Jr.	Finn-Jr.					STATISTICS NOT AVAILABLE										
2001-02	HPK Jr.	Finn-Jr.	31	15	12	4	1877	103	2	3.29	7	3	4	440	31	0	4.23

TOSKALA, Vesa (TAWS-kah-lah, VEH-sa) S.J.

Goaltender. Catches left. 5'10", 190 lbs. Born, Tampere, Finland, May 20, 1977.
(San Jose's 4th choice, 90th overall, in 1995 Entry Draft).

			Regular Season								Playoffs						
Season	Club	Lea	GP	W	L	T	Mins	GA	SO	Avg	GP	W	L	Mins	GA	SO	Avg
1993-94	Ilves Tampere-2	Finn-Jr.					STATISTICS NOT AVAILABLE										
1994-95	Ilves Tampere	Finn-Jr.	17	....	....	....	956	36	....	2.26	....	....	....	....	....	....	....
1995-96	Ilves Tampere	Finn-Jr.	3	....	....	....	180	3	....	1.00	....	....	....	....	....	....	....
	KooVee Tampere	Finland-2	2	....	....	....	119	5	....	2.51	....	....	....	....	....	....	....
	Ilves Tampere	Finland	37	....	....	....	2073	109	1	3.16	2	....	....	78	11	....	8.49
1996-97	Ilves Tampere	Finland	40	22	12	5	2270	108	0	2.85	8	3	5	479	29	0	3.63
1997-98	Ilves Tampere	Finland	43	*26	13	3	2555	118	1	2.77	*9	6	3	519	18	1	2.08
1998-99	Ilves Tampere	Finland	33	21	12	0	1966	70	*5	2.14	4	1	3	248	14	0	3.39
99-2000	Farjestad	Sweden	44	....	....	....	2652	118	3	2.67	7	....	....	439	19	0	2.60
2000-01	Kentucky	AHL	44	22	13	5	2466	114	2	2.77	3	0	3	197	8	0	2.43
2001-02	**San Jose**	**NHL**	**1**	**0**	**0**	**0**	**10**	**0**	**0**	**0.00**	....	....	....	....	....	....	....
	Cleveland Barons	AHL	*62	19	33	7	*3574	178	3	2.99	....	....	....	....	....	....	....
	NHL Totals		**1**	**0**	**0**	**0**	**10**	**0**	**0**	**0.00**	....	....	....	....	....	....	....

TUGNUTT, Ron (TUHG-nuht, RAWN) DAL.

Goaltender. Catches left. 5'11", 160 lbs. Born, Scarborough, Ont., October 22, 1967.
(Quebec's 4th choice, 81st overall, in 1986 Entry Draft).

			Regular Season								Playoffs						
Season	Club	Lea	GP	W	L	T	Mins	GA	SO	Avg	GP	W	L	Mins	GA	SO	Avg
1983-84	Tor. Young Nats	MTHL	34	....	....	....	1690	91	3	2.67	....	....	....	....	....	....	....
	Weston Dukes	MTJHL	1	0	0	0	20	2	0	6.00	....	....	....	....	....	....	....
1984-85	Peterborough	OHL	18	7	4	2	938	59	0	3.77	....	....	....	....	....	....	....
1985-86	Peterborough	OHL	26	18	7	0	1543	74	1	2.88	3	2	0	133	6	0	2.71
1986-87	Peterborough	OHL	31	21	7	2	1891	88	2	*2.79	6	3	3	374	21	1	3.37
1987-88	**Quebec**	**NHL**	**6**	**2**	**3**	**0**	**284**	**16**	**0**	**3.38**	....	....	....	....	....	....	....
	Fredericton	AHL	34	20	9	4	1964	118	1	3.60	4	1	2	204	11	0	3.24
1988-89	**Quebec**	**NHL**	**26**	**10**	**10**	**3**	**1367**	**82**	**0**	**3.60**	....	....	....	....	....	....	....
	Halifax Citadels	AHL	24	14	7	2	1368	79	1	3.46	....	....	....	....	....	....	....
1989-90	**Quebec**	**NHL**	**35**	**5**	**24**	**3**	**1978**	**152**	**0**	**4.61**	....	....	....	....	....	....	....
	Halifax Citadels	AHL	6	1	5	0	366	23	0	3.77	....	....	....	....	....	....	....
1990-91	**Quebec**	**NHL**	**56**	**12**	**29**	**10**	**3144**	**212**	**0**	**4.05**	....	....	....	....	....	....	....
	Halifax Citadels	AHL	2	0	1	0	100	8	0	4.80	....	....	....	....	....	....	....
1991-92	**Quebec**	**NHL**	**30**	**6**	**17**	**3**	**1583**	**106**	**1**	**4.02**	....	....	....	....	....	....	....
	Halifax Citadels	AHL	8	3	3	1	447	30	0	4.03	....	....	....	....	....	....	....
	Edmonton	**NHL**	**3**	**1**	**1**	**0**	**124**	**10**	**0**	**4.84**	**2**	**0**	**0**	**60**	**3**	**0**	**3.00**
1992-93	**Edmonton**	**NHL**	**26**	**9**	**12**	**2**	**1338**	**93**	**0**	**4.17**	....	....	....	....	....	....	....
1993-94	**Anaheim**	**NHL**	**28**	**10**	**15**	**1**	**1520**	**76**	**1**	**3.00**	....	....	....	....	....	....	....
	Montreal	**NHL**	**8**	**2**	**3**	**1**	**378**	**24**	**0**	**3.81**	**1**	**0**	**1**	**59**	**5**	**0**	**5.08**
1994-95	**Montreal**	**NHL**	**7**	**1**	**3**	**1**	**346**	**18**	**0**	**3.12**	....	....	....	....	....	....	....
1995-96	Portland Pirates	AHL	58	21	23	6	3068	171	2	3.34	13	7	6	782	36	1	2.76
1996-97	**Ottawa**	**NHL**	**37**	**17**	**15**	**1**	**1991**	**93**	**3**	**2.80**	**7**	**3**	**4**	**425**	**14**	**1**	**1.98**
1997-98	**Ottawa**	**NHL**	**42**	**15**	**14**	**8**	**2236**	**84**	**3**	**2.25**	**2**	**0**	**1**	**74**	**6**	**0**	**4.86**
1998-99	**Ottawa**	**NHL**	**43**	**22**	**10**	**8**	**2508**	**75**	**3**	***1.79**	**2**	**0**	**2**	**118**	**6**	**0**	**3.05**
99-2000	**Ottawa**	**NHL**	**44**	**18**	**12**	**8**	**2435**	**103**	**4**	**2.54**	....	....	....	....	....	....	....
	Pittsburgh	**NHL**	**7**	**4**	**2**	**0**	**374**	**15**	**0**	**2.41**	**11**	**6**	**5**	**746**	**22**	**2**	**1.77**
2000-01	**Columbus**	**NHL**	**53**	**22**	**25**	**5**	**3129**	**127**	**4**	**2.44**	....	....	....	....	....	....	....
2001-02	**Columbus**	**NHL**	**44**	**12**	**27**	**3**	**2502**	**119**	**2**	**2.85**	....	....	....	....	....	....	....
	NHL Totals		**495**	**168**	**222**	**57**	**27237**	**1405**	**21**	**3.10**	**25**	**9**	**13**	**1482**	**56**	**3**	**2.27**

OHL First All-Star Team (1987) • Played in NHL All-Star Game (1999)

Traded to **Edmonton** by **Quebec** with Brad Zavisha for Martin Rucinsky, March 10, 1992. Claimed by **Anaheim** from **Edmonton** in Expansion Draft, June 24, 1993. Traded to **Montreal** by **Anaheim** for Stephan Lebeau, February 20, 1994. Signed as a free agent by **Washington**, September 25, 1995. Signed as a free agent by **Ottawa**, August 14, 1996. Traded to **Pittsburgh** by **Ottawa** with Janne Laukkanen for Tom Barrasso, March 14, 2000. Signed as a free agent by **Columbus**, July 4, 2000. Traded to **Dallas** by **Columbus** with Columbus' 2nd round choice (Janos Vas) in 2002 Entry Draft for New Jersey's 1st round choice (previously acquired, later traded to Buffalo - Buffalo selected Dan Paille) in 2002 Entry Draft, June 18, 2002.

TURCO, Marty (TUHR-koh, MAHR-tee) DAL.

Goaltender. Catches left. 5'11", 183 lbs. Born, Sault Ste. Marie, Ont., August 13, 1975.
(Dallas' 4th choice, 124th overall, in 1994 Entry Draft).

			Regular Season								Playoffs						
Season	Club	Lea	GP	W	L	T	Mins	GA	SO	Avg	GP	W	L	Mins	GA	SO	Avg
1993-94	Cambridge	OJHL-B	34	19	10	3	1973	114	0	3.47	....	....	....	....	....	....	....
1994-95	U. of Michigan	CCHA	37	*27	7	1	2063	95	1	2.76	....	....	....	....	....	....	....
1995-96	U. of Michigan	CCHA	*42	*34	7	1	*2335	84	*5	*2.16	....	....	....	....	....	....	....
1996-97	U. of Michigan	CCHA	*41	*33	4	4	*2296	87	*4	*2.27	....	....	....	....	....	....	....
1997-98	U. of Michigan	CCHA	*45	*33	10	1	*2640	95	4	2.16	....	....	....	....	....	....	....
1998-99	Michigan K-Wings	IHL	54	24	17	10	3127	136	1	2.61	5	2	3	300	14	0	2.80
99-2000	Michigan K-Wings	IHL	60	23	27	*7	3399	139	*7	2.45	....	....	....	....	....	....	....
2000-01	**Dallas**	**NHL**	**26**	**13**	**6**	**1**	**1266**	**40**	**3**	***1.90**	....	....	....	....	....	....	....
2001-02	**Dallas**	**NHL**	**31**	**15**	**6**	**2**	**1519**	**53**	**2**	**2.09**	....	....	....	....	....	....	....
	NHL Totals		**57**	**28**	**12**	**3**	**2785**	**93**	**5**	**2.00**	....	....	....	....	....	....	....

CCHA Rookie of the Year (1995) • NCAA Championship All-Tournament Team (1996, 1998) • CCHA First All-Star Team (1997) • NCAA West First All-American Team (1997) • CCHA Second All-Star Team (1998) • NCAA Championship Tournament MVP (1998) • Won Garry F. Longman Memorial Trophy (Top Rookie - IHL) (1999) • Won MBNA Roger Crozier Saving Grace Award (2001)

TUREK, Roman (TOOR-ehk, ROH-muhn) CGY.

Goaltender. Catches right. 6'3", 215 lbs. Born, Strakonice, Czech., May 21, 1970.
(Minnesota's 6th choice, 113th overall, in 1990 Entry Draft).

			Regular Season								Playoffs						
Season	Club	Lea	GP	W	L	T	Mins	GA	SO	Avg	GP	W	L	Mins	GA	SO	Avg
1990-91	Ceske Budejovice	Czech	26	....	....	....	1244	98	0	4.73	....	....	....	....	....	....	....
1991-92	Ceske Budejovice	Czech-2					STATISTICS NOT AVAILABLE										
1992-93	Ceske Budejovice	Czech	43	....	....	....	2555	121	....	2.84	....	....	....	....	....	....	....
1993-94	Ceske Budejovice	Czech	44	....	....	....	2584	111	....	2.58	3	....	....	180	12	0	4.00
	Czech Republic	Olympics	2	2	0	0	120	3	0	1.50	....	....	....	....	....	....	....
1994-95	Ceske Budejovice	Czech	44	....	....	....	2587	119	....	2.76	9	....	....	498	25	....	3.01
1995-96	Nurnberg	Germany	48	....	....	....	2787	154	....	3.32	5	....	....	338	14	....	2.48
1996-97	**Dallas**	**NHL**	**6**	**3**	**1**	**0**	**263**	**9**	**0**	**2.05**	....	....	....	....	....	....	....
	Michigan K-Wings	IHL	29	8	13	4	1555	77	0	2.97	....	....	....	....	....	....	....
1997-98	**Dallas**	**NHL**	**23**	**11**	**10**	**1**	**1324**	**49**	**1**	**2.22**	....	....	....	....	....	....	....
	Michigan K-Wings	IHL	2	1	1	0	119	5	0	2.52	....	....	....	....	....	....	....
1998-99♦	**Dallas**	**NHL**	**26**	**16**	**3**	**3**	**1382**	**48**	**1**	**2.08**	....	....	....	....	....	....	....
99-2000	**St. Louis**	**NHL**	**67**	**42**	**15**	**9**	**3960**	**129**	***7**	**1.95**	**7**	**3**	**4**	**415**	**19**	**0**	**2.75**
2000-01	**St. Louis**	**NHL**	**54**	**24**	**18**	**10**	**3232**	**123**	**6**	**2.28**	**14**	**9**	**5**	**908**	**31**	**0**	**2.05**
2001-02	**Calgary**	**NHL**	**69**	**30**	**28**	**11**	**4081**	**172**	**5**	**2.53**	....	....	....	....	....	....	....
	NHL Totals		**245**	**126**	**75**	**34**	**14242**	**530**	**20**	**2.23**	**21**	**12**	**9**	**1323**	**50**	**0**	**2.27**

Shared William M. Jennings Trophy with Ed Belfour (1999) • NHL Second All-Star Team (2000) • Won William M. Jennings Trophy (2000) • Played in NHL All-Star Game (2000)

Rights transferred to **Dallas** after **Minnesota** franchise relocated, June 9, 1993. Traded to **St. Louis** by **Dallas** for St. Louis' compensatory 2nd round choice (Dan Jancevski) in 1999 Entry Draft, June 20, 1999. Traded to **Calgary** by **St. Louis** with St. Louis' 4th round choice (Yegor Shastin) in 2001 Entry Draft for Fred Brathwaite, Daniel Tkaczuk, Sergei Varlamov and Calgary's 9th round choice (Grant Jacobsen) in 2001 Entry Draft, June 23, 2001.

UNDERHILL, Matt (UHN-duhr-hihl, MAT) CGY.

Goaltender. Catches left. 6'2", 195 lbs. Born, Merritt, B.C., September 16, 1979.
(Calgary's 8th choice, 170th overall, in 1999 Entry Draft).

			Regular Season								Playoffs						
Season	Club	Lea	GP	W	L	T	Mins	GA	SO	Avg	GP	W	L	Mins	GA	SO	Avg
1997-98	Notre Dame	SJHL	43	18	22	3	2573	132	2	3.07	....	....	....	....	....	....	....
1998-99	Cornell Big Red	ECAC	25	7	10	4	1320	65	1	2.95	....	....	....	....	....	....	....
99-2000	Cornell Big Red	ECAC	18	8	7	1	912	44	1	2.89	....	....	....	....	....	....	....
2000-01	Cornell Big Red	ECAC	25	13	8	3	1504	47	1	1.88	....	....	....	....	....	....	....
2001-02	Cornell Big Red	ECAC	21	14	6	1	1334	40	3	1.80	....	....	....	....	....	....	....

ECAC First All-Star Team (2002)

VALIQUETTE, Stephen (val-ih-KEHT, STEEV-uhn) NYI

Goaltender. Catches left. 6'5", 190 lbs. Born, Etobicoke, Ont., August 20, 1977.
(Los Angeles' 8th choice, 190th overall, in 1996 Entry Draft).

			Regular Season								Playoffs						
Season	Club	Lea	GP	W	L	T	Mins	GA	SO	Avg	GP	W	L	Mins	GA	SO	Avg
1993-94	Burlington	OPJHL	30	....	....	....	1663	112	1	4.04	....	....	....	....	....	....	....
1994-95	Rayside-Balfour	NOJHA	2	0	2	0	89	12	0	8.09	....	....	....	....	....	....	....
	Smiths Falls	OCJHL	21	10	8	3	1275	75	0	3.53	....	....	....	....	....	....	....
	Sudbury Wolves	OHL	4	2	0	0	138	6	0	2.61	....	....	....	....	....	....	....
1995-96	Sudbury Wolves	OHL	39	13	16	2	1887	123	0	3.91	....	....	....	....	....	....	....
1996-97	Sudbury Wolves	OHL	*61	21	29	7	3311	232	1	4.20	....	....	....	....	....	....	....
	Dayton Bombers	ECHL	3	1	0	0	89	6	0	4.03	2	1	1	118	5	0	2.54
1997-98	Sudbury Wolves	OHL	14	5	7	1	807	50	0	3.72	....	....	....	....	....	....	....
	Erie Otters	OHL	28	16	7	3	1525	65	3	2.56	7	3	4	467	15	1	1.93
1998-99	Hampton Roads	ECHL	31	18	7	3	1713	84	1	2.94	2	0	1	60	7	0	7.00
	Lowell	AHL	1	0	1	0	59	3	0	3.05	....	....	....	....	....	....	....
99-2000	**NY Islanders**	**NHL**	**6**	**2**	**0**	**0**	**193**	**6**	**0**	**1.87**	....	....	....	....	....	....	....
	Lowell	AHL	14	8	5	0	727	36	0	2.97	....	....	....	....	....	....	....
	Providence	AHL	1	1	0	0	60	3	0	3.00	....	....	....	....	....	....	....
	Trenton Titans	ECHL	12	5	6	1	692	36	1	3.12	....	....	....	....	....	....	....
2000-01	Springfield	AHL	20	7	10	1	1066	54	0	3.04	....	....	....	....	....	....	....
2001-02	Bridgeport	AHL	20	10	5	1	1071	45	2	2.52	1	0	0	18	1	0	3.30
	NHL Totals		**6**	**2**	**0**	**0**	**193**	**6**	**0**	**1.87**	....	....	....	....	....	....	....

Traded to **Windsor** (OHL) by **Sudbury** (OHL) with Paul Mara for Glenn Crawford, Kip Brennan and future considerations, December 16, 1997. Traded to **Erie** (OHL) by **Windsor** (OHL) with Jeff Zehr for Craig Jolbert, Jason Polera and Jeff Jkapitanchuk, December 17, 1997. Signed as a free agent by **NY Islanders**, August 18, 1998.

VANBIESBROUCK, John (van-BEES-bruhk, JAWN)

Goaltender. Catches left. 5'8", 176 lbs. Born, Detroit, MI, September 4, 1963.
(NY Rangers' 5th choice, 72nd overall, in 1981 Entry Draft).

			Regular Season								Playoffs						
Season	Club	Lea	GP	W	L	T	Mins	GA	SO	Avg	GP	W	L	Mins	GA	SO	Avg
1979-80	Bishop Gallagher	H.S.					STATISTICS NOT AVAILABLE										
1980-81	Sault Ste. Marie	OMJHL	56	31	16	1	2941	203	0	4.14	11	3	3	457	24	1	3.15
1981-82	Sault Ste. Marie	OHL	31	12	12	2	1686	102	0	3.62	7	1	4	276	20	0	4.35
	NY Rangers	**NHL**	**1**	**1**	**0**	**0**	**60**	**1**	**0**	**1.00**	....	....	....	....	....	....	....
1982-83	Sault Ste. Marie	OHL	*62	39	21	1	3471	209	0	3.61	16	7	6	944	56	*1	3.56
1983-84	**NY Rangers**	**NHL**	**3**	**2**	**1**	**0**	**180**	**10**	**0**	**3.33**	**1**	**0**	**0**	**1**	**0**	**0**	**0.00**
	Tulsa Oilers	CHL	37	20	13	2	2153	124	*3	3.46	4	4	0	240	10	0	*2.50
1984-85	**NY Rangers**	**NHL**	**42**	**12**	**24**	**3**	**2358**	**166**	**1**	**4.22**	**1**	**0**	**0**	**20**	**0**	**0**	**0.00**
1985-86	**NY Rangers**	**NHL**	**61**	***31**	**21**	**5**	**3326**	**184**	**3**	**3.32**	**16**	**8**	**8**	**899**	**49**	***1**	**3.27**
1986-87	**NY Rangers**	**NHL**	**50**	**18**	**20**	**5**	**2656**	**161**	**0**	**3.64**	**4**	**1**	**3**	**195**	**11**	**1**	**3.38**
1987-88	**NY Rangers**	**NHL**	**56**	**27**	**22**	**7**	**3319**	**187**	**2**	**3.38**	....	....	....	....	....	....	....
1988-89	**NY Rangers**	**NHL**	**56**	**28**	**21**	**4**	**3207**	**197**	**0**	**3.69**	**2**	**0**	**1**	**107**	**6**	**0**	**3.36**
1989-90	**NY Rangers**	**NHL**	**47**	**19**	**19**	**7**	**2734**	**154**	**1**	**3.38**	**6**	**2**	**3**	**298**	**15**	**0**	**3.02**
1990-91	**NY Rangers**	**NHL**	**40**	**15**	**18**	**6**	**2257**	**126**	**3**	**3.35**	**1**	**0**	**0**	**52**	**1**	**0**	**1.15**
1991-92	**NY Rangers**	**NHL**	**45**	**27**	**13**	**3**	**2526**	**120**	**2**	**2.85**	**7**	**2**	**5**	**368**	**23**	**0**	**3.75**
1992-93	**NY Rangers**	**NHL**	**48**	**20**	**18**	**7**	**2757**	**152**	**4**	**3.31**	....	....	....	....	....	....	....
1993-94	**Florida**	**NHL**	**57**	**21**	**25**	**11**	**3440**	**145**	**1**	**2.53**	....	....	....	....	....	....	....
1994-95	**Florida**	**NHL**	**37**	**14**	**15**	**4**	**2087**	**86**	**4**	**2.47**	....	....	....	....	....	....	....
1995-96	**Florida**	**NHL**	**57**	**26**	**20**	**7**	**3178**	**142**	**2**	**2.68**	***22**	**12**	**10**	**1332**	**50**	**1**	**2.25**
1996-97	**Florida**	**NHL**	**57**	**27**	**19**	**10**	**3347**	**128**	**2**	**2.29**	**5**	**1**	**4**	**328**	**13**	**1**	**2.38**
1997-98	**Florida**	**NHL**	**60**	**18**	**29**	**11**	**3451**	**165**	**4**	**2.87**	....	....	....	....	....	....	....
	United States	Olympics	1	0	0	0	1	0	0	0.00	....	....	....	....	....	....	....
1998-99	**Philadelphia**	**NHL**	**62**	**27**	**18**	**15**	**3712**	**135**	**6**	**2.18**	**6**	**2**	**4**	**369**	**9**	**1**	**1.46**
99-2000	**Philadelphia**	**NHL**	**50**	**25**	**15**	**9**	**2950**	**108**	**3**	**2.20**	....	....	....	....	....	....	....
2000-01	**NY Islanders**	**NHL**	**44**	**10**	**25**	**5**	**2390**	**120**	**1**	**3.01**	....	....	....	....	....	....	....
	New Jersey	**NHL**	**4**	**4**	**0**	**0**	**240**	**6**	**1**	**1.50**	....	....	....	....	....	....	....
2001-02	**New Jersey**	**NHL**	**5**	**2**	**3**	**0**	**300**	**10**	**0**	**2.00**	....	....	....	....	....	....	....
	NHL Totals		**882**	**374**	**346**	**119**	**50475**	**2503**	**40**	**2.98**	**71**	**28**	**38**	**3969**	**177**	**5**	**2.68**

OHL Second All-Star Team (1983) • CHL First All-Star Team (1984) • Shared Terry Sawchuk Trophy (fewest goals against - CHL) with Ron Scott (1984) • Shared Tommy Ivan Trophy (MVP - CHL) with Bruce Affleck (1984) • NHL First All-Star Team (1986) • Won Vezina Trophy (1986) • NHL Second All-Star Team (1994) • Played in NHL All-Star Game (1994, 1996, 1997)

Traded to **Vancouver** by **NY Rangers** for future considerations (Doug Lidster, June 25, 1993), June 20, 1993. Claimed by **Florida** from **Vancouver** in Expansion Draft, June 24, 1993. Signed as a free agent by **Philadelphia**, July 16, 1998. Traded to **NY Islanders** by **Philadelphia** for NY Islanders' 4th round choice (later traded to Nashville - Nashville selected Jordin Tootoo) in 2001 Entry Draft, June 25, 2000. Traded to **New Jersey** by **NY Islanders** for Chris Terreri and New Jersey's 9th round choice (Juha-Pekka Ketola) in 2001 Entry Draft, March 12, 2001. • Officially announced retirement, June 10, 2001. Signed as a free agent by **New Jersey**, February 4, 2002. • Officially announced retirement, May 22, 2002.

VERNON, Mike (VUHR-nuhn, MIGHK)

Goaltender. Catches left. 5'9", 180 lbs. Born, Calgary, Alta., February 24, 1963.
(Calgary's 2nd choice, 56th overall, in 1981 Entry Draft).

			Regular Season								Playoffs						
Season	Club	Lea	GP	W	L	T	Mins	GA	SO	Avg	GP	W	L	Mins	GA	SO	Avg
1979-80	Calgary Canucks	AJHL	31	21	7	0	1796	88	0	2.95	7	3	4	399	22	0	3.30
1980-81	Calgary	WHL	59	33	17	1	3154	198	1	3.77	22	14	8	1271	82	1	3.87
1981-82	Calgary	WHL	42	22	14	2	2329	143	3	3.68	9	5	4	527	30	0	3.42
	Oklahoma City	CHL	....	....	....	....	....	....	....	....	1	0	1	70	4	0	3.43
1982-83	Calgary	WHL	50	29	18	2	2856	155	*3	*3.26	16	9	7	925	60	0	3.89
	Calgary	**NHL**	**2**	**0**	**2**	**0**	**100**	**11**	**0**	**6.59**	....	....	....	....	....	....	....
1983-84	**Calgary**	**NHL**	**1**	**0**	**1**	**0**	**11**	**4**	**0**	**22.22**	....	....	....	....	....	....	....
	Colorado Flames	CHL	46	30	13	2	2648	148	1	*3.35	6	2	4	347	21	0	3.63
1984-85	Moncton	AHL	41	10	20	4	2050	134	0	3.92	....	....	....	....	....	....	....
1985-86	**Calgary**	**NHL**	**18**	**9**	**3**	**3**	**921**	**52**	**1**	**3.39**	***21**	**12**	***9**	***1229**	**60**	**0**	**2.93**
	Moncton	AHL	6	3	1	2	374	21	0	3.37	....	....	....	....	....	....	....
	Salt Lake	IHL	10	6	4	0	600	34	1	3.40	....	....	....	....	....	....	....
1986-87	**Calgary**	**NHL**	**54**	**30**	**21**	**1**	**2957**	**178**	**1**	**3.61**	**5**	**2**	**3**	**263**	**16**	**0**	**3.65**
1987-88	**Calgary**	**NHL**	**64**	**39**	**16**	**7**	**3565**	**210**	**1**	**3.53**	**9**	**4**	**4**	**515**	**34**	**0**	**3.96**
1988-89 ♦	**Calgary**	**NHL**	**52**	***37**	**6**	**5**	**2938**	**130**	**0**	**2.65**	***22**	***16**	**5**	***1381**	**52**	***3**	**2.26**
1989-90	**Calgary**	**NHL**	**47**	**23**	**14**	**9**	**2795**	**146**	**0**	**3.13**	**6**	**2**	**3**	**342**	**19**	**0**	**3.33**
1990-91	**Calgary**	**NHL**	**54**	**31**	**19**	**3**	**3121**	**172**	**1**	**3.31**	**7**	**3**	**4**	**427**	**21**	**0**	**2.95**
1991-92	**Calgary**	**NHL**	**63**	**24**	**30**	**9**	**3640**	**217**	**0**	**3.58**	....	....	....	....	....	....	....
1992-93	**Calgary**	**NHL**	**64**	**29**	**26**	**9**	**3732**	**203**	**2**	**3.26**	**4**	**1**	**1**	**150**	**15**	**0**	**6.00**
1993-94	**Calgary**	**NHL**	**48**	**26**	**17**	**5**	**2798**	**131**	**3**	**2.81**	**7**	**3**	**4**	**466**	**23**	**0**	**2.96**
1994-95	**Detroit**	**NHL**	**30**	**19**	**6**	**4**	**1807**	**76**	**1**	**2.52**	**18**	**12**	**6**	**1063**	**41**	**1**	**2.31**
1995-96	**Detroit**	**NHL**	**32**	**21**	**7**	**2**	**1855**	**70**	**3**	**2.26**	**4**	**2**	**2**	**243**	**11**	**0**	**2.72**
1996-97 ♦	**Detroit**	**NHL**	**33**	**13**	**11**	**8**	**1952**	**79**	**0**	**2.43**	***20**	***16**	**4**	***1229**	**36**	**1**	**1.76**
1997-98	**San Jose**	**NHL**	**62**	**30**	**22**	**8**	**3564**	**146**	**5**	**2.46**	**6**	**2**	**4**	**348**	**14**	**1**	**2.41**
1998-99	**San Jose**	**NHL**	**49**	**16**	**22**	**10**	**2831**	**107**	**4**	**2.27**	**5**	**2**	**3**	**321**	**13**	**0**	**2.43**
99-2000	**San Jose**	**NHL**	**15**	**6**	**5**	**1**	**772**	**32**	**0**	**2.49**	....	....	....	....	....	....	....
	Florida	**NHL**	**34**	**18**	**13**	**2**	**2019**	**83**	**1**	**2.47**	**4**	**0**	**4**	**237**	**12**	**0**	**3.04**
2000-01	**Calgary**	**NHL**	**41**	**12**	**23**	**5**	**2246**	**121**	**3**	**3.23**	....	....	....	....	....	....	....
2001-02	**Calgary**	**NHL**	**18**	**2**	**9**	**1**	**825**	**38**	**1**	**2.76**	....	....	....	....	....	....	....
	NHL Totals		**781**	**385**	**273**	**92**	**44449**	**2206**	**27**	**2.98**	**138**	**77**	**56**	**8214**	**367**	**6**	**2.68**

WHL First All-Star Team (1982, 1983) • WHL MVP (1982, 1983) • Won Hap Emms Memorial Trophy (Memorial Cup Tournament Top Goaltender) (1983) • CHL Second All-Star Team (1984) • NHL Second All-Star Team (1989) • Shared William M. Jennings Trophy with Chris Osgood (1996) • Won Conn Smythe Trophy (1997) • Played in NHL All-Star Game (1988, 1989, 1990, 1991, 1993)

Traded to **Detroit** by **Calgary** for Steve Chiasson, June 29, 1994. Traded to **San Jose** by **Detroit** with Detroit's 5th round choice (later traded back to Detroit - Detroit selected Andrei Maximenko) in 1999 Entry Draft for San Jose's 2nd round choice (later traded to St. Louis - St. Louis selected Maxim Linnik) in 1998 Entry Draft and San Jose's 2nd round choice (later traded to Tampa Bay - Tampa Bay selected Sheldon Keefe) in 1999 Entry Draft, August 18, 1997. Traded to **Florida** by **San Jose** with San Jose's 3rd round choice (Sean O'Connor) in 2000 Entry Draft for Radek Dvorak, December 30, 1999. Selected by **Minnesota** from **Florida** in Expansion Draft, June 23, 2000. Traded to **Calgary** by **Minnesota** for the rights to Dan Cavanaugh and Calgary's 8th round choice (Jake Riddle) in 2001 Entry Draft, June 23, 2000.

VOKOUN, Tomas (voh-KOHN, TAW-mas) NSH.

Goaltender. Catches right. 6', 195 lbs. Born, Karlovy Vary, Czech., July 2, 1976.
(Montreal's 11th choice, 226th overall, in 1994 Entry Draft).

			Regular Season								Playoffs						
Season	Club	Lea	GP	W	L	T	Mins	GA	SO	Avg	GP	W	L	Mins	GA	SO	Avg
1993-94	HC Kladno	Czech	1	0	0	0	20	2	0	6.01	....	....	....	....	....	....	....
1994-95	HC Kladno	Czech	26	....	....	....	1368	70	....	3.07	5	....	....	240	19	....	4.75
1995-96	Wheeling	ECHL	35	20	10	2	1912	117	0	3.67	7	4	3	436	19	0	2.61
	Fredericton	AHL	....	....	....	....	....	....	....	....	1	0	1	59	4	0	4.09
1996-97	**Montreal**	**NHL**	**1**	**0**	**0**	**0**	**20**	**4**	**0**	**12.00**	....	....	....	....	....	....	....
	Fredericton	AHL	47	12	26	7	2645	154	2	3.49	....	....	....	....	....	....	....
1997-98	Fredericton	AHL	31	13	13	2	1735	90	0	3.11	....	....	....	....	....	....	....
1998-99	**Nashville**	**NHL**	**37**	**12**	**18**	**4**	**1954**	**96**	**1**	**2.95**	....	....	....	....	....	....	....
	Milwaukee	IHL	9	3	2	4	539	22	1	2.45	2	0	2	149	8	0	3.22
99-2000	**Nashville**	**NHL**	**33**	**9**	**20**	**1**	**1879**	**87**	**1**	**2.78**	....	....	....	....	....	....	....
	Milwaukee	IHL	7	5	2	0	364	17	0	2.80	....	....	....	....	....	....	....
2000-01	**Nashville**	**NHL**	**37**	**13**	**17**	**5**	**2088**	**85**	**2**	**2.44**	....	....	....	....	....	....	....
2001-02	**Nashville**	**NHL**	**29**	**5**	**14**	**4**	**1471**	**66**	**2**	**2.69**	....	....	....	....	....	....	....
	NHL Totals		**137**	**39**	**69**	**14**	**7412**	**338**	**6**	**2.74**	....	....	....	....	....	....	....

Claimed by **Nashville** from **Montreal** in Expansion Draft, June 26, 1998.

VOLKOV, Alexei (VOHL-kawf, al-EHX-ay) L.A.

Goaltender. Catches left. 6'1", 195 lbs. Born, Yekaterinburg, USSR, March 15, 1980.
(Los Angeles' 3rd choice, 76th overall, in 1998 Entry Draft).

			Regular Season								Playoffs						
Season	Club	Lea	GP	W	L	T	Mins	GA	SO	Avg	GP	W	L	Mins	GA	SO	Avg
1995-96	SKA Yekaterin.	Russia-2	42	....	....	....	2520	78	....	1.87	....	....	....	....	....	....	....
1996-97	Kryl. Sovetov Jr.	Russia-Jr.	8	....	....	....	48	9	....	1.25	....	....	....	....	....	....	....
	Yekaterinburg 2	Russia-3	34	....	....	....	2040	66	....	1.94	....	....	....	....	....	....	....
1997-98	Krylja Sovetov 2	Russia-3	27	....	....	....	1620	72	....	2.67	....	....	....	....	....	....	....
1998-99	Halifax	QMJHL	39	25	9	3	2332	105	2	2.70	5	1	4	282	21	0	4.47
99-2000	Halifax	QMJHL	40	23	13	2	2222	124	1	3.35	8	3	4	417	29	0	4.18
2000-01	New Orleans	ECHL	29	12	9	5	1577	81	1	3.08	....	....	....	....	....	....	....
	Lowell	AHL	5	1	1	0	202	16	0	4.75	....	....	....	....	....	....	....
2001-02	Ufa	Russia	29	....	....	....	1710	62	3	2.18	....	....	....	....	....	....	....

WANDLER, Bryce (WAND-luhr, BRIGHS) NYR

Goaltender. Catches left. 6', 180 lbs. Born, Lacombe, Alta., February 25, 1979.

			Regular Season								Playoffs						
Season	Club	Lea	GP	W	L	T	Mins	GA	SO	Avg	GP	W	L	Mins	GA	SO	Avg
1996-97	Kamloops Blazers	WHL	1	0	0	0	34	3	0	5.29	....	....	....	....	....	....	....
	Edmonton Ice	WHL	19	2	11	1	936	84	0	5.38	....	....	....	....	....	....	....
1997-98	Edmonton Ice	WHL	47	12	27	4	2576	180	1	4.19	....	....	....	....	....	....	....
1998-99	Swift Current	WHL	51	23	20	4	2882	123	3	2.56	6	2	4	364	17	1	2.80
99-2000	Swift Current	WHL	56	*37	15	2	3255	112	6	*2.06	10	5	5	597	29	0	2.91
2000-01	New Haven	UHL	26	10	13	1	1429	71	0	2.98	2	0	1	71	6	0	5.04
	Hartford	AHL	1	0	0	0	29	1	0	2.04	....	....	....	....	....	....	....
	Charlotte	ECHL	4	2	2	0	240	18	0	4.50	....	....	....	....	....	....	....
2001-02	Hartford	AHL	1	0	0	0	2	0	0	0.00	....	....	....	....	....	....	....
	Charlotte	ECHL	31	16	5	5	1687	76	1	2.70	1	0	1	59	4	0	4.10

Signed as a free agent by **NY Rangers**, March 15, 2000.

WARD, Blake (WOHRD, BLAYK) COL.

Goaltender. Catches left. 6'3", 200 lbs. Born, Lloydminster, Alta., January 18, 1982.
(Colorado's 14th choice, 285th overall, in 2000 Entry Draft).

			Regular Season								Playoffs						
Season	Club	Lea	GP	W	L	T	Mins	GA	SO	Avg	GP	W	L	Mins	GA	SO	Avg
1997-98	Lloydminster	SAHA	35	....	....	....	2100	106	1	2.90	....	....	....	....	....	....	....
1998-99	Tri-City	WHL	21	5	5	2	844	34	0	2.42	1	0	0	20	1	0	3.00
99-2000	Tri-City	WHL	37	14	12	2	1853	111	1	3.59	4	0	3	187	12	0	3.85
2000-01	Tri-City	WHL	8	2	3	3	470	30	0	3.83	....	....	....	....	....	....	....
	Lethbridge	WHL	30	10	15	3	1676	79	2	2.83	4	1	3	238	20	0	5.04
2001-02	Lethbridge	WHL	36	11	18	4	1949	117	0	3.60	....	....	....	....	....	....	....

Traded to **Lethbridge** (WHL) by **Tri-City** (WHL) for future considerations, October 25, 2000.

WARD, Cam (WOHRD, KAM) CAR.

Goaltender. Catches left. 6', 176 lbs. Born, Sherwood Park, Alta., February 29, 1984.
(Carolina's 1st choice, 25th overall, in 2002 Entry Draft).

			Regular Season								Playoffs						
Season	Club	Lea	GP	W	L	T	Mins	GA	SO	Avg	GP	W	L	Mins	GA	SO	Avg
1998-99	Sherwood Park	ABHL	24	13	7	4	1403	85	0	3.64	....	....	....	....	....	....	....
99-2000	Sherwood Park	AMHL	20	9	5	1	1194	71	0	3.57	7	4	3	262	22	0	3.57
2000-01	Sherwood Park	AMHL	25	14	6	3	1449	70	0	2.90	....	....	....	....	....	....	....
	Red Deer Rebels	WHL	1	1	0	0	60	0	1	0.00	....	....	....	....	....	....	....
2001-02	Red Deer Rebels	WHL	46	30	11	4	2694	102	1	*2.27	*23	14	9	*1502	53	*2	2.12

Won Bill Ranford Trophy (Top Goaltender - AMHL) (2001) • WHL East First All-Star Team (2002)

WEEKES, Kevin (WEEKS, KEH-vihn) CAR.

Goaltender. Catches left. 6', 195 lbs. Born, Toronto, Ont., April 4, 1975.
(Florida's 2nd choice, 41st overall, in 1993 Entry Draft).

			Regular Season								Playoffs						
Season	Club	Lea	GP	W	L	T	Mins	GA	SO	Avg	GP	W	L	Mins	GA	SO	Avg
1990-91	Tor. Red Wings	MTHL	STATISTICS NOT AVAILABLE														
	St. Michael's B	OJHL-B	1	0	0	0	41	1	0	1.46	....	....	....	....	....	....	....
1991-92	Tor. Red Wings	MTHL	35	....	....	....	1575	68	4	1.94	....	....	....	....	....	....	....
	St. Michael's B	OJHL-B	2	0	1	1	127	11	0	5.20	4	1	2	214	15	1	4.21
1992-93	Owen Sound	OHL	29	9	12	5	1645	143	0	5.22	1	0	0	26	5	0	11.50
1993-94	Owen Sound	OHL	34	13	19	1	1974	158	0	4.80	....	....	....	....	....	....	....
1994-95	Ottawa 67's	OHL	41	13	23	4	2266	153	1	4.05	....	....	....	....	....	....	....
1995-96	Carolina	AHL	60	24	25	8	3404	229	2	4.04	....	....	....	....	....	....	....
1996-97	Carolina	AHL	51	17	28	4	2899	172	1	3.56	....	....	....	....	....	....	....
1997-98	**Florida**	**NHL**	**11**	**0**	**5**	**1**	**485**	**32**	**0**	**3.96**	....	....	....	....	....	....	....
	Fort Wayne	IHL	12	9	2	1	719	34	1	2.84	....	....	....	....	....	....	....
1998-99	**Vancouver**	**NHL**	**11**	**0**	**8**	**1**	**532**	**34**	**0**	**3.83**	....	....	....	....	....	....	....
	Detroit Vipers	IHL	33	19	5	7	1857	64	*4	*2.07	....	....	....	....	....	....	....
99-2000	**Vancouver**	**NHL**	**20**	**6**	**7**	**4**	**987**	**47**	**1**	**2.86**	....	....	....	....	....	....	....
	NY Islanders	**NHL**	**36**	**10**	**20**	**4**	**2026**	**115**	**1**	**3.41**	....	....	....	....	....	....	....
2000-01	**Tampa Bay**	**NHL**	**61**	**20**	**33**	**3**	**3378**	**177**	**4**	**3.14**	....	....	....	....	....	....	....
2001-02	**Tampa Bay**	**NHL**	**19**	**3**	**9**	**0**	**830**	**40**	**2**	**2.89**	....	....	....	....	....	....	....
	Carolina	**NHL**	**2**	**2**	**0**	**0**	**120**	**3**	**0**	**1.50**	**8**	**3**	**2**	**408**	**11**	**2**	**1.62**
	NHL Totals		**160**	**41**	**82**	**13**	**8358**	**448**	**8**	**3.22**	**8**	**3**	**2**	**408**	**11**	**2**	**1.62**

Shared James Norris Memorial Trophy (fewest goals against - IHL) with Andrei Trefilov (1999)

Traded to **Vancouver** by **Florida** with Ed Jovanovski, Dave Gagner, Mike Brown and Florida's 1st round choice (Nathan Smith) in 2000 Entry Draft for Pavel Bure, Bret Hedican, Brad Ference and Vancouver's 3rd round choice (Robert Fried) in 2000 Entry Draft, January 17, 1999. Traded to **NY Islanders** by **Vancouver** with Dave Scatchard and Bill Muckalt for Felix Potvin and NY Islanders' compensatory 2nd (later traded to New Jersey - New Jersey selected Teemu Laine) and 3rd (Thatcher Bell) round choices in 2000 Entry Draft, December 19, 1999. Traded to **Tampa Bay** by **NY Islanders** with the rights to Kristian Kudroc and NY Islanders' 2nd round choice (later traded to Phoenix - Phoenix selected Matthew Spiller) in 2001 Entry Draft for Tampa Bay's 1st round choice (Raffi Torres) in 2000 Entry Draft, Calgary's 4th round choice (previously acquired, NY Islanders selected Vladimir Gorbunov) in 2000 Entry Draft and NY Islanders' 7th round choice (previously acquired, NY Islanders selected Ryan Caldwell) in 2000 Entry Draft, June 24, 2000. Traded to **Carolina** by **Tampa Bay** for Shane Willis and Chris Dingman, March 5, 2002.

WEIMAN, Tyler (WIGH-muhn, TIGH-luhr) COL.

Goaltender. Catches left. 5'11", 160 lbs. Born, Saskatoon, Sask., June 5, 1984.
(Colorado's 6th choice, 164th overall, in 2002 Entry Draft).

			Regular Season								Playoffs						
Season	Club	Lea	GP	W	L	T	Mins	GA	SO	Avg	GP	W	L	Mins	GA	SO	Avg
99-2000	Ft-Saskatchewan	AMBHL	21	15	4	2	1239	60	0	2.91	....	....	....	....	....	....	....
2000-01	Tri-City	WHL	44	10	25	4	2464	155	0	3.77	....	....	....	....	....	....	....
2001-02	Tri-City	WHL	47	18	17	5	2492	149	2	3.59	5	1	4	300	14	0	2.80

WHITMORE, Kay (WHIHT-mohr, KAY) CGY.

Goaltender. Catches left. 5'11", 175 lbs. Born, Sudbury, Ont., April 10, 1967.
(Hartford's 2nd choice, 26th overall, in 1985 Entry Draft).

			Regular Season								Playoffs						
Season	Club	Lea	GP	W	L	T	Mins	GA	SO	Avg	GP	W	L	Mins	GA	SO	Avg
1982-83	Sudbury Legion	NOJHA	43	....	....	....	2580	108	4	2.51	....	....	....	....	....	....	....
1983-84	Peterborough	OHL	29	17	8	0	1471	110	0	4.49	....	....	....	....	....	....	....
1984-85	Peterborough	OHL	*53	*35	16	2	*3077	172	*2	3.35	17	10	4	1020	58	0	3.41
1985-86	Peterborough	OHL	41	27	12	2	2467	114	*3	*2.77	14	8	5	837	40	0	2.87
1986-87	Peterborough	OHL	36	14	17	5	2159	118	1	3.28	7	3	3	366	17	1	2.79
1987-88	Binghamton	AHL	38	17	15	4	2137	121	*3	3.40	2	0	2	118	10	0	5.08
1988-89	**Hartford**	**NHL**	**3**	**2**	**1**	**0**	**180**	**10**	**0**	**3.33**	**2**	**0**	**2**	**135**	**10**	**0**	**4.44**
	Binghamton	AHL	*56	21	29	4	*3200	241	1	4.52	....	....	....	....	....	....	....
1989-90	**Hartford**	**NHL**	**9**	**4**	**2**	**1**	**442**	**26**	**0**	**3.53**	....	....	....	....	....	....	....
	Binghamton	AHL	24	3	19	2	1386	109	0	4.72	....	....	....	....	....	....	....
1990-91	**Hartford**	**NHL**	**18**	**3**	**9**	**3**	**850**	**52**	**0**	**3.67**	....	....	....	....	....	....	....
	Springfield	AHL	33	22	9	1	1916	98	1	3.07	*15	*11	4	*926	37	0	*2.40
1991-92	**Hartford**	**NHL**	**45**	**14**	**21**	**6**	**2567**	**155**	**3**	**3.62**	**1**	**0**	**0**	**19**	**1**	**0**	**3.16**
1992-93	**Vancouver**	**NHL**	**31**	**18**	**8**	**4**	**1817**	**94**	**1**	**3.10**	....	....	....	....	....	....	....
1993-94	**Vancouver**	**NHL**	**32**	**18**	**14**	**0**	**1921**	**113**	**0**	**3.53**	....	....	....	....	....	....	....
1994-95	**Vancouver**	**NHL**	**11**	**0**	**6**	**2**	**558**	**37**	**0**	**3.98**	**1**	**0**	**0**	**20**	**2**	**0**	**6.00**
1995-96	Detroit Vipers	IHL	10	3	5	0	501	33	0	3.95	....	....	....	....	....	....	....
	Los Angeles	IHL	30	10	9	7	1563	99	1	3.80	....	....	....	....	....	....	....
	Syracuse Crunch	AHL	11	6	4	1	663	37	0	3.35	....	....	....	....	....	....	....
	Binghamton	AHL	....	....	....	....	....	....	....	....	2	0	2	127	9	0	4.27
1996-97	Sodertalje SK	Sweden	25	....	....	....	1320	85	0	3.86	....	....	....	....	....	....	....
1997-98	Long Beach	IHL	46	28	12	3	2516	109	3	2.60	14	9	5	838	43	0	3.08
1998-99	Milwaukee	IHL	23	10	6	4	1304	64	0	2.94	....	....	....	....	....	....	....
	Hartford	AHL	18	8	8	2	1080	47	0	2.61	....	....	....	....	....	....	....
99-2000	Providence	AHL	43	17	19	3	2393	127	1	3.18	1	0	1	59	2	0	2.04
2000-01	**Boston**	**NHL**	**5**	**1**	**2**	**0**	**203**	**18**	**0**	**5.32**	....	....	....	....	....	....	....
	Providence	AHL	26	13	8	2	1460	65	2	2.67	....	....	....	....	....	....	....
2001-02	**Calgary**	**NHL**	**1**	**0**	**1**	**0**	**58**	**3**	**0**	**3.10**	....	....	....	....	....	....	....
	Saint John	AHL	36	10	16	8	2001	83	0	2.49	....	....	....	....	....	....	....
	NHL Totals		**155**	**60**	**64**	**16**	**8596**	**508**	**4**	**3.55**	**4**	**0**	**2**	**174**	**13**	**0**	**4.48**

OHL First All-Star Team (1986) • Won Jack A. Butterfield Trophy (Playoff MVP - AHL) (1991) • Shared James Norris Memorial Trophy (fewest goals against - IHL) with Mike Buzak (1998)

Traded to **Vancouver** by **Hartford** for Corrie D'Alessio and cash, October 1, 1992. Traded to **NY Rangers** by **Vancouver** for Joe Kocur, March 20, 1996. Signed as a free agent by **San Jose**, September 10, 1997. Traded to **Buffalo** by **San Jose** with Colorado's 2nd round choice (previously acquired, Buffalo selected Jaroslav Kristek) in 1998 Entry Draft and San Jose's 5th round choice (later traded to Columbus - Columbus selected Tyler Kolarik) in 2000 Entry Draft for Steve Shields and Buffalo's 4th round choice (Miroslav Zalesak) in 1998 Entry Draft, June 18, 1998. Signed as a free agent by **NY Rangers**, August 17, 1998. Signed as a free agent by **Boston**, August 25, 1999. Traded to **Edmonton** by **Boston** for Mike Matteucci, December 28, 1999. Traded to **Boston** by **Edmonton** for future considerations, July 20, 2000. Signed as a free agent by **Calgary**, July 9, 2001.

YEATS, Matthew (YAYTS, MA-thew) L.A.

Goaltender. Catches left. 5'11", 165 lbs. Born, Montreal, Que., April 6, 1979.
(Los Angeles' 9th choice, 248th overall, in 1998 Entry Draft).

			Regular Season								Playoffs						
Season	**Club**	**Lea**	**GP**	**W**	**L**	**T**	**Mins**	**GA**	**SO**	**Avg**	**GP**	**W**	**L**	**Mins**	**GA**	**SO**	**Avg**
1995-96	Lethbridge	WHL	1	0	0	0	20	3	0	9.00	....	....	....	....	....	....	....
1996-97	Olds Grizzlies	AJHL	32	....	....	....	1678	95	1	3.41	....	....	....	....	....	....	....
1997-98	Olds Grizzlies	AJHL	26	12	12	1	1498	96	0	3.85	....	....	....	....	....	....	....
1998-99	U. of Maine	H-East	DID NOT PLAY														
99-2000	U. of Maine	H-East	32	20	6	4	1821	79	0	2.60	....	....	....	....	....	....	....
2000-01	U. of Maine	H-East	33	18	9	4	1897	76	2	2.40	....	....	....	....	....	....	....
2001-02	U. of Maine	H-East	20	6	8	3	1048	54	0	3.09	....	....	....	....	....	....	....

• Ruled ineligible to play 1998-99 season by NCAA due to appearance with **Lethbridge** (WHL) in 1995-96.

YEREMEYEV, Vitali (yehr-eh-MAY-ehv, VIH-tal-ee)

Goaltender. Catches left. 5'10", 167 lbs. Born, Ust-Kamenogorsk, USSR, September 23, 1975.
(NY Rangers' 11th choice, 209th overall, in 1994 Entry Draft).

			Regular Season								Playoffs						
Season	**Club**	**Lea**	**GP**	**W**	**L**	**T**	**Mins**	**GA**	**SO**	**Avg**	**GP**	**W**	**L**	**Mins**	**GA**	**SO**	**Avg**
1993-94	Ust-Kamenogorsk	CIS	19	....	....	....	1015	38	....	2.24	....	....	....	....	....	....	....
1994-95	CSKA Moscow	CIS	49	....	....	....	2733	97	....	2.13	2	....	....	120	8	....	4.00
1995-96	CSKA Moscow	CIS	25	....	....	....	1339	37	5	1.66	3	....	....	179	7	....	2.34
1996-97	HC CSKA	Russia	14	....	....	....	635	35	0	3.31	1	....	....	59	3	0	3.05
1997-98	Yaroslavl	Russia	17	....	....	....	979	19	3	*1.16	....	....	....	....	....	....	....
	Kazakhstan	Olympics	*7	1	3	1	292	28	0	5.76	....	....	....	....	....	....	....
1998-99	HC CSKA	Russia-2	19	....	....	....	1100	33	....	1.80	....	....	....	....	....	....	....
99-2000	Dynamo Moscow	Russia	26	....	....	....	1564	32	*7	*1.23	*17	....	....	*1039	22	*4	*1.27
2000-01	**NY Rangers**	**NHL**	**4**	**0**	**4**	**0**	**212**	**16**	**0**	**4.53**	....	....	....	....	....	....	....
	Hartford	AHL	36	16	15	3	1977	98	2	2.97	....	....	....	....	....	....	....
	Charlotte	ECHL	5	3	2	0	298	21	0	4.23	....	....	....	....	....	....	....
2001-02	Charlotte	ECHL	3	1	2	0	179	9	0	3.01	....	....	....	....	....	....	....
	Dynamo Moscow	Russia	20	....	....	....	1219	33	3	1.62	3	....	....	177	11	0	3.73
	NHL Totals		**4**	**0**	**4**	**0**	**212**	**16**	**0**	**4.53**	....	....	....	....	....	....	....

AHL All-Rookie Team (2001)

ZEPP, Rob (ZEHP, RAWB) CAR.

Goaltender. Catches left. 6'1", 181 lbs. Born, Scarborough, Ont., September 7, 1981.
(Carolina's 4th choice, 110th overall, in 2001 Entry Draft).

			Regular Season								Playoffs						
Season	**Club**	**Lea**	**GP**	**W**	**L**	**T**	**Mins**	**GA**	**SO**	**Avg**	**GP**	**W**	**L**	**Mins**	**GA**	**SO**	**Avg**
1997-98	Newmarket	OPJHL	3	....	....	....	181	13	0	4.31	....	....	....	....	....	....	....
1998-99	Plymouth Whalers	OHL	31	19	3	4	1662	76	3	2.74	3	1	0	100	10	0	6.00
99-2000	Plymouth Whalers	OHL	53	*36	11	3	3005	119	3	*2.38	*23	*15	8	*1374	52	2	2.27
2000-01	Plymouth Whalers	OHL	55	*34	18	3	3246	122	*4	*2.26	19	14	5	1139	51	2	2.69
2001-02	Florida	ECHL	13	6	5	2	739	41	0	3.33	....	....	....	....	....	....	....

• Re-entered NHL Entry Draft. Originally Atlanta's 5th choice, 99th overall, in 1999 Entry Draft.

Canadian Major Junior Scholastic Player of the Year (1999) • OHL Second All-Star Team (2000, 2001)

• Missed majority of 2001-02 season recovering from groin injury suffered in practice, January 3, 2002.

ZULIANELLO, Colin (zoo-lee-awh-NEHL-loh, COH-lihn) PHX.

Goaltender. Catches left. 6'1", 189 lbs. Born, Thunder Bay, Ont., July 8, 1978.

			Regular Season								Playoffs						
Season	**Club**	**Lea**	**GP**	**W**	**L**	**T**	**Mins**	**GA**	**SO**	**Avg**	**GP**	**W**	**L**	**Mins**	**GA**	**SO**	**Avg**
1996-97	Thunder Bay	USHL	32	11	19	0	1793	134	0	4.49	....	....	....	....	....	....	....
1997-98	Colorado College	WCHA	22	12	6	1	1096	52	1	2.84	....	....	....	....	....	....	....
1998-99	Colorado College	WCHA	9	6	0	0	419	20	0	2.87	....	....	....	....	....	....	....
99-2000	Colorado College	WCHA	12	4	4	2	656	28	1	2.56	....	....	....	....	....	....	....
2000-01	Colorado College	WCHA	17	9	7	1	950	35	0	2.21	....	....	....	....	....	....	....
2001-02	Springfield	AHL	3	0	2	0	162	9	0	3.16	....	....	....	....	....	....	....
	Mississippi	ECHL	33	18	11	1	1898	98	0	3.34	1	0	0	20	2	0	6.00

Signed as a free agent by **Springfield** (AHL), October 3, 2001. Signed as a free agent by **Phoenix**, August 9, 2002.

Retired NHL Player Index

Keith Allen

Mike Allison

Glenn Anderson

Mikael Anderson

Abbreviations: Teams/Cities: – **Ana.** – Anaheim; **Atl.** – Atlanta; **Bos.** – Boston; **Bro.** – Brooklyn; **Buf.** – Buffalo; **Cal.** – California; **Cgy.** – Calgary; **Cle.** – Cleveland; **Col.** – Colorado; **CBJ** – Columbus; **Dal.** – Dallas; **Det.** – Detroit; **Edm.** – Edmonton; **Fla.** – Florida; **Ham.** – Hamilton; **Hfd.** – Hartford;
K.C. – Kansas City; **L.A.** – Los Angeles; **Min.** – Minnesota; **Mtl.** – Montreal; **Mtl.M.** – Montreal Maroons; **Mtl.W.** – Montreal Wanderers; **N.J.** – New Jersey; **NYA** – NY Americans; **NYI** – NY Islanders; **NYR** – New York Rangers; – **Oak.** – Oakland; **Ott.** – Ottawa; – **Phi.** – Philadelphia; **Phx.** – Phoenix; **Pit.** – Pittsburgh; **Que.** – Quebec; **St.L.** – St. Louis; **S.J.** – San Jose; **T.B.** – Tampa Bay; **Tor.** – Toronto; **Van.** – Vancouver; **Wpg.** – Winnipeg; **Wsh.** – Washington

Total seasons are rounded off to the nearest full season. **A** – assists; **G** – goals; **GP** – games played; **PIM** – penalties in minutes; **TP** – total points.
● – deceased. Assists not recorded during 1917-18 season ‡ – Remains active in other leagues.

			Regular Schedule					Playoffs							
Name	NHL Teams	NHL Seasons	GP	G	A	TP	PIM	GP	G	A	TP	PIM	NHL Cup Wins	First NHL Season	Last NHL Season
A															
Abbott, Reg	Mtl.	1	3	0	0	0	0	....	....	....	....	....	....	1952-53	1952-53
● Abel, Clarence	NYR, Chi.	8	333	19	18	37	359	38	1	1	2	58	2	1926-27	1933-34
Abel, Gerry	Det.	1	1	0	0	0	0	....	....	....	....	....	....	1966-67	1966-67
● Abel, Sid	Det., Chi.	14	612	189	283	472	376	97	28	30	58	79	3	1938-39	1953-54
Abgrall, Dennis	L.A.	1	13	0	2	2	4	....	....	....	....	....	....	1975-76	1975-76
Abrahamsson, Thommy	Hfd.	1	32	6	11	17	16	....	....	....	....	....	....	1980-81	1980-81
Achtymichuk, Gene	Mtl., Det.	4	32	3	5	8	2	....	....	....	....	....	....	1951-52	1958-59
Acomb, Doug	Tor.	1	2	0	1	1	0	....	....	....	....	....	....	1969-70	1969-70
Acton, Keith	Mtl., Min., Edm., Phi., Wsh., NYI	15	1023	226	358	584	1172	66	12	21	33	88	1	1979-80	1993-94
Adam, Douglas	NYR	1	4	0	1	1	0	....	....	....	....	....	....	1949-50	1949-50
Adam, Russ	Tor.	1	8	1	2	3	11	....	....	....	....	....	....	1982-83	1982-83
Adams, Greg	Phi., Hfd., Wsh., Edm., Van., Que., Det.	10	545	84	143	227	1173	43	2	11	13	153	....	1980-81	1989-90
Adams, Greg	N.J., Van., Dal., Phx., Fla.	17	1056	355	388	743	326	81	20	22	42	16	....	1984-85	2000-01
● Adams, Jack	Tor., Ott.	7	173	83	32	115	366	10	1	0	1	13	2	1917-18	1926-27
Adams, John	Mtl.	1	42	6	12	18	11	3	0	0	0	0	....	1940-41	1940-41
● Adams, Stew	Chi., Tor.	4	95	9	26	35	60	11	3	3	6	14	....	1929-30	1932-33
Adduono, Rick	Bos., Atl.	2	4	0	0	0	2	....	....	....	....	....	....	1975-76	1979-80
Affleck, Bruce	St.L., Van., NYI	7	280	14	66	80	86	8	0	0	0	0	....	1974-75	1983-84
Agnew, Jim	Van., Hfd.	6	81	0	1	1	257	4	0	0	0	6	....	1986-87	1992-93
Ahern, Fred	Cal., Cle., Col.	4	146	31	30	61	130	2	0	1	1	2	....	1974-75	1977-78
Ahlin, Tony	Chi.	1	1	0	0	0	0	....	....	....	....	....	....	1937-38	1937-38
‡ Ahola, Peter	L.A., Pit., S.J., Cgy.	3	123	10	17	27	137	6	0	0	0	2	....	1991-92	1993-94
Ahrens, Chris	Min.	6	52	0	3	3	84	1	0	0	0	0	....	1972-73	1977-78
Ailsby, Lloyd	NYR	1	3	0	0	0	2	....	....	....	....	....	....	1951-52	1951-52
Aitken, Brad	Pit., Edm.	2	14	1	3	4	25	....	....	....	....	....	....	1987-88	1990-91
‡ Aivazoff, Micah	Det., Edm., NYI	3	92	4	6	10	46	....	....	....	....	....	....	1993-94	1995-96
● Albright, Clint	NYR	1	59	14	5	19	19	....	....	....	....	....	....	1948-49	1948-49
Aldcorn, Gary	Tor., Det., Bos.	5	226	41	56	97	78	6	1	2	3	4	....	1956-57	1960-61
‡ Aldridge, Keith	Dal.	1	4	0	0	0	0	....	....	....	....	....	....	1999-00	1999-00
Alexander, Claire	Tor., Van.	4	155	18	47	65	36	16	2	4	6	4	....	1974-75	1977-78
● Alexandre, Art	Mtl.	2	11	0	2	2	8	4	0	0	0	0	....	1931-32	1932-33
Allan, Jeff	Cle.	1	4	0	0	0	2	....	....	....	....	....	....	1977-78	1977-78
‡ Allen, Chris	Fla.	2	2	0	0	0	2	....	....	....	....	....	....	1997-98	1998-99
Allen, George	NYR, Chi., Mtl.	8	339	82	115	197	179	41	9	10	19	32	....	1938-39	1946-47
Allen, Keith	Det.	2	28	0	4	4	8	5	0	0	0	0	1	1953-54	1954-55
‡ Allen, Peter	Pit.	1	8	0	0	0	8	....	....	....	....	....	....	1995-96	1995-96
● Allen, Vivian	NYA	1	6	0	1	1	0	....	....	....	....	....	....	1940-41	1940-41
Alley, Steve	Hfd.	2	15	3	3	6	11	3	0	1	1	0	....	1979-80	1980-81
Allison, Dave	Mtl.	1	3	0	0	0	12	....	....	....	....	....	....	1983-84	1983-84
Allison, Mike	NYR, Tor., L.A.	10	499	102	166	268	630	82	9	17	26	135	....	1980-81	1989-90
Allison, Ray	Hfd., Phi.	7	238	64	93	157	223	12	2	3	5	20	....	1979-80	1986-87
● Allum, Bill	NYR	1	1	0	1	1	0	....	....	....	....	....	....	1940-41	1940-41
● Amadio, Dave	Det., L.A.	3	125	5	11	16	163	16	1	2	3	18	....	1957-58	1968-69
‡ Ambroziak, Peter	Buf.	1	12	0	1	1	0	....	....	....	....	....	....	1994-95	1994-95
Amodeo, Mike	Wpg.	1	19	0	0	0	2	....	....	....	....	....	....	1979-80	1979-80
● Anderson, Bill	Bos.	1	....	....	....	....	....	1	0	0	0	0	....	1942-43	1942-43
Anderson, Dale	Det.	1	13	0	0	0	6	2	0	0	0	0	....	1956-57	1956-57
Anderson, Doug	Mtl.	1	....	....	....	....	....	2	0	0	0	0	1	1952-53	1952-53
Anderson, Earl	Det., Bos.	3	109	19	19	38	22	5	0	1	1	0	....	1974-75	1976-77
Anderson, Glenn	Edm., Tor., NYR, St.L.	16	1129	498	601	1099	1120	225	93	121	214	442	6	1980-81	1995-96
Anderson, Jim	L.A.	1	7	1	2	3	2	....	....	....	....	....	....	1967-68	1967-68
Anderson, John	Tor., Que., Hfd.	12	814	282	349	631	263	37	9	18	27	2	....	1977-78	1988-89
Anderson, Murray	Wsh.	1	40	0	1	1	68	....	....	....	....	....	....	1974-75	1974-75
Anderson, Perry	St.L., N.J., S.J.	10	400	50	59	109	1051	36	2	1	3	161	....	1981-82	1991-92
Anderson, Ron	Det., L.A., St.L., Buf.	5	251	28	30	58	146	5	0	0	0	4	....	1967-68	1971-72
Anderson, Ron	Wsh.	1	28	9	7	16	8	....	....	....	....	....	....	1974-75	1974-75
Anderson, Russ	Pit., Hfd., L.A.	9	519	22	99	121	1086	10	0	3	3	28	....	1976-77	1984-85
‡ Anderson, Shawn	Buf., Que., Wsh., Phi.	8	255	11	51	62	117	19	1	1	2	16	....	1986-87	1994-95
● Anderson, Tom	Det., NYA, Bro.	8	319	62	127	189	180	16	2	7	9	8	....	1934-35	1941-42
‡ Andersson, Erik	Cgy.	1	12	2	1	3	8	....	....	....	....	....	....	1997-98	1997-98
Andersson, Kent-Erik	Min., NYR	7	456	72	103	175	78	50	4	11	15	4	....	1977-78	1983-84
‡ Andersson, Mikael	Buf., Hfd., T.B., Phi., NYI	15	761	95	169	264	134	25	2	7	9	10	....	1985-86	1999-00
Andersson, Peter	Wsh., Que.	3	172	10	41	51	81	7	0	2	2	2	....	1983-84	1985-86
‡ Andersson, Peter	NYR, Fla.	2	47	6	13	19	20	....	....	....	....	....	....	1992-93	1993-94
Andrascik, Steve	NYR	1	....	....	....	....	....	1	0	0	0	0	....	1971-72	1971-72
Andrea, Paul	NYR, Pit., Cal., Buf.	4	150	31	49	80	10	....	....	....	....	....	....	1965-66	1970-71
● Andrews, Lloyd	Tor.	4	53	8	5	13	10	2	0	0	0	0	1	1921-22	1924-25
‡ Andrievski, Alexander	Chi.	1	1	0	0	0	0	....	....	....	....	....	....	1992-93	1992-93
Andruff, Ron	Mtl., Col.	5	153	19	36	55	54	2	0	0	0	0	....	1974-75	1978-79
‡ Andrusak, Greg	Pit., Tor.	5	28	0	6	6	16	15	1	0	1	8	....	1993-94	1999-00
Angotti, Lou	NYR, Chi., Phi., Pit., St.L.	10	653	103	186	289	228	65	8	8	16	17	....	1964-65	1973-74
Anholt, Darrel	Chi.	1	1	0	0	0	0	....	....	....	....	....	....	1983-84	1983-84
Anslow, Hub	NYR	1	2	0	0	0	0	....	....	....	....	....	....	1947-48	1947-48
Antonovich, Mike	Min., Hfd., N.J.	5	87	10	15	25	37	....	....	....	....	....	....	1975-76	1983-84
Antoski, Shawn	Van., Phi., Pit., Ana.	8	183	3	5	8	599	36	1	3	4	74	....	1990-91	1997-98
● Apps, Syl	Tor.	10	423	201	231	432	56	69	25	29	54	8	3	1936-37	1947-48
Apps Jr., Syl	NYR, Pit., L.A.	10	727	183	423	606	311	23	5	5	10	23	....	1970-71	1979-80
Arbour, Al	Det., Chi., Tor., St.L.	16	626	12	58	70	617	86	1	8	9	92	4	1953-54	1970-71
● Arbour, Amos	Mtl., Ham., Tor.	6	113	52	20	72	77	....	....	....	....	....	....	1918-19	1923-24
● Arbour, Jack	Det., Tor.	2	47	5	1	6	56	....	....	....	....	....	....	1926-27	1928-29
Arbour, John	Bos., Pit., Van., St.L.	5	106	1	9	10	149	5	0	0	0	0	....	1965-66	1971-72
● Arbour, Ty	Pit., Chi.	5	207	28	28	56	112	11	2	0	2	6	....	1926-27	1930-31
Archambault, Michel	Chi.	1	3	0	0	0	0	....	....	....	....	....	....	1976-77	1976-77
Archibald, Dave	Min., NYR, Ott., NYI	8	323	57	67	124	139	5	0	1	1	0	....	1987-88	1996-97
Archibald, Jim	Min.	3	16	1	2	3	45	....	....	....	....	....	....	1984-85	1986-87
Areshenkoff, Ron	Edm.	1	4	0	0	0	0	....	....	....	....	....	....	1979-80	1979-80
Armstrong, Bill	Phi.	1	1	0	1	1	0	....	....	....	....	....	....	1990-91	1990-91
● Armstrong, Bob	Bos.	12	542	13	86	99	671	42	1	7	8	28	....	1950-51	1961-62
Armstrong, George	Tor.	21	1187	296	417	713	721	110	26	34	60	52	4	1949-50	1970-71
Armstrong, Murray	Tor., NYA, Bro., Det.	8	270	67	121	188	72	30	4	6	10	2	....	1937-38	1945-46
● Armstrong, Norm	Tor.	1	7	1	1	2	2	....	....	....	....	....	....	1962-63	1962-63
Armstrong, Tim	Tor.	1	11	1	0	1	6	....	....	....	....	....	....	1988-89	1988-89
Arnason, Chuck	Mtl., Atl., Pit., K.C., Col., Cle., Min., Wsh.	8	401	109	90	199	122	9	2	4	6	4	....	1971-72	1978-79
Arniel, Scott	Wpg., Buf., Bos.	11	730	149	189	338	599	34	3	3	6	39	....	1981-82	1991-92
Arthur, Fred	Hfd., Phi.	3	80	1	8	9	49	4	0	0	0	2	....	1980-81	1982-83
Arundel, John	Tor.	1	3	0	0	0	9	....	....	....	....	....	....	1949-50	1949-50
● Ashbee, Barry	Bos., Phi.	5	284	15	70	85	291	17	0	4	4	22	1	1965-66	1973-74
● Ashby, Don	Tor., Col., Edm.	6	188	40	56	96	40	12	1	0	1	4	....	1975-76	1980-81
Ashton, Brent	Van., Col., N.J., Min., Que., Det., Wpg., Bos., Cgy.	14	998	284	345	629	635	85	24	25	49	70	....	1979-80	1992-93
Ashworth, Frank	Chi.	1	18	5	4	9	2	....	....	....	....	....	....	1946-47	1946-47
Asmundson, Oscar	NYR, Det., St.L., NYA, Mtl.	5	111	11	23	34	30	9	0	2	2	4	1	1932-33	1937-38

Name	NHL Teams	NHL Seasons	Regular Schedule GP	G	A	TP	PIM	Playoffs GP	G	A	TP	PIM	NHL Cup Wins	First NHL Season	Last NHL Season
‡ Astley, Mark	Buf.	3	75	4	19	23	92	2	0	0	0	0	....	1993-94	1995-96
• Atanas, Walt	NYR	1	49	13	8	21	40	....	....	....	....	....	....	1944-45	1944-45
Atcheynum, Blair	Ott., St.L., Nsh., Chi.	5	196	27	33	60	36	23	1	3	4	8	....	1992-93	2000-01
Atkinson, Steve	Bos., Buf., Wsh.	6	302	60	51	111	104	1	0	0	0	0	....	1968-69	1974-75
Attwell, Bob	Col.	2	22	1	5	6	0	....	....	....	....	....	....	1979-80	1980-81
Attwell, Ron	St.L., NYR	1	22	1	7	8	8	....	....	....	....	....	....	1967-68	1967-68
Aubin, Norm	Tor.	2	69	18	13	31	30	1	0	0	0	0	....	1981-82	1982-83
Aubry, Pierre	Que., Det.	5	202	24	26	50	133	20	1	1	2	32	....	1980-81	1984-85
Aubuchon, Ossie	Bos., NYR	2	50	20	12	32	4	6	1	0	1	0	....	1942-43	1943-44
‡ Audet, Philippe	Det.	1	4	0	0	0	0	....	....	....	....	....	....	1998-99	1998-99
Auge, Les	Col.	1	6	0	3	3	4	....	....	....	....	....	....	1980-81	1980-81
‡ Augusta, Patrik	Tor., Wsh.	2	4	0	0	0	0	....	....	....	....	....	....	1993-94	1998-99
• Aurie, Larry	Det.	12	489	147	129	276	279	24	6	9	15	10	2	1927-28	1938-39
Awrey, Don	Bos., St.L., Mtl., Pit., NYR, Col.	16	979	31	158	189	1065	71	0	18	18	150	3	1963-64	1978-79
• Ayres, Vern	NYA, Mtl.M., St.L., NYR	6	211	6	11	17	350	....	....	....	....	....	....	1930-31	1935-36

B

Name	NHL Teams	NHL Seasons	Regular Schedule GP	G	A	TP	PIM	Playoffs GP	G	A	TP	PIM	NHL Cup Wins	First NHL Season	Last NHL Season
Babando, Pete	Bos., Det., Chi., NYR	6	351	86	73	159	194	17	3	3	6	6	1	1947-48	1952-53
Babcock, Bobby	Wsh.	2	2	0	0	0	2	....	....	....	....	....	....	1990-91	1992-93
Babe, Warren	Min.	3	21	2	5	7	23	2	0	0	0	0	....	1987-88	1990-91
Babin, Mitch	St.L.	1	8	0	0	0	0	....	....	....	....	....	....	1975-76	1975-76
Baby, John	Cle., Min.	2	26	2	8	10	26	....	....	....	....	....	....	1977-78	1978-79
Babych, Dave	Wpg., Hfd., Van., Phi., L.A.	19	1195	142	581	723	970	114	21	41	62	113	....	1980-81	1998-99
Babych, Wayne	St.L., Pit., Que., Hfd.	9	519	192	246	438	498	41	7	9	16	24	....	1978-79	1986-87
‡ Baca, Jergus	Hfd.	2	10	0	2	2	14	....	....	....	....	....	....	1990-91	1991-92
Backman, Mike	NYR	3	18	1	6	7	18	10	2	2	4	2	....	1981-82	1983-84
• Backor, Pete	Tor.	1	36	4	5	9	6	....	....	....	....	....	1	1944-45	1944-45
Backstrom, Ralph	Mtl., L.A., Chi.	17	1032	278	361	639	386	116	27	32	59	68	6	1956-57	1972-73
• Bailey, Ace	Tor.	8	313	111	82	193	472	21	3	4	7	12	1	1926-27	1933-34
Bailey, Bob	Tor., Det., Chi.	5	150	15	21	36	207	15	0	4	4	22	....	1953-54	1957-58
• Bailey, Garnet	Bos., Det., St.L., Wsh.	10	568	107	171	278	633	15	2	4	6	28	2	1968-69	1977-78
Bailey, Reid	Phi., Tor., Hfd.	4	40	1	3	4	105	16	0	2	2	25	....	1980-81	1983-84
Baillargeon, Joel	Wpg., Que.	3	20	0	2	2	31	....	....	....	....	....	....	1986-87	1988-89
Baird, Ken	Cal.	1	10	0	2	2	15	....	....	....	....	....	....	1971-72	1971-72
Baker, Bill	Mtl., Col., St.L., NYR	3	143	7	25	32	175	6	0	0	0	0	....	1980-81	1982-83
Baker, Jamie	Que., Ott., S.J., Tor.	10	404	71	79	150	271	25	5	4	9	42	....	1989-90	1998-99
Bakovic, Peter	Van.	1	10	2	0	2	48	....	....	....	....	....	....	1987-88	1987-88
Balderis, Helmut	Min.	1	26	3	6	9	2	....	....	....	....	....	....	1989-90	1989-90
Baldwin, Doug	Tor., Det., Chi.	3	24	0	1	1	8	....	....	....	....	....	....	1945-46	1947-48
Balfour, Earl	Tor., Chi.	7	288	30	22	52	78	26	0	3	3	4	1	1951-52	1960-61
• Balfour, Murray	Mtl., Chi., Bos.	8	306	67	90	157	393	40	9	10	19	45	1	1956-57	1964-65
Ball, Terry	Phi., Buf.	4	74	7	19	26	26	....	....	....	....	....	....	1967-68	1971-72
Balon, Dave	NYR, Mtl., Min., Van.	14	776	192	222	414	607	78	14	21	35	109	2	1959-60	1972-73
Baltimore, Bryon	Edm.	1	2	0	0	0	4	....	....	....	....	....	....	1979-80	1979-80
Baluik, Stan	Bos.	1	7	0	0	0	2	....	....	....	....	....	....	1959-60	1959-60
Bandura, Jeff	NYR	1	2	0	1	1	0	....	....	....	....	....	....	1980-81	1980-81
‡ Banham, Frank	Ana.	3	27	9	2	11	14	....	....	....	....	....	....	1996-97	1999-00
‡ Banks, Darren	Bos.	2	20	2	2	4	73	....	....	....	....	....	....	1992-93	1993-94
‡ Barahona, Ralph	Bos.	2	6	2	2	4	0	....	....	....	....	....	....	1990-91	1991-92
Barbe, Andy	Tor.	1	1	0	0	0	2	....	....	....	....	....	....	1950-51	1950-51
Barber, Bill	Phi.	14	903	420	463	883	623	129	53	55	108	109	2	1972-73	1983-84
Barber, Don	Min., Wpg., Que., S.J.	4	115	25	32	57	64	11	4	4	8	10	....	1988-89	1991-92
• Barilko, Bill	Tor.	5	252	26	36	62	456	47	5	7	12	104	4	1946-47	1950-51
Barkley, Doug	Chi., Det.	6	253	24	80	104	382	30	0	9	9	63	....	1957-58	1965-66
Barlow, Bob	Min.	2	77	16	17	33	10	6	2	2	4	6	....	1969-70	1970-71
Barnes, Blair	L.A.	1	1	0	0	0	0	....	....	....	....	....	....	1982-83	1982-83
Barnes, Norm	Phi., Hfd.	5	156	6	38	44	178	12	0	0	0	8	....	1976-77	1981-82
Baron, Normand	Mtl., St.L.	2	27	2	0	2	51	3	0	0	0	22	....	1983-84	1985-86
Barr, Dave	Bos., NYR, St.L., Hfd., Det., N.J., Dal.	13	614	128	204	332	520	71	12	10	22	70	....	1981-82	1993-94
Barrault, Doug	Min., Fla.	2	4	0	0	0	2	....	....	....	....	....	....	1992-93	1993-94
Barrett, Fred	Min., L.A.	13	745	25	123	148	671	44	0	2	2	60	....	1970-71	1983-84
Barrett, John	Det., Wsh., Min.	8	488	20	77	97	604	16	2	2	4	50	....	1980-81	1987-88
Barrie, Doug	Pit., Buf., L.A.	3	158	10	42	52	268	....	....	....	....	....	....	1968-69	1971-72
Barrie, Len	Phi., Fla., Pit., L.A.	7	184	19	45	64	290	8	1	0	1	8	....	1989-90	2000-01
Barry, Ed	Bos.	1	19	1	3	4	2	....	....	....	....	....	....	1946-47	1946-47
• Barry, Marty	NYA, Bos., Det., Mtl.	12	509	195	192	387	231	43	15	18	33	34	2	1927-28	1939-40
Barry, Ray	Bos.	1	18	1	2	3	6	....	....	....	....	....	....	1951-52	1951-52
Bartel, Robin	Cgy., Van.	2	41	0	1	1	14	6	0	0	0	16	....	1985-86	1986-87
Bartlett, Jim	Mtl., NYR, Bos.	5	191	34	23	57	273	2	0	0	0	0	....	1954-55	1960-61
• Barton, Cliff	Pit., Phi., NYR	3	85	10	9	19	22	....	....	....	....	....	....	1929-30	1939-40
Bassen, Bob	NYI, Chi., St.L., Que., Dal., Cgy.	15	765	88	144	232	1004	93	9	15	24	134	....	1985-86	1999-00
Bathe, Frank	Det., Phi.	9	224	3	28	31	542	27	1	3	4	42	....	1974-75	1983-84
Bathgate, Andy	NYR, Tor., Det., Pit.	17	1069	349	624	973	624	54	21	14	35	76	1	1952-53	1970-71
Bathgate, Frank	NYR	1	2	0	0	0	2	....	....	....	....	....	....	1952-53	1952-53
• Batters, Jeff	St.L.	2	16	0	0	0	28	....	....	....	....	....	....	1993-94	1994-95
‡ Batyrshin, Ruslan	L.A.	1	2	0	0	0	6	....	....	....	....	....	....	1995-96	1995-96
• Bauer, Bobby	Bos.	9	327	123	137	260	36	48	11	8	19	6	2	1936-37	1951-52
Baumgartner, Ken	L.A., NYI, Tor., Ana., Bos.	12	696	13	41	54	2244	51	1	2	3	106	....	1987-88	1998-99
Baumgartner, Mike	K.C.	1	17	0	0	0	0	....	....	....	....	....	....	1974-75	1974-75
Baun, Bob	Tor., Oak., Det.	17	964	37	187	224	1493	96	3	12	15	171	4	1956-57	1972-73
‡ Bautin, Sergei	Wpg., Det., S.J.	3	132	5	25	30	176	6	0	0	0	2	....	1992-93	1995-96
Bawa, Robin	Wsh., Van., S.J., Ana.	4	61	6	1	7	60	1	0	0	0	0	....	1989-90	1993-94
Baxter, Paul	Que., Pit., Cgy.	8	472	48	121	169	1564	40	0	5	5	162	....	1979-80	1986-87
Beadle, Sandy	Wpg.	1	6	1	0	1	2	....	....	....	....	....	....	1980-81	1980-81
Beaton, Frank	NYR	2	25	1	1	2	43	....	....	....	....	....	....	1978-79	1979-80
• Beattie, Red	Bos., Det., NYA	9	334	62	85	147	137	24	4	2	6	8	....	1930-31	1938-39
Beaudin, Norm	St.L., Min.	2	25	1	2	3	4	....	....	....	....	....	....	1967-68	1970-71
Beaudoin, Serge	Atl.	1	3	0	0	0	0	....	....	....	....	....	....	1979-80	1979-80
‡ Beaudoin, Yves	Wsh.	3	11	0	0	0	5	....	....	....	....	....	....	1985-86	1987-88
Beck, Barry	Col., NYR, L.A.	10	615	104	251	355	1016	51	10	23	33	77	....	1977-78	1989-90
Beckett, Bob	Bos.	4	68	7	6	13	18	....	....	....	....	....	....	1956-57	1963-64
Bedard, James	Chi.	2	22	1	1	2	8	....	....	....	....	....	....	1949-50	1950-51
‡ Beddoes, Clayton	Bos.	2	60	2	8	10	57	....	....	....	....	....	....	1995-96	1996-97
Bednarski, John	NYR, Edm.	4	100	2	18	20	114	1	0	0	0	17	....	1974-75	1979-80
Beers, Bob	Bos., T.B., Edm., NYI	8	258	28	79	107	225	21	1	1	2	22	....	1989-90	1996-97
Beers, Eddy	Cgy., St.L.	5	250	94	116	210	256	41	7	10	17	47	....	1981-82	1985-86
• Behling, Dick	Det.	2	5	1	0	1	2	....	....	....	....	....	....	1940-41	1942-43
Beisler, Frank	NYA	2	2	0	0	0	0	....	....	....	....	....	....	1936-37	1939-40
Belanger, Alain	Tor.	1	9	0	1	1	6	....	....	....	....	....	....	1977-78	1977-78
‡ Belanger, Jesse	Mtl., Fla., Van., Edm., NYI	8	246	59	76	135	56	12	0	3	3	2	1	1991-92	2000-01
Belanger, Roger	Pit.	1	44	3	5	8	32	....	....	....	....	....	....	1984-85	1984-85
Belisle, Danny	NYR	1	4	2	0	2	0	....	....	....	....	....	....	1960-61	1960-61
Beliveau, Jean	Mtl.	20	1125	507	712	1219	1029	162	79	97	176	211	10	1950-51	1970-71
• Bell, Billy	Mtl.W., Mtl., Ott.	6	72	4	2	6	14	5	0	0	0	0	1	1917-18	1923-24
Bell, Bruce	Que., St.L., NYR, Edm.	5	209	12	64	76	113	34	3	5	8	41	....	1984-85	1989-90
Bell, Harry	NYR	1	1	0	1	1	0	....	....	....	....	....	....	1946-47	1946-47
Bell, Joe	NYR	2	62	8	9	17	18	....	....	....	....	....	....	1942-43	1946-47
Belland, Neil	Van., Pit.	6	109	13	32	45	54	21	2	9	11	23	....	1981-82	1986-87
• Bellefeuille, Pete	Tor., Det.	4	92	26	4	30	58	....	....	....	....	....	....	1925-26	1929-30
• Bellemer, Andy	Mtl.M.	1	15	0	0	0	0	....	....	....	....	....	....	1932-33	1932-33
Bellows, Brian	Min., Mtl., T.B., Ana., Wsh.	17	1188	485	537	1022	718	143	51	71	122	143	1	1982-83	1998-99
• Bend, Lin	NYR	1	8	3	1	4	2	....	....	....	....	....	....	1942-43	1942-43
‡ Benda, Jan	Wsh.	1	9	0	3	3	6	....	....	....	....	....	....	1997-98	1997-98
Bennett, Adam	Chi., Edm.	3	69	3	8	11	69	....	....	....	....	....	....	1991-92	1993-94
Bennett, Bill	Bos., Hfd.	2	31	4	7	11	65	....	....	....	....	....	....	1978-79	1979-80
Bennett, Curt	St.L., NYR, Atl.	10	580	152	182	334	347	21	1	1	2	57	....	1970-71	1979-80
Bennett, Frank	Det.	1	7	0	1	1	2	....	....	....	....	....	....	1943-44	1943-44
Bennett, Harvey	Pit., Wsh., Phi., Min., St.L.	5	268	44	46	90	347	4	0	0	0	2	....	1974-75	1978-79
• Bennett, Max	Mtl.	1	1	0	0	0	0	....	....	....	....	....	....	1935-36	1935-36
Bennett, Rick	NYR	3	15	1	1	2	13	....	....	....	....	....	....	1989-90	1991-92
Benning, Brian	St.L., L.A., Phi., Edm., Fla.	11	568	63	233	296	963	48	3	20	23	74	....	1984-85	1994-95
Benning, Jim	Tor., Van.	9	605	52	191	243	461	7	1	1	2	2	....	1981-82	1989-90
• Benoit, Joe	Mtl.	5	185	75	69	144	94	11	6	3	9	11	1	1940-41	1946-47
Benson, Bill	NYA, Bro.	2	67	11	25	36	35	....	....	....	....	....	....	1940-41	1941-42
• Benson, Bobby	Bos.	1	8	0	1	1	4	....	....	....	....	....	....	1924-25	1924-25

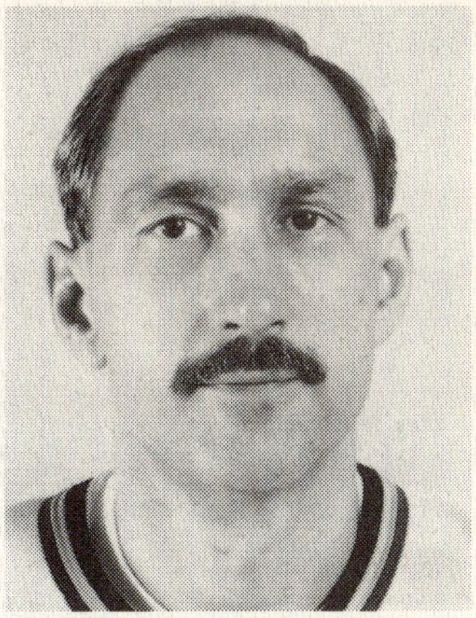
Helmut Balderis

Dave Barr

Bob Baun

Bob Beers

Mike Blaisdell

Mike Bossy

Butch Bouchard

Raymond Bourque

Name	NHL Teams	NHL Seasons	Regular Schedule GP	G	A	TP	PIM	Playoffs GP	G	A	TP	PIM	NHL Cup Wins	First NHL Season	Last NHL Season
• Bentley, Doug	Chi., NYR	13	566	219	324	543	217	23	9	8	17	12	....	1939-40	1953-54
• Bentley, Max	Chi., Tor., NYR	12	646	245	299	544	179	51	18	27	45	14	3	1940-41	1953-54
Bentley, Reg	Chi.	1	11	1	2	3	2	....	....	....	....	....	....	1942-43	1942-43
‡ Beraldo, Paul	Bos.	2	10	0	0	0	4	....	....	....	....	....	....	1987-88	1988-89
Berenson, Red	Mtl., NYR, St.L., Det.	17	987	261	397	658	305	85	23	14	37	49	1	1961-62	1977-78
Berezan, Perry	Cgy., Min., S.J.	9	378	61	75	136	279	31	4	7	11	34	....	1984-85	1992-93
Berg, Bill	NYI, Tor., NYR, Ott.	10	546	55	67	122	488	61	3	4	7	34	....	1988-89	1998-99
• Bergdinon, Fred	Bos.	1	2	0	0	0	0	....	....	....	....	....	....	1925-26	1925-26
Bergen, Todd	Phi.	1	14	11	5	16	4	17	4	9	13	8	....	1984-85	1984-85
Berger, Mike	Min.	2	30	3	1	4	67	....	....	....	....	....	....	1987-88	1988-89
Bergeron, Michel	Det., NYI, Wsh.	5	229	80	58	138	165	....	....	....	....	....	....	1974-75	1978-79
Bergeron, Yves	Pit.	2	3	0	0	0	0	....	....	....	....	....	....	1974-75	1976-77
‡ Bergkvist, Stefan	Pit.	2	7	0	0	0	9	4	0	0	0	2	....	1995-96	1996-97
Bergland, Tim	Wsh., T.B.	5	182	17	26	43	75	26	2	2	4	22	....	1989-90	1993-94
Bergloff, Bob	Min.	1	2	0	0	0	5	....	....	....	....	....	....	1982-83	1982-83
Berglund, Bo	Que., Min., Phi.	3	130	28	39	67	40	9	2	0	2	6	....	1983-84	1985-86
• Bergman, Gary	Det., Min., K.C.	12	838	68	299	367	1249	21	0	5	5	20	....	1964-65	1975-76
Bergman, Thommie	Det.	6	246	21	44	65	243	7	0	2	2	2	....	1972-73	1979-80
Bergqvist, Jonas	Cgy.	1	22	2	5	7	10	....	....	....	....	....	....	1989-90	1989-90
• Berlinquette, Louis	Mtl., Mtl.M., Pit.	8	193	45	33	78	129	11	0	4	4	9	....	1917-18	1925-26
Bernier, Serge	Phi., L.A., Que.	7	302	78	119	197	234	5	1	1	2	0	....	1968-69	1980-81
Berry, Bob	Mtl., L.A.	8	541	159	191	350	344	26	2	6	8	6	....	1968-69	1976-77
Berry, Brad	Wpg., Min., Dal.	8	241	4	28	32	323	13	0	1	1	16	....	1985-86	1993-94
Berry, Doug	Col.	2	121	10	33	43	25	....	....	....	....	....	....	1979-80	1980-81
Berry, Fred	Det.	1	3	0	0	0	0	....	....	....	....	....	....	1976-77	1976-77
Berry, Ken	Edm., Van.	4	55	8	10	18	30	....	....	....	....	....	....	1981-82	1988-89
‡ Bertrand, Eric	N.J., Atl., Mtl.	2	15	0	0	0	4	....	....	....	....	....	....	1999-00	2000-01
• Besler, Phil	Bos., Chi., Det.	2	30	1	4	5	18	....	....	....	....	....	....	1935-36	1938-39
• Bessone, Pete	Det.	1	6	0	1	1	6	....	....	....	....	....	....	1937-38	1937-38
Bethel, John	Wpg.	1	17	0	2	2	4	....	....	....	....	....	....	1979-80	1979-80
‡ Betik, Karel	T.B.	1	3	0	2	2	2	....	....	....	....	....	....	1998-99	1998-99
‡ Bets, Maxim	Ana.	1	3	0	0	0	0	....	....	....	....	....	....	1993-94	1993-94
Bettio, Sam	Bos.	1	44	9	12	21	32	....	....	....	....	....	....	1949-50	1949-50
Beukeboom, Jeff	Edm., NYR	14	804	30	129	159	1890	99	3	16	19	197	4	1985-86	1998-99
Beverley, Nick	Bos., Pit., NYR, Min., L.A., Col.	11	502	18	94	112	156	7	0	1	1	0	....	1966-67	1979-80
Bialowas, Dwight	Atl., Min.	4	164	11	46	57	46	....	....	....	....	....	....	1973-74	1976-77
Bialowas, Frank	Tor.	1	3	0	0	0	12	....	....	....	....	....	....	1993-94	1993-94
Bianchin, Wayne	Pit., Edm.	7	276	68	41	109	137	3	0	1	1	6	....	1973-74	1979-80
‡ Bidner, Todd	Wsh.	1	12	2	1	3	7	....	....	....	....	....	....	1981-82	1981-82
Biggs, Don	Min., Phi.	2	12	2	0	2	8	....	....	....	....	....	....	1984-85	1989-90
Bignell, Larry	Pit.	2	20	0	3	3	2	3	0	0	0	2	....	1973-74	1974-75
Bilodeau, Gilles	Que.	1	9	0	1	1	25	....	....	....	....	....	....	1979-80	1979-80
• Bionda, Jack	Tor., Bos.	4	93	3	9	12	113	11	0	1	1	14	....	1955-56	1958-59
‡ Bissett, Tom	Det.	1	5	0	0	0	0	....	....	....	....	....	....	1990-91	1990-91
Bjugstad, Scott	Min., Pit., L.A.	9	317	76	68	144	144	9	0	1	1	2	....	1983-84	1991-92
Black, Stephen	Det., Chi.	2	113	11	20	31	77	13	0	0	0	13	1	1949-50	1950-51
Blackburn, Bob	NYR, Pit.	3	135	8	12	20	105	6	0	0	0	4	....	1968-69	1970-71
Blackburn, Don	Bos., Phi., NYR, NYI, Min.	6	185	23	44	67	87	12	3	0	3	10	....	1962-63	1972-73
Blade, Hank	Chi.	2	24	2	3	5	2	....	....	....	....	....	....	1946-47	1947-48
Bladon, Tom	Phi., Pit., Edm., Wpg., Det.	9	610	73	197	270	392	86	8	29	37	70	2	1972-73	1980-81
• Blaine, Garry	Mtl.	1	1	0	0	0	0	....	....	....	....	....	....	1954-55	1954-55
• Blair, Andy	Tor., Chi.	9	402	74	86	160	323	38	6	6	12	32	1	1928-29	1936-37
Blair, Chuck	Tor.	1	1	0	0	0	0	....	....	....	....	....	....	1948-49	1948-49
Blair, Dusty	Tor.	1	2	0	0	0	0	....	....	....	....	....	....	1950-51	1950-51
Blaisdell, Mike	Det., NYR, Pit., Tor.	9	343	70	84	154	166	6	1	2	3	10	....	1980-81	1988-89
Blake, Bob	Bos.	1	12	0	0	0	0	....	....	....	....	....	....	1935-36	1935-36
• Blake, Mickey	Mtl.M., St.L., Tor.	3	10	1	1	2	4	....	....	....	....	....	....	1932-33	1935-36
• Blake, Toe	Mtl.M., Mtl.	14	577	235	292	527	272	58	25	37	62	23	3	1934-35	1947-48
Blight, Rick	Van., L.A.	7	326	96	125	221	170	5	0	5	5	2	....	1975-76	1982-83
• Blinco, Russ	Mtl.M., Chi.	6	268	59	66	125	24	19	3	3	6	4	1	1933-34	1938-39
Block, Ken	Van.	1	1	0	0	0	0	....	....	....	....	....	....	1970-71	1970-71
Bloemberg, Jeff	NYR	4	43	3	6	9	25	7	0	3	3	5	....	1988-89	1991-92
Blomqvist, Timo	Wsh., N.J.	5	243	4	53	57	293	13	0	0	0	24	....	1981-82	1986-87
Blomsten, Arto	Wpg., L.A.	3	25	0	4	4	8	....	....	....	....	....	....	1993-94	1995-96
Bloom, Mike	Wsh., Det.	3	201	30	47	77	215	....	....	....	....	....	....	1974-75	1976-77
Blum, John	Edm., Bos., Wsh., Det.	8	250	7	34	41	610	20	0	2	2	27	....	1982-83	1989-90
Bodak, Bob	Cgy., Hfd.	2	4	0	0	0	29	....	....	....	....	....	....	1987-88	1989-90
Boddy, Gregg	Van.	5	273	23	44	67	263	3	0	0	0	0	....	1971-72	1975-76
‡ Bodger, Doug	Pit., Buf., S.J., N.J., L.A., Van.	16	1071	106	422	528	1007	47	6	18	24	25	....	1984-85	1999-00
Bodnar, Gus	Tor., Chi., Bos.	12	667	142	254	396	207	32	4	3	7	10	2	1943-44	1954-55
Boehm, Ron	Oak.	1	16	2	1	3	10	....	....	....	....	....	....	1967-68	1967-68
• Boesch, Garth	Tor.	4	197	9	28	37	205	34	2	5	7	18	3	1946-47	1949-50
Boh, Rick	Min.	1	8	2	1	3	4	....	....	....	....	....	....	1987-88	1987-88
Boileau, Marc	Det.	1	54	5	6	11	8	....	....	....	....	....	....	1961-62	1961-62
Boileau, Rene	NYA	1	7	0	0	0	0	....	....	....	....	....	....	1925-26	1925-26
Boimistruck, Fred	Tor.	2	83	4	14	18	45	....	....	....	....	....	....	1981-82	1982-83
Boisvert, Serge	Tor., Mtl.	5	46	5	7	12	8	23	3	7	10	4	1	1982-83	1987-88
Boivin, Claude	Phi., Ott.	4	132	12	19	31	364	....	....	....	....	....	....	1991-92	1994-95
Boivin, Leo	Tor., Bos., Det., Pit., Min.	19	1150	72	250	322	1192	54	3	10	13	59	....	1951-52	1969-70
Boland, Mike	Phi.	1	2	0	0	0	0	....	....	....	....	....	....	1974-75	1974-75
Boland, Mike J.	K.C., Buf.	2	23	1	2	3	29	3	1	0	1	2	....	1974-75	1978-79
Boldirev, Ivan	Bos., Cal., Chi., Atl., Van., Det.	15	1052	361	505	866	507	48	13	20	33	14	....	1970-71	1984-85
Bolduc, Danny	Det., Cgy.	3	102	22	19	41	33	1	0	0	0	0	....	1978-79	1983-84
Bolduc, Michel	Que.	2	10	0	0	0	6	....	....	....	....	....	....	1981-82	1982-83
• Boll, Buzz	Tor., NYA, Bro., Bos.	12	437	133	130	263	148	31	7	3	10	13	....	1932-33	1943-44
Bolonchuk, Larry	Van., Wsh.	4	74	3	9	12	97	....	....	....	....	....	....	1972-73	1977-78
• Bolton, Hugh	Tor.	8	235	10	51	61	221	17	0	5	5	14	1	1949-50	1956-57
Bonar, Dan	L.A.	3	170	25	39	64	208	14	3	4	7	22	....	1980-81	1982-83
Bonin, Marcel	Det., Bos., Mtl.	9	454	97	175	272	336	50	11	14	25	51	4	1952-53	1961-62
Bonsignore, Jason	Edm., T.B.	4	79	3	13	16	34	....	....	....	....	....	....	1994-95	1998-99
Boo, Jim	Min.	1	6	0	0	0	22	....	....	....	....	....	....	1977-78	1977-78
• Boone, Buddy	Bos.	2	34	5	3	8	28	22	2	1	3	25	....	1956-57	1957-58
Boothman, George	Tor.	2	58	17	19	36	18	5	2	1	3	2	....	1942-43	1943-44
Bordeleau, Christian	Mtl., St.L., Chi.	4	205	38	65	103	82	19	4	7	11	17	1	1968-69	1971-72
Bordeleau, J.P.	Chi.	10	519	97	126	223	143	48	3	6	9	12	....	1969-70	1979-80
Bordeleau, Paulin	Van.	3	183	33	56	89	47	5	2	1	3	0	....	1973-74	1975-76
Borotsik, Jack	St.L.	1	1	0	0	0	0	....	....	....	....	....	....	1974-75	1974-75
‡ Borsato, Luciano	Wpg.	5	203	35	55	90	113	7	1	0	1	4	....	1990-91	1994-95
Borschevsky, Nikolai	Tor., Cgy., Dal.	4	162	49	73	122	44	31	4	9	13	4	....	1992-93	1995-96
Boschman, Laurie	Tor., Edm., Wpg., N.J., Ott.	14	1009	229	348	577	2265	57	8	13	21	140	....	1979-80	1992-93
Bossy, Mike	NYI	10	752	573	553	1126	210	129	85	75	160	38	4	1977-78	1986-87
• Bostrum, Helge	Chi.	4	96	3	3	6	58	13	0	0	0	16	....	1929-30	1932-33
Botell, Mark	Phi.	1	32	4	10	14	31	....	....	....	....	....	....	1981-82	1981-82
Bothwell, Tim	NYR, St.L., Hfd.	12	502	28	93	121	382	49	0	3	3	56	....	1978-79	1988-89
Botting, Cam	Atl.	1	2	0	1	1	0	....	....	....	....	....	....	1975-76	1975-76
Boucha, Henry	Det., Min., K.C., Col.	6	247	53	49	102	157	....	....	....	....	....	....	1971-72	1976-77
Bouchard, Butch	Mtl.	15	785	49	144	193	863	113	11	21	32	121	4	1941-42	1955-56
Bouchard, Dick	NYR	1	1	0	0	0	0	....	....	....	....	....	....	1954-55	1954-55
• Bouchard, Edmond	Mtl., Ham., NYA, Pit.	8	211	19	21	40	117	....	....	....	....	....	....	1921-22	1928-29
Bouchard, Pierre	Mtl., Wsh.	12	595	24	82	106	433	76	3	10	13	56	5	1970-71	1981-82
• Boucher, Billy	Mtl., Bos., NYA	7	213	93	38	131	409	14	3	0	3	17	1	1921-22	1927-28
• Boucher, Bobby	Mtl.	1	11	1	0	1	0	2	0	0	0	0	1	1923-24	1923-24
• Boucher, Clarence	NYA	2	47	2	2	4	133	....	....	....	....	....	....	1926-27	1927-28
• Boucher, Frank	Ott., NYR	14	557	160	263	423	119	55	16	20	36	12	2	1921-22	1943-44
• Boucher, Georges	Ott., Mtl.M., Chi.	15	449	117	87	204	838	28	5	3	8	88	4	1917-18	1931-32
Boudreau, Bruce	Tor., Chi.	8	141	28	42	70	46	9	2	0	2	0	....	1976-77	1985-86
Boudrias, Andre	Mtl., Min., Chi., St.L., Van.	12	662	151	340	491	216	34	6	10	16	12	....	1963-64	1975-76
Boughner, Barry	Oak., Cal.	2	20	0	0	0	11	....	....	....	....	....	....	1969-70	1970-71
Bourbonnais, Dan	Hfd.	2	59	3	25	28	11	....	....	....	....	....	....	1981-82	1983-84
Bourbonnais, Rick	St.L.	3	71	9	15	24	29	4	0	1	1	0	....	1975-76	1977-78
• Bourcier, Conrad	Mtl.	1	6	0	0	0	0	....	....	....	....	....	....	1935-36	1935-36
Bourcier, Jean	Mtl.	1	9	0	1	1	0	....	....	....	....	....	....	1935-36	1935-36
• Bourgeault, Leo	Tor., NYR, Ott., Mtl.	8	307	24	20	44	269	24	1	1	2	18	1	1926-27	1934-35
Bourgeois, Charlie	Cgy., St.L., Hfd.	7	290	16	54	70	788	40	2	3	5	194	....	1981-82	1987-88
Bourne, Bob	NYI, L.A.	14	964	258	324	582	605	139	40	56	96	108	4	1974-75	1987-88
Bourque, Phil	Pit., NYR, Ott.	12	477	88	111	199	516	56	13	12	25	107	2	1983-84	1995-96

			Regular Schedule					Playoffs							
Name	NHL Teams	NHL Seasons	GP	G	A	TP	PIM	GP	G	A	TP	PIM	NHL Cup Wins	First NHL Season	Last NHL Season
Bourque, Raymond	Bos., Col.	22	1612	410	1169	1579	1141	214	41	139	180	171	1	1979-80	2000-01
Boutette, Pat	Tor., Hfd., Pit.	10	756	171	282	453	1354	46	10	14	24	109	....	1975-76	1984-85
Boutilier, Paul	NYI, Bos., Min., NYR, Wpg.	8	288	27	83	110	358	41	1	9	10	45	1	1981-82	1988-89
‡ Bowen, Jason	Phi., Edm.	6	77	2	6	8	109	....	....	....	....	....	....	1992-93	1997-98
Bowman, Kirk	Chi.	3	88	11	17	28	19	7	1	0	1	0	....	1976-77	1978-79
• Bowman, Ralph	Ott., St.L., Det.	7	274	8	17	25	260	22	2	2	4	6	2	1933-34	1939-40
Bownass, Jack	Mtl., NYR	4	80	3	8	11	58	....	....	....	....	....	....	1957-58	1961-62
Bowness, Rick	Atl., Det., St.L., Wpg.	7	173	18	37	55	191	5	0	0	0	2	....	1975-76	1981-82
• Boyd, Bill	NYR, NYA	4	138	15	7	22	72	10	0	0	0	4	1	1926-27	1929-30
Boyd, Irvin	Bos., Det.	4	96	10	10	20	30	5	0	1	1	4	....	1931-32	1943-44
Boyd, Randy	Pit., Chi., NYI, Van.	8	257	20	67	87	328	13	0	2	2	26	....	1981-82	1988-89
Boyer, Wally	Tor., Chi., Oak., Pit.	7	365	54	105	159	163	15	1	3	4	0	....	1965-66	1971-72
‡ Boyer, Zac	Dal.	2	3	0	0	0	0	2	0	0	0	0	....	1994-95	1995-96
Boyko, Darren	Wpg.	1	1	0	0	0	0	....	....	....	....	....	....	1988-89	1988-89
Bozek, Steve	L.A., Cgy., St.L., Van., S.J.	11	641	164	167	331	309	58	12	11	23	69	....	1981-82	1991-92
‡ Bozon, Philippe	St.L.	4	144	16	25	41	101	19	2	0	2	31	....	1991-92	1994-95
• Brackenborough, John	Bos.	1	7	0	0	0	0	....	....	....	....	....	....	1925-26	1925-26
Brackenbury, Curt	Que., Edm., St.L.	4	141	9	17	26	226	2	0	0	0	0	....	1979-80	1982-83
Bradley, Bart	Bos.	1	1	0	0	0	0	....	....	....	....	....	....	1949-50	1949-50
Bradley, Brian	Cgy., Van., Tor., T.B.	13	651	182	321	503	528	13	3	7	10	16	....	1985-86	1997-98
Bradley, Lyle	Cal., Cle.	2	6	1	0	1	2	....	....	....	....	....	....	1973-74	1976-77
‡ Brady, Neil	N.J., Ott., Dal.	5	89	9	22	31	95	....	....	....	....	....	....	1989-90	1993-94
Bragnalo, Rick	Wsh.	4	145	15	35	50	46	....	....	....	....	....	....	1975-76	1978-79
• Branigan, Andy	NYA, Bro.	2	27	1	2	3	31	....	....	....	....	....	....	1940-41	1941-42
Brasar, Per-Olov	Min., Van.	5	348	64	142	206	33	13	1	2	3	0	....	1977-78	1981-82
• Brayshaw, Russ	Chi.	1	43	5	9	14	24	....	....	....	....	....	....	1944-45	1944-45
Breault, Francois	L.A.	3	27	2	4	6	42	....	....	....	....	....	....	1990-91	1992-93
Breitenbach, Ken	Buf.	3	68	1	13	14	49	8	0	1	1	4	....	1975-76	1978-79
Brennan, Dan	L.A.	2	8	0	1	1	9	....	....	....	....	....	....	1983-84	1985-86
Brennan, Doug	NYR	3	123	9	7	16	152	16	1	0	1	21	1	1931-32	1933-34
Brennan, Tom	Bos.	2	12	2	2	4	2	....	....	....	....	....	....	1943-44	1944-45
Brenneman, John	Chi., NYR, Tor., Det., Oak.	5	152	21	19	40	46	....	....	....	....	....	1	1964-65	1968-69
Bretto, Joe	Chi.	1	3	0	0	0	4	....	....	....	....	....	....	1944-45	1944-45
• Brewer, Carl	Tor., Det., St.L.	12	604	25	198	223	1037	72	3	17	20	146	3	1957-58	1979-80
Brickley, Andy	Phi., Pit., N.J., Bos., Wpg.	11	385	82	140	222	81	17	1	4	5	4	....	1982-83	1993-94
• Briden, Archie	Bos., Det., Pit.	2	71	9	5	14	56	....	....	....	....	....	....	1926-27	1929-30
Bridgman, Mel	Phi., Cgy., N.J., Det., Van.	14	977	252	449	701	1625	125	28	39	67	298	....	1975-76	1988-89
• Briere, Michel	Pit.	1	76	12	32	44	20	10	5	3	8	17	....	1969-70	1969-70
Brindley, Doug	Tor.	1	3	0	0	0	0	....	....	....	....	....	....	1970-71	1970-71
• Brink, Milt	Chi.	1	5	0	0	0	0	....	....	....	....	....	....	1936-37	1936-37
Brisson, Gerry	Mtl.	1	4	0	2	2	4	....	....	....	....	....	....	1962-63	1962-63
Britz, Greg	Tor., Hfd.	3	8	0	0	0	4	....	....	....	....	....	....	1983-84	1986-87
• Broadbent, Punch	Ott., Mtl.M., NYA	11	303	121	51	172	564	23	4	5	9	50	4	1918-19	1928-29
‡ Brochu, Stephane	NYR	1	1	0	0	0	0	....	....	....	....	....	....	1988-89	1988-89
Broden, Connie	Mtl.	3	6	2	1	3	2	7	0	1	1	0	2	1955-56	1957-58
Brooke, Bob	NYR, Min., N.J.	7	447	69	97	166	520	34	9	9	18	59	....	1983-84	1989-90
Brooks, Gord	St.L., Wsh.	3	70	7	18	25	37	....	....	....	....	....	....	1971-72	1974-75
• Brophy, Bernie	Mtl.M., Det.	3	62	4	4	8	25	2	0	0	0	2	1	1925-26	1929-30
Brossart, Willie	Phi., Tor., Wsh.	6	129	1	14	15	88	1	0	0	0	0	....	1970-71	1975-76
Broten, Aaron	Col., N.J., Min., Que., Tor., Wpg.	12	748	186	329	515	441	34	7	18	25	40	....	1980-81	1991-92
Broten, Neal	Min., Dal., N.J., L.A.	17	1099	289	634	923	569	135	35	63	98	77	1	1980-81	1996-97
Broten, Paul	NYR, Dal., St.L.	7	322	46	55	101	264	38	4	6	10	18	....	1989-90	1995-96
• Brown, Adam	Det., Chi., Bos.	10	391	104	113	217	378	26	2	4	6	14	1	1941-42	1951-52
Brown, Arnie	Tor., NYR, Det., NYI, Atl.	12	681	44	141	185	738	22	0	6	6	23	....	1961-62	1973-74
‡ Brown, Cam	Van.	1	1	0	0	0	7	....	....	....	....	....	....	1990-91	1990-91
• Brown, Connie	Det.	5	73	15	24	39	12	14	2	3	5	0	1	1938-39	1942-43
Brown, Dave	Phi., Edm., S.J.	14	729	45	52	97	1789	80	2	3	5	209	1	1982-83	1995-96
Brown, Doug	N.J., Pit., Det.	15	854	160	214	374	210	109	23	23	46	26	2	1986-87	2000-01
• Brown, Fred	Mtl.M.	1	19	1	0	1	0	9	0	0	0	0	....	1927-28	1927-28
Brown, George	Mtl.	3	79	6	22	28	34	7	0	0	0	2	....	1936-37	1938-39
• Brown, Gerry	Det.	2	23	4	5	9	2	12	2	1	3	4	....	1941-42	1945-46
‡ Brown, Greg	Buf., Pit., Wpg.	4	94	4	14	18	86	6	0	1	1	4	....	1990-91	1994-95
Brown, Harold	NYR	1	13	2	1	3	2	....	....	....	....	....	....	1945-46	1945-46
Brown, Jeff	Que., St.L., Van., Hfd., Car., Tor., Wsh.	13	747	154	430	584	498	87	20	45	65	59	....	1985-86	1997-98
Brown, Jim	L.A.	1	3	0	1	1	5	....	....	....	....	....	....	1982-83	1982-83
Brown, Keith	Chi., Fla.	16	876	68	274	342	916	103	4	32	36	184	....	1979-80	1994-95
Brown, Larry	NYR, Det., Phi., L.A.	9	455	7	53	60	180	35	0	4	4	10	....	1969-70	1977-78
• Brown, Stan	NYR, Det.	2	48	8	2	10	18	2	0	0	0	0	....	1926-27	1927-28
Brown, Wayne	Bos.	1	....	....	....	....	....	4	0	0	0	2	....	1953-54	1953-54
• Browne, Cecil	Chi.	1	13	2	0	2	4	....	....	....	....	....	....	1927-28	1927-28
Brownschidle, Jack	St.L., Hfd.	9	494	39	162	201	151	26	0	5	5	18	....	1977-78	1985-86
Brownschidle, Jeff	Hfd.	2	7	0	1	1	2	....	....	....	....	....	....	1981-82	1982-83
Brubaker, Jeff	Hfd., Mtl., Cgy., Tor., Edm., NYR, Det.	8	178	16	9	25	512	2	0	0	0	27	....	1979-80	1988-89
Bruce, David	Van., St.L., S.J.	8	234	48	39	87	338	3	0	0	0	2	....	1985-86	1993-94
• Bruce, Gordie	Bos.	3	28	4	9	13	13	7	2	3	5	4	....	1940-41	1945-46
• Bruce, Morley	Ott.	4	71	8	3	11	27	3	0	0	0	2	2	1917-18	1921-22
Brumwell, Murray	Min., N.J.	7	128	12	31	43	70	2	0	0	0	2	....	1980-81	1987-88
• Bruneteau, Eddie	Det.	7	180	40	42	82	35	31	7	6	13	0	....	1940-41	1948-49
• Bruneteau, Mud	Det.	11	411	139	138	277	80	77	23	14	37	22	3	1935-36	1945-46
• Brydge, Bill	Tor., Det., NYA	9	368	26	52	78	506	2	0	0	0	4	....	1926-27	1935-36
Brydges, Paul	Buf.	1	15	2	2	4	6	....	....	....	....	....	....	1986-87	1986-87
• Brydson, Glenn	Mtl.M., St.L., NYR, Chi.	8	299	56	79	135	203	11	0	0	0	8	....	1930-31	1937-38
• Brydson, Gord	Tor.	1	8	2	0	2	8	....	....	....	....	....	....	1929-30	1929-30
Bubla, Jiri	Van.	5	256	17	101	118	202	6	0	0	0	7	....	1981-82	1985-86
• Buchanan, Al	Tor.	2	4	0	1	1	2	....	....	....	....	....	....	1948-49	1949-50
Buchanan, Bucky	NYR	1	2	0	0	0	0	....	....	....	....	....	....	1948-49	1948-49
Buchanan, Jeff	Col.	1	6	0	0	0	6	....	....	....	....	....	....	1998-99	1998-99
Buchanan, Mike	Chi.	1	1	0	0	0	0	....	....	....	....	....	....	1951-52	1951-52
Buchanan, Ron	Bos., St.L.	2	5	0	0	0	0	....	....	....	....	....	....	1966-67	1969-70
Bucyk, John	Det., Bos.	23	1540	556	813	1369	497	124	41	62	103	42	2	1955-56	1977-78
Bucyk, Randy	Mtl., Cgy.	2	19	4	2	6	8	2	0	0	0	0	....	1985-86	1987-88
Buhr, Doug	K.C.	1	6	0	2	2	4	....	....	....	....	....	....	1974-75	1974-75
Bukovich, Tony	Det.	2	17	7	3	10	6	6	0	1	1	0	....	1943-44	1944-45
‡ Bullard, Mike	Pit., Cgy., St.L., Phi., Tor.	11	727	329	345	674	703	40	11	18	29	44	....	1980-81	1991-92
• Buller, Hy	Det., NYR	5	188	22	58	80	215	....	....	....	....	....	....	1943-44	1953-54
Bulley, Ted	Chi., Wsh., Pit.	8	414	101	113	214	704	29	5	5	10	24	....	1976-77	1983-84
‡ Burakovsky, Robert	Ott.	1	23	2	3	5	6	....	....	....	....	....	....	1993-94	1993-94
• Burch, Billy	Ham., NYA, Bos., Chi.	11	390	137	61	198	255	2	0	0	0	0	....	1922-23	1932-33
• Burchell, Fred	Mtl.	2	4	0	0	0	2	....	....	....	....	....	....	1950-51	1953-54
Burdon, Glen	K.C.	1	11	0	2	2	0	....	....	....	....	....	....	1974-75	1974-75
Bureau, Marc	Cgy., Min., T.B., Mtl., Phi.	11	567	55	83	138	327	50	5	7	12	46	....	1989-90	1999-00
Burega, Bill	Tor.	1	4	0	1	1	4	....	....	....	....	....	....	1955-56	1955-56
• Burke, Eddie	Bos., NYA	4	106	29	20	49	55	....	....	....	....	....	....	1931-32	1934-35
• Burke, Marty	Mtl., Pit., Ott., Chi.	11	494	19	47	66	560	31	2	4	6	44	2	1927-28	1937-38
• Burmeister, Roy	NYA	3	67	4	3	7	2	....	....	....	....	....	....	1929-30	1931-32
Burnett, Kelly	NYR	1	3	1	0	1	0	....	....	....	....	....	....	1952-53	1952-53
• Burns, Bobby	Chi.	3	20	1	0	1	8	....	....	....	....	....	....	1927-28	1929-30
Burns, Charlie	Det., Bos., Oak., Pit., Min.	11	749	106	198	304	252	31	5	4	9	6	....	1958-59	1972-73
Burns, Gary	NYR	2	11	2	2	4	18	5	0	0	0	2	....	1980-81	1981-82
• Burns, Norm	NYR	1	11	0	4	4	2	....	....	....	....	....	....	1941-42	1941-42
Burns, Robin	Pit., K.C.	5	190	31	38	69	139	....	....	....	....	....	....	1970-71	1975-76
Burr, Shawn	Det., T.B., S.J.	16	878	181	259	440	1069	91	16	19	35	95	....	1984-85	1999-00
Burridge, Randy	Bos., Wsh., L.A., Buf.	13	706	199	251	450	458	107	18	34	52	103	....	1985-86	1997-98
Burrows, Dave	Pit., Tor.	10	724	29	135	164	373	29	1	5	6	25	....	1971-72	1980-81
• Burry, Bert	Ott.	1	4	0	0	0	0	....	....	....	....	....	....	1932-33	1932-33
Burt, Adam	Hfd., Car., Phi., Atl.	13	737	37	115	152	961	21	0	1	1	8	....	1988-89	2000-01
Burton, Cummy	Det.	3	43	0	2	2	21	3	0	0	0	0	....	1955-56	1958-59
Burton, Nelson	Wsh.	2	8	1	0	1	21	....	....	....	....	....	....	1977-78	1978-79
• Bush, Eddie	Det.	2	26	4	6	10	40	11	1	6	7	23	....	1938-39	1941-42
Buskas, Rod	Pit., Van., L.A., Chi.	11	556	19	63	82	1294	18	0	3	3	45	....	1982-83	1992-93
Busniuk, Mike	Phi.	2	143	3	23	26	297	25	2	5	7	34	....	1979-80	1980-81
Busniuk, Ron	Buf.	2	6	0	3	3	13	....	....	....	....	....	....	1972-73	1973-74
• Buswell, Walt	Det., Mtl.	8	368	10	40	50	164	24	2	1	3	10	....	1932-33	1939-40
Butcher, Garth	Van., St.L., Que., Tor.	14	897	48	158	206	2302	50	6	5	11	122	....	1981-82	1994-95
Butler, Dick	Chi.	1	7	2	0	2	0	....	....	....	....	....	....	1947-48	1947-48
Butler, Jerry	NYR, St.L., Tor., Van., Wpg.	11	641	99	120	219	515	48	3	3	6	79	....	1972-73	1982-83

Carl Brewer

Greg Brown

David Bruce

Garth Butcher

John Byce

Billy Carroll

Bruce Cassidy

Frantisek Cernik

			Regular Schedule					Playoffs							
Name	NHL Teams	NHL Seasons	GP	G	A	TP	PIM	GP	G	A	TP	PIM	NHL Cup Wins	First NHL Season	Last NHL Season
‡ Butsayev, Viacheslav	Phi., S.J., Ana., Fla., Ott., T.B.	6	132	17	26	43	133	....	....	....	....	....	....	1992-93	1999-00
Butters, Bill	Min.	2	72	1	4	5	77	....	....	....	....	....	....	1977-78	1978-79
Buttrey, Gord	Chi.	1	10	0	0	0	0	....	....	....	....	....	....	1943-44	1943-44
Buynak, Gord	St.L.	1	4	0	0	0	2	....	....	....	....	....	....	1974-75	1974-75
‡ Byakin, Ilja	Edm., S.J.	2	57	8	25	33	44	....	....	....	....	....	....	1993-94	1994-95
Byce, John	Bos.	3	21	2	3	5	6	8	2	0	2	2	....	1989-90	1991-92
Byers, Gord	Bos.	1	1	0	1	1	0	....	....	....	....	....	....	1949-50	1949-50
Byers, Jerry	Min., Atl., NYR	4	43	3	4	7	15	....	....	....	....	....	....	1972-73	1977-78
Byers, Lyndon	Bos., S.J.	10	279	28	43	71	1081	37	2	2	4	96	....	1983-84	1992-93
Byers, Mike	Tor., Phi., L.A., Buf.	4	166	42	34	76	39	4	0	1	1	0	....	1967-68	1971-72
‡ Byram, Shawn	NYI, Chi.	2	5	0	0	0	14	....	....	....	....	....	....	1990-91	1991-92
C															
• Caffery, Jack	Tor., Bos.	3	57	3	2	5	22	10	1	0	1	4	....	1954-55	1957-58
Caffery, Terry	Chi., Min.	2	14	0	0	0	0	1	0	0	0	0	....	1969-70	1970-71
• Cahan, Larry	Tor., NYR, Oak., L.A.	13	666	38	92	130	700	29	1	1	2	38	....	1954-55	1970-71
• Cahill, Charles	Bos.	2	32	0	1	1	4	....	....	....	....	....	....	1925-26	1926-27
• Cain, Francis	Mtl.M., Tor.	2	61	4	0	4	35	....	....	....	....	....	....	1924-25	1925-26
• Cain, Herb	Mtl.M., Mtl., Bos.	13	570	206	194	400	178	67	16	13	29	13	2	1933-34	1945-46
Cairns, Don	K.C., Col.	2	9	0	1	1	2	....	....	....	....	....	....	1975-76	1976-77
Calder, Eric	Wsh.	2	2	0	0	0	0	....	....	....	....	....	....	1981-82	1982-83
• Calladine, Norm	Bos.	3	63	19	29	48	8	....	....	....	....	....	....	1942-43	1944-45
Callander, Drew	Phi., Van.	4	39	6	2	8	7	....	....	....	....	....	....	1976-77	1979-80
Callander, Jock	Pit., T.B.	5	109	22	29	51	116	22	3	8	11	12	1	1987-88	1992-93
Callighen, Brett	Edm.	3	160	56	89	145	132	14	4	6	10	8	....	1979-80	1981-82
Callighen, Patsy	NYR	1	36	0	0	0	32	9	0	0	0	0	1	1927-28	1927-28
‡ Camazzola, James	Chi.	2	3	0	0	0	0	....	....	....	....	....	....	1983-84	1986-87
Camazzola, Tony	Wsh.	1	3	0	0	0	4	....	....	....	....	....	....	1981-82	1981-82
Cameron, Al	Det., Wpg.	6	282	11	44	55	356	7	0	1	1	2	....	1975-76	1980-81
• Cameron, Billy	Mtl., NYA	2	39	0	0	0	2	2	0	0	0	0	1	1923-24	1925-26
Cameron, Craig	Det., St.L., Min., NYI	9	552	87	65	152	196	27	3	1	4	17	....	1966-67	1975-76
Cameron, Dave	Col., N.J.	3	168	25	28	53	238	....	....	....	....	....	....	1981-82	1983-84
• Cameron, Harry	Tor., Ott., Mtl.	6	128	88	51	139	189	11	5	4	9	16	2	1917-18	1922-23
• Cameron, Scotty	NYR	1	35	8	11	19	0	....	....	....	....	....	....	1942-43	1942-43
Campbell, Bryan	L.A., Chi.	5	260	35	71	106	74	22	3	4	7	2	....	1967-68	1971-72
Campbell, Colin	Pit., Col., Edm., Van., Det.	11	636	25	103	128	1292	45	4	10	14	181	....	1974-75	1984-85
• Campbell, Dave	Mtl.	1	2	0	0	0	0	....	....	....	....	....	....	1920-21	1920-21
Campbell, Don	Chi.	1	17	1	3	4	8	....	....	....	....	....	....	1943-44	1943-44
• Campbell, Earl	Ott., NYA	3	76	6	3	9	14	1	0	0	0	6	....	1923-24	1925-26
Campbell, Scott	Wpg., St.L.	3	80	4	21	25	243	....	....	....	....	....	....	1979-80	1981-82
Campbell, Wade	Wpg., Bos.	6	213	9	27	36	305	10	0	0	0	20	....	1982-83	1987-88
Campeau, Tod	Mtl.	3	42	5	9	14	16	1	0	0	0	0	....	1943-44	1948-49
Campedelli, Dom	Mtl.	1	2	0	0	0	0	....	....	....	....	....	....	1985-86	1985-86
Capuano, Dave	Pit., Van., T.B., S.J.	4	104	17	38	55	56	6	1	1	2	5	....	1989-90	1993-94
Capuano, Jack	Tor., Van., Bos.	3	6	0	0	0	0	....	....	....	....	....	....	1989-90	1991-92
Carbol, Leo	Chi.	1	6	0	1	1	4	....	....	....	....	....	....	1942-43	1942-43
Carbonneau, Guy	Mtl., St.L., Dal.	19	1318	260	403	663	820	231	38	55	93	161	3	1980-81	1999-00
Cardin, Claude	St.L.	1	1	0	0	0	0	....	....	....	....	....	....	1967-68	1967-68
Cardwell, Steve	Pit.	3	53	9	11	20	35	4	0	0	0	2	....	1970-71	1972-73
• Carey, George	Que., Ham., Tor.	5	72	21	12	33	20	....	....	....	....	....	....	1919-20	1923-24
Carkner, Terry	NYR, Que., Phi., Det., Fla.	13	858	42	188	230	1588	54	1	9	10	48	....	1986-87	1998-99
Carleton, Wayne	Tor., Bos., Cal.	7	278	55	73	128	172	18	2	4	6	14	1	1965-66	1971-72
Carlin, Brian	L.A.	1	5	1	0	1	0	....	....	....	....	....	....	1971-72	1971-72
Carlson, Jack	Min., St.L.	6	236	30	15	45	417	25	1	2	3	72	....	1978-79	1986-87
Carlson, Kent	Mtl., St.L., Wsh.	5	113	7	11	18	148	8	0	0	0	13	....	1983-84	1988-89
Carlson, Steve	L.A.	1	52	9	12	21	23	4	1	1	2	7	....	1979-80	1979-80
‡ Carlsson, Anders	N.J.	3	104	7	26	33	34	3	1	0	1	2	....	1986-87	1988-89
Carlyle, Randy	Tor., Pit., Wpg.	17	1055	148	499	647	1400	69	9	24	33	120	....	1976-77	1992-93
‡ Carnback, Patrik	Mtl., Ana.	4	154	24	38	62	122	....	....	....	....	....	....	1992-93	1995-96
• Caron, Alain	Oak., Mtl.	2	60	9	13	22	18	....	....	....	....	....	....	1967-68	1968-69
Carpenter, Bob	Wsh., NYR, L.A., Bos., N.J.	19	1178	320	408	728	919	140	21	38	59	136	1	1981-82	1998-99
• Carpenter, Ed	Que., Ham.	2	45	10	5	15	41	....	....	....	....	....	....	1919-20	1920-21
Carr, Gene	St.L., NYR, L.A., Pit., Atl.	8	465	79	136	215	365	35	5	8	13	66	....	1971-72	1978-79
Carr, Lorne	NYR, NYA, Tor.	13	580	204	222	426	132	53	10	9	19	13	2	1933-34	1945-46
Carr, Red	Tor.	1	5	0	1	1	2	....	....	....	....	....	....	1943-44	1943-44
Carriere, Larry	Buf., Atl., Van., L.A., Tor.	7	367	16	74	90	462	27	0	3	3	42	....	1972-73	1979-80
• Carrigan, Gene	NYR, Det., St.L.	3	37	2	1	3	13	4	0	0	0	0	....	1930-31	1934-35
Carroll, Billy	NYI, Edm., Det.	7	322	30	54	84	113	71	6	12	18	18	4	1980-81	1986-87
• Carroll, George	Mtl.M., Bos.	1	16	0	0	0	11	....	....	....	....	....	....	1924-25	1924-25
Carroll, Greg	Wsh., Det., Hfd.	2	131	20	34	54	44	....	....	....	....	....	....	1978-79	1979-80
Carruthers, Dwight	Det., Phi.	2	2	0	0	0	0	....	....	....	....	....	....	1965-66	1967-68
• Carse, Bill	NYR, Chi.	4	124	28	43	71	38	13	3	2	5	0	....	1938-39	1941-42
• Carse, Bob	Chi., Mtl.	5	167	32	55	87	52	10	0	2	2	2	....	1939-40	1947-48
• Carson, Bill	Tor., Bos.	4	159	54	24	78	156	11	3	0	3	14	1	1926-27	1929-30
• Carson, Frank	Mtl.M., NYA, Det.	7	248	42	48	90	166	27	0	2	2	9	1	1925-26	1933-34
• Carson, Gerry	Mtl., NYR, Mtl.M.	6	261	12	11	23	205	22	0	0	0	12	1	1928-29	1936-37
Carson, Jimmy	L.A., Edm., Det., Van., Hfd.	10	626	275	286	561	254	55	17	15	32	22	....	1986-87	1995-96
Carson, Lindsay	Phi., Hfd.	7	373	66	80	146	524	49	4	10	14	56	....	1981-82	1987-88
Carter, Billy	Mtl., Bos.	3	16	0	0	0	6	....	....	....	....	....	....	1957-58	1961-62
Carter, John	Bos., S.J.	8	244	40	50	90	201	31	7	5	12	51	....	1985-86	1992-93
Carter, Ron	Edm.	1	2	0	0	0	0	....	....	....	....	....	....	1979-80	1979-80
• Carveth, Joe	Det., Bos., Mtl.	11	504	150	189	339	81	69	21	16	37	28	2	1940-41	1950-51
Cashman, Wayne	Bos.	17	1027	277	516	793	1041	145	31	57	88	250	2	1964-65	1982-83
‡ Casselman, Mike	Fla.	1	3	0	0	0	0	....	....	....	....	....	....	1995-96	1995-96
Cassidy, Bruce	Chi.	7	36	4	13	17	10	1	0	0	0	0	....	1983-84	1989-90
Cassidy, Tom	Pit.	1	26	3	4	7	15	....	....	....	....	....	....	1977-78	1977-78
Cassolato, Tony	Wsh.	3	23	1	6	7	4	....	....	....	....	....	....	1979-80	1981-82
Caufield, Jay	NYR, Min., Pit.	7	208	5	8	13	759	17	0	0	0	42	2	1986-87	1992-93
‡ Cavallini, Gino	Cgy., St.L., Que.	9	593	114	159	273	507	74	14	19	33	66	....	1984-85	1992-93
Cavallini, Paul	Wsh., St.L., Dal.	10	564	56	177	233	750	69	8	27	35	114	....	1986-87	1995-96
Ceresino, Ray	Tor.	1	12	1	1	2	2	....	....	....	....	....	....	1948-49	1948-49
Cernik, Frantisek	Det.	1	49	5	4	9	13	....	....	....	....	....	....	1984-85	1984-85
‡ Chabot, John	Mtl., Pit., Det.	8	508	84	228	312	85	33	6	20	26	2	....	1983-84	1990-91
• Chad, John	Chi.	3	80	15	22	37	29	10	0	1	1	2	....	1939-40	1945-46
• Chalmers, Chick	NYR	1	1	0	0	0	0	....	....	....	....	....	....	1953-54	1953-54
Chalupa, Milan	Det.	1	14	0	5	5	6	....	....	....	....	....	....	1984-85	1984-85
• Chamberlain, Murph	Tor., Mtl., Bro., Bos.	12	510	100	175	275	769	66	14	17	31	96	2	1937-38	1948-49
Chambers, Shawn	Min., Wsh., T.B., N.J., Dal.	13	625	50	185	235	364	94	7	26	33	72	2	1987-88	1999-00
Champagne, Andre	Tor.	1	2	0	0	0	0	....	....	....	....	....	....	1962-63	1962-63
‡ Chapdelaine, Rene	L.A.	3	32	0	2	2	32	....	....	....	....	....	....	1990-91	1992-93
• Chapman, Art	Bos., NYA	10	438	62	176	238	140	26	1	5	6	9	....	1930-31	1939-40
Chapman, Blair	Pit., St.L.	7	402	106	125	231	158	25	4	6	10	15	....	1976-77	1982-83
‡ Chapman, Brian	Hfd.	1	3	0	0	0	29	....	....	....	....	....	....	1990-91	1990-91
‡ Charbonneau, Jose	Mtl., Van.	4	71	9	13	22	67	11	1	0	1	8	....	1987-88	1994-95
Charbonneau, Stephane	Que.	1	2	0	0	0	0	....	....	....	....	....	....	1991-92	1991-92
Charlebois, Bob	Min.	1	7	1	0	1	0	....	....	....	....	....	....	1967-68	1967-68
Charlesworth, Todd	Pit., NYR	6	93	3	9	12	47	....	....	....	....	....	....	1983-84	1989-90
‡ Charron, Eric	Mtl., T.B., Wsh., Cgy.	8	130	2	7	9	127	6	0	0	0	8	....	1992-93	1999-00
Charron, Guy	Mtl., Det., K.C., Wsh.	15	734	221	309	530	146	....	....	....	....	....	....	1969-70	1980-81
Chartier, Dave	Wpg.	1	1	0	0	0	0	....	....	....	....	....	....	1980-81	1980-81
Chartraw, Rick	Mtl., L.A., NYR, Edm.	10	420	28	64	92	399	75	7	9	16	80	4	1974-75	1983-84
Chase, Kelly	St.L., Hfd., Tor.	11	458	17	36	53	2017	27	1	1	2	100	....	1989-90	1999-00
‡ Chasse, Denis	St.L., Wsh., Wpg., Ott.	4	132	11	14	25	292	7	1	7	8	23	....	1993-94	1996-97
Check, Lude	Det., Chi.	2	27	6	2	8	4	....	....	....	....	....	....	1943-44	1944-45
Chernoff, Mike	Min.	1	1	0	0	0	0	....	....	....	....	....	....	1968-69	1968-69
Chernomaz, Rich	Col., N.J., Cgy.	7	51	9	7	16	18	....	....	....	....	....	....	1981-82	1991-92
Cherry, Dick	Bos., Phi.	3	145	12	10	22	45	4	1	0	1	4	....	1956-57	1969-70
Cherry, Don	Bos.	1	....	....	....	....	....	1	0	0	0	0	....	1954-55	1954-55
‡ Chervyakov, Denis	Bos.	1	2	0	0	0	2	....	....	....	....	....	....	1992-93	1992-93
• Chevrefils, Real	Bos., Det.	8	387	104	97	201	185	30	5	4	9	20	....	1951-52	1958-59
• Chiasson, Steve	Det., Cgy., Hfd., Car.	13	751	93	305	398	1107	63	16	19	35	119	....	1986-87	1998-99
‡ Chibirev, Igor	Hfd.	2	45	7	12	19	2	....	....	....	....	....	....	1993-94	1994-95
Chicoine, Dan	Cle., Min.	3	31	1	2	3	12	1	0	0	0	0	....	1977-78	1979-80
Chinnick, Rick	Min.	2	4	0	2	2	0	....	....	....	....	....	....	1973-74	1974-75
Chipperfield, Ron	Edm., Que.	2	83	22	24	46	34	....	....	....	....	....	....	1979-80	1980-81

Name	NHL Teams	NHL Seasons	Regular Schedule GP	G	A	TP	PIM	Playoffs GP	G	A	TP	PIM	NHL Cup Wins	First NHL Season	Last NHL Season
Chisholm, Art	Bos.	1	3	0	0	0	0	....	....	....	....	....	....	1960-61	1960-61
Chisholm, Colin	Min.	1	1	0	0	0	0	....	....	....	....	....	....	1986-87	1986-87
• Chisholm, Lex	Tor.	2	54	10	8	18	19	3	1	0	1	0	....	1939-40	1940-41
Chorney, Marc	Pit., L.A.	4	210	8	27	35	209	7	0	1	1	2	....	1980-81	1983-84
Chorske, Tom	Mtl., N.J., Ott., NYI, Wsh., Cgy., Pit.	11	596	115	122	237	225	50	5	12	17	10	1	1989-90	1999-00
• Chouinard, Gene	Ott.	1	8	0	0	0	0	....	....	....	....	....	....	1927-28	1927-28
Chouinard, Guy	Atl., Cgy., St.L.	10	578	205	370	575	120	46	9	28	37	12	....	1974-75	1983-84
Christian, Dave	Wpg., Wsh., Bos., St.L., Chi.	15	1009	340	433	773	284	102	32	25	57	27	....	1979-80	1993-94
Christie, Mike	Cal., Cle., Col., Van.	7	412	15	101	116	550	2	0	0	0	0	....	1974-75	1980-81
Christoff, Steve	Min., Cgy., L.A.	5	248	77	64	141	108	35	16	12	28	25	....	1979-80	1983-84
Chrystal, Bob	NYR	2	132	11	14	25	112	....	....	....	....	....	....	1953-54	1954-55
• Church, Jack	Tor., Bro., Bos.	5	130	4	19	23	154	25	1	1	2	18	....	1938-39	1945-46
Churla, Shane	Hfd., Cgy., Min., Dal., L.A., NYR	11	488	26	45	71	2301	78	5	7	12	282	....	1986-87	1996-97
Chychrun, Jeff	Phi., L.A., Pit., Edm.	8	262	3	22	25	744	19	0	2	2	65	1	1986-87	1993-94
Chynoweth, Dean	NYI, Bos.	9	241	4	18	22	667	6	0	0	0	26	....	1988-89	1997-98
‡ Chyzowski, Dave	NYI, Chi.	6	126	15	16	31	144	2	0	0	0	0	....	1989-90	1996-97
Ciavaglia, Peter	Buf.	2	5	0	0	0	0	....	....	....	....	....	....	1991-92	1992-93
Ciccarelli, Dino	Min., Wsh., Det., T.B., Fla.	19	1232	608	592	1200	1425	141	73	45	118	211	....	1980-81	1998-99
‡ Ciccone, Enrico	Min., Wsh., T.B., Chi., Car., Van., Mtl.	9	374	10	18	28	1469	13	1	0	1	48	....	1991-92	2000-01
Cichocki, Chris	Det., N.J.	4	68	11	12	23	27	....	....	....	....	....	....	1985-86	1988-89
‡ Cierny, Jozef	Edm.	1	1	0	0	0	0	....	....	....	....	....	....	1993-94	1993-94
• Ciesla, Hank	Chi., NYR	4	269	26	51	77	87	6	0	2	2	0	....	1955-56	1958-59
Cimellaro, Tony	Ott.	1	2	0	0	0	0	....	....	....	....	....	....	1992-93	1992-93
Cimetta, Rob	Bos., Tor.	4	103	16	16	32	66	1	0	0	0	15	....	1988-89	1991-92
Cirella, Joe	Col., N.J., Que., NYR, Fla., Ott.	15	828	64	211	275	1446	38	0	13	13	98	....	1981-82	1995-96
‡ Cirone, Jason	Wpg.	1	3	0	0	0	2	....	....	....	....	....	....	1991-92	1991-92
Clackson, Kim	Pit., Que.	2	106	0	8	8	370	8	0	0	0	70	....	1979-80	1980-81
• Clancy, King	Ott., Tor.	16	592	136	147	283	914	55	8	8	16	92	3	1921-22	1936-37
Clancy, Terry	Oak., Tor.	4	93	6	6	12	39	....	....	....	....	....	....	1967-68	1972-73
• Clapper, Dit	Bos.	20	833	228	246	474	462	82	13	17	30	50	3	1927-28	1946-47
Clark, Dan	NYR	1	4	0	1	1	6	....	....	....	....	....	....	1978-79	1978-79
Clark, Dean	Edm.	1	1	0	0	0	0	....	....	....	....	....	....	1983-84	1983-84
Clark, Gordie	Bos.	2	8	0	1	1	0	1	0	0	0	0	....	1974-75	1975-76
• Clark, Nobby	Bos.	1	5	0	0	0	0	....	....	....	....	....	....	1927-28	1927-28
Clark, Wendel	Tor., Que., NYI, T.B., Det., Chi.	15	793	330	234	564	1690	95	37	32	69	201	....	1985-86	1999-00
Clarke, Bobby	Phi.	15	1144	358	852	1210	1453	136	42	77	119	152	2	1969-70	1983-84
• Cleghorn, Odie	Mtl., Pit.	10	181	95	34	129	142	12	7	1	8	2	1	1918-19	1927-28
• Cleghorn, Sprague	Ott., Tor., Mtl., Bos.	10	259	83	55	138	538	21	4	2	6	28	2	1918-19	1927-28
Clement, Bill	Phi., Wsh., Atl., Cgy.	11	719	148	208	356	383	50	5	3	8	26	2	1971-72	1981-82
Cline, Bruce	NYR	1	30	2	3	5	10	....	....	....	....	....	....	1956-57	1956-57
Clippingdale, Steve	L.A., Wsh.	2	19	1	2	3	9	1	0	0	0	0	....	1976-77	1979-80
Cloutier, Real	Que., Buf.	6	317	146	198	344	119	25	7	5	12	20	....	1979-80	1984-85
Cloutier, Rejean	Det.	2	5	0	2	2	2	....	....	....	....	....	....	1979-80	1981-82
Cloutier, Roland	Det., Que.	3	34	8	9	17	2	....	....	....	....	....	....	1977-78	1979-80
‡ Cloutier, Sylvain	Chi.	1	7	0	0	0	0	....	....	....	....	....	....	1998-99	1998-99
• Clune, Wally	Mtl.	1	5	0	0	0	6	....	....	....	....	....	....	1955-56	1955-56
Coalter, Gary	Cal., K.C.	2	34	2	4	6	2	....	....	....	....	....	....	1973-74	1974-75
Coates, Steve	Det.	1	5	1	0	1	24	....	....	....	....	....	....	1976-77	1976-77
Cochrane, Glen	Phi., Van., Chi., Edm.	10	411	17	72	89	1556	18	1	1	2	31	....	1978-79	1988-89
‡ Coffey, Paul	Edm., Pit., L.A., Det., Hfd., Phi., Chi., Car., Bos.	21	1409	396	1135	1531	1802	194	59	137	196	264	4	1980-81	2000-01
Coflin, Hugh	Chi.	1	31	0	3	3	33	....	....	....	....	....	....	1950-51	1950-51
Cole, Danton	Wpg., T.B., N.J., NYI, Chi.	7	318	58	60	118	125	1	0	0	0	0	1	1989-90	1995-96
Colley, Tom	Min.	1	1	0	0	0	2	....	....	....	....	....	....	1974-75	1974-75
Collings, Norm	Mtl.	1	1	0	1	1	0	....	....	....	....	....	....	1934-35	1934-35
Collins, Bill	Min., Mtl., Det., St.L., NYR, Phi., Wsh.	11	768	157	154	311	415	18	3	5	8	12	....	1967-68	1977-78
Collins, Gary	Tor.	1	....	....	....	....	....	2	0	0	0	0	....	1958-59	1958-59
Collyard, Bob	St.L.	1	10	1	3	4	4	....	....	....	....	....	....	1973-74	1973-74
• Colman, Michael	S.J.	1	15	0	1	1	32	....	....	....	....	....	....	1991-92	1991-92
Colville, Mac	NYR	9	353	71	104	175	130	40	9	10	19	14	1	1935-36	1946-47
• Colville, Neil	NYR	12	464	99	166	265	213	46	7	19	26	32	1	1935-36	1948-49
Colwill, Les	NYR	1	69	7	6	13	16	....	....	....	....	....	....	1958-59	1958-59
Comeau, Rey	Mtl., Atl., Col.	9	564	98	141	239	175	9	2	1	3	8	....	1971-72	1979-80
‡ Comrie, Paul	Edm.	1	15	1	2	3	4	....	....	....	....	....	....	1999-00	1999-00
Conacher, Brian	Tor., Det.	5	155	28	28	56	84	12	3	2	5	21	1	1961-62	1971-72
• Conacher, Charlie	Tor., Det., NYA	12	459	225	173	398	523	49	17	18	35	49	1	1929-30	1940-41
Conacher, Jim	Det., Chi., NYR	8	328	85	117	202	91	19	5	2	7	4	....	1945-46	1952-53
• Conacher, Lionel	Pit., NYA, Mtl.M., Chi.	12	498	80	105	185	882	35	2	2	4	34	2	1925-26	1936-37
‡ Conacher, Pat	NYR, Edm., N.J., L.A., Cgy., NYI	13	521	63	76	139	235	66	11	10	21	40	1	1979-80	1995-96
Conacher, Pete	Chi., NYR, Tor.	6	229	47	39	86	57	7	0	0	0	0	....	1951-52	1957-58
• Conacher, Roy	Bos., Det., Chi.	11	490	226	200	426	90	42	15	15	30	14	2	1938-39	1951-52
• Conn, Red	NYA	2	96	9	28	37	22	....	....	....	....	....	....	1933-34	1934-35
Conn, Rob	Chi., Buf.	2	30	2	5	7	20	....	....	....	....	....	....	1991-92	1995-96
• Connelly, Bert	NYR, Chi.	3	87	13	15	28	37	14	1	0	1	0	1	1934-35	1937-38
Connelly, Wayne	Mtl., Bos., Min., Det., St.L., Van.	10	543	133	174	307	156	24	11	7	18	4	....	1960-61	1971-72
Connor, Cam	Mtl., Edm., NYR	5	89	9	22	31	256	20	5	0	5	6	1	1978-79	1982-83
• Connor, Harry	Bos., NYA, Ott.	4	134	16	5	21	149	10	0	0	0	2	....	1927-28	1930-31
• Connors, Bob	NYA, Det.	3	78	17	10	27	110	2	0	0	0	10	....	1926-27	1929-30
‡ Conroy, Al	Phi.	3	114	9	14	23	156	....	....	....	....	....	....	1991-92	1993-94
Contini, Joe	Col., Min.	3	68	17	21	38	34	2	0	0	0	0	....	1977-78	1980-81
‡ Convery, Brandon	Tor., Van., L.A.	4	72	9	19	28	36	5	0	0	0	2	....	1995-96	1998-99
• Convey, Eddie	NYA	3	36	1	1	2	33	....	....	....	....	....	....	1930-31	1932-33
• Cook, Bill	NYR	11	474	229	138	367	386	46	13	11	24	68	2	1926-27	1936-37
• Cook, Bob	Van., Det., NYI, Min.	4	72	13	9	22	22	....	....	....	....	....	....	1970-71	1974-75
• Cook, Bud	Bos., Ott., St.L.	3	50	5	4	9	22	....	....	....	....	....	....	1931-32	1934-35
• Cook, Bun	NYR, Bos.	11	473	158	144	302	444	46	15	3	18	50	2	1926-27	1936-37
• Cook, Lloyd	Bos.	1	4	1	0	1	0	....	....	....	....	....	....	1924-25	1924-25
• Cook, Tom	Chi., Mtl.M.	9	349	77	98	175	184	24	2	4	6	19	1	1929-30	1937-38
• Cooper, Carson	Bos., Mtl., Det.	8	294	110	57	167	111	7	0	0	0	2	....	1924-25	1931-32
Cooper, Ed	Col.	2	49	8	7	15	46	....	....	....	....	....	....	1980-81	1981-82
• Cooper, Hal	NYR	1	8	0	0	0	2	....	....	....	....	....	....	1944-45	1944-45
• Cooper, Joe	NYR, Chi.	11	420	30	66	96	442	35	3	5	8	58	....	1935-36	1946-47
Copp, Bob	Tor.	2	40	3	9	12	26	....	....	....	....	....	....	1942-43	1950-51
• Corbeau, Bert	Mtl., Ham., Tor.	10	258	63	49	112	629	9	2	2	4	38	....	1917-18	1926-27
Corbet, Rene	Que., Col., Cgy., Pit.	8	362	58	74	132	420	53	7	6	13	52	1	1993-94	2000-01
Corbett, Mike	L.A.	1	....	....	....	....	....	2	0	1	1	2	....	1967-68	1967-68
Corcoran, Norm	Bos., Det., Chi.	4	29	1	3	4	21	4	0	0	0	6	....	1949-50	1955-56
• Cormier, Roger	Mtl.	1	1	0	0	0	0	....	....	....	....	....	....	1925-26	1925-26
Cornforth, Mark	Bos.	1	6	0	0	0	4	....	....	....	....	....	....	1995-96	1995-96
Corrigan, Chuck	Tor., NYA	2	19	2	2	4	2	....	....	....	....	....	....	1937-38	1940-41
Corrigan, Mike	L.A., Van., Pit.	10	594	152	195	347	698	17	2	3	5	20	....	1967-68	1977-78
• Corriveau, Andre	Mtl.	1	3	0	1	1	0	....	....	....	....	....	....	1953-54	1953-54
‡ Corriveau, Yvon	Wsh., Hfd., S.J.	9	280	48	40	88	310	29	5	7	12	50	....	1985-86	1993-94
Cory, Ross	Wpg.	2	51	2	10	12	41	....	....	....	....	....	....	1979-80	1980-81
Cossette, Jacques	Pit.	3	64	8	6	14	29	3	0	1	1	4	....	1975-76	1978-79
Costello, Les	Tor.	3	15	2	3	5	11	6	2	2	4	2	1	1947-48	1949-50
Costello, Murray	Chi., Bos., Det.	4	162	13	19	32	54	5	0	0	0	2	....	1953-54	1956-57
Costello, Rich	Tor.	2	12	2	2	4	2	....	....	....	....	....	....	1983-84	1985-86
• Cotch, Charlie	Ham., Tor.	1	12	1	0	1	0	....	....	....	....	....	....	1924-25	1924-25
Cote, Alain	Que.	10	696	103	190	293	383	67	9	15	24	44	....	1979-80	1988-89
‡ Cote, Alain	Bos., Wsh., Mtl., T.B., Que.	9	119	2	18	20	124	11	0	2	2	26	....	1985-86	1993-94
‡ Cote, Patrick	Dal., Nsh., Edm.	6	105	1	2	3	377	....	....	....	....	....	....	1995-96	2000-01
Cote, Ray	Edm.	3	15	0	0	0	4	14	3	2	5	0	....	1982-83	1984-85
• Cotton, Baldy	Pit., Tor., NYA	12	503	101	103	204	419	43	4	9	13	46	1	1925-26	1936-37
• Coughlin, Jack	Tor., Que., Mtl., Ham.	3	19	2	0	2	3	....	....	....	....	....	1	1917-18	1920-21
Coulis, Tim	Wsh., Min.	4	47	4	5	9	138	3	1	0	1	2	....	1979-80	1985-86
Coulson, D'arcy	Phi.	1	28	0	0	0	103	....	....	....	....	....	....	1930-31	1930-31
• Coulter, Art	Chi., NYR	11	465	30	82	112	543	49	4	5	9	61	2	1931-32	1941-42
Coulter, Neal	NYI	3	26	5	5	10	11	....	....	....	....	....	....	1985-86	1987-88
Cournoyer, Yvan	Mtl.	16	968	428	435	863	255	147	64	63	127	47	10	1963-64	1978-79
Courteau, Yves	Cgy., Hfd.	3	22	2	5	7	4	1	0	0	0	0	....	1984-85	1986-87
‡ Courtenay, Ed	S.J.	2	44	7	13	20	10	....	....	....	....	....	....	1991-92	1992-93
Courtnall, Geoff	Bos., Edm., Wsh., St.L., Van.	17	1048	367	432	799	1465	156	39	70	109	262	1	1983-84	1999-00
Courtnall, Russ	Tor., Mtl., Min., Dal., Van., NYR, L.A.	16	1029	297	447	744	557	129	39	44	83	83	....	1983-84	1998-99
‡ Courville, Larry	Van.	3	33	1	2	3	16	....	....	....	....	....	....	1995-96	1997-98
• Coutu, Billy	Mtl., Ham., Bos.	10	244	33	21	54	478	19	1	1	2	35	1	1917-18	1926-27

Enrico Ciccone

John Chabot

Paul Coffey

Neil Colville

Lionel Conacher

Paul Comrie

Rob Cowie

Craig Coxe

Name	NHL Teams	NHL Seasons	Regular Schedule GP	G	A	TP	PIM	Playoffs GP	G	A	TP	PIM	NHL Cup Wins	First NHL Season	Last NHL Season
• Couture, Gerry	Det., Mtl., Chi.	10	385	86	70	156	89	45	9	7	16	4	1	1944-45	1953-54
• Couture, Rosie	Chi., Mtl.	8	309	48	56	104	184	23	1	5	6	15	1	1928-29	1935-36
‡ Couturier, Sylvain	L.A.	3	33	4	5	9	4	....	....	....	....	....	....	1988-89	1991-92
Cowick, Bruce	Phi., Wsh., St.L.	3	70	5	6	11	43	8	0	0	0	9	1	1973-74	1975-76
‡ Cowie, Rob	L.A.	2	78	7	12	19	52	....	....	....	....	....	....	1994-95	1995-96
• Cowley, Bill	St.L., Bos.	13	549	195	353	548	143	64	12	34	46	22	2	1934-35	1946-47
• Cox, Danny	Tor., Ott., Det., NYR	8	319	47	49	96	128	10	0	1	1	6	....	1926-27	1933-34
Coxe, Craig	Van., Cgy., St.L., S.J.	8	235	14	31	45	713	5	1	0	1	18	....	1984-85	1991-92
‡ Craighead, John	Tor.	1	5	0	0	0	10	....	....	....	....	....	....	1996-97	1996-97
‡ Craigwell, Dale	S.J.	3	98	11	18	29	28	....	....	....	....	....	....	1991-92	1993-94
Crashley, Bart	Det., K.C., L.A.	6	140	7	36	43	50	....	....	....	....	....	....	1965-66	1975-76
Craven, Murray	Det., Phi., Hfd., Van., Chi., S.J.	18	1071	266	493	759	524	118	27	43	70	64	....	1982-83	1999-00
Crawford, Bob	St.L., Hfd., NYR, Wsh.	7	246	71	71	142	72	11	0	1	1	8	....	1979-80	1986-87
Crawford, Bobby	Col., Det.	2	16	1	3	4	6	....	....	....	....	....	....	1980-81	1982-83
• Crawford, Jack	Bos.	13	548	38	140	178	202	66	3	13	16	36	2	1937-38	1949-50
Crawford, Lou	Bos.	2	26	2	1	3	29	1	0	0	0	0	....	1989-90	1991-92
Crawford, Marc	Van.	6	176	19	31	50	229	20	1	2	3	44	....	1981-82	1986-87
• Crawford, Rusty	Ott., Tor.	2	38	10	8	18	117	2	2	1	3	9	1	1917-18	1918-19
Creighton, Adam	Buf., Chi., NYI, T.B., St.L.	14	708	187	216	403	1077	61	11	14	25	137	....	1983-84	1996-97
Creighton, Dave	Bos., Tor., Chi., NYR	12	616	140	174	314	223	51	11	13	24	20	....	1948-49	1959-60
• Creighton, Jimmy	Det.	1	11	1	0	1	2	....	....	....	....	....	....	1930-31	1930-31
Cressman, Dave	Min.	2	85	6	8	14	37	....	....	....	....	....	....	1974-75	1975-76
Cressman, Glen	Mtl.	1	4	0	0	0	2	....	....	....	....	....	....	1956-57	1956-57
Crisp, Terry	Bos., St.L., NYI, Phi.	11	536	67	134	201	135	110	15	28	43	40	2	1965-66	1976-77
Cristofoli, Ed	Mtl.	1	9	0	1	1	4	....	....	....	....	....	....	1989-90	1989-90
• Croghan, Maurice	Mtl.M.	1	16	0	0	0	4	....	....	....	....	....	....	1937-38	1937-38
Crombeen, Mike	Cle., St.L., Hfd.	8	475	55	68	123	218	27	6	2	8	32	....	1977-78	1984-85
Cronin, Shawn	Wsh., Wpg., Phi., S.J.	7	292	3	18	21	877	32	1	0	1	38	....	1988-89	1994-95
• Crossett, Stan	Phi.	1	21	0	0	0	10	....	....	....	....	....	....	1930-31	1930-31
Crossman, Doug	Chi., Phi., L.A., NYI, Hfd., Det., T.B., St.L.	14	914	105	359	464	534	97	12	39	51	105	....	1980-81	1993-94
Croteau, Gary	L.A., Det., Cal., K.C., Col.	12	684	144	175	319	143	11	3	2	5	8	....	1968-69	1979-80
Crowder, Bruce	Bos., Pit.	4	243	47	51	98	156	31	8	4	12	41	....	1981-82	1984-85
Crowder, Keith	Bos., L.A.	10	662	223	271	494	1354	85	14	22	36	218	....	1980-81	1989-90
Crowder, Troy	N.J., Det., L.A., Van.	7	150	9	7	16	433	4	0	0	0	22	....	1987-88	1996-97
‡ Crowe, Phil	L.A., Phi., Ott., Nsh.	6	94	4	5	9	173	3	0	0	0	16	....	1993-94	1999-00
‡ Crowley, Ted	Hfd., Col., NYI	2	34	2	4	6	12	....	....	....	....	....	....	1993-94	1998-99
Crozier, Joe	Tor.	1	5	0	3	3	2	....	....	....	....	....	....	1959-60	1959-60
• Crutchfield, Nels	Mtl.	1	41	5	5	10	20	2	0	1	1	22	....	1934-35	1934-35
Culhane, Jim	Hfd.	1	6	0	1	1	4	....	....	....	....	....	....	1989-90	1989-90
Cullen, Barry	Tor., Det.	5	219	32	52	84	111	6	0	0	0	2	....	1955-56	1959-60
Cullen, Brian	Tor., NYR	7	326	56	100	156	92	19	3	0	3	2	....	1954-55	1960-61
Cullen, John	Pit., Hfd., Tor., T.B.	11	621	187	363	550	898	53	12	22	34	58	....	1988-89	1998-99
Cullen, Ray	NYR, Det., Min., Van.	6	313	92	123	215	120	20	3	10	13	2	....	1965-66	1970-71
Cummins, Barry	Cal.	1	36	1	2	3	39	....	....	....	....	....	....	1973-74	1973-74
Cunneyworth, Randy	Buf., Pit., Wpg., Hfd., Chi., Ott.	16	866	189	225	414	1280	45	7	7	14	61	....	1980-81	1998-99
Cunningham, Bob	NYR	2	4	0	1	1	0	....	....	....	....	....	....	1960-61	1961-62
Cunningham, Jim	Phi.	1	1	0	0	0	4	....	....	....	....	....	....	1977-78	1977-78
• Cunningham, Les	NYA, Chi.	2	60	7	19	26	21	1	0	0	0	0	....	1936-37	1939-40
Cupolo, Bill	Bos.	1	47	11	13	24	10	7	1	2	3	0	....	1944-45	1944-45
Curran, Brian	Bos., NYI, Tor., Buf., Wsh.	10	381	7	33	40	1461	24	0	1	1	122	....	1983-84	1993-94
‡ Currie, Dan	Edm., L.A.	4	22	2	1	3	4	....	....	....	....	....	....	1990-91	1993-94
Currie, Glen	Wsh., L.A.	8	326	39	79	118	100	12	1	3	4	4	....	1979-80	1987-88
Currie, Hugh	Mtl.	1	1	0	0	0	0	....	....	....	....	....	....	1950-51	1950-51
Currie, Tony	St.L., Van., Hfd.	8	290	92	119	211	83	16	4	12	16	14	....	1977-78	1984-85
Curry, Floyd	Mtl.	11	601	105	99	204	147	91	23	17	40	38	4	1947-48	1957-58
Curtale, Tony	Cgy.	1	2	0	0	0	0	....	....	....	....	....	....	1980-81	1980-81
Curtis, Paul	Mtl., L.A., St.L.	4	185	3	34	37	161	5	0	0	0	2	....	1969-70	1972-73
Cushenan, Ian	Chi., Mtl., NYR, Det.	5	129	3	11	14	134	....	....	....	....	....	1	1956-57	1963-64
Cusson, Jean	Oak.	1	2	0	0	0	0	....	....	....	....	....	....	1967-68	1967-68
Cyr, Denis	Cgy., Chi., St.L.	6	193	41	43	84	36	4	0	0	0	0	....	1980-81	1985-86
Cyr, Paul	Buf., NYR, Hfd.	9	470	101	140	241	623	24	4	6	10	31	....	1982-83	1991-92
D															
Dahlin, Kjell	Mtl.	3	166	57	59	116	10	35	6	11	17	6	1	1985-86	1987-88
Dahlquist, Chris	Pit., Min., Cgy., Ott.	11	532	19	71	90	488	39	4	7	11	30	....	1985-86	1995-96
• Dahlstrom, Cully	Chi.	8	342	88	118	206	58	29	6	8	14	4	1	1937-38	1944-45
Daigle, Alain	Chi.	6	389	56	50	106	122	17	0	1	1	0	....	1974-75	1979-80
‡ Daigneault, J.J.	Van., Phi., Mtl., St.L., Pit., Ana., NYI, Nsh., Phx., Min.	16	899	53	197	250	687	99	5	26	31	100	1	1984-85	2000-01
Dailey, Bob	Van., Phi.	9	561	94	231	325	814	63	12	34	46	105	....	1973-74	1981-82
• Daley, Frank	Det.	1	5	0	0	0	0	2	0	0	0	0	....	1928-29	1928-29
Daley, Pat	Wpg.	2	12	1	0	1	13	....	....	....	....	....	....	1979-80	1980-81
Dalgarno, Brad	NYI	10	321	49	71	120	332	27	2	4	6	37	....	1985-86	1995-96
Dallman, Marty	Tor.	2	6	0	1	1	0	....	....	....	....	....	....	1987-88	1988-89
Dallman, Rod	NYI, Phi.	4	6	1	0	1	26	1	0	1	1	0	....	1987-88	1991-92
Dame, Bunny	Mtl.	1	34	2	5	7	4	....	....	....	....	....	....	1941-42	1941-42
Damore, Hank	NYR	1	4	1	0	1	2	....	....	....	....	....	....	1943-44	1943-44
‡ Daniels, Kimbi	Phi.	2	27	1	2	3	4	....	....	....	....	....	....	1990-91	1991-92
Daniels, Scott	Hfd., Phi., N.J.	6	149	8	12	20	667	1	0	0	0	0	....	1992-93	1998-99
Daoust, Dan	Mtl., Tor.	8	522	87	167	254	544	32	7	5	12	83	....	1982-83	1989-90
Dark, Michael	St.L.	2	43	5	6	11	14	....	....	....	....	....	....	1986-87	1987-88
• Darragh, Harold	Pit., Phi., Bos., Tor.	8	308	68	49	117	50	16	1	3	4	4	1	1925-26	1932-33
• Darragh, Jack	Ott.	6	121	66	46	112	113	11	3	0	3	9	3	1917-18	1923-24
David, Richard	Que.	3	31	4	4	8	10	1	0	0	0	0	....	1979-80	1982-83
• Davidson, Bob	Tor.	12	491	94	160	254	398	79	5	17	22	76	2	1934-35	1945-46
• Davidson, Gord	NYR	2	51	3	6	9	8	....	....	....	....	....	....	1942-43	1943-44
• Davie, Bob	Bos.	3	41	0	1	1	25	....	....	....	....	....	....	1933-34	1935-36
Davies, Buck	NYR	1	....	....	....	....	....	1	0	0	0	0	....	1947-48	1947-48
• Davis, Bob	Det.	1	3	0	0	0	0	....	....	....	....	....	....	1932-33	1932-33
Davis, Kim	Pit., Tor.	4	36	5	7	12	51	4	0	0	0	0	....	1977-78	1980-81
Davis, Lorne	Mtl., Chi., Det., Bos.	6	95	8	12	20	20	18	3	1	4	10	1	1951-52	1959-60
Davis, Mal	Det., Buf.	6	100	31	22	53	34	7	1	0	1	0	....	1978-79	1985-86
• Davison, Murray	Bos.	1	1	0	0	0	0	....	....	....	....	....	....	1965-66	1965-66
‡ Davydov, Evgeny	Wpg., Fla., Ott.	4	155	40	39	79	120	11	2	2	4	2	....	1991-92	1994-95
Dawes, Bob	Tor., Mtl.	4	32	2	7	9	6	10	0	0	0	2	1	1946-47	1950-51
• Day, Hap	Tor., NYA	14	581	86	116	202	601	53	4	7	11	56	1	1924-25	1937-38
‡ Day, Joe	Hfd., NYI	3	72	1	10	11	87	....	....	....	....	....	....	1991-92	1993-94
Dea, Billy	NYR, Det., Chi., Pit.	8	397	67	54	121	44	11	2	1	3	6	....	1953-54	1970-71
• Deacon, Don	Det.	3	30	6	4	10	6	2	2	1	3	0	....	1936-37	1939-40
Deadmarsh, Butch	Buf., Atl., K.C.	5	137	12	5	17	155	4	0	0	0	17	....	1970-71	1974-75
Dean, Barry	Col., Phi.	3	165	25	56	81	146	....	....	....	....	....	....	1976-77	1978-79
Debenedet, Nelson	Det., Pit.	2	46	10	4	14	13	....	....	....	....	....	....	1973-74	1974-75
DeBlois, Lucien	NYR, Col., Wpg., Mtl., Que., Tor.	15	993	249	276	525	814	52	7	6	13	38	1	1977-78	1991-92
Debol, Dave	Hfd.	2	92	26	26	52	4	3	0	0	0	0	....	1979-80	1980-81
Defazio, Dean	Pit.	1	22	0	2	2	28	....	....	....	....	....	....	1983-84	1983-84
DeGray, Dale	Cgy., Tor., L.A., Buf.	5	153	18	47	65	195	13	1	3	4	28	....	1985-86	1989-90
• Delmonte, Armand	Bos.	1	1	0	0	0	0	....	....	....	....	....	....	1945-46	1945-46
Delorme, Gilbert	Mtl., St.L., Que., Det., Pit.	9	541	31	92	123	520	56	1	9	10	56	....	1981-82	1989-90
Delorme, Ron	Col., Van.	9	524	83	83	166	667	25	1	2	3	59	....	1976-77	1984-85
Delory, Val	NYR	1	1	0	0	0	0	....	....	....	....	....	....	1948-49	1948-49
Delparte, Guy	Col.	1	48	1	8	9	18	....	....	....	....	....	....	1976-77	1976-77
Delvecchio, Alex	Det.	24	1549	456	825	1281	383	121	35	69	104	29	3	1950-51	1973-74
• DeMarco, Ab	Chi., Tor., Bos., NYR	7	209	72	93	165	53	11	3	0	3	2	....	1938-39	1946-47
DeMarco, Ab Jr.	NYR, St.L., Pit., Van., L.A., Bos.	9	344	44	80	124	75	25	1	2	3	17	....	1969-70	1978-79
• Demers, Tony	Mtl., NYR	6	83	20	22	42	23	2	0	0	0	0	....	1937-38	1943-44
Denis, Jean-Paul	NYR	2	10	0	2	2	2	....	....	....	....	....	....	1946-47	1949-50
Denis, Lulu	Mtl.	2	3	0	1	1	0	....	....	....	....	....	....	1949-50	1950-51
• Denneny, Corb	Tor., Ham., Chi.	9	176	103	42	145	148	6	1	0	1	4	2	1917-18	1927-28
• Denneny, Cy	Ott., Bos.	12	328	248	85	333	301	25	16	2	18	17	5	1917-18	1928-29
Dennis, Norm	St.L.	4	12	3	0	3	11	5	0	0	0	2	....	1968-69	1971-72
• Denoird, Gerry	Tor.	1	17	0	1	1	0	....	....	....	....	....	....	1922-23	1922-23
DePalma, Larry	Min., S.J., Pit.	7	148	21	20	41	408	3	0	0	0	6	....	1985-86	1993-94
Derlago, Bill	Van., Tor., Bos., Wpg., Que.	9	555	189	227	416	247	13	5	0	5	8	....	1978-79	1986-87
• Desaulniers, Gerard	Mtl.	3	8	0	2	2	4	....	....	....	....	....	....	1950-51	1953-54

Name	NHL Teams	NHL Seasons	Regular Schedule GP	G	A	TP	PIM	Playoffs GP	G	A	TP	PIM	NHL Cup Wins	First NHL Season	Last NHL Season
• Desilets, Joffre	Mtl., Chi.	5	192	37	45	82	57	7	1	0	1	7	....	1935-36	1939-40
Desjardins, Martin	Mtl.	1	8	0	2	2	2	....	....	....	....	....	....	1989-90	1989-90
• Desjardins, Vic	Chi., NYR	2	87	6	15	21	27	16	0	0	0	0	....	1930-31	1931-32
Deslauriers, Jacques	Mtl.	1	2	0	0	0	0	....	....	....	....	....	....	1955-56	1955-56
Devine, Kevin	NYI	1	2	0	1	1	8	....	....	....	....	....	....	1982-83	1982-83
• Dewar, Tom	NYR	1	9	0	2	2	4	....	....	....	....	....	....	1943-44	1943-44
Dewsbury, Al	Det., Chi.	9	347	30	78	108	365	14	1	5	6	16	1	1946-47	1955-56
Deziel, Michel	Buf.	1	....	....	....	....	....	1	0	0	0	0	....	1974-75	1974-75
Dheere, Marcel	Mtl.	1	11	1	2	3	2	5	0	0	0	6	....	1942-43	1942-43
‡ Di Pietro, Paul	Mtl., Tor., L.A.	6	192	31	49	80	96	31	11	10	21	10	1	1991-92	1996-97
Diachuk, Edward	Det.	1	12	0	0	0	19	....	....	....	....	....	....	1960-61	1960-61
Dick, Harry	Chi.	1	12	0	1	1	12	....	....	....	....	....	....	1946-47	1946-47
Dickens, Ernie	Tor., Chi.	6	278	12	44	56	98	13	0	0	0	4	1	1941-42	1950-51
Dickenson, Herb	NYR	2	48	18	17	35	10	....	....	....	....	....	....	1951-52	1952-53
Diduck, Gerald	NYI, Mtl., Van., Chi., Hfd., Phx., Tor., Dal.	17	932	56	156	212	1612	114	8	16	24	212	....	1984-85	2000-01
Dietrich, Don	Chi., N.J.	2	28	0	7	7	10	....	....	....	....	....	....	1983-84	1985-86
• Dill, Bob	NYR	2	76	15	15	30	135	....	....	....	....	....	....	1943-44	1944-45
• Dillabough, Bob	Det., Bos., Pit., Oak.	9	283	32	54	86	76	17	3	0	3	0	....	1961-62	1969-70
• Dillon, Cecil	NYR, Det.	10	453	167	131	298	105	43	14	9	23	14	1	1930-31	1939-40
Dillon, Gary	Col.	1	13	1	1	2	29	....	....	....	....	....	....	1980-81	1980-81
Dillon, Wayne	NYR, Wpg.	4	229	43	66	109	60	3	0	1	1	0	....	1975-76	1979-80
Dineen, Bill	Det., Chi.	5	323	51	44	95	122	37	1	1	2	18	2	1953-54	1957-58
Dineen, Gary	Min.	1	4	0	1	1	0	....	....	....	....	....	....	1968-69	1968-69
Dineen, Gord	NYI, Min., Pit., Ott.	13	528	16	90	106	695	40	1	7	8	68	....	1982-83	1994-95
Dineen, Peter	L.A., Det.	2	13	0	2	2	13	....	....	....	....	....	....	1986-87	1989-90
• Dinsmore, Chuck	Mtl.M.	4	100	6	2	8	50	8	1	0	1	2	1	1924-25	1929-30
‡ Dionne, Gilbert	Mtl., Phi., Fla.	6	223	61	79	140	108	39	10	12	22	34	1	1990-91	1995-96
Dionne, Marcel	Det., L.A., NYR	18	1348	731	1040	1771	600	49	21	24	45	17	....	1971-72	1988-89
Dirk, Robert	St.L., Van., Chi., Ana., Mtl.	9	402	13	29	42	786	39	0	1	1	56	....	1987-88	1995-96
‡ Djoos, Per	Det., NYR	3	82	2	31	33	58	....	....	....	....	....	....	1990-91	1992-93
Doak, Gary	Det., Bos., Van., NYR	16	789	23	107	130	908	78	2	4	6	121	1	1965-66	1980-81
Dobbin, Brian	Phi., Bos.	5	63	7	8	15	61	2	0	0	0	17	....	1986-87	1991-92
Dobson, Jim	Min., Col., Que.	4	12	0	0	0	6	....	....	....	....	....	....	1979-80	1983-84
• Doherty, Fred	Mtl.	1	1	0	0	0	0	....	....	....	....	....	....	1918-19	1918-19
‡ Dollas, Bobby	Wpg., Que., Det., Ana., Edm., Pit., Ott., Cgy., S.J.	16	646	42	96	138	467	47	2	1	3	41	....	1983-84	2000-01
Donaldson, Gary	Chi.	1	1	0	0	0	0	....	....	....	....	....	....	1973-74	1973-74
Donatelli, Clark	Min., Bos.	2	35	3	4	7	39	2	0	0	0	0	....	1989-90	1991-92
• Donnelly, Babe	Mtl.M.	1	34	0	1	1	14	2	0	0	0	0	....	1926-27	1926-27
Donnelly, Dave	Bos., Chi., Edm.	5	137	15	24	39	150	5	0	0	0	0	....	1983-84	1987-88
Donnelly, Gord	Que., Wpg., Buf., Dal.	12	554	28	41	69	2069	26	0	2	2	61	....	1983-84	1994-95
Donnelly, Mike	NYR, Buf., L.A., Dal., NYI	11	465	114	121	235	255	47	12	12	24	30	....	1986-87	1996-97
• Doran, John	NYA, Det., Mtl.	5	98	5	10	15	110	3	0	0	0	0	....	1933-34	1939-40
Doran, Lloyd	Det.	1	24	3	2	5	10	....	....	....	....	....	....	1946-47	1946-47
• Doraty, Ken	Chi., Tor., Det.	5	103	15	26	41	24	15	7	2	9	2	....	1926-27	1937-38
Dore, Andre	NYR, St.L., Que.	7	257	14	81	95	261	23	1	2	3	32	....	1978-79	1984-85
Dore, Daniel	Que.	2	17	2	3	5	59	....	....	....	....	....	....	1989-90	1990-91
Dorey, Jim	Tor., NYR	4	232	25	74	99	553	11	0	2	2	40	....	1968-69	1971-72
Dorion, Dan	N.J.	2	4	1	1	2	2	....	....	....	....	....	....	1985-86	1987-88
Dornhoefer, Gary	Bos., Phi.	14	787	214	328	542	1291	80	17	19	36	203	2	1963-64	1977-78
Dorohoy, Eddie	Mtl.	1	16	0	0	0	6	....	....	....	....	....	....	1948-49	1948-49
Douglas, Jordy	Hfd., Min., Wpg.	6	268	76	62	138	160	6	0	0	0	4	....	1979-80	1984-85
Douglas, Kent	Tor., Oak., Det.	7	428	33	115	148	631	19	1	3	4	33	3	1962-63	1968-69
Douglas, Les	Det.	4	52	6	12	18	8	10	3	2	5	2	1	1940-41	1946-47
‡ Douris, Peter	Wpg., Bos., Ana., Dal.	11	321	54	67	121	80	27	3	5	8	14	....	1985-86	1997-98
• Downie, Dave	Tor.	1	11	0	1	1	2	....	....	....	....	....	....	1932-33	1932-33
‡ Doyon, Mario	Chi., Que.	3	28	3	4	7	16	....	....	....	....	....	....	1988-89	1990-91
• Draper, Bruce	Tor.	1	1	0	0	0	0	....	....	....	....	....	....	1962-63	1962-63
• Drillon, Gordie	Tor., Mtl.	7	311	155	139	294	56	50	26	15	41	10	1	1936-37	1942-43
Driscoll, Peter	Edm.	2	60	3	8	11	97	3	0	0	0	0	....	1979-80	1980-81
Driver, Bruce	N.J., NYR	15	922	96	390	486	670	108	10	40	50	64	1	1983-84	1997-98
Drolet, Rene	Phi., Det.	2	2	0	0	0	0	....	....	....	....	....	....	1971-72	1974-75
‡ Droppa, Ivan	Chi.	2	19	0	1	1	14	....	....	....	....	....	....	1993-94	1995-96
• Drouillard, Clarence	Det.	1	10	0	1	1	0	....	....	....	....	....	....	1937-38	1937-38
Drouin, Jude	Mtl., Min., NYI, Wpg.	12	666	151	305	456	346	72	27	41	68	33	....	1968-69	1980-81
‡ Drouin, P.C.	Bos.	1	3	0	0	0	0	....	....	....	....	....	....	1996-97	1996-97
• Drouin, Polly	Mtl.	7	160	23	50	73	80	5	0	1	1	5	....	1934-35	1940-41
Druce, John	Wsh., Wpg., L.A., Phi.	10	531	113	126	239	347	53	17	6	23	38	....	1988-89	1997-98
Drulia, Stan	T.B.	3	126	15	27	42	52	....	....	....	....	....	....	1992-93	2000-01
• Drummond, Jim	NYR	1	2	0	0	0	0	....	....	....	....	....	....	1944-45	1944-45
• Drury, Herb	Pit., Phi.	6	213	24	13	37	203	4	1	1	2	0	....	1925-26	1930-31
‡ Dube, Christian	NYR	2	33	1	1	2	4	3	0	0	0	0	....	1996-97	1998-99
Dube, Gilles	Mtl., Det.	2	12	1	2	3	2	2	0	0	0	0	1	1949-50	1953-54
Dube, Norm	K.C.	2	57	8	10	18	54	....	....	....	....	....	....	1974-75	1975-76
Duberman, Justin	Pit.	1	4	0	0	0	0	....	....	....	....	....	....	1993-94	1993-94
Duchesne, Gaetan	Wsh., Que., Min., S.J., Fla.	14	1028	179	254	433	617	84	14	13	27	97	....	1981-82	1994-95
Dudley, Rick	Buf., Wpg.	6	309	75	99	174	292	25	7	2	9	69	....	1972-73	1980-81
‡ Duerden, Dave	Fla.	1	2	0	0	0	0	....	....	....	....	....	....	1999-00	1999-00
Duff, Dick	Tor., NYR, Mtl., L.A., Buf.	18	1030	283	289	572	743	114	30	49	79	78	6	1954-55	1971-72
Dufour, Luc	Bos., Que., St.L.	3	167	23	21	44	199	18	1	0	1	32	....	1982-83	1984-85
Dufour, Marc	NYR, L.A.	3	14	1	0	1	2	....	....	....	....	....	....	1963-64	1968-69
Dufresne, Donald	Mtl., T.B., L.A., St.L., Edm.	9	268	6	36	42	258	34	1	3	4	47	1	1988-89	1996-97
• Duggan, John	Ott.	1	27	0	0	0	0	2	0	0	0	0	....	1925-26	1925-26
Duggan, Ken	Min.	1	1	0	0	0	0	....	....	....	....	....	....	1987-88	1987-88
Duguay, Ron	NYR, Det., Pit., L.A.	12	864	274	346	620	582	89	31	22	53	118	....	1977-78	1988-89
• Duguid, Lorne	Mtl.M., Det., Bos.	6	135	9	15	24	57	4	1	0	1	6	....	1931-32	1936-37
• Dukowski, Duke	Chi., NYA, NYR	5	200	16	30	46	172	6	0	0	0	6	....	1926-27	1933-34
• Dumart, Woody	Bos.	16	772	211	218	429	99	88	12	15	27	23	2	1935-36	1953-54
Dunbar, Dale	Van., Bos.	2	2	0	0	0	2	....	....	....	....	....	....	1985-86	1988-89
• Duncan, Art	Det., Tor.	5	156	18	16	34	225	5	0	0	0	4	....	1926-27	1930-31
Duncan, Iain	Wpg.	4	127	34	55	89	149	11	0	3	3	6	....	1986-87	1990-91
Duncanson, Craig	L.A., Wpg., NYR	7	38	5	4	9	61	....	....	....	....	....	....	1985-86	1992-93
Dundas, Rocky	Tor.	1	5	0	0	0	14	....	....	....	....	....	....	1989-90	1989-90
• Dunlap, Frank	Tor.	1	15	0	1	1	2	....	....	....	....	....	....	1943-44	1943-44
Dunlop, Blake	Min., Phi., St.L., Det.	11	550	130	274	404	172	40	4	10	14	18	....	1973-74	1983-84
Dunn, Dave	Van., Tor.	3	184	14	41	55	313	10	1	1	2	41	....	1973-74	1975-76
Dunn, Richie	Buf., Cgy., Hfd.	12	483	36	140	176	314	36	3	15	18	24	....	1977-78	1988-89
Dupere, Denis	Tor., Wsh., St.L., K.C., Col.	8	421	80	99	179	66	16	1	0	1	0	....	1970-71	1977-78
Dupont, Andre	NYR, St.L., Phi., Que.	13	800	59	185	244	1986	140	14	18	32	352	2	1970-71	1982-83
Dupont, Jerome	Chi., Tor.	6	214	7	29	36	468	20	0	2	2	56	....	1981-82	1986-87
Dupont, Norm	Mtl., Wpg., Hfd.	5	256	55	85	140	52	13	4	2	6	0	....	1979-80	1983-84
• Dupre, Yanick	Phi.	3	35	2	0	2	16	....	....	....	....	....	....	1991-92	1995-96
Durbano, Steve	St.L., Pit., K.C., Col.	6	220	13	60	73	1127	5	0	2	2	8	....	1972-73	1978-79
Duris, Vitezslav	Tor.	2	89	3	20	23	62	3	0	1	1	2	....	1980-81	1982-83
Dussault, Norm	Mtl.	4	206	31	62	93	47	7	3	1	4	0	....	1947-48	1950-51
• Dutton, Red	Mtl.M., NYA	10	449	29	67	96	871	18	1	0	1	33	....	1926-27	1935-36
Dvorak, Miroslav	Phi.	3	193	11	74	85	51	18	0	2	2	6	....	1982-83	1984-85
Dwyer, Mike	Col., Cgy.	4	31	2	6	8	25	1	1	0	1	0	....	1978-79	1981-82
• Dyck, Henry	NYR	1	1	0	0	0	0	....	....	....	....	....	....	1943-44	1943-44
• Dye, Babe	Tor., Ham., Chi., NYA	11	271	201	47	248	221	10	2	0	2	11	1	1919-20	1930-31
Dykstra, Steve	Buf., Edm., Pit., Hfd.	5	217	8	32	40	545	1	0	0	0	2	....	1985-86	1989-90
Dyte, Jack	Chi.	1	27	1	0	1	31	....	....	....	....	....	....	1943-44	1943-44
Dziedzic, Joe	Pit., Phx.	3	130	14	14	28	131	21	1	3	4	23	....	1995-96	1998-99

E

Name	NHL Teams	NHL Seasons	Regular Schedule GP	G	A	TP	PIM	Playoffs GP	G	A	TP	PIM	NHL Cup Wins	First NHL Season	Last NHL Season
Eagles, Mike	Que., Chi., Wpg., Wsh.	16	853	74	122	196	928	44	2	6	8	34	....	1982-83	1999-00
Eakin, Bruce	Cgy., Det.	4	13	2	2	4	4	....	....	....	....	....	....	1981-82	1985-86
Eatough, Jeff	Buf.	1	1	0	0	0	0	....	....	....	....	....	....	1981-82	1981-82
Eaves, Mike	Min., Cgy.	8	324	83	143	226	80	43	7	10	17	14	....	1978-79	1985-86
Eaves, Murray	Wpg., Det.	8	57	4	13	17	9	4	0	1	1	2	....	1980-81	1989-90
Ecclestone, Tim	St.L., Det., Tor., Atl.	11	692	126	233	359	344	48	6	11	17	76	....	1967-68	1977-78
Edberg, Rolf	Wsh.	3	184	45	58	103	24	....	....	....	....	....	....	1978-79	1980-81
• Eddolls, Frank	Mtl., NYR, Chi.	9	317	23	43	66	114	31	0	2	2	10	1	1944-45	1954-55
Edestrand, Darryl	St.L., Phi., Pit., Bos., L.A.	10	455	34	90	124	404	42	3	9	12	57	....	1967-68	1978-79

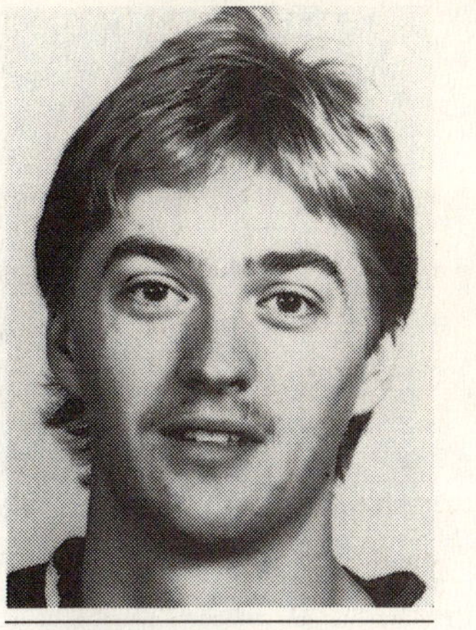
Kjell Dahlin

Kimbi Daniels

Gary Dornhoefer

Gord Dineen

Ivan Droppa

Richie Dunn

Jack Egers

Paul Evans

			Regular Schedule					Playoffs							
Name	NHL Teams	NHL Seasons	GP	G	A	TP	PIM	GP	G	A	TP	PIM	NHL Cup Wins	First NHL Season	Last NHL Season
Edmundson, Garry	Mtl., Tor.	3	43	4	6	10	49	11	0	1	1	8	....	1951-52	1960-61
Edur, Tom	Col., Pit.	2	158	17	70	87	67	....	....	....	....	....	....	1976-77	1977-78
Egan, Pat	NYA, Bro., Det., Bos., NYR	11	554	77	153	230	776	46	9	4	13	48	....	1939-40	1950-51
‡ Egeland, Allan	T.B.	3	17	0	0	0	16	....	....	....	....	....	....	1995-96	1997-98
Egers, Jack	NYR, St.L., Wsh.	7	284	64	69	133	154	32	5	6	11	32	....	1969-70	1975-76
Ehman, Gerry	Bos., Det., Tor., Oak., Cal.	9	429	96	118	214	100	41	10	10	20	12	1	1957-58	1970-71
‡ Eisenhut, Neil	Van., Cgy.	2	16	1	3	4	21	....	....	....	....	....	....	1993-94	1994-95
Eklund, Pelle	Phi., Dal.	9	594	120	335	455	109	66	10	36	46	8	....	1985-86	1993-94
Eldebrink, Anders	Van., Que.	2	55	3	11	14	29	14	0	0	0	10	....	1981-82	1982-83
Elik, Bo	Det.	1	3	0	0	0	0	....	....	....	....	....	....	1962-63	1962-63
‡ Elik, Todd	L.A., Min., Edm., S.J., St.L., Bos.	8	448	110	219	329	453	52	15	27	42	48	....	1989-90	1996-97
Ellett, Dave	Wpg., Tor., N.J., Bos., St.L.	16	1129	153	415	568	985	116	11	46	57	87	....	1984-85	1999-00
● Elliott, Fred	Ott.	1	43	2	0	2	6	....	....	....	....	....	....	1928-29	1928-29
Ellis, Ron	Tor.	16	1034	332	308	640	207	70	18	8	26	20	1	1963-64	1980-81
Eloranta, Kari	Cgy., St.L.	5	267	13	103	116	155	26	1	7	8	19	....	1981-82	1986-87
Elynuik, Pat	Wpg., Wsh., T.B., Ott.	9	506	154	188	342	459	20	6	9	15	25	....	1987-88	1995-96
Emberg, Eddie	Mtl.	1	....	....	....	....	....	2	1	0	1	0	....	1944-45	1944-45
Emmons, Gary	S.J.	1	3	1	0	1	0	....	....	....	....	....	....	1993-94	1993-94
● Emms, Hap	Mtl.M., NYA, Det., Bos.	10	320	36	53	89	311	14	0	0	0	12	....	1926-27	1937-38
Endean, Craig	Wpg.	1	2	0	1	1	0	....	....	....	....	....	....	1986-87	1986-87
Engblom, Brian	Mtl., Wsh., L.A., Buf., Cgy.	11	659	29	177	206	599	48	3	9	12	43	3	1976-77	1986-87
Engele, Jerry	Min.	3	100	2	13	15	162	2	0	1	1	0	....	1975-76	1977-78
English, John	L.A.	1	3	1	3	4	4	1	0	0	0	0	....	1987-88	1987-88
Ennis, Jim	Edm.	1	5	1	0	1	10	....	....	....	....	....	....	1987-88	1987-88
Erickson, Aut	Bos., Chi., Tor., Oak.	7	226	7	24	31	182	7	0	0	0	2	1	1959-60	1969-70
Erickson, Bryan	Wsh., L.A., Pit., Wpg.	9	351	80	125	205	141	14	3	4	7	7	....	1983-84	1993-94
Erickson, Grant	Bos., Min.	2	6	1	0	1	0	....	....	....	....	....	....	1968-69	1969-70
Eriksson, Peter	Edm.	1	20	3	3	6	24	....	....	....	....	....	....	1989-90	1989-90
Eriksson, Roland	Min., Van.	3	193	48	95	143	26	2	1	0	1	0	....	1976-77	1978-79
Eriksson, Thomas	Phi.	5	208	22	76	98	107	19	0	3	3	12	....	1980-81	1985-86
Erixon, Jan	NYR	10	556	57	159	216	167	58	7	7	14	16	....	1983-84	1992-93
Errey, Bob	Pit., Buf., S.J., Det., Dal., NYR	15	895	170	212	382	1005	99	13	16	29	109	2	1983-84	1997-98
‡ Esau, Len	Tor., Que., Cgy., Edm.	4	27	0	10	10	24	....	....	....	....	....	....	1991-92	1994-95
Esposito, Phil	Chi., Bos., NYR	18	1282	717	873	1590	910	130	61	76	137	138	2	1963-64	1980-81
● Evans, Chris	Tor., Buf., St.L., Det., K.C.	5	241	19	42	61	143	12	1	1	2	8	....	1969-70	1974-75
Evans, Daryl	L.A., Wsh., Tor.	6	113	22	30	52	25	11	5	8	13	12	....	1981-82	1986-87
Evans, Doug	St.L., Wpg., Phi.	8	355	48	87	135	502	22	3	4	7	38	....	1985-86	1992-93
● Evans, Jack	NYR, Chi.	14	752	19	80	99	989	56	2	2	4	97	1	1948-49	1962-63
Evans, John Paul	Phi.	3	103	14	25	39	34	1	0	0	0	0	....	1978-79	1982-83
Evans, Kevin	Min., S.J.	2	9	0	1	1	44	....	....	....	....	....	....	1990-91	1991-92
Evans, Paul	Tor.	2	11	1	1	2	21	2	0	0	0	0	....	1976-77	1977-78
Evans, Shawn	St.L., NYI	2	9	1	0	1	2	....	....	....	....	....	....	1985-86	1989-90
● Evans, Stewart	Det., Mtl.M., Mtl.	8	367	28	49	77	425	26	0	0	0	20	1	1930-31	1938-39
Evason, Dean	Wsh., Hfd., S.J., Dal., Cgy.	13	803	139	233	372	1002	55	9	20	29	132	....	1983-84	1995-96
Ewen, Todd	St.L., Mtl., Ana., S.J.	11	518	36	40	76	1911	26	0	0	0	87	1	1986-87	1996-97
Ezinicki, Bill	Tor., Bos., NYR	9	368	79	105	184	713	40	5	8	13	87	3	1944-45	1954-55

F

Name	NHL Teams	NHL Seasons	GP	G	A	TP	PIM	GP	G	A	TP	PIM	NHL Cup Wins	First NHL Season	Last NHL Season
Fahey, Trevor	NYR	1	1	0	0	0	0	....	....	....	....	....	....	1964-65	1964-65
Fairbairn, Bill	NYR, Min., St.L.	11	658	162	261	423	173	54	13	22	35	42	....	1968-69	1978-79
Falkenberg, Bob	Det.	5	54	1	5	6	26	....	....	....	....	....	....	1966-67	1971-72
‡ Falloon, Pat	S.J., Phi., Ott., Edm., Pit.	9	575	143	179	322	141	66	11	7	18	16	....	1991-92	1999-00
Farrant, Walt	Chi.	1	1	0	0	0	0	....	....	....	....	....	....	1943-44	1943-44
Farrish, Dave	NYR, Que., Tor.	7	430	17	110	127	440	14	0	2	2	24	....	1976-77	1983-84
Fashoway, Gordie	Chi.	1	13	3	2	5	14	....	....	....	....	....	....	1950-51	1950-51
Faubert, Mario	Pit.	7	231	21	90	111	292	10	2	2	4	6	....	1974-75	1981-82
Faulkner, Alex	Tor., Det.	3	101	15	17	32	15	12	5	0	5	2	....	1961-62	1963-64
Fauss, Ted	Tor.	2	28	0	2	2	15	....	....	....	....	....	....	1986-87	1987-88
‡ Faust, Andre	Phi.	2	47	10	7	17	14	....	....	....	....	....	....	1992-93	1993-94
Feamster, Dave	Chi.	4	169	13	24	37	154	33	3	5	8	61	....	1981-82	1984-85
‡ Featherstone, Glen	St.L., Bos., NYR, Hfd., Cgy.	9	384	19	61	80	939	28	0	2	2	103	....	1988-89	1996-97
Featherstone, Tony	Oak., Cal., Min.	3	130	17	21	38	65	2	0	0	0	0	....	1969-70	1973-74
Federko, Bernie	St.L., Det.	14	1000	369	761	1130	487	91	35	66	101	83	....	1976-77	1989-90
‡ Fedotov, Anatoli	Wpg., Ana.	2	4	0	2	2	0	....	....	....	....	....	....	1992-93	1993-94
Fedyk, Brent	Det., Phi., Dal., NYR	10	470	97	112	209	308	16	3	2	5	12	....	1987-88	1998-99
Felix, Chris	Wsh.	4	35	1	12	13	10	2	0	1	1	0	....	1987-88	1990-91
‡ Felsner, Brian	Chi.	1	12	1	3	4	12	....	....	....	....	....	....	1997-98	1997-98
‡ Felsner, Denny	St.L.	4	18	1	4	5	6	10	2	3	5	2	....	1991-92	1994-95
Feltrin, Tony	Pit., NYR	4	48	3	3	6	65	....	....	....	....	....	....	1980-81	1985-86
Fenton, Paul	Hfd., NYR, L.A., Wpg., Tor., Cgy., S.J.	8	411	100	83	183	198	17	4	1	5	27	....	1984-85	1991-92
Fenyves, David	Buf., Phi.	9	206	3	32	35	119	11	0	0	0	9	....	1982-83	1990-91
Fergus, Tom	Bos., Tor., Van.	12	726	235	346	581	499	65	21	17	38	48	....	1981-82	1992-93
‡ Ferguson, Craig	Mtl., Cgy., Fla.	5	27	1	1	2	6	....	....	....	....	....	....	1993-94	1999-00
Ferguson, George	Tor., Pit., Min.	12	797	160	238	398	431	86	14	23	37	44	....	1972-73	1983-84
Ferguson, John	Mtl.	8	500	145	158	303	1214	85	20	18	38	260	5	1963-64	1970-71
Ferguson, Lorne	Bos., Det., Chi.	8	422	82	80	162	193	31	6	3	9	24	....	1949-50	1958-59
Ferguson, Norm	Oak., Cal.	4	279	73	66	139	72	10	1	4	5	7	....	1968-69	1971-72
‡ Ferner, Mark	Buf., Wsh., Ana., Det.	6	91	3	10	13	51	....	....	....	....	....	....	1986-87	1994-95
Fetisov, Viacheslav	N.J., Det.	9	546	36	192	228	656	116	2	26	28	147	2	1989-90	1997-98
Fidler, Mike	Cle., Min., Hfd., Chi.	7	271	84	97	181	124	....	....	....	....	....	....	1976-77	1982-83
● Field, Wilf	NYA, Bro., Mtl., Chi.	6	219	17	25	42	151	2	0	0	0	2	....	1936-37	1944-45
Fielder, Guyle	Chi., Det., Bos.	4	9	0	0	0	2	6	0	0	0	2	....	1950-51	1957-58
‡ Filimonov, Dmitri	Ott.	1	30	1	4	5	18	....	....	....	....	....	....	1993-94	1993-94
Fillion, Bob	Mtl.	7	327	42	61	103	84	33	7	4	11	10	2	1943-44	1949-50
● Fillion, Marcel	Bos.	1	1	0	0	0	0	....	....	....	....	....	....	1944-45	1944-45
● Filmore, Tommy	Det., NYA, Bos.	4	117	15	12	27	33	....	....	....	....	....	....	1930-31	1933-34
● Finkbeiner, Lloyd	NYA	1	2	0	0	0	0	....	....	....	....	....	....	1940-41	1940-41
Finn, Steven	Que., T.B., L.A.	12	725	34	78	112	1724	23	0	4	4	39	....	1985-86	1996-97
Finney, Sid	Chi.	3	59	10	7	17	4	7	0	2	2	0	....	1951-52	1953-54
● Finnigan, Ed	St.L., Bos.	2	15	1	1	2	2	....	....	....	....	....	....	1934-35	1935-36
● Finnigan, Frank	Ott., Tor., St.L.	14	553	115	88	203	407	38	6	9	15	22	2	1923-24	1936-37
Fiorentino, Peter	NYR	1	1	0	0	0	0	....	....	....	....	....	....	1991-92	1991-92
Fischer, Ron	Buf.	2	18	0	7	7	6	....	....	....	....	....	....	1981-82	1982-83
● Fisher, Alvin	Tor.	1	9	1	0	1	4	....	....	....	....	....	....	1924-25	1924-25
Fisher, Craig	Phi., Wpg., Fla.	4	12	0	0	0	2	....	....	....	....	....	....	1989-90	1996-97
Fisher, Dunc	NYR, Bos., Det.	7	275	45	70	115	104	21	4	4	8	14	....	1947-48	1958-59
● Fisher, Joe	Det.	4	65	8	12	20	13	12	2	1	3	6	1	1939-40	1942-43
Fitchner, Bob	Que.	2	78	12	20	32	59	3	0	0	0	10	....	1979-80	1980-81
‡ Fitzgerald, Rusty	Pit.	2	25	2	2	4	12	5	0	0	0	4	....	1994-95	1995-96
Fitzpatrick, Ross	Phi.	4	20	5	2	7	0	....	....	....	....	....	....	1982-83	1985-86
Fitzpatrick, Sandy	NYR, Min.	2	22	3	6	9	8	12	0	0	0	0	....	1964-65	1967-68
Flaman, Fern	Bos., Tor.	17	910	34	174	208	1370	63	4	8	12	93	1	1944-45	1960-61
Flatley, Pat	NYI, NYR	14	780	170	340	510	686	70	18	15	33	75	....	1983-84	1996-97
Fleming, Gerry	Mtl.	2	11	0	0	0	42	....	....	....	....	....	....	1993-94	1994-95
Fleming, Reggie	Mtl., Chi., Bos., NYR, Phi., Buf.	12	749	108	132	240	1468	50	3	6	9	106	1	1959-60	1970-71
Flesch, John	Min., Pit., Col.	4	124	18	23	41	117	....	....	....	....	....	....	1974-75	1979-80
Fletcher, Steven	Mtl., Wpg.	2	3	0	0	0	5	1	0	0	0	5	....	1987-88	1988-89
● Flett, Bill	L.A., Phi., Tor., Atl., Edm.	11	689	202	215	417	501	52	7	16	23	42	1	1967-68	1979-80
Flichel, Todd	Wpg.	3	6	0	1	1	4	....	....	....	....	....	....	1987-88	1989-90
Flockhart, Rob	Van., Min.	5	55	2	5	7	14	1	1	0	1	2	....	1976-77	1980-81
Flockhart, Ron	Phi., Pit., Mtl., St.L., Bos.	9	453	145	183	328	208	19	4	6	10	14	....	1980-81	1988-89
Floyd, Larry	N.J.	2	12	2	3	5	9	....	....	....	....	....	....	1982-83	1983-84
● Fogarty, Bryan	Que., Pit., Mtl.	6	156	22	52	74	119	....	....	....	....	....	....	1989-90	1994-95
● Fogolin, Lee	Det., Chi.	9	427	10	48	58	575	28	0	2	2	30	1	1947-48	1955-56
Fogolin, Lee Jr.	Buf., Edm.	13	924	44	195	239	1318	108	5	19	24	173	2	1974-75	1986-87
Folco, Peter	Van.	1	2	0	0	0	0	....	....	....	....	....	....	1973-74	1973-74
Foley, Gerry	Tor., NYR, L.A.	4	142	9	14	23	99	9	0	1	1	2	....	1954-55	1968-69
Foley, Rick	Chi., Phi., Det.	3	67	11	26	37	180	4	0	1	1	4	....	1970-71	1973-74
Foligno, Mike	Det., Buf., Tor., Fla.	15	1018	355	372	727	2049	57	15	17	32	185	....	1979-80	1993-94
Folk, Bill	Det.	2	12	0	0	0	4	....	....	....	....	....	....	1951-52	1952-53
Fontaine, Len	Det.	2	46	8	11	19	10	....	....	....	....	....	....	1972-73	1973-74
Fontas, Jon	Min.	2	2	0	0	0	0	....	....	....	....	....	....	1979-80	1980-81
Fonteyne, Val	Det., NYR, Pit.	13	820	75	154	229	26	59	3	10	13	8	....	1959-60	1971-72
Fontinato, Lou	NYR, Mtl.	9	535	26	78	104	1247	21	0	2	2	42	....	1954-55	1962-63
Forbes, Dave	Bos., Wsh.	6	363	64	64	128	341	45	1	4	5	13	....	1973-74	1978-79

Name	NHL Teams	NHL Seasons	Regular Schedule					Playoffs					NHL Cup Wins	First NHL Season	Last NHL Season
			GP	G	A	TP	PIM	GP	G	A	TP	PIM			
Forbes, Mike	Bos., Edm.	3	50	1	11	12	41	....	....	....	....	....	....	1977-78	1981-82
Forey, Connie	St.L.	1	4	0	0	0	2	....	....	....	....	....	....	1973-74	1973-74
• Forsey, Jack	Tor.	1	19	7	9	16	10	3	0	1	1	0	....	1942-43	1942-43
• Forslund, Gus	Ott.	1	48	4	9	13	2	....	....	....	....	....	....	1932-33	1932-33
‡ Forslund, Tomas	Cgy.	2	44	5	11	16	12	....	....	....	....	....	....	1991-92	1992-93
Forsyth, Alex	Wsh.	1	1	0	0	0	0	....	....	....	....	....	....	1976-77	1976-77
Fortier, Dave	Tor., NYI, Van.	4	205	8	21	29	335	20	0	2	2	33	....	1972-73	1976-77
‡ Fortier, Marc	Que., Ott., L.A.	6	212	42	60	102	135	....	....	....	....	....	....	1987-88	1992-93
Fortin, Ray	St.L.	3	92	2	6	8	33	6	0	0	0	8	....	1967-68	1969-70
‡ Foster, Corey	N.J., Phi., Pit., NYI	4	45	5	6	11	24	3	0	0	0	4	....	1988-89	1996-97
Foster, Dwight	Bos., Col., N.J., Det.	10	541	111	163	274	420	35	5	12	17	4	....	1977-78	1986-87
Foster, Herb	NYR	2	6	1	0	1	5	....	....	....	....	....	....	1940-41	1947-48
• Foster, Yip	NYR, Bos., Det.	4	83	3	2	5	32	....	....	....	....	....	....	1929-30	1934-35
Fotiu, Nick	NYR, Hfd., Cgy., Phi., Edm.	13	646	60	77	137	1362	38	0	4	4	67	....	1976-77	1988-89
• Fowler, Jimmy	Tor.	3	135	18	29	47	39	18	0	3	3	2	....	1936-37	1938-39
Fowler, Tom	Chi.	1	24	0	1	1	18	....	....	....	....	....	....	1946-47	1946-47
Fox, Greg	Atl., Chi., Pit.	8	494	14	92	106	637	44	1	9	10	67	....	1977-78	1984-85
Fox, Jim	L.A.	9	578	186	293	479	143	22	4	8	12	0	....	1980-81	1989-90
• Foyston, Frank	Det.	2	64	17	7	24	32	....	....	....	....	....	....	1926-27	1927-28
Frampton, Bob	Mtl.	1	2	0	0	0	0	3	0	0	0	0	....	1949-50	1949-50
Franceschetti, Lou	Wsh., Tor., Buf.	10	459	59	81	140	747	44	3	2	5	111	....	1981-82	1991-92
Francis, Bobby	Det.	1	14	2	0	2	0	....	....	....	....	....	....	1982-83	1982-83
• Fraser, Archie	NYR	1	3	0	1	1	0	....	....	....	....	....	....	1943-44	1943-44
• Fraser, Charles	Ham.	1	1	0	0	0	0	....	....	....	....	....	....	1923-24	1923-24
Fraser, Curt	Van., Chi., Min.	12	704	193	240	433	1306	65	15	18	33	198	....	1978-79	1989-90
• Fraser, Gord	Chi., Det., Mtl., Pit., Phi.	5	144	24	12	36	224	2	1	0	1	6	....	1926-27	1930-31
Fraser, Harvey	Chi.	1	21	5	4	9	0	....	....	....	....	....	....	1944-45	1944-45
‡ Fraser, Iain	NYI, Que., Dal., Edm., Wpg., S.J.	5	94	23	23	46	31	4	0	0	0	0	....	1992-93	1996-97
Fraser, Scott	Mtl., Edm., NYR	3	72	16	15	31	24	11	1	1	2	0	....	1995-96	1998-99
Frawley, Dan	Chi., Pit.	6	273	37	40	77	674	1	0	0	0	0	....	1983-84	1988-89
• Fredrickson, Frank	Det., Bos., Pit.	5	161	39	34	73	206	10	2	3	5	24	....	1926-27	1930-31
‡ Freer, Mark	Phi., Ott., Cgy.	7	124	16	23	39	61	....	....	....	....	....	....	1986-87	1993-94
• Frew, Irv	Mtl.M., St.L., Mtl.	3	96	2	5	7	146	4	0	0	0	6	....	1933-34	1935-36
Friday, Tim	Det.	1	23	0	3	3	6	....	....	....	....	....	....	1985-86	1985-86
Fridgen, Dan	Hfd.	2	13	2	3	5	2	....	....	....	....	....	....	1981-82	1982-83
Friedman, Doug	Edm., Nsh.	2	18	0	1	1	34	....	....	....	....	....	....	1997-98	1998-99
Friest, Ron	Min.	3	64	7	7	14	191	6	1	0	1	7	....	1980-81	1982-83
Frig, Len	Chi., Cal., Cle., St.L.	7	311	13	51	64	479	14	2	1	3	0	....	1972-73	1979-80
Frost, Harry	Bos.	1	4	0	0	0	0	1	0	0	0	0	1	1938-39	1938-39
Frycer, Miroslav	Que., Tor., Det., Edm.	8	415	147	183	330	486	17	3	8	11	16	....	1981-82	1988-89
Fryday, Bob	Mtl.	2	5	1	0	1	0	....	....	....	....	....	....	1949-50	1951-52
Ftorek, Robbie	Det., Que., NYR	8	334	77	150	227	262	19	9	6	15	28	....	1972-73	1984-85
Fullan, Larry	Wsh.	1	4	1	0	1	0	....	....	....	....	....	....	1974-75	1974-75
Fusco, Mark	Hfd.	2	80	3	12	15	42	....	....	....	....	....	....	1983-84	1984-85

G

Name	NHL Teams	NHL Seasons	GP	G	A	TP	PIM	GP	G	A	TP	PIM	NHL Cup Wins	First NHL Season	Last NHL Season
Gadsby, Bill	Chi., NYR, Det.	20	1248	130	438	568	1539	67	4	23	27	92	....	1946-47	1965-66
‡ Gaetz, Link	Min., S.J.	3	65	6	8	14	412	....	....	....	....	....	....	1988-89	1991-92
Gage, Jody	Det., Buf.	6	68	14	15	29	26	....	....	....	....	....	....	1980-81	1991-92
• Gagne, Art	Mtl., Bos., Ott., Det.	6	228	67	33	100	257	11	2	1	3	20	....	1926-27	1931-32
‡ Gagne, Paul	Col., N.J., Tor., NYI	8	390	110	101	211	127	....	....	....	....	....	....	1980-81	1989-90
Gagne, Pierre	Bos.	1	2	0	0	0	0	....	....	....	....	....	....	1959-60	1959-60
Gagner, Dave	NYR, Min., Dal., Tor., Cgy., Fla., Van.	15	946	318	401	719	1018	57	22	26	48	64	....	1984-85	1998-99
Gagnon, Germain	Mtl., NYI, Chi., K.C.	5	259	40	101	141	72	19	2	3	5	2	....	1971-72	1975-76
• Gagnon, Johnny	Mtl., Bos., NYA	10	454	120	141	261	295	32	12	12	24	37	1	1930-31	1939-40
Gainey, Bob	Mtl.	16	1160	239	262	501	585	182	25	48	73	151	5	1973-74	1988-89
• Gainor, Dutch	Bos., NYR, Ott., Mtl.M.	7	246	51	56	107	129	22	2	1	3	14	2	1927-28	1934-35
Galarneau, Michel	Hfd.	3	78	7	10	17	34	....	....	....	....	....	....	1980-81	1982-83
• Galbraith, Percy	Bos., Ott.	8	347	29	31	60	224	31	4	7	11	24	1	1926-27	1933-34
• Gallagher, John	Mtl.M., Det., NYA	7	205	14	19	33	153	24	2	3	5	27	1	1930-31	1938-39
Gallant, Gerard	Det., T.B.	11	615	211	269	480	1674	58	18	21	39	178	....	1984-85	1994-95
Galley, Garry	L.A., Wsh., Bos., Phi., Buf., NYI	17	1149	125	475	600	1218	89	7	23	30	119	....	1984-85	2000-01
Gallimore, Jamie	Min.	1	2	0	0	0	0	....	....	....	....	....	....	1977-78	1977-78
• Gallinger, Don	Bos.	5	222	65	88	153	89	23	5	5	10	19	....	1942-43	1947-48
Gamble, Dick	Mtl., Chi., Tor.	8	195	41	41	82	66	14	1	2	3	4	1	1950-51	1966-67
Gambucci, Gary	Min.	2	51	2	7	9	9	....	....	....	....	....	....	1971-72	1973-74
Ganchar, Perry	St.L., Mtl., Pit.	4	42	3	7	10	36	7	3	1	4	0	....	1983-84	1988-89
Gans, Dave	L.A.	2	6	0	0	0	2	....	....	....	....	....	....	1982-83	1985-86
• Gardiner, Herb	Mtl., Chi.	3	108	10	9	19	52	9	0	1	1	16	....	1926-27	1928-29
Gardner, Bill	Chi., Hfd.	9	380	73	115	188	68	45	3	8	11	17	....	1980-81	1988-89
• Gardner, Cal	NYR, Tor., Chi., Bos.	12	696	154	238	392	517	61	7	10	17	20	2	1945-46	1956-57
Gardner, Dave	Mtl., St.L., Cal., Cle., Phi.	7	350	75	115	190	41	....	....	....	....	....	....	1972-73	1979-80
Gardner, Paul	Col., Tor., Pit., Wsh., Buf.	10	447	201	201	402	207	16	2	6	8	14	....	1976-77	1985-86
Gare, Danny	Buf., Det., Edm.	13	827	354	331	685	1285	64	25	21	46	195	....	1974-75	1986-87
Gariepy, Ray	Bos., Tor.	2	36	1	6	7	43	....	....	....	....	....	....	1953-54	1955-56
• Garland, Scott	Tor., L.A.	3	91	13	24	37	115	7	1	2	3	35	....	1975-76	1978-79
Garner, Rob	Pit.	1	1	0	0	0	0	....	....	....	....	....	....	1982-83	1982-83
Garpenlov, Johan	Det., S.J., Fla., Atl.	10	609	114	197	311	276	44	10	9	19	22	....	1990-91	1999-00
• Garrett, Red	NYR	1	23	1	1	2	18	....	....	....	....	....	....	1942-43	1942-43
Gartner, Mike	Wsh., Min., NYR, Tor., Phx.	19	1432	708	627	1335	1159	122	43	50	93	125	....	1979-80	1997-98
• Gassoff, Bob	St.L.	4	245	11	47	58	866	9	0	1	1	16	....	1973-74	1976-77
Gassoff, Brad	Van.	4	122	19	17	36	163	3	0	0	0	0	....	1975-76	1978-79
Gatzos, Steve	Pit.	4	89	15	20	35	83	1	0	0	0	0	....	1981-82	1984-85
Gaudreau, Rob	S.J., Ott.	4	231	51	54	105	69	14	2	0	2	0	....	1992-93	1995-96
Gaudreault, Armand	Bos.	1	44	15	9	24	27	7	0	2	2	8	....	1944-45	1944-45
• Gaudreault, Leo	Mtl.	3	67	8	4	12	30	....	....	....	....	....	....	1927-28	1932-33
Gaulin, Jean-Marc	Que.	4	26	4	3	7	8	1	0	0	0	0	....	1982-83	1985-86
Gaume, Dallas	Hfd.	1	4	1	1	2	0	....	....	....	....	....	....	1988-89	1988-89
• Gauthier, Art	Mtl.	1	13	0	0	0	0	1	0	0	0	0	....	1926-27	1926-27
‡ Gauthier, Daniel	Chi.	1	5	0	0	0	0	....	....	....	....	....	....	1994-95	1994-95
• Gauthier, Fern	NYR, Mtl., Det.	6	229	46	50	96	35	22	5	1	6	7	....	1943-44	1948-49
Gauthier, Jean	Mtl., Phi., Bos.	10	166	6	29	35	150	14	1	3	4	22	1	1960-61	1969-70
Gauthier, Luc	Mtl.	1	3	0	0	0	2	....	....	....	....	....	....	1990-91	1990-91
Gauvreau, Jocelyn	Mtl.	1	2	0	0	0	0	....	....	....	....	....	....	1983-84	1983-84
Gavin, Stew	Tor., Hfd., Min.	13	768	130	155	285	584	66	14	20	34	75	....	1980-81	1992-93
Geale, Bob	Pit.	1	1	0	0	0	2	....	....	....	....	....	....	1984-85	1984-85
• Gee, George	Chi., Det.	9	551	135	183	318	345	41	6	13	19	32	1	1945-46	1953-54
Geldart, Gary	Min.	1	4	0	0	0	5	....	....	....	....	....	....	1970-71	1970-71
Gendron, Jean-Guy	NYR, Bos., Mtl., Phi.	14	863	182	201	383	701	42	7	4	11	47	....	1955-56	1971-72
‡ Gendron, Martin	Wsh., Chi.	3	30	4	2	6	10	....	....	....	....	....	....	1994-95	1997-98
Geoffrion, Bernie	Mtl., NYR	16	883	393	429	822	689	132	58	60	118	88	6	1950-51	1967-68
Geoffrion, Danny	Mtl., Wpg.	3	111	20	32	52	99	2	0	0	0	7	....	1979-80	1981-82
• Geran, Gerry	Mtl.W., Bos.	2	37	5	1	6	6	....	....	....	....	....	....	1917-18	1925-26
• Gerard, Eddie	Ott.	6	128	50	48	98	108	16	4	0	4	61	3	1917-18	1922-23
Germain, Eric	L.A.	1	4	0	1	1	13	1	0	0	0	4	....	1987-88	1987-88
• Getliffe, Ray	Bos., Mtl.	10	393	136	137	273	250	45	9	10	19	30	2	1935-36	1944-45
Giallonardo, Mario	Col.	2	23	0	3	3	6	....	....	....	....	....	....	1979-80	1980-81
Gibbs, Barry	Bos., Min., Atl., St.L., L.A.	13	797	58	224	282	945	36	4	2	6	67	....	1967-68	1979-80
Gibson, Don	Van.	1	14	0	3	3	20	....	....	....	....	....	....	1990-91	1990-91
Gibson, Doug	Bos., Wsh.	3	63	9	19	28	0	1	0	0	0	0	....	1973-74	1977-78
Gibson, John	L.A., Tor., Wpg.	3	48	0	2	2	120	....	....	....	....	....	....	1980-81	1983-84
Giesebrecht, Gus	Det.	4	135	27	51	78	13	17	2	3	5	0	....	1938-39	1941-42
Giffin, Lee	Pit.	2	27	1	3	4	9	....	....	....	....	....	....	1986-87	1987-88
Gilbert, Ed	K.C., Pit.	3	166	21	31	52	22	....	....	....	....	....	....	1974-75	1976-77
Gilbert, Greg	NYI, Chi., NYR, St.L.	15	837	150	228	378	576	133	17	33	50	162	3	1981-82	1995-96
Gilbert, Jeannot	Bos.	2	9	0	1	1	4	....	....	....	....	....	....	1962-63	1964-65
Gilbert, Rod	NYR	18	1065	406	615	1021	508	79	34	33	67	43	....	1960-61	1977-78
Gilbertson, Stan	Cal., St.L., Wsh., Pit.	6	428	85	89	174	148	3	1	1	2	2	....	1971-72	1976-77
Giles, Curt	Min., NYR, St.L.	14	895	43	199	242	733	103	6	16	22	118	....	1979-80	1992-93
Gilhen, Randy	Hfd., Wpg., Pit., L.A., NYR, T.B., Fla.	11	457	55	60	115	314	33	3	2	5	26	1	1982-83	1995-96
Gillen, Don	Phi., Hfd.	2	35	2	4	6	22	....	....	....	....	....	....	1979-80	1981-82
• Gillie, Farrand	Det.	1	1	0	0	0	0	....	....	....	....	....	....	1928-29	1928-29
Gillies, Clark	NYI, Buf.	14	958	319	378	697	1023	164	47	47	94	287	4	1974-75	1987-88
Gillis, Jere	Van., NYR, Que., Buf., Phi.	9	386	78	95	173	230	19	4	7	11	9	....	1977-78	1986-87
Gillis, Mike	Col., Bos.	6	246	33	43	76	186	27	2	5	7	10	....	1978-79	1983-84

Dave Feamster

Bernie Federko

Steven Finn

Reggie Fleming

Garry Galley

Bernie Geoffrion

Clark Gillies

Gaston Gingras

Name	NHL Teams	NHL Seasons	Regular Schedule GP	G	A	TP	PIM	Playoffs GP	G	A	TP	PIM	NHL Cup Wins	First NHL Season	Last NHL Season
Gillis, Paul	Que., Chi., Hfd.	11	624	88	154	242	1498	42	3	14	17	156	....	1982-83	1992-93
Gingras, Gaston	Mtl., Tor., St.L.	10	476	61	174	235	161	52	6	18	24	20	1	1979-80	1988-89
Girard, Bob	Cal., Cle., Wsh.	5	305	45	69	114	140	....	....	....	....	....	....	1975-76	1979-80
Girard, Kenny	Tor.	3	7	0	1	1	2	....	....	....	....	....	....	1956-57	1959-60
● Giroux, Art	Mtl., Bos., Det.	3	54	6	4	10	14	2	0	0	0	0	....	1932-33	1935-36
Giroux, Larry	St.L., K.C., Det., Hfd.	7	274	15	74	89	333	5	0	0	0	4	....	1973-74	1979-80
Giroux, Pierre	L.A.	1	6	1	0	1	17	....	....	....	....	....	....	1982-83	1982-83
Gladney, Bob	L.A., Pit.	2	14	1	5	6	4	....	....	....	....	....	....	1982-83	1983-84
Gladu, Jean-Paul	Bos.	1	40	6	14	20	2	7	2	2	4	0	....	1944-45	1944-45
Glennie, Brian	Tor., L.A.	10	572	14	100	114	621	32	0	1	1	66	....	1969-70	1978-79
Glennon, Matt	Bos.	1	3	0	0	0	2	....	....	....	....	....	....	1991-92	1991-92
Gloeckner, Lorry	Det.	1	13	0	2	2	6	....	....	....	....	....	....	1978-79	1978-79
Gloor, Dan	Van.	1	2	0	0	0	0	....	....	....	....	....	....	1973-74	1973-74
● Glover, Fred	Det., Chi.	5	92	13	11	24	62	8	0	0	0	0	1	1948-49	1952-53
Glover, Howie	Chi., Det., NYR, Mtl.	5	144	29	17	46	101	11	1	2	3	2	....	1958-59	1968-69
Glynn, Brian	Cgy., Min., Edm., Ott., Van., Hfd.	10	431	25	79	104	410	57	6	10	16	40	....	1987-88	1996-97
Godden, Ernie	Tor.	1	5	1	1	2	6	....	....	....	....	....	....	1981-82	1981-82
● Godfrey, Warren	Bos., Det.	16	786	32	125	157	752	52	1	4	5	42	....	1952-53	1967-68
Godin, Eddy	Wsh.	2	27	3	6	9	12	....	....	....	....	....	....	1977-78	1978-79
● Godin, Sam	Ott., Mtl.	3	83	4	3	7	36	....	....	....	....	....	....	1927-28	1933-34
‡ Godynyuk, Alexander	Tor., Cgy., Fla., Hfd.	7	223	10	39	49	224	....	....	....	....	....	....	1990-91	1996-97
Goegan, Pete	Det., NYR, Min.	11	383	19	67	86	365	33	1	3	4	61	....	1957-58	1967-68
Goertz, Dave	Pit.	1	2	0	0	0	2	....	....	....	....	....	....	1987-88	1987-88
● Goldham, Bob	Tor., Chi., Det.	12	650	28	143	171	400	66	3	14	17	53	5	1941-42	1955-56
● Goldsworthy, Bill	Bos., Min., NYR	14	771	283	258	541	793	40	18	19	37	30	....	1964-65	1977-78
● Goldsworthy, Leroy	NYR, Det., Chi., Mtl., Bos., NYA	10	336	66	57	123	79	24	1	0	1	4	1	1928-29	1938-39
Goldup, Glenn	Mtl., L.A.	9	291	52	67	119	303	16	4	3	7	22	....	1973-74	1981-82
Goldup, Hank	Tor., NYR	6	202	63	80	143	97	26	5	1	6	6	1	1939-40	1945-46
‡ Goneau, Daniel	NYR	3	53	12	3	15	14	....	....	....	....	....	....	1996-97	1999-00
● Gooden, Bill	NYR	2	53	9	11	20	15	....	....	....	....	....	....	1942-43	1943-44
Goodenough, Larry	Phi., Van.	6	242	22	77	99	179	22	3	15	18	10	1	1974-75	1979-80
● Goodfellow, Ebbie	Det.	14	557	134	190	324	511	45	8	8	16	65	3	1929-30	1942-43
‡ Gordiouk, Viktor	Buf.	2	26	3	8	11	0	....	....	....	....	....	....	1992-93	1994-95
Gordon, Fred	Det., Bos.	2	81	8	7	15	68	2	0	0	0	0	....	1926-27	1927-28
Gordon, Jack	NYR	3	36	3	10	13	0	9	1	1	2	7	....	1948-49	1950-51
Gordon, Robb	Van.	1	4	0	0	0	2	....	....	....	....	....	....	1998-99	1998-99
Gorence, Tom	Phi., Edm.	6	303	58	53	111	89	37	9	6	15	47	....	1978-79	1983-84
Goring, Butch	L.A., NYI, Bos.	16	1107	375	513	888	102	134	38	50	88	32	4	1969-70	1984-85
Gorman, Dave	Atl.	1	3	0	0	0	0	....	....	....	....	....	....	1979-80	1979-80
● Gorman, Ed	Ott., Tor.	4	111	14	6	20	108	8	0	0	0	2	1	1924-25	1927-28
Gosselin, Benoit	NYR	1	7	0	0	0	33	....	....	....	....	....	....	1977-78	1977-78
Gosselin, Guy	Wpg.	1	5	0	0	0	6	....	....	....	....	....	....	1987-88	1987-88
Gotaas, Steve	Pit., Min.	3	49	6	9	15	53	3	0	1	1	5	....	1987-88	1990-91
● Gottselig, Johnny	Chi.	16	589	176	195	371	203	43	13	13	26	18	2	1928-29	1944-45
Gould, Bobby	Atl., Cgy., Wsh., Bos.	11	697	145	159	304	572	78	15	13	28	58	....	1979-80	1989-90
Gould, John	Buf., Van., Atl.	9	504	131	138	269	113	14	3	2	5	4	....	1971-72	1979-80
Gould, Larry	Van.	1	2	0	0	0	0	....	....	....	....	....	....	1973-74	1973-74
Goulet, Michel	Que., Chi.	15	1089	548	604	1152	825	92	39	39	78	110	....	1979-80	1993-94
Goupille, Red	Mtl.	8	222	12	28	40	256	8	2	0	2	6	....	1935-36	1942-43
‡ Govedaris, Chris	Hfd., Tor.	4	45	4	6	10	24	4	0	0	0	2	....	1989-90	1993-94
Goyer, Gerry	Chi.	1	40	1	2	3	4	3	0	0	0	2	....	1967-68	1967-68
Goyette, Phil	Mtl., NYR, St.L., Buf.	16	941	207	467	674	131	94	17	29	46	26	4	1956-57	1971-72
Graboski, Tony	Mtl.	3	66	6	10	16	24	3	0	0	0	6	....	1940-41	1942-43
● Gracie, Bob	Tor., Bos., NYA, Mtl.M., Mtl., Chi.	9	379	82	109	191	205	33	4	7	11	4	2	1930-31	1938-39
Gradin, Thomas	Van., Bos.	9	677	209	384	593	298	42	17	25	42	20	....	1978-79	1986-87
Graham, Dirk	Min., Chi.	12	772	219	270	489	917	90	17	27	44	92	....	1983-84	1994-95
● Graham, Leth	Ott., Ham.	6	27	3	0	3	0	1	0	0	0	0	1	1920-21	1925-26
Graham, Pat	Pit., Tor.	3	103	11	17	28	136	4	0	0	0	2	....	1981-82	1983-84
Graham, Rod	Bos.	1	14	2	1	3	7	....	....	....	....	....	....	1974-75	1974-75
● Graham, Ted	Chi., Mtl.M., Det., St.L., Bos., NYA	9	346	14	25	39	300	24	3	1	4	30	....	1927-28	1936-37
‡ Granato, Tony	NYR, L.A., S.J.	13	773	248	244	492	1425	79	16	27	43	141	....	1988-89	2000-01
Grant, Danny	Mtl., Min., Det., L.A.	13	736	263	273	536	239	43	10	14	24	19	1	1965-66	1978-79
Gratton, Dan	L.A.	1	7	1	0	1	5	....	....	....	....	....	....	1987-88	1987-88
Gratton, Norm	NYR, Atl., Buf., Min.	5	201	39	44	83	64	6	0	1	1	2	....	1971-72	1975-76
Gravelle, Leo	Mtl., Det.	5	223	44	34	78	42	17	4	1	5	2	....	1946-47	1950-51
Graves, Hilliard	Cal., Atl., Van., Wpg.	9	556	118	163	281	209	2	0	0	0	0	....	1970-71	1979-80
Graves, Steve	Edm.	3	35	5	4	9	10	....	....	....	....	....	....	1983-84	1987-88
● Gray, Alex	NYR, Tor.	2	50	7	0	7	32	13	1	0	1	0	1	1927-28	1928-29
Gray, Terry	Bos., Mtl., L.A., St.L.	6	147	26	28	54	64	35	5	5	10	22	....	1961-62	1970-71
● Green, Red	Ham., NYA, Bos., Det.	6	195	59	26	85	290	1	0	0	0	0	1	1923-24	1928-29
Green, Rick	Wsh., Mtl., Det., NYI	15	845	43	220	263	588	100	3	16	19	73	1	1976-77	1991-92
● Green, Shorty	Ham., NYA	4	103	33	20	53	151	....	....	....	....	....	....	1923-24	1926-27
Green, Ted	Bos.	11	620	48	206	254	1029	31	4	8	12	54	1	1960-61	1971-72
Greenlaw, Jeff	Wsh., Fla.	6	57	3	6	9	108	2	0	0	0	21	....	1986-87	1993-94
Gregg, Randy	Edm., Van.	10	474	41	152	193	333	137	13	38	51	127	5	1981-82	1991-92
Greig, Bruce	Cal.	2	9	0	1	1	46	....	....	....	....	....	....	1973-74	1974-75
Grenier, Lucien	Mtl., L.A.	4	151	14	14	28	18	2	0	0	0	0	1	1968-69	1971-72
Grenier, Richard	NYI	1	10	1	1	2	2	....	....	....	....	....	....	1972-73	1972-73
Greschner, Ron	NYR	16	982	179	431	610	1226	84	17	32	49	106	....	1974-75	1989-90
‡ Gretzky, Brent	T.B.	2	13	1	3	4	2	....	....	....	....	....	....	1993-94	1994-95
Gretzky, Wayne	Edm., L.A., St.L., NYR	20	1487	894	1963	2857	577	208	122	260	382	66	4	1979-80	1998-99
Grieve, Brent	NYI, Edm., Chi., L.A.	4	97	20	16	36	87	....	....	....	....	....	....	1993-94	1996-97
Grigor, George	Chi.	1	2	1	0	1	0	1	0	0	0	0	....	1943-44	1943-44
Grisdale, John	Tor., Van.	6	250	4	39	43	346	10	0	1	1	15	....	1972-73	1978-79
‡ Groleau, Francois	Mtl.	3	8	0	1	1	6	....	....	....	....	....	....	1995-96	1997-98
‡ Gronman, Tuomas	Chi., Pit.	2	38	1	3	4	38	1	0	0	0	0	....	1996-97	1997-98
Gronsdahl, Lloyd	Bos.	1	10	1	2	3	0	....	....	....	....	....	....	1941-42	1941-42
Gronstrand, Jari	Min., NYR, Que., NYI	5	185	8	26	34	135	3	0	0	0	4	....	1986-87	1990-91
Gross, Lloyd	Tor., NYA, Bos., Det.	3	62	11	5	16	20	1	0	0	0	0	....	1926-27	1934-35
● Grosso, Don	Det., Chi., Bos.	9	336	87	117	204	90	48	15	14	29	63	1	1938-39	1946-47
Grosvenor, Len	Ott., NYA, Mtl.	6	149	9	11	20	78	4	0	0	0	2	....	1927-28	1932-33
Groulx, Wayne	Que.	1	1	0	0	0	0	....	....	....	....	....	....	1984-85	1984-85
Gruen, Danny	Det., Col.	3	49	9	13	22	19	....	....	....	....	....	....	1972-73	1976-77
Gruhl, Scott	L.A., Pit.	3	20	3	3	6	6	....	....	....	....	....	....	1981-82	1987-88
Gryp, Bob	Bos., Wsh.	3	74	11	13	24	33	....	....	....	....	....	....	1973-74	1975-76
‡ Guay, Francois	Buf.	1	1	0	0	0	0	....	....	....	....	....	....	1989-90	1989-90
Guay, Paul	Phi., L.A., Bos., NYI	7	117	11	23	34	92	9	0	1	1	12	....	1983-84	1990-91
‡ Guerard, Daniel	Ott.	1	2	0	0	0	0	....	....	....	....	....	....	1994-95	1994-95
Guerard, Stephane	Que.	2	34	0	0	0	40	....	....	....	....	....	....	1987-88	1989-90
Guevremont, Jocelyn	Van., Buf., NYR	9	571	84	223	307	319	40	4	17	21	18	....	1971-72	1979-80
Guidolin, Aldo	NYR	4	182	9	15	24	117	....	....	....	....	....	....	1952-53	1955-56
Guidolin, Bep	Bos., Det., Chi.	9	519	107	171	278	606	24	5	7	12	35	....	1942-43	1951-52
Guindon, Bobby	Wpg.	1	6	0	1	1	0	....	....	....	....	....	....	1979-80	1979-80
Gusarov, Alexei	Que., Col., NYR, St.L.	11	607	39	128	167	313	68	0	14	14	38	1	1990-91	2000-01
Gustafsson, Bengt-Ake	Wsh.	9	629	196	359	555	196	32	9	19	28	16	....	1979-80	1988-89
‡ Gustafsson, Per	Fla., Tor., Ott.	2	89	8	27	35	38	1	0	0	0	0	....	1996-97	1997-98
Gustavsson, Peter	Col.	1	2	0	0	0	0	....	....	....	....	....	....	1981-82	1981-82
Guy, Kevan	Cgy., Van.	6	156	5	20	25	138	5	0	1	1	23	....	1986-87	1991-92

H

Name	NHL Teams	NHL Seasons	Regular Schedule GP	G	A	TP	PIM	Playoffs GP	G	A	TP	PIM	NHL Cup Wins	First NHL Season	Last NHL Season
Haanpaa, Ari	NYI	3	60	6	11	17	37	6	0	0	0	10	....	1985-86	1987-88
‡ Haas, David	Edm., Cgy.	2	7	2	1	3	7	....	....	....	....	....	....	1990-91	1993-94
Habscheid, Marc	Edm., Min., Det., Cgy.	11	345	72	91	163	171	12	1	3	4	13	....	1981-82	1991-92
‡ Hachborn, Len	Phi., L.A.	3	102	20	39	59	29	7	0	3	3	7	....	1983-84	1985-86
Haddon, Lloyd	Det.	1	8	0	0	0	2	1	0	0	0	0	....	1959-60	1959-60
Hadfield, Vic	NYR, Pit.	16	1002	323	389	712	1154	73	27	21	48	117	....	1961-62	1976-77
● Haggarty, Jim	Mtl.	1	5	1	1	2	0	3	2	1	3	0	....	1941-42	1941-42
● Hagglund, Roger	Que.	1	3	0	0	0	0	....	....	....	....	....	....	1984-85	1984-85
Hagman, Matti	Bos., Edm.	4	237	56	89	145	36	20	5	2	7	6	....	1976-77	1981-82
Haidy, Gord	Det.	1	....	....	....	....	....	1	0	0	0	0	1	1949-50	1949-50
Hajdu, Richard	Buf.	2	5	0	0	0	4	....	....	....	....	....	....	1985-86	1986-87
Hajt, Bill	Buf.	14	854	42	202	244	433	80	2	16	18	70	....	1973-74	1986-87
Hakansson, Anders	Min., Pit., L.A.	5	330	52	46	98	141	6	0	0	0	2	....	1981-82	1985-86
● Halderson, Harold	Det., Tor.	1	44	3	2	5	65	....	....	....	....	....	....	1926-27	1926-27
Hale, Larry	Phi.	4	196	5	37	42	90	8	0	0	0	12	....	1968-69	1971-72

Name	NHL Teams	NHL Seasons	Regular Schedule					Playoffs					NHL Cup Wins	First NHL Season	Last NHL Season
			GP	G	A	TP	PIM	GP	G	A	TP	PIM			
Haley, Len	Det.	2	30	2	2	4	14	6	1	3	4	6	....	1959-60	1960-61
‡ Halkidis, Bob	Buf., L.A., Tor., Det., T.B., NYI	11	256	8	32	40	825	20	0	1	1	51	....	1984-85	1995-96
● Hall, Bob	NYA	1	8	0	0	0	0	....	....	....	....	....	....	1925-26	1925-26
Hall, Del	Cal.	3	9	2	0	2	2	....	....	....	....	....	....	1971-72	1973-74
● Hall, Joe	Mtl.	2	38	15	8	23	189	7	0	1	1	29	....	1917-18	1918-19
Hall, Murray	Chi., Det., Min., Van.	9	164	35	48	83	46	6	0	0	0	0	....	1961-62	1971-72
Hall, Taylor	Van., Bos.	5	41	7	9	16	29	....	....	....	....	....	....	1983-84	1987-88
Hall, Wayne	NYR	1	4	0	0	0	0	....	....	....	....	....	....	1960-61	1960-61
● Halliday, Milt	Ott.	3	67	1	0	1	4	6	0	0	0	0	1	1926-27	1928-29
Hallin, Mats	NYI, Min.	5	152	17	14	31	193	15	1	0	1	13	1	1982-83	1986-87
Halverson, Trevor	Wsh.	1	17	0	4	4	28	....	....	....	....	....	....	1998-99	1998-99
Halward, Doug	Bos., L.A., Van., Det., Edm.	14	653	69	224	293	774	47	7	10	17	113	....	1975-76	1988-89
Hamel, Gilles	Buf., Wpg., L.A.	9	519	127	147	274	276	27	4	5	9	10	....	1980-81	1988-89
● Hamel, Herb	Tor.	1	2	0	0	0	4	....	....	....	....	....	....	1930-31	1930-31
Hamel, Jean	St.L., Det., Que., Mtl.	12	699	26	95	121	766	33	0	2	2	44	....	1972-73	1983-84
● Hamill, Red	Bos., Chi.	12	419	128	94	222	160	24	1	2	3	20	1	1937-38	1950-51
Hamilton, Al	NYR, Buf., Edm.	7	257	10	78	88	258	7	0	0	0	2	....	1965-66	1979-80
Hamilton, Chuck	Mtl., St.L.	2	4	0	2	2	2	....	....	....	....	....	....	1961-62	1972-73
● Hamilton, Jack	Tor.	3	102	28	32	60	20	11	2	1	3	0	....	1942-43	1945-46
Hamilton, Jim	Pit.	8	95	14	18	32	28	6	3	0	3	0	....	1977-78	1984-85
● Hamilton, Reg	Tor., Chi.	12	424	21	87	108	412	64	3	8	11	46	2	1935-36	1946-47
Hammarstrom, Inge	Tor., St.L.	6	427	116	123	239	86	13	2	3	5	4	....	1973-74	1978-79
Hammond, Ken	L.A., Edm., NYR, Tor., Bos., S.J., Van., Ott.	8	193	18	29	47	290	15	0	0	0	24	....	1984-85	1992-93
Hampson, Gord	Cgy.	1	4	0	0	0	5	....	....	....	....	....	....	1982-83	1982-83
Hampson, Ted	Tor., NYR, Det., Oak., Cal., Min.	12	676	108	245	353	94	35	7	10	17	2	....	1959-60	1971-72
Hampton, Rick	Cal., Cle., L.A.	6	337	59	113	172	147	2	0	0	0	0	....	1974-75	1979-80
‡ Hamr, Radek	Ott.	2	11	0	0	0	0	....	....	....	....	....	....	1992-93	1993-94
Hamway, Mark	NYI	3	53	5	13	18	9	1	0	0	0	0	....	1984-85	1986-87
Handy, Ron	NYI, St.L.	2	14	0	3	3	0	....	....	....	....	....	....	1984-85	1987-88
Hangsleben, Al	Hfd., Wsh., L.A.	3	185	21	48	69	396	....	....	....	....	....	....	1979-80	1981-82
Hankinson, Ben	N.J., T.B.	3	43	3	3	6	45	2	1	0	1	4	....	1992-93	1994-95
Hanna, John	NYR, Mtl., Phi.	5	198	6	26	32	206	....	....	....	....	....	....	1958-59	1967-68
Hannan, Dave	Pit., Edm., Tor., Buf., Col., Ott.	16	841	114	191	305	942	63	6	7	13	46	2	1981-82	1996-97
● Hannigan, Gord	Tor.	4	161	29	31	60	117	9	2	0	2	8	....	1952-53	1955-56
Hannigan, Pat	Tor., NYR, Phi.	5	182	30	39	69	116	11	1	2	3	11	....	1959-60	1968-69
Hannigan, Ray	Tor.	1	3	0	0	0	2	....	....	....	....	....	....	1948-49	1948-49
Hansen, Richie	NYI, St.L.	4	20	2	8	10	4	....	....	....	....	....	....	1976-77	1981-82
Hanson, Dave	Det., Min.	2	33	1	1	2	65	....	....	....	....	....	....	1978-79	1979-80
● Hanson, Emil	Det.	1	7	0	0	0	6	....	....	....	....	....	....	1932-33	1932-33
Hanson, Keith	Cgy.	1	25	0	2	2	77	....	....	....	....	....	....	1983-84	1983-84
● Hanson, Oscar	Chi.	1	8	0	0	0	0	....	....	....	....	....	....	1937-38	1937-38
Harbaruk, Nick	Pit., St.L.	5	364	45	75	120	273	14	3	1	4	20	....	1969-70	1973-74
Harding, Jeff	Phi.	2	15	0	0	0	47	....	....	....	....	....	....	1988-89	1989-90
Hardy, Joe	Oak., Cal.	2	63	9	14	23	51	4	0	0	0	0	....	1969-70	1970-71
Hardy, Mark	L.A., NYR, Min.	15	915	62	306	368	1293	67	5	16	21	158	....	1979-80	1993-94
Hargreaves, Jim	Van.	2	66	1	7	8	105	....	....	....	....	....	....	1970-71	1972-73
‡ Harkins, Todd	Cgy., Hfd.	3	48	3	3	6	78	....	....	....	....	....	....	1991-92	1993-94
Harlow, Scott	St.L.	1	1	0	1	1	0	....	....	....	....	....	....	1987-88	1987-88
Harmon, Glen	Mtl.	9	452	50	96	146	334	53	5	10	15	37	2	1942-43	1950-51
Harms, John	Chi.	2	44	5	5	10	21	4	3	0	3	2	....	1943-44	1944-45
● Harnott, Walter	Bos.	1	6	0	0	0	2	....	....	....	....	....	....	1933-34	1933-34
Harper, Terry	Mtl., L.A., Det., St.L., Col.	19	1066	35	221	256	1362	112	4	13	17	140	5	1962-63	1980-81
Harrer, Tim	Cgy.	1	3	0	0	0	2	....	....	....	....	....	....	1982-83	1982-83
● Harrington, Hago	Bos., Mtl.	3	72	9	3	12	15	4	1	0	1	2	....	1925-26	1932-33
● Harris, Billy	Tor., Det., Oak., Pit.	13	769	126	219	345	205	62	8	10	18	30	3	1955-56	1968-69
Harris, Billy	NYI, L.A., Tor.	12	897	231	327	558	394	71	19	19	38	48	....	1972-73	1983-84
Harris, Duke	Min., Tor.	1	26	1	4	5	4	....	....	....	....	....	....	1967-68	1967-68
● Harris, Henry	Bos.	1	32	2	4	6	20	....	....	....	....	....	....	1930-31	1930-31
Harris, Hugh	Buf.	1	60	12	26	38	17	3	0	0	0	0	....	1972-73	1972-73
Harris, Ron	Det., Oak., Atl., NYR	11	476	20	91	111	474	28	4	3	7	33	....	1962-63	1975-76
● Harris, Smokey	Bos.	1	6	3	1	4	8	....	....	....	....	....	....	1924-25	1924-25
Harris, Ted	Mtl., Min., Det., St.L., Phi.	12	788	30	168	198	1000	100	1	22	23	230	5	1963-64	1974-75
Harrison, Ed	Bos., NYR	4	194	27	24	51	53	9	1	0	1	2	....	1947-48	1950-51
Harrison, Jim	Bos., Tor., Chi., Edm.	8	324	67	86	153	435	13	1	1	2	43	....	1968-69	1979-80
Hart, Gerry	Det., NYI, Que., St.L.	15	730	29	150	179	1240	78	3	12	15	175	....	1968-69	1982-83
● Hart, Gizzy	Det., Mtl.	3	104	6	8	14	12	8	0	1	1	0	....	1926-27	1932-33
‡ Hartman, Mike	Buf., Wpg., T.B., NYR	9	397	43	35	78	1388	21	0	0	0	106	1	1986-87	1994-95
Hartsburg, Craig	Min.	10	570	98	315	413	818	61	15	27	42	70	....	1979-80	1988-89
Harvey, Buster	Min., Atl., K.C., Det.	7	407	90	118	208	131	14	0	2	2	8	....	1970-71	1976-77
● Harvey, Doug	Mtl., NYR, Det., St.L.	20	1113	88	452	540	1216	137	8	64	72	152	6	1947-48	1968-69
Harvey, Hugh	K.C.	2	18	1	1	2	4	....	....	....	....	....	....	1974-75	1975-76
Hassard, Bob	Tor., Chi.	5	126	9	28	37	22	....	....	....	....	....	1	1949-50	1954-55
Hatcher, Kevin	Wsh., Dal., Pit., NYR, Car.	17	1157	227	450	677	1392	118	22	37	59	252	....	1984-85	2000-01
Hatoum, Ed	Det., Van.	3	47	3	6	9	25	....	....	....	....	....	....	1968-69	1970-71
Hawerchuk, Dale	Wpg., Buf., St.L., Phi.	16	1188	518	891	1409	730	97	30	69	99	67	....	1981-82	1996-97
‡ Hawkins, Todd	Van., Tor.	3	10	0	0	0	15	....	....	....	....	....	....	1988-89	1991-92
Haworth, Alan	Buf., Wsh., Que.	8	524	189	211	400	425	42	12	16	28	28	....	1980-81	1987-88
Haworth, Gord	NYR	1	2	0	1	1	0	....	....	....	....	....	....	1952-53	1952-53
Hawryliw, Neil	NYI	1	1	0	0	0	0	....	....	....	....	....	....	1981-82	1981-82
Hay, Bill	Chi.	8	506	113	273	386	244	67	15	21	36	62	1	1959-60	1966-67
● Hay, George	Chi., Det.	7	239	74	60	134	84	8	2	3	5	2	....	1926-27	1933-34
Hay, Jim	Det.	3	75	1	5	6	22	9	1	0	1	2	1	1952-53	1954-55
Hayek, Peter	Min.	1	1	0	0	0	0	....	....	....	....	....	....	1981-82	1981-82
Hayes, Chris	Bos.	1	....	....	....	....	....	1	0	0	0	0	1	1971-72	1971-72
● Haynes, Paul	Mtl.M., Bos., Mtl.	11	391	61	134	195	164	24	2	8	10	13	....	1930-31	1940-41
‡ Hayward, Rick	L.A.	1	4	0	0	0	5	....	....	....	....	....	....	1990-91	1990-91
Hazlett, Steve	Van.	1	1	0	0	0	0	....	....	....	....	....	....	1979-80	1979-80
Head, Galen	Det.	1	1	0	0	0	0	....	....	....	....	....	....	1967-68	1967-68
● Headley, Fern	Bos., Mtl.	1	30	1	3	4	10	1	0	0	0	0	....	1924-25	1924-25
Healey, Rich	Det.	1	1	0	0	0	2	....	....	....	....	....	....	1960-61	1960-61
‡ Heaphy, Shawn	Cgy.	1	1	0	0	0	0	....	....	....	....	....	....	1992-93	1992-93
Heaslip, Mark	NYR, L.A.	3	117	10	19	29	110	5	0	0	0	2	....	1976-77	1978-79
Heath, Randy	NYR	2	13	2	4	6	15	....	....	....	....	....	....	1984-85	1985-86
Hebenton, Andy	NYR, Bos.	9	630	189	202	391	83	22	6	5	11	8	....	1955-56	1963-64
Hedberg, Anders	NYR	7	465	172	225	397	144	58	22	24	46	31	....	1978-79	1984-85
● Heffernan, Frank	Tor.	1	19	0	1	1	10	....	....	....	....	....	....	1919-20	1919-20
Heffernan, Gerry	Mtl.	3	83	33	35	68	27	11	3	3	6	8	1	1941-42	1943-44
Heidt, Mike	L.A.	1	6	0	1	1	7	....	....	....	....	....	....	1983-84	1983-84
● Heindl, Bill	Min., NYR	3	18	2	1	3	0	....	....	....	....	....	....	1970-71	1972-73
Heinrich, Lionel	Bos.	1	35	1	1	2	33	....	....	....	....	....	....	1955-56	1955-56
Heiskala, Earl	Phi.	3	127	13	11	24	294	....	....	....	....	....	....	1968-69	1970-71
Helander, Peter	L.A.	1	7	0	1	1	0	....	....	....	....	....	....	1982-83	1982-83
● Heller, Ott	NYR	15	647	55	176	231	465	61	6	8	14	61	2	1931-32	1945-46
Helman, Harry	Ott.	3	44	1	0	1	7	2	0	0	0	0	1	1922-23	1924-25
‡ Helminen, Raimo	NYR, Min., NYI	3	117	13	46	59	16	2	0	0	0	0	....	1985-86	1988-89
● Hemmerling, Tony	NYA	2	22	3	3	6	4	....	....	....	....	....	....	1935-36	1936-37
Henderson, Archie	Wsh., Min., Hfd.	3	23	3	1	4	92	....	....	....	....	....	....	1980-81	1982-83
Henderson, Murray	Bos.	8	405	24	62	86	305	41	2	3	5	23	....	1944-45	1951-52
Henderson, Paul	Det., Tor., Atl.	13	707	236	241	477	304	56	11	14	25	28	....	1962-63	1979-80
Hendrickson, John	Det.	3	5	0	0	0	4	....	....	....	....	....	....	1957-58	1961-62
Henning, Lorne	NYI	9	544	73	111	184	102	81	7	7	14	8	2	1972-73	1980-81
● Henry, Camille	NYR, Chi., St.L.	14	727	279	249	528	88	47	6	12	18	7	....	1953-54	1969-70
‡ Henry, Dale	NYI	6	132	13	26	39	263	14	1	0	1	19	....	1984-85	1989-90
Hepple, Alan	N.J.	3	3	0	0	0	7	....	....	....	....	....	....	1983-84	1985-86
‡ Herbers, Ian	Edm., T.B., NYI	2	65	0	5	5	79	....	....	....	....	....	....	1993-94	1999-00
● Herberts, Jimmy	Bos., Tor., Det.	6	206	83	31	114	253	9	3	0	3	10	....	1924-25	1929-30
Herchenratter, Art	Det.	1	10	1	2	3	2	....	....	....	....	....	....	1940-41	1940-41
Hergert, Fred	NYA	2	20	2	4	6	2	....	....	....	....	....	....	1934-35	1935-36
Hergesheimer, Phil	Chi., Bos.	4	125	21	41	62	19	6	0	0	0	2	....	1939-40	1942-43
Hergesheimer, Wally	NYR, Chi.	7	351	114	85	199	106	5	1	0	1	0	....	1951-52	1958-59
● Heron, Red	Tor., Bro., Mtl.	4	106	21	19	40	38	21	2	2	4	6	....	1938-39	1941-42
Heroux, Yves	Que.	1	1	0	0	0	0	....	....	....	....	....	....	1986-87	1986-87
‡ Herter, Jason	NYI	1	1	0	1	1	0	....	....	....	....	....	....	1995-96	1995-96
Hervey, Matt	Wpg., Bos., T.B.	3	35	0	5	5	97	5	0	0	0	6	....	1988-89	1992-93
Hess, Bob	St.L., Buf., Hfd.	8	329	27	95	122	178	4	1	1	2	2	....	1974-75	1983-84

Ron Greschner

Bep Guidolin

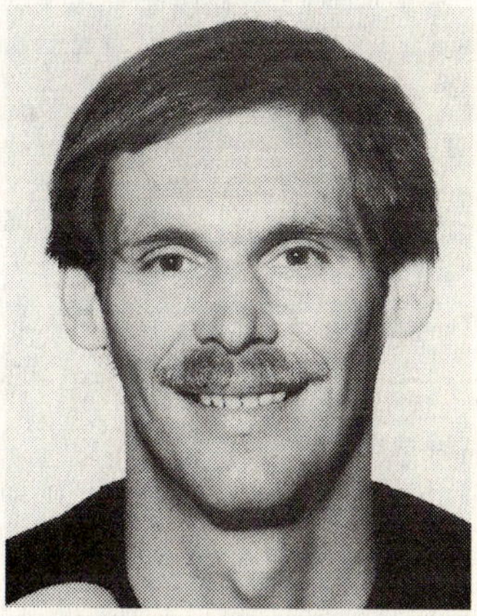

Bill Hajt

Dave Hannan

Billy Harris

Todd Harkins

Bryan Hextall

Tim Higgins

			Regular Schedule					Playoffs							
Name	NHL Teams	NHL Seasons	GP	G	A	TP	PIM	GP	G	A	TP	PIM	NHL Cup Wins	First NHL Season	Last NHL Season
• Heximer, Obs	NYR, Bos., NYA	3	84	13	7	20	16	5	0	0	0	2	....	1929-30	1934-35
• Hextall, Bryan	NYR	11	449	187	175	362	227	37	8	9	17	19	1	1936-37	1947-48
Hextall, Bryan Jr.	NYR, Pit., Atl., Det., Min.	8	549	99	161	260	738	18	0	4	4	59	....	1962-63	1975-76
Hextall, Dennis	NYR, L.A., Cal., Min., Det., Wsh.	13	681	153	350	503	1398	22	3	3	6	45	....	1967-68	1979-80
Heyliger, Vic	Chi.	2	33	2	3	5	2	....	....	....	....	....	....	1937-38	1943-44
Hicke, Bill	Mtl., NYR, Oak., Cal., Pit.	14	729	168	234	402	395	42	3	10	13	41	2	1958-59	1971-72
Hicke, Ernie	Cal., Atl., NYI, Min., L.A.	8	520	132	140	272	407	2	1	0	1	0	....	1970-71	1977-78
Hickey, Greg	NYR	1	1	0	0	0	0	....	....	....	....	....	....	1977-78	1977-78
Hickey, Pat	NYR, Col., Tor., Que., St.L.	10	646	192	212	404	351	55	5	11	16	37	....	1975-76	1984-85
‡ Hicks, Alex	Ana., Pit., S.J., Fla.	5	258	25	54	79	247	15	0	2	2	8	....	1995-96	1999-00
Hicks, Doug	Min., Chi., Edm., Wsh.	9	561	37	131	168	442	18	2	1	3	15	....	1974-75	1982-83
Hicks, Glenn	Det.	2	108	6	12	18	127	....	....	....	....	....	....	1979-80	1980-81
• Hicks, Henry	Mtl.M., Det.	3	96	7	2	9	72	....	....	....	....	....	....	1928-29	1930-31
Hicks, Wayne	Chi., Bos., Mtl., Phi., Pit.	5	115	13	23	36	22	2	0	1	1	2	1	1959-60	1967-68
Hidi, Andre	Wsh.	2	7	2	1	3	9	2	0	0	0	0	....	1983-84	1984-85
Hiemer, Uli	N.J.	3	143	19	54	73	176	....	....	....	....	....	....	1984-85	1986-87
Higgins, Paul	Tor.	2	25	0	0	0	152	1	0	0	0	0	....	1981-82	1982-83
Higgins, Tim	Chi., N.J., Det.	11	706	154	198	352	719	65	5	8	13	77	....	1978-79	1988-89
Hildebrand, Ike	NYR, Chi.	2	41	7	11	18	16	....	....	....	....	....	....	1953-54	1954-55
Hill, Al	Phi.	8	221	40	55	95	227	51	8	11	19	43	....	1976-77	1987-88
Hill, Brian	Hfd.	1	19	1	1	2	4	....	....	....	....	....	....	1979-80	1979-80
• Hill, Mel	Bos., Bro., Tor.	9	324	89	109	198	128	43	12	7	19	18	3	1937-38	1945-46
Hiller, Dutch	NYR, Det., Bos., Mtl.	9	383	91	113	204	163	48	9	8	17	21	2	1937-38	1945-46
‡ Hiller, Jim	L.A., Det., NYR	2	63	8	12	20	116	2	0	0	0	4	....	1992-93	1993-94
Hillier, Randy	Bos., Pit., NYI, Buf.	11	543	16	110	126	906	28	0	2	2	93	1	1981-82	1991-92
Hillman, Floyd	Bos.	1	6	0	0	0	10	....	....	....	....	....	....	1956-57	1956-57
Hillman, Larry	Det., Bos., Tor., Min., Mtl., Phi., L.A., Buf.	19	790	36	196	232	579	74	2	9	11	30	6	1954-55	1972-73
• Hillman, Wayne	Chi., NYR, Min., Phi.	13	691	18	86	104	534	28	0	3	3	19	1	1960-61	1972-73
Hilworth, John	Det.	3	57	1	1	2	89	....	....	....	....	....	....	1977-78	1979-80
• Himes, Normie	NYA	9	402	106	113	219	127	2	0	0	0	0	....	1926-27	1934-35
Hindmarch, Dave	Cgy.	4	99	21	17	38	25	10	0	0	0	6	....	1980-81	1983-84
Hinse, Andre	Tor.	1	4	0	0	0	0	....	....	....	....	....	....	1967-68	1967-68
Hinton, Dan	Chi.	1	14	0	0	0	16	....	....	....	....	....	....	1976-77	1976-77
Hirsch, Tom	Min.	3	31	1	7	8	30	12	0	0	0	6	....	1983-84	1987-88
• Hirschfeld, Bert	Mtl.	2	33	1	4	5	2	5	1	0	1	0	....	1949-50	1950-51
Hislop, Jamie	Que., Cgy.	5	345	75	103	178	86	28	3	2	5	11	....	1979-80	1983-84
• Hitchman, Lionel	Ott., Bos.	12	417	28	34	62	523	35	3	1	4	73	2	1922-23	1933-34
Hlinka, Ivan	Van.	2	137	42	81	123	28	16	3	10	13	8	....	1981-82	1982-83
‡ Hlushko, Todd	Phi., Cgy., Pit.	6	79	8	13	21	84	3	0	0	0	2	....	1993-94	1998-99
‡ Hocking, Justin	L.A.	1	1	0	0	0	0	....	....	....	....	....	....	1993-94	1993-94
Hodge, Ken	Chi., Bos., NYR	14	881	328	472	800	779	97	34	47	81	120	2	1964-65	1977-78
Hodge, Ken	Min., Bos., T.B.	4	142	39	48	87	32	15	4	6	10	6	....	1988-89	1992-93
‡ Hodgson, Dan	Tor., Van.	4	114	29	45	74	64	....	....	....	....	....	....	1985-86	1988-89
Hodgson, Rick	Hfd.	1	6	0	0	0	6	1	0	0	0	0	....	1979-80	1979-80
Hodgson, Ted	Bos.	1	4	0	0	0	0	....	....	....	....	....	....	1966-67	1966-67
Hoekstra, Cec	Mtl.	1	4	0	0	0	0	....	....	....	....	....	....	1959-60	1959-60
Hoekstra, Ed	Phi.	1	70	15	21	36	6	7	0	1	1	0	....	1967-68	1967-68
Hoene, Phil	L.A.	3	37	2	4	6	22	....	....	....	....	....	....	1972-73	1974-75
Hoffinger, Val	Chi.	2	28	0	1	1	30	....	....	....	....	....	....	1927-28	1928-29
Hoffman, Mike	Hfd.	3	9	1	3	4	2	....	....	....	....	....	....	1982-83	1985-86
Hoffmeyer, Bob	Chi., Phi., N.J.	6	198	14	52	66	325	3	0	1	1	25	....	1977-78	1984-85
Hofford, Jim	Buf., L.A.	3	18	0	0	0	47	....	....	....	....	....	....	1985-86	1988-89
Hogaboam, Bill	Atl., Det., Min.	8	332	80	109	189	100	2	0	0	0	0	....	1972-73	1979-80
Hoganson, Dale	L.A., Mtl., Que.	7	343	13	77	90	186	11	0	3	3	12	....	1969-70	1981-82
Holan, Milos	Phi., Ana.	3	49	5	11	16	42	....	....	....	....	....	....	1993-94	1995-96
Holbrook, Terry	Min.	2	43	3	6	9	4	6	0	0	0	0	....	1972-73	1973-74
Holland, Jerry	NYR	2	37	8	4	12	6	....	....	....	....	....	....	1974-75	1975-76
• Hollett, Flash	Tor., Ott., Bos., Det.	13	562	132	181	313	358	79	8	26	34	38	2	1933-34	1945-46
‡ Hollinger, Terry	St.L.	2	7	0	0	0	2	....	....	....	....	....	....	1993-94	1994-95
• Hollingworth, Gord	Chi., Det.	4	163	4	14	18	201	3	0	0	0	2	....	1954-55	1957-58
Holloway, Bruce	Van.	1	2	0	0	0	0	....	....	....	....	....	....	1984-85	1984-85
• Holmes, Bill	Mtl., NYA	3	52	6	4	10	35	....	....	....	....	....	....	1925-26	1929-30
Holmes, Chuck	Det.	2	23	1	3	4	10	....	....	....	....	....	....	1958-59	1961-62
Holmes, Lou	Chi.	2	59	1	4	5	6	2	0	0	0	2	....	1931-32	1932-33
Holmes, Warren	L.A.	3	45	8	18	26	7	....	....	....	....	....	....	1981-82	1983-84
Holmgren, Paul	Phi., Min.	10	527	144	179	323	1684	82	19	32	51	195	....	1975-76	1984-85
• Holota, John	Det.	2	15	2	0	2	0	....	....	....	....	....	....	1942-43	1945-46
Holst, Greg	NYR	3	11	0	0	0	0	....	....	....	....	....	....	1975-76	1977-78
Holt, Gary	Cal., Cle., St.L.	5	101	13	11	24	133	....	....	....	....	....	....	1973-74	1977-78
Holt, Randy	Chi., Cle., Van., L.A., Cgy., Wsh., Phi.	10	395	4	37	41	1438	21	2	3	5	83	....	1974-75	1983-84
• Holway, Albert	Tor., Mtl.M., Pit.	5	112	7	2	9	48	6	0	0	0	0	1	1923-24	1928-29
Homenuke, Ron	Van.	1	1	0	0	0	0	....	....	....	....	....	....	1972-73	1972-73
Hoover, Ron	Bos., St.L.	3	18	4	0	4	31	8	0	0	0	18	....	1989-90	1991-92
Hopkins, Dean	L.A., Edm., Que.	6	223	23	51	74	306	18	1	5	6	29	....	1979-80	1988-89
Hopkins, Larry	Tor., Wpg.	4	60	13	16	29	26	6	0	0	0	2	....	1977-78	1982-83
Horacek, Tony	Phi., Chi.	5	154	10	19	29	316	2	1	0	1	2	....	1989-90	1994-95
Horava, Miloslav	NYR	3	80	5	17	22	38	2	0	1	1	0	....	1988-89	1990-91
Horbul, Doug	K.C.	1	4	1	0	1	2	....	....	....	....	....	....	1974-75	1974-75
Hordy, Mike	NYI	2	11	0	0	0	7	....	....	....	....	....	....	1978-79	1979-80
Horeck, Pete	Chi., Det., Bos.	8	426	106	118	224	340	34	6	8	14	43	....	1944-45	1951-52
• Horne, George	Mtl.M., Tor.	3	54	9	3	12	34	4	0	0	0	4	1	1925-26	1928-29
Horner, Red	Tor.	12	490	42	110	152	1254	71	7	10	17	170	1	1928-29	1939-40
• Hornung, Larry	St.L.	2	48	2	9	11	10	11	0	2	2	2	....	1970-71	1971-72
• Horton, Tim	Tor., NYR, Pit., Buf.	24	1446	115	403	518	1611	126	11	39	50	183	4	1949-50	1973-74
Horvath, Bronco	NYR, Mtl., Bos., Chi., Tor., Min.	9	434	141	185	326	319	36	12	9	21	18	....	1955-56	1967-68
Hospodar, Ed	NYR, Hfd., Phi., Min., Buf.	9	450	17	51	68	1314	44	4	1	5	208	....	1979-80	1987-88
‡ Hostak, Martin	Phi.	2	55	3	11	14	24	....	....	....	....	....	....	1990-91	1991-92
Hotham, Greg	Tor., Pit.	6	230	15	74	89	139	5	0	3	3	6	....	1979-80	1984-85
Houck, Paul	Min.	3	16	1	2	3	2	....	....	....	....	....	....	1985-86	1987-88
Houde, Claude	K.C.	2	59	3	6	9	40	....	....	....	....	....	....	1974-75	1975-76
‡ Houde, Eric	Mtl.	3	30	2	3	5	4	....	....	....	....	....	....	1996-97	1998-99
‡ Hough, Mike	Que., Fla., NYI	13	707	100	156	256	675	42	5	5	10	38	....	1986-87	1998-99
Houle, Rejean	Mtl.	11	635	161	247	408	395	90	14	34	48	66	5	1969-70	1982-83
Houston, Ken	Atl., Cgy., Wsh., L.A.	9	570	161	167	328	624	35	10	9	19	66	....	1975-76	1983-84
Howard, Jack	Tor.	1	2	0	0	0	0	....	....	....	....	....	....	1936-37	1936-37
Howatt, Garry	NYI, Hfd., N.J.	12	720	112	156	268	1836	87	12	14	26	289	2	1972-73	1983-84
Howe, Gordie	Det., Hfd.	26	1767	801	1049	1850	1685	157	68	92	160	220	4	1946-47	1979-80
Howe, Mark	Hfd., Phi., Det.	16	929	197	545	742	455	101	10	51	61	34	....	1979-80	1994-95
Howe, Marty	Hfd., Bos.	6	197	2	29	31	99	15	1	2	3	9	....	1979-80	1984-85
• Howe, Syd	Ott., Phi., Tor., St.L., Det.	17	698	237	291	528	212	70	17	27	44	10	3	1929-30	1945-46
Howe, Vic	NYR	3	33	3	4	7	10	....	....	....	....	....	....	1950-51	1954-55
Howell, Harry	NYR, Oak., Cal., L.A.	21	1411	94	324	418	1298	38	3	3	6	32	....	1952-53	1972-73
• Howell, Ron	NYR	2	4	0	0	0	0	....	....	....	....	....	....	1954-55	1955-56
Howse, Don	L.A.	1	33	2	5	7	6	2	0	0	0	0	....	1979-80	1979-80
Howson, Scott	NYI	2	18	5	3	8	4	....	....	....	....	....	....	1984-85	1985-86
Hoyda, Dave	Phi., Wpg.	4	132	6	17	23	299	12	0	0	0	17	....	1977-78	1980-81
Hrdina, Jiri	Cgy., Pit.	5	250	45	85	130	92	46	2	5	7	24	3	1987-88	1991-92
Hrechkosy, Dave	Cal., St.L.	4	140	42	24	66	41	3	1	0	1	2	....	1973-74	1976-77
Hrycuik, Jim	Wsh.	1	21	5	5	10	12	....	....	....	....	....	....	1974-75	1974-75
Hrymnak, Steve	Chi., Det.	2	18	2	1	3	4	2	0	0	0	0	....	1951-52	1952-53
Hrynewich, Tim	Pit.	2	55	6	8	14	82	....	....	....	....	....	....	1982-83	1983-84
Huard, Bill	Bos., Ott., Que., Dal., Edm., L.A.	8	223	16	18	34	594	5	0	0	0	2	....	1992-93	1999-00
Huard, Rolly	Tor.	1	1	1	0	1	0	....	....	....	....	....	....	1930-31	1930-31
Huber, Willie	Det., NYR, Van., Phi.	10	655	104	217	321	950	33	5	5	10	35	....	1978-79	1987-88
Hubick, Greg	Tor., Van.	2	77	6	9	15	10	....	....	....	....	....	....	1975-76	1979-80
Huck, Fran	Mtl., St.L.	3	94	24	30	54	38	11	3	4	7	2	....	1969-70	1972-73
Hucul, Fred	Chi., St.L.	5	164	11	30	41	113	6	1	0	1	10	....	1950-51	1967-68
Huddy, Charlie	Edm., L.A., Buf., St.L.	17	1017	99	354	453	785	183	19	66	85	135	5	1980-81	1996-97
Hudson, Dave	NYI, K.C., Col.	6	409	59	124	183	89	2	1	1	2	0	....	1972-73	1977-78
Hudson, Lex	Pit.	1	2	0	0	0	0	2	0	0	0	0	....	1978-79	1978-79
Hudson, Mike	Chi., Edm., NYR, Pit., Tor., St.L., Phx.	9	416	49	87	136	414	49	4	10	14	64	1	1988-89	1996-97
Hudson, Ron	Det.	2	33	5	2	7	2	....	....	....	....	....	....	1937-38	1939-40
Huffman, Kerry	Phi., Que., Ott.	10	401	37	108	145	361	11	0	0	0	2	....	1986-87	1995-96
Huggins, Al	Mtl.M.	1	20	1	1	2	2	....	....	....	....	....	....	1930-31	1930-31
Hughes, Albert	NYA	2	60	6	8	14	22	....	....	....	....	....	....	1930-31	1931-32

Name	NHL Teams	NHL Seasons	Regular Schedule					Playoffs					NHL Cup Wins	First NHL Season	Last NHL Season
			GP	G	A	TP	PIM	GP	G	A	TP	PIM			
Hughes, Brent	L.A., Phi., St.L., Det., K.C.	8	435	15	117	132	440	22	1	3	4	53	....	1967-68	1974-75
Hughes, Brent	Wpg., Bos., Buf., NYI	8	357	41	39	80	831	29	4	1	5	53	....	1988-89	1996-97
Hughes, Frank	Cal.	1	5	0	0	0	0	....	....	....	....	....	....	1971-72	1971-72
Hughes, Howie	L.A.	3	168	25	32	57	30	14	2	0	2	2	....	1967-68	1969-70
Hughes, Jack	Col.	2	46	2	5	7	104	....	....	....	....	....	....	1980-81	1981-82
Hughes, James	Det.	1	40	0	1	1	48	....	....	....	....	....	....	1929-30	1929-30
Hughes, John	Van., Edm., NYR	2	70	2	14	16	211	7	0	1	1	16	....	1979-80	1980-81
Hughes, Pat	Mtl., Pit., Edm., Buf., St.L., Hfd.	10	573	130	128	258	646	71	8	25	33	77	3	1977-78	1986-87
Hughes, Ryan	Bos.	1	3	0	0	0	0	....	....	....	....	....	....	1995-96	1995-96
Hull, Bobby	Chi., Wpg., Hfd.	16	1063	610	560	1170	640	119	62	67	129	102	1	1957-58	1979-80
Hull, Dennis	Chi., Det.	14	959	303	351	654	261	104	33	34	67	30	....	1964-65	1977-78
• Hunt, Fred	NYA, NYR	2	59	15	14	29	6	....	....	....	....	....	....	1940-41	1944-45
Hunter, Dale	Que., Wsh., Col.	19	1407	323	697	1020	3565	186	42	76	118	729	....	1980-81	1998-99
Hunter, Dave	Edm., Pit., Wpg.	10	746	133	190	323	918	105	16	24	40	211	3	1979-80	1988-89
Hunter, Mark	Mtl., St.L., Cgy., Hfd., Wsh.	12	628	213	171	384	1426	79	18	20	38	230	1	1981-82	1992-93
Hunter, Tim	Cgy., Que., Van., S.J.	16	815	62	76	138	3146	132	5	7	12	426	1	1981-82	1996-97
‡ Huras, Larry	NYR	1	2	0	0	0	0	....	....	....	....	....	....	1976-77	1976-77
Hurlburt, Bob	Van.	1	1	0	0	0	2	....	....	....	....	....	....	1974-75	1974-75
Hurley, Paul	Bos.	1	1	0	1	1	0	....	....	....	....	....	....	1968-69	1968-69
Hurst, Ron	Tor.	2	64	9	7	16	70	3	0	2	2	4	....	1955-56	1956-57
Huscroft, Jamie	N.J., Bos., Cgy., T.B., Van., Phx., Wsh.	10	352	5	33	38	1065	21	0	1	1	46	....	1988-89	1999-00
Huska, Ryan	Chi.	1	1	0	0	0	0	....	....	....	....	....	....	1997-98	1997-98
Huston, Ron	Cal.	2	79	15	31	46	8	....	....	....	....	....	....	1973-74	1974-75
Hutchinson, Ron	NYR	1	9	0	0	0	0	....	....	....	....	....	....	1960-61	1960-61
Hutchison, Dave	L.A., Tor., Chi., N.J.	10	584	19	97	116	1550	48	2	12	14	149	....	1974-75	1983-84
• Hutton, Bill	Bos., Ott., Phi.	2	64	3	2	5	8	2	0	0	0	0	....	1929-30	1930-31
• Hyland, Harry	Mtl.W., Ott.	1	17	14	2	16	65	....	....	....	....	....	....	1917-18	1917-18
Hynes, Dave	Bos.	2	22	4	0	4	2	....	....	....	....	....	....	1973-74	1974-75
‡ Hynes, Gord	Bos., Phi.	2	52	3	9	12	22	12	1	2	3	6	....	1991-92	1992-93

I

Name	NHL Teams	NHL Seasons	GP	G	A	TP	PIM	GP	G	A	TP	PIM	NHL Cup Wins	First NHL Season	Last NHL Season
Iafrate, Al	Tor., Wsh., Bos., S.J.	12	799	152	311	463	1301	71	19	16	35	77	....	1984-85	1997-98
‡ Ihnacak, Miroslav	Tor., Det.	3	56	8	9	17	39	1	0	0	0	0	....	1985-86	1988-89
Ihnacak, Peter	Tor.	8	417	102	165	267	175	28	4	10	14	25	....	1982-83	1989-90
Imlach, Brent	Tor.	2	3	0	0	0	0	....	....	....	....	....	....	1965-66	1966-67
Ingarfield, Earl	NYR, Pit., Oak., Cal.	13	746	179	226	405	239	21	9	8	17	10	....	1958-59	1970-71
Ingarfield, Earl Jr.	Atl., Cgy., Det.	2	39	4	4	8	22	2	0	1	1	0	....	1979-80	1980-81
Inglis, Billy	L.A., Buf.	3	36	1	3	4	4	11	1	2	3	4	....	1967-68	1970-71
• Ingoldsby, Johnny	Tor.	2	29	5	1	6	15	....	....	....	....	....	....	1942-43	1943-44
Ingram, Frank	Chi.	3	101	24	16	40	69	11	0	1	1	2	....	1929-30	1931-32
• Ingram, John J.	Bos.	1	1	0	0	0	0	....	....	....	....	....	....	1924-25	1924-25
Ingram, Ron	Chi., Det., NYR	4	114	5	15	20	81	2	0	0	0	0	....	1956-57	1964-65
‡ Intranuovo, Ralph	Edm., Tor.	3	22	2	4	6	4	....	....	....	....	....	....	1994-95	1996-97
• Irvin, Dick	Chi.	3	94	29	23	52	78	2	2	0	2	4	....	1926-27	1928-29
Irvine, Ted	Bos., L.A., NYR, St.L.	11	724	154	177	331	657	83	16	24	40	115	....	1963-64	1976-77
Irwin, Ivan	Mtl., NYR	5	155	2	27	29	214	5	0	0	0	8	....	1952-53	1957-58
Isaksson, Ulf	L.A.	1	50	7	15	22	10	....	....	....	....	....	....	1982-83	1982-83
Issel, Kim	Edm.	1	4	0	0	0	0	....	....	....	....	....	....	1988-89	1988-89

J

Name	NHL Teams	NHL Seasons	GP	G	A	TP	PIM	GP	G	A	TP	PIM	NHL Cup Wins	First NHL Season	Last NHL Season
• Jackson, Art	Tor., Bos., NYA	11	468	123	178	301	144	52	8	12	20	29	2	1934-35	1944-45
• Jackson, Busher	Tor., NYA, Bos.	15	633	241	234	475	437	71	18	12	30	53	1	1929-30	1943-44
Jackson, Don	Min., Edm., NYR	10	311	16	52	68	640	53	4	5	9	147	2	1977-78	1986-87
• Jackson, Harold	Chi., Det.	8	219	17	34	51	208	31	1	2	3	33	2	1936-37	1946-47
Jackson, Jack	Chi.	1	48	2	5	7	38	....	....	....	....	....	....	1946-47	1946-47
Jackson, Jeff	Tor., NYR, Que., Chi.	8	263	38	48	86	313	6	1	1	2	16	....	1984-85	1991-92
Jackson, Jim	Cgy., Buf.	4	112	17	30	47	20	14	3	2	5	6	....	1982-83	1987-88
• Jackson, Lloyd	NYA	1	14	1	1	2	0	....	....	....	....	....	....	1936-37	1936-37
• Jackson, Stan	Tor., Bos., Ott.	5	86	9	6	15	75	....	....	....	....	....	1	1921-22	1926-27
Jackson, Walter	NYA, Bos.	4	84	16	11	27	18	....	....	....	....	....	....	1932-33	1935-36
• Jacobs, Paul	Tor.	1	1	0	0	0	0	....	....	....	....	....	....	1918-19	1918-19
Jacobs, Tim	Cal.	1	46	0	10	10	35	....	....	....	....	....	....	1975-76	1975-76
Jalo, Risto	Edm.	1	3	0	3	3	0	....	....	....	....	....	....	1985-86	1985-86
Jalonen, Kari	Cgy., Edm.	2	37	9	6	15	4	5	1	0	1	0	....	1982-83	1983-84
James, Gerry	Tor.	5	149	14	26	40	257	15	1	0	1	8	....	1954-55	1959-60
James, Val	Buf., Tor.	2	11	0	0	0	30	....	....	....	....	....	....	1981-82	1986-87
Jamieson, Jim	NYR	1	1	0	1	1	0	....	....	....	....	....	....	1943-44	1943-44
Jankowski, Lou	Det., Chi.	4	127	19	18	37	15	1	0	0	0	0	....	1950-51	1954-55
Janney, Craig	Bos., St.L., S.J., Wpg., Phx., T.B., NYI	12	760	188	563	751	170	120	24	86	110	53	....	1987-88	1998-99
Janssens, Mark	NYR, Min., Hfd., Ana., NYI, Phx., Chi.	14	711	40	73	113	1422	27	5	1	6	33	....	1987-88	2000-01
‡ Jantunen, Marko	Cgy.	1	3	0	0	0	0	....	....	....	....	....	....	1996-97	1996-97
Jarrett, Doug	Chi., NYR	13	775	38	182	220	631	99	7	16	23	82	....	1964-65	1976-77
Jarrett, Gary	Tor., Det., Oak., Cal.	7	341	72	92	164	131	11	3	1	4	9	....	1960-61	1971-72
Jarry, Pierre	NYR, Tor., Det., Min.	7	344	88	117	205	142	5	0	1	1	0	....	1971-72	1977-78
Jarvenpaa, Hannu	Wpg.	3	114	11	26	37	83	....	....	....	....	....	....	1986-87	1988-89
Jarvi, Iiro	Que.	2	116	18	43	61	58	....	....	....	....	....	....	1988-89	1989-90
Jarvis, Doug	Mtl., Wsh., Hfd.	13	964	139	264	403	263	105	14	27	41	42	4	1975-76	1987-88
• Jarvis, James	Pit., Phi., Tor.	3	112	17	15	32	62	....	....	....	....	....	....	1929-30	1936-37
Jarvis, Wes	Wsh., Min., L.A., Tor.	9	237	31	55	86	98	2	0	0	0	2	....	1979-80	1987-88
Javanainen, Arto	Pit.	1	14	4	1	5	2	....	....	....	....	....	....	1984-85	1984-85
‡ Jay, Bob	L.A.	1	3	0	1	1	0	....	....	....	....	....	....	1993-94	1993-94
Jeffrey, Larry	Det., Tor., NYR	8	368	39	62	101	293	38	4	10	14	42	1	1961-62	1968-69
Jelinek, Tomas	Ott.	1	49	7	6	13	52	....	....	....	....	....	....	1992-93	1992-93
Jenkins, Dean	L.A.	1	5	0	0	0	2	....	....	....	....	....	....	1983-84	1983-84
Jenkins, Roger	Chi., Tor., Mtl., Bos., Mtl.M., NYA	8	325	15	39	54	253	25	1	7	8	12	2	1930-31	1938-39
Jennings, Bill	Det., Bos.	5	108	32	33	65	45	20	4	4	8	6	....	1940-41	1944-45
Jennings, Grant	Wsh., Hfd., Pit., Tor., Buf.	9	389	14	43	57	804	54	2	1	3	68	2	1987-88	1995-96
Jensen, Chris	NYR, Phi.	6	74	9	12	21	27	....	....	....	....	....	....	1985-86	1991-92
Jensen, David	Min.	3	18	0	2	2	11	....	....	....	....	....	....	1983-84	1985-86
Jensen, David	Hfd., Wsh.	4	69	9	13	22	22	11	0	0	0	2	....	1984-85	1987-88
Jensen, Steve	Min., L.A.	7	438	113	107	220	318	12	0	3	3	9	....	1975-76	1981-82
• Jeremiah, Ed	NYA, Bos.	1	15	0	1	1	0	....	....	....	....	....	....	1931-32	1931-32
Jerrard, Paul	Min.	1	5	0	0	0	4	....	....	....	....	....	....	1988-89	1988-89
• Jerwa, Frank	Bos., St.L.	4	81	11	16	27	53	....	....	....	....	....	....	1931-32	1934-35
• Jerwa, Joe	NYR, Bos., NYA	7	234	29	58	87	309	17	2	3	5	16	....	1930-31	1938-39
Jirik, Jaroslav	St.L.	1	3	0	0	0	0	....	....	....	....	....	....	1969-70	1969-70
• Joanette, Rosario	Mtl.	1	2	0	1	1	4	....	....	....	....	....	....	1944-45	1944-45
Jodzio, Rick	Col., Cle.	1	70	2	8	10	71	....	....	....	....	....	....	1977-78	1977-78
Johannesen, Glenn	NYI	1	2	0	0	0	0	....	....	....	....	....	....	1985-86	1985-86
Johannson, John	N.J.	1	5	0	0	0	0	....	....	....	....	....	....	1983-84	1983-84
Johansen, Bill	Tor.	1	1	0	0	0	0	....	....	....	....	....	....	1949-50	1949-50
Johansen, Trevor	Tor., Col., L.A.	5	286	11	46	57	282	13	0	3	3	21	....	1977-78	1981-82
Johansson, Bjorn	Cle.	2	15	1	1	2	10	....	....	....	....	....	....	1976-77	1977-78
‡ Johansson, Roger	Cgy., Chi.	4	161	9	34	43	163	5	0	1	1	2	....	1989-90	1994-95
Johns, Don	NYR, Mtl., Min.	6	153	2	21	23	76	....	....	....	....	....	....	1960-61	1967-68
Johnson, Al	Mtl., Det.	4	105	21	28	49	30	11	2	2	4	6	....	1956-57	1962-63
Johnson, Brian	Det.	1	3	0	0	0	5	....	....	....	....	....	....	1983-84	1983-84
• Johnson, Ching	NYR, NYA	12	436	38	48	86	808	61	5	2	7	161	2	1926-27	1937-38
• Johnson, Danny	Tor., Van., Det.	3	121	18	19	37	24	....	....	....	....	....	....	1969-70	1971-72
Johnson, Earl	Det.	1	1	0	0	0	0	....	....	....	....	....	....	1953-54	1953-54
Johnson, Jim	NYR, Phi., L.A.	8	302	75	111	186	73	7	0	2	2	2	....	1964-65	1971-72
Johnson, Jim	Pit., Min., Dal., Wsh., Phx.	13	829	29	166	195	1197	51	1	11	12	132	....	1985-86	1997-98
Johnson, Mark	Pit., Min., Hfd., St.L., N.J.	11	669	203	305	508	260	37	16	12	28	10	....	1979-80	1989-90
Johnson, Norm	Bos., Chi.	3	61	5	20	25	41	14	4	0	4	6	....	1957-58	1959-60
Johnson, Terry	Que., St.L., Cgy., Tor.	9	285	3	24	27	580	38	0	4	4	118	....	1979-80	1987-88
Johnson, Tom	Mtl., Bos.	17	978	51	213	264	960	111	8	15	23	109	6	1947-48	1964-65
• Johnson, Virgil	Chi.	3	75	1	11	12	27	19	0	3	3	4	1	1937-38	1944-45
Johnston, Bernie	Hfd.	2	57	12	24	36	16	3	0	1	1	0	....	1979-80	1980-81
Johnston, George	Chi.	4	58	20	12	32	2	....	....	....	....	....	....	1941-42	1946-47
‡ Johnston, Greg	Bos., Tor.	9	187	26	29	55	124	22	2	1	3	12	....	1983-84	1991-92
Johnston, Jay	Wsh.	2	8	0	0	0	13	....	....	....	....	....	....	1980-81	1981-82
Johnston, Joey	Min., Cal., Chi.	6	331	85	106	191	320	....	....	....	....	....	....	1968-69	1975-76
Johnston, Larry	L.A., Det., K.C., Col.	7	320	9	64	73	580	....	....	....	....	....	....	1967-68	1976-77
Johnston, Marshall	Min., Cal.	7	251	14	52	66	58	6	0	0	0	2	....	1967-68	1973-74

Bob Hoffmeyer

Randy Holt

Jiri Hrdina

Tim Hunter

Kim Issel

Doug Jarvis

Bob Jay

Steve Jensen

			Regular Schedule					Playoffs							
Name	NHL Teams	NHL Seasons	GP	G	A	TP	PIM	GP	G	A	TP	PIM	NHL Cup Wins	First NHL Season	Last NHL Season
Johnston, Randy	NYI	1	4	0	0	0	4	....	....	....	....	....	....	1979-80	1979-80
Johnstone, Eddie	NYR, Det.	10	426	122	136	258	375	55	13	10	23	83	....	1975-76	1986-87
Johnstone, Ross	Tor.	2	42	5	4	9	14	3	0	0	0	0	1	1943-44	1944-45
• Joliat, Aurel	Mtl.	16	655	270	190	460	771	46	9	13	22	66	3	1922-23	1937-38
• Joliat, Rene	Mtl.	1	1	0	0	0	0	....	....	....	....	....	....	1924-25	1924-25
Joly, Greg	Wsh., Det.	9	365	21	76	97	250	5	0	0	0	8	....	1974-75	1982-83
Joly, Yvan	Mtl.	3	2	0	0	0	0	1	0	0	0	0	....	1979-80	1982-83
‡ Jomphe, Jean-Francois	Ana., Phx., Mtl.	4	111	10	29	39	102	....	....	....	....	....	....	1995-96	1998-99
Jonathan, Stan	Bos., Pit.	8	411	91	110	201	751	63	8	4	12	137	....	1975-76	1982-83
Jones, Bob	NYR	1	2	0	0	0	0	....	....	....	....	....	....	1968-69	1968-69
Jones, Brad	Wpg., L.A., Phi.	6	148	25	31	56	122	9	1	1	2	2	....	1986-87	1991-92
Jones, Buck	Det., Tor.	4	50	2	2	4	36	12	0	1	1	18	....	1938-39	1942-43
Jones, Jim	Cal.	1	2	0	0	0	0	....	....	....	....	....	....	1971-72	1971-72
Jones, Jimmy	Tor.	3	148	13	18	31	68	19	1	5	6	11	....	1977-78	1979-80
Jones, Keith	Wsh., Col., Phi.	9	491	117	141	258	765	63	12	12	24	120	....	1992-93	2000-01
Jones, Ron	Bos., Pit., Wsh.	5	54	1	4	5	31	....	....	....	....	....	....	1971-72	1975-76
Jonsson, Tomas	NYI, Edm.	8	552	85	259	344	482	80	11	26	37	97	2	1981-82	1988-89
Joseph, Tony	Wpg.	1	2	1	0	1	0	....	....	....	....	....	....	1988-89	1988-89
Joyal, Eddie	Det., Tor., L.A., Phi.	9	466	128	134	262	103	50	11	8	19	18	....	1962-63	1971-72
Joyce, Bob	Bos., Wsh., Wpg.	6	158	34	49	83	90	46	15	9	24	29	....	1987-88	1992-93
Joyce, Duane	Dal.	1	3	0	0	0	0	....	....	....	....	....	....	1993-94	1993-94
• Juckes, Bing	NYR	2	16	2	1	3	6	....	....	....	....	....	....	1947-48	1949-50
‡ Juhlin, Patrik	Phi.	2	56	7	6	13	23	13	1	0	1	2	....	1994-95	1995-96
Julien, Claude	Que.	2	14	0	1	1	25	....	....	....	....	....	....	1984-85	1985-86
‡ Junker, Steve	NYI	2	5	0	0	0	0	3	0	1	1	0	....	1992-93	1993-94
Jutila, Timo	Buf.	1	10	1	5	6	13	....	....	....	....	....	....	1984-85	1984-85
Juzda, Bill	NYR, Tor.	9	398	14	54	68	398	42	0	3	3	46	2	1940-41	1951-52

K

Name	NHL Teams	NHL Seasons	GP	G	A	TP	PIM	GP	G	A	TP	PIM	NHL Cup Wins	First NHL Season	Last NHL Season
Kabel, Bob	NYR	2	48	5	13	18	34	....	....	....	....	....	....	1959-60	1960-61
Kachowski, Mark	Pit.	3	64	6	5	11	209	....	....	....	....	....	....	1987-88	1989-90
Kachur, Ed	Chi.	2	96	10	14	24	35	....	....	....	....	....	....	1956-57	1957-58
Kaese, Trent	Buf.	1	1	0	0	0	0	....	....	....	....	....	....	1988-89	1988-89
Kaiser, Vern	Mtl.	1	50	7	5	12	33	2	0	0	0	0	....	1950-51	1950-51
• Kalbfleish, Walter	Ott., St.L., NYA, Bos.	4	36	0	4	4	32	5	0	0	0	2	....	1933-34	1936-37
• Kaleta, Alex	Chi., NYR	7	387	92	121	213	190	17	1	6	7	2	....	1941-42	1950-51
Kallur, Anders	NYI	6	383	101	110	211	149	78	12	23	35	32	4	1979-80	1984-85
Kaminski, Kevin	Min., Que., Wsh.	7	139	3	10	13	528	8	0	0	0	52	....	1988-89	1996-97
• Kaminsky, Max	Ott., Bos., St.L., Mtl.M.	4	130	22	34	56	38	4	0	0	0	0	....	1933-34	1936-37
Kaminsky, Yan	Wpg., NYI	2	26	3	2	5	4	2	0	0	0	4	....	1993-94	1994-95
• Kampman, Bingo	Tor.	5	189	14	30	44	287	47	1	4	5	38	1	1937-38	1941-42
Kane, Francis	Det.	1	2	0	0	0	0	....	....	....	....	....	....	1943-44	1943-44
Kannegiesser, Gord	St.L.	2	23	0	1	1	15	....	....	....	....	....	....	1967-68	1971-72
Kannegiesser, Sheldon	Pit., NYR, L.A., Van.	8	366	14	67	81	292	18	0	2	2	10	....	1970-71	1977-78
‡ Karabin, Ladislav	Pit.	1	9	0	0	0	2	....	....	....	....	....	....	1993-94	1993-94
‡ Karamnov, Vitali	St.L.	3	92	12	20	32	65	2	0	0	0	2	....	1992-93	1994-95
‡ Karjalainen, Kyosti	L.A.	1	28	1	8	9	12	3	0	1	1	2	....	1991-92	1991-92
Karlander, Al	Det.	4	212	36	56	92	70	4	0	1	1	0	....	1969-70	1972-73
‡ Karpov, Valeri	Ana.	3	76	14	15	29	32	....	....	....	....	....	....	1994-95	1996-97
Kasatonov, Alexei	N.J., Ana., St.L., Bos.	7	383	38	122	160	326	33	4	7	11	40	....	1989-90	1995-96
Kasper, Steve	Bos., L.A., Phi., T.B.	13	821	177	291	468	554	94	20	28	48	82	....	1980-81	1992-93
Kastelic, Ed	Wsh., Hfd.	7	220	11	10	21	719	8	1	0	1	32	....	1985-86	1991-92
Kaszycki, Mike	NYI, Wsh., Tor.	5	226	42	80	122	108	19	2	6	8	10	....	1977-78	1982-83
• Kea, Ed	Atl., St.L.	10	583	30	145	175	508	32	2	4	6	39	....	1973-74	1982-83
Kearns, Dennis	Van.	10	677	31	290	321	386	11	1	2	3	8	....	1971-72	1980-81
• Keating, Jack	Det.	2	11	3	0	3	4	....	....	....	....	....	....	1938-39	1939-40
• Keating, John	NYA	2	35	5	5	10	17	....	....	....	....	....	....	1931-32	1932-33
Keating, Mike	NYR	1	1	0	0	0	0	....	....	....	....	....	....	1977-78	1977-78
• Keats, Duke	Bos., Det., Chi.	3	82	30	19	49	113	....	....	....	....	....	....	1926-27	1928-29
Keczmer, Dan	Min., Hfd., Cgy., Dal., Nsh.	10	235	8	38	46	212	12	0	1	1	8	....	1990-91	1999-00
• Keeling, Butch	Tor., NYR	12	525	157	63	220	331	47	11	11	22	34	1	1926-27	1937-38
Keenan, Larry	Tor., St.L., Buf., Phi.	6	233	38	64	102	28	46	15	16	31	12	....	1961-62	1971-72
Kehoe, Rick	Tor., Pit.	14	906	371	396	767	120	39	4	17	21	4	....	1971-72	1984-85
Kekalainen, Jarmo	Bos., Ott.	3	55	5	8	13	28	....	....	....	....	....	....	1989-90	1993-94
Keller, Ralph	NYR	1	3	1	0	1	6	....	....	....	....	....	....	1962-63	1962-63
Kellgren, Christer	Col.	1	5	0	0	0	0	....	....	....	....	....	....	1981-82	1981-82
Kelly, Bob	Phi., Wsh.	12	837	154	208	362	1454	101	9	14	23	172	2	1970-71	1981-82
Kelly, Bob	St.L., Pit., Chi.	6	425	87	109	196	687	23	6	3	9	40	....	1973-74	1978-79
Kelly, Dave	Det.	1	16	2	0	2	4	....	....	....	....	....	....	1976-77	1976-77
Kelly, John Paul	L.A.	7	400	54	70	124	366	18	1	1	2	41	....	1979-80	1985-86
• Kelly, Pep	Tor., Chi., Bro.	8	288	74	53	127	105	38	7	6	13	10	....	1934-35	1941-42
Kelly, Pete	St.L., Det., NYA, Bro.	7	177	21	38	59	68	19	3	1	4	2	2	1934-35	1941-42
Kelly, Red	Det., Tor.	20	1316	281	542	823	327	164	33	59	92	51	8	1947-48	1966-67
• Kemp, Kevin	Hfd.	1	3	0	0	0	4	....	....	....	....	....	....	1980-81	1980-81
Kemp, Stan	Tor.	1	1	0	0	0	2	....	....	....	....	....	....	1948-49	1948-49
• Kendall, Bill	Chi., Tor.	5	131	16	10	26	28	6	0	0	0	0	1	1933-34	1937-38
Kennedy, Dean	L.A., NYR, Buf., Wpg., Edm.	12	717	26	108	134	1118	36	1	7	8	59	....	1982-83	1994-95
Kennedy, Forbes	Chi., Det., Bos., Phi., Tor.	11	603	70	108	178	988	12	2	4	6	64	....	1956-57	1968-69
‡ Kennedy, Mike	Dal., Tor., NYI	5	145	16	36	52	112	5	0	0	0	9	....	1994-95	1998-99
Kennedy, Sheldon	Det., Cgy., Bos.	8	310	49	58	107	233	24	6	4	10	20	....	1989-90	1996-97
Kennedy, Ted	Tor.	14	696	231	329	560	432	78	29	31	60	32	5	1942-43	1956-57
• Kenny, Ernest	NYR, Chi.	2	10	0	0	0	18	....	....	....	....	....	....	1930-31	1934-35
Keon, Dave	Tor., Hfd.	18	1296	396	590	986	117	92	32	36	68	6	4	1960-61	1981-82
‡ Kerch, Alexander	Edm.	1	5	0	0	0	2	....	....	....	....	....	....	1993-94	1993-94
Kerr, Alan	NYI, Det., Wpg.	9	391	72	94	166	826	38	5	4	9	70	....	1984-85	1992-93
Kerr, Reg	Cle., Chi., Edm.	6	263	66	94	160	169	7	1	0	1	7	....	1977-78	1983-84
Kerr, Tim	Phi., NYR, Hfd.	13	655	370	304	674	596	81	40	31	71	58	....	1980-81	1992-93
‡ Kesa, Dan	Van., Dal., Pit., T.B.	4	139	8	22	30	66	13	1	0	1	0	....	1993-94	1999-00
Kessell, Rick	Pit., Cal.	5	135	4	24	28	6	....	....	....	....	....	....	1969-70	1973-74
Ketola, Veli-Pekka	Col.	1	44	9	5	14	4	....	....	....	....	....	....	1981-82	1981-82
Ketter, Kerry	Atl.	1	41	0	2	2	58	....	....	....	....	....	....	1972-73	1972-73
‡ Kharin, Sergei	Wpg.	1	7	2	3	5	2	....	....	....	....	....	....	1990-91	1990-91
Khmylev, Yuri	Buf., St.L.	5	263	64	88	152	133	26	8	6	14	24	....	1992-93	1996-97
Kidd, Ian	Van.	2	20	4	7	11	25	....	....	....	....	....	....	1987-88	1988-89
Kiessling, Udo	Min.	1	1	0	0	0	2	....	....	....	....	....	....	1981-82	1981-82
Kilrea, Brian	Det., L.A.	2	26	3	5	8	12	....	....	....	....	....	....	1957-58	1967-68
• Kilrea, Hec	Ott., Det., Tor.	15	633	167	129	296	438	48	8	7	15	18	3	1925-26	1939-40
• Kilrea, Ken	Det.	5	91	16	23	39	8	15	2	2	4	4	....	1938-39	1943-44
• Kilrea, Wally	Ott., Phi., NYA, Mtl.M., Det.	9	329	35	58	93	87	25	2	4	6	6	2	1929-30	1937-38
‡ Kimble, Darin	Que., St.L., Bos., Chi.	7	311	23	20	43	1082	23	0	0	0	52	....	1988-89	1994-95
Kindrachuk, Orest	Phi., Pit., Wsh.	10	508	118	261	379	648	76	20	20	40	53	2	1972-73	1981-82
King, Frank	Mtl.	1	10	1	0	1	2	....	....	....	....	....	....	1950-51	1950-51
‡ King, Kris	Det., NYR, Wpg., Phx., Tor., Chi.	14	849	66	85	151	2030	67	8	5	13	142	....	1987-88	2000-01
King, Steven	NYR, Ana.	3	67	17	8	25	75	....	....	....	....	....	....	1992-93	1995-96
King, Wayne	Cal.	3	73	5	18	23	34	....	....	....	....	....	....	1973-74	1975-76
Kinnear, Geordie	Atl.	1	4	0	0	0	13	....	....	....	....	....	....	1999-00	1999-00
Kinsella, Brian	Wsh.	2	10	0	1	1	0	....	....	....	....	....	....	1975-76	1976-77
• Kinsella, Ray	Ott.	1	14	0	0	0	0	....	....	....	....	....	....	1930-31	1930-31
• Kirk, Bobby	NYR	1	39	4	8	12	14	....	....	....	....	....	....	1937-38	1937-38
Kirkpatrick, Bob	NYR	1	49	12	12	24	6	....	....	....	....	....	....	1942-43	1942-43
Kirton, Mark	Tor., Det., Van.	6	266	57	56	113	121	4	1	2	3	7	....	1979-80	1984-85
Kisio, Kelly	Det., NYR, S.J., Cgy.	13	761	229	429	658	768	39	6	15	21	52	....	1982-83	1994-95
Kitchen, Bill	Mtl., Tor.	4	41	1	4	5	40	3	0	1	1	0	....	1981-82	1984-85
• Kitchen, Hobie	Mtl.M., Det.	2	47	5	4	9	58	....	....	....	....	....	1	1925-26	1926-27
Kitchen, Mike	Col., N.J.	8	474	12	62	74	370	2	0	0	0	2	....	1976-77	1983-84
Klassen, Ralph	Cal., Cle., Col., St.L.	9	497	52	93	145	120	26	4	2	6	12	....	1975-76	1983-84
• Klein, Lloyd	Bos., NYA	8	164	30	24	54	68	5	0	0	0	2	1	1928-29	1937-38
Kleinendorst, Scot	NYR, Hfd., Wsh.	8	281	12	46	58	452	26	2	7	9	40	....	1982-83	1989-90
‡ Klima, Petr	Det., Edm., T.B., L.A., Pit.	13	786	313	260	573	671	95	28	24	52	83	1	1985-86	1998-99
‡ Klimovich, Sergei	Chi.	1	1	0	0	0	2	....	....	....	....	....	....	1996-97	1996-97
Klingbeil, Ike	Chi.	1	5	1	2	3	2	....	....	....	....	....	....	1936-37	1936-37
Klukay, Joe	Tor., Bos.	11	566	109	127	236	189	71	13	10	23	23	4	1942-43	1955-56
Kluzak, Gord	Bos.	7	299	25	98	123	543	46	6	13	19	129	....	1982-83	1990-91
Knibbs, Bill	Bos.	1	53	7	10	17	4	....	....	....	....	....	....	1964-65	1964-65
Knipscheer, Fred	Bos., St.L.	3	28	6	3	9	18	16	2	1	3	6	....	1993-94	1995-96

Name	NHL Teams	NHL Seasons	Regular Schedule GP	G	A	TP	PIM	Playoffs GP	G	A	TP	PIM	NHL Cup Wins	First NHL Season	Last NHL Season
• Knott, Nick	Bro.	1	14	3	1	4	9	....	....	....	....	....	....	1941-42	1941-42
Knox, Paul	Tor.	1	1	0	0	0	0	....	....	....	....	....	....	1954-55	1954-55
Kocur, Joe	Det., NYR, Van.	15	820	80	82	162	2519	118	10	12	22	231	3	1984-85	1998-99
‡ Kolesar, Mark	Tor.	2	28	2	2	4	14	3	1	0	1	2	....	1995-96	1996-97
Kolstad, Dean	Min., S.J.	3	40	1	7	8	69	....	....	....	....	....	....	1988-89	1992-93
Komadoski, Neil	L.A., St.L.	8	502	16	76	92	632	23	0	2	2	47	....	1972-73	1979-80
Konik, George	Pit.	1	52	7	8	15	26	....	....	....	....	....	....	1967-68	1967-68
Konroyd, Steve	Cgy., NYI, Chi., Hfd., Det., Ott.	15	895	41	195	236	863	97	10	15	25	99	....	1980-81	1994-95
Konstantinov, Vladimir	Det.	6	446	47	128	175	838	82	5	14	19	107	1	1991-92	1996-97
Kontos, Chris	NYR, Pit., L.A., T.B.	8	230	54	69	123	103	20	11	0	11	12	....	1982-83	1992-93
• Kopak, Russ	Bos.	1	24	7	9	16	0	....	....	....	....	....	....	1943-44	1943-44
Korab, Jerry	Chi., Van., Buf., L.A.	15	975	114	341	455	1629	93	8	18	26	201	....	1970-71	1984-85
Kordic, Dan	Phi.	6	197	4	8	12	584	12	1	0	1	22	....	1991-92	1998-99
• Kordic, John	Mtl., Tor., Wsh., Que.	7	244	17	18	35	997	41	4	3	7	131	1	1985-86	1991-92
Korn, Jim	Det., Tor., Buf., N.J., Cgy.	10	597	66	122	188	1801	16	1	2	3	109	....	1979-80	1989-90
Korney, Mike	Det., NYR	4	77	9	10	19	59	....	....	....	....	....	....	1973-74	1978-79
Koroll, Cliff	Chi.	11	814	208	254	462	376	85	19	29	48	67	....	1969-70	1979-80
Kortko, Roger	NYI	2	79	7	17	24	28	10	0	3	3	17	....	1984-85	1985-86
Kostynski, Doug	Bos.	2	15	3	1	4	4	....	....	....	....	....	....	1983-84	1984-85
Kotanen, Dick	NYR	1	1	0	0	0	0	....	....	....	....	....	....	1950-51	1950-51
Kotsopoulos, Chris	NYR, Hfd., Tor., Det.	10	479	44	109	153	827	31	1	3	4	91	....	1980-81	1989-90
Kowal, Joe	Buf.	2	22	0	5	5	13	2	0	0	0	0	....	1976-77	1977-78
Kozak, Don	L.A., Van.	7	437	96	86	182	480	29	7	2	9	69	....	1972-73	1978-79
Kozak, Les	Tor.	1	12	1	0	1	2	....	....	....	....	....	....	1961-62	1961-62
• Kraftcheck, Stephen	Bos., NYR, Tor.	4	157	11	18	29	83	6	0	0	0	7	....	1950-51	1958-59
Krake, Skip	Bos., L.A., Buf.	7	249	23	40	63	182	10	1	0	1	17	....	1963-64	1970-71
‡ Kravets, Mikhail	S.J.	2	2	0	0	0	0	....	....	....	....	....	....	1991-92	1992-93
Krentz, Dale	Det.	3	30	5	3	8	9	2	0	0	0	0	....	1986-87	1988-89
• Krol, Joe	NYR, Bro.	3	26	10	4	14	8	....	....	....	....	....	....	1936-37	1941-42
Kromm, Richard	Cgy., NYI	9	372	70	103	173	138	36	2	6	8	22	....	1983-84	1992-93
Krook, Kevin	Col.	1	3	0	0	0	2	....	....	....	....	....	....	1978-79	1978-79
‡ Kroupa, Vlastimil	S.J., N.J.	5	105	4	19	23	66	20	1	2	3	25	....	1993-94	1997-98
Krulicki, Jim	NYR, Det.	1	41	0	3	3	6	....	....	....	....	....	....	1970-71	1970-71
Kruppke, Gord	Det.	3	23	0	0	0	32	....	....	....	....	....	....	1990-91	1993-94
‡ Kruse, Paul	Cgy., NYI, Buf., S.J.	11	423	38	33	71	1074	28	5	2	7	36	....	1990-91	2000-01
Krushelnyski, Mike	Bos., Edm., L.A., Tor., Det.	14	897	241	328	569	699	139	29	43	72	106	3	1981-82	1994-95
Krutov, Vladimir	Van.	1	61	11	23	34	20	....	....	....	....	....	....	1989-90	1989-90
Krygier, Todd	Hfd., Wsh., Ana.	9	543	100	143	243	533	48	10	7	17	40	....	1989-90	1997-98
Kryskow, Dave	Chi., Wsh., Det., Atl.	4	231	33	56	89	174	12	2	0	2	4	....	1972-73	1975-76
Kryzanowski, Ed	Bos., Chi.	5	237	15	22	37	65	18	0	1	1	4	....	1948-49	1952-53
‡ Kudashov, Alexei	Tor.	1	25	1	0	1	4	....	....	....	....	....	....	1993-94	1993-94
Kudelski, Bob	L.A., Ott., Fla.	9	442	139	102	241	218	22	4	4	8	4	....	1987-88	1995-96
• Kuhn, Gord	NYA	1	12	1	1	2	4	....	....	....	....	....	....	1932-33	1932-33
Kukulowicz, Aggie	NYR	2	4	1	0	1	0	....	....	....	....	....	....	1952-53	1953-54
Kulak, Stu	Van., Edm., NYR, Que., Wpg.	4	90	8	4	12	130	3	0	0	0	2	....	1982-83	1988-89
• Kullman, Arnie	Bos.	2	13	0	1	1	11	....	....	....	....	....	....	1947-48	1949-50
• Kullman, Eddie	NYR	6	343	56	70	126	298	6	1	0	1	2	....	1947-48	1953-54
Kumpel, Mark	Que., Det., Wpg.	6	288	38	46	84	113	39	6	4	10	14	....	1984-85	1990-91
• Kuntz, Alan	NYR	2	45	10	12	22	12	6	1	0	1	2	....	1941-42	1945-46
Kuntz, Murray	St.L.	1	7	1	2	3	0	....	....	....	....	....	....	1974-75	1974-75
Kurri, Jari	Edm., L.A., NYR, Ana., Col.	17	1251	601	797	1398	545	200	106	127	233	123	5	1980-81	1997-98
Kurtenbach, Orland	NYR, Bos., Tor., Van.	13	639	119	213	332	628	19	2	4	6	70	....	1960-61	1973-74
Kurvers, Tom	Mtl., Buf., N.J., Tor., Van., NYI, Ana.	11	659	93	328	421	350	57	8	22	30	68	1	1984-85	1994-95
Kuryluk, Merv	Chi.	1	....	....	....	....	....	2	0	0	0	0	....	1961-62	1961-62
Kushner, Dale	NYI, Phi.	3	84	10	13	23	215	....	....	....	....	....	....	1989-90	1991-92
Kuzyk, Ken	Cle.	2	41	5	9	14	8	....	....	....	....	....	....	1976-77	1977-78
‡ Kvartalnov, Dmitri	Bos.	2	112	42	49	91	26	4	0	0	0	0	....	1992-93	1993-94
Kwong, Larry	NYR	1	1	0	0	0	0	....	....	....	....	....	....	1947-48	1947-48
• Kyle, Bill	NYR	2	3	0	3	3	0	....	....	....	....	....	....	1949-50	1950-51
• Kyle, Gus	NYR, Bos.	3	203	6	20	26	362	14	1	2	3	34	....	1949-50	1951-52
Kyllonen, Markku	Wpg.	1	9	0	2	2	2	....	....	....	....	....	....	1988-89	1988-89
Kypreos, Nick	Wsh., Hfd., NYR, Tor.	8	442	46	44	90	1210	34	1	3	4	65	1	1989-90	1996-97
Kyte, Jim	Wpg., Pit., Cgy., Ott., S.J.	13	598	17	49	66	1342	42	0	6	6	94	....	1982-83	1995-96

L

Name	NHL Teams	NHL Seasons	Regular Schedule GP	G	A	TP	PIM	Playoffs GP	G	A	TP	PIM	NHL Cup Wins	First NHL Season	Last NHL Season
Labadie, Mike	NYR	1	3	0	0	0	0	....	....	....	....	....	....	1952-53	1952-53
Labatte, Neil	St.L.	2	26	0	2	2	19	....	....	....	....	....	....	1978-79	1981-82
L'Abbe, Moe	Chi.	1	5	0	1	1	0	....	....	....	....	....	....	1972-73	1972-73
Labelle, Marc	Dal.	1	9	0	0	0	46	....	....	....	....	....	....	1996-97	1996-97
Labine, Leo	Bos., Det.	11	643	128	193	321	730	60	12	11	23	82	....	1951-52	1961-62
Labossiere, Gord	NYR, L.A., Min.	6	215	44	62	106	75	10	2	3	5	28	....	1963-64	1971-72
Labovitch, Max	NYR	1	5	0	0	0	4	....	....	....	....	....	....	1943-44	1943-44
Labraaten, Dan	Det., Cgy.	4	268	71	73	144	47	8	1	0	1	4	....	1978-79	1981-82
Labre, Yvon	Pit., Wsh.	9	371	14	87	101	788	....	....	....	....	....	....	1970-71	1980-81
Labrie, Guy	Bos., NYR	2	42	4	9	13	16	....	....	....	....	....	....	1943-44	1944-45
Lach, Elmer	Mtl.	14	664	215	408	623	478	76	19	45	64	36	3	1940-41	1953-54
Lachance, Michel	Col.	1	21	0	4	4	22	....	....	....	....	....	....	1978-79	1978-79
Lacombe, Francois	Oak., Buf., Que.	4	78	2	17	19	54	3	1	0	1	0	....	1968-69	1979-80
Lacombe, Normand	Buf., Edm., Phi.	7	319	53	62	115	196	26	5	1	6	49	1	1984-85	1990-91
Lacroix, Andre	Phi., Chi., Hfd.	6	325	79	119	198	44	16	2	5	7	0	....	1967-68	1979-80
‡ Lacroix, Daniel	NYR, Bos., Phi., Edm., NYI	7	188	11	7	18	379	16	0	1	1	26	....	1993-94	1999-00
Lacroix, Eric	Tor., L.A., Col., NYR, Ott.	8	472	67	70	137	361	30	1	5	6	25	....	1993-94	2000-01
Lacroix, Pierre	Que., Hfd.	4	274	24	108	132	197	8	0	2	2	10	....	1979-80	1982-83
Ladouceur, Randy	Det., Hfd., Ana.	14	930	30	126	156	1322	40	5	8	13	59	....	1982-83	1995-96
LaFayette, Nathan	St.L., Van., NYR, L.A.	6	187	17	20	37	103	32	2	7	9	8	....	1993-94	1998-99
Lafleur, Guy	Mtl., NYR, Que.	17	1126	560	793	1353	399	128	58	76	134	67	5	1971-72	1990-91
• Lafleur, Roland	Mtl.	1	1	0	0	0	0	....	....	....	....	....	....	1924-25	1924-25
LaFontaine, Pat	NYI, Buf., NYR	15	865	468	545	1013	552	69	26	36	62	36	....	1983-84	1997-98
Laforce, Ernie	Mtl.	1	1	0	0	0	0	....	....	....	....	....	....	1942-43	1942-43
LaForest, Bob	L.A.	1	5	1	0	1	2	....	....	....	....	....	....	1983-84	1983-84
Laforge, Claude	Mtl., Det., Phi.	8	193	24	33	57	82	5	1	2	3	15	....	1957-58	1968-69
‡ Laforge, Marc	Hfd., Edm.	2	14	0	0	0	64	....	....	....	....	....	....	1989-90	1993-94
Laframboise, Pete	Cal., Wsh., Pit.	4	227	33	55	88	70	9	1	0	1	0	....	1971-72	1974-75
Lafrance, Adie	Mtl.	1	3	0	0	0	2	2	0	0	0	0	....	1933-34	1933-34
• Lafrance, Leo	Mtl., Chi.	2	33	2	0	2	6	....	....	....	....	....	....	1926-27	1927-28
‡ Lafreniere, Jason	Que., NYR, T.B.	5	146	34	53	87	22	15	1	5	6	19	....	1986-87	1993-94
Lafreniere, Roger	Det., St.L.	2	13	0	0	0	4	....	....	....	....	....	....	1962-63	1972-73
Lagace, Jean-Guy	Pit., Buf., K.C.	6	197	9	39	48	251	....	....	....	....	....	....	1968-69	1975-76
Laidlaw, Tom	NYR, L.A.	10	705	25	139	164	717	69	4	17	21	78	....	1980-81	1989-90
Laird, Robbie	Min.	1	1	0	0	0	0	....	....	....	....	....	....	1979-80	1979-80
Lajeunesse, Serge	Det., Phi.	5	103	1	4	5	103	....	....	....	....	....	....	1970-71	1974-75
Lalande, Hec	Chi., Det.	4	151	21	39	60	120	....	....	....	....	....	....	1953-54	1957-58
Lalonde, Bobby	Van., Atl., Bos., Cgy.	11	641	124	210	334	298	16	4	2	6	6	....	1971-72	1981-82
• Lalonde, Newsy	Mtl., NYA	6	99	124	41	165	183	7	15	4	19	23	....	1917-18	1926-27
Lalonde, Ron	Pit., Wsh.	7	397	45	78	123	106	....	....	....	....	....	....	1972-73	1978-79
Lalor, Mike	Mtl., St.L., Wsh., Wpg., S.J., Dal.	12	687	17	88	105	677	92	5	10	15	167	1	1985-86	1996-97
• Lamb, Joe	Mtl.M., Ott., NYA, Bos., Mtl., St.L., Det.	11	443	108	101	209	601	18	1	1	2	51	....	1927-28	1937-38
Lamb, Mark	Cgy., Det., Edm., Ott., Phi., Mtl.	11	403	46	100	146	291	70	7	19	26	51	1	1985-86	1995-96
‡ Lambert, Dan	Que.	2	29	6	9	15	22	....	....	....	....	....	....	1990-91	1991-92
‡ Lambert, Lane	Det., NYR, Que.	6	283	58	66	124	521	17	2	4	6	40	....	1983-84	1988-89
Lambert, Yvon	Mtl., Buf.	10	683	206	273	479	340	90	27	22	49	67	4	1972-73	1981-82
Lamby, Dick	St.L.	3	22	0	5	5	22	....	....	....	....	....	....	1978-79	1980-81
• Lamirande, Jean-Paul	NYR, Mtl.	4	49	5	5	10	26	8	0	0	0	4	....	1946-47	1954-55
Lammens, Hank	Ott.	1	27	1	2	3	22	....	....	....	....	....	....	1993-94	1993-94
• Lamoureux, Leo	Mtl.	6	235	19	79	98	175	28	1	6	7	16	2	1941-42	1946-47
Lamoureux, Mitch	Pit., Phi.	3	73	11	9	20	59	....	....	....	....	....	....	1983-84	1987-88
Lampman, Mike	St.L., Van., Wsh.	4	96	17	20	37	34	....	....	....	....	....	....	1972-73	1976-77
Lancien, Jack	NYR	4	63	1	5	6	35	6	0	1	1	2	....	1946-47	1950-51
Landon, Larry	Mtl., Tor.	2	9	0	0	0	2	....	....	....	....	....	....	1983-84	1984-85
Lane, Gord	Wsh., NYI	10	539	19	94	113	1228	75	3	14	17	214	4	1975-76	1984-85
• Lane, Myles	NYR, Bos.	3	71	4	1	5	41	11	0	0	0	0	1	1928-29	1933-34
Langdon, Steve	Bos.	3	7	0	1	1	2	4	0	0	0	0	....	1974-75	1977-78
Langelle, Pete	Tor.	4	136	22	51	73	11	41	5	9	14	4	1	1938-39	1941-42
Langevin, Chris	Buf.	2	22	3	1	4	22	....	....	....	....	....	....	1983-84	1985-86
Langevin, Dave	NYI, Min., L.A.	8	513	12	107	119	530	87	2	17	19	106	4	1979-80	1986-87

Jarmo Kekalainen

Red Kelly

Forbes Kennedy

Kris King

Mark Kumpel

Steve Larmer

Pierre Larouche

Reggie Leach

			Regular Schedule					Playoffs							
Name	NHL Teams	NHL Seasons	GP	G	A	TP	PIM	GP	G	A	TP	PIM	NHL Cup Wins	First NHL Season	Last NHL Season
Langlais, Alain	Min.	2	25	4	4	8	10	....	....	....	....	....	....	1973-74	1974-75
Langlois, Albert	Mtl., NYR, Det., Bos.	9	497	21	91	112	488	53	1	5	6	50	3	1957-58	1965-66
• Langlois, Charlie	Ham., NYA, Pit., Mtl.	4	151	22	5	27	189	2	0	0	0	0	....	1924-25	1927-28
Langway, Rod	Mtl., Wsh.	15	994	51	278	329	849	104	5	22	27	97	1	1978-79	1992-93
Lank, Jeff	Phi.	1	2	0	0	0	2	....	....	....	....	....	....	1999-00	1999-00
Lanthier, Jean-Marc	Van.	4	105	16	16	32	29	....	....	....	....	....	....	1983-84	1987-88
Lanyon, Ted	Pit.	1	5	0	0	0	4	....	....	....	....	....	....	1967-68	1967-68
Lanz, Rick	Van., Tor., Chi.	10	569	65	221	286	448	28	3	8	11	35	....	1980-81	1991-92
Laperriere, Daniel	St.L., Ott.	4	48	2	5	7	27	....	....	....	....	....	....	1992-93	1995-96
Laperriere, Jacques	Mtl.	12	691	40	242	282	674	88	9	22	31	101	6	1962-63	1973-74
Lapointe, Guy	Mtl., St.L., Bos.	16	884	171	451	622	893	123	26	44	70	138	6	1968-69	1983-84
• Lapointe, Rick	Det., Phi., St.L., Que., L.A.	11	664	44	176	220	831	46	2	7	9	64	....	1975-76	1985-86
Lappin, Peter	Min., S.J.	2	7	0	0	0	2	....	....	....	....	....	....	1989-90	1991-92
Laprade, Edgar	NYR	10	500	108	172	280	42	18	4	9	13	4	....	1945-46	1954-55
LaPrairie, Benjamin	Chi.	1	7	0	0	0	0	....	....	....	....	....	....	1936-37	1936-37
Lariviere, Garry	Que., Edm.	4	219	6	57	63	167	14	0	5	5	8	....	1979-80	1982-83
Larmer, Jeff	Col., N.J., Chi.	5	158	37	51	88	57	5	1	0	1	2	....	1981-82	1985-86
Larmer, Steve	Chi., NYR	15	1006	441	571	1012	532	140	56	75	131	89	1	1980-81	1994-95
• Larochelle, Wildor	Mtl., Chi.	12	474	92	74	166	211	34	6	4	10	24	2	1925-26	1936-37
Larocque, Denis	L.A.	1	8	0	1	1	18	....	....	....	....	....	....	1987-88	1987-88
• Larose, Bonner	Bos.	1	6	0	0	0	0	....	....	....	....	....	....	1925-26	1925-26
Larose, Claude	Mtl., Min., St.L.	16	943	226	257	483	887	97	14	18	32	143	5	1962-63	1977-78
Larose, Claude	NYR	2	25	4	7	11	2	2	0	0	0	0	....	1979-80	1981-82
‡ Larose, Guy	Wpg., Tor., Cgy., Bos.	6	70	10	9	19	63	4	0	0	0	0	....	1988-89	1994-95
Larouche, Pierre	Pit., Mtl., Hfd., NYR	14	812	395	427	822	237	64	20	34	54	16	2	1974-75	1987-88
‡ Larouche, Steve	Ott., NYR, L.A.	2	26	9	9	18	10	....	....	....	....	....	....	1994-95	1995-96
• Larson, Norm	NYA, Bro., NYR	3	89	25	18	43	12	....	....	....	....	....	....	1940-41	1946-47
Larson, Reed	Det., Bos., Edm., NYI, Min., Buf.	14	904	222	463	685	1391	32	4	7	11	63	....	1976-77	1989-90
Larter, Tyler	Wsh.	1	1	0	0	0	0	....	....	....	....	....	....	1989-90	1989-90
Latal, Jiri	Phi.	3	92	12	36	48	24	....	....	....	....	....	....	1989-90	1991-92
Latos, James	NYR	1	1	0	0	0	0	....	....	....	....	....	....	1988-89	1988-89
Latreille, Phil	NYR	1	4	0	0	0	2	....	....	....	....	....	....	1960-61	1960-61
Latta, David	Que.	4	36	4	8	12	4	....	....	....	....	....	....	1985-86	1990-91
Lauder, Martin	Bos.	1	3	0	0	0	2	....	....	....	....	....	....	1927-28	1927-28
Lauen, Mike	Wpg.	1	4	0	1	1	0	....	....	....	....	....	....	1983-84	1983-84
‡ Lauer, Brad	NYI, Chi., Ott., Pit.	9	323	44	67	111	218	34	7	5	12	24	....	1986-87	1995-96
Laughlin, Craig	Mtl., Wsh., L.A., Tor.	8	549	136	205	341	364	33	6	6	12	20	....	1981-82	1988-89
Laughton, Mike	Oak., Cal.	4	189	39	48	87	101	11	2	4	6	0	....	1967-68	1970-71
Laurence, Don	Atl., St.L.	2	79	15	22	37	14	....	....	....	....	....	....	1978-79	1979-80
LaVallee, Kevin	Cgy., L.A., St.L., Pit.	7	366	110	125	235	85	32	5	8	13	21	....	1980-81	1986-87
LaVarre, Mark	Chi.	3	78	9	16	25	58	1	0	0	0	2	....	1985-86	1987-88
Lavender, Brian	St.L., NYI, Det., Cal.	4	184	16	26	42	174	3	0	0	0	2	....	1971-72	1974-75
‡ Lavigne, Eric	L.A.	1	1	0	0	0	0	....	....	....	....	....	....	1994-95	1994-95
• Laviolette, Jack	Mtl.	1	18	2	1	3	6	2	0	0	0	0	....	1917-18	1917-18
Laviolette, Peter	NYR	1	12	0	0	0	6	....	....	....	....	....	....	1988-89	1988-89
‡ Lavoie, Dominic	St.L., Ott., Bos., L.A.	6	38	5	8	13	32	....	....	....	....	....	....	1988-89	1993-94
Lawless, Paul	Hfd., Phi., Van., Tor.	7	239	49	77	126	54	3	0	2	2	2	....	1982-83	1989-90
‡ Lawrence, Mark	Dal., NYI	6	142	18	26	44	115	....	....	....	....	....	....	1994-95	2000-01
Lawson, Danny	Det., Min., Buf.	5	219	28	29	57	61	16	0	1	1	2	....	1967-68	1971-72
Lawton, Brian	Min., NYR, Hfd., Que., Bos., S.J.	9	483	112	154	266	401	11	1	1	2	12	....	1983-84	1992-93
‡ Laxdal, Derek	Tor., NYI	6	67	12	7	19	88	1	0	2	2	2	....	1984-85	1990-91
• Laycoe, Hal	NYR, Mtl., Bos.	11	531	25	77	102	292	40	2	5	7	39	....	1945-46	1955-56
‡ Lazaro, Jeff	Bos., Ott.	3	102	14	23	37	114	28	3	3	6	32	....	1990-91	1992-93
‡ Leach, Jamie	Pit., Hfd., Fla.	5	81	11	9	20	12	....	....	....	....	....	1	1989-90	1993-94
Leach, Larry	Bos.	3	126	13	29	42	91	7	1	1	2	8	....	1958-59	1961-62
Leach, Reggie	Bos., Cal., Phi., Det.	13	934	381	285	666	387	94	47	22	69	22	1	1970-71	1982-83
Leach, Stephen	Wsh., Bos., St.L., Car., Ott., Phx., Pit.	15	702	130	153	283	978	92	15	11	26	87	....	1985-86	1999-00
Leavins, Jim	Det., NYR	2	41	2	12	14	30	....	....	....	....	....	....	1985-86	1986-87
‡ Lebeau, Patrick	Mtl., Cgy., Fla., Pit.	4	15	3	2	5	6	....	....	....	....	....	....	1990-91	1998-99
‡ Lebeau, Stephan	Mtl., Ana.	7	373	118	159	277	105	30	9	7	16	12	1	1988-89	1994-95
LeBlanc, Fern	Det.	3	34	5	6	11	0	....	....	....	....	....	....	1976-77	1978-79
LeBlanc, J.P.	Chi., Det.	5	153	14	30	44	87	2	0	0	0	0	....	1968-69	1978-79
‡ LeBlanc, John	Van., Edm., Wpg.	7	83	26	13	39	28	1	0	0	0	0	....	1986-87	1994-95
‡ LeBoutillier, Peter	Ana.	2	35	2	1	3	176	....	....	....	....	....	....	1996-97	1997-98
LeBrun, Al	NYR	2	6	0	2	2	4	....	....	....	....	....	....	1960-61	1965-66
Lecaine, Bill	Pit.	1	4	0	0	0	0	....	....	....	....	....	....	1968-69	1968-69
Leclair, Jackie	Mtl.	3	160	20	40	60	56	20	6	1	7	6	2	1954-55	1956-57
Leclerc, Rene	Det.	2	87	10	11	21	105	....	....	....	....	....	....	1968-69	1970-71
Lecuyer, Doug	Chi., Wpg., Pit.	4	126	11	31	42	178	7	4	0	4	15	....	1978-79	1982-83
Ledingham, Walt	Chi., NYI	3	15	0	2	2	4	....	....	....	....	....	....	1972-73	1976-77
• Leduc, Albert	Mtl., Ott., NYR	10	383	57	35	92	614	28	5	6	11	32	2	1925-26	1934-35
LeDuc, Rich	Bos., Que.	4	130	28	38	66	69	5	0	0	0	9	....	1972-73	1980-81
• Lee, Bobby	Mtl.	1	1	0	0	0	0	....	....	....	....	....	....	1942-43	1942-43
Lee, Edward	Que.	1	2	0	0	0	5	....	....	....	....	....	....	1984-85	1984-85
Lee, Peter	Pit.	6	431	114	131	245	257	19	0	8	8	4	....	1977-78	1982-83
Leeman, Gary	Tor., Cgy., Mtl., Van., St.L.	14	667	199	267	466	531	36	8	16	24	36	1	1982-83	1996-97
‡ Lefebvre, Patrice	Wsh.	1	3	0	0	0	2	....	....	....	....	....	....	1998-99	1998-99
• Lefley, Bryan	NYI, K.C., Col.	5	228	7	29	36	101	2	0	0	0	0	....	1972-73	1977-78
Lefley, Chuck	Mtl., St.L.	9	407	128	164	292	137	29	5	8	13	10	2	1970-71	1980-81
• Leger, Roger	NYR, Mtl.	5	187	18	53	71	71	20	0	7	7	14	....	1943-44	1949-50
Legge, Barry	Que., Wpg.	3	107	1	11	12	144	....	....	....	....	....	....	1979-80	1981-82
Legge, Randy	NYR	1	12	0	2	2	2	....	....	....	....	....	....	1972-73	1972-73
Lehman, Tommy	Bos., Edm.	3	36	5	5	10	16	....	....	....	....	....	....	1987-88	1989-90
Lehto, Petteri	Pit.	1	6	0	0	0	4	....	....	....	....	....	....	1984-85	1984-85
Lehtonen, Antero	Wsh.	1	65	9	12	21	14	....	....	....	....	....	....	1979-80	1979-80
Lehvonen, Henri	K.C.	1	4	0	0	0	0	....	....	....	....	....	....	1974-75	1974-75
Leier, Edward	Chi.	2	16	2	1	3	2	....	....	....	....	....	....	1949-50	1950-51
Leinonen, Mikko	NYR, Wsh.	4	162	31	78	109	71	20	2	11	13	28	....	1981-82	1984-85
Leiter, Bobby	Bos., Pit., Atl.	10	447	98	126	224	144	8	3	0	3	2	....	1962-63	1975-76
Leiter, Ken	NYI, Min.	5	143	14	36	50	62	15	0	6	6	8	....	1984-85	1989-90
Lemaire, Jacques	Mtl.	12	853	366	469	835	217	145	61	78	139	63	8	1967-68	1978-79
Lemay, Moe	Van., Edm., Bos., Wpg.	8	317	72	94	166	442	28	6	3	9	55	1	1981-82	1988-89
Lemelin, Roger	K.C., Col.	4	36	1	2	3	27	....	....	....	....	....	....	1974-75	1977-78
‡ Lemieux, Alain	St.L., Que., Pit.	6	119	28	44	72	38	19	4	6	10	0	....	1981-82	1986-87
Lemieux, Bob	Oak.	1	19	0	1	1	12	....	....	....	....	....	....	1967-68	1967-68
Lemieux, Jacques	L.A.	3	19	0	4	4	8	1	0	0	0	0	....	1967-68	1969-70
Lemieux, Jean	Atl., Wsh.	5	204	23	63	86	39	3	1	1	2	0	....	1973-74	1977-78
Lemieux, Jocelyn	St.L., Mtl., Chi., Hfd., N.J., Cgy., Phx.	12	598	80	84	164	740	60	5	10	15	88	....	1986-87	1997-98
• Lemieux, Real	Det., L.A., NYR, Buf.	8	456	51	104	155	262	18	2	4	6	10	....	1966-67	1973-74
Lemieux, Rich	Van., K.C., Atl.	5	274	39	82	121	132	2	0	0	0	0	....	1971-72	1975-76
Lenardon, Tim	N.J., Van.	2	15	2	1	3	4	....	....	....	....	....	....	1986-87	1989-90
• Lepine, Hec	Mtl.	1	33	5	2	7	2	....	....	....	....	....	....	1925-26	1925-26
• Lepine, Pit	Mtl.	13	526	143	98	241	392	41	7	5	12	26	2	1925-26	1937-38
Leroux, Gaston	Mtl.	1	2	0	0	0	0	....	....	....	....	....	....	1935-36	1935-36
Lesieur, Art	Mtl., Chi.	4	100	4	2	6	50	14	0	0	0	4	1	1928-29	1935-36
Lessard, Rick	Cgy., S.J.	3	15	0	4	4	18	....	....	....	....	....	....	1988-89	1991-92
Lesuk, Bill	Bos., Phi., L.A., Wsh., Wpg.	8	388	44	63	107	368	9	1	0	1	12	1	1968-69	1979-80
• Leswick, Jack	Chi.	1	37	1	7	8	16	....	....	....	....	....	1	1933-34	1933-34
Leswick, Pete	NYA, Bos.	2	3	1	0	1	0	....	....	....	....	....	....	1936-37	1944-45
• Leswick, Tony	NYR, Det., Chi.	12	740	165	159	324	900	59	13	10	23	91	3	1945-46	1957-58
Levandoski, Joe	NYR	1	8	1	1	2	0	....	....	....	....	....	....	1946-47	1946-47
Leveille, Normand	Bos.	2	75	17	25	42	49	....	....	....	....	....	....	1981-82	1982-83
Leveque, Guy	L.A.	2	17	2	2	4	21	....	....	....	....	....	....	1992-93	1993-94
Lever, Don	Van., Atl., Cgy., Col., N.J., Buf.	15	1020	313	367	680	593	30	7	10	17	26	....	1972-73	1986-87
Levie, Craig	Wpg., Min., St.L., Van.	6	183	22	53	75	177	16	2	3	5	32	....	1981-82	1986-87
‡ Levins, Scott	Wpg., Fla., Ott., Phx.	5	124	13	20	33	316	....	....	....	....	....	....	1992-93	1997-98
• Levinsky, Alex	Tor., NYR, Chi.	9	367	19	49	68	307	37	2	1	3	26	2	1930-31	1938-39
Levo, Tapio	Col., N.J.	2	107	16	53	69	36	....	....	....	....	....	....	1981-82	1982-83
Lewicki, Danny	Tor., NYR, Chi.	9	461	105	135	240	177	28	0	4	4	8	1	1950-51	1958-59
Lewis, Dale	NYR	1	8	0	0	0	0	....	....	....	....	....	....	1975-76	1975-76
Lewis, Dave	NYI, L.A., N.J., Det.	15	1008	36	187	223	953	91	1	20	21	143	....	1973-74	1987-88
• Lewis, Doug	Mtl.	1	3	0	0	0	0	....	....	....	....	....	....	1946-47	1946-47
• Lewis, Herbie	Det.	11	483	148	161	309	248	38	13	10	23	6	2	1928-29	1938-39
Ley, Rick	Tor., Hfd.	6	310	12	72	84	528	14	0	2	2	20	....	1968-69	1980-81
Liba, Igor	NYR, L.A.	1	37	7	18	25	36	2	0	0	0	2	....	1988-89	1988-89
Libby, Jeff	NYI	1	1	0	0	0	0	....	....	....	....	....	....	1997-98	1997-98

	Name	NHL Teams	NHL Seasons	Regular Schedule GP	G	A	TP	PIM	Playoffs GP	G	A	TP	PIM	NHL Cup Wins	First NHL Season	Last NHL Season
	Libett, Nick	Det., Pit.	14	982	237	268	505	472	16	6	2	8	2	....	1967-68	1980-81
	Licari, Tony	Det.	1	9	0	1	1	0	....	....	....	....	....	....	1946-47	1946-47
	Liddington, Bob	Tor.	1	11	0	1	1	2	....	....	....	....	....	....	1970-71	1970-71
	Lidster, Doug	Van., NYR, St.L., Dal.	16	897	75	268	343	679	80	6	15	21	64	1	1983-84	1998-99
‡	Lilley, John	Ana.	3	23	3	8	11	13	....	....	....	....	....	....	1993-94	1995-96
‡	Lindberg, Chris	Cgy., Que.	3	116	17	25	42	47	2	0	1	1	2	....	1991-92	1993-94
‡	Lindbom, Johan	NYR	1	38	1	3	4	28	....	....	....	....	....	....	1997-98	1997-98
	Linden, Jamie	Fla.	1	4	0	0	0	17	....	....	....	....	....	....	1994-95	1994-95
	Lindgren, Lars	Van., Min.	6	394	25	113	138	325	40	5	6	11	20	....	1978-79	1983-84
‡	Lindholm, Mikael	L.A.	1	18	2	2	4	2	....	....	....	....	....	....	1989-90	1989-90
‡	Lindquist, Fredrik	Edm.	1	8	0	0	0	2	....	....	....	....	....	....	1998-99	1998-99
	Lindros, Brett	NYI	2	51	2	5	7	147	....	....	....	....	....	....	1994-95	1995-96
	Lindsay, Ted	Det., Chi.	17	1068	379	472	851	1808	133	47	49	96	194	4	1944-45	1964-65
	Lindstrom, Willy	Wpg., Edm., Pit.	8	582	161	162	323	200	57	14	18	32	24	2	1979-80	1986-87
	Linseman, Ken	Phi., Edm., Bos., Tor.	14	860	256	551	807	1727	113	43	77	120	325	1	1978-79	1991-92
‡	Lipuma, Chris	T.B., S.J.	5	72	0	9	9	146	....	....	....	....	....	....	1992-93	1996-97
	Liscombe, Carl	Det.	9	373	137	140	277	117	59	22	19	41	20	1	1937-38	1945-46
	Litzenberger, Ed	Mtl., Chi., Det., Tor.	12	618	178	238	416	283	40	5	13	18	34	4	1952-53	1963-64
‡	Loach, Lonnie	Ott., L.A., Ana.	2	56	10	13	23	29	1	0	0	0	0	....	1992-93	1993-94
●	Locas, Jacques	Mtl.	2	59	7	8	15	66	....	....	....	....	....	....	1947-48	1948-49
‡	Lochead, Bill	Det., Col., NYR	6	330	69	62	131	180	7	3	0	3	6	....	1974-75	1979-80
●	Locking, Norm	Chi.	2	48	2	6	8	26	....	....	....	....	....	....	1934-35	1935-36
‡	Loewen, Darcy	Buf., Ott.	5	135	4	8	12	211	....	....	....	....	....	....	1989-90	1993-94
	Lofthouse, Mark	Wsh., Det.	6	181	42	38	80	73	....	....	....	....	....	....	1977-78	1982-83
	Logan, Dave	Chi., Van.	6	218	5	29	34	470	12	0	0	0	10	....	1975-76	1980-81
	Logan, Robert	Buf., L.A.	3	42	10	5	15	0	....	....	....	....	....	....	1986-87	1988-89
	Loiselle, Claude	Det., N.J., Que., Tor., NYI	13	616	92	117	209	1149	41	4	11	15	58	....	1981-82	1993-94
	Lomakin, Andrei	Phi., Fla.	4	215	42	62	104	92	....	....	....	....	....	....	1991-92	1994-95
‡	Loney, Brian	Van.	1	12	2	3	5	6	....	....	....	....	....	....	1995-96	1995-96
	Loney, Troy	Pit., Ana., NYI, NYR	12	624	87	110	197	1091	67	8	14	22	97	2	1983-84	1994-95
	Long, Barry	L.A., Det., Wpg.	5	280	11	68	79	250	5	0	1	1	18	....	1972-73	1981-82
●	Long, Stan	Mtl.	1	....	....	....	....	....	3	0	0	0	0	....	1951-52	1951-52
	Lonsberry, Ross	Bos., L.A., Phi., Pit.	15	968	256	310	566	806	100	21	25	46	87	2	1966-67	1980-81
	Loob, Hakan	Cgy.	6	450	193	236	429	189	73	26	28	54	16	1	1983-84	1988-89
	Loob, Peter	Que.	1	8	1	2	3	0	....	....	....	....	....	....	1984-85	1984-85
	Lorentz, Jim	Bos., St.L., NYR, Buf.	10	659	161	238	399	208	54	12	10	22	30	1	1968-69	1977-78
	Lorimer, Bob	NYI, Col., N.J.	10	529	22	90	112	431	49	3	10	13	83	2	1976-77	1985-86
●	Lorrain, Rod	Mtl.	6	179	28	39	67	30	11	0	3	3	0	....	1935-36	1941-42
●	Loughlin, Clem	Det., Chi.	3	101	8	6	14	77	....	....	....	....	....	....	1926-27	1928-29
●	Loughlin, Wilf	Tor.	1	14	0	0	0	2	....	....	....	....	....	....	1923-24	1923-24
	Lovsin, Ken	Wsh.	1	1	0	0	0	0	....	....	....	....	....	....	1990-91	1990-91
	Lowdermilk, Dwayne	Wsh.	1	2	0	1	1	2	....	....	....	....	....	....	1980-81	1980-81
	Lowe, Darren	Pit.	1	8	1	2	3	0	....	....	....	....	....	....	1983-84	1983-84
	Lowe, Kevin	Edm., NYR	19	1254	84	347	431	1498	214	10	48	58	192	6	1979-80	1997-98
	Lowe, Odie	NYR	1	4	1	1	2	0	....	....	....	....	....	....	1949-50	1949-50
●	Lowe, Ross	Bos., Mtl.	3	77	6	8	14	82	2	0	0	0	0	....	1949-50	1951-52
●	Lowrey, Ed	Ott., Ham.	3	27	2	2	4	6	....	....	....	....	....	....	1917-18	1920-21
●	Lowrey, Fred	Mtl.M., Pit.	2	53	1	1	2	10	2	0	0	0	....	....	1924-25	1925-26
●	Lowrey, Gerry	Tor., Pit., Phi., Chi., Ott.	6	211	48	48	96	148	2	1	0	1	2	....	1927-28	1932-33
	Lucas, Danny	Phi.	1	6	1	0	1	0	....	....	....	....	....	....	1978-79	1978-79
	Lucas, Dave	Det.	1	1	0	0	0	0	....	....	....	....	....	....	1962-63	1962-63
	Luce, Don	NYR, Det., Buf., L.A., Tor.	13	894	225	329	554	364	71	17	22	39	52	....	1969-70	1981-82
	Ludvig, Jan	N.J., Buf.	7	314	54	87	141	418	....	....	....	....	....	....	1982-83	1988-89
	Ludwig, Craig	Mtl., NYI, Min., Dal.	17	1256	38	184	222	1437	177	4	25	29	244	2	1982-83	1998-99
	Ludzik, Steve	Chi., Buf.	9	424	46	93	139	333	44	4	8	12	70	....	1981-82	1989-90
	Luhning, Warren	NYI, Dal.	3	29	0	1	1	21	....	....	....	....	....	....	1997-98	1999-00
	Lukowich, Bernie	Pit., St.L.	2	79	13	15	28	34	2	0	0	0	0	....	1973-74	1974-75
	Lukowich, Morris	Wpg., Bos., L.A.	8	582	199	219	418	584	11	0	2	2	24	....	1979-80	1986-87
	Luksa, Charlie	Hfd.	1	8	0	1	1	4	....	....	....	....	....	....	1979-80	1979-80
	Lumley, Dave	Mtl., Edm., Hfd.	9	437	98	160	258	680	61	6	8	14	131	2	1978-79	1986-87
	Lund, Pentti	Bos., NYR	7	259	44	55	99	40	19	7	5	12	0	....	1946-47	1952-53
	Lundberg, Brian	Pit.	1	1	0	0	0	2	....	....	....	....	....	....	1982-83	1982-83
	Lunde, Len	Det., Chi., Min., Van.	8	321	39	83	122	75	20	3	2	5	2	....	1958-59	1970-71
	Lundholm, Bengt	Wpg.	5	275	48	95	143	72	14	3	4	7	14	....	1981-82	1985-86
	Lundrigan, Joe	Tor., Wsh.	2	52	2	8	10	22	....	....	....	....	....	....	1972-73	1974-75
	Lundstrom, Tord	Det.	1	11	1	1	2	0	....	....	....	....	....	....	1973-74	1973-74
●	Lundy, Pat	Det., Chi.	5	150	37	32	69	31	16	2	2	4	2	....	1945-46	1950-51
‡	Luongo, Chris	Det., Ott., NYI	5	218	8	23	31	176	....	....	....	....	....	....	1990-91	1995-96
	Lupien, Gilles	Mtl., Pit., Hfd.	5	226	5	25	30	416	25	0	0	0	21	2	1977-78	1981-82
	Lupul, Gary	Van.	7	293	70	75	145	243	25	4	7	11	11	....	1979-80	1985-86
	Lyle, George	Det., Hfd.	4	99	24	38	62	51	....	....	....	....	....	....	1979-80	1982-83
	Lynch, Jack	Pit., Det., Wsh.	7	382	24	106	130	336	....	....	....	....	....	....	1972-73	1978-79
	Lynn, Vic	NYR, Det., Mtl., Tor., Bos., Chi.	11	327	49	76	125	274	47	7	10	17	46	3	1942-43	1953-54
	Lyon, Steve	Pit.	1	3	0	0	0	2	....	....	....	....	....	....	1976-77	1976-77
	Lyons, Ron	Bos., Phi.	1	36	2	4	6	27	5	0	0	0	0	....	1930-31	1930-31
	Lysiak, Tom	Atl., Chi.	13	919	292	551	843	567	76	25	38	63	49	....	1973-74	1985-86

M

	Name	NHL Teams	NHL Seasons	Regular Schedule GP	G	A	TP	PIM	Playoffs GP	G	A	TP	PIM	NHL Cup Wins	First NHL Season	Last NHL Season
	MacAdam, Al	Phi., Cal., Cle., Min., Van.	12	864	240	351	591	509	64	20	24	44	21	1	1973-74	1984-85
	MacDermid, Paul	Hfd., Wpg., Wsh., Que.	14	690	116	142	258	1303	43	5	11	16	116	....	1981-82	1994-95
	MacDonald, Blair	Edm., Van.	4	219	91	100	191	65	11	0	6	6	2	....	1979-80	1982-83
	MacDonald, Brett	Van.	1	1	0	0	0	0	....	....	....	....	....	....	1987-88	1987-88
‡	MacDonald, Doug	Buf.	3	11	1	0	1	2	....	....	....	....	....	....	1992-93	1994-95
	MacDonald, Kevin	Ott.	1	1	0	0	0	2	....	....	....	....	....	....	1993-94	1993-94
●	MacDonald, Kilby	NYR	4	151	36	34	70	47	15	1	2	3	4	1	1939-40	1944-45
	MacDonald, Lowell	Det., L.A., Pit.	13	506	180	210	390	92	30	11	11	22	12	....	1961-62	1977-78
	MacDonald, Parker	Tor., NYR, Det., Bos., Min.	14	676	144	179	323	253	75	14	14	28	20	....	1952-53	1968-69
	MacDougall, Kim	Min.	1	1	0	0	0	0	....	....	....	....	....	....	1974-75	1974-75
	MacEachern, Shane	St.L.	1	1	0	0	0	0	....	....	....	....	....	....	1987-88	1987-88
	Macey, Hub	NYR, Mtl.	3	30	6	9	15	0	8	0	0	0	0	....	1941-42	1946-47
	MacGregor, Bruce	Det., NYR	14	893	213	257	470	217	107	19	28	47	44	....	1960-61	1973-74
	MacGregor, Randy	Hfd.	1	2	1	1	2	2	....	....	....	....	....	....	1981-82	1981-82
	MacGuigan, Garth	NYI	2	5	0	1	1	2	....	....	....	....	....	....	1979-80	1983-84
	MacIntosh, Ian	NYR	1	4	0	0	0	4	....	....	....	....	....	....	1952-53	1952-53
	MacIver, Don	Wpg.	1	6	0	0	0	2	....	....	....	....	....	....	1979-80	1979-80
	Maciver, Norm	NYR, Hfd., Edm., Ott., Pit., Wpg., Phx.	12	500	55	230	285	350	56	3	11	14	32	....	1986-87	1997-98
	MacKasey, Blair	Tor.	1	1	0	0	0	2	....	....	....	....	....	....	1976-77	1976-77
●	MacKay, Calum	Det., Mtl.	8	237	50	55	105	214	38	5	13	18	20	1	1946-47	1954-55
	MacKay, Dave	Chi.	1	29	3	0	3	26	5	0	1	1	2	....	1940-41	1940-41
●	Mackay, Mickey	Chi., Pit., Bos.	4	147	44	19	63	79	11	0	0	0	6	1	1926-27	1929-30
●	MacKay, Murdo	Mtl.	4	19	0	3	3	0	15	1	2	3	0	....	1945-46	1948-49
	Mackell, Fleming	Tor., Bos.	13	665	149	220	369	562	80	22	41	63	75	2	1947-48	1959-60
●	MacKell, Jack	Ott.	2	45	4	2	6	59	2	0	0	0	0	2	1919-20	1920-21
	MacKenzie, Barry	Min.	1	6	0	1	1	6	....	....	....	....	....	....	1968-69	1968-69
●	MacKenzie, Bill	Chi., Mtl.M., NYR, Mtl.	7	264	15	14	29	145	21	1	1	2	11	1	1932-33	1939-40
	Mackey, David	Chi., Min., St.L.	6	126	8	12	20	305	3	0	0	0	2	....	1987-88	1993-94
●	Mackey, Reg	NYR	1	34	0	0	0	16	1	0	0	0	0	....	1926-27	1926-27
●	Mackie, Howie	Det.	2	20	1	0	1	4	8	0	0	0	0	1	1936-37	1937-38
	MacKinnon, Paul	Wsh.	5	147	5	23	28	91	....	....	....	....	....	....	1979-80	1983-84
	MacLean, Paul	St.L., Wpg., Det.	11	719	324	349	673	968	53	21	14	35	110	....	1980-81	1990-91
	MacLeish, Rick	Phi., Hfd., Pit., Det.	14	846	349	410	759	434	114	54	53	107	38	2	1970-71	1983-84
	MacLellan, Brian	L.A., NYR, Min., Cgy., Det.	10	606	172	241	413	551	47	5	9	14	42	1	1982-83	1991-92
	MacLeod, Pat	Min., S.J., Dal.	4	53	5	13	18	14	....	....	....	....	....	....	1990-91	1995-96
	MacMillan, Billy	Tor., Atl., NYI	7	446	74	77	151	184	53	6	6	12	40	....	1970-71	1976-77
	MacMillan, Bob	NYR, St.L., Atl., Cgy., Col., N.J., Chi.	11	753	228	349	577	260	31	8	11	19	16	....	1974-75	1984-85
	MacMillan, John	Tor., Det.	5	104	5	10	15	32	12	0	1	1	2	2	1960-61	1964-65
	MacNeil, Al	Tor., Mtl., Chi., NYR, Pit.	11	524	17	75	92	617	37	0	4	4	67	....	1955-56	1967-68
	MacNeil, Bernie	St.L.	1	4	0	0	0	0	....	....	....	....	....	....	1973-74	1973-74
	Macoun, Jamie	Cgy., Tor., Det.	16	1128	76	282	358	1208	159	10	32	42	169	2	1982-83	1998-99
●	MacPherson, Bud	Mtl.	7	259	5	33	38	233	29	0	3	3	21	1	1948-49	1956-57
●	MacSweyn, Ralph	Phi.	5	47	0	5	5	10	8	0	0	0	6	....	1967-68	1971-72
	MacTavish, Craig	Bos., Edm., NYR, Phi., St.L.	17	1093	213	267	480	891	193	20	38	58	218	4	1979-80	1996-97
	MacWilliam, Mike	NYI	1	6	0	0	0	14	....	....	....	....	....	....	1995-96	1995-96
	Madigan, Connie	St.L.	1	20	0	3	3	25	5	0	0	0	4	....	1972-73	1972-73
	Madill, Jeff	N.J.	1	14	4	0	4	46	7	0	2	2	8	....	1990-91	1990-91

Stephan Lebeau

Gary Leeman

Tony Leswick

Rick Ley

Pentti Lund

Dan Mandich

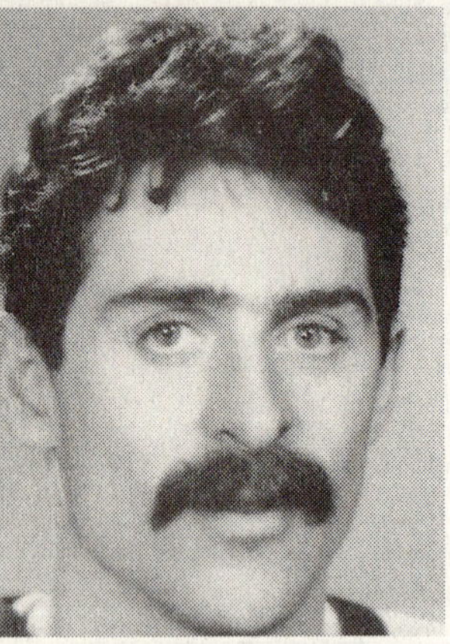

Moe Mantha

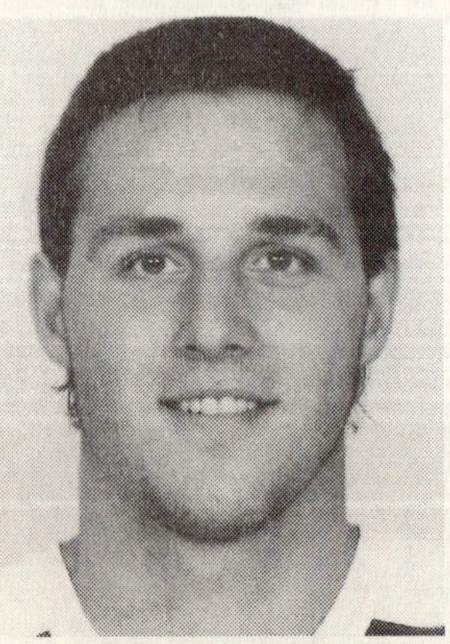

Daniel Marois

Name	NHL Teams	NHL Seasons	Regular Schedule GP	G	A	TP	PIM	Playoffs GP	G	A	TP	PIM	NHL Cup Wins	First NHL Season	Last NHL Season
Magee, Dean	Min.	1	7	0	0	0	4	....	....	....	....	....	....	1977-78	1977-78
Maggs, Daryl	Chi., Cal., Tor.	3	135	14	19	33	54	4	0	0	0	0	....	1971-72	1979-80
Magnan, Marc	Tor.	1	4	0	1	1	5	....	....	....	....	....	....	1982-83	1982-83
Magnuson, Keith	Chi.	11	589	14	125	139	1442	68	3	9	12	164	....	1969-70	1979-80
Maguire, Kevin	Tor., Buf., Phi.	6	260	29	30	59	782	11	0	0	0	86	....	1986-87	1991-92
Mahaffy, John	Mtl., NYR	3	37	11	25	36	4	1	0	1	1	0	....	1942-43	1944-45
Mahovlich, Frank	Tor., Det., Mtl.	18	1181	533	570	1103	1056	137	51	67	118	163	6	1956-57	1973-74
Mahovlich, Pete	Det., Mtl., Pit.	16	884	288	485	773	916	88	30	42	72	134	4	1965-66	1980-81
Mailhot, Jacques	Que.	1	5	0	0	0	33	....	....	....	....	....	....	1988-89	1988-89
Mailley, Frank	Mtl.	1	1	0	0	0	0	....	....	....	....	....	....	1942-43	1942-43
Mair, Jim	Phi., NYI, Van.	5	76	4	15	19	49	3	1	2	3	4	....	1970-71	1974-75
• Majeau, Fern	Mtl.	2	56	22	24	46	43	1	0	0	0	0	1	1943-44	1944-45
Major, Bruce	Que.	1	4	0	0	0	0	....	....	....	....	....	....	1990-91	1990-91
‡ Major, Mark	Det.	1	2	0	0	0	5	....	....	....	....	....	....	1996-97	1996-97
Makarov, Sergei	Cgy., S.J., Dal.	7	424	134	250	384	317	34	12	11	23	8	....	1989-90	1996-97
Makela, Mikko	NYI, L.A., Buf., Bos.	7	423	118	147	265	139	18	3	8	11	14	....	1985-86	1994-95
Maki, Chico	Chi.	15	841	143	292	435	345	113	17	36	53	43	1	1960-61	1975-76
• Maki, Wayne	Chi., St.L., Van.	6	246	57	79	136	184	2	1	0	1	2	....	1967-68	1972-73
Makkonen, Kari	Edm.	1	9	2	2	4	0	....	....	....	....	....	....	1979-80	1979-80
Maley, David	Mtl., N.J., Edm., S.J., NYI	9	466	43	81	124	1043	46	5	5	10	111	1	1985-86	1993-94
Malinowski, Merlin	Col., N.J., Hfd.	5	282	54	111	165	121	....	....	....	....	....	....	1978-79	1982-83
Malkoc, Dean	Van., Bos., NYI	4	116	1	3	4	299	....	....	....	....	....	....	1995-96	1998-99
Mallette, Troy	NYR, Edm., N.J., Ott., Bos., T.B.	9	456	51	68	119	1226	15	2	2	4	99	....	1989-90	1997-98
Malone, Cliff	Mtl.	1	3	0	0	0	0	....	....	....	....	....	....	1951-52	1951-52
Malone, Greg	Pit., Hfd., Que.	11	704	191	310	501	661	20	3	5	8	32	....	1976-77	1986-87
• Malone, Joe	Mtl., Que., Ham.	7	126	143	32	175	57	9	6	0	6	3	1	1917-18	1923-24
Maloney, Dan	Chi., L.A., Det., Tor.	11	737	192	259	451	1489	40	4	7	11	35	....	1970-71	1981-82
Maloney, Dave	NYR, Buf.	11	657	71	246	317	1154	49	7	17	24	91	....	1974-75	1984-85
Maloney, Don	NYR, Hfd., NYI	13	765	214	350	564	815	94	22	35	57	101	....	1978-79	1990-91
Maloney, Phil	Bos., Tor., Chi.	5	158	28	43	71	16	6	0	0	0	0	....	1949-50	1959-60
Maluta, Ray	Bos.	2	25	2	3	5	6	2	0	0	0	0	....	1975-76	1976-77
Manastersky, Tom	Mtl.	1	6	0	0	0	11	....	....	....	....	....	....	1950-51	1950-51
Mancuso, Gus	Mtl., NYR	4	42	7	9	16	17	....	....	....	....	....	....	1937-38	1942-43
Mandich, Dan	Min.	4	111	5	11	16	303	7	0	0	0	2	....	1982-83	1985-86
‡ Maneluk, Mike	Phi., Chi., NYR, CBJ	3	85	11	10	21	57	....	....	....	....	....	....	1998-99	2000-01
Manery, Kris	Cle., Min., Van., Wpg.	4	250	63	64	127	91	....	....	....	....	....	....	1977-78	1980-81
Manery, Randy	Det., Atl., L.A.	10	582	50	206	256	415	13	0	2	2	12	....	1970-71	1979-80
Mann, Jack	NYR	2	9	3	4	7	0	....	....	....	....	....	....	1943-44	1944-45
Mann, Jimmy	Wpg., Que., Pit.	8	293	10	20	30	895	22	0	0	0	89	....	1979-80	1987-88
Mann, Ken	Det.	1	1	0	0	0	0	....	....	....	....	....	....	1975-76	1975-76
• Mann, Norm	Tor.	3	31	0	3	3	4	2	0	0	0	0	....	1935-36	1940-41
• Manners, Rennison	Pit., Phi.	2	37	3	2	5	14	....	....	....	....	....	....	1929-30	1930-31
Manno, Bob	Van., Tor., Det.	8	371	41	131	172	274	17	2	4	6	12	....	1976-77	1984-85
Manson, Ray	Bos., NYR	2	2	0	1	1	0	....	....	....	....	....	....	1947-48	1948-49
• Mantha, Georges	Mtl.	13	488	89	102	191	148	36	6	2	8	24	2	1928-29	1940-41
Mantha, Moe	Wpg., Pit., Edm., Min., Phi.	12	656	81	289	370	501	17	5	10	15	18	....	1980-81	1991-92
• Mantha, Sylvio	Mtl., Bos.	14	542	63	78	141	671	39	5	5	10	64	3	1923-24	1936-37
• Maracle, Bud	NYR	1	11	1	3	4	4	4	0	0	0	0	....	1930-31	1930-31
Marcetta, Milan	Tor., Min.	3	54	7	15	22	10	17	7	7	14	4	1	1966-67	1968-69
• March, Mush	Chi.	17	759	153	230	383	540	45	12	15	27	41	2	1928-29	1944-45
Marchinko, Brian	Tor., NYI	4	47	2	6	8	0	....	....	....	....	....	....	1970-71	1973-74
Marcinyshyn, Dave	N.J., Que., NYR	3	16	0	1	1	49	....	....	....	....	....	....	1990-91	1992-93
Marcon, Lou	Det.	3	60	0	4	4	42	....	....	....	....	....	....	1958-59	1962-63
Marcotte, Don	Bos.	15	868	230	254	484	317	132	34	27	61	81	2	1965-66	1981-82
Marini, Hector	NYI, N.J.	5	154	27	46	73	246	10	3	6	9	14	2	1978-79	1983-84
‡ Marinucci, Chris	NYI, L.A.	2	13	1	4	5	2	....	....	....	....	....	....	1994-95	1996-97
• Mario, Frank	Bos.	2	53	9	19	28	24	....	....	....	....	....	....	1941-42	1944-45
• Mariucci, John	Chi.	5	223	11	34	45	308	12	0	3	3	26	....	1940-41	1947-48
Mark, Gordon	N.J., Edm.	4	85	3	10	13	187	....	....	....	....	....	....	1986-87	1994-95
Markell, John	Wpg., St.L., Min.	4	55	11	10	21	36	....	....	....	....	....	....	1979-80	1984-85
• Marker, Gus	Det., Mtl.M., Tor., Bro.	10	322	64	69	133	133	46	5	7	12	36	1	1932-33	1941-42
Markham, Ray	NYR	1	14	1	1	2	21	7	1	0	1	24	....	1979-80	1979-80
• Markle, Jack	Tor.	1	8	0	1	1	0	....	....	....	....	....	....	1935-36	1935-36
• Marks, Jack	Mtl.W., Tor., Que.	2	7	0	0	0	4	....	....	....	....	....	1	1917-18	1919-20
Marks, John	Chi.	10	657	112	163	275	330	57	5	9	14	60	....	1972-73	1981-82
Markwart, Nevin	Bos., Cgy.	8	309	41	68	109	794	19	1	0	1	33	....	1983-84	1991-92
‡ Marois, Daniel	Tor., NYI, Bos., Dal.	8	350	117	93	210	419	19	3	3	6	28	....	1987-88	1995-96
Marois, Mario	NYR, Van., Que., Wpg., St.L.	15	955	76	357	433	1746	100	4	34	38	182	....	1977-78	1991-92
Marotte, Gilles	Bos., Chi., L.A., NYR, St.L.	12	808	56	265	321	919	29	3	3	6	26	....	1965-66	1976-77
Marquess, Mark	Bos.	1	27	5	4	9	6	4	0	0	0	0	....	1946-47	1946-47
Marsh, Brad	Atl., Cgy., Phi., Tor., Det., Ott.	15	1086	23	175	198	1241	97	6	18	24	124	....	1978-79	1992-93
Marsh, Gary	Det., Tor.	2	7	1	3	4	4	....	....	....	....	....	....	1967-68	1968-69
Marsh, Peter	Wpg., Chi.	5	278	48	71	119	224	26	1	5	6	33	....	1979-80	1983-84
Marshall, Bert	Det., Oak., Cal., NYR, NYI	14	868	17	181	198	926	72	4	22	26	99	....	1965-66	1978-79
Marshall, Don	Mtl., NYR, Buf., Tor.	19	1176	265	324	589	127	94	8	15	23	14	5	1951-52	1971-72
Marshall, Paul	Pit., Tor., Hfd.	4	95	15	18	33	17	1	0	0	0	0	....	1979-80	1982-83
Marshall, Willie	Tor.	4	33	1	5	6	2	....	....	....	....	....	....	1952-53	1958-59
Marson, Mike	Wsh., L.A.	6	196	24	24	48	233	....	....	....	....	....	....	1974-75	1979-80
• Martin, Clare	Bos., Det., Chi., NYR	6	237	12	28	40	78	27	0	2	2	6	1	1941-42	1951-52
‡ Martin, Craig	Wpg., Fla.	2	21	0	1	1	24	....	....	....	....	....	....	1994-95	1996-97
Martin, Frank	Bos., Chi.	6	282	11	46	57	122	10	0	2	2	2	....	1952-53	1957-58
Martin, Grant	Van., Wsh.	4	44	0	4	4	55	1	1	0	1	2	....	1983-84	1986-87
Martin, Jack	Tor.	1	1	0	0	0	0	....	....	....	....	....	....	1960-61	1960-61
‡ Martin, Matt	Tor.	4	76	0	5	5	71	....	....	....	....	....	....	1993-94	1996-97
Martin, Pit	Det., Bos., Chi., Van.	17	1101	324	485	809	609	100	27	31	58	56	....	1961-62	1978-79
Martin, Rick	Buf., L.A.	11	685	384	317	701	477	63	24	29	53	74	....	1971-72	1981-82
• Martin, Ron	NYA	2	94	13	16	29	36	....	....	....	....	....	....	1932-33	1933-34
Martin, Terry	Buf., Que., Tor., Edm., Min.	10	479	104	101	205	202	21	4	2	6	26	....	1975-76	1984-85
Martin, Tom	Tor.	1	3	1	0	1	0	....	....	....	....	....	....	1967-68	1967-68
Martin, Tom	Wpg., Hfd., Min.	6	92	12	11	23	249	4	0	0	0	6	....	1984-85	1989-90
Martineau, Don	Atl., Min., Det.	4	90	6	10	16	63	....	....	....	....	....	....	1973-74	1976-77
‡ Martini, Darcy	Edm.	1	2	0	0	0	0	....	....	....	....	....	....	1993-94	1993-94
Martinson, Steve	Det., Mtl., Min.	4	49	2	1	3	244	1	0	0	0	10	....	1987-88	1991-92
Maruk, Dennis	Cal., Cle., Min., Wsh.	14	888	356	522	878	761	34	14	22	36	26	....	1975-76	1988-89
Masnick, Paul	Mtl., Chi., Tor.	6	232	18	41	59	139	33	4	5	9	27	1	1950-51	1957-58
• Mason, Charley	NYR, NYA, Det., Chi.	4	95	7	18	25	44	4	0	1	1	0	....	1934-35	1938-39
• Massecar, George	NYA	3	100	12	11	23	46	....	....	....	....	....	....	1929-30	1931-32
Masters, Jamie	St.L.	3	33	1	13	14	2	2	0	0	0	0	....	1975-76	1978-79
• Masterton, Bill	Min.	1	38	4	8	12	4	....	....	....	....	....	....	1967-68	1967-68
Mathers, Frank	Tor.	3	23	1	3	4	4	....	....	....	....	....	....	1948-49	1951-52
Mathiasen, Dwight	Pit.	3	33	1	7	8	18	....	....	....	....	....	....	1985-86	1987-88
Mathieson, Jim	Wsh.	1	2	0	0	0	4	....	....	....	....	....	....	1989-90	1989-90
‡ Mathieu, Marquis	Bos.	3	16	0	2	2	14	....	....	....	....	....	....	1998-99	2000-01
• Matte, Joe	Tor., Ham., Bos., Mtl.	4	68	17	15	32	54	....	....	....	....	....	....	1919-20	1925-26
• Matte, Joe	Det., Chi.	2	24	0	3	3	8	....	....	....	....	....	....	1929-30	1942-43
Mattiussi, Dick	Pit., Oak., Cal.	4	200	8	31	39	124	8	0	1	1	6	....	1967-68	1970-71
• Matz, Johnny	Mtl.	1	30	2	3	5	0	1	0	0	0	0	....	1924-25	1924-25
Maxner, Wayne	Bos.	2	62	8	9	17	48	....	....	....	....	....	....	1964-65	1965-66
Maxwell, Brad	Min., Que., Tor., Van., NYR	10	612	98	270	368	1292	79	12	49	61	178	....	1977-78	1986-87
Maxwell, Bryan	Min., St.L., Wpg., Pit.	8	331	18	77	95	745	15	1	1	2	86	....	1977-78	1984-85
Maxwell, Kevin	Min., Col., N.J.	3	66	6	15	21	61	16	3	4	7	24	....	1980-81	1983-84
Maxwell, Wally	Tor.	1	2	0	0	0	0	....	....	....	....	....	....	1952-53	1952-53
May, Alan	Bos., Edm., Wsh., Dal., Cgy.	8	393	31	45	76	1348	40	1	2	3	80	....	1987-88	1994-95
‡ Mayer, Derek	Ott.	1	17	2	2	4	8	....	....	....	....	....	....	1993-94	1993-94
Mayer, Jim	NYR	1	4	0	0	0	0	....	....	....	....	....	....	1979-80	1979-80
Mayer, Pat	Pit.	1	1	0	0	0	4	....	....	....	....	....	....	1987-88	1987-88
Mayer, Shep	Tor.	1	12	1	2	3	4	....	....	....	....	....	....	1942-43	1942-43
• Mazur, Eddie	Mtl., Chi.	6	107	8	20	28	120	25	4	5	9	22	1	1950-51	1956-57
‡ Mazur, Jay	Van.	4	47	11	7	18	20	6	0	1	1	8	....	1988-89	1991-92
McAdam, Gary	Buf., Pit., Det., Cgy., Wsh., N.J., Tor.	11	534	96	132	228	243	30	6	5	11	16	....	1975-76	1985-86
• McAdam, Sam	NYR	1	5	0	0	0	0	....	....	....	....	....	....	1930-31	1930-31
• McAndrew, Hazen	Bro.	1	7	0	1	1	6	....	....	....	....	....	....	1941-42	1941-42
McAneeley, Ted	Cal.	3	158	8	35	43	141	....	....	....	....	....	....	1972-73	1974-75
McAtee, Jud	Det.	3	46	15	13	28	6	14	2	1	3	0	....	1942-43	1944-45
McAtee, Norm	Bos.	1	13	0	1	1	0	....	....	....	....	....	....	1946-47	1946-47
• McAvoy, George	Mtl.	1	....	....	....	....	....	4	0	0	0	0	....	1954-55	1954-55

Name	NHL Teams	NHL Seasons	Regular Schedule					Playoffs					NHL Cup Wins	First NHL Season	Last NHL Season
			GP	G	A	TP	PIM	GP	G	A	TP	PIM			
McBain, Andrew	Wpg., Pit., Van., Ott.	11	608	129	172	301	633	24	5	7	12	39	....	1983-84	1993-94
‡ McBain, Jason	Hfd.	2	9	0	0	0	0	....	....	....	....	....	....	1995-96	1996-97
‡ McBain, Mike	T.B.	2	64	0	7	7	22	....	....	....	....	....	....	1997-98	1998-99
McBean, Wayne	L.A., NYI, Wpg.	6	211	10	39	49	168	2	1	1	2	0	....	1987-88	1993-94
McBride, Cliff	Mtl.M., Tor.	2	2	0	0	0	0	....	....	....	....	....	....	1928-29	1929-30
McBurney, Jim	Chi.	1	1	0	1	1	0	....	....	....	....	....	....	1952-53	1952-53
• McCabe, Stan	Det., Mtl.M.	4	78	9	4	13	49	....	....	....	....	....	....	1929-30	1933-34
• McCaffrey, Bert	Tor., Pit., Mtl.	7	260	43	30	73	202	8	2	1	3	10	1	1924-25	1930-31
McCahill, John	Col.	1	1	0	0	0	0	....	....	....	....	....	....	1977-78	1977-78
• McCaig, Doug	Det., Chi.	7	263	8	21	29	255	7	0	1	1	10	....	1941-42	1950-51
• McCallum, Dunc	NYR, Pit.	5	187	14	35	49	230	10	1	2	3	12	....	1965-66	1970-71
• McCalmon, Eddie	Chi., Phi.	2	39	5	0	5	14	....	....	....	....	....	....	1927-28	1930-31
McCann, Rick	Det.	6	43	1	4	5	6	....	....	....	....	....	....	1967-68	1974-75
McCarthy, Dan	NYR	1	5	4	0	4	4	....	....	....	....	....	....	1980-81	1980-81
McCarthy, Kevin	Phi., Van., Pit.	10	537	67	191	258	527	21	2	3	5	20	....	1977-78	1986-87
• McCarthy, Thomas	Que., Ham.	2	35	22	7	29	10	....	....	....	....	....	....	1919-20	1920-21
McCarthy, Tom	Det., Bos.	4	60	8	9	17	8	....	....	....	....	....	....	1956-57	1960-61
McCarthy, Tom	Min., Bos.	9	460	178	221	399	330	68	12	26	38	67	....	1979-80	1987-88
• McCartney, Walt	Mtl.	1	2	0	0	0	0	....	....	....	....	....	....	1932-33	1932-33
McCaskill, Ted	Min.	1	4	0	2	2	0	....	....	....	....	....	....	1967-68	1967-68
McClanahan, Rob	Buf., Hfd., NYR	5	224	38	63	101	126	34	4	12	16	31	....	1979-80	1983-84
McCleary, Trent	Ott., Bos., Mtl.	4	192	8	15	23	134	....	....	....	....	....	....	1995-96	1999-00
McClelland, Kevin	Pit., Edm., Det., Tor., Wpg.	12	588	68	112	180	1672	98	11	18	29	281	4	1981-82	1993-94
McCord, Bob	Bos., Det., Min., St.L.	7	316	10	58	68	262	14	2	5	7	10	....	1963-64	1972-73
McCord, Dennis	Van.	1	3	0	0	0	6	....	....	....	....	....	....	1973-74	1973-74
McCormack, John	Tor., Mtl., Chi.	8	311	25	49	74	35	22	1	1	2	0	2	1947-48	1954-55
McCosh, Shawn	L.A., NYR	2	9	1	0	1	6	....	....	....	....	....	....	1991-92	1994-95
McCourt, Dale	Det., Buf., Tor.	7	532	194	284	478	124	21	9	7	16	6	....	1977-78	1983-84
McCreary, Bill	NYR, Det., Mtl., St.L.	8	309	53	62	115	108	48	6	16	22	14	....	1953-54	1970-71
McCreary, Keith	Mtl., Pit., Atl.	10	532	131	112	243	294	16	0	4	4	6	....	1961-62	1974-75
McCreary Jr., Bill	Tor.	1	12	1	0	1	4	....	....	....	....	....	....	1980-81	1980-81
• McCreedy, John	Tor.	2	64	17	12	29	25	21	4	3	7	16	2	1941-42	1944-45
McCrimmon, Brad	Bos., Phi., Cgy., Det., Hfd., Phx.	18	1222	81	322	403	1416	116	11	18	29	176	1	1979-80	1996-97
McCrimmon, Jim	St.L.	1	2	0	0	0	0	....	....	....	....	....	....	1974-75	1974-75
McCulley, Bob	Mtl.	1	1	0	0	0	0	....	....	....	....	....	....	1934-35	1934-35
• McCurry, Duke	Pit.	4	148	21	11	32	119	4	0	2	2	4	....	1925-26	1928-29
McCutcheon, Brian	Det.	3	37	3	1	4	7	....	....	....	....	....	....	1974-75	1976-77
McCutcheon, Darwin	Tor.	1	1	0	0	0	2	....	....	....	....	....	....	1981-82	1981-82
McDill, Jeff	Chi.	1	1	0	0	0	0	....	....	....	....	....	....	1976-77	1976-77
McDonagh, Bill	NYR	1	4	0	0	0	2	....	....	....	....	....	....	1949-50	1949-50
McDonald, Ab	Mtl., Chi., Bos., Det., Pit., St.L.	15	762	182	248	430	200	84	21	29	50	42	4	1957-58	1971-72
McDonald, Brian	Chi., Buf.	2	12	0	0	0	29	8	0	0	0	2	....	1967-68	1970-71
• McDonald, Bucko	Det., Tor., NYR	11	446	35	88	123	206	50	6	1	7	24	3	1934-35	1944-45
McDonald, Butch	Det., Chi.	2	66	8	20	28	2	5	0	2	2	10	....	1939-40	1944-45
McDonald, Gerry	Hfd.	2	8	0	0	0	4	....	....	....	....	....	....	1981-82	1983-84
• McDonald, Jack	Mtl.W., Mtl., Que., Tor.	5	69	26	14	40	30	7	1	1	2	3	....	1917-18	1921-22
McDonald, Jack	NYR	1	43	10	9	19	6	....	....	....	....	....	....	1943-44	1943-44
McDonald, Lanny	Tor., Col., Cgy.	16	1111	500	506	1006	899	117	44	40	84	120	1	1973-74	1988-89
McDonald, Robert	NYR	1	1	0	0	0	0	....	....	....	....	....	....	1943-44	1943-44
McDonald, Terry	K.C.	1	8	0	1	1	6	....	....	....	....	....	....	1975-76	1975-76
McDonnell, Joe	Van., Pit.	3	50	2	10	12	34	....	....	....	....	....	....	1981-82	1985-86
• McDonnell, Moylan	Ham.	1	22	1	2	3	2	....	....	....	....	....	....	1920-21	1920-21
McDonough, Al	L.A., Pit., Atl., Det.	5	237	73	88	161	73	8	0	1	1	2	....	1970-71	1977-78
‡ McDonough, Hubie	L.A., NYI, S.J.	5	195	40	26	66	67	5	1	0	1	4	....	1988-89	1992-93
McDougal, Mike	NYR, Hfd.	4	61	8	10	18	43	....	....	....	....	....	....	1978-79	1982-83
‡ McDougall, Bill	Det., Edm., T.B.	3	28	5	5	10	12	1	0	0	0	0	....	1990-91	1993-94
McElmury, Jim	Min., K.C., Col.	5	180	14	47	61	49	....	....	....	....	....	....	1972-73	1977-78
McEwen, Mike	NYR, Col., NYI, L.A., Wsh., Det., Hfd.	12	716	108	296	404	460	78	12	36	48	48	3	1976-77	1987-88
McFadden, Jim	Det., Chi.	8	412	100	126	226	89	49	10	9	19	30	1	1946-47	1953-54
McFadyen, Don	Chi.	4	179	12	33	45	77	11	2	2	4	5	1	1932-33	1935-36
McFall, Dan	Wpg.	2	9	0	1	1	0	....	....	....	....	....	....	1984-85	1985-86
• McFarlane, Gord	Chi.	1	2	0	0	0	0	....	....	....	....	....	....	1926-27	1926-27
‡ McGeough, Jim	Wsh., Pit.	4	57	7	10	17	32	....	....	....	....	....	....	1981-82	1986-87
• McGibbon, Irv	Mtl.	1	1	0	0	0	2	....	....	....	....	....	....	1942-43	1942-43
McGill, Bob	Tor., Chi., S.J., Det., NYI, Hfd.	13	705	17	55	72	1766	49	0	0	0	88	....	1981-82	1993-94
• McGill, Jack	Mtl.	3	134	27	10	37	71	3	2	0	2	0	....	1934-35	1936-37
• McGill, Jack	Bos.	4	97	23	36	59	42	27	7	4	11	17	....	1941-42	1946-47
McGill, Ryan	Chi., Phi., Edm.	4	151	4	15	19	391	....	....	....	....	....	....	1991-92	1994-95
McGregor, Sandy	NYR	1	2	0	0	0	2	....	....	....	....	....	....	1963-64	1963-64
• McGuire, Mickey	Pit.	2	36	3	0	3	6	....	....	....	....	....	....	1926-27	1927-28
McHugh, Mike	Min., S.J.	4	20	1	0	1	16	....	....	....	....	....	....	1988-89	1991-92
McIlhargey, Jack	Phi., Van., Hfd.	8	393	11	36	47	1102	27	0	3	3	68	....	1974-75	1981-82
• McInenly, Bert	Det., NYA, Ott., Bos.	6	166	19	15	34	144	4	0	0	0	2	....	1930-31	1935-36
McIntosh, Bruce	Min.	1	2	0	0	0	0	....	....	....	....	....	....	1972-73	1972-73
McIntosh, Paul	Buf.	2	48	0	2	2	66	2	0	0	0	7	....	1974-75	1975-76
• McIntyre, Jack	Bos., Chi., Det.	11	499	109	102	211	173	29	7	6	13	4	....	1949-50	1959-60
McIntyre, John	Tor., L.A., NYR, Van.	6	351	24	54	78	516	44	0	6	6	54	....	1989-90	1994-95
McIntyre, Larry	Tor.	2	41	0	3	3	26	....	....	....	....	....	....	1969-70	1972-73
McKay, Doug	Det.	1	....	....	....	....	....	1	0	0	0	0	1	1949-50	1949-50
McKay, Ray	Chi., Buf., Cal.	6	140	2	16	18	102	....	....	....	....	....	....	1968-69	1973-74
McKay, Scott	Ana.	1	1	0	0	0	0	....	....	....	....	....	....	1993-94	1993-94
McKechnie, Walt	Min., Cal., Bos., Det., Wsh., Cle., Tor., Col.	16	955	214	392	606	469	15	7	5	12	7	....	1967-68	1982-83
McKee, Mike	Que.	1	48	3	12	15	41	....	....	....	....	....	....	1993-94	1993-94
McKegney, Ian	Chi.	1	3	0	0	0	2	....	....	....	....	....	....	1976-77	1976-77
McKegney, Tony	Buf., Que., Min., NYR, St.L., Det., Chi.	13	912	320	319	639	517	79	24	23	47	56	....	1978-79	1990-91
McKendry, Alex	NYI, Cgy.	4	46	3	6	9	21	6	2	2	4	0	1	1977-78	1980-81
McKenna, Sean	Buf., L.A., Tor.	9	414	82	80	162	181	15	1	2	3	2	....	1981-82	1989-90
McKenney, Don	Bos., NYR, Tor., Det., St.L.	13	798	237	345	582	211	58	18	29	47	10	1	1954-55	1967-68
McKenny, Jim	Tor., Min.	14	604	82	247	329	294	37	7	9	16	10	....	1965-66	1978-79
McKenzie, Brian	Pit.	1	6	1	1	2	4	....	....	....	....	....	....	1971-72	1971-72
McKenzie, John	Chi., Det., NYR, Bos.	12	691	206	268	474	917	69	15	32	47	133	2	1958-59	1971-72
‡ McKim, Andrew	Bos., Det.	3	38	1	4	5	6	....	....	....	....	....	....	1992-93	1994-95
• McKinnon, Alex	Ham., NYA, Chi.	5	193	19	11	30	237	....	....	....	....	....	....	1924-25	1928-29
• McKinnon, John	Mtl., Pit., Phi.	6	208	28	11	39	224	2	0	0	0	4	....	1925-26	1930-31
McLean, Don	Wsh.	1	9	0	0	0	6	....	....	....	....	....	....	1975-76	1975-76
• McLean, Fred	Que., Ham.	2	8	0	0	0	2	....	....	....	....	....	....	1919-20	1920-21
McLean, Jack	Tor.	3	67	14	24	38	76	13	2	2	4	8	1	1942-43	1944-45
McLean, Jeff	S.J.	1	6	1	0	1	0	....	....	....	....	....	....	1993-94	1993-94
• McLellan, John	Tor.	1	2	0	0	0	0	....	....	....	....	....	....	1951-52	1951-52
McLellan, Scott	Bos.	1	2	0	0	0	0	....	....	....	....	....	....	1982-83	1982-83
‡ McLellan, Todd	NYI	1	5	1	1	2	0	....	....	....	....	....	....	1987-88	1987-88
• McLenahan, Rollie	Det.	1	9	2	1	3	10	2	0	0	0	0	....	1945-46	1945-46
McLeod, Al	Det.	1	26	2	2	4	24	....	....	....	....	....	....	1973-74	1973-74
McLeod, Jackie	NYR	5	106	14	23	37	12	7	0	0	0	0	....	1949-50	1954-55
‡ McLlwain, Dave	Pit., Wpg., Buf., NYI, Tor., Ott.	10	501	100	107	207	292	20	0	2	2	2	....	1987-88	1996-97
• McMahon, Mike	Mtl., Bos.	3	57	7	18	25	102	13	1	2	3	30	1	1942-43	1945-46
McMahon, Mike	NYR, Min., Chi., Det., Pit., Buf.	8	224	15	68	83	171	14	3	7	10	4	....	1963-64	1971-72
McManama, Bob	Pit.	3	99	11	25	36	28	8	0	1	1	6	....	1973-74	1975-76
• McManus, Sammy	Mtl.M., Bos.	2	26	0	1	1	8	1	0	0	0	0	1	1934-35	1936-37
McMurchy, Tom	Chi., Edm.	4	55	8	4	12	65	....	....	....	....	....	....	1983-84	1987-88
McNab, Max	Det.	4	128	16	19	35	24	25	1	0	1	4	1	1947-48	1950-51
McNab, Peter	Buf., Bos., Van., N.J.	14	954	363	450	813	179	107	40	42	82	20	....	1973-74	1986-87
• McNabney, Sid	Mtl.	1	....	....	....	....	....	5	0	1	1	2	....	1950-51	1950-51
• McNamara, Howard	Mtl.	1	10	1	0	1	4	....	....	....	....	....	....	1919-20	1919-20
• McNaughton, George	Que.	1	1	0	0	0	0	....	....	....	....	....	....	1919-20	1919-20
McNeill, Billy	Det.	6	257	21	46	67	142	4	1	1	2	4	....	1956-57	1963-64
McNeill, Mike	Chi., Que.	2	63	5	11	16	18	....	....	....	....	....	....	1990-91	1991-92
McNeill, Stu	Det.	3	10	1	1	2	2	....	....	....	....	....	....	1957-58	1959-60
McPhee, George	NYR, N.J.	7	115	24	25	49	257	29	5	3	8	69	....	1982-83	1988-89
McPhee, Mike	Mtl., Min., Dal.	11	744	200	199	399	661	134	28	27	55	193	1	1983-84	1993-94
McRae, Basil	Que., Tor., Det., Min., T.B., St.L., Chi.	16	576	53	83	136	2457	78	8	4	12	349	....	1981-82	1996-97
McRae, Chris	Tor., Det.	3	21	1	0	1	122	....	....	....	....	....	....	1987-88	1989-90
McRae, Ken	Que., Tor.	7	137	14	21	35	364	6	0	0	0	4	....	1987-88	1993-94
• McReavy, Pat	Bos., Det.	4	55	5	10	15	4	22	3	3	6	9	1	1938-39	1941-42
McReynolds, Brian	Wpg., NYR, L.A.	3	30	1	5	6	8	....	....	....	....	....	....	1989-90	1993-94

Andrew McBain

Rob McClanahan

Kevin McClelland

Mike McEwen

Marty McSorley

Larry Melnyk

Max Middendorf

Bob Mongrain

			Regular Schedule					Playoffs							
Name	NHL Teams	NHL Seasons	GP	G	A	TP	PIM	GP	G	A	TP	PIM	NHL Cup Wins	First NHL Season	Last NHL Season
McSheffrey, Bryan	Van., Buf.	3	90	13	7	20	44	....	....	....	....	....	....	1972-73	1974-75
‡ McSorley, Marty	Pit., Edm., L.A., NYR, S.J., Bos.	17	961	108	251	359	3381	115	10	19	29	374	2	1983-84	1999-00
McSween, Don	Buf., Ana.	5	47	3	10	13	55	....	....	....	....	....	....	1987-88	1995-96
McTaggart, Jim	Wsh.	2	71	3	10	13	205	....	....	....	....	....	....	1980-81	1981-82
‡ McTavish, Dale	Cgy.	1	9	1	2	3	2	....	....	....	....	....	....	1996-97	1996-97
McTavish, Gord	St.L., Wpg.	2	11	1	3	4	2	....	....	....	....	....	....	1978-79	1979-80
• McVeigh, Charley	Chi., NYA	9	397	84	88	172	138	4	0	0	0	2	....	1926-27	1934-35
• McVicar, Jack	Mtl.M.	2	88	2	4	6	63	6	0	0	0	2	....	1930-31	1931-32
Meagher, Rick	Mtl., Hfd., N.J., St.L.	12	691	144	165	309	383	62	8	7	15	41	....	1979-80	1990-91
Meehan, Gerry	Tor., Phi., Buf., Van., Atl., Wsh.	10	670	180	243	423	111	10	0	1	1	0	....	1968-69	1978-79
Meeke, Brent	Cal., Cle.	5	75	9	22	31	8	....	....	....	....	....	....	1972-73	1976-77
Meeker, Howie	Tor.	8	346	83	102	185	329	42	6	9	15	50	4	1946-47	1953-54
Meeker, Mike	Pit.	1	4	0	0	0	5	....	....	....	....	....	....	1978-79	1978-79
• Meeking, Harry	Tor., Det., Bos.	3	64	18	12	30	66	9	3	0	3	6	....	1917-18	1926-27
Meger, Paul	Mtl.	6	212	39	52	91	118	35	3	8	11	16	1	1949-50	1954-55
Meighan, Ron	Min., Pit.	2	48	3	7	10	18	....	....	....	....	....	....	1981-82	1982-83
Meissner, Barrie	Min.	2	6	0	1	1	4	....	....	....	....	....	....	1967-68	1968-69
Meissner, Dick	Bos., NYR	5	171	11	15	26	37	....	....	....	....	....	....	1959-60	1964-65
Melametsa, Anssi	Wpg.	1	27	0	3	3	2	....	....	....	....	....	....	1985-86	1985-86
Melin, Roger	Min.	2	3	0	0	0	0	....	....	....	....	....	....	1980-81	1981-82
Mellor, Tom	Det.	2	26	2	4	6	25	....	....	....	....	....	....	1973-74	1974-75
• Melnyk, Gerry	Det., Chi., St.L.	6	269	39	77	116	34	53	6	6	12	6	....	1955-56	1967-68
Melnyk, Larry	Bos., Edm., NYR, Van.	10	432	11	63	74	686	66	2	9	11	127	2	1980-81	1989-90
Melrose, Barry	Wpg., Tor., Det.	6	300	10	23	33	728	7	0	2	2	38	....	1979-80	1985-86
Menard, Hillary	Chi.	1	1	0	0	0	0	....	....	....	....	....	....	1953-54	1953-54
Menard, Howie	Det., L.A., Chi., Oak.	4	151	23	42	65	87	19	3	7	10	36	....	1963-64	1969-70
Mercredi, Vic	Atl.	1	2	0	0	0	0	....	....	....	....	....	....	1974-75	1974-75
Meredith, Greg	Cgy.	2	38	6	4	10	8	5	3	1	4	4	....	1980-81	1982-83
Merkosky, Glenn	Hfd., N.J., Det.	5	66	5	12	17	22	....	....	....	....	....	....	1981-82	1989-90
• Meronek, Bill	Mtl.	2	19	5	8	13	0	1	0	0	0	0	....	1939-40	1942-43
Merrick, Wayne	St.L., Cal., Cle., NYI	12	774	191	265	456	303	102	19	30	49	30	4	1972-73	1983-84
• Merrill, Horace	Ott.	2	8	0	0	0	3	....	....	....	....	....	1	1917-18	1919-20
‡ Mertzig, Jan	NYR	1	23	0	2	2	8	....	....	....	....	....	....	1998-99	1998-99
Messier, Joby	NYR	3	25	0	4	4	24	....	....	....	....	....	....	1992-93	1994-95
Messier, Mitch	Min.	4	20	0	2	2	11	....	....	....	....	....	....	1987-88	1990-91
Messier, Paul	Col.	1	9	0	0	0	4	....	....	....	....	....	....	1978-79	1978-79
‡ Metcalfe, Scott	Edm., Buf.	3	19	1	2	3	18	....	....	....	....	....	....	1987-88	1989-90
Metz, Don	Tor.	9	172	20	35	55	42	42	7	8	15	12	5	1938-39	1948-49
• Metz, Nick	Tor.	12	518	131	119	250	149	76	19	20	39	31	4	1934-35	1947-48
Michaluk, Art	Chi.	1	5	0	0	0	0	....	....	....	....	....	....	1947-48	1947-48
Michaluk, John	Chi.	1	1	0	0	0	0	....	....	....	....	....	....	1950-51	1950-51
Michayluk, Dave	Phi., Pit.	3	14	2	6	8	8	7	1	1	2	0	1	1981-82	1991-92
Micheletti, Joe	St.L., Col.	3	158	11	60	71	114	11	1	11	12	10	....	1979-80	1981-82
Micheletti, Pat	Min.	1	12	2	0	2	8	....	....	....	....	....	....	1987-88	1987-88
• Mickey, Larry	Chi., NYR, Tor., Mtl., L.A., Phi., Buf.	11	292	39	53	92	160	9	1	0	1	10	....	1964-65	1974-75
• Mickoski, Nick	NYR, Chi., Det., Bos.	13	703	158	185	343	319	18	1	6	7	6	....	1947-48	1959-60
Middendorf, Max	Que., Edm.	4	13	2	4	6	6	....	....	....	....	....	....	1986-87	1990-91
Middleton, Rick	NYR, Bos.	14	1005	448	540	988	157	114	45	55	100	19	....	1974-75	1987-88
‡ Miehm, Kevin	St.L.	2	22	1	4	5	8	2	0	1	1	0	....	1992-93	1993-94
Migay, Rudy	Tor.	10	418	59	92	151	293	15	1	0	1	20	....	1949-50	1959-60
Mikita, Stan	Chi.	22	1394	541	926	1467	1270	155	59	91	150	169	1	1958-59	1979-80
Mikkelson, Bill	L.A., NYI, Wsh.	4	147	4	18	22	105	....	....	....	....	....	....	1971-72	1976-77
Mikol, Jim	Tor., NYR	2	34	1	4	5	8	....	....	....	....	....	....	1962-63	1964-65
‡ Mikulchik, Oleg	Wpg., Ana.	3	37	0	3	3	33	....	....	....	....	....	....	1993-94	1995-96
Milbury, Mike	Bos.	12	754	49	189	238	1552	86	4	24	28	219	....	1975-76	1986-87
• Milks, Hib	Pit., Phi., NYR, Ott.	8	317	87	41	128	179	11	0	0	0	2	....	1925-26	1932-33
• Millar, Hugh	Det.	1	4	0	0	0	0	1	0	0	0	0	....	1946-47	1946-47
‡ Millar, Mike	Hfd., Wsh., Bos., Tor.	5	78	18	18	36	12	....	....	....	....	....	....	1986-87	1990-91
‡ Millen, Corey	NYR, L.A., N.J., Dal., Cgy.	8	335	90	119	209	236	47	5	7	12	22	....	1989-90	1996-97
• Miller, Bill	Mtl.M., Mtl.	3	95	7	3	10	16	12	0	0	0	0	1	1934-35	1936-37
Miller, Bob	Bos., Col., L.A.	6	404	75	119	194	220	36	4	7	11	27	....	1977-78	1984-85
Miller, Brad	Buf., Ott., Cgy.	6	82	1	5	6	321	....	....	....	....	....	....	1988-89	1993-94
• Miller, Earl	Chi., Tor.	5	109	19	14	33	124	10	1	0	1	6	1	1927-28	1931-32
Miller, Jack	Chi.	2	17	0	0	0	4	....	....	....	....	....	....	1949-50	1950-51
‡ Miller, Jason	N.J.	3	6	0	0	0	0	....	....	....	....	....	....	1990-91	1992-93
Miller, Jay	Bos., L.A.	7	446	40	44	84	1723	48	2	3	5	243	....	1985-86	1991-92
‡ Miller, Kelly	NYR, Wsh.	15	1057	181	282	463	512	119	20	34	54	65	....	1984-85	1998-99
Miller, Paul	Col.	1	3	0	3	3	0	....	....	....	....	....	....	1981-82	1981-82
Miller, Perry	Det.	4	217	10	51	61	387	....	....	....	....	....	....	1977-78	1980-81
Miller, Tom	Det., NYI	4	118	16	25	41	34	....	....	....	....	....	....	1970-71	1974-75
Miller, Warren	NYR, Hfd.	4	262	40	50	90	137	6	1	0	1	0	....	1979-80	1982-83
Miner, John	Edm.	1	14	2	3	5	16	....	....	....	....	....	....	1987-88	1987-88
Minor, Gerry	Van.	5	140	11	21	32	173	12	1	3	4	25	....	1979-80	1983-84
Miszuk, John	Det., Chi., Phi., Min.	6	237	7	39	46	232	19	0	3	3	19	....	1963-64	1969-70
Mitchell, Bill	Det.	1	1	0	0	0	0	....	....	....	....	....	....	1963-64	1963-64
• Mitchell, Herb	Bos.	2	44	6	0	6	36	....	....	....	....	....	....	1924-25	1925-26
‡ Mitchell, Jeff	Dal.	1	7	0	0	0	7	....	....	....	....	....	....	1997-98	1997-98
Mitchell, Red	Chi.	3	83	4	5	9	67	....	....	....	....	....	....	1941-42	1944-45
‡ Mitchell, Roy	Min.	1	3	0	0	0	0	....	....	....	....	....	....	1992-93	1992-93
• Moe, Bill	NYR	5	261	11	42	53	163	1	0	0	0	0	....	1944-45	1948-49
Moffat, Lyle	Tor., Wpg.	3	97	12	16	28	51	....	....	....	....	....	....	1972-73	1979-80
• Moffatt, Ron	Det.	3	37	1	1	2	8	7	0	0	0	0	....	1932-33	1934-35
‡ Moger, Sandy	Bos., L.A.	5	236	41	38	79	212	5	2	2	4	12	....	1994-95	1998-99
Moher, Mike	N.J.	1	9	0	1	1	28	....	....	....	....	....	....	1982-83	1982-83
Mohns, Doug	Bos., Chi., Min., Atl., Wsh.	22	1390	248	462	710	1250	94	14	36	50	122	....	1953-54	1974-75
Mohns, Lloyd	NYR	1	1	0	0	0	0	....	....	....	....	....	....	1943-44	1943-44
Mokosak, Carl	Cgy., L.A., Phi., Pit., Bos.	6	83	11	15	26	170	1	0	0	0	0	....	1981-82	1988-89
Mokosak, John	Det.	2	41	0	2	2	96	....	....	....	....	....	....	1988-89	1989-90
Molin, Lars	Van.	3	172	33	65	98	37	19	2	9	11	7	....	1981-82	1983-84
Moller, Mike	Buf., Edm.	7	134	15	28	43	41	3	0	1	1	0	....	1980-81	1986-87
Moller, Randy	Que., NYR, Buf., Fla.	14	815	45	180	225	1692	78	6	16	22	197	....	1981-82	1994-95
Molloy, Mitch	Buf.	1	2	0	0	0	10	....	....	....	....	....	....	1989-90	1989-90
• Molyneaux, Larry	NYR	2	45	0	1	1	20	10	0	0	0	8	....	1937-38	1938-39
‡ Momesso, Sergio	Mtl., St.L., Van., Tor., NYR	13	710	152	193	345	1557	119	18	26	44	311	....	1983-84	1996-97
Monahan, Garry	Mtl., Det., L.A., Tor., Van.	12	748	116	169	285	484	22	3	1	4	13	....	1967-68	1978-79
Monahan, Hartland	Cal., NYR, Wsh., Pit., L.A., St.L.	7	334	61	80	141	163	6	0	0	0	4	....	1973-74	1980-81
• Mondou, Armand	Mtl.	12	386	47	71	118	99	32	3	5	8	12	2	1928-29	1939-40
Mondou, Pierre	Mtl.	9	548	194	262	456	179	69	17	28	45	26	3	1976-77	1984-85
‡ Mongeau, Michel	St.L., T.B.	4	54	6	19	25	10	2	0	1	1	0	....	1989-90	1992-93
Mongrain, Bob	Buf., L.A.	6	81	13	14	27	14	11	1	2	3	2	....	1979-80	1985-86
Monteith, Hank	Det.	3	77	5	12	17	6	4	0	0	0	0	....	1968-69	1970-71
Moore, Dickie	Mtl., Tor., St.L.	14	719	261	347	608	652	135	46	64	110	122	6	1951-52	1967-68
• Moran, Amby	Mtl., Chi.	2	35	1	1	2	24	....	....	....	....	....	....	1926-27	1927-28
More, Jay	NYR, Min., S.J., Phx., Chi., Nsh.	10	406	18	54	72	702	31	0	6	6	45	....	1988-89	1998-99
• Morenz, Howie	Mtl., Chi., NYR	14	550	271	201	472	546	39	13	9	22	58	3	1923-24	1936-37
Moretto, Angelo	Cle.	1	5	1	2	3	2	....	....	....	....	....	....	1976-77	1976-77
• Morin, Pete	Mtl.	1	31	10	12	22	7	1	0	0	0	0	....	1941-42	1941-42
• Morin, Stephane	Que., Van.	5	90	16	39	55	52	....	....	....	....	....	....	1989-90	1993-94
Morissette, Dave	Mtl.	2	11	0	0	0	57	....	....	....	....	....	....	1998-99	1999-00
• Morris, Bernie	Bos.	1	6	1	0	1	0	....	....	....	....	....	....	1924-25	1924-25
Morris, Jon	N.J., S.J., Bos.	6	103	16	33	49	47	11	1	7	8	25	....	1988-89	1993-94
• Morris, Moe	Tor., NYR	4	135	13	29	42	58	18	4	2	6	16	1	1943-44	1948-49
Morrison, Dave	L.A., Van.	4	39	3	3	6	4	....	....	....	....	....	....	1980-81	1984-85
Morrison, Don	Det., Chi.	3	112	18	28	46	12	3	0	1	1	0	....	1947-48	1950-51
Morrison, Doug	Bos.	4	23	7	3	10	15	....	....	....	....	....	....	1979-80	1984-85
Morrison, Gary	Phi.	3	43	1	15	16	70	5	0	1	1	2	....	1979-80	1981-82
Morrison, George	St.L.	2	115	17	21	38	13	3	0	0	0	0	....	1970-71	1971-72
Morrison, Jim	Bos., Tor., Det., NYR, Pit.	12	704	40	160	200	542	36	0	12	12	38	....	1951-52	1970-71
• Morrison, John	NYA	1	18	0	0	0	0	....	....	....	....	....	....	1925-26	1925-26
Morrison, Kevin	Col.	1	41	4	11	15	23	....	....	....	....	....	....	1979-80	1979-80
Morrison, Lew	Phi., Atl., Wsh., Pit.	9	564	39	52	91	107	17	0	0	0	2	....	1969-70	1977-78
‡ Morrison, Mark	NYR	2	10	1	1	2	0	....	....	....	....	....	....	1981-82	1983-84
• Morrison, Rod	Det.	1	34	8	7	15	4	3	0	0	0	0	....	1947-48	1947-48
Morrow, Ken	NYI	10	550	17	88	105	309	127	11	22	33	97	4	1979-80	1988-89
‡ Morrow, Scott	Cgy.	1	4	0	0	0	0	....	....	....	....	....	....	1994-95	1994-95
Morton, Dean	Det.	1	1	1	0	1	2	....	....	....	....	....	....	1989-90	1989-90

Name	NHL Teams	NHL Seasons	Regular Schedule GP	G	A	TP	PIM	Playoffs GP	G	A	TP	PIM	NHL Cup Wins	First NHL Season	Last NHL Season
Mortson, Gus	Tor., Chi., Det.	13	797	46	152	198	1380	54	5	8	13	68	4	1946-47	1958-59
Mosdell, Ken	Bro., Mtl., Chi.	16	693	141	168	309	475	80	16	13	29	48	4	1941-42	1958-59
• Mosienko, Bill	Chi.	14	711	258	282	540	121	22	10	4	14	15	....	1941-42	1954-55
Mott, Morris	Cal.	3	199	18	32	50	49	....	....	....	....	....	....	1972-73	1974-75
• Motter, Alex	Bos., Det.	8	255	39	64	103	135	41	3	9	12	41	1	1934-35	1942-43
Moxey, Jim	Cal., Cle., L.A.	3	127	22	27	49	59	....	....	....	....	....	....	1974-75	1976-77
Mulhern, Richard	Atl., L.A., Tor., Wpg.	6	303	27	93	120	217	7	0	3	3	5	....	1975-76	1980-81
‡ Mulhern, Ryan	Wsh.	1	3	0	0	0	0	....	....	....	....	....	....	1997-98	1997-98
Mullen, Brian	Wpg., NYR, S.J., NYI	11	832	260	362	622	414	62	12	18	30	30	....	1982-83	1992-93
Mullen, Joe	St.L., Cgy., Pit., Bos.	17	1062	502	561	1063	241	143	60	46	106	42	3	1979-80	1996-97
Muloin, Wayne	Det., Oak., Cal., Min.	3	147	3	21	24	93	11	0	0	0	2	....	1963-64	1970-71
Mulvenna, Glenn	Pit., Phi.	2	2	0	0	0	4	....	....	....	....	....	....	1991-92	1992-93
Mulvey, Grant	Chi., N.J.	10	586	149	135	284	816	42	10	5	15	70	....	1974-75	1983-84
Mulvey, Paul	Wsh., Pit., L.A.	4	225	30	51	81	613	....	....	....	....	....	....	1978-79	1981-82
• Mummery, Harry	Tor., Que., Mtl., Ham.	6	106	33	19	52	226	2	1	1	2	17	1	1917-18	1922-23
Muni, Craig	Tor., Edm., Chi., Buf., Wpg., Pit., Dal.	16	819	28	119	147	775	113	0	17	17	108	3	1981-82	1997-98
• Munro, Dunc	Mtl.M., Mtl.	8	239	28	18	46	172	21	2	2	4	18	1	1924-25	1931-32
• Munro, Gerry	Mtl.M., Tor.	2	34	1	0	1	37	....	....	....	....	....	....	1924-25	1925-26
Murdoch, Bob	Mtl., L.A., Atl., Cgy.	12	757	60	218	278	764	69	4	18	22	92	2	1970-71	1981-82
Murdoch, Bob	Cal., Cle., St.L.	4	260	72	85	157	127	....	....	....	....	....	....	1975-76	1978-79
Murdoch, Don	NYR, Edm., Det.	6	320	121	117	238	155	24	10	8	18	16	....	1976-77	1981-82
• Murdoch, Murray	NYR	11	508	84	108	192	197	55	9	12	21	28	2	1926-27	1936-37
Murphy, Brian	Det.	1	1	0	0	0	0	....	....	....	....	....	....	1974-75	1974-75
Murphy, Joe	Det., Edm., Chi., St.L., S.J., Bos., Wsh.	15	779	233	295	528	810	120	34	43	77	185	1	1986-87	2000-01
Murphy, Larry	L.A., Wsh., Min., Pit., Tor., Det.	21	1615	287	929	1216	1084	215	37	115	152	201	4	1980-81	2000-01
Murphy, Mike	St.L., NYR, L.A.	12	831	238	318	556	514	66	13	23	36	54	....	1971-72	1982-83
‡ Murphy, Rob	Van., Ott., L.A.	7	125	9	12	21	152	4	0	0	0	2	....	1987-88	1993-94
Murphy, Ron	NYR, Chi., Det., Bos.	18	889	205	274	479	460	53	7	8	15	26	1	1952-53	1969-70
Murray, Allan	NYA	7	271	5	9	14	163	14	0	0	0	10	....	1933-34	1939-40
Murray, Bob	Atl., Van.	4	194	6	16	22	98	10	1	1	2	15	....	1973-74	1976-77
Murray, Bob	Chi.	15	1008	132	382	514	873	112	19	37	56	106	....	1975-76	1989-90
‡ Murray, Chris	Mtl., Hfd., Car., Ott., Chi., Dal.	6	242	16	18	34	550	15	1	0	1	12	....	1994-95	1999-00
Murray, Jim	L.A.	1	30	0	2	2	14	....	....	....	....	....	....	1967-68	1967-68
Murray, Ken	Tor., NYI, Det., K.C.	5	106	1	10	11	135	....	....	....	....	....	....	1969-70	1975-76
• Murray, Leo	Mtl.	1	6	0	0	0	2	....	....	....	....	....	....	1932-33	1932-33
‡ Murray, Mike	Phi.	1	1	0	0	0	0	....	....	....	....	....	....	1987-88	1987-88
Murray, Pat	Phi.	2	25	3	1	4	15	....	....	....	....	....	....	1990-91	1991-92
Murray, Randy	Tor.	1	3	0	0	0	2	....	....	....	....	....	....	1969-70	1969-70
Murray, Terry	Cal., Phi., Det., Wsh.	8	302	4	76	80	199	18	2	2	4	10	....	1972-73	1981-82
Murray, Troy	Chi., Wpg., Ott., Pit., Col.	15	915	230	354	584	875	113	17	26	43	145	1	1981-82	1995-96
Murzyn, Dana	Hfd., Cgy., Van.	14	838	52	152	204	1571	82	9	10	19	166	1	1985-86	1998-99
Myers, Hap	Buf.	1	13	0	0	0	6	....	....	....	....	....	....	1970-71	1970-71
‡ Myhres, Brantt	T.B., Phi., S.J., Nsh., Wsh.	6	153	6	2	8	656	....	....	....	....	....	....	1994-95	2000-01
Myles, Vic	NYR	1	45	6	9	15	57	....	....	....	....	....	....	1942-43	1942-43

N

Name	NHL Teams	NHL Seasons	Regular Schedule GP	G	A	TP	PIM	Playoffs GP	G	A	TP	PIM	NHL Cup Wins	First NHL Season	Last NHL Season
Nachbaur, Don	Hfd., Edm., Phi.	8	223	23	46	69	465	11	1	1	2	24	....	1980-81	1989-90
Nahrgang, Jim	Det.	3	57	5	12	17	34	....	....	....	....	....	....	1974-75	1976-77
Nanne, Lou	Min.	11	635	68	157	225	356	32	4	10	14	8	....	1967-68	1977-78
Nantais, Rich	Min.	3	63	5	4	9	79	....	....	....	....	....	....	1974-75	1976-77
Napier, Mark	Mtl., Min., Edm., Buf.	11	767	235	306	541	157	82	18	24	42	11	2	1978-79	1988-89
Naslund, Mats	Mtl., Bos.	9	651	251	383	634	111	102	35	57	92	33	1	1982-83	1994-95
Nattrass, Ralph	Chi.	4	223	18	38	56	308	....	....	....	....	....	....	1946-47	1949-50
Nattress, Ric	Mtl., St.L., Cgy., Tor., Phi.	11	536	29	135	164	377	67	5	10	15	60	1	1982-83	1992-93
Natyshak, Mike	Que.	1	4	0	0	0	0	....	....	....	....	....	....	1987-88	1987-88
‡ Neaton, Pat	Pit.	1	9	1	1	2	12	....	....	....	....	....	....	1993-94	1993-94
Nechayev, Viktor	L.A.	1	3	1	0	1	0	....	....	....	....	....	....	1982-83	1982-83
Nedomansky, Vaclav	Det., NYR, St.L.	6	421	122	156	278	88	7	3	5	8	0	....	1977-78	1982-83
‡ Nedved, Zdenek	Tor.	3	31	4	6	10	14	....	....	....	....	....	....	1994-95	1996-97
Needham, Mike	Pit., Dal.	3	86	9	5	14	16	14	2	0	2	4	1	1991-92	1993-94
Neely, Bob	Tor., Col.	5	283	39	59	98	266	26	5	7	12	15	....	1973-74	1977-78
Neely, Cam	Van., Bos.	13	726	395	299	694	1241	93	57	32	89	168	....	1983-84	1995-96
Neilson, Jim	NYR, Cal., Cle.	16	1023	69	299	368	904	65	1	17	18	61	....	1962-63	1977-78
Nelson, Gordie	Tor.	1	3	0	0	0	11	....	....	....	....	....	....	1969-70	1969-70
‡ Nelson, Todd	Pit., Wsh.	2	3	1	0	1	2	4	0	0	0	0	....	1991-92	1993-94
Nemeth, Steve	NYR	1	12	2	0	2	2	....	....	....	....	....	....	1987-88	1987-88
Nesterenko, Eric	Tor., Chi.	21	1219	250	324	574	1273	124	13	24	37	127	1	1951-52	1971-72
Nethery, Lance	NYR, Edm.	2	41	11	14	25	14	14	5	3	8	9	....	1980-81	1981-82
Neufeld, Ray	Hfd., Wpg., Bos.	11	595	157	200	357	816	28	8	6	14	55	....	1979-80	1989-90
• Neville, Mike	Tor., NYA	3	65	5	5	10	14	2	0	0	0	0	....	1924-25	1930-31
Nevin, Bob	Tor., NYR, Min., L.A.	18	1128	307	419	726	211	84	16	18	34	24	2	1957-58	1975-76
Newberry, John	Mtl., Hfd.	4	22	0	4	4	6	2	0	0	0	0	....	1982-83	1985-86
Newell, Rick	Det.	2	6	0	0	0	0	....	....	....	....	....	....	1972-73	1973-74
Newman, Dan	NYR, Mtl., Edm.	4	126	17	24	41	63	3	0	0	0	4	....	1976-77	1979-80
• Newman, John	Det.	1	8	1	1	2	0	....	....	....	....	....	....	1930-31	1930-31
Nicholls, Bernie	L.A., NYR, Edm., N.J., Chi., S.J.	18	1127	475	734	1209	1292	118	42	72	114	164	....	1981-82	1998-99
Nicholson, Al	Bos.	2	19	0	1	1	4	....	....	....	....	....	....	1955-56	1956-57
• Nicholson, Ed	Det.	1	1	0	0	0	0	....	....	....	....	....	....	1947-48	1947-48
• Nicholson, Hickey	Chi.	1	2	1	0	1	0	....	....	....	....	....	....	1937-38	1937-38
Nicholson, Neil	Oak., NYI	4	39	3	1	4	23	2	0	0	0	0	....	1969-70	1977-78
Nicholson, Paul	Wsh.	3	62	4	8	12	18	....	....	....	....	....	....	1974-75	1976-77
Nicolson, Graeme	Bos., Col., NYR	3	52	2	7	9	60	....	....	....	....	....	....	1978-79	1982-83
‡ Nieckar, Barry	Hfd., Cgy., Ana.	4	8	0	0	0	21	....	....	....	....	....	....	1992-93	1997-98
Niekamp, Jim	Det.	2	29	0	2	2	37	....	....	....	....	....	....	1970-71	1971-72
‡ Nielsen, Jeff	NYR, Ana., Min.	5	252	20	27	47	70	4	0	0	0	2	....	1996-97	2000-01
Nielsen, Kirk	Bos.	1	6	0	0	0	0	....	....	....	....	....	....	1997-98	1997-98
‡ Nienhuis, Kraig	Bos.	3	87	20	16	36	39	2	0	0	0	14	....	1985-86	1987-88
• Nighbor, Frank	Ott., Tor.	13	349	139	98	237	249	20	4	9	13	13	4	1917-18	1929-30
Nigro, Frank	Tor.	2	68	8	18	26	39	3	0	0	0	2	....	1982-83	1983-84
‡ Nikulin, Igor	Ana.	1	....	....	....	....	....	1	0	0	0	0	....	1996-97	1996-97
Nilan, Chris	Mtl., NYR, Bos.	13	688	110	115	225	3043	111	8	9	17	541	1	1979-80	1991-92
Nill, Jim	St.L., Van., Bos., Wpg., Det.	9	524	58	87	145	854	59	10	5	15	203	....	1981-82	1989-90
Nilsson, Kent	Atl., Cgy., Min., Edm.	9	553	264	422	686	116	59	11	41	52	14	1	1979-80	1994-95
Nilsson, Ulf	NYR	4	170	57	112	169	85	25	8	14	22	27	....	1978-79	1982-83
Nistico, Lou	Col.	1	3	0	0	0	0	....	....	....	....	....	....	1977-78	1977-78
• Noble, Reg	Tor., Mtl.M., Det.	16	510	168	106	274	916	18	2	2	4	33	3	1917-18	1932-33
Noel, Claude	Wsh.	1	7	0	0	0	0	....	....	....	....	....	....	1979-80	1979-80
• Nolan, Paddy	Tor.	1	2	0	0	0	0	....	....	....	....	....	....	1921-22	1921-22
Nolan, Ted	Det., Pit.	3	78	6	16	22	105	....	....	....	....	....	....	1981-82	1985-86
Nolet, Simon	Phi., K.C., Pit., Col.	10	562	150	182	332	187	34	6	3	9	8	1	1967-68	1976-77
Noonan, Brian	Chi., NYR, St.L., Van., Phx.	12	629	116	159	275	518	71	17	19	36	77	1	1987-88	1998-99
‡ Nordmark, Robert	St.L., Van.	4	236	13	70	83	254	7	3	2	5	8	....	1987-88	1990-91
Noris, Joe	Pit., St.L., Buf.	3	55	2	5	7	22	....	....	....	....	....	....	1971-72	1973-74
‡ Norris, Dwayne	Que., Ana.	3	20	2	4	6	8	....	....	....	....	....	....	1993-94	1995-96
Norrish, Rod	Min.	2	21	3	3	6	2	....	....	....	....	....	....	1973-74	1974-75
• Northcott, Baldy	Mtl.M., Chi.	11	446	133	112	245	273	31	8	5	13	14	1	1928-29	1938-39
Norwich, Craig	Wpg., St.L., Col.	2	104	17	58	75	60	....	....	....	....	....	....	1979-80	1980-81
Norwood, Lee	Que., Wsh., St.L., Det., N.J., Hfd., Cgy.	12	503	58	153	211	1099	65	6	22	28	171	....	1980-81	1993-94
Novy, Milan	Wsh.	1	73	18	30	48	16	2	0	0	0	0	....	1982-83	1982-83
Nowak, Hank	Pit., Det., Bos.	4	180	26	29	55	161	13	1	0	1	8	....	1973-74	1976-77
Nykoluk, Mike	Tor.	1	32	3	1	4	20	....	....	....	....	....	....	1956-57	1956-57
Nylund, Gary	Tor., Chi., NYI	11	608	32	139	171	1235	24	0	6	6	63	....	1982-83	1992-93
• Nyrop, Bill	Mtl., Min.	4	207	12	51	63	101	35	1	7	8	22	3	1975-76	1981-82
Nystrom, Bob	NYI	14	900	235	278	513	1248	157	39	44	83	236	4	1972-73	1985-86

O

Name	NHL Teams	NHL Seasons	Regular Schedule GP	G	A	TP	PIM	Playoffs GP	G	A	TP	PIM	NHL Cup Wins	First NHL Season	Last NHL Season
• Oatman, Russell	Det., Mtl.M., NYR	3	120	20	9	29	100	15	1	0	1	18	....	1926-27	1928-29
O'Brien, Dennis	Min., Col., Cle., Bos.	10	592	31	91	122	1017	34	1	2	3	101	....	1970-71	1979-80
O'Brien, Ellard	Bos.	1	2	0	0	0	0	....	....	....	....	....	....	1955-56	1955-56
O'Callahan, Jack	Chi., N.J.	7	389	27	104	131	541	32	4	11	15	41	....	1982-83	1988-89
O'Connell, Mike	Chi., Bos., Det.	13	860	105	334	439	605	82	8	24	32	64	....	1977-78	1989-90
• O'Connor, Buddy	Mtl., NYR	10	509	140	257	397	34	53	15	21	36	6	2	1941-42	1950-51
O'Connor, Myles	N.J., Ana.	4	43	3	4	7	69	....	....	....	....	....	....	1990-91	1993-94
Oddleifson, Chris	Bos., Van.	9	524	95	191	286	464	14	1	6	7	8	....	1972-73	1980-81

Dickie Moore

Bill Mosienko

Rob Murphy

Frank Nighbor

Ulf Nillson

Gary Nylund

Gates Orlando

Bobby Orr

Name	NHL Teams	NHL Seasons	Regular Schedule GP	G	A	TP	PIM	Playoffs GP	G	A	TP	PIM	NHL Cup Wins	First NHL Season	Last NHL Season
Odelein, Selmar	Edm.	3	18	0	2	2	35	....	....	....	....	....	....	1985-86	1988-89
O'Donnell, Fred	Bos.	2	115	15	11	26	98	5	0	1	1	5	....	1972-73	1973-74
O'Donoghue, Don	Oak., Cal.	3	125	18	17	35	35	3	0	0	0	0	....	1969-70	1971-72
Odrowski, Gerry	Det., Oak., St.L.	6	309	12	19	31	111	30	0	1	1	16	....	1960-61	1971-72
O'Dwyer, Bill	L.A., Bos.	5	120	9	13	22	108	10	0	0	0	2	....	1983-84	1989-90
O'Flaherty, Gerry	Tor., Van., Atl.	8	438	99	95	194	168	7	2	2	4	6	....	1971-72	1978-79
O'Flaherty, Peanuts	NYA, Bro.	2	21	5	1	6	0	....	....	....	....	....	....	1940-41	1941-42
Ogilvie, Brian	Chi., St.L.	6	90	15	21	36	29	....	....	....	....	....	....	1972-73	1978-79
● O'Grady, George	Mtl.W.	1	4	0	0	0	0	....	....	....	....	....	....	1917-18	1917-18
Ogrodnick, John	Det., Que., NYR	14	928	402	425	827	260	41	18	8	26	6	....	1979-80	1992-93
‡ Ojanen, Janne	N.J.	4	98	21	23	44	28	3	0	2	2	0	....	1988-89	1992-93
Okerlund, Todd	NYI	1	4	0	0	0	2	....	....	....	....	....	....	1987-88	1987-88
Oksiuta, Roman	Edm., Van., Ana., Pit.	4	153	46	41	87	100	10	2	3	5	0	....	1993-94	1996-97
Olczyk, Ed	Chi., Tor., Wpg., NYR, L.A., Pit.	16	1031	342	452	794	874	57	19	15	34	57	1	1984-85	1999-00
● Oliver, Harry	Bos., NYA	11	463	127	85	212	147	35	10	6	16	24	1	1926-27	1936-37
Oliver, Murray	Det., Bos., Tor., Min.	17	1127	274	454	728	320	35	9	16	25	10	....	1957-58	1974-75
Olmstead, Bert	Chi., Mtl., Tor.	14	848	181	421	602	884	115	16	43	59	101	5	1948-49	1961-62
Olsen, Darryl	Cgy.	1	1	0	0	0	0	....	....	....	....	....	....	1991-92	1991-92
Olson, Dennis	Det.	1	4	0	0	0	0	....	....	....	....	....	....	1957-58	1957-58
‡ Olsson, Christer	St.L., Ott.	2	56	4	12	16	24	3	0	0	0	0	....	1995-96	1996-97
● O'Neil, Jim	Bos., Mtl.	6	156	6	30	36	109	9	1	1	2	13	....	1933-34	1941-42
O'Neil, Paul	Van., Bos.	2	6	0	0	0	0	....	....	....	....	....	....	1973-74	1975-76
● O'Neill, Tom	Tor.	2	66	10	12	22	53	4	0	0	0	6	1	1943-44	1944-45
Orban, Bill	Chi., Min.	3	114	8	15	23	67	3	0	0	0	0	....	1967-68	1969-70
O'Ree, Willie	Bos.	2	45	4	10	14	26	....	....	....	....	....	....	1957-58	1960-61
O'Regan, Tom	Pit.	3	61	5	12	17	10	....	....	....	....	....	....	1983-84	1985-86
O'Reilly, Terry	Bos.	14	891	204	402	606	2095	108	25	42	67	335	....	1971-72	1984-85
Orlando, Gates	Buf.	3	98	18	26	44	51	5	0	4	4	14	....	1984-85	1986-87
● Orlando, Jimmy	Det.	6	199	6	25	31	375	36	0	9	9	105	1	1936-37	1942-43
Orleski, Dave	Mtl.	2	2	0	0	0	0	....	....	....	....	....	....	1980-81	1981-82
Orr, Bobby	Bos., Chi.	12	657	270	645	915	953	74	26	66	92	107	2	1966-67	1978-79
Osborne, Keith	St.L., T.B.	2	16	1	3	4	16	....	....	....	....	....	....	1989-90	1992-93
Osborne, Mark	Det., NYR, Tor., Wpg.	14	919	212	319	531	1152	87	12	16	28	141	....	1981-82	1994-95
Osburn, Randy	Tor., Phi.	2	27	0	2	2	0	....	....	....	....	....	....	1972-73	1974-75
O'Shea, Danny	Min., Chi., St.L.	5	369	64	115	179	265	39	3	7	10	61	....	1968-69	1972-73
O'Shea, Kevin	Buf., St.L.	3	134	13	18	31	85	12	2	1	3	10	....	1970-71	1972-73
Osiecki, Mark	Cgy., Ott., Wpg., Min.	2	93	3	11	14	43	....	....	....	....	....	....	1991-92	1992-93
Otevrel, Jaroslav	S.J.	2	16	3	4	7	2	....	....	....	....	....	....	1992-93	1993-94
Otto, Joel	Cgy., Phi.	14	943	195	313	508	1934	122	27	47	74	207	1	1984-85	1997-98
Ouellette, Eddie	Chi.	1	43	3	2	5	11	1	0	0	0	0	....	1935-36	1935-36
Ouellette, Gerry	Bos.	1	34	5	4	9	0	....	....	....	....	....	....	1960-61	1960-61
Owchar, Dennis	Pit., Col.	6	288	30	85	115	200	10	1	1	2	8	....	1974-75	1979-80
● Owen, George	Bos.	5	183	44	33	77	151	21	2	5	7	25	1	1928-29	1932-33

P

Name	NHL Teams	NHL Seasons	Regular Schedule GP	G	A	TP	PIM	Playoffs GP	G	A	TP	PIM	NHL Cup Wins	First NHL Season	Last NHL Season
Pachal, Clayton	Bos., Col.	3	35	2	3	5	95	....	....	....	....	....	....	1976-77	1978-79
Paddock, John	Wsh., Phi., Que.	5	87	8	14	22	86	5	2	0	2	0	....	1975-76	1982-83
‡ Paek, Jim	Pit., L.A., Ott.	5	217	5	29	34	155	27	1	4	5	8	2	1990-91	1994-95
Paiement, Rosaire	Phi., Van.	5	190	48	52	100	343	3	3	0	3	0	....	1967-68	1971-72
Paiement, Wilf	K.C., Col., Tor., Que., NYR, Buf., Pit.	14	946	356	458	814	1757	69	18	17	35	185	....	1974-75	1987-88
Palangio, Pete	Mtl., Det., Chi.	5	71	13	10	23	28	7	0	0	0	0	1	1926-27	1937-38
Palazzari, Aldo	Bos., NYR	1	35	8	3	11	4	....	....	....	....	....	....	1943-44	1943-44
Palazzari, Doug	St.L.	4	108	18	20	38	23	2	0	0	0	0	....	1974-75	1978-79
Palmer, Brad	Min., Bos.	3	168	32	38	70	58	29	9	5	14	16	....	1980-81	1982-83
Palmer, Rob	Chi.	3	16	0	3	3	2	....	....	....	....	....	....	1973-74	1975-76
Palmer, Robert	L.A., N.J.	7	320	9	101	110	115	8	1	2	3	6	....	1977-78	1983-84
● Panagabko, Ed	Bos.	2	29	0	3	3	38	....	....	....	....	....	....	1955-56	1956-57
‡ Panteleev, Grigori	Bos., NYI	4	54	8	6	14	12	....	....	....	....	....	....	1992-93	1995-96
● Papike, Joe	Chi.	3	20	3	3	6	4	5	0	2	2	0	....	1940-41	1944-45
Pappin, Jim	Tor., Chi., Cal., Cle.	14	767	278	295	573	667	92	33	34	67	101	2	1963-64	1976-77
Paradise, Bob	Min., Atl., Pit., Wsh.	8	368	8	54	62	393	12	0	1	1	19	....	1971-72	1978-79
Pargeter, George	Mtl.	1	4	0	0	0	0	....	....	....	....	....	....	1946-47	1946-47
Parise, Jean-Paul	Bos., Tor., Min., NYI, Cle.	14	890	238	356	594	706	86	27	31	58	87	....	1965-66	1978-79
Parizeau, Michel	St.L., Phi.	1	58	3	14	17	18	....	....	....	....	....	....	1971-72	1971-72
Park, Brad	NYR, Bos., Det.	17	1113	213	683	896	1429	161	35	90	125	217	....	1968-69	1984-85
Parker, Jeff	Buf., Hfd.	5	141	16	19	35	163	5	0	0	0	26	....	1986-87	1990-91
● Parkes, Ernie	Mtl.M.	1	17	0	0	0	2	....	....	....	....	....	....	1924-25	1924-25
‡ Parks, Greg	NYI	3	23	1	2	3	6	2	0	0	0	0	....	1990-91	1992-93
● Parsons, George	Tor.	3	78	12	13	25	20	7	3	2	5	11	....	1936-37	1938-39
● Pasek, Dusan	Min.	1	48	4	10	14	30	2	1	0	1	0	....	1988-89	1988-89
Pasin, Dave	Bos., L.A.	2	76	18	19	37	50	3	0	1	1	0	....	1985-86	1988-89
Paslawski, Greg	Mtl., St.L., Wpg., Buf., Que., Phi., Cgy.	11	650	187	185	372	169	60	19	13	32	25	....	1983-84	1993-94
Paterson, Joe	Det., Phi., L.A., NYR	9	291	19	37	56	829	22	3	4	7	77	....	1980-81	1988-89
Paterson, Mark	Hfd.	4	29	3	3	6	33	....	....	....	....	....	....	1982-83	1985-86
Paterson, Rick	Chi.	9	430	50	43	93	136	61	7	10	17	51	....	1978-79	1986-87
Patey, Doug	Wsh.	3	45	4	2	6	8	....	....	....	....	....	....	1976-77	1978-79
Patey, Larry	Cal., St.L., NYR	12	717	153	163	316	631	40	8	10	18	57	....	1973-74	1984-85
Patrick, Craig	Cal., St.L., K.C., Wsh.	8	401	72	91	163	61	2	0	1	1	0	....	1971-72	1978-79
Patrick, Glenn	St.L., Cal., Cle.	4	38	2	3	5	72	....	....	....	....	....	....	1973-74	1976-77
● Patrick, Lester	NYR	1	1	0	0	0	2	....	....	....	....	....	....	1926-27	1926-27
● Patrick, Lynn	NYR	10	455	145	190	335	240	44	10	6	16	22	1	1934-35	1945-46
● Patrick, Muzz	NYR	5	166	5	26	31	133	25	4	0	4	34	1	1937-38	1945-46
Patrick, Steve	Buf., NYR, Que.	6	250	40	68	108	242	12	0	1	1	12	....	1980-81	1985-86
Patterson, Colin	Cgy., Buf.	10	504	96	109	205	239	85	12	17	29	57	1	1983-84	1992-93
Patterson, Dennis	K.C., Phi.	3	138	6	22	28	67	....	....	....	....	....	....	1974-75	1979-80
‡ Patterson, Ed	Pit.	3	68	3	3	6	56	....	....	....	....	....	....	1993-94	1996-97
● Patterson, George	Tor., Mtl., NYA, Bos., Det., St.L.	9	284	51	27	78	218	3	0	0	0	2	....	1926-27	1934-35
● Paul, Butch	Det.	1	3	0	0	0	0	....	....	....	....	....	....	1964-65	1964-65
● Paulhus, Rollie	Mtl.	1	33	0	0	0	0	....	....	....	....	....	....	1925-26	1925-26
Pavelich, Mark	NYR, Min., S.J.	7	355	137	192	329	340	23	7	17	24	14	....	1981-82	1991-92
Pavelich, Marty	Det.	10	634	93	159	252	454	91	13	15	28	74	4	1947-48	1956-57
Pavese, Jim	St.L., NYR, Det., Hfd.	8	328	13	44	57	689	34	0	6	6	81	....	1981-82	1988-89
● Payer, Evariste	Mtl.	1	1	0	0	0	0	....	....	....	....	....	....	1917-18	1917-18
Payne, Davis	Bos.	2	22	0	1	1	14	....	....	....	....	....	....	1995-96	1996-97
Payne, Steve	Min.	10	613	228	238	466	435	71	35	35	70	60	....	1978-79	1987-88
Paynter, Kent	Chi., Wsh., Wpg., Ott.	7	37	1	3	4	69	4	0	0	0	10	....	1987-88	1993-94
Peake, Pat	Wsh.	5	134	28	41	69	105	13	2	2	4	20	....	1993-94	1997-98
● Pearson, Mel	NYR, Pit.	5	38	2	6	8	25	....	....	....	....	....	....	1959-60	1967-68
‡ Pearson, Rob	Tor., Wsh., St.L.	6	269	56	54	110	645	33	4	2	6	94	....	1991-92	1996-97
Pearson, Scott	Tor., Que., Edm., Buf., NYI	10	292	56	42	98	615	10	2	0	2	14	....	1988-89	1999-00
Pedersen, Allen	Bos., Min., Hfd.	8	428	5	36	41	487	64	0	0	0	91	....	1986-87	1993-94
Pederson, Barry	Bos., Van., Pit., Hfd.	12	701	238	416	654	472	34	22	30	52	25	1	1980-81	1991-92
‡ Pederson, Mark	Mtl., Phi., S.J., Det.	5	169	35	50	85	77	2	0	0	0	0	....	1989-90	1993-94
Pederson, Tom	S.J., Tor.	5	240	20	49	69	142	24	1	11	12	10	....	1992-93	1996-97
● Peer, Bert	Det.	1	1	0	0	0	0	....	....	....	....	....	....	1939-40	1939-40
Peirson, Johnny	Bos.	11	545	153	173	326	315	49	10	16	26	26	....	1946-47	1957-58
Pelensky, Perry	Chi.	1	4	0	0	0	5	....	....	....	....	....	....	1983-84	1983-84
Pelletier, Roger	Phi.	1	1	0	0	0	0	....	....	....	....	....	....	1967-68	1967-68
Peloffy, Andre	Wsh.	1	9	0	0	0	0	....	....	....	....	....	....	1974-75	1974-75
Peluso, Mike	Chi., Ott., N.J., St.L., Cgy.	9	458	38	52	90	1951	62	3	4	7	107	1	1989-90	1997-98
Pelyk, Mike	Tor.	9	441	26	88	114	566	40	0	3	3	41	....	1967-68	1977-78
Penney, Chad	Ott.	1	3	0	0	0	2	....	....	....	....	....	....	1993-94	1993-94
Pennington, Cliff	Mtl., Bos.	3	101	17	42	59	6	....	....	....	....	....	....	1960-61	1962-63
Peplinski, Jim	Cgy.	11	711	161	263	424	1467	99	15	31	46	382	1	1980-81	1994-95
Perlini, Fred	Tor.	2	8	2	3	5	0	....	....	....	....	....	....	1981-82	1983-84
Perreault, Fern	NYR	2	3	0	0	0	0	....	....	....	....	....	....	1947-48	1949-50
Perreault, Gilbert	Buf.	17	1191	512	814	1326	500	90	33	70	103	44	....	1970-71	1986-87
Perry, Brian	Oak., Buf.	3	96	16	29	45	24	8	1	1	2	4	....	1968-69	1970-71
Persson, Stefan	NYI	9	622	52	317	369	574	102	7	50	57	69	4	1977-78	1985-86
Pesut, George	Cal.	2	92	3	22	25	130	....	....	....	....	....	....	1974-75	1975-76
● Peters, Frank	NYR	1	43	0	0	0	59	4	0	0	0	2	....	1930-31	1930-31
Peters, Garry	Mtl., NYR, Phi., Bos.	8	311	34	34	68	261	9	2	2	4	31	1	1964-65	1971-72
Peters, Jimmy	Mtl., Bos., Det., Chi.	9	574	125	150	275	186	60	5	9	14	22	3	1945-46	1953-54
Peters, Jimmy Jr.	Det., L.A.	9	309	37	36	73	48	11	0	2	2	2	....	1964-65	1974-75
Peters, Steve	Col.	1	2	0	1	1	0	....	....	....	....	....	....	1979-80	1979-80

Name	NHL Teams	NHL Seasons	Regular Schedule					Playoffs					NHL Cup Wins	First NHL Season	Last NHL Season
			GP	G	A	TP	PIM	GP	G	A	TP	PIM			
Peterson, Brent	Det., Buf., Van., Hfd.	11	620	72	141	213	484	31	4	4	8	65	....	1978-79	1988-89
‡ Petit, Michel	Van., NYR, Que., Tor., Cgy., L.A., T.B., Edm., Phi., Phx.	16	827	90	238	328	1839	19	0	2	2	61	....	1982-83	1997-98
‡ Petrenko, Sergei	Buf.	1	14	0	4	4	0	....	....	....	....	....	....	1993-94	1993-94
Pettersson, Jorgen	St.L., Hfd., Wsh.	6	435	174	192	366	117	44	15	12	27	4	....	1980-81	1985-86
• Pettinger, Eric	Bos., Tor., Ott.	3	98	7	12	19	83	4	1	0	1	8	....	1928-29	1930-31
• Pettinger, Gord	NYR, Det., Bos.	8	292	42	74	116	77	47	4	5	9	11	4	1932-33	1939-40
Phair, Lyle	L.A.	3	48	6	7	13	12	1	0	0	0	0	....	1985-86	1987-88
Phillipoff, Harold	Atl., Chi.	3	141	26	57	83	267	6	0	2	2	9	....	1977-78	1979-80
• Phillips, Bill	Mtl.M.	1	27	1	1	2	6	4	0	0	0	2	....	1929-30	1929-30
Phillips, Charlie	Mtl.	1	17	0	0	0	6	....	....	....	....	....	....	1942-43	1942-43
• Phillips, Merlyn	Mtl.M., NYA	8	302	52	31	83	232	24	5	1	6	19	1	1925-26	1932-33
Picard, Noel	Mtl., St.L., Atl.	7	335	12	63	75	616	50	2	11	13	167	1	1964-65	1972-73
Picard, Robert	Wsh., Tor., Mtl., Wpg., Que., Det.	13	899	104	319	423	1025	36	5	15	20	39	....	1977-78	1989-90
Picard, Roger	St.L.	1	15	2	2	4	21	....	....	....	....	....	....	1967-68	1967-68
Pichette, Dave	Que., St.L., N.J., NYR	7	322	41	140	181	348	28	3	7	10	54	....	1980-81	1987-88
Picketts, Hal	NYA	1	48	3	1	4	32	....	....	....	....	....	....	1933-34	1933-34
Pidhirny, Harry	Bos.	1	2	0	0	0	0	....	....	....	....	....	....	1957-58	1957-58
Pierce, Randy	Col., N.J., Hfd.	8	277	62	76	138	223	2	0	0	0	0	....	1977-78	1984-85
Pike, Alf	NYR	6	234	42	77	119	145	21	4	2	6	12	1	1939-40	1946-47
Pilote, Pierre	Chi., Tor.	14	890	80	418	498	1251	86	8	53	61	102	1	1955-56	1968-69
Pinder, Gerry	Chi., Cal.	3	223	55	69	124	135	17	0	4	4	6	....	1969-70	1971-72
Pirus, Alex	Min., Det.	4	159	30	28	58	94	2	0	1	1	2	....	1976-77	1979-80
• Pitre, Didier	Mtl.	6	127	64	34	98	84	9	2	4	6	16	....	1917-18	1922-23
Pivonka, Michal	Wsh.	13	825	181	418	599	478	95	19	36	55	86	....	1986-87	1998-99
• Plager, Barclay	St.L.	10	614	44	187	231	1115	68	3	20	23	182	....	1967-68	1976-77
Plager, Bill	Min., St.L., Atl.	9	263	4	34	38	294	31	0	2	2	26	....	1967-68	1975-76
Plager, Bob	NYR, St.L.	14	644	20	126	146	802	74	2	17	19	195	....	1964-65	1977-78
Plamondon, Gerry	Mtl.	5	74	7	13	20	10	11	5	2	7	2	1	1945-46	1950-51
Plante, Cam	Tor.	1	2	0	0	0	0	....	....	....	....	....	....	1984-85	1984-85
Plante, Pierre	Phi., St.L., Chi., NYR, Que.	9	599	125	172	297	599	33	2	6	8	51	....	1971-72	1979-80
Plantery, Mark	Wpg.	1	25	1	5	6	14	....	....	....	....	....	....	1980-81	1980-81
‡ Plavsic, Adrien	St.L., Van., T.B., Ana.	8	214	16	56	72	161	13	1	7	8	4	....	1989-90	1996-97
• Plaxton, Hugh	Mtl.M.	1	15	1	2	3	4	....	....	....	....	....	....	1932-33	1932-33
Playfair, Jim	Edm., Chi.	3	21	2	4	6	51	....	....	....	....	....	....	1983-84	1988-89
Playfair, Larry	Buf., L.A.	12	688	26	94	120	1812	43	0	6	6	111	....	1978-79	1989-90
Pleau, Larry	Mtl.	3	94	9	15	24	27	4	0	0	0	0	....	1969-70	1971-72
• Pletsch, Charles	Ham.	1	1	0	0	0	0	....	....	....	....	....	....	1920-21	1920-21
Plett, Willi	Atl., Cgy., Min., Bos.	13	834	222	215	437	2572	83	24	22	46	466	....	1975-76	1987-88
Plumb, Rob	Det.	2	14	3	2	5	2	....	....	....	....	....	....	1977-78	1978-79
Plumb, Ron	Hfd.	1	26	3	4	7	14	....	....	....	....	....	....	1979-80	1979-80
Pocza, Harvie	Wsh.	2	3	0	0	0	2	....	....	....	....	....	....	1979-80	1981-82
Poddubny, Walt	Edm., Tor., NYR, Que., N.J.	11	468	184	238	422	454	19	7	2	9	12	....	1981-82	1991-92
‡ Podloski, Ray	Bos.	1	8	0	1	1	17	....	....	....	....	....	....	1988-89	1988-89
Podolsky, Nels	Det.	1	1	0	0	0	0	7	0	0	0	4	....	1948-49	1948-49
Poeschek, Rudy	NYR, Wpg., T.B., St.L.	12	364	6	25	31	817	5	0	0	0	18	....	1987-88	1999-00
Poeta, Tony	Chi.	1	1	0	0	0	0	....	....	....	....	....	....	1951-52	1951-52
Poile, Bud	Tor., Chi., Det., NYR, Bos.	7	311	107	122	229	91	23	4	5	9	8	1	1942-43	1949-50
Poile, Don	Det.	2	66	7	9	16	12	4	0	0	0	0	....	1954-55	1957-58
Poirier, Gordie	Mtl.	1	10	0	0	0	0	....	....	....	....	....	....	1939-40	1939-40
Polanic, Tom	Min.	2	19	0	2	2	53	5	1	1	2	4	....	1969-70	1970-71
• Polich, John	NYR	2	3	0	1	1	0	....	....	....	....	....	....	1939-40	1940-41
Polich, Mike	Mtl., Min.	5	226	24	29	53	57	23	2	1	3	2	1	1976-77	1980-81
Polis, Greg	Pit., St.L., NYR, Wsh.	10	615	174	169	343	391	7	0	2	2	6	....	1970-71	1979-80
Poliziani, Dan	Bos.	1	1	0	0	0	0	3	0	0	0	0	....	1958-59	1958-59
Polonich, Dennis	Det.	8	390	59	82	141	1242	7	1	0	1	19	....	1974-75	1982-83
Pooley, Paul	Wpg.	2	15	0	3	3	0	....	....	....	....	....	....	1984-85	1985-86
Popein, Larry	NYR, Oak.	8	449	80	141	221	162	16	1	4	5	6	....	1954-55	1967-68
Popiel, Poul	Bos., L.A., Det., Van., Edm.	7	224	13	41	54	210	4	1	0	1	4	....	1965-66	1979-80
• Portland, Jack	Mtl., Bos., Chi.	10	381	15	56	71	323	33	1	3	4	25	1	1933-34	1942-43
Porvari, Jukka	Col., N.J.	2	39	3	9	12	4	....	....	....	....	....	....	1981-82	1982-83
Posa, Victor	Chi.	1	2	0	0	0	2	....	....	....	....	....	....	1985-86	1985-86
Posavad, Mike	St.L.	2	8	0	0	0	0	....	....	....	....	....	....	1985-86	1986-87
‡ Potomski, Barry	L.A., S.J.	3	68	6	5	11	227	....	....	....	....	....	....	1995-96	1997-98
Potvin, Denis	NYI	15	1060	310	742	1052	1356	185	56	108	164	253	4	1973-74	1987-88
Potvin, Jean	L.A., Phi., NYI, Cle., Min.	11	613	63	224	287	478	39	2	9	11	17	1	1970-71	1980-81
Potvin, Marc	Det., L.A., Hfd., Bos.	6	121	3	5	8	456	13	0	1	1	50	....	1990-91	1995-96
‡ Poudrier, Daniel	Que.	3	25	1	5	6	10	....	....	....	....	....	....	1985-86	1987-88
Poulin, Daniel	Min.	1	3	1	1	2	2	....	....	....	....	....	....	1981-82	1981-82
Poulin, Dave	Phi., Bos., Wsh.	13	724	205	325	530	482	129	31	42	73	132	....	1982-83	1994-95
Pouzar, Jaroslav	Edm.	4	186	34	48	82	135	29	6	4	10	16	3	1982-83	1986-87
• Powell, Ray	Chi.	1	31	7	15	22	2	....	....	....	....	....	....	1950-51	1950-51
Powis, Geoff	Chi.	1	2	0	0	0	0	....	....	....	....	....	....	1967-68	1967-68
Powis, Lynn	Chi., K.C.	2	130	19	33	52	25	1	0	0	0	0	....	1973-74	1974-75
Prajsler, Petr	L.A., Bos.	4	46	3	10	13	51	4	0	0	0	0	....	1987-88	1991-92
• Pratt, Babe	NYR, Tor., Bos.	12	517	83	209	292	463	63	12	17	29	90	2	1935-36	1946-47
Pratt, Jack	Bos.	2	37	2	0	2	42	4	0	0	0	0	....	1930-31	1931-32
Pratt, Kelly	Pit.	1	22	0	6	6	15	....	....	....	....	....	....	1974-75	1974-75
Pratt, Tracy	Oak., Pit., Buf., Van., Col., Tor.	10	580	17	97	114	1026	25	0	1	1	62	....	1967-68	1976-77
Prentice, Dean	NYR, Bos., Det., Pit., Min.	22	1378	391	469	860	484	54	13	17	30	38	....	1952-53	1973-74
Prentice, Eric	Tor.	1	5	0	0	0	4	....	....	....	....	....	....	1943-44	1943-44
Presley, Wayne	Chi., S.J., Buf., NYR, Tor.	12	684	155	147	302	953	83	26	17	43	142	....	1984-85	1995-96
Preston, Rich	Chi., N.J.	8	580	127	164	291	348	47	4	18	22	56	....	1979-80	1986-87
Preston, Yves	Phi.	2	28	7	3	10	4	....	....	....	....	....	....	1978-79	1980-81
Priakin, Sergei	Cgy.	3	46	3	8	11	2	1	0	0	0	0	....	1988-89	1990-91
Price, Jack	Chi.	3	57	4	6	10	24	4	0	0	0	0	....	1951-52	1953-54
Price, Noel	Tor., NYR, Det., Mtl., Pit., L.A., Atl.	14	499	14	114	128	333	12	0	1	1	8	1	1957-58	1975-76
Price, Pat	NYI, Edm., Pit., Que., NYR, Min.	13	726	43	218	261	1456	74	2	10	12	195	....	1975-76	1987-88
Price, Tom	Cal., Cle., Pit.	5	29	0	2	2	12	....	....	....	....	....	....	1974-75	1978-79
Priestlay, Ken	Buf., Pit.	6	168	27	34	61	63	14	0	0	0	21	1	1986-87	1991-92
• Primeau, Joe	Tor.	9	310	66	177	243	105	38	5	18	23	12	1	1927-28	1935-36
Primeau, Kevin	Van.	1	2	0	0	0	4	....	....	....	....	....	....	1980-81	1980-81
• Pringle, Ellie	NYA	1	6	0	0	0	0	....	....	....	....	....	....	1930-31	1930-31
• Prodgers, Goldie	Tor., Ham.	6	111	63	29	92	39	....	....	....	....	....	....	1919-20	1924-25
‡ Prokhorov, Vitali	St.L.	3	83	19	11	30	35	4	0	0	0	0	....	1992-93	1994-95
Prokopec, Mike	Chi.	2	15	0	0	0	11	....	....	....	....	....	....	1995-96	1996-97
Pronovost, Andre	Mtl., Bos., Det., Min.	10	556	94	104	198	408	70	11	11	22	58	4	1956-57	1967-68
Pronovost, Jean	Pit., Atl., Wsh.	14	998	391	383	774	413	35	11	9	20	14	....	1968-69	1981-82
Pronovost, Marcel	Det., Tor.	21	1206	88	257	345	851	134	8	23	31	104	5	1949-50	1969-70
Propp, Brian	Phi., Bos., Min., Hfd.	15	1016	425	579	1004	830	160	64	84	148	151	....	1979-80	1993-94
‡ Proulx, Christian	Mtl.	1	7	1	2	3	20	....	....	....	....	....	....	1993-94	1993-94
• Provost, Claude	Mtl.	15	1005	254	335	589	469	126	25	38	63	86	9	1955-56	1969-70
Pryor, Chris	Min., NYI	6	82	1	4	5	122	....	....	....	....	....	....	1984-85	1989-90
Prystai, Metro	Chi., Det.	11	674	151	179	330	231	43	12	14	26	8	2	1947-48	1957-58
• Pudas, Al	Tor.	1	4	0	0	0	0	....	....	....	....	....	....	1926-27	1926-27
Pulford, Bob	Tor., L.A.	16	1079	281	362	643	792	89	25	26	51	126	4	1956-57	1971-72
Pulkkinen, Dave	NYI	1	2	0	0	0	0	....	....	....	....	....	....	1972-73	1972-73
• Purpur, Fido	St.L., Chi., Det.	5	144	25	35	60	46	16	1	2	3	4	....	1934-35	1944-45
‡ Purves, John	Wsh.	1	7	1	0	1	0	....	....	....	....	....	....	1990-91	1990-91
• Pusie, Jean	Mtl., NYR, Bos.	5	61	1	4	5	28	7	0	0	0	0	1	1930-31	1935-36
Pyatt, Nelson	Det., Wsh., Col.	7	296	71	63	134	69	....	....	....	....	....	....	1973-74	1979-80

Q

Name	NHL Teams	NHL Seasons	GP	G	A	TP	PIM	GP	G	A	TP	PIM	NHL Cup Wins	First NHL Season	Last NHL Season
• Quackenbush, Bill	Det., Bos.	14	774	62	222	284	95	80	2	19	21	8	....	1942-43	1955-56
Quackenbush, Max	Bos., Chi.	2	61	4	7	11	30	6	0	0	0	4	....	1950-51	1951-52
Quenneville, Joel	Tor., Col., N.J., Hfd., Wsh.	13	803	54	136	190	705	32	0	8	8	22	....	1978-79	1990-91
• Quenneville, Leo	NYR	1	25	0	3	3	10	3	0	0	0	0	....	1929-30	1929-30
• Quilty, John	Mtl., Bos.	4	125	36	34	70	81	13	3	5	8	9	....	1940-41	1947-48
Quinn, Dan	Cgy., Pit., Van., St.L., Phi., Min., Ott., L.A.	14	805	266	419	685	533	65	22	26	48	62	....	1983-84	1996-97
Quinn, Pat	Tor., Van., Atl.	9	606	18	113	131	950	11	0	1	1	21	....	1968-69	1976-77
‡ Quinney, Ken	Que.	3	59	7	13	20	23	....	....	....	....	....	....	1986-87	1990-91
‡ Quintin, Jean-Francois	S.J.	2	22	5	5	10	4	....	....	....	....	....	....	1991-92	1992-93

Brad Park

Muzz Patrick

George Parsons

Wilf Paiment

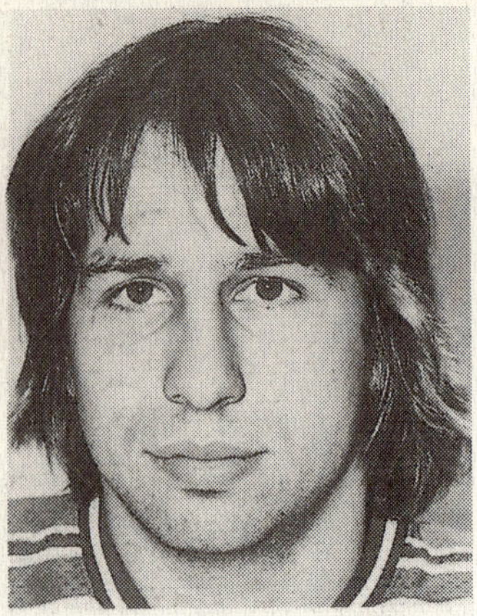
Mark Pavelich

Mark Pederson

J.F. Quintin

Don Raleigh

Name	NHL Teams	NHL Seasons	Regular Schedule GP	G	A	TP	PIM	Playoffs GP	G	A	TP	PIM	NHL Cup Wins	First NHL Season	Last NHL Season
R															
‡ Racine, Yves	Det., Phi., Mtl., S.J., Cgy., T.B.	9	508	37	194	231	439	25	5	4	9	37	....	1989-90	1997-98
• Radley, Yip	NYA, Mtl.M.	2	18	0	1	1	13	....	....	....	....	....	....	1930-31	1936-37
Raglan, Herb	St.L., Que., T.B., Ott.	9	343	33	56	89	775	32	3	6	9	50	....	1985-86	1993-94
Raglan, Rags	Det., Chi.	3	100	4	9	13	52	3	0	0	0	0	....	1950-51	1952-53
Raleigh, Don	NYR	10	535	101	219	320	96	18	6	5	11	6	....	1943-44	1955-56
Ramage, Rob	Col., St.L., Cgy., Tor., Min., T.B., Mtl., Phi.	15	1044	139	425	564	2226	84	8	42	50	218	2	1979-80	1993-94
• Ramsay, Beattie	Tor.	1	43	0	2	2	10	....	....	....	....	....	....	1927-28	1927-28
Ramsay, Craig	Buf.	14	1070	252	420	672	201	89	17	31	48	27	....	1971-72	1984-85
Ramsay, Les	Chi.	1	11	2	2	4	2	....	....	....	....	....	....	1944-45	1944-45
Ramsey, Mike	Buf., Pit., Det.	18	1070	79	266	345	1012	115	8	29	37	176	....	1979-80	1996-97
Ramsey, Wayne	Buf.	1	2	0	0	0	0	....	....	....	....	....	....	1977-78	1977-78
• Randall, Ken	Tor., Ham., NYA	10	218	68	50	118	533	6	2	1	3	27	2	1917-18	1926-27
Ranieri, George	Bos.	1	2	0	0	0	0	....	....	....	....	....	....	1956-57	1956-57
Ratelle, Jean	NYR, Bos.	21	1281	491	776	1267	276	123	32	66	98	24	....	1960-61	1980-81
Rathwell, Jake	Bos.	1	1	0	0	0	0	....	....	....	....	....	....	1974-75	1974-75
‡ Ratushny, Dan	Van.	1	1	0	1	1	2	....	....	....	....	....	....	1992-93	1992-93
Rausse, Errol	Wsh.	3	31	7	3	10	0	....	....	....	....	....	....	1979-80	1981-82
Rautakallio, Pekka	Atl., Cgy.	3	235	33	121	154	122	23	2	5	7	8	....	1979-80	1981-82
Ravlich, Matt	Bos., Chi., Det., L.A.	10	410	12	78	90	364	24	1	5	6	16	....	1962-63	1972-73
• Raymond, Armand	Mtl.	2	22	0	2	2	10	....	....	....	....	....	....	1937-38	1939-40
• Raymond, Paul	Mtl.	4	76	2	3	5	6	5	0	0	0	2	....	1932-33	1938-39
Read, Mel	NYR	1	1	0	0	0	0	....	....	....	....	....	....	1946-47	1946-47
Reardon, Ken	Mtl.	7	341	26	96	122	604	31	2	5	7	62	1	1940-41	1949-50
• Reardon, Terry	Bos., Mtl.	7	193	47	53	100	73	30	8	10	18	12	1	1938-39	1946-47
Reaume, Marc	Tor., Det., Mtl., Van.	9	344	8	43	51	273	21	0	2	2	8	....	1954-55	1970-71
Reay, Billy	Det., Mtl.	10	479	105	162	267	202	63	13	16	29	43	2	1943-44	1952-53
Redahl, Gord	Bos.	1	18	0	1	1	2	....	....	....	....	....	....	1958-59	1958-59
• Redding, George	Bos.	2	55	3	2	5	23	....	....	....	....	....	....	1924-25	1925-26
Redmond, Craig	L.A., Edm.	5	191	16	68	84	134	3	1	0	1	2	....	1984-85	1988-89
Redmond, Dick	Min., Cal., Chi., St.L., Atl., Bos.	13	771	133	312	445	504	66	9	22	31	27	....	1969-70	1981-82
Redmond, Keith	L.A.	1	12	1	0	1	20	....	....	....	....	....	....	1993-94	1993-94
Redmond, Mickey	Mtl., Det.	9	538	233	195	428	219	16	2	3	5	2	2	1967-68	1975-76
Reeds, Mark	St.L., Hfd.	8	365	45	114	159	135	53	8	9	17	23	....	1981-82	1988-89
• Regan, Bill	NYR, NYA	3	67	3	2	5	67	8	0	0	0	2	....	1929-30	1932-33
Regan, Larry	Bos., Tor.	5	280	41	95	136	71	42	7	14	21	18	....	1956-57	1960-61
Regier, Darcy	Cle., NYI	3	26	0	2	2	35	....	....	....	....	....	....	1977-78	1983-84
Reibel, Dutch	Det., Chi., Bos.	6	409	84	161	245	75	39	6	14	20	4	2	1953-54	1958-59
• Reid, Dave	Tor.	3	7	0	0	0	0	....	....	....	....	....	....	1952-53	1955-56
Reid, Dave	Bos., Tor., Dal., Col.	18	961	165	204	369	253	118	9	26	35	34	2	1983-84	2000-01
Reid, Gerry	Det.	1	....	....	....	....	....	2	0	0	0	2	....	1948-49	1948-49
Reid, Gord	NYA	1	1	0	0	0	2	....	....	....	....	....	....	1936-37	1936-37
• Reid, Reg	Tor.	2	39	1	0	1	4	2	0	0	0	0	....	1924-25	1925-26
Reid, Tom	Chi., Min.	11	701	17	113	130	654	42	1	13	14	49	....	1967-68	1977-78
Reierson, Dave	Cgy.	1	2	0	0	0	2	....	....	....	....	....	....	1988-89	1988-89
Reigle, Ed	Bos.	1	17	0	2	2	25	....	....	....	....	....	....	1950-51	1950-51
Reinhart, Paul	Atl., Cgy., Van.	11	648	133	426	559	277	83	23	54	77	42	....	1979-80	1989-90
• Reinikka, Ollie	NYR	1	16	0	0	0	0	....	....	....	....	....	....	1926-27	1926-27
• Reise, Leo	Ham., NYA, NYR	8	241	43	43	86	187	6	0	0	0	16	....	1920-21	1929-30
Reise Jr., Leo	Chi., Det., NYR	9	494	28	81	109	399	52	8	5	13	68	2	1945-46	1953-54
Renaud, Mark	Hfd., Buf.	5	152	6	50	56	86	....	....	....	....	....	....	1979-80	1983-84
‡ Reynolds, Bobby	Tor.	1	7	1	1	2	0	....	....	....	....	....	....	1989-90	1989-90
Ribble, Pat	Atl., Chi., Tor., Wsh., Cgy.	8	349	19	60	79	365	8	0	1	1	12	....	1975-76	1982-83
Rice, Steven	NYR, Edm., Hfd., Car.	8	329	64	61	125	275	2	2	1	3	6	....	1990-91	1997-98
Richard, Henri	Mtl.	20	1256	358	688	1046	928	180	49	80	129	181	11	1955-56	1974-75
Richard, Jacques	Atl., Buf., Que.	10	556	160	187	347	307	35	5	5	10	34	....	1972-73	1982-83
‡ Richard, Jean-Marc	Que.	2	5	2	1	3	2	....	....	....	....	....	....	1987-88	1989-90
• Richard, Maurice	Mtl.	18	978	544	421	965	1285	133	82	44	126	188	8	1942-43	1959-60
‡ Richard, Mike	Wsh.	2	7	0	2	2	0	....	....	....	....	....	....	1987-88	1989-90
‡ Richards, Todd	Hfd.	2	8	0	4	4	4	11	0	3	3	6	....	1990-91	1991-92
Richardson, Dave	NYR, Chi., Det.	4	45	3	2	5	27	....	....	....	....	....	....	1963-64	1967-68
Richardson, Glen	Van.	1	24	3	6	9	19	....	....	....	....	....	....	1975-76	1975-76
Richardson, Ken	St.L.	3	49	8	13	21	16	....	....	....	....	....	....	1974-75	1978-79
Richer, Bob	Buf.	1	3	0	0	0	0	....	....	....	....	....	....	1972-73	1972-73
‡ Richer, Stephane	T.B., Bos., Fla.	3	27	1	5	6	20	3	0	0	0	0	....	1992-93	1994-95
Richmond, Steve	NYR, Det., N.J., L.A.	5	159	4	23	27	514	4	0	0	0	12	....	1983-84	1988-89
Richter, Dave	Min., Phi., Van., St.L.	9	365	9	40	49	1030	22	1	0	1	80	....	1981-82	1989-90
Ridley, Mike	NYR, Wsh., Tor., Van.	12	866	292	466	758	424	104	28	50	78	70	....	1985-86	1996-97
Riley, Bill	Wsh., Wpg.	5	139	31	30	61	320	....	....	....	....	....	....	1974-75	1979-80
Riley, Jack	Det., Mtl., Bos.	4	104	10	22	32	8	4	0	3	3	0	....	1932-33	1935-36
• Riley, Jim	Chi., Det.	1	9	0	2	2	14	....	....	....	....	....	....	1926-27	1926-27
Riopelle, Rip	Mtl.	3	169	27	16	43	73	8	1	1	2	2	....	1947-48	1949-50
Rioux, Gerry	Wpg.	1	8	0	0	0	6	....	....	....	....	....	....	1979-80	1979-80
‡ Rioux, Pierre	Cgy.	1	14	1	2	3	4	....	....	....	....	....	....	1982-83	1982-83
• Ripley, Vic	Chi., Bos., NYR, St.L.	7	278	51	49	100	173	20	4	1	5	10	....	1928-29	1934-35
Risebrough, Doug	Mtl., Cgy.	13	740	185	286	471	1542	124	21	37	58	238	4	1974-75	1986-87
Rissling, Gary	Wsh., Pit.	7	221	23	30	53	1008	5	0	1	1	4	....	1978-79	1984-85
Ritchie, Bob	Phi., Det.	2	29	8	4	12	10	....	....	....	....	....	....	1976-77	1977-78
• Ritchie, Dave	Mtl.W., Ott., Tor., Que., Mtl.	6	58	15	6	21	50	1	0	0	0	0	....	1917-18	1925-26
Ritson, Alex	NYR	1	1	0	0	0	0	....	....	....	....	....	....	1944-45	1944-45
Rittinger, Alan	Bos.	1	19	3	7	10	0	....	....	....	....	....	....	1943-44	1943-44
Rivard, Bob	Pit.	1	27	5	12	17	4	....	....	....	....	....	....	1967-68	1967-68
• Rivers, Gus	Mtl.	3	88	4	5	9	12	16	2	0	2	2	2	1929-30	1931-32
Rivers, Shawn	T.B.	1	4	0	2	2	2	....	....	....	....	....	....	1992-93	1992-93
Rivers, Wayne	Det., Bos., St.L., NYR	7	108	15	30	45	94	....	....	....	....	....	....	1961-62	1968-69
Rizzuto, Garth	Van.	1	37	3	4	7	16	....	....	....	....	....	....	1970-71	1970-71
• Roach, Mickey	Tor., Ham., NYA	8	211	77	34	111	54	....	....	....	....	....	....	1919-20	1926-27
‡ Roberge, Mario	Mtl.	5	112	7	7	14	314	15	0	0	0	24	1	1990-91	1994-95
‡ Roberge, Serge	Que.	1	9	0	0	0	24	....	....	....	....	....	....	1990-91	1990-91
Robert, Claude	Mtl.	1	23	1	0	1	9	....	....	....	....	....	....	1950-51	1950-51
Robert, Rene	Tor., Pit., Buf., Col.	12	744	284	418	702	597	50	22	19	41	73	....	1970-71	1981-82
Roberto, Phil	Mtl., St.L., Det., K.C., Col., Cle.	8	385	75	106	181	464	31	9	8	17	69	1	1969-70	1976-77
‡ Roberts, David	St.L., Edm., Van.	5	125	20	33	53	85	9	0	0	0	16	....	1993-94	1997-98
Roberts, Doug	Det., Oak., Cal., Bos.	10	419	43	104	147	342	16	2	3	5	46	....	1965-66	1974-75
Roberts, Gordie	Hfd., Min., Phi., St.L., Pit., Bos.	15	1097	61	359	420	1582	153	10	47	57	273	2	1979-80	1993-94
Roberts, Jim	Min.	3	106	17	23	40	33	2	0	0	0	0	....	1976-77	1978-79
Roberts, Jimmy	Mtl., St.L.	15	1006	126	194	320	621	153	20	16	36	160	5	1963-64	1977-78
• Robertson, Fred	Tor., Det.	2	34	1	0	1	35	7	0	0	0	0	1	1931-32	1933-34
Robertson, Geordie	Buf.	1	5	1	2	3	7	....	....	....	....	....	....	1982-83	1982-83
Robertson, George	Mtl.	2	31	2	5	7	6	....	....	....	....	....	....	1947-48	1948-49
Robertson, Torrie	Wsh., Hfd., Det.	10	442	49	99	148	1751	22	2	1	3	90	....	1980-81	1989-90
Robidoux, Florent	Chi.	3	52	7	4	11	75	....	....	....	....	....	....	1980-81	1983-84
Robinson, Doug	Chi., NYR, L.A.	7	239	44	67	111	34	11	4	3	7	0	....	1963-64	1970-71
• Robinson, Earl	Mtl.M., Chi., Mtl.	11	417	83	98	181	133	25	5	4	9	0	1	1928-29	1939-40
Robinson, Larry	Mtl., L.A.	20	1384	208	750	958	793	227	28	116	144	211	6	1972-73	1991-92
Robinson, Moe	Mtl.	1	1	0	0	0	0	....	....	....	....	....	....	1979-80	1979-80
Robinson, Rob	St.L.	1	22	0	1	1	8	....	....	....	....	....	....	1991-92	1991-92
Robinson, Scott	Min.	1	1	0	0	0	2	....	....	....	....	....	....	1989-90	1989-90
Robitaille, Mike	NYR, Det., Buf., Van.	8	382	23	105	128	280	13	0	1	1	4	....	1969-70	1976-77
• Roche, Des	Mtl.M., Ott., St.L., Mtl., Det.	4	113	20	18	38	44	....	....	....	....	....	....	1930-31	1934-35
• Roche, Earl	Mtl.M., Bos., Ott., St.L., Det.	4	147	25	27	52	48	2	0	0	0	0	....	1930-31	1934-35
Roche, Ernie	Mtl.	1	4	0	0	0	2	....	....	....	....	....	....	1950-51	1950-51
Rochefort, Dave	Det.	1	1	0	0	0	0	....	....	....	....	....	....	1966-67	1966-67
Rochefort, Leon	NYR, Mtl., Phi., L.A., Det., Atl., Van.	15	617	121	147	268	93	39	4	4	8	16	2	1960-61	1975-76
Rochefort, Normand	Que., NYR, T.B.	13	598	39	119	158	570	69	7	5	12	82	....	1980-81	1993-94
• Rockburn, Harvey	Det., Ott.	3	94	4	2	6	254	....	....	....	....	....	....	1929-30	1932-33
• Rodden, Eddie	Chi., Tor., Bos., NYR	4	97	6	14	20	60	2	0	1	1	0	....	1926-27	1930-31
‡ Rodgers, Marc	Det.	1	21	1	1	2	10	....	....	....	....	....	....	1999-00	1999-00
Rogers, John	Min.	2	14	2	4	6	0	....	....	....	....	....	....	1973-74	1974-75
Rogers, Mike	Hfd., NYR, Edm.	7	484	202	317	519	184	17	1	13	14	6	....	1979-80	1985-86
Rohlicek, Jeff	Van.	2	9	0	0	0	8	....	....	....	....	....	....	1987-88	1988-89
‡ Rohlin, Leif	Van.	2	96	8	24	32	40	5	0	0	0	0	....	1995-96	1996-97
Rohloff, Jon	Bos.	3	150	7	25	32	129	10	1	2	3	8	....	1994-95	1996-97
Rolfe, Dale	Bos., L.A., Det., NYR	9	509	25	125	150	556	71	5	24	29	89	....	1959-60	1974-75

			Regular Schedule					Playoffs					NHL	First	Last
Name	NHL Teams	NHL Seasons	GP	G	A	TP	PIM	GP	G	A	TP	PIM	Cup Wins	NHL Season	NHL Season
Romanchych, Larry	Chi., Atl.	6	298	68	97	165	102	7	2	2	4	4	....	1970-71	1976-77
‡ Romaniuk, Russell	Wpg., Phi.	5	102	13	14	27	63	2	0	0	0	0	....	1991-92	1995-96
Rombough, Doug	Buf., NYI, Min.	4	150	24	27	51	80	....	....	....	....	....	....	1972-73	1975-76
Rominski, Dale	T.B.	1	3	0	1	1	2	....	....	....	....	....	....	1999-00	1999-00
• Romnes, Doc	Chi., Tor., NYA	10	360	68	136	204	42	45	7	18	25	4	2	1930-31	1939-40
Ronan, Ed	Mtl., Wpg., Buf.	6	182	13	23	36	101	27	4	3	7	16	1	1991-92	1996-97
• Ronan, Skene	Ott.	1	11	0	0	0	6	....	....	....	....	....	....	1918-19	1918-19
Ronson, Len	NYR, Oak.	2	18	2	1	3	10	....	....	....	....	....	....	1960-61	1968-69
Ronty, Paul	Bos., NYR, Mtl.	8	488	101	211	312	103	21	1	7	8	6	....	1947-48	1954-55
Rooney, Steve	Mtl., Wpg., N.J.	5	154	15	13	28	496	25	3	2	5	86	1	1984-85	1988-89
Root, Bill	Mtl., Tor., St.L., Phi.	6	247	11	23	34	180	22	1	2	3	25	....	1982-83	1987-88
• Ross, Art	Mtl.W.	1	3	1	0	1	12	....	....	....	....	....	....	1917-18	1917-18
Ross, Jim	NYR	2	62	2	11	13	29	....	....	....	....	....	....	1951-52	1952-53
Rossignol, Roly	Det., Mtl.	3	14	3	5	8	6	1	0	0	0	2	....	1943-44	1945-46
Rota, Darcy	Chi., Atl., Van.	11	794	256	239	495	973	60	14	7	21	147	....	1973-74	1983-84
Rota, Randy	Mtl., L.A., K.C., Col.	5	212	38	39	77	60	5	0	1	1	0	....	1972-73	1976-77
• Rothschild, Sam	Mtl.M., Pit., NYA	4	100	8	6	14	25	6	0	0	0	0	1	1924-25	1927-28
• Roulston, Rolly	Det.	3	24	0	6	6	10	....	....	....	....	....	1	1935-36	1937-38
Roulston, Tom	Edm., Pit.	5	195	47	49	96	74	21	2	2	4	2	....	1980-81	1985-86
Roupe, Magnus	Phi.	2	40	3	5	8	42	....	....	....	....	....	....	1987-88	1988-89
Rouse, Bob	Min., Wsh., Tor., Det., S.J.	17	1061	37	181	218	1559	136	7	21	28	198	2	1983-84	1999-00
Rousseau, Bobby	Mtl., Min., NYR	15	942	245	458	703	359	128	27	57	84	69	4	1960-61	1974-75
Rousseau, Guy	Mtl.	2	4	0	1	1	0	....	....	....	....	....	....	1954-55	1956-57
Rousseau, Roland	Mtl.	1	2	0	0	0	0	....	....	....	....	....	....	1952-53	1952-53
Routhier, Jean-Marc	Que.	1	8	0	0	0	9	....	....	....	....	....	....	1989-90	1989-90
• Rowe, Bobby	Bos.	1	4	1	0	1	0	....	....	....	....	....	....	1924-25	1924-25
Rowe, Mike	Pit.	3	11	0	0	0	11	....	....	....	....	....	....	1984-85	1986-87
Rowe, Ron	NYR	1	5	1	0	1	0	....	....	....	....	....	....	1947-48	1947-48
Rowe, Tom	Wsh., Hfd., Det.	7	357	85	100	185	615	3	2	0	2	0	....	1976-77	1982-83
‡ Roy, Jean-Yves	NYR, Ott., Bos.	4	61	12	16	28	26	....	....	....	....	....	....	1994-95	1997-98
‡ Roy, Stephane	Min.	1	12	1	0	1	0	....	....	....	....	....	....	1987-88	1987-88
‡ Royer, Remi	Chi.	1	18	0	0	0	67	....	....	....	....	....	....	1998-99	1998-99
Rozzini, Gino	Bos.	1	31	5	10	15	20	6	1	2	3	6	....	1944-45	1944-45
Rucinski, Mike	Chi.	2	1	0	0	0	0	2	0	0	0	0	....	1987-88	1988-89
Ruelle, Bernie	Det.	1	2	1	0	1	0	....	....	....	....	....	....	1943-44	1943-44
‡ Ruff, Jason	St.L., T.B.	2	14	3	3	6	10	....	....	....	....	....	....	1992-93	1993-94
Ruff, Lindy	Buf., NYR	12	691	105	195	300	1264	52	11	13	24	193	....	1979-80	1990-91
Ruhnke, Kent	Bos.	1	2	0	1	1	0	....	....	....	....	....	....	1975-76	1975-76
Rundqvist, Thomas	Mtl.	1	2	0	1	1	0	....	....	....	....	....	....	1984-85	1984-85
• Runge, Paul	Bos., Mtl.M., Mtl.	7	140	18	22	40	57	7	0	0	0	6	....	1930-31	1937-38
Ruotsalainen, Reijo	NYR, Edm., N.J.	7	446	107	237	344	180	86	15	32	47	44	2	1981-82	1989-90
Rupp, Duane	NYR, Tor., Min., Pit.	10	374	24	93	117	220	10	2	2	4	8	....	1962-63	1972-73
Ruskowski, Terry	Chi., L.A., Pit., Min.	10	630	113	313	426	1354	21	1	6	7	86	....	1979-80	1988-89
Russell, Cam	Chi., Col.	10	396	9	21	30	872	44	0	5	5	16	....	1989-90	1998-99
• Russell, Church	NYR	3	90	20	16	36	12	....	....	....	....	....	....	1945-46	1947-48
Russell, Phil	Chi., Atl., Cgy., N.J., Buf.	15	1016	99	325	424	2038	73	4	22	26	202	....	1972-73	1986-87
Ruuttu, Christian	Buf., Chi., Van.	9	621	134	298	432	714	42	4	9	13	49	....	1986-87	1994-95
Ruzicka, Vladimir	Edm., Bos., Ott.	5	233	82	85	167	129	30	4	14	18	2	....	1989-90	1993-94
‡ Ryan, Terry	Mtl.	3	8	0	0	0	36	....	....	....	....	....	....	1996-97	1998-99
‡ Rychel, Warren	Chi., L.A., Tor., Col., Ana.	9	406	38	39	77	1422	70	8	13	21	121	1	1988-89	1998-99
Rymsha, Andy	Que.	1	6	0	0	0	23	....	....	....	....	....	....	1991-92	1991-92
S															
Saarinen, Simo	NYR	1	8	0	0	0	0	....	....	....	....	....	....	1984-85	1984-85
Sabol, Shaun	Phi.	1	2	0	0	0	0	....	....	....	....	....	....	1989-90	1989-90
Sabourin, Bob	Tor.	1	1	0	0	0	2	....	....	....	....	....	....	1951-52	1951-52
Sabourin, Gary	St.L., Tor., Cal., Cle.	10	627	169	188	357	397	62	19	11	30	58	....	1967-68	1976-77
Sabourin, Ken	Cgy., Wsh.	4	74	2	8	10	201	12	0	0	0	34	....	1988-89	1991-92
Sacco, David	Tor., Ana.	3	35	5	13	18	22	....	....	....	....	....	....	1993-94	1995-96
Sacharuk, Larry	NYR, St.L.	5	151	29	33	62	42	2	1	1	2	2	....	1972-73	1976-77
Saganiuk, Rocky	Tor., Pit.	6	259	57	65	122	201	6	1	0	1	15	....	1978-79	1983-84
Saleski, Don	Phi., Col.	9	543	128	125	253	629	82	13	17	30	131	2	1971-72	1979-80
Salming, Borje	Tor., Det.	17	1148	150	637	787	1344	81	12	37	49	91	....	1973-74	1989-90
Salovaara, Barry	Det.	2	90	2	13	15	70	....	....	....	....	....	....	1974-75	1975-76
Salvian, Dave	NYI	1	....	....	....	....	....	1	0	1	1	2	....	1976-77	1976-77
Samis, Phil	Tor.	2	2	0	0	0	0	5	0	1	1	2	1	1947-48	1949-50
Sampson, Gary	Wsh.	4	105	13	22	35	25	12	1	0	1	0	....	1983-84	1986-87
Samuelsson, Kjell	NYR, Phi., Pit., T.B.	14	813	48	138	186	1225	123	4	20	24	178	1	1985-86	1998-99
Samuelsson, Ulf	Hfd., Pit., NYR, Det., Phi.	16	1080	57	275	332	2453	132	7	27	34	272	2	1984-85	1999-00
Sandelin, Scott	Mtl., Phi., Min.	4	25	0	4	4	2	....	....	....	....	....	....	1986-87	1991-92
Sanderson, Derek	Bos., NYR, St.L., Van., Pit.	13	598	202	250	452	911	56	18	12	30	187	2	1965-66	1977-78
Sandford, Ed	Bos., Det., Chi.	9	502	106	145	251	355	42	13	11	24	27	....	1947-48	1955-56
Sandlak, Jim	Van., Hfd.	11	549	110	119	229	821	33	7	10	17	30	....	1985-86	1995-96
• Sands, Charlie	Tor., Bos., Mtl., NYR	12	427	99	109	208	58	34	6	6	12	4	1	1932-33	1943-44
‡ Sandstrom, Tomas	NYR, L.A., Pit., Det., Ana.	15	983	394	462	856	1193	139	32	49	81	183	1	1984-85	1998-99
‡ Sandwith, Terran	Edm.	1	8	0	0	0	6	....	....	....	....	....	....	1997-98	1997-98
Sanipass, Everett	Chi., Que.	5	164	25	34	59	358	5	2	0	2	4	....	1986-87	1990-91
Sargent, Gary	L.A., Min.	8	402	61	161	222	273	20	5	7	12	8	....	1975-76	1982-83
Sarner, Craig	Bos.	1	7	0	0	0	0	....	....	....	....	....	....	1974-75	1974-75
Sarrazin, Dick	Phi.	3	100	20	35	55	22	4	0	0	0	0	....	1968-69	1971-72
Sasakamoose, Fred	Chi.	1	11	0	0	0	6	....	....	....	....	....	....	1953-54	1953-54
Sasser, Grant	Pit.	1	3	0	0	0	0	....	....	....	....	....	....	1983-84	1983-84
Sather, Glen	Bos., Pit., NYR, St.L., Mtl., Min.	10	658	80	113	193	724	72	1	5	6	86	....	1966-67	1975-76
Saunders, Bernie	Que.	2	10	0	1	1	8	....	....	....	....	....	....	1979-80	1980-81
Saunders, David	Van.	1	56	7	13	20	10	....	....	....	....	....	....	1987-88	1987-88
• Saunders, Ted	Ott.	1	18	1	3	4	4	....	....	....	....	....	....	1933-34	1933-34
Sauve, Jean-Francois	Buf., Que.	7	290	65	138	203	114	36	9	12	21	10	....	1980-81	1986-87
‡ Savage, Joel	Buf.	1	3	0	1	1	0	....	....	....	....	....	....	1990-91	1990-91
• Savage, Tony	Bos., Mtl.	1	49	1	5	6	6	2	0	0	0	0	....	1934-35	1934-35
Savard, Andre	Bos., Buf., Que.	12	790	211	271	482	411	85	13	18	31	77	....	1973-74	1984-85
Savard, Denis	Chi., Mtl., T.B.	17	1196	473	865	1338	1336	169	66	109	175	256	1	1980-81	1996-97
Savard, Jean	Chi., Hfd.	3	43	7	12	19	29	....	....	....	....	....	....	1977-78	1979-80
Savard, Serge	Mtl., Wpg.	17	1040	106	333	439	592	130	19	49	68	88	8	1966-67	1982-83
‡ Savoia, Ryan	Pit.	1	3	0	0	0	0	....	....	....	....	....	....	1998-99	1998-99
Scamurra, Peter	Wsh.	4	132	8	25	33	59	....	....	....	....	....	....	1975-76	1979-80
Sceviour, Darin	Chi.	1	1	0	0	0	0	....	....	....	....	....	....	1986-87	1986-87
Schaeffer, Butch	Chi.	1	5	0	0	0	6	....	....	....	....	....	....	1936-37	1936-37
Schamehorn, Kevin	Det., L.A.	3	10	0	0	0	17	....	....	....	....	....	....	1976-77	1980-81
Schella, John	Van.	2	115	2	18	20	224	....	....	....	....	....	....	1970-71	1971-72
Scherza, Chuck	Bos., NYR	2	36	6	6	12	35	....	....	....	....	....	....	1943-44	1944-45
Schinkel, Ken	NYR, Pit.	12	636	127	198	325	163	19	7	2	9	4	....	1959-60	1972-73
‡ Schlegel, Brad	Wsh., Cgy.	3	48	1	8	9	10	7	0	1	1	2	....	1991-92	1993-94
Schliebener, Andy	Van.	3	84	2	11	13	74	6	0	0	0	0	....	1981-82	1984-85
Schmautz, Bobby	Chi., Van., Bos., Edm., Col.	13	764	271	286	557	988	84	28	33	61	92	....	1967-68	1980-81
• Schmautz, Cliff	Buf., Phi.	1	56	13	19	32	33	....	....	....	....	....	....	1970-71	1970-71
• Schmidt, Clarence	Bos.	1	7	1	0	1	2	....	....	....	....	....	....	1943-44	1943-44
Schmidt, Jackie	Bos.	1	45	6	7	13	6	5	0	0	0	0	....	1942-43	1942-43
Schmidt, Milt	Bos.	16	776	229	346	575	466	86	24	25	49	60	2	1936-37	1954-55
Schmidt, Norm	Pit.	4	125	23	33	56	73	....	....	....	....	....	....	1983-84	1987-88
Schmidt, Otto	Bos.	1	2	0	0	0	0	....	....	....	....	....	....	1943-44	1943-44
• Schnarr, Werner	Bos.	2	26	0	0	0	0	....	....	....	....	....	....	1924-25	1925-26
‡ Schneider, Andy	Ott.	1	10	0	0	0	15	....	....	....	....	....	....	1993-94	1993-94
Schock, Danny	Bos., Phi.	2	20	1	2	3	0	1	0	0	0	0	1	1969-70	1970-71
Schock, Ron	Bos., St.L., Pit., Buf.	15	909	166	351	517	260	55	4	16	20	29	....	1963-64	1977-78
Schoenfeld, Jim	Buf., Det., Bos.	13	719	51	204	255	1132	75	3	13	16	151	....	1972-73	1984-85
Schofield, Dwight	Det., Mtl., St.L., Wsh., Pit., Wpg.	7	211	8	22	30	631	9	0	0	0	55	....	1976-77	1987-88
Schreiber, Wally	Min.	2	41	8	10	18	12	....	....	....	....	....	....	1987-88	1988-89
• Schriner, Sweeney	NYA, Tor.	11	484	201	204	405	148	59	18	11	29	54	2	1934-35	1945-46
‡ Schulte, Paxton	Que., Cgy.	2	2	0	0	0	4	....	....	....	....	....	....	1993-94	1996-97
Schultz, Dave	Phi., L.A., Pit., Buf.	9	535	79	121	200	2294	73	8	12	20	412	2	1971-72	1979-80
Schurman, Maynard	Hfd.	1	7	0	0	0	0	....	....	....	....	....	....	1979-80	1979-80
Schutt, Rod	Mtl., Pit., Tor.	8	286	77	92	169	177	22	8	6	14	26	....	1977-78	1985-86
Scissons, Scott	NYI	3	2	0	0	0	0	1	0	0	0	0	....	1990-91	1993-94
Sclisizzi, Enio	Det., Chi.	6	81	12	11	23	26	13	0	0	0	6	....	1946-47	1952-53
• Scott, Ganton	Tor., Ham., Mtl.M.	3	57	1	1	2	0	....	....	....	....	....	....	1922-23	1924-25

Rob Ramage

Mark Reeds

Dutch Reibel

Doug Risebrough

Tom Rowe

Reijo Ruotsalainen

Derek Sanderson

Serge Savard

Name	NHL Teams	NHL Seasons	Regular Schedule GP	G	A	TP	PIM	Playoffs GP	G	A	TP	PIM	NHL Cup Wins	First NHL Season	Last NHL Season
• Scott, Laurie	NYA, NYR	2	62	6	3	9	28	....	....	....	....	....	....	1926-27	1927-28
‡ Scremin, Claudio	S.J.	2	17	0	1	1	29	....	....	....	....	....	....	1991-92	1992-93
Scruton, Howard	L.A.	1	4	0	4	4	9	....	....	....	....	....	....	1982-83	1982-83
Seabrooke, Glen	Phi.	3	19	1	6	7	4	....	....	....	....	....	....	1986-87	1988-89
Secord, Al	Bos., Chi., Tor., Phi.	12	766	273	222	495	2093	102	21	34	55	382	....	1978-79	1989-90
Sedlbauer, Ron	Van., Chi., Tor.	7	430	143	86	229	210	19	1	3	4	27	....	1974-75	1980-81
Seftel, Steve	Wsh.	1	4	0	0	0	2	....	....	....	....	....	....	1990-91	1990-91
Seguin, Dan	Min., Van.	2	37	2	6	8	50	....	....	....	....	....	....	1970-71	1973-74
Seguin, Steve	L.A.	1	5	0	0	0	9	....	....	....	....	....	....	1984-85	1984-85
• Seibert, Earl	NYR, Chi., Det.	15	645	89	187	276	746	66	11	8	19	76	2	1931-32	1945-46
Seiling, Ric	Buf., Det.	10	738	179	208	387	573	62	14	14	28	36	....	1977-78	1986-87
Seiling, Rod	Tor., NYR, Wsh., St.L., Atl.	17	979	62	269	331	601	77	4	8	12	55	....	1962-63	1978-79
‡ Sejba, Jiri	Buf.	1	11	0	2	2	8	....	....	....	....	....	....	1990-91	1990-91
Selby, Brit	Tor., Phi., St.L.	8	350	55	62	117	163	16	1	1	2	8	....	1964-65	1971-72
Self, Steve	Wsh.	1	3	0	0	0	0	....	....	....	....	....	....	1976-77	1976-77
Selwood, Brad	Tor., L.A.	3	163	7	40	47	153	6	0	0	0	4	....	1970-71	1979-80
‡ Semak, Alexander	N.J., T.B., NYI, Van.	6	289	83	91	174	187	8	1	1	2	0	....	1991-92	1996-97
Semchuk, Brandy	L.A.	1	1	0	0	0	2	....	....	....	....	....	....	1992-93	1992-93
Semenko, Dave	Edm., Hfd., Tor.	9	575	65	88	153	1175	73	6	6	12	208	2	1979-80	1987-88
Semenov, Anatoli	Edm., T.B., Van., Ana., Phi., Buf.	8	362	68	126	194	122	49	9	13	22	12	....	1989-90	1996-97
Senick, George	NYR	1	13	2	3	5	8	....	....	....	....	....	....	1952-53	1952-53
Seppa, Jyrki	Wpg.	1	13	0	2	2	6	....	....	....	....	....	....	1983-84	1983-84
Serafini, Ron	Cal.	1	2	0	0	0	2	....	....	....	....	....	....	1973-74	1973-74
Serowik, Jeff	Tor., Bos., Pit.	4	28	0	6	6	16	....	....	....	....	....	....	1990-91	1999-00
Servinis, George	Min.	1	5	0	0	0	0	....	....	....	....	....	....	1987-88	1987-88
‡ Sevcik, Jaroslav	Que.	1	13	0	2	2	2	....	....	....	....	....	....	1989-90	1989-90
Severyn, Brent	Que., Fla., NYI, Col., Ana., Dal.	7	328	10	30	40	825	8	0	0	0	12	....	1989-90	1998-99
‡ Sevigny, Pierre	Mtl., NYR	4	78	4	5	9	64	3	0	1	1	0	....	1993-94	1997-98
Shack, Eddie	NYR, Tor., Bos., L.A., Buf., Pit.	17	1047	239	226	465	1437	74	6	7	13	151	4	1958-59	1974-75
• Shack, Joe	NYR	2	70	9	27	36	20	....	....	....	....	....	....	1942-43	1944-45
‡ Shafranov, Konstantin	St.L.	1	5	2	1	3	0	....	....	....	....	....	....	1996-97	1996-97
Shakes, Paul	Cal.	1	21	0	4	4	12	....	....	....	....	....	....	1973-74	1973-74
Shaldybin, Yevgeny	Bos.	1	3	1	0	1	0	....	....	....	....	....	....	1996-97	1996-97
Shanahan, Sean	Mtl., Col., Bos.	3	40	1	3	4	47	....	....	....	....	....	....	1975-76	1977-78
Shand, Dave	Atl., Tor., Wsh.	8	421	19	84	103	544	26	1	2	3	83	....	1976-77	1984-85
‡ Shank, Daniel	Det., Hfd.	3	77	13	14	27	175	5	0	0	0	22	....	1989-90	1991-92
Shannon, Chuck	NYA	1	4	0	0	0	2	....	....	....	....	....	....	1939-40	1939-40
Shannon, Darrin	Buf., Wpg., Phx.	10	506	87	163	250	344	45	7	10	17	38	....	1988-89	1997-98
• Shannon, Gerry	Ott., St.L., Bos., Mtl.M.	5	180	23	29	52	80	9	0	1	1	2	....	1933-34	1937-38
Sharples, Jeff	Det.	3	105	14	35	49	70	7	0	3	3	6	....	1986-87	1988-89
Sharpley, Glen	Min., Chi.	6	389	117	161	278	199	27	7	11	18	24	....	1976-77	1981-82
Shaunessy, Scott	Que.	2	7	0	0	0	23	....	....	....	....	....	....	1986-87	1988-89
Shaw, Brad	Hfd., Ott., Wsh., St.L.	11	377	22	137	159	208	23	4	8	12	6	....	1985-86	1998-99
‡ Shaw, David	Que., NYR, Edm., Min., Bos., T.B.	16	769	41	153	194	906	45	3	9	12	81	....	1982-83	1997-98
• Shay, Norm	Bos., Tor.	2	53	5	3	8	34	....	....	....	....	....	....	1924-25	1925-26
• Shea, Pat	Chi.	1	10	1	0	1	0	....	....	....	....	....	....	1931-32	1931-32
Shedden, Doug	Pit., Det., Que., Tor.	8	416	139	186	325	176	....	....	....	....	....	....	1981-82	1990-91
Sheehan, Bobby	Mtl., Cal., Chi., Det., NYR, Col., L.A.	9	310	48	63	111	40	25	4	3	7	8	1	1969-70	1981-82
Sheehy, Neil	Cgy., Hfd., Wsh.	9	379	18	47	65	1311	54	0	3	3	241	....	1983-84	1991-92
Sheehy, Tim	Det., Hfd.	2	27	2	1	3	0	....	....	....	....	....	....	1977-78	1979-80
Shelton, Doug	Chi.	1	5	0	1	1	2	....	....	....	....	....	....	1967-68	1967-68
• Sheppard, Frank	Det.	1	8	1	1	2	0	....	....	....	....	....	....	1927-28	1927-28
Sheppard, Gregg	Bos., Pit.	10	657	205	293	498	243	82	32	40	72	31	....	1972-73	1981-82
• Sheppard, Johnny	Det., NYA, Bos., Chi.	8	308	68	58	126	224	10	0	0	0	0	1	1926-27	1933-34
Sheppard, Ray	Buf., NYR, Det., S.J., Fla., Car.	13	817	357	300	657	212	81	30	20	50	21	....	1987-88	1999-00
• Sherf, John	Det.	5	19	0	0	0	8	8	0	1	1	2	1	1935-36	1943-44
• Shero, Fred	NYR	3	145	6	14	20	137	13	0	2	2	8	....	1947-48	1949-50
Sherritt, Gordon	Det.	1	8	0	0	0	12	....	....	....	....	....	....	1943-44	1943-44
Sherven, Gord	Edm., Min., Hfd.	5	97	13	22	35	33	3	0	0	0	0	....	1983-84	1987-88
‡ Shevalier, Jeff	L.A., T.B.	3	32	5	9	14	8	....	....	....	....	....	....	1994-95	1999-00
• Shewchuck, Jack	Bos.	6	187	9	19	28	160	20	0	1	1	19	1	1938-39	1944-45
Shibicky, Alex	NYR	8	324	110	91	201	161	39	12	12	24	12	1	1935-36	1945-46
• Shields, Al	Ott., Phi., NYA, Mtl.M., Bos.	11	459	42	46	88	637	17	0	1	1	14	1	1927-28	1937-38
Shill, Bill	Bos.	3	79	21	13	34	18	7	1	2	3	2	....	1942-43	1946-47
• Shill, Jack	Tor., Bos., NYA, Chi.	6	160	15	20	35	70	25	1	6	7	23	1	1933-34	1938-39
Shinske, Rick	Cle., St.L.	3	63	5	16	21	10	....	....	....	....	....	....	1976-77	1978-79
Shires, Jim	Det., St.L., Pit.	3	56	3	6	9	32	....	....	....	....	....	....	1970-71	1972-73
Shmyr, Paul	Chi., Cal., Min., Hfd.	7	343	13	72	85	528	34	3	3	6	44	....	1968-69	1981-82
Shoebottom, Bruce	Bos.	4	35	1	4	5	53	14	1	2	3	77	....	1987-88	1990-91
• Shore, Eddie	Bos., NYA	14	550	105	179	284	1047	55	6	13	19	181	2	1926-27	1939-40
• Shore, Hamby	Ott.	1	18	3	8	11	51	....	....	....	....	....	....	1917-18	1917-18
Short, Steve	L.A., Det.	2	6	0	0	0	2	....	....	....	....	....	....	1977-78	1978-79
‡ Shuchuk, Gary	Det., L.A.	5	142	13	26	39	70	20	2	2	4	12	....	1990-91	1995-96
‡ Shudra, Ron	Edm.	1	10	0	5	5	6	....	....	....	....	....	....	1987-88	1987-88
Shutt, Steve	Mtl., L.A.	13	930	424	393	817	410	99	50	48	98	65	5	1972-73	1984-85
• Siebert, Babe	Mtl.M., NYR, Bos., Mtl.	14	592	140	156	296	982	49	7	5	12	62	2	1925-26	1938-39
Silk, Dave	NYR, Bos., Det., Wpg.	7	249	54	59	113	271	13	2	4	6	13	....	1979-80	1985-86
Siltala, Mike	Wsh., NYR	3	7	1	0	1	2	....	....	....	....	....	....	1981-82	1987-88
Siltanen, Risto	Edm., Hfd., Que.	8	562	90	265	355	266	32	6	12	18	30	....	1979-80	1986-87
Sim, Trevor	Edm.	1	3	0	1	1	2	....	....	....	....	....	....	1989-90	1989-90
Simard, Martin	Cgy., T.B.	3	44	1	5	6	183	....	....	....	....	....	....	1990-91	1992-93
Simmer, Charlie	Cal., Cle., L.A., Bos., Pit.	14	712	342	369	711	544	24	9	9	18	32	....	1974-75	1987-88
Simmons, Al	Cal., Bos.	3	11	0	1	1	21	1	0	0	0	0	....	1971-72	1975-76
• Simon, Cully	Det., Chi.	3	130	4	11	15	121	14	1	0	1	6	1	1942-43	1944-45
‡ Simon, Jason	NYI, Phx.	2	5	0	0	0	34	....	....	....	....	....	....	1993-94	1996-97
Simon, Thain	Det.	1	3	0	0	0	0	....	....	....	....	....	....	1946-47	1946-47
‡ Simon, Todd	Buf.	1	15	0	1	1	0	5	1	0	1	0	....	1993-94	1993-94
Simonetti, Frank	Bos.	4	115	5	8	13	76	12	0	1	1	8	....	1984-85	1987-88
Simpson, Bobby	Atl., St.L., Pit.	5	175	35	29	64	98	6	0	1	1	2	....	1976-77	1982-83
• Simpson, Cliff	Det.	2	6	0	1	1	0	2	0	0	0	2	....	1946-47	1947-48
Simpson, Craig	Pit., Edm., Buf.	10	634	247	250	497	659	67	36	32	68	56	2	1985-86	1994-95
• Simpson, Joe	NYA	6	228	21	19	40	156	2	0	0	0	0	....	1925-26	1930-31
Sims, Al	Bos., Hfd., L.A.	10	475	49	116	165	286	41	0	2	2	14	....	1973-74	1982-83
Sinclair, Reg	NYR, Det.	3	208	49	43	92	139	3	1	0	1	0	....	1950-51	1952-53
• Singbush, Alex	Mtl.	1	32	0	5	5	15	3	0	0	0	4	....	1940-41	1940-41
Sinisalo, Ilkka	Phi., Min., L.A.	11	582	204	222	426	208	68	21	11	32	6	....	1981-82	1991-92
Siren, Ville	Pit., Min.	5	290	14	68	82	276	7	0	0	0	6	....	1985-86	1989-90
Sirois, Bob	Phi., Wsh.	6	286	92	120	212	42	....	....	....	....	....	....	1974-75	1979-80
Sittler, Darryl	Tor., Phi., Det.	15	1096	484	637	1121	948	76	29	45	74	137	....	1970-71	1984-85
• Sjoberg, Lars-Erik	Wpg.	1	79	7	27	34	48	....	....	....	....	....	....	1979-80	1979-80
‡ Sjodin, Tommy	Min., Dal., Que.	2	106	8	40	48	52	....	....	....	....	....	....	1992-93	1993-94
• Skaare, Bjorn	Det.	1	1	0	0	0	0	....	....	....	....	....	....	1978-79	1978-79
Skarda, Randy	St.L.	2	26	0	5	5	11	....	....	....	....	....	....	1989-90	1991-92
• Skilton, Raymie	Mtl.W.	1	1	0	0	0	0	....	....	....	....	....	....	1917-18	1917-18
• Skinner, Alf	Tor., Bos., Mtl.M., Pit.	4	71	26	10	36	87	2	0	1	1	9	1	1917-18	1925-26
Skinner, Larry	Col.	4	47	10	12	22	8	2	0	0	0	0	....	1976-77	1979-80
Skov, Glen	Det., Chi., Mtl.	12	650	106	136	242	413	53	7	7	14	48	3	1949-50	1960-61
Skriko, Petri	Van., Bos., Wpg., S.J.	9	541	183	222	405	246	28	5	9	14	4	....	1984-85	1992-93
Skrudland, Brian	Mtl., Cgy., Fla., NYR, Dal.	15	881	124	219	343	1107	164	15	46	61	323	2	1985-86	1999-00
Sleaver, John	Chi.	2	13	1	0	1	6	....	....	....	....	....	....	1953-54	1956-57
Sleigher, Louis	Que., Bos.	6	194	46	53	99	146	17	1	1	2	64	....	1979-80	1985-86
Sloan, Tod	Tor., Chi.	13	745	220	262	482	831	47	9	12	21	47	2	1947-48	1960-61
• Slobodian, Peter	NYA	1	41	3	2	5	54	....	....	....	....	....	....	1940-41	1940-41
• Slowinski, Ed	NYR	6	291	58	74	132	63	16	2	6	8	6	....	1947-48	1952-53
Sly, Darryl	Tor., Min., Van.	4	79	1	2	3	20	....	....	....	....	....	....	1965-66	1970-71
Smail, Doug	Wpg., Min., Que., Ott.	13	845	210	249	459	602	42	9	2	11	49	....	1980-81	1992-93
Smart, Alex	Mtl.	1	8	5	2	7	0	....	....	....	....	....	....	1942-43	1942-43
Smedsmo, Dale	Tor.	1	4	0	0	0	0	....	....	....	....	....	....	1972-73	1972-73
Smillie, Don	Bos.	1	12	2	2	4	4	....	....	....	....	....	....	1933-34	1933-34
• Smith, Alex	Ott., Det., Bos., NYA	11	443	41	50	91	645	19	0	2	2	28	1	1924-25	1934-35
• Smith, Art	Tor., Ott.	4	144	15	10	25	249	4	1	1	2	8	....	1927-28	1930-31
Smith, Barry	Bos., Col.	3	114	7	7	14	10	....	....	....	....	....	....	1975-76	1980-81
Smith, Bobby	Min., Mtl.	15	1077	357	679	1036	917	184	64	96	160	245	1	1978-79	1992-93
Smith, Brad	Van., Atl., Cgy., Det., Tor.	9	222	28	34	62	591	20	3	3	6	49	....	1978-79	1986-87
Smith, Brian	Det.	3	61	2	8	10	12	5	0	0	0	0	....	1957-58	1960-61
• Smith, Brian	L.A., Min.	2	67	10	10	20	33	7	0	0	0	0	....	1967-68	1968-69

Name	NHL Teams	NHL Seasons	Regular Schedule GP	G	A	TP	PIM	Playoffs GP	G	A	TP	PIM	NHL Cup Wins	First NHL Season	Last NHL Season
• Smith, Carl	Det.	1	7	1	1	2	2	....	....	....	....	....	....	1943-44	1943-44
Smith, Clint	NYR, Chi.	11	483	161	236	397	24	42	10	14	24	2	1	1936-37	1946-47
Smith, Dallas	Bos., NYR	16	890	55	252	307	959	86	3	29	32	128	2	1959-60	1977-78
‡ Smith, Dan	Col.	2	15	0	0	0	9	....	....	....	....	....	....	1998-99	1999-00
Smith, Dennis	Wsh., L.A.	2	8	0	0	0	4	....	....	....	....	....	....	1989-90	1990-91
Smith, Derek	Buf., Det.	8	335	78	116	194	60	30	9	14	23	13	....	1975-76	1982-83
Smith, Derrick	Phi., Min., Dal.	10	537	82	92	174	373	82	14	11	25	79	....	1984-85	1993-94
• Smith, Des	Mtl.M., Mtl., Chi., Bos.	5	196	22	25	47	236	25	1	4	5	18	1	1937-38	1941-42
• Smith, Don	Mtl.	1	12	1	0	1	6	....	....	....	....	....	....	1919-20	1919-20
Smith, Don	NYR	1	11	1	1	2	0	1	0	0	0	0	....	1949-50	1949-50
Smith, Doug	L.A., Buf., Edm., Van., Pit.	9	535	115	138	253	624	18	4	2	6	21	....	1981-82	1989-90
Smith, Floyd	Bos., NYR, Det., Tor., Buf.	13	616	129	178	307	207	48	12	11	23	16	....	1954-55	1971-72
Smith, Geoff	Edm., Fla., NYR	10	462	18	73	91	282	13	0	1	1	8	1	1989-90	1998-99
Smith, Glen	Chi.	1	2	0	0	0	0	....	....	....	....	....	....	1950-51	1950-51
• Smith, Glenn	Tor.	1	9	0	0	0	0	....	....	....	....	....	....	1921-22	1921-22
Smith, Gord	Wsh., Wpg.	6	299	9	30	39	284	....	....	....	....	....	....	1974-75	1979-80
Smith, Greg	Cal., Cle., Min., Det., Wsh.	13	829	56	232	288	1110	63	4	7	11	106	....	1975-76	1987-88
• Smith, Hooley	Ott., Mtl.M., Bos., NYA	17	715	200	225	425	1013	54	11	8	19	109	2	1924-25	1940-41
• Smith, Ken	Bos.	7	331	78	93	171	49	30	8	13	21	6	....	1944-45	1950-51
Smith, Nakina	Det.	1	10	1	2	3	0	....	....	....	....	....	....	1943-44	1943-44
Smith, Randy	Min.	2	3	0	0	0	0	....	....	....	....	....	....	1985-86	1986-87
Smith, Rick	Bos., Cal., St.L., Det., Wsh.	11	687	52	167	219	560	78	3	23	26	73	1	1968-69	1980-81
• Smith, Rodger	Pit., Phi.	6	210	20	4	24	172	4	3	0	3	0	....	1925-26	1930-31
Smith, Ron	NYI	1	11	1	1	2	14	....	....	....	....	....	....	1972-73	1972-73
Smith, Sid	Tor.	12	601	186	183	369	94	44	17	10	27	2	3	1946-47	1957-58
Smith, Stan	NYR	2	9	2	1	3	0	1	0	0	0	0	1	1939-40	1940-41
Smith, Steve	Phi., Buf.	6	18	0	1	1	15	....	....	....	....	....	....	1981-82	1988-89
‡ Smith, Steve	Edm., Chi., Cgy.	16	804	72	303	375	2139	134	11	41	52	288	3	1984-85	2000-01
Smith, Stu	Mtl.	2	4	2	2	4	2	1	0	0	0	0	....	1940-41	1941-42
Smith, Stu	Hfd.	4	77	2	10	12	95	....	....	....	....	....	....	1979-80	1982-83
• Smith, Tommy	Que.	1	10	0	1	1	11	....	....	....	....	....	....	1919-20	1919-20
Smith, Vern	NYI	1	1	0	0	0	0	....	....	....	....	....	....	1984-85	1984-85
Smith, Wayne	Chi.	1	2	1	1	2	2	1	0	0	0	0	....	1966-67	1966-67
Smrke, John	St.L., Que.	3	103	11	17	28	33	....	....	....	....	....	....	1977-78	1979-80
• Smrke, Stan	Mtl.	2	9	0	3	3	0	....	....	....	....	....	....	1956-57	1957-58
Smyl, Stan	Van.	13	896	262	411	673	1556	41	16	17	33	64	....	1978-79	1990-91
• Smylie, Rod	Tor., Ott.	6	74	4	2	6	12	4	0	0	0	2	1	1920-21	1925-26
Smyth, Greg	Phi., Que., Cgy., Fla., Tor., Chi.	10	229	4	16	20	783	12	0	0	0	40	....	1986-87	1996-97
Smyth, Kevin	Hfd.	3	58	6	8	14	31	....	....	....	....	....	....	1993-94	1995-96
‡ Snell, Chris	Tor., L.A.	2	34	2	7	9	24	....	....	....	....	....	....	1993-94	1994-95
Snell, Ron	Pit.	2	7	3	2	5	6	....	....	....	....	....	....	1968-69	1969-70
Snell, Ted	Pit., K.C., Det.	2	104	7	18	25	22	....	....	....	....	....	....	1973-74	1974-75
Snepsts, Harold	Van., Min., Det., St.L.	17	1033	38	195	233	2009	93	1	14	15	231	....	1974-75	1990-91
Snow, Sandy	Det.	1	3	0	0	0	2	....	....	....	....	....	....	1968-69	1968-69
Snuggerud, Dave	Buf., S.J., Phi.	4	265	30	54	84	127	12	1	3	4	6	....	1989-90	1992-93
Sobchuk, Dennis	Det., Que.	2	35	5	6	11	2	....	....	....	....	....	....	1979-80	1982-83
• Sobchuk, Gene	Van.	1	1	0	0	0	0	....	....	....	....	....	....	1973-74	1973-74
Solheim, Ken	Chi., Min., Det., Edm.	5	135	19	20	39	34	3	1	1	2	2	....	1980-81	1985-86
Solinger, Bob	Tor., Det.	5	99	10	11	21	19	....	....	....	....	....	....	1951-52	1959-60
• Somers, Art	Chi., NYR	6	222	33	56	89	189	30	1	5	6	20	1	1929-30	1934-35
Sommer, Roy	Edm.	1	3	1	0	1	7	....	....	....	....	....	....	1980-81	1980-81
Songin, Tom	Bos.	3	43	5	5	10	22	....	....	....	....	....	....	1978-79	1980-81
Sonmor, Glen	NYR	2	28	2	0	2	21	....	....	....	....	....	....	1953-54	1954-55
‡ Sorochan, Lee	Cgy.	2	3	0	0	0	0	....	....	....	....	....	....	1998-99	1999-00
• Sorrell, John	Det., NYA	11	490	127	119	246	100	42	12	15	27	10	2	1930-31	1940-41
• Sparrow, Emory	Bos.	1	8	0	0	0	4	....	....	....	....	....	....	1924-25	1924-25
Speck, Fred	Det., Van.	3	28	1	2	3	2	....	....	....	....	....	....	1968-69	1971-72
• Speer, Bill	Pit., Bos.	4	130	5	20	25	79	8	1	0	1	4	1	1967-68	1970-71
Speers, Ted	Det.	1	4	1	1	2	0	....	....	....	....	....	....	1985-86	1985-86
• Spence, Gordon	Tor.	1	3	0	0	0	0	....	....	....	....	....	....	1925-26	1925-26
• Spencer, Brian	Tor., NYI, Buf., Pit.	10	553	80	143	223	634	37	1	5	6	29	....	1969-70	1978-79
Spencer, Irv	NYR, Bos., Det.	8	230	12	38	50	127	16	0	0	0	8	....	1959-60	1967-68
• Speyer, Chris	Tor., NYA	3	14	0	0	0	0	....	....	....	....	....	....	1923-24	1933-34
‡ Spring, Corey	T.B.	2	16	1	1	2	12	....	....	....	....	....	....	1997-98	1998-99
Spring, Don	Wpg.	4	259	1	54	55	80	6	0	0	0	10	....	1980-81	1983-84
Spring, Frank	Bos., St.L., Cal., Cle.	5	61	14	20	34	12	....	....	....	....	....	....	1969-70	1976-77
• Spring, Jesse	Ham., Pit., Tor., NYA	6	133	11	4	15	74	2	0	2	2	2	....	1923-24	1929-30
Spruce, Andy	Van., Col.	3	172	31	42	73	111	2	0	2	2	0	....	1976-77	1978-79
‡ Srsen, Tomas	Edm.	1	2	0	0	0	0	....	....	....	....	....	....	1990-91	1990-91
‡ St. Amour, Martin	Ott.	1	1	0	0	0	2	....	....	....	....	....	....	1992-93	1992-93
St. Laurent, Andre	NYI, Det., L.A., Pit.	11	644	129	187	316	749	59	8	12	20	48	....	1973-74	1983-84
St. Marseille, Frank	St.L., L.A.	10	707	140	285	425	242	88	20	25	45	18	....	1967-68	1976-77
St. Sauveur, Claude	Atl.	1	79	24	24	48	23	2	0	0	0	0	....	1975-76	1975-76
Stackhouse, Ron	Cal., Det., Pit.	12	889	87	372	459	824	32	5	8	13	38	....	1970-71	1981-82
• Stackhouse, Ted	Tor.	1	13	0	0	0	2	1	0	0	0	0	1	1921-22	1921-22
• Stahan, Butch	Mtl.	1	....	....	....	....	....	3	0	1	1	2	....	1944-45	1944-45
‡ Stajduhar, Nick	Edm.	1	2	0	0	0	4	....	....	....	....	....	....	1995-96	1995-96
Staley, Al	NYR	1	1	0	1	1	0	....	....	....	....	....	....	1948-49	1948-49
Stamler, Lorne	L.A., Tor., Wpg.	4	116	14	11	25	16	....	....	....	....	....	....	1976-77	1979-80
Standing, George	Min.	1	2	0	0	0	0	....	....	....	....	....	....	1967-68	1967-68
Stanfield, Fred	Chi., Bos., Min., Buf.	14	914	211	405	616	134	106	21	35	56	10	2	1964-65	1977-78
Stanfield, Jack	Chi.	1	....	....	....	....	....	1	0	0	0	0	....	1965-66	1965-66
Stanfield, Jim	L.A.	3	7	0	1	1	0	....	....	....	....	....	....	1969-70	1971-72
Stankiewicz, Ed	Det.	2	6	0	0	0	2	....	....	....	....	....	....	1953-54	1955-56
Stankiewicz, Myron	St.L., Phi.	1	35	0	7	7	36	1	0	0	0	0	....	1968-69	1968-69
Stanley, Allan	NYR, Chi., Bos., Tor., Phi.	21	1244	100	333	433	792	109	7	36	43	80	4	1948-49	1968-69
• Stanley, Barney	Chi.	1	1	0	0	0	0	....	....	....	....	....	....	1927-28	1927-28
Stanley, Daryl	Phi., Van.	6	189	8	17	25	408	17	0	0	0	30	....	1983-84	1989-90
Stanowski, Wally	Tor., NYR	10	428	23	88	111	160	60	3	14	17	13	4	1939-40	1950-51
‡ Stanton, Paul	Pit., Bos., NYI	5	295	14	49	63	262	44	2	10	12	66	2	1990-91	1994-95
Stapleton, Brian	Wsh.	1	1	0	0	0	0	....	....	....	....	....	....	1975-76	1975-76
Stapleton, Pat	Bos., Chi.	10	635	43	294	337	353	65	10	39	49	38	....	1961-62	1972-73
Starikov, Sergei	N.J.	1	16	0	1	1	8	....	....	....	....	....	....	1989-90	1989-90
• Starr, Harold	Ott., Mtl.M., Mtl., NYR	7	205	6	5	11	186	15	1	0	1	4	....	1929-30	1935-36
Starr, Wilf	NYA, Det.	4	87	8	6	14	25	7	0	2	2	2	....	1932-33	1935-36
Stasiuk, Vic	Chi., Det., Bos.	14	745	183	254	437	669	69	16	18	34	40	3	1949-50	1962-63
Stastny, Anton	Que.	9	650	252	384	636	150	66	20	32	52	31	....	1980-81	1988-89
Stastny, Marian	Que., Tor.	5	322	121	173	294	110	32	5	17	22	7	....	1981-82	1985-86
Stastny, Peter	Que., N.J., St.L.	15	977	450	789	1239	824	93	33	72	105	123	....	1980-81	1994-95
Staszak, Ray	Det.	1	4	0	1	1	7	....	....	....	....	....	....	1985-86	1985-86
• Steele, Frank	Det.	1	1	0	0	0	0	....	....	....	....	....	....	1930-31	1930-31
Steen, Anders	Wpg.	1	42	5	11	16	22	....	....	....	....	....	....	1980-81	1980-81
Steen, Thomas	Wpg.	14	950	264	553	817	753	56	12	32	44	62	....	1981-82	1994-95
Stefaniw, Morris	Atl.	1	13	1	1	2	2	....	....	....	....	....	....	1972-73	1972-73
Stefanski, Bud	NYR	1	1	0	0	0	0	....	....	....	....	....	....	1977-78	1977-78
Stemkowski, Pete	Tor., Det., NYR, L.A.	15	967	206	349	555	866	83	25	29	54	136	1	1963-64	1977-78
Stenlund, Vern	Cle.	1	4	0	0	0	0	....	....	....	....	....	....	1976-77	1976-77
Stephenson, Bob	Hfd., Tor.	1	18	2	3	5	4	....	....	....	....	....	....	1979-80	1979-80
Stern, Ron	Van., Cgy., S.J.	12	638	75	86	161	2077	43	7	7	14	119	....	1987-88	1999-00
Sterner, Ulf	NYR	1	4	0	0	0	0	....	....	....	....	....	....	1964-65	1964-65
Stevens, John	Phi., Hfd.	5	53	0	10	10	48	....	....	....	....	....	....	1986-87	1993-94
‡ Stevens, Mike	Van., Bos., NYI, Tor.	4	23	1	4	5	29	....	....	....	....	....	....	1984-85	1989-90
• Stevens, Phil	Mtl.W., Mtl., Bos.	3	25	1	0	1	3	....	....	....	....	....	....	1917-18	1925-26
‡ Stevenson, Shayne	Bos., T.B.	3	27	0	2	2	35	....	....	....	....	....	....	1990-91	1992-93
Stewart, Allan	N.J., Bos.	6	64	6	4	10	243	....	....	....	....	....	....	1985-86	1991-92
Stewart, Bill	Buf., St.L., Tor., Min.	8	261	7	64	71	424	13	1	3	4	11	....	1977-78	1985-86
Stewart, Blair	Det., Wsh., Que.	7	229	34	44	78	326	....	....	....	....	....	....	1973-74	1979-80
Stewart, Bob	Bos., Cal., Cle., St.L., Pit.	9	575	27	101	128	809	5	1	1	2	2	....	1971-72	1979-80
‡ Stewart, Cam	Bos., Fla., Min.	7	202	16	23	39	120	13	1	3	4	9	....	1993-94	2001-02
Stewart, Gaye	Tor., Chi., Det., NYR, Mtl.	11	502	185	159	344	274	25	2	9	11	16	2	1941-42	1953-54
• Stewart, Jack	Det., Chi.	12	565	31	84	115	765	80	5	14	19	143	2	1938-39	1951-52
Stewart, John	Pit., Atl., Cal.	5	258	58	60	118	158	4	0	0	0	10	....	1970-71	1974-75
Stewart, John	Que.	1	2	0	0	0	0	....	....	....	....	....	....	1979-80	1979-80
Stewart, Ken	Chi.	1	6	1	1	2	2	....	....	....	....	....	....	1941-42	1941-42
• Stewart, Nels	Mtl.M., Bos., NYA	15	650	324	191	515	953	50	9	12	21	47	1	1925-26	1939-40
Stewart, Paul	Que.	1	21	2	0	2	74	....	....	....	....	....	....	1979-80	1979-80

Eddie Shore

Steve Smith

Peter Stastny

David Struch

Raimo Summanen

Mats Thelin

Esa Tikkanen

Mark Tinordi

Name	NHL Teams	NHL Seasons	Regular Schedule GP	G	A	TP	PIM	Playoffs GP	G	A	TP	PIM	NHL Cup Wins	First NHL Season	Last NHL Season
Stewart, Ralph	Van., NYI	7	252	57	73	130	28	19	4	4	8	2	....	1970-71	1977-78
Stewart, Ron	Tor., Bos., St.L., NYR, Van., NYI	21	1353	276	253	529	560	119	14	21	35	60	3	1952-53	1972-73
Stewart, Ryan	Wpg.	1	3	1	0	1	0	....	....	....	....	....	....	1985-86	1985-86
Stienburg, Trevor	Que.	4	71	8	4	12	161	1	0	0	0	0	....	1985-86	1988-89
Stiles, Tony	Cgy.	1	30	2	7	9	20	....	....	....	....	....	....	1983-84	1983-84
St-Laurent, Dollard	Mtl., Chi.	12	652	29	133	162	496	92	2	22	24	87	5	1950-51	1961-62
Stoddard, Jack	NYR	2	80	16	15	31	31	....	....	....	....	....	....	1951-52	1952-53
‡ Stojanov, Alek	Van., Pit.	3	107	2	5	7	222	14	0	0	0	21	....	1994-95	1996-97
Stoltz, Roland	Wsh.	1	14	2	2	4	14	....	....	....	....	....	....	1981-82	1981-82
Stone, Steve	Van.	1	2	0	0	0	0	....	....	....	....	....	....	1973-74	1973-74
Storm, Jim	Hfd., Dal.	3	84	7	15	22	44	....	....	....	....	....	....	1993-94	1995-96
Stothers, Mike	Phi., Tor.	4	30	0	2	2	65	5	0	0	0	11	....	1984-85	1987-88
Stoughton, Blaine	Pit., Tor., Hfd., NYR	8	526	258	191	449	204	8	4	2	6	2	....	1973-74	1983-84
Stoyanovich, Steve	Hfd.	1	23	3	5	8	11	....	....	....	....	....	....	1983-84	1983-84
• Strain, Neil	NYR	1	52	11	13	24	12	....	....	....	....	....	....	1952-53	1952-53
Strate, Gord	Det.	3	61	0	0	0	34	....	....	....	....	....	....	1956-57	1958-59
Stratton, Art	NYR, Det., Chi., Pit., Phi.	4	95	18	33	51	24	5	0	0	0	0	....	1959-60	1967-68
Strobel, Art	NYR	1	7	0	0	0	0	....	....	....	....	....	....	1943-44	1943-44
Strong, Ken	Tor.	3	15	2	2	4	6	....	....	....	....	....	....	1982-83	1984-85
‡ Struch, David	Cgy.	1	4	0	0	0	4	....	....	....	....	....	....	1993-94	1993-94
Strueby, Todd	Edm.	3	5	0	1	1	2	....	....	....	....	....	....	1981-82	1983-84
• Stuart, Billy	Tor., Bos.	7	195	30	20	50	151	12	1	1	2	6	1	1920-21	1926-27
Stumpf, Bob	St.L., Pit.	1	10	1	1	2	20	....	....	....	....	....	....	1974-75	1974-75
Sturgeon, Peter	Col.	2	6	0	1	1	2	....	....	....	....	....	....	1979-80	1980-81
Suikkanen, Kai	Buf.	2	2	0	0	0	0	....	....	....	....	....	....	1981-82	1982-83
Sulliman, Doug	NYR, Hfd., N.J., Phi.	11	631	160	168	328	175	16	1	3	4	2	....	1979-80	1989-90
Sullivan, Barry	Det.	1	1	0	0	0	0	....	....	....	....	....	....	1947-48	1947-48
Sullivan, Bob	Hfd.	1	62	18	19	37	18	....	....	....	....	....	....	1982-83	1982-83
Sullivan, Brian	N.J.	1	2	0	1	1	0	....	....	....	....	....	....	1992-93	1992-93
Sullivan, Frank	Tor., Chi.	4	8	0	0	0	2	....	....	....	....	....	....	1949-50	1955-56
Sullivan, Peter	Wpg.	2	126	28	54	82	40	....	....	....	....	....	....	1979-80	1980-81
Sullivan, Red	Bos., Chi., NYR	11	557	107	239	346	441	18	1	2	3	6	....	1949-50	1960-61
Summanen, Raimo	Edm., Van.	5	151	36	40	76	35	10	2	5	7	0	....	1983-84	1987-88
• Summerhill, Bill	Mtl., Bro.	4	72	14	17	31	70	3	0	0	0	2	....	1937-38	1941-42
‡ Sundblad, Niklas	Cgy.	1	2	0	0	0	0	....	....	....	....	....	....	1995-96	1995-96
‡ Sundin, Ronnie	NYR	1	1	0	0	0	0	....	....	....	....	....	....	1997-98	1997-98
Sundstrom, Patrik	Van., N.J.	10	679	219	369	588	349	37	9	17	26	25	....	1982-83	1991-92
Sundstrom, Peter	NYR, Wsh., N.J.	6	338	61	83	144	120	23	3	3	6	8	....	1983-84	1989-90
• Suomi, Al	Chi.	1	5	0	0	0	0	....	....	....	....	....	....	1936-37	1936-37
Sutherland, Bill	Mtl., Phi., Tor., St.L., Det.	6	250	70	58	128	99	14	2	4	6	0	....	1962-63	1971-72
• Sutherland, Max	Bos.	1	2	0	0	0	0	....	....	....	....	....	....	1931-32	1931-32
Sutter, Brent	NYI, Chi.	18	1111	363	466	829	1054	144	30	44	74	164	2	1980-81	1997-98
Sutter, Brian	St.L.	12	779	303	333	636	1786	65	21	21	42	249	....	1976-77	1987-88
Sutter, Darryl	Chi.	8	406	161	118	279	288	51	24	19	43	26	....	1979-80	1986-87
Sutter, Duane	NYI, Chi.	11	731	139	203	342	1333	161	26	32	58	405	4	1979-80	1989-90
Sutter, Rich	Pit., Phi., Van., St.L., Chi., T.B., Tor.	13	874	149	166	315	1411	78	13	5	18	133	....	1982-83	1994-95
Sutter, Ron	Phi., St.L., Que., NYI, Bos., S.J., Cgy.	19	1093	205	329	534	1352	104	8	32	40	193	....	1982-83	2000-01
Suzor, Mark	Phi., Col.	2	64	4	16	20	60	....	....	....	....	....	....	1976-77	1977-78
‡ Svejkovsky, Jaroslav	Wsh., T.B.	4	113	23	19	42	56	1	0	0	0	2	....	1996-97	1999-00
Svensson, Leif	Wsh.	2	121	6	40	46	49	....	....	....	....	....	....	1978-79	1979-80
‡ Svensson, Magnus	Fla.	2	46	4	14	18	31	....	....	....	....	....	....	1994-95	1995-96
‡ Svoboda, Petr	Mtl., Buf., Phi., T.B.	17	1028	58	341	399	1605	127	4	45	49	140	1	1984-85	2000-01
Swain, Garry	Pit.	1	9	1	1	2	0	....	....	....	....	....	....	1968-69	1968-69
Swarbrick, George	Oak., Pit., Phi.	4	132	17	25	42	173	....	....	....	....	....	....	1967-68	1970-71
• Sweeney, Bill	NYR	1	4	1	0	1	0	....	....	....	....	....	....	1959-60	1959-60
‡ Sweeney, Bob	Bos., Buf., NYI, Cgy.	10	639	125	163	288	799	103	15	18	33	197	....	1986-87	1995-96
Sweeney, Tim	Cgy., Bos., Ana., NYR	8	291	55	83	138	123	4	0	0	0	2	....	1990-91	1997-98
Sykes, Bob	Tor.	1	2	0	0	0	0	....	....	....	....	....	....	1974-75	1974-75
Sykes, Phil	L.A., Wpg.	10	456	79	85	164	519	26	0	3	3	29	....	1982-83	1991-92
Sylvester, Dean	Buf., Atl.	3	96	21	16	37	32	4	0	0	0	0	....	1998-99	2000-01
Szura, Joe	Oak.	2	90	10	15	25	30	7	2	3	5	2	....	1967-68	1968-69

T

Name	NHL Teams	NHL Seasons	Regular Schedule GP	G	A	TP	PIM	Playoffs GP	G	A	TP	PIM	NHL Cup Wins	First NHL Season	Last NHL Season
Taft, John	Det.	1	15	0	2	2	4	....	....	....	....	....	....	1978-79	1978-79
Taglianetti, Peter	Wpg., Min., Pit., T.B.	11	451	18	74	92	1106	53	2	8	10	103	2	1984-85	1994-95
Talafous, Dean	Atl., Min., NYR	8	497	104	154	258	163	21	4	7	11	11	....	1974-75	1981-82
Talakoski, Ron	NYR	2	9	0	1	1	33	....	....	....	....	....	....	1986-87	1987-88
Talbot, Jean-Guy	Mtl., Min., Det., St.L., Buf.	17	1056	43	242	285	1006	150	4	26	30	142	7	1954-55	1970-71
Tallon, Dale	Van., Chi., Pit.	10	642	98	238	336	568	33	2	10	12	45	....	1970-71	1979-80
Tambellini, Steve	NYI, Col., N.J., Cgy., Van.	10	553	160	150	310	105	2	0	1	1	0	1	1978-79	1987-88
‡ Tancill, Chris	Hfd., Det., Dal., S.J.	8	134	17	32	49	54	11	1	1	2	8	....	1990-91	1997-98
Tanguay, Christian	Que.	1	2	0	0	0	0	....	....	....	....	....	....	1981-82	1981-82
Tannahill, Don	Van.	2	111	30	33	63	25	....	....	....	....	....	....	1972-73	1973-74
Tanti, Tony	Chi., Van., Pit., Buf.	11	697	287	273	560	661	30	3	12	15	27	....	1981-82	1991-92
Tardif, Marc	Mtl., Que.	8	517	194	207	401	443	62	13	15	28	75	2	1969-70	1982-83
‡ Tardif, Patrice	St.L., L.A.	2	65	7	11	18	78	....	....	....	....	....	....	1994-95	1995-96
Tatarinov, Mikhail	Wsh., Que., Bos.	4	161	21	48	69	184	....	....	....	....	....	....	1990-91	1993-94
Tatchell, Spence	NYR	1	1	0	0	0	0	....	....	....	....	....	....	1942-43	1942-43
• Taylor, Billy	Tor., Det., Bos., NYR	7	323	87	180	267	120	33	6	18	24	13	1	1939-40	1947-48
• Taylor, Bob	Bos.	1	8	0	0	0	6	....	....	....	....	....	....	1929-30	1929-30
Taylor, Dave	L.A.	17	1111	431	638	1069	1589	92	26	33	59	145	....	1977-78	1993-94
Taylor, Harry	Tor., Chi.	3	66	5	10	15	30	1	0	0	0	0	1	1946-47	1951-52
Taylor, Mark	Phi., Pit., Wsh.	5	209	42	68	110	73	6	0	0	0	0	....	1981-82	1985-86
• Taylor, Ralph	Chi., NYR	3	99	4	1	5	169	4	0	0	0	10	....	1927-28	1929-30
Taylor, Ted	NYR, Det., Min., Van.	6	166	23	35	58	181	....	....	....	....	....	....	1964-65	1971-72
Taylor Jr., Billy	NYR	1	2	0	0	0	0	....	....	....	....	....	....	1964-65	1964-65
Teal, Jeff	Mtl.	1	6	0	1	1	0	....	....	....	....	....	....	1984-85	1984-85
Teal, Skip	Bos.	1	1	0	0	0	0	....	....	....	....	....	....	1954-55	1954-55
Teal, Vic	NYI	1	1	0	0	0	0	....	....	....	....	....	....	1973-74	1973-74
Tebbutt, Greg	Que., Pit.	2	26	0	3	3	35	....	....	....	....	....	....	1979-80	1983-84
Tepper, Stephen	Chi.	1	1	0	0	0	0	....	....	....	....	....	....	1992-93	1992-93
Terbenche, Paul	Chi., Buf.	5	189	5	26	31	28	12	0	0	0	0	....	1967-68	1973-74
Terrion, Greg	L.A., Tor.	8	561	93	150	243	339	35	2	9	11	41	....	1980-81	1987-88
Terry, Bill	Min.	1	5	0	0	0	0	....	....	....	....	....	....	1987-88	1987-88
• Tertyshny, Dmitri	Phi.	1	62	2	8	10	30	1	0	0	0	0	....	1998-99	1998-99
Tessier, Orval	Mtl., Bos.	3	59	5	7	12	6	....	....	....	....	....	....	1954-55	1960-61
Theberge, Greg	Wsh.	5	153	15	63	78	73	4	0	1	1	0	....	1979-80	1983-84
Thelin, Mats	Bos.	3	163	8	19	27	107	5	0	0	0	6	....	1984-85	1986-87
Thelven, Michael	Bos.	5	207	20	80	100	217	34	4	10	14	34	....	1985-86	1989-90
Therrien, Gaston	Que.	3	22	0	8	8	12	9	0	1	1	4	....	1980-81	1982-83
‡ Thibaudeau, Gilles	Mtl., NYI, Tor.	5	119	25	37	62	40	8	3	3	6	2	....	1986-87	1990-91
Thibeault, Lorrain	Det., Mtl.	2	5	0	2	2	2	....	....	....	....	....	....	1944-45	1945-46
Thiffault, Leo	Min.	1	....	....	....	....	....	5	0	0	0	0	....	1967-68	1967-68
Thomas, Cy	Chi., Tor.	1	14	2	2	4	12	....	....	....	....	....	....	1947-48	1947-48
Thomas, Reg	Que.	1	39	9	7	16	6	....	....	....	....	....	....	1979-80	1979-80
Thomlinson, Dave	St.L., Bos., L.A.	5	42	1	3	4	50	9	3	1	4	4	....	1989-90	1994-95
Thompson, Cliff	Bos.	2	13	0	1	1	2	....	....	....	....	....	....	1941-42	1948-49
Thompson, Errol	Tor., Det., Pit.	10	599	208	185	393	184	34	7	5	12	11	....	1970-71	1980-81
• Thompson, Ken	Mtl.W.	1	1	0	0	0	0	....	....	....	....	....	....	1917-18	1917-18
• Thompson, Paul	NYR, Chi.	13	582	153	179	332	336	48	11	11	22	54	3	1926-27	1938-39
• Thoms, Bill	Tor., Chi., Bos.	13	548	135	206	341	154	44	6	10	16	6	....	1932-33	1944-45
• Thomson, Bill	Det.	2	9	2	2	4	0	2	0	0	0	0	....	1938-39	1943-44
Thomson, Floyd	St.L.	8	411	56	97	153	341	10	0	2	2	6	....	1971-72	1979-80
Thomson, Jim	Wsh., Hfd., N.J., L.A., Ott., Ana.	7	115	4	3	7	416	1	0	0	0	0	....	1986-87	1993-94
• Thomson, Jimmy	Tor., Chi.	13	787	19	215	234	920	63	2	13	15	135	4	1945-46	1957-58
• Thomson, Rhys	Mtl., Tor.	2	25	0	2	2	38	....	....	....	....	....	....	1939-40	1942-43
Thornbury, Tom	Pit.	1	14	1	8	9	16	....	....	....	....	....	....	1983-84	1983-84
• Thorsteinson, Joe	NYA	1	4	0	0	0	0	....	....	....	....	....	....	1932-33	1932-33
• Thurier, Fred	NYA, Bro., NYR	3	80	25	27	52	18	....	....	....	....	....	....	1940-41	1944-45
Thurlby, Tom	Oak.	1	20	1	1	2	4	....	....	....	....	....	....	1967-68	1967-68
Thyer, Mario	Min.	1	5	0	0	0	0	1	0	0	0	2	....	1989-90	1989-90
Tichy, Milan	Chi., NYI	3	23	0	5	5	40	....	....	....	....	....	....	1992-93	1995-96
Tidey, Alex	Buf., Edm.	3	9	0	0	0	8	2	0	0	0	0	....	1976-77	1979-80
‡ Tikkanen, Esa	Edm., NYR, St.L., N.J., Van., Fla., Wsh.	15	877	244	386	630	1077	186	72	60	132	275	5	1984-85	1998-99
‡ Tilley, Tom	St.L.	4	174	4	38	42	89	14	1	3	4	19	....	1988-89	1993-94

Name	NHL Teams	NHL Seasons	Regular Schedule					Playoffs					NHL Cup Wins	First NHL Season	Last NHL Season
			GP	G	A	TP	PIM	GP	G	A	TP	PIM			
● Timgren, Ray	Tor., Chi.	6	251	14	44	58	70	30	3	9	12	6	2	1948-49	1954-55
Tinordi, Mark	NYR, Min., Dal., Wsh.	12	663	52	148	200	1514	70	7	11	18	165	....	1987-88	1998-99
Tippett, Dave	Hfd., Wsh., Pit., Phi.	11	721	93	169	262	317	62	6	16	22	34	....	1983-84	1993-94
Titanic, Morris	Buf.	2	19	0	0	0	0	....	....	....	....	....	....	1974-75	1975-76
Tkaczuk, Walt	NYR	14	945	227	451	678	556	93	19	32	51	119	....	1967-68	1980-81
Toal, Mike	Edm.	1	3	0	0	0	0	....	....	....	....	....	....	1979-80	1979-80
Todd, Kevin	N.J., Edm., Chi., L.A., Ana.	9	383	70	133	203	225	12	3	2	5	16	....	1988-89	1997-98
Tomalty, Glenn	Wpg.	1	1	0	0	0	0	....	....	....	....	....	....	1979-80	1979-80
Tomlak, Mike	Hfd.	4	141	15	22	37	103	10	0	1	1	4	....	1989-90	1993-94
‡ Tomlinson, Dave	Tor., Wpg., Fla.	4	42	1	3	4	28	....	....	....	....	....	....	1991-92	1994-95
Tomlinson, Kirk	Min.	1	1	0	0	0	0	....	....	....	....	....	....	1987-88	1987-88
Tomson, Jack	NYA	3	15	1	1	2	0	2	0	0	0	0	....	1938-39	1940-41
Tonelli, John	NYI, Cgy., L.A., Chi., Que.	14	1028	325	511	836	911	172	40	75	115	200	4	1978-79	1991-92
Tookey, Tim	Wsh., Que., Pit., Phi., L.A.	7	106	22	36	58	71	10	1	3	4	2	....	1980-81	1988-89
Toomey, Sean	Min.	1	1	0	0	0	0	....	....	....	....	....	....	1986-87	1986-87
‡ Toporowski, Shayne	Tor.	1	3	0	0	0	7	....	....	....	....	....	....	1996-97	1996-97
Toppazzini, Jerry	Bos., Chi., Det.	12	783	163	244	407	436	40	13	9	22	13	....	1952-53	1963-64
● Toppazzini, Zellio	Bos., NYR, Chi.	5	123	21	22	43	49	2	0	0	0	0	....	1948-49	1956-57
Torgaev, Pavel	Cgy., T.B.	2	55	6	14	20	20	1	0	0	0	0	....	1995-96	1999-00
‡ Torkki, Jari	Chi.	1	4	1	0	1	0	....	....	....	....	....	....	1988-89	1988-89
‡ Tormanen, Antti	Ott.	1	50	7	8	15	28	....	....	....	....	....	....	1995-96	1995-96
● Touhey, Bill	Mtl.M., Ott., Bos.	7	280	65	40	105	107	2	1	0	1	0	....	1927-28	1933-34
● Toupin, Jacques	Chi.	1	8	1	2	3	0	4	0	0	0	0	....	1943-44	1943-44
● Townsend, Art	Chi.	1	5	0	0	0	0	....	....	....	....	....	....	1926-27	1926-27
Townshend, Graeme	Bos., NYI, Ott.	5	45	3	7	10	28	....	....	....	....	....	....	1989-90	1993-94
Trader, Larry	Det., St.L., Mtl.	4	91	5	13	18	74	3	0	0	0	0	....	1982-83	1987-88
● Trainor, Wes	NYR	1	17	1	2	3	6	....	....	....	....	....	....	1948-49	1948-49
● Trapp, Bob	Chi.	2	82	4	4	8	129	2	0	0	0	4	....	1926-27	1927-28
Trapp, Doug	Buf.	1	2	0	0	0	0	....	....	....	....	....	....	1986-87	1986-87
● Traub, Percy	Chi., Det.	3	130	3	3	6	217	4	0	0	0	6	....	1926-27	1928-29
Tredway, Brock	L.A.	1	....	....	....	....	....	1	0	0	0	0	....	1981-82	1981-82
Tremblay, Brent	Wsh.	2	10	1	0	1	6	....	....	....	....	....	....	1978-79	1979-80
Tremblay, Gilles	Mtl.	9	509	168	162	330	161	48	9	14	23	4	3	1960-61	1968-69
● Tremblay, J.C.	Mtl.	13	794	57	306	363	204	108	14	51	65	58	5	1959-60	1971-72
Tremblay, Marcel	Mtl.	1	10	0	2	2	0	....	....	....	....	....	....	1938-39	1938-39
Tremblay, Mario	Mtl.	12	852	258	326	584	1043	101	20	29	49	187	5	1974-75	1985-86
● Tremblay, Nils	Mtl.	2	3	0	1	1	0	2	0	0	0	0	....	1944-45	1945-46
Trimper, Tim	Chi., Wpg., Min.	6	190	30	36	66	153	2	0	0	0	2	....	1979-80	1984-85
Trottier, Bryan	NYI, Pit.	18	1279	524	901	1425	912	221	71	113	184	277	6	1975-76	1993-94
● Trottier, Dave	Mtl.M., Det.	11	446	121	113	234	517	31	4	3	7	39	1	1928-29	1938-39
Trottier, Guy	NYR, Tor.	3	115	28	17	45	37	9	1	0	1	16	....	1968-69	1971-72
Trottier, Rocky	N.J.	2	38	6	4	10	2	....	....	....	....	....	....	1983-84	1984-85
● Trudel, Lou	Chi., Mtl.	8	306	49	69	118	122	24	1	3	4	4	2	1933-34	1940-41
Trudell, Rene	NYR	3	129	24	28	52	72	5	0	0	0	2	....	1945-46	1947-48
‡ Tsulygin, Nikolai	Ana.	1	22	0	1	1	8	....	....	....	....	....	....	1996-97	1996-97
‡ Tsygurov, Denis	Buf., L.A.	3	51	1	5	6	45	....	....	....	....	....	....	1993-94	1995-96
Tucker, John	Buf., Wsh., NYI, T.B.	12	656	177	259	436	285	31	10	18	28	24	....	1983-84	1995-96
● Tudin, Connie	Mtl.	1	4	0	1	1	4	....	....	....	....	....	....	1941-42	1941-42
Tudor, Rob	Van., St.L.	3	28	4	4	8	19	3	0	0	0	0	....	1978-79	1982-83
Tuer, Allan	L.A., Min., Hfd.	4	57	1	1	2	208	....	....	....	....	....	....	1985-86	1989-90
Turcotte, Alfie	Mtl., Wpg., Wsh.	7	112	17	29	46	49	5	0	0	0	0	....	1983-84	1990-91
‡ Turgeon, Sylvain	Hfd., N.J., Mtl., Ott.	12	669	269	226	495	691	36	4	7	11	22	....	1983-84	1994-95
Turlick, Gord	Bos.	1	2	0	0	0	2	....	....	....	....	....	....	1959-60	1959-60
Turnbull, Ian	Tor., L.A., Pit.	10	628	123	317	440	736	55	13	32	45	94	....	1973-74	1982-83
Turnbull, Perry	St.L., Mtl., Wpg.	9	608	188	163	351	1245	34	6	7	13	86	....	1979-80	1987-88
Turnbull, Randy	Cgy.	1	1	0	0	0	2	....	....	....	....	....	....	1981-82	1981-82
Turner, Bob	Mtl., Chi.	8	478	19	51	70	307	68	1	4	5	44	5	1955-56	1962-63
Turner, Brad	NYI	1	3	0	0	0	0	....	....	....	....	....	....	1991-92	1991-92
Turner, Dean	NYR, Col., L.A.	4	35	1	0	1	59	....	....	....	....	....	....	1978-79	1982-83
● Tustin, Norm	NYR	1	18	2	4	6	0	....	....	....	....	....	....	1941-42	1941-42
Tuten, Aud	Chi.	2	39	4	8	12	48	....	....	....	....	....	....	1941-42	1942-43
Tutt, Brian	Wsh.	1	7	1	0	1	2	....	....	....	....	....	....	1989-90	1989-90
Tuttle, Steve	St.L.	3	144	28	28	56	12	17	1	6	7	2	....	1988-89	1990-91
Twist, Tony	St.L., Que.	10	445	10	18	28	1121	18	1	1	2	22	....	1989-90	1998-99

U V

Name	NHL Teams	NHL Seasons	GP	G	A	TP	PIM	GP	G	A	TP	PIM	NHL Cup Wins	First NHL Season	Last NHL Season
Ubriaco, Gene	Pit., Oak., Chi.	3	177	39	35	74	50	11	2	0	2	4	....	1967-68	1969-70
Ullman, Norm	Det., Tor.	20	1410	490	739	1229	712	106	30	53	83	67	....	1955-56	1974-75
Unger, Garry	Tor., Det., St.L., Atl., L.A., Edm.	16	1105	413	391	804	1075	52	12	18	30	105	....	1967-68	1982-83
‡ Ustorf, Stefan	Wsh.	2	54	7	10	17	16	5	0	0	0	0	....	1995-96	1996-97
Vachon, Nick	NYI	1	1	0	0	0	0	....	....	....	....	....	....	1996-97	1996-97
Vadnais, Carol	Mtl., Oak., Cal., Bos., NYR, N.J.	17	1087	169	418	587	1813	106	10	40	50	185	2	1966-67	1982-83
‡ Vaic, Lubomir	Van.	2	9	1	1	2	2	....	....	....	....	....	....	1997-98	1999-00
Vail, Eric	Atl., Cgy., Det.	9	591	216	260	476	281	20	5	6	11	6	....	1973-74	1981-82
● Vail, Sparky	NYR	2	50	4	1	5	18	10	0	0	0	2	....	1928-29	1929-30
Vaive, Rick	Van., Tor., Chi., Buf.	13	876	441	347	788	1445	54	27	16	43	111	....	1979-80	1991-92
Valentine, Chris	Wsh.	3	105	43	52	95	127	2	0	0	0	4	....	1981-82	1983-84
Valiquette, Jack	Tor., Col.	7	350	84	134	218	79	23	3	6	9	4	....	1974-75	1980-81
Vallis, Lindsay	Mtl.	1	1	0	0	0	0	....	....	....	....	....	....	1993-94	1993-94
Van Boxmeer, John	Mtl., Col., Buf., Que.	11	588	84	274	358	465	38	5	15	20	37	1	1973-74	1983-84
Van Dorp, Wayne	Edm., Pit., Chi., Que.	6	125	12	12	24	565	27	0	1	1	42	....	1986-87	1991-92
‡ Van Drunen, David	Ott.	1	1	0	0	0	0	....	....	....	....	....	....	1999-00	1999-00
Van Impe, Ed	Chi., Phi., Pit.	11	700	27	126	153	1025	66	1	12	13	131	2	1966-67	1976-77
‡ Varis, Petri	Chi.	1	1	0	0	0	0	....	....	....	....	....	....	1997-98	1997-98
‡ Varvio, Jarkko	Dal.	2	13	3	4	7	4	....	....	....	....	....	....	1993-94	1994-95
‡ Vasilevski, Alexander	St.L.	2	4	0	0	0	2	....	....	....	....	....	....	1995-96	1996-97
‡ Vasilyev, Andrei	NYI, Phx.	4	16	2	5	7	6	....	....	....	....	....	....	1994-95	1998-99
Vaske, Dennis	NYI, Bos.	9	235	5	41	46	253	22	0	7	7	16	....	1990-91	1998-99
● Vasko, Moose	Chi., Min.	13	786	34	166	200	719	78	2	7	9	73	1	1956-57	1969-70
Vasko, Rick	Det.	3	31	3	7	10	29	....	....	....	....	....	....	1977-78	1980-81
Vautour, Yvon	NYI, Col., N.J., Que.	6	204	26	33	59	401	....	....	....	....	....	....	1979-80	1984-85
Vaydik, Greg	Chi.	1	5	0	0	0	0	....	....	....	....	....	....	1976-77	1976-77
Veitch, Darren	Wsh., Det., Tor.	10	511	48	209	257	296	33	4	11	15	33	....	1980-81	1990-91
Velischek, Randy	Min., N.J., Que.	10	509	21	76	97	401	44	2	5	7	32	....	1982-83	1991-92
Vellucci, Mike	Hfd.	1	2	0	0	0	11	....	....	....	....	....	....	1987-88	1987-88
Venasky, Vic	L.A.	7	430	61	101	162	66	21	1	5	6	12	....	1972-73	1978-79
Veneruzzo, Gary	St.L.	2	7	1	1	2	0	9	0	2	2	2	....	1967-68	1971-72
Vermette, Mark	Que.	4	67	5	13	18	33	....	....	....	....	....	....	1988-89	1991-92
Verret, Claude	Buf.	2	14	2	5	7	2	....	....	....	....	....	....	1983-84	1984-85
Verstraete, Leigh	Tor.	3	8	0	1	1	14	....	....	....	....	....	....	1982-83	1987-88
Ververgaert, Dennis	Van., Phi., Wsh.	8	583	176	216	392	247	8	1	2	3	6	....	1973-74	1980-81
Vesey, Jim	St.L., Bos.	3	15	1	2	3	7	....	....	....	....	....	....	1988-89	1991-92
Veysey, Sid	Van.	1	1	0	0	0	0	....	....	....	....	....	....	1977-78	1977-78
‡ Vial, Dennis	NYR, Det., Ott.	8	242	4	15	19	794	....	....	....	....	....	....	1990-91	1997-98
Vickers, Steve	NYR	10	698	246	340	586	330	68	24	25	49	58	....	1972-73	1981-82
Vigneault, Alain	St.L.	2	42	2	5	7	82	4	0	1	1	26	....	1981-82	1982-83
‡ Viitakoski, Vesa	Cgy.	3	23	2	4	6	8	....	....	....	....	....	....	1993-94	1995-96
‡ Vilgrain, Claude	Van., N.J., Phi.	5	89	21	32	53	78	11	1	1	2	17	....	1987-88	1993-94
Vincelette, Dan	Chi., Que.	6	193	20	22	42	351	12	0	0	0	4	....	1986-87	1991-92
Vipond, Pete	Cal.	1	3	0	0	0	0	....	....	....	....	....	....	1972-73	1972-73
Virta, Hannu	Buf.	5	245	25	101	126	66	17	1	3	4	6	....	1981-82	1985-86
Visheau, Mark	Wpg., L.A.	2	29	1	3	4	107	....	....	....	....	....	....	1993-94	1998-99
‡ Vitolinsh, Harijs	Wpg.	1	8	0	0	0	4	....	....	....	....	....	....	1993-94	1993-94
‡ Viveiros, Emanuel	Min.	3	29	1	11	12	6	....	....	....	....	....	....	1985-86	1987-88
● Vokes, Ed	Chi.	1	5	0	0	0	0	....	....	....	....	....	....	1930-31	1930-31
Volcan, Mickey	Hfd., Cgy.	4	162	8	33	41	146	....	....	....	....	....	....	1980-81	1983-84
‡ Volchkov, Alexandre	Wsh.	1	3	0	0	0	0	....	....	....	....	....	....	1999-00	1999-00
Volek, David	NYI	6	396	95	154	249	201	15	5	5	10	2	....	1988-89	1993-94
Volmar, Doug	Det., L.A.	4	62	13	8	21	26	2	1	0	1	0	....	1969-70	1972-73
‡ Von Stefenelli, Phil	Bos., Ott.	2	33	0	5	5	23	....	....	....	....	....	....	1995-96	1996-97
Vopat, Jan	L.A., Nsh.	5	126	11	20	31	70	2	0	1	1	2	....	1995-96	1999-00
‡ Vopat, Roman	St.L., L.A., Chi., Phi.	4	133	6	14	20	253	....	....	....	....	....	....	1995-96	1998-99
‡ Vorobiev, Vladimir	NYR, Edm.	3	33	9	7	16	14	1	0	0	0	0	....	1996-97	1998-99
● Voss, Carl	Tor., NYR, Det., Ott., St.L., NYA, Mtl.M., Chi.	8	261	34	70	104	50	24	5	3	8	0	1	1926-27	1937-38

Zelio Toppazzini

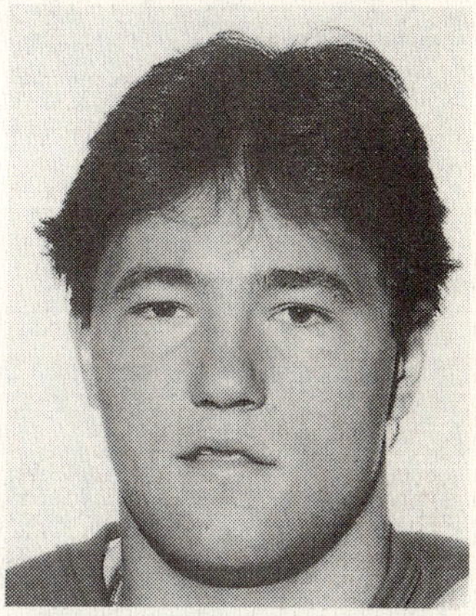

John Tucker

Garry Unger

Ed Van Impe

Ryan Walter

Cooney Weiland

Jay Wells

Ross Yates

			Regular Schedule					Playoffs							
Name	NHL Teams	NHL Seasons	GP	G	A	TP	PIM	GP	G	A	TP	PIM	NHL Cup Wins	First NHL Season	Last NHL Season
Vukota, Mick	NYI, T.B., Mtl.	11	574	17	29	46	2071	23	0	0	0	73	....	1987-88	1997-98
Vyazmikin, Igor	Edm.	1	4	1	0	1	0	....	....	....	....	....	....	1990-91	1990-91
W															
Waddell, Don	L.A.	1	1	0	0	0	0	....	....	....	....	....	....	1980-81	1980-81
• Waite, Frank	NYR	1	17	1	3	4	4	....	....	....	....	....	....	1930-31	1930-31
Walker, Gord	NYR, L.A.	4	31	3	4	7	23	....	....	....	....	....	....	1986-87	1989-90
Walker, Howard	Wsh., Cgy.	3	83	2	13	15	133	....	....	....	....	....	....	1980-81	1982-83
• Walker, Jack	Det.	2	80	5	8	13	18	....	....	....	....	....	....	1926-27	1927-28
Walker, Kurt	Tor.	3	71	4	5	9	142	16	0	0	0	34	....	1975-76	1977-78
Walker, Russ	L.A.	2	17	1	0	1	41	....	....	....	....	....	....	1976-77	1977-78
Wall, Bob	Det., L.A., St.L.	8	322	30	55	85	155	22	0	3	3	2	....	1964-65	1971-72
Wallin, Peter	NYR	2	52	3	14	17	14	14	2	6	8	6	....	1980-81	1981-82
Walsh, Jim	Buf.	1	4	0	1	1	4	....	....	....	....	....	....	1981-82	1981-82
Walsh, Mike	NYI	2	14	2	0	2	4	....	....	....	....	....	....	1987-88	1988-89
Walter, Ryan	Wsh., Mtl., Van.	15	1003	264	382	646	946	113	16	35	51	62	1	1978-79	1992-93
• Walton, Bobby	Mtl.	1	4	0	0	0	0	....	....	....	....	....	....	1943-44	1943-44
Walton, Mike	Tor., Bos., Van., St.L., Chi.	12	588	201	247	448	357	47	14	10	24	45	2	1965-66	1978-79
Wappel, Gord	Atl., Cgy.	3	20	1	1	2	10	2	0	0	0	4	....	1979-80	1981-82
Ward, Don	Chi., Bos.	2	34	0	1	1	16	....	....	....	....	....	....	1957-58	1959-60
• Ward, Jimmy	Mtl.M., Mtl.	12	527	147	127	274	455	36	4	4	8	26	1	1927-28	1938-39
Ward, Joe	Col.	1	4	0	0	0	2	....	....	....	....	....	....	1980-81	1980-81
Ward, Ron	Tor., Van.	2	89	2	5	7	6	....	....	....	....	....	....	1969-70	1971-72
‡ Ware, Jeff	Tor., Fla.	3	21	0	1	1	12	....	....	....	....	....	....	1996-97	1998-99
‡ Ware, Michael	Edm.	2	5	0	1	1	15	....	....	....	....	....	....	1988-89	1989-90
• Wares, Eddie	NYR, Det., Chi.	9	321	60	102	162	161	45	5	7	12	34	1	1936-37	1946-47
Warner, Bob	Tor.	2	10	1	1	2	4	4	0	0	0	0	....	1975-76	1976-77
Warner, Jim	Hfd.	1	32	0	3	3	10	....	....	....	....	....	....	1979-80	1979-80
Warwick, Bill	NYR	2	14	3	3	6	16	....	....	....	....	....	....	1942-43	1943-44
• Warwick, Grant	NYR, Bos., Mtl.	9	395	147	142	289	220	16	2	4	6	6	....	1941-42	1949-50
• Wasnie, Nick	Chi., Mtl., NYA, Ott., St.L.	7	248	57	34	91	176	20	6	3	9	20	2	1927-28	1934-35
Watson, Bill	Chi.	4	115	23	36	59	12	6	0	2	2	0	....	1985-86	1988-89
Watson, Bryan	Mtl., Det., Oak., Pit., St.L., Wsh.	16	878	17	135	152	2212	32	2	0	2	70	....	1963-64	1978-79
Watson, Dave	Col.	2	18	0	1	1	10	....	....	....	....	....	....	1979-80	1980-81
Watson, Harry	Bro., Det., Tor., Chi.	14	809	236	207	443	150	62	16	9	25	27	5	1941-42	1956-57
Watson, Jim	Det., Buf.	8	221	4	19	23	345	....	....	....	....	....	....	1963-64	1971-72
Watson, Jimmy	Phi.	10	613	38	148	186	492	101	5	34	39	89	2	1972-73	1981-82
Watson, Joe	Bos., Phi., Col.	14	835	38	178	216	447	84	3	12	15	82	2	1964-65	1978-79
• Watson, Phil	NYR, Mtl.	13	590	144	265	409	532	54	10	25	35	67	2	1935-36	1947-48
Watters, Tim	Wpg., L.A.	14	741	26	151	177	1289	82	1	5	6	115	....	1981-82	1994-95
Watts, Brian	Det.	1	4	0	0	0	0	....	....	....	....	....	....	1975-76	1975-76
• Webster, Aubrey	Phi., Mtl.M.	2	5	0	0	0	0	....	....	....	....	....	....	1930-31	1934-35
• Webster, Don	Tor.	1	27	7	6	13	28	5	0	0	0	12	....	1943-44	1943-44
Webster, John	NYR	1	14	0	0	0	4	....	....	....	....	....	....	1949-50	1949-50
Webster, Tom	Bos., Det., Cal.	5	102	33	42	75	61	1	0	0	0	0	....	1968-69	1979-80
• Weiland, Cooney	Bos., Ott., Det.	11	509	173	160	333	147	45	12	10	22	12	2	1928-29	1938-39
Weir, Stan	Cal., Tor., Edm., Col., Det.	10	642	139	207	346	183	37	6	5	11	4	....	1972-73	1982-83
Weir, Wally	Que., Hfd., Pit.	6	320	21	45	66	625	23	0	1	1	96	....	1979-80	1984-85
• Wellington, Alex	Que.	1	1	0	0	0	0	....	....	....	....	....	....	1919-20	1919-20
‡ Wells, Chris	Pit., Fla.	5	195	9	20	29	193	3	0	0	0	0	....	1995-96	1999-00
Wells, Jay	L.A., Phi., Buf., NYR, St.L., T.B.	18	1098	47	216	263	2359	114	3	14	17	213	1	1979-80	1996-97
Wensink, John	St.L., Bos., Que., Col., N.J.	8	403	70	68	138	840	43	2	6	8	86	....	1973-74	1982-83
• Wentworth, Cy	Chi., Mtl.M., Mtl.	13	575	39	68	107	355	35	5	6	11	20	1	1927-28	1939-40
Werenka, Brad	Edm., Que., Chi., Pit., Cgy.	7	320	19	61	80	299	19	2	1	3	14	....	1992-93	2000-01
‡ Wesenberg, Brian	Phi.	1	1	0	0	0	5	....	....	....	....	....	....	1998-99	1998-99
Wesley, Blake	Phi., Hfd., Que., Tor.	7	298	18	46	64	486	19	2	2	4	30	....	1979-80	1985-86
Westfall, Ed	Bos., NYI	18	1220	231	394	625	544	95	22	37	59	41	2	1961-62	1978-79
Wharram, Kenny	Chi.	14	766	252	281	533	222	80	16	27	43	38	1	1951-52	1968-69
Wharton, Len	NYR	1	1	0	0	0	0	....	....	....	....	....	....	1944-45	1944-45
‡ Wheeldon, Simon	NYR, Wpg.	3	15	0	2	2	10	....	....	....	....	....	....	1987-88	1990-91
• Wheldon, Don	St.L.	1	2	0	0	0	0	....	....	....	....	....	....	1974-75	1974-75
Whelton, Bill	Wpg.	1	2	0	0	0	0	....	....	....	....	....	....	1980-81	1980-81
Whistle, Rob	NYR, St.L.	2	51	7	5	12	16	4	0	0	0	2	....	1985-86	1987-88
White, Bill	L.A., Chi.	9	604	50	215	265	495	91	7	32	39	76	....	1967-68	1975-76
White, Moe	Mtl.	1	4	0	1	1	2	....	....	....	....	....	....	1945-46	1945-46
• White, Sherman	NYR	2	4	0	2	2	0	....	....	....	....	....	....	1946-47	1949-50
• White, Tex	Pit., NYA, Phi.	6	203	33	12	45	141	4	0	0	0	4	....	1925-26	1930-31
White, Tony	Wsh., Min.	5	164	37	28	65	104	....	....	....	....	....	....	1974-75	1979-80
Whitelaw, Bob	Det.	2	32	0	2	2	2	8	0	0	0	0	....	1940-41	1941-42
Whitlock, Bob	Min.	1	1	0	0	0	0	....	....	....	....	....	....	1969-70	1969-70
‡ Whyte, Sean	L.A.	2	21	0	2	2	12	....	....	....	....	....	....	1991-92	1992-93
• Wickenheiser, Doug	Mtl., St.L., Van., NYR, Wsh.	10	556	111	165	276	286	41	4	7	11	18	....	1980-81	1989-90
• Widing, Juha	NYR, L.A., Cle.	8	575	144	226	370	208	8	1	2	3	2	....	1969-70	1976-77
Widmer, Jason	NYI, S.J.	3	7	0	1	1	7	....	....	....	....	....	....	1994-95	1996-97
• Wiebe, Art	Chi.	11	414	14	27	41	201	31	1	3	4	10	1	1932-33	1943-44
Wiemer, Jim	Buf., NYR, Edm., L.A., Bos.	11	325	29	72	101	378	62	5	8	13	63	....	1982-83	1993-94
• Wilcox, Archie	Mtl.M., Bos., St.L.	6	208	8	14	22	158	12	1	0	1	8	....	1929-30	1934-35
Wilcox, Barry	Van.	2	33	3	2	5	15	....	....	....	....	....	....	1972-73	1974-75
Wilder, Arch	Det.	1	18	0	2	2	2	....	....	....	....	....	....	1940-41	1940-41
Wiley, Jim	Pit., Van.	5	63	4	10	14	8	....	....	....	....	....	....	1972-73	1976-77
Wilkie, Bob	Det., Phi.	2	18	2	5	7	10	....	....	....	....	....	....	1990-91	1993-94
Wilkins, Barry	Bos., Van., Pit.	9	418	27	125	152	663	6	0	1	1	4	....	1966-67	1975-76
• Wilkinson, John	Bos.	1	9	0	0	0	6	....	....	....	....	....	....	1943-44	1943-44
Wilkinson, Neil	Min., S.J., Chi., Wpg., Pit.	10	460	16	67	83	813	53	3	6	9	41	....	1989-90	1998-99
Wilks, Brian	L.A.	4	48	4	8	12	27	....	....	....	....	....	....	1984-85	1988-89
Willard, Rod	Tor.	1	1	0	0	0	0	....	....	....	....	....	....	1982-83	1982-83
• Williams, Burr	Det., St.L., Bos.	3	19	0	1	1	28	7	0	0	0	8	....	1933-34	1936-37
Williams, Butch	St.L., Cal.	3	108	14	35	49	131	....	....	....	....	....	....	1973-74	1975-76
Williams, Darryl	L.A.	1	2	0	0	0	10	....	....	....	....	....	....	1992-93	1992-93
Williams, David	S.J., Ana.	4	173	11	53	64	157	....	....	....	....	....	....	1991-92	1994-95
Williams, Fred	Det.	1	44	2	5	7	10	....	....	....	....	....	....	1976-77	1976-77
Williams, Gord	Phi.	2	2	0	0	0	2	....	....	....	....	....	....	1981-82	1982-83
Williams, Sean	Chi.	1	2	0	0	0	4	....	....	....	....	....	....	1991-92	1991-92
Williams, Tiger	Tor., Van., Det., L.A., Hfd.	14	962	241	272	513	3966	83	12	23	35	455	....	1974-75	1987-88
Williams, Tom	NYR, L.A.	8	397	115	138	253	73	29	8	7	15	4	....	1971-72	1978-79
• Williams, Tommy	Bos., Min., Cal., Wsh.	13	663	161	269	430	177	10	2	5	7	2	....	1961-62	1975-76
Willson, Don	Mtl.	2	22	2	7	9	0	3	0	0	0	0	....	1937-38	1938-39
Wilson, Behn	Phi., Chi.	9	601	98	260	358	1480	67	12	29	41	190	....	1978-79	1987-88
• Wilson, Bert	NYR, St.L., L.A., Cgy.	8	478	37	44	81	646	21	0	2	2	42	....	1973-74	1980-81
Wilson, Bob	Chi.	1	1	0	0	0	0	....	....	....	....	....	....	1953-54	1953-54
Wilson, Carey	Cgy., Hfd., NYR	10	552	169	258	427	314	52	11	13	24	14	....	1983-84	1992-93
• Wilson, Cully	Tor., Mtl., Ham., Chi.	5	127	59	28	87	243	2	1	0	1	6	....	1919-20	1926-27
Wilson, Doug	Chi., S.J.	16	1024	237	590	827	830	95	19	61	80	88	....	1977-78	1992-93
Wilson, Gord	Bos.	1	....	....	....	....	....	2	0	0	0	0	....	1954-55	1954-55
Wilson, Hub	NYA	1	2	0	0	0	0	....	....	....	....	....	....	1931-32	1931-32
Wilson, Jerry	Mtl.	1	3	0	0	0	2	....	....	....	....	....	....	1956-57	1956-57
Wilson, Johnny	Det., Chi., Tor., NYR	13	688	161	171	332	190	66	14	13	27	11	4	1949-50	1961-62
• Wilson, Larry	Det., Chi.	6	152	21	48	69	75	4	0	0	0	0	1	1949-50	1955-56
Wilson, Mitch	N.J., Pit.	2	26	2	3	5	104	....	....	....	....	....	....	1984-85	1986-87
Wilson, Murray	Mtl., L.A.	7	386	94	95	189	162	53	5	14	19	32	4	1972-73	1978-79
Wilson, Rick	Mtl., St.L., Det.	4	239	6	26	32	165	3	0	0	0	0	....	1973-74	1976-77
Wilson, Rik	St.L., Cgy., Chi.	6	251	25	65	90	220	22	0	4	4	23	....	1981-82	1987-88
Wilson, Roger	Chi.	1	7	0	2	2	6	....	....	....	....	....	....	1974-75	1974-75
Wilson, Ron	Tor., Min.	7	177	26	67	93	68	20	4	13	17	8	....	1977-78	1987-88
Wilson, Ron	Wpg., St.L., Mtl.	14	832	110	216	326	415	63	10	12	22	64	....	1979-80	1993-94
Wilson, Wally	Bos.	1	53	11	8	19	18	1	0	0	0	0	....	1947-48	1947-48
Wing, Murray	Det.	1	1	0	1	1	0	....	....	....	....	....	....	1973-74	1973-74
‡ Winnes, Chris	Bos., Phi.	4	33	1	6	7	6	1	0	0	0	0	....	1990-91	1993-94
• Wiseman, Brian	Tor.	1	3	0	0	0	0	....	....	....	....	....	....	1996-97	1996-97
• Wiseman, Eddie	Det., NYA, Bos.	10	456	115	165	280	136	43	10	10	20	16	1	1932-33	1941-42
Wiste, Jim	Chi., Van.	3	52	1	10	11	8	....	....	....	....	....	....	1968-69	1970-71
Witherspoon, Jim	L.A.	1	2	0	0	0	2	....	....	....	....	....	....	1975-76	1975-76
Witiuk, Steve	Chi.	1	33	3	8	11	14	....	....	....	....	....	....	1951-52	1951-52
Woit, Benny	Det., Chi.	7	334	7	26	33	170	41	2	6	8	18	3	1950-51	1956-57
Wojciechowski, Steve	Det.	2	54	19	20	39	17	6	0	1	1	0	....	1944-45	1946-47

Name	NHL Teams	NHL Seasons	Regular Schedule GP	G	A	TP	PIM	Playoffs GP	G	A	TP	PIM	NHL Cup Wins	First NHL Season	Last NHL Season
Wolanin, Craig	N.J., Que., Col., T.B., Tor.	13	695	40	133	173	894	35	4	6	10	67	1	1985-86	1997-98
Wolf, Bennett	Pit.	3	30	0	1	1	133	....	....	....	....	....	....	1980-81	1982-83
Wong, Mike	Det.	1	22	1	1	2	12	....	....	....	....	....	....	1975-76	1975-76
‡ Wood, Dody	S.J.	5	106	8	10	18	471	....	....	....	....	....	....	1992-93	1997-98
Wood, Randy	NYI, Buf., Tor., Dal.	11	741	175	159	334	603	51	8	9	17	40	....	1986-87	1996-97
Wood, Robert	NYR	1	1	0	0	0	0	....	....	....	....	....	....	1950-51	1950-51
Woodley, Dan	Van.	1	5	2	0	2	17	....	....	....	....	....	....	1987-88	1987-88
Woods, Paul	Det.	7	501	72	124	196	276	7	0	5	5	4	....	1977-78	1983-84
‡ Wortman, Kevin	Cgy.	1	5	0	0	0	2	....	....	....	....	....	....	1993-94	1993-94
• Woytowich, Bob	Bos., Min., Pit., L.A.	8	503	32	126	158	352	24	1	3	4	20	....	1964-65	1971-72
Wright, John	Van., St.L., K.C.	3	127	16	36	52	67	....	....	....	....	....	....	1972-73	1974-75
Wright, Keith	Phi.	1	1	0	0	0	0	....	....	....	....	....	....	1967-68	1967-68
Wright, Larry	Phi., Cal., Det.	5	106	4	8	12	19	....	....	....	....	....	....	1971-72	1977-78
Wycherley, Ralph	NYA, Bro.	2	28	4	7	11	6	....	....	....	....	....	....	1940-41	1941-42
• Wylie, Bill	NYR	1	1	0	0	0	0	....	....	....	....	....	....	1950-51	1950-51
Wylie, Duane	Chi.	2	14	3	3	6	2	....	....	....	....	....	....	1974-75	1976-77
Wyrozub, Randy	Buf.	4	100	8	10	18	10	....	....	....	....	....	....	1970-71	1973-74

Y Z

Name	NHL Teams	NHL Seasons	Regular Schedule GP	G	A	TP	PIM	Playoffs GP	G	A	TP	PIM	NHL Cup Wins	First NHL Season	Last NHL Season
• Yackel, Ken	Bos.	1	6	0	0	0	2	2	0	0	0	2	....	1958-59	1958-59
Yaremchuk, Gary	Tor.	4	34	1	4	5	28	....	....	....	....	....	....	1981-82	1984-85
Yaremchuk, Ken	Chi., Tor.	6	235	36	56	92	106	31	6	8	14	49	....	1983-84	1988-89
Yates, Ross	Hfd.	1	7	1	1	2	4	....	....	....	....	....	....	1983-84	1983-84
Yawney, Trent	Chi., Cgy., St.L.	12	593	27	102	129	783	60	9	17	26	81	....	1987-88	1998-99
‡ Yegorov, Alexei	S.J.	2	11	3	3	6	2	....	....	....	....	....	....	1995-96	1996-97
‡ York, Harry	St.L., NYR, Pit., Van.	4	244	29	46	75	99	5	0	0	0	2	....	1996-97	1999-00
‡ Young, B.J.	Det.	1	1	0	0	0	0	....	....	....	....	....	....	1999-00	1999-00
Young, Brian	Chi.	1	8	0	2	2	6	....	....	....	....	....	....	1980-81	1980-81
Young, C.J.	Cgy., Bos.	1	43	7	7	14	32	....	....	....	....	....	....	1992-93	1992-93
• Young, Doug	Det., Mtl.	10	388	35	45	80	303	28	1	5	6	16	2	1931-32	1940-41
• Young, Howie	Det., Chi., Van.	8	336	12	62	74	851	19	2	4	6	46	....	1960-61	1970-71
Young, Tim	Min., Wpg., Phi.	10	628	195	341	536	438	36	7	24	31	27	....	1975-76	1984-85
Young, Warren	Min., Pit., Det.	7	236	72	77	149	472	....	....	....	....	....	....	1981-82	1987-88
Younghans, Tom	Min., NYR	6	429	44	41	85	373	24	2	1	3	21	....	1976-77	1981-82
Ysebaert, Paul	N.J., Det., Wpg., Chi., T.B.	11	532	149	187	336	217	30	4	3	7	20	....	1988-89	1998-99
‡ Zabransky, Libor	St.L.	2	40	1	6	7	50	....	....	....	....	....	....	1996-97	1997-98
Zaharko, Miles	Atl., Chi.	4	129	5	32	37	84	3	0	0	0	0	....	1977-78	1981-82
Zaine, Rod	Pit., Buf.	2	61	10	6	16	25	....	....	....	....	....	....	1970-71	1971-72
‡ Zalapski, Zarley	Pit., Hfd., Cgy., Mtl., Phi.	12	637	99	285	384	684	48	4	23	27	47	....	1987-88	1999-00
Zanussi, Joe	NYR, Bos., St.L.	3	87	1	13	14	46	4	0	1	1	2	....	1974-75	1976-77
Zanussi, Ron	Min., Tor.	5	299	52	83	135	373	17	0	4	4	17	....	1977-78	1981-82
Zavisha, Brad	Edm.	1	2	0	0	0	0	....	....	....	....	....	....	1993-94	1993-94
‡ Zehr, Jeff	Bos.	1	4	0	0	0	2	....	....	....	....	....	....	1999-00	1999-00
Zeidel, Larry	Det., Chi., Phi.	5	158	3	16	19	198	12	0	1	1	12	1	1951-52	1968-69
Zemlak, Richard	Que., Min., Pit., Cgy.	5	132	2	12	14	587	1	0	0	0	10	....	1986-87	1991-92
Zeniuk, Ed	Det.	1	2	0	0	0	0	....	....	....	....	....	....	1954-55	1954-55
Zent, Jason	Ott., Phi.	3	27	3	3	6	13	....	....	....	....	....	....	1996-97	1998-99
Zetterstrom, Lars	Van.	1	14	0	1	1	2	....	....	....	....	....	....	1978-79	1978-79
Zezel, Peter	Phi., St.L., Wsh., Tor., Dal., N.J., Van.	15	873	219	389	608	435	131	25	39	64	83	....	1984-85	1998-99
‡ Ziegler, Thomas	T.B.	1	5	0	0	0	0	....	....	....	....	....	....	2000-01	2000-01
Zmolek, Doug	S.J., Dal., L.A., Chi.	8	467	11	53	64	905	14	0	1	1	16	....	1992-93	1999-00
Zoborosky, Marty	Chi.	1	1	0	0	0	2	....	....	....	....	....	....	1944-45	1944-45
Zombo, Rick	Det., St.L., Bos.	12	652	24	130	154	728	60	1	11	12	127	....	1984-85	1995-96
Zuke, Mike	St.L., Hfd.	8	455	86	196	282	220	26	6	6	12	12	....	1978-79	1985-86
Zunich, Rudy	Det.	1	2	0	0	0	2	....	....	....	....	....	....	1943-44	1943-44

Lars Zetterstrom

Rick Zombo

Retired Players, Goaltenders and Coaches Research Project

Throughout the Retired Players and Retired Goaltenders sections of this book, you will notice many players with a bullet (•) by their names. These players, according to our records, are deceased. The editors recognize that our information on the death dates of NHLers is incomplete. If you have documented information on the passing of any player not marked with a bullet (•) in this edition, we would like to hear from you. We also welcome information on deceased NHL head coaches. Please send this information to:

Retired Player Research Project
c/o NHL Publishing
194 Dovercourt Road
Toronto, Ontario
M6J 3C8 Canada
Fax: 416/531-3939

Many thanks to the following contributors in 2001-02:

Tim Bateman, Corey Bryant, Paul R. Carroll, Jr., Bob Duff, Peter Fillman, Ernie Fitzsimmons, Mel Foster, Glen Goodhand, John Halligan, Gary J. Pearce, Ed Sweeney, Drew "Whitey" White.

Retired NHL Goaltender Index

Abbreviations: Teams/Cities: – **Ana.** – Anaheim; **Atl.** – Atlanta; **Bos.** – Boston; **Bro.** – Brooklyn; **Buf.** – Buffalo; **Cal.** – California; **Cgy.** – Calgary; **Cle.** – Cleveland; **Col.** – Colorado; **CBJ** – Columbus; **Dal.** – Dallas; **Det.** – Detroit; **Edm.** – Edmonton; **Fla.** – Florida; **Ham.** – Hamilton; **Hfd.** – Hartford; **K.C.** – Kansas City; **L.A.** – Los Angeles; **Min.** – Minnesota; **Mtl.** – Montreal; **Mtl.M.** – Montreal Maroons; **Mtl.W.** – Montreal Wanderers; **N.J.** – New Jersey; **NYA** – NY Americans; **NYI** – NY Islanders; **NYR** – New York Rangers; – **Oak.** – Oakland; **Ott.** – Ottawa; – **Phi.** – Philadelphia; **Phx.** – Phoenix; **Pit.** – Pittsburgh; **Que.** – Quebec; **St.L.** – St. Louis; **S.J.** – San Jose; **T.B.** – Tampa Bay; **Tor.** – Toronto; **Van.** – Vancouver; **Wpg.** – Winnipeg; **Wsh.** – Washington

Avg. – goals against per 60 minutes played; **GA** – goals agains; **GP** – games played; **Mins** – minutes played; **SO** – shutouts.
• – deceased. § – Forward, defenseman or coach who appeared in goal. For complete career, see Retired Player Index. ‡ – Remains active in other leagues.

			Regular Schedule								Playoffs										
Name	NHL Teams	NHL Seasons	GP	W	L	T	Mins	GA	SO	Avg	GP	W	L	T	Mins	GA	SO	Avg	NHL Cup Wins	First NHL Season	Last NHL Season
Abbott, George	Bos.	1	1	0	1	0	60	7	0	7.00	....	....	....	....	....	....	....	....		1943-44	1943-44
Adams, John	Bos., Wsh.	2	22	9	10	1	1180	85	1	4.32	....	....	....	....	....	....	....	....		1972-73	1974-75
Aiken, Don	Mtl.	1	1	0	1	0	34	6	0	10.59	....	....	....	....	....	....	....	....		1957-58	1957-58
• Aitkenhead, Andy	NYR	3	106	47	43	16	6570	257	11	2.35	10	6	2	2	608	15	3	1.48	1	1932-33	1934-35
• Almas, Red	Det., Chi.	3	3	0	2	1	180	13	0	4.33	5	1	3	....	263	13	0	2.97		1946-47	1952-53
• Anderson, Lorne	NYR	1	3	1	2	0	180	18	0	6.00	....	....	....	....	....	....	....	....		1951-52	1951-52
Astrom, Hardy	NYR, Col.	3	83	17	44	12	4456	278	0	3.74	....	....	....	....	....	....	....	....		1977-78	1980-81
‡ Bach, Ryan	L.A.	1	3	0	3	0	108	8	0	4.44	....	....	....	....	....	....	....	....		1998-99	1998-99
‡ Bailey, Scott	Bos.	2	19	6	6	2	965	55	0	3.42	....	....	....	....	....	....	....	....		1995-96	1996-97
Baker, Steve	NYR	4	57	20	20	11	3081	190	3	3.70	14	7	7	....	826	55	0	4.00		1979-80	1982-83
‡ Bales, Mike	Bos., Ott.	4	23	2	15	1	1120	77	0	4.13	....	....	....	....	....	....	....	....		1992-93	1996-97
Bannerman, Murray	Van., Chi.	8	289	116	125	33	16470	1051	8	3.83	40	20	18	....	2322	165	0	4.26		1977-78	1986-87
Baron, Marco	Bos., L.A., Edm.	6	86	34	38	9	4822	292	1	3.63	1	0	1	....	20	3	0	9.00		1979-80	1984-85
Bassen, Hank	Chi., Det., Pit.	9	156	46	66	31	8759	434	5	2.97	5	1	3	....	274	11	0	2.41		1954-55	1967-68
• Bastien, Baz	Tor.	1	5	0	4	1	300	20	0	4.00	....	....	....	....	....	....	....	....		1945-46	1945-46
Bauman, Gary	Mtl., Min.	3	35	6	18	6	1718	102	0	3.56	....	....	....	....	....	....	....	....		1966-67	1968-69
Beaupre, Don	Min., Wsh., Ott., Tor.	17	667	268	277	75	37396	2151	17	3.45	72	33	31	....	3943	220	3	3.35		1980-81	1996-97
Beauregard, Stephane	Wpg., Phi.	5	90	19	39	11	4402	268	2	3.65	4	1	3	....	238	12	0	3.03		1989-90	1993-94
Bedard, Jim	Wsh.	2	73	17	40	13	4232	278	1	3.94	....	....	....	....	....	....	....	....		1977-78	1978-79
Behrend, Marc	Wpg.	3	39	12	19	3	1991	160	1	4.82	7	1	3	0	312	19	0	3.65		1983-84	1985-86
Belanger, Yves	St.L., Atl., Bos.	6	78	29	33	6	4134	259	2	3.76	....	....	....	....	....	....	....	....		1974-75	1979-80
Belhumeur, Michel	Phi., Wsh.	3	65	9	36	7	3306	254	0	4.61	1	0	0	....	10	1	0	6.00		1972-73	1975-76
• Bell, Gordie	Tor., NYR	2	8	3	5	0	480	31	0	3.88	2	1	1	....	120	9	0	4.50		1945-46	1955-56
• Benedict, Clint	Ott., Mtl.M.	13	362	190	143	28	22367	863	58	2.32	28	12	11	3	1707	53	9	1.86	4	1917-18	1929-30
Bennett, Harvey	Bos.	1	25	10	12	2	1470	103	0	4.20	....	....	....	....	....	....	....	....		1944-45	1944-45
Bergeron, Jean-Claude	Mtl., T.B., L.A.	6	72	21	33	7	3772	232	1	3.69	....	....	....	....	....	....	....	....		1990-91	1996-97
Bernhardt, Tim	Cgy., Tor.	4	67	17	36	7	3748	267	0	4.27	....	....	....	....	....	....	....	....		1982-83	1986-87
‡ Berthiaume, Daniel	Wpg., Min., L.A., Bos., Ott.	9	215	81	90	21	11662	714	5	3.67	14	5	9	....	807	50	0	3.72		1985-86	1993-94
Bester, Allan	Tor., Det., Dal.	10	219	73	99	17	11773	786	7	4.01	11	2	6	....	508	37	0	4.37		1983-84	1995-96
• Beveridge, Bill	Det., Ott., St.L., Mtl.M., NYR	9	297	87	166	42	18375	879	18	2.87	5	2	3	....	300	11	0	2.20		1929-30	1942-43
• Bibeault, Paul	Mtl., Tor., Bos., Chi.	7	214	81	107	25	12890	785	10	3.65	20	6	14	....	1237	71	2	3.44		1940-41	1946-47
Binette, Andre	Mtl.	1	1	1	0	0	60	4	0	4.00	....	....	....	....	....	....	....	....		1954-55	1954-55
Binkley, Les	Pit.	5	196	58	94	34	11046	575	11	3.12	7	5	2	....	428	15	0	2.10		1967-68	1971-72
Bittner, Richard	Bos.	1	1	0	0	1	60	3	0	3.00	....	....	....	....	....	....	....	....		1949-50	1949-50
Blake, Mike	L.A.	3	40	13	15	5	2117	150	0	4.25	....	....	....	....	....	....	....	....		1981-82	1983-84
Blue, John	Bos., Buf.	3	46	16	18	7	2521	126	1	3.00	2	0	1	....	96	5	0	3.13		1992-93	1995-96
Boisvert, Gilles	Det.	1	3	0	3	0	180	9	0	3.00	....	....	....	....	....	....	....	....		1959-60	1959-60
Bouchard, Dan	Atl., Cgy., Que., Wpg.	14	655	286	232	113	37919	2061	27	3.26	43	13	30	....	2549	147	1	3.46		1972-73	1985-86
• Bourque, Claude	Mtl., Det.	2	62	16	38	8	3830	193	4	3.02	3	1	2	....	188	8	1	2.55		1938-39	1939-40
Boutin, Rollie	Wsh.	3	22	7	10	1	1137	75	0	3.96	....	....	....	....	....	....	....	....		1978-79	1980-81
• Bouvrette, Lionel	NYR	1	1	0	1	0	60	6	0	6.00	....	....	....	....	....	....	....	....		1942-43	1942-43
Bower, Johnny	NYR, Tor.	15	552	250	195	90	32016	1340	37	2.51	74	35	34	....	4378	180	5	2.47	4	1953-54	1969-70
§ • Branigan, Andy	NYA	1	1	0	0	0	7	0	0	0.00	....	....	....	....	....	....	....	....		1940-41	1940-41
• Brimsek, Frank	Bos., Chi.	10	514	252	182	80	31210	1404	40	2.70	68	32	36	....	4395	186	2	2.54	2	1938-39	1949-50
• Broda, Turk	Tor.	14	629	302	224	101	38167	1609	62	2.53	101	60	39	....	6389	211	13	1.98	5	1936-37	1951-52
Broderick, Ken	Min., Bos.	3	27	11	12	1	1464	74	1	3.03	....	....	....	....	....	....	....	....		1969-70	1974-75
Broderick, Len	Mtl.	1	1	1	0	0	60	2	0	2.00	....	....	....	....	....	....	....	....		1957-58	1957-58
Brodeur, Richard	NYI, Van., Hfd.	9	385	131	175	62	21968	1410	6	3.85	33	13	20	....	2009	111	1	3.32		1979-80	1987-88
Bromley, Gary	Buf., Van.	6	136	54	44	28	7427	425	7	3.43	7	2	5	....	360	25	0	4.17		1973-74	1980-81
• Brooks, Art	Tor.	1	4	2	2	0	220	23	0	6.27	....	....	....	....	....	....	....	....		1917-18	1917-18
Brooks, Ross	Bos.	3	54	37	7	6	3047	134	4	2.64	1	0	0	....	20	3	0	9.00		1972-73	1974-75
• Brophy, Frank	Que.	1	21	3	18	0	1249	148	0	7.11	....	....	....	....	....	....	....	....		1919-20	1919-20
Brown, Andy	Det., Pit.	3	62	22	26	9	3373	213	1	3.79	....	....	....	....	....	....	....	....		1971-72	1973-74
Brown, Ken	Chi.	1	1	0	0	0	18	1	0	3.33	....	....	....	....	....	....	....	....		1970-71	1970-71
‡ Brunetta, Mario	Que.	3	40	12	17	1	1967	128	0	3.90	....	....	....	....	....	....	....	....		1987-88	1989-90
Bullock, Bruce	Van.	3	16	3	9	3	927	74	0	4.79	....	....	....	....	....	....	....	....		1972-73	1976-77
• Buzinski, Steve	NYR	1	9	2	6	1	560	55	0	5.89	....	....	....	....	....	....	....	....		1942-43	1942-43
Caley, Don	St.L.	1	1	0	0	0	30	3	0	6.00	....	....	....	....	....	....	....	....		1967-68	1967-68
Caprice, Frank	Van.	6	102	31	46	11	5589	391	1	4.20	....	....	....	....	....	....	....	....		1982-83	1987-88
Carey, Jim	Wsh., Bos., St.L.	5	172	79	65	16	9668	416	16	2.58	10	2	5	....	455	35	0	4.62		1994-95	1998-99
Caron, Jacques	L.A., St.L., Van.	5	72	24	29	11	3846	211	2	3.29	12	4	7	....	639	34	0	3.19		1967-68	1973-74
Carter, Lyle	Cal.	1	15	4	7	0	721	50	0	4.16	....	....	....	....	....	....	....	....		1971-72	1971-72
Casey, Jon	Min., Bos., St.L.	12	425	170	157	55	23255	1246	16	3.21	66	32	31	....	3743	192	3	3.08		1983-84	1996-97
‡ Chabot, Frederic	Mtl., Phi., L.A.	5	32	4	8	4	1262	62	0	2.95	....	....	....	....	....	....	....	....		1990-91	1998-99
• Chabot, Lorne	NYR, Tor., Mtl., Chi., Mtl.M., NYA	11	411	201	148	62	25307	860	73	2.04	37	13	17	6	2498	64	5	1.54	2	1926-27	1936-37
Chadwick, Ed	Tor., Bos.	6	184	57	92	35	11040	541	14	2.94	....	....	....	....	....	....	....	....		1955-56	1961-62
Champoux, Bob	Det., Cal.	2	17	2	11	3	923	80	0	5.20	1	1	0	....	55	4	0	4.36		1963-64	1973-74
Cheevers, Gerry	Tor., Bos.	13	418	230	102	74	24394	1174	26	2.89	88	53	34	....	5396	242	8	2.69	2	1961-62	1979-80
Cheveldae, Tim	Det., Wpg., Bos.	9	340	149	136	37	19172	1116	10	3.49	25	9	15	....	1418	71	2	3.00		1988-89	1996-97
Chevrier, Alain	N.J., Wpg., Chi., Pit., Det.	6	234	91	100	14	12202	845	2	4.16	16	9	7	....	1013	44	0	2.61		1985-86	1990-91
§ • Clancy, King	Ott., Tor.	2	2	0	0	0	3	1	0	20.00	....	....	....	....	....	....	....	....		1924-25	1931-32
§ • Cleghorn, Odie	Pit.	1	1	1	0	0	60	2	0	2.00	....	....	....	....	....	....	....	....		1925-26	1925-26
§ • Cleghorn, Sprague	Ott., Mtl.	2	2	0	0	0	5	0	0	0.00	....	....	....	....	....	....	....	....		1918-19	1921-22
Clifford, Chris	Chi.	2	2	0	0	0	24	0	0	0.00	....	....	....	....	....	....	....	....		1984-85	1988-89
Cloutier, Jacques	Buf., Chi., Que.	12	255	82	102	24	12826	778	3	3.64	8	1	5	....	413	18	1	2.62		1981-82	1993-94
Colvin, Les	Bos.	1	1	0	1	0	60	4	0	4.00	....	....	....	....	....	....	....	....		1948-49	1948-49
§ • Conacher, Charlie	Tor., Det.	3	4	0	0	0	10	0	0	0.00	....	....	....	....	....	....	....	....		1932-33	1938-39
• Connell, Alex	Ott., Det., NYA, Mtl.M.	12	417	193	156	67	26050	830	81	1.91	21	8	5	8	1309	26	4	1.19	2	1924-25	1936-37
Corsi, Jim	Edm.	1	26	8	14	3	1366	83	0	3.65	....	....	....	....	....	....	....	....		1979-80	1979-80
Courteau, Maurice	Bos.	1	6	2	4	0	360	33	0	5.50	....	....	....	....	....	....	....	....		1943-44	1943-44
‡ Cousineau, Marcel	Tor., NYI, L.A.	4	26	4	10	1	1047	51	1	2.92	....	....	....	....	....	....	....	....		1996-97	1999-00
Cowley, Wayne	Edm.	1	1	0	1	0	57	3	0	3.16	....	....	....	....	....	....	....	....		1993-94	1993-94
Cox, Abbie	Mtl.M., NYA, Det., Mtl.	3	5	1	1	2	263	11	0	2.51	....	....	....	....	....	....	....	....		1929-30	1935-36
Craig, Jim	Atl., Bos., Min.	3	30	11	10	7	1588	100	0	3.78	....	....	....	....	....	....	....	....		1979-80	1983-84
Crha, Jiri	Tor.	2	69	28	27	11	3942	261	0	3.97	5	0	4	....	186	21	0	6.77		1979-80	1980-81
• Crozier, Roger	Det., Buf., Wsh.	14	518	206	197	70	28567	1446	30	3.04	32	14	16	....	1789	82	1	2.75		1963-64	1976-77
• Cude, Wilf	Phi., Bos., Chi., Mtl., Det.	10	282	100	132	49	17586	798	24	2.72	19	7	11	1	1257	51	1	2.43		1930-31	1940-41
Cutts, Don	Edm.	1	6	1	2	1	269	16	0	3.57	....	....	....	....	....	....	....	....		1979-80	1979-80
• Cyr, Claude	Mtl.	1	1	0	0	0	20	1	0	3.00	....	....	....	....	....	....	....	....		1958-59	1958-59
Dadswell, Doug	Cgy.	2	27	8	8	3	1346	99	0	4.41	....	....	....	....	....	....	....	....		1986-87	1987-88
D'Alessio, Corrie	Hfd.	1	1	0	0	0	11	0	0	0.00	....	....	....	....	....	....	....	....		1992-93	1992-93
Daley, Joe	Pit., Buf., Det.	4	105	34	44	19	5836	326	3	3.35	....	....	....	....	....	....	....	....		1968-69	1971-72
Damore, Nick	Bos.	1	1	1	0	0	60	3	0	3.00	....	....	....	....	....	....	....	....		1941-42	1941-42
D'Amour, Marc	Cgy., Phi.	2	16	2	4	2	579	32	0	3.32	....	....	....	....	....	....	....	....		1985-86	1988-89
Daskalakis, Cleon	Bos.	3	12	3	4	1	506	41	0	4.86	....	....	....	....	....	....	....	....		1984-85	1986-87
Davidson, John	St.L., NYR	10	301	123	124	39	17109	1004	7	3.52	31	16	14	....	1862	77	1	2.48		1973-74	1982-83
Decourcy, Bob	NYR	1	1	0	1	0	29	6	0	12.41	....	....	....	....	....	....	....	....		1947-48	1947-48

			Regular Schedule								Playoffs										
Name	NHL Teams	NHL Seasons	GP	W	L	T	Mins	GA	SO	Avg	GP	W	L	T	Mins	GA	SO	Avg	NHL Cup Wins	First NHL Season	Last NHL Season
Defelice, Norm	Bos.	1	10	3	5	2	600	30	0	3.00	....	....	....	....	....	....	....	....		1956-57	1956-57
DeJordy, Denis	Chi., L.A., Mtl., Det.	12	316	124	128	51	17798	929	15	3.13	18	6	9	....	946	55	0	3.49	1	1960-61	1973-74
DelGuidice, Matt	Bos.	2	11	2	5	1	434	28	0	3.87	....	....	....	....	....	....	....	....		1990-91	1991-92
‡ DeRouville, Philippe	Pit.	2	3	1	2	0	171	9	0	3.16	....	....	....	....	....	....	....	....		1994-95	1996-97
Desjardins, Gerry	L.A., Chi., NYI, Buf.	10	331	122	153	44	19014	1042	12	3.29	35	15	15	....	1874	108	0	3.46		1968-69	1977-78
• Dickie, Bill	Chi.	1	1	1	0	0	60	3	0	3.00	....	....	....	....	....	....	....	....		1941-42	1941-42
Dion, Connie	Det.	2	38	23	11	4	2280	119	1	3.13	5	1	4	....	300	17	0	3.40		1943-44	1944-45
Dion, Michel	Que., Wpg., Pit.	6	227	60	118	32	12695	898	2	4.24	5	2	3	....	304	22	0	4.34		1979-80	1984-85
Dolson, Dolly	Det.	3	93	35	41	17	5820	192	16	1.98	2	0	2	0	120	7	0	3.50		1928-29	1930-31
‡ Dopson, Rob	Pit.	1	2	0	0	0	45	3	0	4.00	....	....	....	....	....	....	....	....		1993-94	1993-94
Dowie, Bruce	Tor.	1	2	0	1	0	72	4	0	3.33	....	....	....	....	....	....	....	....		1983-84	1983-84
‡ Draper, Tom	Wpg., Buf., NYI	6	53	19	23	5	2807	173	1	3.70	7	3	4	....	433	19	1	2.63		1988-89	1995-96
Dryden, Dave	NYR, Chi., Buf., Edm.	9	203	66	76	31	10424	555	9	3.19	3	0	2	....	133	9	0	4.06		1961-62	1979-80
Dryden, Ken	Mtl.	8	397	258	57	74	23352	870	46	2.24	112	80	32	....	6846	274	10	2.40	6	1970-71	1978-79
‡ Duffus, Parris	Phx.	1	1	0	0	0	29	1	0	2.07	....	....	....	....	....	....	....	....		1996-97	1996-97
Dumas, Michel	Chi.	3	8	2	1	2	362	24	0	3.98	1	0	0	....	19	1	0	3.16		1974-75	1976-77
Dupuis, Bob	Edm.	1	1	0	1	0	60	4	0	4.00	....	....	....	....	....	....	....	....		1979-80	1979-80
• Durnan, Bill	Mtl.	7	383	208	112	62	22945	901	34	2.36	45	27	18	....	2871	99	2	2.07	2	1943-44	1949-50
Dyck, Ed	Van.	3	49	8	28	5	2453	178	1	4.35	....	....	....	....	....	....	....	....		1971-72	1973-74
Edwards, Don	Buf., Cgy., Tor.	10	459	208	155	74	26181	1449	16	3.32	42	16	21	....	2302	132	1	3.44		1976-77	1985-86
Edwards, Gary	St.L., L.A., Cle., Min., Edm., Pit.	13	286	88	125	51	16002	973	10	3.65	11	5	4	....	537	34	0	3.80		1968-69	1981-82
Edwards, Marv	Pit., Tor., Cal.	4	61	15	34	7	3467	218	2	3.77	....	....	....	....	....	....	....	....		1968-69	1973-74
• Edwards, Roy	Det., Pit.	7	236	97	88	38	13109	637	12	2.92	4	0	3	....	206	11	0	3.20		1967-68	1973-74
Eliot, Darren	L.A., Det., Buf.	5	89	25	41	12	4931	377	1	4.59	1	0	0	....	40	7	0	10.50		1984-85	1988-89
Ellacott, Ken	Van.	1	12	2	3	4	555	41	0	4.43	....	....	....	....	....	....	....	....		1982-83	1982-83
Erickson, Chad	N.J.	1	2	1	1	0	120	9	0	4.50	....	....	....	....	....	....	....	....		1991-92	1991-92
Esposito, Tony	Mtl., Chi.	16	886	423	306	151	52585	2563	76	2.92	99	45	53	....	6017	308	6	3.07	1	1968-69	1983-84
• Evans, Claude	Mtl., Bos.	2	5	1	2	1	260	16	0	3.69	....	....	....	....	....	....	....	....		1954-55	1957-58
Exelby, Randy	Mtl., Edm.	2	2	0	1	0	63	5	0	4.76	....	....	....	....	....	....	....	....		1988-89	1989-90
Farr, Rocky	Buf.	3	19	2	6	3	722	42	0	3.49	....	....	....	....	....	....	....	....		1972-73	1974-75
Favell, Doug	Phi., Tor., Col.	12	373	123	153	69	20771	1096	18	3.17	21	6	15	....	1270	66	1	3.12		1967-68	1978-79
Fichaud, Eric	NYI, Nsh., Car., Mtl.	6	95	22	47	10	4799	251	2	3.14	....	....	....	....	....	....	....	....		1995-96	2000-01
Fitzpatrick, Mark	L.A., NYI, Fla., T.B., Chi., Car.	12	329	113	136	49	18329	953	8	3.12	9	0	3	....	289	23	0	4.78		1988-89	1999-00
• Forbes, Jake	Tor., Ham., NYA, Phi.	13	210	85	114	11	12922	594	19	2.76	2	0	2	0	120	7	0	3.50		1919-20	1932-33
Ford, Brian	Que., Pit.	2	11	3	7	0	580	61	0	6.31	....	....	....	....	....	....	....	....		1983-84	1984-85
Foster, Norm	Bos., Edm.	2	13	7	4	0	623	34	0	3.27	....	....	....	....	....	....	....	....		1990-91	1991-92
Fowler, Hec	Bos.	1	7	1	6	0	409	42	0	6.16	....	....	....	....	....	....	....	....		1924-25	1924-25
Francis, Emile	Chi., NYR	6	95	31	52	11	5660	355	1	3.76	....	....	....	....	....	....	....	....		1946-47	1951-52
• Franks, Jimmy	Det., NYR, Bos.	4	42	12	23	7	2520	181	1	4.31	1	0	1	....	30	2	0	4.00	1	1936-37	1943-44
Frederick, Ray	Chi.	1	5	0	4	1	300	22	0	4.40	....	....	....	....	....	....	....	....		1954-55	1954-55
Friesen, Karl	N.J.	1	4	0	2	1	130	16	0	7.38	....	....	....	....	....	....	....	....		1986-87	1986-87
Froese, Bob	Phi., NYR	8	242	128	72	20	13451	694	13	3.10	18	3	9	....	830	55	0	3.98		1982-83	1989-90
Fuhr, Grant	Edm., Tor., Buf., L.A., St.L., Cgy.	19	868	403	295	114	48945	2756	25	3.38	150	92	50	....	8834	430	6	2.92	5	1981-82	1999-00
‡ Gage, Joaquin	Edm.	3	23	4	12	1	1076	67	0	3.74	....	....	....	....	....	....	....	....		1994-95	2000-01
Gagnon, David	Det.	1	2	0	1	0	35	6	0	10.29	....	....	....	....	....	....	....	....		1990-91	1990-91
• Gamble, Bruce	NYR, Bos., Tor., Phi.	10	327	110	150	46	18442	988	22	3.21	5	0	4	....	206	25	0	7.28	1	1958-59	1971-72
Gamble, Troy	Van.	4	72	22	29	9	3804	229	1	3.61	4	1	3	....	249	16	0	3.86		1986-87	1991-92
• Gardiner, Bert	NYR, Mtl., Chi., Bos.	6	144	49	68	27	8760	554	4	3.79	9	4	5	....	647	20	0	1.85		1935-36	1943-44
• Gardiner, Charlie	Chi.	7	316	112	152	52	19687	664	42	2.02	21	12	6	3	1472	35	5	1.43	1	1927-28	1933-34
Gardner, George	Det., Van.	5	66	16	30	6	3313	207	0	3.75	....	....	....	....	....	....	....	....		1965-66	1971-72
Garrett, John	Hfd., Que., Van.	6	207	68	91	37	11763	837	1	4.27	9	4	3	....	461	33	0	4.30		1979-80	1984-85
Gatherum, Dave	Det.	1	3	2	0	1	180	3	1	1.00	....	....	....	....	....	....	....	....	1	1953-54	1953-54
Gauthier, Paul	Mtl.	1	1	0	0	1	70	2	0	1.71	....	....	....	....	....	....	....	....		1937-38	1937-38
‡ Gauthier, Sean	S.J.	1	1	0	0	0	3	0	0	0.00	....	....	....	....	....	....	....	....		1998-99	1998-99
• Gelineau, Jack	Bos., Chi.	4	143	46	64	33	8580	447	7	3.13	4	1	2	....	260	7	1	1.62		1948-49	1953-54
Giacomin, Ed	NYR, Det.	13	610	289	208	97	35693	1675	54	2.82	65	29	35	....	3838	180	1	2.81		1965-66	1977-78
Gilbert, Gilles	Min., Bos., Det.	14	416	192	143	60	23677	1290	18	3.27	32	17	15	....	1919	97	3	3.03		1969-70	1982-83
Gill, Andre	Bos.	1	5	3	2	0	270	13	1	2.89	....	....	....	....	....	....	....	....		1967-68	1967-68
• Goodman, Paul	Chi.	3	52	23	20	9	3240	117	6	2.17	3	0	3	....	187	10	0	3.21	1	1937-38	1940-41
Gordon, Scott	Que.	2	23	2	16	0	1082	101	0	5.60	....	....	....	....	....	....	....	....		1989-90	1990-91
Gosselin, Mario	Que., L.A., Hfd.	9	241	91	107	14	12857	801	6	3.74	32	16	15	....	1816	99	0	3.27		1983-84	1993-94
‡ Goverde, David	L.A.	3	5	1	4	0	278	29	0	6.26	....	....	....	....	....	....	....	....		1991-92	1993-94
Grahame, Ron	Bos., L.A., Que.	4	114	50	43	15	6472	409	5	3.79	4	2	1	....	202	7	0	2.08		1977-78	1980-81
• Grant, Benny	Tor., NYA, Bos.	6	50	17	26	4	2990	187	4	3.75	....	....	....	....	....	....	....	....		1928-29	1943-44
Grant, Doug	Det., St.L.	7	77	27	34	8	4199	280	2	4.00	....	....	....	....	....	....	....	....		1973-74	1979-80
Gratton, Gilles	St.L., NYR	2	47	13	18	9	2299	154	0	4.02	....	....	....	....	....	....	....	....		1975-76	1976-77
Gray, Gerry	Det., NYI	2	8	1	5	1	440	35	0	4.77	....	....	....	....	....	....	....	....		1970-71	1972-73
Gray, Harrison	Det.	1	1	0	1	0	40	5	0	7.50	....	....	....	....	....	....	....	....		1963-64	1963-64
Greenlay, Mike	Edm.	1	2	0	0	0	20	4	0	12.00	....	....	....	....	....	....	....	....		1989-90	1989-90
Guenette, Steve	Pit., Cgy.	5	35	19	16	0	1958	122	1	3.74	....	....	....	....	....	....	....	....		1986-87	1990-91
• Hainsworth, George	Mtl., Tor.	11	465	246	145	74	29087	937	94	1.93	52	22	25	5	3486	112	8	1.93	2	1926-27	1936-37
Hall, Glenn	Det., Chi., St.L.	19	906	407	326	163	53484	2222	84	2.49	115	49	65	....	6899	320	6	2.78	2	1951-52	1970-71
Hamel, Pierre	Tor., Wpg.	4	69	13	41	7	3766	276	0	4.40	....	....	....	....	....	....	....	....		1974-75	1980-81
Hanlon, Glen	Van., St.L., NYR, Det.	14	477	167	202	61	26037	1561	13	3.60	35	11	15	....	1756	92	4	3.14		1977-78	1990-91
Harrison, Paul	Min., Tor., Pit., Buf.	7	109	28	59	9	5806	408	2	4.22	4	0	1	....	157	9	0	3.44		1975-76	1981-82
Hayward, Brian	Wpg., Mtl., Min., S.J.	11	357	143	156	37	20025	1242	8	3.72	37	11	18	....	1803	104	0	3.46		1982-83	1992-93
Head, Don	Bos.	1	38	9	26	3	2280	158	2	4.16	....	....	....	....	....	....	....	....		1961-62	1961-62
Healy, Glenn	L.A., NYI, NYR, Tor.	15	437	166	190	47	24256	1361	13	3.37	37	13	15	....	1930	108	0	3.36	1	1985-86	2000-01
Hebert, Guy	St.L., Ana., NYR	10	491	191	222	56	27889	1307	28	2.81	14	4	7	....	744	33	1	2.66		1991-92	2000-01
• Hebert, Sammy	Tor., Ott.	2	4	2	1	0	200	19	0	5.70	....	....	....	....	....	....	....	....	1	1917-18	1923-24
Heinz, Rick	St.L., Van.	5	49	14	19	5	2356	159	2	4.05	1	0	0	....	8	1	0	7.50		1980-81	1984-85
Henderson, John	Bos.	2	46	15	15	15	2688	113	5	2.52	2	0	2	....	120	8	0	4.00		1954-55	1955-56
• Henry, Gord	Bos.	4	3	1	2	0	180	5	1	1.67	5	0	4	....	283	21	0	4.45		1948-49	1952-53
Henry, Jim	NYR, Chi., Bos.	9	406	161	173	70	24355	1166	27	2.87	29	11	18	....	1741	81	2	2.79		1941-42	1954-55
Herron, Denis	Pit., K.C., Mtl.	14	462	146	203	76	25608	1579	10	3.70	15	5	10	....	901	50	0	3.33		1972-73	1985-86
Hextall, Ron	Phi., Que., NYI	13	608	296	214	69	34750	1723	23	2.97	93	47	43	....	5456	276	2	3.04		1986-87	1998-99
Highton, Hec	Chi.	1	24	10	14	0	1440	108	0	4.50	....	....	....	....	....	....	....	....		1943-44	1943-44
§ • Himes, Normie	NYA	2	2	0	1	0	79	3	0	2.28	....	....	....	....	....	....	....	....		1927-28	1928-29
Hodge, Charlie	Mtl., Oak., Van.	14	358	151	124	61	20593	925	24	2.70	16	7	8	....	804	32	2	2.39	5	1954-55	1970-71
Hoffort, Bruce	Phi.	2	9	4	0	3	368	22	0	3.59	....	....	....	....	....	....	....	....		1989-90	1990-91
Hoganson, Paul	Pit.	1	2	0	1	0	57	7	0	7.37	....	....	....	....	....	....	....	....		1970-71	1970-71
Hogosta, Goran	NYI, Que.	2	22	5	12	3	1208	83	1	4.12	....	....	....	....	....	....	....	....		1977-78	1979-80
Holden, Mark	Mtl., Wpg.	4	8	2	2	1	372	25	0	4.03	....	....	....	....	....	....	....	....		1981-82	1984-85
Holland, Ken	Hfd., Det.	2	4	0	2	1	206	17	0	4.95	....	....	....	....	....	....	....	....		1980-81	1983-84
Holland, Robbie	Pit.	2	44	11	22	9	2513	171	1	4.08	....	....	....	....	....	....	....	....		1979-80	1980-81
• Holmes, Hap	Tor., Det.	4	103	39	54	10	6510	264	17	2.43	2	1	1	0	120	7	0	3.50	1	1917-18	1927-28
§ Horner, Red	Tor.	1	1	0	0	0	1	1	0	60.00	....	....	....	....	....	....	....	....		1931-32	1931-32
‡ Hrivnak, Jim	Wsh., Wpg., St.L.	5	85	34	30	3	4217	262	0	3.73	....	....	....	....	....	....	....	....		1989-90	1993-94
Hrudey, Kelly	NYI, L.A., S.J.	15	677	271	265	88	38084	2174	17	3.43	85	36	46	....	5163	283	0	3.29		1983-84	1997-98
Ing, Peter	Tor., Edm., Det.	4	74	20	37	9	3941	266	1	4.05	....	....	....	....	....	....	....	....		1989-90	1993-94
Inness, Gary	Pit., Phi., Wsh.	7	162	58	61	27	8710	494	2	3.40	9	5	4	....	540	24	0	2.67		1973-74	1980-81
Ireland, Randy	Buf.	1	2	0	0	0	30	3	0	6.00	....	....	....	....	....	....	....	....		1978-79	1978-79
Irons, Robbie	St.L.	1	1	0	0	0	3	0	0	0.00	....	....	....	....	....	....	....	....		1968-69	1968-69
• Ironstone, Joe	Ott., NYA, Tor.	3	2	0	0	1	110	3	1	1.64	....	....	....	....	....	....	....	....		1924-25	1927-28
Jablonski, Pat	St.L., T.B., Mtl., Phx., Car.	8	128	28	62	18	6634	413	1	3.74	4	0	0	....	139	6	0	2.59		1989-90	1997-98
Jackson, Doug	Chi.	1	6	2	3	1	360	42	0	7.00	....	....	....	....	....	....	....	....		1947-48	1947-48
Jackson, Percy	Bos., NYA, NYR	4	7	1	3	1	392	26	0	3.98	....	....	....	....	....	....	....	....		1931-32	1935-36
‡ Jaks, Pauli	L.A.	1	1	0	0	0	40	2	0	3.00	....	....	....	....	....	....	....	....		1994-95	1994-95
Janaszak, Steve	Min., Col.	2	3	0	1	1	160	15	0	5.63	....	....	....	....	....	....	....	....		1979-80	1981-82

Name	NHL Teams	NHL Seasons	Regular Schedule GP	W	L	T	Mins	GA	SO	Avg	Playoffs GP	W	L	T	Mins	GA	SO	Avg	NHL Cup Wins	First NHL Season	Last NHL Season
Janecyk, Bob	Chi., L.A.	6	110	43	47	13	6250	432	2	4.15	3	0	3	....	184	10	0	3.26		1983-84	1988-89
§ Jenkins, Roger	NYA	1	1	0	1	0	30	7	0	14.00	....	....	....	....	....	....	....	....		1938-39	1938-39
Jensen, Al	Det., Wsh., L.A.	7	179	95	53	18	9974	557	8	3.35	12	5	5	....	598	32	0	3.21		1980-81	1986-87
Jensen, Darren	Phi.	2	30	15	10	1	1496	95	2	3.81	....	....	....	....	....	....	....	....		1984-85	1985-86
Johnson, Bob	St.L., Pit.	2	24	9	9	1	1059	66	0	3.74	....	....	....	....	....	....	....	....		1972-73	1974-75
Johnston, Eddie	Bos., Tor., St.L., Chi.	16	592	234	257	80	34216	1852	32	3.25	18	7	10	....	1023	57	1	3.34	2	1962-63	1977-78
Junkin, Joe	Bos.	1	1	0	0	0	8	0	0	0.00	....	....	....	....	....	....	....	....		1968-69	1968-69
Kaarela, Jari	Col.	1	5	2	2	0	220	22	0	6.00	....	....	....	....	....	....	....	....		1980-81	1980-81
Kamppuri, Hannu	N.J.	1	13	1	10	1	645	54	0	5.02	....	....	....	....	....	....	....	....		1984-85	1984-85
• Karakas, Mike	Chi., Mtl.	8	336	114	169	53	20616	1002	28	2.92	23	11	12	0	1434	72	3	3.01	1	1935-36	1945-46
Keans, Doug	L.A., Bos.	9	210	96	64	26	11388	666	4	3.51	9	2	6	....	432	34	0	4.72		1979-80	1987-88
Keenan, Don	Bos.	1	1	0	1	0	60	4	0	4.00	....	....	....	....	....	....	....	....		1958-59	1958-59
• Kerr, Dave	Mtl.M., NYA, NYR	11	427	203	148	75	26639	954	51	2.15	40	18	19	3	2616	76	8	1.74	1	1930-31	1940-41
King, Scott	Det.	2	2	0	0	0	61	3	0	2.95	....	....	....	....	....	....	....	....		1990-91	1991-92
Kleisinger, Terry	NYR	1	4	0	2	0	191	14	0	4.40	....	....	....	....	....	....	....	....		1985-86	1985-86
Klymkiw, Julian	NYR	1	1	0	0	0	19	2	0	6.32	....	....	....	....	....	....	....	....		1958-59	1958-59
Knickle, Rick	L.A.	2	14	7	6	0	706	44	0	3.74	....	....	....	....	....	....	....	....		1992-93	1993-94
Kuntar, Les	Mtl.	1	6	2	2	0	302	16	0	3.18	....	....	....	....	....	....	....	....		1993-94	1993-94
Kurt, Gary	Cal.	1	16	1	7	5	838	60	0	4.30	....	....	....	....	....	....	....	....		1971-72	1971-72
‡ Labrecque, Patrick	Mtl.	1	2	0	1	0	98	7	0	4.29	....	....	....	....	....	....	....	....		1995-96	1995-96
Lacher, Blaine	Bos.	2	47	22	16	4	2636	123	4	2.80	5	1	4	....	283	12	0	2.54		1994-95	1995-96
• Lacroix, Frenchy	Mtl.	2	5	1	4	0	280	16	0	3.43	....	....	....	....	....	....	....	....		1925-26	1926-27
LaFerriere, Rick	Col.	1	1	0	0	0	20	1	0	3.00	....	....	....	....	....	....	....	....		1981-82	1981-82
LaForest, Mark	Det., Phi., Tor., Ott.	6	103	25	54	4	5032	354	2	4.22	2	1	0	....	48	1	0	1.25		1985-86	1993-94
• Larocque, Michel	Mtl., Tor., Phi., St.L.	11	312	160	89	45	17615	978	17	3.33	14	6	6	....	759	37	1	2.92	4	1973-74	1983-84
‡ Larocque, Michel	Chi.	1	3	0	2	0	152	9	0	3.55	....	....	....	....	....	....	....	....		2000-01	2000-01
Laskowski, Gary	L.A.	2	59	19	27	5	2942	228	0	4.65	....	....	....	....	....	....	....	....		1982-83	1983-84
Laxton, Gord	Pit.	4	17	4	9	0	800	74	0	5.55	....	....	....	....	....	....	....	....		1975-76	1978-79
LeBlanc, Ray	Chi.	1	1	1	0	0	60	1	0	1.00	....	....	....	....	....	....	....	....		1991-92	1991-92
§ • Leduc, Albert	Mtl.	1	1	0	0	0	2	1	0	30.00	....	....	....	....	....	....	....	....		1931-32	1931-32
Legris, Claude	Det.	2	4	0	1	1	91	4	0	2.64	....	....	....	....	....	....	....	....		1980-81	1981-82
• Lehman, Hugh	Chi.	2	48	20	24	4	3047	136	6	2.68	2	0	1	1	120	10	0	5.00		1926-27	1927-28
Lemelin, Reggie	Atl., Cgy., Bos.	15	507	236	162	63	28006	1613	12	3.46	59	23	25	....	3119	186	2	3.58		1978-79	1992-93
Lenarduzzi, Mike	Hfd.	2	4	1	1	1	189	10	0	3.17	....	....	....	....	....	....	....	....		1992-93	1993-94
Lessard, Mario	L.A.	6	240	92	97	39	13529	843	9	3.74	20	6	12	....	1136	83	0	4.38		1978-79	1983-84
Levasseur, Jean-Louis	Min.	1	1	0	1	0	60	7	0	7.00	....	....	....	....	....	....	....	....		1979-80	1979-80
§ • Levinsky, Alex	Tor.	1	1	0	0	0	1	1	0	60.00	....	....	....	....	....	....	....	....		1931-32	1931-32
• Lindbergh, Pelle	Phi.	5	157	87	49	15	9150	503	7	3.30	23	12	10	....	1214	63	3	3.11		1981-82	1985-86
• Lindsay, Bert	Mtl.W., Tor.	2	20	6	14	0	1238	118	0	5.72	....	....	....	....	....	....	....	....		1917-18	1918-19
Littman, David	Buf., T.B.	3	3	0	2	0	141	14	0	5.96	....	....	....	....	....	....	....	....		1990-91	1992-93
Liut, Mike	St.L., Hfd., Wsh.	13	664	294	271	74	38215	2221	25	3.49	67	29	32	....	3814	215	2	3.38		1979-80	1991-92
Lockett, Ken	Van.	2	55	13	15	8	2348	131	2	3.35	1	0	1	....	60	6	0	6.00		1974-75	1975-76
• Lockhart, Howard	Tor., Que., Ham., Bos.	5	59	16	41	0	3413	287	1	5.05	....	....	....	....	....	....	....	....		1919-20	1924-25
LoPresti, Pete	Min., Edm.	6	175	43	102	20	9858	668	5	4.07	2	0	2	....	77	6	0	4.68		1974-75	1980-81
• LoPresti, Sam	Chi.	2	74	30	38	6	4530	236	4	3.13	8	3	5	....	530	17	1	1.92		1940-41	1941-42
‡ Lorenz, Danny	NYI	3	8	1	5	0	357	25	0	4.20	....	....	....	....	....	....	....	....		1990-91	1992-93
Loustel, Ron	Wpg.	1	1	0	1	0	60	10	0	10.00	....	....	....	....	....	....	....	....		1980-81	1980-81
Low, Ron	Tor., Wsh., Det., Que., Edm., N.J.	11	382	102	203	38	20502	1463	4	4.28	7	1	6	....	452	29	0	3.85		1972-73	1984-85
Lozinski, Larry	Det.	1	30	6	11	7	1459	105	0	4.32	....	....	....	....	....	....	....	....		1980-81	1980-81
• Lumley, Harry	Det., NYR, Chi., Tor., Bos.	16	803	330	329	142	48044	2206	71	2.75	76	29	47	....	4778	198	7	2.49	1	1943-44	1959-60
MacKenzie, Shawn	N.J.	1	4	0	1	0	130	15	0	6.92	....	....	....	....	....	....	....	....		1982-83	1982-83
Madeley, Darrin	Ott.	3	39	4	23	5	1928	140	0	4.36	....	....	....	....	....	....	....	....		1992-93	1994-95
Malarchuk, Clint	Que., Wsh., Buf.	10	338	141	130	45	19030	1100	12	3.47	15	2	9	0	781	56	0	4.30		1981-82	1991-92
Maneluk, George	NYI	1	4	1	1	0	140	15	0	6.43	....	....	....	....	....	....	....	....		1990-91	1990-91
Maniago, Cesare	Tor., Mtl., NYR, Min., Van.	15	568	189	259	96	32570	1773	30	3.27	36	15	21	....	2245	100	3	2.67		1960-61	1977-78
Marois, Jean	Tor., Chi.	2	3	1	2	0	180	15	0	5.00	....	....	....	....	....	....	....	....		1943-44	1953-54
Martin, Seth	St.L.	1	30	8	10	7	1552	67	1	2.59	2	0	0	....	73	5	0	4.11		1967-68	1967-68
Mason, Bob	Wsh., Chi., Que., Van.	8	145	55	65	16	7988	500	1	3.76	5	2	3	....	369	12	1	1.95		1983-84	1990-91
Mattsson, Markus	Wpg., Min., L.A.	4	92	21	46	14	5007	343	6	4.11	....	....	....	....	....	....	....	....		1979-80	1983-84
May, Darrell	St.L.	2	6	1	5	0	364	31	0	5.11	....	....	....	....	....	....	....	....		1985-86	1987-88
Mayer, Gilles	Tor.	4	9	2	6	1	540	24	0	2.67	....	....	....	....	....	....	....	....		1949-50	1955-56
• McAuley, Ken	NYR	2	96	17	64	15	5740	537	1	5.61	....	....	....	....	....	....	....	....		1943-44	1944-45
McCartan, Jack	NYR	2	12	2	7	3	680	42	1	3.71	....	....	....	....	....	....	....	....		1959-60	1960-61
• McCool, Frank	Tor.	2	72	34	31	7	4320	242	4	3.36	13	8	5	....	807	30	4	2.23	1	1944-45	1945-46
McDuffe, Peter	St.L., NYR, K.C., Det.	5	57	11	36	6	3207	218	0	4.08	1	0	1	....	60	7	0	7.00		1971-72	1975-76
McGrattan, Tom	Det.	1	1	0	0	0	8	1	0	7.50	....	....	....	....	....	....	....	....		1947-48	1947-48
McKay, Ross	Hfd.	1	1	0	0	0	35	3	0	5.14	....	....	....	....	....	....	....	....		1990-91	1990-91
McKenzie, Bill	Det., K.C., Col.	6	91	18	49	13	4776	326	2	4.10	....	....	....	....	....	....	....	....		1973-74	1979-80
McKichan, Steve	Van.	1	1	0	0	0	20	2	0	6.00	....	....	....	....	....	....	....	....		1990-91	1990-91
McLachlan, Murray	Tor.	1	2	0	1	0	25	4	0	9.60	....	....	....	....	....	....	....	....		1970-71	1970-71
McLean, Kirk	N.J., Van., Car., Fla., NYR	16	612	245	262	72	35090	1904	22	3.26	68	34	34	....	4189	198	6	2.84		1985-86	2000-01
McLelland, Dave	Van.	1	2	1	1	0	120	10	0	5.00	....	....	....	....	....	....	....	....		1972-73	1972-73
McLeod, Don	Det., Phi.	2	18	3	10	1	879	74	0	5.05	....	....	....	....	....	....	....	....		1970-71	1971-72
McLeod, Jim	St.L.	1	16	6	6	4	880	44	0	3.00	....	....	....	....	....	....	....	....		1971-72	1971-72
McNamara, Gerry	Tor.	2	7	2	2	1	323	14	0	2.60	....	....	....	....	....	....	....	....		1960-61	1969-70
McNeil, Gerry	Mtl.	8	276	119	105	52	16535	649	28	2.36	35	17	18	....	2284	72	5	1.89	3	1947-48	1957-58
McRae, Gord	Tor.	5	71	30	22	10	3799	221	1	3.49	8	2	5	....	454	22	0	2.91		1972-73	1977-78
Melanson, Rollie	NYI, Min., L.A., N.J., Mtl.	11	291	129	106	33	16452	995	6	3.63	23	4	9	....	801	59	0	4.42	3	1980-81	1991-92
Meloche, Gilles	Chi., Cal., Cle., Min., Pit.	18	788	270	351	131	45401	2756	20	3.64	45	21	19	....	2464	143	2	3.48		1970-71	1987-88
‡ Micalef, Corrado	Det.	5	113	26	59	15	5794	409	2	4.24	3	0	0	....	49	8	0	9.80		1981-82	1985-86
Middlebrook, Lindsay	Wpg., Min., N.J., Edm.	4	37	3	23	6	1845	152	0	4.94	....	....	....	....	....	....	....	....		1979-80	1982-83
• Millar, Al	Bos.	1	6	1	4	1	360	25	0	4.17	....	....	....	....	....	....	....	....		1957-58	1957-58
Millen, Greg	Pit., Hfd., St.L., Que., Chi., Det.	14	604	215	284	89	35377	2281	17	3.87	59	27	29	....	3383	193	0	3.42		1978-79	1991-92
• Miller, Joe	NYA, NYR, Pit., Phi.	4	127	24	87	16	7871	383	16	2.92	3	2	1	0	180	3	1	1.00	1	1927-28	1930-31
Mio, Eddie	Edm., NYR, Det.	7	192	64	73	30	10428	705	4	4.06	17	9	7	....	986	63	0	3.83		1979-80	1985-86
• Mitchell, Ivan	Tor.	3	22	10	9	0	1190	88	0	4.44	....	....	....	....	....	....	....	....	1	1919-20	1921-22
Moffat, Mike	Bos.	3	19	7	7	2	979	70	0	4.29	11	6	5	....	663	38	0	3.44		1981-82	1983-84
Moog, Andy	Edm., Bos., Dal., Mtl.	18	713	372	209	88	40151	2097	28	3.13	132	68	57	....	7452	377	4	3.04	3	1980-81	1997-98
Moore, Alfie	NYA, Chi., Det.	4	21	7	14	0	1290	81	1	3.77	3	1	2	....	180	7	0	2.33	1	1936-37	1939-40
Moore, Robbie	Phi., Wsh.	2	6	3	1	1	257	8	2	1.87	5	3	2	....	268	18	0	4.03		1978-79	1982-83
Morissette, Jean-Guy	Mtl.	1	1	0	1	0	36	4	0	6.67	....	....	....	....	....	....	....	....		1963-64	1963-64
• Mowers, Johnny	Det.	4	152	65	61	26	9350	399	15	2.56	32	19	13	....	2000	85	2	2.55	1	1940-41	1946-47
Mrazek, Jerome	Phi.	1	1	0	0	0	6	1	0	10.00	....	....	....	....	....	....	....	....		1975-76	1975-76
§ • Mummery, Harry	Que., Ham.	2	4	2	1	0	192	20	0	6.25	....	....	....	....	....	....	....	....		1919-20	1921-22
§ • Munro, Dunc	Mtl.M.	1	1	0	0	0	2	0	0	0.00	....	....	....	....	....	....	....	....		1924-25	1924-25
• Murphy, Hal	Mtl.	1	1	1	0	0	60	4	0	4.00	....	....	....	....	....	....	....	....		1952-53	1952-53
• Murray, Mickey	Mtl.	1	1	0	1	0	60	4	0	4.00	....	....	....	....	....	....	....	....		1929-30	1929-30
‡ Muzzatti, Jason	Cgy., Hfd., NYR, S.J.	5	62	13	25	10	3014	167	1	3.32	....	....	....	....	....	....	....	....		1993-94	1997-98
‡ Myllys, Jarmo	Min., S.J.	4	39	4	27	1	1846	161	0	5.23	....	....	....	....	....	....	....	....		1988-89	1991-92
Mylnikov, Sergei	Que.	1	10	1	7	2	568	47	0	4.96	....	....	....	....	....	....	....	....		1989-90	1989-90
Myre, Phil	Mtl., Atl., St.L., Phi., Col., Buf.	14	439	149	198	76	25220	1482	14	3.53	12	6	5	....	747	41	1	3.29	1	1969-70	1982-83
Newton, Cam	Pit.	2	16	4	7	1	814	51	0	3.76	....	....	....	....	....	....	....	....		1970-71	1972-73
Norris, Jack	Bos., Chi., L.A.	4	58	20	25	4	3119	202	2	3.89	....	....	....	....	....	....	....	....		1964-65	1970-71
Oleschuk, Bill	K.C., Col.	4	55	7	28	10	2835	188	1	3.98	....	....	....	....	....	....	....	....		1975-76	1979-80
• Olesevich, Dan	NYR	1	1	0	0	1	29	2	0	4.14	....	....	....	....	....	....	....	....		1961-62	1961-62
‡ O'Neill, Mike	Wpg., Ana.	4	21	0	9	2	855	61	0	4.28	....	....	....	....	....	....	....	....		1991-92	1996-97
Ouimet, Ted	St.L.	1	1	0	1	0	60	2	0	2.00	....	....	....	....	....	....	....	....		1968-69	1968-69
Pageau, Paul	L.A.	1	1	0	1	0	60	8	0	8.00	....	....	....	....	....	....	....	....		1980-81	1980-81
Paille, Marcel	NYR	7	107	32	52	22	6342	362	2	3.42	....	....	....	....	....	....	....	....		1957-58	1964-65

Name	NHL Teams	NHL Seasons	Regular Schedule GP	W	L	T	Mins	GA	SO	Avg	Playoffs GP	W	L	T	Mins	GA	SO	Avg	NHL Cup Wins	First NHL Season	Last NHL Season
Palmateer, Mike	Tor., Wsh.	8	356	149	138	52	20131	1183	17	3.53	29	12	17	....	1765	89	2	3.03		1976-77	1983-84
Pang, Darren	Chi.	3	81	27	35	7	4252	287	0	4.05	6	1	3	....	250	18	0	4.32		1984-85	1988-89
Parent, Bernie	Bos., Phi., Tor.	13	608	271	198	121	35136	1493	54	2.55	71	38	33	....	4302	174	6	2.43	2	1965-66	1978-79
Parent, Bob	Tor.	2	3	0	2	0	160	15	0	5.63	....	....	....	....	....	....	....	....		1981-82	1982-83
‡ Parent, Rich	St.L., T.B., Pit.	4	32	7	11	5	1561	82	1	3.15	....	....	....	....	....	....	....	....		1997-98	2000-01
Parro, Dave	Wsh.	4	77	21	36	10	4015	274	2	4.09	....	....	....	....	....	....	....	....		1980-81	1983-84
§ ● Patrick, Lester	NYR	1	....	....	....	....	....	....	....	....	1	1	0	0	46	1	0	1.30	1	1927-28	1927-28
Peeters, Pete	Phi., Bos., Wsh.	13	489	246	155	51	27699	1424	21	3.08	71	35	35	....	4200	232	2	3.31		1978-79	1990-91
Pelletier, Marcel	Chi., NYR	2	8	1	6	0	395	32	0	4.86	....	....	....	....	....	....	....	....		1950-51	1962-63
Penney, Steve	Mtl., Wpg.	5	91	35	38	12	5194	313	1	3.62	27	15	12	....	1604	72	4	2.69		1983-84	1987-88
● Perreault, Bob	Mtl., Det., Bos.	3	31	8	16	7	1827	103	3	3.38	....	....	....	....	....	....	....	....		1955-56	1962-63
Pettie, Jim	Bos.	3	21	9	7	2	1157	71	1	3.68	....	....	....	....	....	....	....	....		1976-77	1978-79
Pietrangelo, Frank	Pit., Hfd.	7	141	46	59	6	7141	490	1	4.12	12	7	5	....	713	34	1	2.86	1	1987-88	1993-94
● Plante, Jacques	Mtl., NYR, St.L., Tor., Bos.	18	837	435	247	145	49493	1964	82	2.38	112	71	36	....	6651	237	14	2.14	6	1952-53	1972-73
Plasse, Michel	St.L., Mtl., K.C., Pit., Col., Que.	11	299	92	136	54	16760	1058	2	3.79	4	1	2	....	195	9	1	2.77	1	1970-71	1981-82
§ ● Plaxton, Hugh	Mtl.M.	1	1	0	1	0	57	5	0	5.26	....	....	....	....	....	....	....	....		1932-33	1932-33
Pronovost, Claude	Bos., Mtl.	2	3	1	1	0	120	7	1	3.50	....	....	....	....	....	....	....	....		1955-56	1958-59
Puppa, Daren	Buf., Tor., T.B.	15	429	179	161	54	23819	1204	19	3.03	16	4	9	....	786	51	0	3.89		1985-86	1999-00
Pusey, Chris	Det.	1	1	0	0	0	40	3	0	4.50	....	....	....	....	....	....	....	....		1985-86	1985-86
‡ Racicot, Andre	Mtl.	5	68	26	23	8	3357	196	2	3.50	4	0	1	....	31	4	0	7.74	1	1989-90	1993-94
‡ Racine, Bruce	St.L.	1	11	0	3	0	230	12	0	3.13	1	0	0	....	1	0	0	0.00		1995-96	1995-96
‡ Ram, Jamie	NYR	1	1	0	0	0	27	0	0	0.00	....	....	....	....	....	....	....	....		1995-96	1995-96
Ranford, Bill	Bos., Edm., Wsh., T.B., Det.	15	647	240	279	76	35936	2042	15	3.41	53	28	25	....	3110	159	4	3.07	2	1985-86	1999-00
Raymond, Alain	Wsh.	1	1	0	1	0	40	2	0	3.00	....	....	....	....	....	....	....	....		1987-88	1987-88
Rayner, Chuck	NYA, Bro., NYR	10	424	138	208	77	25491	1294	25	3.05	18	9	9	....	1135	46	1	2.43		1940-41	1952-53
Reaugh, Daryl	Edm., Hfd.	3	27	8	9	1	1246	72	1	3.47	....	....	....	....	....	....	....	....		1984-85	1990-91
‡ Reddick, Pokey	Wpg., Edm., Fla.	6	132	46	58	16	7162	443	0	3.71	4	0	2	....	168	10	0	3.57	1	1986-87	1993-94
● Redding, George	Bos.	1	1	0	0	0	11	1	0	5.45	....	....	....	....	....	....	....	....		1924-25	1924-25
Redquest, Greg	Pit.	1	1	0	0	0	13	3	0	13.85	....	....	....	....	....	....	....	....		1977-78	1977-78
Reece, Dave	Bos.	1	14	7	5	2	777	43	2	3.32	....	....	....	....	....	....	....	....		1975-76	1975-76
Reese, Jeff	Tor., Cgy., Hfd., T.B., N.J.	11	174	53	65	17	8667	529	5	3.66	11	3	5	....	515	35	0	4.08		1987-88	1998-99
Resch, Glenn	NYI, Col., N.J., Phi.	14	571	231	224	82	32279	1761	26	3.27	41	17	17	....	2044	85	2	2.50	1	1973-74	1986-87
● Rheaume, Herb	Mtl.	1	31	10	20	1	1889	92	0	2.92	....	....	....	....	....	....	....	....		1925-26	1925-26
Ricci, Nick	Pit.	4	19	7	12	0	1087	79	0	4.36	....	....	....	....	....	....	....	....		1979-80	1982-83
Richardson, Terry	Det., St.L.	5	20	3	11	0	906	85	0	5.63	....	....	....	....	....	....	....	....		1973-74	1978-79
Ridley, Curt	NYR, Van., Tor.	6	104	27	47	16	5498	355	1	3.87	2	0	2	....	120	8	0	4.00		1974-75	1980-81
‡ Riendeau, Vincent	Mtl., St.L., Det., Bos.	8	184	85	65	20	10423	573	5	3.30	25	11	12	....	1277	71	1	3.34		1987-88	1994-95
Riggin, Dennis	Det.	2	18	6	10	2	999	52	1	3.12	....	....	....	....	....	....	....	....		1959-60	1962-63
Riggin, Pat	Atl., Cgy., Wsh., Bos., Pit.	9	350	153	120	52	19872	1135	11	3.43	25	8	13	....	1336	72	0	3.23		1979-80	1987-88
Ring, Bob	Bos.	1	1	0	0	0	33	4	0	7.27	....	....	....	....	....	....	....	....		1965-66	1965-66
Rivard, Fern	Min.	4	55	9	26	11	2865	190	2	3.98	....	....	....	....	....	....	....	....		1968-69	1974-75
● Roach, John Ross	Tor., NYR, Det.	14	492	219	204	68	30444	1246	58	2.46	29	12	14	3	1901	60	7	1.89	1	1921-22	1934-35
● Roberts, Moe	Bos., NYA, Chi.	4	10	3	5	0	501	31	0	3.71	....	....	....	....	....	....	....	....		1925-26	1951-52
● Robertson, Earl	Det., NYA, Bro.	6	190	60	95	34	11820	575	16	2.92	15	7	7	....	995	29	2	1.75	1	1936-37	1941-42
● Rollins, Al	Tor., Chi., NYR	9	430	141	205	83	25723	1192	28	2.78	13	6	7	....	755	30	0	2.38	1	1949-50	1959-60
Romano, Roberto	Pit., Bos.	6	126	46	63	8	7111	471	4	3.97	....	....	....	....	....	....	....	....		1982-83	1993-94
‡ Rosati, Mike	Wsh.	1	1	1	0	0	28	0	0	0.00	....	....	....	....	....	....	....	....		1998-99	1998-99
‡ Roussel, Dominic	Phi., Wpg., Ana., Edm.	8	205	77	70	23	10665	555	7	3.12	1	0	0	....	23	0	0	0.00		1991-92	2000-01
Rupp, Pat	Det.	1	1	0	1	0	60	4	0	4.00	....	....	....	....	....	....	....	....		1963-64	1963-64
Rutherford, Jim	Det., Pit., Tor., L.A.	13	457	151	227	59	25895	1576	14	3.65	8	2	5	....	440	28	0	3.82		1970-71	1982-83
Rutledge, Wayne	L.A.	3	82	28	37	9	4325	241	2	3.34	8	2	4	....	378	20	0	3.17		1967-68	1969-70
St. Croix, Rick	Phi., Tor.	8	130	49	54	18	7295	451	2	3.71	11	4	6	....	562	29	1	3.10		1977-78	1984-85
St. Laurent, Sam	N.J., Det.	5	34	7	12	4	1572	92	1	3.51	1	0	0	....	10	1	0	6.00		1985-86	1989-90
§ ● Sands, Charlie	Mtl.	1	1	0	0	0	25	5	0	12.00	....	....	....	....	....	....	....	....		1939-40	1939-40
Sands, Mike	Min.	2	6	0	5	0	302	26	0	5.17	....	....	....	....	....	....	....	....		1984-85	1986-87
‡ Sarjeant, Geoff	St.L., S.J.	2	8	1	2	1	291	20	0	4.12	....	....	....	....	....	....	....	....		1994-95	1995-96
Sauve, Bob	Buf., Det., Chi., N.J.	13	420	182	154	54	23711	1377	8	3.48	34	15	16	....	1850	95	4	3.08		1976-77	1988-89
● Sawchuk, Terry	Det., Bos., Tor., L.A., NYR	21	971	447	330	172	57194	2389	103	2.51	106	54	48	....	6290	266	12	2.54	4	1949-50	1969-70
● Schaefer, Joe	NYR	2	2	0	2	0	86	8	0	5.58	....	....	....	....	....	....	....	....		1959-60	1960-61
‡ Schafer, Paxton	Bos.	1	3	0	0	0	77	6	0	4.68	....	....	....	....	....	....	....	....		1996-97	1996-97
Scott, Ron	NYR, L.A.	5	28	8	13	4	1450	91	0	3.77	1	0	0	....	32	4	0	7.50		1983-84	1989-90
Sevigny, Richard	Mtl., Que.	9	176	80	54	20	9485	507	5	3.21	4	0	3	....	208	13	0	3.75	1	1978-79	1986-87
Sharples, Scott	Cgy.	1	1	0	0	1	65	4	0	3.69	....	....	....	....	....	....	....	....		1991-92	1991-92
§ ● Shields, Al	NYA	1	2	0	0	0	41	9	0	13.17	....	....	....	....	....	....	....	....		1931-32	1931-32
‡ Shtalenkov, Mikhail	Ana., Edm., Phx., Fla.	7	190	62	82	19	9966	480	8	2.89	4	0	3	....	211	10	0	2.84		1993-94	1999-00
‡ Shulmistra, Richard	N.J., Fla.	2	2	1	1	0	122	3	0	1.48	....	....	....	....	....	....	....	....		1997-98	1999-00
Sidorkiewicz, Peter	Hfd., Ott., N.J.	8	246	79	128	27	13884	832	8	3.60	15	5	10	....	912	55	0	3.62		1987-88	1997-98
Simmons, Don	Bos., Tor., NYR	11	248	101	100	41	14495	698	20	2.89	24	13	11	....	1436	62	3	2.59	3	1956-57	1968-69
Simmons, Gary	Cal., Cle., L.A.	4	107	30	57	15	6162	366	5	3.56	1	0	0	....	20	1	0	3.00		1974-75	1977-78
Skidmore, Paul	St.L.	1	2	1	1	0	120	6	0	3.00	....	....	....	....	....	....	....	....		1981-82	1981-82
Skorodenski, Warren	Chi., Edm.	5	35	12	11	4	1732	100	2	3.46	2	0	0	....	33	6	0	10.91		1981-82	1987-88
Smith, Al	Tor., Pit., Det., Buf., Hfd., Col.	10	233	74	99	36	12752	735	10	3.46	6	1	4	....	317	21	0	3.97		1965-66	1980-81
Smith, Billy	L.A., NYI	18	680	305	233	105	38431	2031	22	3.17	132	88	36	....	7645	348	5	2.73	4	1971-72	1988-89
Smith, Gary	Tor., Oak., Cal., Chi., Van., Min., Wsh., Wpg.	14	532	173	261	74	29619	1675	26	3.39	20	5	13	....	1153	62	1	3.23		1965-66	1979-80
● Smith, Normie	Mtl.M., Det.	8	199	81	83	35	12357	479	17	2.33	12	9	2	0	820	18	3	1.32	2	1931-32	1944-45
Sneddon, Bob	Cal.	1	5	0	2	0	225	21	0	5.60	....	....	....	....	....	....	....	....		1970-71	1970-71
Soderstrom, Tommy	Phi., NYI	5	156	45	69	19	8189	496	10	3.63	....	....	....	....	....	....	....	....		1992-93	1996-97
Soetaert, Doug	NYR, Wpg., Mtl.	12	284	110	104	42	15583	1030	6	3.97	5	1	2	....	180	14	0	4.67	1	1975-76	1986-87
‡ Soucy, Christian	Chi.	1	1	0	0	0	3	0	0	0.00	....	....	....	....	....	....	....	....		1993-94	1993-94
● Spooner, Red	Pit.	1	1	0	1	0	60	6	0	6.00	....	....	....	....	....	....	....	....		1929-30	1929-30
§ ● Spring, Jesse	Ham.	1	1	0	0	0	2	0	0	0.00	....	....	....	....	....	....	....	....		1924-25	1924-25
Staniowski, Ed	St.L., Wpg., Hfd.	10	219	67	104	21	12075	818	2	4.06	8	1	6	....	428	28	0	3.93		1975-76	1984-85
§ ● Starr, Harold	Mtl.M.	1	1	0	0	0	3	0	0	0.00	....	....	....	....	....	....	....	....		1931-32	1931-32
Stauber, Robb	L.A., Buf.	4	62	21	23	9	3295	209	1	3.81	4	3	1	....	240	16	0	4.00		1989-90	1994-95
Stefan, Greg	Det.	9	299	115	127	30	16333	1068	5	3.92	30	12	17	....	1681	99	1	3.53		1981-82	1989-90
Stein, Phil	Tor.	1	1	0	0	1	70	2	0	1.71	....	....	....	....	....	....	....	....		1939-40	1939-40
Stephenson, Wayne	St.L., Phi., Wsh.	10	328	146	103	49	18343	937	14	3.06	26	11	12	....	1522	79	2	3.11	1	1971-72	1980-81
Stevenson, Doug	Chi., NYR	3	8	2	6	0	480	39	0	4.88	....	....	....	....	....	....	....	....		1942-43	1945-46
Stewart, Charles	Bos.	3	77	30	41	5	4742	194	10	2.45	....	....	....	....	....	....	....	....		1924-25	1926-27
Stewart, Jim	Bos.	1	1	0	1	0	20	5	0	15.00	....	....	....	....	....	....	....	....		1979-80	1979-80
● Stuart, Herb	Det.	1	3	1	2	0	180	5	0	1.67	....	....	....	....	....	....	....	....		1926-27	1926-27
Sylvestri, Don	Bos.	1	3	0	0	2	102	6	0	3.53	....	....	....	....	....	....	....	....		1984-85	1984-85
Tabaracci, Rick	Pit., Wpg., Wsh., Cgy., T.B., Atl., Col.	11	286	93	125	30	15255	760	15	2.99	17	4	12	....	1025	53	0	3.10		1988-89	1999-00
Takko, Kari	Min., Edm.	6	142	37	71	14	7317	475	1	3.90	4	0	1	....	109	7	0	3.85		1985-86	1990-91
Tanner, John	Que.	3	21	2	11	5	1084	65	1	3.60	....	....	....	....	....	....	....	....		1989-90	1991-92
Tataryn, Dave	NYR	1	2	1	1	0	80	10	0	7.50	....	....	....	....	....	....	....	....		1976-77	1976-77
Taylor, Bobby	Phi., Pit.	5	46	15	17	6	2268	155	0	4.10	....	....	....	....	....	....	....	....	1	1971-72	1975-76
● Teno, Harvey	Det.	1	5	2	3	0	300	15	0	3.00	....	....	....	....	....	....	....	....		1938-39	1938-39
‡ Terreri, Chris	N.J., S.J., Chi., NYI	14	406	151	172	43	22369	1143	9	3.07	29	12	12	....	1523	86	0	3.39	2	1986-87	2000-01
Thomas, Wayne	Mtl., Tor., NYR	9	243	103	93	34	13768	766	10	3.34	15	6	8	....	849	50	1	3.53		1972-73	1980-81
● Thompson, Tiny	Bos., Det.	12	553	284	194	75	34175	1183	81	2.08	44	20	24	0	2974	93	7	1.88	1	1928-29	1939-40
§ Toppazzini, Jerry	Bos.	1	1	0	0	0	1	0	0	0.00	....	....	....	....	....	....	....	....		1960-61	1960-61
‡ Torchia, Mike	Dal.	1	6	3	2	1	327	18	0	3.30	....	....	....	....	....	....	....	....		1994-95	1994-95
‡ Trefilov, Andrei	Cgy., Buf., Chi.	7	54	12	25	4	2663	153	2	3.45	1	0	0	....	5	0	0	0.00		1992-93	1998-99
Tremblay, Vincent	Tor., Pit.	5	58	12	26	8	2785	223	1	4.80	....	....	....	....	....	....	....	....		1979-80	1983-84
Tucker, Ted	Cal.	1	5	1	1	1	177	10	0	3.39	....	....	....	....	....	....	....	....		1973-74	1973-74
● Turner, Joe	Det.	1	1	0	0	1	70	3	0	2.57	....	....	....	....	....	....	....	....		1941-42	1941-42
Vachon, Rogie	Mtl., L.A., Det., Bos.	16	795	355	291	127	46298	2310	51	2.99	48	23	23	....	2876	133	2	2.77	3	1966-67	1981-82

Name	NHL Teams	NHL Seasons	Regular Schedule GP	W	L	T	Mins	GA	SO	Avg	Playoffs GP	W	L	T	Mins	GA	SO	Avg	NHL Cup Wins	First NHL Season	Last NHL Season
Veisor, Mike	Chi., Hfd., Wpg.	10	139	41	62	26	7806	532	5	4.09	4	0	2	....	180	15	0	5.00		1973-74	1983-84
• Vezina, Georges	Mtl.	9	190	103	81	5	11592	633	13	3.28	13	9	4	1	780	36	2	2.77	1	1917-18	1925-26
Villemure, Gilles	NYR, Chi.	10	205	100	64	29	11581	542	13	2.81	14	5	5	....	656	32	0	2.93		1963-64	1976-77
‡ Waite, Jimmy	Chi., S.J., Phx.	11	106	28	41	12	5253	293	4	3.35	6	0	3	....	211	14	0	3.98		1988-89	1998-99
Wakaluk, Darcy	Buf., Min., Dal., Phx.	8	191	67	75	21	9756	524	9	3.22	8	4	2	....	364	18	0	2.97		1988-89	1996-97
Wakely, Ernie	Mtl., St.L.	5	113	41	42	17	6244	290	8	2.79	10	2	6	....	509	37	1	4.36		1962-63	1971-72
• Walsh, Flat	Mtl.M., NYA	7	108	48	43	16	6641	256	12	2.31	8	2	4	2	570	16	2	1.68		1926-27	1932-33
Wamsley, Rick	Mtl., St.L., Cgy., Tor.	13	407	204	131	46	23123	1287	12	3.34	27	7	18	....	1397	81	0	3.48	1	1980-81	1992-93
Watt, Jim	St.L.	1	1	0	0	0	20	2	0	6.00	....	....	....	....	....	....	....	....		1973-74	1973-74
Weeks, Steve	NYR, Hfd., Van., NYI, L.A., Ott.	18	290	111	119	33	15879	989	5	3.74	12	3	5	....	486	27	0	3.33		1980-81	1992-93
Wetzel, Carl	Det., Min.	2	7	1	3	1	301	22	0	4.39	....	....	....	....	....	....	....	....		1964-65	1967-68
‡ Wilkinson, Derek	T.B.	4	22	3	12	3	933	57	0	3.67	....	....	....	....	....	....	....	....		1995-96	1998-99
‡ Willis, Jordan	Dal.	1	1	0	1	0	19	1	0	3.16	....	....	....	....	....	....	....	....		1995-96	1995-96
Wilson, Dunc	Phi., Van., Tor., NYR, Pit.	10	287	80	150	33	15851	988	8	3.74	....	....	....	....	....	....	....	....		1969-70	1978-79
Wilson, Lefty	Det., Tor., Bos.	3	3	0	0	1	81	1	0	0.74	....	....	....	....	....	....	....	....		1953-54	1957-58
• Winkler, Hal	NYR, Bos.	2	75	35	26	14	4739	126	21	1.60	10	2	3	5	640	18	2	1.69		1926-27	1927-28
Wolfe, Bernie	Wsh.	4	120	20	61	21	6104	424	1	4.17	....	....	....	....	....	....	....	....		1975-76	1978-79
Wood, Alex	NYA	1	1	0	1	0	70	3	0	2.57	....	....	....	....	....	....	....	....		1936-37	1936-37
Worsley, Gump	NYR, Mtl., Min.	21	861	335	352	150	50183	2407	43	2.88	70	40	26	....	4084	189	5	2.78	4	1952-53	1973-74
• Worters, Roy	Pit., NYA, Mtl.	12	484	171	229	83	30175	1143	67	2.27	11	3	6	2	690	24	3	2.09		1925-26	1936-37
Worthy, Chris	Oak., Cal.	3	26	5	10	4	1326	98	0	4.43	....	....	....	....	....	....	....	....		1968-69	1970-71
‡ Wregget, Ken	Tor., Phi., Pit., Cgy., Det.	17	575	225	248	53	31663	1917	9	3.63	56	28	25	....	3341	160	3	2.87	1	1983-84	1999-00
§ • Young, Doug	Det.	1	1	0	0	0	21	1	0	2.86	....	....	....	....	....	....	....	....		1933-34	1933-34
‡ Young, Wendell	Van., Phi., Pit., T.B.	10	187	59	86	12	9410	618	2	3.94	2	0	1	....	99	6	0	3.64	2	1985-86	1994-95
Zanier, Mike	Edm.	1	3	1	1	1	185	12	0	3.89	....	....	....	....	....	....	....	....		1984-85	1984-85

Yves Belanger

Bruce Gamble

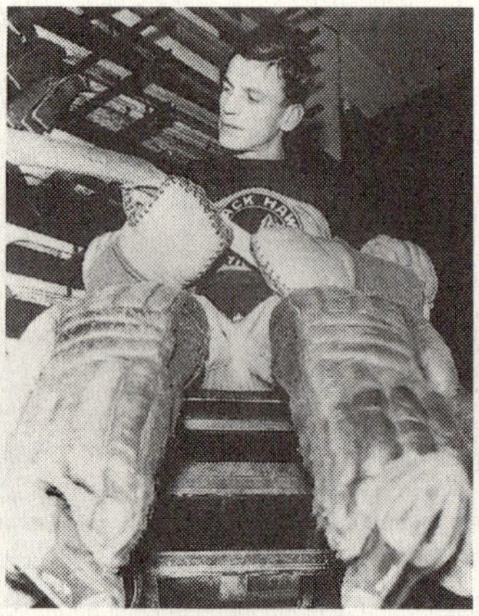

Mike Karakas

Dennis Riggin

Ken Broderick

Ed Giacomin

Rollie Melanson

Chris Terreri

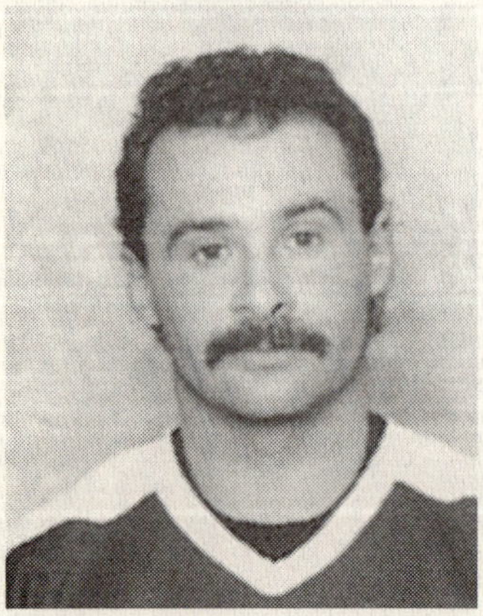

Grant Fuhr

Ron Hextall

Pete Peeters

Roy Worters

2001-02 NHL Player Transactions

(listed in chronological order)

August, 2001

22 – New Jersey trades **Chris Ferraro** to Washington for future considerations.

29 – Dallas trades **Grant Marshall** to Columbus for Columbus' 2nd round pick in 2003.

September, 2001

28 – **NHL Waiver Draft**

Pos.	Player	Claimed By	Claimed From
G	**Chris Osgood**	NY Islanders	Detroit
C	**Sebastien Bordeleau**	Minnesota	St. Louis
LW	**Kirk Muller**	Columbus	Dallas
D	**Jamie Allison**	Calgary	Chicago
D	**Phil Housley**	Chicago	Calgary
LW	**P.J. Stock**	Boston	NY Rangers
RW	**Glen Metropolit**	Tampa Bay	Washington
C	**Josh Holden**	Carolina	Vancouver
G	**Martin Brochu**	Vancouver	Minnesota

28 – Columbus trades **Kirk Muller** to Dallas for the rights to **Evgeny Petrochinin**.

October, 2001

1 – Chicago trades **Marty Wilford** to Toronto for the rights to **Shawn Thornton**.

9 – Chicago trades **Nathan Perrott** to Nashville for a conditional pick in 2003.

24 – Boston trades the rights to **Jason Allison** and **Mikko Eloranta** to Los Angeles for **Jozef Stumpel** and **Glen Murray**.

November, 2001

1 – Anaheim trades **Petr Tenkrat** to Nashville for **Patric Kjellberg**.

1 – Minnesota trades **Sergei Krivokrasov** to Anaheim for Anaheim's 7th round pick in 2002 (**Niklas Eckerblom**) and a conditional pick in 2003.

9 – Tampa Bay trades **Andrei Zyuzin** to New Jersey for **Josef Boumedienne**, **Sascha Goc** and the rights to **Anton But**.

10 – Washington trades **Trevor Linden** and NY Islanders 2nd round pick in 2002 (previously acquired, **Denis Grot**) to Vancouver for Vancouver's 1st round pick in 2002 (**Boyd Gordon**) and 3rd round pick in 2003.

21 – Montreal trades **Martin Rucinsky** and **Benoit Brunet** to Dallas for **Donald Audette** and **Shaun Van Allen**.

21 – Dallas trades **Jyrki Lumme** to Toronto for **Dave Manson**.

December, 2001

4 – Nashville trades **Bert Robertsson** to Anaheim for **Jay Legault**.

5 – Carolina trades **Steven Halko** and its 4th round pick in 2002 (later traded) to St. Louis for **Sean Hill**.

12 – NY Rangers trade **Zdeno Ciger** to Tampa Bay for **Matthew Barnaby**.

17 – Vancouver trades **Donald Brashear** and its 6th round pick (later traded) in 2002 to Philadelphia for **Jan Hlavac** and Tampa Bay's 3rd round pick (previously acquired, **Brett Skinner**) in 2002.

18 – Calgary trades **Jeff Cowan** and the rights to **Kurtis Foster** to Atlanta for **Petr Buzek** and a conditional pick in 2003.

19 – Los Angeles trades **Brett Hauer** to Nashville for **Rich Brennan**.

28 – Tampa Bay trades **Kaspars Astashenko** to Carolina for **Harlan Pratt**.

28 – Phoenix trades **Todd Warriner**, **Tyler Bouck**, **Trevor Letowski** and its 3rd round pick in 2003 to Vancouver for **Drake Berehowsky** and **Denis Pederson**.

January, 2002

4 – Minnesota trades **Sebastien Bordeleau** to Phoenix for **David Cullen**.

7 – Florida trades **Dan Boyle** to Tampa Bay for Tampa Bay's 5th round pick in 2003.

11 – Nashville trades **Yves Sarault** and a conditional pick in 2003 to Philadelphia for **Petr Hubacek** and **Jason Beckett**.

12 – Dallas trades **Benoit Hogue** to Boston for future considerations.

14 – Anaheim trades **Jim Cummins** to NY Islanders for **Dave Roche**.

16 – Carolina trades **Sandis Ozolinsh** and **Byron Ritchie** to Florida for **Bret Hedican**, **Kevyn Adams**, the rights to **Tomas Malec** and a conditional pick in 2003.

16 – Dallas trades **Valeri Kamensky** and a conditional pick in 2003 to New Jersey for **Andre Lakos** and a conditional pick in 2003.

17 – Washington trades **Joe Reekie** to Chicago for Chicago's 4th round pick (**Petr Dvorak**) in 2002.

22 – Minnesota trades **Andy Sutton** to Atlanta for **Hnat Domenichelli**.

24 – Colorado trades **Frederic Cassivi** to Atlanta for **Brett Clark**.

25 – Boston trades **Andrei Nazarov** to Phoenix for Phoenix's 5th round pick (**Peter Hamerlik**) in 2002.

25 – Montreal trades **Brian Savage**, its 3rd round pick in 2002 (**Matt Jones**) and a conditional draft pick to Phoenix for **Sergei Berezin**.

February, 2002

6 – Chicago trades **Steve Dubinsky** to Nashville for future considerations.

11 – Colorado trades **Shjon Podein** to St. Louis for **Mike Keane**.

13 – Carolina trades **Greg Koehler** to Philadelphia for **Jesse Boulerice**.

March, 2002

1 – Nashville trades **Marc Moro** to Toronto for **D.J. Smith** and **Marty Wilford**.

1 – Nashville trades **D.J. Smith** to Colorado for Tampa Bay's 9th round choice (previously acquired, **Matt Davis**) in 2002.

4 – Carolina trades **Mike Rucinski** to New Jersey for **Ted Drury**.

5 – Anaheim trades **Marty McInnis** to Boston for Boston's 3rd round pick (later traded) in 2002.

5 – Philadelphia trades **Joe Dipenta** to Atlanta for **Jarrod Skalde**.

5 – Tampa Bay trades **Kevin Weekes** to Carolina for **Chris Dingman** and **Shane Willis**.

8 – Anaheim trades **Bert Robertsson** to Pittsburgh for **Mark Moore**.

12 – Dallas trades **Martin Rucinsky** and **Roman Lyashenko** to NY Rangers for **Manny Malhotra** and **Barrett Heisten**.

13 – Nashville trades **Tom Fitzgerald** to Chicago for a 4th round pick in 2003 and future considerations.

15 – Carolina trades **Tom Barrasso** to Toronto for a 4th round pick in 2003.

15 – Tampa Bay trades **Juha Ylonen** to Ottawa for **Andre Roy** and a 6th round pick (**Paul Ranger**) in 2002.

15 – Philadelphia trades **Francis Lessard** to Atlanta for **David Harlock** and Atlanta's 3rd and 7th round picks in 2003.

15 – Columbus trades **Jamie Pushor** to Pittsburgh for Pittsburgh's 4th round pick in 2003.

16 – Nashville trades **Cliff Ronning** to Los Angeles for **Jere Karalahti** and a conditional pick in 2003.

16 – Dallas trades **Benoit Brunet** to Ottawa for a conditional pick in 2003.

17 – Calgary trades **Jukka Hentunen** to Nashville for a conditional pick in 2003.

17 – Pittsburgh trades **Billy Tibbetts** to Philadelphia for **Kent Manderville**.

18 – Florida trades **Pavel Bure** and its 2nd round pick (**Lee Falardeau**) in 2002 to NY Rangers for **Igor Ulanov**, the rights to **Filip Novak**, the Rangers' 1st (later traded) and 2nd (**Rob Globke**) round picks in 2002 and 4th round pick in 2003.

18 – Atlanta trades **Ray Ferraro** to St. Louis for Carolina's 4th round pick (previously acquired, **Lane Manson**) in 2002.

19 – Anaheim trades **Dave Roche** to NY Islanders for **Ben Guite** and the rights to **Bjorn Melin**.

19 – Atlanta trades **Jiri Slegr** to Detroit for **Yuri Butsayev** and Detroit's 3rd round pick (later traded) in 2002.

19 – Atlanta trades **Darcy Hordichuk** and Atlanta's 4th (**Lance Monych**) and 5th (**John Zeiler**) round picks in 2002 to Phoenix for **Kirill Safronov**, the rights to **Ruslan Zainullin** and Phoenix's 4th round pick (**Patrick Dwyer**) in 2002.

19 – Atlanta trades **Bob Corkum** to Buffalo for Buffalo's 5th round pick (**Paul Flache**) in 2002.

19 – Boston trades **Greg Crozier** to Minnesota for **Darryl Laplante**.

19 – Boston trades **Bobby Allen** to Edmonton for **Sean Brown**.

19 – Chicago trades **Jaroslav Spacek** and Chicago's 2nd round pick in 2003 to Columbus for **Lyle Odelein**.

19 – Columbus trades **Blake Sloan** to Calgary for **Jamie Allison**.

19 – Colorado trades **Rick Berry** and **Ville Nieminen** to Pittsburgh for **Darius Kasparaitis**.

19 – Dallas trades **Joe Nieuwendyk** and **Jamie Langenbrunner** to New Jersey for **Jason Arnott**, **Randy McKay** and New Jersey's 1st round pick (later traded) in 2002.

19 – Edmonton trades **Tom Poti** and **Rem Murray** to NY Rangers for **Mike York** and the Rangers' 4th round pick (**Ivan Koltsov**) in 2002.

19 – Florida trades **Darren Van Impe** to NY Islanders for the Islanders' 5th round pick in 2003.

19 – Florida trades **Jeff Norton** to Boston for Boston's 6th round pick (**Mikael Vuorio**) in 2002.

19 – Los Angeles trades **Stephane Fiset** to Montreal for future considerations.

19 – Nashville trades **Richard Lintner** to NY Rangers for **Peter Smrek**.

19 – Pittsburgh trades **Stephane Richer** to New Jersey for a conditional pick in 2003.

19 – Washington trades **Adam Oates** to Philadelphia for **Maxime Ouellet** and Philadelphia's 1st (later traded), 2nd (**Maxime Daigneault**) and 3rd (**Derek Krestanovich**) round picks in 2002.

May, 2002

13 – Boston trades **Richard Jackman** to Toronto for the rights to **Kris Vernarsky**.

15 – Pittsburgh trades **Steve Parsons** to Nashville for future considerations.

25 – Montreal trades the rights to **Chris Dyment** to Minnesota for Minnesota's 5th round pick (later traded) in 2002.

June, 2002

12 – Phoenix trades **Michal Handzus** and **Robert Esche** to Philadelphia for **Brian Boucher** and Nashville's 3rd round pick (previously acquired, **Joe Callahan**) in 2002.

18 – Columbus trades **Ron Tugnutt** and its 2nd round pick in 2002 (**Janos Vas**) to Dallas for New Jersey's 1st round choice (previously acquired, later traded) in 2002.

19 – Philadelphia trades **Jiri Dopita** to Edmonton for Edmonton's 3rd round pick in 2003 and a conditional 5th round pick in 2004.

21 – Philadelphia traded **Ruslan Fedotenko**, Tampa Bay's 2nd round pick in 2002 (previously acquired, later traded) and Phoenix's 2nd round pick (previously acquired, later traded) to Tampa Bay for Tampa Bay's 1st round pick (**Joni Pitkanen**) in 2002.

22 – Florida trades its 1st round pick in 2002 (**Rick Nash**) to Columbus for Columbus' 1st round pick (**Jay Bouwmeester**) in 2002 and the right to exchange 1st round picks in 2003, at Florida's option.

22 – Calgary trades its 1st round pick in 2002 (**Petr Taticek**) to Florida for the New York Rangers' 1st round pick in 2002 (previously acquired, **Eric Nystrom**) and Florida's 4th round pick (later traded) in 2002.

22 – NY Islanders trade **Mariusz Czerkawski** to Montreal for **Arron Asham** and Montreal's 5th round pick (**Markus Pahlsson**) in 2002.

22 – Columbus trades **Mattias Timander** to NY Islanders for the Islanders' 4th round pick (**Jekabs Redlihs**) in 2002.

22 – Edmonton trades its 1st round pick in 2002 (**Christopher Higgins**) to Montreal for Montreal's 1st round pick (**Jesse Niinimaki**) and 8th round pick (**Tomas Micka**) in 2002.

22 – Columbus trades New Jersey's 1st round pick in 2002 (previously acquired, **Dan Paille**) to Buffalo for Detroit's 1st round pick in 2002 (later traded) and the rights to **Mike Pandolfo**.

22 – Buffalo trades **Vyacheslav Kozlov** and Buffalo's 2nd round pick in 2002 (later traded) to Atlanta for Atlanta's 2nd round pick (later traded) and Florida's 3rd round pick (previously acquired, **John Adams**) in 2002.

22 – Columbus trades Detroit's 1st round pick in 2002 (previously acquired, **Jim Slater**) to Atlanta for Buffalo's 2nd round pick in 2002 (previously acquired, **Joakim Lindstrom**) and Detroit's 3rd round pick (previously acquired, **Jeff Genovy**) in 2002.

22 – Buffalo trades its 3rd round compensatory pick in 2002 (later traded) and its 2nd round pick in 2003 to Nashville for Nashville's 2nd round pick (later traded) in 2002.

22 – Anaheim trades Boston's 3rd round pick in 2002 (previously acquired, later traded) to Nashville for future considerations.

22 – Florida trades Vancouver's 3rd round pick in 2002 (previously acquired, later traded) and the NY Rangers' 4th round pick in 2003 (previously acquired) to Atlanta for future considerations.

22 – Dallas trades **Brad Lukowich** and its 7th round pick in 2003 to Tampa Bay for Tampa Bay's 2nd round pick (previously acquired, **Tobias Stephan**) in 2002.

22 – Edmonton trades Jochen Hecht to Buffalo for Atlanta's 2nd round pick in 2002 (previously acquired, **Jeff Deslauriers**) and Nashville's 2nd round pick (previously acquired, **Jarret Stoll**) in 2002.

22 – Minnesota trades Jamie McLennan to Calgary for Calgary's 9th round pick (**Mika Hannula**) in 2002.

23 – Vancouver trades **Josh Holden** to Toronto for **Jeff Farkas**.

23 – Tampa Bay trades its 4th round pick in 2003 to Carolina for Carolina's 6th round compensatory pick (**Karri Aakkanen**), Carolina's 8th round pick (**Ryan Craig**), Philadelphia's 8th round pick (previously acquired, **Darren Reid**) and Carolina's 9th round pick (**Alexei Glukhov**) in 2002.

23 – Pittsburgh trades **Krzysztof Oliwa** to NY Rangers for future considerations.

23 – Philadelphia trades Vancouver's 6th round pick (previously acquired, **Jaroslav Balastik**) and Philadelphia's 7th round pick (**Stephen Goertzen**) in 2002 to Columbus for Columbus's 5th round pick in 2003.

23 – San Jose trades its 6th round pick (**Kim Hirschovits**) in 2002 to NY Rangers for the Rangers' 6th round pick in 2003 and future considerations (**Theoren Fleury**, June 26, 2002).

23 – Tampa Bay trades the rights to **Josef Boumedienne** to Ottawa for Ottawa's 7th round pick in 2002 (**Fredrik Norrena**).

23 – Atlanta trades New Jersey's 7th round pick in 2002 (previously acquired, **Tim Conboy**) to San Jose for San Jose's 8th round pick in 2002 (**Pauli Levokari**) and 7th round pick in 2003.

23 – Florida trades its 9th round pick in 2002 (**Sergei Mozyakin**) to Columbus for Columbus' 9th round pick in 2003.

23 – Minnesota trades its 4th round pick in 2002 (previously acquired, **Aaron Rome**) to Los Angeles for the rights to **Cliff Ronning**.

23 – Nashville trades the rights to **Peter Sykora** to Washington for a conditional pick in 2003.

23 – Tampa Bay trades Phoenix's 2nd round pick in 2002 (previously acquired, **Dan Spang**) to San Jose for San Jose's 2nd round pick (**Adam Henrich**) and 5th round compensatory pick (**Gerard Dicaire**) in 2002.

23 – St. Louis trades its 2nd round pick in 2002 (**Vladislav Yevseyev**)

to Boston for Boston's 2nd round pick (**Andrei Mikhnov**) and 5th round pick (**Justin Maiser**) in 2002.

23 – Florida trades **Eric Godard** to NY Islanders for Florida's 3rd round pick (previously acquired, **Gregory Campbell**) in 2002.

23 – Philadelphia trades Carolina's 3rd round pick in 2002 (previously acquired, **Jesse Lane**) to Carolina for Carolina's 6th round pick in 2002 (**Nikita Korovkin**) and 3rd round pick in 2003.

23 – Calgary trades its 3rd round pick in 2002 (**Todd Ford**) to Toronto for Toronto's 3rd round pick (**Matthew Lombardi**) and 5th round pick (**Kristofer Persson**) in 2002.

23 – NY Rangers trade their 3rd round pick in 2002 (**Arttu Luttinen**) to Ottawa for Ottawa's 3rd round pick (**Marcus Jonasen**) and 4th round compensatory pick (**Nathan Guenin**) in 2002.

23 – Chicago trades its 3rd round pick in 2002 (**Jonas Fidler**) to San Jose for San Jose's 3rd round pick (**Alexander Kozhevnikov**) and 4th round pick (**Matt Ellison**) in 2002.

23 – Nashville trades Buffalo's 3rd round compensatory pick in 2002 (previously acquired, **Dominic D'Amour**) to Toronto for a conditional pick in 2003.

23 – Nashville trades Boston's 3rd round pick in 2002 (previously acquired, **Valtteri Filppula**) to Detroit for Detroit's 3rd round pick in 2003.

25 – Anaheim trades **Steve Shields** to Boston for a 3rd round pick in 2003.

25 – Vancouver trades **Ryan Bonni** to Toronto for future considerations.

26 – NY Rangers trade **Theoren Fleury** to San Jose to complete the trade of June 23, 2002.

29 – Ottawa trades **Shawn McEachern** and a 6th round pick in 2004 to Atlanta for **Brian Pothier**.

29 – Dallas trades **Ed Belfour** and **Cameron Mann** to Nashville for **David Gosselin** and a 5th round choice in 2003.

30 – Montreal trades **Sergei Berezin** to Chicago for a 4th round pick in 2004.

30 – Toronto trades **Curtis Joseph** to Calgary for an 8th round pick in 2004 and future considerations.

30 – Toronto trades **Tie Domi** to Nashville for an 8th round choice in 2003.

30 – NY Rangers trade **Mike Richter** to Edmonton for future considerations.

July, 2002

3 – Florida trades **Jason Wiemer** to NY Islanders for **Branislav Mezei**.

6 – New Jersey trades **Petr Sykora**, **Mike Commodore**, **Jean-Francois Damphousse** and **Igor Pohanka** to Anaheim for **Jeff Friesen**, **Oleg Tverdovsky** and **Maxim Balmochnykh**.

18 – Toronto trades **Dmitry Yushkevich** to Florida for **Robert Svelha**.

24 – Buffalo trades **Erik Rasmussen** to Los Angeles for **Adam Mair** and a 5th round draft choice in 2003.

Trades and free agent signings that occurred after August 23, 2002 are listed on page 337.